KIRK-OTHMER

CONCISE
ENCYCLOPEDIA
OF CHEMICAL
TECHNOLOGY

KIRK-OTHMER

CONCISE ENCYCLOPEDIA
OF CHEMICAL TECHNOLOGY

A WILEY-INTERSCIENCE PUBLICATION

John Wiley & Sons

NEW YORK · CHICHESTER · BRISBANE · TORONTO · SINGAPORE

Library of Congress Cataloging in Publication Data:

Main entry under title:

Kirk-Othmer Concise encyclopedia of chemical technology.

"A Wiley-Interscience publication."
Abridged version of the 24 volume Encyclopedia of
chemical technology. 3rd ed. New York: Wiley, c1978–
c1984.
Includes index.
Executive editor, Martin Grayson; associate editor,
David Eckroth.
1. Chemistry, Technical—Dictionaries. I. Kirk,
Raymond E. (Raymond Eller), 1890–1957. II. Othmer,
Donald F. (Donald Frederick), 1904–   . III. Grayson,
Martin. IV. Eckroth, David. V. Encyclopedia of chemical
technology. VI. Title: Concise encyclopedia of chemical
technology.
TP9.K54 1985   660'.03'21   84-22006
ISBN 0-471-86977-5
ISBN 0-471-51700-3 (pbk.)

Printed in the United States of America

10 9 8 7 6 5

PREFACE

This compact desk reference contains all of the subjects covered in the 24 main volumes and the Supplement Volume of the world-renowned Third Edition of the *Kirk-Othmer Encyclopedia of Chemical Technology*. The articles have been condensed by professional science writers, reviewed for accuracy by the original authors or their colleagues, and updated where necessary. This distillation, skillfully prepared to retain the key data, tables, and factual matter of the original, is a complete and self-contained encyclopedia. It is designed to serve as a ready-reference guide for students, scientists, engineers, and technologists seeking answers to questions on any aspect of chemical technology.

This nearly two million word version, like the fifteen million word larger work, provides both SI and common units, carefully selected key references for each article, and hundreds of tables, charts, figures, and graphs. Coverage includes every important industrial sector, such as: agricultural chemicals; chemical engineering; coatings and inks; composite materials; drugs, cosmetics, and biomaterials; dyes, pigments, and brighteners; ecology and industrial hygiene; energy conversion and technology; fats and waxes; fermentation and enzyme technology; fibers, textiles, and leather; food and animal nutrition; fossil fuels and derivatives; glass, ceramics, and cement; industrial inorganic chemicals; industrial organic chemicals; metals, metallurgy, and metal alloys; plastics and elastomers; semiconductors and electronic materials; surfactants, detergents, and emulsion technology; water supply, purification, and reuse; wood, paper, and industrial carbohydrates; and miscellaneous topics from computers, instrumentation and control, information retrieval, patents, trademarks and copyright to transportation, operations management, and product and process topics.

Information from the Supplement Volume has been included alongside the main-volume articles. However, where a specific subject is repeated for updating purposes, the new version has been inserted at the end of this volume in the addendum. Although specific reference to the articles in the original work is not always made, it should be understood that further details, specific bibliographic citations, and much wider coverage of any subject may be obtained by referring to the 26-volume edition.

The editors have also carefully preserved the tradition of citing related articles in the text (*see* and *see also* citations), as well as secondary entries or cross references which cite the synonym or entry term where a subject can be located. An Index of Key Terms provides further access to the contents where desired.

The complete Third Edition of the *Kirk-Othmer Encyclopedia* has been called "the Bible of Chemical Technology." This concise version presents the essence of this monumental work in a useful daily tool comparable to a handbook or dictionary from which the comprehensive, authoritative, and lucidly written data become instantly available. This new encyclopedia has taken three years of intensive effort by the writers, authors, and Wiley editorial staff. Every effort has been made by the editors to provide a work that is unsurpassed in quality and accuracy. We hope we have succeeded and that this work will serve your reference needs for years to come.

MARTIN GRAYSON
Executive Editor and Publisher

EDITORIAL STAFF

CONTRIBUTORS AND REVIEWERS

V.E. Archer, *Vulcan Materials Co.,* Detinning under Tin and tin alloys

Wesley L. Archer, *Dow Chemical U.S.A.,* Survey; Other chloroethanes both under Chlorocarbons and chlorohydrocarbons

Barry Arkles, *Petrarch Systems, Inc.,* Silanes; Silicon ethers and esters; both under Silicon compounds

Herbert G. Arlt, Jr., *Arizona Chemical Company,* Tall oil

George T. Armstrong, *National Bureau of Standards,* Calorimetry

J.P. Arrington, *Dow Chemical U.S.A.,* Amines, aromatic, methylenedianiline

Gary Astrologes, *Halocarbon Products Corporation,* Bromotrifluoroethylene under Fluorine compounds, organic; Fluorinated acetic acids; Fluoroethanols under Fluorine compounds, organic

Thomas A. Augurt, *Propper Manufacturing Co., Inc.,* Sterilization techniques

Joseph P. Ausikaitis, *Union Carbide Technical Center,* Drying agents

John M. Avento, *M & T Chemicals, Inc.,* Antimony and other inorganic compounds under Flame retardants

Oliver Axtell, *Celanese Chemical Company,* Economic evaluation

Ricardo Bach, *Lithium Corporation of America,* Lithium and lithium compounds

R.B. Badachhape, *Rice University,* Direct fluorination under Fluorine compounds, organic

Kim Badenhop, *B.W. Dyer & Co.,* Sugar economics under Sugar

David G. Bailey, *United States Department of Agriculture,* Leather

Robert S. Bailey, *Lilly Industrial Coatings, Inc.,* Stains, industrial

Richard C. Bailie, *West Virginia University,* Incinerators

J.K. Baird, *Kelco Division of Merck & Co., Inc.,* Gums

Malcom H.I. Baird, *McMaster University,* Extraction, liquid – liquid

A.J. Baker, *United States Department of Agriculture,* Wood

Philip J. Baker, Jr., *International Minerals and Chemical Corporation,* Nitroparaffins

Robert Bakish, *Consultant,* Water, supply and desalination

Marilyn Bakker, *Technology Forecast,* Engineering plastics

S.M. Balaban, *Monsanto Co.,* Sorbic acid

F.P. Baldwin, *Exxon Chemical Co.,* Butyl rubber under Elastomers, synthetic

John Ballance, *Brush Wellman Inc.,* Beryllium and beryllium alloys

Bryan Ballantyne, *Union Carbide Corporation,* Toxicology

Nicolaas J. Ballintyn, *Union Carbide Corp.,* Polysulfone resins under Polymers containing sulfur

W. Robert Ballou, *C & I Girdler Incorporated,* Carbon dioxide

Derek Bannister, *CIBA-GEIGY Corporation,* Triphenylmethane and related dyes

D.W. Bannister, *CIBA-GEIGY Corp.,* Dyes and dye intermediates; Indole

Frank J. Barborek, *Gulf Oil Company,* Film and sheeting materials

Hans C. Barfoed, *Nova Industri A/S,* Enzyme detergents

Colin Barker, *University of Tulsa,* Nomenclature; Origin of petroleum both under Petroleum

Robert R. Barnhart, *Uniroyal Chemical,* Rubber compounding

Edmund F. Baroch, *International Titanium, Inc.,* Vanadium and vanadium alloys

Robert Q. Barr, *Climax Molybdenum Company,* Molybdenum and molybdenum alloys

H.F. Barry, *Climax Molybdenum of Michigan,* Molybdenum compounds

Charles M. Bartish, *Air Products and Chemicals, Inc.,* Carbon monoxide

P. Bass, *Mobay Chemical Corp.,* Dyes, application and evaluation

W.E. Bastian, *CPS Chemical Company,* Naphthenic acids

John H. Bateman, *CIBA-GEIGY Corporation,* Hydantoin and derivatives

Roger G. Bates, *University of Florida,* Hydrogen-ion activity

Harinath B. Bathina, *Armak Company,* Amides, fatty acid; Fatty amines both under Amines

Sushil K. Batra, *GIMRET International,* Power generation

O.A. Battista, *Research Services Corp.,* Cellulose

R.G. Bauer, *The Goodyear Tire & Rubber Company,* Styrene – butadiene rubber under Elastomers, synthetic

William Bauer, Jr., *Rohm and Haas Company,* Acrylic acid and derivatives

Burton M. Baum, *FMC Corporation,* Bleaching agents

G.A. Baum, *The Institute of Paper Chemistry,* Paper

Jon A. Baumgarten, *Paskus, Gordon & Hyman,* Trademarks and copyrights

William F. Baxter, Jr., *Eastman Chemical Products, Inc.,* Crotonaldehyde; Crotonic acid

David L. Beach, *Gulf Research & Development Company,* Polymers of higher olefins under Olefin polymers

Harry H. Beacham, *FMC Corporation,* Allyl compounds

Donald A. Becker, *National Bureau of Standards,* Oil under Recycling

George M. Begun, *Oak Ridge National Laboratory,* Isotope separation under Nuclear reactors

Elmer B. Bell, *E.I. du Pont de Nemours & Co., Inc.,* Sulfamic acid and sulfamates

Alan G. Bemis, *Amoco Chemicals Corporation,* Phthalic acids and other benzenepolycarboxylic acids

B.A. Bendtsen, *United States Department of Agriculture,* Wood

R.F. Benenati, *Polytechnic Institute of New York,* Programmable pocket computers

Friederich Benesovsky,‡ *Metallwerke Plansee A.G.,* Survey and Industrial heavy-metal carbides, both under Carbides; Nitrides

Gregory B. Bennett, *Sandoz Inc.,* Hypnotics, sedatives, anticonvulsants

Douglas N. Bennion, *University of California, Los Angeles,* Introduction; Primary cells, both under Batteries and electric cells, primary

A.M. Benson, *Shell Development Company,* Fluid mechanics

Frederic R. Benson, *ICI United States, Inc.,* Sugar alcohols under Alcohols, polyhydric

Harriet Benson, *ALZA Corporation,* Pharmaceuticals, controlled release

Julius Berger, *Hoffmann-La Roche Inc.,* Phenazines under Antibiotics

S. Edmund Berger, *Allied Chemical Corporation,* Hydroxy dicarboxylic acids

Eric M. Bergtraun, *National Semiconductor,* Maintenance

Cicero A. Bernales, *Allied Chemical Corp.,* Oxalic acid

Dwight H. Berquist, *Henningsen Foods, Inc.,* Eggs

R.V. Berthold, *Union Carbide Corporation,* Ethylene glycol and propylene glycol under Glycols

E.R. Bertozzi, *Thiokol Corporation,* Polysulfides under Polymers containing sulfur

Helmut Beschke, *Degussa Company,* Vitamins, nicotinamide and nicotinic acid (B_3)

Fred W. Billmeyer, Jr., *Rensselaer Polytechnic Institute,* Color, Polymers

Jasjit S. Bindra, *Pfizer, Inc.,* Memory-enhancing agents and antiaging drugs

C.D. Binger, *Dow Chemical U.S.A.,* Chlorine and sodium hydroxide under Alkali and chlorine products

Tim Bitler, *Aluminum Corporation of America,* Aluminum compounds; aluminum halides and aluminum nitrate

D.W. Bixby, *The Sulphur Institute,* Sulfur

L.M. Blair, *E.I. du Pont de Nemours & Co., Inc.,* Acetal resins

Alan Bleier, *Massachusetts Institute of Technology,* Colloids

Jan K. Bliek, *Badger, B.V.,* Acrylonitrile

John W. Blieszner, *Amoco Chemicals Corporation,* Propylene

Alexander Bloch, *Roswell-Park Memorial Institute,* Nucleosides under Antibiotics

J.O. Blomeke, *Oak Ridge National Laboratory,* Waste management under Nuclear reactors

Jordan J. Bloomfield, *Monsanto Company,* Photochemical technology

Bernard R. Bluestein, *Witco Chemical Corporation,* Abherents

John G. Blumberg, *PQ Corporation,* Fillers

John H. Blumbergs, *FMC Corporation,* Bleaching agents

Bob Blumenthal, *Noah Industries,* Molybdenum compounds

I.A. Boening, *Monsanto Company,* Phosphorus compounds

D.D. Boesiger, *Phillips Petroleum Co.,* Butanes under Hydrocarbons $C_1 – C_6$

R.T. Bogan, *Tennessee Eastman Co.,* Cellulose derivatives, esters

Henry I. Bolker, *Pulp and Paper Research Institute of Canada,* Çellulose

G.N. Bollenback, *The Sugar Foundation,* Special sugars under Sugar

Allen F. Bollmeier, *International Minerals and Chemical Corp.,* Alkanolamines from nitro alcohols under Alkanolamines; Nitro alcohols; Nitroparaffins

E.R. Booser, *General Electric Company,* Bearing materials; Lubrication and lubricants

George L. Booth, *The Black Clawson Company,* Coating processes

James H. Boothe, *Lederle Laboratories Division, American Cyanamid Company,* Tetracyclines under Antibiotics

E.L. Borg, *Uniroyal Chemical Co.,* Ethylene – propylene rubber under Elastomers, synthetic

W.F.H. Borman, *General Electric Company,* Polyesters, thermoplastic

Sebastian R. Borrell, *Texas Instruments Incorporated,* Photodetectors

W.S. Boston, *CISRO,* Wool

J.H. Bostwick, *E.I. du Pont de Nemours & Co., Inc.,* Sulfamic acid and sulfamates

N.L. Bottone, *Union Carbide Corp.,* Carbon and artificial graphite under Carbon

Edward F. Bouchard, *Pfizer Inc.,* Citric acid.

Max M. Boudakian, *Olin Chemicals,* Fluorinated aromatic compounds under Fluorine compounds, organic

Gilbert Bourcier, *Reynolds Metal Co.,* Nonferrous metals under Recycling

H. Kent Bowen, *Massachusetts Institute of Technology,* Ceramics as electrical materials

Joseph G. Bower, *U.S. Borax & Chemical Corp.,* Boron, elemental

David C. Boyd, *Corning Glass Works,* Glass

R.S. Boynton, *Consultant,* Lime and limestone

Robert L. Boysen, *Union Carbide Corporation,* High pressure (low and intermediate density) polyethylene under Olefin polymers

D.G. Brady, *Phillips Petroleum Company,* Poly(phenylene sulfide) under Polymers containing sulfur

C. Raymond Brandt, *Honeywell Inc.,* Pressure measurement

Donald E. Brasure, *E.I. du Pont de Nemours & Co., Inc.,* Poly(vinyl fluoride) under Fluorine compounds, organic

David B. Braun, *Union Carbide Corporation,* Ethylene oxide polymers under Polyethers

J.P. Brazel, *General Electric Company,* Ablative materials

D.W. Breck, *Union Carbide Corporation,* Molecular sieves

J.J. Brenden, *United States Department of Agriculture,* Wood

Dieter Bretzinger, *Dynamit Nobel A.G.,* Alkoxides, metal

R.J. Brewer, *Tennessee Eastman Company,* Cellulose derivatives, esters

Frank Brigano, *Olin Chemicals,* Water, treatment of swimming pools, spas, and hot tubs

George L. Brode, *Union Carbide Corporation,* Phenolic resins

Lester A. Brooks, *R.T. Vanderbilt Co.,* Antioxidants and antizonants

J.R. Brotherton, *U.S. Borax Research Corp.,* Boric acid esters under Boron compounds

D.B. Broughton, *UOP Process Division, UOP Inc.,* Introduction and Liquids under Adsorptive separation

Allan Brown, *Beecham Research Laboratories,* Antibiotics, β-lactams clavulanic acid, thienamycin, and others

B.F. Brown, *The American University,* Corrosion and corrosion inhibitors

E.S. Brown, *Union Carbide Corporation,* Ethylene glycol and propylene glycol under Glycols

Herbert C. Brown, *Purdue University,* Hydroboration

D.E. Brownlee, *University of Washington,* Space chemistry

Robert D. Bruce, *Hoover Keith & Bruce, Inc.,* Noise pollution

Michael H. Bruno, *Consultant,* Printing processes

D.R. Buchanan, *North Carolina State University,* Olefin fibers

Gordon Buchi, *CIBA-GEIGY Corporation,* Epoxy resins

Bernard Bucholz, *Pennwalt Corporation,* Thiophene

F.A.M. Buck, *King, Buck & Associates,* Feedstocks

Robert H. Buckenmayer, *Airwick Industries, Inc.,* Odor modification

R.A. Budenholzer, *Illinois Institute of Technology,* Power generation

A.M. Bueche,‡ *General Electric Co.,* Research management

Peter R. Buechler, *United States Department of Agriculture,* Leather

Kees A. Bueters, *Combustion Engineering Inc.,* Furnaces, fuel-fired; Steam

Rointan F. Bunshah, *University of California, Los Angeles,* Refractory coatings

Frederick J. Buono, *Tenneco Inc.,* Driers and metallic soaps

Bohdan V. Burachinsky, *Inmont Corporation,* Inks

Joseph V. Burakevich, *FMC Corporation,* Cyanuric and isocyanuric acids

Thomas H. Burgess, *Fischer-Porter Company,* Flow measurement

H.O. Burrus, *E.I. du Pont de Nemours & Co., Inc.,* Chlorosulfuric acid

Steven E. Bushman, *Allied Chemical Corp.,* Oxalic acid

R.M. Bushong, *Union Carbide Corp.,* Carbon and artificial graphite under Carbon

W.R. Busler, *Nalco Chemical Company,* Aluminates (under Aluminum compounds)

James W. Butler, *Naval Research Laboratory,* Ion-implantation

E.J. Butterfield, *Boyce Thompson Institute for Plant Research, Inc.,* Fungicides, agricultural

Israel Cabasso, *Gulf South Research Institute,* Hollow-fiber membranes

Arno Cahn, *Arno Cahn Consulting Services, Inc.,* Surfactants and detersive systems

Elton J. Cairns, *General Motors Research Laboratories,* Fuels cells under Batteries and electric cells, primary

C.R. Campbell, *Monsanto Chemical Intermediates Company,* Adipic acid

M.L. Campbell, *Exxon Chemical Company,* Cyclohexane under Hydrocarbons $C_1 - C_6$

R.W. Campbell, *General Electric Company,* Polyesters, thermoplastic

Thomas F. Canning, *Kerr-McGee Chemical Corporation,* Sodium sulfates under Sodium compounds

C.E. Capes, *National Research Council of Canada,* Size enlargement

S.C. Carapella, Jr., *ASARCO Inc.,* Arsenic and arsenic alloys; Bismuth and bismuth alloys; Cadmium and cadmium alloys

D.A. Carlson, *University of Florida,* Repellents

Frank G. Carpenter, *United States Department of Agriculture,* Sugar analysis; Cane sugar both under Sugar

Dodd S. Carr, *International Lead Zinc Research Organization, Inc.,* Lead salts under Lead compounds

Linda C. Carrico, *U.S. Bureau of Mines,* Mercury

Patrick E. Cassidy, *Southwest Texas University,* Polyimides

Wade Thomas Cathey, *University of Colorado,* Holography

Sigmund C. Catino, *GAF Corp.,* Azo dyes

John A. Caughlan, *Mallinckrodt, Inc.,* Fine chemicals

James N. Cawse, *Union Carbide Corporation,* Ethylene oxide

Dennis J. Cecchini, *Exxon Research and Engineering Company,* Pilot plants and microplants

Peter Cervoni, *Lederle Laboratories,* Cardiovascular agents

Ared Cezairliyan, *National Bureau of Standards,* Calorimetry

D.H. Chadwick, *Mobay Chemical Corporation,* Isocyanates, organic

A.M. Chakrabarty, *University of Illinois at the Medical Center,* Genetic engineering

C.C. Chamis, *National Aeronautics and Space Administration,* Laminated and reinforced metals

Chuan C. Chang, *Bell Laboratories,* Semiconductors, fabrication and characterization

Bernard Chant, *International Flavors & Fragrances,* Perfumes

A.T. Chase, *Struthers Wells Corporation,* Heat transfer under Heat exchange technology

Wai-Kai Chen, *Ohio University,* Dimensional analysis

William B. Chess, *Stauffer Chemical Co.,* Chemical leavening agents under Bakery processes and leavening agents

Chang W. Chi, *W.R. Grace & Co., Davison Chemical Division,* Gases under Adsorptive separation

G.Y. Chin, *Bell Laboratories,* Magnetic materials, bulk; Magnetic materials, thin film

L.J. Chinn, *G.D. Searle & Co.,* Contraceptive drugs

Michael S. Cholod, *Rohm and Haas Company,* Cyanohydrins

D.A. Chung, *The Goodyear Tire & Rubber Company,* Thermoplastic elastomers under Elastomers, synthetic

Rack Hun Chung, *General Electric Co.,* Dyes, anthraquinone

R.H. Chung,‡ *GAF Corporation,* Anthraquinone; Anthraquinone derivatives

Christopher Cimarusti, *Squibb Institute for Medical Research,* Antibiotics, β-lactams, monobactams

David L. Cincera, *Air Products and Chemicals, Inc.,* Poly(vinyl alcohol) under Vinyl polymers

F.P. Civardi, *Inmont Corporation,* Leatherlike materials

T.W. Clapper, *Kerr-McGee Corporation,* Chloric acid and chlorates under Chlorine oxygen acids and salts

David A. Clark, *Pfizer Central Research,* Opioids, endogenous

G.T. Clark, *Eastman Kodak Co.,* Disperse dyes under Thiazole dyes

J. Peter Clark, *Virginia Polytechnic Institute and State University,* Chemurgy

R.K. Clark, *Shell Development Company,* Drilling fluids under Petroleum

Margaret A. Clarke, *Cane Sugar Refining Research Project, Inc.,* Bagasse

George D. Clayton, *Clayton Environmental Consultants, Inc.,* Industrial hygiene and toxicology

Howard P. Clemens, *University of Oklahoma,* Aquaculture

J.T. Clemens, *Bell Telephone Laboratories, Incorporated,* Integrated circuits

T.H. Cleveland, *Mobay Chemical Corporation,* Isocyanates, organic

W.P. Clinton, *General Foods Corp.,* Coffee

David L. Coffen, *Hoffmann-La Roche,* Vitamins, pyridoxine (B_6)

J.G. Cohn, *Englehard Minerals & Chemicals Corp.,* Gold and gold compounds

G. Collin, *Rütgerswerke AG,* Coal chemicals and feedstocks, carbonization and coking

Ward Collins, *Dow Corning Corporation,* Silicon halides under Silicon compounds

James G. Colson, *Hooker Chemicals & Plastics Corp.,* Benzene hexachloride under Chlorocarbons and chlorohydrocarbons

F.B. Colton, *G.D. Searle & Co.,* Contraceptive drugs

J. Ross Colvin, *National Research Council,* Cellulose

J. Clyde Conger, Sr., *American Cyanamid Company,* Azine dyes

James H. Conklin, *E.I. du Pont de Nemours & Co., Inc.,* Exhaust control, industrial

Michael Conway, *University of Oklahoma,* Aquaculture

Conway Publications, Inc., Plant location

David A. Cooney, *Division of Cancer Treatment, National Cancer Institute, National Institutes of Health,* Enzymes, therapeutic

C.M. Cooper, *Michigan State University,* Solvent recovery

Glenn D. Cooper,‡ *General Electric Company,* Polyethers, aromatic under Polyethers

Kenneth W. Cooper, *York Division, Borg-Warner Corp.,* Air conditioning; Refrigeration

Murray C. Cooperman, *CasChem, Inc.,* Castor oil

H.W. Coover, Jr., *Tennessee Eastman Company, Division of Eastman Kodak Company,* 2-Cyanoacrylic ester polymers under Acrylic ester polymers

Edmund S. Copeland, *National Institutes of Health,* Radioprotective agents

A.N. Copp, *Basic, Inc.,* Magnesium compounds

Jim Corbin, *University of Illinois,* Pet and livestock feeds

Geoffrey A. Cordell, *College of Pharmacy, University of Illinois,* Alkaloids

Charles L. Cormany, *PPG Industries,* Drycleaning under Drycleaning and laundering

B. Cornils, *Ruhrchemie AG,* Coal chemicals and feedstocks, gasification

I.W. Cottrell, *Kelco Division of Merck & Co., Inc.,* Gums

W.B. Cottrell, *Oak Ridge National Laboratory,* Safety in nuclear facilities under Nuclear reactors

Charles C. Coutant, *Oak Ridge National Laboratory,* Thermal pollution by power plants

John C. Cowan, *Bradley University,* Drying oils

Richard D. Cowell, *Witco Chemical Corporation,* Abherents

J.A. Cowfer, *BFGoodrich Co.,* Vinyl chloride under Vinyl polymers

Charles M. Cox, *Pravel, Gambrell, Hewitt, Kirk, Kimball & Dodge,* Patents, practice and management

T.D. Coyle, *National Bureau of Standards,* Introduction under Silica

Graham C. Crawley, *Imperial Chemical Industries Ltd.,* Nonsteroidal estrogens under Hormones

Lamberto Crescentini, *Allied Corporation,* Caprolactam under Polyamides

Giovanni Crespi, *Montedison SpA,* Propylene under Olefin polymers

W. Raymond Cribb, *Brush Wellman, Inc.,* Beryllium and beryllium alloys

Burton B. Crocker, *Monsanto Co.,* Air pollution control methods; Fans and blowers; Gas cleaning; Incinerators

L.E. Cross, *The Pennsylvania State University,* Ferroelectrics

E. Lea Crump, *Gulf Oil Chemical Company,* Film and sheeting materials

M.M. Crutchfield, *Monsanto Company,* Phosphorus compounds

William P. Cummings, *W.R. Grace & Co., Davison Chemical Division,* Gases under Adsorptive separation

James A. Cusumano, *Catalytica Associates, Inc.,* Catalysis

Glenn H. Dale, *Phillips Petroleum Co.,* Hydrocarbons, survey; Hexanes under Hydrocarbons $C_1 - C_6$

J. Daley, *Celanese Chemical Company,* Other polyhydric alcohols under Alcohols, polyhydric

W.B. Dancy, *International Minerals & Chemical Corporation,* Potassium compounds

James R. Daniel, *Purdue University,* Starch

Peter J.L. Daniels, *Schering-Plough Corporation,* Aminoglycosides under Antibiotics

Wiley Daniels, *Air Products and Chemicals, Inc.,* Poly(vinyl acetate) under Vinyl polymers

Paul Danielson, *Corning Glass Works,* Vitreous silica under Silica

Richard D. Danielson, *Minnesota Mining & Manufacturing Co.,* Fluoro ethers and amines under Fluorine compounds, organic

D.E. Danley, *Monsanto Chemical Intermediates Co.,* Adipic acid; Electrochemical processing, organic

Eli M. Dannenberg, *Cabot Corporation,* Carbon black under Carbon

K.V. Darragh, *Stauffer Chemical Company,* Aluminum sulfate and alums under Aluminum compounds

John A. Davidson, *BFGoodrich Co.,* Poly(vinyl chloride) under Vinyl polymers

Reg Davies, *E.I. du Pont de Nemours & Co., Inc.,* Sampling

Gerald W. Davis, *Fiber Industries, Inc.,* Polyester fibers

John Davis, *NASA,* Composites, high performance

John H. Davis, *Eastman Chemical Products, Inc.,* Other glycols under Glycols

Robert F. Davis, *North Carolina State University,* Scope under Ceramics

W.T. Davis, *Ethyl Corporation,* Synthetic processes under Alcohols, higher aliphatic

Jesse H. Day, *Ohio University,* Electrochromic and thermochromic under Chromogenic materials

Aldo DeBenedictis, *Shell Chemical Company,* Allyl chloride under Chlorocarbons and chlorohydrocarbons

Ines V. de Gruy, *United States Department of Agriculture,* Cotton

Eckehard V. Dehmlow, *Technischen Universität Berlin,* Catalysis, phasetransfer

Pierre de la Bretèque, *Alusuisse France S.A.,* Gallium and gallium compounds

D.J. DeLong, *Union Carbide Corporation,* Ethylene oxide polymers under Polyethers

Richard DeMarco, *Clairol, Inc.,* Hair preparations

Anthony DeMaria, *Sybron Corp.,* Dye carriers

D.G. Demianiw, *Gulf Oil Chemicals,* Olefins, higher

T. John Dempsey, *The Dow Chemical Company,* Propylene oxide

John M. Derfer, *SCM Corporation,* Terpenoids

Marian M. Derfer, *SCM Corporation,* Terpenoids

J.D. Desai, *General Electric Company,* Tool materials

Robert Desch, *Armstrong World Industries, Inc.,* Plastic building products

H.D. DeShon, *Dow Chemical U.S.A.,* Carbon tetrachloride; Chloroform both under Chlorocarbons and chlorohydrocarbons

C.M. Detz, *Union Carbide Corporation,* Physical and chemical properties and explosive behavior under Acetylene

Robert H. Dewey, *International Minerals & Chemical Corp.,* Alkanolamines from nitro alcohols under Alkanolamines; Nitro alcohols

Martin Dexter, *CIBA-GEIGY Corporation,* Uv stabilizers

François d'Heurle, *IBM,* Electromigration

Philip F. Dickson, *Colorado School of Mines,* Oil shale

Donald R. Diehl, *Eastman Kodak Company,* Polymethine dyes

V.G.DiFate, *Monsanto Co.,* Sorbic acid

John A. Dindorf, *Amoco Chemicals Corporation,* Phthalic acids and other benzenepolycarboxylic acids

Frederick Disque, *Alpha Metals, Inc.,* Solder and brazing alloys

G.O. Doak, *North Carolina State University,* Antimony compounds; Arsenic compounds; Bismuth compounds

James B. Doe, *ESB Technology Co.,* Introduction; Secondary cells, lead-acid; Other cells all under Batteries and electric cells, secondary

Richard L. Doerr, *Olin Corporation,* Chlorine dioxide, chlorous acid, and chlorites under Chlorine oxygen acids and salts

Julius E. Dohany, *Pennwalt Corporation,* Poly(vinylidene fluoride) under Fluorine compounds, organic

Michael J. Dolan, *Monsanto Company,* Phosphoric acids and phosphates

James A. Doncheck, *Bio-Technical Resources, Inc.,* Malts and malting

J.R. Donovan, *Monsanto Enviro-Chem Systems, Inc.,* Sulfuric acid and sulfur trioxide

Daniel J. Doonan, *U.S. Borax Research Corp.,* Boron oxides, boric acid, and borates under Boron compounds

Leonard Doub, *Warner Lambert-Parke Davis Research Laboratories,* Sulfonamides under Antibacterial agents, synthetic

P.A. Dougall, *Kennecott Copper Corporation,* Copper

R.B. Dougherty, *American Cyanamid,* Veterinary drugs

J.H. Downing, *Union Carbide Corporation,* Manganese and manganese alloys

George L. Drake, Jr., *United States Department of Agriculture,* Flame retardants for textiles

Harold J. Drake, *U.S. Bureau of Mines,* Mercury

L.E. Drehman, *Phillips Petroleum Co.,* Hexanes under Hydrocarbons $C_1 - C_6$

Arthur Drelich, *Chicopee Division, Johnson and Johnson,* Nonwoven textile fabrics, staple fibers

Hans Dressler, *Koppers Company,* Chlorinated naphthalenes under Chlorocarbons and chlorohydrocarbons; Naphthalene derivatives (Polyhydroxy)benzenes

L.J. Drew, *U.S. Geological Survey,* Resources under Petroleum

M.P. Dreyfuss, *BFGoodrich Co.,* Tetrahydrofuran and oxetane polymers under Polyethers

P. Dreyfuss, *The University of Akron,* Tetrahydrofuran and oxetane polymers under Polyethers

R.W. Drisko, *U.S. Naval Civil Engineering Laboratory,* Coatings, marine

Gerald M. Drissel, *Air Products and Chemicals, Inc.,* Carbon monoxide

Robert E. Droegkamp, *Fansteel, Inc.,* Tantalum and tantalum compounds

Charles Drum, *PPG Industries, Inc.,* Sodium sulfides under Sodium compounds

Paul Duby, *Columbia University,* Extractive metallurgy

H.S. Dugal, *The Institute of Paper Chemistry,* Paper

Charles B. Duke, *Xerox Corporation,* Polymers, conductive

W.G. Dukek, *Exxon Research and Engineering Company,* Aviation and other gas turbine fuels

Dennis A. Duncan, *Institute of Gas Technology,* From hydrocarbons (under Acetylene)

Frank Duneczky, *CasChem, Inc.,* Castor oil

Gary B. Dunks, *Union Carbide Corp.,* Boron hydrides, heteroboranes, and their metalloderivatives (commercial aspects) under Boron compounds

K.L. Dunlap, *Mobay Chemical Corporation,* Nitrobenzene and nitrotoluenes

Lawrence H. Dunlap, *Consultant,* Plastic building products

Hugh Dunn, *Inmont Corporation,* Inks

Donald B. DuPré, *University of Louisville,* Liquid crystals

Donald F. Durso, *Johnson & Johnson,* Cellulose

Richard A. Durst, *National Bureau of Standards,* Hydrogen-ion activity

Sean G. Dwyer, *S.C. Johnson & Son. Inc.,* Polishes

H.W. Earhart, *Koch Chemical Company,* Polymethylbenzenes

James Early, *National Bureau of Standards,* Ferrous metals under Recycling

G. Yale Eastman, *Thermacore, Inc.,* Heat pipe under Heat exchange technology

Nathan Eastman, *Union Carbide Corporation,* Cellulose

D.B. Easty, *The Institute of Paper Chemistry,* Paper

Claudio L. Eberling, *BASF Aktiengesellschaft, Ludwigshafen, Federal Republic of Germany,* Formamide; Dimethylformamide both under For-

mic acid and derivatives

Frank F. Ebetino, *Norwich Pharmacal Company, Division of Morton-Norwich Products, Inc.,* Nitrofurans under Antibacterial agents, synthetic

Thomas E. Eble, *The Upjohn Company,* Lincosaminides under Antibiotics

Takeshi Egami, *Institut für Physik, Max Planck Institut für Metallforschung,* Glassy metals

Richard Ehrhardt, *The University of North Carolina at Greensboro,* Operations planning

John Ehrlich, *Detroit Institute of Technology,* Chloramphenicol and analogues under Antibiotics

Richard E. Eibeck, *Allied Chemical Corporation,* Sulfur under Fluorine compounds, inorganic; Perfluoroalkylsulfur fluorides under Fluorine compounds, organic

Joseph Eichberg, *American Lecithin Company,* Lecithin

John J. Eisch, *SUNY Binghamton,* Organometallics, metal-complexes

Edward U. Elam, *Tennessee Eastman Company,* Esters, organic

E.M. Elkin, *Noranda Mines Limited,* Selenium and selenium compounds; Tellurium and tellurium compounds

J. Elks, *Glaxo Labs, Glaxochem, Ltd.,* Antibiotics, β-lactams, cephalosporins and penems

Herbert Ellern, *Consultant,* Matches

S.M. Ellerstein, *Thiokol Corporation,* Polysulfides under Polymers containing sulfur

Edward J. Elliott, *FMC Corporation,* Bleaching agents

John Elliott, *CIBA-GEIGY Corporation,* Dyes, reactive; Triphenylmethane and related dyes

J.J. Elliott, *Exxon Research and Engineering Company,* Composition under Petroleum

James K. Ely, *Inmont Corporation,* Inks

Richard B. Engdahl, *Battelle Memorial Institute,* Air pollution

Ralph W. Engstrom, *Consultant, RCA,* Photomultiplier tubes

Mary G. Enig, *University of Maryland,* Mineral nutrients

Stephen H. Erickson, *The Dow Chemical Company,* Salicylic acid and related compounds

Donald M. Ernst, *Thermacore, Inc.,* Heat pipe under Heat exchange technology

Richard A. Erth, *York Division, Borg-Warner Corp.,* Air conditioning

W.E. Eslyn, *United States Department of Agriculture,* Wood

Herbert H. Espy, *Hercules Incorporated,* Papermaking additives

Robert H. Essenhigh, *The Ohio State University,* Furnaces, fuel-fired

P. Ettmayer, *Technical University, Vienna,* Cemented carbides, Nitrides

James V. Evans, *Amoco Oil Company,* Asphalt

Alfred L. Everett, *United States Department of Agriculture,* Leather

James Fair, *University of Texas,* Sprays

J. Falbe, *Ruhrchemie AG,* Coal chemicals and feedstocks, gasification

James S. Falcone, Jr., *The PQ Corporation,* Fillers; Synthetic inorganic silicates under Silicon compounds

John C.C. Fan, *Massachusetts Institute of Technology,* Photovoltaic cells

J.P. Fanaritis, *Struthers Wells Corporation,* Heat transfer under Heat exchange technology

L.W. Fannin, *Texas Alkyls, Inc.,* Organometallics — σ-bonded alkyls and aryls

Hugh Farber, *Dow Chemical Company,* Solvents, industrial

D.F. Farkas, *University of Delaware,* Food processing

Russell E. Farris, *GAF Corporation, and Sandoz Colors & Chemicals,* Aminophenols; Azo dyes; Direct dyes under Thiazole dyes; Dyes, anthraquinone and Dyes, natural; Stilbene dyes; Xanthene dyes

B.P. Faulkner, *Allis Chalmers Corporation,* Size reduction

J. Philip Faust, *Olin Chemicals,* Water, treatment of swimming pools, spas, and hot tubs

Newton C. Fawcett, *University of Southern Mississippi,* Polyimides

R.H. Fay, *Dow Chemical, U.S.A.,* Antifreezes and deicing fluids

Stephen H. Feairheller, *United States Department of Agriculture,* Leather

M. Fefer, *Exxon Chemical Company,* Cyclopentadiene and dicyclopentadiene

Raymond Feinland, *Clairol, Inc.,* Hair preparations

J. Feinman, *United States Steel Corporation,* Iron by direct reduction

Peter W. Feit, *Leo Pharmaceutical Products,* Diuretics

M.J. Feldman, *Oak Ridge National Laboratory,* Special engineering for radiochemical plants under Nuclear reactors

Martin L. Feldman, *Tenneco Inc.,* Driers and metallic soaps

Kelvin H. Ferber, *Allied Chemical Corp.,* Benzidine and related biphenyldiamines

H.L. Fike, *The Sulphur Institute,* Sulfur compounds

A.F. Finelli, *The Goodyear Tire & Rubber Company,* Thermoplastic elastomers under Elastomers, synthetic

Joseph H. Finley, *FMC Corporation,* Bleaching agents

K. Thomas Finley, *State University College at Brockport,* Quinones

E.J. Finnegan, *Hershey Foods Corporation,* Chocolate and cocoa

F. Fischetti, Jr., *Fritzsche Dodge and Olcott, Inc.,* Flavors and spices

J.G. Fisher, *Eastman Kodak Co.,* Disperse dyes under Thiazole dyes

William B. Fisher, *Allied Corporation,* Caprolactam under Polyamides; Cyclohexanol and cyclohexanone

R.W. Fleming, *Warner-Lambert Company,* Histamine and histamine antagonists

Michael Flickinger, *National Cancer Institute,* Antibiotics, peptides; Fermentation

John B. Flynn, *The Gillette Company,* Aerosols

R.T. Foley, *The American University,* Corrosion and corrosion inhibitors

Robert L. Fowlkes, *Air Products and Chemicals, Inc.,* Lower aliphatic amines under Amines

D.W. Fox, *General Electric Company,* Polycarbonates

Stephen J. Fraenkel, *Technology Services, Inc.,* Packaging materials, industrial

Martin S. Frant, *Foxboro Analytical, a division of the Foxboro Corporation,* Ion-selective electrodes

Alan D. Freas, *Consultant,* Laminated wood-based composites

Leon D. Freedman, *North Carolina State University,* Antimony compounds; Arsenic compounds; Bismuth compounds

M.P. Freeman, *Dorr-Oliver Incorporated,* Filtration

I. Freimanis, *Crane Packing Company,* Packing materials

E.R. Freiter, *Dow Chemical U.S.A.,* Chlorophenols; Halogenated derivatives under Acetic acid

William G. French, *3M Company,* Fiber optics

A.L. Friedberg, *University of Illinois,* Enamels, porcelain or vitreous

Heinz Friedrich, *Degussa Company Wolfgang 6451,* Vitamins, nicotinamide and nicotinic acid (B_3)

Wilhelm Friedrich, *Universität Hamburg, Hamburg,* Vitamins, vitamin B_{12}

C.D. Frohning, *Ruhrchemie AG,* Coal chemicals and feedstocks gasification

H.H. Froning, *Amoco Production Company,* Enhanced oil recovery under Petroleum

Peter W. Fryth, *Hoffmann-La Roche, Inc.,* Vitamins, survey

Thomas Furia, *Intermark Corp.,* Food additives

D.D. Fussell, *Amoco Production Company,* Enhanced oil recovery under Petroleum

Milton Fytelson, *Sandoz Colors & Chemicals,* Pigments, organic

John F. Gall, *Philadelphia College of Textiles and Science,* Aluminum; Calcium; Hydrogen under Fluorine compounds, inorganic

James B. Gambrell, *Pravel, Gambrell, Hewitt, Kirk, Kimball & Dodge,* Patents, practice and management

Ted Gammon, *Diamond Shamrock Chemical Co.,* Defoamers

S.V. Gangal, *E.I. du Pont de Nemours & Co., Inc.,* Polytetrafluoroethylene; Fluorinated ethylene–propylene copolymers both under Fluorine compounds, organic

Ashit K. Ganguly, *Schering Corp.,* Oligosaccharides under Antibiotics

John Gannon, *CIBA-GEIGY Corporation,* Epoxy resins

Keith L. Gardner, *BFGoodrich Co.,* Poly(vinyl chloride) under Vinyl polymers

R.C. Gasman, *GAF Corporation,* Vinyl ether monomers and polymers under Vinyl polymers

M.F. Gautreaux, *Ethyl Corporation,* Synthetic processes under Alcohols, higher aliphatic

R.M. Gaydos, *Koppers Company, Inc.,* Naphthalene

J.E. Gearien, *University of Illinois at the Medical Center,* Cholinesterase inhibitors

Charles G. Gebelein, *Chemistry Department, Youngstown State University, and Pharmacology Department, Northeastern Ohio Universities, College of Medicine,* Prosthetic and biomedical devices

Samuel Gelfand, *Hooker Chemicals & Plastics Corp.,* Benzyl chloride, benzal chloride, benzotrichloride; Ring-chlorinated toluenes both under Chlorocarbons and chlorohydrocarbons

R.B. Gengelbach, *Celanese Chemical Company, Inc.,* Methanol

H.R. Gerberich, *Celanese Chemical Company, Inc.,* Formaldehyde

William Germain, *Vulcan Materials Co.,* Detinning under Tin and tin alloys

Robert Geyer, *Harvard University,* Blood-replacement preparations

Dale S. Gibbs, *Dow Chemical U.S.A.,* Vinylidene chloride and poly(vinylidene chloride)

Harry W. Gibson, *Xerox Corporation,* Polymers, conductive

George E. Gifford, *University of Florida,* Chemotherapeutics, antiviral

Kevin E. Gilbert, *Air Products and Chemicals, Inc.,* Diaminotoluenes under Amines — Aromatic amines

Melvin H. Gitlitz, *M&T Chemicals, Inc.,* Tin compounds

Alfred C. Glatz, *General Foods Corporation,* Thermoelectric energy conversion

Gerald L. Goe, *Reilly Tar & Chemical Corporation,* Pyridine and pyridine derivatives

D.W. Goheen, *Crown Zellerbach,* Lignin

Leon Goldman, *Lederle Laboratories, American Cyanamid Co.,* Cardiovascular agents

K.P. Goodboy, *Aluminum Company of America,* Aluminum oxide (Alumina) under Aluminum compounds

Murray Goodman, *University of California, San Diego,* Polypeptides

Richard M. Goodman, *American Cyanamid Co.,* Dispersants

Karl F. Graff, *The Ohio State University,* Welding

Harold Graham, *Thomas J. Lipton, Inc.,* Tea

Margaret H. Graham, *Exxon Research and Engineering Company,* Information retrieval

James R. Grant, *Mytec, Inc.,* Paper under Recycling

Allan M. Green, *New England Nuclear Corporation,* Radioactive drugs

N.R. Greening, *Portland Cement Association,* Cement

John S. Greer, *MSA Research Corporation,* Potassium

G.K. Greminger, Jr., *Dow Chemical U.S.A.,* Cellulose derivatives, ethers

Warren R. Griffin, *Air Force Materials Laboratory,* Perfluoroalkylene triazines under Fluorine compounds, organic

William C. Griffin, *ICI United States, Inc.,* Emulsions

Charles O. Grigsby, *Los Alamos Scientific Laboratory,* Geothermal energy

Ralph E. Grim, *University of Illinois,* Uses under Clays

J.M. Grisar, *Merrell-National Laboratories,* Histamine and histamine antagonists

Louis T. Groet, *Badger, B.V.,* Acrylonitrile

Irwin Gruverman, *New England Nuclear Corporation,* Radioactive drugs

Karl Gschneidner, *Iowa State University,* Rare-earth elements

Hiram Gu, *University of California, Los Angeles,* Primary cells under Batteries and electric cells, primary

Richard A. Guenthner, *3M Company,* Fluorinated higher carboxylic acids; Perfluoroalkane sulfonic acid under Fluorine compounds, organic

R.A. Guest, *James Robinson & Company, Ltd.,* Sulfur dyes

William Gump, *Consultant,* Disinfectants and antiseptics

Zeev Gur-Arieh, *Monsanto Textiles Company,* Antistatic agents

P.O. Haddad, *Dow Chemical U.S.A.,* Magnesium and magnesium alloys

Robert Haddon, *Bell Laboratories,* Semiconductors, organic

Malcom R. Hadler, *Sorex Limited,* Poisons, economic

K.H. Häerdtl, *Philips Forschungslaboratorium Aachen GmbH,* Ferroelectrics

Earl R. Hafslund, *E.I. du Pont de Nemours & Co., Inc.,* Distillation

H.J. Hagemeyer, *Texas Eastman Company,* Acetaldehyde

C. Hagopian, *The Badger Co., Inc.,* Styrene

Gerald J. Hahn, *General Electric Company,* Design of experiments

Robert W. Haisty, *Texas Instruments, Inc.,* Digital displays

Dan Halacy, *Solar Energy Research Institute,* Solar energy

H.S. Halbedel, *The Harshaw Chemical Company,* Ammonium; Boronfluoroboric acid and other fluoroborates; Lithium; Magnesium; Potassium under Fluorine compounds, inorganic

Carl W. Hall, *Washington State University*, Milk and milk products

Richard E. Hall, *FMC Corporation*, Peroxides and peroxy compounds, inorganic

W.L. Hallbauer, *Celanese Chemical Company*, Methanol

Gerald M. Halpern, *Exxon Research and Engineering Company*, Fusion energy

F. Halverson, *American Cyanamid Company*, Flocculating agents

G.E. Ham, *Dow Chemical U.S.A.*, Imines, cyclic

M. Hamell, *General Foods Corp.*, Coffee

Wolf Hamm, *Unilever Limited*, Extraction, liquid–solid; Vegetables oils

John V. Hamme, *North Carolina State University*, Raw materials under Ceramics

Paul A. Hammes, *Merck and Co.*, Antioxidants and antiozonants

W.F. Hamner, *Monsanto Company*, Recreational surfaces

S.E. Handman, *The M.W. Kellog Company*, Piping systems

Bruce B. Hardman, *General Electric Company*, Silicones under Silicon compounds

Alexander P. Hardt, *Lockheed Missiles and Space Company, Inc.*, Pyrotechnics

Edgar E. Hardy, *California State Polytechnic University*, Phosgene

Thomas M. Hare, *North Carolina State University*, Thermal treatment; Properties both under Ceramics

Barbara Harley, *ALZA Corporation*, Pharmaceuticals, controlled release

Charles A. Harper, *Westinghouse Electric Corporation*, Embedding

Phillip V. Harper, Jr., *Texaco Chemical Company*, Amines, cyclic

Robert J. Harper, Jr., *United States Department of Agriculture*, Cyanoethylation

W. James Harper, *New Zealand Dairy Research Institute*, Synthetic and imitation dairy products

B.L. Harris, *Edgewood Arsenal*, Chemicals in war

Guy H. Harris, *Consultant*, Xanthates

J.F. Harris, *United States Department of Agriculture*, Wood

Thomas J. Harrison, *International Business Machines Corporation*, Computers

Winslow H. Hartford, *Belmont Abbey College*, Chromium compounds

Robert H. Hasek, *Tennessee Eastman Company*, Ketenes and related substances

D.M. Haskell, *Phillips Petroleum Co.*, Pentanes under Hydrocarbons C_1–C_6

Roger E. Hatton, *Monsanto Company*, Chlorinated biphenyls and related compounds under Chlorocarbons and chlorohydrocarbons

R.H. Hauge, *Rice University*, Direct fluorination under Fluorine compounds, organic

W.E. Haupin, *Aluminum Company of America*, Aluminum and aluminum alloys

William F. Hauschildt, *Amoco Chemicals Corp.*, Butylenes

C.F. Hauser, *Union Carbide Corporation*, Ethylene glycol and propylene glycol under Glycols

Mason Hayek, *E.I. du Pont de Nemours & Co., Inc.*, Waterproofing and water/oil repellency

Joseph S. Hayes, Jr., *American Kynol, Inc.*, Novoloid fibers

R.E. Hebeda, *CPC International*, Syrups

E.W. Heffern, *Amoco Production Company*, Enhanced oil recovery under Petroleum

Heinz Hefti, *CIBA-GEIGY Ltd.*, Brighteners, fluorescent

Charles M. Heinen, *Chrysler Corporation*, Exhaust control, automotive

W. Heise, *Mobay Chemical Corp.*, Dyes, application and evaluation

John F. Heiss, *Diamond Crystal Salt Co.*, Sodium chloride under Sodium compounds, sodium halides

C.W. Heitsch, *Monsanto Company*, Phosphorus compounds

C.S. Helling, *Agricultural Research Service, USDA*, Soil chemistry of pesticides

Richard A. Helmuth, *Portland Cement Association*, Cement

Paul J. Henkels, *United States Gypsum Co.*, Calcium sulfate under Calcium compounds

Joseph P. Henry, *Union Carbide Corporation*, Ethylene oxide

Frank N. Hepburn, *U.S. Department of Agriculture*, Yeast-raised products under Bakery processes and leavening agents

William F. Herbes, *American Cyanamid Co.*, Amino resins and plastics

David C. Herting, *Eastman Kodak Company*, Vitamins, vitamin E

L.G. Hess, *Union Carbide Corporation*, Acrolein and derivatives

C.S. Hickey, *Monsanto Co.*, Sorbic acid

K.E. Hickman, *York Division, Borg-Warner Corporation*, Refrigeration

Eric S. Hill, *Fiber Industries, Inc.*, Polyester fibers

H. Wayne Hill, Jr., *Phillips Petroleum Company*, Poly(phenylene sulfide) under Polymers containing sulfur

John H. Hillard, *Mountain Fuel Supply Company*, Gas, natural

William B. Hillig, *General Electric Co.*, Composite materials; High temperature composites

Arnold L. Hirsch, *A. L. Laboratories, Inc.*, Vitamins, vitamin D

Joseph J. Hlavka, *Lederle Laboratories Division, American Cyanamid Company*, Tetracyclines under Antibiotics

James P. Hoare, *General Motors Research Laboratories*, Electrolytic machining methods

Charles C. Hobbs, *Celanese Chemical Company, Inc.*, Hydrocarbon oxidation

P.H. Hobson, *Monsanto Triangle Park Development Center, Inc.*, Acrylic and modacrylic fibers

Seymore Hochberg, *E.I. du Pont de Nemours & Co., Inc.*, Coatings, industrial

Melvern C. Hoff, *Amoco Chemicals Corporation*, Butylenes; Toluene

Harold L. Hoffman, *Gulf Publishing Company*, Petroleum products under Petroleum

Stanley Hoffman, *Union Carbide Corporation*, Transportation

John E. Hogan, *The Okonite Co.*, Wire and cable coverings under Insulation, electric

J. Paul Hogan, *Phillips Petroleum Company*, Linear (high density) polyethylene under Olefin polymers

R.L. Hoglund, *Union Carbide Nuclear Division*, Diffusion separation methods

Edward Hohmann, *California State Polytechnic Institute*, Heat exchange technology, network synthesis

Arnold J. Hoiberg,[‡] *Manville Corporation*, Roofing materials

Allan G. Holcomb, *3M Company*, Fluorinated elastomers under Elastomers, synthetic

H. Stuart Holden, *Diamond Shamrock Technologies, S.A.*, Metal anodes

M.L. Hollander, *ASARCO Incorporated*, Cadmium and cadmium alloys

J.F. Holohan, Jr., *Hercules Incorporated*, Hydrocarbons resins

Samuel N. Holter, *Koppers Company*, (Polyhydroxy)benzenes; Quinolines and isoquinolines

John R.E. Hoover, *Smith, Kline & French Laboratories*, β-Lactams under Antibiotics

John M. Hopkins, *Union Oil Company of California*, Oil shale

W.C. Hopkins, *Celanese Chemical Company, Inc.*, Formaldehyde

Alan S. Horn, *University of Groningen*, Neuroregulators

S.E. Horne, Jr., *The BFGoodrich Company*, Polyisoprene under Elastomers, synthetic

Samuel E. Horne, Jr., *Polysar Incorporated*, Elastomers, synthetic, ethylene–propylene rubber; Isoprene

Emanuel Horowitz, *The Johns Hopkins University*, Plastics testing

W.D. Horst, *Hoffmann-LaRoche & Co., Inc.*, Psychopharmacological agents

Eugene V. Hort, *GAF Corporaton*, Acetylene-derived chemicals; Pyrrole and pyrrole derivatives; Vinyl ether monomers and polymers; *N*-Vinyl monomers and polymers both under Vinyl polymers

Mack Horton, *Union Oil Company of California*, Oil shale

Brian Horwood, *Amoco Chemicals Corporation*, Phthalic acids and other benzenepolycarboxylic acids

Fred H. Hoskins, *Louisiana State University*, Food toxicants, naturally occurring

William J. Houlihan, *Sandoz Inc.*, Hypnotics, sedatives, anticonvulsants

J.L. Howard, *United States Department of Agriculture*, Wood

William L. Howard, *Dow Chemical U.S.A.*, Chelating agents

H.E. Howe,[‡] *ASARCO Inc.*, Bismuth and bismuth alloys; Lead

E.D. Howell, *The Carborundum Co.*, Silicon carbides under Carbides

W.R. Howell, *Dow Chemical U.S.A.*, Epoxy resins

Timothy E. Howson, *Columbia University*, Nickel and nickel alloys

H.E. Høyrup, *Bryggeriforeningen*, Beer

C.H. Hoyt, *Crown Zellerbach,* Lignin

Judith K. Hruschka, *Ohio Northern University,* Blood, coagulants and anticoagulants

Robert B. Hudson, *Monsanto Company,* Phosphoric acids and phosphates

Jack Huebler,‡ *Institute of Gas Technology,* Gaseous fuels under Fuels, synthetic

E.O. Huffman, *Consultant,* Fertilizers

David W. Hughes, *Dow Chemical U.S.A.,* Malonic acid and derivatives

Benjamin C. Hui, *Alfa Products,* Thallium and thallium compounds

R.H. Hutchison, *Armstrong Cork Co.,* Cork

G. Frederick Hutter, *Inmont Corporation,* Leatherlike materials

John A. Hyatt, *Tennessee Eastman Co.,* Ketenes and related substances

Un K. Im, *Amoco Chemicals Corporation,* Butylenes

Muthiah N. Inbasekaran, *Ohio State University,* Pharmaceuticals, optically active

E.J. Inchalik, *Exxon Research and Engineering Company,* Oxo process

James W. Ingalls, *Arnold & Marie Schwartz College of Pharmacy and Health Sciences,* Chemotherapeutics, anthelmintic

M.P. Ingham, *Shell International Chemical Co., Ltd.,* Trialkylacetic acids under Carboxylic acids

May Inscoe, *USDA,* Herbicides

Harry Isacoff, *International Flavors and Fragrances Inc.,* Cosmetics

Otto Isler, *F. Hoffmann La Roche & Co.,* Vitamins, vitamin A

Larry A. Jackman, *Special Metals Corporation,* Metal treatments

M.G. Jacko, *Bendix Corporation,* Brake linings and clutch facings

Henry I. Jacoby, *McNeil Laboratories,* Gastrointestinal agents

Gerald M. Jaffe, *Hoffmann-LaRoche, Inc.,* Vitamins, ascorbic acid

V.A. Jagede, *Lederle Laboratories,* Vaccine technology

C.E. Jahnig, *Consultant,* Refinery processes, survey under petroleum

John C. Janka, *Institute of Gas Technology,* Gaseous fuels under Fuels, synthetic

Alfons Jankowski, *Ruhrkohle Oel und Gas GmbH,* Coal chemicals and feedstocks from hydrogenation

Donald B.G. Jaquiss, *General Electric Company,* Polyesters, thermoplastic

Ernest Jaworski, *Monsanto Company,* Genetic engineering

H.N. Jayaram, *Division of Cancer Treatment, National Cancer Institute, National Institutes of Health,* Enzymes, therapeutic

Ronald Jenkins, *Philips Electronics Instruments, Inc.,* X-ray technology

William R. Jenks, *E.I. du Pont de Nemours & Co., Inc.,* Hydrogen cyanide; Alkali metal cyanides; Alkaline earth metal cyanides all under Cyanides

James S. Johnson, Jr., *Oak Ridge National Laboratory,* Reverse osmosis

M. Ross Johnson, *Pfizer Central Research,* Opioids, endogenous

Paul R. Johnson, *E.I. du Pont de Nemours & Co., Inc.,* Chloroprene under Chlorocarbons and chlorohydrocarbons; Chlorosulfonated polyethylene; Neoprene both under Elastomers, synthetic

Richard L. Johnson, *E.I. du Pont de Nemours & Co., Inc.,* Tetrafluoroethylene co-polymers with ethylene; Tetrafluoroethylene co-polymers with perfluorovinyl ethers both under Fluorine compounds, organic

Robert W. Johnson, Jr., *Union Camp Corporation,* Branched-chain acids; Fatty acid from tall oil both under Carboxylic acids

Keith Johnston, *University of Texas, Austin,* Supercritical fluids

Charles R. Jokel, *Bolt, Beranek & Newman,* Insulation, acoustic

Otakar Jonas, *Westinghouse Electric Co.,* Steam

I.C. Jones, Jr., *Dow Chemical U.S.A.,* Chlorine and sodium hydroxide under Alkali and chlorine products

Martha B. Jones, *Royal Crown Cola Company,* Carbonated beverages

Walter Josten, *Dynamit Nobel A.G.,* Alkoxides, metal

B.R. Joyce, *Union Carbide Corp.,* Carbon and artifical graphite under Carbon

Thomas H. Jukes, *University of California, Berkeley,* Choline

P.W. Juneau, Jr., *General Electric Company,* Ablative materials

Charles D. Kalfadelis, *Exxon Research and Engineering Company,* Liquid fuels under Fuels, synthetic

Herbert S. Kalish, *Adamas Carbide Corporation,* Cemented carbides

Vasanth Kamath, *Pennwalt Corporation,* Initiators

Karl Kamman, *Keil Chemical Division, Ferro Corporation,* Sulfurizaton and sulfur-chlorination

Che-I-Kao, *Dow Chemical U.S.A.,* Chlorinated benzenes under Chlorocarbons and chlorohydrocarbons

Martin L. Kaplan, *Bell Laboratories,* Semiconductors, organic

Reuben H. Karol, *Rutgers University,* Chemical grouts

Joseph J. Katz, *Argonne National Laboratory,* Deuterium and tritium

Leonard I. Katzin, *Consultant,* Thorium and thorium compounds

Otto S. Kauder, *Argus Chemical Corporation,* Thioglycolic acid

Percy P. Kavasmaneck, *Union Carbide Corportion,* Ethanol

Antoine Kawam, *The Gillette Company,* Aerosols

P.C. Kearney, *Agricultural Research Service, USDA,* Soil chemistry of pesticides

Donald E. Keeley, *General Electric Co.,* Ethers

Henry J. Kehe, *The BFGoodrich Company,* Diarylamines under Amines —Aromatic amines

S.L. Keil, *Dow Chemical U.S.A.,* Tetrachloroethylene under Chlorocarbons and chlorohydrocarbons

Fred A. Keimel, *Bell Telephone Laboratories,* Adhesives

F.W. Keith, Jr., *Pennwalt Corp.,* Centrifugal separation

C.S. Keller, *Monsanto Co.,* Sorbic acid

W.D. Keller, *University of Missouri,* Survey under Clays

W.J. Kelly, *The Goodyear Tire & Rubber Company,* Polybutadiene under Elastomers, synthetic

Joyce C. Kern, *Glycerine Producers Association,* Glycerol

Melvin H. Keyes, *Owen-Illinois, Inc.,* Enzymes, immobilized

Guy Kieckhefer, *Borg Textile Corporation,* Furs, synthetic

Richard Kieffer,‡ *Technical University, Vienna,* Survey; Industrial heavy-metal carbides, Cemented carbides all under Carbides; Nitrides

Frank Kienzle, *F. Hoffmann La Roche & Co.,* Vitamins, vitamin A

T.K. Kim, *GTE Products Corporation,* Tungsten compounds

Y.K. Kim, *Dow Corning Corporation,* Polyfluorosilicones under Fluorine compounds, organic

Edward L. Kimmel, *Day-Glo Color Corp.,* Luminescent materials, fluorescent pigments (daylight)

Benjamin B. Kine, *Rohm and Haas Company,* Methacrylic polymers; Survey under Acrylic ester polymers

J.C. King, *Mobay Chemical Corp.,* Dyes, application and evaluation

Robert J. King, *United States Steel Corporation,* Steel

Lawrence S. Kirch, *Rohm and Haas Company,* Methacrylic acid and derivatives

J.R. Kirchner, *E.I. du Pont de Nemours & Co., Inc.,* Hydrogen peroxide

Bradley S. Kirk, *AIRCO, Inc.,* Helium-group gases

Richard O. Kirk, *The Dow Chemical Company,* Propylene oxide

William S. Kirk, *U.S. Bureau of Mines,* Cobalt and cobalt alloys

Truman Kirkpatrick, *The Sherwin-Williams Company,* Barium compounds

Isidor Kirshenbaum, *Exxon Research and Engineering Co.,* Butadiene; Oxo process

Yuri V. Kissin, *Gulf Oil Chemical Company,* Polymers of higher olefins under Olefin polymers

Donald L. Klass, *Institute of Gas Technology,* Alcohol fuels; Fuels from biomass

Daniel L. Klayman, *Walter Reed Army Institute of Research,* Radioprotective agents

Axel Kleemann, *Degussa Company,* Vitamins, nicotinamide and nicotinic acid (B_3)

A. Klein, *Lehigh University,* Latex technology

H.G. Klein, *Klein Medical International,* Survey under Antibiotics

Theodor N. Kleinert, *Pulp and Paper Research Institute of Canada,* Cellulose

P. Klinkowski, *Dorr-Oliver, Inc.,* Ultrafiltration

E.A. Knaggs, *Stepan Chemical Company,* Sulfonaton and sulfation

W.A. Knepper, *United States Steel Corp.,* Iron

Ludwig Kniel, *Combustion Engineering, Inc.,* Ethylene

Donald Knittel, *Cabot Corporation,* Titanium and titanium alloys

P. Koch, *The Badger Co., Inc.,* Styrene

R.K. Kochar, *Gulf Oil Chemical Company,* Polymers of higher olefins under Olefin polymers

J.J. Koenig, *Aluminum Company of America,* Aluminum oxide (Alumina) under Aluminum compounds

Truman P. Kohman, *Carnegie-Mellon University,* Radioactivity, natural

E.D. Kolb, *Bell Laboratories,* Synthetic quartz crystals under Silica

James M. Kolb, *Diamond Shamrock Corporation,* Metal anodes

Georg Kölling, *Bergbau-Forschung GmbH,* Coal chemicals and feedstocks, introduction and from hydrogenation

R. Komanduri, *General Electric Company,* Tool materials

Ernest I. Korchak, *Scientific Design Company, Inc., a division of the The Halcon SD Group, Inc.,* Process research and development

Gabriel Kornis, *The Upjohn Company,* Pyrazoles, pyrazolines, pyrazolones

T.M. Korzekwa, *The Carborundum Co.,* Silicon carbides under Carbides

S.G. Kosinski, *AT & T Laboratories,* Fiber optics

K.J. Kowal, *Lederle Laboratories, American Cyanamid Co.,* Vaccine technology

John Kraljic, *Allied Chemical Corp.,* Oxalic acid; Sodium nitrite under Sodium compounds

Hans Krassig, *Lenzing, A.G.,* Cellulose

Paul E. Krieger, *Pravel, Gambrell, Hewitt, Kirk, Kimball & Dodge,* Patents, practice and management

John H. Kroehling, *E.I. du Pont de Nemours & Co., Inc.,* Exhaust control, industrial

Thaddeus Kroplinski, *CasChem. Inc.,* Castor oil

Gerald A. Krulik, *Borg-Warner Corporation,* Electroless plating

R.C. Krutenat, *Exxon Research and Engineering Company,* Survey under Metallic coatings

Thomas J. Kucera, *Consultant,* Reprography

R. Kuehni, *Mobay Chemical Corp.,* Dyes, application and evaluation

Eugene J. Kuhajek, *Morton Salt,* Sodium chloride under Sodium compounds, sodium halides

Karel Kulp, *American Institute of Baking,* Yeast-raised products (under Bakery processes and leavening agents)

Charles J. Kunesh, *Pfizer Inc.,* Barium; Calcium and calcium alloys

S.M. Kuntz, *The Carborundum Co.,* Silicon carbides under Carbides

C.M. Kuo, *Tennessee Eastman Company,* Cellulose derivatives, esters

A.N. Kurtz, *Union Carbide Corporation,* Acrolein and derivatives

Roger N. Kust, *Kennecott Copper Corporation,* Copper compounds

L.J. Kuzma, *The Goodyear Tire & Rubber Company,* Polybutadiene under Elastomers, synthetic

J.A. Kwas, *Struthers Wells Corporation,* Heat transfer under Heat exchange technology

Mitchell A. LaBoda, *General Motors Research Laboratories,* Electrolytic machining methods

Martin Laborde, *CODELCO,* Sodium nitrate under Sodium compounds

E.F. Labuda, *Bell Telephone Laboratories, Incorporated,* Integrated circuits

R.J. Lagow, *University of Texas,* Direct fluorination under Fluorine compounds, organic

Fong M. Lai, *Lederle Laboratories,* Cardiovascular agents

Leonard Lamberson, *Wayne State University,* Materials reliability

John B. Lambert, *Fansteel, Inc.,* Tantalum and tantalum compounds

Alexis B. Lamy, *Essochem Europe, Inc.,* Information retrieval

Richard D. Lanam, *Engelhard Corporation,* Platinum-group metals

John C. Lane, *Ethyl Corporation,* Gasoline and other motor fuels

George Langford, *Drexel University,* Hardness

Josef Langhoff, *Ruhrkohle Oel und Gas GmbH,* Coal chemicals and feedstocks from hydrogenation

H.J. Lanson, *Poly-Chem Resin Corporation,* Alkyd resins

E.R. Larsen, *Dow Chemical U.S.A.,* Halogenated flame retardants under Flame retardants

Leonard Laub, *Vision Three Inc.,* Recording disks

R.A. Laudise, *Bell Laboratories,* Synthetic quartz crystals under Silica

A.C. Lavanchy, *Pennwalt Corporation,* Centrifugal separation

Edward Lavin, *Monsanto Company,* Poly(vinyl acetal)s under Vinyl polymers

B.C. Lawes, *E.I. du Pont de Nemours & Co., Inc.,* Acetic acid derivatives, dimethylacetamide; Sulfuric and sulfurous esters

Barbara Lawrence, *Exxon Corporation,* Information retrieval

Robert W. Layer, *The BFGoodrich Company,* Diarylamines; Phenylenediamines both under Amines – Aromatic amines

Joseph R. Leblanc, Jr., *Pullman Kellogg,* Ammonia

J.J. Leddy, *Dow Chemical U.S.A.,* Chlorine and sodium hydroxide under Alkali and chlorine products

C. Michael Lederer, *Lawrence Berkeley Laboratory, University of California, Berkeley,* Isotopes

Stuart M. Lee, *Ford Aerospace & Communications Corp.,* Film deposition techniques; Xylylene polymers

Edward C. Leonard, *Humko Sheffield,* Dimer acids

Edward F. Leonard, *Columbia University,* Dialysis

L.J. Lefevre, *Dow Chemical U.S.A.,* Ion exchange

David J. Leggit, *Dow Chemical U.S.A.,* Chelating agents

Pedro A. Lehmann F., *Centro de Investigación y de Estudios Avanzados del I.P.N.,* Thyroid and antithyroid preparations

H.D. Leigh, *C-E Basic, Inc.,* Refractories

Charles H. Lemke, *University of Delaware,* Sodium and sodium alloys

George R. Lenz, *Searle Laboratories,* Steroids

Robert W. Lenz, *University of Massachusetts,* Polymerization mechanisms and processes

Richard H. Lepley, *Pfizer, Inc.,* Calcium carbonate under Calcium compounds

George Y. Lesher, *Sterling-Winthrop Research Institute,* Nalidixic acid and other quinolone carboxylic acids under Antibacterial agents, synthetic

Gerd Leston, *Koppers Company, Inc.,* Alkylphenols

C. Scott Letcher, *Petrolite Corporation,* Waxes

Edward K. Letzer, *Eastman Kodak Company,* Optical filters

Betty Lewis, *Cornell University,* Dietary fiber

P.J. Lewis, *The Badger Co., Inc.,* Styrene

W. Bennett Lewis, *Queen's University,* Introduction under Nuclear reactors

Choh Hao Li, *University of California, San Francisco,* Anterior-pituitary hormones under Hormones

Irwin Lichtman, *Diamond Shamrock Chemical Co.,* Defoamers

W.G. Lidman, *Kawecki Berylco Industries, Inc.,* Rubidium and rubidium compounds

Charles H. Lieb, *Paskus, Gordon & Hyman,* Trademarks and copyrights

Hillel Lieberman, *Betz Laboratories,* Technical service

Thomas A. Liederbach, *Eltech Systems Corporation,* Metal anodes

Eric Lifshin, *General Electric Co.,* Analytical methods

J.J. Ligi, *Texas Alkyls, Inc.,* Organometallics — σ-bonded alkyls and aryls

W. Lin, *Lederle Laboratories, American Cyanamid Co.,* Vaccine technology

Charles B. Lindahl, *Ozark-Mahoning Company, a subsidiary of the Pennwalt Corp.,* Introduction; Antimony; Arsenic; Barium; Germanium; Phosphorus; Tin; Titanium; Zinc under Fluorine compounds, inorganic

Victor Lindner, *ARRADCOM,* Explosives and propellants

Merle Lindstrom, *Phillips Research Center,* Sulfolanes and sulfones

John H. Litchfield, *Battelle Columbus Laboratories,* Foods, nonconventional

J.D. Litvay, *The Institute of Paper Chemistry,* Paper

Charles D. Livengood, *North Carolina State University,* Silk

Thomas B. Lloyd, *Gulf and Western National Resources Group,* Zinc and zinc alloys; Zinc compounds

Teh C. Lo, *Hoffmann-LaRoche Inc.,* Extraction, liquid – liquid

David J. Locker, *Kodak Research Laboratories,* Photography

Haines B. Lockhart, Jr., *Eastman Kodak Company,* Silver compounds

L.F. Lockwood, *Dow Chemical U.S.A.,* Magnesium and magnesium alloys

John W. Loeding, *Institute of Gas Technology,* Fuels, synthetic, gaseous fuels

Kurt L. Loening, *Chemical Abstracts Service,* Hydrocarbons, nomenclature; Nomenclature

G. Lohnert, *Ruetgers-Nease Chemical Company,* Coal chemicals and feedstocks, carbonization and coking

Ernest G. Long, *Manville Corporation,* Roofing materials

G. Gilbert Long, *North Carolina State University,* Antimony compounds; Arsenic compounds; Bismuth compounds

George M. Long, *Institute of Gas Technology*, Pipelines

J.C. Long, *Union Carbide Corp.*, Carbon and artificial graphite under Carbon

John W. Lotz, *IFI/Plenum Data Company*, Patents, literature

R.P. Louthan, *Phillips Petroleum Co.*, Thiols

Frederick A. Lowenheim,‡ *Consultant*, Electroplating

Loren D. Lower, *U.S. Borax Research Corp.*, Boron oxides, boric acid, and borates under Boron compounds

B.S. Lowry, *Dow Chemical U.S.A.*, Chlorine and sodium hydroxide under Alkali and chlorine products

F.E. Luborsky, *General Electric Co.*, Amorphous magnetic materials

Luciano Luciani, *Montedison SpA*, Propylene under Olefin polymers

Peter Luckie, *Penn State University*, Size separation

Robert Lukens, *American Society for Testing and Materials*, Conversion factors, abbreviations, and unit symbols; Units and conversion factors

John Lundberg, *Georgia Institute of Technology*, Rayon

Robert D. Lundberg, *Exxon Research and Engineering Company*, Ionomers

John T. Lutz, *Rohm and Haas Co.*, Epoxidation

Anthony M. Luxeder, *The BFGoodrich Co., Chemical Divsion*, Antioxidants and antiozonants

Jesse L. Lynn, Jr., *Lever Brothers Company*, Surfactants and detersive systems

W.L. Lyon, *Westinghouse Nuclear Fuels Division*, Fuel-element fabrication under Nuclear reactors

John W. Lyons, *National Bureau of Standards*, Flame retardants, an overview

John B. MacChesney, *Bell Laboratories, Inc.*, Fiber optics

W.S. MacGregor, *Crown Zellerbach Corp.*, Sulfoxides

M.B. MacInnis, *GTE Products Corporation*, Tungsten compounds

D.C. MacWilliams, *Dow Chemical U.S.A.*, Acrylamide

George MacZura, *Aluminum Company of America*, Aluminum oxide (Alumina) Aluminum halides and aluminum nitrate under Aluminum compounds

John H. Madaus, *MSA Research Corporation*, Potassium

S. Madhaven, *Pullman Kellogg*, Ammonia

Orville L. Mageli, *Lucidol Division, Pennwalt Corporation*, Peroxides and peroxy compounds, organic

A.J. Magistro, *BFGoodrich Co.*, Vinyl chloride under Vinyl polymers

D.S. Maisel, *Exxon Chemical Company*, Methane, ethane, and propane under Hydrocarbons C_1–C_6

Joseph Makhlouf, *PPG Industries*, Polyesters, unsaturated

Earl C. Makin, *Monsanto Chemical Intermediates Co.*, Clathration

E.W. Malcom, *The Institute of Paper Chemistry*, Paper

Alan K. Mallams, *Schering-Plough Corp.*, Macrolides and Polyenes, both under Antibiotics

Walter R. Mallarnee, *The Sherwin-Williams Company*, Paint and varnish removers

D.B. Malpass, *Texas Alkyls, Inc.*, Organometallics — σ-bonded alkyls and aryls

K.H. Mancy, *University of Michigan*, Water, analysis

B.G. Mandelik, *Pullman Kellogg*, Hydrogen

R. St. John Manley, *McGill University*, Cellulose

Robert M. Manyik, *Union Carbide Corporation*, Economic aspects, analysis, and uses under Acetylene

Merritt G. Marbach, *Shell Chemical Company*, Feedstocks

John L. Margrave, *Rice University*, Direct fluorination under Fluorine compounds, organic

Herman F. Mark, *Polytechnic Institute of New York*, Biopolymers

Paul Marsh, *Marsh-Eco-Service Co., Inc.*, Glass under Recycling

R.A. Marshall, *The Goodyear Tire & Rubber Company*, Thermoplastic elastomers under Elastomers, synthetic

Charles M. Marstiller, *Aluminum Company of America*, Aluminum halides and aluminum nitate under Aluminum compounds

Donald R. Martin, *University of Texas*, Boron under Fluorine compounds, inorganic

Edgar J. Martin, *Food and Drug Administration*, Chemotherapeutics, antiprotozoal

G.Q. Martin, *Shell Development Company*, Fluid mechanics

James Martin, *William Zinsser & Co.*, Shellac

Murray J. Martin, *Oak Ridge National Laboratory*, Radioisotopes

Charles J. Masur,‡ *Lederle Laboratories, American Cyanamid Company*, Chemotherapeutcs, antimitotic

L.R. Matricardi, *Union Carbide Corporation*, Manganese and manganese alloys

Christina Matter-Müller, *Swiss Federal Institute of Technology*, Water, properties

J.W. Mausteller, *MSA Research Corporation*, Oxygen-generating systems; Potassium

Ivo Mavrovic, *Consultant*, Urea

R.F. Mawson, *Colorado State University*, Meat products

Donald R. May, *Cyanamid Canada Inc.*, Cyanamides

Samuel Maya, *Beecham Products*, Sodium nitrate under Sodium compounds

C. Glen Mayhall, *Medical College of Virginia*, Chemotherapeutics, antimycotic and antirickettsial

D.J. Maykuth, *Tin Research Institute, Inc.*, Tin and tin alloys

Charles J. Mazac, *PPG Industries*, Iodine and iodine compounds

Robert Mazur, *GD Searle and Co.*, Sweeteners

James V. McArdle, *University of Maryland*, Iron compounds

W. B. McCormack, *E.I. du Pont de Nemours & Co., Inc.*, Organolead compounds under Lead compounds; Sulfuric and sulfurous esters

Paul Y. McCormick, *E.I. du Pont de Nemours & Co., Inc.*, Drying

A.L. McCrary, *Dow Chemical U.S.A.*, Chelating agents

Bruce McDuffie, *State University of New York at Binghamton*, Trace and residue analysis

Thomas E. McEntee, *Arapahoe Chemicals, Inc.*, Grignard reaction

Frederick J. McGarry, *Massachusetts Institute of Technology*, Laminated and reinforced plastics

R.A. McGinnis, *Consultant*, Beet sugar under Sugar

Vincent D. McGinnis, *Battelle Columbus Laboratory*, Radiation curing

John N. McGovern, *The University of Wisconsin*, Fibers, vegetable

James E. McGrath, *Virginia Polytechnic Institute and State University*, Survey under Elastomers, synthetic

W.F. McIlhenny, *Dow Chemical U.S.A*, Ocean raw materials

J.M. McIntire, *Tennessee Eastman Company, Division of Eastman Kodak Company*, 2-Cyanoacrylic ester polymers under Acrylic ester polymers

Ron E. McKeighen, *Krautkramer-Branson, Inc.*, Ultrasonics, low power

W.J. McKillip, *The Quaker Oats Company*, Furan derivatives

John R. McLean, *Warner-Lambert/Parke-Davis*, Epinephrine and norepinephrine

Joseph H. McMakin, *Air Products and Chemicals, Inc.*, Lower aliphatic amines under Amines

Robert C. McMaster, *The Ohio State University*, Nondestructive testing

Donald McNeil, *Technical consultant*, Tar and pitch

W.C. McNeill, Jr., *Dow Chemical U.S.A.*, Trichloroethylene under Chlorocarbons and chlorohydrocarbons

A.L. McPeters, *Monsanto Triangle Park Development Center, Inc.*, Acrylic and modacrylic fibers

James B. McPherson, *American Cyanamid Company*, Sutures

Wayne A. McRae, *Research Ionics, Inc.*, Electrodialysis

Whitney Mears, *Allied Chemical Corporation*, Sulfur-sulfur fluorides under Fluorine compounds, inorganic

J.T. Meers, *Union Carbide Corp.*, Carbon and artificial graphite under Carbon

David Meidar, *University of Southern California, Los Angeles*, Friedel-Crafts reactions

M.T. Melchior, *Exxon Research and Engineering Company*, Composition under Petroleum

P.H. Merrell, *Mallinckrodt, Inc.*, Sodium iodide under Sodium compounds, sodium halides

Edward G. Merritt, *Pfizer Inc.*, Citric acid

Dayal T. Meshri, *Ozark-Mahoning Company, a subsidiary of the Pennwalt Corp.*, Introduction; Antimony; Cobalt; Copper; Iron; Lead; Mercury; Nickel; Silver; Tin; Titanium; Zirconium under Fluorine compounds, inorganic

Robert L. Metcalf, *University of Illinois at Urbana-Champaign*, Insect control technology

Walter C. Meuly, *Rhodia Inc.*, Coumarin

Gerald Meyer, *U.S. Geological Survey*, Water, sources and utilization

William F. Michne, *Sterling-Winthrop Research Institute*, Analgesics, antipyretics, and anti-inflammatory agents

Olaf Mickelsen, *Consultant*, Wheat and other cereal grains

William J. Middleton, *E.I. du Pont de Nemours & Co., Inc.*, Hexafluoroacetone under Fluorine compounds, organic

Craig S. Miller, *Krautkramer-Branson, Inc.*, Ultrasonics, low power

Duane D. Miller, *Ohio State University*, Anti-asthmatic agents

F.M. Miller, *Portland Cement Association*, Cement

H.C. Miller, *Super-Cut, Inc.*, Diamond, natural under Carbon

R.B. Miller, *United States Department of Agriculture*, Wood

Richard M. Miller, *Wellman Thermal Systems Corp.*, Resistance furnaces under Furnaces, electric

W.C. Miller, *Manville Corporation*, Refractory fibers

Norman Milleron, *Neven Corp.*, Vacuum technology

Barton Milligan, *Air Products and Chemicals, Inc.*, Diaminotoluenes under Amines — Aromatic amines

G. Alex Mills, *University of Delaware*, Catalysis

E.F. Milner, *Cominco Ltd.*, Indium and indium compounds

James Minor, *U.S. Forest Products Laboratory*, Pulp

Paul E. Minton, *Union Carbide Corporation*, Heat transfer media other than water under Heat exchange technology

Frank P. Missell, *Universidade de São Paulo*, Superconducting materials

Thomas F. Mitchell, *Darling and Company*, Glue

Tom Mix, *Merix Corporation*, Separations, low energy

Kenneth Mjos, *Jefferson Chemical Co., Inc.*, Cyclic amines (under Amines)

Michel F. Molaire, *Eastman Kodak Company*, Photoreactive polymers

D.P. Montgomery, *Phillips Petroleum Co.*, Hydrocarbons, survey

George E. Moore, *Consultant*, Burner technology

Ralph Moore, *E.I. du Pont de Nemours & Co., Inc.*, Organolead compounds under Lead compounds

Sewell T. Moore, *American Cyanamid Co.*, Amino resins and plastics

William M. Moore, *Dow Chemical U.S.A.*, Methylenedianiline (under Amines — Aromatic amines)

Marguerite K. Moran, *M&T Chemicals, Inc.*, Tin compounds

Herbert Morawetz, *Polytechnic Institute of New York*, Polyelectrolytes

G. Morel, *Université de Paris-Nord*, Membrane technology

Theresa A. Moretti, *American Hoechst Corporation*, Acetamide; Acetyl chloride both under Acetic acid

Booker Morey, *Waste Energy Technical Corp.*, Dewatering

Paul Morgan, *Consultant*, Dicarboxylic acids

G.A. Morneau, *BorgWarner Chemicals*, ABS resins under Acrylonitrile polymers

F.R. Morral, *Consultant*, Cobalt compounds

J.D. Morris, *The Dow Chemical Company*, Acrylamide polymers

Thomas E. Morris, *Dow Chemical U.S.A.*, Ethyl chloride under Chloro carbons and chlorohydrocarbons

Douglas R. Morton, Jr., *The Upjohn Company*, Prostaglandins

Frank H. Moser, *Consultant*, Phthalocyanine compounds

R.A. Mount, *Monsanto Company*, Maleic anhydride, maleic acid, and fumaric acid

James C. Moyer, *New York State Agricultural Experiment Station*, Fruit juices

Paul E. Muehlberg, *Dow Chemical U.S.A.*, Chemicals from brine

Edward E. Mueller, *Alfred University*, Colorants for ceramics

James A. Mullendore, *GTE Product Corp.*, Tungsten and tungsten alloys

J.W. Mullin, *University College, London*, Crystallization

Richard M. Mullins, *The Dow Chemical Company*, Alkanolamines from olefin oxides and ammonia under alkanolamines; Hydroxybenzaldehydes

H.T. Mulryan, *Cyprus Industrial Minerals*, Talc

Charles G. Munger, *Consultant*, Coatings, resistant

Peter Murray, *Westinghouse Electric Corporation*, Fast breeder reactors under Nuclear reactors

Richard J. Nadolsky, *Armak Company*, Amine oxides

David Nagel, *Naval Research Laboratory*, Plasma technology

J.J. Nahm, *Shell Development*, Drilling fluids under Petroleum

Fred Naider, *The College of Staten Island*, Polypeptides

Thomas E. Nappier, *The Harshaw Chemical Company*, Ammonium; Boron-fluoroboric acid and fluoroborates; Lithium; Magnesium; Potassium under Fluorine compounds, inorganic

Claude H. Nash, *Smith, Kline & French Laboratories*, β-Lactams under Antibiotics

Leonard I. Nass, *Consultant*, Heat stabilizers

Frank Naughton, *NL Industries*, Castor oil

Carl Nebel, *PCI Ozone Corp.*, Ozone

Kenneth R. Neel, *American Gilsonite Co.*, Gilsonite

Richard Neerken, *The Ralph M. Parsons Company*, Pumps

R.H. Neisel, *Johns-Manville Sales Corporation*, Insulation, thermal

D.L. Nelson, *The Dow Chemical Company*, Acetone

G.D. Nelson, *Monsanto Industrial Chemicals Co.*, Chloramines and bromamines

Joseph W. Nemec, *Rohm and Haas Company*, Acrylic acid and derivatives; Methacrylic acid and derivatives

Nelson L. Nemerow, *Consulting engineer*, Wastes, industrial

S.P. Nemphos, *Monsanto Commercial Products Co.*, Barrier polymers

David I. Netting, *ARCO Chemical Co.*, Fillers

E.L. Neu, *Grefco Inc.*, Diatomite

John B. Newkirk, *University of Denver*, Cobalt and cobalt alloys

Daniel J. Newman, *Barnard and Burk, Inc.*, Nitric acid

David S. Newsome, *Pullman Kellogg*, Hydrogen

C.L. Newton, *Air Products and Chemicals, Inc.*, Cryogenics

Robert A. Newton, *Dow Chemical U.S.A.*, Propylene oxide polymers and higher 1,2-epoxides under Polyethers

William E. Newton, *Charles F. Kettering Research Laboratory*, Nitrogen fixation

Paul P. Nicholas, *The B.F. Goodrich Co., Corporate Research*, Antioxidants and antiozonants

Louis G. Nickell, *Velsicol Chemical Corporation*, Plant-growth substances

Henry Nielsen, *Teledyne Wah Chang Albany*, Zirconium and zirconium compounds

Ralph H. Nielsen, *Teledyne Wah Chang Albany*, Hafnium and hafnium compounds; Zirconium and zirconium compounds

R.H. Nielsen, *Phillips Petroleum Co.*, Butanes under Hydrocarbons C_1–C_6

A.H. Nishikawa, *Hoffmann-LaRoche Inc.*, Chromatography, affinity

A.A. Nishimura, *Ampex Corporation*, Magnetic tape

Manfred G. Noack, *Olin Corporation*, Chlorine dioxide, chlorous acid, and chlorites under Chlorine oxygen acids and salts

John Norell, *Phillips Petroleum Co.*, Thiols

Jean Northcott, *Allied Chemical Corp.*, Aniline and its derivatives (under Amines — Aromatic amines)

D.A. Novak, *Monsanto Co.*, Air pollution control methods

Ronald W. Novak, *Rohm and Haas Company*, Survey under Acrylic ester polymers; Methacrylic polymers

Milton Nowak, *Troy Chemical Corporation*, Mercury compounds

M.L. Nussbaum, *Stepan Chemical Company*, Sulfonation and sulfation

Dan Oakland, *Whiting Corporation*, Arc furnaces under Furnaces, electric

John Oberteuffer, *Sala Magnetics, Inc.*, Magnetic separation

Robert E. O'Brien, *New England Nuclear Corporation*, Radioactive tracers

Cynthia O'Callaghan, *Glaxo Labs, Glaxochem, Ltd.*, Antibiotics, β-lactams, cephalosporins and penems

T.R. O'Connor, *Portland Cement Association*, Cement

Heribert Offermanns, *Degussa Company*, Vitamins, nicotinamide and nicotinic acid (B_3)

Eilert A. Ofstead, *The Goodyear Tire & Rubber Company*, Polypentenamers under Elastomers, synthetic

Robert Ohm, *R.T. Vanderbilt Co.*, Poly(bicycloheptene) and related polymers

Yoshikazu Oka, *Takeda Chemical Industries, Ltd., Japan*, Vitamins, thiamine (B_1)

Olagoke Olabisi, *Union Carbide Corporation*, Polyblends

George A. Olah, *University of Southern California, Los Angeles*, Friedel-Crafts reactions

James Y. Oldshue, *Mixing Equipment Co., Inc.*, Mixing and blending

A.D. Olin, *Toms River Chemical Corporation,* Dyes and dye intermediates

J. O'M. Brockris, *Texas A&M University,* Hydrogen energy

J.V. Orle, *Crown Zellerbach Corporation,* Sulfoxides

T.A. Orofino, *Monsanto Company,* Recreational surfaces

Clyde Orr, *Micrometrics Instrument Corp.,* Size measurement of particles

John M. Osepchuk, *Raytheon Company,* Microwave technology

Frederick S. Osmer, *Lever Brothers Co., Inc.,* Soap

Donald F. Othmer, *Polytechnic Institute of New York,* Azeotropic and extractive distillation; Engineering and chemical data correlation

Dennis C. Owsley, *Monsanto Company,* Photochemical technology

George W. Packowski, *Joseph E. Seagram & Sons, Inc.,* Beverage spirits, distilled

G.C. Paffenberger, *American Dental Association, Health Foundation Research Unit, National Bureau of Standards,* Dental materials

Richard A. Paluzzi, *Exxon Research and Engineering Company,* Pilot plants and microplants

Morton B. Panish, *Bell Laboratories,* Light-emitting diodes and semiconductor lasers

H.P. Panzer, *American Cyanamid Company,* Flocculating agents

Anthony J. Papa, *Union Carbide Corporation,* Isopropyl alcohol under Propyl alcohols; Ketones

Christos G. Papadopoulos, *Amoco Chemicals Corporation,* Propylene

Harold Papkoff, *University of California,* Anterior-pituitarylike hormones under Hormones

G.D. Parfitt, *Carnegie-Mellon University,* Dispersion of powders in liquids

J.W. Parker, *Shell International Chemical Co., Ltd.,* Trialkylacetic acids under Carboxylic acids

K.J. Parker, *Tate & Lyle, Ltd.,* Sugar derivatives under Sugar

P.D. Parker, *AMAX Base Metal Research and Development, Inc.,* Cadmium compounds

Clyde F. Parrish, *Indiana State University,* Solvents, industrial

Edward Paschke, *Hoechst AG, Frankfurt,* Ziegler process

Nancy R. Passow, *C-E Lummus,* Regulatory agencies

A.R. Patel, *National Cancer Institute,* Anesthetics

John A. Patterson, *International Atomic Energy Agency,* Nuclear fuel reserves under Nuclear reactors

Paul Patterson, *Cyanamid Canada Inc.,* Cyanamides; Guanidine and guanidine salts

Donald R. Paul, *University of Texas,* Membrane technology

John Paul, *Pedco Environmental, Inc.,* Rubber under Recycling

W.A. Pavelich, *Borg-Warner Chemicals,* ABS resins under Acrylonitrile polymers

Patrick H. Payton, *TRW,* Niobium and niobium compounds

William Pearl, *Lederle Laboratories, American Cyanamid Company,* Chemotherapeutics, antimitotic

W.L. Pearl, *NWT Corporation,* Water chemistry of light-water reactors under Nuclear reactors

Eugene Pearlman, *ESB Technology Co.,* Secondary cells, alkaline under Batteries and electric cells, secondary

A. David Pearson, *Bell Laboratories,* Fiber optics

Evaristo Peggion, *University of Padua,* Polypeptides

Peter L. Pellett, *University of Massachusetts,* Proteins

Fred M. Peng, *Monsanto Co.,* Survey and styrene-acrylonitrile copolymers under Acrylonitrile polymers

J.Y. Penn, *Hercules Incorporated,* Hydrocarbon resins

P. Andrew Penz, *Texas Instruments Inc.,* Digital displays

R.J. Penzenstadler, *The Dow Chemical Company,* Acrylamide polymers

Michael Perch, *Koppers Co., Inc.,* Carbonization under Coal conversion processes

Edward G. Perkins, *University of Illinois,* Analysis and standards under Carboxylic acids

David Perlman,‡ *University of Wisconsin,* Fermentation; Peptides under Antibiotics

Theodore P. Perros, *George Washington University,* Forensic chemistry

Harry Perry, *Resources for the Future,* Fuels, survey

R.H. Perry, *Ampex Corporation,* Magnetic tape

K.J. Persak, *E.I. du Pont de Nemours & Co., Inc.,* Acetal resins

E.M. Peters, *Mallinckrodt, Inc.,* Sodium iodide under Sodium compounds, sodium halides

Edward N. Peters, *General Electric,* Inorganic high polymers

Richard A. Peters, *The Procter & Gamble Company,* Survey and Natural alcohols manufacture under Alcohols, higher aliphatic

Timothy V. Peters, *Consultant,* Fibers, elastomeric

D.D. Peterson, *Nuclepore Corporation,* Particle-track etching

David T. Peterson, *Iowa State University,* Electromigration

William R. Peterson, Jr., *Petrarch Systems, Inc.,* Silanes under Silicon compounds

David M. Petrick, *American Cyanamid Co.,* Veterinary drugs

Vladimir Petrow, *Consultant,* Adrenal-cortical hormones; Sex hormones both under Hormones

R.C. Pettersen, *United States Department of Agriculture,* Wood

James R. Pfaffin, *Consultant,* Water, reuse; Water, sewage

R.E. Phillips, *Mobay Chemical Corp.,* Dyes, application and evaluation

Odgen R. Pierce, *Dow Corning Corporation,* Fluorine compounds, organic, introduction

Percy E. Pierce, *PPG Industries, Inc.,* Rheological measurements

W.J. Pierce, *3M Center,* Electroplating

William G. Pinkstone, *Consultant,* Abrasives

E.L. Piper, *Union Carbide Corporation,* Carbon and artificial graphite under Carbon

Robert K. Pitman, *U.S. Department of Energy,* Nuclear reactors, nuclear fuel reserves

James J. Pitts, *Olin Corporation,* Industrial antimicrobial agents

F. Planinsek, *University of Denver,* Cobalt and cobalt alloys

Chester A. Plants, *Union Carbide Corporation,* Heat transfer media other than water under Heat exchange technology

Frank E. Platko, *Clairol, Inc.,* Hair preparations

A.E. Platt, *The Dow Chemical Co.,* Styrene plastics

J.R. Plimmer, *United States Department of Agriculture,* Herbicides; Soil chemistry of pesticides

Edwin P. Plueddemann, *Dow Corning Corporation,* Silylating agents under Silicon compounds

Andrew Pocalyko, *E.I. du Pont de Nemours & Co., Inc.,* Explosively clad metals under Metallic coatings

Noland Poffenberger, *Dow Chemical U.S.A.,* Chlorinated benzenes under Chlorocarbons and chlorohydrocarbons

Bing T. Poon, *Walter Reed Army Institute of Research,* Chemotherapeutics, antiprotozoal

K. Porter, *ICI Fibres,* Fibers, multicomponent; Nonwoven fabrics, spunbonded

M.C. Porter, *Nuclepore Corporation,* Particle-track etching

Mary K. Porter, *Consultant,* Felts

Raymond E. Porter, *Pullman Kellogg,* Ammonia

W. Portz, *Hoechst AG,* Coal chemicals and feedstocks, carbide production

Anthony F. Posteraro, *College of Dentistry, New York University,* Dentifrices

N.P. Potter, *Shell International Chemical Co., Ltd.,* Carboxylic acids, trialkylacetic acid

James E. Potts, *Union Carbide Corporation,* Plastics, environmentally degradable

Ralph H. Potts,‡ *Armak Company,* Manufacture under Carboxylic acids

R. David Prengaman, *RSR Corporation,* Lead alloys

J. Preston, *Monsanto Triangle Park Development Center, Inc.,* Aramid fibers, Heat-resistant polymers

George Prokopakis, *Columbia University,* Azeotropic and extractive distillation

Richard W. Prugh, *E.I. du Pont de Nemours & Co., Inc.,* Plant safety

Everett H. Pryde, *United States Department of Agriculture,* Introduction; Economic aspects both under Carboxylic acids

William P. Purcell, *Union Oil Company of California,* Benzene

Imre Puskas, *Amoco Chemical Corporation,* Butylenes

Stearns T. Putnam, *Hercules Incorporated,* Papermaking additives

Richard E. Putscher, *E.I. du Pont de Nemours & Co., Inc.,* Polyamides, general

Muthiah Ramanathan, *Roy F. Weston, Inc.,* Water, pollution

Francis J. Randall, *S.C. Johnson & Son, Inc.,* Polishes

Derek L. Ransley, *Chevron Research Company,* BTX processing; Xylenes and ethylbenzene

Michael M. Rauhut, *American Cyanamid Co.,* Chemiluminescence

Terence Rave, *Hercules, Incorporated,* Pulp, synthetic

John F. Ready, *Honeywell Corporate Technology Center,* Lasers

B.C. Ream, *Union Carbide Corporation,* Ethylene glycol and propylene glycol under Glycols

Ludwig Rebenfeld, *Textile Research Institute,* Survey under Textiles

Richard A. Reck, *Armak Company,* Amides, fatty acid; Fatty amines (under Amines) Quaternary ammonium compounds

R.L. Reddy, *Union Carbide Corp.,* Carbon and artificial graphite under Carbon

Gerald Reed, *Amber Laboratories, Inc.,* Yeasts

H.W.B. Reed, *Imperial Chemical Industries Limited,* Alkylphenols

Glenn H. Rees, *Brush Wellman Inc.,* Beryllium compounds

Miguel F. Refojo, *Eye Research Institute of Retina Foundation,* Contact lenses

Donald A. Reich, *PPG Industries,* Drycleaning under Drycleaning and laundering

Arno H. Reidies, *Carus Chemical Company,* Manganese compounds

Curt W. Reimann, *National Bureau of Standards,* Analytical methods

Charles E. Reineke, *Dow Chemical U.S.A.,* Bromine

Paul R. Resnick, *E.I. du Pont de Nemours & Co., Inc.,* Perfluoroepoxides under Fluorine compounds, organic

Bertie J. Rueben, *Monsanto Textiles Company,* Antistatic agents

Thomas G. Reynolds, III, *Ferroxcube Corporation,* Ferrites

S.K. Rhee, *Bendix Corporation,* Brake linings and clutch facings

R.C. Rhees, *Pacific Engineering and Production Co.,* Perchloric acid and perchlorates under Chlorine oxygen acids and salts

William H. Rhodes, *BASF Wyandotte,* Phthalocyanine compounds

L.H. Rice, *Hanford Engineering Development Laboratory,* Fuel-element fabrication under Nuclear reactors

Paul N. Richardson, *E.I. du Pont de Nemours & Co., Inc.,* Plastics processing

D.S. Richart, *The Polymer Corporation,* Powder coatings

Ralph E. Ricksecker, *Chase Brass & Copper Co., Inc.,* Wrought copper and wrought copper alloys; Cast copper alloys both under Copper alloys

Gregor H. Riesser, *Shell Development Company,* Chlorohydrins

H.W. Rimmer, *Allis Chalmers Corporation,* Size reduction

William F. Ringk, *Chemical Consultant,* Benzyl alcohol and β-phenethyl alcohol; Cinnamic acid, cinnamaldehyde, and cinnamyl alcohol

M. Ritchey, *Lederle Laboratories, American Cyanamid Co.,* Vaccine technology

Catherine Rivier, *Salk Institute,* Survey; Brain oligopeptides both under Hormones

Jean Rivier, *Salk Institute,* Survey; Brain oligopeptides both under Hormones

Lester E. Robb, *Pennwalt Corporation,* Poly(vinylidene fluoride) under Fluorine compounds, organic

John A. Roberts, *Arco Ventures Co.,* Metal fibers

William J. Roberts, *Consultant,* Fibers, chemical

A.S. Robertson, *Allied Chemical Corporation,* Sodium carbonate under Alkali and chlorine products

J.M. Robertson, *Celanese Chemical Company,* Economic evaluation

Peter M. Robertson, *Eidgenössische Technische Hochschule,* Electrochemical processing, introduction and inorganic

James H. Robins, *Mytec, Inc.,* Paper under Recycling

H.W. Robinson, *Uniroyal, Inc.,* Nitrile rubber under Elastomers, synthetic

W.D. Robinson, *Monsanto Company,* Maleic anhydride, maleic acid, and fumaric acid

William P. Roe, *ASARCO,* Lead

L.G. Roettger, *Borg-Warner Chemicals,* ABS resins, under Acrylonitrile polymers

James A. Rogers, Jr., *Fritzsche Dodge and Olcott, Inc.,* Flavors and spices; Oils, essential

Kempton H. Roll, *Metal Powder Industries Federation, American Powder Metallurgy Institute,* Powder metallurgy

Carl L. Rollinson, *University of Maryland,* Introduction under Aluminum compounds; Survey under Calcium compounds; Mineral nutrients

Christian S. Rondestvedt, Jr., *E.I. du Pont de Nemours & Co., Inc.,* Titanium compounds, organic

R.O. Rooke, *Monsanto Co.,* Herbicides

Joseph Rose, *Drexel University,* Ultrasonics, low power

Joe B. Rosenbaum, *Consultant,* Vanadium compounds

David S. Rosenberg, *Hooker Chemical Co.,* Hydrogen chloride

R.H. Rosenwald, *Consultant, UOP, Inc.,* Alkylation

Stephen T. Ross, *Smith Kline & French Laboratories,* Appetite-suppressing agents

Sydney Ross, *Rensselaer Polytechnic Institute,* Foams

Philip B. Roth, *American Cyanamid Co.,* Amino resins and plastics

L.R. Rothrock, *Union Carbide Corporation,* Gems, synthetic

J.W. Rowe, *United States Department of Agriculture,* Wood

R.M. Rowell, *United States Department of Agriculture,* Wood

Ralph W. Rudolph,‡ *University of Michigan,* Boron hydrides, heteroboranes, and their metalloderivatives under Boron compounds

Walter Runyan, *Texas Instruments, Inc.,* Pure silicon under Silicon and silicon alloys

N.W. Rupp, *American Dental Association, Health Foundation Research Unit, National Bureau of Standards,* Dental materials

R. Russell, *Union Carbide Corp.,* Carbon and artificial graphite under Carbon

M. Salame, *Monsanto Commercial Products Co.,* Barrier polymers

Alvin J. Salkind, *ESB Technology Co.,* Secondary cells, alkaline under Batteries and electric cells, secondary

J.M. Salomone, *Monsanto Enviro-Chem Systems, Inc.,* Sulfuric acid and sulfur trioxide

William M. Saltman, *The Goodyear Tire & Rubber Company,* Isoprene

Cecelia Samans, *Amoco Chemicals Corporation,* Phthalic acids and other benzenepolycarboxylic acids

R.N. Sampson, *Westinghouse Electric Corporation,* Properties and materials under Insulation, electric

J.R. Sanders, *Celanese Fibers Company,* Cellulose acetate and triacetate fibers

Robert L. Sandridge, *Mobay Chemical Corp.,* Amines by reduction

Charles A. Sandy, *E.I. du Pont de Nemours & Co., Inc.,* Organolead compounds under lead compounds

H.B. Sargent, *Union Carbide Corporation,* Physical and chemical properties and explosive behavior under Acetylene

J.H. Saunders, *Monsanto Company,* Polyamide fibers under Polyamides

S.G. Sawochka, *NWT Corporation,* Water chemistry of light-water reactors under Nuclear reactors

R.H. Schatz, *Exxon Chemical Co.,* Butyl rubber under Elastomers, synthetic

B.A. Schenker, *Diamond Shamrock Corp.,* Chlorinated paraffins under Chlorocarbons and chlorohydrocarbons

P.M. Scherer, *Union Carbide Corp.,* Carbon and artificial graphite under Carbon

L. McDonald Schetky, *International Copper Research Association,* Shape-memory alloys

Henry W. Schiessl, *Olin Research Center,* Hydrazine and its derivatives

C.E. Schildknecht, *Gettysburg College,* Allyl monomers and polymers

James H. Schlewitz, *Teledyne Wah Chang Albany,* Zirconium and zirconium compounds

Morris R. Schoenberg, *Amoco Chemicals Corporation,* Propylene

Clifford K. Schoff, *PPG Industries, Inc.,* Rheological measurements

Fred Scholer, *FMC Corporation,* Bleaching agents

W.A. Scholle, *Monsanto Co.,* Air pollution control methods

Glenn R. Schmidt, *Colorado State University,* Meat products

T.W. Schmidt, *Phillips Petroleum Co.,* Pentanes under Hydrocarbons, C_1-C_6

Edward E. Schmitt, *ALZA Corporation,* Pharmaceuticals, controlled release

G.L. Schneberger, *General Motors Institute,* Cleaning, pickling, and related processes; Chemical and electrochemical conversion treatments both under Metal surface treatments

Ronald W. Schroeder, *Union Carbide Corporation,* Nitrogen

D.N. Schulz, *Firestone Tire & Rubber Co.,* Copolymers

G.G. Schurr, *The Sherwin-Williams Company,* Paint

Mortimer Schussler, *Fansteel, Inc.,* Tantalum and tantalum compounds

Brian B. Schwartz, *Brooklyn College, New York,* Superconducting materials

S.A. Schwarz, *Bell Laboratories, Inc.,* Semiconductors, theory and application

Albert E. Schweizer, *Air Products and Chemicals, Inc.,* Lower aliphatic amines under Amines

Don Scott, *Fermco Biochemics Inc.,* Enzymes, industrial

Glenn T. Seaborg, *University of California, Berkeley,* Actinides and transactinides

J.D. Seader, *University of Utah,* Separation systems synthesis

J.K. Sears, *Monsanto Company,* Plasticizers

Oldrich K. Sebek, *The Upjohn Company,* Microbial transformations

Sherwood B. Seeley, *Consultant,* Natural graphite under Carbon

F.E. Selim, *Phillips Petroleum Company,* Liquefied petroleum gas

Joseph Senackerib, *H. Kohnstamm & Co., Inc.,* Colorants for foods, drugs, and cosmetics

Robert E. Sequeira, *Amstar Corporation,* Properties of sucrose under Sugar

George A. Serad, *Celanese Fibers Company,* Cellulose acetate and triacetate fibers

Raymond B. Seymour, *The University of Southern Mississippi,* Sealants

Michael R. Sfat, *Bio-Technical Resources, Inc.,* Malts and malting

J. Shacter, *Union Carbide Nuclear Division,* Diffusion separation methods

Smith Shadomy, *Medical College of Virginia,* Chemotherapeutics, antimycotic and antirickettsial

R.P. Shaffer, *Linde Division, Union Carbide Corp.,* Manufacture from carbide under Acetylene

Morris H. Shamos, *Technicon Instruments Corporation,* Biomedical automated instrumentation

E. Shanty, *Edgewood Arsenal,* Chemicals in war

S.R. Shatynski, *Rensselaer Polytechnic Institute,* High temperature alloys

Henry Shaw, *Exxon Research and Engineering Company,* Liquid fuels under Fuels, synthetic

Walker L. Shearer, *Dow Chemical U.S.A.,* Calcium chloride under Calcium compounds

John E. Sheats, *Rider College,* Metal-containing polymers

O.C. Sheese, *Wellman Furnaces, Inc.,* Furnaces, electric resistance

Robert C. Sheik, *CIBA-GEIGY Corporation,* Pigments, inorganic

J.E. Shelton, *U.S. Bureau of Mines,* Sulfur compounds

Chester S. Sheppard, *Pennwalt Corporation,* Initiators; Peroxides and peroxy compounds, organic

E. Sherman, *The Quaker Oats Company,* Furan derivatives

Paul Dwight Sherman, Jr., *Union Carbide Corporation,* Aldehydes; Amyl alcohols; Butyl alcohols; Butyraldehyde; Ethanol; Ketones

Stanley Sherman, *CIBA-GEIGY Corporation,* Epoxy resins

Joseph C. Sherrill, *Sherrill Associates,* Laundering under Drycleaning and laundering

Noel B. Shine, *Shawinigan Products Dept., Gulf Oil Chemicals,* Calcium carbide under Carbides

A. Ray Shirley, Jr., *Applied Chemical Technology,* Urea

Andrew Shoh, *Branson Sonic Power Company,* Ultrasonics, high power

James N. Short, *Phillips Petroleum Company,* Low pressure linear (low density) polyethylene

Walter Showak, *Gulf and Western National Resources Group,* Zinc and zinc alloys

A. Shultz, *Stepan Chemical Company,* Sulfonation and sulfation; Sulfonic acids

J.C. Siegle, *E.I. du Pont de Nemours & Co., Inc.,* Dimethylacetamide under Acetic acid

Ronald F. Silver, *Elkem Metals Company,* Metallurgical under Silicon and silicon alloys

Walter L. Silvernail, *Consultant,* Cerium and cerium compounds

P.B. Simmons, *Dow Chemical U.S.A.,* Diphenyl and terphenyls

W.T. Simpson, *United States Department of Agriculture,* Wood

Robert W. Singelton, *University of Connecticut,* Testing under Textiles

John J. Singer, *BASF Wyandotte Corporation,* Pigments, dispersed

William Singer, *Troy Chemical Corporation,* Mercury compounds

J.E. Singly, *Environmental Science and Engineering, Inc.,* Water, municipal water treatment

W.E. Sisco, *CPS Chemical Company,* Naphthenic acids

George Sistare, *Consultant,* Silver and silver alloys; Solders and brazing alloys

A.H.P. Skelland, *Georgia Institute of Technology,* Mass transfer

R.E. Skochdopole, *The Dow Chemical Company,* Foamed plastics

Leonard Skolnik, *BFGoodrich Co.,* Tire cords

Morey E. Slodki, *U.S. Department of Agriculture,* Microbial polysaccharides

A.B. Small, *Exxon Chemical Company,* Cyclopentadiene and dicyclopentadiene

Bruce E. Smart, *E.I. du Pont de Nemours & Co., Inc.,* Fluorinated aliphatic compounds under Fluorine compounds, organic

R.B. Smart, *West Virginia University,* Water analysis

Leonard H. Smiley, *PQ Corporation,* Fillers

Robert A. Smiley, *E.I. du Pont de Nemours & Co., Inc.,* Nitriles

H.D. Smith Jr., *Virginia Polytechnic Institute and State University,* Organic boron nitrogen compounds under Boron compounds

Keith Smith, *University College of Swansea, Wales, United Kingdom,* Hydroboration

Mark Smith, *Kay Fries,* Alkoxides, metal

R.A. Smith, *U.S. Borax Research Corporation,* Boron compounds, boron oxides, boric acid, and borates

Richard A. Smoak, *The Carborundum Company,* Silicon carbide under Carbides

James A. Snelgrove, *Monsanto Company,* Poly(vinyl acetyl)s under Vinyl polymers

Lloyd R. Snyder, *Technicon Instruments Corporation,* Biomedical automated instrumentation

R.W. Soffel, *Union Carbide Corp.,* Carbon and artificial graphite under Carbon

Irvine J. Solomon, *IIT Research Institute,* Oxygen under Fluorine compounds, inorganic

J.S. Son, *Shell Development Company,* Fluid mechanics

P.N. Son, *BFGoodrich Company,* Rubber chemicals

Henry E. Sostman, *Yellow Springs Instrument Co., Inc,* Temperature measurement

Thomas F. Soules, *General Electric Company,* Luminescent materials, phosphors

Donald M. Sowards, *E.I. du Pont de Nemours & Co., Inc.,* Exhaust control, industrial

Robert M. Sowers, *Ford Motor Company,* Laminated materials, glass

Ian L. Spain, *Colorado State University,* High pressure technology

Robert E. Sparks, *Washington University,* Microencapsulation

Arno F. Spatola, *University of Louisville,* Posterior-pituitary hormones under Hormones

F.H. Spedding, *Ames Laboratory, Iowa State University,* Rare-earth elements

Gavin G. Spence, *Hercules Incorporated,* Papermaking additives

Herbert E. Spiegel, *Hoffmann-LaRoche Inc.,* Medical diagnostic reagents

F.W. Spillers, *Dow Chemical U.S.A.,* Chlorine and sodium hydroxide under Alkali and chlorine products

L. Spinicelli, *Celanese Chemical Co.,* n-Propyl alcohol under Propyl alcohols

R.D. Spitz, *Dow Chemical U.S.A.,* Diamines and higher amines, aliphatic

Harold B. Staley, *Mobay Chemical Corp.,* Amines by reduction

Edgel Stambaugh, *Battelle Memorial Institute,* Desulfurization under Coal conversion processes

Paul Stamberger, *Crusader Chemical Co.,* Electrodecantation

Ferris C. Standiford, *W.L. Badger Associates, Inc.,* Evaporation

D.B. Stanton, *Union Carbide Corporation,* Acrolein and derivatives

Chuck Starace, *Novo Laboratories, Inc.,* Enzyme detergents

E. Eugene Stauffer, *The Ansul Company,* Fire-extinguishing agents

A.L. Stautzenberger, *Celanese Chemical Co., Inc.,* Formaldehyde

David R. St. Cyr, *Goodyear Tire & Rubber Co.,* Rubber, natural

J.M. Steele, *Dow Chemical U.S.A.,* Methyl chloride under Chlorocarbons and chlorohydrocarbons

L.W. Steele, *General Electric Co.,* Research management

A. Stefanucci, *General Foods Corp.,* Coffee

Claude Stein, *CdF Chimie S.A.,* Poly(bicycloheptene) and related polymers

Lawrence Stein, *Argonne National Laboratory,* Helium-group gases, compounds

S. Steingiser, *Monsanto Commercial Products Co.,* Barrier polymers

Dan Steinmeyer, *Monsanto Inc.,* Process energy conservation

V.A. Stenger, *Dow Chemical U.S.A.,* Bromine compounds; Sodium bromide under Sodium compounds, sodium halides

George Stergis, *Division of Cancer Treatment, National Cancer Institute, National Institutes of Health,* Enzymes, therapeutic

Eric W. Stern, *Englehard Minerals and Chemicals Corporation,* Gold and gold compounds; Platinum-group metals, compounds

L.H. Stembach, *Hoffmann-LaRoche & Co., Inc.,* Psychopharmacological agents

C.V. Sternling, *Shell Development Company,* Fluid mechanics

James E. Stevens, *Hooker Chemicals & Plastics Corp.,* Chlorinated derivatives of cyclopentadiene under Chlorocarbons and chlorohydrocarbons

Violete L. Stevens, *Dow Chemical Company,* Dichloroethylenes under Chlorocarbons and chlorohydrocarbons

Clare A. Stewart, Jr., *E.I. du Pont de Nemours & Co., Inc.,* Chlorocarbons and chlorohydrocarbons, chloroprene

Daniel R. Stewart, *Owens-Illinois,* Glass-ceramics

Philip B. Sticksel, *Battelle Memorial Institute,* Air pollution

H.A. Stingle, *Toms River Chemical Corporation,* Dyes and dye intermediates

N.S. Stoloff, *Rensselaer Polytechnic Institute,* High temperature alloys

A. James Stonehouse, *Brush Wellman Inc.,* Beryllium and beryllium alloys

Robert F. Stoops, *North Carolina State University,* Forming processes under Ceramics

E.R. Stover, *General Electric Company,* Ablative materials

William C. Streib, *Johns-Manville Corporation,* Asbestos

Lorraine Y. Stroumtsos, *Exxon Research and Engineering Company,* Information retrieval

Martin H. Stryker, *The New York Blood Center,* Blood fractionation

David M. Stuart, *Ohio Northern University,* Blood, coagulants and anticoagulants

Werner Stumm, *Swiss Federal Institute of Technology,* Water, properties

David M. Sturmer, *Eastman Kodak Company,* Cyanine dyes; Dyes, sensitizing; Polymethine dyes

K.W. Suh, *The Dow Chemical Company,* Foamed plastics

Edward A. Sullivan, *Thiokol/Ventron Division,* Hydrides

Kishin H. Surtani, *Texas Instruments Inc.,* Digital displays

S. Sussman,‡ *Olin Water Services,* Water, industrial water treatment

Ladislav Svarovsky, *University of Bradford,* Sedimentation

James W. Swaine, Jr., *Allied Corp.,* Thiosulfates

J.W. Swanson, *The Institute of Paper Chemistry,* Paper

Michael W. Swartzlander, *Union Carbide Corporation,* Ethylene oxide

Ray Sweeny, *Brush Wellman Inc.,* Beryyllium and berllium alloys

C. Richard Swenson, *CasChem, Inc.,* Castor oil

Richard Sykes, *Squibb Institute for Medical Research,* Antibiotics, β-lactams, monobactams

L.G. Sylvester, *CIBA-GEIGY Corporation,* Uric acid

Robert M. Talcott, *Dorr-Oliver Incorporated,* Filtration

Mario Tama, *Ajax Magnethermic Corporation,* Introduction; Induction both under Furnaces, electric

Herbert Tanner, *Degussa Company,* Vitamins, nicotinamide and nicotinic acid (B₃)

Barry L. Tarmy, *Exxon Research and Engineering Company,* Reactor technology

Edward Tarnell, *Roger Williams Technical & Economic Services, Inc.,* Market and marketing research

William Tasto, *Dow Chemical U.S.A.,* Ethyl chloride under Chlorocarbons and chlorohydrocarbons

Bryce E. Tate, *Pfizer Inc.,* Itaconic acid and derivatives

David P. Tate, *Firestone Tire and Rubber Co.,* Copolymers; Fluoroalkoxyphosphazenes under Fluorine compounds, organic

Alfred H. Taylor, *Airco Industrial Gases, a division of Airco, Inc.,* Helium-group gases; Oxygen

Donald F. Taylor, *Fansteel, Inc.,* Tantalum and tantalum compounds

Ray Taylor, *BFGoodrich Company,* Rubber chemicals

Robert C. Taylor, *University of Michigan,* Boron compounds, boron hydrides, Heteroboranes, and their metallo derivatives

William I. Taylor, *International Flavors & Fragrances,* Perfumes

H. Teich, *Mobay Chemical Corp.,* Dyes, application and evaluation

A.S. Teot, *Dow Chemical U.S.A.,* Resins, water-soluble

Gary TerHaar, *Ethyl Corporation,* Industrial toxicology under Lead compounds

Jefferson Tester, *MIT,* Geothermal energy

Fred N. Teumac, *Uniroyal Inc.,* Coated fabrics

Ernst T. Theimer, *Consultant,* Benzyl alcohol and β-phenethyl alcohol

John R. Thirtle, *Eastman Kodak Company,* Color photography

David A. Thompson, *Corning Glass Works,* Glass

Q.E. Thomson, *Monsanto Co.,* Diphenyl and terphenyls

Carl Thurman, *Dow Chemical U.S.A.,* Phenol

David Tiemeier, *Monsanto Company,* Genetic engineering

John K. Tien, *Columbia University,* Nickel and nickel alloys

David A. Tillman, *Consultant,* Fuels from waste

Robert W. Timmerman, *FMC Corporation,* Carbon disulfide

Bruce Tippin, *Great Salt Lake Minerals and Chemicals Corp.,* Chemicals from brine

David B. Todd, *Baker Perkins, Inc.,* Mixing and blending

J.R. Tonry, *Portland Cement Association,* Cement

D.C. Torgeson, *Boyce Thompson Institute for Plant Research, Inc.,* Fungicides, agricultural

Arnold Torkelson, *General Electric Company,* Silicones under Silicon compounds

N.W. Touchette, *Monsanto Company,* Plasticizers

IrvingTouval, *M & T Chemicals, Inc.,* Antimony and other inorganic compounds under Flame retardants

Donald Towson, *Petro-Canada,* Tar sands

E.D. Travis, *Ethyl Corporation,* Synthetic processes under Alcohols, higher aliphatic

Paul M. Treichel, *University of Wisconsin,* Rhenium and rhenium compounds

R.O. Tribolet, *Linde Division, Union Carbide Corp.,* Handling of acetylene (under Acetylene)

D.J. Triggle, *State University of New York at Buffalo,* Pharmacodynamics

Samuel I. Trotz, *Olin Corporation,* Industrial antimicrobial agents

L. Trumbley, *Celanese Chemical Company, Inc.,* Methanol

Minoru Tsutsui,‡ *Texas A & M University,* Organometallics — π metal complexes

Harold Tucker, *The BFGoodrich Company,* Polyisoprene under Elastomers, synthetic

W.M. Tuddenham, *Kennecott Copper Corporation,* Copper

Albin F. Turbak, *ITT Rayonier Inc.,* Cellulose; *Georgia Institute of Technology, Atlanta, Georgia,* Rayon

R.J. Turner, *Shell International Chemical Co., Ltd.,* Trialkylacetic acids under Carboxylic acids

Samuel M. Tuthill, *Mallinckrodt, Inc.,* Fine chemicals

G.F. Tuwiler, *McNeil Laboratories,* Insulin and other antidiabetic agents

F. Tweedle, *Mobay Chemical Corp.,* Dyes, application and evaluation

Aaron Twerski, *Hofstra Law School,* Product liability

Henri Ulrich, *The Upjohn Company,* Urethane polymers

W.E. Unger, *Oak Ridge National Laboratory,* Chemical reprocessing under Nuclear reactors

J.D. Unruh, *Celanese Chemical Company,* n-Propyl alcohol under Propyl alcohols

Ivor H. Updegraff, *American Cyanamid Co.,* Amino resins and plastics

Milan R. Uskoković, *Hoffmann-LaRoche, Inc.,* Vitamins, biotin

Sidney L. Vail, *Southern Regional Research Center, United States Department of Agriculture,* Finishing under Textiles

E.J. Vandenberg, *Hercules Incorporated,* Polyethers under Elastomers, synthetic

Gus van Loveren, *International Flavors & Fragrances,* Perfumes

Hendrick C. Van Ness, *Rensselaer Polytechnic Institute,* Thermodynamics

J.H. Van Ness, *Monsanto Company,* Hydroxy carboxylic acids; Vanillin

Jan F. Van Peppen, *Allied Chemical Corp.,* Cyclohexanol and cyclohexanone

Raymond Van Sweringen, *Exxon Research and Engineering Company,* Pilot plants and microplants

John R. Van Wazer, *Vanderbilt University,* Phosphorus and the phosphides

Burt Van Zelt, *Museum of Fine Arts, Boston,* Fine art examination and conservation

J. Varagnat, *Rhône-Poulenc,* Hydroquinone, resorcinol, and catechol

Joseph J. Varco, *Clairol, Inc.,* Hair preparations

Rance A. Velapoldi, *National Bureau of Standards,* Analytical methods

Guy E. Verdino, *Keil Chemical Division, Ferro Corporation,* Sulfurization and sulfur-chlorination

Theodore Vermeulen, *University of California, Berkeley,* Introduction (under Adsorptive separation)

I.D. Verschoor, *Manville Corp.,* Insulation, thermal; insulation, acoustic

Francis A. Via, *Stauffer Chemical Company,* Aluminum compounds, aluminum sulfate and alums

T.M. Vial, *American Cyanamid Company,* Acrylic elastomers under Elastomers, synthetic

Felix Viro, *Kind & Knox, Division of Knox Gelatine, Inc.,* Gelatin

H.F. Volk, *Union Carbide Corp.,* Carbon and artificial graphite under Carbon

Paul T. von Bramer, *Eastman Chemical Products, Inc.,* Other glycols under Glycols

E. Von Halle, *Union Carbide Nuclear Division,* Diffusion separation methods

Urs von Stockar, *University of California, Berkeley,* Absorption

Karl S. Vorres, *Institute of Gas Technology,* Coal; Lignite and brown coal

W.A. Vredenburgh, *Hercules Incorporated,* Hydrocarbon resins

Kenneth Wachter, *Olin Corporation,* Sodium under Fluorine compounds, inorganic

L.E. Wade, *Celanese Chemical Company, Inc.,* Methanol

Robert C. Wade, *Thiokol/Ventron Division,* Hydrides

Parvez H. Wadia, *Union Carbide Corporation,* Ethylene oxide

Frank S. Wagner, *Strem Chemicals, Inc.,* Carbonyls

Frank S. Wagner, Jr., *Nandina Corporation,* Acetic acid; Acetic anhydride both under Acetic acid; Formic acid under Formic acid and derivatives; 1,3 Butylene glycol under Glycols

D. Wahren, *The Institute of Paper Chemistry,* Paper

Alan A. Waldman, *The New York Blood Center,* Blood fractionation

T.C. Wallace, *The Dow Chemical Co.,* Styrene plastics

George Wallis, *P.R. Mallory & Co., Inc.,* Electrostatic sealing

Kenneth Walsh, *Brush Wellman Inc.,* Beryllium and beryllium alloys; Beryllium compounds

Vivian K. Walworth, *Polaroid Corporation,* Color photography, instant

Robert Wannemacher, *Tantatex Chemical Co.,* Dye carriers

Dennis J. Ward, *UOP Inc.,* Cumene

Richard A. Ward, *Day-Glo Color Corporation,* Luminescent materials, fluorescent pigments (daylight)

R. Wardle, *Basis Inc.,* Magnesium compounds

Glenn H. Warner, *Union Carbide Corporation,* Aluminum carboxylates under Aluminum compounds

Clifton Warren, *AGA Infrared System AB,* Infrared technology

J.W. Wasson, *Lithium Corporation of America,* Lithium and lithium compounds

A.F. Waterland, *Waterland, Viar & Associates, Inc.,* Energy management

B.H. Waxman, *GAF Corporation,* N-Vinyl monomers and polymers under Vinyl polymers

Stewart Way, *MHD Consultant,* Magnetohydrodynamics under Coal conversion processes

John C. Weaver, *Case Western Reserve University,* Resins, natural

W.C. Weaver, *Dow Chemical U.S.A.,* Diphenyl and terphenyls

A.D. Webb, *University of California, Davis,* Vinegar

B.P. Webb, *The Dow Chemical Company,* Acetone

Thomas G. Webber, *Consultant,* Colorants for plastics

J. Weber, *Celanese Canada Ltd.,* Other polyhydric alcohols under Alcohols, polyhydric

O.W. Webster, *E.I. du Pont de Nemours & Co., Inc.,* Cyanocarbons

Ionel Wechsler, *Sala Magnetics, Inc.,* Magnetic separation

Theodore H. Wegner, *United States Department of Agriculture,* Wood

E.G. Weierich, *CPS Chemical Company,* Naphthenic acids

Fritz Weigel, *University of Munich,* Plutonium and plutonium compounds; Uranium and uranium compounds

John W. Weigl,[‡] *Xerox Corporation,* Electrophotography

Edward D. Weil, *Stauffer Chemical Co.,* Phosphorus compounds under Flame retardants; Sulfur compounds

R.J. Welgos, *Allied Corporation,* Polyamide plastics under Polyamides

William W. Wells, *Michigan State University,* Vitamins, inositol

Joseph Welsh, *Hayward Baker Co.,* Chemical grouts

William J. Welstead, Jr., *A.H. Robins Company, Inc.,* Expectorants, antitussives, and related agents; Stimulants

Robert J. Wenk, *United States Gypsum Co.,* Calcium sulfate under Calcium compounds

R.H. Wentorf, Jr., *General Electric Company,* Refractory boron compounds under Boron compounds; Diamond, synthetic under Carbon

Leonard A. Wenzel, *Lehigh University,* Simultaneous heat and mass transfer

Alan J. Werner, *Corning Glass Works,* Optical filters

J.H. Wernick, *Bell Laboratories,* Magnetic materials, bulk; Magnetic materials, thin film

R.A. Wessling, *Dow Chemical U.S.A.,* Vinylidene chloride and poly(vinylidene) chloride

Arthur C. West, *3M Company,* Fluorinated elastomers under Elastomers, synthetic; Polychlorotrifluoroethylene under Fluorine compounds, organic

James R. West, *Texasgulf Inc.,* Sulfur recovery

J.H. Westbrook, *General Electric Company,* Chromium and chromium alloys; Materials standards and specifications

Roger Westland, *Parke Davis Co.,* Antibiotics, chloramphenicol and its analogues

John W. Westley, *Hoffmann-LaRoche Inc.,* Polyethers under Antibiotics

Thomas P. Whaley, *Institute of Gas Technology,* Pipelines

R.M. Wheaton, *Dow Chemical U.S.A.,* Ion exchange

Roy L. Whistler, *Purdue University,* Carbohydrates; Starch

C.E.T. White, *Indium Corporation of America,* Indium and indium compounds

Dwain M. White, *General Electric Company,* Aromatic polyethers under Polyethers

Fredrick B. White, Jr., *Kawecki Berylco Industries, Inc.,* Rubidium and rubidium compounds

Lucien White, *Clairol Inc.,* Hair preparations

J. Whitehead, *Tioxide Group PLC,* Titanium compounds, inorganic

Robert C. Whitehead, Jr., *Honeywell,* Liquid-level measurement

Thaddeus E. Whyte, Jr., *Air Products and Chemicals, Inc.,* Lower aliphatic amines under Amines

T.K. Wiewiorowski, *Freeport Minerals Co.,* Sulfur

William R. Wilcox, *Clarkson College of Technology,* Zone refining

C.R. Wilke, *University of California, Berkeley,* Absorption

Joan D. Willey, *University of North Carolina at Wilmington,* Amorphous silica under Silica

Arnold E. Williams, *Kalama Chemical Inc.,* Benzaldehyde; Benzoic acid

C.T. Williams, *Tantalum Mining Corp.,* Cesium and cesium compounds

Elizabeth A. Williams, *General Electric Co.,* Analytical methods

G.P.L. Williams, *Atomic Energy of Canada,* Nuclear reactors, introduction

Jack L. Williams, *Eastman Kodak Co.,* Photoreactive polymers

Ken W. Williams, *Dow Chemical U.S.A.,* Diamines and higher amines, aliphatic; Acetone

Ralph Williams, *Phillips Research Center,* Sulfolanes and sulfones

Theodore J. Williams, *Purdue University,* Instrumentation and control;

Simulation and process design

J. George Wills,[‡] *Mobil Oil Corporation,* Hydraulic fluids

Charles Willus, *Dorr-Oliver Incorporated,* Filtration

H.P. Wilson, *Vulcan Materials Co.,* Detinning under Tin and tin alloys

Leon O. Windstrom, *Consultant,* Succinic acid and succinic anhydride

R.E. Wing, *Dow Chemical U.S.A.,* Chlorine and sodium hydroxide under Alkali and chlorine products

L.L. Winter, *Union Carbide Corp.,* Carbon and artificial graphite under carbon

Olaf Winter, *The Lummus Company,* Ethylene

W.J. Wiseman, *Edgewood Arsenal,* Chemicals in war

John S. Wishnok, *Massachusetts Institute of Technology,* N-Nitrosamines

Romeo R. Witherspoon, *General Motors Research Laboratories,* Fuel cells under Batteries and electric cells, primary

John A. Wojtowicz, *Olin Corporation,* Chlorine monoxide, hypochlorous acid, and hypochlorites under Chlorine oxygen acids and salts; Water, treatment of swimming pools, spas, and hot tubs

Walter J. Wolf, *U.S. Department of Agriculture,* Soybeans and other oilseed proteins

Nikolaus E. Wolff, *Consultant,* Electrophotography

Leszek J. Wolfram, *Clairol, Inc.,* Hair preparations

Stewart Wong, *Boehringer-Ingelheim Ltd.,* Immunotherapeutic agents

W.E. Wood, *James Robinson & Company Ltd.,* Sulfur dyes

J.H. Woode, *Shell International Chemical Co., Ltd.,* Trialkylacetic acids under Carboxylic acids

Henry L. Wooten, *FMC Corporation,* Bleaching agents

Andrew J. Woytek, *Air Products and Chemicals, Inc.,* Fluorine; and Halogens; Molybdenum; Nitrogen; Rhenium; Tantalum; Tungsten under Fluorine, compounds, inorganic

Fred Wudl, *Bell Laboratories,* Semiconductors, organic

Masaaki Yamabe, *Asaki Glass Company, Ltd.,* Perfluorinated ionomer membranes

A. Yamamoto, *Kyowa Hakko Kogyo, Ltd.,* Survey under Amino acids

J.M. Yatabe, *Hanford Engineering Development Laboratory,* Fuel-element fabrication under Nuclear reactors

Patrick P. Yeung, *Toms River Chemical Corporation,* Dyes, reactive

Fumio Yoneda, *Kumamoto University,* Vitamins, riboflavin (B_2)

Toichio Yoshida, *Ajinomoto Co., Inc.,* L-Monosodium glutamate under Amino acids

Clyde T. Young, *North Carolina State University,* Nuts

Ronald D. Young, *Tennessee Valley Authority,* Ammonium compounds

John A. Youngquist, *United States Department of Agriculture,* Laminated wood-based composites

Paul Zanowiak, *Temple University,* Pharmaceuticals

Howard C. Zell, *Food and Drug Administration,* Chemotherapeutics, antiprotozoal

Andrew F. Zeller, *FMC Corporation,* Strontium and strontium compounds

Frederick A. Zentz, *Consultant,* Fluidization

E.G. Zey, *Celanese Corporation,* Esterification

Robert J. Ziegler, *Betz Laboratories,* Technical service

Stanley C. Zink, *The Black Clawson Co.,* Coating processes

D.F. Zinkel, *United States Department of Agriculture,* Wood

Alvin B. Zlobik, *U.S. Bureau of Mines,* Micas, natural and synthetic

B.L. Zoumas, *Hershey Foods Corporation,* Chocolate and cocoa

Samuel Zuckerman, *H. Kohnstamm & Co., Inc.,* Colorants for foods, drugs, and cosmetics

Reinhard Zweidler, *CIBA-GEIGY Ltd.,* Brighteners, fluorescent

Daan M. Zwick, *Eastman Kodak Company,* Color photography

Edward D. Zysk, *Engelhard Minerals and Chemicals Corporation,* Platinum-group metals

John R. Zysk, *Purdue University,* Carbohydrates

[‡] Deceased.

CONVERSION FACTORS, ABBREVIATIONS AND UNIT SYMBOLS

SI Units (Adopted 1960)

Quantity	Unit	Symbol	Acceptable equivalent
BASE UNITS			
length	meter[†]	m	
mass[‡]	kilogram	kg	
time	second	s	
electric current	ampere	A	
thermodynamic temperature[§]	kelvin	K	
amount of substance	mole	mol	
luminous intensity	candela	cd	
SUPPLEMENTARY UNITS			
plane angle	radian	rad	
solid angle	steradian	sr	
DERIVED UNITS AND OTHER ACCEPTABLE UNITS			
*absorbed dose	gray	Gy	J/kg
acceleration	meter per second squared	m/s^2	
*activity (of ionizing radiation source)	becquerel	Bq	l/s
area	square kilometer	km^2	
	square hectometer	hm^2	ha (hectare)
	square meter	m^2	
*capacitance	farad	F	C/V
concentration (of amount of substance)	mole per cubic meter	mol/m^3	
*conductance	siemens	S	A/V
current density	ampere per square meter	A/m^2	
density, mass density	kilogram per cubic meter	kg/m^3	$g/L; mg/cm^3$
dipole moment (quantity)	coulomb meter	$C \cdot m$	

[†] The spellings "metre" and "litre" are preferred by ASTM; however "er-" is used in the Encyclopedia.

[‡] "Weight" is the commonly used term for "mass."

[§] Wide use is made of "Celsius temperature" (t) defined by

$$t = T - T_0$$

where T is the thermodynamic temperature, expressed in kelvins, and $T_0 = 273.15$ by definition. A temperature interval may be expressed in degrees Celsius as well as in kelvins.

*electric charge, quantity of electricity	coulomb	C	A·s
electric charge density	coulomb per cubic meter	C/m^3	
electric field strength	volt per meter	V/m	
electric flux density	coulomb per square meter	C/m^2	
*electric potential, potential difference, electromotive force	volt	V	W/A
*electric resistance	ohm	Ω	V/A
*energy, work, quantity of heat	megajoule	MJ	
	kilojoule	kJ	
	joule	J	N·m
	electron volt†	eV†	
	kilowatt hour†	kW·h†	
energy density	joule per cubic meter	J/m^3	
*force	kilonewton	kN	
	newton	N	kg·m/s^2
*frequency	megahertz	MHz	
	hertz	Hz	1/s
heat capacity, entropy	joule per kelvin	J/K	
heat capacity (specific), specific entropy	joule per kilogram kelvin	J/(kg·K)	
heat transfer coefficient	watt per square meter kelvin	W/(m^2·K)	
*illuminance	lux	lx	lm/m^2
*inductance	henry	H	Wb/A
linear density	kilogram per meter	kg/m	
luminance	candela per square meter	cd/m^2	
*luminous flux	lumen	lm	cd·sr
magnetic field strength	ampere per meter	A/m	
*magnetic flux	weber	Wb	V·s
*magnetic flux density	tesla	T	Wb/m^2
molar energy	joule per mole	J/mol	
molar entropy, molar heat capacity	joule per mole kelvin	J/(mol·K)	
moment of force, torque	newton meter	N·m	
momentum	kilogram meter per second	kg·m/s	
permeability	henry per meter	H/m	
permittivity	farad per meter	F/m	
*power, heat flow rate, radiant flux	kilowatt	kW	
	watt	W	J/s
power density, heat flux density, irradiance	watt per square meter	W/m^2	
*pressure, stress	megapascal	MPa	
	kilopascal	kPa	
	pascal	Pa	N/m^2
sound level	decibel	dB	
specific energy	joule per kilogram	J/kg	
specific volume	cubic meter per kilogram	m^3/kg	
surface tension	newton per meter	N/m	
thermal conductivity	watt per meter kelvin	W/(m·K)	
velocity	meter per second	m/s	
	kilometer per hour	km/h	
viscosity, dynamic	pascal second	Pa·s	
	millipascal second	mPa·s	
viscosity, kinematic	square meter per second	m^2/s	
	square millimeter per second	mm^2/s	

† This non-SI unit is recognized by the CIPM as having to be retained because of practical importance or use in specialized fields (1).

volume	cubic meter	m^3	
	cubic decimeter	dm^3	L(liter)
	cubic centimeter	cm^3	mL
wave number	1 per meter	m^{-1}	
	1 per centimeter	cm^{-1}	

In addition, there are 16 prefixes used to indicate order of magnitude, as follows:

Multiplication factor	Prefix	Symbol	Note
10^{18}	exa	E	
10^{15}	peta	P	
10^{12}	tera	T	
10^{9}	giga	G	
10^{6}	mega	M	
10^{3}	kilo	k	
10^{2}	hecto	h[a]	
10	deka	da[a]	[a]Although hecto, deka, deci, and centi
10^{-1}	deci	d[a]	are SI prefixes, their use should be
10^{-2}	centi	c[a]	avoided except for SI unit-mul-
10^{-3}	milli	m	tiples for area and volume and
10^{-6}	micro	μ	nontechnical use of centimeter,
10^{-9}	nano	n	as for body and clothing
10^{-12}	pico	p	measurement.
10^{-15}	femto	f	
10^{-18}	atto	a	

Conversion Factors to SI Units

To convert from	To	Multiply by
acre	square meter (m^2)	4.047×10^3
angstrom	meter (m)	1.0×10^{-10}[†]
are	square meter (m^2)	1.0×10^{2}[†]
astronomical unit	meter (m)	1.496×10^{11}
atmosphere	pascal (Pa)	1.013×10^5
bar	pascal (Pa)	1.0×10^{5}[†]
barn	square meter (m^2)	1.0×10^{-28}[†]
barrel (42 U.S. liquid gallons)	cubic meter (m^3)	0.1590
Bohr magneton (μ_β)	J/T	9.274×10^{-24}
Btu (International Table)	joule (J)	1.055×10^3
Btu (mean)	joule (J)	1.056×10^3
Btu (thermochemical)	joule (J)	1.054×10^3
bushel	cubic meter (m^3)	3.524×10^{-2}
calorie (International Table)	joule (J)	4.187
calorie (mean)	joule (J)	4.190
calorie (thermochemical)	joule (J)	4.184[†]
centipoise	pascal second (Pa·s)	1.0×10^{-3}[†]
centistokes	square millimeter per second (mm^2/s)	1.0[†]
cfm (cubic foot per minute)	cubic meter per second(m^3/s)	4.72×10^{-4}
cubic inch	cubic meter (m^3)	1.639×10^{-5}
cubic foot	cubic meter (m^3)	2.832×10^{-2}
cubic yard	cubic meter ($m^)$)	0.7646
curie	becquerel (Bq)	3.70×10^{10}[†]
debye	coulomb·meter (C·m)	3.336×10^{-30}
degree (angle)	radian (rad)	1.745×10^{-2}
denier (international)	kilogram per meter (kg/m)	1.111×10^{-7}
	tex[‡]	0.1111
dram (apothecaries')	kilogram (kg)	3.888×10^{-3}
dram (avoirdupois)	kilogram (kg)	1.772×10^{-3}
dram (U.S. fluid)	cubic meter (m^3)	3.697×10^{-6}
dyne	newton (N)	1.0×10^{-5}[†]
dyne/cm	newton per meter (N/m)	1.0×10^{-3}[†]
electron volt	joule (J)	1.602×10^{-19}
erg	joule (J)	1.0×10^{-7}[†]

[†] Exact.
[‡] See footnote on p. xxvi.

fathom	meter (m)	1.829
fluid ounce (U.S.)	cubic meter (m^3)	2.957×10^{-5}
foot	meter (m)	0.3048†
footcandle	lux (lx)	10.76
furlong	meter (m)	2.012×10^{-2}
gal	meter per second squared (m/s^2)	$1.0 \times 10^{-2\dagger}$
gallon (U.S. dry)	cubic meter (m^3)	4.405×10^{-3}
gallon (U.S. liquid)	cubic meter (m^3)	3.785×10^{-3}
gallon per minute (gpm)	cubic meter per second (m^3/s)	6.308×10^{-5}
	cubic meter per hour (m^3/h)	0.2271
gauss	tesla (T)	1.0×10^{-4}
gilbert	ampere (A)	0.7958
gill (U.S.)	cubic meter (m^3)	1.183×10^{-4}
grad	radian	1.571×10^{-2}
grain	kilogram (kg)	6.480×10^{-5}
gram-force per denier	newton per tex (N/tex)	8.826×10^{-2}
hectare	square meter (m^2)	$1.0 \times 10^{4\dagger}$
horsepower (550 ft·lbf/s)	watt (W)	7.457×10^2
horsepower (boiler)	watt (W)	9.810×10^3
horsepower (electric)	watt (W)	$7.46 \times 10^{2\dagger}$
hundredweight (long)	kilogram (kg)	50.80
hundredweight (short)	kilogram (kg)	45.36
inch	meter (m)	$2.54 \times 10^{-2\dagger}$
inch of mercury (32°F)	pascal (Pa)	3.386×10^3
inch of water (39.2°F)	pascal (Pa)	2.491×10^2
kilogram-force	newton (N)	9.807
kilowatt hour	megajoule (MJ)	3.6†
kip	newton (N)	4.48×10^3
knot (international)	meter per second (m/s)	0.5144
lambert	candela per square meter (cd/m^2)	3.183×10^3
league (British nautical)	meter (m)	5.559×10^3
league (statute)	meter (m)	4.828×10^3
light year	meter (m)	9.461×10^{15}
liter (for fluids only)	cubic meter (m^3)	$1.0 \times 10^{-3\dagger}$
maxwell	weber (Wb)	$1.0 \times 10^{-8\dagger}$
micron	meter (m)	$1.0 \times 10^{-6\dagger}$
mil	meter (m)	$2.54 \times 10^{-5\dagger}$
mile (statute)	meter (m)	1.609×10^3
mile (U.S. nautical)	meter (m)	$1.852 \times 10^{3\dagger}$
mile per hour	meter per second (m/s)	0.4470
millibar	pascal (Pa)	1.0×10^2
millimeter of mercury (0°C)	pascal (Pa)	$1.333 \times 10^{2\dagger}$
minute (angular)	radian	2.909×10^{-4}
myriagram	kilogram (kg)	10
myriameter	kilometer (km)	10
oersted	ampere per meter (A/m)	79.58
ounce (avoirdupois)	kilogram (kg)	2.835×10^{-2}
ounce (troy)	kilogram (kg)	3.110×10^{-2}
ounce (U.S. fluid)	cubic meter (m^3)	2.957×10^{-5}
ounce-force	newton (N)	0.2780
peck (U.S.)	cubic meter (m^3)	8.810×10^{-3}
pennyweight	kilogram (kg)	1.555×10^{-3}
pint (U.S. dry)	cubic meter (m^3)	5.506×10^{-4}
pint (U.S. liquid)	cubic meter (m^3)	4.732×10^{-4}
poise (absolute viscosity)	pascal second (Pa·s)	0.10†
pound (avoirdupois)	kilogram (kg)	0.4536
pound (troy)	kilogram (kg)	0.3732
poundal	newton (N)	0.1383
pound-force	newton (N)	4.448
pound-force per square inch (psi)	pascal (Pa)	6.895×10^3
quart (U.S. dry)	cubic meter (m^3)	1.101×10^{-3}
quart (U.S. liquid)	cubic meter (m^3)	9.464×10^{-4}
quintal	kilogram (kg)	$1.0 \times 10^{2\dagger}$
rad	gray (Gy)	$1.0 \times 10^{-2\dagger}$
rod	meter (m)	5.029

† Exact.

roentgen	coulomb per kilogram (C/kg)	2.58×10^{-4}
second (angle)	radian (rad)	4.848×10^{-6}
section	square meter (m^2)	2.590×10^6
slug	kilogram (kg)	14.59
spherical candle power	lumen (lm)	12.57
square inch	square meter (m^2)	6.452×10^{-4}
square foot	square meter (m^2)	9.290×10^{-2}
square mile	square meter (m^2)	2.590×10^6
square yard	square meter (m^2)	0.8361
stere	cubic meter (m^3)	$1.0^{\dagger}$
stokes (kinematic viscosity)	square meter per second (m^2/s)	$1.0 \times 10^{-4\dagger}$
tex	kilogram per meter (kg/m)	$1.0 \times 10^{-6\dagger}$
ton (long, 2240 pounds)	kilogram (kg)	1.016×10^3
ton (metric)	kilogram (kg)	$1.0 \times 10^{3\dagger}$
ton (short, 2000 pounds)	kilogram (kg)	9.072×10^2
torr	pascal (Pa)	1.333×10^2
unit pole	weber (Wb)	1.257×10^{-7}
yard	meter (m)	$0.9144^{\dagger}$

ABBREVIATIONS AND UNITS

A	ampere	*ar*-	aromatic
A	anion (eg, H*A*); mass number	*as*-	asymmetric(al)
a	atto (prefix for 10^{-18})	ASH- RAE	American Society of Heating, Refrigerating, and Air Conditioning Engineers
AATCC	American Association of Textile Chemists and Colorists		
		ASM	American Society for Metals
ABS	acrylonitrile–butadiene–styrene	ASME	American Society of Mechanical Engineers
abs	absolute	ASTM	American Society for Testing and Materials
ac	alternating current, *n.*		
a-c	alternating current, *adj.*	at no.	atomic number
ac-	alicyclic	at wt	atomic weight
acac	acetylacetonate	av(g)	average
ACGIH	American Conference of Governmental Industrial Hygienists	AWS	American Welding Society
		b	bonding orbital
		bbl	barrel
ACS	American Chemical Society	bcc	body-centered cubic
AGA	American Gas Association	bct	body-centered tetragonal
Ah	ampere hour	Bé	Baumé
AIChE	American Institute of Chemical Engineers	BET	Brunauer-Emmett-Teller (adsorption equation)
AIME	American Institute of Mining, Metallurgical, and Petroleum Engineers	bid	twice daily
		Boc	*t*-butyloxycarbonyl
		BOD	biochemical (biological) oxygen demand
AIP	American Institute of Physics		
AISI	American Iron and Steel Institute	bp	boiling point
		Bq	becquerel
alc	alcohol(ic)	C	coulomb
Alk	alkyl	°C	degree Celsius
alk	alkaline (not aikali)	*C*-	denoting attachment to carbon
amt	amount		
amu	atomic mass unit	c	centi (prefix for 10^{-2})
ANSI	American National Standards Institute	*c*	critical
		ca	circa (approximately)
AO	atomic orbital	cd	candela; current density; circular dichroism
AOAC	Association of Official Analytical Chemists		
		CFR	Code of Federal Regulations
AOCS	American Oil Chemists' Society		
		cgs	centimeter–gram–second
APHA	American Public Health Association	CI	Color Index
		cis-	isomer in which substituted groups are on same side of double bond between C atoms
API	American Petroleum Institute		
aq	aqueous		
Ar	aryl	cl	carload

† Exact.

cm	centimeter	estn	estimation
cmil	circular mil	esu	electrostatic unit
cmpd	compound	exp	experiment, experimental
CNS	central nervous system	ext(d)	extract(ed)
CoA	coenzyme A	F	farad (capacitance)
COD	chemical oxygen demand	F	faraday (96,487 C)
coml	commercial(ly)	f	femto (prefix for 10^{-15})
cp	chemically pure	FAO	Food and Agriculture
cph	close-packed hexagonal		Organization (United
CPSC	Consumer Product Safety		Nations)
	Commission	fcc	face-centered cubic
cryst	crystalline	FDA	Food and Drug Administration
cub	cubic	FEA	Federal Energy
D	debye		Administration
D-	denoting configurational	FHSA	Federal Hazardous
	relationship		Substances Act
d	differential operator	fob	free on board
d-	dextro-, dextrorotatory	fp	freezing point
da	deka (prefix for 10^1)	FPC	Federal Power Commission
dB	decibel	FRB	Federal Reserve Board
dc	direct current, *n.*	frz	freezing
d-c	direct current, *adj.*	G	giga (prefix for 10^9)
dec	decompose	G	gravitational constant =
detd	determined		$6.67 \times 10^{11}\,\text{N}\cdot\text{m}^2/\text{kg}^2$
detn	determination	g	gram
Di	didymium, a mixture of all	(g)	gas, only as in $H_2O(g)$
	lanthanons	g	gravitational acceleration
dia	diameter	gc	gas chromatography
dil	dilute	*gem-*	geminal
DIN	Deutsche Industrie Normen	glc	gas-liquid chromatography
dl-; DL-	racemic	g-mol wt;	gram-molecular
DMA	dimethylacetamide	gmw	weight
DMF	dimethylformamide	GNP	gross national product
DMG	dimethyl glyoxime	gpc	gel-permeation
DMSO	dimethyl sulfoxide		chromatography
DOD	Department of Defense	GRAS	Generally Recognized as Safe
DOE	Department of Energy	grd	ground
DOT	Department of	Gy	gray
	Transportation	H	henry
DP	degree of polymerization	h	hour; hecto (prefix for 10^2)
dp	dew point	ha	hectare
DPH	diamond pyramid hardness	HB	Brinell hardness number
dstl(d)	distill(ed)	Hb	hemoglobin
dta	differential thermal	hcp	hexagonal close-packed
	analysis	hex	hexagonal
(*E*)-	entgegen; opposed	HK	Knoop hardness number
ε	dielectric constant (unitless	hplc	high pressure liquid
	number)		chromatography
e	electron	HRC	Rockwell hardness (C scale)
ECU	electrochemical unit	HV	Vickers hardness number
ed.	edited, edition, editor	hyd	hydrated, hydrous
ED	effective dose	hyg	hygroscopic
EDTA	ethylenediaminetetraacetic	Hz	hertz
	acid	i(eg, Pri)	iso (eg, isopropyl)
emf	electromotive force	*i-*	inactive (eg, *i*-methionine)
emu	electromagnetic unit	IACS	International Annealed
en	ethylene diamine		Copper Standard
eng	engineering	ibp	initial boiling point
EPA	Environmental Protection	IC	inhibitory concentration
	Agency	ICC	Interstate Commerce
epr	electron paramagnetic		Commission
	resonance	ICT	International Critical Table
eq.	equation	ID	inside diameter; infective dose
esca	electron-spectroscopy for	ip	intraperitoneal
	chemical analysis	IPS	iron pipe size
esp	especially	IPTS	International Practical
esr	electron-spin resonance		Temperature Scale (NBS)
est(d)	estimate(d)	ir	infrared

IRLG	Interagency Regulatory Liaison Group	μ	micro (prefix for 10^{-6})
ISO	International Organization for Standardization	N	newton (force)
		N	normal (concentration); neutron number
IU	International Unit	N-	denoting attachment to nitrogen
IUPAC	International Union of Pure and Applied Chemistry	n (as n_D^{20})	index of refraction (for 20° C and sodium light)
IV	iodine value	^{n} (as Bu^n), n-	normal (straight-chain structure)
iv	intravenous		
J	joule	n	neutron
K	kelvin	n	nano (prefix for 10^9)
k	kilo (prefix for 10^3)	na	not available
kg	kilogram	NAS	National Academy of Sciences
L	denoting configurational relationship	NASA	National Aeronautics and Space Administration
L	liter (for fluids only)		
l-	*levo*-, levorotatory	nat	natural
(l)	liquid, only as in NH$_3$(l)	NBS	National Bureau of Standards
LC$_{50}$	conc lethal to 50% of the animals tested		
		neg	negative
LCAO	linear combination of atomic orbitals	NF	*National Formulary*
		NIH	National Institutes of Health
LCD	liquid crystal display		
lcl	less than carload lots	NIOSH	National Institute of Occupational Safety and Health
LD$_{50}$	dose lethal to 50% of the animals tested		
LED	light-emitting diode	nmr	nuclear magnetic resonance
liq	liquid	NND	New and Nonofficial Drugs (AMA)
lm	lumen		
ln	logarithm (natural)	no.	number
LNG	liquefied natural gas	NOI- (BN)	not otherwise indexed (by name)
log	logarithm (common)		
LPG	liquefied petroleum gas	NOS	not otherwise specified
ltl	less than truckload lots	nqr	nuclear quadruple resonance
lx	lux	NRC	Nuclear Regulatory Commission; National Research Council
M	mega (prefix for 10^6); metal (as in MA)		
M	molar; actual mass	NRI	New Ring Index
$\overline{M}_w$	weight-average mol wt	NSF	National Science Foundation
$\overline{M}_n$	number-average mol wt	NTA	nitrilotriacetic acid
m	meter; milli (prefix for 10^{-3})	NTP	normal temperature and pressure (25°C and 101.3 kPa or 1 atm)
m	molal		
m-	meta		
max	maximum	NTSB	National Transportation Safety Board
MCA	Chemical Manufacturers' Association (was Manufacturing Chemists Association)		
		O-	denoting attachment to oxygen
		o-	ortho
MEK	methyl ethyl ketone	OD	outside diameter
meq	milliequivalent	OPEC	Organization of Petroleum Exporting Countries
mfd	manufactured		
mfg	manufacturing		
mfr	manufacturer	o-phen	o-phenanthridine
MIBC	methyl isobutyl carbinol	OSHA	Occupational Safety and Health Administration
MIBK	methyl isobutyl ketone		
MIC	minimum inhibiting concentration	owf	on weight of fiber
		Ω	ohm
min	minute; minimum	P	peta (prefix for 10^{15})
mL	milliliter	p	pico (prefix for 10^{-12})
MLD	minimum lethal dose	p-	para
MO	molecular orbital	p	proton
mo	month	p.	page
mol	mole	Pa	pascal (pressure)
mol wt	molecular weight	pd	potential difference
mp	melting point	pH	negative logarithm of the effective hydrogen ion concentration
MR	molar refraction		
ms	mass spectrum		
mxt	mixture		

phr	parts per hundred of resin (rubber)	sl sol	slightly soluble
p-i-n	positive-intrinsic-negative	sol	soluble
pmr	proton magnetic resonance	soln	solution
p-n	positive-negative	soly	solubility
po	per os (oral)	sp	specific; species
POP	polyoxypropylene	sp gr	specific gravity
pos	positive	sr	steradian
pp.	pages	std	standard
ppb	parts per billion (10^9)	STP	standard temperature and pressure (0°C and 101.3 kPa)
ppm	parts per million (10^6)		
ppmv	parts per million by volume	sub	sublime(s)
ppmwt	parts per million by weight	SUs	Saybolt Universal seconds
PPO	poly(phenyl oxide)		
ppt(d)	precipitate(d)	syn	synthetic
pptn	precipitation	^{t}(eg,	tertiary (eg, tertiary butyl)
Pr (no.)	foreign prototype (number)	Bu^t),	
pt	point; part	*t-,*	
PVC	poly(vinyl chloride)	*tert-*	
pwd	powder	T	tera (prefix for 10^{12}); tesla (magnetic flux density)
py	pyridine		
qv	quod vide (which see)	t	metric ton (tonne); temperature
R	univalent hydrocarbon radical		
		TAPPI	Technical Association of the Pulp and Paper Industry
(*R*)-	rectus (clockwise configuration)		
		tex	tex (linear density)
r	precision of data	T_g	glass-transition temperature
rad	radian; radius	tga	thermogravimetric analysis
rds	rate determining step	THF	tetrahydrofuran
ref	reference	tlc	thin layer chromatography
rf	radio frequency, *n.*	TLV	threshold limit value
r-f	radio frequency, *adj.*	*trans-*	isomer in which substituted groups are on opposite sides of double bond between C atoms
rh	relative humidity		
RI	Ring Index		
rms	root-mean-square		
rpm	rotations per minute		
rps	revolutions per second	TSCA	Toxic Substance Control Act
RT	room temperature	TWA	time-weighted average
^{s}(eg,	secondary (eg, secondary butyl)	Twad	Twaddell
Bu^s);		UL	Underwriters' Laboratory
sec-		USDA	United States Department of Agriculture
S	siemens		
(*S*)-	sinister (counterclockwise configuration)	USP	*United States Pharmacopeia*
		uv	ultraviolet
S-	denoting attachment to sulfur	V	volt (emf)
		var	variable
s-	symmetric(al)	*vic-*	vicinal
s	second	vol	volume (not volatile)
(s)	solid, only as in H$_2$O(s)	vs	versus
SAE	Society of Automotive Engineers	v sol	very soluble
		W	watt
SAN	styrene–acrylonitrile	Wb	weber
sat(d)	saturate(d)	Wh	watt hour
satn	saturation	WHO	World Health Organization (United Nations)
SBS	styrene–butadiene–styrene		
sc	subcutaneouss	wk	week
SCF	self-consistent field; standard cubic feet	yr	year
		(*Z*)-	zusammen; together; atomic number
Sch	Schultz number		
SFs	Saybolt Furol seconds		
SI	Le Système International d'Unités (International System of Units)		

A

ABACA FIBER. See Fibers, vegetable.

ABHERENTS

An abherent is a substance that prevents or reduces the adhesion of a material to itself or to another material. Also known as release agents, parting agents, antiblocking agents, slip aids, and external lubricants, abherents function as the antitheses of adhesives.

The primary requirement of a good abherent is chemical inertness towards at least one of the materials whose adhesion is to be prevented. Factors such as chemical inertness, wetting properties, the temperature range of use, interference with subsequent processes, and level of toxicity are evaluated in selection. Health and safety factors are considered from both a manufacturing and consumer standpoint. A number of abherents, such as silicone, waxes, poly(vinyl alcohol), and stearates, have FDA clearance.

Applications

Abherents have wide industrial and commercial uses. In casting, they allow the casting to separate from the mold during cooling and are selected on the basis of the temperatures used in the casting operation. Due to their temperature stability, silicones have become dominant as abherents in compression; transfer die casting; injection, blow, and rotational molding; reaction-injection molding (RIM); and liquid-injection molding (LIM). Abherents are used in forming operations such as extrusion, melt spinning, vacuum forming, drawing, hobbing, and stamping. They have a secondary function as lubricants in many forming operations. Abherents are essential in industrial operations involving pressure-sensitive tapes or labels, the process of transfer coating, the application of hot stamping foils, and the printing industry. They are used as release surfaces on conveyers, bins, chutes, tanks, and reactors; on release papers; and on release fabrics. They protect processing equipment, such as backup rolls in coating or printing operations, pumps, mills, mixing equipment, etc, thereby reducing cleanup time, product transfer to auxiliary equipment, and product buildup in undesired areas, leading to increased production and improved yields. Abherents are particularly important when the finished product is intractable, eg, cured polymers.

Abherents are applied by standard coating methods (see Coating processes) such as spraying, brushing, dusting, dipping, electrostatic powder coating, and plasma-arc powder coating. They are incorporated into one of the materials, a process particularly employed in film or molded polymeric systems.

Abherent Materials

Abherent materials include silicone, the best known and commercially most important abherent, favored because of its low critical tension and temperature stability; waxes, including hydrocarbon, natural ester and synthetic waxes, used as release agents; stearates, including zinc, calcium, magnesium, and lithium stearate, all relatively inexpensive; release papers; cellulose derivatives, such as cellophane, which are generally now replaced by poly(ethylene terephthalate)); polyolefin abherents, accepted as nontoxic and used in shipping or storage of food products, uncured elastomers, and pressure-sensitive tapes (see Olefin polymers); vinyl compounds, such as poly(vinyl acetate) and poly(vinyl alcohol), which are nontoxic, easily removed, and which do not interfere with post-finishing operations; fluorocarbon polymers, noted for their high heat resistance, high chemical inertness, and low surface tension; natural products, such as flour, confectioner's sugar, rice flour, rice paper, sodium, potassium and calcium alginates; and inorganic abherents such as talc, mica, and fumed silica (see Silica), used as dusting agents in the printing

and rubber industries and in the extrusion of films and tubing to prevent blocking.

RICHARD D. COWELL
BERNARD R. BLUESTEIN
Witco Chemical Corporation

J. Brandrup and E.H. Immergut, *Polymer Handbook*, 2nd ed., John Wiley & Sons, Inc., New York, 1975.

L. Goldberg, ed., *Fluorocarbons*, Uniscience, New York, 1976.

K.S. Markley, *Fatty Acids, Their Chemistry, Properties, Production and Uses*, 2nd ed., Pt. 2, John Wiley & Sons, Inc., New York, 1961.

Modern Plastics Encyclopedia, McGraw-Hill, New York, Annual Volumes, 1975–1982.

ABIETIC ACID. See Terpenoids.

ABLATIVE MATERIALS

Ablation describes the erosion and disintegration of meteors resulting from the intense heat generated by passage through the atmosphere at high velocities. The development and study of materials that protect the payload and structure of ballistic missiles, earth satellites, or space probes from damage during ablation is a relatively new area of chemical technology. The same principles of thermal protection also apply to other applications, such as the protection of rocket nozzles from attrition by propellant gases and protection from laser beams.

Attempts to simulate the more severe environments for laboratory tests in almost all cases lack some aspect of the real environment; the mechanisms of ablation are studied to predict reliably performance under conditions that cannot be adequately simulated.

Owing to the diversity in environments, many types of materials are utilized where ablation is involved. During reentry, a protective heat shield must perform as an aerodynamic body, a predictably decomposing ablative material, a structural component, and an insulator.

In missile nose tips, where shape change may affect the trajectory, a material such as graphite may be selected primarily to minimize recession under very high heat fluxes. The element carbon, owing to its high heat capacity per unit weight, high energy of vaporization, and high temperature and pressure required for melting, has the highest heat of ablation, ie, energy absorbed per mass lost, of any material, provided that mechanical removal of particulates does not occur.

If the environment involves erosion by snow or ice particles, a refractory metal may be considered. Beryllium, copper, tungsten, molybdenum, and steel have functioned as a heat sink for protection against heat pulses of relatively short duration. Refractory carbides, borides, nitrides, and oxides have also been considered because of high melting point, high thermal conductivity, and high specific heat.

On the other hand, ability to insulate or protect the payload from the heat generated, with minimum weight, is ordinarily the primary consideration. A variety of ablative plastics and fiber-reinforced plastic composites have been developed. Polytetrafluoroethylene or Teflon was used in the design of a low weight heat shield for a Venus entry probe. The phenolic resins (qv) among the common plastics that give the highest yield of carbon during pyrolysis have been widely used as surface-charring ablative materials. Since a char is relatively weak, and removed mechanically by high shear forces associated with reentry, fibers of carbon, silicon dioxide, refractory oxides, mineral asbestos, or even glass have been added to assist the char retention.

When fiber-reinforced plastic ablators are exposed to ablative environments, they first act as heat sinks (see also Laminated and reinforced plastics). As heating progresses, the outer layer of polymer may become viscous and then begins to degrade, forming a foaming carbonaceous mass and ultimately a porous carbon char. The char is a thermal insulation; the interior is cooled by volatile material percolating through it from the decomposing polymer The resinous component of a typical reinforced plastic ablator comprises ca 35% of the total weight. Since this part provides the essential gas for transpiration cooling of the char, the decomposition products are important.

A variety of resins, including the phenolic resins, a polyimide resin, polybenzimidazole (PBI), poly(p-diphenylene oxide) (DPO), polyphenylenes, p-phenylphenols, and substituted heterocyclic polymers, are utilized for ablative composites (see Polyimides; Heat-resistant polymers; Composite materials).

The mechanical and thermal properties of the ablative materials can limit their application, owing to structural requirements or stresses resulting from the thermal gradients. The plastics and plastic composites are relatively simple to attach to structures because the back face remains at low temperatures. However, the fiber reinforcement must provide good mechanical properties. Laminated composites of the cloth-reinforced phenolic plates have usually been made as heat shields by tape-wrapping cloth impregnated with the resin, molding, and curing. The most critical property affecting thermal stresses is the thermal expansion.

The thermal-stress limitations of industrial graphites have been overcome without significant degradation in ablation performance by incorporating 40–50 vol% of high strength graphite fibers as a three-dimensional reinforcement (see Carbon).

The mechanical properties of the antenna-window portion of a reentry thermal shield have been improved by incorporating fiber reinforcements; either high purity silicon dioxide or boron nitride is used to avoid forming a conducting char.

E.R. STOVER
P.W. JUNEAU, JR.
J.P. BRAZEL
General Electric Company

G.F. D'Alelio and J.A. Parker, eds., *Ablative Plastics*, Marcel Dekker, Inc., New York, 1971, pp. 1–39, 287–313.

H.B. Palmer and M. Shelef in P.L. Walder, Jr., ed., *Chemistry and Physics of Carbon*, Vol. 4, Marcel Dekker, Inc., New York, 1968, pp. 85–135.

J.A. Segletes, *J. Spacecr. Rockets* **12**, 251 (1975).

H.L. Moody and co-workers, *J. Spacecr. Rockets* **13**, 746 (1976).

ABRASIVES

Abrasiveness is defined by measuring and comparing two abrasive substances according to four properties: hardness, toughness, degree of chemical inertness, and, in some cases, resistance to heat.

Hardness is measured either by the Mohs scale, with hardness increasing from 1 to 10, or the Knoop scale, based on the resistance of a material to penetration by diamond, on of the two hardest substances known. The Mohs scale is unsatisfactory for modern abrasives because it is not linear, ie, a hardness of 2 is not twice as hard as a hardness of 1 (see Table 1). The Knoop scale actually measures resistance to plastic flow (see Hardness).

Toughness is the ability of an abrasive to withstand the forces brought to bear on it when in use. It is measured as the inverse proportion of the degree to which particles are broken down or reduced in size when the abrasive grains are shot against a barrier or ball-milled.

Chemical activity is measured since, generally, the best abrasive for a given operation is one that is most nearly inert to the material to be ground. Abrasion produces heat, and thus greatly accelerates any chemical activity between the abrasive and the abraded article; in some cases it is important to measure resistance to heat.

The term abrasive engineering refers to the various processes covering the three main areas of abrasive applications in industry: grinding, abrasive machining, and mechanical finishing.

Materials

Abrasive materials consist of granular particles (classified by a numerical scale) commonly called grits or grains. Each particle is a cutting, grinding, or polishing tool. Abrasive materials are either natural or manufactured. Natural abrasives include:

Diamond (see Carbon), which is carbon in crystalline form, and one of the two hardest substances known. (The other is cubic boron nitride.) Industrial diamonds fall into four classes: three naturally occurring stones (bort, ballast, and carbonado) and a synthetic version developed by General Electric.

Corundum (see Aluminum compounds), a naturally occurring massive crystalline form of aluminum oxide and an impure form of ruby and sapphire, employed as a loose abrasive in grinding and polishing lenses.

Emery, an intimate mixture of corundum (Al_2O_3) and magnetite (Fe_3O_4), with or without hematite (Fe_2O_3), which varies in hardness and toughness according to the iron oxide present. Emery is employed predominantly in the construction of nonskid concrete floors and pavements.

Garnet, a group of seven species of silicates that possess similar physical properties and crystal forms but dissimilar chemical compositions. Their general formula is $(MO)_3M'_2O_3 \cdot (SiO_2)_3$, the divalent element (M) being calcium, magnesium, iron, or manganese and the trivalent element (M′) being aluminum, iron, chromium, or titanium. Garnet is used largely for coated paper and cloth products and in the grinding of optical lenses.

Table 1. Scales of Hardness[a]

Mohs scale		Ridgway's extension of Mohs scale		Knoop scale hardness numbers, at a 100 g-load (K-100) average, kgf/mm^2	
talc	1				
gypsum	2				
calcite	3				
fluorite	4				
apatite	5				
feldspar	6	orthoclase or periclase	6		
quartz	7	vitreous pure silica	7	quartz	820
topaz	8	quartz	8	topaz	1350
corundum	9	topaz	9	corundum	2000
aluminum oxide	9	garnet	10	fused alumina	2050
silicon carbide	9.50	fused zirconia	11	silicon carbide	2500
boron carbide	9.75	fused alumina	12	boron carbide	2800
Borazon (boron nitride)	10	silicon carbide	13		
diamond	10	boron carbide	14	Borazon (boron nitride)	4700
		diamond	15	diamond	8350 ?

[a] Boron carbide, Borazon, silicon carbide, and aluminum oxide are manufactured abrasives. Borazon and boron carbide are relatively recent developments and the above hardness ratings are subject to some differences of opinion and differing laboratory test results.

Pumice and pumicite, forms of cellular, glassy volcanic lava used for dressing wood and metal surfaces of furniture and musical instruments, and in a variety of cleaning and surface-material preparation processes.

Tripoli (not to be confused with Tripolite, a diatomaceous earth found in Tripoli, Libya), the general name for a number of fine grained, lightweight, friable and porous forms of decomposed siliceous rock used in buffing compounds and in optical polishing.

Synthetic abrasives include:

Silicon carbide, made from silicon dioxide, finely ground petroleum coke, salt, and sawdust (see Carbides), available in two basic types: a gray or black type used on cast iron, nonferrous metals, and nonmetallic materials; and a green type used for grinding tungsten carbides.

Aluminum oxide, Al_2O_3, made from bauxite with titanium oxide (TiO_2) added to produce varying degrees of toughness or friability (see Aluminum compounds). It is available in colors ranging from deep brown to pink to white; the brown abrasive is used in most general grinding operations, whereas the pink, red, and white are suited to grinding high-alloy steel tools.

Zirconia alumina, a true alloy of zirconium oxide and aluminum oxide, commercially used for heavy duty grinding and in lighter applications such as cutting-off, portable grinding, and coated-abrasive operations. These abrasives are characterized by extreme strength, hardness, and sharpness.

Synthetic diamond, produced by General Electric.

Boron nitride, a black, brown, or dark-red substance almost as hard as diamond but possessing better heat stability (see Boron compounds), and used in bonded abrasive products.

Tungsten carbide (see Carbides), almost exclusively used as a single-point cutting tool.

Metallic abrasives, eg, malleable cast-iron shot, cast-steel shot, steel wire cut into shot pieces; used extensively in sandblasting, cleaning exteriors of stone buildings, cutting stone, and descaling metals.

Steel wool, made from a variety of steels, used for finishing wood and soft metals, as well as for scouring and cleaning.

Metal oxides, precipitated chemically into a very fine grain size, used as polishing agents.

Forms

Abrasives are used in three basic forms, ie, loose, granular or powdered particles; particles bonded with various agents into wheel, segmental, or stick shapes; and particles coated on paper or cloth with glue or synthetic resins.

Coated abrasives. Coated abrasives consist of abrasive grains, such as aluminum oxide, silicon carbide, garnet, flint, emery, or crocus, deposited through a film of adhesive to a flexible backing, eg, paper, cloth, vulcanized fiber, or a combination of these materials. Manufacturing processes first apply a "make coat" of adhesive, such as hide glue, phenolic resin, synthetic varnish, or a combination of one of these with a mineral filler; the thickness of the make coat is determined by the grit size of the abrasive being applied. The abrasive is then deposited on the coated backing by mechanical or electrostatic means, and a "size coat" is produced which adheres to the make coat and reinforces the abrasive. Finishing or converting steps, such as flexing, may be added to the process to permit the product to bend.

Coated abrasives are used on a myriad of machines, ranging from portable grinders to giant surface grinders and roll grinders. Radial-flap abrasive wheels using coated abrasives can grind and polish contours that are impossible with bonded abrasives. Coated abrasive disks are widely used on a variety of disk grinders.

Bonded abrasives. Vitrified and resinoid bonds are used to bond abrasive materials. Low speed grinding wheels require ceramic (vitrified) bonds, whereas high speed wheels require resinoid bonds. 80–90% of all abrasive products use one of these two types of bonds, with the remaining 10–20% using bonds made of special-purpose rubber, shellac, and magnesium oxychloride.

The basic ceramic-bond materials are feldspar, frit, and high grade clay. Aluminum oxide products are bonded with a glass produced from these materials. Silicon carbide is sufficiently reactive to decompose in glass at elevated temperatures, and therefore it is bonded with a porcelain less reactive than glass.

Resins (see Phenolic resins) consist of phenol and formaldehyde products which are thermosetting in nature and particularly suitable for tough grinding and cutting-off operations at high speeds. The abrasives do not react with the bonds because of the low processing temperature. When compounds such as potassium fluoroborate (KBF_4), potassium sulfate, and sodium chloride are used in resinoid bonds, they lubricate the surface of the metal being ground. Resinoid-bonded wheels are used in the quick removal of large amounts of metal (as in steel-mill and foundry snagging), in grinding with abrasive disks, and in many dry-cutting operations.

Natural and synthetic rubbers are used for bonding purposes owing to their strength and flexibility. Combined with resin bonds, they maintain the heat-resistant character of the resin bond and have the strength and flexibility of the rubber. Only 8–11% of total bonds used today in the U.S. for synthetic wheels are rubber; these bonds are mainly used in grinding bearing races, feed wheels for centerless grinding, and wet cutting-off operations.

Shellac is considered a cool grinding bond since it generates much less heat than other types. Sodium silicate (see Silicates) and magnesium oxychloride are bonds that are of relatively minor use today.

In terms of production volume, the grinding wheel is the most important bonded abrasive product. It is made up of abrasive grains held in a bond, or matrix, with pores, or void spaces. The wheel is identified according to a standard marking system (see Fig. 1). The letters or number symbols may vary in products of the various manufacturers.

Aluminum oxide bonded grinding wheels are used for grinding various types of steel; the so-called softer steels are ground with a tough, titania-containing aluminum oxide or zirconia aluminum oxide. The harder wear-resistant and stainless steels require a purer, more friable aluminum oxide, and the extremely heat-resistant, wear-resistant, and exotic steels use a white or pink pure, very friable aluminum oxide. Silicon carbide wheels are used to grind cast iron, nonferrous metal, and nonmetallic articles. Diamonds and borazon grains grind tungsten carbide, cut or slice precious stones, marble, granite, and slate, and are used to slot the expansion joints in concrete runways and roads.

Performance properties of the grinding wheel depend upon the relative volume of grains, the bond, and the open pores in the body. The grade, ie, tenacity with which the bonding material clings to the abrasive grains during stress to prevent break-out (depends on the volume percent of the bond), and structure, ie, the spacing of the abrasive grains indicated by a numeral following the grade letter from 1 (closed) to 15 (open), determine the system. Because the amounts of abrasive, bond, and voids within a wheel are interrelated in the three-component system, a change in one of the components must affect at least one other factor.

In addition to grinding wheels, disks, segments, and sticks, bonded abrasive products also take special forms such as honing sticks, super-

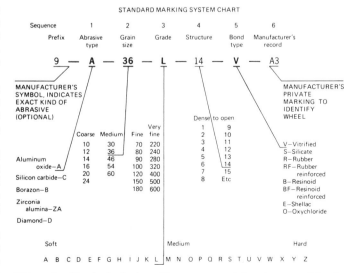

Figure 1. Standard wheel markings recommended and approved by the Standards Committee of The Grinding Wheel Institute.

finishing sticks, diamond wheels, pulpstone wheels, crush-form grinding, and reinforced wheels.

Loose abrasives. In addition to uses in bonded and coated products, abrasives are used in their loose, particulate form in a number of operations including polishing, buffing, pressure, blasting, wire sawing, barrel finishing, in antislip materials, lithography work, as the chemical and metallurgical raw materials, and as electrical materials. Metal polishing usually follows grinding and precedes buffing in the finishing operations; it serves to reduce surface irregularities caused by grinding or machining by producing uniform surfaces of specific characteristics.

Buffing generates a highly lustrous and reflective finish when a 240-grit (240 mesh or 60 μm) or finer abrasive is employed. Lapping, an abrading method for refining the surface and promoting geometrical accuracy on flat, cylindrical, spherical, and formed surfaces, uses natural emerys, aluminum oxide, silicon carbide, boron carbide, and diamonds as abrasives. In pressure blasting, a process for cleaning, altering, or preparing surfaces, a stream of abrasive material is forcibly propelled against the surface of the part with air or liquid. Abrasive materials such as sand, metal shot, aluminum oxide, and silicon carbide are used. In lithography (see Printing processes), loose abrasives are used to produce a surface roughness which promotes better wetting of the plates.

Grinding fluids, eg, water, soluble oils, and petroleum oils dispersed in water, are used to cool and lubricate the work in wet abrasive operations. Air or carbon dioxide forced through jets to the points of contact of wheel and workpiece are also used as coolants, as well as wax, sulfur, and rosin incorporated into grinding wheels to both cool and lubricate. Stearates and grease in stick form are employed with coated abrasives for these purposes. It is of the utmost importance not to use a coolant that would affect the wheel bond: alkalies attack resinoid, shellac, and silicate bonds; oils attack rubber; all coolants may be used with vitrified bonds. Grinding fluids must be kept clean to produce the desired finish on the piece.

WILLIAM G. PINKSTONE
Consultant

W.G. Pinkstone, *The Abrasive Ages*, Sutter House, Lititz, Pa., 1974, 124 pp.

Coated Abrasives, Modern Tool of Industry, Coated Abrasives Manufacturers Institute, Keith Bldg., Cleveland, Ohio, 1965.

W.F. Schleicher, *The Grinding Wheel*, 3rd ed., The Grinding Wheel Institute, Cleveland, Ohio, 1976.

ABSORPTION

Absorption, or gas absorption, is a unit operation in the chemical industry to separate gases by washing or scrubbing a gas mixture with a suitable liquid. One or more of the constituents of the gas mixture will dissolve in the liquid and can be removed. This gaseous constituent reacts to form a physical solution with the liquid or solvent, or reacts with the liquid chemically.

Scrubbing operations are used for gas purification (eg, removal of air pollutants from exhaust gases), product recovery, and production of solutions of gases for various purposes.

Gas absorption is carried out in vertical countercurrent columns, ie, the solvent is fed at the top of the absorber and the gas mixture enters from the bottom. The absorbed substance is washed out by the solvent and leaves at the bottom as a liquid solution. The solvent is often recovered in a subsequent stripped or desorption operation, which is essentially the reverse of absorption. The absorber may be a packed, plate, or simple spray column.

The packed column is a shell filled with packing materials designed to disperse the liquid and bring it into close contact with the rising gas. Packed columns offer simpler and cheaper construction and are preferred for corrosive gases because they can be made, eg, from ceramics. They are used in vacuum applications because the pressure drop is usually less than through plate columns.

In plate towers, liquid flows from plate to plate in cascade fashion, and gases bubble through the flowing liquid at each plate through a multitude of dispersers or through a cascade of liquid as in a shower deck tray. These absorbers are used where tall columns are required. Tall packed towers are subject to channeling and maldistribution of the liquid streams; plate towers can be cleaned more easily. Since coils are more easily installed in plate towers, they are used when large heat effects require internal coils and can be designed for large liquid holdup.

The fundamental physical principles underlying gas absorption are the solubility of the absorbed gas and the rate of mass transfer. Both must be considered when sizing the equipment for a given application (see Distillation; Mass transfer; Heat-exchange technology).

A general way to express solubilities is through the vapor-liquid equilibrium constant m defined by $y_A = mx_A$. The value of m is widely employed to represent hydrocarbon vapor-liquid equilibria in absorption and distillation calculations. At constant pressure and temperature (equivalent to constant m), eg, where Raoult's or Henry's laws for ideal or near-ideal solutions may apply, a plot of y vs x for a given solute is linear from the origin. In other cases, the $y - x$ plot may be approximated by a linear relationship over limited regions. Generally, for nonideal solutions and nonisothermal conditions, y is a curving function of x and must be determined from experimental data or more rigorous theoretical relationships. The $y - x$ plot when applied to absorber design is commonly called the equilibrium line.

Mass-Transfer Concepts

Mass-transfer coefficients and driving forces. In order to determine the size of the equipment necessary to absorb a given amount of solvent per unit time, the equilibrium solubility of the solute in the solvent, and the rate at which equilibrium is established must be known (see Mass transfer). An essentially stable gas-liquid interface to describe this process has been proposed. Large fluid motions are presumed to exist at a certain distance from this interface, distributing all material rapidly and equally in the bulk of the fluid such that no concentration gradients are developed. Closer to this interface, however, the fluid motions are impaired and the slow process of molecular diffusion becomes more important as a mechanism of mass transfer. The rate-governing step in gas absorption may be viewed, therefore, as the transfer of solute through two thin gas and liquid films adjacent to the phase interface. Transfer of material through the interface itself is normally presumed to take place instantaneously so that equilibrium exists between these two films precisely at the interface. Although this assumption has been confirmed in numerous experiments with many systems and different types of phase interfaces, interfacial resistance can develop in some situations.

Film theory. This classical model, put forth to explain and to correlate experimentally measured mass-transfer coefficients, proposes to approximate the situation by "effective" gas and liquid films. Fluid motion may be assumed to be completely laminar within these effective films with a sharp change to totally turbulent flow where the film is in contact with the bulk of the fluid. As a result, mass is transferred through the effective films only by steady-state molecular diffusion, and the concentration profile is governed by Fick's law. Specific equations can be developed for situations involving equimolar counterdiffusion in binary gases, unidirectional diffusion through a stagnant medium, and multicomponent diffusion. Also, rate equations can be derived with concentration-independent mass-transfer coefficients in terms of the constant coefficients for equimolar counterdiffusion. This leads to rate equations with constant mass-transfer coefficients. Similarly, the overall mass-transfer coefficients may be made independent of the effect of bulk flux through the films and thus nearly concentration independent for straight equilibrium lines.

Absorption with chemical reaction. In instances where the solute gas is absorbed into a liquid or a solution with which it is able to undergo chemical reaction, the driving force becomes far more complex. The solute will not only diffuse through the liquid film at a rate determined by the gradient of concentration, but at the same time will react with the liquid at a rate determined by the concentrations of both the solute and the solvent at the point of interest. With addition of a term for the chemical reaction to Fick's second law of diffusion and integration, the concentration profiles through the liquid film can be computed based on a particular mass-transfer model. The results are often expressed as an enhancement factor ϕ, defined as the fractional increase of the liquid film mass-transfer coefficient due to the chemical reaction (k_L^r/k_L^o). The solutions that have been developed in this manner based on film, penetration, and surface-renewal theories are quite similar for a given type of reaction.

Design of Packed Absorption Columns

The concepts and procedures involved in designing packed gas absorption are here confined to "normal" gas absorption processes without complication: isothermal absorption of a solute from an inert gas into a nonvolatile solvent without chemical reaction. The height and diameter of the column required to reduce the concentration of the solute in the gas stream from an original value of $y_{A,1}$ to a residual level $y_{A,2}$ must be determined.

Standard absorber design methods. As the gas mixture travels up through a gas-absorption tower (Fig. 1), the solute A is gradually washed out. The liquid accumulates solute on its way down through the column, and x increases from the top to the bottom of the column. The steady state concentrations y and x at any given point in the column are interrelated through a mass balance either around the upper or lower part of the column (eq. 2), whereas the four concentrations in the streams entering and leaving the system are interrelated by an overall material balance.

Since the total gas and liquid flow rates per unit cross-sectional area vary through the tower, as shown in Figure 1, rigorous material balances should be based on the constant inert-gas and solvent-flow rates, G_M' and L_M' and expressed in terms of mole ratios, Y' and X'. A balance around the upper part of the tower yields

$$G_M' Y' + L_M' X_2' = G_M' Y_2 + L_M' X' \tag{1}$$

which may be rearranged to give

$$Y' = \frac{L_M'}{G_M'}\left(X' - X_{A,2}'\right) + Y_{A,2} \tag{2}$$

where G_M', L_M' = inert gas and solvent flow rate, kg mol/(h·m²) or lb mol/(h·ft²); and $Y' = y/1 - y$ and $X' = x/1 - x$. The overall material balance is obtained by substituting $Y' = T_1'$ and $X' = X_1'$. For

dilute gases, the total molar gas and liquid flows may be assumed constant, and a similar mass balance yields

$$Y = \frac{L_M}{G_M}(x - x_2) + y_2 \tag{3}$$

A plot of either equation 2 or 3 is called the operating line of the process. As indicated by equation 3, the line will be straight with a slope given by L_M/G_M for dilute gases. It is always straight when plotted in $Y' - X'$ coordinates. Together with the equilibrium line, the operating line permits the evaluation of the driving forces for gas absorption along the column. The farther apart the equilibrium and operating lines, the larger the driving forces will become and the faster absorption will occur, resulting in the need for a shorter column.

To place the operating line, the flows, composition of the entering gas y_1, entering liquid x_2, and desired degree of absorption y_2 are usually specified. The specification of the actual liquid rate used for a given gas flow (the L_M/G_M ratio) usually depends upon an economical optimization since the slope of the operating line may be seen to have a drastic effect on the driving force. For example, use of a very high liquid rate will result in a short column and a low absorber cost, but at the expense of a high cost for solvent circulation and subsequent recovery of the solute from the relatively dilute solution. On the other hand, liquid rate near the theoretical minimum, which is the rate at which the operating line just touches the equilibrium line, will require a very tall tower because the driving force becomes very small at its bottom. Use of a liquid rate on the order of one and one-half times the theoretical minimum is not unusual. In the absence of a detailed cost analysis, the L_M/G_M ratio is often specified at one and two-fifths times the slope of the equilibrium line.

The packed height of the tower required to effect a desired composition change may be calculated by combining point values of the mass-transfer rate with a differential material balance for the absorbed component, referring to a slice of the absorber. The equation for h is

$$h = \int_{y_2}^{y_1} \frac{G_M}{k_G' aP} \frac{y_{BM}\, dy}{(1 - y)(y - y_i)} \tag{4}$$

where P = pressure, h_G = gas-phase mass-transfer coefficient; $y_{BM} = (1 - y_A) - (1 - y_{A1})/\ln(1 - y_A/1 - y_{Ai})$ = logarithmic mean of inert gas concentration between the phase interface and the bulk of the gas phase; and a is the interfacial area present per unit volume of packing. Equation 4 may be integrated numerically or graphically evaluating its component terms at a series of points on the operating line. y_1 is found by placing tie lines from each of these points, with slopes given by

$$\frac{y_A - y_{Ai}}{x_A - x_{Ai}} = \frac{k_{LP}}{k_G P}$$

Equation 4 is thus a general expression determining the column height required to effect a given reduction in y_A' and can often be simplified by adopting the concept of a mass-transfer unit. The group $G_M/k_G' aP$ has the dimension of length, or height, and is thus designated the gas-phase height of one transfer unit H_G. The integral is dimensionless and indicates show many of these transfer units it takes to make up the whole tower. Consequently, it is called the number of gas-phase transfer units N_G, expressed as

$$h = (H_G)(N_G) \tag{5}$$

A similar treatment is possible in terms of an overall gas-phase driving force

$$h = (H_{OG})(N_{OG}) \tag{6}$$

H_{OG} and N_{OG} are called the overall gas-phase height of a transfer unit and the number of overall gas-phase transfer units, respectively. Use of equation 6 is especially convenient because N_{OG}' as opposed to N_G can be evaluated without solving for the interfacial concentrations. In all other cases, H_{OG} must be retained under the integral, and its value calculated from H_G and H_L at different points of the equilibrium line as follows:

$$H_{OG} = \frac{y_{BM}}{y_{BM}^*}H_G + \frac{mG_M}{L_M}\frac{x_{BM}}{y_{BM}^*}H_L \tag{7}$$

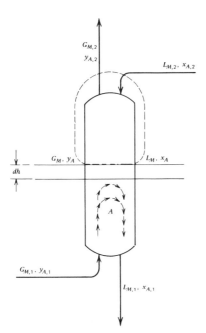

Figure 1. Mass balance in gas absorption columns. The curved arrows labeled A indicate the travel path of the solute A. The upper broken curve delineates the envelope for the material balance, equation 2.

To make use of this concept and the related equations, the heights of the transfer units or the mass-transfer coefficients $k_G'a$ and k_L must be known. Transfer data for packed columns are often measured and reported directly in terms of H_G and H_L and correlated in this form against G_M and L_M. The use of absorption and stripping factors in the Kremser-Brown method allows rapid design for absorption calculations. One of the simplifying assumptions made in the development of this Kremser-Brown method involves the use of the absorption factor.

For relatively dilute systems, the ratios involving $y_{BM} > y_{BM}^*$ ($y_{BM} = (1 - y_A)(1 - y_A^*) \ln[(1 - y_A/1 - y_A^*)]$ (ie, the logarithmic mean of inert gas concentration between the equilibrium concentration and the bulk of the gas phase) and $1 - y$ approach unity so that the computation of H_{OG}, equation 7, and of N_{OG} may be simplified.

$$H_{OG} = H_G + \left(\frac{mG_M}{L_M}\right)H_L \tag{8}$$

$$N_{OG} \approx N_T = \int_{y_2}^{y_1} \frac{dy}{y - y^*} \tag{9}$$

It may be shown that equation 9 is a rigorous expression for the number of overall transfer units for equimolar counterdiffusion, that is, for instance, in distillation columns.

For cases in which the equilibrium and operating lines may be assumed straight, with slopes L_M/G_M and m, respectively, an algebraic expression for the integral of equation 9 has been developed:

$$N_{OG} \approx N_T = \frac{\ln\left[\left(1 - \frac{mG_m}{L_M}\right)\left(\frac{y_1 - mx_2}{y_2 - mx_2}\right) + \frac{mG_M}{L_M}\right]}{1 - \frac{mG_M}{L_M}} \tag{10}$$

The required tower height may thus be easily calculated from the product of H_{OG} and N_{OG}, where H_{OG} is given by equation 8 and N_{OG} by equation 10.

If the operating or the equilibrium line is not straight, equation 10 is of little use because mG_M/L_M will assume a range of values over the tower. The substitution of effective average values for m and for L_M/G_M into equations 10 and 7 obviates lengthy graphic or numerical integrations and leads to a quick, approximate solution for the required tower height. A convenient approach for purposes of approximate design is to define a correction term ΔN_{OG} which can be added to equation 9. Thus,

$$N_{OG} = N_T + \Delta N_{OG} \tag{11}$$

For cases in which y_{BM}^* may be represented by an arithmetic mean,

$$\Delta N_{OG} = \frac{1}{2}\ln\frac{1 - y_2}{1 - y} \tag{12}$$

Equation 12 will be sufficiently accurate for most situations. The required tower height is finally calculated by means of equation 6.

Generalized design equations for gas absorption and distillation. Where ideal equimolar counterdiffusion prevails in distillation columns, the total molar flows L_M and G_M are constant, and the mass balance is given by equation 3. Both unidirectional diffusion through stagnant media and equimolar diffusion are idealizations that are usually violated in real processes. In gas absorption, slight solvent evaporation may provide some counterdiffusion, and in distillation, counterdiffusion may not be equimolar for a number of reasons. This is particularly true for multicomponent operation.

A simple treatment is still possible if it may be assumed that the flux of the component of interest A through the interface stays in a constant proportion to the total molar transfer through the interface over the entire tower. The equations developed are very general and encompass both the usual equations for gas absorption and distillation as well as situations with any degree of counterdiffusion.

Experimental mass-transfer coefficients. Despite the wealth of experimental and theoretical studies of mass-transfer coefficients in packed columns, no comprehensive theory has been developed; most generalizations are based on empirical or semiempirical equations. For some simple systems that have been studied extensively, mass-transfer data is available in the literature; for new systems, prediction is poor.

An experiment that has been applied by numerous authors to measure k_G separately from other factors consists of drying packings previously saturated with a liquid using a gas stream. Since the surface of the packing is normally known in these experiments, k_G can be computed. Application of these kinds of data to gas absorption are, however, difficult because of the different, unknown effective interfacial area when two phases are flowing through the packing.

Bubble-Tray Absorption Columns

General design procedure. Bubble-tray absorbers may be designed graphically, based on a so-called McCabe-Thiele diagram. An operating line and an equilibrium line are plotted in $y - x$ or $Y' - X'$ coordinates using the principles for packed absorbers outlined above (Fig. 2). A minimum number of plates required for a specified recovery may be computed by assuming that equilibrium is reached between the two phases on each bubble tray, so that the gas and the liquid leaving a tray are at equilibrium. A hypothetical tray capable of equilibrating the phase streams is termed a theoretical plate. Starting the calculation at the bottom of the tower, where the concentrations are y_{N+1} and x_N, the concentration leaving the lowest theoretical plate y_N may be found on the design diagram by moving from the operating line vertically to the equilibrium line, because y_N is at equilibrium with x_N. Since the concentrations between two plates are always related by the operating line, x_{N-1} may be found from y_n by moving horizontally to the operating line. By repeating this sequence of steps until the desired residual gas concentration y_1 is reached, the number of theoretical plates can be counted.

The required number of actual plates N_P is larger than the number of theoretical plates N_{TP}, because it would take an infinite contacting time at each stage to establish the equilibrium. The ratio N_{TP}/N_P is called the overall column efficiency. This parameter, however, is difficult to predict from theoretical considerations or to correct for new systems and operating conditions. It is therefore customary to characterize the single plate by the so-called Murphree Vapor Plate Efficiency, E_{MV}.

$$E_{MV} \equiv \frac{y_n - y_{n+1}}{y_N^* - y_{n+1}} \tag{13}$$

The Murphree Vapor Plate Efficiency indicates the fractional approach to equilibrium achieved by the plate. An efficiency of 80% means that the reduction in solute gas concentration effected by the plate is 80% of the reduction obtained from a theoretical plate. Corresponding actual plates may therefore be stepped off by moving from the operating line vertically only 80% of the distance between operating and equilibrium lines (Fig. 2). In some special cases with negligible resistance in the gas phase, E_{MV} values may become unreasonably small. It is then more logical to define a Murphree Liquid Plate Efficiency E_{ML} simply by reversing the role of liquid and gas and by focusing on the change in liquid composition across the plate with respect to an equilibrium given by the leaving vapors. A straight operating line occurs when the concentrations are low and L_M and G_M are essentially constant. The material balance is obtained from the value for N_{OG} (approximately given in eq. 9). In cases where the equilibrium line is straight as well, the number of theoretical

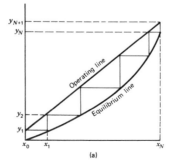
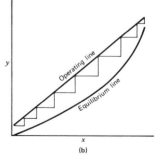

Figure 2. McCabe-Thiele diagram. (**a**) Number of theoretical plates: 5. (**b**) Number of actual plates: 8.

plates can be computed analytically using a special form of the Kremser equations.

C.R. WILKE
URS VON STOCKAR
University of California, Berkeley

T.K. Sherwood, R.L. Pigford, and C.R. Wilke, *Mass Transfer*, McGraw-Hill Book Co., Inc., New York, 1975.

R.E. Treybal, *Mass Transfer Operations*, McGraw-Hill Book Co., Inc., New York, 1955.

C.R. Wilke and U.V. Stockar, *Ind. Eng. Chem. Fundam.* **16**(2), 88 (1977).

R.H. Perry and C.H. Chilton, eds., *Chemical Engineer's Handbook*, 6th ed., McGraw-Hill, Inc., New York, 1983.

ACACIA. See Diuretics; Gums and mucilages.

ACAROID, ACCROIDES. See Resins, natural.

ACETALDEHYDE

Acetaldehyde, CH_3CHO, is a colorless, mobile liquid with a pungent odor that is somewhat fruity and pleasant in dilute concentrations. It is an important intermediate in the production of acetic acid, acetic anhydride, ethyl acetate, peracetic acid, pentaerythritol, chloral, glyoxal, alkylamines, and pyridines, and was used extensively during World War I as an intermediate for making acetone from acetic acid.

Acetaldehyde occurs in traces in all ripe fruits that have a tart taste before ripening; it is an intermediate product of alcoholic fermentation but is reduced almost immediately to ethanol (see Fermentation). An intermediate product in the decomposition of sugars in the body, it occurs in traces in the blood. Acetaldehyde is a product of most hydrocarbon oxidations.

Physical Properties

Acetaldehyde is miscible in all proportions with water, and most organic solvents such as acetone, benzene, ethyl alcohol, ethyl ether, gasoline, paraldehyde, toluene, xylenes, turpentine, and acetic acid.

Table 1 gives some of the physical properties of acetaldehyde.

Chemical Reactions

Acetaldehyde is a highly reactive compound exhibiting the general reactions of aldehydes (qv). Acetaldehyde decomposes at temperatures above 400°C to form methane and carbon monoxide. In aqueous solutions, it exists in equilibrium with the hydrate, $CH_3CH(OH)_2$. The enol form, vinyl alcohol ($CH_2{=}CHOH$), exists in equilibrium with acetaldehyde to the extent of approximately one molecule per 30,000. Acetaldehyde enol has been acetylated with ketene to form vinyl acetate.

Depending on the reaction conditions, it is readily oxidized with oxygen or air to form acetic acid, acetic anhydride, and peracetic acid (see Acetic acid and derivatives). Peracetic acid is formed commercially, either through low temperature oxidation of acetaldehyde in the presence of metal salts, ultraviolet irradiation, or ozone; or directly by liquid-phase oxidation with a cobalt salt catalyst (see Peroxides and peroxy compounds). The nitric acid oxidation of acetaldehyde yields glyoxal.

Acetaldehyde is readily reduced to ethanol (qv), using catalysts such as nickel and copper oxide to support the vapor-phase hydrogenation. When a mineral acid, such as sulfuric, phosphoric, or hydrochloric acid, is added to it, paraldehyde, 2,4,6-trioxane, a cyclic trimer, is formed. Metaldehyde, a cyclic tetramer, is formed at temperatures below 0°C in the presence of dry hydrogen chloride or pyridine-hydrogen bromide. Polyacetaldehyde, an unstable, rubbery polymer with an acetal structure, forms in a few days from acetaldehyde; polyacetaldehyde has no practical significance (see Acetal resins).

The base-catalyzed condensation of acetaldehyde leads to the dimer acetaldol, which can be hydrogenated to form 1,3-butanediol, or dehydrated to form crotonaldehyde, which may also be made directly by the vapor-phase condensation of acetaldehyde over a catalyst. The base-catalyzed reaction of acetaldehyde with excess formaldehyde is the commercial route to pentaerythritol (see Alcohols, polyhydric). The vapor-phase reaction of acetaldehyde and formaldehyde at 475°C over a catalyst gives acrolein. Ethyl acetate is commercially produced by the Tischenko condensation of acetaldehyde with an aluminum ethoxide catalyst; with isobutyraldehyde, this yields a mixture of ethyl acetate, isobutyl acetate, and isobutyl isobutyrate. Butadiene (qv) can be made by the reaction of acetaldehyde and ethyl alcohol. The Strecker amino acid synthesis gives alanine (see Amino acids). Grignard reagents give secondary alcohols.

Manufacture

Acetaldehyde is produced commercially by the oxidation or dehydrogenation of ethanol, the addition of water to acetylene, partial oxidation of hydrocarbons, and the direct oxidation of ethylene. The direct liquid-phase oxidation of ethylene (the Wacker-Hoechst process) is the process of choice; in 1976, 29 companies with $> 82\%$ of the world's 2.3×10^6 t per year plant capacity used this process. A catalytic solution of palladium chloride is used to form acetaldehyde:

$$CH_2{=}CH_2 + PdCl_2 + H_2O \rightarrow CH_3CHO + Pd + 2\ HCl$$

The metallic palladium is reoxidized to $PdCl_2$ with $CuCl_2$, and the cuprous chloride formed is reoxidized with oxygen or air. The net result is a process in which ethylene is oxidized continuously through a series of oxidation–reduction reactions.

There are two variations for the production of acetaldehyde by the oxidation of ethylene: the two-stage process developed by Wacker-Chemie and the one-stage process developed by Farbwerke Hoechst. In the two-stage process, ethylene and oxygen (air) react in the liquid phase in two stages. In the first, the ethylene is almost completely converted to acetaldehyde in one pass in a tubular plug-flow reactor made of titanium. Acetaldehyde produced in the first reactor is removed by adiabatic flashing in a tower. The catalyst solution is recycled from the flash-tower base to the second stage, or oxidation, reactor, where the cuprous salt is oxidized to the cupric state with air. In the one-stage process, ethylene, oxygen, and recycled gas are directed to a vertical reactor for contact with the catalyst solution.

Acetaldehyde is produced commercially by the catalytic oxidation of ethyl alcohol, carried out by passing alcohol vapors and preheated air over a silver catalyst at 480°C. Older commercial processes produced acetaldehyde by the hydration of acetylene under pressure in a vertical reactor containing a mercury catalyst dissolved in sulfuric acid. Acetaldehyde may also be made from methyl vinyl ether and ethylidene diacetate, both of which can be made from acetylene. It is also formed as

Table 1. Physical Properties of Acetaldehyde

Properties	Values
melting point	−123.5
boiling point at 101.3 kPa (1 atm), °C	20.16
density, g/cm^3	0.8045
refractive index, n_D^{20}	1.33113
vapor density (air = 1)	1.52
specific heat at 0°C, $J/(g \cdot K)^a$	2.18
at 25°C	1.41
latent heat of vaporization, kJ/mol^a	25.71
heat of solution in water at 0°C, kJ/mol^a	−8.20
at 25°C	−6.82
flash point, closed cup, °C	−38
ignition temperature in air, °C	165
explosive limits of mixtures with air, vol % acetaldehyde	4.5–60.5

^{a}To convert J to cal, divide by 4.184.

a coproduct in the vapor-phase oxidation of saturated hydrocarbons, eg, butane or mixtures containing butane, with air, or, in higher yield, oxygen (see Hydrocarbon oxidation). Unlike the acetylene route, this method has almost no chance to be used as a main process. Acetaldehyde is produced in a rhodium-catalyzed process capable of converting synthesis gas ($CO + H_2$) directly into acetaldehyde in a single step. In the years 1985 and beyond, interest in the use of synthesis gas from coal gasification as a raw material for acetaldehyde production will increase (see Fuels, synthetic).

Health and Safety Factors

Although in normal industrial operation there is no health hazard in handling acetaldehyde if normal precautions are taken, it may be harmful in the following ways: it paralyzes the respiratory muscles, causes panic, has a general narcotic action which prevents coughing, causes irritation of the eyes and mucus membranes, accelerates heart action, and causes headache and sore throat when breathed in high concentrations. Carbon dioxide solutions in acetaldehyde have a weakened odor and are particularly pernicious. Prolonged exposure causes a decrease in red and white blood cells and a sustained rise in blood pressure. The maximum allowable concentration of acetaldehyde in air is 200 ppm. Mixtures of acetaldehyde vapor with air are flammable and explosive if the concentrations of aldehyde and oxygen rise above 4 and 9%, respectively.

Economic Aspects and Uses

Future growth will depend on the development of a lower cost process based on synthesis gas, and an increased demand for processes based on acetaldehyde activation techniques and peracetic acid. Since 1940, the production pattern has changed from hydration of acetylene and oxidation of ethyl alcohol to liquid-phase oxidation of ethylene using the Wacker-Hoechst processes. Acetaldehyde production is linked with the demand for acetic acid, acetic anhydride, cellulose acetate, vinyl acetate resins, acetate esters, pentaerythritol, synthetic pyridine derivatives, terephthalic acid, and peracetic acid. No new acetaldehyde-based plants are expected due to the rising costs of ethylene as feedstock costs increase. Methanol carbonylation is the preferred route to acetic acid and will continue for the next 10–15 years (see Acetic acid).

H.J. HAGEMAYER
Texas Eastman Company

R. Jira and W. Freiesleben, "Olefin Oxidation and Related Reactions with Group VIII Noble Metal Salts," in E. Becker and M. Tsutsui, eds., *Organometallic Reactions*, Vol. 3, Interscience Publishers, a division of John Wiley & Sons, Inc., New York, 1972.

J. Smidt and co-workers, *Angew. Chem.* **71**, 176 (1959).

U.S. Pat. 3,240,803 (Mar. 15, 1966), B. Thompson and S.D. Neeley (to Eastman Kodak Co.).

U.S. Pat. 3,714,236 (Jan. 30, 1973), H.N. Wright and H.J. Hagemayer, Jr. (to Eastman Kodak Co.).

ACETAL RESINS

Acetal resins, also known as polyoxymethylenes, polyacetals, and aldehyde resins, are engineering thermoplastics that have broad use as replacements for metals in a wide variety of applications (see Engineering plastics). They are high molecular weight polymers with a basic molecular structure of repeating carbon–oxygen links.

This alternating oxymethylene structure, OCH_2, gives the resins a chemistry similar to the simple acetals. Homopolymers consist solely of this carbon-oxygen backbone, whereas copolymers have the oxymethylene structure occasionally interrupted by a comonomer unit such as an ethylene linkage from ethylene oxide.

Properties vary depending on the number-average molecular weight, the type of additives or reinforcing agents, and whether the resin is a copolymer or a homopolymer. For example, additives, eg, glass fibers in the copolymer, can improve tensile strength and stiffness (see Composites; Laminated and reinforced plastics). The resins possess high stiffness, excellent dimensional stability, high tensile and impact strength, good abrasion resistance, and a low coefficient of friction. They resist attack by many organic compounds, but are unsuitable in contact with strong acids. The copolymer offers superior resistance to strong bases. Acetal resins are covered by Federal Specifications L-P-392a and by ASTM Specification D 2133 which identify the resin by specific gravity, melting point, and melt flow rate.

Acetal resins are frequently used to reduce cost and improve performance in applications where one acetal part replaces several metal parts. (see Engineering plastics).

Trade names and producers of commercial products include Delrin (DuPont), Celcon (Celanese), Hostaform (Celanese—Hoechst), Duraon (Celanese—Diacel), and Ultraform (BASF-Degussa).

Manufacture

Homopolymer. Commercial homopolymer production is based on the polymerization of formaldehyde through monomer purification, polymerization, end-capping with alkyl or acyl groups, and finishing. The concentration of chain-transfer agents must be reduced, although high molecular weight polymers have been produced with as much as 0.9% water present. The monomer is purified through scrubbing with an inert liquid, partial polymerization to remove chain-transfer impurities, conversion of formaldehyde to a hemiformal and subsequent decomposition to purified formaldehyde, and treatment with P_2O_5.

There are two main polymerization reactions: the first is addition to the double bond of the monomer. The second is growth of a completely crystalline polymer in aqueous or alcohol medium by stepwise equilibrium with solvated monomer using crystallization as the driving force. Anionic initiators include amines, phosphines, ammonium and sulfonium salts, amides, amidines, etc.

$$\sim CH_2O^- + CH_2{=}O \rightarrow \sim CH_2OCH_2O^- \text{ (anionic)}$$
$$\sim OCH_2^+ + O{=}CH_2 \rightarrow \sim OCH_2OCH_2^+ \text{ (cationic)}$$

Cationic initiators include sulfuric or phosphoric acid, Lewis acids, and stable cations. Ionizing radiation is also effective. The polymer is end-capped for stabilization by reaction with acetic anhydride. It is finished with the addition of antioxidants, typically of the amide or amine type.

Copolymer. Commercial copolymers are made by the trimerization of formaldehyde to trioxane, purification of trioxane, copolymerization, stabilization by alkaline hydrolysis, and finishing. The copolymerization of trioxane, like formaldehyde polymerization, is by cationic initiation using, in this case, boron fluoride etherate or other boron fluoride coordination complexes with organic compounds. Cationic polymerization is relatively complicated and may involve the oxonium ion as well as the carbocation. Transacetalization can occur as a side reaction. The polymerization is carried out in the presence of 0.1–15 mol% of the comonomer, yielding a polymer having 85–99.9% oxymethylene units ($—OCH_2—$); thus, the polymer is still predominantly a polyoxymethylene, allowing its classification as an acetal resin.

Health and Safety Factors

During the processing of acetal resins in molding or extrusion equipment, some decomposition can occur with evolution of formaldehyde. Formaldehyde, as an air contaminant, is regulated under OSHA. The National Sanitation Foundation has listed a number of acetal compositions for use in contact with potable water. The FDA regulations 21 CFR 177.2470 and 21 CFR 177.2480 cover acetals repeatedly used in contact with food.

K.J. PERSAK
L.M. BLAIR
E.I. du Pont de Nemours & Co., Inc.

M.M. Wilson, "Acetal Copolymer" in *Modern Plastics Encyclopedia*, Vol. 58, McGraw-Hill, Inc., New York, 1981–1982 ed., p. 7.

K.J. Persak and S.I. Wilson, "Acetal Homopolymer" in *Modern Plastics Encyclopedia*, Vol. 58, McGraw-Hill, Inc., New York, 1981–1982 ed., p. 5.

M. Sittig, *Polyacetal Resins*, Gulf Publishing Co., Houston, Texas, 1963.

ACETAMIDE. See Acetic acid.

ACETANILIDE. See Amines, aromatic—Aniline and its derivatives.

ACETARSONE. See Arsenic compounds; Chemotherapeutics, protozoal.

ACETATE AND TRIACETATE FIBERS. See Cellulose acetate and triacetate fibers.

ACETIC ACID AND DERIVATIVES

ACETIC ACID

Acetic acid, CH_3COOH, is a colorless, waterlike liquid that has a piercingly sharp, vinegary odor and a burning taste. Although the formula weight of acetic acid is 60.06, its apparent molecular weight varies both with the temperature and the other associating substances present. At pressures below saturation in the temperature range 25–120°C, the vapor is well described as a mixture of the dimer and the monomer in equilibrium. Dimerization, represented by the equation below, accounts for many of the apparent anomalies found in its equation of state, thermodynamic functions, transport properties, etc. At the atmospheric boiling point, acetic acid dehydrates to give traces of acetic anhydride and, at high temperatures in the presence of a catalyst, yields commercially useful amounts of anhydride.

$$\log K_{dimer} = \frac{3166}{T} - 10.4205$$

The properties of acetic acid are summarized in Table 1.

Acetic acid has important properties as a solvent for acids and bases; it protonates bases too weak to detect in water and is often used in their analysis. Other acids also exhibit very powerful acidity properties in acetic acid solutions. Most of the esters of acetic acid are prepared by direct reaction with the corresponding alcohol. A notable exception is ethyl acetate, prepared by the Tischenko reaction of acetaldehyde:

$$2\ CH_3CHO \rightarrow CH_3C(O)OC_2H_5$$

Some ethyl acetate is isolated as a by-product of acetic acid manufacture from liquid-phase hydrocarbon oxidation (qv). Noncatalytic esterification of alcohols is accomplished with a two- to fivefold excess of acetic acid; water, which substantially slows the reaction, is usually removed azeotropically. Catalytic esterification of alcohols uses a mineral acid (see Esterification). Olefins react with anhydrous acetic acid to give esters of secondary or tertiary alcohols. Ethylene, oxygen, and acetic acid passed over a palladium–lithium acetate (with copper) catalyst produce vinyl acetate, $CH_3C(O)OCH{=}CH_2$ (see Vinyl polymers).

Ethylene oxidation in acetic acid gives glycol diacetate, which may be thermally decomposed to vinyl acetate. Acetylene adds to give vinyl acetate (see Acetylene-derived chemicals). Many metals, their oxides, or their carbonates, dissolve in acetic acid to give simple salts. Alkali metal acetates are prepared from corresponding hydroxides or carbonates. Transition-metal acetates can be made from free metal (see Carboxylic acids).

Manufacture

Virtually all acetic acid commercially produced is made by one of three routes: acetaldehyde oxidation, involving direct air or oxygen oxidation of liquid acetaldehyde in the presence of manganese acetate, cobalt acetate, or copper acetate (see Acetaldehyde); liquid-phase oxidation of butane or naphtha, with butane yielding more acetic acid than the higher paraffins, but naphtha having the chief advantages of greater accessibility, initial lower costs, and more ready oxidation (see Hydrocarbon oxidation); methanol carbonylation using a variety of techniques. Cobalt and several other carbonyl-forming metals were studied but found less active than nickel iodide catalysts at 215°C and 14 MPa (138 atm) pressure. BASF built and operated a plant in 1960 that yielded 3600 metric tons of acid per year. BASF technology was used in the Borden Chemical plant in Louisiana, which now has a reported capacity of > 135,000 t/yr.

Monsanto developed a different approach based on rhodium complexes with phosphines, under conditions of 150–200°C at 33–66 MPa (33–65 atm) pressure. At 100% conversion, methanol selectivity to acetic acid is nearly 99%. A vapor-phase process has also been explored. Either Hastelloy C or titanium is required as the material of construction.

Economic Aspects

Acetic acid has a place in the organic chemical industry comparable to sulfuric acid in the inorganic chemical industry. In 1981, production of acetic acid in the United States was 2,706,120 metric tons. Acetic acid is used for the production of vinyl acetate, an important monomer. Acetic acid is used in the fermentation industry (see Fermentation) in large-scale amino acids (qv) production and as the substrate of choice for the production of many dietary supplements, eg, glutamic acid, citric acid, and lysine. Cellulose acetate (qv) was formerly the foremost of synthetic polymers and is still an important thermoplastic and fiber constituent, used in the manufacture of plastic sheeting film and in the formulation of lacquers (see Film and sheeting materials).

Acetic acid is used as the solvent in the liquid-phase oxidation of *p*-xylene to terephthalic acid. The principal use for acetate esters, from methyl through amyl, is as solvents for lacquers and coating. These solvents are attractive because of their great convenience and low price. All acetate solvents may be made obsolete by air-pollution legislation, although these laws have not been promulgated as rapidly as first expected.

Acetic acid is an effective fungicide (see Fungicides) used to control the fungi that cause ropiness in bread; to preserve high moisture grain; and, in combination with propionic acid, to combat *Aspergillus flavus*, the organism responsible for producing aflatoxin in grain.

Table 1. Properties of Pure Acetic Acid

Property	Value
melting point, °C	16.635 ± 0.002
boiling point, °C	117.87
vapor pressure, kPa[a]	$\log p = 6.68206 - 1642.54/(233.386 + t)$
thermal conductivity at 20°C, W/m · K	0.158
heat of vaporization at bp, J/g[a]	
specific heat, J/(g · K)[b]	394.5
of vapor at 124°C	5.029
of solid at 100 K	0.837
density at 20.0°C, g/cm³	1.0493
refractive index, n_D^{25}	1.36965
viscosity of liquid, mPa · s (= cP)	
at 20°C	11.83
at 25°C	10.97
at 40°C	8.18
dielectric constant	
solid at −10.0°C	2.665
liquid at 20°C	6.170
surface tension at 20.1°C, mN/m	
(= dyn/cm)	27.57
flash point, open cup, °C	57

[a] To convert kPa to mm Hg, multiply by 7.5.
[b] To convert J to cal, multiply by 0.239.

Health and Safety Factors

Any solution containing more than half acetic acid in water or organic solvents is considered corrosive. These solutions cause damage to skin, eyes, nose, and mouth. There is no burning sensation upon application to unbroken skin, although symptoms appear as blisters within 30 min to 4 h. There is little or no pain experienced, unless sensory nerve receptors are attacked. Medical care should be sought immediately. Particular care is needed when handling acetic acid to avoid spilling it on the skin or clothing and to avoid breathing the vapors. When exposed, wash acetic acid away with large amounts of water or dilute sodium bicarbonate solution. Washing blisters that have already appeared will not alleviate the pain. Leave areas where the odor of acetic acid vapors is noticeable.

Glacial acetic acid causes severe gastric difficulties if swallowed, but the precise toxic dose for humans is not known. OSHA standard for air in work places is a time-weighted average of 10 ppm.

FRANK S. WAGNER, JR.
Celanese Chemical Company

A.I. Popov in J.J. Lagowski, ed., *The Chemistry of Nonaqueous Solvents*, Vol. 3, Academic Press, Inc., New York, 1966–1976, Chapt. 5.

ACETAMIDE

Acetamide, CH_3CONH_2, mol wt 59.07, is a colorless, deliquescent substance composed of rhombohedral crystals. Also known as acetic acid amide, ethanamide, and methane-carboxamide, it is odorless when pure; otherwise, it has a distinctive odor due to an unknown impurity. It is an excellent dipolar solvent for organic and inorganic compounds; it is compatible with hydrocarbons, ketones, esters, organic acids, alcohols, water, ammonium salts, and ammonia and its derivatives. Its high dielectric constant ($\varepsilon = 59$) makes it a suitable solvent for inorganic compounds.

Physical and Chemical Properties

Acetamide has mp 81.0°C, bp 221.2°C, $n_D^{78.3}$ 1.4274, d_{20}^{20} 1.159 g/cm³, and vapor pressure (at 105°C) 1.33 kPa (10 mm Hg). It is chemically neutral in the classical sense but basic enough to accept a proton from a strong acid. It reacts with alkali metals to give the corresponding metal derivatives, eg, $CH_3CONHNa$. It is useful in the bromination of acid-sensitive compounds since it forms a stable complex with hydrogen bromide $(CH_3CONH_2)_2 \cdot HBr$ which is insoluble in common bromination solvents. Acetamide is dehydrated to acetonitrile when heated either with phosphorus pentoxide or acetic anhydride. It is saponified by hot alkali and hydrolyzed by hot acid. It can be converted to acetic acid by heating with mineral acids and by reaction with nitrous acid.

Preparation

One of the most widely used methods of manufacture is classical synthesis from acetic acid and ammonium carbonate:

$$CH_3C(O)OH \xrightarrow{(NH_4)_2CO_3} CH_3C(O)ONH_4 \xrightarrow{\Delta} CH_3CONH_2 + HOH$$

Widely used laboratory methods are

$$CH_3COCl + 2\,NH_3 \rightarrow CH_3CONH_2 + NH_4Cl$$
$$(CH_3CO)_2O + 2\,NH_3 \rightarrow CH_3CONH_2 + CH_3COONH_4$$
$$CH_3COOC_2H_5 + NH_3 \rightarrow CH_3CONH_2 + C_2H_5OH$$

The most efficient method is distillation of ammonium acetate.

Acetamide is produced in technical and reagent grades, the latter obtained by further recrystallization. Acetamide is shipped in fiber cartons and in 500-g packages, and should be stored in a cool, dry area. It is analyzed by hydrolysis to the acid, or by esterification.

Acetamide is used as general solvent and in the textile industry; it also has applications as an accelerator in the estimation of bilirubin; as an antidote against monofluoroacetamide poisoning; as an antacid in the lacquer, explosives, and cosmetic industries; as a plasticizer in leather, cloth, films, and coatings; as a humectant for paper; as a stabilizer in peroxides; as a raw material in organic synthesis; and in the preparation of hypnotics, insecticides, and a variety of plastics.

THERESA A. MORETTI
American Hoechst Corporation

J. Zabicky, *The Chemistry of Amides*, Interscience Publishers, a division of John Wiley & Sons, Inc., New York, 1970.

ACETIC ANHYDRIDE

Acetic anhydride, $(CH_3CO)_2O$ (often abbreviated Ac_2O), is a mobile, colorless liquid with a penetrating, choking odor, closely related to acetic acid. It is the largest commercially produced acid anhydride; current U.S. production ranges from 500 to 750 thousand metric tons annually. It is used in industry for acetylation reactions and as a solvent in laboratory applications. The greatest single application is in the manufacture of cellulose esters, chiefly cellulose acetate and triacetate fibers; Cellulose derivatives, esters). It is also used in starch acetylation to make textile sizing agents, in salicylic acid acetylation to aspirin (see Salicylic acid and related compounds), and in electrolytic polishing of metals, especially aluminum, and semiconductor processing. An undetermined quantity goes into the manufacture of chloroacetic acid (see Acetic acid, halogenated derivatives) and acetyl chloride (qv). It is peculiarly useful for oxidation reactions, not only of metals but also for organic compounds.

Physical and Chemical Properties

The physical properties of acetic anhydride are given in Table 1.

The chief impurity in acetic anhydride is acetic acid, which is troublesome to remove completely. Purity of the product is best gauged by the refractive index or the specific conductivity. Traces of acetic acid are detected by near-ir spectroscopy. Acetic acid can be estimated by difference when the anhydride is assayed with aniline.

Thermal decomposition of acetic anhydride yields ketene, in a unimolecular dissociation process between 280 and 650°C, making this a convenient way to prepare ketene in the laboratory (see Ketenes and related substances). Photolysis of acetic anhydride generates ethane and carbon monoxide in approximately equal amounts, and carbon dioxide in smaller quantities. Acetic anhydride appears to ionize to acetylium and acetate ions although it is practically an electrical insulator when salt free. The acetylium ion, CH_3CO^+, occurs in strongly acid solutions. Potassium acetate, rubidium acetate, and cesium acetate are very soluble in acetic anhydride, but sodium acetate is only slightly soluble. Barium acetate is the only soluble alkaline-earth acetate. Heavy-metal acetates are poorly soluble.

Table 1. Physical Properties of Acetic Anhydride

Property	Value
melting point, °C	−74.13
boiling point, °C at 101.3 kPa (1 atm)	138.63
vapor pressure equation, kPa (A = 6.2514) (mm Hg (A = 7.1265)	$\log p = A - \dfrac{1427.8}{(198 + t)}$
thermal conductivity, liquid, mW/(m · K), at 30°C	136
specific heat, J/kg[a], at 20°C	1817
heat of vaporization at bp, J/g[a]	406.6
density, d_4^{20}, g/cm³	1.0820
refractive index, n_D^{20}	1.3892
surface tension, mN/m (= dyn/cm), at 20°C	32.65
viscosity, mPa · s (= cP), at 20°C	0.901
dielectric constant, ε^{20}	22.1
dipole moment, C · m[b]	9.3×10^{-30}

[a] To convert J to cal, divide by 4.184.
[b] To convert C · m to debye, divide by 3.336×10^{-30}.

Acetaldehyde adds to acetic anhydride to furnish ethylidene diacetate, $CH_3CH(O_2CCH_3)_2$. The higher carboxylic acid anhydrides may be prepared conveniently by a double decomposition reaction with acetic anhydride and the higher acid. Since acetic acid boils lower than the other fatty acids, it can be distilled and the higher acid anhydride recovered.

Manufacture

There are three main routes to acetic anhydride that are thoroughly understood from an engineering standpoint; the processes depend fundamentally upon their distinctive raw materials, ie, acetic acid, acetone, and acetaldehyde, for anhydride manufacture.

The acetic acid process involves the endothermic decomposition of acetic acid to ketene and the addition of ketene to acetic acid in the liquid phase to produce acetic anhydride.

$$CH_3C(O)OH + CH_2{=}C{=}O \rightarrow$$
$$(CH_3CO)_2O + 62.8 \text{ kJ/mol (15 kcal/mol)}$$

The acetone process involves a similar thermal decomposition of acetone to ketene in a reaction that is less endothermic. Pyrolysis gases are quenched in a liquid recycle mixture composed of acetic acid and anhydride; the quenched gases consist of ketene, acetic acid, acetic anhydride, unconverted acetone, and the permanent gases. During condensation, practically all the ketene reacts with acetic acid to form acetic anhydride. Three distillations recover the acetone, a mixture of acetic acid and anhydride (which is used for quenching), and pure acetic anhydride.

Liquid-phase acetaldehyde oxidation yields acetic acid (as a by-product) and acetaldehyde monoperacetate, which decomposes into acetic anhydride and water. The product may be removed as a liquid or a vapor; as a vapor, the concentration of acetic anhydride is higher than that of acetic acid; air is used to entrain the acetic anhydride and water as fast as they are made. Purification of the liquid products is comparatively easy; low boiling substances are removed with ethyl acetate or acetic acid before the anhydride is distilled.

A number of other routes to acetic anhydride manufacture investigated include methyl acetate, a by-product of methanol carbonylation (see Acetic acid), pyrolyzed to ketene in a system similar to that used for acetone or carbonylated to acetic anhydride in the presence of metal catalyst; and sodium acetate and carbon dioxide reactions form sodium bicarbonate and acetic anhydride in aqueous solutions.

Health and Safety Factors

Acetic anhydride has a powerful effect on the human body. The odor threshold is 0.49 mg/m^3, and the eyes are affected by as little as 0.36 mg/m^3. Concentrations > 0.03 mg/m^3 throughout a regular workday are inadvisable. Acetic anhydride penetrates the skin quickly and can form painful burns and blisters that are slow to heal. It is particularly dangerous to the tissues of the eyes, nose, and mouth; rubber gloves, masks, and goggles are recommended.

FRANK S. WAGNER, JR.
Celanese Chemical Company

G.V. Jeffreys, *The Manufacture of Acetic Anhydride*, 2nd ed., the Institution of Chemical Engineers, London, 1964.

ACETYL CHLORIDE

Acetyl chloride (ethanoyl chloride), CH_3COCl, is a colorless, mobile liquid with a pungent odor. It fumes in moist air, and its vapors are extremely irritating to the eyes, nose, and throat.

Acetyl chloride is used as a catalyst in the chlorination of acetic acid; for this use it is made *in situ* from acetic anhydride and by-product hydrogen chloride. It is used as a powerful acetylating agent in organic synthesis, particularly with a Lewis acid catalyst; it will acetylate many compounds, eg, benzene, that cannot be acetylated with acetic acid or acetic anhydride. Acetyl chloride is used to synthesize pharmaceuticals and dyes; in the preparation of acetanilide, acetophenone, and other acetyl derivatives; and in the laboratory to determine hydroxyl groups; in qualitative organic analyses for distinguishing tertiary from primary and secondary amines. It is also used to form phenolic compound derivatives; as a chlorinating agent for inorganic compounds; as a catalyst in esterification and halogenation of aliphatic acids; and in the nitration of thiophenes and other compounds. Acetylacetone has been prepared in 80% yield from acetyl chloride by a three-stage process involving an addition compound of acetyl chloride and aluminum chloride. Specialty applications include its use as a catalyst in the esterification of cellulose by acetic anhydride; to accelerate the reaction and improve the wetfastness of the resultant yarns; in the manufacture of a pour-point depressant for lubricating grease; as a stabilizer against polymerization of organic isocyanates or diisocyanates; and as a rubber antiscorch agent.

Physical and Chemical Properties

Acetyl chloride has mp $-112.86°C$, bp $51.8°C$, flash pt $4.5°C$, d_4^{20} 1.1051 g/cm^3, n_D^{20} 1.38708, surface tension 26.7 ± 1.0 mN/m ($=$ dyn/cm) at $14.8°C$ and 21.9 mN/m at $46.2°C$, latent heat of fusion $\Delta(H_f)$ (liquid) 275.3 kJ/mol (65.8 kcal/mol), and bond dissociation energy of carbon–chlorine bond 320.9 kJ (76.7 kcal).

It is soluble in acetone, glacial acetic acid, ethyl ether, benzene, toluene, chloroform, and carbon disulfide, and is used as a solvent. Acetyl chloride decomposes vigorously in water to form acetic acid and hydrochloric acid. It reacts with alcohols and amines. Acetamide is formed by treating acetyl chloride with anhydrous ammonia or with concentrated aqueous ammoniacal solutions. Potassium sulfide and potassium hydrogen sulfide react with acetyl chloride to form acetyl sulfide and thioacetic acid. Acetyl chloride reacts with acetylene at $15°C$ in the presence of aluminum chloride to give methyl vinyl ketone. Acetic anhydride is formed in the reaction of acetyl chloride with sodium acetate. It decomposes at $400°C$ over a nickel catalyst into carbon monoxide (62%), hydrogen (32%), hydrogen chloride, ethylene, and carbon dioxide. It reacts with zinc, aluminum, and other metals to form the chloride of the metal and a variety of organic compounds.

Preparation

Large scale *in situ* industrial production of acetyl chloride uses the action of hydrogen chloride at $85–90°C$ on acetic anhydride, to produce a practically quantitative yield. Commercially, it is prepared by treating sulfur dioxide and chlorine with sodium acetate followed by distillation. Laboratory manufacture involves the reaction of phosphorus trichloride with acetic acid. Industrial equipment for acetyl chloride is usually made of glass, glass-lined steel, Teflon, Teflon-lined steel, or heavy-wall polyethylene.

Acetyl chloride is shipped in 220-L (58-gal) polyethylene-lined drums and in 100-g and 500-g packages. It carries the U.S. DOT White Label designation for a corrosive substance. It is produced in both a practical and a reagent grade.

Health and Safety Factors

Acetyl chloride violently decomposes with water and alcohol and is easily hydrolyzed into acetic acid and hydrochloric acid by body moisture. It has a toxic effect of its own because it reacts with the SH group of protein molecules. It is highly toxic and corrosive and a strong irritant. It must be used with adequate ventilation, and proper protective clothing should be worn. Its vapor forms explosive mixtures with air. Spills or leakage of acetyl chloride should be treated with sodium bicarbonate.

THERESA A. MORETTI
American Hoechst Corporation

G.A. Olah, *Friedel-Crafts and Related Reactions*, Interscience Publishers, a division of John Wiley & Sons, Inc., New York, 1963.

DMAC

Dimethylacetamide (DMAC), $CH_3CON(CH_3)_2$, is a colorless, high-boiling polar solvent for a wide range of organic and inorganic compounds. Like the related amide solvent dimethylformamide, it is completely miscible with water, ethers, esters, ketones, and aromatic compounds; it dissolves unsaturated more readily that saturated aliphatics; and it acts as a combined solvent and reaction catalyst, often producing high yields and pure products in a short time. DMAC influences many reactions by solvating cations, minimizing electrostatic attraction between cations and the reactive anions.

DMAC has high solvency for many natural and synthetic resins, eg, functional vinyl polymers and copolymers, acrylates, cellulose derivatives, acrylonitrile polymers, polyimides, and many others, making possible their use in film and fiber processes. Cellulose can be regenerated from DMAC–lithium chloride solution.

Selected physical properties of DMAC are bp 166.1°C; mp −20°C; vapor pressure at 25°C 0.27 kPa (2 mm Hg); density at 15.6°C 0.945 g/cm^3; viscosity at 25°C 0.92 mPa·s (= cP); surface tension at 30°C 32.43 mN/m (= dyn/cm); refractive index n_D^{25} 1.4356; heat of vaporization at 166°C 43.1 kJ/mol (10.3 kcal/mol); flash point (Tag closed cup) 63°C; ignition temperature 490°C; flammability limits in air ca 1.8–11.5 vol%.

DMAC has a low rate of hydrolysis, but this increases somewhat in the presence of acids or bases. The chemical reactions of DMAC are typical of those of disubstituted amides. It undergoes saponification in the presence of strong bases, and alcoholysis in the presence of hydrogen ions.

DMAC can be produced by the reaction of acetic acid and dimethylamine:

$$CH_3C(O)OH + (CH_3)_2NH \rightarrow CH_3CON(CH_3)_2 + H_2O$$

The reaction product is purified by distillation. Solid caustic soda can be used to remove DMAC–acetic acid azeotrope impurity.

The U.S. DOT classified DMAC as a combustible liquid. No DOT label is required for 208-L (55 gal) drums. Because DMAC is a good solvent for many resins, flange gaskets and pump and valve packing should be limited to white asbestos or Teflon fluorocarbon resins.

DMAC can be measured in air by passing a known amount of sample through water in a gas-scrubbing vessel and analyzing the solution chemically or by gas chromotography.

Although it has low acute toxicity from brief exposure, DMAC can produce systemic injury when inhaled or absorbed through the skin in sufficient quantities over a prolonged period of time. DMAC has shown embryotoxic properties in test animals. The OSHA limit for DMAC vapors in air is 10 ppm, or 35 mg/m^3, time-weighted average basis, with a notation to avoid skin contact.

J.C. Siegle
E.I. du Pont de Nemours & Co., Inc.

Reviewed by
Bernard C. Lawes
E.I. du Pont de Nemours & Co., Inc.

HALOGENATED DERIVATIVES

Chlorine Halogenated Derivatives

The most important of the halogenated acetic acids is chloroacetic acid, $ClCH_2CO_2H$, a colorless, white deliquescent solid isolated in three or possibly four crystal modifications (α, fp 63°C; β, fp 55–56°C; γ, fp 50°C; δ, fp 42.75°C). Commercial chloroacetic acid consists of the α form. The boiling point of all forms is 180°C at atmospheric pressure. Density is 1.404$_4^{40}$ g/cm^3; heat capacity (100°C) 181.0 J/(mol·K) (cal/(mol·K)); heat of sublimation (25°C) 88.1 kJ/mol (21.1 kcal/mol); vapor pressure (25°C) 8.68 Pa (0.065 mm Hg), (100°C) 3.24 kPa (24.3 mm Hg); viscosity (100°C) 1.29 mPa·s (= cP); refractive index (65°C) 1.43;

surface tension (100°C) 35.17 mN/m (= dyn/cm); and $K_a = 1.4 \times 10^{-3}$. Solubility of chloroacetic acid (g/100 g-solvent): 257, acetone; 26, benzene; 2.75, carbon tetrachloride; 614, water; 51, methylene chloride; and 1.67, carbon disulfide. Chloroacetic acid can be recrystallized from a variety of chlorinated hydrocarbons such as trichloroethylene, perchloroethylene, and carbon tetrachloride.

Chloroacetic acid is bifunctional and undergoes a variety of chemical reactions. Sodium chloroacetate is made by treating chloroacetic acid in water with aqueous sodium hydroxide. Chloroacetic acid readily undergoes a variety of chlorine substitutions of commercial importance such as reaction with cyanide ion to yield cyanoacetic acid; reaction with 2,4-dichlorophenol and 2,4,5-trichlorophenol to give the well-known herbicides 2,4-D and 2,4,5-T, respectively (see Herbicides); a similar reaction with bisphenol A to give the corresponding diacid, 2,2-bis[p-(carboxymethyleneoxy)phenyl]propane; and with polysaccharides to yield carboxymethyl cellulose, carboxymethyl guar, and carboxymethyl starch, all of which have wide industrial applications. Thioglycolic acid is prepared commercially by treating chloroacetic acid with sodium hydrosulfide. The sodium and ammonium salts of thioglycolic acid have been used for cold-waving hair and the calcium salt as a depilating agent (see Cosmetics; Hair preparations).

Chloroacetic acid is manufactured by the chlorination of acetic acid and by the hydrolysis of trichloroethylene with sulfuric acid. Chlorination of acetic acid cannot be well controlled and gives a mixture containing impurities of acetic acid, monochloroacetic acid, and dichloroacetic acid. The hydrolysis of trichloroethylene yields high purity chloroacetic acid but utilizes a relatively more expensive starting material. Hydrolysis of chloroacetyl chloride is another potential industrial route.

Owing to its corrosive nature, chloroacetic acid is manufactured and stored in glass-lined steel or other special materials. Gas chromatography or liquid chromatography are used for analysis of impurities.

Extreme caution must be used when handling molten or strong aqueous solutions of chloroacetic acid, which in these forms is readily absorbed through the skin in toxic amounts. Rubber gloves, boots, and protective clothing must be worn; exposed skin areas should be immediately washed with large amounts of water, and medical help should be obtained at once. Oral LD$_{50}$ for chloroacetic acid is 76 mg/kg.

Dichloroacetic acid, Cl_2CHCO_2H, is a stronger acid than chloroacetic acid ($K_a = 5.14 \times 10^{-2}$). It has a mp of 13.5°C, bp 194°C, d$_4^{20}$ 1.564 g/cm^3, and n_D^{20} 1.4658. The liquid is infinitely soluble in water, ethyl alcohol, and ether. Most chemical reactions are similar to chloroacetic acid, although the chlorines are much more stable to hydrolysis. The oral toxicity is low (LD$_{50}$ rats, 4.48 g/kg), but it can cause caustic burns of the skin and eyes, and the vapors, if breather, are irritating. It has viricidal and fungicidal activity and is active against several staphylococci.

Trichloroacetic acid, CCl_3CO_2H, forms white deliquescent crystals and has a characteristic odor. Other physical properties include d$_4^{25}$ 1.629 g/cm^3, mp (γ form) 58°C, (β form) 49.6°C, bp at 101 kPa (1 atm) 197.5°C, n_D^{61} 1.4603. It is soluble to the extent of 1306 g/100 g in water, with decreasing solubility in nonpolar organic solvents. It is as strong as hydrochloric acid ($K_a = 0.2159$). It undergoes decarboxylation when heated to yield chloroform. Trichloroacetic acid is manufactured via the exhaustive chlorination of acetic acid. Two alternative methods of synthesis are the nitric acid oxidation of chloral and the hydrolytic oxidation of tetrachloroethene. Trichloroacetic acid has been found as a metabolite in the urine of workers exposed to trichloroethylene vapors. Chloral hydrate is immediately oxidized to trichloroacetic acid in humans. The oral toxicity of sodium trichloroacetate, which is used as a herbicide for various grasses and cattails, is quite low (LD$_{50}$ rats, 5.0 g/kg). Although very corrosive to skin, it does not have the skin-absorption toxicity found with chloroacetic acid.

Bromine Derivatives

Bromoacetic acid, $BrCH_2CO_2H$, occurs as hexagonal or rhomboidal hygroscopic crystals, mp 50°C, bp 208°C, d$_4^{50}$ 1.934 g/cm^3, n_D^{50} 1.4804. Solubility at 25°C, in g/100 g: water, 896; methanol, 896; ethyl ether, 735; and carbon tetrachloride, 26. Bromoacetic acid undergoes many of the same reactions as chloroacetic acid under milder condition. It is not used extensively because of the greater cost in comparison to chloroacetic acid. Esters of bromoacetic acid are the reagents of choice in the Reformatsky reaction which is used to prepare β-hydroxyacids or α,β-unsaturated acids. Chloroacetate esters, however, often react slowly or not at all. Bromoacetic acid can be prepared in good yields via the bromination of acetic acid in the presence of acetic anhydride and a trace of pyridine (80–85%) or by the well-known Hell-Volhard-Zelinsky bromination catalyzed by phosphorus. Bromoacetic acid is corrosive to the skin and is readily absorbed.

Dibromoacetic acid, Br_2CHCO_2H, is a white deliquescent crystal with mp 48°C, bp 232–234°C (decomposition) and is very soluble in water and ethyl alcohol. Tribromoacetic acid, Br_3CCO_2H, mol wt 298.9, occurs as monoclinic crystals with mp 135°C, bp 245°C (decomposition), and is very soluble in water, ethyl alcohol, and diethyl ether, but only slightly soluble in petroleum ether. Both should be handled with great care.

Iodine Derivatives

Iodoacetic acid, ICH_2CO_2H, occurs as colorless, white crystals (platelets out of petroleum ether), with mp 82–83°C which are unstable to heat. It has a K_a of 7.1×10^{-4}. Iodoacetic acid is soluble in hot water and alcohol and slightly soluble in ethyl ether. It cannot be prepared by the direct iodination of acetic acid, but by the direct iodination of acetic anhydride in the presence of sulfuric or nitric acids. Teratogenic studies on mice indicate that feeding iodoacetic acid during days 11–14 of gestation results in a high incidence of cleft palates. The acute oral LD_{50} in rats is 60 mg/kg. It is thought to produce its acute toxicity by a different mechanism than either chloroacetic or fluoroacetic acids and should be handled with care.

Diiodoacetic acid, I_2CHCO_2H, mp 110°C (or 96°C), occurs as white needles and is soluble in water, ethyl alcohol, and benzene; it is not commercially available. Triiodoacetic acid, I_3CCO_2H, molecular weight 437.74, mp 150°C, (decomposition) is soluble in water, ethyl alcohol, and ethyl ether. Solutions of triiodoacetic acid are as a rule unstable, as evidenced by the formation of iodine.

E.R. FREITER
Dow Chemical U.S.A.

N.D. Cheronis and K.H. Spitzmueller, *J. Org. Chem.* **6**, 349 (1941).

E. Ott, *Cellulose and Cellulose Derivatives*, 2nd ed., Interscience Publishers, New York, 1955.

Neth. Pat. 109,768 (Oct. 15, 1964), G. van Messel (to N.V. Koninklijke Nederlandse Zoutindustrie).

O.E. Fancher and co-workers, *Toxicol. Appl. Pharmacol.* **26**, 58 (1973).

ACETIC ACID — TRIALKYLACETIC ACID.   See Carboxylic acids.

ACETIC ANHYDRIDE.   See Acetic acid.

ACETINS (ACETATES OF GLYCEROL).   See Glycerol.

ACETOACET-o-CHLORANILIDE.   See Amines, aromatic—Aniline and its derivatives.

ACETOACETIC ACID AND ESTER.   See Ketenes and related substances.

ACETOACET-o-TOLUIDINE.   See Amines, aromatic—Aniline and its derivatives.

ACETONE

Acetone, CH_3COCH_3, is the simplest and most important of the ketones. It is a colorless, mobile, flammable liquid with a mildly pungent and somewhat aromatic odor. It is miscible in all proportions with water and with organic solvents such as diethyl ether, methanol, ethyl alcohol, and esters. It is used as a solvent for cellulose acetate and nitrocellulose, as a carrier for acetylene, and as a raw material for the chemical synthesis of a wide range of products, eg, ketene, methyl methacrylate, bisphenol A, diacetone alcohol, mesityl oxide, methyl isobutyl ketone, hexylene glycol (2-methyl-2,4-pentanediol), and isophorone.

Physical Properties

Acetone has the following physical properties: melting point, -94.6°C; bp, 56.1°C at 101.3 kPa (1 atm); n_D^{20}, 1.3588; specific heat of liquid, 2.6 /g (0.62 cal/g) at 20°C; heat of vaporization, 29.1 kJ/mol (6.9 kcal/mol) at 56.1°C and 101.3 kPa; specific heat of vapor, 92.1 J/(mol·K) (22.0 cal/mol·K) at 102°C; electrical conductivity, 5.5×10^{-8} S/cm ($= \Omega^{-1} \cdot cm^{-1}$) at 25°C; and heat of combustion of liquid, 1787 kJ/mol (427 kcal/mol); heat of formation at 25°C: gas, -216.5 kJ/mol (-51.7 kcal/mol) and liquid, -248 kJ/mol (59.3 kcal/mol).

Chemical Properties

Acetone has a closed-cup flash point of -18°C and an autoignition temperature of 538°C. The explosive limits of acetone–air mixtures lie between 2.15 and 13.0 vol% acetone in air at 25°C. Acetone shows the typical reactions of saturated aliphatic ketones (qv). It forms crystalline compounds with alkali bisulfites. Reducing agents convert acetone to isopropyl alcohol and pinacol, and propane is the product of the Clemmensen reduction. Isopropylamine is produced by reductive ammonolysis of acetone. Acetone undergoes many condensation reactions: with amines it yields Schiff bases; with orthoformic esters, acetals are formed; with hydrogen sulfide, thioketone is formed; with mercaptans, it yields thioketals; various esters condense readily with acetone in the presence of an amine or ammonia; in the presence of sodium alkoxides and sodium amide, β-diketones are formed (also by reaction with carboxylic anhydrides in the presence of boron trifluoride).

Acetone is stable to many of the usual oxidants, but it can be oxidized with some of the stronger oxidants as well as by an alkali metal hypohalite or by a halogen in the presence of a base to form a haloforom and an alkali metal acetate.

Manufacture

Acetone is mainly produced by cumene hydroperoxide cleavage and by dehydrogenation of isopropyl alcohol; an undefined minor portion of the acetone from isopropyl alcohol is produced by catalytic oxidation.

Over 90% of the phenol produced in this country is via the cumene peroxidation route. Benzene is alkylated to cumene (qv), which is oxidized to cumene hydroperoxide, which in turn, is cleaved to phenol and acetone:

$$C_6H_5CH(CH_3)_2 \xrightarrow{O_2} C_6H_5C(CH_3)_2OOH \xrightarrow[\Delta]{acid} C_6H_5OH + CH_3C(O)CH_3$$

One kilogram of phenol production will result in 0.6 kg of acetone as a coproduct. Several key patents describe the technology of the process which includes the following steps: oxidation of cumene to a concentrated hydroperoxide; cleavage of the hydroperoxide; neutralization of the cleaved products; and distillation to recover acetone.

Catalytic dehydrogenation of isopropyl alcohol is an endothermic reaction:

$$CH_3CHOHCH_3 + 66.5 \text{ kJ/mol (at 327°C)} \xrightarrow{catalyst} CH_3C(O)CH_3 + H_2O$$

A large number of catalysts, including copper, silver, palladium metal and platinum, as well as sulfides of transition metals of Groups IV, V, and VI of the periodic table are used at 400–600°C. Lower-temperature reactions (315–482°C) have used zinc oxide–zirconium oxide combinations and combinations of copper and silicon dioxide. It is usual practice to raise the temperature of the reactor as time proceeds to compensate for the loss of catalyst activity. The dehydrogenation is carried out in a tubular reactor with conversions in the range of 75–95 mol%. A process described by Sheel International Research is a useful two-stage reaction to attain high conversion with lower energy cost and lower capital costs. Although the selectivity of isopropyl alcohol to acetone via vapor-phase dehydrogenation is high, a number of by-products must be removed from the acetone by scrubbing, fractional distillation, and caustic treatment.

Acetone was a product of a former process of hydrocarbon oxidation (qv). The catalytic oxidation of isopropyl alcohol mixed with air and fed to reactors maintained at 400–600°C is used in the production of acetone, but to a lesser extent than catalytic dehydrogenation of isopropyl alcohol.

Acetone is also a product of the Shell glycerol process (qv). It is a by-product of the propylene oxide process (see Propylene) and also a by-product of the diisopropylbenzene process for hydroquinone (see Hydroquinone, resorcinol and catechol).

Analysis

The chief impurity in acetone is usually a few tenths of one percent of water. In current industrial practice, gas chromotography is used for analysis; treatment with hydroxylamine hydrochloride and titration of the liberated hydrochloric acid is also used for analysis. Acetone contains no oxidizable impurities, and the color of a few drops of permanganate is retained for several hours.

Health and Safety

Acetone is a solvent of comparatively low acute and chronic toxicity. High vapor concentrations produce anesthesia and are irritating to the nose, eyes, and throat. The odor is disagreeable. Direct contact with eyes may produce irritation and transient corneal injury. Prolonged or repeated contact with the liquid may dry or defat the skin and cause dermatitis. Industry has generally accepted 1000 ppm as the TLV in air for the normal 8 h work day; at higher levels, respiratory protection should be used.

Uses

Acetone is primarily used as a chemical intermediate for methacrylates, as a solvent for various coatings, and for production of methyl isobutyl ketone. All the U.S. production of methyl methacrylate (MMA), methacrylic acid, and higher molecular weight acrylates is based on the acetone cyanohydrin process (see Methacrylic acid and derivatives; Methacrylic polymers). Acetone derivative solvents are diacetone alcohol, mesityl oxide, methyl isobutyl ketone, methylisobutylcarbinol, hexylene glycol, and isophorone (see Ketones).

Bisphenol A is an important, growing outlet for acetone. Bisphenol A is prepared by the reaction of 2 mol of phenol and 1 mol of acetone in the presence of an acid catalyst (see Alkylphenols). It is used in the manufacture of epoxy resins (qv), polycarbonates (qv), corrosion-resistant polyester resins (see Polyesters), vinyl-ester resins, and other resins.

Acetone is used as a solvent in the processing of cellulose acetate (qv) fibers in sheets, in saturating absorbent packing material in acetylene (qv), and isopropylamines (see Amines).

D.L. Nelson
B.P. Webb
The Dow Chemical Company

Reviewed by
K.W. Williams
Dow Chemical U.S.A.

U.S. Pat. 2,632,774 (Mar. 24, 1953), J.C. Conner, Jr., and A.D. Lohr (to Hercules, Inc.).

Brit. Pat. 999,441 (July 28, 1965), (to Allied Chemical).

Brit. Pat. 665,376 (Jan. 23, 1952), (to Standard Oil Development).

ACETONITRILE. See Nitriles.

ACETONYLACETONE. See Ketones.

ACETOPHENETIDIN. See Analgesics; Antipyretics and anti-inflammatory agents.

ACETOPHENONE. See Phenyl ketones.

ACETYL CHLORIDE. See Acetic acid.

ACETYLENE

PHYSICAL AND CHEMICAL PROPERTIES AND EXPLOSIVE BEHAVIOR

Physical Properties

The triple point of acetylene is at -80.55°C and 128 kPa (1.26 atm). The temperature of the solid under its vapor at 101 kPa (1 atm) is -83.8°C. The vapor pressure of the liquid at 20°C is 4406 kPa (43.5 atm). The critical temperature and pressure are 35.2°C and 6.19 MPa (61.1 atm). The density of the gas at 20°C and 101 kPa is 1.0896 g/L. The specific heats of the gas C_p and C_v (at 20°C and 101 kPa) are 43.91 and 35.45 J/(mol·°C) (10.49 and 8.47 cal/(mol·°C)), respectively. The heat of formation ΔH_f^0 at 0°C is 227.1 kJ/mol (54.3 kcal/mol). The solubility in water is 16.6 g/L at 20°C and 1520 kPa (15.0 atm) and 1.23 g/L at 20°C and 101 kPa. Acetylene forms a hydrate of approximate stoichiometry $C_2H_2 \cdot 6H_2O$. The dissociation pressure of the hydrate is 582 kPa (5.75 atm) at 0°C and 3343 kPa (33 atm) at 15°C. Its heat of formation at 0°C is 64.4 kJ/mol (15.4 kcal/mol). Acetylene is soluble to the extent of 200–300 g/L at room temperature and 1520 kPa (15 atm) pressure in polar solvents such as acetone.

Chemical Properties

Acetylene is highly reactive owing to its triple bond and high positive free energy of formation. It is a linear molecule in which two of the atomic orbitals on the carbon are *sp* hybridized and two are involved in π bonds. Acetylene is more susceptible to nucleophilic attack than is ethylene. The electron withdrawing power of the triple bond polarizes the C—H bond and makes the proton acidic with a pK_a of 25; the acidic nature of acetylene accounts for its strong interaction with basic solvents in which acetylene is highly soluble.

Important reactions involving acetylene are hydrogen replacements, additions to the triple bond, and additions by acetylene to other unsaturated systems. Acetylene undergoes linear polymerization and cyclization reactions to benzene and other aromatics. The formation of a metal acetylide is an example of hydrogen replacement, and hydrogenation, halogenation, hydrohalogenation, hydration, and vinylation are important addition reaction. In the ethynylation reaction, acetylene adds to a carbonyl group to give acetylenic alcohols and glycols (see Acetylene-derived chemicals).

Explosive Behavior

Gaseous acetylene. Commercially pure acetylene can decompose explosively (principally into carbon and hydrogen) under certain conditions of pressure and container size. It can be ignited, ie, a self-propagating decomposition flame can be established, by contact with a hot body, by an electrostatic spark, or by compression (shock) heating. Ignition is more likely the higher the pressure and the larger the cross section of the container. When the wall of an acetylene container is heated, ignition occurs at a temperature that depends on the material of the wall and the composition of any foreign particles that may be present. For example, in clean steel pipe, acetylene at 235–2530 kPa (2.3–25 atm) ignites at 425–450°C, whereas particles such as rust, charcoal, alumina, or silica lower the ignition temperature as much as 150°C, and copper oxide as much as 200°C.

Once a decomposition flame has formed, its propagation through acetylene in a pipe is favored by a large diameter and high pressure. The pressure developed by decomposition of acetylene in a closed container depends not only on the initial pressure, but also on whether the flame propagates as a deflagration or a detonation, and on the length of the container. Deflagration flames in vessels and relatively short pipes usually propagate at an increasing velocity without becoming detonation waves, and develop pressures about 10 times the initial pressure. In long pipes these flames usually become detonation waves that exert at the side wall at least ca 20 times the initial pressure, and at the end of the pipe at least ca 50 times the initial pressure. Pressures much higher than these can develop at the end of a pipe in which a detonation wave travels through acetylene that has been compressed substantially while flame has moved at subsonic velocity toward it. The calculated detonation velocity in room temperature acetylene at 810 kPa (8.0 atm) is 2053 m/s. The predetonation distance (the distance the decomposition flame travels before it becomes a detonation) depends primarily on the pressure and pipe diameter when acetylene in a long pipe is ignited by a thermal, nonshock source.

Propagation of a decomposition flame through acetylene in a piping system (either by deflagration or detonation) can be stopped by a hydraulic back-pressure valve in which the acetylene is bubbled through water. It can also be stopped by filling the pipe with parallel tubes of smaller diameter, or randomly oriented Raschig rings.

Acetylene–air and acetylene–oxygen mixtures. The flammability range for acetylene–air at atmospheric pressure is ca 2.5–80% acetylene in tubes wider than 50 mm. The range narrows to ca 8–10% as the diameter is reduced to 0.8 mm. Ignition temperatures as low as 300°C have been reported for 30–75% acetylene mixtures with air and for 70–90% mixtures with oxygen. Ignition energies are lower for the mixtures than for pure acetylene. In acetylene–air mixtures, the normal mode of burning is deflagration in relatively short containers and detonation in pipes. In oxygen mixtures, detonation easily develops in both short and long containers. The explosion pressure ratios for deflagrations and detonations in the most powerfully explosive acetylene–air mixtures are about the same as for acetylene alone. For equimolar acetylene–oxygen mixture in a pipe, a detonation wave develops a pressure ca 45 times the initial pressure at the side wall and ca 110 times the initial pressure at the closed end.

Liquid and solid acetylene. Both the liquid and the solid have the properties of a high explosive when initiated by detonators or by detonation of adjoining gaseous acetylene. At temperatures near the freezing point, neither form is easily made to explode by heat, impact, friction, or electrostatic spark, but initiation becomes easier as the temperature of the liquid is raised. At liquid temperatures approaching room temperature, violent explosions result from exposure to mild thermal sources.

C.M. DETZ
H.B. SARGENT
Union Carbide Corporation

S.A. Miller, *Acetylene—Its Properties, Manufacture and Uses*, Academic Press, Inc., New York, 1965, Vol. 1.

A. Williams and D.B. Smith, *Chem. Rev.* **70**, 267 (1970).

Acetylene Transmission for Chemical Synthesis, Pamphlet G 1.3, Compressed Gas Association, New York.

HANDLING AND MANUFACTURE FROM CALCIUM CARBIDE

Handling

The design of equipment for handling and using acetylene must consider the factors of pressure, temperature, source of ignition, and ultimate pressures that may result from an acetylene decomposition.

Decompositions do not occur spontaneously but must have a source of ignition. Theoretically, during constant-volume deflagration at ca 100–500 kPa (ca 1–5 atm) without loss of heat, the pressure rises to 11.5–12 times the initial pressure. The minimum pressure at which a deflagration flame can propagate throughout a long tube of any diameter has been determined experimentally. In long pipelines, cascading may occur, an effect that develops maximum pressure several hundred times the initial pressure.

Acetylene cylinders are entirely filled with a monolithic, highly porous (up to 92%) mass designed to stabilize acetylene and permit safe shipment. The mass fills the interior yet occupies only 10% of the total cylinder volume. Acetylene is held in an acetone solution in one cylinder; the solubility of acetylene in acetone increases with rising pressure and diminishing temperature. The pressure gauge attached to the cylinder measures the solution pressure, not the amount of acetylene contained. Unlike oxygen or nitrogen cylinders, the contents of an acetylene cylinder are not accurately determined by pressure-gauge readings alone but must be measured by weight converted into standard volume (atmospheric pressure at 21°C) by the factor of 0.906 L/kg (14.7 ft³/lb).

Manufacture, filling and shipping of acetylene cylinders are subject to DOT regulations. Manufacture and shipping specifications are supplied in CFR Title 49. DOT-8 and DOT-8AL specify the requirements for the manufacture and testing of the steel shell, porous mass, and quantity of acetone. Basic industry guidelines are set forth in an NFPA pamphlet, *Acetylene Cylinder Charging Plants*.

Manufacture from Calcium Carbide

Acetylene is generated by the chemical reaction between calcium carbide and water. $CaC_2 + 2 H_2O \rightarrow Ca(OH)_2 + C_2H_2 + 134$ kJ/mol (900 Btu/lb) of pure calcium carbide. Because of the exothermic reaction and the evolution of gas, the most important safety considerations in acetylene generator design are the avoidance of excessively high temperatures and high pressures. The heat of reaction must be dissipated rapidly and efficiently to avoid local overheating of the calcium carbide which, in the absence of sufficient water, may become incandescent and cause progressive decomposition of the acetylene and development of explosive pressures. Maintaining temperatures < 150°C minimizes polymerization of acetylene and other side reactions that form undesirable contaminants. For protection against high pressures, industrial acetylene generators are equipped with pressure-relief devices that keep pressure below 103 kPa (15 psig), the pressure commonly accepted as a safe upper limit for generator operation.

Most carbide acetylene processes are wet processes from which hydrated lime, $Ca(OH)_2$, is a by-product. The thickened hydrated lime is marketed for industrial wastewater treatment, neutralization of spent pickling acids, as a soil conditioner in road construction, and in the production of sand-lime bricks.

Carbide-to-water generation is the most widely used process in the United States to produce acetylene from calcium carbide. Two classes of generators exist, ie, one that operates below 7 kPa (15.7 psia) (low pressure) and a medium pressure generator which operates between 7 and 10.3 kPa (29.7 psia). The commercially available carbide-to-water acetylene generators consist of a water vessel (or reaction chamber), a carbide feed mechanism, and a carbide storage container that empties into the feed mechanism, with the water vessel equipped to fill with water and drain the lime slurry.

Water-to-carbide generation is a method of limited use mainly employed in small acetylene generators, eg, portable lights or lamps where the generation rate is slow and the mass of carbide is small. A dry generator, water-to-carbide method is used in some large-scale operations and requires ca one kilogram of water per kilogram of carbide and dissipates the heat of reaction through vaporizaton of the water. The dry lime by-product is considered to be favorable compared with the wet lime by-product.

The type of generator employed and the quality of the carbide largely determine the purity of carbide acetylene. Carbide acetylene quality may be affected by four main impurities, ie, phosphine, ammonia, hydrogen sulfide, and organic sulfides. Purification involves oxidation of phosphine to phosphoric acid, the neutralization and absorption of ammonia, and the oxidation of hydrogen sulfide and organic sulfur compounds. Wet and dry processes for purification range from simply passing the gas over purifying media to multistep chemical treatment; oxidizing agents such as chromic acid or chromates, hypochlorite, permanganate, and ferric salts are used. Because of the high materials and labor requirements, dry purification is not practiced where large volumes of gas must be treated.

The maximum amount of impurities allowed in U.S. Grade B acetylene (carbon-generated acetylene) is 2 wt% on a dry basis. This gas meets commercial requirements for acetylene used in cutting and welding (qv). Production of U.S. Grade A acetylene used in sensitive chemical reactions requires further purification to reduce impurities to 0.5 wt%.

R.O. TRIBOLET,
Handling of Acetylene

R.P. SHAFFER,
Manufacture from Calcium Carbide
Linde division, Union Carbide Corp.

Acetylene Transmission for Chemical Synthesis, publication of International Acetylene Association, New York.

Code of Federal Regulations 49, items 173.303, 173.306, 178.59, and 179.60, Department of Transportation, Oct. 1, 1973.

Underwriters Laboratories Inc. Standards for Acetylene Generators, No. 297, Portable Medium Pressure, May 1973; *No. 408, Stationary Medium Pressure*, May 1973.

Federal Specification, Acetylene Technical Dissolved BB-A-106B.

W.E. Alexander, *Purification and Drying of Acetylene for Chemical Use*, OTS report, PB 35211, U.S. Department of Commerce, 1944.

FROM HYDROCARBONS

Modern thermal cracking processes for the manufacture of acetylene from hydrocarbons developed from the 1920s research program at Badische Anilin- und Soda Fabrik (BASF). From the first plant in Germany in 1940, expansion was rapid until the mid- to late 1960s, when acetylene was gradually supplanted by cheaper ethylene as the main petrochemical intermediate. The decline of the acetylene industry has not changed with the advent of OPEC and fuels from coal conversion (see Coal; Feedstocks; Fuels, synthetic).

Theory

Thermal acetylene processes differ essentially only in the manner in which the necessary energy for the reaction is supplied. They all depend on the decrease in free energy of acetylene at higher temperatures. At ≥ 1600 K, acetylene is more stable than any other hydrocarbon but less stable than its component elements; therefore, the contact time at this elevated temperature must be short in order to preserve the acetylene formed in the cracking process.

The reaction based on methane can be expressed as

$$2\,CH_4 + 184\,MJ\,(174{,}500\,Btu) \xrightarrow{1250°C} C_2H_2 + 3\,H_2$$

Acetylene begins to form at ca 1500 K. The competing decomposition of

hydrocarbons to carbon and hydrogen begins at 950 K, and the reaction time is on the order of milliseconds.

The temperature required decreases with increasing molecular weight of the hydrocarbon feed and, at most temperatures, most of the hydrocarbons can decompose to the elements.

Process Technology

The processes for making acetylene from hydrocarbons are classified by the means employed for providing the required high heat of reaction (electric arc, flame or partial combustion, or pyrolysis); the feedstocks used; and the composition of the reactor effluent gas (see Table 1). Some processes make acetylene as an essential product whereas others, such as pyrolysis and the two-stage partial combustion processes, make substantial amounts of ethylene at the same time.

Commercial Acetylene Processing

Electric-discharge processes can supply the necessary energy very rapidly and convert more of the hydrocarbons to acetylene than in regenerative or partial combustion processes. The electric arc provides energy at a very high flux density so that the reaction time can be kept to a minimum (see Fig. 1).

Generalized electric-arc reactor. There have been many variations in the design of electric-arc reactors, but only three have been commercialized, the most important installation being the one at Huels.

A considerable amount of carbon is formed in the reactor in an arc process. Plasma processes can reduce this by using, as a heat carrier, an auxiliary gas such as hydrogen which has the ability to dissociate into very mobile reactive atoms. Plasma processes, developed to industrial

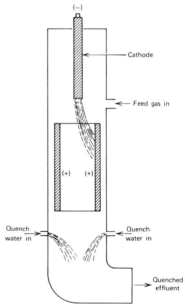

Figure 1. Generalized electric-arc reactor.

Table 1. Commercial Acetylene Processing

Main process category	Particular process type	Usual feedstock	Main energy source	Typical cracked gas concentrations, mol %		
				Acetylene	Ethylene	CO
arc	electric arc	methane	electricity	15	0.9	0.6
	arc plasma	methane to distillates	electricity	14	7	0.6
flame	partial combustion (1-stage)	methane	oxygen plus feed	8	0.2	26
	2-stage combustion	methane to distillates	oxygen plus feed	11	15	14
	submerged combustion	crude oil and distillates	oxygen plus feed	6	6	43
pyrolysis	regenerative furnace	methane to distillates	air plus feed	14	8	7

scale by Farbwerke Hoechst A.G. (Hoechst WLP process) and Chemische Werke Huels (Huels Plasma Process), have the important feature of being able to produce acetylene from heavy feedstocks, even from crude oil, without the excessive carbon formation of a straight arc process.

Flame or partial combustion processes are the predominant processes for manufacturing acetylene from hydrocarbons. They impart the necessary energy to the feedstock by the partial combustion of hydrocarbon feed (one-stage process), or by the combustion of residual gas, or any other suitable fuel, and subsequent injection of the cracking stock into the hot combustion gases (two-stage process).

The design of the burner is of considerable importance since combustion must be as brief and as uniform as possible across the reaction chamber. Preignition, stability, and blow-off of the flame, the possibility of backfiring through the ports of the burner head, and the deposition of carbon on the burner walls depend on the burner design and velocities of the gas and the flame. The feasibility of partial combustion processes results from the high rates of reaction together with the relatively slow rate of decomposition of acetylene and hydrocarbon to carbon and hydrogen. The negative features of partial combustion processes are the cost of oxygen and the dilution of the cracked gases with the combustion products. Flame stability is always a problem. However, these features are more than offset by the inherent simplicity of the operation.

There are several commercial versions of the partial combustion technique including the widely used BASF process (formerly called the Sachsse process); a single-stage and a two-stage process developed by the Societe Belge de l'Azote et des Produits Chimiques de Marly (SBA); the Montecatini process, which operates at a higher temperature than the BASF and SBA processes; a two-stage high temperature pyrolysis (HTP) process operated until recently by Farbwerke Hoechst in Germany; and a BASF submerged-flame process capable of making acetylene from a wide range of feedstocks including crude oil.

Regenerative-furnace processes supply the necessary energy for the cracking reaction by heat exchange with a solid refractory material. Regenerative-pyrolysis processing is very versatile; it can handle varied feedstocks and produce a wide range of ethylene to acetylene. The acetylene content of the cracked gases is high and this assists purification. However, the plant is relatively expensive and requires considerable maintenance because of wear and tear of cyclic operation on the refractory.

In the regenerative-furnace processes, an alternating cycle of operation is used; the hydrocarbon feed is heated by the hot refractory mass to produce acetylene. Carbon and tars, deposited on the refractories during this period, are then removed during a combustion step in which the refractory mass is heated in an oxidizing atmosphere. The refractories must resist both reducing and oxidizing atmospheres at ca 1200°C, as well as withstand the frequent and rapid heating and cooling cycles and abrasion (in the case of moving refractory beds). The regenerative technique is best exemplified by the Wulff process, in operation since 1952, licensed by Union Carbide Corp.

Pyrolysis of hydrocarbon in direct-fired tubes with steam dilution is practiced extensively to make ethylene. This technique may be applicable to the manufacture of acetylene if the requisite high cracking temperature needed can be reached without a high cracking severity and without excessively hot reactor walls which produce catastrophic coking rates. A bench-scale pyrolysis unit based on this principle has been successfully operated using methane feedstock; scale-up to commercial operation is feasible.

Separation and Purification of Hydrogen-Derived Acetylene

The pyrolysis of methane results in a cracked gas that is relatively low in acetylene content and that contains predominantly a mixture of hydrogen, nitrogen, carbon monoxide, unreacted hydrocarbons, acetylene, and higher homologues of acetylene. When the feedstock used is a higher hydrocarbon than methane, the converter effluent also contains olefins, aromatics, and miscellaneous higher hydrocarbons. Most acetylene processes produce significant amounts of carbon black and tars which have to be removed before the separation of acetylene from the gas mixture.

Acetylene's explosive, unstable nature imposes certain restrictions in the use of efficient separation techniques developed for other hydro-

carbon systems. All commercial processes for the recovery of hydrocarbon-derived acetylene are based on absorption-desorption techniques using one or more selective solvents, particularly water (Huels), anhydrous ammonia (SBA), chilled methanol (Montecatini), and N-methylpyrrolidinone (BASF), butyrolactone, acetone, dimethylformamide, and hydrocarbon fractions. Higher acetylenes in the pyrolysis gas are removed by scrubbing with small amounts of a suitable mineral oil or other organic solvent (SBA, Wulff), or by low temperature fractionation (Huels). Carbon black (soot) produced in the partial combustion and electric-discharge processes are removed by electrostatic units, combined with water scrubbers, moving coke beds, and bag filters. The bulk of the carbon in the reactor effluent is removed by a water scrubber (quencher) (see Fig. 2).

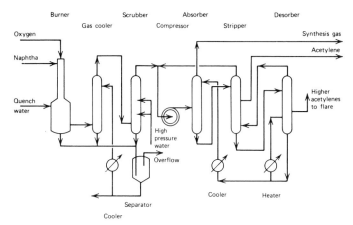

Figure 2. BASF cracking and purification process. Courtesy of Gulf Publishing Co., Houston, Tex.

By-Product Acetylene

A small yield of acetylene is obtained in the commercial fired-tube thermal cracking of hydrocarbons to ethylene (qv). The yield depends principally on the feedstock composition and the severity of cracking. Although this yield has markedly increased as the ethylene manufacturing industry turns to heavier feedstocks, it would be impractical for existing ethylene plants already operating at high severity to raise the cracking temperature and increase the acetylene yield.

Acetylene from Coal

If the price of crude oil continues to rise, the decline in acetylene technology may be reversed by coal-based acetylene becoming competitive with oil-based ethylene. Acetylene has been made traditionally from coal (coke) via the calcium carbide process; laboratory and bench-scale experimental work has investigated the production of acetylene from coal using the widely employed cracking technique of an arc-generated plasma.

One coal-to-acetylene process that could go beyond the laboratory or bench-scale is the Avco arc-coal process, developed under sponsorship to the Office of Coal Research. The process was evaluated by both Stone and Webster Engineering Corporation and Blaw-Knox Chemical Plants, Inc. with the conclusion that the process is practical. Blaw-Knox compared the Avco arc-coal process with partial oxidation of methane to acetylene and concluded that it is the cheapest way to produce acetylene; that it can compete eventually with Gulf Coast ethylene; that economics require location close to the consumer and near a large power plant; and that the effectiveness of the cathode–anode system must be established.

DENNIS A. DUNCAN
Institute of Gas Technology

S.A. Miller, *Acetylene, Its Properties, Manufacture and Uses*, Vol. 1, Academic Press, Inc., New York, 1965.

R.F. Goldstein and A.L. Waddams, *The Petroleum Chemicals Industry*, 3rd ed., E.& F.N. Spon Ltd., London, 1967.

ECONOMIC ASPECTS, ANALYSIS, AND USES

Economic Aspects

In chemical uses, acetylene production as an unavoidable by-product in ethylene manufacture (see Ethylene) or from hydrocarbon-based processes is usually more economical than production from calcium carbide. In special situations, where electrical energy is available cheaply, or where the product must be shipped to relatively small consumers, the carbide route is preferred. Calcium carbide can be shipped over long distances to the point of use where acetylene is required. Acetylene outlets have declined as alternative materials have become available at lower prices; this situation is not expected to change unless there is a change in raw material costs or more by-product acetylene is recovered. By-product acetylene from ethylene production has a potential production of 227,000 t annually in the United States; however, each 450,000 t/yr world-scale ethylene plant only produces 7000–9000 t/yr acetylene, a small volume for an economically scaled derivatives unit.

Analysis

Traces of acetylene can be detected by passing the gas through Ilosvay's solution which contains a cuprous salt in ammoniacal solution; the presence of acetylene is indicated by the pink or red coloration caused by the formation of cuprous acetylide, Cu_2C_2. The preferred quantitative determination is by gas chromatography or infrared and mass spectroscopy (see Analytical methods). Acetylene can be determined volumetrically by absorption in fuming sulfuric acid or by reaction with silver nitrate in solution and titration of the nitric acid formed. Phosphorus, sulfur, and nitrogen compounds, always present in crude gas, are analyzed in acetylene derived from calcium carbide. The presence of unsaturated hydrocarbons and inert gases is determined in acetylene derived from hydrocarbon-based processes. Gas chromatography is the most widely used analytical tool in industrial applications.

Uses

In the United States, 80% of acetylene usage is as a raw material for the synthesis of various organic chemicals (see Acetylene-derived chemicals), with most of the other usage for metal welding or cutting. Chemical markets shrank drastically from 1965 to 1975 as lower cost olefins and and paraffins were substituted for acetylene. Olefins-based processes were adopted for acrylic esters, acrylonitrile, chlorinated ethylenes, 2-chlorobutadiene, vinyl acetate, and vinyl chloride, the principal markets. Markets are now anticipated to be stable in the United States at about 136,000 t/yr for chemical use.

The main chemical markets for acetylene at present are its uses in the preparation of vinyl chloride and vinyl acetate used as monomers for the preparation of a wide variety of polymers. Vinyl chloride was once the principal market for acetylene (see Vinyl polymers). Acetylene is condensed with carbonyl compounds to give a wide variety of products, some of which are the substrates for the preparation of families of derivatives; the most commercially significant is the condensation of acetylene with formaldehyde. The largest growing market for acetylene is the production of 1,4-butanediol by this chemistry.

Small-scale uses include acetylene as an intermediate in the synthesis of vitamins (see Vitamins); in the preparation of vinyl ethers from alcohols; in the preparation of acetylene black used in batteries; and in metal working with the oxyacetylene flame, which continues to consume about 45,000 t/yr of acetylene.

ROBERT M. MANYIK
Union Carbide Corporation

S.A. Miller, *Acetylene*, Vols. I and II, Ernst Benn Ltd., London, 1966.

ACETYLENE BLACK.   See Carbon and artificial graphite.

ACETYLENE-DERIVED CHEMICALS

Acetylene is an extremely reactive hydrocarbon, primarily used as an intermediate (see Acetylene). Because of its thermodynamic instability, it cannot be easily or economically shipped long distances; for large-scale operations, the acetylene consumer must be near the place of acetylene manufacture.

Owing to its high reactivity, acetylene is used in the preparation of a long list of other chemicals, but there are continued efforts to replace acetylene in applications with cheaper raw materials. Ethylene has displaced acetylene as a raw material for acetaldehyde and acetic acid, and in the manufacture of vinyl acetate, vinyl chloride, and chlorinated solvents. Propylene has displaced acetylene as a raw material for acrylonitrile and acrylic acid.

Reaction Products

At one time, acetaldehyde (CH_3CHO) was manufactured principally by hydration of acetylene, but now it is more economically produced by the oxidation of ethylene or other hydrocarbons.

Acrylic acid and acrylates are prepared by the reaction of acetylene, carbon monoxide, and water or an alcohol, using nickel carbonyl as catalyst. Newer processes are based on oxidation of propylene (see Acrylic acid; Acrylate esters).

Tetramerization of acetylene to cyclooctatetraene (C_8H_8), although interesting, apparently has not been used commercially.

Hydrocarbon cracking is used to purify ethylene contaminated by acetylene (see Ethylene).

Successive chlorination and dehydrochlorination of acetylene have furnished commercially important solvents, eg, trichloroethylene ($CHCl=CCl_2$) and perchloroethylene ($CCl_2=CCl_2$), until displaced by chlorination and dehydrochlorination of ethylene or vinyl chloride. Recently, oxychlorination of ethane or other two-carbon raw materials has become attractive (see Chlorocarbons and chlorohydrocarbons).

Addition of acetic acid to acetylene gives vinyl acetate ($CH_3CO_2-CH=CH_2$), but the prominence of this technology in vinyl acetate production has been reduced by newer technology based on oxidative addition of acetic acid to ethylene (see Vinyl polymers).

In the presence of cuprous salt solutions, acetylene dimerizes to vinyl acetylene. With cuprous chloride catalyst, this adds hydrogen chloride to give 2-chloro-1,3-butadiene, chloroprene, the monomer for neoprene rubber (see Elastomers, synthetic).

In the presence of mercuric salts, hydrogen chloride adds to acetylene to give vinyl chloride ($CH_2=CHCl$). Once the principal route to vinyl chloride, this reaction has now been displaced by dehydrochlorination of ethylene dichloride. In similar fashion, vinylidene chloride ($CH_2=CCl_2$) has been prepared from vinyl chloride by successive chlorination and dehydrochlorination steps (see Vinylidene polymers).

Ethynylation Reaction Products

Starting with the ethynylation of formaldehyde, a series of catalytic reactions illustrates the manufacture of Reppe ethynylation products. (See Pyrrole and pyrrole derivatives for products not discussed below.)

$$HC{\equiv}CH + (1-2)\ HCHO \rightarrow HC{\equiv}CCH_2OH + HOCH_2C{\equiv}CCH_2OH$$
propargyl alcohol   2-butyne-1,4-diol

$$HOCH_2C{\equiv}CCH_2OH + H_2 \rightarrow HOCH_2CH{=}CHCH_2OH$$
2-butene-1,4-diol

$$HOCH_2C{\equiv}CCH_2OH + 2\ H_2 \rightarrow HOCH_2CH_2CH_2CH_2OH$$
1,4-butanediol

$$HOCH_2CH_2CH_2CH_2OH \rightarrow \overset{\frown}{CH_2CH_2CH_2\overset{O}{C}O} + H_2$$
γ-butyrolactone

$$\text{(lactone)} + RNH_2 \rightarrow \overset{\frown}{CH_2CH_2CH_2\overset{NR}{C}O} + H_2O$$
R = H   2-pyrrolidinone
R = CH₃   N-methyl-2-pyrrolidinone

$$\text{(pyrrolidinone)} + HC{\equiv}CH \rightarrow \overset{\frown}{CH_2CH_2CH_2\overset{NCH=CH_2}{C}O}$$
N-vinyl-2-pyrrolidinone

Propargyl alcohol, 2-propyn-1-ol, the only commercially available acetylenic primary alcohol, is a colorless, volatile liquid with a mild geraniumlike odor. It is miscible with water and many organic solvents (see Table 1). It has three reactive sites, ie, a primary hydroxyl group, a triple bond, and an acetylenic hydrogen, making it an extremely versatile chemical intermediate. It can be esterified in a normal manner with acid chlorides, anhydrides, and carboxylic acids. It reacts with aldehydes or vinyl ethers in the presence of acid catalysts to form acetals. Using low temperature conditions, oxidation with chromic acid gives propynal ($HC\equiv CCHO$) or propynoic acid ($HC\equiv CCOOH$), which may also be prepared by electrolytic oxidation. Various halogenating agents have been used for replacement of hydroxyl with chlorine or bromine. Hydrogenation gives allyl alcohol, its isomer propionaldehyde, or propanol. In the presence of acidic mercuric salts, water adds to form acetol (1-hydroxyacetone). Under similar conditions, alcohols give ketals which hydrolyze to acetol. Halogens add stepwise to give almost exclusively dihaloallyl alcohol ($CHX=CXCH_2OH$). Propargyl alcohol combines with aldehydes in the presence of copper acetylide catalysts to give acetylenic glycols ($RCH(OH)C\equiv CCH_2OH$). In the presence of dialkylamines and formaldehyde (methylolamines), dialkylaminobutynols ($R_2NCH_2C\equiv CCH_2OH$) are formed. Two equivalents of organomagnesium halide give a Grignard reagent of propargyl alcohol capable of further reactions. Cuprous salts catalyze the oxidative dimerization of propargyl alcohol to 2,4-hexadiyne-1,6-diol.

Propargyl alcohol is a by-product of butynediol manufacture. In the usual high pressure butynediol process, ca 5% of the product is propargyl alcohol. The commercial product is specified as 97% minimum purity, determined by acetylation or gas chromatography. It is shipped in unlined steel containers and handled in standard steel pipe or braided steel hose; rubber is not recommended.

Although propargyl alcohol is stable, violent reactions can take place in the presence of contaminants, particularly at elevated temperatures. Avoid heating with alkaline or strongly acidic catalysts. Weak acids, eg, carboxylic acids, are used as stabilizers. The usual precautions against ignition of vapors must be observed.

Propargyl alcohol is a primary skin irritant and a severe eye irritant; it is toxic and requires good ventilation when being used.

Propargyl alcohol is a component in oil-well acidizing compositions. Corrosion and hydrogen embrittlement of mild steel in mineral acids are inhibited (see Corrosion inhibitors). It is employed in metal pickling and plating operations, and is used as an intermediate in the preparation of the miticide Omite, and sulfadiazine.

Butynediol. 2-Butyne-1,4-diol, is available commercially as a crystalline solid or a 35% aqueous solution manufactured by the ethynylation of formaldehyde (see Table 1 for physical and chemical properties).

Butynediol undergoes the usual reactions of primary alcohols. Because of its linear structure, reactions forming cyclic compounds from cis-1,2-butenediol or butanediol give only polymers with butynediol.

Both hydroxyl groups can be esterified normally. Monoesters can be prepared as mixtures with unesterified butynediol and diesters, but they disproportionate readily and must be purified carefully. The hydroxyl groups can be alkylated in the usual way. Butynediol appears to be more difficult to polymerize than propargyl alcohol, but it does cyclize to hexamethylolbenzene in the presence of catalysts. Heated with acidic oxide catalysts, mixtures of butynediol and ammonia or amines give pyrroles (see Pyrrole and pyrrole derivatives).

The three large manufacturers of butynediol, ie, BASF, GAF, and DuPont, all use formaldehyde ethynylation processes for production.

Although butynediol is stable, in the presence of certain contaminants violent reactions can take place, particularly at elevated temperatures. In the presence of certain heavy metal salts, butynediol can decompose violently. Heating with strongly alkaline materials should be avoided.

Butynediol is a primary skin irritant, requiring appropriate precautions. Acute oral toxicity is relatively high.

Most butynediol produced is consumed domestically for manufacturing butanediol and butenediol. Small amounts are also converted to ethers with ethylene oxide. Butynediol is used to produce the wild-oat herbicide Carbyne and in plating and picking baths.

Butenediol. 2-Butene-1,4-diol, the only commercially available olefinic diol with primary hydroxyl groups, is prepared by partial hydrogenation of butynediol; it consists almost entirely of cis isomer (see Table 1 for physical and chemical properties).

In addition to the usual reactions of primary hydroxyl groups and double bonds, cis-2-butene-1,4-diol undergoes a number of cyclization reactions.

The hydroxyl groups can be esterified normally to prepare the interesting monomer cis-butene-1,4-diol dimethacrylate and the biologically active butenediol haloacetates. Polyesters capable of being cross-linked or suitable for use as soft segments in polyurethanes have been prepared with dibasic acids. Heating with acidic catalysts dehydrates cis-butenediol to 2,5-dihydrofuran. Cupric or mercuric salts give 2,5-divinyl-1,4-dioxane, presumably via 3-butene-1,2-diol. Mixtures of butenediol and ammonia or amines cyclize to pyrrolines upon heating with acidic catalysts (see Pyrrole). Halogens readily add to give 2,3-dihalo-1,4-butanediol. Hydrogen halides cause substitution of the hydroxyl groups. When butenediol is treated with dichromate in acidic solution, oxidation and dehydration give furan (see Furan derivatives). Butenediol does not undergo free-radical polymerization. Butenediol rearranges over various catalysts to 4-hydroxybutyraldehyde.

Butenediol is manufactured by the partial hydrogenation of butynediol. The commercial product is predominantly cis-2-butene-1,4-diol. The heat of hydrogenation (liquid phase) of butynediol to butenediol is 156.0 kJ/mol (37.28 kcal/mol).

In addition to the patents describing supported copper, zinc sponge, and iron catalysts for high pressure hydrogenation, other patents claim more active systems at comparatively low pressure, eg, Raney nickel and supported palladium catalysts.

Table 1. Physical and Chemical Properties of Acetylene-Derived Chemicals

Property	Propargyl alcohol	Butynediol	cis-2-Butene-1,4-diol	Butyrolactone	Butanediol
mp, °C	−52	58			
bp, °C kPa[a]					
0.133		101	84	35	86
1.33		141	122	77	123
13.3		194	176	134	171
101.3	114	248	234	204	228
specific gravity					
d_{15}^{25}			1.070		
d_{4}^{60}		248			
d_{4}^{20}	0.948			1.129	1.017
d_{4}^{25}				1.125	1.015
refractive index					
n_D^{25}		$\alpha = 1.450 \pm 0.002$ $\beta = 1.528 \pm 0.002$	1.4770	1.4348	1.4445
n_D^{20}	1.4310			1.4362	1.4461
viscosity, mPa·s (= cP)					
at 20°C	1.65		22		84.9
at 25°C				1.75	
at 38°C			10.8		
at 99°C			2.5		
dielectric constant					
at 20°C	24.5			39.1	31.5
flash point, open cup, °C	36	152	128	98	121
heat of combustion at constant vol, kJ/mol[b]	1731	2204			
specific heat C_P^{20}, J/(g·K)[b]	2.577				
heat of vaporization at 112°C, kJ/mol[b]	42.09				

[a] To convert kPa to mm Hg, multiply by 7.5.
[b] To convert J to cal, divide by 4.184.

Purity of technical-grade butenediol is determined by gas chromatography. The principal impurities are butanediol, butynediol, and the 4-hydroxybutyraldehyde acetal of butenediol.

Butenediol is stable under normal handling conditions and noncorrosive. It is a primary skin irritant but not a skin sensitizer. Contact with eyes and skin should be avoided. Its low vapor pressure minimizes inhalation exposure. It is much less toxic than butynediol.

Butenediol is used to manufacture the insecticide Endosulfan (Thiodan) and other agricultural chemicals, and pyridoxine hydrochloride, vitamin B$_6$ (see Vitamins).

Butanediol. 1,4-Butanediol, tetramethylene glycol, 1,4-butylene glycol, is produced by the hydrogenation of butynediol. It is miscible in water, ethanol, and acetone, and soluble to the extent of 3.1 g/100 g in ethyl ether, 0.3 g/100 g in benzene, and 0.1 g/100 g in hexane.

The chemistry of butanediol is determined by the two primary hydroxyls. Esterification is normal. For transesterification a nonacidic catalyst is used since strongly acidic catalysts promote cyclic dehydration. With excess phosgene, bischloroformate ester can be prepared. Ethers are formed in the usual way. The bischloromethyl ether is obtained with formaldehyde and hydrogen chloride. The formation of polyurethanes is a commercially important reaction. Carbon monoxide and a nickel carbonyl catalyst give good yields of adipic acid. Heating with acidic catalysts cyclizes butanediol to tetrahydrofuran (see Furan derivatives).

The three main butynediol producers use most of their production for butanediol. Additional small quantities are also produced by other processes. The principal producers are BASF, DuPont, GAF, and Chemische Werke-Huels.

The purity of butanediol is determined by gas chromatography (99% min). The typical product freezes at ca 19°C.

Butanediol is much less toxic than its unsaturated analogues. It is neither a primary skin irritant nor a skin sensitizer.

Substantial amounts of butanediol are used in the manufacture of tetrahydrofuran, butyrolactone, and polyurethanes. It is increasingly being used in the manufacture of polybutylene terephthalate, PBT (see Polyesters, thermoplastic).

Butyrolactone. γ-Butyrolactone, dihydro-2(3H)-furanone, is produced by the dehydrogenation of butanediol and by hydrogenation of maleic anhydric to tetrahydrofuran and butyrolactone. Butyrolactone is completely miscible with water and the common organic solvents. It is only slightly soluble in aliphatic hydrocarbons and is a good solvent for many gases and polymers.

Butyrolactone undergoes the usual reactions of γ-lactones. Particularly characteristic are ring openings and reactions wherein oxygen is replaced by another ring heteroatom. There is also marked reactivity of the alpha hydrogen atoms.

Under neutral conditions, hydrolysis proceeds very slowly at room temperature and comparatively rapidly with elevated temperatures and acidic conditions. The hydrolysis-esterification reaction is reversible. Under alkaline conditions hydrolysis is rapid and irreversible. Heating of the alkaline hydrolysis product at 200–250°C gives 4,4′-oxydibutyric acid after acidification. With alcohols and acid catalysts, esters of 4-hydroxybutyric acid are formed rapidly even at room temperatures. As with water, equilibrium strongly favors the lactone. Attempts to distill the mixtures ordinarily result in complete reversal to butyrolactone and alcohol. The esters can be separated by a quick flash distillation at high vacuum. When butyrolactone and alcohols are heated for a long time at higher temperatures in the presence of acidic catalysts, 4-alkoxybutyric esters are formed. With sodium alkoxides or phenoxides, sodium 4-alkoxybutyrates are obtained. Butyrolactone and anhydrous hydrogen halides give high yields of 4-halobutyric acids.

Benzaldehyde gives α-benzylidenebutyrolactone and ethyl acetate gives α-acetobutyrolactone. The active α-methylene groups add across double bonds; eg, 1-decene reacts with butyrolactone to give α-decylbutyrolactone. Butyrolactone condenses with aromatic hydrocarbons under the influence of Friedel-Crafts catalysts. Condensation with benzene, for example, gives either or both 4-phenylbutyric acid and 1-tetralone, depending on the experimental conditions. Carbonylation of butyrolactone using nickel or cobalt gives high yields of glutaric acid.

Butyrolactone is hygroscopic and should be protected from moisture. When shipping, plain steel, stainless steel, and nickel are suitable for handling and storage but not rubber or phenolic and epoxy resins.

Purity, determined by gas chromatography, is 98% min.

Because of its high boiling point, butyrolactone does not ordinarily present a vapor hazard. It is neither a primary irritant nor a strong skin sensitizer.

Butyrolactone is used as an intermediate with ammonia and methylamine to manufacture 2-pyrrolidinone and N-methyl-2-pyrrolidinone, respectively (see Vinyl polymers); it is also a solvent for agricultural chemicals and polymers and for dyeing and printing.

Other alcohols and diols. Secondary acetylenic alcohols are prepared by ethynylation of higher aldehydes. The commercial products are prepared in alkaline systems, although copper acetylide catalysts may be used.

The ethynylation of ketones is not catalyzed by copper acetylide, but potassium hydroxide is effective, presumably with at least stoichiometric levels of potassium hydroxide. Alcohols are obtained at lower temperatures and glycols at higher temperatures. Other alkaline catalytic systems yield alcohols, but not glycols.

Acetylenic glycols are soluble in the usual organic solvents, but only slightly soluble in aliphatic hydrocarbons. Methylbutynol (2-methyl-3-butyn-2-ol) is the most important of this group of compounds. Manufacture is generally carried out in batch reactors at atmospheric pressure, in a rapid and exothermic reaction. At temperatures < 5°C, the product is almost exclusively the alcohol; at 25–30°C, glycol predominates. A continuous catalytic process for methylbutynol uses a homogeneous aqueous potassium hydroxide–liquid ammonia system.

Acetylenic alcohols and glycols are stable and free of decomposition hazard under normal conditions. The more volatile alcohols present a fire hazard. The glycols are relatively low in toxicity; the alcohols are toxic orally and through skin absorption and vapor inhalation. In general, the tertiary alkynols are less toxic than the secondary.

The secondary acetylenic alcohols, hexynol and ethyloctynol, are used as corrosion inhibitors in oil-well acidizing compositions. The tertiary alcohols, methylbutynol and methylpentynol, are used in the manufacture of vitamin A (see Vitamins) and other products and in metal pickling and plating operations (see Metal surface treatments). Dimethylhexynediol can be used in the manufacture of fragrance chemicals and of peroxide ester catalysts. Dimethyloctynediol and tetramethyldecynediol are antifoaming wetting agents (Surfynols) (see Defoamers). Ethoxylated acetylenic glycols are used as electroplating (qv) additives. Methylbutynol is used as an intermediate for large-scale manufacture of isoprene (qv).

Catalytic Vinylation Products

Unlike ethynylation in which acetylene adds across a carbonyl group and the triple bond is retained, in vinylation a labile hydrogen compound adds across acetylene, forming a double bond.

Catalytic vinylation is applicable to a wide range of alcohols, phenols, thiols, carboxylic acids, and certain amines and amides. Synthetic routes to vinyl esters based on ethylene are rapidly eliminating acetylene as a starting material for vinyl acetate. The manufacture or vinyl ethers and vinylpyrrolidinone still depends upon acetylene.

Vinyl Ethers

Methyl vinyl ether, ethyl vinyl ether, and N-butyl vinyl ether are the principal commercially available vinyl ethers. Others, such as the isobutyl vinyl ether, methoxyethyl, isopropyl hydroxybutyl, decyl, hexadecyl, and octadecyl analogues, are offered as development chemicals (see Ethers). Lower vinyl ethers are miscible with almost all organic solvents.

Union Carbide consumes its entire vinyl ether production for the manufacture of glutaraldehyde. BASF and GAF consume part of their production as monomers. Vinyl ethers are used as comonomers in a number of specialty polymers and as synthetic intermediates.

EUGENE V. HORT
GAF Corporation

S.A. Miller, *Acetylene, Its Properties, Manufacture and Uses*, Vol. 1, Academic Press, Inc., New York, 1965.

W. Reppe and co-workers, *Ann.* **596**, 2 (1955).

W. Reppe and co-workers, *Ann.* **601**, 81 (1956).

ACETYLENE DICHLORIDE (1,2-DICHLOROETHYLENE). See Chlorocarbons and chlorohydrocarbons.

ACETYLENE TETRABROMIDE (1,1,2,2-TETRABROMOETHANE). See Bromine compounds.

ACETYLENE TETRACHLORIDE (1,1,2,2-TETRACHLOROETHANE). See Chlorocarbons and chlorohydrocarbons.

ACETYLENIC ALCOHOLS.   See Acetylene-derived chemicals.

ACETYLENIC GLYCOLS.   See Acetylene-derived chemicals.

ACETYL FLUORIDE.   See Fluorine compounds, organic.

ACETYLIDES.   See Acetylene; Carbides.

ACETYL PEROXIDE.   See Peroxides and peroxy compounds, organic.

ACETYLSALICYLIC ACID.   See Analgesics, antipyretics, and anti-inflammatory agents; Salicylic acid and related compounds.

ACIDS, CARBOXYLIC.   See Carboxylic acids.

ACONITIC ACID.   See Citric acid.

ACRIDINE.   See Pyridine and pyridine derivatives.

ACRIDINE DYES.   See Dyes and dye intermediates.

ACROLEIN AND DERIVATIVES

Acrolein, a colorless, volatile liquid, soluble in many organic liquids, is the simplest member of the class of unsaturated aldehydes. 2-Substituted acroleins, CH_2=CRCHO, are known where R represents alkyl or aryl groups. Extremely reactive owing to its carbonyl double bond, acrolein is an important building block for a large number of derivatives used in a variety of applications. The physical and chemical properties are summarized in Table 1.

Reactions and Derivatives of Acrolein

Acrolein behaves as both diene and dienophile to form the cyclic dimer, 2-formyl-3,4-dihydro-2H-pyran. Acrolein dimer is used to prepare many useful organic intermediates, by reactions of the aldehyde and enol ether groups, eg, hydroxyadipaldehyde by hydrolysis, hexanetriol by hydrogenation, and lysine by oxidation. Condensation yields either a product used in slow-release fertilizers, or 3,4-dihydro-2H-pyran-2,2-

Table 1. Physical and Chemical Properties of Acrolein and Methacrolein

Property	Acrolein	Methacrolein
specific gravity, 20/20 C	0.8427	0.8474
boiling point, °C		
101.3 kPa[a]	53	69
1.33 kPa[a]	−36	−25
melting point, °C	−87.0	−81.0
vapor pressure at 20°C, kPa[a]	29.3	16.1
heat of vaporization at 101.3 kPa (1 atm), kJ/kg[b]	93	76.6
solubility at 20°C, wt%		
in water	20.6	5.9
water in	6.8	1.7
refractive index, n_D^{20}	1.4013	1.4169
viscosity at 20°C, mPa·s (= cP)	0.35	0.49
flash point, open cup, °C	−18	2
closed cup, °C	−26	
heat of polymerization (vinyl), kJ/mol (kcal/mol)[b]	71.1–79.5(17–19)	
heat of condensation (aldol), kJ/mol (kcal/mol)[b]	41.8(10)	

[a] To convert kPa to mm Hg, multiply by 7.5.

[b] To convert J to cal, divide by 4.184.

dimethanol, which is used in polyurethanes (see Urethane polymers). Acrolein dimer is also used to make cotton creaseproof and to harden image-receiving layers for the azo-dye-developer transfer process.

Acrolein is reactive as a dienophile in the Diels-Alder reaction because the carbon–carbon double bond is electron deficient. For example, butadiene gives high yields of 3-cyclohexenecarboxaldehyde.

Acrolein behaves as a 1,3-diene to give dihydropyrans with ethylenic dienophiles; dienophiles with substituents that increase electron density in the carbon–carbon double bond readily add, eg, olefins generally give poor yields and vinyl ethers give good yields.

Acrolein undergoes acid- and base-catalyzed additions. Water, alcohols, and organic acids add readily to give $ROCH_2CH_2CHO$, where R = H, alkyl or acyl. D-Methionine, an essential amino acid used as a feed supplement for poultry and cattle, is one of the growing markets for acrolein. Under basic conditions, acrolein adds active methylenes; this process is used in the synthesis of lysine (see Amino acids).

Acrolein gives the Tischenko reaction to produce allyl acrylate. Acetals form readily under mild acidic conditions. With 1,2- or 1,3-glycols, the cyclic acetal may be the exclusive product.

Acetals of acrolein are of interest because the vinyl group is retained. Hard enamels are produced from such acetals and other vinyl monomers (see Alcohols, polyhydric).

Formaldehyde condenses with acrolein under stoichiometric base conditions to give high yields of pentaerythritol (see Alcohols, polyhydric). It forms a stable adduct with sodium bisulfite, disodium 1-hydroxypropane-1,3-disulfonic acid. At 350–400°C and with acidic catalysts, the amination of acrolein leads to pyridine and 3-picoline as major products. The oxidation of the latter to nicotinic acid (niacin), the antipellagra vitamin, represents an important commercial use for acrolein (see Vitamins).

Reduction of the aldehyde group, as well as vapor-phase hydrogenation, yields allyl alcohol; epoxidation of allyl alcohols followed by hydrolysis is a commercially used route to synthetic glycerol (qv).

Oxidation of the aldehyde in either liquid or vapor phase produces acrylic acid (qv). Vapor-phase oxidation is the basis for large-scale commercial manufacture of acrylic acid which can be converted by direct esterification to various acrylic esters (see Esterification).

In the absence of inhibitors, acrolein polymerizes to an insoluble, highly cross-linked solid—referred to as disacryl in the early literature—which is of no known use. The aldehyde group (1,2-linkage) and vinyl group (3,4-linkage) may polymerize individually or simultaneously by one of two routes (aldehyde-vinyl or 1,4-linkage), but 1,4-polymerization is rare. Polyacrolein, its acetals, and bisulfite derivative have very similar thermoplastic properties, good resistance to aqueous acid but poor resistance to water. Reaction of polyacrolein with formaldehyde improves stability in water and alkali solutions.

Manufacture

Production is based on the propylene oxidation process. Proprietary processes have been described. Products of the reaction include acrolein, acrylic acid, acetic acid, acetaldehyde, and carbon oxides. Acrylic acid can be recovered or discarded, depending upon the economics governing the situation.

Typically, the purity of acrolein marketed approaches 97%; propionaldehyde and acetone are the principal carbonyl impurities. Purity is determined by gas chromatography.

Handling

Acrolein rapidly polymerizes with light or heat and is usually inhibited with hydroquinone. Other highly exothermic reactions are catalyzed by mineral acids, strong alkalies, or amines. Common polymerization inhibitors are not effective in preventing these reactions, but injection of buffer solution is effective to control the reactions. To dispose of acrolein or arrest a runaway reaction, 20 or more volumes of water must be added to completely solubilize the acrolein and absorb the heat of reaction. Acrolein should be stored under an inert atmosphere at reasonably cool temperatures. Adequate ventilation is needed for tanks, processing vessels, pumps, and enclosed areas.

Health and Safety

Acrolein is highly toxic but lacrimatory properties serve as an exposure warning. The TLV for safe, continued 8-h exposure is 0.1 ppm by volume in air. Direct acrolein exposure may result in severe injury, and even dilute solutions can cause residual damage to the eyes. Some individuals are superficially insensitive and may be injured from unsensed overexposure. Acrolein is poisonous by absorption through the skin, and the liquid can cause severe chemical burns. Swallowing produces acute gastrointestinal distress with pulmonary congestion and edema.

Direct Uses of Acrolein

The microbiological activity of acrolein is utilized in the subsurface injection of wastewaters where the addition of acrolein controls the growth of microbes in the feed lines (see Wastes); in protecting liquid fuels against microorganisms; in controlling the growth of algae, aquatic weeds, and mollusks in the recirculating process water systems; and as a slimicide (see Industrial antimicrobial agents). Acrolein is used in biomedical applications as a tissue-fixative; and with its polymers to immobilize enzymes (see Enzymes, immobilized); and in leather tanning (see Leather).

Methacrolein

Methacrolein, methacrylaldehyde, can be commercially prepared by direct oxidation of isobutylene in a manner analogous to that employed for the manufacture of acrolein, by cross-condensation of propionaldehyde and formaldehyde, or by the dehydrogenation of isobutyraldehyde. Despite research activity, no commercial supplier of methacrolein is known.

All safety precautions recommended for acrolein apply to methacrolein. Methacrolein is flammable and highly reactive, forming peroxides and acids on exposure to air (oxygen). Storage under a nitrogen atmosphere is recommended, with at least 0.1 wt% hydroquinone inhibitor added to prevent polymerization.

Methacrolein undergoes most of the reactions discussed for acrolein. However, the higher cost of methacrolein, problems in long-term storage, and the lack of unique properties for its derivatives have discouraged large-scale use. Its principal use is as an intermediate in the production of methacrylonitrile or methacrylic acid (qv).

L.G. HESS
A.N. KURTZ
D.B. STANTON
Union Carbide Corporation

S.A. Ballard, H. de V. Finch, B.P. Geyer, G.W. Hearne, C.W. Smith, and R.R. Whetstone, *World Petroleum Congress*, *Proceedings of the 4th Congress*, *Rome, 1955*, Sect. 4, Part C, pp. 141–154.

C.W. Smith, ed., *Acrolein*, John Wiley & Sons, Inc., New York, 1962.

ACRYLAMIDE

Acrylamide, $H_2C{=}CHCONH_2$, is the simplest and most important member of the series of acrylic and methacrylic amides. Selected physical properties are mol wt 71.08, mp $84.5 \pm 0.3°C$, bp 87°C at 0.27 kPa (2 mm Hg), heat of polymerization -82.8 kJ/mol (-19.8 kcal/mol), and density 1.122 g/cm^3 at 30°C. Acrylamide is soluble to the extent of 215.5 g/100 mL in water, 86.2 g/100 mL in ethanol, 155 g/100 mL in methanol, and 63.1 g/100 mL in acetone, and it has < 1 g/100 ml solubility in hydrocarbon solvents.

Acrylamide contains two reactive centers. The amide group undergoes the reactions characteristic of an aliphatic amide. It exhibits weakly acidic and basic properties. Acrylamide forms a sulfate salt with sulfuric acid and the sulfate reacts with alcohols to yield esters, $CH_2{=}CHCOOR$. Acrylamide is readily hydrolyzed to yield acrylate ion. Acrylamide may be dehydrated by conventional procedures to give acrylonitrile (qv). The reaction with formaldehyde gives N-methylolacrylamide, $CH_2{=}CHCONHCH_2OH$, which is highly significant from an industrial standpoint. Upon acidification, N-methylolacrylamide reacts with additional acrylamide to yield N,N'-methylenebisacrylamide, $(CH_2{=}CHCONH)_2CH_2$, which is useful as a cross-linking agent. The methylol compound may be capped with alcohols to provide the less reactive methylol ethers, $CH_2{=}CHCONHCH_2OR$. Provided the system is not basic, the methylol derivative may also be condensed with phenols, carbamate esters, secondary amines, or phosphines without involving the double bond.

The double bond of acrylamide is electron deficient and gives Michael-type addition reactions, many of which are reversible. In general, the stronger the nucleophilic character of the attacking species, the faster the reactions proceed. Hydroxy compounds such as alcohols and phenols add readily to acrylamide in the presence of a base to yield the corresponding ethers, $ROCH_2CH_2CONH_2$. Polymers having pendant hydroxyl groups, eg, cellulose and starch, may be carbamoylethylated using this reaction. Primary amines give a mono or bis adduct, and secondary amines give only a monoadduct; ammonia gives, 3,3',3''-nitrilotrispropionamide, $N(CH_2CH_2CONH_2)_3$. Tertiary amine salts yield quaternary ammonium adducts. The reaction of sodium sulfite or bisulfite yields sodium β-sulfopropionamide, $NaO_3SCH_2CH_2CONH_2$ and is used to scavenge acrylamide monomer. Electrolytic hydrodimerization of acrylamide to adipamide is an unexpected reaction in that a tail-to-tail dimer, rather than a polymer, is formed, $(H_2NCOCH_2CH_2)_2$. The most important reactions are those that produce vinyl addition polymers. The reaction is initiated by almost all conceivable free-radical sources. Polyacrylamide is predominantly a head-to-tail structure (see Acrylamide polymers). Copolymers with acrylamide are also prepared with ease although the molecular weights are consistently lower than that of polyacrylamide prepared under similar conditions.

Manufacture

Acrylamide production is based on the hydration of inexpensive and readily available acrylonitrile (qv). Until recently, the hydration was done exclusively with sulfuric acid monohydrate. Methacrylamide may be obtained by using methacrylonitrile. N-Alkylacylamides are prepared similarly by running the reaction in the presence of a selected olefin or alcohol. The acrylonitrile must be mixed slowly with the acid and water as the reaction is very exothermic (approximately -130 kJ/mol or -31 kcal/mol, including heats of mixing). Reactant ratios must be controlled very closely, and inhibitors such as copper and copper salts are used to suppress side reactions. The expensive and difficult step in the process is the recovery of the acrylamide from the reaction mixture.

A variety of patents exist on catalyst compositions that eliminate the by-product sulfate. All commercial catalytic plants are believed to be using copper-based catalysts at this time. Acrylamide and its derivatives have been prepared by other noncommercial routes such as the reaction of acryloyl chloride and acrylic anhydride with ammonia.

Specifications

Acrylamide is available in solid and liquid form, and in pellets of 98% and 95% purity. Dry acrylamide is shipped in bags and drums fitted with polyethylene liners. Aqueous acrylamide is the preferred form since it eliminates solids handling and lowers manufacturing cost. The 50% aqueous form is applicable to nearly all the uses of acrylamide when volume is taken into account. The inhibitor system and the pH must be checked periodically and maintained according to the manufacturer's instructions.

Health and Safety

Acrylamide and its principal derivatives are moderately toxic in single doses; by ingestion, inhalation of vapor, dust, or aerosol; and by skin absorption. Most cases of toxic human exposure result from gross and repeated dermal contact. Solutions of acrylamide in organic solvents and acrylamide sulfate are considered more hazardous than aqueous solutions. Acrylamide poisoning signs and symptoms are erythemia and peeling of the hands, numbness, dizziness, and in severe cases, ataxia which impairs the ability to grasp and stand. Acrylamide and N-methylolacrylamide are neurotoxins. Polyacrylamides exhibit very low toxicity. (See Table 1 for a summary of LD_{50} data.)

The maximum exposure for industrial handling is limited to 0.05 mg/kg per day. This value has been translated by the ACGIH to give a TLV of 0.3 mg/m^3 with a reduction in value if any skin contact is involved. The general populace should not be exposed to more than 0.0005 mg/kg per day. This value forms the basis for the approvals granted in the United States for polyacrylamide used in potable water.

Uses

The principal use of acrylamide is in water-soluble polymers (see Acrylamide polymers) which are used as additives for enhanced oil recovery (see Petroleum), flocculants (see Flocculating agents), papermaking aids (see Paper), and thickeners. Smaller quantities are used to introduce hydrophilic centers into essentially oleophilic polymers for promoting adhesion, increasing the softening point and solvent resistance of resins, and introducing a center for dye acceptance. Acrylamide is used frequently as a component of photopolymerizable systems (see Photoreactive polymers). In vinyl polymers, the reactivity of the amide group may be utilized in cross-linking reactions.

D.C. MacWilliams
Dow Chemical U.S.A.

H. Feuer and U.E. Lynch, *J. Am. Chem. Soc.* **75**, 5027 (1953).

Ger. Pat. 1,618,230 (July 5, 1973), H. Enders and G. Pusch (to Chemische Fabrik Pfersee).

Ger. Offen. 2,320,060 (Nov. 7, 1974), T. Dockner and R. Platz (to BASF A.G.).

ACRYLAMIDE POLYMERS

Polyacrylamides are versatile synthetic polymers used worldwide to improve commercial products and processes. They are readily water soluble over a broad range of conditions. Initially designed for applications involving separation and clarification of liquid–solid phases, the polymers are now used for binding, thickening, lubrication, and film formation.

Acrylamide, CH_2=$CHCONH_2$, is unique in its ability to produce very high molecular weight linear polymers which exhibit strong hydrogen bonding. Homopolymers of acrylamide have very low ionic character and limited utility. Of the many derivatives of the parent compound, some possess the anionic character of the carboxylic acid derived from amide hydrolysis or copolymerization with acrylic acid, and there are a variety of cationic copolymers easily made via copolymerization with other monomers.

Physical Properties

Solid polyacrylamide is a hard glassy polymer. It is hygroscopic, the rate of moisture absorption increasing with increasing ionic character of the derivative. Polyacrylamide has good thermal stability when compared with other polyelectrolytes; however, degradation occurs upon prolonged heating at elevated temperatures. Dry polyacrylamide is available in several forms and particle-size distributions, depending on the type of polymerization, drying, and grinding processes employed.

Polyacrylamide is infinitely soluble in water, forming clear solutions, but insoluble in most organic solvents. The viscosity of polyacrylamide in water increases dramatically with molecular weight and decreases with increasing temperature. The viscosity developed by nonionic polyacrylamide is relatively insensitive to electrolyte and hydrogen-ion concentrations. Aqueous solutions of polyacrylamide are quite shear sensitive and at relatively low shear rates are pseudoplastic. At high shear rates, the molecular weight can be degraded by rupturing polymer chains. The solution slowly degrades with time, as measured by loss of viscosity. The stability can be improved by the addition of sodium thiocyanate, thiourea, sodium nitrite, and nonsolvents such as methanol. Degradation can be accelerated by additions of ferrous salts along with a ferric chelating agent and by shear, light, ultrasonic waves, free radicals, and heat.

Chemical Properties

Polyacrylamide has commercial value partly owing to its ability to undergo chemical reactions forming ionic derivatives. Polyacrylamide solutions undergo the general reactions of the aliphatic amide group.

The hydrolysis of polyacrylamide under basic conditions is quite rapid, and a reaction carried out under these conditions will contain some acid groups as well as the derivatives of choice and unreacted amide. Hydroxides and carbonates are used as hydrolyzing agents. Heating a polyacrylamide solution in the presence of excess sodium carbonate or sodium hydroxide will cause up to 30 mol% or 70 mol% hydrolysis, respectively.

Polyacrylamide undergoes methylolation in the presence of formaldehyde under acidic and basic conditions. A secondary reaction with an

Table 1. Toxicity of Acrylamide

Monomer	Species	Oral single dose LD_{50}, g/kg	Eye irritant	Skin irritant	Neurotoxic
acrylamide	mouse	0.17	moderate	moderate	yes
N-methylolacrylamide	mouse	0.42	transient	some	yes
N,N'-methylenebis-acrylamide	rat	0.39		none	no
diacetone acrylamide	rat	2–5	none	none	
acrylamide 50% aq, as is basis	various	0.490–0.565	moderate	minor LD_{50} 2.250 g/kg by absorption	yes

amine produces an aminomethylated polyacrylamide with cationic character. Polyacrylamide will undergo the Hoffmann degradation in the presence of sodium hypochlorite or sodium hypobromite to form a polyamine after neutralization. Polyacrylamide may be converted to a sulfomethyl derivative by condensation with formaldehyde and sodium bisulfite. This polymer will contain significant carboxylate groups (up to 50%) owing to the rapid rate of hydrolysis under the reaction conditions. Polyacrylamide may be cross-linked by imide formation at acid pH and by reaction with hydrated polyvalent metal ions.

Manufacturing and Processing

The most common commercial manufacturing process is aqueous free-radical solution polymerization. Polymers with a bulk solution viscosity less than 20,000 mPa·s (= cP) at 20% solids are generally sold as aqueous solutions. Two main drying techniques are used to prepare solid high molecular weight polymers from aqueous solutions: thermal sheet drying and solvent precipitation. The type of drying system selected depends upon the molecular weight of the polyacrylamide, on the reactions taking place during drying, and on the heat sensitivity of the specific polymer being produced.

Other noncommercial techniques used for aqueous solution polymerization include exposure to heat, high energy radiation, ultrasonic waves, ultraviolet radiation, and ionic polymerization catalysts to produce water-soluble or swellable polymers; base-catalyzed hydrogen-transfer polymerization of acrylamide yields water-insoluble polymer which is soluble in certain hot organic solvents.

In mixed-solvent solution polymerization, the acrylamide and other ingredients are dissolved in a mixture of water and organic cosolvent such as alcohols. The choice of organic solvent and the ratio of organic solvent to water determines the maximum molecular weight of the polymer obtained; polymers prepared in mixed-solvent solution generally have a more narrow molecular weight distribution than those prepared in aqueous solution.

In dispersed-phase polymerization, the aqueous reaction mass is dispersed in an inert organic carrier such as xylene, paraffin oil, fuel oil, or tetrachloroethylene. This produces a final product that is either a beadlike solid or a dispersion. The liquid form of the dispersed-phase product is becoming more popular owing to its ease of handling for low volume uses. Settling, solution rate, and freeze–thaw stability can create application difficulties if the dispersion is not carefully designed. Dispersed-phase systems may be dried by solvent precipitation or azeotropic distillation (qv).

Derivatives

Most uses for polyacrylamide require some ionic functionality, and as a result the solution properties of the polyacrylamide will vary with the electrolyte concentration and hydrogen-ion concentrations. Most analysis procedures for ionic derivatives include the evaluation of polymer viscosity in an electrolyte solution.

Commercial derivatives are produced by copolymerization and by post-polymerization reactions of the amide group. Copolymers are easy to produce using commercial processes, require little additional equipment or process time, and can be used in both solution and dispersed-phase processes. However, the cost of using vinyl monomers can be quite high, since it is difficult to achieve ultrahigh molecular weight, there is a potential for unreacted functional comonomer toxicity, and additional storage is required for the comonomer. The distribution of functional groups may vary from random to block distribution depending on the copolymerization parameters.

The post-polymerization reactions are less expensive than some functional comonomers, they achieve ultrahigh molecular weight derivatives, and there is no potential for unreactive functional comonomer. However, these reactions require additional equipment and processing time, it is difficult to mix reagents into viscous gels or dispersed particles, dispersion instability may occur, there is possible unreacted reagent toxicity, and post-reaction chemicals may require special materials of construction.

Analytical and Test Methods

Typically, the most important consideration for a polyacrylamide is its activity in the intended use, and customers routinely determine activity via their own test methods.

Health and Safety

Polyacrylamides represent little hazard to public health or the environment. The polymers are nonirritating to skin under normal use conditions; however, gross contaminations could cause skin irritations. They are mild eye irritants, and suitable eye protection should be used. The polymers have low single-dose oral toxicity and should present no problem from ingestion under normal conditions of industrial use.

Acrylamide monomer is a severe neurotoxin and a cumulative toxicological hazard. Residual monomer in the polymer appears to be the main toxic concern and is generally controlled.

Some of the cationic polyacrylamides may be more irritating to the skin and eyes and can kill fish if released in natural-water sources.

Uses

Most of the uses of polyacrylamide involve aqueous solutions. The largest application is liquid–solid separation, where the polymer is a flocculant and aids in processing of minerals in mining, waste treatment (see Wastes), and water treatment (see Water) for municipalities or industry (see Flocculating agents). Polyacrylamides are extensively used as additives or processing aids in paper manufacturing. Anionic polymers are used for enhanced oil recovery (see Petroleum).

Polyacrylamides are widely used in industry as adhesives, coagulants, dispersants, friction-reducing agents, suspending agents, and thickeners.

J.D. MORRIS
R.J. PENZENSTADLER
The Dow Chemical Co.

American Cyanamid Co., *Chemistry of Acrylamide*, Wayne, N.J., 1966.

D.C. MacWilliams, "Acrylamide and Other Amides," in R.H. Yocum and E.B. Nyquist, eds., *Functional Monomers*, Vol. 1, Marcel Dekker, Inc., New York, 1973, p. 1.

ACRYLIC ACID AND DERIVATIVES

Acrylic acid ($CH_2=CHCO_2H$) (propenoic acid), a moderately strong carboxylic acid, is a colorless liquid with an acrid odor. Table 1 lists several physical properties of acrylic acid and representative derivatives. Acrylates are derivatives of both acrylic and methacrylic acid ($CH_2=CH(CH_3)CO_2H$) (see Methacrylic acid and derivatives). They are used primarily to prepare emulsion and solution polymers with wide industrial applications: acrylate polymer emulsions are used in coatings, finishes, and binders for leather, textiles, and paper, as well as in the preparation of paints, floor polishes, and adhesives. Solution polymers are used in the preparation of industrial coatings. The polymeric products vary widely in physical properties, depending upon the controlled variation in the ratios of monomers used in their preparation, cross-linking, and molecular weight. All the polymeric products are highly resistant to chemical and environmental attack and have excellent clarity and strength (see Acrylic ester polymers). Table 2 gives properties for commercially important acrylic esters.

Reactions

Acrylic acid and its esters may be viewed as derivatives of ethylene in which one of the hydrogen atoms has been replaced with a carboxyl or carboalkoxy group. These compounds react readily with electrophilic, free-radical, and nucleophilic agents.

Polymerization is avoided by conducting the desired reaction under mild conditions and in the presence of polymerization inhibitors. Acrylic

Table 1. Physical Properties of Acrylic Acid Derivatives

Property	Acrylic acid	Acrolein	Acrylic anhydride	Acryloyl chloride	Acrylamide
melting point, °C	13.5	−88			84.5
boiling point, °C (kPa^a)	141 (101)	52.5 (101)	38 (0.27)	75 (101)	125 (16.6)
refractive index, n_D (°C)	1.4185 (25)	1.4017 (20)	1.4487 (20)	1.4337 (20)	
flash point, Cleveland open cup, °C	68				
density, g/cm³ (°C)	1.045 (25)	0.838 (20)		1.113 (20)	1.122 (30)

^{a}To convert kPa to atm, divide by 101.

Table 2. Properties of Commercially Important Acrylate Esters

Acrylate	Heat of polymerization, kJ/mol^a	Bp, °C	d^{25}, g/cm³	Water solubility, g/100 g H$_2$O	Specific heat, J/(g·K)^a
methyl	78.7	79–81	0.950	5	2.01
ethyl	77.8	99–100	0.917	1.5	1.97
n-butyl	77.4	144–149	0.894	0.2	1.92
isobutyl		61–63^b	0.884	0.2	1.92
tert-butyl		120	0.879	0.2	
2-ethylhexyl		214–220	0.880	0.01	1.92

^{a}To convert J to cal, divide by 4.184.
^{b}At 6.7 kPa (= 50 mm Hg).

acid undergoes the reactions of carboxylic acids and can easily be converted to salts, by reaction with an appropriate base in aqueous medium; acrylic anhydride, by treatment with acetic anhydride or by reaction of acrylate salts with acryloyl chloride; acryloyl chloride, by reaction with phosphorus oxychloride, or benzoyl or thonyl chlorides; esters, by direct esterification with alcohols in the presence of a strong acid catalyst such as sulfuric acid, a soluble sulfonic acid or sulfonic acid resins, addition to alkylene oxide to give hydroxyalkyl esters, or addition to the double bond of olefins such as ethylene in the presence of strong anyhydrous acid catalyst to give ethyl acrylate; and amides, by reaction with ammonia or primary or secondary amines. Most acrylic acid is used in the form of its methyl, ethyl, and butyl esters. Acrylates can also be obtained by carrying out acrylonitrile hydrolysis in the presence of alcohol.

Acrylic esters may be saponified, converted to other esters or converted to amides by aminolysis.

Free-radical initiated polymerization of the double bond is the most common reaction and presents one of the more troublesome aspects of monomer manufacture and purification (see Acrylic ester polymers).

Halogens, HX, and HCN readily add to acrylic acid to give the 2,3-dihalo- and 3-halo- or cyanopropionates, respectively.

Michael condensations are catalyzed by various bases to give β-substituted propionates. These addition reactions are frequently reversible at high temperatures.

Addition of mercaptans with alkaline catalysts gives 3-alkylthiopropionates. In the case of hydrogen sulfide, the initially formed 3-mercaptopropionate reacts with a second molecule of acrylate to give a 3,3'-thiodipropionate, S(CH$_2$CH$_2$CO$_2$R)$_2$.

Ammonia and amines add to acrylates to form β-aminopropionates, H$_2$NCH$_2$CH$_2$CO$_2$R, from NH$_3$, etc, which add easily to excess acrylate to give tertiary amines. The reactions are reversible.

Manufacture

Various methods for the manufacture of acrylates are summarized in Figure 1, which shows their dependence on specific hydrocarbon raw materials. For a route to be commercially attractive, the raw material costs and utilization must be low, plant investment and operating costs must not be excessive, and the waste-disposal charges must be minimal. The favorable supply and cost projections for propylene suggest that all new acrylate plants will employ propylene oxidation technology for at least the next two decades. This two-stage process gives first acrolein and then acrylic acid (see Acrolein and derivatives).

$$CH_2{=}CHCH_3 + O_2 \rightarrow CH_2{=}CHCHO + H_2O$$

$$CH_2{=}CHCHO + 1/2\,O_2 \rightarrow CH_2{=}CHCO_2H$$

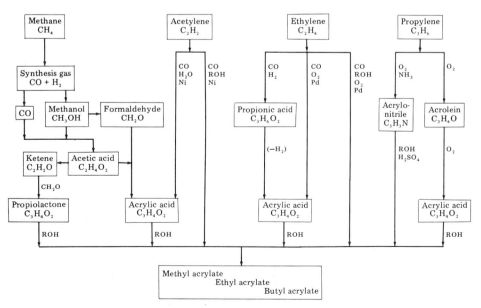

Figure 1. Acrylate manufacturing process.

The most important factor in the oxidation process is catalyst performance. The overall yield of acrylic acid in the oxidation reaction steps of this process is in the range of 76–86%, depending on the catalysts and conditions employed. In the separations step, ca 95% of the acrylic acid is extracted from the absorber effluent with a solvent chosen for high selectivity for acrylic acid and low solubility for water and by-products. Acrylic acid esters are produced in minimum purity of at least 98.5–99% (see Esterification).

Processes based on acetylene—the high pressure Reppe process (BASF) and the modified Reppe process (Rohm and Haas)—or on acrylonitrile were still being used in the late 1970s for the production of acrylic acid and esters. The Reppe reaction uses nickel carbonyl with acetylene and water or alcohols to give acrylic acid or esters (see Acetylene-derived chemicals). The original Reppe reaction requires a stoichiometric ratio of nickel carbonyl to acetylene. The Rohm and Haas modified or semicatalytic process provides 60–80% of the carbon monoxide from a separate carbon monoxide feed and the remainder from nickel carbonyl. The reactions for the synthesis of ethyl acrylate are as follows:

$$4\,C_2H_2 + 4\,C_2H_5OH + Ni(CO)_4 + 2\,HCl \rightarrow$$
$$4\,CH_2{=}CHCO_2C_2H_5 + H_2 + NiCl_2$$

$$C_2H_2 + C_2H_5OH + 0.05\,Ni(CO)_4 + 0.8\,CO + 0.1\,HCl \rightarrow$$
$$CH_2{=}CHCO_2C_2H_5 + 0.05\,NiCl_2 + 0.05\,H_2$$

The stoichiometric and the catalytic reactions occur simultaneously, but the catalytic predominates. The process is started with stoichiometric amounts, but afterward, carbon monoxide, acetylene, and excess alcohol give most of the acrylate ester by the catalytic reaction. The nickel chloride is recovered and recycled to the nickel carbonyl synthesis step. The main by-product is ethyl propionate, which is difficult to separate from ethyl acrylate. However, by proper control of the feeds and reaction conditions, it is possible to keep the ethyl propionate content below 1%. Even so, this is significantly higher than the propionate content of the esters from the propylene oxidation route.

BASF used a high pressure, catalytic route based on the Reppe process at 200°C and 13.9 MPa (2016 psi) in the presence of tetrahydrofuran as an inert solvent. This process gives acrylic acid directly. Nickel carbonyl creates a safety and pollution control problem since it is volatile, has little odor, and is extremely toxic.

The acrylonitrile route, based on the hydrolysis of acrylonitrile, is also a propylene route since acrylonitrile is produced by the catalytic vapor-phase ammoxidation of propylene:

$$CH_3CH{=}CH_2 + NH_3 + 3/2\,O_2 \rightarrow CH_2{=}CHCN + 3\,H_2O$$

The yield of acrylonitrile based on propylene is similar to the yield of acrylic acid based on the direct oxidation of propylene.

More than one-third of the 1977 worldwide manufacturing capacity for acrylic acid and esters was based on the acetylene process; however, raw material supply and cost trends indicate that this process has been phased out in favor of propylene-based plants. All new plants use propylene oxidation. The acrylonitrile route is commercially unattractive because in addition to high raw material costs, large amounts of sulfuric acid–ammonium sulfate wastes are produced; these wastes can be treated in a waste-acid plant for recycling (see Sulfur recovery); however, the investment is relatively high compared with the cost of the rest of the process.

A ketene process, based on the high temperature pyrolysis of acetic acid or acetone is no longer commercially used, and an ethylene cyanohydrin process developed during World War I has been replaced by more economical ones.

Specialty Acrylic Esters

Higher alkyl acrylates and alkyl-functional esters are important in copolymer products, in conventional emulsion applications for coatings and adhesives, and as reactants in radiation-cured coatings and inks. In general, they are produced in direct or transesterification batch processes because of their relatively low volume. The most important higher alkyl acrylate is 2-ethylhexyl acrylate prepared from the available oxo alcohol 2-ethyl-1-hexanol (see Alcohols, higher aliphatic).

Analytical Methods

Chemical assay is preferably performed by gas-liquid chromatography (glc) or by the conventional methods of determination of unsaturation.

Storage and Handling

Acrylic acid and esters are stabilized with minimum amounts of inhibitors such as MEHQ (monomethyl ether of hydroquinone) consistent with stability and safety. The acrylic esters may be stored in mild or stainless steel or aluminum. Acrylic acid is corrosive to many metals and can be stored only in glass, stainless steel, aluminum, or polyethylene-lined equipment. The relatively low flash points of some acrylates create a fire hazard, and the ease of polymerization must be borne in mind in all operations. The lower and upper explosive limits for methyl acrylate are 2.8 and 25 vol%, respectively. Corresponding limits for ethyl acrylate are 1.8 vol% and saturation, respectively. All possible sources of ignition of monomers must be eliminated.

Health and Safety Factors

The toxicity of common acrylic monomers ranges from moderate to slight. They can be handled safely by established practices. The TWA for all is 10 ppm in air over an 8-h shift.

The cornea is particularly sensitive. The lower acrylate esters can be skin sensitizers. Allergic reactions of sensitive individuals may include smarting of the eyes, headache, and skin eruptions. Threshold limit values (TLVs) for an 8-h day have been recommended by the ACGIH as 10 and 25 ppm for methyl and ethyl acrylates, respectively. Vapors of the higher acrylates are appreciably less irritating than those of methyl and ethyl acrylates.

Acrylic acid is strongly corrosive to the skin and eyes. Its oral toxicity is similar to that of methyl acrylate, but it may also cause severe intestinal burns and damage to the gastric tract. It is also moderately toxic in skin absorption tests; superficial destruction of tissues may occur, but the skin subsequently heals. High vapor concentrations are very irritating to the eyes and nasal passages. There are few toxicity data on functional acrylate monomers. Without specific information, precautions should be taken to prevent contact with the liquid or any undue exposure to the vapor of any acrylic ester. Furthermore, some derivatives, such as the alkyl-α-chloroacrylates, are powerful vesicants and can cause serious eye injuries. Thus, although the toxicities of commonly available commercial acrylates are moderate to slight, this should not be assumed to be the case for compounds with chemically different functional groups (see Industrial hygiene and toxicology).

JOSEPH W. NEMEC
WILLIAM BAUER, JR.
Rohm and Haas Company

Storage and Handling of Acrylic and Methacrylic Esters and Acids, Bulletin CM17, Rohm and Haas Co., 1975; *Acrylic and Methacrylic Monomers—Typical Properties and Specifications*, Bulletin CM-16, Rohm and Haas Co., 1972.

E.H. Riddle, *Monomeric Acrylic Esters*, Reinhold Publishing Co., New York, 1954.

M. Sittig, *Vinyl Monomers and Polymers*, Noyes Development Corp., Park Ridge, N.J., 1966.

S. Sakuyama, T. Ohara, N. Shimizu, and K. Kubota, *Chem. Technol.*, 350 (June 1973).

M. Salkind, E.H. Riddle, and R.W. Keefer, *Ind. Eng. Chem.* **51**, 1232, 1328 (1959).

ACRYLIC AND MODACRYLIC FIBERS

Acrylic fiber is the generic name for a manufactured fiber in which the fiber-forming substance is any long-chain synthetic polymer composed of at least 85 wt% acrylonitrile units. Modacrylic fiber is the generic name for a manufactured fiber composed of at least 35 and less than 85 wt% acrylonitrile units. Commercially, the only modacrylics marketed have 25–60% of monomers such as vinyl chloride or vinylidene chloride and consequently possess a high degree of flame resistance.

Acrylic Fibers

Many of the properties of acrylic fibers are determined by the inherent chemical nature or physical behavior of the long-chain polyacrylonitrile (PAN) molecules in oriented structures (see Acrylonitrile polymers).

Acrylic fibers are sold in the form of staple, tow, "tops" (in which the tow has been stretch-broken or cut to form a sliver), and a small amount of continuous filament. Staple lengths vary from 25 to 150 mm depending on the spinning system used to convert the fiber into spun yarn. Fiber size is expressed in denier (the weight in grams of 9000 meters) or tex (the weight in grams of 1000 meters). Specific stress on the fiber during extension or at break has been written as g/den or fg/tex and may be converted to the International System unit for specific stress (modulus and tenacity) expressed in newtons per tex (N/tex) via a formula: 1 N/tex = 102 gf/tex = 11.3 g/den. The tensile properties of acrylic fibers vary according to use. Physical properties of selected acrylic fibers are given in Table 1. Boiling-water shrinkage of standard acrylic fibers is low, but may be much higher for special types. The density of most fibers is about 1.17 g/cm^3 and increases with the incorporation of halogens. The tensile properties of acrylic fibers are little affected by water at normal temperatures. Like other synthetic fibers, their resistance to tension (modulus) decreases at higher temperatures and the elongation increases.

Acrylic fibers are used in some uses requiring flame-retardance. The general flame-resistance test for carpets in the United States specifies a maximum burn distance after ignition from a methenamine (hexamethylenetetramine) pill. Most acrylic fibers produced specifically for carpet use contain sufficient amounts of halogen to pass this test. Where higher degrees of flame retardancy are required, blends of acrylic and modacrylic fibers are used (see Flame-retardant textiles; Flame retardants). Acrylic fibers have outstanding resistance to sunlight and to attack by insects or microorganisms in the soil, properties which make them particularly suitable for outdoor uses (see Recreational surfaces). They are resistant to most ordinary chemicals such as inorganic acids, alkalies, organic solvents, oxidizing agents, and dry-cleaning solvents.

Fiber analysis and identification. Acrylics can be distinguished from other chemical fibers by their solubility in hot dimethylformamide and their density of 1.12–1.20 g/cm^3.

Manufacture. In addition to acrylonitrile, most polymers for acrylic fibers contain minor amounts of various monomers. Neutral monomers such as methyl acrylate, methyl methacrylate, and vinyl acetate are copolymerized with acrylonitrile to increase solubility of the polymer in various solvents and the diffusion rate of dyes into the fiber. Small amounts of ionic monomers are often also included to enhance dyeability. Halogen-containing monomers are added to improve flame resistance.

Although acrylonitrile is moderately soluble in water, the polymer is insoluble in both acrylonitrile and water. The copolymers are generally made by aqueous heterogenous or solution polymerization, using both batch and continuous processes. In a typical continuous process, monomer, water, and initiator are fed to a continuously stirred, overflow reactor at atmospheric pressure and a temperature of 30–70°C. The slurry of polymer, water, and unreacted monomer is filtered; the polymer is washed and dried; and the monomer is recovered from the filtrate and returned to the reactor. Advantages of this process are simplicity, control, and applicability to a variety of polymer compositions. Initiators are usually water-soluble redox combinations such as persulfate–bisulfite, activated by a small amount of iron, or chlorate–sulfite (see Initiators).

Solution polymerization is used to prepare acrylic polymers directly in a form suitable for wet or dry spinning. Solvents include dimethyl sulfoxide, dimethylformamide, and aqueous solutions of zinc chloride or various thiocyanates. Preferred initiators are various organic peroxides and azo compounds. Ultraviolet light may also be used as an initiator. The principal advantage of solution polymerization is reduced manufacturing cost through elimination of process steps.

Mass or bulk polymerization can also be used to prepare acrylic polymers when precautions are taken to prevent autocatalytic polymerization, which can become explosive. A key requirement described in recent patents is an initiator system that produces an easily stirrable monomer-polymer slurry from which heat can be readily removed. Typical conditions listed for continuous polymerization using a cumene hydroperoxide–sodium methyl sulfite redox initiator are a reaction temperature of 30–60°C and a one hour residence time.

Acrylonitrile copolymers used for acrylic fibers are white, easy-flowing powders of viscosity average molecular weight commonly in the range of 100,000–150,000. The polymers decompose before melting; above 180°C, exothermic decomposition becomes rapid. This characteristic, undesirable for acrylic fiber because it causes discoloration, is used advantageously for the manufacture of graphite fibers (see Carbon).

Seven solvents for polyacrylonitrile are used commercially. Dimethylformamide is the principal solvent since it is used in both dry- and wet-spinning processes, followed in importance by dimethylacetamide. Aqueous solutions of sodium thiocyanate and nitric acid are the most important inorganic solvents, along with $ZnCl_2$, dimethyl sulfoxide, and ethylene carbonate. Fiber production from sulfuric acid solution has been reported, but no commercial processes are known. Effective solvents for polyacrylonitrile are highly polar, aprotic compounds which break the intermolecular nitrile–nitrile attraction. Solubility and solution properties are affected by the polymer composition and molecular weight. The presence of 3–10% neutral comonomers increases acrylic polymer solubility and decreases solution viscosity at a given concentration. Molecular weight, composition, spinning concentration, and coagulation conditions are adjusted to give acceptable fiber formation and good tensile properties.

Table 1. Physical Properties of Selected Acrylic Fibers^a

Fiber	Acrilan B-16	Orlon 42	Euro-acril	Cour-telle	Badische type 500^b	Acrilan B-96	Zefran 253A
producer	Monsanto	Du Pont	ANIC	Courtaulds	Badische	Monsanto	Badische
use	textile	textile	textile	textile	industrial	carpet	carpet
tex	0.13	0.38	0.34	0.50	0.67	1.67	1.67
(denier)	(1.2)	(3.4)	(3.1)	(4.5)	(6.0)	(15.0)	(15.0)
tenacity, N/tex	0.32	0.26	0.24	0.22	0.31	0.19	0.20
(g/den)	(3.6)	(3.0)	(2.7)	(2.5)	(3.5)	(2.2)	(2.3)
elongation, %	42	33	35	64	33	56	62
relative knot tenacity, %	90	81	91	93		80	82
initial modulus, N/tex	3.9	4.0	3.6	3.3		1.9	1.8
(g/den)	(44)	(45)	(41)	(37)		(21)	(20)
boiling water shrinkage, %	1.0	0.7	0.7	1.2	< 1.0	1.0	1.0

^{a}Tensile properties measured at 65% rh and 21°C. Extension rate, 100% per minute.
^{b}Source: *Dow Badische Technical Bulletin A-10.*

Fiber production. Acrylic fibers are produced either by wet spinning (78%) or dry spinning (22%) to optimize fiber structure and properties; increase production rate as determined by linear speed, filament tex, and tow size; and improve process stability and fiber uniformity. The combination of process conditions to achieve all these features simultaneously is strongly influenced by the desired filament size or tex. In all spinning systems, fiber tex is controlled by the material balance:

$$\text{tex/filament} = 1.075 \times 10^4 \, (pW_2Q/SV_1)$$

where tex = g/km; p = solution density, g/L; W_2 = weight fraction of polymer in the spinning solution; Q = flow rate per spinnerette hole in L/s; V_1 = first godet (take-out roll) velocity (after coagulation bath or below dry tower) in m/s; and S = overall stretch ratio after first godet, including relaxation shrinkage. These parameters also affect fiber structure and properties.

The main difference between wet- and dry-spinning occurs in the initial fiber-formation stage. In wet spinning, fine streams of viscous polymer solution are extruded through spinnerette holes into a liquid bath containing a coagulant that is miscible with the solvent. Fiber formation occurs rapidly as the polymer comes out of solution and is drawn onto the first godet or take-out roll. Wet spinning is a three-component system (polymer–solvent–nonsolvent) in which two transitions may occur; gelation (the gradual and continuous transition of the fluid solution into a homogeneous elastic gel) and phase separation. Three modified wet-spinning processes have also been described for the production of acrylic fibers. In dry spinning, fiber formation occurs as solvent is evaporated from streams of polymer solution drawn downward through heated gas. Fiber formation in the polymer–solvent systems used for dry spinning results only from gelation. Other transitions do not occur because the systems are miscible, temperatures are high, and the solvent in the fibers lowers their transition temperatures. Gelation occurs gradually as the solvent evaporates and the concentration of polymer in the fiber increases.

Tows from the different wet- and dry-spinning systems all undergo the essential steps of washing, orientation or stretching, finish application, and drying accompanied by collapse, relaxation, and crimping, but the sequence of these steps varies.

Bicomponent acrylic fibers are composed of two polymers in separate areas of the cross section. Each polymer is continuous along the fiber length and the two are permanently joined at an interface. The polymers are selected to shrink or swell differently in response to heat or moisture. As a result, when the fibers are properly treated, differential shrinkage will occur and a spiral or helical crimp will form. Bicomponent fibers may be produced in any of the spinning processes described earlier. Special equipment is required prior to the spinnerette to channel two separate solution streams into each hole.

Acrylic fibers are dyed mainly with cationic dyes which achieve bright, fast colors. A small quantity of acrylic fiber spun from special polymers containing basic groups is dyed with anionic dyes. An oversimplified but useful view of the mechanism of dyeing with cationic dyes has three steps: surface adsorption, diffusion into the fiber, and dye–fiber bonding. Dyeing in nonaqueous liquids (solvent dyeing) is a useful process for coloring acrylics (see Dye carriers). This process requires less energy than aqueous dyeing and greatly reduces water pollution (see Dyes, application).

Uses. Acrylic fibers are used extensively in the apparel market for sweaters, single- and double-knit jersey fabrics, hosiery and craft yarns, and pile and fleece goods. They are used in the home furnishing market for carpets, blankets, curtains, and drapes. A small amount goes into industrial uses.

Modacrylic Fibers

Since modacrylic fibers usually contain more than 50 mol% acrylonitrile, their properties are related to those of acrylic fibers. They differ importantly in the lower maximum temperatures at which the modacrylics are dimensionally stable, a property resulting from the presence of the halogen-containing monomers. Halogen also affects other fiber properties, such as moisture regain, light stability, and resistance to degradation on heating.

The most important property of modacrylic fibers is their flame retardance. The commercial significance of this property has increased rapidly with consumer concern and government regulations (see Flame-retardants in textiles). However, conclusions about the flame retardancy of the modacrylic fibers, and any new fiber, can be drawn only in relation to specific products and tests. Results are dependent upon several factors in addition to fiber composition.

Like the acrylics, modacrylics are highly resistant to biological agents. They are also stable to aqueous solutions of most salts, acids, and bases, and are generally resistant to dry-cleaning solvents although some will be dissolved by acetone and other ketones. Their lightfastness is good but not equal to that of the acrylics.

Modacrylic fibers are sold as staple and tow. In 1976, there were nine known producers in five countries (see Table 2). Textile fibers range from 0.13 to 1.67 tex (1.2–15 den). The pigment-like particles in Orlon 775F, Verel, and S-06 may be antimony oxide. Some producers offer modacrylics with and without antimony. Physical properties of seven modacrylic fibers are shown in Table 2. Most modacrylic fibers are quite lustrous.

Manufacture. All modacrylic fibers of commercial importance are made from copolymers of acrylonitrile and a halogen-containing monomer, with or without additional monomers to enhance dyeability and other physical properties. The two most commonly used monomers are vinyl chloride and vinylidene chloride; the more expensive vinyl bromide, vinylidene bromide, and 2,3-dibromopropyl acrylate are also used. As with acrylic fibers, monomers incorporated to increase ease of dyeing with basic dyes usually contain sulfonic acid groups.

Table 2. Physical Properties of Modacrylic Fibers[a, b]

Fiber	Verel	Kanekalon SE	SEF S-06	Teklan	Orlon 775F	Dralon MA	S-32
Producer	Tennessee Eastman	Kanega-fuchi	Mon-santo	Court-aulds	Du Pont	Bayer	Mon-santo
denier	3.0	3.2	2.9	3.8	3.2	2.9	50.0
tex	(0.33)	(0.36)	(0.32)	(0.42)	(0.36)	(0.32)	(5.6)
tenacity, N/tex	0.24	0.24	0.19	0.22	0.19	0.22	0.22
(g/den)	(2.7)	(2.7)	(2.2)	(2.5)	(2.2)	(2.5)	(2.5)
elongation, %	35	32	52	40	47	35	28
boiling water shrinkage, %	34.3	2.0	1.4	5.2	1.4		2.8
dry heat shrinkage temperature for 10%, °C	187	173	206	175	184	181	
moisture regain, %	2.7	1.2	2.5	0.9	1.2		
density, g/cm^3	1.36	1.25	1.35	1.36	1.24		

[a] Tensile properties measured at 65% rh and 21°C. Extension rate, 100% per minute.
[b] Other products not included in this table are Vonnel H-704 (Mitsubishi Rayon), Toraylon Unfla (Toray Industries), and Saniv (Moscow Textile Institute).

Polymers for modacrylic fibers can be made by emulsion, heterogeneous, and solution polymerization, both batch and continuous. Initiation is usually with the persulfate–bisulfite–iron redox system. Polymer separation, washing, and drying follow standard procedures. In solution polymerization, with dimethylformamide or dimethyl sulfoxide as solvents, azo and peroxy initiations can be used, as well as light. After removing unreacted monomers, these solutions are spun directly into fibers by conventional wet- and dry-spinning techniques. The equipment, factors involved in process development and product optimization, and the sequence of process steps are generally the same as those discussed under spinning of acrylic fibers.

Uses. As a result of flammability legislation and the improvement in textile performance, modacrylics are now used in a wide range of fabrics. The largest volume sales are in the apparel market, particularly in children's sleepwear and pile fabrics, and home furnishings, notably for draperies and curtains that must pass flame-retardant standards. They are also used in a number of industrial applications where their flame retardancy, chemical resistance, and other properties are desirable, such as for paint rollers, battery separators, and protective uniforms. In addition, they are used to manufacture wigs.

P.H. HOBSON
A.L. MCPETERS
Monsanto Triangle Park Development Center, Inc.

P.H. Hobson, *Text. Chem. Color* **4**, 232 (1972).

D.J. Poynton, "Acrylic Fibers", and J. Atkinson, "Modacrylic Fibers," *Text. Prog.* **8**(1), 51 (1976).

F. McNeirney, *Mod. Text.*, 24 (Sept. 1976).

E. Cernia, "Acrylic Fibers" and R.K. Kennedy, "Modacrylic Fibers," in H.F. Mark, S.M. Atlas, and E. Cernia, eds., *Man-Made Fibers Science and Technology*, Vol. 3, John Wiley & Sons, Inc., New York, pp. 135–243.

ACRYLIC ESTER POLYMERS

SURVEY

Acrylic esters are represented by the generic formula $CH_2=CHC(O)OR$. The nature of the R group determines the properties of each ester and the polymers it forms. Polymers of this class are distinguished by their water-clear color and their stability on aging. Acrylic monomers are extremely versatile building blocks. They are relatively moderate to high boiling liquids which readily polymerize or copolymerize with a variety of other monomers. Thus, polymers designed to fit specific application requirements can be tailored from these versatile monomers. Although the acrylics have been higher in cost than many other common monomers, they find use in high quality products where their unique characteristics and efficiency offset the higher cost (see also Methacrylic polymers).

Physical Properties

To a large extent, the properties of acrylic ester polymers depend on the nature of the alcohol radical and the molecular weight of the polymer. As is typical of polymeric systems, the mechanical properties of acrylic polymers improve as molecular weight is increased; however, beyond a critical molecular weight, which often is about 100,000–200,000 for amorphous polymers, the improvement is slight and levels off asymptotically.

Owing to their comparatively low glass temperatures, acrylics in copolymers tend to serve as permanent plasticizers for harder monomers. Table 1 summarizes some of the physical properties of acrylic polymers. In practice, cross-linking of acrylic polymers is used to decrease the thermoplasticity and solubility and increase resilience. In some cases, cross-linking moieties are used in reaction of a polymer with a substrate.

Mechanical and thermal properties. Poly(methyl acrylate) has little or no tackiness at room temperature; it is a tough, rubbery, and moderately hard polymer. Poly(ethyl acrylate) is more rubberlike, con-

Table 1. Physical Properties of Acrylic Polymers

Polymer	T_g, °C	Density, g/cm³	Solubility parameter, $(J/cm^3)^{1/2\,a}$	Refractive index, n_D^{25}
methyl acrylate	6	1.22	20.7	1.479
ethyl acrylate	−24	1.12	19.3	1.464
propyl acrylate	−45		18.4	
isopropyl acrylate	−3	1.08		
n-butyl acrylate	−55	1.08	18.0	1.474
sec-butyl acrylate	−20	1.05		
isobutyl acrylate	−43			
tert-butyl acrylate	43	1.00		
hexyl acrylate	−57 (brittle pt)			
heptyl acrylate	−60			
2-heptyl acrylate	−38			
2-ethylhexyl acrylate	−50			
2-ethylbutyl acrylate	−50			
dodecyl acrylate	−30 (brittle pt)			
hexadecyl acrylate	35			
2-ethoxyethyl acrylate	−50			
isobornyl acrylate	94			
cyclohexyl acrylate	16			

^{a} To convert $(J/cm^3)^{1/2}$ to $(cal/cm^3)^{1/2}$, divide by 2.045.

Table 2. Mechanical Properties of Acrylic Polymers

Polyacrylate	Elongation, %	Tensile strength, kPa (psi)
methyl	750	6895 (1000)
ethyl	1800	228 (33)
butyl	2000	21 (3)

siderably softer, and more extensible, whereas poly(butyl acrylate) is softer still and much tackier. Table 2 summarizes these data.

In the *n*-alkyl acrylate series, the softness decreases through *n*-octyl acrylate. As the chain length is increased beyond *n*-octyl, side-chain crystallization occurs and the materials become brittle.

Unless subjected to extreme conditions, acrylic polymers are durable and degrade slowly. In contrast to methacrylate polymers, which depolymerize on strong heating, poly(methyl acrylate) when pyrolyzed *in vacuo* from 292 to 399°C yields only a small quantity of monomer. Oxidative degradation of acrylic polymers can occur by the combination of oxygen with free radicals generated in the polymer to form hydroperoxides. The rate of oxidation is quite slow unless extreme conditions with oxygen under pressure and high temperatures are employed.

Solution properties. Acrylate polymers that contain short side chains are relatively polar and are soluble in polar solvents such as ketones, esters, or ether alcohols. As the side chain is increased in length, the polymers become less polar and dissolve in relatively nonpolar solvents such as aromatics or aliphatic hydrocarbons.

Chemical Properties of Acrylic Ester Polymers

Acrylic polymers and copolymers have a greater resistance to both acidic and alkaline hydrolyses than poly(vinyl acetate) and vinyl acetate copolymers. Even poly(methyl acrylate), the most readily hydrolyzed polymer of the series, is more resistant to alkali than poly(vinyl acetate). The initial rate of hydrolysis of acrylic copolymer emulsions is considerably lower than that of a vinyl acetate copolymer. Butyl acrylate copolymers are more hydrolytically stable than ethyl acrylate copolymers.

Acrylic polymers are not sensitive to normal uv degradation since the primary uv absorption of acrylics occurs below the solar spectrum of 290 nm. The incorporation of absorbers, such as *o*-hydroxybenzophenone, improves the uv stability. Under use conditions, acrylic polymers have superior resistance to degradation and show remarkable retention of their original properties.

Manufacture of Acrylic Ester Monomers

The two principal processes used for the manufacture of monomeric acrylic esters are the catalytic Reppe process and the newer propylene oxidation process, which is preferred because of economy and safety.

The Reppe process is based on the stoichiometric reaction described by equation 1. This reaction using nickel carbonyl as the source of carbon monoxide proceeds readily under mild conditions. A high temperature, high pressure catalytic reaction using nickel salts is also known (eq. 2).

$$4 \, HC \equiv CH + 4 \, ROH + 2 \, H^+ + Ni(CO)_4 \rightarrow$$
$$4 \, CH_2 = CHC(O)OR + Ni^{2+} + 2 \, (H) \quad (1)$$

$$CO + HC \equiv CH + ROH \xrightarrow[H^+, H_2O]{Ni(II)} CH_2 = CHC(O)OR \quad (2)$$

In the semicatalytic Reppe process, the stoichiometric process is first established, then the catalytic process is initiated by feeding carefully controlled portions of carbon monoxide, acetylene, water, alcohol, and acid. This procedure retains the mild conditions of the stoichiometric process and makes carbon monoxide a starting material, and the more expensive nickel carbonyl need only be present in catalytic amounts. Nickel carbonyl is highly toxic, and great care is taken to limit exposure to this material.

In the newer propylene oxidation process, acrolein is first formed by the catalytic oxidation of propylene vapor at high temperature in the presence of steam (eq. 3). The acrolein is then oxidized to acrylic acid (eq. 4).

$$CH_2 = CHCH_3 + O_2 \xrightarrow{catalyst} CH_2 = CHCHO + H_2O \quad (3)$$

$$2 \, CH_2 = CHCHO + O_2 \xrightarrow{catalyst} 2 \, CH_2 = CHCOOH \quad (4)$$

Both one-step and two-step oxidation processes are known. A number of catalyst systems are known, and most use a molybdenum compound as the main component. The acrylic acid is esterified with alcohol to the desired acrylic ester in a separate process (see Acrylic acid and derivatives).

Handling of Acrylic Ester Monomers

See Acrylic acid and derivatives for information. Inhibitors such as hydroquinone (HQ) or the monomethyl ether of hydroquinone (MEHQ) are added to acrylic monomers to stabilize them during shipment and storage. The common acrylic ester monomers are combustible liquids. Copper or copper alloys must not be allowed to contact acrylic monomers intended for use in polymerization because copper is an inhibitor.

Manufacture of Acrylic Ester Polymers

The vast majority of all commercially prepared acrylic polymers are copolymers of an acrylic ester monomer with one or more different monomers. Copolymerization greatly increases the range of available polymer properties and has led to the development of many different resins suitable for a broad variety of applications.

Bulk and solution polymerization of monomeric acrylic esters. The acrylic monomers do not readily polymerize in solution or in bulk when heat is applied without other initiating species. Generally, organic radical initiators such as azo compounds or peroxides are used to initiate the polymerization of acrylic monomers (see Initiators). Both photochemical and radiation-initiated polymerizations have been reported.

The initial rates of several acrylic monomers initiated with 2,2-azobisisobutyronitrile (AIBN) have been determined at Rohm and Haas Co. (see Table 3). As indicated by the heat of polymerization values, considerable heat is liberated during acrylate polymerizations. The bulk polymerization of acrylic monomers is commercially of limited importance. Bulk polymerizations are characterized by a rapid acceleration in the rate and the formation of a cross-linked insoluble network polymer at low conversion.

The solution polymerization of acrylic monomers to form soluble acrylic polymers is an important commercial process. In general, the

Table 3.   Initial Rates and Heats of Polymerization of Acrylic Monomers

Monomer	K^a		Heat of polymerization,^{b}	
	44.1°C	60°C	kJ/mol	kJ/g
methyl acrylate	250^c	1480^d	78.7	0.912
ethyl acrylate	313^e	1730^f	77.8	0.778
isopropyl acrylate	347^g			
butyl acrylate	324^h		77.4	0.602
isobutyl acrylate	228^c			
tert-butyl acrylate	310^c			
2-ethylhexyl acrylate			60.7	0.331

$^a K = L^{1/2}/(mol^{1/2} \cdot h)$. To calculate the initial rate of polymerization use the following equation: initial rate, %/h, = $K = [AIBN]^{1/2}$. ^{b}To convert J to cal, divide by 4.184. ^{c}3 M in methyl propionate. ^{d}2.5 M in methyl propionate. ^{e}3 M in benzene. ^{f}2.5 M in benzene. ^{g}6 M in methyl propionate. ^{h}1.5 M in toluene.

polyacrylate esters of the lower alcohols are soluble in aromatic hydrocarbons, esters, ketones, and chlorohydrocarbons. They are insoluble or only slightly soluble in aliphatic hydrocarbons, ethers, and alcohols. The higher poly(alkylacrylates) are generally insoluble in the oxygenated organic solvents and soluble in both aliphatic and aromatic hydrocarbons and in chlorocarbons.

The solution viscosity of a polymer prepared from a given monomer under a constant set of conditions varies with the nature of the solvent, a result of the varying degrees of chain-transfer activity associated with solvents of different chemical structures.

The type of initiator utilized for a solution polymerization depends upon several factors including the solubility of the initiator employed, the rate of decomposition of the initiator, and the end use of the polymeric product. The amount of initiator used may vary from a few hundredths to several percent of the monomer weight. As the amount of initiator is decreased, the molecular weight of the polymer is increased as a result of initiating less polymer chains per weight of the monomer. The initiator concentration is often used to control the molecular weight of the final product. Organic peroxides, hydroperoxides, and azo compounds are the initiators of choice for the preparation of most acrylic polymers.

The molecular weight of a polymer also can be controlled by a chain-transfer agent which terminates the growing radical chains and initiates new polymer chains. Chlorinated aliphatic compounds and thiols are particularly effective in regulating the molecular weight of acrylic polymers.

Since acrylic polymerizations liberate considerable heat, violent or runaway reactions are avoided by gradual addition of the reactants to the kettle. A supply of inhibitor is kept on hand to stop polymerization. Since oxygen is often an inhibitor of acrylic polymerizations, its presence is undesirable. A blanket of inert gas is maintained over the polymerization mixture.

Emulsion polymerization of acrylic esters is the most important industrial method of preparation. The primary constituents are water, monomer, a water-soluble initiator, and a surfactant which helps to form micelles and an emulsion of monomer droplets in water. The final product is a milky white dispersion of particle size 0.1 μm to about 1.0 μm at concentrations of 30–60 wt%. The emulsion process gives high molecular weight polymers in a system of relatively low viscosity. It eliminates difficulties in agitation, heat transfer, and transfer of materials that are often encountered in the handling of viscous polymer solutions. In addition, the safety hazards and expense of flammable solvents are eliminated. Water-soluble peroxide salts, such as ammonium or sodium persulfate, are normally used as initiators for the emulsion polymerizations of acrylates. The initiating species is the sulfate radical anion generated from either the thermal or redox cleavage of the persulfate anion. With redox initiator systems, rapid polymerizations are possible at much lower temperatures (25–60°C) than are practical with a thermally initiated system (75–90°C). Industrial acrylic emulsion polymerizations are usually conducted by batch processes in jacketed stainless steel or glass-lined kettles designed to withstand an internal pressure of at least 446 kPa (65 psi).

Suspension polymerization is a special case of bulk polymerization. The monomer is suspended in water using protective colloids and other suspending agents; polymerization is initiated by a monomer-soluble initiator and takes place in the monomer droplets. The aqueous phase serves both as the dispersion medium and to remove the heat of polymerization. Particle size is controlled primarily by the rate of agitation and the concentration of suspending agents. The polymer is obtained as small spherical beads of about 1–5 mm in diameter which are isolated by filtration or centrifugation. Useful initiators for the suspension polymerization of acrylic monomers are organic peroxides or azo compounds which are soluble in the monomer phase and insoluble in the water phase. The amount of initiator influences both the polymerization rate and the molecular weight of the product.

Health and Safety Factors

Acrylic esters are nontoxic. There is a danger of violent polymerization during manufacture.

Uses

Diverse applications include coatings (see Coatings, industrial; Paint); textiles (see Textiles; Nonwoven textiles fabrics); paper (see Paper); leather finishing; cement (qv); ceramic production (see Ceramics); caulks and sealants (qv); film (see Film and sheeting materials); and polishes (see Polishes). Market distribution for use of acrylic monomers is coatings 45%, textiles 25%, export 6%, miscellaneous 6%, fibers 5%, polishes 5%, paper 5%, and leather 3%.

BENJAMIN B. KINE
RONALD W. NOVAK
Rohm and Haas Company

E.H. Riddle, *Monomeric Acrylic Esters*, Reinhold Publishing Corp., New York, 1956.

M.B. Horn, *Acrylic Resins*, Reinhold Publishing Corp., New York, 1960.

L.S. Luskin in E.C. Leonard, ed., *Vinyl and Diene Monomers*, Pt. I, Interscience Publishers, a division of John Wiley & Sons, Inc., New York, 1971.

2-CYANOACRYLIC ESTER POLYMERS

2-Cyanoacrylic polymers provide excellent adhesive bonds between a wide variety of substrates. These bonds are formed rapidly at room temperature without the addition of a catalyst by spreading a monomeric cyanoacrylic ester, CH_2=C(CN)C(O)OR, into a thin film between two adherents.

Physical Properties

The industrially important 2-cyanoacrylic ester adhesives are usually prepared from methyl 2-cyanoacrylate and ethyl 2-cyanoacrylate.

The physical properties of the bonds can be affected by moisture, heat, and solvents. Moisture deteriorates these bonds but not rapidly enough to detract from their usefulness in most applications. Bonds between rubber–rubber, rubber–metal, and rubber–plastic substrates have useful strengths after weathering for a number of years. However, bonds between metals, rigid plastic, and glass are deteriorated by moisture.

Chemical Properties

The most important chemical property of these esters is their ability rapidly to form strong adhesive bonds. This property is derived from the highly electronegative nitrile (—CN) and alkoxycarbonyl (—COOR) substituents which enables ready addition to the polarized carbon–carbon double bond by weak bases derived from water or alcohols, and the resonance stabilization of the subsequent anion which becomes the propagating species to give a high molecular weight polymer.

Manufacture

The most widely used synthesis involves condensing an alkyl cyanoacetate with formaldehyde in the presence of base catalyst to yield a low molecular weight cyanoacrylic ester polymer. Depolymerization at a high temperature gives the 2-cyanoacrylic ester. Other synthetic routes to cyanoacrylic ester have also been reported but are not in commercial use.

Although 2-cyanoacrylic esters are useful in many applications as the unmodified monomer, they are often formulated with inhibitors, thickeners, plasticizers (qv), and colorants to improve their utility as adhesives.

Health and Safety Factors

2-Cyanoacrylic polymers have pungent, unpleasant odors. They are mildly lacrimatory. The TLV for methyl 2-cyanoacrylate is 2 ppm (8 mg/m³). (Good ventilation is effective in maintaining concentrations of the vapor below this level.) Eye and skin contact should be prevented. Both the liquids and resultant polymers are combustible.

Uses

2-Cyanoacrylic esters have excellent adhesion to skin. They are used in surgical adhesives as replacements for sutures as hemostatic agents and by morticians to seal eyes and lips. They are also used in the assembly of a diverse range of products (such as golf clubs, tools, digital watches, models, optical lenses), and in the repair of many rubber, plastic and metal items (see Adhesives).

H.W. COOVER
J.M. MCINTIRE
Tennessee Eastman Company
Division of Eastman Kodak Company

H.W. Coover, Jr. and J.M. McIntire in T. Matsumoto, ed., *Tissue Adhesives in Surgery*, Medical Examination Publishing Company, Inc., Flushing, N.Y., 1972, pp. 154–188.

H.W. Coover, Jr. and J.M. McIntire in I. Skeist, ed., *Handbook of Adhesives*, 2nd ed., Van Nostrand Reinhold Company, New York, 1977, pp. 569–580.

ACRYLONITRILE

Acrylonitrile (ACN), propenenitrile, vinyl cyanide, CH_2=CHC≡N, is a colorless, mobile liquid with a characteristic odor resembling that of peach seeds. It is one of the building blocks of the chemical industry (see Acrylic and modacrylic fibers; Elastomers, synthetic; Acrylonitrile polymers). Table 1 lists the physical properties. Acrylonitrile is miscible with the normal organic solvents.

Table 1. Physical Properties of Acrylonitrile

Property	Value	
boiling point, °C, at 101.3 kPa[a]	77.3	
freezing point, °C	-83.55 ± 0.05	
density, at 20°C, g/cm³	0.806	
viscosity, at 25°C, mPa·s (= cP)	0.34	
refractive index, n_D^{25}	1.3888	
surface tension, at 24°C, mN/m (= dyn/cm)	27.3	
vapor pressure	kPa	°C
	6.7	8.7
	13.3	23.6
	33.3	45.5
	66.7	64.7
	101.3	77.3
water solubility at 20°C		
water, % in acrylonitrile	3.1	
acrylonitrile, wt% in water	7.35	
flash point, °C, tag open cup	-5	
ignition temperature, °C	481	
explosive limits, in air, at 25°C, vol%	$3.05–17.0 \pm 0.5$	
heat of vaporization, 25°C, kJ/mol[b]	32.65	
molar heat capacity, liquid, kJ/(kg·K)[b]	2.09	

[a] To convert kPa to mm Hg, multiply by 7.5.
[b] To convert kJ to kcal, divide by 4.184.

Chemical Reactions

Owing to the presence of both a nitrile group and a double bond, acrylonitrile is very reactive. Pure acrylonitrile polymerizes readily, especially under the influence of light. Storage requires the addition of polymerization inhibitors.

The nitrile group undergoes partial hydrolysis to acrylamide (qv) and complete hydrolysis to acrylic acid. Hydration and hydrolysis reactions catalyzed by hydrochloric acid give good yields of 3-chloropropionamide or 3-chloropropionic acid. Acrylonitrile is converted to acrylic esters with primary alcohols in the presence of sulfuric acid. Imido ethers are formed by reaction with alcohols in the presence of anhydrous halides.

Acrylonitrile reacts with olefins in the presence of concentrated sulfuric acid to give N-substituted acrylamides.

The activated double bond acts as a dienophile in the Diels-Alder reaction, to form cyclic compounds. Polymerization is commercially by far the most important reaction. The homopolymerization and copolymerization of acrylonitrile occur rapidly in the presence of free-radical or anionic initiators in the vapor, liquid, or solid phase, in solution, and in various two-phase systems. Acrylonitrile is hydrogenated to propionitrile with the use of copper, rhodium, or nickel catalysts. A molecule of halogen can be added to acrylonitrile to give 2,3-dihalopropionitriles. With an alkaline catalyst, cyanoethylation (qv) occurs by the addition of an active hydrogen compound (RH) to the double bond to give RCH_2CH_2CN. The reductive dimerization of acrylonitrile can be carried out chemically or electrochemically (see Adipic acid; Electrochemical processing). Hydroformylation or oxo synthesis can be carried out using $CO_2(CO)_8$ in solvents and a mixture of H_2 and CO (see Oxo process).

Manufacture and Processing

Acrylonitrile can be prepared in the laboratory from acetonitrile and formaldehyde, from ethylene cyanohydrin or acrylamide, using phosphorus pentoxide, or from lactonitrile, etc.

Nearly all commercially produced acrylonitrile is made via the Sohio Process, based on the vapor-phase catalytic air oxidation of low cost propylene and ammonia, known as ammoxidation: $CH_2=CHCH_3 + NH_3 + 3/2\ O_2 \rightarrow CH_2=CHCN + 3\ H_2O$. The main by-products formed are hydrogen cyanide, acetonitrile, and carbon oxides. In some plants hydrogen cyanide and acetonitrile are recovered, but in most cases these are disposed of by incineration.

The BP(Distillers)-Ugine Process is a two-step process involving the oxidation of propylene to acrolein, using a Se–CuO catalyst, followed by the reaction of acrolein with ammonia and air in the presence of an MoO_3 catalyst. High acrylonitrile yields are obtained; the main by-products are CO_2, HCN, and acetonitrile. Other catalysts have been used for direct ammoxidation of propylene in processes by various firms.

Other processes. Until the early 1960s, the main process for acrylonitrile manufacture was the addition of hydrogen cyanide to acetylene. The first commercial U.S. production of acrylonitrile was by dehydration of ethylene cyanohydrin, $HOCH_2CH_2CN$.

Other processes include the nitrozation of propylene with nitric oxide catalyst, the dehydrogenation of propionitrile, the reaction of ammonia with propionaldehyde at high temperatures, as well as the two-stage reaction with HCN involving the dehydration of lactonitrile. Ethylene, propane, and butane reactions with HCN are being considered, as well as reactions of propane or butane with oxygen and ammonia. Power Gas Ltd., ICI, and Monsanto indicate that ammoxidation of propane could be competitive with propylene.

Health and Safety

Acrylonitrile is toxic by ingestion, inhalation, or absorption of liquid through skin. OSHA concludes that acrylonitrile poses a carcinogenic risk to workers. Maximum permissible exposure in air is 2 ppm time-weighted average. Danger of violent polymerization in the presence of strong bases exists. Acrylonitrile must be stored in closed systems. Flammable-liquid label and poison label are required.

Uses

U.S.-produced acrylonitrile includes 40% for production of acrylic fibers (see Acrylic and modacrylic fibers); 23% for export; 20% for copolymer resins (acrylonitrile–butadiene–styrene and styrene–acrylonitrile) used in engineering related markets as replacements for traditional materials (see Engineering plastics); and 17% for other miscellaneous uses including adiponitrile, acrylamide, nitrile rubber (see Elastomers, synthetic), and barrier resins (see Barrier polymers). Of the miscellaneous applications, adiponitrile is the major use (see Polyamides; Nitriles).

Louis T. Groet
Badger, B.V.

Reviewed by
Jan K. Bliek
Badger, B.V.

R.B. Stobaugh, S.G. McH Clark, and G.D. Camirand, *Hydrocarbon Process.* **1**, 109 (1971).

M. Sittig, *Chemical Process Monograph No. 14, Acrylonitrile*, Noyes Development Corporation, Park Ridge, N.J., 1965.

"OSHA: Acrylonitrile Standard," *Fed. Regist.* **43**(192), 4510 (Oct. 1978).

ACRYLONITRILE POLYMERS

SURVEY AND STYRENE – ACRYLONITRILE COPOLYMERS

Acrylonitrile (qv) has been established as one of the most important building blocks of the plastics industry. Its use in a family of plastics—the styrene–acrylonitrile copolymers (SAN) and the acrylonitrile–butadiene–styrene terpolymers (ABS)—developed in the late 1950s; these plastics derive superior properties of toughness, rigidity, and resistance to chemicals and solvents from acrylonitrile. In the late 1960s, the acrylonitrile polymers were introduced into the packaging industry in a new class of plastics called barrier resins (see Barrier polymers).

Acrylonitrile homopolymer has little application beyond fibers because the high melting point, poor thermal stability, and high melt viscosity make melt processing very difficult. However, these deficiencies are tempered through copolymerization. Polyacrylonitrile can contribute properties of hardness, heat resistance, slow burning, resistance to most solvents and chemicals and resistance to sunlight, and the ability to form oriented fibers and films to copolymers.

SAN Properties and Test Methods

The incorporation of acrylonitrile into polymeric systems imparts some important properties (see Table 1).

SAN Manufacture

Because the reactivities of acrylonitrile and styrene radicals towards their monomers are not the same, SAN copolymers can vary in compositions from their monomer feeds.

SAN copolymers are manufactured by emulsion, suspension, and continuous-mass processes.

Both batch and continuous emulsion copolymerization processes are used. Generally, the copolymerizing system contains emulsifier, initiator, chain-transfer agent, monomers, and water. The copolymerization is carried out in the temperature range of 70–100°C, and to a conversion of 97% or higher. With redox catalysis systems, the copolymerization can be carried out at temperatures as low as 38°C. The copolymer latex may be used directly to blend with SAN grafted-rubber latex to make ABS, or it may be coagulated, washed, and dried to recover the SAN copolymers.

The reaction system in the suspension process contains monomers, chain-transfer agent, initiator, suspending agent, and water. Co-

Table 1. Properties of SAN Copolymers

Properties	Acrylonitrile content, wt%						
	5.5	9.8	14.0	21.0	27.0	23–24	27–28
Physical							
tensile strength, MPa[a]	42.3	54.6	57.4	63.8	72.5		
elongation, %	1.6	2.1	2.2	2.5	3.2		
notched impact, J/cm[b]	0.67	0.65	0.67	0.67	0.67		
Chemical							
water absorption, %							
at 60°C, 2 weeks						+0.54[c]	+0.57
at RT, 15 days in							
10% NaOH						+0.42	+0.53
40% NaOH						+0.07	+0.05
50% H$_2$SO$_4$						+0.01	+0.01
Solvent resistance							
soluble in	acetone, chloroform, dioxane, methyl ethyl ketone, pyridine						
swells in	benzene, ether, toluene						
insoluble in	carbon tetrachloride, ethyl alcohol, gasoline, lubricating oil, Solvesso (trade name of solvents)						

[a] To convert MPa to psi, multiply by 145.
[b] To convert J/cm to ftlbf·ft/in, divide by 5338 (see ASTM D 256).
[c] As change in weight.

polymerization is carried out from temperatures as low as 60°C to as high as 150°C. A typical recipe is 70 styrene, 30 acrylonitrile, 1.2 dipentene, 0.02 di-*tert*-butyl peroxide, 0.03 acrylic acid–2-ethylhexyl acrylate (90 : 10), and 100 parts water. The amount of suspending agent is quite low, in contrast to the emulsion process.

SAN continuous-mass copolymerization is conceptually simple but practically complicated. The polymerization can be initiated either thermally or catalytically and a chain-transfer agent may be used. Copolymerization is carried out between 100 and 200°C. Solvents can be used to reduce the viscosity or the copolymerization can be conducted at a low conversion level (40–70%). The devolatized polymeric melt is then fed through a strand die, cooled, and pelletized. Because of the high viscosity of the copolymerization medium, it requires complex equipment designed to handle the highly viscous material; to remove the heat of copolymerization, the monomers and solvent; and to keep the composition uniform.

The continuous-mass process has a number of advantages. It is a self-contained system without waste treatment or environmental problems. It does not use a large quantity of water. It is very time and space efficient. It does not require a polymer-drying operation and therefore it consumes less energy and is more economical than the other processes. However, the process operation and equipment design are complicated owing to a number of problems including a long time for the process to reach steady state (which results in the production of some off-grade materials); the production of undesirable intermediate-grade materials during transition periods of product-grade change over; and the difficulty in achieving good mixing and adequate heat removal owing to the highly viscous melt.

Health and Toxicology

Acrylonitrile monomer is a toxic and flammable substance (see Acrylonitrile). The styrene monomer has a relatively low toxicity, but it is irritating to the eyes and respiratory tract. Inhalation of high concentrations produces an anesthetic effect. Exposure to 10,000 ppm for 30–60 min, or 2500 ppm for 8 h is dangerous to life (see Styrene). The SAN copolymers are not considered toxic, but in food-contact applications the amounts of extractable acrylonitrile monomer must not exceed the minimum set forth by FDA regulations.

Uses

A major portion of SAN produced is incorporated in ABS (see Acrylonitrite polymers, ABS resins). The general uses of SAN copolymers are in houseware applications, appliances, packaging, industrial and automotive use, and for general applications including aerosol nozzles, camera parts, sporting goods, telephone parts, terminal boxes, etc.

FRED M. PENG
Monsanto Co.

Brit. Pat. 971,214 (Sept. 30, 1964), to (Monsanto Co.).

U.S. Pat. 3,198,775 (Aug. 3, 1965), R.E. Delacretaz, S.P. Nemphos, and R.L. Walter (to Monsanto Co.).

U.S. Pat. 3,258,453 (June 28, 1966), H.K. Chi (to Monsanto Co.).

U.S. Pat. 3,287,331 (Nov. 22, 1966), Y.C. Lee and L.P. Paradis (to Monsanto Co.).

U.S. Pat. 3,681,310 (Aug. 1, 1972), K. Moriyama and T. Moriwaki (to Daicel Ltd.).

ABS RESINS

When butadiene rubber is grafted with SAN, the resulting rubber and thermoplastic composite is known as ABS, based on its common monomers: acrylonitrile–butadiene–styrene. The amount of rubber in ABS varies from about 5 to 30 wt%, the remainder being SAN copolymer. ABS has the useful properties of SAN, eg, rigidity and resistance to chemicals and solvents, etc, and its rubber additive imparts toughness (see Styrene plastics; Styrene).

The physical properties of ABS plastics vary somewhat with their method of manufacture but more so with their composition. Emulsion processes are used to make materials of higher impact strength, and bulk or suspension processes are preferred for materials with less impact strength. Materials having higher impact strength usually have increased rubber content. For these materials, tensile strength, modulus, hardness, and deflection temperature are generally lower. Elongation, specific gravity, and coefficient of expansion also vary directly with rubber content. Lower impact grades are the most easily used; higher impact grades are somewhat more difficult because their higher rubber content makes them more viscous. Heat-resistant grades, designed to be stiff at elevated temperatures, require somewhat higher processing temperatures and pressures.

Physiochemical Aspects

ABS can be considered a blend of a rubbery domain and a glassy copolymer which can be varied in composition and becomes tougher as the acrylonitrile content is increased. Simple blending of rubber with this glassy copolymer does not lead to optimal impact properties. Free rubber phases tend to separate into large aggregates that are inefficient impact modifiers. Rubbery copolymers that are compatible with the styrene-acrylonitrile copolymer matrix can improve impact resistance, but properties such as tensile strength, hardness, and melt flow are then adversely influenced. Such compatibility allows the formation of uniformly distributed small rubbery domains. Compatibility with the glassy matrix is achieved by grafting styrene–acrylonitrile to the rubber molecules.

Because the two phases are interacting, the rubbery phase modifies the melt flow of the free copolymer. Increasing the amount of rubber raises the impact strength but decreases the hardness. The impact resistance is also influenced by the particle size distribution of the rubbery phase.

Manufacture

Emulsion, suspension, and bulk polymerization processes are used for the commercial production of ABS plastics. Bulk polymerization does not proceed in water and thus has two advantages which are commercially important: wastewater treatment is minimal, and less energy per kilogram of product is consumed since dewatering, drying, and compounding steps are not necessary. The process disadvantages include less product flexibility, greater mechanical complexity, and less complete conversion of monomer to polymer. This means that most ABS materials made in bulk require devolatization to remove residual monomers prior to compounding of the final product.

The emulsion process consists of three distinct polymerizations: A polybutadiene substrate latex is prepared, generally in emulsion batch reactions either as the homopolymer or as a copolymer with up to 35% styrene or acrylonitrile. In the next step, styrene and acrylonitrile are grafted into the polybutadiene substrate using free-radical initiators such as potassium persulfate and chain-transfer agents such as terpinolene. The amount of substrate used in the graft reaction is determined by the physical properties desired in the final polymer, and usually ranges 10–60 wt% of the total polymer. The graft reaction may either be a batch process or semi-batch. Concurrent with the graft reaction is the formulation of styrene–acrylonitrile copolymer. Emulsion ABS resins are recovered from latex by coagulation with dilute salt or acid solutions. Coagulation is carried out at elevated temperatures (80–100°C) to promote agglomeration of the resin particles. The resin particle size can be controlled by salt or acid concentration, temperature, and slurry concentration. The slurry is then dewatered and dried.

In contrast to the emulsion process, the suspension process begins with a lightly cross-linked polybutadiene rubber which is soluble in the monomers. The polybutadiene must be recovered and dried for use in this process. A flow sheet for the ABS suspension process is given in Figure 1. The reaction is carried out at 80–120°C for 6–8 h, until 25–35% conversion of monomer to polymer is achieved. The polymer syrup is

dispersed in water in the suspension reactor, chain-transfer agents are added, and the reactor is heated to 100–170°C, depending upon the initiator used, for 6–8 h.

The first step in the bulk ABS process is essentially the same as in the suspension process. A monomer-soluble polybutadiene rubber or butadiene copolymer, dissolved in the styrene and acrylonitrile with initiators and modifiers, is polymerized through phase inversion to approximately 30% conversion under sufficient shearing to prevent cross-linking. The syrup is then pumped into a specially designed bulk polymerizer where conversion is taken to 50–80%. Several types of these polymerizers may be used interchangeably for high impact polystyrene of ABS. Other more recently described processes are specifically designed for ABS polymerization.

The dry resin made by either the emulsion or suspension process is usually compounded into pellet form before being sold to a plastic processor. High shear devices such as Banbury mixers, single-screw extruders, or twin-screw extruders may be used to create sufficient frictional heat to flux the polymer and disperse pigments, lubricants, stabilizers, or other additives (150–250°C).

ABS polymers made by the emulsion process can be blended with polymers made by the suspension or bulk processes to achieve a material with some of the desirable characteristics of each. The properties of key importance are impact strength, melt flow, tensile strength, and residual monomer content.

Uses

In 1981, the largest single market was electrical and electronic, followed by pipe and fittings, automotive, consumer and institutional, export, and all others, including packaging, luggage, furniture, and recreation.

G. A. MORNEAU
W.A. PAVELICH
L.G. ROETTGER
Borg-Warner Chemicals

R.P. Kambour, *J. Polym. Sci. Macromol. Revs.* **1**, 1 (1973).

L.F. Albright, *Processes for Major Addition-Type Plastics and Their Monomers*, McGraw-Hill, New York, 1974, pp. 365–370.

W.M. Smith, *Manufacture of Plastics*, Vol. I, Reinhold, New York, 1964, pp. 446–454.

C.F. Parsons and E.L. Suck, Jr., *Adv. Chem. Ser.* **99**, 340 (1971).

ACTINIDES AND TRANSACTINIDES

ACTINIDES

The actinides are a group of chemically similar elements with atomic numbers 90 through 103 (see Table 1). Each element has a number of isotopes which are produced by neutron or charged-particle induced transmutations. Each isotope is radioactive and some can be obtained in isotopically pure form (see Radioactivity, natural; Thorium; Uranium; Plutonium; Nuclear reactors; Radioisotopes).

Prior to 1944, the location of the heaviest elements in the periodic table had been a matter of question, and the elements thorium, protactinium, and uranium were commonly placed immediately below the elements hafnium, tantalum, and tungsten. In 1944, on the basis of earlier chemical studies of neptunium and plutonium, the similarity between the actinide and the lanthanide elements was recognized. The intensive study of the heaviest elements shows a series of elements similar to the lanthanide series beginning with actinium. Corresponding pairs of elements show resemblances in spectroscopic and magnetic behavior which arise because of the similarity of electronic configurations for the ions of the homologous elements in the same state of oxidation, and in crystallographic properties, owing to the near matching of ionic radii for ions of the same charge. The two series are not, however, entirely comparable. A difference lies in the oxidation states.

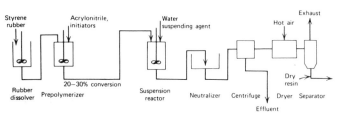

Figure 1. Suspension ABS process.

Table 1. The Actinide and Transactinide Elements

Atomic no.	Element	Approximate no. of isotopes	Symbol	Atomic wt[a]	Oxidation state[b]	Actinide metals Mp, °C	Bp, °C	Density (phase), g/cm³ at 25°C (unless otherwise noted)
89	actinium	25	Ac	227	**3**	1100 ± 50		10.07
90	thorium	25	Th	232	(3), **4**	1750	3850	11.724 (α)
91	protactinium	20	Pa	231	(3), 4, **5**	1575		15.37 (α)
92	uranium	15	U	238	3, 4, 5, **6**	1132	3818	18.97 (α)
								18.11 (β)[c]
								18.06 (γ)[d]
93	neptunium	15	Np	237	3, 4, **5**, 6, 7	637 ± 2	3900	20.45 (α)
								19.36 (β)[e]
								18.04 (γ)[f]
94	plutonium	15	Pu	242	3, **4**, 5, 6, 7	639.5	3235	19.86 (α)[g]
								17.70 (β)[h]
								17.13 (γ)[i]
								15.92 (δ)[j]
								16.01 (δ′)[k]
								16.48 (ε)[l]
95	americium	15	Am	243	(2), **3**, 4, 5, 6	1176	2011	13.67 (α)[m]
								13.65 (β)[m]
96	curium	15	Cm	248	**3**, 4	1340 ± 40	3110	13.51 (α)
								12.66 (β)
97	berkelium	10	Bk	249	**3**, 4	986 ± 25		14.78 (α)
								13.25 (β)
98	californium	17	Cf	249	(2), **3**, (4)	900 ± 30		15.1 (α)
								13.7 (β)
								8.70 (γ)
99	einsteinium	15	Es	254	(2), **3**	860 ± 30		
100	fermium	20	Fm	257	2, **3**			
101	mendelevium	10	Md	258	2, **3**			
102	nobelium	8	No	259	**2**, 3			
103	lawrencium	7	Lr	260	**3**			
104	rutherfordium (USA)	5	Rf	261				
104	kurchatovium (USSR)		Ku					
105	hahnium (USA)	5	Ha	262				
105	nielsbohrium (USSR)		Ns					
106				263				
107				262				

[a] Mass number of longest-lived or more available isotope.

[b] The most stable states are designated by boldface type and those which are very unstable are indicated by parentheses. These latter states do not exist in aqueous solution and have been produced only in solid compounds.

[c] At 720°C; [d] At 805°C; [e] At 313°C; [f] At 600°C; [g] At 21°C; [h] At 190°C; [i] At 235°C; [j] At 320°C; [k] At 460°C; [l] At 490°C; [m] At 20°C.

The tripositive-state characteristic of lanthanide elements does not appear in aqueous solutions of thorium and protactinium and does not become the most stable oxidation state in aqueous solution until americium is reached. The elements uranium through americium have several oxidation states unlike the lanthanides. These differences can be interpreted as resulting from the proximity in the energy of the $7s$, $6d$, and $5f$ electronic levels.

Only the members of the actinide group through Pu have been found to occur in nature. With the exceptions of uranium and thorium, the actinide elements are synthetic in origin for all practical purposes, ie, they are products of nuclear reactions. High neutron fluxes are available in modern nuclear reactors, and the most feasible method for preparing actinium, protactinium, and most of the actinide elements of high atomic number.

For example, actinium is prepared by the transmutation of radium 226 via neutron capture, and protoactinium is formed by nuclear reaction from thorium. Kilogram amounts of neptunium (^{237}Np) have been isolated in wastes as a by-product of the large-scale synthesis of plutonium in nuclear reactors that utilize uranium as fuel (see Diffusion separation methods). Plutonium-237, an important isotope, is prepared in ton

quantities in nuclear reactors (see Plutonium); the plutonium usually contains isotopes of higher mass number and requires a variety of precipitation, solvent extraction, and ion-exchange techniques to recover and purify it. The isotope ^{238}Pu is produced in kilogram quantities from neptunium for use as a fuel for isotopically powered energy sources for terrestrial and extraterrestrial applications (see Thermoelectric energy conversion). Kilogram quantities of americium as ^{241}Am can be obtained by processing reactor-produced plutonium. The elements 95 to 100 are produced in increasing quantities by nuclear reactors.

The key to the discovery of the transuranium elements has been ion exchange (qv), a selective and rapid technique for the separation and chemical identification of curium and higher elements. The elution order and approximate peak position for the undiscovered elements were predicted with considerable confidence: thus, the first experimental observation of the chemical behavior of a new actinide element was often its ion-exchange behavior—an observation coincident with its identification.

There are many similarities in the chemical properties of the lanthanide and actinide elements, particularly with elements in the same oxidation state (see Rare-earth elements). The ion-exchange behavior of each strikingly resembles the others. Actinide ions of the III, IV, and VI oxidation states can be adsorbed by cation-exchange resins and, in general, can be desorbed by elution with chloride, nitrate, citrate, lactate, α-hydroxyisobutyrate, ethylenediaminetetraacetate, and other anions. Ion-exchange separations can also be made by the use of a polymer with exchangeable anions such as Dowex-1 (a copolymer of styrene and divinylbenzene with quaternary ammonium groups) and Amberlite IRA-400 (a quaternary ammonium polystyrene). In this case, the lanthanide or actinide elements must be initially present as complex ions.

Another useful method for separating the actinide elements from each other is extraction chromatography, a process in which the organic extractant is adsorbed on the surface of a fine porous powder placed in a column (see Extraction). Bis(2-ethylhexyl) phosphoric acid, mono(2-ethylhexyl) phenylphosphonic acid ester, and tri-n-butyl phosphate are useful cation-extraction agents. Tertiary amines such as tricaprylamine or trilaurylamine, or quaternary amines such as tricaprylmethylammonium chloride (nitrate, thiocyanate) are excellent anion-extracting agents.

It is possible to prepare very heavy elements in thermonuclear explosions, owing to the very intense, although brief (microsecond) neutron flux furnished by the explosion. Einsteinium and fermium were first produced this way in the fallout materials from the first thermonuclear explosion in 1952. The process of neutron irradiation in high flux reactors cannot be used to prepare the elements beyond fermium (^{257}Fm), except at extremely high neutron fluxes, because some of the intermediate isotopes that must capture neutrons have very short half-lives which precludes the necessary concentrations. Transfermium elements are prepared by bombardment with heavy ions accelerated by cyclotrons or linear accelerators. The chemical properties of the elements have been studied by the tracer technique.

Properties

There is a close chemical resemblance between the actinide elements. See Table 1 for a summary of their oxidation states. The actinides exhibit uniformity in ionic types. In acidic aqueous solution, there are four types of cations, M^{3+}, M^{4+}, MO_2^+, MO_2^{2+}, and MO_5^{3-}, with a wide variety in colors characteristic of the transition series of elements. Reduction potentials have been reported.

The degree of hydrolysis or complex-ion formation decreases in the order $M^{4+} > MO_2^{2+} > M^{3+} > MO_2^+$. On the basis of increasing charge and decreasing ionic size, it could be expected that the degree of hydrolysis for each ionic type would increase with increasing atomic number. For the ions M^{4+} and M^{3+}, beginning at about uranium, such a regularity of hydrolytic behavior is observed, but for the remaining two ions, MO_2^+ and MO_2^{2+}, the degree of hydrolysis decreases with increasing atomic number, thus indicating more complicated factors than simple size and charge.

Actinide ions form complex ions with a large number of organic substances. Their extractibility by these substances varies from element to element and depends markedly on oxidation state. A number of important separation procedures are based on this property. Solvents that behave in this way are tributyl phosphate, diethyl ether, ketones such as diisopropyl ketone or methyl isobutyl ketone, and several glycol ether-type solvents such as diethyl Cellosolve (ethylene glycol diethyl ether) or dibutyl Carbitol (diethylene glycol dibutyl ether).

A number of organic compounds, eg, acetylacetone and cupferron, form compounds with aqueous actinide ions (IV state for reagents mentioned) that can be extracted from aqueous solution by organic solvents. The chelate complexes are especially noteworthy and, among these, the ones formed with diketones, such as 2-thenoyltrifluoroacetone ($C_4H_3SCOCH_2COCF_3$), are of importance in separation procedures for plutonium.

Actinide metals. The actinide metals, like the lanthanide metals, are highly electropositive. They can be prepared by the electrolysis of molten salts or by the reduction of a halide with an electropositive metal, such as calcium or barium. See Table 1 for physical properties. With respect to chemical reactivity, the actinide metals resemble the lanthanide metals more than metals of the 5d elements, such as tanta-

lum, tungsten, rhenium, osmium, and iridium. A wide range of intermetallic compounds has been observed and characterized including compounds or alloys with members of groups IB, IIA, IIIA, IVA, VIII, VA, and VIB-chalcogenides.

Solid compounds. The tripositive actinide ions resemble tripositive lanthanide ions in their precipitation reactions. Thousands of compounds of the actinide elements have been prepared. Hundreds of actinide organic derivatives are known including organometallic compounds of the π-bonded type (see Organometallics).

Crystal structure and ionic radii. Crystal structure data have provided the basis for the ionic radii. For both M^{3+} and M^{4+} ions there is an actinide contraction, analogous to the lanthanide contraction, with increasing positive charge on the nucleus. As a consequence of the ionic character of most actinide compounds and of the similarity of the ionic radii for a given oxidation state, analogous compounds are generally isostructural. The actinide contraction and the isostructural nature of the compounds constitute some of the best evidence for the transition character of this group of elements.

Absorption and Fluorescence Spectra

The absorption spectra of actinide and lanthanide ions in aqueous solution and in crystalline form contain narrow bands in the visible, near-ultraviolet, and near-infrared regions of the spectrum. Much evidence indicates that these bands arise from electronic transitions within the 4f and 5f shells, in which the $4f^n$ and $5f^n$ configurations are preserved in the upper and lower states for a particular ion. In general, the absorption bands of the actinide ions are some ten times more intense than those of the lanthanide ions. Fluorescence, for example, is observed in the trichlorides of uranium, neptunium, americium, and curium, diluted with lanthanum chloride.

Uses

Thorium, uranium, and plutonium are well known for their role as the basic fuels (or sources of fuel) for the release of nuclear energy. The importance of the remainder of the actinide group lies at present, for the most part, in the realm of pure research, but a number of practical applications are also known. The actinides present a storage-life problem in nuclear waste disposal and consideration is being given to separation methods for their recovery prior to disposal (see Nuclear reactors, waste management). ^{238}Pu is used to provide energy for small thermoelectric-power units such as the thermoelectric generator left on the moon, and the similar generator which powers the instrumentation for the Viking Mars explorer. Americium has a predominant gamma-ray energy of 60 keV and a long half-life of 433 years for decay by the emission of alpha particles, which makes it particularly useful for a wide range of industrial gauging applications and the diagnosis of thyroid disorders. When mixed with beryllium it generates neutrons at the rate of 1.0×10^7 neutrons per (s·g ^{241}Am). The mixture is designated ^{241}Am–Be and a large number of such sources are in worldwide daily use in oil-well logging operations to find how much oil a well is producing in a given time span. Californium-252 is an intense neutron source: 1 gram emits 2.4×10^{12} neutrons per second. This isotope is being tested for applications in neutron activation analysis, neutron radiography, and portable sources for field use in mineral prospecting and oil-well logging. Both ^{238}Pu and ^{252}Cf are being studied for possible medical applications: the former as a heat source for use in heart pacemakers and heart pumps and the latter as a neutron source for irradiation of certain tumors for which gamma-ray treatment is relatively ineffective.

TRANSACTINIDES

The elements beyond the actinides in the periodic table are the transactinides. They begin with atomic number 104. Only four transactinides are definitely known, 104 through 107.

Study of the chemical properties of element 104 has confirmed that it is homologous to hafnium as demanded by its position in the periodic table. No meaningful chemical studies have been made for element 105, and no chemical studies have been made for elements 106 and 107. Such studies are difficult because the longest lived isotope of 104 (^{261}104) has a half-life of about 40 s, 106 (^{263}106) has a half-life of about 1 s, and 107 (^{262}107) has a half-life of about 5 ms. On the basis of the simplest projections, it is expected that the half-lives of the elements beyond element 107 will become shorter and shorter as the atomic number is increased, and this is true even for the isotopes with the longest half-life for each element.

GLENN T. SEABORG
University of California, Berkeley

C. Keller, *The Chemistry of the Transuranium Elements*, Verlag Chemie GmbH, 1971.

J.J. Katz and G.T. Seaborg, *The Chemistry of the Actinide Elements*, Methuen & Co., Ltd., London, and John Wiley & Sons, Inc., New York, 1957; second edition, in 1983.

G.T. Seaborg, ed., *Transuranium Elements: Products of Modern Alchemy*, Dowden, Hutchinson & Ross, Inc., Stroudsberg, Pa., 1978.

ACTIVATED SLUDGE.   See Water, sewage.

ACTIVATION ANALYSIS.   See Analytical methods.

ADAMANTANE.   See Chemotherapeutics, antiviral.

ADHESION.   See Adhesives.

ADHESIVES

Adhesives, often referred to as cements, glues, or pastes, are defined as substances "capable of holding materials together by surface attachment". Adhesive technologists do not all agree as to what is and what is not an adhesive. Substances may attach to surfaces and develop the internal or cohesive strength necessary to hold materials together while cooling from liquid to solid state, while losing solvent, or during chemical reaction. Pressure-sensitive adhesives do not undergo a phase change in order to hold materials together.

Many of the substances designated as adhesives may be called paints, finishes, or coatings when applied in thin films to only one surface, or may be called caulking, potting, casting, or encapsulating compounds when employed in thick masses (see Chemical grouts; Sealants; Embedding).

To be termed an adhesive, a substance must be a liquid or tacky semisolid, at least for an instant to contact and wet a surface, and be used in relatively thin layer to form a useful joint capable of transmitting stresses from one substrate to another. Adhesives and sealants include cement (qv), glue (qv), paste, and waxes (qv).

Theory

Adhesion is an interfacial phenomenon: wetting of substrates is essential. The electrical theory presumes that the adhesive and the substrates are like the plates of a capacitor that become charged due to the contact of two substances. The theory fails to predict the strong joints that result when a layer of water is frozen to join two blocks of ice, or when an epoxy adhesive is used to join two previously cured blocks of cast epoxy. Diffusion theory presumes the penetration of the substrates by the adhesive prior to its solidification; this is easily applied to many porous plastics; however, not to metal, glass, or glazed ceramic. Adsorption theory specifies the concept of forces, such as van der Waals forces, acting across the space between molecules in a material. Rheological theory suggests that the removal of weak boundary layers in plastics leaves the joint's mechanical properties determined by the mechanical properties of the materials making up the joint and local stresses. In the absence of weak boundary layers, joint failure must be cohesive within the bondline of the substrates.

Types and Application

Adhesives have been categorized by suitability for bonding various substrates by physical form, by method of application, and by temperature-resistance recommendations for many substrates when using a particular type of bonding agent.

Much of the equipment used to apply adhesives is the same as used to apply surface coatings such as roller, flow, curtain, and knife coaters (see Coating processes).

Setting and Curing

Only the reactive adhesives truly cure. For example, acrylics, unsaturated polyesters, and other monomer adhesives containing ethylenic unsaturation cure by formation of free radicals when catalyzed by peroxides and accelerated by metal-ion donors. Epoxies cure by addition mechanisms (see Epoxy resins). Polysulfides react with active oxidizing agents to form rubbery polymers particularly useful in joining low expansion glasses to metals (see Polymers containing sulfur). Many urethane and silicone adhesives cure in the presence of moisture. Most of the reactive adhesive cures are accelerated by heat. Cyanoacrylate adhesives cure only in thin bondlines, ie, 1.0 μm to 0.1 mm, without the addition of a catalyst or hardener. The basicity present on many substrates initiates cure when inhibitors present in the adhesive are overcome (see Acrylic ester polymers).

Uses

Adhesives are used widely in a variety of industries including automotive; construction; electronic bonding, with special applications to computer memories and both active and inactive micro- and macroelectronic elements; packaging; plastics (see Laminated and reinforced wood; Leatherlike materials); and in the textile products and apparel industry (see Nonwoven textiles fabrics; Coated fabrics).

Health and Safety

Application may expose the user to hazards such a strong acid or base solutions and contact with 4,4'-methylenebis(2-chloroaniline), a chemical carcinogen used for cross-linking urethane resins.

FRED A. KEIMEL
Bell Telephone Laboratories

J. Shields, *Adhesives Handbook*, CRC Press, Cleveland, Ohio, 1970.

D.J. Alner, ed., *Aspects of Adhesion*, in seven volumes, University of London Press Ltd., 1965–1973.

C.V. Cagle, *Handbook of Adhesive Bonding*, McGraw-Hill, Inc., New York, 1973.

I. Skeist, ed., *Handbook of Adhesives*, 2nd ed., Reinhold, New York, 1977.

ADIPIC ACID

Adipic acid (hexanedioic acid) (1,4-butanedicarboxylic acid), HCOO-(CH$_2$)$_4$COOH, is a white crystalline solid with a melting point of 153.0°C. From a commercial viewpoint, it is the most important of all the aliphatic dicarboxylic acids; primarily, it is used in the manufacture of nylon-6,6, (see Polyamides).

Physical and Chemical Properties

The physical and chemical properties are summarizes in Table 1.

Adipic acid undergoes the usual reactions of the aliphatic dicarboxylic acids, including salt formation, esterification, amidation, and acid halide and anhydride formation, as well as those reactions characteristic of the methylene group alpha to carboxyl groups. Its greatest utility stems from its capability to undergo condensation reactions with difunctional compounds to form polymers.

Table 1. Physical and Chemical Properties of Adipic Acid

Property		Value	
melting point, °C		153.0–153.1	
specific gravity of solid at 18°C		1.344	
specific heat of solid, J/kg·K[a]		1590	
heat of fusion, kJ/mol[a]		16.7	
dissociation constants in water	$Temp.$, °C	$k_1 \cdot 10^5$	$k_2 \cdot 10^6$
	25	3.70	3.86
flash point, Cleveland open cup, °C		210	

[a]To convert J to cal, divide by 4.184.

Esterification with acid catalysts produces both mono- and diesters, although the reaction proceeds readily in the absence of a catalyst at elevated temperatures if water of reaction is removed or if excess alcohol is employed. Reactions with glycols leads to the formation of polyesters (qv). Adipic esters can be reduced by catalytic hydrogenation to hexamethylene glycol. Adipic acid reacts with ammonia or amines on heating to form the corresponding amide. Diammonium adipate readily forms adipamide on heating at 200°C in the presence of ammonia. On further heating in the presence of a dehydration catalyst, adipamide is converted to adiponitrile, $NC(CH_2)_4CN$, which on hydrogenation yields hexamethylenediamine, $H_2N(CH_2)_6NH_2$, the amine component of nylon-6,6.

Adipic acid is converted into a linear polymeric anhydride by treatment with acetic anhydride or acetyl chloride at reflux. The cyclic anhydride is obtained by distillation on heating the linear anhydride at 210°C but is unstable and reverts to the polymeric form at 100°C. Cyclopentanone is formed at high temperatures by the decomposition of an alkali-metal salt. Dimethyl adipate undergoes the acyloin condensation to yield adipoin (2-hydroxycyclohexanone) in 55–57% yield on treatment with metallic sodium followed by acidification of the sodium enolate. Adipoyl chloride is formed by the action of thionyl chloride.

Manufacture

The predominant commercial route to adipic acid is via oxidation of cyclohexane with air to form a mixture of cyclohexanone and cyclohexanol (termed ketone–alcohol, KA, or ol-one). This is followed by oxidation of the KA mixture with nitric acid to produce adipic acid. The balance of adipic acid is made from phenol by hydrogenation to cyclohexanol followed by similar nitric acid oxidation.

Other routes to adipic acid include air oxidation of KA, nitric acid oxidation or air oxidation of cyclohexane to adipic acid in one step, and carbonylation of butadiene in the presence of methanol to form dimethyl adipate (or the unsaturated diester), which can readily be converted to adipic acid.

Health and Safety

Adipic acid is a food additive (qv) approved by FDA. Danger of dust explosions exists in concentrations of 0.010–0.015 g/L. Irritation along wrists, neck, and ankles and to mucous membranes occurs upon dust exposure.

Uses

Large quantities used as esters and polyesters in plasticizers (qv) and high performance lubricants (see Lubrication; Hydraulic fluids; Esters, organic). Higher esters are valuable as plasticizers for poly(vinyl chloride) (PVC) and copolymers, natural and synthetic rubbers, polystyrene, and cellulose derivatives. Significant quantities of diesters are used in PVC for food packaging, films, electrical insulation, and coated fabrics (see also Nonwoven textiles). It is used in production of complex linear polyesters and polymeric plasticizers consisting mostly of adipic acid polyesters, particularly urethane elastomers (see Urethane polymers). Adipic acid is approved for use in all nonstandardized foods as stipulated by the FDA.

D.E. Danly
C.R. Campbell
Monsanto Chemical Intermediates Co.

T.L. Vesel'chakova, V.A. Preobrazhenskiy, A.M. Gol'dman, and M.S. Furman, *Sov. Chem. Ind.* **3**, 223 (1971).

W. Rosler and H. Lunkwitz, *Chem. Tech. Leipzig* **27**, 345 (1975).

W.J. van Asselt and D.W. van Krevelen, *Chem. Eng. Sci.* **18**, 471 (1963).

ADRENALINE. See Epinephrine; Hormones; Psychopharmacological agents.

ADSORPTION, INDUSTRIAL. See Adsorptive separation, gases; Adsorptive separation, liquids.

ADSORPTION, THEORETICAL. See Adsorptive separation.

ADSORPTIVE SEPARATION

INTRODUCTION

Adsorption is the selective collection and concentration, onto solid surfaces, of particular types of molecules contained in a liquid or gas. Solid adsorbents are usually porous granular particles (≤ 5 mm dia) used in fixed beds and in fluidized beds (≥ 0.05 mm dia), with interior pore diameters of 0.01 μm or less (see Adsorptive separation, gases) and with total volume approaching 50% of the whole particle.

Adsorption is selected as the most economical method for treating a fluid stream on an industrial scale owing to one of the following characteristics: high concentrating power of the absorbent, related to selectivity; chemical instability of the adsorbate (solute), restricting it to temperatures unsuited for other separations; or fluctuating or intermittent supply of the fluid feed (see also Ion exchange; Analytical methods).

Adsorption, as a separation operation, commonly includes the use of other solid separating agents, such as aluminosilicate molecular sieves (qv), certain anhydrous salts (such as anhydrous $CaSO_4$), and homogeneous organic polymers of amorphous gel type, such as polystyrene or dextran, all of which function in the same overall manner as true adsorbents.

The physical properties for the primary classes of adsorbent materials are listed in Table 1, along with the uses for which each material is particularly suitable.

Equilibrium Relations

An isotherm for a solute-adsorbent system at a designated temperature is an isothermal contour for the equilibrium, measured between the concentration or partial pressure of adsorbate in the external fluid (liquid or gaseous) and the concentration of adsorbate held by the particular adsorbent. A polytherm is a single curve showing the gassorbent equilibria over a range of temperatures for a specific solute or even for a homologous series of solutes. Isotherms (and polytherms) are generally used in a semitheoretical or an empirical form where the constants are fitted empirically to experimental data and their algebraic form is theoretically derived.

Table 1. Principal Types of Adsorbents

Composition	Internal porosity, %	External void-fraction, %	Bulk dry density, kg/m³	Surface area, m²/g	Uses[a]
acid-treated clay	ca 30	ca 40	560–880	100–300	a
activated alumina and bauxite	30–40	40–50	720–880	200–300	a,j,k
aluminosilicate (molecular sieves)	45–55	ca 35	660–705	600–700	a,k,o,p
bone char	50–55	18–20	640	ca 100	d,h
carbons	55–75	35–40	160–480	600–1400	nearly all
fuller's earth	50–55	ca 40	480–640	130–250	a,c
iron oxide (Fe₂O₃)	22	37	1440	20	o
magnesia (MgO)	ca 75	ca 45	ca 400	200	a
silica gel	ca 70	ca 40	ca 400	ca 320	j,k,l,n

[a]See list of uses a–p under "Uses."

Adsorption from a liquid is normally less selective than from a gas. The total volumetric uptake is nearly constant, so that separation is often based upon exchange adsorption analogous to ion-exchange processes. The composition of liquid held by the porous solid is the sum of the layer next to the pore surface (which reflects nearly the same selectivity as a gas) and of the liquid filling the remaining pore space (which approaches the composition of the bulk liquid outside the particle).

Adsorption into a homogeneous solid surface of varying quantities of solute up to the limit of a full monolayer, in the absence of lateral interactions between adsorption sites, was first treated by Langmuir who termed his result the "hyperbolic isotherm." With Q the monolayer capacity of the surface (eg, in moles per unit weight of solid), of the actual quantity adsorbed at equilibrium, K the equilibrium constant, and p the partial pressure of a single solute, the Langmuir isotherm is

$$\theta = \frac{q}{Q} = \frac{Kp}{1 + Kp}$$

where Kp may be replaced by $K_c c$, K_c is another constant, and c the concentration of solute in the bulk fluid phase. The quantity θ denotes the fraction of the monolayer coverage that has been attained. Relative to a reference set of p and q, usually the largest values reached under given conditions, the concentration on the solid may be expressed as $y = q/q_{ref}$, and in solution as $x = p/p_{ref}$.

Multilayer systems can be explained by the Brunauer, Emmett, and Teller relationship. Adsorptive energies are related by the well-known Freundlich isotherms

$$\frac{q}{Q} = Ap^\beta; \quad y = x^\beta$$

with A, a, and β all constant at a given temperature. Other permutations of these equations are discussed in the literature. The potential-theory model of Polanyi is useful to extend an isotherm to cover different conditions of temperature and molecular volume.

Process Arrangement

Two main classes of treatment are encountered in adsorption: contact filtration, in which the fluid and adsorbent follow identical paths through the processes with equal residence times without any segregation occurring between the two phases; the segregation of flows is carried out in two different types of systems. Countercurrent contact between solid and gas or liquid is operated in a manner analogous to gas–liquid and liquid–liquid separations, with side feed as in distillation columns or bottom feed as in absorption columns. Fixed beds usually require a larger inventory of adsorbent but are often more economic overall than moving beds (see Absorption; Distillation; Extraction).

Regeneration, the purging of solutes held in the bed following one production step and preceding the next, is necessary in every type of system where the sorbent is to be reused. Some adsorbents must be removed from the system and processed in a separate system to restore their activity. Activated carbon, for instance, is often regenerated by controlled oxidation at 600–900°C in a rotary kiln (see Carbon and artificial graphite). For many uses, carbon or other adsorbents can be regenerated in place, and reverse-flow regeneration is often advantageous when the regeneration conditions are mild.

Uses

For liquid-phase treatment, the following types of processes predominate; (a) decolorizing, drying, or degumming of petroleum fractions (solvents, fuels, lubricants, waxes); (b) odor, taste, and color removal from municipal water supplies; (c) decolorizing of vegetable and animal oils; (d) decolorizing of crude sugar syrups; (e) "clarification" of beverages and pharmaceutical preparations; (f) recovery of vitamins and other products from fermentation mixtures; (g) purification of process effluents for control of water pollution (includes ion exchange); (h) removal of salt or ash from process streams (includes demineralization by ion exchange, ion retardation, ion exclusion): (i) separation of aromatic from paraffinic hydrocarbons; and for gas-phase application; (j) solvent recovery from air leaving a chamber where an evaporative process occurs such as paint drying, newspaper printing, textile dry cleaning, or rayon spinning; (k) dehydration of gases (including in-package desiccation); (l) odor removal and toxic gas removal in ventilation systems or from vent gases for air pollution control; (m) separation of rare gases (krypton, xenon) at low temperatures; (n) impurity removal from air feed to low temperature fractionation; (o) odor removal from municipal illuminating-gas supplies; and (p) gas-phase separation of low molecular weight hydrocarbon gases (an alternative to rectified adsorption or to low temperature distillation).

Theodore Vermeulen
University of California, Berkeley

Reviewed by
D.B. Broughton
UOP Process Division

W.A. Adamson, *Physical Chemistry of Surfaces*, 4th ed., Wiley-Interscience, New York, 1982, Chapts. 8 and 13.

V. Ponec, Z. Knor, and S. Cerny, *Adsorption on Solids*, Butterworths, London, 1974.

T. Vermeulen, G. Klein, and N.K. Hiester in R.H. Perry and C.H. Chilton, eds., *Chemical Engineers' Handbook*, 5th ed., McGraw-Hill, New York, 1973, Sect. 16.

GASES

Adsorptive separation can be divided into bulk separation, involving a separation of less than 20–50% of the process stream, and purification, involving the separation of less than 3–5% of the impurities from a process stream. Bulk separations include the separation of straight-chain hydrocarbons from iso-compounds, unsaturated-hydrocarbon separation, and the separation of nitrogen from air. Purifications include dehydration and removal of hydrogen sulfide, carbon dioxide and mercaptans from sour natural gas, air plant feed drying and CO_2 removal, and cracked-gas drying.

The success of adsorptive separation depends on choosing the right adsorbents, such as silica gel, silica-base beads, activated bauxite and alumina, activated carbon, and molecular sieves, and on optimizing process variables (see Molecular sieves; Aluminum compounds—alumina; Carbon and artificial graphite; Silica).

Molecular sieves have a high capacity at a low adsorbate concentration, the ability to maintain much of the capacity at moderately elevated temperature, and selectivity based on size, configuration, and affinity. The molecular sieves possess thermal, hydrothermal and chemical stability, and inertness. Formed molecular sieves (eg, beads, extrudates, granules, etc) should have a network of macropores to allow rapid diffusion of molecules to inner crystallites. The control of both size and distribution of the macropores is important for efficient adsorption and regeneration. There are two types of zeolites of particular interest for adsorptive separation: type A, the most common, has a uniform, effective pore opening of about 0.3 to 0.5 nm (0.3 nm for KA; 0.4 nm for NaA; 0.5 nm for CaA) and will selectively adsorb molecules with critical diameters of less than the nominal openings; and type X, which has an effective pore opening in the range of 0.9–1.0 nm.

With adsorptive separation, contaminants can be removed to nearly undetectable levels with minimal capital and operating costs, and with design flexibility and high reliability. The most highly polar or more easily polarizable molecules will be preferentially adsorbed, sometimes to the virtual exclusion of other molecules which in the absence of competition would be firmly held. In addition, the degree of unsaturation of organic molecules influences their degree of adsorption, the most unsaturated being held most strongly. Within homologous groups of common types of molecules, those having the highest molecular weight are the most readily adsorbed.

Characteristics of Adsorptive Operations

Dynamic adsorption. The most common mode of adsorptive separation process employs a fixed bed, cyclic operation. As the wet process stream enters the fresh molecular-sieve bed, the water vapor is adsorbed in a finite length of bed called a mass-transfer zone (MTZ), the bed length through which the concentration of the adsorbate is reduced from essentially inlet to outlet conditions. As wet gas continues to flow, the bed may be divided into three distinct zones: the dynamic equilibrium zone with inlet gas, the MTZ, and the remaining active zone. In most molecular-adsorption systems, the MTZ becomes stabilized because of its favorable isotherm and moves at a constant speed. The bed is exhausted when the leading edge of the MTZ reaches the outlet end of the bed. High effective capacity results from high dynamic equilibrium capacity and a short MTZ.

Regeneration of the exhausted bed removes the adsorbate in preparation for the next step. This is accomplished by the thermal swing, whereby the bed is heated to a temperature at which the adsorptive capacity is reduced to a low level so that the adsorbate leaves the molecular-sieve surface and is easily removed by a stream of pure gas; the pressure swing which depends on reducing the adsorptive capacity by lowering the pressure at essentially constant temperature; inert-purge stripping which removes the adsorbate without changing the temperature or pressure by passage off a fluid (liquid or gas) containing no adsorbable molecules, and in which the adsorbate is soluble or miscible; and displacement desorption which depends on passage of fluid containing a high concentration of an adsorbable molecule or a more strongly adsorbed molecule. Because of this high concentration, these molecules are able to displace material previously adsorbed by mass action.

System variables influencing the performance of adsorptive-separation processes include temperature, pressure, fluid velocity, adsorbate concentration, phase, particle size, degree of molecular-sieve activation, and contaminants.

Bulk Separations

Normal-paraffin separation involves the selective adsorption of *n*-paraffins by 0.5 nm (CaA) molecular sieves from a variety of petroleum fractions on the basis of molecular dimension and configuration. The high purity *n*-paraffins are recovered for the production of biodegradable detergents, specialty solvents, plasticizers, fatty acids, and synthetic protein; their removal enhances the octane value of low boiling gasoline fractions and reduces the pour point of lube oils.

The main commercial processes include the UOP Molex process, BP Process, Shell process, Texaco Selective Finishing (TSF) process, Exxon Ensorb process, UCC IsoSiv process, and VEB Leuna Werke's Parex process. All processes, except the UOP Molex, employ a vapor-phase, fixed-bed operation including adsorption, purging, and desorption steps. The residual branched-chain hydrocarbons and aromatics present in the bed void space and on the nonselective surfaces are removed during the purging step to improve product purity. The purging step is especially desirable for liquid-phase operations. The processes differ in method of desorption, the desorbent, and the purging agent employed as well as operating pressure, temperature, and cycle time.

Unsaturated-hydrocarbon separation is based on the degree of saturation of the molecules, eg, separation of acetylene from ethylene or methane oxidation mixtures and ethylene recovery from other gas mixtures.

Oxygen enrichment is accomplished with molecular sieves used to produce high purity oxygen from air or oxygen-rich air. A pressure-swing adsorption process is used (see Oxygen; Cryogenics).

Purification

New natural-gas processes are designed to purify or alter the natural-gas composition or to extract salable components, and encompass such operations as dehydration, carbon dioxide removal, sulfur and mercaptan removal (sweetening), and hydrocarbon separation (see Gas, natural).

Although adsorption processes have long been used for dehydration and separation, it is only in the last decade that molecular sieves have become a dominating factor in natural-gas processing (see Molecular sieves). Molecular-sieve dehydration, CO_2 removal, and sweetening may be applied to any size gas stream from 10^2 to 10^7 m^3/d (ca 4×10^3–10^8 ft^3/d). In addition, the performance of molecular sieves is not affected by the degree of saturation of the feed. Therefore, they can be used with good results on saturated gas streams or on those that have been given prior well-head treatment.

In larger installations, where the concentrated sour regeneration gas from the molecular-sieve beds cannot be used as fuel or otherwise disposed of, it may be further treated in a small liquid-process system (see Alkanolamines).

Cracked gas can mean any complex gas mixture containing olefins produced for ethylene manufacture from a variety of feedstocks. The compositions depend on the acetylene-removal systems. Historically, these mixed-olefin feeds have been dried by a variety of desiccants: glycols, silica gels, and aluminas (see Drying agents). Current methods employ molecular sieves and are more effective.

Air drying and purification for liquefaction is accomplished by molecular sieves, silica gels, and activated alumina. For normal drying, silica gels and activated alumina are quite adequate. For extreme drying, 0.4 nm (NaA) molecular sieves are cost-effective. Molecular sieves are also suited for drying air, such as in cryogenic applications, in which water and carbon dioxide have to be removed prior to air liquefaction to avoid heat-exchanger freeze-up.

Other applications include removal of air pollutants, and adsorbents for glass insulation systems.

Design Methods

Isothermal adsorption models, adiabatic systems, and a multicomponent model have been proposed to explain adsorption phenomena. In any application, preliminary adsorption designs should provide adsorbent selection and quantity, vessel size and configuration, cycle time, pressure drop, and regeneration requirements. The aging factor based on varied plant experience must also be applied.

Most common adsorption design problems utilize the concepts of equilibrium zone(s), coadsorption zone(s), and the MTZ. The equilibrium and coadsorption section are the portions of the bed in which the adsorbent is loaded to its dynamic equilibrium capacity for one or more components and can be estimated from the adsorption isotherms. The thermodynamic properties expressed in isotherms, isobars, and isosteres are essential design elements for determining capacity, ultimate degree of separation, driving force, and heat required for regeneration.

For removal of a single component from an inert fluid, the bed may be considered to have an equilibrium section and an unused bed section (or lost bed height) owing to finite transfer rates. The length of the unused bed (LUB) is approximately one half of the MTZ length and can be determined from the breakthrough data. The change in adsorbate concentration in the fluid phase through the MTZ has been considered to be from 5 to 95% of the entering concentration. In Figure 1, area A is proportional to useful (break) capacity and area B is proportional to lost capacity. The total dynamic equilibrium capacity is proportional to the sum of areas A and B. The total dynamic equilibrium value is used for estimating the length of the dynamic equilibrium section. And MTZ is used for estimating LBH. If the dynamic capacity is not known, the value for the static capacity can be used.

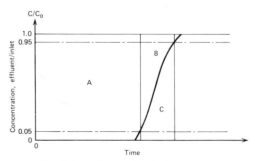

Figure 1. Breakthrough curve for single component removal.

For a multicomponent system where coadsorption takes place, the design almost totally depends on experimental and field performance data. In general, the bed behaves as a chromatographic column because of differences in the affinities of various adsorbates. A detailed design method for the bed has been discussed.

During regeneration of the bed, the reverse of adsorption takes place; ie, there is a progressive broadening of the MTZ front. Thermal swing is the most common method of regeneration. The outlet air temperature rises rapidly at the start and then remains fairly constant for the greater part of the regeneration cycle. As the molecular sieve approaches complete regeneration, the outlet temperature starts to rise again. When the effluent temperature reaches 90% of the temperature differential between the regenerant inlet and the initial bed, the water is practically all removed and the adsorbent is considered regenerated. After the bed has been heated for regeneration, a cooling period is used to reduce the temperature to within 20°C of that of the stream to be processed.

Thermal requirements for adequate regeneration include the following components: heat for the bed; heat of desorption of adsorbate; and heat for the vessel walls and bed supports. The heated gas is used to increase the bed temperature and also serves to carry away the adsorbates. The gas rate required for regeneration depends on total heat load, specific heat of the regenerating gas, dynamics of heat and mass transfer, and the time available for regeneration. Countercurrent heating is usually preferred in order to minimize the residual water content at the bed exit.

The quantity of gas necessary for regeneration can be estimated by standard heat-transfer methods incorporating the properties of the adsorbent, adsorbate, and heating gas. Another empirical approach developed for type NaA molecular sieves is analogous to fixed-bed heat transfer. The correlation is linear in terms of dimensionless bed height and dimensionless desorption time.

CHANG W. CHI
WILLIAM P. CUMMINGS
W.R. Grace & Co., Davison Chemical Division

D.W. Breck, *Zeolite Molecular Sieves*, John Wiley & Sons, Inc., New York, 1974.

E.N. Lightfoot, R.J. Sanchez-Palma, and D.O. Edwards, *Interscience Libr. Chem. Eng. Process.* **2**, 135 (1962).

LIQUIDS

Adsorption is used commercially for the recovery of major components of feed streams as pure products. Adsorption from the liquid phase has long been used for removing contaminants present at low concentrations in process streams. Common adsorbents are either polar or hydrophilic types, or nonpolar, hydrophobic types. The polar adsorbents include silica gel, activated alumina, molecular sieves, and various clay minerals, such as rutile, bentonite, bauxite, diatomaceous earth, attapulgite, and fuller's earth, and are generally employed when the materials to be removed are more polar than the process liquid. Nonpolar adsorbents, such as coal-derived carbons and activated carbons, are generally used to remove less polar contaminants from polar bulk streams (see Wastes, industrial; Water, industrial water treatment). Polar-type solvents are remarkably similar to each other qualitatively in their selectivities for components of some mixtures and cannot accomplish some of the separations that are possible with adsorbents.

Equilibrium

An adsorbent can be visualized as a porous solid with certain characteristics: When immersed in a liquid mixture, the pores of the solid fill with liquid which at equilibrium differs in composition from that of the liquid surrounding the particles. These compositions can be related to each other by enrichment factors analogous to relative volatility in distillations. The adsorbent is selective for the component that is more concentrated in the pores than in the surrounding liquid.

The choice of separation method to be applied to a particular system depends largely on the phase relations that can be developed by use of various separative agents. Adsorption is usually considered to be a more complex operation than the use of selective solvents in liquid–liquid extraction, extractive distillation, or azeotropic distillation (see Azeotropic and extractive distillation; Extraction). Consequently, adsorption is employed where it achieves higher selectivities than those obtained with solvents.

Adsorbents can be produced to differ widely from each other and from polar solvents in selectivity, particularly those based on the synthetic crystalline zeolites; a wide variety of selectivities is obtained by variations in silica–alumina ratio, crystalline type, and nature of the replaceable cations (see Molecular sieves). Commercial operations with sieves include *n*-paraffin separation; olefin–paraffin separation in wide-boiling mixtures; and separation of C_8-aromatic isomers. In addition, the technology and theory are equally applicable to suitably selective nonsieve adsorbents such as those of the polar type, eg, silica gel and activated alumina, and those of the nonpolar type, eg, activated carbon. The order of affinity for various chemical species with activated carbon is saturated hydrocarbons < aromatic hydrocarbons = halogenated hydrocarbons < ethers = esters = ketones < amines = alcohols < carboxylic acids. With some exception, the less polar component of a mixture is selectively adsorbed; eg, paraffins are adsorbed selectively relative to olefins of the same carbon number, but bicyclic aromatics are adsorbed selectively relative to monocyclic aromatics of the same carbon number.

Commercial Operation

Commercial development of large-scale bulk separations from the liquid phase has been accomplished by use of a flow scheme simulating the continuous countercurrent flow of adsorbent and process liquid, without actual movement of the adsorbent. The processes are UOP Sorbex; in its application to *n*-paraffin separation, licensed as UOP Molex; for olefin–paraffin separation as UOP Olex; for *p*-xylene separation as UOP Parex; and for ethylbenzene separation as UOP Ebex. The Toray Aromax process is a variant of this flow process used for *p*-xylene separation. These separations involve continuous countercurrent operations and moving-bed operations. The theoretical performance of the commercial simulated moving-bed operation is practically identical to that of a system in which solids flow continuously as a dense bed countercurrent to liquid. A model in which the flows of solid and liquid are continuous, as shown in Figure 1, is adequate.

The operation is modeled in terms of theoretical equilibrium trays having the same significance as in fractionating columns (see Distillation). Solid and liquid are assumed to flow continuously through hypothetical well-mixed trays in which equilibrium is attained. The number of theoretical trays has no relation to the number of bed segments in the actual operation, each segment can be equivalent to many theoretical trays.

Bulk Separations

Commercial operations include the following specific separations: *n*-paraffins from naphtha, kerosine and gas oils (UOP Molex); olefins from olefin-paraffin mixtures (UOP Olex); *p*-xylene from other C_8-aromatics

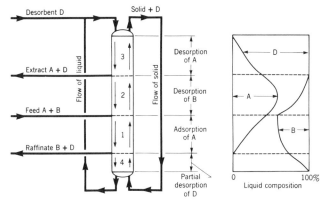

Figure 1. Adsorptive separation with moving bed.

and nonaromatic hydrocarbons (UOP Parex) (Toray Industries Aromax); p-diethylbenzene from mixed diethylbenzene isomers (UOP p-DEB Separation).

Linear paraffins in the C_{10}–C_{15} range are used for the production of alcohols and plasticizers and biodegradable detergents of the linear alkylbenzenesulfonate and nonionic types (see Alcohols; Plasticizers; Surfactants and detersive systems). Recovery of n-paraffins in the C_{10}–C_{23} range for protein production provides a new outlet for these hydrocarbons (see Foods, nonconventional). Recovery is 93.5% at a product purity of 99.5%. It is characteristic of liquid-phase operations that such wide-boiling stocks be handled in one operation. Desorbents used in these operations are generally n-paraffins of lower boiling range than the feedstocks.

Catalytic dehydrogenation of n-paraffins is a route to commercial production of linear olefins. Conversion is incomplete, owing to limitations imposed by equilibrium and side reactions. To obtain a concentrated olefin product, the olefins must be separated from the reactor effluents and the unreacted n-paraffins recycled to the catalytic reactor for further conversion. The feedstock includes C_{11}–C_{14} components. Extraction of olefins is 93.9% and the olefin product contains 1.1% residual n-paraffins. Essentially similar results have been obtained in commercial operations on C_8–C_{10} and C_{15}–C_{18} feedstocks. The desorbents used are generally hydrocarbon mixtures of lower boiling range than the feed components.

p-Xylene is widely used as a precursor for the production of polyester fibers and plastics. Before the advent of adsorptive techniques, it was commonly separated from hydrocarbon mixtures by crystallization at a rate of 55–60% recovery. Extraction is 99.7% at a product purity of 99.3% by current adsorptive techniques (see Xylenes and ethylbenzene).

Ethylbenzene is used in the production of styrene (qv). It is difficult to separate from mixed C_8-aromatics by fractionation. No commercial adsorptive unit is at present installed although the operation has been successfully performed in pilot plants with about 99% of the ethylbenzene in the feed recovered at a purity of 99.7% (UOP Ebex process).

Fructose–dextrose separation is an example of the application of adsorption to nonhydrocarbon systems. An aqueous solution of the isomeric monosaccharide sugars, fructose and dextrose, accompanied by minor quantities of polysaccharides, is produced commercially under the designation of "high-fructose corn syrup" by enzymatic conversion of cornstarch (see Sugar; Sweeteners).

Aromatics are partially removed from kerosenes and jet fuels on a commercial basis (as sometimes required to improve smoke point and burning characteristics) by liquid–liquid extraction with solvents such as furfural or by adsorptive separation. The extent of extraction does not vary greatly for each of the various species of aromatics present. Silica gel tends to extract all aromatics from nonaromatics.

Naphthalene and its higher homologues can be extracted from petroleum fractions, followed by conversion of the mixture to naphthalene by hydrodealkylation. In contrast to the operation with silica gel, naphthalene homologues are extracted to a much greater extent than are other aromatic types. Activated carbon tends to be specifically selective for the more condensed aromatic structures.

D.B. BROUGHTON
UOP Process Division, UOP, Inc.

T. Vermeulen, N.K. Hiester, and G. Klein, *Perry's Chemical Engineers' Handbook*, 6th ed., MacGraw-Hill, Inc., New York, 1983, Sect. 16.

D.B. Broughton, H.J. Bieser, and M.C. Anderson, *Pet. Int. (Milan)* **23**(3), 91 (1976), (in Eng.).

D.B. Broughton, H.J. Bieser, and R.A. Persak, *Pet. Int. (Milan)* **23**(5), 36 (1976), (in Eng.).

D.B. Broughton, R.W. Heuzil, J.M. Pharis, and C.S. Brearley, *Chem. Eng. Prog.* **66**(9), 70 (1969).

AEROSOLS

The term aerosol refers to products packaged in pressurized, sealed containers with liquefied or compressed gases, which enable the product to self-dispense by merely opening a valve. The significant attributes of the aerosol package are convenience, prevention of product contamination, good distribution, and in some cases, use efficiency.

The three components of an aerosol package are the product being dispensed, the propellant system, and the hardware necessary for containment and delivery.

The dispensed product may take the form of a wet or fine liquid or powder spray, foam, or paste, depending upon the application. The composition of the formulated product in the pressurized container may be solution, emulsion, or suspension.

Propellants

Propellants can be liquefied or compressed gases with a vapor pressure greater than 101.3 kPa (14.7 psia) at a temperature of 41°C. Pressures are often commonly expressed either in absolute units (psia) or gauge units (psig). The system can be approximated by the ideal gas equation, $PV = nRT$. The chief advantage of liquefied propellants is that they keep the pressure constant in the aerosol can until the contents are exhausted. They are classified into two chemical categories: halocarbons (fluorochlorohydrocarbons, and chlorinated hydrocarbons) and hydrocarbons. The fluorochlorocarbons are given numbers based on their chemical composition; in general, these numbers are preceded by the trademark of the manufacturer. In the numbering system, the first digit on the right denotes the number of fluorine atoms in the compound; the second digit from the right indicates the number of hydrogen atoms plus 1; the third digit from the right indicates the number of carbon atoms less 1. In the case of isomers, each has the same number. The most symmetrical is indicated by the number without any letter following it. As the isomers become more unsymmetrical, the letters a, b, and c are appended. If a molecule is cyclic, the number is preceded by C. For example, Propellant 11 is trichloromonofluoromethane, CCl_3F. The hydrocarbon propellants are liquefied petroleum gases such as propane, butane, and isobutane. They are also given numbers like 31, 46, etc, which refer to their vapor pressures (in psig), eg, aerosol-grade propane at 21°C consisting of 10% isobutane and 90% Propellants 11 and 12 in a 1:1 ratio. The chief problem of liquefied hydrocarbon propellants is flammability.

The compressed-gas propellants CO_2, N_2O, and N_2 are not liquid in conventional aerosol containers. They are nontoxic, nonflammable, low in cost, and very inert. However, the vapor pressure in the container drops as the contents are depleted, possibly causing changes in the rate and characteristics of the spray. This situation can be considerably improved when the contents are materials in which the compressed propellant is somewhat soluble; therefore, CO_2 and N_2O are preferred. For cosmetic use, ethanol appears to be the best material. In other applications, like insecticide products and spray paints, acetone, petroleum distillates, or acetate esters can be used in the formulations. Finally, in the case of liquefied propellants, the headspace plays no part in the system design, since pressure is constant throughout; however, in designing a CO_2-propelled system, the headspace becomes a variable since the distribution of gas within the can is important. Nitrous oxide is at present used in food items.

The EPA and other Federal agencies determined that fluorochlorocarbons cause depletion of the ozone layer and present a public hazard. The fluorochlorocarbons, when released on earth, can take up to 10 yr to reach the stratosphere where exposure to ultraviolet radiation causes decomposition. Ozone in the stratosphere acts as a screen for ultraviolet radiation which is harmful to living systems (see Ozone; Air pollution). Some controversy outside the agency surrounds the extent of ozone-layer depletion that would occur from continuing use of fluorochlorocarbons. Many other countries have not followed the United States' lead in this regulatory area. The EPA banned fluorochlorocarbon-propelled aerosols on December 15, 1978 in the U.S. The ban covers, for any aerosol propellant use, manufacturing of fluorochlorocarbons, importing fluorochlorocarbons, or products containing fluorochlorocarbons, or processing fluorochlorocarbons into an aerosol propellant article for domestic use or for export. The impact of this regulatory structure is to make the distribution in commerce of finished aerosol propellant products containing fluorochlorocarbons virtually impossible (very minor exceptions exist). Hence, since the ban, the propellants currently in use in the U.S. have been the liquefied hydrocarbons, CO_2 and N_2O.

Hardware

The hardware of the aerosol system consists primarily of pressure containers and dispensing valves. The container materials include steel, aluminum, glass, plastic, and their linings such as epoxy resins (qv). The heart of the aerosol system is the dispensing valve and actuator, consisting of a mounting cup, housing, stem, gasket, spring, dip tube, and actuator. The largest volume of aerosol valves is used to produce sprays, although foam valves, metering, filtered, codispensing, automatic, tip-seal, and refillable valves exist.

Leading areas of production are for hair sprays, deodorants-antiperspirants, perfumes, household products, insect sprays, and paints.

ANTOINE KAWAM
JOHN B. FLYNN
The Gillette Company

Aerosol Guide, 7th ed., Aerosol Division, Chemical Specialties Manufacturers Association, Inc., New York, Mar. 1981.

P.A. Sander, *Principles of Aerosol Technology*, Van Nostrand Reinhold Company, New York, 1970; an excellent general reference on technology and product design.

A. Herzka, *International Encyclopedia of Pressurized Packaging*, Pergamon Press, Inc., Elmsford, N.Y., 1966.

AGAR. See Gums.

AGAVE. See Fibers, vegetable.

AGRICULTURAL CHEMICALS. See Fertilizers; Fungicides; Herbicides; Insect control technology; Soil chemistry of pesticides.

AIR CONDITIONING

Basic Principles

Thermodynamic principles govern all air-conditioning processes (see Thermodynamics). The Carnot cycle, formulated directly from the second law of thermodynamics, is a perfectly reversible, adiabatic cycle consisting of two constant entropies and constant-temperature processes. It defines the ultimate efficiency for any process operating between two temperatures. The coefficient of performance (COP) of the reverse Carnot cycle (refrigerator) is expressed as:

$$\text{COP} = \frac{\text{useful effect}}{\text{work input}} = \frac{T_1}{T_2 - T_1}$$

where T_1 = absolute temperature of the cold source and T_2 = absolute temperature of the hot source (see Refrigeration).

In some applications, large quantities of waste or low cost heat are generated, and the absorption cycle can be directly powered from such heat.

There is a need to minimize the temperature difference between the heat source and sink to increase the efficiency of the process. This implies rejecting heat to the sink with the lowest possible temperature and obtaining heat from the source at the highest possible temperature. This also suggests using the largest heat exchangers consistent with economy to minimize temperature differences. Other considerations resulting from the second law include minimizing fluid-pressure losses by using the largest practical duct work and piping, and utilizing the most efficient mechanical devices obtainable to minimize system losses. Lack of maintenance, especially in regard to heat exchangers, significantly increases the temperature differentials and the energy consumption.

Psychrometrics is the branch of thermodynamics that deals specifically with moist air, a binary mixture of dry air and water vapor. The following properties of moist air are presented on psychometric charts: dry-bulb temperature, dew-point temperature, enthalpy, humidity ratio, relative humidity, saturation temperature, sensible-heat factor, specific volume, and wet-bulb temperature. This chart may be used to determine the properties of air required for a condition or process. Air-conditioning processes, except those involving substantial pressure changes, can be plotted on the chart although the process may not always be a straight line.

Design Conditions

Fundamental to the design of any air-conditioning system is the selection of appropriate conditions for the process requirements, such as temperature, humidity, and worker comfort. Specific design conditions are required in industrial applications to control temperature and humidity so that: constant temperature is maintained for machining, close-tolerance measuring, gauging, or grinding operations and to minimize expansion and contraction of machine parts; increases in relative humidity are controlled below the critical point at which metals may etch, nonhygroscopic materials may be affected, or the electrical resistance of insulating materials is significantly decreased; surface moisture films are maintained; comfort conditions consistent with the operator's expected level of activity are maintained; relative humidity and temperature are controlled to maintain the strength, pliability, and regain of hygroscopic materials such as textiles and papers, and the rates of chemical or biochemical reactions (such as varnish drying, application of sugar coatings, etc) are regulated. If the process does not require specific conditions, air conditioning is chosen to maintain worker comfort.

Determining the proper size of air-conditioning and heating equipment and yearly energy requirements using computer or manual methods requires detailed study and calculation.

Air-Conditioning and Humidification Systems

Air-conditioning and humidification systems heat or cool air, humidify or dry it, and control chemical impurities to maintain the desired space conditions. Two broad categories of air-conditioning systems exist: unitary and central-station systems. Frequently, systems are "off the shelf" self-contained units which use electricity for cooling and may use electricity, natural gas, fuel oil, or propane for heating. Heat rejected during the cooling cycle is dissipated to the outdoor air. Multiple unitary systems may be used to provide greater overall reliability and to permit individual control of various sections of the plant.

Central-station equipment provides more flexibility. It is normally used to condition a relatively large area of the plant as part of a "field erected" system. Outdoor air for ventilation or cooling (economizer cycle) is drawn through a preconditioning or preheat coil and mixed with air returned from the conditioned space. Dampers are used to control the relative amounts for temperature control. The air is filtered before passing into the conditioning section. Cooling and dehumidifying coils, air washers for humidity control, and heating coils are present in this section. The refrigerating effect is normally provided by refrigerating machines ("chillers"). Although most systems are electrically powered, low pressure steam, hot water, process streams and, more recently, solar-heated water may provide the motive force (see Solar energy).

For winter operation or special process requirements, humidification may be required. Humidification may be effected by use of an air washer which employs direct water sprays, low pressure steam injection, capillary-type humidifiers, or pan humidifiers.

Dehumidification may be accomplished in several ways (see also Drying). Moderate changes in humidity can be made by exposing the air stream to water or to a surface whose temperature is below the dew point of air. The air is cooled and releases a portion of its moisture. Direct contact between air and cold water using "open circuit" equipment achieves moderate dehumidification. In many industrial situations, dehydration equipment is more effective, using solid sorbents such as silica (qv), alumina (see Aluminum compounds), and molecular sieves (qv) or liquid sorbents such as brines of calcium chloride, lithium chloride, lithium bromide, calcium bromide, triethylene glycol, diethylene glycol, or ethylene glycol (see Drying agents; Glycols).

The removal or neutralization of noxious components from large volumes of air is achieved through the use of spray-type air washers employing appropriate reagents (eg, sodium hydroxide solution when the air contains acid gases) sprayed into the washer to purify the air (see Air Pollution control methods).

Evaporative air cooling applies the process of evaporating water into an air stream in situations where a high sensible heat load exists and inside design relative humidities of 50% rh or higher are required. Basic evaporative cooling devices include spray air washers, cell washers, and wetted-media air coolers (see Evaporation). Uninterrupted performance of any evaporative cooling device depends largely on a regular cleaning and inspection schedule.

Air conditioning controls range from a single thermostat controlling one unit to complex systems (see Instrumentation and control).

To conserve energy, knowledge of the building, its operating schedule, and the systems being installed is needed. To reduce energy consumption, use equipment only when needed; supply heating and cooling from the most efficient source; sequence heating and cooling; and provide only the heating or cooling actually needed.

Uses

Many industrial processes require accurate environmental control such as those in the synthetic textile industry, represented by rayon (qv) and nylon (see Polyamides; see also Textiles); the pharmaceutical industry for processing requirements of pharmaceutical materials (see Pharmaceuticals); the rubber industry (see Rubber compounding and fabrication); photographic processing (see Photography); the munitions industry (see Explosives); brewing (see Beer); food processing (qv); the metal industry (see Metal treatments; Powder metallurgy); ceramics (see Ceramics); and laboratories to maintain constant conditions.

KENNETH W. COOPER
RICHARD A. ERTH
York Division, Borg-Warner Corp.

ASHRAE handbooks, American Society of Heating, Refrigerating, and Air Conditioning Engineers, Inc., Publications Department, New York, four volumes: *Fundamentals, Equipment, Systems,* and *Applications.* One volume is revised each year.

J.P. Holman, *Thermodynamics,* McGraw-Hill, Inc., New York, 1974.

ASHRAE Standards, American Society of Heating, Refrigerating, and Air Conditioning Engineers, Publications Department, New York. Standards are upgraded on a regular basis.

Comprehensive Bibliography of Available Computer Programs in the General Area of Heating, Refrigerating and Air Conditioning and Ventilating, American Society of Heating, Refrigerating, and Air Conditioning Engineers, Atlanta, 1982.

AIR, LIQUID.   See Cryogenics.

AIR POLLUTION

Air pollution is defined as the "presence in the atmosphere (or ambient air) of one or more contaminants of such quantity and duration as may be injurious to human, plant, or animal life, property, or conduct of business." Air pollution suggests that the objectionable atmospheric concentration could have been prevented (see Air pollution control methods). The pollution may be toxic, irritate eyes and nose, cause difficulties in breathing, or degrade visibility. Ambient air quality standards are designed to protect the atmosphere from objectionable air pollution (see Industrial hygiene and toxicology).

Air pollution has three components: the emitting source, atmospheric transport and dispersal, and the receptor. It is characterized by a time-concentration relationship and may be controlled by reducing the concentration, the exposure time, or both. Standards for air pollutant concentrations, and some emission standards as well, specify an averaging period.

Pollutants can be emitted as gases (eg, sulfur oxides and nitrogen oxides) or as aerosols (eg, mist, smoke, soot, fume, and dust). Once in the atmosphere, the primary pollutant may undergo transformations into a secondary pollutant (eg, sulfur dioxide to sulfates). To make field measurements on the quantity and characteristics of pollutant emissions, emission factors are used. These statistical averages estimate the rate at which a pollutant is released to the atmosphere as a result of an activity such as raw material consumed, finished material produced, quantity of fuel burned, or vehicle miles traveled.

Atmospheric transport and dispersal control the destination of emitted pollutants. Factors that determine atmospheric transport include wind strength, wind direction, and atmospheric stability. Atmospheric stability is the propensity of the atmosphere to either enhance (instability) or retard (stability) an initial horizontal or vertical movement of an air parcel.

Predictions of air pollutant concentrations are generally made with the Gaussian diffusion equation. This approach assumes the pollutant concentrations in the cross section of a plume resemble the normal or Gaussian distribution in both the horizontal and vertical dimensions. For predicting ground-level concentrations, the Gaussian diffusion equation has the form:

$$\chi_{(x,y,z=0)} = \frac{Q}{\pi \sigma_y \sigma_z u} \exp\left[-1/2 \left(\frac{y}{\sigma_y} \right)^2 \right] \exp\left[-1/2 \left(\frac{H_e}{\sigma_z} \right)^2 \right]$$

where χ is concentration at the receptor (g/m^3); Q is the emission rate (g/s); u is the wind speed (m/s); x, y, and z are distances locating the receptor with respect to the base of the source $0, 0, 0$ (m); H_e is the effective stack height of the source, ie, physical height plus plume rise (m); σ_y, σ_z are the crosswind and vertical dispersion coefficients (m). The dispersion coefficients σ_y and σ_z are functions of atmospheric stability and distance downwind from the source.

Receptors are humans, animals, vegetation, and building materials. Based on criteria studies of the effects of pollutants on receptors, EPA has established primary and secondary National Ambient Air Quality Standards which specify maximum concentrations that are not to be exceeded more than once per year. Primary standards are directed toward protecting human health and take into account the range of human sensitivity by including a margin of safety. The stricter secondary standards are aimed at achieving an even cleaner atmosphere, a goal which would promote the public welfare by such means as improving visibility or protecting sensitive ornamental plants. See Table 1.

Air Pollutants

Air pollutants are divided into three groups by the Clean Air Act Amendments of 1970: criteria, hazardous, and designated. Criteria pollutants have an adverse effect on public health or welfare and result from diverse sources. Hazardous pollutants cause an increase in mortality or an increase in serious illness, eg, asbestos, mercury, beryllium, and vinyl chloride. Designated pollutants are those whose emissions are neither covered by the ambient air quality standards nor considered as hazardous, but which may be harmful to public health or welfare.

Pollutants fall into three main categories: particulates, gases, and odors.

Atmospheric particulates encompass both solid- and liquid-dispersed matter ranging in diameter from individual aggregates of about 0.0002 μm (just larger than single small molecules), to particles such as coarse sand which at 500 μm are visible. Both industrial and natural particulates are formed by one of four mechanisms: (1) condensation or sublimation of combustion or volatilization products (eg, soot particles and metallic fumes); (2) disintegration or attrition of material by natural forces, such as wind, or by mechanical processes such as grinding and drilling (eg, sea salt and milled flour); (3) coagulation of small particles to form large particles (eg, soot and fume particle agglomerates); and (4) chemical reactions involving gases in the atmosphere through the action of radiation, heat, or humidity (eg, oxidation of terpenes from trees which produces the haze in the Smoky Mountains). Particles formed by the first three mechanisms are sometimes designated as primary particles since they are injected directly into the atmosphere. Secondary particles, predominantly sulfates, nitrates, and oxyhydrocarbons, are formed within the atmosphere.

Table 1. United States National Ambient Air Quality Standards

| Pollutant | Averaging time | Maximum permissible concentration[a] | | Measurement method |
		Primary	Secondary	
sulfur oxides	annual arithmetic mean	80 μg/m^3 (0.03 ppmv)		West-Gaeke pararosaniline
	24 h max	365 μg/m^3 (0.14 ppmv)		
	3 h max		1300 μg/m^3 (0.5 ppmv)	
particulates	annual geometric mean	75 μg/m^3	60 μg/m^3	gravimetric 24 h hi-vol sample
carbon monoxide	8 h max	10 mg/m^3	same as primary	nondispersive infrared analyzer
photochemical oxidants	1 h max	235 μg/m^3 (0.12 ppmv as O$_3$)	same as primary	gas-phase chemiluminescence analyzer
hydrocarbons	3 h max (6–9:00 AM)	160 μg/m^3 (0.24 ppmv as CH$_4$)	same as primary	flame ionization detector
nitrogen oxides	annual arithmetic mean	100 μg/m^3 (0.05 ppmv as NO$_2$)	same as primary	chemiluminescence analyzer

[a]Standards for periods shorter than annual average may be exceeded once per year.

Sulfur oxides (SO$_x$) as air pollutants include sulfur dioxide (SO$_2$), sulfur trioxide (SO$_3$), plus the corresponding acids (H$_2$SO$_3$ and H$_2$SO$_4$) and salts (sulfites and sulfates) (see Sulfur compounds; Sulfuric acid and sulfur trioxide). Combustion of sulfur-containing solid and liquid fossil fuels (coal or oil) in power plants accounts for a major source of these chemical pollutants.

Once in the atmosphere, SO$_2$ is further oxidized to sulfuric acid and sulfates (generally between 0.2 and 2.0 μm in diameter). This typically accounts for 5–20% of the suspended particulates in urban areas. Increases in atmospheric humidity, as well as higher percentages of relative humidity, result in increases in the ratio of sulfuric acid to sulfur dioxide and in larger sizes of sulfate particles and sulfuric acid droplets. Sulfur dioxide alone is only a mild respiratory irritant; most of the deleterious effects of the sulfur oxides are caused by a combination of sulfur oxides and particulates, or by sulfuric acid and sulfates.

Carbon monoxide (qv) emissions to the atmosphere exceed the combined total of the other four major pollutants. To accommodate the higher levels, ambient concentrations of carbon monoxide are frequently expressed in μg/m^3 as for other pollutants. In the United States, approximately 75–95% of the CO emissions are produced by transportation sources. The remainder of CO emissions are due to incomplete combustion of carbonaceous materials other than gasoline. Potential sinks include migration to the upper atmosphere, metabolization by plants and soil microorganisms, absorption in the oceans, and adsorption on surfaces. The mean atmospheric residence time for CO has been estimated as five years. Carbon monoxide is absorbed by the lungs. It reacts with hemoglobin to form carboxyhemoglobin (COHb) which reduces the oxygen carrying capacity of the blood. The basis for U.S. CO standards is the protection against carboxyhemoglobin levels above 2% in nonsmokers.

Nitrogen oxides (NO$_x$) refer to two gaseous oxides of nitrogen, nitrogen dioxide (NO$_2$), and nitric oxide (NO). When measurements of nitric oxide concentrations are made, it is expressed as nitrogen dioxide. Most of the NO$_x$ is nitric oxide, created by fertilizers, by natural-gas-fueled turbines driving compressors at gas plants and along oil and gas pipelines, and by the high temperature combustion of coal, oil, or natural gas in power plants and of gasoline in internal combustion engines. NO$_2$ is the only major air pollutant gas that is visible, appearing as a brownish color of the smog layer. The residence time of nitrogen dioxide in the atmosphere is only a few days. Average annual concentrations range from less than 5 μg/m^3 as a natural background to 150 μg/m^3 in urban areas.

Hydrocarbons. The primary objective in controlling hydrocarbons is to reduce photochemical oxidant concentrations by limiting emissions of nonmethane hydrocarbons, especially olefins and aromatics. Most of the polluting nonmethane hydrocarbons are discharged into the air during incomplete combustion, especially from gasoline-powered automobile engines, and by evaporation from various sources (see Hydrocarbons).

Photochemical oxidants are atmospheric substances formed by sunlight-stimulated chain reactions. Typical oxidants are ozone (the most prevalent in polluted atmospheres), peroxyacetyl nitrate and nitrite (PAN), and nitrogen dioxide. Ozone exists in large quantities in the stratosphere where it is created by photochemical processes involving the absorption of ultraviolet solar radiation by upper atmospheric oxygen, and by lightning discharges. As pollutants, ozone and other oxidants are primarily produced in the lowest atmospheric layer by photochemical reactions between solar radiation and the oxidant precursors (nitrogen oxides and hydrocarbons). Nitric oxide is the principal supplier of oxygen in the photochemical reaction process. Although there is no correlation between high oxidant concentrations and increased mortality or morbidity, oxidants do have significant deleterious effects on vegetation, materials, and human comfort. The combined effect of O$_3$ and SO$_2$ or of O$_3$, SO$_2$, and NO$_2$ can cause more vegetation damage than any of the pollutants in the same concentrations individually. Many organic polymers undergo chemical alterations and rubber cracks when exposed to ozone. When ambient oxidant concentrations reach 200 μg/m^3 (0.10 ppm), eye irritation occurs (see also Photochemical technology).

Asbestos. There are five major sources of asbestos as an air pollutant: mining and milling of asbestos; manufacture and fabrication of asbestos-containing materials; abrasion of asbestos products; demolition of building where asbestos has been used in construction; and use of sprays containing asbestos materials. Occupational exposure to asbestos fibers causes asbestosis, a crippling and sometimes fatal lung disease, and mesothelioma, a cancer of the stomach or lung lining. Effects of asbestos inhalation may be cumulative.

Mercury (qv) is emitted from mercury-ore processing facilities and mercury-cell chloralkali plants (see Alkali and chlorine products). If mercury-saturated gases present in these operations are ventilated without treatment, they carry mercury vapor into the atmosphere. Particulate mercury emissions occur as elemental mercury mist, solid mercury compounds, and mercury adsorbed on soot.

Lead. Approximately 50% of the atmospheric lead in urban areas results from gasoline combustion. Alkyl lead vapors produced during the manufacture of antiknock compounds for gasoline account for 10% of atmospheric lead. The 1966–1967 U.S. National Air Sampling Network indicated average lead concentrations of 0.02–19 μg/m^3. Natural atmospheric cleansing processes of sedimentation and raindrop scavenging remove about one-half the lead-containing particulate matter from automotive exhausts within one hundred meters of roadways. Mean residence time of lead in the atmosphere is 7–30 d. Ingestion of food which has been exposed to high concentrations of airborne lead causes an elevated concentration of lead in the blood.

Polycyclic organic matter (POM) in air is associated exclusively with particulate matter, especially soot. The size of the POM particles is generally $< 5\ \mu m$ dia. Two classes of POMs are the azaarenes, identified as animal carcinogens (see Azine dyes; Pyridine; Naphthalene), and the polycyclic aromatic hydrocarbons, particularly benzo[a]pyrene (BaP) which results from cigarette smoke, coal-fired furnaces, and motor vehicles without emission controls.

Fluoride compounds (qv) are released through the mining and processing of minerals containing fluorides and the use of fluorine compounds as catalysts or fluxes. Phosphate fertilizer and aluminum industries have problems related to the emission of fluorides (see Phosphorus and phosphides; Fertilizers; Aluminum; Phosphoric acid). Other atmospheric fluorides are produced in the steel and glass industries. Fluoride emissions cause concern because of adsorption by nearby vegetation; injury to ornamental plants, and accumulation in the tissue of forage crops which are ingested by farm animals. This high fluoride forage can cause lameness and other harmful effects.

Carbon dioxide (qv) has increased in average concentrations from 290 ppm before 1900 to 330 ppm in the 1950s. Very little visible solar radiation is absorbed by atmospheric carbon dioxide. However, CO_2 is opaque to certain bands of the Earth's long-wave radiation. Thus increases in atmospheric CO_2 decrease the amount of energy Earth losses to space resulting in a higher Earth surface temperature.

Chlorofluorocarbons, CCl_3F and CCl_2F_2 (Propellants 11 and 12), man-made chlorofluorocarbons, are used widely as refrigerants and as propellants for aerosol cans (see Aerosols; Fluorine compounds). These inert compounds dissociate to release chlorine atoms in the stratosphere. The chlorine then catalyzes a chain reaction resulting in partial depletion of the ozone in the stratosphere. Depletion of the ozone layer would permit increased penetration of ultraviolet radiation at higher wavelengths (290–320 nm), a range predicted to produce a greater degree of skin cancer. As a result of the slow rate at which the fluorocarbons rise to the stratosphere (on the order of a year) and their long residence time there, it has been predicted that ozone destruction would continue until 1990 even if fluorocarbon emissions were curtailed in the late 1970s.

Odors. Two notable sources of objectionable odors are the meat rendering and kraft pulping industries. Threshold concentrations (the concentration at which all members of a group of people recognize an odor) can range over several orders of magnitude for different compounds. Among the criteria pollutants, nitrogen dioxide, sulfur dioxide, and some hydrocarbons, especially the aromatics and halogenated hydrocarbons, have detectable odors at concentrations found in the air. Hydrogen sulfide and the various mercaptans (eg, methylmercaptan, CH_3SH) are objectionable. Human perception of odor intensity is not directly proportional to the pollutant concentration, but to the logarithm of the concentration.

PHILIP R. STICKSEL
RICHARD B. ENGDAHL
Battelle Memorial Institute

Compilation Air Pollutant Emission Factors, U.S. Environmental Protection Agency, Research Triangle Park, N.C., 1973 (twelve supplements 1973–1981).

A.C. Stern, ed., *Air Pollution*, ed., Vols. 1–5, Academic Press, Inc., New York, 1976.

M. Smith, ed., *Recommended Guide for the Prediction of the Dispersion of Airborne Effluents*, 2nd ed., The American Society of Mechanical Engineers, New York, 1973.

AIR POLLUTION CONTROL METHODS

The choice of pollution control methods should be based on the need to control ambient air quality, to achieve compliance with standards for criteria or hazardous pollutants, or to protect human health and vegetation from numerous nonregulated contaminants. Three elements are necessary for a pollution problem: a pollution source, a receptor affected by pollutants, and transport of the contaminants from source to receptor. Elimination or modification of one of these elements can eliminate or change the nature of the pollution problem.

Pollutants fall into three main categories: gases, particulates (which may be either liquids or solids, or both in combination), and odors. Odors can be present in gaseous or particulate form and can be controlled in the same manner as gases and particulates in general, but are discussed separately owing to the different sensing and measurement methods used.

Measurement and Sampling

Pollution measurements are either ambient or source. Air samples are often measured in the ppmv to ppbv range; source concentrations range from tenths of a percent to a few hundred ppmv.

Air sampling may fulfill one or more of the following objectives: establishing and operating a pollution-alert network; monitoring the effect of an emission source; predicting the effect of a proposed installation; establishing seasonal or yearly trends; locating the source of an undesirable pollutant; obtaining permanent sampling records for legal action; and correlation of pollutant dispersion with meterological, climatological, and topographic data and societal activities.

Source sampling typically fulfills one or more of the following objectives: demonstrating compliance with regulations; providing emission data; measuring product loss or optimizing process variables; obtaining engineering data for design; acceptance testing of control equipment; and determining the need for maintenance of process or control equipment.

Gaseous pollutants are detected by their chemical nature (see Analytical methods; Trace and residue analysis). Suspended particulate concentration is frequently determined gravimetrically by filtering a sample. Various methods for particle-size measurements, such as microscopy, sieve analysis, air and liquid sedimentation, centrifugal classification, and electrical and optical counters, may be used to determine the size in particulate samples extracted from a source or collected from ambient air by filtration, electrostatic or thermal precipitation, or impaction (see Size measurement of particles).

Selection of Control Equipment

Table 1 summarizes the basics of an engineering approach to selection of control equipment.

Control of Gaseous Emissions

Five methods are available for control of gaseous emissions: absorption, adsorption, condensation, chemical reaction, and incineration. Atmospheric dispersion from a tall stack is sometimes considered a sixth method. Absorption is particularly attractive for pollutants present in appreciable concentration, although it is quite applicable to dilute, highly water-soluble gases (see Absorption). Nonaqueous liquids may be used for gases with low water solubility.

Absorption. Packed columns, open spray chambers and towers, cyclonic spray towers, and combinations of sprayed and packed cham-

Table 1. Engineering Requirements for Design and Selection of Control Equipment

know properties of pollutant: chemical species, physical state, particle size, concentration, quantity of conveying gas

know effects of pollutant on surrounding environment

design for likely future collection requirements

determine advantages of alternative collection techniques: collection efficiency; ease of reuse or disposal of recovered material; ability of collector to handle variations in gas flow and loads at required collection efficiencies, equipment reliability and freedom from operational and maintenance attention; initial investment and operating cost

try to recover or convert contaminant into a saleable product

apply known engineering principles even in areas of extremely dilute concentration

bers are most used for air contaminants, whereas countercurrent packed towers are used for particulate-free gas. The cross-flow packed scrubber (Fig. 1) is a newer device used as a pollutant absorber. Typical design parameters for gas absorption only are gas-flow rate $G = 2.44 \text{ kg/(s·m}^2)$ and liquid-flow rate $L = 2.03 \text{ kg/(s·m}^2)$. With particulates, sprays are added upstream directed at the bed-retaining grillwork. Most of the solids are impacted on the first 150 mm of packing in the gas-flow direction. To remove deposited solids, the liquid rate over the first 300 mm of the packing is increased to $L = 13.56 \text{ kg/(s·m}^2)$. Solids loadings up to 11 g/m^3 have been handled successfully. A single transfer unit has been achieved in a gas-flow depth of 200 mm when absorbing HF in water. Open horizontal spray chambers and vertical spray towers are desirable when solids are present, but cyclonic spray towers provide slightly better scrubbing. The number of transfer units attainable in a single countercurrent spray tower is limited by loss of countercurrency caused by spray droplet entrainment with the gas.

Water, the most common absorption liquid, is often used for removing acidic gases such as HCl, HF, and SiF_4, especially if the last contact is with water of alkaline pH. Problems can arise in the initial absorption stages when contacting concentrated gases and neutralizing agents. Vapor-phase reactions can produce a submicrometer smoke which is difficult to wet and collect. Less soluble acidic gases, such as SO_2, Cl_2, and H_2S, can be absorbed more readily in an alkaline solution. Scrubbing with alkaline ammonium salt solutions, such as the Cominco SO_2-absorption process, is practiced.

Disposal of recovered gaseous pollutants may be a problem. Precipitation of the pollutant as an insoluble sludge may be possible by the addition of lime or other reagents. Rather than disposal in streams or landfill, conversion of the pollutants to a usable form is preferable (see Wastes).

Adsorption is desirable for removal of contaminant gases to extremely low levels (less than 1 ppmv) and handling large volumes of gases with quite dilute contaminants. It is always exothermic so that gas precooling is practiced to prevent capacity loss. Pretreatment to remove or reduce competing adsorbable molecules should also be considered. Water vapor is a typical material that can compete for adsorption sites. Carbon is one of the few adsorbents that can work in a humid gas stream, even though water vapor is adsorbed equally well and reduces the bed adsorptive capacity. Other adsorbents typically are simple or complex metal oxides (see Molecular sieves).

After the pollutants have been adsorbed, they may be disposed of by discarding the saturated adsorbent or by adsorbent regeneration. Disposal is attractive when the quantity of material to be adsorbed is small or occurs only infrequently, and when the cost of fresh adsorbent is insignificant compared to the cost or inconvenience of regeneration. When pollutant quantities are appreciable, regeneration is usually practiced. Desorption can occur through bed heating, evacuating, stripping with an inert gas, or displacement with a more adsorbable material, or combinations. Desorption is never complete so excess adsorption capacity must be provided. Some materials undergo chemisorption in which they combine chemically with the adsorbent. Typical pollution-adsorption applications include odor control from food processing; chemical and process manufacturing; odor control from foundries, animal laboratories, painting, and coating operations; and control of solvent and hydrocarbon emissions as well as radioactive gases from nuclear activities.

Condensation is best for vapors with reasonably high vapor pressure. This method is most feasible for organic compounds. When the gas must be cooled more than 40–50°C to achieve adequate condensation, condensate fog may form. When fog formation is unavoidable, it may be removed by a high efficiency mist collector such as those using Brownian diffusion (see Particle filtration below). Elimination of noncondensable diluents is beneficial where refrigeration is needed for the final condensation step.

Chemical reaction of gaseous pollutants can make them more easily collectable. Nitrogen oxides can be decomposed to N_2 and O_2 in a catalytic device. Many odors and organic compounds can be destroyed by strong oxidants such as $KMnO_4$ and HNO_3, and hypochlorite solutions.

Incineration (suitable only for combustibles) is used for control of organic vapors and destruction of small quantities of H_2S and NH_3. Dilute gases, not more concentrated than 25% of the lower explosive limit, are burned in a gas incinerator. Gases sufficiently concentrated to support combustion are burned in waste-heat boilers, flares, or used for fuel (see Incinerators).

Specific problem gases. Sulfur dioxide, nitrogen oxides, and automobile exhaust (containing hydrocarbons, carbon monoxide, and nitrogen oxide pollutants) are widespread gaseous pollutants (see Air pollution). Sulfur oxide emissions from combustion are controlled by the use of low sulfur fuel, fuel desulfurization, and SO_2 removal from flue gas (see Coal; Petroleum; Sulfur recovery). Control of SO_2 emissions from sulfuric acid manufacture, from nonferrous smelters, from pulp and paper manufacture, as well as general control methods are discussed in the literature (see Sulfuric acid; Pulp; Paper). Control of nitrogen oxide emission is discussed as well (see Nitric acid; Nitration; Burner technology; Furnaces, fuel-fired; Exhaust control, industrial and automotive).

The removal of particles (liquids, solids, or mixtures) from a gas stream requires their deposition and attachment to a surface. The surface may be continuous, such as the wall and cone of a cyclone or the collecting plates of an electrostatic precipitator, or discontinuous, such as spray water droplets in a scrubbing tower. Once deposited on a surface, a means must be provided to remove the collected particle at intervals without appreciable reentrainment in the gas stream. Seven physical principles are commonly employed, singly or in combination, to move the particles from the bulk of the gas stream to the collecting surface: gravity settling; centrifugal deposition; flow-line interception; inertial impaction; diffusional deposition; electrostatic precipitation; and thermal precipitation.

Gravity settling in a chamber is one of the oldest forms of gas–solid separation. It can be a large room with well-distributed gas entry at one end and exit at the other. Although mechanical devices might minimize labor costs, gravity settlers are disappearing because of bulky size and low collection efficiency (impractical for particles smaller than 40–50 μm) (see Gravity concentration).

Cyclonic collection utilizes a stationary device with no moving parts, which converts the entering gas stream to a vortex. Centrifugal force acting on the suspended particles causes them to migrate to the outside wall where they are collected by inertial impingement. Since the centrifugal force developed can be many times that of gravity, very small particles can be collected in a cyclone. Figure 2 illustrates four common types of cyclone design.

In the conventional cyclone (Fig. 2(**a**)), the gas enters the cylinder tangentially, and spins in a vortex as it proceeds down the cylinder. A cone section causes the vortex diameter to decrease until the gas reverses on itself and spins up the center to the outlet pipe or vortex finder. A cone causes flow reversal to occur sooner and makes the cyclone more compact. Dust particles are centrifuged toward the wall and collected by inertial impingement. The collected dust flows down in the gas-boundary layer to the cone apex where it is discharged through an air lock or into a dust hopper serving a number of parallel cyclones.

A grade-efficiency curve for a particular cyclone design is required to predict overall dust collection efficiency. Geometric design affects the grade-efficiency curve.

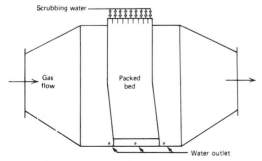

Figure 1. Cross-flow packed scrubber.

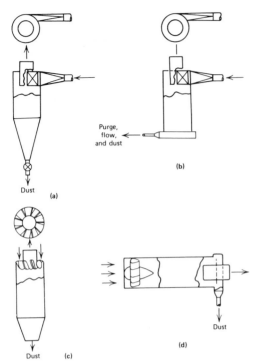

Figure 2. Cyclone types commonly used; (**a**) type A tangential inlet axial discharge; (**b**) type B tangential inlet peripheral discharge; (**c**) type C axial inlet and discharge; (**d**) type D axial inlet peripheral discharge. Courtesy of American Industrial Hygiene Association.

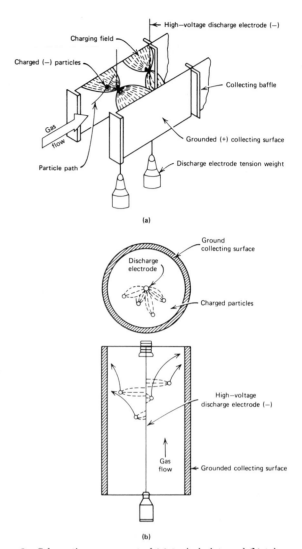

Figure 3. Schematic arrangement of (**a**) typical plate and (**b**) tube precipitators.

Cyclone application problems include parallel cyclone distribution, fouling and caking, and erosion. Cyclone efficiency pressure-drop increases with flow up to the particle pickup velocity. Increase in dust loading increases efficiency and decreases pressure drop. Thus, poor gas or dust distribution between parallel cyclones can result in poorer overall efficiency for the group. Unequal pressure drop can result in gas backflow through cyclone dust outlets connected to a common dust hopper. Fouling occurs as buildup on cyclone walls from sticky, hygroscopic particulates or moisture condensation on cold cyclone walls. Erosion can be a severe problem when handling a heavy loading of abrasive dust.

Inertial centrifugal forces may be utilized alone or in combination with cyclones in which mechanically driven parts similar to centrifugal fans accelerate the gas and particles.

Electrostatic precipitation can collect either solid or liquid particulates, or combinations. Simultaneous gas absorption is also possible in a wet precipitator. In an electrical precipitator, particles are charged in an electric field and move to a surface of opposite charge where they are deposited. For particles smaller than 2 μm, the electrical forces developed can be more powerful than any other collectional force and precipitators can have the highest energy-utilization efficiency. Precipitators have low pressure drops (often no more than 250 Pa or 1 in. of water), low consumption of electric power, and low operating costs. A precipitator recovers particulates in agglomerated form, a frequent advantage rendering them more easily collectable in case of reentrainment. Offsetting the desirable features is the fact that precipitators are the most capital-intensive of all control devices. Further, mechanical, electrical, or process problems can cause poor on-stream time and reliability.

Precipitators may be single stage or two stage. In single-stage precipitators, used for industrial gases, the particles are charged and collected in the same electrical field. A complete precipitator consists of discharge electrodes, collecting surfaces (plates or tubes), a suspension and tensioning system for the discharge electrodes, a rapping system to remove collected dust from the tubes, dust hoppers and dust-discharge system, gas-distribution and precipitator housing, and power supply and control system. Figure 3 shows typical plate and tube type arrangements.

Principal problems encountered with dry-dust precipitators are dust reentrainment, dust resistivity, gas distribution, and electrical problems of arcing and control. Water-irrigated wet precipitators using sprays or overflow weirs with plates, either flat or concentric, are suited for fine particulates and eliminate dust resistivity problems. However, they introduce other problems such as corrosion (requiring plastic or expensive alloy parts where carbon steel would be satisfactory if dry), scaling, water treatment, and waste-handling problems. Wet precipitators are used to treat mixtures of gaseous and submicrometer particulates such as aluminum-potline and carbon-anode baking fumes, fiber-glass fume control, coke-oven and metallurgical fumes, and phosphate fertilizer emissions.

In two-stage precipitators, the particles are charged in an ionizing section and precipitated in a separate collection section. Two-stage units operate at lower voltages, with less hazard of electrical sparking, and are used for air purification, ventilation, and air conditioning (qv). Figure 4 illustrates the operating principles of two-stage precipitation. The most frequent use of the two-stage precipitator is for air cleaning in ventilation and air conditioning. A principal advantage of the unit is its much lower cost because of mass production in modules. However, it has a low dust-holding capacity which has precluded its use for heavily loaded industrial gases. Collection of liquids is possible, but if the liquids are conductive, shorting of the spacer insulators is a problem. These units have recently been used for hydrocarbon mists where the collected liquid is nonconductive.

A new area for improved fine-particle collection is the combination of electrostatic forces with devices using other collection mechanisms. Much research has been devoted to optimizing hardware design.

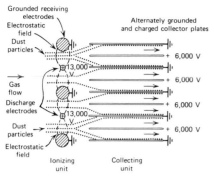

Figure 4. Operating principle of two-stage electrical precipitator. Courtesy of McGraw-Hill, Inc.

Particle filtration devices are divided into three main categories: cloth filters using either woven or felted fabrics in a bag or envelope; paper and mat filters; and in-depth aggregate bed filters.

The cloth or bag dust filter is the oldest and often the most reliable of the many methods for removing dusts from an air stream. Advantages are high collection efficiency, often ≥ 99%, moderate pressure drop and power consumption, recovery of the dust in a dry and often reusable form, and no introduction of water in the exhaust gases as with a wet scrubber. They also have numerous disadvantages: maintenance cost due to bag replacement can be quite high; they are suitable only for low to moderate temperature use; they cannot be used in situations where liquid condensation can occur; and they are bulky and require considerable installation space.

Bag filters have been used for controlling low temperature dusts from milling and screening operations, drying, packaging, loading, and unloading, as well as material conveying; they are being extended to control fine particles from large volume hot-gas streams such as power boilers.

In-depth fiber-bed filters are used for the collection of liquid droplets, fogs, and mists. Horizontal pads of knitted metal wire (or plastic fibers such as fluorocarbon resins) are used with upflow of gas. The pressure drop is 250–500 Pa (1–2 in. water). Characteristics of the pads vary slightly with mesh density, but a typical void space is 97–99% of total volume. Collection is by inertial impaction and direct impingement; thus, efficiency will be low at low (below 2.3 m/s) superficial velocity.

Three types of randomly oriented fiber beds, 25–75 mm thick and using chemically resistant glass, polypropylene, or fluorocarbon fibers, are used for collection of fine mist particles: a high velocity filter which collects by inertial impaction; a high efficiency mist eliminator which is custom designed for submicrometer mist and collects by Brownian diffusion; and a lower efficiency spray catcher type of unit similar to the high velocity unit but designed for smaller pressure losses (138–276 Pa or 0.5–1.0 in. water). Although fiber mist eliminators are designed primarily for collection of liquid particles, they can be used for efficient collection of water-soluble solid particulates if the particle loading is not too high and the bed is flushed with water.

Some effluents contain submicrometer oily or tar-like particulates. A dry, glass-fiber disposable-mat filter can handle such particulates successfully.

Wet scrubbing can be highly effective for both particulate collection and gas absorption. The scrubber purchase cost may be quite reasonable but addition of water treatment and waste disposal costs may make the total cost as great as other collection devices. Scrubbers automatically provide cooling of hot gases but the water-saturated effluent may produce offensive plume condensation in cold weather. Many effluents become more corrosive in a wet environment, solids accumulation may occur at wet–dry interfaces, and icing problems can occur in winter. For fine particles, energy consumption may exceed that of dry collectors by an appreciable amount.

It is difficult to classify a scrubber by collection mechanism since more than one mechanism is frequently present. However, at least for low energy scrubbers, inertial impaction is frequently predominant. Where this is so, a "contacting power" principle has been proposed which states that the efficiency of collection is proportional to power expended and that more energy is required to capture particles as their size decreases.

Many different types of scrubbers have been developed. An open spray tower is the simplest. A number of commercial scrubbers such as type N and R Rotoclones and the Microdyne use a combination of water atomization and centrifugal force for capture. Sieve-plate towers and those with other types of trays have been investigated and found effective for particles down to 1 μm. Ejector-venturi fume scrubbers are reasonably effective on particles larger than 2 μm. Venturi scrubbers can be operated at 2.5 kPa (0.36 psi) to collect particles coarser than 1 μm; for smaller particles, a pressure dip of 7.5–10.0 kPa (1.09–1.45 psi) is usually required.

Particle growth through vapor condensation can enhance wet scrubber performance above normally expected levels. Collection of hydrophobic dust particles can be enhanced with the addition of nonfoaming surfactants to the water.

Control of Odors

Odor is the subjective perception of the sense of smell. Odor control describes any process that gives a more acceptable perception of smell as a result of dilution, removal of the offending substance(s), or by counteraction or masking (see Odor counteractants; Perfumes). Odor measurement methods are static or dynamic, depending on the presentation method. The objective is to measure odor intensity by determining the dilution necessary to make the odor imperceptible or doubtful to a human test panel. The detection threshold is the lower concentration at which an odor stimulus is detected. The recognition threshold is a higher value at which the chemical entity is recognized. An odor unit (o.u.) has been widely defined as 0.0283 m³ (1 ft³) of air at the odor threshold. It is a dimensionless unit representing the quantity of odor which, when dispersed in 28.3 L (1 ft³) of odor-free air, produces a positive response by 50% of the panel members. Odor concentration is the number of cubic meters that one cubic meter of odorous gas will occupy when diluted to the odor threshold.

In static dilution methods, a known volume of odorous sample is diluted with a known amount of odor-free air, mixed and presented statically or quiescently to a test panel. The ASTM D 1391 syringe dilution technique is the best known of these methods.

In dynamic dilution methods, odor dilution is achieved by continuous flow which promotes accurate results, simplicity, reproducibility, and speed. Dynamic olfactometer control, the flow of both odorous and pure diluent air, permit ration adjustment to give desired dilutions and present multiple, continuous samples for test-panel observers at ports beneath odor hoods.

Absorption, adsorption, and incineration are typical control methods for gaseous odors. Odorous particulates are controlled by the usual particulate-control methods. For oxidizable odors, chemical reaction with oxidizers such as H_2O_2, ozone (qv), and $KMnO_4$ is also practiced occasionally; catalytic oxidation has also been employed (see Exhaust control). Industry-oriented odor control examples include rendering plants, spent grain dryers, pharmaceutical plants, and cellulose pulping.

B.B. CROCKER
D.A. NOVAK
W.A. SCHOLLE
Monsanto Company

A.C. Stern, ed., *Air Pollution*, 3rd ed., Vols, 1–5, Academic Press, Inc., New York, 1967–1977.

W. Straus, *Industrial Gas Cleaning*, 2nd ed., Pergamon Press, Oxford, UK, 1975.

R.H. Perry, ed., *Engineering Manual*, 3rd ed., Sect. 10, McGraw-Hill, Inc., New York, 1976.

H.E. Hesketh, *Air Pollution Control*, Ann Arbor Science Publishers, Ann Arbor, Mich., 1979.

ALCOHOLATES.   See Alkoxides, metal.

ALCOHOL FUELS

Although modern commercial marketing of gasohol, a blend of 90 vol% unleaded gasoline and 10 vol% ethanol, began in the United States in 1979, many of the problems associated with alcohol cost and performance have not been solved. The alcohol used with motor fuel, either as a gasoline extender or neat fuel, is usually ethanol. Ready availability, good combustion characteristics, and octane enhancement are the great advantages offered by ethanol. In addition, laws permitting the production of tax-free ethanol for industrial applications were passed in many countries, including the United States. State and Federal excise tax subsidies also help to promote ethanol-fuel usage. Synthetic methanol was introduced in the United States at about half the price of wood alcohol (methanol), and was considered a possible alternative for gasoline in the 1930s (see Coal chemicals and feedstocks). Little has been published about the use of higher alcohols as motor fuels. But considerable experience has been accumulated with methanol, which is expected to be a strong competitor for ethanol, especially if tax subsidies are given to methanol similar to those for ethanol.

In the United States, several oil companies are marketing ethanol–unleaded gasoline blends and blends of unleaded gasoline with methanol and cosolvent. In Brazil, up to 20% ethanol–gasoline blends sold are distributed on a large scale by the government-sponsored Proalcol program.

Properties

Methanol and ethanol have ca 50 and 66%, respectively, of the volumetric heating value of gasoline (see Table 1). The alcohols used as liquid-fuel extenders exhibit specific boiling points, whereas commercial hydrocarbon fuels consist of mixtures of paraffinic, aromatic, and cycloparaffinic compounds that exhibit a boiling range. Furthermore, methanol and ethanol have much lower stoichiometric air : fuel ratios for combustion than gasoline, which is the reason that alcohol–air mixtures supplied by a gasoline-set carburetor are too lean for combustion to occur.

The high octane numbers of alcohol compared to those of unleaded regular gasoline suggest greater volumetric efficiencies of neat alcohol than of gasoline, provided the compression ratios of the engine are high enough to take advantage of the higher octane values. For ethanol and especially for methanol, the latent heats of vaporization are sufficiently large that the fuel does not completely evaporate during the suction stroke and continues to evaporate during the compression stroke. The

higher octane values of methanol and ethanol and the higher operational engine compression ratios imply higher latent heats of vaporization and the resulting increases in fuel-air cooling, density, and mass flow, as well as the more favorable molar ratio of combustion products to charge. These factors suggest greater efficiency of alcohols than of hydrocarbon fuels in terms of mileage per unit energy expended.

Only anhydrous ethanol forms solutions with diesel fuels and up to ca 30 vol% anhydrous ethanol can be added without engine modification. However, contamination by 0.5 vol% water causes phase separation of ethanol–diesel fuel blends at 0°C; severe knocking can also occur. Diesel fuel cannot be completely replaced by methanol or ethanol because of the low cetane numbers and high autoignition temperatures of the alcohols.

Combustion. The relatively high volume of air needed for methanol and ethanol to burn stoichiometrically results in the blend-leaning effect. The increased thermal efficiency of low molecular weight alcohols is owing to higher pressure in the combustion chamber and higher power output.

The heating values for alcohols, methyl *tert*-butyl ether, isooctane, and gasoline are in a very narrow range (ca 3.72 MJ/m^3 = ca 100 Btu/ft^3 at 20°C). In theory, a properly aspirated and timed spark-ignition engine could thus be expected to deliver the same power outputs, independent of which fuel was used. However, many differences cause deviation from idealized behavior.

Octane numbers. The octane numbers of neat alcohols are high (see Table 1) and, as expected, the alcohols enhance the octane numbers of gasolines. Many petroleum refiners are marketing regular and premium grades of gasoline, which contain alcohols as octane-improving substitutes for tetraethyllead and not as gasoline extenders.

Manufacture

Ethanol. Ethanol is manufactured from natural feedstocks by fermentation or from ethylene by direct or indirect hydration (see Ethanol; Fuels from biomass). Alcohol fermentation under anaerobic conditions can be conducted on sugar crops and sugar-containing by-products; starchy crops, eg, corn, and grains; and lignocellulosic materials, eg, wood, sulfite waste liquor from paper pulp, and crop residues. Pretreatment of the feedstock such as hydrolysis may be necessary to facilitate alcohol fermentation.

Synthetic ethanol refers to ethanol made by nonfermentation routes. Of these, direct ethylene hydration is the primary production method. It is carried out in the vapor phase with liquid or solid catalysts, steam,

Table 1.   Comparison of Some Properties for Several Liquid Fuels and Gasoline Additives

Property	Methanol	Ethanol	Unleaded regular gasoline	Diesel fuel
formula	CH_3OH	CH_3CH_2OH	C_4–C_{12}	C_{14}–C_{19}
element, wt%				
C	37.48	52.13	85–88	85–88
H	12.58	13.12	12–15	12–15
O	49.94	34.72	0	0
density (at 20°C), g/cm^3	0.7914	0.7893	0.69–0.80	0.82–0.86
boiling point, °C	65.0	78.5	27–225	240–360
latent heat of vaporization (at 20°C), MJ/L^a	0.931	0.662	0.251	0.237
flashpoint, °C	11.1	12.8	−43	38
autoignition temperature, °C	464	423	495	260
flammability limits, vol% in air	6.7–36.0	4.3–19.0	1.4–7.6	
lower heating value (at 20°C), MJ/L^a	15.76	21.09	32.16	35.4
stoichiometric air : fuel mass ratio	6.45	8.97	14.73	15.00
stoichiometric air : fuel volumetric ratio	7.16	14.32	57.28	
water solubility, %	infinite	infinite	0.009	nil
research octane number (RON)	112	111	91–93	
motor octane number (MON)	91	92	82–84	
cetane number	3	8	8–14	40–60
Reid vapor pressure, kPa^b	32	17	48–103	nil

^{a}To convert MJ/L to Btu/gal, multiply by 3590.
^{b}To convert kPa to psi, multiply by 0.145.

and ethylene at temperatures and pressures that minimize ethylene polymerization, ethanol dehydration, and formation of by-products. Catalysts include tungsten oxides, phosphoric acid, and ion-exchange resins. Indirect hydration involves sulfation of ethylene with concentrated sulfuric acid. Dilution of the reaction mixture with water and hydrolysis of the sulfates yield ethanol, diethyl ether, and dilute sulfuric acid, which is concentrated and recycled. The products are separated by distillation.

Because of the sensitivity of gasohol-type formulations to phase separation on contamination with water, anhydrous ethanol is generally used for gasoline blending.

Other routes to synthetic ethanol include direct oxidation of ethane and higher hydrocarbons, hydrolysis of ethyl esters, reduction of diethyl ether or acetaldehyde, conversion of synthesis gas, and homologation. Many of these methods are not economically feasible or are in the research stage.

Methanol. Methanol (qv) can be produced by wood pyrolysis, non-catalytic oxidation of hydrocarbons, as a by-product in the Fischer-Tropsch synthesis, and by reduction of carbon monoxide. Today, methanol is almost entirely manufactured from synthesis gas, a mixture of carbon monoxide and hydrogen, at pressures of 5–10 MPa (50–100 atm) and at temperatures of 220–270°C.

$$CO_2 + 3 H_2 \rightarrow CH_3OH + H_2O$$

Conversion catalysts are based on Cu/ZnO systems.

Methane is oxidized to methanol by nitrous oxide at selectivities up to 60% in the presence of water over silica-supported molybdenum oxide at 0.1 MPa (1 atm) and 560–600°C in a continuous-flow reactor.

Fermentation methods for methanol have not been developed.

Research

Fermentation ethanol. Research has concentrated on minimizing the energy inputs, increasing yields, reducing residence times, and improving the pretreatment processes to convert cellulosic materials to fermentation feedstocks (see Fermentation).

Other improvements include the use of reduced pressure to remove ethanol as fast as it forms; use of bacteria instead of yeasts to shorten fermentation times; continuous fermentation; simultaneous saccharification and fermentation of low grade cellulosic material with enzymes and yeasts; thermophilic anaerobes for one-step hydrolysis and fermentation; packed columns containing live, immobilized yeast cells or both enzymes and yeast cells through which glucose is passed; and recombinant-DNA techniques to develop new microbial strains for rapid conversion of starch to sugars.

Thermochemical ethanol. Catalytic conversion of synthesis gas to mixed alcohols can yield ethanol as a main product. A typical distribution over a catalyst containing a mixture of Cu and Co oxides, alkali-metal salts, and oxides or Cr, Fe, V, or Mn, is methanol 20%, ethanol 35%, 1-propanol 21%, 2-propanol 3%, and 1-butanol 17%.

Thermochemical methanol. In the Chem System liquid-phase catalytic process, an inert hydrocarbon liquid in the presence of a heterogeneous catalyst produces a high conversion of synthesis gas. The liquid controls the temperature more efficiently than vapor and allows a closer approach to equilibrium and better recovery of the reaction heat as steam. The methanol product stream is claimed to be suitable for direct use as fuel without distillation.

In another system, three different catalysts are employed in successive beds that are intercooled for removal of the heat of reaction. The crude product can be used as fuel without distillation. It contains ca 97% methanol, 2% higher alcohols, and not more than 1% water. Overall thermal efficiency from coal-derived synthesis gas is 58%. The presence of higher alcohols can reduce or prevent phase separation when methanol fuel–gasoline blends are contaminated with water.

Legal Blending Limitations

Ethanol has the advantage over methanol of more lenient tax-forgiveness and fewer controls on blending. Up to 10 vol% can be blended in unleaded gasoline. In the United States, neat methanol is presently limited in unleaded gasoline to a maximum concentration of 0.3 vol%. On the other hand, the Permanent Bureau of Europoean Automobile Manufacturers Association accepts gasoline blends with 3 vol% methanol and has approved the proprietary formulation Petrocoal, a blend of methanol and higher alcohol cosolvent, for use in unleaded gasoline at methanol concentrations up to 12 vol%. Such methanol formulations can also be used in the United States.

Storage and Handling

Long-term storage of alcohol-gasoline blends increases gum formation and loosens deposits of rust and sediments. In addition, evaporative emissions, phase separation, a wide flammability range, low odor intensity, flame luminosity, and the ability to form explosive mixtures in closed containers can present storage and handling problems.

Operating Problems

Water contamination can cause separation of alcohol-gasoline blends into a lower water-rich layer and an upper hydrocarbon-rich layer. The amount of water that the blend can tolerate before separation occurs depends on the alcohol structure and concentration and the gasoline composition. Phase separation causes engine-operating difficulties, corrosion, and plugging of small orifices with gums and deposits. Addition of cosolvents, eg, higher alcohols, appears to be the best solution. Phase separation does not occur in neat-alcohol fuels since water is essentially miscible with methanol and ethanol.

Emissions

Methanol and ethanol form azeotropes with many of the hydrocarbons present in gasoline. In blends, the boiling points of many components are reduced and, at lower temperatures, the vapor pressure of the blend is higher than that of the gasoline. At concentrations of 10%, methanol and ethanol can increase evaporative emissions by as much a 130–220% and 49–62%, respectively. This can have an adverse effect on the volatility balance of the fuel and result in vapor lock and stoppage of the fuel flow to the engine. These problems can be overcome by adjusting the concentrations of the lower boiling fractions of gasolines used for the blends, particularly the butanes and pentanes.

Driveability

The term driveability refers to the response of the vehicle to manipulations by the driver. Modern automobiles powered with gasoline-type fuels generally respond satisfactorily to throttle inputs. Older vehicles may suffer from excessive blend-leaning effects, particularly when operating on fuels containing methanol, and may require manual adjustments of the carburetion system. Vapor lock and difficulty with starting at high ambient or underhood temperatures with alcohol–gasoline blends seems to be vehicle-dependent with ethanol blends but common with methanol blends when no volatility adjustments are made. The increased volatility of the blends, however, facilitates cold starting.

Materials Compatibility

Cars powered by neat ethanol require special materials of construction in certain areas of the fuel system because of possible corrosion. For example, electroless nickel plating instead of terneplate is used for fuel tanks, and steel fuel lines are replaced with nylon-7 or polyethylene. Methanol is more corrosive than ethanol, especially in the presence of water.

Fuel Economy

Brazilian automobiles equipped with engines designed specifically for neat ethanol consume ca 15–20 vol% more fuel than corresponding gasoline-fueled cars. The Ford Motor Company reports that Escorts equipped with alcohol engines operated at ca 64 vol% (methanol fuel) and 85 vol% (ethanol fuel) of a gasoline-fueled Escort. Thus, on a distance basis, 1.56 L (0.41 gal) methanol and 1.18 L (0.31 gal) ethanol are equivalent to 1.0 L (0.26 gal) gasoline. At the same thermal efficiencies, the equivalency should be 2.04 L (0.54 gal) methanol and 1.53 L (0.40 gal) ethanol per liter of gasoline. Therefore, thermal efficiencies and the distance traveled for each energy unit expended in the alcohol-fueled cars are considerably higher than in the gasoline-fueled cars.

Toxicology

Ingestion of small amounts of methanol can cause blindness and death; ethanol can be as toxic to animals as methanol. These materials should be handled with caution. At present, no data are available on the toxic effects of gasohol-type mixtures, but their widespread use and lack of problems indicate no serious toxicological conditions.

Uses

Alcohols are being developed as intermediates in the production of other fuels and for nonvehicular fuel applications. In the Mobil Oil Company process, methanol is converted to gasoline over zeolite catalysts, which are crystalline aluminosilicates. Their unusual shape-selective properties seem to be responsible for the high yields of liquid hydrocarbons. The catalytic conversion of ethanol to hydrocarbon fuels is potentially competitive with distillation routes because of the energy consumed in the latter by the removal of water. Cost estimates of $0.33/L ($1.26/gal) gasoline have been made.

Alcohols can also be utilized by first reforming them in catalytic reactors with exhaust heat from the engine, and then operating the engine on the hydrogen and carbon monoxide in the reformer gas. The strongly endothermic dissociation effectively increases the energy content of the alcohol in the fuel tank through energy capture and reuse.

Methanol is being evaluated as a substitute for fossil fuels in stationary applications, eg, boilers. Turbines fueled by methanol appear to offer long-range benefits to electric utilities.

Although the emphasis on alcohol fuels has been devoted almost entirely to methanol and ethanol, *tert*-butyl alcohol is used as a cosolvent for methanol in blends with gasoline, and 1-butanol and acetone in gasoline blends have been reported to perform well in spark-ignition engines.

DONALD L. KLASS
Institute of Gas Technology

D.L. Klass, *Energy Topics*, Institute of Gas Technology, Chicago, Ill., April 14, 1980, 8 pp.

Alcohols—A Technical Assessment of Their Applications as Fuels, Publication No. 4261, American Petroleum Institute, New York, July 1976.

J.L. Keller, G.M. Nakaguchi, and J.C. Ware, *Methanol Fuel Modification for Highway Vehicle Use*, HCP W3683-18, final report for U.S. Department of Energy, Washington, D.C., July, 1978.

ALCOHOLS, HIGHER ALIPHATIC

SURVEY AND NATURAL ALCOHOLS MANUFACTURE

The monohydric aliphatic alcohols of six or more carbon atoms are generally referred to as higher alcohols, historically derived from the natural fats, oils, and waxes, and are known as fatty alcohols. The higher alcohols are characterized by their use, rather than their derivation from natural or synthetic stocks, because this determines the manufacturing process, feedstock, and chemical and physical properties required.

The higher alcohols can be separated into a plasticizer range, alcohols of generally six to eleven carbon atoms (C_6–C_{11}), and a detergent range, ie, alcohols of twelve or more carbon atoms (C_{12}–C_{18}), although there are some alcohols that do not fit comfortably into either category.

Commercial descriptions of plasticizer-range alcohols are rather confusing, but in general, a commercially pure material is called "-anol" (as in 2-ethylhexanol), and mixtures are called "-yl alcohols" or "iso...yl alcohol" (as in hexyl alcohol and isoctyl alcohol) (see Plasticizers).

Natural or synthetic detergent-range alcohols are usually described as middle cut or heavy cut. Only a small percentage of these alcohols is sold as pure materials (see Surfactants and detersive systems).

Physical Properties

Table 1 provides physical property data for selected pure alcohols. The homologous series of primary normal alcohols exhibits definite trends in physical properties. For every added CH_2 unit, the normal boiling point increases by about 20°C and the specific gravity (20°C) increases by about 0.003 units. The melting point increases by about 10°C in the lower end of the range and about 4°C in the upper end for each additional CH_2 unit. The water solubility decreases with increasing molecular weight and the oil solubility increases. In general, the higher alcohols are soluble in lower alcohols such as ethanol and methanol and in diethyl ether and petroleum ether. The solubility of water in 1-hexanol and 1-octanol is appreciable, but it drops off rapidly with increased molecular weight; however, enough solubility remains even in 1-octadecanol to make it slightly hygroscopic. Below C_{12} the normal alcohols are colorless, oily liquids with light, rather fruity odors. Pure 1-dodecanol solidifies slightly below room temperature to soft, crystalline platelets. The physical form of higher molecular weight alcohols progresses from these soft platelets to crystalline waxes. Although 1-dodecanol has a slight odor, the higher homologues are essentially odorless. The secondary and branched primary alcohols are oily liquids at room temperature and have light, fruity odors. They are soluble in alcohol solvents and diethyl ether, and they generally show less affinity for water as their molecular weights increase. The members of this group do not have well-defined freezing points, but rather they set to a glass at a very low temperature. It is difficult to determine physical properties because of the difficulty of obtaining pure samples.

Chemical Properties

Higher alcohols undergo the same chemical reactions as other primary or secondary monohydric alcohols. In general, the reactivity decreases with increasing molecular weight and with increasing chain branching. The lower reactivity and reduced solubility of higher alcohols in water and other solvents compared to shorter-chain alcohols generally require modification of reaction conditions. The following reactions are typical of higher alcohols.

Esterification

$$ROH + R'COOH \rightarrow R'COOR + HOH$$

Sulfation

$$ROH + ClSO_3H \rightarrow ROSO_3H + HCl$$
$$\text{alkyl sulfuric acid}$$
$$ROH + SO_3 \rightarrow ROSO_3H$$
$$ROSO_3H + NaOH \rightarrow ROSO_3Na + HOH$$
$$\text{sodium alkyl sulfate}$$

Etherification

$$ROH + nH_2C\overset{\diagdown O \diagup}{-}CH_2 \rightarrow R(OCH_2CH_2)_nOH$$
$$\text{polyethoxylated alcohol}$$

$$ROH + H_2C\overset{\diagdown O \diagup}{-}CHCH_2Cl \rightarrow ROCH_2\underset{OH}{CH}CH_2Cl$$
$$\text{alkyl chlorohydrin ether}$$

Halogenation

$$3 ROH + PCl_3 \rightarrow 3 RCl + P(OH)_3$$

Dehydration

$$RCH_2CH_2OH \rightarrow RCH{=}CH_2 + HOH$$

Typical properties of several commercial plasticizer-range alcohols are presented in Table 2. Because in most cases these are mixtures of several carbon chain lengths or isomers, the properties of a particular material can vary somewhat among manufacturers.

Health and Safety

Higher alcohols are generally nontoxic and cause no primary skin irritation. Sustained breathing of alcohol vapor mist should be avoided.

Table 1. Physical Properties of Pure Alcohols

IUPAC name	Molecular formula	Other common names	Specific gravity, 20/20°C[a]	Refractive index, 20°C[a]	Bp, °C, 101.3 kPa[b]	Mp, °C	Viscosity, mPa·s[a,c]	Solubility, % by wt in water	Solubility, % by wt of water	Solubility, % by wt in other solvents
Primary normal aliphatic										
1-hexanol	$C_6H_{14}O$	*n*-hexyl alcohol	0.8212	1.4181	157	−44	5.9	0.59[20]	7.2	petroleum ether, ethanol
1-heptanol	$C_7H_{16}O$	*n*-heptyl alcohol	0.8238	1.4242	176	−35	7.4	0.10[18]		
1-octanol	$C_8H_{18}O$	*n*-octyl alcohol	0.8273	1.4296	195	−15.5	8.4	0.06[25]	4.5	ethanol, petroleum ether
1-nonanol	$C_9H_{20}O$	*n*-nonyl alcohol	0.8295	1.4338	213	−5	11.7			
1-decanol	$C_{10}H_{22}O$	*n*-decyl alcohol	0.8312	1.4371	230	7	13.8		2.8	glacial acetic acid, benzene, ethanol, petroleum ether
1-undecanol	$C_{11}H_{21}O$	*n*-undecyl alcohol	0.8339	1.4402	243	16	17.2	<0.02		
1-dodecanol	$C_{12}H_{26}O$	*n*-dodecyl alcohol lauryl alcohol	0.8306[25]	1.4428	138[1.33]	24	18.8	i	1.3	petroleum ether, ethanol
1-tridecanol	$C_{13}H_{28}O$	*n*-tridecyl alcohol	0.8238[31]		155[2.0]	30.5				
1-tetradecanol	$C_{14}H_{30}O$	*n*-tetradecyl alcohol myristyl alcohol	0.8165[50]	1.4358[50]	158[1.33]	38		<0.02	nil	petroleum ether, ethanol
1-pentadecanol	$C_{15}H_{32}O$	*n*-pentadecyl alcohol		1.4408[50]		44				
1-hexadecanol	$C_{16}H_{34}O$	cetyl alcohol palmityl alcohol	0.8157[60]	1.4392[60]	177[1.33]	49	53[75]	0.06[20]	nil	ethanol, methanol, diethyl ether, benzene
1-heptadecanol	$C_{17}H_{36}O$	margaryl alcohol	0.8167[60]	1.4392[60]		54				
1-octadecanol	$C_{18}H_{38}O$	stearyl alcohol *n*-octadecyl alcohol	0.8137[60]	1.4388[60]	203[1.33]	58		i	nil	
1-nonadecanol	$C_{19}H_{40}O$	*n*-nonadecyl alcohol				62				
1-eicosanol	$C_{20}H_{42}O$	eicosyl alcohol arachidyl alcohol			251[1.33]	66		i	nil	benzene, ethanol, petroleum ether
1-hexacosanol	$C_{26}H_{54}O$	ceryl alcohol			305[2.67]	79.5		i		ethanol, ether
1-hentriacontanol	$C_{31}H_{64}O$	melissyl alcohol myricyl alcohol	0.7784[95]			87		nil		
9-hexadecen-1-ol	$C_{16}H_{32}O$	palmitoleyl alcohol			205–210[2.0]					
9-octadecen-1-ol	$C_{18}H_{36}O$	oleyl alcohol	0.8504[58]	1.4473[60]						ethanol, diethyl ether
10-eicosen-1-ol	$C_{20}H_{40}O$	eicosenyl alcohol								
Primary branched aliphatic										
2-methyl-1-pentanol	$C_6H_{14}O$	2-methylpentyl alcohol	0.8254	1.4190	148		6.6	0.31	5.4	
2-ethyl-1-butanol	$C_6H_{14}O$	2-ethylbutyl alcohol	0.8348	1.4224	146.5	−114				
2-ethyl-1-hexanol	$C_8H_{18}O$	2-ethylhexyl alcohol	0.8340	1.4316	184	−70	9.8	0.07	2.6	ethanol, diethyl ether
3,5-dimethyl-1-hexanol	$C_8H_{18}O$		0.8297	1.4250	182.5					
2,2,4-trimethyl-1-pentanol	$C_8H_{18}O$		0.839	1.4300	168	−70				ethanol
Secondary aliphatic										
4-methyl-2-pentanol	$C_6H_{14}O$	methyl amyl alcohol methylisobutyl carbinol	0.8083	1.4112	132	−90	5.2	1.7	5.8	ethanol, diethyl ether
2-octanol	$C_8H_{18}O$	capryl alcohol	0.835[15/4]	1.4256	178–179	−38	8.2	0.096[25]		ethanol, petroleum ether
2,6-dimethyl-4-heptanol	$C_9H_{20}O$	diisobutyl carbinol	0.8121	1.4231	178	−65	14.3	0.06	0.99	ethanol, diethyl ether
2,6,8-trimethyl-4-nonanol	$C_{12}H_{26}O$		0.8193	1.4345	225	−60	21	<0.02	0.60	

[a]20°C if not otherwise noted. [b]Unless otherwise noted (superscript pressures also in kPa). To convert kPa to mm Hg, multiply by 7.50. [c]mPa · s = cP.

Table 2. Typical Properties of Commercial Plasticizer-Range Alcohols

Descriptive name	Alcohol, wt%	Hydroxyl value	Carbonyl number	Specific gravity[a]	Boiling range, °C	Moisture, wt%	Flash point, Tag open cup, °C
hexyl alcohol	98.9	543	0.05	0.820	151–157	0.03	62
2-ethylhexanol	99.5	429	0.06	0.834	182–185	0.03	82
isooctyl alcohol	99.3	428	0.07	0.832	184–191	0.08	82
isodecyl alcohol	99.1	351	0.05	0.838	216–221	0.06	102
hexyl-decyl alcohols (Epal[b] 610)		420	< 0.15	0.825[25/25]	175–240	0.08	88[c]
octanol	99.2	430		0.824[25/25]	190–193	0.03	
decanol	99.0	351		0.829[25/25]	223–229	0.04	
4-methyl-2-pentanol	98.5	547		0.808	131–133	0.06	41

[a]20/20°C if not otherwise noted.
[b]Registered trademark for Ethyl Corporation alcohols.
[c]Cleveland open cup.

Uses

Detergent-range alcohols are used for their surface-active properties or as a means of introducing a long-chain moiety into a chemical compound. Most are used as derivatives, eg, poly(oxyalkylene) ethers, or esters of acids, eg, sulfuric acid. Surfactants derived from detergent range alcohols are used for emulsification, dispersion, wetting, or detergent properties (see Surfactants and detersive systems; Emulsions); stabilizers for fire-extinguishing foams (see Fire-extinguishing agents); bar soaps; and antifoam agents (see Defoamers). Hexadecanol and octadecanol are used extensively in the cosmetics and pharmaceuticals industries (see Cosmetics); lubricants (see Lubrication); and miscellaneous uses.

Plasticizer-range alcohols are used primarily in plasticizers for poly(vinyl chloride) (PVC) and other plastics; in the plastics industry as additives for heat stabilizers (qv) and uv absorbers (qv); flame retardants (qv); antioxidants (qv); lubricants (see Hydraulic fluids); agriculture, surfactants (see Flotation); and breaking foams (see Defoamers).

Fats, oils, and waxes from a number of animal, vegetable, and marine sources are the feedstocks for the manufacture of natural higher alcohols. Most of these materials consist of fatty triglycerides, ie, glycerol esterified with three moles of a fatty acid, and the alcohol is manufactured by reduction of the fatty acid functional group. A small amount of natural alcohol is obtained commercially by saponification of natural wax esters of the higher alcohols.

The carbon chain lengths of the fatty acids available from natural fats and oils range from six through twenty-four and higher, although a given material has a narrower range. Each triglyceride material has a random distribution of fatty acid chain lengths and unsaturation, but the proportion of the various fatty acids is fairly uniform for fats and oils from a common source. Any fatty triglyceride or fatty acid may be used as a raw material for the manufacture of fatty alcohols, but the commonly used fats and oils are coconut oil, palm kernel oil, lard, tallow, and palm oil, and to a lesser extent sperm oil, soybean oil, rapeseed oil, castor oil, and menhaden and other fish oils (see Fats and fatty oils; Vegetable oils). Coconut and palm kernel oil are the sources of dodecanol and tetradecanol; tallow and palm oil are the sources of hexadecanol and octadecanol.

Natural fatty alcohols are produced by hydrogenolysis of esters or acids in the presence of a heterogeneous catalyst, usually a complex mixture of copper(II) oxide and copper(II) chromite in the form of a finely divided powder. Reaction proceeds at 20.7–31 MPa (3000–4500 psi) and about 300°C in conversions of 90–95%. A higher conversion can be achieved with more rigorous conditions, but it is accompanied by a significant amount of hydrocarbon production; at 330°C the hydrocarbon production increases noticeably.

$$RCOOR' + 2 H_2 \xrightarrow[\text{catalyst}]{\text{high pressure}} RCH_2OH + R'OH$$

By-product fatty esters are formed by ester interchange of the product alcohol (RCH_2OH) with the starting material in the hydrogenolysis reactors; these esters are separated from the product alcohol. A nonselective copper chromite catalyst always gives completely saturated alcohol, even if double bonds are present. To produce an unsaturated alcohol, the catalyst and reaction conditions must be tailored for this purpose (see below). Fatty acids can be used for the production of fatty alcohols, but this process requires the use of corrosion-resistant process equipment whereas the ester process uses carbon-steel equipment. Most processes use a low molecular weight ester of the fatty acid, typically the methyl ester. Other fatty acid esters could be used, including the triglyceride ester, but the methyl ester process is preferred because it allows easy separation of the low molecular weight alcohol ($R'OH$) formed in the reaction. Use of the triglyceride would also lead to decomposition of by-product glycerol at the high temperatures of reaction.

The fats and oils used for fatty alcohol production first undergo alkali or steam refining to remove free fatty acid that would interfere with subsequent alcoholysis. The alcoholysis (ester interchange) is performed at atmospheric pressure in carbon-steel equipment. Sodium methylate is used as a catalyst (see Alkoxides, metal). It can be prepared in the same reactor used for the alcoholysis with methanol by adding sodium before the triglyceride is charged.

Production of Unsaturated Alcohols

Unsaturated higher alcohols may be produced by saponification, by sodium reduction, or by hydrogenolysis of unsaturated acids or esters. Hydrogenolysis of unsaturated fatty acids or esters to alcohol without loss of the double bond requires the use of a specially designed catalyst with a reaction temperature somewhat lower than that for production of saturated alcohols. Cadmium-modified catalysts have been widely claimed to be effective. A zinc–aluminum catalyst reportedly avoids isomerization of the cis double bond during the hydrogenolysis of the methyl esters of octadecanoic acid, soybean fatty acid, and linseed fatty acid. Ashland Chemical Company (U.S.), Henkel GmbH (FRG), and the New Japan Chemical Company (Japan) practice the known commercial hydrogenolysis processes.

The sodium reduction process may be considered a variant of hydrogenolysis in which hydrogen is obtained from the reaction of metallic sodium with a reducing alcohol.

$$RCOOR' + 4 Na + 2 R''OH \rightarrow RCH_2ONa + R'ONa + 2 R''ONa$$

$$RCH_2ONa + R'ONa + 2 R''ONa + 4 H_2O \rightarrow$$

$$RCH_2OH + R'OH + 2 R''OH + 4 NaOH$$

Because of high manufacturing cost owing to the usage of sodium (4 mol/mol alcohol), the process is no longer used for large-scale production; today it probably is used only for specialty alcohols.

Minor quantities of fatty alcohols occur in many oilseeds and in marine sources, but significant quantities as wax esters are found in sperm oil.

RICHARD A. PETERS
The Procter & Gamble Company

J.A. Monick, *Alcohols, Their Chemistry, Properties and Manufacture*, Reinhold Book Corp., New York, 1968, pp. 519–579.

SYNTHETIC PROCESSES

Ziegler Process

The Ziegler process, invented by Dr. Karl Ziegler of the Max Planck Institute in 1953, produces predominantly linear, primary alcohols with an even number of carbon atoms. Alcohol linearity is a function of the olefin feedstock and process used.

Four basic steps are used to synthesize alcohols from aluminum alkyls (see Organometallics):

1. Triethylaluminum preparation by a one- or two-stage process.

$$Al + 3/2 H_2 + 2 Al(C_2H_5)_3 \rightarrow 3 Al(C_2H_5)_2H$$

$$3 Al(C_2H_5)_2H + 3 C_2H_4 \rightarrow 3 Al(C_2H_5)_3$$

2. Chain growth as triethylaluminum reacts with ethylene in a controlled, highly exothermic polymerization to produce a spectrum of higher molecular weight alkyls of even carbon number (see Olefin polymers).

$$Al(C_2H_5)_3 + 3x C_2H_4 \rightarrow AlR_3$$

3. Oxidation with dry air above atmospheric pressure in a fast, highly exothermic reaction using a solvent to help avoid localized overheating and to decrease the viscosity of the solution. By-products formed include paraffins, aldehydes, olefins, esters, and alcohols which, along with the solvent, must be removed before hydrolysis to prevent contamination (this can be done by high temperature vacuum flashing or stripping).

$$AlR_3 + 3/2 O_2 \rightarrow Al(OR)_3$$

4. Hydrolysis with sulfuric acid or other acids, bases, and water, usually around 100°C. Neutral hydrolysis gives marketable aluminum oxide. Sulfuric acid gives marketable alum. The crude alcohols are washed with a dilute caustic solution and water, then fractionated.

$$Al(OR)_3 + 3/2 H_2SO_4 \rightarrow 3 ROH + 1/2 Al_2(SO_4)_3$$

The Ziegler process typically produces a distribution of about 50% C_6–C_{10} and 50% C_{12}–C_{18}. In a modified version of the Ziegler process, Ethyl Corporation separates the mixture of aluminum alkyls from the chain-growth step into a shorter-chain fraction (predominantly C_2–C_{10}) and a longer-chain fraction (predominantly C_{12}–C_{18}). The former is recycled for additional chain growth, and the latter is sent to the oxidation step. The resulting product distribution is about 15–25% C_6–C_{10} and 75–85% C_{12}–C_{18}.

Environmental problems in Ziegler-chemistry alcohol processes are not severe. A small quantity of aluminum alkyl wastes is usually produced and represents the most significant waste-disposal problem.

Oxo Process

The oxo process, or hydroformylation reaction, discovered by Roelen in Germany in 1938, was first used on a commercial scale by the Enjay Chemical Company (now Exxon) in 1948 (see Oxo process). In a typical oxo process, primary alcohols are produced from monoolefins in two steps. In the first, the olefin, hydrogen, and carbon monoxide react in the presence of a cobalt catalyst to form an aldehyde, which is hydrogenated to the alcohol.

$$RCH=CH_2 + CO + H_2 \xrightarrow{\text{catalyst}} RCH_2CH_2CHO$$

$$RCH_2CH_2CHO + H_2 \xrightarrow{\text{catalyst}} RCH_2CH_2CH_2OH$$

The oxo catalyst may be modified to function as a hydrogenation catalyst as well:

$$RCH=CH_2 + CO + 2 H_2 \xrightarrow{\text{catalyst}} RCH_2CH_2CH_2OH$$

These reactions are applicable to most monoolefins and are used to obtain a large number of products.

The technology of the process involves either a cobalt catalyst, two-step high pressure process, or a modified cobalt catalyst, one-step low pressure process, as well as the aldol process in which 2-ethylhexanol is manufactured from n-butyraldehyde by aldol condensation in an alkaline medium followed by catalytic hydrogenation.

Feedstocks currently used include propylene (qv) to produce 2-ethylhexanol. Purified commercial paraffin waxes (mp 50–60°C) cracked with steam produce C_2–C_{20} linear, primarily alpha olefins (Kuhlmann, BASF, ICI, Shell, Mitsubishi Chemical, and Nissan use this process). Refinery propylene and butenes polymerized with a phosphoric acid catalyst yield a mixture of branched olefins up to C_{15} (used primarily in producing plasticizer alcohols) (Exxon, Kuhlmann, BASF, Ruhrchemie, ICI, Nissan, Getty Oil, and U.S. Steel Chemicals use this olefin source). With linear olefin feedstock, the Shell oxo process produces alcohols with 15–20% branching, and other oxo processes produce alcohols with ca 50% branching. Dimerization of isobutene or the codimerization of isobutene and n-butene produce olefins for plasticizer alcohols. Propylene and butylene feedstocks produce hexenes, heptenes, and octenes via the Dimersol process (French Petroleum Institute). Refinery streams are used to isolate detergent-range n-paraffins by molecular-sieve processes (see Adsorptive separation) (these are converted to olefins either by the Shell process, involving chlorination–dehydrochlorination to produce a linear, random, primarily internal olefin, or by the process developed by Universal Oil Products in which an n-paraffin of the desired chain length is dehydrogenated, using the Pacol process in a catalytic fixed-bed reactor in the presence of excess hydrogen at low pressure and moderately high temperature, to produce a linear, random, primarily internal olefin). Ethylene is used to produce higher linear alpha olefins via chain growth on triethylaluminum (the products are used as oxo feedstock in both the plasticizer and the detergent range) (Monsanto, ICI, and Mitsubishi Chemical). Ethylene is used as a feedstock to produce C_4–C_{30} and higher alpha olefins via catalytic oligomerization over a nickel chelate catalyst in a solvent in conjunction with disproportionation (this process, used by Shell Chemical, produces an olefin and subsequently an alcohol which has essentially the same characteristics as that derived from n-paraffins).

Although cobalt compounds are used in almost all present commercial oxo operations, rhodium, ruthenium, and iridium are reported to be effective as oxo catalysts. Plants using rhodium catalyst are operated by Union Carbide Corporation.

M.F. GAUTREAUX
W.T. DAVIS
E.D. TRAVIS
Ethyl Corporation

P.A. Lobo, D.C. Coldiron, L.N. Vernon, and A.T. Ashton, *Chem. Eng. Prog.* **58**(5), 85 (May 1962).

G.U. Ferguson, *Chem. Ind.* (11), 451 (1965).

E.J. Wickson and H.P. Dengler, *Hydrocarbon Process.* **51**(11), 69 (1972).

ALCOHOLS, POLYHYDRIC

SUGAR ALCOHOLS

The sugar alcohols bear a close relationship to the simple sugars from which they are formed by reduction and from which their names are often derived (see Carbohydrates). Most of the sugar alcohols have the general formula, $HOCH_2(CHOH)_nCH_2OH$, $n = 2$–5, and are classified according to the number of hydroxyl groups as tetritols, pentitols, hexitols, and heptitols. Polyols from aldoses are sometimes called alditols. Each class contains stereoisomers. Counting meso and optically active forms, there are three tetritols, four pentitols, ten hexitols, and sixteen heptitols, all of which are known either from natural occurrence or through synthesis. Of the straight-chain polyols, sorbitol and mannitol have the greatest industrial significance.

Physical Properties

In general, these polyols are water-soluble crystalline compounds with small optical rotations in water and have a slightly sweet to very sweet taste. Selected physical properties of many of the sugar alcohols are listed in Table 1.

Polymorphism has been observed for both D-mannitol and D sorbitol with three different forms for each hexitol. Bond lengths of crystalline pentitols and hexitols are all similar with an average distance for C—C

Table 1. Physical Properties of the Sugar Alcohols

Sugar alcohol	Melting point, °C	Optical activity in H_2O, $[\alpha]_D^{20-25}$	Solubility g/100 g H_2O^a	Heat of solution in water, kJ/mol[b]	Heat of combustion, constant vol, kJ/mol[b]
Tetritols					
erythritol	120	meso	61.5	23.3	−2091.6
threitol					
D-threitol	88.5–90	+4.3	very sol		
L-threitol	88.5–90	−4.3			
D,L-threitol	69–70				
Pentitols					
ribitol	102	meso	very sol		
arabinitol					
D-arabinitol	103	+131[c]	very sol		
L-arabinitol	102–103	−130[c]			−2559.4
D,L-arabinitol	105				
xylitol	61–61.5 (metastable) 93–94.5 (stable)	meso	179		−2584.5
Hexitols					
allitol	155	meso	very sol		
dulcitol (galactitol)	189	meso	3.2 (15°C)	29.7	−3013.7
glucitol					
sorbitol (D-glucitol)	93 (metastable) 97.7 (stable)	−1.985	235	20.2	−3025.5
L-glucitol	89–91	+1.7			
D,L-glucitol	135–137				
D-mannitol	166	−0.4	22	22.0	−3017.1
L-mannitol	162–163				
D,L-mannitol	168				
altritol					
D-altritol	88–89	+3.2	very sol		
L-altritol	87–88	−2.9			
D,L-altritol	95–96				
iditol					
D-iditol	73.5–75.0	+3.5			
L-iditol	75.7–76.7	−3.5	449		
Disaccharide alcohols					
maltitol (4-O-α-D-glucopyranosyl-D-glucitol)		+90			
lactitol (4-O-β-D-galactopyranosyl-D-glucitol)	146	+14			

^{a}At 25°C unless otherwise indicated.
^{b}To convert J to cal, divide by 4.184.
^{c}In aqueous molybdic acid.

of 0.152 nm and C—O of 0.143 nm. Conformations in the crystal structures of sugar alcohols are rationalized by Jeffrey's rule that "the carbon chain adopts the extended, planar zigzag form when the configurations at alternate carbon centers are different, and is bent and nonplanar when they are the same." Conformations are adopted which avoid parallel C—O bonds on alternate carbon atoms. Very little, if any, intramolecular hydrogen bonding exists in the crystalline sugar alcohols, but extensive intermolecular hydrogen bonds are found, usually with involvement of each hydroxyl group in two hydrogen bonds, one as a donor and one as an acceptor.

The small optical rotations of the alditols arise from the low energy barrier for rotation about C—C bonds permitting easy interconversion and the existence of mixtures of rotational isomers (rotamers) in solution.

In aqueous solution, sugar alcohols influence the structure of water presumably by hydration of the polyol hydroxyl groups through hydrogen bonding, as indicated by effects on solution compressibility, vapor pressure, enthalpies of solution, dielectric constant, and Ag–AgCl electrode potential. Compressibility measurements indicate that mannitol in aqueous solution is hydrated with two molecules of water at 25°C. Osmotic coefficients are related to the number of hydrophilic groups per molecule, those of sorbitol being larger than those of erythritol.

The weakly acidic character of acyclic polyhydric alcohols increases with the number of hydroxyl groups as indicated by the pK_a values in aqueous solution at 18°C.

Alcohol	pK_a
glycerol	14.16
erythritol	13.90
xylitol	13.73
sorbitol	13.57
mannitol	13.50
dulcitol	13.46

Chemical Properties

The sugar alcohols undergo the expected typical reaction for polyhydric compounds such as: anhydrization to give cyclic ethers (monoanhydro internal ethers of the hexitols are called hexitans, and the dianhydro derivatives are called hexides); esterification to produce both partial and complete esters; etherification, for example, by alkyl halides with sugar polyols in the presence of aqueous alkaline reagents, and reaction of olefin oxides (epoxides) to produce poly(oxoalkylene) ether derivates is of great commercial importance (see Uses); acetal formation; oxidation; reduction; formation of metal complexes (see also Alkoxides, metal; Chelating agents); and isomerization.

Manufacture of Sorbitol and Mannitol

Sorbitol is manufactured by catalytic hydrogenation of glucose using either batch or continuous hydrogenation procedures. Corn sugar, hydrolyzed starch, and other sources of glucose may be used. Both supported nickel and Raney nickel are used as catalysts.

When invert sugar is used as a starting material, sorbitol and mannitol are produced simultaneously. Mannitol crystallizes from solution after hydrogenation owing to its lower solubility in water. Mannitol may also be extracted from seaweed.

Health and Safety

Mannitol and sorbitol, with a relatively harmless toxicity classification, are accepted for use in foods.

Uses

The hexitols and their derivatives are used in many diverse fields including foods, pharmaceuticals, cosmetics, textiles, and polymers. In foods, sweetness is an important characteristic of sugar alcohols. Sorbitol is about 60% as sweet as sucrose, and mannitol, D-arabinitol, ribitol, maltitol, and lactitol are generally comparable to sorbitol (see Sweeteners). Sorbitol is used to add body and texture as well as sweetness (see Food additives) and has the property of reducing the undesirable aftertaste of saccharin. Sorbitol is extensively used in the food industry. In

the pharmaceutical industry, mannitol is used as a base for vitamins and other pharmaceuticals (see Pharmaceuticals); sorbitol is used as a bodying agent, as an ointment vehicle, in skin treatment, in emulsions, in solubilization of oil-soluble vitamins A and D, and in the manufacture of vitamin C. Cosmetic applications include sorbitol as a humectant (see Cosmetics) and plasticizer. In textiles, sorbitol is used in bleaching or scouring solutions (see Textiles), as a plasticizer, in emulsifiers, and in dry-cleaning detergents (see Dry cleaning and laundering). In polymer applications, sorbitol is used as a stabilizer against heat and light (see Heat stabilizers; Uv absorbers; Rubber compounding; Olefin polymers); as a polyol component of alkyd resins and rosin esters for protective coatings (see Alkyd resins); and in coatings and elastomers (see Urethane polymers). Sorbitan fatty esters are used as plasticizers, lubricants, and antifog agents for vinyl resins and other polymers.

FREDERIC R. BENSON
ICI Americas
formerly
ICI United States, Inc.

G.A. Jeffrey, *Carbonhydr. Res.* **28**, 233 (1973).

M.A. Phillips, *Br. Chem. Eng.* **8**, 767 (1963).

Chemical Economics Handbook 583.7800 A, Stanford Research Institute, Menlo Park, Calif., Apr. 1979.

A.E. Waltking and co-workers, *J. Am. Oil Chem. Soc.* **50**, 353 (1973).

OTHER POLYHYDRIC ALCOHOLS

Pentaerythritol

Pentaerythritol (2,2-bis(hydroxymethyl)-1,2-propanediol or tetramethylolmethane), a tetrahydric neopentyl alcohol, is an odorless, white crystalline compound. It is nonhygroscopic, practically nonvolatile, and stable in air. Pure pentaerythritol melts at 261–262°C, sublimes slowly on heating, and boils at 276°C at 4 kPa (30 mm Hg). Its density is 1.396 g/cm³. It is moderately soluble in cold water and quite soluble in hot water (solubility 7.23 g/100 g water at 25°C, 772. g/100 g water at 97°C). It is only slightly soluble in alcohols and other organic liquids.

Pentaerythritol usually crystallizes in tetragonal form. At temperatures above 180°C it undergoes a transition from its usual form to the cubic lattice structure.

The heat of solution is approximately 21 kJ/mol (5 kcal/mol); the heat of combustion is 2765 kJ/mol (661 kcal/mol) at constant volume and 2769 kJ/mol (662 kcal/mol) at constant pressure; the heat of formation is 948 kJ/mol (226.6 kcal/mol); the heat of vaporization is approximately 92 kJ/mol (22 kcal/mol); the heat of sublimation is 131.5 kJ/mol (31.4 kcal/mol). The specific heat is 254.6 J/mol (60.85 cal/mol) at 100°C and 432.5 J/mol (103.4 cal/mol) at 190°C.

Reactions. Pentaerythritol undergoes the usual reactions of polyhydric alcohols. Nitrates are formed by reaction with HNO_3 without the addition of a catalyst, such as H_2SO_4 (see Explosives). It is selectively oxidized to trimethylolacetic acid (2,2-bis(hydroxymethyl)hydracrylic acid) in aqueous solution with molecular oxygen in the presence of a platinum catalyst or with chlorine. It forms complexes with many metals (eg, reacts with $Ni(NO_3)_2$ and boric acid to give a nickel–borate complex).

Manufacture. Pentaerythritol is manufactured by the reaction of acetaldehyde with formaldehyde in alkaline medium such as sodium or calcium hydroxide. First, the α-hydrogen atoms of the acetaldehyde condense with the formaldehyde in three sequential aldol reactions to form pentaerythrose. These reactions are fast, reversible, and base-catalyzed. The pentaerythrose is reduced to pentaerythritol in a crossed Cannizzaro reaction with formaldehyde. This reaction is irreversible and consumes a stoichiometric amount of the base, driving the overall reaction to completion and allowing the commercial manufacture of pentaerythritol in high yield.

Various competing side reactions result in the formation of by-products (especially at $\geq 80°C$) such as polypentaerythritols, methyl ethers, and formals of pentaerythritol gums and formose sugars.

Uses. Pentaerythritol is used in fast-drying surface-coating compositions, particularly in manufacture of alkyd resins (qv); *in situ* varnishes; caulking and putty compounds; and fire-retardant coatings (see Flame retardants). The esters are used as base stock in high performance lubricants for jet, turbine, and automotive engines; additives to mineral oil lubricants (see Lubrication and lubricants); ashless dispersants; and plasticizers for poly(vinyl chloride) (see Vinyl polymers). Reaction with epoxides yields compounds useful as surfactants; as reactants in radiation-cured coatings; and as thermosetting resins. Nitrates are used as a filling in detonating-fuse and priming compositions and in the treatment of angina pectoris (see Cardiovascular agents).

Health and safety. Pentaerythritol is nontoxic. When finely powdered, it can form explosive dust clouds (min explosive dust concentration in air is 30 g/m³; min temperature for ignition is 450°C).

Dipentaerythritol

Dipentaerythritol (2,2-[oxybis(methylene)]-bis[2-(hydroxymethyl)-1,3-propanediol]) is a white, nonhygroscopic crystalline compound. Heat of combustion is 7737 kJ/mol (1849 kcal/mol). It is less soluble in most solvents than pentaerythritol. It is considered nontoxic. Dust explosions occur at a concentration in air of 30 g/m³ at a minimum ignition temperature of 400°C. It is obtained as a by-product in the manufacture of pentaerythritol and it is separated by fractional crystallization. (See physical properties in Table 1.)

Tripentaerythritol

Tripentaerythritol (2,2-bis[[3-hydroxy-2,2-bis(hydroxymethyl)propoxy)methyl]-1,3 propanediol) is a white solid with heat of combustion 12,850 kJ/mol (3071 kcal/mol). It is considered nontoxic.

Trimethylolethane

Trimethylolethane (2-hydroxymethyl-2-methyl-1,3-propanediol), a white, odorless, crystalline trihydric alcohol, is produced by the reaction of formaldehyde with propionaldehyde. It is used in the manufacture of specialty alkyd resins; polyurethane foams and coatings; and in the preparation of epoxy resins and explosives.

Trimethylolpropane

Trimethylolpropane (2-ethyl-2-hydroxymethyl-1,2-propanediol) is a white solid, trihydric alcohol. Essentially nontoxic, it is formed by the condensation of *n*-butyraldehyde with formaldehyde in the presence of an alkaline condensing agent. It is used in the production of polyester and polyether urethane foams; alkyd coatings; synthetic lubricants; and

radiation-curing of urethane polyesters, epoxy resins, and polyether resins containing acrylates.

Anhydroenneaheptitol

Anhydroenneaheptitol (3,3,5,5-(4*H*,6*H*)-tetramethanol-4-hydroxy-2*H*-pyran or 3,3,5,5-tetrakis(hydroxymethyl)-4-hydroxytetrahydropyran) is a hygroscopic crystalline compound. Considered nontoxic, it is manufactured by exhaustive hydroxymethylation of acetone in a base-catalyzed aldol reaction with formaldehyde.

J. WEBER
Celanese Canada Ltd.

J. DALEY
Celanese Chemical Company

E. Berlow, R.H. Barth, and E.J. Snow, *The Pentaerythritols*, Reinhold Publishing Corp., New York, 1958.

R.R. Suchanec, *Anal. Chem.* **37**, 1361 (1965).

I. Mellan, *Polyhydric Alcohols*, Spartan Books, Washington, D.C., 1962.

Anhydroenneaheptitol, technical bulletin, Proctor Chemical Co., Cincinnati, Ohio.

Trimethylolpropane, product bulletin, Celanese Corp., New York.

ALCOHOLS, UNSATURATED. See Acetylene-derived chemicals.

ALDEHYDE RESINS. See Acetal resins.

ALDEHYDES

Aldehydes are carbonyl-containing organic compounds, similar to ketones, in which the carbonyl group is at a terminal carbon (see Ketones). They are useful as intermediates in synthetic organic chemistry since the carbonyl group can easily be transformed by oxidation or reduction to yield an acid or alcohol function, respectively, or condensation. Aldehyde nomenclature is derived from the base name of the related acids (see Carboxylic acids) or uses the base IUPAC acyclic parent name combined with the suffix "-al".

Physical Properties

The one- and two-carbon aliphatic aldehydes, formaldehyde and acetaldehyde, are gases at ambient conditions. The aldehydes propanal (C_3) through undecanal (C_{11}) are liquids, and the higher aldehydes are solids. The lower aldehydes have pungent, penetrating, unpleasant odors attributed to the presence of the corresponding acids formed by air oxidation. Above C_8, aldehydes have more pleasant odors, and some

Table 1. Physical Properties of Polyhydric Alcohols

	Dipenta-erythritol	Tripenta-erythritol	Tri-methylol-ethane	Tri-methylol-propane	Anhydro-ennea-heptitol	1,2,4-Butane-triol	1,2,6-Hexane-triol
melting point, °C	221–222.5	248–250	202	58.8	156	supercools	−32.8
boiling point, °C			283	295		131/20 Pa[a]	178/660 Pa[a]
density, g/cm³	1.356	1.30				1.184	1.1063
refractive index, n_D^t						1.4758[17]	1.4771[20]
solubility[b] in							
water	s	sl	s	s	s	s	s
alcohols	s	ins	s	s	s	s	s
ketones	sl	ins	sl	s	sl	sl	s
esters	sl	ins	sl	sl	ins		ins
hydrocarbons		ins	ins	ins	ins		ins

[a] To convert Pa to mm Hg, multiply by 0.0075.
[b] s = soluble; sl = slightly soluble; ins = insoluble.

higher aldehydes are used in the perfume and flavoring industry (see Perfumes). See physical properties in Table 1.

Reactions

The hydrogenation of aldehydes, using a platinum or nickel catalyst, results in the formation of the corresponding alcohol:

$$CH_3CH_2CH_2CHO \xrightarrow{H_2} CH_3CH_2CH_2CH_2OH$$

With an unsaturated aldehyde, special reduction techniques such as the Meerwein-Pondorf-Verley reduction, also used for ketones, must be employed because of the reactive nature of the olefin function. The aldehyde function is easily oxidized in air, or with oxidizing agents such as chromic acid or ammoniacal silver oxide, to form the corresponding carboxylic acid:

$$CH_3CH_2CHO \xrightarrow{[O]} CH_3CH_2COOH$$

The Tischenko reaction produces an ester. The acid- or base-catalyzed condensation of aldehydes, or the aldol reaction, forms β-hydroxy aldehyde which readily dehydrates to form an α,β-unsaturated aldehyde. The overall transformation of acetaldehyde to 2-butenal has been used commercially for the formation of butanol after hydrogenation. The Claisen reaction, carried out by combining an aromatic aldehyde and an ester with a catalyst, is useful for obtaining α,β-unsaturated esters. The Perkin reaction, carried out by combining an aromatic aldehyde with an acid aldehydride and a suitable base, produces the corresponding α,β-unsaturated acid.

Methods of Preparation

Formaldehyde is commercially produced via oxidation of methanol. Acetaldehyde (qv) is derived by the direct oxidation of ethylene. Olefin hydroformylation (see Oxo process) is now used to produce many of the other industrially important aldehydes. A mixture of carbon monoxide, hydrogen, and olefin is heated under pressure in the presence of an appropriate catalyst (Rh, Co) to form aldehyde:

$$CH_3CH{=}CH_2 + CO + H_2 \rightarrow \underset{90\%}{CH_3CH_2CH_2CHO} + \underset{10\%}{(CH_3)_2CHCHO}$$

Aldehydes are produced as well by the catalytic dehydrogenation of primary alcohols at high temperature in the presence of a copper or a copper–chromite catalyst. An olefin can be oxidized to form an unsaturated aldehyde; eg, propylene is oxidized with air at high temperatures to form acrolein (qv). Alkenes will also fragment to form aldehydes or ketones by treatment with ozone followed by decomposition of the resultant ozonide by water or catalytic hydrogenation.

Small-scale preparative procedures for aldehydes include oxidation of the corresponding alcohol with manganese dioxide or a sulfuric acid solution of potassium dichromate to form the lower aldehydes; the Rosenmund reduction, consisting of the reaction of an acid chloride with hydrogen in the presence of an appropriate catalyst to form an aldehyde and hydrogen chloride; the pyrolysis of the calcium salt of a carboxylic acid in the presence of calcium formate produces an easily distilled mixture of ketones and aldehydes; and the formation of an acetal by the reaction of an organomagnesium compound (Grignard reagent) with an alkyl orthoformate, which is then hydrolyzed to the corresponding aldehyde with dilute acid.

Health and Safety

Aldehydes are considered to be toxic by ingestion, and the low molecular weight aldehydes (with high volatility) are dangerous (see Acetaldehyde; Formaldehyde).

Uses

Aldehydes are widely used as chemical intermediates in the production of corresponding alcohols; as derivatives for the plasticizer industry (see Plasticizers); and in the manufacture of solvents (alcohols, esters, and ethers), resins, and dyes. They are used in perfume and flavoring applications (see Perfumes; see also Acetaldehyde; Benzaldehyde; Butyraldehyde, Crotonaldehyde; Formaldehyde).

P.D. Sherman
Union Carbide Corporation

C.D. Gutsche, *The Chemistry of Carbonyl Compounds*, Prentice-Hall, Inc., Englewood Cliffs, N.J., 1967.

S. Patai, *The Chemistry of the Carbonyl Group*, Interscience Publishers, a division of John Wiley & Sons, Inc., New York, 1970.

I.T. Harrison and S. Harrison, *Compendium of Organic Synthetic Methods*, Interscience Publishers, a division of John Wiley & Sons, Inc., New York, Vol. 1, 1971, pp. 132–176; Vol. 2, 1974, pp. 53–69.

J. Falbe, *Carbon Monoxide in Organic Synthesis*, Springer-Verlag, New York, Inc., New York, 1970.

ALDOL CONDENSATION.   See Aldehydes; Ketones.

Table 1.   Properties of Aldehydes

Aldehyde	Formula	Mp, °C	Bp, °C	Solubility, g/100 g water
formaldehyde (methanal)	HCHO	−92	−21	v sol
acetaldehyde (ethanal)	CH_3CHO	−121	20	
propionaldehyde (propanal)	CH_3CH_2CHO	−81	49	16
butanal (*n*-butyraldehyde)	$CH_3CH_2CH_2CHO$	−99	74.6	7
2-methylpropanal (isobutyraldehyde)	$(CH_3)_2CHCHO$	−65.9	64	11
pentanal (*n*-valeraldehyde)	$CH_3(CH_2)_3CHO$	−91	103	sl s
3-methylbutanal (isovaleraldehyde)	$(CH_3)_2CHCH_2CHO$	−51	93	sl s
hexanal (caproaldehyde)	$CH_3(CH_2)_4CHO$		131	sl s
heptanal (heptaldehyde)	$CH_3(CH_2)_5CHO$	−42	155	0.1
octanal (caprylaldehyde)	$CH_3(CH_2)_6CHO$		163.4	sl s
phenylacetaldehyde	$C_6H_5CH_2CHO$		194	sl s
benzaldehyde	C_6H_5CHO	−26	179	0.3
o-tolualdehyde	$(o\text{-}CH_3C_6H_4)CHO$		202	
m-tolualdehyde	$(m\text{-}CH_3C_6H_4)CHO$		199	
p-tolualdehyde	$(p\text{-}CH_3C_6H_4)CHO$		205	
salicylaldehyde (*o*-hydroxy benzaldehyde)	$(o\text{-}HOC_6H_4)CHO$	2	197	1.7
p-hydroxybenzaldehyde (4-formylphenol)	$(p\text{-}HOC_6H_4)CHO$	116		1.4
p-anisaldehyde (*p*-methoxy benzaldehyde)	$(p\text{-}CH_3OC_6H_4)CHO$	0	248	0.2

ALDOSES. See Carbohydrates; Sugar.

ALE. See Beer.

ALGAL CULTURES. See Foods, nonconventional.

ALGIN. See Gums.

ALIZARIN. See Anthraquinone derivatives; Dyes and dye intermediates.

ALKALI AND CHLORINE PRODUCTS

CHLORINE AND SODIUM HYDROXIDE

Production and Manufacture

The primary products of the alkali and chlorine industries are sodium hydroxide (caustic soda), chlorine, sodium carbonate (soda ash), potassium hydroxide (caustic potash) (see Potassium compounds), and hydrochloric acid (muriatic acid) (see Hydrochloric acid).

Advances in the chlor-alkali industry include more efficient solid-state rectifiers, larger and higher capacity electrolytic cells, dimensionally stable anodes to replace graphite (see Metal anodes), and cation-exchange membranes (still under development) to replace traditional asbestos diaphragms.

All of the caustic soda and caustic potash produced in the United States and most of the chlorine is made by the electrolysis of sodium or potassium chloride.

Electrolytic-cell operating characteristics. Three types of electrolytic cells are in commercial use for the production of chlor-alkali: mercury cell, diaphragm cell, and membrane cell (Fig. 1). In each, the objective is the isolation of the separate reaction products.

The mercury cell process actually involves two cells, one a primary electrolyzer (or brine cell) in which purified, saturated brine containing ca 25 wt% sodium chloride is decomposed, and a decomposer (denuder or soda) cell.

In the electrolyzer, the brine is decomposed as it passes through the narrow space between the electrodes, and chlorine gas is liberated at the anode with sodium metal liberated at the cathode. The chlorine gas is accumulated above the anode assembly and discharged to the purification process. The sodium liberated at the cathode immediately forms an amalgam, which essentially eliminates the reaction with water in the brine to form caustic soda and hydrogen, and reacts with dissolved chlorine.

$$Na^+ + Cl^- \xrightarrow{Hg} Na-Hg + 1/2\,Cl_2 \uparrow$$

From the electrolyzer, the dilute amalgam is fed to a separate packed-bed reactor, the denuder or soda cell, where it reacts with water.

$$Na-Hg + H_2O \rightarrow Na^+ + OH^- + 1/2\,H_2 \uparrow + Hg$$

Thermodynamic potential. The voltage required to drive the reaction is the thermodynamic potential (often referred to as the decomposition voltage), plus the voltage required to overcome cell resistance and electrode overvoltage (or more strictly, overpotential), which is the additional voltage needed to drive the reaction at an acceptable rate. The thermodynamic potential E^0 is equated with the standard free energy change ΔG^0 which is defined by the overall chemical reaction:

$$\Delta G^0 = -nFE^0$$

where n is the number of moles of electrons involved in the primary electrode reaction, and F is Faraday's constant, expressed in ampere-hour equivalents. The thermodynamic potential at the actual cell temperature may be calculated from the value of ΔG^0 at that temperature, which is, in turn, derived from G^0, ΔH^0, and S^0 at 25°C and the appropriate heat capacity data. Table 1 lists the thermodynamic potentials for both diaphragm and mercury cells. The overall consumption of electrical energy is defined by energy (kWh) = cell potential (V) × current (kA) × time (h).

Energy efficiency. In the overall electrolytic process the energy efficiency is determined as the ratio of the theoretical and actual energies. It is the usual practice of an operating plant to express energy consumption in terms of chlorine produced:

$$\frac{kW \cdot h \text{ (of dc)}}{\text{metric ton of } Cl_2} = \frac{75,600 \times \text{cell voltage}}{\% \text{ chlorine current efficiency}}$$

Thus, energy consumption is a direct, linear function of the cell working voltage and an inverse, linear function of the percent chlorine current efficiency.

The mercury cell depends upon the higher overpotential of hydrogen vs mercury to achieve the preferential release of sodium rather than hydrogen without the need for a diaphragm. However, impurities that can appear on the mercury surface may lack this high overvoltage protection and can cause localized release of hydrogen into the chlorine that is being collected from the anode. The presence of trace amounts of certain metals, eg, vanadium, can cause the release of dangerous amounts of hydrogen.

Chlor-alkali diaphragm cells produce chlorine and sodium (or potassium) hydroxide by electrolysis of saturated brine. A diaphragm is employed to separate the chlorine liberated at the anode, and the hydrogen and caustic soda produced at the cathode. Without the diaphragm, the caustic soda and chlorine would react to form sodium hypochlorite, with further reaction to produce sodium chlorate. (Elec-

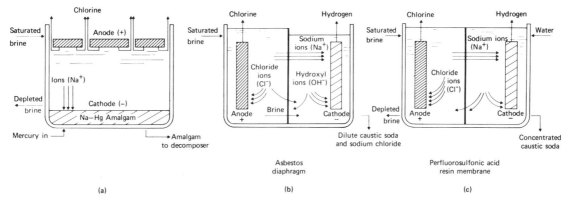

Figure 1. Chlorine electrolysis cells: (**a**) mercury cell; (**b**) ordinary diaphragm cell; (**c**) new membrane cell. Reprinted with permission of Mc-Graw Hill, Inc.

Table 1.   Thermodynamic Potentials for Diaphragm and Mercury Cells at 25°C

Potentials	Diaphragm cell	Mercury cell
anode reaction	$2\,Cl^- \rightarrow Cl_2 + 2\,e$	$2\,Cl^- \rightarrow Cl_2 + 2\,e$
anode potential[a]	-1.360 V	-1.360 V
cathode reaction	$2\,H_2O + 2\,e \rightarrow H_2 + 2\,OH^-$	$Na^+ + e \xrightarrow{Hg}$ (Na amalgam)
cathode potential[a]	-0.828 V	-2.77 V
overall cell reaction	$2\,H_2O + 2\,Cl^- \rightarrow Cl_2 + H_2 + 2\,OH^-$	$2\,Na^+ + 2\,Cl^- \xrightarrow{Hg} Cl_2 + 2\,Na$ (amalgam)
cell potential	-2.188 V	-4.13 V

[a]Standard reduction potential.

Table 2.   Mercury Cell and Diaphragm Cell Comparison

Factor	Mercury cell	Diaphragm cell
diaphragm	none	asbestos or ion-exchange membrane
anode	graphite or titanium-based	graphite or titanium-based
cathode	mercury on steel	mild steel
cathode product	sodium amalgam	caustic-brine[a] cell effluent and hydrogen
decomposer product	low-salt 50% caustic and hydrogen	none
evaporator product	none	high-salt[c] 50% caustic and solid salt
steam consumption	none	ca 0.5 kg of 1 MPa (145 psia) steam per kg of evaporated H_2O
cell voltage	4–5 V	3–4 V
electrical load	$7–10\ kA/m^2$	$1.5–4\ kA/m^2$

[a]Ion-exchange membrane cells can produce cell effluent at a higher NaOH concentration containing very little salt, compared with conventional diaphragm cells.

trolysis without a diaphragm is used commercially to produce sodium chlorate.)

Modern diaphragm cells employ vertical electrodes and a cathode-supported diaphragm. Asbestos is vacuum-deposited on the cathode to form the diaphragm, which separates the feed brine, or anolyte, from the caustic-containing catholyte. Purified brine enters the anode compartment and percolates through the diaphragm into the cathode chamber.

In the diaphragm cell, saturated brine (ca 25 wt% NaCl) is decomposed to approximately half of its original concentration in a pass through the electrolyzer as compared to a 16% decomposition of salt per pass in mercury cells. Electrolytic cells are designed to operate at specific values of voltage and current, which vary among cells from different manufacturers. Various commercially employed diaphragm cells include the Hooker cell, the Diamond-Shamrock cell, the Glanor electrolyzer, and the Dow cell.

In the membrane cell, no brine passes through the membrane into the water-filled cathode compartment. Only the hydrated positive ions (Na^+ and H_3O^+) migrate through the cation-exchange membrane into the cathode compartment, where the water of hydration is decomposed into hydrogen gas and OH^- ions. Saturated brine enters the anode compartment, and depleted brine is discharged from the same compartment. As in the standard diaphragm cell, chloride ions are oxidized at the anode and discharged from the cell as a gas.

Dimensionally stable anodes have revolutionized the chlor-alkali industry. Over half the chlorine in the United States is now produced with these anodes, and the replacement of graphite anodes by DSAs will continue. DSAs, consisting of a coating of ruthenium dioxide and titanium applied to an expanded titanium metal base, have advantages of reduced anodic overpotential; dimensional stability; ease of perforation for gas release; long cell life; light weight; and versatility in fabrication for unique cell design.

Although diaphragms and membranes each serve a similar function in physically separating the anolyte and catholyte in their respective cells, they operate on different mechanisms (Table 2). In the diaphragm cell, brine flows through the asbestos diaphragm at a carefully controlled rate that is sufficient to prevent the back-migration of most of the hydroxyl

ions. On the other hand, the cation-exchange membrane permits the passage of sodium ions into the catholyte, but blocks the passage of chloride ions. However, because the membrane is very thin, some chloride ion transfer occurs. Pure water may be added to the cathode compartment to maintain a predetermined and constant concentration.

Other chlorine production processes include electrolysis to produce potassium hydroxide; coproduct chlorine from sodium metal production; and chlorine recovered from HCl (see Hydrogen chloride) by electrolysis and oxidation. In the electrolytic decomposition of HCl, $2\,HCl \rightarrow H_2 + Cl_2$. Three companies—DeNora, Krebs-BASF, and Uhde—offer electrolytic cells for the decomposition of hydrogen chloride. Oxidation includes the Deacon, Shell, and Kel-Chlor processes. The Deacon process employs a copper oxychloride catalyst, the Shell process uses a chloride catalyst in a fluidized bed, and the Kel-Chlor process is the only commercial process based on the air oxidation of hydrogen chloride to yield chlorine using nitrogen oxides as catalysts.

Chlorine plant auxiliaries. Brine purification and resaturation, chlorine cooling and drying, hydrogen recovery, and direct-current electric power supply are common auxiliaries to all three cell technologies. Brine purification removes contaminants such as calcium, magnesium and sodium sulfates, and trace metals such as iron, titanium, molybdenum, chromium, vanadium, and tungsten. A significant portion (5–10%) of overall chlor-alkali production costs is in removal of contaminants. Hot chlorine contains appreciable amounts of water which is removed by cooling in titanium heat exchangers or scrubbing with cold water; final drying is achieved in countercurrent sulfuric acid contact towers. Chlorinated organics, often referred to as taffy, consisting of decomposition products from the graphite electrodes, may need to be removed by washing the gas with liquid chlorine in a countercurrent packed tower. Although almost all of the chlorine produced in the U.S. is dried and purified, only about half is compressed, stored, and shipped; the remainder is transported as a gas by pipeline. Hydrogen from diaphragm cells is relatively pure; drying is accomplished with refrigeration systems and, if necessary, with chemical absorbents. Hydrogen from mercury cells, in addition to entrained casutic soda, also contains significant amounts of mercury which is removed by scrubbing with sodium hypochlorite or sodium persulfate. The major portion of hydrogen produced in chlor-alkali cells is simply burned as a source of low energy fuel (see Hydrogen; Hydrogen energy). The operation of a chlor-alkali plant is heavily dependent upon the availability of huge quantities of direct-current electric power, usually obtained from a high voltage source of alternating current.

Chlorine

Binary chlorides are discussed in the article dealing with compounds of other elements; eg, for aluminum chloride, see Aluminum compounds; for hydrogen chloride, see Hydrochloric acid (see also Chlorine oxygen acids and salts; Chlorocarbons and chlorohydrocarbons).

Chlorine, atomic number 17, is a member of the halogen family. Its two stable isotopes, of atomic weight 35 and 37, have a natural abundance of 75.53 and 24.47%, respectively. At normal pressures and temperatures, chlorine exists as a diatomic gas (Cl_2) with a yellowish-green color and a characteristic pungent odor. Physical properties are given in Table 3. Properties of chlorine, such as gas and liquid density, vapor pressure, and pressure enthalpy, which vary with changing conditions, are also of interest to industry.

Chlorine is used as a strong oxidizer and as a chlorinating agent, through addition reactions to double bonds in aliphatic compounds, or through substitution reactions in either aliphatic or aromatic compounds. In its binary compounds with hydrogen and the metals, chlorine has an oxidation number of -1. In hypochlorous acid and hypochlorite, however, it exhibits the oxidation number $+1$. The oxidation number is $+3$, $+5$, and $+7$ in the compounds sodium chlorite ($NaClO_2$), sodium chlorate ($NaClO_3$), and sodium perchlorate ($NaClO_4$), respectively.

Chlorine, gas or liquid, is neither explosive nor flammable; however like oxygen, it is capable of supporting combustion. It reacts with most elements under specific conditions, often with extreme rapidity. It reacts with sulfur, phosphorus, iodine, bromine, and fluorine, as well as with nearly all metals, under suitable conditions, to form soluble metal chlorides.

Chlorine is very slightly soluble in water (maximum solubility ca 1% at 9.6°C, decreases with rising temperature). When dissolved it reacts reversibly to form hypochlorous and hydrochloric acids; this causes corrosion of metals ordinarily unaffected by dry chlorine. Below 9.6°C, chlorine hydrate crystals ($Cl_2.8H_2O$) may form in solution; this compound can cause blockages in chlorine-handling equipment.

Chlorine does not react directly with nitrogen or oxygen. Depending upon conditions, chlorine may react with aqueous ammonia to form monochloramine, dichloramine, or nitrogen trichloride, a highly explosive oily liquid. Several oxides of chlorine, chlorine monoxide (Cl_2O), chlorine dioxide (ClO_2), chlorine trioxide (Cl_2O_3), chlorine hexoxide (Cl_2O_6), and chlorine heptoxide (Cl_2O_7) have been prepared by indirect means. Chlorine has a powerful affinity for hydrogen: $H_2S + Cl_2 \rightarrow 2$ $HCl + S$. The preparation of sodium hypochlorite is a typical reaction of chlorine with the alkali- and alkaline-earth metal hydroxides. The hypochlorite formed is a powerful oxidizing agent (see Bleaching agents): $2 NaOH + Cl_2 \rightarrow NaOCl + NaCl + H_2O$. Sodium hypochlorite, manu-

factured *in situ*, reacts with ammonia to form chloramines, and subsequently hydrazine (N_2H_4) which is a component of rocket fuel (see Explosives and propellants).

Chlorine is also used as a chlorinating agent for some inorganic chemicals.

Health and safety. Chlorine is a respiratory irritant. Breathing difficulty may result in death by suffocation. Physiological response to various concentrations of chlorine gas is given in Table 4. Penetrating odor and greenish-yellow color give ample indication of presence. No attempt should be made to handle chlorine for any purpose without a thorough understanding of its properties and the hazards involved.

Sodium Hydroxide (Caustic Soda)

Pure anhydrous sodium hydroxide (NaOH) is a white, somewhat crystalline solid known as caustic soda because even in moderate concentration it is highly corrosive to skin. Physical properties are given in Table 5.

Although caustic soda is produced and shipped in the anhydrous state in the form of a solid, flakes, or beads, very little is actually consumed as such. Because it is used almost exclusively in water solutions of less than 50% concentration, the properties of its aqueous solutions are most important to industry.

In water solution, depending on concentration, caustic soda can form five hydrates containing 1, 2, 3, 5, and 7 molecules of water. Hydrate formation is exothermic. With concentrations of $\geq 40\%$, the heat generated can raise the temperature above the boiling point, resulting in sporadic dangerous eruptions of the solution. Any dilution of caustic soda from concentrations $> 25\%$ should be done cautiously.

Chemical properties. Aqueous solutions of sodium hydroxide are highly basic and especially useful in reaction with weakly acidic materials where weaker bases such as sodium carbonate are ineffective.

Sodium hydroxide will neither burn nor support combustion, although in its reaction with amphoteric metals, such as aluminum, tin, and zinc, hydrogen gas is generated which may form an explosive mixture. A principal use of sodium hydroxide is to form sodium salts, thus neutralizing strong acids and solubilizing water-insoluble chemicals through the

Table 4. Physiological Response to Chlorine

Effect	Parts of chlorine per million (10^6) parts of air (ppmv)
least amount required to produce slight symptoms after several hours exposure	1
least detectable odor	3
maximum amount that can be inhaled for one hour without serious disturbances	4
least amount required to cause irritation of throat	15
least amount required to cause coughing	30
amount dangerous in 30–60 min	40–60
amount that kills most animals in very short time	1000

Table 3. Physical Constants of Chlorine

Property	Value
melting point, °C	-100.98
boiling point at 101.33 kPa (760 mm Hg), °C	-34.05
density of dry gas at 0°C and 101.33 kPa, g/L	3.209
density of saturated gas at 0°C and 366.49 kPa (53.155 psia), g/L	12.07
density of liquid at 0°C and 366.49 kPa, g/L	1468.4
critical density, g/L	573
critical pressure, MPa[a]	7.71
critical temperature, °C	1144
critical volume, L/g	0.001745
latent heat of vaporization, J/g (cal/g)	287.4 (68.7)

[a] To convert MPa to psi, multiply by 145.

Table 5. Physical Constants of Pure Sodium Hydroxide

specific gravity, 20°/4°C	2.130
melting point, °C	318
boiling point at 101.3 kPa (760 mm Hg), °C	1390
index of refraction	1.3576
latent heat of fusion, J/g[a]	167.4
heat of transition, alpha to beta, J/g[a]	103.3
heat of formation from the elements:	
alpha form, kJ/mol[a]	422.46
beta form, kJ/mol[a]	426.60
transition temperature, °C	299.6
solubility at 20°C, g/100 g water	109

[a] To convert J to cal, divide by 4.184.

formation of the sodium salt. Sodium hydroxide is also useful in the precipitation of heavy metals as their hydroxides and in the control of acidity of aqueous solutions.

Caustic soda reacts with all mineral acids to form the corresponding salts. It also reacts with weak-acid gases, such as hydrogen sulfide, sulfur dioxide, and carbon dioxide. Typical reactions are

$$H_2S + 2\,NaOH \rightarrow Na_2S + 2\,H_2O$$

$$SO_2 + 2\,NaOH \rightarrow Na_2SO_3 + H_2O$$

$$CO_2 + 2\,NaOH \rightarrow Na_2CO_3 + H_2O$$

With appropriate equipment and operating conditions, the stronger acid gases can be selectively separated from carbon dioxide with a reasonable degree of efficiency, such as in the preferential removal of hydrogen sulfide from natural gas that contains carbon dioxide (see Sulfur recovery).

Caustic soda reacts with amphoteric metals and their oxides to form soluble salts. For example, hydrated alumina forms sodium aluminate:

$$Al(OH)_3 + NaOH \rightarrow NaAlO_2 + 2\,H_2O$$

This reaction is the basis for the extraction of alumina from bauxite by the Bayer process (see Aluminum compounds). The precipitation of metals as their hydroxides is becoming increasingly important for pollution control. Nearly complete metal removal can be accomplished by precipitation and settling, providing an acceptable effluent for disposal.

All organic acids also react with sodium hydroxide to form soluble salts. However, saponification of esters to form the corresponding salt of the organic acid and an alcohol is of greater industrial importance. The reaction of caustic soda with fatty acid triglycerides to form soap (qv) and glycerol is typical:

$$
\begin{array}{l}
RCOOCH_2 \\
\quad| \\
RCOOCH \\
\quad| \\
RCOOCH_2
\end{array}
+ 3\,NaOH \rightarrow
\begin{array}{l}
HOCH_2 \\
\quad| \\
HOCH \\
\quad| \\
HOCH_2
\end{array}
+ 3\,RCOONa
$$

Another common reaction of caustic soda is dehydrochlorination, eg, the reaction of propylene chlorohydrin and sodium hydroxide (often referred to as an internal coupling) to form propylene oxide (qv).

Other coupling reactions include reactions of cellulose, caustic, and monochloroacetic acid to produce the sodium salt of carboxymethylcellulose. The first step in this reaction is similar to that used in mercerizing cotton (qv) and in the preparation of rayon (qv) via cellulose xanthate.

Other processes for producing caustic soda include the lime-soda process, no longer used to produce caustic soda for sale but still employed in industry.

Health and safety. Sodium hydroxide is a highly hazardous chemical. No attempt should be made to handle it without review of properties and safety precautions. It gives no warning signs of its danger. In contact with skin, immediate damage is caused, but not immediate pain. If contact with eyes occurs, wash within 10 s of contact and continue for at least 30 min. In the case of skin contact, wash with water immediately to prevent slow-healing chemical burns.

Economic data. Markets for chlorine include vinyl chloride monomer, propylene oxide (qv), and methylene chloride (see Chlorocarbons and chlorohydrocarbons). Other markets are chlorinated ethanes, inorganic chemicals, wastewater treatment, pulp and paper, miscellaneous organics, and fluorocarbons. Strong markets for caustic soda are organic chemicals, inorganic chemicals, particularly sodium hypochloride, petroleum (qu), and food processing (1%). Other markets include pulp and paper, soap and detergents, miscellaneous uses, exports, alumina, textiles, and rayon and cellulose.

Industry problems include raw energy requirements; balance of markets for chlorinated chemicals and caustic; and control by regulatory agencies.

J.J. LEDDY
I.C. JONES, JR.
B.S. LOWRY
F.W. SPILLERS
R.E. WING
C.D. BINGER
Dow Chemical U.S.A.

J.S. Sconce, *Chlorine; Its Manufacture, Properties and Uses*, Reinhold Publishing Co., New York, 1962.

N.R. Iammartino, *Chem. Eng. N.Y.* **83**, 86 (June 21, 1976).

Chlor-Alkali: World Survey, C.H. Klind & Co., Inc., Fairfield, N.J., 1977.

V. Gutmann, *Halogen Chemistry*, Vols. I–III, Academic Press, Inc. (London) Ltd., London, UK, 1967.

SODIUM CARBONATE

Sodium carbonate, Na_2CO_3, is a white, crystalline, hygroscopic powder known in the chemical trade as ash, soda ash, soda, and calcined soda. Sodium carbonate is moderately soluble in cold water and soluble to approximately 30 wt% in hot water. The solution is strongly alkaline. Melting point is 851°C; heat capacity at 25°C, 1043.01 J/(kg·K) (249.3 cal/(kg·K); density at 20°C, 2.533 g/cm³.

Manufacture

Most of the world's soda ash is produced by the ammonia-soda or Solvay process. The raw materials are common salt and limestone; ammonia enters the process but is not consumed. The overall reaction for the entire process is

$$CaCO_3 + 2\,NaCl \rightarrow Na_2CO_3 + CaCl_2$$

which does not take place directly but in a number of steps, shown in Figure 1.

It is convenient to consider as the first step the calcination of limestone with fuel in a kiln to produce carbon dioxide and lime (CaO). The lime is discharged from the kiln and may be slaked with excess water to form a thick milk of lime $(Ca(OH)_2)$.

Common salt, in the form of near-saturated, purified brine, is treated with ammonia and carbon dioxide to produce ammonium chloride and precipitate sodium bicarbonate, which is filtered and thermally decomposed, at about 200°C, to the carbonate. The filtrate contains ammonium chloride, some unreacted sodium chloride, and the excesses

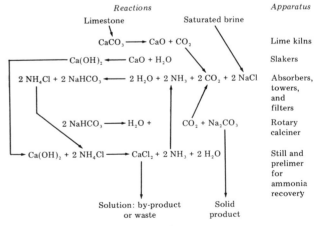

Figure 1. Reaction scheme of the Solvay ammonia-soda process.

of both ammonia and carbon dioxide; the latter probably exists as bicarbonate ions, although the ammonia is not fully bicarbonated. All of the comparatively expensive ammonia is recovered in two steps. Heating alone suffices to drive off the "free" ammonia corresponding to the bicarbonate and hydroxide ions. The hot solution, containing only ammonium chloride and unreacted salt, is then treated with the milk of lime to recover "fixed" ammonia. The reactions in Figure 1 are reversible. The bicarbonation of the sodium salt is never complete, and the reaction is carried only as far as economically feasible.

Process steps. Brine is usually prepared by solution mining and contains low concentrations of impurities (see Chemicals from brine). This requires purification to prevent scaling of processing equipment and contamination of the product. The brine is then saturated with ammonia at slightly less than atmospheric pressure in a countercurrent absorber. The ammoniated brine from the absorber coolers is pumped to the top of one countercurrent column in a block of columns used to precipitate bicarbonate. The first column is the cleaning column and those remaining are the making or crystallizing columns. Lime kiln gas and bicarbonate calciner gas are the sources of carbon dioxide fed to the crystallizing columns.

This process precipitates sodium bicarbonate, and considerable heat is involved. Crystals formed during the carbonation step gradually foul the heat-exchange surfaces, and thus a crystallizing column must alternately be the cleaning column. Vented gases are recycled from the cleaning column to the absorber. The slurry, collected from the crystallizing towers, is filtered. The filter cake, known as crude bicarbonate or ammonia soda, is made up of sodium bicarbonate and small amounts of ammonium bicarbonate. Free carbon dioxide and ammonia is recovered in the distiller by decomposing the filtrate with heat and steam. The stripped solution is usually treated with lime to give more ammonia gas, and the residue, known as distiller waste, contains calcium chloride. In some plants, this calcium chloride is recovered. The crude filtered bicarbonate is calcined continuously by indirect heat, usually at 175–225°C to provide a gas of 95% carbon dioxide or higher. The hot soda ash, or light ash, is cooled, screened, and packaged for shipment; otherwise, it is converted to dense ash, by a hydration and dehydration process, for use in the glass industry. Demand and price of soda ash fluctuate with price and availability of competing alkali, and sodium hydroxide (caustic soda) (see Sodium hydroxide).

Natural Soda Ash

Natural deposits containing sodium carbonate are scattered throughout the world. The most abundant form of sodium carbonate in these deposits is as trona ($Na_2CO_3.NaHCO_3.2H_2O$). A sodium carbonate deposit is, normally, a massive brown matrix of trona containing other alkaline minerals as well as shales.

In general, two types of process, simpler than the Solvay process, are used to refine trona ore. In the first, a monohydrate process, the trona ore is calcined to impure soda ash which is then purified. In the second, a sodium sesquicarbonate process, the soda ash is produced by calcination of sodium sesquicarbonate which had previously been purified.

Soda ash is also produced in the United States from the natural brine at Searles Lake, Calif. More complicated processes are required than for trona because of the complex nature of the brines: in one of the processes, brine is evaporated to give sodium carbonate in the form of burkeite ($Na_2CO_3.2Na_2SO_4$); lithium compounds are separated and the brine processed for other by-products (see Chemicals from brine). In another process, brine is first treated with carbon dioxide in carbonation towers, washed, and calcined to convert sodium bicarbonate to soda ash.

Health and Safety

Although low in toxicity, ingestion can be harmful. Product dust may produce irritation of nose, throat, and lungs. Soda ash may irritate or burn eyes.

Uses

Primary use is in the glass industry. Sodium carbonate is used in the production of chemicals, particularly sodium silicates and sodium phosphates. It is used in pulp (qv) and paper (qv) industries, detergents and cleaners, and water treatment. By-products include calcium chloride (see Calcium compounds); ammonium chloride used in dry cells and fluxing agents (see Ammonium compounds) and fertilizers; and refined bicarbonate, which is frequently manufactured concurrently and used for baking powders and for medicinal purposes.

A.S. ROBERTSON
Allied Corporation

ALKALI METALS. See Cesium; Lithium; Potassium; Rubidium; Sodium.

ALKALINE EARTH METALS. See Barium; Calcium; Strontium.

ALKALOIDS

Alkaloids are naturally occurring substances with a particularly wide range of structures and pharmacologic activities. They may be conveniently divided into three main categories: the true alkaloids, the protoalkaloids, and the pseudoalkaloids.

The true alkaloids have the following characteristics: they show a wide range of physiological activity, are usually basic, normally contain nitrogen in a heterocyclic ring, are biosynthesized from amino acids, are of limited taxonomic distribution, and occur in the plant as the salt of an organic acid. Exceptions are colchicine, aristolochic acid, and the quaternary alkaloids. The protoalkaloids, also known as the biological amines, include mescaline and N,N-dimethyltryptamine. They are simple amines synthesized from amino acids in which the nitrogen is not in a heterocyclic ring. The pseudoalkaloids, those not derived from amino acids, include two major series of compounds: the steroidal and terpenoid alkaloids (eg, conessine), and the purines (eg, caffeine). Most alkaloids occur in the Angiosperms, the flowering plants, but they are also found in animals, insects, marine organisms, microorganisms, and the lower plants. The only common characteristic of alkaloid names is that they end in "ine" (except camptothecin).

Alkaloids are preferably grouped by their biosynthetic origin from amino acids (eg, ornithine, lysine, phenylalanine, tryptophan, histidine and anthranilic acid) rather than their heterocyclic nucleus. The pseudoalkaloids, which are not derived from amino acids and clearly cannot be classified this way, are best organized in terms of their parent terpenoid class, eg, diterpenoid and steroidal alkaloids, or as purines, ansamacrolides, etc (see also Antibiotics; Amino acids; Steroids).

Physical Properties

Most alkaloids are colorless, crystalline solids with a defined melting point or decomposition range, eg, vindoline and morphine. Some alkaloids are amorphous gums and some are liquids, eg, nicotine and coniine, and some are colored (eg, berberine is yellow and betanidine is red).

The free base of the alkaloid is normally soluble in an organic solvent; however, the quaternary bases are only water soluble, and some of the pseudo- and protoalkaloids show substantial solubility in water. The salts of most alkaloids are soluble in water.

The solubility of alkaloids and their salts is of considerable significance in the pharmaceutical industry, both in the extraction of the alkaloid from plant or fungus and in the formulation of the final pharmaceutical preparation. Solubility is also of considerable significance in the clinical distribution of an alkaloidal drug.

Chemical Properties

Most alkaloids are basic, which makes them extremely susceptible to decomposition, particularly by heat and light.

Ornithine-Derived Alkaloids

Ornithine-derived alkaloids include the tropanes (atropine, *l*-hyoscyamine, *l*-scopolamine, and cocaine), the *Senecio* alkaloids, and nicotine.

Tropane alkaloids are derived from plants in the Solanaceae, Erythroxylaceae, and Convolvulaceae families. These alkaloids comprise two parts: an organic acid and an alcohol (normally a tropan-3α-ol). The pharmacologically active members of this group include atropine, the optically inactive form of *l*-hyoscyamine which is isolated from deadly nightshade (*Atropa belladonna*); *l*-hyoscyamine and *l*-scopolamine, which are found in the leaves of *Duboisia metel* L., *D. meteloides* L., and *D. fastuosa* var *alba* leaves, are also manufactured.

The tropane alkaloids are parasympathetic inhibitors. For example, atropine acts through antagonism of muscarinic receptors, the receptors responsible for the slowing of the heart, constriction of the eye pupil, vasodilation, and stimulation of secretions. Atropine prevents secretions (eg, sweat, saliva, tears, pancreas) and dilates the pupil. Atropine is used to reduce pain of renal and intestinal cholic and other gastrointestinal tract disorders, to prolong mydriasis when necessary, and as an antidote to poisoning by cholinesterase inhibitors. Small doses produce respiratory and myocardial stimulation and decrease nasal secretion, and the drug has little local anesthetic action (see Cardiovascular agents; Veterinary drugs). Hyoscyamine and scopolamine have mydriatic effects. They are also used in combination as sedatives, in antimotion sickness drugs, and in antiperspirant preparations.

Cocaine is a potent central nervous system (CNS) stimulant and adrenergic blocking agent. It is extracted from South American coca leaves, or prepared by converting ester alkaloids to ecgonine followed by methylation and benzoylation. It is too toxic to be used as an anesthetic by injection, but the hydrochloride is used as a topical anesthetic. It has served as a model for a tremendous synthetic effort to produce an anesthetic of increased stability and reduced toxicity.

cocaine

Senecio alkaloids possess a pyrrolizidine nucleus and occur in the genera *Senecio* (Compositae), *Heliotropium* and *Trichodesma*, and *Crotalaria*. They are biosynthetically derived from two units of ornithine in a manner similar to some of the lupin alkaloids. Certain of the alkaloids having an unsaturated nucleus are potent hepatotoxins.

Nicotine is toxic, soluble in water, and a constituent of tobacco. The lethal human dose is ca 40 mg/kg. Pharmacologically, there is an initial stimulation followed by depression and paralysis of the autonomic ganglia. The biosynthesis of nicotine is well established.

nicotine

Lysine-Derived Alkaloids

Lysine-derived alkaloids contain the pyridine nucleus or its reduced form, piperidine. They include alkaloids derived from the Areca or Betel nut, *Lobelia* alkaloids, and those derived from pomegranate or club mosses.

Arecoline, a colorless, liquid alkaloid which has a pronounced stimulant action (in large doses, paralysis may occur), is found in the Areca or Betel nut. As the hydrobromide, it is used as a diaphoretic and anthelmintic (see Chemotherapeutics, anthelmintic).

Lobelia inflata, known as Indian tobacco, contains lobeline, which is similar to—but less potent than—nicotine in pharmacologic action and is used as an emetic; the sulfate salt is used in antismoking tablets.

The root of *Punica granatum* contains alkaloids such as pelletierine and pseudopelletierine, which are formed from lysine and acetate. Pelletierine is toxic to tapeworms and is used as an anthelmintic.

The club mosses, *Lycopodium* species, produce polycyclic alkaloids such as lycopodine, whereas *Hydrangea* species yield febrifugine, an active antimalarial agent (see Chemotherapeutics, antiprotozoal). Anabasine is found in *Haloxylon persicum* Bunge; this alkaloid has antismoking and respiratory muscle stimulation action similar to lobeline, and is also used as a metal anticorrosive.

A host of complex alkaloids such as lupinine, sparteine, cytisine, and matrine are found in the lupins, a large plant family of the *Leguminosae*. Sparteine paralyzes motor nerve endings and sympathetic ganglia. The sulfate is used as an oxytoxic and the adenylate derivative is used to treat cardiac insufficiencies. Cytisine, found in the seeds of the highly toxic plant *Cytisus laburnum* L., is a strongly basic alkaloid which produces convulsions and death by respiratory failure.

pelletierine   lupinine

Anthranilic Acid-Derived Alkaloids

Anthranilic acid-derived alkaloids exhibit great structural diversity. This group includes dictamnine, platydesmine, vasicine, cusparine, and rutecarpine. Vasicine has oxytocic activity

Phenylalanine- and Tyrosine-Derived Alkaloids

Phenylalanine- and tyrosine-derived alkaloids are by far the most numerous group of alkaloids, ranging from simple phenethylamines to the very complex dimeric benzylisoquinolines and the highly rearranged *Cephalotaxus* alkaloids.

Ephedrine, from the Chinese drug Ma Huang, is soluble in water, alcohol, chloroform, and diethyl ether. It melts over a range of 33–42°C, depending on the water content. Little of the ephedrine of U.S. commerce is obtained from natural sources. Ephedrine is produced commercially through a biosynthetic process. Ephedrine has mydriatic effects, and demonstrates adrenaline-like activity. It causes a rise in arterial blood pressure, increased secretions, and dilated pupils; it is a monamine oxidase inhibitor and is used in nasal decongestants (see Cardiovascular agents; Epinephrine).

Peyote, the small cactus of the Indians of north central Mexico, contains over 60 constituents. It is used as a hallucinogen in religious ceremonies and for medicinal purposes. A principal constituent is mescaline, a simple trimethoxyphenethylamine (see Psychopharmacological agents).

mescaline

Ipecac, derived from *Cephaelis ipecacuanha* (native to Brazil), contains emetine and cephaeline, both of commercial significance. Emetine exhibits profound pharmacologic effects including clinical antiviral activity and is used in the treatment of amebic dysentery. Its side effects are cardiotoxicity, muscle weakness, and gastrointestinal problems including diarrhea, nausea, and vomiting. Two synthetic routes to emetine are of commercial importance: the Roche synthesis, which produces dehydro-emetine, and the Burroughs-Wellcome synthesis. In handling emetine and its products, exposure should be limited as it can cause severe conjunctivitis, epidermal inflammation, and asthma attacks in suscepti-ble individuals.

Isoquinoline and related alkaloids are the largest group of alkaloids derived from phenylalanine (or its hydroxylated derivatives) and corre-sponding β-phenylacetaldehydes. The alkaloids are prevalent in the plant families Fumariaceae, Papaveraceae, Ranunculaceae, Rutaceae, and Berberidaceae. There are many structural types, including the simple tetrahydroisoquinolines; the benzylisoquinolines; the bisbenzyl-isoquinolines, such as the dl- and d-isomers of tetrandrine, which exhibit anticancer activity; the proaporphines such as glaziovine, which is an antidepressant; the aporphines such as glaucine; the aporphine-benzyl-isoquinoline dimers such as thalicarpine, which shows cytotoxic and antitumor activity (see Chemotherapeutics, antimitotic); the oxoa-porphines; the protoberberines, a group of over 70 alkaloids such as xylopinine, berberine, canadine, and corydaline, known for such diverse pharmaceutical activities as tranquilizers, CNS depressants, anti-bacterial and antiprotozoal agents, anticancer agents, and alpha adren-ergic blockers; the benzophenanthridines, a group of thirty alkaloids such as fagaronine and nitidine, which exhibit antitumor properties; the protopines; the phthalideisoquinolines, a group that includes narcotine, which possesses antitussive activity, depresses smooth muscles and is not narcotic, and hydrastine, which is used as an astringent in mucous membrane inflammation; and the homoaporphines.

The opium alkaloids number over 25, some of which are of commercial importance and major significance.

Opium is the air-dried milky exudate from incised, unripe capsules of *Papaver somniferum* L. or *P. albumen* Mill (Papaveraceae) (see Analges-ics, antipyretics and antiinflammatory agents; Hypnotics, sedative and anticonvulsants).

Notable opium-derived alkaloids include morphine, codeine, thebaine, noscapine, and papaverine.

Morphine is the most important alkaloid. It is isolated from opium. Along with its salts, it is classified as a narcotic analgesic and is strongly hypnotic. Side effects include constipation, nausea, and vomiting in addition to habituation, reduced power of concentration, and reduction in fear and anxiety. Respiration is also deepened (see also Opioids, endogenous).

Codeine is the methyl ether of morphine, and thebaine is one of the methyl enol ethers of codeinone. Codeine pharmacologically resembles morphine but is weaker, less toxic, and exhibits less depressant action (it does not depress respiration in normal therapy). It is used in the treatment of minor pain and as an antitussive. Codeine is available as a free base or the sulfate or phosphate salts. Thebaine is a convulsant poison rather than a narcotic.

morphine R = H
codeine R = CH₃

Papaverine, a smooth-muscle relaxant which is neither narcotic nor addictive, occurs in *P. somniferum* to the extent of 0.8–1.0%, commonly accompanied by noiscopine (narcotine). Most papaverine used is synthet-ically produced. The glyoxylate salt is used in treating arterial and venous disorders.

Amaryllidaceae alkaloids include galanthamine, margetine, and narciprimine. Galanthamine is a water-insoluble crystalline alkaloid which exhibits powerful cholinergic activity and analgesic activity com-parable to morphine. It has been used to treat diseases of the nervous

system. Its derivatives show anticholinesterase, antibacterial, and CNS depressant activity. Narciclasine, margetine, and narciprimine exhibit anticancer activity.

Colchicine, also known as hermodactyl, surinjan, and ephemeron, has some of the most unusual solubility characteristics of any alkaloid: it is soluble in water, alcohol, and chloroform, but only slightly soluble in ether or petroleum ether. Colchicine-type alkaloids are present in ten other genera of the *Liliaceae* and 19 species of *Colchicum*. Reviews of the chemistry of colchicine and related compounds, history, and phar-macology are available. Colchicine has the ability to artificially induce polyploidy or multiple chromosome groups. It is also used to suppress gout.

Cephalotaxus alkaloids are found in the Japanese plum yews, *Cepha-lotaxus* species. The esters, such as harringtonine and homohar-ringtonine, are potent antileukemic agents. The absolute configuration of the ester moiety has been determined. The α-hydroxy ester is essential for *in vivo* antileukemic activity.

cephalotaxine R = H
harringtonine R = —COC(CH₂)₂C(CH₃)₂
CH₂CO₂CH₃
deoxyharringtonine R = —COC(CH₂)₂CH(CH₃)₂
CH₂CO₂CH₃

Securinine, isolated from *Securinega suffruticosa* Rehd., is similar to strychnine in action, but exhibits lower toxicity, stimulates respiration, raises blood pressure, and increases cardiac output. The chemistry, pharmacology, and biosynthesis of securinine and related compounds have been reviewed.

Tryptophan-Derived Alkaloids

Tryptophan-derived alkaloids, which occur in the families Apo-cynaceae, Rubiaceae, and Loganiaceae, have recently become of great interest, particularly those derived from tryptamine and a monoterpene unit. There are many diverse structural types in this group. Attempts to determine the important details of the biosynthetic interconversions of these compounds have been reported.

The simplest indole alkaloids are derived from tryptamine itself. These include indole-3-acetic acid, a potent plant-growth stimulator; serotonin (5-hydroxytryptamine), a vital mammalian product which inhibits or stimulates smooth muscles and nerves; N-acetyl-5-methoxy-tryptamine (melotonin), a constituent of the pineal gland with melanophase-stimulating properties; 5-methoxy-N,N-dimethyltrypta-mine, a constituent of the hallucinogenic *Virola* snuffs; psilocybine, a hallucinogenic found in the mushroom *Psilocybe mexicana* Heim.

The harmala alkaloids, such as harmine and harmaline, are powerful monoamine oxidase inhibitors, previously used in the treatment of Parkinsonism. Harmine is the active ingredient of the narcotic drug yage (see Hypnotics, sedatives and anticonvulsants).

Ellipticine and derivatives show anticancer activity.

Several alkaloids of *Calycanthus* species are the products of dimeriza-tion or trimerization of simple tryptamine residues, such as folicanthine.

The main indole alkaloid skeletons are derived from tryptamine and a C₁₀ unit. Examples include corynanthine; yohimbine, which has hypo-tensive and cardiostimulant properties and is used to treat rheumatic disease; ajmaline, which has coronary dilating and antiarhythmic prop-erties; decarbomethoxydihydrovobasine, which shows vasodilating and hypotensive activities whereas related compounds exhibit antiviral activ-ity; akuammicine; tabersonine; catharanthine; rhynchophylline; vindo-line; dihydrovobasine; 10-methoxyibogamine, whose acyl derivatives ex-hibit analgesic and antiinflammatory activity; and strychnine. This structural diversity has been the source of intense biosynthetic interest.

Physostigmine, found in the perennial West African woody climber, *Physostigma venenosum* Balfour, is pharmacologically similar to pilo-carpine and is a reversible cholinesterase inhibitor used to treat glaucoma.

The **ergot alkaloids** are obtained from ergot, the dried sclerotium of the fungus *Claviceps purpurea* (Fries) Tulsane (Hypocreaceae). Ergot alkaloids are produced by isolation from the crude drug grown in the field; by extraction from saprophytic cultures; and by partial and total synthesis.

The ergot alkaloids act pharmacologically to produce peripheral, neurohormonal, and adrenergic blockage, and produce smooth-muscle contraction as well. The two medicinally important ergot alkaloids are ergotamine and ergonovine (see Neuroregulators).

lysergic acid

Three main groups of ergot alkaloids exist. (*1*) The clavine type, a group of over 20 alkaloids, is water insoluble and does not give lysergic acid on hydrolysis. This group includes elymoclavine, agroclavine (a potent uterine stimulant) and chanoclavine-I. (*2*) The aqueous lysergic acid derivatives such as ergonovine, which in its maleate salt (or as methyl ergonivine maleate) is the drug of choice to treat post-partum hemorrhage. (*3*) The "peptide" ergot alkaloids, a group of water-insoluble lysergic acid derivatives. This group includes ergotamine, ergocornine, and ergocryptine. Ergotoxine, a mixture of three peptide ergot alkaloids, possesses strong sympatholytic action and is used as a peripheral vasodilator and antihypertensive. Dihydroergotoxine is used for vascular disorders in the aged (see Cardiovascular agents).

Some ergot alkaloids (eg, 2-bromo-α-ergocryptine) stimulate prolactin release and are being evaluated for treatment of breast cancer.

Catharanthus and Vinca alkaloids, usually discussed together, are quite distinct. The most important alkaloids of the *Catharanthus* genus are vincaleukoblastine, leurosine, leurocristine, and leurosine, all antileukemic agents. Vincaleukoblastine and leurocristine are used clinically. The most important alkaloid of *Vinca* is vincamine, used to treat hypertension, angina and migraine headaches. Alkaloids of this type produce marked hypotensive effects and curare-like action. The ethers of vincaminol are potent muscle relaxants.

Rauwolfia alkaloids include reserpine, the first tranquilizer, rescinnamine, and deserpidine. Reserpine is a sedative and tranquilizer useful in treating hypertension. It is also used as a rodenticide (see Poisons, economic).

Strychnine, from the seeds of many *Strychnos* species, is a widely known poison (although in fact it is only moderately toxic). Pharmacologically, strychnine excites all portions of the CNS; it is a powerful convulsant and death results from asphyxia. It has no therapeutic uses in Western medicine, although its nitrate is used in treating chronic aplastic anemia.

Cinchona alkaloids, derived from the dried stem or root bark of various *Cinchona* species, include quinine and quinidine. These alkaloids are bitter tasting, white, crystalline solids sparingly soluble in water. Quinine is toxic to many bacteria and other unicellular organisms, and was the only specific antimalarial remedy until the second World War. It is a local anesthetic of considerable duration. Quinine is commonly used as the sulfate and dihydrochloride. Quinidine, produced by the isomerization of quinine or found in *Cuprea* bark, is more effective on cardiac muscle than quinine and is used to prevent or abolish certain cardiac arrhythmias.

Camptothecine, isolated from the Chinese tree *Camptotheca acuminata* Decsne, is used to treat cancer in the People's Republic of China.

Histidine-Derived Alkaloids

Histidine-derived alkaloids include pilocarpine and saxitoxin. Pilocarpine stimulates parasympathetic nerve-endings and is used to treat glaucoma. The main commercial source of pilocarpine is *Pilocarpus*

microphyllus Stapf., known as Marnham jaborandi. Saxitoxin is an extremely toxic neuromuscular blocking agent found in the so-called coastal red tides of North America.

Monoterpenoid Alkaloids

Monoterpenoid alkaloids include chaksine, a guanidine alkaloid from *Cassia lispidula* Vahl which induces respiratory paralysis in mice; β-skytanthine, which is tremorigenic; cantleyine, derived from a monoterpene before loganin; and those derived from secologanin such as gentianine, which exhibits hypotensive, antiinflammatory and muscle-relaxant actions, gentioflavine, gentiatibetine, pedicularine, and actinidine, a potent feline attractant.

Diterpene Alkaloids

Diterpene alkaloids are not of commercial or therapeutic significance, but some have potent pharmacological activity, eg, aconitine and *Erythrophleum* alkaloids.

Steroidal and Triterpene Alkaloids

Steroidal and triterpene alkaloids are found in the plant families Solanaceae, Liliaceae, Apocynaceae, and Buxaceae (see Steroids). There are four main groups based on the botanical source: the *Veratrum*, *Solanum*, *Holarrhena* and *Funtumia*, and *Buxus* alkaloids.

The *Veratrum* alkaloids include jervine, protoveratrine A, and protoveratrine B; the latter two produce pronounced bradycardia and a fall of blood pressure by stimulation of vagal afferents. The *Solanum* alkaloids are of interest as potential sources of steroids. Examples of these alkaloids are tomatidine and solanidine. Some *Solanum* alkaloids exhibit fungistatic activity (see Chemotherapeutics, antibacterial and antimycotic). Biosynthetically, the alkaloids are derived from acetate and mevalonate.

tomatidine   solanidine

Purine Alkaloids

Purine alkaloids are derivatives of the xanthine nucleus and include caffeine, theophylline, and theobromine, the principal constituents of plants used throughout the world as stimulating beverages (see Stimulants).

Caffeine has the structure 1,3,7-trimethylxanthine. It is derived from cola, coffee (qv), tea (qv), guarana, and maté. Theophylline, 1,3-dimethylxanthine, is found in tea. Theobromine, 3,7-dimethylxanthine, is found in cocoa and tea (see Chocolate and cocoa). The xanthine derivatives have pharmacological properties in common: central nervous system (CNS) and respiratory stimulation; skeleton-muscle stimulation; diuresis; cardiac stimulation; and smooth-muscle relaxation. Caffeine is used to increase CNS activity; it acts on the cortex to produce clear thought and to reduce drowsiness and fatigue (see Analgesics, antipyretics, and anti-inflammatory agents). Theophylline is used in smooth-muscle relaxants. Theobromine, as the ethylenediamine salt, is used in preference to caffeine in cardiac edema and in angina pectoris.

Miscellaneous Alkaloids

Coniine is an extremely toxic alkaloid which induces paralysis of the motor nerve endings and is the primary toxic constituent of poison hemlock. It was the first alkaloid to be synthesized. Carpaine, a crystalline macrocyclic alkaloid which induces bradycardia, depresses the CNS and is a potent amoebicide. Alkaloids are found in the poisonous *Amanita* species of mushrooms, such as α- and β-amanita-toxins, ibotenic acid, muscimol, and muscazone (see Psychopharmacological agents). Maytansine and related ansamacrolides are potent antileukemic agents (see

Antibiotics, ansamacrolides). Surugatoxin, found in the carnivorous gastropod *Babylonia japonica*, produces a pronounced mydriatic effect sometimes resulting in death.

GEOFFREY A. CORDELL
College of Pharmacy
University of Illinois

G.A. Cordell, *Introduction to Alkaloids, a Biogenetic Approach*, Wiley-Interscience, a division of John Wiley & Sons, Inc., New York, 1981.

J.S. Glasby, *Encyclopedia of the Alkaloids*, Vols. 1 and 2, Plenum Press, New York, 1975; Vol. 3, 1977; Vol. 4, 1983.

R.H.F. Manske, *Alkaloids, Chemistry and Physiology*, Vols. 1–22, Academic Press, Inc., New York, 1950–1983.

Specialist Periodical Reports, Vols. 1–13, Royal Society of Chemistry, London, 1971–1983.

ALKANOLAMINES

FROM OLEFIN OXIDES AND AMMONIA

Alkanolamines are prepared from ammonia and ethylene oxide (qv), propylene oxide (qv), or butylene oxide (see Butylenes) leading to three major classes of compounds listed in Table 1. The first series, the ethanolamines, include mono-, di-, and triethanolamine, $(HOC_2H_4)_nNH_{3-n}$ (where $n = 1, 2,$ or 3). The second series, the isopropanolamines, include the three products obtained from the reaction of ammonia and propylene oxide, mono-, di-, and triisopropanolamine, $(CH_3CHOHCH_2)_nNH_{3-n}$ (where $n = 1, 2,$ or 3). The third series, the secondary butanolamines, prepared from ammonia and butylene oxide, include mono-, di-, and tri-*sec*-butanolamine, $(C_2H_5CHOHCH_2)_nNH_{3-n}$ (where $n = 1, 2,$ or 3). Other substituted alkanolamines have gained some commercial importance, particularly aminoethylethanolamine, diethylethanolamine, and dimethylethanolamine, prepared from amines and ethylene oxide.

Table 1. Physical Properties of Alkanolamines

Name	Formula	Freezing point, °C	Boiling point, °C
Prepared from ammonia and olefin oxides			
monoethanolamine	$NH_2C_2H_4OH$	10	170
diethanolamine	$NH(C_2H_4OH)_2$	27.5	270[a]
triethanolamine	$N(C_2H_4OH)_3$	17.9	360
monoisopropanolamine	$NH_2CH_2CHOHCH_3$	−2	160
diisopropanolamine	$NH(CH_2CHOHCH_3)_2$	47	248
triisopropanolamine	$N(CH_2CHOHCH_3)_3$	58	300
mono-*sec*-butanolamine	$NH_2CH_2CHOHC_2H_5$	3	169
di-*sec*-butanolamine	$NH(CH_2CHOHC_2H_5)_2$	68–70	256
tri-*sec*-butanolamine	$N(CH_2CHOHC_2H_5)_3$	41–47	310
Prepared from nitro alcohols			
2-amino-2-methyl-1-propanol (AMP)	$NH_2C(CH_3)_2CH_2OH$	30–31	165[b]
2-amino-2-ethyl-1,3-propanediol (AEPD)[c]	$NH_2C(CH_2OH)_2C_2H_5$	37.5–38.5	152–153[d]
2-dimethylamino-2-methyl-1-propanol (DMAMP)	$(CH_3)_2NC(CH_3)_2CH_2OH$	19	160[b]
tris(hydroxymethyl)aminomethane; 2-amino-2-(hydroxymethyl)-1,3-propanediol (Tris Amino)[c]	$NH_2C(CH_2OH)_3$	171–172	219–220[d]

[a] At 20/20°C.
[b] At 101.3 kPa (1 atm).
[c] Registered trademark of IMC Chem.
[d] At 1.3 kPa (10 mm Hg).

Physical Properties

Monoisopropanolamine, mono-*sec*-butanolamine, and all of the ethanolamines are colorless liquids at or near room temperature. Di- and triisopropanolamines, and di- and tri-*sec*-butanolamines are white solids. The monoalkanolamines have a basicity about equal to aqueous ammonia, whereas dialkanolamines and trialkanolamines are weaker bases than aqueous ammonia. In general, basicity decreases in the order mono > di > tri.

With the exception of tri-*sec*-butanolamine, all of the alkanolamines are infinitely soluble in water, have rather limited solubility in heptane, and are either completely miscible with or very soluble in polar solvents but not in nonpolar solvents. Exceptions are triisopropanolamine and di-*sec*-butanolamine, which are slightly soluble in *n*-heptane, and tri-*sec*-butanolamine, which is very soluble in *n*-heptane.

Chemical Properties

Under anhydrous conditions, mono- and diethanolamines and isopropanolamines form carbamates with carbon dioxide:

$$2\ HOCHRCH_2NH_2 + CO_2 \rightarrow$$
$$HOCHRCH_2NHCOOH.H_2NCH_2CHROH$$

Alkanolamines in aqueous solution react with carbon dioxide and hydrogen sulfide to yield salts, which dissociate on heating.

$$HOCHRCH_2NH_2 + CO_2 + H_2O \underset{\Delta}{\rightleftharpoons} HOCHRCH_2NH_2.H_2CO_3$$
$$HOCHRCH_2NH_2 + H_2S \underset{\Delta}{\rightleftharpoons} HOCHRCH_2NH_2.H_2S$$

These reactions form the basis of an important industrial application, ie, the "sweetening" of natural gas (see Uses below). Salts are formed with other acids. Heating with sulfuric acid gives bisulfates, H_2NCH_2-$CHROSO_3H$, which on further heating form 2-alkylaziridines (see Imines, cyclic). In a similar manner, diethanolamine and diisopropanolamine give morpholines (see Amines). Alkanolamines and long-chain fatty acids react at room temperature to give neutral alkanolamine soaps, which are waxy, noncrystalline materials with widespread commercial application as emulsifiers (qv). At elevated temperatures, *N*-alkanolamides are the main products. Reaction of dialkanolamines with fatty acids in a 2:1 ratio at 140–160°C gives a second type of alkanolamide, which consists of water-soluble, complex mixtures of *N*-alkanolamides, amine esters and diesters, amide esters and diesters, and a considerable amount of unreacted dialkanolamine (accounting for the aqueous solubility of the product). Both the 1:1 and 2:1 alkanolamides are of commercial importance (see Surfactants).

At ordinary temperatures, acyl halides and acid anhydrides give amides (qv). On warming, or in the presence of alkyl anhydrides, esters are formed. Alkyl halides form *N*-alkyl derivatives of alkanolamines. Polyester polyamides can be formed from dicarboxylic acids. Carbon disulfide gives 2-mercaptothiazolines. Ethyleneurea (2-imidazolidinone) can be prepared by heating a mixture of monoethanolamine and urea for several hours:

$$HOCHRCH_2NH_2 + NH_2CONH_2 \rightarrow HN\overset{\underset{\displaystyle\|}{O}}{\diagup}\hspace{-0.5em}NH + NH_3 + H_2O$$

Ethylene or propylene carbonates give carbamates (see Carbamic acid). Owing to their multifunctional character, the alkanolamines are chelating agents for metal ions (see Chelating agents).

Manufacture

Alkanolamines are principally manufactured from alkylene oxide and excess ammonia. For example, the ethanolamines are prepared from ethylene oxide and aqueous ammonia at high temperature and pressure. The resulting mixture of mono-, di-, and triethanolamine with excess ammonia and water is separated by distillation.

$$NH_3 + RCH\overset{O}{\overset{\triangle}{-}}CH_2 \rightarrow NH_2CH_2CHROH$$

$$NH_2CH_2CHROH + RCH\overset{O}{\overset{\triangle}{-}}CH_2 \rightarrow NH(CH_2CHROH)_2$$

$$NH(CH_2CHROH)_2 + RCH\overset{O}{\overset{\triangle}{-}}CH_2 \rightarrow N(CH_2CHROH)_3$$

where R equals H, CH_3, or C_2H_5. The reactions are exothermic and are usually carried out at 50–100°C. The product ratio is generally controlled by the ratio of ammonia to alkylene oxide. A high ammonia-to-alkylene oxide ratio is used when mono- and dialkanolamines are desired. A recycle technique can be used when di- and triethanolamine are desired; excess monoalkanolamine is added to suppress its further formation. This product ratio flexibility is of great value when demands shift from one product to another. A variety of substituted alkanolamines can be made from amines and alkylene oxides, such as the commercially important aminoethylethanolamine (from ethylenediamine and ethylene oxide), dimethylethanolamine (from dimethylamine and ethylene oxide), and diethylethanolamine (from diethylamine and ethylene oxide).

Health and Safety

No hazard results from normal use. Alkanolamines are toxic when swallowed. Excessive vapor concentrations of monoethanolamine, diethanolamine, triethanolamine, monoisopropanolamine, diisopropanolamine, triisopropanolamine, and mixed isopropanolamines occur on heating and are eye and nose irritants.

Uses

There is extensive industrial use where a nonvolatile water-soluble amine is desired (eg, removal of acid gases; anionic surfactants; nonionic surfactants; chemical intermediates; algicides) (see Surfactants). Other uses include the cement and concrete industry (see Cement); corrosion inhibitors (see Corrosion and corrosion inhibitors); raw material in cosmetics (see Cosmetics); detergents (see Surfactants and detersive systems; Emulsions); gas purification or "sweetening" (see Gas, natural; see also Carbon dioxide; Sulfur recovery); textile production (see Dyes, application; see also Textiles); electroplating (see Electroplating; Electroless plating); cross-linking agents in polyurethanes and chelating agents (see Metal surface treatments; Dispersants); secondary vulcanization accelerators in the rubber industry (see Rubber; Elastomers, synthetic). Aminoethylethanolamine and its derivatives are used in textiles, chelating agents, petroleum products, agricultural chemicals, detergents, emulsifiers, mining chemicals, corrosion inhibitors, and other products. Dimethylethanolamine, diethylethanolamine, and their derivatives are used in pesticides, corrosion inhibition, drugs and pharmaceuticals, emulsification, paints and coatings, metal fabrication and finishing, petroleum and petrochemical products, and plastics and resins.

RICHARD M. MULLINS
The Dow Chemical Company

F.A. Lowenheim and M.K. Moran, eds., *Industrial Chemicals*, 4th ed., 1975, pp. 339–344.

Alkanolamines Handbook, The Dow Chemical Company, Midland, Mich., 1968.

FROM NITRO ALCOHOLS

The nitro alcohols (see Nitro alcohols) obtained by the condensation of nitroparaffins (see Nitroparaffins) with formaldehyde may be reduced to a unique series of alkanolamines (β-amino alcohols):

$$RCH_2NO_2 + CH_2O \longrightarrow \underset{\underset{NO_2}{|}}{RCHCH_2OH} \overset{H}{\longrightarrow} \underset{\underset{NH_2}{|}}{RCHCH_2OH}$$

The condensation may occur one to three times, giving rise to amino alcohols with one to three hydroxy groups. 2-Amino-2-ethyl-1,3-propanediol (AEPD), 2-amino-2-methyl-1-propanol (AMP), 2-dimethylamino-2-methyl-1-propanol (DMAMP), tris(hydroxymethyl)-aminomethane, and 2-amino-2-hydroxymethyl-1,3-propanediol (Tris Amino) are of commercial importance.

Physical Properties

Table 1 lists the properties of the alkanolamines. These compounds are highly soluble in water, very soluble in alcohol, slightly soluble in aromatic hydrocarbons, and nearly insoluble in aliphatic hydrocarbons (tris(hydroxymethyl)aminomethane is appreciably soluble only in water and methanol). The two compounds, DMAMP ($pK_a = 10.20$ at 25°C) and AMP ($pK_a = 9.72$), are among the strongest alkanolamines commercially available, whereas Tris Amino (see Table 1) ($pK_a = 8.03$) is one of the weaker amines. All alkanolamines have a mild amine odor in the liquid state; the solids are nearly odorless.

Table 1. Physical Properties

Compound	Boiling point, °C	Melting point, °C	Specific gravity	pH of 0.1 M aqueous solution	Solubility, in water at 20°C g/100 mL
2-amino-2-methyl-1-propanol (AMP)	165[a]	30–31	0.934[b]	11.3	completely miscible
2-amino-2-ethyl-1,3-propanediol (AEPD)[c]	152–153[d]	37.5–38.5	1.099[b]	10.8	completely miscible
2-dimethylamino-2-methyl-1-propanol (DMAMP)	160[a]	19	0.90[e]	11.6	completely miscible
tris(hydroxymethyl)-aminomethane; 2-amino-2-(hydroxymethyl)-1,3-propanediol (Tris Amino)[c]	219–220[d]	171–172		10.4	80

[a] At 101.3 kPa (1 atm).
[b] At 20/20°C.
[c] Registered trademark of IMC Chemical Group, Inc.
[d] At 1.3 kPa (10 mm Hg).
[e] At 25/25°C.

Chemical Properties

The alkanolamines discussed here attack copper, brass, and aluminum but not steel or iron. With mineral acids they form ammonium salts which hydrolyze readily in the presence of water and dissociate on heating. Fatty acids give soaps which are highly efficient emulsifying agents with important industrial uses (see Emulsions; Surfactants).

On heating, alkanolamine soaps first dehydrate to amides, then to oxazolines (on further heating) which have cationic surface-active properties and are emulsifying agents of the water-in-oil type. Formaldehyde reacts with the hydrogens on the α-carbon in the fatty acid from which the oxazoline was formed to yield a vinyl monomer which can be polymerized. These products are useful for the modification of alkyd resins, the preparation of paint vehicles, and copolymerization with other monomers. Alkanolamines react with nitro alcohols to form nitro-hydroxyamines which show antibacterial activity. Mercaptothiazolines are obtained from the corresponding sulfate esters and carbon disulfide. Aldehydes react with monohydric alkanolamines to give monocyclic oxazolidines or with polyhydric alkanolamines to give bicyclic oxazolidines. These can be hydrogenated to N,N-substituted alkanolamines. Oxidation of the hydroxy group gives amino acids, eg, oxidation of 2-amino-2-methyl-1-propanol to 2-methylalanine, $(CH_3)_2CNH_2COOH$.

Manufacture

Reduction of nitro alcohols to alkanolamines proceeds smoothly, even at room temperature, by conventional chemical or catalytic methods. Both AMP and AEPD are purified by distillation. Tris Amino is purified by recrystallization. 2-Dimethylamino-2-methyl-1-propanol is manufactured from AMP by hydrogenation in the presence of formaldehyde and purified by distillation.

Health and Safety

Alkanolamines are slightly toxic by ingestion. Undiluted DMAMP and AMP cause eye burns and severe skin irritation. AEPD is a severe eye irritant. Stearate soaps are nonhazardous.

Uses

Uses of alkanolamines include emulsions; pigment dispersants for water-based paints (see Pigments); resin solubilizers (see Resins, water-soluble); catalysts for textile resins, coatings resins, adhesives, etc; corrosion protection for steam-boiler condensate-return lines (see Corrosion and corrosion inhibitors); formaldehyde scavenging; and synthetic applications (see Alkyd resins; Coatings, resistant). Tris Amino is used pharmaceutically to counteract CO_2-induced acidosis.

ROBERT H. DEWEY
ALLEN F. BOLLMEIER, JR.
International Minerals & Chemical Corp.

H.B. Hass and E.F. Riley, *Chem. Rev.* **32**, 373 (1943).

M. Senkus, *Ind. Eng. Chem.* **40**, 506 (1948).

J.A. Frump, *Chem. Rev.* **71**, 483 (1971).

ALKOXIDES, METAL

Metal alkoxides, also known as metal alcoholates, are compounds in which a metal is attached to one or more alkyl groups by an oxygen atom. Alkoxides are derived from alcohols by the replacement of the hydroxyl hydrogen by a metal. Alkoxides of nonmetals can be found under the corresponding compounds (eg, Boron or Silicon compounds). Metal alkyls, compounds in which the alkyl group is bound directly to the metal, are discussed under the corresponding elements (eg, Aluminum compounds; and Organometallics).

Physical Properties

Properties depend on the metal, and secondarily on the alkyl group. Many alkoxides are strongly associated depending on the nature and shape of the alkyl groups; methoxides are usually solid and nondistillable compounds, but with a larger number of methyl groups and a small atom, they become sublimable and even distillable. Many metal alkoxides are soluble in the corresponding alcohols, but magnesium alkoxides are practically insoluble. Only the distillable alkoxides, like those of aluminum, titanium, and zirconium, are soluble in weakly polar solvents. The simple alkali alkoxides have an ionic lattice and a layer-like structure, but alkaline-earth alkoxides and aluminum oxides show more covalent character. Structures are highly varied among the transition metals. Metal alkoxides are colored when the corresponding metal ions are colored.

Chemical Properties

The outstanding property of metal alkoxides is the ease of irreversible hydrolysis. Uranium hexa-*tert*-butoxide is an exception, and does not react with water.

Alcohols give equilibrium reactions:

$$MOR + R'OH \rightleftharpoons MOR' + ROH$$
$$M(OR)_x + R'OH \rightleftharpoons M(OR)_{x-1}(OR') + ROH$$
$$M(OR)(OR')_{x-1} + R'OH \rightleftharpoons M(OR')_x + ROH$$

Dihydric alcohols give addition compounds or cyclic esters. Amino alcohols react similarly.

Metal alkoxides and carboxylic acids give salts. Titanium tetraalkoxides react with only three equivalents of acid. In some cases, the products are unstable and eliminate an ester. Metal alkoxides and phenol usually form phenolates smoothly. Enols and alkoxides give chelates, with elimination of alcohol. For example, all four alkoxide groups attached to zirconium can be replaced, but only two of the four attached to titanium (zirconium has coordination number 8, titanium 6).

Metal alkoxides catalyze the Tishchenko condensation of aldehydes, the transesterification of carboxylic esters, the Meerwein-Ponndorf reaction, and other enolization and condensation reactions.

Alkoxides often react to give double alkoxides, eg, $MgAl(OR)_4$, $LiZr(OR)_9$, $UAL_4(OR)_{16}$, and $LaAl_3(OR)_{12}$. Alkaline-earth metal alkoxides decompose to carbonates, olefins, hydrogen, and methane; calcium alkoxides give ketones.

Aluminum alkoxides give ethers, alcohols and olefins. Zirconium alkoxides behave similarly. Many metal alkoxides decompose at higher temperatures to lower valency compounds, in some cases to metal.

Preparation

From metals and alcohol. Alkali, alkaline earth metals, and aluminum react with alcohols to give metal alkoxides and hydrogen as co-product.

$$M + ROH \rightarrow MOR + 1/2 H_2$$

The speed of the reaction depends on both the metal and the alcohol, and increases with increasing electropositivity and decreases with length and branching of the chain. Thus, sodium reacts strongly with ethanol, but slowly with tertiary butanol. Reaction solvents for alkali metals are ether, benzene, or xylene.

Reactions from metal oxides and hydroxides. This reaction is usually an equilibrium in which the water must be removed:

$$NaOH + C_2H_5OH \rightleftharpoons NaOC_2H_5 + H_2O$$

(also with K, Tl; not with Li);

$$V_2O_5 + 6 ROH \rightarrow 2 VO(OR)_3 + 3 H_2O$$

(also with Sb_2O_3).

From metal halides. The reaction of metal chlorides with alcohols can give mixed metal chloride alkoxides such as:

$$TiCl_2(OC_2H_5)_2 \cdot C_2H_5OH$$

The HCl formed can lead to secondary reactions, especially with unsaturated or tertiary alcohols. Metal alkoxides can be used to achieve complete replacement of halogen.

By alcoholysis and transesterification. This method is used for metal alkoxides of higher, unsaturated or branched alcohols.

From metal amides. Dimethyl and diethyl amides of some metals react smoothly to give good yields of certain metal alkoxides, such as $V(OR)_4$, that are otherwise difficult to obtain.

Other methods. Other methods include alcoholysis of organometallic compounds; alcoholysis of carbides; oxidation of organometallic compounds, an important reaction in the manufacture of long-chain alcohols by means of hydrolysis of the aluminum oxide; oxidation of metal alkoxides reduction of esters; the alcoholysis of sulfides; and the Meerwein-Ponndorf reaction:

$$Ti[OCH(CH_3)_2]_4 + 4 CCl_3CHO \rightarrow Ti(OCH_2CCl_3)_4 + 4 CH_3COCH_3$$

Commercial Alkoxides

Commercial alkoxides are generally caustic, hygroscopic powders (see Table 1 for properties of selected compounds).

Health and Safety

Metal alkoxides are strongly caustic. They are decomposed by humid air, skin moisture, and require the use of protective glasses and gloves. The health hazard reflects the toxicity of the metals they contain and of the metallic hydroxides and alcohols they form on hydrolysis.

Table 1. Commercial Alkoxides

Alkoxide	d_4^{20}, g/cm^3	Solubility
sodium methylate	0.45	33% in CH_3OH at 20°C; insol most solvents
sodium ethylate	0.25	28% in ethanol; insol most solvents
potassium methylate	0.75	33% at 20°C in methanol; insol hydrocarbons
potassium ethylate	0.65	28% at 20°C in ethanol; insol hydrocarbons
magnesium ethylate	0.48	
aluminum isopropylate	1.0346	sol aromatic and chlorinated hydrocarbons
tetraisopropyl titanate	0.97	sol org solvents
zirconium tetra-n-butylate	1.07	sol hydrocarbons

Uses

Metal alkoxides are used as drying agents or intermediates, as catalysts, as hardening and cross-linking agents and fireproofing agents.

DIETER BRETZINGER
WALTER JOSTEN
Dynamit Nobel, A.G.

D.C. Bradley, in W.L. Jolly, ed., *Preparative Inorganic Reactions*, Vol. 2, John Wiley & Sons, Inc., New York, 1965, pp. 169–186.

F. Schmidt in E. Müller and co-eds., *Houben-Weyl, Methoden der organischen Chemie*, 4th ed., Vol. 6, Part 2, G. Thieme, Stuttgart, FRG, 1963.

J.H. Harwood, *Industrial Application of the Organometallic Compounds*, Reinhold Publishing Corp., New York, 1963, pp. 199–329.

ALKYD RESINS

Alkyd resins are the polymerization products of polyhydric alcohols and polybasic acids modified with monobasic fatty acids (see Polyesters). Nonoil or oil-free alkyds, best described as saturated or hydroxylated reactive polyesters, are formed by the reaction of polybasic acids with excess polyhydric alcohols and are used in the coating industry (see Coatings, industrial; Paint).

Alkyd resins are classified by alkyd ratio or polyhydric alcohol–phthalate ratio, oil length or percent oil for alkyds containing glycerol as the only polyol, and percent phthalic anhydride. Alkyds may be roughly classified into four main types: short (30–42% fatty acid content, 38–46% phthalic anhydride content); medium (43–54% fatty acid content, 30–37% phthalic anhydride content); long (55–68% fatty acid content, 20–30% phthalic anhydride content); and very long (> 68% fatty acid content, < 20% phthalic anhydride content). The percentage of fatty acid modification greatly influences the properties of alkyd resins.

Alkyds are characterized by rapid drying, good adhesion, flexibility, mar resistance, and durability. Their principal weakness is the ease of alkaline hydrolysis of the ester groups, which form so large a part of the molecules. Through the use of special polyols or of acids containing sterically hindered hydroxyl groups, it is possible to produce alkyds that improve resistance to hydrolysis.

A wide variety of reactive chemicals and other polymeric materials modify the properties of alkyd resins. In a long-oil resin, the large number of long-chain fatty acid groups impart a nonpolar character to the molecule. In general, compatibility of alkyd resins depends upon polarity, and it is inversely proportional to the degree of polymerization. In a short-oil alkyd, the high proportion of hydroxyl groups imparts polarity and provides centers of potential reactivity with many hydroxyl-reactive materials. The aromatic ring in phthalic anhydride and the ester linkages contribute polarity, and the unsaturation in the fatty acid groups allows interpolymerization with many vinyl monomers, epoxidation, and many of the other reactions of double bonds.

Functionality Theory

In a polyester formulation, such as glycerol phthalate, trifunctional components can lead to a gelled polymer long before completion of the esterification. The conditions for the gelation of such condensation polymers has been treated mathematically by Carothers. The simplified form of the Carothers equation states that as the molecular weight becomes infinite at the gel point, $p = 2/f$, where p = extent of reaction, and f = the average degree of functionality or the average number of functional groups in the reacting molecules, considering only stoichiometric equivalents of interacting functional groups. The Carothers equation, when applied to bifunctional reactants, indicates that no gelation will occur even at 100% esterification. Thus, to avoid gelation, the average functionality of the reactant must be approximately two. If glycerol (3 hydroxyl groups) and phthalic anhydride (2 carboxyl groups) are present in equimolecular proportions, the actual functionality of the glycerol is limited to two since that is the number of carboxyl groups available for reaction. If all the primary hydroxyl groups of glycerol react first, a linear polymer will result. Therefore, the use of more glycerol than the stoichiometric quantity required by the carboxyl groups present reduces the functionality. The use of excess polyol is an important method of reducing functionality. An additional method for controlling functionality is the basis for the formation of oil-modified polyesters alkyds as they are commonly known. This involves conversion of trifunctional polyol to a bifunctional polyol by forming a monoglyceride and then conversion to a fatty acid-modified linear polymer.

Fatty acids and oils. The monobasic acid modifies the properties of the resin by its capacity to control functionality (and so allow control of polymer growth) and by its physical and chemical properties. Most of the monobasic acids used in alkyd resins are derived from natural oils (see Carboxylic acids). Drying and nondrying oils are used in the formation of alkyd resins (see Drying oils).

The degree and nature of unsaturation in the fatty acid groups dictate the susceptibility of a drying oil to absorb oxygen and cross-link to a solid film. When the fatty acids present in the alkyd are derived from semidrying or drying oils, the alkyd can undergo autoxidation at ambient temperature, the oxygen attacking the unsaturated area of the fatty acid molecule. The general mechanism of film formation is similar to that for glyceride drying oils (see Driers and metallic soaps).

Alkyds based on soybean oil give good drying rates and color retention. Alkyds based on linseed oil give faster drying and less stringent color retention. Alkyds based on dehydrated castor oil are color-retentive. Alkyds based on tung oil and oiticica oil, in admixture with other oils, dry faster and harden earlier. Alkyds based on nonoxidizing castor and coconut oils are color-retentive plasticizer resins. Fish, sunflower, walnut, and safflower seed oils are used for special properties. Tall oil fatty acids in alkyds impart slower air-drying than soybean fatty acids, but give slightly better color retention and are especially useful for baking finishes (see Carboxylic acids). Monobasic fatty acids, such as lauric, pelargonic, isononanoic, and isodecanoic acids, are used in plasticizing alkyds.

Rosin, in conjunction with oil acids, improves the alkyd's hardness, apparent drying time, gloss, and water resistance, but decreases exterior durability if much above 5–6% of the total alkyd resin. Aromatic monobasic acids stoichiometrically replace a portion of the monobasic fatty acids to impart faster drying, better color, greater hardness, and improved gloss retention on exterior exposure.

Polyhydric alcohols. Glycerol (qv) is the workhorse polyol for alkyds, closely followed by pentaerythritol (see Alcohols, polyhydric). Mixtures of pentaerythritol and ethylene glycol are used to prepare medium- and short-oil alkyds with good compatibility properties, gloss retention, and durability. Polyols such as sorbitol and diethylene glycol are used in smaller quantity.

Polybasic acids. The most important polybasic acids for alkyd resins are phthalic acid and isophthalic acid. Properly formulated isophthalic alkyds have fast drying, tough, flexible films and show superior thermal stability. The anhydrides of maleic acid and fumaric acid form polyesters as well as Diels-Alder and other adducts with unsaturated acids in drying oils. Nonconjugated linoleic acids and even oleic acid react to give noncyclic adducts. The total functionality of the resin system is increased by the reaction of these acids with the unsaturation in fatty acids; eg, replacing phthalic anhydride by maleic anhydride produces a more complex molecular structure with high viscosity, and gelation may occur.

Other dibasic acids used in alkyds to impart special properties are adipic acid, azelaic acid, and sebacic acid to impart flexibility in the alkyd structure for alkyds used as plasticizers; and tetrachlorophthalic anhydride and chlorendic anhydride to impart fire-retardant properties to the resin system (see Flame retardants).

Functionality and Formulation

In alkyd formulation, the polymer chemist generally seeks to prepare the resin with the highest molecular weight at low acid numbers without encountering gelation. Patton has derived an alkyd constant based on an adaptation of the Carothers functionality equation. His proposed alkyd constant is the ratio of m_o/e_A and equal to unity, for the following:

$$K = \frac{m_o}{e_A} = 1$$

where m_o = total number of acid and hydroxyl-bearing molecules in a given alkyd composition at the beginning of a condensation between carboxyl and hydroxyl groups, and e_A = total number of acid equivalents in the reaction. Patton showed that those alkyds based on phthalic anhydride tend to have low alkyd-constant values (1.005 ± 0.014), whereas those based on isophthalic acid tend to have higher values (1.05 ± 0.008). An alkyd constant lower than 0.05 unit from the above values indicates a probability of encountering premature gelation short of 100% reaction.

Synthesis

The fatty acid method is the simultaneous direct esterification of all ingredients at 220–260°C. Any polyhydric alcohol or polyhydric alcohol blend, fatty acids not available as glycerides, and specialty fatty acids that have been segregated and refined for specific alkyd performance can be used.

The alcoholysis or monoglyceride method is the most common method now used. It starts with an oil, a polyol, and a dibasic acid or anhydride and catalysts such as litharge (PbO), calcium hydroxide, lithium carbonate, or their soaps. The degree of alcoholysis has an important bearing on the properties of the resulting resin. The monoglyceride is formed at 225–250°C by reaction of the triglyceride oil and glycerol or other polyol. When a dibasic acid is added to the defunctionalized glyceride, a homogeneous resin will result.

In general, alcoholysis gives a slower rate of esterification at somewhat higher acid numbers than the fatty acid method; bodying and gelation occur at slightly higher acid numbers; air-dry set time is somewhat slower; and the resin tolerates more aliphatic thinner.

Other methods include the fatty acid-triglyceride oil process, acidolysis process with polybasic acids, and the fusion (absence of solvent) or solvent or azeotropic process.

Uses

95% of the alkyds produced are used in the coatings industry (see Coatings, industrial; Paint). With modifications they are used to improve gloss, adhesion, durability, cold-check resistance of lacquers, drying, and curing speed.

H.J. LANSON
Poly-Chem Resin Corporation
(present affiliation
LanChem Corporation)

T.C. Patton, *Alkyd Resin Technology*, Interscience Publishers, a division of John Wiley & Sons, Inc., New York, 1962.

H.J. Lanson in J.K. Craver and R.W. Tess, eds. *Applied Polymer Science*, American Chemical Society, Washington, D.C., 1975, Chapt. 37.

D.H. Solomon, *The Chemistry of Organic Film Formers*, John Wiley & Sons, Inc., New York, 1967.

D.T. Moore, *Ind. Eng. Chem.* **43**, 2348 (1951).

ALKYLATION

Alkylation is the addition or insertion of an alkyl group into a molecule to produce significant changes in chemical and physical properties.

The prefix iso is used to denote one methyl group on the next-to-terminal carbon atom and to mean a branched chain.

Alkylation of Paraffins

Alkylation of paraffins can be carried out either thermally or catalytically. Thermal or noncatalytic alkylation of a paraffin with an olefin is not commercially practiced. It is carried out at high temperatures and pressures of 2–41 MPa (3000–6000 psi). In a free-radical mechanism, instigated by chain initiators, the free radical derived from the paraffin adds to the olefin and the cycle is completed by a chain-transfer reaction with the paraffin. Halogen-containing reaction modifiers can increase the yield of alkylate.

Catalytic alkylation of paraffins involves the addition of an isoparaffin containing a tertiary hydrogen to an olefin using conventional protonic and Lewis catalysts. The process is used by the petroleum industry to prepare highly branched paraffins, mainly in the C_7 to C_9 range, that are high quality fuels for spark-ignition engines. The overall process is a composite of complex reactions and requires rigorous control of the operating conditions and of the catalyst to assure predictable results. The process conditions and the product composition depend upon the particular hydrocarbons involved (see Catalysis).

Large-scale alkylation of ethylene is not carried out since ethylene is too valuable a chemical to convert to a fuel. In alkylation with ethylene, aluminum chloride is used, promoted by either hydrogen chloride or ethyl chloride. The sequence involves the formation of the *tert*-butyl cation; the addition of the cation to ethylene; and rearrangement followed by transfer of hydride (H^-) from isobutane. The same mechanism applies in the alkylation of higher olefins but with more pronounced isomerization and side reactions.

Propene is alkylated commercially as a component of a C_3–C_4 fraction (see Propylene). It can be brought into reaction with an isoparaffin in the presence of either concentrated sulfuric acid or hydrogen fluoride. Aluminum chloride, unless modified in activity, gives undue side reactions. The heptanes produced by alkylation of isobutane are mainly 2,3,- and 2,4-dimethylpentane. Sulfuric acid and hydrogen fluoride are suitable catalysts for alkylation with 1- and 2-butenes.

Alkylation in the petroleum industry. Isoparaffin-olefin alkylation is used to prepare branched paraffins that distill in the gasoline range (up to ca 200°C) for use as fuels with high antiknock properties. Polymerization processes produce fuels with lower octane ratings (particularly in the case of leaded fuels) (see Petroleum).

Commercial alkylation is carried out with either sulfuric acid or hydrogen fluoride catalysts. The process requires isobutane. The alkylation reaction is a two-phase system with a low solubility of the isobutane reactant in the catalyst phase. Efficient mixing with fine subdivision must be provided.

Alkylation of Aromatic Hydrocarbons

Alkylation can be carried out with olefins, alcohols, halides, and ethers. The attacking carbonium ion in these electrophilic substitution reactions is formed from the olefin by addition of a proton or a Friedel-Crafts type of catalyst (Friedel-Crafts reaction). The overall reaction consists of the formation of the carbonium ion from the olefin and the preferential attack by the carbonium ion of those nuclear positions on the benzene ring where electrons are most available.

The alkylation of benzene with ethylene is the main source of ethylbenzene, a commercially important intermediate in the manufacture of styrene (qv) (see Xylenes and ethylbenzene).

Yields greater than 98% are produced by commercial-scale processes: (1) aluminum chloride is used as catalyst at a temperature of ca 100°C and a pressure near atmospheric; (2) aluminum chloride is used in a homogeneous system; (3) a solid catalyst promoted by BF_3 is used in the Alkar process for alkylation and transalkylation; (4) a molecular sieve (qv) solid catalyst is used.

The alkylation of benzene with propylene and aluminum chloride or a solid phosphoric acid as catalyst produces cumene (qv) (isopropylbenzene) which is used in phenol manufacture (see Phenol; Acetone). The alkylation of benzene with higher molecular weight olefins or alkyl chlorides in the 10 to 18 carbon range yields detergent alkylate which is then sulfonated and converted to a detergent (see Surfactants).

Alkylation is dependent upon the activating influence of substituents; eg, in alkylbenzenes, such as toluene, the alkyl group increases the ease of substitution. Clean-cut alkylation to specific disubstituted compounds is rare, and separation of pure isomers is not easily accomplished.

Dealkylation becomes appreciable at elevated temperatures, particularly with tertiary alkyl groups. Selective hydrogenolysis is used to dealkylate aromatics to form benzene and naphthalene in 60–90% conversions. The use of the dehydroalkylation process is a result of the imbalance of benzene, toluene, and xylenes (BTX) production by catalytic naphtha reforming and extraction (see BTX processing).

Carbanion reactions, in the presence of alkali metals or their derivatives or hydrides as catalysts, offer potential new methods to prepare alkylbenzenes.

Alkylation of Phenols

Alkylation of phenols takes place with any catalyst that converts the olefin into the carbonium ion. The alkylation of p-cresol with isobutylene uses sulfuric acid and yields 2,6-di-t-butyl-4-methylphenol, an antioxidant (see Antioxidants). With thiophenol the reaction gives alkyl phenyl sulfides.

The alkylation of phenol with straight-chain α-olefins can be conducted thermally. Alkylation of phenols with olefins can be accomplished using metal salts of the phenol, particularly zinc, aluminum, magnesium, and calcium (see Alkylphenols). The ease of alkylation depends on the olefin and proceeds with the greatest ease from isobutylene to ethylene.

Nitrogen Alkylation

Alcohols react with ammonia (qv) or amines to give alkyl derivatives (see Amines; Ammonium compounds). The methanol–ammonia reaction, for example, is the basis for the manufacture of methylamines, and ethyl and butyl alcohols are used to give corresponding amines.

Commercial preparation of amyl amines from secondary amyl halides in an alcohol solution yields a product consisting of three parts primary amine and two parts secondary amine.

Alkylamines can be prepared by direct addition of an olefin to ammonia or an amine using alkali metals as catalysts. Ammonia and ethylene yield ethylamine; higher olefins yield amines in moderate yields. These are formed according to Markownikov's rule, ie, the NH_2 attaches to that carbon atom bearing the fewest hydrogens.

The addition of hydrogen cyanide to an olefin followed by hydrolysis (Ritter reaction) is an indirect route to alkylamines. tert-Alkylamines with C_8, C_9, and higher groups are commercially manufactured by use of butylene polymers or propylene polymers.

High yields of any alkyl derivative can be obtained by reductive alkylation, a process of controlling the reaction of either ammonia or an amine with a carbonyl compound in the presence of a proper hydrogenation catalyst. The commercially useful process is based on ketones, although aldehydes can be used (see Amines by reduction). The further alkylation of a tertiary amine leads to formation of quaternary ammonium compounds (qv).

Oxygen and Sulfur Alkylation

Ethers are produced by the reaction of an alkyl halide with an alkali metal salt of either an alcohol or phenol. Ethers of cellulose are prepared by an alkylation of alkali cellulose with an alkyl halide (see Cellulose derivatives).

The addition of an alcohol or phenol to an olefin can be used to produce secondary or tertiary alkyl ethers.

Dimethyl sulfate and diethyl sulfate are common alkylating reagents for the preparation of ethers from both alcohols and phenols. The reaction is run in the presence of a base, usually a metal hydroxide.

Dehydration of alcohols, or of an alcohol and phenol, produces ethers in a reaction catalyzed by acid-type catalysts or metal oxides at high temperature.

R.H. ROSENWALD
Consultant, UOP Inc.

J.O. Iverson and L. Schmerling, *Advances in Petroleum Chemistry and Refining*, Vol. 1, Interscience Publishers, Inc., New York, 1958.

M. Siskin and co-workers, *Symposium on New Hydrocarbon Chemistry, Division of Petroleum Chemistry, ACS Meeting*, Aug. 29–Sept. 3, 1976.

L. Schmerling, *The Chemistry of Petroleum Hydrocarbons*, Vol. 3, Reinhold Publishing New York, 1955.

ALKYLBENZENES. See BTX processing; Cumene; Toluene; Xylenes and ethylbenzene.

ALKYL HALIDES. See Bromine compounds; Chlorocarbons and chlorohydrocarbons; Fluorine compounds, organic; Iodine and iodine compounds.

ALKYLPHENOLS

Alkylphenols are normally manufactured by alkylation of phenol or its homologues with olefins or equivalent alcohols or chlorides. Most alkylphenols are used as antioxidants.

Physical Properties

The solubility in water and the density of alkylphenols or their sodium salts decreases and the solubility in hydrocarbon solvents increases with the number and size of the alkyl groups. Crypto or partially hindered phenols are soluble in Claisen solution (6.24 M KOH in CH_3OH solution), although they are insoluble in aqueous alkali. Hindered phenols give only alkali-metal salts by reaction with solid alkali hydroxide and azeotropic removal of water.

The boiling points of alkylphenols are reduced when the ortho positions are substituted.

Chemical Properties

Their chemistry is similar in type to that of phenol with respect to chlorination, bromination, sulfonation and nitration.

Bulky alkyl groups ortho to the hydroxyl group drastically decrease reactivity.

Commercially important reactions include ether formation with an alkyl halide, sulfate, etc, in the presence of alkali. The most important etherification reaction is polycondensation with ethylene oxide to give alkylphenol ethoxylates (see Surfactants).

Aldehydes condense readily under both acidic and basic conditions with phenols unsubstituted in positions ortho and/or para to the hydroxyl group to give initially hydroxyalkyl derivatives. These are usually very unstable compounds and immediately react further to give polymers or diphenylmethanes, depending on the nature of the phenol used (see Phenolic resins). Dialkylphenols containing one unsubstituted ortho or para position give diphenylmethane derivatives on reaction with aldehydes which are important rubber antioxidants (see Antioxidants). Condensation of alkylphenols with formaldehyde and dimethylamine readily gives the expected Mannich bases.

Alkylphenols condense under mild alkaline conditions with acrylate esters to give para-substituted propionates. Mild conditions are necessary for the nitration and halogenation of most phenols containing tertiary alkyl groups ortho or para to the hydroxyl group; otherwise, replacement of these groups or 2,5-cyclohexadienone formation can take place. Alkylphenols react readily with sulfur mono- and dichlorides giving, respectively, bisphenol disulfides and bisphenol monosulfides which are used as lubricating-oil detergents and antioxidants. 2,4,6-Trialkylphenols and halogens or nitric acid react to give the 2,5-cyclohexadienone derivatives.

The only oxidation products that have been considered commercially, largely as antioxidants, are based on the coupling of 2,6-dialkylphenols to diphenoquinones by chemical oxidation or catalyzed air oxidation (see Polyethers).

Manufacture

The olefins used are propylene, 1- and 2-butene, isobutene, isopentene, isoheptene, diisobutylene, nonenes, dodecenes and styrene. The alkylation reaction takes place at or near atmospheric pressure in the presence of protonated acids or metal phenoxide catalysts. Alkylations may also be performed over solid acidic clay or molecular-sieve catalysts. The nature of the products obtained in alkylation of substituted phenols, such as cresols and xylenols, depends to a large extent on the positions already occupied and the sizes of the alkyl groups as well as the size of the entering alkyl group.

The ease of ring alkylation alkyl phenols follow the order tertiary > secondary > primary > alkyl. Ease of alkylation on the ring position is ortho > para > meta. On the other hand, the thermodynamic stability is just about the reverse of the above rates, ie, primary > secondary > tertiary and meta > para > ortho.

Thus, ortho t-alkylation takes place readily even with mild metal phenoxide-type catalysts. More drastic conditions, eg, higher temperatures, stronger catalyst, and longer reaction time convert ortho-t-alkylphenols to the paraisomers and finally lead either to de-t-alkylation or isomerization to the meta-isomer. On the other end of the spectrum, one finds that meta-primary-alkylphenols are the most stable. Thus both coal tar (1000°C) with aluminum chloride catalysis (RT) produces a 2:1 meta/para cresol isomer ratio.

Alkylation-dealkylation with isobutylene was used to separate meta- from para-cresol. This is based on the fact that p-cresol will di-t-butylate in the ortho positions giving the lower boiling derivative whereas m-cresol yields the higher boiling 2,4-di-t-butyl-5-methylphenol. (The 6-position does not t-butylate owing to excessive steric hindrance). Some examples are shown in Table 1. Secondary alkyl groups, being more stable, are more difficult to remove from the ring than tert-alkyl groups.

Acid catalysts used are the phosphoric acid on Fuller's earth, sulfuric acid, or oleum and aluminum chloride. 3-tert-Butylphenol is prepared by direct high temperature (190°C), reaction of phenol and isobutene, and by isomerization of the 2- or 4-isomers or mixtures.

H.W.B. REED
Imperial Chemical Industries Limited

Reviewed by
GERD LESTON
Koppers Co., Inc.

S.H. Patinkin and B.S. Friedman in G.A. Olah, ed., *Friedel-Crafts and Related Reactions*, Vol. 2, Part I, Interscience Publishers, a division of John Wiley & Sons, Inc., New York, 1964, pp. 75–97.

R. Stroh and co-workers in W. Foerst, ed., *Newer Methods of Preparative Organic Chemistry*, Vol. 2, Academic Press, Inc., New York, 1963, pp. 337–359.

M.W. Ranney, *Antioxidants: Recent Developments*, Noyes Data Corporation, Park Ridge, N.J., 1979.

Table 1. Properties and Specifications of Important Alkylphenols

Alkylphenol	Mp, °C	Bp, °C/kPa[a]	Solubility Water	Solubility Hydrocarbons
2-isopropyl-5-methyl-phenol(thymol)	50.5	231.8/101.3	insol	sol
4-tert-butylphenol	99	240/101.3	insol	sl sol
5-methyl-2-tert-butylphenol	22	244/101.3, 171/13.3	sl sol[c]	sol
4-methyl-2-tert-butylphenol	51	241/101.3	sl sol	sol
2-phenylphenol	56–57	283–286/101.3	insol	sol
4-phenylphenol	159–163		insol	sl sol
2,4-dimethyl-6-tert-butylphenol	22	250/101.3, 174/13.3	insol	sol
2,4-di-tert-butylphenol	52	264/101.3, 190/13.3	insol	sl sol
2,6-di-tert-butylphenol	37	253/101.3, 161/6.7	insol	sol
4-octylphenol	81–83	280–302/101.3	insol	sl sol
4,4'-isopropylidenediphenol (bisphenol A)	153	220/0.53	insol	sl sol
2,4-dimethyl-6-(1-methylcyclohexyl)-phenol		180/3.0	insol	sol
nonylphenol		295–320/101.3	sl sol	sl sol
4-methyl-2,6-di-tert-butylphenol	70	265/101.3, 190/13.3	insol	sol
dodecylphenol		322–335/101.3	insol	sl sol
2,2'-thiobis(4-methyl-6-tert-butylphenol)	84–87		insol	sol
4,4'-thiobis(5-methyl-2-tert-butylphenol)	161		insol	sol
2,2'-methylenebis(4-methyl-6-tert-butylphenol)	130–132.5		insol	sl sol
2-methyl-4,6-di-tert-nonylphenol			insol	sol
2,2'-methylenebis(4-methyl-6-(1-methylcyclohexyl)phenol)	130		insol	insol
4,4'-methylenebis(2,6-di-tert-butylphenol)	155	217/0.13	insol	sol
octadecyl 3-(3,5-di-tert-butyl-4-hydroxyphenyl)propionate	49–54		insol	sol
1,1,3-tris(2-methyl-4-hydroxy-5-tert-butylphenyl)butane	183		insol	sl sol
tris(3,5-di-tert-butyl-4-hydroxybenzyl) isocyanurate	221		insol	sl sol
1,3,5-trimethyl-2,4,6-tris(3,5-di-tert-butyl-4-hydroxybenzyl)benzene	244		insol	sl sol
pentaerythrityl tetrakis[3-(3,5-di-tert-butyl-4-hydroxyphenyl)propionate]	110–125		insol	sl sol

[a]To convert kPa to mm Hg multiply by 7.50.
[b]Forms hydrate $C_{11}H_{16}O \cdot \frac{1}{4} H_2O$, mp, 37°C.

ALKYL PHOSPHATES.   See Phosphoric acids and phosphates.

ALKYL SULFATES.   See Sulfuric acid and sulfurous esters.

ALLANTOIN.   See Uric acid.

ALLENE.   See Hydrocarbons.

ALLERGENS.   See Industrial hygiene and toxicology; Anti-asthmatic agents.

ALLITOL.   See Alcohols, polyhydric.

ALLOSE, $C_6H_{12}O_6$.   See Carbohydrates; Sugar.

ALLULOSE, $C_6H_{12}O_6$. See Carbohydrates; Sugars; Syrups.

ALLYL COMPOUNDS

Allyl compounds occur widely in nature although they are main constituents of only a few natural substances, such as oil of mustard, oil of cloves, and oil of sassafras. Large commercial volumes are derived synthetically. Most of these are used as intermediate in chemical processes or as polymerization monomers (see Allyl monomers and polymers).

Reactions

The term allylic is applied generally to reactions involving the hydrogen atoms on carbons adjacent to double bonds. These hydrogens are especially labile towards radical attack because of the stabilizing resonance of the allyl radical.

$$\left(-CH{=}CH\dot{C}H- \leftrightarrow -\dot{C}HCH{=}CH- \right) \text{ or } \overline{-CH{\cdots}CH{\cdots}CH-}$$

The allylic hydrogens are similar to benzylic hydrogen which is labilized by ring interaction.

The oxidation of allylic unsaturation is complex. Peroxides may form at the allylic carbon, or the reaction may proceed to give products such as acrolein (qv) and acrylic acids (qv). Epoxidation (qv) of allylic alcohol gives glycidol, a commercially important intermediate. The double bond undergoes additions typical of olefins; eg, allyl chloride reacts with hypochlorous acid to give 70% 2,3- and 30% 1,3-glycerol dichlorohydrins (see Chlorohydrins). In displacement reactions, allyl chloride reacts faster than alkyl halides to give allyl alcohol and amines and ammonium salts with ammonia and amines. A common reaction of allyl groups is the allylic rearrangement which can occur in isomerizations, additions, or displacement reactions via probably the 3-carbon cation $(CH_2{=}CH{=}CH-)^+$.

Table 1 lists the physical properties of important allyl compounds. Allyl chloride (3-chloropropene) is a colorless, pungent, mobile liquid prepared by noncatalytic low pressure, high temperature chlorination of propylene via an allyl radical intermediate. Allyl alcohol (2-propen-1-ol) is a colorless, pungent, mobile liquid and potent lacrimator. It is prepared by the alkaline hydrolysis of allyl chloride; oxidation of propylene to acrolein, which in turn reacts with a secondary alcohol to yield allyl alcohol and a ketone; and isomerization of propylene oxide over a lithium phosphate catalyst. Important allyl esters are allyl acetate, diethylene glycol bis(allyl carbonate), diallyl maleate, diallyl phthalate, and triallyl cyanurate manufactured by direct esterification of acids and anhydrides, from acid chlorides, by transesterification, and from allyl chloride (see Esterification; also Esters).

Regardless of the manufacturing process, the bulk of allyl alcohol is consumed in the manufacture of glycerol (qv).

Table 1. Physical Properties of Allyl Chloride and Allyl Alcohol

	Allyl chloride	Allyl alcohol
molecular weight	76.53	58.08
boiling point, °C	44.96	96.9
freezing point, °C	−134.5	−129
critical temperature, °C	ca 240.7	272
density, d_4^{20}, g/cm³	0.9382	0.8520
refractive index, n_D^{20}	1.4160	1.4127
vapor pressure at 20°C, kPa (mm Hg)	39.4 (295.5)	2.5 (19)
latent heat of vaporization at bp, kJ/mol (cal/mol)	29.0 (6940)	40.0 (9550)
specific heat J/(kg·°C) [cal/(g·°C)]		
vapor 1000°C	962 [0.230]	3095 [0.740]
liquid 40°C	1652 [0.395]	2217 [0.530]
viscosity at 20°C, mPa·s (= cP)	3.36	1.37
solubility in water at 20°C, %	0.36	∞

Health and Safety

Allyl compounds are generally toxic, irritating to the respiratory tract and eyes. The effects of exposure may be cumulative and delayed. Allyl esters are less toxic than the alcohols and chloride. Skin absorption is dangerous. Allyl compounds are normally stable at room temperature. Prolonged exposure to air can present a hazard. Contamination with either peroxides or catalytically active transition metals can result in strong, rapid exothermic polymerization.

Uses

Allyl chloride is used in the preparation of glycerol and epoxy resins (qv) (see also Alkyd resins). Other derivatives, such as esters, are used in coatings (see Electroplating), latex paints (see Paint), bonding, and in various resins.

HARRY H. BEACHAM
FMC Corporation

R.H. DeWolfe and W.G. Young, *Chem. Rev.* **56**, 753 (1956).

P. de Mayo, ed., *Molecular Rearrangements*, Pt. I, Interscience Publishers, a division of John Wiley & Sons, Inc., New York, 1978.

C.E. Schildknecht, *Allyl Compounds and Their Polymers*, Vol. 28, John Wiley & Sons, Inc., New York, 1973.

ALLYL MONOMERS AND POLYMERS

Allyl compounds (qv) comprise a large group of ethylenic compounds with unique reactivities and uses often contrasting with those of typical vinyl-type compounds (eg, styrenes, acrylics, vinyl esters).

Reactivity and Properties

Most monoallyl compounds, such as allyl acetate, give only viscous liquid polymers. This is explained by the low reactivity of the allyl double bond, together with the prevalence of chain termination through reaction of the allylic H atoms:

$$-CH_2\dot{C}H \quad + \quad CH_2{=}CH \quad \rightarrow -CH_2CH_2 \quad + \quad CH_2{=}CH$$
$$\quad | \qquad\qquad\qquad | \qquad\qquad\qquad\qquad | \qquad\qquad\qquad\qquad |$$
$$CH_2O(O)CCH_3 \quad CH_2O(O)CCH_3 \quad CH_2O(O)CCH_3 \quad \dot{C}HO(O)CCH_3$$

Thus, a polymer molecule of relatively low molecular weight is formed, as well as a new free radical. This radical can begin to form another polymer molecule and therefore has the tendency to undergo ready chain transfer.

With few exceptions, such as the poly-1-alkenes by Ziegler catalysts (see Olefin polymers), few monoallyl monomers have yet been polymerized as useful, well-characterized products of high molecular weight. In special cases, allyl compounds such as diallyl (1,5-hexadiene) and diallyl ether can form high polymers by addition of active hydrogen atoms (eg, polymerization of diallyls with dimercaptans).

The most commercially important polymer is based on diallyl diglycol carbonate or diethylene glycol bis(allyl carbonate), designated as CR-39 monomer, obtained from the reaction of diethylene glycol bis(chloroformate) with allyl alcohol in the presence of alkali. Purified CR-39 is a colorless liquid of mild odor, which is miscible with higher alcohols, ketones, esters, hydrocarbons, and allyl halides, but insoluble in water (see Table 1 for physical properties). Unlike styrene and many acrylic monomers, CR-39 does not polymerize readily by heating without an added initiator. Properties of the typical CR-39 polymer castings are specific gravity (25°C), 1.32; refractive index (20°C), 1.50; light transmission (0.62 cm%), 89–92; Rockwell M hardness, 95–100; tensile strength, 34.5–41.4 MPa (5000–6000 psi).

The physical properties of the diallyl phthalate monomers, allyl acrylic monomers, diallyl esters of maleic, fumaric and adipic acids, and triallyl cyanurate monomers are given in Table 1. Four additional polyfunctional allyl ester monomers recently available are diallyl itaconate, diallyl chlorendate, triallyl citrate, and triallyl trimellitate.

Table 1. Properties of Some Allyl Monomers

Monomer	Bp, °C	Density °C, g/cm³	Refractive index °C	Viscosity at 25°C, mPa·s (= cP)
Diallyl carbonate				
ethylene glycol bis(allyl carbonate)	122	1.114[20]	1.443[20]	
CR-39, diethylene glycol bis(allyl carbonate)	160	1.143[20]	1.4503[20]	9
triethylene glycol bis(allyl carbonate)	poly-merized	1.135[20]	1.4520[20]	
ethylene glycol bis(meth-allyl carbonate)	142	1.110[20]	1.4490[20]	
Diallyl phthalate				
diallyl-*o*-phthalate (DAP)	161[a]	1.117[25]	1.518[25]	12
diallyl isophthalate (DAIP)	181[a]	1.124[20]	1.521[25]	17
Diallyl esters of maleic, fumaric, and adipic acids				
diallyl maleate	112[a]	1.070[25]	1.4664[20]	4.3
diallyl fumarate	140[b]	1.0516[25]	1.4670[25]	3.0
diallyl adipate	140[a]	1.0235[20]	1.4544[20]	4.1
Allyl acrylic esters				
allyl acrylate	65[c]	0.935[25]	1.429[25]	
allyl methacrylate	150[c]	0.930[25]	1.453[25]	13
Triallyl cyanurates				
triallyl cyanurate (TAC)	140[d]	1.1133[30]	1.5049[25]	12.6[e]
triallyl isocyanurate (TAIC)	126[f]	1.1720[30]	1.5115[25]	ca 100

[a] At 0.53 kPa (4.0 mm Hg).   [d] At 67 Pa (0.5 mm Hg).
[b] At 0.40 kPa (3.0 mm Hg).   [e] At 30°C.
[c] At 13.3 kPa (100 mm Hg).   [f] At 40 Pa (0.3 mm Hg).

Diallyl *o*-phthalate (DAP) is made by esterification of phthalic anhydride using an excess of allyl alcohol. Complete polymerization of DAP requires higher concentrations of free-radical initiator than typical vinyl polymerizations. Diallyl isophthalate (DAIP) polymerizes on heating with peroxides somewhat faster than the ortho isomer, undergoes less cyclization, and yields cured polymers of higher heat resistance.

Diallyl succinate, adipate, and sebacate are commercially important allyl esters. Diallyl maleate and diallyl fumarate are strong smelling, high boiling liquid monomers prepared by reactions of allyl alcohol with maleic anhydride or with respective acids.

Allyl methacrylate is made by alcoholysis of methyl methacrylate by allyl alcohol in the presence of sodium methylate and a polymerization inhibitor. Both esters are made by esterification of the acids by allyl alcohol.

Triallyl cyanurate (TAC), 2,4,6-tri(allyloxy)-*s*-triazine, is a colorless monomer prepared by the gradual addition of cyanuric chloride to an excess of allyl alcohol in the presence of concentrated aqueous alkali (see Cyanuric and isocyanuric acids). TAC is a relatively stable enol isomer of the more stable *N*-allyl imide, triallyl isocyanurate (TAIC) (see Table 1).

Uses

CR-39 monomer is used to produce cast sheets, lenses, and shapes with outstanding scratch resistance and optical and mechanical properties; these are the most important clear, organic optical materials shaped by casting and machining processes. DAP polymers are used in thermosetting molding plastics with solvent and heat resistance, electrical properties, and dimensional stability. Diallyl esters are used as copolymers to cross-link for heat and solvent resistance properties in plastics, coatings, and adhesives; curing is promoted by heating with small concentrations of peroxide or by uv or high energy radiation. Properties of the allyl monomers can be modified by minor proportions of comonomers.

C.E. Schildknecht
Gettysburg College

C.E. Schildknecht, *Allyl Compounds and Their Polymers (Including Polyolefins)*, Interscience Publishers, a division of John Wiley & Sons, Inc., New York, 1973.

G.B. Butler and R.J. Angelo, *J. Am. Chem. Soc.* **79**, 3128 (1957).

C.E. Schildknecht in *Polymerization Processes, High Polymers*, Vol. 29, Interscience Publishers, a division of John Wiley & Sons, Inc., New York, 1977, Chapt. 2.

ALUMINUM AND ALUMINUM ALLOYS

Aluminum, the most abundant metallic element on the surfaces of the earth and moon, is a silver-white metallic element in group III of the periodic table. Naturally occurring aluminum consists of a single isotope having a mass number of 27. The electronic configuration of the atom is $1s^2 2s^2 sp^6 3s^2 3p^1$. Its many desirable physical, chemical, and metallurgical properties make aluminum the most widely used nonferrous metal.

Physical Properties

Properties vary with purity and alloying. Table 1 lists physical properties for aluminum of 99.99% min purity. Naturally occurring aluminum consists of a single stable isotope, ^{27}Al, but a number of radioactive isotopes have been artificially produced (see Radioisotopes).

Thermal and electrical conductivities are related by the equation: $K = 2.1\lambda T \times 10^{-6} + 12.55$, where K = thermal conductivity in W/(m·K), λ = electrical conductivity in S/m, and T = temperature, K. The electrical resistivity of aluminum of 99.996% purity is 2.6548×10^{-8} $\Omega \cdot m$ at 20°C. It becomes superconductive at temperatures below 1.187 K. Resistivity at very low temperatures is strongly increased by impurities. Aluminum is cathodic to magnesium and anodic to zinc, cadmium, iron, nickel, and copper. Aluminum electrode potentials are highly irreversible in aqueous solutions and vary significantly with pH. Hydrogen is the only gas known to be appreciably soluble in solid or molten aluminum.

Chemical Properties

Aluminum exhibits a valence of $+3$ in most compounds, except a few high temperature monovalent and divalent gaseous species. It reacts with oxygen with a heat of formation of 1675 kJ/mol (400 kcal/mol): $2 Al + 3/2 O_2 \rightarrow Al_2O_3$. In dry air at room temperature, this reaction is self-limiting, producing a highly impervious film about 5 nm-thick, self-healing if scratched, which provides stability in ambient temperature exposures and resistance to corrosion by seawater and other aqueous and chemical solutions. Thicker oxide films are formed at elevated temperatures and other conditions of exposure. Molten aluminum is protected by an oxide film, and oxidation of the liquid proceeds very slowly in the absence of agitations.

At high temperatures, aluminum reduces many compounds containing oxygen, particularly metal oxides. These reactions are used in the manu-

Table 1. Physical Properties of Aluminum

Property	Value
melting temperature, °C	660.2
boiling point, °C	2494
crystal structure	fcc
density at 25°C, g/cm³	2.698
thermal conductivity at 25°C, W/(m·K)	234.3
latent heat of fusion, J/g (cal/g)	395 (94)
latent heat of vaporization at bp, kJ/g (kcal/g)	10,777 (2576)
electrical conductivity	65% IACS[a]
electrical resistivity at 20°C, Ω·m	2.6548×10^{-8}
temperature coefficient of electrical resistivity	0.0043/°C
electrochemical equivalent, mg/C	0.0932
electrode potential, V	-1.66
magnetic susceptibility	0.6276×10^{-6}/g
Young's modulus, MPa (psi)	65,000 (9.4×10^6)
tensile strength, MPa (psi)	50 (7100)

[a] International Annealed Copper Standard.

facture of certain metals and alloys, as well as in the thermite welding process. The reaction is of the type: $3 MO + 2 Al \rightarrow Al_2O_3 + 3 M$. Molten aluminum reacts violently with water and should not be allowed to touch damp tools or containers.

At high temperatures and in the absence of O_2, aluminum reacts with nitrogen, sulfur, and carbon. Aluminum is rapidly attacked by solutions of alkali hydroxides and reacts vigorously with halides as well as with chlorinated hydrocarbons in the presence of water. Aluminum hydroxide and aluminum chloride do not ionize appreciably in solution but behave in some respects as covalent compounds. The aluminum ion has a coordination number of six and in solution binds six molecules of water as $[Al(H_2O)_6]^{3+}$. On addition of a base, the hydroxyl ion substitutes for the water molecule until the normal hydroxide results and precipitation is observed. Dehydration is essentially complete at pH 7.

Other than halogenated derivatives (see above), aluminum is chemically stable in the presence of most organic compounds. Compounds Al—X—R exist where X may be oxygen, nitrogen, or sulfur, and R is an organic group (see Alkoxides). Compounds with an aluminum-to-carbon-bond include certain polymers derived from vinyl and aluminum halides. Aluminum alkyls are also aluminum–carbon-bond compounds (see Organometallics). They are used as catalysts in the production of poly-olefins and synthetic elastomers (qv).

Raw Materials

Aluminum is usually combined with silicon and oxygen to form silicate minerals, which when subjected to tropical weathering form aluminum hydroxide minerals. Bauxite, a mineral of this type containing 40–60% Al_2O_3, is the raw material for almost all production of alumina (see Aluminum compounds, aluminum oxide). Bauxite ranges in characteristics from an earthy, dark-brown ferruginous material to cream or light-pink-colored layers of hard crystalline gibbsite. Under favorable hydrological conditions, bauxite can originate from almost any alumina-containing rocks. Principal world bauxite deposits exist in Australia, Guinea, Brazil, Jamaica, Cameroon, Greece, and Surinam.

Other potential sources of alumina are numerous clays; igneous rocks rich in plagioclase feldspar; shales; alunite; aluminum phosphates; fly ash; and the kaynite–sillimanite mineral group. Cryolite, Na_3AlF_6, a main constituent of the Hall-Héroult cell electrolyte, is produced by the reaction of sodium aluminate from the Bayer process with hydrofluoric acid. The aluminum industry in the U.S. uses about 15 kg of fluoride ion per metric ton aluminum. Fluoride ion comes largely from widespread and abundant reserves of fluorspar.

Manufacture of Alumina

The Bayer process, the most economical method of manufacture, takes advantage of the reaction of aluminum trihydroxide and aluminum oxide hydroxide with aqueous caustic soda to form sodium aluminate:

$$Al(OH)_3 + NaOH \rightleftharpoons NaAlO_2 + 2 H_2O$$

$$AlO(OH) + NaOH \rightleftharpoons NaAlO_2 + H_2O$$

The reaction equilibria move to the right with increases in caustic soda concentration and temperature. The following operations are performed: (1) dissolution of the alumina at high temperature; (2) separation and washing of the insoluble impurities of bauxite (red muds) to recover the soluble alumina and caustic soda; (3) partial hydrolysis of sodium aluminate at a lower temperature to precipitate aluminum trihydroxide; (4) regeneration of the solutions for recycle to step (1) by evaporation of the water introduced by the washings; and (5) transformation of the trihydroxide to anhydrous alumina by calcination at 1450 K. Figure 1 shows the flow sheet of the Bayer process.

Variations of lime-soda sintering processes are used in treatment of high-silica bauxites. Numerous sintering and acid-extraction processes have been investigated for kaolin and other clays, as well as from dawsonite and coal ash.

Electrolysis of Alumina

Nearly all aluminum is produced by the electrolysis of alumina (Al_2O_3) dissolved in a molten cryolite-based bath, the Hall and Héroult process. The aluminum is deposited onto a carbon cathode which also serves as a

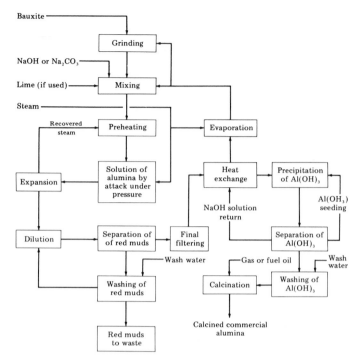

Figure 1.   Flow sheet of the Bayer process.

melt container. Simultaneously, oxygen is deposited on and consumes the cell's carbon anode(s). Figure 2 gives liquidus temperatures in the system Na_3AlF_6–AlF_3–Al_2O_3.

Cryolite ionizes to form sodium (Na^+) and hexafluoroaluminate (AlF_6^{3-}) ions. The latter dissociates to form tetrafluoroaluminate (AlF_4^-) and fluoride (F^-) ions. Alumina dissolves at low concentrations by forming oxyfluoride ions with a 2:1 ratio of aluminum to oxygen ($Al_2OF_{2n}^{(4-2n)}$); at higher alumina concentrations, oxyfluoride ions with a 1:1 ratio of aluminum to oxygen ($Al_2OF_{2n}^{(4-2n)}$) are formed. Cells are generally operated with a 2–6 wt% Al_2O_3 in the electrolyte. Saturation ranges between 7–12% Al_2O_3, depending upon electrolyte composition. Ion transport measurements indicate that Na^+ ions carry most of the current; however, aluminum is deposited. Most probably, a charge transfer occurs at the cathode interface and hexafluoroaluminate ions are discharged, forming aluminum and F^- ions to neutralize the charge of the current carrying Na^+ ions.

$$12 Na^+ + 4 AlF_6^{3-} + 12 e \rightarrow 12 (Na^+ + F^-) + 4 Al + 12 F^-$$

Oxyfluoride ions discharge on the anode, forming carbon dioxide and AlF_6^{3-} ions.

$$3 Al_2O_2F_4^2 + {}^-3 C + 24 F^- \rightarrow 3 CO_2 + 6 AlF_6^{3-} + 12 e$$

The addition of anode and cathode equations, plus solution of alumina gives the overall reaction:

$$2 Al_2O_3 + 3 C \rightarrow 4 Al + 3 CO_2$$

According to Faraday's law, one Faraday (26.8 A·h) theoretically should deposit one gram-equivalent of aluminum (8.994 g). In practice, only 85–95% of this amount is obtained.

A modern alumina-smelting cell consists of a rectangular steel sheet lined with refractory insulation surrounding an inner lining of baked carbon which can withstand the combined corrosive action of molten fluorides and molten aluminum (see Carbon and artificial graphite). Steel collector bars are joined to the carbon cathode at the bottom to conduct electric current from the cell. Current enters through prebaked carbon anodes (see Fig. 3) or through a continuous Soderberg anode.

In spite of its industrial dominance, the Hall-Héroult process has several inherent disadvantages: it requires a large capital investment; it requires expensive electric power rather than cheap thermal power; most producing countries must import alumina or bauxite; and petroleum coke supplies for anodes are limited.

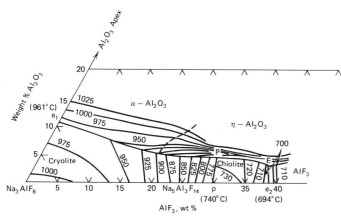

Figure 2. The system Na_3AlF_6–AlF_3–Al_2O_3. P, ternary peritectic point (28.3% AlF_3–4.4% Al_2O_3–67.3% Na_3AlF_6, 723°C); E, ternary eutectic point (37.3% AlF_3–3.2% Al_2O_3–59.5% Na_3AlF_6, 684°C); p, binary peritectic point; e_1, e_2, binary eutectic points.

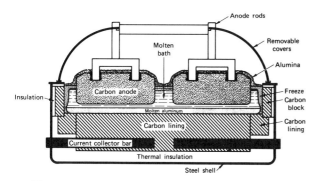

Figure 3. Aluminum electrolyzing cell with prebaked anode.

Alternative processes for obtaining aluminum include metallothermic, carbothermic, and electrolytic reduction processes. As high as 67% yields using direct carbothermic reduction of alumina can be obtained. A new closed-system Alcoa smelting process claims to use 30% less electrical power to produce aluminum than the best Hall-Héroult cells and to be environmentally more attractive.

The Hall-Héroult process cannot ensure aluminum purity higher than 99.9%. To produce metal of higher purity, other techniques are used, including electrolytic refining, fractional crystallization (see Zone refining), and fractional distillation (to produce aluminum of > 99.9995% purity).

Environmental Considerations

Fluoride emitted from aluminum-smelting cells is removed by dry scrubbers that catch particulates and sorb HF on alumina which is subsequently fed to the cells. Hydrocarbon fumes evolved during anode baking are disposed of by burning. Alumina and coke dust are collected by hoods and exhaust systems and are separated by cyclones, electrostatic precipitators, and filter bags, and recycled to the process (see Air pollution control methods). Fumeless chlorine-fluxing procedures remove hydrogen and undesirable metallic impurities from aluminum.

Aluminum Alloys

Primary metal as produced contains iron (0.05–0.6%) and silicon (0.04–0.3%) as impurities, as well as small amounts of Cu, Mn, Ni, Zn, Ti, V, Na, and Ga. Aluminum alloys consist of a matrix (a solid solution of other elements in aluminum) in which are distributed particles of intermetallic phases. Binary alloys show three types of reaction between liquid alloy, aluminum solid solution, and other phases: eutectic, peritectic, and monotectic. Table 2 shows representative data for these reactions in binary metal systems.

Table 2. Phase Transformations in Binary Aluminum Alloys

Al-M system	Reaction Type	Temperature, °C	Solubility of M, % Solid	Solubility of M, % Liquid	Other phase
Al-Si	eutectic	577	1.65	12.60	Si
Al-Cr	peritectic	651	0.77	0.41	$CrAl_7$
Al-Pb	monotectic	326.8			Pb

The ability to supersaturate and precipitate at low temperatures, which occurs in Al-Cu systems, for example, results in a variety of types of high strength alloys. Al-Si systems exhibit good fluidity and castability, and many important alloys with magnesium reach desired levels of strengthening by solid-solution hardening.

Almost all commercial alloys are of ternary or greater complexity. The alloy type is defined by the nature of the principal alloying additions, and phase reactions in several classes can be described by ternary phase diagrams. Minor alloying additions may have a powerful influence on alloy properties but a small effect on phase reactions among the major elements.

Iron and silicon are present in primary aluminum but may also be added to produce enriched alloys for specific purposes.

Silicon enhances fluidity, increases resistance to hot-cracking of other alloys, and improves pressure tightness. Copper and magnesium, added to binary-silicon alloys, yields higher strengths through heat treatment. Foundry alloys may contain a small count of titanium for grain refinement, as well as manganese, chromium, or nickel. Commercially important alloys are based on the Al-Mg-Si system. They exhibit good age-hardening capabilities and corrosion resistance. Selective compositions in the Al-Cu-Mg and Al-Mg-Zn systems have age-hardening characteristics. Silicon is sometimes added to Al-Cu-Mg alloys to modify aging response and properties. The highest strength alloys are of the Al-Cu-Mg-Zn type, containing more zinc than copper or magnesium.

The alkali and some alkaline-earth metals often have detrimental effects on the alloys (eg, sodium embrittles Al-Mg or Al-Cu-Mg alloys).

Alloys for the production of wrought products, chosen for their flexibility as well as physical, chemical, and mechanical properties, are less highly alloyed than those for foundry use and contain less iron and silicon. They may also contain small amounts of chromium, titanium, zirconium, vanadium, or manganese for added strength, grain size control, etc. The mechanical properties of wrought aluminum alloys vary with alloys, fabricated shape, and processing.

Generally, prealloyed powders for precision, strong, lightweight parts resemble wrought alloys in composition; however, elemental powder mixes omit elements that do not sinter well in powder blends.

Aluminum alloys are thermally treated during manufacture to assist fabrication or to control properties. Ingots and billets are usually homogenized prior to rolling, forging, or extrusion to improve the hot working characteristics of the alloys and increase plasticity, resistance to corrosion, and mechanical properties. Annealing is used to increase the plasticity of the metal during fabrication and to produce a highly formable material. Precipitation hardening is achieved by solution heat treatment, followed by quenching and precipitation heat treatment to obtain the desired strength, ductility, and stable mechanical properties.

Aluminum alloys are commercially available in a wide range of wrought forms produced by rolling, extrusion, drawing, or forging, and in compacted and sintered shapes made from powders.

Aluminum and aluminum alloys are often chosen because of their ability to resist corrosion. Wrought alloys of the Al, Al-Mn, Al-Mg, and Al-Mg-Si types have excellent corrosion resistance in most weathering exposures; alloys based on additions of copper, or copper, magnesium, and zinc, have significantly lower resistance to corrosion. Atmospheric corrosion of aluminum is electrochemical and depends on the flow of current between areas of anodic and cathodic behavior. Aluminum generally has a high resistance to corrosion by chemical solutions in the pH range of 4.5–8.5 (see Corrosion).

Finishes for aluminum are decorative and protective, and include anodic oxides, chemical conversion coating, electroplating, porcelain enameling, and painting (see Metal treatments).

Uses

Aluminum is the most widely used nonferrous metal. It is used in transportation and structural applications where weight saving is important; in electrical transmission lines because of its high electrical conductivity and lightness; in the construction and building industry for siding, roofing, doors, windows, and sandwich structures for its corrosion resistance. In the transportation industry, it is used for commercial and military aircraft, automobiles, mobile homes, rail freight and tank cars, engine parts, satellites, and moon rockets; in the packaging industry for cans and household foil; in the home as cooking utensils, refrigerators, other appliances, sporting equipment, and tools; in the chemical and petrochemical industries for piping and tanks; and in explosives and rocket fuels.

W.A. ANDERSON
W.E. HAUPIN
Aluminum Company of America

G. Gerard, ed., *Extractive Metallurgy of Aluminum*, Vol. 1, Interscience Publishers, a division of John Wiley & Sons, Inc., New York, 1963.

K. Grojotheim, C. Krohn, M. Malinovský, K. Matiasovský, and J. Thonstad, *Aluminum Electrolysis, The Chemistry of the Hall-Heroult Process*, Aluminium-Verlag GmbH, Düsseldorf, FRG, 1977.

K.R. Van Horn, *Aluminum*, Vols. 1–3, American Society for Metals, Metals Park, Ohio, 1967.

ALUMINUM COMPOUNDS

INTRODUCTION

Aluminum, Al, is found in many rare and common minerals and ores including feldspars; micas; kaolin, $H_4Al_2Si_2O_9$; magnesium silicates; cryolite, $Na_3(AlF_6)$; bauxite, $Al_2O_3 \cdot xH_2O$; spinel, $MgAl_2O_4$; chrysoberyl, $BeAl_2O_4$; garnet, $3CaO \cdot Al_2O_3 \cdot 3SiO_2$; beryl and emerald, $3BeO \cdot Al_2O_3 \cdot 6SiO_2$; and turquoise, $Al_2(OH)_3PO_4 \cdot H_2O$ (colored blue or green with copper compounds). Impure mixtures of hydrated aluminum oxides (gibbsite, boehmite, or diaspore, colloidal $Al(OH)_3$) are produced by leaching (under extreme conditions) of silicon and alkalies from the clay minerals (aluminum silicates) that are principal constituents of ordinary soil. The composition of bauxite ranges from $AlO(OH)$ to $Al(OH)_3$, depending on the source. The most important aluminum hydroxides are $Al(OH)_3$, as either gibbsite (α-$Al(OH)_3$), bayerite (β-$Al(OH)_3$), or nordstrandite; and $AlO(OH)$ as boehmite, the principal constituent of European bauxites, or diaspore.

Chemical Behavior

Aluminum is amphoteric, forming both aluminum salts and aluminate. Hydrated aluminum ions tend to hydrolyze which complicates many reactions of aluminum compounds in aqueous solution. Aluminum does not react with water under ordinary conditions but reacts with steam with production of hydrogen; amalgamated aluminum foil reacts with water to produce the hydrous oxide. The metal reacts with common strong acids (except nitric, in which it becomes passive) and with solutions of strong bases.

Aluminum forms binary compounds with nonmetallic elements, eg, AlX_3 (X = H, F, Cl, Br, I), Al_2Y_3 (Y = S, Se, Te), AlZ (Z = N, As, Sb, P), and Al_4C_3.

All except the nitride and fluoride are hydrolyzed by water, some incompletely to hydroxy-aquo Al^{3+} ions, and others completely to hydrated aluminum hydroxide and the volatile hydride of the other element. The phosphide, arsenide, and antimonide have semiconducting properties.

Anhydrous aluminum halides, covalent compounds with low melting points and poor electrical conductivity, cannot be prepared by heating the crystalline hydrates because hydrolysis occurs (see Aluminum halides). The chemical and physical properties of anhydrous AlF_3 differ markedly from those of the other aluminum halides: it sublimes without melting at ca 1300°C to produce discrete AlF_3 molecules; it is inert, and

it is decomposed only slowly by water. Anhydrous $AlCl_3$ is produced for use in industrial Friedel-Crafts reactions (eg, alkylation of aromatic compounds), and aluminum fluoride is used in the Hall aluminum process.

The anhydrous oxide corundum, Al_2O_3, one of the hardest substances known, resists hydration and attack by acids; it is used as an abrasive (see Abrasives). Several precious stones are forms of Al_2O_3 with traces of metals.

Aluminum is used as a reducing agent in the Goldschmidt (thermite) process to obtain other metals (eg, manganese, chromium, vanadium) from their compounds, particularly the oxides. Thermite, a mixture of aluminum and one of the iron oxides, has been used in welding (qv) operations.

Higher order compounds include hydrated aluminum salts (see Aluminum carboxylates; Aluminum sulfate and alums; Aluminum halides and aluminum nitrate); coordination compounds, eg, cryolite, $Na_3(AlF_6)$, and analogous salts such as $(NH_4)_3(AlF_6)$ (see Chelating agents; Coordination compounds); silicates and clays; and organometallic compounds (see Organometallics). Several aluminum compounds are useful in organic synthesis.

C.L. ROLLINSON
University of Maryland

R. Wade and A.J. Banister, "Aluminium, Gallium, Indium, and Thallium" in *Comprehensive Inorganic Chemistry*, Vol. 1, Pergamon Press, Oxford, UK, 1973, pp. 993–1064.

R. Norris Shreve, *Chemical Process Industries*, McGraw-Hill Book Company, New York, 1967.

A.J. Carty, "Aluminium-Oxygen Compounds" in *MTP International Review of Science, Inorganic Chemistry*, Series Two, Vol. 1, Butterworths, London, 1975.

ALUMINATES

Physical and Chemical Properties

Pure sodium aluminate, $Na_2Al_2O_3$ or $NaAlO_2$, is a hygroscopic white crystalline solid. It is biaxial-negative with refractive indexes: $a = 1.566$, $b = 1.575$, and $c = 1.580$, all ± 0.003. Commercial grades of sodium aluminate contain water of hydration and excess sodium hydroxide. In solutions, high pH retards the reversion of sodium aluminate to insoluble aluminum hydroxide.

Manufacture

The form of sodium aluminate produced depends upon the manufacturing process used. Small amounts of sodium aluminate are prepared by fusion of equimolar quantities of sodium carbonate and aluminum acetate at 800°C. Commercial quantities are made from alumina via the reaction of sodium hydroxide with amorphous alumina or aluminum metal. Alumina used in the manufacture of sodium aluminate is either in the monohydroxide or trihydroxide form, the latter occurring naturally as gibbsite, hydrargillite, or as a product of the Bayer process. High yields of sodium aluminate are formed by the digestion of aluminum trihydroxide in aqueous caustic at atmospheric pressure and near boiling temperature.

Uses

Aluminates are used to treat industrial or municipal water supplies; in lime softening; as a coagulant aid; in the papermaking industry to improve size and filler retention; to solubilize binders; in preparation of of alumina-based catalysts; as adsorbents; and in insulating materials.

W.R. BUSLER
Nalco Chemical Company

J.R. Glastonbury, *Chem. Ind. (London)*, 121 (Feb. 1969).

P.L. Hayden and A.J. Rubin in A.J. Rubin, ed., *Aqueous-Environmental Chemistry of Metals*, Ann Arbor Science, Ann Arbor, Mich., 1976, pp. 317–381.

ALUMINUM CARBOXYLATES

The aluminum salts of carboxylic acids are derived from aluminum hydroxide, $Al(OH)_3$, by successive replacement of hydroxyl groups by carboxylate anions. For example, with formic acid the aluminum formate family would include dibasic aluminum formate, $(HO)_2Al(OOCH)$; monobasic aluminum formate, $(HO)Al(OOCH)_2$; and normal aluminum formate, $Al(OOCH)_3$.

General methods of preparation for the mono- and dibasic aluminum carboxylates include direct reaction of aluminum as the catalytic electrode with an aqueous solution of the acid in an electrolytic process; reaction of the metal with the acid in the presence of catalytic amounts of mercury or mercuric chloride; and reaction by double replacement between aluminum alkoxide and the sodium salt of the organic acid. Normal aluminum cabroxylates, eg, tris(formato)aluminum, are prepared by the direct action of the acid with aluminum chloride in an organic solvent or by a double displacement reaction using a soluble aluminum salt and sodium salt of the organic acid. (Aluminum distearate is the largest selling aluminum carboxylate.)

Uses

About 15–20 of the aluminum carboxylates are industrially important compounds with commercial applications. Commercial applications fall into three general areas: pharmaceutical preparations, particularly aluminum acetate and aluminum formate, based on the very low toxicity of the salts and upon either their basic or astringent properties (see Gastrointestinal agents; Antiobiotics, β-lactams); gelling agents, particularly aluminum mono-, di-, and tristearates, aluminum palmitate, and aluminum octanoate (2-ethylhexanoate); the manufacture of cosmetics, coatings, and rocket fuels (see Cosmetics; Chemicals in war); and aluminum carboxylates, which are used in the textile industry as finishing agents for waterproofing fabric and as mordants in the dyeing process.

1984 Directory of Chemical Producers, Chemical Information Service, Stanford Research Institute, Menlo Park, Calif., 1984.

The United States Pharmacopeia XX, (*USP XX–NFXV*) The United States Pharmacopeial Convention, Inc., Rockville, Md., 1980.

ALUMINUM HALIDES AND ALUMINUM NITRATE

All halogens form binary trihalides with aluminum. Those halides that are commercially important are formed from aluminum chloride and fluoride (see Fluorine compounds, inorganic). In addition, aluminum nitrate is a stable compound of minor commercial significance.

Aluminum Chloride

Aluminum chloride is marketed as anhydrous $AlCl_3$; as aluminum chloride hexahydrate, $AlCl_3 \cdot 6H_2O$; or as a 27.82%, 1.28 g/cm^3 (32°Be) aqueous solution. Anhydrous aluminum chloride is a highly hygroscopic white or light yellow material if pure, green or light gray if impure; it has either an amorphous or hexagonal crystalline structure; density (25°C)2.44 g/cm^3; heat of formation (25°C) − 705.6 kJ/mol; heat of solution (20°C) − 325.1 kJ/mol. In the gaseous states below 300°C, aluminum chloride exists primarily as the dimer, Al_2Cl_6. Above 300°C, appreciable dissociation occurs with the chloride and bromide. Aluminum chloride is soluble in hydrochloric acid and organic solvents. The monochloride, $AlCl$, is a gas which is stable only at elevated temperatures.

Preparation. Anhydrous aluminum chloride is produced commercially by the action of dry hydrogen chloride or chlorine on aluminum metal at temperatures of about 150°C.

The Toth process at 900°C produces aluminum, from clay, brown coal, or lignite and a small amount of sodium chloride, as well as aluminum free of manganese and gaseous manganese chloride. A high purity aluminum chloride is prepared for electrolytic reduction to aluminum by the Alcoa smelting process. This process produces an unusually pure, free-flowing powder with aluminum oxide, aluminum oxychloride, and any hydrogenous material kept to a minimum.

Uses. Anhydrous aluminum chloride is used as a catalyst in organic reactions such as reforming hydrocarbons and polymerization of low

molecular weight hydrocarbons (see Hydrocarbon resins; Petroleum); in Friedel-Crafts reactions; and in metallurgical and metal-finishing operations.

Aluminum Chloride Hexahydrate

Aluminum chloride hexahydrate is usually made by dissolving aluminum hydroxide, $Al(OH)_3$, in concentrated hydrochloric acid. It is used primarily in deodorant, antiperspirant, and fungicidal preparations; for water treatment (see Flocculating agents; Water); in the manufacture of alumina and alumina–silica refractories; and in textile finishing. It may be toxic on ingestion.

Aluminum Bromide and Aluminum Iodide

Aluminum bromide and aluminum iodide are produced commercially only in small quantities. Anhydrous aluminum bromide, $AlBr_3$, (density (25°C)3.01 g/cm^3; melting point 97°C; boiling point 256°C) is used as a catalyst in Friedel-Crafts reactions. Anhydrous aluminum iodide, AlI_3, (density (25°C)3.98 g/cm^3; melting point = 191°C) has limited use as a catalyst in the laboratory.

Aluminum Nitrate

Aluminum nitrate, $Al(NO_3)_3 \cdot 9H_2O$, is a white crystalline material, mp 73.5°C, soluble in cold water, alcohols, and acetone; decomposes to nitric acid and basic aluminum nitrates, $Al(OH)_x(NO_3)_y$ ($x + y = 3$); and dissociates to aluminum oxide and oxides of nitrogen above 500°C. It is commercially produced from aluminous materials such as bauxite and nitric acid. Hydrated aluminum nitrate is used primarily as a salting-out agent in the extraction of actinides.

CHARLES M. MARSTILLER[‡]
Aluminum Company of America

Reviewed by
GEORGE MacZURA
Aluminum Company of America (Alcoa)

ALUMINUM OXIDE (ALUMINA)

Aluminum oxide (alumina) occurs abundantly in nature as impure hydroxides in bauxites and laterites from which numerous purer grades of alumina are manufactured (see Aluminum and aluminum alloys).

Aluminum Hydroxides

Alumina monohydrate and hydrated alumina do not contain water of hydration but only hydroxide and oxide groups. Preparations of aluminum hydroxide obtained from solutions of aluminum salts, alkaline aluminates, or aluminum alcoholates that are mainly amorphous are called gels (see Activated Aluminas). They are amphoteric, with low solubility in aqueous solutions of intermediate pH. Colloidal aluminum hydroxide is hydrophilic. Depending on the method of preparation, not all gelatinous aluminas are amorphous.

α-Alumina trihydrate is obtained by the Bayer process (see Aluminum) from crystallized alumina trihydroxide (known as gibbsite or hydrargillite, the principal constituent of tropical bauxites). Grains of α-trihydrate precipitated by the Bayer process are aggregates of relatively spherical shape, 50–100 μm in size. Solubility of α-trihydrate and heat of solution in various bases and acids is important in connection with the manufacture of aluminum chemicals. It contains about 35% chemically bound water. Typical Bayer hydrate contains 0.4 wt% of sodium ions in the crystal lattice; along with other alumina hydrates, it chemisorbs various anions effectively.

β-Alumina trihydrate is not a product of the Bayer process but is prepared by aging of gels, gassing of sodium aluminate liquor with carbon dioxide, and calcination of aluminum oxychloride. Nordstandite, a new trihydrate, is not yet commercially produced.

α-Alumina oxide hydroxide, a well-crystallized modification of AlOOH, also known as α-alumina monohydrate or boehmite, is a principal constituent of many bauxites of the Mediterranean type. Crystallized

[‡] Deceased.

boehmite is prepared by hydrothermal digestion of aqueous slurries of trihydroxides or by solid-state reaction when gibbsite is heated in air at 383–573 K. Boehmite usually contains water in excess of the theoretical 15%. Another well-crystallized modification of AlOOH, β-alumina oxide hydroxide or diaspore, has been produced in the laboratory.

Activated Aluminas

These chemicals are obtained from various hydrated aluminas by controlled heating which eliminates most of the water of constitution. Their crystal structure is that of the transition aluminas: γ, η, χ or ρ alumina. Structural properties of the transition aluminas are given in Table 1. These transition aluminas are known as activated alumina or active alumina.

Preparation.   Type 1, the oldest commercial form still widely used, is made from Bayer α-trihydrate. Processes produce various activated aluminas in cylindrical or spherical forms, but the pore distribution is not broad and few pores are greater than several tens of nanometers in diameter. Surface areas of these products are 300–500 m^2/g. Type 2, activated bauxites with properties similar to Type 1, is obtained by thermal activation of bauxite containing alumina in the form of gibbsite. Type 3, another activated alumina, is obtained by very rapid activation of Bayer hydrate at 673–1073 K and has a pore surface area of 300–350 m^2/g. Type 4 is derived from alumina gels, principally those prepared from solutions of $Al_2(SO_4)_3$ and NH_3, from $NaAlO_2$ and an acid, or from $NaAlO_2$ and $Al_2(SO_4)_3$, and produce corresponding by-product salts. Surface areas of pores are 300–600 m^2/g.

Calcined Alumina

These are generally obtained from Bayer hydroxide.

Commercial calcined aluminas are heated to obtain α-Al_2O_3, the stable form of anhydrous alumina known also as the naturally occurring substance corundum. α-Al_2O_3 also results from solidification of molten alumina (see Gems, synthetic) or by sintering processes. Although α-Al_2O_3 has a mp of 2326 K, the relatively fine hexagonal crystal plates obtained during calcination permit sintering to occur at much lower temperatures for ceramic-product manufacture. Higher sintering temperatures improve thermal stability for refractory applications, and sintered and fused α-Al_2O_3 products made from calcined aluminas exhibit extreme hardness, resistance to wear and abrasion, chemical inertness, outstanding electrical and electronic properties, good thermal shock resistance and dimensional stability, and high mechanical strength at elevated temperatures (see also Abrasives; Ceramics; Glass; Enamels).

Preparation.   Calcination is performed in rotary kilns and/or fluid calciners (see Fluidization) similar to those used for preparing alumina for the manufacture of aluminum by electrolysis. The particle size of the unground calcined Bayer aluminas is controlled, and the resulting porous agglomerates of alumina crystals (< 1–10 μm) are nominally 44–149 μm 100–325 mesh. Physical properties begin to exhibit the effect of crystal size above ca 2–3 μm. Bulk density decreases, angle of repose increases, attrition rate and dustiness increase, and bulk handling characteristics worsen with increasing crystal size. Special Bayer calcined aluminas can be broadly categorized according to sodium content and crystal size.

Dense sintered alumina containing large (typically 50–300 μm) tablet-like crystals of α-Al_2O_3 is commonly known as tabular alumina. Fused alumina is available as larger tonnage brown alumina and higher quality white alumina. Calcined sodium β-Al_2O_3, in the forms of $Na_2O.11Al_2O_3$ and $Na_2O.5Al_2O_3$, is also available.

Uses

About 90% of alumina is used in the production of aluminum metal. The remainder, including Bayer-source alumina, chemical-grade bauxite, and calcined abrasive and refractory-grade bauxite, is consumed in other applications such as flame-retardant fillers, preparation of aluminum compounds, pigments, adsorbents, catalysts, ceramics, refractories, and abrasives (see Abrasives; Batteries and electric cells, secondary; Adsorptive separation; Drying agents; Sulfur recovery; Petroleum; Exhaust control).

Geörge MacZura
K.P. Goodboy
J.J. Koenig
Aluminum Company of America

K. Wefers and G.M. Bell, *Oxides and Hydroxides of Aluminum*, *Technical Paper No. 19*, Aluminum Company of America, Pittsburgh, Pa., 1972.

W.H. Gitzen, *Alumina as a Ceramic Material*, *Special Publication No. 4*, American Ceramic Society, Inc., Columbus, Ohio, 1970.

G. MacZura, T.L. Francis, and R.E. Roesel, *Interceram* **25**(3), 200 (1976).

ALUMINUM SULFATE AND ALUMS

Aluminum Sulfate

Aluminum sulfate hydrate, $Al_2(SO_4)_3.18H_2O$, and its aqueous solutions are known as papermakers alum or alum.

Properties.   Over 50 acidic, basic, and neutral hydrates of aluminum sulfate are formed by crystallization from aqueous solution. Only a few of these species are well characterized.

Aqueous solutions of aluminum sulfate hydrolyze at elevated temperatures to yield $3Al_2O_3.4SO_3.9H_2O$, which is structurally related to alunite, $K_2Al_6(SO_4)_4(OH)_{12}$. Clear acidified aluminum sulfate solutions can be supercooled without immediate nucleation. Aluminum sulfate dehydrates in two distinct stages: 15 molecules of water are lost between 40 and 250°C; three are lost between 250 and 420°C. The formula weight for the octadecahydrate is 648.1; specific gravity is 1.69 at 17°C.

Manufacture.   Commercial grade aluminum sulfate is produced by direct reaction of bauxite or kaolin clay with sulfuric acid (30–60%). Most alum producers use raw materials that are naturally low in iron and potassium to avoid difficult removal of iron. The iron-free grade (< 0.005% Fe_2O_3 max) is produced by using pure alumina trihydrate, $Al_2O_3.3H_2O$, in place of the bauxite or clay.

Uses.   Aluminum sulfate is used in the paper industry to clarify process waters and to control the pH of pulp slurries and setting dyes. It is also used in potable and wastewater treatment, cellulosic insulation, the manufacture of chemicals, catalysis, pharmaceuticals, dyeing operations, soaps and greases, fire-extinguishing solutions, tanning, foods, and modifying concrete (see Paper; Water).

Alums

The term alum is used in a number of different senses. Alum or papermakers alum refers to commercial aluminum sulfate. Common alum or potash alum refers to potassium aluminum sulfate, $KAl(SO_4)_2.12H_2O$ or $K_2SO_4.Al_2(SO_4)_3.24H_2O$, whereas ammonia alum is ammonium aluminum sulfate. The term also applies to a series of crystallized double

Table 1.   Structural Properties of Transition Aluminas

Form	Crystal system	Molecules per unit cell	Unit axis length, nm			Density, g/cm^3
			a	b	c	
γ	tetragonal		0.562	0.780		3.2[a]
δ	orthorhombic	12	0.425	1.275	1.021	3.2[a]
	tetragonal		0.796		2.34	
η	cubic (spinel)[b]	10	0.790			2.5–3.6
θ	monoclinic[c]	4	1.124	0.572	1.174	3.56
χ	cubic	10	0.795			3.0[a]
	hexagonal		0.556		1.344	
	hexagonal		0.557		0.864	
κ	hexagonal	28	0.971		1.786	3.1–3.3
	hexagonal		0.970		1.786	
	hexagonal		1.678		1.786	
ι	orthohombic	4	0.773	0.778	0.292	3.71[a]
	orthorhombic[d]	3	0.759	0.767	0.287	3.0

[a] Estimated.
[b] Space group O_h^7.
[c] Space group C_{2h}^3; angle of crystal, 103°20′.
[d] Space group D_{2h}^9 or C_{2v}^8.

sulfates $(M(I)M'(III)(SO_4)_2.12H_2O)$ with the same crystal structure as the common alums. Sodium and other univalent metals may replace the potassium or ammonium, and other metals may replace the aluminum; even the sulfate radical may be replaced. Examples of alums are cesium alum, $CsAl(SO_4)_2.12H_2O$; iron alum, $KFe(SO_4)_2.12H_2O$; chrome alum, $KCr(SO_4)_2.12H_2O$; and chromoselenic alum, $KCr(SeO_4)_2.12H_2O$. Pseudoalums are a series of double sulfates, such as $FeSO_4.Al_2(SO_4)_3 24H_2O$, containing a bivalent element in place of the univalent element of ordinary alums, and have different crystal structures.

Potassium aluminum sulfate (potassium alum), $KAl(SO_4)_2.12H_2O$, is a white crystal with mp 92.5°C; sp gr, 1.757; and solubility of 11.4 g/100 cm^3 H_2O at 20°C. Dehydration takes place at ca 200°C to yield a porous potassium alum, $KAl(SO_4)_2$. Potassium alum occurs naturally in alunite, $K_2Al_6(SO_4)_4(OH)_{12}$, sp gr 2.58–2.75, and kalinite, $KAl(SO_4)_2.12H_2O$, sp gr 1.75. It is commercially produced by treating bauxite with sulfuric acid and then potassium sulfate. It is used in medicines, dyeing, water purification, paper sizing, and dressing of skins.

Sodium aluminum sulfate (sodium alum), $NaAl(SO_4)_2.12H_2O$, is a colorless crystal, sp gr 1.675, mp at 61°C. Solubility is 110 g anhydrous salt per 100 cm^3 at 15°C. Sodium alum occurs naturally as mendozite. It is commercially produced from aluminum sulfate by the addition of a clear solution of sodium sulfate. The largest single use of sodium alum is as leavening acid in baking (see Bakery processes and leavening agents).

K.V. DARRAGH[‡]
Stauffer Chemical Company

Reviewed by
FRANCIS A. VIA
Stauffer Chemical Company

J.W. Mellor, *Inorganic and Theoretical Chemistry*, Vol. V, Longmans, Green and Co., New York, 1946, pp. 332–339.

AMALGAMATION PROCESS. See Mercury; Gold and gold compounds.

AMALGAMS. See Mercury.

AMATOL. See Explosives.

AMBER. See Resins, natural.

AMBERGRIS. See Perfumes.

AMERICIUM. See Actinides.

AMIDES, FATTY ACID

Amides are produced by the reaction of a fatty acid (or its acid chloride or ester) with ammonia or an amine. The simplest fatty acid amides are $RCONH_2$ (R = a saturated or unsaturated alkyl chain derived from a fatty acid moiety). Substituted amides are formed when one or both of the hydrogen atoms in $RCONH_2$ are replaced by an alkyl, aryl, or other radical.

[‡] Deceased.

Physical Properties

Many anomalous physical properties of fatty acid amides have been explained on the basis of tautomeric structures. Simple amides have unusually high melting points and low solubilities in most solvents as compared with the substituted amides, which show remarkable differences in their melting points and solubility data. All fatty amides are essentially insoluble in water and slightly soluble in nonpolar and polar solvents (solubility decreases with increasing chain length). In general, fatty amides are stable towards air oxidation, heat, and dilute acids or bases.

Chemical Reactions

An important group of fatty amides is the *N*-methylol amides, the reaction products of fatty acid amides with formaldehyde or its polymer in the presence of a catalyst such as boric acid, sodium hydroxide, or potassium carbonate, eg, $RCONHCH_2OH$. Methylenebisamides are formed when the condensation is carried out in the presence of an acid catalyst. Zelan-type water repellents are formed by the reaction of a fatty acid amide with formaldehyde and pyridine in the presence of HCl: $(RCONHCH_2(NC_5H_5))^+Cl^-$. Amides also undergo acylation, and nonionic surfactants are produced by addition of ethylene oxide (see Surfactants).

Manufacture

Simple amides are manufactured by the reaction of a fatty acid with anhydrous ammonia in a batch or continuous process using boric acid, Al_2O_3, or Ti and Zn alkoxides catalysts; from hexadecanoic acid, octadecanoic acid, and mixed fatty acids from coconut oil treated with urea and purified; or by hydrogenolysis of hydroxamic acids and hydrazides.

Fatty acids, esters, acid halides, and simple amides can be converted to substituted amides by treating with primary- and secondary alkyl- and arylamines, polyamines, and hydroxyalkylamines under suitable conditions. Industrially important products include fatty acid amides of the alkanolamide type, $RCON[(CH_2)_xOH]_y$, used as surface-active agents; substituted amides of the Lamepon type, $RCON(CH_3)CH_2COONa$, and Igepon type, $RCON(CH_3)CH_2CH_2SO_3Na$, prepared by an acylation reaction using fatty acid chlorides; and fatty acid bisamines.

Uses

Mainly used as antislip and antiblock additives for polyethylene films, fatty acid amides are also used as mold-release agents (see Abherents); in the production of water repellents (see Waterproofing and water/oil repellency); and in lubricants (see Lubrication and lubricants). The alkanolamide derivatives are used in detergents.

HARINATH B. BATHINA
RICHARD A. RECK
Armak Company

R.A. Reck, *J. Am. Oil Chem. Soc.* **39**, 461 (1962).
R.H. Potts and G.W. McBride, *Chem. Eng.* **57**(2), 124 (1950).

AMINE OXIDES

Amine oxides are *N*-oxides of tertiary amines, either aromatic (actually heteroaromatic) or aliphatic, depending upon whether the tertiary nitrogen is part of an aromatic ring system.

Physical and Chemical Properties

Amine oxides are hygroscopic solids which are sparingly soluble in nonpolar solvents but readily soluble in water and lower alcohols. Their properties vary considerably. Amine oxides are usually depicted with a coordinate-covalent bond between nitrogen and oxygen with the greater electron density residing on oxygen; dipole moments are $13-17 \times 10^{-30}$

C·m (4–5 D). Aliphatic oxides have a tetrahedral nitrogen configuration. In aromatic amine oxides, the trigonal nitrogen forces the oxygen into the same plane as the aromatic ring and permits resonance structures involving the formally nonbonded electrons on oxygen. This accounts for their added stability and special properties. Amine oxides are weaker bases than the amines from which they are derived. For unsubstituted aromatic amine oxides, pK_as generally range from 0.8–1.6. Aliphatic and mixed aromatic amine oxide pK_as range from 4.0 to 5.0. Aliphatic amine oxides behave as typical surfactants in aqueous systems. Below the critical micelle concentration (CMC), dimethyldodecylamine oxide exists as a simple basic molecule. In the presence of an anionic surfactant, any protonated amine oxide forms an insoluble salt which raises the pH of the solution. Recently, amine oxides from amidoamines, derived by amidation of fatty acids with N,N-dimethylethylenediamine or N,N-dimethyltrimethylenediamines, have become important. For all commercial oxides, physical data are usually available only as industrial specifications for solutions of these materials. Some studies of phase behavior, chain mobility in mesomorphic phases, and thermodynamic data are also available. The amine oxides from dimethyldodecyl through dimethyoctadecyl are better detergents than sodium dodecylbenzenesulfonate; in addition they are foam stabilizers.

Reactions

Most amine oxides decompose between 100 and 200°C with aliphatics giving olefin and a dialkylhydroxylamine (Cope elimination) and aromatics giving the parent amine. If there is no β-hydrogen, an alkyl group may migrate from nitrogen to oxygen (Meisenheimer rearrangement). Decomposition is catalyzed by Fe(II), generally by a free-radical mechanism. O-Acylated aliphatic amine oxides decompose to aldehydes and the acyl derivative of secondary amines (the Polonovski reaction).

Aliphatic amine oxides, in contrast to aromatic amine oxides, are readily reduced to tertiary amines by sulfurous acid at room temperature. The oxides can also be reduced by catalytic hydrogenation, or metals and acids.

Manufacture and Preparation

The most widely used method of commercial manufacture is the oxidation of aliphatic tertiary amines with hydrogen peroxide at 60–80°C using water, lower alcohols, acetone or acetic acid solvents:

$$R_3N + H_2O_2 \rightarrow R_3N \rightarrow O + H_2O$$

Aromatic oxides cannot be formed directly by H_2O_2 oxidation, but can be obtained by oxidation of the corresponding amine with perbenzoic or other peracid, but the low temperatures and large volumes of volatile solvents required are impractical. A procedure better suited to large-scale use consists in the addition of aqueous H_2O_2 to a solution of the amine in glacial acetic acid at 80–95°C. Industrial amine oxides are derived from alkyldimethylamines, alkylbis(2-hydroxyethyl)amines, and amido-amines. Because hydrogen peroxide is sensitive to heavy metals and amine oxides are readily decomposed by iron salts, sequestering agents (eg, trisodium EDTA) are generally added to the reaction mixture, particularly when water is the solvent. Total reaction time is 4–12 h. The only precaution required for storage or transport of amine oxides is that the temperature not exceed ca 60°C.

Procter and Gamble is the largest manufacturer of amine oxides and the only known producer utilizing α-olefins as raw material. However, their production is solely for captive use.

Health and Safety

Aliphatic amine oxides such as alkydimethylamine oxides and alkylbis(2-hydroxyethyl) amine oxides are practically nontoxic to slightly toxic. Commercial-concentrated products are primary skin irritants. Aromatics such as 4-nitroquinoline N-oxide, 4-hydroxylaminoquinoline N-oxide, 2-methyl, 2-ethyl, and 6-chloro derivatives of 4-nitroquinoline N-oxide are carcinogenic (see also N-Nitrosamines).

Uses

The bulk of aliphatic amine oxides is used as surfactants, detergents, and shampoos. Aromatic amine oxides are used in pharmaceuticals.

RICHARD J. NADOLSKY
Armak Company
present affiliation
Miranol Chemical Company, Inc.

P.A.S. Smith, *The Chemistry of Open-Chain Organic Nitrogen Compounds*, Vol. II, W.A. Benjamin, Inc., New York, 1966, pp. 21–28.

E. Orchiai, *Aromatic Amine Oxides*, Elsevier Publishing Co., Amsterdam, The Netherlands, 1967 (transl. by D.U. Mizoguchi).

Amine Oxides, A Literature Survey, Electrochemicals Dept., E.I. du Pont de Nemours & Co., Inc., Wilmington, Del., 1963.

AMINES

LOWER ALIPHATIC AMINES

Lower aliphatic amines are derivatives of ammonia where one, two, or three of the hydrogen atoms are replaced by alkyl groups of six carbons or less. Amines with higher alkyl groups are known as fatty amines (qv).

Physical and Chemical Properties

The important commercial amines (production > 1000 t/yr) are the following (bp °C): monomethylamine (−6.3); dimethylamine (6.9); trimethylamine (2.9); monoethylamine (16.6); diethylamine (55.9); triethylamine (88.8); dipropylamine (109.3); isopropylamine (32.4); N-butylamine (77.8); dibutylamine (159.6); diisobutylamine (139.5); cyclohexylamine (134.5); and dicyclohexylamine (256).

Commercially important lower amines are toxic, colorless gases or liquids which are highly flammable and have strong odors. The amines with lower alkyl groups are water-soluble and sold as aqueous solutions and in pure form. Amines react with water and acids to form bases and salts, respectively, analogous to ammonia. The base strengths in water of the primary, secondary, and tertiary amines and of ammonia are essentially the same. Primary and secondary amines can also react as very weak acids ($k_a = 10^{-33}$), eg, with acyl halides, anhydrides, and esters (as does ammonia). With carbon disulfide and carbon dioxide, they form alkyl ammonium salts of dithiocarbamic and carbamic acid, respectively, and with isocyanic acid and alkyl or aryl isocyanates, substituted ureas. Primary amines give alcohols with nitrous acid; secondary amines give highly toxic nitrosamines (see N-Nitrosamines). Tertiary amines give complex products with nitrous acid. On cooling the aqueous solutions, crystalline hydrates are formed. Methylamine solutions are good solvents for any inorganic and organic compounds.

Manufacture

In order of importance, methods of commercial production are as follows: (*1*) Ammonia and the alcohol are passed continuously over a dehydration catalyst such as ammonia, silica–alumina, silica, titania, tungstic oxides, clays, or various metal phosphates to produce yields of 80% or higher of amines. The reaction produces a mixture of primary, secondary, and tertiary amines which undergo separation by distillation and extraction. (*2*) Ammonia and the alcohol passed continuously over a dehydrogenation catalyst such as metallic silver, nickel, or copper, with hydrogen maintaining catalyst activity. These conditions give high conversion with good selectivity of amines; the mixtures are separated by distillations and extractions. (*3*) Ammonia and an aldehyde or ketone over a hydrogenation catalyst is an alternative in time of shortages or when a specific amine is available only by this route (see Amines by reduction). (*4*) Addition of HCN to a solution of an olefin, such as isobutylene, in an acidic medium (Ritter reaction) to produce primary amines of the type R_3CNH_2. (*5*) Amination of alkyl halides, instead of alcohols.

Health and Safety

Alkylamines are toxic; liquids and vapors can cause severe irritations to mucous membranes, eyes, and skin. Amines are flammable, with low flash points. Alkylamines give a rapid and exothermic reaction.

Uses

Lower aliphatic amines are widely used as intermediates in the manufacture of medicinal, agricultural, textile, rubber, and plastic chemicals.

ALBERT E. SCHWEIZER
ROBERT L. FOWLKES
JOSEPH H. McMAKIN
THADDEUS E. WHYTE, JR.
Air Products and Chemicals, Inc.

D.R. Stull, E.F. Westrum, and G.C. Sinke, *The Chemical Thermodynamics of Organic Compounds*, John Wiley & Sons, Inc., New York, 1969, pp. 457–467.

J. Pasek, P. Kondelik, and P. Richter, *Ind. Eng. Che. Prod. Res. Dev.* **11**, 333 (1972).

B.C. Challis and S.A. Kyrtopoulos, *J. Chem. Soc. Chem. Commun.*, 877 (1976).

FATTY AMINES

Fatty amines are classified as primary, secondary, and tertiary depending on whether one, two, or three of the hydrogen atoms in ammonia are replaced by alkyl groups. Commercial fatty amines contain mixed alkyl chain lengths based on the fatty acid occurring in nature (see Fats and fatty oils). Typical trade names for primary, secondary, and tertiary amines are Adogen (Ashland), Jetamine (Jetco), Armeen (Armak), Alamine (General Mills), and Kemamine (Humko). Trade names for diamines are Duomeen O (Armak), Jetamine D-O (Jetco), Diam 21 (General Mills), and Kemamine D-997 (Humko).

Chemical Reactions and Physical Properties

Fatty amines are essentially insoluble in water, weakly basic with ionization constants indicating that the secondary amines are somewhat stronger than the corresponding primary amines.

Fatty amines undergo reactions characteristic of aliphatic amines such as formation of cationic salts, which are used as corrosion inhibitors and in ore flotation; reactions of amines with alkyl and aryl halides to form surface-active agents, the quaternary ammonium salts; condensation with ethylene oxide and oxidation by peroxides and peracids to form nonionic surfactants (qv) such as ethoxylates and amine N-oxides, respectively; addition to unsaturated compounds; and reaction with chloroacetic acid to form substituted betaines, eg, $R(CH_3)_2 \overset{+}{N}CH_2COO^-$, used in shampoo bases (see Cosmetics).

Manufacture

Amines are most commonly prepared by the conversion of a fatty acid to a nitrile by treatment with ammonia, followed by catalytic hydrogenation of the nitrile to primary, secondary, or tertiary amine with suitable adjustment in the reaction conditions (see Amines by reduction). Primary amines are made in continuous processes using both Raney cobalt and nickel catalysts. Sometimes ammonia is added to favor the formation of primary amines and suppress secondary amines. Fatty acids can be converted directly to primary amines by catalytic hydroammonolysis at high temperatures and pressures in the presence of nickel and/or cobalt or zinc and/or chromium or rhenium. Similarly, the methyl esters and fatty acid glycerides can be directly converted to amines. Fatty aldehydes and ketones are converted to primary amines by reductive catalytic ammonolysis.

Secondary amines are made from either nitriles or primary amines by continuous removal of ammonia during the hydrogenation reaction of a nitrile using catalysts such as cobalt or nickel. Primary amines react with alkyl halides to produce secondary or tertiary amines.

Most of the industrially useful fatty tertiary amines are either symmetrical amines, dimethylalkylamines (used in manufacture of quaternaries and other derivatives), or those derived by the reaction of primary or secondary amines with ethylene oxide. The Leuckart reaction uses formaldehyde and formic acid for the alkylation. At high temperature and pressure in the presence of hydrogen and catalyst, formaldehyde can be used to produce dimethylalkylamines. Alkylation of long-chain amines to high molecular weight tertiary amines is achieved by treating the amine with a long-chain alcohol under hydrogenation conditions with Raney nickel or Cu-Cr catalyst and by treating with alkyl halide under pressure without a catalyst, or with a catalyst such as Cu. Secondary amines can be treated with olefins, carbon monoxide, and hydrogen in the presence of group VIII metal catalysts to yield tertiary amines.

Uses

Commercially important derivatives include polyethoxylated fatty amines, eg, bis(2-hydroxyethyl)cocoamine, polyoxypropylene derivatives, a large variety of quaternary ammonium compounds, amphoteric surface-active agents, N-alkylamino acids, acrylic acid ester derivatives, as well as salts of amines and diamines. Fatty amines and derivatives are used mainly in fabric softeners (30%), chemical intermediates, asphalt emulsifiers (18%), petroleum additives (13.5%), ore-flotation agents (11.5%), and textile chemicals (7%).

HARINATH B. BATHINA
RICHARD A. RECK
Armak Company

U.S. Pat. 2,287,219 (June 23, 1943), H.P. Young and C.W. Christensen (to Armour & Co.).

U.S. Pat. 2,355,356 (Aug. 8, 1944), H.P. Young, Jr. (to Armour & Co.).

Brit. Pat. 860,922 (Feb. 15, 1961), S.H. Shapiro and F. Pilch (to Armour & Co.).

N.O.V. Sonntag in K.S. Markley, ed., *Fatty Acids*, Part 5, John Wiley & Sons, Inc., New York, 1964, p. 1674.

CYCLIC AMINES

Morpholine, $\overset{\frown}{OCH_2CH_2\,NHCH_2CH_2}$, and piperazine, $\overset{\frown}{NHCH_2CH_2}$-$\overline{NHCH_2CH_2}$, are the most important cyclic amines in commercial use.

Physical and Chemical Properties

Morpholine is a colorless liquid; fp, $-4.9°C$; bp, 128.9°C; flash point (tag closed cup), 35°C; density (20°C), 0.999 g/cm³; and viscosity (20°C), 2.23 mPa·s (= cP). Piperazine is a white solid; bp, 148.5°C; fp, 109.6°C; viscosity (138°C), 0.666 mm²/s (= cSt); and flash point (Cleveland open cup), 107°C. Both are hygroscopic with a characteristic amine odor.

Reactions of morpholine and piperazine are typical of those of a secondary amine. With inorganic acids, they form salts, and with organic acids, either a salt or an amide. Both amines can be alkylated on the nitrogen atom by the usual methods using alkyl halides, dialkyl sulfates, and a combination of aldehyde and formic acid. Piperazine generally gives the N,N'-disubstituted compound unless one nitrogen is blocked first. Ethylene oxide and propylene oxide react to form the corresponding amine alcohols, which can be further alkoxylated (see Alkanolamines). Substituted ureas are prepared by reaction of piperazine with isocyanates or isothiocyanates.

Manufacture

Morpholine is prepared by reaction of diethylene glycol with ammonia in the presence of hydrogen and a hydrogenation catalyst (eg, containing copper, nickel, cobalt, chromium, molybdenum, etc). Excess ammonia is removed from the crude reaction by a stripping operation, and the morpholine is recovered by fractional distillation. It has also been produced by a method based upon the dehydration of diethanolamine by a strong acid using oleum, concentrated sulfuric acid, or concentrated hydrochloric acid.

Piperazine can be prepared from ethylenediamine, aminoethylethanolamine, diethylenetriamine, monoethanolamine, diethanolamine, or triethanolamine. It is often a coproduct with other desired amines such as ethylenediamine and diethylenetriamine, also prepared by continuously passing ammonia and monoethanolamine over a catalyst. High yields of piperazine have been produced by the reaction of excess ammonia with diethanolamine in the presence of a Raney nickel catalyst; by the reaction of an alkylenediamine with an alkanolamine at elevated temperatures and superatmospheric pressure in the presence of hydrogen and a hydrogenation catalyst; and from diethylenetriamine in contact with a Raney nickel catalyst.

Health and Safety

Morpholin is not toxic when swallowed in small doses; it is slightly toxic in appreciable quantities; it exhibits a moderately high degree of hazard by skin contact. Anhydrous piperazine has low acute oral toxicity; however, it causes severe eye damage because of its alkalinity.

Uses

Morpholine is used as an intermediate in the manufacture of many commercial chemicals such as rubber accelerators (see Rubber chemicals); as water and metal-corrosion inhibitors; wax emulsifiers; and optical brighteners. Piperazine is used as an anthelmintic.

KENNETH MJOS
Jefferson Chemical Co., Inc.

Reviewed by
PHILLIP V. HARPER
Texaco

AMINES, AROMATIC

ANILINE AND ITS DERIVATIVES

Aniline ($C_6H_5NH_2$) is the simplest of the primary aromatic amines.

Physical Properties

Aniline is miscible with acetone, alcohol, benzene, and ether, soluble in most other organic solvents, and soluble to the extent of 3.5% at 25°C in H_2O. Bp 184.4°C; mp $-6.15°C$; d $\pm$ 1.022 g/cm^3 (at 20/20°C); n_D^{20} 1.5863; viscosity mPa·s (= cP) 4.423–4.435; flash point (closed cup) 76°C.

Chemical Properties

Aromatic amines are weaker bases than aliphatic amines but resemble them in many reactions. A characteristic difference between the two is the behavior toward nitrous acid. Primary aromatic amines give diazo compounds which are important intermediates for commercial dyes (see Azo dyes).

Aniline can be alkylated with various agents such as alkanols or olefins to give N-alkyl and N,N-dialkyl derivatives; treatment with metallic sodium and reaction with an olefin gives N-ethyl or N,N-diethylaniline. Diphenylamine can be obtained from aniline at high temperatures in the presence of a catalyst. Ring substitution occurs under conditions of elevated temperatures and the presence of a catalyst such as aluminum anilide or other aniline salts.

A wide variety of important amides are formed by reactions with acids, acid chlorides, anhydrides, and esters. Depending on conditions, reaction of aniline with phosgene can give various products; in particular, vapor-phase reaction with equimolecular proportions gives phenyl isocyanate, C_6H_5NCO. Aromatic primary diamines give important isocyanates (see Isocyanates, organic). 2-Mercaptobenzothiazole is obtained by heating aniline with carbon disulfide and sulfur under pressure (see Sulfur compounds).

Under certain conditions, the reaction of formaldehyde gives 4,4'-methylenedianiline, an important intermediate for 4,4'-methylenebis (phenyl isocyanate). Polymeric products are obtained by the reaction of aniline hydrochloride and formaldehyde.

Numerous N-heterocyclic compounds can be obtained from aniline; for example, the Skraup synthesis yields quinoline by condensation of aniline with glycerol or acrolein. Aniline undergoes a variety of oxidation reactions depending on the oxidizing agents and conditions. The first synthetic dyestuffs, eg, Perkin's mauve, were formed by oxidation of impure aniline. Air oxidation at room temperature yields azobenzene. Bichromate or chlorate oxidation in the presence of metal yields aniline black.

Important halogen derivatives include o-chloroaniline by vapor-phase chlorination: 2,4,6-trichloroaniline by chlorination in organic solvents, and chloranil (2,3,5,6-tetrachlorobenzoquinone) by chlorination in hot sulfuric acid.

Sulfanilic acid (4-aminobenzenesulfonic acid) is formed from sulfuric acid or sulfur trioxide. Fuming sulfuric acid gives all three isomeric monosulfonic acids. Mononitroanilines are formed at $-20°C$ with nitric acid in sulfuric acid. The meta derivative can be obtained by nitration of an aniline–BF_3 complex, whereas nitration of acetanilide (N-phenylacetamide) yields the 4-nitro compounds.

Manufacture

Aniline is produced commercially by catalytic reduction of nitrobenzene at temperatures below 350°C using catalysts such as copper oxide, sulfides of nickel, molybdenum, or tungsten, and palladium–vanadium/lithium–aluminum spinels (see Amines by reduction). Important derivatives of aniline include the toluidines, xylidines, anisidines, phenetidines, the chloro-, nitro-, and sulfonic acids, and the N-allyl, N-aryl, and N-acyl derivatives.

Health and Safety

Aniline is highly toxic, readily absorbed through the skin, and fatal if swallowed or vapors are inhaled. Chronic poisoning can occur by repeated exposure to low concentrations. Aniline is classified by the ICC as a class B poison.

Uses

Principal uses of aniline are in the following industries: polymer (45%), mainly for manufacture of isocyanates to make polyurethanes (43%); rubber (30%), in the manufacture of antioxidants, antidegradants, and vulcanization accelerators; agricultural (10%), in herbicides, fungicides, insecticides, animal repellants, and defoliants; and dyes.

JEAN NORTHCOTT
Allied Chemical Corporation

U.S. Pat. 3,649,693 (Mar. 14, 1972), J.P. Napolitano (to Ethyl Corp.).

K. Venkataraman, *Chemistry of Synthetic Dyes*, Vol. II, Academic Press, Inc., New York, 1952.

M. Gans, *Hydrocarbon Process.* **55**, 145 (1976).

DIAMINOTOLUENES

Toluene-2,4-diamine (diaminotoluene, TDA), an important industrial chemical, has six possible isomers which are separated into two commercial synthetic mixtures, one containing meta-isomers and the other ortho-isomers. Dinitration of toluene produces a mixture of ca 78% 2,4-DNT, 18% 2,6-DNT, 3.5% 2,3-DNT and 3,4-DNT, 0.5% 2,5-DNT, and traces of 3,5-DNT.

Physical Properties and Reactions

2,4-Toluenediamine has a density of 1.047 g/cm^3 at 100°C. The commercial m-isomer mixture has mp 90°C and bp 283°C.

The most important reaction of toluenediamine is with phosgene to give toluene diisocyanate, TDI (see Isocyanates, organic). The procedure for making TDI from TDA involves a cold phosgenation in an inert solvent with excess phosgene, followed by heating with excess phosgene. By-products are substituted ureas and benzimidiazolones, formed from ortho-substituted diamines and phosgene.

Manufacture

Dinitrotoluene can be catalytically hydrogenated to TDA under a wide variety of temperatures, pressures, and solvents. The catalyst used may be a supported noble metal (eg, Pd/C) or nickel, either supported or Raney type. Most commercial processes use continuous, stirred-flow, liquid-phase reactors.

Health and Safety

Toluenediamines are toxic, like other aromatic amines. 2,4-Toluene-diamine is carcinogenic to laboratory animals.

Uses

The main commercial use for TDA is the manufacture of toluene diisocyanate (TDI). The function of toluenediamines in most applications can be classified as monomer, chain extender or cross-linker, or intermediate for dyes and heterocyclic compounds.

BARTON MILLIGAN
KEVEN E. GILBERT
Air Products and Chemical, Inc.

R.J. Lindsay, *Compr. Org. Chem.* 2, 131 (1979).

DIARYLAMINES

Diarylamines are compounds with two of the hydrogens of ammonia replaced by aryl groups. The simplest is diphenylamine (DPA), $(C_6H_5)_2NH$, also called *N*-phenylbenzeneamine.

Physical and Chemical Properties

Solid diphenylamine has mp 53.0°C; bp 302°C; and d_4^{25} 1.159 g/cm^3. N,N'-Diphenyl-*p*-phenylenediamine, $C_6H_5NHC_6H_4NHC_6H_5$, has mp 152°C; bp 200°C; and d_4 1.18 g/cm^3. The solubility of DPA in water is given by:

$$\log c = 1.5786 - 1571 T^{-1}$$

where c is the concentration of DPA in mol/L. Solubility at 10, 15, 20, 25, and 30°C is 1.06, 1.35, 1.75, 2.11, and 2.61×10^{-4}, respectively.

The simple diarylamines are weaker bases than primary aromatic amines but stronger than triarylamines. Dilute acids separate primary aromatic amines from diarylamines. The hydrogen atom attached to the nitrogen in diarylamines can be replaced by alkali metals. Reaction of DPA with metallic aluminum leads to aluminum diphenylamide, which may be prepared more easily from *N*-sodium diphenylamide and aluminum chloride.

Diarylamines do not react with carbon disulfide. Formaldehyde reacts to give N,N,N',N'-tetraphenylmethylenediamine. Using catalysts, diarylamines are readily alkylated in the para position with isobutylene, styrene, and α-methylstyrene. Depending upon conditions, acetone reacts with diphenylamine to form a variety of products such as 9,9-dimethylacridan. Diphenylamine is easily dehydrogenated to carbazole. Similarly, heating of alkyl derivatives of diphenylamines gives acridines. Reaction with sulfur produces phenothiazine.

The diarylamines are easily halogenated and nitrated. Hydrogenation leads to dialicyclic amines. Oxidation of unsubstituted diarylamines with various oxidizing agents gives aminyl radicals, eg, $(Ar)_2N \cdot$, which cannot be isolated.

Manufacture

Diarylamines are manufactured by the self-condensation of a primary aromatic amine in the presence of an acid, or the reaction of an arylamine with a phenol. They readily donate their hydrogen atom to terminate various free-radical reactions.

Toxicity

N-Phenyl-2-naphthylamine (PBNA) is metabolized to form 2-naphthylamine, a carcinogen. Rats fed DPA and sodium nitrite form *N*-nitrosodiphenylamine (see Nitrosamines).

Uses

Diarylamines are used mainly as stabilizers for nitrocellulose propellants, as antioxidants for various polymers and elastomers, and for numerous small-scale applications. Typical rubber antioxidants include diphenyl-*p*-phenylenediamines (Age-Rite HP), the octylated diphenylamines, such as Age-Rite Stalite, Stalite S, Octamine, and Nonox OD, and the acetone-diphenylamine condensation products (Age-Rite Superflex and BLE). Antioxidants that are also antiozonants are paraphenylenediamines such as Antozite, Flexzone, Santoflex, and Wingstay.

ROBERT W. LAYER
HENRY J. KEHE
The B.F. Goodrich Company

W.C. Danen and F.A. Neugebauer, *Angew. Chem., Int. Ed. Engl.* 14(2), 783 (1975).
Rubber World 174, 35 (Sept. 1976).

METHYLENEDIANILINE

Synonyms and abbreviations for 4,4'-methylenedianiline (MDA) (*p*-$H_2NC_6H_4)_2CH_2$, include *p*,*p*'-methylenedianiline, methylenedianiline, 4,4'-diaminodiphenylmethane, bis(4-aminophenyl)methane, dianilinomethane, 4,4'-methylenebisaniline, 4,4'-methylenebisbenzeneamine, DMA, DADPM, and DAPM.

Physical Properties

Purified 4,4'-methylenedianiline is a light brown, crystalline solid with a faint amine-like odor. It oxidizes slowly in air which results in darker colored crystals. It has bp range = 262–268°C at 4.66 kPa (35 mm Hg) 5–95%; fp 89.0°C; density at 100°/4°C 1.056; viscosity at 100°C, mPa·s (or cP) 8.04; flash point 221.1°C. It is soluble to the extent of 273.0 g/100 g in acetone, 9.0 g/100 g in benzene, 9.5 g/100 g in ethyl ether, 143.0 g/100 g in methanol, and 0.1 g/100 g in water.

Chemical Properties

4,4'-Methylenedianiline reacts similarly to other aromatic amines. Most reactions involve substitution of the amine hydrogens or the aromatic hydrogens. In some cases, the functional groups are totally altered, eg, the reaction with phosgene yields diisocyanate. Polymers are formed when the amine is condensed with multifunctional reactants.

A bisacetamide is formed by reaction with acetic acid and acetic anhydride. With maleic anhydride, both hydrogens are replaced, and the bismaleimide can be homopolymerized through the double bonds or can add to nucleophilic groups, eg, the primary amine. The rings can be brominated or nitrated in different positions depending on the conditions (see Aniline). Other reactions include: diazotization; reaction with epoxides; conversion to isocyanate derivatives, such as the commercially important diisocyanate MDI, $(OCNC_6H_4)_2CH_2$, formed by reaction with phosgene (see Isocyanates).

Manufacture

Commercial production of 4,4'-methylenedianiline is by the acid catalyzed reaction of formaldehyde with aniline. The reaction proceeds through several intermediates to yield predominantly the 4,4'-isomer. Although the reaction is complex, production can be carried out in one reactor by either a continuous or batch process. Longer reaction times, eg, 6–12 h, enhance further alkylation of the diamine.

Health and Safety

MDA is a dangerous chemical, particularly when handled in vapor or dust form. Oral ingestion, inhalation, or skin absorption can result in toxic effects on the liver. MDA has a moderate single-dose oral toxicity. Repeated ingestion of small doses may present a higher degree of toxicity. Toxic effects on the liver can result from oral ingestion, inhalation or skin absorption. The vapor or dust forms of MDA appear to pose the greatest hazard. MDA has been reported to produce tumors in animal studies (rats and mice, drinking water).

Uses

MDA's largest use is as an intermediate for isocyanates. Also it is used as an intermediate in the preparation of epoxy resins, polyurethanes, rubber chemicals and polymers. It is also used in manufacture of rigid polyurethane foams and urethane products.

WILLIAM M. MOORE
Dow Chemical USA

Reviewed by
J.P. ARRINGTON
Dow Chemical USA

D.D. McGill and J.D. Motto, *N. Eng. J. Med.* **291**, 278 (1974).
S.V. Williams and co-workers, *N. Eng. J. Med.* **291**, 1256 (1974).

PHENYLENEDIAMINES

Phenylenediamines, $C_6H_4(NH_2)_2$, are aromatic amines with two amino groups attached to benzene.

Physical Properties

There are three isomeric phenylenediamines: ortho (mp 102°C; bp 256–258°C); meta (mp 63°C; bp 287°C); and para (mp 140°C; bp 280°C). All are white solids that darken upon standing in air, and all are completely soluble in hot water. Certain anils of *p*-phenylenediamines, eg, *N,N′*-bis[*p*-(octadecyloxy)benzylidene]-*p*-phenylenediamine, are liquid crystals (qv).

Reactions

o-Phenylenediamines undergo many reactions to heterocyclic compounds. Anhydrides, acids, amides, and esters give benzimidazoles; 1,2-dicarbonyl compounds give quinoxalines. *o*-Phenylenediamine gives triazole when treated with nitrous acid, whereas *m*-phenylenediamine gives an azo dye, Bismarck Brown G. Under special conditions, it can be tetrazotized. *p*-Phenylenediamine can be diazotized and tetrazotized giving intermediates for various azo dyes.

p-Phenylenediamines are oxidized more readily than the ortho or meta isomers, which are oxidized only slightly more easily than aniline. *p*-Phenylenediamines are readily oxidized by free radicals, or free-radical sources such as bromine, chlorine, oxygen, or peroxides, to give radical cations called Würster salts, which are stable in water–ethanol solutions at pH 3; they can be isolated as the hydrochloride. The unique reactivity of *p*-phenylenediamines explains why they are antioxidants, polymerization inhibitors, and coupling agents in color photography. Oxidation with silver oxide in ether solution gives *p*-benzoquinonediamines, colorless compounds which are very unstable, polymerize easily, and are readily hydrolyzed to ammonia or amines and *p*-benzoquinone.

Color photography utilizes the one-electron oxidation of *N,N*-dialkyl-*p*-phenylenediamines (developing agents) by light-sensitized silver salts. These radicals react with various coupling agents, eg, phenols, to give indoaniline dyes. Both *meta*- and *para*-phenylenediamines are used to manufacture sulfur dyes either by refluxing in aqueous sodium polysulfide or by heating with elementary sulfur at 330°C to give the leuco form of the dye. These dyes are polymeric, high molecular weight compounds, soluble in base. The color is developed by oxidation on the fabric. 2,4-Toluenediamine and sulfur give Sulfur Orange 1.

Phenylenediamines are difunctional and react with other difunctional compounds, eg, dianhydrides, diacyl chlorides, dicarboxylic acids, and disulfonyl chlorides, to give polyamines. With epoxides, diols, diacetals, diisocyanates, and diesters, they give potential commercially important aromatic polymers. Both *m*- and *p*-phenylenediamine react with benzenedicarbonyl dichlorides to give linear, fully aromatic polyamides called aramids (see Aramid fibers).

Health and Safety

Phenylenediamines are toxic, affecting kidney, liver, and blood and causing convulsions when taken internally. A common toxic effect is dermatitis. 2-Naphthylamine, obtained in small amounts in the production of *N,N′*-di-2-naphthyl-*p*-phenylenediamine, is a potent carcinogen.

Uses

Toluenediamines are used to manufacture diisocyanates for polyurethanes. *m*-Phenylenediamines are used in azo dyes. *p*-Phenylenediamines are used as antioxidants, as antiozonants, in color photography, in hair dyes, and to make polymers (aramids).

ROBERT LAYER
The BF Goodrich Company

W.W. Wright, "The Development of Heat-Resistant Organic Polymers," in G. Geuskens, ed., *Degradation and Stabilization of Polymers*, John Wiley & Sons, Inc., New York, 1975, Chapt. 3.
N.V. Sidgwick, *The Organic Chemistry of Nitrogen*, The Clarendon Press, Oxford, UK, 1937, pp. 45–104.

AMINATION BY REDUCTION

Amines (qv) are derivatives of ammonia in which one or more of the hydrogens are replaced by an alkyl, aryl, cycloalkyl, or heterocyclic group. They are classified as primary, secondary, or tertiary, depending on the number of replaced hydrogens. Primary amines, for example, can be obtained from reduction of nitro, nitroso, azo, azoxy, hydrazo, oxime, amide, and nitrile compounds. These reductions can be carried out by a variety of reducing agents.

Catalytic Hydrogenation

Commercial large volume aromatic and aliphatic amines are obtained by continuous, catalyzed high pressure hydrogenation of nitro compounds or nitriles. In the reduction of nitro compounds, the oxygen in the —NO_2 group is progressively replaced by hydrogen. In the production of aniline, for example, hydrogenation can be carried out in the vapor phase by the Cyanamid process in which nitrobenzene is vaporized and mixed with excess hydrogen and passed through a fluidized bed of copper-on-silicon catalyst at 270°C and 239 kPa (20 psig). Allied Corp. developed a similar process in which nitrobenzene is reduced in the vapor phase using a fixed bed of nickel sulfide on alumina.

Several large manufacturers use liquid-phase hydrogenation with Raney nickel or supported precious-metal catalysts. The nitro compound in a suitable solvent (eg, methanol), slurried with the catalyst and excess hydrogen, is passed through a series of reactors. This process is also used to reduce dinitrotoluene to toluenediamine.

Liquid-phase catalytic hydrogenation is also used to reduce aniline to cyclohexylamine, large quantities of which are used for corrosion inhibitors, insecticides, and rubber chemicals, and as a starting material for the sweetener Sucaryl (see Sweeteners).

Another catalytic hydrogenation process is used to produce aniline-based polyamines. Aniline reacts with formaldehyde in the presence of an acidic catalyst to produce a mixture of diaminodiphenylmethane (MDA) and higher homologues of the general structure:

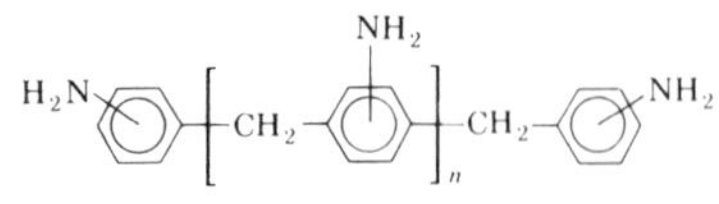

These are phosgenated to produce isocyanates useful for polyurethane products. The rings can be reduced to produce bis(4-aminocyclohexyl)methane; the trans-trans isomer is used to make a nylon-type fiber, and the other isomers are converted to isocyanates for light-stable coatings.

Nitriles are also reduced to the amines by liquid-phase catalytic hydrogenation. Thus, adiponitrile, hydrogen, and liquid ammonia are passed through a fixed-bed catalyst to produce hexamethylenediamine, which is used in nylon fibers and is converted to hexamethylene diisocyanate for light-stable urethane polymers.

Nitro alcohols are reduced to a variety of alkanolamines (qv) by continuous hydrogenation in the presence of Raney nickel. For example, 2-amino-2-methyl-1-propanol is produced from the corresponding nitro alcohol.

Metals in Acid Solution

The discovery in 1854 that nitrobenzene could be reduced by iron and acetic acid resulted in the Béchamp process for the manufacture of aniline. Subsequently, it was found that HCl could be used advantageously. Clean iron turnings, sometimes with the addition of ferrous salt, are used to catalyze the reaction. Less iron reacts in this system than is predicted by the overall equations representing the reactions.

Toluidines and large quantities of p-phenylenediamine have been produced as dyestuff intermediates by iron reduction of nitrotoluenes and p-nitroaniline, respectively.

Because of the efficiency of catalytic hydrogenation, it has been assumed that the Béchamp process is obsolete. However, interest has remained strong since the iron oxide produced by this process is in great demand as a pigment. Mobay Chemical Corp. has an iron oxide pigment plant based on the Béchamp process from which aniline is the by-product rather than the principal product.

Other metals have been used in acid solutions to prepare amines. Zinc and HCl, for example, reduce nitrobenzene to aniline. By a proper selection of the reducing agent and careful control of the pH of the solution, the reduction can be stopped at intermediate stages and valuable products other than amines obtained. The use of zinc metal and acid is useful for preparing amines that are insoluble at low pH, thus making it easier to separate the amine from the zinc salts.

Tin metal and stannous salts with HCl have also been used to selectively reduce one nitro group in compounds containing two nitro groups. Stannous chloride can also be employed for the reduction of nitroparaffins. In general, this procedure yields both the hydroxylamine and the primary amine.

Metals in Alkaline Solutions

Nitrobenzene, nitrotoluenes, and homologues are reduced to the corresponding hydrazo compounds in the presence of iron or zinc in alkaline solutions. The reduction proceeds stepwise through nitroso, hydroxylamine, azoxy, azo to the hydrazo stage. If the NaOH concentration is too low at the start of the reaction, the reduction may go to the amine. Dianisidine has been prepared on a commercial scale by I.G. Farben starting with o-nitroanisole using zinc dust and caustic soda. Hydrazobenzene has been produced by reduction of nitrobenzene using iron borings and caustic soda. Azoxybenzene is formed from nitrobenzene in the presence of zinc dust and strong alkali, but in faintly alkaline solutions, phenylhydroxylamine is produced. Zinc and ammonium chloride reduce nitroparaffins to alkylhydroxylamines.

Sulfides in Alkaline Solution

Reductions with alkali or metal sulfides, although more expensive than iron and acid or catalytic reduction, are useful for the reduction of dinitro compounds to nitroamines, nitrophenols, and nitroanthraquinones and for the preparation of aminoazo compounds. Thus, m-nitroaniline is prepared from m-dinitrobenzene using sodium sulfide, and 2-amino-4-nitrophenol is prepared from 2,4-dinitrophenol using ferrous sulfide. Likewise, 2-amino-7-chloroanthraquinone can be prepared from 2-nitro-7-chloroanthraquinone using alkaline sodium sulfide. The nitroazo compound obtained by diazotization and coupling of p-nitroaniline with an amine or phenol can be reduced to 4-amino-4'-hydrazobenzene using sodium sulfide in weakly alkaline solution.

Miscellaneous Reducing Agents

Sodium hydrosulfite in alkaline solution is used to reduce anthraquinone and indigoid derivatives to the leuco compounds, and for the preparation of o-aminophenol from the corresponding nitro compound.

Sodium sulfite reduces nitro compounds to coproduce o- and p-aminosulfonic acids by a procedure known as the Piria reduction.

Hydrazine reduces nitro compounds to amines only in the presence of catalysts such as Raney nickel and platinum-group metals.

Electrolytic reduction, because of the high energy cost, has not been applied to high volume production of amines. In general, such reductions can be readily controlled and few by-products form, which may be advantageous in the reduction of sensitive organic compounds. Some nitroalcohols have been reduced to alkanolamines by this method.

Lithium aluminum hydride reduces aliphatic nitro compounds to amines, but aromatic nitro compounds in general are reduced only to the azo compounds.

Sodium borohydride reduces aromatic nitro and nitroso compounds to amines when catalyzed by palladium on charcoal.

Robert L. Sandridge
Harold B. Staley
Mobay Chemical Corp.

P.H. Groggins, ed., *Unit Processes in Organic Synthesis*, 5th ed., McGraw-Hill Book Company, New York, 1958, pp. 129–203.

J. Werner, *Ind. Eng. Chem.* **50**(9), 1329 (Sept. 1950); **51**(9), 1065 (Sept. 1959); **53**(9), 77 (Jan. 1961).

M.S. Gibson, "The Introduction of the Amino Group," in S. Patai, ed., *The Chemistry of the Amino Group*, Interscience Publishers, a division of John Wiley & Sons, Inc., New York, 1968.

R. Schroter in E. Muller, ed., *Houben-Weyl Methoden der Organischen Chemie*, 4th ed., Vol. II, Pt. 1, Georg Thieme Verlag, Stuttgart, FRG, 1957, pp. 341–781.

AMINOACETIC ACID, NH₂CH₂COOH.   See Amino acids (Glycine).

AMINO ACIDS

SURVEY

Approximately 20 amino acids are the main components of proteins, the elementary nutrients of living organisms (see Table 1). There are eight, and possibly ten, amino acids that are essential for human existence and must be ingested through food.

Physical Properties

Every amino acid has equal electric charges of opposite sign caused by the amino and carboxyl groups at the asymmetric α-carbon atom. The shapes of crystals in each amino acid vary widely. The dielectric constants of solution are very high; their ionic dipolar structure confers on them special vibrational spectra (Raman, ir) as well as characteristic properties (specific volumes, specific heats, electrostriction). In aqueous solution, amino acids undergo a pH-dependent dissociation. The melting points are poorly defined, between 200–300°C, and the amino acids frequently decompose below their melting points. With the exception of glycine, all α-amino acids contain at least one asymmetric carbon atom, and all are characterized by their ability to rotate the plane of polarized light depending on the solvent and the degree of ionization.

Chemical Properties

α-Amino acids are ampholytic compounds, with chemical properties classified according to their carboxyl, amino and side-chain groups, and whether reversible or irreversible reactions are involved.

All α-amino acids form salts in alkaline and acidic aqueous solutions.

Esters of α-amino acids can be prepared by refluxing anhydrous alcoholic suspensions of α-amino acids saturated with dry HCl. Diketopiperazines are formed by heating the alcoholic solution of the α-amino acid ester, and decompose by hydrolysis under drastic conditions. Peptides are produced by hydrolysis under mild conditions.

Table 1. Amino Acids

Common name	Abbreviation	Formula	Common name	Abbreviation	Formula
Aliphatic, monoamino monocarboxylic			*Aliphatic, monoamino dicarboxylic*		
glycine	Gly	H_2NCH_2COOH	aspartic acid	Asp	$HOOCCH_2CH(NH_2)COOH$
α-alanine	Ala	$CH_3CH(NH_2)COOH$	glutamic acid	Glu	$HOOCCH_2CH_2CH(NH_2)COOH$
valine[a]	Val	$CH_3CH(CH_3)CH(NH_2)COOH$	*Aliphatic, diamino monocarboxylic*		
leucine[a]	Leu	$CH_3CH(CH_3)CH_2CH(NH_2)COOH$	lysine[a]	Lys	$H_2N(CH_2)_4CH(NH_2)COOH$
isoleucine[a]	Ileu	$CH_3CH_2CH(CH_3)CH(NH_2)COOH$	arginine	Arg	H_2N $CNH(CH_2)_3CH(NH_2)COOH$ HN
serine	Ser	$CH_2OHCH(NH_2)COOH$	*Aromatic*		
threonine[a]	Thr	$CH_3CH(OH)CH(NH_2)COOH$	phenylalanine[a]	Phe	$C_6H_5CH_2CH(NH_2)COOH$
cysteine	Cys	$HSCH_2CH(NH_2)COOH$	tyrosine	Tyr	HO—⬡—$CH_2CH(NH_2)COOH$
cystine	Cys₂	$SCH_2CH(NH_2)COOH$ $\mid$ $SCH_2CH(NH_2)COOH$	*Heterocyclic* proline	Pro	⬠ NH $COOH$
methionine[a]	Met	$CH_3SCH_2CH_2CH(NH_2)COOH$	hydroxyproline	Hypro	HO—⬠ NH $COOH$
asparagine	Asn	$H_2NCOCH_2CH(NH_2)COOH$	histidine	His	N ⬠ NH $CH_2CH(NH_2)COOH$
glutamine	Gln	$H_2NCOCH_2CH_2CH(NH_2)COOH$	tryptophan[a]	Trp	⬡⬠ N H $CH_2CH(NH_2)COOH$

[a] Essential for human nutrition.

The α-amino moiety of amino acids produces the Schiff base with aldehydes in a neutral or weakly alkaline aqueous solution. Important reactions between amino acids and carbonyl compounds include the Strecker degradation, and Maillard reaction, or nonenzymatic browning reaction; and the Ninhydrin color reaction.

In general, a peptide, which is an amino acid polymer, is synthesized by stepwise elongation. This process involves: (1) protection of amino and carboxyl groups; (2) formation of a peptide bond; (3) selective removal of the amino-terminal protecting group; (4) elongation of the peptide chain by repeating reactions (2) and (3); and removal of all protecting groups (see Polypeptides).

Manufacture and Processing

Amino acids are commonly produced by fermentation (qv) or enzymatically using microbes. Glycine, alanine and methionine are chemically synthesized; hydroxyproline and cysteine are obtained by separation from protein hydrolysates.

In the fermentative production of amino acids (see Fig. 1), all raw materials are natural or biologically available substances with no harmful by-products yet found.

Excretion of excess amino acids is a problem for the microorganism since in natural selection the microorganism that overproduces amino acids is at a disadvantage; regulation or control is essential. Industrial production of L-glutamic acid uses both glucose, sugarcane waste molasses, or acetic acid as the carbon source, together with urea or ammonia as nitrogen sources, phosphates and other ordinary salts. The carbon source can be converted to glutamic acid in 50% yield under optimum aerobic conditions. The success of lysine production brought about a new production method for amino acids which uses artificially-derived auxotrophic or regulatory mutants as the strains resistant to growth inhibition by the structural analogues of the desired amino acid. Complex regulatory systems controlling biosynthesis can be artificially by-passed in these mutants to allow overproduction (see Microbiological transformations). See Figure 2 for biosynthetic pathways for amino acid production.

L-Amino acids are produced by enzymatic resolution of racemic derivatives. Most L-amino acids are now produced directly by fermentation. Some amino acids—such as L-alanine, L-asparatic acid, L-tryptophane, and L-lysine—are more economical to produce by enzymatic methods than fermentation.

Glycine, DL-alanine, and DL-methionine are produced by chemical synthesis. L-DOPA is produced by asymmetric synthesis (see

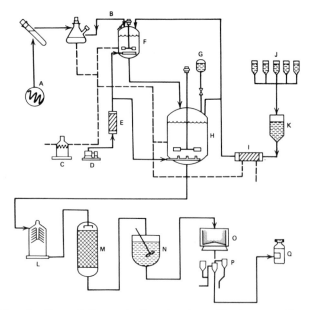

Figure 1. Fermentative production of amino acids. A, pure culture; B, inoculation; C, boiler; D, air compresser; E, air filter; F, seed tank; G, pH control medium; H, fermenter; I, sterilizer; J, culture media; K, preparation tank; L, centrifugal separator; M, ion-exchange column; N, crystallizing tank; O, crystal separator; P, dryer; Q, amino acid production.

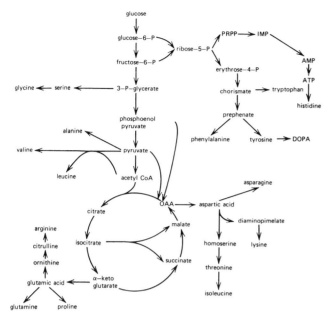

Figure 2. Amino acid biosynthesis. AMP, adenosine monophosphate; ATP, adenosine triphosphate; CoA, coenzyme A; DOPA, 3(3,4-dihydroxyphenyl)alanine; IMP, inosine monophosphate; OAA, oxaloacetic acid; P, phosphate; PRPP, 5-phosphoribosyl-1-pyrophosphate.

Pharmaceuticals, optically active). DL-Methionine is produced in the largest amount synthetically of the group. L-Arginine is produced directly by fermentation.

Cystine and cysteine are produced by isolating them from hydrolyzed keratin protein, eg, hair and feathers. Proline and hydroxyproline are precipitated from gelatin hydrolysates.

Health and Safety

In some studies, monosodium glutamate (MSG) was implicated in brain injury in mice; however, FDA has classified MSG as safe. Naturally occurring new amino acids such as those from some legumes, which can be classified as proteinaceous or nonproteinaceous, are toxic.

Uses

Amino acids are used to improve the nutritive value of proteins (qv). The enrichment of required essential amino acids significantly promotes the growth of animals. Some amino acids, particularly L-glutamic acid, are used to enhance flavor (see Flavors and spices). Some have been found to have special tastes. In medicine, they are largely used in transfusion and for pharmaceuticals. Amino acid polymers have been developed as raw materials for artificial leathers.

A. YAMAMOTO
Kyowa Hakko Kogyo Co., Ltd.

T. Kaneko and co-workers, *Synthetic Production and Utilization of Amino Acids*, Kodansha, Tokyo, Japan; John Wiley & Sons, Inc., New York, 1974.

J.P. Greenstein and M. Winitz, *Chemistry of the Amino Acids*, Vol. 1, John Wiley & Sons, Inc., New York, 1961.

R. Walter, ed., *Peptides—Chemistry, Structure, and Biology*, Ann Arbor Science Publishers, Inc., Ann Arbor, Michigan, 1975.

L-MONOSODIUM GLUTAMATE (MSG)

Physical Properties

Monosodium L-glutamate (MSG), HOOCCH(NH$_2$)(CH$_2$)$_2$CCOONa. H$_2$O found in stable crystals, is soluble in water and normally forms rhombic prisms when crystallized from water. Commercially preferable crystals are grown in the presence of amino acid contaminants. Below $-0.8°C$, MSG crystallizes as the pentahydrate. The filtered pentahydrate loses its water of crystallization by exposure to air to become the extremely porous monohydrate.

Solubilities of MSG are given by the equation:

$$S = 35.30 + 0.098t + 0.0012t^2$$

where S is % of C$_5$H$_8$NO$_4$Na (anhydrous) in a saturated solution at $t°C$ and the remainder at equilibrium is the monohydrate.

The α-carbon of MSG is asymmetric and both the natural L-form, which possesses taste-modifying characteristics, and the unnatural D-forms are known. L-Glutamic acid does not racemize in neutral solution, even at 100°C. But intramolecular dehydration is noticeable under acidic or alkaline conditions. Microbial racemase (eg, found in *Lactobacillus fermenti*) brings about racemization. L-Glutamic acid is split into α-ketoglutaric acid (2-oxoglutaric acid) and into ammonia by glutamate dehydrogenase, and this reaction may be reversed. L-Glutamate is the only amino acid in mammalian tissues that undergoes oxidative deamination at an appreciable rate. The formation of ammonia from α-amino groups requires their conversion to the α-amino nitrogen of L-glutamate. Thus, L-glutamic acid is a key substance in the nitrogen metabolism of amino acids.

Manufacture

MSG is biosynthetically produced by fermentation processes. Glutamic acid-producing microorganisms (*Micrococcus, Brevibacterium, Corynebacterium, Arthrobacter,* and *Microbacterium*) have intense glutamate dehydrogenase activity and oxidative degradability to both L-glutamic acid and α-ketoglutaric acid. The carbon sources for biosynthesis include acetic acid and the commonly used carbohydrate sources, as well as sugar molasses processed with penicillin to depress excessive biotin activity (see Vitamins). Gaseous ammonia or a solution of urea are convenient nitrogen sources for the fermentation, not only as feedstock but also to maintain pH of the culture media at 7–8. Other important culture conditions include regulation of aeration and stirring; an optimum temperature of 30–37°C depending on the microorganisms used; and sterilized and aseptic air to prevent bacterial and bacteriophage contamination. Industrial-scale fermenters are pressure-tight, stainless-steel containers, built to hold several hundred kiloliters of cultivating medium and equipped with automatic aeration, stirring and control devices. Fermentation takes 35–45 h.

MSG is also produced by chemical synthesis. But the product is racemic. Therefore, it requires a process to separate D-acid and recycle it.

Health and Safety

MSG has been toxicologically cleared for use in food, with the exception of use in baby food for infants under 12 weeks of age. At ordinary dose levels used for flavor enhancement, MSG presents no health problem.

Uses

Plain MSG is used as a flavor enhancer (see Flavors and spices; Flavor characterization). In Japan, MSG mixed with Na salts of ribonucleotides has a wider domestic market.

TOICHI YOSHIDA
Ajinomoto Co., Inc.

U.S. Pat. 2,834,805 (May 13, 1958), J.L. Purvis and V. Bassel (to International Minerals & Chemical Corp.).

Jpn. Pat. 178,486 (Apr. 16, 1948), T. Ogawa (to Ajinomoto Co., Inc.).

T. Ogawa, *J. Chem. Soc. Japan, Ind. Chem. Sec.* **52,** 69 (1949).

H.A. Harper, *Review of Physiol. Chem.*, 13th ed., Lange Medical Publication, 1971, p. 30.

N. Katsuya and co-workers, *J. Biochem. (Japan)* **53,** 333 (1963).

T. Tsunoda and co-workers, *Agr. Biol. Chem. (Japan)* **27,** 858 (1963).

AMINO ALCOHOLS. See Alkanolamines.

AMINOGLYCOSIDES (AMINOCYCLITOL) AMINOLYSIS. See Antibiotics—Aminoglycosides.

AMINONAPHTHALENES. See Naphthalene derivatives.

AMINONAPHTHOLS. See Naphthalene derivatives.

AMINOPHENOLS

Aminophenols and their derivatives are amphoteric compounds possessing properties as both weak acids and weak bases. The unsubstituted or monosubstituted aminophenols are unstable to air, oxidizing readily to brown or black substances; in the case of ortho isomers, to phenoxazines.

Health and Safety

In general, these compounds are irritating, and some cause dermatitis and allergies.

Uses

Most of the aminophenols are used in the dyestuff industry as intermediates or in mordant or acid dyes, as antioxidants and in photography (see Table 1).

Table 1. Properties and Uses of Aminophenols

Name	Mp, °C	Solubility at 0°C	Use
2-amino-6-chloro-4-nitro-phenol	160	sol in ethanol and ether	intermediate in mordant dyes
2-amino-4,6-dinitrophenol	169	sol in H_2O, alcohol, and benzene; sl sol in chloroform and ether	indicator dye
2-amino-4-nitrophenol	142–143	sol in ethanol and ether; sl sol in H_2O	intermediate in mordant and acid dyes
2-aminophenol	174	59 parts H_2O; 23 parts ethanol; sol in ether; sl sol in benzene	mfg of azo and sulfur dyes
3-amino-2-hydroxybenzoic acid			
3-dimethylaminophenol	87	sol in ethanol, ether, benzene, acetone, alkalies, mineral acids; practically insol in H_2O	dye intermediate
3-(diethylamino)phenol	78		
4-aminophenol	189.5–190	90 parts H_2O; 22 parts ethanol; sl sol in ether; insol in chloroform and benzene	
5-amino-2-hydroxybenzoic acid	283	insol in ethanol; sol in HCl	
4-(2-naphthylamino)phenol	135		
4-(2,4-dinitroanilino)phenol	195–196		
6-amino-4-chloro-1-phenol-2-sulfonic acid			
2-amino-6-nitro-1-phenol-4-sulfonic acid		sol in water; sl sol in ethanol	

RUSSELL E. FARRIS
GAF Corporation

Colour Index, 3rd ed., The Society of Dyers and Colorists, Bradford, Yorkshire, UK, Vol. 4, 1971, pp. 4001–4689, 4691–4863; Vol. 6, 1975, pp. 6391–6404, 6407–6410.

"General-Formelregister," *Beilsteins Handbuch der Organischen Chemie*, Vierte Auflage, Springer-Verlag, Berlin, 1956.

M. Windholz, ed. *The Merck Index*, 10th ed., Merck & Co., Rahway, N.J., 1983.

Dictionary of Organic Compounds, 5th ed., Vols. 1–5, Oxford University Press, New York, 1983.

AMINO RESINS AND PLASTICS

The term amino resin applies to a broad class of materials regardless of application, whereas the term aminoplast or aminoplastic is more commonly used for thermosetting molding compounds based on amino resins.

Most amino resins are based on the reaction of formaldehyde with urea (qv) or melamine (1) (see Cyanamides). Formaldehyde links two molecules and is hence difunctional. Urea and melamine react polyfunctionally with formaldehyde to form three-dimensional, crosslinked polymers. Compounds with a single amino group, such as aniline or toluenesulfonamide, can react with formaldehyde to form only linear polymer chains. This is true under mild conditions, but in the presence of an acid catalyst at higher temperatures, the aromatic ring of aniline, for example, may react with formaldehyde to produce a cross-linked polymer.

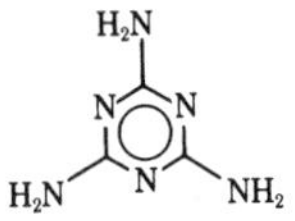

melamine

Raw Materials

Urea, melamine, and formaldehyde are the basic materials. Benzoguanamine (2) and acetoguanamine may be used in place of melamine for greater solubility in organic solvents and greater chemical resistance.

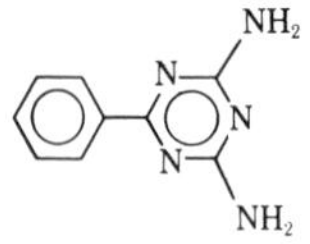

benzoguanamine

Urea (carbamide) is the most important building block for amino resins; urea–formaldehyde is the largest-selling amino resin, and urea is the raw material for melamine, the amino compound used in the next largest-selling type of amino resin. Urea is soluble in water, and the crystalline materials is somewhat hygroscopic. It is made by reaction of carbon dioxide and ammonia at high temperature and pressure to yield a mixture of urea and ammonium carbamate which is recycled.

Melamine (cyanurotriamide, 2,4,6-triamino-s-triazine) is a white crystalline solid, with melting point ca 350°C with vaporization; it is only slightly soluble in water and is made from urea (see also Cyanuric and isocyanuric acids). Essentially all of the melamine produced is used for making amino resins and plastics.

Formaldehyde (qv) is a colorless, pungent-smelling reactive gas. The commercial product is handled either as a solid polymer, as paraformaldehyde, or in aqueous or alcoholic solutions.

Aniline and toluenesulfonamide react with formaldehyde to form resins that are used as plasticizers. Acrylamide is an interesting monomer for use with amino resins (see Acrylamide polymers).

Chemistry of Resin Formation

Successful manufacture and utilization of amino resins depend largely on the precise control of two reactions: methylolation and methylene bridge formation. The first step, methylolation or hydroxylation, is the addition of formaldehyde, using acid or base catalysts, to introduce the hydroxymethyl group, eg, $RNH_2 + HCHO \rightarrow RNHCH_2OH$. The second step, methylene bridge formation (also known as polymerization, resinification, or simply cure), involves condensation of the monomer unit using an acid catalyst, liberating water to form a dimer, a polymer chain, or a vast network.

In addition to these two main reactions, methylolation and condensation, a number of other reactions important for the manufacture and uses of amino resins include the following:

Two methylol groups may combine to produce a dimethylene ether linkage and liberate a molecule of water; a more stable and more soluble (in organic solvents) compound results from the replacement of the hydrogen of the methylol compound with an alkyl group, a reaction catalyzed by acids and carried out in the presence of excess alcohol to suppress the competing self-condensation reaction.

The mechanism of the alkylation reaction is similar to curing. The methylol group becomes protonated and dissociates to form a carbonium-ion intermediate which may react with alcohol to produce an alkoxymethyl group or with water to revert to the starting material. Desired compatibility with organic solvents can be achieved by employing an amino compound with an organic solubilizing group in the molecule, such as benzoguanamine. Displacement of a volatile with a nonvolatile alcohol is an important technique in curing paint films with amino cross-linkers and amino resins on textile fabrics or paper.

Manufacture

Precise control of the course, speed, and extent of the reaction is essential for successful manufacture. Important factors are mole ratio of reactants; catalyst (pH of reaction mixture); and reaction time and temperature. Amino resins are usually made by a batch process. The formaldehyde and other reactants are charged to a kettle, the pH adjusted, and the charge heated. Often the pH of the formaldehyde is adjusted before adding the other reactants. Aqueous formaldehyde is the most convenient to handle and lowest in cost. In the manufacture of amino resins, every attempt is made to recover and recycle raw materials. Figure 1 illustrates the manufacture of amino resin syrups, cellulose-filled molding compounds, and spray-dried resins.

Types of Resins

Ethyleneurea, a resin based on dimethylolethyleneurea (1,3-bis(hydroxymethyl)-2-imidazolidinone), is prepared from urea, ethylenediamine, and formaldehyde. Propyleneurea–formaldehyde resin, 1,3-bis(hydroxymethyl)-tetrahydro-2(1H)-pyrimidinone, is prepared from urea, 1,3-diaminopropane, and formaldehyde. Triazone resin, is made from urea, formaldehyde, and a primary aliphatic amine, usually hydroxyethylamine (see Alkanolamines). Uron resins are mixtures of a minor amount of melamine resin and so-called uron (predominantly N,N'-bis(methoxymethyl)uron plus 15–25% methylated urea–formaldehyde resins). Glyoxal resins made from urea, glyoxal, and formaldehyde, are based on dimethyloldihydroxyethyleneurea (DMDHEU), in which methanol groups are attached to each nitrogen. Melamine–formaldehyde resins are the most versatile textile-finishing resins, and include the dimethyl ether of trimethylolmelamine. Methylol carbamate resins, derivatives made from urea and an alcohol (R can vary from a methyl to a monoalkyl ether of ethylene glycol), are second to melamine–formaldehyde and urea–formaldehyde resins as textile finishers. Other amino resins used in the textile industry include methylol derivatives of acrylamide, hydantoin, and dicyandiamide. DMDHEU is used as a wrinkle-recovery, wash-and-wear, durable-press agent in the textile industry.

$$H_2NCONH_2 + OHCCHO + 2\ HCHO \longrightarrow HOCH_2N \underset{OH\ \ OH}{\overset{O}{\underset{\bigg|\ \ \bigg|}{\parallel\ \ C}}} NCH_2OH$$

DMDHEU

Textiles are finished with amino resins in four steps. The fabric is (1) passed through a solution containing the chemicals; (2) passed through squeeze rolls (padding) to remove excess solution; (3) dried; and (4)

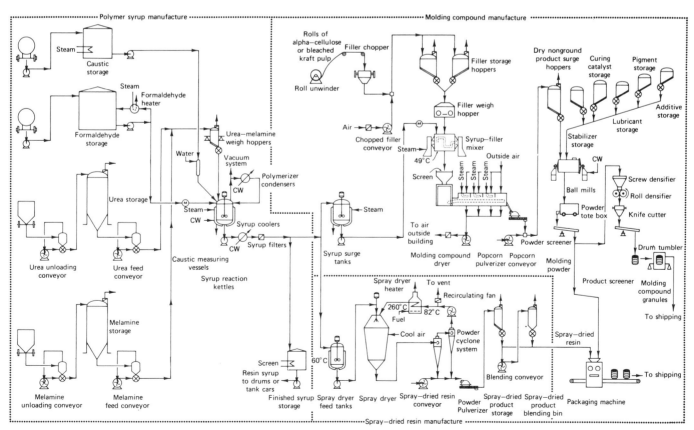

Figure 1. Urea–formaldehyde and melamine–formaldehyde resin manufacture. CW = cold water. Courtesy of Stanford Research Institute.

heated (cured) to bond the chemicals with the cellulose or to polymerize them on the fabric surface.

Health and Safety

Urea– and melamine–formaldehyde resins are substantially nontoxic, but there have been some problems with the release of formaldehyde from urea–formaldehyde foam used as insulation. Also, some urea–formaldehyde-bonded chipboard can release formaldehyde. Off-gassing can be alleviated by finishing the board with coatings or overlays that act as sealants, or by storing the boards for longer periods of time before shipping to users.

Uses

Major uses of resins are adhesives for bonding and laminating; in molding compounds; paper treating and coatings; protective surface coatings; and textile finishes.

Ivor Updegraff
Sewell Moore
William Herbes
Philip Roth
American Cyanamid Company

C.P. Vale and W.G.K. Taylor, *Amino Plastics*, Iliffe Books, Ltd., London, 1964.

AMMINES.   See Coordination compounds.

AMMONIA

Ammonia is a colorless, alkaline gas, lighter than air, and possessing a unique, penetrating odor.

Physical Properties

Bp, $-33.35°C$; fp, $-77.7°C$; critical temperature, $133.0°C$; specific heat (at $0°C$), 2097.2; soluble to the extent of 33.1 wt% in H_2O at $20°C$; specific gravity (anhydrous ammonia) at $-40°C$, 0.690; and density of aqueous ammonia is 0.618 g/cm^3 ($15°C$). The flammable limits of ammonia in air are 16 and 25 vol%; in oxygen the range is 15–79 vol%. Such mixtures can explode, although ammonia–air mixtures are quite difficult to ignite (ignition temperature ca $650°C$). Ammonia is readily absorbed in water to make ammonia liquor. Approximately 2180 kJ (521 kcal) of heat is evolved when 1 kg of ammonia gas dissolves in water.

The alkali and alkaline-earth metals (except beryllium) are readily soluble in ammonia, metallic magnesium only slightly. Iodine, sulfur, and phosphorus dissolve in ammonia; ammonia readily attacks copper in the presence of oxygen. Most fluorides are insoluble in liquid ammonia. Potassium, silver, and uranium are only slightly soluble. Both ammonium and beryllium chloride are very soluble, whereas most other metallic chlorides are slightly soluble or insoluble. Bromides are in general more soluble in ammonia than chlorides, and most of the iodides are soluble. Oxides, hydroxides, sulfates, sulfites, and carbonates are insoluble. Nitrates, eg, ammonium nitrate, and urea are soluble in anhydrous and aqueous ammonia (this property makes the production of fertilizer nitrogen solutions possible). Many organic compounds, eg, amines, nitro compounds, and aromatic sulfonic acids, also dissolve in liquid ammonia.

Chemical Properties

Ammonia is comparatively stable at ordinary temperatures but decomposes into hydrogen and nitrogen at elevated temperature. The rate of decomposition is greatly affected by the nature of the surfaces with which the gas comes into contact. Whereas glass is very inactive, porcelain

and pumice have a distinctly accelerating effect, and metals, such as iron, nickel, osmium, zinc, and uranium, even more.

At atmospheric pressure, dissociation of ammonia begins at about 450–500°C, whereas in the presence of catalysts, it begins as low as 300°C and is nearly complete at 500–600°C. However, at 1000°C a trace of ammonia remains.

Ammonia reacts readily with a large variety of substances (see Ammonium compounds; Amines by reduction; Amines).

One of the more important reactions is oxidation giving nitrogen and water. Gaseous ammonia is oxidized by many oxides of the less positive metals (eg, cupric oxide) when heated to a relatively high temperature. Powerful oxidizing agents react similarly at ordinary temperatures, eg, potassium permanganate:

$$2\,NH_3 + 2\,KMnO_4 \rightarrow 2\,KOH + 2\,MnO_2 + 2\,H_2O + N_2$$

The action of chlorine on ammonia can also be regarded as an oxidation reaction:

$$8\,NH_3 + 3\,Cl_2 \rightarrow N_2 + 6\,NH_4Cl$$

With a platinum–rhodium catalyst, ammonia can be oxidized to nitric oxide and water which is a means of producing nitric acid. Base metal catalysts are also used to promote ammonia oxidation. Reducing agents normally do not react.

The neutralization of acids is of commercial importance, eg, three major fertilizers, ammonium nitrate, ammonium sulfate, and ammonium phosphate, are made from ammonia.

The reaction between ammonia and water is reversible:

$$NH_3 + H_2O \rightleftarrows [NH_4OH] \rightleftarrows NH_4^+ + OH^-$$

With increasing temperature, the solubility of ammonia decreases rapidly.

Ammonia is a comparatively weak base and ionizes in water much less than sodium hydroxide. Aqueous ammonia also acts as a base in precipitating metallic hydroxides from solutions of their salts and forming complex ions in excess ammonia solution. For example, with copper sulfate solution the cupric hydroxide, which is at first precipitated, redissolves in excess ammonia solution owing to the formation of the complex cuprammonium, tetramminecopper(II) ion, $[Cu(NH_3)_4]^{2+}$. Both potassium and sodium metals dissolve to give the amides, eg, $NaNH_2$. Other metals, such as magnesium calcium, strontium, and barium, give the nitrides. Reactions with halogens give unstable trihalides.

Ammonia reacts with chlorine in dilute solution to give chloramines (qv), an important reaction in water purification (see Bleaching agents; Water). Ammonia reacts with phosphorus vapor at red heat to give nitrogen and phosphine. Sulfur also reacts with liquid anhydrous ammonia to produce nitrogen sulfide, N_4S_4. Ammonia and carbon at red heat give ammonium cyanide.

Ammonia forms a great variety of addition or coordination compounds (qv), also called ammoniates, in analogy with hydrates. Thus $CaCl_2 \cdot 6NH_3$ and $CuSO_4 \cdot 4NH_3$ are comparable to $CaCl_2 \cdot 6H_2O$ and $CuSO_4 \cdot 4H_2O$, respectively. Such compounds when regarded as coordination compounds are called amines and are written as complexes, eg, $[Cu(NH_3)_4]SO_4$. The solubility in water of such compounds is often quite different from the solubility of the salts themselves, eg, silver chloride, $AgCl$, is almost insoluble in water, whereas $[Ag(NH_3)_2]Cl$ is readily soluble. Thus silver chloride dissolves in aqueous ammonia. Similar reactions take place with other insoluble salts such as silver phosphate and cuprous chloride.

The reaction of ammonia and carbon dioxide via ammonium carbamate to urea is of considerable industrial importance.

Manufacture

In 1913, Haber and Bosch in Germany developed the first commercial process for the direct synthesis of ammonia (see Fig. 1). In the mid 1960s, the ammonia industry grew very rapidly owing to improvements in the energy cycle, including recovery of waste heat to generate high pressure steam, and technical advances in compressors, reformers, and converters.

Ammonia is synthesized by the reversible reaction of hydrogen and nitrogen:

$$N_2 + 3\,H_2 \rightleftarrows 2\,NH_3$$

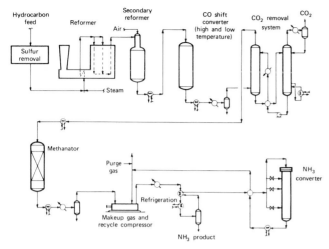

Figure 1. Haber-Bosch process.

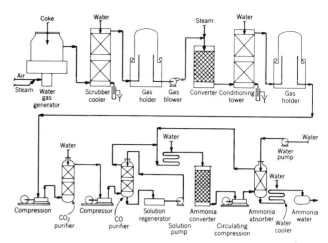

Figure 2. Flow sheet for a high capacity single-train ammonia plant.

In the design of synthesis facilities, the rate at which the ammonia is formed has to be considered.

The reaction of hydrogen and nitrogen to form ammonia on an industrial scale is always carried out on a catalytic surface based on metallic iron (mostly magnetite) that has been promoted with other oxides. Oxygen and oxygen-containing compounds have a temporary poisoning effect on the catalyst and sulfur, arsenic, phosphorus, chlorine, and heavy hydrocarbons cause permanent poisoning.

The following variables affect ammonia synthesis: (1) Synthesis pressure, an increase in pressure increases the equilibrium percent rate of ammonia and the reaction rate. (2) Synthesis temperature, higher temperatures increases reaction rates but decreases the equilibrium amount of ammonia and increases the thermal degradation of the catalyst. (3) Space velocity (the ratio of the volumetric rate of gas at standard conditions to the volume of catalysts), the percentage of ammonia in the exiting gas decreases with increasing space velocity. However, the same volume of catalyst at increased space velocities can produce more ammonia (normal space velocities for commercial operations are 8,000–60,000 h^{-1}). (4) Inlet gas composition, effects vary depending on the concentration of inert materials, ammonia concentration, and hydrogen to nitrogen ratio. Most commercial facilities operate at a hydrogen to nitrogen ratio of 3 : 1; however, the optimum may be something less than that. (5) Catalyst particle size, catalyst activity increases with smaller particles (most common size 6–10 mm).

The manufacture of synthetic anhydrous ammonia consists of three basic steps: synthesis-gas preparation, purification, and ammonia synthesis. The first two involve the generation of hydrogen (qv), the introduction of nitrogen in the stoichiometric synthesis proportion, and the removal of catalyst poisons (namely, carbon dioxide, carbon monoxide, and water). Ammonia synthesis includes the catalytic fixation of nitrogen at elevated temperatures and pressures and the recovery of the ammonia. Although conditions vary greatly, the chemistry of ammonia synthesis is basic to all commercial processes.

A typical flow sheet for a high capacity single-train ammonia plant is shown in Figure 2.

By and large, the raw-material sources in modern ammonia plants are coal, petroleum fractions, and natural gas (see Feedstocks; Gas, natural). In general, the most economical feedstock has the highest hydrogen to carbon ratio.

The two basic generation techniques for processing these raw materials are partial oxidation and steam reforming. Steam reforming is most often used in commercial plants, but partial oxidation processes are employed where steam-reformable feeds are not available.

The heart of a coal-based partial-oxidation process is the gasification step. The two important commercial processes are the Lurgi, employing a fixed-bed reactor using oxygen and steam at ca 2000–3000 kPa (20–30 atm) and temperatures ranging from 560 to 620°C, and the Koppers-Totzek, in which gasification takes place at low pressures and much

higher temperatures (ca 1480°C). In recent years, much attention has been given to production of synthetic gas via coal-gasification processes (see Fuels, synthetic).

Partial oxidation of hydrocarbon feedstocks for the production of synthetic gas uses the Shell and Texaco processes with operating conditions in the gas generator varying from 1200 to 1370°C and 3200–8370 kPa gauge (465–1215 psig) using heavy oils as hydrocarbon feed.

The bulk (75–80%) of worldwide ammonia production uses hydrocarbon steam-reforming operations, approximately 60–65% use; natural gas feed. The process involves the following: (1) Desulfurization of the feedstock by removal of sulfur via adsorption on activated carbon at ca 15–50°C, or by reaction with zinc oxide to remove H_2S, mercaptans, and chlorine. (2) Primary and secondary reforming of hydrocarbon feedstock, carried out in two catalytic stages in the presence of steam: the first producing a partially reformed gas which is then further processed in a secondary reformer to achieve the low methane content desired. (3) Shift conversion, for conversion of carbon dioxide to hydrogen and carbon dioxide. (4) Carbon dioxide removal using reaction systems, eg, MEA (20% monoethanolamine), promoted MEA (monoethanolamine plus Amine Guard), Vetrocoke (K_2CO_3 plus As_2O_3–glycine), Carsol (K_2CO_3 plus additives), Catacarb (K_2CO_3 plus additives), Benfield (K_2CO_3 plus diethanolamine and additives), and Lurgi (K_2CO_3 plus additives); combination reaction–physical systems, eg, Sulfinol (sulfolane and 1,1′-iminobis-2-propanol) and TEA-MEA (triethanolamine and MEA); and physical absorption systems that use various solvents, eg, Purisol (NMP) (*N*-methyl-2-pyrrolidinone), Rectisol (methanol), Fluor solvent (propylene carbonate), and Selexol (dimethyl ether of polyethylene glycol). (5) Synthesis gas purification, to remove or convert to inert species residual carbon monoxide and carbon dioxide, either by methanation or cryogenic purification. (6) Ammonia synthesis and recovery, the actual step in which the ammonia is made. Currently, the operating pressure for most modern synthesis loops fall in the range of 13,785 to 34,475 kPa (2000–5000 psi). Plants designed before 1964 tended to favor the region above 31,030 kPa (4500 psi), whereas new plants favor the lower range. The economics of the more recent large tonnage ammonia plants is more favorable at these pressures.

The question of what is the right synthesis facility varies from project to project with the synthesis pressure the most important consideration, followed by loop configuration, makeup gas-recycle compression, and the converter. The design of modern ammonia plants reflects a high degree of integration between the process and energy systems. Waste process heat is used to provide energy for boiler-feedwater heating and steam generation. Energy recovery in the convection section of the primary reformer includes satisfying similar utility demands of boiler-feedwater heating, and steam generation and superheating.

Health and Safety

Effects of ammonia can range from a mild irritation to severe corrosion of sensitive membranes of the eyes, nose, throat, and lungs, depending on the concentration. Because of its great affinity for water, it is particularly irritating to moist skin surfaces.

Uses

The largest market for ammonia is the fertilizer industry. Other important markets include commercial explosives (nitric acid, initially derived from ammonia, is used in manufacture of explosives) (see Explosives and propellants); and fibers and plastics.

Joseph LeBlanc, Jr.
S. Madhavan
Raymond E. Porter
M.W. Kellogg

M. Appl, *Nitrogen* **100**, 47 (Mar.–Apr. 1976).

O.J. Quartulli, *J. Hungarian Chem. Soc.* **30**(8), 404 (1975).

AMMONIUM COMPOUNDS

The ammonium compounds consist of a large number of salts, many of which are of considerable industrial importance, eg, ammonium sulfate, ammonium chloride, and ammonium nitrate. The ammonium salts are much like salts of the alkali metals in such properties as solubility, but they differ in being considerably less stable. Except for complexes containing metallic radicals, they are completely volatile on heating or ashing. They are more like the salts of potassium than those of sodium in some ways, eg, in the lesser solubility of the ammonium and potassium perchlorates and chloroplatinates.

Ammonium Chloride

NH_4Cl, known as sal ammoniac, occurs naturally in crevices in the vicinity of volcanoes as a sublimation product. It has specific heat of 1.55 J/g (0.371 cal/g) in the temperature range 1–55°C. It is soluble to the extent of 29.4 g/100 g H_2O at 0°C. Vapor pressure at 250°C 6.5 kPa; at 320°C 60.9 kPa; and at 338°C 101.1 kPa. The notable high vapor pressure of ammonium chloride at elevated temperatures causes it to sublime readily into a vapor consisting of equal volumes of ammonia and hydrogen chloride. Like other ammonium salts of strong acids, it has an acid reaction in aqueous solution and the solid tends to lose ammonia and become more acid on exposure and in storage. Its aqueous solutions have a notable tendency to attack ferrous metal and other metal surfaces.

Ammonium chloride manufacture is based on metathesis, or double decomposition reactions of sodium chloride; the ammonium salt is used directly or formed *in situ*. Ammonium chloride is obtained as a by-product of the classic ammonia-soda or Solvay process. Double decomposition using the ammonium sulfate–sodium chloride process is another method used. Preparation by direct neutralization of hydrochloric acid with ammonia is simple but not as attractive economically.

Ammonium Nitrate

NH_4NO_3 does not occur in nature and was first made in 1659 by Glauber, who called it *nitrum flammans* because of the difference of its yellow flame from that of potassium nitrate. It is the most important of the ammonium compounds from the standpoints of volume of production and major uses.

Ammonium nitrate is a white crystalline solid, d_4^{20} 1.725 g/cm³. It is extremely soluble in water, and therefore very hygroscopic (see Table 1). In the solid state, it occurs in five different crystallographic modifications.

The change from one crystal form to another at various transition points can be detected by time–temperature cooling curves.

The solid salt has a specific heat of 1.70 J/g (0.406 cal/g) between 0 and 31°C. It has a negative heat of solution in water and can be effectively used to prepare freezing mixtures.

Ammonium nitrate decomposes under extreme shock or at elevated temperatures in two widely divergent manners

$$NH_4NO_3 \rightarrow N_2O + 2 H_2O$$

$$2 NH_4NO_3 \rightarrow 2 N_2 + 4 H_2O + O_2$$

Table 1. Solubility of Ammonium Nitrate

Temperature, °C	Soly of NH_4NO_3, g	
	in 100 g water	in 100 g soln
0	118	54.2
20	187	65.2
40	297	74.8
60	410	80.4
80	576	85.2
100	843	89.4

The first reaction, which occurs when ammonium nitrate is heated to temperatures from 200 to 260°C, can be carried out safely and can be controlled even when rapid. The second reaction takes place with great rapidity and violence when ammonium nitrate detonates (see Explosives).

Ammonium nitrate acts as an oxidizing agent in many reactions at ordinary temperatures, and in aqueous solutions it is reduced by various metals.

Ammonium nitrate manufacturing processes depend almost entirely on the neutralization of nitric acid with ammonia in liquid or gaseous form and involve three essential steps: neutralization, evaporation of the neutralized solution, and control of the particle size and characteristics of the dry product.

Ammonium Sulfate

$(NH_4)_2SO_4$ is a white crystalline solid, whose crystals are of rhombic structure, d_4^{20} 1.769 g/cm³. The melting point as determined in a closed system is 513 ± 2°C. Upon heating in an open system, the sulfate begins to decompose at 100°C and yields ammonium bisulfate, NH_4HSO_4, which has a melting point of 146.9°C. The solubility of ammonium sulfate in 100 g water is 70.6 g at 0°C and 103.8 g at 100°C. It is insoluble in alcohol and acetone.

Ammonium sulfate is produced by reaction of by-product ammonia from coke ovens with sulfuric acid. Additional amounts have increasingly been produced by the reaction of synthetic ammonia and sulfuric acid; from the process for production of caprolactam; and via a method using gypsum, $CaSO_4.2H_2O$.

Health and Safety

Ammonium nitrate as commonly handled in small amounts is properly considered safe. However, it is a potential high explosive when three conditions are present: priming by a high velocity explosive; confinement at elevated temperatures; and the presence of an oxidizable substance.

Uses

Ammonium acetate is used in pharmaceuticals as a diaphoretic and diuretic. Ammonium bicarbonate is used as a leavening agent in the production of certain baked foods. Ammonium chloride is used in the manufacture of dry cells; as a metal cleaner in soldering; as a flux in tinning and galvanizing; and in pharmaceuticals. Ammonium nitrate is used as a fertilizer and as an industrial explosive, as well as in the manufacture of nitrous oxide. Ammonium sulfate is used as a fertilizer, particularly for rice; and as an additive to supply nitrogen in fermentation processes.

Ronald D. Young
Tennessee Valley Authority

G. Feick and R.M. Hainer, *J. Am. Chem. Soc.* **76**, 5860 (1954).

J.W. Mellor, *Comprehensive Treatise on Inorganic and Theoretical Chemistry*, Vol. 2, Longmans, Green and Co., New York, 1927, pp. 837 and 843.

T.P. Hou in R. Furnas, ed., *Manual of Industrial Chemistry*, 6th ed., Vol. 1, D. Van Nostrand Co., Inc., New York, 1942, pp. 402–459.

G. Drake, "Production of Ammonium Nitrate," *The Fertilizer Society of London Proceedings No. 136*, Oct. 25, 1973.

A.V. Slack, *Fertilizer Developments and Trends*, Noyes Development Corporation, Park Ridge, N.J., 1968, pp. 146–150.

AMMONOLYSIS. See Amines.

AMORPHOUS MAGNETIC MATERIALS

Amorphous magnetic materials, such as common glasses and plastics, contain no long-range atomic order. They are usually prepared by the direct solidification from the melt since the cooling rates required to inhibit crystallization are easily attainable. They are composed of long chains of silicates or polymers, and they typically act as insulators.

Various techniques are used to determine whether a particular metallic alloy is amorphous, eg, x-ray, electron, or neutron-scattering experiments (see Analytical methods). There is disagreement over what evidence is necessary to characterize a particular sample as definitely amorphous. These alloys typically show no contrast when examined in the electron microscope at resolutions of the order of 0.02 nm; they produce only one broad, strong diffraction line followed by a number of weak, broad lines; on heating, rapid changes in resistivity, heat evolution, magnetic coercive force, mechanical hardness, and other properties all occur at the temperature corresponding to the onset of crystallization.

Preparation

All amorphous alloys are in a metastable state. Their preparation and stability at room temperature depend on various kinetic barriers to growth of crystal nuclei, if nuclei are present, or to nucleation barriers that hinder formation of stable crystal nuclei. During quenching, the alloy must pass too rapidly through the temperature range where nucleation occurs. The following methods are used to prepare amorphous alloys:

1. Vacuum deposition for elements such as Co, Cr, Fe, Mn, and Ni, carried out at very low temperatures. In particular, the rare earth–transition metal (RE–TM) alloys are of interest for magnetic bubble applications. The criterion for the formation of an amorphous phase by deposition can be described in terms of whether the added atom is prevented from diffusion more than an atomic distance before it is fixed in position by deposition of additional atoms.

2. Sputter deposition, carried out in a partial pressure of an inert gas, such as argon at 13.3 Pa (10^{-1} torr), using an r-f power of about 200 W at 14 MHz. For bubble memories which require thicknesses of ca 1 μm, the deposition takes about 30 min.

3. Quenching from the melt, with the alloy cooling through the temperature range T_m to T_g, the melting temperature to the glass-transition temperature, without crystallizing. The factors controlling T_g and crystallization are both structural and kinetic. A low eutectic temperature compared to the melting point of the metallic element coupled with a metal-rich eutectic composition favors the formation of the amorphous phase.

Preparation of amorphous metals in general, and of amorphous magnetic metals in particular, by direct solidification from the melt is of great interest because of the wide variety of alloys that can be made as well as the potential low cost of the method.

Magnetic Properties

The intrinsic properties of amorphous ferromagnets are still not quantitatively understood. The basic problem is to incorporate into the existing theoretical approaches the structural distribution contained, eg, in the radial distribution function. In a qualitative sense, the intrinsic properties (magnetic moment M_s, Curie temperature T_c, and magnetostriction λ) and the extrinsic properties (coercive force H_c, remanence-to-saturation ratio M_r/M_s, and the permeability and losses W as a function of frequency) can all be understood using the same underlying physical principles applicable to crystalline alloys with appropriate modifications to take into account the differences in structure (see Magnetic materials). In the case of the differences in the intrinsic properties, the changes in average interatomic distances and their distribution are considered, but in many cases it is the presence of the metalloids alone, not the absence of long-range order, that makes the amorphous alloys different from crystalline alloys.

Magnetic moment. The saturation magnetization of a material at a temperature of 0 K is one of its fundamental properties. Measurements are usually expressed as average moment per magnetic atom in units of the Bohr magneton, μ_B, or as saturation magnetization for the alloy M_s in units of emu/cm^3. The simplest case to analyze would be for the amorphous pure elements Fe, Ni, and Co.

Curie temperature and temperature dependence of M_s. In spite of their chemical and structural disorder, amorphous ferromagnets have a well-defined magnetic ordering Curie temperature T_c reflecting the retention of short-range order. This has been confirmed from magnetization temperature and Mossbauer and specific-heat measurements. The T_c is a measure of the strength of the magnetic interactions between atoms.

Magnetic anisotropy. Amorphous solids have been assumed to have no long-range order; thus they should be isotropic on a macroscale. However, anisotropic magnetic behavior is observed, with its origin varied and not completely understood: in many cases, it reflects the existence of short-range order in the amorphous alloy.

Like crystalline alloys, the amorphous alloys order under the influence of a magnetic field or stress applied at temperatures below the Curie temperature. This results in a uniaxial anisotropy arising from the ordering of the magnetic and nonmagnetic atoms. In the case of magnetic ordering, the magnitude of the directional order anisotropy K_u at the measurement temperature and at 0 K is expected to be a function of the annealing temperature, of the magentization at the annealing temperature, and of the concentrations of the ordering atoms. These relations for amorphous Fe–Ni–B alloys have been studied in detail.

Anisotropic microstructures can arise in amorphous alloys prepared by any of the techniques described. These may involve density or compositional fluctuations that produce variations in magnetization. The resultant anisotropy field can then be no greater than that for long rods, namely, $4\pi M_s$. In some amorphous alloys, anisotropy fields greater than $4\pi M_s$ are observed even with no external stresses. The magnitude of K_u depends on the composition and conditions of preparation.

Extrinsic magnetic properties. The magnetic anisotropy of amorphous alloys may arise from field or stress annealing, nonuniform strains, or from chemical inhomogeneities. The magnitude of the total anisotropy K depends on the alloy, its preparation, and thermal-mechanical treatments. This anisotropy, and its local variations, together with the sample shape then determine the domain structure. The motion of these domains under the influence of the internal material parameters and the externally applied fields determine the static and dynamic magnetic properties. The domain configurations can be observed directly. The direction of magnetization is determined only by the field applied during the annealing and by the sample geometry. The coercivity is dominated by pinning of the domain walls at the ribbon surface and is thus proportional to the reciprocal of the ribbon thickness. The losses of amorphous alloys are extremely low owing to the reduction of hysteresis loss.

Resistivity

One of the advantages of amorphous metallic alloys for application in devices is their higher resistivity compared to conventional metallic materials. This results in smaller eddy-current contributions to the permeability and losses. Resistivities for some of the commercially available amorphous alloys are two to four times larger than those of the corresponding transition-metal alloys without the glass formers.

Uses

Magnetic-bubble materials are used to produce devices with memory storage densities of greater than 1.5×10^5 bit/cm^2 and with data rates greater than 1 MHz (see Computers; Magnetic materials). Conventional soft magnetic materials are used in a wide variety of applications ranging from large power transformers and generators requiring many thousands of kilograms of Si–Fe to very small transformers and inductors in hybrid electronic circuits requiring only a few grams of one of the many available grades of Fe–Ni, Fe–Co, or ferrites. For application in small electronic devices, the amorphous alloys have losses and permeabilities that are somewhat poorer than those of conventional Fe–Ni alloys but are much better than those of Fe–Co and Fe–Si alloys. Where design optimization requires the lower cost of the amorphous alloys, higher induction than the Fe–Ni alloys, or lower losses at high frequency than Fe–Co, Fe–Si, and Fe–No, all will favor the use of amorphous alloys.

F.E. LUBORSKY
General Electric Company

J.J. Gilman, *Phys. Today*, 46 (May 1975).

G.S. Cargill, III, *Solid State Phys.* **30**, 227 (1975).

F.E. Luborsky, "Amorphous Ferromagnets," in E.P. Wohlfarth, ed., *Ferromagnetic Materials*, Vol. 1, North Holland Publishing Co., Amsterdam, The Netherlands, 1980, Chapt. 6.

AMOSITE.   See Asbestos.

AMPHIBOLE.   See Asbestos.

AMYL ACETATE, $CH_3COOC_5H_{11}$.   See Esters, organic.

AMYL ALCOHOLS

The term amyl alcohol describes any five-carbon saturated alcohol. Four of the alcohols are primary and three are secondary. Three are pentanols, four are substituted butanols, and one is a disubstituted propanol.

Physical and Chemical Properties

Amyl alcohols, $C_5H_{11}OH$, occur in eight isometric forms, seven of which are clear, colorless, mobile liquids at ambient conditions whereas the eighth isomer, 2,2-dimethylpropanol (neopentyl alcohol), is a solid that melts at 53°C. The amyl alcohols include: 1-pentanol, CH_3-$(CH_2)_4OH$, bp 137.8°C; 2-pentanol, $CH_3(CH_2)_2CH(OH)CH_3$, bp 119.3°C; 3-pentanol, $(CH_3CH_2)_2CHOH$, bp 115.6°C; 2-methyl-1-butanol, $CH_3CH_2CH(CH_3)CH_2OH$, bp 128.0°C; 3-methyl-1-butanol, $(CH_3)_2$$CH(CH_2)_2OH$, bp 128.0°C; 2-methyl-2-butanol, $CH_3CH_2C\text{-}(OH)(CH_3)_2$, bp 101.8°C; 3-methyl-2-butanol, $(CH_3)_2CHCH(OH)CH_3$, bp 112.0°C; and 2,2-dimethyl-1-propanol, $(CH_3)_3CCH_2OH_3$, bp 113–114°C.

The amyl alcohols have pungent odors characteristic of alcohols, are only slightly soluble in water, but are miscible with most organic solvents including mineral and vegetable oils.

The amyl alcohols undergo reactions typical of alcohols of the corresponding class and isomeric structure. There are differences in reactivity of the primary, secondary, and tertiary alcohol groups as a consequence of the hydrocarbon radical and the degree of branching.

Manufacture

Amyl alcohols are now produced almost exclusively via the oxo process, also known as the hydroformylation reaction (see Oxo process; Alcohols, higher aliphatic, synthetic processes). The alcohol fermentation process is an alternative means of manufacture.

Health and Safety

Amyl alcohols are generally narcotic, lethal to animals in high dosage and should not be taken internally or inhaled. They are volatile and flammable.

Uses

Amyl alcohols are used as solvents for resins and gums; as lubricants; frothers in ore-flotation processes; plasticizers and solvents; corrosion inhibitors and antioxidants.

PAUL DWIGHT SHERMAN, JR.
Union Carbide Corporation

J. Falbe, translated by C.R. Adams, *Carbon Monoxide in Organic Synthesis*, Spring-Verlag, New York, 1970, pp. 3–75.

J.A. Monick, *Alcohols*, Reinhold Book Corp. New York, 1968, pp. 136–146.

AMYLAMINES.   See Amines, Lower aliphatic.

AMYLASES.   See Enzymes.

AMYLODEXTRINS.   See Starch.

AMYLOPECTINS.   See Starch.

AMYLOSES.   See Starch.

ANALEPTICS.   See Stimulants.

ANALGESICS, ANTIPYRETICS, AND ANTI-INFLAMMATORY AGENTS

Analgesics are substances that relieve pain without decreasing sensibility or consciousness.

Those that act within the central nervous system are used for relief of moderate to severe pain of surgery, major trauma, terminal illness, etc. These include morphine (**1**) and morphine-like agonists such as heroin (**2**) and codeine (**3**), as well as derivatives and simpler synthetic analogues such as phenazocine (Prinadol), levorphanol (Dromoran), meperidine (Demerol hydrochloride), and *d*-propyphene (Darvon).

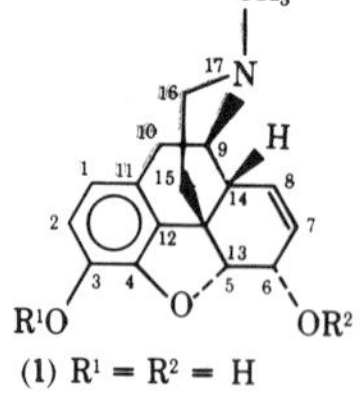

(1) $R^1 = R^2 = H$
(2) $R^1 = R^2 = COCH_3$
(3) $R^1 = CH_3, R^2 = H$

Morphine-like agonists used for analgesia may cause drowsiness, constipation, respiratory depression, tolerance (decrease in analgesic effect following repeated administration), and physical dependence.

Nalorphine (*N*-allylnormorphine)like narcotic antagonists are included in the class of centrally acting analgesics. These substances reverse some of the effects of the morphinelike agonists, and precipitate withdrawal syndrome when administered to individuals dependent on narcotics. Some halorphinelike antagonists, eg, pentazocine (Talwin), butorphanol, and buprenorphine (made from thebaine), are potent analgesics that generally are less likely to produce tolerance and physical dependence than the morphine agonists.

The pure antagonists naloxone (4) (Narcan), prepared from an oxidation product of thebaine and naltrexone are devoid of analgesic effect and are useful as antidotes for acute narcotic overdose. They are also being studied for their ability to prevent post-addicts' return to narcotics use. Naltrexone blocks a heroin challenge in volunteer subjects and is currently the best candidate for opiate-antagonist treatment of heroin dependence. Naloxone has proved to be a powerful tool for research on the biochemical mechanisms of analgesia, tolerance, and physical dependence. Radiolabeled naloxone was used in the first direct biochemical demonstration of opiate receptors in nervous tissue and in determining the regional distribution of these receptors in monkey and human brains. Opiate receptors have been characterized with respect to their time, temperature, and pH dependence, and the binding affinities of a number of agonists and antagonists have been found to parallel their pharmacological activities.

(4) R = $CH_2CH=CH_2$

Aspirinlike analgesics are used for relief of mild to moderate pain of headache, muscle ache, arthralgia, etc. These include the salicylates such as acetylsalicyclic acid (5) (aspirin), and methyl salicylate (oil of wintergreen); derivatives of *p*-aminophenol, eg, acetaminophen; pyrazolones such as phenylbutazone (used to treat acute inflammatory episodes of rheumatoid arthritis); indomethacin (6) (Indocin); ibuprofen (Motrin); and ketoprofen.

(5)

(6)

They produce their analgesic effect by acting primarily at the site of origin of the pain, and may be antipyretic (lowering body temperature to normal) and/or anti-inflammatory. Aspirinlike analgesics do not produce drowsiness, constipation, or respiratory depression, nor do they give rise to tolerance and physical dependence.

A possible mechanism of action for aspirinlike compounds was suggested by Vane involving inhibition of prostaglandin synthesis as the basis of their pharmacological action (see Prostaglandins).

WILLIAM F. MICHNE
Sterling-Winthrop Research Institute

J.H. Jaffe and W.R. Martin in L.S. Goodman and A. Gilman, eds., *The Pharmacological Basis of Therapeutics*, 5th ed., Macmillan Publishing Co., Inc., New York, 1975, pp. 245–283.

R.K.S. Lim, F. Guzman, D.W. Rogers, K. Goto, C. Braun, G.D. Dickerson, and R.J. Engle, *Arch. Int. Pharmacodyn. Ther.* **152**, 25 (1964).

J.R. Vane, *Nature (London) New Biol.* **231**, 233 (1971).

ANALYTICAL METHODS

Analytical chemistry is concerned with either the qualitative determination of elemental and molecular constituents of a selected specimen of matter, or the quantitative measurement of the fractional distribution of those constituents. It is an extremely broad discipline essential to many industries, particularly those concerned with the processing and manufacture of petroleum, chemical, semiconductor, metal, plastic, pharmaceutical, ceramic, paper, textile, and glass products. It is also fundamental to environmental monitoring and product qualification (see Trace and residue analysis). The professional analytical chemist has the responsibility of determining what procedures should be used for a given problem or, if required, developing new ones.

Analytical Strategy

In order to describe specific procedures for a given analytical problem, it is important to consider first the following series of questions directed at determining both sample requirements and the appropriateness of conventionally available methods: How should the materials be sampled to ensure that the analysis is truly representative? What is the physical state of the substance? Is it a solid, liquid, or gas? Does the material consist of more than one chemically or physically distinct phase? Will separation be required prior to analysis? What sample size will be needed for the various methods under consideration? What degree of accuracy is necessary and what degree is possible? Is the material organic or inorganic? Is it crystalline or amorphous? Is an elemental or molecular analysis required? Does the measurement have to be nondestructive or can the sample be consumed? (see Nondestructive testing); and is a representative bulk analysis required or is the region of interest limited to a discrete area such as a thin surface film or fine internal precipitate?

The analysis of a completely unknown sample requires more effort compared to one where some background information is available. If a general elemental survey analysis is required, some techniques like spark-source mass spectrometry, optical emission, or x-ray fluorescence are relatively fast and provide semiquantitative information for a range of elements in a time period of a few minutes to several hours. Other methods required for maximum accuracy in quantitative analysis include a variety of traditional wet chemical techniques which often involve totally different procedures for each element and can take many hours to set up. In cases where extremely low elemental concentrations are needed, entirely different methods are often used and special precaution must be taken to avoid interference effects. For example, an instrumental technique like atomic absorption, known for its sensitivity, requires that special optical-emission source tubes be available for each element to be examined.

For convenience, a general survey of some of the more common techniques is given in Table 1.

ERIC LIFSHIN
ELIZABETH A. WILLIAMS
General Electric Company

G.W. Ewing, *Instrumental Methods of Chemical Analysis*, McGraw-Hill, Inc., New York, 1975.

D.A. Skoog and D.M. West, *Fundamentals of Analytical Chemistry*, Holt, Rinehart and Winston, New York, 1976.

H.A. Strobel, *Chemical Instrumentation: A Systematic Approach*, Addison-Wesley, Reading, Pa., 1973.

D.G. Peters, J.M. Hayes, and G.M. Hieftje, *A Brief Introduction to Modern Chemical Analysis*, Saunders, Philadelphia, Pa., 1976.

ANALYTICAL METHODS. See also Supplement.

ANDROGENS. See Hormones.

Table 1. Common Analytical Methods

Method	Principal applications	Molecular phenomenon	Advantages in qual. analysis	Advantages in quant. analysis	Average sample desired for qual. anal.[b]	Method limitations	Sample limitations
Molecular identification and analysis[a]							
infrared spectroscopy	structure determination and identity of organic and inorganic compounds general quant. analysis	excitation of molecular vibrations by light absorption	identification of functional groups largest file of reference spectra available virtually no sample limitations impurity detection	widely applicable	3 mg	medium sensitivity[c] no direct information about size of molecule	avoid aqueous solutions
Raman spectroscopy	structure determination and identity of organic compounds symmetry of molecular groups in solid state	excitation of molecular vibrations by light scattering	identification of functional groups (usually different from those identified by ir) water solutions	(special applications)	0.01 mg	low sensitivity[c] no direct information about size of molecule	sample must not fluoresce avoid turbid materials some restrictions on colored material
mass spectrometry	structure determination and identity of organic compounds analysis of trace volatiles in nonvolatiles	ionization of molecule and cracking of molecule into fragment ions	precision molecular wt (molecular ion) masses of integral parts of molecule (fragment ions) very high sensitivity impurity detection	high sensitivity wide applicability to volatile materials	0.1 mg	does not detect functional groups directly comparatively slow	$\geqslant 0.133$ Pa (10^{-3} torr) vapor pressure at sample inlet temperature unless special ionization techniques are available
nuclear magnetic resonance	structure determination and identity of organic compounds molecular conformation	reorientation of magnetic nuclei in a magnetic field	determination of chemical type and number of atoms molecular configuration and conformation impurity determination applicable to water solutions	standards not required	10 mg (^{1}H) 300 mg (^{13}C)	medium sensitivity applicable only to nuclei with a nuclear spin	liquid or soluble solid (wide variety of solvent choices)
ultraviolet and visible spectrophotometry and colorimetry	quantitative analysis, esp. as end methods in chemical analysis schemes	excitation of loosely bonded electrons	(special applications)	high precision high sensitivity simplicity	0.01 mg	low specificity; little information on molecular structure	soluble in uv-transparent solvent (wide variety of choices)
gas chromatography	general multicomponent quantitative analysis of volatile organics highly efficient separation technique	partitioning between vapor phase and substrate	separates materials for examination by other techniques	generality—widely applicable to volatile materials multicomponent analyses high sensitivity in special cases	1 mg	identifies materials only in special cases not applicable to materials of low volatility	$\lesssim 133$ Pa (1 torr) vapor pressure at sample inlet temperature
combined gas chromatography-mass spectrometry	identification and analysis of trace organic materials	combines separation efficiency of gc with sensitivity and specificity of mass spec	applicable to identity of sub-ppm components in mixtures	specific identification of gc peaks being determined	1 mg	not applicable to materials of low volatility	>133 Pa (1 torr) vapor pressure at sample inlet temperature
liquid chromatography (including ion exchange and thin layer)	separation technique for less volatile materials multicomponent quant. analysis	partitioning between liquid solution and substrate	separates materials for examination by other techniques	multicomponent analyses of less volatile materials	300 mg	resolution rather poor compared to other chromatographic methods method development is time-consuming	(none)
gel permeation chromatography	separation on basis of molecular weight determination of mol wt distribution in polymers	separation by variation of penetration into cross-linked gel structure	separates materials for examination by other techniques	determines molecular weight distribution	500 mg	requires extensive calibration time-consuming	material must be soluble in narrow choice of solvents
x-ray	identification of crystalline	diffraction of x-	high specificity for	applicable to inorganic	0.1 mg	limited structural	crystalline solids,

Table 1 (continued)

Method	Principal applications	Molecular phenomenon	Advantages in qual. anaysis	Advantages in quant. analysis	Average sample desired for qual. anal.[b]	Method limitations	Sample limitations
diffraction	substances, especially inorganic determination of crystallinity esp. polymers	rays from crystal planes	crystalline solids, esp. inorganics large file of reference patterns available	crystalline solids		information—essentially a "fingerprint" method	partly crystalline polymers
chemical reaction methods (classical analysis)	variety of specialized quantitative analysis applications	stoichiometry of chemical reactions	(special applications)	high precision for assay analyses absolute calibration	1000 mg	time-consuming interferences often a problem	(none)
Atomic identification and analysis							
atomic emission spectroscopy	general qual. and semi-quant. survey of all metallic elements trace metal analysis	light emission from excited electronic states of atoms	general for all metallic elements simultaneous analysis of all metallic elements	general for all metallic elements high sensitivity in many cases	10 mg[d]	detects the volatile elements (nonmetals) only with difficulty calibration required for precision quant. analysis	applicable principally to nonvolatile materials[e]
atomic absorption spectroscopy	precision quant. analysis for a given metal trace analysis for a given metal	absorption of atomic resonance line	(not applicable)	fast, reliable analysis for a given element high sensitivity in some cases simplicity	100 mg	metals analyzed individually not simultaneously usually not applicable to nonmetallic elements	element being analyzed must be in a solution (many solvent choices)
x-ray fluorescence	general qual. and semi-quant. survey of all elements atomic no. <11 (Na) precision quant. analysis of elements esp heavier nonmetals (P, S, Cl, Br, I) trace analysis	re-emission of x-rays from excited atoms	general for all elements atomic no. ≥11 (Na) minimum sample preparation	general for all elements atomic no. ≥14 (Si) high sensitivity in some cases simplicity minimum sample preparation	500 mg (nondestructive)	nonsensitive to elements of atomic no. ≥11 (Na) precision limited by nonuniformity of spl.	applicable principally to solids and nonvolatile liquids
neutron activation	precision quant. analysis of most elements trace and ultratrace element analysis general qual. analysis of most elements	counting of radio-active species produced by neutron reactions	minimum sample preparation	highest sensitivity for many elements high confidence level only general instrumental method capable of N, O, and F analysis	100 mg (nondestructive)	sensitivity varies considerably among elements (but sensitive to amounts <1 μg for most elements) multicomponent analyses present some problems	applicable to solids and liquids
chemical reaction methods (classical analysis)	variety of specialized quantitative analysis applications	stoichiometry of chemical reactions	(special applications)	high precision for assay analysis absolute calibration	1000 mg	time-consuming	(none)

[a] This table compares the more widely used techniques. Other more specialized techniques are also discussed in this article.

[b] The amount of sample listed in this column is an estimate of the average minimum sample required, usually a larger sample is preferred in order to do the best possible analysis. On the other hand successful identifications can often be done with much smaller amounts.

[c] "Sensitivity" as used here indicates ability to determine a small amount of one material in the presence of large amounts of other materials.

[d] The amount of sample listed in this column is only a rough index. Sample required will vary enormously depending on whether the problem is identity of a major elemental component or precision determination of a trace element.

[e] None of these methods handles gases easily and conveniently (although it could be done in special cases). For elemental analysis in gases mass spectrometry is the best general choice.

ANDROSTERONE, $C_{19}H_{30}O_2$. See Hormones.

ANESTHETICS

Anesthetics are substances that affect vital functions of all types of cells, especially those of nervous tissue. General anesthetics produce a hiatus in consciousness and thus abolish the sensitivity to pain. Local anesthetics are applied directly to restricted areas to abolish pain. Some local anesthetics are injected directly into the spinal fluid to produce block anesthesia.

Theories of General Anesthesia

A unified theory of narcosis is precluded by the varied physiochemical properties and physiological effects of the general anesthetics. Several theories have been advanced including: the colloid theory; the permeability theory; the surface tension or absorption theory; the lipid theory; a variety of neurophysiological theories; a variety of physical theories; and biochemical theories. A recent review discusses in detail various aspects such as structure–activity relationships, distribution, and metabolism of experimental and currently used inhalation and intravenous anesthetics.

General Anesthetics

The most commonly used volatile and gaseous general anesthetics include: nitrous oxide (N_2O), a water-soluble, nonirritating, colorless, sweet-smelling, inert gas which exerts its influence on the central nervous system through the blood; cyclopropane (C_3H_6), a colorless, flammable and explosive gas with a slightly unpleasant odor which relaxes skeletal muscles; and halothane (Fluothane), 1-bromo-1-chloro-2,2,2-trifluoroethane, a nonflammable, nonexplosive and nonirriatating gas. Less commonly used anesthetics in this category include diethyl ether $[(C_2H_5)_2O]$; fluroxene (Fluoromar), 2,2,2-trifluoroethyl vinyl ether; methoxyflurane (Penthrane), 2,2-dichloro-1,1-difluoroethyl methyl ether; enflurane (Ethrane), 2-chloro-1,1,2-trifluoroethyl difluoromethyl ether; and isoflurane (Forane), 1-chloro-2,2,2-trifluoroethyl difluoromethyl ether. Ethylene, chloroform, ethyl chloride, halopropane (3-bromo-1,1,2,2-tetrafluoropropane), trichloroethylene, and divinyl ether, inhalation anesthetics previously used for induction or maintenance of general anesthesia, have been supplanted by nonflammable anesthetics with more desirable properties.

Fixed general anesthetics include ultrashort-acting barbiturates, which are poor analgesics and generally used for induction and for short surgical procedures. When injected intravenously, they produce a quick onset of unconsciousness and rapid recovery after small doses. Anesthetics in this category include Thiopental (Pentothal), 5-ethyl-5-(1-methylbutyl)-2-thiobarbituric acid; Methohexital (Brevital), 5-allyl-5-(1-methyl-2-pentynyl)-1-methylbarbituric acid; Propanidid (Epontol), propyl 3-methoxy-4-(N,N-diethylacetamidoxy)phenylacetate; the new drug Etomidate, ethyl 1-methylbenzylimidazole-5-carboxylate; and less often used drugs such as ketamine, 2-(o-chlorophenyl)-2-methylaminocyclohexanone; the neuroleptanalgetic anesthetic Innovar, a mixture of droperidol and fentanyl; and Althesin, a mixture of two steroids, 3α-hydroxy-5-α-pregnane-11,20-dione (Alphaxalone); and 21-acetoxy-3α-hydroxy-5α-pregnane-11,20-dione (Alphadolone).

Spinal anesthesia is achieved by injecting a local anesthetic such as procaine, tetracaine, dibucaine, and lidocaine (see Local Anesthetics) into the subarachnoid space. This technique is useful for abdominal, pelvic and lower-extremity surgery.

Local Anesthetics

Local anesthetics produce loss of sensation and motor activity in a restricted area of the body by decreasing the permeability of the cell membrane to sodium ions and thereby reversibly blocking conduction in nerve fibers. A good local anesthetic should be nonirritating to the tissue to which it is applied; confine its action to nerve tissue; have a short latency for the onset of anesthesia; have a sufficiently long duration of action; be of low systemic toxicity; be heat stable; be soluble in saline and water; be compatible with vasoconstrictor agents; and be effective whether used topically or parenterally. Local anesthetics are metabolized primarily in the liver and the plasma with toxicity due to high concentration in the blood.

Among others, commonly used local anesthetics include: benzocaine (Anesthesin), $C_9H_{11}NO_2$, topically applied; bupivacaine (Marcaine), $C_{18}H_{28}N_2$, applied via infiltration, nerve block, caudal, and epidural; cocaine, $C_{17}H_{21}NO_4$, applied topically; dibucaine (Nupercaine), $C_{20}H_{29}N_3O_2$, applied topically and spinally; dimethisoquin (Quotane), $C_{17}H_{24}N_2O$, applied topically; dyclonine (Dyclone), $C_{18}H_{27}NO_2$, applied topically; lidocaine (Xylocaine), $C_{14}H_{22}N_2O$, applied topically, via infiltration, nerve block, peridurally, regional intravenous; pramoxine (Tronothane), $C_{17}H_{27}NO_3$, applied topically; procaine (Novocaine), $C_{13}H_{20}N_2O_2$, applied via infiltration, nerve block, caudally, epidurally and spinally; and tetracaine (Pontocaine), $C_{15}H_{24}N_2O_2$, applied topically, via infiltration, nerve block, caudally. and spinally.

A.R. PATEL
National Cancer Institute

M.E. Wolff, ed., *Burger's Medicinal Chemistry*, 4th ed., Pt. III, John Wiley & Sons, Inc., New York, 1981, Chapts. 50, p. 623, and 51, p. 645.

ANETHOLE, $CH_3CH{=}CHC_6H_4OCH_3$.   See Ethers.

ANILINE AND ITS DERIVATIVES.   See Amines, aromatic—Aniline and its derivatives.

ANSAMACROLIDES.   See Antibiotics—Ansamacrolides.

ANTACIDS, GASTRIC.   See Gastrointestinal agents.

ANTHELMINTICS.   See Chemotherapeutics, anthelmintic.

ANTHRACENE.   See Hydrocarbons.

ANTHRANILIC ACID, o-$NH_2C_6H_4COOH$.   See Benzoic acid.

ANTHRAQUINONE

Anthraquinone, also known as 9,10-dihydro-9,10-diketoanthracene, 9,10-dihydro-9,10-dioxoanthracene, and 9,10-anthracenedione, is not found in nature. Positions 1, 4, 5, and 8 are frequently referred to as the α-positions, and 2, 3, 6, and 7 as β. The 9 and 10 positions are sometimes designated as meso or ms.

Physical Properties

Anthraquinone is a pale yellow, crystalline solid with mp 286°C, bp 379–381°C under atmospheric pressure, easily sublimed but not nearly as volatile as most p-quinones. It begins to decompose appreciably at about 450°C and has low solubility in solvents such as the aliphatic alcohols or the usual aromatic hydrocarbons. It is best recrystallized from acetic acid or a high boiling solvent such as nitrobenzene or di- or trichlorobenzene.

Chemical Properties

Anthraquinone is remarkably stable to heat and oxidizing agents, which attack it with great difficulty, yielding phthalic acid. It behaves very little like an ordinary quinone or a true ketone. It is soluble in concentrated sulfuric acid forming oxonium salts. It is reduced by zinc and ammonia to yield anthracene, an important method in synthesis of dyestuff intermediates, and yields 9,10-dihydroxyanthracene (important to the dyestuff chemist) by reduction with various other agents such as $Na_2S_2O_4$.

Manufacture

The oxidation of anthracene to remove contaminants yields anthraquinone. Many processes for the purification of anthracene involve the distillation of coal tar followed by recrystallization from a variety of solvents, eg, pyridine. Dichromate is an important industrial oxidant. Several other oxidation processes include the use of potassium chlorate, sodium chlorate or nitrate, or catalytic oxidation at a temperature ranging from 350–450°C using catalysts containing vanadium.

The preferred manufacturing method is the reaction of phthalic anhydride with benzene in the presence of aluminum chloride to give o-benzoylbenzoic acid followed by acid cyclization, either by a solvent or ball-mill process.

Health and Safety

Anthraquinone has low systemic toxicity, but local contact may cause skin irritation and sensitization.

Uses

Anthraquinone is used mainly in the production of various intermediates in the manufacture of dyestuffs.

R.H. CHUNG[‡]
General Electric Company

N. Rabjohn, ed., *Organic Synthesis IV*, John Wiley & Sons, New York, 1963, p. 665. U. S. Pat. 3,655,741 (April 11, 1972), H. Sturm, H. Nienburg, and W. Gisfield (to Badische Aniline and Soda Fabrik).

ANTHRAQUINONE AND RELATED QUINONOID DYES. See Dyes, anthraquinone.

ANTHRAQUINONE DERIVATIVES

Homologues of Anthraquinone

The 2-alkylanthraquinones, derivatives of anthraquinone with alkyl chains ranging from one to five carbon atoms, are of great technical importance. They are pale-yellow solids with melting points lower than anthraquinone, soluble in benzene, and are commonly recrystallized from ethanol or acetic acid.

Properties. 1-Chloro-2-methylanthraquinone is made by chlorination of 2-methylanthraquinone either with chlorine in the presence of iodine, or sulfuryl chloride in nitrobenzene at about 100°C. 2-Methylanthraquinone can be nitrated to give exclusively 1-nitro-2-methylanthraquinone.

Alkylanthraquinone and 2-ethylanthraquinone are hydrogenated in the continuous production of hydrogen peroxide using solvents such as alkynaphthalenes, aromatic alcohols and their esters, alkylphenyl methyl ketones, and esters of dibasic acids using catalysts, eg, Raney nickel, palladium, and platinum.

Preparation. The 2-alkylanthraquinones are manufactured by the Friedel-Crafts reaction (qv) using phthalic anhydride and the corresponding substituted alkylbenzene; or by the Diels-Alder process, using 2-substituted butadiene and 1,4-naphthoquinone.

Halogenated Anthraquinones

These are yellow needles, plates or crystals with sharp melting points; soluble in nitrobenzene, acetic acid, benzene, pyridine, chloroform, and concentrated sulfuric acid; and are recrystallized from nitrobenzene, chlorinated benzenes, anisole, acetic acid, and ligroin.

Properties. α-Halogen atoms are more easily replaced than β-halogens by hydroxy, alkoxy, amino, amide, and sulfonic acid groups. Certain anthraquinonesulfonic acids are not readily prepared by the usual sulfonation reactions, but by reaction of the halogenated anthraquinone with sodium sulfite; eg, the reaction of 1,4-diamino-2,3-dichloroanthraquinone with sodium sulfite produces the 2,3-disulfonic acid derivative.

α-Chloroanthraquinones, such as 1,5-dichloroanthraquinone, react with sodium nitrate in dimethylformamide to give the corresponding dinitroanthraquinones. Some α-halogen derivatives can be converted to the β-analogues by heating. Substituted chloroanthraquinones, not readily accessible by other methods, can be recovered from the chlorosulfonic acids. Depending on reaction conditions, various sulfonic acids are obtained. Chloroanthraquinones, such as 2-chloro-, 1,6-dichloro, and 1,7-dichloroanthraquinone are reduced to anthracene derivatives by zinc and ammonium hydroxide.

Preparation. Direct halogenation of anthraquinone itself is rather difficult. Amino groups are usually quite readily replaced by halogens by

treating the diazonium salt with cuprous halide in the usual manner. Since aminoanthraquinones are easily halogenated, the reaction with subsequent removal of the amino groups by diazotization affords a method of obtaining certain chloroanthraquinone derivatives not readily produced by other methods. Hydroxyl groups can also be replaced by chlorine, by treatment with phosphorus trichloride, phosphorus pentachloride, or phosphorus oxychloride, eg, 1-chloro-4-hydroxy-anthraquinone is converted to the 1,4-dichloroanthraquinone in good yield. Other methods of preparing haloanthraquinones include replacement of nitro groups either in the α or β position using chlorine. Sulfonic acid groups in either the α or β position also are readily replaced by chlorine or bromine. Chloroanthraquinone is synthesized by condensation of phthalic anhydride with chlorobenzene.

Two main methods of producing haloanthraquinones employ (a) the replacement of the sulfonic acid groups in anthraquinonesulfonic acids by halogen, eg, to yield 1-chloroanthraquinone and 1,5- and 1,8-dichloroanthraquinones, and (b) ring closure by Friedel-Crafts synthesis from phthalic anhydride and halogenated benzenes, eg, to yield 2-chloroanthraquinone. Less important methods include replacement of nitro, amino, or hydroxy groups by halogen or direct halogenation of anthraquinone.

Nitroanthraquinones

Nitroanthraquinones are light to deep yellow solids with relatively high melting points; are sparingly soluble in most organic solvents; and commonly recrystallized from dichlorobenzene, nitrobenzene, or pyridine.

Properties. The nitro groups are easily replaced by nucleophiles, the α-nitro groups being much more reactive than the β-derivatives. In typical replacement reactions, treatment with alkaline methanol gives methoxy derivatives, reaction with phenol gives the phenoxy product, reaction with sodium sulfite gives the sulfonic acid, replacement by ammonia or amine gives amino derivatives, and hydrolysis with pyridine gives the hydroxy derivatives. Nitroanthraquinones are readily reduced to aminoanthraquinones by reagents such as sodium sulfide, sodium hydrosulfide, sodium hydrosulfite, and by catalytic hydrogenation.

Preparation. Nitroanthraquinones are manufactured by the direct nitration of anthraquinone, which does not pose the pollution problems associated with the manufacture of 1-aminoanthraquinone from 1-anthraquinonesulfonic acid. It is also produced by the Diels-Alder reaction of 5-nitro-1,4-naphthoquinone and 1,3-butadiene, followed by oxidation of the tetrahydroanthraquinone.

Aminoanthraquinones

The unsubstituted aminoanthraquinones are orange-colored, high melting solids of poor solubility, requiring high boiling solvents such as nitrobenzene, various halogenated benzenes, and some heterocyclic amines.

Properties. Primary aminoanthraquinones are extremely weak bases; basicity increases upon alkylation. They all form salts with strong inorganic acids that are readily dissociated by water. Stable diazonium compounds can be obtained by the reaction of sodium nitrite in strong sulfuric or phosphoric acid with α- and β-aminoanthraquinones. Aminoanthraquinones can be acylated by treatment with an aliphatic or aromatic acid chloride, or their respective anhydrides; or by heating the aminoanthraquinone with a free acid in a high-boiling inert solvent such as nitrobenzene or naphthalene. The reaction of thionyl chloride with aminoanthraquinones affords the corresponding thionylamines which can be hydrolyzed to the aminoanthraquinones. Boric acid is frequently used for protecting the α-amino group in many reactions of aminoanthraquinones. The primary aminoanthraquinones are converted to secondary and tertiary derivatives by reaction with alkyl halides.

[‡] Deceased.

Arylation is accomplished by heating with an aryl halide in the presence of a catalyst. Primary aminoanthraquinones are significantly more stable than other primary aromatic amines, and can often be nitrated without protecting the amino group. However, a complicating factor in nitration reactions is the tendency for the aminoanthraquinones to form nitroamines, unstable somewhat explosive compounds. The halogenation of aminoanthraquinones is as complicated as their nitration, and bromination is affected by the nature of the acyl group in 1- and 2-acylaminoanthraquinones. Alkaline fusion of 2-aminoanthraquinone is an important reaction giving indanthrone, a blue vat dye, together with alizarin (1,2-dihydroxyanthraquinone) a yellow dye. α-Aminoanthraquinone is readily hydroxylated with formaldehyde.

Preparation.   The replacement of sulfonic acid groups by amino groups is the most important method for commercial production of aminoanthraquinones, since a large number of sulfonic acids are readily obtainable by sulfonating the anthraquinones with or without the addition of a mercury catalyst. In this high temperature reaction, the sodium or potassium sulfite formed can attack the anthraquinone nucleus unless it is destroyed, or rendered inactive as rapidly as it appears by the addition of various oxidizing agents. Highly pure 1-aminoanthraquinone may be obtained by ammonolysis of 1-anthraquinonesulfonic acid in the presence of boric acid. Aminoanthraquinones can be prepared from the amides of the anthraquinone carboxylic acids by a Hofmann reaction, eg, treatment with hypochlorite or hypobromite.

The manufacture of α-aminoanthraquinones by ammonolysis of sulfonic acids presents pollution problems. The production of α-aminoanthraquinones by reduction of nitroanthraquinones with amines has become very important.

Anthraquinonesulfonic Acids

Anthraquinonesulfonic acids are pale yellowish crystals which generally occur as hydrates with rather irregular melting points. They are strong acids; readily soluble in water with the differences in solubility of their various salts used for their separation. The potassium salt of 1-anthraquinonesulfonic acid is named gold salt and the sodium salt of the 2-acid is called silver salt because of their physical appearance.

Properties.   These acids are easily desulfonated by hydrolysis, the sulfonic acid group in the position being easier to remove than in the β position. They are converted into the sulfonyl chlorides by treatment with phosphorus pentachloride or phosphorus oxychloride. Their chemical properties are similar to those of other sulfonyl chlorides. The important reactions of the anthraquinonesulfonic acids include replacement of the sulfonic acid group by amino, alkoxy, chloro, hydroxy, or aryloxy groups.

Preparation.   Direct sulfonation of anthraquinone is the most important method of preparing anthraquinonesulfonic acids. In contrast to benzene and naphthalene compounds, it is difficult to sulfonate anthraquinone even with concentrated sulfuric acid because destructively high temperatures are required. The advantage is that the very slow reaction allows the use of concentrated sulfuric acid as a solvent in other anthraquinone reactions. At lower temperature, sulfonation is accomplished by using fuming sulfuric acid when two sulfonic acid groups are introduced.

Hydroxyanthraquinones

Unsubstituted hydroxyanthraquinones are high melting solids varying in color from yellow through orange and red. They are sparingly soluble in lower-boiling aromatic hydrocarbons, but soluble in strong alkali and concentrated sulfuric acid. α-Hydroxyanthraquinone is only sparingly soluble in weak alkali because of the hydrogen bonding effect. Upon addition of boric acid to a sulfuric acid solution of hydroxyanthraquinones, distinctive color changes occur, which may be used for qualitative analysis purposes. Hydroxyanthraquinones are recrystallized from pyridine, di- or trichlorobenzenes, nitrobenzene, or acetic acid. Many of them may be sublimed as a means of purification.

Chemical properties.   The hydroxyanthraquinones undergo the usual reactions of phenols. Owing to difficulties in acylation and alkylation, both alkyl and aryl ethers of α-hydroxyanthraquinones are obtained by the replacement of nitro, chloro, or sulfonic acid substituents. Alkylation

is used to protect the hydroxyl groups in nitrations, whereas aryl ethers, being more stable, are used in sulfonation. p-Hydroxyanthraquinones are easily oxidized. The hydroxyanthraquinones react with formaldehyde to give hydroxyanthraquinonyl carbinols. Boric acid takes part in many reactions of hydroxyanthraquinones, facilitating transformations and improving yields, eg, boracetic anhydride and α-hydroxyanthraquinones give a boracetic ester; however, with β-hydroxyanthraquinones, a stable acetylanthraquinone is obtained. α-Hydroxyanthraquinones and oleum give sulfonic esters of anthraquinones. Hydroxyanthraquinones can be easily nitrated without protecting the hydroxyl groups. The chlorination of quinizarin leads to various products, depending upon the reaction conditions employed.

Preparation.   An oxidizing agent, eg, air, alkaline nitrate, or chlorate in alkaline or acid solution, is used to introduce hydroxyl groups into the anthraquinone molecule. Oxidation of hydroxyanthraquinones is carried out in concentrated sulfuric acid with an oxidizing agents such as nitric acid, manganese dioxide, arsenic acid, ammonium persulfate, etc (eg, manganese dioxide oxidizes quinalizarin to 1,2,4,5,8-pentahydroxy-anthraquinone, alizarin cyanine R).

quinalizarin

Further hydroxylation is best accomplished by addition of excess boric acid which controls the oxidation and prevents ring cleavage. Nitric acid in concentrated sulfuric acid can act either as a nitrating or oxidizing agent or both. Hydroxyanthraquinones can also be obtained by removing one or more hydroxyl groups from the more extensively hydroxylated compounds, eg, reduction of purpurin to xanthopurpurin.

purpurin

Nitro groups are replaced by hydroxyl groups by heating the nitro compounds in pyridine or quinoline; eg, α-nitroanthraquinone gives 1-hydroxyanthraquinone. Anthrarufin is obtained by condensation of 3-hydroxybenzoic acid; condensation of phthalic anhydride with

quinizarin

p-chlorophenol or hydroquinone yields quinizarin; purpurin is produced by condensation of 4-chlororesorcinol with phthalic anhydride. Chlorohydroxyanthraquinones are prepared by direct chlorination of hydroxyanthraquinones or by Friedel-Crafts reaction of chlorophenol and phthalic anhydride. Aminohydroxyanthraquinones can be produced by nitration of hydroxyanthraquinones followed by reduction. Alizarin is commercially produced by replacement of a sulfonic acid group from the sodium salt of 2-anthraquinonesulfonic acid with caustic alkali.

Benzanthrone

Benzanthrone crystallizes in pale yellow needles, melting at 170°C. It is soluble in hot aromatic solvents and recrystallized from alcohol and xylene; sublimation is the best method of purification. A solution of benzanthrone in concentrated sulfuric acid produces an orange color with green fluorescence.

benzanthrone

Properties. The 3-position is the most reactive towards electrophilic agents, eg, benzanthrone halogenates with molecular or nascent halogen to yield the 3-halo derivative, which further halogenates to yield mainly the 3,9-dihalogen compound along with some 3,11-derivative. Nitration of benzanthrone in nitrobenzene yields 3-nitrobenzanthrone. Sulfonation produces mainly 9-benzanthronesulfonic acid. Benzanthrone is readily oxidized with chromic acid in acetic acid to yield 1-anthraquinone-carboxylic acid. Chloromethylation with bis-(chloromethyl) ether in sulfuric acid gives 3,9-bis-(chloromethyl)benzanthrone. Dibenzanthrone (violanthrone) and isodibenzanthrone (isoviolanthrone) are obtained by alkaline fusion of benzanthrone.

Preparation. Benzanthrone is formed by treating an anthraquinone with glycerole and a dehydrating agent, eg, sulfuric acid, in the presence of a metal catalyst, via an aldol-type condensation followed by dehydration, giving an anthrone derivative which cyclizes to benzanthrone. It is also prepared by the 1,4-addition to the β-unsaturated carbonyl of acrolein by anthranone to afford an aldehyde intermediate; dehydration and dehydrogenation, followed by intramolecular cyclization, yield benzanthrone. Substituted benzanthrones are frequently synthesized by the condensation of anthrone with a substituted acrolein.

Anthrimides

Anthrimides, also known as anthraquinonylamines or iminoanthraquinones, are secondary amine derivatives of at least two molecules of anthraquinone joined by an amine bridge. They are very high melting colored substances, sparingly soluble in organic solvents. Solutions in concentration sulfuric acid are green to greenish blue. They are recrystallized from nitrobenzene or aniline.

Properties. They behave much like secondary aryl aminoanthraquinones with the nitro groups para to the imino group; further nitration leads to a tetranitro or pentanitro derivative. The most important reaction is the formation of carbazole derivatives which may be converted into a vat dye having fast brown shades on cotton.

Uses. 2-Alkylanthraquinones are used for the manufacture of hydrogen peroxide and dye intermediates, and possibly as photosensitizers (see Plastics, environmentally degradable). Nitroanthraquinones are important in the manufacture of various dyestuff intermediates. Aminoanthraquinones are important intermediates in the preparation of anthraquinone dyes, and as light sensitizers for photodegradable polystyrene resin. Anthraquinonesulfonic acids are used in the manufacture of various dye intermediates; in cellulose pulp production; as hydrogenation catalyst for nitroaromatic compounds to amines; as antifoggants; and in the synthesis of photosensitive glass. Hydroxyanthraquinones are used in the manufacture of dye intermediates. They are important dyes themselves. Benzanthrone is used to manufacture dibenzanthrone (Vat Blue 20), an important dye; as a luminophore; and as sensitizers in light-degradable thermoplastics. Anthrimides are used as intermediates for the manufacture of dyes.

R.H. CHUNG[‡]
General Electric Co.

U.S. Pat 3,767,779 (Oct. 23, 1973), M. Coingt (to Oxy Synthese).

U.S. Pat 3,766,222 (Oct. 16, 1973), E. Hartwig, O. Ackermann, and H. Eilingsfeld (to BASG A.G.).

ANTIANEMIA PREPARATIONS. See Iron compounds; Pharmaceuticals; Vitamins.

[‡] Deceased.

ANTI-ASTHMATIC AGENTS

Nearly 9×10^6 people are afflicted with bronchial asthma, a hypersensitivity reaction of the lower respiratory system which produces bronchoconstriction, mucosal edema and excess mucus, and results in bronchial narrowing.

Clinical asthma is categorized as extrinsic, sometimes called allergic or atopic asthma, and intrinsic, also termed idiopathic or nonatopic, asthma. The onset of extrinsic asthma is normally during childhood or adolescence with observed elevated levels of specific antibodies, eg, immunoglobulin E (IgE). Intrinsic asthma often begins in adulthood with attacks seemingly related to infection, exercise, and/or psychological stress.

The fundamental causes of asthma are still unknown. Management is usually thought of in terms of removing the trigger mechanism if possible. There is no known cure for the fundamental abnormality in asthma, and drugs are used for the prevention or relief of symptoms. In some cases, hyposensitization or densensitization is successful. A schematic diagram of the possible sites of action of anti-asthmatic drugs is shown in Figure 1.

Adrenergic Stimulants

Adrenergic stimulants can bring about symptomatic relief in acute asthmatic attacks (see Epinephrine; Hormones). All members of this class of drugs are structural modifications of norepinephrine, a neurotransmitter in the adrenergic nervous system. The adrenergic nervous system is one portion of the autonomic nervous system which regulates the so-called vegetative or involuntary functions of the body, eg, enervating principal organs such as the heart and lungs. The cholinergic nervous system is the other main segment of the autonomic nervous system and has acetylcholine as a neurotransmitter.

The primary receptors upon which adrenergic stimulants act have been divided into α- and β-adrenergic receptors. β-adrenergic receptors are subdivided into β_1 (cardiac stimulation and lipolysis) and β_2 (bronchodilation and vasodilation). Considerable effort has been spent to develop selective β_2 adrenergic stimulants which would be devoid of the cardiovascular, gastrointestinal, and central nervous side effects produced by compounds possessing significant α- and β_1 adrenergic receptor stimulation.

Among the theories of the mechanism of asthmatic attack are those that implicate an abnormality in the autonomic nervous system, eg, an inherited or acquired deficiency of β-adrenergic receptors (adenyl cyclase) giving rise to low levels of cyclic-3′,5′-adenosine monophosphate (cyclic-3′,5′-AMP) and no bronchodilation when exposed to noxious stimuli. Another theory proposes that under stressful circumstances asthmatics

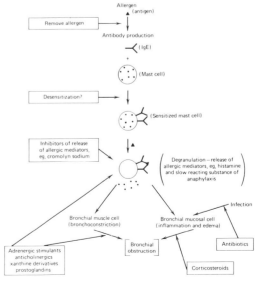

Figure 1. Schematic diagram of possible sites of action of anti-asthmatic drugs.

have a decreased circulating level of epinephrine, thus again leading to low levels of cyclic-3′,5′-AMP needed for bronchodilation.

The adrenergic stimulant, isoproterenol, administered by inhalation, and the catecholamine epinephrine have both been used to treat asthmatics. Both are synthesized by similar processes. Since both are catecholamines, they are susceptible to oxidation.

Ephedrine, a natural product found in Ma Huang, is a popular agent used in a host of products for the treatment of bronchial asthma. It is much less potent than epinephrine, but has the advantage of oral activity and a longer duration of action (3–4 h).

Therapeutically, it would be desirable to have an orally active, direct acting β_2-adrenergic agonist available that has rapid onset and long duration of action. Three general types of modifications of the catecholamine structure have been attempted to produce this substance: (1) replacing or altering the catechol ring system, eg, replacing the catechol ring system with a resorcinol ring system as in metaproternol, terbutaline and fenoterol, replacing the 3-hydoxyl group to form saligenin derivatives such as salbutamol and salmefamol, clenbuterol (in which a chlorine is added at the 2, 5, or 6 positions of isoproterenol), quinterenol (in which the catechol group is replaced with an 8-hydroxyquinoline derivative), trimetoquinol (cyclized phenethylamine derivative), isoetharine (an α-ethyl analogue of isoproternol); (2) varying the type of substituent attached to the nitrogen, eg, hexaprenaline (in which a hexane moiety is placed between two molecules of norepinephrine); and (3) altering the phenethylamine side chain.

Anticholinergics

The role that anticholinergics play in treating asthma is not well defined at present, although Asthmador, a product containing the naturally occurring stamonium and belladonna alkaloids, has been successfully used for over 100 yr in the treatment of asthma.

Inhibitors of the Release of Allergic Mediators

A relatively new means of treatment is the use of cromolyn sodium, commercially called Intal, a prophylactic drug which must be administered before the antigen challenge to be effective. The introduction of cromolyn sodium has led toward thinking in terms of prophylaxis rather than crisis therapy of asthma. Compounds similar in activity to cromolyn sodium have been discussed. Other drugs that inhibit allergic reactions by a similar mechanism include bufrolin (also called ICI-74,917); doxantrazole; and PR-D-92-EA.

Adrenergic stimulants have an added action besides bronchodilation—the ability to inhibit the release of allergic mediators in lung tissue. Drugs such as isoproterenol have the ability to block the release of chemical mediators of anaphylaxis. Xanthine derivatives are also reported to have this action by increasing the level of cyclic-3′,5′-AMP which in turn inhibits the release of chemical mediators of anaphylaxis such as histamine and the slow-reacting substance of anaphylaxis.

Xanthine Derivatives

Theophylline, one of the naturally occurring xanthines, has been most effective in treating asthma (see Alkaloids). Like adrenergic stimulants, it brings about bronchial smooth-muscle relaxation. Theophylline and related xanthines, including theobromine and caffeine, produce bronchial smooth-muscle relaxation by increasing cellular levels of cyclic-3′,5′-AMP. Theophylline acts by a different mechanism in increasing cyclic-3′,5′-AMP than do the adrenergic stimulants. A derivative of theophylline, dyphylline (dihydroxypropyltheophylline) is also used in treatment of asthma.

Prostaglandins

The prostaglandins are a family of cyclic, oxygenated, C_{20} fatty acids biosynthesized in human tissue (see Prostaglandins). They are quite potent in their ability to alter smooth-muscle tone in the lung, and although further work needs to be done on their possible role, they are promising candidates for treating bronchial asthma. The prostaglandins work by a different mechanism than the adrenergic stimulants; they increase intracellular cyclic-3′,5′-AMP and prevent the release of chemical mediators of anaphylaxis.

Other Drugs

In addition to the above classes of anti-asthmatics, other classes of drugs used in treating asthma include: corticosteroids such as corticotropin; antihistamines; tranquilizers, sedatives, mucolytics, and antibiotics. An important new area of research will be finding inhibitors of the pharmacological effects of slow-reacting substances of anaphylaxis (SRS-A) which are thought to be attributed to the direct action of the active components (leukotrienes) (see Prostaglandins).

DUANE D. MILLER
Ohio State University

F.F. Austen and L.M. Lichtenstein, *Asthma Physiology, Immunopharmacology, and Treatment*, Academic Press, New York, 1973.

J.P. Devlin, *Annu. Rep. Med. Chem.* **16**, 61 (1981).

ANTIBACTERIAL AGENTS, SYNTHETIC

NALIDIXIC ACID AND OTHER QUINOLONE CARBOXYLIC ACIDS

Quinolones, which refer to quinolone carboxylic acids, are synthetic compounds useful as antibacterial agents. They include nalidixic acid (Betaxina, Dixiben, Kusnarin, Nalidixin, Nalix, Negabatt, NeGram, Negram, Nevigramon, Nicelate, Nogram, Specifin, Uralgin, Uriclar, Urodixin, Urogram, Uroneg, Valuren, Wintomylon), piromidic acid (Panacid), pipemidic acid (Pipram), and oxolinic acid (Ossian, Pietil, Prodoxal, Vritrate, Urotrate, Utibid), cinoxacin (Cinabec) and rosoxacin (Eradacil). Two more recently discovered derivatives that appear to be much more potent under study are norfloxacin and perfloxacin.

nalidixic acid

All have a range of gram-negative antibacterial activity and the ability to concentrate in the urine after oral administration.

The obvious common structural feature of these compounds is a fused 1-alkyl-1,4-dihydro-4-oxo-3-pyridinecarboxylic acid (except for cinoxacin). They are chemically stable and readily soluble in aqueous base.

Preparation

The preparation of all these agents involves a portion of the Gould-Jacobs synthetic sequence. Nalidixic acid is produced from diethyl [[(6-methyl-2-pyridinyl)amino]methylene]propanedoicacid, which is thermally cyclized, subsequently alkylated, and hydrolyzed to nalidixic acid. Oxolinic acid is prepared in a similar manner from 1,3-benzodioxol-5-amine.

Resistance

Nalidixic acid is susceptible to chromosomal resistance (the bacteria's genetic ability to circumvent the lethal effects of the drug and to survive and multiply) but significantly not to R-factor resistance (where the organism contains a plasmid that carries the codes for resistance to one or more antibacterial agents).

Mode of Action

Studies indicate that nalidixic acid selectively inhibits the functioning of bacterial DNA.

GEORGE Y. LESHER
Sterling-Winthrop Research Institute

G.C. Crumplain, J.M. Midgley, and J.T. Smith in P.G. Sammes, ed., *Topics in Antibiotic Chemistry*, Vol. 3, John Wiley & Sons, Inc., New York, 1980, pp. 9–38.

M. Gellert in P.D. Boyer, ed., *The Enzymes*, 3rd ed., Vol. 14, Academic Press, Inc., New York, 1981, pp. 345–366.

NITROFURANS

Nitrofurans are a class of synthetic compounds with antibacterial activity, the class characterized by the presence of the 5-nitrofuran-2-yl group

$$O_2N \underset{O}{\boxed{}} R$$

In the most prominent members each R substituent includes an azomethine (—CH=N—) linkage attached to the furan ring.

Properties

Most nitrofurans are relatively stable, crystalline solids which darken on prolonged exposure to strong alkali and light; they are relatively insoluble in water; and have characteristic ultraviolet absorption. Derivatives undergo chemical reactions typical of the side chain. Chemical reduction leads to aminofurans and furan ring-opened products. The nitro group can be displaced by methoxy, halogen, azido, alkylmercapto, and phenylsulfonyl groups. Photochemical hydroxylation of 5-nitro-2-furancarboxaldehyde leads to 4-hydroxy derivatives.

Most commercial nitrofurans are derivatives of 5-nitro-2-furancarboxaldehyde or its diacetate. Representative antibacterial nitrofurans include: nitrofurazone, a topical antibacterial agent and a systemic agent for bacterial disease in poultry and swine; furazolidone, used for GI and vaginal infections in humans, as well as a wide variety of bacterial and protozoal infections in poultry and swine; nitrofurantoin, used in treating bacterial urinary trace infections; nifuroxime, active against many bacteria and fungi; Z-Furan, 3-(5-nitrofuran-2-yl)-2-propenamide, used as a food preservative in Japan; nitrofurfuryl methyl ether, a topical drug used against dermatomycoses; nitrovin, useful for its bacteriostatic and bactericidal action; furalazine, used as systemic bacterial agent; furium, active against cavilli and pathogenic enterobacteria; furazolium chloride, an effective broad spectrum antibacterial agent; and nitrofurathiazide, used to treat acute or chronic otitis externa of dogs and cats.

FRANK F. EBETINO
Norwich Eaton Pharmaceuticals, Inc.

K. Miura and H.K. Reckendorf, *Progress in Medicinal Chemistry*, Vol. 5, Plenum Press, New York, 1967, p. 320.

R.E. Chamberlain, *J. Antimicrob. Chemother.* **2**, 325 (1976).

H.E. Paul and M.F. Paul in R.J. Schnitzer and F. Hawking, eds., *Experimental Chemotherapy*, Vol. 2, Academic Press, Inc., New York, 1964, pp. 207–370; Vol. 4, 1966, pp. 521–536.

K. Miura and H.K. Reckendorf in G.P. Ellis and G.B. West, eds., *Progress in Medicinal Chemistry*, Vol. 5, Plenum Press, New York, 1967, pp. 320–381.

SULFONAMIDES

Sulfa drugs are sulfonamides derived from sulfanilamide (*p*-aminobenzenesulfonamide). The therapeutically active derivatives usually are N^1-substituted (referring to substitution on the sulfonamide nitrogen rather than substitution on the *p*-amino group, referred to as N^4-substitution). The structures of sulfanilamide and sulfadiazine illustrate these relations.

$$H_2N \underset{(N^4)}{\boxed{}} \underset{(N^1)}{SO_2NH_2} \qquad H_2N \boxed{} SO_2NH\text{—}$$

Important sulfa drugs are listed in Table 1. The fact that sulfa drugs are antibacterial agents separates them sharply from the diuretic, hypoglycemic, and uricosuric sulfonamide drugs.

Table 1. Selected Sulfa Drugs and Related Compounds

Generic or common name	Mp, °C	Solubility in H_2O at 25°C, mg/100 mL	Approximate half-life	pK_a
sulfanilamide	164.5–166.5	750	6.8	10.08
sulfadiazine	252–256	8	16.7	7.25
sulfamethazine	176	ca 100	7.0	7.37
sulfamethizole	208	25	short	5.45
sulfisomidine	243	ca 200	7.4	7.57
sulfisoxazole	194	350 at pH 6	6.0	5.0
sulfacytine	167–168	109 at pH 5	4	6.9
sulfacetamide	182–184	670	7.0	5.78
sulfaguanadine	190–193	100		(base)
dapsone (4,6′-diamino-diphenylsulfone)S	175–176	(soluble)		(base)
Prontosil (sulfamido-chrysoidine)		(soluble)		

Sulfa drugs are antibacterial agents that act primarily by holding back growth (bacteriostatic) rather than by destroying the pathogen (bactericidal); this is thought to arise from a derangement of the metabolic conversion of *p*-aminobenzoic acid (PABA) to the folic acid-related vitamins (see Vitamins).

Physical and Chemical Properties

The sulfa drugs are white or nearly white powders melting above 150°C. The azo and nitro derivatives may be yellow. These compounds are not very soluble in cold water; somewhat soluble in acetone or alcohol; sulfanilamide, sulfaguanidine, sulfacetamide, and mafenide are fairly soluble in hot water The drugs are weak acids; the more important ones generally have a pK_a in the range of 5–8. They are generally soluble in basic aqueous solutions. As the pH is lowered, the solubility of the N^1-substituted sulfonamides reaches a minimum, usually in the pH range 3–5. The sulfa drugs are generally stable in strong alkali and strong acid, but there are exceptions. The binding of the drugs to blood proteins is important in theoretical interpretations of their systemic activity. The amino group is readily diazotized with nitrous acid, the basis for its assay. Aldehydes react with the amino group to form anils and other condensation products. The yellow product formed with 4-(dimethylamino)benzaldehyde is useful for the visualization of sulfa drugs with free amino groups in thin-layer and paper chromatography. Acylation of the amino group occurs readily with agents such as acetic or phthalic anhydride.

Manufacture

The most usual method of preparation is by the reaction of *N*-acetylsufanilyl chloride (ASC) with the appropriate amine. Extra amine or a suitable base is used to neutralize the hydrochloric acid freed in the reaction.

$$CH_3CONHC_6H_4SO_2Cl + RNH_2 \xrightarrow{\text{base}} CH_3CONHC_6H_4SO_2NHR$$

The resulting acetyl product is usually hydrolyzed with aqueous alkali to the free amino compound, eg, sulfanilamide is made by the reaction with excess concentrated aqueous ammonia and hydrolysis of the product. On occasion, certain other *N*-acylsulfanilyl chlorides may be used, eg, *p*-nitrobenzenesulfonyl chloride. Occasionally, N^1-heterocyclic sulfanilamides are prepared by an alternative condensation.

Health and Safety

A small percentage of those treated with sulfa drugs develop symptoms of toxicity, including the Stevens-Johnson syndrome (an extremely severe dermatologic reaction). Acidic sulfa drugs may be potentially damaging to the kidney.

Uses

Sulfa drugs have been effective in internal infections caused by *Streptococci*, pneumococci, *Staphylococci*, meningococci gonococci, *Hemophilis influenza*, and the fungus-related *Nocardia*. They have largely been displaced by antibiotics (qv) in treatment of systemic disease, although they are still important substances for the treatment of urinary-tract infections. Related sulfones, eg, Dapsone, are used to treat leprosy (see Chemotheraputics, antibacterial and antimycotic).

LEONARD DOUB
Warner Lambert-Parke Davis Research Laboratories

E.H. Northey, *The Sulfonamides and Allied Compounds*, Reinhold Publishing Corp., New York, 1948.

J.K. Seydel, *J. Pharm. Sci.* **57**, 1455 (1968).

L. Weinstein, M.A. Madoff, and C.M. Samet, *N. Engl. J. Med.* **263**, 793 (1960).

ANTIBIOTICS

SURVEY

Antibiotics are chemical substances produced as intermediates or end products of metabolism by various species of microorganisms and other living systems that are capable in small concentrations of either inhibiting the growth of or killing bacteria and other microorganisms. A substance can be classified as an antibiotic agent although it is without effect *in vivo* (within the body) or is too toxic to permit its use in the body.

History

In 1877, Pasteur noted the antagonism of some growing organisms for other groups when he studied the rate of growth of different species of bacteria. In 1889, Vuillemin coined the term antibiosois to denote antagonism between living creatures. Emmerich first attempted the use of antibiotics therapeutically against anthrax, diphtheria, typhoid fever, and bubonic plague bacilli in 1898. In 1928, Alexander Fleming discovered penicillin; Chain and Florey developed the antibiotic preparation for human use, and in 1941, it became commercially available. Waksman isolated actinomycin, streptothricin, streptomycin, and neomycin, and introduced the word, antibiotic.

Antibiotics are identified by: (*1*) the conventional chemical name or a name descriptive of the chemical structure of the compound based on the rules of standard nomenclature; (*2*) the generic name, a shorter, established name which is commonly used in the scientific literature; and (*3*) the trade or brand name given to the drug by the manufacturer.

Classification

Chemically, the antibiotics are low molecular weight compounds exhibiting a variety of chemical structures, elemental composition, and physical–chemical properties. They have been classified according to chemical structure, microbial source, and mechanism of action. Following the classification of Garrod, Lambert, and O'Grady, based on the general similarity of chemical structure, those that are manufactured today can be divided into the following groups:

Penicillin and related antibiotics, all of which have a β-lactam ring in their structure, include the natural penicillins, the semisynthetic penicillins, the cephalosporins, the clavulanic acids and thienamycins and the monolactoms.

Aminoglycoside antibiotics, which have amino sugars in glycosidic linkage, include the streptomycins, neomycin, kanamycin, paromomycin, gentamicin, tobramycin, and amikacin.

Macrolide antibiotics, which consist of a macrocyclic lactone ring to which sugars are attached, comprise erythromycin, oleandomycin, and spiromycin.

Tetracycline antibiotics, which are derivatives of polycyclic naphthacenecarboxamide, consist of tetracycline, chlortetracycline, demeclocycline, oxytetracycline, and minocycline.

Chloramphenicol, an antibiotic in a class of itself, is a nitrobenzene derivative of dichloroacetic acid.

Peptide antibiotics form a large group but have few therapeutic applications. These antibiotics are composed of peptide-linked amino acids which commonly include both D- and L-forms. This group includes bacitracin, gramicidin, and the polymyxins.

Antifungal antibiotics include polyenes, which are a group of over 50 compounds including nystatin and amphotericin B, and other antifungal antibiotics including 5-fluorocytosine, clotrimazole, and griseofulvin (see Fungicides; Industrial antimicrobial agents).

Unclassified antibiotics have varied structures, including cycloserin, fusidic acid, novobiocin, prasinomycin, spectinomycin, and vancomycin.

Production

All but a few antibiotics are the complete products of microbial synthesis. The exceptions are the semisynthetic penicillins, the cephalosporins, and the tetracyclines, in addition to clindamycin and dihydrostreptomycin, which are made by chemical manipulation of microbially produced antibiotics. Chloramphenicol and cycloserine are manufactured entirely by chemical synthesis.

Almost all antibiotics are made by aerobic fermentation processes, involving growth of the microorganism using a fermentation medium designed to stimulate maximum antibiotic production, with some produced from intermediates.

In the United States, antibiotics as a group are the single most important bulk medicinal chemical produced. Leading U.S. producers are Pfizer, Eli Lilly, Bristol Myers, Upjohn, Schering-Plough, American Cyanamid, American Home Products, Abbott, Squibb, Warner-Lambert, and SKF. Combined shipments represent 44% of the world's market.

Health and Safety

The FDA requires companies to undertake extensive tests for approval.

Uses

The primary use is in the prevention and treatment of diseases in humans and animals (see Veterinary drugs).

H.G. KLEIN
Klein Medical International

AMINOGLYCOSIDES

Aminoglycosides, among the oldest known antibiotics, all contain an aminocyclitol unit as well as being aminoglycosides, and are more accurately known as aminocyclitol or aminoglycoside–aminocyclitol antibiotics. With the exception of dihydrostreptomycin, made by the chemical reduction of streptomycin, all clinically useful compounds in this class were naturally occurring until 1973. Important new entries are the semisynthetic compounds, such as dibekacin, amikacin, and netilmicin. These compounds are listed in Table 1.

The aminoglycoside–aminocyclitol antibiotics are relatively small, basic, water-soluble molecules which form stable acid addition salts; they are biosynthesized from carbohydrate components of their fermentation media. The aminoglycoside–aminocyclitol antibiotics are broad spectrum, active against gram-positive and particularly against gram-negative bacteria, eg, *Escherichia coli*, *Klebsiella*, *Proteus*, and *Enterobacter*. None are absorbed well from the alimentary trace or when applied topically and must be administered parenterally for systemic use. None are effective against anaerobic bacteria or aerobic organisms growing under anaerobic conditions. Some are active against *Pseudomonas aeruginosa* which constitutes an important use for this class of antibiotics. None are effective when used alone against streptococci, and they have varying activities against myobacterium tuberculosis.

Their action is bactericidal, rather than bacteriostatic, involving binding to bacterial ribosomes with inhibition of protein synthesis.

All antibiotics of this group encounter some problems of bacterial resistance following several years of extensive use.

Table 1.   Aminoglycoside-Aminocyclitol Antibiotics Used in Therapy

Antibiotic	Trade names[a]	Producing organism	Aminocyclitol derivation	Active against
streptomycin	Streptosulfat, Ampistrep	*Streptomyces griseus*	streptidine	broad spectrum including Mycobacterium tuberculosis; inactive against *P. aeruginosa*
dihydrostreptomycin	Vibriomycin, Didromycin	*S. humidus* *S. fradiae*	streptidine [b]	same as streptomycin
neomycin (framycetin) (neomycin B)	Neocin, Neobiotic, Myciguent, Soframycin, Framygen			broad spectrum but not active against *P. aeruginosa*; used topically because of systemic toxicity
paromomycin (aminosidin)	Pargonyl, Humatin, Gabbromycin	*S. rimosus* forma *paramomycinus*, *S. pulufraceus*	[b]	*Entamoeba histolytica*
kanamycin (kanamycin A)	Kantrex, Kanacyn	*S. kanamyceticus*	[c]	broad spectrum but inactive against *P. aeruginosa*; many strains have acquired resistance
aminodeoxykanamycin (kanamycin B)	Kanendomycin, Nekacyn, Bekanacyn	*S. kanamyceticus*	[c]	
gentamicin (gentamicins C_1, C_2, and C_{1a})	Garamycin, Gentacin, Cidomycin, Refobacin	*Micromonospora purpurea*	[c]	broad spectrum, active against *Pseudomonas aeroguinosa* and gram-neg. bacteria which have acquired resistance to kanamycin
sisomicin	Extramycin, Pathomycin	*M. inyoensis*	[c]	same as above
ribostamycin	Vistamycin	*S. ribosidificus*	[b]	broad spectrum but inactive *P. aeruginosa* and strains which have acquired resistance to aminoglycosides
lividomycin (lividomycin A)	Livalline	*S. liuidus*	[b]	same as above
tobramycin	Nebcin, Obracin, Tobrasix	*S. tenebrarius*		broad spectrum, active against *P. aeruginosa* and strains which have acquired resistance to kanamycin
dibekacin (DKB)	Panimycin, Dibekacin	semisynthetic		same as above
amikacin (BB-K8)	Biklin, Briclin, Novamin	semisynthetic		broad spectrum including *P. aeruginosa*; active against many bacterial strains resistant to other aminoglycosides
netilmicin	Netromycin	semisynthetic		broad spectrum including *P. aeruginosa* and certain organisms resistant to other aminoglycosides; less toxic than other aminoglycosides in laboratory animals

[a] A number of other antibiotics in this list have many trade names; consequently, this list is not comprehensive.
[b] Contains a 4,5-disubstituted 2-deoxystreptamine.
[c] Contains a 4,6-disubstituted 2-deoxystreptamine.

Chemical Aspects

This is a structurally diverse group of colorless, water-soluble polyhydroxy–polyamino compounds with molecular weights in the 300–800 range. Although some compounds in this series have been obtained crystalline, in general they are hydrated, amorphous solids without characteristic melting points, uv or ir absorption. They are most easily characterized by their ^{13}C nmr spectra.

Streptomycin, the first antibiotic in this class to be discovered, contains the aminocyclitol streptidine. The largest group, however, is that containing 2-deoxystreptamine, a group further divided into those antibiotics in which the deoxystreptamine is substituted on two adjacent hydroxyl groups (4,5-disubstituted 2-deoxystreptamine group); those in which the deoxystreptamine is substituted on two nonadjacent hydroxyl groups (4,6-disubstituted 2-deoxystreptamine group); and those having a monosubstituted deoxystreptamine unit.

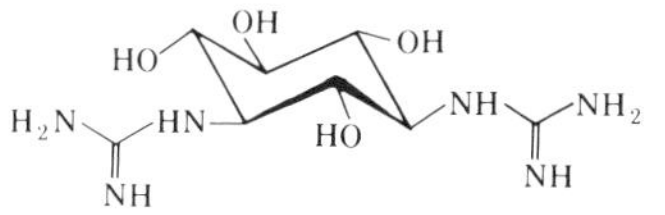

streptidine

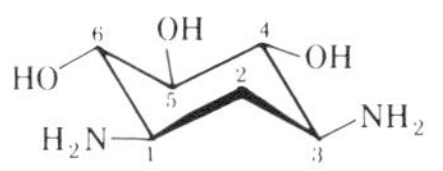

2-deoxystreptamine

Biosynthesis

Aminoglycoside biosynthesis has been comprehensively reviewed. Glucose provides the carbon skeletons of all subunits of the antibiotics. It has been shown that, whereas the isolated, underivatized sugar components of the antibiotics usually cannot serve directly as precursors for their parent antibiotics, the isolated aminocyclitol units can. This finding has allowed workers at the University of Illinois to devise an important method of biosynthetically creating new antibiotics containing novel aminocyclitol units.

This process, known as mutasynthesis, has been used to produce novel compounds related to streptomycin, neomycin, ribostamycin, butirosin, kanamycin, paromomycin, spectinomycin, gentamicin, and sisomicin.

Manufacture

All aminoglycoside–aminocyclitol antibiotics are produced by submerged aerobic fermentation of the appropriate producing organism in an aqueous nutrient medium containing assimilable carbon and nitrogen sources, inorganic ions, and occasionally some form of pH control (see Fermentation). The precise conditions of fermentation differ in detail for different antibiotics and manufacturers, and specialized ingredients may be added to germination or fermentation stages. Vigorous aeration and agitation are generally necessary, as well as temperature control in the range of 25–38°C. Isolation of the antibiotic is accomplished by filtration of the acidified broth followed by resin extraction and elution with acid or base to yield the crude antibiotic which is then further purified.

Antibacterial Activity and Uses

Typical antimicrobial activities of the more important compounds in the series are shown in Tables 2 and 3.

The particular usefulness of aminogylcosides lie in their rapid bactericidal activity against most gram-negative bacteria, especially *Pseudomonas aeruginosa*. They have no useful activity against anerobic bacteria or fungi. Because of their relative toxicity, they are used only for serious infections, usually in hospitalized patients. Aminoglycosides show useful synergy with penicillins against streptococi and are used in combination for treatment of streptococcal endocarditis.

During the last decade gentamicin, particularly, has been used widely because of its broad-spectrum activity including efficacy against many organisms that have developed resistance to kanamycin. Tobramycin and sisomicin, which were introduced into clinical practice considerably later than gentamicin, show similar activity but are somewhat more potent against pseudomonads. Both antibiotics are being used with increasing frequency. Dibekacin, also similar in its activity to gentamicin, is the most widely used compound in Japan where it was discovered. Amikacin, a semisynthetic compound, has important activity against organisms that have become resistant to gentamicin and tobramycin. In some countries the use of amikacin has become widespread, although in the United States its use has been more conservative, being reserved for resistant strains in most hospitals. During the last few years netilmicin has been introduced into therapy in a number of countries. This semisynthetic antibiotic is active against a number of gentamicin-resistant bacteria but, more importantly, showed substantially reduced toxicity in animals compared to standard aminoglycosides. Evidence is now becoming available that this safety factor has clinical relevance. Streptomycin was once widely used for tuberculosis, but its use has declined.

Bacterial Resistance

Three general mechanisms are known by which bacteria acquire resistance to aminoglycosides: (1) the organism mutates, leading to altered ribosomes which no longer bind the drug. (This type of resistance is important only to streptomycin-type compounds); (2) the bacterium becomes less permeable to the antibiotic; and (3) most commonly, bacterial enzymes, coded for by extra-chromosomal loops of DNA

Table 2. Mean Minimum Inhibitory Concentrations (MIC) (µg/mL) of Aminoglycoside Antibiotics[a]

Organism	No. of strains	Streptomycin	Neomycin	Kanamycin	Paromomycin	Gentamicin
Staphylococcus aureus	29	2	0.5	1	1	0.125
Streptococcus faecali	32	64	64	32	64	8
Escherichia coli	22	8	8	4	8	1
Klebsiella sp	20	4	2	2	2	1
Aerobacter sp	10	4	2	2	2	0.5
Proteus mirabilis	6	8	8	4	8	2
Proteus vulgaris	6	4	4	4	4	1
Proteus morganii	10	8	8	4	4	1
Proteus rettgeri	7	4	8	2	4	1
Pseudomonas aeruginosa	31	32	32	128	512	4
Salmonella sp	14	16	2	2	2	1
Shigella sp	17	8	8	4	8	2

[a]Tests mainly on recent isolates at Hammersmith Hospital by plate dilution method with two-fold differences. Means are of $\log_2$ of MIC to the nearest $\log_2$. In the series, tested strains showing a clearly abnormal degree of resistance (sometimes following treatment with the antibiotic) were omitted from these calculations.

Table 3. Median MIC (µg/mL) of Aminoglycoside Antibiotic[a]

Organism	No. of strains	Gentamicin	Tobramycin	Amikacin	Netilmicin
Escherichia coli	9	0.25	0.5	1.0	0.25
Enterobacter sp	7	0.25	0.5	1.0	0.25
Klebsiella pneumoniae	8	0.25	0.25	1.0	0.25
Providencia sp	9	4.0	4.0	1.0	8.0
Pseudomonas aeruginosa	21	2.0	0.5	4.0	4.0
Serratia marcescens	9	0.5	2.0	2.0	1.0
Proteus, indole pos	9	1.0	1.0	2.0	1.0
Proteus, indole neg	3	1.0	2.0	1.0	2.0

[a]Aminoglycoside-sensitive strains only were used. Values were determined in Mueller-Hinton agar by standard dilution methods.

(plasmids) which inactivate the drug. This latter mechanism is by far the most common, accounting for perhaps 80% of all resistant clinical isolates.

Health and Safety

All aminoglycosides have some potential for producing nephrotoxicity and ototoxicity. This generally limits their use to the treatment of serious infections in hospitalized patients who can be monitored closely.

Semisynthetic Aminoglycoside–Aminocyclitols

Many semisynthetic compounds have been prepared in attempts to circumvent enzymatic inactivation. Amikacin, dibekacin, and netilmicin are such compounds which are already in clinical usage.

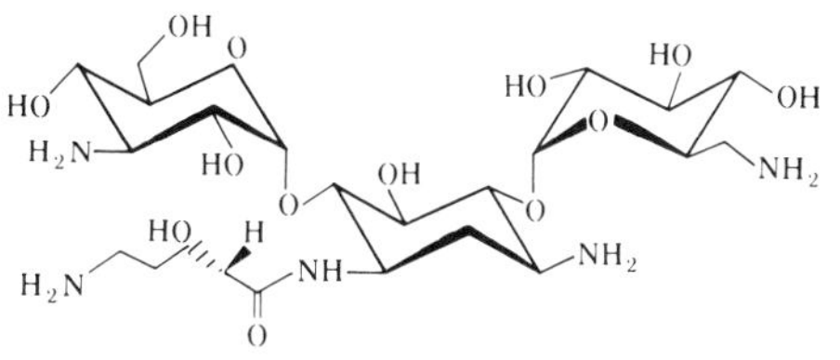

amikacin

Related Compounds

A number of related antibiotics, some of which are not strictly aminoglycoside–aminocyclitols, are spectinomycin (Trobicin), a bacteriostatic compound, produced by *Streptomyces spectabilis* which is used to treat gonorrhea; kasugamycin, produced by *S. kasugaensis*, which is useful against *Periculariai oryzae*, the causitive organism of rice blast disease; minosaminomycin, isolated from Streptomyces No. MA514-Al; hygromycin A, isolated from *Streptomyces hygroscopicus*; hygromycin B; destomycin; validamycin A and E; and fortimicin A and B.

PETER J.L. DANIELS
Schering-Plough Corporation

"Proceedings of the 5th Anniversary Symposium of the Institute of Bioorganic Chemistry," *J. Antibiot.* **32**, Suppl. (1979).

K.L. Rinehart and T. Suami, eds., *Aminocyclitol Antibiotics, ACS Symposium Series*, American Chemical Society, Washington, D.C., 1980.

A. Whelton and H.C. Neu, eds., *The Aminoglycosides*, Marcel Dekker, Inc., New York, 1982.

ANSAMACROLIDES

The ansamycins are a family of antibiotics characterized by an aliphatic *ansa*-bridge which connects two nonadjacent positions of the aromatic nucleus. The ansamacrolides are divided into the naphthoquinoid ansamycins, which include the streptovaricins, rifamycins, tolypomycins, halomicins, and naphthomycin, and the benzoquinoid ansamacrolides, which include geldanamycin and the maytansinoids. Table 1 outlines the biological activities and producing organisms of this group of antibiotics.

Naphthoquinoid Ansamacrolides

Streptovaricins. The streptovaricins undergo thermal isomerization to the corresponding atropisostreptovaricin. In the natural streptovaricins, the *ansa*-bridge lies above the aromatic nucleus, but in the atropisostreptovaricins it lies below. All of the streptovaricins except streptovaricin D react with one mole of sodium periodate to yield the corresponding streptovals. Basic methanolysis yields damavaricins C and D and the corresponding atropisodamavaricins and lactonized damavaricins. Very few derivatives have been prepared. The order of antibacterial activity is streptovaricins A and G > streptovaricins B and C > streptovaricins D and J > streptovaricin E ≫ streptovaricin F_G. The damavaricins are not as active as the corresponding streptovaricins, and the streptovals are inactive as antibacterial agents. The structures of individual streptovaricins are shown in Figure 1.

Table 1. Biological Activities of Ansamacrolides

Name	Biological activity	Producing organism
streptovaricins	antibacterial (gram-positive and mycobacterial) antiviral inhibitors of reverse transcriptase	*Streptomyces spectabilis*
rifamycins	antibacterial (gram-positive, gram-negative, and mycobacteria) antiviral inhibitors of reverse transcriptase	*Nocardia mediterranei*
tolypomycins	antibacterial (gram-positive)	*Streptomyces tocypophorus*
halomicins	antibacterial (gram positive)	*Micromonospora halophytica*
naphthomycin	antibacterial (gram-positive) vitamin K antagonist	*Streptomyces collinus*
geldanamycins	antiprotozoal inhibitors of reverse transcriptase	*Streptomyces hyeroscopicus*
maytansinoids	antileukaemic, antitumor	*Maytenus buchananii* (and others)

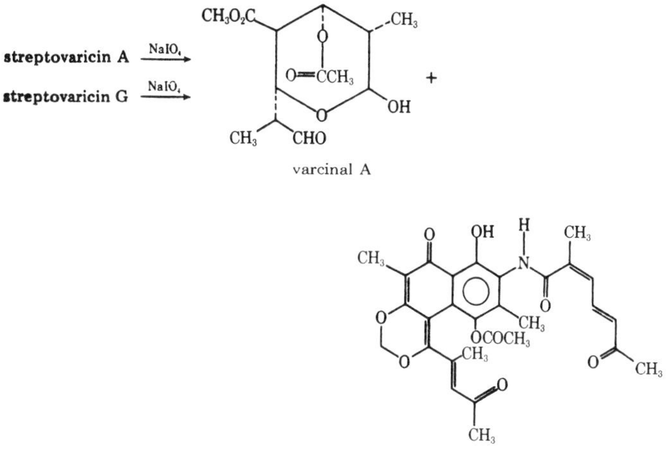

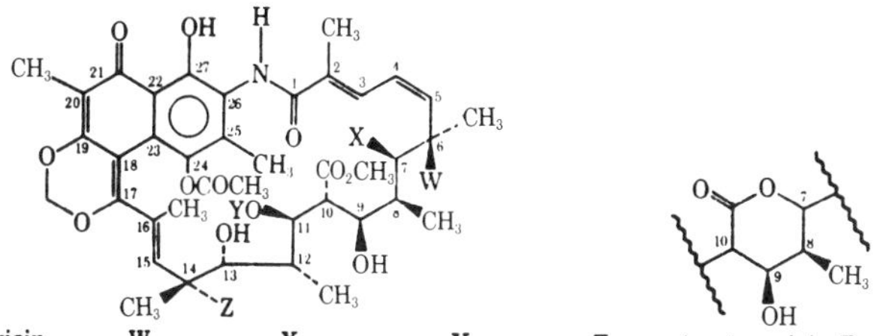

varcinal A

prestreptovarone

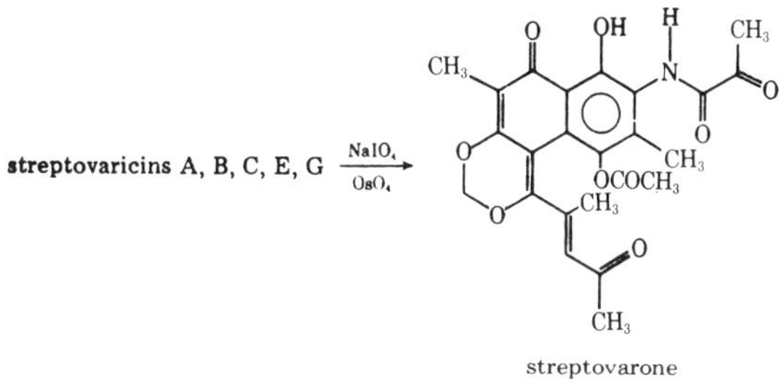

streptovarone

streptovaricin	W	X	Y	Z
A	OH	OH	COCH₃	OH
B	H	OH	COCH₃	OH
C	H	OH	OH	OH
D	H	OH	H	H
E	H	O=	H	OH
G	OH	OH	H	OH
J	H	OCOCH₃	H	OH
K	OH	OCOCH₃	H	OH

streptovaricin F_G
W = OH
Y = H
Z = OH

Figure 1. Structures of the streptovaricins.

Rifamycins. At least five substances having biological activity are designated rifamycins A through E. Thousands of derivatives have been prepared in the attempt to obtain a broader-spectrum antibiotic as well as one with good oral absorption. Rifamycins B, O, and S have served as starting materials for a large group of derivatives.

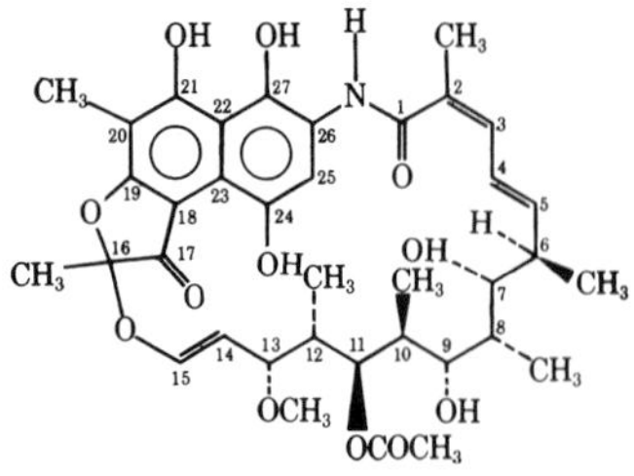

rifamycin SV

Treatment of rifamycin B with amines, hydrazines, and alcohols yields the corresponding amides, hydrazides, and esters, respectively. Rifamycin O reacts with a variety of aromatic amines, hydrazides, amidrazones, and aminoguanidines to give quinonimine derivatives. Reaction of rifamycin S with a variety of o-phenylenediamines and o-aminophenols produces a series of phenazines and phenoxazines, respectively; rifazine is the simplest of the phenazines. Rifamycin S also undergoes conjugate addition reactions to the quinone ring by a variety of nucleophiles to give the C-25 substituted derivatives of rifamycin SV, many of which show excellent antibacterial properties. Rifampicin is a therapeutically useful derivative of a rifamycin S aminomethyl derivative and is active against a variety of gram-negative and gram-positive bacteria and is used in the treatment of tuberculosis. In addition, a number of oxime derivatives of rifaldehyde have been prepared.

Tolypomycins. Mild acid hydrolysis of tolypomycin Y yields tolypomycinone and tolyposamine.

Halomicins. Halomicins are a group of four antibiotics (A, B, C, D) particularly active against bacterial strains resistant to penicillin G.

Naphthomycin. A vitamin K antagonist; its activity against gram-positive bacteria is reversed by the addition of cysteine.

Benzoquinoid Ansamacrolides

Geldanamycin. Geldanamycin is active against protozoa and fungi. It undergoes reaction with o-phenylenediamines and o-aminophenols to give compounds similar to rifazine.

Maytansinoids and maytanasides. Unlike other ansamacrolides, the matansinoids are isolated after repeated column chromatography and preparative thin-layer chromatography from ethanol extracts of plants obtained in Ethiopia and Kenya. Maytansinoids are those compounds similar to maytansine; maytansides are maytansinoids possessing the macrocyclic ring system but lacking the ester moiety. The maytansides (maysin, normaysine, maysenine and maytansinol) lack antitumor activity, indicating that the ester side chain is a requirement for activity.

Mode of action. The ansamacrolides inhibit bacterial growth by inhibiting RNA synthesis. The antiviral activity is owing to inhibition of the assembly of virus particles. Commercially available ansamacrolides are Rifamide (Rifocin M), Rifamysin SV (Rifocin), and Rifampicin (Rimactane; Rifadine).

FREDERICK J. ANTOSZ
The Upjohn Company

K.L. Rinehart, Jr., and L.S. Shield, *Fortschr. Chem. Org. Naturst.* **33**, 231 (1976).

W. Wehrli and M. Staehelin in J.W. Corcoran and F.E. Hahn, eds. *Antibiotics*, Vol. 3, Springer-Verlag Inc., New York, 1975, p. 252.

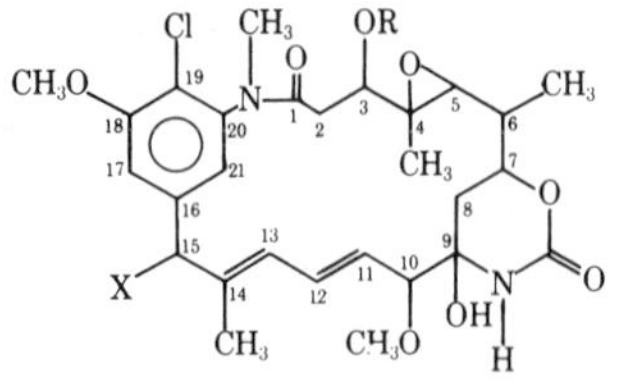

β-LACTAMS

The β-lactam antibiotics comprise two principal groups of therapeutic agents, ie, the penicillins and the cephalosporins (including the cephamycins and oxacephalosporins), and several nonclassical types ob-

maytanvaline X = H; R = COCHCH$_3$NCH$_3$COCH$_2$CH(CH$_3$)$_2$
maytansine X = H; R = COCHCH$_3$NCH$_3$COCH$_3$
maytanprine X = H; R = COCHCH$_3$NCH$_3$COCH$_2$CH$_3$
maytanbutine X = H; R = COCHCH$_3$NCH$_3$COCH(CH$_3$)$_2$
maytanacine X = H; R = COCH$_3$
maytansinol X = R = H
colubrinol X = OH; R = COCHCH$_3$HCH$_3$COCH(CH$_3$)$_2$
colubrinol acetate X = OCOCH$_3$; R = COCHCH$_3$NCH$_3$COCH(CH$_3$)$_2$

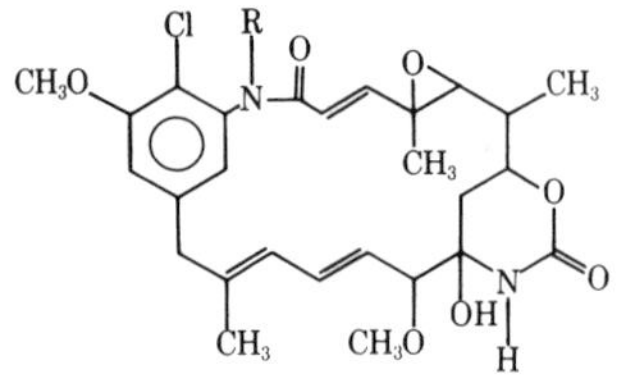

maysine R = CH$_3$
normaysine R = H

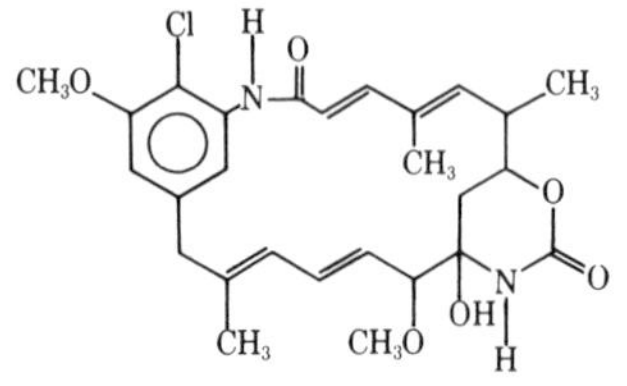

maysenine

Figure 2. Structures of the maytansinoids and maytansides.

tained more recently by fermentation and (semi)synthesis. The latter include the nocardicins, carbapenems (thienamycins and olivanic acids), penems, monobactams, and clavulanic acids. Figure 1 illustrates the essential structural features of these antibiotics. These antibiotics have in their chemical structures a four-membered lactam. With the exception of the nocardicins and monobactams, the β-lactam is fused through the nitrogen and the adjacent tetrahedral carbon atom to a second ring, ie, a five-membered thiazolidine, pyrroline, thiazoline, or oxazolidine for the penicillins, carbapenems, penems, and clavulanic acid, respectively, or a six-membered dihydrothiazine or dihydrooxazine for the cephalosporins, cephamycins, and oxacephalosporins. These antibiotics carry a variety of substituents which contribute to their biological properties. A structural feature common to most members of this group is the carboxyl group on the carbon adjacent to the lactam nitrogen, although this group has been chemically converted to a tetrazole with retention of biological activity, and the monobactams have a directly linked sulfonic acid group (Fig. 1). The penicillins, cephalosporics, nocardicins, and monobactams also possess a functionalized amino group on the carbon atom opposite the nitrogen of the β-lactam. The recently discovered carbapenems and clavulanic acids are not functionalized in this way.

The β-lactam antibiotics inhibit bacteria, exhibiting activities that differ in pattern and intensity by interfering with the synthesis of essential structural components of the bacterial cell wall. They tend to be irreversible inhibitors of cell-wall synthesis, and are bactericidal at concentrations close to their bacteriostatic levels. They are regarded as highly effective antibiotics with low toxicity, used for treating a wide variety of bacterial infections. Some of the β-lactams are inhibitors of the β-lactamases, which have become important factors in the therapeutic use of these antibiotics.

Nomenclature

Chemical Abstracts indexes most penicillins as 4-thia-1-azabicyclo[3.2.0]heptanes, cephalosporins and cephamycins as 5-thia-1-azabicyclo[4.2.0]oct-2-enes, and clavulanic acids as 4-oxa-1-azabicyclo[3.2.0]-heptanes. The unsubstituted bicyclic β-lactam systems are commonly designated, for example, as penam, carbapenam, cepham, oxacepham, and clavam. Thus, the penicillins are generally 6-acylamino-2,2-dimethylpenam-3-carboxylic acids, and the cephalosporins are 3-acetoxymethyl-7-acylamino-3-cepham-4-carboxylic acids. For penicillins, a straightforward naming system widely used relies on the fact that nearly all of the variations of the penicillins' structures are on the acyl group: the carbonyl of the acyl group is included in the basic moiety name of penicillin, eg, penicillin G is named benzylpenicillin.

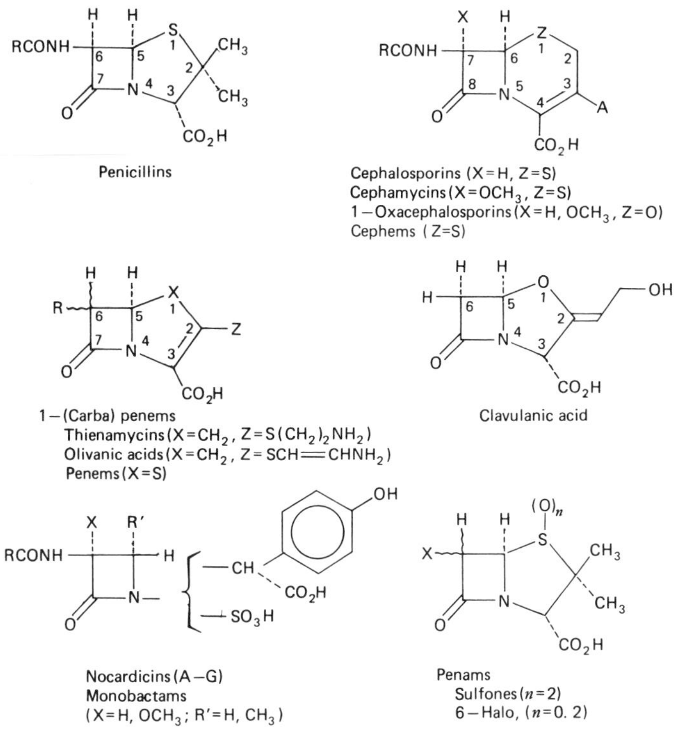

Figure 1. β-Lactam antibiotics.

Types

Biosynthetic penicillins include, among others, benzylpenicillin ($C_6H_5CH_2—$), or penicillin G; D-4-amino-4-carboxybutylpenicillin [D-$HO_2CCH(NH_2)(CH_2)_3—$], or penicillin N; and phenoxymethylpenicillin ($C_6H_5OCH_2—$), or penicillin V. Naturally occurring cephalosporins and cephamycins include cephalosporin C, deacetylcephalosporin C, deacetoxycephalosporin C, carbamyldeacetylcephalosporin C, 3′-methyl thiodeacetoxycephalosporin C, cephamycin A, cephamycin B, Takeda C2801X, cephamycin C, and 7-methoxycephalosporin C.

At least 18 carbapenems, designated as (epi)thienamycins and/or olivanic acids, have been isolated from fermentation broths (d). Natural nocardicins include nocardicin A-G. Natural monobactams include (iso)sulfazecins, SQ26180, and related structures. Several naturally occurring clavulanic acid-related structures have also been described. The oxacephalosporins and penems do not occur naturally but are obtained via (semi)synthesis. β-Lactams of potential clinical importance are included in Table 1 which lists analogues that have been studied in humans. Not all of these antibiotics have been marketed.

Physical Properties

For the most part, when obtained in reasonably pure form these antibiotics are white, off-white, tan, or yellow solids that are usually amorphous but sometimes crystalline. They do not have sharp melting points, but decompose upon heating to elevated temperatures. Most natural members have a free carboxyl group; as their salts, they are soluble in polar solvents, especially water. When other ionizable substituents are absent, they are soluble in organic solvents as their free acids. See Table 2 for physical properties.

Table 1. Important β-Lactam Antibiotics

A. Penicillins

Biosynthetic	Penicillinase-resistant	α-Aminopenicillins
benzylpenicillin	methicillin	ampicillin
phenoxymethylpenicillin	nafcillin	bacampicillin
phenethicillin	oxacillin	pivampicillin
propicillin	cloxacillin	talampicillin
clometocillin	dicloxacillin	hetacillin
azidocillin	flucloxacillin	metampicillin
		amoxicillin

α-Carboxy and α-sulfopenicillins	N-Acyl ampicillins	epicillin
carbenicillin	azlocillin	cyclacillin
carfecillin	mezlocillin	Amidiniopenicillanic acids
carindacillin	furazlocillin	mecillinam
ticarcillin	piperacillin	promecillinam
sulbenicillin	apalcillin	
suncillin	timoxicillin	
	pirbenicillin	

B. Cephalosporins/Cephamycins/Oxacephalosporins

β-Lactamase-sensitive cephalosporins		β-Lactamase-resistant cephalosporins
Parenteral	Oral	Parenteral
cephalothin	cephaloglycin	cefamandole
cephapirin	cephalexin	cefonicid
cephacetrile	cephradine	cefotaxime
cephaloridine	cefroxadine	cefmenoxime
cefazolin	cefadroxil	cefodizime
ceftezole	cefaclor	cefsulodin
cefazedone	cefatrizine	ceftazidime
cefotiam	FR10612	cefuroxime
cefazaflur	RMI19592	ceftizoxime
ceforanide	SCE100	ceftriaxone
		cefoperazone
Cephamycins		Oxacephalosporin
cefoxitin		moxalactam
cefmetazole		
cefotetan		
SQ14359		

C. Nonclassical

Carbapenams	Clavulanates	Monobactams
thienamycin	clavulanic acid	azthreonam
MK0787		SQ26180

Table 2. Physical Properties of Some β-Lactam Antibiotics

Name	Melting point	pK_a[a]	β-Lactam ir stretching frequency, cm^{-1}[b]	Ring system uv absorbance nm[c]	Nmr resonance[d] δ(H$_A$)	δ(H$_B$)	JAB	Optical rotation[e] Temp, °C	[α]$_D^t$
6-APA	209–210 (dec)	2.29, 4.90	1775		4.64d	5.54d	4.0	31	+ 273 (c, 1.2)[f]
benzylpenicillin sodium salt	215 (dec)	2.76	1775		5.56d	5.44d	4.0	24.8	+ 301 (c, 2.0)
ampicillin	199–202 (dec)	2.53, 7.24	1770		5.51s	5.51s		23	+ 287.9
7-ACA	> 200 (dec)	1.75, 4.63	1806	261 (ε8500)	50% 5.53d 50% 4.83d	5.13d	4.5	20	+ 114 (c, 1)
cephalosporin C		< 2.6, 3.1, 9.8	1780	260 ($E_{1cm}^{1\%}$ 200)	5.66d	5.75d	4.7	20	+ 103[g]
cephalothin sodium			1760	265 ($E_{1cm}^{1\%}$ 204)	5.70d	5.14d	4.5		+ 130 (c, 5)
cephalexin		5.2, 7.3	1775	260 (ε7750)	6.10d	5.45d	4.201-01.153 (c, 1)	+ 153 (c, 1)	
cephamycin C		4.2, 5.6, 10.4[h]	1770	264, 242 (ε6900, 5700)		5.19s			
thienamycin		3.08[i]	1765[j]	296.5 (ε7900)	3.39dd	4.25m	2.9	27	+ 82.7 (c, 0.1)
nocardicin A	234–235 (dec)	3.2, 4.5, 10.0, 11.6, 12.7[n]	1730[k]	270 (ε14900)	4.96m[l]	3.83t (4α)			− 135 (c, 1)
clavulanic acid		2.4–2.7			3.54dd (6α)	5.72dd (5α)	2.8	20	+ 47[m]
methyl ester			1800					22	+ 38.0[n]

[a] Except where noted pK_a values are for aqueous solutions, determined directly or extrapolated from mixed solvents.
[b] In most cases from Nujol mull or natural film.
[c] In aqueous solution except nocardicin A (ethanol–water).
[d] In D$_2$O except where noted: s, singlet; d, doublet; m, multiplet; δ, ppm relative to TMS; J, coupling constant in Hz; H$_A$, α proton on penultimate carbon from β-lactam nitrogen; H$_B$, α proton on carbon adjacent to β-lactam nitrogen.
[e] In water unless indicated; [f] In 0.1 N HCl; [g] For dihydrate; [h] In 66% DMF; [i] In phosphate buffer; [j] Methyl ester 1779 cm^{-1}; [k] In 50% DMSO; [l] In DMSO-d$_6$; [m] Solvent unreported; [n] Amine salt of tetrahydrate.

Chemical Properties

Determinations of basic chemical properties, degradation and structure elucidations, as well as descriptions of total syntheses, interconversions, modifications, and semisynthesis are in the literature.

A number of chemical properties pertaining to the manufacture and use of these antibiotics and to those manipulations that have been carried out to create new derivatives and analogues of the natural antibiotics with configurations of the β-lactams are indicated in the structures in Figure 1. Much of the chemical reactivity of these molecules is associated with the β-lactam ring. For example, the level of reactivity ranges from the nocardicins, which are stable over a fairly wide pH range, through the penicillins, which are only moderately stable in solutions that are near to neutrality (pH 5–8), to thienamycin, which is extremely unstable outside of a very narrow pH range (pH 6–7) and requires the presence of phosphate buffer or its equivalent for reasonable stability even within this pH range; it undergoes self-condensation at concentrations > 1.0 mg/mL. Penicillins are destroyed by aqueous acids via initial interaction of the side-chain carbonyl group with the lactam. The cephalosporins are more resistant to ring opening than the penicillins; eg, although alcohols readily attack the penicillin lactam, cephalosporins are sufficiently stable to permit the use of ethanol as a recrystallization solvent.

Penicillin to Cephalosporin Conversions

The penicillin sulfoxides have provided a means for chemically correlating the penicillin and cephalosporin structures, and commercial processes for obtaining cephalosporins and oxacephalosporins from the less expensive penicillins. The myriad reactions and mechanism studies which this work has generated has been extensively reviewed. Reviews of new β-lactam-containing structures are also available.

Biological Properties

The many structure–activity studies that have been published suggest the need to confer an appropriate amount of chemical reactivity on the β-lactam ring. The fused-ring structures increase reactivity through ring

strain. The monobactam ring appears to be activated by the attached sulfonic acid group. The penicillins and cephalosporins (3-cephems, ie, 5-thia-1-azabicyclo [4.2.0] oct-3-ene-4-carboxylic acid) appear to be optimal in this respect, equipping the β-lactam with a high degree of chemical reactivity, but retaining sufficient stability to survive in the environment of the host. For the penicillins, changes in the 6-acyl group have generally been the most profitable. Acyl groups that tend to confer the highest gram-positive activity on the penicillin structure are acetic acids substituted by an aromatic moiety (usually a phenyl group or a simple heterocycle such as thiophene) attached directly or through a hetero atom. In cephalosporins, there is more latitude for structural variation. The phenyl and phenoxyacetyl side chains are less effective, whereas acetic acid side chains carrying heterocyclic rings, notably thiophene, pyridine, tetrazole, furan, sydnone, and aminothiazole are more effective. Additional substituents (eg, hydroxyl, amino alkoxy-imino) on the β-carbon of the acyl group frequently provide beneficial effects on activity and resistance to inactivation by β-lactamases. Many other changes on the penicillin and cephalosporin structures give products with reduced antibacterial activities, eg, sulfoxides, sulfones, 6-alkylpenicillins, 2-substituted cephalosporins, 2-cephams, etc.

Bacterial resistance to β-lactam antibiotics can be natural or acquired. Resistance usually results from the production, by bacteria, of a β-lactamase enzyme which opens the β-lactam ring. The concentration of the β-lactamase produced is controlled by mutation, induction and the acquisition by the bacteria of R plasmids. The transfer of resistance between strains and even species has enhanced the problems of β-lactam-antibiotic resistance. The inherent permeability of the antibiotic determines the amount of the antibiotic that reaches the β-lactamase. Some clavulanic acids, olivanic acids, and penam sulfones are powerful inhibitors of β-lactamases.

Manufacture

All of the β-lactam antibiotics are derived from fermentation processes. Penicillins G and V are obtained by direct fermentation; the other penicillins and all of the cephalosporins are produced by a combination of fermentation and subsequent chemical manipulation of the fermentation product.

The manufacturing process consists of four stages:

(1) Fermentation under optimal physical and nutritional conditions to provide the desired microbial product (see Fermentation). Organisms used for commercial production of penicillins and cephalosporins are mutants of *Penicillium chrysogenum* and *Cephalosporium acremonium*, both true fungi. Cephamycins are produced by certain species of procaryotic *Streptomyces*. The commercial penicillin and cephalosporin fermentations can be divided into three interrelated phases: rapid growth during the first phase (ca 30 h for penicillins and 48 h for cephalosporins) which produces most of the mycelia for the fermentation; second phase, in which the antibiotic is produced at a rapid rate (5–7 d) with the growth rate rapidly reduced during this period; third phase, characterized by depletion of the carbon and nitrogen sources, followed by a termination of the antibiotic synthesis. The cells lyse, releasing ammonia, and the pH rises.

(2) Isolation, relying primarily on solubility, adsorption, and ionic characteristics of the antibiotic to separate it from the large number of other substances present in the fermentation mixture. The penicillins are monobasic carboxylic acids which lend themselves to solvent extraction techniques. The separation of amphoteric cephalosporin C is more complex, usually using a combination of ion exchange and precipitation procedures or adsorption on neutral macroporous resins and precipitations.

(3) Chemical modification, since only benzylpenicillin and phenoxymethylpenicillin are used in their original chemical form. The semisynthetic penicillins are derived from 6-aminopenicillanic acid, obtained, in turn, by removal of the 6-acyl group from a fermentation-processed penicillin. Likewise, 7-aminocephalosporanic acid, obtained by removal of the α-aminoadipic acid side chain of cephalosporin C, is a key intermediate for the cephalosporins that have a 3-acetoxymethyl group, or a functionalized 3-substituent derived therefrom. On the other hand, the cephalosporins with a 3-methyl substituent (deacetoxycephalosporanic acids) are derived from the penicillins. A considerable number of modifications to each position on the penicillin and cephalosporin rings has been attempted in the search for biologically useful altered products.

(4) Finishing, with the final isolation and purification of these antibiotics controlled by the nature of the chemical substance and the Federal certification controls indicated.

Of the β-lactam antibiotics, those of importance in commerce are the many penicillins and cephalosporins and the one oxacephalosporin in clinical use. Several additional penicillins, cephalosporins, a thienamycin, and a monobactam are now under development. Table 3 lists some of the main β-lactam antibiotics of commercial importance and their primary manufacturers.

Health and Safety

Large-scale handling of these agents associated with their manufacture and formulation entails some risk. Certain manufacturing intermediates produce contact dermatitis. In rare cases, individuals sensitized to penicillins may exhibit sensitivity reactions that range from mild to severe skin rashes and life-threatening anaphylaxis.

Uses

β-Lactam antibiotics are used for treating infectious diseases of bacterial origin in humans and animals. The clinical effectiveness of the penicillins and caphalosporins depends on a number of properties associated with the individual antibiotic such as its *in vitro* activity and toxicity properties, so that it may achieve concentrations in the host (serum and tissue levels) great than those required to inhibit the pathogenic organism (this capacity is controlled by pharmacokinetic properties, eg, efficiency and rate of absorption, rate of metabolism and excretion, tissue distribution, serum binding, serum and tissue inactivation, etc).

Benzylpenicillin is the most widely used of the biosynthetic penicillins and is effective against gram-positive bacteria. Semisynthetic penicillins, eg, methicillin, oxacillin, cloxacillin, dicloxacillin, and nafcillin, are used to treat benzylpenicillin-resistant staphylococci. Ampicillin is active against gram-positive and some gram-negative bacteria. For the most part, penicillins and cephalosporins are ineffective against *Pseudomona aeruginosa* (except carbenicillin), its related analogues, and several of the newer β-lactams (eg, third generation cephalosporins, oxacepha-

Table 3. Primary Manufacturers and Bulk Suppliers of Penicillins in the United States and Cephalosporins in the United States and Elsewhere

Chemical	Company	Chemical	Company
penicillin G	Wyeth Laboratories	amoxicillin	Beecham, Inc.
	E. R. Squibb & Sons		Bristol Laboratories
	Pfizer, Inc.		Roche Laboratories
	Bristol Laboratories	carbenicillin	Beecham, Inc.
	Eli Lilly & Co.		Roerig Laboratories
penicillin V	Eli Lilly & Co.	cephalothin	Eli Lilly & Co.
	E. R. Squibb & Sons		Glaxo Laboratories, Ltd. (UK)
	Bristol Laboratories	cephapirin	Bristol Laboratories
	Wyeth Laboratories	cephacetrile	Ciba-Geigy, Ltd. (Switz.)
	Abbott Laboratories	cephaloridine	Glaxo Laboratories, Ltd. (UK)
phenethicillin	Bristol Laboratories		Eli Lilly & Co.
	Pfizer, Inc.	cefazolin	Fujisawa Pharmaceutical Corp. (Jpn.)
methicillin	Beecham, Inc.		Eli Lilly & Co.
	Bristol Laboratories		Smith Kline & French Laboratories
oxacillin	Beecham Laboratories	cephaloglycin	Eli Lilly & Co.
	Bristol Laboratories	cephalexin	Eli Lilly & Co.
cloxacillin	Beecham Laboratories		Glaxo Laboratories, Ltd. (UK)
	Bristol Laboratories	cephradine	Smith Kline & French Laboratories
dicloxacillin	Beecham Laboratories		E. R. Squibb & Sons
	Bristol Laboratories	cephalosporin C	Glaxo Laboratories, Ltd. (UK)
	Ayerst Laboratories		Eli Lilly & Co.
	Wyeth Laboratories		Fujisawa Pharmaceutical Corp. (Jpn.)
nafcillin	Wyeth Laboratories		Montedison Pharmaceutical Div. (Italy)
ampicillin	Beecham Laboratories		Antibioticos S.A. (Spain)
	Biocraft Laboratories, Inc.	7-ADCA	Gist-Brocades N.V. (Neth.)
	Bristol Laboratories		Glaxo Laboratories, Ltd. (UK)
	E. R. Squibb & Sons		Eli Lilly & Co.
	Wyeth Laboratories		Montedison Pharmaceutical Div. (Italy)
	Ayerst Laboratories		Antibioticos S.A. (Spain)
hetacillin	Bristol Laboratories		

losporins and tienamycins). All the cephalosporins are highly active against a large number of gram-positive and gram-negative organisms, including penicillin-resistant staphylococci.

JOHN R.E. HOOVER
CLAUDE H. NASH
Smith Kline & French Laboratories

J.R.E. Hoover and G.L. Dunn in M. Wolff, ed., *Burger's Medicinal Chemistry*, 4th ed., Pt. II, John Wiley & Sons, Inc., New York, 1979, pp. 83–172.

P.G. Sammes, ed., *Topics in Antibiotic Chemistry*, Vol. 3, Halsted Press, a division of John Wiley & Sons, Inc., New York, 1979.

J. Elks, ed., *Recent Advances in the Chemistry of β-Lactam Antibiotics*, The Chemical Society, Special Publication 28, Burlington House, London, 1977.

J.R.E. Hoover in A. Demain, ed., *The β-Lactam Antibiotics*, Springer-Verlag, Heidelberg, FRG, 1983.

M.J. Basker, R.J. Boon, and P.A. Hunter, *J. Antibiot.* **33**, 878 (1980).

R.B.Morin and M. Gorman, eds., *Chemistry and Biology of β-Lactam Antibiotics*, Vols. 1, 2, and 3, Academic Press, New York, 1982.

β-LACTAMS (CEPHALOSPORINS AND PENEMS)

Cephalosporins

The principal deficiency of the older cephalosporins was their lack of resistance to β-lactamases, and one very important property of the newer compounds is their resistance to these enzymes. Recent research has given 11 new injectable compounds with varying amounts of β-lactamase resistance; some examples are given below. The close relationship between enzyme susceptibility and antibacterial activity is much less clearcut with the new compounds.

(1) cefoxitin   (2) cefuroxime

(3) cefotaxime   (4) moxalactam

Some compounds with improved stability to β-lactamases were discovered in the fermentation liquors of some *Streptomyces sp*, although the antibacterial activity of these naturally occurring 7α-methoxycephalosporins was weak. They enabled a new class of compounds to be synthesized which had good activity against species which heretofore had been resistant to cephalosporins, notably *Bacteroides sp*, and many β-lactamase-producing gram-negative organisms, eg, cefoxitin (1).

Several methods have been devised for the stereoselective introduction of a 7α-methoxy group into a cephalosporin derivative. Usually, a double bond is introduced between C-7 and the attached N group, and methanol addition occurs from the α-side of the molecule. All 7α-methoxycephalosporins show excellent resistance to β-lactamases, but they have different levels of antibacterial activity.

The first compound in which the sulfur atom in the cephalosporin ring was replaced by oxygen was 1-oxadethiacephalothin. A later, improved compound, stabilized by the introduction of a 7α-methoxy group, is moxalactam (4), which has excellent stability to β-lactamases and very high activity.

The high potency of cephalosporins with a 7-aminothiazolyl group was first demonstrated in 1977 with cefotiam. The β-lactamase resistance conferred by a methoximino group at the α-carbon atom in the 7-acyl group was first demonstrated by cefuroxime. Several cephalosporins have now appeared that have combined these structural features, and all

have very high antibacterial activity and excellent enzyme resistance, eg, cefotaxime, (3), cefmenoxime, ceftizoxime and ceftriaxon. This last compound is notable for its unusually long serum half life.

A feature of the newer compounds is their promise of activity against *Pseudomonas*, but even so, most of them are less effective than cefsulodin and cefoperazone. Ceftazidime is now the most active cephalosporin against *Pseudomonas*.

The pharmacokinetic properties of the cephalosporins depend to a large extent on the substituent at position 3. There is a tendency for compounds with a 1-*N*-substituted-tetrazol-5-yl-mercaptomethyl group in this position to be excreted via the bile, especially cefoperazone. The presence of comparatively large amounts of a potent antibiotic in the gastrointestinal tract may give rise to digestive upset. Acidic 3-substituents appear to be associated with a very high level of serum protein binding.

Orally Active Cephalosporins

In the search for orally active cephalosporins, cefatrizine, cefadroxil, cefroxadine and cefaclor were synthesised. Cefroxadine and cefadroxil have virtually the same antibacterial activity and enzyme stability as cephalexin and are absorbed orally to a similar extent. Cefaclor and cefatrizine are more active *in vitro* than cephalexin, but the oral absorption is lower and they are unstable in serum.

Intrinsic Activity and Biochemical Targets

All β-lactam antibiotics act by binding to and inhibiting the action of members of the enzyme complex involved in the formation of cell-wall peptidoglycan. These penicillin-binding proteins (PBPs) are species specific; seven have been identified in gram-negative organisms, but only four in *S. aureus*. Although all the cephalosporins act in the same general way, as shown by competitive binding experiments, they differ in their avidities for the various PBPs. Preferential binding to PBP 1B produces rapid lysis with little malformation of the cells. If PBP 2 is the most susceptible, then large, round, misshapen forms are produced which lyse slowly. Inhibition of PBP 3 gives long filamentous forms which ultimately lyse.

The intrinsic activity of any compound depends largely on the concentration needed to saturate the most susceptible PBP. However, bacterial growth proceeds uninterruptedly if the cell wall can prevent entry of the antibiotic. Proteins in the outer cell wall of bacteria appear to be connected with ability of antibiotics to enter the cell. Cells deficient in some of these proteins, called porins, appear to be much more impermeable than cells rich in porins.

Biogenesis

The biosynthesis of cephalosporins and penicillins starts from the same aminoacids which form the common intermediate δ-(L-α-aminoadipyl)-L-cysteinyl-D-valine, the LLD-tripeptide, but its role was difficult to prove as it does not penetrate into intact cells.

Penems

These compounds were first prepared by an intramolecular Wittig reaction of compounds of type (5). Compound (6) was found to have antibacterial activity resembling that of a cephalosporin rather than a penicillin. The work was followed up to provide a variety of 2-substituted ethyl penem-3-carboxylic acids, with hydroxy, alkoxy, acyloxy, amino, acylamino, alkylureido, and aryl and acylthio groups as the substituent. The antibacterial activities of these compounds were similar. The *in vivo* activity of some early compounds was disappointing

(5)   (6)

because of a lack of activity in the presence of serum proteins. A new compound, Sch 29482, is undergoing clinical trials.

CYNTHIA O'CALLAGHAN
J. ELKS
Glaxo Pharmaceuticals Ltd.
Glaxo Group Research Ltd.

"Penicillin Fifty Years After Fleming," *Phil. Trans. R. Soc. Lond.* **B289** (1980).

J. Elks, ed., *Recent Advances in the Chemistry of β-lactam Antibiotics*, The Chemical Society, London, UK.

β-LACTAMS (CLAVULANIC ACID)

The discovery of the cephamycins about 15 years ago led to a resurgence of interest in β-lactam antibiotics. The detection of novel β-lactam systems has resulted from the examination of actinomycetes isolated from soil samples obtained from a wide range of localities throughout the world. So far, of these novel β-lactams, only clavulanic acid has been used widely chemotherapeutically. Clavulanic acid (1) formulated with amoxyllin in various proportions is in clinical use in Europe, the United States, and several other countries. The structures of clavulanic acid, its derivatives, and related naturally occurring compounds are shown in Figure 1.

In clavam, and carbapenem nomenclature, the numbering of the atoms of the fused bicyclic-β-lactam ring system follows the sequence used for the penicillins and cephalosporins. On the other hand, with the systematic IUPAC system, the numbering of the azabicyclo[3.2.0]heptane ring commences at the nitrogen. Olivanic acid (7) and its derivatives are carbapenems that differ from thienamycin in the stereochemistry of C-8. By analogy with carbapenem (6a) and penem (6b), (6c) is designated as clavem.

Clavulanic Acid

In the search for β-lactamase inhibitors, investigation of fermentation broths revealed that a cephamycin-producing culture, ie, *S. clavuligerus*, produced a β-lactamase inhibitory metabolite with chromatographic properties distinctly different from the known cephamycins and the β-lactamase inhibitory compounds of *S. olivaceus*. The metabolite responsible for the inhibitory activity was called clavulanic acid.

Augmentin, a formulation of amoxicillin trihydrate (250 mg) and potassium clavulanate (125 mg) has been developed by Beecham and is marketed in the United Kingdom. Clinical investigations are underway in the United States and several other countries.

Properties. Salts of clavulanic acid are colorless crystalline solids with indefinite melting points. Clavulanic acid is an unstable viscous gum with pK_a of 2.3–2.7. The ir spectra of clavulanic acid salts and derivatives contain the characteristic β-lactam carbonyl stretching frequency in the range of 1790–1815 cm^{-1} in addition to a band at 1680–1695 cm^{-1} assigned to the strained exocyclic double bond. The nmr spectra of clavulanic acid and its derivatives show features characteristic of the fused β-lactam ring system.

The structure of clavulanic acid was deduced from spectroscopic data, although confirmation was obtained by x-ray analysis of the *p*-nitrobenzyl and *p*-bromobenzyl esters. Clavulanic acid was found to be a novel fused β-lactam which differed dramatically from the known penicillins and cephalosporins. The molecule contained oxygen instead of sulfur at C-1, ie, possessed an oxazolidine ring instead of a thiazolidine ring; it lacked an acylamino side chain at C-6, characteristic of active penicillins and cephalosporins, and at C-2 there was an exocyclic double bond in the form of a β-hydroxyethylidene substituent.

Clavulanic acid and its salts are readily converted into esters, many of which have been used for further transformations. Reaction of the esters with certain bases and treatment with carbonyl compounds give C-6 substituted clavams. Photochemical isomerization of the esters produces isoclavulanic acid derivatives (14), which are also sometimes isolated as by-products.

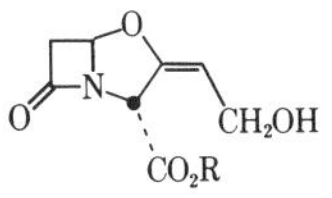

(14)

Biological properties. As a potent progressive inhibitor of β-lactamases derived from gram-positive and gram-negative bacteria, clavulanic acid differs from the β-lactams of the penicillin and cephalosporin series. The β-lactamases readily inhibited are mainly of the penicillinase type. Clavulanic acid is highly effective against plasmid-mediated β-lactamases with limited activity against those that are chromosomally mediated. In conjunction with certain penicillins and cephalosporins, clavulanic acid produces a pronounced synergistic effect against clinically important β-lactamase-producing organisms.

Clavulanic acid is well absorbed in animals orally and by the subcutaneous or intramuscular route. No compound other than clavulanic acid itself has been detected as a bioactive, ie, β-lactamase-inhibitory, metabolite in serum or urine. In combination with amoxicillin or penicillin G, it is highly effective in mice against infections caused by β-lactamase-producing organisms where both clavulanic acid and penicillin alone were ineffective. Clavulanic acid is well absorbed by the oral route in man; a dose of 250 mg gives a maximum concentration of ca 5.5 µg/mL after 1 h with a urinary recovery of 30–40%. Clavulanic acid alone or with amoxycillin is well tolerated. The half-life of clavulanic acid is ca 1 h, whereas that of amoxycillin is ca 1.3 h with urinary recoveries of 30 and 54%, respectively.

The proposed degradative metabolic pathway of clavulanic acid is shown in Figure 2.

Synthesis. Clavulanic acid has been obtained by two synthetic routes, which were extensions of earlier procedures to yield the clavem and clavam ring systems.

The biosynthesis has been studied using ^{13}C- and ^{14}C-labeled precursors. Evidence indicates that C-2 and C-3 of, and C-8, C-9, and C-10 on, the oxazolidine ring of clavulanic acid are derived from α-amino-δ-hydroxyvalerate via glutamate and 2-oxoglutarate. As all these pre-

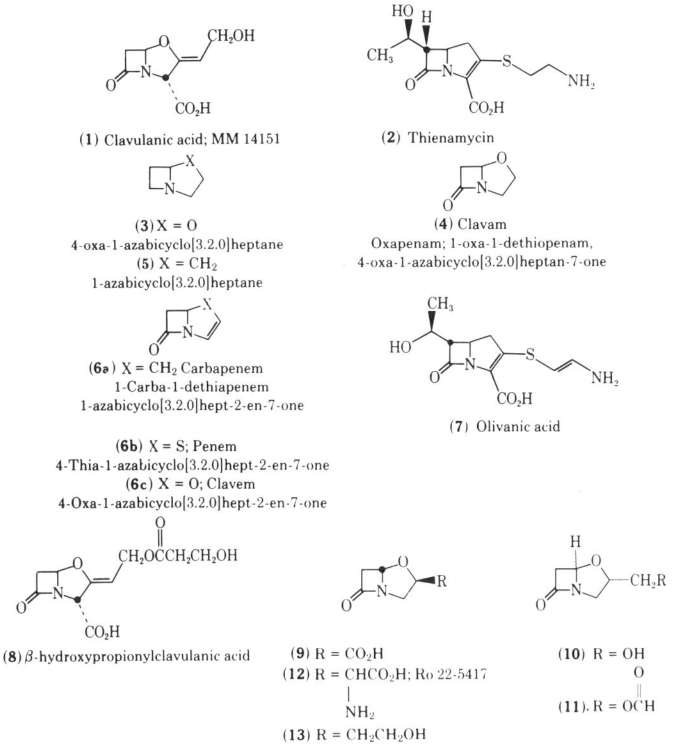

Figure 1. Clavulanic acid, thienamycin, and olivanic acid and naturally occurring clavam derivatives.

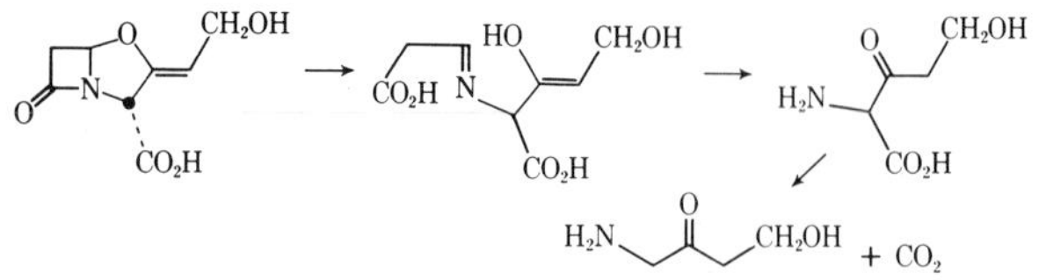

Figure 2. Clavulanic acid, postulated metabolic pathway.

cursors can be metabolized via the tricarboxylic acid cycle, the label appears in oxazolidine and azetidinone fragments of the fused β-lactam or clavam ring system when *S. clavuligerus* is grown in a medium based on triglyceride.

Manufacture and processing. Like penicillin G and cephalosporin C, clavulanic acid is produced by submerged fermentation. Stock cultures of *S. clavuligerus*, preserved in lyophilized form in ampuls or desiccated soil, are used to prepare working slant cultures on agar. Production media in large-scale fermenters are generally based on soybean flour. A typical medium consists of 1.5% soybean flour, 1.0% glycerol, 0.1% KH_2PO_4, and 0.2% (v/v) of 10% Pluronic L81 antifoam in soybean oil.

Broth, clarified by filtration or centrifugation, can be treated in two ways to provide a clavulanic acid salt for further purification: solvent extraction at acid pH or adsorption on a basic amine ion-exchange resin.

Assay. The β-lactamase-inhibitor properties are utilized in microbiological assays. The hole-in-plate synergy assay with penicillin G and *Klebsiella aerogenes* (KAG assay) is used for biological fluids. It can be used in the presence of other β-lactams such as amoxycillin or ticarcillin. An assay method based on the absorbance at 312 nm of the product formed by the reaction of clavulanic acid with imidazole has been described. High performance liquid chromatography is another method.

Thienamycin and Other Carbapenem Antibiotics

Until recently, *S. cattleya* was the only organism known to yield thienamycin (**2**); however, *S. penemifaciens* is claimed to produce (**2**) without penicillin N and cephamycin C, which were both coproduced by *S. cattleya*.

To date, approximately 38 carbapenem have been reported. The non-β-lactam metabolites cyclopentenedione (**15**) and the azabicyclo-[3.3.0]octa-5.7-dien-4-one (**16**) derivatives have been isolated from *S. cattleya* and *S. olivaceus*, respectively.

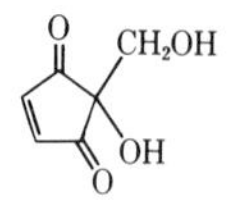

(**15**)

G2201-C;
2-hydroxy-2-(hydroxymethyl)
cyclopent-4-ene-1,3-dione

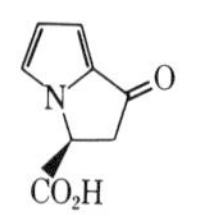

(**16**)

MM 22702;
(2S)-4-oxo-1-azabicyclo[3.3.0]
octa-5,7-diene-2-carboxylic acid

Properties. Thienamycin is a colorless hygroscopic solid, freely soluble in water but sparingly so in methanol. There are distinct differences in uv absorption maxima of carbapenems, depending on the saturation of the substituent on the sulfur at *C*-2 and whether the sulfur atom is at the thiol or the sulfoxide level of oxidation. Maximum stability of thienamycin is obtained at pH 6–7; it is comparable to that of penicillin G at pH 2–5; whereas degradation above pH 7 is rapid. Concentrated solutions (> 10 mg/L) decompose rapidly, yielding yellow and brown solutions with a considerable loss in potency. This creates a problem in the final stages of purification. Kinetic studies indicate that the β-lactam of thienamycin is degraded by intramolecular aminolysis involving the primary amino function of a second molecule. A dimer may be formed, suggesting a mechanism analogous to the degradation of 6-aminopenicillanic acid (6-APA) and ampicillin.

The structure of thienamycin, the olivanic acid derivatives and other carbapenems was established with the help of nmr spectra and mass spectra. The mass spectra of the carbapenem derivatives gave fragmentation patterns analogous to those seen with cephalosporins and penicillins but with a carbon substituent at *C*-6 rather than an acylamino

group. Thienamycin, the first reported example of a naturally occurring trans-fused β-lactam; compared to a penicillin or cephalosporin, it contains a methylene unit instead of sulfur at *C*-1.

The unsaturated *C*-2 side chain present in a number of carbapenems has always been found to occur naturally with a trans double bond. *N*-acetyl derivatives of thienamycin are readily obtained under Schotten-Bauman conditions. Esterified *N*-phenoxyacetylthienamycin can be cleaved reductively to afford the 6-(hydroxyethyl)carbapenem, called descysteaminylthienamycin. *N*-Alkylation of thienamycin gives a mixture of secondary, tertiary, and quaternary amines. The *O*-methyl and *O*-acyl derivatives have been prepared after protection of the amine and carboxyl functions. In general, thienamycin esters are unstable but give stable Schiff bases with salicylaldehyde. Esters of the *N*-acyl derivatives are stable.

The naturally occurring carbapenems can be transformed into analogues in which the alkyl or alkenyl substituent on the *C*-2 sulfur has been modified.

Both chemical and enzymatic methods have been used for the deacetylation of *N*-acetyl carbapenems, Enzymatic deacylation of epithienamycin A, epithienamycin C, and *N*-acetylthienamycin was achieved by treatment with an amidohydrolase derived from a strain of *Protaminobacter ruber* isolated from soil. Epithienamycin A is also deacetylated by an enzyme prepared from hog kidney. Conversion of the amino group in (**2**) into a more basic but less nucleophilic substituent dramatically increases the stability and results in the retention of the antipseudomonal activity.

The most interesting derivatives of thienamycin are the amidines (eg, MK-0787, *N*-formimidoylthienamycin (**17**)), derived by reaction with an imidate ester at pH 8 or by reaction of a dialkyliminium chloride with trimethylsilyl thienamycin. At high concentrations, the amidines are five to ten times more stable than thienamycin.

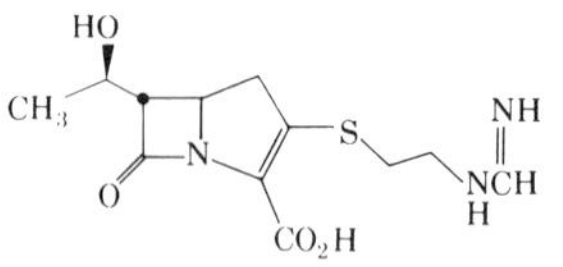

(**17**)   MK-0787

Other naturally occurring carbapenems, eg, the epithienamycins, are also broad-spectrum antibiotics but are generally not as potent as thienamycin and all lack significant activity against *Pseudomonas*. Thienamycin affords good *in vivo* protection against a wide range of organisms, although not via the oral route. Of all the naturally occurring carbapenems, it is the most effective in animal models. Detailed studies show that carbapenems are extensively degraded in the kidney, resulting in low and variable urinary recoveries. Many carbapenems are highly effective β-lactamase inhibitors.

Synthesis. The synthesis of the carbapenem nucleus and of thienamycin and its analogues is the aim of many groups throughout the world. 1-Carba-2-penem-3-carboxylic acid was synthesized before its isolation from *Serratia* and *Erwinia* spp. The racemic form was obtained as the sodium salt in solution by hydrolysis of various esters. The Merck workers prefer the synthetic approach to large quantities of thienamycin rather than the fermentation route from *S. cattleya*. This is because of the difficulty in producing large titers by strain improvement, a problem encountered with all metabolites in the carbapenem series. A stereocontrolled synthesis of thienamycin was achieved at Merck from an amino acid, and similarly, 6-APA was converted to thienamycin. Recent routes involve synthesis by way of intermediates derived from glucose.

The literature gives no detailed information on the biosynthesis of thienamycin or other carbapenems. A preliminary report, however, indicates that glutamic acid is a key building block.

Manufacture. Many species of *Streptomyces* have been reported to produce carbapenem antibiotics; however, only two species, *S. cattleya* and *S. penemifaciens*, are capable of producing thienamycin. The carbapenems are obtained as complex mixtures and in some cases are coproduced with penicillin N and cephamycin C. For thienamycin the filtered fermentation broth is absorbed on a Dowex 50 Na$^+$ column. Elution with 2% pyridine, reabsorption, elution with a lutidine-based volatile buffer, desalting on XAD-2, and finally lyophilization gives thienamycin in an overall yield of < 2% and a purity of 94%. Another procedure employs liquid ion exchange using O,O-dinonylnaphthalenesulfonic acid in butanol.

In the production of sulfated olivanic acids, S. olivaceus produces several very closely related compounds. It is possible to influence their proportions by using sulfate-free media with certain strains or developing mutants blocked in the synthesis of the sulfated compounds.

Ion-pair extraction has been of great value in the isolation of carbapenems containing sulfate ester groups. Anion-exchange processes to isolate carbapenems have utilized acrylic resins containing quaternary and tertiary ammonium groups.

Assay. Thienamycin is assayed by microbiological disk diffusion using *Staphyloccocus aureus*. A differential spectrophotometric procedure involves reaction of the antibiotic in buffered solution with hydroxylamine at pH 7 for 30 min at 23°C; cysteine can be used instead of hydroxylamine. The half-lives of thienamycin to hydroxylamine and cysteine were 5.7 and 2.5 min, respectively, which is considerably shorter than for penicillin G.

Penicillanic Acid and Related Compounds

A number of β-lactam derivatives have been prepared as possible alternative inactivators of β-lactamase, eg, penicillanic acid sulfone (**18**).

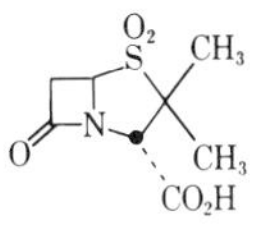

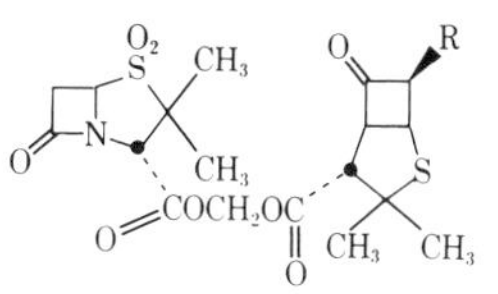

It is obtained by the oxidation of penicillanic acid or the hydrogenolysis of 6,6-dibromo- and 6α-bromopenicillanic acid sulfones. It is an irreversible inhibitor of a wide range of β-lactamase types, but was found to be less effective than clavulanic acid in its capacity to inhibit plasmid-mediated β-lactamases of *E. coli* and *K. aerogenes*. Esters such as sultamicillin (**19**) are well absorbed from the gastrointestinal tract and are subsequently hydrolyzed to liberate antibiotic and inhibitor.

ALLAN BROWN
Beecham Research Laboratories

A.G. Brown, *J. Antimicrob. Chemother.* **7**, 15 (1981).

P.C. Cherry and C.E. Newall, "Clavulanic Acid," in R.B. Morin and M. Gorman, eds., *Chemistry and Biology of β-Lactam Antibiotics*, Academic Press, Inc., New York, 1982, Chapt. 6.

R.W. Ratcliffe and G. Albers-Schönberg, in R.B. Morin and M. Gorman, eds., *Chemistry and Biology of β-Lactam Antibiotics*, Academic Press, Inc., New York, 1982, Chapt. 4.

β-LACTAMS (MONOBACTAMS)

The monobactams are characterized by the 2-oxoazetidine-1-sulfonic acid moiety. They are the only family of β-lactam antibiotics produced solely by gram-negative bacteria. Some of the naturally occurring monobactams are given in Table 1.

The structurally simple monobactams, SQ 26.180 (**1**), sulfazecin (**2**), and isosulfazecin (**3**) were the first to be isolated and characterized (see Fig. 1).

The naturally occurring monobactams correspond in configuration to the cephalosporins and cephamycins. Monobactams have been encountered in nature with amino side chains ranging from acetyl groups polypeptides.

As with bicyclic β-lactams, monobactams of absolute configuration opposite to those encountered in nature are inactive.

In both the penicillins and the cephalosporins, a 6α- or 7α-methoxyl substituent, respectively, increases chemical stability, but methoxylated

Table 1. Naturally Occurring Monobactams

Name	Producing organism	Structure
SQ 26.180[a]	*Chromobacterium violaceum*	(**1**)
sulfazecin	*Pseudomonas acidophila*	(**2**)
SQ 26.445	*Gluconobacter oxydans*	(**2**)
isosulfazecin	*Pseudomonas mesoacidophila*	(**3**)
EM 5400	*Agrobacterium radiobacter*	

[a]3-(Acetylamino)-3-methoxy-2-oxo-1-azetidinesulfonic acid, monopotassium salt.

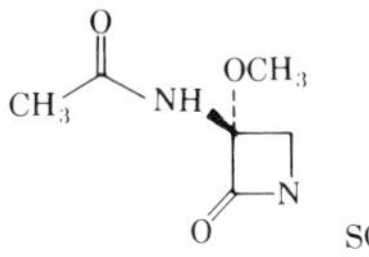

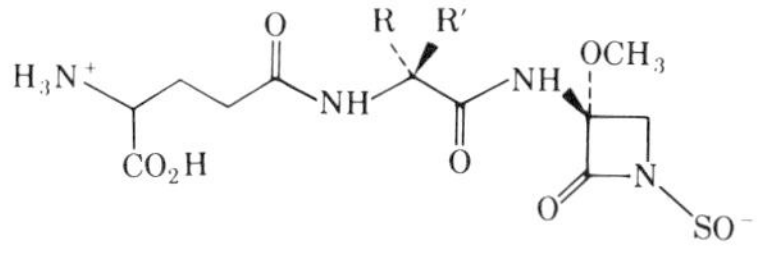

Structure no.	R	R′	Name
(2)	CH$_3$	H	sulfazecin, SQ 26,445
(3)	H	CH$_3$	isosulfazecin

Figure 1. Structures of naturally occurring monobactams.

monobactams with simple side chains are much less stable than the nonmethoxylated compounds. However, methoxylated monobactams are more β-lactamase stable than their unsubstituted analogues. Increased β-lactamase stability is thus displayed by the chemically less stable compound, a remarkable example of the importance of enzymatic fit relative to inherent reactivity.

Mode of Action

The monobactams, like the penicillins and cephalosporins, interfere with the synthesis of bacterial cell-wall peptidoglycan. The enzymatic steps involved in the insertion of newly synthesized cell-wall polymers into expanding peptidoglycan may play a role in the action of penicillin and other β-lactam antibiotics.

In accord with their poor antibacterial activity, naturally occurring monobactams exhibit a lack of affinity for any essential penicillin-binding proteins (PBPs) of *Escherichia coli* or *Staphylococcus aureus*.

The appropriate modifications of the monobactam nucleus result in compounds with potent antibacterial activity. A large number of monobactams have been prepared synthetically, that carry penicillin- or cephalosporin-like side chains. Compounds with the penicillin-G chain are predominantly active against gram-positive organisms.

Biosynthesis and Synthesis

The monocyclic ring of the naturally occurring monobactams is derived from serine, whereas the methyl moiety of the methoxyl group in 3α-methoxylated monobactams is derived from methionine.

The failure to degrade naturally occurring monobactams to intermediates led to the development of two totally synthetic routes. In one synthesis, the key step is the sulfonation of an N − 1 unsubstituted azetidinone. More than 100 monobactam side-chain analogues have been made by the sulfonation of the N − 1 unsubstituted derivatives. The other route starts with L-threonine and leads via cyclizaton to 3S-(*trans*)-3-amino-4-methylmonobactamic acid.

High intrinsic activity and excellent β-latamase stability are exhibited by monobactams that contain aminothiazole oxime side chains and 4-alkyl groups.

Aztreonam. Aztreonam is a totally synthetic β-lactam antibiotic that exhibits potent and specific activity against a wide range of β-lactamase-producing and nonproducing gram-negative bacteria, and displays minimal inhibition against gram-positive organisms and anaerobes. Aztreonam interacts poorly with penicillin-binding proteins (PBPs) derived from these latter organisms. Its MBC (minimal bactericidal concentration), like that of other β-lactam antibiotics, is either the same as or double the MIC.

In experimental animals, aztreonam shows excellent pharmacokinetic properties and activity *in vivo* in animal model infection with aerobic gram-negative bacteria. These promising results have translated directly to humans: extensive clinical trials have demonstrated the safety and efficacy of aztreonam in the treatment of infections caused by gram-negative bacteria.

CHRISTOPHER M. CIMARUSTI
RICHARD B. SYKES
Squibb Institute for Medical Research

C.M. Cimarusti and R.B. Sykes, *Chem. Brit.* **19**, 302 (April, 1983).
R.B. Sykes, D.P. Bonner, N.H. Georgopapadakou, and J.S. Wells, *J. Antimicrob. Chemother. (Suppl. E)* **8**, 1 (1981).

CHLORAMPHENICOL AND ITS ANALOGUES

Chloramphenicol (**1**) is the first of the so-called broad-spectrum medicinal antibiotics. A neutral, crystalline substance with a bitter taste, it is moderately soluble (4.4 mg/mL at 28°C) in water and freely soluble in many organic solvents. Originally isolated from cultures of *Streptomyces venezuelae*, it is manufactured by chemical synthesis and commercially available as the free compound and its esters.

(1)

Physical and Chemical Properties

Chloramphenicol, $C_{11}H_{12}Cl_2N_2O_5$, has mp 150.5–151.5°C; is soluble at 25°C to the extent of 2.5 mg/ml in H_2O; 150.8 mg/ml in $HOCH_2$-CH_2CH_2OH; very soluble in CH_3OH, C_2H_5OH, *n*-BuOH, CH_3COCH_3, ethyl and amyl acetate, and *N,N*-dimethylacetamide, and soluble in C_6H_6 and petroleum ether.

The stereochemical configuration of the 2-acylamino propanediol side chain, containing two asymmetric carbon atoms, confers antimicrobial properties. The aromatic part of the molecule is less specific for antimicrobial activity.

Esterification of the primary alcohol group of the propanediol side chain with palmitic acid readily yields chloramphenicol palmitate, a white crystalline substance only slightly soluble in water and petroleum ether but freely soluble in ethyl alcohol, chloroform, ether, and benzene, and is hydrolyzed in the duodenum to active chloramphenicol; with succinic acid it yields chloramphenicol acid succinate. The free acid ester is converted to chloramphenicol succinate sodium salt, anhydrous, a faintly yellow crystalline substance, freely soluble in water which is inactive *in vitro* but which, upon parenteral injection, is hydrolyzed by tissue enzymes, liberating chloramphenicol within the body.

Manufacture

Chloramphenicol was originally prepared by fermentation of *Streptomyces venezuelae*, and is now manufactured synthetically. Several different synthetic processes have been described. One involves the addition of benzaldehyde to β-nitroethanol to yield 2-nitro-1-phenyl-1,3-propanediol; another uses acetophenone, or preferably *p*-nitroacetophenone as the starting material; a third uses a cinnamyl alcohol conversion compound; and a fourth procedure uses phenylserine derivatives.

Chloramphenicol is marketed for human and veterinary medical use in product forms suitable for oral, parenteral, and topical administration.

Health and Safety

In humans, therapeutic doses are generally well tolerated. Aplastic anemia is the most serious toxicity associated with chloramphenicol and may not occur after exclusively parenteral use. Hypoplastic anemia, thrombocytopenia, and granulocytopenia have been associated with administration of chloramphenicol.

Uses

Chloramphenicol is antimicrobially active against a wide range of gram-positive and gram-negative bacteria, rickettsiae, the lymphogranuloma–psittacosis group, and *Vibrio cholerae*, *Salmonella typhi*, and *Hemophilus influenzae*. The chloramphenicol esters are essentially inactive microbiologically until hydrolyzed *in vivo* to the free antibiotic. It is bacteriostatic at low concentrations, inhibiting protein synthesis in bacteria and in cell-free systems.

JOHN EHRLICH
Detroit Institute of Technology

Reviewed by
ROGER WESTLAND
Parke Davis Co.

J. Ehrlich, Q.R. Bartz, R.M. Smith, D.A. Joslyn, and P.R. Burkholder, *Science* **106**, 417 (1947).
J.E. Smadel and E.B. Jackson, *Science* **106**, 418 (1947).
D. Szulczewski and F. Eng, *Anal. Profiles Drug Subst.* **4**, 48 (1975).
E.F. Gale, E. Cundliffe, P.E. Reynolds, M.H. Richmond, and M.J. Waring, *The Molecular Basis for Antibiotic Action*, John Wiley & Sons, Inc., New York, 1972, pp. 325–331.

LINCOSAMINIDES

Lincosaminide antibiotics are characterized by an alkyl 6-amino-6,8-dideoxy-1-thio-α-D-*galacto*-octopyranoside joined with a proline moiety by an amide linkage.

Lincomycin, $C_{18}H_{34}N_2O_6S \cdot H_2O$, is produced by *Streptomyces lincolnensis* and other Streptomyces SPP as a white, colorless, needlelike crystalline hydrochloride salt, it is commercially available as lincomycin hydrochloride (Lincocin). Lincomycin hydrochloride is readily soluble in water, moderately soluble in methanol and ethanol, and relatively insoluble in less polar organic solvents. It is available in oral dosage forms and in sterile forms for injection.

Celesticetin, $C_{24}H_{36}N_2O_9S$, a related antibiotic produced by *Streptomyces caelestis*, and its antibacterially active alkaline hydrolysis product, desalicetin, $C_{17}H_{32}N_2O_7S$, do not have clinical applications. 4'-Depropyl-4'-ethyllincomycin, $C_{17}H_{32}N_2O_6S$, a homologue of lincomycin co-produced in lincomycin fermentations, possesses one-tenth the activity of lincomycin.

A series of modified lincomycins is obtained by manipulating fermentation conditions, adding presumed precursors, biochemical antagonists and fragments of the lincomycin molecule to *Streptomyces lincolnensis*. These include 1-demlethylthio-1-ethylthiolincomycin, $C_{19}H_{36}N_2O_6S$; 1'-demethyllincomycin, $C_{17}H_{32}N_2O_6S$, which occurs under the influence of methylation inhibitors; the 1-demethylthio-1-ethylthio-1'-demethyl-1'-ethylthiolincomycin analogue, $C_{20}H_{38}N_2O_6S$; lincomycin sulfoxide, $C_{18}H_{34}N_2O_7S$; and 1-demethylthio-1-hydroxylincomycin, $C_{17}H_{32}N_2O_7$.

Clindamycin is produced by the successful reaction of lincomycin with thionyl chloride. It is isolated as the crystalline hydrochloride hydrate. The 2-palmitate ester of clindamycin (Cleocin Pediatric), is tasteless and is readily hydrolyzed *in vivo* to active clindamycin. Clindamycin 2-phosphate (Cleocin Phosphate), antibacterially inactive per se, is rapidly hydrolyzed after injection to active clindamycin.

Uses

Lincomycin is used to treat infections caused by staphylococci, pneumococci, and streptococci (other than enterococci), diphtheria, and a variety of anaerobic infections. Clindamycin is used to treat common infections caused by gram-positive cocci, and anaerobic infections.

THOMAS E. EBLE
The Upjohn Company

R.J. Fass, "Lincomycin and Clindamycin," in E.M. Kagan, ed., *Antimicrobial Therapy*, 2nd ed., W.B. Saunders Company, Philadelphia, Pa., 1974, p. 83.

R.D. Birkenmeyer, B.J. Magerlein, and F. Kagan, *Abstr. 5th Intersci. Conf. Antimicrob. Agents and Chemother.* 1965, p. 18.

A.A. Sinkula, W. Morozowich, and E.L. Rowe, *J. Pharm. Sci.* **62**, 1106 (1973).

MACROLIDES

Macrolide antibiotics are produced as secondary metabolites of soil microorganisms, and the majority have been produced by various strains of *Streptomyces*. Picromycin was the first to be isolated.

As a class, they are characterized by having excellent antibacterial activity, particularly against gram-positive bacteria. They are classified according to the size of the macrocyclic lactone ring forming the aglycone, either as 12-, 14-, or 16-membered ring macrolides.

Macrolides are polyfunctional molecules, the majority containing at least one aminosugar moiety, and, are basic compounds. However, neutral macrolides, containing only neutral sugar moieties attached to the aglycone, are also known.

The macrolide antibiotics are invariably produced as a complex mixture of closely related components by submerged aerobic fermentation of suitable cultures at 20–40°C in an aqueous nutrient medium containing a variety of carbohydrate and nitrogen sources. The optimum yields are usually obtained after 3–4 d. The antibiotic complex is then isolated by adjusting the whole broth to pH 9.5 and extracting with suitable solvents such as ethyl acetate, chloroform, or methylene chloride.

Erythromycins (Erthrocin, E-Mycin, Robimycin, Ilotycin) and their derivatives are examples of the 14-membered ring macrolides. They are produced by *Streptomyces erythreus* NRRL 2338.

cladinose (R_1 = H, R_2 = CH_3)
mycarose (R_1 = R_2 = H)

erythromycin A	R_1 = R_3 = H, R_2 = CH_3, R_4 = OH
erythromycin B	R_1 = R_3 = R_4 = H, R_2 = CH_3
erythromycin C	R_1 = R_2 = R_3 = H, R_4 = OH
erythromycin D	R_1 = R_2 = R_3 = R_4 = H

Derivatives include ethyl succinate (Erythrocin Ethyl Succinate; Pediamycin); lactobionate (Erythrocin Lactobionate); stearate (Erythrocin Stearate, Bristamycin, Ethril, Erypar, SK-Erythromycin, Pfizer-E); estolate (Ilosone); and gluceptate (Ilotycin Gluceptate).

Another example of the 14-membered ring macrolides is oleandomycin, produced by *Streptomyces antibioticus*. A triacetyl derivative (TAO or Troleandomycin) is manufactured by Roerig.

oleandrose (R = CH_3)
olivose (R = H)

oleandomycin	R = CH_3
O-demethyloleandomycin	R = H

Tylosin (Tylan), produced by *Streptomyces fradiae*, is an example of a 16-membered ring group.

R = CHO   tylosin

Activity

Macrolide antibiotics are medium- or narrow-spectrum antibiotics. Some possess activity against a wide range of gram-positive bacteria, but have only limited activity against gram-negative bacteria. Macrolides have also been shown to be active against *Rickettsia* strains, spirochetes, large viruses, protozoa, amoebae, schistosomes, and mycoplasma. Erythronolide B has no antibacterial activity, but has hypocholesterolemic activity as well as activity against schistosomes. In general, many macrolide-resistant clinical isolates of bacteria exhibit cross-resistance to other macrolides, as well as to lincomycin and chloramphenicol.

The mode of action of macrolide antibiotics involves inhibition of protein synthesis owing to binding of the antibiotic to the 50S subunit of the ribosomes of the bacteria, although the specific step in the protein synthesis that is affected has not yet been identified.

Macrolide antibiotics used extensively in human medicine include erythromycin, oleandomycin, kitasamycin, and josamycin. The major clinical use is against group A beta-hemolytic streptococcal infections as well as staphylococcal and pneumococcal infections; erythromycin is also used to treat diptheria, erythrasma, and intestinal amebiasis. Tylosin (Tylan) is used to treat mycoplasma in animals and in food preservation.

In general, the macrolide antibiotics have low toxicity, severe adverse reactions being rare and usually resulting from patient hypersensitivity, the form of the drug used, or the use of unusually large doses of the drug.

ALAN K. MALLAMS
Schering-Plough Corporation

S. Mitsuhashi, ed., *Macrolide Antibiotics and Lincomycin*, Vol. 1 of *Drug Action and Drug Resistance in Bacteria*, University Park Press, Tokyo, Japan, 1971.

Z. Vanek and J. Majer in *Biosynthesis*, Vol. 2 of D. Gottlieb and P.D. Shaw, eds., *Antibiotics*, Springer-Verlag, New York, 1967, p. 154.

W. Keller-Schierlein, *Progress in the Chemistry of Organic Natural Products*, Vol. 30, Springer-Verlag, Berlin, 1973, p. 313.

NUCLEOSIDES

The group of nucleoside antibiotics comprises compounds that consist of a heterocycle (aglycone base) linked by a carbon–nitrogen or a carbon–carbon bond to a carbohydrate moiety. The compounds are obtained predominantly from microbial sources, and they have the capacity to interfere with the growth or function of various other biological systems. The agents function essentially as structural analogues of the natural nucleosides or their metabolic derivatives, and a few of them (eg, aminoacyl and peptidyl nucleosides) act by virtue of their structural resemblance to end-portions of aminoacyl or peptidyl RNA (see Biopolymers). The heterocyclic moiety comprises variously substituted purine and pyrimidine rings and other ring systems including imidazole, oxazine, pyrazole, pyrrole, pyrrolopyrimidine and pyrazolopyrimidine. The carbohydrate portion comprises furanose and pyranose rings and various derivatives thereof.

Table 1. Selected Properties of Nucleosides

Compound/ structure no.	Mp, °C	$[\alpha]_D^{(t)a}$, c[b]	Uv max, nm, $(E_{1cm}^{1\%})$ solvent
9-arabinofuran-osyladenine (Ara-A) (1)	257	−5° (27), 0.25 in H$_2$O	257.5, (12,700), pH 1; 259, (14,000), pH 13
5-azacytidine (2)	228–230	+39° (25), 1.0 in H$_2$O	241, (8,767); H$_2$O; 249, (3,077), 0.01 N HCl; 223, (24,200), 0.01 N KOH
gougerotin (3)	200–215 dec	+45° (21), 1.0 in H$_2$O	267, (9,400), H$_2$O; 235, (9,300), H$_2$O; 275, (13,600), 0.1 N HCl; 267, (9,800), 0.1 N NaOH
pentopyranine A (4)	258 dec	+31.5° (21), 2.0 in H$_2$O	278, (13,100), 0.1 N HCl; 270, (8,850), 0.1 N NaOH
puromycin (5)	175–177	−11° (25), 1.0 in C$_2$H$_5$OH	267.5, (19,500), 0.1 N HCl; 275, (20,300), 0.1 N NaOH
tubercidin (6)	247–248 dec	−67° (17), 1.0 in 50% CH$_3$COOH	272, (12,200), 0.01 N HCl; 227, na, 0.01 N HCl; 270, (12,100), 0.01 N NaOH

[a] Temperature (°C) is shown parenthetically.
[b] c = concentration by volume, g/100 mL solvent.

Table 1 lists selected properties of some nucleoside analogues.

Nucleoside antibiotics containing structural changes in the heterocycle include 5-azacytidine (2), oxazinomycin, pyrazofurin, showdomycin, crotonoside (isoguanosine), coformycin, 2′-deoxycoformycin, formycin, formycin B, oxoformycin B, nebularine, tubercidin (6), toyocamycin, and sangivamycin.

Nucleoside antibiotics with structural modifications in the carbohydrate moiety include 3-acetamido-3′-deoxyadenosine, 2′-amino-2′-deoxyguanosine, 3′-amino-3′-deoxyguanosine, arabinofuranosyl adenine (Ara A) (1), aristeromycin, cordycepin, decoyinine, hikizimycin, nucleocidin, octosyl acids, pentopyranines (4), psicofuranine, spongothymidine, and spongouridine.

Nucleosides with aminoacyl or peptidyl residues comprise amicetin, bamicetin, blasticidin S, gougerotin (3), homocitrullylaminoadenosine, lysylamino adenosine, oxamycetin, plicacetin, polyoxins, puromycin (5), and septacidin.

Nucleosides with modifications in both the heterocyclic and carbohydrate moieties include the streptothricins and the endotoxin of *Bacillus thuringiensis*.

Uses

The observed biological activity of these agents includes inhibition of bacterial growth, interference with viral replication, and interruption of parasitic and fungal infections in experimental animals (see Chemotherapeutics; Fungicides). Some of the agents have demonstrated anticancer activity (see Chemotherapeutics, antimitotic) and a number of them have displayed cardiovascular (see Cardiovascular agents), immunosuppressive (see Immunotherapeutic agents) and cytokinin effects.

ALEXANDER BLOCH
Roswell Park Memorial Institute

R.J. Suhadolnik, *Nucleoside Antibiotics*, Interscience Publishers, a division of John Wiley & Sons, Inc., New York, 1970.

J.J. Fox, K.A. Watanabe, and A. Bloch, *Progr. Nucleic Acid Res. Mol. Biol.* **5**, 251 (1966).

A. Bloch in E.J. Ariens, ed., *Drug Design*, Academic Press, Inc., New York, 1973.

OLIGOSACCHARIDES

The four members of the oligosaccharide group of antibiotics are everninomicins, curamycins, avilamycins, and flambamycins. They represent very complex structures and possess many asymmetric centers.

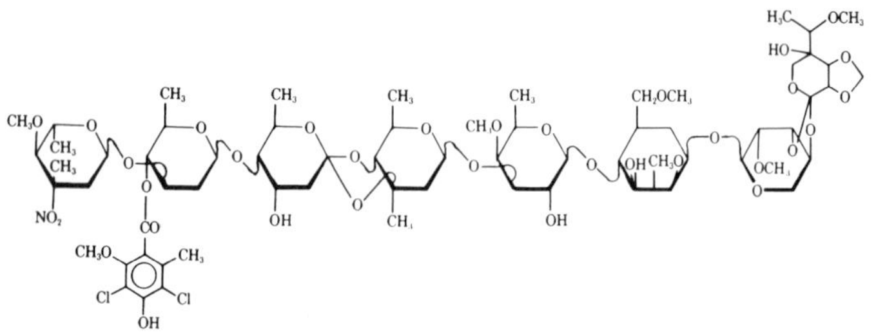

everninomicin D

Everninomicins are active against a wide variety of gram-positive aerobic and anaerobic bacteria, neisseria and mycobacteria, mycoplasma, and a small group of anaerobes. They are bactericidal for group A streptococci and bacteriostatic for other organisms and lack cross-resistance with other antimicrobials.

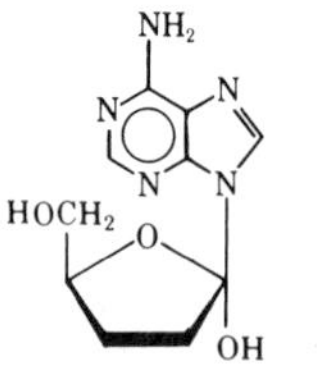

(1) arabinofuranosyladenine (Ara-A)

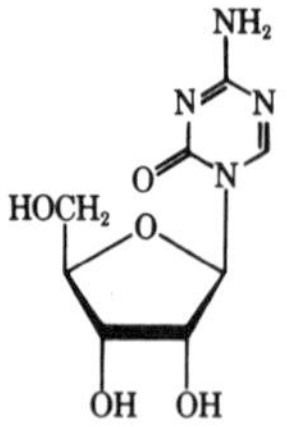

(2) 5-azacytidine

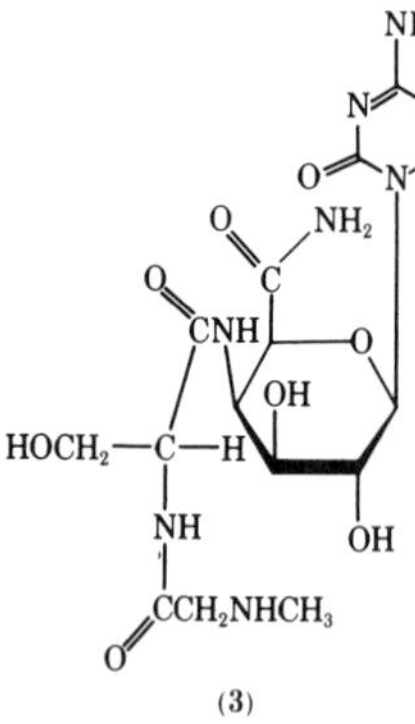

(3) gougerotin

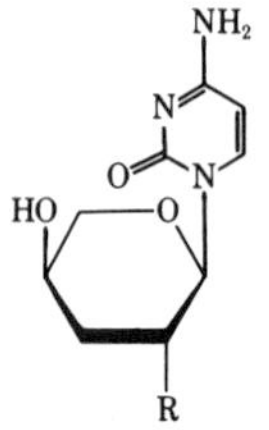

(4) R = H pentopyranine A

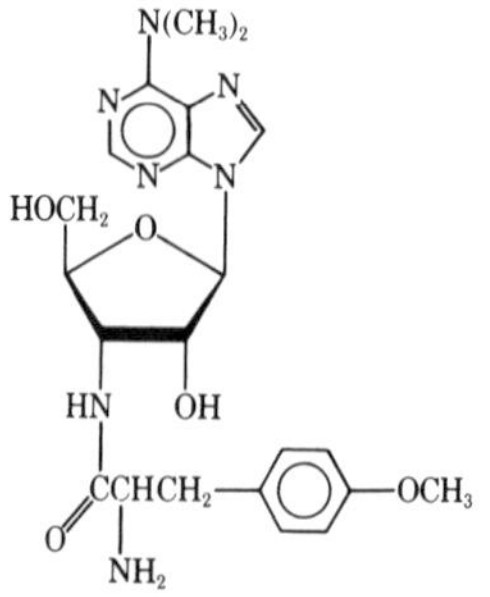

(5) puromycin

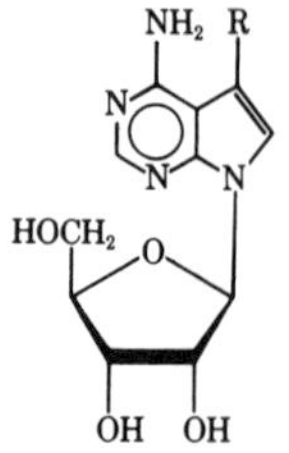

(6) R = H tubercidin

Chemical Properties

Everninomicin D (**1**) is hydrolyzed with aqueous acid to everninomicin D₁; on treatment with diazomethane it undergoes smooth cleavage to the methyl ether of everninonitrose pyranosyl (1 → 4) digitolactone and olgose, both colorless crystalline solids. Flambamycin, produced by *Streptomyces hygroscopius* DS 2320 is active in vitro against gram-positive bacteria and neisseria.

ASHIT K. GANGULY
Schering Corp.

M.J. Weinstein, G.M. Luedemann, E.M. Oden, and G.H. Wagman, *Antimicrob. Agents Chemother.*, 24 (1964).

O.L. Galmarini and V. Deulofeu, *Tetrahedron* **15**, 76 (1961).

Ger. Pat. 1,116,864 (Nov. 9, 1961), E. Gaeumann, V. Prelog, and E. Vischer (to Ciba, Ltd.).

L. Ninet, F. Benazet, Y. Charpentie, M. Dubost, J. Flovent, J. Lunel, D. Mancy, and J. Preud'homme, *Experientia* **30**, 1720 (1974).

A.K. Ganguly, O.Z. Sarre, D. Greeves, and J. Morton, *J. Am. Chem. Soc.* **97**, 1982 (1975).

PEPTIDES

Relatively few natural antibiotics are linear peptides. The exceptions are the gramicidins A, B, and C, and amphomycin. Although certain synthetic polypeptides have significant antimicrobial activity, eg, poly-L-lysine, it is usually low in comparison with antibiotics from microorganisms.

Most antibiotic peptides contain cyclic structures distinct from those involving the disulfide bridges, eg, those found in insulin and oxytocin. In cyclic antibiotic peptides, the terminal carboxyl group of an otherwise linear peptide participates in the cyclization by condensation with an amino or a hydroxyl function at some positions along the peptide chain. In a homomeric cyclic peptide, the ring is composed entirely of atoms from amino acids (eg, depsipeptides, enniantins, and valinomycin); these are either homodetic, eg, amino acids joined only through amide bonds, or heterodetic, that is two or more amino acids joined through other than amide bonds.

Peptide antibiotics differ in the following ways from proteins and peptides having hormonal or other functions: they have low molecular weights, ranging from 500–1500; the usual amino acids, with most peptide antibiotics having between 6 and 12 amino acid residues, some including methionine and histidine; they have unusual amino acids are absent; they have lipids and other moieties; they have D-amino acid residues; they resist hydrolysis by proteolytic enzymes; they are biosynthesized by enzymes rather than ribosomes; they are biosynthesized in families of peptide antibiotics; and substitution by related amino acids does not change the antibiotic activity, eg, N-methyl-L-alanine and N-methyl-L-isoleucine can be substituted for N-methyl-L-valine in the actinomycins.

Table 1. Source and Biological Activity of Peptide Antibiotics Currently Manufactured on an Industrial Scale for Therapeutic and Agricultural Use[a]

Compound	Microbial source	Antibiotic activity[b]	Some uses
amphomycin	*Streptomyces canus*	G +	topical antibacterial
bacitracin	*Bacillus subtilis*	G +	topical antibacterial, animal feed supplement
bleomycins	*Streptomyces verticillus*	G + , G − , and TB; antitumor	antitumor
cactinomycin	*Streptomyces chrysomallus*	G + ; antitumor	antitumor
capreomycin	*Streptomyces capreolus*	TB	anti-TB
colistin	*Bacillus colistinus*	G −	systemic antibacterial
dactinomycin	*Streptomyces antibioticus*	G + ; antitumor	antitumor
enduracidin	*Streptomyces fungicidus*	G + and TB	animal feed supplement
gramicidin A	*Bacillus brevis*	G +	topical antibacterial
gramicidin J(S)	*Bacillus brevis*	G + and G −	topical antibacterial
mikamycins	*Streptomyces mitakaensis*	G +	animal feed supplement
polymyxins	*Bacillus polymyxa*	G −	systemic and topical antibacterial
pristinamycins	*Streptomyces pristinae spiralis*	G +	systemic antibacterial
siomycin A	*Streptomyces sioyaensis*	G +	veterinary
stendomycin	*Streptomyces endus*	G + and TB; antifungal	agricultural (withdrawn)
thiopeptin	*Streptomyces tateyamensis*	G + and TB	animal feed supplement
thiostrepton	*Streptomyces azureus*	G +	veterinary
tyrocidine	*Bacillus brevis*	G +	topical antibacterial
tyrothricin	*Bacillus brevis*	G + and G −	topical antibacterial
viomycin	*Streptomyces floridae*	TB	anti-TB
virginiamycin	*Streptomyces virginiae*	G +	systemic antibacterial, animal feed supplement

[a]Cactinomycin is also known as actinomycin C; dactinomycin is also known as actinomycin D; colistin is a member of the polymyxin group; capreomycin and viomycin are members of the same group; mikamycin, pristinamycin, and virginiamycin are all members of the same group; tyrothricin is a mixture of gramicidin A and tyrocidine; siomycin and thiostrepton are members of the same group.

[b]G + = gram-positive bacteria; G − = gram-negative bacteria; TB = tubercle bacillus bacteria.

Preparation

More than 250 compounds designated as antibiotic peptides exist. Many have recently had their structures elucidated for clinical or agricultural use. Those produced by fermentation on a large scale are listed in Table 1 (with the exception of dactinomycin and catinomycin, produced on the kilogram scale) (see Fermentation).

Peptide antibiotics produced by Bacillus species. The polymyxins are basic polypeptides whose basicities are associated with the uncommon basic amino acid, α,γ-diaminobutyric acid (see Table 2). Polymyxins form water-soluble salts with mineral acids; only phosphates are isolated in crystalline form. Sulfates and hydrochlorides are normally presented as amorphous solids for pharmaceutical use. The naphthalene-2-sulfonates and azobenzene-4-sulfonates are insoluble in water, crystalline forms are obtained from aqueous alcohols; the picrates, reineckates, helianthates, Polar Yellow and other dyestuff salts, long-chain alkyl sulfates, etc, are very soluble in water. Polymyxins are inactivated in alkaline solution, or by treatment with acetic anhydride or other acylating agents.

Table 2. The Structures of the Members of the Polymyxin Group[a]

Polymyxin	R	X	Y	Z
A₁ (= M1)	MOA	D-Dab	D-Leu	Thr
A₂ (= M2)	IOA	D-Dab	D-Leu	Thr
B₁	MOA	Dab	D-Phe	Leu
B₂	IOA	Dab	D-Phe	Leu
D₁	MOA	D-Ser	D-Leu	Thr
D₂	IOA	D-Ser	D-Leu	Thr
E₁ (colistin A)	MOA	Dab	D-Leu	Leu
E₂ (colistin B)	IOA	Dab	D-Leu	Lue
circulin A	MOA	Dab	D-Leu	Ile

[a] Dab = α,γ-diaminobutyric acid; MOA = (+)-6-Methyloctanoic acid; IOA = isooctanoic acid.

The polymyxins inhibit growth of gram-negative organisms, eg, *Pseudomonas*, *Escherichia*, *Klebsiella*, *Enterobacter*, but not *Proteus* and gram-positive bacteria. A wide range of antibiotic formulations are marketed: preparation of sulfates of polymyxin B and of colistin (polymyxin E) are used for local, topical, oral and intravenous medication; the sodium *N*-sulfomethyl derivatives are used for intramuscular and intrathecal administration.

Bacitracin. Commercial bacitracin is a white amorphous powder which is soluble in water, ethanol, methanol, and *n*-butanol, and insoluble in acetone, ethyl ether, and chloroform. Commercial-grade bacitracin has a potency of about 50–100 units/mg (the USP unit of bacitracin is defined as the bacitracin activity given by 26 μg of the FDA master standard). The bacitracins are usually stable at temperatures between 25–27°C when the moisture content is less than 1%; aqueous solutions deteriorate rapidly at room temperature. They inhibit the growth of gram-positive bacteria and are used in combination with other antibiotics in topical preparations. Intramuscular injections can cause renal damage.

Bacitracin A

Tyrothricin. Commercial tyrothricin is a white amorphous powder which is an unresolved mixture of purified thyrothricin precipitates with a gramicidin content of about 20%. It is practically insoluble in water, and soluble in pyridine, methanol, ethanol, acetic acid, and ethylene glycol. Although effective against many gram-positive pathogens, its hemolytic action prevents its use in systemic chemotherapy.

Tyrocidine is a crystalline, nonhomogeneous material of three closely related compounds, tyrocidines A, B, and C.

Gramicidin, the insoluble neutral fraction isolated by treating tyrothricin with an acetone–ethyl ether mixture, can be crystallized from acetone or dioxan. Gramicidin A, B, C, and D are linear pentadecapeptides, not cyclic peptides. The valine and isoleucine gramicidins A have been prepared by chemical synthesis.

Gramicidin S

Gramicidin S, structurally related to tyrocidine, forms colorless needles which are insoluble in water and soluble in ethanol and methanol.

Peptide antibiotics from Streptomyces species. Over 150 peptide antibiotics have been isolated from different species of *Streptomyces*; however, only a limited number have value as therapeutic agents.

Viomycin, and the related capreomycin, are tuberculostatic antibiotics. They are strongly water soluble, readily forming highly colored crystalline salts, eg, the hydrochloride, the sulfide, or the naphthalene-2-sulfonate. They are stable in acid solution; somewhat unstable under alkaline conditions. Owing to their toxic manifestations, including nephrotoxicity, ototoxicity, abnormal liver function, leukocytosis, leukopenia, and hypersensitivity, they are not utilized unless other therapies have been tried and failed.

The bleomycins are water and methanol soluble antibiotics that inhibit the growth of gram-positive and gram-negative bacteria. They also cause induction of bacteriophage in a lysogenic strain and inhibit vaccinia virus replication in infected HeLa cells and in mice. They are insoluble in organic solvents. Bleomycin sulfate is a palliative treatment and/or adjuvant to surgery and radiation therapy for squamous cell carcinoma. Among the side effects are pulmonary fibrosis (in 50% of patents), skin changes, and renal, hepatic, and central nervous system toxicity which may occur at any time during treatment.

Other *Streptomyces*-derived antibiotics include amphomycin, a water-stable mixture of closely related peptides used in topical preparations, that inhibits gram-positive bacterial growth (causes hemolysis when injected); thiostrepton, insoluble in water and lower alcohols, and highly potent against gram-positive bacteria; thiopeptins, active against gram-positive bacteria and *Mycoplasma laidlawii*; the enduracidins, a group of chlorine-containing antibiotics; stendomycin, an antifungal peptide antibiotic; neocarzinostatin, an antitumor compound that is an acidic single-chain molecule.

The antibiotic cyclodepsipeptides. The cyclodepsipeptides or peptolides are often produced by microorganisms as families of chemically related compounds. They can be divided into two groups.

(1) Peptide lactones are one group, having the lactone linkage through a hydroxy amino acid (eg, triostins and quinomycins). This group includes actinomycins, peptide derivatives of a chromophore, actinocin, a phenoxazonecarboxylic acid liberated from all actinomycins under mild hydrolytic conditions (it is the drug of choice for Wilm's tumor); the synergistic family of antibiotics, separated into groups A and B, whose total complexes are more active microbiologically against gram-positive bacteria than the individual components. Group A are macrolides and group B are cyclic polypeptides having a lactone structure. Synergistic antibiotic complexes include streptogramin, staphylomycin (virginiamycin), ostreogrycin, mikamycin, pristinamycin, vernamycin, and synergistin.

(2) True cyclodepsipeptides contain both amino and hydroxy acids. Two classes can be distinguished: those with alternating hydroxy and amino acid residues, and those consisting largely of amino acids with the lactone linkage through hydroxy acids or hydroxy amino acids. Several

have unique chelation properties and serve as special agents in ion-selective sensors (eg, valinomycin is used in potassium electrodes). The true cyclodepsipeptides include angolide, enniatin A and B, beauvericin, destruxin A and B, insariin, peptidolipine, pithomycolide, serratamolide, sporidesmolide I, II, and III, surfactin, valinomycin, and viscosin.

Antimetabolites. An antimetabolite is a structural analogue of an essential metabolite, vitamin, hormone, or amino acid, etc, that is able to cause signs of the deficiency of the essential metabolite. Some 75 microbially produced antimetabolites have been catalogued. Theoretically, antimetabolites act as isosteric enzyme inhibitors and reversible inhibition would be relieved by the addition of either the production or an excess of substrate. The potential utility of microbial peptides that act as antimetabolites has rarely been demonstrated.

D. PERLMAN‡
University of Wisconsin

Reviewed by
MICHAEL FLICKINGER
National Cancer Institute

H.J. Peppler and D. Perlman, eds., *Microbial Technology and Microbial Processes*, Vol. 1, Academic Press, Inc., New York, 1979.

L.C. Vining in G.T. Tsao, M.C. Flickinger, and R.K. Finn, eds., *Annual Reports on Fermentation Processes*, Vol. 47, *Peptides*, Academic Press, Inc., New York, 1980.

T.L. Zimmer, O. Froyshov, and S.G. Laland in A.H. Rose, ed., *Economic Biology*, Vol. 3, *Secondary Products of Metabolism*, Academic Press, Inc., London, 1979.

J.M. Cassady and J.D. Douros, eds., *Agents Based On Natural Product Models*, Academic Press, Inc., New York, 1980.

PHENAZINES

As early as 1860, Fordos reported the production of the dark-blue pigment pyocyanin, and Guignard and Sauvageau in 1894 described the production of the pigment chlororaphin. Since then, over 30 variously substituted phenazines have been isolated from microorganisms. The phenazine derivatives have been reported to be produced by bacteria belonging to the genera *Pseudomonas* and *Brevibacterium*, by myxobacteria of the genus *Sorangium*, by actinomycetes of the genera *Streptomyces*, *Streptosporangium*, and *Microbispora (Waksmania)*, as well as by members of a group of *Nocardiaceae* (reclassified as *Actinomadura dassonvillei*). There is considerable overlap in the production pattern, several microorganisms producing the same compounds, and in other cases, the same microorganism produce 3–10 phenazine derivatives.

The phenazine antibiotics bear the following common skeleton:

Chemical Abstracts Service Numbering System   older numbering system (Beilstein)

C or O substituents often are found at the 1, 4, 6, and 9 positions with identical substituents often attached at the diagonally opposed positions, 1, 6 or 4, 9.

With the exception of internal salts, eg, aeruginosins A and B, phenazines can generally be extracted from fermentation broths with water-immiscible solvents such as chloroform. Many of the isolated phenazines possess varying degrees of antibiotic properties, a feature that has been related to their interaction with deoxyribonucleic acid, presumably by intercalation of the planar aromatic ring system. Inhibitory activities of phenazine antibiotics have been described against gram-positive and gram-negative bacteria, yeasts, fungi, animal tumors, helminths, selected

protozoa such as *Trichomonas*, amoeba, noxious plants, acarids, and HeLa cells.

Myxin, mp 130–135°C dec, is a potent, broad-spectrum antibiotic prepared by chemical methylation of iodinin. The cupric complex of myxin has been used for mastitis in cattle and for topical bacterial, yeast, fungal, and protozoan infections in mammals. Because of their extreme sensitivity to violent decomposition when exposed to heat, high friction, or static charges, myxin and copper myxin are handled with great care in manufacturing. For example, because of the explosion hazard, conventional means of deaggregation cannot be used to achieve the necessary initial size distribution of copper myxin complex particles in pharmaceutical products. However, ultrasonic deaggregation can be used safely to reduce copper-complex aggregates to their micronized form in the necessary 5–20 μm range. Sales of the copper myxin Unitop (Hoffmann-LaRoche) continue, but at a very low level.

Lomofungin (Lomondomycin), mp 320°C dec, is a useful biochemical tool in the study of yeast nucleic acids. *In vivo*, lomofungin is quite toxic, the LD$_{50}$ is 10 mg/kg ip.

Iodinin, mp 230°C dec, has direct practical application as an antibiotic except as an intermediate in the manufacture of myxin and copper myxin. On methylation with dimethyl sulfate in alkali or with diazomethane, it yields the antibiotic myxin. Iodinin is also claimed to be effective as an antihypertensive when administered daily to patients at dosages of ca 0.1–300 mg/kg of body weight, preferably orally in divided dosages three to four times daily. Since the oral LD$_{50}$ of iodinin is > 1 mg/kg body weight, a therapeutic index of safety was claimed.

Other phenazines include aeruginosin A ($C_{14}H_{11}N_3O_2$); aeruginosin B ($C_{14}H_{11}N_3O_5S$); chlororaphin ($C_{13}H_{11}N_3O$); griseoluteic acid ($C_{15}H_{12}N_2O_4$); griseolutein A ($C_{17}H_{14}N_2O_6$), mp 193°C, LD$_{50}$ > 500 mg/kg sc; griseolutein B ($C_{17}H_{16}N_2O_6$), mp 220°C dec, LD$_{50}$ (mg/kg) 200 iv, 450 sc, > 800 po; hemipyocyanin ($C_{12}H_8N_2O$), mp 158°C, less toxic in animals than pyocyanin; 9-hydroxy-6-(hydroxymethyl)-1-phenazine-carboxylic acid (antibiotic T-4138) ($C_{14}H_{10}N_2O_4$); 1-hydroxymethyl-6-carboxyphenazine ($C_{14}H_{10}N_2O_4$); 1-methoxy-4-methyl-9-carboxyphenazine ($C_{13}H_{10}N_2O_2$), mp 124–126°C; 6-methoxy-1-phenazinol ($C_{13}H_{10}N_2O_2$), mp 192°C; oxychlororaphin ($C_{13}H_9N_3O$), mp 243°C, tolerated 500 mg/kg ip in olive oil; 1,6-phenazinediol ($C_{12}H_8N_2O_3$), mp 248–250°C dec; pyocyanin ($C_{13}H_{10}N_2O$), mp 133°C, minimum lethal dose 100 mg/kg ip; tubermycin A ($C_{17}H_{16}N_2O_2$), mp 174°C, LD$_{50}$ 160 mg/kg iv; and tubermycin B ($C_{13}H_8N_2O_2$), mp 238°C, LD$_{50}$ 400 mg/kg iv.

JULIUS BERGER
Hoffmann-LaRoche, Inc.

N.N. Gerber in A.I. Laskin and H.A. Lechavalier, eds., *Handbook of Microbiology*, Vol. 3, CRC Press, Inc., Cleveland, Ohio, 1973, pp. 329–1332.

POLYENES

The observation in 1950 that *Streptomyces noursei* produced a novel antifungal agent led to the discovery of nystatin, the first of the polyene macrolide antifungal antibiotics. Most of the polyenes isolated to date (1977) have been produced from soil actinomycetes, mainly of the genus *Streptomyces*, and are formed as a complex mixture of both polyenic and nonpolyenic products by submerged aerobic fermentation of suitable cultures at 20–40°C in an aqueous nutrient medium. The bulk of the polyene is contained in the mycelium which usually is separated from the whole broth and extracted with polar solvents such as water-saturated *n*-butanol. A crude polyene complex is obtained as a yellow-orange solid that is soluble in very polar solvents such as aqueous alcohols, dimethyl-formamide, and dimethyl sulfoxide. The various components of the complex usually are separated by countercurrent distribution, column chromatography, or high pressure liquid chromatography.

The polyenes have a large lactone ring bearing a number of hydroxy substituents and a conjugated polyene chromophore of 3–7 double bonds. The compounds are classified according to the number of double bonds present in the principal chromophore either as trienes, tetraenes, pentaenes, hexaenes, or heptaenes.

‡Deceased.

Trienes have been reported in the literature, but little is known about their chemical structures, and some may have to be reclassified in the future.

Tetraenes. The nystatin complex produced by *Streptomyces noursei* consists of three parts: a main component, nystatin A₁ (**1**), and two minor components, nystatin A₂ and nystatin A₃. It is the most widely used polyene in clinical applications. Recently, *Streptomyces noursei* var. *polifungini* ATCC 21581 has been shown to produce nystatin A₁, nystatin A₂, nystatin A₃, and a new tetraene polifungin B. Unlike the nystatin group, which has a 38-membered macrocyclic lactone, many of the tetraenes have a 26-membered lactone, including pimaricin (natamycin) from *Streptomyces natalensis*, tennecetin from *Streptomyces chattanoogensis*, lucensomycin from *Streptomyces lucensis*, tetrin A and tetrin B from a *Streptomyces* species, arenomycin B from *Actinomyces tumemacerans Krass.*, Kov., 1962, var. *griseoarenicolor* var. nov., and rimocidin from *Streptomyces rimosus*.

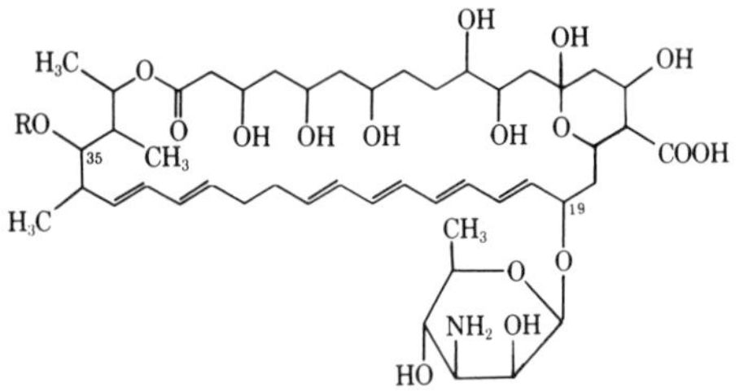

(**1**)   R = H   nystatin A₁

Pentaenes may be divided into three subgroups. Subgroup one possesses a pentaene chromophore having a terminal methyl substituent and may either be neutral or contain an aminosugar moiety. Included in this group are filipin from *Streptomyces filipinensis*, fungichromin from *Streptomyces cellulosae*, lagosin from a *Streptomyces* species, pentamycin from *Streptomyces pentaticus*, moldcidin B from *Streptomyces griseofuscus* (identical to pentamycin), cogomycin from *Streptomyces fradiae*, chainin from a *Chainia* species, and aurenin (**2a**) and homoaurenin (**2b**) from *Actinomyces aureorectus* sp. nov.

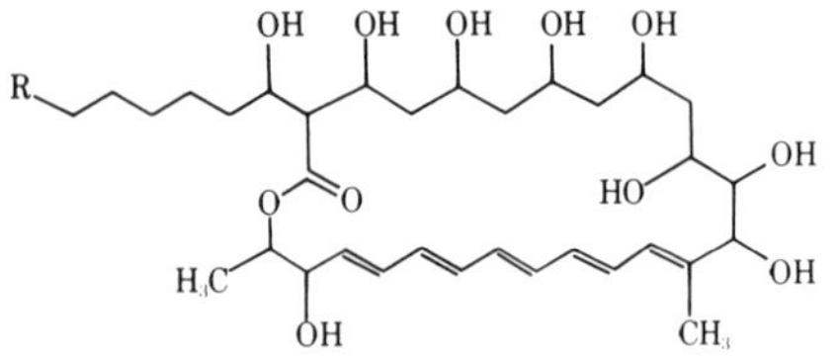

(**2a**) R = H aurenin
(**2b**) R = CH₃ homoaurenin
(**2**)

The second subgroup of pentaenes is characterized by a conjugated lactone chromophore. Polyenes in this group include mycoticin A and B from *Streptomyces ruber*, flavofungin A and B from *Streptomyces flavofungini*, and flavomycoin from *Streptomyces roseflavus* ARA1 1951, var. *jenesis* nov. var. JA 5068.

The third subgroup of pentaenes comprises the classical polyene structure of an unsubstituted, isolated pentaene chromophore. This group includes eurocidin A and B from *Streptomyces albireticuli* and rectilavendomycin from *Actinomyces rectilavendulae* var. *pentaenicus* var. nov.

Hexaenes. Although little is known about this group of polyenes, two structures have been assigned to Dermostatin A and B from *Streptomyces viridogriseus*. The dermostatins possess a novel 36-membered macrocyclic conjugated lactone.

Heptaenes have been used in clinical practice to treat systemic fungal infections and may be divided into two groups, ie, nonaromatics, which include amphotericin B (**3**) from *Streptomyces nodosus*, candidin from *Streptomyces viridoflavus*, mycoheptin from *Streptomyces netropsis*, and aureofungin B from *Streptomyces cinnamomeus* var. *terricola*; and aromatics which include aureofungin A from *Streptomyces cinnamomeus* var. *terricola*; trichomycin A, B, and C from *Streptomyces hachijoensis*; DJ-400 B₁ and DJ-400 B₂ from *Streptomyces surinam*; Vacidin A from *Streptomyces aureofaciens*; hamycin and hamycin X from *Streptomyces pimprina* Thirum; and perimycin from *Streptomyces coelicolor* var. *aminophilus*.

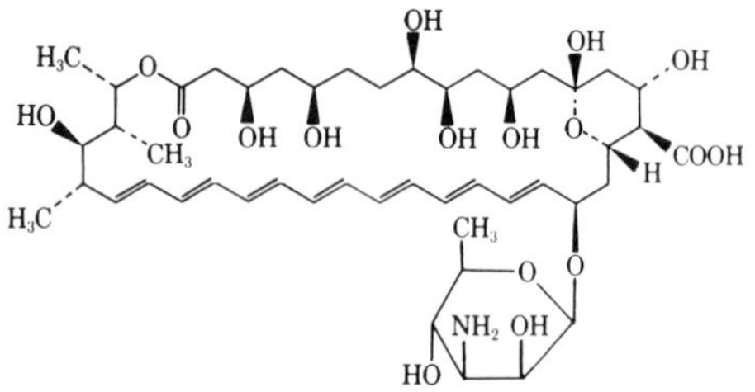

(**3**)   amphotericin B

Derivatives of Polyenes

The poor solubility characteristics of the polyene macrolides greatly restrict their parenteral use, and consequently, a variety of derivatives has been prepared with a view to increasing the solubility in aqueous solutions, while retaining the antifungal activity. These include complexes between polyenes such as amphotericin A and amphotericin B and calcium ions, a borate complex of amphotericin B, *N*-glycosyl derivatives of several polyenes, *N*-methylglucamine salts of polyenes containing carboxylic acid groups, and methyl ester derivatives of several polyenes. Irradiation of levorin A₂ with uv light has been shown to produce a cis-isomer, isolevorin A₂, which has greater antifungal activity than levorin A₂.

In general, the polyene macrolides are essentially devoid of antibacterial activity, but possess excellent activity against a wide range of

Table 1. MIC^a (µg/mL) after 72 h Incubation

	Pimaricin	Nystatin	Amphotericin B	Candicidin
Candida albicans Burke	7.5	0.75	0.3	<0.01
Candida albicans Collins	7.5	0.75	0.055	<0.01
Candida albicans Sparks	7.5	0.3	0.3	<0.01
Candida albicans Wisconsin	7.5	0.75	0.055	<0.01
Candida tropicalis	3.0	0.75	0.3	0.3
Candida stellatoidea	<0.01	0.3	<0.01	<0.01
Cryptococcus neoformans	3.0	0.3	0.3	<0.01
Pityrosporum ovale	<0.01	0.3	0.75	0.3
Rhodotorula rubra	7.5	0.3	0.3	<0.01
Torulopsis glabrata	17.5	3.0	0.3	0.055
Trichophyton rubrum	7.5	0.3	0.3	0.3
Microsporum canis	3.0	0.75	0.055	3.0
Microsporum gypseum	>50	0.75	0.3	3.0
Monosporium apiospermum	3.0	3.0	0.3	0.3
Aspergillus niger	>50	>50	>50	>50
Geotrichum candidum	0.75	3.0	>50	>50

^a MIC = minimum inhibitory concentration.

yeasts and fungi (see Table 1). In addition, many of the polyenes exhibit good antiprotozoal activity against organisms such as *Trichomonas vaginalis*, *Entamoeba histolytica*, and a number of *Leishmania* strains. Trienin also has been reported to exhibit antitumor activity. Heptaene macrolides have been reported to be useful in treating prostatic hypertrophy, in reducing cholesterol absorption, and in oral treatment of acne.

The polyene macrolides have a high degree of potential toxicity associated with their use. The magnitude of toxicity varies according to the route of administration; it is greatest when the polyene is administered intravenously or intraperitoneally. Amphotericin B (intravenous) remains the most widely used polyene for the treatment of systemic infections. However, adverse reactions are invariably produced by systemic therapy. They include chills, fever, headache, anorexia, nausea, vomiting, and anemia. Nephrotoxicity is a considerable problem with amphotericin B and can be controlled successfully only by minimizing exposure of the kidneys to the drug. The aromatic heptaenes exhibit the greatest toxicity. The polyenes show very low toxicity when administered orally owing to the lack of absorption during passage through the gastrointestinal tract. Candicidin and nystatin are used topically to treat vaginal candidiasis. Nystatin is also used in other topical as well as oral treatments.

Polyene macrolide antibiotics form complexes with sterols in the plasma membrane of various cells which result in morphological and permeability changes in the cells. It is believed that this produces pore formation and causes lysis and death of the cell. This is thought to be the reason for the susceptibility of fungi to polyenes, whereas bacteria with no sterols in their cell membranes are not susceptible.

Commercial antifungal polyenes in the United States are listed in Table 2.

Table 2. Polyene Macrolide Antifungals Marketed in the United States

Polyene macrolide	Trade name	Manufacturer
amphotericin B[a]	Fungizone	E.R. Squibb and Sons
candicidin[a]	Candeptin	Schmid Laboratories, Inc.
	Vanobid	Merrell-National Laboratories (Div. Richardson-Merrell, Inc.)
nystatin[a]	Nilstat	Lederle Laboratories (Div. American Cyanimid Company)
	Mycostatin	E.R. Squibb and Sons
	Korostatin	Holland-Rantos Company, Inc.
	Nystaform	Dome Laboratories (Div. Miles Laboratories, Inc.)
	Terrastatin[b]	Pfizer Laboratories (Div. Pfizer, Inc.)

[a] Produced by fermentation.
[b] In combination with oxytetracycline.

ALAN K. MALLAMS
Schering-Plough Corporation

J.F. Martin and L.E. McDaniel, *Adv. Appl. Microbiol.* **21**, 1 (1977).

D. Pappagianis in P.D. Hoeprich, ed., *Infectious Diseases*, Harper & Row, Publishers, Inc., New York, 1972.

S.M. Hammond in G.P. Ellis and G.B. West, eds., *Progress in Medicinal Chemistry*, Vol. 14, North Holland Publishing Co., Amsterdam, The Netherlands, 1977.

POLYETHERS

Most of the known polyether antibiotics are isolated from the *Streptomyces* genus of microorganisms. *S. albus* and *S. hygroscopicus* are the most common sources of the eighty reported polyethers. The exceptions are antibiotics from *Streptoverticillium*, *Actinomadura*, *Dactylosporangium*, and *Nocardia*. Polyethers exhibit good *in vitro* activity against gram-positive and mycobacteria but do not inhibit gram-negative microorganisms. Owing to their high parenteral toxicity, polyethers have no use as clinical antibacterial agents but are playing an increasing role in veterinary medicine as coccidiostats in poultry and growth promotants in ruminant such a cattle and sheep (see Veterinary drugs; Chemotherapeutics, antiprotozoal).

These antibiotics are all acids, but unlike simple carboxylic acids, they cannot be extracted from organic solvents by aqueous sodium carbonate or bicarbonate. Under these conditions, the polyethers, like the crown ethers and other ionophores, extract alkali–metal cations (see Table 1) like sodium into the organic phase from which neutral antibiotic–salt complexes can be isolated by evaporation and crystallization (see also Chelating agents).

Table 1. Cation Selectivity[a] of Polyether Antibiotics as Determined by Either Fluorimetric or Two-Phase Distribution Studies

Antibiotic	Mol wt	Cation selectivity
lasalocid	590	$Ba^{2+} \gg Cs^+ > Rb^+, K^+ > Na^+, Ca^{2+}, Mg^{2+} > Li^+$
monensin	671	$Na^+ \gg K^+ > Rb^+ > Li^+ > Cs^+$
nigericin	724	$K^+ > Rb^+ > Na^+ > Cs^+ > Li^+$
salinomycin	750	$K^+ > Na^+ > Cs^+ \gg Ca^{2+}$
dianemycin	866	$Na^+, K^+ > Rb^+, Cs^+ > Li^+$
antibiotic X-206	870	$K^+ > Rb^+ > Na^+ > Cs^+ > Li^+$

[a] Lyotropic series (with ionic radii in nm) are: Cs^+ (0.169) > Rb^+ (0.148) > K^+ (0.133) > Na^+ (0.095) > Li^+ (0.060) and Ba^{2+} (0.135) > Sr^{2+} (0.113) > Ca^{2+} (0.099) > Mg^{2+} (0.065).

The structure of a typical polyether antibiotic, lenoremycin, is illustrated below. All of these antibiotics exist in cyclic conformation in the crystalline and solution states, owing to a hydrogen bond between the acid and terminal hydroxyl groups.

lenoremycin

The *in vitro* activity of some polyether antibiotics is summarized in the Table 2.

Commercially important polyethers include lasalocid (as Avatec) and monensin (as Coban), which are sold as coccidiostats for poultry. The same two antibiotics, as Bovatec and Rumensin, respectively, are sold as enhancers of feed utilization in cattle.

JOHN W. WESTLEY
Hoffmann-LaRoche, Inc.

J.W. Westley, *Polyether Antibiotics: Naturally Occurring Acid Ionophores*, Vols. 1 and 2, Marcel Dekker, Inc., New York, 1982–1983.

TETRACYCLINES

The tetracyclines are a group of antibiotics having an identical 4-ring carbocyclic structure as a basic skeleton and differing from each other chemically only by substituent variation. In general, the tetracyclines are yellow crystalline compounds with amphoteric properties. They are soluble in both aqueous acid and aqueous base. The acid salts tend to be soluble in organic solvents such as 1-butanol, dioxane, and 2-ethoxyethanol.

The first tetracycline discovered, produced by a soil organism, *Streptomyces aureofaciens*, now known as chlortetracycline (7-chlorotetracycline), was marketed by Lederle Laboratories in 1948. This compound ushered in a new era in antibacterial chemotherapy since it was not only effective orally, but was also effective against a much wider range of gram-positive and gram-negative bacteria. Other members of the family were produced in the next decade, including oxytetracycline (5-hydroxytetracycline) from *Streptomyces rimosus*, tetracycline, and 6-demethylchlortetracycline (6-demethyl-7-chlorotetracycline). The three

Table 2. *In Vitro* Antimicrobial Activity and Toxicity of the Polyether Antibiotics

Polyether antibiotic	Minimum inhibitory concentration, μg/mL[a]									Toxicity in mice, LD$_{50}$, mg/kg[c]	
	Staph. aureus ATCC No. 6538P	Sarcina lutea 9341	Bacillus sp E 27859	Bacillus subtilis 558[b]	Bacillus megaterium 8011	Bacillus sp TA 27860	Mycobacterium phlei 355	Streptomyces cellulosae 3313	Paecilomyces varioti 26820	ip	po
monensin (A3823)	3.1	12.5	0.4	1.6	3.1	1.6	12.5	6.3	d	16.8	43.8
nigericin (X 464)	0.2	0.1	0.004	0.1	0.1	0.02	0.4	0.4	0.8	65	190
grisorixin	0.4	0.2	0.1	0.2	0.1	0.4	0.2	0.2	3.1		
salinomycin	3.1	3.1	0.2	0.8	0.2	0.8	6.3	3.1	3.1	18	50
narasin (A 28086A)	0.8	1.6	0.2	0.9	0.4	0.4	3.1	3.1		12	45
lonomycin (DE 3936)	1.6	12.5	1.6	3.1	0.8	1.6	12.5	6.3	50	13	45.8
X 206	0.2	0.8	0.02	0.4	0.2	0.2	0.2	0.8	0.8	1.2	17
dianemycin	1.6	3.1	0.2	3.1	3.1	1.6	6.3	6.3	6.3	9	150
lenoremycin (Ro 21-6150)	0.2	0.2	0.02	0.4	0.1	0.4	0.2	1.6	0.8		55
septamycin (A28695A)	0.8	1.6	0.006	1.6	1.6	0.2	1.6	3.1	6.3		41.1
A 204A	3.1	12.5	0.8	6.3	12.5	1.6	12.5	12.5	25		8
CP 38952	12.5	6.3	1.6	6.3	3.1	3.1	3.1	12.5	12.5		
lasalocid (X 537A)	1.6	3.1	0.2	1.6	3.1	1.6	12.5	6.3		64	146
iso-lasalocid	12.6	6.3	1.6	1.6	1.6	3.1	6.3	25	12.5	250	1000
lysocellin	0.8	0.4	0.03	0.2	0.4	0.4	3.1	1.6	1.6	65	350
A 23187	0.2	0.04	0.2	0.1	0.8	0.04	1.6	6.3		10	
BL 580α	0.8	1.6	0.2	0.4	0.4	0.4	3.1	3.1		14	220
ionomycin		12.5	12.5	25	6.3					12	650

[a] Lowest two-fold dilution giving zone of inhibition in agar-well diffusion assay.
[b] NRRL collection number.
[c] 24 h acute toxicities unless reference is given.
[d] Indicates no activity up to 50 μg/mL.

tetracyclines most recently marketed, ie, methacycline (6-demethyl-6-deoxy-5-hydroxy-6-methylenetetracycline), doxycycline (6α-deoxy-5-hydroxytetracycline), and minocycline (6-demethyl-6-deoxy-7-dimethylaminotetracycline), were made by a semisynthetic pathway. Such chemical modifications of the tetracycline molecule have been difficult historically because of a sensitive chemical structure. Chemical pathways had to be devised that would bring about the necessary transformation, yet preserve the BCD ring chromophore and its antibacterial properties. The lability of the 6-hydroxy group to acid and base degradation plus the ease of epimerization at the carbon atom at position 4 contribute to chemical instability under many reaction conditions (see Pharmacodynamics).

tetracycline

The very large number of prescriptions written for tetracyclines is proof of their importance to modern medicine. Certain of these antibiotics have been used in large quantities as additives to the food of meat-producing animals where they improve the efficiency of converting animal feeds to meat protein (see Pet and livestock feeds). A rather novel use of a radioactive tetracycline molecule exploits the property that tetracycline localizes in tumor tissue. Radioactive 7-iodo-6-demethyl-6-deoxytetracycline was administered to dogs with well-developed mammory tumors. The level of radioactivity was determined in different areas by external scanning. As was expected, the liver, kidney, spleen, and blood contained the highest level of radioactivity. However, the tumors contained twice as much radioactivity as adjacent healthy tissue.

The manufacture of tetracyclines begins with the cultivated growth of certain selected strains of *Streptomyces* in a chosen medium that produces optimum growth and maximum production of the antibiotic. The choice of the strain of microorganism is one of the important variables in the process. Strains to be used in manufacture are mutants of the original producer which are chosen as the result of a planned program of mutant selection. The choice depends on the strain's ability to produce large amounts of the proper antibiotic in a reasonable time from ingredients that are economically feasible.

All steps of manufacture are monitored for antibiotic potency, purity, stability, and bioavailability by quality-control methods. In addition, each batch of bulk powder and each lot of finished pharmaceutical dosage form must be certified by the FDA. During all stages of manufacture, the FDA requires compliance with specified standards of manufacture.

Evaluation of the biological activities of the various tetracyclines usually begins with an *in vitro* turbidimetric assay against *Staphylococcus aureus*. *In vivo* evaluation in mice usually is against a wide range of bacterial infections, eg, staphylococci, streptococci, and pneumococci. The drug is administered either in the diet or as a single oral dose. The ED$_{50}$ (effective dose that prevents death in 50% of test animals) and LD$_{50}$ are determined in mice following oral, intraperitoneal, and intravenous doses. The mode of action of the tetracyclines is reported to be inhibition of protein synthesis of sensitive bacteria which interferes with a variety of biochemical systems, eg, cell-wall synthesis, biosynthesis of bacterial respiratory systems, and similar systems.

Since the introduction of tetracyclines into clinical practice, a number of microorganisms have developed a resistance to these drugs. The changes leading to this resistance probably result from spontaneously occurring mutations that modify biochemical processes in the resistant microorganisms. Minocycline, however, has been shown to be effective orally against tetracycline-resistant staphylococcal infections in mice.

In the United States, the manufacturers of fermentation-derived tetracyclines are Lederle Laboratories, Division of American Cyanamid, Charles Pfizer, Bristol Laboratories, and Rachelle Laboratories. Pfizer's Doxycycline and Lederle's Minocycline are the only members of this group whose sales continue to increase.

James H. Booth
Joseph J. Hlavka
Lederle Laboratories Div.,
American Cyanamid Company

L.A. Mitscher, *The Chemistry of the Tetracyclines*, Marcel Dekker, Inc., New York., 1978

ANTIBIOTICS, ANTHRACYCLINES.   See Chemotherapeutics, antimitotic.

ANTIBLOCKING AGENTS.   See Abherents.

ANTIFREEZES AND DEICING FLUIDS

An antifreeze is a substance that is added to a liquid, usually water, to lower its freezing point. The largest single use of antifreeze is to protect liquid-cooled internal-combustion engines against freezing and consequent damage to the engine water jacket and the radiator.

Because of its universal availability, low cost, and good heat-transfer properties, water was selected early as the coolant for internal-combustion engines despite its disadvantages of corrosiveness to metals, relatively high freezing point and 9% volume expansion on freezing, and relatively low boiling point.

Prior to 1920, the most frequently used antifreezes were methanol, denatured ethanol, glycerol, and corrosive calcium chloride or salt brines. Sugar and honey solutions were used to a limited extent but required such high concentrations to obtain even moderate freezing-point depression that the viscosity of the coolant was too high at low temperatures, which was also true of glycerol.

During the next decade, denatured ethanol became the most important antifreeze material because of its efficiency, low cost, and availability. Ethylene glycol, which became the most widely used antifreeze, was introduced in 1925. It became known as a permanent antifreeze because its high boiling point reduced coolant loss when an engine overheated, and glycol antifreeze increased its share of the market to > 90% by 1960 as automobile companies began installing ethylene glycol as a year-round antifreeze-coolant during assembly.

Principal U.S. producers include Union Carbide Corp., Dow Chemical U.S.A., BASF-Wyandotte Corp., Continental Oil Co., Northern Petrochemical Co., Shell Oil Co., and Texaco Chemical Co.

Antifreeze Requirements and Test Methods

An acceptable antifreeze must satisfy many requirements. The most essential of these are the ability to lower the freezing point of water to the lowest winter operating temperatures likely to be encountered; satisfactory chemical stability and service and the ability to protect the cooling-system metals from corrosion or deposits; the absence of undesirable effects on engine cooling and heat transfer; minimal effect on elastomers; low cost; minimum effect on surface finishes; acceptably low viscosity at low temperatures; low coefficient of expansion; usefulness for at least one winter season; and ability to be easily checked for freezing point. Although ethylene glycol is the most popular antifreeze today, several other substances are acceptable for special application.

Freezing Protection

Antifreeze solutions, unlike water, do not solidify when exposed to temperatures many degrees below their freezing points. As ice crystals form in the antifreeze solution, the liquid portion becomes more concentrated, and its freezing point is thus lowered. Consequently, instead of forming a solid as does water, the frozen antifreeze is a thick slush that permits the coolant to flow toward the engine headspace as it expands. However, if the slushes form and are unable to circulate through the radiator, overheating, boiling, and engine damage can result. For this reason, antifreeze-protection charts always show the temperatures at which the first ice crystals form.

For dilute ideal solutions of nonelectrolytes, the depression of the freezing point brought about by addition of a solute is proportional to the molal concentration of the solute and is independent of its nature: $\Delta T_f = K_f \cdot m$, where T_f is the freezing-point depression, K_f is the molal freezing-point lowering of the solvent (1.86°C for water), and m is the molality of the solution (moles of solute per 1000 g of water). Although this equation is strictly applicable only to dilute ideal solutions, fair results are obtained when it is applied to nonelectrolytic antifreezes.

The exact determination of the freezing points of antifreeze solutions must be done experimentally. Field testers are usually based on hydrometer-thermometers determining specific gravity. The refractive-index type is more accurate and precise and is of increasing use.

The most important property of an antifreeze formulation, next to heat-transfer and freeze-depressing characteristics, is it ability to prevent corrosion in an automotive cooling system. All antifreezes contain a combination of inhibitors designed to eliminate or at least minimize corrosive attack on a variety of metals. These inhibitors may include organic, inorganic, and silicone compounds. The choice of individual components and their relative proportions affect the performance of the antifreeze, and most inhibitor formulations are patented or kept a trade secret (see Corrosion and corrosion inhibitors).

The service period over which the antifreeze solution will provide satisfactory corrosion protection varies greatly, depending on the mechanical condition of the cooling system, the type of operation to which the vehicle is subject, and the initial corrosion protection of the antifreeze. Factors that materially reduce inhibitor life include presence of rust or residual contamination in the cooling system at the time the antifreeze is installed, leakage of combustion gases into the coolant, excessive aeration, and engine overheating due to a failure of the fan, coolant pump, or thermostat. The general recommendation is that the cooling system be flushed and refilled with fresh coolant annually.

There are many areas in addition to automotive cooling systems where freeze-retarding substances are needed. Stationary engines used in pipeline service or other industrial applications need both freeze protection and corrosion prevention for the cooling system. Refrigeration systems, snow-melting systems, hot-water heating systems, air-conditioning systems, solar-energy units, freezing and freeze-drying units, and hydraulic systems often contain antifreeze.

Deicing fluids are used chiefly for removing ice and frost from parked aircraft and from windows. A heated chemical breaks the surface bonding of the ice which then can be readily removed by mechanical means, often by the force of the sprayed chemicals.

The toxicity of antifreeze is largely determined by the antifreeze base. For humans, a lethal dose of ethylene glycol antifreeze is ca 1.4 g/kg or 100 mL for an adult. Skin or eye exposure is not hazardous, but repeated or prolonged exposure to mists or heated vapors should be avoided. More than a liter of propylene glycol is the estimated lethal dose for humans over 45 kg. Small animals such as dogs and cats are attracted by the sweetish taste of glycols. Solutions should be properly disposed of to prevent accidental ingestion.

Both ethylene and propylene glycol may enter the water environment but, except in local high concentration, pose little threat owing to their rapid and complete biodegradation and low toxicity to wildlife and plants.

R.H. FAY
Dow Chemical, U.S.A.

Maintenance of Automotive Engine Cooling Systems, Society of Automotive Engineers (SAE) Booklet TR-40, Warrendale, Pa.

L.C. Rowe, *Corrosion (Houston)* **13**, 750t (1957).

J.D. Jackson and co-workers, *Corrosion of Metals by Ethylene Glycol-Water, DMIC Report 215,* Defense Metals Information Center, Battelle Memorial Institute, Columbus, Ohio, May 1965.

ANTIMALARIALS.   See Chemotherapeutics, antiprotozoal.

ANTIMICROBIAL AGENTS.   See Industrial antimicrobial agents.

ANTIMONY AND ANTIMONY ALLOYS

Antimony, Sb, mp 630.7°C, belongs to Group VA of the periodic table, may exhibit a valence of +3, +5, or −3 (see Antimony compounds), and is classified as a nonmetal or metalloid, although it has metallic characteristics in the trivalent state. It is a silvery white, brittle, crystalline solid that is a poor conductor for electricity and heat, and is ordinarily quite stable, ie, not readily attacked by air or moisture. The two stable isotopes of antimony are abundant and have masses of 121 (57.25%) and 123 (42.75%). Antimony ore bodies are small and scattered throughout the world with crustal abundance estimated at 0.2 g/t.

The antimony content of commercial ores is 5–60% and determines the method of recovery. In general, the lowest grades of sulfide ores, 5–25% antimony, are volatilized as oxides; 25–40% antimony ores are smelted in a blast furnace; and 45–60% antimony ores are liquated or treated by iron precipitation. The blast furnace is generally used for mixed sulfide and oxide ores and for oxidized ores containing up to ca 40% antimony; direct reduction is used for rich oxide ores, and complex ores are treated by leaching followed by electrolysis.

Antimony is often found associated with lead ores. The smelting and refining of these ores yield antimony-bearing flue, baghouse and Cottrell dusts, drosses, and slags. These materials may be treated to recover elemental antimony or an antimonial lead from which antimony oxide or sodium antimonate may be produced. Secondary antimony is obtained from the treatment of scrap antimony-bearing lead and tin materials such as battery plates, type metal, bearing metal, antimonial lead, etc. The metal produced by a simple pyrometallurgical reduction, however, is normally not pure enough for a commercial product and must be refined. The matting and fluxing technique yields 85–90% refined antimony.

The most important metallurgical use for antimony is as an alloying ingredient for imparting hardness and stiffness to lead and other metals. Antimonial lead used as a grid-metal alloy in the lead acid-storage battery is the largest area of application. Antimony additives to gray cast iron act as a powerful pearlite stabilizer. High purity antimony (> 99.999%) has only a limited, although important, application in the manufacture of semiconductor devices.

The 8-h average exposure limit for antimony and its compounds expressed as Sb is 0.5 mg/m³, and that for stibine (SbH₃) is 0.5 mg/m³ or 0.1 ppm. Although metallic antimony may be handled freely without danger, it is recommended that direct skin contact with antimony and its compounds be avoided. In operations creating dust or fumes, a properly designed exhaust ventilation system is required, or an approved respirator when proper ventilation is not technically feasible. Stibine is a highly toxic gas, and exposure to it should be avoided. Any sign of illness involving possible exposure to stibine should receive immediate medical evaluation.

Estimated world production of antimony is ca 65,500 metric tons. Metallic antimony is sold as ingots and in broken lumps and in cake form (25 × 25 × 6.9 cm) weighing ca 22–26 kg. Normally, it is sold under brand names. U.S. primary producers of antimony are ASARCO, Inc., Anzon, Sunshine Mining Co., and U.S. Antimony Corp. Under depressed demand and price situations, metallic antimony may be produced by these companies on an intermittent basis.

The United States is heavily dependent on imports of antimony ore and metal. Principal suppliers are Bolivia, South Africa, China, France, and Mexico.

S.C. CARAPELLA, JR.
ASARCO, Inc.

Code of Federal Regulations, Title 29, Pt. 1910, OSHA, Washington, D.C., May 21, 1971.

ANTIMONY COMPOUNDS

Antimony is the fourth member of the nitrogen family and, like its congeners, the antimony atom has five electrons in its outermost shell ($5s^2, 5p^3$). The utilization of these orbitals and, in some cases, of one or two $5d$ orbitals permits compounds in which the antimony atom forms three, four, five, or six covalent bonds.

The valence-bond theory predicts that trivalent compounds of antimony should have pyramidal structures derived from the $5p$ orbitals and that the $5s$ electrons should act as an inert pair. Many trivalent derivatives of antimony, however, have intervalency angles significantly larger than the 90° angle predicted by this model, suggesting that sp^3 hybridized antimony orbitals are being employed and that the lone pair occupies one of the tetrahedral positions. The fact that the bond angles are often considerably less than the regular tetrahedral value of 109.05° may be ascribed to repulsion by the lone pair. Pentacoordinate compounds of antimony usually exhibit trigonal bipyramidal geometry corresponding to the sp^3d hybridized antimony orbitals of valence-bond theory.

Inorganic Compounds of Antimony

Stibine (SbH₃, mp −88°C, bp −17°C) is a colorless, poisonous gas with a disagreeable odor. High purity stibine is used as an *n*-type gas-phase dopant for Si in semiconductors.

Metallic antimonides.   Numerous binary compounds of antimony with metallic elements are known. The most important of these are indium antimonide (InS), gallium antimonide (GaSb), and aluminum antimonide (AlSb), which have extensive use as semiconductors (qv).

Antimony trioxide.   Antimony(III) oxide (antimony sesquioxide), Sb₂O₃, is dimorphic, existing in an orthorhombic modification, valentinite, and a cubic form, senarmontite. It is used as a flame retardant in fabrics (see Flame retardants), as a catalyst, and as an opacifier in glass (qv), ceramics (qv), and vitreous enamels (qv).

Antimony tetraoxide.   Antimony(III, V) oxide, antimony dioxide, SbO₂ or Sb₂O₄, occurs in two modifications. Orthorhombic antimony tetroxide has long been known as the mineral cervantite, α-Sb₂O₄, but in recent years a monoclinic modification, β-Sb₂O₄, has been recognized. Antimony tetroxide is used as an oxidation catalyst, particularly for oxidative dehydration of olefins.

Antimony pentoxide hydrates.   Antimonic acid, antimonic(V) acid, antimony(V) oxide, Sb₂O₅·nH₂O. Antimonic acid has been used as an ion-exchange material for a number of cations in acidic solution. Many oxidation and polymerization catalysts are listed as containing Sb₂O₅.

Antimony trifluoride.   Antimony(III) fluoride, SbF₅ (mp 280 ± 1°C, bp 346 ± 10°C), is a white, crystalline, orthorhombic solid and is used as a fluorinating agent.

Antimony trichloride.   Antimony(III) chloride, SbCl₃ (mp 73.4°C, bp 222.6°C), is a colorless, crystalline solid that is used as a catalyst or as a component of catalysts to polymerize hydrocarbons and to chlorinate olefins; in hydrocracking of coal and heavy hydrocarbons; as an analytical reagent for chloral, aromatic hydrocarbons, and vitamin A; and in the microscopic identification of drugs (see Vitamin A). Liquid SbCl₃ is used as a nonaqueous solvent.

Antimony tribromide/antimony triiodide.   Antimony(III) bromide, SbBr₃ (mp 96.0 ± 0.5°C), and antimony(III) iodide, SbI₃ (mp 170.5°C, bp 401°C). The compounds antimony bromide sulfide (SbBrS), antimony iodide sulfide (SbIS), and antimony iodide selenide are of interest with respect to their solid-state properties, ferroelectricity, pyroelectricity, photoconduction, and dielectric polarization.

Antimony pentafluoride.   Antimony(V) fluoride, SbF₅ (mp 6°C, bp 150°C), is a colorless, hygroscopic, viscous liquid used principally as a fluorinating agent.

Antimony pentachloride.   Antimony(V) chloride, SbCl₅ (mp 3.2 ± 0.1°C, bp 69°C at 1.82 kPa (14 mm Hg)), is a colorless, hygroscopic, oily liquid. It is a strong Lewis acid and a useful chlorine carrier.

Derivatives of antimony pentabromide and pentaiodide.   The existence of SbBr₅ and SbI₅ is in doubt, but several derivatives of these compounds are known, including the 1 : 1 adduct SbBr₅·O(C₂H₅)₂ and a number of complex bromoantimony compounds with alkali metal and organic bases.

Antimony trisulfide.   Antimony(III) sulfide (antimony sesquisulfide), Sb₂S₃, exists as a black crystalline solid, stibnite, and as an amorphous red to yellow-orange powder. It is used in fireworks (see Pyrotechnics), in certain types of matches (qv), as a pigment, and in the manufacture of ruby glass (see Pigments, inorganic).

Antimony pentasulfide.   Antimony pentasulfide is a yellow to orange to red amorphous solid of indefinite composition used in vulcanization to produce a red variety of rubber, as a pigment, and in fireworks.

Antimony(III) salts. Concentrated acids dissolve trivalent antimony compounds. From the resulting solutions, it is possible to crystallize normal and basic salts, eg, antimony(III) sulfate.

Compounds containing Sb—O—C or Sb—S—C linkages. A large number of compounds have been prepared in which the antimony atom is linked to carbon through an oxygen or sulfur atom. The best known compound of this type is tartar emetic (antimony potassium tartrate) used as an antiparasitic agent in medicine (see Chemotherapeutics, antiprotozoal), an insecticide, and a mordant in the textile and leather industries.

Organoantimony Compounds

Antimony forms a wider variety of compounds containing the Sb—C bond. Organoantimony compounds can be broadly divided into Sb(III) and Sb(V) compounds. The former may contain from one to four organic groups, and the Sb(V) compounds may contain from one to six organic groups.

Organoantimony compounds include primary, secondary, and tertiary stibines; halostibines, dihalostibines, and related compounds; antimono compounds and distibines; antimonins; stibonic and stibinic acids; pentacovalent antimony halides; stibonium ylids, and related compounds; and pentaalkyl- and pentaarylantimony compounds.

Antimony Compounds Used in Medicine

The introduction of antimonides for the treatment of parasitic diseases is regarded as one of the important milestones in the history of therapeutics (see Chemotherapeutics, anthelmintic). Medical antimony compounds include antimony potassium tartrate (tartar emetic; potassium antimonyl tartrate); antimony sodium tartrate (sodium antimonyl tartrate); stibophen (sodium antimony(III) bis(pyrocatechol-3,5-disulfonate); antimony(III) sodium gluconate (triostam; T.S.A.G.); stibocaptate (sodium antimony(III) dimercaptosuccinate; astiban); ethylstibamine (neostibosan); and antimony(V) sodium gluconate (pentostam, solustibosan).

Leon D. Freedman
G.O. Doak
G. Gilbert Long
North Carolina State University

P.J. Durrant and B. Durrant, *Introduction to Advanced Inorganic Chemistry*, 2nd ed., John Wiley & Sons, Inc., New York, 1970, pp. 756–770.

J.D. Smith in A.F. Trotman-Dickenson, ed., *Comprehensive Inorganic Chemistry*, Vol. 2, Pergamon Press, Oxford, UK, 1973, pp. 643–644.

G.O. Doak and L.D. Freedman, *Organometallic Compounds of Arsenic, Antimony, and Bismuth*, John Wiley & Sons, Inc., New York, 1970.

ANTINAUSEA AGENTS. See Hypnotics, sedatives, and anticonvulsants.

ANTIOXIDANTS AND ANTIOZONANTS

Antioxidants

Antioxidants are substances that retard oxidation by atmospheric oxygen at moderate temperatures (autoxidation). Autoxidation is a free-radical chain reaction and, therefore, can be inhibited at the initiation and propagation steps. Antioxidants are used widely in polymers, petroleum products, and food.

Two main classes of antioxidants inhibit the initiation step in thermal autoxidation. The peroxide decomposers function by decomposing hydroperoxides through polar reactions. Metal deactivators are strong metal-ion complexing agents that inhibit catalyzed initiation through reduction and oxidation of hydroperoxides.

Autoxidation often has a long kinetic chain length. Therefore, agents that interrupt the propagation step markedly reduce the oxidation rate.

The most important commercial propagation inhibitors are hindered phenols and secondary alkylaryl- and diarylamines.

Combinations of certain antioxidants sometimes provide synergistic protection. The most common synergistic combinations are mixtures of antioxidants operating by different mechanisms. For example, combinations of peroxide decomposers with propagation inhibitors are commonly used in polyolefins. Similarly, combinations of metal chelating agents with propagation inhibitors are used in certain elastomers. Synergistic combinations of structurally similar antioxidants are also known, particularly combinations of phenols.

Hydroperoxides are important initiators in photooxidation. They are decomposed by solar radiation both photochemically and thermally. Agents that protect substrates from photooxidation include those that reduce the amount of damaging radiation entering the substrate (uv absorbers) and those that deactivate photoexcited chromophores by energy transfer (quenching agents). However, despite the importance of radical-chain autoxidation in photodegradation processes, most hindered phenols alone provide little protection, probably because they are rapidly decomposed under uv irradiation. Nevertheless, combinations of uv absorbers and phenolic antioxidants can be synergistic and are commercially important (see Uv stabilizers).

Many different commercial antioxidants have been developed for specific applications:

Amines. The amine-type antioxidants usually are the most effective in rubber and are used in the largest volume (see Amines, aromatic). They can be classified further as ketone–amine condensation products, diaryldiamines, diarylamines, and ketone–diarylamine condensation products. Both solid and liquid products are marketed. Most are discoloring and staining and are used in applications where this property can be ignored.

Phenols. Phenolic antioxidants are less discoloring than amines. They are used in applications where these properties are necessary, and are of increasing commercial importance. Most of the newer commercial antioxidants are of this type, such as alkylated hydroquinones and phenols. In high temperature applications, polynuclear phenols generally are preferred over monophenols because of their lower sublimation rates (see Phenols).

Phosphites. Certain phosphites are used to protect rubber during manufacture and in storage. For plastics, they generally are used in combination with other antioxidants, particularly phenols. Tri (mixed nonyl- and dinonylphenyl) phosphite is used in the largest volume (see Phosphorus compounds).

Sulfides. Dilauryl thiodipropionate and distearyl thiodipropionate are the most important commercial antioxidants in this class. They are used with phenols in plastics to give synergistic combinations (see Sulfur compounds).

Metal salts of dithioacids. These substances act as hydroperoxide decomposers and propagation inhibitors, and are used in conjunction with other antioxidants, particularly phenols for rubber, petroleum products, and plastics.

Bound antioxidants. Recently, antioxidants have been developed that are copolymerized into the elastomer chain. The main advantage of such a system is low antioxidant extractability in applications where the elastomer is in contact with solvents capable of extracting conventional antioxidants.

Most phenolic antioxidants are readily prepared by alkylating either hydroquinones, phenols, or cresols. All aromatic amine antioxidants or antiozonants are made from aniline or substituted anilines. Diphenylamine and *p*-phenylenediamine likewise are derived from aniline and are the basic raw materials for other products. Unsymmetrical alkylphenyl-substituted phenylenediamines are best prepared from diphenylamine. Phosphites and dithiophosphates are made from alcohols or phenols, or their mixtures, whereas dithiocarbamates and xanthates are prepared from carbon disulfide (qv).

Antioxidants are used in the production of raw, cured and synthetic rubbers, plastics, gasoline and other petroleum products, adhesives, food, and animal feed. The rubber market is by far the largest consumer of antioxidants.

Antioxidants are considered to be nonhazardous, although the only direct additives allowed for food are those with FDA acceptance. Also, some hazard to workers can be created by the physical form of an anitoxidant, such as vapor, dust, or aerosol. Personal hygiene practices to avoid accidental ingestion, skin contact, and inhalation are essential.

Antiozonants

Many highly unsaturated elastomers are susceptible to crack development when stressed in an ozone-containing atmosphere (see Ozone). These cracks propagate normal to the stress direction when certain minimum stress levels are exceeded. Cracks also occur at low atmospheric ozone concentrations and can be a major cause of rubber failure in outdoor applications.

Two broad groups of additives are used commercially to inhibit ozonolysis of unsaturated elastomers. One group includes relatively unreactive, film-forming waxes (qv). The other comprises certain reagents that react rapidly with ozone and possibly ozonolysis products. These reactive reagents are called antiozonants. In order to be effective, waxes must quickly diffuse to the elastomer surface to form a protective film, or so-called bloom. Wax protection alone is inadequate for products, such as tires, that flex in service causing the film to break, and antiozonants are essential when highly unsaturated elastomers are used in these applications.

In general, the N,N'-disubstituted p-phenylenediamines are the most effective commercial antiozonants, and continue to enjoy widespread growth, especially for tire applications. Certain dihydroquinolines, thioureas, metal dithiocarbamates, and N,N'-diaryl-substituted p-phenylenediamines, such as N,N'-diphenyl-p-phenylenediamine, also have weak antiozonant activity.

The di-sec-alkyl-p-phenylenediamines are prepared from ketones and p-phenylenediamine. Condensation gives the dianil which is catalytically hydrogenated to the desired product. The N-sec-alkyl-N'-phenyl-p-phenylenediamines are made in the same way from p-aminodiphenylamine. Condensation with the ketone gives the monoanil which is then hydrogenated.

Antiozonants are incorporated into rubber on a mill or in a Banbury mixer. They are often added to unsaturated rubbers such as polyisoprene, polybutadiene, and SBR, particularly when used in outdoor applications such as tires. Important product applications for antiozonants and waxes include tires, power-transmission fluids, conveyor belts, hoses, gaskets, and other molded articles.

The p-phenylenediamines can be toxic and may cause dermatitis when the N-substituents are small groups, such as sec-butyl. However, the use of higher homologues has largely eliminated this problem. Nevertheless, the general care outlined for antioxidants should be followed.

PAUL P. NICHOLAS
The BF Goodrich Co., Corporate Research

ANTHONY M. LUXEDER
The B.F. Goodrich Co., Chemical Division

LESTER A. BROOKS
R.T. Vanderbilt Co.

PAUL A. HAMMES
Merck and Co.

G. Scott, ed., *Developments in Polymer Stabilization*, Vols. 1–6, Applied Science Publishers, Ltd., London, 1979–1983.

G. Scott, *Atmospheric Oxidation and Antioxidants*, Elsevier Publishing Co., New York, 1965.

B. Ranby and J.F. Rabek, *Photodegradation, Photo-oxidation and Photostabilization of Polymers*, John Wiley & Sons, Inc., New York, 1975.

R. Lattimer, R.W. Layer, and C.K. Rhee in J. Kroschwitz, ed., *Encyclopedia of Polymer Science and Engineering*, John Wiley & Sons, Inc., New York, 1985.

ANTIPYRETICS. See Analgesics, antipyretics, and anti-inflammatory agents.

ANTISEPTICS. See Disinfectants and antiseptics.

ANTISTATIC AGENTS

The increased processing speeds of textiles, paper and plastics often are accompanied by significant static-electricity problems. Accumulation of electrical charges on textiles, plastics, and other materials of an insulating character causes a large number of problems which range from general nuisance and discomfort to processing difficulties, loss of productivity, poor product quality, and often serious fire and explosion hazards in processing as well as in use. In general, static difficulties occur with greater frequency as the relative humidity decreases, particularly during winter months in heated rooms.

The amount of electrostatic charge developed on a material represents a balance between the rate of generation and the rate of dissipation. Good conductors disperse a static charge quickly; however, textiles and plastics have a high surface resistivity and charge decay can only occur at a low rate.

Static charges can be reduced either by reducing the rate of or preventing generation or by increasing the rate of dissipation. Reducing friction generally results in reduction of static, but friction is essential for many processes and often cannot be excessively reduced. Thus, all practical remedial measures are based on increasing the rate of charge dissipation or leakage.

Static charges on polymers dissipate by surface and volume conductivity of charges through the substrate and by loss of charge by radiation to the air. Shashoua represented this process by the equation:

$$V_t = V_o e^{-(K_1 + K_2)t}$$

where: V_o = initial voltage, V_t = voltage at time t, t = time (s), K_1 = conduction rate constant, and K_2 = radiation rate constant. The radiation process has a significant effect on charge decay, and in some polymers, eg, nylon-6,6, the evidence strongly suggests that the loss of charge by radiation is greater than by conductivity.

The main methods for increasing the rate of static dissipation are grounding, ie, using a grounded bar, rod, or roll to discharge static charge buildup on a material; increasing the electrical conductivity of the surrounding air by using static eliminators such as silent (corona) discharge eliminators or radioactive eliminators, which ionize the atmosphere; increasing the electrical conductivity of the material by increasing its electrolytic (ionic) or electronic conductivity through such means as humidification, application of internal or external antistatic agents, surface modification to increase hygroscopic properties, grafting of functional groups leading to ionic configurations or greater moisture sensitivity, coating fibers with a thin layer of a conductor such as silver or carbon black, blending fine metal fibers with static-prone fibers in textiles, or incorporating carbon black in plastics; and humidification of the air to control and help dissipate static electricity in processing.

Humidification within a tolerable range is the most widely used antistatic process in handling materials of limited static propensity such as cotton and rayon, but this does not suffice in processing other materials, such as wool, synthetic fibers, plastics, etc, and the use of suitable antistatic agents is essential. Most antistatic agents, however, also rely on atmospheric moisture for their effectiveness, and humidity control is also required.

In general, most antistatic agents belong to one or another of the following classes: nitrogen compounds such as long-chain amines, amides and quaternary ammonium salts; esters of fatty acids and their derivatives; sulfonic acids and alkyl aryl sulfonates; polyoxyethylene derivatives; polyglycols and their derivatives; polyhydric alcohols and their derivatives; phosphoric acid derivatives; solutions of electrolytes in liquids with high dielectric constants; molten salts; metals; carbon black or semiconductors; and liquids with high dielectric constant, such as water, which are usually volatile and have a temporary effect only.

Antistatic Agents for Textile Materials

Effective antistatic agents must act at a relative humidity below 40%, preferably below 25%, and must have a low electrical resistivity even at low temperature and humidity conditions. Some examples of commercial antistatics are given in Table 1. The agent must form a film on various surfaces and be applied from a solution or dispersion in water or other inexpensive solvent. The antistat must not interfere with subsequent processing of the product or impair the hand, color, odor, appearance, and performance properties of the substrate. It should be nontoxic and nonflammable.

Nondurable antistatic agents are suitable for textile processing aids in such operations as spinning, winding, weaving, and fiber manufacturing, but cannot provide the material with antistatic protection in consumer use because nondurable finishes are easily removed by washing. For consumer use, an effective antistatic finish must be durable and capable of withstanding repeated laundering and dry-cleaning cycles. Only a small number of such durable agents are available. They are polyelectrolytes and involve variations of cross-linked polyamines containing polyethoxy segments.

Aging is known to improve the resistance to laundering of durable antistatic finishes that have been incompletely cross-linked by curing, but a gradual reduction of static protection on aging has been observed with some nondurable antistats owing to migration of the agent from the surface to the interior of the polymer. Extraction and reapplication with a solvent can restore the static protection.

Antistatic finishes are commonly applied to textile materials by padding (dipping and squeezing off excess finish), exhausting (absorption from solution owing to affinity of the antistat for the textile material), and by spraying and coating. Durable antistatic finishes require cross-linking of the resin usually achieved by subjecting the treated, dried material to heat curing. Without special measures, the amount of antistat needed for protection of textile fabrics is in the range of 0.1%, depending on the efficiency of the antistat, relative humidity, temperature, diameter of fiber, fabric structure, and the degree of antistatic protection desired.

Efforts to develop fibers that are inherently static-free have involved incorporation of internal antistatic additives in the bulk of the polymer before spinning to increase ionic conductivity, and coating the fiber with conducting metals or carbon black and incorporating fine metal fibers in the textile material which increases electronic conductivity. Both methods increase the conductivity of the whole material rather than the surface alone.

Most antistatic fibers that contain internal antistatic agents are polyamides (qv) such as Antistat Celon (Courtaulds), Perlon Antistatic (Bayer), and Ecstatic (Enkalon), which are all nylon-6 type, and Counterstat (ICI) which is a nylon-6,6. Examples of carpet antistat fibers are Anso 4 (Allied), Antron (Dupont), Ultron (Monsanto), and Enkalon (Enka).

Antistatic agents for plastics. Added ionic materials or ionic impurities on the surface of a plastic provide antistatic properties. In addition, antistatics for plastics are generally hydrophilic, and they attract moisture and other ionic impurities. They may be applied by dipping, wiping or spraying the surface for nondurable protection, or by incorporating the agent directly in the bulk of the polymer from which it continually migrates to the surface and forms a permanent, antistatic film.

The main groups of antistatic agents include ionic compounds such as quaternary ammonium salts and amines and hydrophilic compounds such as polyglycols and ethylene oxide derivatives, which increase the surface ionic conductivity. The high electrical conductivity of carbon black in the absence of moisture is used where its color is not objectionable and its reinforcing actions also can be used (see Carbon, carbon black; Fillers).

Several new antistatic agents for plastics have been introduced since 1975. they are geared toward higher levels of performance and convenience, broader materials applicability, and an effective lower cost to users. Among the most significant in terms of potential market growth are, Armak's Armostat 375 P for polyolefins, which has FDA acceptance, and Armostat 450S for styrene polymers. Sandoz Colors and Chemicals has introduced Sandin PU and VU, which are intended for use in poly(vinyl chloride) and reportedly meet the National Fire Protection Associations standards for use in hospitals. Continued developments in antistats for plastics in contact with food are typified by Ashland's Verstat K-22 and T-22 antistats which have FDA acceptance for use in polyolefin films (see Food additives).

The two large-volume applications of antistatic agents are in the textile and plastics industries, although antistatic compositions are being sold for application during household laundering. It has been estimated that approximately 250 t/yr of antistats is used in textile-mill applications.

ZEEV GUR-ARIEH
BERTIE J. REUBEN
Monsanto Textiles Company

V.E. Shashoua, *J. Polym. Sci.* **33**, 65 (1958); **1A**, 169 (1963).

N.R. Cansdale and J.M. Heaps, *Plastics*, 68 (June 1964).

J.L. Rogers, *Soc. Plast. Eng. J.* **29**, 52 (1973).

Table 1. Some Examples of Antistatic Agents for Textile Materials

Trade name	Manufacturer	Chemical type	Remarks
Aerotex Antistatic D[a]	American Cyanamid		durable antistat
Aston 123[a]	Millmaster Onyx	polyamine resin	durable antistat
Aston MS[b]	Millmaster Onyx	quaternery fatty amine condensate	cationic antistat
Avitex DN[c]	DuPont	cationic compound	antistatic softener
Birocostat 31	Virkler Chemical	cationic compound	
Cirrasol PT[a]	ICI		durable antistat
Igepal Co 430	GAF	nonyl phenol ethylene oxide condensate	
Larostat 192	Jordan Chemical	cationic compound	
Nonax 1166[a]	Henkel	polyamine resin	durable antistat
Stanax 1166[a]	Standard Chemical Products	polyamine resin	durable antistat
Tomostat DPW	Piedmont Chemical Industries	organic amine derivative	
Zelec DP	DuPont	cationic compound	
Zero C	Lutex Chemical	nitrogeneous polymer	

[a] Durable antistat.
[b] Cationic.
[c] Antistatic softner.

ANTISYPHILITICS. See Chemotherapeutics, antibacterial and antimycotic.

ANTITUSSIVES. See Expectorants, antitussives, and related agents.

APPETITE-SUPPRESSING AGENTS

Appetite-suppressing agents belong to the class of sympathomimetic amines. The sympathetic nervous system is responsible for such excitatory reactions as the fight or flight response. In the natural state, activation of the sympathetic nervous system is mediated by release of the neurohormones epinephrine (qv), norepinephrine, and dopamine (see Psychopharmacological agents). Drugs with a similar effect to these three natural substances or which cause their release or alter their metabolism frequently have similar chemical structures. Thus, amphetamine (**1**) bears the phenethylamine moiety and all but one of the currently used appetite suppressants, mazindol (**2**), are structurally related to amphetamine.

Aside from their role as appetite suppressants, amphetamine and related agents are used to prevent and reverse fatigue and in the treatment of narcolepsy, hyperkinetic syndromes related to minimal brain dysfunction, epilepsy and postencephalitic parkinsonism.

(1) amphetamine

d-Amphetamine enhances the turnover rate of both norepinephrine and dopamine and depletes norepinephrine centrally. These actions are thought to be critical in the behavioral state of appetite suppression. Fenfluramine (2) has little effect on norepinephrine or dopamine levels or turnover, but does cause central release of 5-hydroxytryptamine (serotonin) at therapeutic doses suggesting a distinctly different mechanism of action. Mazindol (3) blocks neuronal uptake of norepinephrine like amphetamine, but the latter alters the brain norepinephrine release rate, levels, and synthesis—effects not shared by mazindol.

(2) fenfluramine

(3) mazindol

The most serious drawback of amphetamine and similar appetite-suppressing agents is dependency and abuse. On account of this, these compounds are under control by the Bureau of Narcotics and Dangerous Drugs (BNDD). Amphetamine, methamphetamine, and phenmetrazine are in BNDD Schedule II which is highly restrictive, whereas the others are in Schedules III and IV which are succeedingly less restrictive but still indicative of abuse liability. Except for one compound, all of these drugs may cause mood elevation at the dose level that suppresses appetite. Tolerance to both effects develops over a period of weeks. However, the mood elevation may be retained with steadily increasing doses, leading to psychological and physical dependency. The massive dose required to support severe dependency finally cause irreversible peripheral and central nervous system damage. Overdosage causes restlessness, dizziness, tremor, insomnia, hyperactive reflexes, talkativeness, tenseness, weakness and euphoria. Fatigue and general depression follow these symptoms of overstimulation. Fatal overdosage usually terminates in convulsions or coma, and pathology discloses mainly cerebral hemorrhage.

This group of medicinals cannot currently be considered a pharmaceutical growth market. The need for the development of an effective appetite-suppressing agent that is free from abuse liability and tolerance development remains great since recently introduced agents, with the possible exception of mazindol, have been disappointing.

STEPHEN T. ROSS
Smith Kline & French Laboratories

A.G. Gilman, L.S. Goodman, and A. Gilman, *The Pharmacological Basis of Therapeutics*, 6th ed., MacMillan Co., New York, 1980.

A.C. Sullivan, H.W. Baruth, and L. Cheng in H.-J. Hess, ed., *Recent Advances in the Design and Development of Antiobesity Agents in Annual Reports in Medicinal Chemistry*, Vol. 15, Academic Press, Inc., New York, 1980, pp. 172–181.

AQUACULTURE

By 1985, total world production of fish through aquaculture should reach 12×10^6 metric tons, roughly twice the annual production of the mid-1970s. Of the 6×10^6 t then produced annually was 66% is finfish, 16.2% molluscs, 17.5% seaweeds, and 0.3% crustaceans.

A number of phases of the industry are highly mechanized, and include the use of bulldozers and scrapers for building ponds, pullers for seines, tractors for feeding, transportation equipment, machines for removal of scales and bones, refrigerators for storing, and a great array of milling processes for preparing fish food.

Culture ponds are usually turbid; therefore, the culturist must indirectly assess the condition of the fish. Ponds should be prepared by drying followed by filling and immediate stocking. When drying time for drained ponds is unavailable, they may be sterilized to rid them of trash fish, insects, crayfish, tadpoles, and all competitors and predators of fry. Calcium hypochlorite at the rate of 10–50 ppm, hydrated lime at 1000 kg/ha, or 5 ppm sodium cyanide are commonly used for sterilization (see Water). Trichlorfon at 0.25–0.5 ppm removes insects and crayfish; rotenone and antimycin A are good piscicides. Aquatic vegetation is removed by a variety of herbicides (qv) or by overstocking with weed eaters, but encroachment is best prevented biologically by proper stocking rates or the inclusion of a herbivorous species where law permits.

Disease outbreaks in fish usually are the result of stress caused by poor water quality, nutritional deficiencies, and overcrowding. The two recurring causes of stress, anoxia and poor water quality caused by the buildup of waste products, are controlled by running fresh water into the pond. Anoxia, characteristic of phytoplankton blooms, also can be controlled by algicides such as cupric sulfate. Uncontrolled blooms can cause the loss of an entire fish population overnight. If warning signs, such as a decrease in food consumption, rapid change of pond color, or fish piping (gasping) at the surface at daybreak, are noticed, potassium permanganate often can be used to reduce the high chemical oxygen demand (COD) caused by dead phytoplankton, although pumping large quantities of fresh water into the pond is still essential. Recirculation of water and the use of aerators are also effective.

Disease outbreaks can occur at any time during the year. Infestations of common pathogenic protozoa such as *Ichthyophthurius*, and *Trichodina* can be treated prophylactically with formalin, copper sulfate, or malachite green (see also Poisons, economic). Bacterial infections such as *Aeromonas* and *Pseudomonas* can be treated prophylactically with medicated feeds containing oxytetracycline or a furan derivative though such feeds are ineffective for treatment of fish already sick (see Antibiotics; Antibacterial agents, synthetic). Potassium permanganate at 2–3 ppm, however, is an effective pond treatment although variations in the COD make dose-level determinations extremely difficult.

Three types of advanced culture approaches have been developed: natural bodies of water where fish are cultured in high density cages; raceways, constantly supplied with large quantities (thousands of liters per minute) of fresh water, and more recently the use of biofilters to maintain extremely high densities of fish (15–80 kg/m^3); and shallow culture ponds, the most common approach, which produces 2000–3000 kg/ha of fish. Table 1 lists selected examples of aquacultural yields.

In the simplest form of aquaculture, fish are stocked in densities low enough that their growth to a marketable size is sustained on the natural food in the pond. Production is low, about 50–200 kg/ha, depending on the natural fertility of the soil. Where fish are produced at rates of 2000–3000 kg/ha, feeding must be intensive. The culturist brings the food to the fish to eliminate loss of energy in hunting. Pelletized food facilitates handling and maintains a high nutritional level. The fish should be trained to take the pellet as it hits the water so that it does not dissolve into fine particles unavailable to the fish. Highest production is achieved when food and fertilizers are combined in polyculture to allow two to five species of fish occupying different ecological niches to be raised in the same pond.

In harvesting fish, the usual practice is to drain and seine, or simply seine. Seining, when properly executed, is 50–85% effective. Since the holding, handling, and transport of live fish is a highly stressful period, treatments with anesthetics (qv) such as tricaine, antibacterials such as oxytetracycline, and parasiticides such as formalin are necessary to quiet fish and control bacteria and protozoa.

The fate of all chemical agents, with respect to the fish, the consumer, and the environment, must be considered. Chronic toxicities in a culture

Table 1. Selected Examples of Aquacultural Yields Arranged by Ascending Intensity of Cultural Methods

Culture method	Species	Location	Yield, kg/(ha·yr)
stocking and rearing in fertilized enclosures, no feeding	carp	Java[a]	62,500–125,000
stocking and rearing with fertilization and feeding	channel catfish	U.S.	3,000
	carp, mullet	Israel	2,100
	tilapia	Cambodia	8,000–12,000
	carp and related sp (in polyculture)	China, Hong Kong, Malaysia	3,000–5,000
	Clarias	Thailand	97,000
intensive cultivation in running water; feeding	rainbow trout	U.S.	170 kg/(L·s)
	carp	Japan	ca 100 kg/(L·s)
	shrimp	Japan	6,000
intensive cultivation of sessile	oysters	Japan, Inland Sea	20,000
organisms, molluscs and algae	oysters,	U.S.	5,000 (best yields)
	mussels	Spain	300,000
	Porphyra	Japan	7,500
	Undaria	Japan	47,500

[a]Sewage stream 25–50% of water area used.

unit may be manifested by poor growth and susceptibility to disease. The persistence of an agent after it has accomplished its purpose can result in residues in the fish, concentration in the pond bottom, and distribution into the surrounding watershed by the effluent. For the protection of the consumer and the environment, the culturist is required to use only chemicals approved by and registered with the EPA and FDA.

HOWARD P. CLEMENS
MICHAEL CONWAY
University of Oklahoma

F.W. Wheaton, *Aquacultural Engineering*, John Wiley & Sons, Inc., New York, 1977.

R.R. Stickney, *Principles of Warmwater Aquaculture*, John Wiley & Sons, Inc., New York, 1979.

J.E. Bardach, J.H. Ryther, and W.O. McLarney, *Aquaculture, The Farming and Husbandry of Freshwater and Marine Organisms*, John Wiley & Sons, Inc., New York, 1972, p. 6.

ARAMID FIBERS

Aromatic polyamides are formed by reactions that produce amide linkages between aromatic rings. In practice, this generally means the reaction of aromatic diamines and aromatic diacid chlorides in an amide solvent. From solutions of these polymers, it is possible to produce fibers of exceptional heat and flame resistance and fibers of good to quite remarkable tensile strength and modulus. Because the physical property differences between fibers of aromatic aliphatic polyamides are greater than those between other existing generic classes of fibers, a new generic term for fibers from aromatic polyamides was requested by the DuPont Company in 1971. Subsequently, the generic term aramid was adopted (1974) by the U.S. Federal Trade Commission for designating fibers of the aromatic polyamide type: aramid—"a manufactured fiber in which the fiber-forming substance is a long-chain synthetic polyamide in which at least 85% of the amide (—CONH—) linkages are attached directly to two aromatic rings."

There is no recognized systematic nomenclature for aromatic polyamides as exists for aliphatic polyamides. However, acronmyms or the initial letters of the various monomers often are combined, the initials of the diamine moieties preceding those of the diacid moieties. The letters T and I have the commonly accepted significance of standing for,

respectively, terephthalamide and isophthalamide. The first aramid fiber to be developed, known experimentally as HT-1 and almost certainly based on poly(m-phenyleneisophthalamide), MPD-I, was commercialized under the trademark Nomex by DuPont in 1967.

Aramid fibers do not melt in the conventional sense because decomposition generally occurs simultaneously. An endothermic peak in the differential thermal analysis (dta) thermogram for aramid fibers can be obtained; the values are generally > 400°C and can be as high as 550°C. Glass transitions range from ca 250 to > 400°C. Weight loss, as determined by thermogravimetric analysis (tga) in an inert gas, begins at ca 425°C for most aramids although some of the rodlike polymers do not lose substantial weight to 550°C.

Aramid fibers are characterized by medium to ultra-high tensile strength, medium to low elongation, and moderately high modulus to ultra-high modulus. Most of these fibers have been reported to be either highly crystalline or crystallizable, and densities for crystalline fibers are ca 1.35–1.45 g/cm^3. Poly(m-phenyleneisophthalamide) fiber of low orientation has a density of ca 1.35 g/cm^3, and hot-drawn fiber has a density of ca 1.38 g/cm^3; Kevlar, poly(p-phenyleneterephthalamide), has a density of ca 1.45 g/cm^3.

Since most of the work on aramid fibers has been directed toward the development of heat and flame-resistant fibers and ultra-high strength, high modulus fibers, it is possible to categorize them in these terms. In general, polymers useful for the former contain a high portion of meta-oriented phenylene rings, and polymers useful for the latter fibers contain principally para-oriented phenylene rings. The meta-oriented polymers are generally considered to be chain-folding polymers in contrast to the rodlike para-oriented polymers which show no evidence of chain-folding.

Aramid fibers have tensile strengths at 250°C that are characteristic of conventional textile fibers at room temperature and have useful tenacities to > 300°C. In contrast, nylon-6,6 fiber loses almost all of its strength at ca 205°C. The tensile moduli of the aramid fibers fall off considerably at 300–350°C; the modulus values are still substantial, however, even at these temperatures. Surprisingly, the elongations of these fibers tend to be less at 300°C than at room temperature. Evaluation of tensile properties after heat-aging at 300°C indicates that aramid fibers retain useful tensile properties for 1–2 weeks under such conditions.

Aramid fibers characteristically burn only with difficulty and do not melt like nylon-6,6 or polyester fibers. Upon burning, the aramid fibers produce a thick char which acts as a thermal barrier and prevents serious burns to the skin. Some aramid fibers, such as Nomex, shrink away from a flame or a high heat source. Durette fabrics, based on Nomex and treated with hot chlorine gas or other chemical reagents to promote surface cross-linking which stabilizes the fibers, were developed for greater dimensional stability.

Wholly aromatic polyamides have high volume resistivities and high dielectric strengths, and significantly, they retain these properties at elevated temperatures to a high degree. Accordingly, they have considerable potential as high temperature dielectrics, particularly for use in motors and transformers. The high temperature electrical properties of aramid fibers are exemplified by breakdown voltages of 76 V/mm at temperatures up to 180°C. By comparison, the breakdown voltage for polytetrafluoroethylene is only 57 V/mm at 150°C (see Insulation, electric).

Paper of Nomex has almost twice the dielectric strength of high quality rag paper and retains its useful electrical properties at higher operating temperatures than the upper limit for the rag paper (ca 105°C) (see Pulp, synthetic). In the early years of the commercialization of Nomex, it is almost certain that the production of Nomex in the form of paper equaled or outstripped the production of Nomex in the form of fiber.

Although aramid fibers are much more resistant to acid than nylon-6,6 fibers, they are not as acid resistant as polyester fibers, except at elevated temperatures, particularly in the range of pH 4–8. Their resistance to strong base is comparable to that of nylon-6,6 fibers, and hydrolytic stability of the aramids is superior to that of polyester and comparable to nylon-6,6. Resistance of aramids to ionizing radiation is greatly superior to that of nylon-6,6.

Aramid fibers, like their aliphatic counterparts, are susceptible to degradation by uv light. The mechanism of photodegradation of the aramids is quite different from that of the aliphatic polyamides and presents greater problems regarding the development of color.

Aramid fibers are exceedingly difficult to dye, probably as a consequence of their high glass-transition temperatures. One procedure for dyeing Nomex makes use of a dye carrier (qv) at an elevated temperature (ca 120°C) in pressurized beam and jet-dyeing machines. Cationic dyes, which are used exclusively, must be selected carefully in order to achieve reasonable lightfastness.

Fibers from the rodlike polymers have rather low elongations, and they lose elongation rapidly after being heated at an elevated temperature, thereby becoming too brittle to be useful. However, for very short-term exposure, retention of strength at elevated temperatures is outstanding. Flame resistance of aramid fibers from rodlike polymers also is high, but a flame tends to propagate along a vertically held fiber of this class because of the grid formed by the residual char. Incorporation of phosphorus has been used to raise the flame resistance to very high levels for fiber from poly(p-phenyleneterephthalamide) (PPD-T) which has not been highly drawn during spinning. The high flame resistance of PPD-T fabrics containing phosphorus coupled with their inherent high resistance to shrinkage can contribute significantly to providing insulation and protection from burns.

Two important features of the ultra-high strength, high modulus fibers are their fine diameter and their production from rodlike polymers to give extremely high intrinsic and inherent viscosity values and moderate molecular weights. The situation regarding fine linear densities is not unlike that for glass fiber, and stress–strain curves for ultra-high strength, high modulus fibers also show a strong similarity to those for glass and steel.

Most wholly aromatic polyamide fibers are spun from solutions of the polymers. Both dry-spinning and wet-spinning techniques appear to be equally satisfactory for fiber formation from polymers that are soluble in organic solvents, although properties may vary with the method of spinning. The molecular weight of a given polymer generally should be higher when spun by the wet method and should be $\geq 60,000$ for the best balance of initial tensile properties and retention of tensile properties after the fiber is placed in service. Almost all wholly aromatic polyamides require the presence of a salt, either organic or inorganic, in the amide solvents to dissolve the polymers if they are prepared by interfacial polymerization, or to prevent them from precipitating if prepared by solution polycondensation. The salt, however, must be extracted since any residual salt in the fiber will lead to poorer thermal stability and poorer electrical properties. The yarn must be dried and is usually hot-drawn whether spun by a wet or dry process. An important exception to the need for a hot-drawing step is for high modulus fibers spun from an anisotropic solution for which hot drawing can result in more than a twofold increase in initial modulus, but at a sacrifice in elongation-to-break without an attendant appreciable increase in tenacity.

The two types of aramid fibers, ie, heat-resistant fibers (eg, Nomex and Conex) and ultra-high strength, high modulus fibers (eg, Kevlar and Arenka), are relatively expensive compared to other synthetic fibers such as nylon and polyester because of the relatively low volume production of both types of aramid fibers, the high cost of monomers employed, and high processing costs.

Typical applications for aramid fibers are listed according to the distinctive properties of these fibers: (1) heat resistance: filter bags for hot stack gases (see Air pollution control methods), press cloths for industrial presses, home ironing-board covers, sewing thread for high speed sewing, insulation paper for electrical motors and transformers (see Insulation, electric), braided tubing for insulation of wires, papermakers' dryer belts; (2) flame resistance: industrial protective clothing, fire-department turnout coats, pants, and shirts, flight suits for military pilots, jump suits for forest-fire fighters, pajamas and robes, mailbags, carpets, upholstery, drapes, boat covers, and tents (see Flame retardants in textiles); (3) dimensional stability: reinforcement of fire hose and V belts by aramid fibers of moderately high modulus (eg, Nomex), conveyor belts; (4) ultra-high strength, high modulus: tire cord for use in tire carcasses and as the belt in bias-belted and radial-belted tires (see Tire cords), V belts, cables, parachutes, body armor, rigid reinforced plastics, radomes and antenna components, circuit boards, filament-wound vessels, fan blades, sporting goods; (5) electrical resistivity: electrical insulation (paper and fiber); (6) chemical inertness: use in filtration (qv); (7) permoselective properties: hollow-fiber permeation separation membranes to purify seawater or brackish water or to make separation of numerous types of salts and water (see Hollow-fiber membranes).

In recent inhalation studies, hexamethylphosphoric triamide (HPT) used in the preparation of rodlike polyamides has been found to be carcinogenic in rats. Accordingly, any work with HPT should be carried out in an efficient hood to avoid the inhalation of vapors. Since HPT is absorbed by the skin, care should be taken to avoid contact with the substance. Despite the fact that the HPT solvent is washed from the PPD-T prior to drying the polymer and that the fibers spun from sulfuric acid are vigorously washed with water, some HPT solvent apparently persists in the fiber.

No adverse mechanical or chemical skin effects are to be expected from the usual industrial handling of either Kevlar or Nomex. For Nomex, and especially for Kevlar, inhalation of the "fly" generated in certain textile-handling operations should be avoided. Also, because of its unusual high strength and cutting resistance, caution should be exercised in splicing, handling, or cutting Kevlar. Manual cutting and splicing should be attempted only with stationary yarns to avoid possible injury from entanglement in moving yarn or fabric.

J. PRESTON
Monsanto Triangle Park Development Center, Inc.

W.B. Black and J. Preston in H.F. Mark, S.M. Atlas, and F. Cernia, eds., *Man-Made Fibers: Science and Technology*, Vol. 2, Interscience Publishers, a division of John Wiley & Sons, Inc., New York, 1968, pp. 297–364.

J. Preston in A. Blumstein, ed., *Liquid Crystalline Order in Polymers*, Academic Press, Inc., New York, 1978.

ARGON. See Helium-group gases.

ARSENIC AND ARSENIC ALLOYS

Arsenic, although often referred to as a metal, is classified chemically as a nonmetal or metalloid and belongs to Group VA of the periodic table. The principal valences of arsenic are $+3$, $+5$, and -3. Only one stable isotope of arsenic with mass 75 (100% natural abundance) has been observed. Elemental arsenic normally exists in the α-crystalline metallic form (see Table 1 for physical properties), which is steel gray in appearance and brittle, and in the β-form, a dark-gray amorphous solid. Arsenic is widely distributed about the earth and has a terrestrial abundance of ca 5 g/t.

Metallic arsenic can be obtained by the direct smelting of the minerals arsenopyrite or loellingite. The metal also can be prepared commercially

Table 1. Physical Properties of Arsenic

atomic weight (^{12}C = 12.0000)	74.9216
mp at 3.91 MPa (38.6 atm), °C	816
bp, °C	615, sublimes
density at 26°C, g/cm^3	5.778
latent heat of fusion, J/(mol·K)^a	27,740
specific heat at 25°C, J/(mol·K)^a	24.6
electrical resistivity at 0°C, $\mu\Omega\cdot$cm	26
crystal system	hexagonal (rhombohedral)
lattice constants at 26 C, nm	a = 0.376, c = 1.0548
hardness, Mohs scale	3.5

^{a}To convert J to cal, divide by 4.184.

by the reduction of arsenic trioxide with charcoal. Since the demand for metallic arsenic is limited, however, arsenic usually is marketed in the form of the trioxide, referred to as white arsenic, arsenious oxide, arsenious acid anhydride, and also by the misnomer arsenic. Arsenic trioxide is volatilized during the smelting of copper and lead concentrates, and is therefore concentrated with the flue dust.

Commercial arsenic metal is sold at a typical purity of 99% and in lump or fragmented form (5–7.5 cm). Both As_2O_3 and metallic arsenic are produced by ASARCO in the United States.

The main use of arsenic in the United States is in agricultural chemicals with wide application in cotton farming. Refined arsenic trioxide is used as a decolorizer and fining agent in the production of glassware. Arsenic acid is used as a feed additive for poultry. Sodium arsenite solutions have been used in cattle and sheep dips, for debarking trees, and for aquatic weed control. Arsenicals also are used in preparations for medicinal veterinary use. A limited but important demand for arsenic metal of exceedingly high purity ($\geq 99.999\%$) exists for semiconductor-device applications. The main use of arsenic metal is in alloys in combination with lead and to a lesser extent with copper.

OSHA's standard for employee exposure, in effect since Aug. 1, 1978, is $10 \ \mu g/m^3$.

S.C. CARAPELLA, JR.
ASARCO, Inc.

American Institute of Physics Handbook, McGraw-Hill Book Co., Inc., New York, 1957, Sect. 7, Chapt. 9.

S. deLajarte, *Arsenic in Glass*, Arsenic Development Committee, Rue LaFayette, Paris, France, Mar. 1969.

D.D. Nicholas, ed., *Preservative and Preservative Systems*, Vol. II, Syracuse University Press, Syracuse, N.Y., 1973, pp. 66–84.

Chem. Eng. News, 6 (May 8, 1978).

ARSENIC COMPOUNDS

Arsenic is the third member of the nitrogen family of elements and hence possesses five electrons in the outermost shell: $4s^2, 4p^3$. The most common oxidation states of arsenic are -3, $+3$, and $+5$, although compounds containing the simple As^{3-}, As^{3+}, and As^{5+} ions are unknown. In most arsenic compounds, the arsenic atom is in the tetrahedral hybridized valence state. Compounds in which the arsenic atom is three-coordinate are assumed to contain the tetrahedrally hybridized arsenic atom with a lone pair of electrons in one of the hybrid orbitals. Since the valence shell of the arsenic atom contains d orbitals, compounds are known in which the arsenic atom adopts trigonal-bipyramidal or octahedral valence states.

Inorganic Compounds

Arsenic hydrides. Although there are occasionally reports in the literature of other arsenic hydrides, the only well-characterized binary compound of arsenic and hydrogen is arsine, AsH_3, mp $-116°C$, bp $-62.4°C$, a colorless, exceedingly poisonous gas with an unpleasant garlic-like odor. Arsine may be formed accidentally by the reaction of arsenic impurities in commercial acids stored in metal tanks, so that a test should be made for arsine before entry is made into such vessels. It is injurious in 1 : 20,000 dilution, and a few inhalations may cause death from anoxia or pulmonary edema. Although arsine is toxic to higher living forms, some strains of bacteria and fungi are capable of producing arsine or arsine derivatives.

Other arsenic hydrides include diarsine, As_2H_4, a by-product in the preparation of arsine by treatment of a magnesium aluminum arsenide alloy with dilute sulfuric acid; arsenic monohydride, As_2H_2 or AsH, a brown amorphous powder; and hydrogen diarsenide, As_4H_2.

Arsenic halides. Arsenic forms a complete series of trihalides, but arsenic pentafluoride is the only simple pentahalide known. All of the arsenic halides are covalent compounds that hydrolyze in the presence of water: arsenic trifluoride, AsF_3, mp $-8.5°C$, bp $60.4°C$, a colorless liquid;

arsenic pentafluoride, AsF_5, mp $-88.7°C$, bp $-58.2°C$, a colorless gas; arsenic trichloride, $AsCl_3$, mp $-13°C$, bp $130.2°C$, a colorless liquid; arsenic tribromide, $AsBr_3$, mp $31°C$, bp $221°C$, a yellow solid; arsenic triiodide, AsI_3, mp $140.7°C$, bp ca $400°C$, a red solid; and arsenic diiodide, As_2I_4, mp $130°C$, a red solid.

Arsenic oxides and acids. The only oxides of commercial importance are the trioxide and the pentoxide: arsenic trioxide (arsenic(III) oxide, arsenic sesquioxide, arsenous oxide, white arsenic, arsenic), As_2O_3, the most important arsenic compound of commerce: arsenous acids and arsenites; arsenic pentoxide (arsenic oxide, arsenic(V) oxide), As_2O_5, mp ca $300°C$; arsenic acid and arsenates (arsenates of calcium or lead are used as insecticides; sodium arsenate is used in printing inks and as a mordant); and arsenous arsenate (arsenic dioxide, arsenic tetroxide), As_2O_4.

Arsenic sulfides. Arsenous sulfide (arsenic(III) sulfide, arsenic sesquisulfide, arsenic red), As_4S_6, mp $320°C$, bp $707°C$, is a yellow solid; arsenic sulfide, As_4S_4, mp $307°C$, bp $565°C$, occurs naturally as the mineral realgar and is a red or orange solid; arsenic pentasulfide, As_4S_{10}, is a yellow solid.

Other arsenic compounds include arsenic trisulfate, $As_2(SO_4)_3$, arsenyl sulfate, $(AsO)_2SO_4$, and arsenic triacetate, $As(C_2H_3O_2)_3$, mp $82°C$, bp $165–170°C$.

Organic Compounds

Since arsenic is a metalloid (electronegativity 2.0 on the Pauling scale), it combines readily with carbon to form a wide variety of compounds containing the As–C bond. Such compounds are now largely synthesized for studies of their structure and stereochemistry or as ligands in coordination chemistry, although they also have considerable commercial use as medicinal agents, herbicides, and catalysts in polymerization processes. Organoarsenic compounds can be divided roughly into those derived from trivalent and those derived from pentavalent arsenic. In general, the former are considerably more toxic than the latter.

Arsines and their derivatives. The two primary arsines methylarsine, CH_3AsH_2, and trifluoromethylarsine, CF_3AsH_2, and the secondary arsine bis(trifluoromethyl)arsine, $(CF_3)_2AsH$, are gases at room temperature; all other primary and secondary arsines are liquids or solids. Tertiary arsines, R_3As, are widely used as ligands in coordination chemistry (see Coordination compounds).

Haloarsines and dihaloarsines, R_2AsX and $RAsX_2$, are generally liquids or low melting solids with irritating effects on the eyes and mucous membranes of the mouth and nose.

Haloarsoranes. Halides of the types $RAsX_4$, R_2AsX_3, R_3AsX_2, and R_4AsX are known.

Arseno compounds, $(RAs)_n$, and diarsines, $(R_2As)_2$, contain As–As bonds. The aromatic arseno compounds at one time had widespread use in medicine as antisyphilitic agents.

Other compounds include arsenins, arsonous and arsinous acids, arsonic and arsinic acids (used as corrosion inhibitors for steel, additives to motor fuel, agricultural bactericides, herbicides, and fungicides), arsine oxides, arsonium salts (used as analytical reagents for anions such as perchlorate, perrhenate, and permanganate) and ylids and pentaalkyl- and pentaarylarsoranes, R_5As.

Arsenic compounds have numerous practical applications as tanning agents, corrosion inhibitors, catalysts, animal-feed additives, veterinary medicines, wood preservatives, and pharmaceuticals. Though once considered a powerful therapy against syphilis, arsenicals today are considered obsolete in the treatment of that disease. They are still used, however, against amebic dysentery and are indispensable for the treatment of the late neurological stage of African trypanosomiasis. Melarsoprol [2-[4-(4,6-diamino-1,3,5-triazin-2-yl)amino]phenyl]-1,3,2-dithiarsolane-4-methanol; Mel B; arsobal) is the drug of choice for human trypanosomiasis caused by *Trypanosoma gambiense* or *T. rhodesiense* and has been employed successfully for the treatment of tryparsamide-resistant infections of the nervous system. It often shows toxic effects and should be used only with strict medical supervision. Trypanosomes are becoming resistant to this drug. Toxic reactions caused by arsenicals often can be treated with the antidote, 2,3-dimercapto-1-propanol (dimercaprol; BAL).

Most arsenic compounds are highly toxic. According to OSHA, excess exposure may cause dermatitis, acute or chronic poisoning, and perhaps cancer. Workers exposed to inorganic arsenicals include those in the copper, lead, and zinc smelting industries, and in factories producing arsenical pesticides and wood preservatives.

G.O. DOAK
G. GILBERT LONG
LEON D. FREEDMAN
North Carolina State University

P.J. Durrant and B. Durrant, *Introduction to Advanced Inorganic Chemistry*, 2nd ed., John Wiley & Sons, Inc., New York, 1970, p. 756.

Office of Toxic Substances, *Summary Characterizations of Selected Chemicals of Near-Term Interest*, U.S. EPA, Washington, D.C., 1976, pp. 1–2.

Chem. Eng. News, 4 (June 28, 1976).

G.O. Doak and L.D. Freedman, *Organometallic Compounds of Arsenic, Antimony and Bismuth*, John Wiley & Sons, Inc., New York, 1970.

ASBESTOS

Asbestos is a generic term describing a variety of naturally formed hydrated silicates that, upon mechanical processing, separate into mineral fibers. There are two fundamental varieties of asbestos: serpentine and amphiboles. Serpentine asbestos is known as chrysotile, and the amphiboles include five species identified as anthophyllite, amosite, crocidolite, actinolite, and tremolite. The serpentine group all have the approximate chemical composition of $Mg_3(Si_2O_5)(OH)_4$. The structure of all the amphiboles consists of two chains or ribbons based on Si_4O_{11} units separated by a band of cations. Of mineral origin, asbestos does not burn, does not rot, and dependent on variety, possesses extremely high tensile strength as well as resistance to acids, bases, and heat. Similarly, when processed into long, thin fibers, asbestos is sufficiently soft and flexible to be woven into fire-resistant fabrics. (See Table 1 for physical properties.)

Imbedded asbestos fibers are removed from the ore by a repeated series of crushing, fiberizing, screening, aspirating, and grading operations. In general, a 5% recovery of milled asbestos fiber from ore is typical. The chrysotile mines in California are notable exceptions where 50% recoveries are common. The fibers, however, are short and normally sold as reinforcing fillers (qv).

Most of the world's chrysotile is produced by Canada and the USSR. The main producing area of the important amphiboles, crocidolite, and amosite is Africa.

The largest use of asbestos is in asbestos cement for products such as pipes, ducts, and flat and corrugated sheets. Other uses include fire-resistant textiles, friction materials (see Brake linings), underlayment and roofing paper, and floor tiles.

Three diseases are associated with inhalation of asbestos: asbestosis, bronchogenic carcinoma, and mesothelioma. Reduction of asbestos dust exposure is the only known method of preventing disease among asbestos industry workmen. There are governmental regulations (OSHA) that describe the allowable airborne fiber levels in work areas (see Regulatory agencies).

WILLIAM C. STREIB
Manville Corp.

ASBESTOS-CEMENT PRODUCTS. See Cement.

ASBESTOSIS. See Industrial hygiene and toxicology.

ASCORBIC ACID. See Vitamins.

Table 1. Properties of Asbestos Fibers

	Chrysotile	Anthophyllite	Amosite (ferroanthophyllite)	Crocidolite	Tremolite	Actinolite
structure	in veins of serpentine, etc	lamellar, fibrous asbestiform	lamellar, coarse to fine fibrous and	fibrous in iron-stones	long, prismatic and fibrous aggregates	reticulated long prismatic crystals
color	white, gray, green, yellow	gray–white, brown, gray, or green	ash gray, green, or brown	lavender, blue	gray–white, green, yellow, blue	green
luster	silky	vitreous to pearly	vitreous, somewhat pearly	silky to dull	silky	silky
hardness, Mohs	2.5–4.0	5.5–6.0	5.5–6.0	4	5.5	ca 6
specific gravity	2.4–2.6	2.85–3.1	3.1–3.25	3.2–3.3	2.9–3.2	3.0–3.2
cleavage	010 perfect	110 perfect	110 perfect	110 perfect	110 perfect	110 perfect
optical properties	biaxial positive extinction parallel	biaxial positive extinction parallel	biaxial positive extinction parallel	biaxial extinction inclined	biaxial negative extinction inclined	biaxial negative extinction inclined
refractive index	1.50–1.55	ca 1.61	ca 1.64	1.7 pleochroic	1.61	1.63 weakly pleochroic
fusibility, Seger cones	fusible at 6, 1190–1230°C	infusible or difficult to fuse	fusible at 6, loses water at moderate temperatures	fusible at 3, 1145–1170°C	fusible at 4, 1165–1190°C	fusible at 4, 1165–1190°C
flexibility	very flexible	very brittle, nonflexible	good, less than chrysotile	fair to good	generally brittle, sometimes flexible	brittle and nonflexible
length	short to long	short	5–28 cm	short to long	short to long	short to long
texture	soft to harsh, also silky	harsh	coarse but somewhat pliable	soft to harsh	generally harsh, sometimes soft	harsh
acid resistance	soluble up to approximately 57%	fairly resistant to acids	fairly resistant to acids	fairly resistant to acids	fairly resistant to acids	relatively insoluble in HCl
spinnability	best	poor	fair	fair	generally poor, some are spinnable	poor
specific heat, J/(kg·K) [or Btu/(lb·°F)]	1113 [0.266]	879 [0.210]	908 [0.217]	841 [0.201]	887 [0.212]	908 [0.217]

ASPARAGINE. See Amino acids.

ASPARTIC ACID. See Amino acids.

ASPHALT

Asphalt is a dark brown to black cementitious material in which the predominating constituents are bitumens that occur in nature or are obtained in petroleum processing. Asphalts characteristically contain very high molecular weight hydrocarbons called asphaltenes and are essentially soluble in carbon disulfide, and aromatic and chlorinated hydrocarbons. Bitumen is a generic term for a class of black or dark-colored (solid, semisolid, or viscous) cementitious substances, natural or manufactured, composed principally of high molecular weight hydrocarbons, of which asphalts, tars, pitches, and asphaltites are typical.

Prior to 1907, most of the asphalt used occurred naturally and included native asphalts, rock asphalts, and lake asphalts. Since the early 1900s, however, most asphalts have been produced from the refining of petroleum and used primarily in paving and roofing applications. Unlike native asphalts, petroleum asphalts are organic with only trace amounts of inorganic materials.

At normal service temperatures, asphalt is viscoelastic; at higher temperatures, it becomes viscous. The disperse phase is a micelle of asphaltenes and the higher molecular weight aromatic components of the petrolenes. Determination of the components of asphalts has always presented a challenge because of the complexity and high molecular weights of the hydrocarbons present. The component of highest carbon content is the fraction termed carboids, which is insoluble in carbon disulfide. This fraction, although organic, is nonasphaltic. The so-called carbenes are insoluble in carbon tetrachloride and soluble in carbon disulfide. Both carboids and carbenes, if present, occur in small amounts. Asphaltenes have a great influence on the viscosity of asphalt. They seem to be relatively constant in composition in residual asphalts, despite the source, as determined by carbon–hydrogen analysis. The nonasphaltene components of asphalt are called maltenes or petrolenes. Properties of asphalts appear in Table 1.

Asphalts are used as protective films, adhesives, and binders because of their waterproof and weather-resistant properties. Some movement without fracture can occur because of their viscous (sol) nature. They have long and continuous satisfactory service because of their slow rate of hardening from heat, oxidation, fatigue, and weathering. Exposed asphalt films harden partially from a loss of volatile oils and to a greater extent from the formation of additional asphaltene fractions and loss of maltenes through oxidation. Such chemical change undoubtedly is catalyzed by uv irradiation. Recent studies have indicated that asphalt stiffness can be used in optimizing performance, although fundamental measures of mechanical properties are preferable. A stiffer asphalt, under uniform loading conditions, could reduce pavement deflection, extend fatigue life, and allow less flow deformation. A softer material would normally allow a longer weathering life before the maltene–asphaltene composition becomes critical in service. Usually, the softest material allowed by initial service needs is selected.

The water resistance of asphalt films is also a manifestation of durability. Asphalts that have a low content of soluble salts show a low water absorption. The pickup of water is primarily a surface manifestation; it softens the film and can cause blistering. Even with a high rate of absorption, asphalt films show little loss of bond to surfaces on continued immersion in water and continue to protect metals from corrosion for long periods of time. Bacteria and fungi can attack the very low molecular weight portion of bituminous materials.

Mineral fillers often are added to asphalts to influence their flow properties and reduce costs. They are used commonly as stabilizers in roofing coatings at concentrations up to 60 wt%. Mineral-filled films show improved resistance to flow at elevated temperatures, improved impact resistance, and better flame-spread resistance. Fillers may increase the water absorption of asphalts. Mineral fillers commonly used are ground limestone, slate flours, finely divided silicas, trap rocks, and mica; they often are produced as by-products in rock-crushing operations. Opaque fillers offer protection from weathering. Asbestos filler has special properties because of its fibrous structure, high resistance to flow, and toughness. It has been used in asphalt paving mixes to increase the

Table 1. Properties of Asphalts

Property	Straight-reduced, residual	Thermal	Air-blown residual
softening point (ring and ball), °C	46	113	93
penetration of 100 g at 25°C for 5 s, mm/10	90	0	20
ductility at 25°C, 5 cm/min, cm	≥ 150	too hard	3.2
specific gravity, 15.6/15.6°C	1.03	1.12	1.05
mean coefficient of cubical expansion/°C			
15.6–65.6°C	0.00063	0.00058	0.00063
15.6–232°C	0.00068	0.00063	0.00068
specific heat, J/(kg·K)[a]			
4.4°C	1675	1549	1633
93.3°C	1968	1842	1926
204.4°C	2345	2177	2303
thermal conductivity at 26.7°C, W/(m·K)	0.16	0.16	0.16
permeability constant at 25°C, kg·m/(m²·s·Pa)[b]			
water vapor	0.62–1.93×10^{-15}	1.1×10^{-15}	1.25–2.4×10^{-15}
oxygen			0.08×10^{-15}
water absorption of 10-mil films on aluminum panels, wt%			
50 weeks			1.5–10
100 weeks			2.5–16.5
surface tension, mN/m (= dyn/cm)			
25°C	34		32
100°C	27		28
dielectric strength, spherical electrodes, V/m	11–45×10^6	36×10^6	30–35×10^6
dielectric constant, 50 Hz at 20°C	2.7	3.0	2.7

[a] To convert J to cal, divide by 4.184.
[b] To convert Pa to mm Hg, multiply by 0.0075.

resistance to movement under traffic and in roofing materials for fire-retardant purposes (see Fillers).

Petroleum-derived asphalt, which represents > 99% of total asphalt and asphalt products sold in the United States, is manufactured by the following methods:

Straight reduction. Crude oil at 340–400°C is injected into a fractionating column. The lighter fractions are separated as overhead products, and the residuum is straightreduced asphalt. Crude oil containing ca 30% or more of asphalt can be refined completely in an atmospheric unit to an asphalt cement product. However, most crude oil cannot be distilled at atmospheric pressure because of high percentages of high boiling fractions. As a supplement to the atmospheric process, a second fractionating tower (a vacuum tower) is added. This two-stage process is particularly applicable to crude oils containing 18–30% asphalt. Straight-reduced asphalts are used mainly in pavements, where they serve primarily as binders in paving mixes. The most important recent technical innovation in asphalt paving has been to use asphalt throughout the entire pavement structure (termed total asphalt) to provide more efficient and economical distribution of traffic stresses to the subgrade and provide better protection of the base from intrusion of outside materials, eg, water, soil, etc.

Air-blowing. Asphalt stock (flux) is converted to a harder product by air contact at 200–275°C. Air-blown asphalts are generally more resistant to weather and changes in temperature than straight-reduced asphalts and are produced by batch and continuous methods. Air-blown asphalts of diverse viscosities and flow properties with added fillers, polymers, solvents, and in water emulsions provide products for many applications in roofing and other industries. Air-blowing is also used to produce the harder paving-asphalt grades when the crudes available have a low asphalt content and cannot be reduced directly to grade.

Propane deasphalting involves the precipitation of asphalt from a residuum stock by treatment with propane under controlled conditions. The petroleum stock is usually atmospheric-reduced residue from a primary distillation tower. Propane usually is used in this process although propane–butane mixtures and pentane have been used with some variation in process conditions and hardness of the product. Propane deasphalting is used primarily for crude oils of relatively low asphalt content, generally ≤ 12%. Asphalt produced from this process is blended with other asphaltic residua for making paving asphalt.

Thermal asphalts differ from other asphalts in that they are products of a cracking process. They have relatively high specific gravity, low viscosity, and high temperature susceptibility, and they contain cokelike bodies (carbenes) as indicated by the spot test. Thermal asphalts are used principally as saturants for cellulosic building products such as insulation boards, brick-finish siding, and fiber soil pipes. Currently, their use in road asphalts is rare. Thermal asphalt actually is in very short supply because of changes in cracking methods, and there is little likelihood that it will ever become commonly available.

Blended asphalts may be produced when a refinery stocks two grades of asphalt, one at each end of the viscosity spectrum of the entire product grade requirements. Intermediate grades are prepared by blending (proportioning) the extremes.

Emulsions are immiscible liquids dispersed in one another in the form of very fine droplets from ca 1–25 μm and an average of 5 μm dia. In the most common asphalt emulsion, ie, the oil-in-water type, the asphalt is the dispersed (internal) phase, and water is the continuous (external) phase (see Emulsions). Colloid mills are most commonly used for the manufacture of road emulsions in the United States. A colloid mill usually consists of a rapidly revolving conical disk (rotor). The asphalt, water, and emulsifying agent are forced through the narrow clearance between the rotor and the stator (stationary section).

Industrial emulsions have applications outside the road-building industry. They are made with harder grades of asphalt and contain clays, casein, gelatin, or blood albumin as peptizing agents. Certain clays, such as bentonite, are good emulsion dispersants and impart a buttery consistency to the emulsion. These emulsions have a wide variety of applications, such as in surface coating of asphalt pavements, for built-up roofs, and for other weather coverings.

The large demands for asphalt as a building material were created primarily by mass production of the automobile and the development of asphalt roofing materials (qv) for home construction. The use of asphalt in pavement base-course construction (instead of untreated aggregates), hydraulics, rapid growth in home construction, and the interstate road system have greatly increased its use.

In recent years, the paving market has consumed ca 80% of the product. Asphalt has been used to surface 94% of the United States' highways. The roofing industry typically has accounted for ca 15% of the asphalt market, and miscellaneous industrial asphalts make up the remaining 5%. These products usually are classified only by types (Bureau of Mines), eg, liquid, solid, emulsion; or by use, eg, laminates, pipe coating, automotive asphalts, etc.

The petroleum industry can produce larger quantities of asphalt by adjusting the use of the residual product from refining processes. The integrated refineries have alternative uses for crude-oil residua, ie, coke and residual fuel oil. Very recently, technology improvements have allowed the use of crude residua in existing catalytic cracking units which, in the absence of sufficient distillates normally used for this purpose, places a high alternative value on residua without the attendant need for capital to expand coking facilities. The value of residua for coking and residual fuel establishes a basis for asphalt prices. Asphalt is the preferred product from high sulfur crude stocks because it is a construction material and does not require desulfurization for use as a fuel. Asphalt yields from three crudes are shown in Figure 1.

No significant air-pollution problems are associated with emissions from hot paving operations using several asphalt cements. Concentrations of gaseous substances and emissions from paving-asphalt cement have been found to be very low and within existing EPA and OSHA standards, even when the ambient air sampling was done under confined conditions. Asphalt's very minor content of high molecular weight polynuclear aromatic constituents, however, has been studied as a possible health hazard. The conclusion from these studies suggests that unlike tars, asphalt can be classified in the same manner as particulate dust, but surveillance should be continued although asphalt has not been shown to be a material of significant hazard.

Steps to minimize potential safety hazards in the handling of asphalt are set forth by the American Petroleum Institute and the Asphalt Institute. Hazards include sudden pressure increases from hot asphalt in contact with moisture in enclosed tanks or transports, exposure to air at

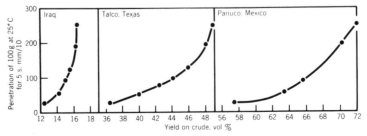

Figure 1. Asphalt yield from three crudes.

$\geq 150°C$, local overheating above heating coils, flashing of asphalt volatiles in the presence of an ignition source or possible auto-ignition, and hydrogen sulfide from high temperature operations.

JAMES V. EVANS
Amoco Oil Co.

H. Abraham, *Asphalts and Allied Substances*, 6th ed., Vol. 1, D. Van Nostrand Co., Inc., Princeton, N.J., 1960, Chapt. 2.

R.N. Traxler, *Asphalt*, Reinhold Publishing Corp., New York, 1961.

A Brief Introduction to Asphalt and Some of Its Uses, Manual Series No. 5, The Asphalt Institute, College Park, Md., 1975.

A.J. Hoiberg, *Bituminous Materials: Asphalts, Tars, and Pitches*, Vols. 1, 2, and 3, Interscience Publishers, a division of John Wiley & Sons, Inc., New York, 1964–1966.

ASPHALT TILE.   See Plastic building products.

ASPIRIN, o-$CH_3COOC_6H_4COOH$.   See Analgesics, antipyretics, and anti-inflammatory agents; Salicylic acid and related compounds.

ASTATINE.   See Radioactivity, natural.

AUTOMATION.   See Instrumentation and control.

AUTOMOBILE EXHAUST CONTROL.   See Exhaust control.

AUTOXIDATION.   See Antioxidants and antiozonants.

AUXINS.   See Plant growth substances.

AVIATION AND OTHER GAS-TURBINE FUELS

The first fuels burned in a turbine combustor were petroleum liquids, ie, kerosene in Whittle's prototype jet engine, and diesel in the Brown-Boveri industrial turbine, selected because of their availability and convenience.

Gas-Turbine Products

Because the jet aircraft is a weight-limited vehicle, a high premium is assigned to hydrocarbon fuels with a maximum gravimetric heat content or hydrogen-to-carbon ratio. A high hydrogen content in hydrocarbons is consonant with straight-chain paraffins. However, because the density of these fuels is low, the heat content on a volumetric basis is less than that of nonparaffinic fuels of the same boiling range. Fuels are measured and sold on a volumetric basis. Hence, the aircraft user obtains more energy for each cubic meter of a less paraffinic fuel. Kerosene, distillation range 150–300°C, is the best compromise to combine maximum mass and heat content with other desirable properties.

The ground-turbine user is not concerned with weight limitations and seeks maximum energy per unit of cost. This usually means that the heavier nonparaffinic fuels with high heat content per unit of volume are more economical for ground turbines.

Liquid fuels for ground-based gas turbines are best defined by ASTM Specification D 2880, which is the basis for an international standard. For aircraft fuels, specifications tend to be industry standards issued by a government body or a consensus organization like ASTM rather than manufacturer's requirements. Civil aviation fuels are defined by ASTM Specification D 1655.

Because of specification limitations on olefins and aromatics, aviation turbine fuels are blended from straight-run distillates rather than cracked stocks. Fuels for ground turbines are equivalent to their fuel-oil counterparts and are manufactured as dual-purpose products.

In a typical refinery processing sequence, the basic processes after distillation are catalytic reforming for gasoline, catalytic cracking of gas oil, and hydrogen processing of naphtha, kerosene, and gas oil using the hydrogen from reforming. This refinery sequence maximizes gasoline rather than distillates and is less energy intensive than one that converts heavy gas oil and residuals to lighter products by hydrocracking or coking processes (see Petroleum).

The distillation cut-points must be closely controlled to yield a product that meets the requirements of flash point on the one hand and the freezing point on the other. In practical terms, a kerosene for jet aircraft requires a fraction of ca 160°C initial boiling point and final boiling point not exceeding 300°C.

The distilled fractions from the crude are apt to contain mercaptans or organic acids in excess of the specification limits. A common process step in the refinery is caustic washing to remove mercaptans and acidity. This usually is followed by either a water-washing step or clay-bed absorption of residual mercaptides and soaps that have not settled. In some cases, a sweetening process is used to convert mercaptans into disulfides rather than caustic washing to remove them. The newest process, developed for gasoline and applied to turbine fuels, involves catalytic oxidation of mercaptans with dissolved air over a fixed bed. Mercaptan oxidation, Merox, has the advantage of simplicity and minimum waste-disposal problems, but it does not lower the sulfur content of the product. Most of the crudes available in greatest quantity today, however, are high in sulfur and yield products that must be desulfurized to meet specifications or environmental standards. The process most widely used as a replacement for acid treating for desulfurization is mild catalytic hydrogenation. Some of the oxygen-containing species that are removed by hydrogen treating act as natural inhibitors of hydrocarbon oxidation by trapping peroxy radicals. Recognition of this by-product of hydrodesulfurization has led refiners to add antioxidants (qv) to hydrotreated fractions.

Preserving the quality of gas-turbine fuels between the refinery and the point of use is difficult because of the complicated distribution systems of multiproduct pipelines and tankers that move fuel and sometimes introduce contaminants such as water, corrosion products, metal salts, microorganisms, and other extraneous materials into the fuel. The principal means of removing contaminants are tank settling and filtration.

Quality-control limits must be met by third parties who distribute and handle fuel. Tests on receipt at terminals include appearance, distillation, flash point (or vapor pressure), density, freezing point, smoke point, corrosion, existing gum, water reaction, and water separation. Tests on delivery to the aircraft include appearance, particulates, membrane color, free water, and conductivity.

The extensive processing and cleanup steps carried out on gas-turbine fuels produce a purified liquid dielectric of high resistivity which is capable of retaining electrical charges long enough for buildup of large surface voltages. Several approaches have been taken to reduce the risk of static discharge during delivery which may result in tank explosions: introduction of an additive to increase the electrical conductivity of the fuel to a safe level; modification of filters to substitute screens for filter media, thereby lowering the amount of charge and allowing more residence time for the charge to disperse; and reduction in flow rate through filters to reduce the current carried by the fuel (see also Antistatic agents).

Performance

The primary reaction carried out in the gas-turbine combustion chamber is oxidation of a fuel to release its heat content at constant pressure. Atomized fuel mixed with enough air to form a close-to-stoichiometric mixture is continuously fed into a primary zone. There its heat of formation is released at a flame temperature determined by the pressure in the combustor. Because the atomization process is so important for subsequent mixing and burning, fuel-injector design is as critical as fuel properties.

Aromatics readily form soot in the fuel-rich spray core as their hydrogens are stripped, leaving a carbon-rich benzenoid structure to condense into large molecular aggregates. Multi-ring aromatics form soot more rapidly than single-ring aromatics and exhibit greater smoking tendency (unburned soot) and greater flame luminosity. Luminous flames raise temperatures of the metal in the combustion liner and turbine inlet by radiation and result in shortened life of these parts. At the other extreme, n-paraffins undergo little cyclization in the fuel-rich spray core, holding soot formation and flame radiation at a minimum. A combustor design that causes proper mixing is critical for ensuring smoke-free exhaust products. Nevertheless, in any given combustor, there is a relationship between smoke output, liner temperature, and the hydrogen content of the fuel.

Exhaust emissions are a reflection of combustion conditions rather than fuel properties except in the case of sulfur. The SO_2 amounts of in exhaust emissions are directly proportional to the content of bound sulfur in the fuel. Sulfur values in fuel are determined by crude type and desulfurization processes. Specifications for aircraft fuels impose limits of 3000–4000 ppm, but average amounts are half of these values. For heavier gas-turbine fuels, sulfur limits are determined by legal limits on stack emissions from fuel oils in general (see Exhaust control).

Aviation fuel is exposed to a wide range of thermal environments, and these greatly influence required fuel properties; eg, in the tank of a long-range subsonic aircraft, fuel temperatures can drop so low that ice crystals, wax formation, or viscosity increases may affect the plane's performance. On its way to the engine, the fuel then absorbs heat from airframe or engine components and is sometimes used as a coolant for engine lubricant. The inlet manifold is blanketed in hot compressed air, and in the injection nozzle itself the fuel is mixed with primary atomizing air at a temperature $\geq 300°C$. At that point, fuel temperatures increase and fuel stability assumes primary importance. Fuel stability is not as important in ground-based turbines because fuel lines can be insulated better than in aircraft.

The volatility of aircraft gas-turbine fuel is controlled primarily by the aircraft itself and its operating environment. For example, limits on the vapor pressure of military jet fuels were dictated by the vapor and liquid entrainment losses that could occur in an aircraft capable of climbing to an altitude (ca 12,000 m) where the vapor pressure of warm fuel exceeds atmospheric pressure (21 kPa or 3 psi). The volatility of kerosene fuel is measured by its temperature at the point where its vapors just prove to be flammable, ie, the flash point or lean flammability temperature. Since vaporization promotes engine startup, a minimum volatility frequently is specified to assure adequate vaporization under low temperature conditions.

The effects of water in fuel inside a tank, particularly an aircraft tank, are important because of the demonstrated proclivity of free water to form undrainable pools where microorganisms can flourish. It is common practice to curb growth of organisms by biocidal treatment (see Industrial antimicrobial agents).

Boundary lubrication of rubbing surfaces, such as those in pumps and fuel controls, has been related to the presence in fuel of species that form a chemisorbed film that reduces wear and friction. Fuels that have been highly refined lack natural lubricity agents. Corrosion inhibitors are used to improve fuel lubricity when field problems are encountered.

Trends

Demand for aviation gas-turbine fuels expressed as a share of total crude will reach 9% by 2000. Some of the measures taken in 1974 to cope with the drastic rise in OPEC oil prices and the shortage of jet fuel, eg, aromatics up to 25%, have become permanent specification changes because it is increasingly clear that aircraft gas turbines will have to be operated on a wider cut than the 160–260°C fraction of crude. Synthetic fuels derived from shale or coal also will have to supplement domestic supplies of petroleum by 2000. Some have proposed that nuclear energy may become the key to production of cheap hydrogen for advanced air transports in the next century. Should this occur, a revolution could take place in all forms of transportation (see Hydrogen energy).

The expansion under way in electric-power generation by coal-burning plants or nuclear reactors is accompanied by growing requirements for peak power generation best satisfied by standby gas turbines requiring either synthetic gas made from coal by gasification or a No. 2-GT fuel made by a coal-liquefaction process. In the latter case, the fuel will be deficient in hydrogen compared with the petroleum product and will present a formidable challenge to the designers of advanced combustors (see Fuels, synthetic; Burner technology).

W.G. DUKEK
Exxon Research and Engineering Company

M. Smith, *Aviation Fuels*, G.T. Foulis & Co., Ltd., Henley-on-Thames, Oxfordshire, England.

J.K. Appeldoorn and W.G. Dukek, *Lubricity of Jet Fuels*, SAE 660712, Society of Automotive Engineers, Warrendale, Pa.

K.M. Brown, *Commercial Results with UOP MEROX Process for Mercaptan Extraction in U.S. and Canada*, UOP Booklet 267, UOP, Inc., Des Plaines, Ill., 1960.

AZELAIC ACID, $HOO(CH_2)_7COOH$.   See Dicarboxylic acids.

AZEOTROPIC AND EXTRACTIVE DISTILLATION

Azeotropic Distillation

In industry, it is often difficult or impossible to separate solutions of two or more liquids by ordinary distillation and rectification because of the closeness of boiling points or because, by themselves, they form an azeotrope. Azeotropic distillation may be used to separate some such liquid mixtures. Water, for example, frequently must be separated from another volatile liquid. When ordinary rectification fails, a possible method for some cases may use a distillation with a third liquid, ie, an entrainer (also called a withdrawing agent). This is a liquid that effectively codistills with water to give a constant composition of vapors over a wide range of liquid composition. These vapors are those of the heterogeneous azeotrope, which has a lower boiling point than that of any single component present. Figure 1 is a flow sheet of a typical azeotropic distillation for water removal with an entrainer.

One of the simplest examples of azeotropic distillation was developed in 1927 to lessen the large quantities of butanol in decantation storage because of the low settling rate following its use in removing water from nitrocellulose and to reduce the cost of the large number of redistillations (Fig. 1). In this system, butanol, saturated with water, was fed to the top of the dehydrating column by line A. The water distilled overhead as the azeotrope, with the butanol acting itself as the entrainer.

Water also may be added as an entrainer for pairs of organic liquids with which it forms an azeotrope, and the converse of the following operation would be used for the separation of such a pair. The aqueous binary mixture enters at some point between line A at the top of the dehydrating column and line B at an intermediate point. If a third liquid is to be used as an entrainer, it may enter the system through line C prior to start-up and during operation for makeup. In the dehydrating column, the water–entrainer mixture is rectified to give the vapors overhead in almost the theoretical proportions of the azeotropic mixture as the lowest-boiling component of the liquid system. The condensate, which may be one or more phases, flows to a decanter where the two liquids are separated. The entrainer layer, saturated with water, is passed back to the top plate of the dehydrating column to serve as reflux. Here, distillation occurs again or more water is entrained. The liquid being dehydrated passes down in the column and is discharged practically water-free at the base, which is heated by closed steam or another heat source. The water layer, saturated with the entrainer, passes from the decanter to the water column where the entrainer is stripped to give the azeotropic vapor mixture with water having the

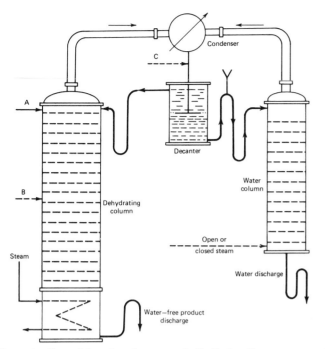

Figure 1. General flow sheet for azeotropic distillation. For water removal, aqueous feed is through line A or B. Water column may have open steam if water is discharged or closed steam if other liquid is discharged. If it is necessary to separate the condensate into two phases, water is supplied through line C and in some cases, an extractor may replace the decanter.

same composition as the vapors from the dehydrating column. Thus, vapors from both columns are passed to the same condenser to give, ideally, an easily separated two-phase condensate which is passed to the decanter. The process continues with the entrainer recycling around the upper part of each column, the condenser, and the decanter. The entrainer acts merely as a carrier that brings over the water entrained in the heterogeneous azeotrope. This is constant in composition as long as enough material is present to allow rectification to continue. The water contains less and less entrainer as it descends the water column and finally passes out of the bottom.

Many processes have been developed during the last twenty years to dry hydrocarbons of their relatively small amount of dissolved water. One related recent development is the azeotropic removal of water from emulsions with crudes from the processing of tar sands stabilized by ultrafine solid particles so that for practical purposes the three phases are dispersed permanently.

Dehydration of acetic acid. The industrial use of acetic acid increased greatly in the late 1920s as a result of expanded manufacture of volatile acetates for auto lacquers, cellulose acetate for film and rayon, and other important materials. Manufacturing of acetic acid by wood pyrolysis and by fermentation of carbohydrates gave 6–7% aqueous solutions from which the water was removed by azeotropic distillation. During World War II, large amounts of the powerful explosive RDX (cyclonite) were produced by acetylation of the dinitrate of hexamethylenetetramine, yield large quantities of dilute acid. Many millions (10^6) of tons of acetic acid have been recovered in such processes for recycle using azeotropic distillation.

In dehydrating acetic acid, an entrainer such as ethylene dichloride, propyl acetate, or butyl acetate may be added to bring over the water. The entrainer is chosen to decrease effectively the boiling point or to increase the vapor pressure of one of the components, in this case water. The choice of an entrainer depends on its boiling temperature and that of its azeotrope which control the ratio to water in the azeotropic mixture; its mutual solubility with water which controls the efficiency of the decantation; its partition coefficient of acetic acid with water which

controls its ability, in refluxing, to extract and to carry down the acetic acid; and the concentration of the dilute acid to be handled.

Absolute alcohol. The preparation of absolute or anhydrous ethanol is difficult because most possible entrainers form binary azeotropes with both water and alcohol and, usually, a ternary azeotrope of all three compounds. Up to about World War I, absolute alcohol was obtained by chemically removing the water. However, the need for absolute ethanol made alcohol the first liquid to be dried by azeotropic distillation according to Young's basic batch-distillation process, which is somewhat more complicated than Figure 1. The modernization of the chemical engineering design of the original Young process gives a superior product of 100% (200 proof) spirits from the 95% alcohol feed at a much lower heat cost. Most anhydrous alcohol made is now used as the solution of ca 5% in gasoline, called gasohol, for use in motor fuel. Any water present would cause the separation into phases, but most of the volatile impurities are retained.

Separation of formic acid, acetic acid, and water. Vapors from boiling solutions of acetic and formic acids have very nearly the same compositions as those of the liquid. Separation of these two acids, which are often obtained together and with more or less water in industrial processing, is difficult. No entrainer has been found to remove formic acid from acetic acid that does not also form a heterogeneous azeotrope also with water; with some, a ternary azeotrope of the entrainer, water, and formic acid is formed. A satisfactory entrainer for distilling with both formic acid and the small amount of water usually present is chloroform. No azeotrope of chloroform and acetic acid exists.

Separation of hydrocarbons. Hydrocarbons constitute by far the largest volumes of liquids to be separated industrially for use as fuels and lubricants and as chemicals or intermediates. Fuels and lubricants are seldom pure compounds but are fractions separated by boiling range. By adding an entrainer for a whole series of hydrocarbons within the boiling range of the fraction, it may be possible to separate the compounds of one series from others by azeotropic distillation.

Some chemical reactions give water as one product. Another liquid of suitable volatility and capable of forming an azeotrope with water may be used to distill with the water as fast as it is formed in the chemical reaction. Thus, the reaction is expedited greatly in continuous processes through this azeotropic distillation.

Extractive Distillation

In an extractive distillation, such as may be used as still another process for the separation of water from acetic acid, a relatively high boiling solvent for acetic acid is used that is immiscible with water. This solvent is added to the top plate of a distilling column, which thus distills nearly pure water over the top while the solvent dissolves and carries the higher boiling acetic acid to the base. The solvent acts as a liquid reflux in the column and goes to the very bottom because of its high boiling point. Since it extracts the acetic acid from the rising vapor stream, the mixture of solvent and acetic acid discharges from the column base and is passed to a second column for separation.

To compare the two systems, the added liquid in an extractive distillation lowers the relative volatility of one of the constituents, ie, the acetic acid, so as to lower its boiling point and bring it out the base; whereas in an azeotropic distillation, the added liquid increases the relative volatility of one constituent, ie, the water, and brings it over the head of the column. Most of the industrial mixtures of liquids to be separated in recent decades have been derived from petroleum crudes or gases and many extractive distillation processes have been developed for the separations necessary in the petroleum and petrochemical industries. During this same period, mathematical modeling of the physical processes with the assistance of computers allows the analysis and optimization of the separation of azeotropic and close-boiling mixtures of liquids. These and other techniques have been productive in the prediction and selection of suitable added liquids and of process and plant design for extractive distillation.

DONALD F. OTHMER
Polytechnic Institute of New York

D.F. Othmer, *Chem. Metall. Eng.* **48**, 91 (June 1941).

L.H. Horsely, *Azeotropic Data*, 1952; L.H. Horsely and W.S. Tamplin, in R.F. Gould, ed., *Azeotropic Data II*, American Chemical Society, Washington, D.C., 1963.

AZEOTROPIC AND EXTRACTIVE DISTILLATION.   See also Supplement.

AZINE DYES

Azine dyes, including the related oxazine and thiazine dyes, range in color from yellow to deep blue. The chromogens in these dyes are of the phenazinium, phenazonium, and phenazthionium groups. Substitution in the 3,7 positions with amino, alkylamino, or aryl amino groups produces a large auxochromic effect, modifying the intensity of color and the shade of the dye. In some cases, a hydroxy group at 3 and a substituted amino at 7 lead to mordant dyes as exemplified by Resorcin Blue, the simplest dye of this type. Selective physical properties are given for methylene blue, nigrosines, induline base, safranine T, and indanthrone in Table 1.

Table 1.   Physical Properties of Selected Azine Dyes

| | uv max, nm | mp, °C | Fastness, % | | |
			light	heat, °C	water
safranine T	515–520	dec 150–160	poor	good	good
induline base[a]	565–570	dec 175	fair	dec 175	exc
methylene blue	665	dec 100–110	good	dec 100–110	good
indanthrone	580[b]	dec 500	exc	very high	exc
nigrosine spirit					
soluble blue[a]	565–570	dec 175	good	dec 150–175	exc
nigrosine base[a]	565–570	dec 175 ignites	good	dec 170–175	exc

[a] These three types varies in solubility because the percent of isomers varies from lot to lot. The nigrosines have appreciable (10–30%) solubility in a 50–50 mixture of xylene and ethanol.

[b] Leuco dye solution.

Methylene Blue is used for dyeing paper. Nigrosines are used as solvent dyes for coloring phenolic plastics, in shoe polish, spirit lacquers, ball-pen inks, wood stains, carbon papers, and many related formulas and applications. When converted to sulfonic acids, they are used for dyeing paper, wood, silk, and in coloring solutions where deep blue-blacks are needed. Indulines are used for coloring transparent or translucent plastics, but are poor in phenolic plastics. Safranine T, sometimes called safranine Y or safranine GF, a yellowish-red dye, is used mainly in coloring paper products and sometimes in coloring leather and silk.

Of the chemicals used in making azine dyes, sodium dichromate, aminoazobenzene, and sodium nitrite are suspected carcinogens. Aniline, dimethylaniline, *o*-toluidine, nitrobenzene, and in-process, intermediate chemicals derived from them cause cyanosis by inhalation of vapors, ingestion, or skin contact. Many azine dyes create special fire and explosion hazards.

J. CLYDE CONGER, SR.
American Cyanamid Co.

The Colour Index, 3rd ed., Vol. 4, published jointly by the Society of Dyes and Colourists, Yorkshire, UK and the American Association of Textile Chemists and Colorists, Lowell, Mass., 1971, pp. 4445–4473.

G.A. Swan and D.G.I. Felton, *Phenazines*, *Chemistry of Heterocyclic Compounds*, Interscience Publishers, New York, 1957.

K. Venkataraman, *The Chemistry of Synthetic Dyes and Pigments*, Vol. II, Academic Press, Inc., New York, 1952, pp. 773–775.

AZINES.   See Azine dyes.

AZO DYES

Azo dyes are the largest and most versatile class of dyes. This well-defined group of compounds is characterized by the presence of one or more azo groups (—N=N—). Peter Griess synthesized the first azo dye soon after his discovery of the diazotization reaction in 1858. The two reactions that form the basis for azo dye chemistry are diazotization and coupling. The diazotization reaction may be stated as:

$$ArNH_2 + 2\,HX + NaNO_2 \rightarrow ArN_2^+\ X^- + NaX + 2\,H_2O \qquad (1)$$

where X = Cl, Br, NO_3, HSO_4, BF_4, etc. Coupling proceeds according to the equation:

$$ArN_2^+\ X^- + HR \rightarrow ArN{=}NR + HX \qquad (2)$$

where R is alkyl or aryl.

Practically every aromatic primary amine is a potential diazo component. The value of an amine is determined chiefly by the properties of the dye prepared from it. Cost of amine, ease of diazotization, stability of the diazonium salt, and final cost of the dye are factors that influence the selection of an amine.

Diazotization involves treating a primary aryl amine with nitrous acid to form a diazonium salt (eq. 1). Since the mechanism of diazotization involves attack by the nitrosating species on the free amine and not the amine salt, it often is difficult to choose the acidity of the reaction medium to suit the basicity of the amine and to promote the formation of an active nitrosating species. The reaction medium should also provide stabilization of the diazonium salt by solvation effects.

The coupling reaction between an aromatic diazo compound and a coupling component is the single most important synthetic route to azo dyes. For the preparation of azo compounds in which either or both Ar and R are heterocyclic groups, apart from the normal azo coupling reaction, oxidative coupling can be employed.

All coupling components used to prepare azo dyes possess one common feature, ie, an active hydrogen atom bound to a carbon atom. The following compounds can be used as azo coupling components: compounds that possess phenolic hydroxyl groups, such as phenols and naphthols; aromatic amines; compounds that possess enolizable ketone groups of an aliphatic character, ie, compounds that have active methylene groups,

$$XCH_2COY \rightleftarrows XCH = C(OH)Y$$

where X is an electron-attracting group of the type —COR, —COOH, —CN (R being alkyl or aryl), and Y, for the most part, is a substituted or unsubstituted amino group; and heterocyclic compounds, such as those containing pyrrole, indole, and similar ring systems, eg, 5-pyrazolones.

Analogous to aromatic halogenation, nitration, and sulfonation, the azo coupling reaction is an electrophilic aromatic substitution. The effect on the reaction rate of substituents on both the diazo and coupler components is in agreement with this mechanism. Thus, the reaction is facilitated by electron-attracting groups in the diazo component, and by electron-donating groups in phenol and aromatic amine couplers.

The phenoxide anion is important in coupling to phenol; thus, any contribution by the keto form can be ignored. Electron-attracting α substituents are needed to increase the reactivity of methylene compounds. On the other hand, the presence of groups such as nitro and sulfonate depress the coupling capability of phenols and aromatic amines. Electron-attracting groups promote the shift of the phenol-phenolate anion equilibrium to the side of the reactive phenolate species; however, the lowering of electron density on the aromatic ring at the reaction site is the overriding factor. Thus, nitronaphthols in which both substituents are in the same nucleus do not couple or do so only sluggishly. This is true especially of nitrophenols. A technically important example is salicylic acid, in which the effect of the carboxyl group on the π-electrons of the ring and the lower acidity of the phenolic proton due to hydrogen bonding are responsible for depressing the ability to couple. On the other hand, since electron-donating groups such as alkyl, alkoxy, and amino residues facilitate reaction, the second hydroxyl group of resorcinol enables this component to be used for syntheses that could not be carried out with phenol.

Normally, coupling occurs at the position para to the hydroxyl group. If this is occupied by a substituent not readily eliminated, as in *p*-cresol or β-naphthol, the diazo component attacks the ortho position. When both ortho and para positions are free, two aryl azo residues can be introduced, but with phenols the second coupling occurs with great difficulty.

In coupling there are always two competing reactions:

$$RN_2^+ \ X^- + HR' \rightarrow RN{=}NR' + HX \qquad (3)$$

and

$$RN_2^+ \ X^- + HOH \rightarrow ROH + N_2 + HX \qquad (4)$$

where R is an aryl radical, R'H a coupling component, and X^- a halide ion. Factors that increase the rate of the desired reaction (eq. 3) also increase the rate of diazo decomposition (eq. 4). Since the rates of the two reactions do not parallel each other, one can generally maximize the yield of equation 3 by a judicious choice of conditions. Coupling reactions should be carried out in a medium in which the acid–base equilibria of the diazo and coupling components favor as much as possible the diazonium ions and the phenolate ions or the free amine, respectively.

An optimum pH exists for each azo coupling reaction which is limited by acidities, expressed in pH units, that correspond to the pK_a value of the coupling component or the constants of the diazonium–diazotate equilibria. Increasing the temperature of the coupling reaction usually has an unfavorable effect since the diazo decomposition reactions possess greater activation energies and, therefore, a larger temperature gradient than azo coupling. For every 10°C rise in temperature, the coupling rate increases 2.0–2.4 times, whereas the diazo decomposition rate rises 3.1–5.3 times.

A few broad rules for the activity of coupling components: (*1*) Diazo coupling follows the rules of orientation of substituents in aromatic systems in accordance with the mechanism of electrophilic aromatic substitution and the concept of resonance. (*2*) Generally, phenols (as the phenolate anion) couple more readily than amines, and members of the naphthalene series more readily than members of the benzene series. (*3*) Electron-attracting substituents in the coupling component such as halogen, nitro, sulfo, carboxyl, and carbonyl are deactivating and tend to retard coupling. (*4*) A lower alkyl or alkoxy group substituted in the ortho or meta position of an aminoarye group may promote coupling. Good couplers are obtained from dimethylaniline when lower alkyl, lower alkoxy, or both groups are present in the 2 and 5 positions. (*5*) It is possible for diazo compounds to attack both the ortho and para positions of hydroxyl and amino coupling components when these positions are not already occupied. Technologically, the most important examples of such couplers are 1-naphthylamine, 1-naphthol, and sulfonic acid derivatives of 1-naphthol. Of great importance in the dyestuff industry are derivatives of 1-naphthol-3-sulfonic acid, such as H acid (8-amino-1-naphthol-3,6-disulfonic acid), J acid (6-amino-1-naphthol-3-sulfonic acid), and gamma acid (7-amino-1-naphthol-3-sulfonic acid).

The ortho/para isomer ratio depends on factors such as the nature of the diazo component, the solvent system, the pH of the medium, the temperature of the coupling reaction, presence of catalysts, and the position of the substituents.

The important dye Eriochrome Black A (**1**) (R = H) (CI Mordant Black 5; CI No. 26695) is a mixture of ortho and para isomers whereas Eriochrome Black T (**1**) (R = NO_2) (CI Mordant Black 11; CI No. 14645) is purported to be exclusively ortho.

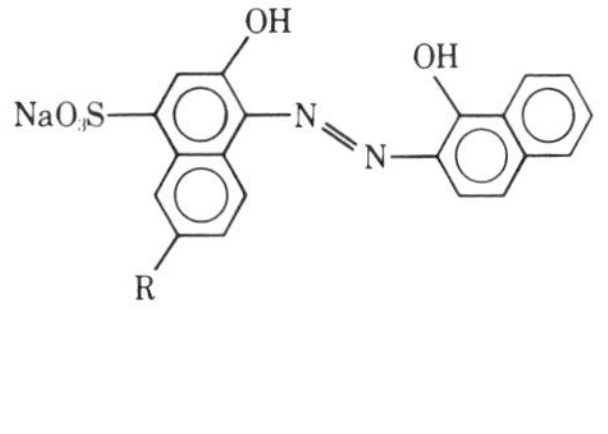

Oxidative coupling of heterocyclic hydrazones opened the way to the preparation of azo derivatives of diazo compounds unobtainable by other means, ie, heterocyclic compounds in which the diazotizable amino group is conjugated with the heterocyclic nitrogen atom as in 2- and 4-aminopyridine, compounds which do not normally yield stable diazonium salts. The most suitable oxidizing agent is potassium ferricyanide, but ferric chloride, hydrogen peroxide in the presence of ferrous salts, ammonium persulfate, lead dioxide, lead tetraacetate or chromate, or silver and cupric salts may be useful. Water mixed, eg, with methanol, dimethylformamide, or glycol ethers is employed as reaction medium.

In the *Colour Index* a dual classification system is employed to group dyes according to area of usage and chemical constitution. Because of the ease of synthesis of azo, disazo, and polyazo dyes, and their wide range of applications, azo dyes comprise the largest chemical class in numbers, monetary value, and tonnage produced. There are more than two thousand chemical structures of azo dyes disclosed in Volume 4 of the *Colour Index*.

Chemically, the azo class is subdivided according to the number of azo groups present into mono-, dis-, tris-, tetrakis-, etc. Mono- and disazo dyes are essentially equal in importance. Trisazo and tetrakisazo dyes are less important. Thus, substances with more than three azo linkages are included under the heading of polyazo dyes.

The second method of subdivision distinguishes between primary and secondary types of disazo dyes. The former covers compounds made from two molecules of a diazo derivative and one molecule of a bifunctional coupling component. In both cases, the monofunctional reagent may consist of two molecules of one compound or one molecule of each of two substances used stepwise, the first alternative yielding symmetrical products. In stepwise reactions it is the rule to carry out the coupling occurring with greater difficulty first, since components containing one or more azo groups become progressively more inert. Secondary disazo dyes are made by diazotizing an aminoazo compound, the amino group of which derives from the original coupling component, and coupling it to a suitable intermediate. A third group of miscellaneous disazo dyes is prepared by condensation of two identical or different aminoazo compounds, commonly with phosgene, cyanuric chloride, or fumaryl dichloride.

Of all classes of dyestuffs, azo dyes have attained the widest range of usage because variations in chemical structure are readily synthesized and methods of application generally are not complex. There are azo dyes for dyeing all natural substrates such as cotton, paper, silk, leather, and wool as well as the synthetics: polyamides, polyesters, acrylics, polyolefins, viscose rayon, cellulose acetate, etc; for the coloring of paints, varnishes, plastics, printing inks, rubber, foods, drugs, and cosmetics; for staining polished and absorbed surfaces; and for use in diazo printing, and color photography (qv). Types of dyes include the following:

Acid dyes are capable of coulombic interactions between the dye molecule and the fiber. Acid dyes themselves are not generally acids (ie, dye–SO_3H) but are sulfonate salts (dye–SO_3Na) of strong sulfonic acid groups. Commercial acid dyes contain one or more sulfonate groups, thereby providing solubility in aqueous media. These dyes are applied in the presence of organic or mineral acids. Such acids protonate any available cationic sites on the fiber, thereby making possible bonding between the fiber and the dye molecule. There are three general classifications of acid dyes: acid dyes that dye directly from the dyebath, mordant dyes that are capable of forming metallic lakes on the fiber when aftertreated with metallic salts, and premetallized dyes.

Azoic dyes (known also as ice colors and ingrain colors) are water-insoluble azo pigments, free from solubilizing groups, formed on the fiber by reaction of a diazo component with a coupling component (a Naphthol AS compound, such as an arylide of 3-hydroxy-2-naphthoic acid). The discovery that 3-hydroxy-2-naphthoic acid arylides have greater substantivity for cotton than 2-naphthol led to the introduction of an extensive line of Naphthol AS derivatives, fast color salts, rapid fast dyes, and the rapidogen dyes from diazoamino compounds.

Fast color salts, diazonium salts in the form of stable dry powders were introduced to simplify the work of the dyer. When dissolved in water, they react like ordinary diazo compounds. These diazonium salts, derived from amines free from solubilizing groups, are prepared by the usual method and are salted out from the solutions as the sulfates, the metallic double salts, or the aromatic sulfonate. The isolated diazonium salt is sold in admixture with anhydrous salts such as sodium sulfate or magnesium sulfate. Commercially, the most important fast color bases and salts include Azoic Diazo Component 3 (2,5-dichloroaniline, produced by the reduction of 1.4-dichloro-2-nitrobenzene); Azoic Diazo Component 5 (4-nitro-o-anisidine, produced by nitration of O-acetanisidide with subsequent hydrolysis); Azoic Diazo Component 8 (2-nitro-p-toluidine produced by nitration of o-acetotoluidide in H_2SO_4 and subsequent saponification); Azoic Diazo Component 13 (5-nitro-o-anisidine, produced either by partial reduction of 2,4-dinitroanisole or nitration of o-anisidine); Azoic Diazo Component 20 (4'-amino-2',5'-diethoxybenzanilide, produced by condensation of 2,5-diethoxyaniline with benzoyl chloride, followed by nitration and reduction); Azoic Diazo Component 32 (5-chloro-o-toluidine, produced by nitration of p-chlorotoluene followed by reduction of the resulting 4-chloro-2-nitrotoluene); and Azoic Diazo Component 34 (4-nitro-o-toluidine, produced by nitration of o-benzenesulfonotoluidide followed by hydrolysis).

Disperse azo dyes are coloring substances having very low aqueous solubility which are applied to hydrophobic fibers from an aqueous system in which the dye is present in a highly dispersed state. The sharp increase in the importance of disperse dyes in the last 20 yr can be attributed directly to the emergence of polyester and nylon as the main man-made fibers. Azo and anthraquinone compounds comprise the two principal structural types that are used as disperse dyes. Other compounds used to a much lesser extent include methines, cyanostyryls, hydroxyquinophthalones, and nitrodiarylamines.

Disperse dyes are used mainly in the coloring of polyester, cellulose acetate, and triacetate fibers in textile applications. Another important use is in the dyeing of nylon, especially carpets and hosiery, in which disperse dyes effectively cover barré defects in the fiber caused by processing inconsistencies. Disperse dyes also are used for acrylics, modacrylics, and cellulose acetate.

Oil-soluble azo dyes dissolve in oils, fats, waxes, etc. Generally yellow, orange, red, and brown oil colors are azo structures and greens, blues, and violets are primarily anthraquinones. Blacks are nigrosines and indulines of the azine type. Uses include the coloring of hydrocarbons, waxes, oils, candles, etc. Substitution by chloro, nitro, and similar groups increases the molecular weight and improves sublimation fastness, but lowers the oil solubility of this group of dyes.

Spirit-soluble azo dyes dissolve in polar solvents such as alcohol and acetone, and are applied in the coloring of lacquers, plastics, printing inks, and ball-point pen inks. Of the two principal types of azo structures used, the most important are the insoluble salts of azo dyes containing sulfo groups and relatively organic amines.

Azo pigments are an important class of organic colorants used in the mass coloration of man-made fibers, in textile printing in the textile field, as well as in the nontextile area, and plastics. A pigment is insoluble in the medium in which it is used. The physical properties of pigments are of great significance since the coloring process does not involve solution of the colorant. Azo pigments can be grouped as metal toners, metal chelates, and metal-free azo pigments. Because these pigments are organic in nature, they tend to bleed in resins and solvents. Increasing the molecular weight often reduces this tendency.

SIGMUND C. CATINO
RUSSELL E. FARRIS
GAF Corporation

K.H. Saunders, *The Atomatic Diazo Compounds*, 2nd ed., Edward Arnold and Co., London, 1949.

H. Zollinger, *Diazo and Azo Chemistry*, transl. from the German by H.E. Nursten, Interscience Publishers, Inc., New York, 1961.

K. Venkataraman, ed., *The Chemistry of Synthetic Dyes*, Vol. III, Academic Press, Inc., New York, 1970.

B

BABBIT METALS. See Bearing materials.

BACITRACINS. See Antibiotics, peptides.

BACTERIAL, RICKETTSIAL, AND MYCOTIC INFECTIONS, CHEMOTHERAPY. See Chemotherapeutics, bacterial and mycotic.

BAGASSE

Bagasse is the fibrous residue from sugarcane after the cane is harvested and crushed to extract the juice. Quantity and quality of bagasse vary with variety and age of the cane and nature of the harvesting and milling process. Yield on cane is ca 25% by weight of mill-run bagasse (or 50% moisture). Bagasse has two main components, ie, pith and fiber, derived from the respective inner and outer areas of the cane stalk. Pith (ca 30% dry wt of cane) is the nonfibrous parenchyma-cell portion of the inner stalk where the thin-walled cells contain most of the juice. The true fiber is the hard-walled cylindrical cell rind of the stalk, composing 50% dry wt of the cane. When cane is ground, the pith and fiber portions are intimately mixed and must be separated before the fiber can be used for pulp, paper, board, or other such applications.

The traditional method of bagasse storage has been as bales, stored outdoors under cover, with well-arranged ventilation to encourage rapid fermentation and sugar destruction. In recent years, bulk storage of moist bagasse has become popular. The piles are kept wet with water or a bacteriological liquor that prevents deterioration of fibrous material by controlling pH and maintaining a high concentration of lactic acid bacteria and other organisms.

The most widespread use for bagasse has been as fuel for steam production in sugar factories. Excess steam is increasingly produced to be sold as electricity to the grid. Aside from that main function, bagasse also is used to make briquettes for open fires, in paper production, for dissolving pulp for the rayon industry, as a source for various chemicals, as a raw material for thermosetting plastics, in animal feed, as litter and mulch, and as a cellulose source for microbial protein production. The current use of greatest potential is as a fermentation substrate for alcohol fuel production.

Margaret A. Clarke
Cane Sugar Refining Research Project, Inc.
present affiliation
Sugar Processing Research

J.M. Paturau, *By-Products of the Cane Sugar Industry*, 2nd ed., Elsevier Publishing Co., Amsterdam, The Netherlands, 1982.

G.P. Meade and J.C.P. Chen, *Cane Sugar Handbook*, 10th ed., John Wiley & Sons, Inc., New York, 1977.

BAKERY PROCESSES AND LEAVENING AGENTS

YEAST-RAISED PRODUCTS

Bakery products are highly perishable foods. Much of their production is done in small plants scattered about the country. From these, distribution is made within a relatively limited area unless the product is frozen. According to the 1977 Census of Manufactures, however, only slightly over one quarter of the dollar volume of the baking industry was attributed to these small plants which account for almost 75% of the baking establishments in the United States.

Dry baked goods (biscuits, cookies, crackers, pretzels, and ice cream cones), which have a long shelf life, can be distributed over a wide area from a single large bakery. In 1977, the Census of Manufactures reported there were 24 plants manufacturing these dry baked goods with a combined output of $\$2.8 \times 10^9$. The total value of the entire baking industry approaches $\$11 \times 10^9$.

Yeast-Raised Products

The leavening action of yeast depends upon the ability of yeast to generate carbon dioxide and alcohol by breaking down fermentable sugars and upon the unique ability of the protein in wheat flour to trap evolved gases (see Fermentation; Beer; Yeasts). Product variety is obtained by the incorporation of special ingredients or of unusually large amounts of a particular ingredient, or by shaping of the dough prior to baking, or by the method of baking. Of the total dollar volume of the baking industry, yeast-raised goods constitute 66.4%.

Chemically Leavened Products

Chemical leaveners (baking powders) are exclusively used in cakes, cookies, refrigerated dough, and quick bread manufacturing. These products constitute 33.6% of the bakery industry output (see Bakery processes and leavening agents, chemical leavening agents).

Ingredients

Ingredients of bakery foods include the following: flour; fresh compressed or active dry yeast; yeast foods; sugar, including corn syrups, sucrose, and dextrose; animal and vegetable fats and oils; emulsifiers; high-heat milk and milk substitutes; eggs; salt; enzymes; mold inhibitors (antimycotics); flavorings; and enriching ingredients.

Procedures and Equipment

Larger bakeries purchase flour shipped in bulk from the mill in specially designed railroad cars or tracks. The flour is transferred pneumatically from the cars or trucks to bins in the plant, from which it is conveyed to the mixing room as needed. A tote bin system of bulk handling is available to smaller users. Flour used in small quantities or by bakers with no bulk handling facilities is delivered in multiwall paper bags and is handled in the bakery on skids with forklift trucks. Much of the sugar used by bakers is in the form of syrup shipped in tank cars and piped directly to the mixing room where it may be metered into the mixing bowl. Shortening can also be handled in bulk; plastic shortening is heated and pumped through heated pipes. Yeast and other perishable ingredients are stored under refrigeration. Water is metered rather than weighed. All minor ingredients may be weighed directly from the shipping containers or transferred to small covered bins from which they are weighed.

Most bread is made by the sponge-dough process which has four steps: a sponge of flour, yeast, yeast food, water, and enzymes is mixed and fermented, then mixed again with the remaining ingredients to develop the gluten in the dough; after a rest period (floor time) the dough is cut, shaped, and panned in the makeup step; the panned dough is permitted to rise (proof) and then is baked; the baked bread is cooled, sliced, and wrapped. Instead of solid sponges, liquid sponges (brews containing different amounts of flour) are commonly used.

Continuous breadmaking is carried out in equipment that continuously mixes and deposits dough in the pans. Unlike the conventional method, fermentation is accomplished in a liquid before enough flour to make dough is added. The ferment contains water, yeast, sugar, enrichment nutrients, and yeast food. This process is less widely used than 30 years ago.

Brown 'n serve. These baked products are made in the same way as their fully baked counterparts except that they are baked long enough to gelatinize the starch and set the dough so the cooled product retains its shape, but not long enough to brown the crust. After purchase, the consumer completes the baking at normal baking temperatures.

Cake. Because cakes are not made by fermentation, they must be leavened with air whipped into the batter or by combinations of air and baking powder. Today, cakes are made by a single-stage mixing process. Continuous mixing is common in large plants.

Biscuits and crackers. These products have a long shelf life and may be distributed over wide areas. For this reason, biscuit and cracker plants are larger and contain more intricate mechanical equipment than those in which bread and cake are produced. The former are the most highly automated in the baking industry.

Frozen bakery products. Appreciable amounts of all types of bakery products are marketed frozen. In addition, a substantial volume of frozen doughs is being sold to the consumer or shipped from a central bakery to supermarkets for baking off within the stores. For certain baked foods containing perishable cream or custard filling, distribution in the frozen state safeguards against the growth of food-poisoning organisms.

Standards of Identity for Baked Foods

Nearly all breads and rolls conform to the Federal standards of identity which, as of Jan. 1, 1978, permit the addition of ingredients that are safe and suitable and require that each ingredient be listed in descending order of prominence. Since 1975, the FDA has required that all enriched bread-type products carry nutrition labeling.

KAREL KULP
American Institute of Baking

FRANK N. HEPBURN
U.S. Department of Agriculture

S.A. Matz, *Bakery Technology and Engineering*, 2nd ed., Avi Publishing Co., Inc., Westport, Conn., 1972.

D.K. Tressler and W.J. Sultan, *Food Products Formulary*, Vol. 2 of *Cereal, Bakery Goods, Dairy and Egg Products*, Avi Publishing Co., Inc., Westport, Conn., 1975.

S.A. Matz, *Cookie and Cracker Technology*, Avi Publishing Co., Inc., Westport, Conn., 1968.

E.J. Pyler, *Baking Science and Technology*, Vol. 2, Siebel Publishing Co., Chicago, Ill., 1973.

CHEMICAL LEAVENING AGENTS

Baking powder has rapidly declined in home use since 1940. However, the chemical leavening components are used in an increasing array of modern food products. These products include prepared cake, pancake, cookie, and doughnut mixes, self-rising flour and cornmeal, and refrigerated or frozen doughs and batters.

Composition of Chemical Leavening Systems

Sodium bicarbonate (baking soda) is an ingredient common to most leavening systems. Powdered soda is preferred for self-rising flour and prepared biscuit mixes. A granular soda is required for stability for baking powders and prepared cake mixes or any product subjected to storage for ≥ 6 mo.

Because of flavor effects, the amount of sodium bicarbonate must be balanced in such proportions with the baking acid that few unchanged reactants remain in the finished product, resulting in an essentially neutral product. The concept of neutralizing value was developed to ensure that leavening ingredients would be used in the proper ratio. The neutralizing value, NV, of a baking acid is defined as the number of parts by weight of sodium bicarbonate that will result in a neutral baked product when used with 100 parts of the baking acid.

Baking Powder

Modern baking powder consists of a mixture of sodium bicarbonate, one or more acid ingredients, and an inert ingredient. The most prevalent type of household baking powder contains sodium aluminum sulfate (SAS or soda alum). A small amount of monocalcium phosphate monohydrate (MCP) is used in order to preform the gas cells during the makeup of the dough or batter so that uniform and efficient expansion occurs in the oven. This is needed since SAS is almost completely unreactive until heat is applied. Other double-acting type baking powders, usually used in commercial applications are sodium acid pyrophosphate

(SAPP), sodium aluminum phosphate (SALP), and MCP in various combinations.

The blending of baking powder, although simply a physical mixing, must be done in the proper sequence. The inert ingredient serves to keep the reactive components physically separated to minimize premature reaction in the dry mixture. It should be preblended with the least stable component to ensure thorough coating before introducing other reactive ingredients into the mixer. The powder is usually packaged in airtight metal or fiber cans. However, the package must not be hermetically sealed because of the potential build up of high internal pressures.

Leavening Agents for Preleavened Mixes and Doughs

By far the greatest use of leavening in the home is in preleavened mixes. For many of these products, leavening is the most critical component. The baking acids and soda are incorporated into the mix separately. The type, quantity, combination and order of incorporation are all tailored to each product to give maximum stability, tolerance, performance, and quality. Most of the baking acids used in the mix industry are of the phosphate type. Monocalcium phosphate, sodium acid pyrophosphate and sodium aluminum phosphate are acids, primarily in use and with some dicalcium phosphate dihydrate and glucono-δ-lactone used for special purposes.

WILLIAM B. CHESS
Stauffer Chemical Co.

S. Mendelsohn, *Baking Powders*, Chemical Publishing Co., Inc., New York, 1939.

U.S. Pat. 3,501,314 (Mar. 17, 1970), T.P. Kichline and N.E. Stuhlhebec (to Monsanto Chemical Co.).

U.S. Pat. 3,736,151 (May 29, 1973), L.B. Post, H.J. Rosen, and J.H. Zeh (to Stauffer Chemical Co.).

BARBITURIC ACID AND BARBITURATES.   See Hypnotics, sedatives, and anticonvulsants.

BARIUM

Barium, a soft, volatile, silvery-white metal, is a member of Group IIA of the periodic table and is classed as an alkaline-earth metal (see Table 1). Barium metal does not occur free in nature; however, its compounds occur in small but widely distributed amounts in the earth's crust, especially in igneous rocks, sandstone, and shale. The principal barium

Table 1.   Physical Properties of Barium

Property	Value
density at 20°C, g/cm^3	3.51
melting point, °C	729
boiling point, °C	1640
heat of fusion, kJ/mol[a]	7.66
heat of vaporization, kJ/mol[a]	149.20
specific heat at 20°C, J/(g·K)[a]	0.192
coefficient of thermal expansion, m/(m·°C)	1.8×10^{-5}
electrical resistivity, μΩ·cm	
commercial purity at 0°C	34.8
ultrahigh purity at 0°C	29.4
liquid at mp	314
electron work function, eV	2.11
crystal structure	bcc
lattice constant, nm	0.5025
superconducting temp. at 8.8×10^6 kPa[b], K	3.05

[a] To convert J to cal, divide by 4.184.
[b] Increases with increasing pressure at the rate of 1.3×10^{-7} K/kPa. To convert kPa to atm, divide by 101.3.

minerals are barytes (barium sulfate) and witherite (barium carbonate), also known as heavy spar. Barium has a valence electron configuration of $6s^2$ and characteristically forms divalent compounds. Finely divided barium is susceptible to rapid, violent combination with atmospheric oxygen. Barium powder must be stored under dry argon or helium to avoid the possibility of violent explosions.

Owing to its high reactivity, production of barium by electrolysis of aqueous solutions or high temperature carbon reduction is impossible. At present, barium is prepared by the thermal reduction of barium oxide with aluminum. Pfizer, Inc. is the only U.S. producer of barium metal.

Barium metal and most barium compounds, with the exception of barium sulfate, are highly poisonous. Barium ion acts as a muscle stimulant and can cause death through ventricular fibrillation of the heart. Workers must wear respirators approved for toxic airborne particles, goggles, gloves, and protective clothing at all times. Barium is classified as a flammable solid and is nonmailable. It must be kept dry and preferably sealed in shipping containers. Principal uses of barium are as a getter for removing residual gases from vacuum systems and as a deoxidizer for steel and other metals.

CHARLES J. KUNESH
Pfizer, Inc.

F. Emley in D.M. Considine, ed., *Chemical and Process Technology Encyclopedia*, McGraw-Hill, Inc., New York, 1974, p. 151.

C.A. Hampel in C.A. Hampel, ed., *The Encyclopedia of the Chemical Elements*, Reinhold Book Corp., New York, 1968, pp. 45–46.

C.L. Mantell in C.A. Hampel, ed., *Rare Metals Handbook*, 2nd ed., Reinhold Publishing Corp., London, 1961, p. 26.

BARIUM COMPOUNDS

All barium compounds, except barium sulfate, are manufactured commercially from barite, usually starting with the high temperature reduction of barium sulfate to water-soluble barium sulfide.

Water- and acid-soluble barium salts, such as the carbonate, chloride, hydroxide, nitrate, acetate, and sulfide, are quite toxic; however, barium sulfate is so insoluble as to be nontoxic. The lethal adult dose for most barium salts is 1–15 g. The barium ion stimulates muscular tissue and causes a depression in serum potassium. Death may occur from respiratory failure owing to paralysis of the respiratory muscles, or from cardiac arrest or fibrillation. Barium poisoning should be treated by oral administration of magnesium sulfate solution, followed by gastric lavage, respiratory assistance if necessary, administration of atropine sulfate, and intravenous administration of potassium salts, accompanied by monitoring the serum potassium level.

Barium acetate, $Ba(CH_3CO_2)_2$, crystallizes from water as a trihydrate, sp gr 2.02, below 24.7°C; as a monohydrate between 24.7 and 41°C; and as the anhydrous salt above 41°C. Solubility is 58.8 g $Ba(C_2H_3O_2)_2/100$ g H_2O at 0°C.

Barium borate. Barium metaborate monohydrate, $BaB_2O_4 \cdot H_2O$ (mp, > 900°C) has been used in the United States as a paint pigment since 1959. The commercial product is a white powder with 0.3% solubility in water. In paint, barium metaborate confers a degree of resistance to mold, corrosion, and fire. Two other barium borates, $BaO \cdot 2B_2O_3$ and $2 BaO \cdot 3B_2O_3$, are also reported.

Barium carbonate. Precipitated barium carbonate, $BaCO_3$, is the most widely used barium chemical, with the exception of natural barite. It is a fine, white powder with an absolute density of 4.25 and a bulk density varying from 0.80 to 2.16 g/cm³ (50–135 lb/ft³), depending on the grade. Production rates of barium carbonate have stabilized after several years of decline. The number of U.S. producers declined from seven in 1961 to two in 1981. Prices increased steadily. The U.S. Bureau of Mines estimates that demand for barium carbonate and other barium chemicals will increase, possibly reaching three times its present level by the year 2000. Barium carbonate is used in the brick, ceramic, oil-well drilling, photographic, glass, and chemical manufacturing industries.

Barium formate, $Ba(HCO_2)_2$, forms white rhombic crystals; sp gr, 3.21; solubility, 26.2 g/100 g H_2O at 0°C.

Halides

Barium chloride is produced commercially as barium chloride dihydrate, $BaCl_2 \cdot 2H_2O$, and also as the anhydrous salt. It crystallizes from aqueous solution as the dihydrate, fine white monoclinic crystals; sp gr, 3.097; solubility, 31.7 g/100 g H_2O at 0°C. It is sold in moistureproof bags and steel or fiber drums and is used as a raw material in the manufacture of barium colors, as an ingredient in case-hardening and heat-treating baths (see Metal surface treatment) and as a flux in the manufacture of magnesium metal.

Barium bromide may be crystallized from aqueous solution as white crystals of barium bromide dihydrate, $BaBr_2 \cdot 2H_2O$. Solubility is 92.2 g $BaBr_2/100$ g H_2O at 0°C.

Barium iodide is crystallized from hot aqueous solution as barium iodide dihydrate, $BaI_2 \cdot 2 H_2O$; solubility, 169.4 g $BaI_2/100$ g H_2O at 0°C.

Barium hydroxide. Barium hydroxide octahydrate, $Ba(OH)_2 \cdot 8H_2O$, occurs as white monoclinic crystals; sp gr, 2.18. It melts in its water of crystallization at 77.9°C. Solubility is 1.67 g $Ba(OH)_2/100$ g H_2O at 0°C. Barium hydroxide is used in lubricating oils and greases; in plastics stabilizers; as a papermaking additive; as an ingredient in sealing compositions, vulcanization accelerators, and antirust-paper formulations; as a pigment dispersant; in a formula for self-extinguishing polyurethane foams; and in the protection of objects made of limestone from deterioration (see Fine art examination and conservation).

Barium nitrate, $Ba(NO_3)_2$, occurs as colorless crystals; solubility, 5.0 g/100 g H_2O at 0°C; mp, 592°C; sp gr, 3.24. The compound is used in pyrotechnics (qv), in green flares, in tracer bullets, in primers, in detonators, and as a source of barium oxide in enamels.

Barium nitrite, $Ba(NO_2)_2$, crystallizes from aqueous solution as barium nitrite monohydrate, $Ba(NO_2)_2 \cdot H_2O$, yellowish hexagonal crystals; sp gr 3.173; solubility, 54.8 g $Ba(NO_2)_2/100$ g H_2O at 0°C.

Barium oxide, BaO, occurs as colorless cubic or hexagonal crystals; mp, 1923°C; bp, ca 2000°C; sp gr (cubic), 5.72. Commercial barium oxide is a white or gray powder with minor impurities of barium silicate, barium carbonate, and carbon, originating from the manufacturing process. At present, barium oxide is not being manufactured in the United States, and the market is supplied by imports. Accumulation of barium oxide (and peroxide) dust presents a definite fire hazard.

Barium peroxide. When heated in air or oxygen to 500°C, barium oxide is converted readily to barium peroxide, BaO_2.

Barium silicate. When barium oxide, carbonate, or sulfate is heated with silica to a white heat, barium silicates are formed. The two silicates identified are monobarium silicate, $BaO \cdot SiO_2$ or $BaSiO_3$, and tribarium silicate, $3BaO \cdot SiO_2$ or Ba_3SiO_5.

Barium sulfate, $BaSO_4$, occurs as colorless rhombic crystals; mp, 1580°C (dec); sp gr, 4.50; sol, 0.00285 g/100 g H_2O at 30°C. In medicine, barium sulfate is used widely as an x-ray contrast medium (see X-ray technology). Commercially, barium sulfate is sold both as natural barite ore and as a precipitated product. Drilling muds consume 94% of all barite produced. Also, barite is used as a filler or extender in paints, as a filler in plastics and rubber products, in floor mats and carpet backings made of polyurethane foam, in the glass and ceramic industry, and as the starting point in the manufacture of other barium compounds. Precipitated barium sulfate is used for medical purposes and to form photographic paper.

Barium Sulfides

Barium sulfide, BaS, barium hydrosulfide, $Ba(SH)_2$, and several polysulfides are known; the polysulfides include dibarium trisulfide, Ba_2S_3; barium disulfide, BaS_2; barium trisulfide, BaS_3; barium tetrasulfide monohydrate, $BaS_4 \cdot H_2O$; and barium pentasulfide, BaS_5. Barium sulfide is the largest-volume barium compound manufactured, although little of it is sold. Commercial barium compounds are almost all made by first manufacturing barium sulfide by high temperature solid-phase reduction of barium sulfate with a carbonaceous reducing agent, and then converting it into the desired substance. Barium sulfide is handled in the form of an aqueous solution and only rarely is separated as a solid.

Barium sulfite, $BaSO_3$, occurs as colorless cubic (or hexagonal) crystals, with solubility 0.02 g/100 g H_2O at 0°C.

Barium titanate, $BaTiO_3$ (mp, ca 1625°C), has both ferroelectric and piezoelectric properties and is used in sonar equipment, in phonograph cartridges, capacitators, and other electronic equipment.

TRUMAN KIRKPATRICK
The Sherwin-Williams Company

F.B. Fulkerson, "Barium," preprint from Bureau of Mines Bulletin 667, *Mineral Facts and Problems*, U.S. Department of the Interior, Bureau of Mines, Superintendent of Documents, Washington, D.C., 1975.

S.K. Haines, "Barite," preprint from Bureau of Mines *Minerals Yearbook 1976*, U.S. Department of the Interior, Bureau of Mines, Superintendent of Documents, Washington, D.C.

R.B. Reznik and H.D. Toy, Jr., *Source Assessment: Barium Chemicals*, EPA MRC-DA-530, Contract No. 68-02-1874, Office of Research and Development, U.S. Environmental Protection Agency, Washington, D.C., Feb. 1976.

D.A. Brobst, "Barite" in D.A. Brobst and W.P. Pratt, eds., *United States Mineral Resources*, U.S. Geological Survey Professional Paper 820, U.S. Government Printing Office, Washington, D.C., 1973, pp. 75–84.

BARRIER POLYMERS

The general theory of permeation of a gas or liquid through a polymer matrix states that the permeation rate is the product of a diffusion term and a solubility constant of the gas–liquid in the polymer matrix, each of which is often independent of the other. The process of permeation through a polymeric barrier involves four steps: absorption of the permeating species into the polymer wall; solubility in the polymer matrix; diffusion through the wall along a concentration gradient; and desorption from the outer wall. In order to be a good barrier polymer, the material must have some degree of polarity such as contributed by the nitrile, ester, chlorine, fluorine, or acrylic functional groups; high chain stiffness; inertness; close chain-to-chain packing by symmetry, order crystallinity, or orientation; some bonding or attraction between chains; high glass-transition temperature T_g. Permeability also is affected by fillers and additives, moisture content, temperature, thickness, and molecular structure of permeating gas or liquid.

Measurement of Barrier Properties

The most common method of measuring gas permeation uses a Dow Permeation Cell. Water permeation through a polymer generally is measured by gravimetric weight loss of a sealed water-filled container made from the test polymer or by gravimetric weight loss of a special metal cup (such as the Paine cup) which uses the test polymer as a lid. Organic-liquid permeation usually is measured by using a filled molded container made of the test polymer and noting gravimetric weight loss. In all measurements of gas or liquid permeation, it is necessary to allow time for equilibrium rates to become established or erroneous values will be obtained.

A high barrier polymer can be defined as one that exhibits a high resistance to molecular flow of a permeating agent or agents through the polymer matrix. To qualify as a truly high barrier polymer, the following limits of permeation should apply based on studies of the packaging of products that are sensitive to gases, liquids, or organic-vapor diffusion: a gas transmission of not more than 10 cm^3 of oxygen per 0.025 nm of polymer per 645 cm^2 (10 in.²) of surface per day per 101 kPa (760 mm Hg) driving force at 23°C and humidity conditions of use; water permeation of not more than 7 g per 0.025 mm per 645 cm^2 per day in direct contact with water at 38°C and with low rh air circulating on the downstream side of the barrier; and less than 5% loss of an organic substance by absorption and/or diffusion from a solution of the substance in contact with the polymer for a period of at least 6 months at 23°C (or equivalent).

Table 1 lists currently available polymers meeting the requirements of high barrier polymers and compares their permeation rates for oxygen, carbon dioxide, water, and organic compounds. There are also several well-known polymers that come close to meeting the limits set for gas and water permeation and that can be considered as moderate barrier polymers: nylon-6; nylon-6,6; Delrin, Penton, poly(vinyl fluoride), poly(methyl methacrylate); nylon-11; and XT Polymer. Some polymers show excellent gas barrier properties but poor water barrier rates, eg, poly(vinyl alcohol) (dry) and cellophane (dry-uncoated); and others are poor gas barriers but good water barriers, eg, high density polyethylene, polypropylene, Teflon (polytetrafluoroethylene), polybutene, low density polyethylene, Surlyn ionomer, and butyl rubber.

Absorption from Dilute Solutions

A property related to barrier properties, but more subtle, is that of the absorption by the polymer of the molecules of a solution in contact with it. In many cases, these can be large bulky organic molecules, and the actual diffusion through the polymer matrix can be slow. But, because of the depletion of some molecules from the packaged solution, the properties of the product are altered as in the case of direct permeation of the product through the polymer. This is especially true if the absorbed molecules are related to taste, odor, or flavor of the contacting food or beverage. This phenomenon is directly proportional to the barrier properties of the polymer in most cases.

The main use of high barrier polymers is packaging, especially for foods and beverages, as a replacement for glass and metal containers. Light weight, nonshatterability, ease of disposal by incineration, and potentially lower costs are the forces behind the increasing popularity of barrier plastics. To be a successful food-and-beverage packaging material, the polymer must resist absorption from dilute solutions, retain carbon dioxide, protect food from oxygen, be durable, and have good creep strength, clarity, packaging processability, antistatic properties, and general chemical resistance (see Table 2).

The high nitrile polymers are the most interesting of the barrier polymers to be introduced. The permeation of any high nitrile polymer depends upon the level of nitrile, the type of nitrile, and the amount and type of comonomer and the presence of additives. Although the amount of nitrile is controlling, the comonomer can have significant barrier effects. For instance, whereas a 70:30 acrylonitrile–styrene copolymer

Table 1. Permeability Properties of High Barrier Polymers

Polymer	Polymer class	Permeation rates		
		Oxygen[a]	Carbon dioxide[a]	Water[b]
poly(vinylidene chloride)	halogenated	0.4	1.2	7.9
Lopac[c]	nitrile	3.9	12	200
Barex	nitrile	4.3	12	240
Cycopac[d]	nitrile	4.3	16	200
Saran wrap	halogenated	5.1	18	20
epoxy (bisphenol A; amine cure)	thermoset	12	35	160
Kel-F (polychlorotrifluoro-ethylene)	halogenated	13	47	12
Trogamid T	polyamide	18	47	205
Kynar [poly(vinylidene fluoride)]	halogenated	18	59	39
poly(ethylene terephthalate)	polyester	20–39[e]	47–79[e]	80–160[e]
nylon-6,9; nylon-6,10	polyamide	22	47	240
phenoxy [poly(phenylene oxide)]	aromatics	28	59	180
poly(vinyl chloride)	halogenated	31–59[f]	79–157[f]	80–120[f]

[a] At 23°C (100% rh), $(m^3 \cdot m)/(m^2 \cdot d \cdot PPa)$. To convert PPa to bar, multiply by 10^{10}.

[b] kg·cm/(im²·d) at 38°C (100% rh).

[c] Acrylonitrile (70%)–styrene copolymer used for manufacturing Monsanto Cycle-Safe container.

[d] A similar polymer, Vicobar (DuPont), is no longer made. Vicobar had similar barrier properties.

[e] Depends on exact level of crystallinity and orientation.

[f] Depends on exact compound formulation.

Table 2. Properties of True High-Barrier Polymers

Polymer	Tensile strength, MPa[a]	Drop impact (toughness)	Creep (cold flow)	Heat-distortion temperature, °C	Optical properties	Process-ability
Lopac, oriented	103	good	low	>90	clear	fair
Lopac, unoriented	76	poor	low	>90	clear	fair
Cycopac	69	good	low	90	clear	fair
Barex 210	55	good	moderate	71	clear	fair
Kel-F	41	good	moderate	>150	clear	poor
Trogamid T	62	fair–good	high	>90	clear	fair
Kynar [poly(vinyl-idene fluoride)]	41	fair	low	>90	opaque	poor
poly(ethylene terephthalate), unoriented	48	poor	low	>90	clear/opaque	poor
poly(ethylene terephthalate), oriented	>70	good	low	>90	clear	fair
nylon-6,10	55	good	moderate	>90	translucent	fair
phenoxy resin	62	fair	low	>120	clear	fair–poor
poly(vinyl chloride) (rigid)	52	fair–good	high	66	clear	fair

[a] To convert MPa to psi, multiply by 145.

has an oxygen permeation rate of 4.0, the corresponding acrylo-nitrile–methyl acrylate copolymer has an oxygen permeation of 3.2. This improvement in gas barrier is related to the polarity of the acrylate.

Poly(ethylene terephthalate) (PET), commonly called polyester, has a balance of properties and economics that make it useful for 1-L (34-oz) and 2-L (68-oz) soft-drink containers. Characteristics of the polyester bottle that have led to its commercialization fall into three categories: physical properties (clear, glossy, light, strong, and shatterproof); chemical properties (resistant to mineral and organic acids and weak alkalies at room temperature and to bleaches, oxidizing agents, and most common solvents and oils: protects taste of the beverage); and properties relevant to government regulations of PET as a food additive (*Title 21, CFR, 121 (F)*).

Restrictive packaging and antilitter legislation is at various stages of preparation or consideration in over 30 different state legislatures. Their passage could conceivably force a policy of returnable and refillable bottles. Currently produced PET bottles cannot be refilled.

S. STEINGISER
S.P. NEMPHOS
M. SALAME
Monsanto Company

M. Salame, *Polym. Sci. Technol.* **6**, 275 (1974).

M. Salame and E.J. Temple, *Adv. Chem. Ser.* (135), 61 (1974).

M. Salame, *Package Eng.* **17**(8), 61 (1972).

BATTERIES AND ELECTRIC CELLS, PRIMARY

INTRODUCTION

Devices that convert chemical energy into electrical energy are commonly referred to as batteries. In these devices, electrical energy is the result of chemical reactions that give products with lower energy content. When the higher energy compounds are formed by putting electrical energy into the battery and electrical energy is later withdrawn with formation of lower energy compounds, the battery acts as an electrical energy storage device, a secondary battery. Primary batteries are devices initially assembled with high energy chemical compounds, and stored chemical energy is withdrawn as electrical energy at some later time. Batteries sold in their charged state and discarded without recharge are called primary batteries. Fuel cells are a special class of primary batteries in which the high energy reactants are fed continuously into the battery and the low energy reaction products are removed continuously.

Today, a battery may be one or many individual cells. A single electrochemical cell is composed of a negative electrode and a positive electrode separated by an electrolytic solution. When the cell is discharging, converting chemical energy to electrical energy, an oxidation reaction occurs at the negative electrode or anode: $Zn \rightarrow Zn^{2+} + 2\ e$. At the positive electrode or cathode during discharging, a reduction reaction occurs: $Cl_2 + 2\ e \rightarrow 2\ Cl^-$.

In electrode reactions of any compatible pair of anode and cathode processes, often called an electrochemical couple, electrons pass through the external circuit from the anode to the cathode. The circuit is completed by ionic species transferring across the cell through the intervening electrolyte. The change from electronic conduction to ionic conduction occurs at the electrodes and involves an electrochemical reaction or Faradaic reaction. Electrons cannot pass from the positive to the negative electrode through the electrolyte. If that occurs, an electronic short exists, and the cell will self-discharge. This can cause cell overheating, pressure buildup, and rupture. This factor must be considered in battery cell design.

Stoichiometric reactions shown above consist of a sequence of more elementary steps that may occur at slightly different locations on a microscopic scale. However, the reactants must approach within molecular distances of one another and products must be removed on a continuing basis for a cell to operate properly. These steps imply that electrons from the external circuit must reach or leave the reaction sites. Reaction sites must, therefore, be electronically connected to the external circuit. Usually, ionic species in the electrolyte solution must move to or away from the reaction site. The solution-phase transport causes concentration variations near reaction sites affecting the rate of reaction. There may also be ion transport through solid phases. The energy losses associated with concentration variations result in lowered cell potential called concentration overpotential. Associated with concentration overpotential is the heat produced by the energy necessary to drive ions through the solution to carry the electrical current. This energy is often called the resistance loss or ohmic loss which appears as a reduction in the cell potential, called ohmic overpotential. In addition, energy is necessary to drive the reactions taking place. This driving energy comes from conversion of chemical energy which then degrades to heat rather than producing electrical energy. Losses associated with driving the transfer of charge at electrode reaction sites result in a reduction in the cell potential called surface overpotential. The irreversible heat generated by each loss process is the appropriate overpotential times the electrical current. Understanding these loss processes and minimizing them is an important factor in battery design. There is also a reversible heat associated with cell operation. The irreversible heats are always positive, adding heat to the cell. The reversible heat can be positive or negative. Besides voltage losses, losses in overall useful energy occur because of many types of unwanted side reactions, often called corrosion reactions.

The driving force pushing electrons through an external circuit is the change in the free energy: $G = H - TS$, where H is the enthalpy or heat content, S is the entropy, and T is the absolute temperature. If there are no corrosion reactions or other unwanted or unknown side reactions, the open circuit cell potential U is related to the free energy change as follows: $\Delta G = -nFU$, where ΔG is the free energy change for the overall cell reaction based on the reversible transfer of n equivalents of electrons through the external circuit and F is the Faraday constant (96, 487 C/equiv).

DOUGLAS N. BENNION
University of California, Los Angeles
present affiliation
Brigham Young University

J. Newman, *Electrochemical Systems*, Prentice-Hall, Inc., Englewood Cliffs, N.J., 1973.

P. Delahay, *Double Layer and Electrode Kinetics*, John Wiley & Sons, Inc., New York, 1965.

N.C. Cahoon and G.W. Heise, eds., *The Primary Battery*, Vols. 1 and 2, John Wiley & Sons, Inc., New York, 1976.

PRIMARY CELLS

The Leclanché cell, which uses an amalgamated zinc anode, an electrolyte of ammonium and zinc chlorides dissolved in water, and a manganese dioxide cathode, has dominated the primary battery market since its introduction in the 1860s. In present-day dry cells, the cathode is MnO_2 with ca 10–30 wt% acetylene black to improve the matrix conductivity. The anode is usually the zinc can or a zinc sheet. Amalgamation of zinc with mercuric chloride minimizes corrosion and associated hydrogen evolution. The electrolyte layer in the dry cell is also a separator which prevents a short circuit of the cathode and anode. Common forms of the separator are an electrolyte paste of flour and starch; the paste plus a thin kraft paper; and a layer of methyl cellulose bonded to one side of a paper sheet.

Effective sealing of the Leclanché dry cell is important in preventing loss of moisture from the battery during storage as well as access of oxygen which increases corrosion at the zinc electrode. Various waxes, asphalts, gilsonite (qv), and plastic materials have been used to close the top of the cell.

The two general categories for the Leclanché dry cell are the round and the flat cells whose difference is chiefly physical, not chemical. The cell has an open-circuit potential of 1.60 V when fresh. Discharge efficiency of the cell generally varies inversely with the discharge current density; shelf life is an important factor and can be controlling at low discharge rates. Service life also is dependent on the relative time of operation and recuperation period. The popularity of Leclanché dry cells is based in part on their relatively low cost, availability in many voltages and sizes, and suitability to intermittent and light-to-medium current-drain capability.

A recent innovation in Leclanché cell fabrication allows production of bipolar cells in a thin, flat, sandwich configuration. These batteries are used to operate automatic cameras and are packaged with the photographic film (see Color photography, instant). The cost of the fabrication machinery is high. However, production costs are low on a per-battery basis. This technique for battery manufacture shows considerable promise and can probably be applied to other chemical systems such as zinc alkaline cells. Principal manufacturers of Leclanché cells in the United States are Bright Star Industries, Burgess Battery, ESB-Polaroid, Marathon Battery, Ray-O-Vac, and Union Carbide.

The zinc chloride cell is better than the conventional Leclanché cell in several ways: higher discharge efficiency during heavy continuous use; longer life of the working voltage level; superior performance at low temperatures; and water is consumed during the discharge. There is probably no leakage because the cell is dry when fully discharged. Without the ammonium chloride, corrosion owing to oxygen is increased. A better sealing technique is needed for the zinc chloride battery so that its shelf life is equivalent to that of a Leclanché cell.

Alkaline cells use sodium or potassium hydroxide as the electrolyte. The first commercially successful alkaline cell using the zinc–copper oxide couple was developed in the 1880s. It has applications in railroad signal and track circuits where high capacity, low cost, and relatively constant operating voltage are required. Zinc is favored for the anode because of its low polarization losses, high exchange-current density, low equivalent weight, and strong reducing potential. In the group of cathodic materials, oxygen, oxides of silver, oxides of manganese, and oxides of copper are commercially significant. Alkaline cells generally perform better than Leclanché cells. The alkaline cells have a stable electrolyte which can be used over a wide temperature range. Zinc corrosion is slight, especially when the zinc is amalgamated. Very little electrolyte solution is required, and the cell can be very compact. The alkaline cells have higher energy densities and a more nearly flat voltage-discharge curve than the Leclanché cells.

The alkaline manganese dioxide–zinc dry cell has the same sloping discharge curve as the Leclanché cell. However, at high current drain, it sustains the operating voltage much better than the Leclanché system. It also has a good low temperature performance and a long shelf life. The cell system, MnO_2–KOH–Zn, is essentially the same as the Leclanché system except for the electrolyte. Commercially, the flat or crown and the cylindrical cell construction is available. The alkaline manganese dioxide–zinc cells can be used as rechargeable units with some changes in structural details (see Batteries, secondary).

Alkaline manganese cells are most suitable for use in movie-camera cranking, electronic flash devices, tape recorders, toys, shavers, and any heavy-discharge use. Principal manufacturers of primary alkaline manganese cells are P.R. Mallory, Ray-O-Vac, and Union Carbide.

The alkaline copper oxide–zinc (Lalande) cell, represented by CuO–$NaOH_{aq}$–Zn(Hg), is mainly used in railroad applications in track circuits, signals, and lighting. Despite a relatively low operating voltage (0.6–0.7 V), its high discharge capacity, flat discharge curve, and stability have made the Lalande cell an important commercial unit. The open-circuit potential of the cell is ca 1.1 V. Principal manufacturers are Carbone and T.A. Edison.

The mercuric oxide–zinc, or RM, cell has a relatively flat discharge curve, good high temperature performance, and a high ratio of discharge capacity to volume. It is used in transistorized equipment, walkie-talkies, photoelectric exposure devices, hearing aids, electronic watches, cardiac pacemakers, and as a secondary voltage standard. The RM cell, Zn(Hg)–KOH–HgO, is available in cylindrical, flat, or crown forms and as a rechargeable battery. Manufacturers include Burgess Battery, P.R. Mallory, Ray-O-Vac, and Union Carbide.

The mercuric oxide–cadmium cell, produced by Elca Battery, is similar to the mercuric oxide-zinc cell. Cadmium is more expensive than zinc and yields a lower cell voltage and theoretical or thermodynamic energy density. However, the cadmium mercuric oxide combination has shelf-life advantages, a large operating- and storage-temperature range, and voltage stability.

The silver oxide–zinc cell, Ag_2O–KOH (or NaOH)–Zn, has a high energy output per unit weight or volume and a fairly constant voltage level during discharge. Because of the high cost of silver and its compounds, the silver oxide–zinc cell has mainly been used in military applications. The button-shaped commercial unit is usually a miniature dry cell and has uses in hearing aids, electronic watches, photoelectric exposure devices, and as reference-voltage sources.

Monovalent silver oxide (Ag_2O) is usually used in commercial cells, although the development of the divalent silver oxide cell has been active, especially since the electrical capacity per gram of silver is twice as great in AgO as in Ag_2O. The open-circuit voltage of a silver oxide–zinc battery is either 1.85 V with AgO cathode or 1.60 V with Ag_2O cathode.

Sometimes NaOH is substituted for KOH in applications where service conditions are uniform, such as the operation of electric wristwatches. For use in hearing aids, where the relatively rapid loss in power as the cell approaches exhaustion is undesirable, the cathode contains a mixture of silver oxide and manganese oxide to give a gradual drop in working voltage. The cell's capacity is slightly over 100 mA·h, which is equivalent to 40–65-h service in a hearing aid or one year's operation of an electric watch. Manufacturers include Union Carbide, Mallory, and ESB.

The alkaline air–zinc cell, like the Lalande cell, has been used widely in railway applications. A small version also has been used in hearing aids. The cell has a large capacity, a fairly constant voltage with moderately large current, and a long storage life. Its voltage and capacity (W·h) are roughly double that of the Lalande cell, and it has a free cathodic reactant. Recent advances in zinc–air cells have yielded improved small batteries for long life, low power applications such as watches, flashing barricade lights, and navigation aids. A dry-cell version of the air-depolarized zinc cell is under development. The immediate objective is to produce miniature units that can be used in hearing aids and electronic watches to replace mercuric or silver oxide batteries. Main manufacturers of air–zinc cells are Carbone, T.A. Edison, Gould, and Union Carbide.

Lithium batteries. Because of its low equivalent weight and strong reducing potential, lithium is an attractive material to use as the reducing agent in producing high energy density primary batteries. Lithium batteries with iodine positive electrodes are being produced by at least four companies for use in cardiac pacemakers. Values as high as 0.73–0.95 (W·h)/cm^3, exceptionally high energy densities, are claimed for the $SOCl_2$ batteries. Further development of lithium batteries appears likely because of their inherent possibilities for high energy density, high reliability, and long shelf life.

A primary lithium battery can be built by pressing an inert wire screen of nickel or iron against a sheet of lithium with a backing plate. A lithium hydroxide–water solution is pumped around the wires of the screen between the backing plate and the lithium. When stored in a dry, inert atmosphere, such as argon or helium, lithium–water batteries can have indefinite storage life. They are activated by introducing LiOH–water solution into the flow channels. Besides use as a reserve battery, the lithium–water cell appears promising for underwater and other applications.

Sulfur dioxide, thionyl chloride, and phosphoryl chloride are all soluble positive-electrode reactants. The positive electrode is typically a paste of carbon powder mixed with a Teflon binder spread on a nickel grid. An electrolyte such as $LiAlCl_4$, $LiBCl_4$, or LiBr is dissolved in the liquid $SOCl_2$ or $POCl_3$. In the case of SO_2, nonaqueous solvents such as propylene carbonate and acetonitrile are added to decrease the SO_2 vapor pressure and increase electrolyte solubility and ionic conductivity. The negative electrode is typically sheet lithium pressed into a nickel screen or grid. Often, the positive and negative electrodes are made in the form of long, narrow sheets. A key feature of the cell design is the glass-to-metal seal (G/M seal) through which the positive lead passes without contacting the case which is at the potential of the negative electrode.

Considerable hazards are involved with nonaqueous batteries. Whenever high energy density storage is achieved, as in these batteries, the possibility for uncontrolled release of the energy exists. Nonaqueous batteries must be used with care, and safe handling procedures must be well developed and carefully followed to prevent explosion and fire. In considering explosion protection of these systems, it must be remembered that $SOCl_2$ and $POCl_3$ are highly corrosive to many materials including most electronic devices. Extensive venting in a closed space may severely damage the device the battery is designed to power. Furthermore, these materials are quite toxic.

A large number of nonaqueous, room temperature or near room temperature, battery systems have been investigated in recent years. Many of these batteries use lithium as the reducing agent at the negative electrode. A few of the systems have been produced on a limited basis for cardiac pacemakers, watches, small electronic devices, and special military uses. Some of the systems are designed for high energy density, high power, short life applications, eg, a power supply for a guidance system in a military weapon. Others are designed for very long life and low drain rates as required in a cardiac pacemaker. All must have high reliability. Achieving reliability has been one obstacle preventing wider use of nonaqueous batteries. Cells of the type Li–LiI–I_2 are gaining wide acceptance in cardiac pacemakers.

Fused-salt cells have been used for many years as reserve batteries in military weapons, including nuclear weapons. They are reserve batteries in the sense that they are assembled in the solid state at room temperature and activated by heating at the time of use.

The development of pellet technology has reduced assembly cost, extended operating life of fused-salt cells to an hour, and has allowed a simpler, more easily characterized battery. The basic battery assembly is made by stacking three types of pellets in an insulated casing. The cell is safe and easy to handle in the pelletizing and assembly steps, and all materials are assembled in dry rooms in which people work quite comfortably with an atmosphere containing < 150 ppm water (0.5% rh at 22°C). Although these pellet batteries have been used only for specialized military applications to date, that type of assembly may eventually be used in longer life, more general-purpose batteries.

Hiram Gu
University of California, Los Angeles
present affiliation
General Motors

Douglas N. Bennion
University of California, Los Angeles
present affiliation
Brigham Young University

K. Fester, R.L. Doty, and R.V. Velden, *Sixteenth Annual ASME Symposium on Energy Alternatives*, Albuquerque, New Mexico, Feb. 26, 1976.

B.H. Van Domelen and R.D. Wehrle, *9th Intersociety Energy Conversion Engineering Conference*, 1974.

FUEL CELLS

Fuel cells carry out the electrochemical reaction of conventional fuels with oxygen, preferably from air. A fuel cell thus is a continuous-feed electrochemical cell in which energy from such a reaction is converted directly and usefully into (low voltage) direct-current electrical energy. This definition excludes metal fuels and oxidants other than oxygen.

Reformed Hydrocarbon–Air Fuel Cells

The most practical approach to the use of hydrocarbons in fuel cells is the conversion of the hydrocarbon fuel to a hydrogen-rich mixture of gases via the reforming reaction with steam, and the electrochemical conversion of the hydrogen-rich fuel stream in the fuel cell.

The phosphoric acid electrolyte fuel-cell system has given rise to the largest fuel-cell system ever built and is probably the most highly developed of all fuel-cell systems. The primary reasons for the use of phosphoric acid (85–95 wt%) are ability to operate with ambient air containing CO_2, ability to operate with a thin matrix electrolyte (no liquid circulation), chemical stability of the electrolyte over a fairly wide temperature range, and relatively low cost. Systems of this type have been developed by the Power Systems Division of the United Technology Corporation in a series of progressively larger power plants which have reached a 4.8-MW size. The fuel-cell system has a high thermal efficiency and is capable of a much lower level of pollutants than conventional power-generation systems.

The present accumulated operating time of fuel-cell stacks is reported to be in the range of 6000 h for a 40-kW unit. In large-scale units for dispersed power generation (26-MW size), the goal is a stack-life of 40,000 h.

Currently, there are two reformed hydrocarbon–phosphoric acid–air fuel-cell systems undergoing further development. The smaller unit, rated at 40 kW, is designed for use with apartment complexes and small shopping centers. The other is 26.4 MW and intended for use in dispersed power-generation systems. The inverter is an important part of the entire system since its efficiency influences the fuel-cell system efficiency. Fuel cells for apartment complexes and shopping centers supply electrical energy for common needs, but the waste heat energy can also be used to supply hot water; heat pumps can be used for space heating and cooling to provide maximum overall energy efficiency (35–41%). Compared to more conventional approaches to heating and cooling, considerable energy savings are possible.

Molten Carbonate Electrolyte Fuel Cells

The overall reaction that takes place in the molten carbonate fuel-cell system is $H_2 + 1/2\ O_2 \rightarrow H_2O$. Since the operating temperature of this fuel-cell system is 500–700°C, the electrode reactions are fast enough so that high specific area catalytic electrodes like those for lower temperature operation are not needed. A common electrode material for both electrodes is porous nickel. Many of the cell components are either nickel or stainless steel. The main exceptions are the electrolyte filler, which can be one of several ceramic compositions, and the electrolyte, which consists of mixtures of alkali-metal carbonates with low melting points.

Overall efficiency of the molten carbonate fuel-cell system with reformed hydrocarbon fuel depends not only on the reformer efficiency but also on the ability of the fuel from the fuel stream. Endurance has always been a problem with the molten carbonate cell. One of the more serious causes for short life has been the instability of the tiles that contain the electrolyte. Another problem has been the loss of the active area and electrode porosity, as well as poor contact between the electrodes and the past electrolyte. Significant progress has been made in the areas of stable electrolyte matrices, more sinter-resistant electrodes, and generally more stable cell structures. Provided that cell durability at higher power densities can be improved, and the electrolyte loss problem can be solved, it is quite probable that the molten carbonate fuel-cell system with reformed hydrocarbon fuel will represent the next step in improved fuel-cell systems for dispersed power generation.

Solid Oxide Electrolyte Fuel Cells

This type of fuel cell makes use of ceramic electrolytes which consist of zirconium or cerium oxides doped with alkaline-earth or rare-earth oxides to enhance the ionic conductivity (see Ceramics). Since the oxide electrolyte does not require the addition of carbon dioxide to the cell (as is the case for molten carbonate cells), a simple overall fuel-cell system for reformed hydrocarbons as well as other fuels might be expected, but the advantages have not materialized because of flaws and electronic conduction in the electrolyte.

Ammonia Fuel Cells

Ammonia is an easily stored, relatively inexpensive liquid fuel free of problems associated with petroleum-based fuels. There have been two approaches in the use of ammonia in fuel cells: directly feeding ammonia to the fuel-cell anode and feeding the ammonia to a catalytic reactor where it is decomposed to a mixture of hydrogen and nitrogen.

Research on direct ammonia fuel cells shows current densities approaching 1 A/cm^2 at the peak power density of 175 mW/cm^2, all at 140°C with 54% KOH electrolyte. These high current densities allow the cell to be started at room temperature and self-heated to at least 140°C although overvoltages are high.

The indirect ammonia fuel cell allows the use of an alkaline-electrolyte H_2-air fuel cell in a straightforward manner, but the catalytic decomposer adds complexity and requires energy.

Compact Fuel Cells: The Dynamic Mass-Transport Concept

The Alsthom fuel cell has thin electrocatalyst layers bonded by polymers to the faces of thin goffered-metal bipolar current collectors and a thin porous polymeric separator. These two elements are stacked to form a fuel battery with a cell thickness of 0.5 mm and an active area of 100 cm^2. This type of cell has been operated most successfully using liquid reactants that are miscible with aqueous KOH electrolyte. Because of the very small interelectrode distance, relatively dilute electrolyte can be used without suffering unreasonably large resistance losses and severe corrosion problems. A complete Alsthom fuel-cell system includes the stacks of bipolar cells, pumps for fuel and oxidant stream circulation, gas–liquid separators for removing gaseous products such as N_2, heat exchangers for cooling, provision for injecting fuel and oxidant into the recirculating streams, and storage tanks.

Complete N_2H_4–KOH–H_2O_2 systems of power ratings of 1–100 kW have been built and tested. A number of applications including special-purpose vehicles, standby power, and undersea vehicles have been considered. This system appears to be very attractive for powering special-purpose submersibles because it is near-neutral in buoyancy and is nearly unaffected by the pressure of the sea. Extension of this approach to the use of other fuels and oxidants has been less successful.

Fuel-Cell Powered Vehicles

The first fuel-cell powered road vehicles were the U.S. Army M-37 (0.68 t or 3/4 short ton) truck, a fuel-cell van, and an automobile. The M-37 truck was powered by a hydrazine–air fuel battery (provided by Monsanto) made of four cell modules of porous nickel electrodes separated by a porous mat and stacked with external connections. The fuel usage was 2.1 km/L (4.9 mi/gal). When the 20-kW fuel battery was supplemented by a 20-kW lead–acid power battery, the truck accelerated to 72.5 km/h (45 mph) in 25 s, slightly faster than for the 70 kW (94 hp) gasoline engine that was replaced. The emission of NH_3 (with a $CuSO_4$ scrubber) was too high for continuous exposure (150–2500 ppm, depending on location on the vehicle, vs 100 ppm maximum allowable). Thus, the hydrazine fuel cells have an acceptable specific power for vehicle propulsion, but their cost is too high, lifetime too short, fuel too costly, and NH_3 emissions too high for practical application.

The hydrogen–oxygen powered van was constructed and tested by General Motors, making use of Union Carbide fuel cells and cryogenically stored hydrogen and oxygen (see Cryogenics). The peak power capability of the fuel battery was 160 kW, and the continuous rating was 32 kW. The system's open-circuit voltage was 550 V; the maximum

current was set at 400 A. The electric drive system had a peak power of 93 kW and a top speed of 13,700 rpm.

The most recent fuel-cell powered vehicle to be reported was a converted Austin A40 4-passenger automobile with a hybrid powerplant of Union Carbide H_2–KOH–air fuel battery with a lead–acid battery for peak power. The total system delivered 36.4 kW·h and had a peak power of 22 kW, providing a top speed of 88 km/h (55 mph).

The technical feasibility of fuel-battery powered vehicles has been demonstrated, but problems remain in system weight, volume, cost, and lifetime.

For wide acceptance of fuel cells, it is necessary to develop low cost catalyst systems and substrates to replace the expensive platinum–carbon systems.

ELTON J. CAIRNS
ROMEO R. WITHERSPOON
General Motors Research Laboratories

H.A. Liebhafsky and E.J. Cairns, *Fuel Cells and Fuel Batteries*, John Wiley & Sons, Inc., New York, 1968.

G. Sandstede, ed., *From Electrocatalysis to Fuel Cells*, University of Washington Press, Seattle, Wash., 1972.

BATTERIES AND ELECTRIC CELLS, SECONDARY

INTRODUCTION

The chemical, electrochemical, and physical properties associated with the primary battery are similar to those utilized in the secondary battery system. The nature of the electrochemically active material is the main difference between these systems. For example, the zinc–manganese primary battery has a zinc metal can as the anode, whereas the zinc–silver oxide secondary battery uses a zinc powder anode.

When primary batteries are completely discharged, they are discarded. However, the secondary battery can discharge its energy and be recharged as many as several thousand times depending on the use and operating conditions. Recharging is accomplished by forcing an external current through the battery in a direction opposite to the current flow during discharge in order to restore the electrochemically active materials to their original charged condition. This is the chief advantage of the secondary battery for which a penalty in energy density or initial cost usually prevails.

There is a limit to how many times a battery can be recharged before it no longer has a reasonable recharge efficiency (A-h) and/or becomes unchargeable. Reasons for this failure are crystal growth, decrease in porosity, and reduction of active material particle size; corrosion of conductive plate grids and cell or battery connectors; loss of active material on the electrode or plate by mechanical means (such as vibration, shock, electrostatic stresses, expansion, and contraction), electrolyte erosion of active material, dissolution and nonredeposition of active material, and formation of an electrically nonconducting species; separator deterioration by active material (oxidant) and temperature, and perforation by metallic dendritic growths.

The term secondary battery as used here means more than one secondary cell connected to another for purposes of increasing voltage or energy capacity.

The largest segment of the U.S. battery industry is the lead–acid battery for automotive starting, lighting, and ignition (SLI). In most cases, the suitability of the application of secondary batteries (lead–acid (SLI, industrial, and sealed) and nickel–cadmium (sintered plate, pocket plate, and sealed)) is more important than the cost of the battery. The principal uses for the sealed lead–acid batteries are in security and alarm systems and emergency lighting. The sintered-plate nickel–cadmium battery is used in extremely high discharge rate functions, such as starting small jet and helicopter engines used in military and business applications and starting auxiliary power units in large military jets and helicopters and in commercial aircraft. A much broader range of applications for pocket-plate nickel–cadmium batteries includes emergency lighting, electrical switchgear, security and alarm systems, railroad track-and-signal applications, heavy-duty engine cranking, emergency control and lighting batteries for rapid-transit cars, railroad-car lighting, and standby emergency power systems. The sealed nickel–cadmium battery has extensive use in cordless tools and pocket calculators.

Electrochemistry

When a secondary-battery reaction occurs spontaneously, there is an accompanying decrease in the energy content of the system. This energy change (Gibbs free energy) is calculated by the equation:

$$\Delta G = -nFE$$

where ΔG = Gibbs free energy corresponding to the appropriate secondary-battery reaction, J/equiv; n = number of equivalents of electrons transferred in the stoichiometric reaction; F = Faraday's constant, 96,487 C/equiv; E = reversible whole cell potential (electromotive force, emf), V.

Cell and Battery Design

Design parameters. The basic battery building block is the cell. Its constituent parts are the negative and positive electrodes or plates, an ionically conducting electrolyte, and usually a separator which prevents direct contact between the negative and positive plates and retains the electrolyte within its structure. The enclosure of these elements in a case constitutes a completed cell. The cell and battery configurations are available in prismatic, cylindrical, and button-type units.

Cell-Battery Operation

Cell-battery discharge characteristics. The initial discharge period during which the cell voltage decays below the starting voltage may have a combination of causes. Surface or concentration overpotential may be developing, concentrations of products may be changing (causing a shift in the reversible potential), or resistive films or layers may be forming. Occasionally, at high currents, these phenomena occur so rapidly they are not observed. When all overpotential phenomena have reached a steady state, the discharge plateau begins. The last period is reached when the cell is almost exhausted. Decrease in voltage may result from a decrease in the quantity of reactant in the solution or on the surface of the electrode as these reactants are changed to the undischarged state. This decrease in quantity of reactants causes an increase in current density at the remaining surface of the charged-state materials and an increase in both concentration and surface overpotentials. Voltage losses also can be caused in part by the increase of ohmic resistance in the cell.

Cell-battery charging characteristics. The first period of cell-battery charging involves the establishment of overpotential, the second period the plateau voltage, and the third period the exhaustion zone. The causes of the observed phenomena are similar to those mentioned for discharge.

Heat Generation

Regardless of the type of secondary battery, heat will be generated by $T\Delta S(\Delta H - \Delta G)$, Joule heating ($I^2R$ from intercell-battery connectors) and overpotential (resistance, concentration, and surface). The latter two represent irreversible heat rejections. The necessity to remove this heat depends on the battery application and the cell-battery design; otherwise a deterioration in battery performance (eg, capacity, retention, voltage, output, charge efficiency) and battery components may result. Depending on the battery temperature, the heat removal can be achieved by radiation, free or natural convection, forced convection, or forced liquid cooling (see Heat-exchange technology).

An expression used to calculate the rate of heat generation Q under reversible conditions and at constant pressure (except where gases are directly involved, the effects of pressure change tend to be small) is

$$Q = I \int_{t_1}^{t_2} \Delta E \, dt + \phi T \Delta S$$

where Q = total heat evolved ($-$) or absorbed ($+$), W·h; I = current, A; $\Delta E = (E_{ocv} - E_{ccv})$ = overpotential, V; E_{ocv} = battery open circuit voltage V; E_{ccv} = battery closed circuit voltage, V; t, dt = time, h; ϕ = equivalents; T = temperature, K; ΔS = reaction entropy, J/K. The expression is valid when the temperature is constant and when the current is held constant and the voltage allowed to vary, and it expresses the rate of heat evolution or absorption resulting from any secondary battery's electrochemical process as a function of the voltage at the battery, the current through the battery, and the entropy change for the electrochemical process.

James B. Doe
ESB Technology Co.

E. Yeagen and A.J. Salkind, eds., *Techniques of Electrochemistry*, Vol. 3, Wiley-Interscience, a division of John Wiley & Sons, Inc., New York, 1978.

ALKALINE CELLS

Active materials for the positive electrodes of alkaline secondary cells have been fabricated from the hydroxides or oxides of copper, nickel, silver, mercury, and manganese, and for the negative electrodes from various geometric forms of iron, cadmium, and zinc. In addition, hydrogen has been investigated as a negative electrode-active material and the nickel hydroxide-hydrogen battery is in limited-scale production.

Nickel Positive Electrode Systems

Nickel–cadmium cells. A number of different types of nickel oxide electrodes have been used in fundamental studies and practical systems. (The term nickel oxide is a common usage.) Actually, the active materials are hydrated hydroxides at oxidation state $2 +$ for the discharged condition and oxidation state $3 +$ in the charged condition. The many varieties of practical nickel electrodes can be divided into two main types. In one, nickel hydroxide active material is blended, admixed or layered with an electronically conductive material. The so-called pocket, tubular, and most button-cell nickel electrodes fall into this category. The other main type of nickel electrode includes those constructions in which the active material is not chemically prepared away from the electrode but is deposited *in situ* on or in the pores as well as some new types in which nickel hydroxide is electrodeposited on a conducting nickel sheet or a metallic fiber. Almost all the variations indicated for the nickel electrode have been used to fabricate cadmium electrodes.

A fundamental understanding of the behavior of the nickel electrode is not complete. A charge–discharge mechanism can be rewritten in simplified form as:

$$2 \ \beta\text{-NiOOH}^* + \text{Cd} + 2 \ \text{H}_2\text{O} \underset{\text{charge}}{\overset{\text{discharge}}{\rightleftharpoons}} 2 \ \text{Ni(OH)}_2^* + \text{Cd(OH)}_2$$

The asterisks represent differing amounts of hydrated or absorbed water and/or electrolyte.

The chemistry, electrochemistry, and crystal structure of the cadmium electrode is much simpler than that of the nickel electrode. The overall reaction is recognized as:

$$\text{Cd} + 2 \ \text{OH}^- \underset{\text{charge}}{\overset{\text{discharge}}{\rightleftharpoons}} \text{Cd(OH)}_2 + 2 \ e$$

At higher temperatures, the nickel–cadmium cell has a slightly lower open-circuit voltage and this, combined with a reduction of internal resistance at higher temperatures, can result in a reduced back emf in charging. Therefore, one must take special precautions in charging under constant-voltage conditions to avoid the so-called runaway condition; ie, an increase in charge current because of a reduction in back emf which results in a subsequent increase in temperature and a new increase in charge current. Cells can destroy themselves in this runaway condition unless current limiters are provided in the charge circuit.

The maximum capacity that can be achieved normally from nickel active material is 0.30 A·h/g calculated on the basis of Ni(OH)_2. The capacity of negative active material is ca 0.37 A·h/g based on Cd(OH)_2.

Pocket electrode nickel–cadmium cells. These cells are made in two main process steps: manufacture of the electrodes; and assembly of the cell. Originally, nickel-plated steel was the predominant material used for cell containers but, in the last few years, plastic containers have been used for a considerable proportion of pocket cells. Polyethylene, high-impact polystyrene, and a copolymer of propylene and ethylene have been the most widely used plastics (see Olefin polymers; Acrylonitrile polymers).

The cells usually are filled with an electrolyte solution of potassium hydroxide of density 1.18–1.23 g/cm³ which may also contain lithium hydroxide to increase the life of the positive pocket electrodes in cycling operation. Individual cells are precycled before assembling into batteries. These early charge–discharge cycles, often called formation cycles, improve the capacity of the cell by increasing the surface area of the active material and effecting crystal structure changes.

Tubular nickel electrode. Although the tubular nickel electrode invented by Edison is almost always combined with an iron negative electrode, a small quantity of cells is produced in which it is used with a pocket cadmium electrode. This construction is used for low operating temperature environments, where iron electrodes do not perform well, or where charging current must be limited.

Sintered-electrode batteries. The fabrication of these batteries can be divided into five major operations: preparation of sintering-grade nickel powder; preparation of the sintered-nickel plaque; impregnation of the plaque with active material; assembly of the impregnated plaques (often called plates) into electrode groups and into cells; and assembly of cells into vented or sealed batteries. By far the majority of sealed cells is of the small, cylindrical self-supporting type in the familiar AA, C, and D commercial sizes.

The button cell. The button cell, used for calculators and other electronic devices, is commonly made using a pressed-powder nickel electrode mixed with graphite similar to a pocket electrode. The cadmium electrode is made in a similar manner.

Lower cost and lower weight cylindrical cells have been made using plastic-bound or pasted active material pressed into a metal screen.

Nickel–cadmium cells represent over 7% of the market of all storage batteries (including lead–acid) manufactured in the noncommunist world. Their uses are divided mainly into three categories. Pocket cells are used in emergency lighting, diesel starting, and stationary and traction applications where the reliability, long life, medium high rate capability, and low temperature performance characteristics warrant the extra cost over lead-acid storage batteries. Sintered, vented cells are used in extremely high rate applications, such as jet-engine starting and large diesel-engine starting. Sealed cells (sintered and button types) are used in portable tools, electronic devices, calculators, cordless razors, toothbrushes, carving knives, flashguns, and in space applications. Nickel–cadmium is optimum because it can be recharged a great number of cycles and given prolonged trickle overcharge. Cells of this category are made in sizes comparable to conventional dry cells.

Nickel–iron batteries. Nickel–iron batteries constitute only a minor part of the commercial use of alkaline storage batteries. There is some renewed interest in this system because of the plentiful nature of the components for long-life electric-vehicle batteries. A simplified equation representing the cell's charge–discharge cycle can be given as

$$2\ NiOOH^* + Fe + 2\ H_2O \underset{\text{charge}}{\overset{\text{discharge}}{\rightleftharpoons}} 2\ Ni(OH)_2^* + Fe(OH)_2$$

The asterisks indicate adsorbed water and KOH.

Lithium hydroxide is present in the electrolyte of nickel–iron cells and has been shown to have a beneficial effect on the life cycle. Cobalt is usually present in the nickel active material to aid in charge efficiency; and sulfur as FeS was incorporated by Edison to minimize oxidation and passivation of the iron electrode.

The construction methods usually employed in assembling nickel-iron cells are intended to give a product of extreme ruggedness and long life (10–25 yr). The electrodes are not close, and the cells are not optimized for low internal resistance. As a consequence, there is little danger of thermal runaway on charge. The iron electrode does not charge efficiently, and 30–80% overcharge is required to maintain cell capacity.

This results in more loss of electrolyte by electrolysis than with any other alkaline storage battery, ie, greater rewatering and maintenance are required. However, most of these maintenance problems have been eliminated by self-watering systems.

Nickel–iron cells are always designed to be limited in capacity by the amount of active materials in the positive electrodes. Nonetheless, modern cells using sintered nickel and sintered iron electrodes intended for electric vehicles exceed 40 W·h/kg even at high discharge rates (1 h).

The classic tubular battery designed by Edison was manufactured in three main process steps: chemical preparation of active materials and nickel flake; manufacture of electrodes; and assembly of cells and batteries. Several other constructions of nickel–iron cells have been tested recently as prototypes. All of them aim to achieve higher energy densities than the classical Edison construction and are intended for electric-vehicle applications. The approaches are toward closer plate spacing, higher electrolyte concentration, lighter-weight supporting structures; and larger active material-to-support structure ratio.

Nickel–zinc cells. Nickel–zinc cells with energy densities in the range of 40–60 W·h/kg have successfully been demonstrated. The overall reactions in the nickel–zinc cell can be represented by

$$2\ NiOOH + Zn + 2\ H_2O \underset{\text{charge}}{\overset{\text{discharge}}{\rightleftharpoons}} 2\ Ni(OH)_2 + Zn(OH)_2$$

$$E_0 = +1.73\ V$$

State-of-the-art nickel–zinc batteries are housed in molded plastic cell jars of styrene, SAN or ABS material for maximum weight savings. Nickel electrodes can be of the sintered or pocket type; however, both of these types are not cost effective. In recent years, several different types of plastic-bonded nickel electrodes have been developed. Negative electrodes are fabricated of zinc oxide and separators are both of the organic and inorganic type.

The occurrence of the oil embargo in 1973 and the realization of the ultimate limitation of fossil-fuel resources has created further interest in electric-vehicle development. At present there are several private companies as well as government agencies engaged in this effort which is considered the application most likely to result in a commercial nickel–zinc battery. If the problems of limited life and high installation cost ($100–150/(kW·h)) are solved, a nickel–zinc EV (electric vehicle) battery could provide twice the driving range of available lead–acid batteries. However, at present there is no commercial production of nickel–zinc batteries.

Nickel–hydrogen cells. In the mid-1970s, nickel–hydrogen cells were developed in the United States and in Europe to overcome some of the problems associated with deep cycle nickel–cadmium, long-life cells for space-satellite use. The memory effect of the cadmium electrode was eliminated and gravimetric energy density improved. The overall electrochemical reaction is simple:

$$Ni(OH)_2 \underset{\text{discharge}}{\overset{\text{charge}}{\rightleftharpoons}} NiOOH + \tfrac{1}{2}\ H_2$$

However, the generation and migration of water in the half-cell reactions must be considered in the cell design. The cells are designed as pressure vessels and are cylindrical in shape with hemispherical caps. Electric terminals are made with ceramic-to-metal seals in the end caps. The positive electrodes are of a conventional sintered type. The hydrogen-reacting (negative) electrode usually consists of a Teflon-bonded platinum black layer and a porous Teflon layer pressed into a fine-mesh nickel screen.

Silver Positive-Electrode Systems

The silver–zinc battery. The silver–zinc battery has the highest attainable energy density of any rechargeable system in use today. In addition, it has an extremely high rate capability coupled with a very flat voltage-discharge characteristic. Its use at present is limited almost exclusively to military applications such as various aerospace applications, submarine and torpedo propulsion applications and some limited military portable-communications applications. The main drawback of these cells is the limited life of the silver–zinc system. Its life is normally less than 200 cycles with a total wet life of no more than about 2 yr. This along with the very high cost of the silver–zinc system makes its use justified only in applications where cost is a minor factor. Cellophane or its derivatives are used in most silver–zinc batteries as the basic separator. The electrolyte is KOH of 30–45% concentration. Cell jars are constructed almost exclusively of injection-molded plastics which are resistant to the strong alkali electrolyte used in silver–zinc cells. The most often-used materials are modified styrenes or copolymers of styrene and acrylonitrile (SAN). Charging of the cells most often is done with the constant-current method which consists of a single-rate current usually equivalent to a full input within the 12–16 h period.

There is still considerable difference of opinion on the specific cell reactions that occur in the silver–zinc battery. Equations that are readily acceptable are the following:

$$2\ AgO + Zn + H_2O \underset{\text{charge}}{\overset{\text{discharge}}{\rightleftharpoons}} Ag_2O + \begin{matrix} Zn(OH)_2 \\ (ZnO + H_2O) \end{matrix} \qquad E_0 = 1.85\ V$$

$$Ag_2O + Zn + H_2O \underset{\text{charge}}{\overset{\text{discharge}}{\rightleftharpoons}} 2\ Ag + \begin{matrix} Zn(OH)_2 \\ (ZnO + H_2O) \end{matrix} \qquad E_0 = 1.59\ V$$

Performance of silver–zinc cells is considered to be adequate in the temperature range of 10–38°C.

Silver–cadmium cells. In the late 1950s, interest was shown in silver–cadmium cells for use in appliances, power tools, and scientific satellites. It was hoped that silver–cadmium batteries could offer an energy density close to silver–zinc batteries and a life characteristic approaching nickel–cadmium batteries. In these applications, the energy density of the silver–cadmium system was attractive; and in the satellite applications the nonmagnetic property of the battery was of utmost importance; magnetometers were used on satellites to measure radiation and the effects of magnetic fields of energetic particles. In addition the aerospace application required sealed batteries. Another recent application for silver–cadmium batteries is propulsion power for submarine simulator target drones.

The overall reactions are

$$2\ AgO + Cd + H_2O \underset{\text{charge}}{\overset{\text{discharge}}{\rightleftharpoons}} Ag_2O + Cd(OH)_2 \qquad E_0 = 1.38\ V$$

$$Ag_2O + Cd + H_2O \underset{\text{charge}}{\overset{\text{discharge}}{\rightleftharpoons}} 2\ Ag + Cd(OH)_2 \qquad E_0 = 1.16\ V$$

The positive plates are sintered silver on a silver grid and the negative plates are fabricated from a mixture of cadmium oxide powder, silver powder, and a binder pressed onto a silver grid. The main separator is four or five layers of cellophane with one or two layers of woven nylon on the positive plate. The electrolyte is aq (50 wt%) KOH. In the aerospace applications, the plastic cases were encapsulated in epoxy resins (see Embedding). Energy densities of sealed batteries are 26–31 W·h/kg.

Silver–iron battery. One of the latest unusual combinations of electrodes to receive attention is the silver–iron battery. This system combines the advantages of the high rate capability of the silver electrode with the cycling characteristics of the iron electrode. The commercial development has been undertaken by Westinghouse Electric in conjunction with Yardney Electric to solve problems associated with deep cycling of high power batteries for ocean-systems operations.

Cells consist of porous sintered silver electrodes and high rate iron electrodes. The latter are enclosed with a seven-layered, controlled-porosity polypropylene bag which serves as the separator. The electrolyte contains 30% KOH and 1.5% LiOH. At the 3-h rate, the energy density is 100 W·h/kg, or about 3.5 times the energy density of comparable lead–acid batteries.

Applications have been found for the batteries in emergency power applications for telecommunications systems in tethered balloons. Because of the high cost of the silver electrode, applications are sought where raw materials can be recovered. In Sweden, O. Lindstrom has produced small silver–iron sealed button cells for long-life operation of hearing aids, calculators and electric razors.

Manganese, Mercury and Other Systems

Alkaline MnO₂–zinc secondary cells. These cells have been manufactured in three common dry cell sizes, D, F, and G. However, they have not been marketed as individual cells but as nominal 12-V batteries assembled in a metal case. Under controlled conditions, 50–60 cycles are obtained.

Mercuric oxide–cadmium battery. The mercuric oxide–cadmium battery, although of a lower potential than the mercuric oxide–zinc battery, is capable of long storage life at elevated temperatures. The overall cell reaction is

$$HgO + H_2O + Cd \underset{\text{charge}}{\overset{\text{discharge}}{\rightleftharpoons}} Hg + Cd(OH)_2 \qquad E_0 = +0.907\ V$$

Both of these mercury batteries are produced as sealed button cells.

Cadmium–Cadmium Coulometer

Ampere-hour meter. A great deal of effort has been devoted to developing charge-control equipment for nickel–cadmium cells and batteries. This development has been aimed especially at devices capable of recharging sealed-cell batteries in less than 1 h, such as the cadmium–cadmium coulometer, developed at the Canadian Defense Research Establishment.

Recent work has been conducted on iron–air and zinc–air secondary batteries in many countries. Both stationary electrolyte and pumped electrolyte systems have been studied, and long cycle life at low discharge rates has been achieved. Prototype batteries achieved energy densities of over 100 W·h/kg. However, the electrical turnaround efficiency is very low, and the electrolyte becomes contaminated with the CO₂ to form a carbonate. Nevertheless, these cells are being considered for some hybrid vehicles as high power batteries and in some electric vans where high discharge rates are not required.

In selecting the optimum battery system for a particular application, important considerations are energy density, discharge characteristics, charge characteristics, A·h and W·h efficiency, charge retention, cycle life, mechanical properties, cost and availability, and maintenance requirements.

The classic fabrication techniques are being modified to take into account environmental and personnel safety as well as consumer protection needs.

ALVIN J. SALKIND
EUGENE PEARLMAN
ESB Technology Co.

Yardney Elect. Corp., *Design and Cost Study Zinc–Nickel Oxide Battery for Electric Vehicle Propulsion in Final Report*, ANL Contract No-109-38-3543, Oct. 1976.

D.P. Boden and E. Pearlman in D.H. Collins, ed., *Power Sources 4, Processings of 8th International Symposium*, 1972.

A.J. Salkind and G.W. Bodamer in D.H. Collins, ed., *Proc. 4th Int. Battery Symp. Brighton, Eng.*, 1964.

LEAD–ACID

The lead–acid battery consists of a number of cells in a container. These cells contain positive (PbO₂) and negative (Pb) electrodes or plates, separators to keep the plates apart, and sulfuric acid electrolyte. The electrochemical system is highly reversible and can be discharged and charged repeatedly before failure of some sort causes the charge-cycle to be impractical. The most widely used secondary battery is the lead–acid type. This battery is available in many sizes and capacities, and the weight can vary from ca 100 g to several metric tons. There are three principal categories: *Automotive*: SLI (starting, lighting, and ignition) for cranking of internal combustion engines is the main application of the lead–acid battery; *Industrial*: heavy-duty applications such as motive power and standby power; *Consumer*: emergency lighting and security-alarm systems, cordless convenience devices and power tools, and small-engine starting.

Recently, two types of batteries utilizing immobilized-electrolyte systems have come into use. These batteries are for consumer applications, eg, emergency lighting and power and alarm-security systems, and use either a gel or a highly absorbent separator to immobilize the electrolyte.

Because of the gelled electrolyte, these batteries can work in any position, whereas batteries with free electrolyte can work only in the upright position.

In spite of the numerous studies about charging and discharging the lead electrodes in sulfuric acid solution, there are still many doubts regarding the exact mechanism. The double sulfate theory has been confirmed by a number of methods as the only reaction consistent with the thermodynamics of the system. The following reactions are based on the two separate electrode reactions that combine to become the cell reaction known as the double sulfate theory:

Positive electrode (plate)

$$PbO_2 + 4 H^+ + SO_4^{2-} + 2 e \underset{charge}{\overset{discharge}{\rightleftharpoons}} 2 H_2O + PbSO_4$$

Negative electrode (plate)

$$Pb + SO_4^{2-} \underset{charge}{\overset{discharge}{\rightleftharpoons}} PbSO_4 + 2 e$$

The components of a lead–acid battery are the container, which includes case, cover, and vent plugs; and the cell, which includes plates (positive, lead dioxide (PbO_2) on lead grid; negative, lead (Pb) on lead grid), separators, and sulfuric acid solution.

The materials used depend on the application, eg, polypropylene and vulcanized rubber for automotive batteries, polystyrene for stationary batteries, polycarbonate for a large single cell, and ABS plastic and drawn steel cans for sealed lead cells. All plastic and rubber containers are molded. Physical qualities required of the container are to resist attack by sulfuric acid solution and the battery components, to be nonporous, to be nonreactive, and to withstand extremes of heat and cold, as well as shock and vibration.

The lead alloy grid is the mechanical framework of support for the active material (PbO_2 or Pb) of the plates and conducts the current to and from these active materials. The grid must possess sufficient stiffness to prevent damage or distortion during the casting, plate pasting, and battery fabrication operations. The required grid strength is reported to be ca 40 MPa (5802 psi) UTS (ultimate tensile strength). The grid also must be corrosion-resistant. Currently, most battery grids are cast, but mechanically formed grids are beginning to make their appearance in specialized applications. In general, the positive and negative grids are of the same design (except for the tubular positive grid which uses an automotive-type grid counterpart). New lightweight, economical grids for the negative plate made of ABS plastic grids coated with lead and a grid frame structure of polystyrene interwoven with lead strands are being produced. Both examples incorporate lead terminal connectors in their design for easily soldering to one another or to current collectors.

The lead alloy containing antimony (2–12 wt%) has wide acceptance as grid material (see Antimony). Antimony hardens the lead and improves its castability. Other additives include calcium and strontium (for maintenance-free applications) and arsenic, tellurium, and tin. Lead-calcium alloys are being used for the new mechanically formed grids. If these grids gain wide acceptance, a revolutionary high speed manufacturing processes with lead–acid battery designs and battery performance characteristics will be possible. Recyclers of lead scrap that supply many battery manufacturers, however, say that the shift to lead-calcium alloys could drastically disrupt the lead scrap business because of the difficulties of segregating calcium and antimony alloys from the lead. Thus, improved or new techniques must be developed to reprocess the lead alloy scrap so it will be accepted for use by the battery industry (see Recycling).

The active materials for the positive (PbO_2) and negative (Pb) plates are prepared from lead oxides (formed by tumbling high purity lead pigs or balls in a water-cooled ball mill or by the Barton pot method) in combination with finely divided metallic lead. The lead oxide (PbO) used in battery plates exists in two crystalline forms, the yellow orthorhombic form and the red tetragonal form. The red oxide Pb_3O_4 is sometimes used in making battery plates, but its use is declining.

The positive plate (PbO_2) is formed from lead oxide (PbO) that is mixed in some cases with ≤ 20 wt% Pb_3O_4, red lead (to aid in the formation step), sulfuric acid (40–42 wt%), fibers (to facilitate handling of the plate after pasting), and water which is added until the paste has the proper consistency for application to the grids. The negative plate (Pb) is made of the same ingredients except that the red lead is omitted, and expanders such as lamp black (ca 0.2 wt%), certain organic compounds (eg, ligninsulfonic acid), and barium sulfate (≤ 3 wt%) are added. The grids are then pasted and cured before undergoing formation (or charging), which converts the inert lead oxide sulfate paste into active plates. In this process, the cured plates are electrically oxidized or reduced in a forming tank containing dilute sulfuric acid solution (< 20 wt%). The positive plates are the anode and the negative plates are the cathode.

To prevent contact of the positive and negative plate, a separator is placed between them. The separators are usually in sheet form and are commonly made of paper, rubber, glass, or plastic. Since sulfuric acid must permeate the separator, it must be microporous. The separators are ribbed on the side placed toward the positive plate. This procedure prevents excessive separator contact with the oxidizing plate material, and it allows a greater volume of sulfuric acid to be used by the positive plate during discharge.

The dilution of sulfuric acid must be done with adequate cooling because the dilution process is highly exothermic. Each manufacturer selects a particular sulfuric acid concentration that is a compromise between lead grid corrosion and battery performance. An example of immobile electrolyte is made by mixing sodium silicate, demineralized water, and dilute sulfuric acid to obtain a thixotropic gel.

After formation, the dry plates are assembled into cells. The individual cell consists of several positive plates and several negative plates with separators sandwiched between the plates. The positive and negative groups are assembled by burning the individual grid lugs to their respective lead alloy strap and terminal post via a jig fixture. The negative group always has one more plate than its positive counterpart so that when the two groups are meshed, each positive plate is located between two negative plates. This configuration prevents distortion or buckling of the positive plates. After assembly of the cell elements, the container is sealed. If the plates have been stabilized, the battery can be stored in the dry condition until activated. The final step for a dry-charge battery is to fill it with the desired sulfuric acid solution.

The finished battery is tested for voltage, capacity, charge rate acceptance, cycle life, accelerated life tests, storage, overcharge, normal temperature operation, low temperature cranking, and shock and vibration.

Automotive batteries have the lion's share of the market, but other types of lead–acid batteries, such as sealed and small maintenance-free varieties, are making inroads into areas where other types of batteries traditionally have been dominant. The possibility of using lead–acid batteries for utility load leveling and peaking also has been reported. Since the passage of *Public Law 94-413* requiring the assessment of the need for electric-powered or electric hybrid vehicles, there has been a renewed interest in this field in which most of the operational data have been obtained with lead–acid batteries.

JAMES B. DOE
ESB Technology Co.

M.A. Dasoyan and J.A. Aguf, *Current Theory of Lead Acid Batteries*, Technicopy Limited, UK, 1979.

M. Barak, *Electrochemical Power Sources*, Peter Peregrinus, Ltd., Stevenage, UK, 1980.

J. Power Sources **2**(1), (Oct. 1977).

OTHER CELLS

Considerable effort is being directed toward development of advanced secondary battery systems for utility load-leveling and electric vehicles. U.S. organizations working on advanced secondary batteries include EDA and Gould Inc., who are working on a zinc–chlorine system; Battelle (Geneva Research Center) and NASA (Lewis Lab.), who are working on a redox flow system; Dow Chemical Co., ESB Technology Co., Ford Motor Co., and General Electric Corp., who are working on a sodium–sulfur system; and Argonne National Lab., Catalyst Research Corp., Eagle-Pitcher Industries, Inc., General Motors Corp., Gould Inc., and Rockwell International Corp. (Atomics International Div.), who are working on a lithium–iron sulfide system.

Zinc – chlorine. This unique battery uses chlorine gas in a solid form, chlorine hydrate ($Cl_2.6H_2O$), for a safe, practical means of chlorine storage. The electrochemical reaction is the following:

$$Zn + Cl_2 \underset{\text{charge}}{\overset{\text{discharge}}{\rightleftharpoons}} ZnCl_2 \qquad E_{cell} = 2.12 \text{ V at } 25°C$$

The zinc (negative) electrode is electroplated on a conductive but inactive carbon substrate, and the chlorine (positive) electrode is a porous titanium substrate treated with ruthenium. The electrolyte is ca 10 wt% aqueous zinc chloride and contains a special amine compound that enhances the electrodeposition of zinc by minimizing zinc dendrites and nodules.

Redox flow is a developing technology for storing bulk quantities of electrical power for the utility industry. Basically, the energy storage system consists of two electrolyte storage tanks connected to a flow cell. One tank contains the anolyte or negative fluid, and the other tank contains the catholyte or positive fluid. The flow cell has two compartments, one for each electrolyte, containing inert carbon electrodes separated by an anion-permeable, selective ion-exchange membrane. The ion-exchange membrane permits the passage of chloride ions from one compartment to the other to maintain electroneutrality and prevents the mixing of cathode and anode electrolytes. The system is recharged by reversing the direction of the current flow.

Sodium – sulfur. This battery uses a solid electrolyte membrane (β-alumina ceramic) or borate glass fibers to provide sodium ion transport and to separate the two liquid electrodes: sodium (negative) and sulfur (positive). The electrochemical reaction is

$$2 \text{ Na} + x \text{ S} \underset{\text{charge}}{\overset{\text{discharge}}{\rightleftharpoons}} Na_2S_x \qquad E_{cell} = \text{ca } 3.22 \text{ V}$$

The Battery Energy Storage Test facility is expected to develop a 100 MW·h sodium–sulfur system for utility peak load-leveling by 1985–1987.

Lithium – iron sulfide. This battery utilizes a molten salt eutectic of lithium chloride–potassium chloride at a temperature of 400–500°C as electrolyte. The positive electrode is either iron sulfide (FeS) or iron disulfide (FeS_2), and the negative electrode is a lithium alloy (lithium–aluminum ($LiAl_x$) or lithium–silicon ($LiSi_x$). A representative electrochemical reaction would be

$$2 \text{ LiAl}_x + \text{FeS} \underset{\text{charge}}{\overset{\text{discharge}}{\rightleftharpoons}} \text{Fe} + Li_2S + 2x \text{ Al} \qquad E_{cell} = \text{ca } 1.34 \text{ V}$$

Other cell systems that have been investigated include zinc–bromine, sodium–chlorine, sodium–antimony trichloride, and lithium–titanium disulfide.

James B. Doe
ESB Technology Co.

R. Robert, *Status of the DOE Battery and Electrochemical Technology Program III*, contract No. DE-ACO3-76SF00098, The MITRE Corporation, McLean, Va., Feb. 1982.

BEARING MATERIALS

The selection of materials for sliding and rolling contacts depends on an optimum matching of material properties with demands for low friction and wear in the machinery involved.

Properties of the lubricant combined with mating bearing materials determine the level of frictional forces and the nature of wear processes. For compatibility in sliding against steel, good bearing metals must have (1) atomic diameters of at least 0.285 nm (15% greater than iron) to minimize atomic junctions and solubility between the sliding surfaces and (2) be members of groups of the periodic table of elements which implies covalent atomic bonds at any junction rather than more tenacious metallic bonds, eg, silver, cadmium, indium, lead, and bismuth. Hardness and modulus of elasticity of oil-film and boundary-lubricated bearing materials should be as low as possible while providing sufficient strength to carry the applied load. Ability to absorb foreign dirt particles helps to prevent scoring and wear. For sliding bearings, materials of intermediate compressive strength are usually desirable. Fatigue strength is important in applications where load changes direction, such as in reciprocating engines. Selection also should be limited to materials that are not readily attacked by lubricants or any other corrosive media contacting the bearing.

Oil-Film Bearing Materials

The high lead and tin alloys known as babbitts offer an almost unsurpassed combination of compatibility, conformability, and embeddability for use as bearing surfaces. Even under severe operating conditions, ie, high loads, fatigue problems, or high temperatures, babbitts are used as a thin surface coating because of their good rubbing characteristics. Babbitts are used in electric motors, turbines, blowers, and industrial equipment, as well as in automotive engines and fractional horsepower motors.

Copper–lead alloys consist of a copper matrix with 20–40% lead dispersed in pockets and are used extensively in reciprocating engines where high fatigue strength and improved high temperature performance are required, eg, connecting-rod and main bearings in internal-combustion engines for automobiles, aircraft, commercial vehicles, trucks, and diesels.

Bronzes are used in large volumes for cast bushings since they combine economy with generally adequate bearing properties, good compressive and fatigue strength, and excellent casting and machining characteristics. Bronzes can be grouped into leaded, tin, and high strength bronzes, each having successively higher hardness and strength (see Copper alloys). Leaded bronzes are used for bearings in machine tools, home appliances, farm machinery, and pumps. Tin bronzes are used in high load, low speed applications such as trunnion bearings, gear bushings for farm equipment, earthmoving machines, and rolling-mill bearings. Aluminum bronzes of very high strength are best suited for heavy-duty, low speed applications with plentiful lubrication because of their poor ease of embedding, compatibility, and conformability.

Bearings of aluminum alloys have high fatigue strength, excellent corrosion resistance, and high thermal conductivity, and they are low in cost. They are used extensively in connecting-rod and main bearings in automobile and diesel engines, reciprocating compressors, and aircraft equipment. Because of their relatively poor compatibility and embeddability, aluminum bearings require sufficient lubrication, good surface finish, and shaft Rockwell hardness of ca HRB 85.

Cast iron and steel can be used as inexpensive bearing materials for light loads at low speed. Other metals, such as cadmium, silver, and zinc have limited uses as bearing materials. Silver bearings have given excellent heavy-duty service in reciprocating aircraft engines and diesels, and cadmium alloys, with high temperature fatigue properties somewhat superior to babbitt, have been used in some passenger car and truck engines.

Boundary-Lubricated Bearing Materials

A variety of bearing materials have been introduced on a wide scale in the past several decades for use under dry or sparsely lubricated conditions. These materials often allow design simplification, give freedom from regular maintenance, and reduce contamination problems. They provide superior performance at low speed and for intermittent operation.

Porous metal bearings are easily shaped to final dimensions, are low in cost, and contain their own lubricating oil. They are popular in home appliances, small motors, machine tools, aircraft and automotive accessories, business machines, and farm and construction equipment.

Nearly the whole range of commercial resins has at least occasionally been employed in forming sliding surfaces. Phenolic bearings, reinforced commonly with cotton fabric, other reinforcing fibers, cellulose, or rigid fillers (qv), have replaced wood and metals in small instruments, clocks, home appliances, ship-propeller shafts, rolling mills, conveyors, and earthmoving machines. Nylon and polyacetal usually provide the simplest and cheapest injection-molded small bearings and are used respectively for home appliances, toys, office machines, textile and food machinery, and automotive, appliance, and industrial application. Polytetrafluoroethylene, often woven with a secondary fiber into a fabric, is used in automotive ball-and-socket joints, aircraft controls and accessories, bridge bearings, and electrical switch gear (see Composite materials). Polyimides, polycarbonates, and polysulfones are among the newer high temperature moldable plastics with excellent resistance to both chemical attack and burning, and polypropylene, polyurethane, epoxies, and various copolymers give satisfactory performance in many applications involving slow speed sliding at light loads.

Because most polymers exhibit some viscoelastic properties, an appreciable portion of sliding friction arises from elastic hysteresis which varies with load and especially with speed. Wear rates are usually related to the *PV* factor which also is used to define the load speed limit. Tests have shown repeatedly that the total volume of material worn away is approximately proportional to the total normal load multiplied by the distance traveled in a length of time. Since the friction and wear values vary, suppliers should be checked for their best estimates for specific applications. Model trials under moderately accelerated speed and load conditions are desirable.

Bearings of graphite are being used extensively with water and other low viscosity fluids and in dry operation at temperatures up to 400°C. Common applications are in high temperature furnaces and kilns where normal lubrication is impossible; in food, drug, and textile equipment where contamination with oil and grease must be avoided; and in water, gasoline, and chemical pumps where the boundary-lubrication properties permit operation on low viscosity, nonlubricating fluids. Certain precautions should be observed in applying graphite, however, since the material chips and cracks easily when struck on an edge or a corner or when subjected to high thermal, tensile, or bending stresses.

Rubber bearings are used for ships as main-shaft bearings and pumps including those handling sand and slurries. They also are used in vertical turbines and process equipment such as gravel washers, thickeners, and classifiers.

Though displaced from many of their applications by plastics, porous metals, and rubber, bearings of lignum vitae and oil-impregnated maple and oak are useful at temperatures up to 70°C (see Wood). They offer good self-lubricating properties, low cost, cleanliness, and resistance to oils, acids, and chemicals. Wood bearings are applied with water and related fluids at speeds up to several hundred rpm for ship propeller shafts and hydraulic turbines, as well as in pumps, conveyors, and agitators in chemical, food, and beverage plants.

High Temperature Bearing Materials

Engineering innovations have created a need for materials with improved tolerance for higher temperature in such fields as nuclear systems, supersonic aircraft, guided missiles, gas turbines, and compact compressors. No ideal bearing materials have come into general use for this high temperature range. However, selected graphites (see above) and solid-film coatings are the usual choice up to ca 400–500°C. Hard metals and superalloys, such as Mo alloys (TZM), tool steels, nitrided steels, Hastelloy C, Stellite 6, Stellite Star J, Inconel X, Stellite 19, Rene 41, and Tribaloy T-400 are employed up to ca 800°C. For even higher temperatures, ceramics such as a-Al_2O_3, B_4C, Si_3N_4–SiC, Haynes LT1B (19% Al_2O_3, 59% Cr, 20% Mo, 2% Ti), Kl62B (64% TiC, 25% Ni, 5% Mo), and graphite are used.

Bearing Materials with Liquid Metals

The bearing system using liquid metals as lubricants must meet the requirements of application in addition to mechanical strength, dimensional stability, differential thermal expansion, oxidation resistance, corrosion resistance to the liquid metal, solubility of the bearing components in the liquid metal, and compatibility of materials. Some materials that have shown promise as potential bearing materials operating in the liquid metal NaK are aluminum, copper, iron, and cobalt alloys; cermets; and ceramics.

Rolling-Element Bearings

Ball- and roller-bearing materials are commonly selected to provide a minimum hardness of HRC 58 at load-carrying contacts. The more common bearing steels are AISI 3115, AISI 4320, AISI 4615, AISI 4820, AISI 5120, AISI 6120, AISI 6195, AISI 8620, AISI 52100, AISI 420 and AISI 440C./ For higher temperatures and for use in special atmospheres, the following alloys are used: MHT, Ml tool steel, M2 tool steel, M10 tool steel, M50 tool steel, MV-1 tool steel, Halmo (VM), T1 tool steel, Stellite 3, Stellite 19, Stellite 25, Stellite Star J, and K Monel.

E.R. BOOSER
General Electric Co.

Mach. Des., Mechanical Drives Reference Issue, published annually.
J.K. Lancaster, *Tribology*, 219 (Dec. 1973).
G.C. Pratt, *Int. Metall. Rev.* **18**, 62 (1973).

BEER

Beer is generally defined as an alcoholic beverage made by fermentation of a farinaceous extract that is obtained from a starchy raw material, barley, in the form of malt. Although it is possible to replace some part of the barley with other starchy materials (eg, corn, rice, wheat, oats, or potatoes), barley is usually the main constituent of brewing materials.

Types of Beer

Bottom-fermented beers. Lager beers are stored or lagered after fermentation in cold cellars or storehouses for clarification and maturing. The various types of lager beer include Pilsner, Dortmund, and Munich, named after the cities where they originate.

Top-fermented beers. Ale is pale with a pronounced hop taste and aroma and 4.0–4.5 wt% alcohol. In Burton-on-Trent, UK, where the best ales are made, the brewing water is rich in calcium sulfate; elsewhere, it is usual to burtonize the water by adding calcium sulfate. In addition to the amount of hops used in the brewing, more dry hops are added during storage in order to have a distinct hop aroma in the finished product. Other kinds of top-fermented beers are porter, stout, and lambic beer.

Attempts have been made to make beers with a low alcohol content. The difficulty in making such a beer is its pronounced wort taste. The following measures can be used in making such beers: interruption of the fermenting process; scalding beer in the copper kettle; vacuum or distillation; and reverse osmosis.

A beer with normal alcohol content, but much lower in caloric content, can be made with the help of extra enzymes that are added during mashing to achieve a further breakdown of malt substances.

Properties of Beer

The properties of the finished beer vary with the type of beer and place of origin. Most beers have an alcohol content of 3–5 wt%, a protein content of 0.3–0.6%, and a pH of ca 4. Ales have a slightly higher alcohol content and a protein content of ca 0.6–0.8 wt%. The quality of a beer is judged by the following attributes: aroma, taste, appearance, and formation and stability of foam.

The caloric content of beer is significant, but not especially high. A 355 mL (12 oz) bottle of average beer yields ca 600 kJ (143 kcal). The calories are provided by the unfermented residues and by the alcohol. The metabolic role of the latter is not fully understood, but it replaces carbohydrates, fats, and proteins, so that there may be a gain in body weight. Besides its caloric value, beer also contributes to the mineral requirements of the body and supplies useful quantities of B-complex vitamins (see Vitamins).

A special use of beer is for the control of sodium intake in the treatment of diseases such as congestive heart failure, high blood pressure, and certain kidney and liver ailments. Beer cannot, because of its low pH, harbor any pathogenic germs. The content of nourishing components is all in dissolved form. Beer is free from fat, it acts as a diuretic (see Diuretics), and it promotes the formation of gastric acid, thus acting as an appetite inducer. The alcohol in beer is effective to the amount and the concentration (ie, the speed at which it is drunk).

Beer defects. Among beer defects, turbidity is of primary importance. It may be of either biological or physiochemical nature. The former occurs only in unpasteurized beer and is caused by growth of microorganisms, such as brewers yeast, wild yeast, or certain bacteria. The latter may be induced by oxalic acid (from the barley), metals, chilling, and heating (pasteurization). Chill haze is rich in sulfur (1.8–2.0%) indicating a rather high content of active su!fhydryl groups (—SH) which are readily subject to oxidation. When oxidation occurs, the —SH groups combine to form large molecules with sulfur to sulfur linkage (—S—S—). The oxidized compounds are insoluble, thus producing an oxidation haze with its consequent deleterious effect on the flavor and shelf life of the beer. The effect of sunlight in the sunstruck flavor in beer is caused by formation of mercaptans. Wild or gushing beer is a defect observed as a rather violent over-foaming from the bottle immediately after opening; this defect, however, does not affect the taste of the beer. The fundamental cause of gushing is attributed to the formation of microbubbles, which are caused by certain substances in barley or malt that has been molding.

Besides the culture yeasts, there are many wild yeasts, molds, and bacteria that may contaminate the beer. By constant cleaning, brewers try to keep to a minimum the number of harmful microorganisms. The infecting organisms are nonpathogenic, but they increase the acidity which makes the beer less palatable, though not toxic. Foreign organisms capable of growing in beer are most often introduced through contamination of innoculation yeast.

Brewing Materials

Barley has technical advantages that make it superior for malting and brewing. Barley differs from the other common cereals in that the husk adheres to the kernel after threshing. This makes the process of malting, and subsequent extraction of the wort, much easier than with wheat or other grains.

During the malting process, the raw, hard, flat-tasting barley is changed into a crisp, mellow, sweet-tasting malt. The aim of malting is to bring forth substances, ie, enzymes, that will break down (hydrolyze) starch and proteins to less complex, water-soluble compounds, ie, amino acids, fermentable sugars, and small peptides. When these compounds are dissolved in water, the resulting liquid is known as wort. Brewers are especially interested in the starch-splitting amylases. They are formed in barley during germination, but they react during mashing. The protein-splitting enzymes, proteases, are formed and react during the germination so that some of the proteins of the barley are split into compounds soluble in water. The hemicellulases, another group of enzymes, split the cell tissue, thereby leaving the starch open to attack by amylases.

The character of the malt plays a large role in the resulting beer. The two extremes in malt character are reflected in Pilsner and Munich beers. For Pilsner, a pale malt with no pronounced aroma must be used; Munich beer demands malts with a pronounced aroma. Although the water content of the final malt is 3–4%, the interplay of water content and temperature during the drying is of fundamental importance for the character of the malt. The pale malt must be quickly dried to reach the final temperature; the dark malt must have a suitable high water content as the temperature rises. Special malts such as caramel and black are used for the darkest beers such as porter and stout. These malts have pronounced color and aroma but little or no diastatic power (see Malts and malting).

Adjuncts in the form of cereals (especially corn) other than barley are often part of the brewing materials when the demand is for a stable, nonsatiating, sparkling beer. Sugar can be added during various steps of the brewing instead of forming fermentable sugar through the splitting

of starch during mashing. Since adjuncts are essentially starch with very little protein, they are a source of additional alcohol, but contribute little to the color, taste, aroma, or protein content of the beer.

Originally, the principal aim of adding hops was to compensate for the insipid, sweet taste of the unhopped beer with the characteristic bitter taste and aroma of hops. Other benefits of adding hops include increasing the biological stability of the beer and improving the head retention and body of the beer. The amount of hops added varies from 0.4 to 4.0 g/L.

Yeast for brewing is propagated from single-cell cultures and is recovered after fermentation for use throughout many generations. There are two types of yeast in brewing: top yeast, which forms spores, ferments vigorously at elevated temperatures, and tends to float on top of the beer; and bottom yeast (weakly attenuating Saaz and strongly attenuating Frohberg), which does not usually form spores, is well adapted to slow fermentation at low temperatures, and settles to the bottom of the tank at the end of fermentation (see Yeasts). *Saccharomyces carlsbergensis* is considered the type of species for bottom fermenters, and certain strains of *S. cervisiae* for the top fermenter.

Perhaps the greatest preoccupation in relation to brewing yeasts is with stability, ie, whether a strain or mixture of strains remains unchanged and uncontaminated with prolonged use and whether the range of performance of the constant strain or mixture of strains is small enough to provide adequate constancy of beer quality despite inevitable minor variation in wort composition. Various strains of yeast have individual flavor characteristics; thus, brewers yeasts are not selected solely on the basis of fermenting power, but decidedly on the flavor they give beer.

Water has a great influence on the character of the beer, and the hardness of the water manifests itself by the extent of its reaction (alkalinity) with the weak acids of the mash. Certain ions are harmful to brewing; nitrates slow down fermentation, iron destroys the colloidal stability of beer, and calcium ions give beer a purer flavor than magnesium or sodium ions. Brewing water plays so large a role that some of the world's best known beer types such as Pilsner, Munich, Dortmunder, and Burton Pale Ale are special because of the properties of water used in their production. Water with a large carbonate content, as found in Munich and Copenhagen, is best suited for dark types of beer. The water in Pilsen is poor in minerals, it is best suited for pale beer made from malt dried at low temperatures and with methods of mashing that do not favor a rise in acidity.

If the production of a pale beer with a fine hop taste is desired and the water is rich in carbonates, one of the following methods of treatment may be used: addition of gypsum (burtonizing); addition of acid; decarbonation; and demineralizing by ion exchange. These methods are easily employed, but there is a chance that salt may remain in the water, which is undesirable. Today, it is possible to treat brewing water through double ion exchange, a process in which all salts are completely removed.

Production

The most important phases during brewing are mashing, wort boiling, fermenting, lagering, filtering, and bottling. During mashing, a mixture of finely crushed malt and warm water is exposed to enzymatic activity, thus converting starch into miscellaneous sugars and protein into peptides and amino acids. Adjuncts must be pretreated to be accessible to attack by the malt enzymes, usually by boiling before addition to the mash. The dissolved product from mashing is called wort and the insoluble remainder (mostly husks of malt) is called spent grain, which is sold as cattle fodder. The wort is boiled with hops, and during boiling, the enzymes are destroyed while bitter substances are extracted from the hops. Boiling of the wort also causes a certain amount of unconverted protein to coagulate and flocculate. This flocculation appears in the boiled wort as flakes; the brewer says that the wort has a fine break. After separation from the wort, it is called sludge.

After cooling of the wort to ca 10°C, the yeast is added in order to convert the sugar into alcohol and carbon dioxide during fermentation. After fermentation, most of the yeast is separated from the beer followed by a slow after-fermentation and maturing of the green beer in lagering tanks at ca 0°C. By the end of the lagering there is a sediment in the tank, called draff, consisting of the remaining yeast together with precipitated proteins and tannins. The beer is then filtered into pressure tanks and transferred to the bottling machine. It is then pasteurized to keep its clear color for longer periods of time. The finished product is completed in 2 mo for Pilsner beer and 4–6 mo for the stronger beers.

Much has changed in the technology of beer making; however, the methods of transforming barley into beer have not undergone profound modifications because of the lack of fundamental knowledge, a lack that is common to the whole food industry. Today, attempts are being made to avoid the malting process by brewing directly from barley with enzymes extracted from microorganisms. Industrial enzymes (qv), such as bacterial or fungal α-amylases, amyloglucosidases, proteases, and pullulanase, mainly of microbial origin, are being considered for use. Using these enzymes will yield hitherto unknown degrees of freedom in selecting conditions of the mashing and, consequently, the control of wort composition.

Although there is controversy over the principles of continuous brewing, with a large setback in the United States and a success in Spain, more cautious approaches in other parts of the world, particularly the UK, are proving their worth. Continuous fermentation is the most recently studied development in breweries. Principally, the primary fermentation is continuous, and the secondary fermentation and lagering is conventionally done. Because great demands are put on the operative yeast, the right yeast must be found for the system to work at an optimum. Use of the right yeast eliminates the need for the following lagering step.

Two systems have been described, and both are in commercial operation with good results. The open system depends on the wort being continually supplied to the system and the beer removed simultaneously. The partially closed system is built on the principle that a high concentration of yeast is artificially held back in the system.

The degree of automation varies widely. Newer plants have the ultimate in automatic wort production where one man per shift, per brewhouse unit, makes twelve brews per day. An important consideration in current brewhouse operations is to increase the utilization of all vessels by the rapid batch process. A rapid filtration of grains is required and achieved with the use of a lauter tun, mash filter, or strainmaster. The separation of hops from the boiled wort has been accelerated by the use of hop powder pellets or omitted when hop extract is used. High density brewing involves the production of worts with high initial density and the introduction of a substantial part of water at as late a stage in the conventional brewing process as possible, usually after preliminary beer filtration. A further development in the process is the use of universal refrigerant-cooled tanks, designed to ferment, age, and finish beer in a single tank without the usual transfers.

The production of concentrated beer is basically a crystallization process in which water, constituting ca 90% of a normal beer, forms pure ice crystals at temperatures below the freezing point of water which are separated from the beer concentrate. After filtering, the beer concentrate is ready for reconstitution by the addition of water and carbon dioxide. In the concentration and subsequent filtering step, some of the constituents that contribute to the instability of finished beer are removed. Since these processes can be accomplished in a matter of hours, the resulting beer can be ready for consumption shortly after fermentation is completed, rather than weeks later as is the case with present lagering techniques. Also, in recent years, automatic regulation of processes has steadily gained a foothold in the brewing industry. The aim has been to produce beer with a better and more even quality and at lower costs.

Cleaning and disinfection of brewery equipment at every stage of the process is necessary. It has been the object of much research, and more effective chemicals are constantly marketed. Physiological, chemical, and bacteriological cleanliness all are important, though bacteriological cleanliness through disinfection is the most important. In the breweries, both chemical and bacteriological degrees of cleanliness are nearly always obtained.

Environmental Problems

The awakening of public interest toward environmental pollution means that brewing companies must anticipate future legislation. Recent changes in state laws have placed severe restrictions on breweries in terms of amounts of suspended solids and biological oxygen demand (BOD) that can be discharged with wastewaters. Most breweries have an average BOD of 1.0–1.5 g/L; however, with strict controls and collection of waste streams, this can be decreased to as low as 0.35 g/L BOD. Cost of pollution control also is climbing rapidly. Municipalities that charged ca 5¢/kg of BOD a few years ago are now charging 10¢/kg, and some are contemplating increases to 30¢/kg.

The impact on the environment is not caused solely by effluents; it is also a matter of disposal of one-way bottles and cans. Local authorities know that their problem is growing, and in many places in Europe, authorities have begun to impose extra duties on firms using one-way packages to be assured of payment for their disposal.

H.E. HØYRUP
Bryggeriforeningen, Denmark

R.H.M. Langer and G.D. Hill, *Agricultural Plants*, Cambridge University Press, New York, 1982.

H.M. Broderic, ed., *The Practical Brewer*, Master Brewers Association of America (MBAA), Madison, Wisc., 1978.

BENTONITE.   See Clays.

BENZAL CHLORIDE, $C_6H_5CHCl_2$.   See Chlorocarbons and chlorohydrocarbons.

BENZALDEHYDE

Benzaldehyde, C_6H_5CHO, mp 26°C, bp at 101.3 kPa (1 atm) 179°C, is the simplest member of the series of aromatic aldehydes. It is a strongly refractive liquid with a characteristic odor of bitter almonds, used primarily as a flavoring agent, odorant, and intermediate in organic synthesis. Benzaldehyde is a colorless liquid, readily miscible with organic solvents. It is a versatile intermediate because of its reactive aldehyde group.

In the United States, benzaldehyde is mainly produced as a co-product with benzoic acid by oxidation of toluene (see Toluene). The main United States producer is Kalama Chemical.

Although benzaldehyde is nontoxic, its vapors have a mild narcotic effect. Contact with the skin or eyes should be avoided. The fatal ingested dose is estimated at 50–60 g.

Benzaldehyde usually is handled in phenolic resin-lined drums. Since the flash point is 64.5°C, all containers must be marked benzaldehyde and combustible liquid.

Derivatives

Derivatives based upon the reactivity of the aldehyde group include the following: benzoin, β-hydroxy-β-phenylacetophenone, $C_6H_5CH(OH)COC_6H_5$, mp 133–137°C, bp 343–344°C at 101.7 kPa (763 mm Hg), used on a small scale as a photopolymerization catalyst in polyester resin manufacture; α-(trichloromethyl)benzyl alcohol, $C_6H_5CH(OH)CCl_3$, mp 37°C, bp 154–155°C at 333 Pa (25 mm Hg); and benzylidene sorbitols, suggested for use as gelation agents.

Nuclear-substitution derivatives of benzaldehyde such as the chloro derivatives (chlorobenzaldehydes, ClC_6H_4CHO, and dichlorobenzaldehydes, $Cl_2C_6H_3CHO$) and the nitro derivatives (nitrobenzaldehydes, $NO_2C_6H_4CHO$), are of limited use as intermediates in various syntheses. The following hydroxy and related derivatives are used in flavoring materials: salicylaldehyde (2-hydroxybenzaldehyde), p-anisaldehyde (4-methoxybenzaldehyde), veratraldehyde (3,4-dimethoxybenzaldehyde), vanillin (4-hydroxy-3-methoxybenzaldehyde), ethyl vanillin (3-ethoxy-4-hydroxybenzaldehyde), and piperonal (3,4-methylenedioxybenzaldehyde). Also included are aminobenzaldehydes, $NH_2C_6H_4CHO$, used as dyestuff intermediates; o-formylbenzenesulfonic acid, $HO_3SC_6H_4CHO$, used in dyestuff syntheses; *p*-tolualdehyde, CH₃C₆H₄CHO, bp 204°C, used in synthetic apricot, plum, and cherry essences; *p*-isopropylbenzaldehyde (cuminal), (CH₃)₂CHC₆H₄CHO, bp 103°C at 1.3 kPa (10 mm Hg); and phenylacetaldehyde (*α*-tolualdehyde, hyacinthin), C₆H₅CH₂CHO, bp 193–194°C, which has a very strong hyacinth like odor, and is used in perfumes.

ARNOLD E. WILLIAMS
Kalama Chemical Inc.

U.S. Pharmacopeia XX (USP XX-NFXV), The United States Pharmacopeial Convention, Rockville, Md., 1980, pp. 1211, 2051.

A.H. Blatt, *Organic Synthesis*, Collective Vol. 2, John Wiley & Sons, Inc., New York, 1943, p. 441.

BENZENE

Benzene, C₆H₆, is a volatile, colorless, and flammable liquid aromatic hydrocarbon that possesses a characteristic odor. It is used primarily as a chemical raw material in the synthesis of styrene (polystyrene plastics and synthetic rubber), phenol (phenolic resins), cyclohexane (nylon), aniline, maleic anhydride (polyester resins), alkylbenzenes (detergents), chlorobenzenes, and other products used in the production of drugs, dyes, insecticides, and plastics, Benzene, along with other light, high octane aromatic hydrocarbons, such as toluene and xylenes, is a component of motor gasoline. Benzene is also used as a solvent, but it has been replaced by safer solvents in most applications. Representative properties of benzene are shown in Table 1.

The reactions of benzene are of three distinct types: substitution, addition, and cleavage or rupture of the ring. All three types are used industrially, but substitutions are by far the most important.

Under suitable conditions, one or more hydrogen atoms of benzene can be replaced by atoms such as halogens or by groups such as the nitro and sulfonic acid groups. The number of isomers possible depends on the number of substituents, eg, with two substituents, ortho(1,2), meta(1,3), and para(1,4) isomers are possible. With three substituents, up to ten isomers are possible.

The presence of benzene in practically all crude oils has been recognized almost from the beginning of the petroleum industry. The amount of benzene present in most crude oils is small, and its separation into a relatively pure state has ordinarily been uneconomical. However, when the benzene concentration in a petroleum fraction is increased by a catalytic or thermal reaction, the recovery may become attractive. Re-

cently, benzene has been produced from toluene by transalkylation, also called toluene disproportionation. Toluene is treated catalytically to produce benzene and xylenes. The high value of *p*-xylene has at times made this process more profitable than dealkylation. Toluene dealkylation may be economical when benzene is four to six cents a kilogram more valuable than toluene. However, requirements for lower lead levels in motor gasolines have increased the need for toluene in gasoline.

The production of ethylene by the pyrolysis of naphtha or gas oil feedstocks provides an important new source of aromatics in the United States. Most ethylene in the U.S. is produced from natural gas. Aromatics are recovered from the liquid by-product, which is called pyrolysis gasoline or dripolene, by partial hydrogenation and extractive distillation; by hydrogenation, desulfurization, and solvent extraction; or by partial hydrogenation, desulfurization, hydrocracking, hydrodealkylation, and distillation. Solvent extraction processes have increased benzene production. Many different solvents are now used to extract aromatics selectively. The aromatics are then distilled from the extract (see Azeotropic and extractive distillation; Adsorptive separation).

Benzene is one of the principal components of the light oil recovered from coal carbonization gases (see Coal carbonization). It also has been recovered from coal tar and as a by-product of water carburetion. New coking, liquefaction, and gasification processes for coal are potential sources of benzene. The distant future may bring processes for the conversion of limestone (calcium carbonate) and water to acetylene that can then be converted to benzene (see Feedstocks, coal; Coal chemicals and feedstocks).

The United States is the principal world producer of benzene. The top five U.S. producers of benzene from petroleum are Commonwealth Petrochemicals, Inc., Puerto Rico, Gulf Oil, Exxon, and Shell Chemical. The main U.S. producer of benzene from coal is U.S. Steel.

In defining specifications for the various grades of benzene, a number of test methods are important. The ASTM designations of the test methods are listed under each specified test in Table 2.

Benzene, because of its flammability, volatility, and toxicity, is very heavily regulated by several agencies in the U.S. government, including OSHA, the EPA, the DOT, the CPSC, and NIOSH. Although publications of the appropriate Federal agency must be consulted for specific regulations, the basic safety practices regarding benzene are the same as those for any flammable, volatile, and toxic material. Benzene is stored in steel containers. Adequate ventilation, labeling, and electric grounding are required. A benzene fire must be extinguished by smothering, using CO₂ or dry-chemical fire extinguishers.

Benzene is a poisonous substance with acute and chronic toxic effects. Most serious industrial benzene exposures now occur as a result of malfunction or during maintenance. The current U.S. workroom benzene limit is 32 mg/m³ for an 8-h TWA. In Jan. 1977, OSHA issued a nonmandatory guideline recommendation that benzene exposure limits be reduced to 3.2 mg/m³ (1.0 ppm) for an 8-h TWA. This recommendation is likely to become mandatory.

The short term effects of the inhalation, ingestion, or dermal contact of benzene are immediately apparent, but the effects of chronic exposure

Table 1. Properties of Benzene

Property	Value
formula weight	78.11
mp, °C	5.533
bp, °C	80.100
density, at −3.77°C, kg/m³	873.7
vapor pressure at 26.075°C, kPa (mm Hg)	13.33 (100)
refractive index, η_0^{25}	1.49792
viscosity (absolute) at 20°C, mPa·s (= cP)	0.6468
surface tension at 25°C, mN/m (= dyn/cm)	28.18
critical temperature, °C	289.45
critical pressure, kPa (atm)	4,924.4 (48.6)
critical density, kg/m³	300.0
flash point (closed cup), °C	−11.1
ignition temp in air, °C	538
flammability limits in air, vol %	1.5–8.0
heat of fusion, kJ/(kg·mol) [kcal/(kg·mol)]	9,847 [2,353]
heat of vaporization at 80.100°C, kJ/(kg·mol) [kcal/(kg·mol)]	33,871 [8,095]
heat of combustion at constant pressure and 25°C (liquid C₆H₆ to liquid H₂O and gaseous CO₂), kJ/g (kcal/g)	41.836 (9.999)
solubility in water at 25°C, g/100 g water	0.180
solubility of water in benzene at 25°C, g/100 g benzene	0.05

Table 2. Specifications for Commercial Grades of Benzene

Specification	Refined benzene-535 (ASTM D2359-69)	Refined benzene-485 nitration-grade (ASTM D835-71)	Industrial-grade benzene (ASTM D836-71)
sp gr, 15.56/15.56°C (ASTM D891)	0.8820–0.8860	0.8820–0.8860	0.875–0.886
color (ASTM D1209)	no darker than 20 max on the platinum cobalt scale		
distillation range, 101.3 kPa (1 atm) (ASTM D850)	not more than 1°C including 80.1°C (on dried sample)	not more than 1°C including 80.1°C	not more than 2°C including 80.1°C
solidifying point (ASTM D852)	5.35°C min (dry basis)	4.85°C min (anhydrous basis)	
acid wash color (ASTM D848)	no. 1 max	no. 2 max	no. 3 max
acidity (ASTM D847)	nil	no free acid (no evidence of acidity)	
sulfur compounds (ASTM D853)	free of H₂S and SO₂	free of H₂S and SO₂	free of H₂S and SO₂
thiophene (ASTM D1685)	1 ppm max		
copper corrosion (ASTM D849)	copper strip shall not show iridescence nor a gray or black deposit or discoloration		
nonaromatics (ASTM D2360)	0.15% max		

to lower levels of benzene are not. Chronic exposure to benzene, however, has been associated with leukemia. Periodic blood tests should be required for those regularly exposed to benzene. Benzene should be handled only in closed systems and used as a solvent only when no substitutes are possible. When it is used, thorough ventilation is a necessity. Safety goggles or plastic shields and PVC gloves should be used.

Inhalation of air with a benzene concentration of 64,000 mg/m³ is fatal in 5–10 min. A benzene concentration of 8000 mg/m³ produces toxic effects in 60 min. Treatment for cases of acute benzene poisoning requires immediate removal of the subject from the benzene vapor, artificial respiration, oxygen, and respiratory stimulus.

WILLIAM P. PURCELL
Union Oil Company of California

C.R. Noller, *Chemistry of Organic Compounds*, W.B. Saunders Company, New York, 1965.

M. Sittig, *Aromatic Hydrocarbons, Manufacture and Technology*, Noyes Data Corp., Park Ridge, N.J., 1976.

A.M. Brownstein, *Trends in Petrochemical Technology*, Petroleum Publishing Company, Tulsa, Okla., 1976.

BENZENESULFONIC ACIDS.   See Sulfonic acids.

BENZIDINE AND RELATED BIPHENYLDIAMINES

Benzidine (**1**) (mp 128°C, bp 400–401°C), *o*-tolidine (**2**) (mp 129°C), *o*-dianisidine (**3**) (mp 137–138°C), 3,3′-dichlorobenzidine (**4**) (mp 133°C), and several other derivatives of biphenyl-4,4′-diamine are of commercial value as intermediates in the manufacture of dyes and pigments. Some are used as analytical reagents or indicators in forensic and clinical medicine and in sanitation control (see Forensic chemistry). A number of main producers of benzidine have ceased manufacture, and use of the substance is rapidly diminishing because of its carcinogenic nature.

(**1**) R = H
(**2**) R = CH₃
(**3**) R = OCH₃
(**4**) R = Cl

Benzidine undergoes chemical reactions characteristic of primary arylamines (see Amines) as well as those related to its difunctional character such as oxidation to quinonoid structures and copolymerization. Most important commercially is the reaction of benzidine with nitrous acid to form the bisdiazonium (or tetrazonium) salt which may be coupled with assorted naphthylamine derivatives and other reactive compounds to produce a wide variety of useful azo dyes (qv). Most derivatives of benzidine undergo similar reactions.

Commercial manufacture of benzidine involves the alkaline reduction of nitrobenzene in either one stage to hydrazobenzene, or stepwise, changing conditions at the azoxybenzene or azobenzene stages. U.S. manufacture of benzidine dyes continues only by *in situ* conversion of hydrazobenzene to benzidine tetrazonium salt.

Benzidine and its salts are potent human bladder carcinogens capable of being absorbed through the skin as well as by inhalation. Several benzidine dyes have been found to be carcinogenic in test animals. Industrial consumers of benzidine dyes thus are restricting their use and seeking alternatives.

Derivatives

The only commercially important alkoxy derivative of benzidine is *o*-dianisidine used mainly in preparation of direct azo dyes. Tolidine

dyes have increasing use as substitutes for benzidine dyes. However, *o*-tolidine has been found to be carcinogenic in rodents, and its metabolism resembles that of benzidine, so its value as a substitute is open to question. *o*-Tolidine also is used as an indicator for the detection of residual chlorine in potable and recreational waters. 3,3′-Dichlorobenzidine is sold as dihydrochloride and is used in the manufacture of yellow, orange, and red pigments. It has also been found to be carcinogenic in rats, hamsters, and dogs.

KELVIN H. FERBER
Allied Chemical Corporation
present affiliation
Buffalo Color Corporation

G.W.H. Cheeseman and P.F.G. Praill in G. Wilson, ed., *Rodd's Chemistry of Carbon Compounds*, 2nd ed., Vol. III, Pt. F, Elsevier Scientific Publishing Co., Amsterdam, The Netherlands, 1974, pp. 1–33.

13-Week Subchronic Toxicity Studies. Direct Blue 6, Direct Black 38, and Direct Brown 95 Dyes, DHEW Publication No. (NIH) 78-1358, U.S. Dept. of Health, Education, and Welfare, Washington, D.C., 1978.

WHO International Agency for Research on Cancer, Monographs, Supplement 1, Lyon, France, Sept. 1979, pp. 45–46.

BENZINE.   See Petroleum (petroleum ether).

BENZOIC ACID

Benzoic acid (benzenecarboxylic acid, phenylformic acid, BzOH), C₆H₅COOH, mp, 122.4°C, bp, 249.2°C, crystallizes in lustrous, white, monoclinic leaflets or needles. In the free state, or in the form of simple derivatives, such as salts, esters, and amide, benzoic acid is widely distributed in nature. Gum benzoin (from styrax benzoin) (see Resins, natural) may contain as much as 20% benzoic acid. Acaroid resin (from *Xanthorrhoca hasilis*) contains 4.5–7% benzoic acid. Smaller amounts of the free acid are found in natural products of the most diverse character, including the scent glands of the beaver, the bark of the wild black cherry tree, cranberries, prunes, ripe cloves, and oil of anise seed. In its chemical behavior benzoic acid shows few exceptional properties; the reactions of the carboxyl group are normal, and ring substitutions take place as would be predicted.

Manufacture

Benzoic acid is manufactured in the United States by liquid-phase oxidation of toluene employing various cobalt catalysts. Under current air-emission standards, it is likely that most manufacturers are employing adsorption on activated carbon rather than scrubbers to treat their vent gases.

Benzoic acid sales have shown steady growth over the past quarter century. In addition, two processes were introduced in the early 1960s, Snia Viscosa's caprolactam and Dow's toluene to benzoic acid to phenol process. Outside of the larger captive requirements, the production of sodium benzoate continues to be a major market for benzoic acid since it is such an important perservative in food products (see Food additives). Conversion of benzoic acid into plasticizers and benzoyl chloride consume most of benzoic acid produced. The main United States producers of benzoic acid are Kalama Chemical, Pfizer and Velsicol.

Many uses have been found for benzoic acid and its salts and esters, such as medicinals, veterinary medicines, food and industrial preservatives, cosmetics, resin preparations, plasticizers, dyestuffs, synthetic fibers, and intermediates. Sodium benzoate has been used as a corrosion inhibitor and is used in substantial quantities to improve the properties of various alkyd resin coating formulations, where it tends to improve gloss, adhesion, hardness, and chemical resistance (see Alkyd resins; Coatings). A new use for benzoic acid is as a temporary plugging agent in subterranean formations and it is of considerable use in oil production.

The systemic toxicity of the benzoates is low. Symptoms generally resemble those produced by salicylates.

Derivatives

Benzoyl chloride, C_6H_5COCl, mp, $-1°C$; bp, $197.2°C$, is an important benzoylating agent. Other main uses are in the production of benzoyl peroxide, benzophenone, and derivatives employed in the fields of dyes, resins, perfumes, pharmaceuticals, and as polymerization catalysts.

Benzoic anhydride, $(C_6H_5CO)_2O$, mp, $42°C$, bp, $360°C$, is used as a benzoylating agent in special situations where benzoyl chloride is unsatisfactory because of excessive reactivity or the generation of hydrochloric acid.

Esters, especially butyl, have a large-volume use as dye carriers for polyester fibers. Other well-established uses of benzoic acid esters are in the fields of perfumes, insect repellents, medicinals, and plastics. Esters include benzyl benzoate, $C_6H_5COOCH_2C_6H_5$, mp, $21°C$, bp, $323-324°C$; methyl benzoate, $C_6H_5COOCH_c$, bp $198-200°C$; butyl benzoate, $C_6H_5COOC_4H_9$, mp $-22°C$, bp $250°C$; glycol benzoates, such as esters of diethylene glycol and dipropylene glycol; phenyl benzoate, $C_6H_5COOC_6H_5$, mp $70-71°C$, bp $314°C$; and vinyl benzoate, $C_6H_5COOCH=CH_2$.

Salts. Sodium benzoate, $NaC_7H_5O_2$, is the only benzoic acid salt produced in large quantities, although other salts, such as ammonium benzoate ($C_6H_5COONH_4$), do have uses in many of the same fields. Main uses of sodium benzoate correspond to those of benzoic acid.

Ring-substituted derivatives. A large number of benzoic acid derivatives are known, but only a small portion of these are manufactured in any significant volume. The main end uses of these derivatives are in the dye industry, medicine, insecticides, and the alkyd resin field. They include o-aminobenzoic acid (anthranilic acid), o-$NH_2C_6H_4COOH$, mp, $144-146°C$; m-aminobenzoic acid, mp, $174°C$; p-aminobenzoic acid (PABA), mp, $187-187.5°C$; o-benzoylbenzoic acid (benzophenone-o-carboxylic acid) o-$C_6H_5COC_6H_4COOH$; and p-$tert$-butylbenzoic acid, p-$(CH_3)_3CC_6H_4COOH$.

Halogen derivatives. o-Bromobenzoic acid has been mentioned repeatedly in the literature as a plant growth regulator, and p-bromobenzoic acid is used as a vat-dye intermediate. All three monochlorobenzoic acids, ClC_6H_4COOH, are commercial products. o-Chlorobenzoic acid is used as a preservative for glues and paints and as an intermediate in dye and fungicide manufacture. p-Chlorobenzoic acid is used as a dye carrier and in plasticizers as well as a fungicide and dye intermediate. 2,4-Dichlorobenzoic acid is used as an intermediate, especially in the synthesis of the antimalarial quinacrine hydrochloride (atabrin) (see Chemotherapeutics, antiprotozoal. Many chlorinated benzoic acid derivatives act as plant-growth regulators. 3-Amino-2,5-dichlorobenzoic acid (Amiben), an important herbicide, is a benzoic acid derivative.

Nitro derivatives. Of the nitrobenzoic acids, the meta and para are used in the manufacture of various intermediates.

ARNOLD E. WILLIAMS
Kalama Chemical Inc.

U.S. Pharmacopeia XX, United States Pharmacopeial Convention, Rockville, Md., 1980, pp. 73, 1051.

U.S. Pat. 3,797,575 (Mar. 14, 1974), R. Walter, J.T. Chatterji, and J.A. Knox (to Halliburton Co.).

BENZOIN, $C_6H_5CH(OH)COC_6H_5$. See Benzaldehyde; Resins, natural.

BENZOL. See Benzene.

BENZOPHENONE. See Ketones.

BENZOTRICHLORIDE, $C_6H_5CCl_3$. See Chlorocarbons and chlorohydrocarbons.

BENZOYL CHLORIDE, C_6H_5COCl. See Benzoic acid.

BENZOYL PEROXIDE, $(C_6H_5CO)_2O_2$. See Peroxides and peroxy compounds, organic.

BENZYL ACETATE, $CH_3COOCH_2C_6H_5$. See Esters, organic.

BENZYL ALCOHOL AND β-PHENETHYL ALCOHOL

Benzyl alcohol (1) and β-phenethyl alcohol (2) are the simplest of the aromatic alcohols. They are similar in chemical structure and have somewhat similar uses.

Benzyl alcohol (1) (phenylmethanol), mp $-15°C$, bp $205.4-205.7°C$ at 101.3 kPa (1 atm), is a colorless liquid with a mild aromatic odor and occurs either free or as an ester in oils of jasmine, castoreum, gardenia, and ylang-ylang, and in balsams of peru and tolu. It undergoes the characteristic reactions of a primary alcohol forming esters, ethers, halides, etc, and additionally forms ring-substituted derivatives. It forms a large number of azeotropes that are useful in separating it from mixtures. It is miscible with ethyl ether and chloroform.

Benzyl alcohol is manufactured on a commercial scale from benzyl chloride (see Chlorocarbons and chlorohydrocarbons) and sodium carbonate. Recent developments have necessitated the production of special grades of benzyl alcohol that must meet stringent specifications.

Because of the new OSHA regulations concerning the safety of chemicals, most manufacturers furnish customers with product safety information bulletins. In the case of benzyl alcohol, the bulletins discuss chemical reactivity, stability, fire hazards, fire-fighting procedures, health hazards, first-aid procedures, threshold limit values, warning properties, spill or leak control, and special precautions in handling and storage.

The aliphatic esters of benzyl alcohol are used in the soap, perfume, and flavor industries. Benzyl benzoate, a scabicide, is used in lotion form for delousing. Benzyl salicylate (see Esters, organic) is used in sunscreen lotions and ointments. Photo-grade benzyl alcohol is used in a developing bath for color motion pictures and for other film-processing procedures. The textile grade is used as a dyeing assistant for wool and nylon. The NF and reagent grades of benzyl alcohol are used for the preservation of aqueous and oily parenteral drugs as well as for the anesthetic, antiseptic, and solvent properties of the alcohol. Technical grade is used as a degreasing agent in rug cleaners, in leather dyeing, in ballpoint inks, as an extractive distillation solvent for m- and p-xylenes and m- and p-cresols, and in a cleaner for soldering. It is also used in insect repellents and in the polymer field.

β-Phenethyl alcohol (2) (β-phenylethanol), mp $-25°C$, bp $219.5-220°C$ at 101.3 kPa (1 atm), has only one significant commercial use, ie, as a fragrance material, but in that field no product is more important. The alcohol, as well as a number of its esters, has been isolated from a great variety of natural sources including orange juice, *Plumeria acutifolia* oil, Japanese tobacco leaf oil, beer (8–15 ppm), cigarette smoke, ylang-ylang, narcissus, hyacinth, lily, tea leaves, and otto of rose. It possesses a remarkably pervasive rose odor that is not especially apparent in concentrated form. Traces of impurities markedly influence the flowery odor and render the alcohol without value as a perfume material. It forms a number of azeotropes. The compound also undergoes the usual chemical reactions of alcohol or aromatic compounds.

At present, the Friedel-Crafts (qv) process produces most of the phenethyl alcohol today, although hydrogenation of styrene oxide also is used.

The use of β-phenethyl alcohol presents no health problems. It has GRAS status and has been approved for food use by the FDA. The only by-product is hydrated aluminum chloride that is used as a flocculating agent in municipal sewage-treatment installations.

WILLIAM RINGK (Benzyl Alcohol)
Consultant

ERNST T. THEIMER (β-Phenthyl Alcohol)
Consultant

L.H. Horsley, *Azeotropic Data, Advances in Chemistry Series No. 6*, American Chemical Society, Washington, D.C., 1952.

W.T. Hansen, Jr., and W.I. Kisner, *Soc. Motion Pict. Telev. Eng.* **61**, 667 (1953).

U.S. Pat. 2,304,925 (Dec. 15, 1942), E.E. Jelley (to Eastman Kodak Co.).

BENZYL CHLORIDE, $C_6H_5CH_2Cl$. See Chlorocarbons and chlorohydrocarbons.

BERKELIUM, BK. See Actinides and transactinides.

BERYLLIDES. See Beryllium and beryllium alloys.

BERYLLIUM AND BERYLLIUM ALLOYS

Beryllium has a specific gravity of 1.846. It is the only light metal with a high melting point. Natural beryllium consists of 100% of the ^{9}Be isotope; four unstable isotopes with masses 6, 7, 8, and 10 have been made artificially. Only two industrial beryllium extraction plants are in operation today, one based on beryl ore and the other bertrandite. Both yield beryllium hydroxide as an intermediate product. The hydroxide is the starting material for the manufacture of beryllium, beryllia, and alloys. Beryllium is extracted from bertrandite ore by a leaching process. Two processes, sulfate extraction and a fluoride process, are used to extract beryllium from beryl ore.

More than two thirds of beryllium producion is used in alloys, principally copper–beryllium alloys. In the early 1970s, nearly 20% of the extracted beryllium was used as the metal in heat sinks and specialized structural elements for air- and spacecraft. Beryllium oxide ceramic products represent ca 10% of the usage of beryllium.

In the form of beryl, aquamarine, and emerald, beryllium compounds have been known for centuries as gemstones. The precious gemstone beryl approaches a pure beryllium–aluminum–silicate composition, $3BeO.Al_2O_3.6SiO_2$. The beryllium content of the earth's surface rocks has been estimated at 4–6 ppm. Of the 45 beryllium minerals identified, only beryl and bertrandite have achieved commercial significance.

A summary of the physical constants of beryllium is given in Table 1.

Most methods for reducing beryllium compounds to the metal are based on chloride or fluoride electrolysis, which are not too complicated in the laboratory. Electrowinning of beryllium on a commercial scale has never been economically competitive with the thermal reduction of beryllium fluoride with metallic magnesium. Since the demand for beryllium metal has decreased, only one plant continues to operate.

The specification on commercially available electrorefined metal indicates maximum impurities as follows: iron, 33 ppm; aluminum, 100 ppm; silicon, 100 ppm; carbon, 300 ppm; nickel, 200 ppm; magnesium, 60 ppm; and copper, 50 ppm.

Most beryllium hardware is produced by powder metallurgy (qv) techniques in order to achieve fine-grained structures with a nearly random crystallographic orientation. For some specialized applications, sheet has been rolled from cast beryllium ingot, such material exhibiting an average grain size of 50–100 μm as compared to the typical 12 μm or less of the powder metallurgy products. Small quantities of foil and wire also are processed from ingot. Working of beryllium results in the

Table 1. Physical Properties of Beryllium

mp, °C	1287–1292 ± 3
bp, °C	2970
crystal structure	
α-Be	
close-packed hexagonal, P6$_3$/mmc	a = 22.856 nm, c = 35.832 nm
	c/a = 1.5677
β-Be (above 1250°C)	
body-centered cubic	a = 25.51 nm
latent heat of fusion, kJ/mol^a	11.7 ± 2
latent heat of sublimation, kJ/mol^a	320–330
latent heat of vaporization, kJ/mol^a	230–310
contraction on solidification, %	3
specific heat, kJ/(kg·K)^a, 25°C	1970
thermal conductivity, kW/(m·K), 25°C	190
coefficient of linear expansion, K^{-1},	
25–800°C	16.5×10^{-6}
electrical resistivity, Ω·m, 25°C	4.266×10^{-8}

^{a}To convert J to cal, divide by 4.184.

establishment of a high degree of preferred orientation, generally enhancing the properties in the working direction, but impairing properties normal to it. The preferred orientation problem has limited the use of wrought beryllium; shapes are usually machined from billets where standard metallurgical practice would involve rolled, extruded, or forged components with other metals. Axial or isostatic cold pressing followed by vacuum sintering, as well as hot isostatic pressing also are used commercially, particularly in the production of a large number of small components (< 1 kg).

Health and Safety Aspects

Skin problems are associated with contact by the water-soluble salts of beryllium, notably the fluoride. Acute and chronic pulmonary disease entities can result from exposure to respirable, beryllium-containing material. The former is due exclusively to inhalation of soluble beryllium salts and is not caused by exposure to the oxide or the metal. Beryllium in massive metal form presents no hazard.

U.S. Government exposure limits. *Occupational*: (1) No person shall be exposed to atmospheric concentrations of beryllium greater than 25 micrograms per cubic meter at any time; (2) no person shall be exposed to atmospheric concentrations of beryllium averaged over an 8-h day in excess of 2 μg/m^3. *Nonoccupational*: The concentration of beryllium in the ambient air surrounding a beryllium facility should not exceed 0.01 μg/m^3 as a monthly average. Under these limits, a worker could be exposed to any combination of levels < 25 μg/m^3 so long as the time-weighted average does not exceed 8 h × 2 μg/m^3. Workers in operations that exceed either of such levels must be provided with approved respirators or other protective equipment. Along with above limits, periodic physical examinations must be given emphasizing respiratory systems and changes in body weight. Chest x-ray is mandatory. Clothing contaminated with beryllium dust or chips must be washed separately from other clothing.

Beryllium is used extensively as a window material for x-ray tubes because of its ability to transmit low energy x rays. It also is used in nuclear weapons, as a neutron reflector in high flux test reactors, as a heat-sink material in low weight, high performance military aircraft brakes, in inertial guidance components, and in space optics. Beryllium oxide articles take advantage of the high thermal conductivity, high melting point, very low electrical conductivity, and high transparency to microwaves in microelectronic substrate applications (see Microwave technology).

Alloys

The most important beryllium alloys are with copper. Beryllium copper is produced by calcination of Be(OH)$_2$ to BeO and arc reduction to produce a master alloy which is used to produce billets for conventional hot and cold rolling processes. The two general categories of copper–beryllium alloys are the high strength alloys containing 1.6–2.0

wt% Be and 0.25 wt% Co [used for small electrical contacts, springs, clips, switches, bellows, Bourdon pressure gauges, and dies for plastics (see Plastics technology)], and the high conductivity alloys (used for electrical applications such as springs, switches, contacts, and welding disks and tips).

Nickel–beryllium alloys find limited applications as electrical connectors and in the glass industry. Small additions of beryllium are made to magnesium and aluminum melts to improve casting fluidity and decrease oxidation losses.

JOHN BALLANCE
A. JAMES STONEHOUSE
RAY SWEENEY
KENNETH WALSH
Brush Wellman Inc.

Standards Handbook, Wrought Copper and Copper Alloys Mill Products, Part II, Alloyed Data, Copper Development Association, New York, 1973.

BERYLLIUM COMPOUNDS

Beryllium compounds include beryllium carbide, Be_2C; beryllium carbonates, such as $BeCO_3 \cdot 4H_2O$; beryllium salts of organic acids which can be divided into normal beryllium carboxylates, $Be(RCOO)_2$, and beryllium oxide carboxylates, $Be_4O(RCOO)_6$; beryllium fluoride, BeF_2, an ionic compound, and other halides, beryllium chloride, $BeCl_2$, beryllium bromide $BeBr_2$, and beryllium iodide, BeI_2, which are covalent; beryllium hydroxide, $Be(OH)_2$, which exists in three forms, beryllium intermetallic compounds (beryllides) which are formed with most metals and that may exhibit excellent oxidation resistance, high strength at elevated temperature, good thermal conductivity, and low densities as compared with refractory metals and many ceramics (used in turbine engines); beryllium nitrate, $Be(NO_3)_2$; beryllium nitride, Be_3N_2; beryllium oxalate, BeC_2O_4; and beryllium sulfate, $BeSO_4$, used in the production of beryllium oxide powder for ceramics.

The most important high purity commercial beryllium chemical is beryllium oxide, BeO, the availability of which has spawned a new industry in the last few decades. Beryllium oxide can be fabricated in classical ceramic-forming processes such as dry pressing, isostatic pressing, extrusion, and slip casting (see Ceramics). Beryllia ceramics offer the advantages of a unique combination of high thermal conductivity and heat capacity with high electrical resistivity, and are frequently used in electronic and microelectronic applications requiring thermal dissipation.

KENNETH WALSH
GLENN H. REES
Brush Wellman Inc.

D.A. Everest, "Beryllium" in J.C. Bailar and co-eds., *Comprehensive Inorganic Chemistry,* Pergamon Press Ltd., Oxford, UK, 1973.

A.J. Stonehouse, R.M. Paine, and W.W. Beaver, "Physical and Mechanical Properties of Beryllides" in J.T. Weber and co-eds., *Compounds of Interest in Nuclear Reactor Technology,* American Institute of Mining, Metallurgical and Petroleum Engineers, Inc., New York, 1964.

BETA-LACTAM ANTIBIOTICS.   See Antibiotics, β-lactams.

BEVERAGE SPIRITS, DISTILLED

The early methods of distillation utilized the alembic, a simple closed container to which heat was applied. The vapors were transferred through a tube to a cooling chamber in which they were condensed. Alembics made of copper, iron, or tin were preferred. From this devel-oped stills consisting of clay or brick fireboxes into which the copper pots of the still were fitted. Direct heat was applied to the body or cucurbit containing the fermented mixture, and the vapors rose into the head, passed through a pipe (the crane neck) to the worm tub, and condensed by a copper coil immersed in a barrel of cool water. Variations and improvements of this technique resulted in pot stills, some of which are still in use in the production of malt whiskies, notably in Scotland, and of brandies in France.

In 1830 Aeneas Coffey, in Dublin, developed his continuous still. The Coffey still is composed of two columns; in one, the fermented mash is stripped of its alcohol, and in the other, the vapors are rectified to a high proof (94–96%). Almost all of the fundamentals of distillation had been recognized and incorporated into the Coffey still. Later developments do not differ fundamentally from the stills of Cellier-Blumenthal and Coffey.

The distilling industry always has been taxed heavily, not only in the United States, but throughout the world. For example, in 1981, the total revenue for Federal, state, and local government from excise taxes was 6808 million (10^6) dollars. Total liquor consumption in the United States for 1981 was 1701.8 million liters.

Since most governments had an economic interest because of the great source of revenue realized, they established standards for distilled beverage spirits, generally in keeping with the customs prevailing in their country. As a result, geographical identification has become accepted, and each country respects the identity and exclusiveness of the other's products.

Within each category of product, wide variations in flavor can be caused by types of materials and their proportions, methods of preparation, selection of yeast types, fermentation conditions, distillation processes, maturation techniques, and blending.

Since the alcohol and water components are insignificant factors in flavor intensity or palatability, distillers are primarily interested in the more flavorful constituents, ie, the so-called congeners, substances that are generated with the alcohol during the fermentation process and in the course of maturation. In order to produce a palatable product, it is therefore necessary to select the proper configuration of these constituents (congeners). This cannot be accomplished by production techniques such as listed above, and therefore, the majority of beverage spirits are blended to provide uniformity, balanced bouquet, and palatability.

Whiskies

Whiskies are by far the leading distilled beverage spirits, with those from Canada, Scotland, and the United States accounting for most of the sales. Irish whiskey, a distinctive product of Ireland, although not enjoying a large volume of sales, does have good distribution as a specialty item. The Irish use the spelling whiskey, the Scots and Canadians use whisky, and the Americans use both, with whiskey predominating.

Canadian. Canadian whisky is manufactured in compliance with the laws of Canada regulating the manufacture of whisky for consumption in Canada and containing no distilled spirits less than three years old. Canadian whiskies are premium products usually bottled at six years of age or more, and since they are blended, they are not designated as straight whiskies. They are light-bodied, and though delicate in flavor, they nevertheless retain a distinct positive character. The Canadian government exercises the customary rigid controls in matters pertaining to labeling and in collection of the excise tax. However, it sets no limitations as to grain formulas, distilling proofs, or special types of cooperage for the maturation of whisky.

The principal cereal grains, ie, corn, rye, and barley malt, are used, and their proportions in the mashing formula remain a distiller's trade secret; otherwise, the process is substantially the same as is found in the main distilleries of the United States.

White oak casks (189 L, 50 U.S. gal) are used in the maturation process.

Scotch. Scotch whisky is a distinctive product of Scotland, manufactured in Scotland in compliance with the laws of the UK regulating the manufacture of Scotch whisky for consumption in the UK, and containing no distilled spirits less than three years old. Most Scotch brands are blends of grain whiskies and numerous distinctive malt whiskies produced by over 100 distilleries in four main areas of Scotland.

The outstanding taste characteristic of Scotch, its subtle smoky flavor, results from the techniques used in the production of malt whiskies. Malted barley, dried over peat fires, is the only grain ingredient in the mash. The kind and the amount of peat used in the fires determines the intensity of flavor in the final product. The aroma of the burning peat, known as peat reek, is absorbed by the barley malt. This smoky flavor is carried through to the distillate and becomes a characteristic of single-malt whisky. Peat is heather, fern, and evergreen that have been subjected to aging and compression processes over the centuries.

The dried malted barley is ground to a grist, gelatinized in a mash tub, and after conversion is completed, the liquid portion or wort is drained, cooled, and placed in a fermenter.

After fermentation is complete, the separation of malt whisky from the fermented wort takes place in a batch distillation system, a copper kettle with a "worm" or spiral tube leading from its head. The size and shape of these pot stills exert a definite influence on the character of the whiskies. Another critical factor is the selection of the product during that portion of the distillation cycle that will produce the desired flavors. The first portion in the cycle is referred to as foreshots and the last as feints (heads and tails in the United States). The middle portion, after further distillation, becomes high wines and is subsequently reduced to maturation proof for storage in oak casks. The final distillation proof is in the 140–160° (70–80%) range.

The grain whiskies used in Scotch brands are produced in a manner similar to the production techniques used in Canada and the United States. Corn (referred to in the UK as maize) and barley malt are the ingredients. The proportions again depend on the individual distiller. Since delicate flavors are desired, the distillation proof is ca 180–186° (90–93%).

The grain whiskies are generally aged in matured oak casks of 190 L capacities not unlike American and Canadian barrels. Some malt whiskies acquire other distinctive qualities by being matured in oak casks that were previously used for sherry.

Irish.   Irish whiskey contains no distilled spirits less than three years old. Like Scotch, Irish whiskeys are blends of grain and malt whiskeys. Unlike Scotch, Irish whiskey does not have the unique smoky taste because the barley malt is not dried over peat fires. The production techniques are closely related to those used in Scotland. One variation, however is the use of some small grain, mostly barley, in addition to barley malt in the production of malt whiskey. Irish whiskey brands, although flavorful, are distinctive and reflect the process variations unique to Irish distillers.

United States.   Distilled spirits for beverage purposes in the United States are characterized specifically as to type, materials, composition, distillation proofs, maturation proofs, storage containers, and the extent of the aging period. The Federal government also requires that a detailed statement of the production process be filed for each type, and any subsequent improvements or changes must be filed and approved before being placed into operation. In addition, a generalized application of the regulations is made to establish the identity of products where the intensity of flavor may not conform to an arbitrary organoleptic evaluation based on chemical analysis.

Title 27, Code of Federal Regulations, Subpart C, lists several classes of distilled spirits: *grain neutral spirits*, an alcohol distillate from a fermented mash of grain distilled at or above 190° proof (95%); *grain spirits*, neutral spirits distilled from a fermented mash of grain stored in oak containers and bottled at not less than 80° proof (40%); *light whiskey*, distilled at more than 160° proof (80%) and less than 190° proof (95%) and stored in used or uncharred new oak containers (permitted to be entered for storage at proofs higher than 125° proof (62.5%)); *blended light whiskey*, if light whiskey is mixed with < 20 vol% of 100° proof (50%) straight whiskey; *bourbon/rye*, bourbon whiskey or rye whiskey which has been distilled at ≤ 160° proof (80%) from a fermented mash of ≥ 51% corn or rye grain, respectively, and stored at ≤ 125° proof (62.5%) in charred new oak containers. They also include mixtures of such whiskeys where the mixture consists exclusively of whiskeys of the same type. *Corn whiskey* has the same distillation proof limitation, but

it must be produced from a fermented mash of ≥ 80% corn grain and stored in uncharred or reused charred oak containers. *Straight whiskey*, any of the whiskey types mentioned above further qualify as straight whiskey by complying with the following: withdrawn from the distillery at not more than 125° proof (62.5%) and not less than 80° proof (40%) (for maturation purposes), and aged for not less than 24 months. A straight whiskey may further be identified as bottled in bond, provided it is at least four years of age, bottled at 100° proof (50%), and distilled at one plant by the same proprietor. *Blended whiskey*, a mixture that contains at least 20 vol% of 100° proof (50%) straight whiskey and, separately or in combination, whiskey, grain spirits, or grain neutral spirits. As many as 40 whiskeys may be included in a blend.

Gin

Gin is a product obtained by original distillation from mash or by the redistillation of distilled spirits, by mixing neutral spirits with or over juniper and other aromatic berries, or with or over extracts derived from infusion, percolations, or maceration of such materials, and includes mixtures of gin and neutral spirits. Gin derives its main characteristic flavor from juniper berries. In addition to juniper berries, other botanicals may be used, including angelica root, anise, coriander, caraway seeds, lime, lemon and orange peel, licorice, calamus, cardamon, cassia bark, orris root, and bitter almonds. The use and proportion of any of these botanicals in the gin formula is left to the producer, and the character and quality of the gin depends to a great extent on the skill of the craftsman in formulating his recipe. The more skilled producers formulate their aromatic ingredients on the basis of the essential oil content in the raw materials to assure a greater degree of product uniformity.

Brandy

Brandy is a distillate or a mixture of distillates obtained solely from the fermented juice, mash, or wine of fruit, or from the residue thereof, distilled at < 190° proof (95%) in such a manner as to produce the taste, aroma, and characteristics generally attributed to it. The most important category of brandy is fruit brandy. Cognac is a brandy produced in the Cognac region of France, and it enjoys an exclusive identity under which no other brandy may be labeled. Brandy derived exclusively from one variety of fruit is so designated. However, a fruit brandy derived exclusively from grapes may be designated as brandy without further qualifications. A minimum of two years' maturation in oak casks is required; otherwise, the term immature must be included in the designation. Blended applejack is accorded a special classification as a mixture that contains at least 20% apple brandy (applejack) on a proof basis, stored in oak containers for not less than two years, and not more than 80% neutral spirits on a proof gallon basis and bottled at not less than 80° proof (40%). Another class of beverage spirits, flavored brandy, is a brandy to which natural flavoring materials have been added with or without addition of sugar; it is bottled at no less than 70° proof (35%).

Rum

Rum is produced from the fermented juice of sugar cane, its syrup, molasses, or other by-products distilled at less than 190° proof (95%) and bottled at not less than 80° proof (40%). Blackstrap molasses is the most common raw material for the manufacture of rum. Rums are characterized as light-bodied, of which the Puerto Ricans are the best known, and full-bodied, which come from Jamaica and certain other islands of the West Indies.

Tequila

Tequila, a distinctive product of Mexico, is produced in Mexico from the fermented juice of the heads of Agave Tequilana Weber (blue variety), with or without additional fermentable substances. This is Mexico's most popular distilled spirits drink. Tequila, as consumed in Mexico, is unaged and usually bottled at 80–86° proof (40–43%). Tequila aged one year is identified as Añejo; aged as much as 2–4 years, it is identified as Muy Añejo.

Cordials and Liqueurs

Cordials and liqueurs are the same, with the former term being American and latter European. They are obtained by mixing or redistilling neutral spirits, brandy, gin, or other distilled spirits with or over fruits, flowers, plants, or pure juices therefrom, or other natural flavoring materials, or with extracts derived from infusions, percolations, or maceration of such materials. Cordials must contain a minimum of 2.5 wt% of sugar or dextrose, or a combination of both. If the added sugar and dextrose are < 10 wt%, then it may be designated dry. Three basic processes, namely, maceration, percolation, and distillation, or any combination thereof, are used to make cordials. Synthetic or imitation flavoring materials cannot be included in U.S. cordials.

Vodka

Vodka, produced on a multicolumn distillation system from fermented mash of grain at or above 190° proof (95%), must be further treated with charcoal or activated carbon or further refined by distillation in such a manner as to be without character, aroma, or taste. If any flavoring material is added to the distillate, the vodka is so characterized.

The Manufacturing Process

Any material rich in carbohydrates is a potential source of ethyl alcohol, which for industrial purposes is obtained by the fermentation of materials containing sugar (molasses), or a substance convertible into sugar, such as the starches. In the production of distilled spirits for beverage purposes, however, cereal grains are the principal types of raw material used. Any reference to alcohol in beverages is always to ethyl alcohol, C_2H_5OH.

The chemical composition of grain varies considerably and depends to a large extent on environmental factors such as climatic conditions and the nature of the soil. Another variable is the malt (sprouted or germinated grain) used. Malt is generally understood to be germinated barley, unless it is further qualified as rye malt, wheat malt, etc. The purpose of malting is to develop the amylases, the active ingredients in malt. Amylases are enzymes of organic origin that change grain starch into sugar (maltose). Besides providing the means of converting the grain starch into sugar, the malt also contributes to the final flavor and aroma of the distillate.

Milling breaks the outer cellulose protective wall around the kernel and exposes the starch to the cooking and conversion processes. Distillers require an even, coarse meal without flour. The mashing process consists of cooking, ie, gelatinization of starch, and conversion (saccharification), ie, changing starch to grain sugar (maltose). Distillers vary mashing procedures, but generally conform to basic principles, especially in the maintenance of sanitary conditions. The cooking and conversion equipment is provided with direct or indirect steam, propeller or rake-type agitation, and cooling coils or a barometric condenser.

In fermentation (qv), the grain sugars (largely maltose), produced by the action of malt enzymes (amylases) on gelatinized starch are converted into nearly equal parts of ethyl alcohol and carbon dioxide. This is accomplished by zymase, which is produced by yeasts. Although yeasts of several genera are capable of some degree of fermentation, *Saccharomyces cerevisiae* is almost exclusively used by the distilling industry. A great variety of strains exist, and the characteristics of each strain are evidenced by the type and amount of congeners the yeast is capable of producing Alcohol fermentation is represented by the following reactions:

$$C_{12}H_{22}O_{11} \xrightarrow[\text{H}_2\text{O}]{\text{maltase}} C_6H_{12}O_6 \xrightarrow{\text{zymase}} 4\,C_2H_5OH + 4\,CO_2$$

Distillation separates, selects, and concentrates the alcohol and flavor constituents produced by yeast fermentation from the fermented grain mash, sometimes referred to as fermented wort or distillers beer (see Beer; Distillation). In addition to the alcohol and the desirable secondary products (congeners), the fermented mash contains solid grain particles, yeast cells, water-soluble proteins, mineral salts, lactic acid, fatty acids, and traces of glycerol and succinic acid. The most common distillation systems used in the United States are the continuous whisky separating column, the continuous multicolumn system used for the production of grain neutral spirits, and the batch column and kettle unit used primarily in the production of grain neutral spirits. The discharge from the base of the whisky column, called stillage, contains substances derived from grain and is processed into distillers feeds which are used to fortify dairy, poultry, and swine formula feeds.

In the United States, the final phase of the whisky production process, called maturation, is the storage of beverage spirits in new, white oak barrels whose staves and headings are charred. The duration of storage in the barrel depends on the time it takes a particular whisky to attain the desirable ripeness or maturity. The thickness of the stave, depth of the char, temperature and humidity, entry proof, and length of storage impart definite and intended changes in the aromatic and taste characteristics of a whisky. These changes are caused by three types of reactions occurring in the barrel: extraction of complex wood constituents by the liquid; oxidation of components originally in the liquid and material extracted from the wood; and reaction between the various organic substances present in the liquid, leading to the formation of new congeners.

Federal regulations prescribe and limit the standards of fill (size of containers). The most common packaging operation used is the straight-line system with a line speed of 150–300 bottles per minute.

GEORGE W. PACKOWSKI
Joseph E. Seagram & Sons, Inc.

Annual Statistical Review (1982), Distilled Spirits Council of the U.S., Inc., Washington, D.C., 1983.

Regulations under the Federal Alcohol Administration Act, (27 CFR) ATF P 51005 (3-79), Dept. of the Treasury, U.S. Government Printing Office, Washington, D.C., 1979.

BILE CONSTITUENTS. See Meat products.

BIOASSAY. See Biomedical instrumentation; Diagnostic chemicals.

BIOCIDES. See Industrial antimicrobial agents.

BIODEGRADABLE POLYMERS. See Plastics, environmentally degradable.

BIOMEDICAL AUTOMATED INSTRUMENTATION

Automated clinical laboratory techniques have been in use for only a quarter century. In the United States, about 8×10^8 tests were performed in clinical laboratories in 1970, and about 1.6×10^9 in 1975. Thereafter, the annual growth rate diminished from about 15 percent to 5–6 percent per year.

The primary benefits of automation are speed, precision and reliability, in addition to the economy achieved by greatly increased productivity. Automated systems and their manual counterparts are generally identical with regard to the basic steps. The main elements of an automated system include the sampling system, the transport system that moves the samples through the chemical processing to the readout stage, the chemical processing and separation system, the temperature control system, the measuring system (colorimetric nephelometric, turbidimetric, or potentiometric, depending on the chemical reaction employed), and the readout (signal handling) system.

The desired features of automated biochemical analyzers include speed (high sample throughput and short lag time), large dynamic range, low operating costs, reliability, ease of operation, long-term stability, test specificity, precision, accuracy, and verifiability (standards).

A completely automated system for clinical analysis consists of the instrument *per se* (hardware), the reagents, and the experimental conditions (time, temperature, etc) required for each determination. With the exception of assays for various electrolytes (Na, K, Cl, CO_2), determination is normally by photometric means at wavelengths in the uv and visible region. Other means of assay are required only for more difficult assays, particularly those of serum or urine constituents at concentrations below 100 μg/L. Assay methods for various biochemical species can be classified according to end-point vs kinetic methods; blanking schemes; and reaction principle and type of reagents. Specificity, sensitivity, speed, and adequate range are essential. The three general categories of reactions are chemical, biochemical, and immunological.

Discrete Biochemical Analyzers

In discrete systems, sample and reagents are transported by essentially mechanical means as opposed to hydraulic means. Discrete analyzers can be further subdivided according to their degree of automation (either discrete-discontinuous, requiring operator intervention, or discrete-continuous systems), their flexibility toward test selection for a given assay, and the manner in which the final sample comes in contact with the detector. Table 1 lists a number of currently available systems.

Continuous-Flow Analyzers

In discrete analyzers, sample and reagents are manipulated mechanically. Consequently, the instrumentation is complex and often less reliable than desired. The continuous-flow analyzers have the advantages of simple, economical and reliable mechanical devices (the pump and sampler) as well as essentially foolproof calibration or standardization, since both samples and standards are treated identically in these systems. A typical continuous-flow analyzer, such as the modular single-channel Technicon Autoanalyzer II system, includes a sampler; a multichannel peristaltic pump that moves sample and wash fluid into the system and aspirates the reagents used; a reaction cartridge which houses a number of separate operations such as precise admixture of fixed proportions of sample and reagents; controlled incubation of the reaction mixture; continuous dialysis to free the compound to be assayed from various interferences such as protein; a colorimeter; and a recorder. A major advance in clinical analysis came about with the design of multichannel continuous-flow systems that provided a standard series of tests or profile on each sample analyzed.

The principal demands in the future will be for greater sensitivity and specificity of analytical methodology, thus allowing the determination of biochemical constituents present in lower concentration in body fluids. Greater emphasis will be placed on such special fields as enzyme chem-

istry and immunochemistry. Techniques such as fluorimetry, electrochemistry and chemiluminescence (qv) will replace optical absorption in some cases. Separation techniques such as gas chromatography, liquid chromatography, and mass spectrometry will be used to enhance detection specificity. In the future a greatly increased use of ion selective electrodes (qv) as well as other electrochemical techniques (eg, coulometry and polarography) should occur.

MORRIS H. SHAMOS
LLOYD R. SNYDER
Technicon Instruments Corp.

B. Jacobson and J.G. Webster, *Medicine and Clinical Engineering*, Prentice-Hall, Inc., Englewood Cliffs, N.J., 1977.

W.L. White, M.M. Erickson, and S.C. Stevens, *Practical Automation for the Clinical Laboratory*, The C.V. Mosby Co., St. Louis, Mo., 1972.

BIOPOLYMERS

The term biopolymers refers to naturally occurring, large, polymeric molecules, such as proteins, nucleic acids, and polysaccharides, that are essential components of all living systems. The same term also is used to describe synthetic polymers prepared from identical or similar monomers or subunits to those that make up the natural biopolymers.

The three main classes of natural macromolecules, ie, proteins, nucleic acids, and polysaccharides, are all formed by the condensation of monomeric units. The condensation reaction conceptually involves the elimination of a molecule of water from monomeric units, resulting in covalent bonding; repetition of the process results in long chains or polymers. The sequence of the monomeric units along such chains represents the primary structure. Secondary structure results from the interaction of the monomeric units with each other along the chain by, for example, hydrogen bonding or hydrophobic interactions. The most common types of secondary structure for proteins are the α-helix in which hydrogen bonding between groups along the same chain results in helical arrays about the axis of the chain (Fig. 1); and β-sheet patterns which arise from hydrogen bonding between groups on adjacent chains, resulting in sheetlike arrays of side-by-side, fully extended, or pleated chains (Fig. 2). Chains may also assume tertiary structures in which the chains are folded into complex, 3-dimensional forms stabilized by intermolecular forces such as hydrogen bonding, polar and hydrophobic interactions, and even covalent bonds.

Table 1. Automated Systems and Their Characteristics

Manufacturer and system	Transport system[a]	Throughput Samples/h	Throughput Tests/h	Sample size microliters	Numbers of tests available	Test methods[b]	Selectivity	Operating temp, °C	Positive sample identification[c]	No. of channels
American Monitor										
Parallel	DC	240	7200	30	10–80	R, E	random	37	yes	28
Boehringer Mannheim										
Diagnostics 8700	DC	120	3000	25	7.5–95	R, E, ISE	random	37	yes	25
Coulter DACOS	DC		450	24	2–20	R, E, ISE	random	30, 37	yes	na
DuPont ACA III	DC		100	60	20–600	R, E, ISE	random	37	yes	na
Hitachi 705	DC		180	19	5–20	R, E, ISE	random	25, 30, 37	no	na
Kodak Ektachem 400	DC		300–500	16	10	E, ISE	random	37	no	2
Technicon RA-1000										
System	DC		240	20	3–30	R, E, ISE	random	30, 37	yes	na
Technicon SMAC II										
System	CF	150	3450	23	600	R, E, ISE	random	37	yes	23
Olympus DEMAND	DC		400	20	5–27	R, E, ISE	random	37	no	na

[a]DC = discrete-continuous; CF = continuous flow.
[b]R = rate reactions; E = end points; ISE = ion-selective electrode.
[c]Such as ID (identification) number.

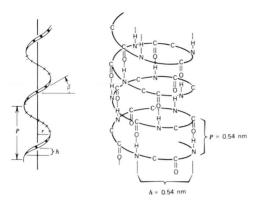

Figure 1. Schematic representations of a simple helix and the α-helix. P refers to the helix pitch, r, the radius of the helix; h, the axial translation per residue; and β, the pitch angle. Courtesy of Gordon & Breach, Science Publishers, Inc.

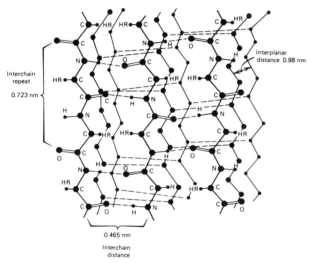

Figure 2. Schematic representation of the polypeptide chain in the β conformation. Courtesy of Gordon & Breach, Science Publishers, Inc.

Proteins are the most abundant of the natural biopolymers, constituting ≥ 50% of the dry weight of cells. Their functions are enormously diverse. For example, some serve structural roles; others have catalytic function (enzymes (qv)) and accelerate or facilitate specific metabolic reactions; other proteins are important in transport or storage; and still others act as hormones (qv) and regulate specific processes (see Antibiotics, peptides; Proteins).

The basic monomeric units of proteins are amino acids (qv), of which ca 20 commonly occur in natural proteins: glycine, alanine, valine, leucine, isoleucine, serine, threonine, lysine, hydroxylysine, histidine, arginine, phenylalanine, tyrosine, tryptophan, aspartic acid, asparagine, glutamic acid, glutamine, cysteine, cystine, methionine, proline, and hydroxyproline. Many amino acids (usually more than a hundred) are linked in the polypeptide, and a protein may consist of one or more polypeptide chains. Although the backbone of polypeptide chains is repetitious, each amino acid residue has a unique side-chain group whose chemical nature, size, and charge play an important role in determining the final conformation of the protein and hence its interaction with other substances and the role it will play in the living system.

Nucleic acids are of fundamental importance in controlling the metabolism, reproduction, and growth of living systems. These biopolymers are made of repeating sequences of nucleotides joined by phosphodiester linkages. Nucleotides themselves are composed of pyrimidine or purine bases attached at the 1′-position of the sugar, ie, β-D-ribose, the sugar residues being joined via their 3′- and 5′-hydroxyl groups in the phosphodiester linkage of the polynucleotide chain.

Nucleic acids may be divided into two classes depending upon the nature of their sugar residues; ribonucleic acids (RNAs) contain β-D-ribose, and deoxyribonucleic acids (DNAs) contain β-2-deoxy-D-ribose. Both classes share similar chemical and physical properties, although they serve different functions within the cell. The primary function of DNA is the storage of genetic information. The function of RNAs is to assist in the conversion of the genetic information into the synthesis of the corresponding protein. In some DNAs, molecular weights of polynucleotide chains in nucleic acids are typically in the millions (10^6) or even billions.

Polysaccharides are important for their structural and functional roles in the cell walls of bacteria and other plants, in the cell membranes of animals, and also as structural components of the fibrous and woody tissue of plants. In addition, some polysaccharides are important forms of fuel storage in plants and animals (see Carbohydrates; Cellulose; Sugar). The monomeric units of polysaccharides are simple sugars or monosaccharides which are polyhydroxy-containing molecules. The monomeric sugar molecules are linked to form polysaccharides by condensation involving removal of a molecule of water from hydroxyl groups on different sugar molecules. Most of the polysaccharides serving skeletal, storage, and gel-forming functions are long-chain homopolymers or alternating copolymers of the type $(A\text{-}B)_n$. These polysaccharides are built up mainly from a handful of monomer units such as D-glucose, D-mannose, D-galactose, D-xylose, and some of their substituted derivatives.

Starch (qv) is the most common fuel storage material in plants. It occurs in two forms, ie, α-amylose and amylopectin, the majority of starches containing 15–39% of amylose. Glycogen is the storage polysaccharide found in animal tissues, being especially abundant in liver and muscle. Like amylose and amylopectin, glycogen is a homopolysaccharide of D-glucose with α-(1 → 4) linkages. Cellulose is the simplest and most abundant structural and cell-wall polysaccharide, constituting > 50% of the total organic carbon in the biosphere. Chains of cellulose contain D-glucose with β-(1 → 4) linkages. The polysaccharide chitin is a principal structural element in the hard exoskeletons of insects and crustaceans. It is a homopolymer of N-acetylglucosamine, which is structurally related to cellulose. Other, more complex polysaccharides of importance include peptidoglycan or murein which is the amine component of bacterial cell walls.

Because of their simplified primary structure, synthetic analogues of proteins (polypeptides) and nucleic acids (polynucleotides) have been used successfully to help elucidate the secondary structures of these biopolymers. Structural and conformational studies provide valuable insight into the function of polynucleotides, and therefore nucleic acids, at the molecular level. Nucleic acids interact with other biologically important molecules, eg, with proteins in histones and chromosomes, and also with drug molecules. In the latter case, a number of drugs such as antibiotics are thought to act via inhibition of the replication and transcription of nucleic acids. Most recently, a number of anticancer drugs, such as daunomycin and adriamycin also have been shown to intercalate within the DNA helix to inhibit replication (see Antibiotics; Chemotherapeutics, antimitotic).

In the area of synthetic polymers that have been used successfully to replace biopolymers, plasma expanders, such as fluorinated hydrocarbons, polypeptides [poly(hydroxypropyl glutamine) and poly(hydroxyethylaspartamide)], polymers derived from hydroxyalkylmethacrylates, linear polyvinylpyrrolidines, and dextrans in the molecular weight range of 40,000–80,000 have constituted a prime area of research, as have polymers that simulate the catalytic effects of certain enzymes. Another area where synthetic polymers are being used is in the design of matrices for drug delivery. Synthetic and natural polymers that have potential for this use include vinyl and acrylic polymers, polypeptides, polysaccharides, proteins, and DNA. Studies also suggest that the properties of the polymer may be used to control the biological activity of the drug. Some researchers are attempting to direct drugs to specific sites through the use of "messenger" molecules that are attached to the polymer together with the drug. Polymeric derivatives of food additives (qv) such as dyes, antioxidants, and sweeteners (qv) are under investigation as a possible approach to producing additives that will pass through the gastrointestinal tract without being absorbed (See Pharmaceuticals, controlled-release).

Another interesting application of synthetic polymers involves the manipulation of the biological activity of molecules such as enzymes. Some immobilized enzymes have proved so stable that they are now being used in a variety of industrial processes. A related area, known as affinity chromatography, takes advantage of the specificity of biological interactions to effect highly efficient purification of proteins, enzymes, etc (see Chromatography, affinity).

Another important application of synthetic polymers in medicine is in the area of surgical implantation of, for example, artificial joints, heart valves, arterial replacements, etc, to improve the biocompatibility of the implants (See Prosthetic and biomedical devices).

HERMAN F. MARK
Polytechnic Institute of New York

R.G. Dickerson and I. Geis, *The Structure and Action of Proteins*, Harper and Row, New York, 1969.

G.N. Ramachandran, *Conformation of Biopolymers*, Academic Press, Inc., New York, 1967.

H. Ringsdorf, "Pharmacologically Active Polymers," *J. Polym. Sci. Symp.* **51**, 135 (1975).

L.G. Donaruma, "Synthetic Biologically Active Polymers," *Progr. Polym. Sci.* **4**, 1 (1975).

BIOTIN. See Vitamins.

BIPHENYL, $C_6H_5C_6H_5$. See Diphenyl and terphenyls.

BISMUTH AND BISMUTH ALLOYS

Bismuth, Bi, mp 271.4°C, bp 1564°C, is a member of Group VA of the periodic table in the same subgroup as arsenic and antimony. Its valences are +3 and +5, and the only natural isotope is ^{209}Bi. Bismuth will fracture as a brittle, crystalline metal having a high metallic luster with a pinkish tinge. Its crustal abundance is estimated to be 0.000431 g/t. The important bismuth-containing minerals are bismite (Bi_2O_3), bismuthinite (Bi_2S_3), and bismutite (($BiO)_2CO_3 \cdot H_2O$). The principal source of bismuth is as a by-product from the treatment of lead and copper ores. Some bismuth also is found associated with molybdenum, gold, silver, tin, tungsten, and zinc ores.

Bismuth is the most diamagnetic of all the metals with a mass susceptibility of -1.35×10^6, and it displays the greatest increase in resistance when influenced by a magnetic field (Hall effect). However, its thermal conductivity decreases in such a field. Bismuth and gallium are the only metals that increase in volume on solidification, and mercury is the only metal with a lower thermal conductivity than bismuth.

The United States is dependent on imports since the domestic supply of bismuth is insufficient to satisfy annual requirements of ca 1000–1500 metric tons per year. ASARCO is the only primary producer of bismuth in the United States. The metal is produced and sold at purities of 99.99 and 99.999%.

The principal uses of bismuth are in the manufacture of pharmaceuticals (see Bismuth compounds), fusible alloys (low melting), and metallurgical additives to improve machinability of steel and aluminum in the manufacture of malleable iron. Because of its low absorption cross section for thermal neutrons, liquid bismuth has been used in the design of a liquid-metal fission reactor. Bismuth also is used as a window material for neutron transparency in irradiation devices used in medicine. In wire form, it is useful for Hall effect and thermoelectric applications (see Thermoelectric energy conversion). A thin film of bismuth on a plastic film substrate has been patented by the Bell Laboratories for microfilming information supplied by a pulsed laser.

Alloys

Bismuth forms a number of binary, ternary, quaternary, and quinary alloys melting in the range of 47–262°C. The more common alloys are referred to as fusible alloys and are sold under the trade names of AsarcoLo or Cerro alloys. The low melting point of these alloys is not the only property that affords them unique uses. Alloys of bismuth often exhibit unusual expansion and contraction characteristics. Various applications make use of these characteristics for grip tools, punches, and parts to be machined.

One of the earliest applications for fusible alloys was sprinkler heads that spray water automatically through a valve at a predetermined temperature. Other safety devices include holding fire-protection doors open until the temperature reaches a predetermined point, and a fusible alloy plug for tanks containing flammable liquids or gases. Research into the unusual properties of fusible alloys has suggested many ingenious labor- and time-saving applications in areas such as soldering and sealing, spotting fixtures, anchoring parts, bending thin sections, foundry applications, molding, lens blocks, heat transfer, and metallurgical uses.

No industrial poisoning from bismuth has been reported. However, precautions should be taken against the careless handling of bismuth and its compounds; ingestion and inhalation of dusts and fumes should be avoided.

S.C. CARAPELLA, JR.
H.E. HOWE[‡]
ASARCO Inc.

Metals Handbook, American Society for Metals, Cleveland, Ohio, 1948.

M. Hansen and K. Anderko, *Constitution of Binary Alloys*, 2nd ed., McGraw-Hill Book Co., New York, 1958.

E. Browning, *Toxicity of Industrial Metals*, Butterworths, London, 1961.

BISMUTH COMPOUNDS

Bismuth is the fifth member of the nitrogen family of elements and, like its congeners, possesses five electrons in its outermost shell: $6s^2, 6p^3$. In many compounds, the bismuth atom utilizes only the three $6p$ electrons in bond formation and retains the two $6s$ electrons as an inert pair. Compounds are also known, however, where bismuth is bonded to four, five, or six other atoms. Many bismuth compounds do not have simple molecular structures and exist in the solid state as polymeric chains or sheets. The +3 oxidation state is exhibited by bismuth in the vast majority of its compounds.

It is difficult to generalize about the toxicity of bismuth compounds. Some of these substances, eg, bismuth oxychloride, appear to have no significant toxicity, whereas others like bismuth subnitrate and subgallate can be quite harmful.

Inorganic Compounds

Bismuthine, BiH_3, is a colorless gas, unstable at room temperature but isolable as a colorless liquid at lower temperatures.

Bismuthides are intermetallic compounds of bismuth with alkali metals and alkaline-earth metals, many of which have the expected formulas M_3Bi and M_3Bi_2, respectively. Both alkali and alkaline-earth metals form another series of alloylike bismuth compounds that become superconducting at low temperatures. The MBi_2 compounds are particularly noteworthy as having extremely short bond distances between the alkali metal atoms.

Bismuth halides, of which the trihalides are the best known, include bismuth trifluoride, BiF_3, mp 649°C, bp 900 ± 10°C; bismuth trichloride, $BiCl_3$, mp 233.5°C, bp 441°C; bismuth tribromide, $BiBr_3$, mp 219°C, bp 462°C; bismuth triiodide, BiI_3, mp 408.6°C, bp 542°C; and bismuth pentafluoride, BiF_5, mp 151°C, bp 230°C.

[‡]Deceased.

Bismuth oxide halides, include bismuth oxychloride, BiOCl, used in fingernail polishes, lipsticks, and face powders to give a nacreous effect, and bismuth oxybromide, BiOBr, used in the manufacture of dry-cell cathodes.

Bismuth oxides and bismuthates, eg, bismuth trioxide, Bi_2O_3, are used in ceramics and glasses, in soldering and brazing, as a ceramic dielectric, in ceramic varistors, and as a component in a number of catalysts.

Higher oxides of bismuth and related compounds are used in products in which much of the bismuth is in the $+5$ oxidation state as a result of oxidation of either a fused mixture of sodium oxide and bismuth trioxide or a suspension of bismuth trioxide in 40–50% sodium hydroxide solution.

Sulfides and related compounds. Bismuth trisulfide, Bi_2S_3, mp 850°C, is used as a high temperature lubricant and for its photo- and thermoelectric properties. Bismuth triselenide, Bi_2Se_3, and bismuth tritelluride, Bi_2Te_3, both are of interest for their semiconducting properties.

Bismuth salts include bismuth triperchlorate pentahydrate, $Bi(ClO_4)_3 \cdot 5H_2O$, bismuth trintrate pentahydrate, $Bi(NO_3)_3 \cdot 5H_2O$, and bismuth trisulfate, $Bi_2(SO_4)_3$.

Organobismuth Compounds

In a similar manner to the other Group-V elements, the bismuth atom can be either tri- or pentacovalent. However, organobismuth compounds are less stable thermally than the corresponding phosphorus, arsenic, or antimony compounds, and there are fewer types of organobismuth compounds.

Bismuthines. The only known primary and secondary bismuthines are methylbismuthine, CH_3BiH_2, and dimethylbismuthine. The tertiary bismuthines, R_3Bi (R is alkyl or aryl), are better known and probably constitute the largest known group of organobismuth compounds.

Halo- and dihalobismuthines and related compounds. The organobismuth halides and dihalides are crystalline solids with high melting points (eg, phenylbismuth dibromide, $C_6H_5BiBr_2$, mp 205–206°C; and diphenylbismuth chloride, $(C_6H_5)_2BiCl$, mp 184–185°C).

Bismuth compounds and compounds with bismuth bonded to other metals. A few compounds are known in which bismuth is linked to Group-IV metals: triethylgermyldiethylbismuth, $(C_2H_5)_2BiGe(C_2H_5)_3$, tris(triphenylsilyl)bismuthine, $[(C_6H_5)_3Si]_3Bi$, and tris(triphenylgermyl)bismuthine, $[(C_6H_5)_3Ge]_3Bi$.

Bismin. Bismabenzene, C_5H_5Bi, has not been isolated but shown to exist as a reaction intermediate.

Trialkyl- and triarylbismuth dihalides and related compounds. After the tertiary bismuthines, the triarylbismuth dihalides constitute the most important class of organobismuth compounds and are by far the largest class of compounds containing pentacovalent bismuth. Only a few trialkylbismuth dihalides are known, eg, *cis*- and *trans*-tripropenylbismuth dibromide. The triarylbismuth dihalides are stable crystalline solids with high melting points. Several triarylbismuth dihalides and other compounds related to the dihalides have limited industrial use. Triphenylbismuth dichloride, triphenylbismuth dibromide, carbonate, and hydroxychloride, and related compounds have been patented as bactericides and fungicides for the treatment of plastics and fibrous materials. Compounds of the type $(C_6H_5)_3Bi(O_2CR)_2$, eg, triphenylbismuth dimethacrylate, $(C_6H_5)_3Bi(O_2CC(CH_3)=CH_2)_2$, and triphenylbismuth di(*p*-vinylbenzoate), $(C_6H_5)_3Bi(O_2CC_6H_4CH=CH_2-p)_2$, are claimed to be effective agents against *Staphylococcus aureus* infections.

Other compounds. Quaternary bismuth compounds, eg, tetraphenylbismuth bromide, tetraphenylbismuthonium tetrafluoroborate, and tetraphenylbismuthonium hexafluorophosphate, are generally unstable with the exception of the perchlorates, tetrafluoroborates, and hexafluorophosphates. The chemistry of pentaarylbismuth compounds, such as pentaphenylbismuth, has not been explored thoroughly.

Bismuth Compounds Used in Medicine

Milk of bismuth and bismuth subnitrate, both used in the management of diarrhea and intestinal inflammation, are the only bismuth preparations listed as therapeutic agents in the latest edition of the USP. However, bismuth subcarbonate is probably used more than any other bismuth compound for the treatment of gastric disturbances. Tripotassium dicitratobismuthate (De-Nol) is a new drug said to be effective for the treatment of gastric and duodenal ulcers. A closely related preparation referred to as bismuth ammonium citrate (Ulcerine) has also been reported to give excellent results in treating peptic ulcers.

G. Gilbert Long
Leon D. Freedman
G.O. Doak
North Carolina State University

P.J. Durrant and B. Durrant, *Introduction to Advanced Inorganic Chemistry*, 2nd ed., John Wiley & Sons, Inc., New York, 1970.

G.P. Smith in D.B. Sowerby, ed., *MTP International Review of Science, Series Two, Inorganic Chemistry*, Vol. 2, Butterworths, London, 1975.

L.D. Freedman and G.O. Doak, *Chem. Rev.* **82**, 15 (1982).

BISPHENOL A, $(HOC_6H_4)_2C(CH_3)_2$.

See Alkylphenols (4,4'-Isopropylidenediphenol).

BITUMENS.

See Asphalt; Roofing materials.

BLANC FIXE, $BaSO_4$.

See Barium compounds (Barium sulfate).

BLEACHING.

See Pulp.

BLEACHING AGENTS

A bleaching agent is a material that lightens or whitens a substrate through chemical action. This action can involve either oxidative or reductive processes that make color bodies in the substrate more soluble and more easily removed during processing. Bleaching can also involve chemical processes that change the ability of color bodies to absorb light by changing their degree of unsaturation (see Chloramines; Chlorine oxygen acids and salts; Hydrogen peroxide; Peroxides).

Chlorine-containing bleaching compounds. Three classes of chlorine-containing compounds are used as bleaching agents: chlorine, hypochlorites and *N*-chloro compounds, and chlorite and chlorine dioxide. The first two classes, termed available-chlorine compounds, produce hypochlorous acid and hypochlorite anion in bleaching baths. Chlorine dioxide (ClO_2) is a gas that is extremely explosive in high concentrations and must be stored and shipped under refrigeration as a stable octahydrate. It is used extensively in pulp bleaching. Sodium chlorite ($NaClO_2$) is a relatively stable solid source of chlorine dioxide which is used mainly outside of the United States for textile bleaching.

Peroxygen compounds. Peroxygen or active oxygen compounds always contain a peroxide linkage ($—O—O—$) in which one oxygen atom is active. Hydrogen peroxide (qv) is one of the most common bleaching agents used in textile, pulp and paper, and home laundry applications. Pure hydrogen peroxide has an active oxygen content of 47%. It is the least expensive source of active oxygen commercially available. Moreover, it is a liquid, making it convenient for many bleaching applications. Hydrogen peroxide reacts with many compounds, such as borates, pyrophosphates, sulfates, silicates, and several organic carboxylic acids to give peroxy compounds or peroxyhydrates. Usually these are stable solids that hydrolyze readily to give hydrogen peroxide in solution. These compounds, such as borates (the most widely used solid peroxygen compound), sodium carbonate peroxyhydrate, phosphate peroxyhydrates, and potassium permonosulfate, therefore serve as useful solid sources of hydrogen peroxide. One advantage of hydrogen peroxide is that it is nonpolluting, ultimately decomposing to oxygen and water.

Reducing bleaches. The reducing agents generally used in bleaching include sulfur dioxide, sulfurous acid, bisulfites, sulfites, hydrosulfites (dithionites), sodium formaldehyde sulfoxylate, and sodium borohydride. These materials are used mainly in pulp and textile bleaching (see Sulfur compounds; Boron compounds).

The Mechanism of Bleaching

Bleaching is essentially a whitening or decolorizing process. The color producing agents in fibers are often organic compounds that contain conjugated chains, that is, alternating single and double bonds. These groups are called chromophores.

Decolorization often can be brought about by destroying one or more of the double bonds in the conjugated systems. This can be accomplished by addition to the double bond or its cleavage. the reactions of chlorine-containing bleaches can be classified into three types: chlorhydrination, chlorination (by addition or substitution), and oxidation; all three types occur irreversibly. Hydrogen peroxide and other peroxygens can destroy double bonds by epoxidation (qv).

Applications

The principal home laundry bleach product in the United States is liquid chlorine bleach, sodium hypochlorite. However, some synthetic fabrics, some wash and wear fabrics, and some colored fabrics are attacked by chlorine bleaches and must be bleached with solid safety-bleach formulations based on sodium perborate. In Europe, perborate is widely used in detergent formulations.

In industrial and institutional bleaching, both liquid chlorine and dry chlorine bleaches are used.

Scouring cleansers, designed to clean hard surfaces, are composed of a finely ground abrasive, a synthetic detergent, auxiliaries, and bleaches. The most commonly used materials are chlorine bleaches such as sodium or potassium salts of chlorinated isocyanuric acid and chlorinated tri-sodium phosphate (see Cyanuric acid and isocyanuric acids).

Textile bleaching. Industrially, textile bleaching is carried out on fabrics containing such natural fibers as cotton or wool that require a high degree of brightness or that will be dyed or printed in a light shade. The cotton and cotton–polyester fabrics used extensively today are the two fabrics most often bleached. Hydrogen peroxide is the main bleaching agent used for textiles in the United States. Sodium hypochlorite, sodium chlorite, and sodium hydrosulfite are used to a much smaller extent. Sodium chlorite is used to bleach alkali-sensitive dyed goods and fabrics containing nylon, polyester, and acrylic fibers. Its use has severe drawbacks in that solutions are corrosive and require special materials of construction (titanium). Moreover, sodium chlorite presents toxicity problems and explosion hazards.

Pulp bleaching. There are three types of pulp: chemical, semichemical, and mechanical. Most of the chemical pulp produced in the United States is kraft pulp which is dark in color and has to be bleached for many applications. A variety of bleaching agents including chlorine, chlorine dioxide, hypochlorite, peroxide, and oxygen are used on chemical pulp. Bleaching usually requires the use of more than one of these chemicals in sequence. Hydrogen peroxide also is being used as a total or partial replacement for hypochlorite. This results in a stronger bleached pulp, lower energy use, and a reduced chloride level in the bleach-plant effluent.

New developments in chemical pulp bleaching. Hooker has developed a series of antipollution sequences (APS) that can be applied to the conventional kraft mill. In the APS-1 process, which is being used commercially, half of the available chlorine in the chlorination stage is replaced by chlorine dioxide in a sequential process. Sequential chlorination is followed by a hypochlorite stage and buffered at a high pH, instead of the conventional caustic-extraction stage. These changes are claimed to reduce the color and BOD and COD levels of bleach plant effluent. In the closed-cycle bleaching process, all process streams including the ones from the bleach plant are recycled. The plant requires a higher capital investment because a salt-recovery process is required, but it is claimed to be cheaper than a conventional mill with external effluent treatment.

Another process that has gained interest because of current more stringent pollution controls is oxygen bleaching. The use of oxygen instead of chlorine-based chemicals allows the recycling of the water from the oxygen bleaching stage to brown-stock washing. Moreover, any organic material in the effluent can be burned in the recovery process. The use of an oxygen bleaching sequence can lead to reduced BOD and color in the effluent and to lower operating costs than for plants using more conventional processes.

Displacement bleaching, wherein the bleaching times can be dramatically reduced by passing the concentrated bleaching chemical through a layer of pulp and displacing the liquid already present, and gas-phase bleaching, a sequential process using gaseous chlorine (C_g) and gaseous chlorine dioxide (D_g) at atmospheric pressure, also are being studied.

Bleaching of mechanical pulp often entails a two-stage, sequential process using hydrogen peroxide and sodium hydrosulfite. Hydrogen peroxide is the most satisfactory bleaching agent for human hair, and fur is usually bleached using hydrogen peroxide stabilized with sodium silicate.

Sulfur dioxide is used to preserve grapes, wine (qv), and apples. Flour can be bleached with a variety of chemicals including chlorine, chlorine dioxide, oxides of nitrogen, and benzoyl peroxide. Bleaching agents such as chlorine dioxide or sodium dichromate are used in the processing of nonedible fats and fatty oils for the oxidation of pigments to colorless forms (see Food processing; Vegetable oils).

BURTON M. BAUM
JOSEPH H. FINLEY
JOHN H. BLUMBERGS
EDWARD J. ELLIOTT
FRED SCHOLER
HENRY L. WOOTEN
FMC Corp.

B.K. Easton, *Ciba Geigy Rev.* **3**, 3 (1971).

The Pulping of Wood, Vol. 1 of R.G. MacDonald, ed., *Pulp and Paper Manufacture*, 2nd ed., McGraw-Hill Book Co., New York, 1969.

V.B. Bodenheimer and J.O.E. Patchen, *South. Pulp Pap. Manuf.* **139**, 29 (Mar. 1976); **139**, 30 (Apr. 1976).

BLENDING.   See Mixing and blending.

BLOOD, ANIMAL.   See Meat products.

BLOOD, COAGULANTS AND ANTICOAGULANTS

Hemostasis or stoppage of blood flow can be shown as the disturbance of a delicately poised system of two processes—coagulation and fibrinolysis. Under normal circumstances blood remains fluid, but if vascular damage occurs or if certain abnormal physiological states develop, steady states in one or both of the processes are disturbed and hemostasis results.

Coagulants

Biochemically, blood coagulation is the result of polymerization of a polypeptide monomer (modified fibrinogen) into a cross-linked mesh of fibrils, an insoluble gel called fibrin. The theory accounting for the coagulation of blood has evolved from a two-step mechanism to the present model; this depicts coagulation as a multistep cascade of activations of protein factors which culminates in fibrin formation. Roman numerals have been assigned to each participating substance. It is common practice to refer to factors V through XIII by Roman numerals, and the subscript (a) is used to indicate the activated form. Factors I–IV are generally referred to by common names. Table 1 shows a list of synonyms of factors that have been used. All but one of the factors (III,

Table 1. Nomenclature of Coagulation Factors

Factor number	Common names
I	fibrinogen
II	prothrombin; activated form: thrombin
III	tissue factor, tissue thrombokinase, extrinsic thromboplastin
IV	calcium ion (Ca^{2+})
V	proaccelerin, labile factor, plasma accelerator globulin, plasma Ac-globulin, accelerator factor, plasma prothrombin convertin factor, thrombogene; activated form: accelerin
VII	proconvertin, stable factor, prothrombinogen, autoprothrombin I; activated form: serum prothrombin, conversion accelerator (SPCA), convertin
VIII	antihemophilic factor (AHF), antihemophilic globulin (AHG), hemophilic factor A, platelet cofactor 1, thromboplastinogen, thrombocytolysin
IX	Christmas factor, antihemophilic factor B, autoprothrombin II, plasma thromboplastin component (PTC), platelet cofactor 2, thromboplastinogen B; activated form: prephase accelerator (PPA)
X	Stuart factor, Prower factor, Stuart-Prower factor, autoprothrombin III; activated form: plasma thromboplastin, autoprothrombin C, thrombokinase
XI	plasma thromboplastin antecedent (PTA), antihemophilic factor C, activated form: third prothromboplastic factor
XII	Hageman factor, surface factor, contact factor, clot-promoting factor
XIII	fibrin stabilizing factor (FSF), fibrin stabilizing enzyme, Laki-Lorand factor, fibrinase, plasma transglutaminase, fibrinoligase
XIV[a]	Protein C[a]; activated form: autoprothrombin II-A

[a] Factor XIV usage has been recently proposed; it is not in common usage

tissue factor) circulate in the blood plasma. It should be noted that most of these factors exist as zymogens (factor precursors) and must be modified by hydrolytic cleavage to achieve full activity. Cleavage is accomplished by the factor immediately preceding in the coagulation cascade; consequently, most of the activated factors are endopeptidases (XIIa, XIa, Xa, IXa, VIIa, IIa).

It has been known for many years that vitamin K exerts a procoagulant effect *in vivo*; a deficiency of vitamin K results in a bleeding tendency. This vitamin plays a role in the biosynthesis of factors VII, X, IX, and prothrombin (the vitamin K-dependent coagulation factors), factors involved in reactions that occur in lipoprotein complexes.

The traditional treatment of hemophilia and other blood disorders due to the lack of activity of coagulation factors has been plasma transfusion. A more recent development is the use of plasma concentrates; because 90% of congenital hemorrhagic states are caused by deficiencies in activity of factors VIII (classical hemophilia, von Willebrand's disease) and IX (Christmas disease), concentrates containing these factors have received the most attention. AHF (factor VIII) concentrates can be made locally at blood banks by cryoprecipitation (see Blood fractionation), and both factor VIII and factor IX concentrates are available commercially. Fibrinogen concentrates, also available commercially, are used in treatment of the congenital hemorrhagic states afibrinogenemia and dysfibrinogenemia, and in acquired hypofibrinogenemia.

Miscellaneous hemostatics include carbazochrome salicylate, the sodium salicylate complex of the oxidation product of epinephrine–adrenochrome (Adrenosem Salicylate, Beecham Labs), which is used to reduce oozing and bleeding after some surgical procedures. Its effect is not on the coagulation process; rather, it reduces capillary permeability, thus controlling capillary bleeding. Other products are used locally to control oozing: oxidized cellulose (Oxycel, Parke Davis and Surgicel, Johnson & Johnson), a specially treated gelatin product (Gelfilm, Upjohn), and microfibrillar collagen, a powder derived from bovine corium collagen (Avitene, Avicon).

Anticoagulants

Substances that remove Ca^{2+} from plasma have long been used to prevent the coagulation of blood samples, eg, ethylenediaminetetraacetic acid (EDTA) (Sequestrene); dilithium oxalate, $Li_2C_2O_4$; disodium oxalate, $Na_2C_2O_4$; dipotassium oxalate monohydrate, $K_2C_2O_4 \cdot H_2O$; sodium fluoride, NaF; sodium polyanetholesulfonate (Liquoid, Hoffmann-La Roche, Inc.); and heparin.

Several plasma proteins have been shown to inhibit the activity of thrombin and other serine proteases. Among these are Antithrombin III (AT-III, antithrombin–heparin cofactor), α_2-macroglobulin, α_1-antitrypsin, and C-1 esterase inhibitor. A main portion of the antithrombin activity of plasma (70–90%) has been shown to be due to AT-III.

Therapeutic anticoagulants. Heparin, a glycosaminoglucuronan (acid mucopolysaccharide) which occurs in most tissues, interferes with the coagulation mechanism, enhances fibrinolysis, and can potentiate or inhibit platelet aggregation and release reaction. It has been particularly useful in immediate treatment of deep vein thrombosis and pulmonary and systemic embolism, and recently has been shown efficacious in low doses for prophylaxis of post-operative thrombo-embolism. Low dosage regimens also may be useful for control of coronary heart disease and for prophylactic control of deep vein thrombosis after myocardial infarction.

The parent compound for synthesis of coumarin substances is 4-hydroxycoumarin. These anticoagulants, in contrast to heparin, are active *in vivo*, but not *in vitro*. They produce their effect by a competitive inhibition of vitamin K which leads to a reduction in the activity, in blood, of factors VII, IX, X, and prothrombin through interference with the synthesis of these agents in the liver.

Anticoagulant antagonists. Heparin, owing to its strongly acidic character, is readily precipitated by strongly basic agents; thus, protamine sulfate is used widely as a heparin antagonist. The substance has no effect, however, on the coagulation time of normal individuals. The antidote for overdosage of the coumarin drugs is vitamin K.

Fibrinolysis

Like coagulation, fibrinolysis involves a number of components, many of which are proteases. The culminating event of the fibrinolytic system is the dissolution of a clot by the degradation of fibrin to water-soluble fibrin degradation products (FDP). This hydrolysis is accomplished by a serine protease, plasmin, which is generated from its zymogen, plasminogen, by substances termed plasminogen activators, most of which also are proteases. Activators responsible for the generation of plasmin from plasminogen by proteolytic cleavage are present in many body tissues and fluids. Factor XII plays a role in initiating fibrinolysis and recent studies show that this activity may be mediated by kallikrein.

Thrombolytic agents. Various substances have been used in an attempt to hydrolyze existing clots and to prevent formation of thrombi. These agents supplement the role of therapeutic anticoagulants, although they are aimed not at preventing coagulation, but at enhancing fibrinolysis and thrombolysis. They have not yet achieved the status of the anticoagulants in management of thrombotic disorders. In view of the current concept of *in vivo* fibrinolysis, the ideal thrombolytic agent would be one that activates thrombibound (gel-phase) plasminogen. It would, therefore, be classified as a plasminogen activator and would have a high gel-phase–soluble-phase plasminogen activation ratio, eg, urokinase, purified from urine, which is available commercially as Win-Kinase (Winthrop) and Abbokinase (Abbott). Alternatively, the agent could act in a less direct manner and stimulate the release of plasminogen activator *in vivo*. One of the anabolic steroids, Stanazol, has shown promise as a useful therapeutic agent of this kind.

Inhibitors of fibrinolysis. There are two types of inhibitors, those that inhibit plasminogen activation and those that inhibit plasmin directly. Of the inhibitors, only the synthetic amino acids have had any wide clinical use. They enhance formation of clots by preventing lysis of incipient fibrin deposits. Agents commercially available include ε-aminocaproic acid, 6-aminohexanoic acid (Amicar, Lederle); tranexamin acid, *trans*-4-(aminomethyl)cyclohexanecarboxylic acid (Amstat, Lederle); and aprotinin, trypsin inhibitor, trasylol (Trasylol, Delbay). They may be indicated in pathological conditions where hemostasis does not occur. In addition, because of possible inhibitory effects on the complement system, they may be useful in immunologically mediated phenomena such as organ transplants, asthma, and blood-transfusion induced hemolytic states.

DAVID M. STUART
JUDITH K. HRUSCHKA
Ohio Northern University

J.W. Suttie and C.M. Jackson, *Physiol. Rev.* **57**, 1 (1977).

V.V. Kakkar and D.P. Thomas, eds., *Heparin, Chemistry and Clinical Uses*, Academic Press, London, 1976.

C.R.M. Prentice and J.F. Davidson in D.W. Ribbon and K. Brew, eds., *Proteolysis and Physiological Regulation*, Academic Press, Inc., New York, 1976.

BLOOD FRACTIONATION

In the United States, all facilities that draw blood or prepare blood components are licensed by or registered with the Federal government and are regulated by the Bureau of Biologics of the FDA. Strict guidelines for donor selection, as well as for collection and handling of blood and blood components, are contained in the Code of Federal Regulations (CFR), which is published annually.

Current procedures for large-scale (at least one unit) fractionation of human blood into plasma and cellular components are based on one of two principles: sedimentation or adhesion. Sedimentation techniques take advantage of differences in density between plasma and suspended cells and of differences in size and density among the cell types. Methods utilizing centrifugation to enhance sedimentation are the most common. Such methods can separate whole blood into four fractions, ie, plasma, platelet concentrate, white cell (leukocyte) concentrate, and leukocyte-poor packed red cells. Adhesion techniques take advantage of variations in the ability of different cell types to bind to solid supports. Methods based on these techniques have been developed for removing leukocytes from packed red cells and for selectively isolating granulocytes from whole blood. The following components must be isolated within 6 h of collection of the source blood: platelets, leukocytes, and plasma that is to be used for the preparation of labile coagulation factors.

The conditions and maximum length of storage for the main blood components and derivatives are presented in Table 1 and, except in the case of leukocyte concentrates, are mandated by the CFR.

Plasma Fractionation

Plasma is fractionated to produce therapeutic materials containing one or more of the plasma proteins in concentrated and purified form in order to achieve optimal clinical usefulness. The main therapeutic fractions currently produced are albumin (in several degrees of purity), immune serum globulin (ISG) (both normal and specific), antihemophilic factor (AHF) (factor VIII), and prothrombin complex (PTC) (factors II, VII, IX, and X). A fractionation method was developed during the 1940s by a group headed by Cohn at the Harvard Medical School. The methods developed during that period, with some modifications, are still the most popular methods for preparation of albumin and ISG.

Cold ethanol methods for albumin production. The monumental paper of Cohn and co-workers describes methods for the separation and purification of the protein and lipoprotein components of human plasma. In each of these methods, there was an initial separation of the protein components of plasma into a small number of fractions in which the main components were separated, and then into a large number of subfractions into which these components were further concentrated and purified. The methods involved lowering the solubility of proteins by reducing the dielectric constant of the solution by the addition of ethanol. Thus, separations could be carried out in the range of low ionic strengths at which the interactions of proteins with electrolytes differ from each other markedly.

Cold ethanol methods for immune serum globulin production. The procedure developed by Oncley and co-workers has become the classic method for the production of ISG. All ISG for therapeutic use is prepared from large pools of plasma from many donors so that the final product will contain a broad spectrum of antibodies.

Alternative methods for the production of albumin and ISG include heat denaturation of the nonalbumin components of plasma, using polyethylene glycol (PEG) as a protein precipitant, adsorption chromatography (for the purification of ISG), a large scale method for the production of albumin using PEG, adsorption chromatography, and gel chromatography, continuous preparative electrophoresis, polarization chromatography, and isotachopheresis and isoelectric focusing.

Table 1. Blood Components: Use and Storage

Component	Contents	Indications for use	Shelf life	Storage temperature
red cells	red cells, some plasma, some WBC, and platelets or their degradation products	increase patient red-cell mass	35 d closed system	1–6°C
leukocyte-poor red cells	red cells, some plasma, and few WBC	prevent febrile reaction from leukoagglutinins	21 d closed system; 24 h open system	1–6°C; 1–6°C
frozen red cells	red cells, no plasma, minimal WBC and platelets	increase red-cell mass, prevent tissue antigen sensitization, prevent febrile or anaphylactic IgA[a] reaction, provide rare bloods	3 yr frozen; 24 h thawed	frozen, −65°C or −196°C; thawed, 1–6°C
leukocyte concentrate	WBC, few platelets, and some RBC	agranulocytosis	24 h	1–6°C
platelet concentrate	platelets, few WBC, and some plasma	bleeding due to thrombocytopenia	5 d	20–24°C or 1–6°C
single-donor plasma	plasma with no labile coagulation factors	blood-volume expansion	5 yr	frozen, −18°C
fresh frozen plasma	plasma with all coagulation factors and no platelets	treatment of coagulation disorders	1 yr frozen; 2 h thawed	frozen, −18°C; thawed (not above 37°C)
cryoprecipitate	coagulation factors I and VIII	hemophilia A and von Willebrand's disease, fibrinogen deficiency	1 yr	frozen, −18°C
antihemophilic factor	factor VIII	hemophilia A	up to 2 yr	lyophilized, 1–6°C
prothrombin complex	factors II, VII, IX, and X	hemophilia B (Christmas disease)	1 yr	lyophilized, 1–6°C
albumin	albumin	blood-volume expansion, replacement of protein	3 yr	room temperature (not above 37°C)
plasma protein fraction	albumin, α- and β-globulin	blood-volume expansion	3 yr	room temperature (not above 30°C)
immune serum globulin	γ-globulin	disease prophylaxis or attenuation, agammaglobulinemia	3 yr	1–6°C
$Rh_0(D)$ immune globulin	γ-globulin from sensitized donors	prevention of $Rh_0(D)$ sensitization	6 mo	1–6°C

[a] IgA is one of the γ-globulins.

Methods of antihemophilic factor production. Cryoprecipitation is the first step in most, if not all, methods in use today for the large scale production of AHF concentrates. Fresh frozen plasma is pooled and then thawed at 2°C, and the precipitate collected in continuous-flow centrifuges. The cryoprecipitate thus obtained is then processed by a variety of methods all of which have as their last steps sterile filtration and lyophilization. High purity concentrates of AHF are produced from cryoprecipitate extracts by fractional precipitation with PEG, PEG and glycine, and ethanol. A good approximation of the recovery of AHF from plasma is 200–400 units per liter for intermediate purity concentrates and 100–250 units per liter for high purity concentrates.

Methods for prothrombin complex production. Coagulation factors II, VII, IX, and X are usually isolated together as a fraction called prothrombin complex. Two prothrombin-complex preparations are produced commercially in the United States. One preparation starts from cold ethanol fraction III that is suspended in saline and treated with calcium phosphate. The prothrombin complex is then eluted with citrate and freed of lipoproteins by cold ethanol fractionation. Precipitation with PEG is used to purify and concentrate the coagulation factors.

The second commercial preparation performed on a large scale starts from cold ethanol fraction I supernatant. The prothrombin complex is adsorbed onto DEAE-Sephadex and eluted in several fractions.

Procedures, Equipment, and Reagents for Plasma Fractionation

Plasma fractionation is carried out in large tanks with jackets to allow circulation of fluid for precise control of temperature. The tanks are made of stainless steel and constructed so that they may be cleaned in place or easily disassembled for thorough cleaning. Some tanks may be equipped with temperature control coils inside the vessel. Most fractionation procedures in general use, such as the cold ethanol methods and the AHF methods requiring cryoprecipitation, call for low temperatures. The requirements for cooling make refrigeration one of the most important and expensive factors in plasma fractionation (see Refrigeration).

Reagents must be added slowly, with efficient mixing, to avoid local excesses of precipitant concentration, temperature, pH, and ionic strength. This often is accomplished by adding the reagents as liquid through a device with multiple narrow-bore openings directly beneath the propeller blades of the stirrer.

Liquid–solid separations are critical for the efficiency of a plasma fractionation procedure. Filtration and centrifugation are the two methods generally used. Lyophilization is a standard procedure in plasma fractionation for removal of organic solvents and for drying. Several alternatives to lyophilization have been devised for the removal of ethanol from plasma protein fractions. Vacuum distillation, molecular-sieve chromatography, and semipermeable-membrane filtration have been used successfully.

There has been considerable recent work on methods for converting the cold ethanol batch process to a semicontinuous-flow stream. A continuous-flow precipitating system with automated controls and rapid mixing devices has been shown to be capable of producing PPF equal in purity and yield to that obtained by the comparable batch method. It is hoped that continuous-flow systems with automated controls will allow for more reproducible, efficient, and sanitary operation of plasma fractionation.

Reagents used in plasma fractionation must be completely removed from the final products or must be demonstrated to be safe. One of the problems in plasma fractionation is contamination of fractions by pyogens, eg, bacterial endotoxins. Water for processing and washing as well as all chemical reagents are regularly tested for pyrogenic contamination. Water is distilled and then stored at high temperatures to inhibit bacterial growth. Chemical reagents are generally USP grade. Sampling and testing are performed at each step of the fractionation process to provide information for further processing and to assure the quality of the product. Extensive testing of products is performed to assure their safety, potency, and efficacy.

The main hazards in producing fractions from large pools of plasma is the transmission of hepatitis and, probably, AIDS (auto-immune deficiency syndrome). It is now required in the United States that all donors of blood or plasma be tested for the presence of hepatitis B surface antigen by radioimmunoassay or assays of equivalent sensitivity, though this screening does not identify non-A, non-B hepatitis. Plasma protein fraction (PPF) has occasionally been associated with hypotensive reations in recipients. This has led to a contraindication for the use of PPF in patients on cardiopulmonary bypass and a recommendation that it not be infused at a rate greater than 10 mL/min. Another hazard of plasma fractionation is the partial denaturation of some fractions such as ISG, caused by fractionation methods. These denatured proteins may have toxic effects or may be immunogenic in the recipients.

Therapeutic uses of plasma fractions. Albumin and PPF as 5% solutions are used primarily as blood volume expanders. Immune serum globulin is used for the treatment of congenital and acquired agammaglobulinemias. It also has been found useful for the prevention of poliomyelitis and for the prevention and modification of measles and hepatitis A. ISG is used as a general prophylactic against various infectious diseases. ISG suitable for intravenous administration is a recent development. Antihemophilic factor is used for the treatment of hemophilia A. Prothrombin complex is indicated for the treatment of hemophilia B.

At present in the United States, most single-donor blood products are produced and distributed by regional blood centers. Plasma fractionation is controlled to a great extent by several large firms, most of which are divisions of pharmaceutical corporations.

Martin H. Stryker
Alan A. Waldman
The New York Blood Center

E.J. Cohn and co-workers, *J. Am. Chem. Soc.* **68**, 459 (1946).

J.L. Oncley and co-workers, *J. Am. Chem. Soc.* **71**, 541 (1949).

B. Blomback and A.L. Hanson, eds., *Plasma Proteins*, John Wiley & Sons Ltd., Chichester, UK, 1979.

J.M. Curling, ed., *Methods of Plasma Protein Fractionation*, Academic Press, London, 1980.

Seminars in Thrombosis and Hemostasis **6** (1979–1980).

A.A. Waldman and M.H. Stryker, *Laboratory Management* **19**, 43 (1981).

R. Gallo and co-workers, "HTLV-III," *Science* **224**, 497 (May 4, 1984).

BLUEPRINTING.   See Printing processes.

BOILERS.   See Steam power generation.

BORDEAUX MIXTURE.   See Fungicides.

BLOOD-REPLACEMENT PREPARATIONS

Oxygen in solution in F-tributylamine (F = perfluoro) is biologically available and makes it possible to replace the blood of rats, which survive and thrive. Fluosol-DA, a mixture of F-decalin and F-tripropylamine, is the only commercially available product that can be used in humans. Certain U.S. clinics have developed protocols for specific uses with FDA approval.

The perfluoro compounds are colorless and odorless organic compounds in which all hydrogen atoms have been replaced with fluorine. The number of possible compounds is enormous, but the number suitable for blood replacement is limited (see Fluorine compounds, organic).

Properties

Surface tensions are very low, and kinematic viscosities are lower than those of the corresponding hydrocarbons. Perfluoro compounds are insoluble in water and other polar liquids. They dissolve in halogenated liquids and usually in each other; refractive indexes are extremely low. Vapor pressure affects application. Above 3.3 kPa (25 mm Hg), a lethal condition, known as bloated lungs, occurs and normal respiration is impossible. The stability of the C—F bond renders perfluoro compounds chemically inert.

Manufacturing and Processing

In electrochemical fluorination (Simon process) the starting material is placed in anhydrous HF in an electrolytic cell equipped with Ni and Fe electrodes and cooling; F-tributylamine, F-tripropylamine and F-2-butyltetrahydrofuran are prepared this way. Both cis- and trans-F-decalin are prepared by fluorination with fluorine gas at elevated temperatures with CoF_3 as catalyst.

For emulsification, ultrasonic or high pressure homogenization equipment is employed. Properly purified perfluoro compounds are not toxic in tissue-culture assay systems. Both stationary and suspension cultures of human and animal cells can be used.

Administration

A small amount of blood is removed first, followed by administration of the same volume of artificial product. The blood can also be removed at the same rate as the artificial material is administered.

Health and Safety Factors

There are few risks associated with purified perfluoro compounds or other ingredients of blood-replacement preparations. They should, however, not be given in successive doses more frequently than the rate of loss from the body indicates.

Uses

Perfluoro compounds have many potential uses in biology and medicine. In addition, they may be used in industry where gas transport is important. Fluosol-DA has been used clinically in a wide variety of hemorrhaging and nonhemorrhaging patients and is being investigated clinically in stroke and heart disease patients. Oxypherol is designed for animal experimentation. The compositions of these two preparations are given in Table 1.

Table 1.   Composition of Perfluoro Preparations, w / v%[a]

Constituent	Oxypherol	Fluosol-DA, 20%
perfluorotributylamine	20.0	
perfluorodecalin		14.0
perfluorotripropylamine		6.0
Pluronic F-68	2.56	2.7
yolk phospholipids		0.4
potassium oleate		0.032
glycerol		0.8
NaCl	0.60	0.60
KCl	0.034	0.034
$MgCl_2$	0.020	0.020
$CaCl_2$	0.028	0.028
$NaHCO_3$	0.21	0.210
glucose	0.180	0.180
hydroxyethyl starch	3.0	3.0

[a] Water added to a total of 100 mL.

ROBERT GEYER
Harvard University

Technical Information Service No. 4, Green Cross Corp., Osaka, Japan, Dec. 26, 1976.

P. Menasche and co-workers in R.P. Geyer, G. Nemo, and R. Bolin, eds., Progress in Clinical and Biological Research, Vol. 122: Advances in Blood Substitute Research, Alan R. Liss Inc., New York, 1983, pp. 363–373.

BORON, ELEMENTAL

Boron, the fifth element in the periodic table, is composed of two stable isotopes with mass numbers of 10 and 11. Although widespread in nature, it has been estimated to constitute only 0.0001% of the earth's crust, usually occurring as alkali or alkaline-earth borates, or as boric acid (see Boron compounds, boron oxides). Pure forms of the element are difficult to prepare; the most common technique involves vapor deposition from a boron halide, usually in admixtures with hydrogen. Research continues for methods of obtaining commercial quantities of the pure element, particularly as filamentary reinforcement for advanced composites. In nuclear technology, thin films of boron are used for neutron counters, and dispersions of powdered boron in poly(vinyl chloride) or polyethylene castings are effective for shielding against thermal neutrons (see Nuclear reactors). Table 1 lists physical properties of boron.

Table 1.   Physical Properties of Boron

Property	Value
atomic weight	10.811 ± 0.003
mp, °C	2190 ± 20
bp, °C	3660
coefficient of thermal expansion per °C	
from 25 to 1050°C	$(5-7) \times 10^{-6}$
hardness	
Knoop, HK	2110–2580
Mohs, modified scale[a]	11
Vickers, HV	5000
density	
liquid[b]	2.08
α-rhombohedral crystals	2.46
Filamentary boron	
tensile strength, MPa (psi)	3450–4830 (500,000–700,000)
Young's modulus, MPa (psi)	3040–3330 (440,000–480,000)
Structural modifications[c]	
preparation temp, °C	
amorphous	800
α-rhombohedral	800–1100
α-tetragonal	1100–1300
β-rhombohedral	1300
density, g/cm³	
amorphous	2.3
α-rhombohedral	2.46
α-tetragonal	2.31
β-rhombohedral	2.35

[a] Diamond = 15.
[b] Just above melting point.
[c] The crystalline forms not listed here are designated tetragonal II (β-tetragonal), tetragonal III, and hexagonal.

JOSEPH G. BOWER
U.S. Borax & Chemical Corp.

J.L. Hoard and R.E. Hughes, "Elemental Boron and Compounds of High Boron Content," in E.L. Muetteries, ed., The Chemistry of Boron and Its Compounds, John Wiley & Sons, Inc., 1967.

J.G. Bower in R.J. Brotherton and H. Steinberg, eds., Progress in Boron Chemistry, Vol. 2, Oxford, 1970, pp. 231–271.

BORON COMPOUNDS

BORON OXIDES, BORIC ACID, AND BORATES

At present, borax (tincal), colemanite, probertite, ulexite, and szaibelyite are the only borate minerals of commercial importance. Borax and colemanite are the most important. Present borate production comes mostly from five countries: the United States, Turkey, the USSR, Argentina, and the People's Republic of China.

Boron Oxides

Boric oxide, B_2O_3, is the only commercially important oxide; however, one high oxide and several suboxides have been reported. Boric oxide, also known as diboron trioxide, boric anhydride, or anhydrous boric acid, normally is encountered in the vitreous state. This colorless, glassy solid usually is prepared by dehydration of boric acid at elevated temperatures. It is quite hygroscopic at room temperature, and the commercially available material contains ca 1% moisture as a surface layer of boric acid (see Table 1).

vitreous boric oxide

A high purity grade of vitreous boric oxide (99% B_2O_3) is produced by fusing refined, granular boric acid in a glass furnace fired by oil or gas. Principal uses of boric oxide relate to its behavior as a flux, an acid catalyst, or a chemical intermediate.

Boric Acid

The name boric acid usually is associated with orthoboric acid, which is the only commercially important compound. Three forms of metaboric acid also exist. Orthoboric acid, H_3BO_3, crystallizes from aqueous solutions as white, waxy plates, mp 170.9°C. When heated slowly, it loses water to form metaboric acid, HBO_2.

Most boric acid is produced by the reaction of inorganic borates with sulfuric acid in an aqueous medium. Sodium borates are the principal raw material in the United States. Boric acid serves as a source of B_2O_3 in many fused products such as textile fiberglass and other borosilicate glasses. An important new use of boric acid is as a fire retardant in cellulosic materials. It also serves as a component of fluxes for welding and brazing (see Solders; Welding). A number of boron chemicals are prepared directly from boric acid. It also catalyzes the air oxidation of hydrocarbons and increases the yield of alcohol by forming esters that prevent further oxidation of hydroxyl groups to ketones and carboxylic acids (see Hydrocarbon oxidation). The bacteriostatic and fungicidal properties of boric acid, although weak, have led to its use as a preservative in natural products. NF-grade boric acid serves as a mild, nonirritating antiseptic. Boric acid also is quite poisonous to many insects and has been used to control cockroaches and to protect wood against insect damage (see Insect control technology). A special quality grade of boric acid is added to nuclear-reactor cooling water.

Solutions of Boric Acid and Borates

Boric acid is essentially monomeric in dilute aqueous solutions, but polymeric species may form at concentrations $> 0.1\ M$. The conjugate base of boric acid in water is the tetrahydroxyborate anion, $B(OH)_4^-$. This species also is the principal anion in solutions of alkali-metal (1:1) borates such as $Na_2O.B_2O_3.4H_2O$. Mixtures of $B(OH)_3$ and $B(OH)_4^-$ would appear to form classical buffer systems where the solution pH is governed primarily by the acid–salt ratio. This relationship is nearly correct for solutions of sodium or potassium (1:2) borates, where the mole ratio $B(OH)_3 : B(OH)_4^- \cong 1$, and the pH remains near 9.0 over a wide range of concentrations. However, for solutions that have pH values much greater or less than 9.0, the pH changes greatly on dilution. This anomalous pH behavior is due to the presence of polyborates which dissociate into $B(OH)_3$ and $B(OH)_4^-$ as the solutions are diluted. Formation of polyborates also greatly enhances the mutual solubilities of boric acid and alkali borates.

From a series of very accurate pH studies, Ingri calculated equilibrium constants involving the species $B(OH)_3$ and $B(OH)_4^-$, and the polyions $B_3O_3(OH)_5^{2-}$, $B_3O_3(OH)_4^-$, $B_5O_6(OH)_4^-$, and $B_4O_5(OH)_4^{2-}$. The ratio between the total anionic charge and the number of borons per ion increases with increasing pH.

borate ion in borax decahydrate and pentahydrate

borate polyion in kernite

borate ion in sodium metaborate

Sodium Borates

Disodium tetraborate decahydrate (borax decahydrate), $Na_2O.2B_2O_3.10H_2O$ [formula wt 381.43, monoclinic, sp gr 1.71, specific heat 1.611 kJ/(kg·K) (0.385 cal/g·°C)) at 25–50°C, heat of formation -6.2643 MJ/mol (-1497.2 kcal/mol)] exists in nature as the mineral borax. Disodium tetraborate pentahydrate (borax pentahydrate), $Na_2O.2B_2O_3.5H_2O$ [formula wt 291.35, trigonal, sp gr 1.88, specific heat 1.32 kJ/(kg·K) (0.316 cal/(g·°C)), heat of formation -4.7844 MJ/mol (-1143.5 kcal/mol)] is found in nature as a fine-grained deposit formed by dehydration of borax. Disodium tetraborate tetrahydrate, $Na_2O.2B_2O_3.4H_2O$ [formula wt 273.34, monoclinic, sp gr 1.91, specific heat ca 1.2 kJ/(kg·K) (0.287 cal/(g·°C)), heat of formation -4.4890 MJ/mol (-1072.9 kcal/mol)] exists in nature as the mineral kernite. Disodium tetraborate (anhydrous borax), $Na_2O:2B_2O_3$ [formula wt 201.27, sp gr (glass) 2.367, heat of formation (glass) -3.2566 MJ/mol (-778.4 kcal/mol)] exists in several crystalline forms as well as a glassy form. Sodium pentaborate decahydrate, $Na_2O.5B_2O_3.10H_2O$ [formula wt 590.34, monoclinic, sp gr 1.71] exists in nature as the mineral sborgite. Sodium metaborate octahydrate, $Na_2O.B_2O_3.8H_2O$ [formula wt 295.76, triclinic, sp gr 1.74] and sodium metaborate tetrahydrate, $Na_2O.B_2O_3.4H_2O$ [formula wt 203.68, triclinic, sp gr 191] can also be prepared.

Table 1. Physical Properties of Vitreous Boric Oxide

Property	Value
vapor pressure[a], 1331–1808 K	$\log P_{kPa} = 5.849 - \dfrac{16960}{T}$
heat of vaporization[b], ΔH, kJ/mol, 1500 K	390.4
298 K	431.4
boiling point, °C extrapolated	2316
viscosity, $\log \eta$, mPa·s $(= cP)$	
1000°C	4.00
density, g/cm³, 0°C	1.8766
18–25°C,	1.844
well-annealed	
1000°C	1.528
index of refraction, 14.4°C	1.463
heat capacity (specific)[b], J/(kg·K)	
1000 K	131.38
heat of formation, ΔH_f, kJ, 298.15 K[b]	
for 2 B(s) + 3/2 O_2(g) = B_2O_3 (glass)	-1252.2 ± 1.7

[a] To convert kPa to torr, multiply by 7.5.
[b] To convert J to cal, divide by 4.184.

Borax decahydrate and pentahydrate are produced from sodium borate ores, from dry lake brines, from colemanite, and from magnesium borate ores. U.S. Borax's open-pit borax mine and refinery in Boron, Calif., represents the largest single source of borate chemicals in the world. Anhydrous borax is produced from its hydrated forms by fusion. Primary producers in North America are U.S. Borax and Kerr-McGee. In general, the production of fused materials is much more energy intensive than that of hydrated products, and this difference is reflected in their prices. Polybor, a proprietary product of U.S. Borax, has the approximate composition of disodium octaborate tetrahydrate, $Na_2O.4B_2O_3.4H_2O$. The material is produced by spray-drying mixtures of borax and boric acid.

The bulk borate products, borax decahydrate and pentahydrate, anhydrous borax, boric acid and oxide, and upgraded colemanite and ulexite, account in both tonnage and monetary terms for > 99% of sales of the boron primary products industry. A large increase in demand in the United States has been related to the use of borates in energy-conserving products (ie, insulation). The industry historically has been dominated by the United States, but Turkey has recently become an important producer.

Poisonings by boric acid have been reported following its use over large areas of burned or denuded skin. The handling of borax or boric acid is generally not considered dangerous, however.

In the United States, borates are used mainly for glass manufacture. They also are used as fluxing agents for porcelain enamels and ceramic glazes, in soap and cleaning compositions, in agriculture as a fertilizer and herbicide, as a catalyst in the air oxidation of hydrocarbons, in the manufacture of alloys and refractories, as flux in metallurgy, as fire retardants, and as neutron absorbers in nuclear reactors.

Alkali-Metal and Ammonium Borates

These include dipotassium tetraborate tetrahydrate, $K_2O.2B_2O_3.4H_2O$ [formula wt 305.51 orthorhombic, sp gr 1.92]; potassium pentaborate octahydrate, $K_2O.5 B_2O_3.8 H_2O$ [formula wt 586.42, orthorhombic, sp gr 1.74, heat capacity 329.0 J/(mol·K) (78.6 cal/(mol·K)) at 296.6 K]; and diammonium tetraborate tetrahydrate, $(NH_4)_2O.2B_2O_3.4H_2O$ [formula wt 263.38, tetragonal, sp gr 1.58]. Ammonium pentaborate octahydrate, $(NH_4)_2.5B_2O_3.8 H_2O$ [formula wt 544.4, sp gr 1.58, heat capacity 359.4 J/(mol·K) (85.9 cal/(mol·K)) at 301.2 K] exists in two crystalline forms, orthorhombic (α) and monoclinic (β).

Potassium tetraborate tetrahydrate may be prepared from an aqueous solution of KOH and boric acid with a $B_2O_3 : K_2O$ mole ratio of ca 2.0, or by separation from KCl–borax solution. Potassium pentaborate is prepared in a manner analogous to that used for the tetraborate, but the strong liquor has a $B_2O_3 : K_2O$ mole ratio near 5. Ammonium tetraborate tetrahydrate is prepared with a $B_2O_3 : (NH_4)_2O$ mole ratio of 1.8–2.1. Ammonium pentaborate is similarly produced from an aqueous solution of boric acid and ammonia with a $B_2O_3 : (NH_4)_2O$ mole ratio of 5.

The potassium and ammonium borates are low volume products with production figures of hundreds of metric tons per year for the tetra- and pentaborates. Dipotassium tetraborate tetrahydrate is used to replace borax in applications where an alkali-metal borate is needed but sodium salts cannot be used, or where a more soluble form is required. The potassium compound is used as a component in lubricants, as a solvent for casein, as a constituent in welding fluxes, and as a component in diazo-type developer solutions. Potassium pentaborate octahydrate is used in fluxes for welding and brazing of stainless steels and nonferrous metals. Diammonium tetraborate tetrahydrate is used when a highly soluble borate is desired, but alkali metals cannot be tolerated. Ammonium pentaborate octahydrate is used as a component in electrolytes for electrolytic capacitors, as an ingredient in flameproofing formulations, and in paper coating.

Calcium-Containing Borates

Dicalcium hexaborate pentahydrate, $2CaO.3B_2O_3.5H_2O$ [formula wt 411.16, monoclinic, sp gr 2.42, heat of formation -3469 kJ/mol (-0.38 kcal/mol)] exists in nature as the mineral colemanite. Sodium calcium

pentaborate hexadecahydrate, $Na_2O.2CaO.5B_2O_3.16H_2O$ [formula wt 810.60, triclinic, sp gr 1.95] exists in nature as the mineral ulexite. Sodium calcium pentaborate decahydrate, $Na_2O.2CaO.5B_2O_3.10H_2O$ [formula wt 702.50, monoclinic, sp gr 2.15] exists in nature as the mineral probertite.

The alkaline-earth metal borates of primary commercial importance are colemanite and ulexite. Both of these borates are sold as impure ore concentrates. United States production is in Death Valley, Calif. Late in 1976, the American Borate Corp. purchased the rights to the Death Valley colemanite and ulexite reserves which had been held by Tenneco Inc.; the main world producer is Turkey.

Colemanite is used in the production of boric acid and borax, as well as for the manufacture of the "E" glass used in textile glass fibers and plastic reinforcements (where sodium cannot be tolerated). It also has limited application as a slagging material in steel manufacture and as a precursor to some boron alloys. Ulexite and probertite are used in the production of insulation fiber-glass and borosilicate glass as well as in the manufacture of other borates.

Borate Melts and Glasses

Most of the interest in metal borate glasses has centered on reports that indicated the existence of maxima and minima in some of the physical properties of the glasses with increasing metal oxide content. This phenomenon has been called the boron oxide anomaly. Modern theory on borate glass structure, however, indicates that the changes in the properties with alkali content are not anomalous, but are the result of well-defined structural changes in the glass at the molecular level. The borate glass compounds of commercial importance are boric oxide and disodium tetraborate pentahydrate.

Other Metal Borates

Borate salts or complexes of virtually every metal have been prepared. For most metals, a series of hydrated and anhydrous compounds may be obtained by varying the starting materials and/or reactions conditions. In general, hydrated borates of heavy metals are prepared by mixing aqueous solution or suspensions of the metal oxides, sulfates, or halides with boric acid or alkali-metal borates (eg, borax). Anhydrous metal borates may be prepared by heating the hydrated salts to 300–500°C or by direct fusion of the metal oxide with boric acid or B_2O_3.

Barium metaborate, $BaO.B_2O_3$, is used as an additive to impart fire-retardant and mildew-resistant properties to latex paints, plastics, textiles, and paper products, and as a preservative in protein-based glues (see Glue). A material described as technical grade cobalt tetraborate, $CoO.2B_2O_3.4H_2O$, is marketed as an acid catalyst by the Shepherd Chemical Co. This material may actually be a mixture of metaborates and hexaborates. Hydrated copper metaborate, $CuO.B_2O_3.2H_2O$, has been used as a fungicide for treatment of lumber and other cellulose materials. The anhydrous salt, $CuO.B_2O_3$, is used as an oil pigment.

Two grades of manganese borates are marketed as $MnO.2Br_2O_3.xH_2O$ by the General Metallic Oxides Co. Both materials are used as printing ink driers.

Zinc borates. A series of hydrated zinc borates has been developed over the past fifty years for use as fire-retardant additives in coatings and polymers. Worldwide consumption of these zinc salts is currently several thousand metric tons per year. A substantial portion of this total is used in vinyl plastics where zinc borates are added alone or in combination with other fire retardants such as antimony oxide. The most commonly encountered zinc borate is $2 ZnO.3B_2O_3.3.5H_2O$, which is formed when boric acid is added to solutions of soluble zinc salts.

Boron Phosphate

BPO_4 is a white, infusible solid that vaporizes slowly above 1450°C, without apparent decomposition. It normally is prepared by dehydrating mixtures of boric acid and phosphoric acid at temperatures up to 1200°C. The principal application of boron phosphate is as a heterogeneous acid catalyst, and a high degree of chemical purity is often unnecessary.

DANIEL J. DOONAN
LOREN D. LOWER
U.S. Borax Research Corp.

Reviewed by
ROBERT A. SMITH
U.S. Borax Research Corp.

N.P. Nies and G.W. Campbell, "Inorganic Boron-Oxygen Chemistry" in R.M. Adams, ed., *Boron, Metallo-Boron Compounds, and Boranes*, Interscience Publishers, a division of John Wiley & Sons, Inc., New York, 1964.

J.W. Mellor, "Boron-Oxygen Compounds," Vol. V, Supplement, Part A of *Inorganic and Theoretic Chemistry*, Longman, Inc., New York, 1980.

BORIC ACID ESTERS

The term boric acid esters refers to compounds with the general formula $B(OR)_3$. Much of the chemistry of these compounds is related to the electrophilic nature of boron. A series of related compounds can be formed when electron-deficient boron accepts a fourth nucleophilic substituent leading to tetrahedral boron structures such as $NaB(OR)_4$.

Nomenclature in the boric acid ester series is confusing at best. The IUPAC committee on boron chemistry has suggested the use of trialkoxy- and triaryloxyboranes for compounds widely referred to in current literature as boric acid esters, trialkyl (or aryl) borates, trialkyl (or aryl) orthoborates, alkyl (or aryl) borates, and alkyl (or aryl) orthoborates, and in the older literature as boron alkoxides and aryloxides.

Properties

Trialkoxy (or aryloxy) boranes range from colorless low boiling liquids such as trimethoxyborane (commonly referred to as methyl borate) to high melting soilds. Trialkoxy (or aryloxy) boranes are typically monomeric, soluble in most organic solvents, and dissolve in water with accompanying hydrolysis to boric acid and the corresponding alcohol or phenol. This hydrolysis occurs rapidly except in cases where the boron atom is protected by bulky alkyl or aryl substituent groups. The boron atom in trialkoxy (or aryloxy) boranes is in a trigonal coplanar state with sp^2 bond hybridization, and has a vacant p orbital along the threefold symmetry axis perpendicular to the BO_3 plane. This vacant orbital is readily available for acceptance of nucleophiles, such as water or alcohols. Structural studies of trialkoxy- and triaryloxyboranes have confirmed this structure having the angle between C—O bonds at 120° except in those cases where the angles are slightly distorted by bulky substituent groups. The susceptibility of the boron atom to attack by nucleophiles is similar to that of the carbonyl carbon in carboxylic acid esters and leads to the analogy of referring to trialkoxy (or aryloxy) boranes as boric acid esters.

Trialkoxyboranes from straight-chain alcohols and triaryloxyboranes are stable to relatively high temperatures. Attempts to use this potentially useful property in borate ester-based high temperature lubricants and heat-transfer media have met with limited success because of the reactivity of these compounds to water and oxygen.

Preparation. The most common preparative method for trialkoxy- and triaryloxyboranes is the reaction of the appropriate alcohol or phenol with boric acid, an inexpensive and readily available boron source. The boric acid ester prepared in the largest quantities is methyl borate, and most of it is used captively in the production of sodium borohydride by the Ventron Division of Thiokol Corporation. Other sources of methyl borate are Anderson Development Co., May and Baker Ltd., and Manchem Ltd. Other borate esters that have been offered commercially include ethyl borate, *n*-propyl borate, isopropyl borate, *n*-butyl borate, cresyl borate (from a mixture of *meta* and *para* cresols), and tri(hexylene glycol) diborate.

Uses

The principal application of methyl borate is as an intermediate in the commercial production of sodium borohydride. Another commercial outlet is the use of methyl borate azeotrope as gaseous flux for welding (qv) and brazing (see Solders). Smaller quantities of borate esters are finding use as epoxy resin curing agents, and specific types of borate esters are used as gasoline additives to reduce engine knocking and engine deposits. Uses of borate esters in hydraulic fluids (qv) and lubricants and as a variety of polymer additives have been described in numerous patents (see Lubrication). Glycol borate esters also constitute the active ingredients in biocides used in jet and diesel fuels.

R.J. BROTHERTON
U.S. Borax Research Corp.

H. Steinberg, *Boron-Oxygen and Boron Sulfur Compounds*, Vol. 1 of *Organoboron Chemistry*, Interscience Publishers, a division of John Wiley & Sons, Inc., New York, 1964.

REFRACTORY BORON COMPOUNDS

Borides have metallic characteristics, with high electrical conductivity and positive coefficients of electrical resistivity. Many of them have high melting points, great hardness, low coefficients of thermal expansion, and good chemical stability, particularly the borides of metals of Subgroups IVA, VA, and VIA, the MB_6 compounds of Groups II and III, and the borides of aluminum and silicon.

Borides are inert toward nonoxidizing acids; however, a few, such as Be_2B and MgB_2, react with aqueous acids to form boron hydrides. Most borides dissolve in oxidizing acids such as nitric or hot sulfuric acid. They are also readily attacked by hot alkaline salt melts or fused alkali peroxides and more stable borates are formed. In dry air, where a protective oxide film can be preserved, borides are relatively resistant to oxidation.

The simplest method of preparation is a combination of the elements at a suitable temperature, usually in the range of 1100–2000°C. On a commercial scale, borides are prepared by the reduction of mixtures of metallic and boron oxides with aluminum, magnesium, carbon, boron, or boron carbide followed by purification. In spite of their unique properties, there are few commercial applications for monolithic shapes of borides. They are used for resistance-heated boats (with boron nitride), for aluminum evaporation, and for sliding electrical contacts.

Boron and carbon form one compound, boron carbide, B_4C, mp about 2400°C. It has a rhombohedral structure consisting of an array of nearly regular icosahedra, each with twelve boron atoms at the vertices and three carbon atoms in a linear chain outside the icosahedra.

Hot-pressed boron carbide is used as wear parts, sandblast nozzles, seals, and ceramic armor plates. Boron carbide is used in the shielding and control of nuclear reactors.

Boron and nitrogen form one compound, boron nitride, BN, mp 3000°C, that may exist in a cubic zinc blende form, or a hexagonal graphite-like form with a layered structure and planar six-membered rings of alternating boron and nitrogen atoms. It is colorless and a good electrical insulator when pure; traces of impurities add color and make it a semiconductor.

Hot-pressed hexagonal boron nitride is useful for high temperature electrical or thermal insulation, vessels, etc, especially in inert or reducing atmospheres. The greatest use of cubic boron nitride is as the abrasive Borazon.

R.H. WENTORF, JR.
General Electric Co.

R. Thompson, *Endeavour*, 34 (Jan. 1970).

N.N. Greenwood, "Boron" in J.C. Bailar and co-eds., *Comprehensive In-

organic Chemistry, Vol. 1, Pergamon Press, New York, 1973, pp. 665–993.

R.H. Wentorf, Jr., *J. Chem. Phys.* **34**, 809 (1961).

BORON HALIDES

The important physical and thermochemical properties of the boron trihalides are given in Table 1. The thermal conductivity of BBr_3 at 20°C is 0.112 W/(m·K). The boron trihalides are planar (sp^2) molecules with X–B–X angles of 120°C. Orbital energy assignments have been made for the trihalides based on their photoelectron spectra.

Table 1.   Physical Properties of the Boron Trihalides

Property	BCl_3	BBr_3	BI_3
mp, °C	−107	−46	−49.9
bp, °C	12.5	91.3	210
density ρ^a (liq), g/mL	1.434_4^0	2.643_4^{18}	3.35
	1.349_4^{11}		
crit. temp, °C	178.8	300	
crit. pressure, kPa^b	3901.0		
ΔH_f^0, kJ/mol gas^c	−403	−206	+18
ΔH_{vap}, kJ/mol^c	23.8	34.3	
C_ρ, J/(mol·°C), for gas at 25°C^c	62.8	67.78	
C_ρ, J/(mol·°C), for liquid at 25°C^c	121	128	
ΔH_{hydrol} kJ/mol liquid at 25°C^c	−289	−351	
ΔH_{fusion}, J/g at mp^c	18		
B—X bond energy, kJ/mol^c	443.9	368.2	266.5
B—X distance, nm	0.173	0.187	0.210

^{a}For BCl_3: ρ (g/mL) = 1.3730–2.159 × 10^{-3}°C; −8.377 × 10^{-7}°C; −44 to 5°C.

For BBr_3: ρ (g/mL) = 2.698–2.996 × 10^{-3}°C; −20 to 90°C.

^{b}To convert kPa to mm Hg, multiply by 7.50.

^{c}To convert J to cal, divide by 4.184.

The only United States producer of commercial quantities of BCl_3 and BBr_3 is the Kerr-McGee Chemical Corporation. Their processes are proprietary, but they almost certainly produce both compounds by the reaction of B_4C with the halogens. Approximately 96% of United States production of boron trichloride is used in the manufacture of boron fibers, the remainder as catalysts in various organic reactions. About 95% of the tribromide production is used for catalysis, the remainder for semiconductors.

Boron trihalides are highly toxic and contact should be avoided. They react vigorously (sometimes explosively) with water to yield hydrogen halides. At high temperatures they decompose to yield toxic halogen-containing fumes.

LOREN D. LOWER
U.S. Borax Research Corp.

Technical Bulletin 0211, Ker-McGee Chemical Corp., Oklahoma City, Okla., 1973.

G. Urry in E. Muetterties, ed., *The Chemistry of Boron and Its Compounds*, John Wiley & Sons, Inc., New York, 1967, p. 325.

BORON HYDRIDES, HETEROBORANES, AND THEIR METALLO DERIVATIVES

Through nearly a quarter century of classic synthetic work, Stock fathered a family of toxic, air-sensitive, and volatile hydrides of general composition B_nH_{n+4} and B_nH_{n+6}. This fact is most remarkable because it required that he and his collaborators first develop basic techniques of vacuum-line manipulation, synthesis, and characterization. Hydroboration (qv) procedures, now so important to synthetic organic chemists, were developed by Brown. Significant interest was generated also by the high energy fuel projects HEF and ZIP (see Explosives and propellants). The Nobel Prizes in Chemistry were awarded to Lipscomb in 1976 and in 1979 to Brown for their definitive work in borane chemistry.

The emergence of a theoretical understanding of boranes and the residual momentum of the high energy fuel program spawned a recent

(since ca 1960) rapid rise in the discovery of significant new boranes and the elaboration of their chemistry.

Structural Systematics

The polyhedral skeletons described here can be accurately described as deltahedra (all faces triangular) or deltahedral fragments. The left column in Figure 1 illustrates the deltahedra from $n = 4$–12 vertices. In their regular form, all of these idealized structures are convex deltahedra except for the octadecahedron, which is not a regular polyhedron. The left column of Figure 1 also constitutes the class of deltahedral *closo* molecules from which all the other idealized structures can be generated systematically. Any *nido* or *arachno* cluster can be generated from the appropriate deltahedron by ascending a diagonal from left to right. This progression generates the *nido* structure (center column) by removing the most highly connected vertex of the deltahedron and *arachno* structure (right column) by removal of the most highly connected atom of the open face of the *nido* cluster.

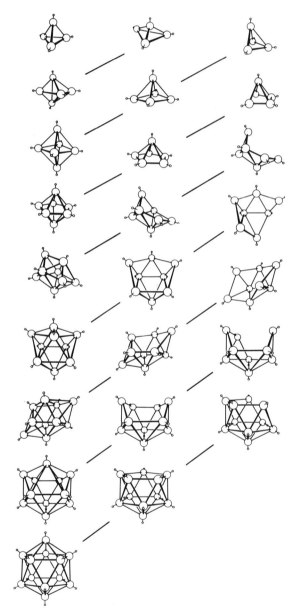

Figure 1. Idealized deltahedra and deltahedral fragments for *closo*, *nido*, and *arachno* boranes and heteroboranes. From left to right, the vertical columns give the basic *closo*, *nido*, and *arachno* frameworks; bridge hydrogens and BH_2 groups are not shown, but when appropriate they are placed around the open face of the framework. The diagonal progression are the known members of the *hypho* class.

The nomenclature *closo, nido, arachno,* and *hypho* implies closed, nestlike, weblike, and netlike structures, respectively. In addition to boranes, the classifications *closo, nido, arachno,* and *hypho* apply to heteroboranes and their metallo analogues and are intimately connected to a quantity called the framework electron count. The partitioning of electrons into exopolyhedral and framework classes allows the accurate prediction of structure in most cases, even though these systematics are not concerned explicitly with the assignment of localized bonds within the skeleton. Proposal of a structure from Figure 1 for a given borane or heteroborane proceeds by selecting the row that corresponds to the number of framework atoms n and determining the number of electrons that can be reasonably assigned to the skeleton as opposed to exopolyhedral electrons. (Counts of $2n + 2$, $2n + 4$, and $2n + 6$ framework electrons give a *closo, nido,* or *arachno* classification, respectively, and suggest the structure corresponding to the appropriate column of Fig. 1.) The systematics also emphasize the oxidation–reduction nature of *closo-nido-arachno* interconversions for frameworks of the same size.

Metalloboranes and metalloheteroboranes. The correlation of skeletal electron count with structure appears to be applicable to metalloborane derivatives and other cluster molecules such as carbonium ions and metal clusters. The method has been termed the PERC approach (Paradigm for the Electron Requirements of Clusters). Metalloboranes formed from a main group metal and a borane or heteroborane can be treated as heteroboranes by the PERC method.

In addition to the framework electron requirements of the cage, transition element metalloboranes and metalloheteroboranes generally adhere to the rule of eighteen, and therefore, require a somewhat different PERC treatment. If, for instance, the metal vertex uses only three orbitals in cluster bonding, then 12 of the 18 electrons at a metal vertex are not involved in cluster bonding. The d-electrons in effect are not included as framework electrons. Such premises to give the number of skeletal electrons per metal vertex as $v + x - 12$, where v equals the number of valence electrons of the metal, and x is the number of electrons donated by exocluster ligands and substituents. In this formalism, moieties such as $Fe(CO)_3$ and $Co(\eta\text{-}C_5H_5)$ are analogous to a BH vertex and $Ni(\eta\text{-}C_5H_5)$ is effectively a CH vertex.

Heteroatom placement. Obviously, many of the deltahedra and deltahedral fragments in Figure 1 have two or more nonequivalent vertices. Heteroatoms exhibit a positional preference that can be deduced on the basis of the electron richness of the heteroatom relative to boron and the order of the polyhedral vertex. Electron-rich heteroatom groupings contribute more framework electrons than a :B—H moiety (2 framework electrons) and seem to prefer low-order vertices. Transition metal moieties appear to occur predominantly at high-order vertices.

Placement of extra hydrogens. Some empirical rules of hydrogen placement in boranes are evident: bridging occurs only between boron atoms, although in metallo boranes, hydrogen sometimes bridges between boron and the metal; when possible, the bridge termini are the low-order vertices of the open face; and there is only one bridge per edge.

Numerous metalloboranes and metalloheteroboranes now are known to contain hydrogens bridging between the metal and skeletal borons, but complexes containing covalently bound $[BH_4]^-$ constitute the prototypical class.

Of course, exceptions to these structural systematics do occur.

Bonding

Since there are more valence orbitals than valence electrons available in the boron hydrides, they have been termed electron-deficient molecules. This so-called electron deficiency is responsible, at least in part, for the great interest surrounding borane chemistry and molecular structure. The structure of diborane, B_2H_6, led to the description of a new bond type, the three-center bond, in which one electron pair is shared by three atomic centers.

$$\text{(1)}$$

The delocalization of a bonding electron pair over a three-center bond allows for the utilization of all the available orbitals in an electron

deficient system. Of course, normal two-center bonds (B—B and B—H) also are used.

Lipscomb has described precisely the possible number of valence structures for a given boron hydride using three general equations of balance. For a borane of composition $[B_pH_{q+p+c}]^c$, where c is the charge, the equations are

$$s + x = q + c$$
$$s + t = p + c$$
$$t + y = p - c - 1/2q$$

where p equals the number of BH units; s is the number of B—H—B bonds; t is the number of B—B—B bonds; y is the number of B—B bonds; and x is the number of BH_2 groups. Usually, there are several possible solutions to the equations of balance that are differentiated by a so-called styx number, a four-digit number that gives the respective values of s, t, y, and x. The digits of the styx number give one half of the number of electrons involved in framework bonding. Therefore, after using PERC to arrive at a framework structure, a valence-bond description of the localized bonds can be deduced from the styx formalism.

In addition to the localized bond descriptions just described, Lipscomb and co-workers have developed molecular orbital (MO) descriptions of bonding in the boranes and carbaboranes. In fact, their early work on boranes developed one of the most widely applicable approximate MO methods, the extended Hückel method. Molecular orbital descriptions are particularly useful for *closo* molecules where localized bond descriptions become cumbersome because of the large number of resonance structures that do not accurately reflect molecular symmetry.

Boranes

Nido and *arachno* boranes are generally much more reactive and thermally less stable than *closo* boranes. Table 1 lists some of the physical properties of these boranes.

Closo boranes contain some very stable molecules that make their chemistry atypical among the boranes. The molecules that have been characterized constitute a series of $[B_nH_n]^{2-}$ ($n = 6$–12) deltahedral anions as shown in Figure 1. Unlike their *nido* and *arachno* counterparts with bridge hydrogens, for practical purposes, the abstraction of H^+ does not occur in *closo* borane chemistry. Instead, acid catalysis plays an important role in their substitution chemistry.

Tetrahydroborates

Tetrahydroborates of virtually every metal in the periodic chart except the actinides (qv) have been reported. However, the important commercial ones are the alkali metal tetrahydroborates. Some of their properties are given in Table 2.

The tetrahydroborates have been used as reducing agents for a variety of inorganic reductions. Several such reactions have been used in quantitative analytical procedures. However, the use of hydroborates (and boranes) for organic processes has proven to be even more significant since the reduction reactions are highly selective and nearly quantitative. The reducing characteristics of the hydroborate may be varied by changing the associated cation and by changing the solvent.

Heteroboranes

Here, a heteroelement is classified as a nonmetal. The heteroatoms known to form part of a borane polyhedron include C, N, Si, P, S, As, Se,

Table 1. Physical Properties of Some Boranes

Compound and Name		mp, °C	bp, °C	ΔH_f°, kJ/mol[a]	ΔG_f°, kJ/mol[a]	S_{298}° m J/(K·mol)[a]
B_2H_6	diborane(6)	−164.9	−92.6	35.5	86.6	232.0
B_4H_{10}	tetraborane(10)	−120	18	66.1		
B_5H_9	pentaborane(9)	−46.6	48	73.2	174.0	275.8
B_5H_{11}	pentaborane(11)	−123	63	103.0		
B_6H_{10}	hexaborane(10)	−62.3	108	94.6		
$B_{10}H_{14}$	decaborane(14)	99.7	213	31.5	216.1	353.0

[a]To convert J to cal, divide by 4.184.

Table 2. Some Properties of the Alkali Metal Tetrahydroborates

Property	Li[BH$_4$]	Na[BH$_4$]	K[BH$_4$]	Rb[BH$_4$]	Cs[BH$_4$]
mp, °C	268	505	585		
decomp. temp, °C	380	315	584	600	600
density, g/cm^3	0.68	1.08	1.17	1.71	2.40
refractive index		1.547	1.490	1.487	1.498
lattice energy, kJ/mol^a	792.0	697.5	657	648	630.1
ΔH_f^0, kJ/mol^a	−184	−183	−243	−246	−264
S_{298}^0, J/(mol·K)^a	−128.7	−126.3	−161	−179	−192

^{a} To convert J to cal, divide by 4.184.

and Sb, either singly or in combination. Those heteroboranes with the greatest demonstrated scope and flexibility of chemistry are the carbaboranes and thiaboranes. Both carbaboranes and thiaboranes are available in practical amounts.

Carbaboranes. The heteroborane that ushered in a new era of borane chemistry is 1,2-dicarba-*closo*-dodecaborane,1,2-C$_2$B$_{10}$H$_{12}$, which often is termed *ortho*-carborane. A large variety of C-substituted *o*-carboranes can be obtained by using substituted acetylenes, RC≡CR′. Two *arachno* dicarbaboranes have been reported: 6,8-C$_2$B$_7$H$_{13}$ and 6,9-C$_2$B$_8$H$_{14}$. The most readily accessible carbaborane known is *nido*-7-(NH$_3$)-7-CB$_{10}$H$_{12}$.

Heteroboranes other than carbaboranes. In principle, most heteroboranes could have a wide range of skeletal sizes; however, with the important exception of the carboboranes, such a series is emerging only for the thiaboranes. Most of the other known heteroboranes have a 12- or 11-atom framework. The thiaboranes [*arachno*-6-SB$_9$H$_{12}$]$^-$ and *nido*-6-SB$_9$H$_{11}$ were the first reported heteroboranes (other than carbaboranes) with 10-atom framework.

Metallo Analogues

Seemingly limitless varieties of metals and heteroatoms are capable of combination with boron in cluster compounds:

13- and 14-Atom clusters. Polyhedral expansion (reduction followed by metal insertion) of C$_2$B$_{10}$H$_{12}$ leads to supraicosahedral metallodicarbaboranes such as *closo*-13-Cp-13,11,7-CoC$_2$B$_{10}$H$_{12}$. Further expansion of 13-vertex-species leads to the 14-vertex cluster (CpCo)$_2$C$_2$B$_{10}$H$_{12}$.

12-Atom clusters. The first demonstration of the insertion of a transition metal into an open face of a borane was the reaction of iron(II) chloride with [7,8-C$_2$B$_9$H$_{11}$]$^{2-}$, which has sometimes been termed the dicarbollide ion. The product was readily air-oxidized to the complex containing a formal Fe(III), [(7,8-C$_2$B$_9$H$_{11}$)$_2$Fe(III)].

11-Atom clusters. A variety of heteroboranes are readily reduced from *closo* to *nido* molecules with a concomitant opening of the deltahedron (left column of Fig. 1). The open, nontriangular faces of the resulting *nido* anions appear to be ideally suited to coordination to metals. A variety of *nido* 11-atom clusters are known, including both main-group and transition metals: 7,7-(CH$_3$)$_2$-7-MB$_{10}$H$_{12}$ (M = Ge, Sn) and [7,7-(CH$_3$)$_2$-7-TlB$_{10}$H$_{12}$]$^-$; [M(B$_{10}$H$_{12}$)$_2$]]$^{2-}$ (M = Zn, Cd, Hg, Co, Ni, Pd, Pt); L = PR$_3$; and [L$_3$M(B$_{10}$H$_{12}$)]$^-$ (M = Co, Rh, Ir; L = Co, PR$_3$).

10-Atom clusters. The precursor to several *closo* 10-atom cobaltadicarbaboranes is [*arachno*-6,8-C$_2$B$_7$H$_{11}$]$^{2-}$. When treated with excess CoCl$_2$ and cyclopentadienide ion, [6,8-C$_2$B$_7$H$_{11}$]$^{2-}$ gives *closo*-CpCo (C$_2$B$_7$H$_9$). Other interesting 10-atom *closo* species include *closo*-10-(CpNi)-1-CB$_8$H$_9$ and *closo*-[1-(CpNi)B$_9$H$_9$]$^-$, in which cases the metal is bound to a B$_4$-face. *Nido* ten-atom metalloboranes appear to have a boatlike framework with the metal at a prow, 6- or 9-position. However, the nature of the bonding depends upon the metal. Examples include 6,6′-[(C$_2$H$_5$)$_3$P]$_2$-9-L-6-PtB$_9$H$_{11}$ (L = phosphine, amine, nitrile, or sulfide) and 9,9′-(L)$_2$-6,9-SPtB$_8$H$_{10}$ (L = P(C$_6$H$_5$)$_3$, P(CH$_3$)$_2$C$_6$H$_5$, P(C$_2$H$_5$)$_3$).

5, 6, 7, 8, and 9-Atom clusters. The expected tricapped trigonal prism structure of a 9-atom *closo* cluster is observed for [2-(CO)$_3$-1,2,6-

CMnCB$_6$H$_8$]$^-$, (CpNi)$_2$C$_2$B$_5$H$_7$, and 6,8-(CH$_3$)$_2$-1,1-[(CH$_3$)$_2$-1,6,8-PtC$_2$B$_6$H$_6$. The smaller boranes and dicarbaboranes have also been shown to form a wide variety of metalloderivatives in which the metal occupies a bridging position between two boron atoms, not a polyhedral vertex. An extensively studied system is μ-R$_3$MB$_5$H$_8$, where R = H, CH$_3$, C$_2$H$_5$, halogen, and M = Si, Ge, Sn, Pb.

Exopolyhedral metalloboranes. Metalloboranes with an exopolyhedral M—B sigma bond have been prepared by nucleophilic displacement reactions and oxidative addition of B—H and B—Br bonds to the metal center. Examples of oxidative addition to the B—H of *closo* heteroboranes include 3-[[P(C$_6$H$_5$)$_3$]$_2$IrHCl]-1,2-C$_2$B$_{10}$H$_{11}$ and 2-[[P(C$_6$H$_5$)$_3$]$_2$IrHCl]-1-SB$_9$H$_8$. It also has been shown that metals can be bound to polyhedral boranes through exo B—H—M bonds as in the case of [[(C$_6$H$_5$)$_3$P]$_2$Cu]$_2$ μ-B$_{10}$H$_{10}$.

RALPH W. RUDOLPH[‡]
The University of Michigan

Reviewed by
ROBERT C. TAYLOR
The University of Michigan

E.L. Muetterties and W.H. Knoth, *Polyhedral Boranes*, Marcel Dekker, Inc., New York, 1968.

R.E. Williams, *Adv. Inorg. Chem. Radiochem.* **18**, 67 (1976).

BORON HYDRIDES, HETEROBORANES, AND THEIR METALLODERIVATIVES, COMMERCIAL ASPECTS

The boranes, or B—H-containing compounds, of proven commercial value include the following: Tetrahydroborate, BH$_4^-$, more commonly called sodium borohydride, is the only boron hydride used routinely on a commercially significant scale. One of the earliest and presently important uses of sodium tetrahydroborate is as a specific reducing agent in the synthesis of pharmaceuticals, vitamins, antibiotics, flavors, and other fine chemicals. A large and rapidly growing market for sodium tetrahydroborate is its use in wood-pulp bleaching, clay bleaching, and textile-dye reductions. Applications reported for the octahydrotriborate ion include its use as an ignition aid in solid propellants, as a fogging agent in photographic films, and as a specialty reducing agent in organic synthesis. One of the most intriguing, albeit commercially insignificant, uses of boranes is in neutron-capture therapy for the sterilization of malignant tissue in the treatment of brain tumors. Several boranes, eg, B$_{12}$H$_{11}$SH^{2-}, have been studied in neutron-capture therapy with some positive results.

Although many amine boranes are available in research quantities, their major commercial application is the use of an estimated several thousand kilograms per year of dimethylamineborane, (CH$_3$)$_2$NHBH$_3$, as the reducing agent in electroless plating (qv) baths in the deposition of nickel, copper, and cobalt on the surface of metals and plastics.

Heteroboranes that have carbon and boron atoms in the polyhedral structure are called carboranes. The largest member of the series, C$_2$B$_{10}$H$_{12}$, has been incorporated into polymeric systems including polyvinyls, polyamides, polyesters, and carborane siloxanes. At present, carborane siloxane polymers that incorporate the *meta*-C$_2$B$_{10}$H$_{12}$ moiety are marketed under the trade name Dexsil and are commercially available for use as the stationary phase in gas–liquid chromatography columns (see Analytical methods). Derivatives of the C$_2$B$_{10}$H$_{12}$ carborane have been reported useful in optical switching devices, as a defoliant-desiccant that aids in harvesting crops, and as propellant additives.

Metalloboranes have not been shown to have commercial value although some potential uses have been cited.

GARY B. DUNKS
Union Carbide Corp.

[‡]Deceased.

M.A. Kaplan, J.H. Lannon, and F.H. Buckwalter, *Appl. Micro.* **13**, 505 (1965).

H. Hatanaka, *J. Neurol.* **209**, 81 (1975).

U.S. Pat. 3,667,991 (June 6, 1972), G.A. Miller (to Texas Instruments Inc.).

R.N. Grimes, *Carboranes*, Academic Press, Inc., New York, 1970.

ORGANIC BORON – NITROGEN COMPOUNDS

Organic boron–nitrogen compounds include all compounds in which there is a direct boron–nitrogen bond with at least one conventional organic group or moiety attached to boron, nitrogen, or both. There are three types of boron–nitrogen bonds: (*1*) a coordinate covalent bond in which the nitrogen atom supplies both electrons in the B—N bond, generally represented by the formula R_3N—BR_3, sometimes by $R_3N \rightarrow BR_3$ or $R_3\overset{+}{N} - \overset{-}{B}R_3$ to show bond polarity (amine boranes); (*2*) a normal covalent bond in which an electron from each atom is shared. In such systems the hybridization of both boron and nitrogen is sp^2 resulting in a planar moiety capable of π-interaction by utilizing the nitrogen's free electron pair and boron's vacant p orbital; $R_2\overset{+}{N} - \overset{-}{B}R_2$ (aminoboranes); (*3*) formal analogues of aromatic carbon compounds in which the isoelectronic B—N linkage replaces the C—C bond.

A fourth distinct clas might be compounds in which an organonitrogen moiety is attached to a boron atom in a boron cluster compound (substituted boron hydrides, carboranes, metalloboranes, etc) via either a sigma or dative bond. The distinguishing feature of this class is that the boron atom of interest has a coordination number greater than four.

Amine Boranes

The amine adducts of borane, BH_3, are relatively stable when pure and many are commercially available. The derivatives of haloboranes are quite sensitive to hydrolysis and are best prepared *in situ*. Amine boranes are molecular species in which the strength of the B—N interaction is a function of the nature of the groups attached to boron and nitrogen. Conductivity studies show amine boranes to be essentially nonconductors. They are monomeric even in polar solvents. Molecular association, owing to dipole–dipole interactions, is observed in benzene. Vibrational spectral data indicate that the B—N stretching frequency of amine boranes lies in the region 700–800 cm^{-1}. Most amine boranes are thermally stable when pure, exhibiting sharp melting points. Appreciable dissociation is observed in the vapor phase.

Aminoboranes

Aminoboranes are classified as mono-, bis- or trisaminoboranes depending on the number of amine groups attached to the boron atom. Monoaminoboranes readily undergo association, the extent of which depends principally on the steric requirements of the groups attached to boron and nitrogen. The monomers are generally liquids or low melting solids whereas the dimers and trimers are crystalline solids. The monomers are readily hydrolyzed but the oligomers show appreciable hydrolytic stability except at high temperatures. Monoaminoboranes undergo extensive dissociation at elevated temperatures. This fact probably accounts for the hydrolytic instability at higher temperatures since nucleophilic attack on the tricoordinate boron is possible. Monoaminoboranes are soluble in a variety of organic solvents. The degree of solubility decreases in the order monomer > dimer > trimer. Bisaminoboranes and trisaminoboranes are known only as monomers. They are less sensitive to hydrolysis than the monomeric monoaminoboranes. They are generally high-boiling liquids or crystalline solids.

Borazines

The largest and most extensively studied family of boron–nitrogen compounds is the borazines, characterized by a six-membered ring system containing alternating boron and nitrogen atoms. In contrast to the trimeric aminoboranes, $(—R'_2N—BR_2—)_3$, borazines are tricoordinate, planar, and have B—N bond distances intermediate between calculated single- and double-bond distances. Because borazines are isostructural and isoelectronic with benzenes, an inevitable comparison has evolved. The physical properties of borazines tend to confirm a resonance structure quite similar to that of benzenes. However, the chemical evidence indicates that the reactions of borazines are dominated by polarization of the B—N bonds. Symmetrically substituted borazines are generally crystalline solids and unsymmetrically substituted borazines are generally liquids or low melting solids. Borazines are planar molecules, although some puckering may be observed in highly substituted borazines owing to steric interactions.

Other B—N Ring Systems

Several unusual ring systems containing only B—N linkages have been reported. These include the eight-membered borazocines and the four-membered diazaboretanes as well as systems containing B_2N_3, B_3N_2, BN_4, B_2N_4, B_4N_2, etc, frameworks.

Manufacture and Uses

Organic boron–nitrogen compounds do not have extensive usage and, therefore, very few are manufactured on a large scale. Callery Chemical Company appears to be the largest manufacture of amine boranes and borazines.

Amine boranes are principally used as reagents to produce synthetic intermediates in organic syntheses. The amine adducts of BH_3 are valuable reducing agents, both in organic chemistry and metal plating (see Electroless plating). Borazines and aminoboranes have limited use as polymerization catalysts or oxidation stabilizers (see Boron hydrides, commercial aspects).

H.D. SMITH, JR.
Virginia Polytechnic Institute and State University

H. Steinberg and R.J. Brotherton, *Organoboron Chemistry*, Vol. 2, John Wiley & Sons, Inc., New York, 1966.

K. Niedenze and J.W. Dawson in E.L. Muetterties, ed., *The Chemistry of Boron and Its Compounds*, John Wiley & Sons, Inc., New York, 1967.

N.N. Greenwood and B.S. Thomas in A.F. Trotman-Dickerson, ed., *Comprehensive Inorganic Chemistry*, Vol. 1, Pergamon Press Ltd., Oxford, UK, 1973, p. 919.

BRAKE FLUIDS. See Hydraulic fluids.

BRAKE LININGS AND CLUTCH FACINGS

Brakes and clutches operate dry or wet. In dry friction couples, the heat is removed by conduction to the surrounding air and structural members. Wet friction couples operate within a fluid (usually an oil) which absorbs the heat and maintains the couple at relatively low temperature (below 200°C). The fluid also traps the wear debris (see Hydraulic fluids).

Friction materials serve in a variety of ways to control the acceleration and deceleration of a variety of vehicles and machines that may be as small as a clutch in a business machine or a brake on a bicycle to as large as jumbo-aircraft brakes. A qualitative analysis of a friction couple suggests that a frictional force is likely to consist of several components such as adhesion tearing, ploughing (or abrasion), elastic and plastic deformation, and asperity interlocking, all occurring at the sliding interface. These mechanisms presumably depend upon the temperature as well as the normal load and sliding speed, since material properties are known to be dependent upon these variables. In the case of automotive friction materials, the coefficient of friction is usually found to decrease with increasing unit pressure and sliding speed at a given temperature, contrary to Amonton's law. This decrease in friction is controlled by the composition and microstructure of friction materials.

For a fixed amount of braking, the amount of wear of automotive friction materials tends to increase slightly or remain practically constant with respect to brake temperature, but once the brake rotor temperature reaches about 204°C the wear of resin-bonded materials increases exponentially with increasing temperature. This exponential wear is owing to thermal degradation of organic components. At low temperatures, the practically constant wear rate is primarily controlled by abrasion and adhesion. The wear, W, of friction materials can best be described by the following wear equation: $W = KP^aV^bt^c$ where K is the wear coefficient; P, the normal load; V, the sliding speed; t, the sliding time; and a, b, and c are a set of parameters for a given friction material–rotor pair at a given temperature.

Wear is an economic consideration. Wear resistance generally is inversely related to friction and other desirable performance characteristics within any class of friction material. The formulator's objective is to provide the highest degree of wear resistance in the normal use temperature range, a controlled moderate increase at elevated temperatures, and a return to the original lower wear rate when temperatures again return to normal. Maximum wear life does not require maximum physical hardness.

Cermet or carbon friction materials operate at substantially higher temperatures than normal automotive or truck friction materials. One unique feature of these two materials is the formation of a glazed friction layer which reduces the wear rate. Without this glazed layer, the wear rate is very high.

Types of Friction Materials

The most common type of friction materials used in brakes and clutches for normal duty is termed organic. These materials usually contain about 30–40 wt% of organic components. The main constituent of practically all organic friction materials is asbestos fiber, although small quantities of other fibrous reinforcement may be used. Since asbestos alone does not offer all of the desired properties, other materials called property modifiers, either abrasive or nonabrasive, are added to provide desired amounts of friction, wear, fade, recovery, noise, and rotor compatibility. A resin binder, such as phenolic or cresylic resin, holds the other materials together. This binder is not completely inert and makes contributions to the frictional characteristics of the composite.

Gray cast iron is of reasonably low cost, provides good wear resistance and damping characteristics, and has been in long time use as a brake drum or disk material for passenger cars and trucks.

Semimetallics were introduced in the late 1960s and gained widespread usage in the mid 1970s. These materials usually contain more than 50 wt% metallic components. They are primarily used as disk pads and blocks for heavy-duty operation. The main constituent of practically all semimetallics is iron powder in conjunction with steel fiber and contains no asbestos.

Heavy weights and high landing speeds of modern aircraft or high speed trains require friction materials that are extremely stable such as cermet friction materials which are metal-bonded ceramic compositions. Currently, several companies are actively developing and field-testing lighter brake materials based on carbon fiber-reinforced carbon-matrix composites to be used primarily for aircraft brakes (see Carbon and artificial graphite; Ablative materials).

An important relation exists between composition, performance requirements, and ease of manufacture. Organic linings that must bend also require higher resin contents and longer fibers. Heavy-duty materials with reduced resin loading for improved performance require molding-to-shape. Sintered and carbon friction materials require high pressure forming and high temperature treatment in inert atmospheres. Woven and some clutch materials require special fiber-forming methods.

Environment and Health

Organic friction materials have been in increasing jeopardy because OSHA regulations have limited the exposure of workers to airborne asbestos fibers to a time-weighted average of 2 fibers/cm³, and that limit is expected to be reduced further to 0.5 to 0.1 fibers/cm³. Lead may also be present in some friction materials.

Friction material and rotor emissions are generated by wear. However, a study of brake emissions adjacent to a city freeway exit ramp on the downwind side indicated that the asbestos emissions were so low they could not be distinguished from the background on the upwind side.

The trend toward more energy-efficient passenger cars and trucks will put an increased demand on friction materials. Ventilated disk brakes may be replaced by solid disk brakes for weight savings. Most passenger car and truck manufacturers are requiring asbestos-free friction materials. More sintered friction materials will appear in the heavy-vehicle brake and clutch markets. At the same time, aircraft will move toward light carbon brakes.

M.G. JACKO
S.K. RHEE
Bendix Corp.

S.K. Rhee, *SAE Trans.* **83**, 1575 (1974).
S.K. Rhee, *Wear* **28**, 277 (1974).
W.R. Tarr and S.K. Rhee, *Wear* **33**, 373 (1975).

BRANDY. See Beverage spirits, distilled.

BRASS. See Copper alloys.

BREAD. See Bakery processes and leavening agents.

BRIGHTENERS, FLUORESCENT

The operation of bleaching or brightening is concerned with the preparation of fabrics whose commercial value is dependent on the highest possible whiteness. With the aid of optical brighteners, also referred to as fluorescent whitening agents (FWAs) or fluorescent brightening agents, optical compensation of the yellow cast of substrates such as textiles, paper, detergents, etc, may be obtained. The yellow cast is produced by the absorption of short-wavelength light (violet-to-blue). With optical brighteners this lost light is in part replaced; thus, a complete white is attained without loss of light. This additional light is produced by the brightener by means of fluorescence. Optical brightening agents absorb the invisible ultraviolet portion of the daylight spectrum, ie, into blue-to-blue-violet light (see Fig. 1). Optical brightening, therefore, is based on the addition of light.

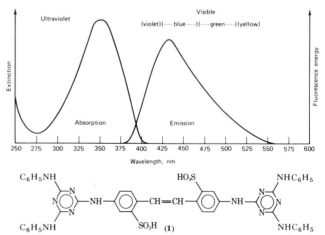

Figure 1. Absorption and emission spectra in solution of compound (**1**).

Two requirements are indispensable for an optical brightener; it should be optically colorless on the substrate, and it should not absorb in the visible part of the spectrum. In the application of optical brighteners, it is possible not only to replace the light lost through absorption, thereby attaining a neutral, complete white but, through the use of excess brightener, to convert still more uv radiation into visible light so that the whitest white is made still more sparkling. Since the fluorescent light of an optical brightener is itself colored, ie, blue-to-violet, the use of excess brightener always gives either a blue-to-violet or, sometimes, a bluish green cast.

Collectively, these fluorescent compounds belong to the aromatic or heterocyclic series; many of them contain condensed ring systems. An important feature of these compounds is the presence of an uninterrupted chain of conjugated double bonds, the number of which is dependent on substituents as well as the planarity of the fluorescent part of the molecule. Almost all of these compounds are derivatives of stilbene or 4,4'-diaminostilbene, biphenyl, 5-membered heterocycles (triazoles, oxazoles, imidazoles, etc) or 6-membered heterocycles (coumarins, naphthalimide, s-triazine, etc).

Stilbene derivatives include 4,4'-bis(triazin-2-ylamino)stilbene-2,2'-disulfonic acids (1), mono(azol-2-yl) stilbenes, and bis(azol-2-yl) stilbenes. They are used for brightening synthetic fibers and cellulosics. Pyrazolines include 1,3-diphenyl-2-pyrazolines and are used for brightening synthetic fibers. Bis(benzazol-2-yl) derivatives, such as bis(benzoxyazol-2-yl) derivatives, bis(benzimidazol-2-yl) derivatives, and 2-(benzofuran-2-yl)benzimidazoles, are suitable for brightening of plastics, synthetic fibers, and cotton. Coumarins, such as 7-hydroxy and 7-(substituted amino)coumarins, and carbostyrils are useful for brightening wool, nylon, polyamides, cellulose acetates, plastics, and other synthetic fibers. Other heterocyclic systems include naphthalimides for brightening synthetic fibers, derivatives of dibenzothiophene-5,5-dioxide, pyrene derivatives for polyester fibers, and pyridotriazoles for acrylic fibers.

Initially, fluorescent whitening agents (FWAs) were used exclusively in textile finishing. The detergent and paper industries followed thereafter, and today these products are in widespread use in fiber-spinning masses, plastics, and paints.

Measurement of Whiteness

The Ciba-Geigy Plastic White Scale is effective in the visual assessment of white effects. Its whiteness formula makes use of instrumental measurement of whiteness levels. As a complement to the whiteness level, shade is determined from the Ciba-Geigy Shade Formula.

REINHARD ZWEIDLER
HEINZ HEFTI
Ciba-Geigy Ltd.

"Fluorescent Whitening Agents" in F. Coulston and F. Korte, eds., *Environmental Quality and Safety*, Suppl. Vol. 4, Georg Thieme Verlag, Stuttgart, FRG, and Academic Press, Inc., New York, 1975.

BRONZE DISEASE.   See Fine art examination and conservation.

BUBBLE MEMORY.   See Magnetic material.

BROMINE

Bromine, a nonmetallic element of the halogen family, is a dark red-brown liquid at ordinary temperatures and pressures. It vaporizes readily and has a sharp, penetrating odor. The diatomic nature of bromine persists throughout the solid, liquid, and gaseous phases. Bromine's properties are between those of chlorine and iodine. Many reactions of bromine are a result of its strong oxidizing properties. The most common oxidation states of bromine are -1 and $+5$, but positive valences of 1, 3, and 7 are observed. Table 1 summarizes the physical properties of the element.

Bromine does not occur in nature as the free element, but is found only as the bromide. The most readily recoverable form of bromine occurs in the hydrosphere as soluble bromide salts. The currently accepted value of 65 mg/L for the bromide concentration in seawater was reported in 1871. Other extractable sources of bromine occur in salt lakes, saline, and inland seas, as well as brine wells.

The key factors of bromine isolation processes are the selective separation of bromide from chloride and the removal of small concentrations of bromine from large volumes of aqueous solution. The selective oxidation of bromide in the presence of large amounts of chloride is possible because of the difference in their reduction potentials.

The four principal steps in bromine production are (1) oxidation of bromide to bromine; (2) bromine vapor removal from solution; (3) isolation of bromine from the vapor; and (4) purification. For step (1) the use of chlorine has been mentioned, but other methods have been explored including oxidation by manganese dioxide, chlorates, bromates, hypochlorite, and electrochemical methods.

It is in the second step that the two most significant developments in bromine manufacture have been made. These are the steaming-out process as modified by Kubierschky in 1909 and the blowing-out process developed by Dow in 1889. Steam is suitable when the raw brine contains more than one gram of bromine per liter; however, air is more economical when the source is as dilute as ocean water. When steam is used the vapor may be condensed directly; otherwise, the bromine must be trapped in an alkaline or reducing solution. In either case, purification is necessary to remove chlorine and possible other impurities that may vary with the brine source.

Uses

Elemental bromine is used primarily in the manufacture of bromine compounds (qv) which are characterized mainly by their chemical or biological activity, high density, or fire-retarding and extinguishing ability (see Flame retardants). Bromine compounds are well represented in such use areas as gasoline additives, agricultural chemicals, flame retardants, dyes, photographic chemicals, pharmaceuticals, and others. The high density characteristics of bromine compounds are advantageously applied in hydraulic fluids (qv), gauge fluids, and ore flotation. The largest industrial use of bromine, however, is as ethylene dibromide (EDB), a lead scavenger additive used in gasoline with tetraethyllead and tetramethyllead to reduce engine knocking. A decrease in EDB production to 66% of total bromine usage in 1974 was attributed partly to regulations issued in 1973 by the EPA to reduce the lead content of gasoline.

Sanitary preparations include the use of bromine in water disinfection in swimming pools (see Water) and cooling towers for control of bacteria, algae, odors, and other uses (see Industrial antimicrobial agents). Bromine

Table 1.   Physical Properties of Bromine

Property	Value
stable isotopes	Br^{79}, 50.54%
	Br^{81}, 49.46%
freezing point, °C	-7.25
boiling point, °C	58.8
density at 25°C, g/cm^3	3.1055
vapor density at 101.3 kPa^a and 0°C, g/L	7.139
refractive index at 25°C	1.6475
surface tension at 25°C, mN/m (= dyn/cm)	40.9
dielectric constant at 25°C, 10^5 Hz	3.33
expansion coefficient at 20–30 °C, per °C	0.0011
heat capacity at 15 K, J/mol^b	7.217
solid, at 265.9 K	61.64
liquid, at 265.9 K	77.735

^{a} To convert kPa to mm Hg, multiply by 7.50.
^{b} To convert J to cal, divide by 4.184.

also is used, either directly or indirectly, in the desizing of cotton, in bleaching of pulp and paper, air-conditioning absorption fluids, and hair-waving compositions (see Bleaching agents; Hair preparations).

Health and Safety Factors

Liquid bromine or its vapors attack the skin and other tissues to produce irritation and necrosis. The maximum time-weighted average concentration of bromine considered safe for repeated 8-h exposure is 0.1 ppm. This is the limit specified by OSHA. Production facilities where bromine is manufactured or used should be designed to dispose rapidly of liquid bromine spills. Full body protection constructed of resistant materials should be worn when handling bromine in significant quantities. Major manufacturers of bromine will furnish information on the safe handling procedures for bromine.

CHARLES E. REINEKE
Dow Chemical U.S.A.

Z.E. Jolles, ed., *Bromine and Its Compounds*, Academic Press, Inc., New York, 1966.

A.J. Downs and C.J. Adams, "Chlorine, Bromine, Iodine and Astatine," in A.F. Trotman-Dickenson, ed., *Comprehensive Inorganic Chemistry*, Vol. 2, Chapt. 26, Pergamon Press, New York, 1973, pp. 1107–1573.

V. Gutmann, ed., *Halogen Chemistry*, Vol. 1, Academic Press, Inc., New York, 1967.

Minerals Yearbook, U.S. Dept. of Interior, Bureau of Mines, U.S. Government Printing Office, Washington, D.C.

BROMINE COMPOUNDS

The most important compound of bromine is 1,2-dibromoethane (ethylene bromide), used primarily in antiknock gasoline but also to some extent as a fumigant. Both uses have been declining in the United States. Estimates based on Bureau of Mines data indicate that of domestic bromine production, ca 37% went into ethylene dibromide, 16% to flame retardants, 10% to methyl bromide, 5–10% to other fumigants and bromoorganics, and 27–32% to inorganics (calcium bromide makes up the largest part of the latter).

Inorganic Compounds

Bromides. Considerable amounts of sodium bromide and potassium bromide are employed in the preparation of light-sensitive silver bromide emulsions for photography. Various inorganic bromides, chiefly those of the alkalies, alkaline-earths, and ammonium ion, are prescribed in medicine. Their sedative action is of value in the treatment of nervous disorders. Lithium bromide and calcium bromide are effective desiccants used in the industrial drying of air (see Drying agents). Recent applications of calcium and zinc bromides is in dense packing fluids for oil-well completion (see Drilling fluids). Anhydrous aluminum bromide is a catalyst for some types of bromination reactions.

Hydrogen bromide, HBr (hydrobromic acid), is a very irritating, colorless gas, mp $-86°C$, bp $-67°C$. It resembles hydrochloric acid but is a more effective solvent for some ore minerals because of its higher boiling point and stronger reducing action. Its toxic effects are similar to those of hydrochloric acid, although they may be less severe. A fair amount of hydrobromic acid is consumed in the manufacture of inorganic bromides, as well as in the synthesis of alkyl bromides from alcohols. The acid can also be used to hydrobrominate olefinic linkages directly. In the petroleum industry, hydrogen bromide can serve as an alkylation catalyst. Applications of HBr with NH_4Br or with H_2S and HCl as promoters in the dehydrogenation of butene to butadiene have been described, and either HBr or HCl can be used in the vapor-phase ortho methylation of phenol with methanol over alumina. An important reaction of HBr in organic syntheses is the replacement of aliphatic chlorine by bromine in the presence of an aluminum catalyst.

Bromine halides. Bromine chloride, BrCl, iodine bromide, IBr, and a tribromo complex, IBr_3, are soluble in carbon tetrachloride or acetic acid and are chiefly of interest as halogenating agents for organic substances. Bromine chloride is an efficient disinfectant for wastewater.

Fluorine reacts violently with bromine forming various fluorides (BrF, BrF_3, and BrF_5). The tri- and pentafluorides are available commercially. As strong fluorinating agents, they are useful in organic syntheses and in forming uranium fluorides, for both isotope enrichment and fuel-element reprocessing.

Bromine oxides, acids, and salts. At least some of the oxides of bromine may be considered as anhydrides of the oxygen-containing bromo acids described below. All are unstable at ordinary temperature: bromine monoxide, Br_2O, a dark brown solid, stable below $-40°C$ and melting with decomposition at $-17.5°C$, and several higher oxides have been reported.

The oxygen acids of bromine, such as hypobromous acid, HOBr, bromous acid, $HBrO_2$, and bromic acid, $HBrO_3$, are unstable strong oxidants capable of existing at ordinary temperatures only in solution. Hypobromites, the salts of hypobromous acid, are strong bleaching agents, similar to hypochlorites, and have been incorporated into scouring formulations. Sodium bromite is used to limited extent as a desizing agent in the textile industry.

Bromates are stable under ordinary conditions and have various applications based upon their oxidizing properties. Potassium bromate, $KBrO_3$, is used in flour treatment to improve baking characteristics. It also serves as a primary standard and a brominating agent in analytical chemistry. An important outlet for sodium bromate, $NaBrO_3$, is as a neutralizer or oxidizer in some hair-wave preparations. Perbromates can be prepared only with great difficulty and are chiefly of academic interest.

Organic Compounds

Organic compounds of bromine usually resemble their chlorine analogues but have higher densities and lower vapor pressures. The bromo compounds are more reactive toward alkalies and metals; brominated solvents should be kept from contact with active metals such as aluminum. On the other hand, they present less fire hazard. Bromine compounds that have commercial application are the following:

Ethylene dibromide, $BrCH_2CH_2Br$ [ethylene bromide, 1,2-dibromoethane], EDB mp 9.9°C, bp 131.4°C, is a clear, colorless liquid with a characteristic sweet odor. In the manufacture of ethylene dibromide, gaseous ethylene is brought into contact with bromine by various methods, allowing for dissipation of the heat of reaction. The vapor of ethylene dibromide is toxic, and exposure to a time-weighted average of 20 ppm by volume in air should be avoided. That ethylene dibromide may be a carcinogen has been reported from feeding studies of rodents. Ethylene dibromide is one of the cheapest organic compounds of bromine. Its principal use as an additive in leaded gasoline is declining because of regulation of the amount of lead in automotive fuel. Other uses as an ingredient of soil fumigants against wireworms and nematodes, and of grain-fumigant formulations for insect control are also being diminished by tightened regulations.

Methyl bromide, CH_3Br (bromomethane), mp $-93.7°C$, bp 3.56°C, is a colorless liquid or gas with practically no odor. Commercial and laboratory methods of manufacturing methyl bromide are based primarily upon the reaction of hydrobromic acid with methanol. Methyl bromide, sold as a pure compound or in formulations, is classified as a poison, class B, and requires a poison label. The upper safe limit for daily 8-h exposure to the vapor in air is considered as 15 ppm by volume, or ca 0.06 mg/L. The main use for methyl bromide is in the extermination of insect and rodent pests (see Poisons, economic). The material is suitable for the fumigation of food commodities and the facilities in which they are processed or stored, as well as for tobacco and many kinds of nursery stock.

Other bromomethanes. The main outlet for bromochloromethane, CH_2BrCl, bp 68.1°C, is as a fire-extinguisher fluid (see Fire-extinguishing agents). Dibromomethane, CH_2Br_2 (methylene bromide), mp 52.7°C, bp 96.9°C, has limited uses in syntheses, as a solvent, and in gauge fluids. Tribromomethane, $CHBr_3$ (bromoform), mp 7.7°C, bp 149.5°C, is usually sold mixed with 3–4% ethanol as a stabilizer. Tetrabromomethane, CBr_4 (carbon tetrabromide), mp 90.1°C, bp 189°C, is capable of direct addition to olefins. It also is light-sensitive and may have uses in photography and photoduplicating systems.

Various bromofluoromethanes have been proposed for use as fire-extinguishing agents (qv); among these are CBr_2F_2 and $CBrF_3$.

Ethyl bromide, CH_3CH_2Br (bromoethane), mp $-119.3°C$, bp $38.4°C$, is prepared by refluxing ethanol with hydrobromic acid, or with an alkali bromide and sulfuric acid. It is used mainly as an ethylating agent in synthesis, particularly of pharmaceuticals.

Bromochloropropanes. 1-Bromo-3-chloropropane, $CH_2BrCH_2CH_2Cl$ (trimethylene chlorobromide), mp $-58.9°C$, bp $143.4°C$, is used for the synthesis of cyclopropane. 1,2-Dibromo-3-chloropropane (DBCP), mp $6°C$, had become a useful pesticide for control of nematodes, root-knot disease, etc, until sterility problems were encountered among male employees in production facilities. The compound also has been reported as carcinogenic in feeding studies. Its use has been banned in the United States except in Hawaiian pineapple fields.

Dyes and indicators. Among the dyes that contain bromine, those of the indigo group have been in greatest demand. Compounds with from one to eight atoms of bromine per molecule can be prepared. Other bromo compounds are employed as intermediates for the preparation of dyes. Bromine as a substituent in dye or indicator molecules causes absorption of light at longer wavelengths.

Pharmaceuticals. Bromine may be significant in medicinal chemistry in several ways: bromide released from a compound may have a sedative effect; the presence of one or more bromine atoms in a molecule may impart desirable physical or chemical properties; and bromine compounds may be useful synthetic intermediates.

In general, exposure to bromine compounds should be avoided. In case of skin contact, the affected parts should be washed thoroughly, first with water, then soap and water. Medical treatment should be obtained promptly following any significant exposure.

V.A. STENGER
Dow Chemical, U.S.A.

M.W. Kluwe, "Chemical Modulation of 1,2-Dibromo-3-chloropropane Toxicity," *Toxicology* **27**(3–4), 287 (1983).

R.D. White, A.J. Gandolfi, G.T. Bowen, and I.G. Sipes, "Deuterium Isotope Effect on the Metabolism and Toxicity of 1,2-Dibromoethane," *Toxicol. Appl. Pharmacol.* **69**(2), 170 (1983).

A.G. Sharpe and co-workers in J.W. Mellor, ed., *Comprehensive Treatise on Inorganic and Theoretical Chemistry*, Vol. II, Supplement II, John Wiley & Sons, Inc., New York, 1956, pp. 689–812.

A.J. Downs and C.J. Adams in A.F. Trotman-Dickenson, ed., *Comprehensive Inorganic Chemistry*, Vol. II, Pergamon Press, New York, 1973, pp. 1107–1594.

BTX PROCESSING

Benzene (B), toluene (T), and xylenes (X) are familiar chemicals to those even slightly acquainted with the chemical industry; BTX, however, is less familiar as the main contributor to high performance gasolines. Processing for benzene, toluene, xylene, and the enormous gasoline industry are inextricably linked. If benzene, toluene, or xylenes are removed as basic chemicals, the gasoline reservoir or pool has to be compensated for the loss of its more important high octane components (see Benzene; Gasoline; Petroleum; Toluene; Xylenes and ethylbenzene).

A refinery must maintain a pool of gasoline with sufficiently high octane values to supply transportation demands. Removal of BTX depletes the gasoline pool. Addition of tetraethyl- or tetramethyllead raises the octane rating; however, government regulations restrict the amount of lead that may be used. Thus, additional processing is needed to replace those aromatics removed for chemical purposes, resulting in higher costs.

The crude oil received by the refiner contains a fraction in the 65–175°C boiling range. It is separated by distillation but is of little use in gasoline or for chemical purposes because the concentration of aromatics and thus the octane value is low (see Feedstocks). The process

that increases the aromatics concentration is called reforming. The balance of the crude oil is further distilled to recover fuels for diesel or jet engines or for heating purposes. The even higher-boiling fractions are broken down or cracked to bring them into the more useful boiling range.

The key to BTX processing is the catalytic reforming step, which simplified may be considered paraffin → cycloparaffin → aromatics conversion. The octane rating of the various classes of reformer feeds is in the order of aromatics > cycloparaffins > isoparaffins > n-paraffins. Thus, it is the objective of the gasoline refiner and BTX producer to increase the proportion of aromatics. The main reactions occurring in a reformer are shown in Figure 1; most are reversible, indicating the importance of reaction equilibrium. Reformer feeds of high aromatics-plus-cycloparaffins content produce high octane reformates using only the cycloparaffins-to-aromatics reactions; cyclization of paraffins is generally more difficult.

Most benzene is formed from cyclohexane and methylcyclopentane. In order to obtain a good yield, the severity must be raised, and if possible, low pressures should be used. The use of broad-range feeds is not advisable because of higher catalyst coking rates. Toluene is less valuable as a chemical but is an important octane contributor. Reforming for xylene is common. A sharply cut 90–190°C fraction may contain only 40–50% C_8 hydrocarbons because of boiling point overlap with C_7 and C_9 compounds. A reformate from such a feed will contain only 30% C_8 aromatics.

If the heart-cut reformate contains appreciable nonaromatics, extraction may be necessary to prevent the accumulation of nonaromatics in the processing loop. Some isomerization processes are capable of cracking the paraffins but at the expense of additional hydrogen consumption. These processes do, however, eliminate the need for extraction of the xylenes and heavier compounds. The most important extraction processes use sulfolane or glycols. Recovery of benzene and toluene is usually $\geq 99\%$; of C_8 aromatics, 97%; and of $C_{>9}$ aromatics, 75–90%.

Dealkylation of toluene provides benzene. Temperatures in the 540–810°C range are required, and an excess of hydrogen is used. Transalkylation of toluene or toluene and $C_{>9}$ aromatics has not been widely used in the past but is likely to become more important as feed preparation becomes more expensive. Transalkylation of toluene alone leads mainly to benzene and xylenes, whereas transalkylation of toluene and $C_{>9}$ aromatics gives mainly xylenes. p-Xylene separation is accomplished by crystallization or adsorption. Isomerization of xylenes or the C_8 aromatics increases the recovery of the required product from a given amount of feed. The processes may isomerize xylenes only; isomerize ethylbenzene to xylenes; or isomerize xylenes and selectively crack ethylbenzene (recently developed by Mobil).

BTX by Pyrolysis

Side products of ethylene production may include large quantities of BTX, depending on the feed. Low octane stocks, often in excess in a refinery, are favored feeds to an ethylene cracker (see Ethylene). Pyrolysis gasoline contains a large proportion of olefins which must be hydrogenated before BTX recovery. The C_5 and $C_{>9}$ fractions are usually separated by distillation prior to hydrogenation in order to save hydrogen. The C_6–C_8 heart cut is hydrotreated to desulfurize and saturate the olefins. The aromatics are recovered by extraction as discussed earlier. The raffinate from the aromatic-extraction step has a high cycloparaffin content and is an excellent reformer feed.

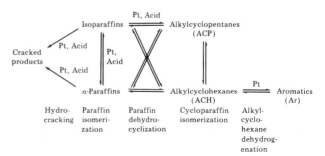

Figure 1. Main reactions of catalytic reforming.

The Role of the Computer in BTX Processing

Most refineries have an overall computer program, including models of the various units in the refinery, that provides a material balance. By inserting operating costs of each unit, values for each stream, and specific product requirements, an optimum economic operating mode can be reached.

Environmental Considerations

The petroleum industry has to operate in a highly competitive area within the framework of a continuously changing set of rules, many of which are intended to protect the environment. Limits are imposed on the permissible level of sulfur in heavy fuel oils. The BTX industry uses fuel oil that must now be either treated or replaced by an acceptable and more expensive fuel. Finally, there is an ever-present problem in handling large volumes of low boiling materials.

DEREK L. RANSLEY
Chevron Research Company

B.C. Gates, J.R. Katzes, and G.C.A. Schmit, *Chemistry of Catalytic Processes*, McGraw-Hill, Inc., New York, 1979, pp. 184–324.

BURNER AND COMBUSTION TECHNOLOGY

Most familiar and important combustion processes involve diffusion flames in which fuel and oxidant are separately introduced, and in which the rate of overall process is determined mainly by mixing, eg, by laminar or turbulent diffusion, or by some other physical process. Flames of candles, matches, gaseous fuel jets, oil sprays, and large fires, accidental or otherwise, are essentially of this type; but many others of practical interest are premixed flames, propagating in homogeneous mixtures of reactants.

Definitions, Terminology, and Basic Ideas

The overall composition of a mixture (whether uniform or not) of fuel with air or other oxidant is often specified as the ratio of fuel to oxidant: the weight ratio (m_f/m_o), or for gaseous reactants, the molar or the volume ratio (v_f/v_o). In either case, it can be normalized to the stoichiometric composition by defining the equivalence ratio:

$$\phi = (m_f/m_o)/(m_f/m_o)_{stoich} = (v_f/v_o)/(v_f/v_o)_{stoich}$$

The denominator describes the composition that would yield just CO_2, H_2O, and N_2 on complete combustion, eg, in a calorimeter, if the initial mixture contained only C, H, O, and N (sometimes called perfect combustion). A mixture of $\phi < 1$ is said to be lean, and its burned gas contains unreacted or excess O_2; if $\phi > 1$, some of the C and H necessarily appears as CO and H_2 in the burned gas, and the mixture is said to be rich.

In mixtures of a given fuel (gas or vapor) with oxidant, there usually are two values of ϕ: the lean or lower limit (ϕ_L) and the rich or upper limit (ϕ_U) of flammability between which the mixtures are said to be flammable (ie, explosive), or can propagate a flame as a steady wave from some local ignition source, such as a spark, at a velocity that depends on ϕ; it is normally a maximum near $\phi = 1$ and near the limits tends to lower values. The limits most often measured and quoted refer to fuel–air mixtures initially at ambient conditions ignited with a spark or small flame at the bottom of a vertical tube of diameter ≥ 10 cm.

There is an oxygen content of the atmosphere, sometimes called the oxygen index (OI) of the fuel, below which no mixture with the fuel is flammable. Near the OI, both ϕ_U and ϕ_L approach unity, and only an approximately stoichiometric mixture will be flammable. ϕ_L and OI may be considered roughly equivalent as measures of flammability, eg, in hazard evaluation; both decrease slowly with increasing mixture temperature (ca 10% for 100°C increase) while ϕ_U increases. The flammability region widens with increasing temperature, as it usually does also with

elevated pressure, though the effect of pressure is less predictable. As rough rules, for most hydrocarbons, the oxygen index is in the range of 12–14%, the lower limit in mixtures with air is at ϕ_L ca 0.5, and the flame temperatures of their limit mixtures are about the same (1500–1700 K).

A closely related measure of liquid flammability is the flash point. It is primarily a measure of volatility and is ideally the temperature at which the vapor pressure of the liquid is such that the mole fraction of its vapor in the atmosphere above the liquid ($x_f \simeq$ vapor pressure in atmospheres) is that of the lower limit mixture of the fuel in air at that temperature; ie, it will just propagate a flame.

Thermochemical quantities. The higher heating value of a fuel is its heat of combustion at constant pressure and temperature (usually ambient), from calorimetric measurements in which any water formed by combustion is completely condensed. The lower heating value is the similarly measured or defined heat of combustion if there was no condensation of water. The lower value is most often used in combustion and flame calculations, since the water is steam in most processes of practical interest; the combustion efficiency of the process is then the ratio of the heat actually released (enthalpy of the burned gas) to this ideal heat release.

Adiabatic flame temperatures. If the products of any combustion process or of any chemical reaction are at equilibrium and uniform with respect to composition, pressure, temperature (whether or not the reactant mixture is uniform), and if there are no heat losses to the surroundings in the process, the thermodynamically attainable temperature is calculable as the adiabatic flame temperature T_a. In addition, if the process occurs at constant pressure, the overall enthalpy change is zero and T_a is the temperature that satisfies the condition.

Burning velocity of premixed flames. In the propagation of a flame or combustion wave in a premixture, the chemical reactions occur mainly at or near T_a. In most of the transit time of a gas element through the very thin wave, typically in ms, it is simply being heated to T_a. An observer moving with the wave would see a steady laminar flow of unburned gas at a uniform velocity, S_u^o, into the stationary wave or flame. S_u^o is thus defined as the normal burning velocity of the mixture. If its unburned density is ρ_o, the mass velocity, $\rho_o S_u^o$, is then the rate of mass consumption of the mixture per unit area of the wave. S_u^o as conventionally used implies measurement at the ambient ρ_o [101.3 kPa (1 atm), 298 K]; though it is given, eg, as cm/s, it actually refers to a mass velocity, with that value of ρ_o and should be so understood (see Mass transfer).

With increasing pressure, S_u^o decreases (for some usually low S_u^o) mixtures and for some it varies little if at all; it increases with pressure for most fast-burning mixtures, eg, H_2 or C_2H_2 in air or oxygen, or generally those mixtures in which a flame may become a detonation. Often loosely applied to a variety of explosion processes, the term detonation should refer to a specific, constant velocity and violent mode of flame propagation that usually takes place only in mixtures of high S_u^o.

The quenching distance d_q is the limiting dimension of passage, containing an essentially quiescent flammable mixture, through which a flame can propagate. d_q is roughly related to the heat loss rate that a flame can experience without extinction. The concept is applied in a flame arrester, exemplified by the grid in the Davy safety lamp, which is essentially a porous or perforated barrier that allows the passage of a gas mixture that may be explosive, but quenches a propagating flame should the mixture be ignited on one side of the arrester. For a wide range of fuels and mixture compositions, d_q can be correlated with S_u^o and the thermal diffusivity α of the mixture; $S_u^o d_q/\alpha \simeq$ constant; d_q varies approximately inversely with absolute pressure as does α.

Autoignition

If a homogeneous-fuel-oxidant mixture is uniformly and instantaneously heated to some temperature (≥ 500 K), the mixture may spontaneously self-ignite after a time delay or induction period γ_{ig}. Ideally, the temperature at which such self-ignition is observed is called the autoignition temperature T_{ig}, corresponding to that value of the time delay or ignition. In the range of T_i usually of interest (500–1000 K), τ_{ig} (the inverse of T_g) is typically 10^{-4}–10 s.

Although thermal ignition has been exhaustively studied, the complex nature of the process permits only a few useful generalities. The ignition lag decreases with increasing pressure, varying about as p^{-1} or p^{-2}, though the dependence appears to vary with the range of temperature. The τ_{ig} varies slowly with O_2 content of the mixture and is usually somewhat less in O_2 than in air at the same equivalence ratio. It varies with mixture composition in a complicated way, usually decreasing with increasing fuel concentration. There are no definite composition limits for thermal ignition, but in any case the range is much wider than the range of flammability at ambient temperature. By preheating to a sufficiently high temperature, any fuel–air mixture will, of course, react exothermically in some characteristic time.

Thermal Ignition and Flame Chemistry

Early in the ignition or induction period, the attack of fuel by O_2 is initially slow, but generates free radicals (OH, H, O, HO_2 hydrocarbons) and other intermediate species (CO, H_2, and partial oxidation and decomposition products of hydrocarbons, if present). For some time, there may be little or no temperature rise, the energy essentially is stored in the free radicals to be released later. When the partial pressure of radicals becomes high, their homogeneous recombination becomes fast, the heat evolution much exceeds losses, eg, to any walls, and the temperature rise accelerates any uncompleted consumption of fuel to produce more radicals. Around the maximum temperature, recombination exhausts the radical supply and the heat evolution rate may not compensate radiation losses.

Although most of the chemistry summarized very briefly here has evolved from study of flames and explosions in premixtures, it also applies at least semiquantitatively to many nonpremixed systems as well. It is consistent with results obtained in studies of such systems by the quite different approach of the well-stirred reactor, as well as observations made by shock-heating of premixtures. Combustion in the well-stirred reactor resembles that in some types of combustion chambers in common use, and some aspects of combustion of practical as well as fundamental interest can be experimentally examined with such a device.

Flame Chemistry and Emissions

From combustion with air of fuels containing only C, H, and O (clean fuels), the usual air pollutants or emissions of interest are carbon monoxide, unburned hydrocarbons, and oxides of nitrogen (NO_x); the interaction of the last two in the atmosphere produces photochemical smog (see Air pollution). NO_2, the sum of NO and NO_2, is formed almost entirely as NO in the products of flames; typically 5 or 10% of NO is subsequently converted to NO_2 at low temperatures. Occasionally, conditions in a combustion system may lead to a much larger fraction of NO_2 and undesirable visibility of, eg, a very large exhaust plume.

Control of thermal NO_x involves reduction of the maximum attainable temperature, the residence time at high temperature, or both.

Heat abstraction or cooling of the flame must always occur to some extent by radiation from the highly luminous flames of pulverized coal or oil, typical of boilers and similar furnaces. When the rejected heat is taken away, eg, by a boiler fluid, and is not returned or recuperated to the unburned mixture, the maximum temperature and thermal NO_x formation is reduced by the heat transfer. An extension of this effect has been applied to achieve low NO_x emissions in some furnaces and boilers in which combustion occurs in a very rich, relatively low temperature primary stage, followed by heat abstraction by convection as well as radiation to reduce the gas enthalpy.

A different and often more serious source of NO_x is chemically bound nitrogen or fuel-N in any form because some of it is almost certain to appear as NO. If combustion can be effected in two stages, with or without the intermediate heat rejection for thermal NO_x control discussed above, the conversion of fuel-N to NO can be largely circumvented by a primary stage at ϕ ca 1.5–2 with a modest residence time to allow formation of N_2 in the hot primary products, followed by

rapid addition of secondary air to complete the combustion at an effective $\phi \simeq 0.8$. Of course, there will be a maximum in the temperature near $\phi = 1$ in the course of the secondary air addition, but if time at that condition is minimized, the production of thermal NO is also minimized.

When present in the fuel as inorganic sulfides or organic compounds, sulfur is always converted practically quantitatively to SO_2 in the products of complete combustion. There are no known techniques for prevention of this conversion process in flames, and emission control measures necessarily involve either desulfurization of the fuel or removal of the SO_2 in the exhaust or stack gases.

Emitted smoke from clean (ash-free) fuels consists of unoxidized and aggregated particles of soot (sometimes referred to as carbon, though it actually is a hydrocarbon). Typically, the particles are of submicrometer size and are initially formed by pyrolysis or partial oxidation of hydrocarbons in very rich but hot regions of hydrocarbon flames; conditions that cause smoke will usually also tend to produce unburned hydrocarbons and their potential contribution to smog formation. Both may be objectionable, though for different reasons, at concentration levels equivalent to only 0.01–0.1% of the initial fuel.

It may be expected that compromises are inevitable if NO_x, UHC, CO, and smoke are to be simultaneously minimized; experience with emissions control in various types of furnaces, engines, etc, has generally confirmed that those measures or techniques effective for NO_x tend to make control of one or more of the other three more difficult. The magnitude of the problem varies with other constraints of the system such as size limitations, variability of operating conditions, etc.

A more intractable source of particulate matter or smoke is the inorganic or ash content of some fuels. Like sulfur oxides, smoke of this kind is not susceptible to control through combustion chemistry and usually can be reduced only by control of the original fuel composition or by its removal from the exhaust or stack gas.

Stabilization of Premixed Flames

The Bunsen flame. In any useful burner or torch, some mechanism must exist, or some device (a flameholder or pilot) must be provided to stabilize the flame against a variable flow of unburned mixture and to fix the position of the flame at the burner port. The volume flow rate of mixture V_{mix} into the steady flame must in general be such that the average mixture velocity is many times S_u^o. Although burners vary greatly in form and complexity, in many the stabilization mechanism is fundamentally the same; in some small fraction of the mixture flow, its linear velocity is maintained $< S_u^o$ to form a steady pilot flame from which flame spreads to consume the main flow at much higher velocity. The basic idea may be illustrated by the simple Bunsen flame on a tube of circular cross section in which the stabilization depends on the inevitable velocity variation in the flow emerging from the tube and the low gas velocity near the wall.

Atmospheric or induced-air-flow burners. Many gas–air premixed flame burners are basically laminar-flow Bunsen burners. Most of these are atmospheric, ie, the primary air is induced from the atmosphere by the fuel flow with which it mixes in the burner passage or pipe leading to the burner port or ports where the mixture is ignited and the flame stabilized. The induced air flow is determined by the fuel flow through momentum exchange and by the position of a shutter or throttle at the air inlet. Among burners of this type are many multiple-port burners for domestic furnaces, heaters, and stoves, as well as for industrial use. The Meker burner also is a multiple-port burner in which the ports are the square openings in a grid.

Burners with independent air supply. The mixture may also be prepared from separately metered or controlled supplies of fuel and air if they are available at somewhat elevated pressures. Torches and burners that require such supplies are usually intended for much higher mixture velocity or heating intensity as is often the case in industrial applications.

Interchangeability and substitution of gaseous fuels. It often is desired to substitute directly some more readily available fuel for the gas for which a premixed burner or torch and its associated feed system were designed. Satisfactory behavior with respect to flashback, blowoff, and heating capability, or the local enthalpy flux to the work, generally requires reproduction as nearly as possible of the maximum temperature and velocity of the burned gas and of the shape or height of the flame cone. A satisfactory substitution can often be obtained by adjustment of the fuel composition to reproduce the Wobbe Index (= heating value/$\sqrt{\text{density}}$) of the fuel gas.

Other types of flame stabilizers. Flames in high velocity, highly turbulent streams of premixtures in large ducts, etc, are often stabilized with obstructions or flameholders of various kinds, such as bars, grids, rods, V-gutters, abrupt changes in duct cross section, etc. To some extent, local reduction in velocity and establishment of a piloting region analogous to that of the Bunsen flame may be involved.

Combustion of Nonflammable Mixtures (Catalytic Burners)

The effluent gas from process equipment, engines, etc, often contains environmentally objectionable organic vapor too dilute to be either economically recoverable or flammable per se. The most straightforward solution is incineration in a conventional flame in the stack or exhaust of the process if fuel can be added to render the mixture flammable if it contains sufficient oxygen (see Exhaust control). Alternatively, it may be possible to consume the mixture directly by passing it through a bed of catalyst (eg, Pt or Pd on a porous support).

Combustion of Particle Suspensions of Liquids and Solids

In the design of efficient, high intensity burners for liquid and solid fuels, a basic objective is to minimize limitations on the overall combustion rate imposed by surface-limited processes. The most important features of practical burner and combustion design are the mechanisms provided for particle size reduction or maximizing the surface : mass ratio of the fuel, and for its suspension, rapid evaporation (of liquids), and mixing of the suspension with additional air or with hot burned gas, eg, for distillate fuels of moderate viscosity (ca 30 mm^2/s or 30 cSt) at ordinary temperatures, simple pressure atomization with some type of spray nozzle is most commonly used.

GEORGE E. MOORE
Consultant

B. Lewis and G. von Elbe, *Combustion, Flames and Explosions of Gases*, 2nd ed., Academic Press, Inc., New York.

H.B. Palmer and J.M. Beér, eds., *Combustion Technology, Some Modern Developments*, Academic Press, Inc., New York, 1974.

Gas Engineers' Handbook, Industrial Press, New York, 1966.

BUTADIENE

Butadiene (1,3-butadiene), CH_2=$CHCH$=CH_2, is a major commodity product of the petrochemical industry. Annual U.S. production has been in the range of $1.2–1.7 \times 10^6$ metric tons in recent years. A principal use is the manufacture of SBR (butadiene–styrene copolymer) elastomer, of which 60% or more is used for tires. 1,3-Butadiene is a colorless gas, slightly soluble in water, somewhat more soluble in methanol and ethanol, but readily soluble in common organic solvents. It forms azeotropes with ammonia, methylamine, acetaldehyde, n-butene, and 2-butene. Its physical properties are given in Table 1.

Reactions

The versatility of butadiene as a chemical raw material is reflected in a large variety of reactions, especially 1,4 additions with formation of a double bond in the 2,3 position. Butadiene undergoes the Diels-Alder reaction with a wide variety of dienophiles to form six-membered ring compounds. It can be converted into several different cyclic or open-chain dimers and trimers depending upon the reaction conditions and catalyst used. An efficient route to 1,5,9-cyclodecatriene led to the large-scale manufacture of dodecandioic acid used in the manufacture of DuPont's Qiana. Butadiene also is readily polymerized under a variety of conditions to yield structures corresponding to 1,4 and 1,2 addition of the

monomer. Several synthetic elastomers are based on butadiene, the three most important being styrene–butadiene (SBR), polybutadiene (BR), and acrylonitrile–butadiene (NBR) (see Elastomers, synthetic). The most widely used is SBR, and its manufacture accounts for close to half of the butadiene consumption in the United States. Most SBR is made by an emulsion process with a free-radical initiator giving a product with a styrene–butadiene ratio of ca 23 : 77.

A variety of heterogeneous catalysts are used for the hydrogenation of butadiene to give butenes and butane. The reaction of butadiene with an active hydrogen compound in the presence of a palladium catalyst, eg, a tertiary phosphine complex, yields a 2,7-octadienyl derivative. Butadiene also reacts easily and rapidly with both electrophilic and free-radical reagents via 1,2 and 1,4 mechanisms, depending upon reagent and reaction conditions. The vapor-phase air oxidation in the presence of vanadium or molybdenum oxide catalysts at 250–400°C produces maleic anhydride (qv) or maleic acid. Butadiene and oxygen give polymeric peroxides, and oxidation by a peroxy acid (eg, perbenzoic acid) yields butadiene monoxide. Finally, the replacement of a hydrogen in butadiene by a substituent is readily effected by a number of reagents.

Commercial Aspects

The principal processes now used in the United States for the manufacture of butadiene are coproduct production in the steam cracking of naphtha and gas–oil fractions; catalytic dehydrogenation of n-butane; and oxidative dehydrogenation of n-butene. In the early 1980s, coproduct butadiene accounted for ca 65% of U.S. production. This fraction has been increasing as additional steam-cracking plants are built to meet a growing demand for ethylene and other lower olefins. The hydrocarbon conversion processes yield a crude C_4 fraction containing butadiene and other closely boiling hydrocarbons (see Table 2). For synthetic rubber

Table 1. Physical Properties of 1,3-Butadiene

Constant	Value
bp, °C	−4.41
freezing point, °C	−108.9
critical temperature, °C	152
critical pressure, MPa (psi)	4.32 (626)
critical volume, mL/mol	221
critical density, g/cm^3	0.245
density at 20°C, g/cm^3	0.6211
heat of fusion, J/g (cal/g)	147.6 (35.28)
heat of vaporization, J/g (cal/g)	
25°C	389 (93)
bp	418 (100)
heat of formation at 25°C, kJ/mol (kcal/mol)	
gas	110.2 (26.33)
liquid	88.7 (21.21)
free energy of formation at 25°C, kJ/mol (kcal/mol)	
gas	150.7 (36.01)
explosive limits, vol% butadiene in air	
lower	2.0
upper	11.5

Table 2. Composition^a of a Crude Butadiene Fraction

Component	bp °C	Vol %
C_3 hydrocarbons		0.9
isobutylene	−6.9	27.7
1-butene	−6.3	17.2
1,3-butadiene	−4.4	39.1
n-butane	−0.5	4.1
trans-2-butene	+0.9	6.0
cis-2-butene	+3.7	4.5
C_4 acetylenes	+5.1	0.2
1,2-butadiene	+10.9	<0.1
C_5 hydrocarbons		0.1

^{a} Varies depending on process and conditions.

manufacture, a 99.0 wt% min purity butadiene is needed with acetylenes in the ppm range. Acetylenes are particularly undesirable because they can polymerize, contributing to equipment fouling and foaming problems.

Commercially, two basic methods have been used for the separation and purification of 1,3-butadiene. One is selective extraction with aqueous cuprous ammonium acetate, the CAA process, which produces high purity butadiene with a recovery higher than 98%. In the other method, butadiene is extractively distilled with selective solvents, including acetonitrile, furfural, dimethylformamide, N,N-dimethylacetamide, N-methylpyrrolidinone, β-methoxypropionitrile, and others. Both single-stage and two-stage extractive distillation processes are employed.

Butadiene can dimerize during handling and storage. The rate of dimerization is not apparently affected by the presence of peroxides but is a function of temperature. Consequently, butadiene should be stored at low temperature. The dimer, vinylcyclohexene, is miscible with butadiene in all proportions and can be removed by distillation. Butadiene is shipped as liquid via tank cars, tank trucks, and in DOT approved cylinders. It must contain an inhibitor and have a red gas label. In handling butadiene, the liquid should not be permitted to come into contact with the skin or clothing. Rapid evaporation can cause a burn or frostbite. Exposure at high concentration of 1,3-butadiene has a narcotic effect. It also is irritating to skin, eyes, and upper respiratory passages.

The principal hazards of butadiene arise from its high flammability and chemical reactivity. The explosive limits are 2–11.5% by volume of butadiene in air; the autoignition temperature is 450°C. Explosive peroxide formation is best prevented by the exclusion of air or by the addition of an inhibitor, eg, t-butylcatechol, hydroquinone, di-n-butylamine, etc. The inhibitor has no effect on butadiene in contact with oxygen in the vapor phase; moreover, the inhibitor can be used up by frequent or prolonged exposure to air, and its content should be checked at appropriate intervals.

The main use for butadiene is in the manufacture of synthetic elastomers, principally styrene–butadiene rubber and poly(cis-1,4-butadiene). It also is the raw material for many important chemicals, such as adiponitrile (for conversion to hexamethylenediamine, a nylon-6,6 precursor); 1,4-hexadiene and ethylidenenorbornene (monomers for ethylene–propylene terpolymer, EPDM); 1,5,9-cyclodecatriene (converted to dodecanedioic acid); and others.

ISIDOR KIRSHENBAUM
Exxon Research and Engineering Co.

C.L. Yaws, Chem. Eng. 83(5), 107 (Mar. 1, 1976).

W.J. Bailey in E.C. Leonard, ed., Vinyl and Diene Monomers, Pt. II, Wiley-Interscience, a division of John Wiley & Sons, Inc., New York, 1971, Chapt. 4.

T. Reis, Chem. Proc. Eng. 51(3), 65 (1970).

BUTANEDIOLS, $C_4H_8(OH)_2$.   See Glycols.

BUTANES.   See Hydrocarbons, C_1–C_6.

BUTTER.   See Milk and milk products.

BUTYL ACETATE, $CH_3COOC_4H_9$.   See Esters, organic.

BUTYL ALCOHOLS

There are four isomeric, four-carbon alcohols of molecular formula $C_4H_{10}O$; two are primary, one is an unsymmetrical secondary alcohol, and one is tertiary (Table 1).

Manufacture

The most widespread means of producing n-butanol today is the oxo process (qv); isobutyl alcohol (2-methyl-1-propanol) is the main by-product. Hydration, in the presence of sulfuric acid, of 1-butene and isobutylene derived from the cracking of petroleum leads to the production of 2-butanol and $tert$-butyl alcohol respectively.

Table 1.   Physical Properties of the Butyl Alcohols

Properties	1-Butanol	2-Methyl-1-propanol	2-Butanol	2-Methyl-2-propanol
alternative name	n-butyl alcohol	isobutyl alcohol	sec-butyl alcohol	tert-butyl alcohol
classification	primary	primary	secondary	tertiary
mp, °C	−90.2	−108	−114.7	25.5
bp, °C	117.7	108.1	99.5	82.5
density, d_4^t, g/cm^3	0.8134[15]	0.8058[15]	0.8109[15]	0.7762[30]
refractive index, n_D^t	1.39711[20]	1.39768[15]	1.39446[15]	1.3811[25]
flash pt, °C	35.0	27.5	24.4	8.9
viscosity, mPa·s^t (= cP^t)	33.79[15]	47.03[15]	42.10[15]	33.16[30]
heat of vaporization, J/g^a	591.2	578.4	562.4	535.4
sp heat, J/(g·K)^a	2.33[20]	2.38[20]	2.73[20]	3.04[27]
heat of fusion, J/g^a	125			91.6
heat of combustion, J/g^a	2674	2670		2633
critical temp, °C	287	265	265	235
critical pressure, kPa^b	4890	4850		
vapor pressure, kPa^b	0.628[20]	1.173[20]	4.132[32]	4.079[20]
electrical conductivity, S/cm	9.12 × 10^{-9}	8 × 10^{-8}		
dipole moment, C·m^c	1.66 × 10^{-18}	5.97 × 10^{-30}		5.54 × 10^{-30}
dielectric constant, ϵ^t	17.7[17.2]	17.95[25]	15.5[19]	11.4[19]
solubility at 30°C, wt%				
in water	7.08	7.5	18	miscible
of water	20.62	17.3	36.5	miscible

[a] To convert J to cal, divide by 4.184.
[b] To convert kPa to mm Hg, multiply by 7.5.
[c] To convert C·m to debye (D), divide by 3.336 × 10^{-30}.

The largest use for butanol and its derivatives is in the coatings industry for the formulation of nitrocellulose lacquers. An emerging market is latexes.

The histological effects of the butyl alcohols are similar; they cause fatty accumulations in the liver, heart, and kidneys of experimental animals.

PAUL DWIGHT SHERMAN, JR.
Union Carbide

J. Falbe, *Carbon Monoxide in Organic Synthesis*, Springer-Verlag, New York, 1970, pp. 4–14.

J.A. Monick, *Alcohols, Their Chemistry, Properties, and Manufacture*, Reinhold Book Corp., New York, 1968, pp. 125–136.

BUTYLENE OXIDES, C_4H_8OO.   See Epoxides.

BUTYLENES

The four C_4H_8 mono-olefins, ie, 1-butene, *cis*-2-butene, *trans*-2-butene, and 2-methylpropene (isobutylene), collectively are called butylenes. These four isomers are treated as a group because, with only minor exceptions, they are obtained as a mixture from the C_4 fraction from processes that crack petroleum fractions and natural gas. This C_4 fraction is commonly known as the B–B stream because it contains butanes as well as butylenes. The structures and important physical properties for the four butylenes are summarized in Table 1.

Butylenes behave as typical olefins. In addition to the reactions listed in the literature under the individual butylene isomers, the many reactions studied with the more easily handled C_5–C_{10} olefins may be expected to apply to butylenes. The double bonds are centers of high electron density, and reactions take place by electrophilic, metallation, and free-radical mechanisms. The main reactions exhibited by butylenes are addition reactions, isomerizations, and polymerization.

Almost all commercially produced butylenes are obtained as by-products from two principal processes: catalytic or thermal cracking, refinery processes that upgrade high boiling petroleum fractions to gasoline; and steam cracking, which produces light olefins for chemical feedstocks by pyrolysis of saturated hydrocarbons derived from natural gas or crude oil. Catalytic cracking accounts for ca 95% of total production in the United States.

Other commercial processes that are sometimes used to produce specific isomers or mixtures of butenes, or both, either directly or as by-products, include the Oxirane process for making propylene oxide, the dehydrogenation of butane and isobutane, the disproportionation of olefins, and the oligomerization of ethylene.

Butylenes are not regarded as toxic. However, because their physiological properties are not adequately known, they should be handled with caution. Some of the isomers may have narcotizing action if inhaled. All of them are asphyxiants.

The butylenes are extremely flammable. Fire hazards should be minimized by appropriate preventive measures. In case of fire, both water spray and carbon dioxide extinguishers are applicable. Butylenes also form explosive mixtures with air and oxygen. In air, the explosive limits are 1.7–9.7% butylene.

Fuel Uses

Methyl *tert*-butyl ether has been manufactured since 1979 as an octane enhancer for motor fuels.

Chemical Uses

Butadiene manufacture is the single largest consumer of butylenes for chemical use. *sec*-Butyl alcohol is almost always converted to methyl ethyl ketone (MEK) by catalytic dehydrogenation. The dimer of isobutylene, a mixture of mainly 2,4,4-trimethylpentene-1 and -2, is used for alkylation of phenol, an intermediate for surfactants. Butylated phenols and cresols are used primarily as oxidation inhibitors and chain terminators. Heptenes from mixed butylenes and propylene, and dimer octenes are used for the preparation of isooctyl and isopropyl alcohols by hydroformylation (see Oxo process). *tert*-Butylamine is used as an intermediate in the manufacture of lubricating oil additives and miscella-

Table 1. Physical Properties of the Butylenes

Properties	1-Butene	*cis*-2-Butene	*trans*-2-Butene	Isobutylene
mp, °C	−185.35	−138.922	−105.533	−140.337
bp, °C	−6.25	+3.718	+0.88	−6.896
density of liquid at 25°C, g/L	588.8	615.4	598.4	587.9
vapor pressure (Antoine equation constants)[a, b]				
temp range, °C	−82 to +13	−73.4 to +23	−76 to +20	−82 to +12
A	7.7180	7.74436	7.74462	7.71644
B	926.1	960.1	960.8	932.2
C	240.0	237.0	240.0	240.0
heat of vaporization, J/g[c]				
at 25°C	359.21	359.08	380.87	367.46
at boiling point	391.18	416.74	406.18	394.78
critical temperature, °C	146.4	162.43	155.48	144.75
critical pressure, kPa[b]	4022.6	4205.0	4103.6	4000.3
critical volume, L/mol	0.240	0.234	0.238	0.239
heat of combustion at constant pressures and 25°C, J/mol[c]	2.524×10^6	2.515×10^6	2.512×10^6	2.507×10^6
isobaric specific heat at 25°C, J/(kg·K)[c]				
gas in ideal state	1528.9	1408.7	1567.5	1590.5
liquid at 101.3 kPa[b]	2300.3	2251.7	2277.3	2338.0
surface tension at 20°C, mN/m (= dyn/cm)	0.0125	0.10507	0.01343	0.01242

[a] $\log_{10} P = A - B/(t + C)$, where P is in kPa and t in °C.
[b] To convert kPa to mm Hg, multiply by 7.5.
[c] To convert J to cal, divide by 4.184.

neous chemicals. *tert*-Butyl mercaptan is used primarily as an odorant at 300 ppm for natural gas so that leaks can be readily detected. The main use of the amyl alcohols is as esters such as acetates for solvents. Triisobutylaluminum is used to manufacture plasticizer (C_6–C_{10}) and detergent-range (C_{16}–C_{22}) alcohols (see Alcohols, synthetic), Diisobutylaluminum chloride and diisobutylaluminum hydroxide are used as cocatalysts for Ziegler polymerization systems. A major use of butylene oxide is as an acid scavenger for chlorine-containing materials such as trichloroethylene. *p-tert*-Butyltoluene is an intermediate in the production of *p-tert*-butylbenzoic acid. Neopentanoic (pivalic) acids are converted to peroxy esters for use as polymerization initiators. The metal salts are used as driers in paint formulations (see Driers). Methallyl chloride is a chemical intermediate for various specialty products. Butylated hydroxy anisole (BHA) is an oxidation inhibitor and has been accepted for use in foods where the use of butylated hydroxytoluene (BHT) is restricted (see Food additives).

Polymer Use

Polybutylenes are used most often in the manufacture of additives for motor oils. Other uses are in formulations of sealants, caulks, coatings, adhesives, and laminating agents; in high voltage electrical cables as impregnating oils and pipe oils, and as industrial lubricants. Polyisobutylenes are also used as adhesives, caulks, sealants, and polymer additives. Very high molecular weight polyisobutylenes are used as viscosity index improvers for lubricating oils. Exxon Chemical and Lubrizol are the domestic producers. Butyl elastomers can be used for the manufacture of automotive inner tubes and inner liners. In recent years, small volumes of lower molecular weight butyl rubbers also have been used for caulk and sealant applications. Poly(1-butene) products are particularly suitable for pipe fabrication. Butene-1 is also used in high density polyethylene (HDPE) and linear low density polyethylene (L-LDPE).

UN K. IM
WILLIAM F. HAUSCHILDT
IMRE PUSKAS
Amoco Chemicals Corp.

Technical Data Book—Petroleum Refining, American Petroleum Institute, New York, 1970.

R.R. Dreisbach, *Physical Properties of Chemical Compounds II*, American Chemical Society, Washington,D.C., 1959.

Butylenes, Vol. II of *World Petrochemicals*, SRI International, Menlo Park, Calif., Jan. 1982.

Butylenes, Vol. II of *Chemical Economics Handbook, Primary Petrochemicals*, SRI International, Menlo Park, Calif., Sept. 1982.

L.F. Hatch and S. Matar, *Hydrocarbon Process.*, 153 (Aug. 1978).

A.L. Waddams, *Chemicals from Petroleum*, John Murray, London, 1978.

BUTYL RUBBER. See Elastomers, synthetic.

BUTYRALDEHYDE

Both *n*-butyraldehyde and isobutyraldehyde occur naturally in small quantities. The former has been isolated in small quantities in the essential oils of several plants. It also has been detected in oil of lavender, the oil of the *Eucalyptus globulus* of California, in tobacco smoke, in tea leaves, and in other leaves. Isobutyraldehyde is found in trace quantities in Jeffrey pine oil and tea leaves.

n-Butyraldehyde, $CH_3CH_2CH_2CHO$, is a colorless, flammable liquid with a characteristic aldehyde odor. It is used chiefly as an intermediate in the production of synthetic resins, rubber accelerators, solvents, and plasticizers. Because of the large number of condensation and addition reactions it can undergo, it is a useful starting material in the production of a wide variety of compounds containing at least 6 to 8 carbon atoms.

Isobutyraldehyde, $(CH_3)_2CHCHO$, is also a colorless liquid with a pungent odor and is available commercially. Like *n*-butyraldehyde, it can be used in the manufacture of resins and rubber chemicals as well as in the synthesis of isobutyric acid, acetals, mercaptals, and derivatives for use as corrosion inhibitors, insecticides, and amino acids.

The four-carbon aldehydes are highly flammable liquids with physical properties as shown in Table 1. They are colorless and have pungent odors that are characteristic of the lower carbon aldehydes. Both butyraldehyde and isobutyraldehyde are slightly soluble in water and miscible with most organic solvents, eg, ethanol, ethyl ether, benzene, toluene, and acetone. The butyraldehydes contain a terminal carbonyl group and undergo reactions that are characteristic of aldehydes (qv), ie, oxidation, reduction, and condensation.

Table 1. Physical Properties of Butyraldehydes

Properties	Butanal (butyraldehyde)	2-Methyl-propanal (isobutyral-dehyde)
mp, °C	−99.0	−65.9
bp, °C	75.7	64.5
density, d_{20}^{20}, g/cm^3	0.8048	0.7938
vapor density (air = 1)	2.48	
refractive index, n_D^{20}	1.3843	1.3730
flash point, °C[a]	−9.4	−10.6
viscosity (20°C), mPa·s (= cP)	0.433	
heat of formation, kJ/mol[b]	240.3	
specific heat, J/(kg·K)[b]	2121	2544
heat of vaporization at bp, J/g[a]	436	409
heat of combustion, kJ/mol[b]	2478.7	2510.0
dipole moment (vapor), C·m[c]	9.07×10^{-30}	
surface tension, mN/m (= dyn/cm), 24°C	29.9	
vapor pressure, kPa[d] (20°C)	12.2	18.4

[a] Tag open cup, ASTM D 1310.
[b] To convert J to cal, divide by 4.184.
[c] To convert C·m to debye, divide by 3.336×10^{-30}.
[d] To convert kPa to mm Hg, multiply by 7.5.

The most widely used manufacturing technique for both butyraldehyde and isobutyraldehyde is the oxo process (qv), in which propylene, carbon monoxide, and hydrogen are combined with a suitable catalyst, usually a cobalt or rhodium compound, at about 130–160°C and 1–20 MPa (10–200 atm) pressure.

Although tests have shown that *n*-butyraldehyde exhibits some adverse physiological effects, normal manufacturing practice is to minimize exposure.

PAUL DWIGHT SHERMAN, JR.
Union Carbide Corp.

S. Patai, ed., *The Chemistry of Functional Groups*, Vol. 2, Interscience Publishers, a division of John Wiley & Sons, New York, 1966.

H.J. Hagemayer, Jr., and G.C. DeCroes, *The Chemistry of Isobutyraldehyde and its Derivatives*, Eastman Kodak Co., 1953.

J. Falbe, *Carbon Monoxide in Organic Synthesis*, Springer-Verlag, New York, 1970, pp. 3–77.

BUTYRIC ACID AND BUTYRIC ANHYDRIDE. See Carboxylic acids.

BUTYROLACTONE. See Acetylene-derived chemicals.

C

CABLE COVERINGS. See Insulation, electric.

CACAO. See Chocolate and cocoa.

CADMIUM AND CADMIUM ALLOYS

Cadmium, Cd, a Group-IIB element (between zinc and mercury), is a soft, ductile, silvery-white metal with a distorted hexagonal close-packed structure ($a = 0.29793$ nm, $c = 0.56181$ nm). The crustal abundance of cadmium is somewhere between 0.1 and 0.5 ppm, and although several cadmium minerals have been identified—the most common one being greenockite, CdS—the element is generally encountered in zinc ores, zinc-bearing lead ores, or complex copper–lead–zinc ores, where it forms an isomorphic impurity in the zinc mineral sphalerite, ZnS, usually in concentrations of 0.1–0.5% cadmium. For this reason, cadmium is almost invariably recovered as a by-product from the processing of zinc, lead, and copper ores.

Some typical physical properties of cadmium are listed in Table 1. Its electronic structure is $1s^2 2s^2 2p^6 3s^2 3p^6 3d^{10} 4s^2 4p^6 4d^{10} 5s^2$, and its oxidation state in almost all of its compounds is $+2$, although a few compounds have been reported in which cadmium exists in the oxidation state of $+1$.

Cadmium occurs primarily as sulfide minerals in zinc, lead–zinc, and copper–lead–zinc ores. Beneficiation of these minerals, usually by flotation (qv) or heavy-media separation, yields concentrates that are then processed for the recovery of the contained metal values. Cadmium follows the zinc with which it is so closely associated.

Air-pollution problems and labor costs have led to the closing of older pyrometallurgical zinc plants, and to increased electrolytic production. On a worldwide basis, 56% of total zinc production in 1970 was by the electrolytic process. In electrolytic zinc plants, the calcined material is dissolved in aqueous sulfuric acid, usually spent electrolyte from the electrolytic cells. The filtered leach solution is treated with zinc dust to remove cadmium and other impurities. Recent U.S. production has dropped sharply owing to a cutback in zinc production. The outlook is

influenced by reduced cadmium levels of new ore sources and changes in environmental regulations.

Cadmium is classified as a toxic metal. Most cases of cadmium poisoning result from the inhalation of dust or fumes affecting the respiratory tract. The current OSHA atmospheric limit (8-h time-weighted average standard) is 100 $\mu g/m^3$ of air for cadmium fume and 200 $\mu g/m^3$ of air for cadmium dust.

Elemental cadmium is used principally as an electroplated coating on fabricated steel and cast iron parts for corrosion protection (see Corrosion). Another significant application of elemental cadmium is in NiCd rechargeable batteries.

Alloys

Cadmium is an important component in brazing and low melting alloys, used in bearings, solders, and nuclear-reactor control rods, and as a hardener for copper (see Bearing materials). Of interest are two brazing alloys, 20% Ag–45% Cu–30% Zn–5% Cd (mp 615°C; flow point 815°C), and ASTM Ag 2 which is 35% Ag–26% Cu–21% Zn–18% Cd (mp 607°C; flow point 702°C). Other useful brazing compositions are also available (see Solders and brazing alloys).

The commonly used low melting or fusible alloys are AsarcoLo 158 or Cerrobend, AsarcoLo 158-190 or Cerrosafe, AsarcoLo 117 or Cerrolow 117. For soldering aluminum, combinations of cadmium and zinc are used widely. Additions of cadmium to copper raise the recrystallization temperature and improve the mechanical properties especially in cold-worked conditions.

The annual consumption of cadmium during the past few years averaged about 4300 metric tons.

M.L. HOLLANDER
S.C. CARAPELLA, JR.
ASARCO Incorporated

I.M. Kolthoff and P.J. Elving, *Treatise on Analytical Chemistry*, Pt. II, Vol. 3, Interscience Publishers, New York, 1961.

A. Butts and C.D. Coxe, eds., *Silver Economics, Metallurgy and Use*, D. Van Nostrand Co., Inc., Princeton, N.J., 1967, p. 387.

Brazing Alloy Handbook, ASARCO Incorporated, New York, 1968.

CADMIUM COMPOUNDS

The only naturally occurring cadmium compound of significance, the sulfide greenockite, CdS, which is fairly rare, is almost always associated with the polymetallic sulfide ores of zinc, lead, and to a lesser extent copper, at grades up to ca 0.5%. Cadmium compounds are prepared from metallic cadmium obtained from lead–zinc production. The cadmium can be converted to the oxide which is a more convenient starting material for many compounds.

Cadmium, a member of Group IIB (Hg, Cd, Zn) of the periodic table, exhibits almost entirely a $+2$ valence state in its compounds. Cadmium shows a marked tendency to form aqueous complexes in which it binds from one to four ligands. The most important are the cyanide, amine, and various halide complexes.

Technically and commercially important cadmium compounds include the oxide, sulfide, selenide, chloride, sulfate, nitrate, hydroxide, and various organic cadmium salts such as the stearate and benzoate. Some of the main areas of application are metal finishing, pigment manufacture, batteries, plastic stabilizers, electronics, and catalysts.

In 1977, NIOSH proposed standards that limit the 8-h exposure level to 40 $\mu g/m^3$ for all forms of airborne cadmium. The ceiling concentration for dust is 200 $\mu g/m^3$, and for fumes, it is 100 $\mu g/m^3$. Cadmium is toxic and poisoning occurs through inhalation and ingestion.

Inorganic compounds include cadmium arsenides, Cd_3As_2 (mp 721°C) and $CdAs_2$ (mp 621°C); cadmium antimonide, CdSb (mp 456°C); cadmium phosphides, Cd_3P_2, CdP_2, and CdP_4; cadmium borates, $nCdO \cdot mB_2O_3$; cadmium carbonate, $CdCO_3$ (decomposes at 332°C); cadmium complexes (eg, chloride, $CdCl_3^-$, cyanide, $Cd(CN)_4^{2-}$, ammine, $Cd(NH_3)_4^{2+}$, bromide, $CdBr_3^-$, and thiocyanate, $Cd(SCN)_4^{2-}$); cadmium

---

Table 1. Physical Properties of Cadmium

Property	Value
melting point, °C	321.1
boiling point, °C	767
latent heat of fusion, kJ/mol[a]	6.2
latent heat of vaporization, kJ/mol[a]	99.7
specific heat at 20°C, J/(mol·K)[a]	25.9
coefficient of linear expansion at 20°C, $\mu cm/(cm \cdot °C)$	31.3
electrical resistivity at 22°C, $\mu\Omega \cdot cm$	7.27
electrical conductivity, % IACS[b]	25
density at 26°C, g/cm³	8.642
volume change on fusion, % increase	4.74
thermal conductivity at 0°C, W/(m·K)	98
vapor pressure at 382°C, kPa[c]	0.1013
surface tension at 330°C, mN/m (= dyn/cm)	564
viscosity at 340°C, mPa·s (= cP)	2.37
thermal neutron capture cross section at 2200 m/s, $\times 10^{-28}$ (= barns)	2450 ± 50

[a] To convert J to cal, divide by 4.184.
[b] International Annealed Copper Standard.
[c] To convert kPa to mm Hg, multiply by 7.5.

Table 1.   Properties of Dialkyl Cadmium Compounds

Compound	Formula	mp, °C	bp, °C (kPa)[a]	Density g/cm³
dimethylcadmium[b]	$(CH_3)_2Cd$	− 4.5	105.5 (101.3)	1.9846
diethylcadmium	$(C_2H_5)_2Cd$	− 21	64 (2.6)	1.6564
dipropylcadmium	$(C_3H_7)_2Cd$	− 83	84 (2.8)	1.4184
dibutylcadmium	$(C_4H_9)_2Cd$	− 48	103.5 (1.6)	1.3054
diisobutylcadmium	$(C_4H_9)_2Cd$	− 37	90.5 (2.6)	1.2674
diisoamylcadmium	$(C_5H_{11})_2Cd$	− 115	121.5 (2.0)	1.2184

[a] To convert kPa to mm Hg, multiply by 7.5.
[b] $\Delta H^\circ_{f,298}$ = 63.6 kJ/mol (15.2 kcal/mol), $\Delta G^\circ_{f,298}$ = 139.3 kJ/mol (33.3 kcal/mol), and S°_{298} = 201.88 J/(mol·K) [48.25 cal/(mol·K)].

ferrite, $CdFe_2O_4$; cadmium fluoride, CdF_2 (mp 1110°C); cadmium chloride, $CdCl_2$ (mp 568°C); cadmium bromide, $CdBr_2$ (mp 568°C); cadmium iodide, CdI_2 (mp 387°C); cadmium hydroxide, $Cd(OH)_2$ (decomposes at 150°C); cadmium nitrate, $Cd(NO_3)_2$ (mp 350°C); cadmium nitride, $Cd(N_3)_2$; cadmium oxide, CdO (subl 1540°C); cadmium selenide, CdSe (dissoc 680°C); cadmium telluride, CdTe (mp 1045°C); cadmium silicate, Cd_2SiO_4 (mp 1246°C); cadmium metasilicate, $CdSiO_3$ (mp 1252°C); cadmium sulfate, $CdSO_4$ (mp 1000°C); cadmium sulfide, CdS (subl 980°C); and cadmium tungstate.

Organic Compounds

Alkyl and aryl cadmiums are now employed as polymerization catalysts, and cadmium salts of organic acids are used as heat and light stabilizers in plastics (see Heat stabilizers; Uv stabilizers). Table 1 lists the properties of dialkyl cadmium compounds.

Cadmium forms acetates $Cd(CH_3COO)_2 \cdot nH_2O$ (n = 1–3); anhydrous cadmium acetate, d = 2.34 g/cm³, melts at 256°C. Cadmium acetate is the starting material for cadmium halides and is a colorant in glass, ceramics, and textiles (see Colorants for ceramics).

Cadmium salts of organic acids have widespread use as heat and light stabilizers in plastics such as poly(vinyl chloride). Cadmium stabilizers only prevent early discoloration and generally are used in combination with barium organic soaps which help in preventing long-term yellowing. Cadmium soaps are marketed as solid and liquid. The former include the cadmium salts of lauric, stearic, myristic, and palmitic acids. Most stabilizers are used in liquid form including cadmium octanate, phenolate, decanoate, naphthenate, and benzoate. The liquid stabilizers impart better physical characteristics to PVC and are more economical. The use of Ba–Cd stabilizers in plastics that come in contact with food is not permitted by the FDA because of the high toxicity of cadmium.

P.D. PARKER
AMAX Base Metals Research & Development, Inc.

B.J. Aylett in J.C. Bailar and A.F. Trotman-Dickenson, eds., *Comprehensive Inorganic Chemistry*, Pergamon Press, Oxford, UK, 1973, pp. 187, 190, and 258–272.

D.M. Chizhikov, *Cadmium*, trans. D.E. Hayler, Pergamon Press, Oxford, UK, 1966.

L. Friberg and co-workers, *Cadmium in the Environment*, 2nd ed., CRC Press, Cleveland, Ohio, 1974, Chapt. 2.

CAFFEINE. See Alkaloids.

CALCIUM AND CALCIUM ALLOYS

Calcium, Ca, a member of Group II of the periodic table (between magnesium and strontium) is classified, together with barium and strontium, as an alkaline-earth metal and is the lightest of the three. Calcium metal does not occur free in nature; however, in the form of numerous compounds, it is the fifth most abundant element constituting 3.63% of the earth's crust.

Calcium is mainly used as a reducing agent for many reactive, less common metals; to remove bismuth from lead; as a desulfurizer and deoxidizer for ferrous metals and alloys; and as an alloying agent for aluminum, silicon, and lead. Smaller amounts are used as a dehydrating agent for organic solvents, and as a purifying agent for removal of nitrogen and other impurities from argon and other rare gases.

Some of the more important physical properties of calcium are given in Table 1.

Table 1.   Physical Properties of Calcium

Property	Value		
density at 20°C, g/cm³	1.55		
melting point, °C	839 ± 2		
boiling point, °C	1484		
heat of fusion, kJ/mol[a]	9.2		
heat of vaporization, kJ/mol[a]	161.5		
heat of combustion, kJ/mol[a]	634.3		
vapor pressure data			
pressure, kPa[b]	0.133	13.3	101.3
temperature, °C	800	1200	1484
specific heat at 25°C, J/(g·K)[a]	0.653		
electrical resistivity at 0°C, μΩ·cm	3.91		

[a] To convert J to cal, divide by 4.184.
[b] To convert kPa to mm Hg, multiply by 7.5.

Calcium, a bright silvery-white metal, has a valence-electron configuration of $4s^2$ and characteristically forms divalent compounds. It is very reactive and reacts vigorously with water, liberating hydrogen and forming calcium hydroxide, $Ca(OH)_2$. Calcium is an excellent reducing agent and is widely used for this purpose.

At present, all calcium metal is produced by high temperature vacuum reduction of calcium oxide with aluminum. For certain applications (especially those involving reduction of other metal compounds), a better than 99% purity is required. This can be achieved by redistillation.

Because of its extreme chemical reactivity, calcium metal must be carefully packaged for shipment and storage. Calcium is classed as a flammable solid and is nonmailable. Calcium metal is produced by Pfizer Inc. Calcium-containing alloys are produced by Union Carbide and Foote Mineral Company.

Calcium metal and most calcium compounds are nontoxic. Care must be taken, however, to avoid contact with water owing to the exothermic liberation of hydrogen and the resulting explosion hazard.

Calcium alloys are produced commercially by various techniques such as direct alloying of the pure metals and chemical reduction of one or more of the components. Alloys include calcium–aluminum, calcium–lithium, calcium–magnesium and calcium–beryllium, trilead calcium, and calcium disilicide.

CHARLES J. KUNESH
Pfizer Inc.

C.L. Mantell, "The Alkaline Earth Metals—Calcium, Barium, and Strontium" in C.A. Hampel, ed., *Rare Metals Handbook*, 2nd ed., Reinhold Publishing Corp., London, 1961, pp. 15–25.

C.L. Mantell and C. Hardy, *Calcium Metallurgy and Technology*, Reinhold Publishing Corp., New York, 1945.

C.L. Mantell, "Calcium" in C.A. Hampel, ed., *The Encyclopedia of Chemical Elements*, Reinhold Book Corp., New York, 1968, pp. 94–103.

CALCIUM-CHANNEL BLOCKING AGENTS. See Cardiovascular agents.

CALCIUM COMPOUNDS

SURVEY

Calcium is the fifth most abundant element and the third most abundant metal, amounting to about 4% of the earth's crust. About 7% of the earth's crust is calcite, $CaCO_3$, in the forms of limestone, marble, and chalk.

Inorganic Compounds

Calcium carbonate. Limestone is the most widely used of all rocks since calcium carbonate is an indispensable chemical in industry, either as such, or as the precursor of lime and hydrated lime.

Lime (CaO) and hydrated lime ($Ca(OH)_2$). In industrial countries, lime is second only to sulfuric acid in tonnage consumed. Lime is one of the few chemical compounds having a negative temperature coefficient of solubility. It is used in mortar, metallurgy, treatment of industrial wastes, and treatment of municipal and industrial water supplies (see Lime and limestone).

Halogen compounds. Halogen compounds include calcium halides and hypochlorites (used as bleaches).

Sulfates and sulfites. Calcium sulfate occurs in large deposits as $CaSO_4$ (anhydrite and $CaSO_4 \cdot 2H_2O$ (gypsum)). Calcium sulfite and acid sulfite may be prepared by reaction of SO_2 with hydrated lime or limestone.

Phosphates. Important industrial phosphates are the phosphate fertilizers, phosphoric acid, and monocalcium phosphate used in baking powder (see Phosphoric acid; Fertilizers).

Silicates. Silicates include ordinary glass, portland cement, whitewares, and calcium silicate brick.

Coordination Chemistry of Calcium

Calcium ion is essential in the sequence of reactions resulting in coagulation of blood (see Blood, coagulants). Blood samples can thus be protected against coagulation by substances that mask Ca^{2+}. One of the most effective agents, citrate, reduces the Ca^{2+} concentration by coordination to a value below that needed for coagulation (see also Chelating agents).

Organic Chemistry of Calcium

Probably the most important organic calcium compound is calcium carbide. Derivatives of this compound include acetylene, calcium cyanamide (used as a fertilizer), and calcium cyanide. Salts of organic acids include calcium sucrate and lactate. Reagents in synthesis include calcium borohydride, $Ca(BH_4)_2$, hexammine calcium, $Ca(NH_3)_6$, calcium hydride, CaH_2, and anhydrous calcium sulfate (Drierite), $CaSO_4$.

Physiological Role of Calcium

Calcium accounts for about 2% of body weight, about 99% is found in bones and teeth. The major mineral component of these structures is a hydroxyapatite, $Ca(OH)_2 \cdot 3Ca_3(PO_4)_2$. In addition to its major functions of formation and maintenance of the skeletal structure, calcium is also essential in the following ways: blood clotting; control of muscle and nerve cell response; influencing permeability of cell membranes; functioning of certain enzyme systems; and milk formation.

CARL L. ROLLINSON
University of Maryland

R.D. Goodenough and V.A. Stenger, "Magnesium, Calcium, Strontium, Barium and Radium" in *Comprehensive Inorganic Chemistry*, Vol. 1, Pergamon Press, Oxford, UK, 1973, pp. 591–664.

Limestone Purifies Water, National Limestone Institute, Inc., Fairfax, Va., May, 1977; valuable up-to-date discussion of limestone including statistics of production and use, application in agriculture, neutralization of acidic wastes and acidic rain, effects on natural waters, etc (92 references).

CALCIUM CARBONATE

Calcium carbonate, $CaCO_3$, mol wt 100.09, is the major constituent of limestone which is used in the manufacture of quicklime or hydrated lime; these in turn are the sources of most calcium compounds including precipitated calcium carbonate. Calcium carbonate occurs naturally in the form of marble, chalk, and coral. Powdered calcium carbonate is produced by either chemical methods or by the mechanical treatment of the natural materials. The term precipitated calcium carbonate applies to the commercial types of the compound produced chemically in a precipitation process. The precipitated products are generally finer, have a more uniform particle size distribution and a higher degree of chemical purity. Products averaging as fine as 0.05 μm are now routinely produced. However, by controlling the process, a wide range of product sizes and particle shapes can be produced.

Precipitated calcium carbonate is one of the most versatile mineral fillers and is consumed in a wide range of products including paper, paint, plastics, rubber, textiles, putties, caulks, sealants (qv), adhesives (qv), and printing ink (see Fillers). USP grades are used in dentifrices (qv), cosmetics, foods, and pharmaceuticals. Precipitated calcium carbonate is produced in a number of grades for these applications.

Calcium carbonate occurs naturally in two crystal structures, calcite and aragonite. Calcite is thermodynamically stable at all investigated pressures and temperatures. The aragonite polymorph is metastable and irreversibly changes to calcite when heated in dry air to about 400°C, the rate increasing with temperature. The transformation is much more rapid when in contact with water or solutions containing calcium carbonate and may take place at room temperature. The crystal forms of calcite are in the hexagonal system. Aragonite is in the orthorhombic system.

Precipitated calcium carbonate can be produced by several methods but only the carbonation process is used commercially in the United States today. This is the simplest and most direct process, using the most readily available and lowest cost raw materials. Current U.S. producers of precipitated calcium carbonate include the Minerals, Pigments and Metals Division of Pfizer Inc., and the Mississippi Lime Company.

There are minimal environmental concerns in the handling of precipitated calcium carbonate.

RICHARD H. LEPLEY
Pfizer Inc.

C. Palache, H. Berman, and C. Frondel, *Dana's System of Mineralogy*, 7th ed., Vol. II, John Wiley & Sons, Inc., New York, 1951, p. 151.

CALCIUM CHLORIDE

Calcium chloride is an extremely soluble salt that forms many hydrates with properties as shown in Table 1 on p. 197. At present the greatest volume of calcium chloride is derived from evaporation of underground brines (see Chemicals from brine). The major markets for calcium chloride are in deicing, dust control, road stabilization, and production of concrete products, as well as in oil-well drilling and completion operations. Most calcium chloride environmental concerns involve its effects on automobiles, concrete and asphalt pavements, vegetation, and drinking-water supplies.

WALKER L. SHEARER
Dow Chemical, U.S.A.

Chem. Process (Chicago) **39**(5), 71 (1976).

CALCIUM SULFATE

Mineral calcium sulfate is commonly called anhydrite, and occurs in many parts of the world. The mineral gypsum, calcium sulfate dihydrate, is widely distributed and is of much more economic importance.

Table 1. Properties of Calcium Chloride Hydrates

Property	$CaCl_2 \cdot 6H_2O$	$CaCl_2 \cdot 4H_2O$	$CaCl_2 \cdot 2H_2O$	$CaCl_2 \cdot H_2O$	$CaCl_2$
composition, wt% $CaCl_2$	50.66	60.63	75.49	86.03	100.00
mol wt	219.09	183.05	147.02	129.00	110.99
mp[a], °C	29.9	45.3	176	187	772
bp, °C			175[b]	181[b]	1935
density, d_4^{25}, g/cm³	1.71	1.83	1.85	2.24	2.16
heat of fusion, J/g (Btu/lb)	209 (90)	163 (70)	88 (38)	134 (58)	257 (111)
heat of soln (to infinite diln)[c] in H_2O, J/g (Btu/lb)	72 (31)	−59.4 (−25.6)	−304.6 (−131.1)	−405 (−174.3)	−737.2 (−317.2)
heat capacity, J/(g·K)[d], at 25°C	1.4	1.4	1.2	0.84	0.67

[a] Incongruent mp for hydrates.
[b] Temperature where dissociation pressure is 101.3 kPa (1 atm).
[c] Negative sign means heat is evolved (exothermicity).
[d] To convert J to cal, divide by 4.184.

About 55 million (10⁶) metric tons of gypsum are consumed annually; about half is processed to calcium sulfate hemihydrate. The hemihydrate is also called plaster of Paris. Smaller amounts are dehydrated to hexagonal calcium sulfate, or soluble anhydrite, and orthorhombic calcium sulfate, identical to the mineral anhydrite. The latter is called dead-burnt gypsum. Both anhydrite and gypsum are produced in large quantities as a by-product of various chemical operations. Pure anhydrite and the hydrates have the following percent compositions: anhydrite ($CaSO_4$), 41.2% CaO, 58.8% SO_3; gypsum ($CaSO_4 \cdot 2H_2O$), 32.6% CaO, 46.5% SO_3, 20.9% H_2O; and hemihydrate ($CaSO_4 \cdot 1/2H_2O$), 38.6% CaO, 55.2 SO_3, 6.2% H_2O. Major uses of gypsum are in construction, portland cement and agriculture.

Gypsum, $CaSO_4 \cdot 2H_2O$, is the most useful form of $CaSO_4$. It is useful primarily because a controlled, modest amount of heat will convert gypsum to hemihydrate $CaSO_4 \cdot 1/2H_2O$. This intermediate, relatively stable phase of calcium sulfate is the basis for over 90% of the commercial value of all calcium sulfate products in the United States.

There are two types of gypsum: a natural mineral and a synthetic product of chemical reaction from a variety of industries. The natural mineral is quarried or mined in many areas of North America. Major producing areas are Nova Scotia, Mexico, Michigan, Texas, California, Iowa, and Oklahoma. France, the USSR, the UK, Spain, Italy, and the FRG also have significant deposits of natural gypsum. Deposits vary in both ease of extraction and purity. Natural gypsum is seldom found in the pure form. The anhydrous form and the dihydrate are commonly found together. Most gypsum that is commercially used is a minimum of 80% pure, although some deposits require beneficiation or selective mining practices to achieve this degree of purity. Some physical properties are shown in Table 1.

calcination in the gypsum industry, is used to prepare hemihydrate or anhydrite. Kettle calcination continues to be the most commonly used method of producing beta hemihydrate. The kettle can be operated on either a batch or continuous basis. In addition to this method, soluble anhydrite is commercially manufactured in a variety of forms, from fine powders to granules 4.76 mm (4 mesh) in size. Insoluble anhydrite is manufactured commercially by several methods. Where large rock gypsum is the starting material, beehive kilns are used and 24-h processing time is not unusual. Rotary calciners or traveling grates are often used for small rock feed. Fine ground gypsum is calcined to the insoluble form in flash calciners. Temperature control is somewhat critical in all methods; low temperatures result in soluble anydrite being present and high temperatures dissociate the $CaSO_4$ into CaO and oxides of sulfur. Phosphogypsum, the by-product of phosphoric acid manufacture, is the major source of synthetic gypsum. The $CaSO_4$ can be produced in either the dihydrate or hemihydrate form. Because these synthetic gypsums are by-products, most contain objectionable quantities of impurities that relate to the process and raw materials from which they originate.

Crude gypsum rock is the only form of calcium sulfate that has a significant movement in international trade, although certain special products move across international borders. Widespread location of gypsum deposits plus the relatively low ratio of manufacturing cost to product weight (shipping cost) are the reasons for the limited trade. Canada is the largest exporter of gypsum rock with most of it being shipped to eastern seaboard facilities in the United States. The United States imports more gypsum than any other nation.

ROBERT J. WENK
PAUL L. HENKELS
United States Gypsum Co.

K.K. Kelly, J.C. Southard, and C.T. Anderson, *U. S. Bur. Min. Tech. Papers* **625**, (1941).

Mineral Industry Surveys, U.S. Department of Interior, Bureau of Mines, Mar. 1977.

Table 1. Physical Properties of Calcium Sulfate

Physical properties	Gypsum	Hemihydrate	Anhydrite
mol wt	172.17	145.15	136.14
mp, °C	128 (−1 1/2 H_2O) 163 (−2 H_2O)	163 (−1/2 H_2O)	1360 (dec)
sp gr	2.32		2.96
Mohs hardness	1.5–2.0		3.0–3.5
soly in 100 g H_2O at 25°C	0.24	0.30	0.20

CALIFORNIUM. See Actinides and transactinides.

Manufacture

Gypsum rock from the mine or quarry is crushed and sized to meet the requirements of future processing or direct marketing of the dihydrate. The dehydration of gypsum (dihydrate), commonly referred to as

CAULKING AND SEALING COMPOSITIONS. See Sealants; Chemical grouts.

CALORIMETRY

With the development of the first law of thermodynamics in the nineteenth century, as heat became more clearly defined, calorimetry assumed its present meaning: a procedure for establishing increments in the internal energy of a system by work done on the system plus energy transferred into it as a result of temperature gradients (heat flow) (see Thermodynamics). From the time of the earliest quantitative heat measurements, the reference heat effect usually adopted was the amount of heat needed to raise the temperature of unit mass of water one temperature unit, leading to a heat scale on which the specific heat capacity of water is unity. This custom, which was a practical approach before the equivalence of thermal and mechanical energies was established by Joule, led to established units of heat, the calorie (cal) and the British thermal unit (Btu), different from those of other forms of energy. The unit of energy derived from mechanical work, the newton-meter, was named the joule. This SI unit is properly applicable to heat.

In thermodynamic terminology (first law), heat is observable only in transfer, the transfer being the result of a temperature gradient. A material has a greater internal energy when hot than when cold. However, it is not correct to say that a substance contains more heat when molten than when solid, for the solid and liquid can coexist in contact indefinitely without any flow of heat between them if they are at the same temperature. Increase of temperature of a substance, melting of a solid, or vaporization of a condensed phase can be caused by doing work on the system as well as by the flow of heat into the system. Work can easily be done on the system mechanically, as by stirring, or by electrical work done on a resistor imbedded in the system. Thus, measurements of heat and work are complementary; internal energy is a proper descriptor of the form of the energy that heat and work change as they enter or leave the system. A calorimeter is an appropriate device for measuring these changes.

Principles

First law of thermodynamics. At equilibrium at a given temperature, pressure, composition, and values of other intensive and extensive variables, the value of the internal energy U is unique for a system and independent of the path by which the state was reached. The first law states that increments in U are the sum of heat kq flowing into the system and work w done on the system:

$$\Delta U = q + w \tag{1}$$

For an isobaric system,

$$\Delta H = q + w' \tag{2}$$

where w' is work other than that due to $P\Delta V$.

Both q and w' (or w) can be measured. Mechanical or electrical work is a measure of w. Aside from pressure–volume work ($P\Delta V$ in an isobaric system), mechanical stirring and other frictional effects driven from outside the system are included in mechanical work done on the system. Electrical work, w_{el}, is easily done and easily measured.

$$w_{el} = EIt = I^2 Rt = \frac{E^2}{R} t \tag{3}$$

E is the emf that drives a current I through a resistance R in the system for an elapsed time t. Because w_{el} is easily measured it is customary to relate calorimetric experiments ultimately to electrical measurements of work.

Calorimeter isolation. To control or measure q, two extreme procedures are used in practice. In one extreme procedure, the calorimeter is isolated as fully as possible from its surroundings so that heat transfer is minimized; work added, or energy converted by a chemical process, causes a change in temperature of the calorimeter and its contents. Electrical work and the thermodynamic properties of the materials (heat capacity, enthalpy, and enthalpy or internal energy of reaction) are related to one another in terms of the temperature changes that occur in the system and the effective heat capacity of the calorimeter system. Heat flow is measured only in order to make small corrections. In the extreme case of ideal adiabatic calorimetry q is zero. The second extreme procedure has a good, well-defined path for heat flow between the calorimeter system and its surroundings, and a means by which q is measured accurately. This is a conduction-type calorimeter, and q must be measured with the full accuracy expected of the experiment.

Temperature measurement. As a dominant parameter of calorimetry, temperature is measured for several purposes: to establish the temperature with which the process or property measured is to be associated; to measure a temperature increment caused by work done or by a chemical reaction in an isolated calorimeter or by heat flow to or from the calorimeter; to measure temperature differentials that cause heat flow.

The thermodynamic properties of materials are functions of temperature, and their values are unambiguous only when temperatures are assigned to them. The appropriate temperature scale is the thermodynamic temperature scale, for which the kelvin (K) is now internationally accepted as the unit of temperature or temperature difference.

Measurement of work done. In a typical calorimetric experiment, work is done on the calorimeter electrically. The electrical energy to establish the energy equivalent of the calorimeter is provided by a d-c power supply. The electrical power that is dissipated in the calorimeter is measured potentiometrically by measuring the emf across two standard resistors. The emf across a 0.1-Ω resistor is used to measure the heater current; the emf across a 10-Ω resistor, which forms part of a voltage divider of ratio about 1 : 1000 in parallel with the heater, is used to determine the emf across the heater terminals. The time of heating is measured electronically by counting cycles of a reference frequency, using the appearance and disappearance of the power supply voltage itself to trigger the electronic counter.

Applications — Types of Calorimeters

Calorimetry is used to determine the thermodynamic properties of materials and also to measure thermal effects that can be derived from more or less complex devices or physical and chemical processes without special regard to the properties of materials.

From a technological point of view, these properties are valuable in choosing practical manufacturing processes in the chemical-process and metallurgical industries, predicting and optimizing yields of reaction products, making energy balances, choosing materials for particular applications, controlling or monitoring effluent concentrations, and in other ways. Calorimetry also is used directly to determine technologically important properties of complex poorly defined materials. Of principal importance in this area are the heating values of fuels, coal and coke, petroleum products, gaseous fuels, and others. In recent years, the heating values of incinerator refuse and refuse-derived fuels have become of interest because of their potential use in integrated utilities systems (see Fuels from waste). Calorimetry of hazardous and explosive materials has extremely important industrial consequences for the safe handling and transport of chemicals.

Interest in the use of thermal effects as a diagnostic tool has caused increasing development of procedures allowing rapid determinations, such as differential thermal analysis (dta) and differential scanning calorimetry (dsc), with some loss of precision and definition. Calorimetry also is used for purity determination of substances, measurement of laser power and energy, and in investigation of nuclear phenomena. The major growth of commercial instrumentation for calorimetry has occurred where applications in routine analysis and in rapid characterization of materials have been found. Almost all the commercially available calorimeters deal with reacting systems or with physical properties of nonchemical systems as described above. There are few, if any, commercially available instruments for accurately determining thermal properties of nonreacting substances over a wide temperature range.

Nonreacting and reacting systems are treated differently in determination of the properties of materials. The properties determined are often differentiated as thermodynamic (or thermophysical) and thermochemical properties. The boundary between nonreacting and reacting systems may not be sharp. Calorimetry of a typical nonreacting system involves determining its heat capacity and enthalpy over a range of temperatures. This may include heating a substance through several phase transitions. The changes are usually reversible and the substance is still the same substance identified as a single component according to Gibbs' phase rule. The calorimetry of reacting systems is most unambiguous when a chemical reaction occurs in an irreversible way, leading to products that are not readily regained merely by reversing the temperature path. Reaction processes include mixing of gases, mixing of liquids, dissolving gases, liquids, or solids in a liquid, etc.

Calorimetric procedures are not uniquely different or suited for only one kind of process or the other, though certain types of instruments and procedures have been specifically developed and optimized in design to carry out a certain type of process.

Calorimeters for nonreacting systems include isothermal calorimeters, isoperibol calorimeters, adiabatic calorimeters, receiving calorimeters, levitation calorimeters, pulse calorimeters, and modulation calorimeters. Calorimeters for reacting systems include bomb combustion calorimeters, gas-flow combustion calorimeters, solution calorimeters, and microcalorimeters.

Trends in Calorimetry

Differential scanning calorimeters and similar devices are becoming more accurate. They are supplemented by other dynamic procedures that use rapid data-acquisition devices and convey the benefits of the rapid determination of properties. The availability of continuous or very rapid data acquisition to procedures combined with programmed temperature control is encouraging attempts to measure kinetic as well as thermodynamic information in the same procedure. Emphasis on biological materials and processes as subjects of calorimetric study is increasing. Finally, it is reasonable to expect that a revitalization of calorimetric studies will be one response to an increasing awareness of the need for thrift in the use of energy, the desirability of developing new convenient forms of fuels, and an increasing need to utilize material resources that are either completely different or are of lower quality than have been used in the past (see Fuels).

Nomenclature

E = emf
H = enthalpy
I = current
k = Boltzmann constant
  $(1.38 \times 10^{-23} \text{ J/K})$

q = heat flow
t = time
U = internal energy
w = work
w_{el} = electrical work

GEORGE T. ARMSTRONG
ARED CEZAIRLIYAN
National Bureau of Standards

G.T. Armstrong, *J. Chem. Ed.* **41**, 297 (1964).

Standard for Metric Practice, ASTM Designation E 380-82, American Society for Testing and Materials, Philadelphia, Pa., 1982.

C.G. Hyde and M.W. Jones, *Gas Calorimetry*, 2nd ed., Ernest Benn, Ltd., London, UK, 1960, p. 18.

CARBAMIC ACID

Carbamic acid, NH_2COOH, is the hydrated form of isocyanic acid, $HN{=}C{=}O$. It is not known in the free state; hydrolysis rapidly gives ammonia and carbon dioxide. Carbamic acid is the monoamide of carbonic acid; the diamide is the well-known compound urea, $(NH_2)_2CO$, also called carbamide (see Urea). Salts of *N*-substituted dithiocarbamic acids are used as fungicides (qv) and rubber vulcanization accelerators (see Rubber chemicals).

CARBIDES

SURVEY

Although the element carbon is comparatively inert at room temperature, at higher temperatures it forms carbides with most other elements, particularly metals and metal-like elements. Some of these carbides are extremely important in technology; for instance, calcium carbide, CaC_2, the abrasives silicon carbide, SiC, and boron carbide, B_4C and tungsten carbide, WC, titanium carbide, TiC, and tantalum (niobium) carbide, TaC (NbC), the basic carbides of modern cemented carbides. Cementite, Fe_3C, should also be mentioned, as well as the numerous complexes such as $(Co,W)_6C$, $(Cr,Fe,Mo)_{23}C_6$, and $(Cr,Fe)_7C_3$. These are responsible for hardness, wear resistance, and excellent cutting performance in tool steels and Stellite-type alloys (see Boron Compounds; Tungsten and tungsten alloys; Tantalum and tantalum compounds; Niobum and niobium compounds; Titanium and titanium compounds). Carbides are divided into four main groups: the saltlike, metallic, diamondlike, and volatile compounds of carbon (ie, those with hydrogen, nitrogen, oxygen, and sulfur). The important industrial carbides are all stable at high temperatures and thus can be prepared by the direct reaction of carbon with metals or oxides at high temperatures.

Saltlike carbides include almost all carbides of Groups I, II, and III of the periodic system. Metallic carbides include the interstitial carbides of the transition metals of Groups IVB, VB, and VIB, and the carbides of the Group VIIB and Group VIII metals. The metalloconductive carbides P_2C_6 and As_2C_6 also are included. Silicon and boron carbides belong to the diamondlike carbide group; beryllium carbide, with a high degree of hardness, can be included.

RICHARD KIEFFER[‡]
Technical University, Vienna

FRIEDERICH BENESOVSKY[‡]
Metallwerke Plansee A.G.,
Reutle/Tyrol

H. Nowotny and A. Neckel, *J. Inst. Met.* **97**, 1961 (1969).

F. Binder, *Radex Rdsch.*, 531 (1975).

CEMENTED CARBIDES

The term cemented carbides denotes powder-metallurgical products consisting of carbides of Group IVB to VIB metals in a matrix of a metal, usually cobalt or nickel. The preparation of cemented carbides involves the following stages: preparation of the starting materials (metals, oxides, carbon black); preparation of the carbides, and of the carbide solid solutions; solvent ball-milling of the carbides with cobalt powder; pressing the carbide–cobalt mixtures with hydraulic or mechanical presses or by cold isostatic pressing; forming of presintered shapes (double-sintering process) or pressing to the finished shape; high temperature sintering under a protective gas or *in vacuo*; hot-pressing of finished powder batches or isostatic hot-pressing to remove the last traces of porosity for applications where perfection is essential and pores are unacceptable, such as in drawing and Sendzimir mill rolls; and inspection.

The classification of cemented carbides is a controversial subject. It is based on applications, ie, cutting (including machining steel, nonferrous metals or even wood), and wear parts (including mining tools).

The standard grades of cemented carbides account for about 99% of the total, but the special grades fulfill a small, yet important function, and have considerable growth potential. The acid resistance of the WC–Co carbides is unsatisfactory for some applications, and WC can be replaced by TiC or Cr_3C_2, and Co by Ni–Cr, Co–Cr, or Pt.

Recycling of cemented carbide scrap is an important operation. Recently, an ingenious method was initiated by the U.S. Bureau of Mines and developed by Teledyne, Wah Chang Division. The cleaned scrap is heated with molten zinc that attacks and embrittles the cobalt phase. The zinc is distilled *in vacuo* and reclaimed. The treated carbide pieces have the strength of a presintered block and can easily be broken and ball-milled. The resulting reclaimed powder is comparable to virgin powder mixtures (zinc content is less than 20 ppm).

About 50% of the production is used for cutting tools (30% for long-chip, 20% for shot-chip materials) and 50% for wear-resisting

[‡] Deceased.

materials, comprising 25% mining tools and the balance wear parts, dies, and forming tools.

Toxic effect by inhalation of dust can occur with extremely fine carbide and cobalt powders. Therefore, efficient exhaust devices, dust filters, and protective masks are essential; regular medical examinations of workers are recommended.

RICHARD KIEFFER[‡]
Technical University, Vienna

FRIEDRICH BENESOVSKY[‡]
Metallwerke Plansae A.G.,
Reutle Tyrol

Reviewed by
HERBERT S. KALISH
Adamas Carbide Corporation

P. ETTMAYER
Technical University, Vienna

K.J.A. Brookes, *World Directory and Handbook of Hardmetals*, 3rd ed., Engineers Digest Limited and International Carbide Data, Surrey KT22 8AH, UK, 1982.

H.S. Kalish, "Status Report: Cutting Tool Materials," *Met. Prog.* (Nov. 1983).

P. Schwarzkopf and R. Kieffer, *Cemented Carbides*, Macmillan Co., New York, 1960.

INDUSTRIAL HEAVY-METAL CARBIDES

The four most important carbides for the production of hard metals are tungsten carbide, WC, titanium carbide, TiC, tantalum carbide, TaC, and niobium carbide, NbC. In general, the carbides of metals of Groups IVB–VIB are prepared by the action of elementary carbon and hydrocarbons on metals and metal compounds at sufficiently high temperatures. The process may be carried out in the presence of a protective gas, under vacuum, or in the presence of an auxiliary metal (menstruum). Methods include fusion, carburization, reduction, and chemical separation.

Tungsten carbide, mol wt 195.87, mp 2720°C, and the fused W_2C–WC eutectic are important in the mining industry and for wear-resistant welding (eg, hardfacing of oil drills).

Titanium carbide, TiC, mol wt 59.91, mp 2940°C, is an important component of sintered cemented carbides because of its hardness and solvent action for other carbides. Standard cemented carbides for steel cutting contains 5–25%, and the special grades, 30–85% TiC. Next to Cr_3C_2, TiC is the major component for heat- and oxidation-resistant cemented carbides. TiC-based boats, containing AlN, BN, and TiB_2, have been found satisfactory for the evaporation of metals.

Tantalum carbide, TaC, mol wt 192.96, mp 3825°C, is used in small quantities in high pressure lamps and as a barrier between tungsten and graphite in rocket jet nozzles. WC–TiCCo steel grades for long-chip materials have largely been replaced by WC–TiC–TaC(NbC)–Co grades that show higher hot-hardness and better cutting performance. This, as well as the addition of small quantities (0.5–2%) of TaC to straight WC–Co alloys to prevent grain growth, has created an increasing market for TaC.

Niobium carbide has a mol wt of 104.92 and a melting point of 3613°C. The grayish-brown NbC powder is used in increasing quantities in sintered cemented carbides in order to replace TaC by the less expensive NbC. TaC–NbC solid solutions with TiC and WC are used (see under Solid Solutions).

Chromium carbide, Cr_3C_2, mol wt 180.05, mp 1810°C, is used in small quantities in oxidation-resistant hard alloys based on TiC (1–5%). As the main component (60–85% Cr_3C_2) with Ni or Ni–Cu binders, it forms acid- and wear-resistant alloys that have found application in the chemical industry in valve parts. Cast Cr_7C_3 or the eutectic Cr_7C_3–Cr_3C_2 powders are used in rods for hard facing.

Molybdenum carbide, Mo_2C, mol wt 203.91, mp 2520°C, is used in special tungsten-free alloys based on TiC–Mo_2C–Ni(Mo) or in the newly developed spinodal alloys based on $(Ti,Mo)(C,N)_{1-x}$–Ni(Mo). During sintering, the latter alloy shows a spinodal decomposition to an intimate mixture of the phases $(Ti,Mo)C_{1-x}$ and $Ti(C,N)_{1-x}$.

Although vanadium carbide, VC, mol wt 62.96, mp 2684°C, is very hard, it is very brittle and has, therefore, been used only in special cemented carbides. Addition of hafnium carbide, HfC, mol wt 190.51, mp 3820°C, as NbC–HfC or TaC–HfC solid solutions to WC–TiC–Co alloys, improves the properties. Zirconium carbide, ZrC, mol wt 103.23, mp 3420°C, is important as an intermediate formed during the preparation of zirconium from its ores.

Since World War II, the carbides of ^{235}U and thorium have gained importance as nuclear fuels and breeder materials for gas-cooled, graphite-moderated reactors. The actinide carbides include UC, mol wt 250.08, mp 2560°C; UC_2, mol wt 262.09, mp ca 2500°C; ThC, mol wt 244.06, mp 2625°C; and ThC_2, mol wt 256.07, mp 2655°C.

The carbides of iron, Fe_3C (mol wt 179.56, mp 1650°C), nickel, Ni_3C, cobalt, Co_3C, and manganese, Mn_3C, are not classified with the hard metallic materials as they have lower melting points and hardness, and different structures. Nonetheless, these carbides, particularly iron carbide and the double carbides with other transition metals, are of great technical importance as hardening components of alloy steels and cast iron.

Solid Solutions

Pure WC is of paramount importance in straight WC–Co alloys for short-chip materials. Fine-grain-size straight WC inserts perform better than alloy carbides when cutting superalloys even though the latter produce long chips. In all other grades of cemented carbides, binary and ternary solid solutions of the type WC–TiC, WC–TaC(NbC), TiC–TaC(NbC), WC–TiC–TaC(NbC), WC–MoC, and (W,Mo)(C,N) occur in addition to the carbide component WC. In tungsten-free alloys, the solid solutions $(Ti,Mo)C_{1-x}$, $(Ti,Mo,V,Nb)C_{1-x}$, and $(Ti,Mo)(C,N)_{1-x}$ are of growing interest.

The solid solutions of the carbides are generally prepared like pure carbides employing the following methods which can also be modified or combined: carburization of metal oxide mixtures with carbon black, separation of solid solutions of carbides from metal melts (metallic menstruum), and reduction of oxides in the presence of a metal or carbide.

Mixed Phases with Nitrogen, Boron, and Silicon

The chemistry of carbonitrides is attracting increasing attention. Ti(C,N) coatings appear to compete with TiC, TiN, HfN, and Al_2O_3 coatings. The Ni–Mo cemented carbonitrides Ti(C,N) and $(Ti,Mo,V,Nb,Ta)(C,N)_{1-x}$ show good cutting properties. The spinoidal $(Ti,Mo)(C,N)_{1-x}$ carbonitrides are superior to the $(Ti,Mo)C_{1-x}$ based alloys. Solid solutions of (Mo,W)C and (Mo,W)(C,N) may partially replace pure WC.

Complex Carbides

Complex carbides are ternary or quaternary intermetallic phases containing carbon and two or more metals. One metal can be a refractory transition metal and the second also a transition metal or from the iron or A-groups. Nonmetals also can be incorporated.

RICHARD KIEFFER[‡]
Technical University, Vienna

FRIEDRICH BENESOVSKY[‡]
Metallwerke Plansee A.G., Reutle, Tyrol

CALCIUM CARBIDE

Calcium carbide is a transparent colorless solid. The pure material can be prepared only by very special laboratory techniques. Commercial calcium carbide is not a pure chemical compound but is composed of

---

[‡] Deceased.

[‡] Deceased.

calcium carbide (CaC_2), lime (CaO), and other impurities occurring in the coke and lime used in its manufacture. Its calcium carbide content varies, and the commercial material is sold based upon a minimum acetylene yield as specified by U.S. Government regulations. Industrial-grade calcium carbide, sold for generation of acetylene gas, contains about 80% calcium carbide, the remainder being CaO with 2–5% other impurities.

The use of calcium carbide-based acetylene is now almost completely confined to the oxyacetylene welding and metal-cutting markets. Calcium carbide, which is safe and convenient to store and ship, provides an economical supply of acetylene to the metal-working industry. Minor quantities of calcium carbide are used for the desulfurization of steel and iron, a use expected to grow rapidly over the next few years. Table 1 lists the more important physical properties of calcium carbide.

The highly exothermic reaction of calcium carbide and water to give acetylene (qv) is the basis of the most important industrial use of calcium carbide. The second important industrial use of calcium carbide is the production of calcium cyanamide (see Cyanamides). Another use of calcium carbide is developing in the iron and steel industry where it has been found to be a very effective desulfurizing agent for blast furnace iron.

Calcium carbide is produced commercially by the reaction of high purity quicklime with coke in an electric furnace at 2000–2200°C. Material requirements per ton of carbide vary within moderate limits. On the basis of 95% available CaO in the lime and 88% fixed carbon in the coke, the coke-to-lime ratio required technically to produce calcium carbide (80%) is about 0.57, or about 865 kg of lime and 494 kg of coke per metric ton of carbide. Theoretical power requirements per metric ton of calcium carbide are about 2200 kW·h but because of heat losses, about 2800–3100 kW·h is required.

No special health or safety factors are involved in the manufacture of calcium carbide. The major environmental problem is prevention of particulate dust emission which can be handled with cloth filtration equipment (see Air pollution control methods). Treatment of the furnace gas stream is complicated by the high temperature of the gas, its explosiveness, toxicity, the dust concentration, and particle size. Filtra-tion of taphole fumes consisting entirely of submicrometer particles, which rapidly clog the filtration media, is difficult and expensive.

NOEL B. SHINE
Shawinigan Products Dept.
Gulf Oil Chemicals

S.A. Miller, *Acetylene, Its Properties, Manufacture and Uses*, Vol. 1, Academic Press, New York, 1965, p. 475.

J.W. Frye, *50 M.W. Calcium Carbide Furnace Operation*, Electric Furnace Proceedings, Airco Alloys & Carbide, Louisville, Ky., 1970.

Kaess and Vogel, *Chem. Ing. Tech.* **28**, 759 (1956).

SILICON CARBIDE

Silicon carbide, SiC, is a crystalline material, with a color that varies from nearly clear through pale yellow or green to black, depending upon the impurities. It occurs naturally only as the mineral moissanite in the meteoric iron of Canyon Diablo, Arizona.

The commercial product, which is made in an electric resistance furnace, is usually obtained as an aggregate of iridescent crystals. It takes 6–12 kW·h to produce 1 kg of crude, depending on the grade and recovery from furnaces. The loose black or green grain of commerce is prepared from the manufactured product by crushing, purification treatments, and grading for size.

Table 1. Physical Properties of Silicon Carbide

Property	Value	
mol wt	40.10	
decomposition temp,[a] °C		
α-form	2825 ± 40	
β-form	2985	
sp gr, g/cm³ at 20°C		
β-form	3.210	
6H polytype[b]	3.211 (3.208)	
commercial	3.1	
refractive index		
β-form	2.48	
α-form	ε	ω
4H[b]	2.712	2.659
6H	2.69	2.647
15R	2.687	2.650
free energy of formation, ΔG^0, kJ/mol[c]		
α-form	504.1	
β-form	506.2	
heat of formation, ΔH_{298}^0, kJ/mol[c]		
α-form	−25.73 ± 0.63	
β-form	−28.03 ± 2	
thermal conductivity,[d] W/(m·K)		
commercial, high density	4.60	
80% density refractory	24.3	
coefficient of thermal expansion,[e] per °C		
25–200°C	2.97×10^{-6}	
25–600°C	4.27×10^{-6}	
700–1500°C	6.08×10^{-6}	
elasticity		
Young's modulus, GPa[f]		
α-form, hot-pressed	480	
α-form, sintered	410	
β-form, sintered	410	
shear modulus, GPa[f]		
reaction-bonded	167.3	
α-form, sintered	177	
β-form, sintered	140–190	
sublimed	19	

[a] The decomposition products are Si, Si_2C, Si_2, SiC, and Si_3.
[b] H = hexagonal. R = rhombohedral.
[c] To convert J to cal. divide by 4.184.
[d] To convert W/(m·K) to cal/(s·m·°C) divide by 4.184.
[e] Sintered α-form.
[f] To convert GPa to psi, multiply by 145,000.

Table 1. Physical Properties of Calcium Carbide

Property	Value
molecular weight	64.10
melting point, °C	2300
crystal structure	
phase I	face centered tetragonal, 25–447°C
phase II	triclinic, below 25°C
phase III	monoclinic, metastable
phase IV	fcc, above 450°C
commercial	grain structure, 7–120 μm
specific gravity, coml grade	
at 15°C	2.34
at 2000°C (liquid)	1.84
electrical conductivity, tech grade, $(S/cm)^{-1}$	
at 25°C	3000–10,000[a]
at 1000°C	200–1000
at 1700°C (liquid)	0.36–0.47
at 1900°C (liquid)	0.075–0.078
viscosity at 1900°C, Pa·s (P)	
CaC_2 50%	6.0 (60)
CaC_2 87%	1.7 (17)
specific heat,[b] 0–2000°C, J/(mol·K)[c]	74.9
heat of formation, ΔH_{298}, kJ/mol[c]	−59 ± 8
latent heat of fusion, H_f, kJ/mol[c]	32

[a] Depending on impurities.
[b] 100% CaC_2.
[c] To convert J to cal, divide by 4.184.

Silicon carbide may crystallize in the cubic, hexagonal or rhombohedral structure. The properties of silicon carbide depend upon purity, polytype, and method of formation. It is well known as a hard material, and it has a Knoop hardness of 2700 to 2900. Because of its high thermal conductivity and low thermal expansion, silicon carbide is more resistant to thermal shock than other refractory materials. Sax has described silicon carbide as having a slight toxicity by acute or chronic inhalation and an unknown toxicity with respect to acute or chronic systemic reaction. Table 1 (on p. 201) lists the properties of fully compacted, high purity silicon carbide.

In 1977, six firms in the United States and Canada were producing crude silicon carbide under various trade names, including American Metallurgical Products Co., Inc.; The Carborundum Company; Electro-Refractories and Abrasives, Ltd.; The Exolon Co.; General Abrasive Co.; and Norton Co.

The metallurgical, abrasive, and refractory industries are the largest users of silicon carbide. It is also used for heating elements in electric furnaces, in electronic devices, and in applications where its resistance to nuclear radiation damage is advantageous.

Recent developments indicate a growing market for pressureless or reaction-sintered SiC shapes, mostly for high temperature applications. High chemical and wear resistance, high thermal conductivity, moderate electrical resistivity and low thermal expansion, all at high temperatures, make sintered SiC suitable for seals, heat exchangers, components for advanced heat engines, etc.

RICHARD H. SMOAK
S.M. KUNZ
T.M. KORZEKWA
E.D. HOWELL
The Carborundum Co.

R.C. Marshall and co-workers, eds., *Silicon Carbide–1973*, University South Carolina Press, Columbia, S.C., 1974.

E.H. Kraft, *Carborundum Company Internal Report*, The Carborundum Co., Niagara Falls, N.Y., 1977.

CARBOHYDRATES

Carbohydrates are the most abundant class of organic compounds found in living matter. They constitute three-fourths of the dry weight of the plant world and are widely distributed in other life forms. In plants and animals, carbohydrates mainly serve as structural elements and food reserves. Plant carbohydrates, in particular, represent an enormous store of energy, either as human and animal food or after transformation in the geological past, as coal and peat. Large industries process carbohydrates such as sucrose, starch, cellulose, pectin, and certain seaweed polysaccharides. Some carbohydrates and their derivatives have been examined as chemotherapeutic drugs for various pathological conditions such as cancer (see Antibiotics; Chemotherapeutics). Derivatives have been successfully used in biochemical analysis (see Biopolymers; Medical diagnostic reagents).

Although it is not possible to give a simple, yet comprehensive definition of such a broad group of compounds, one fairly good definition describes them as compounds of carbon, hydrogen, and oxygen containing the saccharose group, or its first reaction product, and which usually contain hydrogen and oxygen in the ratio found in water. All monosaccharides and many oligosaccharides are called sugars. Frequently the monosaccharides are called simple sugars. All sugars are readily soluble in water but vary greatly in sweetness.

$$-\underset{\underset{OH}{|}}{C}H-\underset{\underset{O}{\|}}{C}-$$

Structure and Classification

Monosaccharides are the simplest carbohydrates. They are polyhydroxy aldehydes, ketones, or derivatives. An important characteristic of monosaccharides is their ability to undergo glycoside formation in order to yield a mixed acetal. The attached group, termed aglycone, may be a sugar or nonsugar. Treatment of D-glucose with methanol and acid catalyst results in the formation of methyl glycosides. These are commonly methyl D-glucofuranosides or methyl D-glucopyranosides.

Methyl β-D-glucofuranoside   Methyl β-D-glucopyranoside

Instead of reacting with a monohydroxylic compound to form a glycoside, a sugar hemiacetal may react with a hydroxyl group of another sugar to form a disaccharide, and the reaction can be repeated to yield polysaccharides. Disaccharides may be composed of two identical monosaccharides as in maltose, or may be composed of two dissimilar monosaccharides as in sucrose or in lactose (milk sugar), which is made up of a D-galactose and a D-glucose unit.

Oligosaccharides are water-soluble polymers consisting of 2–10 monosaccharide units. Such polymers can be further classified as homopolymers containing only one type of monosaccharide, or heteropolymers containing several different kinds of monosaccharides. Examples of oligosaccharides are stachyose, a naturally occurring tetrasaccharide, maltopentaose, a five-monosaccharide-unit oligosaccharide, and cyclomaltohexaose, a cyclic polysugar (Schardinger dextrin), consisting of six monosaccharide units.

Polysaccharides are large polymers of monosaccharides in a branched or unbranched chain. Cellulose, amylose, and amylopectin are examples of homoglycans, polysaccharides that contain one kind of monosaccharide. Polysaccharides with more than one kind of monosaccharide are known as heteroglycans, those with two different monosaccharide units are called diheteroglycans, and those with three different kinds of sugar units are known as triheteroglycans. Not more than six different monosaccharide units have been found in natural heteroglycans. The main chain of a natural heteroglycan is composed of no more than two different monosaccharide units. Whenever more than two monosaccharide units is found in a polysaccharide, the carbohydrate is branched and often bushlike in structure.

Hydroxyl groups of carbohydrates are readily oxidized to acids, aldehydes, and ketones. The mechanism of oxidation is quite complex and only with lead tetraacetate and periodic acid is the reaction specific. Reduction of aldoses by hydrogen and nickel converts these sugars to polyhydroxy alcohols. Reduction of glucose in this manner yields sorbitol, an effective bacteriostat and food additive (see Alcohols, polyhydric).

Industrial Sugars

Sucrose is the most important industrial sugar, followed by D-glucose and other much less important sugars, eg, lactose, maltose, and fructose. A few sugars are used in the synthesis of medicinal products (see Sugar).

Sucrose is obtained from sugar cane and sugar beets. Other than cane and beet sugar, industrial sugars include liquid sugar, invert sugars, lactose, maltose, and fructose.

Water-Soluble Polysaccharide Gums

Gums (qv), in the form of natural, biosynthetic, or modified polysaccharides, are industrially used in tremendous quantities. Over 500,000 t is consumed annually within the United States where the growth of gum usage exceeds 8–10% per year. Gums seldom constitute an entire finished product. They are mainly used as additives to improve or control the properties of a commodity. The extent of their use results from the low cost of many gums and the important properties they contribute to products in low concentrations. Commercial gums are water-soluble or water-dispersible hydrocolloids. Their aqueous dispersions usually possess suspending, dispersion, and stabilizing properties; or the gums may act as emulsifiers, have gelling characteristics, or be either adhesive or mucilaginous. Some act as coagulants, binders, lubricants, and film formers. The term gum as used in industry refers to plant or microbial polysaccharides or to their derivatives that are dispersible in either cold or hot water to produce viscous mixtures or solutions.

Cellulose

Cellulose, the most abundant polysaccharide, constitutes approximately one-third the weight of annual plants and one-half the weight of perennial plants (see Cellulose). It is a high molecular weight, stereoregular, linear polymer of repeating β-D-glucopyranosyl units. Owing to its availability from rich, continuously replenished sources and its stereoregularity, celluose is a relatively low cost polymer with valuable physical and chemical properties. Its largest use, approximately 45×10^6 t annually, is as wood fiber in the manufacture of paper and paper products, including paperboard (see Paper). Over 400,000 t is used annually in textile fibers. In spite of the increasing use of synthetic fibers, cellulose rayons (see Rayon) and cotton (qv) account for over 70% of textile production.

The purest natural cellulose is the cotton fiber which (on a dry basis) consists of about 98% cellulose, 1% protein, 0.45% wax, 0.65% pectic substance, and 0.15% mineral matter. However, as techniques for isolation and purification have improved, it has been recognized that chemical celluloses derived from wood and other sources can be fully equivalent to cotton cellulose in purity and performance.

Derivatives. Since all D-glucosyl units in the chain, except a single unit at each end, have hydroxyl groups available at carbons C-2, C-3, and C-6, cellulose is trifunctional in alcohol groups that may be etherified or esterified (see Cellulose derivatives). Derivatives include cellulose xanthate, cellulose acetate, cellulose acetate butyrate, cellulose trinitrate, cellulose sulfate, and carboxymethyl cellulose.

In recent years, efforts have been made to modify the properties of cellulose by attaching (grafting) chains of synthetic polymers (see Copolymers). Usually a cellulose is mixed with a monomer, and chain growth is initiated in the mixture by additives that decompose into free radicals, radiation that produces free radicals, or by redox agents. The most successful of the latter have been ceric ions, which abstract hydrogen atoms from hydroxyl groups of cellulose molecules, producing free-radical sites. Initiation of the chain reaction leads to attachment of monomers that continue to grow into long-chain polymer molecules.

Hydrolysis. Cellulose molecules are hydrolyzed under acidic conditions with rapid depolymerization occurring in strong mineral acids. Heterogeneous acid hydrolysis results in a rapid reduction in tensile strength and viscosity because of preferential attacks on those cellulose chains in the more accessible amorphous regions. Following this initial rapid reaction, the chain length approaches a leveling or limiting value. At this stage the resulting microcrystalline particles can be isolated and sold as nonmetabolized food additives. Recently, efforts have been made to hydrolyze cellulose enzymatically with microbial cellulases. Such hydrolysis to D-glucose, if economically practical, could represent a new, large volume method of processing cellulose from woods or annual plants (see Enzymes).

Oxidation. Many oxidants attack the cellulose chain to produce chain cleavage or to insert carbonyl functions.

Starches

Starches are distributed widely throughout the plant world as reserve polysaccharides. They are stored principally in seeds, fruits, tubers, roots, and stem pith (see Starch). They occur as discrete particles or granules of 2–150 µm in diameter. The physical appearance and properties of granules vary widely from one plant to another and may be used to classify original starches. Some granules are round, some elliptical, and some polygonal; many have a spot termed the hilum that is the intersection of two or more lines or creases and is the center from which the granule grew. The granules are anisotropic and show strong birefrin-

gence. Two dark extinction lines extend from edge to edge of the granule intersecting at the hilum. Some granules show a series of striations arranged concentrically around the hilum. Although starches hydrolyze only to D-glucose, they are not single substances. Except in very rare instances, they are mixtures of two structurally different glucans. Most starches contain 22–26% amylose, a linear molecule, and 74–78% amylopectin, a bush-shaped molecule.

Corn starch is the most important of starches manufactured in the United States. Approximately 10.6×10^6 m³ (300×10^6 bu) of corn is processed annually. Most waxy starches are entirely free of amylose and, therefore, consist solely of branched amylopectin molecules. Starch of excellent quality can be prepared from white potatoes. Commercial wheat starch usually is prepared in small quantity in the United States but in large quantities elsewhere. Starches from other sources (cassava, tapioca, etc) are prepared in a manner similar to that for corn or potato starch.

One of the most important properties of starch granules is their behavior on heating with water. Water is at first slowly and reversibly taken up and limited swelling occurs without any perceptible changes in viscosity or birefringence. At a temperature characteristic for the type of starch, the granules undergo an irreversible rapid swelling, losing their birefringence, and with rapid increase in the viscosity of the suspension. Finally, at higher temperatures starch diffuses from some granules and others are ruptured leaving formless sacs. Swelling can be induced at room temperature by numerous chemicals such as formamide, formic acid, chloral, strong bases, and metallic salts. Sodium sulfate and to a lesser degree sodium chloride impede gelatinization.

For many industrial applications the properties of natural starches are changed by various treatments including the action of enzymes, acids, or oxidizing agents on an aqueous suspension of the starch, or by heating essentially dry starch with or without small quantities of acids or alkalies.

Hemicelluloses

Hemicelluloses are a large group of well-characterized polysaccharides found in the primary and secondary cell walls of all land and freshwater plants, and in some seaweeds. Hemicelluloses are made up of a relatively limited number of sugar residues; the principal ones are D-xylose, D-mannose, D-glucose, D-galactose, L-arabinose, D-glucuronic acid, 4-O-methyl-D-glucuronic acid, D-galacturonic acid, and to a lesser extent L-rhamnose, L-fucose, and various O-methylated neutral sugars. Hemicelluloses have many properties similar to the exudate gums. Because of the great abundance of hemicelluloses and the continued price escalation of exudate gums, forced by increases in labor rates for harvesting, hemicelluloses are expected to enter the industrial market in growing volume.

ROY L. WHISTLER
JOHN R. ZYSK
Purdue University

R.L. Whistler, E.F. Paschal, and J.N. BeMiller, eds., *Starch: Chemistry and Technology*, Academic Press, Inc., New York, 1983.

R.L. Whistler and J.N. BeMiller, eds. *Industrial Gums*, Academic Press, Inc., New York, 1973.

CARBON

CARBON AND ARTIFICIAL GRAPHITE

Structure, Terminology, History, and Applications

Carbon occurs widely in its elemental form as crystalline or amorphous solids. Coal, lignite, and gilsonite are examples of amorphous forms; graphite and diamond are crystalline forms. Carbon forms chemical bonds with other elements, but is also capable of forming compounds in which carbon atoms are bound to carbon atoms and, as such, is the chemical element that is the basis of organic chemistry. Carbon and graphite can be manufactured in a wide variety of products with exceptional electrical, thermal, and physical properties. They are a unique family of materials in that properties of the product can be controlled by changes in the manufacturing processes and by raw materials selection.

Elemental carbon exists in nature as two crystalline allotropes: diamond and graphite. There are six electrons in the carbon atom with four electrons in the outer shell available for chemical bonding. Three of the four electrons form strong covalent bonds with adjacent in-plane carbon atoms. The fourth electron forms a less-strong bond of the van der Waals type between the planes. The weak forces between planes account for such properties of graphite as good electrical conductivity, lubricity, and the ability to form interstitial compounds.

Although the terms carbon and graphite are used interchangeably frequently, the two are not synonymous. The terms carbon, manufactured carbon, amorphous carbon, or baked carbon refer to products that result from the process of mixing solid particulates (such as petroleum coke, carbon blacks, or anthracite coal), called fillers, with binder materials of coal tar or petroleum pitch, forming these mixtures by molding or extrusion, and baking the mixtures in furnaces at temperatures from 800–1400°C. The filler is material that makes up the body of the finished product (see Fillers). Green carbon refers to formed carbonaceous material that has not been baked.

Graphite, also called synthetic or artificial graphite, electrographite, manufactured graphite, or graphitized carbon, refers to a carbon product that has been further heat-treated at a temperature exceeding 2400°C and preferably 2800–3000°C.

With the development of nuclear and aerospace technologies, several new forms of carbon and graphite are being produced commercially. Products that are deposited on a heated graphite substrate by vapor-phase decomposition of gaseous hydrocarbons, usually methane, at 1800–2300°C, are termed pyrolytic carbons. Chemical vapor deposited (CVD) carbon or pyrocarbon are other terms used to designate pyrolytic carbons. Pyrolytic graphite is a product resulting from high temperature annealing and has a crystallite interlayer spacing similar to that of ideal graphite (see Ablative materials; Nuclear reactors).

The term polymeric carbon is a generic term for products that result when high polymers with some degree of cross-linking are heated in an inert atmosphere. Vitreous or glassy carbon prepared by carbonization of cellulose and phenolic or polyfurfuryl resins is an example of a polymeric carbon.

There is no standard industry-wide system for designating the various grades of carbon and graphite that are commercially available. The data cited in the following sections are representative average values for commercially available materials.

The most significant event in the development of the carbon and graphite industry was the invention by E.G. Acheson in 1896 of an electric furnace capable of reaching approximately 3000°C, the temperature necessary for graphitization. This permitted the development of improved products which were used in the production of alkalies, chlorine, aluminum, calcium, and silicon carbide, and for electric furnace production of steel and ferroalloys. In 1942, a new application for graphite was found when it was used as a moderator by E. Fermi in the first man-made self-sustaining nuclear chain reaction. This nuclear application and subsequent use in the developing aerospace industries opened new fields of research and new markets for carbon and graphite. Carbon and graphite fibers are an example of a new form and a new industry. A list of main applications is shown in Table 1.

J.C. Long
R.M. Bushong
R. Russell
B.R. Joyce
P.M. Scherer
N.L. Bottone
R.L. Reddy
Union Carbide Corporation

P.L. Walker, Jr., ed., *Chemistry and Physics of Carbon; A Series of Advances*, Vol. 1, Marcel Dekker, New York, 1965.

L.C.F. Blackman, ed., *Modern Aspects of Graphite Technology*, Academic Press, Inc., New York, 1970.

C.L. Mantell, *Carbon and Graphite Handbook*, 3rd ed., Interscience Publishers, a division of John Wiley & Sons, Inc., New York, 1968.

Activated Carbon

Activated carbon, a microcrystalline, nongraphitic form of carbon, has been processed to develop internal porosity. Commercial grades of activated carbon are designated as either gas-phase or liquid-phase adsorbents. Liquid-phase carbons are generally powdered or granular in form; gas-phase, vapor-adsorbent carbons are hard granules or hard, relatively dust-free pellets. Activated carbons are used widely to remove impurities from liquids and gases and to recover valuable substances from gas streams (see Adsorptive separation).

Surface area is the most important physical property of activated carbon. For specific applications, the surface area available for adsorption depends upon the molecular size of the adsorbate and the pore diameter of the activated carbon. Generally, liquid-phase carbons are characterized as having a majority of pores of 3 nm dia and larger, whereas most of the pores of gas-phase adsorbents are 3 nm in diameter and smaller. The most important chemical properties of activated carbon are the ash content, ash composition, and pH of the carbon.

For economic reasons, lignite, coal, bones, wood, peat, and paper-mill waste (lignin) are most often used for the manufacture of liquid-phase or decolorizing carbons, and coconut shells, coal, and petroleum residues are used for the manufacture of gas-adsorbent carbons.

Activation of the raw material is accomplished by two basic processes, depending upon the starting material and whether a low or high density, powdered or granular carbon is desired: (1) Chemical activation depends upon the action of inorganic chemical compounds, either naturally present or added to the raw material, to degrade or dehydrate the organic molecules during carbonization or calcination. (2) Gas activation depends upon selective oxidation of the carbonaceous matter with air at low temperature, or steam, carbon dioxide, or flue gas at high temperature. The oxidation is preceded usually by a primary carbonization of the raw material.

Antipollution laws have increased the sales of activated carbon for control of air and water pollution (see Air pollution control methods). Pollution-control regulations also affect the manufacture of activated carbons. Chemical activating agents have a large potential for the

Table 1. Applications for Manufactured Carbon and Graphite[a]

Aerospace	Metallurgical
nozzles	electric furnace electrodes for
nose cones	the production of iron and
motor cases	steel, ferroalloys, and
leading edges	nonferrous metals
control vanes	furnace linings for blast
blast tubes	furnaces, ferroalloy
exit cones	furnaces, and cupolas
thermal insulation	aluminum pot liners and
Chemical	extrusion tables
heat exchangers and centrifugal	run-out troughs
pumps	for molten iron
electrolytic anodes for the	from blast furnaces
production of chlorine,	and cupolas
aluminum, and other	metal fluxing and inoculation
electrochemical products	tubes for aluminum and ferrous furnaces
electric furnace electrodes for	ingot molds for steel, iron,
making elemental phosphorus	copper, and brass
activated carbon	extrusion dies for copper and
porous carbon and graphite	aluminum
reaction towers and accessories	Nuclear
Electrical	moderators
brushes for electrical motors and	reflectors
generators	thermal columns
anodes, grids, and baffles for mercury	shields
arc power rectifiers	control rods
electronic tube anodes and parts	fuel elements
telephone equipment products	Other
rheostat disks and plates	motion picture projector carbons
welding and gouging carbons	turbine and compressor packing and seal rings
electrodes in fuel cells and batteries	spectroscopic electrodes and powders for
contacts for circuit breakers and	spectrographic analyses
relays	structural members in applications requiring high
electric discharge machining	strength-to-weight ratios

emission of corrosive acid gases during the activation process. If the emissions from chemical activation processes cannot be economically controlled, alternative selective oxidation methods will dominate. Selection of raw materials favoring low-sulfur materials will depend upon the local standards for permissible emission levels of sulfur oxides.

Regeneration of activated carbons is a major factor in the cost effectiveness of the use of carbon. Gas-phase activated carbons used for recovery of valuable process gases and solvents are responsible for the continued cost effectiveness of many industrial processes, particularly solvent-based fiber and tape manufacture, dry cleaning, and rotogravure printing.

To avoid contamination and possible loss of properties, gas-phase activated carbon should not be exposed to vapors or moisture. Although activated carbon is not an easily combustible material, stored carbon should be kept away from heat, electricity, and flames. Use of activated carbon in protective masks does not protect the wearer from an oxygen-depleted atmosphere. An asphyxiation hazard also exists inside activated-carbon vessels because of adsorption of oxygen from the air onto the carbon. Proper ventilation and breathing equipment should be used by persons entering activated-carbon vessels.

Approximately 60% of the activated carbon manufactured for liquid-phase applications is used in powdered form. The two principal uses of powdered activated carbon are removal from solution of color, odor, taste, or other objectionable impurities such as those causing foaming or retarding crystallization, and concentration or recovery from solution of a solute. Miscellaneous uses for liquid-phase carbons include laboratory uses, reaction catalysis, and aquarium water filters.

The largest single application for gas-phase carbon is in gasoline-vapor emission-control cannisters on automobiles. It also is used in purification and separation of gases, in air conditioning (qv) systems to remove industrial odors and irritants as well as body, tobacco, and cooking odors from building air, in nuclear-reactor systems to adsorb radioactive gases in carrier or coolant gases and from emergency exhaust systems, for recovery of volatile organic compounds from process air streams, in cigarette filters to adsorb some of the harmful components of tobacco, and as the catalyst in the manufacturing of phosgene (qv).

R.W. SOFFEL
Union Carbide Corporation

J.W. Hassler, *Purification with Activated Carbon*, Chemical Publishing Co., Inc., New York, 1974.

M. Smisek and S. Cerny, *Active Carbon*, American Elsevier Publishing Co., Inc., New York, 1970.

P.N. Cheremisinoff and F. Ellerbusch, eds., *Carbon Adsorption Handbook*, Ann Arbor Science Publishers, Inc., Ann Arbor, Mich., 1978.

Baked and Graphitized Carbon

The raw materials used in the production of manufactured carbon and graphite largely determine the ultimate properties and practical applications of the finished products. This dependence can be attributed to the nature of carbonization and graphitization processes. Throughout the entire process of the thermal conversion of organic materials to carbon and graphite, the natural chemical driving forces cause the growth of larger and larger fused-ring aromatic systems, and ultimately result in the formation of the stable hexagonal carbon network of graphite. Differences in the final materials depend upon the ease and extent of completion of these overall chemical and physical ordering processes.

In order to produce useful carbon and graphite bodies, filler and binder materials are mixed, formed by molding or extrusion, and finally baked, or baked and graphitized to yield the desired shaped-carbon or graphite bodies. More than 30 different raw materials are used in the manufacture of carbon and graphite products. The primary materials, in terms of tonnage consumption, are the petroleum coke or anthracite coal fillers and the coal-tar or petroleum-pitch binders and impregnants. Other materials, called additives, are often included to improve processing conditions or to modify certain properties in the finished products.

Coke and pitch purchased by carbon companies from suppliers must meet rigid specifications. Some materials, such as pitch and natural graphite, may be used as received from suppliers; other materials, such as raw coke and anthracite, require calcining, a thermal treatment to temperatures above 1200°C. Calcining consists of heating raw filler to remove volatiles and to shrink the filler to produce a strong, dense particle.

L.L. WINTER
Union Carbide Corporation

J.M. Hutcheon in L.C.F. Blackman, ed., *Modern Aspects of Graphite Technology*, Academic Press, Inc., New York, 1970, Chapt. 2.

M.L. Deviney and T.M. O'Grady, eds., *Petroleum Derived Carbons*, American Chemical Society, Washington, D.C., 1976.

Processing of Baked and Graphitized Carbon

In the first step of artificial-graphite production, the run-of-kiln coke is crushed, sized, and milled to prepare it for the subsequent processing steps. The degree to which the coke is broken down depends on the grade of graphite to be made. Roll crushers commonly are used to reduce the incoming coke to particles which are classified in a screening operation.

Generally, the guiding principle in designing carbon mixes is the selection of the particle sizes, the flour content, and their relative proportions in such a way that the intergranular void space is minimized. If this condition is met, the volume remaining for binder pitch and the volatile matter generated in baking are also minimized. The volatile evolution is often responsible for structural and property deterioration in the graphite product. In practice, most carbon mixes are developed empirically with the aim of minimizing binder demand and making use of all the coke passed through the first step of the system. Typically, a coarse-grained mix may contain a large particle (eg, 6 nm dia), a small particle half this size or smaller, and flour. In this formulation, approximately 25 kg of binder pitch would be used for each 100 kg of coke.

The manufacturing process begins with the mixing operation, the purpose of which is to blend the coke filler materials and to melt and distribute the pitch binder over the surfaces of the filler grains and structural integrity of the graphite. Thus, the more uniform the binder distribution is throughout the filler components, the greater the likelihood for a structurally sound product.

Following the mixing operation, the hot mix must be cooled to a temperature slightly above the softening point of the binder pitch at which point it is ready to be charged into an extrusion press or mold. One purpose of the forming operation is to compress the mix into a dense mass so that the pitch-coated filler particles and flour are in intimate contact. Another purpose of the step is to produce a shape and size as near that of the finished product as possible.

The next stage is the baking operation during which the product is fired to 800–1000°C. One function of this step is to convert the thermoplastic pitch binder to solid coke. Another function of baking is to remove most of the shrinkage in the product associated with pyrolysis of the pitch binder at a slow heating rate. This procedure avoids cracking during subsequent graphitization where very fast firing rates are used. The conversion of pitch to coke is accompanied by marked physical and chemical changes in the binder phase, which if conducted too rapidly, can lead to serious quality deficiencies in the finished product. For this reason, baking is regarded as the most critical operation in the production of carbon and graphite.

In some applications, the baked product is taken directly to the graphitizing facility for heat treatment to 3000°C. However, for many high performance applications of graphite, the properties of stock processed in this way are inadequate. The method used to improve those properties is impregnation with coal-tar or petroleum pitches. Upon carbonization the impregnant deposits additional coke in the open pores of the baked stock, thereby improving properties of the graphite product. Many nuclear and aerospace graphites are multiply pitch-treated to achieve the greatest possible assurance of high performance.

Graphitization, an electrical heat treatment of the product to ca 3000°C, causes the carbon atoms in the petroleum coke filler and pitch coke binder to orient into the graphite lattice configuration. This ordering process produces graphite with the intermetallic properties that make it useful in many applications.

In the temperature range of 1500–2000°C, most petroleum cokes undergo an irreversible volume increase known as puffing. This effect has been associated with thermal removal of sulfur from coke and increases with increasing sulfur content. Because of the recent emphasis on the use of low sulfur fuels, many of the sweet crudes that have been used as coker feeds are now being processed as fuels. Desulfurization of the sour crudes available for coking is possible but expensive. The result is an upward trend in the sulfur content of many petroleum cokes, leading to greater criticality in heating rate in the puffing-temperature range during graphitization.

E.L. PIPER
Union Carbide Corporation

W.P. Eatherly and E.L. Piper in R.E. Nightingale, ed., *Nuclear Graphite*, Academic Press, Inc., New York, 1962, Chapt. 2.

Properties of Manufactured Graphite

The graphite crystal, the fundamental building block for manufactured graphite, is one of the most anisotropic bodies known. Properties of graphite single crystals illustrating this anisotropy are shown in Table 1. Anisotropy is the direct result of the layered structure with extremely strong carbon–carbon bonds in the basal plane and weak bonds between planes. The anisotropy of the single crystal is carried over in the properties of commercial graphite, although not nearly to the same degree. By the selection of raw materials and processing conditions, graphites can be manufactured with a wide range of properties and degree of anisotropy.

Table 1. Room-Temperature Properties of Graphite Crystals

Property	Value in basal plane	Value across basal plane
resistivity, $\Omega \cdot m$	40×10^{-4}	ca 6000×10^{-4}
elastic modulus[a], TPa	0.965	0.034
tensile strength (est)[a], TPa	0.096	0.034
thermal conductivity, W/(m·K)	ca 2000	10
thermal expansion, °C^{-1}	-0.5×10^{-6}	27×10^{-6}

[a] To convert TPa to psi, multiply by 1.45×10^8.

Manufactured graphite is a composite of coke aggregate (filler particles), binder carbon, and pores. Most graphites have a porosity of 20–30%, although special graphites can be made that have porosity well outside this range. Manufactured graphite is a highly refractory material that has been thermally stabilized to as high as 3000°C. At atmospheric pressure, graphite has no melting point but sublimes at 3850°C, the triple point being approximately 3850°C and 12.2 MPa (120 atm). The strength of graphite increases with temperature to 2200–2500°C; above 2200°C, graphite becomes plastic and exhibits viscoelastic creep under load. Graphite is a valuable structural material because it has high resistance to thermal shock, does not melt, and possesses structural strength at temperatures well above the melting points of most metals and alloys. For many applications of graphite, one or more of the following characteristics is important: density, elastic modulus, mechanical strength, electrical and thermal conductivity, and thermal expansion.

J.T. MEERS
Union Carbide Corporation

B.T. Kelley, *Physics of Graphite*, Applied Science Publishers, Englewood, N.J., 1981.

R.E. Nightingale, H.H. Yoshikawa, and H.W. Losty in R.E. Nightingale, ed., *Nuclear Graphite*, Academic Press, Inc., New York, 1962, Chapt. 6.

Carbon Fibers and Fabrics

Carbon fibers are filamentary forms (fiber dia 5–15 μm) of carbon (carbon content exceeding 92 wt%) and are characterized by flexibility, electrical conductivity, chemical inertness except to oxidation, refractoriness, and in their high performance varieties, high Young's modulus and high strength. Considerable confusion exists over the terms carbon and graphite fibers. The term graphite fibers should be restricted to materials with the three-dimensional order characteristic of polycrystalline graphite; essentially all commercial fibers are carbon fibers.

Most desirable properties of carbon fibers depend on the degree of preferred orientation. In a high modulus fiber, the carbon-layer planes are predominantly parallel to the fiber axis; however, when viewed in cross section, the layers in most carbon fibers are oriented in all directions, although mesophase pitch-based fibers with concentric layers (onionskin structure) and with layers radiating from the fiber axis (radial structure) have been made. The high strength, high modulus properties are found only along the fiber axis. The electrical and thermal conductivity also increases with increased orientation. Tensile strength of highly oriented carbon fibers should, in theory, be approximately 5% of the Young's modulus. The useful tensile strength of commercial carbon yarn rarely exceeds 4.5 GPa (650,000 psi). Interlamellar shear strength of carbon-fiber composites tends to decrease with increasing fiber modulus.

Rayon-based carbon fibers. The production process for rayon-based carbon yarn and cloth involves three distinct steps: preparation and heat treating, carbonization, and optional high temperature heat treatment. These carbon fibers are used usually in the form of cloth for aerospace ablative applications. Through 1975, the volume of rayon-based, low modulus yarn and cloth probably exceeded that of all other carbon fibers combined.

PAN-based carbon fibers. Production of PAN-based (polyacrylonitrile) carbon fibers also involves the three steps of low temperature heat treatment, carbonization, and optional high temperature heat treatment. The principal differences from the rayon-based fiber process are that a well-oriented ladder polymer structure is developed during oxidative heat treatment under tension, and the orientation is essentially maintained through carbonization.

Although official production statistics are unavailable, it is estimated that the combined production in the three principal producing countries (United States, Japan, and United Kingdom) may have risen to 1000 t in 1981 (see Novoloid fibers). Most PAN-based carbon fibers are used in epoxy "prepregs" from which structures are fabricated. Composites are used in major flight-critical aircraft structures, in golf clubs, fishing rods, tennis rackets, bows and arrows, skis, and sailboat masts and spars. Applications for textile and computer machinery, automotive and general transportation, and musical instruments are emerging rapidly (see Composite materials). Carbon fibers, like other forms of carbon, are inert. For this reason they are now used in medical research for body implant studies such as pacemaker leads, replacement of tendons, etc (see Prosthetic and biomedical devices).

Mesophase pitch-based fibers. The process for these fibers starts with commercial coal-tar or petroleum pitch that is converted through heat treatment into a mesophase or liquid-crystal state. Since the mesophase is highly anisotropic, the shear forces acting on the pitch during spinning and drawing result in a highly oriented fiber, with the essentially flat aromatic polymer molecules oriented parallel to the fiber axis. The spun yarn is then thermoset in an oxidizing atmosphere which renders the fibers infusible and amenable to rapid carbonization. The original preferred orientation is maintained during thermosetting.

Mesophase pitch-based fibers are produced in three forms: low modulus mat used for nonstructural applications, eg, to provide electrical conductivity and/or good friction and wear characteristics to plastic parts; carbon-fiber fabrics that present an attractive, low cost alternative to rayon-based cloth; and continuous filament yarn used in stiffness-critical applications.

H.F. VOLK
Union Carbide Corporation

M. Sittig, ed., *Carbon and Graphite Fibers; Manufacture and Applications*, Noyes Data Corporation, Park Ridge, N.J., 1980.

J. Delmonte, *Technology of Carbon and Graphite Fiber Composites*, Van Nostrand Reinhold Co., New York, 1981.

G. Lubin, ed., *Handbook of Composites*, Van Nostrand Reinhold Co., New York, 1982.

Other Forms of Carbon and Graphite

Carbon and graphite foams have been produced from resinous foams of phenolic or urethane base by careful pyrolysis to preserve the foamed cell structure in the carbonized state. The foams are attractive as high temperature insulating packaging material in the aerospace field and as insulation for high temperature furnaces (see Insulation, thermal). Variations of the resinous-based foams include the syntactic foams where cellular polymers or hollow carbon spheres comprise the major volume of the material bonded and carbonized in a resin matrix.

The carbon commercially produced by the chemical vapor deposition (CVD) process is referred to as pyrolytic graphite. The material is not true graphite in the crystallographic sense, and wide variations in properties occur as a result of deposition methods and conditions. Commercial applications for pyrolytic graphite include rocket nozzle parts, nose cones, laboratory ware, and pipe liners for smoking tobacco. Pyrolytic graphite coated on surfaces or infiltrated into porous materials also is used in other applications such as nuclear-fuel particles and prosthetic devices (qv).

A special form of pyrolytic graphite is produced by annealing under pressure at temperatures above 3000°C. This pressure-annealed pyrolytic graphite exhibits the theoretical density of single-crystal graphite; and although the material is polycrystalline, the properties of the material are close to single-crystal properties. The highly reflective, flat faces of pressure-annealed pyrolytic graphite have made the material valuable as an x-ray monochromator (see X-ray technology).

When carbon is produced from certain nongraphitizable carbonaceous materials, the material resembles a black glass in appearance and brittleness; hence the terms glassy carbon or vitreous carbon. Nongraphitizing carbons are obtained from polymers that have some degree of cross-linking, and it is believed that the presence of these cross-linkages inhibits the formation of crystallites during subsequent heat treatments. By pyrolyzing polymers such as cellulosics, phenol-formaldehyde resins, and poly(furfuryl alcohol) under closely controlled conditions, glassy carbon is produced. These carbons are composed of random crystallites, of the order of 5.0 nm across, and are not significantly altered by ordinary graphitization heat treatment to 2700°C.

The properties of glassy carbon are quite different from those of conventional carbons and graphites. The density is low (1.4–1.5 g/cm³), but the porosity and permeability also are quite low. The extreme chemical inertness of glassy carbon, along with its impermeability, makes it a useful material for chemical laboratory glassware, crucibles, and other vessels. It has found use as a susceptor for epitaxial growth of silicon crystals and as crucibles for growth of single crystals.

Carbon and graphite have been produced for many years in roughly spherical shape in small sizes (1–2 mm dia). Globular carbon of this type was used in telephone receivers in the mid-1930s. Carbon of nearly spherical shape can be produced by mechanically tumbling or rolling a mixture of finely divided carbon or graphite with a resinous binder, screening to select the desired size balls, and carefully carbonizing the selected material. Usage of these nearly spherical carbon balls has been limited to specialized applications where flowability was at a premium or uniform contact area was important. Shim particles for use in nuclear reactor fuel applications are examples.

A new type of carbon sphere, usually hollow, has been made from pitch. The spheres are of uniform, very nearly spherical shape and are graphitizable. The particle density is usually 0.20–0.25 g/cm³, and consequently, attention has been given to application in syntactic carbon foam for high temperature insulation and for lightweight composite structures. The spheres can be activated to produce an adsorptive material of high flow-through capacity.

R.M. BUSHONG
Union Carbide Corporation

G.M. Jenkins and K. Kannenmura, *Polymeric Carbons—Carbon Fibre, Glass, and Char*, Cambridge University Press, New York, 1976.

A.W. Moore, *Chem. Phys. Carbon* **17**, 233 (1981).

CARBON BLACK

Carbon black is an important member of the family of industrial carbons. Its various uses depend on chemical composition, pigment properties, state of subdivision, adsorption activity, and other colloidal properties. The basic process for manufacturing carbon black has been known since antiquity. The combustion of fuels with insufficient air produces a black smoke containing extremely small carbon-black particles which, when separated from the combustion gases, comprise a fluffy powder of intense blackness. The term carbon black refers to a wide range of such products made by partial combustion or thermal decomposition of hydrocarbons in the vapor phase, in contrast to cokes and chars that are formed by the pyrolysis of solids. Whether partial combustion or thermal decomposition methods are used, the basic reaction is represented by:

$$C_xH_y \xrightarrow{\Delta} xC + \frac{y}{2}H_2$$

Carbon exists in two crystalline forms and numerous so-called amorphous, less-ordered forms. The crystalline forms are diamond and graphite, and the less-ordered forms are mainly cokes and chars. All forms of industrial carbons other than diamond and graphite, including carbon black, can be classified as amorphous carbons characterized by degenerate or imperfect graphitic structures.

Carbon blacks differ in particle size or surface area, average aggregate mass, particle and aggregate mass distributions, morphology or structure, and chemical composition. The form of these products, loose or pelleted, is another feature of some special grades. The ultimate colloidal units of carbon black, or the smallest dispersible entities in the elastomer, plastic, and fluid systems are called aggregates, fused assemblies of particles. The particle size is related to the surface area of the aggregates. The oil-furnace process produces carbon blacks with particle diameters of 10–250 nm and the thermal black process, 120–500 nm.

A primary aggregate is further characterized by its size, the volume of carbon comprising the aggregate, and its morphology. Aggregates have a variety of structural forms. Some are clustered like a bunch of grapes, whereas others are more open, branched, and filamentous. The three most important properties used to identify and classify carbon blacks are surface area, structure, and tinting strength.

Most commercial rubber-grade carbon blacks contain over 97% elemental carbon. A few special pigment grades have carbon contents below 90%. In addition to chemically combined surface oxygen, carbon blacks contain varying amounts of moisture, solvent-extractable hydrocarbons, sulfur, hydrogen, and inorganic salts. Hygroscopicity is increased by surface activity, high surface area, and the presence of salts. Extractable hydrocarbons result from the adsorption of small amounts of incompletely burned hydrocarbons. The combined sulfur content of carbon black has its origin in the sulfur content of the feedstocks. Most of the inorganic salt content comes from the water used for quenching and pelletizing.

Manufacture

The discovery of rubber reinforcement by carbon black about 1912 and the growth of the automobile and tire industries transformed carbon black from a small volume specialty product to a large volume basic industrial raw material. Today's carbon black production is almost entirely by the oil-furnace black process, which was introduced in the United States during World War II. However, the unique relationship of manufacturing process to special performance features of products has prevented the total abandonment of the older processes. The thermal, lamp black, channel black, and acetylene black processes account for less than 4% of the world's production. The easy sea and land transportation of feedstocks for the oil-furnace process has made possible worldwide manufacturing facilities convenient to major consumers. The remarkable success of the oil-furnace process was due to its improved yield performance in the range of 50–70%, higher capacities, and its ability to produce a wide range of products.

The feedstocks (qv) used by the carbon-black industry are viscous, residual aromatic hydrocarbons consisting of branched polynuclear aromatic types mixed with smaller quantities of paraffins and unsaturates. Preferred feedstocks are high in aromaticity, free of coke or other gritty materials, and contain low levels of asphaltenes, sulfur, and alkali metals. As a result of the introduction of more efficient petroleum-cracking catalysts, decant oil from gasoline production has now become the most important feedstock in the United States, displacing residual thermal tars and catalytic cycle-stock extracts.

The reactor feeds are preheated feedstock, preheated air, and gas. Carbon-black properties, yield, and production rate can be adjusted by controlling reactor feed variables. The carbon-containing combustion products are quenched with water sprays and pass through heat exchangers that preheat the primary combustion air. The production stream is again cooled by a secondary water quench in vertical towers. Carbon black in light fluffy form is separated in bag filters and conveyed to micropulverizers discharging into a surge tank. From the surge tank, the carbon black is fed to dry drums for dry pelletization or pin-type wet pelletizers, followed by dryers, to produce pelleted products (see Size enlargement). Some of the special carbon-black grades used for pigmentation are lightly densified and shipped as "loose" or "fluffy" products.

Most of the carbon black used by the rubber industry is wet pelletized. Dry pelletization is preferred for some rubber applications and for pigmentation. In the dry process the slightly compacted loose black is conveyed to large rotating steel drums whose rolling motion causes a steady increase in bulk density and the formation of pellets. A fraction of the pelletized product is recirculated and added to the loose black at the drum entry to increase the density of the bed and provide seed material to increase the formation rate.

At least two different types of reactors are required to make all the furnace grades for the rubber industry: one for the reinforcing and another for the less reinforcing grades. Additional types are used to produce special pigment grades. During the last decade, reinforcing furnace-black reactors have undergone substantial design changes to give higher gas velocities and turbulence, more rapid mixing of reactant gases and feedstock, and increased capacity. Some of the latest reactor designs have refractory-lined metal construction, heat-recovery chambers for preheating combustion air, increased thermal efficiency, more efficient feedstock atomization, and improved quenching.

During the early 1970s, the industry adopted improved reactor and burner designs that provided more rapid and uniform feedstock atomization, more rapid mixing, and shorter residence times. Higher capacities and increases of 6–8% in yields for products rated at equivalent roadwear levels have been reported. During the late 1970s, the demands of the automobile industry for lower rolling resistance tires have resulted in the development of new carbon-black grades for tire treads. These grades are used to make lower hysteresis tread compounds which reduce tire rolling resistance without sacrificing tread-wear performance. The new tread grades have broader aggregate size distributions and lower tinting strengths and give evidence of higher surface activity.

Carbon black manufacturing processes. The high temperature decomposition of hydrocarbons in the absence of air or flames is the basis for the manufacture of thermal blacks and acetylene black. Thermal black is made by a strongly endothermic reaction requiring a large heat-energy input, whereas the acetylene-black decomposition reaction is strongly exothermic. Thermal-black production has been drastically curtailed in recent years, and thermal black now has only limited use as a functional filler for specially engineered rubber and plastic products (see Fillers).

Acetylene black is made by a continuous decomposition process at 800–1000°C in water-cooled metal retorts lined with refractory. The process is started by burning acetylene and air to heat the retort to reaction temperature, followed by shutting off the air supply to allow the acetylene to decompose to carbon and hydrogen in the absence of air. A major use for acetylene black is in dry-cell batteries because it contributes low electrical resistance and high capacity. In rubber it gives electrically conductive properties to heater pads, heater tapes, antistatic belt drives, conveyor belts, and shoe soles. It also is used in electrically conductive plastics. Many of these electrical applications are now also served by newly developed electrically conductive grades of furnace blacks. Some applications of acetylene black in rubber depend on its contribution to improved thermal conductivity, such as rubber curing bags for tire manufacture.

Early lampblack processes used large open shallow pans from 0.5 to 2 m in diameter and 16 cm deep for burning various oils in an enclosure with a restricted air supply. The smoke from the burning pans was allowed to pass at low velocities through settling chambers from which it was cleared by motor-driven ploughs. Today lampblack substitutes are made by the furnace process. Traditional lampblack manufacture is still practiced, but the quantities produced are small.

The following trends in carbon-black manufacture can be expected: increasing use of by-product pyrolysis tars from ethylene manufacture for feedstocks; replacement of natural gas by liquid hydrocarbons for the primary heat source; increased sensible-heat recovery from the combustion stream; increased use of tail gas for fuel in the primary combustion zone; use of higher sulfur content feedstocks; and increasing use of automatic computer control.

Carbon-black consumption by the rubber industry has grown rapidly from the time of the discovery of its outstanding reinforcing properties by S.C. Motte in Great Britain about 1912.

Carbon is a relatively stable, unreactive element, insoluble in organic or other solvents. Carbon in the form of char has been used in the pharmaceutical industry for many years. Recently, vitreous and pyrolytic carbons have been found to be biocompatible and useful for artificial heart valves and other medical applications (see Prosthetic and biomedical devices). There is no evidence of carbon-black toxicity in humans despite the fact that it contains trace amounts of some polynuclear aromatic compounds known to be carcinogenic.

Owing to the low thermal conductivity of bulk carbon black, storage fires may go undetected for some time. Fires may be controlled by purging with carbon dioxide. When this is done, respirators must be used to avoid breathing carbon monoxide. Even in the absence of a fire, an air line or adequate respirator should be used when entering any confined carbon-black storage facilities to avoid breathing possible toxic gases or low oxygen atmosphere.

The combustion of fuels can cause serious problems of atmospheric pollution from the emission of particulate material, sulfur compounds, nitric oxides, hydrocarbons, and other gases. Because of the intense blackness of carbon black, it cannot be allowed to escape to the atmosphere even in minute quantities. Huge bag filters for complete recovery were put into operation over 35 years ago (see Air pollution control methods).

Uses

The general effect of different carbon blacks on rubber properties is dominated mainly by surface area and aggregation or structure. High surface area and small particle size impart higher levels of reinforcement as reflected in tensile strength, tear resistance, and resistance to abrasive wear with resulting higher hysteresis and poorer dynamic performance. Higher structure gives improved extrusion behavior, higher stock viscosities, improved green strength, and higher modulus values. Rubber property changes produced by carbon-black addition depend on loading. Tensile strength and abrasion resistance increase with increased loading to an optimum and then decrease. The optimum is normally in the range of 40–60 phr (parts per hundred or rubber). Increasing the loading level to 35–80 phr produces linear increases in hardness and modulus. The magnitude of these changes increases with structure.

About 8–9% of the total carbon black production consists of special industrial carbon grades for nonrubber applications (special blacks). Of increasing importance in recent years have been applications in plastics where special carbon blacks are required to improve weathering resistance, or to impart antistatic and electrically conductive properties (see Antistatic agents). About 35% of the special blacks is used in printing inks, 8% in paints and lacquers, 33% in plastics, 1% in paper, and 23% in miscellaneous applications. News inks account for most of the printing-ink market. For this use, special N300 series (HAF) grades containing 6% mineral oil are made to give rapid dispersion. Medium and high color blacks, the most expensive grades, are used in enamels, lacquers, and plastics for their intense jetness.

The low thermal conductivity of carbon black makes it an excellent high temperature insulating material. For high temperature insulation up to 3000°C, it must be maintained in an inert atmosphere to prevent oxidation. Thermal-grade carbon black has been most widely used. Carbon black also is a source of pure carbon both for ore reduction and carburizing. Carbon brushes and electrodes are fabricated from carbon black. It is used as well as a pigment in the cement industry, in linoleum, leather coatings, polishes, and plastic tile.

ELI M. DANNENBERG
Cabot Corporation

J.B. Donnet and A. Voet, *Carbon Black Physics, Chemistry and Elastomer Reinforcement*, Marcel Dekker, Inc., New York, 1976.

P.H. Johnson and C.R. Eberline in *Encyclopedia of Chemical Processing and Design*, Vol. 6, Marcel Dekker, Inc., New York, 1977, pp. 187–257.

J. Janzen, *Rubber Chem. Technol.* **55**, 669 (1982).

D. Rivin and R.G. Smith, *Rubber Chem. Technol.* **55**, 707 (1982).

DIAMOND, NATURAL

Diamond, the high pressure allotrope of carbon, changes to graphite, the high temperature allotrope of carbon, when heated above 1500°C *in vacuo*. Diamond is metastable and chemically inert at moderate temperatures. Although diamond is the hardest known substance, it is quite brittle and is readily crushed to grit and powder, the form in which approximately 75% of all industrial diamond is used. These properties make diamond of ever-increasing importance in industry and technology as fixed abrasive in saws, drills, and grinding wheels. In the form of single crystals (stones), it is used in turning and boring tools, wire dies, indenters, tools to shape and dress conventional abrasive wheels, and in drill crowns. Those diamonds of near flawless quality having colors of blue, blue-white, water-white, pink, green, or a slight tinge of yellow are of greatest value as gems. Approximately 20% of mined diamonds are classified as gems.

Diamonds are found in ancient volcanic pipes in relatively soft, dark, basic periotidite rock called blue ground or kimberlite. As a result of erosion of the volcanic pipes, diamonds are also recovered from alluvial gravels and marine terraces.

Diamond is composed of the single element carbon and has a cubic crystal structure of space group 0_h^7. Diamonds are classified into four groups based on their optical and electrical properties as types 1a, 1b, 11a, and 11b. Typical physical properties of natural diamond are listed in Table 1.

H.C. MILLER
Super-Cut, Inc.

E. Burton, *Diamonds*, N.A.G. Press Ltd., London, UK, 1970.

P. Grodzinski, *Diamond Technology*, 2nd ed., N.A.G. Press Ltd., London, UK, 1953.

Table 1. Physical Properties of Natural Diamond

Property	Value
density,[a] g/cm³, 25°C	3.51524 ± 0.00005
thermal conductivity, W/(cm·K), 20°C	
Type 1	9
Type 11a	26
maximum at 190°C	
Type 1	24
Type 11a	120
thermal expansion coefficient at 20°C	$0.8 \pm 0.1 \times 10^{-6}$
specific heat (constant volume) at 20°C, J/(mol·K)[b]	6.184
refractive index at 546.1 nm (Hg green), μm	2.4237
dielectric constant, at 0–3 kHz, 27°C	5.58 ± 0.03
optical transparency	
Type 11a diamond	225 nm to 2.5 μm > 6 μm
Type 1 diamond	340 nm to 2.5 μm > 10 μm
resistivity, 20°C, Ω·cm	
Type 1 and most Type 11a	$> 10^{16}$
Type 11b	$10–10^3$
hardness, Mohs scale (scratch hardness)	10
Knoop scale (indention hardness), kgf/mm²[c]	5,700–10,400
compressive strength, GPa²[d]	
average (flawless octahedral diamond)	8.68
maximum	16.5
modulus of elasticity, GPa[d]	1160
Young's modulus, GPa[d]	1160

[a] Average of 35 diamonds.
[b] To convert J to cal, divide by 4.184.
[c] To convert kgf/mm² to psi, multiply by 1422.
[d] To convert GPa to psi, multiply by 145,000.

DIAMOND, SYNTHETIC

Catalyzed Synthesis

In this process, a mixture of carbon (eg, graphite) and catalyst metal is heated high enough to be melted while the system is at a pressure high enough for diamond to be stable. Graphite is then dissolved by the metal, and diamond is produced from it. Effective catalysts are Cr, Mn, Fe, Co, Ni, Ru, Rh, Pd, Os, Ir, Pt, and Ta, and their alloys and compounds. If the metal is not molten, graphite is obtained instead of diamond even at pressures high enough to produce diamond. (The exception is tantalum which does not have to be molten to be effective.)

Many kinds of apparatus have been devised for simultaneously producing the high pressures and temperatures necessary for diamond synthesis. An early, successful design is the belt apparatus, in which two opposed, conical punches, made of cemented tungsten carbide and carried in strong steel binding rings, are driven into the ends of a short, tapered chamber that also is made of cemented tungsten carbide supported by strong steel rings. A compressible gasket, constructed in a sandwich fashion of stone, usually pyrophyllite, and steel cones, seals the annular gap between punch and chamber, distributes stress, provides lateral support for the punch, and permits axial movement of the punches to compress the chamber contents. The reaction zone, usually a cylinder, is buried in pyrophyllite stone in the chamber. The pyrophyllite, a good thermal and electrical insulator, is easily machined and transmits pressure fairly well. The reaction zone is heated electrically with a heavy current.

Size, shape, color, and impurities are dependent on the conditions of synthesis. Lower temperatures favor dark-colored, less pure crystals; higher temperatures promote paler, purer crystals. Low pressures (about 5 GPa or 50 kbar) favor the development of the cube faces, whereas higher pressures produce octahedral faces. Nucleation and growth rates increase rapidly as the process pressure is raised above the diamond-graphite equilibrium pressure.

If diamond seed crystals are placed in the active diamond-growing zone of a typical graphite-catalyst metal apparatus, new diamond usually forms on the seed crystals. However, the new growth tends to be uneven in the thickness and quality, with gaps or inclusions of foreign material. Such defects probably appear because the main driving force for the nucleation and growth under these conditions is the Gibbs free energy difference between diamond and graphite, which is a function of pressure, temperature, and composition; none of these variables can be sufficiently controlled. However, excellent growth can be obtained if pressure and composition are held relatively constant while the change of composition with temperature is employed as a driving force.

Direct Graphite-to-Diamond Process

In this process, diamond forms from graphite without a catalyst. The refractory nature of carbon demands a fairly high temperature (2500–3500 K) for sufficient atomic mobility for the transformation, and the high temperature in turn demands high pressure (above 12 GPa; 120 kbar) for diamond stability. The combination of high temperature and pressure may be achieved statically, or dynamically using shock waves. During the course of experimentation on this process, a new form of diamond with a hexagonal (wurtzitic) structure was discovered.

A diamond prepared by the direct conversion of well-crystallized graphite, at pressures of about 13 GPa (130 kbar), shows certain unusual reflections in the x-ray-diffraction patterns. They could be explained by assuming a hexagonal diamond structure (related to wurtzite) with $a = 0.252$ and $c = 0.412$ nm, space group $P6_3/mmc\text{-}D_{6h}^4$ with 4 atoms per unit cell. The calculated density would be 3.51 g/cm³, the same as for ordinary cubic diamond, and the distances between nearest-neighbor carbon atoms would be the same in both hexagonal and cubic diamond, 0.154 nm.

Metastable Vapor-Phase Deposition

Metastable growth of diamond takes place at low pressures where graphite is thermodynamically stable. The subject has a long history, and work in the United States and the USSR indicates that diamond may form at moderate pressures during decomposition of gases such as methane.

The Synthetic-Diamond Industry

Soon after the first successful diamond syntheses by the catalyst process, small batches of crystals were prepared in the laboratory. When mounted in abrasive wheels, these first synthetic diamonds performed better than comparable natural diamonds for shaping of ceramics and cemented tungsten carbide. A pilot plant for producing synthetic diamond was established, the efficiency of the operation was increased, production costs declined, and product performance was improved. Today, the price of General Electric man-made diamond is competitive with natural-diamond prices. General Electric is the major producer of synthetic industrial diamonds.

With the exception of the rare, natural type IIb, diamond is normally a good electric insulator. However, semiconducting diamonds are prepared by adding small amounts of boron, beryllium, or aluminum to the growing mixture, or by diffusing boron into crystals at high pressures and temperatures.

Some natural diamonds known as carbonado or ballas occur as tough, polycrystalline masses. The production of synthetic sintered diamond masses of comparable excellent mechanical properties has only been achieved recently. Natural single-diamond crystals and carbonado can now be replaced in many industrial uses by sintered-diamond tool blanks.

R.H. WENTORF, JR.
General Electric Co.

F.P. Bundy, H.M. Strong, and R.H. Wentorf, Jr., in P.L. Walker and P.A. Thrower, eds., *Chemistry and Physics of Carbon*, Vol. 10, Marcel Dekker, Inc., New York, 1973, pp. 213–263.

F.P. Bundy and J.S. Kasper, *J. Chem. Phys.* **46**, 3437 (1967).

L.E. Hibbs, Jr., and R.H. Wentorf, Jr., *High Temp. High Pressures* **6**, 409 (1974).

NATURAL GRAPHITE

Natural graphite, the mineral form of graphitic carbon, occurs worldwide. It differs from the carbon of coal and of diamond in its predominantly lamellar hexagonal crystal structure. The ore usually contains associated silicate minerals that vary in kind and amount with the source. Except for technical terminology, the name natural graphite is seldom used. It may be simply termed graphite or any of several common names such as plumbago, black lead, silver lead, carburet of iron, potelot, crayon noir, carbo-minerals, and reissblei. The macrophysical form depends on geological genesis, whereas the properties depend on both the macrophysical form and associated mineral suite. The commercial value depends on specific characteristics such as form, percentage, and kind of mineral suite, and availability. Graphite occurs in widely distributed places as flakes, lumps, and cryptocrystalline masses referred to commercially as amorphous graphite.

Parallel layers of condensed planar C_6-rings constitute the graphite crystallite. Each carbon atom joins to three neighboring carbon atoms at 120° angles in the plane of the layer. The C—C distance is 0.1414 nm (this bond is 0.1397 nm in benzene); the width of each C_6-ring is 0.246 nm. The hexagonal form of graphite (Fig. 1) contains the most common stacking order, ie, ABABAB.... Grinding graphite to particle sizes smaller than ca 0.1 μm reduces the crystallite size to < 20 nm, at which size two-dimensional ordering of carbon (char) replaces the three-dimensional ordering of graphite. This is depicted in the Franklin-Bacon curve (Fig. 2). The weak van der Waals forces that pin the carbons in adjoining layers are strong enough to hold the three-dimensional ordering of graphite at 0.3354 nm distance, but at 0.344 nm the structure is completely disoriented in the third dimension because the forces are too weak. This accounts in part for the marked anisotropic properties of graphite crystal.

Table 1 lists some of the physical properties of natural graphite (see Composite materials).

Graphite burns slowly in air above 450°C, the rate increasing with temperature and exposed area. The particle size and shape govern the ignition temperature. Flake graphites generally resist oxidation better than granular graphites. Above 800°C graphite reacts with water vapor, carbon monoxide, and carbon dioxide. Chlorine has a negligible effect on graphite, and nitrogen none. Many metals and metal oxides form carbides above 1500°C. These reactions occur with the carbon atom and destroy the graphitic structure (see Carbides). A series of compounds in which the graphite structure is retained, known as graphite compounds, consists of two general kinds: crystalline and covalent compounds.

Graphite can be regenerated from the crystalline compounds because the graphitic structure has not been too greatly altered. These dark crystalline compounds are called intercalation compounds, interstitial compounds, or lamellar compounds because they are formed by reactants that fit between the planar carbon networks. Each interlayer may be occupied, or every other interlayer, or every third interlayer, etc. Thus, the same element, or group, can form a series of distinct compounds. The alkali metals form a variety of such addition compounds; potassium, rubidium, and cesium reactions with graphite are well known.

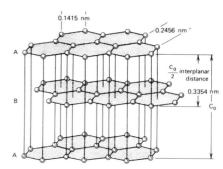

Figure 1. Hexagonal structure of graphite; $C_0/2$ = d-spacing.

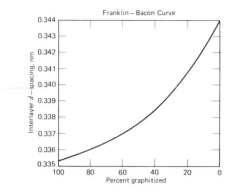

Figure 2. Change in mean interlayer spacing with fraction graphitized.

Table 1. Physical Properties of Natural Graphite

density[a], g/cm^3	
calculated	2.265
experimental, pure Sri Lanka	ca 2.25
compressibility, GPa^{-1}[b], Sri Lanka	
at low pressures	4.5×10^{-2}
at high pressures	2×10^{-2}
average	3.1×10^{-2}
shear modulus, GPa[c]	2.3
Young's modulus, GPa[c]	1.13×10^5
heat of vaporization, kJ/mol[d]	711
sublimation point, K	4000–4015
triple point, K	
graphite–liquid–gas, 10.1 MPa[c]	3900 ± 50
graphite–diamond–liquid, 12–13 GPa[c]	4100–4200
surface energy, J/cm^2[d]	ca 1.2×10^{-5}

[a] The difference between the calculated and experimental values of density is caused by dislocations and imperfections.
[b] To convert GPa^{-1} to cm^2/dyn, multiply by 10^{-10}.
[c] To convert GPa to psi, multiply by 145,000.
[d] To convert J to cal, divide by 4.184.

The compounds of sodium and lithium are less well known. When potassium vapor enters graphite interstitially, it forms a series of intercalation compounds, eg, C_8K, $C_{24}K$, $C_{26}K$, and others, depending on the sequence of interstitial layers filled.

Compounds of graphite with alkali metals or ammonia are electron donors. Compounds with the halogens (except fluorine), metal halides, and sulfuric acid (graphite sulfate) are electron acceptors.

Graphite sulfate forms when graphite is warmed in concentrated sulfuric acid containing a small quantity of an oxidizing agent such as concentrated nitric acid. The graphite swells and becomes blue. Bromine vapor forms C_8Br by direct addition to well-oriented graphite. Some intercalated crystalline compounds find their way into commerce as catalysts for chemical synthesis. Graphite-FeCl$_3$.KCl has been used in the synthesis of ammonia (qv). Acetic acid (qv) has been synthesized in high yield from methanol and carbon monoxide over a graphite-RhCl$_3$-I$_2$ compound.

The covalent compounds of graphite differ markedly from the crystal compounds. They are white or lightly colored electrical insulators, have ill-defined formulas, and occur in only one form, unlike the series typical of the crystal compounds. In the covalent compounds, the carbon network is deformed and the carbon atoms rearrange tetrahedrally as in diamond. Often they are formed with explosive violence.

Graphite oxide (graphitic acid) may explode when heated above 200°C. Below this temperature, it converts to a black powder once known as pyrographitic acid. The composition varies with the heat treatment and the end point; according to x-ray diffraction studies, it is a form of carbon that reconverts to well-ordered graphite on heating to 1800°C. Fluorine forms covalent compounds with graphite, eg, C_4F and CF.

These fluorine compounds of graphite explode on heating. Graphite fluoride continues to be of interest as a high temperature lubricant. Careful temperature control at 627 ± 3°C results in the synthesis of poly(carbon monofluoride), a compound that remains stable in air to ca 600°C and is a superior lubricant under extreme conditions of high temperatures, heavy load, and oxidizing conditions (see Lubrication).

Worked deposits of natural graphite are limited to those of Sri Lanka, Madagascar, Mexico, Brazil, Austria, the Republic of Korea, India, Norway, Canada, the USSR, the People's Republic of China, and Zimbabwe. There has been no recent domestic production in the U.S.; imports are from Mexico (60%), Republic of Korea (12%), Madagascar (5%), People's Republic of China (5%), and other (18%). The three physically different forms of natural graphite, which provide essentially different commodities, are in Sri Lanka (lump), Madagascar (large flake), Norway (small flake), Mexico (amorphous), and the Republic of Korea (amorphous).

Graphite's many useful properties give rise to a wide variety of products: unctuous, dry lubricant; marks readily, writing and drafting pencils; combination of lubricity and electrical conductivity, motor and generator brushes; excellent weathering properties and inertness, industrial paint pigment; solubility in molten iron, carbon raiser for steel; poorly wet by most metals and alloys, foundry-mold facings; and burns slowly, conducts heat, and retains strength over a large temperature range, refractories such as crucibles, retorts, and stopper heads for steel ladles.

Some additional properties of interest include hydrophobicity, formation of water-in-oil emulsions, negative charge, low photoelectric sensitivity, strong diamagnetism, and infrared absorbent. Some minor uses of graphite include coating smokeless powder and gunpowder grains to control burning rate and to prevent static sparking from friction between grains, roofing granules, packings, brake linings, gaskets, stove polish, static eliminator, polish for tea leaves and coffee beans, pipe-joint compounds, boiler compounds, wire drawing, welding-rod coatings, catalysts, oil-well drilling muds, lock lubrication, coatings for eight-track tape cartridges, mechanical mounts in cassettes, mercury and silver dry cells, exfoliated flake for gaskets and packings, aircraft disk brakes, catalyst pellet production, O-rings and oil seals, and interior and exterior coating for cathode-ray tubes.

SHERWOOD B. SEELEY
The Joseph Dixon Crucible Co.

G.R. Henning, *Proceedings of the Second Conference on Carbon*, University of Buffalo, Buffalo, N.Y., 1956.

Graphite, Minerals Yearbook, U.S. Bureau of Mines, Washington, D.C., 1974, p. 8.

Carbon, journal, Pergamon Press, Inc., Elmsford, N.Y.

CARBONATED BEVERAGES

The Federal Standard of Identity refers to soda water as "the class of beverages made by absorbing carbon dioxide in potable water. They either contain no alcohol or only such alcohol, not in excess of 0.5 wt% of the finished beverage, as is contributed by the flavor ingredient used."

The carbonated soft drink of today is composed of carbonated water, a sweetening agent, acid, flavor, color, and a preservative. These ingredients must be combined in the proper ratio in order to make an appealing and refreshing carbonated beverage.

Ingredients

Flavors. Flavors that are alcoholic solutions or extracts, emulsions or fruit juices, are prepared by a specialized industry associated with the beverage industry which also supplies the exact formula to be followed in preparation of the finished syrup. Two of the most common flavorings are caffeine and extract of kola nut.

Acidulants. Acids in carbonated beverages serve several different purposes: to impart a sour or tart taste, often imitating the fruit for which the beverage is named; to modify the sweetness of the sugar present; to act as a preservative in syrups, color solutions, and the finished product; and to catalyze the inversion of sucrose in the syrup or beverage. Citric, phosphoric, malic, and tartaric acids are those most commonly used in carbonated beverages.

Sweetening agents. The sweeteners (qv) used in carbonated beverages fall into two categories: natural, or nutritive, and synthetic, or nonnutritive. The natural sweeteners are granulated sucrose, liquid sucrose, liquid invert sugar, high fructose corn syrup, and dextrose. The only synthetic sweeteners that are approved by the FDA are saccharin and aspartame.

Water. The ingredient contained in greatest quantity in carbonated beverages is water, constituting 86–92% of the beverage. It is evident, therefore, that the water used must be of highest quality.

Municipal water is treated in the bottling plant to remove objectionable characteristics. There are a number of possible treatments, and the exact pattern varies according to local requirements. A popular sequence is chlorination, treatment with lime, coagulation, sedimentation, and sand filtration, followed by treatment with activated carbon (see Carbon).

Color. Visual appeal of a carbonated beverage is largely influenced by its color. Most colors used in carbonated beverages are natural or synthetic (see Colorants for foods, drugs, and cosmetics). The only natural colors present in carbonated beverages are those imparted by the use of natural fruits such as grapes, strawberries, or cherries. Even then, additional synthetic color must be added. A foaming caramel is frequently used in root beer, and a nonfoaming caramel in cola drinks. The FDA permits eight synthetic colors in food use. Only five of the eight colors are recommended for carbonated beverages: FD & C Red No. 40, FD & C Yellow No. 5, FD & C Yellow No. 6, FD & C Blue No. 1, and FD & C Green No. 3 (see Colorants for food, drugs, and cosmetics).

Preservatives. Many carbonated beverages are satisfactorily preserved by the acid used in the beverage and the carbon dioxide content. Beverages containing fruit juices or little carbonation are preserved with 0.05% sodium benzoate.

Carbon dioxide. Carbon dioxide (qv) contributes the characteristic pungent taste or bite associated with carbonated beverages. It inhibits the growth of mold and bacteria, and may destroy bacteria, depending on the extent of carbonation used.

Carbon dioxide for beverages must be odorless and as pure as possible. The carbonator, saturator, or carbo-cooler is the machine used to effect solution of carbon dioxide gas.

Manufacture

The complete mixture of all ingredients required to make up carbonated beverages, with the exception of carbonated water, is referred to as the syrup. In a bottling or canning plant, a room equipped with mixing and storage tanks is set aside for syrup mixing.

The final step in the production of carbonated beverages comprises mixing the syrup with water and carbon dioxide in the proper ratio, and packaging in a bottle or can. Beverages are produced either by the presyrup process or by the premix process.

In the presyrup bottling process, a measured amount of beverage syrup is placed in the freshly washed bottle as it passes through the syruper. The bottles then move to the filler where the carbonated water is added. This method is now used only in very small old bottling plants. In the premix process of filling, the proper volumes of syrup and water are automatically measured by a continuous metering system, mixed, cooled, and carbonated simultaneously by being forced through a special type of carbonator.

Packaging. Seventy-seven percent of carbonated beverages sold are packaged in bottles and cans that range in size from 180 mL to 2 L. Approximately 26% of packaged beverages sold in 1982 were in a returnable glass bottle. Twenty-one percent of the packaged product in 1982 was sold in plastic bottles, primarily of the two-liter size.

Reusable bottles make bottle washing an important part of production. A carbonated beverage bottle must be of good mechanical strength, attractive in appearance, and sterile for each use. Bottles are cleansed and sterilized with a warm alkaline solution, then rinsed with potable water.

A can for packaging soft drinks should be sanitary, easily filled and sealed, and corrosion resistant. It must withstand the maximum internal pressure developed by the product during storage and must not impart off-flavors to the product. Today, aluminum and tin-free steel are the most important metals in the manufacture of carbonated-beverage cans.

Quality control. Certain quality control tests must be made frequently during production of carbonated beverages. The water should be checked for foreign odors, chlorine, and alkalinity. Density tests are made on the beverage syrup, reported in g/cm^3 or degrees Baumé, and on the finished beverage, reported in Brix. Carbonation of the finished beverage also is tested by determining its gas pressure and temperature which can be related by charts to the number of volumes of gas present. Colorimetric tests and titrations are made to confirm that the correct amount of color and acid are present in the finished product. Taste tests are made to confirm the flavor. Microbiological tests also can be run on the product, and bottles and cans should also be tested.

Soft drinks are recognized by all responsible authorities as food products. As such, they are subject to those laws and regulations of Federal and state governments that generally apply to the production, labeling, and sale of foods.

Martha B. Jones
Royal Crown Cola Co.

L.F. Green, *Development in Soft Drinks Technology*, Applied Science Publishers, London, UK, 1978.

M.B. Jacobs, *Manufacture and Analysis of Carbonated Beverages*, Chemical Publishing Co., New York, 1959.

Woodruff and Phillips, *Beverages: Carbonated and Non-Carbonated*, AVI Publishing Co., Inc., Westport, Conn., 1974.

CARBON DIOXIDE

Carbon dioxide, CO_2, is a colorless gas with a faintly pungent odor and acid taste. Today carbon dioxide is a by-product of many commercial processes: synthetic ammonia production, hydrogen production, substitute natural-gas production, fermentation, limestone calcination, certain chemical syntheses involving carbon monoxide, and reaction of sulfuric acid with dolomite. Generally present as one of a mixture of gases, carbon dioxide is separated, recovered, and prepared for commercial use as a solid (dry ice), liquid, or gas.

Carbon dioxide also is found in the products of combustion of all carbonaceous fuels, in naturally occurring gases, as a product of animal metabolism, and in small quantities, about 0.03 vol%, in the atmosphere. Its many applications include beverage carbonation, chemical manufacture, fire fighting, food preservation, foundry-mold preparation, greenhouses, mining operations, oil-well secondary recovery, rubber tumbling, therapeutical work, and welding. Although it is present in the atmosphere and the metabolic processes of animals and plants, carbon dioxide cannot be recovered economically from these sources.

Physical Properties

Sublimation point, $-78.5°C$ at 101 kPa (1 atm); triple point, $-56.5°C$ at 518 kPa (75.1 psi); critical temperature, $31.1°C$; critical pressure, 7.39 MPa (1071 psi); critical density, 467 g/L; latent heat of vaporization, 348 J/g (149.6 Btu/lb) at the triple point and 235 J/g (101.03 Btu/lb) at 0°C; gas density, 1.976 at 0°C and 101 kPa (1 atm); liquid density, 914 g/L at 0°C and 0.759 vol/vol at 25°C and 101 kPa pressure of carbon dioxide; viscosity, 0.015 mPa·s (= cP) at 25°C and 101 kPa pressure; and heat of formation of carbon dioxide, 393.6 kJ/mol (373.4 Btu/mol) at 25°C.

Carbon dioxide, the final oxidation product of carbon, is not very reactive at ordinary temperatures. However, in water solution it forms carbonic acid, H_2CO_3, which forms salts and esters through the typical reactions of a weak acid. The primary ionization constant is 3.5×10^{-7} at 18°C; the secondary is 4.4×10^{-11} at 25°C. The pH of saturated carbon dioxide solutions varies from 3.7 at 101 kPa (1 atm) to 3.2 at 2.37 MPa (23.4 atm). A solid hydrate, $CO_2 \cdot 8H_2O$, separates from aqueous solutions of carbon dioxide that are chilled at elevated pressures.

In addition to the common stable carbon isotope of mass 12, traces of a radioactive carbon isotope of mass 14 with a half-life estimated at 5568 years are present in the atmosphere and in carbon compounds derived from atmospheric carbon dioxide (see Radioactive elements). Formation of radioactive ^{14}C is thought to be caused by cosmic irradiation of atmospheric nitrogen. Procedures have been developed for estimating the age of objects containing carbon or carbon compounds by determining the amount of radioactive ^{14}C present in the material as compared with that present in carbon-containing substances of current botanical origin. Carbon dioxide containing known amounts of ^{14}C has been used as a tracer in studying botanical and biological problems involving carbon and carbon compounds and in organic chemistry to determine the course of various chemical reactions and rearrangements. It also has been used in testing gaseous diffusion theory with mixtures of CO_2 and $^{14}CO_2$ at elevated pressures.

Carbon dioxide plays a vital role in the earth's environment. It is a constituent in the atmosphere and, as such, is a necessary ingredient in the life cycle of animals and plants. The balance between animal and plant life cycles as affected by the solubility of carbon dioxide in the earth's water results in the carbon dioxide content in the atmosphere of about 0.03 vol%. However, carbon dioxide content of the atmosphere seems to be increasing as increased amounts of fossil fuels are burned. There is some evidence that the rate of release of carbon dioxide to the atmosphere may be greater than the earth's ability to assimilate it. Such an increase could result in a warmer temperature at the earth's surface by allowing the short heat waves from the sun to pass through the atmosphere while blocking larger waves that reflect from the earth. If the earth's average temperature were to increase by several degrees, portions of the polar ice caps could melt causing an increase in the level of the oceans, or air-circulation patterns could change, altering rain patterns to make deserts of farmland or vice versa. On the other hand, it has been demonstrated that the addition of CO_2 to greenhouses increases the growth rate of plants so that an increase in the partial pressure of CO_2 in the air could stimulate plant growth making possible shorter growing seasons and increased consumption of carbon dioxide from the air.

Sources of carbon dioxide for commercial carbon dioxide recovery plants are flue gases resulting from the combustion of carbonaceous fuels; synthetic ammonia and hydrogen plants in which methane or other hydrocarbons are converted to carbon dioxide and hydrogen ($CH_4 + 2 H_2O \rightarrow CO_2 + 4 H_2$); fermentation in which a sugar such as dextrose is converted to ethyl alcohol and carbon dioxide ($C_6H_{12}O_6 \rightarrow 2 C_2H_5OH + 2 CO_2$); lime-kiln operation in which carbonates are thermally decomposed ($CaCO_3 \rightarrow CaO + CO_2$); sodium phosphate manufacture ($3 Na_2CO_3 + 2 H_3PO_4 \rightarrow 2 Na_3PO_4 + 3 CO_2 + 3 H_2O$); and natural carbon dioxide gas wells.

There are a number of methods of recovering carbon dioxide from industrial or natural gases. The potassium carbonate and ethanolamine processes are most common. In all these processes, the carbon dioxide-bearing gases are passed countercurrent to a solution that removes the carbon dioxide by absorption and retains it until it is desorbed by heat in separate equipment. All of these processes are in commercial use, and the most suitable choice for a given application depends on individual conditions. Water could be used as the absorbing medium, but this is uncommon because of the relatively low solubility of carbon dioxide in water at normally encountered pressures. The higher solubility in alkali carbonate and ethanolamine is due to a chemical combination of the carbon dioxide with the absorbing medium. Recovery methods include the sodium carbonate, potassium carbonate, Girbotol amine, Sulfinol, Rectisol, Purisol, and Fluor processes.

Although carbon dioxide produced and recovered by the methods mentioned above has a high purity, it may contain traces of hydrogen sulfide and sulfur dioxide which cause a slight odor or taste. The fermentation-gas recovery processes include a purification stage, but carbon dioxide recovered by other methods must be further purified before it is acceptable for beverage, dry ice, or other uses. The most commonly used purification methods are treatment with potassium permanganate, potassium dichromate, and active carbon.

Carbon dioxide may be liquid at any temperature between its triple point (-56.6°C) and its critical point (31°C) by compressing it to the corresponding liquefaction pressure and removing the heat of condensation. Solid carbon dioxide is produced in blocks by hydraulic presses.

Although carbon dioxide is a constituent of exhaled air, high concentrations are hazardous. Five volume percent carbon dioxide in air causes a threefold increase in breathing rate, and prolonged exposure to concentrations higher than 5% may cause unconsciousness and death. Ventilation sufficient to prevent accumulation of dangerous percentages of carbon dioxide must be provided where carbon dioxide gas has been released or dry ice has been used for cooling.

W. Robert Ballou
C & I Girdler Incorporated

R.H. Perry and C.H. Chilton, eds., *Chemical Engineers' Handbook*, 6th ed., McGraw-Hill Book Co., 1983.

CARBON DISULFIDE

Carbon disulfide (carbon bisulfide), CS_2, is a volatile, flammable liquid that is denser than water. It has many useful properties and has been an important industrial chemical since the nineteenth century. Principal uses are in the manufacture of regenerated cellulose fibers and films, and as a raw material for the manufacture of carbon tetrachloride. It also is used in preparation of various organic sulfur compounds and as a solvent.

Commercial manufacture began about 1880. Until about 1950, the principal method was by the reaction of carbon (wood charcoal) and sulfur. That procedure has largely been supplanted by processes employing the reaction of hydrocarbons and sulfur. Worldwide production of carbon disulfide was over one million (10^6) metric tons in the mid-1970s although it has declined significantly since then.

Carbon disulfide is clear and colorless and has a mild ethereal odor; however, minor impurities impart a disagreeable sulfurous odor. It is slightly soluble in water and an excellent solvent for many organic compounds. Selected physical properties are given in Table 1.

Carbon disulfide is highly flammable and is easily ignited by sparks or hot surfaces. The flammability or explosive range of carbon disulfide with air is very wide and depends on conditions. For example, the extreme range is 1.06–50.0 vol% for upward propagation in a glass tube 75 mm in diameter, whereas for downward propagation, the limits are 1.91 and 35.0 vol%. Concentrations of carbon disulfide in air of 4–8% explode with maximum violence, although the thermal energy released in carbon disulfide explosions is low compared to that from most other flammable substances. The maximum absolute pressure developed is reported as 730 kPa (7.2 atm). The equilibrium constant of formation of carbon disulfide increases with temperature, reaching a maximum corresponding to 91% conversion to CS_2 at about 1000 K.

Carbon disulfide is very toxic and is harmful by inhalation of the vapor, skin absorption of the liquid, or ingestion. Carbon disulfide poisoning has disastrous effects upon the nervous system and brain, with symptoms ranging from simple irritability to manic depression. The recommended limit in workroom air is 20 ppm (ca 60 mg/m^3) TWA for an 8-h work day and 40-h work week. The short-term exposure limit is 30 ppm in the United States. OSHA standards are 20 ppm TWA, 30 ppm ceiling limit, and 100 ppm/30 min peak.

The largest single application of carbon disulfide (ca one third of production) is in the manufacture of viscose rayon (see Rayon). It is also used for carbon tetrachloride production (ca 31% and cellophane-film manufacture (ca 13%). There are many other uses including agricultural and pharmaceutical applications with carbon disulfide either as a direct

Table 1. Physical Properties of Carbon Disulfide

Property	Value		
melting point, K	161.11		
latent heat of fusion, kJ/kg[a]	57.7		
boiling point at 101 kPa[b], °C	46.25		
flash point at 101 kPa[b], °C	−30		
ignition temperature in air, °C			
10-s lag time	120		
0.5-s lag time	156		
critical temperature, °C	273		
critical pressure, kPa[b]	7700		
critical density, kg/m^3	378		
solu H$_2$O in CS$_2$ at 10°C, ppm	86		
at 25°C, ppm	142		
Liquid at temperature, °C	*0*	*20*	*46.25*
density, kg/m^3	1293	1263	1224
specific heat, J/(kg·K)[a]	984	1005	1030
latent heat of vaporization, kJ/kg[a]	377	368	355
surface tension, mN/m(=dyn/cm)	35.3	32.3	28.5
thermal conductivity, W/(m·K)		0.161	
viscosity, mPa·s (=cP)	0.429	0.367	0.305
refractive index, n_D	1.6436	1.6276	
solubility in water, g/kg soln	2.42	2.10	0.48
vapor pressure, kPa[b]	16.97	39.66	101.33
Gas at temperature, °C[c]	*46.25*	*200*	*400*
density, kg/m^3	2.97	1.96	1.37
specific heat, J/(kg·K)[a,d]	611	679	730
viscosity, mPa·s (=cP)	0.0111	0.0164	0.0234
thermal conductivity, W/(m·K)	0.0073		
Thermochemical data at 298 K[a]			
heat capacity[a], C_p^o, J/(mol·K)	45.48		
entropy[a], S^o, J/(mol·K)	237.8		
heat of formation[a], H_f^o, kJ/mol	117.1		
free energy of formation[a], G_f^o, kJ/mol	66.9		

[a] To convert J to cal, divide by 4.184.
[b] To convert kPa to atm, divide by 101.3.
[c] At absolute pressure, 101.3 kPa.
[d] $C_p/C_v = 1.21$ at 100°C (1).

reactant, chemical intermediate, or solvent. Dwindling production reflects trends of declining consumption for carbon tetrachloride and regenerated cellulose.

ROBERT W. TIMMERMAN
FMC Corp.

CS$_2$, brochure, Stauffer Chemical Company, Westport, Conn., 1975.

C.M. Thacker, *Hydrocarbon Process.* **49**(4), 124 (1970); **49**(5), 137 (1970).

H.O. Folkins, E. Miller, and H. Hennig, *Ind. Eng. Chem.* **42**, 2202 (1950).

F.A. Patty, *Industrial Hygiene and Toxicology*, 2nd rev. ed., Vol. II, John Wiley & Sons, Inc., New York, 1962, pp. 901–904.

CARBONIC AND CHLOROFORMIC ESTERS

The reaction of phosgene (carbonic dichloride) with alcohols gives two classes of compounds: carbonic esters and chloroformic esters. The carbonic esters (carbonates), ROC(O)OR, are the diesters of carbonic acid. The chloroformic esters (chloroformates, chlorocarbonates), ClC(O)OR, are the esters of hypothetical chloroformic acid, ClCOOH.

The reaction proceeds in stages, first producing a chloroformate, and then a carbonic acid diester. When a different alcohol is used for the second stage, a mixed radical or unsymmetrical carbonate is produced.

Chloroformic Esters

In the literature, chloroformates are also referred to as chlorocarbonates because of the structural parallel with these acids. In general, chloroformates are clear, colorless liquids with low freezing points and high boiling points. They are soluble in most organic solvents, but insoluble in water, although they do hydrolyze slowly. The physical properties of the most widely used chloroformic esters are given in Table 1.

Chloroformates are reactive in intermediates that combine acid, chloride, and ester functions. The ester moiety determines thermal stability, generally in the following order of decreasing stability: aryl > primary alkyl > secondary alkyl > tertiary alkyl.

Reaction with water gives the parent hydroxy compound, HCl and CO$_2$, in addition to the symmetrical carbonate formed by the hydroxy compound and the chloroformate. Esters are the main products in the reaction with carboxylic acids. The reaction with ammonia is the classical method for preparing primary carbamates. Excess ammonia is used to remove the hydrogen chloride as it is evolved (see Carbamic acid). Aryl chloroformates are good acylating agents, reacting with aromatic hydrocarbons under Friedel-Crafts conditions to give the expected aryl esters of the aromatic acid (see Friedel-Crafts reactions). However, alkylation takes place with aliphatic chloroformates under similar conditions.

The DOT (Bureau of Hazardous Materials) and the U.S. Coast Guard classify most chloroformates as flammable or combustible liquids, requiring the "Flammable," or "Combustible Liquid," label. Chloroformates, especially those of low molecular weight, are lacrimators and vesicants and produce effects similar to those of hydrochloric acid or carboxylic acid chlorides.

Carbonic Esters

The physical properties are given in Table 2. The lower alkyl carbonates are neutral, colorless liquids with a mild odor. Aryl carbonates are normally solids.

Diethyl, dimethyl, and dipropyl carbonates are fire hazards. They are mildly irritating to skin, eyes, and mucous membranes. Protective cloth-

Table 1. Physical Properties of Selected Chloroformates

Chloroformate	Density, d_4^{20}, g/cm^3	Refractive index, n_D^{20}	Bp, °C		
			2.67 kPa[a]	13.3 kPa[a]	101.3 kPa[a]
methyl	1.250	1.3864			71
ethyl	1.138	1.3950			94
isopropyl	1.078	1.3974		47	105
n-propyl	1.091	1.4045	25.3	57.5	112.4
allyl	1.1394	1.4223	25	57	
n-butyl	1.0585	1.4106	44	77.6	
sec-butyl	1.0493	1.4560	36	69	
isobutyl	1.0477	1.4079	39	71	
2-ethylhexyl	0.9914	1.4307	98	137	
n-decyl	0.9732	1.4400	122	159	
phenyl	1.2475	1.5115	83.5	121	185
benzyl	1.2166	1.5175	103	123	152
ethylene bis	1.4704	1.4512	108	137	
diethylene glycol bis	1.388	1.4550	148	180	

[a] To convert kPa to mm Hg, multiply by 7.50.

Table 2. Physical Properties of Selected Carbonates

Carbonate	Density, d_4^{20}, g/cm^3	Refractive index, n_D^{20}	bp	
			°C	kPa[a]
dimethyl	1.073	1.3697	90.2	101.31
diethyl	0.975	1.3846	23.8	1.33
			69.7	13.33
			126.8	101.31
di-*n*-propyl	0.941	1.4022	165.5–166.6	101.31
diisopropyl			147.0	101.31
diallyl	0.994	1.4280	97	8.13
			105	13.33
			166	97.31
di-*n*-butyl	0.9244	1.4099	207.7	101.31
di-2-ethylhexyl	0.8974$_4^{20}$	1.4352	173	1.33
diphenyl	1.1215$_4^{87}$		302	101.31
diethylene glycol bis(allyl)	1.143	1.4503	160	0.27
tolyl diglycol	1.189	1.5229	247–248	0.27
ethylene	1.3218$_4^{39}$	1.4158	248	101.31

[a] To convert kPa to mm Hg, multiply by 7.5.

ing, rubber gloves, and goggles should be worn when handling these chemicals.

Commercially, the most important carbonate is the diethyl ester. It is used in many organic syntheses, particularly of pharmaceuticals and pharmaceutical intermediates. It also is used as a solvent for many synthetic and natural resins, and in vacuum tube cathode-fixing lacquers. Diethylene glycol bis(allyl carbonate) polymerizes easily because of its two double bonds and is used for colorless, optically clear castings. Polymerization is catalyzed by the use of diisopropyl peroxydicarbonate. Such polymers are used in the preparation of safety glasses, lightweight prescription lenses, glazing cast sheet, and optical cement (see Allyl monomers and polymers; Polycarbonates).

EDWARD ABRAMS
Chemetron Corporation

M. Matzner, R.P. Kurkjy, and R.J. Cotter, *Chem. Rev.* **64**, 645 (1964).

S. Yura and T. Ono, *J. Soc. Chem. Ind. Japan* **48**, 30 (1945).

U.S. Pat. 3,846,468 (Nov. 5, 1974), E. Perrotti and G. Cipriani (to Snam Progetti S.p.A.).

CARBONIZATION.   See Coal–Coal conversion processes.

CARBON MONOXIDE

Carbon monoxide (CO), a colorless, odorless, flammable, toxic gas, is produced by steam reforming or partial oxidation of carbonaceous materials. It is used as a fuel, a metallurgical reducing agent, and a feedstock in the manufacture of a variety of chemicals, notably methanol, acetic acid, phosgene, and oxo alcohols. Increased usage of carbon monoxide from coal in chemicals and fuels manufacture is likely if economic coal gasification technology evolves (see Fuels, synthetic). A summary of particularly useful physical constants is presented in Table 1. Thermodynamic properties have been reported from 70 to 300 K at pressures from 0.1 to 30 MPa (1–300 atm), and from 0.1 to 121 MPa (1–1200 atm).

The bond energy of 1070 kJ/mol (255.7 kcal/mol) is consistent with the triple bond formulation and is the highest observed bond energy for any diatomic molecule. The fundamental absorption in the infrared spectrum of carbon monoxide is located at 2143 cm^{-1}. The bonding between carbon monoxide and transition-metal atoms is particularly important because transition metals, whether deposited on solid supports or present as discrete complexes, are required as catalysts for the reaction between carbon monoxide and most organic molecules.

Table 1.   Physical Properties of Carbon Monoxide

Property	Value
mp, K	68.09
bp, K	81.65
ΔH, fusion (68 K), kJ/mol[a]	0.837
ΔH, vaporization (81 K), kJ/mol[a]	6.042
density (273 K, 101.3 kPa (1 atm)), g/L	1.2501
sp gr, liquid, 79 K[b]	0.814
relative sp gr, gas, 298 K[c]	0.968
critical temperature, K	132.9
critical pressure, MPa (atm)	3.496 (34.5)
critical density, g/cm^3	0.3010
triple point, K/kPa (°C/torr)	68.1/15.39 (−205/115.4)
bond length, nm	0.11282

[a]To convert J to cal, divide by 4.184.

[b]With respect to water at 277 K.

[c]With respect to air at 298 K.

Industrially Significant Reactions

Carbon monoxide reacts with water over a catalyst to produce hydrogen and carbon dioxide. This reaction is used to prepare high purity hydrogen or synthesis gas with a higher hydrogen-to-carbon monoxide ratio than the feed. Carbon monoxide can be oxidized without a catalyst or at a controlled rate with a catalyst. It readily disproportionates into elemental carbon and carbon dioxide on a catalyst surface as well. The reaction between carbon monoxide and chlorine is catalyzed by activated carbon and gives phosgene in nearly quantitative yield. Methanol is manufactured by the reaction between carbon monoxide and hydrogen at 230–400°C and 5–60 MPa (50–600 atm). Manufacture of acetic acid by methanol carbonylation has become a leading commercial route to acetic acid. Probably the best known catalytic carbonylation reaction is the hydroformylation, or oxo reaction, for producing aldehydes and alcohols from carbon monoxide, hydrogen, and olefins:

$$RCH{=}CH_2 + CO + H_2 \rightarrow RCH_2CH_2CHO \xrightarrow{H_2} RCH_2CH_2CH_2OH$$

About one-third of the acrylic acid and ester produced in the United States is manufactured by the Reppe reaction from acetylene, methanol, and carbon monoxide:

$$HC{\equiv}CH + CH_3OH + CO \rightarrow CH_2{=}CHCOOCH_3$$

Carbon monoxide is catalytically hydrogenated to a mixture of straight-chain aliphatic, olefinic, and oxygenated hydrocarbon molecules in the Fischer-Tropsch reaction (see Fuels, synthetic):

$$n\,CO + 2\,n\,H_2 \rightarrow {-}(CH_2)_{\overline{n}} + n\,H_2O$$

Carbon monoxide is hydrogenated to methane to make substitute natural gas or purify gas streams in which CO cannot be tolerated. The Mond process for nickel purification is based on the formation of volatile nickel carbonyl, $Ni(CO)_4$, which is stable below 60°C but decomposes rapidly and completely into nickel and carbon monoxide at 180°C.

General Reactions

In a liquid-phase high pressure reaction (60 MPa or 600 atm), a rhodium cluster complex catalyzes the direct formation of ethylene glycol, propylene glycol (see Glycols), and glycerol (qv) from synthesis gas. Carbon monoxide reacts with alcohols, ethers, and esters to give carboxylic acids (qv). The sulfuric acid-catalyzed reaction of formaldehyde with carbon monoxide and water to glycolic acid at 200°C and 71 MPa (700 atm) pressure was the first step in an early process to manufacture ethylene glycol. The reaction of unsaturated organic compounds with carbon monoxide and molecules containing an active hydrogen atom leads to a variety of interesting organic products. Under proper conditions, carbon monoxide and oxygen react with organic molecules to form carboxylic acids or esters. In the presence of a catalyst, nitroaromatic compounds can be converted into isocyanates, using carbon monoxide as a reducing agent. The industrial solvent dimethylformamide is manufactured by the reaction between carbon monoxide and dimethylamine. Carbon monoxide reacts with aromatic hydrocarbons or aryl halides to yield aromatic aldehydes (see Aldehydes). It forms metal carbonyls or metal carbonyl derivatives with most transition metals (see Coordination compounds). Carbon monoxide forms copolymers with ethylene and suitable vinyl compounds.

Manufacture and Purification

Carbon monoxide is produced by steam reforming of methane or naphtha, partial oxidation of coal or petroleum, or from blast-furnace gases. It is obtained with variable quantities of hydrogen, carbon dioxide, and water. Shifting the ratio of hydrogen to carbon monoxide, removal of carbon dioxide and water, or isolation of pure CO depends on the intended application.

Two commercial processes to purify carbon monoxide are based on its absorption by copper–salt solutions, the Cosorb process currently being preferred. A third approach uses cryogenic condensation and fractionation. All three routes use similar techniques to remove minor impurities, water, and carbon dioxide.

Health and Safety

Carbon monoxide is the most widely spread gaseous hazard to which man is exposed. The toxicity of carbon monoxide is a result of its reaction with the hemoglobin of blood. The carboxyhemoglobin (COHb) that is formed displaces oxygen and leads to asphyxiation. The recommended NIOSH limit of 35 ppm is the TWA exposure to carbon monoxide based on a carboxyhemoglobin level of 5%. Prevention of carbon monoxide poisoning is best accomplished by providing good ventilation where contamination is a problem.

CHARLES M. BARTISH
GERALD M. DRISSEL
Air Products and Chemicals, Inc.

F.A. Lowenheim and M.K. Moran, *Industrial Chemicals*, 4th ed., Wiley-Interscience, a division of John Wiley & Sons, Inc., New York, 1975.

J. Falbe, *Carbon Monoxide in Organic Synthesis*, Springer-Verlag, New York, 1970.

R.N. Shreve and J.A. Brink, Jr., *Chemical Process Industries*, 4th ed., McGraw-Hill Book Co., New York, 1977.

CARBON MONOXIDE – HYDROGEN REACTIONS. See Fuels, synthetic.

CARBONYLS

Carbon monoxide (qv), the most important π-acceptor ligand, forms a host of neutral, anionic, and cationic transition-metal complexes. There is at least one known type of carbonyl derivative for every transition metal, as well as evidence supporting the existence of the carbonyls of some lanthanides and actinides (see Coordination compounds; Organometallics).

Carbonyls are useful in the preparation of high purity metals, in catalytic applications, and in the synthesis of organic compounds. Metal carbonyls are employed in the preparation of complexes where the carbon monoxide ligand is replaced by halides, hydrogen, group VB and VIB derivatives, arenes, and many chelating ligands (see Chelating agents).

Nearly every metal forming a carbonyl obeys the inert gas rule. One exception is vanadium which forms a hexacarbonyl in which the number of electrons is 35. This carbonyl has a paramagnetism equivalent to one unpaired electron. Many metals with an odd number of electrons achieve the inert configuration by forming a covalent metal–metal bond, using CO molecules to act as bridges between two metals with each metal receiving one electron from each CO molecule, or using a CO molecule to form a bridge among three metal centers.

The accepted view of bonding in metal carbonyls is one in which charge is donated from the ligand to the metal by a sigma bond and electron density from the metal d-orbitals is back-donated into the π^* (unoccupied or antibonding) orbitals of the ligand. The electron density in the π^* orbitals of the ligand is dependent to a certain extent on the charge donation from the ligand to the metal; therefore, the σ and π bonding is synergistic.

The CO molecule coordinates in three ways as shown diagrammatically in Figure 1. Terminal carbonyls are the most common. Bridging carbonyls are common in most polynuclear metal carbonyls. As depicted in Figure 1, metal–metal bonds also play an important role in poly-

nuclear metal carbonyls. The metal atoms in carbonyl complexes show a strong tendency to coordinate all their valence orbitals in forming bonds. These include the nd^5, $(n + 1)s$, and the $(n + 1)p^3$ orbitals. As a result, the inert gas rule is successful in predicting the structure of most metal carbonyls.

Some physical properties of metal carbonyls are presented in Table 1. Most metal carbonyls are volatile solids that can be sublimed under vacuum. Iron and nickel carbonyls are distillable liquids. The volatility of metal carbonyls is an important safety consideration owing to their toxicity.

Table 1. Physical Properties of Metal Carbonyls

Carbonyl formula	Color (solid)	Melting point, °C	Boiling point, °C	Density, g/cm³	M—M distance, nm
$V(CO)_6$	blue–green	50 dec	40–50^{15} subl.		
$Cr(CO)_6$	white	149–155	70–75^{15} subl.	1.77	
$Mo(CO)_6$	white	150–151 dec		1.96	
$W(CO)_6$	white	169–170	50 subl.	2.65	
$Mn_2(CO)_{10}$	yellow	151–155	$50^{0.01}$ subl.	1.81	0.293
$Tc_2(CO)_{10}$	white	159–160	$40^{0.01}$ subl.	2.08	0.3036
$Re_2(CO)_{10}$	white	177	$60^{0.01}$ subl.	2.87	0.302
$Fe(CO)_5$	white	-20	103	1.52	
$Fe_2(CO)_9$	yellow	100 dec		2.08	0.2523
$Fe_3(CO)_{12}$	green–black	140 dec	$60^{0.1}$ subl.	2.00	0.263 (av)
$Ru_3(CO)_{12}$	orange	150 dec		2.75	0.2848
$Os_3(CO)_{12}$	yellow	224		3.48	0.288
$Co_2(CO)_8$	orange	50–51	45^{10} subl.	1.73	0.2542
$Co_4(CO)_{12}$	black	60 dec		2.09	0.249
$Rh_4(CO)_{12}$	red	dec		2.52	0.275 (av)
$Rh_6(CO)_{16}$	black	220 dec		2.87	0.2776
$Ir_4(CO)_{12}$	yellow	210 dec			0.268
$Ni(CO)_4$	white	-25	43	1.32	

Since transition metals even in a finely divided state do not readily combine with CO, various metal salts have been used to synthesize metal carbonyls. Metal salts almost always contain the metal in a higher oxidation state than the resulting carbonyl complex. Therefore, most metal carbonyls result from the reduction of the metal in the starting material. Such a process has been referred to as reductive carbonylation. Although detailed mechanistic studies are lacking, the process probably proceeds through stepwise reduction of the metal with simultaneous coordination of CO.

The dry method for synthesizing metal carbonyls from salts and oxides has proven useful in a number of cases (eg, $Re_2(CO)_{10}$). The metal carbonyl is formed in the presence of suitable reducing agents, such as CO.

Some metal carbonyls of lower molecular weight lose CO on heating or uv irradiation leading to the formation of polynuclear species. In some cases this method is a useful preparative tool (eg, $CO_4(CO)_{12}$, $Rh_6(CO)_{16}$). A few metal carbonyls can be prepared by exchange of CO molecules (eg, $W(CO)_6$). Heteronuclear metal carbonyls are usually synthesized by either metathesis or condensation.

The majority of metal carbonyls are synthesized under high pressures of CO and in organic solvents. Reducing agents are normally active metals such as Na, K, Mg, Zn, or Al. A few metal carbonyls are synthesized in aqueous medium (eg, $Fe_3(CO)_{12}$, $Rh_4(CO)_{12}$). Recently, attention has been directed toward finding methods of preparing metal carbonyls under less drastic conditions. Numerous reports have appeared in the literature concerning low pressure syntheses of metal carbonyls, but the reactions have been restricted primarily to the carbonyls of the transition metals of Groups VIII, IB, and IIB.

The commercial manufacturers of metal carbonyls include Pressure Chemical Company ($Ni(CO)_4$), GAF Corporation ($Fe(CO)_5$), and Strem Chemicals Inc. and Pressure Chemical Company ($W(CO)_6$, $Mo(CO)_6$, and $Cr(CO)_6$).

Many catalytic reactions involving metal carbonyls have important industrial applications. The reactions include carbonylation of olefins,

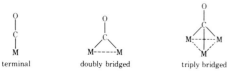

Figure 1. Bonding modes of CO.

carboxylation reaction, olefin isomerization, stabilization of unstable intermediates, coupling and cyclization reactions, formation of functional groups, homogeneous hydrogenation catalysts, and the water-gas shift reaction.

Easily decomposed, volatile metal carbonyls have been used in metal-deposition reactions where heating forms the metal and carbon monoxide. Other products such as metal carbides and carbon may also form, depending on the conditions. The commercially important Mond process depends on the thermal decomposition of $Ni(CO)_4$ to form high purity nickel. Another important use of metal carbonyls is as an antiknock compound in gasoline (qv).

Exposure to metal carbonyls can present a serious health threat. The toxic symptoms from inhalation of nickel carbonyl are believed to be caused by both nickel metal and carbon monoxide. In many acute cases the symptoms are headache, dizziness, nausea, vomiting, fever, and difficulty in breathing. If exposure is continued, unconsciousness follows with subsequent damage to vital organs and death.

When heated to about 60°C, nickel carbonyl explodes. For both iron and nickel carbonyl, suitable fire extinguishers are water, foam, carbon dioxide, or dry chemical. Large amounts of iron pentacarbonyl also have been reported to ignite spontaneously. Solutions of molybdenum carbonyl have been reported to be capable of spontaneous detonation.

FRANK S. WAGNER
Strem Chemicals Inc.

C.F. Powell, J.H. Oxley, and J.M. Blocher, Jr., eds., *Vapor Deposition*, John Wiley & Sons, Inc., New York, 1966.

F.A. Glaski, ed., *The Third International Conference on Chemical Vapor Deposition*, Salt Lake City, Utah, Apr. 24 and 27, 1972.

I. Wender and P. Pino, eds., *Organic Synthesis via Metal Carbonyls*, Wiley-Interscience, a division of John Wiley & Sons, Inc., New York, 1968; Vol. 2, 1977.

E.L. Kugler and F.W. Steffgen, eds., *Adv. Chem. Ser.* **178**, American Chemical Society, Washington, D.C., 1979.

CARBOXYLIC ACIDS

SURVEY

Carboxylic acids from the smallest, formic, to the 22-carbon fatty acids, eg, erucic, are economically important; several million (10^6) metric tons is produced annually. The shorter-chain aliphatic acids are colorless liquids. Each has a characteristic odor ranging from sharp and penetrating (formic and acetic acids) or vinegary (dilute acetic) to the odors of rancid butter (butyric acid) and goat fat (the 6–10-carbon acids). At room temperature, the cis unsaturated acids through C_{18} are liquids and the saturated aliphatic acids from decanoic through the higher acids and trans-unsaturated acids are solids. The latter are higher melting because of their greater symmetry of structure (eg, elaidic acid).

Both odd- and even-numbered alkanoic acids of molecular formula $C_nH_{2n}O_2$ up through hexanoic acid occur naturally. Only the even-numbered higher acids, most often the C_{18} acids, occur naturally. Formic (qv), acetic (qv), propionic, and butyric acids are manufactured in large quantities from petrochemical feedstocks. The higher fatty acids are derived from animal fats, vegetable oils, or fish oils. Some higher saturated fatty acids with significant industrial applications are pelargonic, lauric, myristic, palmitic, and stearic acids.

In the alkenoic series of molecular formula $C_nH_{2n-2}O_2$, acrylic (qv), methacrylic (qv), undecylenic, and oleic acids have important applications. Acrylic and methacrylic acids have a petrochemical origin, and undecylenic and oleic acids have natural origins.

The polyunsaturated aliphatic monocarboxylic acids having industrial significance include sorbic, linoleic, linolenic, eleostearic, and various polyunsaturated fish acids. Of these, only sorbic acid (qv) is made synthetically. The other acids, except those from tall oil, occur naturally as glycerides and are used in this form.

The shorter-chain alkynoic (acetylenic) acids are common in laboratory organic syntheses, and several long-chain acids occur naturally. Many substituted fatty acids, particularly methacrylic, 2-ethylhexanoic, and ricinoleic acids, are commercially significant. Several substituted fatty acids exist naturally. Fatty acids with a methyl group in the penultimate position are called iso acids, and those with a methyl group in the antipenultimate position are called anteiso acids (see Carboxylic acids, branched-chain acids).

Some naturally occurring fatty acids have alicyclic substituents; the cyclopentyl-containing chaulmoogra acids, notable for their use in treating leprosy (see Chemotherapeutics, antimycotic), and the cyclopropenyl and sterculic acids.

The prostaglandins (qv) constitute another class of fatty acids with alicyclic structures. These are of great biological importance and are formed by *in vivo* oxidation of 20-carbon polyunsaturated fatty acids, particularly arachidonic acid. The several prostaglandins, eg, PGE_1, have different degrees of unsaturation and oxidation when compared to the parent compound, prostanoic acid.

Aromatic carboxylic acids are produced annually in amounts of several million (10^6) metric tons. Several aromatic acids occur naturally (eg, benzoic (qv), salicylic (qv), cinnamic (qv), and gallic acids), but those used in commerce are produced synthetically. These acids generally are crystalline solids with relatively high melting points, attributable to the rigid, planar, aromatic nucleus (see also Phthalic acids).

Melting points, boiling points, densities, and refractive indexes for carboxylic acids vary widely depending upon molecular weight, structure, and the presence of unsaturation or other functional groups. Equations for the specific heats for solid and liquid states of palmitic acid are, respectively, (J/g):

$$Cp = 1.604 + 0.00544t \quad (-73 \text{ to } 40°C)$$
$$Cp = 1.936 + 0.00734t \quad (63 \text{ to } 92°C)$$

For stearic acid, the equations are (to convert J to cal, divide by 4.184)

$$Cp = 1.7886 + 0.00754t \quad (-120 \text{ to } 65°C)$$
$$Cp = 1.7861 + 0.00754t \quad (70 \text{ to } 78°C)$$

The viscosity of a mixture of fatty acids depends upon the average chain length $\bar{n}$ and can be calculated from the equations:

$$70°C, \log \eta = -0.602802 + 0.134844\bar{n} - 0.00259(\bar{n})^2$$
$$90°C, \log \eta = -0.510490 + 0.101571\bar{n} - 0.001628(\bar{n})^2$$

where η = viscosity in mPa·s (= cP).

Heats of combustion for liquid alkanoic acids at 25°C are given by the equation:

$$-\Delta H_R^{25} = 654.4n - 430.4 \text{ J/mol} \quad (n \geq 5)$$

Crystallographic properties of solid alkanoic acids significantly affect many of their other properties. For example, heat of crystallization, melting point, and solubility depend upon whether the acid is even- or odd-numbered, and vary alternately in a homologous series. Other physical properties such as boiling point, density (liquid), and refractive index depend upon molecular rather than crystal structure and change in a regular manner according to molecular weight. An important chemical characteristic of unsaturated acids is the iodine value (IV) which indicates the average degree of unsaturation. It is equal to the number of grams of iodine absorbed under standard conditions by 100 g of the unsaturated acid.

The alkanoic acids, with the exception of formic acid, undergo typical reactions of the carboxyl group. Formic acid has reducing properties and does not form an acid chloride or an anhydride. The hydrocarbon chain of alkanoic acids undergoes the usual reactions of hydrocarbons except that the carboxyl group exerts considerable influence on the site and ease of reaction. The alkenoic acids in which the double bond is not conjugated with the carboxyl group show typical reactions of internal olefins. All three types of reactions are industrially important. Reactions of the carboxyl group include salt and acid chloride formation, esterification, pyrolysis, reduction, and amine, nitrile, and amide formation.

EVERETT H. PRYDE
United States Department of Agriculture

K.S. Markley, ed., *Fatty Acids. Their Chemistry, Properties, Production, and Uses*, 2nd rev. ed., 5 Parts, Interscience Publishers, a division of John Wiley & Sons, Inc., New York, 1960–1968.

E.H. Pryde, ed., *Fatty Acids*, The American Oil Chemists' Society, Champaign, Ill., 1979.

MANUFACTURE

This section deals mainly with those acids commonly known as fatty acids that have 6 to 24 carbon atoms, different degrees of unsaturation, and are found in substantial amounts in natural fats and oils.

The C_6–C_{24} acids, or fatty acids, are obtained from animal tallows and greases, vegetable, coconut, palm, and marine oils. They are also produced synthetically from petroleum sources. For many years fatty acids have been produced by oxidation of hydrocarbons. Actual and potential methods for synthesizing fatty acids are listed here.

Catalytic oxidation of paraffinic hydrocarbons. Catalytic air oxidation produces mixtures of odd and even carbon atoms with many impurities; recovery of pure acids is almost impossible.

Oxidation of olefins. Many manufacturing processes are based on the reaction of carbon monoxide, water, and olefins (see Oxo process).

Oxidation or carboxylation of ethylene-growth compounds. Compounds obtained from ethylene by the Ziegler process contain an even number of carbon atoms. They are more easily converted to alcohols and acids than naturally occurring compounds (see Alcohols; Olefin polymers). The cost of producing the acids from the Ziegler process prevents them from competing with natural fatty acids. Furthermore, the cost of ethylene has increased more rapidly than the costs of natural oils and fats.

Oxidation of natural fats. Dibasic and straight-chain monobasic acids are produced by oxidation of unsaturated fatty acids with ozone or other oxidizing agents, eg, oleic acid oxidation produces azelaic, $HO_2C(CH_2)_7$-CO_2H, and pelargonic acids, $CH_3(CH_2)_7CO_2H$.

Alkali fusion of alcohols. Treatment of fatty alcohols with alkalies at 300°C and ca 5.6 MPa (800 psig) produces the corresponding fatty acid plus hydrogen.

Telomerization. Telomerization of ethylene and butadiene with formic and acetic acids, methyl acrylate, or methyl chloride produces acids of an odd number of carbon atoms.

The first step in the manufacture of fatty acids from natural fats and oils is the saponification or hydrolysis of the triglyceride. Saponification of a fat is stepwise, ie, removal of one acid group forms a diglyceride, another acid removal leaves a monoglyceride, and finally, free glycerol is produced.

Sweet water from a continuous countercurrent fat splitter contains ca 12–20% glycerol, emulsified fat, slight amounts of soluble acids and proteinaceous material, and a trace of inorganic salts. Its pH is 4.5–5. The sweet water may be concentrated to some degree with acid-resistant evaporators but it is usually allowed to settle so that the insoluble fatty materials may be skimmed. It is filtered after treatment with lime which precipitates the dissolved fatty acids. The excess lime is removed by soda ash (Na_2CO_3) at pH 8. A second filtration removes $CaCO_3$. This treatment can be carried out in a single step but extreme care is necessary since excess lime causes scale formation in the evaporator tubes. Crude glycerol from saponification (88% glycerol) varies in color from pale yellow to brown. It is distilled or deionized to make the commercial chemical pure (cp) grade or dynamite grade (see Glycerol).

Separation of oleic, stearic, and linoleic acids is usually done by crystallization (qv) either with or without solvents. Cottonseed, soybean, corn oil, and other liquid edible oils are winterized by lengthy treatments at low temperatures to allow the more saturated triglycerides to crystallize and drop out of solution; the resulting slurries are then filtered.

The solid fractions from crystallization must be hydrogenated to obtain saturated acids with an IV of less than one. Hydrogenation is not difficult and is usually accomplished by the batch method.

For distillation (qv) the preferred method is film evaporation which can be achieved by a falling-film evaporator or steam injection into the heating tubes. Present pollution-control laws have caused criticism of the use of large amounts of water in contact condensers (injected steam). The use of steam is unnecessary with the falling-film heater which maintains a thin film over the heating surface. Mechanical means of maintaining a thin film over the heating surface, such as white film evaporators and spinning disks, are used increasingly in the distillation of heat-sensitive substances.

RALPH H. POTTS[‡]
Armak Company

F.D. Gunstone, *An Introduction to the Chemistry and Biochemistry of Fatty Acids and Their Glycerides*, Halsted Press, London, 1975.

K.S. Markley, ed., *Fatty Acids: Their Chemistry and Physical Properties*, 2nd ed., Interscience Publishers, Inc., a division of John Wiley & Sons, Inc., New York, 1960–1968.

E.S. Pattison, ed., *Fatty Acids and Their Industrial Applications*, Marcel Dekker, Inc., New York, 1968.

ANALYSIS AND STANDARDS FOR FATTY ACIDS

The fatty acids most commonly encountered as solids or liquids in natural fats have common names related to their natural occurrence. Most fatty acids arise from processing selected fats and oils such as coconut, soybean, cottonseed, corn, and other vegetable oils, and lard and beef tallow. Such fatty acids are originally present as triesters of glycerol (triglycerides) with other mixed lipid constituents such as sterols (see Vegetable oils; Fats and fatty oils). Standardized methods are available in the *Official and Tentative Methods of the American Oil Chemists' Society* (AOCS) to determine the following chemical and physical properties of fats, oils, and fatty acids: titer, acid and iodine values, color, stability, saponification value, unsaponifiables, and fatty acid composition. Methods for determination of physical and chemical properties of fatty acids are also described by ASTM in the *Standard Methods for Analysis of Fats and Oils* and by the Association of Official Analytical Chemists.

The melting points of fatty acid mixtures vary according to the chain lengths, crystalline structures, and proportions of saturated fatty acids in the mixture. Crsytalline structures are termed polymorphic since they are easily altered by heat treatment. Fatty acid mixtures with the same titer may have varying iodine values. For instance, pure oleic acid has the same iodine value (ca 90) as a 1:1 mixture of stearic (IV = 0) and linoleic acids (IV = 180).

Both acid and saponification values are measures of average molecular weight. However, mixtures of fatty acids can differ widely in composition yet yield very similar values. These values are also influenced by the unsaponifiable content and cannot be relied upon for quantitative information concerning composition.

Gas–liquid chromatography (glc) is now routinely used for quantitative analysis of fatty acid mixtures. Information concerning the exact percentage composition of any saturated fatty acid may be obtained with modern liquid phases such as the cyano silicone Sp2340 in under 30 minutes. Other qualitative and semiquantitative methods for the analysis of fatty acids are paper chromatography, thin layer chromatography (tlc), and high pressure liquid chromatography (hplc).

Fatty acid deterioration is related to the total unsaturation of the acid mixture. The extent and relative speed of deterioration decreases in the series of acids: linoleic, oleic, saturated acids. Such deterioration arises as a result of free radical reactions resulting in polymerization and chain cleavage. Rancid odor and flavor, objectionable in food and cosmetics, are caused by short chain (C_1–C_{10}) aldehydes and ketones. The AOCS

[‡] Deceased.

method td3a-64 measures the color stability of fatty acids after heating under the specified conditions of the test; the specification used is the maximum allowable color of the sample after the test.

EDWARD G. PERKINS
University of Illinois

Official and Tentative Methods of the American Oil Chemists' Society, 3rd ed., American Oil Chemists' Society, Champaign, Ill, 1980.

Standard Methods for the Analysis of Fats and Oils, Fat Commission, IUPAC, Paris, France, 1954.

W. Horowitz, ed., *Official Methods of Analysis of the Association of Official Analytical Chemists*, 12th ed., AOAC, Washington, D.C., 1975.

ECONOMIC ASPECTS

Several aliphatic carboxylic acids are produced on a large scale (1982 production in thousand metric tons): for example, formic (29), acetic (1,250), acetic anhydride (762), propionic (30), acrylic (243), and 2-ethylhexanoic (5.7, as various metal salts). The 1981 production of higher fatty acids from vegetable oils and tall oil, including both saturated and unsaturated acids, totaled more than 550 thousand metric tons.

A principal use for the higher fatty acids and their derivatives is as surface-active agents, synthetic detergents, and soaps. The nonbenzenoid surfactants of the total market, derived in a large part but not entirely from fatty acids, have increased from 34% in 1964 through 57% in 1965 and 72% in 1966–1972 to 75% in 1981. This dramatic turn-around in market pattern reflects environmental concerns and the superior degradability of straight-chain carbon compounds. Straight-chain compounds are available not only from fatty acids and their derivatives but also from the mixture of linear alcohols obtained by telomerization of ethylene (see Soap; Surfactants and detersive systems). Other significant uses for higher fatty acids are in alkyd resins, and as plasticizers. Dodecyl mercaptan is an important polymerization regulator in the manufacture of synthetic rubber (see Elastomers, synthetic). Erucamide is an important antiblock agent for polyethylene film.

Several aromatic carboxylic acids are produced on a large scale (1982 production in thousand metric tons): benzoic (50), phthalic anhydride (313) (see Phthalic Acids), salicyclic (12), and terephthalic as the dimethyl ester (2,197) (see Phthalic Acids). Phthalic anhydride markets are distributed as follows: plasticizers for poly(vinyl chloride), ca 50%; unsaturated polyester resins, ca 22%; and alkyd resins, ca 20%. Dimethyl terephthalate markets are confined mainly to polyester textile fibers (82%) and film (8%) (see Polyester fibers; Polyesters).

Carboxylic acids will continue to be a significant part of the chemical industry and the advanced industrial economy. However, as petroleum resources decline, more emphasis will be placed on the acids based on renewable resources. The extent of such a shift will depend on the relative prices for starting materials derived from fats and oils vs petrochemicals.

EVERETT H. PRYDE
United States Department of Agriculture

Chem. Mark. Rep. (formerly Oil, Paint, and Drug Reporter), published weekly on Monday by Schnell Publishing Company, Inc., 100 Church Street, New York, N.Y. 10007.

Fatty Acid Producer's Council, Soap and Detergent Association, 475 Park Avenue South at 32nd Street, New York, N.Y. 10016.

Synthetic Organic Chemicals, United States Production and Sales, 1982, USITC Publication 1422, United States International Trade Commission, U.S. Government Printing Office, Washington, D.C., 1983.

FATTY ACIDS FROM TALL OIL

Tall oil fatty acids (TOFA) primarily consist of oleic and linoleic acids and are obtained by the distillation of crude tall oil. Crude tall oil, a by-product of the kraft pulping process, is a mixture of fatty acids, rosin acids, and unsaponifiables. These components are separated from one another by a series of distillations. Several grades of TOFA are available depending on rosin, unsaponifiable content, color, and color stability. The most widely growing area of application is intermediate chemicals which include dimer acids (qv) and epoxidized TOFA esters. Other areas of significant use are protective coatings, soaps and detergents, and ore flotation (qv).

ROBERT W. JOHNSON, JR.
Union Camp Corp.

L.G. Zachary, H.W. Bajok, and G.J. Eveline, eds., *Tall Oil and Its Uses*, Pulp Chemicals Association, New York, 1965.

U.S. Pat. 3,216,909 (Nov. 9, 1965), D.F. Bress (to Foster Wheeler Corp.).

Chemical Division, Sales Department, Union Camp Corporation, Wayne, N.J.

BRANCHED-CHAIN ACIDS

Branched-chain acids contain at least one branching alkyl group attached to the carbon chain which causes the acid to have different physical, and in some cases different chemical, properties than their corresponding straight-chain isomers. For example, stearic acid has a melting point of about 69°C, whereas isostearic acid has a melting point of about 58°C. Branched-chain acids have a wide variety of industrial uses as paint driers, vinyl stabilizers, and cosmetic products. The hazards of handling branched-chain acids are similar to those encountered with other aliphatic acids of the same molecular weight. Eye and skin contact as well as inhalation of vapors of the shorter-chain acids should be avoided.

ROBERT W. JOHNSON, JR.
Union Camp Corp.

A. Fisher, *J. Am. Oil Chem. Soc.* **43**, 469 (1966).

M. Fefer and G. Rubin, *Mod. Plast.*, 178 (Apr. 1970).

TRIALKYLACETIC ACIDS

The trialkylacetic acids comprise a range of synthetically produced acids characterized by the structure below. No member of the series has been reported as occurring naturally. Since 1975, manufacture has been concentrated on two more commercially successful products. These are the first members of the series, trimethylacetic acid, or 2,2-dimethylpropanoic acid (pivalic or neopentanoic acid, or Versatic 5), and the C_{10} acid, neodecanoic acid (Versatic 10). Three of the largest individual outlets for trialkylacetic acids are for the manufacture of the vinyl (VeoVa 10), glycidyl (Cardura E10), as well as peroxy esters.

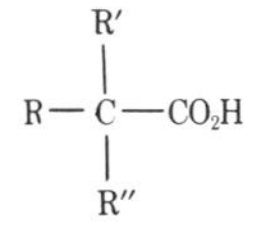

$$R - \overset{\displaystyle R'}{\underset{\displaystyle R''}{\overset{|}{\underset{|}{C}}}} - CO_2H$$

Pivalic acid, mp 33.5°C, bp 163–164°C, $(CH_3)_3CCO_2H$, is a solid at room temperature. Essentially dry pivalic acid can be obtained by drying over molecular sieves immediately before use. The commercially available product is usually more than 99% pure. The reactions of pivalic acid are characterized by steric hindrance at the carboxyl group caused by the alpha substituents. Although reactions proceed less readily than with straight-chain acids, the derivatives prepared are correspondingly more resistant to hydrolysis and oxidation.

The principal handling hazard of pivalic acid is from skin and eye irritation. At elevated temperatures the vapors will be irritant. Any such contact should therefore be avoided.

Pivalic acid is used in the following areas: polymers and resins; pharmaceuticals; photographic chemicals; cosmetic preparations; fuels, lubricants, and transmission fluids; agricultural applications; and metal extractants.

The C_{10} trialkylacetic acids, mp $< -40°C$, are mobile liquids at room temperature. They are a mixture of a large number (at least 27) of highly branched isomers of monocarboxylic acids, at least 98% of which have the tertiary structure.

The C_{10} trialkylacetic acids have toxicities similar to pivalic acid. The same health and safety factors apply. The acids are used in surface-coating industries, as PVC stabilizers, and in the solvent extraction of metals.

The most important reaction of the glycidyl ester, bp 249°C, is with carboxylic acids at 100–150°C, giving diacyl-substituted glycerides, in the production of alkyds (see Alkyd resins) and acrylic resins. That of the vinyl ester, bp 210°C, is the persulfate-initiated emulsion copolymerization with vinyl acetate in paint latex manufacture.

J.W. Parker
M.P. Ingham
R.J. Turner
J.H. Woode
Shell International Chemical Co., Ltd.

Reviewed by

N.P. Potter
Shell International Chemical Co., Ltd.

CARBOXYLIC ACIDS, POLYBASIC ACIDS.   See Dimer acids.

CARBURIZING.   See Metal surface treatments.

CARCINOGENS.   See Industrial hygiene and toxicology.

CARDBOARD.   See Paper.

CARDIOVASCULAR AGENTS

Cardiovascular agents are drugs that affect the circulatory system of the body. This complex network consists of the heart and a system of elastic blood vessels (the aorta, arteries, capillaries, and veins) that carry the blood from the heart to and from the various organs. Cardiovascular agents elicit their responses by actions on the heart and blood vessels, but may in some instances principally affect the autonomic and central nervous systems or the kidneys.

Antihypertensive Agents

Hypertension (diastolic blood pressure > 12.7 kPa (95 mm Hg)) is a disease known to affect 10–15% of the adult population of the United States and is thought to have been identified only in 50–75% of afflicted persons. Therapy by drug treatment, surgery, or both, is aimed at reducing arterial pressure as close to normal as possible (diastolic blood pressure < 11.3 kPa [85 mm Hg]) in order to prevent or postpone cardiovascular complications, ameliorate congestive heart failure, reverse retinopathy, and prolong life. The principal physiological abnormality in essential hypertension is increased total peripheral resistance, yet no defect of the effector organ has been found.

Physiology and Biochemistry of Blood-Pressure Regulation

The level of arterial blood pressure is nearly constant and maintained despite the variation in requirements for blood flow through vascular beds such as kidneys, skeletal muscle and gastrointestinal tract, exercise, and postural changes. The maintenance of mean arterial pressure is a result of homeostatic mechanisms that control cardiac output and peripheral resistance. The heart's pumping action must overcome the resistance of peripheral arterioles (vasomotor tone) to the blood flow into the arterial circuit and to the capillaries. Blood pressure is directly proportional to the product of the flow per unit time and the resistance to flow. Blood pressure is also controlled by cardiac output which is dependent on blood volume. Various metabolic and hormonal factors determine the vasomotor tone: the sympathetic nervous system, the adrenal catecholamines, and angiotension II are known vasoconstricting factors; and the parasympathetic nervous system, the kallikrein–kinin system and the prostaglandins (qv) are known vasodilating factors.

Antihypertensive Drug Therapy

In the stepped-care program, a widely used approach effective in most patients when diastolic pressure is between 13.7–18.7 kPa (103–140 mm Hg), an oral diuretic (usually a thiazide type) is administered in the first step. In more than 30% of patients, blood pressure is normalized to $\le 18.7/12$ kPa (140/90 mm Hg). If this does not occur within two months, addition of the direct-acting vasodilator hydralazine, or a sympathetic depressant, propranolol, methyldopa or reserpine, is taken as the second step. In malignant hypertension (diastolic pressure > 17.3 kPa (130 mm Hg)), this combination of two drugs is the start of therapy. If this two-drug combination therapy is not successful, a third drug is added: if hydralazine was used in the second step, methyldopa is added in the third; if methyldopa or reserpine was used in the second step, hydralazine is added in the third. Individualized therapy is advised in the event that unsatisfactory results are obtained with the triple combination. Neuronal blocking agents, such as guanethidine, are the last drugs to be added to the regimen.

If enough information is known to allow characterization of the nature of the hypertensive disease the patient is suffering from, a rational approach to therapy can be used. For this purpose antihypertensive drugs can be classified into three categories, (1) diuretics, (2) drugs that depress the sympathetic nervous system, and (3) vasodilators. They may act centrally or peripherally.

Antihypertensive Drugs

Diuretics. Three classes of diuretics are used in the treatment of hypertension; (1) thiazides and related nonthiazides (7-sulfamoyl-6-chloro-2H-benzo-1,2,4-thiadiazine 1,1-dioxides (1) and their 3,4-dihydro derivatives, chlorthalidone—a (1-hydroxyl-3-oxo-1-isoindolinyl)benzenesulfonamide, and the 1,2,3,4-tetrahydro-4-oxo-6-quinazolinesulfonamides, metolazone and quinethazone); (2) the loop diuretics, ethacrynic acid and furosemide, along with bumetanide and indapamide (2); and (3) the potassium-sparing diuretics, spironolactone and triamterene (3).

(1) 2H-benzo-1,2,4-thiadiazine 1,1-dioxides

(2) indapamide

(3) triamterene

The thiazides are usually the first type of drug prescribed for the treatment of mild or moderate hypertension. Initially, the antihypertensive effect is a result of a reduction in plasma volume with consequent decrease in cardiac output, probably through an initial negative sodium balance.

The thiazide diuretics, although possessing antihypertensive properties of their own, find greatest use in combination with other antihypertensive drugs, such as hydralazine, methyldopa, and reserpine, by potentiating their action and reducing the dose requirement. This has the beneficial effect of reducing the incidence and severity of adverse side effects. They are well tolerated and effective in lowering blood pressure in both supine and standing positions. Long term antihypertensive effects may result from reduction in vascular reactivity to sympathetic stimulation, thereby preventing compensation for reduction in plasma volume. When given with an antihypertensive drug of the vasodilator type, they counteract secondary fluid retention and may diminish development of drug resistance.

The main side effects of thiazide diuretics are hypokalemia, increase in concentrations of blood uric acid, serum calcium or blood urea, hyperglycemia, weakness, dizziness, fatigue, leg cramps, and gastrointestinal disturbances. Plasma renin activity is often increased and stays elevated.

Loop diuretics are more potent and act more rapidly than the thiazides (see Diuretics). They inhibit the reabsorption of sodium by acting primarily on the medullary portion of the ascending limb of the Henle loop where they block the active transport of sodium. They also inhibit reabsorption of sodium in the proximal and distal tubules. They are used in hypertensive patients who have become refractory to the thiazide diuretics and for patients with severely impaired renal function. With the exception of indapamide, loop diuretics can cause hyperglycemia and hyperuricemia.

Potassium-sparing diuretics interfere with the reabsorption of sodium at the distal exchange sites. Hence sodium excretion is promoted while potassium is conserved. Spironolactone, an antagonist of the corticosteroid aldosterone, has only weak antihypertensive activity whereas triamterene, which interferes directly with electrolyte transport in the distal tubules, has no significant antihypertensive activity. Both drugs are used mainly in conjunction with the thiazide diuretics in order to reduce potassium excretion and minimize alkalosis. Side effects include occasional gastrointestinal disturbances, muscular weakness, and skin rashes.

Drugs that depress the sympathetic nervous system. Depression of the sympathetic nervous system, either centrally or peripherally, is the predominant blood-pressure lowering mechanism of the majority of antihypertensive drugs. Inhibition of any steps involved in the formation and disposition of norepinephrine has potential for antihypertensive action.

Inhibition of adrenergic nerve transmission by preventing impulses from the sympathetic nerve terminals from reaching the effector site (arterioles) is produced by so-called adrenergic neuronal blocking agents. This has the effect of decreasing sympathetic tone (vasodilatation) with consequent decreased peripheral resistance, increased peripheral flow, and reduction in blood pressure. In contrast with the sympathetic nervous system, where norepinephrine is released at the sympathetic postganglionic fibers and into the synaptic cleft, acetylcholine is released at the parasympathetic postganglionic end organs. Blocking of parasympathetic and sympathetic impulses at the first level of ganglia (ganglionic blockade) results in the desired reduction in blood pressure by interference with the adrenergic system, but at the same time produces undesirable side effects because of interference with parasympathetic (cholinergic) control of many visceral functions: urinary retention, constipation, dry mouth, blurred vision, and impotence. Orthostatic hypotension and weakness are adverse side effects that may occur from sympathetic blockade.

Drugs that lower blood pressure mainly by sympatholytic action are methyldopa (3-hydroxy-α-methyl-L-tyrosine) (4), *Rauwolfia serpentina* (the powdered root of a small climbing shrub indigenous to India) and reserpine (5) and related alkaloids, propranolol (6) and other β-adrenergic blocking agents, clonidine, guanethidine, α-adrenergic blocking agents, ganglionic blocking agents, and guanabenz.

(4) methyldopa

Name	R	R′
(5) reserpine	OCH₃	

(6)

Vasodilators. Lowering of peripheral resistance with resultant lowering of blood pressure can occur by direct action of drugs on vascular smooth muscle or by central α-adrenergic stimulation. Since direct-acting vasodilators do not affect sympathetic reflexes, little or no postural or exercise-induced hypotension results. The drugs are given generally with another antihypertensive drug since, alone, they are not effective in lowering blood pressure at doses that are safe. The reflex tachycardia that results from the action of antihypertensive vasodilators requires concomitant administration of a β-adrenergic blocking agent. Vasodilators include hydralazine, minoxidil (7), diazoxide (8), sodium nitroprusside, and prazosin (9).

(7) minoxidil

(8) diazoxide

(9) prazosin

Antagonists of angiotensin II and its formation. Saralasin (sar¹-ala⁸-angiotensin II (sar-sarcosine)), a competitive antagonist of angiotensin II, and teprotide (a modified bradykinin B potentiator and angiotensin I-converting enzyme inhibitor), have been found useful for the diagnosis and treatment of hypertension resulting from high renin levels. Administration of sar¹-ala⁸-angiotension II produces a fall in blood pressure in sodium-depleted patients with renal artery stenosis. Application of these angiotensin II antagonists is limited since they are active only by the iv route. Captopril (SQ 14,225), (S)-1-(3-mercapto-2-methyl-1-oxopropyl)-L-proline, is an orally active angiotensin-converting enzyme (ACE) inhibitor that is indicated for treatment of hypertensive patients who on multidrug regimens have either failed to respond satisfactorily or developed unacceptable side effects. Usually, multidose regimens include combinations of a diuretic, a sympathetic nervous system-active agent (such as a beta-blocker) and a vasodilator. Enalapril, (S)-1-[N-[1-(ethoxycarbonyl)-3-phenylpropyl]-L-alanyl]-L-proline, another orally active ACE inhibitor, has been shown to lower blood pressure in hypertensive patients.

Cardiac Glycosides

Cardiac glycosides, used in modern medicine as cardiotonics, are found in plants of the *Digitalis*, *Strophanthus*, and *Scilla* (Squill) species, and in the skin of certain toads. They are made up of a steroid aglycone (genin) in combination with 1–4 mol sugar. Some of the sugars are acetylated. The genins, which are generally convulsive poisons, have some cardiotonic activity of their own, but the saccharide moieties contribute to their potency and transport. The glycosides are extraordinarily powerful on a weight basis in their ability to contract heart muscle (the average daily oral therapeutic dose of digitoxin in man is 0.05–0.2 mg), but they also have a very low margin of safety.

The beneficial effects that the cardiac glycosides demonstrate in congestive heart failure are the result of increased cardiac efficiency with attendant increase in blood flow to other organs, especially the kidneys. Diuresis is increased with reduction in edema; venous pressure, blood volume, and heart size are decreased. These effects result from increased myocardial contractile force (positive inotropic effect) in the failed heart. The cardiac glycosides also slow the ventricular rate in atrial fibrillation or flutter. Digitalis terminates and prevents the recurrence of supraventricular tachycardia. It is useful for the treatment of sinus tachycardia and supraventricular and ventricular premature beats when primarily caused by congestive heart failure. The mechanism of the positive inotropic action of digitalis is thought to result from its strong inhibition of the Na-K pump at the cellular membrane by binding specifically to Na-K ATPase (the enzyme that mediates the pumping of sodium ions out of the heart muscle cell and potassium ions into the cell).

Chemistry

Medicinally important cardiac glycosides that occur in or are isolated from leaves of *Digitalis Lanata* Ehrd. and *Digitalis purpurea* L. (family Scrophulariacea) are the primary glycosides, lanatosides A, B, and C, and deacetylanatosides A and B. Seeds of *Strophanthus kombé* Oliv. (family Apocynaceae) yield K-strophanthoside, K-strophanthin-β, and cymarin. Ouabain is obtained from seeds of *Strophanthus gratus* (Wall. & Hock.) Baill. (family Apocynaceae), and scillaren A from *Urginea* (*Scilla*) *maritima* (L.) Baker (family Liliceae). These glycosides are degraded to their component parts in stepwise fashion by appropriate sequences of acidic, enzymatic, and basic hydrolysis.

The genins of the cardiac glycosides all have the cyclopentanoperhydrophenanthrene backbone of the steroids and bile acids. They are substituted with a 3β-hydroxyl group (the point of attachment of the sugar moiety) in common with many steroids, but uniquely have a 14β-hydroxyl group with consequent C/D-*cis* ring fusion (eg, digitoxigenin (10)). Each has a 17β-butenolide or 17β-pentadienolide substituent and there may be, additionally, hydroxyl, carbonyl, epoxide, and unsaturation functions in the molecule.

(10) digitoxigenin

Therapeutic Indications

The principal therapeutic indications for cardiac glycosides are congestive heart failure and arrhythmias.

Adverse Reactions

Severe toxicity caused by digitalis is treated by withholding digitalis and diuretics, and treating cardiac failure if present. Potassium salts, either orally or iv, are administered, with the exception that they are not given iv in the presence of A-V block or renal failure. Digitalis-induced arrhythmias are treatment with phenytoin, lidocaine, and propanolol.

Antiarrhythmic Agents

The cardiac cycle consists of contraction (systole) and relaxation (diastole), including a brief period of inactivity before the next cycle. Cardiac arrhythmias can arise from disturbances of impulse formation, disturbances of conduction, or both. Antiarrhythmic drugs, by affecting the electrophysiology of heart muscle, act mainly by (a) diminishing automaticity by reducing the rate of phase 4 diastolic depolarization and, hence, depressing repetitive firing of impulses, and (b) altering conduction velocity and refractory period, thereby inhibiting disorders of reentry and reciprocal excitation. Antiarrhythmic drugs affect conduction velocity or membrane responsiveness in different ways, both qualitatively and quantitatively. Type 1 drugs decrease automaticity and delay conduction velocity, but otherwise differ: quinidine, procainamide hydrochloride (11), bretylium tosylate, and propranolol or other β-adrenergic blocking agents. Type 2 drugs decrease automaticity, but have little effect on or in low concentrations can increase conduction velocity: lidocaine (12) and phenytoin.

(11) procainamide hydrochloride

(12) lidocaine

Other antiarrhythmic drugs include digitalis, the drug of choice for atrial flutter or fibrillation; atropine sulfate hydrate, a blocker of the parasympathetic neurotransmitter; acetylcholine, used to treat bradyarrhythmias resulting from acute myocardial infarction; edrophonium chloride, a rapid acting cholinesterase inhibitor used to terminate supraventricular tachycardia that does not respond to vagal maneuvers; phenylephrine hydrochloride, a potent constrictor of arteries and arterioles used to control supraventricular tachycardia in patients who do not respond to carotid massage; disopyramide, synthesized in an attempt to develop a quinidine-like antiarrhythmic agent and found to be less toxic than quinidine and more effective in reversing arrhythmias following myocardial infarction; verapamil, a coronary vasodilator and clinically effective antianginal agent; canrenoate potassium, an aldosterone antagonist that suppresses frequent ventricular premature beats and ventricular bigeminal and trigeminal rhythms; and mexiletine hydrochloride, which is effective in preventing recurrence of ventricular arrhythmias resulting from mycardial infarction once these arrhythmias have been suppressed by iv lidocaine.

Antiatherosclerotic Agents

The most common cause of death in the United States is a result of the complications stemming from atherosclerosis. These are coronary heart disease (over a million (10^6) heart attacks and 600,000 deaths annually in the 1970s), cerebral vascular disease, and peripheral vascular diseases, and they account for 50–60% of all deaths.

A common occurrence of the aging process is the development from early childhood of atheromatous plaques that are incorporated in the walls of the coronary, aortal, cerebral, vertebral, and renal arteries, and the principal arteries to the legs. These lesions begin as gelatinous elevations or fatty streaks and progress to pearly-white fibrous plaques or atheromatous plaques containing a central core of lipid (mainly cholesterol) overlaid with a fibrous cap of connective tissue (mostly collagen and elastic fibers). Disorders involving lipids are the result of deranged lipoprotein metabolism where there is either an overproduction or faulty catabolism of one or more lipoproteins.

Treatment of hyperlipoproteinemia is almost always initiated with a suitable diet since this often alleviates the condition without the use of potentially toxic drugs. Drugs, if needed, are used in combination with diet for an additive effect. Drug therapy is aimed at correcting specific lipoprotein elevation either by affecting production of the lipoprotein or effecting its removal.

Drugs that decrease lipoprotein production include clofibrate (**13**), an aryloxyisobutyric acid ester, and the vitamin, nicotinic acid, which are useful in the treatment of type III and some cases of types IV and V hyperlipoproteinemias. Drugs that increase lipoprotein catabolism include cholestyramine resin, the bile-sequestering quaternary ammonium salt anion-exchange resin, D-thryoxine sodium (**14**), the sodium salt of the dextrorotatory enantiomer of thyroxine, colestipol, the high molecular weight anion-exchange resin, and the antibiotic neomycin. Tibric acid, a sulfonamidobenzoic acid, and p-aminosalicylic acid also are used. Cholestyramine resin, colestipol, and p-aminosalicylic acid are used to treat type II hyperlipoproteinemia; D-thyroxine sodium is used to treat types II and III; and tibric acid is used to treat type IV.

(**13**) clofibrate   (**14**) D-thyroxine sodium

Antianginal Agents

Angina pectoris is typically a pressing, tight, or burning sensation in the precordial region, sometimes with pain radiating to the left arm or to the neck or back. The pain is characteristically brought on by consumption of heavy meals, exercise, anxiety, or other stress, and is quickly relieved by rest. The pain results when the myocardial oxygen supply is insufficient to meet the metabolic demand and produces myocardial hypoxia. This condition can be alleviated by either increasing the supply of oxygen or reducing the demand. Drugs that have been clearly shown to be of value in alleviating angina pectoris are the sublingually, topically and transdermally administered nitrates and the orally administered β-adrenergic blocking agents. The latter are used for protracted prophylaxis of angina pectoris.

Of the sublingual, topical, and transdermal nitrates, nitroglycerin is the choice drug, since it dramatically and rapidly relieves the pain of an attack of angina pectoris. The introduction of transdermal nitroglycerin adhesive "band-aid" patches has solved the problems of ease of handling controlled dosing (eliminating the peaks and valleys of plasma levels with conventional therapy) and prolonged duration of action (see Supplement article). Nitroglycerin and the other nitrates (isosorbide dinitrate, erythrityl tetranitrate, mannitol hexanitrate, pentaerythritol tetranitrate, and trolnitrate phosphate) are thought to relieve anginal pain by reducing myocardial oxygen requirement through reduction in ventricular volume and myocardial tension and by lowering arterial blood pressure. Nitrates, when administered orally (erythrityl tetranitrate, mannitol hexanitrate, pentaerythritol tetranitrate, trolnitrate phosphate) are claimed to have a long duration of action and, hence, provide long-term prophylaxis.

Because β-adrenergic blocking agents reduce heart rate and contractility resulting from sympathetic stimulation, particularly from exercise, myocardial oxygen consumption is reduced and the pain of angina pectoris is delayed or prevented. Long-term therapy reduces the frequency of anginal attacks, nitroglycerin consumption, and may have the effect of reducing the incidence of myocardial infarction. Combination therapy with sublingual nitrates results in additive effects of the two types of drugs because of their different modes of action.

Other antianginal drugs include amyl nitrite, perhexiline maleate, lidoflazine, nifedipine, verapamil, and the calcium channel blocking agents.

Peripheral Vasodilators

The peripheral circulation is impaired under certain conditions, and vasodilator drugs are prescribed in an attempt to restore circulation to ischemic areas. However, most drugs are nonselective in their action and produce generalized vasodilatation that can prove deleterious by shunting blood away from the affected area to normal areas in the extremities. The most common vascular disorders affect circulation to the skin, skeletal muscle, and cerebrum, but drug treatment has only been well-established in cutaneous disorders.

α-Adrenergic receptors are present in coronary, skin, skeletal, and cerebral arterioles and cause constriction in response to sympathetic nerve stimulation. Vasodilating drugs that act by α-adrenergic blockade are effective in cutaneous disorders. Certain cerebrovascular disorders, such as impaired memory and intellectual function, depression, and psychosis, are reported to respond to vasodilator drug therapy. Although most drugs that have been tried in an attempt to improve intellectual function and memory are ineffective, the dihydrogenated ergot alkaloids and some direct-acting vasodilators are still in use.

The agents that act on sympathetic nerve terminals are reserpine and phenoxybenzamine hydrochloride. The direct-acting vasodilators are nicotinic acid, tolazoline, cyclandelate, isoxsuprine, nylidrin, nicotinyl alcohol, papaverine, dioxyline, naftidrofuryl, ethaverine, and suloctidil, as well as hydergine, a mixture of hydrogenated ergot alkaloids.

LEON GOLDMAN
Lederle Laboratories
American Cyanamid Company

AMA Drug Evaluations, 5th ed., American Medical Association, Chicago, Ill., 1983.

M.E. Wolfe, ed., *Burger's Medicinal Chemistry*, 4th ed., John Wiley & Sons, Inc., New York, 1981.

A.G. Gilman, L.S. Goodman, and A. Gilman, eds., *Goodman and Gilman's The Pharmacological Basis of Therapeutics*, 6th ed., Macmillan Publishing Co., Inc., New York, 1980.

J.E. Knoben and P.O. Anderson, eds., *Handbook of Clinical Drug Data*, 5th ed., Drug Intelligence Publications, Inc., Hamilton, Ill., 1983.

CARDIOVASCULAR AGENTS.   See also Supplement.

CARNALLITE, $KCl \cdot MgCl_2 \cdot 6H_2O$.   See Potassium compounds.

CARNAUBA WAX.   See Wax.

CARNOTITE, $K_2O \cdot 2UO_3 \cdot V_2O_5 \cdot 3H_2O$.   See Uranium and uranium compounds.

CAROTENOIDS.   See Vitamins—Vitamin A.

CASEHARDENING.   See Metal surface treatments.

CASEIN.   See Milk and milk products.

CASTOR OIL

Castor oil is derived from the bean of the castor plant, *Ricinus communis* L., of the family Euphorbiaceae, found in many tropical and subtropical countries. It is also known as ricinus oil, oil of Palma Christi, tangantangan oil, and Neoloid. The oil is a triglyceride of fatty acids which contains 87–90% ricinoleic acid, *cis*-12-hydroxyoctadec-9-enoic acid, $CH_3(CH_2)_5CH(OH)CH_2CH=CH(CH_2)_7COOH$, a rare source of an eighteen carbon hydroxylated fatty acid with one double bond. The oil is pale, yellow, and viscous, with a slightly characteristic odor, and nearly tasteless but familiarly unpleasant through its minor use as a purgative.

Recovery of castor oil from castor beans is commonly by the use of hydraulic pressers or expellers followed by solvent extraction (see Vegetable oils).

Properties

The average fatty acid composition of castor oil contains ricinoleic acid (89.5%), dihydroxystearic acid (0.7%), palmitic acid (1.0%), stearic acid (1.0%), oleic acid (3.0%), linoleic acid (4.2%), linolenic acid (0.3%), and eicosanoic acid (0.3%).

The oil is distinguished from other triglycerides by its high specific gravity (0.957–0.961 at 25/25°C), viscosity = (6.5–8.0) cm²/s hydroxyl value (160–168) and its solubility in alcohol (1:2 in 95% ethanol at 20°C); it is soluble in polar organic solvents, less soluble in aliphatic hydrocarbon solvents, and slightly soluble in petroleum ether.

Chemical Modification

Castor oil serves as an industrial raw material for the manufacture of a number of complex derivatives.

Catalytic dehydration uses as catalysts sulfuric acid and its acid salts, oxides, and activated clay at 230–280°C. This converts castor oil to an excellent drying oil called dehydrated castor oil (see Driers and metallic soaps), used extensively by the coatings industry. In a typical process, castor oil is heated to 230–280°C under vacuum and 3–5% diluted sulfuric acid is added to it at a controlled rate.

In sulfonation, concentrated sulfuric acid is added over a period of several hours while cooling to maintain a temperature of 25–30°C, followed by washing to remove surplus acid, and then neutralized to produce sulfonated castor oil, also known as Turkey red oil. Commercial sulfated castor oil contains ca 8.0–8.5% combined SO_3; the sulfate group imparts excellent wetting, emulsification, and dispersing characteristics. Sulfonation with anhydrous SO_3 (2 mol) at temperatures higher than the sulfuric acid treatment produces a product with better hydrolytic stability and contains less inorganic salts and free fatty acids.

Alkali fusion yields two different sets of products, depending on the reaction conditions. At 180–200°C using one mole of sodium or potassium hydroxide, methyl hexyl ketone and 10-hydroxydecanoic acid result. Two moles of alkali per mole of ricinoleate at 250–275°C and a shorter reaction cycle produce 2-octanol and sebacic acid.

Oxidation produces clear viscous oils resulting from controlled oxidation by intimate mixing (blowing) of castor oil at 80–130°C with air or oxygen with or without the use of a catalyst. The reaction is promoted by transition metals and the properties of the oil are dependent on the reaction conditions.

Pyrolytic decomposition at 340–400°C splits the ricinoleate molecule at the hydroxyl group to form heptaldehyde and undecylenic acids. Heptaldehyde is converted to heptanoic acid by oxidation techniques, and to heptyl alcohol by catalytic hydrogenation.

Hydrogenation can be performed in a number of ways to produce unique derivatives. For example, simple double-bond hydrogenation at 140°C in the presence of Raney nickel catalyst produces glyceryl tris(12-hydroxystearate), mp 86°C. Hydrogenation of ricinoleic acid with a copper cadmium catalyst at 220°C and 26 MPa (ca 250 atm) yields 70% ricinoleyl alcohol. Stearates are prepared by dehydrating hydrogenated castor oil, or by dehydration of castor oil followed by full hydrogenation of the diene intermediate. The hardness, flexibility, mp, and iodine value of the finished products are controlled by the degree of hydrogenation. Esters of castor oil are changed by hydrogenation from fluid products to soft waxes with a mp range of 45–65°C.

Alkoxylation with ethylene oxide produces ethyloxylated derivatives with varied degrees of hydrophobic–hydrophilic properties, and with propylene oxide yields mineral oil soluble derivatives. The reaction is carried out at 120–180°C and 0–405 kPa (0–4 atm) using alkaline catalysts.

In the production of nylon-11, castor oil is transesterified with methyl alcohol to form glycerol and methyl ricinoleate, which undergoes pyrolysis at ca 450–500°C to form methyl 10-undecylenate and heptaldehyde. Rilsan monomer is formed by addition of bromide and then heated to give the polymer by condensation. Nylon-11 has a mp range of 186–190°C which facilitates high speed, uniform processing. It displays excellent chemical resistance and stability in contact with all types of fuels, along with shock and vibration resistance.

Uses. Castor oil is sometimes used as a purgative, but is primarily used as a raw material for the preparation of chemical derivatives used in coatings; urethane derivatives (see Urethane polymers); surfactants (qv) and dispersants; cosmetics; and lubricants. Nylon-11 is used in the automotive industry, in powder coatings (qv), or coatings that require no solvents.

FRANK C. NAUGHTON
FRANK DUNECZKY
C. RICHARD SWENSON
THADDEUS KROPLINSKI
MURRAY C. COOPERMAN
CasChem, Inc.

F.C. Naughton, *J. Am. Oil Chem. Soc.* **51**, 65 (1974).

K.T. Achaya, *J. Am. Oil. Chem. Soc.* **48**, 759 (1971).

M. Brauer and R. Sabia, "Design Consideration, Chemistry and Performance of a Reenterable Polyurethane Encapsulant," *International Wire and Cable Symposium Proceedings*, U.S. Army Electronics Command, Washington, D.C., Nov. 1975, pp. 104–111.

CATALYSIS

A catalyst is a substance that alters the velocity of a chemical reaction without appearing in the products. Catalysts accelerate reactions but do not change equilibria; both forward and backward reactions are catalyzed to the same extent. Increasing the velocity of a desired reaction relative to unwanted reactions maximizes the formation of a desired product.

Catalysts are believed to function through an unstable chemical complex formed between the catalyst and reactant molecules. This complex reacts to produce new compounds with dissociation of the complex and regeneration of the catalyst which can then bring about the transformation of additional reactant molecules (see Enzymes).

Chemical and Physical Aspects

The rate of a chemical reaction is proportional to a rate constant:

$$k = Ae^{-E/RT}$$

Because of the exponential form of the equation, even a small decrease in E, the activation energy, can bring about a large increase in velocity. The catalytic pathway is a series of three chemical reactions: formation of the complex (activated adsorption), surface reaction, and desorption.

Solid surface catalysts invariably adsorb reactant molecules, adsorption and chemical reaction occurring on active centers identified as certain types of lattice defects. The geometric factor in catalysis emphasizes the significance of the reactant molecule's spacial arrangement and the catalyst crystalline lattice.

The activity and selectivity of a catalyst are affected by its pore structure, eg, in a very slow reaction molecules can diffuse through a pore system to the center of a catalyst particle before they react and the entire internal surface area will be used. The fraction of surface available in a given reaction can be calculated. The experimental rate constant, K_{exp}, is compared with

$$K_d = (18/a^2)\rho BVgD$$

where a = catalyst granule size, ρB = bulk density, Vg = porosity, and D = diffusion constant.

Various theories have been applied to catalysis, including quantum mechanical treatment of solids and the electron band, crystal field, resonance valence bond, and molecular orbital ligand field theories.

Techniques

The following factors are considered when choosing a catalyst for a specific reaction: selectivity, or the efficiency in catalyzing a desired transformation; activity, or the overall conversion, expressed as the amount of reactant in contact with the catalyst under a given set of conditions converted to all products; stability, the ability to retain initial activity and selectivity over the catalyst's lifetime; physical suitability; regenerability; and cost.

Heterogeneous catalysts, in contrast to homogeneous catalysts, form a separate phase from reactants and products and are usually solids. They have a microporous structure and a very large internal surface area (it may reach $1000 \ m^2/g$) but these colloidal catalysts are often inactivated by the presence of small amounts of poisons. Poisons act by combining with an active site in the catalyst surface and rendering it inactive.

Most poisons are strongly sorbed and react quantitatively with active sites to form stable inactive surface compounds, eg, acidic catalysts are poisoned by basic nitrogen and alkali-metal ions, metal catalysts by sulfides, arsenic and lead compounds. Some poisons cause loss of selectivity, eg, nickel and copper present in gas oils used in catalytic cracking. In some cases, poisons minimize a desired reaction while permitting an undesired reaction. Reactivation of a catalyst is sometimes accomplished by decomposition or desorption of the poison.

In contrast, small amounts of promoters or cocatalysts may cause striking increases in catalytic activity.

In industrial use, catalysts lose activity and sometimes selectivity. In general, heterogeneous catalysts lose effectiveness owing to overheating and contamination. Contaminants affecting catalytic efficiency are metals entrained in feedstocks; oxygen, nitrogen, or sulfur compounds; and polynuclear aromatic compounds which form coke.

Preparation

Properties of a catalyst are determined by its preparation, with care taken to ensure duplication of characteristics. Colloid chemistry is important, including gel preparation (eg, hydrous oxides and systems mainly of silica or alumina are amenable to gel formation), leaching selected components from solids (eg, Raney nickel catalyst is prepared by leaching aluminum from a nickel–aluminum alloy by means of a caustic alkali), or decomposition of salts or hydrates with gas evolution. Frequently, the support is manufactured first by one of these techniques followed by impregnation with additional chemical components. The product may again require suitable treatment, eg, precipitation, calcination, reduction (eg, for hydrogenation catalysts such as nickel, cobalt, iron, or copper catalysts), etc.

Organic and Inorganic Chemicals

Important catalytic commercial processes include the Deacon process; platinum-catalyzed oxidation of ammonia for nitric acid production; the Haber and Mittasch ammonia synthesis process; hydrogen production employing catalytic reactions such as reforming, shift, and methanation; hydrogenation techniques for hardening fat; synthesis of methanol from carbon monoxide and hydrogen; production of phthalic anhydride via oxidation of naphthalene; oxo reaction to produce aldehydes, which can be catalytically hydrogenated to the corresponding alcohols; catalytic dehydrogenation of butene or butane (see Butadiene); the catalytic cracking, reforming, or hydrodesulfurization of petroleum; polymerization for production of plastics and synthetic rubber; the production of polyurethane plastics.

Transition-Metal Complexes — Homogeneous Catalysis

The distinction between homogeneous and heterogeneous catalysts has been disappearing as fundamental similarities have been recognized. However, certain transition-metal complexes are classified as homogeneous catalysts since they can be molecularly dispersed in solution, eg, hydrocobalt tetracarbonyl, $HCo(CO)_4$. Some homogeneously catalyzed reactions are hydrogenation, hydroformylation, double-bond migration, polymerization, and oxidation.

Organometallic complex catalysts tend to be highly active, specific, and selective, and may resist poisoning better than heterogeneous systems. The catalytic center may have two metal sites. Organometallic catalysts have a central transition-metal atom or ion bonded to molecules known as ligands which form a polyhedron around the metal. The nature of the metal and ligands critically influences the electronic structure of the complex. The coordination bonding properties of the metal are a key feature of the catalytic complexes. Important reactions of catalyst complexes are ligand exchange, oxidative addition, and the insertion reaction.

The most important industrial reactions are the Wacker process (see Acetaldehyde); vinyl acetate (see Acetic acid; Vinyl polymers) the oxo process (see Oxo process); methanol carbonylation (see Acetic acid); and the Ziegler-Natta polymerizations (see Olefin polymers).

The advantage of complex catalysts including the anchored complexes lies in the predictability of their properties, the ease of interpreting reaction mechanisms, and the consequent potential for extending the homogeneous catalysts to many new applications.

Oxidation Catalysis

Classical examples of heterogeneous catalysis include oxidation of CO over copper, of NH_3 over Pt, and of SO_2 over Pt or V_2O_5. Heterogeneous oxidation catalysis is similar to catalysis by metal complexes since both utilize a metal of variable valence.

Commercially important hydrocarbon-oxidation processes include oxidation of benzene to maleic anhydride and toluene to benzoic acid, naphthalene and xylenes to phthalic acids, ethylene and propylene to their respective epoxides (see Hydrocarbon oxidation). The catalytic ammoxidation of propylene, eg, the Sohio process, is the successful process for the manufacture of acrylonitrile (see Acrylonitrile). Ethylene dichloride, the precursor of vinyl chloride, is manufactured by the oxidation of ethylene (see Chlorocarbons; Vinyl polymers).

Catalysis Applied to Fuels

The concepts of acid and basic catalysis, as proposed by Brønsted and Lowry, apply to reactions of industrial importance, particularly hydrocarbon reactions used in large-scale petroleum refining.

Acid catalysts are employed in such diverse areas as Friedel-Crafts reactions (qv) and reactions in the presence of clays. The isomerization of n-butane over an aluminum halide–hydrogen catalyst takes place only in the presence of olefins. Friedel-Crafts catalysts, mineral acids, such as H_2SO_4 and HF, and claylike substances, eg, $H_2O–Al_2O_3–SiO_2$, all act in a similar manner because of their essential acidity. In each case the active catalyst contains a proton; it is a Brønsted acid. For example, hydrogen chloride plus aluminum trichloride give

$$H^+ \begin{bmatrix} Cl & & Cl \\ & Al & \\ Cl & & Cl \end{bmatrix}^-$$

Catalytic activity of acid catalysts is correlated with surface acidity for different reactions. The Brønsted acidity is necessary for successful catalytic cracking, and the distribution of acid strengths is important.

Basic catalysts are employed less than acid catalysts, but are used industrially for polymerization of conjugated dienes as well as for the isomerization of olefins, dehydrogenation of certain diolefinic materials to aromatics, alkylation of arylalkanes, and polymerization of monoolefins. Base catalysis gives different products than acid catalysis, eg, ethylene alkylates toluene in the side-chain with a base catalyst whereas acid catalysis gives ring alkylation.

Zeolites (Molecular Sieves)

Crystalline zeolites are silicates with an open framework, regular structure, and apertures of molecular dimensions based on a tetrahedron of four oxygen ions surrounding a smaller silicon or aluminum ion (see Molecular sieves). Important zeolite catalysts are known as X, Y (similar to the rare natural mineral, faujasite), erionite, and mordentite, and the newer ZSM-5 and ZSM-11. The most significant application of crystalline zeolites has been in catalytic cracking where catalysts containing 10–20% zeolites greatly improve selectivity and activity.

Dual-Function Catalysts

Reforming represented a significant advance in the utilization of the dual-function catalysts, eg, reforming with platinum catalysts is today the second largest industrial catalytic application (see Petroleum). Another example of dual functionality is the conversion of C_6-cycloparaffins (cyclohexane and methylcyclopentane) over single- and dual-function catalysts for converting methylcyclopentane to benzene. For certain reactions, dual-function catalysts can accomplish more than passing the reactants through two reactors in sequence, each filled with a different single-function catalyst.

Multimetallic Catalysis

Bimetallic cluster catalysts are composed of atoms of two different metals in a state of high dispersion on a carrier. Dispersion refers to the fraction of atoms in a metal crystallite that exist on the surface, with the degree of dispersion increasing with decreasing crystallite size. These novel bimetallic systems offer a new range of catalytic properties and are not limited to combinations of metals that form bulk alloys. New metallic catalysts with high activity and unprecedented activity maintenance are employed in the reforming of petroleum naphthas to high-octane gasoline and in isomerization, hydrocracking, and hydrogenation. In the petrochemicals field, improved Pd–Au catalysts are utilized for the synthesis of vinyl acetate, whereas more selective catalysts (eg, Ag–Au, Cu–Au) are used for the partial oxidation of olefins. Bimetallic catalysts show increased activity, improved resistance to poisoning by coke accumulation or deposition, and optimize product selectivity, eg, dehydrocyclization and aromatic hydrogenation are possible with a minimum of cracking reactions (hydrogenolysis) while maintaining very high hydrogenation activity.

Synthetic Fuels

Synthetic fuels from coal, oil shale, and biomass are expected to become of enormous industrial importance with catalysis playing a key role in the conversion of these complex solids to more convenient gaseous and liquid forms, and in the removal of sulfur, nitrogen, and oxygen in addition to inorganic mineral matter (see Fuels, synthetic).

Four general types of coal liquefaction processes are catalytic hydroliquefaction, pyrolysis, solvent extraction, and indirect synthesis (Fischer-Tropsch). The mechanism of direct coal liquefaction is believed to consist of a primary thermal pyrolysis of coal, followed by catalytic hydrogenation of the reactive fragments or asphaltenes formed by pyrolysis. Several direct catalytic coal-hydroliquefaction processes use active cobalt–molybdenum–alumina catalysts developed for petroleum desulfurization, including the H-Coal process (HRI), Synthoil (DOE), SRC and EDS (Exxon), and CCL (Gulf Oil), with others being researched.

In contrast to direct coal hydroliquefaction, in indirect liquefaction coal is gasified to a mixture of $CO + H_2$, followed by catalytic conversion to fuel products (Fischer-Tropsch process). The influence of the catalyst is profound, product distribution obtained varies considerably (see Table 1). The SASOL process is in commercial operation in South Africa. A plant using the Mobil process is being built in New Zealand.

Catalysis is effective in all three reactions involved in coal gasification, ie, gasification, shift, and synthesis.

Methanation is used commercially to convert relatively small amounts of harmful carbon oxides to CH_4, where, as in ammonia synthesis, CO would especially interfere with catalytic conversions of gas mixtures. Methanation also will be used on a large scale in the manufacture of high heat values or pipeline gas, essentially methane.

Chemicals are expected to be derived from the refining of coal hydroliquefaction, especially aromatics, or from gas synthesis.

Table 1. Syngas Process Products

| Product | Process, % of product | | |
| | SASOL-I | | |
	Fixed	Fluid	Mobil
light gas C_1, C_2	7.6	20.0	1.3
LPG C_3, C_4	10.0	23.0	17.8
gasoline C_5—C_{12}	22.5	39.0	80.9
diesel C_{13}—C_{18}	15.0	5.0	0
heavy oil C_{19+}	41.0	6.0	0
oxygen compounds	3.9	7.0	0
aromatics, % of gasoline		5	38.6

Fuel Cells

Catalysts, eg, noble metals on electrically conducting supports, are used at both the anode and cathode in a fuel cell (see Batteries and electric cells, primary; see also Hydrogen energy).

Environmental Control

Catalysts used for the abatement of noxious emissions (see Exhaust control, automotive; Exhaust control, industrial) were developed where the active component and substrate are supported on sturdy ceramic honeycomb structures, called monolithic supports. Platinum-group metal catalysts are used for nitric acid (qv) plant tail-gas cleanup. Platinum metals supported on SiO_2, Al_2O_3, and carbon are used as catalysts to eliminate organic fumes from air pollution and for fire safety (see Air pollution control methods). Hydrocarbon oxidation reactions are promoted via catalysts for emission control. Combustor systems are usually ceramic monolithic substrates coated with a catalytic agent. Catalysts are used in hydrogen sulfide conversion to sulfur (see Sulfur recovery).

About 60% of catalysts is used in the petroleum industry, and the remainder for major chemicals and chemical intermediates.

G. ALEX MILLS
U.S. Department of Energy
present affiliation
University of Delaware

H. Heinemann and J.J. Carberry, eds., *Catalysis Reviews*, Marcel Dekker, Inc., New York, 1968–present.

W.K. Hall, ed., *J. Catal.*, Academic Press, Inc., New York.

"Catalysts: a Chemical Market Poised for Growth," *Chem. Eng. News*, 19 (Dec. 5, 1983).

CATALYSIS, PHASE-TRANSFER

Phase-transfer catalysis (PTC) is a technique by which reactions between substances located in different phases are brought about or accelerated. It is also known as extractive alkylation, catalytic two-phase reaction, and preparative ion-pair extraction. Typically, one or more of the reactants is an organic liquid or solid dissolved in a nonpolar organic solvent and the coreactants are salts or alkali-metal hydroxides as solids or in aqueous solution. In principle, however, one reactant can be a gas, and the extracted species—normally an anion—can be a cation, a metal atom, or a neutral molecule. Catalysts used most extensively are quaternary ammonium or phosphonium salts, and crown ethers and cryptates.

There are two distinct classes of reactions formed under PTC conditions:

(1) Reactions without added bases, eg, displacement reactions such as $RX + Y^- \rightarrow RY + X^-$; oxidation reactions with oxidants such as $KMnO_4$, KO_2, $K_3[Fe(CN)_6]$, $K_2Cr_2O_7$, NaOCl; as well as some reduction reactions.

(2) Reactions in the presence of inorganic bases (conc. aqueous solutions or finely ground solids of NaOH, KOH, K_2CO_3, etc). These

reactions include alkylations of weak C—H acidic compounds up to a pK_a limit of 22–25; alkylations of ambident anions; alkylations of OH, NH, and SH bonds; isomerizations and H—D exchange; additions across C=C and C=O bonds; β-eliminations; formation of carbenes by α-eliminations; hydrolysis and saponifications; Darzens reactions; Horner-Wittig reactions; nucleophilic aromatic substitutions; and many others.

Mechanism of Phase-Transfer Catalysis

A mechanistic picture of class (1) reactions shows that catalyst cation Q^+ extracts the more lipophilic anion Y^- from the aqueous to the nonpolar organic phase where it is present in the form of a poorly solvated ion pair $[Q^+Y^-]$. This then reacts rapidly with RX, and the newly formed ion pair $[Q^+X^-]$ exchanges X^- for Y^- at the interphase for the next cycle. The rate of the reaction is proportional to the catalyst concentration and independent of the stirring speed above the minimum required for efficient mixing. Base-catalyzed reactions are mechanistically diverse: They can be brought about by extraction of deprotonated substrate (Sub^-) from the aqueous phase, by deprotonation at the phase boundary and subsequent transport of $[Q^+Sub^-]$ into the depth of the organic phase, or by extraction of $[Q^+OH^-]$ as reacting base. In addition, many anions behave as bases under the virtually anhydrous condition of PTC, eg, F^-, Cl^-, $CH_3CO_2^-$.

Factors Influencing the Usefulness of Phase-Transfer Catalysis

The best solvents for PTC are nonmiscible with water and nonhydroxylic, eg, $CHCl_3$, hydrocarbons, chlorobenzene. The logarithms of extraction constants for symmetrical tetra-n-alkylammonium salts (log E_{QX}) rise by ca 0.54 per added C atom, but the lipophilicity of phenyl and benzyl groups carrying ammonium salts is much lower than the number of C atoms might suggest. For example E_{QX} of tetra-n-butylammonium salts is about 140 times larger than E_{QX} for tetra-n-propylammonium salts of the same anion in the same solvent–water system. For practical application in mixtures of water–organic solvent, only ammonium and phosphonium salts containing 15 or more C atoms are sufficiently lipophilic unless a "salting-out effect" is operative as in the presence of concentrated sodium hydroxide. Crown ethers are sometimes as effective as the best onium salts in empirical catalyst comparisons; complexing agents advocated include 15-crown-5 and 18-crown-6 and their benzo derivatives or cryptates (see Chelating agents). Benzyltriethylammonium chloride is the most widely used catalyst under strongly basic conditions.

The successful extraction of the necessary anion into the organic phase is crucial for PTC. Often three anions compete for the catalyst cation: the one that is to react; the one formed in the reaction; and the one brought in originally with the catalyst.

Uses

Typical applications include displacement reactions (see Nitriles); formation of esters (see Esterification); formation of ethers (see Ethers); alkylation of carbanions (see Alkylation); Horner-Wittig reactions; oxidations using hypochlorite, permanganate, chromate or other oxidants; carbene reactions for preparing dihalocarbene adducts of olefins. Many other reactions assisted by PTC are described in the general references.

ECKEHARD V. DEHMLOW
Universität Bielefeld
Federal Republic of Germany

E.V. Dehmlow and S.S. Dehmlow, *Phase-Transfer Catalysis*, 2nd ed., Verlag Chemie, Weinheim, FRG, 1982.

A. Brändström, *Preparative Ion Pair Extraction, An Introduction to Theory and Practice*, Apotekarsocieteten/Hässle Läkemedel, Stockholm, Sweden, 1974.

C.M. Starks and C. Liotta, *Phase Transfer Catalysis, Principles and Techniques*, Academic Press, Inc., New York, 1982.

CATECHOL, $o\text{-}C_6H_4(OH)_2$. See Hydroquinone, resorcinol, and catechol.

CATHARTICS. See Gastrointestinal agents.

CAULKING COMPOSITIONS. See Sealants; Chemical grouts.

CAUSTIC SODA. See Alkali and chlorine products.

CELCON. See Acetal resins.

CELLOPHANE. See Film and sheeting materials.

CELLS, ELECTRIC. See Batteries and electric cells.

CELLS, HIGH TEMPERATURE. See Batteries and electric cells, primary.

CELLULOSE

The Raw Material

Cellulose is the principal fiber cell-wall material of green terrestial and marine plants, produced also by a few bacteria, animals and fungi, and thus the most abundant natural material ($\sim$ 40% in wood, over 70% in bast and leaf fibers, 95% in cotton, 70% in the cell wall of the green alga *Valonia ventricosa*; ca 5×10^{11} metric tons biosynthesized yearly). Cellulose is a long linear polymer of anhydroglucose units (Fig. 1), and this is reflected in the thread-like structures of cellulose found in plant cell walls (elementary fibrils, approximately 3.5 nm in width and indefinite length) which are further laterally associated to provide strength (as microfibrils, generally 10–30 nm in breadth). Although the biochemical pathway for the biosynthesis of cellulose from glucose in plants is relatively well understood, the mechanisms of the formation of long chains, of association as microfibrils and of deposition in plant cell walls are still being researched.

Cellulose is partly ordered (crystalline) and partly disordered (amorphous), presumably the result of regions of regularity and nonregularity within the elementary and microfibrils. Accessibility of cellulose is the relative ease by which the hydroxyl groups can be accessed by reactants. The amorphous regions are highly accessible and react readily, whereas the crystalline regions with close packing and hydrogen bonding can be completely inaccessible.

Cellulose also exists in several polymorphs. Native cellulose or cellulose I, is converted to cellulose II when cellulose fibers are regenerated or treated with 12–18% NaOH solution (mercerized), and to cellulose III and cellulose IV upon being subjected to certain chemical treatments or

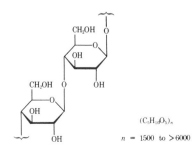

Figure 1. Repeat unit of cellulose (β-cellobiose residue).

heat. X-ray and electron-diffractometric studies, conformational analyses and vibrational spectroscopy have been used to define the crystalline nature of these polymorphs, and several unit-cell structures based on intermolecular hydrogen bonding of cellulose chains have been proposed and refined over the years. Recent work with Raman spectroscopy and solid-state ^{13}C nmr spectroscopy suggests that adjacent anhydroglucose units in cellulose are nonequivalent, leading to the hypothesis that cellulose is a polymer not of anhydroglucose but of anhydrocellobiose. Further, varying abundances of three anhydrocellobiose conformations (two ordered states, K_I and K_{II}, and one disordered state, K_0) result in differing intramolecular order which, with the intermolecular bonding of cellulose chains, make up the various polymorphs.

Cellulose in its differing conformations exhibits differing properties. The degree of crystallinity also depends on cellulose preparation. There is not one cellulose, but a number of celluloses. It is not a compound but a material whose usefulness depends on how it has been modified by all steps prior to its final use.

Cellulose never occurs in pure form, instead it is usually embedded in lignocellulose (an amorphous matrix of hemicellulose and lignin containing ordered cellulose), making up the cell walls of fibers such as found in wood (well-developed matrix) and cotton (matrix of almost vanishing magnitude). The hemicelluloses are polysaccharides, usually branched, of various sugars and some uronic acids, which can usually be extracted from lignocellulosics with alkali. Lignins are highly cross-linked aromatic polymers, of no regular repeating unit because of their formation by free-radical condensation. Industrially useful fibers are the textile fibers: bast or stem fibers (flax, jute, hemp, ramie), leaf fibers (sisal, abaca) and the seed and fruit fibers (cotton, kapok); and the nontextile fibers (chiefly from hardwood and softwood). The geometry of the arrangement of microfibrils in fiber walls has a pronounced effect on the physical properties and thus use of these fibers.

Lignin (qv) removal from wood fibers is the basic process of the chemical pulping industry for the production of papermaking pulp and cellulose. Pulping consists of separating and recovering delignified cellulosic fibers from fibrous plant materials by mechanical and/or chemical methods (see Paper; Pulp).

Bleaching

For many pulp applications, color is important, and one or more stages of reductive or oxidative chemical treatments are used following pulping to provide the desired level of whiteness in the finished product. In one-stage reductive bleaching used for newsprint (with zinc or sodium hydrosulfide for example), color bodies are bleached but not completely removed, and color reversion can take place when such papers are exposed to light and air (see Bleaching agents). At the other extreme, for dissolving pulps, bleaching and purification steps are combined, and as many as 12 steps are required. In oxidative bleaching for papermaking the CEHD semi-bleach, and the CEDED and CEHDED bleach sequences are the most common (C = chlorination, E = caustic extraction to remove oxidized species, H = hypochlorite, D = chlorine dioxide). Peroxide (P) may be used as a final step to provide high brightness and stability. Oxygen and ozone are receiving increased attention, especially in concert with oxygen–alkali pulping in moves towards pollution-free pulping and bleaching systems.

Derivatives

Cellulose derivatives are of the ether and ester type. Preswelling of cellulose is necessary in both etherifications (with alkali) and esterifications (with acid). The most important swelling complexes of cellulose are those with sodium hydroxide, compounds with given stoichiometric relations between alkali and cellulose. The alkali celluloses are important intermediates because they exhibit markedly enhanced reactivity compared to original cellulose. Reagents can penetrate more easily into the swollen cellulose structure and react with hydroxyl groups. Preparation of alkali cellulose (called mercerization after its inventor John Mercer in 1844) is an important step in producing cellulose xanthate, from which regenerated cellulose as viscose fibers and cellophane are produced (see Rayon; Xanthates).

Cellulose ethers (most common are sodium carboxymethyl, methyl, ethyl, hydroxyethyl and hydroxypropyl) are used as gums and thickeners in laundry detergents, water-based paints, oil-well fluids, cement, plaster, foods, textiles and cosmetics.

Cellulose esters (most common are the nitrates and acetates) are used for a variety of applications.

Microcrystalline Cellulose

An increasingly important cellulose product is microcrystalline cellulose. Crystallite zones appear in regenerated, mercerized, and alkali celluloses, differing from those found in native cellulose. By applying a controlled chemical pretreatment to destroy molecular bonds holding these crystallites, followed by mechanical treatment to disperse the crystallites in aqueous phase, smooth colloidal microcrystalline cellulose gels with commercially important functional and rheological properties can be produced. Microcrystalline celluloses have wide use in foods, pharmaceuticals (qv), and cosmetics (qv). In foods, these provide smooth textures, stable viscosities over a wide temperature range, heat stability, ability to thicken with acceptable mouth feel, and flow control (see Food processing). They extend starches, form sugar gels, stabilize foam, and control formation of ice crystals. Applications are in fillings, meringue, chocolate cake sauce, cookie fillings, whipped toppings, and imitation ice cream. In pharmaceuticals for tableting, microcrystalline celluloses assist in flow, lubrication, bonding of tablet ingredients, improve stability of drugs, and rapid disintegration of the tablet in the stomach. In cosmetics, mixtures of microcrystalline cellulose and carboxymethyl cellulose are used in suspensions, shampoos, cream rinses, hair conditioners, hair-coloring products, hair bleaches, and toothpaste (see also Hair preparations; Dentifrices).

Uses

Cellulose can be cross-linked with formaldehyde to effect intermolecular cross-links for improved fabric and sheet wet strength. In textile finishing, the precondensates of formaldehyde with urea and its derivatives react with cellulose to give improved dimensional stability, crease resistance, wash-and-wear performance, and durable-press properties (see Textiles, finishes). Although satisfactory for properties like crease and shrink resistance, cross-linking makes products more rigid and brittle. Other specialty celluloses are the cation- and anion-exchange celluloses, produced by introducing acidic or basic substituents into cellulose. Advantages of cellulose as an ion-exchange base are its inertness and high surface area. Grafting of polymers to cotton, rayon, paper, cellophane and wood, although promising for the production of new cellulose-base materials with desirable properties, has little commercial application.

Cellulose products, however, are most visible in fiber and film forms. Refined cotton linters and high grade wood pulps are used in the production of rayon (qv), cellulose acetate, and triacetate fibers (qv) for textile applications. Rayon is used for the manufacture of staple fibers and filament yarns. These are used for clothing, curtains, carpets, blankets, nonwovens, disposables, and sanitary products (see Nonwoven textile fabrics). Blended with synthetic fibers, rayons contribute water absorptivity for greater wearing comfort. The high tenacity filament yarns and staple fibers are used in tire cords (qv), belts, tapes, hoses, coated fabrics, etc.

Cast as films instead of fibers, and often coated to give desired properties, regenerated celluloses and cellulose derivatives serve product areas where dimensional stability, clarity, wet shrinkage, shrink sealing, and gas or solute permeability are required. Cellophane is receiving competition from polyolefins but still finds use for packaging snacks and cookies, baked goods and candy, tobacco and other nonfood items. Wiener (70% cellulose, 12% glycerol, 18% water) and salami/bologna (hemp-paper reinforced 23% paper, 46% cellulose, 21% glycerol, 10% moisture) sausage casings are the most profitable high volume cellulose products sold today. Other film uses are for packaging of sterile instruments (steam or ethylene oxide can penetrate to kill bacteria, but organisms cannot pass through in storage), for dialysis (qv) in artificial kidneys, and as cellulose acetate reverse-osmosis membranes for water purification, desalination, and food concentration (see Reverse osmosis; Membrane technology).

Special attention is being given to cellulose-based materials which outperform synthetic polymers in such areas as water absorbancy, and visible efforts in this direction are disposable diapers and feminine-hygiene products. Cellulose, in this age of material substitution, is experiencing a rebirth as a natural and plentiful raw material for a variety of commodity and specialty product needs.

ED J. SOLTES
Texas A & M University

D.P. Delmer, "Cellulose Synthesis," in C.C. Black, A. Mitsui, and O.R. Zaborsky, eds., *Biosolar Resources*, Vol. 1 of *Basic Principles*, CRC Press, Boca Raton, Fla., 1981.

J.R. Colvin, *Appl. Polym. Symp.* **37**, 25 (1983).

E.J. Soltes, "Cellulose: Elusive Component of the Plant Cell Wall," in W.A. Cote, Jr., ed., *Biomass Utilization*, Plenum Press, New York, 1983, pp. 271–298.

Proceedings of the 1983 International Dissolving and Specialty Pulps Conference, TAPPI Press, Atlanta, Ga., 1983.

R.J. Atalla, "The Structure of Cellulose: Recent Developments," in E.J. Soltes, ed., *Wood and Agricultural Residues: Research on Use for Feed, Fuels, and Chemicals*, Academic Press, New York, 1983, pp. 59–77.

F.E. Bailey, Jr., ed., *Initiation of Polymerization*, ACS Symp. Series 212, American Chemical Society, Washington, D.C., 1983.

CELLULOSE ACETATE AND TRIACETATE FIBERS

Acetate fiber, also known as secondary acetate, is the generic name for a cellulose acetate fiber which is a partially acetylated cellulose. Triacetate is the generic name for cellulose triacetate fiber, also known as primary acetate, an almost completely acetylated cellulose. Acetate and triacetate differ only moderately in the degree of acetyl substitution on cellulose, yet they have different chemical and physical properties.

Cellulose acetate is the reaction product of cellulose and acetic anhydride (see Cellulose derivatives, esters). The degree of polymerization (DP) of the cellulose used for esterification is ca 1000–1500, whereas the esterification product DP is ca 300. Secondary cellulose acetate is obtained by acid catalyzed hydrolysis of the triacetate to an average degree of substitution of 2.4 acetyl groups per glucose unit. The degree of acetylation is specified by the acetyl value (%) and combined acetic acid (%); the two are always in the ratio of 43:60. The combined acetic content of commercial cellulose triacetate is 61.5%, and of cellulose acetate is 55%. Commercial cellulose acetates contain small amounts of residual-free carboxyl and sulfate groups as well as hydroxyl and acetyl groups. The salt form is ordinarily more thermally stable. Cross-linking between residual sulfate and carboxyl groups can cause an artificially high solution viscosity which limits the concentration of cellulose acetate used in yarns.

Properties

Cellulose acetate fibers have a low degree of crystallinity and orientation even after heat treatment; cellulose triacetate develops considerable crystallinity after heat treatment. Fibers of both have a bright lustrous appearance. Duller, whiter yarns are produced by the addition of pigments (qv). The appropriate azo, anthraquinone, or diphenylamine disperse dye is selected to ensure colorfastness. Fading inhibitors are used to resist the effects of nitrogen oxides and ozone.

The specific gravity values of 1.32 for acetate and 1.30 for triacetate are accepted for fibers of combined acetic acid contents of 55 and 61.5%, respectively.

Triacetate fiber is hydrophobic. An hysteresis effect is noted for both acetate and triacetate fibers over the entire range of relative humidities. Heat-treated triacetate fiber has a lower regain than nonheat-treated fiber, values are in the range of 2.5–3.2%. The average percentages of water-imbibition are acetate, 24%; nonheat-treated triacetate, 16%; heat-treated triacetate, 10%.

Like other thermoplastic fibers, acetate sticks, softens, and even melts when ironed at high temperatures; between 190–205°C acetate softens and sticks, but fuses at ca 260°C. The sticking and glazing temperatures of nonheat-treated triacetate fiber are in the same range, those of

heat-treated triacetate fibers considerably higher (triacetate mp ca 300°C). Mechanical properties are not altered significantly; however, as the temperature is raised the modulus of the acetate and triacetate fibers is reduced, and they extend more readily under stress. Both fibers are weakened by prolonged exposure to elevated temperatures in air; their resistance is lowered upon exposure to direct weathering.

Acetate and triacetate yarns readily develop static charges, and often have antistatic finishes added (see Antistatic agents). The resistivity of cellulose acetate is 1.27 MΩ-cm; of nonheat-treated cellulose triacetate is 3.81; of heat-treated cellulose triacetate is 15.2; and of specially scoured and heat-treated cellulose triacetate 1016.

The physical forms of the fibers and the geometrical construction of the fabric determine the mechanical and aesthetic properties of textile fabrics. Performance criteria include hand, drape, wrinkle resistance and recovery, strength, and flexibility. Fiber mechanical properties are described by stress–strain and recovery behavior under conditions of tensile, torsional, bending, and shear loading. Under standard conditions, acetate and triacetate tenacity ranges from 0.10–0.12 N/tex (1.3 gf/den); loop ranges from 0.09–0.10 N/tex (65% rh, 21°C); and elongation at break is 25–45%.

The ability of a material to resist deformation under applied tensile stress is measured by the modulus of elasticity (Young's modulus). In viscoelastic materials, the apparent modulus of a textile fiber is defined as the ratio of stress to strain in the initial, linear portion of the stress–strain curve. Most stress–strain data are reported under standard conditions of 21°C and 65% rh. Strains of more than 10% are avoided in textile processing so that the dimensional or shape stability of the resultant fabric will be acceptable. The modulus of elasticity can be affected by drawing (elongating) the fiber and by other manufacturing procedures. Values for commercial acetate and triacetate fibers are generally 2.2–4.0 N/tex (25–45 gf/den). The wet modulus of fibers at various temperatures influences the degree of creasing and mussiness caused by laundering. The ability of a fiber to absorb energy during straining is known as toughness, or work of rupture. The ratio of the work recovered to the total work absorbed is known as resilience.

The viscoelastic behavior of a fiber is the elongation of a stretched fiber described as a combination of instantaneous extension and a time-dependent extension or creep. Conversely, recovery of viscoelastic fibers is typically described as a combination of immediate elastic recovery, delayed recovery, and permanent set or secondary creep, which is residual extension which is not recoverable. The bending properties of a fiber generally depend upon the viscoelastic behavior of the material, measured in flex rigidity of $g \cdot m^2$.

Chemical properties. Under slightly acidic or basic conditions at room temperature, acetate and triacetate fibers are very resistant to chlorine bleach at the concentrations normally encountered in laundering. Triacetate fiber is significantly more resistant than acetate to alkalies encountered in normal textile operations. Acetate should be dyed at temperatures no higher than 85°C or in a pH above 9.5; triacetate may be dyed with alkali up to pH 9.5 and temperatures up to 96°C. Acetate and triacetate are unaffected by dilute solutions of weak acids, or by perchloroethylene dry-cleaning solutions. Strong mineral acids cause serious degradation. Triacetate is very resistant to microorganisms.

Manufacture of Cellulose Acetate and Triacetate Flakes

Acetate and triacetate flakes are white amorphous solids produced in granular, flake, powder, or fiber form which are used as raw materials in the preparation of fibers, films, and plastics. Density varies with physical form, ranging from 100–500 kg/m³.

Manufacture of flakes. Acetate and triacetate flakes are prepared by the esterification of high purity chemical cellulose with acetic anhydride. The solution properties of the resultant acetate or triacetate flake may be adversely affected by impurities even at low concentrations; extrusion processes require the polymer to flow freely through small diameter capillaries (eg, 30–80 μm). Wood pulp, which has replaced cotton linters as a source of chemical cellulose, must have the noncellulose impurities (eg, lignin, hemicellulose) removed.

Secondary acetate processes. Three main processes used are: (1) A solution process, the most common, in which acetylation is obtained with acetic anhydride using glacial acetic acid as the solvent, and sulfuric acid as catalyst. Two major variants of this process are high catalyst (10–15% sulfuric acid based on cellulose) and low catalyst (< 7% sulfuric acid) procedures. (2) A solvent process in which methylene chloride is substituted for all or part of the acetic acid and acts as a solvent for the triacetate as it is formed, perchloric acid is used as catalyst. (3) A heterogeneous process (nonsolvent or Schering process) in which an inert organic liquid, such as benzene or ligroin is used as a nonsolvent to prevent the acetylated cellulose from dissolving as it is formed. The cellulose ester produced is never solubilized and its physical form is similar to the original cellulose fiber. To produce a secondary acetate soluble in acetone, it is necessary to acetylate the cellulose during the dissolution step and then hydrolyze while still dissolved to the required acetyl value. This four-step process involves: (1) preparation of the cellulose for acetylation; (2) acetylation at temperatures less than 50°C using a 5–15 wt% excess of acetic anhydride and yielding a 15–25% cellulose acetate concentration; (3) hydrolysis, normally at 50–100°C using steam injection for 1–24 h to reduce the number of acetyl groups present in each anhydroglucose unit at the end of acetylation from slightly less than 3.0 to ca 2.4; (4) recovery of cellulose acetate (via precipitation, washing, and drying) and of solvents.

Batch preparation of triacetate. The batch triacetate process differs from the preparation of secondary acetate in that there is little hydrolysis and the temperature is in the range of 50–100°C. The triacetate hydrolysis step (or desulfation) removes only the sulfate groups from the polymer by slow addition of a dilute acetic acid solution containing sodium or magnesium acetate or triethanolamine to neutralize the liberated sulfuric acid. Cellulose triacetate product has a combined acetic acid content of 61.5%.

Continuous flake-manufacturing processes. The basic processing stages are identical to those of the batch process but some equipment and materials handling techniques differ. For example, the acetylizer is a specialized, high energy input reactor with a range of materials-handling capabilities which receives matted pulp, acetic acid, catalyst, and acetic anhydride and produces a highly viscous, homogeneous solution of cellulose triacetate. The hydrolysis step consists of a series of retention tanks which provide a controlled time–temperature exposure to reach the desired combined acetic acid and residual sulfate level.

Manufacture of Cellulose Acetate and Triacetate Fibers

A four part dry-extrusion process is primarily used to convert a solution of cellulose acetate or triacetate polymer into fiber form: (1) the cellulose acetate is dissolved in a volatile solvent, eg, acetone for secondary acetate (ca 95% acetone and 5% water) and chlorinated hydrocarbons (ie, methylene chloride containing 5–15% methanol) for triacetates; (2) the solution is filtered to remove insoluble matter; (3) the solution is extruded to form fibers by forcing the solvent through the spinneret (which has numerous holes 30–80 mm in diameter) into a cabinet of warm air; (4) the yarn is lubricated and taken up on a suitable package, finish or lubricant giving the extruded yarn the frictional and antistatic properties required for further processing. The solvent used to form the extrusion dope is recovered by adsorption, or condensation.

Acetate and triacetate continuous-filament yarns, staples, and tows are manufactured with varying properties, eg, tex (wt (g) of a 1000-m filament) or denier (wt (g) of a 9000-m filament), cross-sectional shape, and number of filaments. The same basic extrusion technology that produces continuous-filament yarn also produces staple and tow.

Uses. Principal applications are women's apparel and home furnishing fabrics (see Nonwoven textile fabrics; Textiles; Flame-retardant in textiles); and cigarette filters.

GEORGE A. SERAD
J.R. SANDERS
Celanese Fibers Company

L. Segal in N.M. Bikales and L. Segal, eds., *High Polymers Series*, Vol. V, Wiley-Interscience, New York, 1971, Chapt. XVII-A.

C.J. Malm and G.D. Hiatt, *Cellulose and Cellulose Derivatives* in E. Ott, H.M. Spurlin, and M.W. Graffin, eds., *High Polymers Series*, 2nd ed., Vol. V, Pt. II, Interscience Publishers, New York, 1954.

CELLULOSE DERIVATIVES, ESTERS

Organic Esters

Cellulose acetate is by far the most important organic ester of cellulose owing to its broad applications in plastics and fibers. It is prepared with varying degrees of substitution, ranging from water-soluble monoacetate to triacetate. The common commercial celluloses are primary (triacetate) and secondary (acetone-soluble) (see Cellulose acetate and triacetate fibers). Cellulose esters of higher aliphatic acids circumvent to some degree the shortcomings of cellulose acetate, such as restrictions due to moisture sensitivity, compatibility with plasticizers and other resins, and relatively high flow temperatures.

Physical and chemical properties. Commercial cellulose acetates are white, odorless, and nontoxic materials. Their properties are dictated by their combined acetic acid content (acetyl) and molecular weight. The general solubility characteristics are as follows: (a) acetyl content above 43%, sol in dichloromethane, insol in acetone; (b) acetyl content 37–42%, sol in acetone, insol in dichloromethane; (c) acetyl content 24–32%, sol in 2-methoxyethanol, insol in acetone; (d) acetyl content 13–14%, sol in water, insol in 2-methoxyethanol; (e) acetyl content < 13%, insol in all solvents listed above.

Moisture regain and vapor permeability rate increase with decreasing acetyl content. At 50% rh % moisture regain for cellulose 10.8; acetate 2.0; priopionate 0.5; butyrate 0.2. Melting point range of secondary acetate ca 230–250°C; triacetate 270–300°C; propionate 230°C; butyrate 183°C. Bulk density varies with physical form and is 160–481 kg/m³ (10–30 lb/ft³). Tensile strength (MPa) of acetate 71.6; propionate 48.0; and butyrate 30.4. Specific gravity (1.29–1.30), refractive index (1.48), and dielectric strength of most commercial cellulose acetates are very similar. Mechanical properties of fibers, plastics, and film made from cellulose acetate vary with degree of polymerization (DP) and DP distribution, and improve when DP is increased from about 100 to 250. Thermoplastic characteristics improve as acetyl content (degree of substitution or DS) increases from ca 1 to 2.25 max. Cellulose acetates are compatible with many liquid plasticizers including dialkyl phthalates. Some synthetic resins compatible with cellulose acetates are Acryloid A-10, Bakelite BR-3180, and Vinylite AYAA, degree of compatibility varying according to acetyl content and DP. The number of acetyl groups per anhydroglucose unit, acyl chain length, and the degree of polymerization influence the properties, as well as the acyl–acyl (eg, acetyl–propionyl) ratios of the ester. Gas permeability and crystalline properties are similarly influenced.

Manufacture. All commercially important cellulose acetates, except the fibrous triacetate, are manufactured by a solution process using acetic anhydride in the presence of a suitable catalyst and a solvent such as acetic acid. The esterification of cellulose is a heterogeneous topochemical reaction. The triacetate is prepared and then hydrolyzed in solution to the desired degree of substitution. The two main sources of cellulose are cotton (qv) linters and purified wood pulps (see Cellulose; Pulp). Sulfuric acid is the most popular catalyst, other catalysts include zinc chloride, perchloric acid, and others. The course of the acetylation reaction is largely controlled by the rate of diffusion of the reagents into the cellulose fibers. The usual commercial practice is to activate cellulose by treatment with acetic acid containing a part of the total amount of catalyst, as the molecular weight of the cellulose must be reduced to obtain a satisfactory product. Efficiency of activation depends on catalyst and water concentrations, duration, and temperature. Hydrolysis removes some of the acetyl groups and the combined sulfate. Cellulose triacetate is normally hydrolyzed in an aqueous acid solution in the presence of catalyst, at temperatures as low as 38°C and as high as 229°C in an autoclave and at 129°C in a continuous process. For commercial hydrolysis, the water content of the acid dope is usually 5–20%.

Several continuous processes for manufacture of cellulose acetate have been used, including continuous cellulose activation, acetylation, hydrolysis, precipitation, and washing and drying. Other methods have been reported (see Cellulose acetate and triacetate fibers).

Dicarboxylate mixed esters, eg, cellulose acetate phthalate, are prepared by the reaction of the free hydroxy group of a hydrolyzed cellulose acetate with phthalic anhydride. These have commercially significant characteristics, such as alkaline solubility, and excellent film-forming properties. These esters may be prepared by treating a hydrolyzed cellulose acetate with phthalic anhydride in pyridine solution or in acetic acid using sodium acetate catalyst. In general, the solubilities of cellulose acetate phthalate esters in organic solvents approximate those of the simple cellulose acetates having comparable degrees of substitution. Salts of cellulose acetate phthalate are obtained by dispersing the ester in a water–alcohol mixture and adding the proper amounts of ammonium salt, alkali-metal hydroxide, or lower molecular weight amine.

Cellulose propionate and cellulose butyrate may be prepared by esterification of efficiently activated cellulose with propionic or butyric anhydride in the presence of acid catalysts. Cellulose valerate has been synthesized by conventional methods using valeric anhydride and sulfuric acid catalyst, but higher ester homologues cannot normally be prepared by this method. Cellulose esters are customarily hydrolyzed to eliminate small amounts of combined sulfate and modify the properties of the ester by removing a small number of acyl groups with sulfuric acid as catalyst.

Most cellulose esters are stabilized to minimize color development and prevent substantial loss of molecular weight during thermal processing operations. Stabilization is achieved by removing the small amounts of free and combined sulfuric acid present in the ester using various methods, eg, neutralization with alkali-metal salts, boiling water, or treatment with aqueous potassium and cadmium iodide solution.

Inorganic Esters

Cellulose nitrate is the most important and only commercially available inorganic ester of cellulose. Two other classes of inorganic derivatives of possible interest are sulfur-containing cellulose esters, formed by the reaction of cellulose with chlorosulfonic acid (qv) in pyridine solvent; and phosphorus-containing esters, prepared by reaction of cellulose with phosphoric acid in molten urea.

Properties. Cellulose nitrate is odorless, tasteless, and water-white as film. Specific gravity of cast film is 1.58–1.65; tensile strength at 23°C, 50% rh 62–110.3 MPa (10,000–16,000 psi); softening-point range (Parr) 155–220°C. The most important properties related to its use in lacquer coating, inks, and adhesives are solubility; viscosity; compatibility with plasticizers, resins, and pigments; and toughness. Three standard types (RS, AS, and SS) are available: RS-type is soluble in ketones, esters, and ether–alcohol mixtures, a high tolerance for aromatic hydrocarbons. AS-type is soluble in the same types of solvents, but tolerates higher proportions of alcohols in solvent blend. Type SS is soluble in alcohols, and is more thermoplastic than the other types.

Purification. Cellulose nitrate undergoes a purification process consisting of a series of boiling-water and washing treatments at controlled pH which is designed to remove the residual traces of nitrating acids and to eliminate, by hydrolysis, the combined sulfate ester groups which cause instability. For certain applications stressing low color, cellulose nitrate is bleached by potassium permanganate or sodium hypochlorite by procedures similar to those used in bleaching cellulose.

Manufacture. Until recently, cellulose nitrate was manufactured only by a batch process in which a rather small quantity (15 kg) of cellulose was agitated vigorously in a vessel containing ca 800 kg of HNO_3/H_2SO_4 nitrating mixture for 20–30 min, followed by centrifugation, quick drowning in a large volume of water, and purification.

A successful continuous process has been developed in which the cellulose and mixed nitrating acids are fed simultaneously into a vessel where nitration occurs; this mixture is centrifuged for separation of spent acids, and the cellulose nitrate is washed in stages with progressively weaker aqueous acid. Spent acids are recovered. Although the predominant stabilization of cellulose nitrate occurs during manufacture, additional thermal and light stabilization may be required.

Cellulose nitrates undergo a digestion process to produce low viscosity types used in lacquers.

Health and safety. Dry cellulose nitrate is extremely flammable and may explode when subjected to heat or sudden shock. It is therefore shipped and handled wet with water, or most often, alcohol.

Uses. Among organic esters, cellulose acetate is used in textile fibers, plastics, film, sheeting, and lacquers; the film is used in photography and recording tape. Cellulose acetate propionate and butyrate are used as sheeting compositions, molding plastics, film products, and lacquer formulations, and cellulose acetate butyrate is extensively used in film-forming applications such as lacquers. Among inorganic esters, cellulose nitrate is used in lacquer coatings; for explosives and propellants (see Explosives and propellants; Inks; and Adhesives).

R.T. BOGAN
C.M. KUO
R.J. BREWER
Tennessee Eastman Company

N.M. Bikales, ed., *Encyclopedia of Polymer Science and Technology*, Vol. 3, Interscience Publishers, a division of John Wiley & Sons, Inc., New York, 1965.

E. Ott, H.M. Spurlin, and M.W. Grafflin, eds., *Cellulose and Cellulose Derivatives*, Part II of *High Polymers*, Vol. 5, 2nd ed., Interscience Publishers, New York, 1954.

N.M. Bikales and L. Segal, eds., *Cellulose and Cellulose Derivatives*, Part V of *High Polymers*, Vol. 5, 2nd ed., Wiley-Interscience, New York, 1971.

CELLULOSE DERIVATIVES, ETHERS

Cellulose ethers are manufactured from a preformed polymer, cellulose, comprised of linear chains of β-anhydroglucose rings. Either chemical-grade cotton linters or wood pulp can be used to prepare cellulose ethers. An idealized structure of a cellulose ether is shown in Figure 1, where the R-groups can be either single representatives or combinations of the above substituent groups. The chemical nature, quantity, and distribution of the substituent groups govern such properties as solubility, surface activity, thermoplasticity, film characteristics, and biodegradation (see Biopolymers; Carbohydrates; Cellulose).

The degree of substitution (DS), which defines the number of substituted ring sites, is used for process control and specifications; the maximum DS value is equal to 3. The indefinite term molar substitution (MS) is used in the case of alkylene oxides. By dividing MS by DS, the average length of the side chain can be measured.

Cellulose ether commercial products are subdivided into water-soluble ethers and organic-soluble ethers.

Water-soluble ethers include sodium carboxymethyl cellulose, sodium carboxymethyl 2-hydroxyethyl cellulose, 2-hydroxyethyl cellulose, methyl cellulose, 2-hydroxypropyl methyl cellulose, 2-hydroxyethyl methyl cellulose, 2-hydrobutyl methyl cellulose, 2-hydroxyethyl ethyl cellulose, and 2-hydroxypropyl cellulose. Selected properties for these ethers are listed in Table 1.

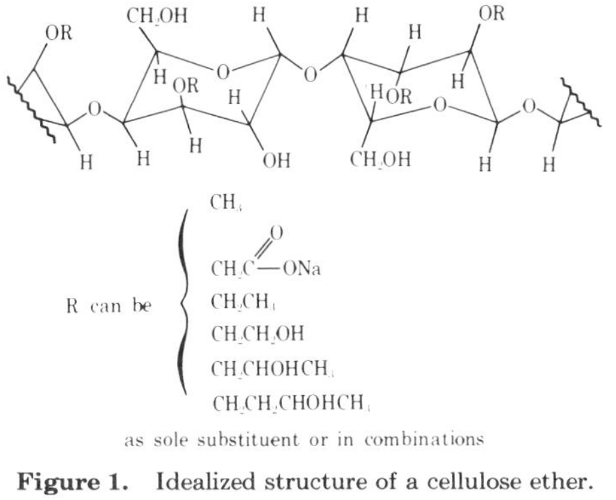

Figure 1. Idealized structure of a cellulose ether.

Table 1. Selected Properties of Water-Soluble Cellulose Ethers

Property	Sodium carboxymethyl cellulose	Methyl cellulose and derivatives	2-Hydroxypropyl cellulose	Ethyl 2-hydroxyethyl cellulose	2-Hydroxyethyl cellulose
apparent density of powder, g/mL		0.25–0.70	0.5		
bulk density, g/mL	0.75			0.4–0.6	0.6
solutions		(sp gr, 20/4°C)	(sp gr, 30°C)		
sp gr, 2% soln, 25°C	1.0068	1.0245^a	1.010^b		1.0033
refractive index, 2% soln	1.3355	1.336	1.337		1.336
surface tension, mN/m (= dyn/cm)	71	47–53	43.6		66.8
films					
refractive index	1.515			1.49	1.51
sp gr	1.59	1.39		1.33	1.34

^{a}Sp gr, 20/4°C.
^{b}Sp gr, 30°C.

Sodium carboxymethyl cellulose, also known as cellulose gum (CMC, formerly called sodium cellulose glycolate) is an anionic cellulose ether manufactured by the reaction of monochloroacetic acid, as either the acid or the sodium salt, and alkali cellulose. Sodium carboxymethyl cellulose can react with metal ions, such as alkaline-earth elements and heavy metal ions to form precipitates. Lower sensitivity to precipitation by salt solutions and acids can be achieved by modification with hydroxyethyl substitution. The commercial products span a DS range of 0.38–1.4. The polyelectrolyte structure of CMC influences viscosity and stability in solution, maximum viscosity and best stability occurring at 7–9 pH. CMC is soluble in both hot and cold water. Since CMC is a polyelectrolyte, metathesis reactions with other cations can occur. CMC is prepared by treating alkali cellulose, prepared batchwise or by continuous process, with sodium chloroacetate. A side reaction, the conversion of sodium chloroacetate to sodium glycolate, occurs simultaneously. In general, the alkylation reaction involves the traditional Williamson ether reaction where alkali cellulose reacts with an alkyl halide:

$$R_{cell}ONa + R'Cl \rightarrow R_{cell}OR' + NaCl$$

Other water-soluble ethers include methyl cellulose and 2-hydroxypropyl methyl cellulose, alkylene oxide modifications, which are nonionic in structure (see Ethylene oxide; Propylene oxide). They are available in powder and granular forms. Water solutions of these products possess the unusual property of thermal gelation. Temperature of gelation and the character and texture of the gel can be controlled by choice of the ratio of substituents on the anhydroglucose ring. Alkali cellulose is prepared by the traditional steeping and pressing method of the viscose industry, the use of screw presses, and the newer techniques of dipping, spraying, or use of slurries for improving the distribution of the sodium hydroxide.

Hydroxyethyl cellulose, HEC, is a water-soluble, nonionic cellulose ether that differs from methyl cellulose and 2-hydroxypropyl methyl cellulose in that it is soluble in both hot and cold water. The nonthermal gelling property increases the tolerance for ionic salts. 2-Hydroxyethyl cellulose does not have a thermal gel point, so the viscosity of solutions continues to decrease with increase of temperature. Manufacturing requires a lower quantity of sodium hydroxide in the preparation of HEC than that required for the alkylation of cellulose.

2-Hydroxyethyl methyl cellulose, HEMC, can also be formed by methyl substitution of HEC, which is used to modify the properties of HEC by limiting side chain formation and producing increased hydroxyethyl substitution along the cellulose chain. HEMC is similar to HEC in tolerance to most salts. HEMC has no thermal gel characteristics in water at boiling temperatures.

2-Hydroxypropyl cellulose results when propylene oxide reacts with alkali cellulose in a modification of the hydroxyethyl cellulose process.

Hydroxypropyl cellulose has a thermal gel point like methyl cellulose, and is also thermoplastic.

Ethyl 2-hydroxyethyl cellulose is a nonionic cellulose, soluble in cold water and insoluble in hot water.

Organic-soluble ethers include ethyl cellulose, ethyl 2-hydroxyethyl cellulose, and 2-cyanoethyl cellulose.

Ethyl cellulose is a white, odorless granular solid, with ethoxyl substitution values of commercial products available in two ranges: ca 2.0–2.3 and ca 2.4–2.6. The ethoxyl content greatly affects thermal behavior. Ethyl cellulose with a DS of 2.4–2.6 dissolves completely in all organic solvents except naphthas, purely aliphatic hydrocarbons, polyhydric alcohols, and a few ethers. Properties of ethyl cellulose in the 2.4–2.6 DS range are bulking value of powder in soln 0.826–0.868 g/mL; softening point of film 152–162°C. The sequence of chemical reactions is similar to that for methylation of cellulose. In commercial practice, sodium hydroxide concentrations of 50% or higher are used at temperatures of 90–150°C and 828 to 965 kPa (120 to 140 psi) for 6–12 h.

Ethyl 2-hydroxyethyl cellulose is a modified ethyl cellulose with a small amount (DS ca 0.3) of hydroxyethyl substitution; it is soluble in mixtures of solvent rich in aliphatics. Cellulose, in the presence of a dilute sodium hydroxide (2%) reacts with acrylonitrile (qv) to form the cyanoethyl ether (see Cyanoethylation).

Health and Safety

Cellulose ethers are of very low toxicity. Several types of cellulose ethers are permitted as food and pharmaceutical additives. Fine dusts can form explosive mixtures with air.

Uses

Each type of cellulose either has its individual spectrum of uses. There is also considerable overlap in some market areas such as latex paint. The water-soluble cellulose ethers are used in drilling fluids, detergents, construction products, foods, pharmaceuticals, coatings, tobacco products, textile auxiliaries, adhesives, protective colloidal polymerization processes, agricultural products, and paper. The water-insoluble cellulose ethers are used in furniture finishes, inks, impact-resistant coatings, hot melts, pharmaceuticals, and thermoplastic foaming.

G.K. GREMINGER, JR.
Dow Chemical, USA

N.M. Bikales and L. Segal, eds., *Investigations of the Structure of Cellulose and Its Derivatives in High Polymers*, 2nd ed., Vol. V, Wiley-Interscience, New York, 1971, pp. 1–325.

A.B. Savage, A.E. Young, and A.T. Maasberg in E. Ott, H.M. Spurlin, and M.W. Graffin, eds., *High Polymers*, 2nd ed., Vol. V, Interscience Publishers, Inc., New York, 1954, pp. 882–958.

R.L. Whistler, *Industrial Gums*, Academic Press, Inc., New York, 1973.

R.L. Davidson, *Handbook of Water-Soluble Gums and Resins*, McGraw-Hill, New York, 1980.

CELLULOSE DERIVATIVES, PLASTICS. See Plastics technology.

CEMENT

The term cement refers to many different kinds of substances that are used as binders or adhesives (qv). Here it refers to inorganic hydraulic cements (principally portland and related cements) which on hydration form relatively insoluble bonded aggregations of considerable strength and dimensional stability.

Hydraulic cements are manufactured by processing and proportioning suitable raw materials, burning (or clinkering at a suitable temperature), and grinding the resulting hard nodules called clinker to the fineness required for an adequate rate of hardening by reaction with water. Portland cement consists mainly of tricalcium silicate and dicalcium silicate. Usually two types of raw materials are required: one rich in calcium, such as limestone, chalk, or marl; the other rich in silica, such as clay or shale.

Clinker Chemistry

Phase equilibria. During burning in the kiln, about 20–30% of liquid forms in the mix at clinkering temperatures. Reactions occur at surfaces of solids and in the liquid. The crystalline silicate phases formed are separated by the interstitial liquid. The interstitial phases formed from the liquid in normal clinkers during cooling are also completely crystalline to x rays, although they may be so finely subdivided as to appear glassy under the microscope.

In the relatively small portland cement zone of the equilibrium ternary system, $CaO–Al_2O_3–SiO_2$, almost all cements fall in the high lime portion (about 65% CaO; 21–25% SiO_2; 3–6% Al_2O_3). Cements of lower lime content tend to be slow in hardening and may show trouble from dusting of the clinker by transformation of the β- to γ-C_2S phase, especially if the clinker cooling is very slow. The zone is limited on the high lime side by the need to keep the uncombined CaO at low enough values to prevent excessive expansion due to hydration of the free lime. On the high alumina side, the zone is limited by excessive liquid-phase formation which prevents proper clinker formation in rotary kilns.

Clinker formation. Portland cements are ordinarily manufactured from raw mixes including components such as calcium carbonate, clay or shale, and sand. As the temperature of the materials increases during their passage through the kiln, the following reactions occur: (1) evaporation of free water; (2) release of combined water from the clay; (3) decomposition of magnesium carbonate; (4) decomposition of calcium carbonate (calcination); and (5) combination of the lime and clay oxides. Cooling is ordinarily too rapid to maintain the phase equilibria.

Phases formed in portland cements. Most clinker compounds take up small amounts of other components to form solid solutions, eg, the C_3S solid solution called alite. Under reducing conditions in the kiln, reduced phases such as FeO and calcium sulfide may be formed. Properties of the more common phases are summarized in Table 1. The major phases all contain impurities which stabilize the structures formed at high temperatures so that decomposition or transformations do not occur during cooling, as does occur with pure compounds. Portland cement clinker structures vary considerably with composition, particle size of raw materials, and burning conditions, resulting in variations of clinker porosity, crystallite sizes and forms, and aggregations of crystallites.

Raw material proportions. The three main considerations in proportioning raw materials for cement clinker are the potential compound composition; the percentage of liquid phase at clinkering temperatures (in the range from 18–25%); and the burnability of the raw mix. Potential compound composition is determined by Bogue calculation or ASTM C150 calculation.

Hydration

Calcium silicates. Lime and silica are frequently used in various proportions with or without portland cement in the manufacture of calcium silicate hydrate products. Typical compositions in cement chemists' notation include tricalcium silicate hydrate, $C_6S_2H_3$; calcio-chondrodite, C_5S_2H; α-dicalcium silicate hydrate, C_2SH; afwillite, $C_3S_2H_3$; foshagite, C_4S_3H; xonotlite, C_5S_5H; ranging in density from 2.6–2.98 g/cm^3. Calcium silicate hydrate is not the only variable in composition, but is very poorly crystallized and is generally referred to as calcium silicate hydrate gel (or tobermorite gel) because of the colloidal sizes of the gel particles. Significant changes in the structure of the gel continue over very long periods, changes often having a positive effect on both strength development and reduction of drying shrinkage. Drying and other chemical processes can have significant effects on this structure, such as loss of hydrate water and collapse of the structure to form more stable aggregations of particles.

Tricalcium aluminate and ferrite. The hydration of C_3A alone and in the presence of gypsum usually produces well-crystallized reaction products, eg, C_3AH_6, the cubic calcium aluminum hydrate, and C_4AH_{19} and $C_3A\overline{CS}H_{12}$ (monosulfate), the hexagonal phases. The formation of highly hydrated trisulfate, ettringite, is responsible for the necessary retardation of hydration of the aluminates in portland cements and the expansion process in expansive cements.

The hydration of the ferrite phase (C_4AF) is of greatest interest in mixtures containing lime and other cement compounds because of the strong tendency to form solid solutions. When the sulfate in solution is very low, solid solutions are formed between the cubic C_3AH_6 and an analogous iron hydrate C_3FH_6. In the presence of water and silica, solid solutions such as $C_3ASH_4.C_3FSH_4$ may be formed.

Other phases. In cements, free lime (CaO) and periclase (MgO) hydrate to the hydroxides. The hydrated silicates formed by portland cements at 100°C are similar to those obtained from lime–silica mixtures or C_3S and C_2S but with a higher C:S ratio of the C–S–N gel.

Hydration at ordinary temperatures. Portland cement is generally used at 5–40°C, being temperature extremes avoided. The initial conditions for the hydration reactions are determined by the concentration of the cement particles in the mixing water and the fineness of the cement. The exact course of the early hydration reactions depends mainly on the C_3A, ferrite, and soluble alkali contents of the clinker and the amount of gypsum in the cement. The overall hydration rate increases with the temperature, the fineness of the cement, and to a lesser extent with the water–cement ratio; measurements of the activation energy indicate that the reaction becomes increasingly diffusion controlled.

Cement Paste Structure and Concrete Properties

The properties of both fresh and hardened mortars and concretes depend mainly on the cement–water paste properties. The engineering properties of the concrete, eg, strength, elastic moduli, permeability to water and aggressive solutions, and frost resistance, depend strongly on the water–cement ratio and degree of hydration of the cement. Under sustained loads hardened cements and concrete creep or deform continuously with time, in addition to the initial elastic deformation. Drying of hardened cements results in shrinkage of the paste structure and of concrete members. The slowness of drying and the penetration of the hardened cement by carbon dioxide or chemically aggressive solutions is a result of the small size of the pores.

Manufacture of Portland Cements

Wet process and dry process plants produce portland cements. Manufacture consists of: quarrying and crushing the rock, including control of the clinker composition by systematic core drillings and selective quarrying; grinding the carefully proportioned materials to high fineness (argillaceous, siliceous, and ferriferous raw-mix components are added to the crushed product; ball mills are used for both wet and dry processes to grind material, although roller mills are used for dry process only); subjecting the raw mix to pyroprocessing in a rotary kiln (a countercurrent heating device—see Clinker Chemistry) when, during the burning process, the high temperatures cause vaporization of the alkalies, sulfur, and halides (kilns are either wet process, dry process, suspension preheaters, or precalciners); and grinding the resulting clinker to a fine powder via open-circuit or closed-circuit grinding.

Table 1. Phases in Portland Cement Clinker

Name	Crystal system	Density, g/cm^3	Moh's hardness	Cement chemists' notation
alite, Ca_3SiO_5	triclinic			C_3S
	monoclinic	3.14–3.25	ca 4	
	trigonal			C_2S
belite, Ca_2SiO_4	hexagonal	3.04		
	orthorhombic	3.40		
	monoclinic	3.28	> 4	
	orthorhombic	2.97		
C_3A, $Ca_3Al_2O_6$	cubic	3.04	< 6	C_3A
ferrite, $Ca_2Al_{<0.7}Fe_{>0.3}O_{3.5}$	orthorhombic	3.74–3.77	ca 5	$C_2A_xF_{1-x}$
free lime, CaO	cubic	3.08–3.32	3–4	C
magnesia, MgO	cubic	3.58	5.5–6	M

Pollution control. The Clean Air Act and its amendments set standards for dust collection on kilns and coolers. Modern equipment collects dust at about 99.8% efficiency. The Federal Water Pollution Control Act Amendments of 1972 established limits for cement-plant effluents.

Special Purpose and Blended Cements

These are manufactured essentially by the same processes as ordinary portland cements but have specific compositional and process differences. For example, white cements are made from raw materials of very low iron content; regulated set cements are made with fluorite (CaF_2) additions which also act as fluxing agents or mineralizers to reduce burning temperatures; expansive cements depend upon aluminate and sulfate phases which result in more ettringite formation during hydration than in normal portland cements (see Hydration); oil-well cements are manufactured for sluggish reactivity and thus levels of C_3A and alkali sulfates are kept low and hydration-retarding additions are employed; blended cements differ in the finish-grinding processes where the cement clinker is interground with pozzolans (natural materials such as diatomaceous earths, opaline cherts, and shales), granulated blast-furnace slag, or a variety of materials; masonry cements are finely interground mixtures, including limestones, hydrated lime, natural cement, pozzolans, clays, and air-entraining agents.

Nonportland Cements

These cements include calcium aluminate cements, manufactured by heating until molten or by sintering a mixture of limestone and bauxite with small amounts of SiO_2, FeO, and TiO_2; supersulfated cements, containing about 80% slag interground with 15% gypsum or anhydrite and 5% portland cement clinker; and hydraulic limes, produced by heating below sintering temperature a limestone containing considerable clay.

Uses

Portland cement is most widely used in concrete for construction of underground, marine, and hydraulic structures. Masonry and oil-well cements are produced for special purposes; calcium aluminate cements are extensively used for refractory concretes (see Refractories). Other cements used are air-entraining cements; expansive cements; regulated set cements; natural cements; calcium aluminate cements; Trief cements; and hydraulic limes.

RICHARD A. HELMUTH
F.M. MILLER
T.R. O'CONNOR
N.R. GREENING
Portland Cement Association

J.F. Young, ed., *Cements Research Progress 1976*, American Ceramic Society, Columbus, Ohio, 1977.

H.F.W. Taylor, ed., *The Chemistry of Cements*, Vols. 1 and 2, Academic Press, Inc. (London) Ltd., London, 1964, Appendix 1.

F.M. Lea, *The Chemistry of Cement and Concrete*, 3rd ed., Edward Arnold (Publishers) Ltd., London, 1971.

W.H. Duda, *Cement Data Book*, Bauverlag G.m.b.H., Wiesbaden, FRG, 1976.

Cement Standards of the World, Cemburear, Paris, France, 1968.

CENTRIFUGAL SEPARATION

Centrifugal separation is a mechanical means of separating the components of a mixture by accelerating the material in a centrifugal field. The field acts upon the mixture in the same manner as a gravitational field, but can also be varied by changes in rotational speed or dimensions of the equipment. Commercial centrifugal equipment can reach accelerations of 20,000 times gravity and laboratory equipment can reach up to 360,000 G. Most centrifugation equipment is intended to separate immiscible or insoluble components from a liquid medium. The ultracentrifuge and gas centrifuge represent special cases that establish separation gradients on a molecular scale. The usual gravitational operations, such as sedimentation or flotation of solids in liquids, drainage of liquids from solid particles, and stratification of liquids according to density, are accomplished more effectively in a centrifugal field (see Flotation; Gravity concentration; Size classification).

The complex flow patterns in the centrifuge bowls have defied satisfactory mathematical modelling. The concept of a theoretical capacity factor for sedimentation depends only upon the characteristics of the equipment and is independent of the system, and the theoretical effect of particle size distribution in a single and multistage centrifugation has been demonstrated.

Centrifugal separation of a mixture of immiscible components makes use of either density differences between the components or drainage of a liquid phase through a packed bed or cake of solid particles.

Separation by Density Difference

A single solid particle or discrete liquid drop settling under the acceleration of gravity in a continuous liquid phase accelerates until a constant terminal velocity is reached. At this point, the force resulting from gravitational acceleration and the opposing force resulting from frictional drag of the surrounding medium are equal in magnitude. The terminal velocity largely determines what is commonly known as the settling velocity of the particle or drop under free-fall or unhindered conditions; for a small spherical particle, it is given by Stokes' law:

$$v_g = \frac{\Delta\delta\, d^2 g}{18\mu}$$

where v_g = the settling velocity of a particle or drop in a gravity field; $\Delta\delta = \delta_s - \delta_L$ = the difference between the true mass density of the solid particle or liquid drop and that of the surrounding liquid medium; d = the diameter of the solid particle or liquid drop; g = the acceleration of gravity; and μ = the viscosity of the surrounding medium.

Stokes' law can readily be extended to the centrifugal field:

$$v_s = \frac{\Delta\delta\, d^2 \omega^2 r}{18\mu} = v_g\left(\frac{\omega^2 r}{g}\right)$$

where v_s = the settling velocity of a particle or drop in a centrifugal field; ω = the angular velocity of the particle in the settling zone; and r = the radius at which settling velocity is determined.

These concepts are used to analyze separations in the bottle centrifuge, the imperforate bowl centrifuge, and the disk centrifuge. Separation by density difference in other types can be analyzed by analogy.

Separation by Drainage

The theory covering drainage in a packed bed of particles is incomplete even for a gravity field, and requires much more development for a centrifugal field.

Equipment

Centrifugal sedimentation equipment is characterized by limiting flow rates and theoretical settling capabilities. The important parameters of centrifugal filtration equipment are screen area, degree of centrifugal acceleration in the final drainage zone, and cake thickness, which affects both residence time and volumetric throughput rate.

Equipment Operating by Density Difference

This type of equipment can be categorized by capacity range and theoretical settling velocities of the particles normally handled.

Bottle centrifuges are designed to handle small batches of material for laboratory separations, testing and control. The basic structure is a motor-driven verticle spindle supporting various heads or rotors (either swinging bucket, fixed-angle head, or small perforate or imperforate baskets for larger quantities of materials); a surrounding cover to reduce windage, to facilitate temperature control, and to provide a safety shield against breakage of the spinning container. Optional accessories include timer, tachometer, and manual or automatic braking. Bench-top bottle centrifuges operate with rheostat control at 500–20,000 rpm with 34,000 G in the high speed units. Larger units operate up to 6000 rpm developing 8000 G with special attachments permitting 40,000 G and may also be equipped with automatic temperature control down to −10°C (see Fig. 1).

Preparative ultracentrifuges are used to process quantities of subcellular particles, viruses, and proteins. Operating speeds range from 20,000 rpm (generating about 40,000 G), to 75,000 rpm (generating ca 500,000 G). Pressure in the casing surrounding the rotor is reduced to ca 0.13 Pa (1 μm Hg) for high speeds and to < 130 kPa (< 1 mm Hg) for lower speeds; temperature is controlled in the range of −15 to +30°C within ±1°C; rotor speed may be automatically programmed for sequential changes in speed.

Analytical ultracentrifuges allow continuous observation inside the cells of the rotor. They operate on particles so small that diffusion must also be considered, and are used to measure sedimentation velocities of macromolecules in solution or to determine the weight of particles or molecules by letting them settle until equilibrium in a concentration gradient has been reached.

Zonal centrifuges use density gradients in the centrifuge rotors to greatly increase the sharpness of separations and the quantities of the materials that can be handled. Low molecular weight solutes are frequently used, and a natural, continuous, or step gradient may be formed.

Tubular centrifuges separate liquid–liquid mixtures or clarify liquid–solid mixtures with less than 1% solids content and fine particles. Liquid is discharged continuously, whereas solids are removed manually when sufficient bowl cake has accumulated. For industrial use, cylindrical bowls 100 to 130 mm in diameter (length : diameter ratios range from 4 to 8) attain speeds up to 15,000 rpm, generating centrifugal accelerations up to 13,000 G at the bowl wall. The laboratory tubular centrifuge operates at speeds of 10,000 to 50,000 rpm, generating 65,000 G at the latter speed in the 4.5 cm diameter bowl. The nominal capacity range is 30 to 60,000 cm³/min. They are used primarily for the purification of contaminated lubricating oils (see Fig. 2).

Disk centrifuges channel feed through a large number of conical disks to facilitate separation; they combine high flow rates with high theoretical capacity factors, depending on the number and the cube of the size of disks. Capacities are maximized with bowls of equal height and diameter. The outstanding feature of the bowl design is a stack of cones (disks) arranged so that the mixture to be clarified must pass through the disk before discharge. The stratification of the liquid medium greatly reduces the sedimenting distance required before a particle reaches a solid surface and may be considered removed from the process stream. The angle of the cones is great enough so that even solid particles deposited on their surfaces slide either individually or as a concentrated phase according to the difference in their density and that of the medium. Disk centrifuges operate at temperatures from −40°C to about 200°C and at pressures up to 1.03 MPa (140 psi). Industrial units operate at speeds from 4500 to 8000 rpm. This type of centrifuge was originally employed for the separation of cream from milk (see Milk products), and is used for purifying fuels and lubricating oils, separating wash water from fats and vegetable or fish oils, and removing moisture from jet fuel.

Variants of the imperforate-bowl disk centrifuge are used in the food and vegetable-oil industries. Continuous-discharge disk centrifuges are equipped with 8 to 24 orifices (nozzles) located on the bowl periphery. These continuous disk centrifuges are used to dewater kaolin clay, to separate fish oil from press liquor, to concentrate starch and gluten, to clarify wet-process phosphoric acid, to concentrate yeast, bacteria, and fungi from growth media in protein synthesis, and to thicken secondary excess activated sludge from biological treatment in industrial and municipal wastewater treatment plants (see Wastes, industrial).

Continuous conveyor-discharge centrifuges collect solids by sedimentation and continuously discharge both liquid and solid material using centrifugal fields lower than in disk or tubular centrifuges because of the conveyor and associated mechanisms. Maximum speeds range from 1600–8500 rpm; operating pressures up to 1.03 MPa (150 psi) and temperatures up to 200°C are standard and 300°C is possible. Bowl designs include countercurrent or concurrent movement of the phases. They are used to dewater or thicken sludges in waste treatment (see Water, sewage), to clarify and recover many types of crystals, meal from fish press liquor, and polymers; they dewater coal and concentrate solids from flue-gas desulfurization sludges.

Liquid cyclones, or hydroclones, are sedimentation-type, imperforate-wall centrifugation equipment. In general, they can economically handle large volumes of feed to produce a ten to twentyfold concentration of feed solids. Normal operating pressure drops are 35–350 kPa (5–50 psi); temperatures are up to 300°C. Optimum cyclone performance has been obtained with fairly consistent designs. They are used for classification, eg, degritting of kaolin clay before finer classification in a conveyor-discharge centrifuge; concentrating ores; desliming tailings; recovering catalysts (see also Air pollution control methods).

Equipment Operating by Filtration

Centrifuges that separate by filtration produce drained solids and are categorized by final moisture, drainage time, G, and physical characteristics of the system.

Basket centrifuges are the simplest and most common form of centrifugal filter. They consist of a cylindrical bowl with a diameter of ca 100–2400 mm (diameter-to-height ratio ranges from 1 to 3). The wall is perforated with a large number of holes, and is lined with a filter medium, eg, a single layer of fabric or metal screen. Three types of basket centrifuge are distinguished by their ways of discharging accumulated solids: (1) those that are stopped for discharge, used if a variety of materials must be filtered in small batches, if equipment must be sterilized between batches, or if the production rate is too low to warrant more automation, and to remove liquid from crystalline materials, to dry bulk materials, and to clarify process liquids and waste streams; (2) those that are decelerated for discharge; (3) those that discharge at full speed, primarily used for materials draining freely in a high centrifugal field, for medium tonnages, and for multiple rinses where nearly complete segregation of the rinses and mother liquor is desired; also used on borax and boric acid and for dewatering various products or slurries.

Basket centrifuges are suitable for applications where the nature of solids may vary widely with time and for collection of soft or fine solids that are difficult to filter, eg, to dewater excess biological and alum wastes.

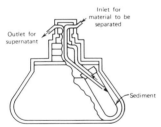

Figure 1.   Tube-type continuous flow rotor. Courtesy of Sorvall.

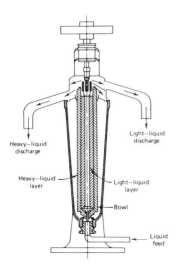

Figure 2.   Tubular centrifuge.

Continuous cylindrical-screen centrifuges. Continuous filtering centrifuges are used for very fast-draining solids that do not require extremely dry final products. Rinsing efficiency varies considerably; power requirements are usually low; initial slurry concentrations can be somewhat more variable and not as high as for the high speed basket. The cylindrical-screen centrifuge may consist of two to six steps with successively larger diameters. They are used on sugar where the high viscosity of the mother liquor causes slow drainage, a high degree of plasticity in the partially drained cake, and poor penetration of wash liquor. Conical-screen centrifuges are those that are self-discharging (used for drying crystalline materials and for separating or dewatering fine fiber in wet-corn milling); those that discharge by means of a helical conveyor (used for dewatering medium and coarse crystals, deoiling proteinaceous solids and for removing solids from fruit and vegetable pulps and other food slurries); and those that apply an axial vibration or oscillation to the bowl or the bowl and casing (used in dewatering potash and other crystalline solids).

A.C. LAVANCHY
F.W. KEITH, JR.
Pennwalt Corp.

C.M. Ambler, *Chem. Eng. Prog.* **48**, 150 (1952).
F.A. Records, *Chem. Eng.* 281 (Jan. 1974).

CEPHALOSPORINS. See Antibiotics—β-Lactams.

CEPHEMS. See Antibiotics—β-Lactams.

CERAMIC COLORS. See Colorants for ceramics.

CERAMIC COMPOSITE ARMOR. See Heat-temperature composites.

CERAMICS

SCOPE

Ceramics comprise all engineering materials or products (or portions thereof) that are chemically inorganic, except metals and alloys, and are usually rendered serviceable through high temperature processing. Ceramic materials are normally composed of both cationic and anionic species; their primary difference from other materials is in the nature of their chemical bonding. They are frequently termed ionic solids, ie, possessing ionic bonding.

Modern ceramics encompass a wide range of materials ranging from single crystals and dense polycrystalline materials through glass-bonded aggregates to insulating foams and wholly vitreous substances. The need for high performance ceramic materials has increased steadily in the last ten to twenty years (see Ceramics as electrical materials; Coal; Solar energy; Boron compounds, refractory carbides and nitrides; Silicon and silicides; Refractones).

As a class of materials, ceramics are better electrical and thermal insulators and more stable in chemical and thermal environments than metals, are appreciably stronger in compression than in tension, and exhibit greater ridigity, hardness, and temperature stability than polymers (see also Glass; Glass ceramics; Enamels).

ROBERT F. DAVIS
North Carolina State University

F.H. Norton, *Elements of Ceramics*, 2nd ed., Addison Wesley Publishing Company, Inc., Reading, Mass., 1974.

Institute of Ceramics, Textbook Series, published by McClaren and Sons, Ltd., London: (**a**) W.E. Worrall, *Raw Materials*, 1964; (**b**) F. Moore, *Rheology of Ceramic Systems*, 1967; (**c**) R.W. Ford, *Drying*, 1964; (**d**) W.F. Ford, *The Effect of Heat on Ceramics*, 1976.
W.D. Kingery, H.K. Bowen, and D.R. Uhlmann, *Introduction to Ceramics*, 2nd ed., John Wiley & Sons, Inc., New York, 1976.

RAW MATERIALS

The principal raw materials of the ceramic industry are clay (including shale and mudstone), silica, and feldspar (see Clays). Other raw materials include a wide variety of rocks, minerals and synthetic compounds used to manufacture diverse products.

Clays

Clay minerals are of three principal families: kaolinite, $Al_4Si_4O_{10}(OH)_8$, montmorillonite, $X_yAl_2(Al_ySi_{4-y})(O_{10})(OH)_2$ where X is usually Na, Mg, or Al, and illite, $K_y(AlFeMg_4Mg)(Al_ySi_{8-y})O_{20}(OH)_4$. Closely associated minerals are gibbsite, $Al(OH)_3$, diaspore, $HAlO_2$, and bauxite (of indefinite composition but usually given as $Al_2O_3.2H_2O$ which is an intermediate between the first two). All clays have as the major constituents one or more of these minerals.

The kaolinite group includes kaolinite, halloysite, dickite, and nacrite. The montmorillonite group includes montmorillonite, nontronite, beidellite, hectorite, and saponite. The illite group, similar to muscovite but containing less potassium, more silica, and more combined water, includes the illites, the hydromicas, phengite, brammallite, glaucomite, and celadonite. There is a wide range of substitutions that occur in each family. All three groups of minerals are classified as sheet silicas (see Silica; Silicon compounds). In addition, most clays will have one or more accessory minerals, eg, quartz, muscovite, biotite, limonite, hydrous micas, feldspar, vermiculite, and/or carbonaceous matter.

Clay terminology. The clays used in the ceramic industry are usually referred to by names that reflect their use. For example: a pottery clay is used to make pottery; a sewer-pipe clay is used for sewer pipe; fire-clays (or refractory clay) are used to manufacture fireclay bricks, crucibles, refractory mortars, etc; china clay is used to manufacture whitewares; slip clays are used for glazing stoneware and porcelain; and brick clays are used to manufacture common brick and face brick. Exceptions to this pattern do occur. For example, kaolins are clays consisting chiefly of kaolinite (see Table 1). The most refractory of all clays, with softening points of 1700–1785°C, it is used extensively in whitewares as the chief ingredient of most porcelain and china bodies. Ball clays are any sedimentary clay of high plasticity. They are composed primarily of kaolinite with varying amounts of impurities and some organic matter. They are used whenever increased workability is needed.

Table 1.   Analyses of Typical Clays

Typical kaolins, %	SiO₂	Al₂O₃	Fe₂O₃	CaO	MgO	Alkalies	H₂O
sedimentary (Fla.)[a]	47.0	36.8	0.8	0.2	0.2	0.2	15.0
crude residual (N.C.)	62.4	26.5	1.4	0.6		1.0	8.8
washed residual (N.C.)[b]	45.8	36.5	0.3	0.5		0.3	13.4
English china clay	48.3	37.6	0.5	0.1		1.6	12.0
Tennessee ball clay, 1	46.9	33.2	2.0	0.3	0.4	0.7	
English ball clay	49.0	32.1	2.3	0.4	0.2	3.3	
New Jersey ball clay[c]	45.6	38.9	1.1		0.1	0.2	

[a] Also 0.2% TiO_2.
[b] Also 11% FeO.
[c] Also 1.3% TiO_2.

Nonclay Minerals

Silica (qv) occurs in the form of quartz, tridymite, cristobalite, vitreous silica, cryptocrystalline forms, hydrated silica, and diatomite. Principal sources of silica used in the ceramic industry are the sandstones, quartzites, and sands. Principal uses of silica in ceramic industry are in glass, fused silica, brick, whiteware bodies, whiteware glazes, and enamels (qv). Diatomite (qv) is used for insulating refractories up to a temperature of ca 900°C; when bonded with clay, it may be shaped and used for insulation up to a temperature of 1370°C.

Feldspars are used chiefly as fluxes and sources of Al_2O_3, SiO_2, alkalies (K_2O, Na_2O), and CaO. The plagioclase feldspars vary in composition from albite, $NaAlSi_3O_8$, to anorthite, $CaAl_2Si_2O_8$, in a continuous series of solid solutions. Orthoclase and microcline feldspar ($KAlSi_3O_8$) are referred to as potash feldspar, used extensively in whiteware bodies. Anorthoclase, $(Na,K)AlSi_3O_8$, is a combination of albite and potash feldspar. High soda feldspars are used in glasses and glazes. All feldspars contain some free silica which is not detrimental as long as it is controlled.

Other nonclay minerals include nepheline syenite, a rock that contains a large percentage of the mineral nephelite, $(Na,K)_2Al_2Si_2O_8$, along with some soda and potash feldspars; lime, derived by calcination of limestone chiefly calcium carbonate, $CaCO_3$, which is used in glazes, enamels, and glasses (eg, whiting is a high purity calcium carbonate) (see Calcium compounds); magnesium silicates such as talc (used as a flux with clays), asbestos (used chiefly for insulating refractories) (see Asbestos), and olivine (used for refractory products; fluxing minerals which lower the vitrification temperature, the melting temperature or the reaction temperature, eg, lithium minerals spodumene, $Li_2Al_2Si_4O_{12}$, lepidolite, $[K_2(Li_3Al_3)-(Al_2Si_6O_{20})(OH,F)_4]$, amblygonite, $Li_2F_2Al_2-P_2O_8$, and petralite, $LiAlSi_4O_{10}$, as well as barium minerals such as barite, $BaSO_4$, and whiterite, $BaCO_3$; refractory minerals such as zirconium minerals (see Zirconium and zirconium compounds), hydrated alumina minerals, titania, TiO_2, thoria, ThO_2, and graphite, one of the most refractory materials available (see Carbon and artificial graphite). More minerals in this group are magnesite, dolomite, gypsum, chromite, the aluminum silicates, and pyrophyllite.

Special Materials

Super refractories include the carbides of elements such as silicon, boron, zirconium, hafnium, tantalum, vanadium, molybdenum, tungsten, and niobium (see Carbides; Boron compounds); nitrides (qv); sulfides (see Cerium and cerium compounds); single and complex oxides.

Nuclear ceramic materials being researched include uranium oxide, UO_2, uranium carbide, nitrides, sulfides, phosphides, and combinations (see Nuclear reactors; Uranium).

Electronic ceramic materials include ferrites (qv) with improved magnetic properties (see Magnetic materials), titanates, principally $BaTiO_3$ (see Ultrasonics; Ceramics as electrical materials). Numerous other materials used in the ceramics industry are covered in *Ceramic Age*.

JOHN V. HAMME
North Carolina State University

W.E. Worrall, *Clays and Ceramic Raw Materials*, Halsted Press, a division of John Wiley & Sons, Inc., New York, 1975.

R.W. Grimshaw, *The Chemistry and Physics of Clays*, 4th ed., John Wiley & Sons, Inc., New York, 171.

Ceram. Age **4** (July 1976).

E.J. Kliff, D.R. Dykes, and J.R. Hickey, *Ceramic Raw Materials*, Charles H. Kline and Co., Fairfield, N.J., 1971.

FORMING PROCESSES

Forming imparts to ceramic ware permanent shape, and temporary size (altered by shrinkage) and strength (which increases during drying and firing). Forming methods used in the production of ceramics are plastic deformation (eg, extrusion, dry pressing and all hot forming methods) and casting (eg, slip and fusion casting).

In preparing materials for forming, optimal particle size and size distribution may be obtained by crushing and grinding the various materials, separating the various size fractions by screening, and then blending the desired size fractions of each material.

Plastic Forming Processes

In cold forming, the ware is made oversize and shrinks during firing and drying. In hot forming (eg, hot pressing, hot extrusion, hot rolling, forging and swaging, hot isostatic pressing, and chemical vapor deposition), the ceramic ware is made close to its final size.

Extrusion processes are used to form ware of any shape that has an axis normal to a fixed cross section. It is widely used in the structural clay products industry and to a lesser extent in the whiteware and refractories industries (see Refractories). Completely nonplastic materials can be extruded with the addition of suitable plasticizers (qv) such as gums, starches, poly(vinyl alcohol), waxes, and wax emulsions. Extrusion usually comprises three stages: pugging, deairing, and extrusion through a die.

Soft plastic forming is usually used to produce shapes with symmetrical circular cross sections from materials with 20–30% water and significant percentages of plastic clays. Common processes include jiggering, used principally in the whiteware industries, and hand molding.

Dry pressing is used when many parts of relatively simple shapes are to be made, eg, in the whiteware, refractories, and abrasive industries and in the production of cermets (ceramic–metal composites). The parts must have no undercuts, be fairly uniform in thickness, and have a length not normally exceeding twice their diameter. The material to be dry pressed is usually in the form of granules, either dry or containing up to 12% water. Isostatic presses and vibratory forming are used to overcome the problems of nonuniform density which arise.

Casting processes. For slip casting, a suspension of the raw materials in a liquid such as water is prepared by blunging or ball milling. This slip is poured into a porous plaster of Paris mold, with the interior surfaces of the mold conforming to the exterior surface of the desired ware. As the plaster absorbs water from the slip, solid particles are deposited on the interior surface of the mold. Slip casting permits the formation of complex shapes, and is widely used in the whiteware industries to make art objects and dinnerware that is not adaptable to jiggering, and in the refractories industry to make crucibles and other special shapes. Slip-cast molds are relatively inexpensive and reusable. Special techniques can be used to form materials without plasticity; thin sheets of ceramics can be made by the doctor-blade process, and complex shapes can be made by injection molding.

ROBERT F. STOOPS
North Carolina State University

W.D. Kingery, H.D. Brown, and D.R. Uhlman, *Introduction to Ceramics*, 2nd ed., John Wiley & Sons, Inc., New York, 1976.

F.H. Norton, *Elements of Ceramics*, 2nd ed., Addison-Wesley Publishing Co., Reading, Mass., 1974.

G.Y. Onoda, Jr. and L.L. Hench, eds., *Ceramic Processing Before Firing*, John Wiley & Sons, Inc., New York, 1978.

THERMAL TREATMENT

Thermal treatment is an essential step in the manufacturing of ceramic products. Materials that are stable at room temperature have to be raised to relatively high temperatures (ranging from 700°C for enamels (qv) to ca 1650°C for alumina ceramics) for reactions to take place.

Methods of Thermal Treatment

Thermal treatment is a smoothly varying time–temperature profile, rather than a particular temperature held for some time period. To determine firing temperature and quality control, thermocouples, radiation pyrometers, and optical pyrometers are used; expendable pyrometric cones are used for a large class of materials, eg, structural clay products, porcelains, sanitary ware (see Temperature measurement). Microstructures and properties of many specialty ceramics vary according to time–temperature profile and often require precise monitoring of thermal gradients and temperature (see Amorphous magnetic materials; Glassy metals).

Extreme temperatures are achieved in protective atmospheres using refractory metals; the upper limit is the temperature at which vaporization occurs, eg, for molybdenum, 1800–2200°C; for tungsten, 3000°C. Above 3000°C, graphite is used; plasma devices are used up to 15,000°C, eg, for dynamic thermal processing of ceramic particulates (see Furnaces, electric; Plasma technology). Atmospheric control of the furnace is required for oxide ceramics to control the volatile components, and for nonoxide ceramics, nitrogen and inert gases are used to exclude oxygen.

Kilns. A tunnel kiln is the most prevalent equipment for thermal treatment; it is used for structural clay products, refractories, tile, porcelain, dinnerware, and electronic components. A periodic kiln is ordinarily used for small-scale production. A rotary kiln is used for manufacturing products such as cement (qv) and calcined lime or dolomite. Arc melting is used for processing materials such as alumina, magnesia, zirconia, and mullite.

Hot-pressing processes, which incorporate forming and thermal treatment in one step, are used to obtain a dense ceramic in chemical systems that cannot be fully densified using conventional firing methods; it is limited to very simple shapes and slow rates of production.

Physical and Chemical Changes During Thermal Treatment

Drying and binder removal occurs between 0–400°C. The ware loses physically held water and organic binders which are essential for the forming operation.

Firing occurs at higher temperatures, and involves these changes: dissociation, the loss of carbon dioxide by carbonate constituents of ceramic compositions; compound formation, the continued heating of clays which results in the formation of mullite, $3Al_2O_3 \cdot 2SiO_2$; polymorphic transformation of materials capable of existing in more than one crystallographic arrangement, eg, silica and zirconia; sintering, the fabrication of a product usually involving densification by pore removal through diffusional mechanisms; and vitrification.

THOMAS M. HARE
North Carolina State University

W.D. Kingery, H.D. Brown, and D.R. Uhlman, *Introduction to Ceramics*, 2nd ed., John Wiley & Sons, Inc., New York, 1976.

PROPERTIES AND APPLICATIONS

Composition and Microstructure

Composition is the chemical and mineralogical makeup of the total substance of the material (including its deliberate and accidental impurities); microstructure is the crystallographic structure of each component or phase and, in particular, the size, shape, and distribution of each of the phases present (including the unoccupied space or porosity). These are determinative parameters fixed by the combined response of the starting materials to all the processing steps.

A ceramic material is a crystalline, vitreous, or mixed type solid which may be monophase (composed of a single substance) or polyphase (containing two or more discrete substances); monophase crystalline ceramics may be monocrystals. Most ceramics are polycrystalline masses with abrupt changes in orientation or composition occurring across each grain or phase boundary. Many properties of ceramics are surface-sensitive, both with respect to mechanical condition and/or chemical and atmospheric environment.

Chemical Properties of Ceramic Materials

Ceramic materials based on oxides, carbides, nitrides, and borides exhibit a high degree of chemical stability, eg, oxides are extremely stable over wide ranges of temperature and environmental conditions. The stability of nonoxide ceramic compounds results from the compactness of the crystal structure, the directed chemical bonding (generally ionic or covalent, or of mixed ionic–covalent character), and of the high field strengths associated with the relatively small, highly charged cations encountered in refractory ceramics. Thermodynamically, the most stable ceramic will be that having the greatest negative free energy of formation from the elements, eg, yttria, thoria, alumina, beryllia, magnesia, and stabilized zirconia (see Refractories). Most refractory ceramics have very low vapor pressures in the most stable valence state, but the suboxides tend to be volatile (SiO, Al_2O, etc). Interactions between dissimilar ceramics, or between ceramics and metals, become increasingly likely as operating temperature is increased, or at low pressure. Even a system of materials with high melting points can form fluid liquids at temperatures much below either melting point; eg, Al_2O_3 melts at 2050°C and CaO at 2500°C, but when in contact they form a reactive, eutectic liquid at 1450°C. The presence of a few percent impurities can significantly lower the temperature of liquid formation.

Optical Properties

Many ceramic substances are optically transparent in single crystal or vitreous forms. Optical properties are closely linked to composition and structure of the ceramic; eg, the degree of polarization is a function of ion size, binding energy, and crystallographic direction, and is measured by the index of refraction. Also, the refractive index of SiO_2 is 1.55; of MgO 1.74; of $ZrSiO_4$ 1.95; of TiO_2 2.71; and for LiF is 1.392. Principal and very practical sources of unfilled energy bands are impurities, or deliberate colorant oxide additives such as cobalt blue and chrome green. The microstructure of a polycrystalline ceramic tends to dominate the useful optical properties of the base material. Polyphase polycrystalline materials may develop additional scattering owing to differences in refraction across boundaries, dispersed phases in particular acting as opacifiers; dispersions of SnO_2, $ZrSiO_4$, TiO_2 and/or Sb_2O_3 are commercially employed as opacifiers and whiteners in enamels (qv), glazes, and glasses. Highly translucent, almost transparent, pore-free polycrystalline ceramics are produced by sintering or the hot-pressing methods, and require careful control of impurities. Control of impurities can minimize internal absorption.

Thermal Properties

The thermal properties of greatest interest include specific heat, thermal conductivity, and thermal expansion. The principal properties improving thermal shock resistance are low thermal expansion, high thermal conductivity, and high strength. Ceramic materials in general do not have particularly good thermal shock resistance because of their brittle nature. They are used in many applications where ductile materials could not be used because of superior high temperature properties. Materials such as fused silica, cordierite, and lithium aluminosilicates have very low thermal expansion and good thermal shock resistance despite being weak when compared to other ceramic materials.

Elastic Properties

Elastic properties of solids arise from the interaction of mechanical distortion, called strain, with the periodicities of the atomic or ionic structure. Elastic moduli measure the force involved in achieving recoverable, small, unit displacements of atoms or ions from their equilibrium positions, and are dependent upon the type of bonding and the electron density configuration of the crystal. For isotropic substances such as glass, and for randomized polycrystalline materials having pseudoisotropic behavior, elastic properties can be adequately expressed in terms of:

$$E = \text{Young's modulus of elasticity} = \frac{\sigma}{\epsilon}$$

$$G = \text{Shear modulus} = \frac{\tau}{\gamma}$$

$$\mu = \text{Poisson's ratio} = \frac{E\gamma}{2G} - 1$$

where σ = normal stress, τ = shear stress, ε = normal strain, γ = shear strain.

Strength

Ceramics are brittle materials, strong in compression but much weaker in transverse bending or in tension. As in glass at room temperature, deformation is entirely elastic up to the point of failure and brittle fractures attributed to stress-induced catastrophic extension of preexisting defects, the well-known Griffith cracks. The theoretical strengths of ceramic materials, derived from atomic bonding consideration, are far larger than the observed tensile strengths which are explained by the presence of flaws. The crack-tip areas, or microflaws from which fractures originate, are produced during the fabrication process and can be identified by careful microscopy; the largest flaw is often the one that affects strength.

At high temperatures, ceramics can deform plastically, a mechanical behavior similar to metals, and in creep. Ceramic material selected for resistance to creep at high temperatures may be rather weak and prone to brittle fracture at low temperatures, and subject to excessive deformation at higher temperatures.

Engineering with ceramics. A minimum service-life guarantee can be determined from knowledge of crack propagation behavior. The value obtained for strength in a test is highly dependent on how the tests are made; the type of flaw present at the surface of the material can dominate if these flaws are larger than those inherent in the bulk materials.

Electrical and Magnetic Properties

Ceramics are increasingly used as electronic and magnetic components (see Ceramics as electrical materials). Most oxide ceramics in their normal valence states are insulators (see Insulation, electric) while most nonoxide ceramics are semiconductors (qv); eg, SiC is much used as a high temperature heating element and as a voltage-dependent resistor to ground. Ceramics containing transition metal ions may display pronounced magnetic effects; in contrast to magnetic metals, they do not act as electrical conductors (see Magnetic materials; Glassy metals; Amorphous magnetic materials; Microwave technology).

Magnetic materials are classified as hard (permanent magnets) or soft (materials that may be magnetized and demagnetized). The main types are the spinel ferrites, rare earth garnets, and hexagonal ferrites.

Composites and Cermets

An important segment of the ceramic industry is concerned with complex materials or components of which the ceramic phase is only a part; another part is metallic or plastic, and the useful properties of the combination are different from those of either phase alone. Composites contain at least one phase which is macroscopic in at least one dimension (fibrous shape). Cermets contain ceramic and metallic phases randomly shaped and intimately dispersed one within the other on a microstructural scale.

Uses

All useful properties of a ceramic must be provided in advance; failure in service will necessitate replacing the piece *in toto*. Many ceramics are produced in volume as standard items, eg, refractory bricks and shapes, crucibles, furnace tubes, and hermetic seals. Ceramic–metal composites and glass–fiber-reinforced plastics are employed for aerospace hardware (see Ablative materials). Ceramic fiber-reinforced metals, metal-impregnated ceramics, and metal–ceramic laminates are being researched (see Composite materials). The largest group of cermets are carbides or cemented carbides (see Carbides).

THOMAS M. HARE
North Carolina State University

W.D. Kingery, H.D. Brown, and D.R. Uhlman, *Introduction to Ceramics*, 2nd ed., John Wiley & Sons, Inc., New York, 1976.

CERAMICS AS ELECTRICAL MATERIALS

The electrical characteristics of ceramic materials vary greatly; oxides, borides, nitrides, and carbides have metallic conductivity in some cases but behave like semiconductors in others. Many ceramics are good insulators even at high temperatures. The atomic processes are very different for the various conduction modes and the transport of current may be caused by the motion of electrons or electron holes, or by ions.

Semiconductor materials are used in many specialized applications such as resistance heating elements; semiconductor devices such as rectifiers, photocells, transistors, thermistors, detectors, and modulators. Ceramics are equally important as electrical insulators; porcelains are used for both low- and high-voltage insulation. Consequently, the entire range of electrical conduction properties should be considered (see Photovoltaic cells; Semiconductors; Insulation, electric). For background information see *Introduction to Ceramics* as well as reviews on electronic and ionic conductors.

Electrical Conduction Phenomena

Materials are usually classed according to the specific conductivity mode, such as insulators which have low concentrations and low mobility of carriers. Metallic conductors such as borides, carbides, some oxides, etc, have a high conductivity value which is not a strong (exponential) function of temperature. Insulators have a low conductivity which increases exponentially with temperature; for example, the conductivity $(\Omega \cdot m)$ at room temperature of SiO_2 glass is $< 10^{-12}$; of low voltage porcelain $10^{-10} - 10^{-12}$; of fire-brick clay 10^{-6}. Semiconductors are intermediate and have an exponential temperature dependence. The conductivity $(\Omega \cdot m)^{-1}$ of dense silicon carbide is 10; of boron carbide 200; of pure germanium 10^4. Note that conductivity is the inverse of resistivity $(\sigma_i = l/\rho_i)$ and the two are used interchangeably. The resistivity in SI units is $\Omega \cdot m$ but much of the literature remains in units of $\Omega \cdot cm$.

The total conductivity is that owing to the net motion of electrons, holes, and each of the ions:

$$\sigma = \sigma_1 + \sigma_2 + \sigma_3 \ldots \sigma_i$$

The fraction of the total conductivity (at a specific temperature, composition, etc) owing to the particle is called the transference number of i: $t_i = \sigma_i/\sigma$.

Ionic Conduction in Ceramics

Ions in compounds or glasses such as oxides and halides constitute an ever-present charge carrier that can contribute to electrical conductivity. Electrical conductivity resulting from ion migration is important in many ceramic materials. Its analysis requires a determination of the concentration and mobility of charge carriers. For an ion to move through the lattice under the driving force of an electric field, it must have sufficient thermal energy to pass over a free-energy barrier, ΔG^+, the intermediate position between lattice sites. The result is a mobility that depends on the ion jumping from site to site. The ion mobility (for the rocksalt crystal structure) is expressed as:

$$\mu_i = \frac{4ez_i a^2 v}{kT} e^{-\Delta G^+/kT}$$

where a is the lattice parameter (ca 3×10^{-8} cm), ez_i is the charge on the ion, and v is the vibrational frequency (ca 10^{13}/s).

When aliovalent (different valence) impurities are added to an ionic solid, the crystal lattice compensates by forming defects that maintain both electrical neutrality and the anion-to-cation ratio of the host lattice. For example, addition of a mol of CaO to ZrO_2 requires the formation of a mol of oxygen vacancies. If this concentration is larger than the oxygen vacancies created by the thermal effects, then the conductivity from the motion of doubly charged oxygen ions is directly proportional to the concentration of CaO.

In polycrystalline materials, ion transport within the grain boundary must also be considered. For oxides with close-packed oxygens, the O-ion almost always diffuses much faster in the grain boundary than in the bulk. Other examples are less clear except that second phases at boundaries often promote higher ion diffusivities.

Several types of compounds show exceptionally high ionic conductivity and have recently become of technological interest. Such phases fall into three broad groups: (1) halides and chalcogenides of silver and copper in which the metal atom is disordered over several alternative sites; (2) oxides with the β-alumina structure in which a monovalent cation is mobile; and (3) oxides of the fluorite structure in which large concentrations of defects are caused by either a variable-valence cation or solid solution with a second cation of lower valence, eg, $CaO-ZrO_2$ or $Y_2O_3-ZrO_2$.

Electron Conduction in Ceramics

The relatively high mobility of conducting electrons or electron holes when present contributes appreciably to electrical conductivity. In some cases, metallic levels of conductivity result; in others, the electronic contribution is extremely small. In all cases, the electrical conductivity can be interpreted in terms of carrier concentrations and carrier mobilities. Metallic conduction occurs in a few cases of transition-metal oxides such as ReO_3, CrO_2, VO, TiO, and ReO_2; the doped perovskite structures $LaTiO_3$, $CaVO_3$, $CaMoO_3$, $BaTiO_3$, $SrCrO_3$, $SrFeO_3$, $LaNiO_3$, $LaFeO_3$, $LaCoO_3$, and $LaCrO_3$; many of the bronzes, Na_xWO_3, La_xWO_3, and $Na_xNb_2O_5$; and some of the spinels, $Li_{0.5}In_{0.5}Cr_2S_4$. It occurs at specific dopant concentrations.

Two types of scattering affect the motion of electrons and holes. Lattice scattering results from thermal vibrations of the lattice and increases with the increasing amplitude of vibrations at higher temperatures, $T^{-3/2}$. A second source of scattering is the presence of impurities which distort the periodicity of the lattice, $T^{+3/2}$. The total mobility is proportional to the sum of these two terms. The temperature dependence of the mobility term for quasi-free electrons and holes is much smaller than that for their concentrations (exponential). As a result, the conductivity has a temperature dependence chiefly determined by the concentration term.

Nonstoichiometric and solute-controlled electronic ceramics. Most oxide semiconductors are either doped with impurities to create extrinsic defects or annealed under conditions in which they become nonstoichiometric. Although these effects have been carefully studied in oxides, the precise nature of the low mobility value is often difficult to measure. Because the impurity atoms introduce new localized energy levels for electrons, intermediate between the valence and conduction bands, impurities strongly influence the properties of semiconductors. If the new energy levels are unoccupied and lie close to the energy of the top of the valence band, electrons are easily excited out of the filled band into these new acceptor levels. This leaves an electron hole in the valence band which can contribute to electrical conductivity. Positive-carried (p-type) oxide conductors commonly arise as a result of nonstoichiometric composition with a decreased metal content (eg, $Cu_{2-x}O$) and are sometimes called deficit semiconductors. If the impurity additions have filled electron levels close to the energy level of the conduction band, electrons may be excited from impurity atoms into the conduction band; these are called donor levels. The electron excited into the conduction band is able to contribute to conductivity. Negative-carried (n-type) oxide conductors commonly result from a nonstoichiometric composition with an excess metal content (eg, $Zn_{1+x}O$) and are sometimes called excess semiconductors.

Silicon carbide can be doped with boron to provide acceptor levels within the band gap (ca 0.3 eV above the valence band), thus making it a p-type conductor, or nitrogen can be added to provide donor levels and n-type conduction (ca 0.07 eV below the conduction band).

Another method of obtaining semiconductors with controlled resistivity which avoids difficulties caused by stoichiometry deviations is the formation of solid solutions of two or more compounds with widely different conductivity. Magnetite, Fe_3O_4, is an excellent conductor with a specific resistance of about 10^{-4} $\Omega \cdot m$, compared to values of about 10^8 $\Omega \cdot m$ for most stoichiometric transition-element oxides. The good electrical conductivity of magnetite is a function of the random location of Fe^{2+} and Fe^{3+} ions on octahedral sites which allows electron transfer from cation to cation.

In general, a condition for appreciable conductivity in the spinel structure is the presence of ions having multiple valence states at equivalent crystallographic sites. Examples of solid-solution semiconductors include $Fe_3O_4-MgCr_2O_4$, $Fe_3O_4-FeAl_2O_4$, and $Fe_3O_4-FeCr_2O_4$. The temperature coefficient or activation energy increases with increasing resistivity. Semiconductor materials of this type with materials like $MgAl_2O_4$, $MgCr_2O_4$, and Zn_2TiO_4 as the nonconducting component can

be prepared with a controlled temperature coefficient of resistivity. Semiconductors made in this way are used as thermistors.

Ceramics with high electronic conductivity or with nonlinear behavior. As previously indicated, many oxides are metallic conductors. The perovskite- and rutile-structure oxides are the most frequently studied. The d- or f-electron states of the transition-metal ions overlap, causing a partially filled band of ca 10^{28} electrons/m^3 for conduction. The mobility of the electrons is usually less than 1 $cm^2/(V \cdot s)$. These materials are technologically significant as electrodes and in microelectronics. The ReO_2 and RuO_2 compounds are used in organic and inorganic pastes as electrical components and are applied as thick films followed by a firing step. The material most often studied is doped barium titanate, $BaTiO_3$, which is used as a barrier layer capacitor, intergranular capacitor, and nonlinear positive-temperature-coefficient (PTC) resistor.

Mixed Conduction in Ceramics

The conductivity of many ceramic materials is significantly influenced by both ionic and electronic defects. Highly resistive ceramics such as MgO and Al_2O_3 fall into this category; however, good electrical conductors may be of a mixed type. The CeO_2-ZrO_2 system is an example of mixed conduction. The ionic transference number varies from 0.03 to 1.0 depending on temperature and composition.

H. Kent Bowen
Massachusetts Institute of Technology

W.D. Kingery, H.K. Bowen, and D.R. Uhlmann, *Introduction to Ceramics*, 2nd ed., John Wiley & Sons, Inc., New York, 1976.

N.M. Tallen, ed., *Electrical Conductivity in Ceramics and Glasses*, Marcel Dekker, Inc., New York, 1974.

R.J. Friauf, "Basic Theory of Ionic Transport Processes," in J. Hladik, ed., *Physics of Electrolytes*, Vol. 1, Academic Press, Inc., New York, 1972.

B.R. Rossing, L.H. Cadoff, and T.K. Gupta, *The Fabrication and Properties of Electrodes Based on Zirconium Dioxide*, Vol. 2, 6th Intl. Conf. MHD Electrical Power Generation, Washington, D.C., June 1975, pp. 105–117.

CERAMICS, CHEMICAL WARE. See Ceramics.

CERMETS. See High temperature composites.

CERIUM AND CERIUM COMPOUNDS

Cerium is a member of Group IIIA of the periodic table, referred to as rare-earth metals, lanthanons, or lanthanides (see Rare-earth elements). Traditionally, the lanthanons have been classified into three groups: cerium, including elements from lanthanum (no. 57) through samarium (no. 62), terbium, including the elements europium (no. 63) through dysprosium (no. 66), and yttrium, including the elements holmium (no. 67) through leutetium (no. 71) (although yttrium, no. 39, is not a member of the lanthanum series, it is always found with these elements in nature).

Cerium has two common oxidation states: +3, cerous, Ce(III), and +4, ceric, Ce(IV). Cerous compounds resemble the other trivalent lanthanons, but ceric compounds are more like those of the elements titanium, zirconium, and thorium. The potential for the reaction $Ce^{3+} \rightleftarrows Ce^{4+} + e$ depends on the nature of the medium. Values of E^0_{298} are reported between -1.28 and -1.70 V. The electrode potential for the reaction $Ce(s) \rightleftarrows Ce^{3+} + 3e$ is $+2.335$ V on the hydrogen scale.

Properties

Cerium occurs primarily (88.48%) as ^{140}Ce isotope. Mp 798°C; bp 3257°C; density 6.773 g/cm^3.

Occurrence

Cerium is the most abundant of the rare-earth metals, comprising about 50% of a natural mixture. Although the lanthanons are widely distributed and there are many minerals containing them, sizable deposits are few in number. The chief source for rare-earth elements is bastnasite, $Ln_2F_3(CO_3)_3$, a fluorocarbonate, which contains more lanthanum and fewer heavier lanthanons than monazite, the second source of commercial importance. The main deposit of bastnasite is in the Mountain Pass area of California (90% of U.S., and 2/3 of world output). Monazite, a thorium–rare-earth orthophosphate, is found in many countries, eg, India, Brazil, Australia, Malaysia, and U.S., occurring primarily as beach placers and concentrates with other heavy minerals such as zircon, cassiterite, ilmenite, and rutile (see also Thorium and thorium compounds). Monazite forms light brown to hyacinth-red monoclinic crystals (sp gr, 4.9–5.5; Mohs hardness, 5–5.5). Thorium oxide content of monazite is 5–9% for commercial uses.

Ore Extraction

Anhydrous rare-earth chlorides may be derived directly from bastnasite or monazite by chlorination of an ore–carbon mixture in a special, electrically heated furnace at 1000–1200°C. Fused anhydrous chlorides are obtained which are suitable for the production of rare-earth metals.

Bastnasite

Bastnasite is extracted by hydrochloric acid or sulfuric acid treatment. Hydrochloric acid treatment involves concentrating the ore to ca 60% Ln_2O_3 by flotation (qv); drying and roasting to drive off CO_2 and oxidize the cerium; and slurrying in water and treating with diluted HCl to solubilize the bulk of the trivalent lanthanons, which are then separated by filtration. Sulfuric acid treatment involves first roasting the ore; dissolving it in dilute sulfuric acid; removing the lanthanon sulfate solution by filtration; and subsequent processing, eg, precipitation as fluorides or treatment with reducing agents, depending upon the products desired. Bastnasite may also be opened by heating the unroasted ore with concentrated sulfuric acid, driving off the fluoride, and leaving lanthanons as anhydrous sulfates which are dissolved in water for separation, and the rare-earth sulfates are then recovered. Heating unroasted bastnasite with sodium hydroxide solution converts the lanthanons to lanthanon hydroxides, which are washed free of soluble impurities, and dissolved in hydrochloric or nitric acid for further separation.

Monazite

Monazite is extracted via sulfuric acid or aqueous sodium hydroxide treatment. Sulfuric acid treatment involves heating a mixture of monazite sand and 98% sulfuric acid to 120–150°C to obtain soluble or insoluble thorium compounds which are converted to water-soluble sulfates. Ores containing a considerable amount of gangue (calcium, silica, etc) have both thorium and lanthanon values solubilized. The lanthanons are precipitated from solution with sodium sulfate; the rare-earth values may be recovered as insoluble oxalates by addition of oxalic acid. Treatment with aqueous sodium hydroxide involves heating monazite with a 65–70% sodium hydroxide solution to convert the rare earth phosphates to hydroxides which are washed with water after leaching the sodium phosphate from the reaction mass. The P_2O_5 concentration finally should be less than 1%. Thorium is separated from the lanthanons by a partial dissolution of hydroxides in hydrochloric acid; the lanthanon chloride solution contains Ce(III). Oxidation with sodium hypochlorite followed by selective precipitation upgrades the cerium oxide to 90% or more.

Cerium Metal

Properties. Cerium metal resembles steel in appearance, a bright silvery luster that tarnishes readily in air forming the oxide Ce_2O_3, or, in the presence of water vapor, $Ce(OH)_3$. At higher temperatures, CeO_2 or $Ce(OH)_4$ occurs. Cerium dissolves in dilute mineral acids and is attacked by alkaline solutions. Above 200°C vigorous reaction with chlorine, bromine, and iodine takes place to form the corresponding cerium trihalide. At high temperatures, cerium reacts directly with carbon, sulfur, nitrogen, and boron to form metalloid compounds; cerium combines with hydrogen at 345°C to form cerium hydride, CeH_2. Cerium metal as well as cerium alloys, eg, mischmetal, are strongly pyrophoric

when sawed or scratched with a file; at elevated temperatures, the metal burns in air forming the stable oxide, CeO_2.

Preparation. Cerium metal and cerium alloys are prepared by metallothermic reduction of the halides using calcium or lithium as the reductant, or by electrolytic reduction of the fused chloride. On a commercial scale, cerium metal and mischmetal are prepared by electrolysis of a melt containing the partially dehydrated chlorides mixed with sodium and calcium chloride in a cast-iron pot which serves as the cathode. Graphite anodes are used and the pot is heated externally until fusion begins; the metal collects at the bottom of the pot and is easily removed and cast into ingots. Cerium and other lanthanons are prepared by electrolysis of the oxides in a fluoride melt, a method which can also be applied to the preparation of cerium alloys.

Cerium Alloys

The most important alloys are those with iron and with other lanthanon metals. Mischmetal, an alloy in which the cerium comprises about 50% of the lanthanon content, is prepared by electrolysis (see above). Mischmetal is used to increase the ductility of iron (nodular iron), usually added as ferro-mischmetal silicide or ferro-cerium silicide (see Iron; Steel).

Cerium Compounds

Cerous salts, derived from the strong base, cerous hydroxide, $Ce(OH)_3$, are only slightly hydrolyzed in solution and like the other trivalent rare earths, the nitrate, chloride, sulfate, acetate, sulfamate, and formate are soluble in water, whereas the carbonate, oxalate, hydroxide, oxide, phosphate, fluoride, and many basic salts are insoluble. Cerous salts are oxidized in acid solutions by ammonium peroxydisulfate with a silver catalyst, sodium bismuthate, etc, and in basic suspensions by oxygen, air, and alkali permanganates and hypochlorites. Cerous salts are colorless. Cerous hydroxide also is colorless (white) when freshly precipitated, but on standing becomes violet and eventually yellow due to oxidation.

Ceric compounds are derived from the weak base ceric oxide, CeO_2, and are more acidic, more highly hydrolyzed, and more susceptible to complex formation than the corresponding cerous compounds. Ceric salts are strongly colored, are strong oxidizing agents, and are easily reduced by halogen acids, oxalic acid, ferrous salts, sulfurous acid, hydrogen peroxide in acid solution, etc.

The only simple carbonate of cerium is cerous carbonate, $Ce_2(CO_3)_3$, which precipitates as the pentahydrate when an alkali bicarbonate solution is added to a solution of a cerous salt. Cerous chloride, $CeCl_3$, is prepared by dissolving cerous carbonate or cerous hydroxide in hydrochloric acid. Anhydrous cerous chloride is made commercially by heating hydrated cerous chloride in cast-iron pots under conditions where access to air is restricted. Ceric chloride exists as chloroceric acid, H_2CeCl_6; in the presence of excess chloride, Ce(IV) is reduced to Ce(III) with the liberation of chlorine. Cerous fluoride tetrahydrate, $CeF_3 \cdot 4H_2O$, is made by addition of hydrofluoric acid to a soluble cerous salt, or by treating cerous oxalate with hydrofluoric acid. Cerium nitrates include cerous nitrate hexahydrate, $Ce(NO_3)_3 \cdot 6H_2O$; and ammonium hexanitratocerate (ceric ammonium nitrate), $(NH_4)_2Ce(NO_3)_6$. Only cerous oxalate non-ahydrate, $Ce_2(C_2O_4)_3 \cdot 9H_2O$, is stable. Cerium oxides and hydroxides include ceric oxide (ceria) CeO_2, a refractory, soluble with difficulty in acids; cerous oxide, Ce_2O_3; and ceric hydroxide (hydrous ceric oxide, cerium hydrate), $CeO_2 \cdot xH_2O$ ($x = 1/2$ to 2). Cerium sulfates include cerous sulfate, $Ce_2(SO_4)_3$; ceric sulfate, $Ce(SO_4)_2$; and ammonium trisulfatocerate dihydrate (ceric ammonium sulfate).

Health and Safety

Cerium and its compounds have low to moderate toxicity ratings, and low oral toxicity is ascribed to poor intestinal absorption. Local injection or inhalation of lanthanon compounds can cause both skin and lung granulomas.

Uses

The greatest use of cerium is in the form of rare-earth mixtures for petroleum cracking catalysts (26%); metallurgical applications (43%); ceramics and glass (31%); and a variety of uses including arc carbons (see

Molecular sieves; Air pollution control methods; Exhaust control). Rare-earth metals are used in air-pollution control; as additives to iron and steel to counteract the influence of various impurities; and in pyrophoric devices. Cerium oxide is used extensively in the glass industry for polishing glass and as a decolorizer (see Glass).

WALTER L. SILVERNAIL
Consultant

T. Moeller, *The Chemistry of the Lanthanons*, Reinhold Publishing Corp., New York, 1963.

L. Eyring, ed., *Progress in the Science and Technology of the Rare Earths*, Vol. 3, Pergamon Press, Inc., New York, 1968.

K.A. Gschneider, ed., *Industrial Applications of Rare-Earth Elements*, ACS Symposium Series 164, American Chemical Society, Washington, D.C., 1981.

CESIUM AND CESIUM COMPOUNDS

Cesium is the most electropositive and least abundant of the naturally occurring alkali metals. Cesium was the first element discovered spectroscopically; the name, after the Latin *caesius* for sky blue, refers to the characteristic blue spectral lines of the element.

Physical Properties

Cesium is a silvery-white, soft ductile metal, mp $28.64 \pm 0.17°C$. Of the five naturally occurring alkali metals, cesium has the highest vapor pressure (solid, from $-23°C$ to mp, $\log_{10} P$ kPa $= -4120/T + 11.321 - 1.0 \log_{10} T$ (to convert kPa to mm Hg, multiply by 7.5); highest density (solid 1.892 g/cm^3 at 18°C); lowest boiling point (670°C); and lowest ionization potential; viscosity at mp 0.686 mPa·s ($=$ CP).

Chemical Properties

Cesium is similar in chemical behavior to potassium and rubidium except that it is oxidized more readily than any of the other alkali metals; when exposed to air, the metal ignites spontaneously and burns vigorously. It reacts with water to give the hydroxide and hydrogen. If exposed to both air and water, a hydrogen explosion usually occurs. Cesium is the most reactive of the alkali metals with nitrogen, carbon, and hydrogen. The hydroxide absorbs carbon dioxide readily and attacks glass relatively rapidly. Cesium forms simple alkyl and aryl compounds and reacts readily with hydrocarbons. A brown solid addition product is formed in reaction with ethylene; triphenylmethylcesium, $(C_6H_5)_3CCs$, is formed by reaction of cesium amalgam with triphenylmethyl chloride solution in anhydrous ether. Cesium ions can be extracted from aqueous solutions in organic phases for ion-pair association complexes. Cesium salts are generally similar chemically to other alkali-metal salts; the salts and simple anions are usually hygroscopic and soluble, but sparingly soluble salts of cesium and complex anions are seldom hydrates and usually are not hygroscopic.

Occurrence

Cesium is widely distributed in the earth's crust at very low concentrations, eg, granites contain about 1 ppm, sedimentary rocks contain ca 4 ppm, and seawater contains about 0.2 ppm cesium. Higher concentrations are found in lepidolite, a lithium mica; in carnallite, a double salt of potassium and magnesium chloride; and beryl, rhodonite, leucite, spodumene, petalite, potash feldspars, and related minerals. By far the most important primary source of cesium has been the mineral pollucite, ideally $Cs_2.Al_2O_3.4SiO_2$, with lepidolite and carnallite as minor commercial sources. Cesium-137 is a by-product of nuclear-fission reactors, and can be recovered from fuel elements in nuclear-fuel reprocessing facilities.

Manufacture of Cesium Compounds

Commercial production is based on pollucite. A process is available, if required, for low grade ores, although there are many high grade ores that do not require processing. Cesium values in pollucite are made soluble by sintering with CaO and $CaCl_2$ at 800–900°C; leaching sinter cake gives an impure, alkaline cesium chloride solution.

Manufacture of Cesium Metal

In thermochemical methods, cesium metal is distilled from a mixture of CsCl and calcium. Magnesium is used to obtain cesium metal from the hydroxide, carbonate, or aluminate. Thermochemical reduction reactions are carried out under vacuum or in an atmosphere of an inert gas then vacuum distillation at low temperatures. Cesium for use in vacuum tubes is usually generated within the evacuated tube by ignition of a pelleted getter charge, eg, cesium chromate and zirconium. Calcium metal, or aluminum, silicon, or ferrosilicon, after lime pretreatment, can be used to reduce cesium thermally from pollucite. Thermal decomposition of cesium azide, CsN_3, made from the sulfate and an azide, which melts at 326°C and decomposes at 390°C.

Electrolytic reduction involves reduction from concentrated aqueous solutions by electrolysis with a mercury cathode, and Cs is recovered by distillation of the amalgam.

Alloys of the Amalgams

Eutectics melting at about -30, -47, and $-40°C$ are formed in the binary system, cesium–sodium at about 9 wt% sodium, and cesium–rubidium at ca 14% rubidium.

Cesium compounds include cesium carbonate, Cs_2CO_3, a colorless crystalline solid. Cesium halides include cesium bromide, CsBr, a colorless crystalline solid; cesium chloride, CsCl, well-defined, colorless, cubic crystals; cesium fluoride, CsF, an extremely hygroscopic, colorless, crystalline solid; and the most soluble of cesium salts, cesium iodide, CsI, a colorless, crystalline solid; cesium hydroxide, CsOH, the strongest base known; cesium nitrate, $CsNO_3$, which crystallizes from aqueous solution in well-defined, glittering, hexagonal prisms; cesium oxides, ranging from the suboxide, Cs_7O, to the superoxide, CsO_2; cesium sulfate, Cs_2SO_4, which forms rhombic or hexagonal crystals; and cesium alum (cesium aluminum sulfate), $Cs_2SO_4.Al_2(SO_4)_3.24H_2O$.

Uses

The principal use of cesium is in developmental research on ion propulsion and thermionic, turboelectric, and magnetohydrodynamic (MHD) electric power generation (see Coal, conversion processes, MHD; Explosives and propellants; Power generation). Principal commercial use is in manufacture of vacuum tubes. Cesium salts are used as antishock reagents in medicine following administration of arsenical drugs. Cesium bromide is used in the manufacture of optical crystals. Cesium chloride is used in biological research. Cesium fluoride is used as a fluorinating agent (see Fluorine compounds, organic). Cesium iodide is used for optical crystals. Cesium hydroxide has been patented for use as an agent in removal of sulfur from heavy oils.

C. T. WILLIAMS
Tantalum Mining Corp.

F.W. Wessel, *Minor Metals and Minerals—Cesium and Rubidium*, in *Minerals Yearbook*, Vol. 1, U.S. Dept. of the Interior, Washington, D.C., 1959–1962.

P.D. Bergman and D. Bienstock, "Economics of a Mined Potassium–Cesium Seeding of a MHD Combustion Plasma," *U.S. Dept. of the Interior, Bureau of Mines Report of Investigation* 7717, 1972, p. 12.

CHELATING AGENTS

A chelating agent (or ligand) is a compound containing donor atoms that can combine by coordinate bonding with a single metal ion to form a cyclic structure called a chelating complex, or simply, a chelate. The descriptive word chelate is derived from the chela, or great claw, of the lobster and symbolizes the chelate ring(s) formed between ligand and metal ion; each ring gives the appearance of a metal ion being held in a pincer formed by the other atoms.

The technological importance of chelation derives from the almost universal presence of metal ions of one kind or another, either naturally or by intentional addition. Chelating agents provide a means of manipulating and controlling metal ions by forming complexes that usually have properties that are markedly different from those of either the original ions or chelants.

Structure and Terminology in Chelate Chemistry

The structural essentials of a chelate are coordinate bonds between a metal ion (M) and two or more atoms in the molecule of the chelating agent, or ligand (see structure (1)). Chelate atoms are electron donors and the metal atom is the electron acceptor. When coordinate bonds are formed between the metal and two donor atoms, the atoms of the ligand that connect the donor atoms complete the ring, thereby giving the structure its chelate character.

(1)

Chelating agents may be bidentate, tridentate, etc, depending on the number of chelate rings in the complex. Molecules with only one donor atom, eg, dimethylamine, are monodentate and form coordination complexes but not chelates. The principal donor atoms in practical use are N, O, and S, but P, As, and Se also form chelates. Coordination numbers of metal atoms give the maximum number of donor atoms to which the metal ions can coordinate, most commonly 4 and 6, less often 2 and 8, and state the actual number of donor atoms bound to the central metal ion in a particular chelate.

If the coordination number of a metal ion M is greater than the number of donor atoms in the ligand L, more than one ligand molecule may combine with the metal to form ML_n. Different chelating molecules can combine with the same metal ion to form species such as L_mML_n. In the same way, a ligand molecule with enough donor atoms in the proper configuration can bind more than one metal ion. A chelate compound may be a neutral molecule, or a positive or negative complex ion associated with appropriate counterions to produce electroneutrality. The donor atoms of most chelating agents are contained in linear or branched-chain structures, separated by suitable numbers of other atoms to allow the formation of the chelate rings.

In macrocyclic structures, the donor atoms of the chelating agent become positioned near the center of the macro ring and have the metal ions coordinated to them. The spacer atoms complete the chelate rings between pairs of coordinating donor atoms, forming a pattern of fused rings centered around the metal, eg, the porphyrins. Crown ethers are a group of cyclic polyethers having donor atoms that are ether functions separated by two or three carbon atoms, and the chelated metal ion is positioned near the center of the macro ring, coordinated to the donor atoms. Sandwich structures are also known (see also Antibiotics, polyethers). Three-dimensional polymacrocyclic chelating agents are formed by joining bridgehead structures with chains that contain properly spaced donor atoms. For example, cryptates (2) are bicyclic structures that result from joining nitrogen bridgeheads with chains of ($-OCH_2$-CH_2-). Synthesis, metal binding, and thermodynamic properties of synthetic multidentate macrocyclic complexes are reviewed.

(2)
2.2.2-cryptate

Incorporating ligand groups into a cross-linked polymer structure gives the chelate-forming resins which perform ion-exchange functions by chelation (see Ion exchange).

The number of known inorganic chelating agents is very small, but their annual consumption exceeds that of all the organic chelating agents combined. For example, polyphosphates are less expensive than organics, but are hydrolytically unstable at high temperature and pH.

Important organic chelating agents include the phosphonic acids, analogous to the amino and hydroxycarboxylic acids (qv). These chelants possess many of the complexing properties of the inorganic polyphosphates, particularly threshold scale inhibition, but are water stable at high temperature and pH (see Dispersants, Phosphorus compounds).

Substances known for chelating properties include polyphosphates, aminocarboxylic acids, 1,3-diketones, hydroxycarboxylic acids, polyamines, aminoalcohols, aromatic heterocyclic bases, phenols, aminophenols, oximes, Schiff bases, tetrapyrroles, thiols, xanthates and other sulfur compounds, polymers such as polyethylenimine, polymethylacryloylacetone, and poly(acrylic acid).

The Chelation Reaction

Chelate formation equilibria. In homogeneous solution the equilibrium constant for the formation of the chelate complex from the solvated metal ion and the ligand in its fully dissociated form is called the formation or stability constant. The ligand displaces the solvent molecules coordinated to the metal ion, but the solvent molecules are not usually shown in equations. When more than one ligand complexes with a metal ion, the reaction usually proceeds in a stepwise fashion. A metal with a coordination number of six and a bidentate chelating agent proceeds through steps giving ML, ML_2, ML_3. The overall stability equation for the reaction $M + 3 L$ is:

$$K = \frac{[ML_3]}{[M][L]^3}$$

The overall stability constant is the product of the step-stability constants, $K = K_1 K_2 K_3$, and is often designated by the Greek letter β. The chelate of a tetracoordinate metal with EDTA is shown (see Structure (3)). Such a reaction is the sum of an acid dissociation reaction, and the reaction of chelate formation from the fully dissociated form of the ligand.

(3)
Chelate or tetracoordinate
metal with hexadentate
ligand EDTA

Because various species in solution are in a formation–dissociation equilibrium, displacement reactions of one metal or ligand by another are possible. The equilibrium constant depends on stability and is a function of the stability constants for the two complexes. If the metals or ligands involved in a displacement reaction form chelates whose type formulas are different, the exchange equilibrium constant will not be the simple ratio of the formation constants of the chelates.

Complexation is almost instantaneous, but the reaction rates of many ions with higher valences are slow enough to measure by ordinary kinetic techniques. Practical use has been made of both lability and inertness, and of stability and instability. For example, a reaction or handling process may be completed during the period of apparent stability that is actually a result of inertness. Many parameters influence stability, including the size and number of the rings, substituents on the rings, and the nature of the metal ions and donor atoms. In the macrocyclic complexes, the degree to which the size of the metal ion fits the space enclosed by the macro rings is a major factor. In any series of chelates, the variability of the stability constants is usually influenced by more than one of the parameters that are known to affect chelate stability.

Protonated chelating agents exhibit titration behavior typical of their acidic groups (eg, carboxyl, phenolic hydroxyl, ammonium, sulfhydryl) if they are titrated with bases whose cations have very weak or no tendency to form chelates. In the presence of a metal ion that coordinates with the donor atoms of one of these acidic groups, hydrogen ions will be displaced by the metal, the acid strength of the group will thus appear to be enhanced, and the hydrogen-ion concentration developed will be higher than in the absence of the metal. The stronger the coordination tendency, the greater will be the apparent acidity of the group.

By choosing chelating agents with appropriate stability constants, the pM can be regulated over a wide range (pM is defined as the negative logarithm of the concentration of uncomplexed metal ions, ie, pM = $[-\log M^{n+}]$). Two or more metals may be selectively buffered at different concentrations by a single chelating agent with different stability constants for the metals. The selective buffering of one metal to a low concentration in the presence of other metals is termed masking. The ability to maintain a nearly constant concentration of metal ions at almost any level of concentration is the basis of many of the commercial uses of chelating agents.

Consider the reaction between a metal ion and an anion X^{n-} to give MX which is insoluble. The presence of a sufficiently strong chelating agent will keep the concentration of free metal ions suppressed so that pM will be larger than the saturation pM given by the solubility product relation. No solid phase will form even in the presence of relatively high concentrations of the anion X^{n-}. In other words, the metal will be sequestered with respect to precipitation by the anion. This phenomenon has application in the prevention of the formation of insoluble soaps by hardness ions in water, and the dissolution of an insoluble salt (see Dispersants).

Applications

Three features of chelation chemistry are fundamental to most applications. The first is the control of the concentration of the free form (aqueous or solvated) of the metal ions by means of the binding-dissociation equilibria. Sequestration (the suppression of certain properties of a metal without removing it from the system or phase), solubilization (causing constituents of a phase that is normally insoluble to dissolve in the medium), and buffering (the effect in which the addition or removal of an appreciable amount of a metal ion produces only a relatively small change in the concentration of that ion in solution) all depend upon the concentration control feature of chelation. For example, sequestration is used to control water hardness; solubilization is used in dissolving boiler scale, scale in heat exchangers, and hardness scale from pipes (see Dispersants; Water, industrial water treatment) and in metal recovery from ores as well as in cleaning up radioactive contamination (see Nuclear reactors). Chelation buffering is useful in supplying micronutrient metal ions to biological growth systems at controlled, very low concentration (see Mineral nutrients) as well as in controlling the activity of redox polymerization catalysts. Some of the chief applications of the preparative feature of chelation depend on the solubility properties, color, or catalytic effects of chelates, eg, dyeing processes that achieve color or fastness by chelation. In catalysis chemistry, chelating ligands are used for special properties of the chelate itself (eg, asymmetric syntheses) (see Pharmaceuticals, optically active; Enzymes). Chelation itself is sometimes useful in directing the course of synthesis. The introduction of the crown ethers and cryptates is owing to their ability to strongly complex the alkali metals.

Extraction and precipitation use chelation to transfer the metal into another phase. A special kind of transfer of metal ions to the solid phase is found in the chelating resins, eg, in the preconcentration of ions from media such as seawater, body fluids, and geological materials in which their original concentration is exceedingly small, so that they may be detected or determined analytically.

Displacements using chelating agents or chelation include treatment of poisoning by lead and other metals with EDTA (ethylenediaminetetracetic acid).

Uses

The main use of STPP (sodium tripolyphosphate) is in cleaning formulations (see Surfactants); citric acid is used as a pH regulator, flavor enhancer, and antioxidant synergist; the aminopolycarboxylic acids are used principally as chelating agents, particularly in cleaning formulations; and gluconic acid and its salts are used primarily in strongly alkaline cleaning preparations.

A.L. McCrary
William L. Howard
Dow Chemical U.S.A.

Reviewed by
David J. Leggett
Dow Chemical U.S.A.

M.M. Jones, *Elementary Coordination Chemistry*, Prentice Hall, Inc., Englewood Cliffs, N.J., 1964.
J.J. Christensen, D.J. Eatough, and R.M. Izatt, *Chem. Rev.* **74**, 351 (1974).

CHEMICAL CLEANING.   See Metal surface treatments.

CHEMICAL GROUTS

A chemical grout is defined as a true solution (as opposed to a suspended solid) which is injected into soil, rock, or concrete as a liquid to alter the physical characteristics of the grouted mass when it changes to a solid. An ideal chemical grout would have the following characteristics: permanence; low viscosity; controllable set or gel time; economical qualities, as well as inexpensive handling, storing, and shipping costs; nontoxic, noncorrosive, nonhazardous, and ecologically safe, and permeability reduction of the substrate or strength increase of that substance by significant amounts, or both.

Sodium silicate-based formulas are the most widely used of all the chemical grouts. The most common silicate systems use an organic ester or an amide as a catalyst, and salts such as sodium bicarbonate, calcium chloride, and sodium aluminate as accelerators. Portland cement can also be used as a catalyst.

Sodium silicate gels are subject to syneresis, and (under certain environmental conditions) to partial dissolution. These characteristics limit the residual permeability value of treated formations to values greater than 10^{-5} cm/s, and may also limit such uses as curtain-wall construction where high hydraulic gradients may exist.

Silicate grouts are relatively inexpensive, and may be formulated to yield high strength, or low viscosity. Most formulations are environmentally nonhazardous.

Acrylamide- and acrylate-based grouts have the lowest permeability of all the chemical grouts, and therefore, the best penetrability into fine cracks and voids. In general, they are not as strong as the silicates, and considerably more expensive. However, they have the best gel-time control of any of the grouts and are easy to handle; thus they have many uses in construction and seepage control. Acrylamide is neurotoxic, and poses a health hazard to users who are continually exposed to it. Acrylates also are toxic, but the degree of toxicity may be as low as 1% of of that of acrylamide. The acrylates (relatively new commercial products) are beginning to replace acrylamide, particularly in the chief area of use (sewer sealing).

Acrylamide-based grouts are commercially available, have excellent gel time control, and their fieldwork concentrations permit penetration of finer materials than those that can be treated with other grouts. These groups are subject to creep and consolidation at high stress points.

Other grouts are low viscosity phenoplasts used for waterproofing and improving strength; easy to handle, limited-use, lignosulfonate derivatives consisting of calcium lignosulfonate and a hexavalent chromium salt (see Pulp); and aminoplasts (urea-formaldehydes) in specific formulations which gel with an acid or salt (see Amino resins). All of these materials pose serious health hazards to applicators.

Health and Safety

Sodium silicate is nonhazardous and nonpolluting. Lignin (qv) in lignosulfonate derivatives can cause skin problems and hexavalent chromium can cause contamination of the environment. Acrylamide in powder and solution form is neurotoxic; gels from acrylamide are totally nontoxic except for minute amounts of unreacted acrylamide they may contain. Phenoplasts and aminoplasts generally release free formaldehyde and ammonia, and are irritants and health hazards.

Use

Grouts are used in construction to rectify unexpected water-infiltration problems or to strengthen a distressed foundation.

REUBEN H. KAROL
Rutgers University

R.H. Karol, *Introduction to Chemical Grouting*, Marcel Dekker, New York, 1983.

CHEMICALS FROM BRINE

Brines are important commercial sources of basic industrial chemicals such as boron compounds, bromine, calcium chloride, iodine, lithium compounds, magnesium compounds, potassium compounds, sodium carbonate, sodium chloride, and sodium sulfate. These chemicals are recovered directly from brines produced by solution mining as well as from the naturally occurring brines. Sources of brine include seawater, inland lake as well as subterranean waters, potash (sylvite), and numerous salt (NaCl) deposits including salt domes and bedded salt. For processing, limestone, dolomite, lime, ammonia, and sulfuric acid are also used as raw materials.

Occurrence and Recovery Processes

The two brine bodies of Searles Lake, Calif., are principal sources for boron compounds, bromine, lithium compounds, sodium carbonate, and sodium sulfate.

Boron compounds are recovered by crystallizing borax pentahydrate, $Na_2B_4O_7.5H_2O$, from plant mother liquor resulting from upstream processes in which sodium and potassium salts are removed.

Bromine production depends upon oxidation of bromide ion to bromine, $Cl_2 + 2 NaBr \rightarrow 2 NaCl + Br_2$, steam stripping of the dissolved elemental bromine, recovery of the bromine from the stripping agent, and purification by distillation.

Calcium chloride is the final constituent recovered in a multiproduct brine-processing operation involving separation of magnesium ion by precipitation of magnesium hydroxide or crystallization of tachyhydrite. Michigan and California natural brines are principal sources, as well as the waste brine resulting from the production of soda ash by the Solvay process (see Alkali and chlorine products; Calcium compounds, calcium chloride).

Iodine is recovered from underground brines in a process involving two concentration steps. A blowing-out process similar to that used for recovering bromine from seawater is used. Japan, the world's leading iodine producer, uses a process similar to the silver process formerly used in the U.S. Lithium compounds are produced from Silver Peak Marsh, Nev. brine, via a process which consists of pumping brines from shallow wells, solar evaporation to concentrate the brine, and the removal of unwanted ions during the final processing in the plant (see Lithium and lithium compounds).

Magnesium compounds are recovered from seawater, the waters of the Great Salt Lake, and brines in Utah, Michigan, and West Texas (see Magnesium compounds). Magnesium hydroxide and magnesium chloride are recovered from natural brines; magnesium hydroxide can be recovered from the brine itself or from intermediate plant liquor. Production of magnesium chloride, recovered as a direct product, is usually economically feasible only where other salable products are also re-

covered. Magnesium metal is produced by the electrolysis of molten $MgCl_2$ by: the Dow seawater process in which $MgCl_2$ liquor is dried to a hydrous cell feed and which requires the addition of make-up chlorine; and another process which recovers $MgCl_2$ instead of $Mg(OH)_2$ from the source brine and forms anhydrous cell feed from which both Mg metal and marketable chlorine are produced.

Potassium compounds are recovered from three brine–potash operations in Utah and a fourth in California. One of the Utah operations employs solution mining (recovery of a water-soluble substance from an underground deposit by dissolving the substance *in situ* and forcing the resulting solution to the surface) (see Fertilizers; Potassium compounds). At the Great Salt Lake, the recovery involves brine collection; concentration by solar evaporation; precipitation and harvesting of mixed crystals; and beneficiation to produce potash.

Sodium carbonate is recovered from nonbrine sources, from Searles Lake, and by the Solvay process. The latter depends economically on solution-mined salt (see Sodium compounds, sodium carbonate).

Sodium sulfate is recovered from terrestrial natural brines and dry lakes. The anhydrous form is recovered from Glauber's salt by submerged combustion evaporation of the liquor formed by remelting Glauber's salt in the Texas brines process.

Sodium chloride is recovered by evaporation of brines (see Sodium compounds, sodium chloride). Along the coast of California, seawater undergoes solar evaporation followed by crystallization.

R. BRUCE TIPPIN
Great Salt Lake Minerals and Chemicals Corp.

PAUL E. MUEHLBERG
Dow Chemical Company

P.E. Muehlberg and co-workers in P.W. Spaite and I.A. Jefcoat, eds., *Industrial Process Profile for Environmental Use*, Rept. No. EPA-600/2-77-0230, Environmental Protection Agency, Cincinnati, Ohio, 1977, Chapt. 15.

J.E. Ryan, *Min. Eng.* **3**, 447 (1951).

W.E. VerPlank and R.F. Heiser, *Salt in California*, Bulletin 175, California Department of Natural Resources, Mar. 1958.

CHEMICALS IN WAR

Toxic chemical agents are chemical substances in gaseous, liquid, or solid state intended to produce casualty effects ranging from harassment, through varying degrees of incapacitation, to death. A few such agents are true gases, but most are solids or liquids that, in actual use, are converted into a gaseous state or disseminated as aerosols. For contamination of terrain or materiel, the agent can also be disseminated in bulk form. Throughout military history there are references to the four major types of chemicals discussed in this article: flame agents, incendiaries, smokes, and toxic chemical agents. A fifth type of chemical is included here, the riot-control agent CS, although it has a limited defensive role in war (see Irritants).

Toxic chemical munitions are unique in contrast to other weapon systems in that they seek out the enemy whether widely dispersed or concentrated in fortifications. In addition, toxic chemicals are minimum-destruction weapons as regard materiel; they are directly effective against personnel but leave cities and industrial facilities intact.

Effects of toxic chemical agents include death, mild incapacitation, blistering, choking, blood poisoning, lacrimation, nerve poisoning, laxation, and various forms of mental and physical disorganization.

Criteria used for selecting a suitable agent are effectiveness in extremely small concentrations; time to onset of symptoms and duration of action; effectiveness through various routes of entry into the body; stability to long-term storage; and ease of dissemination in feasible munitions.

Lethal Agents

Mustard and nerve agents are among the most important lethal agents available.

Mustard and related vesicants. Mustard, bis(2-chloroethyl) sulfide (Chemical Agent Symbol: HD), $ClCH_2CH_2SCH_2CH_2Cl$, is a colorless, oily liquid when pure with a characteristic garlic-like odor. It is a vesicant forming blisters by either liquid or vapor contact, and a systemic poison attacking eyes and lungs. Symptoms do not appear until 4–5 h after exposure. A variety of sulfur analogues of mustard are known vesicants, including: 1,2-bis(2-chloroethylthio)ethane (Chemical Agent Symbol: Q), $ClCH_2CH_2SCH_2CH_2SCH_2CH_2Cl$; and bis(2-chloroethylthioethyl) ether (T), $ClCH_2CH_2SCH_2CH_2OCH_2CH_2SCH_2CH_2Cl$. A number of nitrogen mustards are known. These are colorless liquids when pure, but which turn yellow to amber in storage; they have faintly fishy or soaplike odors, and exhibit vesicant properties.

Physical and chemical properties. The sulfur mustards are slightly soluble in water; nitrogen mustards are slightly soluble at neutral pH but form water-soluble salts under acid conditions; both are extremely soluble in most organic solvents. The soluble portion in water is extremely reactive. Mustards readily alkylate inorganic thiosulfates to form Bunte salts. Sulfur mustard reacts rapidly with chlorine or with bleach, and this reaction is a means of decontamination. Mustards can be oxidized with oxidizing agents, eg, hydrogen peroxide or potassium bichromate in sulfuric acid.

Nerve Agents

Nerve agents are two groups of highly toxic chemical compounds, G- and V-agents, that are generally organic esters of substituted phosphoric acid. These compounds act to inhibit cholinesterase (qv) enzymes. The three most active G-agent compounds are tabun, ethyl phosphorodimethylamidocyanidate (GA), $((CH_3)_2N)P(O)(CN)OC_2H_5$; sarin, isopropyl methylphosphonofluoridate, (GB), $CH_3P(O)(F)OCH(CH_3)_2$; and soman, pinacolyl methylphosphonofluoridate, (GD), $CH_3P(O)(F)OCH(CH_3)C(CH_3)_3$. V- and G-agents are colorless, odorless liquids. They are readily absorbed through the lungs, eyes, intestinal tract and skin, the V-agents absorbing more effectively. G-agents are miscible in both polar and nonpolar solvents; they hydrolyze slowly in water at neutral or slightly acidic pH and more rapidly under strong acid or alkaline conditions. Inhalation of G-agent vapor is immediately incapacitating, with initial symptoms culminating in death. A binary munitions concept may be used to dispense standard nerve agents.

Lethal chemical agents are used in close military engagements where their rapidity of action, area-search, and multiple-casualty capabilities can influence the immediate tactical situation.

Incapacitating Agents

Also known as incapacitants, these compounds render individuals incapable of performing their jobs. One example is agent BZ (3-quinuclidinyl benzilate), an atropine mimetic which acts on the central and peripheral nervous systems. Effects of BZ are like those of an anticholinergic psychotomimetic drug, including disorientation, visual and auditory hallucinations, disturbances of higher integrative functions of memory, problem solving, attention, and comprehension.

By U.S. army criteria, incapacitating agents do not include the following: lethal agents that are incapacitating at sublethal doses; substances that cause permanent or long-lasting injury; medical drugs that exert marked effects on the central nervous system; agents of temporary effectiveness that produce reflex responses interfering with performance of duty; agents that disrupt basic life-sustaining systems of the body and prevent physical activity.

Irritants

Irritant compounds are harassing agents whose effects are reversible and briefly incapacitating, eg, lacrimators, sternutators, and riot control agent CS. CS (o-chlorobenzylidenemalonitrile, $ClC_6H_4CH{=}C(CN)_2$) causes physiological effects that are immediate even in extremely low concentrations. A water-soluble white crystalline solid, CS is disseminated as a spray solution, a cloud of dust or powder, or as an aerosol.

Flame

Flame agents are various hydrocarbons, blends of hydrocarbons, and other readily flammable liquids can be projected or delivered to military targets. In the U.S., their major application is in flame throwers and flame projectors; the fire bomb is rapidly becoming obsolete. Thickened pyrophoric flame agents are deployed in the field as expedients and as the payload for the U.S. Army's flame rocket system. These flame agents do not require an ignition system to function on target and can be prepackaged in warheads, etc. Optimization of several low viscosity metal alkyl formulations has made flame agents relatively independent of temperature variations.

Incendiaries

Incendiary agents are compositions of chemical substances designed for use in the planned destruction of building, property, and materiel by fire. They burn with an intense, localized heat, are very difficult to extinguish, and are capable of setting fire to materials that normally do not ignite and burn readily. The mechanics of starting fires with incendiary agents involve: a source of heat which acts as a "match" to initiate combustion, initiation temperature varying from 200–440°C; combustible material which serves as "kindling"; and fuel. The rate of heat release will vary with the nature of the incendiary agent, and depends on: flash-fire or decomposition temperature; the particle size of the agent after ejection; oxidizing agents blended with the combustible material to increase the rate of heat evolution.

Metal incendiaries include those consisting of magnesium in various forms and powdered or granular aluminum mixed with powdered iron oxide, eg, thermite.

Oil and metal incendiary mixtures include PT1, a complex mixture based on a paste of magnesium dust, magnesium oxide, and carbon, with an adequate amount of petroleum distillate and asphalt, and PTV, an improved oil and metal incendiary.

Smokes

Military smokes are aerosols of gaseous, liquid, or particulate matter that are tactically employed. Traditional battlefield applications include screening and marking. Useful smoke screens are opaque and long-lasting. White smokes, which do not have light-absorbing particles, eg, carbon, have the greatest screening action. Types of screening smoke include: oil smoke; HC, type C, a volatile hygroscopic chloride; white phosphorus (WP); red phosphorus (RP); and sulfur trioxide (SO_3) solution (FS). Signaling smokes must be clearly distinguishable from the smoke incident to battle, must afford good visibility, and be unmistakable in identity. They are colored with volatilized and condensing mixtures of organic dyes, principally azo (qv), anthraquinone (qv), azine (qv), or diphenylmethane dyes.

Defense Against Toxic Agents

Defense measures against toxic chemical agents include: detection and identification of the agent using chemical kits and warning devices employing infrared absorption, the Raman effect, conductivity, and enzyme inhibition; individual and collective protection through the use of protective ("gas") masks, airtight, impermeable clothing, and Chemical Protective Suits impregnated with labile chlorine compounds, or repellents to repel liquid agents and sorbents (collective protection involves the use of shelters); medical defense against organophosphorus (nerve) agents involving the use of atropine and pralidoxime to neutralize the effects of anticholinesterase compounds and to reactivate inhibited enzymes (see Cholinesterase inhibitors) (copious washing of mustard burns is also effective when used early); decontamination of persistent and thickened chemical agents using natural aeration (if possible), chemical neutralizers, eg, supertropical bleach (STB), and DS2 (diethylenetriamine plus caustic).

B.L. HARRIS
F. SHANTY
W.J. WISEMAN
Chemical Systems Laboratory, U.S. Department of Defense

J.H. Rothschild, *Tomorrow's Weapons*, McGraw-Hill Book Co., New York, 1964.

B.H. Liddell-Hart, *A History of the World War, 1914–1918*, Little, Brown & Co., Boston, Mass., 1948.

W.A. Noyes, Jr., ed., "Chemistry," *Science in World. War II*, Little, Brown & Co., Boston, Mass., 1948.

Stockholm International Peace Research Institute (SIPRI), *The Problem of Chemical and Biological Warfare, Volume LL: CB Weapons Today*, Humanities Press, New York, 1973.

CHEMILUMINESCENCE

Chemiluminescence is the emission of visible light, (and also ultraviolet or infrared radiation) from chemical reactions at ordinary temperatures. It is a special case of chemiexcitation wherein the excited state reaction products are luminescent. Chemiluminescent reactions produce a reaction intermediate or product in an electronically excited state; the radiative decay of the excited state is the source of light. For example, when the excited state is a singlet, the emissions are fluorescent, and when a triplet, the emissions are phosphorescent. A significant number of chemiluminescent reactions is known, and a few combine high chemiexcitation efficiency with high fluorescence yield to provide substantial light production.

Mechanism

Most chemiluminescent reactions can be classified into: peroxide decomposition, including bioluminescence and peroxyoxalate chemiluminescence; singlet-oxygen chemiluminescence; and ion-radical or electron-transfer chemiluminescence, which includes electrochemiluminescence.

In principle, one molecule of a chemiluminescent reactant can react to form one electronically excited molecule, which in turn can emit one photon of light. Thus, one mole of reactant can generate Avogadro's number of photons defined as one einstein (ein) with light yields defined in the same terms as chemical product yields (units of einsteins of light emitted per mole of chemiluminescent reactant), but in practice, most Q_c (chemiluminescence quantum yield) values are on the order of 1% or much less.

Efficient chemiluminescence requires a selective reaction producing a key intermediate and efficient conversion of the key intermediate to the singlet excited state of a highly fluorescent product. The yield of the second step is called the excitation yield, and the product of the yields of the first two steps is the yield of excited state. In general, efficient excitation requires a large energy release in a single reaction step and a reaction pathway that promotes crossing of the ground-state potential energy surface to an electronically excited potential energy surface.

Liquid Phase Chemiluminescence

Peroxide decomposition. In many chemiluminescent reactions of peroxides, two carbonyl groups are formed simultaneously by decomposition of an intermediate. In such reactions, the substantial heat of the simultaneous (concerted) formation of the two carbonyl groups meets the energy requirement. It is not necessary for the new carbonyl groups to be a part of the structure of the excited product, only that the excited state be formed synchronously with the two carbonyl groups.

1,2-Dioxetanes. Simple dioxetanes (1) are obtained from an α-halohydroperoxide by treatment with base or reaction of singlet oxygen with an electron-rich olefin such as tetraethoxyethylene.

$$R_2C \underset{\displaystyle |}{\overset{\displaystyle O-O}{\underset{\displaystyle }{|}}} CR_2 \longrightarrow R_2C{=}O + [R_2C{=}O]^*$$

(1)

They decompose thermally near or below room temperature to generate excited states of carbonyl products.

Excitation appears to be general for this reaction but yields of excited products vary substantially with the substituent R; the highest yields reported are from tetramethyl-1,2-dioxetane (TMD), and other tetraalkyldioxetanes behave similarly.

Since fluorescence and phosphorescence radiative yields from simple ketones are very low, chemiluminescence is weak, with a quantum yield of 1×10^{-6} ein/mol; this can be increased significantly by adding a fluorescent acceptor, eg, 9,10-diphenylanthracene for singlets. 1,2-Dioxetanes undergo ready thermal decomposition below 80°C, and are rapidly decomposed by transition metals, amines, electron-donor olefins, and by ultraviolet light. A number of chemiluminescent reactions may proceed through unstable dioxetane intermediates, eg, classical chemiluminescent reactions of lophine, lucigenin, and transannular peroxide decomposition.

$$(C_6H_5)_2C{=}C{=}O + (O_2)^i \longrightarrow (C_6H_5)_2C\underset{\displaystyle |}{\overset{\displaystyle O-O}{\underset{\displaystyle }{|}}}C{=}O + FLR \longrightarrow$$
$$(C_6H_5)_2C{=}O + CO_2 + [FLR]^* \quad (1)$$

α-Peroxylactones (1,2-dioxetanones). Alkyl substituted 1,2-dioxetanones are prepared using low temperature techniques, eg, reaction 1. The α-hydroperoxy acids can be prepared in high yield and cyclized to the dioxetanone with dicyclohexylcarbodiimide in carbon tetrachloride at low temperatures. Dioxetanones decompose near or below room temperature to aldehydes or ketones in weakly chemiluminescent reactions yielding poorly fluorescent reaction products ($Q_c = $ ca 10^{-7} ein/mol). The addition of rubrene substantially increases the quantum yield and decomposition rate, suggesting catalysis. In reaction 1, light is observed using the triphenyl phosphite ozonide complex as a singlet-oxygen source to add O_2 to the ketenes in the presence of the fluorescer 9,10-bis(phenylethynyl)anthracene. More recently, dioxetanones have been isolated from this reaction, which appears to be a relatively simple and general synthetic method for these compounds.

Peroxyoxalate (1,2-dioxetanedione). Peroxyoxalate chemiluminescence is the most efficient nonenzymatic chemiluminescent reaction known. It is illustrated by the reaction of an oxalic ester, such as bis(2,4,6-trichlorophenyl) oxalate, with hydrogen peroxide and a fluorescer, such as 9,10-diphenylanthracene or rubrene, in benzene or dimethyl phthalate. A key aspect of the proposed mechanism for this reaction is catalytic decomposition of the key intermediate, 1,2-dioxetanedione, by the fluorescer in the excitation step, with nonluminescent intermolecular side reactions minimized by selection of reaction conditions. Efficient chemiluminescent oxalates include electronegatively substituted aromatic and aliphatic esters, amides, sulfonamides, O-oxalylhydroxylamine derivatives, mixed oxalic carboxylic anhydrides, and oxalyl chloride. A variety of fluorescers can be used with excitation yields generally decreasing as the excitation energy of the fluorescer increases, and the Q_c for higher energy (blue) fluorescers tends to be relatively low. Most peroxyoxalate chemiluminescent reactions are catalyzed by bases (weak bases are preferred) and the reaction rate, chemiluminescent intensity, and chemiluminescent lifetime can be varied by the selection of the base and its concentration.

Luminol (phthalhydrazide). Chemiluminescence from luminol (3-aminophthalhydrazide) and analogues takes place in aqueous solution with hydrogen peroxide and a supplemental oxidant including ferricyanide, hypochlorite, persulfate, or the hydroxyl radical generated from hydrogen peroxide, and a metal derivative such as hemin. It also takes place with oxygen and a strong base in a dipolar aprotic solvent such as dimethyl sulfoxide. Under both conditions, Q_c is about 1%.

Organometallics. Arylmagnesium halides, especially bromides, react with oxygen in ether to generate light. p-Chlorophenylmagnesium bromide is the most efficient with $Q_c = $ ca 10^{-4} ein/mol, and ArOOMgX is probably an intermediate; free radicals may be involved. The emitting species are brominated biphenyls and, since these are only weakly fluorescent, the excitation yield must be high.

Autooxidation. Liquid-phase oxidation of hydrocarbons, alcohols, and aldehydes by oxygen produces chemiluminescence in quantum yields of 10^{-8} to 10^{-10} ein/mol. The chemiluminescent reaction is important because it provides an easy tool for study of the kinetics and properties of autooxidation reactions, including industrially important processes.

Singlet oxygen. The electronically excited singlet state of oxygen can be produced by passing ground-state oxygen through a microwave discharge, by reaction of hydrogen peroxide with hypochlorite ion, by energy transfer from triplet excited states formed by irradiation to ground-state oxygen, and by low temperature thermal decomposition of the triphenyl phosphite–ozone complex. The intensity of chemiluminescence can be increased 100-fold with the addition of 5×10^{-4} M violanthrone. Chemiluminescence from singlet oxygen and rubrene probably involves the same mechanisms as involved in the presence of violanthrone; singlet oxygen may also be responsible for the red chemiluminescence observed in the reaction of pyrogallol with formaldehyde and hydrogen peroxide in aqueous alkali, as well as the decomposition of secondary dialkyl peroxides and hydroperoxides.

Electron transfer. Electron-transfer reactions appear to be inherently capable of producing excited products when sufficient energy is released. Examples include luminescence from anthracene crystals subjected to alternating electric current, and from oxidation of aromatic radical anions.

Stable anion radicals are easily prepared from aromatic hydrocarbons, eg, 9,10-diphenylanthracene, by electrochemical reduction in acetonitrile or dimethylformamide-containing electrolytes such as tetrabutylammonium perchlorate. Reversal of electrode polarity generates cation radicals from the hydrocarbon, and their reaction with the anion-radical reservoir generates chemiluminescence. More simply, an alternating current may be used so that cation and anion radicals are continuously formed and annihilated to produce light and regenerate the original hydrocarbon. When hydrocarbons with stable ion radicals are used and impurities reactive with ion radicals are rigorously excluded, long-lasting electrochemiluminescence occurs.

Three processes can result in electrochemiluminescence, depending on the energy released by the electron-transfer process and the excitation energy of the aromatic hydrocarbon. In each case, a charge-transfer complex between the oppositely charged radicals is probably formed: efficient luminescence when sufficient energy is available for the complex to dissociate to one ground-state molecule and one excited-singlet molecule; triplet excitation resulting in excited singlets by triplet–triplet annihilation; excimer (excited dimer) emission from the complex itself. In the first two cases, the luminescence spectrum matches the normal fluorescence spectrum of the hydrocarbon, whereas in the third case typical, red-shifted, broad-band excimer emission results.

Gas-Phase Chemiluminescence

Gas-phase chemiluminescence is illustrated by the classic sodium–chlorine cool flame:

$$Na + Cl_2 \rightarrow NaCl + Cl$$

$$Cl + Na_2 \rightarrow NaCl + [Na]^*$$

Intense sodium D-line emission results from excited sodium atoms produced in a highly exothermic step (second equation above). Many gas-phase reactions of alkali metals are chemiluminescent, in part because their low ionization potentials favor electron transfer to produce intermediate charge-transfer complexes such as $[Cl^- . Na_2^+]$. There appears to be an analogy with solution-phase electron-transfer chemiluminescence in such reactions. Atom-transfer reactions in the gas phase can produce chemiluminesence from vibrationally excited ground states, eg, reaction of oxygen atoms with carbon disulfide, acetylene, or methylene.

Combination chemiluminescence is illustrated by afterglow reactions, so named because the luminescence persists following the dissociation of molecules in an electric discharge:

$$N \cdot + N \cdot \rightarrow N_2^* \text{ (nitrogen afterglow)}$$

$$O \cdot + NO \rightarrow NO_2^* \text{ (air afterglow)}$$

Such reactions can also be produced by microwave-discharge dissociation.

White phosphorus oxidation. Emission of green light from the oxidation of elemental white phosphorus in moist air is one of the oldest recorded examples of chemiluminescence. Although this is normally observed from solid phosphorus, the reaction actually occurs primarily just above the surface with gas-phase phosphorus vapor; the reaction mechanism is unknown, but careful spectral analysis of the reaction with water and deuterium oxide vapors indicates that the primary emitting species in the visible spectrum are excited states of $(PO)_2$ and HPO or DPO. Ultraviolet emission from excited PO is also detected.

Solid-Phase Chemiluminescence

Siloxene $(Si_6H_6O_3)_n$ is fluorescent and red chemiluminescence results from oxidation with ceric sulfate, chromic acid, potassium permanganate, nitric acid, and several other strong oxidants. Solid lithium organophosphides are chemiluminescent in reaction with oxygen.

Bioluminescence is characteristic of numerous marine organisms (eg, sponges, worms, crustaceans, corals, snails, squid, clams, shrimp, and jellyfish) and a few land organisms (eg, fungi, centipedes, millipedes, worms, beetles, and fireflies) as well as bacteria and dinoflagellates.

The chemistry of bioluminescence is complex. In general, the reactions involve oxygen, a luciferin, and a luciferase enzyme with other essential reactants. Most studies have been carried out with the American firefly (*Photinus pyralis*), the crustacean *Cypridina hilgendorfi*, the coelenterates *Renilla reformis* (sea pansy) and *Aequorea*, and the genus *Photobacterum*.

American firefly bioluminescence is the most efficient reaction known, with Q_c reported to be 88% and (ϵ-max at 562 nm. In the first step of the reaction, luciferin reacts with adenosine triphosphate (ATP) and the enzyme to give complex (**2**) of the adenylate ester (AMP = adenosine monophosphate). Reaction of the complex with oxygen produces the excited state of a highly fluorescent compound (**3**) and carbon dioxide. In common with the specificity of other enzyme reactions, only the D(—)enantiomorph of luciferin produces light.

(2)

(3)

Chemiluminescence is also obtained by anionic autooxidation of luciferin with oxygen in alkaline dimethyl sulfoxide which provides the same products as the bioluminescent reaction and a quantum yield of 10%.

Applications

Chemiluminescence is studied for several reasons: chemiexcitation relates to fundamental molecular interactions and transformations; efficient chemiluminescence can provide an emergency or portable light source, eg, chemical light-sticks used on aircraft emergency slides; means to detect and measure trace elements and pollutants for environmental control (see Trace and residue analysis), or metabolites for disease detection, eg, glucose in urine; classification of the bioluminescent relationship between different organisms to define their biological relationships and patterns of evolution; examination of the effects of enzyme or substrate structural modifications.

MICHAEL M. RAUHUT
American Cyanamid Co.

F. McCapra in W. Carruthers and J.K. Sutherland, eds., *Progress in Organic Chemistry*, John Wiley & Sons, Inc., New York, 1973, p. 231.

M.J. Cormier, D.M. Hercules, and J. Lee, eds., *Chemiluminescence and Bioluminescence*, Plenum Press, New York, 1973.

CHEMOTHERAPEUTICS, ANTHELMINTIC

Anthelmintic drugs are used to relieve diseases caused by parasitic flatworms or roundworms. The drugs injure or destroy worm parasites, facilitate their removal from the body, or interfere with their protective mechanisms against the natural defenses of the host.

Blood Fluke Disease (Schistosomiasis)

The influence of antischistosomal drugs on the disease depends upon the decrease or arrest of egg production, since it is the movement of eggs through tissues that produces the pathology. The number of pairs of worms in a patient varies from a few to hundreds, and in untreated patients the worms may survive for 5 yr, or even decades. The efficacy and toxicity of drugs used to treat schistosomiases leaves something to be desired.

Antimonials. Antimony potassium tartrate (Tartar emetic), stibophen (**1**) (Fuadin), and stibocaptate (Astiban) inhibit phosphofructokinase, an enzyme involved in the anaerobic metabolism of glucose. The worms lose their energy supply since the enzyme in schistosomes is much more sensitive than the corresponding enzyme in human tissues. However, as drug concentration rises in the host, the important effects cease to be therapeutic and become toxic. Lucanthone (Miracil D) was the first oral drug used to treat a schistosomal disease; hycanthone (Etrenol), an hydroxymethyl analogue recognized as the active metabolic product, was used in mass population treatment by a single intramuscular injection but later withdrawn owing to hepatotoxicity. The initial effect of this drug was to decrease egg production possibly through shifts in the tissue distribution of 5-hydroxytryptamine (serotonin), which acts as a neurotransmitter in schistosomes (see Neuroregulators).

Niridazole (**2**) (Ambilhar) is an oral drug that inhibits glucose uptake by schistosomes, and arrests shell formation. There are many common toxic manifestations to the use of this product, which has also shown carcinogenic, lucigenic, and immunosuppressive effects in laboratory animals. Compared to niridazole, which can be toxic and must be administered for a week, the newer drug praziquantel (Drancit, Pyquiton) can be used orally in a single dose and it appears to be without important side effects.

Fluke (Trematode) Infections in the Lungs, Intestines and Liver

Bithionol (Actamer, Bithin, Lorothidol), a phenolic compound similar to hexachlorophene (see Disinfectants and antiseptics) is used to treat *Paragonimus* (lung fluke) or *Fasciola* (sheep liver fluke) infections. It interferes with the neuromuscular physiology of helminths, impairs egg formation, and may cause the protective cuticle covering the worm to become defective. Hexylresorcinol (Crystoids) is a phenolic compound used against *Fasciolopsis buski* (the giant intestinal fluke). It produces blisters and cuts in the surface of the parasites. Its selective toxicity is presumably derived from the natural coatings of the human gastrointestinal tract; the drug can burn unprotected mucous membranes of the mouth and upper esophagus if it comes into contact with these surfaces by eructation.

Tetrachloroethylene (Nema) has a broad anthelmintic spectrum. It can be used against *F. buski*, and also against two small intestinal flukes, *Heterophyes heterophyes* and *Metagonimus yokogawai*. Alternatively, the two small flukes can be treated with niclosamide (**3**) (Yomesan).

Chloroquine (Aralem), a 4-aminoquinoline, bears the same alkyl side chain as the acridine dye quinacrine (see Dyes) and reduces the egg output of the Chinese liver fluke (*Chlonorchis sinesis*). Chloroquine (**4**), like quinacrine, interrelates with deoxyribonucleic acid (DNA) and this upsets the role of DNA as a template in replication and transcription. However, short-term administration of chloroquine is relatively harmless.

(3) Chloroquine

(4) Niclosamide

(1) Stibophen

(2) Niridazole

Tapeworm (Cestode) Infections

Niclosamide can be used to treat all the adult tapeworms that are natural to the intestine of man: *Diphyllobothrium latum* (the broad fish tapeworm), *Taenia saginata* (the beef tapeworm), *T. solium* (the pork tapeworm), and *Hymenolepis nana* (the dwarf tapeworm).

Quinacrine hydrochloride (Atabrin) which causes tapeworms to lose their hold on the intestinal wall, turn bright yellow, and be passed with the stool, is the second drug of choice against tapeworms. It is more toxic than niclosamide without being more efficacious, and is troublesome to administer.

Paromomycin (Humatin), a broad-spectrum antibiotic, is used to treat tapeworm disease, and for amoebiasis as well as for its antibacterial effects in diarrheas and dysenteries (see Antibiotics, aminoglycosides).

Treatment of Intestinal Roundworm (Nematode) Infections

Pyrantel pamoate (Antiminth) in a single oral dose cures infections with *Enterobius vermicularis* (pinworm) and with *Ascaris lumbricoides* (large roundworm of man). Given for three days, it is used to treat infections with the two hookworms *Ancylostoma duodenale* and *Necator americanus*. Thus, this drug is suitable for use in patients with mixed roundworm infections. In all these worms, muscle is persistently activated resulting in spastic paralysis. Bephenium hydroxynaphthoate (Alcobar) can be used against both species of hookworm and also against *Ascaris*, but it is most effective for *A. duodenale*. It is a quaternary ammonium compound with a structure similar to that of acetylcholine. Hookworms lose their hold on the intestinal mucosa and in a contracted state they are carried along with the fecal stream. Tetrachloroethylene (Nema) works against both species of hookworm, but is more successful in the treatment of *N. americanus*. The worms detach from the mucosa and appear in the stool, alive and motile.

Piperazine citrate (Antebar) is effective against *Ascaris* by blocking the ability of the worm to respond to acetylcholine. The worms are carried out of the body, still alive. Pyrvinium pamoate (Poxan) can be employed against pinworm infections and against *Stronglyoides stercoralis* (threadworm). Pyrvinium is the salt of an asymmetrical cyanine dye with a resonating amidinium system in the molecule. In anaerobic worms, such as pinworm and threadworm, it irreversibly interferes with the absorption of glucose. This causes relative failure of muscular activity and reduced motility, loss of adenosine triphosphate (ATP) and death.

Mebendazole (Vermox), a benzimidazole derivative, is a broad-spectrum anthelmintic which has effects against all the above named roundworms and also is most effective against *Tricharis trichiura* (whipworm). In addition, it can be used to treat taenioid tapeworms. The drug interferes with glucose metabolism of helminths. Glycogen stores in the worms are depleted and ATP supplies fail, the worms die and are lost from the body.

Tissue Roundworm (Nematode) Infections

Thiabendazole (Mintezol) was the first broad-spectrum benzimidazole anthelmintic. It was the first drug reported to have a beneficial influence on trichinosis, the disease in which *Trichinella spiralis* larvae migrate through skeletal muscle. When given to someone who has just eaten infected pork, it is prophylactic against the disease. Thiabendazole can be used against *Dracunculus medinensis* (guineaworm) and *Ancylostoma braziliensis* (cutaneous larva migrans) and *Toxacara* (visceral larva migrans). Intestinal roundworms are also treatable with this drug, which inhibits helminthic fumarate reductase and so depletes supplies of ATP. Other drugs used against the guineaworm, which is a stringlike worm about 1 m long in the skin, are niridazole and metronidazole (Flagyl), a broad-spectrum antiprotozoal agent (see Chemotherapeutics, antiprotozoal).

Diethylcarbamazine citrate (Hetrazan) is the traditional drug for treating filariases including elephantiasis, which is caused by *Wuchereria bancrofti*. The drug kills the microfilariae of this and other filarial species, including *Onchocerca volvulus*. It is a derivative of piperazine and it affects microfilariae so that they become susceptible to phagocytosis by fixed macrophages of the reticulo-endothelial system. For treatment of *Onchocerca*, causative agent for the disease called river blindness, diethylcarbamazine is given in conjunction with suramin (an antitrypanosomal drug). Diethylcarbamazine is one of the anthelmintics that may have pharmacological effects that enhance immunological responses. Another such anthelminthic drug is levamisole (Nemacide). Many anthelmintics have additional pharmacological actions. For example, thiabendazole has antipyretic, analgesic, and antiinflammatory activities that may contribute to its medicinal value.

JAMES W. INGALLS
Arnold & Marie Schwartz College of Pharmacy & Health Sciences
Long Island University

American Medical Association Department of Drugs, "Anthelmintics" in *AMA Drug Evaluations*, 4th ed., American Medical Association Publisher, Chicago, Ill., 1980, pp. 1409–1426.

P.D. Marsden and K.S. Warren, "Helmintic Diseases," in P.B. Beeson, W. McDermott and J.B. Wyngaarden, eds., *Cecil Textbook of Medicine*, 15th ed., W.B. Saunders, Philadelphia, Pa., 1979, pp. 605–638.

I.M. Rollo, "Drugs Used in the Chemotherapy of Helminthiasis," in A.G. Gilman, L.S. Goodman, and A. Gilman, eds., *The Pharmacological Basis of Therapeutics*, 6th ed., Macmillan, New York, 1980, pp. 1013–1037.

CHEMOTHERAPEUTICS, ANTIMITOTIC

The modern era in cancer chemotherapy began with the introduction of polyfunctional alkylating agents, such as nitrogen mustard in malignant lymphoma, in the early 1940s, followed by the use of the folic acid antagonist aminopterin to induce remission in acute lymphocytic leukemia in children. Today, a number of new drugs are designed to kill the cancer cell either directly or by depleting its essential growth elements. They are divided according to their general pharmacologic activity: alkylating agents; antimetabolites; antibiotics; plant alkaloids; miscellaneous agents; and hormones.

Alkylating agents. This diverse group of chemicals is able to form covalent linkages with various substances, including such important moieties as phosphate, amino, sulfhydryl, hydroxyl, carboxyl, and imidazole groups, in biologically vital macromolecules (see Biopolymers). Alkylation of the purine base, guanine, in the nucleic acids of deoxyribonucleic acid (DNA), and similar alkylating reactions usually lead to gene miscoding, serious damage to the DNA molecule, or major disruption in nucleic acid function, and inhibition of a variety of other cell functions such as glycolysis and respiration. Tumors can shrink in 1 or 2 days with intravenous administration. All of these agents are potentially mutagenic, teratogenic, and carcinogenic themselves. Their antineoplastic activity occurs throughout the cell cycle (toxicity is cell-cycle independent).

Alkylating agents include: nitrogen mustards, eg, chlorambucil USP (Leukeran, Burroughs Wellcome), melphalan USP (Alkeran, Burroughs Wellcome), uracil mustard NF (Uracil Mustard, Upjohn), cyclophosphamide USP (Cytoxan, Mead Johnson), mechlorethamine hydrochloride USP (Mustargen hydrochloride, Merck, Sharp, and Dohme), and carmustine (BCNU, Bristol Labs); nitrosoureas, eg, lomustine (C-CNU, Bristol Labs), and streptozocin (investigational drug); triazenes, eg, dacarbazine (DTIC, Dome); and others such as thiotepa NF (Thiotepa, Lederle), busulfan USP (Myleran, Burroughs Wellcome).

Antimetabolites. These compounds compete with and displace the substrate of specific enzymes involved in DNA synthesis, thereby inhibit-

ing cell reproduction. They are classified according to their specific inhibitory action (antagonists of purine, pyrimidine, glutamine, etc). Tumors shrink only after 4–8 wk of treatment. Antimetabolite cytotoxicity is cell-dependent.

Antimetabolites include mercaptopurine USP (Purinethol, Burroughs Wellcome), thioguanine USP (Thioguanine, Burroughs Wellcome), cytarabine USP (Cytosar, Upjohn), fluorouracil USP (Fluorouracil, Roche), and methotrexate USP (Methotrexate, Lederle) (1).

Antibiotics. These chemical substances, produced by certain microorganisms (actinomycetes, fungi, bacteria), suppress the growth of or destroy other microorganisms; they are also antagonists of tumor cells, although not as frequently as alkylators and antimetabolites. The mechanisms of their antineoplastic, bactericidal, and bacteriostatic actions are similar (see Antibiotics, peptides); they bind to DNA, inhibit DNA-dependent ribonucleic acid (RNA) synthesis and consequently the synthesis of proteins required by the cell. Their cytotoxicity is cell-cycle dependent.

Antibiotics used as antimitotic chemotherapeutic agents include: bleomycin sulfate (Blenoxane, Bristol Labs), which can cause chain scission, nicking, and fragmentation of DNA molecules, and inhibits the repair of scission; daunorubicin (investigational drug) (2) and doxorubicin (Adriamycin, Adria) (3) which are representative of an extensive series of anthracycline antibiotics under investigation in antimitotic chemotherapy; mithramycin USP (Mithracin, Pfizer); dactinomycin USP (Cosmegen, Merck, Sharp and Dohme); and mitomycin (Mutamycin, Bristol Labs).

(1)

(2) R = H
(3) R = OH

(4) R = CH₃
(5) R = CHO

Plant alkaloids. The two clinically useful alkaloids are vinblastine sulfate USP (Velban, Lilly) (4) and vincristine sulfate USP (Oncovin, Lilly) (5), both derived from the periwinkle plant (*Vinca rosea*) although notably different in their potency, clinical use, and toxicity. They both abort mitosis in the metaphase portion of the cell cycle, affect cellular movement and phagocytosis, and can cause aberrations of the cell nucleus. These agents are cell-cycle dependent (see Alkaloids). Vincristine is of relatively low toxicity for normal cells, making it useful in the presence of impaired bone-marrow function.

Miscellaneous agents. L-Asparaginase (Elspar, Merck, Sharp, and Dohme), an enzyme used as an antitumor agent, catalyzes the hydrolysis of asparagine to L-aspartate and ammonia, and deprives the malignant cell of an essential amino acid used in protein synthesis without damaging normal tissue. Hydroxyurea USP (Hydrea, Squibb) is representative of a group of compounds (eg, substituted urea, guanazole, thiosemicarbazones) that inhibit the enzyme ribonucleoside diphosphate reductase, an enzyme necessary in the biosynthesis of DNA. Mitotane USP (Lysodren, Calbio) is an adrenocortical suppressant which acts selectively on the normal and neoplastic cells of the adrenal cortex, causing a rapid reduction in blood and urinary levels of the adrenocorticosteroids and their metabolites as the drug suppresses the adrenal tumor. Procarbazine hydrochloride USP (Matulane, Roche) is a methylhydrazine derivative that inhibits DNA, RNA, and protein synthesis.

Hormones. These agents, including estrogens, androgens, progestins, and adrenocorticosteroids, are not specific oncolytic agents, but are employed to manipulate the hormonal environment of endocrine-dependent cancers (eg, breast, ovary, and prostate), thus depriving the tumors of the required hormonal growth factors and altering the neoplastic process (see Hormones). The adrenocorticosteroids are of the greatest value in treating acute lymphocytic leukemia in children, and malignant lymphomas; estrogens and androgens are of value in treating neoplastic diseases in the prostate and the mammary glands. The anabolic effects of many steroidal agents may benefit the patient (see Steroids).

Treatment

Combination therapy is used to treat neoplastic disease, chemotherapy being used as an adjunct to radiation therapy and/or surgery. Multidrug treatment uses simultaneous or sequential administration of synergistic combinations of chemicals that are effective in different phases of the cell cycle. Additive cell kill may be obtained without additive toxicity when equally effective drugs with different mechanisms of toxicity are combined. Multiple drug programs are the rule in neoplastic treatment, eg, cyclophosphamide and doxorubicin in breast cancer, cytarabine and thioguanine in acute myelocytic leukemia, and cyclophosphamide, vincristine, methotrexate and cytarabine in diffuse histiocytic lymphoma.

Immunology

Still in experimental stages, immunotherapy aims to enhance the body's natural ability to suppress tumor growth, produce regression of nodules, and prevent metastases. Bacteria, bacterial adjuvants, or other materials are administered as antigens to amplify cellular or humoral immune responses and activate macrophages. Some immune responses are thought to play an important role in the natural host resistance to malignant tumors; immunosuppression by antineoplastic agents may increase the susceptibility of chemically treated cancer patients to infections. Chemotherapy must be carefully designed and monitored to selectively suppress or enhance the immune responses of the patient.

Drug Toxicity

Anticancer drugs are potent and have severe adverse effects. For example, cycle-dependent antineoplastic drugs often damage normal tissues that proliferate rapidly (eg, bone marrow, gastrointestinal epithelium, hair follicles). Intermittent courses of chemotherapy are used to allow the restoration of any normal cells whose numbers may have been reduced by treatment. Certain drugs are toxic because they reduce body levels of essential metabolites; their adverse effects can be decreased by concomitant administration of the normal metabolite. The toxicity of anticancer drugs may be markedly increased if liver, kidney, or bone-marrow function has been impaired by previous treatment or disease. Cell kinetics and pharmacokinetics are being used to analyze the problem of selective toxicity (see Pharmacodynamics).

Radiation Therapy

Radiation is a selectively cytotoxic agent which destroys certain types of malignant growth *in situ* without simultaneously destroying the normal tissue in which the tumor is growing; it is the primary treatment for certain types of malignancy, eg, Hodgkin's disease, and is used in combination therapy as well (see Radioactive drugs; Radioisotopes; X-ray technology).

Charles J. Masur[‡]
William Pearl
Lederle Laboratories
American Cyanamid Company

J.F. Holland and E. Frei, III, eds., *Cancer Medicine*, Lea and Febiger, Philadelphia, Pa., 1973.

[‡] Deceased.

S.K. Carter and M. Slavik, *Annu. Rev. Pharmacol.* **14**, 157 (1974).

P. Calabresi and R.E. Parks, Jr. in L.S. Goodman and A. Gilman, eds., *The Pharmacological Basis of Therapeutics*, 5th ed., Macmillan Publishing Co., Inc., New York, 1975, pp. 1254–1307.

CHEMOTHERAPEUTICS, ANTIMYCOTIC AND ANTIRICKETTSIAL

Mycotic Infections

Some 50 species out of a total of 40,000–50,000 different fungi produce infections that are either superficial (dermatophytic infections of the keratinized tissues including skin, hair, and nails) or systemic or generalized. The latter are the most important in terms of severity of disease and mortality.

Superficial Mycoses

These are the most numerous and the only fungal infections capable of direct host-to-host transmission. They include six different clinical manifestations depending upon the anatomical site involved: (*1*) tinea (pityriasis) versicolor, superficial asymptomatic infections of smooth skin only, caused by *Malassezia furfur* (*Pityrosporum orbiculare*); (*2–3*) the piedras (black and white piedra), infections of hair only, caused by *Piedraia horta* (black) and *Trichosporon beigelii* (white); (*4*) tinea nigra palmaris, asymptomatic fungus infection of the palmar surfaces of the hand, caused by Exophiala (*Cladosporium*) *werneckii*; (*5*) dermatophytoses, ringworm infections of the skin, nails, and hair, caused by *Epidermophyton floccosum*, *Microsporum* sp., or *Trichophyton* sp.; and (*6*) candidiasis, mucocutaneous infections of the mouth, vagina, etc, caused by *Candida albicans* and other *Candida* sp. Correct diagnosis, based on microscopic examination of samples from infected areas, is crucial for effective chemotherapy.

Systemic and Generalized Mycoses

The deep mycoses include two distinct groups of life-threatening fungal infections in humans: the systemic fungal infections caused by pathogenic fungi in normal hosts (eg, blastomycosis, chromoblastomycosis, cladosporiosis, coccidiodomycosis, cryptococcosis, histoplasmosis, lobomycosis, maduromycosis (mycetoma), paracoccidioidomycosis, and sporotrichosis); and opportunistic infections caused by fungi of low virulence in individuals with compromised resistance factors (eg, aspergillosis, candidiasis, and mucormycosis (phycomycosis)). Diagnosis is based upon epidemiologic fact-finding; demonstration of fungal pathogens in tissues or other specimens; isolation, recovery, and identification of the responsible pathogen; and serologic techniques.

Antibiotic Antifungal Agents

Antibiotics with antifungal properties include polyene and nonpolyene compounds.

Polyene antibiotics are high molecular weight, polyhydroxy compounds belonging to the macrolides but differing from them in possessing conjugated chains of chromophoric double bonds in the macrolactone ring (see Antibiotics, macrolides). Polyenes are chemically unstable, have poor solubility in water, have strong and characteristic uv absorption spectra, and have poor stability to light and temperature. Their potent antifungal activity is related to their ability to produce profound changes in cell-membrane permeability. They have some antiprotozoal activity, but little or no antibacterial activity. Their toxicity and efficacy is determined by the degree of binding and by the specificity and avidity of certain polyenes for specific sterols.

Polyenes include the following list of drugs: nystatin (Mycostatin), an amphoteric tetraene, used for topical treatment of superficial infections caused by *Candida albicans*; amphotericin B, an amphoteric heptaene with a sugar (mycosamine) moiety, used to treat systemic fungal infections such as coccidioidomycosis and cryptococcosis and a highly toxic substance when given parenterally; amphotericin B, methyl ester, the water-soluble methyl ester of amphotericin B which has *in vitro* antifungal properties comparable to the parent but because of toxicity is no longer being actively pursued; candicidin, a heptaene complex produced by *Streptomyces griseus*, which exists as either Complex A or Complex B, used to topically treat vaginal and mucocutaneous candidiasis; pimaricin (Natamycin), a tetraene polyene produced by *Streptomyces natalensis*, limited to topical treatment of mycotic keratitus.

Nonpolyene antifungal antibiotics include griseofulvin, a phenolic benzofuran cyclohexane produced by several species of *Pencillium*, used in oral form (Fulvicin, Grifulvin, Grisactin, Gris-Peg, etc) against dermatophytic species; cycloheximide (Acti-dione), produced by *Streptomyces griseus*, mainly active against nonpathogenic fungi and also *Cryptococcus neoformans*; pyrrolnitrin (not an approved drug in U.S.), a potent antifungal metabolite produced from tryptophan, active against *Trichophyton mentagrophytes*, *T. rubrum*, *Candida albicans*, and systemic fungal pathogens and used for topical treatment of dermatophytic infections; sinefugin, produced by *Streptomycees griseolus*, is active against *C. albicans*; ambruticin, a new class of cyclopropylpyran antibiotic, active at low concentrations *in vitro* (0.025 µg/mL) against systemic and dermatophytic fungal pathogens; orally and topically active against *T. mentagrophytes*, and orally against *Histoplasma capsulatum* and *Coccidiodes immitis*.

Synthetic Antifungal Agents

Nonspecific topical medications were used as early chemotherapy of the mycoses. For example, Whitfield's ointment, comprised of 6% benzoic and 12% salicylic acids combined in petrolatum, was used to treat smooth-skin infections. Potassium permanganate (in soaking solutions, topically, and vaginally) was used to treat cutaneous and mucocutaneous candidiasis. Gentian or crystal violet (hexamethyl-*p*-rosaniline) was used to treat superficial fungus infections, and oral and vaginal candidiasis. Undecylenic acid (10-undecenoic acid) and various salts, eg, zinc undecylenate) have been used to treat smooth-skin ringworm infections (Desenex). Other fatty acids or their derivations used to treat superficial fungal infections have included calcium propionate (Sporonol) and sodium propionate (see Carboxylic acids). Sulfur-containing compounds, eg, sulfur-salicylic ointment, sodium thiosulfate, and salicylic acid, once played an important role in treating superficial fungal infections. Iodochlorhydroxyquin (5-chloro-7-iodo-8-quinolinol, Vioform) is used topically to treat dermatophytic infections and *Candida* vaginal infections. Acrisoncin (9-aminoacridine-4-hexylresorcinolate, Akrinol) is effective in treating tinea versicolor.

Nonpecific systemic medications as antifungal chemotherapeutics began with the iodides. Various sulfonamides have been employed in treatment of human mycoses, including sulfadiazine, to treat South American blastomycosis; an aromatic diamine, stilbamidine, used to treat North American blastomycosis; 2-hydroxystilbamidine isethionate, which replaced stilbamidine, now used against less severe, primary pulmonary infections.

5-Fluorocytosine (Flucytosine, 5-FC) was originally developed as an antimetabolite for use in treatment of leukemia, but instead is used in combination therapy (5-FC plus amphotericin B) in treatment of life-threatening yeast infections and some cases of chromoblastomycosis. There is no justification for the use of 5-FC alone in such infections.

Imidazole compounds represent the most versatile and most valuable source of antifungal compounds. As a group, they have a uniquely broad spectrum of activity which includes bacteria and fungi, as well as protozoa, helminths, and nematodes. Specific activity and spectra of individual compounds depend highly on structure.

The first of the antifungal imidazoles was 1-chlorobenzyl-2-methylbenzimidazole (Myco-Polycid); thiabendazole (Mintezol) was then introduced as an oral antihelminthic for roundworm infections. Clotrimazole is used topically to inhibit isolates of *Candida albicans* and dermatophytic infections. Miconazole has important broad-spectrum antifungal activity against dermatophytic fungi, yeasts, and systemic pathogens such as *C. immitis* and is available as topical miconazole cream (2% miconazole nitrate, Monistat) for treatment of vaginal candidiasis and smooth-skin dermatophytosis. Miconazole also is available in an intravenous preparation; however, it is not considered to be the drug of choice for any of the common systemic mycoses. Ketoconazole (Nizoral) is an orally active imidazole with a spectrum similar to that of micona-

zole. At this time, however, the clinical use of ketoconazole is limited to candidiasis, coccidioidomycosis, histoplasmosis, chronic mucocutaneous candidiasis, oral thrush candidureia, chromoblastomycosis and paracoccioidomycosis.

Econazole nitrate (Spectazole, Ortho) is a synthetic imidazole differing from miconazole only in that it lacks a chlorine atom at the two position of the phenylmethoxy group. It has the same indications as miconazole and is applied topically twice daily as is miconazole. Spectazole is now available in this country. Bifonazole (Mycospor, Bayer) also is an imidazole derivative. However, it differs from all other azole-like compounds in that it contains no halogens. Available (outside this country including many countries in Asia and Europe) in topical preparations only. Bifonazole has the same broad spectrum of clinical utilty as do other topical azoles. The principal advantage claimed for bifonazole is that it need be applied only once daily.

Haloprogin, an analogue of the antibiotic lenamycin, is active *in vitro* against pathogenic yeasts, gram-positive cocci, *Mycobacterium tuberculosis*, and many species of dermatophytic fungi.

Agricultural Use of Antifungal Agents

Antifungal agents are used to control fungal plant pathogens, particularly in Japan (see Fungicides). Griseofulvin is fungistatic for most phytopathogenic fungi; pimaricin is used as a dip to prevent fungal growth in skins of harvested apples; cycloheximide is used to treat downy mildew of onions and shoot blight of the Japanese larch; and kasugamycin, blasticidin S, polyoxin D, and validamycin A are used extensively for agricultural purposes (see Antibiotics, aminoglycosides, and nucleosides).

Rickettsial Infections

Rickettsia share some characteristics of bacteria and viruses; they have cell walls, contain muramic acid, have metabolic enzymes, are capable of independent respiratory activity, and their growth is inhibited by antibiotics (all characteristics of bacteria), but they are unable to propagate outside living host cells. The rickettsial infections are classified into five categories: (*1*) typhus group, eg, epidemic typhus, Brill-Zinsser disease, endemic typhus; (*2*) spotted-fever group, eg, Rocky Mountain spotted fever, rickettsialpox, and tick-borne rickettsioses of the eastern hemisphere; (*3*) scrub typhus; (*4*) Q fever; and (*5*) trench fever.

Treatment

All rickettsial infections respond to treatment with chloramphenicol or tetracycline, the latter being the drug of choice (see Antibiotics, chloramphenicol, and tetracyclines). Other antimicrobial agents or antibiotics used include *para*-aminobenzoic acid and doxycycline. Prevention is attained by elimination of lice, mites, ticks, rodents, and use of the appropriate insecticides such as benzyl benzoate, *N*,*N*-diethyl-*m*-toluamide, dimethyl phthalate, and dibutyl phthalate (see Insect control technology).

SMITH SHADOMY
C. GLEN MAYHALL
Medical College of Virginia

G.S. Kobayashi and G. Medoff, *Ann. Rev. Microbiol.* **31**, 291 (1977).

N.F. Conant and co-workers, *Manual of Clinical Mycology*, W.B. Saunders, Philadelphia, Pa., 1971, 255 pp.

P.B. Beeson and W. McDermott, eds., *Textbook of Medicine*, W.B. Saunders Co., Philadelphia, Pa., 1975.

CHEMOTHERAPEUTICS, ANTIPROTOZOAL

Antiprotozoal chemotherapy utilizes drugs that are either protozoocidal (eradicates the parasite in all its states within the host) or protozoostatic (suppresses the parasite's clinically relevant developmental stages but not certain latent stages, which may then either cause clinical relapse or stimulate immune defenses to cure or control the infection).

Coccidiosis

The life cycle of *Coccidia* includes cysts, sporozoites, schizonts, merozoites, and gametocytes. Medically important *coccidia* species include *Eimeria*, which infects many vertebrates by invading the epithelial lining of the digestive tract, causing diarrhea, sloughing, ulceration of the intestinal lining, and hemorrhage, leading to possible metastatis spread, malabsorption of nutrients and vitamins, metabolic imbalance, anemia, and bacteremia. Eradication of *Eimeria* species is impractical due to toxicological and economic considerations. Anticoccidials are mixed with feeds to minimize the intensity of infection or to reduce clinical symptoms. The efficacy of any given anticoccidial can vary from one species to another, and the safety range of anticoccidials varies widely.

Drugs used as anticoccidials against *Eimeria* include: thiamine competitors, thiamine derivatives (eg, amprolium and aminoalkenyl sulfides), antifolates (eg, sulfonamides, pyrimethamine, and *p*-aminobenzoic acid competitors such as Ethopabate (see Antibacterial agents), antibiotics (eg, monensin and oxytetracycline (see Antibiotics, polyethers; Antibiotics, tetracyclines). Efforts are also underway to develop nitrobenzamide and nitrofuran anticoccidials, eg, Robenidine, bisthiosemicarbazones, Clopidol, quinolones, and quinazolinones.

Toxoplasmosis. Toxoplasma may pass from one host species to another, changing their target organs from host to host. An estimated one-third or more of human populations may have quiescent toxoplasma infections. The relatively rare clinical manifestations are very serious; they are more prevalent in children and immunodeficient individuals. Humans are contaminated either transplacentally or by consuming infected meat or animal material. Antifolates, sulfonamides, pyrimethamine, and spiramycin are useful chemotherapeutics (see Antibiotics, macrolides).

Anaplasmosis

Anaplasma species parasitize red blood cells of cattle, causing death. Useful chemotherapeutics are tetracyclines, dithiosemicarbazones (eg, gloxazone), and Imidocarb.

Babesiasis

Babesia, transmitted by ticks, parasitize the red blood cells of mammals and birds causing mortality rates of more than 10% in cattle; rare cases of human infection have been reported. Effective chemotherapeutics include diamidine and related compounds, pentamidine, amicarbalide, diaminazene aceturate (Berenil), and the toxic, cholinesterase-inhibiting quaternary compounds of the quinuronium class (see Choline; Cholinesterase inhibitors).

Theileriasis

Theileria species are transmitted by ticks to cattle and other ungulates. In east and south Africa the species *T. parva* has as high as a 90% mortality rate in herds of farm animals; the species *T. annulata* has a wide geographic distribution. Various developmental stages of the parasite are in lymphocytes, histiocytes, and erythrocytes. Although chemotherapy with antiparasitic drugs has been unsuccessful, administration of tetracycline during the incubation period provides good chemoprophylaxis.

Trypanosomiasis

Species of *Trypanosoma brucei* cause a variety of diseases. Trypanosomes are transmitted by insect vectors with distinct developmental cycles in the host and in most vectors. Epidemiologic control is difficult since certain wild and domestic animals are disease reservoirs. Available drugs are quite toxic and frequently inadequate for some of the diseases.

African Trypanosomiasis. The African trypanosomes of the brucei type cause African sleeping sickness (*T. br.* Gambiense and *T. br.* rhodesiense) and damage to livestock in endemic areas. In humans, unless treated, they are invariably fatal, progressing from blood to lymphatic tissue and, finally, to the central nervous system. Chem-

otherapy early in the infection, utilizing pentamidine and suramin sodium, is more promising than at the later stages. When parasites have invaded the central nervous system, drugs that penetrate the blood-brain barrier are used. Other drugs useful in the early stages are the aromatic diamidine, Berenil (too toxic for humans); the nitrofuran, nifiurtimox (see Antibacterial agents, nitrofurans); the phenanthridines such as homidium bromide; quinapyramine; and some of their derivatives and antimonials such as ethylstibamine (Stibosamine, Neostiboson) (see Antimony compounds).

Chagas' disease is caused by *T. cruzi* (a collective term comprising regional varieties of physiologically distinct organisms). No effective chemotherapy is known, although drugs being considered are 8-aminoquinolines, nitrofuran, and a nitroimidazole, Ro-7-1051.

Leishmaniasis

Members of the genus *Leishmania* are transmitted by insect vectors, and wild and domestic animals are disease reservoirs. They cause a variety of diseases in man; eg, visceral cutaneous, and mucocutaneous leishmaniases (*L. donovani*, *L. tropica*, and *L. brasiliensis*). The parasites invade the reticulo-endothelial system and their amastigotes penetrate and survive in the host's macrophages, thus reducing the host's immune-defense capabilities and making direct drug contact difficult. Few drugs are selective against leishmania; the highly toxic tartar emetic, antimony potassium tartrate, is one of the oldest. Less toxic organic antimonials are linear polymers containing pentavalent antimony, such as sodium antimonyl gluconate, Pentostam; the antimonate of *N*-methylglucamine, Glucantime; and ethylstibamine. Other effective compounds are amphotericin B, pentamidine isethionate, and hydroxystilbamidine isethionate; some 8-aminoquinolines which are not too toxic for human use seemed promising, as well as cycloguanil embonate and Berberine.

Pneumocystosis

Caused by *Pneumocystis carinii*, an organism that is cosmopolitan among humans and other vertebrates, this is an opportunistic pathogen that proliferates and becomes clinically manifest only in individuals whose immune defenses are impaired. Available drugs are parenteral pentamidine, and the oral antifolate combination of a pyrimidine with a sulfonamide (eg, Fansidar and Bactrim).

Trichomoniasis

Extracellular, flagellated parasites transmitted by direct contact, the trichomonads reside in the superficial layers of infected organs. In humans, *Trichomonas vaginalis* may cause significant disease in both sexes, and symptomless carriers do occur. Various nitroimidazoles are effective, eg, metronidazole and tinidazole. No specific chemotherapy is known for *T. foetus*, which causes abortion in cattle, or *T. equi*, which causes death in horses. *T. gallinae*, which parasitizes a variety of birds, is effectively treated with Enheptin, 2-amino-5-nitrothiazole.

Hexamitosis

Reported in the United Kingdom and the Americas, this disease of birds is caused by the flagellate *Hexamita meleagridis*, and is transmitted by contaminated food. No effective drug is known.

Balantidial Dysentery

The agent of this condition, *Balantidium coli*, occurs in humans, primates and several other vertebrates; it is cosmopolitan and transmitted by food and beverages. Metronidazole, tetracycline and paromomycin are effective drugs (see Antibiotics, aminoglycosides).

Giardiasis

The agent of this condition is the flagellate *Giardia lamblia*, which lives on the epithelial lining of the human duodenum and jejunum, causing a variety of digestive disturbances including malabsorption, although symptom-free carriers are common. The parasite is cosmopolitan, and transmitted by contaminated food and beverages. No specific chemotherapeutic agent is available; antimalarials, eg, quinacrine, and antiamebic drugs, eg, metronidazole, tinidazole, and furazolidone, are effective.

Amebiases

The agents of these conditions are amebae, which are cosmopolitan and transmitted by ingestion of contaminated material. *Entamoeba histolytica*, the common pathogenic species, causes amebic dysentery in humans, although symptomless carriers occur. Humans become infected by ingesting the amebic cysts that descend in the intestinal tract. Amebic trophozoites develop and reside in the caecum, colon, and sigmoid, producing the next generation of infective cysts which are eliminated with the feces. When the trophozoites invade the intestinal wall, they cause ulcerations and clinical intestinal symptoms that vary in severity. Metastatic amebic lesions may develop in various organs. Amebicides include nitroimidazoles, metronidazole, tinidazole, and the more toxic nitro heterocycle niridazole, all effective at all sites. Parenteral emetine and dehydrometine are effective against amebiasis in the bowel wall and liver. Paromomycin, iodoquinoline analogues, and various oral arsenical and bismuth preparations (eg, carbarsone, Milibis and emetine–bismuth iodide) are effective in the lumen of the bowel against either trophozoites or cysts, or both. Conjunctive antibiotic chemotherapy utilizes Fumagillin (see Chemotherapeutics, antimitotic) and tetracyclines (see Antibiotics, tetracyclines) as adjuvants.

Primary amebic meningo-encephalitis is a rare and fatal condition caused by amebae of the genera *Naegleria* and *Hartmanella*; no chemotherapy is available.

The Malarias

Agents of these infections are intracellular parasites of the genus *Plasmodium*, particularly *P. falciparum*, *P. vivax*, *P. ovale*, and *P. malariae*, which cause in humans the historically termed malignant, benign, oval tertian, and quartan malarias, respectively. Mosquito vectors normally transmit the parasites between host individuals. When feeding on an individual with malaria, the mosquito ingurgitates blood containing male and female gametocytes of the plasmodium which then undergo a developmental cycle yielding infective sporozoites. These are transmitted to the new host. In man, the malaria parasite develops through sequential stages: sporozoites; pre-erythrocytic schizonts; merozoites which may develop into the sexual stage (completing the cycle) or into asexual forms; and erythrocytic schizonts, which produce new generations of merozoites. The clinical manifestations of the disease are caused at repeated intervals by the erythrocytic circuit of the developmental cycle in *P. falciparum*. *P. vivax* and *P. ovale* undergo a somewhat more complicated development, releasing merozoites much later and causing clinical relapses even after years of latency. Malaria has also been transmitted from person to person by blood transfusion and needle contamination.

Antimalarial drugs can affect one or more developmental stages of plasmodia species. Those of clinical relevance to man are classified according to their principal target stages: (*1*) Drugs acting on asexual blood forms such as quinine salts, acridines, biguanides, chlorguanide, dihydrotriazines (cycloguanil), pyrimidines (pyrimethamine), sulfones and sulfonamides (dapsone) (see Antibacterial agents, sulfonamides), antibiotics (tetracycline, clindamycin) (see Antibiotics, lincosaminides), and 4-aminoquinolines (chloroquine, amodiaquin), and experimental drugs such as the quinolinemethanols (mefloquin HCl, WR 030,090HCl), and a phenanthrenemethanol (WR 033,063HCl). The antifolate combination drug Fansidar has been useful. Drugs that affect the asexual blood form alone can eradicate susceptible strains of *P. falciparum* and cure these infections, but they can only suppress the relapsing type of infections since they do not affect their exoerythrocytic tissue forms; (*2*) Drugs affecting tissue forms eg, 8-aminoquinolines (pamaquine, and primaquine); and (*3*) Drugs acting on gametocytes, eg, quinacrine, 4-aminoquinolines (affect *P. vivax* and *P. malariae* only), chlorguanide and 8-aminoquinolines (affect *P. vivax* and *P. falciparum*).

Antimalarial drugs are used in chemoprophylaxis; drugs that suppress merozoites and asexual blood forms are commonly used to prevent clinical patency and permit time for immunodefenses to develop. Endemic prophylaxis is also obtained with drugs that antagonize sexual forms.

Regional drug resistance of plasmodium strains poses a serious problem. The oldest antimalarial is quinine, a herbal product (see Alkaloids).

HOWARD C. ZELL
EDGAR J. MARTIN
Food and Drug Administration

N.D. Levine, *Protozoan Parasites of Domestic Animals and Man*, Burgess Publishing Co., Minneapolis, Minn., 1973.

P.E.C. Manson-Bahr and F.I.C. Apted, *Manson's Tropical Diseases*, Balliere Tindall, London, 1982.

M. Windholz, ed., *The Merck Index*, 10th ed., Merck & Co., Inc., Rahway, N.J., 1983.

CHEMOTHERAPEUTICS, ANTIVIRAL

The difficulty in designing and developing effective antiviral substances is caused largely by the very nature of viruses, a diverse group of infectious agents that differ greatly in size, shape, chemical composition, host range, and effects on hosts. Inhibitors of cellular processes will often prevent viral replication, but are also toxic for the host. It should be possible to design inhibitors that have specific effects upon virus-induced functions, eg, it should be possible to find a chemotherapeutic agent which specifically inhibits the function of the unique RNA-dependent viral enzyme, RNA-polymerase (replicase) or the retrovirus, a unique viral enzyme, RNA-dependent DNA polymerase (reverse-transcriptase) (see Antibiotics, ansamacrolides).

Areas of Application

Antiviral chemotherapy and prophylaxis would be useful: (*1*) for viral diseases for which vaccines have not been successful owing to antigenic instability of the virus, eg, influenza A viruses; (*2*) for viral diseases that are antigenically stable, but which have a plethora of viral serotypes that may cause the disease, eg, rhinoviruses; (*3*) for viruses that cause disease even in the presence of circulating antibody, eg, herpes simplex; and (*4*) for individuals who are immunologically compromised, eg, by immunosuppressants following transplantation, and for individuals with immunodeficiency disease.

Antiviral Agents in Humans

Only four of the many hundreds of antiviral compounds synthesized or isolated from nature over the last three decades are approved by the FDA for use: amantadine hydrochloride (**1**), idoxuridine (**2**), arabinosyladenine (Ara-A (**3**)), and acyclovir (**4**). These four are effective in only a few types of viral infections.

Amantadine hydrochloride (1) (1-adamantanamine hydrochloride, aminoadamantane hydrochloride, Symmetrel; DuPont), mp = 360°C, has a protective effect against influenza A in humans, and appears to have *in vitro* activity against rubella, parainfluenza, and vaccina. Replication of arenaviruses in cell cultures has been reduced in the presence of amantadine; the compound appears to act in the early stage of the influenza replication cycle by blocking or slowing the penetration of the virus into the host cell. In 1966, it was licensed by the FDA for use in prevention of respiratory illness owing to influenza A viruses by prophylactic treatment of contact of patients and when index cases appear in the area.

5-Iodo-2'-deoxyuridine (2) (idoxuridine, IDU, IUDR, Stoxil; Smith Kline and French Calbio, HEM Research Inc), mp 240°C, is a halogenated pyrimidine used against acute herpetic keratitis infections, and shown to be effective against varicella-zoster virus, cytomegalovirus, and vaccinia in tissue culture. The drug is limited to topical application.

Arabinosyladenine (3) (9-β-D-arabinofuranosyladenine) (Ara-A, Vidarabine; Parke Davis) is licensed in the U.S. (see Antibiotics, nucleosides). This purine nucleoside analogue is virtually identical in potency and activity to idoxuridine and cytosine arabinoside, but less toxic. It does not suppress the hematopoietic system, nor does it exert an immunosuppressive effect on antibody formation or on cellular immunity;

it produces only minimal systemic symptoms. Adenine arabinoside has been licensed for topical use in ophthalmic ointments, and is the first antiviral drug licensed for systemic treatment of herpetic infections.

(1) Amantadine hydrochloride

(2) Idoxuridine, R = I
Trifluorothymidine, R = CF

(3) Ara-A

(4) Acyclovir

Acyclovir (4) [9-(2-hydroxyethoxymethyl)guanine] (Acycloguanosine, Zo-Virax; Burroughs-Wellcome) is licensed in the U.S. for parenteral use in herpetic infections. Acyclovir is effective in treating established herpes 1 and 2 infections and in preventing recurrent infections in immunosuppressed patients while taking the drug. It does not seem to eliminate latent infections. Acyclovir may be superior to Ara-A since it is more soluble and is less toxic.

Other antiviral agents include thiosemicarbazones, represented by methisazone (N-methylisatin-3-thiosemicarbazone) (Marboran, Burroughs Wellcome; Aldrich), mp 245°C; trifluorothymidine (5′-trifluoromethyl-2′ deoxyuridine) (Heinrich Mack) mp 169–172°C; arabinosylcytosine (1-β-Arabinofuranosyl cytosine) (Ara-C, Upjohn, Merck, Sharp and Dohme) mp 212–213°C; ribavirin (1-β-D-ribofuranosyl-1,2,4-triazole-3-carboxamide) (Virazole; ICN Pharmaceuticals); inosiplex (Isoprinosine; Newport Pharmaceuticals), and Levamisole (Lederle) mp 60–61.5°C (see Chemotherapeutics, antihelmintic).

Interferon, discovered in 1956 by Isaacs and Lindenmann, is a glycoprotein produced by cells in response to most viral infections. A natural defense of the body, it has a broad spectrum of antiviral activity, and can prevent the replication of most viruses in pretreated cells by synthesizing short-lived inactive proteins (2′,5′-oligodenylate (2′,5′-A) synthetase and a protein kinase). Evidence suggests that these proteins, when activated by double-stranded RNA or by virus, activate the synthesis of 2′,5′-A which in turn induces a latent cellular ribonuclease that destroys messenger RNA. The protein kinase inhibits peptide chain initiation by phosphorylating a cellular initiation factor for protein synthesis. Prolonged interferon treatment is well tolerated.

GEORGE E. GIFFORD
University of Florida

R. Saral, W.H. Burns, O.L. Laskin, and co-workers, *New Engl. J. Med.* **305**, 63 (1981).

CHEMURGY

Chemurgy is that branch of chemistry devoted to industrial utilization of organic raw materials, particularly from farm products; it is the use of renewable resources for materials and energy (see also Fuels from biomass).

Chemurgy was really a social movement during the 1920s and 1930s, a time when there was large surpluses of agricultural materials and severe economic problems in farm areas. The idea of using farm commodities as chemical or industrial raw materials was seen a great contribution to solving these economic problems. Probably one of the movement's chief accomplishments was the founding of the regional laboratories of the USDA, laboratories which still exist as major centers of research in the application of agricultural materials.

One broad distinction that can be made is between the use of natural products grown solely for industrial purposes (eg, trees, cotton (qv), flax, starch (qv), proteins from oilseed meals such as soy, peanuts, and cottonseed) and those grown primarily for food. Renewable resources of industrial materials also include waste products or by-products of food processing (eg, soybean-oil foods or soap stock from soybean oil) and wastes from the pulp and paper industry such as lignin (qv) and sugars in the form of sulfite waste liquor. The fermentation industry is based almost exclusively on renewable materials in the form of molasses, starch, and other materials (see Fermentation). An area that seems to have been a common thread through the history of chemurgy is that of generating energy from biomass in some form, eg, ethanol (qv), the oldest chemical made by fermentation (qv) (see Gasoline and other motor fuels; see also Fuels from waste).

Examples Based on Industrially Known Raw Materials

Trees are used for structural purposes, pulp, and paper. Cotton, grown for its fiber, is processed into thread and textiles and also yields protein and oil from the seed. Flax, grown primarily for its seed (linseed oil), also yields fiber used in fine papers. Jojoba, a desert crop which gives a small bean containing ca 50% of a wax, a fatty acid ester with a fatty alcohol, can substitute for sperm whale oil. Guayule, a new crop, is a potential source of natural rubber from its milled leaves. Crambe, another new crop, is an oilseed whose oil is very high in erucic acid which can be used to provide industrial lubricants. Kenaf, a grass fiber crop, has been proposed as an alternative papermaking source. Lesquerelle, an oil-seed crop, is a potential industrial crop.

Examples Based on Food Crops

Crops that are grown primarily for food can be used for industrial purposes when the crops are in surplus or are found to be unfit for human consumption or for their intended purpose. For example, corn and wheat starch-separation processes give starch used in paper sizing and textiles; oil seeds grown primarily for use in salad dressings, etc, also produce an important by-product in the form of high protein meal (see Vegetable oils); by-products from meat animals are commercially significant (see Meat products), yielding leather (qv) from tanned hides, commercial gelatin and animal glue products (see Glue), and inedible animal tallow which is used in soaps and detergents.

Examples Based on Wastes

The trimming and slash from forest operations have been left in the forest but are increasingly being used in much the same way as higher quality timber, primarily for pulping or chipping. Agriculture produces large amounts of wastes in the form of animal manures, branches, stems, stalks, and straws which can be used for fertilizer; animal wastes can be recycled as animal feed. Agricultural product processing into food, fiber, and feeds yields additional wastes which may be considered as raw materials, eg, kraft black liquor, sulfite waste liquor, and other dilute streams from pulp and paper making are potential fermentation substrates because of the dissolved sugar, the very material that is the most serious pollutant. Cellulosic materials, such as farm wastes, may be upgraded for animal feed by contacting with ammonia (see Pet and other livestock feeds).

Potential for Renewable Resources

The 1976 National Academy of Science study conducted on renewable resources for industrial materials concluded that competition between a material made from a renewable or a nonrenewable resource would be resolved by economics. The future of chemurgy is interconnected with the questions of energy, environment, and food. Most likely, the competition of chemurgic materials with synthetic materials will be settled in the marketplace. The problems of chemurgic process may be summarized as those caused by variable and complex raw materials; involving natural, ie, degradable, raw materials; involving reductive rather than oxidative chemistry because of the composition of the raw materials; and involving relatively simple operations.

J. PETER CLARK
Virginia Polytechnic Institute and State University

U.S. Dept. of Agriculture, *Crops in Peace and War*, Government Printing Office, Washington, D.C., 1950.

CHICLE.   See Gums and mucilages.

CHLORAL.   See Hypnotics, sedatives, anticonvulsants.

CHLORAMINES AND BROMAMINES

Chloramine denotes NH_2Cl (the monochlor-), $NHCl_2$ (the dichlor-), and NCl_3 (the trichloramine). Historically, the term included over 1000 compounds containing one or more chlorine atoms attached to a nitrogen atom (chloramines, chlorimines, chloramides, and chlorimides) (see also Cyanuric and isocyanuric acids).

Properties

The N—Cl bond is covalent, but properties differ from those of the covalent C—Cl bond. Chlorine bonded to nitrogen is regarded as positive, and this is also the active-chlorine content of a compound (eg, organic chloramines are better termed *N*-chloramines to indicate attachment of a chlorine atom to a nitrogen so that the chlorine is positive). When the bond is broken, chlorine may be replaced by a hydrogen that is considered positive. Chloramines are similar in some respects to hypochlorous acid and its salts; the N—Cl bond is formed by treating an amine, imine, amide, or imide with hypochlorous acid in the presence of a base.

All compounds containing the N—Cl bond liberate iodine from an acidified iodide solution (this is commonly regarded as a test for *N*-chloro compounds):

$$RR'NCl + 2 HI \rightarrow RR'NH + I_2 + HCl$$

The available-chlorine content is measured by the ability to liberate iodine from an acidified iodine solution, expressed as weight percent (grams of available chlorine per 100 grams of solution). The time needed for a given content of combined available chlorine (chlorine present as chloramines, eg, monochloramine, NH_2Cl) to destroy a specific organism may be from 10 to 20 times that required by the same amount of free available chlorine (chlorine present as hypochlorous acid or hypochlorite or in the form of a more active chloramine).

Hypochlorous acid.   Chlorine dissolved in water gives hypochlorous acid and hydrogen and chloride ions. The oxidizing and bactericidal properties of chloramines and hypochlorites are related to the hypochlorous acid content in their solutions (see Chlorine oxygen acids). The content depends on the pH and the concentration and chloramine hydrolysis constant (eg, $K \times 10^{-4}$ is 1.46 at 0°C; 3.95 at 25°C; 6.05 at 45°C). At pH above 5, the concentration of molecular chlorine is negligible; at 7.5 pH (25°C), the amount of available chlorine present as hypochlorous acid equals that of hypochlorite; strong alkali (> 10 pH) concentrations have their own bactericidal action; below 4.5 pH, acid concentrations inhibit or destroy bacteria.

With hydrolysis constants as high as 10^{-4}, the amount of chloramine present as hypochlorous acid is large in dilute solutions and the action is comparable to that of hypochlorites of the same available chlorine content. Chloramines with constants over 10^{-6} yield bactericidal concentrations of hypochlorous acid, although they may or may not be less effective than the same content of available chlorine in hypochlorites.

Chloramines with constants as low as 10^{-10} may have useful sanitizing value, although they may be bacteriostatic rather than bactericidal.

The amount of available chlorine present as hypochlorous acid tends to be low in strong solutions, but rises in weak solutions. In alkaline solutions, chloramines may react with hydroxide ion to form hypochlorite (see Water, swimming-pool treatment; Bleaching agents).

Reactions.   *N*-Chloro derivatives of aniline rearrange to form new *N*-chloramines. Two chloramines and an amine or other molecules may form diazo and other chromophoric groups. Most chloramines can be precipitated only in acid solutions; they decompose in alkaline solutions, particularly as pH increases from 9 to 11 and as concentrations increase.

Inorganic Chloramines

Formed by the action of hypochlorites or hypochlorous acid on ammonia, or on other nitrogenous materials, such as urea, the product obtained depends on the pH of the solution (eg, monochloramine above pH 9.5; nitrogen trichloride below pH 4.5).

Important inorganic chloramines include monochloramine, NH_2Cl, fp −66°C, a colorless liquid with strong odor, soluble in water, prepared and handled in an ether solution free of water, which may explode at room temperature; dichloramine, $NHCl_2$; and nitrogen trichloride, NCl_3, an explosive, bright-yellow liquid with a powerful odor, limited solubility in H_2O, mp below −40°C, bp 70°C; and others such as *N*-chlorosulfamic acid, *N,N*-dichlorosulfamic acid, sodium *N*-chloroimidodisulfonate, and trichlorimidometaphosphates.

Organic Chloramines

The three main classes are (*1*) chloroisocyanurates (see Cyanuric and isocyanuric acids); (*2*) heterocyclic chloramines with the chlorine attached to the nitrogen in the ring, eg, glycoluril, hydantoins such as 1,3-dichloro-5,5-dimethylhydantoin, and other chloro derivatives, 1,3,5-trichloro-2,4-dioxohexahydrotriazine, and succinchlorimide; (*3*) *N*-chloramine condensation products from cyanamide derivatives, eg, dichloroazodicarbonamide, Chloroazodin, $NH_2C(=NCl)N=NC(=NCl)NH_2$, and melamines (see Cyanamides); (*4*) *N*-chloroanilides (see Amines, aromatic (aniline)); (*5*) *N*-chlorosulfonamides and related compounds such as chloramine-T, obtained from reaction of *p*-toluenesulfonamide treated with sodium hypochlorite, chloramine-B, $C_6H_5SO_2NClCa$, halazone, $HOOCC_6H_4SO_2NCl_2$, and *N*-chloro-*N*-methyl-*p*-toluenesulfonamide.

Other organic chloramines include *N*-chloramines, eg, quinone dichlorimide ($ClN=C_6H_4=NCl$), ethyl *N*-chlorobenzimidate ($C_6H_5C=NClO(C_2H_5)$); and mixed chlor- and bromamines such as *N*-bromo-*N*-chloro-5,5-dimethylhydantoin and sodium *N*-bromo-*N*-chlorocyanurate.

Uses

Chlorinated *s*-triazinetriones are used as swimming pool disinfectants, bleaches, sanitizers, and dishwasher detergent products. Chlorinated hydantoins are used for bleaching.

G.D. NELSON
Monsanto Industrial Chemicals Company

E.M. Smolin and L. Rapaport, *s-Triazines and Derivatives*, Interscience Publishers, New York, 1959.

G.C. White, *Handbook of Chlorination*, Van Nostrand Reinhold, New York, 1972.

CHLORAMPHENICOL.   See Antibiotics—Chloramphenicol.

CHLORINATED BIPHENYLS.   See Diphenyl and terphenyls.

CHLORINE.   See Alkali and chlorine products.

CHLORINE OXYGEN ACIDS AND SALTS

CHLORINE MONOXIDE, HYPOCHLOROUS ACID, AND HYPOCHLORITES

Oxidation States

Chlorine has positive oxidation states in oxychlorine compounds since its appreciable electronegativity (2.83) on the Allred-Rochow scale is exceeded by that of oxygen. All oxychlorine compounds are strong oxidants. Decomposition reactions of chlorine oxides and oxo-acids are always energetic and violent in many cases. Their chemical properties indicate a trend toward greater thermodynamic and kinetic stability with increasing oxidation state. The oxo-anions of chlorine are weaker and more stable oxidants than the corresponding acids.

Chlorine Monoxide (Dichlorine Oxide)

The anhydride of hypochlorous acid, Cl_2O, typifies the chlorine oxides as a highly reactive and explosive compound with strong oxidizing properties; however, it may be handled safely. The two compounds are readily interconvertible via the equilibrium: $Cl_2O + H_2O \rightleftarrows 2 HOCl$.

Physical and chemical properties. Chlorine monoxide is a brownish-yellow gas, bp $2.0°C$, mp $-120.6°C$, soluble in H_2O and carbon tetrachloride. Explosions of gaseous chlorine monoxide are initiated by spark or heat whereas the liquid phase is shock sensitive, and it thermally and photochemically decomposes into Cl_2 and O_2. Thermal decomposition is complete in 12–24 h at $60–100°C$; in only a few minutes at $150°C$; and terminates in explosion above $110°C$. Chlorine monoxide reacts with a variety of inorganic substances (eg, reaction with N_2O_5 is a convenient route to $ClNO_3$).

Preparation. Commercial production involves reaction of Cl_2 with moist sodium carbonate in either a tower or a rotating tubular reactor; it is also prepared by reaction of Cl_2 (diluted with moist air) with activated soda ash; by addition of HgO or moist soda ash to a solution of Cl_2 in CCl_4 followed by filtration; and from concentrated HOCl solution by vacuum distillation, stripping with air, or treatment with anhydrous $Ca(NO_3)_2$.

Hypochlorous Acid

Known primarily in aqueous solution, HOCl is a highly reactive, unstable compound which is also the most stable and strongest of the hypohalous acids, and one of the most powerful oxidants among the chlorine oxyacids.

Physical and chemical properties. Hypochlorous acid is a weak acid (dissociation constant 2.90×10^{-8} at $25°C$) that is stable in cold, dilute, pure solutions. Hypochlorite solutions oxidize numerous inorganic substrates, eg, oxidation of CN^- is an important reaction in wastewater treatment, and proceeds via the intermediate ClCN. Hypochlorous acid reacts with peroxide with the evolution of oxygen. The addition of HOCl to ammonia results in stepwise formation of chloramines. Dichloramine decomposes with regeneration of some HOCl by the overall reaction:

$$2 NHCl_2 + H_2O \rightarrow N_2 + HOCl + 3 HCl$$

This reaction foms the basis of breakpoint chlorination, which is important in the disinfection of municipal water supplies, swimming-pool sanitation, and wastewater treatment (see Chloramines).

Hypochlorous acid undergoes C- and N-chlorination, oxidation, addition, and ester formation reactions. The most important industrial reaction is with olefins to form chlorohydrins via aqueous chlorination, dichlorides and ethers being formed as by-products. Adding HOCl to acetylenic compounds produces dichloroketones (see Acetylene-derived chemicals).

Preparation. Chloride-containing solutions are prepared by the hydrolytic reaction of chlorine with an aqueous base; strong bases (eg, caustic or lime) produce a stepwise reaction having hypochlorite as an intermediate, whereas weak bases (eg, $NaHCO_3$ or $CaCO_3$) do not. Chloride-free solutions of HOCl are obtained in the following ways: from gaseous Cl_2O or solutions in CCl_4 generated from Cl_2 and either HgO or soda ash; from liquid Cl_2O; from chlorination of an aqueous slurry of bismuth oxide to give insoluble bismuth oxychloride; by chlorination of aqueous HgO slurries followed by filtration of the basic mercuric chloride to yield HOCl solutions contaminated with significant amounts of $HgCl_2$. Electrodialysis of aqueous Cl_2 or HOCl salt solutions using semipermeable membranes has yielded aqueous HOCl substantially free of chloride ion. Near quantitative yields of organic solutions of HOCl are prepared by extraction of chloride-containing aqueous solutions of HOCl with polar solvents such as ketones, nitriles, and esters.

Metal Hypochlorites

The only known stable solid neutral hypochlorites are those of lithium, strontium, barium, and calcium (the major commercial solid hypochlorite).

Physical and chemical properties. Li, Na, and Ca hypochlorites are soluble in H_2O ($25°C$) to the extent of 40, 45, and 21.4%, respectively. Hypochlorites yield HOCl when treated with stoichiometric amounts of acid, and are converted to Cl_2 when excess HCl is used. They react quantitatively with iodide in acid media liberating iodine and with hydrogen peroxide liberating oxygen. These two reactions are employed in the analysis of hypochlorites. Hypochlorite is a strong oxidant, capable of oxidizing MnO_4^{2-} to MnO_4^-, IO_3^- to IO_4^-, and Fe^{3+} to FeO_4^{2-}. Its reaction with ammonia to form chloramine is the basis of the manufacture of hydrazine (qv). Ammonia, hydrazine, or amido compounds treated with excess NaOCl are converted to N_2. Anhydrous hypochlorites are oxidized to chlorates by Cl_2O. Dry hypochlorites decompose in the presence of gaseous chlorine. Hypochlorite solutions decompose under the influence of temperature (and light, depending on the concentration), ionic strength, pH, and impurities. Co, Ni, and Cu are powerful catalysts for accelerating decomposition reactions; iridium catalyzes chlorate formation. Calcium hypochlorite exposed to a stream of N_2 during dta undergoes dehydration at ca $65–70°C$; further heating results in exothermic decomposition at ca $200–210°C$. The available chlorine in a hypochlorite is a measure of the oxidizing power of its active chlorine expressed in terms of elemental chlorine.

Hypochlorite acts as a chlorinating and oxidizing agent toward organic compounds. It has numerous and useful synthetic applications, eg, preparation of carboxylic acids by the haloform oxidation and amines by the Hoffmann rearrangement. Aromatically bound methylene groups in acetyl-substituted aromatics are oxidized by NaOCl to carboxylic acids. Aliphatic oximes and primary and secondary nitro compounds are converted to geminal chloro nitro alkanes. Unsaturated aldehydes, ketones, and nitriles are epoxidized in one step in high yields via nucleophilic attack by hypochlorite ion (see Epoxidation). Hypochlorite readily chlorinates phenols to mono-, di-, and tri-substituted compounds.

Preparation of hypochlorite solutions. Sodium hypochlorite, NaOCl (liquid bleach), is usually prepared by chlorination of NaOH, and also electrolytically. Calcium hypochlorite, $Ca(OCl)_2$, (bleach liquor), is a solution of calcium hypochlorite and calcium chloride containing some dissolved lime. It is prepared by chlorination of lime slurry in a process similar to sodium hypochlorite manufacture.

Preparation of solid hypochlorites. Commercial crystalline trisodium phosphate (TSP) and chlorinated trisodium phosphate are complexes of the type $(Na_3PO_4 \cdot x H_2O)_n NaY$ where $n = 4$ to 7, $x = 11$ or 12, and Y is a monovalent anion (see Phosphoric acids and phosphates). A high purity product is obtained by batch-wise crystallization from a liquor with the proper $Na_2O-P_2O_5$ ratio containing an excess of NaOCl. Commercial material usually contains ca 3.65% chlorine.

Dibasic magnesium hypochlorite, a white, fine powdery solid, has been synthesized by addition of either a sodium or calcium hypochlorite solution to an excess of aqueous $MgCl_2$ or $Mg(NO_3)_3$ with available chlorine in the 52—58% range. It is highly stable to moisture, thermally stable, and decomposes endothermically rather than exothermically at $325°C$.

Calcium hypochlorite is similar in composition to ahydrous $Ca(OCl)_2$ except for its higher water content of about 6–12% and a slightly lower chlorine content. 65% calcium hypochlorite contains salt and water as the main diluents along with small amounts of lime, $CaCl_2$, $Ca(ClO_3)_2$, and $CaCO_3$. A high purity, completely hydrated lime of high reactivity is employed in manufacture. Commercial production is via the Olin process (previously known as the Olin-Mathieson process), in which a slurry of hydrated lime in sodium hypochlorite solution is chlorinated and then cooled to yield crystallized hypochlorite in the form of a triple salt; this is separated, mixed with chlorinated lime slurry, and crystals of neutral calcium hypochlorite dihydrate precipitate are filtered and dried. Other commercial processes include the Pennwalt process; the PPG process, based on hypochlorous acid; and various European processes such as the Imperial Chemical Industries process, the Potasse et Produits Chimiques process, and the Thann process. Other processes not based on intermediates employ cocrystallization of $Ca(OCl)_2$ and $NaCl$ and utilize classification for separation of the mixed crystals.

Bleaching powder, $Ca(OCl)_2.CaCl_2.Ca(OH)_2.2 H_2O$, is made by chlorination of slightly moist hydrated lime. Ordinary bleaching powder is a mixture of compounds, produced in a batch or continuous process, which has benefited from numerous improvements over the original chamber process.

Organic and Nonmetal Hypochlorites

Alkyl hypochlorite (esters of hypochlorous acid) are volatile liquids with irritating odors. The primary and secondary hypochlorites are very unstable, decomposing vigorously on warming and explosively when exposed to light, yielding aldehydes and ketones as major initial decomposition products. In contrast, the tertiary hypochlorites are significantly more stable. *t*-Butyl hypochlorite has been the alkyl hypochlorite of choice for experimental work owing to its stability and ease of preparation in high yield and purity. In bright sunlight, it decomposes slowly, giving chiefly acetone and methyl chloride. It is readily prepared by chlorination of an alkaline solution of the alcohol, is a convenient source of positive chlorine, and is soluble in most organic solvents. In contrast, the fluoroalkyl hypochlorites are extremely susceptible to hydrolysis, but are much more thermally stable. They readily react with CO and SO_2 to form the corresponding chloroformates and chlorosulfates in near quantitative yields, and add to olefins to give α-chloroethers. Although perfluoroacyl hypochlorites are thermally unstable and explosive, CF_3CO_2Cl and $C_3F_7CO_2Cl$ are easily handled.

Uses. Chlorine compounds are applied in disinfectants, bleaches, and sanitizers (see Disinfectants and antiseptics; Industrial antimicrobial agents). Chlorine monoxide is used as an intermediate in the manufacture of calcium hypochlorite; in sterilization for space applications (see Sterlization techniques); in preparation of chlorinated solvents; in bleaching of pulp (qv) and textiles. HOCl is used in water purification; chloramine manufacture (see Cyanuric and isocyanuric acids). Sodium hypochlorite is also used in the manufacture of hydrazine (qv), of chlorinated TSP, and in the synthesis of organic chemicals (see Bleaching agents). *t*-Butyl hypochlorite is used in preparation of α-substituted acrylic acid esters and other polymers.

J.A. Wojtowicz
Olin Corporation

J.J. Rennard and H.I. Bolker, *Chem. Rev.* **76**, 487 (1976).

A.J. Downs and C.J. Adams in *Comprehensive Inorganic Chemistry*, Pergamon Press, London, 1973.

Ullmanns Encyklopadie der technischen Chemie, Verlag Chemie, Weinheim FRG, Vol. 9, 1975.

CHLOROUS ACID, CHLORITES, AND CHLORINE DIOXIDE

Sodium chlorite ($NaClO_2$) and chlorine dioxide (ClO_2) are both manufactured in commercial quantities. However, only $NaClO_2$ is sufficiently stable to be an article of commerce whereas ClO_2, an explosive gas, must be made where it is to be consumed. Chlorous acid is an unstable compound even in dilute solutions.

Chlorine Dioxide

Physical and chemical properties. ClO_2 exists almost entirely as the monomeric free radical; mp $-59°C$; bp $11°C$; vapor pressure (at fp) 1.3

kPa (10 mm Hg). Liquid chlorine dioxide density 1.765 g/cm^3 at $-56°C$, and 1.62 g/cm^3 at $11°C$; explosive at temperatures above $-40°C$. The vapor decomposes at partial pressure of ca 10.7 kPa (80 mm Hg) and may produce mild explosions or "puffs"; it may detonate above 40 kPa (300 mm Hg); soluble in H_2O where it forms a yellow solution that is quite stable if kept cool and away from light. Various solid polyhydrates have been described. At 25°C chlorine dioxide is about 23 times more concentrated in the aqueous phase than it is in the gas phase with which it is in equilibrium.

The oxidation potential of chlorine dioxide in aqueous solution is 0.95 V at 4–7 pH. ClO_2 is reduced to chlorite in basic solution by hydrogen peroxide. Other reducing agents convert it to chlorite at neutral pH (eg, potassium iodide, sodium sulfite, sodium arsenite, and plumbous oxide; borohydride, iodide at 1 pH, and sulfurous acid in acidic solution reduce it completely to chloride ion. Oxidation of chlorine dioxide by chlorine or hypochlorous acid proceeds as follows:

$$2 ClO_2 + Cl_2 + 2 H_2O \rightleftarrows 2 ClO_3^- + 2 Cl^- + 4 H^+ \tag{1}$$

$$2 ClO_2 + HOCl + H_2O \rightleftarrows 2 ClO_3^- + Cl^- + 3 H^+ \tag{2}$$

Chlorine dioxide is oxidized readily to chlorate in neutral solutions (eq. 2) but not in acidic solution where the equilibrium (eq. 1) is shifted to the left. In the manufacture of chlorine dioxide from chlorate and chloride, the concentration of sulfuric acid should be greater than $8N$. Oxidation by ozone yields Cl_2O_6.

Only slow minimal decomposition of aqueous solutions occurs in the dark at 0°C; in sunlight, dry chlorine dioxide gas decomposes into chlorine and oxygen radicals with subsequent chain reactions yielding chlorine trioxide in addition to chlorine. Thermal decomposition of aqueous solutions at high levels of acidity and temperature is accelerated by hydrogen and chloride ion.

Chlorine dioxide exhibits a pattern of reactivity toward organic compounds considerably different from that of other oxidants commonly employed in the laboratory. For example, olefins react more rapidly with permanganate than with chlorine dioxide whereas triethylamine is nearly 10^4 times more reactive with chlorine dioxide than with permanganate. The most reactive compounds are the aliphatic tertiary amines, phenols and aromatic amines. Alcohols, carbonyl compounds, and carbohydrates react rather slowly. Phenol and chlorine dioxide have first-order reactions yielding the 2-chloro-1,4-, 2,5-dichloro-1,4-, and 2,6-dichloro-1,4-benzoquinones, 2-chlorophenol, oxalic acid, and maleic acid (see also Lignin). Hydroquinones are readily oxidized to the corresponding quinones without ring chlorination. Small amounts of chlorine dioxide inhibit chlorine oxidation of a number of typical low molecular weight carbohydrates (qv).

Manufacture. Chlorine dioxide is made either from chlorite by oxidation or from chlorate by reduction. Requirements for the chemical dictate the method of manufacture: large consumption calls for reduction of sodium chlorate [from Cl(V) to Cl(IV)] by means of rather complex processes, usually carried out in strongly acidic solutions with reducing agents such as sodium chloride, hydrochloric acid, sulfur dioxide, and methanol. In smaller applications, the less complicated oxidation of sodium chlorite through reaction with chlorine or anodic irritable oxidation is favored. The three commercial processes based on chlorate use sulfuric acid (eg, Erco's ER-3 and Hooker's single-vessel process (SVP)), hydrochloric acid (eg, Kesting process), or sulfur dioxide.

Chlorous Acid and Chlorites

Chlorous acid, $HClO_2$, has no great significance in technology, except that it is thought to be the first in a series of short-lived active species in the bleaching of textiles and in reactions of acidified solutions of chlorites. Most of the physical and chemical properties have been inferred from investigations of acidified solutions of alkali chlorites, its existence is based primarily upon spectroscopic evidence.

Crystallization of sodium chlorite from aqueous solution yields $NaClO_2.3H_2O$ below 37.4°C and anhydrous $NaClO_2$ above 37.4°C. Analytical grade $NaClO_2$ is a colorless crystalline, solid, mp 180–200°C (decomp), density of crystals 2.468 g/cm^3. Technical grade $NaClO_2$ is white flakes or powder, mp 218–219°C, density 2.19 g/cm^3.

Chemical properties. Reported dissociation constants of chlorous acid (25°C) range from 0.49×10^{-2} to 1.10×10^{-2}. Several chlorites explode or detonate when struck or heated, eg, the chlorites of the heavy metal ions Hg^+, Tl^+, Pb^{2+}, Cu^{2+}, and Ag^+, ammonium and tetramethylammonium chlorite; $NaClO_2$ is not shock sensitive unless contaminated with combustibles. The decomposition and disproportionation reactions of sodium chlorite in acidic media are significant in industrial processes, including bleaching and generation of chlorine dioxide. At least three mechanisms have been proposed for acid decomposition reactions of chlorites. Chlorate is formed rather slowly in alkaline solutions. When using $NaClO_2$ to make ClO_2, the reaction conditions are manipulated to maximize the yield of chlorine dioxide, since chlorite converted to chlorate is lost for purposes of chlorine dioxide generation; chlorine dioxide formation is maximized by increasing concentrations of chlorine and sodium chlorite. Common reducing agents are oxidized by chlorous acid and chlorites, although ClO_2^- is a rather weak oxidant in neutral and more so in alkaline solutions.

Mixing solid sodium chlorite with combustibles may result in violent explosions occurring spontaneously or upon grinding, sparking or shock. However, many organic compounds are oxidized only partially and comparatively slowly by ClO_2^- in aqueous solution. Aldehydes are readily oxidized to the corresponding carboxylic acids in weakly acidic and neutral media.

Manufacture. The manufacture of sodium chlorite requires sodium chlorate as the source of chlorine dioxide (see Chloric acid and chlorates; see also Chlorine dioxide, synthesis). Dilute chlorous acid essentially free of other compounds can be obtained by the reaction of barium chlorite with an equivalent of dilute sulfuric acid and separating the insoluble $BaSO_4$. In general, chlorine dioxide is absorbed in caustic soda containing a reducing agent, most commonly hydrogen peroxide.

Health and Safety

Chlorine dioxide is bactericidal and viricidal (see Disinfectants and antiseptics). It is toxic at relatively low levels of concentration in air or water. The NIOSH standard for maximum average concentration of ClO_2 in air is 0.1 ppm on a time-weighted basis. Sodium chlorite induces methemoglobia in warm-blooded animals. A single LD_{50} of 30 g could be lethal to a human adult. No wood, or lumber should be present where $NaClO_2$ is handled; contact with acid results in the release of toxic and explosive ClO_2.

Uses

Chlorine dioxide is used for bleaching (see Pulp), water treatment, and odor control. Sodium chlorite is consumed as a source of chlorine dioxide in applications where required volumes are relatively small; it is used as a disinfectant and oxidant.

MANFRED G. NOACK
RICHARD L. DOERR
Olin Corporation

G. Gordon, R.G. Kieffer, and D.H. Rosenblatt, *Prog. Inorg. Chem.* **15**, 208 (1972).

CHLORIC ACID AND CHLORATES

Chlorates are salts of chloric acid, $HClO_3$, which is fairly stable in cold water solution in concentrations up to ca 30%. Upon heating, chlorine and chlorine dioxide may be evolved, depending on the strength of the solution.

Chloric acid is a strong oxidizing agent. Its oxidizing properties vary somewhat with the pH and temperature of the solution and violent reactions with organic substances may occur in strong solutions. The only industrially important oxide of chlorine is chlorine dioxide, ClO_2 (see Chlorous acid, chlorites and chlorine dioxide). The hexoxide, Cl_2O_6, is a dark red liquid below its boiling point of 203°C, sp gr = 2.023 at fp (3.5°C). The least explosive of the chlorine oxides, it is a powerful oxidizing agent and can exist in either the monomeric, ClO_3, or dimeric, Cl_2O_6, form.

Sodium and Potassium Chlorate

Sodium chlorate, $NaClO_3$, forms cubic crystals, mp 248°C, decomposition point 265°C, sp gr 2.49, n_D^{20} 1.515. The crystals are slightly hygroscopic, a property which limits the use of $NaClO_3$ in some industrial applications.

Potassium chlorate, $KClO_3$, crystallizes in the monoclinic system usually as short prisms, mp 368°C; decomposition pt 400°C, sp gr 2.32, n_D^{20} 1.440, and is nonhygroscopic. Sodium chlorate is much more soluble than potassium chlorate. On thermal decomposition, both may produce the corresponding perchlorate. Mixtures of $KClO_3$ with metal oxide catalysts are used as a laboratory source of oxygen; mixtures of chlorates with organic materials have been employed as explosives.

Manufacture. Chlorates may be prepared by the chlorination of a hypochlorite solution. Most of the chlorate is manufactured by the electrolysis of sodium chloride solution in electrochemical cells without diaphragms (see Electrochemical processing).

In general, the sodium chlorate production system consists of cells that generate active chlorine, and a holding volume usually in a closed loop, which acts as a reactor for further conversion of the product. The cells range from small monopolar cells with an annual capacity of 22–90 metric tons per year of sodium chlorate to large multipolar or bipolar cell assemblies.

A number of changes affect the design of the sodium chlorate cell: development of noble metal-coated titanium anodes for the chlorine and chlorate industry (see Alkali and chlorine products; Metal anodes); new cell design to take advantage of the coated titanium anodes; OSHA requirements for improved working conditions in manufacturing plants; EPA requirements of less air pollution; increased cost and reduced availability of graphite; and increased cost of electric power. The new cells were designed to take advantage of the characteristics of the new anodes, and practically all the new cells have connections for the recovery of evolved hydrogen, necessitated by air-pollution controls and increased costs of fuels (see Hydrogen).

Other Chlorates

Barium chlorate, $Ba(ClO_3)_2 \cdot H_2O$, colorless, monoclinic crystals, mp 120°C ($-H_2O$), sp gr 3.18, is prepared by the reaction of barium chloride and sodium chlorate in solution.

Lithium chlorate, $LiClO_3$, rhombic needles, mp 124–129°C, decomposes on heating to 270°C, is very hygroscopic and one of the most soluble salts known.

Health and Safety

Chlorine hexoxide, like the other oxides of chlorine, is very dangerous at high concentrations and no safe procedures for handling such concentrations are known. Chlorates are strong oxidizing agents and extreme care must be taken to ensure that they do not contact heat, organic materials, phosphorus, ammonium compounds, sulfur compounds, oils, greases or waxes, powdered metals, paint, metal salts (especially copper), and solvents, eg, a mixture of sodium and potassium chlorate with any combustible organic or inorganic matter should be regarded as dangerous.

Uses

Sodium chlorates are mainly used (78%) in the conversion to chlorine dioxide bleach by the pulp and paper industry (see Pulp). It is also used as an intermediate in the production of other chlorates and perchlorates (see Perchloric acid and perchlorates under Chlorine oxygen acids and salts); as a herbicide (qv); as a defoliant; and as an oxidizing agent. Potassium chlorate is used mainly in the manufacture of matches (see Pyrotechnics) and in pharmaceutical preparations.

T.W. CLAPPER
Kerr-McGee Corporation

C.F. Goodeve and F.A. Todd, *Nature (London)* **132**, 514 (1933).

J.C. Schmacher, *ACS Monographs*, No. 146, 1960, p. 77.

S. Ardizzone and co-workers *J. Electrochem. Soc.* **129**(8) 1689, (Aug. 1982).

PERCHLORIC ACID AND PERCHLORATES

Perchloric acid, $HClO_4$, and perchlorate compounds form a large group of relatively stable compounds of chlorine. All of these compounds are essentially made directly or indirectly by electrochemical oxidation of chlorine compounds. Their most useful property is their oxidizing capability: safe, reproducible oxidations using perchlorates can be achieved under controlled conditions. Aqueous perchlorate solutions exhibit little oxidizing power when they are diluted and cold, but are powerful oxidizers when hot and concentrated.

Chlorine Heptoxide (Dichlorine Heptoxide)

The anhydride of perchloric acid, Cl_2O_7, is a colorless, volatile, oily liquid; d 1.82 g/cm^3 at 20°C, mp -91.5°C, bp 85°C at 101 kPa (1 atm) and 0°C at 3.2 kPa (23.7 mm Hg). It explodes violently upon concussion or upon contact with flame or iodine; it does not react with paper, sulfur, or wood when cold. It is prepared by dehydration of perchloric acid with phosphorous pentoxide or by electrolysis of 55–73% $HClO_4$ at a cold anode (0 to -55°C). It decomposes to chlorine and oxygen at low pressures and at temperatures of 100–120°C.

Perchloric Acid

$HClO_4$ is a colorless, explosive and shock-sensitive, strong mineral acid; hygroscopic; mp 112°C; bp 110°C (explosive); d 1.768 g/cm^3. Commercial perchloric acid is an aqueous solution containing usually 60–62% or 70–72% (bp 203°C) $HClO_4$; above this concentration, solutions are hygroscopic and unstable. Reactivity becomes increasingly influenced by the oxidizing properties of the acid as the concentration and temperature increases. Thermal decomposition yields oxygen and Cl_2, HCl, and Cl_2O, with chlorine monoxide, ClO_3, and ClO_4 as decomposition intermediates.

Preparation. Commercial preparation has largely been by the chemical reactions of sodium perchlorate and excess concentrated hydrochloric acid. The sodium chloride formed is removed to give a filtrate containing about 32% $HClO_4$.

An attractive method for direct electrochemical production by electrolysis of chlorine in cold (-5°C) dilute perchloric acid (40%) has been developed. It is also produced by direct electrolytic oxidation of cold (18°C) very dilute (0.5 N) hydrochloric acid using a platinum anode and a silver or copper cathode. Anhydrous perchloric acid is prepared by vacuum distillation from a mixture of 72% $HClO_4$ and fuming sulfuric acid at pressures of ≤ 0.13 kPa (≤ 1 mm Hg).

Ammonium Perchlorate

Ammonium pechlorate, NH_4ClO_4, is highly soluble in ammonia; soluble in H_2O to the extent of 20.02 g/100 g soln (25°C); dissociation constant in ammonia 5.4×10^{-3}; d 1.95 g/cm^3 (20°C). It undergoes reversible crystallographic transition from low temperature orthorhombic to cubic structure at 240°C. Low temperature decomposition (210°C) products are NH_3, $HClO_4$, N_2, O_2, H_2O, NO_2, Cl_2, and ClO_2. Decomposition and sublimation of ammonium perchlorate is suppressed under an atmosphere of ammonia. At 345–350°C a fast reaction begins with a subsequent explosion. Effective inhibitors are NH_4Cl, NH_3, NH_4F, CdF_2, ZnF_2, and $PbCl_2$; urea and dicyandiamide are effective at elevated temperatures.

Perchlorates

Compounds have been identified where perchlorate has been combined with one or more elements of every group in the periodic table except Group O (the inert gases).

Group IA (alkali) perchlorates. These are white or colorless, with the solubility of salts in water decreasing in the order of Na > Li > NH_4 > K > Rb > Cs. The higher solubility of $NaClO_4$ makes it useful as an intermediate for production of all other perchlorates by double metathesis reactions and controlled crystallization. All perchlorates in this category except the lithium salt undergo crystalline-phase transformations from orthorhombic to cubic on heating; only lithium perchlorate appears to have a definite melting point without decomposition in a narrow temperature range around the mp. Representative compounds include potassium perchlorate, $KClO_4$, mp 580–610°C, d 2.5298 g/cm^3; lithium perchlorate, $LiClO_4$, mp 236–247°C, bp 470°C, d 2.43 g/cm^3, highly sol in organic solvents. Commercial ammonium perchlorate is manufactured by the double-exchange reaction of sodium perchlorate and ammonium chloride.

Group IB includes perchlorates of both copper (I), $CuClO_4$, and copper(II), $Cu(ClO_4)_2$. The divalent copper perchlorates form a series of hydrates and a number of complexes with ammonia, ammonia and water, pyridine, and with organic derivatives of these compounds. Copper perchlorate is an effective burning-rate accelerator for solid propellants; silver perchlorate is used to prepare other perchlorates.

Group IIA (alkaline earth) perchlorates form many hydrates, ammoniates, and other solvate compounds. The salts may be prepared in the anhydrous state by heating ammonium perchlorate with the corresponding hydroxide or carbonate; the more basic metals react more rapidly and at lower temperatures. The hydrates may be prepared readily by reaction of the metal oxide, carbonate, or hydroxide with perchloric acid. The strong affinity of anhydrous magnesium perchlorate for water gives it great value as a dehydrating agent; the other anhydrous alkaline-earth perchlorates have absorption properties similar to those of $MgClO_4 \cdot BaClO_4$, which is used as a desiccant (see Drying agents).

Group IIB perchlorates include those of zinc, cadmium, and mercury

Group IIIA includes compounds of each of the elements, eg, boron perchlorates (which occur as double salts with alkali perchlorates); aluminum perchlorate (which forms a series of hydrates having 3, 6, 8, or 15 moles of water per mole of $Al(ClO_4)_3$).

Group IIIB and inner transition metals include yttrium and lanthanum, and the trivalent perchlorate compounds of the lanthanide series of inner transition-metal compounds; also tetravalent cerium perchlorate and uranium perchlorate (see Actinides).

Group IVA includes the large number of organic carbon-containing perchlorates such as amine, diazonium, and oxonium perchlorates, and perchlorate esters. Many decompose violently when heated, contacted by other agents, or subjected to mechanical shock.

Group IVB perchlorates include titanium tetraperchlorate.

Group VA perchlorates are particularly important, eg, the nitrogen perchlorates have been used as oxidizers in rocket propellant systems.

Group VB includes vanadyl perchlorate, $VO(ClO_4)_3$.

Group VIA includes a compound containing sulfur, diperchlorate sulfate, $SO_4(ClO_4)_2$, a strong oxidizing agent.

Group VIB includes both di- and trivalent chromium perchlorate.

Group VIIA includes fluorine perchlorate, $FClO_4$, and perchloryl fluoride, $FClO_3$ (see Fluorine compounds, inorganic).

Group VIIB includes both di- and trivalent manganese perchlorate.

Group VIII, the transition elements, include perchlorate compounds of Fe, Co, Ni, Rh, and Pd, all of which are usually colored. Double salts have been reported as well.

Manufacture of Perchlorates

The commercially important method of production is the electrochemical oxidation of lower valence chlorine-containing compounds. The reaction to form perchlorate at the anode of an electrolytic cell can be written as:

$$ClO_3^- + H_2O \rightarrow ClO_4^- + 2\,H^+ + 2\,e$$

Some oxygen is also produced at the anode and is the main cause of the inefficiency of the process. Hydrogen is produced from reduction of the water at the cathode.

Most metal perchlorate compounds are made by metathesis with sodium perchlorate. Sodium perchlorate is made by electrooxidation of sodium chlorate (see Chloric acid and chlorates). Legendre has outlined the pretreatment of electrolyte feed to cells for various processes and the aftertreatment of the electrolyzed cell product.

Health and Safety

All perchlorate compounds, owing to their high oxygen content, can undergo vigorous and explosive reactions when oxidizable substances are present. There is wide variance in associated hazards: anhydrous perchloric acid should be prepared only in small quantities; perchlorates in which all elements in the compounds are at their highest valence, eg, the alkali and alkaline-earth perchlorates, are very stable. The hazards of these compounds are associated with mixtures containing oxidizable substances. Inorganic perchlorates in which the cation contains elements having a variable valence decompose at lower temperatures and can be strongly exothermic and explosive. Organic perchlorates have both oxidizing and reducing moieties and can easily explode.

Inhalation and ingestion should be avoided. Perchlorates influence the iodine balance of the normal human thyroid gland.

Uses

Perchlorates are used principally as oxidizers in rocket and missile propellants (see Explosives and propellants). Perchloric acid is used in analytical chemistry or for research. Perchlorates are used in explosive and pyrotechnic formulations (see Pyrotechnics).

R.C. RHEES
Pacific Engineering and Production Company

J.C. Schumacher, *Perchlorates, Their Manufacture & Uses*, ACS Monograph 146, Reinhold Publishing Corp., New York, 1960.
A. Legendre, *Chem. Ing. Tech.* **34**, 379 (1962).

CHLORITES.   See Chlorine oxygen acids and salts.

CHLORITES (MINERALS).   See Silica.

CHLOROCARBONS AND CHLOROHYDROCARBONS

SURVEY

Chlorination of a variety of hydrocarbon feedstocks produces many valuable and useful chlorinated solvents, intermediates, and chemical products (see individual articles). The chlorinated derivatives provide an important means of upgrading the value of industrial chlorine.

General Properties of Chlorinated Hydrocarbons

Progressive chlorination of a hydrocarbon molecule yields a sequence of liquids, solids, or both, of increasing nonflammability, density, and viscosity, as well as improved solubility for a large number of inorganic and organic materials. Specific heat, dielectric constant, and water solubility of a solvent exhibit a progressive decrease with increasing chlorine content.

All chlorinated hydrocarbons are susceptible at elevated temperatures to pyrolysis breakdown which liberates hydrogen chloride. Olefinic chlorinated derivatives are oxidized in the presence of ultraviolet light to give hydrogen chloride, phosgene, and chlorinated acetyl chlorides (acid derivatives). Saturated aliphatic chlorine derivatives are usually quite stable to oxidation.

Although many chlorinated hydrocarbons attack aluminum, proprietary organic inhibitors permit commercial use of reactive solvents such as 1,1,1-trichloroethane and trichloroethylene in both cold and hot cleaning of aluminum. Inhibitors that may be classified as antioxidants (qv), acid acceptors, and metal stabilizers are added to minimize stress on the stability of the solvent in commercial use.

Types of Aliphatic Chlorination Reactions

Substitution chlorination is an important commercial process, eg, chlorination of methane yields all four possible chlorinated derivates by a radical-chain mechanism. Addition chlorination, as in the chlorination of olefins, eg, ethylene, can be carried out in a catalytic or vapor-phase or liquid-phase process. Hydrochlorination is the addition of hydrogen chloride to simple olefins in the absence of peroxides via an electrophilic mechanism in accord with Markovnikov's rule. Dehydrochlorination is used in the commercial production of vinyl chloride through the thermal dehydrochlorination of 1,2-dichloroethane via a radical-chain mechanism. Chlorinolysis, with excess chlorine in high temperature chlorinations, cleaves the C—C bonds of a hydrocarbon to give chlorinated derivatives of shorter chain length. Oxychlorination is an important process for the production of 1,2-dichloroethane. Thermal cracking is used in the production of trichloroethylene and tetrachloroethylene.

Chlorinated Aromatic Derivatives

Aromatic compounds may be chlorinated in the presence of a catalyst such as iron, ferric chloride, or other Lewis acids. The halogenation reaction involves the electrophilic displacement of the aromatic hydrogen by halogen. Introduction of a second chlorine atom into the monochloro aromatic structure leads to ortho and para substitution, the presence of a Lewis acid favoring polarization of the chlorine molecule.

Health and Safety

The greatest industrial hazard is excessive vapor inhalation which results in central nervous system depression. The degree of toxicity of a solvent cannot be predicted from its chlorine content or from the structure of the chlorinated derivatives.

WESLEY L. ARCHER
Dow Chemical U.S.A.

W.L. Archer and V.L. Stevens, *Ind. Eng. Chem. Prod. Res. Dev.* **16**, 319 (Dec. 1977).
W.L. Archer, *Ind. Eng. Chem. Prod. Res. Dev.* **21**, 670 (Dec. 1982).

METHYL CHLORIDE

Methyl chloride (monochloromethane), CH_3Cl, at ordinary temperatures and pressures is a colorless gas with an ethereal, nonirritating odor and sweet taste. It is miscible with the principal organic solvents, and slightly soluble in water. The dry liquid is stable and noncorrosive. Gaseous methyl chloride is moderately flammable, and owing to its high vapor pressure can form explosive mixtures with air.

Physical and Chemical Properties

Methyl chloride is the simplest chlorinated hydrocarbon; mp $-97.7\,°C$; bp $-23.73°C$ at 101.3 kPa (1 atm); sp gr of liq ($-23.7°C$) 0.920; n_D of liq ($-23.7°C$) 1.3712. Dry methyl chloride in the absence of air does not decompose at an appreciable rate at temperatures approaching 400°C, even in contact with many metals. Thermal dissociation is virtually complete at 1400°C. Oxidative breakdown of gas requires temperatures of several hundred °C. Methyl chloride is decomposed by an open flame to give hydrogen chloride and carbon dioxide, with possible formation of small amounts of carbon monoxide and phosgene. In the liquid state contact with water or moisture will cause slow hydrolysis to methanol and hydrogen chloride even at ambient temperatures but accelerated by increasing temperature. It is readily hydrolyzed by boiling dilute sodium hydroxide solution. Dry methyl chloride is unreactive with all common metals except the alkali and alkaline-earth metals, magnesium, zinc, and aluminum. In dry ether solution, it reacts with sodium to yield ethane by Wurtz synthesis; condensation with higher chloroparaffins gives propane, butane, etc. It reacts with magnesium to form the Grignard reagent, methylmagnesium chloride, which has been applied to the synthesis of alcohols and to the preparation of intermediates for the formation of silicone polymers. Hoffman reaction occurs with ammonia in alcoholic solution or in the vapor phase where it forms a mixture of hydrochlorides of methylamines, di- and trimethylamine, and tetramethylammonium chloride. It forms quaternary derivatives with amines.

Manufacture

The principal processes for industrial production are: (1) a liquid phase reaction of hydrogen chloride and methanol at 0°C to yield methyl chloride as the sole product, (2) a similar reaction in the gas phase at elevated temperatures over a catalyst, chlorination of methane thermally (in a chain reaction carried out in the absence of light or catalyst), photochemically, or catalytically in a multiple-product process with the other chloromethanes.

Health and Safety

This is a moderately toxic chlorinated hydrocarbon. There is no adequate warning of the presence of harmful concentrations and there is a delay in the development of symptoms. Prolonged exposure to high concentrations of the vapor can produce severe toxic effects; OSHA 1981 Standard places a limit of 50 ppm for an 8-h time-weighted average.

Uses

Methyl chloride is used mainly in the manufacture of silicones, tetramethyllead (TML, antiknock agent), synthetic rubber, and methyl cellulose, and as a general methylating agent; it has secondary uses as a refrigerant and extractant.

J.M. STEELE
R.C. AHLSTROM, JR.
Dow Chemical U.S.A.

N.B. Vargaftik, *Tables on the Thermophysical Properties of Liquids and Gases*, John Wiley & Sons, Inc., New York, 1975, p. 362.

Brit. Pat. 1,230,743 (May 5, 1971) (to the Dow Chemical Company).

N.I. Sax, *Dangerous Properties of Industrial Materials*, 4th ed., Van Nostrand Reinhold Company, New York, 1975, p. 915.

METHYLENE CHLORIDE

Methylene chloride (dichloromethane, methylene dichloride), CH_2Cl_2, is one of the more stable of the chlorinated hydrocarbon solvents. It is a clear, colorless, volatile liquid with a mild ethereal odor; bp 39.8°C at 101.3 kPa; fp −96.7°C; sp gr at 20°C 1.320; slightly soluble in water (13.2 g/kg at 20°C). It is completely miscible with other grades of chlorinated solvents, diethyl ether, and ethyl alcohol in all proportions. It dissolves most other common organic solvents and is an excellent solvent for many resins, waxes, and fats. It has no flash point, but as little as 10 vol% acetone or methyl alcohol is capable of producing one. Initial degradation temperature is 120°C in dry air, decreasing as the moisture content increases. Decomposition produces mainly HCl with trace amounts of phosgene and may be inhibited by adding small quantities of phenolic compounds, amines, or a mixture of nitromethane and 1,4 dioxane; addition of epoxides can inhibit aluminum reactions catalyzed by iron. On prolonged contact with water, it hydrolyzes very slowly forming HCl as the primary product; on prolonged heating with water, it yields formaldehyde and hydrochloric acid (140–170°C), or formic acid, methyl chloride, methanol, hydrochloric acid, and some carbon monoxide (at 180°C).

Dry methylene chloride does not react with the common metals under normal conditions but reactions may be initiated by small amounts of other halogenated or aromatic solvents, eg, aluminum and iron catalyzes this reaction and can be significant in the handling and storage and in the formulation of products (see Friedel-Crafts reactions). In the gas phase, it reacts with nitrogen dioxide at 270°C to yield a mixture of carbon monoxide, nitric oxide, and hydrogen chloride. It is easily reduced to methyl chloride (qv) and methane by alkali-metal ammonium compounds in liquid ammonia. In the presence of chlorination catalysts, it may form chloroform (qv) and carbon tetrachloride (qv).

Manufacture

The predominant method employs a first step as the vapor-phase reaction of hydrogen chloride and methanol with the aid of a catalyst at ca 180°C to give methyl chloride; this is subsequently passed through a converter, fed to reactors and combined with chlorine to produce methylene chloride, chloroform, and carbon tetrachloride (see also Methanol). Methylene chloride is also produced industrially by direct reaction of excess methane (natural gas) with chlorine at high temperature (ca 485–510°C) producing methyl chloride, chloroform, and carbon tetrachloride as coproducts.

Health and Safety

Methylene chloride is one of the least toxic chlorinated methanes. Vapors have an anesthetic action at high levels. There is a wealth of toxicity and environmental data on this solvent, and suppliers should be contacted for more details.

Uses

It is used as an extraction solvent in the decaffeination of coffee; in paint strippers; solvent carrier (see Dye carriers); in chemical processing (see Photoreactive polymers); as blowing agent (see Urethane polymers) and as a vapor-pressure depressant in aerosol formulations.

T. ANTHONY
Dow Chemical U.S.A.

W.L. Archer, *Ind. Eng. Chem. Prod. Res. Dev.* **18**, 131 (1979).

CHLOROFORM

Chloroform (trichloromethane, methenyl trichloride), $CHCl_3$, at normal temperature and pressure is a heavy, water-white, volatile liquid with a pleasant, ethereal, nonirritating odor.

Physical and Chemical Properties

Mp −63.2°C; bp 61.3°C (101 MPa); sp gr (25/4°C) 1.481; slightly soluble in H_2O; miscible with principal organic solvents; dissolves alkaloids, cellulose acetate and benzoate, ethyl cellulose, essential oils, fats, gutta-percha, halogens, methyl methacrylate, mineral oils, many resins, rubbers, tars, vegetable oils, and a wide range of organic compounds. It decomposes at ordinary temperatures in sunlight in the absence of air, and in the dark in the presence of air. Small quantities of ethyl alcohol are added to act as a stabilizer in the storage of most chloroform.

Chloroform forms a series of binary azeotropes; at 25°C it dissolves 3.59 times its volume of carbon dioxide. At 290°C, chloroform vapor is not attacked by oxygen, and it resists thermal decomposition. Pyrolysis of chloroform vapor occurs above 450°C, producing tetrachloroethylene, hydrogen chloride and a number of chlorohydrocarbons in minor amounts. Oxidation with powerful oxidizing agents, eg, chromic acid, results in formation of phosgene and liberation of chlorine. Chloroform can be reduced to methane with zinc dust and aqueous alcohol.

It reacts readily with halogens or halogenating agents; with aluminum bromide to form bromoform, $CHBr_3$ (see Bromine compounds); with aniline and other aromatic and aliphatic primary amines in alcoholic alkaline solution it forms isonitriles (isocyanides, carbylamines); with phenols in alkaline solution it gives hydroxy-aromatic aldehydes (Reimer-Tiemann reaction) (eg, phenol gives chiefly *p*-hydroxybenzaldehyde and some salicyaldehyde). Chloroform cannot be directly fluorinated with elementary fluorine; fluoroform, CHF_3, is produced by reaction with hydrogen fluoride in the presence of a metallic fluoride catalyst. Chloroform combines with the inner anhydride of salicyclic acid to form a well-defined crystalline double compound which readily liberates chloroform when heated; this reaction has been used to produce very pure chloroform (see Salicylic acid).

Manufacture

The most common method of producing chloroform is methane or methyl chloride chlorination (see Methylene chloride). There is a variety of processes. For example, methane, chlorine, and carbon tetrachloride are fed to a fluidized bed at velocities of 6.1–27 cm/s and a temperature of 650–775°C; when the ratios of chlorine and carbon tetrachloride to methane in the feed mixture are between 1.75–2.60 : 1 and 0.75–1.25 : 1, respectively, chloroform is the predominant product. Noncommercial processes include limited reduction of carbon tetrachloride to chloroform, effected by reaction with hydrogen, methane, zinc dust, or ethyl alcohol; decomposition of pentachloroethane with aluminum chloride; electrolysis of alkali-metal or alkaline-earth metal chlorides in aqueous alcohol solution; or monoxide and hydrogen chloride under pressure at ca 400°C in the presence of catalytic oxides.

Health and Safety

The principal industrial hazard is damage to liver and kidneys resulting from inhalation or ingestion. Toxic effects resemble those of carbon tetrachloride. In 1976, the FDA banned chloroform for use in the US as an anesthetic and for other consumer uses. The OSHA-recommended maximum time-weighted average concentration in the workroom atmosphere for an 8-h daily exposure is 50 ppm, but the American Conference of Governmental Industrial Hygienists recommends exposure limits be reduced to 10 ppm per 8 h workday.

Uses

Chloroform is largely used for the manufacture of a fluorocarbon (chlorodifluoromethane). It is used as a refrigerant in closed air-conditioning systems and as a chemical intermediate in the production of thermally stable polymers. Chloroform has minor applications as a solvent to dissolve fats, greases, gums, oils and waxes, and as a chemical intermediate for dye and pesticide preparation. Once widely used as an anesthetic, it has been replaced by safer and more versatile materials with fewer side effects (see Anesthetics).

H.D. DeShon
Dow Chemical U.S.A.

G.D. Clayton and F.E. Clayton, eds. *Patty's Industrial Hygiene and Toxicology*, Vol. 23, Wiley-Interscience, New York, 1983.

CARBON TETRACHLORIDE

Carbon tetrachloride (tetrachloromethane), CCl_4, is one of the most toxic of the chloromethanes and the most unstable upon thermal oxidation. At ordinary temperature and pressure, it is a heavy, colorless, and nonflammable liquid with a characteristic nonirritant odor. CCl_4 contains 92 wt% chlorine.

Physical and Chemical Properties

Mp $-22.92°C$ (101.3 kPa or 1 atm); bp $76.72°C$ (101.3 kPa); sp gr 1.595 (20/4°C); soluble to the extent of 0.08 g/100 g H_2O (25°C); miscible with many common organic liquids, and a powerful solvent for asphalt, benzyl resin (polymerized benzyl chloride), bitumens, chlorinated rubber, ethyl cellulose, fats, gums, resins, and waxes. Carbon tetrachloride forms a large number of binary and several ternary azeotropic mixtures, eg, n-butyl alcohol (bp 77°C), acetic acid (bp 77°C), ethyl nitrate (bp 75°C), ethyl alcohol (bp 65°C), nitromethane (bp 71°C), ethylene dichloride (bp 76°C), and acetone (bp 56°C).

Many polymer films (eg, polyethylene, polyacrylonitrile) are permeable to CCl_4 vapor which also can affect the explosion limits of several gaseous mixtures, eg, air–hydrogen and air–methane. At ca 400°C, thermal decomposition occurs slowly; at 900–1300°C, dissociation is extensive, forming perchloroethylene, and hexachloroethane, and liberating some chlorine. The vapor decomposes to give toxic products such as phosgene when in contact with a flame or a very hot surface. Carbon tetrachloride is the chloromethane least resistant to oxidative breakdown. Dry carbon tetrachloride does not react with most commonly used construction metals, eg, iron and nickel; it reacts very slowly with lead; it is sometimes explosive with aluminum and its alloys. In contact with metallic sodium or potassium, shock may produce an explosion. Carbon tetrachloride can be reduced to chloroform using zinc and acid, and has been used in the dehydrogenation of chloroethanes in the presence of a catalyst. It forms telomers with ethylene and certain vinyl derivatives under pressure. It reacts with benzene to give triphenylchloromethane. At elevated temperatures, CCl_4 is attacked by silica gel forming a silicon oxychloride.

Manufacture

Carbon tetrachloride is produced mainly by chlorinolysis, the chlorination of hydrocarbons, eg, ethylene. The quantity of carbon tetrachloride produced depends on the nature of the hydrocarbon starting material and the conditions of chlorination, eg, the Hüls process uses a 5:1

mixture (by vol) of chlorine and methane at 650°C with perchloroethylene as the principal by-product; when ethylene is substituted for methane, perchloroethylene becomes the main by-product. An alternative process of methane chlorination brings the reactants into contact with a fluid catalyst bed at ca 300°C with a Fuller's earth catalyst to yield a crude product of approximately equal quantities of carbon tetrachloride and perchloroethylene. A number of other processes have been described for producing carbon tetrachloride by chlorination of various hydrocarbons other than methane.

Other processes include the pyrolysis of hexachloroethane at 300–420°C to yield carbon tetrachloride and perchloroethylene.

Crude carbon tetrachloride is customarily purified by neutralization and drying followed by distillation.

Health and Safety

Carbon tetrachloride is one of the most harmful of the common solvents; it is toxic by inhalation of its vapor and oral intake of the liquid. It has an anesthetic effect and may produce organic injury either on a single prolonged exposure, or a series of short-duration exposures. The OSHA maximum safe concentration is a time-weighted average of 10 ppm by vol of CCl_4 for daily 8-h exposure. The ACGIH recommends daily 8-h exposure limits of 5 ppm (maximum).

Uses

Carbon tetrachloride is a raw material for the manufacture of chlorofluorocarbon refrigerants for air conditioning and refrigeration equipment. It is also a carrier of chlorine for chemical process reactions, in which the use of carbon tetrachloride permits close control of the process temperature and eliminates the need for an additional ingredient to serve as a solvent.

It should be noted that carbon tetrachloride is restricted to industrial uses. Use of the compound in products intended for home use is prohibited by two Federal agencies—the FDA and the Consumer Product Safety Commission (CPSC).

H.D. DeShon
Dow Chemical, U.S.A.

W.L. Archer and M.K. Harter, *Corrosion*, **34**(5), 159 (1978).

H.B. Elkins, *The Chemistry of Industrial Toxicology*, John Wiley & Sons, Inc., New York, 1950, p. 132.

ETHYL CHLORIDE

Ethyl chloride (chloroethane), C_2H_5Cl, is a colorless, mobile liquid with a nonirritant ethereal odor and a pleasant taste. It is flammable, burning with a green-edged flame, and producing hydrogen chloride fumes.

Physical and Chemical Properties

Mp 138.3°C; bp 12.4°C; sp gr 0.8970 (20/4°C). At 0°C, 100 g ethyl chloride dissolves in 0.07 g water; 100 g water dissolves in 0.447 g ethyl chloride, increasing sharply with temperature to 0.36 g/100 g H_2O at 50°C. It is miscible with methyl and ethyl alcohols, diethyl ether, ethyl acetate, methylene chloride, chloroform, carbon tetrachloride, and benzene. Ethyl chloride dissolves many organic substances, sulfur, and phosphorus. Three binary azeotropes have been reported. Its thermal stability is similar to that of methyl chloride. Decomposition to ethylene and hydrogen chloride is nearly complete at 500–600°C on pumice or about 300°C with chlorides of nickel, cobalt, iron, and lead; several inorganic salts, metals, and oxides also catalyze the cracking of ethyl chloride. Dehydrochlorination using potash yields ethylene. Ethyl chloride forms regular crystals of a hydrate with water at 0°C. Dry, it can be used in contact with most common metals in the absence of air up to 200°C. Its oxidation and hydrolysis are slow at ordinary temperatures. It yields ethyl alcohol, acetaldehyde, and some ethylene in the presence of steam with various catalysts, eg, titanium dioxide. Chlorination under light forms both ethylidene and ethylene chlorides; chlorination in the presence of antimony pentachloride at 100°C produces ethylene chloride. Photochemical bromination yields a series of bromochloroethanes. Reaction with an alcoholic solution of ammonia yields ethylamine, diethylamine, triethylamine, and tetraethylammonium chloride (see Amines, lower aliphatic).

Manufacture

Ethyl chloride is commercially produced by: hydrochlorination of ethylene, eg, under 0.1–0.3 MPa (1–3 atm) pressure at normal temperature; chlorination of ethane either thermally (eg, at 250–500°C and 207 kPa (30 psi pressure)), catalytically (using metal chlorides or crystalline carbon (graphite)); and reaction of hydrochloric acid with ethanol using a zinc chloride catalyst at 100–140°C. Other processes for production include a proposed one based on diethyl sulfate.

Health and Safety

Ethyl chloride, having a TLV of 1000 ppm, is one of the safest chlorinated hydrocarbons. It is less toxic than many common hydrocarbons such as hexane (TLV 100 ppm), toluene (100), xylene (100), and methyl ethyl ketone (200).

Uses

Ethyl chloride is used in the manufacture of tetraethyllead (TEL) (see Lead compounds); ethyl chloride is used as an ethylating agent for, eg, ethyl cellulose (see Alkylation; Friedel-Crafts reactions).

THOMAS E. MORRIS
WILLIAM D. TASTO
Dow Chemical U.S.A.

Chemical Profiles, Ethyl Chloride, Schnell Publishing Co., New York, Apr. 1977.

G.D. Clayton and F.E. Clayton, eds., *Patty's Industrial Hygiene and Toxicology*, Vol. 23, Wiley-Interscience, New York, 1983.

OTHER CHLOROETHANES

Table 1 lists the physical properties, uses, and safety data for the chloroethanes discussed below.

1,1-Dichloroethane

Ethylidene chloride, ethylidene dichloride, is a colorless liquid with an ethereal odor.

Chemical properties. The compound decomposes at 356–453°C to give vinyl chloride and hydrogen chloride. Dehydrochlorination proceeds on activated alumina, on magnesium sulfate or potassium carbonate, and readily in the presence of anhydrous aluminum chloride. Refluxing with an aluminum coupon leads to complete metal consumption and formation of aluminum chloride and 2,3-dichlorobutane, $CH_3CHClCHClCH_3$. Reaction with iron and zinc is minimal under dry conditions, whereas the wet solvent (7% water phase) increases corrosion rates. Reaction with an amine gives low yields of chloride ions and the dimer 2,3-dichlorobutane.

Manufacture. 1,1-Dichloroethane is produced from hydrogen chloride and vinyl chloride at 20–55°C in the presence of an aluminum, ferric, or zinc chloride catalyst. 1,1-Dichloroethane is usually an intermediate in the production of vinyl chloride and of 1,1,1-trichloroethane by photochlorination.

1,2-Dichloroethane

Ethylene chloride, ethylene dichloride, s-dichloroethane, is a colorless, volatile liquid with a pleasant odor.

Chemical properties. The compound is stable at ordinary temperatures. Pyrolysis in the range of 350–515°C gives vinyl chloride, hydrogen chloride, and traces of acetylene, and decomposition is accelerated by chlorine, bromine, bromotrichloromethane, or carbon tetrachloride. Hydrolysis gives ethylene glycol. Atmospheric oxidation at room or reflux temperatures generates some hydrogen chloride and results in solvent discoloration. Dry refluxing 1,2-dichloroethane completely consumes an aluminum coupon but barely attacks iron and zinc (2). Nucleophilic substitution occurs, eg, the reaction with sodium cyanide in methanol at 50°C gives 3-chloropropionitrile slowly; using dimethyl sulfoxide as the solvent greatly enhances the substitution. Ammonolysis with 50% aqueous ammonia at 100°C is a commercial process for ethylenediamine (see Diamines).

Manufacture. 1-2-Dichloroethane is produced by: catalytic vapor- or liquid-phase (using catalysts, eg, ferric oxide) chlorination at 40–50°C, of ethylene; or by oxychlorination of ethylene, a process usually incorporated into an integrated vinyl chloride plant.

1,1,1-Trichloroethane

Methyl chloroform, CH_3CCl_3, is a colorless, nonflammable liquid with a characteristic ethereal odor. It is among the least toxic of the industrial chlorinated solvents.

Chemical properties. Pyrolysis at 325–425°C yields 1,1-dichloroethylene and hydrogen chloride. Under normal conditions, hydrolysis at 75–160°C under pressure proceeds slowly and yields acetyl chloride, acetic acid, or acetic anhydride. 1,1,1-Trichloroethane is stable to oxidation when compared to olefinic chlorinated solvents. Refluxing uninhibited 1,1,1-trichloroethane reacts vigorously with aluminum to give aluminum chloride, 2,2,3,3-tetrachlorobutane, 1,1-dichloroethylene, and hydrogen chloride, a reaction that can be prevented by adequate metal inhibitors (see Corrosion and corrosion inhibitors). Dehydrochlorination occurs over activated alumina or anhydrous aluminum chloride at 0°C to give rapid hydrogen chloride evolution and formation of polymers; aluminum fluoride is the most active metal fluoride catalyst.

Manufacture. Vinyl chloride is hydrochlorinated to 1,1-dichloroethane which is then thermally or photochemically chlorinated to give 1,1,1-trichloroethane. In another process, hydrogen chloride is added to 1,1-dichloroethylene in the presence of $FeCl_3$ catalyst.

1,1,2-Trichloroethane

Vinyl trichloride, $CH_2ClCHCl_2$, is a colorless, nonflammable liquid with a pleasant odor.

Chemical properties. 1,1,2-Trichloroethane is much slower to hydrolyze, and much more prone to oxidative degradation than 1,1,1-trichloro-

Table 1. Physical Properties of Chloroethanes

Compound	TLV, ppm[a]	Mp °C	Bp °C	d g/mL 20°C	sol, g/100 g H₂O (20°C)[b]
1,1-dichloroethane	200	−96.7	57.3	1.1747	0.55; miscible with organic solvents
1,2-dichloroethane	10; highly toxic	−35.3	83.7	1.2529	0.869; miscible with chlorinated solvents; sol in organic solvents
1,1,1-trichloroethane	350; least toxic	−33.0	74.0	1.3249	0.0095; miscible with chlorinated solvents; sol in common organic solvents
1,1,2-trichloroethane	10; highly toxic	−37.0	113.7	1.4432	0.45; miscible with chlorinated solvents; sol in common organic solvents
1,1,1,2-tetrachloroethane	highly toxic	−68.7	130.5	1.5465	water sol to the extent of 0.0565 in 100 g CCl_3CH_2Cl
1,1,2,2-tetrachloroethane	5; highly toxic	−42.5	146.3	1.593	0.32; miscible with chlorinated solvents; high solvency for natural organic and inorganic compounds and sulfur
pentachloroethane	narcotic effect	−29.0	161.95	1.678	0.05; miscible with common organic solvents
hexachloroethane	1; highly toxic	185.0	186.0	2.094	0.005

[a] Time-weighted average in ppm.
[b] Although many of these chloroethanes are good solvents, their use is discouraged owing to their high toxicity.

ethane. Uninhibited 1,1,2-trichloroethane is corrosive to aluminum, iron, and zinc at reflux temperatures. Addition of a water phase results in complete dissolution of all three metals. It is easily dehydrochlorinated by a number of catalytic reagents to give 1,1- and some 1,2-dichloroethylene. Anhydrous aluminum chloride also gives rapid dehydrochlorination.

Manufacture. The compound is produced directly or indirectly from ethylene, eg: by chlorination of 1,2-dichloroethane, a product from ethylene; by oxidation of ethylene with hydrogen chloride and oxygen to give 1,2-dichloroethane and 1,1,2-trichloroethane along with some higher chlorinated ethanes; as a coproduct in the thermal chlorination of 1,1-dichloroethane to produce 1,1,1-trichloroethane.

1,1,1,2-Tetrachloroethane

1,1,1,2-Tetrachloroethane is produced as an incidental by-product in the manufacture of chlorinated ethanes. Thermal decomposition yields tetrachloroethylene, hydrogen chloride, and trichloroethylene. Oxidation in the presence of ionizing radiation gives dichloroacetyl chloride, $Cl_2CHCOCl$.

Manufacture. The compound is prepared by heating the 1,1,2,2-isomer with anhydrous aluminum chloride or chlorination of 1,1-dichloroethylene at 40°C.

1,1,2,2-Tetrachloroethane

Acetylene tetrachloride, $CHCl_2CHCl_2$, is a heavy, nonflammable liquid with a sweetish odor.

Chemical properties. Thermal degradation with or without a catalytic agent gives trichloroethylene, tetrachloroethylene, and hydrogen chloride. Effective catalysts may include ferric chloride, $FeCl_3$–KCl, activated alumina, aluminum chloride, aluminum chloride in nitrobenzene, calcium hydroxide, molecular sieves containing metal cations, or a Fe–Cr–K oxide catalyst. Thermal cracking gives a 95% conversion to trichloroethylene and tetrachloroethylene. Air oxidation under ionizing radiation gives dichloroacetyl chloride. Contact with strong alkali gives dichloroacetylene and an explosion may result.

Manufacture. 1,1,2,2-Tetrachloroethane is commercially produced by direct chlorination or oxychlorination utilizing ethylene as a feedstock.

Pentachloroethane

$CHCl_2CCl_3$ is a colorless, heavy, nonflammable liquid with a chloroformlike odor.

Chemical properties. Various catalysts promote dehydrochlorination. Chlorination in the presence of anhydrous aluminum chloride gives hexachloroethane. Dry pentachloroethane does not corrode iron at temperatures up to 100°C. It is slowly hydrolyzed by water at normal temperatures and oxidized in the presence of light to give trichloroacetyl chloride.

Manufacture. Pentachloroethane is produced by chlorinating 1,1,2,2-tetrachloroethane under ultraviolet light, or trichloroethylene at 70°C in the presence of ferric chloride, sulfur, or ultraviolet light. Oxychlorination of ethylene gives pentachloroethane as well as lower chlorinated hydrocarbons.

Hexachloroethane

Perchloroethane, CCl_3CCl_3, is a white crystalline solid, which is nonflammable and has a camphorlike odor.

Chemical properties. Thermal cracking in the gaseous phase at 400–500°C gives tetrachloroethylene, carbon tetrachloride, and chlorine. Thermal decomposition occurs by means of a radical chain mechanism and is inhibited by traces of nitric oxide. Hexachloroethane reacts violently with powdered zinc in alcoholic solution to give the metal chloride and tetrachloroethylene; aluminum gives a less violent reaction. It is unreactive with aqueous alkali and acid at moderate temperatures, and decomposes to oxalic acid when heated with solid caustic above 200°C or with alcoholic alkalies at 100°C. Degradation of carbon tetrachloride by photochemical, radiolytic, or ultrasonic energy produces the trichloromethyl free radical which on dimerization gives hexachloroethane.

Manufacture. Hexachloroethane is formed in minor amounts in many industrial chlorination processes designed to produce lower chlorinated hydrocarbons.

WESLEY L. ARCHER
Dow Chemical U.S.A.

W.L. Archer and E.L. Simpson, *Ind. Eng. Chem. Prod. Res. Dev.* **16**, 158 (June, 1977).

W.L. Archer and V.L. Stevens, *Ind. Eng. Chem. Prod. Res. Dev.* **16**, 319 (Dec. 1977).

W.L. Archer, *Ind. Eng. Chem. Prod. Res. Dev.* **21**, 670 (Dec. 1982).

1,2-DICHLOROETHYLENE

1,2-Dichloroethylene (1,2-dichloroethene), also known as acetylene dichloride, dioform, α,β-dichloroethylene, and *sym*-dichloroethylene, exists as a mixture of the trans and cis isomers as a colorless, mobile liquid with a sweet, slightly irritating odor resembling that of chloroform. The trans isomer has mp -49.44°C; bp 47.7°C; sp gr 1.2631^{10}; n_D 1.44620 (20°C); soluble to the extent of 0.63 g/100 g H_2O (25°C). The cis isomer has mp -81.47°C; bp 60.2°C; sp gr 1.2917^{15}; n_D 1.44900 (20°C); soluble to the extent of 0.35 g/100 g H_2O (25°C). The trans isomer is more reactive than the cis isomer in 1,2-addition reactions; both undergo benzyne cycloaddition and dimerize to tetrachlorobutene in the presence of organic peroxides.

Manufacture

It is produced by direct chlorination of acetylene at about 40°C. It is often used as a by-product in the chlorination of chlorinated compounds, and is recycled as an intermediate for the synthesis of more useful chlorinated ethylenes.

Health and Safety

The compound is moderately toxic (TLV of 200 ppm).

Uses

1,2-Dichloroethylene is used as an extraction solvent and as a chemical intermediate in synthesis of other chlorinated solvents and compounds.

VIOLETE L. STEVENS
Dow Chemical Company

R. Ausubel, *Int. J. Chem. Kinet.* **7**, 739 (1975).

U.S. Pat. 3,654,358 (April 4, 1977), J. Gaines (to The Dow Chemical Company).

1977 ACGIH publication, American Conference of Government Industrial Hygienists, Cincinnati, Ohio, p. 15.

TRICHLOROETHYLENE

Trichloroethylene, trichloroethene, $CHCl=CCl_2$, is a colorless, sweet smelling, volatile liquid.

Physical and Chemical Properties

Immiscible with water, trichloroethylene is miscible with many organic liquids and is a powerful solvent for a large number of natural and synthetic substances. Although it does not have a flash or fire point, it exhibits a flammable range when high concentrations of vapor are mixed with air and exposed to high-energy ignition sources. Mp -87.1°C; bp 86.7°C; sp gr of liquid at 20/4°C 1.465, or vapor at 0°C 1.325; n_D of liquid 1.4782 (20°C), of vapor 1.001784 (0°C).

The most important reactions are atmospheric oxidation and degradation catalyzed by aluminum chloride. In the absence of stabilizers, trichloroethylene is decomposed (autoxidized) by air in a reaction catalyzed by free radicals and greatly accelerated by elevated temperature and exposure to light. The degradation proceeds through the intermediates

which finally decompose to dichloroacetic acid and hydrochloric acid, and phosgene, carbon monoxide, and hydrogen chloride, respectively; these products are corrosive (see also Tetrachloroethylene).

In the presence of aluminum, a self-sustaining pathway to solvent decomposition forms via aluminum chloride; sufficient quantities of aluminum can cause violent decomposition. All commercial grades are stabilized against autoxidation and $AlCl_3$-catalyzed degradation. Trichloroethylene is not readily hydrolyzed by water. In the presence of catalysts, it is readily chlorinated to pentachloro- and hexachloroethane.

Manufacture

Formerly made from acetylene, most trichloroethylene is now made from ethylene or dichloroethane. The chlorination of ethylene yields dichloroethane, which can be further chlorinated to trichloroethylene and tetrachloroethylene in an exothermic reaction carried out at 280–450°C using catalysts, eg, potassium chloride, aluminum chloride, graphite, or activated carbon. Trichloroethylene conversion is maximized at a chlorine to dichloroethane ratio of 1.7 : 1; tetrachloroethylene conversion is maximized at a feed ratio of 3.0 : 1.

The oxychlorination of ethylene or dichloroethane, using catalysts (eg, mixtures of potassium and cupric chlorides), at 425°C and 138–207 kPa (20–30 psi) yields a mixture of tetrachloroethylene and trichloroethylene. Below 425°C tetrachloroethane becomes the main product, and above 480°C excessive burning and decomposition reactions occur.

Health and Safety

Trichloroethylene is intrinsically toxic, primarily because of its anesthetic effect on the central nervous system (see Industrial hygiene and toxicology). The OSHA time-weighted-average concentration is 100 ppm for an 8-h exposure.

Uses

Approximately 80% of U.S. production is used in vapor degreasing of fabricated metal parts (see Metal surface treatments) and the remaining 20% is divided equally between exports and miscellaneous applications such as in poly(vinyl chloride) production (see Vinyl polymers).

W.C. McNEILL, Jr.
Dow Chemical U.S.A.

Criteria Document: Recommendations for an Occupational Exposure Standard for Trichloroethylene, NIOSH, Contract No. HSM 73-11025, Washington, D. C. Nov. 1973.

J.L. Blackford, "Trichloroethylene," in *Chemical Economics Handbook*, Stanford Research Institute, Menlo Park, Calif., 697.301A–697.302Y, Nov. 1975.

L.M. Elkin, *Process Economics Program, Chlorinated Solvents Report No. 48*, Stanford Research Institute, Menlo Park, Calif., Feb. 1969.

TETRACHLOROETHYLENE

Tetrachloroethylene, $CCl_2{=}CCl_2$, (tetrachloroethene, perchloroethylene), is known as perc and sold under a variety of trade names. It is a nonflammable liquid with a pleasant, ethereal odor and the most stable of the chlorinated ethanes and ethylenes, requiring only small amounts of stabilizers.

Physical and Chemical Properties

Mp −22.7°C; bp 121.2°C; sp gr (liquid) 1.62260 (20/4°C); viscosity at 25°C 0.839 mPa·s (cP); n_D 1.50547; soluble to the extent of 15 g/100 g

H_2O. Tetrachloroethylene dissolves sulfur, iodine, mercuric chloride, and appreciable amounts of aluminum chloride; it is a solvent for many organic compounds; miscible with chlorinated organic solvents and most other common solvents and forms about sixty binary azeotropes.

Commercially pure, stabilized tetrachloroethylene can be used in contact with the common construction metals at about 140°C, even in the presence of air, water, and light. It resists hydrolysis at temperatures up to 150°C. It is stable to about 500°C in the absence of catalysts, air, or moisture. Decomposition products are mainly hydrogen chloride and phosgene, but products depend upon reaction conditions; an excess of hydrogen in the presence of reduced nickel catalyst at 220°C results in total decomposition to hydrogen chloride and carbon. In the absence of light, tetrachloroethylene is unaffected by oxygen, but under uv radiation in the presence of air or oxygen it undergoes autoxidation to trichloroacetyl chloride via formation of peroxy compound intermediates.

It reacts explosively with butyllithium in petroleum ether solution, and metallic potassium at its mp, but not with sodium.

Manufacture

For many years, tetrachloroethylene was produced almost exclusively from acetylene and chlorine via trichloroethylene (see Acetylene-derived chemicals). Now other hydrocarbons, eg, methane, ethane, propane or higher paraffins or their chlorinated derivatives, are employed as feedstocks in a simultaneous chlorinating and pyrolysis processes. About 40% of current US production is based on ethane and propane. Typical reactions involved, eg, based on propane, are

$$CH_3CH_2CH_3 + 8\,Cl_2 \rightarrow CCl_2{=}CCl_2 + 8\,HCl + CCl_4$$
$$2\,CCl_4 \rightarrow CCl_2{=}CCl_2 + 2\,Cl_2$$

In a single-stage fluid-bed reactor oxychlorination of ethylene dichloride (dichloroethane) with chlorine, at 425°C and 138–207 kPa (20–30 psi), using an inexpensive oxychlorination catalyst (eg, potassium chloride), tetrachloroethylene is a coproduct with trichloroethylene (qv).

Health and Safety

Tetrachloroethylene is a central nervous depressant and can cause death. Concentrations of 1500 ppm cause unconsciousness in less than 30 min. The distinctive odor does not provide adequate warning.

Uses

Estimated pattern of use (1976) was in dry cleaning (66%), textile processing (13%), metal degreasing (13%), fluorocarbon manufacture (see Fluorine compounds) (3%), and miscellaneous and exports (5%).

S.L. Keil
Dow Chemical USA

W.L. Archer and E.L. Simpson, *Ind. Eng. Chem. Prod. Res. Dev.* **16**, 158 (June, 1977).

W.L. Faith, D.B. Keyes, and R.L. Clark, *Industrial Chemicals*, 3rd ed., John Wiley & Sons, Inc., New York, 1965.

L.M. Elkin, *Report No. 48—Chlorinated Solvents*, Stanford Research Institute, Menlo Park, California, Feb. 1969, pp. 137–185.

ALLYL CHLORIDE

Allyl chloride, $CH_2{=}CHCH_2Cl$, is a colorless liquid with a pungent odor. It is the most important of the allyl compounds.

Physical and Chemical Properties

Fp $-134.5°C$; bp at 101 kPa (1 atm) $44.96°C$; sp gr at $20°C$ 0.9392; soluble to the extent of 0.36 wt% in H_2O at $20°C$.

Allyl chloride exhibits reactivity as an olefin and as an organic halide. Reactions can be directed by control of conditions, selection of reagents, and provision of suitable catalysts. Allyl chloride does not polymerize well by free-radical techniques (see Allyl monomers and polymers). It undergoes typical additions to the double bond; eg, chlorine, bromine, and iodine chloride at temperatures below the inception of the substitution reaction produce the 1,2,3-trihalides, and high temperature halogenation by a free-radical mechanism leads to unsaturated dihalides $CH_2{=}CHCHClX$. Simple replacement reactions of chlorine occur, eg, the alkaline hydrolysis of allyl chloride to allyl alcohol (see Allyl compounds) and acidic hydrolysis in the presence of cuprous chloride. Allyl esters are formed by reaction of allyl chloride with sodium salts of appropriate acids under conditions of controlled pH. Mono-, di-, and triallylamines are prepared by reaction with ammonia, the ratio of reagents determining the product distribution; mixed amines are prepared in similar fashion using a substituted amine in place of ammonia. Many syntheses of complex substances benefit from the reactivity of allyl chloride, eg, mixed allyl ethers may be prepared from allyl chloride and the appropriate alkoxide or alcohol–alkali mixture; polyol ethers form resinous polymers (see Allyl monomers and polymers); allyl aryl ethers undergo the Claisen rearrangement to allyl-substituted phenols and are used as starting points for several syntheses; alkylation reactions are known with either the olefin or chloro group involved (eg, reactions with benzene are typical of those involving the double bond); and several allylation reactions are known, frequently using a metallo-organic derivative of the compound being allylated or a strongly electropositive metal in conjunction with the reactants, eg, Grignard reactions.

Manufacture

Allyl chloride is produced mainly through the high temperature ($> 300°C$) substitutive chlorination of propylene. The temperature variable affects the monochloride substitution reaction directly, and pressure, mole ratio, residence time, impurities, and degree of mixing affect side reactions.

Industrial-scale allyl chloride facilities perform three major processing operations: (1) feed preparation, to ensure that both reactor feeds are dry and reasonably pure in order to limit yield losses; propylene purification measures vary with the feedstock (see Feedstocks); (2) reaction, carried out in an adiabatic reactor at 500–510°C and 205 kPa (15 psig); and (3) product recovery, allyl chloride being separated from the product stream by a series of fractionations.

Health and Safety

Allyl chloride is toxic, extremely flammable, and severely irritating; high concentration can cause death. Results of a recently completed study (a 90-d inhalation study in rats and mice) to determine the potential hazard from long-term exposure to the vapors has been evaluated and the report is available from the CMA. A concentration of 1 ppm allyl chloride in air is the threshold limit value (TLV) for worker exposure.

Uses

Allyl chloride is a parent compound for a number of useful derivatives, eg, epichlorohydrin (see Chlorohydrins), glycerol (qv), a variety of resins and polymers (see Allyl monomers and polymers), allyl alcohol, and allylamines.

ALDO DEBENEDICTIS
Shell Chemical Company

H.P.A. Groll and G. Hearne, *Ind. Eng. Chem.* **31**(12), 1530 (1939).

Criteria for a Recommended Standard...Occupational Exposure to Allyl Chloride, HEW Publication No. (NIOSH) 76-204, GPO, 1976.

F. He, J.M. Jacobs, and F. Scarvieli, *Acta Neuropathologica (Berlin)* **55**, 125 (1981).

CHLOROPRENE

Chloroprene, 2-chloro-1,3-butadiene, is a colorless, mobile, volatile liquid with an ethereal odor similar to that of ethyl bromide.

Physical and Chemical Properties

Mp $-130 \pm 2°C$; bp at 101 kPa $59.4°C$; d at $20°C$ 0.9585 g/cm³. It is very slightly soluble in water and miscible with most organic solvents.

The chemical properties are mainly a product of electronic interactions between the conjugated butadiene structure and the chlorine atom in the molecule. The chlorine atom strongly enhances free-radical activity of the molecule but decreases activity toward electrophilic reagents. Thus, chloroprene polymerizes much more readily than either butadiene or isoprene to form high molecular weight elastomeric polymers (see Elastomers, synthetic-neoprene), but it reacts less readily with maleic anhydride. Chloroprene forms cyclic dimers on prolonged standing in the presence of polymerization inhibitors. Dimerization is independent of free-radical catalysts and inhibitors, high pressure accelerating the formation considerably. Many inorganic and organic compounds add to the double bonds. Chloroprene reacts with oxygen to form peroxides. It enters into addition reactions with halogens, hydrogen halides, hypohalous acids, and mercaptans. The general chemistry is reviewed in the literature.

Related compounds of interset are the cis and trans isomers of 1-chloro-1,3-butadiene (bp 66–67°C; d_4^{20} 0.954 g/cm³), frequently referred to as α-chloroprene. These are by-products of most reactions leading to 2-chlorobutadiene. 4-Chloro-1,2-butadiene, or "isochloroprene" (bp 88°C; n_D^{20} 1.4775; d_4^{20} 0.9891 g/cm³) is prepared from hydrochlorination of vinylacetylene in the absence of cuprous chloride or other isomerizing materials. 2,3-Dichloro-1,3-butadiene (bp = 98°C; n_D^{20} = 1.489; d_4^{20} 1.183 g/cm³) is significant because it is very reactive in copolymerization with chloroprene and it is available from by products of both the acetylene and butadiene routes to chloroprene.

Manufacture

Formerly, chloroprene was manufactured from acetylene (see Acetylene-derived chemicals). Most chloroprene is currently produced from butadiene (qv). Three essential steps in the process are: (1) chlorination of butadiene; (2) isomerization using catalysts containing, typically, copper salts and a solubilizing agent; and (3) caustic dehydrochlorina-

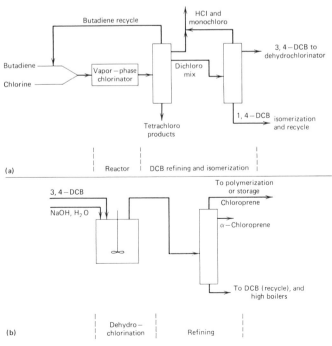

Figure 1. Conversion of butadiene to chloroprene. 3,4-DCB is 3,4-dichloro-1-butene; 1,4-DCB is 1,4-dichloro-2-butene.

tion of 3,4-dichloro-1-butene. These reactions are shown by the following equations:

Chlorination

$$CH_2=CHCH=CH_2 + Cl_2 \rightarrow ClCH_2-CH=CH-CH_2Cl + ClCH-CHClCH=CH_2$$

Isomerization

$$ClCH_2CH=CHCH_2Cl \overset{catalyst}{\rightleftarrows} CH_2=CHCHClCH_2Cl$$

Dehydrohalogenation

$$CH_2=CHCHClCH_2Cl + NaOH \rightarrow CH_2=CHCCl=CH_2 + NaCl + H_2O$$

Figure 1 on p. 267 depicts a general flow diagram of an integrated production plant.

Health and Safety

Chloroprene presents acute hazards because of its tendencies to peroxidize, polymerize, and burn. There are reports of mutagenic, embryotoxic, and teratogenic effects. It is physiologically active as an anesthetic. The TLV in the workplace is 10 ppm.

PAUL R. JOHNSON
E.I. du Pont de Nemours & Co., Inc.

Reviewed by
CLARE A. STEWART, JR.
Wilmington, Delaware

P.S. Bauchwitz, J.B. Finlay, and C.A. Stewart, Jr., in E.C. Leonard, ed., *Vinyl and Diene Monomers*, Part II, John Wiley & Sons, Inc., New York, 1971, pp. 1149–1183.
P.R. Johnson, *Rubber Chem. Technol.* **49**, 650 (1976).

CHLORINATED PARAFFINS

Chlorinated paraffins are classified as chlorinated hydrocarbons that have the general formula $C_xH_{(2x-y+2)}Cl_y$.

Physical and Chemical Properties

Commercial chlorinated paraffins have a 20–70 wt% chlorine content. The bulk of the manufactured products fall within the 40–70 wt% range. The important physical properties include viscosity, solubility, color, and thermal instability (see Table 1). For a given paraffin, increasing chlorine content increases viscosity and specific gravity. Chlorinated paraffins are miscible with many organic solvents, and insoluble in water, glycerol, and the glycols. Water emulsions can be made with the use of proper emulsifying agents. These paraffins are relatively inert materials. Prolonged exposure to heat or light can cause dehydrochlorination catalyzed by the presence of aluminum, zinc, and iron. Stabilizers are normally added for storage purposes.

Manufacture

Chlorinated paraffins are produced in an exothermic reaction by passing chlorine gas into a liquid paraffin; chlorine feed rates and reaction temperatures differ slightly among the different producers, usually falling between 90–100°C. Manufacture of resinous chlorinated paraffins (70% chlorine content) requires the use of a solvent during the chlorination step.

Health and Safety

Chlorinated paraffins are nontoxic.

Uses

Chlorinated paraffins are used as extreme-pressure lubricant additives (50%) in the metal-working industry; in plastics (25%); and in rubber (qv) caulks and sealants (25%).

B.A. SCHENKER
Diamond Shamrock Corp.

Diamond Shamrock Corp, *Technical Data—Chlorowax* & *Delvet*, 1977.
Fed. Reg. (Aug. 12, 1961), et seq.

CHLORINATED DERIVATIVES OF CYCLOPENTADIENE

Hexachlorocyclopentadiene (HCCP) and octachlorocyclopentene (OCCP) are the only known commercially significant cyclic C_5 chlorocarbons produced from cyclopentadiene.

Physical Properties

Hexachlorocyclopentadiene is a nonflammable liquid with a characteristically pungent musty odor. The pure compound is a light lemon-yellow. Mp 11.34°C; bp 239°C; sp gr at 20°C 1.710. Octachlorocyclopentene has a very faint camphoric odor. It freezes to a white solid at room temperature and is quite stable on prolonged storage. At its atomspheric reflux temperature (283°C) it dissociates slightly into hexachlorocyclopentadiene and chlorine. Mp 38.2°C; bp 285°C; sp gr at 20°C 1.816.

Chemical Properties

HCCP has a high order of reactivity with a variety of dienophilic materials in the Diels-Alder reaction as well as a variety of addition and substitution reactions; products are generally 1:1 adducts containing a hexachlorobicyclo[2.2.1]heptene structure and invariably contain the constituent derived from the olefin in the endo position.

Manufacture

A number of manufacturing processes for hexachlorocyclopentadiene have been commercialized. Three alternative chemical routes are the following: (*1*) a process employing cyclic hydrocarbons, eg, cyclopentadiene, mixed with alkaline hypochlorite; (*2*) processes employing aliphatic hydrocarbons, eg, the preparation of a polychlorinated straight- or branched-chain hydrocarbon followed by cyclization to give OCCP or HCCP; (*3*) processes employing chlorination of cyclic hydrocarbons, eg, those used by Shell Development Company, the catalytic Lidov process, and the Lummus process.

Octachlorocyclopentene and hexachlorocyclopentadiene are mutually interconvertible. The dechlorination reaction is accelerated in the presence of metals or metal salts of the transition elements.

Health and Safety

HCCP is toxic; it is included on the EPA's Toxic Pollutant list. It causes skin burns and is readily absorbed through the skin.

Uses

HCCP is used in the production of insecticides, eg, 6,7,8,9,10,10-hexachloro-1,5,5a,6,9,9a-hexahydro-6,9-methano-2,4,3-benzodioxathiepin 3-oxide (Endosulfan, Beosit, Chlortiepin, Insectophene, Thionex). It is also used to produce flame-retardant chemicals, eg, 1,4,5,6,7,7-hexachlorobicyclo[2.2.1]hept-5-ene-2,3-dicarboxylic acid (Het Acid, Chlorendic Acid).

JAMES E. STEVENS
Hooker Chemical and Plastics Corp.

H.E. Ungnade and E.T. McBee, *Chem. Rev.* **58**, 249 (1958).
Hexachlorocyclopentadiene, Hooker Chemical & Plastics Corp., Data Sheet No. 815A, Apr. 1969.
U.S. Pat. 2,714,124 (July 26, 1955), A.H. Maude and D.S. Rosenberg (to Hooker Chemical & Plastics Corp.).
Fed. Reg. **43**(21), 4108 (1978).

Table 1. Physical Properties of Chlorinated Paraffins

Paraffin feedstock	Wax				C_{13}-C_{17}	C_{10}-C_{13}
chlorine content, wt%	39	42	48	70	52	60
density at 25°C, g/cm³	1.12	1.17	1.23	1.65	1.25	1.36
refractive index	1.501	1.505	1.516		1.510	1.516

CHLORINATED BENZENES

Twelve chlorinated benzenes can be formed by replacing some or all of the hydrogen atoms of benzene with chlorine atoms.

Physical and Chemical Properties

Table 1 lists some important physical properties.

The chlorine atom in chlorobenzene is sufficiently labile to be hydrolyzed. The hydrolysis of one of the chlorine atoms in 1,2,4,5-tetrachlorobenzene with sodium hydroxide dissolved in a suitable alcohol solvent produces 2,4,5-trichlorophenol. The reaction must be carried out under extremely restricted conditions to prevent the formation of the very toxic compound, 2,3,7,8-tetrachlorodibenzo-*p*-dioxin (TCDD) (see Industrial hygiene and toxicology). Nitration of chlorobenzenes with nitric acid has wide industrial applications; however, their uses are declining rapidly.

Manufacture

All chlorobenzenes are now produced by chlorination of benzene in the presence of a Friedel-Crafts catalyst, usually ferric chloride (see Friedel-Crafts reactions). Each compound, except hexachlorobenzene, can be further chlorinated; hence the product is always a mixture of chlorinated benzenes. Pure compounds are obtained by distillation and crystallization. The chlorination reaction is exothermic, liberating heat of ca 1.83 kJ/g Cl_2. Heat is removed in some cases by circulating the reaction liquid through a suitable cooler (see Heat exchange technology); in other cases, chlorination occurs at the boiling point and the heat is removed by the vaporizing liquid.

Benzene chlorination proceeds in a batch, single-, or multistage stirred reactor. The ratio of 1,4- to 1,2-dichlorobenzene varies from 1.4 to 5, depending upon the catalyst. Reactors are subject to design and operating hazards: stagnant areas must be avoided since they allow chlorination to tetra- and pentachlorobenzenes, compounds that cause plugging; spontaneous combustion is a hazard, principally in the vapor phase. Since HCl is present in most parts of the equipment, corrosion is always a potential problem.

Health and Safety

All of the chlorobenzenes are less toxic than benzene. The threshold limit value (TLV) (in ppm per volume of air) for human exposure in 8 h is 50 for 1,2-dichlorobenzene, and 75 for chlorobenzene and 1,4-dichlorobenzene.

Uses

Formerly, large quantities of chlorobenzene had been used in the production of phenol, aniline, diphenyl oxide, and DDT, a widely used and now restricted insecticide. Monochlorobenzene, 1,2-di- and 1,2,4-trichlorobenzene are used as solvents.

CHE-I KAO
NOLAND POFFENBERGER
Dow Chemical U.S.A.

Chemical Economics Handbook, SRI International, Menlo Park, Calif., 1980.

BENZENE HEXACHLORIDE

Benzene hexachloride, commonly called BHC (1,2,3,4,5,6-hexachlorocyclohexane, Agrocide, Ambiocide, Benzanex, Benzex, Gammacide, Gammaloid, Gamaspra, Gamtox, Gyben, HCH, Hexdow, Isotox, Lintox, Lexone, Trivex-T), is the fully saturated product formed by light-catalyzed addition of chlorine to benzene. This reaction produces a number of stereoisomeric compounds of the composition $C_6H_6Cl_6$, with varying amounts of both under-chlorinated and over-chlorinated compositions. It is used as technical BHC, fortified BHC, or lindane.

Physical Properties

Technical benzene hexachloride (BHC) is a brownish-to-white solid having a musty odor. It melts around 63°C and can be distilled with no decomposition. The composition of the various isomers in benzene hexachloride depends on conditions of manufacture, but is commonly: (mp 159.2°C, n_D^{20} 1.626) 65%; (mp 311.7°C, n_D^{20} 1.626) = 14%; (mp 218.2 C, n_D^{20} 1.635) = 4%; (mp 154.8 C, n_D^{20} 1.576–1.674) = 10%; and traces of others. The odor of BHC has been attributed to both under- and overchlorinated cyclohexanes as well as chlorocyclohexenes and can be removed by a simple carbon treatment.

Fortified benzene hexachloride (FBHC) is a brown-to-white amorphous powder with a musty odor. This mixture is processed by selective crystallization of technical BHC to a crystalline solid containing 40–48% gamma isomer. FBHC melts near 78°C.

Lindane is a crystalline white, free-flowing solid from technical BHC, ≥ 99% gamma isomer. The pure gamma isomer is the only benzene hexachloride isomer with significant insecticidal activity; it has three modifications, one rhombic, and two monoclinic which are enantiotropic to the rhombic form. The melting points of the two stable forms are very close together. These two are relatively heat and chemically stable.

Chemical Properties

With the exception of the beta isomer, the isomers are dehydrochlorinated by alkali at 60°C to give primarily 1,2,4-trichlorobenzene largely from the alpha isomer. The gamma and delta isomers also yield lesser quantities of 1,2,3- and 1,3,5-trichlorobenzenes. When BHC is chlorinated in carbon tetrachloride, various substitution derivatives are formed; use of liquid chlorine yields higher-substituted derivatives. High temperature chlorination in a chlorobenzene solvent in the presence of ferric chloride also results in quantitative conversion of BHC to hexachlorobenzene. BHC reacts readily with sulfur at 240–290°C to form SPC (Sulphure de Polychlorocyclane), used in Europe as a pest-control agent. With elementary sulfur, *p*-dichlorobenzene and 1,2,4,5-tetrachlorobenzene are formed. To inhibit decomposition, BHC is stored for long times at temperatures above ambient with added 1% sodium thiosulfate. Treatment of BHC with zinc dust in acids dechlorinates the material to benzene. Although generally inert to strong acids, BHC is decomposed to benzene by chlorosulfonic acid or sulfur trioxide.

Manufacture

Technical BHC is produced with a batch or continuous process, in stirred or tubular reactors, by the addition chlorination of benzene in the presence of free-radical initiators such as visible or ultraviolet light, x rays, or gamma rays. A typical method uses reactors at 15–35°C at 101 kPa (1 atm). The gamma-isomer content is significantly increased by the reaction of benzene and chlorine in a nonreactive solvent of high dielectric constant, eg, methylene chloride. No commercial process has yet been found that permits recovery of all of the gamma isomer present in crude benzene hexachloride. Numerous procedures for isolating the highly

Table 1. Physical Properties of Chlorobenzenes

Compound	Mp, °C	Bp 101.3 kPa[a], at °C	Liquid density, g/cm³	Refractive index of liquid, n_D^{25}
chlorobenzene	−45.34	131.7	1.10118	1.5219
1,2-dichlorobenzene	−16.97	180.4	1.3022	1.5492
1,3-dichlorobenzene	−24.76	173.0	1.2828	1.54337
1,4-dichlorobenzene	53.04	174.1	1.2475	1.52849 (55°C)
1,2,3-trichlorobenzene	53.5	218.5		
1,2,4-trichlorobenzene	17.15	213.8	1.44829	1.56933
1,3,5-trichlorobenzene	63.5	208.5		
1,2,3,4-tetrachlorobenzene	46.0	254.9	1.70	
1,2,3,5-tetrachlorobenzene	51	246		
1,2,4,5-tetrachlorobenzene	139.5	248.0	1.833(s)	
pentachlorobenzene	85	276		
hexachlorobenzene	228.7	319.3	1.596	

[a] To convert kPa to mm Hg, multiply by 7.5.

desired gamma isomer have been patented; the two used commercially for lindane preparation are the supersaturation process, and the fluid classification process.

Health and Safety

Lindane is of intermediate acute toxicity. BHC used in plant protection produces considerable residues which are carried on in the food chain; use of technical BHC is discouraged.

Uses

Lindane is a broad-spectrum insecticide and selective acaricide with contact and fumigant action. Its primary use in the United States is in seed treatment.

JAMES G. COLSON
Occidental Chemical Corporation

G.T. Brooks, *Chlorinated Insecticides*, CRC Press, Cleveland, Ohio, 1974.

E. Uhlmann, *Lindane*, Verlag K. Schillinger, Freiburg im Breisgau, Berlin, 1972.

D.S. Rosenberg, *Chlorine*, R.E. Kruger Publishing Co., Huntington, N.Y. 1962.

RING-CHLORINATED TOLUENES

The ring-chlorinated derivatives of toluene form a group of stable, colorless compounds. They are prepared directly by chlorination, and indirectly by routes such as reactions involving the replacement of the amino, chlorosulfonyl, hydroxyl, and nitro groups by chlorine, and the use of the sulfonic acid and amino groups to orient substitution followed by their removal from the ring.

Monochlorotoluenes

These are mobile, colorless liquids, miscible in all proportions with many organic liquids such as aliphatic and aromatic hydrocarbons, chlorinated solvents, lower alcohols, ketones, glacial acetic acid, and di-n-butylamine. They are insoluble in water, ethylene, and diethylene glycols, and triethanolamine. o-Chlorotoluene (1-chloro-2-methylbenzene) has mp $-35.6°C$, bp $159.2°C$, and d_{20} 1.083 g/cm^3. p-Chlorotoluene (1-chloro-4-methyl benzene) has mp 7.5°C, bp 162.4°C, and d_{20} 1.070 g/cm^3. m-Chlorotoluene (1-chloro-3-methylbenzene) has mp $-47.8°C$, bp 161.7°C, and d_{20} 1.072 g/cm^3. o- and p-Chlorotoluenes form binary azeotropes with various organic compounds and stable ionic complexes with antimony pentachloride.

Chemical properties. Monochlorotoluenes are stable to the action of steam, alkalies, amines, and hydrochloric and phosphoric acids at moderate temperatures and pressures. Reactions are divided into (1) reactions of the aromatic ring, eg, ring chlorination of o-chlorotoluene yields a mixture containing all four possible dichlorotoluene isomers; (2) reactions of the methyl group, which relate to most of the uses of monochlorotoluene products, eg, chlorination under free-radical conditions leads successively to the chlorinated benzyl, benzal, and benzotrichlorides (qv); and (3) reactions involving the chlorine substituent, eg, displacement and benzyne mechanisms are involved in the hydrolysis of chlorotoluenes to cresols with aqueous sodium hydroxide.

Preparation. Chlorination of toluene is carried out with a wide variety of chlorinating agents and catalysts and under a variety of reaction conditions. Chlorinations with elemental chlorine yield varying ratios of ortho and para isomers depending on both the chlorinating agent and catalyst used. The para isomer content is of greatest commercial significance. The meta isomer must be prepared by indirect means since only a small amount (usually 1%) is formed by direct chlorination. Pure chlorotoluene isomers (eg, m-chlorotoluene) are available from diazotization of the toluidine isomers followed by reaction with cuprous chloride (Sandmeyer reaction).

Higher Chlorotoluenes

Physical properties of some of the higher chlorotoluenes are listed in Table 1. Of the six dichlorotoluenes, only the 2,4-, 2,5-, and 3,4-isomers

Table 1. Physical Properties of the Higher Chlorotoluenes

Toluene	Mp, °C	Bp, °C	Density at 20°C, g/cm^3
2,4-dichloro	-13.5	201.1	1.250
2,6-dichloro		200.6	1.269
3,4-dichloro	-15.3	208.9	1.256
2,3,4-trichloro	43–44	244	
2,4,5-trichloro	82.4	229–230[a]	
2,3,4,5-tetrachloro	98.1		
2,3,4,6-tetrachloro	92	266–276	
pentachloro	224.5–225.5	301	

[a] At 95.4 kPa (716 mm Hg).

are available from direct chlorination of monochlorotoluenes. 2,3-Dichlorotoluene is best prepared by the Sandmeyer reaction on 3-amino-2-chlorotoluene. The chlorination of toluene and o- and p-chlorotoluenes produces mixtures of 2,3,6- and 2,4,5-trichlorotoluenes containing small amounts of the 2,3,4,- and 2,4,6- isomers. 2,3,4,6-Tetrachlorotoluene is prepared from the Sandmeyer reaction on 3-amino-2,4,6-trichlorotoluene. Pentachlorotoluene is formed by the ferric chloride catalyzed chlorination of toluene in carbon tetrachloride or hexachlorobutadiene.

Health and Safety

Monochlorotoluenes are slightly toxic.

Uses

Major use is in the manufacture of p-chlorobenzotrifluoride. Isomers are widely used as solvents and herbicide intermediates.

SAMUEL GELFAND
Hooker Chemical & Plastics Corp.

P.E. Hoch in J.S. Sconce, ed., *Chlorine: Its Manufacture, Properties and Uses*, ACS Monograph No. 15, Van Nostrand Reinhold Company, New York, 1962, Chapt. 28, pp. 834–863.

P. Kovacic in G.A. Olah, ed., *Friedel-Crafts & Related Reactions*, Vol. IV, Interscience Publishers, a division of John Wiley & Sons, Inc., New York, 1965, Chapt. XLVIII, pp. 111–127.

Houben-Weyl, *Methoden Der Organischer Chemie*, Band V/3, Georg Thieme Verlag, Stuttgart, FRG, 1962, pp. 657–668.

BENZYL CHLORIDE, BENZAL CHLORIDE, AND BENZOTRICHLORIDE

Benzyl Chloride

Benzyl chloride, ((chloromethyl)benzene, α-chlorotoluene), $C_6H_5CH_2Cl$, is a colorless liquid with a pungent odor; fp $-39.2°C$, bp 179.4°C, d_4^4 1.114 g/cm^3. Thirty-six binary and nine ternary azeotropic systems containing benzyl chloride have been reported. Benzyl chloride is insoluble in cold water, decomposes slowly in hot water to give benzyl alcohol, and is miscible in all proportions with most organic solvents at room temperature.

Manufacture. Benzyl chloride is manufactured by the thermal or photochemical chlorination of toluene at 65–100°C. Various materials, eg, phosphorus pentachloride or sulfuryl chloride, are reported to catalyze side-chain chlorination. These compounds and others such as amides also reduce ring chlorination by complexing metallic impurities. Commercial chlorination is carried out either batchwise or continuously in glass-lined or nickel reactors. To reduce the risk of decomposition during distillation, additives such as lactams and amines are used; lime, sodium carbonate, and triethylamine are used as stabilizers during storage and shipment.

Benzal Chloride

Benzal chloride, ((dichloromethyl)benzene, α,α-dichlorotoluene, benzylidene chloride), $C_6H_5CHCl_2$, is a colorless liquid with a pungent, aromatic odor; fp $-16.4°C$, bp 205.2°C, d_{14}^{14} 1.256 g/cm^3. Eight binary azeotropic systems are reported. Benzal chloride is insoluble in water at room temperature and miscible with most organic solvents.

Manufacture. Benzal chloride is manufactured by chlorination with 2.0–2.2 mol of chlorine per mole of toluene, is the benzal chloride purified by distillation. It is also formed by the reaction of dichlorocarbene (CCl_2:) with benzene.

Benzotrichloride

Benzotrichloride, ((trichloromethyl)benzene, α, α, α-trichlorotoluene, phenylchloroform), $C_6H_5CCl_3$, is a colorless, oily liquid with a pungent odor; fp $-4.75°C$, bp $220.6°C$, d_4^{20} 1374 g/cm^3. It is soluble in most organic solvents but reacts with water and alcohol. Three binary azeotropes are reported.

Manufacture. Benzotrichloride is manufactured by exhaustive chlorination of toluene with ultraviolet light catalysis and final temperature of 100–140°C. Additives, eg, phosphorus trichloride, are used to complex metallic impurities. After purging with an inert gas to remove hydrogen chloride, the crude product is utilized or purified by distillation.

Chemical Properties

Reactions of the side chain containing the halogen include hydrolysis, esterification, ether formation, benzylation of a variety of aliphatic, aromatic, and heterocyclic amines, oxidation, and self condensation (eg, benzyl chloride self-condenses to form polymeric oils and solids). Hydrolysis of benzyl chloride proceeds slowly with boiling water and more rapidly at elevated temperature and pressure in the presence of alkalies. Benzal chloride is hydrolyzed to benzaldehyde under both acid and alkaline conditions. Benzotrichloride is hydrolyzed to benzoic acid by hot water, concentrated sulfuric acid, or dilute aqueous alkali. Hydrolysis with an equimolar amount of water gives benzoyl chloride.

Aromatic ring reactions include chlorination (eg, benzyl chloride in the presence of an iodine catalyst chlorinates to a mixture consisting of the ortho and para compounds); nitration (eg, benzotrichloride nitrated below 30°C with a mixture of nitric and sulfur acids gives *m*-nitrobenzotrichloride containing some ortho and para isomers), chlorosulfonation (eg, benzotrichloride with chlorosulfonic acid (qv) or sulfur trioxide gives *m*-chlorosulfonylbenzoyl chloride).

Health and Safety

Benzyl chloride is slightly toxic; the permissible level established is 1 ppm (5 mg/m^3). Vapors of both benzal chloride and benzotrichloride are strongly irritating and lacrimatory. Order of decreasing toxicity is BTC > BAC > benzyl chloride. According to the WHO, there is sufficient evidence that benzotrichloride is carcinogenic in mice.

Uses

Benzyl chloride is used mainly in the manufacture of benzyl butyl phthalate, benzyl alcohol, and benzyl chloride-derived quaternary ammonium compounds; it is also used as an intermediate in triphenylmethane dye manufacture. Benzal chloride is used to manufacture benzaldehyde (qv). Benzotrichloride is used mainly in the manufacture of benzoyl chloride (see Benzoic acid). Lesser amounts are consumed in the manufacture of benzotrifluoride and hydroxybenzophenone ultraviolet-light stabilizers (see Uv stabilizers). Benzyl-derived quaternary ammonium compounds are used as cationic surface-active agents, germicides, fungicides, and sanitizers.

Derivatives

Ring-chlorinated derivatives are produced by the direct side-chain chlorination of the corresponding chlorinated toluenes or by one of several indirect routes if the required chlorotoluene is not readily available. These include *o*-, *m*-, and *p*-chlorobenzyl chloride, chlorobenzal chloride, and chlorobenzotrichloride; 2,4-, 2,6-, and 3,4-dichlorobenzyl chloride; 2,6-, 3,4-, and 2,4-dichlorobenzal chloride; and 2,4- and 3,4-dichlorobenzotrichloride.

Samuel Gelfand
Hooker Chemicals and Plastics Corp.

IARC Monographs on the Evaluation of the Carcinogenic Risk of Chemicals to Humans, Vol. 29, May 1982, pp. 73–82.

P.E. Hoch in J.S. Sconce, ed., *Chlorine and Its Manufacture, Properties and Uses*, ACS Monograph No. 154, Van Nostrand Reinhold Company, 1962, Chapt. 28, pp. 834–863.

Houben-Weyl, *Methoden Der Organischer Chemie*, Band V/3, Georg Thieme Verlag, Stuttgart, FRG, 1962, pp. 735–748.

CHLORINATED NAPHTHALENES

In 1833 Laurent discovered that naphthalene could be chlorinated to obtain waxlike materials, the chloronaphthalenes. There are 77 possible chloronaphthalenes; not all are known or are rigidly identified materials.

Physical Properties

The industrially significant chloronaphthalenes are generally not pure materials, but mixtures of isomers of mono- and/or polychloronaphthalenes. They range from thin liquids to hard waxes to a high melting solid, with melting points from ca -40 to 180°C. Their specific gravities are 1.2–2.0 at 25°C. The solids give melts of low viscosity. Liquid chloronaphthalenes are soluble in almost all organic solvents. Waxy or solid chloronaphthalenes are soluble in chlorinated solvents, aromatic solvents, and petroleum naphthas, and are compatible with petroleum waxes, chlorinated paraffins, polyisobutylenes, and plasticizers such as dioctyl phthalate and tricresyl phosphate. Chloronaphthalenes have very good thermal and chemical stability, and low flammability, volatilities ranging from modest to extremely low (eg, octachloronaphthalene).

Cheimcal Properties

1-Chloronaphthalene has mp $-2.3°C$, bp 260°C, d 1.189^{25} g/cm^3 can be made by ferric chloric-catalyzed chlorination of molten naphthalene at 100–110°C with the neutralized product then distilled. It undergoes the usual electrophilic substitution reactions such as nitration, sulfonation, and chloromethylation; it is converted to 1-naphthol, eg, by aqueous caustic in the presence of Cu catalysts.

2-Chloronaphthalene has mp 59.5–60°C, bp 259°C, d 1.266^{16} g/cm^3 and cannot be isolated easily from the product of the direct chlorination of naphthalene but can be easily made from 2-naphthylamine (a carcinogen) via the diazonium salt by the Sandmeyer reaction. Pure isomers can be prepared also by treating hydroxyl-, sulfonyl chloride-, or nitro-substituted naphthalenes with phosphorus pentachloride.

Manufacture

Commercial chloronaphthalenes are manufactured by the metal halide-catalyzed chlorination of molten naphthalene to the desired chlorination stage at a temperature slightly above the melting point of the desired product.

Health and Safety

Skin contact can cause dermatitis and acneform lesions; inhalation of the vapors or fumes can result in liver disease. Precautions in the industrial handling of chlorinated naphthalenes include strict personal cleanliness during and after work, the use of fat-free barrier creams, protective clothing, periodic changes of clothing, efficient exhaust systems for the work place, handling in closed systems, good housekeeping, and informed employee and supervisory practices. Preemployment and periodic physical examinations of personnel have also been recommended.

Uses

There are no commercial uses for purified di- or polychloronaphthalene isomers with the exception of octachloronaphthalene, which is used in the manufacture of heptachloro-1-naphthol. Monochloronaphthalenes are used for chemical-resistant gauge fluids, dyes, and instrument seals. The tri- and higher-chlorinated naphthalene products are used as impregnants (see Ceramics) and in flame-proofing (see Flame retardants).

Hans Dressler
Koppers Company, Inc.

N. Donaldson, *The Chemistry and Technology of Naphthalene Compounds*, Edward Arnold Publishing, London, 1958, Chapt. V, p. 128.

CHLORINATED BIPHENYLS AND RELATED COMPOUNDS

The term PCB is commonly used as an abbreviation for polychlorinated biphenyl. Biphenyl (diphenyl), terphenyls, higher polyphenyls, or mixtures of these compounds, can be chlorinated to give a wide range of products that have outstanding chemical and thermal stabilities. Industrial grade PCBs are mixtures of various congeners and isomers of chlorinated biphenyl. The most common commercial PCB products contained an average of from 3 to 5 chlorine atoms per molecule.

Physical and Chemical Properties

The individual isomers of the chlorobiphenyls vary from liquids to waxes to crystalline solids. Mixed isomers have properties quite different from the individual isomers. PCBs are considered to be generally chemically inert. However, they will react with certain materials under high temperature conditions, eg, with sodium hydroxide under extreme conditions they yield phenolic materials. They are insoluble in water, glycerol, and glycols, but soluble in most organic solvents. They resist oxidation, are permanently thermoplastic at the higher chlorination levels, and are extremely fire resistant. They have high dielectric constants, high volume resistivities and dielectric strengths, and low power factors.

Health and Safety

Domestic U.S. production of PCBs was voluntarily halted in Oct. 1977 because of these products' tendency to accumulate and persist in the environment (owing to their low degradation rates), and because of some alleged toxic effects in animal tests. Prolonged exposure to PCB vapor at high temperature may lead to systemic toxic effects in certain animal species. Small amounts of PCB can accumulate in the food chain; PCBs are fat soluble and are stored in the lipids of animals where they resist metabolic changes. PCBs are destroyed by incineration at high temperatures ($> 1100°C$) with long residence time in properly designed incinerators. The amount of PCB present in environmental materials can be determined using ATSM D3304-74 "Standard Method for Analysis of Environmental Materials for Chlorinated Biphenyls."

Uses

PCBs are no longer manufactured commercially in the United States, and their use and disposal is totally regulated by the EPA under the TSCA. Former uses were wide-ranging owing to their unique blend of fire resistance, thermal and oxidative stability, electrical characteristics, solvency, inertness, and liquid range. The largest use was in electrical equipment such as transformers and capacitors.

ROGER E. HATTON
Monsanto Company

Reviewed by
JOHN H. CRADDOCK
Monsanto Company

Interdepartmental Task Force on PCBs, *Polychlorinated Biphenyls and the Environment*, COM-72-10419, Washington, D.C., 1972.

O. Hutzinger, S. Safe, and V. Zitko, *The Chemistry of PCBs*, CRC Press, Cleveland, Ohio, 1974; contains reviews on preparation, chemistry, and properties of chlorobisphenyl isomers.

Proceedings of the National Conference on Polychlorinated Biphenyls, Chicago, Ill., 19–21, Nov. 1975, EPA-560/6-75-004, 1976.

Polychlorinated Biphenyls, National Research Council, National Academy of Sciences, Washington, D.C., 1979.

CHLOROCARBONS AND CHLOROHYDROCARBONS — VINYL CHLORIDE.   See Vinyl polymers.

CHLOROCARBONS AND CHLOROHYDROCARBONS — VINYLIDENE POLYMERS.   See Vinylidene chloride.

CHLOROHYDRINS

The chlorohydrins are organic compounds containing one or more chlorine atoms and one or more hydroxyl groups. With the possible exception of Jaconine, they are unknown as natural products and are prepared by synthetic methods.

The commercially important chlorohydrins are epichlorohydrin (1-chloro-2,3-epoxypropane), a chloroether and the only chlorohydrin sold on a large scale; the 1-2-chlorohydrins, eg, propylene α- and β-chlorohydrin; and glycerol chlorohydrins, esters of hydrochloric acid, and glycerol (qv). Chlorohydrins in which the hydroxyl and chlorine groups are not on adjacent carbons (eg, trimethylene chlorohydrin) are not commercially important.

Physical and Chemical Properties

Table 1 lists selected physical properties of the chlorohydrins. These compounds are usually colorless liquids of varying viscosity. With the exception of ethylene chlorohydrin, they are not miscible with water, and are less soluble than the corresponding glycols. They are soluble in ethyl alcohol, acetone, and ethyl ether, and moderately soluble in hydrocarbon solvents. Their boiling points range between those of the corresponding glycols and dichlorides. They are slightly unstable and yellow on standing.

Chemically, they behave either as alcohols or chlorides, and show the typical reactions of these compounds. The chlorine in 1,2-chlorohydrins is very labile; it dehydrohalogenates readily through an intermediate epoxide, which hydrolyzes to glycols (see Epoxidation). Bromohydrins and iodohydrins can be prepared similarly to chlorohydrins, and are often used as intermediates in organic synthesis.

Both the glycerol mono- and dichlorohydrins occur in two isomeric forms; because of their reactivity and their many functional groups, they are important intermediates. Reactions of the glycerol dichlorohydrins can be divided into reactions involving both the chlorine and hydroxyl

Table 1. Selected Physical Properties of Chlorohydrins

Compound	Mp, °C	Bp, °C	Density, d_4^{20}, g/cm^3
ethylene and propylene chlorohydrins			
ethylene chlorohydrin (2-chloro-1-ethanol)	−62.6	128.7[a]	1.2045 (d_{20}^{20})
propylene β-chlorohydrin (2-chloro-1-propanol)		133–134[a]	1.1092
propylene α-chlorohydrin (1-chloro-2-propanol)		127.4[a]	
trimethylene and tetramethylene chlorohydrins			
trimethylene chlorohydrin (3-chloro-1-propanol)		165[a]	1.1318
tetramethylene chlorohydrin (4-chloro-1-butanol)		84–85[b]	1.0083
glycerol mono- and dichlorohydrins			
α-monochlorohydrin (3-chloro-1,2-propanediol)		213[a]	1.3204
β-monochlorohydrin (2-chloro-1,3-propanediol)		146[c]	1.4831
α,γ-dichlorohydrin (1,3-dichloro-2-propanol)		175[a]	1.3645
α,β-dichlorohydrin (2,3-dichloro-1-propanol)		182[a]	1.3607, 1.3616
epichlorohydrin (1-chloro-2,3-epoxypropane)	−57.2	30–32	1.18066

[a]101.3 kPa (to convert kPa to mm Hg, multiply by 7.5).
[b]16 kPa.
[c]2.4 kPa.

groups, eg, usually base catalyzed with epichlorohydrin formed as the intermediate, and reactions involving either the hydroxyl, eg, esterification with organic acids or acid chlorides, or the chlorine, eg, formation of mono- and diazido derivatives.

Epichlorohydrin (chloromethyl oxirane) exists as a racemic mixture containing equal amounts of dextro- and levorotary forms. It can be an intermediate in the synthesis of a wide variety of products, including derivatives of monochlorohydrin, dichlorohydrins, glycidol, and glycerol.

Preparation

Glycerol was the main source for the preparation of glycerol chlorohydrins until the process for direct substitutive chlorination of propylene (qv) to allyl chloride paved the way for the synthesis by chlorohydrination of allyl chloride. Chlorohydrins are generally prepared by chlorohydrination, the addition of hypochlorous acid to olefins (eg, glycerol chlorohydrins), or by reaction of hydrochloric acid with epoxides or glycols. Major side reactions that occur during chlorohydrination are chlorination (formation of dichlorides) and formation of chloro ethers, both of which processes can be inhibited.

Health and Safety

The chlorohydrins are harmful when taken internally and are absorbed through the skin. Ethylene chlorohydrin is toxic; air concentrations should be held below 5 ppm. The glycerol mono- and dichlorohydrins have low vapor pressures at moderate temperatures, but care should be taken to avoid exposure by inhalation. Epichlorohydrin is a toxic, severely irritating compound. OSHA limits exposure to less than 5 ppm. It is flammable, and can form explosive mixtures with air at elevated temperatures. It may polymerize and burst its container when heated. Toxic gases are released when it burns.

Uses

Ethylene chlorohydrin was formerly used as the principal source of ethylene oxide (qv). Propylene chlorohydrin is an intermediate of commercial significance, eg, in the production of polyurethane foams. The glycerol chlorohydrins are available as specialty products. The glycerol dichlorohydrins are produced on a large scale for captive use; they are hydrolyzed to epichlorohydrin. Epichlorohydrin is used to manufacture glycerol (qv) and epoxy resins (qv), with smaller volumes used in a variety of applications such as the synthesis of glycerol and glycidol derivatives used as intermediates.

GREGOR H. RIESSER
Shell Development Company

J. Myszkovski and co-workers, *Chimie et Industrie* **91**, 654 (1964).

G.H. Riesser and J.G. Riesser in *Encyclopedia of Chemical Analysis*, Vol. 14, John Wiley & Sons, Inc., New York, 1971, pp. 153–178.

Houben-Weyl, *Methodern der Organischen Chemie*, Band V/3, Georg Thieme Verlag, 1962, p. 768.

Epichlorohydrin, Shell Chemical Corp., New York, 1959; *Toxicity and Safety Bulletin*, Shell Chemical Corp., Houston, Texas, 1976.

CHLOROPHENOLS

All of the nineteen possible chlorinated phenols are commercially available; 2,4-dichlorophenol, pentachlorophenol, 2,4,5-trichlorophenol, and o- and p-chlorophenol the most important.

Monochlorophenols

Of the three isomers of monochlorophenol, the 2- and 4-monochlorinated phenols are utilized in modest volumes, mainly as dyestuff intermediates and in the manufacture of higher chlorinated phenols. 4-Chlorophenol (mp 40–41°C, bp 219°C) is used as an intermediate in the manufacture of the disinfectant Chlorophene (2-benzyl-4-chlorophenol) (see Disinfectants) and quinizarin 1,4-dihydroxyanthraquinone.

Treatment of phenol with SO_2Cl_2 in the presence of Fe, $FeCl_3$, or $ZnCl_2$ gives primarily the para isomer. Higher concentrations and higher temperatures during the chlorination of phenol with *tert*-butyl hypochlorite and chlorine tend to favor the formation of the ortho isomer. Solvents such as acetonitrile, nitromethane, and bis-2-chloroethyl ether diminish ortho substitution. Monochlorinated phenols have been obtained by chlorination of sodium p-phenol sulfonate and phenyl phosphate; via the hydrolysis of di- and trichlorinated benzenes. Dehalogenations on polysubstituted chlorophenols yield mixtures of monochlorinated phenols. Chlorocumenes (o- and p-) oxidize to form the corresponding peroxides, which are converted to the corresponding chlorophenols. *m*-Chlorophenol is prepared by the oxidation of p-chlorobenzoic acid with stoichiometric quantities of copper(II) oxide as the oxidizing agent.

Dichlorophenols

2,4-Dichlorophenol (mp 43–44°C, bp = 210–211°C) is of commercial interest; it is used in large volumes in the manufacture of 2,4-dichlorophenoxyacetic acid (2,4-D). It is manufactured via the chlorination of phenol. Hydrolysis of 1,2,4-trichlorobenzene with copper, iron, or zinc halide catalysts results in 3,5-dichlorophenol (mp 68°C, bp 233°C). 3,5-Dichlorophenol is also prepared by hydrogenation of polychlorinated phenols over heavy metal catalysts with additives, and via partial dehalogenation of polychlorophenol using group VIII metals and sulfur or sulfide catalysts.

Trichlorophenols

2,4,5-Trichlorophenol (mp 68°C, bp 245–246°C) is used in the manufacture of herbicides and insecticides. It is manufactured via the chlorination of benzene with four moles of chlorine to form 1,2,4,5-tetrachlorobenzene, which is hydrolyzed in base to form the desired product. 2,4,5-trichlorophenol is also prepared during the chlorination of 3,4-dichlorophenol.

Tetrachlorophenols

2,3,4,6-Tetrachlorophenol (mp 69–70°C, bp 64°C (3.0 kPa or 22.5 mm Hg)) is available commercially and is used as a preservative. The chlorination of phenol using (specifically) potassium tellurate, K_2TeO_3, as catalyst gives exclusively tetrachlorinated phenols.

Pentachlorophenol

Pentachlorophenol (Penta, PCP) and its sodium salt are used extensively as antimicrobial agents. Pentachlorophenol (mp 190°C, bp 309–310°C) is manufactured via the chlorinatin of phenol at 100–180°C using catalysts, eg, $AlCl_3$; the reaction is complete in 5–15 h. Catalyst concentration is critical (about 0.0075 mol/mol of phenol). Dehalogenations are usually carried out on pentachlorophenols using alumina granules impregnated with $CuCl_2$.

Health and Safety

Many of the chlorinated phenols are readily absorbed through the skin in toxic amounts. The degree of toxicity displayed by the dioxins can vary from very extreme to nontoxic and is dependent upon the exact positions that are occupied by chlorine atoms on the aromatic rings.

Regarding chlorophenols in the environment: they are much more environmentally stable than the parent unsubstituted phenol; as the number of chlorine atoms increases, the rate of decomposition decreases; compounds containing a meta chlorine are more persistent than compounds lacking a chlorine atom in positions meta to the hydroxyl group. Higher chlorinated phenols will be more persistent in the environment than the lower chlorinated analogues.

E.R. FREITER
Dow Chemical U.S.A.

CHLOROPHYLL. See Dyes, natural.

CHLOROPRENE, $CH_2=CHCCl=CH_2$. See Chlorocarbons and chlorohydrocarbons.

CHLOROSULFURIC ACID

Chlorosulfuric acid, ClSO₃H, is a highly reactive compound containing equimolar quantities of HCl and SO₃. The clear, colorless liquid is actually an equilibrium mixture of chlorosulfuric acid with minor amounts of hydrogen chloride, sulfur trioxide, and some related compounds. The acid reacts violently with water, producing heat and dense white fumes of hydrochloric acid and sulfuric acid. It reacts with almost all organic materials, in some cases with charring.

Physical and Chemical Properties

Fp −81 to −80°C; bp 151–152°C (with dec); d_4^{20} 1.753 g/cm³. Heating chlorosulfuric acid results in the equilibrium formation of sulfuryl chloride, sulfuric acid, pyrosulfuryl mono- and dichlorides, and pyrosulfuric acid. Heating beyond the boiling point results in decomposition into SO_2, Cl_2, and H_2O. Distillation tends to degrade the acid rather than purify it. Chlorosulfuric acid is miscible with sulfur trioxide, 100% sulfuric acid, and pyrosulfuric chloride in all proportions; it is soluble in *sym*-tetrachloroethane ($C_2H_2Cl_4$), chloroform and dichloromethane. Chlorosulfuric acid is insoluble in carbon tetrachloride and carbon disulfide; it is soluble in acetic acid and acetic anhydride, trifluoroacetic acid and its anhydride as well as in sulfuryl chloride. It reacts with alcohols, ketones, diethyl ether, and dimethyl sulfoxide. The solubility of hydrogen chloride in chlorosulfuric acid is ca 2.4 mol/(L·MPa) but decreases rapidly with increasing temperature.

Chlorosulfuric acid is a strong acid containing a relatively weak sulfur–chlorine bond. It is a powerful sulfating and sulfonating agent, a fairly strong dehydrating agent, and a specialized chloridating agent. The reactions are the result of attachment of an —SO₃H group to give a sulfate or sulfonate. In the presence of excess acid, sulfonyl chlorides are formed. Sulfamation forms a —C—N—S— bond, as in $C_6H_{11}NHSO_3Na$, sodium cyclohexylsulfamate. The fluorine analogue of chlorosulfuric acid, fluorosulfuric acid, is considerably mores stable because of the strong fluorine–sulfur bond. Many salts and esters of chlorosulfuric acid are known, most of them relatively unstable or hydrolyzing readily in moist air. Strong dehydrating agents, eg, phosphorus pentoxide or sulfur trioxide, convert chlorosulfuric acid to its anhydride, pyrosulfuryl chloride (disulfuryl dichloride), $S_2O_5Cl_2$. Noncatalytic decomposition at elevated temperature generates sulfuryl chloride, chlorine, sulfur dioxide, and other compounds. The acid is rather slow to react with aliphatic hydrocarbons unless a double bond or some other reactive group is present.

Manufacture

Chlorosulfuric acid is commercially produced in a continuous reaction involving the direct union of sulfur trioxide with dry hydrogen chloride gas at 50–70°C. The sulfur trioxide may be in the form of 100% liquid or gas, or may be present as a dilute gaseous mixture obtained directly from a contact sulfuric acid plant. Two processes for the manufacture of high quality acid are described in a number of patents.

Health and Safety

Principal hazard is contact with the liquid acid which severely burns body tissue. The vapor is hazardous and extremely irritating to the skin, eyes, nose, and throat. When exposed to air, fumes react with moisture of the air and release highly irritating and corrosive hydrochloric acid fumes and sulfuric acid mist which may cause delayed lung damage. Chlorosulfuric acid may cause ignition by contact with combustible material. During storage, excessive pressures caused by liquid thermal expansion can create hazardous acid sprays or pipe rupture.

Uses

Chlorosulfuric acid is used in detergents, pharmaceuticals (eg, sulfa drugs, synthetic sweeteners, diuretics), dyes and pigments, ion-exchange resins, catalysis, and miscellaneous uses (eg, preparation of pesticides, plasticizers, condensing agents).

H.O. BURRUS
E.I. du Pont de Nemours & Co., Inc.

K.E. Jackson, *Chem. Revs.* **25**, 81 (1939).

CHOCOLATE AND COCOA

All cocoa and chocolate products are derived from the cocoa bean. In the United States, chocolate and cocoa are foods standardized by the U.S. Food, Drug and Cosmetic Act.

Cocoa

Worldwide cocoa bean production has increased from about 75,000 metric tons in 1895 to well over 10⁶ t today. The Ivory Coast is by far the largest single producer, followed by Brazil and Ghana. Recent government assistance programs have enabled Brazilian farmers to make significant progress in cocoa production. Because this assistance has emphasized research to increase cocoa bean yields and prevent disease, many experts predict that Brazil will be the leading producer of beans in the future.

Consumption of cocoa beans has, for the most part, paralleled supply. The United States has been the largest single consumer of cocoa beans for many years. At present, about one half of the world supply undergoes some processing in the producing country. Though some of this cocoa is consumed, most is converted to cocoa butter, cocoa cake, cocoa powder, and unsweetened chocolate (chocolate liquor) for export. Brazil is leading the trend by processing as much as 40% of its annual crop.

An International Cocoa Agreement, signed in 1972, governs the marketing of much of the world's cocoa. However, most chocolate and cocoa products imported into the United States are marketed through the New York Cocoa Exchange, Inc.

Cocoa Beans

Prior to shipment from producing countries, most cocoa beans undergo a process known as curing, fermenting, or sweating. These terms are used rather loosely to describe a procedure in which seeds are removed from the pods, fermented, and dried. Unfermented beans, particularly from Haiti and the Dominican Republic, are used in the United States.

Cocoa beans vary widely in quality, necessitating a system of inspection and grading to ensure uniformity. Only recently, however, has a procedure for grading beans been established at an international level. It classifies beans in two major categories according to the fraction of moldy, slaty, flat, germinated, and insect-damaged beans.

Manufacture of Cocoa and Chocolate Products

The cocoa bean is the basic raw ingredient in the manufacture of all cocoa products. The beans are converted to chocolate liquor, the primary ingredient from which all chocolate and cocoa products are made. Figure 1 depicts the conversion of cocoa beans to chocolate liquor, and in turn, to cocoa powder, cocoa butter, and sweet and milk chocolate, the chief chocolate and cocoa products manufactured in the United States.

Most chocolate and cocoa products consist of blends of beans chosen for flavor and color characteristics. Cocoa beans may be blended before or after roasting, or nibs may be blended before grinding. In some cases finished liquors are blended.

Chocolate liquor. Chocolate liquor is the solid or semiplastic food prepared by finely grinding the kernel or nib of the cocoa bean. It is also commonly called chocolate, baking chocolate, or cooking chocolate.

Cocoa powder. When chocolate liquor is exposed to pressures of 35–41 MPa (5000–6000 psig) in a hydraulic press, and part of the fat (cocoa butter) is removed, cocoa cake (powder) is produced. Cocoa powder is produced by grinding cocoa cake. Commercial cocoa powders are produced for various specific uses and many cocoas are alkali treated or "Dutched" to produce distinctive colors and flavors. The alkali process can involve the treatment of nibs, chocolate liquor, or cocoa with a wide variety of alkalizing agents. Cocoa powders not treated with alkali are known as cocoa, natural cocoa or American processed cocoa. Natural cocoa has a pH of ca 5.4–5.8 depending upon the type of cocoa beans used. Alkali-processed cocoa ranges in pH from ca 6–8.

Cocoa butter. Cocoa butter is a unique fat with specific melting characteristics. It is a solid at room temperature, starts to soften around

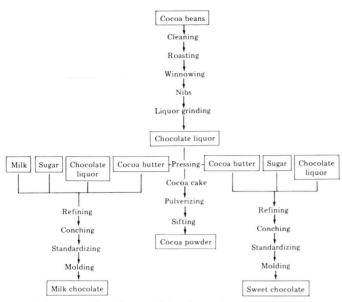

Figure 1. Flow diagram of chocolate and cocoa production.

30°C and melts completely just below body temperature. Its distinct melting characteristic makes cocoa butter the preferred fat for chocolate products. Cocoa butter is composed mainly of glycerides of stearic, palmitic, and oleic acids.

In the past 25 years, many fats have been developed to replace part or all of the added cocoa butter in chocolate-flavored products. These fats fall into two basic categories commonly known as cocoa butter substitutes and cocoa butter equivalents. Neither can be used in the United States in standardized chocolate products, but substitutes of all types enjoy widespread use in the United States chiefly as ingredients in chocolate-flavored products.

Sweet and milk chocolate. Most chocolate consumed in the United States is consumed in the form of sweet chocolate or milk chocolate. Sweet chocolate is chocolate liquor to which sugar and cocoa butter have been added. Milk chocolate contains these same three ingredients and milk or milk solids.

Theobromine and Caffeine

Chocolate and cocoa products, like coffee, tea and cola beverages, contain alkaloids (qv). The predominant alkaloid in cocoa and chocolate products is theobromine, although significant amounts of caffeine may be present, depending upon the origin of the beans. Chocolate and cocoa products also supply proteins, fats, carbohydrates, vitamins, and minerals.

B.L. ZOUMAS
B.J. FINNEGAN
Hershey Foods Corp.

W.T. Clarke, *The Literature of Cacao*, American Chemical Society, Washington, D.C., 1954.

Gill and Duffus, *Cocoa Market Report No. 278*, Dec. 1977.

P. Kalustian, *Candy Snack Industry* **141**(3), (1976).

CHOLINE

Choline (trimethyl(2-hydroxyethyl)ammonium hydroxide), $[(CH_3)_3$-$NCH_2CH_2OH]^+ OH^-$, derives its name from bile (Greek *cholē*), from which it was first obtained. Choline is a colorless, hygroscopic liquid with an odor of trimethylamine. Choline is a constituent of certain phos-

pholipids universally present in protoplasm. It is a dietary requirement for birds, but mammals can synthesize it if methionine, folic acid and vitamin B_{12} are supplied in the diet. If the amounts of any of these three substances are inadequate, choline deficiency may occur, unless choline is supplied in the diet. It is also of great physiological interest because one of its esters, acetylcholine (ACh), appears to be responsible for the mediation of parasympathetic nerve impulses and has been postulated to be essential to the transmission of impulses of all nerves. ACh and other more stable compounds that simulate its action are pharmacologically important because of their powerful effect on the heart and on smooth muscle. Choline is used clinically in liver disorders and as a constituent in animal feeds, especially poultry feeds.

Physical and Chemical Properties

Choline is a strong base. It crystallizes with difficulty and is usually known as a colorless deliquescent syrupy liquid that absorbs carbon dioxide from the atmosphere. Choline is very soluble in water and in absolute alcohol, but insoluble in ether. It is stable in dilute solution but in concentrated solutions tends to decompose at 100°C, giving ethylene glycol, polyethylene glycol, and trimethylamine.

Choline is not usually encountered as the free base but as a salt; the one most commonly used is the chloride $[(CH_3)_3N(CH_2CH_2OH)]^+ Cl^-$. As a quaternary ammonium hydroxide, choline reacts with hydrochloric acid to form the chloride and water, whereas primary, secondary, and tertiary amines combine with hydrochloric acid to form so-called hydrochlorides without the elimination of water (see Amines; ammonium compounds). Choline chloride is a crystalline deliquescent salt, usually with a slight odor of trimethylamine, and with a strongly brackish taste somewhat resembling that of ammonium chloride. It is very soluble in water, freely soluble in alcohol, slightly soluble in acetone and chloroform, and practically insoluble in ether, benzene, and ligroin.

In nutrition, the most important function of choline appears to be the formation of lecithin (qv) and other choline-containing phospholipids.

Choline occurs widely in nature and, prepared synthetically, it is available as an article of commerce. Choline is produced and used for medicinal and nutritional purposes as the chloride, dihydrogen citrate, and bitartrate and as tricholine citrate.

Derivatives

The most important derivatives of choline are acetylcholine (ACh) $[CH_3CO_2CH_2CH_2N(CH_3)_3]^+ OH^-$, the vagus substance produced in the body in small quantities by the stimulation of the cholinergic (parasympathetic) nerves; acetyl-β-methyl choline chloride (methacholine chloride) $[CH_3CO_2CH(CH_3)CH_2N(CH_3)_3]^+ Cl^-$, used as a parasympathetic stimulant and as an antiepinephrine substance; chlorocholine chloride (2-chloroethyltrimethylammonium chloride), $[ClCH_2CH_2N(CH_3)_3]^+$, mp 245°C (dec), a colorless, crystalline deliquescent salt, very soluble in cold water and alcohols, stable in aqueous solution and used as a plant growth regulant; and carbamoylcholine chloride (carbachol), $[H_2NCO_2$-$CH_2CH_2N(CH_3)_3]^+ Cl^-$, used to increase peripheral circulation in cases of vasospasm in peripheral vascular disease.

THOMAS H. JUKES
University of California at Berkeley

C.W. Prince and W.C.M. Lewis, *Trans. Faraday Soc.* **29**, 775 (1933).

E. Kahane and J. Lévy, *Biochemie de la Choline et de ses Dérivés*, Hermann, Paris, 1938.

W.M. Latimer and W.H. Rodebush, *J. Am. Chem. Soc.* **42**, 1419 (1920).

CHOLINESTERASE INHIBITORS

The cholinesterases are a group of hydrolytic enzymes present in both vertebrates and insects. Of these, acetylcholinesterase (AChE) is specifically used for the hydrolysis of the ion acetylcholine (**1**); a neurotransmitter at synapses and at neuroeffector junctions of the cholinergic nervous system.

(1)

Inhibition of AChE is a major cause of increased stimulation of the cholinergic nervous system through increased concentrations of (1) (see Neuroregulators). Cholinesterase (ChE) inhibitors are useful in medicine and as insecticides (see Insect control technology). They have potential application as nerve gases in chemical warfare (see Chemicals in war).

Compounds That Inhibit ChE

A number of compounds inhibit the hydrolysis of ACh by AChE. Most of these are esters that serve as substrates for AChE and are hydrolyzed by it. In all cases they act as inhibitors since their hydrolyses proceed at rates appreciably lower than that of ACh.

Two types of compounds are used as inhibitors of AChE: esters of carbamic acid; and derivatives of the acids of phosphorus. The reduced rate of hydrolysis of carbamate and phosphate esters by AChE results from the regeneration of the esterified enzyme. In the case of carbamate esters, the enzyme is carbamoylated. In the case of the esters of phosphorus, it is phosphorylated. The rate required for the regeneration of the carbamoylated enzyme can be expressed in minutes, whereas that for the acetylated enzyme is measured in microseconds. The rate of hydrolysis of the phosphorylated enzyme is measured in hours.

Because of their prolonged action, the phosphate esters are used only for topical application in medicine but have been widely employed as insecticides. Carbamate esters are employed both in medicine and as insecticides.

Carbamate esters that inhibit AChE and are used in medicine include physostigmine, neostigmine bromide, pyrodostigmine bromide and demecarium bromide. Those used for insecticides include mobam, carbaryl, baygon, aldicarb, and isolan (see Insect control technology).

Derivatives of Phosphorus Oxo Acids

Derivatives of phosphoric acid, pyrophosphoric acid, phosphonic acid, and their thio derivatives are inhibitors of cholinesterases (see Phosphoric acid; Phosphorus compounds). Diisopropyl fluorophosphate (DFP), and echothiopate iodide are applied topically to relieve intraocular tension of glaucoma. Among the compounds used as insecticides are dicapthon, tetraethylpyrophosphate (TEPP), malathion, and ethyl p-nitrophenylphenylphosphonothioate (EPN).

Toxicity of AChE Insecticides

The toxicity of cholinesterase inhibitors varies widely. The relatively high toxicity of some of these compounds coupled with the long duration of action of the phosphorus derivatives poses a hazard to those involved in their application and transportation. This can be reduced by limiting their utilization to those trained properly and by the prompt administration of an effective antidote in case of poisoning.

J.E. GEARIEN
University of Illinois at the Medical Center

M.E. Wolff, ed., *Burger's Medicinal Chemistry*, 4th ed., Part III, Wiley-Interscience, New York, 1981, p. 450.

C.O. Wilson, O. Giswald, and R.F. Doerge, *Textbook of Organic Medicinal Chemistry and Pharmaceutical Chemistry*, 7th ed., J.B. Lippincott Co., Philadelphia, Pa., 1977.

CHROMATOGRAPHY.   See Analytical methods; Chromatography, affinity.

CHROMATOGRAPHY, AFFINITY

The process of affinity chromatography is quite simple. It involves a selective adsorbent that is placed in contact with a solution containing several kinds of substances including the desired species, the ligate. The ligate is selectively adsorbed to the ligand, which is attached via a leash (or tether) to the insoluble support or matrix. The nonbinding species are removed by washing. The ligate is then recovered by eluting with a specific desorbing agent.

Biological Specificity or Affinity

In addition to enzymes and antibody molecules, there are several other classes of biopolymers that, as ligates, exhibit significant affinities to appropriate ligands. The known permutations and combinations are listed in Table 1. In general, most affinity interactions are reversible.

Table 1.   Affinity Systems

Ligand, immobile entity	Ligate, soluble entity
inhibitor, cofactor, prosthetic group, polymeric substrate	enzymes; apoenzymes
enzyme	polymeric inhibitors
nucleic acid, single strand	nucleic acid, complementary strand
hapten; antigen	antibody
antibody (IgG)	proteins; polysaccharides
monosaccharide; polysaccharide	lectins; receptors
lectin	glycoproteins; receptors
small target compounds	binding proteins
binding protein	small target compounds

To prepare an affinity sorbent, the first task is to find a suitable ligand. The preferred ligand compound shows a high structure selectivity for the desired ligate and also possesses a second functional site where immobilization may be effected without adversely affecting ligate binding. Many ligands are selective for a group or class type of ligate species. Such group-selective ligands have been referred to as general affinity ligands. Ligands come in all sizes. Although size does not determine the intrinsic selectivity of a ligand, practical considerations of sorbent preparation are affected by size.

The next task requires an appropriate coupling chemistry to attach the ligand to the carrier support or matrix. In research applications, agarose gels and cross-linked agarose gels have been the most widely used support materials. A variety of covalent coupling methods for attaching ligands and their leashes to supports have been developed; their choice and use, however, depends on the chemical nature of the support itself. For carbohydrates such as agarose or cellulose, the reaction with cyanogen bromide in aqueous alkali has been the most common.

Operational Modes

The most commonly reported mode of using affinity sorbents is in column packing. Batch operation has been employed frequently when column arrangements have suffered flow rate or clogging problems, especially during the sample-loading steps. It is also convenient to work in batch mode when long contact times for ligate and sorbent are needed to obtain complete binding.

Analytical Applications

The bulk of principle affinity applications has been in isolation biopolymers for research purposes. Other areas of application include the study of enzyme reaction mechanisms with immobilized affinity ligands, and medical diagnostics applications of immunosorbents in a rapidly growing technique called radioimmunoassay (RIA), or its variant, enzyme immunoassay (EIA) (see Medical diagnostic reagents; Radioactive traces). Biomedical applications include treatment of jaundice, hypercholesterolemia, systemic lupus erythmatosus, cancer, and removal of hepatitis contaminants from collected human plasma.

A.H. NISHIKAWA
Hoffmann LaRoche Inc.

W.H. Scouten, *Affinity Chromatography*, John Wiley & Sons, Inc., New York, 1981.

J. Turkova, *Affinity Chromatography*, Elsevier, Amsterdam, Netherlands, 1978.

CHROME DYES. See Azo dyes; Dyes—Application and evaluation.

CHROMIUM AND CHROMIUM ALLOYS

Chromium is the 21st element in the earth's crust in relative abundance, ranking with V, Zn, Ni, Cu and W. With atomic number 24, it belongs to Group VIB of the periodic table. The only commercial ore, chromite, has the ideal composition $FeO.Cr_2O_3$, ie, 68 wt% Cr_2O_3, 32 wt% FeO or ca 46 wt% chromium. Actually, the Cr : Fe ratio varies considerably and the ores are better represented as (Fe,Mg)O. $(Cr,Fe,Al)_2O_3$. Chromite deposits occur in olivine and pyroxene-type rocks and their derivatives. Almost all U.S. chromium supplies come from the USSR and the Union of South Africa either as chromite or increasingly as ferrochrome.

The valence states of chromium are $+2$, $+3$, and $+6$, the latter two being the most common. The $+2$ and $+3$ states are basic, whereas the $+6$ is acidic, forming ions of the type $(CrO_4)^{2-}$ (chromates) and $(Cr_2O_7)^{2-}$ (dichromates). The blue-white metal is refractory and very hard. Its properties are listed in Table 1. Perhaps more so than any other common metal, the mechanical properties of chromium depend on purity, history, grain size, strain rate, and surface condition.

Very little chromite, of course, is processed all the way to ductile chromium, since most can be used in various intermediate forms. Ferrochrome is usually made by reduction of chromite with coke in a three-phase electric submerged-arc furnace. The principal pyrometallurgical process for commercial metal is the reduction of Cr_2O_3 by aluminum.

Electrowinning of chromium. Chromium is obtained by electrolysis of either chrome–alum or chromic acid. The Union Carbide Corporation's Metals Division Plant at Marietta, Ohio, is a typical chrome–alum plant with a capacity of 2000 metric tons per year. In this process, high carbon ferrochromium is leached with a hot solution of reduced anolyte plus chrome alum mother liquor and makeup sulfuric acid. The slurry is then cooled to 80°C by the addition of cold mother liquor from the ferrous ammonium sulfate circuit, and the undissolved solids, mostly silica, are separated by filtration. The chromium in the filtrate is then converted to the nonalum form by several hours' conditioning treatment at elevated temperature.

The feed for the alternative electrowinning process, chromic acid, CrO_3, is obtained from sodium dichromate (see Chromium compounds). Small amounts of an ionic catalyst, specifically sulfate, chloride, or fluoride, are essential to this electrolytic process. Fluoride and complex fluoride-catalyzed baths have become especially important in recent years. The product from this process contains less iron and oxygen than that from the chrome–alum electrolysis.

The metal obtained from the two electrolytic processes mentioned above contains considerable oxygen which is believed to cause brittleness at room temperature. For most purposes, the metal as plated is satisfactory. However, if ductile metal is desired, the oxygen can be removed by hydrogen reduction, the iodide process, calcium refining, or melting in a vacuum in the presence of a small amount of carbon. Thereafter, chromium metal may be consolidated by powder-metallurgy techniques or by arc melting in an inert atmosphere. The initial cast structure of arc-melted ingots must be carefully broken down by hot working in order to permit subsequent warm working.

Electroplating, Chromizing, and Other Chromium-Surfacing Processes

Electroplating of chromium on various substrates is practiced in order to realize a more decorative and corrosion- or wear-resistant surface. About 80% of the chromium employed in metal treatment is used for chromium plating; over 50% is for decorative chromium plating. Hard chromium plating differs from decorative plating mostly in terms of thickness. Hard chromium plate may be 10 to several 100 μm thick,

Table 1. Physical Properties of Chromium

Property	Value
atomic weight	51.996
isotopes, %	
50	4.31
52	83.76
53	9.55
54	2.38
crystal structure	bcc
a_0, nm	0.2844–0.2848
density at 20°C, g/cm³	7.19
melting point, °C	1875
boiling point, °C	2680
vapor pressure, 130 Pa[a], °C	1610
specific heat at 25°C, kJ/(mol·K)[b]	23.9 (0.46 kJ/(kg·K))
linear coefficient of thermal expansion at 20°C	6.2×10^{-6}
thermal conductivity at 20°C, W/(m·K)	91
electrical resistivity at 20°C, $\mu\Omega\cdot$m	0.129

[a] To convert Pa to mm Hg, multiply by 0.0075.
[b] To convert J to cal, divide by 4.184.

whereas the chromium layers in a decorative plate may be as little as 0.25 μm or about 2 g Cr/m² of surface. Hard plating is noted for its excellent hardness, wear resistance, and low coefficient of friction. Decorative chromium plating retains its brilliance because air exposure immediately forms a thin, invisible protective oxide film. The chromium is not applied directly to the surface of the metal but rather over a nickel plate that in turn is laid over a copper plate.

Chromizing is the other principal method of obtaining a chromium-rich surface on steel. The material to be treated is embedded in a mixture of ferrochrome powder, chromium halide, alumina, and sometimes NH_4Cl. The chromium is diffused in by a furnace treatment at about 1100°C to produce an effective stainless-steel surface whose mean composition is about 18% Cr and whose thickness is controlled by the time of treatment. This is an economical process for improving the corrosion resistance of steel parts where cut edges and appearance are not important considerations, eg, automotive exhausts, heat exchangers, and silos.

Other surface processes using chromium include sputtering, ion implantation (qv), chemical vapor deposition, metal spraying, cladding, and weld overlayment. Only the latter two have commercial significance. The most popular alloys deposited are the Co–Cr-based Stellites and similar compositions, but chromium carbides and oxides can also be deposited (see Metal surface treatments; Film deposition techniques).

Chromium Alloys

In addition to inorganic compounds of chromium, important as pigments and tanning agents, and certain organic compounds used in greases, catalysts and plastic compounding agents, there are a number of metallic compounds of chromium significant either in their own right or as metallurgical constituents in Cr-bearing alloys. The carbide Cr_3C_2 is important as a wear-resistant gauge material, CrB is used for oil-well drilling, and $Cr_xMn_{2-x}Sb$ as a magnetic material with unique characteristics. The intermetallic compounds Cr_3Al, Cr_3Si, and Cr_2Ti are encountered in developmental oxidation-resistant coatings. $Cr_{23}C_6$, Cr_7C_3, CrFe (σ phase), and $Cr_{12}Fe_{36}Mo_{10}$ (χ phase) are found as constituents in many alloy steels and $CrAl_7$ and CoCr in aluminum and cobalt-based alloys. The chromium-rich interstitial compounds (Cr_2H, Cr_2N, $Cr_{23}C_6$) play an important role in the effect of trace impurities on the properties of unalloyed chromium. The intermetallic and the interstitial compounds of chromium are stabilized by electronic and/or spatial factors and are not to be regarded as simple ionic or covalent compounds.

Alloying has not solved the problem of resistance to gaseous embrittlement. Alloying with yttrium improves the resistance to embrittlement by high temperature exposure to oxygen but not to nitrogen-bearing atmospheres, and a barrier-coating approach must be used. Furthermore, although solid-solution additions can improve high temperature strength by three- to four-fold over unalloyed chromium, these have also resulted in increases in the ductile-to-brittle transition temperature (DBTT). A better combination of properties has been achieved through precipitation or dispersion hardening. The second phases may be oxides, carbides, or borides.

Stainless Steels

The stainless quality is conferred on steels if they contain enough chromium to form a protective surface film. About 12 wt% chromium is required for protection in mild atmospheres or in steam. With 18–20 wt% chromium, sufficient protection is achieved for satisfactory performance in a wide variety of more destructive environments, including those occurring in the chemical, petrochemical, and the power-generating industries. Stainless grades with 25 wt% wt chromium or more and containing other alloying elements such as molybdenum provide even higher corrosion resistance. In certain stainless steels, chromium depresses martensite transformation below room temperature; by thus stabilizing austenite, it permits achievement of desired mechanical properties without loss of corrosion resistance (see Steel).

Other Alloy Steels

In low-alloy steels, chromium contributes more to hardenability, tempering resistance, and toughness than to solid-solution hardening or oxidation resistance. In the high-chromium tool steel compositions, chromium carbides improve the high hot hardness. Wrought alloy steels, alloyed cast irons and steels, and tool steels account for 20–25% of the annual U.S. consumption of chromium.

Nonferrous Alloys

Nonferrous alloys account for only about 2% of the total chromium used in the United States. Nonetheless, some of these applications are unique and constitute a vital role for chromium. For example, chromium confers corrosion and oxidation resistance on the nickel-base superalloys used in jet engines; the familiar electrical resistance heating elements are made of a Ni–Cr alloy; and a variety of Fe–Ni and Ni-based alloys used in a diverse array of applications (especially in the nuclear reactor field) depend on chromium for oxidation and corrosion resistance.

Recovery and Reuse

At the present time only about 15% of chromium consumed in the United States is recycled and this largely from stainless-steel scrap. Steps are being taken to improve recovery of the substantial chromium losses incurred in the past from refractory and foundry applications of chromite grain (see Recycling).

J.H. Westbrook
General Electric Co.

A.H. Sully and E.A. Brandes, *Chromium*, 2nd ed., Plenum Publ. Corp., New York, 1967.

E.R. Parker and co-workers, *Contingency Plan for Chromium Utilization*, NMAB Report 335, 1977.

D. Peckner and I.M. Bernstin, *Handbook of Stainless Steels*, McGraw-Hill, New York, 1977.

CHROMIUM COMPOUNDS

Some of the properties of a group of typical chromium compounds are given in Table 1 (p. 279).

In its compounds, chromium may use any of its six $3d$ and $4s$ electrons and may show any oxidation state from -2 to $+6$. Those of $+2$, $+3$, and $+6$ are the most important, whereas the -1 and -2 oxidation states are of little significance. Commercial applications center about the $+6$ state, with some interest in the $+3$ state. Research interest at present centers around kinetic studies involving the $+3$ state and structural studies of the π-bonded complexes of the lower oxidation states.

Manufacture

The two primary industrial compounds of chromium made directly from chrome ore are sodium chromate and sodium dichromate. Secondary chromium compounds produced in substantial quantity include potassium chromate and dichromate, ammonium dichromate, chromic acid (chromium(VI) oxide), and various formulations of basic chromic sulfate used principally for leather tanning.

Sodium chromate and dichromate. Sodium chromate is made by leaching of the calcine obtained by roasting chrome ore with soda ash to which lime and/or leached calcine may be added. On acidification, the chromate yields sodium dichromate, the principal primary chromium compound.

Other chromates and dichromates. Potassium and ammonium dichromates are generally made from sodium dichromate by a crystallization process involving equivalent amounts of potassium chloride or ammonium sulfate.

Chromate and dichromates are sold in both technical and reagent grades. Chlorides and sulfates are the principal impurities. Both manufacturers' and U.S. GSA specifications exist for the technical grades.

Chromic acid. Chromic acid is produced by the reaction of sulfuric acid with sodium dichromate: $Na_2Cr_2O_7 + 2\ H_2SO_4 \rightarrow 2\ CrO_3 + NaHSO_4 + H_2O$.

Chromic sulfate, basic. Basic chromic sulfate is manufactured as a proprietary product under various trade names for use in leather tanning. It is made generally by reduction of sodium dichromate with either sulfur dioxide or sugar in the presence of sulfuric acid and contains sodium sulfate, small amounts of organic acids if carbohydrate reducing agents are used, plus various additives.

Health, Safety, and Environmental Considerations

From the practical standpoint, the only chromium compounds encountered are those in oxidation stages $+3$ and $+6$. Health considerations are concerned mainly with chromium(VI) compounds. Acute systemic poisoning is rare. Acute effects of chromates are mainly on the skin and mucous membranes. Exposure standards have been set at 0.1 mg CrO_3/m^3, or 0.05 mg $Cr(VI)/m^3$. Standards for soluble chromic salts are set at 0.5 mg Cr/m^3, and for chromium metal and insoluble chromic compounds at 1.0 mg Cr/m^3. Plants have to be designed and operated so that emissions are kept to a practical minimum. However, it is recognized that wastes are unavoidable. Waste chromium can, however, be converted to insoluble, inert chromium(III) compounds.

Chromium in nutrition. Certain evidence for chromium as an essential element in animal and human nutrition was reported in 1969 (see Mineral nutrients). Further work has established improvement in glucose tolerance, especially in the elderly, with supplemental chromium intake of the order of 150 $\mu g/d$. A definite recommended intake has not been established. The U.S. daily average of 80 $\mu g/d$ is probably marginal, as chromium is poorly assimilated from many foods. Chromium plays a role in sugar production in plants and sugar metabolism in animals.

Chromium compounds in the environment. Chromium is widely distributed in the earth's crust, but is concentrated in the basic, ultramafic rocks. At an overall crustal concentration of 125 ppm Cr, it is the twentieth most abundant element.

Uses

Chromium compounds are essential to many industries. The distribution of major uses is given in Table 2 (p. 279).

Winslow H. Hartford
Belmont Abbey College

"Chrom" in *Gmelins Handbuch der Anorganischen Chemie*, 8th ed., System No. 52, Springer-Verlag, N. Y., 1960–1963.

F.A. Cotton and G. Wilkinson, *Advanced Inorganic Chemistry: A Comprehensive Text*, 4th ed., Wiley-Interscience, a division of John Wiley & Sons, Inc., New York, 1980, pp. 721–736.

Table 1. Physical Properties of Typical Chromium Compounds

Compound	Formula	Appearance	Density, g/cm^3_t	Mp, °C	Bp, °C	Solubility
Oxidation state 0						
chromium carbonyl	$Cr(CO)_6$	colorless crystals	1.77_{18}	150 (dec) sealed tube	151 (dec)	sl sol CCl_4; insol H_2O, $(C_2H_5)_2O$, C_2H_5OH, C_6H_6
Oxidation state +1						
bis(biphenyl)chromium(I) iodide	$(C_6H_5C_6H_5)_2CrI$	orange plates	1.617_{16}	178	dec	sol C_2H_5OH, C_5H_5N
Oxidation state +3						
chromic chloride	$CrCl_3$	bright purple plates	2.87_{25}	subl	885	insol H_2O; sol presence Cr^{2+}
chromic acetylacetonate	$Cr(CH_3COCHCOCH_3)_3^3$	red-violet crystals	1.34	208	345	insol H_2O; sol C_6H_6
chromic potassium sulfate (chrome alum)	$KCr(SO_4)_2 \cdot 12H_2O$	deep purple crystals	1.826_{15}	89 (incongruent)		sol H_2O
Oxidation state +4						
chromium(IV) oxide	CrO_2	dark brown or black powder	4.98 (calcd)		dec to Cr_2O_3	sol acids to Cr^{3+} and Cr^{6+}
Oxidation state +5						
barium chromate(V)	$Ba_3(CrO_4)_2$	black-green crystals				sl dec H_2O; sol dil acids to Cr^{3+} and Cr^{6+}
Oxidation state +6						
chromium(VI) oxide	CrO_3	ruby-red crystals	2.7_{25}	197	dec	v sol H_2O; sol CH_3COOH, $(CH_3CO)_2O$
chromyl chloride	CrO_2Cl_2	cherry-red liquid	1.9145_{25}	-96.5	115.8	insol H_2O, hydrolyzes; sol CS_2, CCl_4
ammonium dichromate	$(NH_4)_2Cr_2O_7$	red-orange crystals	2.155_{25}	dec 180		sol H_2O
potassium dichromate	$K_2Cr_2O_7$	orange-red crystals	2.676_{25}	398	dec	sol H_2O
sodium dichromate	$Na_2Cr_2O_7 \cdot 2H_2O$	orange-red crystals	2.348_{25}	84.6 incongruent	dec	v sol H_2O
potassium chromate	K_2CrO_4	yellow crystals	2.732_{18}	971		sol H_2O
sodium chromate	Na_2CrO_4	yellow crystals	2.723_{25}	792		sol H_2O
barium chromate	$BaCrO_4$	pale yellow solid	4.498_{25}	dec		v sl sol H_2O; sol strong acids
strontium chromate	$SrCrO_4$	yellow solid	3.895_{15}	dec		sl sol H_2O; sol dil acids
lead chromate	$PbCrO_4$	yellow solid orange solid red solid	6.12_{15}	844		practically insol H_2O; sol strong acids

Table 2. Distribution of Uses of Chromium Chemicals, 1979

Use	Percent
metal finishing and corrosion control	43.5
pigments and allied products	21.6
leather tanning and textiles	14.9
wood preservation	11.0
drilling muds	3.0
other uses (catalysts, intermediates)	6.0

Division of Medical Sciences, National Research Council, *Chromium*, Report of Panel, A.M. Baetjer, Chairman, National Academy of Sciences, Washington, D.C., 1974.

W.H. Hartford in R. Thompson, ed., *Specialty Inorganic Chemicals*, The Royal Society of Chemistry, London, 1981, pp. 311–345.

CHROMOGENIC MATERIALS

PHOTOCHROMIC

Photochromism is the phenomenon in which absorption of electromagnetic energy by a system has resulted in a change in its color. It can also be defined as the phenomenon in which a single chemical species is caused, by the absorption of electromagnetic radiation, to reversibly change to a state having an absorption spectrum different than the first. The word reversibly does not exclude examples having immeasurably slow reverse rates, merely those in which the state resulting from the photolysis is thermodynamically much more stable than the initial state. The single chemical species may be a molecule, an ion, or even a trapped electron or hole. The definition of photochromism, in terms of a single chemical species, is intended to exclude reversible processes operating through a cyclic series of reactions. Specification of the rates of the forward and reverse reactions is an important part of the characterization of photochromic materials.

Organic Photochromic Materials

Organic photochromic materials have not yet been proven useful although they have been studied extensively. They are all plagued by small spectral shifts between the excited and unexcited species, slow reversal times, fatigue, or any combination of the three. Mechanical constraints imposed for certain applications also eliminate many organic materials from consideration. The most important mechanisms leading to photochromism of organic materials include heterolytic cleavage, homolytic cleavage, cis-trans isomerization, photoinduced tautomerism, and triplet states.

Inorganic Photochromic Systems

Most inorganic photochromic materials are solids that have a band gap in excess of 3 eV and consequently do not absorb in the visible in their unactivated state. Absorption of light having energy equal to or greater than the band gap creates electron–hole pairs. Color centers are formed by the trapping of the electrons and holes at defects in the structure. Photochromism, when it is observed in inorganic solids, is determined by the nature of the defects characteristic of the solid.

Photochromic Glass

Photochromic glass has several advantages over other photochromic materials simply because it is glass. By standard glass melting and forming techniques, it can be made in any desired size or shape. Its transparency and durability under chemical attack, scratching, or moderate heating offer advantages for some applications (see Glass). Most important is that photochromic glass usually shows no fatigue.

There are several families of homogeneous glasses that darken reversibly in response to ultraviolet excitation. Also, photochromic materials have been made by precipitating silver compounds, thallium compounds, or copper compounds from alkali aluminoborosilicate-base glasses. The materials other than the silver and copper halides were characterized by low darkening efficiencies.

Applications

Many proposed applications of photochromic materials are related to their ability to act as a light valve; other applications are based on the change in the material itself. Only in the ophthalmic field have photochromics had commercial success and thus far only photochromic glasses containing silver halides have been successfully used in that application.

ROGER J. ARAUJO
Corning Glass Works

R.J. Araujo in M. Tomozawa and R.H. Doremus, eds., *Treatise on Materials Science and Technology*, Vol. 12, Academic Press, Inc., New York, 1977, p. 91.

G.H. Brown, ed., *Photochromism*, Wiley-Interscience, a division of John Wiley & Sons, Inc., New York, 1971.

ELECTROCHROMIC AND THERMOCHROMIC

Electrochromic Materials

Electrochromism is a color change caused by an electric current. It has some use in determination of dipole moments of excited molecules and in the study of photosynthesis. Many uses are proposed in the patent literature for display devices and duplication of images (see also Chemiluminescence; Digital displays; Electrophotography).

Solid electrochromic materials. An electrochromic cell consists of the electrolyte solution and a pair of transparent electrodes, one of which bears the electrochromic material. The electrodes usually consist of a thin layer of stannic oxide (NESA glass), SnO_2, indium, or cadmium stannate, Cd_2SnO_4, on glass. The electrochromic material is most often tungsten trioxide, WO_3, although the whole range of transition-metal oxides appears in patents. The electrolyte layer may be sulfuric acid with gelling agents and metal salts added to improve conductivity and reduce leakage. The completed cell may be encapsulated to preserve its integrity.

Liquid cells. Electrochemical cells that reduce a soluble organic material to a colored form in cells of similar construction to those discussed above have been investigated using a variety of color-forming materials. These include salts of aromatic-substituted pyridines and bipyridines, pyrazolines, and picrylhydrazyl compounds, as well as solid complexes of diphthalocyanine with lanthanide metals as the negative electrode.

Photosynthesis. In photosynthesis a burst of light causes a shift in the spectrum of the photosensitive pigment molecule whose orientation is fixed by the membrane in which it resides. A study of thin layers of chlorophyll and carotenoids in oriented films has been made using the electrochromic effect.

Thermochromic Materials

Thermochromic compounds change color reversibly with temperature, gradually in solution, sharply on melting. Potential uses are many, but the best established is the irreversible system that leaves a record of heat distribution on an engine part or a reaction vessel.

Reversible color changes of organic compounds. Many compounds that can be designated as overcrowded ethylenes are thermochromic. Structural features in common are at least one ethylene group, a multiplicity of aromatic rings, and a hetero atom, usually nitrogen or oxygen.

Thermochromic materials include spiropyrans (di-β-naphthospiro-pyran), anils, pyridine or quinoline quaternary amines, and naphthothiazoles, low molecular weight polymers with two to six salicylidene units, and pyridinium-N-phenol betaines. Thermochromic materials that are a combination of an electron-donating compound that forms a weak association with an electron-accepting material may form a brightly colored complex in the dissolved or melted states, which vanishes on cooling or freezing. These behave essentially as mixtures in the solid state, but in the liquid state form colored complexes.

Polydiacetylenes alone or in a fluorocarbon polymer change color reversibly. It is useful over a wide range as a temperature indicator.

Reversible color changes of inorganic compounds. A few organic complexes of metal ions are reversibly thermochromic. The nitrogen bases complexed with Cu(II) or Ni(II) are most numerous, their color change being caused by a change from tetrahedral to square-planar geometry or by a change in the number of kinds of ligands present. Very important to the color change are the anions, most often ClO_4^- or BF_4^-, and the choice of solvent. The double salts Ag_2HgI_4 and Cu_2HgI_4 have been extensively studied. Aqueous solutions of some simple metal salts are thermochromic when heated to 230°C, including $FeCl_3$, $CuCl_2$, $CoCl_2$, $NiCl_2$, and UO_2Cl_2.

Fluorescence thermochromism. Monovalent copper halides complexed with nitrogen gases show strong ultraviolet fluorescence. The fluorescence color depends on the temperature and is called fluorescence thermochromism.

Irreversible thermochromism, temperature-indicating paints. Commercially available temperature-indicating paints and crayons change color between 40 and 1300°C, with an accuracy of about ±5%. Manufacturers' literature indicates which of these is suitable for use in the presence of steam, under water, and in various environments. Most have a single color change, but several have up to four distinct color changes at well-defined temperatures. Applications are in electrical, manufacturing, and chemical industries, in aeronautics and space flight, and in the development of various types of engines. Salts and organic complexes of cobalt, chromium, nickel, manganese, iron, vanadium, and copper are used most. The color change is owing to loss of water of hydration, the splitting out of some other small molecule, or decomposition.

JESSE H. DAY
Ohio University

Displays **3**(1), 3 (1982); **3**(2), 67 (1982).

Ind. Eng. Chem. Prod. Res. Dev. **21**(2), 261 (1982).

CINNAMIC ACID, CINNAMALDEHYDE, AND CINNAMYL ALCOHOL

These unsaturated aromatic compounds and their derivatives, characterized by the grouping $C_6H_5CH=CHCH_2-$, are important in the flavors, perfumes, cosmetics, pharmaceuticals, graphic arts, plastics, and polymers industries.

Cinnamic acid, 3-phenyl-2-propenoic acid, $C_6H_5CH=CHCOOH$, exists partly in the free state and partly in the form of esters in ethereal oils, resins, storax (styrax), balsams of Tolu and Peru, and coca leaves. Commercial cinnamic acid occurs as pale yellow to off-white crystals with a characteristic aromatic odor. Its solubility in water is only 0.546 g/L at 25°C. It is more soluble in absolute alcohol and very soluble in ether. Cinnamic acid occurs in cis and trans forms. Ordinary cinnamic acid possesses the trans configuration; mp, 133°C; bp, 300°C; d_4^4, 1.2475; heat of combustion, 4477 kJ/mol (1040 kcal/mol); K_{25}, 3.5×10^{-5}. Cinnamic acid undergoes reactions typical of a carboxyl group, an olefinic double bond, and the benzene nucleus. *trans*-Cinnamic acid is synthesized by the Perkin reaction, by heating benzaldehyde, acetic anhydride, and anhydrous sodium acetate to 180–200°C.

Cinnamaldehyde, 3-phenyl-2-propenal, $C_6H_5CH=CHCHO$, occurs naturally in Chinese cinnamon oil from the leaves and twigs of *Cinnamonum cassia*. It is obtained by steam distillation and further rectified by vacuum distillation. It also occurs in Sri Lanka (Ceylon) cinnamon oil from *C. zeylanicum*. Cinnamaldehyde is a yellow mobile liquid that turns to a dark brown viscous liquid on exposure to light and air. It has the characteristic odor of cinnamon oil and a burning aromatic taste. Cinnamaldehyde is slightly soluble in water, infinitely soluble in ether, and insoluble in petroleum ether; its solubility in 50% alcohol is 1 part in 25 parts, and in 70% alcohol, 1 part in 2–3 parts. It is volatile with steam. Cinnamaldehyde has the following properties: mp, 7.5°C; bp, 252°C at 101 kPa (760 mm Hg); d_{20}^{20}, 1.1102; n_D^{20}, 1.61949. It reacts typically as an aldehyde. Cinnamaldehyde is made commercially by the condensation of benzaldehyde and acetaldehyde in the presence of dilute sodium hydroxide, hydrochloric acid, or sodium ethylate.

Cinnamyl alcohol, 3-phenyl-2-propen-1-ol, styrylcarbinol, cinnamic alcohol, $C_6H_5CH=CHCH_2OH$, does not occur free in nature, but is usually found as an ester, eg, cinnamyl cinnamate in storax and balsam of Peru. It forms colorless needles and has an odor resembling hyacinths. Cinnamyl alcohol is slightly soluble in water and very soluble in alcohol and ether. Cinnamyl alcohol has the following properties: mp, 33°C; bp, 257°C (cor) at 101 kPa (760 mm Hg); d_{35}^{35}, 1.0397; n_D^{33}, 1.57580. Cinnamyl alcohol is made commercially by the Meerwein-Ponndorf reduction of cinnamaldehyde using aluminum isopropylate or ethylate.

W.F. Ringk
Consultant

F.K. Beilstein, *Handbuch der Organischen Chemie*, Vol. 6, 4th ed., Springer, Berlin, 1923, pp. 570–571; 1st Suppl., Vol. 6, 1932, p. 281; 2nd Suppl., Vol. 6, 1949, pp. 525–528; 3rd Suppl., Vol. 6, 1965, pp. 2670–2680; Vol. 7, 1925, pp. 348–359; 1st Suppl., Vol. 7, 1931, pp. 187–190; 2nd Suppl., Vol. 7, 1948, pp. 273–283; 3rd Suppl., Vol. 7, 1968, pp. 1364–1371; Vol. 9, 1926, pp. 572–609; 1st Suppl., Vol. 9, 1932, pp. 225–251; 2nd Suppl., Vol. 9, 1949, pp. 377–407; 3rd Suppl., Vol. 9, 1970, pp. 2401–2403.

J.R. Johnson in R. Adams, ed., *Organic Reactions*, Vol. 1, John Wiley & Sons, Inc., New York, 1942, p. 218.

CINNAMON. See Flavors and spices.

CINNAMYL ALCOHOL, $C_6H_5CH=CHCH_2OH$. See Cinnamic acid, cinnamaldehyde, and cinnamyl alcohol.

CITRAL, $(CH_3)_2C=CH(CH_2)_2C(CH_3)=CHCHO$. See Flavors and spices; Perfumes; Terpenoids.

CITRIC ACID

Citric acid (2-hydroxy-1,2,3-propanetricarboxylic acid), $C_6H_8O_7$, a natural constituent and common metabolite of plants and animals, is the most versatile and widely used organic acid in the field of foods and pharmaceuticals.

Citric acid is also used in many industrial applications to sequester ions, neutralize bases, and act as a buffer. In cosmetics it is used as a buffer to control pH in shampoos, hair rinses, and setting lotions. Many citrates, especially the neutral sodium salt, are used extensively in food and pharmaceutical products and in detergents. Esters of citric acid are used commercially as plasticizers in the preparation of polymer compositions, protective coatings, and adhesives.

Physical Properties

Anhydrous citric acid, mol wt 192.13, crystallizes from hot concentrated aqueous solutions. The crystals are translucent and colorless; they belong to the holohedral class of the monoclinic system. The melting point of the anhydrous form is 153°C; the density is 1.665. A monohydrate may also be formed. Citric acid is optically inactive and manifests no piezoelectric effect.

Citric acid is a relatively strong organic acid as indicated by the first dissociation constant, 8.2×10^{-4} at 18°C. Second and third dissociation constants are 1.77×10^{-5} and 3.9×10^{-7}, respectively. pK values at 25°C are pK_1 3.128, pK_2 4.761, and pK_3 6.396. Citric acid is readily soluble in water and moderately soluble in alcohol.

Chemical Properties

As a tribasic acid, citric acid manifests the usual properties of a polybasic acid. It forms a variety of salts, including those of the alkali metal and alkaline earth metal families (see Salts). In addition, citric acid forms a variety of esters, amides, and acyl chlorides. Mixed compounds such as the salts of the acid esters can also be formed. The anhydride itself cannot be formed, but acyl derivatives of the acid can be dehydrated to form acyl citric anhydrides. The hydroxyl group may form acyl derivatives, ethers, etc. A wide variety of such mixed compounds is possible, and many have been prepared and studied. In aqueous solution, citric acid can be mildly corrosive toward carbon steels and, therefore, should be used with an appropriate inhibitor. It is not corrosive to stainless steels which are most often employed as the material of construction for processes involving citric acid.

At proper pH in aqueous media, citric acid's hydroxyl and carboxylic acid groups act as multidentate ligands forming complexes or chelates with metal ions. These chelating reactions are the basis for many of today's industrial processes, including elimination or control of metal-ion catalysis, lowering of metal oxidation potentials, removal of corrosion products (ie, Fe^{3+}), regeneration of ion-exchange resins, recovery of valuable metals by precipitation of insoluble chelates, decontamination of radioactive materials, quenching reactions, and driving reactions to completion (see Chelating agents).

Occurrence

Citric acid occurs widely in both the plant and animal kingdoms. It is found most abundantly in the fruits of the citrus species such as lemons (4.0–8.0%), grapefruit (1.2–2.1%), tangerines (0.9–1.2%), oranges (0.6–1.0%), and limes (ca 7.0%). The citrate ion occurs in all animal tissues and fluids, eg, in human whole blood (15 ppm). The total circulating citric acid in the serum of humans is approximately 1 mg/kg body weight.

Citric acid occurs in the terminal oxidative metabolic system of all but a few organisms. This system, variously referred to as the Krebs cycle, the tricarboxylic acid cycle, or the citric acid cycle, is a metabolic intermediate cycle involving the terminal steps in the conversion of carbohydrates, fats, or proteins to carbon dioxide and water with concommitant release of energy necessary for growth, movement, luminescence, chemosynthesis, and reproduction. This cycle also provides the carbonaceous materials from which amino acids and fats are synthesized by the cell.

Manufacture and Processing

The microbial production of citric acid on a commercial scale was begun by Pfizer, Inc. in 1923 based primarily on the work of Currie, who found that certain strains of *Aspergillus niger*, when grown on the surface of a sucrose-and-salt solution, produced significant amounts of citric acid. Variations of this surface-culture technique still account for a substantial portion of the world's production of this acid.

In the submerged process using *Aspergillus niger*, the microorganism is grown dispersed through a liquid medium. The fermentation vessel usually consists of a sterilizable tank of several hundred cubic meters (thousands of gallons) capacity equipped with a mechanical agitator and a means of introducing sterile air.

Until about 1969 or 1970 *A. niger* was considered to be the only organism that could be used to produce citric acid in commercial quantities. A patent issued in 1970 challenged this prevailing point of view by demonstrating the production of citric acid by species of yeast (eg, *Candida guilliermondii*) grown submerged in a medium containing either glucose or blackstrap molasses with an equivalent amount of sugar. Fermentation time was shorter than with *A. niger*. A subsequent improvement patent quotes citric acid concentrations of 110 g/L (see Yeast). Another yeast (eg, *Candida lipolytica*) was patented in 1970 for the conversion of C_9 to C_{20} normal paraffins to citric acid.

The energy requirement of the surface-culture method is low. Energy requirements of 1.3–2.6 MJ/m³ (4.8–9.5 Btu/gal) of fermenting medium are sufficient for both cooling and aeration. Submerged culture techniques must provide agitation, aeration, and cooling. Energy demands range from 8–16 MJ/m³ (28.5–57.0 Btu/gal) depending on the type and size of the fermenter.

Citric acid is generally recovered from a fermented aqueous solution by first separating the microorganism (using rotary filtration or centrifugation and then precipitating the citrate ion as the insoluble calcium salt. Classically, the calcium citrate salt has been used to separate fermentation by-products and other impurities from the citrate ion. Acidification with sulfuric acid to convert calcium citrate back to citric acid is then followed by concentration and filtration steps to remove formed calcium sulfate and a series of evaporative crystallizations to separate citric acid from remaining trace impurities.

A newer recovery process consists of selectively transferring citric acid via a solvent from an aqueous solution containing various by-products to another aqueous solution in which the citric acid is more concentrated and contains substantially fewer by-products. The final processing steps begin with a diluent wash of the aqueous solution by the hydrocarbon solvent, followed by passage of the acid solution through granular activated-carbon columns. Effluent from the carbon columns is processed through a conventional sequence of evaporator–crystallizer steps to complete the manufacturing process.

The residual solubles and biomass from the fermentation process contain nutrients that can be recycled usefully as animal feeds and for other agricultural purposes.

Specifications, Standards, and Quality Control

Citric acid for pharmaceutical use in the United States must meet the following USP specifications: must conform to the pyridine–acetic anhydride test for identity and contain not less than 99.5% of citric acid calculated on the anhydrous basis; the anhydrous acid must contain no more than 0.5% water and hydrous form no more than 8.8%; ash may not exceed 0.05%; the limit on heavy metals is 0.001% max, and within the heavy metals group, the maximum level of arsenic tolerated is 0.0003%. Other USP tests include oxalate, sulfate, and readily carbonizable substances.

The same forms of citric acid are commercially available for food and feed applications, and both must meet the specifications of the Food Chemicals Codex (FCC). With two exceptions, the FCC requirements are essentially equivalent to the USP specifications. The FCC lists no limit for sulfate. On the other hand, the FCC does limit the abundance of trace substances having uv absorbance properties in the range of 280–400 nm. The limit also is mandated for food use by FDA regulations.

Specifications are also available for the 50% aqueous technical solution forms of citric acid for industrial applications.

Citric acid, USP, FCC, anhydrous and hydrous, can be stored in dry form without difficulty, however, high humidity conditions and elevated temperatures should be avoided to prevent caking. Materials packaged with desiccants are commercially available.

Health and Safety Factors (Toxicology)

Citric acid is universally accepted as a safe food ingredient. The FDA lists citric acid as a multiple purpose generally-recognized-as-safe (GRAS) food substance and as a GRAS sequestrant limited only by good manufacturing practice. Citric acid also is approved by the Joint FAO/WHO Expert Committee on Food Additives for use in foods without limitation. As might be expected for a compound so widespread in living systems, aquatic toxicity is very low.

Salts

The salts of citric acid of commercial significance include dibasic ammonium citrate, calcium citrate, ferric ammonium citrate, potassium citrate, and sodium citrate. Sodium citrate is the citrate salt of major importance in the food and beverage industries. It is used as an emulsifier, stabilizer, and acidity and flavor modifier. Sodium citrate also is used in pharmaceuticals, cosmetics and toiletries, and in industrial applications and household cleansers.

Esters include triethyl, acetyl triethyl, tri-*n*-butyl, acetyl tri-*n*-butyl, triallyl, acetyl triallyl, isopropyl, tristearyl, and calcium monostearyl citrates.

EDWARD F. BOUCHARD
EDWARD G. MERRITT
Pfizer Inc.

Pfizer Organic Chelating Agents, *Technical Bulletin No. 32*, Pfizer Chemicals Division, New York, 1972.

Food Acidulants, *Technical Bulletin*, Pfizer Inc., New York, 1977.

Evaluation of the Health Aspects of Citric Acid, Sodium Citrate, Potassium Citrate, Calcium Citrate, Triethyl Citrate, Isopropyl Citrate, and Stearyl Citrate as Food Ingredients, PB 280 954, National Technical Information Service, Springfield, Va., 1977.

CITRONELLA OIL. See Oils, essential.

CITRONELLAL, $(CH_3)_2C{=}CH(CH_2)_2CH(CH_3)CH_2CHO$. See Perfumes; Terpenoids.

CLATHRATION

Clathrates or cage compounds are organic inclusion species, characterized by incorporation of "guest" molecules into vacancies in the "host" molecular array without strong bonding between host and guest. In these compounds chemical reactivity between host and guest may be totally lacking as evidenced by the formation of a hydroquinone clathrate with an inert gas such as argon. Clathrates also exist where two different components, differing in molecular size, are accommodated by appropriately sized holes in the lattice structure, such as gas hydrates, where one component may stabilize the lattice structure to form second cavities accommodating a different species (see Gas, natural).

The effect of the clathrate on the encapsulated species is frequently dramatic as in the case of the hydroquinone clathrate of argon. This clathrate is stable at atmospheric pressure and room temperature even though argon has a boiling point of −185.7°C.

Hydroquinone complexes are formed by linking through hydrogen bonds to form infinite three-dimensional complexes of trigonal symmetry of units. The hydroquinone clathrates are very stable at atmospheric pressure and room temperature. They have no smell of the encapsulated species but decompose when heated close to the melting point of hydroquinone or when dissolved in water. The only significant forces between guest molecules in hydroquinone clathrates are those caused by interaction of a guest molecule with the cage compounds.

Carbohydrates provide examples of two types of inclusion complexes. Both amylose and cyclodextrins form channel structures; cyclodextrins also form cagelike structures. Amylose forms clathrates with long-chain fatty acids such as lauric, palmitic, stearic, and oleic acids.

Trimers of phosphonitrilic compounds are a recent addition to the list of clathrating agents available to the investigator for separating, purifying, or analyzing organic compounds with a high degree of specificity. 2,4,6-Tris[1,2-phenylenedioxycyclo]triphosphazene forms clathrates by contact with many organic compounds in liquid or vapor form.

Studies show a complex containing from 2–9 molecules of Dianin's compound (**1**), geometrically aligned by the hydrogen bonding of hydroxyl groups to form the cage, similar to the behavior of hydroquinone in its cage-forming action. Most of these clathrates contain a 6:1 host:guest ratio. Although the cavities in quinol are roughly spherical, ca 0.4 nm, Dianin's compound has only one OH group available for hydrogen bonding, forming columns of independent cages with an hourglass structure about 1.1 nm long and 0.43 nm wide at maximum extension.

(**1**) Dianin's compound

(**2**) tri-*o*-thymotide

A number of adducts of alcohols, alkyl halides, and aliphatic ethers with the compound, tri-*o*-thymotide (**2**), form channel and clathrate-type complexes depending upon the guest dimensions. Adducts with very long-chain alcohols are extended channels, as shown by x ray, with an essentially constant enclosing structure. Cavity-type occlusions are formed by bulkier molecules such as chloroform or benzene. Werner salt clathrate compounds have been investigated in partition and adsorption chromatography. Use of these compounds is limited to temperatures below 90°C; above this temperature serious decomposition may occur. Clathrates serve for storage of volatile materials and are useful analytical tools as well. Optical resolution of tri-*o*-thymotide clathrates has been reported. Clathrates may be used also to dehydrate potable solutions such as fruit juices (qv), coffee (qv) extracts, and solutions such as those of latex that are concentrated in order to reduce shipping costs.

EARL C. MAKIN
Monsanto Chemical Intermediates Co.

Reviewed by
E.J. FULLER
Exxon Research and Engineering Co.

L. Mandelcorn, ed., *Non-Stoichiometric Compounds*, Academic Press, Inc., New York, 1964.

V.M. Bhatnagar, *Clathrate Compounds*, Chemical Publishing Co., New York, 1970.

E.J. Fuller in J.J. McKetta, ed., *Encyclopedia of Processing and Design*, Vol. 8, Marcel Dekker, New York, 1979, p. 333.

CLAYS

SURVEY

Clays as they occur in nature are rocks that may be consolidated or unconsolidated. They are distinctive in at least two properties that render them technologically useful: plasticity, and composition of extremely fine crystals or particles. The very fine particles yield very large specific-surface areas that are physically sorptive and chemically surface-reactive. Many clay-mineral crystals carry an excess negative electric charge owing to internal substitution by lower-valent cations, and thereby increase internal reactivity in chemical combination and ion exchange. Clays may have served as substrates that selectively absorbed and catalyzed amino acids in the origin of life. They apparently catalyze petroleum formation in rocks (see Petroleum).

Because clays (rocks) usually contain more than one mineral and the various clay minerals differ in chemical and physical properties, the term clay may signify entirely different things to different technologists. A broad definition of clay includes the following properties: the predominant content of clay minerals, which are hydrated silicates of aluminum, iron, or magnesium, both crystalline and amorphous; the possible content of hydrated alumina and iron; the extreme fineness of individual clay particles that may be of colloidal size in at least one dimension; the property of thixotropy in various degrees of complexity; the possible content of quartz sand and silt, feldspars, mica, chlorite, opal, volcanic dust, fossil fragments, high density so-called heavy minerals, sulfates, sulfides, carbonate minerals, zeolites, and many other rock and mineral particles ranging upward in size from colloids to pebbles.

Geology and Occurrence

Clays may originate through several processes: hydrolysis and hydration of a silicate by weathering or by hydrothermal alteration; solution of a limestone or other soluble rock containing relatively insoluble clay impurities that are left behind; slaking and weathering of shales (clay-rich sedimentary rocks); replacement of a preexisting host rock by invading guest clay whose constituents are carried in part or wholly by solution; deposition of clay in cavities or veins from solution; bacterial and other organic activity, including the extraction of metal cations as nutrients by plants; action of acid clays, humus, and inorganic acids on primary silicates; alteration of parent material or diagenetic processes following sedimentation in marine and freshwater environments; and resilication of high alumina minerals. Every state in the United States has within its boundaries clays or shales that may be utilized in the manufacture of bricks, tiles, or other heavy clay products. Kaolin is most diverse in physical properties, eg, it is free-slaking, or, diametrically opposite, it is slaking resistant; likewise it has an equally diverse array of technological uses.

Mineralogy

Today, crystalline clay minerals are identified and classified primarily on the basis of crystal structure and the amount of locations of charge (deficit or excess) with respect to the basic lattice. Amorphous (to x ray) clay minerals are poorly organized analogues of crystalline counterparts. Clay minerals are divided into crystalline and paracrystalline groups and a group amorphous to x rays.

Crystalline and paracrystalline groups include kaolins [kaolinite, dickite, nacrite (all $Al_2O_3.2SiO_2.2H_2O$), and halloysite–endellite ($Al_2—O_3.2 SiO_2.2H_2O$ and $Al_2O_3.2 SiO_2.4H_2O$, respectively)]; serpentines which substitute Mg_3 in the kaolin structure ($Mg_3Si_2O_5(OH)_4$); smectites (montmorillonites), which are the 2:1 clay minerals that carry a lattice charge and characteristically expand when solvated with water and alcohols, notably ethylene glycol and glycerol; illites or micas, illites being the general term for the clay mineral constituents of argillaceous sediments belonging to the mica group; glauconite, a green, dioctahedral, micaceous clay rich in ferric iron and potassium; chlorites and vermiculites which are regularly interstratified (1:1) and attributed to the mineral corrensite; attapulgite and sepiolite; and mixed-layer clay minerals that exhibit ordered and random intercalation of sandwiches with one another.

Amorphous and Miscellaneous Groups

Allophane is an amorphous clay that is essentially an amorphous solid solution of silica, alumina, and water. Imogolite is an uncommon, thread-shaped paracrystalline clay mineral assigned a formula 1.1 $SiO_2.Al_2O_3.2.3-2.8H_2O$. Several hydrated alumina minerals should be grouped with the clay minerals because the two types may occur so intimately associated as to be almost inseparable (diaspore and boehmite, both $Al_2O_3.H_2O$, gibbsite, $Al_2O_3.3H_2O$, and cliachite).

W.D. KELLER
University of Missouri at Columbia

C.E. Marshall, *The Colloid Chemistry of the Silicate Minerals*, Academic Press, Inc., New York, 1949.

R.E. Grim, *Clay Mineralogy*, McGraw-Hill Book Co., Inc., New York, 1968.

W.D. Keller, *Clays Clay Miner.* **26**, 1 (1978).

W.D. Keller, *Geol. Soc. Amer. Bull.* **93**, 27 (1982).

USES

Ceramic Products

Ceramic is defined as "relating to the art of making earthenware or to the manufacture of any or all products made from earth by the agency of fire, as glass, enamels, cements." To this list should be added brick, tile, heat-resisting refractory materials, porcelain, pottery, chinaware, and earthenware (see Ceramics). In general, ceramic ware is produced by plasticizing the clay by the addition of water so that it may be shaped or formed by some means into the desired object. A relatively low value for water of plasticity is desired in ceramics and hence kaolinite, illite, and chlorite clays have better plasticity characteristics than attapulgite or montmorillonite. Important properties of clays used for ceramics other than plasticity are firing and drying properties, resistance to thermal shock, attack by slag, and thermal expansion in the case of refractories. For whiteware, translucency, acceptance of glazes, etc, may be extremely important. These properties depend on the clay mineral composition and the method of manufacture, such as the forming procedure and intensity of firing.

Raw material. Almost any clay composition is satisfactory for the manufacture of brick unless it contains a large percentage of coarse stony material that cannot be eliminated or ground to adequate fineness. Roofing and structural tiles are usually made from the same material as face brick. Clays composed of mixtures of clay minerals containing 25–50% fine-grained unsorted quartz are well suited for the manufacture of terracotta, stoneware, sewer pipe, and paving brick. Porcelain and dinnerware are made up of about equal amounts of kaolin, ball clay, flint (ground quartz), feldspar, or some other white-burning fluxing material such as talc and nepheline. The slurry used in enameling is commonly composed of ball clay, frits, and coloring pigments. Refractory products are prepared from a wide variety of naturally occurring materials such as chromite and magnesite, or from clays predominantly composed of kaolinite.

Molding Sands

Molding sands, which are composed essentially of sand and clay, are used extensively in the metallurgical industry for the shaping of metal by the casting process. Using a pattern, a cavity of the desired shape is formed in the sand, and into this molten metal is poured and then allowed to cool. The bentonites, composed essentially of montmorillonite and used extensively in bonding molding sands, are of two types. The type carrying sodium as a principal exchangeable cation is produced largely in Wyoming. The calcium-carrying type is produced in Mississippi and in many countries outside the United States. The third type of clay used in foundries is composed essentially of illite. Green compression strengths in the range of ca 35–75 kPa (5–11 psi) are desired in actual practice.

Drilling Fluids

The suitability of a clay for use in drilling must be measured primarily by (a) the number of cubic meters (barrels) of mud with a given viscosity (usually 15 mPa·s or cP) obtained from one metric ton of clay in fresh water and in salt water; (b) the difference in gel strength taken immediately after agitation and 10 min later; (c) the wall-building properties, as measured by the water loss through a filter paper when a 15-mPa·s (= cP) clay is subjected to a pressure of 690 kPa (100 psi) for 30 min; and (d) the thickness of the filter cake produced in the standard water-loss test. The most widely used clay for drilling fluids is a bentonite from Wyoming composed of montmorillonite carrying sodium as the major exchangeable cation.

Catalysis

Catalysts made from various clay minerals are extensively used in the cracking of heavy petroleum fractions (see Catalysis; Petroleum). These catalysts are produced from halloysites, kaolinites, and bentonites composed of montmorillonite by chemical treatment.

Oil Refining and Decolorization

Clay materials are used widely to decolorize, deodorize, dehydrate, and neutralize mineral, vegetable, and animal oils. Decolorization is generally the major objective of such processes. A wide range of clay materials have been used for decolorization, ranging from fine-grained silts to clays composed of almost pure clay minerals. Clays composed of attapulgite and some montmorillonites possess superior decolorizing powers. Clays used to decolorize edible oils must not impart an obnoxious odor or taste to the product.

Paper

A sheet of cellulose fibers is not well suited to high fidelity printing because of transparency and irregularities of the surface. These deficiencies are corrected by the addition of binding agents such as starch and resin, and by the incorporation into the fiber stock of mineral fillers such as calcium carbonate or sulfate, and especially pure white clay. Pure kaolinite clays are used for filling and coating paper, except when special properties are desired (see Papermaking additives).

Miscellaneous Uses

Clays are used as ingredients in a vast number of products, for example, asphalt, paints, rubber, adhesives, Portland cement, pozzolanas, paint, medicinals and cosmetics, pesticides, plastics, and rubber. They also are used as agents in many processes, such as water purification, radioactive waste disposal, and pelletizing ores, fluxes and fuels. Also, they are a source of raw materials, eg, aluminum.

RALPH E. GRIM
University of Illinois

R.E. Grim, *Applied Clay Mineralogy*, McGraw-Hill Book Co., New York, 1962.

CLUTCH FACINGS. See Brake linings and clutch facings.

CMC. See Cellulose derivatives—Esters (sodium carboxymethylcellulose).

COAGULANTS AND ANTICOAGULANTS. See Blood, coagulants and anticoagulants.

COAL

Coal is a dark burnable solid, usually layered, that resulted from the accumulation and burial of partially decayed plant matter over earlier geological ages. These deposits were converted to coal through biological changes and later effects of pressure and temperature. Composition as well as chemical and physical properties vary widely for the different kinds of coal. Differences in type are caused by variations in the amounts of different plant parts exemplified by common-banded, splint, cannel, and boghead coals. The degree of coalification is referred to as rank. Brown coal and lignite, subbituminous coal, bituminous coal, and anthracite make up a natural series with increasing carbon content. The impurities in coal cause differences in grade (see Lignite and brown coal).

Coal is red-brown black or brownish-black and is sold in a range of sizes. The color, luster, texture, and fracture vary with the type, rank, and grade. Coal is composed chiefly of carbon, hydrogen, and oxygen with small amounts of nitrogen and sulfur, and varying amounts of moisture and ash or mineral impurities.

The formation of coal, the variation in its composition, its microstructure, and its chemical reactions indicate that coal is a mixture of compounds. The organic composition depends on the degree of biochemical change of the original plant material, and on the later coalification by pressure and temperature effects in the deposit. The major organic compounds have resulted from the formation and condensation of polynuclear carbocyclic and heterocyclic ring compounds containing carbon, hydrogen, oxygen, nitrogen, and sulfur. The amount of carbon in aromatic ring structures increases with rank (see Constitution).

Nearly all coal is utilized in combustion and coking. Over 80% is burned directly for generation of electricity (see Coal conversion processes, magnetohydrodynamics), or steam generation for industrial uses, transportation (see Power generation), or residential heating, and metallurgical processes, in the firing of ceramic products, etc. The balance is mostly carbonized to produce coke, coal gas (see Coal chemicals and feedstocks, gasification), ammonia, coal tar, and light oil products from which many chemicals are derived (see Coal conversion processes, carbonization; Coal chemicals and feedstocks, carbonization). Combustible gases can also be produced by the gasification of coal (see Fuels, synthetic). Carbon products are produced by heat treatment (see Carbon). A relatively small amount of coal is used for miscellaneous purposes such as fillers (qv), pigments (qv), foundry material, and water filtration.

Most organic chemicals can be synthesized from coal carbonization products or by different processes utilizing gases derived from coal. Some aliphatic and aromatic carbocyclic acids can be produced by the oxidation of coal. Processes to produce gaseous and liquid fuels, as well as chemicals, have been extensively developed and are being improved in the United States, Europe, and South Africa. Development has been limited by the ability of these products to compete economically with petroleum and natural gas (see Feedstocks).

Coal is one of the world's major consumed commodities. Current annual production averages ca 700×10^6 metric tons in the United States and 3.3×10^9 metric tons for the entire world (2.4×10^9 t coal, and 0.9×10^9 t brown coal and lignite). Reserves of coal are far greater than the known reserves of all other mineral fuels (petroleum, natural gas, oil shale, and tar sands) combined. This is true for the entire world. The reserves of petroleum and natural gas are being depleted more rapidly than coal. Coal consumption has been cyclic and, as energy consumption increases, coal will be used in continually larger amounts. Many new processes for coal conversion and utilization are being developed and will be of significant interest to chemists and chemical engineers (see also Coal chemicals and feedstocks, survey; Gas, natural; Oilshale; Petroleum; Tar and pitch; Tar sands).

Microscopic examination of coal has been carried out using transmitted light through thin sections and reflected light from polished samples. The first studies were made with transmitted light and thin sections. This technique permits the identification of the plant parts that became coal components. Three components are termed anthraxylon, attritus, and fusain from a examination of thin sections of bituminous coals from the United States.

The mineral matter in coal is the result of several separate processes. Part is caused by the inherent material present in all living matter; another type is caused by the detrital minerals that were deposited during and after the time of peat formation; and a third type is caused by secondary minerals that have crystallized from water that has percolated through the coal seams. Clay is the most common detrital mineral; however, other common ones include quartz, feldspar, garnet, apatite, zircon, muscovite, epidote, biotite, augite, kyanite, rutile, staurolite, topaz, and tourmaline. The secondary minerals are generally kaolinite, calcite, and pyrite. Analyses have shown the presence of almost all elements in at least trace quantities in the mineral matter. Certain elements, ie, germanium, beryllium, boron, and antimony are found primarily with the organic matter in coal. Zinc, cadmium manganese, arsenic, molybdenum, and iron are found with the inorganic material in coal. The primary mineral constituents in coal are aluminum, silicon, iron, calcium, magnesium, sodium, and sulfur. The relative concentrations will depend primarily on the geographical location of the coal seam, and will vary from place to place within a given field (see Constitution).

Coal petrology is used to identify the macerals and microlithotypes present in coals. The behavior of coal depends on the properties of the individual constituents, the relative amounts in the coal, and the rank of the coal. The difference in resistance of macerals to breakage together with other properties is used in the Longwy-Burstlein or Sovaco process in the selection of coals for blending in the optimum proportions for producing high quality coke. By careful control of the petrological composition and the rank of a coal blend, its behavior during carbonization can be controlled. The petrological composition of a coal is quantitatively determined on polished surfaces by measuring the volume percentage of the different macerals with a point count method. The rank of the coal is determined microscopically by measuring reflectance of vitrinite with a microphotometer. The coking behavior of a coal or a blend of coals of different ranks can be reliably predicted with these two measurements.

Two types of classification may be identified, ie, scientific and commercial. In the scientific category, the Seyler chart has found the widest acceptance. The commercial classification uses scientific measurements to make the distinctions that occur in the classifications. Both methods are complementary and are used in research. In industry the commercial classification is essential.

In the ASTM classification system, the higher rank coals are specified by fixed carbon (for volatile matter $\leq 31\%$) on a dry, mineral-matter-free basis and lower rank coals are classified by calorific value on the moist, mineral-matter-free basis. The latter parameter depends on two properties: the moisture absorbing capacity and the calorific value of the pure coal matter. Some overlap between bituminous and subbituminous coals occurs, and is resolved on the basis of the agglomerating properties.

The increasing amount of coal in international commerce since ca 1945 indicated a need for an international system of coal classification. The Coal Committee of the European Economic Community considered different national classifications and agreed on a system designated International Classification of Hard Coal by Type. Similar to the other methods, volatile matter and gross calorific value on a moist, ash-free basis are the parameters used. Coals with volatile matter up to 33% are divided into classes 1–5. Coals with volatile matter greater than 33% are divided into classes 6–9. The calorific values are given for a moisture content obtained after equilibrating at 30°C and 96% rh.

The nine classes are divided into four groups determined by caking properties as measured through either the free swelling index or the Roga index. These tests indicate properties observed when the coal is heated rapidly.

The brown coals and lignites have been classified separately and have been defined as those coals with heating values less than 23.85 MJ/kg (10,260 Btu/lb, 5700 kcal/kg) (see Lignite and brown coal).

Constitution

The functional groups of interest contain O, N, or S. The significant oxygen-containing groups found in coals are those of carbonyl, hydroxyl, carboxylic acid, and methoxy. The nitrogen-containing groups include aromatic nitriles, pyridines, and pyrroles. The sulfur is largely found in dialkyl, aryl alkyl thioether, and thiophene groups.

The relative and absolute amounts of these groups vary with coal rank and maceral type. The major oxygen-containing functional groups in vitrinites of mature coals are phenolic hydroxyl and conjugated carbonyl as in quinones. Evidence exists for hydrogen bonding of hydroxyl and carbonyl groups. There are unconjugated carbonyl groups such as ketones in exinites. The infrared absorption bands are displaced from the normal range for simple ketones by the conjugation in vitrinites. The interactions between the carbonyl and hydroxyl groups affect the normal reactions.

The anthracites approach graphite in composition (see Carbon, graphite). As such they have higher rank than bituminous coals and have less oxygen and hydrogen. They are less reactive and insoluble in organic solvents. These characteristics are more pronounced as rank increases in the anthracite group. The anthracites have greater optical and mechanical anisotropy than lower rank coals. The internal pore volume and surface increase with rank after the minimum below about 90 wt% C.

Properties

Most of the physical properties discussed here depend on the direction of measurement compared to the bedding plane of the coal. Additionally, the properties vary according to the history of the piece of coal. They also vary between pieces because of the brittle nature of coal and of the crack and pore structure. The specific electrical conductivity of dry coals is very low (specific resistance 10^{10}–10^{14} $\Omega \cdot cm$), although it increases with rank. Coal has semiconducting properties. The conductivity tends to increase exponentially with increasing temperatures.

The dielectric constant (ϵ) is also affected by structural changes on strong heating. Also it is very rank dependent, exhibiting a minimum at about 88 wt% C, and rises rapidly for carbon contents over 90 wt%. The dielectric constant (in debye units) equals the square of the refractive index only at the minimum value.

Density values of coals differ considerably, even after correcting for the mineral matter, depending on the method of determination. The true density of coal matter is most accurately obtained from measurement of the displacement of helium after the absorbed gases have been removed from the coal sample. Density values increase with carbon content or rank for vitrinites. They are 1.4–1.6 g/cm^3 above 85 wt% carbon where there is a shallow minimum.

The thermal conductivity and thermal diffusivity are also dependent on the pore and crack structure. Thermal conductivities for coals of different ranks at room temperature are in the range of 0.23–0.35 $W/(m \cdot K)$.

The specific heat of coal can be determined by direct measurement or from the ratio of separate measurements of thermal conductivity and thermal diffusivity. The latter method gives values decreasing from 1.25 $J/(g \cdot K)$ [(0.3 cal/(g·K)]) at 20°C to ≤ 0.4 $J/(g \cdot K)$ [(≤ 0.1 cal/(g·K))] at 800°C.

Coal contains an extensive network of ultrafine capillaries that pass in all directions through any particle. The smallest and most extensive passages are caused by the voids from imperfect packing of the large organic molecules. Vapors pass through these passages during adsorption, chemical reaction, or thermal decomposition. The rates of these processes depend on the diameters of the capillaries and any restrictions in them. Most of the inherent moisture in the coal is contained in these capillaries. The porous structure of the coal and products derived from it will have a significant effect on the absorptive properties of these materials. Effectiveness of coal conversion processes depends on rapid contact of gases with the surface. Large internal surfaces are required for satisfactory rates.

Mechanical properties are important for a number of steps in coal preparation from mining through handling, crushing, and grinding. The properties include elasticity and strength as measured by standard laboratory tests and empirical tests for grindability and friability; and indirect measurements based on particle size distributions. A number of efforts to correlate grinding energy input to size reduction have been made.

The Young's modulus for medium rank coals has been found to be ca 4 GPa (4×10^{10} dyn/cm^2) by several workers. Sharp increases in the Young's and shear moduli have been found in vitrinites with increase in carbon content over 92%.

There are indications that compressive strength (measured by compression of a disk) may give useful correlations with the ease of cutting with different kinds of equipment. Studies of the probability of survival of pieces of different size suggest that the breaking stress should be most closely related to the linear size rather than the area or volume of the piece. The effects of rank on both compressive and impact strength have been studied, and usual minima were found at 20–25% dry, ash-free volatile matter (88–90 wt% carbon). Accordingly, the Hardgrove grindability index exhibits maximum values in this area.

Solvent extraction. A wide range of organic solvents can dissolve part of coal samples but dissolution is never complete and usually requires heating to temperatures sufficient for some thermal degradation or reaction with the solvent to take place (ca 400°C). At room temperature the best solvents are primary aliphatic amines, pyridine, and some higher ketones, especially when used with dimethylformamide. Above 300°C large amounts can be dissolved with phenanthrene, 1-naphthol, and some coal-derived high boiling fraction. Dissolution of up to 40 wt% can be achieved near room temperature, and up to 90% near 400°C. coals with 80–85% carbon in the vitrinite give the largest yields of extract. Very little coal above 90 wt% C dissolves.

Solvent extraction work was carried out by a number of organizations in the United States. Heating value of the product solvent refined coal (SRC) is ca 37 MJ/kg (ca 16,000 Btu/lb). Sulfur contents have been reduced from 2–7% initially to 0.9% and possibly less. Ash contents have been reduced from 8–20% to 0.17%. These properties permit compliance with 1976 EPA requirements for SO_2 and particulate emissions. The SRC is primarily intended to be used as a boiler fuel in either a solid or molten form (heated to ca 315°C).

Properties involving utilization. Coal rank is the most important single property for almost every application of coal. The rank sets limits on many properties such as volatile matter, calorific value, and swelling and coking characteristics. Other properties of significance include grindability, ash content and composition, sulfur content, and size.

Most of the mined coal is burned in boilers to produce steam for electric-power generation. The calorific value determines the amount of steam that can be generated. The design and operation of a boiler requires consideration of a number of properties (see Furnaces, fuel-fired; Burner and combustion technology).

In general, high rank coals are more difficult to ignite, requiring supplemental oil firing and slower burning, with large furnaces to complete combustion. The greater reactivity of lower rank coals makes them better suited for cyclone burners that carry out rapid, intense combustion to maximize carbon utilization and minimize smoke emission.

Ash content is important. Ash discharge at high temperature, as molten ash from a slagging boiler, involves substantial amounts of sensible heat. The higher cost of washed coal of lower ash content will not always merit its use. Ash disposal and extra freight costs for high ash coals enter the selection of coal. The current use of continuous mining equipment produces coal with about 25% ash content. The average ash content of steam coal is about 15% in the United States. For some applications, such as chain-grate stokers, a minimum ash content of about 7–10% is needed to protect the metal parts.

For pulverized coal firing, a high Hardgrove index or grindability index is desired. This implies a relatively low energy cost for pulverizing since the coal is easier to grind. The abrasiveness of the coal is also important since this determines the wear rate on pulverizer elements.

Moisture content affects handling characteristics. It is most important for fines smaller than 0.5 mm. The moisture content of peat or brown coal that is briquetted for fuel must be reduced to about 15% for satisfactory briquetting. Mechanical or natural means are used because of the cost of thermal drying.

The sulfur content is important in meeting air quality standards. Volatile matter is important for ease of ignition. High rank coals have low volatile matter and burn more slowly.

Coke production. Coking coals are mainly selected on the basis of the quality and amount of coke that they will produce. Gas yield is also considered. About 65–70% of the coal charged is produced as coke. The gas quality depends on the coal rank and is a maximum (in energy in gas per mass of coal) for coals of about 89% carbon (dry, mineral-matter-free) or 30% volatile matter.

A Clean Coke process is being developed by U.S. Steel Corp., with partial Department of Energy support, to convert coals not normally used for coke production, including high sulfur coal, to a low ash, low sulfur metallurgical coke, chemical feedstocks, and other fuels. The coal is initially carbonized, then pelletized using a binder derived from another part of the process before heating in a continuous coking kiln.

Gasification. A number of gasifiers are either available commercially or in various stages of development. The range of coals that may be used vary from one gasifier type to another. Many of the coal selection criteria for combustion apply to gasification, which is a form of partial oxidation.

Reactions

Mature coals (> 75% C) tend to be built of assemblages of polynuclear ring systems connected by a variety of functional groups. The ring systems themselves contain many functional groups. These so-called molecules differ from each other to some extent in the coal matter. For bituminous coal, a tar-like material fills some of the interstices between the molecules. Generally, the initial volatile constituents of coal are moisture and light hydrocarbons. The volatile matter produced on carbonization reflects decomposition of parts of the molecule and the release of moisture. Rate of heating affects the volatile-matter content. Faster rates give higher volatile-matter yields. Coal is not carbon or a hydrocarbon. The composition will depend on the rank. The empirical composition of a vitrinite of ca 84 wt% carbon content is approx $C_{13}H_{10}O$ with small quantities of N and S.

The surface of coal particles undergoes air oxidation. This process may initiate spontaneous combustion in storage piles or weathering with a loss of heating value and coking value during storage.

Many of the products made by hydrogenation, oxidation, hydrolysis, or fluorination are of industrial interest. The increasing cost of petroleum and natural gas is increasing interest in some of these materials primarily as upgraded fuels to meet air-pollution control requirements as well as to take advantage of the greater ease of handling of the liquid or gaseous material, and to utilize existing facilities such as pipelines and furnaces. Demonstration plants are planned for conversion of coal to methane (substitute natural gas or SNG), industrial fuel gas, ammonia, coal-derived refinery feedstock, and boiler fuel (see Gas, natural; Hydrocarbons). A chemistry based on synthesis-gas conversion to other materials is being developed and may have more extensive application in another decade. At present, the major production of chemicals from coal involves the by-products of coke manufacturing.

Plasticity of heated coals. Coals with a certain range of composition associated with the bend in the Seyler diagram and having 88–90 wt% carbon will soften to a liquid condition when heated. These are known as prime coking coals. The soft condition is somewhat reversible for a time, but does not persist for many hours at 400°C, and is not be observed above ca 550°C if the sample is continuously heated as in a coking process. This is caused by degradation of coal matter, releasing vapors and resulting in polymerization of the remaining material.

Several laboratory tests are used to determine the desirability of a coal or blend of coals for making coke. These are empirical and are carried out under conditions that approach the coking process. The three properties that have been studied are swelling, plasticity, and agglomeration.

Pyrolysis of coal. Most coals decompose below temperatures of about 400°C, characteristic of the onset of plasticity. Moisture is released near 100°C, and traces of oil and gases appear between 100–400°C depending on the coal rank. As the temperature is raised in an inert atmosphere at a rate of 1–2°C/min, the evolution of decomposition products reaches a maximum rate near 450°C, and most of the tar is produced in the range of 400–500°C. Gas evolution begins in the same range but most evolves

about 500°C. If the coal temperature in a single reactor exceeds 900°C, the tars can be cracked, the yields are reduced, and the products are more aromatic. Heating beyond 900°C results in minor additional weight losses but the solid matter changes its structure. The tests for volatile matter indicate the loss in weight at a specified temperature in the range of 875–1050°C from a covered crucible. The weight loss represents the loss of volatile decomposition products rather than volatile components.

The loss of fusible material can be accelerated by reducing the pressure, and can also reduce caking properties. Further reactions of the volatile products are also reduced. Increasing the pressure has little effect. Mild oxidation is used to destroy the caking properties and can eliminate the production of tars.

Mining and Preparation

Coal is obtained by either surface mining of outcrops or seams near the surface, or by underground mining. The method chosen depends on the geological conditions which may vary from thick, flat seams to thin, inclined seams that are folded and need special methods of mining.

Coal preparation normally involves some size reduction of coal particles and the systematic removal of some ash-forming material and very fine coal. The percentage of mined coal that is mechanically cleaned in major coal producing countries has risen during the past 30 yr. There are a number of reasons for this. The most important in the United States is the increased use of continuous mining equipment. The nature of this operation tends to include inorganic foreign matter from the floor and ceiling of the seams, and run-of-mine coal currently includes about 25% ash.

Storage of coal may be necessary at any of the major steps in production or consumption. For utilities, two types of storage are used. A small amount of coal in storage meets daily needs and will be continually turned over. This coal is loaded into storage bins or bunkers. Long-term reserves are carefully piled and left undisturbed except as necessary to sustain production. Coal storage results in some deterioration of the fuel owing to air oxidation. If inadequate care is taken, spontaneous heating and combustion will result. As the rank of coal decreases, it oxidizes more easily and must be piled more carefully. Anthracite does not usually present a problem.

The usual means of transporting coal are railroad, water shipment, truck, and in some instances by conveyor belt from mine to plant.

In 1976, the United States coke exports totaled 1,190,000 metric tons, mainly to Canada; anthracite exports were 558,000 t, mainly to Canada.

Health and Safety Factors

Coal mining has been a relatively dangerous occupation. In the seven years after the passage of the Federal Coal Mine Health and Safety Act of 1969, however, the average fatality rate decreased to 0.58, a reduction of 44.8%. Major causes of fatalities are falling rock from mine roofs and faces, haulage, surface accidents, machinery, and explosions.

A disease called pneumoconiosis, also called black lung, results from breathing coal dust over prolonged periods of time. The coal particles coat the lungs and prevent proper breathing.

KARL S. VORRES
Institute of Gas Technology

M.A. Elliott, ed., *Chemistry of Coal Utilization*, 2nd Suppl. Vol., Wiley-Interscience, a division of John Wiley & Sons, Inc., New York, 1981.

D.W. van Krevelen, *Coal*, Elsevier Scientific Publishing Co., Amsterdam, The Netherlands, 1961.

Annual Book of ASTM Standards, 1977—Part 26, Gaseous Fuels; Coal and Coke; Atmospheric Analysis, ASTM, Philadelphia, Pa., 1977.

G.L. Tingey and J.R. Morrey, *Coal Structure and Reactivity*, Battelle Energy Program Report, Battelle Pacific Northwest Laboratories, Richland, Wash., Dec. 1973.

COAL CHEMICALS AND FEEDSTOCKS. See Supplement.

COAL — GASIFICATION OF COAL.   See Fuels, synthetic; Supplement.

COAL — POWER FROM COAL BY GASIFICATION AND MAGNETOHYDRODYNAMICS.   See Coal—Magnetohydrodynamics.

COAL — SYNTHETIC CRUDE OIL FROM COAL.   See Fuels, synthetic.

COAL CONVERSION PROCESSES

CARBONIZATION

Carbonization of coal is an old established industry by which coal is subjected to destructive distillation in a heated retort in the absence of air. Coke is the residue remaining upon devolatilization of bituminous coal. Coke is the main product, although tar, light oil, ammonia liquor, and coke-oven gas are also produced. Coke is used principally as a fuel and reductant in the blast furnace for ironmaking. The other products are further refined into commodity chemicals such as ammonium sulfate, benzene, toluene, xylene, naphthalene, pyridine, phenanthrene, anthracene, creosote, road tars, roofing pitches, and pipeline enamels. The coke-oven gas is a valuable heating fuel used mainly within steel plants. Hence the carbonization industry is closely tied to the steel industry (see Iron; Steel). In the United States, most carbonization is conducted in the high temperature range of 900–1200°C; medium (750–900°C) and low temperature carbonization is rare. Carbonization operations are conducted almost entirely in slot-type ovens constructed of silica brick; less than about 1% of the total coke in the United States is produced in beehive ovens (see Coal chemicals and feedstocks, Carbonization and coking). The total world production of coke amounts to about 375×10^6 metric tons.

Coals for Coking

It is general practice in the United States to blend two or more types of coking coals. Usually a high-volatile coal is blended with a low-volatile coal at a 90 : 10 to a 60 : 40 blend, respectively. A larger amount of the low-volatile coal, reduced coking rate, and incorporation of 5–15% of an inert material such as anthracite fines or coke breeze are customary in the production of foundry coke. The resultant coke is large and has a high shatter strength and reduced reactivity.

Carbonization pressure is an important consideration in the selection of coals for use in coke ovens. It is important that a coal blend that meets the requirement of producing a satisfactory coke, with an acceptable analysis, also meets the safe-pressure criterion.

Coal Preparation

Coals arrive at the coke plant by rail, ship, or other means, and are unloaded into separate stockpiles according to high-, medium-, or low-volatile classification or are conveyed directly to the coal-blending plant, mixed, crushed, and stored in the storage bunkers near the ovens.

Coking Mechanisms

When coal is charged into a coke oven, the coal particles adjacent to the heated walls melt and fuse, lose much of their volatile materials, and finally solidify into semi-coke. This melting and fusion phase of the coal particles during carbonization is the most important part of the coking of coal. The degree of melting and degree of assimilation of the coal particles into the molten mass, especially the inert particles, determine the characteristics of the coke produced. A coal rich in vitrinite produces a weak spongy coke; a coal deficient in vitrinite has insufficient plastic properties necessary for good coke formation. To produce the strongest coke for a particular coal or coal blend, there is an optimum ratio of reactive to inert entities.

Energy Considerations

The recent awareness of energy availability problems and increased cost of all forms of energy has focused more attention on the amount of energy required for coking. In the past, coke-oven design was based on a requirement of ca 2.4–2.7 MJ/kg (1050–1150 Btu/lb) of wet coal when underfiring with coke-oven gas and carbonizing at a rate of ca 25 mm/h. The change to higher capacity ovens, faster coking rates, and higher temperature coke to minimize emissions as required by air pollution regulations, has resulted in higher underfiring requirements.

Oven Operation

The coal blend is charged into hot ovens according to a planned schedule. The covers on the standpipes are closed and the valves in the collecting mains are opened so that the volatile products from the coal are discharged directly into the collecting mains. The coal is carbonized for a length of time proportional to the flue temperature, and dependent on oven design and characteristics of the coal. Heating is adjusted so that every oven charge is coked out in approximately the same amount of time.

The coke is pushed into a quenching car for transport to a quenching station where it is sprayed with thousands of liters of water to cool it below its ignition temperature. After draining, the quenched coke is deposited on an inclined brick wharf where it is further cooled by ambient air.

Coke Preparation and Properties

The quenched or cooled coke is conveyed from the wharf to the screening station for separation into desired sizes. Generally, the desired blast-furnace size is about 75 × 20 mm. Coke larger than 75 mm may be crushed to the desired size. Coke smaller than ca 20 mm, called breeze, is used for iron-ore sintering, boiler firing, as an "inert" in foundry-coke manufacture, electric smelting, chemicals manufacture, and other purposes. Foundry coke is separated into various double-screened sizes larger than 75 mm. Smaller coke is used for other applications.

Pollution Aspects

Several systems for pollution control have been under development. These attempt to control continuous emissions from doors and lids, intermittent emissions that result from charging the coal into the ovens, and emissions from pushing and quenching the coke. The charging emissions are considered to be the most severe. Conventional charging practice has been improved in some plants by the adoption of a sequential charging practice: coal from the larry hoppers is dropped into the oven in a predetermined sequence, each lid is replaced when a hopper is emptied, as the last hopper is emptied, the levelling bar is put into operation. Other plants have adopted modified systems using either double collecting mains or single mains and jumper pipes.

Recent Technology

The most promising technology will continue to be improved during the next few years. Oven design considerations to promote more uniform heating for ovens of higher capacity will continue. Preheating coal charges and charging them into ovens presents a number of problems, particularly that of coal carry-over. A solution to this problem is imperative for more universal adoption of the technology. In addition, partial briquetting of coal charges will probably be developed further (see Size enlargement). Both preheating and partial briquetting have the potential of using poorly coking coals which may become necessary when premium coals are exhausted. The advancement of formcoke technology should continue and the adoption of one or more processes should be made, particularly in regions where suitable coking coals are scarce.

MICHAEL PERCH
Koppers Co., Inc.

M.A. Elliott, ed., *Chemistry of Coal Utilization*, 2nd Suppl. Vol., John Wiley & Sons, Inc., New York, 1981, Chapts. 14–15.

H.H. Lowry, ed., *Chemistry of Coal Utilization*, Suppl. Vol., John Wiley & Sons, Inc., New York, 1963, Chapts. 10–11.

H.H. Lowry, ed., *Chemistry of Coal Utilization*, John Wiley & Sons, Inc., New York, 1945, Chapts. 25 and 31.

P.J. Wilson and J.H. Wells, *Coal, Coke and Coal Chemicals*, McGraw-Hill Book Co., Inc., New York, 1950.

DESULFURIZATION

Physical and chemical coal cleaning are being studied extensively in an effort to provide a clean source of solid energy from coal and at the same time reduce pollution of the environment by sulfur oxides, nitrogen oxides, and toxic metals. U.S. coals may contain any or all of four petrographic subclasses or lithotypes, eg vitrain, clarain, durain, and fusain.

Impurities in Coal

The three categories of potential air pollutants found in coal are sulfur, ash, and nitrogen (as a source of NO_x). Sulfur and ash are associated with the mineral and organic portions, and nitrogen most likely with the organic matter. Sulfur is an undesirable constituent of all coals. It is present in amounts ranging from traces to > 10 wt%. Sulfur is present in coal as sulfate, pyritic, and organic sulfur. Physical cleaning and chemical cleaning are employed in the desulfurization of coal.

Physical Coal Cleaning

Physical coal cleaning is a proven technology for upgrading raw coal by physical removal of associated impurities. Coal has traditionally been cleaned to meet certain market requirements as to size and ash and moisture contents. However, governmental regulations of atmospheric emission of sulfur oxides from coal combustion have been focused on sulfur reduction. Physical coal-cleaning processes grind coal to release impurities; the fineness of the coal governs the degree to which the impurities are released.

Current industrial and laboratory physical coal-cleaning methods can be divided into four broad categories based on the physical properties that effect the separation: gravity, flotation, magnetic, and electrical methods. Commercial coal cleaning is currently limited to gravity separation (qv) with minor application of froth flotation (qv) methods. Table 1 summarizes the types of processes and equipment used in coal cleaning since 1942.

Conventional coal washing. In a modern physical cleaning plant, coal is typically subjected to: size reduction and screening; separation of impurities; and dewatering and drying. Table 2 shows the beneficiation of the sulfur reductions possible by physical cleaning of coal from various regions of the United States.

Chemical coal cleaning. Chemical coal cleaning is not practiced commercially. Theoretically, however, chemical cleaning of coal to produce an environmentally acceptable solid fuel as a clean source of energy is technically feasible. In chemical cleaning, raw coal is treated with a reagent or reagents that react with and liberate pollutant-forming constituents. Chemical cleaning can be used alone or in combination with physical coal cleaning and has the potential to significantly increase the amount of coal reserves that could be used directly as a solid fuel with little or no pollution control. Chemical cleaning can achieve essentially complete removal of the pyritic sulfur from most coals and up to 40–50 wt% of the organic sulfur from some coals. Several processes that can achieve this degree of cleaning are oxidative desulfurization of coal,

Table 2. Sulfur Reduction by Physical Coal Cleaning

Energy recovery, %	Level 1[a] 99	Level 2[a] 95	Level 3[a] 90	Level 4[a] 85
Pyritic sulfur reduction, %				
Northern Appalachian	10	33	47	54
Southern Appalachian	15	35	44	48
Alabama	10	32	38	39
Eastern Midwest	20	45	54	59
Western Midwest	15	33	41	45
Western	8	30	33	33
U.S. average	*13*	*35*	*46*	*52*
Total sulfur reduction, %				
Northern Appalachian	5	20	28	33
Southern Appalachian	2	6	9	9
Alabama	3	9	12	12
Eastern Midwest	6	22	29	32
Western Midwest	5	21	25	29
Western	2	8	11	11
U.S. average	*6*	*20*	*25*	*29*
Emission, kg SO_2/GJ (lb SO_2/10^6 Btu)				
Northern Appalachian	1.9 (4.5)	1.5 (3.5)	1.3 (3.1)	1.2 (2.9)
Southern Appalachian	0.6 (1.5)	0.6 (1.4)	0.6 (1.3)	0.6 (1.3)
Alabama	0.9 (2.0)	0.9 (2.0)	0.9 (2.0)	0.9 (2.0)
Eastern Midwest	2.5 (5.8)	2.0 (4.7)	1.8 (4.2)	1.7 (4.0)
Western Midwest	3.4 (8.0)	2.9 (6.8)	2.7 (6.2)	2.5 (5.9)
Western	0.5 (1.1)	0.4 (0.9)	0.4 (0.9)	0.4 (0.9)
U.S. average	*2.0 (4.7)*	*1.6 (3.8)*	*1.2 (2.9)*	*1.2 (2.7)*

[a] Beneficiation.

nitrogen dioxide oxidative cleaning, ferric salt leaching, promoted oxidative leaching, microwave desulfurization, wet oxygen leaching, hydrogen peroxide–sulfuric acid leaching, fluidized-bed air oxidation, hydrosulfurization, chlorinalysis, and hydrothermal leaching.

EDGEL STAMBAUGH
Battelle Memorial Institute

A.W. Deurbrock and P.S. Jacobsen, "Coal Cleaning State of the Art," paper presented at *Coal Utilization Symposium—SO₂ Emission Control, National Coal Conference*, Louisville, Kentucky, Oct. 1974.

J.A. Cavallaro, M.T. Johnston, and A.W. Deurbrock, "Sulfur Reduction Potential of the Coals of the United States," *U.S. Bur. Mines Rep. Invest. RI 8118* (1976).

R.R. Oder and co-workers, Bechtel Corporation, San Francisco, Calif., "Technical and Cost Comparison For Chemical Coal Cleaning Processes," *Coal Conference, American Mining Congress*, Pittsburgh, Penn., May 1977.

E.P. Stambaugh, "Review of Hydrothermal Coal Process," *Proceedings of 173rd National Meeting of ACS*, New Orleans, La., March 21–25, 1977, EPA-600 17-78-068, April, 1978.

MAGNETOHYDRODYNAMICS

In the magnetohydrodynamic (MHD) method of electrical power generation an electrically conducting fluid is caused to flow in a duct through a transverse magnetic field. When electrodes are placed along the sides of the duct at appropriate locations, electric power can be extracted from the system. This results from the induced electric field in the moving fluid.

The electrically conducting fluid for MHD generation may be a liquid metal or a hot ionized gas. The latter method is the more direct and efficient for the conversion of thermal to electric energy in commercial application. This article is concerned with generation of power by the MHD process from a stream of hot conducting gas. Such a stream may consist of products of combustion of a fossil fuel, in particular products of combustion of coal or a coal-derived fuel such as char. The hot gas stream might also be a clean gas flowing in a closed loop, with heating provided by an externally fired coal-combustion chamber.

Table 1. Preparation of Coal by Type of Equipment, % of Clean Coal Produced, Year

Washer type	1942	1952	1962	1972
jigs	47.0	42.8	50.2	43.6
dense-medium processes	8.8	13.8	25.3	31.4
concentrating tables	2.2	1.6	11.7	13.7
flotation			1.5	4.4
pneumatic	14.2	8.2	6.9	4.0
classifiers	7.4	8.5	2.1	1.0
launders	13.1	5.2	2.2	1.9

The MHD process in its simplest form is illustrated in Figure 1. The hot ionized gas flows from the chamber at the left through the MHD generator duct, which is provided with a transverse magnetic field of flux density B (tesla, T or 10^4 gauss). With electrodes placed as shown, there is a transverse current density of magnitude j_y (A/m²) which contributes to the total current I flowing to the load circuit. Actually, the generator process is more complex than shown here, and a modification of the electrode arrangement is desirable.

In a power plant the MHD generator may use either a closed system or an open system. In either system, the gases exhausting from the generator are very hot, and it is desirable to recover their heat. A large part of the exhaust heat can be used to preheat the combustion air; then a further utilization can be realized in a bottom plant, which may be either a gas turbine system or a conventional steam power plant. In principle, the closed cycle could draw heat from a nuclear reactor rather than a coal-fired combustor (see also Plasma technology).

There are four important types of MHD generator channels designated as follows: type a, continuous electrodes, $E_x = 0$; type b, segmented Faraday type, $j_x = 0$; type c, Hall type, $E_y = 0$; and type d, diagonal Faraday type, $j_x = 0$.

The MHD-duct flow relations are necessary to describe the behavior of the channel as a whole. The flow through the duct is accompanied, in general, by changes in pressure, temperature, velocity, density, enthalpy, entropy, conductivity, and gas composition. Many of these variables are interrelated. The situation is further complicated by internal friction and thermal transfer. Real MHD channels are also characterized by such phenomena as leakage currents and nonuniformities of plasma properties. A preferred way of dealing with this very complex situation is to carry out an analysis based on assumptions that are reasonably valid, but which, at the same time, lead to simplifications. Special investigations can then be made of aspects ignored in the simplified treatment. A one-dimensional, hydraulic type of analysis or an analysis based on boundary-layer treatment can be used.

Applications in Power Plants

Fuel. The fuels of primary interest for MHD power generation are coal and coal derivatives, including char and gasified coal. Attention centers on coal because of its plentiful supply and moderate cost. Coal and char are fuels admirably suited for MHD power generation.

The factors that determine the suitability of a fuel for MHD application are the capability of giving both good electrical conductivity and high flame temperature. The factor most conducive to high conductivity is a low ratio of hydrogen to carbon. Abundance of hydrogen is disadvantageous for two reasons. First, in the presence of oxygen and alkali atoms, it promotes a higher concentration of potassium or cesium hydroxide, thereby diminishing the concentration of alkali atoms and reducing the numbers of positive ions (K^+ or Cs^+) and free electrons. This reduction of n_e lowers electrical conductivity. The second unfavorable effect of hydrogen is also related to an adverse effect on n_e. The hydrogen tends to promote a greater concentration of OH, ie, OH⁻.

The second favorable fuel characteristic of thermodynamic nature is the high flame temperature. Fuels that give a higher flame temperature for a given air-preheat temperature are more desirable on account of both beneficial effects on conductivity and increase in enthalpy drop attainable in the generator. Char, anthracite, and coke are superior MHD fuels. Next are bituminous and sub-bituminous coals, followed in turn by lignite, fuel oil, and natural gas.

Preheat temperature requirements. The air-preheat temperature required to give a certain flame temperature can be calculated from the following equation:

$$\frac{Q_{add}}{N_c w_c} = \frac{m_{aw}}{m_c}(h_{2aw} - h_{0aw}) - \lambda_1 \theta_c \tag{1}$$

where Q_{add} is the heat added to the system above the reference state per mole of fuel mixture to bring the products to the state (p_3, T_3); N_c is the number of moles of combustible (eg, dry, ash-free coal) per mole of fuel mixture; λ_1 is the fractional heat-loss ratio; w_c = equivalent molecular weight of the combustible; m_{aw}/m_c = ratio of moist air to combustible, weight basis; h_{2aw} = specific enthalpy of moist air at T_0; and h_{0aw} = specific enthalpy of moist air at reference temperature T_0.

The calculation cannot be completed without detailed knowledge of the gas composition. Usually kinetic effects can be neglected and the flame temperature calculation based on the gas composition at chemical equilibrium. Such equilibrium calculations, as well as calculations of molar enthalpies, should be made at the outset of MHD thermodynamic investigations. Use of JANAF tables for basic data is recommended. One completes the calculation of the preheat temperature by first assuming a set of flame temperatures, T_3, at some chosen pressure p_3, and then applying equation 1 to find $h_{2aw} - h_{0aw}$. The preheat temperature, T_2, can be found from the known enthalpy properties of the moist air. The quantity, Q_{add} would have previously been determined for the given fuel and equivalence ratio θ by computerized calculation for many pressures and temperatures.

Oxygen enrichment. Oxygen enrichment may be used to avoid a costly investment in air preheaters. For a given flame temperature, the air preheat temperature can then be reduced, even to the level of no preheat at all if that condition is desired. The penalty is a reduction of generation efficiency because the ratio of fuel burned to mass flow of gas in the MHD generator is increased. For short-burst generators, or generators applied for peaking or emergency service, oxygen enrichment may be attractive to reduce capital cost and the cost of heat development.

The other aspect of oxygen enrichment is the possibility of increasing flame temperature at a given air-preheat temperature. In this case the enthalpy drop of the gases in the MHD duct can be increased, and the efficiency of the plant improved. There are essentially two reasons why one does not invariably resort to oxygen enrichment. First, it is undesirable to operate with the duct materials at a higher temperature than that necessary to give acceptably good performance. Second, the economic penalty of the cost of the oxygen (or of the equipment and energy to produce it) discourages the use of extensive enrichment. Some optimum level of oxygen enrichment may well be found, perhaps at a very low value such as 1–2% of the air, which will be helpful and advantageous.

Selection of parameters. Deciding upon the maximum temperature to use in the MHD power plant involves various trade offs. It is possible to increase plant efficiency by increasing the combustion chamber temperature, but this necessarily implies higher air-preheat temperature and more costly air preheaters. Problems in the combustor, ducting, and MHD generator walls are also magnified by temperature increases. Metal strength considerations set a practical upper limit of 1300–1350 K on the air-preheat temperature. Selection of the maximum temperature leads directly to assignment of a maximum pressure. Selection of the combustion-chamber pressure, p_3, depends on the power level of the plant, the desired generator duct length, and the magnetic field strength.

The assigned exit pressure, after the diffuser at the outlet of the MHD duct, also influences the selection of p_3. The heat exchanger and steam generator system may be designed to run at pressures above or below atmospheric. Thus, diffuser exit pressure p_6 might be either about 1.16 kPa (1.15 atm) or 0.96 kPa (0.95 atm). Still another variation is to use a low pressure system throughout, with a powerful induced-draft blower ahead of the stack. The lower pressure can reduce wall heat fluxes and lower the required magnetic field. A disadvantage is the increase of flow area.

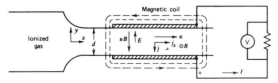

Figure 1. Simple MHD generator, showing E-field due to electrodes; induced field uB; terminal voltage $V = Ed$ and current density components; and the total current, I.

The specification of channel inlet velocity must be made with due consideration of the pressure recovery in the diffuser as well as overall enthalpy drop per unit of channel length. One generally finds that for base-load plants, with maximum temperature around 2700 K, an inlet velocity of 750–800 m/s is appropriate.

The decision on seed concentration is based partly on economic factors and partly on technical considerations. It is advantageous to use enough alkali seed material to combine with any sulfur introduced with the fuel. There is a strong tendency for the formation of K_2SO_4 (or Cs_2SO_4). This has the fortunate consequence of greatly reducing or eliminating SO_2 discharges in the stack gas. It is necessary to recover the seed material in any case, and so doing removes sulfur from the system. It is only necessary to process the K_2SO_4 to separate the sulfur and form K_2CO_3 for reintroduction into the system. The unit processes involved are fairly well known.

Air preheaters.   Air preheaters are classified as follows: (1) regenerative, or storage-type, heaters (direct or indirect-fired); (2) recuperators (direct and indirect-fired radiant type, and direct and indirect-fired convective type).

Direct firing uses the heat of the MHD exhaust stream directly. Indirect firing uses a separate, generally clean, fuel and combustion system to supply heat. Indirect-fired heaters are considered useful chiefly for test facilities and interim installations until the direct-fired heaters can be fully developed. Use of indirect firing in practical MHD power plants tends to render the plant noncompetitive.

Direct-fired regenerative, or storage-type, heaters are receiving a great deal of attention now. The well-known cowper stoves used in the steel industry are an example.

The combustion process.   Although a clean fuel could be made from coal by gasification or solvation refining, coal or char is preferably burned directly. If the combustion process can be carried out with maximum rejection of the ash, the problems in the MHD duct and heat exchange system will be considerably simplified. The goal might be at least 90% ash-rejection, leading to 10% carry-over.

A process well-suited to coal combustion with ash rejection as slag is that used in the cyclone furnace. The wall is protected primarily by a layer of frozen slag that also reduces the heat flux. The vortex flow in the chamber also is conducive to efficient combustion, since continuous ignition is maintained. Strong radial accelerations promote separation of particles and slag droplets and tend to bring the unreacted oxidant (cool air) into contact with cool particles on the chamber wall. There would normally be considerable ash vaporization in a high temperature vortex coal combustor but this can be minimized by use of a two-stage configuration.

MHD power system analysis.   The MHD generator must not only perform in a satisfactory manner but also must be integrated into the complete power-plant system. The bottom plant must be matched to the MHD topping unit. Adequate provision must be made for exhaust-heat utilization, pollution control, and seed recovery. For example, a coal-fired MHD generator could have potassium seeding, a direct-fired air preheater, and no oxygen enrichment.

Inversion to a-c power.   The inversion from d-c power to a-c power for the multielectrode Faraday generator requires, in general, as many inverters as there are electrode pairs. The inverter circuits present no unusual complications. The inverters could be built integrally with the transformers, each inverter supplying power to a portion of the primary winding.

One of the major advances in the area of d-c to a-c inversion or MHD has been the concept, recently proposed by Rosa, of load consolidation. Essentially, his idea is to introduce d-c emf between adjacent electrodes in such a way as to bring leads from two or more adjacent electrodes to the same potential. The power source for the emf would come from a power take-off from the main a-c output circuit, the power in the feedback lines being inductively coupled to rectifiers that produce the d-c incremental voltages. Thus power is circulated in the electrical output circuitry but without dissipation.

Load consolidation can effect great simplification of the MHD plant electrical system and reduce costs. The relative attractiveness of segmented Faraday generators compared to diagonally-conducting wall

(DCW) generators is increased. It should be pointed out, however, that the consolidation technique can also be applied profitably to DCW generators.

Chemical regeneration.   Chemical regeneration, as applied in MHD systems, is the process of recovering energy from the hot exhaust system by storage of chemical enthalpy in fuel components. The types of reactions that might be considered are the following: $CH_4 + Q \rightarrow C + 2 H_2$; $CH_4 + H_2O + Q \rightarrow 3 H_2 + CO$; $C + CO_2 + Q \rightarrow 2 CO$; and $C + H_2O + Q \rightarrow CO + H_2$. The third and fourth reactions represent a simplified formulation of the reaction of coal and the exhaust stream gases. The coal can be rapidly devolatilized with evolution of vaporized fuel constituents, and the resulting char can react with CO_2 or H_2O to yield CO and H_2.

The use of chemical regeneration makes possible a gain of three or more efficiency points in the MHD system without increase of air-preheat temperature or oxygen enrichment. It is therefore viewed as one of the principal avenues for further improvement of the MHD combined cycle.

Control of effluents and environmental effects.   The MHD combined-cycle plant, which may reach efficiencies over 50%, is attractive in that it has low thermal discharge. This reduces heat evolution at the power-plant site and also reduces cost of the heat-dissipation equipment. The discharge of particulate materials is a second important environmental effect. In the coal-fired MHD plant, there are arrangements in which not more than 10% of the ash is carried downstream. This typically is ca 0.1 wt% of the gas flow. The seed material, on the other hand, when converted to K_2SO_4, amounts to about 1.4% of the gas flow. Total particulates are thus ca 1.5% of the flow. By use of a good bag-house or electric-precipitation cleaning system, 99.7% of this particulate material can be captured, leaving discharged particulates of only 0.004% or 40 ppm in the discharged gas flow. The particulate discharge might be even less with superior filtration. Good particulate recovery is essential to minimize cost of seed make-up material. By carrying out the fuel-rich combustion at values less than 0.95 and allowing a sufficiently slow cooling rate at ca 1800 K, the nitric oxide concentration can also be brought to a satisfactorily low value.

Generator construction features.   The duct or channel of an MHD generator for long-term use has four main functions: to provide a safe and secure means of containing and carrying the hot plasma stream; to provide current collection at the electrodes and transmission of these currents to the external load circuits; to provide adequate internal and external electrical insulation; and to ensure good durability of all internal parts in contact with the hot plasma stream. Generators for short-burst operation may be designed on the heat-sink principle which leads to a configuration quite different from that of long-duration channels with cooled walls. These requirements should be met with minimal electrical and thermodynamic losses. The internal construction of a generator includes electrodes, insulators, insulating walls, and cooling means; and the external casing construction provides the main structural support and gas-tight envelope.

The magnetic field for large MHD generators is normally provided by a superconducting (SC) magnet. To maintain the superconducting state in the winding, the temperature should be held close to that of liquid helium, ca 4 K. The winding consists of a composite material of superconductors such as Nb–Ti, Nb_3Sn, and copper. It is important in the generator design to provide for a casing of minimal external dimensions so that the inner warm bore of the magnet can be made as small as possible. Magnetic field levels generally considered are of the order of 6–7 T (60,000 to 70,000 gauss) at the inlet for large central-station generators, 5–6 T for smaller pilot plants.

Alternative Arrangements of the MHD Power Plant

Disk generator.   The disk generator has been investigated experimentally and theoretically by Louis, Klepeis, and others. The basic idea involves a radially symmetrical configuration in which flow takes place radially between flat disk-shaped walls. The magnetic field is in the axial direction. It is generally considered advisable to impart a swirl component to the velocity.

Coal-plex processes.   The Coal-plex concept has been suggested by Squires as a means of more effectively using our coal resources. The coal is processed in a carbonizer, where volatile materials are driven off and the coal is partly pyrolyzed to form a relatively clean gaseous fuel. This off-gas can be used either as a clean fuel or as feedstock for various chemical processes. The nonvolatile char residue can be used as fuel for power generation (see Fuels, synthetic—gaseous).

Gas turbine bottom plants. The MHD generator can be used effectively in conjunction with a gas-turbine or air-turbine plant.

STEWART WAY
Westinghouse Electric Corp.

J. B. Heywood and G.J. Womack, *Open Cycle MHD Power Generation*, Pergamon Press, London, 1969.

R. Bünde, H. Muntenbruch, J. Raeder, R. Volk, G. Zankl, *MHD Power Generation*, Springer, New York, 1975.

M. Petrick, B.Y. Shumyatsky, *Open Cycle Magnetohydrodynamic Electrical Power Generation*, Argonne National Laboratory, Argonne, Ill., 1978.

COAL GAS. See Coal chemicals and feedstocks; Coal conversion processes; Tar and pitch.

COAL TAR. See Coal conversion processes; Tar and pitch.

COATED FABRICS

A coated fabric is a construction that combines the beneficial properties of a textile and a polymer. The textile (fabric) provides tensile strength, tear strength, and elongation control. The coating is chosen to provide protection against the environment in the intended use. A polyurethane might be chosen to protect against abrasion or a polychloroprene (Neoprene) to protect against oil (see Urethane polymers; Elastomers, synthetic).

Textile Component

For many years cotton (qv) and wool (qv) were used as primary textile components, contributing the properties of strength, elongation control, and aesthetics. Polyester, by itself and in combination with cotton, is used extensively in coated fabrics. Nylon is the strongest of the commonly used fibers. Rayon and glass fibers are the least used; rayon has low adhesion and glass is brittle.

There are many choices in textile construction. The original, and still the most commonly used, is the woven fabric. Knitted fabrics are used where moderate strength and considerable elongation are required. Many types of nonwoven fabrics are used as substrates (see Nonwoven textiles fabrics).

The construction that results from either weaving or knitting is called a greige good. Other steps are required before the fabric can be coated: scouring to remove surface impurities; and heat setting to correct width and minimize shrinkage during coating.

Polymer Component

In addition to natural rubber and polychloroprene, other polymers in use include: styrene–butadiene (SBR), polyisoprene, polyisobutylene (Vistanex), isobutylene–isoprene copolymer (Butyl), polysulfides (Thiokol), polyacrylonitrile (Paracril), silicones, chlorosulfonated polyethylene (Hypalon), poly(vinyl butyral), acrylic polymers, polyurethanes, ethylene–propelene copolymer (Royalene), fluorocarbons (Viton), polybutadiene, polyolefins, and many more. Copolymerizations and physical blends make the number available staggering (see Copolymers; Olefin polymers; Polymers containing sulfur; Acrylic ester polymers; Vinyl polymers; Fluorine compounds; Acrylonitrile polymers; Silicon compounds; Urethane polymers).

Processing

Coated fabrics can be prepared by lamination, direct calendering, direct coating or transfer coating (see Coating processes). The basic

problem in coating is to bring the polymer and the textile together without altering undesirably the properties of the textile. Almost any technique in applying polymers to a textile requires having the polymer in a fluid condition, which requires heat. Therefore, damage to the synthetic or thermoplastic fabric may occur.

Coated fabrics can be decorated by printing with an ink. If a textured surface is desired, the coated fabric is heated to soften it and pressure is applied by an engraved embossing roll. The final layer is called the slip. Most coatings are tacky enough to stick to themselves (block) during stacking or rolling. The main purpose of the slip is to prevent blocking (see Abherents).

Health and Safety Factors

Some materials used in coating operations have been identified by the U.S. Government as being hazardous to the workers' health. Coating operations should not be initiated without consulting the *Federal Regulations on Occupational Safety and Health Standards*, Subpart Z, *Toxic and Hazardous Substances*.

FRED N. TEUMAC
Reeves Brothers, Inc.

M. Morten, *Rubber Technology*, Van Nostrand Reinhold Co., New York, 1973.

COATING PROCESSES

Plastic film-forming materials may be applied as coatings in many different processes. Coatings may be applied for either functional purposes (eg, waterproofing, flameproofing, mildewproofing, abrasion resistance, rust resistance, reflection, insulation, adhesion, or a coating impermeable to gases and liquids) or decorative purposes (eg, gold coloration on aluminum foil or wrapping paper).

In addition to coated, shaped rigid articles such as automobile bodies, furniture frames, machinery, etc, the range of coated web or sheet materials includes a wide spectrum of products such as artificial leather, garment interlinings, impermeable products such as shower curtains, book bindings, paper and paperboard, a variety of tapes, coated metal strips for building and furniture applications, hard-surface phenolic and melamine laminations for furniture and insulating uses, floor and wall coverings, food packaging and other coated films, and coated plywood and fiber panels.

Processes

A coating process is the application of liquid to a traveling web or substrate. The primary substrates include paper and paperboard grades, films, foils, nonwovens, and wovens. The coating principles are better suited to continuous webs than to short individual sheets. In general, there is an ideal coater arrangement for any given product; however, most machines are required to produce many different products and coating thicknesses and the machine is therefore usually a compromise for several applications.

All coating machines contain application and metering devices. In order to ensure that a given quantity of coating is retained on a substrate, it is necessary to incorporate the metering principle. Coaters can be tension sensitive and tension insensitive. Tension-sensitive coaters (eg, the metering rod) simply means that the ability of the coating station to maintain coating weight is dependent on the ability of the process line to maintain a constant substrate tension. Tension-insensitive coating stations have the inherent ability to maintain coat weight even when web tensions vary.

Brush-coating methods are best suited to slow-drying coatings such as oleoresinous paints, starches, latexes, or emulsions. They permit the application of a flawless coating, such as required on playing cards, or the uniform coverage and penetration of uneven surfaces, such as required on rough fiber panels. Brush-coating machines have been replaced largely by air-knife coaters. Advantages: smooth application; contour coater; not tension sensitive. Disadvantages: noisy; high maintenance; slow speed; high viscosity; difficult to clean.

The simplest, least expensive, and yet one of the best coating methods is rigid knife coating, either unsupported or supported, with the latter being the more accurate. Advantages: low cost; simple, compact; precision to 1% variation; heavy coat weights, 0.012–2.5 mm wet; high viscosities up to 100,000 imPa·s (= cP); levels rough surfaces; low solvent loss; produces uniform total web cross section. Disadvantages: streaking or scratches; passing slices; messy on web breaks; contacting-type edge dams; coating eight changes with web caliper or profile; speeds above 300 m/min.

The familiar example of sheep wool or synthetic fleece-covered hand-painting rolls is an excellent illustration of the fibrous belt coating method preferred for the application of latex and emulsion coatings.

The metering rod coater (Fig. 1), also known as a Mayer rod coater, utilizes a driven single-roll applicator to coat in either direction and is followed by a driven metering rod assembly to control weight. Advantages: low cost; simple, compact. Disadvantages: Mayer rod wear; rod holder wear; speed above 300 m/min; high viscosity; high cost weight.

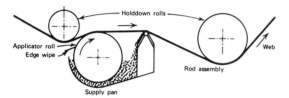

Figure 1. Metering rod coater.

The flexible-blade coater is known as the puddle coater. Its advantages are high speed, low coat weight, and short dwell. Disadvantages include: high coating loss on web breaks; backing roll wear from dikes; poor coat-weight running adjustment; poor operator vantage point; difficult to change doctor blade; difficult to filter coating effectively; difficult to go off-coat.

A second-generation blade coater is known as the flooded-nip, inverta-blade, or roll applicator. The latest blade coater arrangement is the twin-blade concept that consists of two fountain-blade coaters back to back without a backing roll.

One of the simplest coating machines used today on films that have stringent optical requirements utilizes the meniscus principle. Advantages: excellent precision application; produces good optical qualities with minimum pattern; does not scratch substrate or undercoat; does not distort or elongate substrate; low coating weight; low viscosity; clean room qualities. Disadvantages: slow speed; high viscosity; coat weight changes with speed and viscosity; sensitive to room air currents; sensitive to air bubbles and ripples; high coating weight; low speed; different width-back-up roll for each web width.

There are many types of kiss-roll coaters. They are usually tension sensitive and require a postmetering or smoothing device such as a metering rod or air knife.

Size-press coaters are used to apply aqueous sizing materials to fabrics, but more specifically on machines making paper and paperboard. Advantages: simple; compact; easy to maintain; can add size to either or both sides. Disadvantages: film split pattern; poor coating-weight control; reduces caliper; low solids, low coating weight.

Paperboard surface characteristics are improved by adding water boxes to the steel roll calender stack on the board-making machine. Advantages: simple; inexpensive; improves end product; reduces curl; can treat either or both sides. Disadvantages: corrosion of rolls covered by solutions; usually slows machine process down because of additional drying requirements; reduces caliper; increases bulk density.

The transfer-roll coater, also known as the Kimberly-Clark-Mead (KCM), upgrades the performance of a conventional two-roll size press so that size solutions and coatings can be applied in a single station at higher solids and/or coat weights. Advantages: high speed; adjusts coating weight while running; high solids; high viscosity; can apply two different formulas simultaneously; easy to thread. Disadvantages: roll split pattern on coatings; very cumbersome and expensive; roll deflection compensation is required to ensure uniform coating weight across the machine.

Direct gravure coaters are the most accurate of the roll-type coaters. The coating weight is usually controlled by proper selection of the gravure roll etch. Advantages: simple to operate; compact; easy to maintain coating weight; coats a wide range of substrate materials; low solvent or heat loss; not tension sensitive; gives low coating weights. Disadvantages: difficult to change coating weight except on reverse application; backing roll is undercut for substrate width; gravure roll maintenance and life; doctor blade maintenance; high viscosity; film-split pattern; coating weights change with speed.

In offset or indirect gravure, a steel back-up roll is added above the direct-gravure roll arrangement allowing an additional film split before the application to the substrate. This minimizes coating patterns and reduces coat weight. Advantages: low coating weight; can adjust coating weight while running; high speed; minimal film-split gravure-cell patterns; higher viscosity; good coating-weight stability with process speed changes. Disadvantages: high coating weight; hot melt cools on resilient roll.

Both sides of a substrate can be simultaneously coated by a symmetrical coater arrangement. However, it is limited to low viscosities and web-speed applicator rolls that can form a pattern. It is commonly used for the application of primers or precoats to a plastic web.

During the early 1970s, printing multi-or single-color designs on fabrics was an important development. Rather than flexographic printing directly onto the fabric with its shrinkage and register problems, the patterns are heat-transferred from a release sheet. The release sheet consists of a base paper with a release coating or a web of foil.

The vertical or in-line reverse-roll coater utilizes the principles of the reverse-roll kiss coater in addition to a back-up roll above the applicator roll to eliminate the tension-sensitive application. Advantages: open design; simple; wide range of coating weights, process speeds, and viscosity; applies a uniform coating even with substrate caliper variations; not tension sensitive. Disadvantages: solvent evaporation; deterioration of the resilient roll caused by solvents; solids buildup; contamination buildup; scratches on sensitive substrates; new top roll for each web width; promotes air bubbles; strike-through on open webs.

The three-roll nip-fed coater consists of a rubber-covered back-up roll and precision metal-applicator rolls. Advantages: low solvent loss; wide range of coating weights, viscosity, and speeds; improved resilient roll life; maintains coat weight even with caliper variations; not tension sensitive. Disadvantages: end-dam maintenance and wear on coating rolls; coating streaks by particles in the nip; danger of clashing and damaging metering and applicator rolls; poor operator vantage point when adjusting coater; TIR (total indicator run out) maintenance; promotes air bubbles; backing-roll strike-through on open webs; backing-roll maintenance.

Saturators are webs designed to control the amount of residual coating solution evenly distributed through the cross-sectional area of a substrate. Final cost weights are often referred to in percent pick-up. Saturating-grade substrates include porous or unsized papers, creped paper, woven fiberglass, canvas, and nonwoven and cotton linter. Products of saturators include masking tape, counter tops, electric circuit boards, fish poles, and air and oil filters, to name a few. Three basic methods of saturating include dip and scrape, dip and squeeze, and reverse roll.

Many plastic compositions can be softened by heat and formed into self-supporting sheets by squeezing between heated iron calender rolls (Fig. 2).

Film-forming materials can be applied to continuous substrates by cast-coating methods, ie, the process that includes contact of a wet coated surface with a smooth polished drum or metal belt during the drying or fusing operation. This process gives a flat-coated surface for

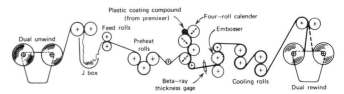

Figure 2. Z calender-coating arrangement.

decorative effects or for subsequent printing operations (see Printing processes). In the precast-coating process, a metered coating is deposited on a metal drum or metal belt followed by contact with the substrate and transfer of the coated material from the substrate, resulting in a smooth coated finish.

Powdered thermoplastic resins (particle size 25–200 μm) may be applied by flame-spray coating methods, principally on metallic substrates. They also may be applied by electrostatic attraction by spray-gun delivery.

Irregularly shaped objects are frequently coated by immersion in a coating composition, and then removed, drained, and dried or baked. Coating thickness is determined by the use. Tears or drops of coating at the bottom of dip-coated articles may be removed by electrostatic attraction as the article is moved along the conveyor.

Plastic coatings may be applied by spraying on irregularly shaped and compound curved or sharp-edged surfaces. The principal distinction in spray-gun design depends on the method used to obtain atomization of the coating liquid.

In pressure atomization or airless spraying, the coating is delivered to the nozzle under very high pressure and atomization results from dispersion in air as it leaves the nozzle. Air atomization spray-nozzle design depends on the impingement of an air stream against the coating to break it into small droplets and carry it to the surface to be coated. Most spray-nozzle operations require the fluid coating to be delivered through the nozzle under pressure rather than by suction from its supply source. Multiple-component spray nozzles may be arranged for internal mixing or external mixing of components. Materials of short pot life such as epoxy resins, polyurethanes, polyesters, etc, may be mixed and applied as coatings in this manner.

Many coating materials of suitable dielectric constant may be electrically polarized so that they are attracted to a grounded or oppositely charged surface. This principle (electrophoresis) has been used for many years in electrostatic spraying equipment. The coating particles may be generated by airless or air-atomized spray nozzles, centrifugal force, or electric attraction. Practical applications extend from paint or plastic solutions and suspensions to hot-melt compositions and powdered resins.

Many monomers can be polymerized directly on a substrate to form a uniform surface coating. Several different methods are used: magnetically contained glow discharge; exposure to an electron beam; and exposure to ultraviolet radiation (see Radiation curing).

In the curtain-coating process, the coating composition is delivered in a falling sheet or curtain to the substrate, which is moved through the curtain at a controlled rate. The principal limitation of this process is that the coating composition must be sufficiently fluid to fall freely and sufficiently cohesive to present a continuous film to the substrate.

Extrusion coating is the process of melting, metering, combining, and cooling a molten homogeneous sheet of thermoplastic material onto a continuous substrate such as paper, paperboard, film, foil, or fabric. Today's expanding food-packaging industry is the direct result of packaging improvements that can be attained by altering the surface and physical characteristics of a flexible web by the extrusion-coating process (see Film and sheeting materials).

Drying Systems

Drying systems are a very important part of the coating process since once a coating has been properly applied to a substrate, it must be conveyed, dried, and cured before the coated side can be contacted again. The drying system is based on the available fuel for the heat source, strength or stability of the coated substrate and the maximum rate at which evaporation and curing can be accomplished without detrimental effect on the end product (see Drying). The most common heat sources for drying include steam, oil, dual gas (natural and propane) and/or oil, electric radiation, ultraviolet light, and high frequency generation. Drying can be contacting or noncontacting.

STANLEY C. ZINK
The Black Clawson Co.

Reviewed by
GEORGE L. BOOTH
The Black Clawson Co.

G.L. Booth, *Coating Equipment and Processes*, Lockwood Publishing, New York, 1970.

Modern Plastics Encyclopedia, McGraw-Hill, New York, 1977.

L.G. Janett, *Drying of Coatings and a Summary of Methods*, Midland Ross Corporation, New Brunswick, N.J.

COATINGS, INDUSTRIAL

Coatings are available as liquid or fusible compositions. Table 1 gives a classification of coatings formulations. The liquids are usually aqueous or organic solutions. The coatings are applied by the user to the substrates, allowed to flow out smoothly, chiefly by forces of surface tension, and then cured to the final solid form. An industrial organic coating usually comprises an organic binder, pigments, a carrier liquid (sometimes omitted), and various additives.

The binder exists in the final film as a polymer of high molecular weight that may or may not be cross-linked. It is primarily responsible for the plastic quality of the film. Binders are grouped into certain overlapping classes such as acrylic, vinyl, alkyd, polyester, etc. The molecular structure of the binder and the forces operating between the molecules largely determine the mechanical properties.

The pigments, which may be organic or inorganic, contribute primarily to opacity and color, in addition to durability, hardness, adhesion, and particular rheological properties of the coating in fluid form.

Table 1. Formulation Possibilities[a] of Synthetic Resins in Coatings Formulation[b]

Vehicle system	Uses
Class 1: Vehicles containing oil-modified alkyds or other polymers containing drying oil	
(1) oxidizing alkyd resins (sometimes mixed with oleoresinous varnishes)	(1) architectural enamels, house paints, interior paints, flat wall paints, baking and air-drying undercoats and enamels for machinery, prefab housing structural units, and other factory products
(2) alkyd and phenoplast alkyd and nitrocellulose alkyd and chlorinated rubber alkyd and polystyrene alkyd and diisocyanate alkyd and vinyl and epoxy	(2) air-drying or low-temperature baking undercoats and enamels (for metal products) that have more plasticlike film properties than is possible to attain with alkyds alone
(3) alkyd and aminoplast alkyd and aminoplast and epoxy alkyd and silicone	(3) similar uses as above, but where a high premium is placed on color retention, and superior chemical and heat resistance
(4) oil-modified epoxy resins and aminoplast	(4) air-drying or baking-type undercoats or enamels; improved baking enamels and undercoats
Class 2: Vehicle systems containing no alkyd or drying oil	
(5) vinyl acetals and/or phenolic allylaminoplast and 2,4,6-trimethylolphenyl ether alkyd epoxy	(5) chemically resistant baking undercoats and enamels
(6) phenoplasts (with or without epoxy, or vinylacetal, or aminoplast)	(6) thermosetting undercoats and/or enamels for high corrosion protection, especially in thin films (not resistant to discoloration); room-temperature setting mastics for corrosion and abrasion protection
(7) polyester and triazine resin allyl polyester silicone thermosetting acrylics complex amino resins some other polyesters	(7) chemical and discoloration resistant, glossy, clear films and pigmented baking enamels for metallic and nonmetallic production goods (thermosetting film formers)
(8) vinyl acetate–chloride, copolymers vinylidene or vinyl chloride–acrylonitrile copolymers butadiene copolymers acrylic copolymers poly(vinyl acetate)	(8) thermoplastic lacquers, baking or air-drying emulsion paints for production finishes on nonmetallic goods such as acoustical board or molded plastics, fire-retardant and corrosion-protective mastics; exterior house paints and interior decorators' paints of emulsion type
(9) nylons some cellulosic esters and ethers polyurethanes polytetrafluoroethylene poly(vinyl acetals) saturated polyesters unsaturated polyesters and styrene epoxy and polyamide copolymers of ethylene or propylene	(9) special type of coatings, potting compounds, mastics, etc, for electrical insulation and corrosion protection

[a] Commercial formulations are generally proprietary.
[b] Courtesy *Chemistry in Canada*.

Coatings are classified as primers that are applied directly to a substrate or as topcoats. The latter are applied over a primer and are usually the last coat. Intermediate coats are called sealers. In some uses, one coat satisfies all requirements.

Industrial coatings are generally applied with specialized machinery representing a large investment. Coating properties must be controlled carefully to ensure continuous and satisfactory application (see Coating processes). Properties to be controlled include rheological characteristics, stability, color, smoothness or gloss, metallic luster, durability, adhesion, permeability, hardness, flexibility, and protection of the substrate against deterioration.

Government Regulations

(1) In 1974, the EPA set national air-quality standards for particulates, sulfur dioxide, carbon monoxide, photochemical oxidants, and hydrocarbons (see Air pollution). The standards for photochemical oxidants and hydrocarbons in the atmosphere are most significant for the coatings industry. The concentration of photochemical oxidants is limited to 160 $\mu g/m^3$ as the maximum average concentration for one hour; this 1-h concentration is not to be exceeded more than once per year. Hydrocarbon standards are similarly set at 100 $\mu g/m^3$. The states are responsible for local regulations that allow the national standards to be met. These limit the content of volatile solvents in coating formulations.

(2) The Water Pollution Control Act amendment of 1972 extends the Federal authority to all United States water. By July 1, 1984 industrial plants must reduce pollutants using the best feasible measures so that progress is made towards elimination of effluent pollutants by 1985 (see Water pollution).

(3) NIOSH has developed stringent standards for maximum permissible concentrations of many materials employed in the organic coatings industry. Asbestos, chromates, lead, cadmium, and mercury are some of the hazardous materials used, and an intensive search is going on for safe substitutes. Residual vinyl monomers in polymers, particularly vinyl chloride and acrylonitrile (qv), are being reduced greatly.

(4) The FDA issues regulations governing materials that may contact food, and a new coating meant for contact with food must pass a series of extraction tests in order to obtain FDA approval.

(5) The Toxic Substances Control ACT of 1977 requires that new manufactured materials undergo adequate testing and must meet toxicity standards (see Industrial hygiene and toxicology; Regulatory agencies).

SEYMORE HOCHBERG
E.I. du Pont de Nemours & Co., Inc.

"Guidance to State and Local Agencies in Preparing Regulations to Control Volatile Organic Compounds From Ten Stationary Categories," U.S. Environmental Protection Agency, *EPA-450/2-79-004*, Sept. 1979.

COATINGS, MARINE

Ships, offshore working platforms, and onshore waterfront structures are damaged by contact with the harsh marine environment. Control of this destructive action is best achieved through a program of (1) selection of the materials most resistant to deterioration, (2) design to minimize conditions favorable to corrosion, and (3) effective utilization of protective coatings and/or cathodic protection (an electrical method of preventing metal corrosion in a conductive medium by placing a charge on the item to be protected) to deter corrosion. Protective (anticorrosive) coatings impart protection to the substrate by forming a barrier to the water, salt, and oxygen which accelerate corrosion. Antifouling paints control the attachment and growth of marine fouling organisms on immersed areas.

The modern synthetic coatings have three common ingredients: solvent, binder, and pigment. Coatings that cure by chemical reaction of two component parts are the most widely used in submerged marine applications. For example, epoxies, coal-tar epoxies, urethanes, and polyesters are durable and resistant to water, solvents, and chemicals.

Modern synthetic marine coatings provide much longer protection than earlier ones, but require both complete cleanliness and surface profile (tooth) in order to obtain adequate bonding of prime coats. High speed abrasive (sand, grit, or shot) blasting by conventional air-pressured equipment or by newer equipment that utilizes centrifugal force to propel the abrasive helps provide an appropriate surface (see Abrasives). Actions taken by personnel to eliminate particulate emissions during abrasive blasting include the use of hard, sharp, and properly sized abrasives that produce minimum emission of particulates, blasting inside a building or under a temporary shroud, and using equipment that automatically moves across a regular surface and picks up and recycles the spent abrasive.

Manufacturers of marine coatings always have printed information available on the use of their products. This information includes recommendations on the equipment to be used, mixing of components, time and temperature requirements, coverage rate at a recommended dry-film thickness, and good application practices.

Protective (anticorrosive) coatings include vinyls, chlorinated rubbers, epoxies, coal-tar epoxies, urethanes, polyesters, inorganic zincs, zinc-rich organics, and specialized coatings (eg, solvent-free epoxies, powder coatings, plastic-coated steel electrical conduits and fittings, and petrolatum-coated tapes).

Currently, the only two biocidal materials that are used extensively in antifouling paints are cuprous oxide and organotin compounds (usually tributyltin oxide or tributyltin fluoride). Most commercial antifouling paints use a vinyl binder, although products with other binders are also marketed. Rosin or some other leaching agent must be added to cuprous oxide formulations to permit its controlled release into seawater where it is lethal to fouling larva forms. Organotins usually do not require leaching agents to dissolve slowly in seawater.

Recent research has lead to the development of organometallic polymers for use in antifouling paints. A sheet material (ca 2 mm thick) of black neoprene rubber (see Elastomers, synthetic, neoprene) impregnated with tributyltin is currently marketed. Because the sheet is so much thicker than an antifouling coating system (usually about 100 μm) it has a larger reservoir of biocide that can result in longer-lasting fouling control.

R.W. DRISKO
U.S. Naval Civil Engineering Lab.

Antifouling Marine Coatings, Noyes Data Corporation, 1973.
Coating Systems Guide for Hull, Deck, and Superstructure, Society of Naval Arch. & Marine Engineering Report No. TR-4-10, 1973.

COATINGS, RESISTANT

Resistant or high performance coatings or linings are specialty products used to give long-term protection under difficult corrosive conditions to industrial structures. This contrasts with paint, which is used for general appearance and shorter-term protection against milder atmospheric conditions, and industrial coatings, ie, coatings that are applied to manufactured products (see Paint; Coatings, industrial; also Coating processes; Coatings, marine).

There are two basic methods by which coatings or linings protect the surface. The first is based on the principle of impermeability. The second method uses anodically active or inhibitive pigments in the primer or in the coating to regulate corrosion (see Pigments). The inert impervious system performs best as a lining where it is subject to continual moist, wet, or immersion conditions and where it is subject to little or no physical abrasion. Inhibitive coatings perform best in areas where the coating is subject to weathering, atmospheric conditions, high humidities, or chemical fumes.

Components of Resistant Coatings and Linings

The similarity to paint extends to the essential ingredients of the coating or linings which include the binder, color-carrying pigments, inert and reinforcing pigments, inhibitive pigments for primer only, and solvents. In general, the same equipment and manufacturing procedures are used for paint (qv) or industrial coatings.

Application of Resistant Coating Systems

The characteristics provided by each part of a coating system are as follows: *primer*—good adhesion to surface, satisfactory bonding surface for next coat, ability to retard the spread of corrosion, enough chemical and weather resistance to protect the surface until application of next coat, chemical resistance equivalent to the remainder of the system; *intermediate coats*—adequate film thickness for the coating or lining system, uniform bond between the primer and the topcoats, a superior barrier to moisture vapor and aggressive chemicals; *finish coats*—must be pleasing in appearance as well as resistant, may serve to provide a nonskid surface, a matrix for antifouling agents or other specialized purposes, must have sufficient weather, chemical, and abrasion resistance in any environment to ensure its remaining intact and providing protection to the substrate. Proper surface preparation and coating or lining application are essential to the performance of the resistant coating.

Resistant coatings or linings represented about 9% of the overall paint or coatings market in 1983. Three health and safety factors are important to consider both during manufacture and application of protective coatings: air pollution, health hazards to the individual and fire and explosion hazards.

New Coating Developments

As a consequence of the new and increasing governmental environmental standards as well as health and safety requirements, there are three avenues of development that will have a strong influence on the future of protective coatings: the use of water-based coatings; the use of solvent-free coatings or 100% solids coatings; and the use of inorganic materials as coatings, eg, organic zinc coatings.

Table 1 allows a quick comparison of the properties and uses of resistant coatings.

CHARLES G. MUNGER
Consultant

Table 1. Properties of Several Resistant Coatings[a]

Uses of resistant coatings	Lacquer coatings		Coreactable coatings				Inorganic coatings	
	Vinyl chloride–acetate copolymer	Chlorinated rubber	Epoxy–amine	Epoxy–poly-amide	Urethane moisture cure	Urethane–aliphatic isocyanate cure	Water-base	Solvent-base
abrasion resistance	G	G	G	G	E	E	VG, metallic	VG, metallic
bacterial and fungal resistance	E	G	G	NR				
chemical resistance	BSR	BSR	G	G	G	G	NR	NR
acid-oxidizing	S or F	S or F	NR	NR	F	S or F	NR	NR
acid-nonoxidizing	S or F	S or F	F	F	F	S or F	NR	NR
acid-organic	I fatty acid	dissolves in fatty acids	Spillage, F	F	F	splash or F	NR	NR
alkali	G	G	G	G	D	S or D	NR	NR
salts-oxidizing	Splash, S	F or D	F	F	NR	NR	NR	NR
salts-nonoxidizing	I, G	I, G	I, G	I, G	splash	splash	G, marine splash, S	G, marine splash, S
solvent-aliphatic	E	G	E	G	VG	E	E	E
solvent-aromatic	swells	dissolves	G	NR	VG	E	E	E
solvent-oxygenated	dissolves	dissolves	NR	NR	NR	NR	E	E
water	VGI	VGI	VGI	VGI		[b]	G	G
moisture	low	low	low	low	medium, low	medium, low		
permeability								
contamination of contacting materials								
food	[c]	water, G	G	G				
chemical	VG	G	G	G	G	VG		
decontamination	VG		VG	VG			E	E
friction resistance (faying surfaces)							coef. of friction 0.47	coef. of friction 0.52
heat resistance, °C								
wet	48	38	48	48		38		
dry	65	60	95	95	120	120	370	315
radiation resistance, Gy[d]	10^6	10^6	10^7	10^8	5×10^6	5×10^6	10^8	10^8
soil resistance							NR	NR
weather and light resistant	G, properly pigmented	G, properly pigmented	heavy chalking	G	G	E, color and gloss retention	E > 20 yr	E > 20 yr
principal hazard application	solvent F	solvent F	dermatitis solvent F	solvent F	solvent F	solvent F	none, water base	solvent F

[a] G, Good; VG, Very Good; E, Excellent; NR, Not Recommended; BSR, Broad-Spectrum Resistance; I, Immersion; S, Spray; F, Fumes; D, Dusts.

[b] G, primer required; critical for immersion.

[c] G, odorless; tasteless; nontoxic.

[d] To convert gray to rad, multiply by 100.

F.L. LeQue, *Marine Corrosion, Causes, and Prevention*, John Wiley & Sons, Inc., New York, 1975, pp. 291–297.

"Good Painting Practices," *Steel Structures Painting Manual*, Vol. 2, Steel Structures Painting Council, Pittsburgh, Pa., 1982.

Federation Series on Coatings Technology, Federation of Societies for Coating Technology, Philadelphia, Pa.

COBALT AND COBALT ALLOYS

Cobalt, a transition-series metal, is a metallic element that is similar to silver in appearance. Cobalt and cobalt compounds have expanded from strict use as coloring agents in glasses and ground coat frits for pottery to drying agents in paints and lacquers, animal and human nutrients, electroplating materials, high temperature alloys, high speed tools, magnetic alloys, alloys used for prosthetics, and uses in radiology. Cobalt is also used as a catalyst for hydrocarbon refining. The most timely use of this last application is the synthesis of heating fuels. The largest and most important use of cobalt is in superalloys (see High temperature alloys).

Cobalt is the 30th most abundant element on earth and comprises approximately 0.0025% of the earth's crust. It occurs in mineral form as arsenides, sulfides, and oxides; trace amounts are also found in other minerals of nickel and iron as substitute ions. Cobalt minerals are commonly associated with ores of copper and nickel, and to a much lesser extent with ores of zinc, lead, gold, and platinum. The largest cobalt reserves are in Zaire, Zambia, New Caledonia, Indonesia, and Cuba. Lateritic ores are becoming increasingly important as a source of nickel having cobalt as a by-product. In the United States, laterites are found in California and Oregon. Deposits also occur in Cuba, Indonesia, New Caledonia, the Philippines, Venezuela, Guatemala, and Australia.

The physical properties of cobalt are listed in Table 1.

Cobalt Alloys

Many of the investigated binary systems are high-temperature alloys (qv), such as Co–Cr, Co–Fe, Co–Mo, Co–Ta, Co–Ti, Co–V, and Co–W.

Table 1. Properties of Cobalt

Property	Value
atomic number	27
atomic weight	58.93
Thermal	
transformation temperature, °C	417
heat of transformation, J/g[a]	251
mp, °C	1493
bp, °C	3100
latent heat of fusion, J/g[a]	259.4
latent heat of vaporization, J/g[a]	6276
specific heat, J/(g·°C)[a]	
15–100°C	0.442
molten	0.560
coefficient of thermal expansion, per °C	
cph room temperature	12.5
fcc at transformation temperature	14.2
thermal conductivity at room temp, W/(m·K)	69.16
thermal neutron absorption, Bohr atom	34.8
Electrical	
resistivity, at 20°C[b]	6.24
Curie temperature, °C	1121
Selected mechanical	
Young's modulus, GPa[c]	211
hardness, diamond pyramid (Vickers)	99.9% Co
at 20°C	225
strength of 99.9% cobalt, MPa[d]	as cast
tensile	237

[a] To convert J to cal, divide by 4.184.
[b] Conductivity = 27.6% of International Annealed Copper Standard.
[c] To convert GPa to psi, multiply by 145,000.
[d] To convert MPa to psi, multiply by 145.

More recently, rare-earth alloys such as $SmCo_5$ have become prominent because of their outstanding permanent magnetic properties.

A group of superalloys is based upon cobalt. They were developed to improve turbo-superchargers on aircraft engines but later found application in steam turbines and gas turbines as well.

Cobalt also is a common constituent of many types of magnetic alloys including precipitation-hardened alloys, quench-hardened steels, alloys with ordered structures, cold-worked alloys, and single-domain powder magnets. Magnetic materials (qv) with cobalt additions characteristically retain their magnetism at high temperatures and have higher coercive force at room temperature. Typical cobalt-containing magnetic alloys include Fe–Co–Mo, Fe–Ni–Al–Co, Fe–Ni–Cu–Co, Co–Pt, and Fe–Ni–Co–Mn, in addition to Sm–Co alloys.

Fe–Ni–Co or Fe–Co–V combination are typical of the soft magnetic alloys. Commercially important alloys are Permendur, Supermendur, Hiperco, and Perminuvar.

Cobalt is added to both molybdenum and tungsten to make high speed cutting tools that remain hard at elevated temperatures, as well as to W–Cr hot-work tool steels to increase toughness and strength. Under service conditions where impact and thermal cycling occur, Co–Cr–W alloys are used as hardfacing and wear-resistant alloys.

F. PLANINSEK
JOHN B. NEWKIRK
University of Denver
Reviewed by
WILLIAM S. KIRK
U.S. Bureau of Mines

R.S. Young, *Cobalt*, American Chemical Monograph Series, No. 149, American Chemical Society, Reinhold Publishing Co., New York, 1960.

C.S. Hurlbut, Jr., *Dana's Manual of Mineralogy*, 17th ed., John Wiley & Sons, Inc., New York, 1966.

J.C. Agarwal, *J. Met.* **28**(4), 24 (1976).

COBALT COMPOUNDS

Cobalt is similar to its neighbors, iron and nickel, in the periodic table. In nearly all its compounds it exhibits a valence of $+2$ or $+3$. The stable divalent form is not subject to appreciable hydrolysis in aqueous solutions whereas trivalent cobalt compounds are powerful oxidizing agents that are mostly unstable. In the complexed state cobalt(II) is relatively unstable and is readily oxidized to cobalt(III) by ordinary oxidants. An extremely large number of complex ions have been identified, most of which are quite stable in aqueous media. Cobalt has a formal valence of $+1$ only in a few complex nitrosyls and carbonyls. Tetravalent cobalt exists solely in fluoride complexes and in one unusual series of binuclear peroxo compounds. However, many oxidation states have been reported. Cobalt salts are usually made from the hydroxide or carbonate or by dissolving fine cobalt powder in acid. Table 1 gives typical analyses of a group of commercial cobalt salts.

Table 1. Typical Analyses of Commercial Cobalt Compounds

Assay, %	Acetate	Carbonate	Hydrate	Oxide Ceramic grade	Oxide Chemical grade	Hydrated sulfate
Analysis						
cobalt	23.50	46.00	61.00	71.70	56.00	21.00
nickel, max	0.10	0.15	0.20	0.40	0.20	0.10
iron, max	0.04	0.10	0.10	0.13	0.20	0.04
manganese, max	0.02	0.05	0.05	0.01	0.05	0.02
silica, max				0.10	0.30	
copper, max	0.02	0.05	0.05	trace	0.04	0.02
sulfur, max					0.50	
calcium oxide, max				0.22	0.10	
moisture at 105°C				trace	31.00	
alkali and alkaline earth sulfates,	0.40	0.80	0.80			0.30
hydrochloric acid-insoluble matter,		0.05	0.05			
water-insoluble matter, max	0.05					0.10
apparent sp gr	0.960	0.835	0.350			1.114

Complex Cobalt Compounds

The innumerable coordination compounds of cobalt(III) exhibit substantial diversity in their coordination number, geometric structure, and stability, and in many aspects of their chemistry. The most common coordination number is 6.

Historically, the ammines of cobalt(III) have dominated the chemistry of cobalt complexes and their influence on the development of chemistry has been substantial. The important donor atoms (in order of decreasing tendency to complex) are nitrogen, carbon in cyanides, oxygen, sulfur, and the halogens. Divalent cobalt exhibits a coordination number of either four or six, whereas that of the trivalent cobalt ion is invariably six.

Ammines. By adding excess ammonia to a cobalt salt and exposing it to air, oxidation occurs and brown solutions form that become pink on boiling. The solutions contain complex cobalt ammines, eg, $[Co(NH_3)_6]Cl_3$, $[Co(NH_3)_5Cl]Cl_2$, and $[Co(NH_3)_5H_2O]Cl_3$. These solutions show none of the reactions of cobalt. The chelates with bis(salicyaldehyde)ethylenediamine derivatives and diamines have unusual oxygen-carrying properties.

Cyanides. If KCN is added to a solution of cobalt salt, reddish-brown cobaltous cyanide, $Co(CN)_2 \cdot 3H_2O$ precipitates. However, with excess cyanide this compound redissolves forming a red solution of potassium cobalt(II) cyanide. If a little HCl or acetic acid is added to this solution and the solution is boiled in the presence of oxygen, oxidation occurs and potassium cobalt(III) cyanide, $K_3(Co(CN)_6)$ forms. Single crystals of this compound have been made in fairly large amounts for use in lasers (qv).

Other complexes. The hexaaquo complex, $(Co(H_2O)_6)^{2+}$, is pink, but introduction of a halide into the coordination sphere to give the complex anion, $(CoX_4)^{2-}$, produces a blue color. Thus, cobalt(II) halides in dilute aqueous solutions are normally pink, but on heating or in concentrated solutions they may turn blue. This behavior is the basis for "sympathetic (invisible) inks" and desiccant indicators.

Health and Safety

Cobalt salts in sufficiently large doses can irritate the gastrointestinal trace and cause nausea, vomiting, and diarrhea. Cobalt salts are used in the treatment of anemia, in some cases together with iron or manganese salts. Externally, cobalt may produce dermatitis. It is also suspected as a carcinogen of the connective tissue and lungs. Cobalt affects the digestive and colonic system, but only in ruminants; nonruminants apparently require no cobalt in their diet.

Uses

Cobalt oxides are a source of metallic cobalt powder which is used to prepare alloys and cemented carbides by powder-metallurgy techniques. Like the other transition elements, cobalt is an effective catalyst for many organic and inorganic reactions. Cobalt compounds also fit into all categories of ceramic pigments. Cobalt soaps have enjoyed a wide use as catalysts to accelerate the drying or oxidation of linseed, soybean, and similar unsaturated oils (see Driers). Cobalt oxides and salts are employed in the vitreous enameling industry to provide color and promote adherence of enamel to steel. Cobalts salts, especially the chlorides, have also proved useful as visual indicators of humidity.

F.R. MORRAL
Consultant

I.C. Smith and B.L. Carson, eds., "Trace Metallic in the Environment," *Cobalt*, Vol. 6, Science Publishers, Ann Arbor, Mich., 1977.

W. Betteridge, *Cobalt and Its Alloys*, John Wiley & Sons, Inc., New York, 1982.

COCHINEAL.   See Colorants for foods, drugs, and cosmetics.

COCOA.   See Chocolate and cocoa.

COCONUT OIL.   See Fats and fatty oils.

COFFEE

Commercial coffees are grown in tropical and subtropical climates at altitudes up to ca 1800 m; the best grades are grown at high elevations. The premium coffee blends contain higher percentages of Colombian and Central American coffees.

Coffee has been commercially significant for about 165 yr. It is second only to petroleum in importance as an article of international trade. The total world exportable production of green coffee in the 1975–1976 growing season was 52.6×10^6 bags.

The coffee plant is a relatively small tree or shrub, often controlled to a height of 3–5 m, belonging to the family Rubiaceae. *Coffea arabica* accounts for 69%; *Coffea robusta*, 30%; and *Coffea liberica* and other, 1% of world production.

The outer portion of the coffee plant fruit is removed by curing; yellowish or light-green seeds, the coffee beans, remain. They are covered with a tough parchment and a silvery skin known as the spermoderm. Curing is effected by either the dry or wet method. The dry method produces so-called natural coffees; the wet method, washed coffees. The latter coffees are usually more uniform and of high quality.

Coffee varies in composition according to the type of plant region from which it comes, altitude, soil, and method of handling the beans. The lower oil, trigonelline, and sucrose contents are typical of robusta beans as is the higher caffeine content (see Alkaloids). Green coffee contains little reducing sugar but a considerable quantity of carbohydrate polymers. The polymers are mainly mannose with varying percentages of glucose, arabinose, and galactose (see Biopolymers; Carbohydrates).

Green coffee has no desirable taste or aroma; these are developed by roasting. Table 1 shows the most significant and well-established chemical changes that occur in green coffee as a result of roasting.

The main processing steps in the manufacture of roasted coffee are blending, roasting, grinding, and packaging. Instant coffee, the dried water extract of roasted, ground coffee, is blended, roasted, and ground as for regular coffee, and then charged into columns called percolators through which hot water is pumped to produce a concentrated coffee extract. The extracted solubles are dried, usually by spray or freeze drying, and the final powder is packaged in glass jars at rates of up to 200 jars per minute.

The basic process for decaffeinating coffee is described in a 1908 patent. Green coffee beans are steamed to increase the moisture content to at least 20%. The additional water and heat separate the caffeine from its natural complexes and aid its transport to the surface of the beans. An organic solvent extracts the caffeine from the wet beans. The beans are again steamed to remove the solvent, and then dried and roasted.

Table 1.   Average Composition of Green and Roasted Coffee

Constituents	Green, % db[a]	Roasted, % db[a,b]
hemicelluloses	23.0	24.0
cellulose	12.7	13.2
lignin	5.6	5.8
fat	11.4	11.9
ash	3.8	4.0
caffeine	1.2	1.3
sucrose	7.3	0.3
chlorogenic acid	7.6	3.5
protein (based on nonalkaloid N)	11.6	3.1
trigonelline	1.1	0.7
reducing sugars	0.7	0.5
unknown	14.0	31.7

[a] Dry basis.

[b] Not corrected for dry-weight roasting loss, which varies from 2 to 5%.

Decaffeination processes cause some changes in the beans that subsequently affect roasting and development of flavor. However, a water extraction process patented in 1943 is supposedly more rapid and efficient and causes less damage to the quality of the coffee flavor than the original process. Consumption of decaffeinated coffees, regular and instant, has increased during recent years and currently accounts for about 30% of all instant coffees sold in the United States and Europe.

A. STEFANUCCI
W.P. CLINTON
M. HAMELL
General Foods Corp.

W.A. Ukers, *All About Coffee*, 2nd ed., Tea and Coffee Trade Journal, New York, 1935, pp. 1–3.
U.S. Pat. 897,763 (Sept. 1, 1908), J.F. Meyer, L. Roselius, and K.H. Wimmer.
U.S. Pat. 2,309,092 (Jan. 26, 1943), N.E. Berry and R.H. Walters (to General Foods Corp.).

COGENERATION.   See Energy management.

COKE.   See Coal, coal conversion processes.

COKE OVEN GAS.   See Coal, coal conversion processes.

COLLIDINES (TRIMETHYLPYRIDINES, ETC).   See Pyridine and pyridine bases.

COLLOIDS

A colloid is a material that exists in a finely divided state. It is usually a solid particle but may also be a liquid droplet or gas bubble. Surface-area to volume ratios are high, which is characteristic of matter in the submicrometer-size range.

Properties

The dimensions of most colloids are between 5 and 1000 nm. If only two dimensions (fibrillar geometry) or only one dimension (laminar geometry) are in this range, the unique properties due to the high surface area may still be observed. For highly dispersed systems that do not exhibit a particularly high surface-area to volume ratio, a dispersion factor expresses the ratio of the number of surface atoms to the total number of atoms in the particle.

Colloids are classified according to their states of subdivision and agglomeration, as well as with respect to the dispersing medium or environment (see Table 1).

General physical properties include diffusion, Brownian motion, electrophoresis, osmosis, rheology, mechanic and mechanical, optical, and electrical properties. In many industries, it is necessary to predict and control suspension rheology, in particular, thixotropy and dilatancy.

Monitoring Techniques

The preparation and destruction of colloids is studied with electron and optical microscopy, light-scattering, and surface tensiometric methods. Agglomeration or coalescence can be followed by these techniques as well as by conductivity, filtration, sedimentation, and electrokinetics.

Formation

Colloid formation involves either nucleation and growth phenomena or subdivision processes. The former require a phase change; the latter pertain to the comminution or atomization of coarse particles (solids) or droplets (liquids). Nucleation and growth processes involve the con-

Table 1.   Classifications of Two-Phase Colloidal Systems

System[a]	Dispersing medium	Dispersed (colloidal) matter
aerosol, fog, mist	gas	liquid
aerosol, smoke		solid
foam	liquid	gas
emulsion		liquid
sol[b], gel, dispersion, suspension		solid
solid foam	solid	gas
gel, solid emulsion		liquid
solid sol, alloy		solid

[a] The appropriate descriptive name depends on the specific properties of the system.
[b] Aqueous sols are sometimes called hydrosols.

densation of vapor either to yield liquid or solid directly or to form colloidal liquid droplets (aerosols) that subsequently solidify, as well as precipitation from liquid or solid solution.

Colloidal liquid aerosols and emulsions are produced by atomizers and nebulizers of various designs, and material with a relatively narrow size distribution can be routinely prepared. Electrostatic classifiers select the desired particle sizes in the final stage of aerosol production. Colloidal powders prepared as liquid suspensions have uniform chemical and phase composition, size, and shape. The preparation of uniform, monosized colloids of complex composition has advanced rapidly, including many inorganic compounds and organic polymers. Monosized powders or monodispersed colloidal solids are used in many applications, eg, pigments (qv), coatings (qv), and pharmaceuticals (qv).

Characterization

Particle size distribution, shape, and morphology are determined by scanning and transmission electron microscopy. Indirect techniques include sedimentation and centrifugation, ultrafiltration, and diffusiometric methods. Surface area is determined by gas or solute adsorption, and permeability by mercury porosimetry. The degree of crystallinity and polycrystallinity is obtained via x-ray diffraction and the interpretation of Bragg reflections. Internal and surface stresses are evaluated by electron microscopy, x-ray and neutron diffraction, and physical property measurements.

Agglomeration

Colloidal systems undergo agglomeration leading to a distribution of aggregate size for solid colloids and droplet size for liquid colloids. Agglomeration is evaluated by ultramicroscopy and optical techniques. Though wetting phenomena and nonwetting colloidal factors may play a significant role, the agglomeration process is induced by particulate collisions arising from diffusion, as in Brownian motion, velocity or shear gradients in a liquid-dispersion medium, and gravitational settling.

Irreversible agglomeration can be quantified using various models for repulsive or attractive electrostatic, London–van der Waals, and steric forces which affect stabilization of aqueous and nonaqueous colloidal systems. A comprehensive model of colloidal stability, the DLVO (Derjagiun-Landau-Verwey-Overbeek) model has proved to be essentially correct regarding the roles of electrolytes, dielectric constant, and other physical quantities in colloidal systems, and is considered a cornerstone of modern colloid science. This theory considers the electrostatic interactions between two identically charged, suspended particles to be repulsive and to arise from the overlap of the electrical double layers associated with each particle. For systems containing soluble polymer or surfactant, molecular arrangement, thickness of the absorbed layer, temperature, and chain or segment solvation are additional critical parameters in determining the effectiveness of a dispersed agent in providing steric stabilization.

If velocity or shear gradients are present and are sufficiently large, the frequency of collisions depends on the volume fraction of solids and the mean velocity gradient as well as the system parameters described above. Assuming that sedimentation is slow compared to the first two collision mechanisms, the overall agglomeration rate is

$$-dN/dt = k_d N^2 + k_s N \tag{1}$$

where N is the particle-number concentration, k_d and k_s are the respective rate constants corresponding to diffusion-controlled and shear-induced collision processes, and the minus sign denotes that the particle-number concentration decreases with time t.

Chemical Properties

Wet chemical analysis or instrumental techniques are suitable for colloids. The available methods include spectrographic techniques and Raman spectroscopy (see Analytical methods).

Hazards

Dust explosions and spontaneous combustion are dangers in the presence of finely divided dry matter. Some colloidal substances may cause health hazards, eg, short-term allergic congestions (asthma, hay fever) or long-term effects (silicosis, asbestosis, black-lung disease).

Applications

Colloids are utilized in many processes, such as mineral beneficiation and the preparation of ceramics, polymers, composites, paper, foods, textiles, photographic materials, cosmetics, and detergents. Diffusion, chemical reactivity, particulate interactions, adhesion/deposition, and electrical, thermal and magnetic behavior are responsible for this wide usage.

As reinforcement agents in metals, ceramics, and polymers, colloidal particles may be spherical, angular, fibrillar, or flake-shaped. Fibrillar fillers are used as discontinuous fibers in metals and plastics. Unidirectional-oriented continuous fibers are incorporated in laminated structures. Ceramics and metals are reinforced by precipitation of colloidal material during thermal treatment of the matrix composition. Other applications include the preparation of rigid, elastic, and thixotropic gels and surface coatings. Colloidal liquids are used in the form of emulsions, eg, as in antifoams, paints, lacquers, varnishes, and insulation. Fluid foams (colloidal gases) are used in foods, shaving cream, and detergents. Solid foams, such as polyurethane foam and pumice, contain dispersed gas bubbles. Colloidal gas bubbles are also essential in solid-foam technology.

ALAN BLEIER
Massachusetts Institute of Technology

E. Matijevic, *Chem. Technol.* **3**, 656 (1973).

P.C. Hiemenz, *Principles of Colloid and Surface Chemistry*, Marcel Dekker, Inc., New York, 1977.

C.R. Veale, *Fine Powders*, John Wiley & Sons, Inc., New York, 1972; J.K. Beddow, *Particulate Science and Technology*, Chemical Publishing, New York, 1980.

COLOR

Color can be defined as that part of the visual experience that deals with aspects of objects perceived other than their si   ape, and surface texture. This definition, purposely vague, emphasizes the perceptual, and therefore highly personal, nature of color: color is what we see. Light (electromagnetic radiation in the visible region, roughly 400–700 nm) either falls directly on the eye, or is seen after it is modified by an object. Neural impulses are created in the eye and sent to the brain. There they are interpreted as color. One cannot know exactly what another sees as color, but through common experience and agreed terminology the basis for developing a science of color is produced.

Interactions among the source of light, the object, and the eye and brain must always be considered in understanding the visual experience. However, color is not the totality of visual experience. Many other perceptions contribute to the appearance of objects, including such properties as size and shape, translucency, gloss, surface uniformity, and metallic character.

The Variables of Perceived Color

In the simplest case only three variables are required for a complete description of color; in a common set are hue, lightness, and saturation. Achromatic colors (whites, grays, blacks) do not exhibit a hue and have zero saturation. They are described by their lightness, judged relative to the lightest achromatic color, white. Chromatic colors do exhibit hues, defined by the common terms red, yellow, green, blue, etc.; their saturation is a measure of the "amount of hue" they contain, and of course they exhibit a lightness as well.

Physical Aspects of Color

Sources of light. Light is one type of electromagnetic radiation, limited to the visible region of the spectrum. Sources of light emit power, and terms related to this power (radiometric terms) and to the response it creates in the eye (photometric terms) are described by the adjectives radiant and luminous. Radiometric quantities are functions of the wavelength of the radiation, and are designated by the adjective spectral. Photometric quantities are obtained by multiplying the corresponding spectral radiant quantity by the eye's response to power, the spectral luminous efficiency function $V(\lambda)$, and integrating the product over the wavelengths of the visible spectrum. The total radiant power emitted by a light source is the radiant flux, measured in watts (W). The corresponding luminous flux is measured in lumens (lm) (683 lm = 1 W).

The effect of a light source on color is best described by its spectral power distribution. For critical visual judgments of color, standard sources are defined and made available; for the equivalent measurements to be described, the corresponding spectral power distributions are tabulated and referred to as standard illuminants.

Interaction of light with matter. Of the many forms of the interaction of light with objects, only three are of major interest for their influence on the colors of objects: absorption, scattering, and reflection. In absorption, radiant power is utilized in molecular transitions and ultimately is converted to heat. Spectrally selective absorption is responsible for color effects but does not detract from the transparency of objects. Opacity and translucency are provided by scattering, the result of the redirection of radiation without change in wavelength. On the macroscopic scale, reflection occurs at surfaces, either specularly (as from a glossy object) or diffusely (as from a mat object), or diffusely from within the body of an object; the latter is the predominant source of color effects in most translucent and opaque objects. Beyond the retina, the neural impulses transmitting color-vision information to the brain appear to traverse complex networks. For many purposes it is convenient to think in terms of derived opponent signals (yellow–blue, red–green, light–dark) instead of the three signals derived from the respective cones. Of major importance is the trichromacy of color vision, ie, the need for only three signals to explain all known facts of photopic vision. About 8% of the population, almost all males, has some degree of color-vision deficiency, characterized by difficulty in recognizing colors or the confusion of dissimilar colors.

Metamerism. If two objects have different spectral reflectance curves which, when combined with the spectral power distribution of a specific source and the spectral responsivities of a specific observer, give rise to the same color, ie, to two colors that match, then the objects form a metameric pair. In general, a change in either the source or the observer characteristics will destroy the color match. The recognition and avoidance of metamerism plays a major role in the use of industrial coloring materials.

Other phenomena are related to metamerism. Color constancy refers to the change in the appearance of a color as the illumination or other conditions change. Color rendering describes the ability of a given test illuminant to produce the same color effects as a reference illuminant. Chromatic adaptation results when the responsivities of the eye are altered so as to preserve the colors of objects as seen under specified conditions. To a large extent these represent aspects of color science that are not yet fully understood.

Colorants. Colorants (qv) are chemical substances added to materials to produce color effects. They are commonly classified as dyes (qv) or pigments (qv). Dyes are substances that are molecularly dispersed, often in aqueous solution, transferred to a material, and bound to it by intermolecular forces to provide color effects by spectrally selective absorption processes. Pigments are larger than molecular-particle size, held in place by their corresponding low mobility, and usually scatter as well as absorb light. Traditionally, dyes were organic in nature and pigments inorganic, but many exceptions exist, notably the widespread use of organic pigments.

Colorant mixing. The mixing of dyes and pigments to produce desired colors is a major objective of color technology, applying to art and design as well as to the coloring of materials in industry. Procedures and considerations include simple-subtractive mixing for the mixing of colorants in transparent systems in which only absorption is important, and complex-subtractive mixing in which both scattering and absorption are important. Additive mixing of colored lights is used in color television, stage lighting, and some color photography.

Color-Order Systems

The major color-order systems provide a substantial body of data against which the results of color measurement and color technology may be tested. Systems include the Munsell System and the Universal Color Language of which it is a part, the OSA Uniform Color Scales System, Natural-Color System, Ostwald System, and a variety of commercial colorant-based systems and single-number color scales.

Colorimetry: The CIE System

Colorimetry, the quantitative science of color, exists in two parts: basic colorimetry, which sets in numerical terms the criteria for the identity of two color stimuli and is concerned with color matching; and advanced colorimetry, which is concerned with color difference, color appearance, and other difficult and largely unsolved problems in color science. Basic colorimetry does not depict color sensations, but only whether two stimuli do or do not match; advanced colorimetry includes methods of assessing the appearance of color stimuli presented to the observer in complicated surroundings as they might occur in everyday life. Colorimetry rests on methodology and standardization introduced by the International Commission on Illumination (Commission Internationale de l'Éclairage—CIE) beginning in 1931. The methodology combines the spectral reflectances of the object with the spectral power of a standard illuminant and standardized spectral color-matching functions representing the average human observer to calculate tristimulus values X, Y, Z defining the color of the object (Fig. 1). Basic colorimetry states that two colors having identical tristimulus values match; advanced colorimetry seeks transformations and combinations of the tristimulus values that correctly predict quantities related to degree of mismatch (color differences, metamerism) and color appearance.

Color Measurement

The purpose of color measurement is to provide a permanent, objective measure of the spectral character and color coordinates of samples. The basic tool is the diffuse-reflectance spectrophotometer, usually computer-interfaced for control and colorimetric calculations. Optical-analogue tristimulus colorimeters have limited but wide use for assessing the color differences between similar and nonmetameric pairs of samples, eg, a production batch and its standard.

Industrial Application of Color Science

Color science is widely applied in many industries including, eg, coatings, plastics, textiles, and paper, for color matching and color control in production. It is also an essential part of color reproduction (photography, television, motion pictures, and graphic arts), lighting, and signalling. It is safe to say that business, science and industry worldwide is dependent on color in a wide variety of ways.

FRED W. BILLMEYER, JR.
Rensselaer Polytechnic Institute

F.W. Billmeyer, Jr., and M. Saltzman, *Principles of Color Technology*, 2nd ed., John Wiley & Sons, Inc., New York, 1981.

R.S. Hunter, *The Measurement of Appearance*, John Wiley & Sons, Inc., New York, 1975.

G. Wyszecki and W.S. Stiles, *Color Science—Concepts and Methods, Quantitative Data and Formulas*, 2nd ed., John Wiley & Sons, Inc., New York, 1982.

D.B. Judd and G. Wyszecki, *Color in Business, Science, and Industry*, 3rd ed., John Wiley & Sons, Inc., New York, 1975.

COLOR AND CONSTITUTION OF ORGANIC DYES. See Cyanine dyes; Color.

COLORANTS FOR CERAMICS

Ceramics colorants can be divided into two broad groups: colorants used in conjunction with clay-based products such as dinnerware, tile, sanitaryware, etc, and colorants used in the manufacture of glass (usually containers) (see Ceramics; Glass). In the manufacture of glass, wherein silica is the major ingredient, the presence of iron as an impurity imparts a decided colored tint under normal conditions. Since this discoloration is generally considered to be undesirable, decolorizers are used as components of the glass batch. In clear glasses, color is produced primarily by transition elements, alone or in combination. In effect, the glass, most often a silicate composition, acts as a solvent for solute coloring ions that produce a color characteristic of the element(s) involved.

Color can also be produced by developing a colloidal suspension in a glass matrix. Metals such as copper (or possibly Cu_2O), gold, and silver, and pigmenting materials such as cadmium sulfoselenides are known to produce strong colors and are used commercially to this end. Translucent or opaque glasses, most often white, can be produced by the precipitation of a crystalline phase during the cooling of a glass, by separation of an immiscible vitreous phase, or by precipitation of a crystalline phase from the glass during a secondary heating.

Both temperature and firing atmosphere can have marked effects on color. Also, chemical composition of the matrix in which the color is developed can have a significant effect on the color produced. The average particle diameter and particle size distribution of ceramic colorants can have a considerable effect on the color of ceramic products as well. Generally, if a colorant is reduced to too small a size, the resultant color will be weakened; the colorant will be more readily attacked by the matrix with a probable total loss of color.

Because of the inherent relatedness of ceramic colorants, it is possible to utilize similar manufacturing methods in their preparation. Close control is exercised in the selection of raw materials, which may be metallic oxides or some salts of appropriate metals, plus normally inert substances (from the point of view of not developing color by themselves) such as silica, alumina, zirconia, calcium carbonate, clays, borax, etc. Certain mineralizers, such as some of the alkali or alkaline-earth halides, may be added to promote the formation of desired crystalline

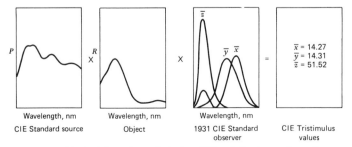

Figure 1. Methodology for calculating CIE tristimulus values. Source: Billmeyer and Saltzman, *Principles of Color Technology*, 2nd ed., John Wiley & Sons, Inc., New York, 1981, p. 45.

species in a subsequent thermal treatment. Depending on the colorant being prepared, both very narrow and relatively broad impurity tolerances are encountered.

Following a weighing operation, the raw materials are thoroughly blended. The end-use coloring capabilities of the raw-material mixtures are developed through a high temperature calcining operation. Following calcination the colorant is ground or milled to the optimum size desired. Finally, colorants may be blended either with other colorants to produce an intermediate hue (eg, the ZrO_2-SiO_2 family) or with inert materials, such as silica, alumina, kaolin, etc, to produce different tints or tones of the same hue. The ultimate control utilized in the manufacture of ceramic colorants is the final appearance in an application.

Ceramic colorants are seldom manufactured by the ultimate user. Owing to the limited market and the variety and complexity of the systems involved, only a small number of manufacturers (eg, Ferro Corp, Fusion Ceramics Inc., General Color and Chemical Co., Harshaw Chemical Co., Hercules Inc. (Drakenfeld), Mason Color and Chemical, Pemco Products Group (Mobay Corp.), and the O. Hommel Co.) produce colorants for the ceramic industry; in addition, some colorants are imported.

There are no serious problems with health or safety in the use and manufacture of ceramic colorants.

Colorant Systems

Black. Most black ceramic colorants are formed by a calcination of oxides (usually) of selected elements that develop a spinel structure when heated.

Blue. Cobalt is the major element used in producing blue colors for the ceramic industry; vanadium, ZrO_2, and SiO_2 are also used. Blue colors in glass, including some glazes, also can be developed by the addition of copper oxide. Characteristically, the resultant color is bluish-green, somewhat between those colors produced by cobalt and chromium oxides when they are used separately.

Brown. A wide range of brown colors can be produced from combinations of chromium, iron, and zinc, most usually from equal parts of the oxide forms, although metal salts of these elements can be used as well.

Gray. Gray colors in ceramics can be obtained by use of one of the black colorants with a suitable inert diluent or a combination of tin and antimony oxides.

Green. Chromium oxide is the main ingredient in most green ceramic colorants and is used to prepare most green-colored glasses.

Pink. Several basic colorant families are available for use in the manufacture of products requiring a pink coloration. These can be enriched, darkened, or lightened through the use of additives. Probably the oldest composition series is a combination of chromium, tin, and calcium to form the so-called chrome-tin pinks. Combinations of chromium and aluminum oxides will also yield a pink-to-ruby color. Another series involves the use of iron in combination with zirconium and silicon. Very delicate pinks can be produced in vitreous systems by colloidal suspensions of gold or selenium.

Red. Iron oxide is one of the oldest red colorants and also one of the most powerful. By far the most prevalent red colorant series for lower temperature applications, ie, over glaze colors, glass enamels, porcelain enamels, and colored glazes, is a family of pigments based on cadmium, selenium, and sulfur.

White. In ceramic bodies white achromatic colors are developed as a result of the use of materials relatively free from impurities. However, most so-called white bodies have a slight cream tint resulting often from minor quantities of iron and titanium dioxide that are often associated with kaolins, ball clays, feldspar, and silica, the usual ingredients in most whiteware products. Iron is also a contaminant in many other materials as well.

Yellow. A number of different compounds are in commercial use to produce yellow colors in ceramics; one of the most widely used is cadmium sulfide. A combination of lead and antimony oxides or metallic salts that is stable to ca $1100°C$ is useful in most ceramic applications in that very strong yellow colors can be achieved. Tin–vanadium yellows are usually prepared from tin oxide and a compound, such as ammonium vanadate, by calcining at $1100-1200°C$. The last system comprises a

combination of zirconium, silicon, and praseodymium. The main colorant for producing a very pure yellow in glass is CdS which is characterized by a sharp absorption edge.

Edward E. Mueller
Alfred University

K. Shaw, *Ceramic Colours and Pottery Decoration*, 2nd ed., Maclanen and Sons, London, UK., 1968.

S.R. Scholes and C.H. Greene, *Modern Glass Practice*, 7th ed., Cahners Publishing Co., Boston, Mass., 1975, pp. 302–329.

N.J. Kreidl in F.V. Tooley, ed., *The Handbook of Glass Manufacture*, Vol. II, Books for Industry, Inc., New York, 1974, pp. 976–987.

C.W. Parmalee and C.G. Harman, *Ceramic Glazes*, 3rd ed., Cahners Books, Boston, Mass., 1973, pp. 460–512.

COLORANTS FOR FOODS, DRUGS, AND COSMETICS

The Federal Food and Drugs Act of June, 1906, brought the use of coloring matter in foods under government supervision under Public Law No. 618. All approved colorants are classified by the FDA as either permanently or provisionally listed. They are further classified as those that require certification and those exempt from certification.

Chemical Classification of Certified Colorants

Certified colorants may be classified into groups according to their chemical structures.

Nitro colorants. External D & C Yellow No. 7 is the only nitro dye certified.

Azo colorants. This group includes the greatest number of colorants on the certified list. They are characterized by the presence of the azo functional group ($-N=N-$) and can be separated into four types: (1) the unsulfonated compounds which are insoluble in water but soluble in aromatic solvents and oils, eg, D & C Red No. 17; (2) the insoluble pigments that contain a sulfonic acid group in the ortho position which is converted by a permissible precipitant into an insoluble metal salt, eg, D & C Red No. 7, 9, and 34; (3) the soluble azo dyes which contain one or more sulfonic acid or carboxy groups to produce their water solubility. The sulfonic acid group is generally in the meta or para position to the azo group, eg, FD & C Red No. 40 and D & C Orange No. 4 (see Azo dyes); and (4) the unsulfonated pigments that are precipitated directly, eg, D & C Orange No. 17 and D & C Red No. 36. These pigments have some solubility in organic solvents.

Triphenylmethane colorants. This is a group of sulfonated dyes (derivative of the corresponding basic dye) containing two or more sulfonic acid groups, eg, FD & C Blue No. 1 and FD & C Green No. 3.

Xanthene (fluoran) colorants. The xanthene group is characterized by the structure:

Anthraquinone Colorants. This group includes D & C Green 5.

Indigoid Colorants. This group includes FD & C Blue 2 and D & C Red 30.

Colorant Regulations

The Code of Federal Regulations of April 1982 contains the most recent version of the colorant regulations. Additional changes will be made in later *Federal Registers*. Some of the regulations important to users of food, drug, and cosmetic color additives are as follows:

General Provisions. CFR Section 70.3 — Definitions. A colorant (color additive) is any material not exempted under section 201 (t) of the Act, that is a dye, pigment, or other substance made by a process of synthesis or similar artifice, or extracted, isolated, or otherwise derived, with or without intermediate or final change of identity from a vegetable, animal, mineral, or other source. When added or applied to a food, drug, or cosmetic or to the human body, the colorant is capable (alone or through reaction with another substance) of imparting a color thereto. Substances capable of imparting a color to a container for foods, drugs, or cosmetics are not color additives unless the customary or reasonably foreseeable handling or use of the container may reasonably be expected to result in the transmittal of the color to the contents of the package or any part thereof. Food ingredients such as cherries, green or red peppers, chocolate, and orange juice that contribute their own natural color when mixed with other foods are not regarded as color additives; but where a food substance such as beet juice is deliberately used as color it is a color additive.

For a material otherwise meeting the definition of colorant to be exempt from section 706 of the Act, on the basis that it is used (or intended to be used) solely for a purpose or purposes other than coloring, the material must be used in a way that any color imparted is clearly unimportant insofar as the appearance, value, marketability, or consumer acceptability is concerned. (It is not enough to warrant exemption if conditions are such that the primary purpose of the material is other than to impart color).

The term safe means that there is convincing evidence that establishes with reasonable certainty that no harm will result from the intended use of the colorant.

The FDA has mandated a cyclic review program for all permanently listed colorants. The future of natural colorants also is uncertain. The FDA has indicated that natural does not necessarily assure the color is safe (see Food toxicants, naturally occurring). The FDA will require toxicology studies and more complete information about the chemistry of these natural colorants (see also Dyes, natural; Food additives).

SAMUEL ZUCKERMAN
JOSEPH SENACKERIB
H. Kohnstamm & Co., Inc.

U.S. Department of HEW, FDA, *Report on Certification of Color Additives*, Fiscal years 1978, 1979, 1980, 1981, 1982.

Code of Federal Regulations, Parts 1–99, April 1, 1982, pp. 217–309.

COLORANTS FOR PLASTICS

There are three general types of colorants for plastics: dyes, and inorganic and organic pigments. Dyes (qv) are soluble under the conditions of use but must be completely dissolved, leaving no color streaks and little or no haze. Pigments are insoluble and consist of particles that must be dispersed by physical means (see Pigments).

Resins (plastics) fall into two categories: thermoplastics and thermosets. For thermoplastics, the polymerization reaction has been completed, the materials are processed at or close to their melting points, and scrap may be reground and remolded, eg, polyethylene, polypropylene, poly(vinyl chloride), acetal, acrylics, ABS, nylons, cellulosics, and polystyrene. In the case of thermoset resins, the chemical reaction is only partially complete when the colorants are added, and is concluded when the resin is molded. The result is a nonmeltable cross-linked resin that cannot be reworked, eg, epoxy resins, urea–formaldehyde, melamine–formaldehyde, phenolics, and thermoset polyesters. The method of coloring differs depending upon whether the resin is a transparent acrylic or an opaque ABS and also upon the desired transparency or opacity of the product.

There is a possibility of a chemical reaction between a plastic and a colorant at processing temperatures. More often, allowance must be made for additives in the plastic, such as antioxidants (qv), flame retardants (qv), ultraviolet light absorbers (see Uv stabilizers), and fillers (qv). Thermal stability is a factor. The final use of the colored resin often dictates the selection of a colorant. Obviously, colorants must be evaluated for the user's processing conditions in view of the subsequent fate of fabricated parts.

Dispersion and Stability

Colorants are chosen for their hues, among other reasons. In matching a shade submitted by a customer, it is desirable to use the same colorant combination so that the match will be valid under all types of illumination (see Color). The process of dispersion of a dry pigment into a liquid vehicle or molten plastic may be visualized as taking place in two steps. These are the breaking up of the pigment agglomerates into the much smaller ultimate particles, and then the displacement of air from the particles to obtain a complete pigment-to-vehicle interface.

Dyes

Dyes should be checked for migration, sublimation, and heat stability before use. These precautions are particularly important for plasticized resins. Commonly used dyes include azo dyes (Solvent Yellow 14, Disperse Yellow 23), azo acid dyes (Metanil Yellow), anthraquinone dyes (Solvent Red 111, Disperse Violet 1, Solvent Blue 56, and Solvent Green 3) (see Dyes, anthraquinone), azine dyes (induline and nigrosines), and miscellaneous dyes such as Brilliant Sulfoflavin, Acid Yellow 7, and Solvent Orange 60.

Solvent Yellow 14

Disperse Yellow 23

Inorganic pigments

This category includes white (titanium dioxide), black (carbon black (qv)), iron oxides, chromium oxide greens, iron blue and chrome green, violet, ultramarine pigments, blue, green, yellow, and brown metal combinations, lead chromates and molybdates, cadmium pigments, titanate pigments, pearlescent pigments, and metallic pigments.

Organic Pigments

Organic pigments include monazo pigments (Hansa Yellows, toluidine reds, naphthol reds, Nickel Azo Yellow, Lake Red C, Permanent Red 2B, Pigment Scarlet 3B Lake), disazo pigments (Benzidine Yellows, AAMX, AAOT, AAOA, and HR Yellow (PY 13, 14, 17, and 43)), disazo condensation pigments (Cromophtal), quinacridone pigments (Pigment Violet 19, Pigment Red 122), dioxazine violet, vat pigments [Flavanthrone Yellow (PY 112), Anthrapyrimidine Yellow (PR 108), Pyranthrone Orange (PO40); Perinone Orange (PO 43), Brominated Anthranthrone Orange (PR 168), Brominated Pyranthrone (PR 197), Anthramide Orange (VO 15); Indanthrone Blue, Red Shade (PB 64) and Isoviolanthrone Violet (PV 31 or PV 33)], perylene pigments (Pigment Red 123 R), thioindigo pigments (Pigment Red 88R, Pigment Red 198R), phthalocyanine pigments (Phthalocyanine Blue, PB 15), tetrachloroisoindolinones, and fluorescent pigments.

Pigment Violet 10, R = H
Pigment Red 122, R = CH$_3$

Pigment Red 88 R = Cl
Pigment Red 198 R = CH$_3$

Pigment Red 123 R = $-\!\langle\bigcirc\rangle\!-OC_2H_5$

The cost of the coloring process should be calculated as soon as a formula is projected, and compared against that of any alternative colorant systems that may be available.

Health and Safety Factors

Dyes may sublime under operating conditions and show up in dust collectors. Migration is tested for by making a sandwich of alternately colored and uncolored pieces under a weight at an elevated temperature. Toxicity of colored plastics is a complex and rapidly changing subject. For food packaging, an elaborate series of extraction tests must be run to demonstrate that neither colorants nor other additives can migrate into the food. Only titanium dioxide, iron oxides, and ultramarine blue are exempt from this requirement at present. A few pigments are certified for use in drugs and cosmetics and a very few dyes and alumina hydrate lakes are certified as food colors. In the past decade, colorant manufacturers have eliminated not only colorants of known toxicity, but also some made from toxic intermediates (see Colorants for food, drugs, and cosmetics). At the same time, it has been shown that some resin monomers are toxic on long-time exposure. Handlers of colorants, additives, and resins should use maximum practical ventilation to protect their workers.

THOMAS G. WEBBER
Consultant

T.G. Webber, ed., *Coloring of Plastics*, Wiley-Interscience, New York, 1979.

COLOR MEASUREMENT. See Dyes, application and evaluation; Color.

COLORIMETRY AND FLUOROMETRY. See Color; Dyes, application and evaluation, Analytical methods.

COLOR PHOTOGRAPHY

Color photography is a process by which light of varying wavelength, intensity, and location, collected and focused by a lens system, can initiate chemical reactions leading to a relatively permanent record in two-dimensional space of these variations. All the practical systems of color photography fulfill at least the following two requirements: the sensitivity of the photographic material must be such that objects producing different colors and different lightnesses form latent images than can be differentiated; the chemical or physical process that makes the latent image visible must maintain these differences and translate them into material images that will modulate some form of general illumination to cause the observer or other sensor to perceive the desired result.

Much of color photography may be called pictorial in that the end result is a picture that appears natural. In such photography, the sensitive materials must in some way approximate the spectral sensitivity of the eye, and the viewing system must provide a gamut of colors similar to those of the observer's experience.

In technical uses, the image frequently need not be natural to convey information. Energy outside the visible spectrum may be involved: eg, ultraviolet or infrared. Colors need not be reproduced in a natural or pleasing fashion. The reproduction may be designed for presentation to the human observer or for presentation to some other sensor such as a photoelectric cell.

The fundamental light-sensitive element in most color-photographic materials, just as in most black-and-white materials, is a silver halide or mixture of silver halides, silver chloride, silver bromide, or silver iodide, dispersed as crystals or grains in a medium such as gelatin. For color photography a silver halide must be sensitive to all regions of the visible spectrum and sometimes beyond.

Experiments with light and color and human vision have established that for full color reproduction the spectrum must be separated into three components, generally in the red, green, and blue light regions. The identity of the separate components must be maintained long enough so that they can be translated into the physical or chemical systems that serve to modulate three components of the viewing illuminant to give the sensation of full color to the observer. One process called the additive color process can be illustrated by current color television systems that use juxtaposed multicolor phosphor elements in the display tubes to generate the gamut of colors.

Most current color processes depend on silver halide grains which can be selectively sensitized to red, green, or blue and placed in separate layers so that the three records can maintain their identity. The rest of the color-image-forming process consists of using some reactions associated with the developing silver grain or the silver grain itself to form, destroy, or modify dyes, generally yellow, magenta, and cyan. The latter dyes are termed *subtractive primaries* in that each absorbs or subtracts one of the colors that is the basis of color vision. Thus yellow absorbs blue but transmits green and red; magenta absorbs green but transmits blue and red; and cyan absorbs red but transmits blue and green. The combination of yellow in superposition with magenta appears red; yellow with cyan appears green; and magenta with cyan appears blue. The three subtractive colors in equivalent amounts reproduce a range of grays through black, depending upon the densities of the three dye images (see also Color).

Current Basis for Color Photography

Sensitizing dyes. The natural sensitivity of silver halide crystals to ultraviolet and to blue light can be extended to green and red by the adsorption of sensitizing dyes to the crystal surfaces. The sensitizing dyes absorb green and red light and transfer the energy to the silver halide substrates. In this way the silver halide can form a latent image of light to which it is otherwise insensitive.

The most widely used sensitizing dyes are of the cyanine class (see Cyanine dyes). These consist of heterocyclic moieties linked by conjugated systems of atoms. An example of this class is:

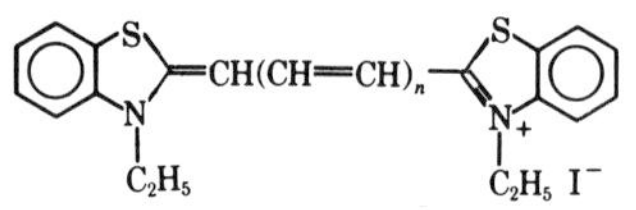

Photographic emulsions. As in black-and-white photography, specially sensitized silver salts are precipitated in gelatin media with which they form emulsions (see Emulsions; Photography). In order to make use of the ability of such silver salts to differentiate a scene into three-color records, they must be prepared in separate emulsions, which are most commonly coated as separate layers, in superposition on a supporting film. A variety of other layers are frequently used; a typical color-photographic material is illustrated in Figure 1.

The color records obtained by exposure of these emulsions may be processed to a color negative, from which a color positive can be obtained by printing onto a similar emulsion package, followed by color processing. Alternatively, the color positive may be obtained by reversal processing.

Color development. The chemistry of color photography involves the conversion of the latent images into dye images. Fischer reported that N,N-dialkyl-p-phenylenediamines are useful developing agents whose oxidation products are capable of forming dyes in proportion to the amount of exposure to a given region of the spectrum. The second component of the dye is the coupler, the choice of which determines the color obtained with a given developing agent. Fischer's proposal is given schematically as follows: exposed silver salt + developer → oxidized developer; oxidized developer + coupler → dye.

Color-Forming Agents

Yellow couplers give dyes with maximum absorption (λ_{max}) between 400 and 500 nm. The compounds suitable for this purpose contain an

```
======== h clear gelatin overcoat
-------- g blue-sensitive silver halide emulsion
======== f blue-absorbing interlayer
-------- e green-sensitive silver halide emulsion
======== d gelatin interlayer
-------- c red-sensitive silver halide emulsion
======== b gelatin interlayer
-------- a antihalation layer
/ / / / / / film base
```

Figure 1. Typical cross-section of a color-photographic product.

active methylene group that is generally not part of a ring. By far the most important class of yellow couplers consists of benzoylacetanilides,

$$X \longrightarrow \text{CCH}_2\text{CNH} \longrightarrow Y$$

Ease of preparation and versatility of substituent effects are features of this system.

Magenta couplers are selected to give dyes whose maximum absorption occurs between 500 and 600 nm. Selection of a suitable magenta coupler is facilitated by the wide variety of types available. In general, the magenta dye formers are similar to the yellows in structure except that they give dyes that are shifted in hue as a result of substitution. Such substitution may involve cyclization, which makes the active methylene group part of a ring structure, or it may involve substituents that extend or intensify the conjugation of the dye structure.

Cyan couplers give indoaniline dyes (see Dyes and dye intermediates) with λ_{max} generally between 600 and 700 nm; in almost all cases considerable absorption is noted beyond 700 nm, outside the visual range. Cyan dyes generally also adsorb a considerable amount of green light and usually some blue light.

Color developers. Important properties of the color-developing agents are their solubilities, their reactivities, and their effect on the hue of the dyes (see below). Most of the simple N,N-dialkyl-1,4-phenylene-diamines are sufficiently soluble in alkaline processing solutions to perform adequately.

Color developers are notoriously allergenic and must be handled with care (see Amines, aromatic, phenylenediamines). However, the incidence of allergenic reactions, resulting in skin irritation similar to that caused by poison ivy, can be reduced by certain substituents. Most prominent among these substituents are the β-hydroxyethyl group and the β-methylsulfonamidoethyl group.

Modification of the Dye Hue

Within each of the classes of yellow, magenta, and cyan dyes, considerable variation of hue is possible and, indeed, necessary in order to produce hues suitable for a given use. Dye hues can be modified by the introduction of substituents into either of the dye-forming components, viz, coupler or developer.

Practical Chromogenic Color Processes

Developer-soluble couplers. The positive silver halide can be developed chromogenically to form a positive dye image without removal of the negative silver image. Two musicians, L. Mannes and L. Godowsky, used this principle to devise a color process that was developed in the Kodachrome film and process marketed by Eastman Kodak Company in 1935. Agfacolor film, like most of the many color films that have succeeded it, used color couplers that were incorporated in the emulsion layers. Kodachrome film, on the other hand, depends on soluble couplers in three different color-developing solutions to obtain the dye images. The complex Kodak process now is used mostly for amateur motion pictures and for 35-mm still transparencies.

Incorporated couplers. A simplification in the color process is effected when the different dye-forming couplers are incorporated into the ap-propriate sensitized layers. Most manufacturers use the type of incorporation exemplified by Kodak Ektachrome film. That type involves couplers that are nondiffusible by virtue of long aliphatic chains or combinations of short aliphatic chains and aromatic systems without hydrophilic groups. Color materials containing incorporated couplers can be processed in a reversal mode or in a negative–positive mode.

Reflection prints. By coating the proper amounts of emulsion and color formers on white reflecting supports instead of on transparent film, it is possible to make reflection-print materials.

Nonchromogenic Color Photography

A number of color-photographic processes depend upon the black-and-white development of exposed silver salts followed by chemical reactions that result in the destruction of a preformed dye or in the transfer of a dye out of the emulsion to a mordanted receiver (see Color photography, instant). The Cibachrome process is an example.

Quality characteristics of color materials that should be considered include color fidelity—masking, dye stability, color balance, grain, speed, and sharpness.

Two-Color Systems

Because of the complexity and cost of arriving at a natural three-color photographic image, several processes were derived that depend on just two color records. Since it is impossible to obtain a full gamut of colors with only two records, a compromise in color quality was made in two-color systems for pictorial photography. Two-color systems were also developed for data-recording uses, where a full range of colors was not needed.

Special applications of color photography include color television and infrared-recording color film.

JOHN R. THIRTLE
DAAN M. ZWICK
Eastman Kodak Co.

P. Kowalski, *Applied Photographic Theory*, John Wiley & Sons, Inc., New York, 1972.

R.W.G. Hunt, *The Reproduction of Colour*, 3rd, ed., Fountain Press, London, UK, 1975.

E.J. Wall, *History of Three-Color Photography*, American Photographic Publishing Co., Boston, Mass., 1925.

COLOR PHOTOGRAPHY, INSTANT

Instant color photography comprises the group of color processes that result in finished color photographs within moments after exposure of the film. Processing is accomplished in a single step that takes place rapidly under ambient conditions. The processes are essentially dry and the reagent is provided as a part of the film unit. In most cases the reagent is applied by mechanical action of the camera, film holder, or other processing device immediately after the film is exposed.

Principles of Instant Photography

Film and process design. The broad requirements of instant photography were described by Land in 1947 when he introduced the first one-step print process. To be practical for handheld cameras, a film requires sufficient photographic speed to permit short exposures at small apertures. For this reason Land worked with a silver halide emulsion, taking advantage of its high sensitivity to light and the enormous amplification provided by the development process. He described a one-step process that entailed the spread of a viscous reagent between two sheets, one bearing an exposed silver halide emulsion and the other an image-receiving layer, as both sheets were drawn out of a camera through a pair of pressure rollers. A sealed pod attached to one of the sheets ruptured to release the viscous reagent which spread to form a thin layer between the two sheets, temporarily bonding them. The action of the reagent produced concomitantly a negative image in the emulsion layer and a positive image in the image-receiving layer. After about a minute, the two sheets were stripped apart to reveal the positive image. Films handled in this way are described as peel-apart materials. An integral print film comprises two sheets permanently secured as a single unit. The image-forming layers are located on the inner surfaces of the two sheets, one of which is transparent.

Reagents for instant photography. In each instant process, an essential component is the viscous reagent. The concept of essentially dry processing works by using a highly viscous fluid reagent and restricting the amount to just that needed to complete the image-forming reactions for one picture. The viscosity of the reagent is increased by the addition of water-soluble polymeric thickeners. Suitable polymers include hydroxyethyl cellulose and the alkali-soluble salts of carboxymethyl hydroxyethyl cellulose. In addition to high molecular weight polymer, reducing agents, and alkali, the viscous reagent may contain reactive species that participate in image formation and stabilization. Fresh reagent in a sealed pod for each picture permits the use of more highly reactive reducing agents and more strongly alkaline conditions than would be feasible in a tank or tray process.

Formation of transfer images. In investigating mechanisms for one-step photography, Land identified within the conventional photographic process an entire family of images resulting from reactions that vary from point to point as a consequence of exposure and development. These images, shown in Table 1, represent possible starting points for transfer processes leading to useful positive images.

In the instant color processes using dye developers—molecules that are both image dyes and photographic developers—image 6 (the unoxidized dye developer) transfers to form a positive dye image in the image-receiving layer; in color development and redox dye-release systems, image 5 (an oxidized developer) forms or releases the final positive dye image. Also, the three color records, whether initially formed in a single layer or in three separate layers, produce the final picture in a single, separate image-receiving layer.

Both additive and subtractive systems have been used in instant color films. The Polavision instant transparency system is based on an additive color process, and the Polacolor, SX-70, and PR-10 instant-print systems use subtractive color processes.

Each of the subtractive instant color films is based on preformed dyes. Polaroid's Polacolor and SX-70 dye developers are initially mobile in alkali and are used in a positive-working sense in conjunction with an emulsion that develops negative silver images. Kodak's PR-10 film uses initially immobile dyes in a negative-working sense in conjunction with an emulsion that forms direct positive silver images.

Commercial Instant Color Processes

The Polacolor process. The Polacolor process produces subtractive multicolor prints comprising positive images in terms of cyan, magenta, and yellow dye developers. The dye developers that form the positive image transfer from a multilayer color negative to an image-receiving layer concomitantly with the formation of negative silver images and negative dye developed images within the layers of the negative. Image formation is based on the immobilization of dye developers by oxidation in areas where exposed silver halide grains are developed, and on diffusion of dye-developers from areas that are unexposed and hence free from oxidation.

Handheld cameras using Polacolor pack films range from inexpensive amateur models to professional models with great flexibility. Pack films

Table 1. Images Present in Silver Halide Emulsion upon Exposure and Treatment with a Developer

1	the exposed grains of the latent image
2	the unexposed grains of the latent image
3	the developed silver
4	the undeveloped silver halide
5	the oxidized developer
6	the developer that is not oxidized
7	the neutralized alkali
8	the alkali that is not neutralized
9	the hardened gelatin
10	the unhardened gelatin

are also used in a variety of cameras and camera backs for professional and industrial applications.

Polaroid integral films. Used in cameras that control exposure automatically and eject each film unit through processing rollers immediately after exposure, the Polaroid integral films provide images that require no timing and no peeling apart. The entire process takes place within the film unit under ambient conditions. The picture unit is an integral multilayer structure both before and after processing (Fig. 1).

The viscous reagent is contained in a small pod concealed within the wide border of the film unit. As an exposed film unit is ejected through the processing rollers and out of the camera, the pod bursts and the viscous reagent forces apart the top negative layer and the image-receiving layer, forming a new layer comprising water, alkali, white pigment, polymer, and other processing addenda. Alkali quickly permeates the layers of the negative. In areas that have been exposed, silver halide is reduced and the associated dye developer is oxidized and immobilized. The auxiliary developing agent 4'-methylphenylhydroquinone acts as an electron-transfer agent or "messenger" developer. In areas that have not been exposed, the dye developers, ionized and solubilized by the action of the alkali, migrate through overlying layers of the negative to reach the image-receiving layer above the new stratum of pigment-containing reagent.

An important aspect of the integral system is that the developing film unit is ejected while still light sensitive. Hence both surfaces must be opaque to ambient light. However, the top surface, through which the camera exposure is made, cannot become opaque until after that exposure has taken place. Protection against further exposure through this surface is provided by the combined effects of opacifying dyes and pigment included in the viscous reagent layer. The opposite surface is protected by the support layer of black polyester. The opacifying dyes are phthalein indicators that have high extinction coefficients at very high pH levels.

The first Polaroid integral film was SX-70, introduced in 1972. This was followed by SX-70 Time Zero, which used new components to provide images that appeared more rapidly. Further innovations led to Polaroid 600 High Speed Film, rated at ASA 600. Cameras using Polaroid integral films include both folding single-lens reflex models and nonfolding models with separate camera and viewing optics. All of the cameras are motorized. Additional features on some cameras include automatic ultrasonic rangefinders and built-in strobe units. Immediately after exposure, the integral picture unit is automatically ejected from the camera. In a number of the cameras, pictures may be taken as frequently as every 1.5 s.

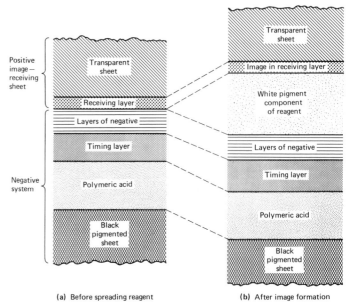

(a) Before spreading reagent

(b) After image formation

Figure 1. Schematic cross sections of SX-70 Time Zero film (**a**) before and (**b**) after development.

Kodak instant processes. These film systems are based on a negative-working dye-release process using preformed dyes. Transfer of dyes is initiated by the oxidized developing agent formed in areas where silver halide grains are undergoing development. The oxidized developing agent reacts with alkali-immobile dye-releasing compounds, which in turn release mobile image dyes that diffuse to the image-receiving layer. A positive print is formed with emulsions that yield positive silver images directly by developing only those grains that have *not* received exposure.

Kodak PR-10 and Kodamatic instant-print films are used in Kodak instant cameras which include battery-operated motorized models and lower cost hand-cranked models. Additional features on some cameras include infrared automatic focusing devices and built-in strobe units. The pictures begin to develop as they are ejected from the camera and development proceeds over several minutes under ambient conditions.

Two-sheet peel-apart print processes using similar chemistry are marketed under the name Ektaflex. These materials are designed for darkroom processing. Both negative-working and positive-working versions are provided.

In 1981, Fuji introduced an integral color film similar to Kodak's. At this writing, the Fuji film is distributed only in Japan.

Polaroid addition color processes. The Polavision and Polachrome systems are based on a very fine additive color screen and an integral silver transfer image. The additive screen comprises a microscopically fine pattern of red, green, and blue filter stripes. To produce lines fine enough to be essentially invisible when projected, Land devised a new process using temporary lenticules. The film base is embossed to form fine lenticules on one surface, and a layer of dichromate-treated gelatin on the opposite surface is exposed through the lenticules to form hardened line images. After washing away the unhardened gelatin, the lines that remain are dyed. The process is repeated to complete the array of alternating red, green, and blue stripes. The lenticules are removed after they have been used to form lines.

Polavision Super-8 film is provided in a sealed cassette, and the film is exposed, processed, viewed, and rewound for further viewing without leaving the cassette.

Polavision cassettes (Polavision Phototape Cassette, Type 608) are used in the Polavision instant movie camera and player.

Domestic camera sales in the field of instant photography were estimated at 55% of total mass-market cameras purchased in 1981.

No toxicological hazards have been associated with the normal use of instant color films. However, the manufacturers caution that direct contact with the highly alkaline processing fluids can cause alkali burns. The fluids are provided inside sealed pods and are not usually handled by the user.

Many laboratory instruments and diagnostic machines now include built-in instant-film camera backs. Specialized pack cameras permit the use of Polacolor film for rapid fabrication of portrait identification cards, widely used for drivers' licenses, student registration, and credit cards. Instant-portrait studios using SX-70 integral color film were introduced in 1977. Large format Polacolor photography using 20×25 cm (8×10 in.) and 51×61 cm (20×24 in.) films was first demonstrated in 1976. One of the most remarkable uses of instant color in large format is the making of full-size Polacolor replicas of paintings and tapestries, a technique of great value in professional museum work. Medical applications include photomicrographs of stained tissue and scintillation-camera records of the brain and other organs in the presence of radioactive isotopes (see Radioactive drugs; Radioactive tracers).

VIVIAN K. WALWORTH
Polaroid Corp.

E.H. Land, H.G. Rogers, and V.K. Walworth in J.M. Sturge, ed., *Neblette's Handbook of Photography and Reprography*, 7th ed., Van Nostrand Reinhold, New York, 1977, pp. 258–330.

S.M. Bloom and co-workers in K. Venkataraman, ed., *The Chemistry of Synthetic Dyes*, Vol. 8, Academic Press, Inc., New York, 1978, pp. 195–213.

COLUMBIUM. See Niobium and niobium compounds.

COMPLEXING AGENTS. See Chelating agents.

COMPOSITE MATERIALS

Composites are combinations of two or more materials present as separate phases and combined to form desired structures so as to take advantage of certain desirable properties of each component. The constituents can be organic, inorganic, or metallic (synthetic or naturally occurring) in the form of particles, rods, fibers, plates, foams, etc. Compared with homogeneous materials, these additional variables often provide greater latitude in optimizing, for a given application, such physically uncorrelated parameters as strength, density, electrical properties, and cost. Furthermore, a composite may be the only effective vehicle for exploiting the unique properties of certain special materials, eg, the high strength of graphite, boron, or aramid fibers (qv).

Some measure of coarseness of the homogeneous constituent structures is needed for a meaningful definition of composite material. The term as used here assumes that the average dimension of the largest single homogeneous geometric feature, in at least one direction, is small relative to the size of the total body in that direction; in addition, it assumes that the dimensions of the minor constituent phase are sufficiently large so that its characteristic properties are substantially the same as if it were present in bulk.

Composite materials consist of a continuous matrix phase that surrounds the reinforcing-phase structures. Possible exceptions are (*a*) a laminated stacking of sheets in which the phases are kept separated, and (*b*) two continuous interpenetrating phases, such as an impregnated sponge structure, in which it is arbitrary as to which phase is designated as the matrix. The relative role of the matrix and reinforcement generally fall into the following categories: (*1*) The reinforcement has high strength and stiffness, and the matrix serves to transfer stress from one fiber to the next to produce a fully dense structure. (*2*) The matrix has many desirable, intrinsic physical, chemical, or processing characteristics, and the reinorcement serves to improve certain other important engineering properties, such as tensile strength, creep resistance, or tear resistance. (*3*) Emphasis is placed on enhancing the economic attractiveness of the matrix, eg, by mixing or diluting it with materials that will improve its appearance, processability, or cost advantages while maintaining adequate performance.

Composites typically are made up of the continuous matrix phase in which are embedded: a three-dimensional distribution of randomly oriented reinforcing elements, eg, a particulate-filled composite; a two-dimensional distribution of randomly oriented elements, eg, chopped fiber mat; an ordered two-dimensional structure of high symmetry in the plane of the structure, eg, an impregnated cloth structure; or a highly aligned array of parallel fibers randomly distributed normal to the fiber directions, eg, a filament-wound structure, or a prepared sheet consisting of parallel rows of fibers impregnated with a matrix.

Certain properties, such as the colligative thermodynamic ones, can be accurately calculated from knowledge of the volume fractions and chemical composition of the constituent phases (see Thermodynamics). Other properties, such as thermal and electrical conductance and the elastic properties, can be estimated from idealized models that closely approximate real composite behavior. Other important properties, such as failure strength and fracture toughness, can only be approximated roughly. Composite properties often are assumed, frequently unjustifiably, to be representable by the rule of mixtures: $P = P_1 V_1 + P_2 V_2 + P_3 V_3 + \cdots$ in which P is the property value for the composite, and P_i and V_i are the property values and volume fractions of the ith phase. The rule of mixtures for strength S of a fiber–matrix composite in the direction of the fiber is often used [$S_{composite} = V_f S_{fiber} + V_m \sigma^*_{matrix}$] where σ^* is the stress in the matrix at the failure strain, and S_{fiber} is the mean fiber strength.

Fabrication Methods

The methods used to make composite materials and structures depend, among other factors, on the type of reinforcement, the matrix, the required performance level, the shape of the article, the number to be made, and the rate of production. Large diameter, single-filament materials, such as boron, silicon carbide or metal wires, are often fed in precisely controlled, parallel arrays to form tapes of sheet materials. In the case of finer filaments, such as fiberglass, carbon fiber, or boron nitride fiber, bundles (tows or roving) of hundreds to tens of thousands of loosely aggregated fibers are handled as an entity. When these fibers are to be incorporated into a polymer matrix composite, it is usually convenient to form a semiprocessed, shapable, intermediate ribbon or sheet product known as prepreg in which the fibers are infiltrated by incompletely cured resin. Another approach is to form dry structures first, such as wire armatures, which are then impregnated with the matrix material. Where larger shapes or lower production volume is needed, the spray-up technique may be useful. Molding compounds useful for injection, transfer, compression, or other similar types of force-flow forming can be compounded by use of chopped fiber or other types of fillers. When the matrix is metallic or ceramic, techniques such as hot pressing or diffusion bonding are required.

Reinforcements

If a reinforcement is to improve the strength of a given matrix, it must be both stronger and stiffer than the matrix, and it must significantly modify the failure mechanism in an advantageous way, or both. The various types of fiber reinforcements include glass (vitreous silica, E glass, S glass), carbonaceous types (derived from polyacrylonitrile, rayon, or pitch) typically exhibiting high strength and high stiffness, polymer (aramid, olefin, nylon, rayon), inorganic (monocrystalline alumina, polycrystal alumina, whisker alumina, alumina silicates, asbestos, tungsten core boron, boron nitride, carbon core silicon carbide, polycrystallene silicon carbide, whisker silicon carbide, whisker silicon nitride, polycrystal zirconia), and metal (beryllium, molybdenum, steel, tungsten).

A variety of reinforcements or fillers of plate-like or particulate geometries are frequently used in nonhigh-performance composites. Minerals, such as clay, talc (qv), sand, mica (qv), and asbestos (qv), are often used because they are inexpensive and impart desired characteristics to the matrix/reinforcement combination during processing and in its use properties or both (see Clays; Silica). These properties include thixotropy, strength, ease of finishing, creep, and heat distortion temperature. The filler materials also function to extend the polymers, especially the more expensive engineering types (see Engineering plastics).

Matrices

Any solid that can be processed so as to embed and adherently grip a reinforcing phase is a potential matrix material. The polymers and metals have been the most successful in this role although cements, glasses, and ceramics have also been used. Thermosetting resins are particularly convenient because they can be applied in a fluid state, which facilitates penetration and wetting in the unpolymerized state, followed by hardening of the system at times and conditions largely controlled by the operator. Exothermicity, shrinkage, and the evolution of volatiles, particularly if the polymerization is of the condensation type, are among the difficulties encountered using resins.

W.B. HILLIG
General Electric Co.

L.J. Broutman and R.H. Krock, eds., *Modern Composite Materials*, Addison-Wesley Publishing Company, Reading, Mass., 1967.

J.E. Gordon, *The New Science of Strong Materials*, Penguin Books Inc., Baltimore, Md., 1968, Chapt. 8.

D. Johnson, ed., *Composites*, a journal published by IPC Science and Technology Press Limited, Surrey, UK; Vol. 1, 1968, and continuing quarterly.

S.W. Tsai, ed., *Journal of Composite Materials*, Technomic Publishing Co., Inc., Westport, Conn.; quarterly.

COMPRESSORS. See Fluid mechanics (transportation); High pressure technology.

COMPOSITES, HIGH PERFORMANCE

High performance composites are formed by combining two or more homogeneous materials in order to obtain properties that are superior to the properties of either single material. Increased strength, stiffness, fatigue life, fracture toughness, environmental resistance, and reduced weight and manufacturing costs are some of the advantages of composites. The most common form is laminated fiber-reinforced plastic, where the fibers in each layer are usually arranged in one of the four configurations shown in Figure 1 (see also Laminated and reinforced plastics).

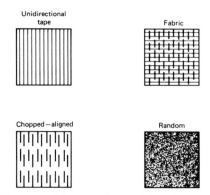

Figure 1.   Schematic of fiber-reinforced concepts.

Reinforcing Fibers

S-glass fibers exhibit higher tensile strength than any other commercial fibers (see Table 1). However, the use of such fibers in structural applications is limited because of their low elastic modulus and high specific gravity.

Boron fibers have approximately eight times the strength of aluminum but are 10% less dense; their modulus is nearly twice that of steel. Although these properties offer great advantages in aircraft-structure applications, the outlook for producing boron fibers for less than $200/kg is not promising, and the development of carbon fibers has greatly reduced the interest in boron fibers.

Carbon fibers are made either from polyacrylonitrile fibers or pitch-based fibers. The former are oxidized and stretched at the same time, whereas the latter do not have to be held under tension during the oxidation process. The low negative coefficient of thermal expansion can be utilized in the design of zero thermal-expansion structures. Most carbon fibers can be woven into fabrics.

Aramid fibers offer attractive strength, stiffness, and density properties and a high degree of toughness. They are marketed under the trade name Kevlar (DuPont). They are made by treating poly(*p*-phenylene-terephthalamide) with strong acid and extruding the resulting fibers from spinnerets.

Plastic Matrix Materials

Polyester, the cheapest matrix material, is easy to process and is the most common material used in glass-reinforced plastics. However, tensile elongation at breaking is low. Therefore, epoxy resins are preferred for high performance carbon-fiber composites. Polybutadiene, vinyl ester polymers, and polyimides are also used.

Properties

The basic building block in high performance fiber-reinforced plastics is a single ply of unidirectional tape. Properties are assumed to equal those of a laminate where all fibers are aligned in one direction. The wide range of properties of unidirectional tapes offers designers the opportunity to tailor strength and stiffness to achieve maximum efficiency.

Table 1. Properties of Reinforcing Fibers

Fiber	Tensile strength, MPa[a]	Tensile modulus, GPa[a]	Specific gravity	Fiber diameter, μm	Coefficient of thermal expansion, μm/(m °C)	Electrical resistivity, $\Omega \cdot$ cm
glass						
E-glass	3450	72	2.54	7.0	5.0	10^{15}
S-glass	4820	85	2.49	10.0	5.6	10^{16}
boron	3551	386	2.45	140.0	5.4	10^{14}
carbon						
Thornel 300	3448	231	1.75	7.0	−0.5	1.8×10^{-4}
AS6	4137	248	1.74	5.5		
GY-70	1860	517	1.96	8.4		6.5×10^{-4}
P-100S	2241	690	2.15	11.0	−1.6	2.5×10^{-4}
aramid						
Kevlar 49	2758	131	1.44	12.0	−2.0	10^{16}

[a] To convert MPa to psi, multiply by 145 (GPa × 145,000).

Fatigue behavior and resistance to environmental degradation are excellent. Glass, boron, and Kevlar composites exhibit high electrical resistivity in both longitudinal and transverse directions. Carbon composites exhibit high electrical resistivity only in the transverse direction.

Processing

In general, fiber and resin are combined to form tape (prepegging); sheets of tape are stacked in the specific orientation; the resulting laminate is bagged and cured in an autoclave; and the cured product is machined and inspected. Precision equipment is used to control the tape thickness (generally 1.5–2.5 mm), fiber alignment and spacing, resin content, and the degree of curing. Automation greatly reduces costs. Inspection methods include visual, radiographic, microwave, infrared, holographic, and eddy-current techniques.

Fiber-reinforced composites can be sawed, drilled, ground, sanded, chamfered, milled, and turned. Carbon-epoxy is the easiest to machine, Kevlar-epoxy the most difficult.

Filament winding is commonly used to fabricate bodies of revolution, eg, rocket-motor cases, high pressure tanks, and cylinders, whereas so-called pultrusion is used for fence posts, beams, and tubing. Molding techniques are also employed.

Design

Fiber-reinforced plastics, although designed with the same principles as metallic structures, do not have the same ductility, and stress considerations are important. Laminates designed to fail by a break in fibers in one or more plies are defined as fiber-dominated composites. Laminates that fail in other modes exhibit poor fatigue behavior and interlaminar stresses are designed to be low. The composite structures may be joined by mechanical fastening and adhesive bonding.

Applications

Composite components in the space shuttle orbiter include carbon–epoxy payload bay doors and panels for the orbital maneuvering system, boron–epoxy reinforced titanium aft-thrust structures, and Kevlar–epoxy pressure vessels. In aircraft applications, composite components can result in weight savings of 25–50%. Kevlar–epoxy covers about 45% of the exterior surface of the Sikorsky S-76 helicopter, and is also used in internal structures.

Aramid (Kevlar) fibers (qv) are used extensively in automotive tires, and Ford is investigating carbon–epoxy to reduce the weight of automobiles.

More than 90% of pleasure boats and approximately 40% of 29 m or shorter fishing trawlers are made of glass-fiber-reinforced plastics. These materials are also used in sporting goods. Health-related applications include wheelchairs and prosthesis equipment.

JOHN DAVIS
NASA

D.V. Rosato in G. Lubin, ed., *Handbook of Composites*, Van Nostrand Reinhold Company, Inc., New York, 1982, pp. 1–14.

C. Zweben, H.T. Hahn, and R.B. Pipes, *Composites Design Guide*, Vol. 1, Mechanical Properties Section 1.7, 1980, pp. 65–71.

M.L. Phelps, *In-Service Inspection Methods for Graphite-Epoxy Structures on Commercial Transport Aircraft*, NASA CR 165746, 1981.

COMPUTERS

Computer Hardware

In the jargon of the computer industry, the term hardware refers to the physical equipment in a computer system—the electronic devices, power supplies, mounting racks and chassis, and electromechanical equipment required to handle cards, magnetic tape, and the other media used with the system. To understand how hardware can perform mathematical and other information-processing operations requires some understanding of information representation and manipulation by electronic circuits.

For numerical problem-solving by humans, the numerical system based on powers of 10, the decimal number system, has evolved into common use. However, the binary number system, based on powers of 2, can be used similarly for the arithmetic operations addition, subtraction, multiplication, and division, and to indicate greater than, less than, and equal. This is the system most commonly used in computers.

Computers also must process alphanumeric information, ie, information consisting of both numerals and alphabetic characters. This is done by assigning a numeric code to each of the symbols to be represented. For example in the decimal system, the word DOG could be coded as 041507 if A = 01, B = 02, C = 03, etc. One of the best known binary codes is the American Standard Code for Information Interchange (ASCII).

Finally, it can be shown that binary-coded information can be manipulated by electronic circuits through the use of logical operations. Logic is a branch of mathematics that deals with two states, true and false, as inputs (variables) and outputs (results). The binary number system also has only two states, 0 and 1, and it is possible to state all of the binary arithmetic and logical operations in terms of formal logic statements. As a result, all of the operations required for information processing in a computer can be reduced to logical equations that are easily realized using electronic circuits.

Digital computers are constructed of many such circuits, from about 2000 circuits in a very small computer to 200,000 or more in a large computer. Present semiconductor technology makes it possible to construct (integrate) thousands of these circuits on a single chip of silicon less than 10 mm². Such a chip is called an integrated circuit (qv). If the number of circuits on the chip is greater than about 100, it is known as a large-scale integrated (LSI) circuit; if greater than about 1000, it is a very large-scale (VLSI) circuit. LSI and VLSI circuits are used extensively in today's computers, and the rapidly increasing density of circuits per chip, mass consumption, and mass production techniques result in a very low cost per circuit.

To implement a digital computer, thousands of circuits must be combined into a structure for processing data. The typical structure for a small general-purpose computer consists of a processing unit, a storage unit, and one or more peripheral units used primarily to transfer data to and from the processing unit and/or storage to input/output (I/O) units such as keyboards and displays. Storage consists of electronic circuits that retain instructions specifying an operation to be performed and the operands on which the instruction is to operate. Both instructions and operands are coded in binary form and represented in storage and other circuits as the presence or absence of a voltage corresponding to each binary digit (bit) in the binary-coded representation. The datum or instruction stored in each location is typically a fixed-length "word" that may be subdivided into "bytes" (fractions of a word). The fact that the datum is of fixed length limits the numerical precision that can be expressed in a single word. For example, in a 16-bit word, the integers 0 to 65,535 ($= 2^{16} − 1$) can be directly expressed. Expression of more significant digits requires the use of more than one word.

The processing unit typically can perform all the basic mathematical and relational operations, such as *add*, *subtract*, and *greater than*. It can also cause transfer of control to be effected; that is, it can cause the computer to execute a sequence of instructions starting at an arbitrary storage address, rather than continuing with the instruction stored in the next sequential location. This concept, called branching, is very important because it provides the capability to choose alternative actions conditionally based on the result of a computation or other operation.

The peripheral units provide the system a means to input or output information or of storing large quantities (millions (10^6) to billions (10^9) of words) of data or instructions. Mass storage units, such as disk files or magnetic tape (qv), also serve as peripheral devices. Although several mass-storage forms are available, magnetic media are most commonly used.

Computer Programming

The purpose of computer programming is to create a sequence of computer instructions that results in solution of a problem such as inverting a matrix or alphabetizing names. The programming activity consists of a sequence of steps: define the problem; design a solution (an algorithm); convert the algorithm into a form for internal representation in storage by using a computer language (this step is known as coding and results in a program); test the program to ensure correct operation; document the program; and install it on the computer (ie, read it into storage for subsequent use).

Computer vendors usually also supply a control program, known as an operating system or executive for the system. The operating system is a program that controls the execution of the user's programs and provides common facilities for use by a number of user programs. Its purpose is to assist in the efficient utilization of the machine and to relieve the user of tedious and routine details of machine operation.

Computer Applications

Information processing, in its most general sense, involves the manipulation of data, both numeric and alphabetic, to provide for their organization according to a defined set of rules (eg, alphabetizing a set of names or ordering a list of quantities by magnitude), or to derive numerical results from a set of numeric data according to some mathematical equation (eg, calculating equilibrium conditions of a chemical mixture), or for empirically finding a solution by comparing many potential solutions to a set of rules or guidelines. Information processing applications increasingly involve manipulation and management of large quantities of information. This information is organized into a series of files which, taken collectively, are termed a data base. For example, a data base might include files of the production records, product production statistics, customer order status, and financial records.

Computers have been used since the late 1950s for controlling and gathering data from industrial processes. Applications range from the control of a single process unit, such as a reformer in an oil refinery, to integrated control of entire industrial complexes—a steel mill from ore carrier to finished steel products.

Recent Trends

The advent of the microprocessor has spawned significant developments in instrumentation and control (qv). It now is possible to incorporate a digital processor in an instrument to provide unit control (eg, controlling the valve sequence in a vapor-phase chromatograph) and to provide some data processing (eg, finding peak values, averages, and deviations). As another example, microprocessors are built into some single-loop proportional-integral-derivative (PID) controllers used in the control of industrial processes. The primary implication of this trend is that some data processing can be performed in the instrument, thereby reducing the load and dependency on the central computer system.

There is increasing emphasis on digital communication methods between instruments and between computer systems. A related development is the current interest in computer networks and distributed processing. In distributed processing a given problem is solved by interconnected computers that may be located close together or distributed

across several countries and interconnected by means of common-carrier telephone facilities or dedicated high bandwidth communication circuits. Typical benefits claimed for distributed processing and networks are simplicity (since each node is dedicated to a single task), lower cost through the use of smaller, relatively simple computers at each node, availability of special computing resources (eg, a Fast Fourier Transform Analyzer) within the network, and faster response time since some tasks are performed at the local node without communication delays.

Thomas J. Harrison
International Business Machines Corporation

H.T. Nagel, Jr., B.D. Carroll, and J.D. Irwin, *An Introduction to Computer Logic*, Prentice-Hall, Englewood Cliffs, N.J., 1975.

C.J. Date, *An Introduction to Data Base Systems*, 3rd ed., Addison-Wesley, Reading, Mass., 1981.

H. Lorin and H.M. Dietel, *Operating Systems*, Addison-Wesley, Reading, Mass., 1981.

R.J. Cypser, *Communications Architecture for Distributed Systems*, Addison-Wesley, Reading, Mass., 1978.

CONCRETE. See Cement.

CONDIMENTS. See Flavors and spices.

CONGO COPAL. See Resins, natural.

CONSTANTAN. See Nickel and nickel alloys.

CONTACT LENSES

Contact lenses are optical devices that are placed over the cornea of the eye in such a manner that the lens remains on the eye's surface throughout blinking. The main purpose of contact lenses is correction of vision deficiencies; in this application they are called cosmetic lenses. Contact lenses can also be used medically for the treatment of certain corneal diseases; in such cases they are called therapeutic, or bandage lenses.

Glass contact lenses were first used at the end of the 18th century, but extensive development of the contact-lens industry occurred only when a plastic that was lighter than glass and had excellent optical properties, poly(methyl methacrylate), became easily available after World War II. The industry in the United States now involves over 300×10^6 annually. It is estimated that over 12×10^6 people wear contact lenses. The newer oxygen-permeable lenses and hydrogel soft lenses offer advantages over the hard poly(methyl methacrylate) lenses, particularly with regard to comfort, a shorter break-in period which accustoms the cornea to the presence of the lens, and an improved physiological environment for the cornea. Currently, the contact-lens industry is undergoing rapid change with new lenses made of a variety of materials appearing with increasing frequency (see Methacrylic acid and derivatives).

Contact lenses made of any material other than poly(methyl methacrylate) are regulated in the United States by the FDA. Before approving a new lens and its auxiliary systems (such as cleaning, disinfecting, and wetting solutions) for sale to the public, the FDA must be convinced of the safety and efficacy of the complete device. The FDA has stringent standards, and obtaining approval for a new contact lens requires several years of preclinical and clinical testing (see Regulatory agencies).

Contact lenses can generally be classified as hard lenses, silicone rubber lenses, and soft hydrogel lenses. In all three types there are lenses that are oxygen permeable and in hard and silicone rubber types there are lenses that are hydrophobic. An essential difference among the three types is the manner in which they fit the eye. Hard lenses and silicone rubber ones require a relatively thick tear film between their posterior surface and the cornea of the eye which is a matter of considerable optical and physiological importance. Soft hydrogel lenses adhere closely to the cornea with a tear film of only capillary thickness between the lens and the cornea surface.

With any kind of contact lens, the cornea's surface must be wet and oxygenated at all times to remain transparent and healthy. Thus, with any type of contact lens, disruption of oxygen supply to the cornea surface must be minimized, either by oxygen-rich tear exchange under the lens or by oxygen permeation through the lens, or both. Therefore, the main challenge in the art of fitting contact lenses is to provide oxygen to the cornea.

Trade literature refers to contact lenses not only by the chemical name of the polymer and its acronym, or by the trade name or registered name, or the manufacturer's or distributor's name, but often by the name of the lens inventor or entrepreneur who developed it.

Hard contact lenses include poly(methyl methacrylate) lenses, lenses made of diverse copolymers of methyl methacrylate with siloxanylalkyl methacrylates, silicone resin lenses, and cellulose acetate butyrate (CAB) lenses. (CAB is a group of plastics that has received wide attention in the last few years as an oxygen-permeable hard contact-lens material.) In order for a contact lens to be comfortable, one of the most important requirements is that its surface must be uniformly wet by the tear film at all times. Good water wettability is also required for good optical performance of the lens and for its ability to remain clean of mucoid deposits in the eyes. To impart the desired surface characteristics to the normally relatively hydrophobic poly(methyl methacrylate) and other poorly wettable silicone-containing lenses, various specially prepared wetting solutions are commercially available. These solutions are reported to make contact lenses more comfortable because a hydrophilic lens is held better than a hydrophobic lens by the capillary action of tear film on the cornea. Temporarily soaking a lens in a wetting solution is reported to improve its wetting properties.

To increase the permanence of improved wettability, certain hydrophilic coatings have been developed. Improved hard-lens wettability has also been attempted by means of copolymerization of hydrophobic and hydrophilic monomers.

The flexible rubber contact lenses are made almost exclusively of diverse silicone rubbers. These are hydrophobic materials, but some of the silicone rubber lenses have hydrophilic surfaces.

Hydrogel contact lenses can be classified according to the chemical composition of the main ingredient in the polymer network regardless of the type or amount of minor components such as cross-linking agents and other by-products or impurities in the main monomer. Hydrogel contact lenses can be classified as (1) 2-hydroxyethyl methacrylate lenses; (2) 2-hydroxyethyl methacrylate–N-vinyl-2-pyrrolidinone lenses; (3) hydrophilic–hydrophobic moiety copolymer lenses (the hydrophilic component is usually N-vinyl-2-pyrrolidone or glyceryl methacrylate, the hydrophobic component is usually methyl methacrylate); and (4) miscellaneous hydrogel lenses, such as lenses with hard optical centers and soft hydrophilic peripheral skirts, and two-layer lenses. Most hydrogels are obtained by bulk polymerization of the monomer mixture with the addition of a free-radical initiator, or in some cases by solution polymerization with a free-radical or redox initiator.

Manufacture

Starting from a polymer blank or "button" (generally made by cutting a rod of cross-linked acrylic resin) or from a molded semifinished lens, most hard contact lenses are made by standard methods of machining and polishing. A widely available hydrogel contact lens is made by a spin-casting technique. Static cast-molded lenses are also currently available. However, the most common system for the production of hydrogel contact lenses is a standard machining technique that is similar in many respects to the technique for manufacturing most hard contact lenses. Silicone lenses are molded. Currently the principal problem with silicone lenses is surface coating rather than edge fabrication.

Oxygenation of the Corneal Surface

With all hard contact lenses and most hydrophobic flexible lenses, oxygenation of the corneal surface is obtained in part by a pumping mechanism that operates in the tear film during blinking. The tear film is normally oxygenated from the air; however, under a contact lens it must ideally be oxygenated across the lens. Thus, oxygen-permeable contact lenses perform better than do oxygen-impermeable lenses. Hydrogel lenses are, in general, fitted closer to the cornea surface than other types of lenses, and they remain practically static on the cornea whereas other types move with the blink. Thus, the corneal surface depends more on the oxygen that passes across the lens with hydrogel lenses than with hard lenses. However, this point is still moot and a mechanism has been postulated for oxygenation of the cornea, at least in part, by means of a hypothetical capillary tear film that pumps tears between the lens and the corneal surface with the blink.

MIGUEL F. REFOJO*
Eye Research Institute of Retina Foundation
Dept. of Ophthalmology
Harvard Medical School

R.B. Mandell, *Contact Lens Practice*, 3rd ed., Charles C. Thomas, Springfield, Ill., 1981.

M. Ruben, ed., *Soft Contact Lenses: Clinical and Applied Technology*, John Wiley & Sons, Inc., New York, 1978.

N.A. Peppas and W.H.M. Yang, *Cont. Intraocular Lens Med. J.* **7**(4), 300 (1981).

CONTAINERS. See Packaging materials, industrial products.

CONTRACEPTIVE DRUGS

Control of fertility continues to be an important issue throughout the world even though the population growth rate has shown a steady decline in many countries, partly owing to the extensive use of oral contraceptives. The first of these products to be marketed was Enovid, a combination of norethynodrel (**1**) (a progestin) and mestranol (**2a**) (an estrogen), originally introduced in 1957 for the treatment of menstrual disorders. In 1960, the FDA approved its use for the cyclic control of ovulation.

(**1**)
norethynodrel

(**2**)
(**a**) R = CH₃   mestranol
(**b**) R = H   ethinyl estradiol

Oral contraceptives used by the majority of women today are each a combination of two steroidal substances, a progestin and an estrogen. The two substances are present in various ratios and act principally by inhibiting ovulation in normally cycling women (see Hormones; Steroids).

Although the combination of a progestin and estrogen is very effective in suppressing ovulation, certain undesirable side effects became apparent on widespread usage. Thromboembolic and related vascular disorders, as well as alterations in carbohydrate and lipid metabolism, have been reported in women taking contraceptive pills, particularly among those who are older. The estrogen component has been implicated in

*Supported by USPHS grant EY-00327 from the National Eye Institute, National Institutes of Health.

these disorders. For this reason, current practice suggests prescribing those products in which each pill contains 50 μg or less of estrogen. Estrogen is required for good cycle control. Reducing its quantity may result in a greater incidence of breakthrough bleeding. Because of the side effects reported with the progestin–estrogen combination pills, other approaches to contraception are being investigated.

Steroids

Ovulation and/or implantation suppression. For the most part, the oral contraceptives that are currently in use contain either mestranol (2a) or ethinyl estradiol (2b) as the estrogen component. The progestin component of the vast majority of oral contraceptives is also an estrane derivative. Norethindrone is the progestin present in most preparations, as for example, in Ortho-Novum, Norinyl, Modicon, Brevicon, Micronor, and Nor-Q.D. Many modifications of the norethindrone structure can be made without adversely affecting the biological activity. Acylation of the 17-hydroxyl group can enhance or prolong the effects of norethindrone. Thus, norethindrone enanthate (Norigest) is utilized as an injectable contraceptive, given once every several months.

Attempts also have been made to impart oral activity to progesterone and to enhance its parenteral activity by introducing various groups into the molecule without alteration of the steroid skeleton. Among the earlier efforts was the insertion of an acyloxy group into the 17α-position, eg, 17α-acetoxyprogesterone is an orally active progestin, and 17α-hydroxyprogesterone is inactive. The corresponding hexanoate has prolonged activity when administered parentally. It has been marketed as Delalutin. As was the case with norethindrone, further enhancement of activity could be achieved by inserting methyl and halogen groups into certain positions of the molecule and by extending the conjugated keto system with an additional double bond.

The potent, orally active progestin dydrogesterone belongs to a class of compounds known as retrosteroids. In these steroids, the methyl group at C-10 and the hydrogen atom at C-9 are oriented α and β, respectively. The backbone configuration of the retrosteroids is *anti-cis-anti-trans*. Dydrogesterone is marketed in Europe as Duphaston for the treatment of menstrual disorders.

At one time it was thought that a high degree of structural and configurational specificity was required for hormonal activity. However, recent findings clearly indicate that the structural and configurational requirements for progestational activity are not as rigorous as they were one thought to be.

Postcoital contraceptives. Agents that prevent implantation are known as interceptives. Generally they act by altering the rate of ovum or zygote transport. They may also act by luteolysis. Estrogens are leutolytic in certain species as evidenced by a reduction in plama progesterone levels or a decrease in basal body temperature following postovulatory administration. Estrogen may also prevent implantation by altering the sensitive hormonal balance that is essential for synchronization of the uterine environment with endometrial implantation. To be effective, the estrogens must be given in relatively high doses. This may result in the production of side effects, such as edema, thrombophlebitis, menstrual irregularities, vomiting, and nausea. If ineffective doses are administered or if treatment is begun after implantation, fetal malformations may occur.

Daughters of women treated with diethylstilbestrol in the first trimester of pregnancy were observed to have a high incidence of vaginal adenosis and adenocarcinoma. Consequently, it has been recommended that termination of pregnancy by surgical means be employed should the interceptive method fail.

Because of the reported side effects, efforts are being made to modify the structures of the estrogens in order to achieve a greater separation of antifertility and estrogenic activity. Analogues of ethinyl estradiol, in which the ethynyl hydrogen has been replaced by either an ethynyl or a trifluoromethyl group, have an enhanced separation of antifertility and estrogenic activity.

Besides the estrogens, progestins such as norethindrone, norgestrel, quingestanol acetate, and R 2323 are currently being studied in the clinic as potential postcoital contraceptives.

Progesterone blockers. The binding of a hormone to its polypeptide receptor in the cytosol initiates the chain of events that results in the biological phenomenon observed. This has been demonstrated for progesterone, the pregnancy hormone, as well as for the estrogens, androgens, and corticoids.

The progesterone–receptor complex formed in the cytosol is translocated into the nucleus where it acts on the chromatin to promote the synthesis of deoxyribonucleic acid (DNA)-dependent mRNA (messenger ribonucleic acid). Through the process of translation, the latter induces the synthesis of proteins that are essential for implantation and maintenance of pregnancy.

Interference with the binding of progesterone to its receptor may terminate pregnancy. The isolation of the progesterone receptor from human uteri, as well as from the uteri of several other species, has stimulated interest in finding compounds that would compete with progesterone in binding to its receptor and that would not support pregnancy.

Suppressions of spermatogenesis. Oral contraceptives for women inhibit ovulation by suppressing the mid-cycle surge of gonadotrophins. Similar suppression of gonadotrophins in males can be attained with steroidal agents. Although spermatogenesis is inhibited as a result, which effect often is accompanied by loss of libido, potency, and secondary sexual characteristics. However, this can be avoided if an androgen is administered with the antigonadotrophic steroid. If the androgen is given at a high dose, a synergistic suppression of spermatogenesis is achieved as well.

Manufacture. The commercial preparation of the progestin and estrogen components of the oral contraceptives can be accomplished by either partial or total synthesis. An ingenious procedure that makes use of carbon dating can be employed to determine which form of synthesis had been utilized to prepare a specific product. The starting material used in partial synthesis is from the plant or animal kingdom and is of recent origin. Total synthesis, on the other hand, uses materials that are most likely derived from fossil fuels, either coal or petroleum. In general, the starting material for partial synthesis is a sapogenin.

In contrast to partial synthesis, which affords a product free of its enantiomer, total synthesis as practiced in the past furnished products or intermediates that were racemic. At some stage in the synthesis, a resolution with a chiral reagent was required to remove the undesired enantiomer which was generally discarded. To avoid the necessity of having to generate the undesired enantiomer, an approach was developed in which a chiral reagent could be employed to react selectively at a particular prochiral center present in an intermediate (or a substrate) to afford an optically active product having predominantly the desired configuration. A variety of microorganisms and chiral chemical reagents were found to be effective in the processes developed. As a result, asymmetric synthesis has become a useful means for the commercial production of optically active steroids (see Pharmaceuticals, optically active).

Until recently, diosgenin obtained from the Mexican species of the dioscorea plant was the chief starting material for the production of steroid drugs. With technical improvements made in microbiological transformations, however, processes involving the fermentation of abundant sterols, such as sitosterol and cholesterol, are now being widely used for the production of contraceptives. Efficiency achieved in total synthesis has not only made it economically possible to prepare steroids with an ethyl group attached to C-13 (eg, d-norgestrel), but also steroids with a methyl group at that position. As a result, production of steroids either involving fermentation or by total synthesis competes favorably with the production of steroids from dioscorea. Indeed, in some cases steroids prepared from the fermentation of sterols or by total synthesis are 30–60% less expensive than the same steroids derived from dioscorea.

Many steroidal contraceptive drugs are listed here, including virtually all those manufactured in the United States: Anovlar, Norigest (Schering AG); Brevicon, Norinyl, Nor-G.D. (Syntex); Deladroxate (Squibb); Demulen, Enovid, Ovluen (Searle); Loestrin, Norlestrin, Ortho-Novum (Parke-Davis); Lo/Ovral, Ovral, Ovrette (Wyeth); Lyndiol (Organon); Micronor, Modicon (Ortho); Ovcon (Mead Johnson); Planor (Roussel-Uclaf); Provest (Upjohn); Riglovis (Warner-Lambert); Zorane (Lederle).

Polypeptides

Luteinizing hormone – releasing hormone (LH – RH) analogues. Since 1971 immunologic studies have led to the development of antisera to LH–RH that inhibit ovulation in estrous-cycling rats and also produce abortion in pregnant rats. Numerous analogues of LH–RH have been synthesized. Some have proved to be agonists having potencies greater than that of the natural hormone, and others have been found to be antagonists. Both agonists and antagonists have been examined for their potential contraceptive usage. Agonists that are more potent than LH–RH either have a greater affinity for the pituitary receptor or their half-life may be longer than that of LH–RH, which has been reported to be about four minutes (as computed from the first exponential portion of the disappearance curve of LH–RH in hymen plasma). LH–RH and stimulating analogues that cause the prolonged secretion of LH and FSH induce ovulation. They are also capable of interfering with pregnancy when administered postcoitally.

Immunologic approach. Active immunization is currently being explored as a means for controlling fertility. Identification of a polypeptide containing 30 amino acid residues, which is unique to the β-subunit of human chorionic gonadotrophin (HCG), has given rise to the hope that antibodies can be raised that would react specifically with HCG. This hormone is a glycoprotein that is normally produced by the placenta. It is excreted in the urine in large amounts during the first trimester of pregnancy. Its main function in the early stages of pregnancy is to prolong the life of the corpus luteum and to stimulate the luteinized cells to produce progesterone which is necessary for the maintenance of pregnancy.

Prostaglandins

The prostaglandins (qv) comprise another class of naturally occurring physiologically active substances that have been studied extensively for their effects on reproduction. 15-Methyl PGF_2 methyl ester has been observed to terminate pregnancy in humans when administered intravaginally. Although 15-methyl PGF_2 methyl ester stimulates uterine contractions, termination of early pregnancy has been attributed in part to its luteolytic effect. The abortifacient activity of the prostaglandins can be ascribed to the effects they exert on the uterus, either directly as a result of smooth-muscle stimulation or indirectly via the ovary or pituitary.

Other Methods of Contraception

The contraceptive drugs considered so far are intended to disrupt the delicate hormonal balance that exists during the onset of ovulation or during pregnancy. These drugs are also intended to act systemically, as they are given either orally or by injection. Other routes of administration are being explored. Medicated intrauterine devices (IUDs) have been developed that will release a small quantity of the drug daily in the uterus (see Pharmaceuticals, controlled-release). The copper IUD is highly effective in controlling fertility as well.

Although the intrauterine devices do not elicit the side effects attributed to the oral contraceptives, they may produce bleeding, be expelled, be associated with ectopic pregnancies, and be implicated in pelvic inflammatory disease.

Investigations are being conducted on male fertility with the view of developing male contraceptive agents, eg, gossypol derived from cotton seed. The approaches under study include disruption of the biochemical events associated with spermatogenesis. Modification of the essential components of the epididymal fluid or other accessory secretions and disruption of epididymal sperm maturation offer additional methods for the control of male fertility. In another approach, specific agents are being sought that will block the protease, acrosin, present in the acrosome of the sperm. Inhibition of acrosin will cause the spermatozoan to lose its capacity to fertilize the ovum.

Concern for safety and the availabilty of abortion on demand in many regions have rekindled an interest in the spermicidal and barrier approach to fertility control. Spermicidal jellies and foams, as well as diaphragms and condoms, are acquiring an increasing number of users despite their double-digit pregnancy failure rates (per 100 woman-years).

L.J. CHINN
F.B. COLTON
G.D. Searle & Co.

M.H. Briggs and M. Briggs, *Biochemical Contraception*, Academic Press, Inc., New York, 1976.
Contraceptives of the Future, The Royal Society of London, London, 1976.
J.P. Bennett, *Chemical Contraception*, Columbia University Press, New York, 1974.
R. Wiechert, *Angew. Chem. Int. Ed. Engl.* **16**, 506 (1977).

COOLANTS. See Heat-transfer media other than water; Refrigeration; Antifreezes and deicing fluids.

CONVEYING

Conveying equipment can be classified as units that feed material into a system at a predetermined and controlled rate, generally referred to as feeders, and units that transport material from one location to another. The latter fall into the general category of conveyors. The factors that must be considered in equipment selection are material characteristics, layout, and economics. Other considerations include environmental conditions such as temperature and corrosiveness.

Units

Feeders include the apron feeder (a series of apron pans attached to a chain or pivotally to each other); belt feeder (an endless rubber or metal belt operating overdrive, tail, or bend terminals); chain-curtain feeder (a power-operated curtain of endless length of chain resting on and retarding the flow of bulk material in an inclined chute); live-roll grizzly (a series of rotating, parallel rolls with fixed spaces between them); reciprocating feeder (a reciprocating driven plate or pan operating under a head of bulk material); roll feeder (a smooth, fluted, or cleated roll or drum rotating to deliver bulk material) (see Fig. 1a); rotary-table feeder (a rotating circular table to which the material flows from a round bin or hopper opening and from which it is discharged by a plow) (see Fig. 1b); screw feeder (a screw revolving in a stationary trough or casing fitted with hangers, trough ends, and other accessories); and vibrating feeder (a trough or tube, usually supported on or suspended from a steel or rubber spring system, which is excited by an electric motor or magnet).

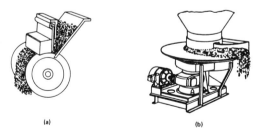

(a)   (b)

Figure 1.   (**a**) Roll feeder. (**b**) Rotary-table feeder.

Conveyors include the apron and the belt conveyors, both of which operate like the apron feeder and belt feeder. The belt conveyor is more flexible, versatile, and economical for more installations than any other conveying medium. Other conveyors include the drag-chain conveyor (one or more endless chains that drag bulk material in a wooden, steel, or concrete trough) (see Fig. 2); en-masse conveyor (a series of skeleton or solid flights on an endless chain, cable, or other linkage operating in a horizontal, inclined, or vertical path within a closely fitted casing); flight conveyor (one or more endless propelling media, such as a chain, to which flights are attached, and a trough through which material is

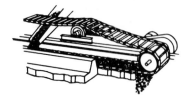

Figure 2. Drag-chain conveyor.

pushed); gravity-discharge conveyor-elevator (buckets mounted between two endless chains that operate through troughs and casings on horizontal, inclined, and vertical paths of overdrive, corner, and takeup terminals) (see Fig. 3); pivoted-bucket carrier (the buckets remain in the carrying position until tipped for discharge); pneumatic conveyor (material is charged into an airtight bucket and propelled by a continuous blast of forced air); screw conveyor (a conveyor screw revolves in a stationary trough or casing fitted with hangers, trough ends, and other accessories); shuttle conveyor (any conveyor mounted on a self-contained mobile structure, generally operating on rails); slat conveyor (one or more endless chains to which nonoverlapping, noninterlocking, spaced slats are attached; sliding-chain conveyor (one or more endless chains slide on tracks); and the vibrating conveyor which operates like the vibrating feeder.

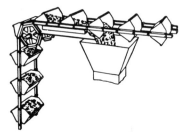

Figure 3. Gravity-discharge conveyor–elevator.

Bucket elevators consist of a series of buckets mounted on a continuous strand of chain or rubber belting which are filled or loaded at the bottom of the elevator casing and lifted to the top of the elevator where they are emptied. The most common styles include the centrifugal-discharge bucket spaced to permit the free discharge of bulk material; the continuous type mounted on a single- or double-chain strand; the super-capacity type, which operates like the continuous type except that the buckets are very large and are hung between two strands of chain; the internal-discharge type designed for loading and discharging along the inner boundary of a closed path; and the mill-duty type which is a centrifugal-discharge elevator that incorporates hooded buckets with a high front lip allowing the buckets to be spaced very closely. In bucket elevators, the speed of a unit has little allowable variation. Operation above or below the recommended speed creates a condition known as backlegging, in which the discharging material does not follow its prescribed path but falls back into the elevator boot where plugging becomes inevitable.

Elevator capacity can be determined by calculating the volume to be handled from the following formula:

$$\frac{m^3}{h} = \frac{t/h}{D}$$

where D = bulk density.

Belt conveyors have an extremely wide speed range from < 0.3 to > 300 m/min; typical speeds are 15–180 m/min. Gravity takeups are desirable on any belt conveyor with pulley centers > 90 m. Because of the inertia of a long belt conveyor, gradual starting is often used to reduce stress. Most manufacturers supply data that show the capacities that can be attained for various belt widths.

Chain conveyors usually operate at lower speeds than belt conveyors. They are designed with maximum speeds of 30–38 m/min for chain pitches up to 23 cm, and somewhat lower speeds (15–23 m/min) for greater pitches.

Operation. A unit serving as a feeder is operated at a lower speed than when used as a conveyor. As the feeder withdraws material from a bin or hopper through a restricted opening, the lower operating speed results in reduced material agitation, increased accuracy and uniformity of the volume withdrawn, less dusting of fines, and longer equipment life.

Power requirement is the sum of the power required to move the empty conveyor, to move the load horizontally, to lift the load, and to overcome friction. It is calculated as follows:

$$\text{power} = \frac{(T_e)(V)}{C}$$

where T_e = force (in N or lbf) developed in overcoming the forces of friction and gravity, V = velocity (m/s or ft/min), and C = constant. This formula applies to most conveyors and feeders with a linear motion. It is not applicable to rotary-motion or reciprocating units.

Pneumatic conveyors. The principle of pneumatic conveying is to fluidize a dry, granular material and direct it through a pipe with a flow of air. Systems, if classified by air flow, are either high pressure or low pressure systems. The former require air to be supplied at normal plant air pressure of 590–690 kPa (85–100 psi) and are equipped with a pressure vessel, called a transporter, into which the material is loaded by gravity. Low pressure systems operate at 21–69 kPa (3–10 psi) and require large volumes of air to develop higher velocities. The positive-displacement blower, which is an integral part of these systems, must be properly sized to suit the installation.

Vibrating conveyors and processing units. Vibrating conveyors are often more than carriers of material or products. In addition to moving, lifting, and feeding, they can be used for sorting, screening, mixing, and tumbling, and to reduce the size of the material. Vibrating conveyors are not affected by heat, dust, impact, and heavy loads, and therefore provide a longer service life, reliable operation, and low maintenance. When vibrating conveyors are installed in elevated structures, dynamic balancing may be required. The spiral elevator and fluid-bed drying or cooling are two recent applications of vibrating conveying.

E.J. ANDERSON
Rexnord Inc.

Unique Materials-Handling with Vibration, Bulletin 16912, Rexnord Inc., Milwaukee, Wisc.

Rexnord Inc. Cat R80, Rexnord Inc., Milwaukee, Wisc.

Material Handling with Pneumatic Conveyors, Whirl-Air-Flow, Engineering and Manufacturing Facilities, Minneapolis, Minn., 1982.

COORDINATION COMPOUNDS

Coordination compounds (or metal complexes) are important throughout chemistry and chemical technology. The chemistry of the metallic elements, which constitute 80% of the periodic table, is predominately coordination chemistry. Coordinating or complexing Lewis bases (electron-pair donors), called ligands, are often used commercially to modify the properties of the metals or metal ions. Ligands interact differently with different metal ions; therefore, analysis of metallic elements is possible through coordination followed by solvent extraction, spectroscopy, gravimetry, electrochemistry, etc (see Analytical methods). Some ligands are used to selectively extract metal ions from biological tissue. Most of these have been employed in analytical applications as well. Conversely, metal ions are sometimes used to alter the properties of commercially important species through coordination; eg, the metallation of azo dyes (qv) to improve permanence or change color, or both, and the coordination of zinc to bactericides to modify their properties.

The coordination compound often has properties entirely different from those of the ligand or the metal ion. Coordinated metallic species serve as catalysts or intermediates in a wide range of organic and biochemical reactions. Biochemists have recently begun to study the large number of metalloenzymes that are important to life (see Enzymes). Conversely, chemists are now using metalloenzymes to produce chemicals that cannot be prepared easily (if at all) by other known routes.

Typically a coordination compound consists of a metal atom, ion, or other Lewis acid plus a number of electron-pair donors (ligands) coordinated to the metal. Ligands with two or more donor atoms coordinated to the same acceptor atom are *chelating ligands*. The number of donor atoms coordinated to the same metal atom determine the dentate number of the ligand (bidentate, terdentate, quadridentate, etc, for 2, 3, 4, etc, donors to the same metal atom). The resultant ionic charge on the complex, if any, depends on the charges of the components. The coordination number is based on the total number of donor atoms that coordinate to the metal atom. Common geometries for coordination numbers from three through nine are shown in Figure 1.

Properties

The thermodynamic stability of coordination compounds in solution has been extensively studied. The equilibrium constants may be reported as stability or formation constants: $M + nL \rightleftarrows ML_n$. In addition to the normal regularities that can be rationalized by electronic considerations, steric factors are important in coordination chemistry. An example of steric selectivity involves the homopoly and heteropoly ions of molybdenum, tungsten, etc. Each molybdenum(VI) and tungsten(VI) ion is octahedrally coordinated to six oxygen (oxo) ligands. Chromium(VI) is too small and forms only the well-known chromate-type species with four oxo ligands. The ability of other cations to participate in stable heteropoly-ion formation is also size related.

Reactions

Coordination species are often categorized in terms of the rate at which they undergo substitution reactions. Complexes that react with other ligands to give equilibrium conditions almost as fast as the reagents can be mixed by conventional techniques are termed labile. Included are most of the complexes of the alkali metals, the alkaline earths, the aluminum family, the lanthanides, the actinides, and some of the transition-metal complexes. On the other hand, numerous transition metal complexes that resist substitution reaction are termed inert. Substitution rates of many complexes are enhanced by irradiation of the low energy d–d transitions ($t_{2g} \rightarrow e_g^*$ in octahedral coordination compounds.

Redox or oxidation–reduction reactions are often governed by the hard–soft rule; eg, a metal in a low oxidation state (relatively soft) can be oxidized more easily if surrounded by hard ligands or a hard solvent. Metals tend toward hard-acid behavior on oxidation. Organometallic coordination species are usually between 16- and 18-electron systems.

Applications

Applications include use in bactericides and fungicides, catalysis, coordination polymers, dyes and pigments, electroplating, petroleum additives, and therapeutic chelates.

RONALD D. ARCHER
University of Massachusetts

F.A. Cotton and G. Wilkinson, *Advanced Inorganic Chemistry*, 4th ed., Interscience Publishers, a division of John Wiley & Sons, Inc., New York, 1972, pp. 61–194, 619–688, 1049–1345, plus sections on individual elements.

T. Moeller, *Inorganic Chemistry, A Modern Introduction*, John Wiley & Sons, Inc., New York, 1982, pp. 358–548, 702–754, 578–584, 606–610, 678–688.

COPOLYMERS

Synthetic polymeric materials have become some of the most useful commodities of the modern world. In the form of elastomers, plastics, and fibers they meet basic human needs (eg, shelter, clothing, transportation, etc), and improve the overall quality of human life. In recent years the emphasis in research, development, and production of synthetic macromolecules has been directed toward preparation of cost-effective multicomponent polymer systems (ie, copolymers, polyblends, and composites), rather than the preparation of new and frequently more expensive homopolymers.

Homopolymers are high molecular weight molecules prepared by linking a large number of smaller molecules called monomers (eq. 1) (see also Polymers).

$$n A \rightarrow -(A)_{\overline{n}} \qquad (1)$$
$$\text{monomer} \quad \text{polymer}$$

Macromolecules in which two or more different monomers (comonomers) are incorporated in the same polymer chain are copolymers (eq. 2).

$$n A + n B \rightarrow -(A—B)_{\overline{n}} \qquad (2)$$
$$\text{comonomers} \quad \text{copolymer}$$

Copolymers can be further described by specifying the number and distribution of monomer units within the copolymer molecule, eg, ABBABABAAB (random), ABABABABA (alternating), AAAABBBB (block), AAAAA (graft), AABA (network), and polyblend, a physical

$$\begin{array}{ll} B & C \\ | & | \\ B & C \\ | & | \\ B & AAABAAA \end{array}$$

or mechanical blend (alloy) of two or more homopolymers or copolymers.

Although there is presently no accepted IUPAC naming scheme for copolymers, the semisystematic system introduced by Ceresa has received the most support. Ceresa's classification method is based upon a homopolymer nomenclature that uses *poly* plus the monomer. Copolymers are distinguished from homopolymers and the various types of copolymers from each other by the use of prefixes. Thus, a random copolymer of butadiene and styrene is named poly(butadiene-*co*-styrene).

Copolymers extend the number and range of available materials, enabling the polymer scientist to achieve combinations of material properties (eg, tensile strength, solubility, solvent resistance, low temperature flexibility, etc) unattainable from the simple constituent homopolymers. As a result, a large number of copolymers have become

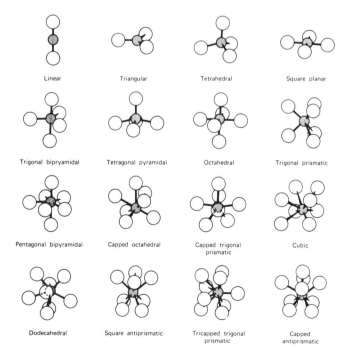

Figure 1. Common geometries for coordination numbers from three through nine. The principal axis orientation is vertical for all as shown for the first diagram.

Linear   Triangular   Tetrahedral   Square planar

Trigonal bipyramidal   Tetragonal pyramidal   Octahedral   Trigonal prismatic

Pentagonal bipyramidal   Capped octahedral   Capped trigonal prismatic   Cubic

Dodecahedral   Square antiprismatic   Tricapped trigonal prismatic   Capped antiprismatic

commercially important such as poly(butadiene-co-styrene)—synthetic rubber GRS, SBR; poly(ethylene-co-propylene)—EPM, EPR rubber; poly(ethylene-co-propylene-co-5-ethylidene-2-norbornene)—EPDM rubber; poly(butadiene-co-acrylonitrile)—NBR rubber; poly(styrene-b-butadiene-b-styrene)—SBS thermoplastic rubber (b = block); poly(iso-butylene-co-isoprene—butyl rubber, GR-1; poly(vinyl chloride-co-vinyl acetate—Vinylite flooring, Tygon tubing, coatings; poly(vinyl chloride-co-acrylonitrile)—Dynel fibers; poly(vinyl chloride-co-vinylidene chloride)—Saran packaging, fibers; poly(acrylonitrile-co-vinyl acetate)—Orlon, Acrilon acrylic fibers; poly(diacid-co-glycol-co-diisocyanate-co-diamine)—Spandex fibers; poly(butadiene-g-styrene-co-acrylonitrile) (g = graft) + poly(styrene-co-acrylonitrile)—ABS plastics; poly(butadiene-g-styrene) + polybutadiene + polystyrene—IPS impact styrene plastics; poly(styrene-co-acrylonitrile)—SAN plastics; and poly(styrene-alt-maleic anhydride) (alt = alternating)—SMA resins.

Copolymerization Reactions

The mutual polymerization of two or more monomers is called copolymerization. In the case of chain-growth copolymerization, growing polymer chains must choose between more than one monomer. Such a choice or relative reactivity has been quantitatively treated by the reactivity ratio and $Q - e$ schemes.

The synthesis of random, alternating, and block (graft) copolymers involves mechanisms and processes (eg, bulk or mass, solution, emulsion, suspension) similar to those used in homopolymerization (see also Polymerization mechanisms and processes). Copolymerization mechanisms can be further divided into categories such as free-radical, anionic, cationic, coordination, ring-opening, condensation, and post-polymerization reactions.

Effects of Monomer Unit Arrangement on Physical Properties

The primary incentive for preparing copolymers is to attain certain properties in the products. The effect of random copolymerization on polymer properties is easily shown by the difference in polymer crystallinity, melting point T_m, glass-transition temperature T_g, and solubility between the copolymer and the corresponding homopolymers. Since random comonomer enchainment tends to reduce symmetry and modify intermolecular forces, it is not surprising that random copolymers have different melting behavior than the corresponding constituent crystalline homopolymers. The solubility of random copolymers is often low in solvents for the respective homopolymers, yet high in solvent pairs. Physical properties of homogeneous block or graft copolymers are similar to those of random copolymers.

By far the most interesting block copolymers are those in which there is little or no mixing of the block phases. Such heterogeneous block copolymers tend to show the properties of the components rather than an averaging of homopolymer properties. Not only do block copolymers have properties different from those of homopolymers, random copolymers, and polyblends, but the properties of block copolymers themselves differ depending upon the arrangement of the blocks.

Characterization of Copolymers

The characterization of copolymers must distinguish copolymers from polyblends and the various types of copolymers from each other. In addition, the exact molecular structure, architecture, purity, and supermolecular structure must be determined.

Economic Importance of Copolymers

Clearly, SBR, a random copolymer, constitutes the bulk of the entire United States elastomer production. Copolymers of ethylene and propylene, and nitrile rubber (a random copolymer of butadiene and acrylonitrile) are manufactured in smaller quantities. Nevertheless, the latter copolymers approach or exceed the synthesis levels for the elastomeric homopolymers of butadiene or isoprene. Copolymers are also a very significant proportion of resin production.

Prediction of the long range usage of copolymers is especially risky at this time. The copolymers of the future will be closely related to society's energy choices, which will greatly affect monomer and polymer availability and price. Furthermore, increased governmental scrutiny of

chemical substances will make it more difficult to bring a new product to market. The choice of comonomers and copolymers may be based partly upon EPA, FDA, OSHA, and TSCA rulings (see Regulatory agencies).

When new copolymers cannot be competitively introduced, new ways must be developed for improving old ones such as blending or processing changes. There will also be a growing need for macromolecules for special requirements. Of special interest will be biomedical materials. Recent applications of polyurethane copolymers to surgical devices, such as endotrachial tubes, synthetic blood vessels, heart valves, and even artificial hearts, suggest that specialty copolymers will play increasingly important roles in the future (see Prosthetic and biomedical devices.)

D.N. SCHULZ
D.P. TATE
Firestone Tire & Rubber Co.

R.J. Ceresa, *Block and Graft Copolymers*, Butterworths Co., Washington, D.C., 1962.

J.P. Kennedy in L.H. Sperling, ed., *Recent Advances in Polymer Blends, Grafts, and Blocks*, *Polymer Science and Technology*, Vol. 4, Plenum Press, New York, 1974.

F.W. Billmeyer, Jr., *Textbook of Polymer Science*, 3rd ed., Wiley-Interscience, a division of John Wiley & Sons, New York, 1984.

D. Braun, H. Cherdron, and W. Kern, *Techniques of Polymer Synthesis and Characterization*, Wiley-Interscience, a division of John Wiley & Sons, Inc., New York, 1971, pp. 88–89.

D.J. Lyman in H.G. Elias, ed., *Trends in Macromolecular Science*, Gordon & Breach Science Publishers, Inc., New York, 1973, p. 55.

COPPER

Copper has been of major importance in the development of civilization. Because of its unique physical and chemical properties and its tendency to concentrate in large ore bodies, copper has retained a position with iron and aluminum as one of the most important metallic elements.

Copper was critical in the development of civilization because it was the only metal found naturally in the metallic state suitable for the production of tools. Furthermore, the very factors that encourage the occurrence of copper metal in nature made it relatively easy for humans to produce copper from naturally occurring minerals by reduction in a wood fire. The relative ease with which it could be reduced from the oxide form to the metal and its tendency to alloy to advantage with other metals naturally present in the ores promoted its broad use by emerging civilizations.

Copper is the first element of subgroup IB of the periodic table. It is classed with silver and gold as a noble metal and like them it can be found in nature in the elemental form. Copper occurs as two natural isotopes, ^{63}Cu and ^{65}Cu.

Today, high electrical and thermal conductivities and corrosion resistance combine with the traditional attributes of easy workability and beauty to give this metal its very wide range of commercial applications.

Introduction of flotation for beneficiation of sulfide ores during the 1920s improved metal recovery and gave impetus to the exploitation of the low grade porphyry deposits in Arizona which thus became the leading copper producing area in the United States. As a result of this development, the United States became the largest copper producer in the world.

Occurrence

The high affinity of copper for sulfur is the main factor in determining the manner of occurrence in the earth's crust. Copper shows a strong tendency to combine with all available sulfur during the crystallization of rocks. Copper–iron sulfides are the last minerals to crystallize and fill the interstices between other minerals in igneous rocks that contain an average of about 60–70 ppm copper. Other copper compounds occurring in nature are oxides and silicates. The strong affinity of copper for sulfur is the prime factor in separating copper from iron in the pyrometallurgical reduction of copper from sulfide ore.

Copper ore minerals are classified as primary, secondary, oxidized, and native copper. Primary minerals are considered to have been concentrated in ore bodies by hydrothermal processes, whereas secondary minerals have been formed when copper leached from surface deposits by weathering and groundwater was reprecipitated near the water level (see Extractive metallurgy). The important copper minerals include sulfide ores, bornite, chalcopyrite, and tetrahedrite-tennantite which are primary minerals, and covellite, chalcocite, and digenite which were formed as secondary deposits. The oxide minerals such as chrysocolla, malachite, and azurite were formed by oxidation of the surface sulfides. Native copper is usually found in the oxidized zone. However, the major native copper deposits in Michigan are considered to be of a primary nature.

Properties

Like silver and gold, copper has an atomic structure that results in outstanding electrical and thermal conductivities and a high degree of malleability. Although the assigned electronic configuration of 2-8-18-1 implies a stable closed shell of 18 electrons, this shell is not inert. Rather, the underlying d orbitals appear to participate in metallic bonding by promotion of at least one d electron into a high energy orbital of the outermost principal quantum level. There this electron is exceptionally available for participation in electrical and thermal conduction. A comparison of the properties of copper, silver, and gold is presented in Table 1.

Recovery and Processing

Most copper today is processed by mining, waste leaching and cementation, concentrating, smelting and refining. Open-pit mining is more common than underground mining and the overburden, or waste, contains some copper. Frequently the waste is leached to extract the copper, which may be recovered by passing the leach solution through a bed of scrap iron, precipitating metallic copper, and dissolving the iron; the last operation is called cementation.

The copper ore from the mine, often containing less than 1% copper, is transported to the concentrator where it is first crushed and then ground with water. The ground ore slurry enters flotation cells, where copper concentrates are collected as froth (see Flotation). Following dewatering, they enter the smelter. In the smelter the sulfide minerals react with oxygen and fluxes to produce impure copper metal, SO_2, and slag.

Smelting occurs in two steps. In the reverberatory furnace, the copper concentrate is melted to produce matte, the mixed sulfides of copper and iron. Next, air is blown through the matte in the converters, producing impure copper plus a slag containing the iron. The oxygen in the impure copper is adjusted. The copper is then cast into anodes and purified by plating onto pure copper in an electrolytic tankhouse.

Other hydrometallurgical processes include the direct leaching of ore followed by recovery of copper by cementation or electrowinning. Recently, there have been serious attempts to develop hydrometallurgical treatment of concentrates in lieu of smelting in order to avoid the high cost of environmental-control facilities required for new smelters.

Effluents from production. In the United States, the EPA sets rules for plant emissions whereas OSHA is concerned with the sulfur dioxide concentrations in working areas. The new source standards of the EPA specify that for plant expansions or new plants 95% of the sulfur in the plant feed must be captured. In locations where sulfur dioxide background levels are higher than specified, new smelters cannot be built. After all existing smelters in an area come into compliance, a new smelter can be built provided it uses the best available technology and the other smelters reduce emissions by an amount equal to the emissions from the new smelter.

Formation of SO_2 in the smelting operation can be controlled by replacement of reverberatory furnaces with equipment that allows the formation of relatively high strength gases that can be treated to produce sulfuric acid in acid plants. Alternatively, the formation of SO_2 can be prevented by radical changes in copper processing methods employing hydrometallurgical techniques for treatment of copper sulfide ores or concentrates.

The Solid Waste Disposal Act, amended in 1976 by the Resources Conservation and Recovery Act, requires solid and hazardous wastes to be managed in accordance with regulations to be adopted by the EPA. Since specific methods for defining hazardous wastes are still being established, the full impact of this government regulation is not yet known.

Present and impending United States government regulations on water discharge require the installation of water-treatment plants for smelter effluents carrying significant levels of heavy metals as well as discharges from the concentrator tailings ponds. It is anticipated that by 1985 all copper concentrators, smelters, and refineries in the United States will be practicing maximum water recycle. Such action will require new approaches to internal water treatment and possible modification of flotation-reagent systems to compensate for buildup in circulating loads of organic and inorganic constituents.

Since total recycle of water from dump-leaching operation is the current practice, actions should be restricted to whatever steps are necessary to keep leach solutions from entering groundwater systems. Uncontrolled runoff from mining operations requires impoundment with appropriate treatment before release into surface systems.

In 1977 energy represented ca 10% of the cost of copper production; yet in the total picture, the copper industry was not a major consumer of energy, using only ca 0.5% of the total energy used in the United States industrial sector. Production of 1 t of copper consumes ca 130 GJ (123×10^6 Btu) distributed among mining (20%), concentrating (40%), smelting (30%), and refining (10%).

The United States is largely self-sufficient with respect to copper, meeting any shortfall by imports from Canada, Chile, and Peru.

The limits of toxicological tolerance for copper in humans are so high that no problems arise in its use for parasite control on crops for which copper salts have been so used for more than 50 years.

Uses

The properties of copper and its alloys that make it a major metal of commerce may be summarized as follows: high electrical conductivity; high thermal conductivity; ease of casting; extrusion, rolling, and drawing to produce wire, tubing, and strip; low corrosion rate of copper when used for food preparation; excellent alloying characteristics; high aesthetic appeal; and low toxicity to humans.

Copper has been employed as a bactericide, molluscicide, and fungicide for a long time and is of importance in the control of schistosomiasis. In this case its addition to lake water acts as an efficient deterrent to

Table 1.   Comparison of the Properties of Pure Copper with Those of Silver and Gold

Property	Copper	Silver	Gold
atomic weight	63.54	107.87	196.97
oxidation states	1, 2, 3	1, 2, 3	1, 2, 3
standard electrode potential, 25°C, V	Cu/Cu^+ = 0.520 Cu/Cu^{2+} = 0.337	Ag/Ag^+ = 0.799	Au/Au^+ = 1.692
density, g/cm^3	8.96	10.49	19.32
crystal structure	fcc	fcc	fcc
{metallic radius, nm	0.1276	0.1442	0.1439
ionic radius, (M +), nm	0.096	0.126	0.137
covalent radius, nm	0.138	0.153	0.150
electronegativity	2.43	2.30	2.88
thermal conductivity, W/(m·K)	394	427	289
electrical resistivity, 20°C, μΩ/cm	1.6730	1.59	2.35
melting point, °C	1083	960.8	1063
heat of fusion, J/kg[a]	212×10^3	102×10^3	67.4×10^3
boiling point, °C	2595	2212	2970
specific heat at 20°C, J/(kg/°C)[a]	384	233	131 (18°C)
linear coefficient of expansion ×10^6 per °C at 20°C	16.5	10.68	14.2
modulus of elasticity, for hard drawn metal, MPa[b]	$10.2{-}12 \times 10^4$	7.75×10^4	7.85×10^4

[a] To convert J to cal, divide by 4.184.
[b] To convert MPa to psi, multiply by 145.

transmittal of the disease by elimination of snails that act as host for the responsible parasite.

Medical uses range from a copper intrauterine device for contraceptive purposes (see Contraceptive drugs) to copper drugs in cancer therapy. Copper sulfate is a powerful emetic and has been used clinically as such in the treatment of intoxications.

The main use of copper is as an electrical conductor, and about 50% of United States demand is for electrical uses.

The resistance to saltwater corrosion of admiralty brass, an alloy with 71% copper, 28% zinc, and 0.75–1.0% impurities, led to extensive use of this alloy in ships and condenser tubing. Since 1963, there has been a steady replacement of admiralty brass by copper–nickel alloys. The resistance to corrosion of copper by food and the nontoxicity of copper in dilute concentrations has encouraged its use for food-preparation equipment. Low corrosion rates coupled with ease of forming, bending, and of soldering resulted in extensive use of copper tubing for domestic water pipe. Copper radiators for automobiles use the high thermal conductivity of copper, and the ease of mechanical working and brazing.

Brass can be cast readily into intricate shapes and is used for many cast products having utilitarian or decorative applications. The ease of chrome plating brass has diversified the use of copper alloys where appearance and resistance to corrosion are major requirements.

W.M. TUDDENHAM
P.A. DOUGALL
Kennecott Copper Corporation

A. Butts, ed., *Copper—The Science and Technology of the Metal, Its Alloys and Compounds*, ACS Monograph 122, Reinhold Publishing Corp., New York, 1954.

J.W. Laist in M.C. Sneed, J.L. Maynard, and R.C. Brasted, eds., *Comprehensive Inorganic Chemistry*, Vol. 2, D. Van Nostrand Company, Inc., New York, 1954, p. 9.

Metals Handbook, 9th ed., Vol. 2, American Society of Metals, Metals Park, Ohio, 1979, pp. 726–732.

COPPER ALLOYS

WROUGHT COPPER AND WROUGHT COPPER ALLOYS

Copper and its alloys form a group of materials of major commercial importance. The general applications of these materials are influenced by properties such as electrical and thermal conductivity, both hot and cold formability, machinability, the ease with which they may be joined by welding, brazing, and soldering, fatigue characteristics, generally good to excellent corrosion resistance, and the fact that the materials are nonmagnetic.

A significant percentage of copper produced is used as the basis of engineering alloys, a variety of high copper content alloys, which include a number of precipitation or age-hardenable alloys, the brasses or the copper–zinc series with or without other designated alloying elements, nickel silvers, cupro-nickels, and the bronzes (copper alloys in which the major alloying element is at least one other than zinc or nickel).

Copper Classifications of Important Alloys

Electrolytic tough pitch (ETP) (the exact state or quality of well reduced and refined copper) has long been the standard type of copper used in the manufacture of rod, wire, sheet, and strip products. Electrolytic copper is refined by electrolytic deposition as a cathode, whereas fire-refined copper uses a furnace process. ETP copper contains small amounts of O_2 present as cuprous oxide.

Tough-pitch copper, deoxidized copper, and oxygen-free copper relate to the characteristic of the copper during melt preparation and casting (see also Copper). Deoxidized copper is produced by introducing an element that reacts with oxygen (CuO) present in a molten bath of copper (copper for deoxidation is usually in the tough-pitch condition). Up to 0.10% of phosphorus does not significantly impair the mechanical properties, but the electrical and thermal properties are seriously affected. Oxygen-free high conductivity copper (OFHC, produced by AMAX) is

not a deoxidized copper, but it is made by melting select copper cathodes in a furnace constructed to exclude air and using a protective gas atmosphere over the molten copper; the copper is protected from any oxidizing condition during melting, casting, and solidification.

Properties

The coppers are soft with relatively low tensile strength and high elongation, and are capable of high reduction in area by cold-working processes, even in the as-cast condition. They are not greatly increased in hardness by a given reduction in cross-sectional area by cold-working as are alloys of lower copper content. Owing to the initial softness and low rate of hardening under deformation, copper is rated as having a high working capacity. It is well suited to forming and shaping by both hot and cold metal-working operations (see Copper). The low rate of work hardening is probably accounted for by copper's fcc crystallization, with plastic deformation taking place by a shearing action along certain planes within a crystal: in the fcc lattices of copper, the (111) planes slip most readily, and the direction of movement is also the direction corresponding to the greatest atomic density.

Copper has no allotropic modifications and thus no critical temperature.

Annealing (recovery, recrystallization, and softening).   Cold-work distorts the copper structure, and results in increased hardness and strength, and decreased ductility and electrical conductivity. Heating the cold-worked material gradually restores these properties to values characteristic of soft or annealed copper. Changes take place at a temperature that more nearly accompanies the appearance of recrystallization by rearrangement of the dislocations into a lower energy level without a very marked decrease in the total number of dislocations (the process is sometimes called polygonization). Severe cold deformation leads to an annealing phenomenon in which stacking faults seem to assume an important function. Recrystallization takes place at a temperature somewhat higher than that of recovery, which results from the rearrangement of atoms of the solid in an entirely new set of crystals.

The recrystallization and grain-growth characteristics of the coppers depend upon the degree of deformation or the residual strain energy in the crystal lattice, temperature, and the heating rate. The electrolytic copper exhibits the lowest recrystallization temperature for a given condition of prior cold-work and heating characteristics.

Silver is added to copper to increase its resistance to softening at elevated temperature and improves its creep characteristic compared to ETP copper.

Electrical and thermal properties.   Copper is second only to silver in its electrical and thermal conductivity. Electrical conductivity is expressed in terms of mass resistivity, namely the resistivity of a wire one meter long weighing one gram (percentage of the International Annealed Copper Standard, IACS); this standard resistance is 0.15328 $(\Omega \cdot g)/m^2$ (875.2 $(\Omega \cdot lb)/mi^2$) at 20°C. High purity fire-refined, chemically refined, OFHC, ETP, and deoxidized low residual phosphorus (DLP) copper are all known as high conductivity coppers. Deoxidized copper having high residual phosphorus (0.02–0.04%) DHP is known as low conductivity copper (80–90% IACS). Electrical conductivity of annealed copper is not affected by grain size, annealing temperature, or crystallographic orientation. Cold-work decreases electrical conductivity only slightly. Elements held in solid solution in copper have a marked effect on electrical conductivity, particularly when used in quantities added by intent for alloying. Small concentrations of impurity elements usually have a significant detrimental effect. The presence of oxygen along with the impurity element often forms compounds with the impurity, removing it from the solid solution and thereby nullifying its effect on the conductivity. The electrical conductivity is not affected by elements that are insoluble in copper or form insoluble intermetallic compounds; it is increased by heat treatment (aging—see Precipitation-Hardening Copper Alloys) which causes precipitation of elements or intermetallic compounds from solid solution; and it is decreased if elements present are in solid solution with copper (especially at conc > 0.01 wt%). The thermal conductivity of copper with an electrical conductivity of 100% IACS is 391 W/(m·K) at 20°C.

Oxidation.   Surface oxidation: When heated in air, Cu develops a cuprous oxide film that exhibits a succession of interference tints, eg, dark brown (film thickness = 37–38 nm) to red (film thickness = 124–126 nm). Black cupric oxide forms over the cuprous oxide layer as film thickness increases beyond the interference color range. At −183°C, copper rapidly reacts with a small amount of oxygen to a thickness of about 2.4 nm which may represent a monolayer of absorbed oxygen; after this, further oxidation is very small.

Internal oxidation: Dilute copper alloys of base metals that are of relatively high solubility and diffusivity for atomic oxygen are subject to internal oxidation (eg, dilute copper alloys containing Al, Zn, Cd, Be, P, Ni, and Si). The oxygen diffusion into copper is dependent upon a second element such as a low concentration of silicon or phosphorus; absolutely pure copper cannot be internally oxidized because oxygen then reacts with the copper atoms to produce a surface layer of copper oxide.

Directional properties. Mechanical directionality (fibering) usually results from the attenuation, in the rolling direction, of second phases such as cuprous oxide or of the grains alone.

The preferred orientation developed by cold-work frequently changes on recrystallization to a very different preferred orientation or directionality.

The most prominent recrystallization texture is the "cubic" texture, found in most face-centered cubic metals that have been annealed following heavy-rolling reductions. The cubic axes in this texture are very well aligned along the strip axes, and its behavior approaches that of a single crystal, hence the directional properties.

Effects of other elements in copper. Impurities: Some of the elements present as impurities in certain instances are added on other occasions as alloying elements, eg, silver, tellurium, and lead. Arsenic increases tensile strength slightly in the cold-worked condition, raises the recrystallization temperature, and improves the working properties of oxygen-bearing copper. Antimony increases tensile strength and endurance limit, and fatigue properties are improved as compared with plain arsenited copper. Bismuth promotes hot shortness, tends to make the material brittle, and cannot be tolerated in even small amounts. Sulfur, selenium, and tellurium increase the ease with which copper may be machined when they are added in quantities up to 1%. These elements do not induce hot- or cold-shortness, and have little effect upon tensile strength although there may be some loss in ductility. All three form compounds with copper. Sulfur significantly affects the recrystallization temperature at low concentrations (up to 10 ppm, the effect decreasing at higher concentrations). Selenium affects the softening of copper. As opposed to lead, tellurium copper is relatively free from fire-cracking and lends itself well to hot-forging operations. Lead is added to copper to be fabricated into rod by the hot-extrusion process to improve the machinability of the basic copper and by increasing the brittleness of the alloy. The presence of oxygen with low lead concentrations restrains hot shortness in hot rod rolling. Iron present in quantities up to 2% in a series of copper–iron alloys both strengthens and hardens copper without significantly reducing ductility. Cobalt's effects are similar to those of iron. Nickel generally slightly hardens copper and increases its strength without reducing its ductility; it has a striking effect upon the color. In concentrations up to 0.05%, nickel has no effect upon recrystallization temperature. The commercially important cupro-nickels are three copper-base nickel alloys containing 10, 20, and 30% nickel dissolved in copper with small amounts of manganese and iron to enhance casting qualities and corrosion resistance. Carbon, sulfur, and phosphorus as impurities in these alloys can have a marked effect upon fabricating and welding characteristics. Silver is one of the most common impurities found in copper; it does not significantly modify the usual mechanical properties of copper, but can significantly increase the softening temperature.

Residual deoxidizing elements include phosphorus, lithium, silicon, boron, and calcium, as well as strontium, aluminum, magnesium, zinc, barium, and carbon. Phosphorus forms copper-rich solid solutions with copper, and has a strong effect upon electrical conductivity in the absence of cuprous oxide. It increases both the tensile and fatigue strength without decreasing ductility significantly. The two commercially important phosphorus-deoxidized coppers available are DLP, high conductivity low residual phosphorus copper, and DHP, low conductiv-

ity high residual phosphorus copper. Silicon in low residual concentrations has little effect upon mechanical properties, but small additions have a pronounced effect on electrical conductivity in oxygen-free copper, effects which are nullified in the presence of cuprous oxide. Although there are a number of silicon bronze alloys available, they differ principally in the silicon content and the nature of the third element. Silicon is a very potent hardener (addition of 3% Si practically doubles the tensile strength of copper).

Controlled alloying elements include aluminum, arsenic, beryllium, cadmium, cobalt, chromium, lead, nickel, oxygen, phosphorus, silicon, silver, sulfur, tellurium, tin, zinc, and zirconium. Oxygen in the form of cuprous oxide is always present in tough-pitch coppers, and under ordinary conditions does not interfere with the use of this material. All copper alloys that contain appreciable amounts of cuprous oxide are susceptible to embrittlement, damage caused by heating under reducing conditions (especially if hydrogen is the reducing agent). Oxygen-free or deoxidized copper is not susceptible upon heating in a hydrogen atmosphere unless it has first been heated in an oxidizing atmosphere. Foreign oxides embedded in copper are reduced by hydrogen at elevated temperatures and a type of embrittlement can result from this cause alone. Tin has a significant effect upon electrical conductivity of oxygen-free copper. Tin bronzes are alloys of copper and tin (commercial wrought bronzes usually contain 11% or less of tin). It is common to deoxidize the bronze melts with phosphorus (phosphor bronzes). In added concentrations of up to 2%, tin produces relatively high strength and high electrical conductivity alloys. Cadmium increases copper's strength and has minimal influence upon electrical conductivity. Added in even small amounts, Cd significantly increases the softening temperature of Cu. Aluminum and copper alloys of commercial importance are the single-phase alpha solid-solution alloys, and the alpha–beta alloys. The commercially important aluminum bronzes usually contain 4–10% Al with or without other metals. See Table 1 for the mechanical properties of aluminum bronzes. Lead is deliberately added to copper to be fabricated into rod by hot-rolled extrusion; it increases the brittleness of the alloy, allowing it to break off readily into short chips when the material is machined. Additions of iron, manganese, silicon, or nickel are made to increase strength and hardness. Zinc is added to form the brasses, the industrially important series containing from a few percent to about 40% zinc.

Other properties of wrought copper alloys. The modulus of elasticity is generally quite similar for almost all copper alloys, ranging from 96 GPa (to convert to psi multiply by 145,000) for leaded brasses (3% Pb 58–62% Cu) to 131 GPa for beryllium copper (98.1% Cu, 1.9% Be). The modulus of rigidity (shear) is generally expressed as 0.38 times the modulus of elasticity in tension. Maximum shear stresses of wrought copper alloys in torsion or simple shear are related to the tensile strength of the material involved. Endurance (fatigue) failure is fracture that occurs as a result of many applications of a stress which, if applied only once, would not cause failure. The fatigue limit is not related to the ductility, is not affected by time, and the number of applications for a given stress is not affected by the frequency of applied stress. Alloying does have an effect upon this property: hard-rolled or drawn silicon copper and the phosphor bronzes have increased fatigue resistance as compared to the copper–zinc alloys. Beryllium copper has excellent fatigue resistance. Magnetic properties of copper alloys, which are generally considered to be nonmagnetic, are exhibited under certain conditions most often associated with the iron concentration of the alloys and the

Table 1. Mechanical Properties of Aluminum Bronzes

Name	UNS designation	Temper	Tensile strength, MPa[c]	Yield strength, MPa[c]	Elongation %, 5 cm
aluminum bronze 5%	C 60800	soft	413.4	186	55
aluminum bronze D	C 61400	soft	523.7–565	227.4–310.1	45–40
	C 61900	soft	634	338	60
		hard[a]	931	829[b]	2
		spring[a]	999.2	931[b]	3

[a] After heat treatment for 1 h at 232°C.
[b] 0.2% Offset.
[c] To convert MPa to psi, multiply by 145.

mode of its occurrence (eg, in solid solution, magnetic or nonmagnetic form) (see Magnetic materials).

Precipitation-Hardening Copper Alloys

The sole requirement for precipitation or age hardening is the characteristic of decreasing solid solubility of some phase, element, or intermetallic compound with decreasing temperature. Precipitation is a decomposition of a solid solution leading to a new phase of different composition: namely, the precipitation of the new phase from the solid solution that comprises the matrix. In such alloy systems, cooling at an appropriately rapid rate (quenching) from a temperature well within the all-alpha field will preserve the alloy as a single solid solution possessing relatively low hardness, strength, and electrical conductivity. A second heat treatment (aging) at a lower temperature will cause precipitation of the unstable phase; this is usually accompanied by an increase in strength, hardness, and electrical conductivity. In some alloy combinations, only a very slight increase in hardness follows aging, but often the increase is significant. In addition to composition and heat-treating conditions, the properties obtained in certain alloys are significantly affected by cold-working after solution treatment. Regardless of the degree of hardness obtained after aging, any effective precipitation invariably results in some increase in electrical conductivity.

Copper–Zinc Alloy Brasses

Copper and zinc melted together in various proportions produce one of the most useful groups of alloys, the brasses. Six distinct phases are formed in the complete range of possible compositions. Alloys having less than ca 58% Cu are limited in commercial application. Those containing > 58% are characterized by high ductility and malleability, good strength, pleasing color, and excellent resistance to most corrosive media. More complex alloy systems containing various other elements in addition to zinc are produced commercially. As the copper content decreases from 100%, the brasses vary in color from rich bronze (90% Cu, 10% Zn), to golden (85% Cu, 15% Zn), to yellow (70% Cu, 30% Zn), to reddish-yellow (60% Cu, 40% Zn). All commercially important brass alloys are cold-workable, and may be hot-worked at certain optimum temperatures.

Two broad classes of copper zinc alloys are commercially important: those containing 65–99% Cu consist of a single phase (alpha brasses), and those containing ca 55–64% Cu contain two phases (alpha–beta brasses). Properties vary with condition or temper and composition, and can be made to vary with the degree of deformation, the deformation temperature, and the time and temperature of annealing. Cold-working has a marked effect upon electrical conductivity (conductivity of compositions 60–100% Cu decrease regularly; compositions 90–100% Cu show slight changes in conductivity; more abrupt changes occur for 65–90% Cu. Other properties such as color and susceptibility to stress corrosion can change rapidly in the 80–90% copper range. Toughness is considered to be high for the entire nonleaded-brass series, increasing with zinc content to a maximum value of 30% Zn. Alloys containing only small amounts of zinc (< ca 2%) are sometimes deoxidized with other elements, eg, phosphorus. Elements added intentionally to improve special properties include lead, tellurium, tin, iron, manganese, phosphorus, aluminum, and nickel. For example, architectural bronze contains Cu 57 wt%, Zn 40.0, lead 3.0; free-cutting brass contains Cu 62 wt%, Zn 34.75, lead 3.25; nickel-leaded commercial bronze contains Cu 90.25 wt%, Zn 6.9, lead 1.75, and 1.0 Ni, 0.10 P; manganese bronze A contains Cu 58.5 wt%, Zn 39.25%, 1 tin, 1 iron, 0.25 manganese; inhibited admiralty type D brass contains Cu 71.5 wt%, Zn 27, phosphorus 0.04; low-tin commercial bronzes contain Cu 90 wt%, Zn 9.5, tin 0.5%; and nickel–silver 65–10 contains Cu 65 wt%, Zn 25, and Ni 10.

Various impurities may be encountered in alloys, such as bismuth, antimony, arsenic, phosphorus, lead, tellurium, manganese, tin, aluminum, nickel, and iron.

Corrosion of Copper and Copper-Base Alloys

Copper itself possesses very high resistance to the effects of the atmosphere, fresh water and seawater, alkaline solutions (except those containing ammonium ion), and many organic chemicals. The behavior of copper in acids depends mainly on the severity of existing oxidizing conditions. Sulfur and sulfides attack copper and high copper alloys vigorously. Brasses with high zinc content offer much better resistance toward sulfur-bearing environments (see Corrosion and corrosion inhibitors).

Many of the alloying elements added to copper improve corrosion resistance of the parent metal and enhance its mechanical properties. Extensive successful application of copper in many service areas has included the following broad classifications: (1) atmospheric exposures, eg, roofing, grillwork; (2) fresh water supply lines; (3) seawater applications; (4) heat exchangers and condensers; and (5) industrial and chemical plant equipment involving exposure to a wide variety of conditions. Copper roofing, in addition to forming a protection from the elements, develops a pleasing green patina consisting essentially of green basic copper sulfate.

Corrosion is classified by the appearance of the corroded material, eg, general thinning, galvanic corrosion, pitting, impingement, fretting, dezincification, parting, and intercrystalline stress-corrosion cracking (caused by ammonia and ammonium compounds or mercury and mercury compounds). Stress, velocity, effects of galvanic coupling, concentration cells, effects of surface condition, and contamination of the surrounding media (eg, fouling by marine organisms), all influence the rate of corrosion reactions. If damage results from pitting, intergranular corrosion, or selective leaching as in dezincification, a measure of corrosion rate based on change in weight may be misleading; measurement in terms of loss in mechanical strength is more meaningful.

Effect of alloy composition. The copper–zinc series (with zinc content as high as 42%) are the most widely used group of copper alloys. The resistance of these alloys to corrosion by aqueous solutions does not change significantly until the zinc content exceeds about 15%, then dezincification can occur. Alloys containing higher zinc concentrations tend to exhibit stress-corrosion cracking, but resist corrosion due to exposure to sulfur compounds. Tin added to certain brasses significantly increases the resistance toward acidic media as well as the tendency to dezincify. Copper alloys containing 5–12% aluminum have excellent resistance to impingement attack and to high temperature oxidation. Aluminum additions to brasses, as well as to copper, tend to form self-healing films on the surface of the alloy, and are fairly resistant toward acid conditions while dissolving readily in alkalies. Aluminum brass is susceptible to dezincification unless inhibited by the addition of a small amount (0.05%) of arsenic. The additions of very small amounts of phosphorus, arsenic, and/or antimony to the alpha-structured brass alloys containing more than 2.0% zinc eliminates or greatly reduces the incidence of damage by dezincification. The phosphor bronze alloys are comparable in respect to the coppers in general corrosion resistance; tin does improve the resistance to most nonoxidized acids except hydrochloric, and the higher tin alloys have high impingement-attack resistance. Silicon enhances resistance in acidic environments. The 30% cupro-nickel has the best general corrosion resistance of the three alloys of copper with nickel additions. The nickel silvers have good resistance to corrosion by fresh and salt water.

Protective coatings. The relatively high general resistance of copper and its alloys to corrosion in many kinds of environments is a result of the protective reaction products developed by converting the surface of the metal or alloy into one or more compounds. The application of various protective coatings, either metallic or organic (see Coatings, industrial; Paint; Metallic coatings) may enhance the natural resistance of the materials. Metallic coatings may make the electropotential relationship of coating to base metal important, especially at uncoated or cut edges.

Behavior of Copper Alloys in Various Media

Copper and copper alloys resist corrosion by industrial, marine, and rural atmospheres, the most widely used copper materials for these environments being copper, commercial bronze, red brass, architectural bronze, and the nickel silvers. Copper has a high resistance to corrosion by soils, the highest rate of corrosion occurring in sulfide soils (rifle peat with sulfide). Copper is used extensively for handling freshwater; the greatest single application of copper tubing is in hot and cold distribution lines. Copper and copper alloys resist attack by pure steam (qv), but if much carbon dioxide, oxygen, or ammonia is present, the condensate becomes corrosive. The cupro-nickels are preferred for high temperatures and pressures. Copper is less resistant than the inhibited admiralty alloys, aluminum brass, or cupro-nickel alloys for saltwater environments. In general, copper alloys are used in contact with nonoxidizing acids (eg, acetic, sulfuric, and phosphoric) if the concentration of oxidizing agents such as dissolved oxygen (air) or ferric or bichromate ions is low. Hydrochloric acid is one of the most corrosive of the nonoxidizing acids in contact with copper alloys, but is successfully handled in dilute solutions (see Hydrogen chloride). Hydrofluoric acid is less corrosive and can be handled by 30% cupro-nickel (see Fluorine compounds, inorganic). Copper and alloys can be used in commercial processes involving exposure to acetic acid (qv) and related chemical compounds. Oleic and stearic acids attack copper alloys to form metallic soaps (see Carboxylic acids). Copper and its alloys resist alkaline solutions, except those containing ammonium hydroxide or compounds that hydrolyze to ammonium hydroxide or cyanides; they also resist a range of organic compounds. The chloride salts are the most corrosive, and copper and copper alloys corrode rapidly in contact with all but dilute solutions of oxidizing salts. Sulfur compounds react freely to form CuS at varying reaction rates. CO_2 and CO in dry form are usually inert to Cu and its alloys; SO_2 attacks copper; moist hydrogen sulfide gas reacts to form copper sulfide. The halogens are noncorrosive in the dry state, but aggressive when moisture is present. Hydrogen can diffuse into copper and react with oxides to form steam, damaging the grain boundaries of the metal, when oxygen-bearing copper is heating in hydrogen. Copper and copper alloy tubes convey oxygen at room temperature, but scaling results at high temperatures.

Alloy Selection

Generally, the members of each group in Table 2 behave similarly in the same corrosive media.

Table 2. Alloy Classification According to Composition

Group 1 coppers oxygen free	Group 3 alloys containing less than 80% copper
electrolytic tough-pitch	low brass 80%
deoxidized low phosphorus	cartridge brass 70%
deoxidized high phosphorus	yellow brass 65%
silver bearing	Muntz metal 60%
cadmium bearing	67 Cu, 32.5 Zn, 0.5 Pb
tellurium	64 Cu, 35.5 Zn, 0.5 Pb
leaded	63 Cu, 37 Zn
sulfur	62 Cu, 35.5 Zn, 2.5 Pb
zirconium	free cutting brass
arsenical	60 Cu, 0.5 Pb nickel silver
beryllium	60 Cu, 2.0 Pb nickel silver
chromium	57 Cu, 3.0 Pb nickel silver
phosphor bronze, 1.25% Sn	naval brass
	manganese brass
Group 2 alloys containing more than 80% copper	**Group 4 special brasses**
gilding 95%	inhibited brasses
commercial bronze	inhibited aluminum brass
87 Cu, 13 Zn	inhibited admiralty brass
red brass 85%	
leaded commercial bronze	
Group 5 phosphor bronzes	**Group 8 cupro–nickel alloys**
95 Cu, 92 Cu, 90 Cu, 5 Sn, 8 Sn, 10 Sn	cupro–nickel (10% Ni)
88 Cu, 444 alloy 4 Sn, 4 Pb, 4 Zn	cupro–nickel (20% Ni)
88 Cu, 6 Sn, 1.5 Pb, 4.5 Zn	cupro–nickel (30% Ni)
87 Cu, 8 Sn, 1.0 Pb, 4.0 Zn	
85 Cu, 5 Sn, 9 Pb, 1 Zn	**Group 9 nickel–silvers**
80 Cu, 10 Sn, 10 Pb	55% Cu, 18% Ni
83 Cu, 7 Sn, 7 Pb, 3 Zn	65% Cu, 18% Ni
	65% Cu, 12% Ni
Group 6 aluminum bronze	
7 Al, 3 Si	
7 Al, 2 Fe	
5 Al	
Group 7 copper–silicon alloys	
3 Si	
1.5 Si	

RALPH E. RICKSECKER
Chase Brass & Copper Co., Inc.

Metals Handbook, American Society for Metals, Metals Park, Ohio.

COPPER ALLOYS, CAST

The chemical compositions of casting alloys have greater latitudes than those of comparable wrought alloys; however, the chemistry of the heat-treatable alloys must be carefully controlled to attain the proper hardening effects, and compositions that do not lend themselves to forming operations are available in cast form for specific applications.

Properties and Characteristics

Table 1 contains some of the physical properties of the various cast copper alloys.

There are two main categories of cast copper alloys. (1) *Single phase alloys*, which are characterized by moderate strength (eg, for the copper–tin–lead alloys, tensile strength may range from 234–303 MPa (34,000–44,000 psi); yield strength ranges from 103–152 MPa (15,000–22,000 psi), high ductility (with the exception of the leaded varieties), moderate hardness and good impact strength. Most of the alloys have a tensile strength two or three times greater than the yield strength. Included in this category are the copper–tin–lead–zinc alloys, the tin–bronzes, and the leaded tin–bronzes. (2) *Polyphase alloys* are characterized by high strength (tensile strength may be 415–825 MPa (60,000–120,000 psi); yield strength may be 170–585 MPa (25,000–85,000 psi), moderate ductility, and moderate impact strength. These have a wide range of properties resulting from variations within structure. The desired balance of ductility and strength can be obtained in age-hardenable alloys, eg, beryllium copper, by controlling the amount of precipitate. The mechanical properties of the manganese–bronzes are dependent upon the relative amounts of alpha and beta phases present and their distribution.

Stress and stress relieving. Nonuniform cooling leads to unbalanced residual stress patterns in the casting which can lead to cracking if exposed to environments containing ammonia, ammoniacal compounds, or mercury. Stress-corrosion cracking does not distinguish between residual or applied stresses. Stress relieving to safe stress levels is achieved by stress-relief annealing, a thermal treatment, effective for alloys such as high copper alloys, red brass, semi-red brass, yellow brass, manganese–

Table 1. Properties of Selected Cast Copper Alloys

Alloy, UNS designation	Melting point, °C	Pouring temperature, °C		Sp gr[a] (density, lb/in³)
		Light castings	Heavy castings	
copper, C 80100	1064–1083	1149–1204	1110–1166	8.94
leaded red brass,				
C 83600	854–1010	1149–1288	1066–1177	8.7–8.9
C 83800	843–1004	1149–1266	1066–1177	8.6–8.7
leaded semi-red brass,				
C 84400	843–1060	1149–1260	1066–1177	8.6–8.8
C 84800	832–954	1149–1260	1066–1177	8.55–8.69
copper-nickel 90 : 10,				
C 96200	1099–1149	1316–1427	1204–1316	8.94
chromium–copper,				
C 81500	1075–1085	1149–1204	1110–1165	8.83
beryllium–copper,				
C 81700	1029–1068	1132–1188	1093–1149	8.75
silicon–brass,				
C 87400	821–916	1093–1177	1038–1066	8.30–8.44
C 87500	821–917	1093–1177	1038–1066	8.30–8.44
copper-nickel 70 : 30,				
C 96400	1171–1238	1371–1483	1316–1399	8.94
leaded nickel-brass, nickel silver 12%				
C 97300	1010–1040	1204–1316	1193–1204	8.8–8.91

[a] To convert sp gr to density (lb/in³), divide by 27.68.

bronze, silicon–bronze, tin–bronze, nickel–bronze and aluminum–bronze alloys.

Machinability. Cast copper alloys are rated in the same general manner as wrought copper alloys. The leaded alloys are considered free-machining. The polyphase alloys with a second phase generally harder than the matrix can be brittle and experience chip breakage. This includes leaded tin–bronze, silicon–bronze, high tin–bronze, aluminum–bronze, and manganese bronze. The most difficult alloys to machine are the high-strength manganese and aluminum–bronze high in iron or nickel content.

Electrical and thermal conductivities. Electrical conductivity is customarily expressed as a percentage of the International Annealed Copper Standards (IACS). Elements in solid solution with copper often have a marked effect on both the electrical and thermal conductivity of the alloy. Alloying elements present in significant concentrations and low concentrations of deoxidized elements decrease both properties. The wrought alloys have higher conductivities than comparable casting alloys because of composition differences. The thermal conductivity of copper having an electrical conductivity of 100% IACS is 391 W/(m·K) at 20°C.

Bearing and wear properties. Bearing alloys containing copper are placed in three groups. (*1*) The phosphor–bronze alloys, containing Cu, Sn or Cu, Sn, Pb, have a residual phosphorus concentration of a few hundredths to 1%; these are high wear resistance, high hardness, moderately high strength alloys. (*2*) The copper–tin–lead alloys are softer, and used as bearing material for lighter loads (below 5.5 MPa or 800 psi) moving at moderate speeds. (*3*) The aluminum–bronze alloys, containing 8–9% Al, are widely used for brushing and bearing in light or high speed applications; alloys containing 11% Al are used for heavy service, eg, valve guides, wear plates. The addition of manganese and silicon increase strength and hardness of the aluminum–bronze alloys (see Bearing materials).

Joining may be accomplished by welding (qv), brazing, and soldering techniques with varying degrees of ease and success (see Solders and brazing alloys).

Mechanical properties. Most alloys containing tin, lead, or zinc have moderate tensile and yield strength, and high elongation. Higher tensile or yield strengths are available through the use of aluminum– and manganese–bronzes, silicon brasses and bronzes, and some nickel alloys. Some alloys are heat-treatable for maximum tensile strength. The entire group of leaded alloys is easiest to cut and machine. Bearing and wear-resistant alloys are phosphor–bronze (Cu–Sn) alloys; copper–tin–lead (low zinc) alloys, which have high resistance to wear, high hardness, and moderately high strength; and manganese–, aluminum–, and silicon–bronze alloys, with the high strength manganese–bronze alloys having high tensile strength, hardness, and resistance to shock.

Production Methods

Sand casting is the most popular, and the cheapest method with a low pattern cost; it achieves the greatest dimensonal tolerances. *Permanent mold casting* is expensive, uses a metal mold and, optionally, a core. It is generally limited to tin–, aluminum–, silicon–, and manganese–bronzes, and to the yellow brasses. *Die casting* is a special form of permanent mold-casting used when a difficult cored section of relatively large surface area compared to weight is required. Yellow brass is the largest-volume alloy to be die cast. *Plaster casting* is a refinement of sand molding, using a match-plate system, cope and drag assembly, cores, and gravity pouring. The process can be largely automatic and continuous and is best suited for alloys such as the aluminum–bronzes, yellow brass, manganese bronze, low nickel bronze, and silicon bronze. Lead in the alloy reacts with plaster to discolor castings. *Centrifugal casting* consists of pouring the melt into a rotating mold which may be in either the horizontal or vertical position. Mechanical properties of centrifugally cast alloys vary with the alloy and the mold material. *Precision investment casting* (*cire perdue*, or *lost wax* process) is a special process for making small castings to close dimensional tolerances by using a pattern of materials such as wax, plastics, a fusible alloy or frozen mercury, and a suitable molding or investment compound (eg, ethyl silicate) cast around the pattern; this is then cured, and the invested pattern is melted out to form the finished mold. *Shell molding* employs synthetic resin to make shells that can be used with or without cores. *Continuous casting methods* are being applied to production with molten metal continuously poured into the top of the water-cooled lubricated mold, and the solid cast shape continuously withdrawn mechanically from the bottom of the mold; the process is useful where soundness and high volume of parts are needed.

Alloy Identification and Chemistry

Effect of various alloying elements. Mechanical properties are a function of alloying elements and their concentration. For example, zinc is added to copper (5–40%) to form the alloy series known as the brasses; zinc increases the tensile strength at a significant rate up to the concentration of ca 20%. The cast-copper zinc alloys are described as red brasses and leaded red brasses, semi-red, silicon, yellow, and high strength yellow brasses. Tin is added to copper (5–20%) to form the tin–bronze series, and leaded tin–bronze. Tin imparts strength and hardness to copper-base alloys, enhancing toughness and wear resistance. Lead, insoluble in the copper-base alloys, is added (up to 40%); because of its low melting point, lead is found distributed in the grain boundaries of the casting. Up to 1.5% concentrations significantly improve machinability. It is undesirable in high-strength manganese–bronze, silicon–bronze, and silicon–brass. Silicon alloys are high strength and tough, with improved corrosion resistance. In the leaded tin–bronze alloys, silicon is a very harmful impurity leading to lead sweat and unsoundness. Aluminum is added to form the aluminum–bronzes, a series of high strength alloys. Present in leaded tin–bronzes, Al promotes unsoundness. Iron adds strength to silicon–, aluminum–, and manganese–bronzes. Phosphorus is used principally as a deoxidizer; the alloy should contain a minimum residual of 0.02% phosphorus. Boron is a commercial deoxidizer. Manganese is added in high strength brasses where it forms compounds with other elements. Nickel is added to whiten the resulting alloy; the cupro–nickel alloys have high corrosion resistance. Nickel in some of the high tin gear bronze alloys enhances wear. Beryllium and chromium both form age- or precipitation-hardenable alloys; the beryllium alloys are the strongest of all known copper-base alloys (see Age-Hardenable Alloys under Copper alloys, wrought). Arsenic, antimony, and phosphorus can be added to inhibit dezincification in the brass alloys.

Health and Safety

Smoke or fumes of certain of the metals formed during melting or pouring can be hazardous to health. EPA, NIOSH and OSHA have set limits on quantities of particulate, eg, zinc, copper, lead, iron, aluminum, manganesium, and silicon, thought permissible in casting shop atmosphere and also in exhaust from collection systems.

Uses

Copper alloy castings are used for their generally superior corrosion resistance, high electrical and thermal conductivities, and good bearing and wear qualities. Some of the alloys are heat-treatable and couple high strength with good electrical and thermal conductivity. The copper and high copper alloys are used as electrical and thermal conductors. The red brass alloys are widely used in fittings, valves, and general hardware. The tin–bronzes are used in values and fittings, bearings and bushings. The manganese–bronze alloys are used in marine fittings, in free-machining, bearings, and brushings. The aluminum–bronze alloys are used in acid resistant pumps, marine equipment, valve guides, and corrosion-resistant parts. The silicon–bronze and silicon–brass alloys are used in bearings, corrosion-resistant casting, propellers, and bearings. The copper–nickel and leaded nickel bronze and brass alloys are used in corrosion-resistant marine applications and hardware.

RALPH E. RICKSECKER
Chase Brass & Copper Co., Inc.

Cast Metals Handbook, American Foundrymen's Society, Des Plaines, Ill., 1957.

G.J. Cook, *Engineered Castings*, McGraw-Hill Book Co., New York, 1961.

COPPER COMPOUNDS

The electronic configuration of copper is $1s^2 2s^2 2p^6 3s^2 3p^6 3d^{10} 4s$ but, along with its compounds, it is usually discussed with the transition metals.

Occurrence

Copper is widely distributed throughout the earth, primarily in sulfidic minerals, eg, chalcopyrite ($CuFeS_2$), and bornite (Cu_5FeS_4), chalcocite (Cu_2S), and covellite (CuS); in deposits of basic carbonates, eg, malachite ($Cu_2CO_3(OH)_2$), azurite ($Cu_3(CO_3)_2$), and the oxide cuprite (Cu_2O); in the silicate chrysocolla ($CuSiO_3.2H_2O$); and in ocean manganese nodule deposits.

Properties

Copper displays four oxidation states: copper(I) and copper(II) compounds are most common and industrially important, eg, acetates, arsenic compounds such as copper(II) acetoarsenite ($3Cu(AsO_2)_2Cu(C_2H_3COO)_2$), copper(II) chromate ($CuCrO_4$), copper(II) ferrate ($CuFe_2O_3$), copper(II) gluconate ($Cu(C_6H_{11}O_7)_2$), the halogen compounds such as copper(I) chloride ($CuCl$), widely used in industry, the hydrides, the oxides and the hydroxides. Copper(II) sulfate, $CuSO_4$, as the pentahydrate, $CuSO_4.5H_2O$, is the most important copper salt in terms of the amount produced. Copper(III) compounds are few but possibly becoming more significant. Several compounds of copper(0) have been claimed.

The metallic state of copper is preferred. The dissolution of metallic copper requires a mild oxidizing agent or a strong complexing agent that stabilizes one of the positive-valent ions (eg, nitric acid dissolves copper with the evolution of nitric oxide). Sulfuric acid dissolves copper only under oxidizing conditions. Cyanide forms very strong complexes with copper(I) and hence dissolves copper in aqueous solution with the evolution of hydrogen. The copper(I) ions are unstable in aqueous solution, and tend to disproportionate to copper(II) ions and copper metal unless a stabilizing ligand is present. The only copper(I) compounds that are stable in water are the insoluble ones, eg, $Cu_2S.CuCN$, and the copper(I) halides. Copper(III) compounds are difficult to prepare and are only of academic interest. As in the case of Cu(I), the Cu(III) species can be stabilized by the presence of certain complexing agents, eg, potassium persulfate or deprotonated amides, eg, tetraglycine. The easy oxidation of Cu(II) to the Cu(III) state with deprotonated amides is particularly important to biochemists for it allows a two-electron oxidation step from Cu(I) to Cu(III) in some enzymatic reactions, hitherto explained by a free-radical mechanism.

Copper(II) sulfate occurs in the form of blue triclinic crystals (blue vitriol), found in the mineral chalcanthite. The anhydrous salt also occurs in the mineral hydrocyanate. Large quantities are used as fungicides, algicides, as a source of copper in animal nutritions, and as a fertilizer. It is a primary source from which other copper compounds can be derived.

Uses

Copper is one of the trace metals essential to life, over 18 cuproteins participating in various metabolic cycles. Those particularly important are the copper-bearing enzymes which participate in oxygen transfer, such as mammalian dopamine hydroxylase and galactose oxidase. Naturally occurring sulfidic copper deposits serve as a main source of copper for plant growth. In large quantities, copper can be poisonous, especially to lower life forms such as fungi. Copper salts are used as fungicides (qv) (eg, the Bordeaux mixture Paris green) (see Insect control technology).

Copper compounds are used as catalysts, herbicides, fungicides, pigments, in antifouling paints, for cloud seeding, as corrosion inhibitors, in electrolysis and electroplating, in fabrics and textiles, in flameproofing, in fuel-oil treatment, in glass, ceramics and cement, in medical, food and drug applications, in metallurgy, in mining and metals, in organic reactions as catalysts, in the paper industries, in pyrotechnics, and in wood preservation.

ROGER N. KUST
Kennecott Copper Corporation

M. Goldstein and co-workers, *Fed. Proc.* **24**, 604 (1965).

J. Peisach, P. Aisen, and W.E. Blumberg, eds., *Biochemistry of Copper*, Academic Press, Inc., New York, 1966.

CORK

Cork is one of the few naturally grown closed-cell foams that has never been duplicated with synthetic material; it is the outer bark of the cork oak, *Quercus suber*, which grows mainly near the Mediterranean Sea on the Iberian Peninsula and along the shores of North Africa. The cork oak is unique in that the outer bark of cork, or the phellen, can be stripped from most of the tree trunk and major limbs once the tree reaches maturity (20–25 yr) without harming the tree.

Cork is nonfibrous, composed of tiny, closely packed cells that are tetrakaidecahedral, or 14-sided; six of the faces are quadrilateral and eight are hexagonal. This shape provides for the optimum packing of the cells without extra void space, and accounts for cork's excellent gasketing and flotation characteristics. The cells range from 0.025–0.050 mm in the longest dimension, the cell walls and resinous binder being highly resistant to water, to most organic liquids, and to all but strong acid and alkali solutions.

Composition

Cork is comprised of, in wt%: fatty acids, 30; miscellaneous organic 17; lignin, 16; other acids, 13; ceroids, 10; tannins, 4; glycerol, 4; cellulose, 3; and inorganic ash, 3.

Uses

Most of the cork harvested from trees ends up in varying types of cork composition. The uses of these range from gasket and packing applications, to insulation, rocket-engine casing, and fuel tanks (see Ablative materials; Packing materials).

R.H. HUTCHINSON
Armstrong Cork Co.

E. Palmgren, *Cork Production and International Cork Trade*, International Institute of Agriculture, F.A.O., 1947.

J. Marcos de Lanuga, *The Cork of Quercus Suber*, Inst. For. Invest. Exper. Madrid **35**, 82 (1964).

CORROSION. See Corrosion and corrosion inhibitors.

CORROSION AND CORROSION INHIBITORS

Corrosion is defined as the destructive attack of a metal by the environment, by chemical or by electrochemical processes (as contrasted with mechanical means such as the erosion of a metal structure by sand in the desert). Corrosion includes conjoint mechanical and chemical action to produce early failure of a load-carrying metal structure, eg, stress-corrosion cracking.

Manifestations of Corrosion

Most of the observed corrosion occurs as one or more of the following five cases: (1) uniform attack, the most common manifestation, in which the entire metal surface is covered with the corrosion product, eg, the rusting of iron in any humid atmosphere, or the tarnishing of silver and copper in sulfur-containing environments; (2) pitting corrosion, in which a pit, or number of pits, may cause considerable damage to a metal structure, or may penetrate the structure without the metal exhibiting any appreciable loss in weight (it is usually seen on metals that are normally passivated with an oxide film, eg, aluminum alloys, or stainless steels of the nickel-chromium type); (3) stress-corrosion cracking, in which a metal part exposed simultaneously to a constant tensile stress and specific corroding agent will crack intergranularly (between metal grains) or transgranularly (across metal grains); when stress is cyclic rather than constant, the failure is termed corrosion fatigue; (4) intergranular attack, often occurring during (3) but also as a result of different potentials between grains and grain boundaries; (5) dezincification, in which certain alloys, such as those containing a reactive metal (Zn) and a more noble metal (Cu), dezincify upon exposure to a particularly corrosive environment; in gold and silver alloys this is called "parting". There are also some special forms of corrosion, such as the dissolution of a metal or alloy in a hot fused-salt bath.

Origin of Corrosion

Most commonly, corrosion originates in the basic thermodynamic tendency for metal to react as expressed in terms of the free energy of reaction (see Thermodynamics). In addition, there are factors that accelerate the corrosion rate.

The free energy of formation of a compound is the free energy for the reaction of the elemental metal with another chemical species also in its elemental state. A negative free energy of formation indicates a tendency for the metal to react; that is, the oxide is stable. A positive free energy of formation indicates that the elemental metal is stable.

Some free energies of formation of compounds important in corrosion are (ΔG^0, kJ/mol): $AlCl_3$ -636.8; $Al_2O_3 \cdot H_2O$ -1820.0; CdO -225.1; CaO -604.2; CuO -127.2; Cu_2O -146.4; Au_2O_3 $+163.2$; Fe_2O_3 -1014.2; $Fe(OH)_2$ -483.5; $Fe(OH)_3$ -694.5; MgO -569.6; NiO -216.7; Ag_2O -10.8; ZnO -318.2; $Zn(OH)_2$ -554.8; $ZnCO_3$ -731.4; Cr_2O_3 -1046.8 (J/4.184 = cal).

Although a large free energy of formation indicates that the formation of the compound is favored thermodynamically, the specific reaction may not go readily because of a high activation energy for the reaction, eg, as in the corrosion of lithium metal by air and its component gases.

A measure of the reaction tendency (free energy) is the electrode potential. A negative potential indicates the strong tendency for the metal to oxidize (corrode). The standard electrode potential refers to the metal in a solution containing its ions at unit activity; but it is altered by the environment (special tables of electrode potentials are required to measure the proper galvanic relationships existing between metals and alloys in a specific environment, eg, seawater); and by polarization or the alteration of the potential as the current flows. In the absence of these complicating effects, the electromotive force of the galvanic cell can be calculated from the expression $\Delta G = -nFE^0$ (ΔG is the free energy change for the reaction in kJ (cal), E^0 is the emf of the cell in volts, n is the number of electrons involved in the oxidation, and F the Faraday, 96.49 kJ/V (23.061 kcal/V)). Imposed on this basic thermodynamic tendency for metals to seek their lowest energy level are other sources that lead to accelerated corrosion, eg, galvanic corrosion, often labeled dissimilar metal corrosion. Corrosion can also be accelerated by differential salt or oxygen concentration cells; by differential temperature cells; by the conjoint action of mechanical forces with chemical ones (eg, fretting corrosion, which arises from the relative slippage of surfaces in contact with each other by vibration); and by corrosion fatigue (a metal is stressed in a cyclic manner in the presence of a corrosion environment).

Thermodynamic basis. The thermodynamic data pertinent to the corrosion of metals in aqueous solutions include the potential, the pH dependence of the metal, metal oxide, and metal hydroxide reactions, and complex ions. This data is illustrated with a Pourbaix diagram in which, along with specific data for a given metal, the potential and pH dependence of the hydrogen and oxygen reaction are superimposed.

Electrochemical basis. Corrosion is an electrochemical process rather than a strictly chemical reaction (see also Electrochemical processing). A corroding metal surface is comprised of a large number of local anodes and a large number of local cathodes whose sites may actually shift as the corrosion reaction ensues. Under certain environmental conditions, notably in the absence of chloride ion, some metals can be anodically protected. If the metal is potentiostatically maintained in the proper potential range, it will resist corrosion indefinitely.

Environmental effects. The important environmental factors are the oxygen concentration in the water of the atmosphere, the pH of the electrolyte or the temperature, and the concentrations of various salts in solution in contact with the metal. Chloride ions, ubiquitous in nature,

play an important role in corrosion of metals. This role in iron corrosion is theoretically analyzed into oxide film properties, absorption, field effect, catalytic effects, and complex formation. Very often the environment is reflected in the composition of corrosion products, eg, the composition of the green patina formed on copper roofs over a period of years.

Metallurgical Factors

These factors include crystallography, grain size and shape, grain heterogeneity, second phases, impurity inclusions, and residual stress owing to cold work. For example, stainless steels undergo sensitization, a condition caused by the precipitation of chromium-rich carbides in the grain boundaries, giving rise to chromium-depleted grain-boundary media. Copper alloys respond to residual stresses left from forming which can cause stress-corrosion cracking; also, the presence of the beta phase can lead to dezincification (see Copper and copper alloys). Wrought high strength aluminum alloys tend to be highly textured which causes the grains of the primary alpha phase to be flattened and elongated, a metallurgical texture that can promote exfoliation corrosion (intergranular corrosion leading to the leafing-off of uncorroded grain bodies). Nickel alloys sensitize similarly to austenitic stainless steels, becoming vulnerable to intergranular corrosion.

Stress-Corrosion Cracking (SCC)

This is a fracturing process that affects alloys, but not pure metals. It is caused by the conjoint action of corrosion and tensile stress (eg, design operating stress, residual stresses from welding, heat treatment, fit-up, cold-forming, or combinations of these). The path of SCC may be either inter- or transgranular and the damage caused can be grossly out of proportion to the amount of corrosion that has occurred (eg, SCC can initiate fatigue cracks in aircraft fuselage panels, initiate brittle fracture of high strength alloy components, or perforate a condenser tube wall permitting cooling water to contaminate a boiler). Theories of SCC are either mechano-electrochemical, film rupture, embrittlement, adsorption, or periodic electrochemical–mechanical, and many types of macroscopic specimens are currently used to evaluate SCC in various systems (eg, smooth specimens such as Brinell impression, Erichsen cup, U-bend, beams, tensile or C-ring). In addition, two fundamentally different tests also characterize the SCC response of alloys: one employs a precracked specimen, and the other is known as the constant extension rate test (or constant strain rate test, or slow strain rate test).

Mitigation (by alloy families). Ammoniacal SCC is the most common form of SCC in copper-base alloys. In general, alloys must be selected with the minimum susceptibility to SCC; in addition, one must avoid designs that permit water to accumulate in contact with aluminum, avoid conditions in which salts can concentrate in contact with the aluminum, and use an alloy clad with an anodic coating.

Corrosion-Resistant Materials

At the lower end of the alloying scale are the low alloy steels (iron-base alloys) in which corrosion resistance is based on the protective nature of the surface film, which in turn, is based on the physical and chemical properties of the oxide film. All of the stainless steels, and several copper alloys, offer exceptional improvement in all sorts of atmospheric conditions; they depend for their corrosion resistance on the formation of a passive film, and are thus susceptible to pitting.

Corrosion Inhibitors

These are materials that interact with the metal surface to prevent corrosion.

Inorganic inhibitors. These inhibitors are subclassified into passivators, those that can function without oxygen, eg, chromate and nitrate, and those that require oxygen, eg, sodium phosphates, silicates, and borates. Inhibitors may also be classified in terms of their mechanisms; those that function by influencing the anodic side of the electrochemical corrosion cell, eg, chromates, nitrites, silicates, phosphates, and borates, and the cathodic inhibitors, eg, calcium polyphosphate. To inhibit corrosion in cooling waters, polyphosphates, nitrites, and chromates have been used. Concentrations required for inhibition depend on the presence or absence of chloride ion, the temperature, and the movement of the corroding solution. Usually, the effective concentration for inorganic inhibition falls in the range of several hundred ppm. With respect to the oxidizing anions, there exists a critical concentration.

Organic inhibitor compounds. These inhibitors must be adsorbed, but the type of adsorption bond varies with the electrochemical configuration of the molecule. The main types involve electrostatic adsorption (eg, aniline, pyridine, and benzoic acid); chemisorption (eg, nitrogen or sulfur heterocycles, benzotriazole and butylamine); and π-bond (delocalized electron) adsorption. The concentration generally needed is substantially higher than that required for the inorganic inhibitors such as the chromates.

Vapor-phase inhibitors. These are volatile compounds containing one or more functional groups capable of inhibiting corrosion. To be effective, the inhibitor must contain certain functional groups, have a vapor pressure above a minimum value, and be adsorbed on the metal surface. Some classes of successful compounds include the amine salts with nitrous or chromic acids; with carbonic, carbamic, acetic and substituted or unsubstituted benzoic acids; organic esters of nitrous, phthalic, or carbonic acids; aliphatic amines; cycloaliphatic and aromatic amines; polymethyleneamines; mixtures of nitrites with urea; nitrobenzene. Dicyclohexylamine nitrate has been used commercially for many years. A large number of commercial inhibitors are available.

Coatings for Protection Against Corrosion

These are characterized by the temperature at which they are applied, (eg, hot dip coatings using molten metal, zinc or aluminum); whether or not they require electrical current for deposition, (eg, cadmium and zinc deposits or nickel electroplating); or whether or not they convert the original surface metal to another chemical compound involving the same metal (eg, phosphate, chromate, and sprayed coatings) (see Coating processes; Coatings).

Corrosion Testing

Corrosion tests are either laboratory tests, which are usually intended to be accelerated tests, or field and service tests, which are designed to reproduce actual conditions.

R.T. FOLEY
B.F. BROWN
The American University

B.F. Brown, *Stress Corrosion Cracking Control Measures*, NBS Monograph 156, National Bureau of Standards, 1977.

C.C. Nathan, *Corrosion Inhibitors*, National Association of Corrosion Engineers, Houston, Texas, 1973.

W.H. Ailor, *Handbook on Corrosion Testing and Evaluation*, John Wiley & Sons, Inc., New York, 1971.

CORUNDUM. See Abrasives; Aluminum compounds.

COSMETICS

Cosmetics are the product of cosmetic chemistry, a science that combines the skills of specialists in chemistry, physics, biology and medicine. The following discussions deal with product groups individually.

Cosmetic Emulsions

Cosmetic lotions and creams are emulsions of water-based and oil-based phases. An emulsion is a two-phase system consisting of two incompletely miscible liquids, the internal, or discontinuous, phase dispersed as finite globules in the other. Special designations have been devised for oil and water emulsions to indicate which is the dispersed and which the continuous phase. Oil-in-water (o/w) emulsions have oil as the dispersed phase in water as the continuous phase. In water-in-oil (w/o) emulsions, water is dispersed in oil, which is the external (continuous) phase (see Emulsions).

Emulsifiers can be classified as ionic or nonionic according to their behavior. An ionic emulsifier is composed of an organic lipophilic group (L) and a hydrophilic group (H). The hydrophilic–lipophilic balance (HLB) is often used to characterize emulsifiers and related surfactant materials. The ionic types may be further divided into anionic and cationic, depending upon the nature of the ion-active group. The lipophilic portion of the molecule is usually considered to be the surface-active portion. Nonionic emulsifiers are completely covalent and show no apparent tendency to ionize. Emulsifiers, being surface-active agents, lower surface and interfacial tensions and increase the tendency of their solution to spread.

O/w emulsifiers include P.E.G. 300 distearate (nonionic) and triethanolamine stearate (anionic). W/o emulsifiers produce emulsions in which the continuous phase is lipophilic in character and the lipophilic portion of the molecules is predominant, eg, lanolin alcohols (nonionic) and propylene glycol monostearate S/E (anionic).

Cosmetic Creams

Materials used in creams may be prepared in o/w or w/o emulsions. The aesthetic effect and degree of emolliency depend to a great extent on the emulsion type as well as on the emulsion composition. O/w emulsions produce a cooling effect on application to the skin owing to water evaporation. W/o emulsions do not produce this effect since water evaporation is slowed by the film of the oil in the continuous phase.

Cold-cream emulsions are w/o, the emulsifier is sodium cerotate formed by reaction of borax and free cerotic acid in beeswax. If the water content is raised to approximately 45% or more, the composition changes to an o/w emulsion. Vanishing cream can be considered to be an emulsion of a free fatty acid (usually stearic acid) in a nonalkaline medium.

Cosmetic Lotions

The oils and waxes in lotions are identical to those of an emollient cream but they are present in lower concentration.

Deodorants and Antiperspirants

Deodorant and antiperspirant products are marketed as aerosols, creams, gels, lotions, powders, soaps, and sticks. Aluminum chloride, once commonly used in deodorants and antiperspirants, may cause fabric damage and skin irritation because of its low pH. This has led to the development of various basic aluminum compounds, eg, aluminum chlorohydroxide (ACH) and basic aluminum hydroxychloride–zirconyl hydroxy oxychloride.

Sunscreens

Sunscreens are of two types: physical, eg, titanium dioxide and zinc oxide, which are opaque materials that block and scatter light, thus acting as mechanical barriers and acting nonselectively over all wavelengths; and chemical, which act by absorbing uv light, offering selective protection against certain uv wave bands, depending on their absorption spectrum, eg, anthranilates, cinnamates, benzyl and homomenthyl salicylate, and *p*-aminobenzoic acid (PABA) and its ester derivatives.

Make-Up Preparations

Color. Straight colorants are referred to in the trade as primary colors (principally red, blue, and yellow which may be combined to produce practically every hue), primary dyes, and primaries. Straight colorants include both primaries and lakes of such primary colorants made by extending the primary colorant on a substratum such as zinc oxide, or a mixture of aluminum hydroxide and barium sulfate.

Inorganic colorants, except for the white pigments such as titanium dioxide, zinc oxide and talcum, are used for their excellent lightfastness and complete insolubility in solvents and aqueous solutions.

Face powder. Face powder, both the loose and the now-popular cake or pressed powder, is a blend of white pigments and inorganic colorants, principally iron oxides. The function of a face powder is to impart a smooth, matte finish to the skin by masking any shine owing to the secretions of the sebaceous and sweat glands.

Lipstick. Lipstick is a solid fatty-base product containing dissolved and suspended colorant materials. A good lipstick must possess a certain maximum and minimum of thixotropy, ie, it must soften enough to yield a smooth, even application with the minimum of pressure. It should not be necessary to apply a lipstick more often than every 4 to 6 h. The applied film, no matter how thin, should to some extent be impervious to the mild abrasion encountered during eating, drinking, or smoking, and the lipstick should not bleed, streak or feather into the surrounding tissue of the mouth. The colorants used in lipsticks are either insoluble lakes or oil-soluble dyes of the eosin group, or both.

The bromo acids, bromo derivatives of fluorescein, are used to produce indelibility in the applied film. Two are dibromo- and tetrabromo-fluorescein. Dibromofluorescein produces a yellow–red color; the tetra compound gives a more purple stain. From 1–5% of these dyes is used in lipsticks; but ordinarily, 2 or 3% is adequate.

Mascara. Mascara ingredients are strictly limited by law because of the proximity of these products to the eye. The colorants are limited to natural dyes and inorganic and carbon pigments.

Eye shadows. Colors are carbon blacks, ultramarine blue, and the various yellows, browns, and reds of the iron oxide pigments. The base into which the colorants are ground usually consists of beeswax, ozokerite, mineral oil, lanolin, and petrolatum.

Nail products. Nail lacquer, or nail polish, usually consists of a resin (nitrocellulose or dinitrocellulose, also known as pyroxylin), plasticizer (eg, high molecular weight esters as well as castor oil), solvents (preferably mixtures of low and medium boiling point alcohols, aromatic hydrocarbons, and aliphatic hydrocarbons), and pigment. Lacquer removers may be made from any of a number of solvents, eg, any of the solvents used in the formulation of enamels can be used such as acetone, the fastest acting. Cuticle removers and cuticle softeners usually consist of a dilute solution of alkali in water with some glycerol or other humectant added.

Hair Preparations

Shampoos. Shampoos usually contain a primary detergent which can be a fatty alcohol sulfate, ether sulfate, sarcosinate, or one of many other anionics (see Hair preparations; Surfactants). Soap shampoos are primarily aqueous solutions of soft soap combined with preservatives, sequestrants, color and perfume. Soapless shampoos are primarily aqueous solutions of sulfonated oils. It has been reported that mild aqueous acids cause antiswelling action on the cuticle scales of the hair. As the cuticle tightens, the hair gains luster because light is more efficiently reflected from the surface of the hair shaft. In the absence of external charges, hair shows its highest strength and resiliency at pH 4.0–6.0. Combinations of triethanolamine and sodium and ammonium lauryl sulfates with amine oxides can be adjusted to low pH with good stability. Amphoteric shampoos are based on the use of imidazoline, betaine, and sulfobetaine surfactants, which are generally assumed to be less irritating to the eyes than other detergent or soap preparations. Shampoos to counteract dandruff, the product of hyperkeratinization, contain ingredients that allow a normal turnover rate of epidermal cells, eg, coal tar, quaternary ammonium compounds, resorcinol, salicylic acid, selenium sulfide, sulfur, undecylenic acid and derivatives, and zinc pyrithione (zinc 2-mercaptopyridine N-oxide salt).

Depilatories. Depilatories are chemical products used for the removal of unwanted hair. The depilatory most widely used today is based on calcium thioglycolate in a strongly alkaline medium, pH ca 12.3.

Bath Products

Bath oils. Bath oils are either those that spread or float (usually an anhydrous system that is oleaginous and floats on the surface of the water), or those that disperse, usually containing a sufficient concentration of surface-active agents to solubilize the oleaginous components through micellar solubilization, or to disperse them as the internal phase of an oil-in-water emulsion. The spreading oils action is facilitated by the inclusion of small amounts of an oil-soluble surfactant. The oleaginous composition may consist of one or more of the commonly accepted hydrophobic components; natural vegetable oils, animal-derived lipids such as lanolin and mink oil; low fatty alcohols such as oleyl and hexadecyl, low melting synthetic glycerides such as glyceryl mono-oleate, and glyceryl monolaurate, and mineral oils. The dispersible bath oil loses some of its oleaginous character as the result of oil-in-water dispersibility, leaving a layer of dilute oil in a water dispersion on the skin. As the water evaporates from this, a hydrophobic residue remains, providing a moisture barrier on the skin surface. Foaming bath oils and foam baths are based on suitable surfactants, either liquid or powder, which are blended with color, perfume, and foam stabilizers. The addition of warm or hot running water produces a fragrant foam.

Soaps

Soaps can be made from natural fats or from the isolated fatty acids. The natural fats, glycerides of fatty acids, are used to make soap where cost is a consideration (see Soap).

Transparent soaps are prepared with sugar, alcohol, and glycerol. About half the soap market is claimed by deodorant soaps, containing either Irgasan DP-300 (Triclosan) or 3,4,4'-trichlorocarbanilide (Triclocarban).

Shaving Preparations

Lather shave creams are soft soaps formulated with potassium hydroxide salts of saturated and unsaturated fatty acids, and containing a small percentage of sodium hydroxide to regulate viscosity, body, and consistency.

Brushless shave creams are similar to vanishing creams, an excess of free fatty acid, usually stearic, emulsified into a nonalkaline base of suitable viscosity or consistency. Their softening effect is not so rapid as that obtained with a lather shave.

Gel shave creams are emollient aqueous gels that are soap based and contain a volatile solvent such as pentane. Expelled as a soft gel from a pressure container, the shave cream is rubbed on the face, and the solvent is released, expanding as it volatilizes and the gel becoming foam.

Health and Safety

Preliminary animal tests are performed to determine whether the material has an acute or chronic toxicity, and if so, the MLD_{50}. In human safety trials of topically applied materials, the principal concern is induced cutaneous reactions. If, however, animal studies indicate significant absorption of components of the formulation, clinical trials must also monitor such absorption by testing blood and urine which may reflect subclinical systemic toxic effects. The cutaneous reactions are of three types: primary irritation; contact sensitization; and reactions that develop following exposure to the sun. The patch test is used to measure the potential of topically applied materials to reduce these reactions.

The CTFA has set up a Cosmetic Ingredient Review (CIR) to evaluate the safety of cosmetic ingredients.

HARRY ISACOFF
International Flavors and Fragrances, Inc.

F.E. Wall, *Origin and Development of Cosmetic Science and Technology*, Interscience Publishers, New York, 1957.

P. Carter *Am. Perfume Arom. Doc. Ed.* **1**, 233 (1960).

G. Barnett, *Emollient Creams and Lotions*, Vol. 1, Wiley-Interscience, New York, 1972, pp. 27–104.

M.G. de Navarre, ed., *Chemistry and Manufacture of Cosmetics*, 2nd ed., Continental Press, Orlando, Fla.

E. Jungermann, *J. Cosmet. Toiletries* **91**, 50 (July 1976).

COTTON

Although the actual origin of cotton is still unknown, there is evidence that it existed in Egypt as early as 3000 BC. Cotton is the most important vegetable fiber used in spinning (see Fibers, vegetable). Its origin, breeding, morphology, and chemistry have been described in innumerable publications. A member of the Malvaceae or mallow family, a plant of the genus *Gossypium*, it is widely grown in warm climates the world over.

The average cotton plant is an herbaceous shrub having a normal height of 1.2–1.8 m, with maximum height for some varieties of 4.5–6.0 m. The most important species in the genus *Gossypium* are *hirsutum*, *barbadense*, *arboreum*, and *herbaceum*. The most favorable growing conditions include a warm climate (17–27°C mean temperature) where fairly moist and loamy, rather than rich, soil is an important factor; seeds planted in dry soil produce fine and strong, but short, fibers of irregular lengths and shapes. Under normal climatic conditions, seeds germinate in 7–10 d, flower buds (known as squares) appearing in 35–45 d followed by open flowers 21–25 d later. The mature cotton boll opens 45–90 d after flowering. Within the boll are 3–5 divisions called locks, each of which normally has 7–9 seeds that are covered with both lint and fuzz fibers (Fig. 1).

Morphology

Cotton is essentially 95% cellulose; noncellulosic materials, mostly waxes, pectinaceous substances and nitrogenous matter, located in the primary wall with small amounts in the lumen. Typical composition (% of dry weight) is cellulose, 94.0; protein (% N × 6.25), 1.3; pectic substances, 1.2; ash, 1.2; wax, 0.6; total sugars, 0.3; pigment, trace; and others, 1.4.

The cotton fiber is tapered for a short length at the tip and along its entire length is twisted frequently; the direction of twist reversing occasionally. These twists, referred to as convolutions, are important in spinning because they contribute to the natural interlocking of fibers in a yarn (Fig. 2).

Physical Properties

Classing of cotton involves describing the quality of cotton in terms of grade and staple length in accordance with the official cotton standards of the United States. About 95% of cotton produced in U.S. is classed by grade and staple length, often supplemented by the inclusion of a Micronaire reading to indicate the fineness.

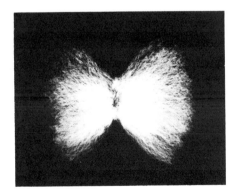

Figure 1. Cotton butterfly with lint and fuzzy fibers.

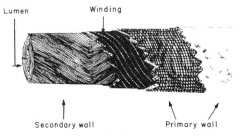

Figure 2. Schematic diagram for cotton fiber.

Length is the most important dimension in the cotton fiber, measured in a representative sample of fiber without regard to quality, or value, at 65% rh and 21°C. Diameter, an inherited characteristic greatly influenced by soil and weather, is best observed in cross sections of bundles of fine, medium, or coarse fibers. Whereas the typical shape resembles that of the kidney bean, the shapes range from circular to elliptical to linear. Diameter may also be derived from wall thickness, lumen axes, or combinations of these. Fineness, referred to also as linear density, is defined as a relative measure of size, diameter, linear density, or weight per unit length expressed as micrograms per inch (μg/in.) [0.394 μg/cm]. Strength is attributed to the cellulose (qv) and to the molecular chain length and orientation of the cellulose (see Biopolymers). Directly associated with strength is the degree of maturity.

Chemical Properties

Because the cotton fiber has such a high cellulose content, the chemical properties are essentially those of the cellulose polymer. The cellulose polymer contains chains of varying lengths. Like other forms of cellulose, the molecular chains of cotton cellulose consist of anhydroglucose units joined by 1–4 linkages (see Carbohydrates). These glycosidic linkages characterize the cotton cellulose as a polysaccharide, and are cleaved during hydrolysis, acetolysis, or oxidation resulting in shorter chains. If degradation is extensive enough, cellobiose or glucose derivatives are produced. In addition to the length of cotton-cellulose chains and to the chain-length distribution, the degree of accessibility of the long, thread-like, polymeric molecules of cellulose within the elementary fibril of cotton characterizes cotton cellulose.

Reactions. One of the earliest known modifications of cotton was mercerization. The process traditionally employed a cold concentrated sodium hydroxide treatment of yarn or woven fabric, followed by washing and a mild acetic acid neutralization with the fabric maintained under tension during the entire procedure to ensure the expected properties. The resultant mercerized cotton of commerce has improved luster and dyeability, and to a lesser extent, improved strength. A recent variation of this procedure substitutes hot sodium hydroxide that is allowed to cool while the cotton remains immersed in the caustic solution. If the cotton is allowed to shrink freely during contact with mercerizing caustic, slack mercerization takes place; this technique produces a product with greatly increased stretch (stretch cotton) which has application in both medical and apparel fields. Effects similar to those from sodium hydroxide mercerization have been produced by exposure of the cotton to volatile primary amines or to ammonia.

The two most common classes of cotton cellulose reactions are esterification and etherification of the cotton cellulose hydroxyls, as well as addition reactions with certain unsaturated compounds to produce a cellulose ether. These reactions result in modification of the cotton to impart special properties. By far the most important commercial modifications of cotton cellulose occur through etherification. Commercial modification of cotton to impart durable-press, smooth drying, or shrinkage-resistance properties involves cross-linking adjacent cellulose chains through amidomethyl ether linkages using reagents such as di- or polyfunctional amidomethylol compounds (see Amino resins). Cross-linking proceeds via either Lewis or Brønsted catalysis, the effects being increased resiliency, manifested in wrinkle resistance, smooth drying properties, and greater shape-holding properties. Base-catalyzed reactions of cotton cellulose with either mono- or diepoxides to form cellulose ethers also result in fabrics with increased resiliency.

Health and Safety

Byssinosis, also called mill fever or brown lung disease, is a pulmonary ailment similar to bagassosis (see Bagasse) or silicosis that is alleged to develop on repeated inhalation of cotton dust. OSHA has issued standards for control of cotton dust.

Uses

Besides its obvious value to the textile industry in the form of cloth and clothing, cotton is used for filters, bandages, as a growth medium, and in other applications where a cellulose fiber can be utilized. The seed is crushed and cottonseed oil is extracted for use in food processes.

B.A. KOTTES ANDREWS
INES V. DE GRUY
United States Department of Agriculture

W.L. Balls, *The Development and Properties of Raw Cotton*, A. & C. Black, Ltd., London, UK, 1915.

C.B. Purves, in E. Ott, H.M. Spurlin, and M.W. Graefflin, eds., *Cellulose and Cellulose Derivatives*, 2nd ed., Pt. 1, Interscience Publishers, Inc., New York, 1954, pp. 29–53.

H.B. Brown and J.O. Ware, *Cotton*, 3rd ed., McGraw-Hill Book Co., New York, 1958.

COTTONSEED.   See Vegetable oils.

COUMARIN

Coumarin, 2*H*-1-benzopyran-2-one, $C_9H_6O_2$ (1), is the odoriferous principle of the tonka bean (*Dipteryx odorata*), sweet clover (*Melilotus officinalis* and *alba*), and woodruff (*Asperula odorata*). Widely distributed in the plant kingdom, coumarin occurs in oil of lavender, oil of cassia, citrus oils, balsam Peru, and in some 60 other species of plants. In many cases the plants are odorless because coumarin occurs as a complex combined with sugars and acids; the resultant glucoside compounds can be split by acids, by natural enzymatic action or by uv irradiation. Coumarin is the parent substance of a very large group of derivatives, many of which occur naturally and some of which are of economic importance.

(1)

Physical Properties

Coumarin is usually sold in the form of colorless shiny leaflets or rhombic prisms. Mp 70.6°C; bp 303°C; it begins to sublime at 100°C; it is soluble to the extent of 0.25 g/100 mL in water (at 25°C), and to the extent of 47.00 g/100 mL in 70% ethanol at 40°C.

Chemical Properties

The lactone is easily hydrolyzed by alkali to the corresponding salt of coumarinic acid, *cis-o*-hydroxycinnamic acid, eg, by treatment with dilute sodium hydroxide or by boiling with potassium carbonate. Fusion of coumarin with caustic alkali yields salicylic and acetic acids. It combines readily with sodium bisulfite solutions to form stable, soluble sodium 3- or 4-hydrosulfonates. Halogenation occurs with treatment in chloroform at room temperature with bromine, yields coumarin-3,4-dibromide or 3-bromo- and 3,6-dicromocoumarin. Sulfonation occurs at water-bath temperature to give coumarin-6-sulfonic acid, and at 150°C to give 3,6-disulfonic acid. Nitration forms mainly 6-nitrocoumarin and 8-nitrocoumarin. Hydrogenation with an active Raney catalyst yields 3,4-dihydrocoumarin and continued hydrogenation yields the saturated *octahydrocoumarin*. Reduction occurs by sodium amalgam, yielding melilotic acid. Coumarin is oxidized with difficulty and is stable to chromic acid, but oxidative biochemical attack occurs frequently at the 7-position. Prolonged exposure to sunlight, or uv irradiation, converts coumarin to a dimer (mp 263°C).

Methods of Preparation

The practical synthetic methods for coumarin use as starting materials either salicylaldehyde (or its equivalent in the form of side-chain chlorinated *o*-cresol) or phenol, from which the pyrone ring is elaborated.

Synthesis from salicylaldehyde proceeds via: the extensively used Perkin reaction, an example of Claisen condensation, in which the sodium salt of salicylaldehyde, or sodium acetate, is heated with acetic anhydride to yield coumarin; the Knoevenagel reaction, a condensation reaction in which the acetic anhydride of the Perkin reaction is replaced by acetic acid derivatives possessing an activated methylene group, and organic amines are used as catalysts. This reaction is used primarily in the synthesis of many substituted coumarins; and the Raschig method, which is based on hydrolysis of the benzal chloride intermediate formed in the dichlorination of the methyl group of esters of *o*-cresol. Processes based on chlorinated *o*-cresol esters have been used to a considerable extent in the commercial production of coumarin.

Synthesis from phenol proceeds via the Pechmann condensation, in which phenols are heated with malic acid in the presence of sulfuric acid, and CO is eliminated. Beta-keto esters in place of malic acid yield coumarins substituted in the pyrone ring. Maleic and fumaric acids can be used in place of malic acid; the next step is vinylation with acrylic esters, which involves the vinylation of the phenol group with methyl acrylate in acidic medium and in the presence of air.

Health and Safety

Since 1954, coumarin has been classed by the FDA as a toxic substance; foods containing coumarin are regarded as adulterated.

Uses

Coumarin is widely used in the perfumery, cosmetic and related industries, and in electroplating (qv) industries.

Derivatives

A few of the derivatives are of economic importance. Some are physiologically active, eg, the alkyl, hydroxy, and methoxy derivatives as well as more complicated condensed systems. A large number of synthetic derivatives are described, some of which are of commercial significance.

Derivatives include: 3,4-dihydrocoumarin, which has a GRAS status and is thus permitted in imitation flavors; 6-methylcoumarin, which also has a GRAS status; umbelliferone (7-hydroxycoumarin), an important coumarin metabolite used in cosmetics and in fluorescent brighteners; 4-hydroxycoumarin, a metabolite occurring in spoiled hay. Dicumarol, a product derived from it, causes sweet clover hemorrhagic disease in cattle. The synthetic version has been introduced as an anticoagulant drug in the therapy of thrombic diseases (see Blood coagulants and anticoagulants). Warfarin is a related synthetic compound with hemorrhagic properties, as is Phenprocoumon. Both are used extensively as rodenticides (see Poisons, economic).

WALTER C. MEULY
Rhodia Inc.

Fed. Reg. **19**, 1239 (Mar. 5, 1954).

A. Seidell, *Solubilities of Organic Compounds*, Vol. 2, Van Nostrand Co., Inc., New York, 1941, p. 623.

COUMARONE INDENE RESINS.   See Hydrocarbon resins.

CRACKING.   See Petroleum.

CREAM.   See Milk products.

CRESIDINE, 2-METHOXY-5-METHYLANILINE.   See Amines—Aromatic amines—Aniline and its derivatives.

CRESOLS. See Alkylphenols.

CRESYLIC ACIDS, SYNTHETIC. See Alkylphenols.

CROTONALDEHYDE

Crotonaldehyde (2-butenal), $CH_3CH{=}CHCHO$, is a water-white liquid which exists as either the cis or trans isomer (the commercial product is > 95% trans).

Physical Properties

Mp −69°C; bp 102.2°C; d_{20}^{20} 0.853 g/cm³. Crotonaldehyde is soluble to extent of 18.1 g/100 g H_2O (at 20°). Like acrolein (qv), crotonaldehyde is very reactive. It resinifies readily, the amount and rate of polymerization depending upon temperature, type of container, etc. It is difficult to store without its becoming discolored. Dicroton or dicrotonaldehyde is the best-known dimer, and one trimer is known. Hydroquinone and water inhibit resinification and oxidation.

Chemical Properties

Reduction occurs selectively and by cyclization by a variety of agents, eg, crotyl alcohol, $CH_3CH{=}CHCH_2OH$, is produced by reduction of the carbonyl group by $LiBH_4$, $NaBH_4$, KBH_4, $LiAlH_4$, H_2 (using various catalysts). Other reduction products include 2-butene, butyraldehyde, butanol, and hydroxytetrahydrofuran.

Autooxidation occurs when an organic solvent and acetates or crotonates of Mn, Co, and Cu are used; crotonic anhydride and crotonic acid are formed. Oxidation by bromine converts —CHO to —COBr in an exothermic reaction.

Halogenation is the direct chlorination of crotonaldehyde at room temperature results in the addition of two chlorine atoms; upon warming to 50°C in the presence of water, the monohydrate, $CH_3CHClCHCl$-$CHO.H_2O$, is formed. Bromination of cooled crotonaldehyde followed by boiling with an excess of 1% HCl in absolute ethanol for 46 h gives 2-bromo-1,1,3-triethoxybutane. The kinetics of the acid-catalyzed addition of Cl_2 and Br_2 to crotonaldehyde in acetic acid solution indicates a nucleophilic mechanism. Thiols generally add at the 2,3 positions. The reaction with hydrogen sulfide gives the thio ether. Ethanol adds to the carbon–carbon double bond, $CH_3CH(OC_2H_5)CH_2CH(OC_2H_5)_2$. Nitromalonic ester and nitroacetic ester add in the presence of a basic catalyst. Crotonaldehyde undergoes the Diels-Alder reaction as both a diene and a dienophile. It forms many acetals; those with unsaturated alcohols undergo rearrangement to form 2-alkenyl crotonaldehydes. With amines, it undergoes characteristic reactions of aldehydes (qv). Metal alkyls react to produce secondary alcohols after hydrolysis (see Organometallics). Reactions with nitroalkanes result in the secondary alcohol. $NaHSO_3$ adds to the carbonyl group, the carbon–carbon double bond, or to both. Formaldehyde reacts to form the methylol derivative. Acetaldehyde reacts to form, after hydrogenation, 1-hexanol, 1-octanol, and 1-decanol. Crotonaldehyde reacts with ketene (qv) to form the β-lactone in the presence of boric acid or zinc salts, or in the presence of strong acid to form the acetoxy compound. Substituted phenol reacts in acetic solution to give a derivative of 2-(substituted phenoxy)-4-methylchroman. Autocondensation yields a variety of products depending on the catalyst and conditions.

Condensation polymers are formed with ethylamines, ethyleneimine, and ammonium thiocyanate as well as hydroxy compounds. Triethlylamine forms a resin with film-forming properties.

Synthesis and Manufacture

The most widely used method for synthesis is the aldol condensation of acetaldehyde, accompanied or followed by dehydration, using various catalysts. Direct oxidation of hydrocarbons offers a possible synthesis (see Hydrocarbon oxidation). Pyrolysis of various hydroxy compounds can lead to crotonaldehyde.

Health and Safety

Crotonaldehyde is more toxic than its saturated analogue; it irritates the nose, pharynx, and larynx.

Uses

The largest use is in the manufacture of *n*-butanol, and in the production of sorbic acid (qv). Despite its reactive nature, it does not have wide usage.

WILLIAM F. BAXTER, JR.
Eastman Chemical Products, Inc.

D.W. Clayton and co-workers, *J. Chem. Soc.*, 581 (1953).

U.S. Pat. 2,413,235 (Dec. 24, 1946), D.J. Kennedy (to Shawinigan Chemicals, Inc.)

M. Hori, *J. Agr. Chem. Soc. Japan*, **18**, 155 (1942).

CROTONIC ACID

Crotonic acid (*trans*-2-butenoic acid), $CH_3CH{=}CHCHO$, is a white crystalline solid, mp 781.4–71.7°C, bp 184.7°C (101.3 kPa or 1 atm), d 1.018 g/cm³. It is soluble to the extent of 65.6 g/100 g H_2 (at 40°C), and 126 g/100 g H_2O (42°C). Crotonic acid crystals are monoclinic needles or prisms. The crotonic and isocrotonic acids are cis-trans stererisomers:

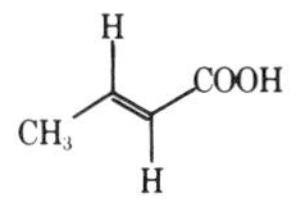

trans-crotonic acid

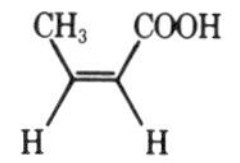

cis-isocrotonic acid

Isocrotonic acid (*cis*-2-butenoic acid) is miscible with water at 25°C. Although its sodium salt is soluble in 4–5 parts of absolute ethyl alcohol, sodium crotonate is practically insoluble. This offers a method of separating the acids.

Chemical Properties

Crotonic acid contains the conjugated system —C=C—C=O, which influences the velocity and direction of many of its reactions. Esterification is much slower than for butyric acid and crotonic esters undergo hydrolysis and alcoholysis more slowly than the corresponding butyric esters. Ethyl crotonate is called a vinylog of ethyl acetate.

Crotonic acid in toluene solution is converted into isocrotonic acid by uv radiation. Concentrated nitric acid oxidizes crotonic acid to acetic and oxalic acids; chromic acid gives acetic acid and acetaldehyde; fusion with potassium hydroxide gives two moles of acetate. Crotonic acid is destructively oxidized by electrolysis; it is oxidized by the liver to acetoacetic acid. Crotonic acid is hydrogenated to butyric acid in the presence of catalysts.

Manufacture

Crotonic acid is prepared commercially by the oxidation of crotonaldehyde (qv) with air or oxygen. Percrotonic acid is believed to be an intermediate. In a continuous process developed by Shawinigan Chemical Ltd (now Gulf Oil of Canada, Ltd), inert organic diluents and catalysts such as copper acetate–cobalt acetate mixtures are used. Batch processes are also used.

Health and Safety

It is nontoxic when injected and a strong irritant when in contact with skin or eyes.

Uses

Crotonic acid is used in the formation of copolymers with vinyl acetate, in coating, paper products (see Paper), resins, as a fungicide, to improve corrosion resistance. Water-soluble resins and copolymers are suitable for pharmaceuticals (see Resins, water, soluble), binders, and films.

Derivatives

These include crotonamide (*trans*-2-butenamide), $CH_3CH=CHCONH_2$, mp 159–160°C; crotonanilide (*trans*-2-butenanilide), $CH_3CH=CHCONHC_6H_5$, mp 115–118°C; crotonic anyhydride (*trans*-2-butenoic anhydride), $(CH_3CH=CHCO)_2O$, bp 247°C; crotonyl chloride, $CH_3CH=CHCOCl$, bp = 124–124°C; and esters, which are all colorless liquids with pleasant odors.

WILLIAM. F. BAXTER, JR.
Eastman Chemical Products, Inc.

D.J.G. Ives, R.P. Linstead, and H.L. Riley, *J. Chem. Soc.*, 561 (1933); E.J. Boorman, R.P. Linstead, and H.N. Rydon, *J. Chem. Soc.*, 568 (1933); D.J.G. Ives, *J. Chem. Soc.*, 86 (1938).

M.A. Matthews, "Manufacture of Crotonaldehyde and Crotonic Acid at I.G. Hoechst," *BIOS*, *Final Rept. 758*, (1946).

CROWN ETHERS.   See Catalysis, phase-transfer; Chelating agents.

CRYOGENICS

Cryogenics is the branch of physics that relates to the attainment and effects of very low temperatures. It is used to produce low cost, high purity gases through fractional condensation and distillation, and to refrigerate materials or to alter their physical properties.

Refrigeration Methods

Refrigeration for cryogenic temperatures is produced within a system by absorbing or extracting heat at low temperatures and rejecting it at higher temperatures to the surroundings (see Refrigeration).

Three methods of producing low temperature refrigeration are

(*1*) Liquid vaporization is a process in which a fluid with the desired physical properties is compressed to the pressure at which it can be condensed by heat exchange with a refrigerant that boils at a higher temperature. The pressure of the condensed liquid phase is reduced (isenthalpic expansion), resulting in formation of a flash vapor fraction which reduces the temperature and enthalpy of the remaining liquid. This liquid is then vaporized by heat exchange, thereby cooling the load stream; the vapor phase is recompressed to complete the cycle.

(*2*) The Joule-Thomson (J-T) cycle starts with a compressed fluid precooled below its inversion temperature and uses isenthalpic expansion through an orifice. Lower temperatures are obtained by incorporating a heat exchanger to cool the fluid further.

(*3*) The expansion-engine cycle approaches isentropic (reversible) expansion with a resultant reduction in fluid enthalpy and temperature. Both reciprocating engines (which handle low volumetric flow rates and high expansion pressure ratios) and centrifugal expansion engines (which handle high flow rates at low ratios) are used in present-day cryogenic plants (see Thermodynamics; Heat-exchange technology).

Applications

Air separation. The practical application of cryogenics began with air separation. Air-separation plants produce oxygen, nitrogen, argon, and, if desired, the rare gases neon, krypton, and xenon. The main switching heat exchanger is the heart of the basic air-separation plant. In the present-day large tonnage air-separation plant which uses a combination of expander and Joule-Thomson refrigeration, air is compressed to 700 kPa (ca 100 psi) in the compressor; the main exchanger precools the air against effluent product streams and extracts carbon dioxide and moisture through condensation and solidification; the expander provides the process refrigeration; high pressure and low pressure columns separate

the oxygen and nitrogen primary components and concentrate the argon; and the crude argon column further separates the argon from oxygen.

Liquified natural gas. The two primary reasons for liquefying natural gas are for "peakshaving", a process in the U.S. for the liquefaction of pipeline natural gas during the low demand periods, and for "baseload supply", a process used in foreign countries for the liquefaction of natural gas at its source for export. The refrigerant cycles used are cascade (J-T expansion), stepwise refrigeration using three vaporization systems, each with a different working fluid, typically propane, ethylene, and methane; (2) expansion-engine refrigeration; and multicomponent refrigerant process, one of the most common cycles that uses a compressor which raises the pressure of the mixed refrigerant to a nominal 2.75 MPa (400 psi), a condenser that partially condenses the mixed refrigerant, and a main heat exchanger.

Other industrial applications. Other applications include hydrogen purificaton, important in the production of benzene by the hydrodealkylation of toluene (qv) or the hydrotreatment pyrolysis of gasoline; ammonia purge-gas recovery; helium recovery; ethane and ethylene recovery; nuclear off-gas systems to minimize the release of gaseous radionuclides from power reactors and nuclear-fuel reprocessing plants (see Nuclear reactors); helium refrigeration and liquefaction; hydrogen liquefaction; and uses for small cryogenic refrigerators, eg, cooling infrared detectors (see Infrared technology).

Cryobiology

The low temperature of liquid nitrogen has many important applications in the biological sciences. Certain biological materials, ranging from single cells to more highly organized structures, can be cooled to extremely low temperatures without loss of life. The first commercial application was in the storage and distribution of semen for dairy-cattle breeding. Liquid nitrogen is also used in the cryogenic preservation of whole blood. Cryosurgury holds promise for painless hemorrhage-free surgery.

Equipment

The two most common types of heat exchangers are the spirally wound, tube-in-shell exchanger, and the plate-fin exchanger. Bulk insulation, rigid foam, vacuum, vacuum-powder, and multilayer insulations are used. Primary criteria in selecting materials for equipment fabrication are ductility, thermal conductivity, coefficient of expansion, and combustibility. Materials selection for pressure vessels, piping, and heat-exchange surface is generally restricted to aluminum, copper, 18-8 stainless steel, and 9% nickel steel below −40°C service.

Health and Safety

The potential for the uncontrolled release of cryogenic liquids must be considered in the design of all facilities that process and handle them. Trace quantities of impurities in the plant may concentrate, and can result in the combination of an oxidant with a flammable cryogen (eg, solid oxygen in liquid hydrogen) or a combustible with an oxidant (eg, acetylene in liquid oxygen).

C.L. NEWTON
Air Products and Chemicals, Inc.

K.D. Timmerhaus and co-workers, eds., *Advances in Cryogenic Engineering*, Vol. 1 (1955)–Vol. 30 (1984), Plenum Press, New York.

C.A. Bailey, *Advanced Cryogenics*, Plenum Press, New York, 1971.

L.K. Lozina-Lozinskii, *Studies In Cryobiology*, John Wiley & Sons, Inc., New York, 1972.

CRYSTALLIZATION

Crystallization is one of the oldest of the chemical engineering unit operations and a primary processing technique in the chemical industry. Vast quantities of crystalline substances are manufactured commercially, eg, sodium chloride, sodium and ammonium sulfates, and sucrose, all with worldwide production rate in excess of 100×10^6 metric tons per year. A high proportion of the products of the pharmaceutical and organic fine-chemicals industries are crystalline. Many organic liquids are purified on a large scale by crystallization, as an alternative to distillation, for the separation of azeotropes and close-boiling mixtures. As shown in Table 1, enthalpies of crystallization are generally much lower than enthalpies of vaporization and crystallization operations are usually carried out much nearer the ambient temperature than are distillation processes.

Crystallization is a very complex operation. The growth of crystals in a crystallizer involves the simultaneous processes of heat and mass transfer in a multiphase, multicomponent system. These conditions alone present complications enough, but the crystallization process is also strongly dependent on fluid and particle mechanics in a system where the size and size distribution of the particulate solids, neither property being capable of unique definition, can vary with time. Furthermore, the solution in which the solids are suspended is thermodynamically unstable, frequently fluctuating between so-called metastable and labile states, and sometimes entering the unsaturated condition, eg, in the vicinity of heat-exchanger surfaces. Traces of impurity, sometimes a few parts per million (10^6), can profoundly affect the nucleation and crystal-growth kinetics. In view of these complexities, it is perhaps understandable why crystallization has been slow to submit to simple analytical procedures.

Nevertheless, crystallizers can be designed and operated successfully despite the uncertainties surrounding fundamental principles, uncertainties that lie in four main areas: solubility and phase relationships; hydrodynamics of crystal suspensions; nucleation characteristics; and crystal-growth rates.

Saturation and Supersaturation

Solubility and crystal yield. Solubility may be expressed most conveniently in units of mass of solute per mass of solvent. An estimate of the crystal yields for a simple cooling or evaporating crystallization may be made from a knowledge of the solubility characteristics of the solution. The general equation may be written:

$$Y = \frac{WR\left[c_o - c_f(1 - V)\right]}{1 - c_f(R - 1)}$$

where c_o = initial solution concentration, kg anhydrous salt/kg water; c_g = final solution concentration, kg anhydrous salt/kg water; W = initial mass of water, kg; V = water lost by evaporation, kg/kg of original water present; R = ratio of molecular weights of hydrates and anhydrous salts; and Y = crystal yield, kg. In practice, the actual yield may differ slightly from that calculated.

Supersaturation. A supersaturated solution contains more dissolved solute than that required for equilibrium saturation. Ostwald (1897) and Miers (1906) suggested that two types of supersaturation could be recognized, namely, the metastable and labile states, respectively. These two states can be represented on a temperature–concentration diagram.

Table 1.   Crystallization and Distillation Energy Requirements

Substance	Melting point, °C	Enthalpy of crystallization, kJ/kg[a]	Boiling point, °C	Enthalpy of vaporization, kJ/kg[a]
o-cresol	31	115	191	410
m-cresol	12	117	203	423
p-cresol	35	110	202	435
o-xylene	−25	128	141	347
m-xylene	−48	109	139	343
p-xylene	13	161	138	340
o-nitrotoluene	−4.1	120	222	344
m-nitrotoluene	15.5	109	233	364
p-nitrotoluene	51.9	113	238	366
water	0	334	100	2260

[a] To convert kJ to kcal, divide by 4.184.

If a solution represented by point A in Figure 1 is cooled without loss of solvent (line ABC), spontaneous nucleation cannot occur until conditions represented by C are reached. Supersaturation can also be achieved by removing some of the solvent from the solution by evaporation. Line ADE represents such an operation carried out at constant temperature. In industrial practice, some combination of cooling and evaporation is most often used.

Crystallization Phenomena

The starting point for any crystallization operation is the state of the supersaturation that may be achieved by cooling, partial evaporation of the solvent, addition of a precipitating agent, as the result of a chemical reaction, etc. However, supersaturation alone is not sufficient to cause crystals to grow. Before growth can commence, there must exist in the system a number of seed crystals which may be formed spontaneously, induced artificially or added deliberately.

Nucleation kinetics. In the industrial crystallizer, the greatest hazard comes not from primary, but from secondary nucleation, the generation of nuclei by the crystals already present in suspension. Metastable zone widths and other nucleation data can be measured in the laboratory with a simple apparatus. Secondary nucleation in industrial crystallizers arises predominantly from crystal–crystal or crystal–equipment (especially the agitator) contact.

Crystal-growth kinetics. For crystallizer design and assessment, rates of crystallization are most conveniently expressed in terms of the supersaturation by empirical relationships. Two aspects of crystal growth are of interest: the overall mass-deposition rates, needed for the design of industrial crystallizers, and growth rates of individual crystal faces under different environmental conditions, helpful for specification of the operating conditions.

A useful pictorial representation, although admittedly a gross simplification, of the crystallization process is given by the diffusion–integration theory in which two separate steps are postulated. The first is a diffusional process whereby solute is transported from the bulk fluid phase through the solution boundary layer close to the crystal surface. This is followed by the integration of adsorbed solute ions or molecules at the crystal surface into the crystal lattice (see Fig. 2).

Habit modification. In general, under different environmental conditions, high index crystal faces grow faster than low ones, and changes in

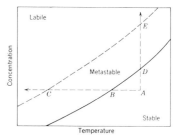

Figure 1.   The solubility–supersolubility diagram. The diagram can be divided into three parts: (*1*) the stable (unsaturated) zone where crystallization is impossible; (*2*) the metastable (supersaturated) zone, between the two curves, where spontaneous nucleation is improbable (although a crystal located in a metastable solution would grow); and (*3*) the unstable or labile (supersaturated) zone where spontaneous nucleation is probable but not inevitable.

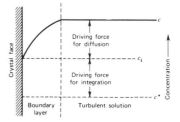

Figure 2.   Concentration driving forces for crystal growth from solution.

the environment can have a profound effect on the individual face-growth rates. Changes in the face-growth rates give rise to habit (shape) changes in the crystals.

The solvent frequently influences the crystal habit. For example, naphthalene crystallizes in the form of needles from cyclohexane and as thin plates from methanol. The possibility of habit change is an important factor when a solvent for crystallization is being chosen. One of the most common causes of habit modification is the presence of impurities in the crystallizing solution.

In nearly every industrial crystallization, some form of habit modification is necessary to control the type of crystal produced. This may be done, eg, by controlling the supersaturation, temperature or pH, or by changing the solvent, or by deliberately adding some impurity which acts as a habit modifier.

Inclusions in crystals. Crystals generally contain inclusions (pockets of solid, liquid, or gaseous impurities) which are a frequent source of trouble in industrial crystallizations. Crystals grown from aqueous solution can contain up to 0.5 wt% of included liquid, which can significantly affect the purity of the product. The main interest lies in finding methods to prevent the formation of inclusions, eg, keeping the crystallizing system clean to avoid dirt, rust, and other debris, avoiding vigorous agitation or boiling, or the application of ultrasonic irradiations to help prevent bubbles or particles adhering to a growing crystal face.

Ripening. Ripening, frequently referred to as Ostwald ripening, can occur in suspensions of very small crystals. Those smaller than some critical size dissolve, despite the fact that the system may apparently be supersaturated, while the larger crystals grow. One reason for ripening is the difference in solubility between crystals of different size, expressed by the Gibbs-Thomson (or Ostwald-Freundlich) equation:

$$\ln\left[\frac{c(r)}{c^*}\right] = \frac{2\gamma v}{\nu k T r}$$

where $c(r)$ is the solubility of a particle of radius r; c^* is the normal equilibrium solubility; γ is the surface energy of the solid in contact with its solution; v the molecular volume of the solid; k the Boltzmann constant; T the absolute temperature; and ν the number of ions in the solute, if it an electrolyte. For nonelectrolytes, $\nu = 1$.

Deliberate temperature cycling has been used successfully as a technique for increasing the mean crystal-size distribution in industrial batch crystallizers.

Industrial Crystallization Processes

Growth of single crystals. There is an ever increasing demand for single crystals of a large number of substances for use in the electronics industry where their dielectric, piezoelectric, semiconductor, and other properties are exploited (see Semiconductors; Zone refining). Single crystals are also in demand for masers, lasers (qv), and gem stones (see Gems, synthetic).

Fractional solidification. In recent years, a considerable number of melt-crystallization techniques have been developed, eg, the Newton-Chambers process for the purification of benzene from a coal–tar benzole fraction using either batch or multistage operation; the Proabd Refiner, a batch process in which a static liquid feedstock is progressively cooled and crystallized and then remelted; and the MWB process which acts as a continuous multistage countercurrent scheme and has applications in the purification of a wide range of organic substances including benzoic acid and caprolactam. Column crystallizers utilizing countercurrent washing are also available, such as that used in the Phillips process for the large-scale production of p-xylene; the Brodie Purifier used for p-dichlorobenzene; and the TNO process which is still in the pilot-scale development stage. Several comprehensive reviews of the theory and practice of column crystallizers have been made.

Salting-out crystallization. This process is widely used for the crystallization of organic substances from water-miscible organic solvents, and for the precipitation of inorganic salts from aqueous solution by an alcohol. A solution can be supersaturated by the addition of a substance, preferably a liquid, that reduces the solubility of the solute in the solvent. Iron-free alums and coarse-grained anhydrous precipitates of normally hydrated salts have been prepared by this method.

Reaction crystallization. The precipitation of a solid phase by the chemical interaction of gases or liquids is a common method of preparing many industrial chemicals. The reactants are usually fed into a zone where intimate mixing takes place quickly. Precipitation occurs when the fluid phase becomes supersaturated with respect to the solute. This is widely employed in industry, eg, sodium bicarbonate can be prepared from flue gases containing 10–20% carbon dioxide by countercurrent contact with brine in packed towers.

Spray crystallization. Solids are deposited from a highly concentrated solution or a melt by spraying droplets into a large chamber where they fall countercurrently to an upflowing stream of hot air. Fertilizer chemicals, particularly ammonium nitrate and urea, are manufactured on a large scale by this method.

Emulsion crystallization. Crystallization is carried out by cooling an aqueous emulsion. The organic substance should be practically insoluble in water and able to melt and solidify in a heterogeneous medium and remain stable.

Extractive crystallization. A binary mixture that forms a eutectic cannot be separated into its pure components by conventional fractional crystallization. However, if a third component, which alters the solid–liquid phase relationships, is added to the system, it is sometimes possible to effect the separation by a stepwise crystallization process. Many hydrocarbon isomers are amenable to this treatment.

Adductive crystallization. Liquid mixtures, especially those of chemical isomers, that are not amenable to conventional crystallization can often be separated by adding another component to the system that causes a solid phase to be deposited. Molecular compound formation can be utilized, eg, in the separation of the eutectic system of m- and p-cresol.

Industrial Crystallization from Solution

Industrial crystallizers may be classified into a number of general categories: for example, batch or continuous; agitated or nonagitated; controlled or uncontrolled (referring to supersaturation control); classifying or nonclassifying (referring to production of a selected product size by classification in a fluidized bed of crystals); circulating liquor (crystals remain in the crystallization zone) or circulating magma (crystals and mother liquor are circulated together). Crystallizers operating continuously with fully mixed suspensions are generally referred to as mixed-suspension, mixed-product moval (MSMPR) crystallizers. Classification of crystallizers according to the method by which supersaturation is achieved is still widely used: eg, evaporating, cooling, vacuum, and reaction crystallizers.

The simplest types of cooling crystallizers use unstirred tanks. When a stirrer is employed in an open-tank crystallizer (agitated vessels), smaller and more uniform crystals are formed and the batch time is reduced. Tank crystallizers, stirred or otherwise, vary in design from shallow pans to large cylindrical tanks, according to the needs of a particular process.

One method for avoiding the use of a conventional heat exchanger is to employ vacuum cooling. Another technique is to use direct-contact cooling (DCC) where supersaturation is achieved by contacting the process liquor with a cold heat-transfer medium.

Design and Operation of Crystallizers

The temperature-solubility relationship between the solute and solvent is of prime importance in the selection of a crystallizer. For solutions that yield appreciable quantities of crystals on cooling, the choice of equipment will generally lie between a simple-cooling or a vacuum-cooling crystallizer. For solutions that change little in composition with a reduction in temperature, an evaporating crystallizer would normally be used, although the method of salting-out could be employed in certain cases.

Continuous crystallizers are generally more economical in operating and labor costs than the batch units, especially for large production rates. The batch units are usually cheaper in initial capital cost. (A complete overview of crystallizer operation may be found in the unabridged edition).

J.W. MULLIN
University College London

J.W. Mullin, *Crystallization*, 2nd ed., Butterworths, London, UK, 1972.

R.C. Bennett in R.H. Perry and C.H. Chilton, eds., *Chemical Engineers' Handbook*, 6th ed., McGraw-Hill, New York, 1983, Section 19.

A.D. Randolph and M.A. Larson, *Theory of Particulate Processes*, Academic Press, Inc., New York, 1971.

CRYSTALS.   See X-ray technology.

CUMENE

Cumene (1-methylethylbenzene, 2-phenylpropane, isopropylbenzene), C_9H_{12}, is normally a liquid, substituted aromatic compound in the benzene (qv), toluene (qv), ethylbenzene series (see BTX processing; Xylenes and ethylbenzene).

Physical and chemical properties. Bp 152.39°C; fp −96.03°C; d 0.8619 g/cm³. The transport and thermodynamic properties are presented by Yaws.

Manufacture. Cumene is manufactured exclusively from propylene and benzene utilizing an acidic catalyst, eg, solid phosphoric acid (see Alkylation; Friedl-Crafts reactions).

Health and safety. Liquid cumene is a primary skin irritant; some data suggest that long exposures may result in cumulative effects.

Uses. It is the major intermediate chemical used worldwide in the production of phenol (qv), acetone and α-methylstyrene, all of which are components in plastic resins (see Styrene plastics). Since the alternative route to acetone utilizes propylene, the cumene route is essentially equivalent to the production of phenol from benzene while obtaining high yields of acetone from propylene. Cumene is also used as a chain initiator (see Initiators) and as a component in aviation gasolines (see Aviation and other gas-turbine fuels; Gasoline).

DENNIS J. WARD
UOP Inc.

C.L. Yaws, *Chem. Eng.* (*N.Y.*) **82**(20), 73 (1975).

P.R. Pujado, J.R. Salazar, and C.V. Berger, *Hydrocarbon Process.* **55**(3), 91 (1976).

U.S. Dept. of Labor, "Toxic Substances," *Fed. Reg.* **40**(196), 77262 (Oct. 8, 1975).

CUMIN OIL.   See Oils, essential; Flavors and spices.

CUPFERRON.   See Zinc.

CUPRITE, Cu₂O.   See Copper.

CURARE AND CURARELIKE DRUGS.   See Alkaloids.

CYANAMIDES

Cyanamide

Cyanamide, (carbamic acid nitrile; carbamodiimide), $H_2NC{\equiv}N$, is a weak acid with a very high solubility in water.

Properties. Cyanamide crystallizes from a variety of solvents as somewhat unstable, colorless, orthorhombic, deliquescent crystals. Dimerization is prevented by traces of acidic stabilizers, eg, monosodium phosphate. Studies of the infrared and Raman spectra support the *N*-cyanoamine structure, $NH_2{-}C{\equiv}N$. It is completely soluble at 43°C, and has a minimum solubility (eutectic) at −15°C. It is highly soluble in polar organic solvents, and less soluble in nonpolar solvents. Mp 46°C; bp 140°C (101.3 kPa); d 1.282 g/cm³; ref. index 1.4418 at 48°C.

Reactions. Reactions are either additions to the nitrile group or substitution at the amino group. Both are involved in the dimerization to dicyandiamide, which occurs readily at pH 8–10. The cyanamide anion is strongly nucleophilic and reacts with most alkylating or acylating reagents; addition to a variety of unsaturated systems occurs readily. In some cases, a cyanamide salt is used; in others base catalysis suffices. Alkylation with a variety of common alkyl halides or sulfates gives stable dialkyl cyanamides. The reaction with formaldehyde produces first an unstable hydroxymethyl derivative that resinifies more or less rapidly, passing through a water-soluble stage which permits various applications. Hydrogen chloride gives a dihydrochloride. Chloroformamidine hydrochloride is a convenient anhydrous form which is easily stored and handled.

Manufacture. The basic process for manufacture comprises four stages: lime is made from high grade limestone (see Lime and limestone); calcium carbide is manufactured from lime and coal or coke (see Carbides); calcium cyanamide is produced by passing gaseous nitrogen through a bed of calcium carbide, which is heated to 1000–1100°C in order to start the reaction (the heat source is then removed as the reaction continues because of its strong exothermic character); and cyanamide is manufactured from calcium cyanamide by continuous carbonation in aqueous medium. For production of commercial 50% solution, and for recovery of crystalline cyanamide, this process is modified to improve purity and concentration, eg, calcium and iron may be removed by ion-exchange treatment.

Other processes include the reaction of lime with hydrogen cyanide; the reaction of limestone with ammonia; and the reaction of lime and urea to form calcium cyanate which is then converted to calcium cyanurate which gives calcium cyanamide at a higher temperature.

Dicyandiamide

Dicyandiamide (**1**) (cyanoguanidine) is the dimer of cyanamide and crystallizes in colorless monoclinic prisms. It is amphoteric, and generally soluble in polar solvents and insoluble in nonpolar solvents. Mp 208°C; bp dec.

(1)

(2)

Reaction. The reactions resemble those of cyanamide. However, cyclizations take place easily and the nitrile group is less reactive. For example, under pressure and in the presence of ammonia, it cyclizes to melamine. Guanamines are obtained when dicyandiamide is heated with alkyl or aryl nitriles in the presence of small amounts of alkali. Reaction with ammonium salts gives biguanide salts which react further with the ammonium salt forming guanidine salts; guanidine nitrate is manufactured by this route. Hydrolysis occurs easily at elevated temperatures. It can be treated with formaldehyde to produce resinous compositions of varying proportions (see Amino resins).

Dicyandiamide is manufactured by the dimerization of cyanamide in aqueous solution, with pH adjusted to 8–9 and held at ca 80°C for 2 h to give complete conversion.

Melamine

Melamine (**2**) (cyanurotriamide, cyanuramide, 2,4,6-triamino-1,3,5-triazine) is a white crystalline material that is insoluble in most organic solvents. It is appreciably soluble in water. Mp 3550°C; d 1.573 g/cm³. Although moderately basic, it is better considered as the triamide of cyanuric acid than as an aromatic amine (see Cyanuric and isocyanuric acids). Its reactivity is poor in nearly all reactions considered typical for amines.

Melamine was formerly manufactured by the conversion of dicyandiamide under heat. The dehydration condensation of urea has displaced the dicyandiamide process (see Urea).

Health and Safety

Manufacture of cyanamide does not present any serious health hazard. Contact or ingestion must be avoided. Dicyandiamide is essentially nontoxic, but may cause dermatitis. Melamine crystal may be handled in ordinary industrial use without special hygienic precautions.

Uses

In Europe, cyanamide and calcium cyanamide are used as fertilizers (qv). Calcium cyanamide is used for steel nitridation and desulfurization (see Steel); It is the raw material for dicyandiamide. Dicyandiamide is used as a raw material for the manufacture of guanamines, biguanide, and guanidine salts, and various resins. Most of the melamine produced is used in the form of melamine-formaldehyde resins (see Amino resins and plastics).

DONALD R. MAY
Cyanamid Canada Inc.

D. Costa and C. Bolis-Cannella, *Ann. Chim. (Rome)* **43**, (1953).
W.H. Fletcher, *J. Chem. Phys.* **39**, 2478 (1963).
B. Bann and S.A. Miller, *Chem. Res.* **58**, 131 (1958).

CYANIDES

HYDROGEN CYANIDE

Hydrogen cyanide (hydrocyanic acid, prussic acid, formonitrile), HCN, is a colorless, poisonous, low viscosity liquid with an odor characteristics of bitter almonds. It was prepared in dilute solution by Scheele in 1782 and as the pure compound by Gay-Lussac in 1815. Its structure is that of a linear, triply-bonded molecule, HC≡N. It is likely that HCN is formed, usually in trace quantities, whenever hydrocarbons are burned in air or when photosynthesis takes place.

Physical Properties

Mp $-13.24°C$; bp $25.70°C$; d_4^t 0.7150 g/cm³ (0°C); sp gr (10% aqueous solution) 0.9838.

Chemical Properties

Hydrogen cyanide is a very weak acid; its ionization constant has the same magnitude as that of the natural amino acids. The reactions of the cyanide ion are discussed under Alkali metal cyanides.

As the nitrile of formic acid, it undergoes many of the typical nitrile reactions (see Nitriles). It can be oxidized by air at 300–650°C over silver or gold catalysts to yield cyanic acid, HOCN, and cyanogen $(CN)_2$. Reaction with chlorine in the liquid phase gives cyanogen chloride (ClCN). This is the basic route to triazines, eg, melamine (see Cyanamides; Urea).

Hydrogen cyanide reacts with phenols in the presence of hydrogen chloride and aluminum chloride to give aromatic aldehydes (the Gattermann synthesis). It is the starting chemical for alanine, phenylalanine, valine, and α-aminobutyric acid. HCN reacts with formaldehyde and aniline to form *N*-phenylglycinonitrile, and with formaldehyde alone to form glycolic nitrile. Cyanohydrins are formed by the reaction of glucose and similar compounds with HCN. It reacts violently with sulfuric acid in strong concentrations. Under certain conditions, it can polymerize to black solid compounds, eg, the homopolymers and the tetramer. Although HCN is a weak acid and not considered corrosive, it has a corrosive effect: in water solutions, which cause crystalline stress-cracking of carbon steels under stress even at room temperature and in dilute solution; and in water solution containing sulfuric acid as a stabilizer,

they severely corrode steel above 40°C and stainless steels above 80°C (see Corrosion).

Manufacture

The dominant commercial production process is based on the classical technology developed by Andrussow. This is the reaction of ammonia, methane (natural gas) and air over platinum metals as catalysts. The stoichiometry of the Andrussow process may be presented by the following:

$$CH_4 + NH_3 + 1.5 O_2 \rightarrow HCN + 3 H_2O \quad 481.9 \text{ kJ/mol}$$
$$(115.2 \text{ kcal/mol})$$

Usually, the overall reaction is carried out adiabatically, with catalyst temperature ca 1100°C.

A second process, called the BMA process, involves the reaction of ammonia with methane in the absence of air. HCN is also obtained as a by-product in the manufacture of acrylonitrile by the ammoxidation of propylene (Sohio technology) (see Acrylonitrile).

Health and Safety

The cyanides are true noncumulative protoplasmic poisons, ie, they can be detoxified readily. They combine with those enzymes at the blood tissue interfaces that regulate oxygen transfer to the cellular tissues. Death results from asphyxia. Hydrogen cyanide can enter the body by inhalation, oral absorption, or skin absorption. The threshold limit is 10 ppm. The body has a mechanism for continuous removal of small amounts of hydrogen cyanide by converting it to thiocyanate which is removed in the urine. Oral absorption is rapidly fatal.

Hydrogen cyanide is classified by the DOT as a Class A Flammable Poison; when working with HCN, never work alone.

Uses

Hydrogen cyanide is the basic chemical building block for sodium cyanide, potassium cyanide, methyl methacrylate, methionine, triazines, iron cyanides, adiponitrile, and various chelating agents.

WILLIAM R. JENKS
E.I. du Pont de Nemours & Co., Inc.

D.M.L. Griffiths and H.A. Standing, *Adv. Chem. Ser.* **55**, 666 (1966).
A. Blumenthal and G. Kiss, *Mitt. Geb. Lebensmittelunters. Hyg.* **61**, 394 (1970).

ALKALI-METAL CYANIDES

Sodium Cyanide

Sodium cyanide, NaCN, is a white cubic crystalline solid commonly called white cyanide. Mp $563.7 \pm 1°C$ (100%); bp 1500°C (extrapolated); d 1.60 g/cm³. The solid phase in contact with a saturated aqueous solution at temperatures above 34.7°C is the anhydrous salt; below 34.7°C the solid phase is the dihydrate. The solubility of the dihydrate in g NaCN/100 g satd soln is 16.01 at $-15°C$; 32.8 at 10°C; and 45 at 34°C. The solubility of the anhydrous salt is less dependent on temperature. At room temperature (20°), it is a crystalline ionic compound with a structure similar to that of sodium chloride or bromide (an fcc crystal lattice). Sodium chloride and sodium cyanide are isomorphous and form an uninterrupted series of mixed crystals.

Chemical properties. When heated in a dry CO_2 atmosphere, sodium cyanide fuses without much decomposition. In the presence of a trace of iron or nickel oxide, rapid oxidation occurs when cyanide is heated in air, first to cyanate and then to carbonate. Case hardening of steels using a sodium cyanide molten bath depends on these reactions where the active carbon and nitrogen are absorbed into the steel surface; hence the names carburizing and nitriding (see Metal surface treatments; Steel). In the presence of oxygen, aqueous sodium cyanide dissolves most metals in the finely divided state, with the exception of lead and platinum, This is the basis of the MacArthur process for the extraction of gold and silver from their ores.

Manufacture. Almost all sodium cyanide is manufactured by the neutralization or so-called wet processes in which hydrogen cyanide reacts with sodium hydroxide solution.

$$HCN + NaOH \rightarrow NaCN + H_2O$$

Most modern high tonnage plants use the reaction of high purity HCN and aqueous sodium hydroxide in a unit system that embodies evaporation of water and crystallization. Control of the system avoids HCN polymer formation.

Health and safety. Environmental aspects and toxicity of cyanides are discussed under Hydrogen cyanide. Molten sodium cyanide reacts with strong oxidizing agents such as nitrates and chlorates with explosive violence.

Use. The largest single market for sodium cyanide has been electroplating (see Electroplating). It is also used in ore extraction and recovery. Chemical uses include dyes, pharmaceuticals, and chelating agents.

Potassium Cyanide

Potassium cyanide, KCN, is a white crystalline, deliquescent solid. Mp 634.5°C (100%); d 1.553 g/cm³ at 20°C. It is soluble to the extent of 71.6 g/ 100 g H_2O at 25°C. Unlike sodium cyanide, potassium cyanide does not form a dihydrate. At room temperature, KCN has an fcc crystal structure. Average KCN crystals inherently have 3 to 4 times the mass or size of NaCN crystals.

Chemical properties. KCN is readily oxidized to potassium cyanate by heating in the presence of oxygen or easily reduced oxides (eg, those of lead or tin, or manganese dioxide), and in aqueous solution by reaction with hypochlorites or hydrogen peroxide. Many reactions can be carried out with organic compounds, the alkalinity of the KCN acting as a catalyst. These reactions, as well as those with sulfur and sulfur compounds are analogous to reactions of sodium cyanide.

Manufacture. KCN is manufactured by the reaction of an aqueous solution of potassium hydroxide with hydrogen cyanide, in a reaction analogous to sodium cyanide synthesis.

Uses. KCN is used primarily for fine silver plating and for dyes and specialty products.

Lithium, Rubidium, and Cesium Cyanides

These are all white or colorless salts, isomorphous with potassium cyanide. In their physical and chemical properties, they closely resemble sodium and potassium cyanide. All are very soluble in water. They have as yet no industrial uses.

Ammonium Cyanide

NH_4CN is a colorless crystalline solid, relatively unstable, and decomposing into ammonia and hydrogen cyanide at 36°C. Because of its unstable nature, it is not shipped or sold commercially.

WILLIAM R. JENKS
E.I. du Pont de Nemours & Co., Inc.

U.S. Pat. 3,619,132 (Nov. 9, 1971), H.J. Mann and co-workers (to Degussa).

U.S. Pat. 3,615,176 (Oct. 26, 1971), W.R. Jenks and O.W. Shannon (to E.I. du Pont de Nemours & Co., Inc.).

U.S. Pat. 3,695,833 (Oct. 3, 1972), O. Wiedeman and co-workers (to American Cyanamid Co.).

ALKALINE–EARTH METAL CYANIDES

Crude calcium cyanide, $Ca(CN)_2$, about 48–50 eq% NaCN (also called black cyanide), is the only commercially important alkaline-earth metal cyanide currently produced. It is marketed in flake form, as a powder, or as cast blocks.

Physical and chemical properties. Because of decomposition, the melting point of calcium cyanide can only be estimated by extrapolation to be 640°C. As the diammoniate, it is formed in liquid ammonia by reaction of calcium hydroxide or nitrate with ammonium cyanide. Deammoniation under heat and high vacuum yields calcium cyanide, a white

powder which is readily hydrolyzed to hydrogen cyanide. The normal humidity content of air is sufficient to cause evolution of hydrogen cyanide.

Aqueous solutions of calcium cyanide prepared even at low temperature turn yellow or brown owing to the formation of the HCN polymer. Calcium cyanide hydrolyzes readily and is decomposed by carbon dioxide, acids, and acidic salts, liberating hydrogen cyanide.

Manufacture. Calcium cyanide is commercially produced by heating crude calcium cyanamide (which contains carbon) in an electric furnace above 1000°C in the presence of sodium chloride. The resulting melt is cooled rapidly to prevent reversion to calcium cyanamide.

Health and safety. This compound is extremely toxic to humans, animals and fish. Precautions similar to those used for sodium cyanide should be used for black cyanide. It is a Class B poison.

Uses. The first, and still largest, use is in the extraction or cyanidation of precious metal ores. The leaching action of the cyanide is due to the formation of soluble cyanide complexes. Calcium cyanide is also used in froth flotation (qv) of minerals.

Other Alkaline-Earth Cyanides

Included in this category are magnesium cyanide, $Mg(CN)_2$, and strontium cyanide and barium cyanide, which are somewhat more stable than calcium and magnesium cyanides.

WILLIAM R. JENKS
E.I. du Pont de Nemours & Co., Inc.

G. Peterson and H.H. Franck, *Z. Anorg. Allgem. Chem.* **237**, 1 (1938).

J.V.N. Dorr and F.L. Bosqui, *Cyanidation and Concentration of Gold and Silver Ores*, McGraw-Hill Book Co., Inc., New York, 1950.

CYANINE DYES

The cyanine dyes, which are among the oldest synthetic dyes known, comprise a large group with a wide variety of colors. Absorption spectra of cyanines range in position from the ultra-violet to the infrared region and cover a wider span of the spectrum than those of any other dye class. In addition, the cyanines are quite useful and efficient as photographic sensitizers; their variety of absorption wavelengths allows sensitization to most of the visible and infrared spectrum, often with an efficiency nearly equal to that of the ultraviolet response of the silver halide itself (see Color photography; Dyes, Sensitizing).

Properties

The most typical cyanine dyes are molecules that are planar, rectangular in overall shape, and 5–9 nm (length) × 0.5–1.0 nm (height) × 0.3 nm (depth). Color and constitution may be understood through consideration of the chromophoric systems, terminal groups and crystal structure. Chromophoric systems in cyanine dyes are primarily of three types: the amidinium-ion system (**A**), the carbonyl-ion system (**B**) and the dipolar amidic system (**C**) (see Fig. 1).

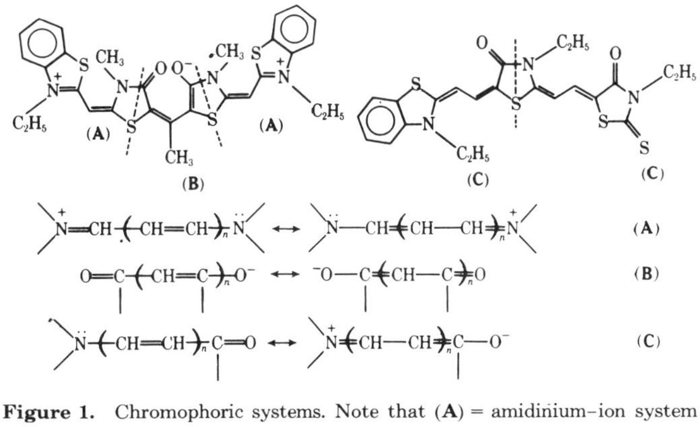

Figure 1. Chromophoric systems. Note that (**A**) = amidinium–ion system (a cyanine), (**B**) = carbonyl–ion system (an oxonol), and (**C**) = dipolar amidic system (a merocyanine).

Dyes derived from the primary chromophores (**B**) and (**C**) are designated oxonols and merocyanines, and neutrocyanine has also been used for (**C**). Dyes that differ only by the number of vinyl groups (—CH= CH—) in the methine chain are termed vinylogous series, eg, see compound (**1**).

$$X = \text{—CH=CH—}$$

$n = 0$, a simple cyanine
$n = 1$, a carbocyanine
$n = 2$, a dicarbocyanine
$n = 3$, a tricarbocyanine

Many cyanines and related dyes are formed by addition–elimination reactions in which a simple terminal group is replaced by more complex ones. The largest classes of useful cyanines are derived from dyes with heterocyclic terminal groups of two main types: basic or electron-donating, eg, 2-thiazole, benzimidazole, benzoxazole and 2-benzothiazole; and acidic or electron-accepting, eg, rhodanine, hydantoin, isoxazolone and pyrazolidinedione (see Table 1).

Table 1.   Examples of Nuclei Occurring in Important Cyanine Dyes

Basic nuclei for cyanines and merocyanines

2-thiazole   2-benzoxazole

Acidic nuclei for merocyanines and oxonols

rhodanine   pyrazolidin-3,5-dione

Crystals of these molecules are highly ordered and resemble a tilted deck of cards or staircase in the arrangement of molecules. These same arrangements lead to very narrow absorption bands which have even more selective absorption than the individual molecules. To the extent that such behavior also takes place when dyes adsorb to photosensitive crystals, the same selectivity can be used to benefit color reproduction in color photographic materials.

In aqueous solution at room temperature, these changes are most evident as the concentration of the dye increases or as dye is concentrated by adsorption to crystal surfaces or oppositely charged polymer sites. Relative to the absorptions for monomeric dye molecules (M-band, Fig. 2), progressive shifts of absorption maxima to shorter wavelength are designated as H-bands (hypsochromic) although the term metachromic has been used as well. Shifts to longer wavelength are less progressive. The new absorption is designated as a J-band (bathochromic). Self-association among dye molecules is termed aggregation. The degree of self-association was determined spectroscopically for several H-aggregates (dimers, trimers, and tetramers) and for J-aggregates. Under ideal circumstances, aggregation in solution involves two-component equilibria (Fig. 2): monomeric dye (curve M) with either dimer (curve D) or J-aggregate (curve J).

Larger H-aggregates, formed at increasing concentration for dye only, shift the absorption peaks progressively to shorter wavelengths (peaks H-3, 4). Most H-bands have been considered to have larger half-band width than the monomer until recently, when an extremely sharp H-band designated H* was noted for two thiacarbocyanines. On the other hand, most J-bands, which are favored also by increased concentrations of certain dyes such as (**2a**) or (**3**) exhibit unusually sharp absorptions.

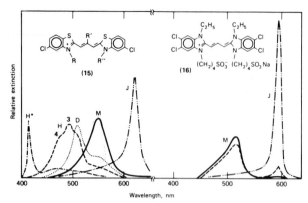

Figure 2.   Relative absorption of aggregates. For (**2a**) $R = C_2H_4CO_2^-$, $R' = CH_3$, $R'' = C_2H_4CO_2H$; (**2b**) $R = (CH_2)_4SO_3^-$, $R' = C_2H_5$, $R'' = (CH_2)_4SO_3Na$; and (**3**).

Manufacture

The manufacture of cyanine, merocyanine, and oxonol dyes for photographic uses is highly specialized, and synthetic routes are selected to maximize both the yield and the purity of the final dyes. Dye-forming reactions may be classified as oxidative or nonoxidative. Nonoxidative syntheses are more versatile; they combine various nucleophilic and electrophilic reagents and have been characterized as condensation reactions in which two intermediates react and eliminate a simple molecule.

Other than variations in either the methine chain lengths or the structures of the heterocycles used for cyanine dyes, significant modifications of the properties of the dyes are accomplished in two other ways: the use of substituted ortho esters $RC(OR)_3$ places a substituent larger than hydrogen on the methine chain (this often affects dye planarity, the tendency to form aggregates and solubility in simple solvents like methanol); and the use of charged substituents attached to the heterocyclic nitrogens such as anionic groups (eg, $(CH_2)_nCO_2^-$), replace the simple alkyl groups and alter the net charge on cyanine dyes from positive (the charge due to the chromophore) to neutral (one anionic group) to negative (two anionic groups).

Chemical and Photochemical Reactions

Certain cyanine dyes are readily protonated in acidic media, a reaction that is favored by highly basic heterocyclic groups (eg, quinoline) or long methine chains (eg, pentamethine). Correlations between dye structure and pKa are well documented. Many other dyes undergo nucleophilic attack, reacting with amines, sulfite, and other reagents to produce new products with distinctly altered absorptions.

Photooxidation of cyanines with various heterocycles and methine chain lengths was sensitized by methylene blue. The rate constants are consistent with singlet-oxygen addition to the chromophore and are generally parallel to electrochemical oxidation potentials determined for the same dyes. Both the sensitized photooxidations and the electrochemical oxidations lead to new chemical products, indicating that the oxidation of cyanine dyes in solution is not a highly reversible process. Electrochemical reductions of a wide variety of cyanines more often exhibit reversibility on the timescale of the elecrochemical measurement.

Direct photoexcitation of cyanine dyes in solution requires a high concentration of another chemical reactant to achieve significant chemical reaction. Cyanine dyes show low quantum-yield photofading reactions under either oxidative conditions (oxygen, reducible metal ions) or reductive conditions (ascorbic acid, gelatin), particularly when compared to the noncyanine classes of dyes such as xanthenes, thiazenes and acridines. The excited-state behavior of cyanines is consistent with these results, because cyanines generally show low singlet-triplet intersystem crossing yields, rapid radiationless deactivation in solution and high fluorescence yields in rigid media.

The main nonfluorescent process in cyanine dyes is the radiationless deactivation $S_1 \rightarrow S_0$. Maximum singlet-triplet interconversion (Φ_{ST}) in methanol for carbocyanines is about 3% (max $\Phi_{ST} \leq 0.03$), and the sum $(\Phi_{Fl} + \Phi_{ST})$ is less than 0.10. In contrast to this, hydrocarbons and other classes of dyes exhibit significant triplet yields ($\Phi_{ST} = 0.2$–0.9), and the sum $(\Phi_{Fl} + \Phi_{ST})$ is close to 1.0.

Uses

Cyanine dyes are used primarily for speciality purposes: photographic sensitizers and desensitizers (see Color photography); laser dyes and certain medicinal applications, eg, Povan and Dithiazanine are useful anthelmintics (see Chemotherapeutics, anthelmintics) and Indocyanine Green is used as an ir-absorbing tracer for blood-dilution studies (see Infrared technology).

DAVID M. STURMER
Eastman Kodak Company

L.G. S. Brooker in C.E.K. Mees and T.H. James, eds., *The Theory of the Photographic Process*, 3rd ed., Macmillan, New York, 1966, p. 198.

F.M. Hamer "The Cyanine Dyes and Related Compounds," in A. Weissberger, ed., *The Chemistry of Heterocyclic Compounds*, Vol. 18, Interscience Publishers, Inc., a division of John Wiley & Sons, Inc., New York, 1964.

D.M. Sturmer in E.C. Taylor and A. Weissberger, eds., *The Chemistry of Heterocyclic Compounds*, Vol. 30, Wiley-Interscience, New York, 1977, p. 441.

A.H. Herz, *Advances in Colloid and Interface Science*, Vol. 8, Elsevier Scientific Publishing Co., Amsterdam, Neth., 1977, pp. 237–298.

CYANOCARBONS

Cyanocarbons are compounds having such a large number of cyano groups that the chemical reactions of the class are essentially new in kind and not shared by analogous compounds free of such groups. The unique reactivity of cyanocarbons was first recognized with the synthesis of tetracyanoethylene (TCNE). The cyano group is a powerful electron-withdrawing group, sufficiently small to present no steric problem.

The most important members of the cyano class are the alkenes: tetracyanoethylene, hexacyanobutadiene, and tetracyanoquinodimethane (TCNQ), $(NC)_2C{=}C_6H_4{=}C(CN)_2$; the alkanes: tetracyanomethane and hexacyanoethane; dicyanoacetylene; hexacyanobenzene; tetracyanoquinone; cyanocarbon acids, eg, cyanoform, methanetricarbonitrile; oxacyanocarbons eg, tetracyanoethylene oxide; thiacyanocarbons; and azacyanocarbons (see below).

Tetracyanoethylene is described because an understanding of its chemistry is helpful in understanding the chemistry of other cyanocarbons.

Tetracyanoethylene

Tetracyanoethylene must be regarded as a highly electron-deficient olefin that should be a strongly electrophilic reagent. It has a high positive heat of formation, but is stable to shock. Mp 200–202°C; bp 223°C; d 1.318 g/cm³. It has good thermal stability, resists oxidation, but once ignited in oxygen it burns with an intensely hot flame.

TCNE forms intensely colored complexes with olefins or aromatic hydrocarbons, eg, benzene solutions are yellow, xylene solutions are orange, and mesitylene solutions are red. The colors arise from complexes of a Lewis acid-base type, with partial transfer of a electron from the aromatic hydrocarbon to TCNE. TCNE is conveniently prepared in the laboratory by debromination of dibromomalononitrile with copper powder:

$$CH_2(CN)_2 + Br_2 \rightarrow Br_2C(CN)_2 \rightarrow (NC)_2C{=}C(CN)_2$$

TCNE undergoes two principal types of reactions: addition to the double bond and replacement of a cyano group. Among reagents that readily add to TCNE are hydrogen, 1,3-dienes, electron-rich alkenes, and ketones as well as nucleophilic radicals.

Although a C—CN bond is normally strong, one or two cyano groups in TCNE can be replaced easily, about as easily as the one in an acyl cyanide. The replacing group can be hydroxyl, alkoxyl, amino, or a nucleophilic aryl group. Thus, hydrolysis of TCNE under neutral or mildly acidic conditions leads to tricyanoethanol, a strong acid isolated only in the form of its salts.

Health and safety. All cyanocarbons should be considered as toxic as sodium cyanide or hydrogen cyanide (see Cyanides), unless specifically tested.

O.W. WEBSTER
E.I. du Pont de Nemours & Co., Inc.

E. Ciganek, W.J. Linn, and O.W. Webster, in Z. Rapport, ed., *The Chemistry of the Cyano Group*, Interscience Publishers, London, UK, 1970, pp. 423–638.

CYANOETHYLATION

Cyanoethylation, the reaction in which a compound possessing an active hydrogen adds across the double bond of acrylonitrile (qv), offers the synthetic chemist a convenient route for incorporating the propionitrile moiety into a molecule. The characteristic feature of compounds undergoing this reaction is their possession of a labile hydrogen atom that, once removed, produces a nucleophilic group which then attacks the most positive position in acrylonitrile. Examples of chemical types that undergo cyanoethylation include hydroxyl compounds, nitrogen compounds, some carbon compounds (eg, malononitrile, cyanoacetic esters, benzyl cyanide, and cyclopentadiene), hydrogen chloride, hydrogen bromide, sodium bisulfite, as well as arsines, boranes, germanes, phosphines, silanes, and stannanes.

Mechanism

The generally accepted mechanism is summarized in equations 1–3 in which an alcohol is used as a reactant with acrylonitrile. This mechanism is believed to apply to all base-catalyzed cyanoethylations, eg, alcohols, mercaptans, phosphines, and activated carbon compounds.

$$ROH + base \rightleftharpoons RO^- + (base{-}H)^+ \tag{1}$$

$$RO^- + CH_2{=}CHC{\equiv}N \rightleftharpoons \begin{cases} [ROCH_2C^-HC{\equiv}N] \\ \updownarrow \\ [ROCH_2CH{=}C{=}\bar{N}] \end{cases} \tag{2}$$

(2)

$$(2) + ROH \rightleftharpoons ROCH_2CH_2C{\equiv}N + RO^- \tag{3}$$

The reaction is most frequently performed with a basic catalyst, such as the hydroxides, oxides, alkoxides, cyanides, hydrides, and amides of the alkali metals or the alkali metals themselves. Acids and metallic compounds have also been used, producing high yields. In general, 1–5% of catalyst, based on the weight of acrylonitrile, is satisfactory.

The reaction is strongly exothermic. Solvents used to control it include benzene, toluene, and pyridine.

Reactions

On carbon, at least one hydrogen atom adjacent or alpha to one or more strong electron-withdrawing groups is required (yield ca 80–90%). On oxygen, in the presence of a strongly basic catalyst, hydroxy compounds react readily with acrylonitrile to produce 3-alkoxypropionitriles. The reaction proceeds rapidly with primary alcohols, somewhat more slowly with secondary alcohols, and requires drastic conditions with tertiary alcohols (yields 70–95%). On nitrogen, compounds readily undergo cyanoethylation. However, since many nitrogen compounds have several active sites, selectivity of reaction to achieve the degree of substitution desired has presented problems. In some cases, reaction proceeds without any catalyst or again, basic, acidic or metallic catalyst may be required (yields ca 60–85%).

Cyanoethylated cotton has been the subject of continuing research, but is not at present of commercial importance. There are no known producers of commercial quantities of cyanoethylated cotton. Cyanoethylation of cellulose products imparts heat and rot resistance properties to these substances.

Uses

Important uses are in organic synthesis, eg, in the preparation of amino acids, chemotherapeutic agents, dyes, blowing agents, and numerous intermediates; for cotton (see Textiles); highly cyanoethylated cellulose is preferred material for embedding phosphors in electroluminescent devices (see Luminescent materials).

ROBERT J. HARPER, JR.
United States Department of Agriculture

J. Compton, W.H. Martin, and D.M. Gagarine, *Text. Res. J.* **40**, 813 (1970).

The Chemistry of Acrylonitrile, 2nd ed., American Cyanamid Co., New York, 1959.

H.A. Bruson in R. Adams, ed., *Organic Reaction*, Vol. 5, "Cyanoethylation," John Wiley & Sons, Inc., New York, 1949.

CYANOHYDRINS

A cyanohydrin is an organic compound that contains both a cyanide and a hydroxy group on an aliphatic section of the molecule. The cyanohydrin category is usually comprised of α-hydroxy nitriles which are the products of addition of hydrogen cyanide to the carbonyl group of aldehydes and ketones.

The outstanding chemical property of cyanohydrins is their ready conversion to α-hydroxy acids and their derivatives, especially α-amino and unsaturated acids. All commercially important cyanohydrins occur as chemical intermediates.

Physical and Chemical Properties

Cyanohydrins are usually colorless to straw-yellow liquids with an objectionable odor akin to that of hydrogen cyanide. Table 1 lists selected properties of some cyanohydrins.

Cyanohydrins can react either at the nitrile group or at the hydroxy group. Hydrolysis of the nitrile group proceeds through the amide to the corresponding carboxylic acid. Owing to the instability of cyanohydrins at high pH, this hydrolysis must be carried out with an acid catalyst. In cases where amide hydrolysis is slower than nitrile hydrolysis, the amide may be isolated. Cyanohydrins react with ammonium carbonate to form hydantoins which yield amino acids upon hydrolysis. Reaction with absolute ethanol in the presence of HCl yields the ethyl ester of α-

Table 1.   Physical Properties of Some Cyanohydrins

	Mp, °C	Bp, °C	Sp gr	n_D^{20}
formaldehyde cyanohydrin, (glycolonitrile, $HOCH_2CN$)	−72	119 at 3.2 kPa[a]	1.104	1.4117
acetaldehyde cyanohydrin, (lactonitrile, $HOCH(CH_3)CN$)	−40	182–184	0.988	1.405
acetone cyanohydrin, (methyllactonitrile, $HOC(CH_3)_2CN$)	−19	82 at 3.1 kPa	0.932	1.3992
cyclohexanone cyanohydrin, ($\widehat{CH_2(CH_2)_4C}(CN)OH$)	29	109–113 at 1.2 kPa[a]	1.032	1.4576
benzaldehyde cyanohydrin, ($HOCH(C_6H_5)CN$)	−10	170	1.117	1.5315
ethylene cyanohydrin, (hydracrylonitrile, $HOCH_2CH_2CN$)	−46.2	228, 116–118 at 2.7 kPa[a]	1.059	1.4256
propylene cyanohydrin, (2-methylhydracrylonitrile, $HOCH(CH_3)CH_2CN$)		207		1.4280

[a] To convert kPa to mm Hg, multiply by 7.5.

hydroxy acids. *N*-Substituted amides can be synthesized by heating a cyanohydrin with an amine in water.

The hydroxy group of cyanohydrins is subject to displacement with other electronegative groups. For example, cyanohydrins react with ammonia to yield amino nitriles, a step in the Strecker synthesis of amino acids.

Preparation

Cyanohydrins can be formed by: the acid- or base-catalyzed reaction of hydrogen cyanide with an aldehyde or ketone; the displacement of bisulfite ion by cyanide ion on the bisulfite addition compounds of aldehydes and ketones; or the exchange of cyanide ion between a ketone cyanohydrin and an aldehyde to give the more stable (usually) aldehyde cyanohydrin.

All aliphatic aldehydes and most ketones react to form cyanohydrins. Ketones are usually less reactive than aldehydes. This behavior has been attributed to a combination of electron-donating effects and increased steric hindrance of the second alkyl group in the ketones. The magnitude of the equilibrium constants for the addition of hydrogen cyanide to a carbonyl group is a measure of the stability of the cyanohydrin relative to a carbonyl compound plus hydrogen cyanide.

Direct combination of hydrogen cyanide and a carbonyl compound is the commercial and most common route to cyanohydrins. Production is accomplished through the base-catalyzed combination of hydrogen cyanide and the carbonyl compound in a solvent, usually the cyanohydrin itself. The reaction is carried out at high dilution of feeds, at 10–15°C and a pH pf 6.5–7.5 with the product continuously removed from the reaction zone, cooled, and then stabilized with a mineral acid.

Health and Safety

Cyanohydrins are highly toxic by inhalation or ingestion, and moderately toxic through skin absorption.

Uses

They are primarily used as intermediates in the production of other chemicals. Manufacture of methyl methacrylate from acetone cyanohydrin is the most economically important cyanohydrin process (see Methacrylic acid). Used also as solvents, anti-knock fuel agents, and avicides.

MICHAEL S. CHOLOD
Rohm and Haas Company

E.C. Wagner and M. Baizer in E.C. Horning, ed., *Organic Syntheses*, Coll. Vol. III, John Wiley & Sons, Inc., New York, 1955, p. 323.

V.P. Belikov and co-workers, *Izv. Akad. Nauk SSSR Ser. Khim.* **8**, 1862 (1967).

CYANURIC AND ISOCYANURIC ACIDS

Although cyanuric acid has been known for two hundred years, it did not achieve commercial importance until the mid-1950s and even then mainly via its derivatives. Nomenclature is based on two of the several possible equilibrating tautomeric species, ie, the trioxo and the trihydroxy forms. The trihydroxy form is variously designated cyanuric acid, s-triazine-2,4,6-triol or 2,4,6-trihydroxy-s-triazine. The trioxo structure, or s-triazine-2,4,6(1H,3H,5H)-trione is the basis for the isocyanuric acid nomenclature. The triazine-based system is preferred in the scientific literature and is used by government regulatory agencies; however, the less cumersome (iso)cyanurate designations persist in commerce. Through common usage, both forms are collectively called cyanuric acid (CA).

isocyanuric acid   cyanuric acid (CA)

Physical and Chemical Properties

(Iso)cyanuric acid is a white, odorless, crystalline solid that does not melt up to 330°C; at higher temperature isocyanic acid (HNCO) has been entrapped as a decomposition product and has been used for further reaction.

Solubilization of (iso)cyanuric acid is a frequently encountered problem in developing applications for the materials, eg, water solubility at 25°C is only 0.2 wt%, and at 90°C is 2.6 wt%. It is less than 0.1 wt% soluble at room temperature in most common organic solvents, but is appreciably soluble in dimethyl sulfoxide (17.4 wt%), and sulfuric acid (14.1 wt%) at 25°C. Solubility is frequently achieved through salt formation.

The chemistry of (iso)cyanuric acid is best interpreted in terms of its cyclic triimide structure. Virtually all reactions find their counterparts in straightforward imide chemistry, eg, salt formation, hydrolysis, N-halogenation, and alkylation.

The chemical reactivity of (iso)cyanuric acid is frequently limited by its low solubility in common organic solvents. As in general imide chemistry, reaction of (iso)cyanuric acid normally occurs by replacement of hydrogens at nitrogen: this, of course, results in derivatives of the isocyanurate series.

The N-chloro derivatives of isocyanuric acid (chlorinated s-triazinetrios) are the most important commercial products derived from (iso)cyanuric acid (see Uses). Properties of some of the more important commercial derivatives are listed in Table 1.

Manufacture

Commercial production of (iso)cyanuric acid involves pyrolysis of urea. Historically, pyrolysis of solid urea in a kiln was the general method of large-volume manufacture; newer plants employ a solvent medium. The solid urea is heated at 200–300°C for several hours and (iso)cyanuric acid is produced according to the following equation:

$$3\ H_2NCOHNH_2 \xrightarrow{200-300°C} CA + 3\ NH_3$$

Table 1. Properties of N-Chlorinated Derivatives of Isocyanuric Acid

Property	NaDCC[a]	NaDCC.2 H$_2$O[a]	KDCC[a]	TCCA[b]
available chlorine (theory), %	64.5	55.4[c]	60.1	91.5
pH of 1% solution	5.8–7.0	5.8–7.0	5.9–6.7	2.0–3.7
solubility in H$_2$O at 25°C,				
g/100 mL solution	22.7	> 22.7[c]	9.0	1.2

[a]DCC = dichloroisocyanuric acid.
[b]TCCA = trichloroisocyanuric acids.
[c]Actual values vary according to hydrate water content.

The equation is deceptively simple. During pyrolysis, urea initially forms a free-flowing melt which subsequently thickens and finally solidifies. Components of the pyrolyzate include urea, biuret ($H_2NCONHCONH_2$) and presumably triuret ($H_2NCONHCONHCONH_2$), all intermediates to (iso)cyanuric acid, and aminotriazines which must be removed by acid digestion.

Health and Safety

These compounds are safe for use in bleaching, sanitizing, and swimming-pool disinfection applications when handled properly.

Uses

Most of the cyanuric acid produced commercially is chlorinated to produce sodium dichloroisocyanurate (NaDCC), sodium dichloroisocyanurate dihydrate (NaDCC.2H$_2$O), potassium dichloroisocyanurate (KCDD), trichloroisocyanuric acid (TCCA), and the mixed complex 4 KDCC : 1 TCCA. These have become standard ingredients for scouring powders, bleaches, sanitizers, cleansers, and swimming-pool disinfectants.

JOSEPH V. BURAKEVICH
FMC Corporation

W. Weigert, G. Düsing, N. Kriebitzsch, and H. Pfleger in E. Bartholomé, E. Biekert, H. Hellmann, H. Ley, W.M. Weigert, H. Buchholz-Meisenheimer, J. Frenzel, and R. Pfefferkorn, eds., *Ullmann's Encyklopädie der technischen Chemie*, Band 9, Verlag Chemie, Gmg H, Weinheim/Bergotr., Ger., 1975, pp. 648–665.

CYCLIZATION.   See Petroleum.

CYCLOHEXANE.   See Hydrocarbons, C$_1$–C$_6$.

CYCLOHEXANOL AND CYCLOHEXANONE

Cyclohexanol

$\overset{\frown}{CH_2(CH_2)_4}CHOH$, is a colorless, viscous liquid with a camphoraceous odor. Mp 25.15°C; bp 161.1°C; n_D^{25} 1.4648; soluble in water (30°C) to the extent of 4.3 g/100 g H$_2$O. It is miscible in all proportions with most organic solvents, including those customarily used in lacquers, and dissolves many oils, waxes, gums and resins.

Cyclohexanol shows most of the typical reactions of secondary alcohols. It is prepared commercially by the boric acid-catalyzed air oxidation of cyclohexane at 140–180°C, or by the nickel-catalyzed hydrogenation of phenol carried out at elevated temperatures and pressure in either the liquid or in the vapor phase. The oxidation of cyclohexane to a mixture of cyclohexanol and cyclohexanone, known as KA-oil (ketone–alcohol, cyclohexanone–cyclohexanol crude mixture), is used for most production.

Cyclohexanone

$\overset{\frown}{CH_2(CH_2)_4}CO$, is a colorless, mobile liquid with an odor suggestive of peppermint and acetone. Mp −37°C; bp 156.7°C; n_D^{20} 1.4507. It is soluble to the extent of 5 g/100 g H$_2$O at 30°C, and is miscible with most organic solvents and alcohols. It dissolves many organic compounds.

Cyclohexanone shows most of the typical reactions of aliphatic ketones, such as the usual addition reactions and various condensation reactions that are typical of ketones having α-methylene groups. Reduction converts cyclohexanone to cyclohexanol or cyclohexane, and oxidation with nitric acid converts cyclohexanone almost quantitatively to adipic acid.

The most efficient route to cyclohexanone is the hydrogenation of phenol, in either a liquid phase, catalyzed by palladium on carbon, or in the vapor phase, catalyzed by palladium on alumina. The liquid-phase catalytic air oxidation of cyclohexane is also important to cyclohexanone. It is also produced by the catalytic dehydrogenation of cyclohexanol, and by the oxidation of cyclohexanol.

Health and Safety

Cyclohexanol is slightly to moderately toxic. The toxic-dose limitation for cyclohexanone inhalation is 75 ppm, with a time-weighted average OSHA standard for air 50 of ppm.

Uses

Cyclohexanol is used chiefly as a chemical intermediate, a stabilizer, and as a homogenizer for various soap and detergent emulsions, and as a solvent for lacquers and varnishes. Cyclohexanone is used chiefly as a chemical intermediate and as solvent for resins, lacquers, dyes, and insecticides.

Williams B. Fisher
Jan F. Van Peppen
Allied Chemical Corporation

"Cyclohexanol and Cyclohexanone: Salient Statistics," *Chemical Economics Handbook*, SRI International, Menlo Park, Calif., June 1976.

Registry of Toxic Effects of Chemical Substances, 1976 ed., U.S. Department of Health, Education, and Welfare, 1976.

U.S. Pat. 4,162,267 (July 24, 1979), W.B. Fisher and J.F. Van Peppen (to Allied Corporation).

CYCLOPENTADIENE AND DICYCLOPENTADIENE

Two well-established chemical building blocks are cyclopentadiene (CP), (1), and its more stable and available form, the dimer dicyclopentadiene (DCP), (2). The discovery in 1951 of stable metal derivatives gave added impetus to study the chemistry of cyclopentadiene (see Organometallics).

Physical and Chemical Properties

DCP exists in two steroisomeric forms, the endo and exo isomers (see below). The commercial product is predominantly the endo isomer. DCP, 4,5-methano-3a,4,7,7a-tetrahydroindene, is the form in which cyclopentadiene is commercially available.

CP   DCP
(1)   (2)

Cyclopentadiene is a colorless liquid with a sweet terpenic odor, bp 41.5°C (101.3 kPa); mp −85°C; d_4^{20} 0.8024 g/cm^3. Dicyclopentadiene forms colorless crystals with a camphoraceous odor, with bp 170°C (101.3 kPa) (depolymerizes at boiling point to form two molecules of cyclopentadiene); mp 33.6°C; d_4^{35} 0.9770 g/cm^3.

CP contains two conjugated double bonds and an active methylene group and can thus undergo a diene addition reaction with a wide variety of unsaturated compounds. The diene-addition reaction involves the addition of an ethylenic group contained in the first reactant, called the dienophile, across the 1,4-position of CP. The Diels-Alder products are usually bicyclo[2.2.1]heptene derivatives. The highly exothermic reaction is carried out by simply bringing the two reactants together in the presence or absence of a solvent at temperatures ranging from essentially room temperature in some cases to ca 200°C.

CP polymerizes spontaneously at ordinary temperature to DCP. At temperatures above 100°C, CP can be made to polymerize noncatalytically to tri-, tetra-, and higher polymers, the reactions consisting of a series of consecutive Diels-Alder reactions. The thermal polymers are crystalline compounds with little odor. In addition to thermal polymerization, CP can be polymerized with inorganic halides as catalyst, eg, with trichloroacetic acid catalyst, deeply colored, blue polymers that conduct electricity in nonpolar solvents are obtained.

The methylene group in CP is extremely reactive and can undergo condensation-type reactions. A large number of derivatives have been prepared in this way, eg, aldehydes and ketones condense with CP in the presence of alkaline condensing agents to produce colored fulvene derivatives.

DCP has been progressively hydrogenated through the dihydro to the tetrahydro derivative, and the higher polymers may also be hydrogenated. Halogen and halogen acids add readily to the unsaturated carbon linkages of the CP molecule; of all of the possible chloro derivatives of CP, only the hexachlorocyclopentadiene has reached commercial status. CP reacts spontaneously with oxygen to form brown, gummy products that usually contain substantial amounts of peroxides.

Manufacture

No commercial process for the sole production of cyclopentadiene exists. It is obtained as a by-product from thermal operations, eg, thermal cracking of hydrocarbons, particularly gas-oil and naphtha in the presence of steam; and ethylene (qv) production from thermal cracking of ethane, propane and other raw materials. CP is recovered from other hydrocarbons in thermal cracking by distilling the total cracked product to recover a distillate consisting of C$_5$ hydrocarbons and lighter components. The disallate is heated to a temperature of ca 100°C to convert monomeric CP to DCP. The DCP boils higher than the unreacted hydrocarbons of the distillate, and is recovered as distillation bottoms. Regardless of its source, only the dimer is available since the monomer spontaneously reacts to form the dimer. Some CP can be recovered from the foreruns of coke-oven benzene refining but the rest is from hydrocarbon cracking operations. In the US and Europe, DCP streams of 70–95 wt% purity are available.

Health and Safety

DCP is a moderately toxic material and, to some extent, an irritant and a narcotic. The 1978 TLV for DCP is 5 ppm by the ACGIH on a 8-h exposure time-weighted average.

Uses

Owing to the unusual reactivity of the DCP molecule, there are a number of wide and varying uses. The primary uses are DCP-based hydrocarbon-type resins; elastomers via a third monomer, 5-ethylidene-2-norbornene; polychlorinated pesticides; and polyhalogenated flame retardants. Products obtained from the diene addition reaction are extremely versatile chemicals suitable as intermediates for the production of plasticizers, pharmaceuticals, pesticides, resins, paint driers, perfumes, etc.

M. Fefer
A.B. Small
Exxon Chemical Company

A.F. Plame and J.M. Terent'eva, *Usp. Khim* **20**, 560 (1951).

M. Moulin, *Bull. Assoc. Fr. Tech. Pet.* **135**, 563 (1959).

Chem. Week **94**, 119 (Apr, 25, 1964); A.W. Gailbraith and co-workers, *Lancet* ii, 113 (1971).

D

DAIRY PRODUCTS. See Milk and milk products.

DAMMAR, DAMAR. See Resins, natural.

DATA INTERPRETATION AND CORRELATION. See Engineering and chemical data correlation; Programmable pocket computers.

DDT (p-ClC$_6$H$_4$)$_2$CHCl$_3$. See Insect control technology.

DECAHYDRONAPHTHALENE, DECALIN. See Naphthalene.

DECARBOXYLASES. See Enzymes.

DEFLOCCULATING AGENTS. See Clays; Dispersants; Flocculating agents.

DEFOAMERS

A defoamer is a material that, on addition in low concentration to a foaming liquid, controls the foam problem. Defoamers are complex blends of many different organic compounds. As process aids, they improve filtration, dewatering, washing, and drainage of many types of suspensions, mixtures, or slurries.

Use Requirements of a Defoamer

To be marketable, a proprietary defoamer must meet five basic requirements in addition to any special criteria for a particular industry or process; cost efficiency; ease of handling; specificity of action; absence of any adverse effect on the product, and environmental safety.

Functional Classifications of Defoamer Components

A defoamer must be reasonably inert in the foaming system, because it must exhibit excellent longevity to be economically justified. It must not be capable of reacting with the finished product or the aqueous system for which it was designed. Almost all defoamers are considered unreactive in the particular process to which they are applied.

Five general functional classifications are used. Many defoamers contain more than five components and some contain only two or three with overlapping of the function of each component.

Primary antifoam agent. The primary antifoam agent, the main active ingredient in the defoamer and the backbone of the defoamer formulation, frequently consists of highly insoluble particulate materials such as hydrophobic (silicone-treated) silica, fatty amides, hydrocarbon waxes, fatty acids and fatty esters. They are usually the most costly ingredients in the product.

Secondary antifoam agents. The secondary agents modify the surface effect of the primary antifoam agent by changing the spreading, solubility, or crystallinity of that material with regard to the carrier and the foaming medium. Some of these are fatty alcohols, fatty esters, silicones, and certain oil-insoluble polymers.

Carriers. Carriers, or vehicles, comprise the bulk of the defoamer formulation. They are usually hydrocarbon oils or water, or fatty alcohols, solvents, or fatty esters.

Emulsifiers. Emulsifiers, or spreading agents, function by introducing the main components (primary and secondary antifoam agents and carrier) into the system. These products dictate the speed of foam decay by spreading the product quickly throughout the foaming medium (see Emulsions).

Coupling or stabilizing agents. These agents are additives that contribute to defoamer stability or shelf life. Examples include red oil (oleic acid), hexylene glycol, fatty alcohols, naphthalenesulfonates, butyl alcohol, and formaldehyde.

Defoamers in Current Use

Blends of fats and oils. Blended products have been used for years in pulp and paper, sugar beet, fertilizer and effluent applications. Typical formulations may consist of solutions of saturated fatty acids and fatty alcohols in mineral oil carriers or blends of polyethylene glycol esters. The disadvantage of this class of defoamer is the titer effect that occurs when dissolved fats or fatty acids come out of solution at low temperatures, with the product losing effectiveness owing to lack of homogeneity.

Paste-type defoamers. Paste defoamers consist of immobile oil-in-water (o/w) emulsions of fatty acid soaps, esters of saturated fats, paraffin waxes, mineral oils, emulsifiers, and the like. These were the first large-volume defoamers. These rather stiff pastes are usually converted to dilute emulsions before introduction to the foamy system. The disadvantages of these defoamers is that they are affected by temperature, they form hard-water soaps, they cream-out owing to lack of agitation, and they form insoluble deposits in the feed system.

Dispersion-type defoamers. Basically, these are dispersions of finely divided particulate matter such as hydrophobic silica or quick-chilled amides in various insoluble vehicles, eg, mineral oils, kerosene, fatty alcohols, silicone oils and the like. They have the advantage of being easy to handle owing to relatively low viscosities and low pour points, and appear to be the most effective and versatile of the types in use.

Water-based defoamers. Water-based defoamers are fluid, as easy-to-handle as the dispersion-type defoamers, and some are more pumpable than the blended fat defoamers at proper temperature. They are considered combination dispersion and o/w emulsion products, composed of fatty alcohols, fatty acid soaps, fatty amides, emulsifiers, ethoxylates, mineral oils, and esters.

Other types of defoamers. Other defoamers include: silicone defoamers, dispersions of finely divided particles of silica dispersed in polydimethylsiloxane or similar silicones; solid defoamers composed of waxes, esters, fatty alcohols, soaps, etc; surfactants (see Surfactants); powdered defoamers; simple compounds such as fatty oils and polyethylene glycol esters; and defoamers for nonaqueous systems, including silicones, and fluorinated compounds.

Defoamer application is a large and very specialized market. Some industries rely heavily on the efficient and economical use of defoamers, eg, the pulp and paper industry, the paint and latex industry, coating processes of all types, the fertilizer industry, the textile industry, fermentation-based processes, the metal-working industry, adhesive manufacture, polymer manufacture, the sugar-beet industry, and various chemical processes.

Recently, the cost of mineral oils has risen to the point where they are no longer cost effective as a component in many applications. Products that make use of different blends and higher concentrations of primary and secondary antifoam agents have taken their place. These compounds are dispensed in water as either o/w or w/o emulsions. Many of these "water-based" defoamers have proven equal or superior to oil-based products on a cost basis.

IRWIN LICHTMAN
TED GAMMON
Diamond Shamrock Chemical Co.

H.T. Kerner, *Foam Control Agents*, Noyes Data Corp., Park Ridge, N.J., 1976.

U.S. Pat. 2,797,198 (June 25, 1957), F.L. Chappell (to Hercules Powder Company).

DEFORMATION RECORDING MEDIA

Deformation recording media are deformable materials used for recording information in the form of ripples or grooves. The deformations, produced by selectively depressing the surface of the recording material, represent relief images that can be projected by optical techniques to give black-and-white or colored images. Images are developed by heat without need for fixing, and processing requires no chemicals; development is dry and fast. The images can be erased and the media can, in principle, be reused.

The depth of deformations is usually very small (0.1–1.1 μm). The grooves cause a retardation of transmitted (or reflected) light and change the phase of light. This change can be detected with a schlieren optical system.

Recording Media

These are classified according to the type of stimulus used for producing the electrostatic image, electron beam or light, and according to the physical state of the deformable component. Practically all recording media consist of at least three components, ie, deformable layer, electrically conductive layer which serves as an electrical ground plane, and substrate.

Electron-beam-sensitive fluid media. These materials form the basis of light-valve television systems capable of producing monochrome and color images projected on large screens. Two systems are at present in use. Whereas the operations of these vary in detail, both employ a scanning electron beam to deposit a raster of electron charges on the surface of a fluid layer. The beam is modulated by electric signals corresponding to the images to be projected. The electrostatic forces associated with deposited charges deform the surface of the fluid—called eidophor (image bearer)—and the resulting deformation images are projected with schlieren optics. The fluids must have appropriate dielectric and physical properties for use in television. Electrical resistivity must be sufficiently high (about 10^{11} $\Omega \cdot$cm at the operating temperature) to permit the buildup of a proper potential for deformation and to prevent loss of resolution of the raster by lateral dissipation of charge.

After the fluid has deformed, the charge has to decay so that self-erasure due to the restoring forces of surface tension can take place within a TV time frame (1/30 s). The fluid should have high resistance to electron radiation, and must be photochemically stable so as not to be degraded by the radiation of the projector light source. The preferred materials are polysiloxanes (see Silicon compounds, silicones) and products made from benzyl chloride and aromatic compounds such as naphthalene and toluene.

Electron-beam-sensitive solid media. Information recording with an electron beam on thermoplastic films was first described as thermoplastic recording. Recording and reading are similar to the eidophor, except that the images are developed by heating and fixed by cooling. The transparent recording medium is a thin film of a dielectric thermoplastic on a conductive tape substrate. A satisfactory material is a blend of a poly(diphenylsiloxane) and a commercially available poly(2,6-dimethyl-1,4-phenylene ether) (PPO) (see Polyethers). The sensitivity of thermoplastic film is ca 10^{-7}–10^{-8} C/cm^2, and resolving power is in excess of 500 lp (line pairs)/mm.

Photosensitive solid media. Photosensitive solid media, designed for photographic or reprographic applications, are basically photoconductors (see Electrophotography). The general recording technique involves sensitizing the film in the dark by an electric charge, exposing it to an optical image, and then heating it to develop the image. When the surface potential exceeds a certain value, the film tends to deform into worm-like light-scattering patterns called frost because of their frosted appearance; this may serve as a natural carrier for continuous tone rendition. Frost images scatter light at large angles and do not require schlieren projection for readout. An ordinary projector with a small aperture, eg, f/16 or smaller, gives satisfactory results.

Single-layer configurations. A single-layer configuration has a photoconductive and thermoplastic layer of three types of compositions: dispersions of insoluble photoconductor particles in a thermoplastic polymer; solid solutions of an organic photoconductor in a thermoplastic polymer; and inherently photoconductive thermoplastic polymers.

Double-layer configuration. The double-layer configuration, a technique also called thermoplastic xerography or frost process, comprises adjacent photoconductor and thermoplastic layers.

Photosensitive elastomeric media. Elastomeric media are similar to the double-layer configuration, except that an elastomer layer is substituted for the thermoplastic (see Elastomers, synthetic). The device, called ruticon, is made of a transparent conductive substrate, thin photoconductive layer, thin elastomer layer, and a deformable electrode. The latter may be a conductive liquid such as mercury or gallium–indium alloy (α-ruticon), a conductive gas such as argon operated in the gas discharge mode (β-ruticon), or a thin metal mirror (γ-ruticon) (this type has been used as a TV projection display).

SIEGFRIED AFTERGUT
General Electric Company

W.E. Good, *IEEE Trans. Broadcast Telev. Rec.* **15**, 21 (1969).

W.E. Glenn, *J. Appl. Phys.* **30**, 1870 (1959).

J. Gaynor, *IEEE Trans. Electron Devices* **19**, 512 (1972).

DEGREASING AGENTS.   See Metal-surface treatments.

DEHUMIDIFICATION.   See Air conditioning.

DELRIN.   See Acetal resins.

DENSITY AND SPECIFIC GRAVITY.   See Analytical methods.

DENTAL MATERIALS

Much of dental therapy involves replacement of hard and soft oral tissues lost through disease with inert materials (metallic, ceramic, and organic) or with composites employing combinations of these three broad classes. Dental restorations and prostheses are made of amalgam, chromium-based alloys, precious-metal alloys, special cements, synthetic polymers, composites and porcelain, all of which must withstand the rigors of the mouth (see also Prosthetic and biomedical devices). The accessory materials needed in the fabrication procedures include synthetic polymers, gums and waxes, both synthetic and natural, hydrocolloids, gypsums, and refractories.

Dental cements. Dental cements are used in ca 50% of all dental restorations as cementing (luting) agents for fixed restorations and orthodontic appliances, as a base under permanent restorations to insulate against thermal and chemical shock, for pulp capping, root-canal sealers, and temporary fillings. The various types include zinc phosphate, silicate, zinc silicate, resin, zinc oxide-eugenol, EBA (zinc oxide-eugenol-o-ethoxybenzoic acid), polycarboxylate and glass–ionomer cements.

Gypsums. Gypsums, separately, and in combination with other substances, are used as impression and dental cast materials, which are used in processing artificial dentures and as a binder in refractories for forming molds for the lost-wax process of casting both precious and base-metal alloys. The latter combination is called a dental investment.

In this specialized form of the lost-wax casting process, the cavity is formed in the mold by burning out an expendable wax or plastic pattern that has been imbedded in the investment. Various investments generally consist of a powdered refractory, a binder, and modifiers. The refractory is usually some crystalline form of silica (SiO_2) (mainly quartz and/or crystobalite) for use with precious-metal casting alloys and for base-metal alloys with relatively low melting ranges. For high melting point gold alloys and most base-metal casting alloys, phosphate bonded investments are used.

Dental porcelain. A fine ceramic powder, pigmented to produce the color and translucency of human teeth, is mixed with water to form a paste which is formed into the desired shape and fused to form a ceramic body that is relatively strong, insoluble in oral fluids, and has excellent aesthetic qualities. One type of dental porcelain is used for artificial teeth, another for jacket crowns and inlays, and a third type, more properly designated as an enamel, is used as a veneer over cast metals (see Enamels).

Porcelain teeth, comprising about 20% of artificial teeth, are generally made of a mixture of fine particles of feldspar, quartz, and kaolin, heated to ca 1250–1500°C. After fusing, this mixture and other lower maturing mixtures, are then quenched in water, causing considerable cracking and fracturing and producing a frit which can be readily ground to a fine powder of almost colloidal dimensions. These porcelains are translucent. Opaquing agents, eg, oxides of zirconium, tin, or titanium, are used to mask the color of dentin, or more frequently the color of the metal to which the porcelain is fused.

Porcelain-to-metal bonded systems. When porcelains or glasses are used in porcelain-to-metal (metal–ceramic) systems, they are called ceramic enamels (see Ceramics). The technique is as follows: a thin metal casting is made to fit the prepared tooth, and the ceramic is then fused as a veneer on the metal casting so that little or no metal is visible. A layer of opaque porcelain is fused against the casting, and then the tooth contour is built by using an overlay of translucent enamel. The final enamel-veneered structure is then cemented on the prepared tooth.

The alloys used for the construction of metal–ceramic restorations have a number of stringent requirements. Both the metal and ceramic must have coefficients of thermal expansion closely matched to prevent undesirable tensile stresses at the interface. The metal should have a high proportional limit and particularly a high modulus of elasticity; these, together with the bulk of the alloy, determine the ability of the casting to resist deformation. Compositions of some veneering porcelains show high silica and alumina contents.

Chemical bonding is the most probable mechanism, and the bond probably occurs through diffusion of oxides from the metal to the porcelain. Manufacturers generally recommend degassing the metal by heating the alloy to ca 1035°C before veneering. Handling the metal with the fingers before fusing produces a volatile contaminant.

Colloids

Agar-based impression materials (reversible hydrocolloids). Agar-based impression materials are thermally reversible, aqueous gels that become viscous fluids in boiling water and set to an elastic gel when cooled below 35°C. About 6–12% agar is used with a 75–85% water content. Filler may include zinc oxides and clays, and small percentages of boron compounds that increase the viscosity, strength, toughness and resilience of the compositions.

The agar-based materials are also used extensively for duplicating casts.

Alginate-based impression materials (irreversible hydrocolloids). Alginate impression materials are usually the potassium or sodium salts of alginic acid, a high molecular weight linear polymer of anhydro-β-d-mannuronic acid derived from specific varieties of kelp. Alginate impression materials are supplied as a dry powder. When the correct proportions of the powder and water are mixed, a viscous but slightly fluid mass is formed. A tough and elastic gel is formed, based upon the conversion of soluble salts to insoluble salts of alginic acid, usually calcium alginate.

Elastomers

Polysulfide impression materials. In 1953, the first nonaqueous elastic dental impression material was introduced. This material was based on the room-temperature conversion of a liquid polymer, a polyfunctional mercaptan (polysulfide), to a strong, tough, dimensionally accurate elastomer. This first elastomeric impression material was supplied as two pastes, usually labeled base and catalyst. The active component (80%) in the base was a low molecular weight polymer with a terminal and a few side-chain mercaptan groups. The limited side chains permitted crosslinking. The most common catalyst is lead peroxide (PbO_2). The polymerization of the mixed polysulfide and catalyst pastes results from oxidation of the mercaptan group producing a polysulfide polymer with the liberation of water as a by-product. Most curing agents produce an initial solid mass characterized by a high degree of plasticity but poor elasticity. The development of adequate elastic properties is so slow that the materials are not suitable for accurate reproduction of severe undercut areas of interest. The polysulfides do have quite low dimensional change if gypsum casts are poured within 30 min. They also have high tear strength and good reproduction of detail.

Silicone impression materials. The silicone impression materials were introduced to overcome some of the problems associated with the polysulfide materials, such as slow setting, odor, staining and distortion upon removal from severe undercuts.

Two types of silicone materials are available, one polymerized by a condensation reaction and the other by an addition reaction. The condensation-reaction materials were introduced first. These are supplied as two-paste, paste–liquid, or putty–liquid systems. The base consists of 60–65% low molecular weight silicone polymer with terminal hydroxy groups, a small amount of ethyl silicate functioning as a cross-linking agent and 30–40% silica reinforcing filler. The catalyst consists of stannous octanoate or butyl tin laurate. The polymerization, after mixing the two ingredients, results from condensation of the terminal hydroxy groups and the ethoxy groups on the ethyl silicate to produce a rubber and ethyl alcohol as a by-product. This release of ethyl alcohol causes severe dimensional changes in the impression if the casts are not poured within 20 minutes after completion of the mix.

This large dimensional change after polymerization was partially overcome by increasing the silica filler content to 75% to make the putty form. This putty, when combined with a thin layer of low viscosity wash, increases the accuracy of the condensation silicone impressions. The other type of silicone that cures by an addition reaction increases the dimensional stability of elastomeric impression materials. This material is also supplied in the two-paste form. One paste contains a low molecular weight silicone with terminal vinyl groups, reinforcing filler, and chloroplatinic acid catalyst. The second paste contains a low molecular weight silicone with terminal silane hydrogens and a reinforcing filler. When the two pastes are mixed, the addition reaction occurs between the hydrogen and vinyl groups, no by-product being formed. The material has a relatively short working time and rather low flexibility. Advantages include excellent elasticity or low permanent deformation, low flow and very low dimensional change after setting.

Polyether impression materials. The polyether impression materials are also supplied in a two-paste system. The base paste contains a low molecular weight polyether with imine terminal groups. The catalyst paste contains an aromatic alkyl sulfonate and a thickening agent. When mixed, the reaction results in cross-linking by cationic polymerization via the imine terminal groups.

The setting time is relatively short and the material has very low flexibility. Advantages of this material include excellent elasticity, low permanent deformation, low flow, and when stored dry, very low dimensional change after setting. The polyether material has water-sorption values of about 14% at equilibrium in water The material should not be in contact with water during setting and storage.

Polymers

Synthetic polymers in dentistry. The use of synthetic polymers in dentistry involves combinations of unique properties. There are overall requirements that have general applications, such as minimal dimensional changes during and subsequent to polymerization, lifelike appearance, color stability, passivity, biocompatibility, and suitability for dental techniques. The polymers that best satisfy the foregoing are the acrylic resins.

Polymerization is induced by heat, chemical means or electromagnetic irradiation. For the polymerization of denture bases, a dough is formed by blending a powder (polymer) and liquid (monomer). This dough is compressed and heated in a gypsum mold. The molecular weight of the resultant denture-base acrylic is as high as 600,000 and such a polymer will have a higher molecular weight if cross-linked. Heat is the most common medium for polymerizing. Benzoyl peroxide is the usual initiator. It decomposes at relatively low temperatures (50–100°C) generating free radicals which are potent initiators of the polymerization. There are several chemicals, eg, tertiary amines, that induce the decomposition of the peroxide.

The presence of impurities impedes the reaction, and some chemicals inhibit the polymerization. Butylated hydroxytoluene is the inhibitor commonly used in the monomer (0.006% or less) to prevent premature polymerization until curing is desired.

Radiopaque denture materials. The detection of dentures or fragments of dentures that have been swallowed, inhaled, or imbedded in the tissues has been difficult because the materials are radiolucent. Radiopaque materials that satisfy current ADA specifications for denture base polymers are not deficient in strength or color stability (see Radiopaques).

Restorative resins. Resins were introduced as tooth-restorative material in the early 1940s. These materials can be classified as unfilled tooth-restorative-resins, composite or filled restorative resins, and pit and fissure sealants.

Unfilled tooth-restorative resins. Chemically-cured resins are used in this direct-filling aesthetic material for conservative restorations. Their physical properties limit the use of unfilled resins to restoring for aesthetics those surfaces of the teeth not subject to load bearing. The color stability of the unfilled resins is unsatisfactory because of the straw-colored reaction products of amine initiators used to decompose the benzoyl peroxide. Some products use a sulfinic acid activator giving marked improvements in color stability. Ultraviolet absorbers are also added to most present-day tooth-restorative resins to lessen discoloration. Weaknesses of these unfilled resins include a high volume shrinkage during curing (6–7 vol%), a coefficient of linear thermal expansion (100 ppm/°C) that is excessively different from that of tooth structure (10 ppm/°C), and low resistance to abrasion.

Composite tooth-restorative materials. The addition of silane-treated inorganic fillers to unfilled resins has improved the properties of the resin restorative materials. The addition reaction product of bisphenol A and glycidyl methacrylate is an adduct of epoxy and methacrylate resins. This BIS–GMA resin, which polymerizes through the methacrylate groups, is not an epoxy resin molecule. Mineral fillers coated with a silane coupling agent were incorporated into this BIS–GMA monomer which was diluted with other methacrylate monomers to make it less viscous and which bonded the powdered inorganic fillers chemically to the resin matrix. In addition to the peroxide–amine mechanism for polymerization at ambient room or mouth temperatures, a photoinitiator such as benzoin methyl ether is decomposed by ultraviolet radiation into free radicals which initiate polymerization. A similar initiating mechanism occurs using camphoroquinone with an activator such as a tertiary amine and visible light.

Pit-and-fissure sealants. The BIS–GMA resin portion of the composite restorative material has been further diluted with methyl methacrylate monomer or other low viscosity monomers and used to seal poorly developed pits and fissures. In incompletely closed or exceptionally deep pits and fissures, decay tends to occur after the teeth erupt. By properly acid-etching the enamel surface within and surrounding the pits and fissures, good adhesion is attained. The low viscosity, modified BIS–GMA flows into the porosities created by the etching and fills and effectively seals the pits and fissures. This sealing is effective clinically in reducing the rate of decay in pits and fissures.

Metals and Alloys

Metallurgical and dental science have given dentistry many excellent alloys for restorative dentistry, prosthetic dentistry, orthodontics, and dental techniques (see Prosthetic and biomedical materials). Examples include amalgams; base-metal alloys (cobalt–chromium casting alloys for partial dentures, chromium-containing casting alloys for crowns and bridges); foil and crystal gold; Au–Ca alloy and gold-casting and wrought alloys; platinum and platinum alloys; gold and silver solders and electrodeposited copper and silver.

Amalgams. Amalgams are one of dentistry's main therapeutic agents which are used extensively in restoring single teeth. Dental amalgam is a novel alloy. Within a few seconds after the start of trituration, the alloy and mercury amalgamate to a smooth plastic mass which is packed into the prepared cavity. After 3–5 minutes, it sets to a carvable mass and remains so for 15 minutes. Within two hours, it develops sufficient strength, hardness, and toughness to resist mild biting and chewing forces and does not tarnish or corrode extensively in the mouth nor react to produce toxic or soluble salts. A satisfactory restoration should have a minimum mercury content. High mercury content drastically reduces values for pertinent properties. The most widely used powdered alloys are commonly referred to as high copper content and contain silver, tin, copper, and sometimes zinc, palladium or indium.

Base-metal alloys. Although inferior to gold-based casting alloys in some dental applications, base-metal casting alloys are sometimes superior, eg, castings for partial dentures, and are gradually replacing precious metal alloys for all types of dental castings. Research on titanium and its alloys for dentistry is being actively pursued.

Cast chromium-containing alloys. These products are used for removable partial dentures owing to their light weight, low cost, stiffness, and general passiveness (see Chromium and chromium alloys). The special materials, devices and techniques that have evolved to fabricate restorations from the chromium-containing alloys have essentially restricted the use of these alloys to commercial dental laboratories. These complex alloys may contain four to nine principal constituents, eg, chromium, cobalt, nickel, molybdenum, and beryllium, plus several minor ingredients, eg, carbon, iron, silicon, manganese, and nitrogen, all of which exert a significant influence on the properties of the alloys.

Since precious metals are expensive, intensive development of alloys based on nickel–chromium with as many as eight modifying elements has been pursued, and these alloys are being used successfully with the advocated techniques.

Stainless steels. Stainless steels of various types are being used in dentistry for orthodontic appliances, space maintainers, crown forms, and a variety of instruments including root-canal files and reamers for endodontic treatment. The detailed properties of 18-8 stainless steels may be obtained by reference to appropriate handbooks.

The alloys most often used for orthodontic appliances are selected from the low carbon austenitic stainless steels, containing ca 17–19% Cr, 7–9% Ni, and 0.08–0.2% C with Si limited to 0.75% max and Mn limited to 0.60% max.

Electrodeposited metals. Electrodeposited, electroplated, or electroformed copper and silver are used for producing accurate metal-clad casts from compound or elastomeric impression materials (see Electroplating).

Gold and gold alloys. Cohesive and noncohesive gold foil and crystal gold are cold-welded into cavities. Dental gold alloys contain gold, silver, and copper often modified with iridium, indium, nickel, tin, iron, and zinc.

Wrought gold alloys are sometimes used for the fabrication of orthodontic and prosthetic dental appliances but have to a great extent been replaced with chromium-containing alloys.

Platinum and platinum alloys. Platinum is used as a foil for the matrix in the construction of fused-porcelain restorations and laminated with gold foil for cold-welded metal restorations.

Dental Impression Compounds

Dental impression compounds are thermoplastic compositions designed for taking impressions of oral structures. The compounds are softened by dry heat or in warm water at 50–60°C, depending upon the type of compound, until they are thoroughly plastic and free of lumps. The material is adapted to a previously prepared metal tray, placed in the mouth where an impression is desired. The assorted compounds are designed with variations in softening temperatures, plastic ranges, plasticity, and body.

Composition. The gums, resins, and waxes of dental impression compounds are essentially natural products. The objective of many manufacturers has been to improve formulas and to take advantage of cleaner, cheaper, more uniform, and more available synthetic products. Frequently, such substitutions are most difficult to achieve without sacrificing some of the more subtle qualities that characterize a particular product. Essentially, the impression-compound formulas include materials from the following classifications: thermoplastic resins, which may include the coumarone–indene resins, shellac; plasticizers, which may be selected from stearic acid, carnauba wax, Japan wax (see Waxes), palm oil, or synthetics; fillers for body, which may include talc and calcium carbonate; and pigments for color, which are selected from the iron oxides, bone black, and titanium dioxide.

Dental Waxes

Waxes and wax compositions have been important in dentistry for many years. The diversity of their usefulness is illustrated by pattern waxes, impression waxes, bite-registration waxes and disclosing waxes; processing waxes, including boxing waxes, sticky waxes, and utility waxes; and study waxes (carving wax).

Two important characteristics of dental waxes are flow and thermal expansion. Dental wax compositions are usually trade secrets.

Abrasives

Dental abrasives range in fineness from those that do not damage tooth dentin to those that are coarse and hard enough to cut tooth enamel or the hardest of the restorative materials (see Abrasives). Dental abrasives can be classified according to their use or according to their degree of abrasivity (see also Dentifrices).

Adhesives

The fundamentals of adhesion have a role in many aspects of dentistry, especially the adhesion of restorative materials to the tooth tissues, enamel, and dentin (see Adhesives). The retention of tooth restorations depends upon either the mechanical locking in cavities prepared with undercuts or microstructural roughness etched in the enamel (created by acid treatment) into which tags of the resin project. Some chemical bonding systems have promise: the chelating effect of carboxyl groups in poly(acrylic acid) to the calcium in the tooth and a bonding system using an acidic mordant solution, a surface-active co-monomer and a coupling agent.

Health and Safety

In addition to specifications for dental materials, it is necessary to have clinical and biological evaluations of any new materials, or of any material that has been in use for a long time if substantial changes are made. Biological and clinical evaluations include toxicity tests for implantation, acute systemic toxicity, mucous-membrane irritation, irritation of the tooth pulp, and efficacy for the capping of exposed pulps.

The Council on Dental Materials, Instruments and Equipment of the American Dental Association has established specifications for dental materials and devices, guidelines for the clinical evaluation for dental materials, and recommended standard practices for biological evaluations of dental materials. The Council also presents status reports on the safety and efficacy of various dental materials and devices. Inquiries should be addressed to the American Dental Association, Chicago, Ill.

The FDA has the responsibility of enforcing the Federal Food, Drug, and Cosmetic Act as amended and the Radiation Control for Health and Safety Act. Dental materials and devices come under the amended Act, so any new material or substantially modified old materials have to have premarket clearance from the FDA prior to introduction in interstate commerce.

G.C. Paffenbarger
N.W. Rupp
American Dental Association Health Foundation
Research Unit, National Bureau of Standards

R.W. Phillips, *Skinner's Science of Dental Materials*, 8th ed., W.B. Saunders Co., Philadelphia, Pa., 1982.

R.G. Craig, ed., *Restorative Dental Materials*, 6th ed., C.V. Mosby Co., St. Louis, 1980.

Index of Dental Literature, American Dental Association, Chicago, Ill. (yearly publication).

Journal of Dental Research, International Association for Dental Research, Washington, D.C.

DENTIFRICES

A dentifrice is a substance or preparation used with a toothbrush to aid mechanical cleaning of the accessible surfaces of the teeth. A typical formulation for a dentifrice paste contains abrasives (generally insoluble inorganic salts), flavoring mixture, humectants (glycerol, sorbitol), thickening agents (carboxymethyl cellulose, carrageenan), foaming agents, and water. Commercial dentifrices are available as pastes, powders, and gels.

From a mechanical standpoint, the most important ingredient is the abrasive agent, eg, several phosphate salts such as dicalcium phosphate, insoluble sodium metaphosphate, calcium pyrophosphate, calcium carbonate, magnesium carbonate, hydrated aluminum oxide, as well as silicates and dehydrated silica gels.

On the basis of clinical studies involving the use of dentifrices containing sodium monofluorophosphate, Na_2PFO_3, as the active ingredient, two dentifrices, Colgate with MFP and Macleans Fluoride, have been accepted by the Council on Dental Therapeutics of the American Dental Association (ADA) as effective in helping to prevent caries. When incorporated in dentifrice formulations at a level of ca 0.76%, it has been shown to be of benefit in 17–38% reduction of dental caries. These findings are in the same general range as those reported for dentifrices already accepted by the ADA that contain 0.4% stannous fluoride (SnF_2) (Crest and Aim).

Mouthwashes

A mouthwash is a solution for rinsing the teeth and mouth and usually serves as an adjunct in cleaning the mouth and as an aid in removing loose debris after brushing. The ADA does not recognize any substantial contribution to oral health in the unsupervised use of medicated mouthwashes by the general public.

Anthony F. Posteraro
College of Dentistry
New York University

A. Osol, ed., *Blakiston's Gould Medical Dictionary*, 3rd ed., McGraw-Hill, New York, 1972.

J. Am. Dent. Assoc. **81**, 1177 (1970).

Accepted Dental Therapeutics, 39th ed., American Dental Association, Chicago, Ill., 1982, pp. 369–378.

DESIGN OF EXPERIMENTS

The main reason for designing an experiment statistically is to obtain unambiguous results at a minimum cost. Although results of a well-planned experiment are often evident from simple graphical analyses, the best statistical analysis cannot rescue a poorly planned experimental design.

Statistically planned experiments are characterized by: the proper consideration of extraneous variables; the fact that primary variables are changed together, rather than one at a time, in order to obtain information about the magnitude and nature of the interactions of interest and to gain improved precision in the final estimates; and built-in procedures for measuring the various sources of random variation and for obtaining a valid measure of experimental error.

Variables

Primary variables are those that created the need for the investigation. Such variables may be quantitative (eg, concentration of catalyst, temperature, or pressure) and are frequently related to the performance variable by some assumed statistical relationship or model; or they may be qualitative. Qualitative variables are either fixed effects (Type 1) whose specific mean effects are to be compared directly, or random effects (Type 2), whose individual contributions to performance variability are to be evaluated (eg, material batch when one is not interested in specific batches *per se*, but in the magnitude of variation in performance caused by differences between batches). When there are two or more variables, they might interact with one another, and an important purpose of a designed experiment is often to obtain information about interactions among primary variables.

Background variables and blocking. In addition to the primary controllable variables, there are variables that cannot be held constant. Such variables arise, for example, when an experiment is to be run over a number of days, or when different machines or operators are to be used. If these background variables do not interact with the primary variables, they may be introduced into the experiment in the form of experimental blocks. An experimental block represents a relatively homogeneous set of conditions within which different conditions of the primary variables are compared. A main reason for running an experiment in blocks is to ensure that the effect of a background variable does not contaminate evaluation of the primary variables.

Uncontrolled variables and randomization. Other variables, such as ambient conditions, can be identified but not controlled. To ensure that such uncontrolled variables do not bias the results, randomization is introduced in various ways into the experiment to the extent that doing so is practical. Randomization means that the sequence of preparing experimental units, assigning treatments, running tests, taking measurements, and so forth, is randomly determined, based, for example, on numbers selected from a random-number table. The total effect of such uncontrolled variables is thus lumped together into experimental error as unaccounted variability.

Variables held constant. Holding a variable constant limits the size and complexity of the experiment, but can also limit the scope of the resulting inferences.

Experimental Environment and Constraints

The operational conditions under which the experiment is to be conducted and the manner in which each of the factors is varied must be clearly spelled out. In many programs, variables are introduced at different operational levels. This gives rise to "split-plot" experiments, in which more precise information is obtained on the low level variables than on the high level variables.

Different types of repeat information. The various ways of obtaining repeat results in the experiment need to be specified, eg, taking replicate measurements on the same experimental unit, cutting a sample in half at the end of the experiment and obtaining a reading on each half, or taking readings on two samples prepared independently of each other. Initial estimates of overall repeatability should be obtained before embarking on any major test program and repeat runs, under supposedly identical conditions, should be included in the experiment.

Running the Experiment in Stages

A statistically planned experiment does not require all testing to be conducted at one time. Instead, the program can be conducted in stages of, eg, 8–20 runs; this permits changes to be made in later tests, if needed, based on early results and allows preliminary results to be reported.

Other Considerations

Other matters that warrant consideration in planning an experiment include: the most meaningful way to express the controllable or independent variables; the proper experimental range for the selected quantitative variables; the statistical model, or equation, to approximate the relationship between the independent variables and each response variable; the desired degree of precision of the final results and conclusions; previous benchmarks of performance; and the statistical techniques required for the analysis.

GERALD J. HAHN
General Electric Co.

G.J. Hahn, *Chem. Technol.* **10**, 36 (Jan. 1980).

V.L. Anderson and R.A. McLean, *Design of Experiments—A Realistic Approach*, Marcel Dekker, Inc., New York, 1974.

G.E.P. Box, W.G. Hunter, and J.S. Hunter, *Statistics for Experimenters*, John Wiley & Sons, Inc., New York, 1978.

DETERGENCY.   See Surfactants and detersive systems.

DETONATING AGENTS.   See Explosives.

DEUTERIUM AND TRITIUM

The element hydrogen has three known isotopes. These hydrogen species are identical in atomic number, ie, they have identical extranuclear electronic configurations ($1s^1$) but they differ in nuclear mass. Over 99.98% of the hydrogen in nature has a nucleus consisting of a single proton, and therefore, has mass 1 (symbol ^{1}H). Two heavier isotopes of hydrogen, present in small amount in nature, of mass 2 and 3 are also known; these have nuclei consisting of one proton and one neutron (deuterium) or one proton and two neutrons (tritium). Whereas ordinary hydrogen and deuterium are stable isotopes, tritium is unstable and its nucleus undergoes radioactive decay (see Radioisotopes). Both isotopically pure deuterium and tritium can now be manufactured on a large industrial scale. Heavy water, D_2O, is the most important compound of deuterium and the only form of deuterium produced and used on a large scale.

DEUTERIUM

Deuterium (symbol ^{2}H or D) occurs in nature in all hydrogen-containing compounds to the extent of about 0.0145 atom%. Small but real differences in the deuterium content of water from various sources (rain, snow, glaciers, freshwater, seawater from different oceans) can readily be detected, and variations in the natural abundance of deuterium resulting from evaporation, precipitation, and molecular exchange make it possible to draw far-reaching conclusions about the genesis and geological history of natural waters.

Deuterium was first isolated in relatively pure form by Urey and co-workers at Columbia University in 1931, and nearly pure D_2O was prepared by G.N. Lewis shortly thereafter by electrolysis. Differences in rates of reaction between corresponding ^{1}H and D compounds are an important tool for the elucidation of organic reaction mechanisms. The recognition in 1940 that deuterium as heavy water has nuclear properties that make it a highly desirable moderator and coolant for nuclear reactors (qv) fueled by uranium (qv) of natural isotopic composition stimulated the development of industrial processes for the manufacture of heavy water.

Physical Properties

Although the chemical and physical properties of all isotopes of an element are qualitatively the same, there are quantitative differences among them. The physical and chemical differences between the hydrogen isotopes are relatively much greater than those among the isotopes of all other elements because of their large relative differences in mass (H : D : T = 1 : 2 : 3).

Some physical properties of liquid H_2 and D_2 at 20.4 K and light and heavy water are presented in Tables 1 and 2.

Deuterium has very desirable properties as a moderator for neutrons. The principal difference in chemical behavior between H and D derives from the generally greater stability of covalent chemical bonds; thus, a C—D bond is less reactive than a comparable C—H bond and a greater amount of energy is required to activate a C—D bond in comparison to a C—H for reaction to occur. The most important factor contributing to the difference in bond energy and the kinetic isotope effect is the lower (5.021–5.272 kJ/mol) (1.2–1.5 kcal/mol) zero-point vibrational energy for D bonds. These effects are very important in the elucidation of organic reaction mechanisms.

Table 1. Some Properties of Liquid H_2 and D_2[a] at 20.4 K

Property	H_2	D_2
density, g/L	70	169
viscosity, mPa·s (= cP)	1.4×10^{-2}	4.0×10^{-2}
thermal conductivity, W/(cm·K)	11.6	12.6

[a] For the equilibriated mixture of ortho and para species at the indicated temperature.

Table 2. Physical Properties of Light and Heavy Water

Property	H_2O	D_2O
molecular weight, ^{12}C scale	18.015	20.028
melting point, T_m, °C	0.00	3.81
triple point, T_{tr}, °C	0.01	3.82
normal boiling point, T_b, °C	100.00	101.42
density at 25°C, g/cm^3	0.99701	1.1044
C_p (liquid at 25°C), J/(mol·K)[a]	75.27	84.35
vapor pressure (liquid at 25°C), kPa[b]	3.166	2.734
length of the hydrogen bond, nm	0.2765	0.2760
refractive index, n_D^{20}	1.3330	1.3283

[a] To convert MPa to atm, multiply by 10.1.

[b] To convert kPa to mm Hg, multiply by 7.5.

Replacement of more than one third of the hydrogen by deuterium in the body fluids of mammals, or two thirds of the hydrogen in higher green plants has catastrophic consequences for the organisms, resulting variously in mammals in sterility, impairment of kidney function, anemia, disturbed carbohydrate metabolism and CNS disturbances, and altered adrenal function among many others. Extensive replacement of H by D in living organisms is not invariably fatal: numerous green and blue-green algae have been grown in which 99.5% of H has been replaced by D. These fully deuterated organisms can be used to start a food chain to provide fully deuterated nutrients for organisms that have more demanding nutritional requirements.

Production of Heavy Water

Chemical exchange processes. Isotope-exchange reactions between H gas and water or ammonia and between water and hydrogen sulfide provide the basis for the most efficient large-scale methods known for the concentration of deuterium. The efficiency of chemical exchange processes can be increased by taking advantage of the difference in the equilibrium constants as a function of temperature in the form of a dual-temperature exchange process.

All of the major installations currently producing highly enriched deuterium in quantities greater than 20 metric tons of D_2O per year use the H_2S/H_2O dual-temperature exchange system. In the dual-temperature H_2O/H_2S process, exchange of deuterium between $H_2O(l)$ and $H_2S(g)$ is carried out at pressures of about 2 MPa (20 atm). At elevated temperatures deuterium tends to displace hydrogen (^1H) in the hydrogen sulfide and thus concentrates in the gas. At lower temperatures the driving force is reversed and the deuterium is extracted from the H_2S and accumulates in the liquid water.

The deuterium-exchange reactions in the H_2S/H_2O process (the GS process) occur in the liquid phase without the necessity for a catalyst. Dual-temperature operation avoids the necessity for an expensive chemical reflux operation which is essential in a single-temperature process.

In producing tonnage quantities of heavy water, large equipment must be used to perform the initial separating because of the low D concentration in the feed the high throughput entails. To produce one metric ton of D_2O, the plant must process 41,000 t of water and must cycle 135,000 t of hydrogen sulfide.

Vacuum distillation of water was the first method used for the large-scale extraction of deuterium, but is now only used for final

enrichment of D_2O. Electrolysis has been the method of choice for the further enrichment of moderately enriched H_2O–D_2O mixtures.

Uses. The only large-scale use of deuterium in industry is as a moderator (in the form of D_2O) for nuclear reactors (see Nuclear reactors). If fast-breeder reactors using plutonium-239 as fuel become important in energy production, the role of D_2O in nuclear technology would likely be adversely affected.

JOSEPH J. KATZ
Argonne National Laboratory

E.C. Arnett and D.R. McKelvey in J.F. Coetzee and C.D. Ritchie, eds., *Solute–Solvent Interactions*, Marcel Dekker, Inc., New York, 1969, pp. 343–398.

C.J. Collins and N.S. Bowman, eds., *Isotope Effects in Chemical Reactions*, American Chemical Society Monograph 167, Van Nostrand Reinhold Co., New York, 1971.

TRITIUM

Tritium is the name given to the hydrogen isotope of mass 3 (symbol ^3_1H or more commonly T). Its isotopic mass on the ^{12}C scale (^{12}C = 12.000000) is 3.0160497. The molecular form is T_2, analogous to that for the other hydrogen isotopes. The tritium nucleus is energetically unstable and decays radioactively by the emission of a low energy β particle. The half-life is relatively short (12.26 yr) and therefore tritium occurs in nature only in equilibrium with amounts produced by cosmic rays or man-made nuclear devices (see Radioactivity, natural; Radioisotopes).

Physical Properties

Tritium has mp 20.62°C; bp 25.04°C; molar density of liquid 42.65 M at 25°K.

As in the case of molecular hydrogen, molecular tritium exhibits nuclear spin isomerism. Below 5 K molecular tritium is 100% para at equilibrium. At high temperatures (110°C) the equilibrium concentrations is 25% para and 75% ortho. In the solid phase the conversion of molecular tritium to a state of ortho-para equilibrium is 210 times as fast as that for molecular hydrogen.

Properties of T_2O. Tritium oxide can be prepared by the catalytic oxidation of T_2 or by reduction of copper oxide with tritium gas. T_2O has bp 101.51°C; d 1.2138 g/cm^3 at 25°C; and molecular weight (^{12}C scale) 22.032. T_2O even of low isotopic abundance (1–19% T) undergoes radiation decomposition to form HT and O_2.

Nuclear properties. Radioactivity: tritium decays by emission, $^3_1T \rightarrow ^3_2He + \beta$. It has a mean β energy 5.7 keV.

Chemical Properties

Most of the chemical properties of tritum are common with those of the other hydrogen isotopes. Notable deviations in chemical behavior result from isotope effects and from enhanced reaction rates resulting from the emissions in tritium systems.

Enhanced reaction kinetics. For reactions involving tritium, the reaction rates are frequently larger than expected because of the ionizing effects of the tritium β-decay. For example, the uncatalyzed reaction $2 T_2 + O_2 \rightarrow 2 T_2O$ can be observed under conditions (25°C) for which the analogous reaction of H_2 or D_2 would be too slow for detection.

Isotopic exchange reactions. The reaction $T_2 + H_2 \rightarrow 2$ HT equilibrates at room temperature even without a catalyst. In 1957, Wilzbach demonstrated that tritium could be introduced into organic compounds by merely exposing them to tritum gas. Since that time hundreds of compounds, of types as simple as methane and as complex as insulin, have been labeled with tritium by this basic method of isotope exchange.

Natural Production and Occurrence

Tritium arises in nature by the action of primary cosmic rays (high-energy protons) or cosmic-ray neutrons on a number of elements. In the upper atmosphere, fast neutrons, protons and deuterons collide with components of the stratosphere to produce tritium:

$$^{14}_{7}N + ^{1}_{0}n \rightarrow ^{3}_{1}T + ^{12}_{6}C; \qquad ^{14}_{7}N + ^{1}_{1}H \rightarrow ^{3}_{1}T + \text{fragments};$$

$$^{2}_{1}D + ^{2}_{1}D \rightarrow ^{3}_{1}T + ^{1}H.$$

The most important of these reactions by far is the $^{14}_{7}N(n, ^{12}_{16}C)^{3}_{1}T$ reaction, in which the energetic tritons so produced are incorporated into water molecules by exchange or oxidation, and the tritium reaches the earth's surface as rainwater.

Manufacture

Tritium is produced on a large scale by neutron irradiation of $^{6}_{3}Li$.

$$^{6}_{3}Li + ^{1}_{0}N \rightarrow ^{3}_{1}T + ^{4}_{2}He$$

The principal U.S. site of production is the Savannah River plant near Aiken, South Carolina. Tritium is produced there in large heavy-water moderated, uranium-fueled reactors, and may be produced either as a primary product by placing target elements of Li–Al alloy in the reactor, or as a secondary product by using Li–Al elements as an absorber for control of the neutron flux.

In the recovery process, the gaseous constituents of the target are evolved, and the hydrogen isotopes separated from other components of the gas mixture.

Isotopic concentration. Low temperature (20–25 K) distillation has been widely used to separate hydrogen and deuterium and also to separate tritium from the other hydrogen isotopes. Concentration by gas chromatography has also been demonstrated.

Health and Safety

Since tritium decays with emission of low energy radiation, it does not constitute an external radiation hazard, but it is a serious hazard through ingestion and subsequent exposure of vital body tissue to internal radiation. For humans, the estimated maximum permissible total body burden is 37 MBq (1 mCi).

Uses

Tritium is a reactant in nuclear fusion reactions and is used as a source of thermonuclear energy in weapons and controlled fusion reactions.

Joseph J. Katz
Argonne National Laboratory

E.A. Evans, *Tritium and Its Compounds*, 2nd ed., John Wiley & Sons, Inc., New York, 1974.

C.M. Lederer and V.S. Shirley, *Table of Isotopes*, 7th ed., John Wiley & Sons, Inc., New York, 1978.

K. Wilzbach, *J. Am. Chem. Soc.* **79**, 1013 (1957).

DEXTROSE AND STARCH SYRUPS. See Syrups.

DEWATERING

Dewatering is the last mechanical process applied to separate a solid from a liquid. In municipal wastewater treatment, dewatering is regarded as the final process used to achieve a water content of 85% or less in the sludge. Other industries would regard a sludge with such a water content as feed to a primary liquid–solid-separation device and would consider a product containing 15% moisture to be dewatered. Thus, dewatering is not defined by moisture content or by the type of separation equipment used.

Table 1 presents most of the liquid–solid separation and dewatering equipment and techniques. Dewatering is mostly concerned with water bound in capillaries in filter and centrifuge cakes. Dewatering processes affect the water and solids ratio by changing the size distribution of

Table 1. Liquid-Solid Separation Techniques

Equipment	Processes and Pretreatments
filters: pressure, vacuum, centrifugal, cartridge, membrane	flocculation, coagulation, sludge conditioning
depth or granular filters	filter thickening
electrofilters	sludge drainage on sand beds
centrifuges	filter and press admix
hydrocyclones	viscosity adjustment
expression presses	electromagnetic dewatering
dewatering screens and strainers: static, kinetic	surface tension and wettability adjustment
thickeners, clarifiers	oil agglomeration, shear flocculation
dissolved air flotation thickeners	oil-phase extraction
	solvent extraction of water
	thermal treatment
	freeze-thaw, freeze crystallization
	thermal drying

capillary radii; reducing adhesion of water to the solids; displacing water from the capillaries; increasing the force applied on the liquid; and reducing the energy required to cause flow in the capillaries.

Industrial Dewatering Practice

Cake dewatering. Centrifuges and centrifugal filters dewater by providing sufficient centrifugal force to expel the liquid from the pores of the cake. The final moisture content of all centrifuge cakes is further lowered by using steam, surfactants, or flocculants, as well as by pretreatments (pelletizing, oil agglomeration, thermal treatment, and freeze-thaw processes).

Filter cakes are dewatered by compression or by displacement of the residual liquid with another phase, usually a gas. Compression reduces the total void space by pumping the slurry into a chamber that includes a filter medium which lets the liquid pass through. Expression (mechanical pressing) is often used with compressible cakes of particle sizes below 50 μm, superflocculated cakes or cakes containing loosely entrapped liquid. Filters that use expression include variable-chamber plate-and-frame filters, belt-filter presses and tube presses.

In displacement, the liquid in the voids of the cake is displaced by a gas. Filters that provide displacement dewatering include vacuum filters, rotary pressure disk, leaf, and drum filters, the Lasta and Larox filter presses, some Nutsche-type filters, and the Alfa-Dyne tube press. Displacement is controlled by the cake properties, fluid properties, interfacial properties, and temperature and pressure gradients.

Expression is also useful for dewatering fibrous materials, eg, paper. Expression is the best mechanical method for processing oilseeds, fruits, and other materials containing liquids in closed cells. Continuous presses include screw, disk (see Fig. 1), and roll presses. Pot and cage presses

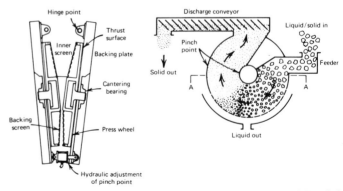

Figure 1. Disk press. Courtesy of Bepex Corp., a subsidiary of Berwind Corp.

process batches. They recover, eg, cocoa butter from cocoa beans using pressures up to 41 MPa (6000 psi). Continuous expression presses do not produce a clear liquid, and the filtrate is frequently passed through a filter press.

Cake dewatering can be improved by viscosity reduction which is controlled by temperature. The viscosity can drop by a factor of three as the temperature rises from 15 to 80°C and the flow rate of the water through the cake is tripled. Heat is applied with low pressure steam. Steaming typically removes 0.7–1.5 kg of water per kilogram of steam applied. The largest use is on nonmagnetic taconite, and the largest potential use is to dewater pipeline coal. Permeability is the critical factor.

Surfactants. Anionic, nonionic or cationic surfactants at a few ppm of the slurry effectively reduce moisture in filter cakes of froth-floated coal, metal sulfide concentrates, and fine iron ores. Surfactant effects appear inconsistent. A number of filter cakes, eg, coal refuse, clay, and hydroxide sludge do not respond to surfactants. The effect of the surfactant on downstream processes and equipment must also be considered. The use of both steam and surfactant on coal and iron ore is additive, giving twice the moisture reduction of either treatment alone.

The main effect of the surfactant is to lower the surface tension and therefore the capillary pressure of water in the filter cake. Surfactants also lower the contact angle. However, a concentration near or above the critical micelle concentration (CMC) can reverse the beneficial effects.

Flocculants. Flocculants traditionally improve the sedimentation rate of solids and the clarity of the legend overflow; however, they are also useful in filtration. Small additions of polyacrylate and polyacrylamide flocculants increase the moisture content of coal and kaolin cakes. However, when air is blown through the cake, the moisture content drops sharply below the moisture level obtained without the flocculant. Flocculants do not impair the effectiveness of surfactants. They have the capability to improve filtration rates up to 100 times, especially those of fine clay, sludges, or tailings. In the processing of municipal sludge, flocculants improve the dewatering rates, but often an increase of the moisture in the final cake occurs.

In sludge conditioning, polymeric flocculants and coagulants like iron chlorides, alum, and lime are added, that react with the sludge and improve the dewatering rates. Fly ash and ash derived from incinerated sludge, as well as bark, newspaper, and diatomaceous earth have also been used to improve dewatering rates (see Flocculating agents).

Less Common Commercial Dewatering Processes

Flocculated, thickened municipal sludge can be dewatered by gentle rolling in a mesh-covered drum which allows water to drain as the sludge is reworked and agglomerated by its own weight. Such a device, called the dual-gravity sludge dewatering unit (DCG) is inexpensive and can be used on many types of sludges. It is, however, difficult to operate since it is sensitive to changes in flocculant addition. The Rotoplug thickener operates similarly. Municipal sludges can be dewatered to 7–17% solids.

A device called the Dehydrum flocculates and pelletizes thickened sludges into compact 3-mm pellets. These units are used to thicken fine refuse from Japanese coal-preparation plants. The product contains 50% moisture, compared to a 3% solids feed.

Similar pelletizing can be obtained with chemicals that make hydrophilic particles hydrophobic. Treated particles agglomerate into tight hydrophobic clusters. In the spherical-agglomeration processes, the added reagents specifically adsorb on only one type of mineral in the slurry (eg, on sulfides, not on quartz). Vigorous stirring of the slurry causes sulfide minerals to agglomerate. By screening, the agglomerates are concentrated and dewatered.

In a new agglomeration process, a chlorofluorocarbon (Freon) is added to a slurry of coal and mineral matter (particle size < 30 μm). The Freon causes the fine coal to agglomerate into large pellets, which are recovered from the slurry on coarse, static screens. The Freon is recovered by evaporation, leaving only a little residual water in the pellets. In the Shell pelletizing separator, heavy oil is added to a dilute soot-in-water slurry and the mixture is agitated in a multistage mixer. The soot agglomerates with the oil to form pellets of 3–5 mm dia, which are easily screened from the water. The pellets contain 5–10% water. Another

modification combines the water exclusion of agglomeration with the ability of froth flotation to separate and achieve thickening.

Extraction processes. The use of very large quantities of an immiscible liquid allows extraction or transfer of particles from one phase to another. The particles remain in a dispersed state in the new phase. To extract solids from water, flotation reagents (long-chain fatty acids, alkylsulfonates, amines, xanthates) provide the required hydrophobic surface. In pigment flushing, paint particles are transferred from the aqueous suspension to a dispersed state in a nonaqueous carrier and water is removed at the same time. The Lummus antisolvent deashing process uses kerosene which is immiscible with the coal liquids, yet has an affinity for agglomerating and extracting the solids from the coal liquid. The two liquids are separated in a settler.

Water contained in a cake or slurry can be separated from the solids by dissolving the water in a nonaqueous solvent. The Institute of Gas Technology and, separately, the Resource Conservation Co. have developed processes, called solvent dewatering, that use solvents like triethanolamine which dissolve or precipitate water depending on temperature. Kerr McGee's critical-solvent deashing process uses the same principle to purify coal liquids using a supercritical solvent.

Thermal processes. Thermal treatment can cause municipal sludge to thicken to 10–15% solids, which can then be filtered to 35–40% solids. Thermal treatment requires heating the sludge to 150–225°C under a pressure of ca 2.4 MPa (350 psi). Peat or lignite can be easily dewatered after thermal treatment at 150–200°C under ca 1.3 MPa (189 psi) for ca 15 min. Higher temperatures and pressures up to 10 MPa (1500 psi) completely dewater lignite, resulting in a stable product with little tendency to reabsorb moisture.

Slow freezing of hydroxide, clay, or municipal sludge affects the water-retention properties of the solids when the frozen slurry is thawed. This technique has been used on municipal sludge in the U.S., Canada, and the UK. Technical and economic feasibilities have been discussed for applying freeze-thawing to Florida phosphate slimes, allowing a typical thickened 13% solids sludge to drain to 45% solids.

Freeze crystallization has been considered for solutions but not for suspensions. In this process, pure ice is formed and screened from the remaining liquor. It has been tested for the concentration of orange juice.

Electromagnetic processes. Applying an electric field to a water suspension can cause the particles to move in one direction. This electrophoresis can prevent a filter cake from forming while allowing water to pass through the medium from the slurry. Alternatively, an electric field can cause ionic solutions to move through a porous sludge where the particles cannot move. Electroosmosis (this effect) is used to dewater soil and fine mineral tailings.

Magnetic forces can replace thickening and filtration in a few applications. The Frantz Ferrofilter is used for removing suspended ferromagnetic impurities from liquids. The principle has been extended to large magnetized volumes (1.7 m³ or 60 ft³) and at high fields (0.15–2.0 T or 1500–20,000 G). Large-scale magnetic separators have been used to remove 90% of the suspended mill-scale solids in steel-mill wastewaters and for cleanup of steam boiler water (see Magnetic separation). Laboratory demonstrations of dewatering using magnetic separation have been made.

BOOKER MOREY
Telic Technical Service

Sludge Dewatering, Manual of Practice 20, C.T. Way, ed., Water Pollution Control Federation, Washington, D.C., 1983.

R.J. Wakeman, *Filtration Post-Treatment Processes,* Elsevier Publishing Co., Amsterdam, 1975.

D.B. Purchas, *Solid/Liquid Separation,* Uplands Press, Croyden, UK, 1981.

DIALYSIS

Dialysis is a membrane transport process in which solute molecules are exchanged between two liquids. The process proceeds in response to differences in chemical potentials between the liquids. Dialysis requires that the membrane separating the liquids permit diffusional exchange between at least some of the molecular species present while effectively preventing any convective exchange between, or commingling of, the solutions. It is used in industry, medicine, and the laboratory as a separation and dilution process (see Separations systems synthesis).

As currently practiced in industry, dialysis is a passive separation process with a low operating cost because it uses no external thermal or chemical energy sources. It requires a high capital investment, essentially because of the large membrane areas required per volume of fluid to be processed per unit of time.

Theory

Three kinds of theory address three kinds of questions about dialysis: theories of intramembrane transport that seek to interrelate different fluxes, concentration gradients across the membrane, and concentration levels within it, and identify how the fluxes are affected both by the chemical composition and physical–chemical state of the membrane, as well as by the nature of the permeating molecules; boundary-layer transport theories designed to elucidate the relationship between fluxes and the fluid-mechanical conditions adjacent to the membrane, often with particular emphasis on the prediction of accumulation of impermeable constituents to an extent that fouling of the membrane or at least a sharp reduction in flux occurs; and theories that show how fluxes and concentrations vary along the fluid pathway and how certain recycle and reflux schemes may be used to alter the molecular sensitivity intrinsic to the membrane.

Intramembrane transport. The basic formulation that is now commonly used to compare specific models of intramembrane transport is irreversible thermodynamics. The relative velocities between pairs of species in the same region, when multiplied by a frictional coefficient, gives rise to a frictional force that characterizes the interaction of the two molecular groups. Thus, in ideal single-solute solutions when there is negligible solvent flux:

$$J_s = \frac{K_s RT}{\Delta X (f_{sw} + f_{sm})} \cdot (C_s^o - C_s^{\Delta x}) \qquad (1)$$

Here J_s is the solute; K_s is the coefficient of distribution of solute between solution and membrane; f_{sw} and f_{sm} are the coefficients, respectively, between solute and solvent and solute and membrane; and C_s^o and $C_s^{\Delta x}$ are the concentrations of solute in solution at the two faces of the membrane. The quantity $K_s/\Delta X (f_{sw} + f_{sm})$ has been termed the permeability coefficient by Katchalsky and Curran, and this quantity multiplied by RT is the membrane permeability K_m used by most authors in treatments of dialyzer design.

Juxtamembrane transport. The theory of intramembrane transport considered above assumes that regardless of the rate at which molecules enter the membrane on one side and leave it on the other, their concentration in the immediately adjacent fluid layers is somehow maintained. Maintaining these concentrations when there is a finite flux through the membrane requires that there be a tandem process in each fluid, matching the membrane flux with a supply process to deliver molecules from the fluid bulk on one side, and a removal process to take molecules from the membrane into the fluid bulk on the other. These necessary components of useful dialysis are termed juxtamembrane transport.

In practice, effective juxtamembrane transport is by convective diffusion, involving convection and diffusion processes, with convection occurring principally parallel to the membrane surface and diffusion being responsible for transverse transport, especially near the membrane surface. Convection is either forced, as by pumping or gravity feed, or free, as induced by density gradients arising from dilution of the feed and concentration of the opposite solution.

Staging, recycling, and refluxing. As noted above, dialysis is used primarily as a separation process rather than a dilution or concentration process, and the flux of one solute must not only be absolutely large enough to keep the membrane area requirement within practical limits but it must also be large relative to that of another solute from which it is to be separated.

Applications

The archetypal industrial application of dialysis has been the recovery of caustic from hemicellulose (see Pulp). Dialysis is under active consideration in conjunction with development of biomass projects (see Fuels from biomass; Foods, nonconventional).

The largest contemporary use is in hemodialysis, the treatment of the blood of persons with end-stage renal disease in which the kidneys are no longer capable of removing products of metabolism from the blood and excreting them. Dialysis here is essential because of two requirements: that the processing be gentle and that the basic separation is between quite large (eg, proteins) and quite small (eg, urea) molecules. The absence of large concentration differences is apparently outweighed by the benefits just cited and is offset by using large areas of membranes to compensate for small fluxes, eg, areas of ca 1 m² are used to process a relatively small flow of ca 200 mL/min.

Probably the most important technical innovation in dialysis technology in support of hemodialysis is the development of hollow-fiber dialysis membranes and devices. As with sheet membranes, most hollow fibers are cellulosic although other materials have also been prepared and evaluated. Hollow-fiber artificial kidneys are mass produced (see Hollow-fiber membranes).

EDWARD F. LEONARD
Columbia University

A. Katchalsky and P.F. Curran, *Non-Equilibrium Thermodynamics in Biophysics*, Harvard University Press, Cambridge, Mass., 1967, p. 124.

DIAMINES AND HIGHER AMINES, ALIPHATIC

Commercially available polyamines are usually slightly viscous liquids with a strong ammoniacal odor. Most are completely miscible in water, except PEHA and TEDA, and completely miscible with alcohol, acetone, benzene, and ethyl ether, but only slightly soluble in heptane. Properties of commercial diamines and higher amines are listed in Table 1 (p. 351).

Chemical Properties

Aliphatic polyamines are strong bases. They undergo reactions typical of compounds containing primary and secondary amines. They react readily with the common inorganic acids to form the corresponding salts. These salts are water-soluble, crystalline materials that react with strong bases to regenerate the original amine. The polyamines react with alkylene oxides, such as ethylene oxide, propylene oxide, or butylene oxide, the mole ratio of the reactants and the reaction conditions controlling the products obtained. Ethyleneamines react exothermically with aliphatic aldehydes to form a variety of adducts, the specific type depending upon the stoichiometry and reaction conditions employed. Reaction of the aldehyde with one of the primary amine groups results in the formation of a Schiff base that undergoes cyclization to a 2-substituted imidazolidine. EDA and its homologues react when heated with an excess of a monobasic carboxylic acid to yield mono- and disubstituted amides. They react with urea (qv) to give 2-imidazolidinones (ethyleneureas). Other compounds that react with aliphatic polyamines include organic halides, eg, polymeric, cross-linked, water-soluble cationic products are obtained with aliphatic dihalides; cyanides and nitriles, eg, EDA reacts with sodium cyanide and formaldehyde to give the tetrasodium salt of ethylenediaminetetraacetic acid; and sulfur and sulfur compounds.

Manufacture

Ethyleneamines. Two processes are currently used to prepare ethyleneamines commercially: the reaction of aqueous ammonia with 1,2-dichloroethane (EDC), and the catalytic amination of monoethanolamine (MEA). The EDC-ammonia process is the major commercial route, yielding the entire family of ethyleneamines produced as their hydrochloride salts. The product distribution obtained in the EDC–NH_3

Table 1. Properties of Commercial Diamines and Higher Amines

Commercial name	Abbreviation	Molecular formula	Freezing point, °C	Boiling point at 101.3 kPa,[a] °C	n_D^{20}	Viscosity at 20°C, mPa·s (= cP)
ethylenediamine	EDA	$H_2N(CH_2)_2HN_2$	10.8	117.0	1.4565	1.6
diethylenetriamine	DETA	$H_2N(CH_2)_2NH(CH_2)_2NH_2$	−35	206.7	1.4859	7.1
triethylenetetramine	TETA	$H_2N(CH_2CH_2NH)_2CH_2CH_2NH_2$	−39	277.4	1.4986	26.7
tetraethylenepentamine	TEPA	$H_2N(CH_2CH_2NH)_3CH_2CH_2NH_2$	< −40	340.3	1.5067	96.2
pentaethylenehexamine	PEHA	$H_2N(CH_2CH_2NH)_4CH_2CH_2NH_2$	−30	180–280[b]		300–400
aminoethylpiperazine	AEP	$NH_2CH_2CH_2N(CH_2CH_2)_2NH$	−17.6	222.0	1.5003	14.1
piperazine	DEDA	$HN(CH_2CH_2)_2NH$	109.6	148.5		
propylenediamine	1,2-PDA	$H_2NCH(CH_3)CH_2NH_2$	−37.2	120.9	1.4455	1.6
1,3-diaminopropane	1,3-PDA	$H_2NCH_2CH_2CH_2NH_2$	−12	139.7	1.4555	2.0
iminobispropylamine	IBPA	$H_2N(CH_2)_3NH(CH_2)_3NH_2$	−15.1	238.3	1.4791	9.6
dimethylaminopropylamine	DMAPA	$(CH_3)_2NCH_2CH_2CH_2NH_2$	−100	134.9	1.4350	1.1
menthanediamine	MDA	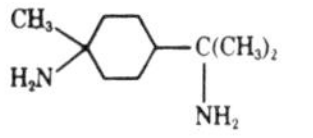 (structure)	−45	107–126[c]	1.479	17.5
triethylenediamine	TEDA	$N(CH_2CH_2)_3N$	158	174		
N,N,N′,N′-tetramethyl-ethylenediamine	TMEDA	$(CH_3)_2NCH_2CH_2N(CH_3)_2$	−55.1	119–122	1.4160	
N,N,N′,N′-tetramethyl-1,3-butanediamine	TMBDA	$(CH_3)_2NCH(CH_3)CH_2CH_2N(CH_3)_2$	< −100	165.1	1.4311	1.0
hexamethylenediamine	HMDA	$H_2N(CH_2)_6NH_2$	40.9	204.5		

[a] 101.3 kPa = 1 atm.
[b] At 0.67 kPa (mm Hg), 10–60% distills in this range.
[c] At 1.3 kPa (mm Hg).

reaction is influenced by the ratio of EDC : NH_3 employed, the reaction time, and the reaction temperature, the ratio being most often used to control product distribution. The catalytic reductive amination of MEA is more selective, producing primarily EDA with small amounts of DEDA, DETA, AEP and other amines coproduced.

Propyleneamine process. The reductive amination of propylene oxide, 1,2-propylene glycol, or monoisopropanolamine are useful procedures for the preparation of 1,2-PDA. 1,2-PDA, and 1,3-PDA can be produced from aqueous ammonia and the appropriate dichloropropanes, through processes similar to the production of EDA from EDC.

Health and Safety

The vapors of these products are painful and irritating to the eyes, nose, throat, and respiratory system. The liquids severely damage the eye and may cause serious burns upon contact with the skin.

Uses

Aliphatic polyamines are used in the manufacture of a variety of derivatives which have a broad spectrum of industrial applications, notably as fungicides (qv), chelating agents (qv), wet-strength resins, epoxy curing agents, polyamide resins, fabric softeners, corrosion inhibitors (qv), lube-oil additives and asphalt emulsifiers.

R.D. SPITZ
K. WILLIAMS
Dow Chemical USA

H.A. Bruson in R. Adams, ed., *Organic Reactions*, Vol. 5, John Wiley & Sons, Inc., New York, 1949, pp. 79–87.

DIAMOND. See Carbon.

DIARYLAMINES. See Amines—Aromatic amines—Diarylamines.

DIASPORE, β-ALUMINA MONOHYDRATE, $Al_2O_3·H_2O$. See Aluminum monohydrate.

DIASTASES. See Enzymes.

DIATOMITE

Diatomite (diatomaceous earth, kieselguhr) is a sedimentary rock of marine or lacustrine deposition. It consists mainly of accumulated shells or frustules of hydrous silica secreted by diatoms, microscopic one-celled flowerless plants of the class Bacillarieae. Diatoms still inhabit fresh, brackish, or seawaters today. The prehistoric diatom plants, which thrived in the waters of the earth from 60 million (10^6) years ago to as late as ca 100,000 years ago, extracted silica from the water and used this substance to form an encasing shell or exoskeleton. At the end of a very brief life, the diatom settled to the bottom of the body of water and the organic matter decomposed, leaving the siliceous skeleton. These fossil skeletons (specifically frustules) are in the shape of the original diatom plant in designs as varied and intricate as snow flakes. Over 10,000 varieties have been classified.

Physical and Chemical Properties

Chemically, diatomite consists primarily of silicon dioxide, and is essentially inert. It is attacked by strong alkalies and by hydrofluoric acid, but it is practically unaffected by other acids. Because of the intricate structure of the diatom skeletons that form diatomite, the silicon dioxide has a very different physical structure from other forms in which it occurs (see Silicon compounds). The chemically combined water content varies from 2–10%. Impurities are other aquatic fossils, sand clay, volcanic ash, calcium carbonate, magnesium carbonate, soluble salts, and organic matter. Typical chemical analyses of diatomite from different deposits are given in Table 1.

The color of pure diatomite is white, or near white, but impurities such as carbonaceous matter, clay, iron oxide, volcanic ash, etc, may

Table 1. Typical Spectrographic Analysis of Various Diatomites (Dry Basis)

	Deposit		
Constituent, %	Lompoc, Calif.	Basalt, Nev.	Sparks, Nev.
SiO_2	88.90	83.13	87.81
Al_2O_3	3.00	4.60	4.51
CaO	0.53	2.50	1.15
MgO	0.56	0.64	0.17
Fe_2O_3	1,69	2.00	1.49
Na_2O	1.44	1.60	0.77
V_2O_5	0.11	0.05	0.77
TiO_2	0.14	0.18	0.77
ignition loss	3.60	5.30	4.10

darken it. The refractive index ranges from 1.41 to 1.48, almost that of opaline silica. Diatomite is isotropic, the apparent density of powdered diatomite being 0.112–0.320 g/cm^3, and reaching 0.960 g/cm^3 for impure lump material. The true specific gravity of diatomite is 2.1–2.2, the same as for opaline silica, or opal. Diatomite has only weak adsorption powers but shows excellent absorption.

Uses

Several hundred diatomite products are available, generally grouped according to use as follows: filter aids, fillers or extenders, thermal insulation, absorbents, catalyst carriers, insecticide carriers and diluents, fertilizer conditioners, and miscellaneous.

E.L. Neu
A.F. Alciatore
Grefco, Inc.

R. Calvert, *Diatomaceous Earth*, American Chemical Society Monograph Series, Chemical Catalog Company, J.J. Little & Ives Co., New York, 1930.

Minerals Yearbook, Vol. 1, U.S. Department of Interior, Bureau of Mines, U.S. Government Printing Office, Washington, D.C., 1974, pp. 539–541.

W.G. Hull and co-workers, *Ind. Eng. Chem.* **45**, 256 (Feb. 1953).

DICARBOXYLIC ACIDS

The aliphatic dicarboxylic acids, $HOOC(CH_2)_nCOOH$, are best known by trivial names for $n = 0$–8. IUPAC names, which are derived from the name of the parent hydrocarbon plus the suffix dioic, are given for all the acids.

Physical and Chemical Properties

All of the diacids are colorless, crystalline solids. As the aliphatic chain becomes longer, the trend of the melting points is downward and the solids tend to become more waxy (see Table 1 for selected physical properties). Many of the dibasic acids exhibit more than one crystalline form. There is an alternation of some physical properties of neighboring members of the series, eg, melting point, decarboxylation temperature, solubility, and index of refraction. Such alternating melting-point effects are known for other related series of compounds.

The chemical reactions of the dicarboxylic acids depend mainly on the terminal carboxyls and their reactive hydroxyl groups. The dibasic acids undergo all the reactions of the monocarboxylic acids (see Acids, carboxylic) but some interactions are peculiar to the presence of two carboxyl groups in proximity. This is demonstrated by the facile elimination of carbon dioxide from malonic acid where the carboxyls are both attached to the same carbon. Succinic and glutaric acids, on heating or removal of water by acetic anhydride or acetyl chloride, produce cyclic anhydrides. Polyanhydrides are obtained from all of the higher diacids (C_6 and above) upon heating with acetic anhydride.

Each carboxyl group in the diacids is subject to the formation of many normal acid derivatives; this capability is the basis for much of the utility of the higher members, particularly in polymers, plasticizers, and lubricants. In recent years, diacid dichlorides have been much used in low temperature synthesis of condensation polymers, primarily polyamides (qv).

Table 1. Melting Points and Boiling Points of Linear Dicarboxylic Acids, $HOOC(CH_2)_nCOOH$

Total number of carbons atoms	IUPAC name	Common name	Mp, °C	Bp, °C
2	ethanedioic	oxalic	187 (dec)	
3	propanedioic	malonic	134–136 (dec)	
4	butanedioic	succinic	187.6–187.9	
5	pentanedioic	glutaric	98–99	$200_{2.7\ kPa}$[a]
6	hexanedioic	adipic	153.0–153.1	$265_{13.3\ kPa}$, $216.5_{2.0\ kPa}$
7	heptanedioic	pimelic	105.7–105.8	$272_{13.3\ kPa}$, $223_{2.0\ kPa}$
8	octanedioic	suberic	143.0–143.3	$279_{13.3\ kPa}$, $230_{2.0\ kPa}$
9	nonanedioic	azelaic	107–108	$286.5_{13.3\ kPa}$, $237_{2.0\ kPa}$
10	decanedioic	sebacic	134.0–134.4	$294.5_{13.3\ kPa}$, $243.5_{2.0\ kPa}$
11	undecanedioic		110.5–112	
12	dodecanedioic		128.7–129.0	$254_{2.0\ kPa}$
13	tridecanedioic	brassylic	114	
14	tetradecanedioic		126.5	
15	pentadecanedioic		114.7	
16	hexadecanedioic	thapsic	125	
17	heptadecanedioic		117–118	
18	octadecanedioic		124.6–124.8	
19	nonadecanedioic		118–119.5	
20	eicosanedioic		124–125	
21	heneicosanedioic	japanic	118–120	

[a] To convert kPa to mm Hg, multiply by 7.5.

Esterification (qv) of the acids with monoalcohols in the presence of acid catalysts has led to a series of commercially important diesters (see Esters, organic). Higher alkyl esters are normally prepared by transesterification of the lower alkyl diesters:

$$2\ ROH + CH_3OOC(CH_2)_2COOCH_3 \rightarrow ROOC(CH_2)_2COOR + 2\ CH_3OH$$

A polyester is made by the reaction of an aliphatic diol with the diacid in the presence of mildly acidic or basic catalysts to accelerate the reaction and minimize dehydration of the diol to olefin (see Polyesters). Wholly aliphatic polyesters have low melting points (many below 100°C). Higher melting polyesters can be prepared from aliphatic dicarboxylic acids and aromatic diols, and are most easily synthesized by low temperature procedures with alkali and diacid chlorides.

Manufacture

A one-source reference to the laboratory preparation of diacids from oxalic to brassylic is the paper by Verkade, Hartman, and Coops. Generally, the acids from succinic to dodecanedioic are prepared by controlled oxidation or other reactions on available cyclic compounds, eg, succinic acid is produced by hydrogenation of maleic anhydride or HNO_3 oxidation of tetrahydrofuran; glutaric acid is produced from cyclopentanol or cyclopentanone oxidation; and dodecanedioic acid is produced from cyclododecene oxidation. Many diacids are synthesized from linear compounds containing a carbon chain similar to that required in the product, eg, oxalic acid from ethylene or propylene via HNO_3–H_2SO_4 oxidation, succinic acid by maleic acid reduction, butanediol or paraffin wax oxidation; suberic acid by HNO_3 oxidation of ricinoleic acid; and azelaic acid by oleic acid ozonolysis–oxidation.

Health and Safety

Toxicity varies markedly throughout the series related in part to acidity, spacing of the carboxyl groups, and solubility. Oxalic and malonic acids are hazardous. Succinic and adipic acids have been used as acidulants in foods. Although the higher dibasic (C_8 and higher) are reported to have low internal toxicity, inhalation of dusts may be irritating to the respiratory system.

PAUL MORGAN
Consultant

E.H. Pryde and J.C. Cowan "Aliphatic Dibasic Acids," in J.K. Stille and T.W. Campbell, eds., *Condensation Monomers*, Wiley-Interscience, New York, 1972, pp. 1–153.

J.A. Dean, ed., *Lange's Handbook of Chemistry*, 11th ed., McGraw-Hill, New York, 1973.

P.E. Verkade, H. Hartman, and J. Coops, *Rec. Trav. Chim. Pays-Bas*, **45**, 380 (1926).

DIENE POLYMERS. See Elastomers, synthetic.

DIESEL FUEL. See Gasoline and other motor fuels; Petroleum products.

DIETARY FIBER

Dietary fiber is defined, at present, as the plant-derived food polysaccharides and lignin that are not digested by enzymes of the human gastrointestinal tract. These polysaccharides include cellulose, hemicelluloses (xylan, arabinoxylan, xyloglucan, glucomannan), pectic substances and associated polysaccharides (arabinan, arabinogalactan, galactan), as well as fructan, galactomannan, and β-glucan. Dietary fiber is a complex mixture composed of cellulose with lesser amounts of other polysaccharides and lignin, the precise composition being related to the plant source and also to the stage of maturity as well as to food processing effects (see Lignin; Cellulose).

A resurgence of interest in dietary fiber was stimulated by epidemiological evidence of differences in colonic disease patterns between certain cultures whose diets contain larger quantities of fiber and western people who consume more highly refined diets.

The primary sources of fiber in the diet are vegetables, cereals, and to a lesser extent, fruits. Currently, certain processed foods and breads are supplemented in their fiber content by incorporation of purified cellulose or cereal bran. In addition, water-soluble cellulose derivatives, seaweed polysaccharides (alginate, carrageenan), seed mucilaginous polysaccharides, highly complex plant-exudate polysaccharides (gum arabic, tragacanth, etc), and microbially-synthesized polysaccharides such as xanthan gum are added to foods and contribute to the dietary-fiber content.

Physiological Properties

Dietary fiber has a profound effect on the quality and appearance of the feces and on the passage time of digesta through the gastrointestinal tract. The particle size and shape, density, and water-holding capacity influence the flow rates. Increased levels of fiber increase the flow rates and frequency of defecation and the fecal weight. High fiber diets play a role in excretion of bile acids and cholesterol. The composition and level of dietary fiber may affect the generation and action of tumorigenic substances in the colon; however, evidence is not yet available to prove this hypothesis.

The colon bacteria possess inducible enzymes capable of degrading and fermenting many of the fiber polysaccharides. Some possess cellulolytic activity. The volatile fatty acids generated by bacterial action are partially absorbed through the colon wall and are available as a supplementary energy source.

Physiochemical Properties

Water-holding capacity, ion-exchange capacity, solution viscosity, density, and molecular interactions are characteristics determined by the chemical structure the crystallinity, and surface area of the component polysaccharides.

Uses

Food use: used to control consistency of various food products; to improve texture and appearance; to enhance stability of food formulations (see Bakery processes); as an additive. Nonfood use: filter aid and binders.

BETTY LEWIS
Cornell University

G.E. Inglett and S.I. Falkehag, eds., *Dietary Fibers: Chemistry and Nutrition*, Academic Press, Inc., New York, 1979, pp. 1–285.

"Symposium on the Role of Dietary Fiber in Health," *Amer. J. Clin. Nutr.* **S31** (1978) pp. S1–S291.

G.A. Spiller and R.J. Amen, eds., *Fiber in Human Nutrition*, Plenum Press, New York, 1976, pp. 1–278.

DIETHANOLAMINE, $NH(C_2H_4OH)_2$. See Alkanolamines.

DIETHYLENE GLYCOL, $HOCH_2CH_2OCH_2.CH_2OH$. See Glycols.

DIFFUSION. See Diffusion separation methods.

DIFFUSION SEPARATION METHODS

Ordinary diffusion involves molecular mixing caused by the random motion of molecules. It is much more pronounced in gases and liquids than in solids. The effects of diffusion in fluids are also greatly affected by convection or turbulence. These phenomena are involved in mass-transfer processes, and therefore, in separation processes (see Mass transfer; Separation systems synthesis).

In chemical engineering, the term diffusional unit operations normally refers to the separation processes in which mass is transferred from one phase to another, often across a fluid interface, and in which diffusion is considered to be the rate-controlling mechanism governing the interphase mass transfer. Thus, the standard unit operations such as distillation (qv), gas absorption, extraction (qv), drying (qv), and the sorption processes, as well as the less conventional separation processes, are usually classified under this heading (see Absorption; Adsorptive separation).

Since the advent of nuclear energy, a number of special processes have been developed for difficult separations such as stable-isotope separation (see Nuclear reactors), eg, the gaseous diffusion process used on a very large scale to separate the isotopes of uranium.

Owing to the importance of the diffusion phenomena in these special processes, that is, because the separation of components is based upon different rates of diffusion, they are often referred to as diffusion separation methods. Since the more traditional unit operations are considered elsewhere (see Distillation; Extraction), emphasis is given here to the process fundamentals and design considerations connected with these more novel diffusional separation methods. Special attention is focused on gaseous, mass and sweep, and thermal diffusion processes, and on gas centrifugation or other pressure or gravity diffusion, particularly for the separation of the uranium isotopes.

Industrial Uranium Enrichment

The most important industrial application of the diffusion separation methods has been for uranium isotope (^{235}U) enrichment—initially for nuclear explosive devices and, in more recent years, for nuclear power reactors. Even moderate (3% ^{235}U) enrichment of uranium fuel lowers the size of "critical mass" assemblies and therefore capital cost requirements for such reactors. Capital costs tend to dominate the economics of nuclear power plants.

The wartime gaseous-diffusion facilities were greatly expanded after World War II in Oak Ridge, Tenn., Paducah, Ky., and Portsmouth, Ohio. They constitute a multibillion (10^9) dollar national investment, and form the basis of the largest business and export venture ever owned and managed by the government.

Even before the end of the war, Union Carbide, under government contract, reduced the unit cost of product from that of the original K-25 plant by about a hundredfold. The quality of the separation membrane or barrier, and improved process design and equipment were among the key improvements.

Since the 1950s, the program was also extended to develop the gas-centrifuge process and, most recently, a laser isotope separation process toward commercial application. Additional contractors, laboratories and companies, became involved in these government-sponsored efforts.

In 1977, the government authorized the construction of GCEP, a gas centrifuge enrichment plant in Portsmouth, Ohio. The $7 billion ($10^9$) facility, if completed, will constitute the biggest single-site industrial construction project in the nation. It is anticipated that ultimate centrifuge performance will lower the plant's unit cost of separative work by a substantial factor. However, in earlier years, a unilateral U.S. moratorium on further uranium-enrichment orders, combined with sharp increases in U.S. costs (TVA power) and prices, had accelerated foreign investment in enrichment plants. A number of foreign customers no longer perceive the U.S. as a reliable supplier of their nuclear fuel.

In addition to the U.S., UK, and USSR plants, new gaseous diffusion capacity was constructed in France as well as new centrifuge facilities in the UK and on the Continent. Other countries are also becoming involved in the development of additional capacity. Delays and cut-backs in nuclear power plants have also contributed to a severe, worldwide glut in enrichment capacity, and to many cancellations of long-term purchase contracts. Consequently, over the last decade, projected demand for U.S. enriched uranium in the year 2000 has dropped by a very large factor.

General Process and Design Selection

In difficult separations, such as isotope separations involving the separation of molecules with very similar physical and chemical properties, the enrichment that can be obtained in a single equilibrium stage or transfer unit of the process is quite small; hence, a large number of these elementary separating units must be connected to form a separation cascade in order to achieve most desired separations. Usually, very large separations systems and amounts of energy are needed. Such processes are expensive.

There are, broadly speaking, three general types of continuous separation processes:

Reversible processes. Distillation (qv) is an example of a theoretically reversible separation process. In fractional distillation, heat is introduced at the bottom stillpot to produce the column upflow in the form of vapor; the vapor is then condensed and turned back down as liquid reflux or column downflow. This system is fed at some intermediate point, and product and waste are withdrawn at the ends. Except for losses, through the column wall, etc, the heat energy spent at the bottom vaporizer can be recovered at the top condenser, but at a lower temperature. Ideally, the energy input of such a process is dependent only on the properties of feed, product, and waste.

Among the diffusion-separation methods discussed here, the centrifuge process (pressure diffusion) constitutes a theoretically reversible separation process.

Partially reversible processes. In a partially reversible type of process, exemplified by chemical exchange, the reflux system is generally derived from a chemical process. The process requires the consumption of chemicals needed to transfer the components from the upflow into the downflow at the top of the cascade, and to accomplish the reverse at the bottom. In such systems, the separation process itself may well be reversible; however, if the reflux is not accomplished reversibly, the entire process can of course not be considered to be reversible.

The total consumption of reflux-producing chemicals is proportional to the interstage flows, or width of the cascade at the widest point (feed), but independent of the number of stages in series, or length of the system.

Irreversible processes. Irreversible processes are the most expensive continuous processes and are used only in special situations. Among them are the cases in which the separation factors of more efficient processes are found to be uneconomically small.

Except for pressure diffusion, the diffusion methods discussed here are essentially irreversible processes: gaseous diffusion, in which gas expands from a region of high pressure to one of low pressure; mass diffusion, in which a vapor flows from a region of high partial pressure to one of low partial pressure; and thermal diffusion, in which heat flows from a high temperature source to a low temperature sink, are all irreversible processes.

In contrast to reversible and partially reversible processes, the energy demand in an irreversible process is distributed over the whole cascade in direct proportion to the distribution of flow.

In gaseous diffusion, the cascade consists of individual stages that are connected in series. In each stage, part of the gaseous feed is forced through a diffusion membrane or barrier. The barrier contains holes which are smaller than the mean free path of the gas. Because of their slightly greater mobility, the lighter components flow preferentially through the barrier. This enriched portion of the feed is transported to a neighboring stage, up the cascade, where the lighter components tend to concentrate. The other portion of the gas that does not pass through the barrier is rejected to a neighboring stage, down the cascade, where the heavier components tend to concentrate. The feed to each stage is thus composed of combined upflow and downflow from neighboring stages.

In mass and sweep diffusion, the lighter components of the gas mixture flow preferentially against a stream of readily condensible vapor. The optional presence of a screen or barrier in these processes serves an auxiliary (fluid flow) role since the pores are much larger than the mean free path of the gas mixture. Again, the effect of the separation is multiplied by cascading.

Finally, in pressure diffusion, a pressure gradient is established by gravity in a centrifugal field. The lighter components concentrate toward the lower-pressure (central) portion of the fluid. Countercurrent flow and cascading extend the separation effect.

Cascade Design

Cascade is the term given to the aggregation of separating units that have been interconnected so as to be able to produce the desired material. The optimum arrangement of the separating units in a separation cascade generally minimizes the unit cost of product. Its design involves a number of considerations that are common to all separation processes.

The separation stage. A fundamental quantity α exists in all stochastic separation processes as an index of the steady-state separation that can be attained in an element of the process equipment. The numerical value of α is developed for each process under consideration in the subsequent sections.

The separation stage or, in a continuous separation process, the transfer unit or equivalent theoretical plate (see Distillation), may be considered as a device separating a feed stream or streams into two effluent streams, often called heads and tails, or product and waste. The quantity relates the concentrations of the components in the two effluent streams. For the case of separation of a binary mixture this relationship is

$$\left(\frac{y}{1-y}\right)\bigg/\left(\frac{x}{1-x}\right) = \alpha \qquad (1)$$

where y is mole fraction of the desired component in the upflowing (heads) stream from the stage and x the mole fraction of the same component in the downflowing (tails) stream from the stage. The quantity α is usually called the stage separation factor. For the case of separating a binary mixture for the processes considered here, α exceeds unity by only a very small fraction and the relationship between the concentrations leaving the stage can be written in the form:

$$y - x = (\alpha - 1)x(1 - x) \tag{2}$$

without appreciable error.

A separation stage or transfer unit operating on a binary mixture is shown schematically in Figure 1.

In a cascade of separating stages, the feed stream can be formed by mixing the downflowing stream from the stage above and the upflowing stream from the stage below. The quantity θ ie, the fraction of the combined stage feed that goes into the stage upflow stream is termed the cut of the stage. In cascades ordinarily designed for difficult separations, the stage cut is normally very nearly equal to one half.

In the case of a theoretical plate or unit in a continuous process, the feed consists of two separate streams, one from above and one from below. In cascades for either stagewise or continuous processes, the upflow rate L and the downflow rate L' [or $L(1 - \theta)/\theta$] are very nearly equal. For continuous process units, the length S is the length of equipment necessary to satisfy the requirement of equation 1 that the streams leaving the unit related by α; it is usually called the height of a transfer unit (HTU) or the height equivalent to a theoretical plate (HETP). Although the HTU and HETP are defined differently and are not precisely equivalent to each other, the difference between them becomes negligible when the value of the quantity $\alpha - 1$ is small.

Comparisons may be made between different separation processes using the separative capacity of the stage, a quantity indicative of the amount of useful separative work that can be done per unit time by a single stage. The separative capacity of the stage by definition is set equal to the increase in separative value it creates.

The separative capacity. The separative capacity of the stage is termed δU (eq. 3)

$$\delta U = Lv(y) + \frac{1 - \theta}{\theta} Lv(x) - \frac{1}{\theta} Lv(z) \tag{3}$$

It is desirable that the separative capacity of the stage be independent of the concentration of the material with which it is operating, and so the terms in the equation involving the concentration are set equal to a constant, taken for convenience to be unity, and the separative capacity of a single stage operating with a cut of one half is seen to be (eq. 4)

$$\delta U = 1/4 L(\alpha - 1)^2 \tag{4}$$

Thus, the separative capacity of a stage is directly proportional to the stage upflow as well as to the square of the separation effected. This is a basic relationship. It is at times overlooked by researchers who forecast great commercial advantages for a process with an appreciable separation factor without giving any consideration to the contribution of throughput of the process to the separative capacity.

The separative capacity of a theoretical plate in a continuous process can be obtained in the same manner. The maximum value of the separative capacity of a theoretical plate in a continuous process is equal to that of a single separation stage when both units have the same value of the $L(\alpha - 1)^2$ product. The definition of a theoretical plate in the continuous process is arbitrary and not required. However, it is a useful concept, permitting both the stagewise and continuous processes to be treated with the same set of equations.

In addition to providing a relatively simple means for estimating the production of separation cascades, separative capacity is useful for solving some basic cascade-design problems. Among them would be the determination of the optimum size of the stripping section and total cost of the operation, based on feed costs and cost per unit of separative work.

If the separative capacity of a single separation element can be determined, then the total number of identical elements required in an ideal cascade to perform a desired separation job is simply the ratio of the needed overall separative capacity to that of the element. The concept of an ideal plant is particularly useful, since moderate departures from ideality do not appreciably affect the results presented above.

Cascade gradient equations. The gradient equations for the cascade are obtained from a combination of the material-balance equations, frequently called the operating-line equations, and the α relationship, usually called the equilibrium-line equation.

At stage n, when the value of $\alpha - 1$ is small, the stage enrichment $x_{n+1} - x_n$ can be approximated by the differential ratio dx/dn without appreciable error. The gradient equation for the enriching section of a simple cascade therefore takes the form:

$$dx/dn = (\alpha - 1)x(1 - x) - (P/L)(y_P - x) \tag{5}$$

Similarly, one obtains a gradient equation for the stripping section that has the form:

$$dx/dn = (\alpha - 1)x(1 - x) - (W/L)(x - x_W) \tag{6}$$

These equations are basic for cascade design. Although they were derived from a consideration of a cascade composed of discrete separation stages, equations of the same form are also obtained for cascade designs based on continuous or differential separation processes.

It is evident from the gradient equations that the enrichment per (fixed) stage decreases as the withdrawal rate (product or waste) increases. Thus, the minimum number of stages required to span a given concentration difference is obtained when no material is withdrawn from the cascade. This mode of operation ($P = W = F = 0$) is frequently called total reflux operation. With a constant $\alpha - 1$, integration of the gradient equation for this case gives:

$$N_{\min} = \frac{1}{\alpha - 1} \ln\left(\frac{x_T}{1 - x_T} \bigg/ \frac{x_B}{1 - x_B}\right) \tag{7}$$

where the concentration range to be spanned is from concentration x_B at the bottom to concentration x_T at the top (eg, at the waste and product ends of the cascade).

With feed(s) and withdrawals from the cascade, it is necessary that the stage upflow from the stage at which the cascade concentration is x must exceed some critical value in order that there be any positive enrichment at that point in the cascade. This critical value is called the minimum stage upflow and is obtained by setting dx/dn equal to zero in the gradient equation. Thus, for any point in the enriching section, the minimum stage upflow is given by:

$$L_{\min} = P(y_P - x)/[(\alpha - 1)x(1 - x)] \tag{8}$$

for the case of enriching ^{235}U to 90 mol % product concentration, the stage upflow at the feed point ($x_F = 0.0072$) must therefore exceed 29,046 times the product withdrawal rate.

It can now be seen from a consideration of the minimum stage upflow that the approximation made in deriving the gradient equation, ie, taking the quantity $(1 - P/L)$ equal to unity, introduces negligible error except possibly in the immediate vicinity of the withdrawal points.

The condition that arises in a cascade at points where the stage upflow approaches the value $L_{\min}$ is commonly called pinching.

A section of cascade composed of identical stages, that is, a number of stages with the same separation factor and the same stage flow, is called

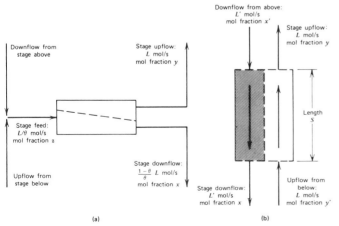

Figure 1. The analogy between (**a**) the separation stage in a stagewise process, and (**b**) the transfer unit or equivalent theoretical plate in a continuous-separation process.

a square section. For sections of this type, the gradient equations are readily integrable.

Some of the preceding concepts can be illustrated graphically by means of a McCabe-Thiele diagram (see Distillation).

The ideal cascade. Of particular interest to design engineers is a continuously tapered cascade (ie, one in which L is a continuously varying function of x, or n) that has the property of minimizing the sum of the stage upflows of all the stages required to achieve a given separation task. In general, the total volume of the equipment required and the total power requirement of the cascade are directly proportional to the sum of the stage upflows. This consideration of the ideal plant requirements often permits a good economic estimate of the unit cost of product to be made without having to resort to the much more painstaking labor of designing a real (as opposed to ideal) cascade to accomplish the separation job. A simple, intuitive approach to the ideal cascade concept in the case of a cascade composed of discrete stages is given below.

For the case of a stagewise enrichment process, the ideal cascade may be defined as one in which there is no mixing of streams of unequal concentrations. Clearly, the mixing of streams of unequal concentrations to form the feed to a separation stage constitutes an inefficiency since it counteracts the process taking place in the stage itself. The no-mixing condition leads directly to the gradient equation for the ideal plant:

$$\frac{dx}{dn} = \frac{\alpha - 1}{2}x(1 - x) \qquad (9)$$

The number of stages required to span a given concentration difference in an ideal plant in which all stages have the same separation factor is therefore equal to $2\,N_{\min}$ (see eq. 7), twice the minimum number of stages required.

The combination of equations 5 and 9 gives the equation for the stage upflow at any point in the enricher of an ideal cascade which is twice the minimum upflow. These equations can be used in conjunction with the corresponding equations for the stripping section to produce an ideal plant profile such as is shown in Figure 2 where L_{ideal} is plotted against N_{ideal}. The graph applies to the example of an ideal cascade to produce one mol of uranium per unit time, enriched to 90 mol % in ^{235}U, from natural feed containing 0.72 mol % ^{235}U, with a waste stream rejected at a concentration of 0.5% ^{235}U. The characteristic lozenge shape of the ideal cascade is evident. No two stages in either the enricher or stripper are the same size. The operating line for an ideal cascade on a McCabe-Thiele diagram is a curved line lying midway between the equilibrium line and the 45° reference line.

Equations for large stage-separation factor. Partly because of the development of the gas-centrifuge process, in recent years, there has been renewed interest in the design of cascades composed of stages with large stage-separation factors. When the stage-separation factor is large, the number of stages required in an ideal cascade in which all stages have the same separation factor is given by equation 10:

$$N_{\text{ideal}} = \frac{2}{\ln \alpha} \ln \left(\frac{y_P}{1 - y_P} \middle/ \frac{x_W}{1 - x_W} \right) - 1 \qquad (10)$$

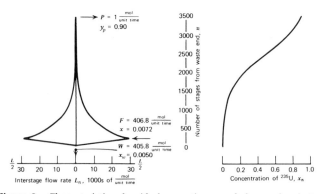

Figure 2. Characteristics of an ideal separation cascade for uranium isotope separation. For this cascade: $\alpha = 1.0043$; $N_T = 3484$ stages; $\Delta U = 153.08$ mol/unit time; $\Sigma L_T = 33.116 \times 10^2$ mol/unit time.

When dealing with cascades composed of stages with large separation factors, it is somewhat more convenient to calculate the sum of all the stage feed flows in the cascade rather than the sum of all the stage upflows as was done in the case when $(\alpha - 1)$ is small. When $(\alpha - 1)$ is small with respect to unity, the stage feed flow is essentially just twice the stage upflow rate.

Real cascades. Although the ideal cascade minimizes the volume of equipment and the energy requirements, it does not generally minimize the cost of the cascade because production economies are realized in the manufacture of the process equipment when a large number of identical units are produced. Thus, a minimum-cost cascade consists of a number of square cascade sections rather than uniformly tapered nonidentical stages. Once the size (length and width) of the individual separating units available is determined, a first approximation to the optimum practical cascade is obtained by fitting the ideal plant shape with square sections.

Time-dependent cascade behavior. The period of time during which a cascade must be operated from start-up until the desired product material can be withdrawn is called the equilibrium time of the cascade. The equilibrium time of cascades utilizing processes with small values of $\alpha - 1$ is a very important quantity, since a cascade with an excessively long equilibrium time may prove to be quite impractical. An estimate of the equilibrium time of a cascade can be obtained from the ratio of the enriched inventory of desired component, at steady state, to the net upward transport of desired component averaged over the entire transient period from start-up to steady state.

The total equilibrium time is directly proportional to the average stage hold-up of process material. Thus, since liquid densities are normally on the order of a thousand times larger than gas densities, a preference for gas-phase processes for difficult separations becomes understandable.

The Gaseous-Diffusion Process

The gaseous diffusion separation process depends on the separation effect arising from the phenomenon of molecular effusion (the flow of gas through small orifices or pores). The equation for the separation factor of this process reflects the relative ease with which the light (versus heavy) molecules can escape through the pores.

It turns out that α^*, the ideal separation factor, is the ratio of the two molecular velocities. Since the kinetic energies, $1/2\,mv^2$, of the two species are the same, α^*, the ratio of the two velocities, is equal also to the square root of the inverse ratio for the two molecular weights. This separation effect was first discovered experimentally by Graham in 1846 and later explained theoretically with the advent of Maxwell's kinetic theory of gases.

Process description. The basic unit of a gaseous-diffusion cascade is the gaseous-diffusion stage (see Fig. 3). Its main components are the converter holding the barrier in tubular form, motors, and compressors moving the gas between stages, a heat exchanger removing the heat of compression introduced by the stage compressors, the interstage piping, and special instruments and controls to maintain the desired pressures and temperatures.

The important parameters in stage design are the stage-separation factor and the size of a stage required to handle the desired stage flows. Both factors depend on the characteristics of the barrier, which must be fine-pored and have many pores per unit area. Preparation and characterization of such a material presents a difficult technological problem. An ideal separation barrier is one that permits flow through it to take place by effusion only, which occurs when the diameter of the pores in the barrier is sufficiently small compared with the mean free path of the gas molecules.

An ideal point-separation factor can be defined on the basis of the separation obtained when a binary mixture flows through an ideal barrier into a region of zero back-pressure. The ideal point-separation factor is equal to:

$$\alpha^* = \frac{y'/(1 - y')}{x'/(1 - x')} = \sqrt{M_B/M_A} \qquad (11)$$

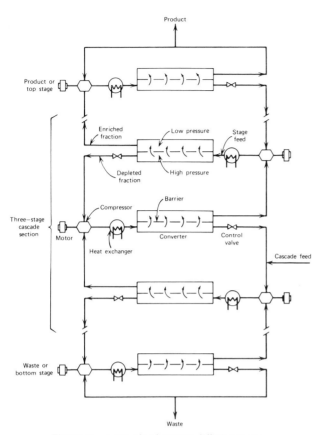

Figure 3. A cascade of gaseous diffusion stages.

which, for example, for the case of uranium isotope separation with UF_6 is equal to 1.00429. This is also the expression for the ratio of the velocity of the light molecules to that of the heavy molecules.

The stage-separation factor differs appreciably from the ideal point-separation factor because of the existence of four efficiency terms that must be taken into account: a barrier efficiency factor, which to a first approximation is equal to the Knudsen flow multiplied by a pressure-dependent term associated with its degree of separation, divided by the total flow; a back-pressure efficiency factor, caused by the tendency for the lighter component to effuse preferentially back through the barrier, equal to a first approximation to $(1-r)$, where r is the pressure ratio (of back-pressure to fore-pressure) p_b/p_f; a mixing-efficiency factor, since as the gas flows along the high pressure side of the diffusion barrier it becomes, as a result of the effusion process, preferentially depleted with respect to the lighter component in the neighborhood immediately adjacent to the barrier and a concentration gradient perpendicular to the barrier is set up on the high pressure side; and a cut-correction factor, which takes into account the difference between the concentrations of streams on both sides of the barrier. In practice, a compromise between the various factors is to be made.

The actual power requirement must also compensate for frictional losses in the cascade piping, compressor inefficiencies, and losses in the power distribution system.

The Sweep and Mass Diffusion Processes

A partial separation of a binary gas mixture can be obtained if the gas mixture is allowed to diffuse through a third auxiliary gas at a constant total pressure. For these processes to be effective, there must exist a difference in the diffusion coefficients of two components of the gas mixture with respect to the auxiliary gas.

A porous screen is occasionally used to divide the process region; however, unlike the gaseous-diffusion barrier, this screen is not essential to the separation and may be dispensed with entirely. If the screen is

included, the process is generally referred to as mass diffusion; if it is not used, the process is call sweep diffusion.

Normally, the power consumption of the mass diffusion plant would be about three times that of a gaseous-diffusion plant of the same capacity, and the capital cost of a mass diffusion plant would be five times that of a gaseous-diffusion plant with its associated power plant.

Process description. The sweep and mass diffusion processes can be carried out either in countercurrent-flow columns containing many equivalent equilibrium stages, or as stagewise processes in which the equilibrium stages are contained in separate and distinct parts of the process equipment.

The sweep diffusion process (Fig. 4) is irreversible, and continuous, since many effective separation stages can be attained in a single column. Thus, it will probably never compete with conventional separation processes such as distillation and extraction, where these more economical processes are applicable (ie, exhibit significant separation factors).

No industrial applications of the sweep or mass diffusion processes are known. Laboratory-scale separations of some gas pairs, including the separation of some light isotopes, have been reported.

The sweep diffusion process is best suited to the separation of light process gas and is most economical if heavy sweep vapors with low latent heats of vaporization are employed. The process gas should be nearly insoluble in the condensed sweep vapor and no chemical reaction should take place between them.

The Thermal Diffusion Process

Thermal diffusion arises when a mixture is subjected to a temperature gradient leading to partial separation of the components. For example, if a homogeneous binary mixture in a closed vessel is heated at one end and cooled at the other, a relative motion of the two components takes place in the mixture; one component tends to concentrate in the hotter region, the other in the colder region.

The separation effect in the thermal diffusion process depends on the flow of heat through the process-gas mixture from the hot to the cold wall. A large amount of power is needed to achieve a given separation. In the column application, the process is continuous.

Thermal diffusion can be classified as a physical rather than a chemical process since its elementary separation mechanism depends on the intermolecular forces between the molecules of the process material.

Process description. Conventional thermal diffusion apparatus are either single-stage cell, or thermogravitational columns (see Fig. 5).

Although single-stage thermal diffusion equipment is quite inefficient for effecting the separation of gas mixtures, experiments of this type are of great theoretical interest. Of the various transport phenomena in gases, thermal diffusion is most sensitive to the nature of molecular

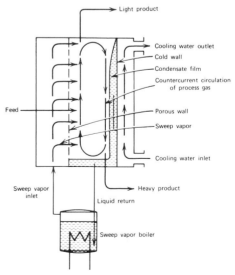

Figure 4. Schematic drawing of a sweep diffusion column.

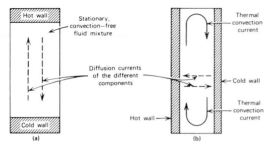

Figure 5. Schematic diagram of conventional thermal diffusion apparatuses. (**a**) Single-stage cell. (**b**) Thermogravitational column.

interactions, and physicists find it to be a valuable tool for the study of intermolecular forces.

In the multistage thermal diffusion apparatus, commonly called a thermal diffusion column or a thermogravitational column, a confined fluid mixture is subjected to a horizontal temperature gradient. Natural thermal convection currents are thereby established. Flow is upward in the neighborhood of the hot wall and downward near the cool wall. This countercurrent flow multiplies the single-stage separation effect, concentrates at the top of the column that component which diffuses preferentially toward the hot and at the bottom of the column that component which diffuses preferentially toward the cold wall.

The separative capacity of a thermal diffusion column operating in an ideal cascade is given by:

$$\delta U = \frac{H^2 Z}{4(K_c + K_d)} \tag{12}$$

Since the process is thermodynamically irreversible and the column throughput is small, it is unlikely to compete with other feasible methods of separation when large amounts of product are desired. However, since thermal diffusion columns are simple and relatively inexpensive, and can run for long periods of time without attention, the process can become competitive when small amounts of highly enriched product are desired. Thermal diffusion columns have been used to achieve a great many separations, particularly separation of gaseous isotopic mixtures.

Pressure Diffusion Processes

The development of the kinetic theory of gases led to the conclusion that a partial separation of the components of a gaseous mixture results when the gas is subjected to a pressure gradient. For example, lighter components would concentrate at the top of a column of gas standing in the earth's gravity field, and heavier components at the bottom. However, steeper pressure gradients are needed.

In order to utilize the pressure diffusion phenomenon in a practical sense, a number of types of equipment have been developed to provide such gradients.

The gas centrifuge. The gas centrifuge is one such device, developing gravitational fields equal to many thousand times that of the earth (see Centrifugal separation).

The construction and operation of a precision, high speed centrifuge in a high vacuum environment presents some formidable mechanical problems, as well as problems associated with flow, separation and critical-speed phenomena.

For isotope separation, the separative efficiency of a gas centrifuge is best defined in terms of separative work. Thus, the separative efficiency E is defined by: (eq 13)

$$E = \frac{\delta U \text{ (experimental)}}{\delta U \text{ (max)}} \tag{13}$$

where U (experimental) is the actual separative work produced per unit time by the centrifuge under consideration and U (max) is the maximum

theoretical separative capacity of the machine. The maximum separative capacity of a gas centrifuge is given by:

$$\delta U \text{ (max)} = \frac{\pi Z c D_{AB}}{2} \left(\frac{\Delta M V^2}{2 RT} \right)^2 \tag{14}$$

where δU is the separative capacity in moles per unit time, Z is the length of the rotor, C is the total molar concentration, mol/L, D_{AB} is the binary diffusion coefficient for the pair AB, L^2/t, M is the difference in the molecular weights of the components being separated, and V is the peripheral velocity of the centrifuge ($V = \omega r_2$).

The ideality efficiency of the gas centrifuge takes into account the difference between the shape of a centrifuge which may be regarded as a square cascade (one in which all the units are equivalent) and that of an ideal cascade.

The vortex chamber. The vortex chamber utilizes pressure diffusion in a rapidly rotating gas, angular acceleration effects providing the pressure gradient.

The separation nozzle. Also referred to as the jet diffusion method process, the separation nozzle was developed in the FRG for enrichment of the light uranium isotope ^{235}U.

A good understanding of the separation phenomenon of the this process is obtained from a very simple model that treats the separation nozzle as a gas centrifuge at steady state. The feed mixture is assumed to traverse the circular flow path at a constant and uniform angular velocity (wheel flow) and the peripheral velocity of the flow is assumed equal to the sonic velocity of the entering feed gas. The separation of the isotopes is effected by pressure diffusion in the pressure gradient resulting from the curved streamlines and the associated centrifugal forces.

Nomenclature

F = feed flow rate to a unit or cascade, mol/t

H = steady-state enriched inventory of desired component in a cascade, moles

K_c = transport coefficient of convective remixing, mol·L/t

K_d = transport coefficient for back-diffusional remixing, mol·L/t

L = upflow rate of process gas in a stage, mol/t

M_A = molecular weight of component A

N = number of stages required to span a given range ov concentrations, dimensionless

P = product flow rate of a unit or cascade, mol

t = time

$v(x)$ = value function, dimensionless

y_P = mol fraction of desired component of a binary mixture in the product stress of a unit or cascade, dimensionless

R.L. Hoglung
J. Shacter
E. Von Halle
Union Carbide Nuclear Division

E. Von Halle, *The Countercurrent Gas Centrifuge for the Enrichment of U-235*, K/OA-4058, Union Carbide Nuclear Div., Oak Ridge, Tenn., Nov., 1977.

P.R. Vanstrum and S.A. Levin, *New Process for Uranium Isotope Separation*, paper presented at IAEA meeting, Salzburg, Austria, 1977. Union Carbide Nuclear Div., Oak Ridge, Tenn., IAEE-CN-36/12 (II.3), 1977.

International Conference on Nuclear Power and Its Fuel Cycle, Salzburg, Austria, May 2–13, 1977, sponsored by the International Atomic Energy Agency; for this conference only a *Book of Abstracts* has been published.

M. Benedict, ed., *Development in Uranium Enrichment*, AICHE Symposium Series, Vol. 73, No. 169, American Institute of Chemical Engineers, New York, ¹1977.

DIGITAL DISPLAYS

The digital display is a device that indicates numerical information to a viewer. The data are usually generated by a semiconductor integrated circuit (IC) (see Integrated circuits; Semiconductors). The IC processes data entered by the viewer or other sources and reproduces the information in an electrical form. The display transforms the data from the electrical to the optical domain.

There are considerable operational differences between light-emitting (active) and light-modulating (passive) displays. Light-emitting displays require a power source of sufficient capacity to create the required luminence. Light-modulating displays use ambient light and therefore require much lower power since they do not have to generate photons. Passive displays are becoming dominant in applications that require battery operation. The power limitation and the requirement that many battery-powered displays be visible in bright sunlight give light-modulating displays a substantial advantage in portable systems. In systems that can be operated from line power, however, light-emitting displays have an important advantage: people like bright lights and color. Since line power generally implies ambient home or office light levels, active displays remain dominant in this area.

The detailed information that is transmitted to the display is in an electrical form. The variety of electrooptic techniques for transforming these signals into visible numbers all rely on the local generation of electrical fields. In a seven-segment format, seven distinct segments generate the ten Arabic numerals when arranged as in Figure 1. This is the dominant type of digital display in use today. A common geometrical feature of these displays is flatness, in contrast to the bulkiness of the cathode ray tube (CRT). This characteristic permits the design of small compact systems.

Figure 1 explains how any single number can be formed by the seven-segment electrode pattern. It follows that any number with n digits can be formed by repeating the pattern n times and addressing $7n + 1$ electrodes. When n is large, eg, 12 as in a scientific calculator, the number of leads to the display for the LSI (large-scale integration) chip becomes too large to handle economically. The reason for this lies in the format of the data supplied by the electronics. Data are read in and out from the driver chip sequentially over a limited number of pins to the chip. The viewer, on the other hand, requiring a parallel data format, must be able to view the whole display simultaneously. The mismatch between the display data requirement and the chip data format becomes intolerable for n greater than four. The solution is to supply data to the display in a serial format and take advantage of the averaging properties of the display or the eye to produce the parallel format. This is accomplished by timesharing leads (multiplexing) of an $N \times M$ matrix which can be addressed by choosing one of N row electrodes and M column electrodes ($N + M$ leads).

Four display technologies are described: LCDs; LEDs; gas-discharge displays; and vacuum fluorescent displays. All four are in volume production for use in the vast majority of digital display applications.

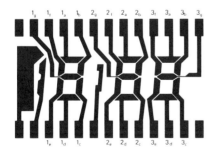

Figure 1. The seven-segment electrode pattern for a $3\frac{1}{2}$-digit watch display. The display can be converted to a 3×7 array by electrically connecting all the a segments for an A lead, b segments for a B lead, etc, and by arranging for each digit to have its own common electrode. One then selects the common electrode for digit one, eg, by raising it from ground to a given potential, and simultaneously applying voltages on the A → G leads to write all of digit one. The second and third digits are then written sequentially, returning to digit one before the eye can detect a flicker of the data on any of the digits. Whether this process supplies enough energy to the electrooptic media to provide a good display is the key question in displaying large amounts of data for a limited number of display connections. The size of the array in this fashion varies from one display technology to the next.

Liquid-Crystal Displays

The term liquid crystal refers to several thermodynamically stable phases of matter that occur over a limited temperature range for certain organic molecules (see Liquid crystals). The medium is anisotropic. One would expect different electrical susceptibilities for an electric field along the average molecular axis (optical axis) than for the field applied transverse to the axis. This is the case for both the low frequency susceptibility (dielectric constant) and the visible frequency response (index of refraction). The anisotropy of the dielectric response enables an applied electric field to reorient the optical axis of a liquid crystal. The optical anisotropy permits detection of this change via the difference in birefringence or scattering. These electrooptical effects are essential for the use of liquid crystals in digital display devices.

The twisted-nematic display. The key to high contrast LCD is to produce a twisted structure in the off-state. The surfaces enclosing the liquid-crystal layer are treated to promote alignment of the nematic director in one direction on the surface. The twisted structure is produced by orienting the surfaces so that the director at the top surface makes an angle of 90° with the bottom surface. In order to conform to these boundary conditions, the director in the bulk of the sample twists 90° in going from the top to the bottom surface.

In the early 1970s, Fergason in the U.S. and Schadt and Helfrich in Switzerland realized independently that the optical characteristics of the twisted-nematic structure offered unique device characteristics. Between crossed polarizers oriented parallel to the director axes on the two plates, the sample would appear transparent, allowing for the insertion loss of the polarizers. The application of a voltage above the threshold tilts the director parallel to the field. This eliminates the twist in the sample and the display appears dark. This device is a high contrast display but the brightness is low owing to the polarizers.

The twisted-nematic device has become the overwhelmingly favorite display for handheld calculators, and has replaced LEDs for watch applications.

Light-Emitting Diodes

The visible light-emitting diode (VLED) display is based on injection electroluminescence (see Light-emitting diodes and semiconductor lasers). The p–n junction represents the most effective way to inject minority charge carriers, and it is the recombination of these carriers with carriers of the opposite type, ie, electrons and holes, that produces the radiation. Nonradiative recombinations by several mechanisms are also possible, and dominate in many materials. The energy-band structure of GaAs, described below, favors the radiative process.

Basically, in the physics of the p–n junction, the application of a forward bias lowers the potential barrier at the junction, allowing electrons to diffuse from the n-region into the p-region, and holes from the p-region to diffuse to the n-region. In GaAs, most of the current is generally carried by electrons, ie, electrons are injected into the p-region, and the light is generated there. One of the important considerations in designing an efficient VLED is minimizing the internal absorption. To this end, the VLEDs or display segments are typically constructed as a thin p-region on an n-substrate.

Although they cannot compete with liquid-crystal displays where current drain is a major factor, VLED displays offer the advantage of appearance, readability, high speed multiplexing, low operating voltage (< 2 V) and self-illumination.

Construction of VLED displays. Most commercial VLED displays are constructed in a layer of GaAsP grown by vapor-phase epitaxy. The growth is accomplished in a furnace of two or more temperature zones in which the gases H_2, HCl, AsH_3, PH_3, and a dopant gas such as H_2S are introduced at the high temperature end, and carefully polished single crystal GaAs (or GaP) substrates are positioned, usually in a rotating holder, at the other end. The HCl–H_2 stream is directed over a Ga reservoir where a volatile chloride is formed and swept by the flow of hydrogen, along with the other gases, across the substrates.

Gas-Discharge Displays

Physics of the discharge. Gas-discharge displays emit light with an orange glow characteristic of neon. They employ a cold cathode from which electrons are emitted by positive ions striking the cathode. The electrons are accelerated by the electric potential (typically 150–250 V). In the immediate vicinity of the cathode, the electrons acquire sufficient energy to ionize the gas. Some ions emit light and others strike the cathode. This negative glow is the most commonly used region for display purposes.

Operating gas-discharge displays. Although a firing voltage for a typical cathode-glow display may be 170–180 V, a swing of only 40–50 V is needed to turn the discharge on or off, so that lower voltage devices can be used to control the display. A typical 10-digit display is designed to be operated in a multiplexed mode where the cathode drive and decoder circuitry are timeshared among all of the digit positions. Corresponding cathode segments for all 10 character positions are bus-connected so that there are only nine cathode connections to the display. Anodes, which are essentially transparent conductors, cover each character. The anode is scanned fast enough to refresh information in less than 3 ms. Characters are displayed by successively applying positive signals to the anodes while negative signals are applied to the appropriate cathode bus line.

Vacuum-Fluorescent Displays

The vacuum-fluorescent display (VF) emits green light which results when electrons from a hot cathode are accelerated into a phosphor (see Luminescent materials). Its extensive usage in handheld calculators is essentially owing to multiplex capability at relatively low drive voltage (ca 25 V). This makes the electronic drivers cheaper for VF displays as compared to gas-discharge displays.

Basic principle and materials. The display is built around the working principle of a triode tube with three basic parts to the display: cathode filament (source of electrons); metal grid (to accelerate and control the electrons); and anode, coated with low voltage luminescent materials, typically $ZnO:Zn$ phosphor powder.

Tube characteristics. Display characteristics are typically: filament current 22 mA; anode grid voltage (on voltage) 24 V; grid off voltage −6 V; brightness (1/9 duty cycle) 514 cd/m^2; and power consumption (per digit) 12 mW. For a large display requiring higher multiplexibility, anode and grid voltages are higher to get peak brightness. A typical 40-digit display is driven at 40–50 V. The lifetime of the VF tube is a primary problem. Although most manufacturers guarantee the minimum life to be 10,000 h, deterioration of the performance is generally seen much earlier. The display is, in fact, no different than a conventional vacuum tube. Further complications result from the aging of the ZnO–Zn phosphor, and mechanical design considerations are complex, restricting its use to relatively small displays.

P. ANDREW PENZ
ROBERT W. HAISTY
KISHIN H. SURTANI
Texas Instruments Incorporated

L.E. Tannas and contributing authors, *Flat Panel Displays*, Van Nostrand Reinhold, New York, 1983.

U.S. Pat. 3,731,986 (May 8, 1973), J. Fergason (to International Liquid Xtal Co.).

M. Schadt and W. Hefrich, *Appl. Phys. Lett.* **18**, 127 (1971).

DIMENSIONAL ANALYSIS

Dimensional analysis is a technique that treats the general forms of equations governing natural phenomena. It provides procedures of judicious grouping of variables associated with a physical phenomenon to form dimensionless products of these variables; therefore, without destroying the generality of the relationship, the equation describing the physical phenomenon may be more easily determined experimentally (see also Design of experiments). The method is particularly valuable when the problems involve a large number of variables.

Units and Dimensions

The concepts used to describe natural phenomena are based on the precise measurement of quantities. The quantitative measure of anything is a number that is found by comparing one magnitude with another of the same type. It is necessary to specify the magnitude of the quantity used in making the comparison if the number is to be meaningful. The statement that "the length of a car is six meters" implies that a length, namely one meter, has been chosen, and that the ratio of the length of the car to the chosen length is six. The chosen magnitudes, such as the meter, are called units of measurement (see Units). To each kind of physical quantity there corresponds an appropriate kind of unit. The physical concepts such as length, area, and time are referred to as dimensions, which are different from units.

Over the years, the number of reference dimensions in physics has evolved from the original three (length l, mass m, and time t), to four, to five, and then gradually downwards to an absolutely necessary one, and then upwards again through an understanding that, though only one is absolutely necessary, a considerable convenience can stem from using three, four, or five reference dimensions depending on the problem at hand. There is nothing sacrosanct about the number of reference dimensions and dimensional analysis is merely a tool that may be manipulated at will. Thus, an important step is the selection of reference dimensions in such a way that others called the secondary or derived dimensions, can be expressed in terms of them. The relation between reference and derived dimensions is generally established either through the fundamental law or equation governing the phenomenon or through definitions. When length, mass, and time are taken to be the reference dimensions, the dimensions of velocity v, for example, are the dimensions of length divided by time, or expressed in symbols, $v = lt^{-1}$. Likewise, through Newton's law of motion which relates force, mass, and acceleration by

$$\text{force} = \text{constant} \cdot \text{mass} \cdot \text{acceleration} \tag{1}$$

the dimensions of force f must be mass-length/time^2 or $f = mlt^{-2}$. If force, length and time are chosen as the reference dimensions, then mass becomes secondary and the constant in Newton's law is dimensionless. However if force, mass, length, and time are chosen as the reference dimensions, the constant is no longer dimensionless and the units are generally selected so that the constant is numerically equal to the standard acceleration of gravity.

To eliminate the ambiguities in the subject of electricity and magnetism, it is convenient to add charge q to the traditional l, m, and t dimensions of mechanics to form the reference dimensions. In many situations, permittivity or permeability is used in lieu of charge. For thermal problems, temperature T is considered as a reference dimension.

Other dimensional systems have been developed for special applications which can be found in the technical literature. In fact, to increase the power of dimensional analysis it is advantageous to differentiate between the lengths in radial and tangential directions. In doing so ambiguities for the concepts of energy and torque, as well as for normal and sheer stress are eliminated.

Dimensional Matrix and Dimensionless Products

An appropriate set of independent reference dimensions may be chosen so that the dimensions of each of the variables involved in a physical phenomenon can be expressed in terms of these reference dimensions. In order to utilize the algebraic approach to dimensional analysis, it is convenient to display the dimensions of the variables by a matrix. The matrix is referred to as the dimensional matrix of the variables and is denoted by the symbol D, with each column denoting a variable under consideration and each row representing a reference dimension.

The validity of the method of dimensional analysis is based on the premise that any equation that correctly describes a physical phenomenon must be dimensionally homogeneous; that is, each term has the same exponents of dimensions. Such an equation is of course independent of the systems of units employed provided the units are compatible with the dimensional system of the equation. It is convenient to represent the exponents of dimensions of a variable by a column vector called dimensional vector represented by the column corresponding to the variable in the dimensional matrix.

WAI-KAI CHEN
University of Illinois at Chicago

J.F. Douglas, *An Introduction to Dimensional Analysis for Engineers*, Sir Isaac Pitman & Sons, London, UK, 1969.

DIMER ACIDS

The dimer acids, C_{21} dicarboxylic acids, 9- and 10-carboxystearic acids, and C_{19} dicarboxylic acids, are products resulting from three different reactions of C_{18} unsaturated fatty acids. These reactions are, respectively, self-condensation, Diels-Alder reaction with acrylic acid, and reaction with carbon monoxide followed by oxidation of the resulting 9- or 10-formylstearic acid (or, alternatively, by hydrocarboxylation of the unsaturated fatty acid).

The starting materials for these reactions have been almost exclusively tall oil fatty acids or oleic acid, although other unsaturated fatty acid feedstocks can be used (see Carboxylic acids, from tall oil; Tall oil).

Physical and Chemical Properties

The physical properties of polymerized fatty acids are influenced by the basestock, by the dimerization conditions, and by the degree to which monomer, dimer, and higher oligomers are separated following the dimerization. Dimer acids are relatively high molecular weight (ca 560) and yet are liquid at 25°C, a liquidity resulting from the many isomers present (see Table 1).

Structure and mechanism of formation. Thermal dimerization of fatty acids has been explained both by a Diels-Alder mechanism, a mechanism which satisfactorily applies to high linoleic acid starting materials, and by a free-radical route involving hydrogen transfer, which best refers to oleic acid polymerization. Clay-catalyzed dimerization of unsaturated fatty acids appears to be a carbonium-ion reaction based on the observed double-bond isomerization, acid catalysis, chain branching, and hydrogen transfer. Different precursors for dimer acid preparation give quite different structures.

Chemical reactions of dimer acids. The most important is polymerization, the greatest tonnage of dimer acids being incorporated into the non-nylon polyamides (qv). Other reactions of commercial importance include polyesterification, hydrogenation, esterification, and conversion of the carboxy groups to various nitrogen-containing functional groups.

Manufacture

The clay-catalyzed intermolecular condensation of oleic and/or linoleic acid on a commercial scale produces approximately a 60:40 mixture of dimer acids (C_{36} and higher polycarboxylic acids) and monomer acids (C_{18} cyclized, aromatized, and isomerized fatty acids). The polycarboxylic acid and monomer fractions are usually separated by wiped-film

Table 1. Properties of Distilled Dimer Acids[a]

Physical characteristics	Hydrogenated	Unhydrogenated
composition, %		
dimer	95	95
trimer	4	4
monomer	1	1
monobasic acids, max	1.5	1.5
neutralization equivalent	289–298	283–289
acid number	188–194	194–198
saponification number	190–197	198–202
viscosity at 25°C, mm²/s (= cSt)	7000–8000	7000–8000
specific gravity, 25/25°C	0.98	0.95
refractive index, 25°C	1.478	1.483

[a] Hystrene series of dimer acids, Humko Sheffield Chemical.

evaporation. The monomer fraction, after hydrogenation, can be fed to a solvent separative process that produces commercial isostearic acid, a complex mixture of saturated fatty acids that is liquid at 10°C. Dimer acids can be further separated into distilled dimer acids and trimer acids.

The production of dimer acids is quite energy-intensive. A standard operating sequence at one manufacturing plant results in the expenditure of ca 18.6 MJ (17,600 Btu) to produce each kilogram of crude dimer and to separate it into a monomer, dimer, and trimer.

There are a number of experimental alternatives to clay-catalyzed or thermal polymerization of dimer acids. However, none has achieved commercial success. These include the use of peroxides, hydrogen fluoride, a sulfonic acid ion-exchange resin, and corona discharge (see Initiators).

Health and Safety

Trimer acids, dimer acids, monomer acids, and distilled dimer acids are classified as nontoxic by ingestion, are not primary skin irritants or corrosive materials, and are not eye irritants. Federal regulations do not permit direct use of dimer acids in food products; however, they may be indirectly used in packaging materials with incidental food contact.

Uses

Nonreactive polyamide resins are the largest-volume commercial application of dimer acids. The largest commercial application is hot-melt adhesives (qv), thermoplastics that have fairly sharp melting ranges. Other applications include printing inks, and surface coatings, eg, in producing thixotropy (see Alkyd resins). Reactive polyamide resins are the second-largest commercial application, particularly in surface coatings, adhesives, potting and casting. Other uses include "downwell" corrosion inhibitors for oil-drilling equipment (see Petroleum); lubricants, and intermediates.

EDWARD C. LEONARD
Humko Sheffield Chemical

E.C. Leonard, "Dimer Acids Applications," *American Oil Chemists' Society Convention, Higher Molecular Weight Dibasic Acids Symposium*, New York, May 12, 1977.

A. Wolfe "Polyamide Resins (Non-Nylon Types)," in *Chemical Economics Handbook*, Stanford Research Institute, Menlo Park, Calif., June 1977, p. 580.1031L.

E.C. Leonard "The Higher Aliphatic Di- and Polycaroxylic Acids," in E.H. Pryde, ed., *Fatty Acids*, American Oil Chemists' Society, Champaign, Ill., in press.

DIPHENYL ETHER, DIPHENYL OXIDE $(C_6H_5)_2O$. See Heat-exchange technology—Heat-transfer media other than water; Ethers.

DIPHENYL AND TERPHENYLS

Diphenyl (biphenyl, phenylbenzene) and terphenyl are the lowest members of a family of polyphenyls in which benzene rings are attached one to another in a chain-like manner, $C_6H_5(C_6H_4)_mC_6H_5$. Although many higher polyphenyls are known, only 1,1'-biphenyl and terphenyl mixed isomers: 1,1':2,1''-terphenyl, 1,1':3,1''-terphenyl, and 1,1':4,1''-terphenyl are of commercial importance.

Physical and Chemical Properties

Diphenyl is a white or slightly yellow crystalline solid. It separates from solvents as plates or monoclinic prismatic crystals. As one of the most stable organic compounds, it resists thermal decomposition and degradation by radiation. Mp 69.2 C; bp 255.2 ± 0.2°C; d_4^{20} 1.041 g/cm³; vapor pressure (200°C) 25.43 kPa (191 mm Hg); liq d (100°C) 0.970 g/mL.

Diphenyl may be regarded as a substituted benzene, and thus exhibits most of the chemistry but it is less reactive than benzene. Partial reduction gives phenylcyclohexane, full hydrogenation gives dicyclohexyl.

Pure terphenyls are white crystalline solids, whereas the commercial grades are yellow or tan. o-Terphenyl has mp 56.2°C; bp 332°C (101.3 kPa); vapor pressure (93°C) 11.72 Pa (87.9 μm Hg); sp grav (93°C) 1.022. m-Terphenyl has mp 87.5°C; bp 365°C (101.3 kPa); vapor pressure (90°C) 1.65 Pa (12.4 μm Hg); sp grav (93°C) 1039. p-Terphenyl has mp 212.7°C and bp 376°C (101.3 kPa). Terphenyls are essentially insoluble in water and dissolve only sparingly in the lower alcohols and glycols whereas aromatic solvents readily dissolve o- and m-terphenyls. The p-isomer is much less soluble.

Terphenyls, like diphenyl, undergo the usual organic reactions of aromatic hydrocarbons. The o- and p-isomers add a nitro group at the 4-position, whereas the m-isomer adds a nitro group at the 4′ position. Halogenation apparently proceeds similarly. All three isomers are attacked most readily at the 4 position by acyl or sulfonyl halides in Friedel-Crafts reactions (qv).

o- and p-Terphenyls can be isomerized to the p-isomer. The ortho isomer, refluxed for a short time with a small amount of aluminum chloride in benzene, gives 94% meta isomer. More drastic isomerization conditions give conversions of up to 84% para isomer. Under strongly dehydrogenating conditions, o-terphenyl cyclizes to triphenylene.

In the presence of certain ether solvents, terphenyls can be metalated by sodium amalgam. The meta isomer reacts least rapidly and adds only one atom of sodium per molecule whereas o- and p-terphenyl give 2:1 adducts. Terphenyl can be reduced in liquid ammonia, and reductively methylated by alkali metals.

Manufacture

Before the early 1970s, the major commercial process for diphenyl and terphenyls was thermal dehydrogenation of benzene. Today, the main source of diphenyl is as a by-product in benzene production by hydrodealkylation of toluene (qv). The by-product stream can be recycled into the process or utilized to enrich fuel oil. For marketing, by-product diphenyl is refined to 93–97% purity by distillation.

When both biphenyl and terphenyls are desired, the thermal dehydrocondensation of benzene remains the commercial route of choice.

A number of other diphenyl processes involve a variety of hydrocarbon feedstocks and conditions, eg, the pyrolysis of coal tar in the presence of Mo-oxide/AlO$_3$ catalyst, hydrocarbon gas pyrolysis tar, crude benzene fractions, and CH$_4$ passed through silica gel at 1000°C.

Terphenyls are obtained from the higher-boiling polyphenyl by-products formed in the pyrolysis process of diphenyl manufacture. After benzene and diphenyl are distilled from the pyrolysis mixture, the high boiling residue consists of approximately 3–8% o-terphenyl, 44% m-terphenyl, 24% p-terphenyl, 1.5% triphenylene, and about 22–27% higher polyphenyls and tars. To obtain high purity m- or p-terphenyl, the appropriate distillation fraction has to be further purified by recrystallizing, zone melting, or other refining technique (see Zone refining).

Health and Safety

Because of its low vapor pressure and low toxicity, diphenyl is not an industrial hazard; dust explosions are a great hazard. A maximum permissible level in working areas of 5 mg/m^3 is recommended for both mixed terphenyls and p-terphenyl.

Uses

In the past, the chief use for diphenyl was heat-transfer fluids, eg, the diphenyl oxide–diphenyl eutectic mixture marketed under the trademarks Dowtherm A and Therminol VP-1 are widely used for heat transfer by the condensation of vapors in the temperature range of 250–360°C (see Heat-transfer technology). The current primary use for biphenyl is in the formulation of dye carriers (qv) for textile dyeing. Diphenyl is also used as a paper impregnant for citrus fruit where it acts as a fungicide. Terphenyls are used almost exclusively in high temperature heat-transfer applications.

W.C. WEAVER
P.B. SIMMONS
Dow Chemical USA
Q.E. THOMPSON
Monsanto Co.

J.P. Stone, C.T. Ewing, and R.R. Miller, *J. Chem. Eng. Data* **7**(1), 519 (1962).

M. McEwen, *Organic Coolant Handbook*, sponsored by U.S. Atomic Energy Commission, published by Monsanto Co., St. Louis, Mo., 1958.

G. Descotes, J. Praly, and M. Lebaupain, *Bull. Soc. Che. France*, 896, 901 (1976).

DISINFECTANTS AND ANTISEPTICS

Antiseptics are generally understood to be agents applied to living tissue and are considered to be drugs. Therefore, they may be inactivated by contact with body fluids, such as blood or serum. Disinfectants are generally understood to be agents suitable for application to inanimate objects. Environmental surfaces are deemed to be contaminated, not infected. Disinfectants kill the growing forms but not necessarily the resistant spore forms of microorganisms.

Definitions

The FDA defines an antimicrobial (active) ingredient as a compound or substance that kills microorganisms or prevents or inhibits their growth and reproduction and that contributes to the claimed effect of the product in which it is included. The FDA defines an antimicrobial preservative (inactive) ingredient as a compound or substance that kills microorganisms, or that prevents or inhibits their growth and reproduction, and that is included in a product formulation only at a concentration sufficient to prevent spoilage or prevent growth of inadvertently added microorganisms, but does not contribute to the claimed effects of the product to which it is added.

The EPA has adopted the following definition for antimicrobial agents: Disinfectants destroy or irreversibly inactivate infectious or other undesirable bacteria, pathogenic fungi, or viruses on surfaces or inanimate objects; sanitizers reduce the number of living bacteria or viable virus particles on inanimate surfaces, in water, or in the air; bacteriostats inhibit the growth of bacteria in the presence of moisture; sterilizers destroy viruses and all living bacteria, fungi, and their spores on inanimate surfaces; and fungicides (qv) and fungistats inhibit the growth of or destroy fungi (including yeasts) pathogenic to man or animals on inanimate surfaces.

Disinfectants

Alcohols. The antibacterial effectiveness of several of the aliphatic alcohols has been known for a long time, particularly the bactericidal action of ethyl alcohol.

Isopropyl alcohol appears to be somewhat more effective as an antimicrobial agent than ethyl alcohol, eg, *Staphylococcus aureus* is reported to be killed by 50–91% isopropyl alcohol in 1 min at 20°C. However, earlier reports claim that 90% isopropyl alcohol fails to kill this organism in 2 h. Isopropyl and ethyl alcohols are essentially similar in tuberculocidal performance when acting upon dried sputum smears.

In the homologous series of primary normal alcohols, the germicidal action increases with increasing molecular weight from methyl to octyl alcohol as shown in tests with *Salmonella typhosa* and *Staphylococcus aureus*. Moreover, there exists a fairly constant ratio between the molecular coefficients of successive members of the series:

$$\text{molecular coefficient} = \frac{\text{mol wt alcohol}}{\text{mol wt phenol}} \times \text{phenol coefficient}$$

Primary alcohols with 6 to 16 carbon atoms inhibit *Mycobacterium tuberculosis* and *Trichophyton gypseum* but not *Staphylococcus aureus*. Relating thermodynamic activity to cytostatic activity accounts satisfactorily for many regularities encountered in homologous series of alcohols.

Vapors of propylene glycol and triethylene glycol can protect animals against airborne bacteria and influenza virus under controlled conditions of temperature and humidity (see Alcohols; Glycols).

Halogens

Chlorine and hypochlorites. The bactericidal action of hypochlorites, eg, calcium hypochlorite (which has replaced chlorinated lime for large-scale usage), and sodium hypochlorite (the active principle of many household products and of some hospital specialties) is caused primarily by the release of hypochlorous acid. However, the hypochlorite ion, OCl^-, may be a contributory factor since even distinctly alkaline hypochlorite solutions show some antibacterial potency. The pH of the hypochlorite solution has a marked effect on its germicidal activity in tests employing various test organisms. The antibacterial action of a number of organic chlorine compounds depends upon their capacity for releasing active chlorine. Chlorine dioxide is more sporicidal than hypochlorous acid. The hypochlorites are subject to gradual deterioration over a period of time, and lose much of their bactericidal potency under practical conditions of disinfection.

N-Chloramines. This group is comprised of derivatives of amines in which one or two valences of trivalent nitrogen are taken up by chlorine. Compared with hypochlorous acid, monochloramine, NH_2Cl, requires more time and a higher concentration to produce sporicidal action. Other chloramines includes Chloramine T (sodium N-chloro-p-toluenesulfonamide), Dichloramine T (N,N-dichloro-p-toluenesulfonamide), Halazone (N,N-dichloro-p-carboxybenzenesulfonamide), and chlorinated cyanuric acid derivatives which consist of dichloroisocyanuric acid, their sodium and potassium salts, and trichloroisocyanuric acids (see Bleaching agents; Cyanuric and isocyanuric acids).

Iodine. Iodine, unlike bromine, is a significant antibacterial agent, particularly in the field of antisepsis. Like chlorine, iodine is a highly reactive substance, combining with proteins partly by chemical reaction and partly by adsorption. Its antimicrobial action is subject to substantial impairment in the presence of organic matter, eg, serum, blood, urine, milk, etc. As a bactericidal antiseptic, iodine is most frequently used in the form of a 2% tincture, but stronger preprations are occasionally employed, eg, for preoperative antisepsis. Iodophors are combinations of iodine with suitable solubilizing organic compounds (usually nonionic surfactants) which are effective owing to the free, or available, iodine they contain. Several disinfectant, antiseptic, and sanitizing agents are formulated on the iodophor principle. Polyvinylpyrrolidinone (PVP) is used as a solubilizing agent and carrier of iodine in external applications on humans (PVP-iodine, Betadine (The Purdue Frederick Co.)); quaternary ammonium compounds are also used as solubilizing agents.

Metals, Their Salts, and Other Compounds

Inorganic mercurials. Mercuric chloride (bichloride of mercury, $HgCl_2$) is one of the earliest antibacterial agents known. Its antibacterial action is primarily inhibitory rather than bactericidal.

Organic mercurials. Organic mercurials are mainly used as antiseptics rather than disinfectants. Here, the mercury is linked directly with carbon and is not released in solution as an ion. If ionization takes place, organomercuric ions, RHg^+, are formed. The inactivation of bacteria is possibly due to a blocking of cellular enzymatic thiol reception by formation of mercaptide bonds and without any other demonstrable cell injury, the inhibition is reversible. The two reactions involved are illustrated below for a mercuric salt:

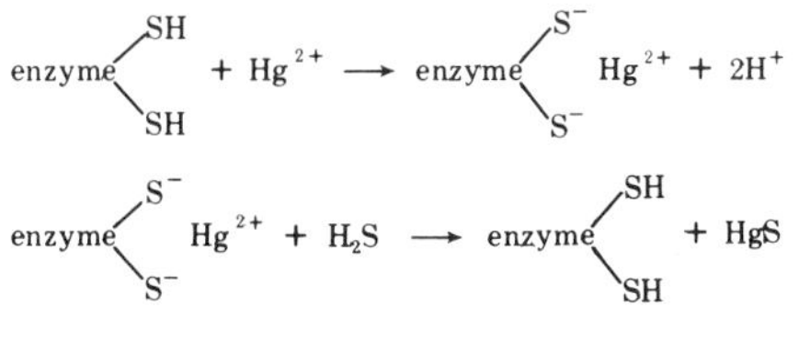

Organic mercurials used as antiseptics include: phenyl mercuric nitrate (Merphenyl nitrate, Merphene (Hamilton)); merbromin (2,7-dibromo-4-hydroxymercurifluorescein, sodium salt; Mercurochrome (Hynson)); nitromersol (4-nitro-3-hydroxymercurio-o-cresol anhydride; Metaphen (Abbott); thimerosal (sodium ethylmercurithiosalicylate; Merthiolate (Lilly)) among others.

Silver compounds. Silver, both free and in salts, produces antimicrobial effects. Colloidal dispersions of metallic silver display oligodynamic action owing to the formation of silver ions which are adsorbed by the bacterial cell with cytotoxic results. Even insoluble silver chloride and

silver iodide show some antibacterial action when dispersed in colloidal form and applied in fairly high concentrations. The degree of antibacterial action depends upon the number of free silver ions present; this is illustrated by the comparatively high effectiveness of silver nitrate as against the low value of an ammoniacal silver complex. Silver nitrate is still used occasionally for routine prophylactic instillation into the eyes of newborn infants, although it has been displaced largely by other more effective and less irritant antimicrobial agents, especially penicillin.

Other metals and their compounds. Metals and compounds other than those of mercury and silver are of lesser importance as antibacterial agents. Several salts of heavy metals are bactericidal under the conditions of the FDA method.

Peroxides and Other Oxidants

The antibacterial potency of some representatives of this group depends primarily upon the reactivity of free OH radicals. Although different microorganisms vary considerably in their susceptibility to oxidizing agents, the obligate anaerobic bacteria are most sensitive to oxidation. Of these, the species capable of forming hydrogen peroxide metabolically, but incapable of producing catalase to decompose it, are damaged and eventually killed by exposure to molecular oxygen (see Peroxides).

Hydrogen peroxide (qv) is used as an antiseptic in a 3% solution; concentrates of 30% or stronger are corrosive to skin and denuded tissue. Solutions containing as little as 0.1 and 0.25% H_2O_2 kill *Salmonella typhosa*, *Escherichia coli*, and *Staphylococcus aureus* in 1 h but not in 5 min. Addition of ferric or cupric ions, potassium dichromate, cobaltous sulfate, or manganous sulfate enhances the bactericidal action of hydrogen peroxide. With urea, hydrogen peroxide forms a solid compound capable of yielding over 35% of H_2O_2.

Zinc peroxide, as used for medical purposes, consists of a mixture of this chemical with zinc carbonate and zinc hydroxide. Especially effective against anaerobic and microaerophilic bacteria, it is applied in the treatment of wound infections in the form of a thick suspension.

Benzoyl peroxide, applied as an antiseptic dressing, is useful in treating wounds infected with gas-forming anaerobes.

Ozone, is actively germicidal in the presence of moisture, and is used for disinfecting swimming-pool water. Ozone generators introduced for the purification of air are not efficient controls of airborne pathogens and may create a toxicological hazard.

Phenolic Compounds

The bactericidal effectiveness of several phenol derivatives has been correlated with their respective partition ratios between oil and water; a high lipophilic character is associated with greater antibacterial activity. However, the surface activity of the phenol derivatives plays an important part in their total antibacterial performance by altering the permeability of the cell wall. In this respect, their behavior resembles that of other antibacterial surfactants.

Coal-tar disinfectants, until recently, constituted the most important category of disinfectants for environmental use. Because of the crude, unrefined character of the coal-tar fractions entering into resin soaps, these disinfectants emit a rather pungent, disagreeable odor (see Soap).

Phenol homologues. The homologous series of alkylchlorophenol derivatives show a regularity like that of the phenol derivatives. The regularities noted in the series of o- and p-chlorophenol, and o- and p-bromophenol derivatives are as follows: Halogen substitution intensifies the microbicidal potency of phenol derivatives; halogen in the para position to the hydroxyl group is more effective than in the ortho position. Introduction of aliphatic or aromatic groups into the aromatic nucleus of halogenated phenols increases the bactericidal potency up to certain limits, depending in the case of alkyl substitution upon the number of carbon atoms present in the substituting group or groups. A normal aliphatic chain with a given number of carbon atoms generally exerts a greater intensifying effect upon the bactericidal potency than that of a branched chain or of two alkyl groups with the same total number of carbon atoms. o-Alkyl derivatives of p-chlorophenol are more actively germicidal than the corresponding p-alkyl derivatives of o-chlorophenol.

Hydroxybenzoic acids. The antibacterial and antifungal properties of the hydrobenzoic acids depend primarily upon the reactivity of their phenolic hydroxy group. It is masked by intramolecular hydrogen bonding in o-hydroxybenzoic (salicyclic) acid which is unimportant as an antimicrobial agent (see Salicyclic acid and related compounds). On the other hand, salicylanilides with halogens as substituents have commercial use as antimicrobials.

p-Hdyroxybenzoate esters (parabens) play an important role in the preservation of different perishable organic materials, eg, those based upon carbohydrates, gums and proteins, against attack and destruction by airborne bacteria, fungi and yeasts.

Bis(hydroxyphenyl) alkanes. The most important member of the series is 2,2'-methylenebis(3,4,6-trichlorophenol), also named bis(3,5,6-trichloro-2-hydroxyphenol)methane (hexachlorophene, G-11, Givaudan Corporation). It has extensive use as an antibacterial soap additive as it does not lose activity in the presence of soap, as most phenolic compounds do. Hexachlorophene has also been incorporated in soapless detergent bases (pHisoHex, Winthrop Laboratories). Because of potential neurotoxicity in humans, the FDA has regulated the use of hexachlorophenes.

Among orthohydroxydiphenyl ethers, 2,4,4'-trichloro-2'-hydroxydiphenyl ether has shown to be active against gram-positive and gram-negative bacteria and against fungi. It has use as an additive for soaps and washing products.

8-Hydroxyquinoline and derivatives. Of the seven isomeric hydroxyquinolines, only 8-hydroxyquinoline exhibits an antimicrobial character (8-quinolinol, oxine). It is used either by itself or as a salt as the active ingredient of several antiseptics whose effect is bacteriostatic and fungistatic rather than microbicidal.

hexachlorophene

Quaternary Ammonium and Related Compounds

Generally, the long-chain alkylammonium and the quaternary ammonium salts, like other cation-active compounds, are incompatible with soaps or other anion-active materials. Mutual precipitation usually occurs when aqueous solutions of the representatives of the cation-active and anion-active classes are brought into contact, except where the molecular weights of the cations and anions are sufficiently low.

Of the numerous quaternary ammonium salts investigated, only a comparatively small number have retained importance as antibacterial agents. Among them are benzalkonium chloride, ie, alkylbenzyldimethylammonium chloride, in which R is a mixture of alkyls from C_8H_{17} to $C_{18}H_{37}$, $C_{12}H_{25}$ predominating; benzethonium chloride, ie, N-benzyl-N,N-dimethyl-N-[2-[2-[p-(1,1,3,3,-tetramethylbutyl)phenoxy]ethoxy]ethyl]ammonium chloride; methylbenzethonium chloride; and alkylisoquinolinium bromide.

Soaps or other anionic surfactants must be removed by thorough rinsing from surfaces to be treated with quaternary ammonium germicides. The hardness of water can also affect the antibacterial performance of the quaternaries. However, within these limitations, the quaternaries render satisfactory service in the sanitization of eating and drinking utensils, the disinfection of equipment used in dairies and in other food-processing plants, the cleaning of eggs, etc.

Pine Oil Compounds

Pine oil is obtained from waste pine wood by destructive distillation or by distillation with superheated steam. It is insoluble in water, but is readily emulsifiable when combined with a soap, a sulfonated oil, or other suitable dispersing agents. Most pine-oil disinfectants contain from 60–65 vol% of pine oil. Diluted with water such preparations yield white, milky emulsions, with a characteristic odor (see Odor counteractants). In recent years, a serious problem has been created in hospitals, owing to the emergence of antibiotic-resistant strains of staphylococci; however,

these are susceptible, eg, to the action of most phenolic disinfectants or combinations of pine oil with phenolic bactericides.

Carbamic Acid and Urea Derivatives

The class of dithiocarbamates was previously studied for its capacity to control fungus diseases of tomatoes, beans, potatoes, etc. (see Fungicides). More recently, some dithiocarbamates were found to exert powerful inhibitory action upon fungi pathogenic for man (see Chemotherapeutics, antibacterial and antimycotic). Tetramethylthiuram disulfide (TMTD) was the most active of the compounds tested against pathogenic fungi as well as against bacteria (see also Rubber chemicals). 3,4-4'-Trichlorocarbanilde (TCC, trichlorocarban, Monsanto Chemical Co), has been introduced commercially as a soap additive for its effective skin degerming and deodorant action. The combination of TCC with hexachlorophene exhibits antibacterial synergism.

Other Disinfectants and Antiseptics

These include nitrofuran derivatives, which possess antiseptic properties (eg, 5-nitro-2-furaldehyde-2-ethylsemicarbazone (nitrofurazone, Furacin, Eaton Laboratories)). Formaldehyde and other aldehydes, including certain saturated dialdehydes and particularly glutaraldehyde, $OCH(CH_2)_3CHO$, which has a mild odor and a low irritation potential for the skin and mucous membranes (see Aldehydes). Ethylene oxide, available as Carboxide R, represents a mixture of 10% ethylene oxide and 90% carbon dioxide. The bactericidal activity of ethylene oxide is thought to be the result of direct ethoxylation of functional groups of bacterial proteins. Propylene oxide was found similar in properties to ethylene oxide, but less volatile and less active biologically. Interest in the compound was revived when the FDA restricted use of ethylene oxide in the food industry because of the toxicity of ethylene glycol, a hydrolytic product found in small amounts. Propylene oxide hydrolyzes to nontoxic propylene glycol (see Glycols; Sterilization techniques).

WILLIAM GUMP
Consultant

P.S. Herman and W.M. Sams, *Soap Photodermatitis and Photosensitivity to Halogenated Salicylanilides*, Charles C Thomas, Springfield, Ill., 1972.

S.S. Block, *Disinfection, Sterilization, and Preservation*, 2nd ed., Lea & Febiger, Philadelphia, Pa., 1977.

C.A. Lawrence, *Surface-Active Quaternary Ammonium Germicides*, Academic Press, Inc., New York, 1950.

DISPERSANTS

Dispersants are a class of materials that are capable of bringing fine solid particles into a state of suspension so as to inhibit or prevent their agglomerating or settling in a fluid medium. Dispersants may break up agglomerates or aggregates of fine particles, bring fine particles into a colloidal solution, or solubilize supersaturated salts to leave a clear solution. In particular, dispersants are useful whenever one wishes to prevent deposition, precipitation, settling, agglomerating, adhering, caking, etc, of solid particles in a fluid medium.

This article deals only with materials capable of dispersing completely aqueous systems. Dispersants can also refer to the large number of proprietary compositions that facilitate the formulation of paints, inks, dyes, etc, especially by the maintenance of uniform consistency. Often the term is applied to surface-active materials that stabilize oil–water emulsions in the manufacture of lotions, lubricating oils, resins, and latexes (see Latex technology), materials used primarily in systems where a second, nonaqueous liquid phase is present (see Emulsions; Surfactants). Dispersants can be contrasted with flocculants or coagulants which facilitate the aggregation of fine particles in aqueous media to improve separations. Dispersants are normally anionic materials of lower molecular weight and higher charge than flocculants (see Flocculating agents).

With aqueous dispersants, there are three primary categories of materials: deflocculants, primarily used to fluidize concentrated slurries to reduce their bulk viscosity or stickiness in processing or handling; antinucleation agents, usually referred to by function as antiscalants or antiprecipitants, used to prevent the deposition of sparingly soluble salts or scale in boilers, recirculating cooling-water systems, or desalination units; and builders, the largest dispersants market, the additives in most cleaning and detergent compositions that facilitate cleaning action by sequestering water-hardness ions (see also Chelating agents).

Physicochemical Principles of Dispersancy

Virtually all dispersants must adsorb on the substrate in order to perform their function. The degree of adsorption is related to solubility, strength of surface interaction, and the presence or absence of coulombic or van der Waals forces. Once adsorbed, deflocculants have the ability to disaggregate insoluble or suspended matter in aqueous media. Scale inhibitors (or antiscalants) are agents that inhibit deposition of adherent, crystalline deposits at surfaces, especially heat-transfer surfaces. Antiprecipitants are materials that hinder precipitation of solids or formation of turbidity in the bulk solution. Antiscalants and antiprecipitants are both classified as antinucleation agents. All currently available commercial scale and precipitation inhibitors are threshold inhibitors, ie, they inhibit nucleation of macroscopic scale or precipitates.

Deflocculation. Deflocculation has been termed "the breaking up from a flocculant state: converting into very fine particles." The key phenomenon of deflocculation is the breaking apart or tearing down of aggregates, ie, reversing the tendency of particles in dense suspensions (not supersaturated solutions) to settle, stick, or adhere, leading to inhibition of flow or increasing bulk viscosity.

Detergent builders. Builders improve surfactant performance by acting as deflocculating agents, buffering agents, and water-softening agents particularly as related to the removal of hardness which retards detergency (see Surfactants and detersive systems).

Dispersant Materials

Condensed phosphates. The most active threshold inhibitors for many applications are amorphous polyphosphate salt glasses which contain long phosphate chains. Today, virtually all condensed phosphates are prepared by calcining the appropriate ratio of Na_2O : P_2O_5 : H_2O under carefully controlled conditions to form the desired compositions directly. They are used as oil-field drilling mud thinner (see Petroleum) deflocculants, detergent builders, and scale inhibitors for recirculating cooling water (see Phosphoric acids and phosphates).

Organic polymers. Polyacrylates are probably the most flexible dispersant products. They are easily produced in a variety of molecular weights and degrees of anionic charge (see Acrylamide polymers; Acrylic acid) and typically cost less than the phosphonates but more than polyphosphates. Polyacrylates are very stable polymers and are active deflocculants. However they provide no corrosion inhibition.

Polymaleates generally show similar properties to polyacrylates but are particularly useful for evaporative technologies (see Maleic anhydride). Natural product-derived dispersants, such as tannins and lignins, are still widely used as drilling-mud thinners (see Petroleum). Other specialty dispersants such as alginates, find use in shipboard boilers where low toxicity is a crucial property.

Uses

Dispersants are used in antinucleation agents in boiler treatments, recirculating water, desalination, and in oil-field applications, eg, prevention of blockage of rock pores by scale formation; as deflocculants in clay deflocculation; as drilling mud thinners and portland cement dispersants; in pigment preparation; in mineral process; in recirculating cooling water; and in handling high solids slurries.

RICHARD M. GOODMAN
American Cyanamid Company

Handbook of Industrial Water Conditioning, 7th ed., Betz Laboratories, Inc., Trevose, Pa., 1976.

G.D. Parfitt, *Dispersion of Powders in Liquids*, 3rd ed., Applied Science Publishers, London, UK, 1981.

J. Am. Oil Chem. Soc. **55**(1) (1978), for a complete review of detergents.

DISPERSINS AGENTS.   See Sulfonic acids.

DISPERSION OF POWDERS IN LIQUIDS

The term dispersion refers to the process of incorporating a powder into a liquid medium in such a manner that the product consists of fine particles distributed throughout the medium. The dispersion of fine particles is normally termed colloidal if at least one dimension of the particles lies between 1 nm and 1 μm. Such dispersions are classified, in terms of the affinity of the colloidal particle for the medium, as lyophobic (possessing aversion to liquid) or lyophilic (possessing affinity for liquid); and as hydrophobic and hydrophilic, respectively, for aqueous media.

A powder containing nonporous particles of colloidal dimensions has a high surface area. A 1-cm cube, initially with an area of 6 cm², when divided into 10^{12} cubes of 1-μm edge length has an area of 6 m², whereas when divided into 10^{18} cubes of 10-nm edge length, it has an area of 600 m² or 0.06 ha (ca 0.15 acre). In many uses of powders, the primary particle is sufficiently small for further subdivision to be unnecessary.

In the early stages of the dispersion process, the solid–air interface is replaced by one between solid and liquid (wetting). The powder consists of aggregates/agglomerates of similar particles, and to disperse these particles into a liquid, mechanical work is performed. The forces that exist at the interface determine the ease with which the process can be brought about.

The dispersion process includes incorporation, wetting, breakdown of particle clusters, and flocculation of the dispersed particles. These stages overlap, but one is likely to be dominant with respect to a particular property of the dispersed system.

Incorporation

The interparticle forces that must be overcome in dispersion may be mechanical forces, surface-tension forces, forces arising from plastic welding, electrostatic van der Waals forces, and solid-bridge forces. All are subject to changes in environment. Different methods of particle production give particles of different roughness, leading to changes in the interlock or friction forces between particles.

Liquid bridges. When appreciable quantities of liquid are adsorbed, the surface film covers the asperities and forms liquid bridges between particles in contact. The curvature of the meniscus determines the vapor pressure above the liquid bridge and the magnitude of the adhesive force owing to surface tension. For two smooth spheres, the total force F_{tot} is

$$F_{tot} = \frac{2\pi r \gamma}{1 + (\tan \phi /2)}$$

where ϕ is the central angle (see Figure 1), r is the particle radius, and γ is the surface tension.

Interaction between surface asperities. Particle surfaces are rough when viewed on a macroscopic level, and it seems reasonable to expect surface asperities of adjacent particles to interlock. When two particles come into contact, however, only relatively small parts of their total surface area are actually touching. Only the tips of the asperities come into contact, and if the material is at all compacted, tremendous pres-

Figure 1. Condensed liquid at the point of contact between two smooth solid spheres, assuming zero contact angle.

sures can be generated over the tiny areas of contact, and plastic flow can occur.

Electrostatic forces. Electrostatic forces may arise from surface charges which are either permanent or acquired from adjacent surfaces. When two solid bodies are in contact, electrons are transferred from one to the other. In theory, this transfer process should continue until an equilibrium is reached when the electron currents (or work function) of the solids become equal. When this equilibrium has been reached, there is a potential difference between the two solids.

Van der Waals force. The derivation of the attractive force between macroscopic bodies assumes that interactions between individual atoms and molecules in the interacting materials are additive. For two equal spheres of radius a, the attractive potential energy V_A is given by

$$V_A = -\frac{A}{6}\left[\frac{2}{s^2 - 4} + \frac{2}{s^2} + \ln\left(\frac{s^2 - 4}{s^2}\right)\right]$$

where A is the Hamaker constant and $s = R/a$; R is the distance between the centers of the spheres.

Solid bridges. The strength of an agglomerate can be considerably enhanced by the formation of solid bridges between particles. Such bridging can be due to chemical reactions at the surface.

A wide variety of mechanisms make it difficult to quantify the effects in any theory of powder strength. However, for practical purposes, the magnitude of the principal forces involved can be estimated (van der Waals and gravity) to illustrate the problem of disagglomeration of the powder during the early stages of incorporation.

Tensile strength. A simple demonstration of the range of cohesiveness in pigments is the measurement of tensile strength, which is the stress required to fracture a specimen in simple tension, as given by Rumpf's equation:

$$T = \frac{9}{8} \cdot \frac{(1 - \epsilon)}{\pi d^2} \cdot cF$$

where T is the tensile strength, d is the diameter of the spherical particles, c is the mean coordination number, F is the mean microscopic force at the contact face, and ϵ is the voidage. The equation was verified using limestone particles. In another approach, it is assumed that individual particles are packed in a somewhat random fashion. Tensile strength is determined by the force which acts between a pair of particles.

Wetting

The initial stage of wetting involves both the external surface of particles and the internal surfaces which exist between the particles in the clusters that make up the dry powder. The principles of wetting external surfaces are well established, but extension to internal surfaces involves geometric factors which, for fine particle systems, are not readily defined. The three distinct types of wetting are adhesional wetting, spreading wetting, and immersional wetting. They are readily defined in terms of changes in interfacial free energy.

Interactions at the Solid–Liquid Interface

To reduce the surface tension of water, ie, the free energy of the liquid–vapor interface, a large number of compounds are available. They are amphipathic, ie, they contain a hydrocarbon group that is expelled by water and a polar group that is water-soluble. These so-called surface-active agents are classified into anionic, cationic, and nonionic types.

Analysis of concentration changes caused by adsorption at the solid–liquid interface is straightforward. Taking m g of a solid in contact with a solution that initially contained n_1^0 mol of component 1 at mol fraction x_1^0, and n_2^0 mol of 2 at x_2^0, and after adsorption equilibrium is reached the solution contains n_1 mol at x_1, and n_2 at x_2, then the change on adsorption is given by

$$n^0\left(x_1^0 - x_1\right)/m = n_1^s x_2 - n_2^s x_1$$

where n_1^s and n_2^s mol of the components are adsorbed per gram of solid and n^0 $(= n_1^0 + n_2^0)$ is the total number of moles of liquid components present in the system.

Adsorption from dilute solution is analytically simpler, since the measured concentration change gives directly the individual isotherm of the solute. Systems of practical interest usually involve adsorption from dilute solution.

Normal long-chain acids adsorb onto oxide surfaces by hydrogen bonding with the surface hydroxyl groups, and in addition, for unsaturated acids there may be interactions between the surface and the alkene residue. The adsorption of some long-chain electrolytes (cationic and anionic) from aqueous solutions on polar and nonpolar adsorbents illustrates the effect of ionic strength and pH and shows further that the adsorption reaches a saturation value at the critical micelle concentration (CMC) of the electrolyte, a phenomenon that is commonly observed.

Polymers are important components of many systems, and their adsorption at the solid–liquid interface may be complex. The time taken to reach equilibrium depends on the molecular weight of the polymer. The amount adsorbed at equilibrium increases with increasing molecular weight. It is usually assumed that the polar, low molecular weight fractions and the hydroxyl number of the adsorbed resin are larger than those of the initial resin.

Breakdown of Agglomerates and Aggregates

In the initial stage of incorporating and wetting, the magnitude of the cohesive forces between the individual particles determines the extent of disagglomeration before the mechanical action is applied. Once the particles are wetted, mechanical energy is required to separate them completely.

Dispersion equipment falls into four classes. Low shear equipment employs slow mixing at high viscosity or high solids concentration. High shear equipment uses impellers moving at high peripheral speeds. In ball mills, the energies transmitted as applied to the dispersion by use of free-moving members irrespective of size, shape, or means of activation. Roll mills include equipment that uses one or more rollers and is designed to carry the millbase between either two rollers or a roller and a blade at a closely controlled gap under pressure.

Stability to Flocculation

Flocculation is the process that reduces the number of particles by collisions under both static (Brownian motion) and dynamic (shear) conditions. The types of forces between particles as they approach collision are the van der Waals force of attraction, the Coulombic force (repulsive or attractive) associated with charged particles, and the repulsive force associated with the interaction between adsorbed layers of polymer on the particle surfaces.

The interplay of the van der Waals and Coulombic forces forms the basis of the classical theory of flocculation of lyophobic dispersions, known as the DLVO (Derjaguin-Landau-Verwey-Overbeek) theory. Application of the DLVO theory to flocculation problems has become common practice despite all limitations. The theory considers the interaction between two charged particles in terms of the overlap of their electric double layers leading to a repulsive force which is combined with the attractive term to give the total potential energy as a function of distance for the system.

Since in a medium containing ions, a charged particle with its electric double layer is electrically neutral, no net Coulombic force exists between charged particles at large distances from each other. As the particles approach, the diffuse parts of the double layers interpenetrate, giving rise to a repulsive force which increases in magnitude as the distance between the particle decreases. The distance at which the repulsive force becomes significant increases with the thickness of the double layer, and the force increases with the surface potential.

The simplest way to calculate the force of repulsion between two interacting double layers is to consider the force as arising from the osmotic pressure of the excess ions in the space where the double layers overlap. Since the double-layer repulsion and the van der Waals attraction operate independently, the total potential energy V_{tot} for the system is given by the sum of the two terms.

For practical systems with high particle concentration, the validity of the DLVO theory is open to question. The use of the theory for two-particle interaction is probably useful as a first approximation for concentrated aqueous dispersions with moderate to high electrolyte concentrations.

Using an alternative approach, the force between two charged spherical particles in a concentrated dispersion in a hydrocarbon medium has been calculated. It was demonstrated that the electric double-layer force is not sufficient to promote stability to flocculation, in contrast to its effect in dilute dispersions; this is borne out in practice.

Interaction between dissimilar particles is termed heteroflocculation. The potential energy of interaction between two dissimilar particles in nonpolar media has been evaluated by applying Maxwell's equations for charged conducting spheres.

Steric stabilization. The term steric stabilization embraces all aspects of the stabilization of colloidal particles by adsorbed nonionic macromolecules, and is equally applicable to aqueous and nonaqueous media. Of particular interest are the thickness of the adsorbed layer, the configuration of the polymer chain, the fraction of segments adsorbed, and the segment-density distribution, both normal and parallel to the surface.

A characteristic feature of sterically stabilized systems is their different responses to temperature change. Entropically stabilized systems flocculate on cooling, enthalpically stabilized systems on heating, and for combined stabilization there is no accessible temperature for flocculation.

Mixing effects are readily interpreted in terms of polymer-solution theory, and apply when the separation between the approaching surfaces is one or two adsorbed layer thicknesses. Since polymer adsorption is usually irreversible, the method of mixing of the components can have a profound effect on the flocculation process. The effect has been demonstrated with silica and aqueous solutions of poly(vinyl alcohol).

Measurement of Dispersion

Perhaps one of the most difficult aspects of the experimental approach to dispersion is the assessment of the state of the dispersion, particularly for particles in the colloidal range where the measurement can disturb the system, eg, in electron microscopy, rheology, etc.

Various electron-microscope techniques have been used for assessing pigment dispersion in coatings. In principle, rheological methods provide a useful way of investigating the state of dispersion although interpretation of the observed phenomena is beset with difficulties (see Rheological measurements). Much of the fundamental work has been carried out with large spheres and at low concentration; such systems usually exhibit Newtonian behavior. With increasing concentration, deviations from ideal behavior are observed, and there is a large number of equations, both empirical and theoretical, that purport to describe experimental viscosity data.

G.D. Parfitt
Carnegie-Mellon University

G.D. Parfitt and K.S.W. Sing, eds., *Characterization of Powder Surfaces*, Academic Press, London, 1976.

G.D. Parfitt, ed., *Dispersion of Powders in Liquids*, 3rd ed., Applied Science Publishers, London, 1981.

G. D. Parfitt and C.H. Rochester, eds., *Adsorption from Solution at the Solid/Liquid Interface*, Academic Press, London, 1983.

B. Vincent and S.G. Whittington in E. Matijevic, ed., *Surface and Colloid Science*, Vol. 12, Plenum Press, New York, 1982, p. 1.

DISTILLATION

Distillation is broadly defined as the separation of more volatile materials from less volatile materials by a process of vaporization and condensation. In engineering terminology, the separation of a liquid from a solid by vaporization is considered evaporation (qv), and the term distillation is reserved for the separation of two or more liquids by vaporization and condensation.

Distillation is the most widely used method of separating liquid mixtures and is at the heart of the separation processes in many chemical and petroleum plants. Some of the many types of distillations are covered in this article, but azeotropic and extractive distillation (qv) are discussed under a separate heading in this *Encyclopedia*.

Vapor-Liquid Equilibria (VLE)

VLE is the relationship of the composition of the vapor phase and the liquid phase when the phases are in physical equilibrium. The driving force for any distillation is a favorable VLE. If the VLE are unfavorable, distillation is impossible. Reliable VLE data are essential for distillation column design and for most operations involving liquid–vapor-phase contacting. This information may be measured or estimated if not directly available from in-house or contract experimental measurements. The most common method to measure VLE data is direct vapor–liquid phase analysis from recycle equilibrium stills and from total pressure measurements. Chromatographic methods are used extensively for initial equilibria screening.

Distillation Processes

Basic distillation involves application of heat to a liquid mixture, vaporization of part of the mixture and removal of heat from the vaporized portion. The resultant condensed liquid, the distillate, is richer in the more volatile components. Most commercial distillations involve some form of multiple staging to obtain a greater enrichment than is possible for a single vaporization and condensation.

Simple distillations. Simple distillations utilize a single equilibrium stage to obtain separation. They may be either batch (also called differential distillation) or continuous (also called flash distillation). Batch distillation may be represented on boiling-point or phase diagrams. In Figure 1, if the batch distillation started with a liquid of composition x_L^A, the initial distillate vapor composition would be y_L^A. As the distillate is removed, the remaining liquid becomes less rich in L (low boiler) and the boiling liquid composition moves to the left. If the distillation was continued until the liquid had a composition of x_L^E, the last vapor distillate would have had a composition of y_L^E. Simple batch distillation is not widely used in industry. Calculation methods are found in most standard distillation texts.

Simple continuous distillation has a continuous feed to an equilibrium stage; the liquid and vapor leaving the stage are in phase equilibrium. A schematic representation is shown in Figure 2. On the boiling-point diagram, Figure 1, the feed is represented by x_L^F, the bottoms liquid by x_L^B, and the equilibrium vapor distillate by y_L^D. The mass balances are:

$$F = D + B \quad \text{(overall balance)}$$

$$x_L^F F = y_L^D D + x_L^B B \quad \text{(component balance)}$$

Flash distillations are widely used where a crude separation is adequate.

Multiple equilibrium staging. Multiple equilibrium staging is used to increase the component separation over that obtainable in simple distillation (see Fig. 3). The feed F enters the column at equilibrium stage $\bar{f}$. The heat $\bar{q}^S$ required for vaporization is added at the base of the column in a reboiler or calandria. The vapors V^T from the top of the

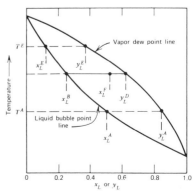

Figure 1. Boiling point diagram (isobaric).

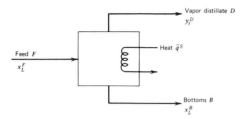

Figure 2. Simple continuous distillation with single equilibrium stage.

column flow to a condenser from which heat $\bar{q}^C$ is removed. The liquid condensate from the condenser splits into two streams. The first, a distillate D, which is the overhead product (also called heads or make), is withdrawn from the system and the second, a reflux R, is returned to the top of the column. A bottoms stream is withdrawn from the boiler. The overall separation is the feed F separating into a distillate D and bottoms B.

Above the feed, a typical equilibrium stage is designated as n; the stage above n is $n + 1$ and the stage below n is $n - 1$. The section of column above the feed is called the rectifying section and the section below the feed the stripping section.

Graphical McCabe-Thiele method. The graphical McCabe-Thiele design method facilitates a visualization of distillation principles. In the following discussion, the subscripts L and H are not used and x and y refer to the low boiler of the binary system. A McCabe-Thiele diagram is given in Figure 4. P, Q, and S are the x^B, x^F, and x^D compositions on the $x = y$, 45° construction line. Line OP is the stripping operating line and line OS is the rectifying operating line.

The McCabe-Thiele method makes the simplifying assumptions that the molal overflows in the stripping and the rectification sections are constant. These assumptions reduce the rectifying and stripping operating line equation to:

$$y^{n-1} = \left(\frac{\bar{L}}{\bar{V}_R}\right)x^n + \left(\frac{D}{\bar{V}_R}\right)x^D \tag{1}$$

$$y^{m-1} = \left(\frac{\bar{L}}{\bar{V}}\right)_S x^m - \left(\frac{B}{\bar{V}_S}\right)x^B \tag{2}$$

$\bar{L}$ and $\bar{V}$ designate the constant molal flow in each section. The McCabe-Thiele assumptions imply that the molal latent heats of the two components are identical, the sensible heat effects are negligible, and the heat of mixing and the heat losses are zero. These simplifying assumptions are closely approximated for many distillations. Equation 1 now represents the straight upper operating line OS and equation 2 represents the straight lower operating line OP. The upper operating line has the slope $(\bar{L}/\bar{V})_R$ and the intercept at x^D ($= y^T$) on the $x = y$ line. Similarly, the lower operating line has a slope of $(\bar{L}/\bar{V})$ and the intercept is at x^B on the $y = x$ line. The line QO from the feed intercept Q to the intersection of the operating lines at O is called the q line.

The equilibrium line gives the vapor–liquid relationship of y^n and x^n above the feed and of y^m and x^m below the feed. The upper operating line gives the relationship of y^{n-1} and x^n and the lower operating line gives the relationship y^{m-1} and x^m. The graphical representation of theoretical equilibrium stages (or steps or plates) n and m is shown. The y^{m-1}, x^m to y^m, x^m to y^m, x_L^{m+1} represents the mass balance and phase equilibrium for theoretical stage m; similarly, y^{n-1}, x^n to y^n, x^n to y^n, x^{n+1} represents theoretical stage n. The total number of theoretical stages in the column can now be stepped off starting either at the x^B and stepping upward or started at x^D and stepping downward.

Condition of feed (q line) is determined by mass and enthalpy balances around the feed plate. The q-line balances are detailed in distillation texts.

The reflux is the liquid returned to the top of the column and ratio R/D, is the external reflux ratio or simply reflux ratio. The ratio $(\bar{L}/\bar{V})_R$, which is the slope of the rectifying operating line, is the internal rectifying reflux ratio. Similarly, the ratio $(\bar{L}/\bar{V})_S$, which is the slope of the stripping operating line, is the internal stripping reflux ratio.

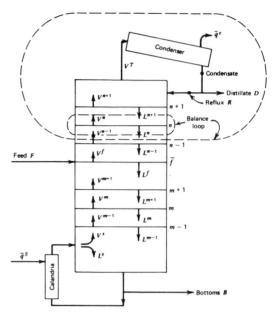

Figure 3. Distillation column with stacked multiple equilibrium stages.

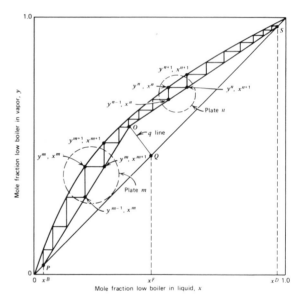

Figure 4. McCabe-Thiele diagram.

As the reflux ratio R/D increases, the internal rectifying reflux ratio increases and numerically approaches 1; similarly, the internal stripping reflux ratio decreases and numerically approaches 1. In the McCabe-Thiele plot, the two operating lines move away from the equilibrium line toward the $y = x$ diagonal; as the reflux ratio increases, the individual theoretical stage steps become larger and fewer theoretical stages are required to make a given separation.

Distillation Columns

Distillation columns (also called towers) provide the contact area between streams of descending liquids and ascending vapor and, thereby, furnish an approach toward physical vapor–liquid equilibrium. Depending upon the type of distillation column which is used, the contact may occur in discrete steps, called plates or trays, or in a continuous differential contacting on the surface of packing. The fundamental requirement of the column is to provide the mass-transfer contacting at a required rate and at a minimum overall cost. Individual column requirements vary from high vacuum to high pressure, from low liquid to high liquid rates, from clean to dirty systems, etc. As a result, a large variety of equipment has been developed to fill these needs.

Plate columns. Plate columns of two general types are in use: the crossflow plate, in which the liquid flows across the plate and from plate to plate, and the counterflow plate, on which the liquid flows downward through the same orifices through which the vapor rises.

The three principal vapor–liquid contacting devices are used in current crossflow plate design: the sieve plate, the valve plate, and the bubble-cap plate. These devices provide the mass-transfer contacting area between the rising vapor and the crossflowing liquid.

The contacting orifices in the conventional sieve plate are holes that measure 3.2–50 mm dia and which exhibit ratios of open perforated area to bubbling area ranging from 1 : 20 to 1 : 5.

The principal advantage of valve plates over sieve plates is their ability to maintain efficient operation over a wider operating range by using variable orifices. The valves open or close to expose the holes for vapor passage.

Bubble-cap plates used in new installations are limited to very low liquid flow rate applications or where the widest possible operating range is desired. Vapor flows through a hole in the plate floor through a riser, reverses direction in the dome of the cap, flows downward in an annular area between the riser and the cap and exits through slots in the cap.

Packed columns. In packed columns, the vapor–liquid contacting takes place in continuous beds of packing rather than in discrete individual plates as in plate columns. The contacting occurs in differential increments across the height of the packing. Mechanically, the packed column is a simple structure as compared to a plate column. The simplest packed column consists of a vertical shell with dumped packing on a packing support and a liquid distributor above the packing that distributes the liquid uniformly across the packed bed. The vapor enters the column below the bottom packed bed and flows upward through the column.

Packed vs plate towers. Packed towers are more useful than plate towers for: small diameter towers (under 1.5 m), severe corrosion environments; vacuum operation where a low pressure drop per theoretical plate is a critical requirement; high liquid rates above 49,000 kg/(h·m²) [10,000 lb/(h·ft²)]; foaming systems; or debottlenecking plate towers having relatively close plate spacing (under 0.5 m).

Steam Distillation

Steam distillation is used to lower the distillation temperatures of high boiling organic compounds that are essentially immiscible with water. In these cases, both the compound and the water will exert their full vapor pressure upon vaporization from the immiscible two-component liquid. At a system pressure of P, the partial pressure would be:

$$P = p_{water} + p_{organic}$$

And since the water and organic compound are immiscible:

$$P = p_{water} + p_{organic} = P^o_{water} + P^o_{organic}$$

Commercial applications of steam distillation include the distilling of turpentine (see Terpenoids) and certain essential oils (see Oils, essential).

Economics of Distillation

The suitability and economics of a distillation separation depends on many factors including: favorable vapor–liquid equilibria, feed composition, number of components to be separated, product purity requirements, the absolute pressure of the distillation, heat sensitivity, corrosivity, and continuous vs batch requirements.

EARL R. HAFSLUND
E.I. du Pont de Nemours & Co., Inc.

R.H. Perry and C.H. Chilton, eds., *Chemical Engineers' Handbook*, 6th ed., McGraw-Hill Book Co., New York, 1983.

R.E. Treybal, *Mass-Transfer Operations*, 3rd ed., McGraw-Hill Book Co., New York, 1980.

M. Van Winkle, *Distillation*, McGraw-Hill Book Co., New York, 1967.

DIURETICS

The word diuretic originally referred solely to the ability of agents to increase the rate of urine formation. In a modern therapeutic sense, it is generally used to describe all drugs that act on the kidney to promote the urinary excretion of water and electrolytes, particularly sodium ion. Additionally, the term saluretic is used for those drugs that exert their diuretic effect by primarily increasing the excretion of sodium chloride. It should be emphasized that these valuable clinical tools are heterogenous, not only in respect to chemical structure but, even more important, in their mode and site of action in the kidney. The main and traditional use of diuretics is in the treatment of edematous states. Diuretic treatment permits, in most cases, satisfactory mobilization and subsequent prevention of edema, ascites, and pleural effusion. Such symptomatic treatment appears obvious, as the pathogenesis of local or general accumulation of extracellular fluid comprises either a secondary or primary increase in renal water and sodium chloride retention. The underlying disease might be of cardiac, hepatic, or renal origin, or the retention might be associated with other clinical situations such as inflammation, hypersensitivity, and premenstrual tension.

During the past decade, diuretics have been increasingly employed in nonedematous conditions, usually unaccompanied by sodium and water retention. Thus, their efficacy in the treatment of hypertension, a well-established use of certain diuretics, is not closely related to their natriuretic effect and the choice of the agent depends on the nature of the hypertension. Powerful diuretics may be used in the treatment of drug overdose to produce an enhanced clearance of salicylates and some barbiturates by forced alkaline diuresis. Even the altered renal handling of calcium ion, which might be considered to be a side effect of diuretic treatment, may be utilized. Certain diuretics exert an inhibitory effect upon renal calcium excretion and thereby decrease the likelihood of renal stone formation, and others show an increased rate of calcium excretion, causing a fall in serum calcium levels valuable in the management of hypercalcemia.

It is remarkable that even now, with few exceptions, significant progress in diuretic research is taking place without a knowledge of the precise biochemical mode of action of diuretic drugs. The net effect of various diuretics on the urinary excretion of H_2O, Na^+, K^+, Cl^-, and HCO_3^- is a consequence of their interference with transport processes and is highly dependent on their site of action in the nephron.

Diuretic Compounds

Logically, diuretics should be classified from a therapeutic point of view, according to their diuretic profile and the maximal natriuretic effect that can be achieved. This reflects their sites of action in the kidney or, more precisely, in the nephron.

However, in order to discuss related chemical structures or structural features together whenever possible, diuretic compounds are divided into the following groups: osmotic diuretics, sulfamoyl diuretics, newer loop diuretics, potassium-sparing diuretics, xanthines, pyrimidines, organomercurials, and aryloxylacetic acids. It is likely that new diuretic drugs will result from the large number of new structures with diuretic activity reported in recent years. For the characterization of diuretic behavior in a general sense, however, a division is made into the following classes.

Osmotic diuretics act at multiple sites, primarily in response to the osmotic load in the tubule. They provide a mild saluretic action and an enhanced urine formation, probably by decreasing the medullary osmolar gradient. Clinical application is concentrated on D-mannitol (**1**) and isosorbide (**2**).

(1) D-mannitol   (2) isosorbide

Carbonic anhydrase inhibitors act in the proximal convoluted tubule by decreasing carbonic anhydrase enzyme activity responsible for sodium and bicarbonate reabsorption in this segment of the nephron. Additionally, hydrogen-ion secretion in the distal tubule is impaired, leading to increased potassium excretion. The maximal obtainable natriuretic effect —their natriuretic efficacy—is highly dependent on the acid–base status of the organism. It is generally low, as in the more distal part of the tubule, and the higher load of sodium results in increased compensatory sodium reabsorption. Acetazolamide (**3**) is an example of one of the most potent compounds.

(**3**) acetazolamide

Thiazide-type diuretics act mainly at the earlier cortical part of the distal tubule. Their action can lead to maximal excretion of ca 8% of the filtered load of sodium chloride. Similarly to the loop diuretics, potassium excretion is enhanced by the distal tubule. This type of diuretic action does not impair the concentrating ability of the kidney and, consequently, provides a urine of relatively high osmolarity. The group name is derived from chlorothiazide, which was the first diuretic of this type. Compounds (**4**) and (**5**) are about ten times more potent.

(**4**) X=Cl hydrochlorothiazide
(**5**) X=CF₃ hydrofl4uorthiazide

Loop diuretics such as (**6**) and (**7**) act at the medullary and cortical thick ascending loop of Henle by decreasing active chloride reabsorption. In providing high maximal saluresis and possessing high ceiling diuretic activity, they are the most efficient diuretic compounds available to date. The particular site of action can lead to an excretion up to ca 20% of the filtered load of sodium chloride. A diminished medullary osmolar gradient owing to a wash-out of the medullary solute results, furthermore, in a dilute urine even in the presence of antidiuretic hormone (ADH). The increased sodium delivery to the distal tubule is responsible for kaliuretic effects.

(**6**) furosemide

(**7**) R=NH(CH₂)₃CH₃
bumetanide

Potassium-sparing diuretics such as spironolactone (**8**) decrease aldosterone-dependent sodium–potassium exchange at the distal site of the nephron, the collecting duct. They act either directly on the tubular cells or indirectly by interfering with aldosterone synthesis or as competitive antagonists of this mineralocorticoid. Natriuretic efficacy is relatively low. In the concurrent administration with more efficient saluretics, this important feature is utilized for counteracting potassium loss of the more proximally acting drugs.

(**8**) spironolactone

Miscellaneous diuretics show a diuretic profile in which the main site of action is not common to any of the above classes; the xanthines are representative of this particular group.

Independent of this classification is, of course, both the duration of action, which is dictated by the pharmacokinetic behavior of the com-

pound in question, and the potency. The words potent and potency are used exclusively in the context of their relationship to activity per unit weight.

PETER W. FEIT
Leo Pharmaceutical Products

D.L. Davis and G.M. Wilson, *Drugs* **9**, 178 (1975).

A.F. Lant and G.M. Wilson, "Diuretics" in D. Black, ed., *Renal Diseases*, 3rd ed., Blackwell Scientific Publications, Oxford, UK, 1972, p. 684.

R.C. Allen, E.H. Blaine, E.J. Cragoe, Jr., and R.L. Smith in E.J. Cragoe, Jr., ed., *Diuretics—Chemistry, Pharmacology, and Medicine*, John Wiley & Sons, Inc., New York, 1983.

DRIERS AND METALLIC SOAPS

Metallic soaps are a group of water-insoluble compounds containing alkaline-earth or heavy metals combined with monobasic carboxylic acids of 7 to 22 carbon atoms. They can be represented by the general formula $(RCOO)_x M$ (neutral soap), where R is an aliphatic or alicyclic radical and M is a metal with valence x. They differ from ordinary soap by composition and their insolubility in water. Their solubility or solvation in a variety of organic solvents accounts for their many and varied uses (see Table 1). Driers are the most important group of metallic soaps promoting or accelerating the drying, curing, or hardening of oxidizable coating vehicles such as paints (see Paint).

Composition and Properties

The acid or anion portion of a metal soap can be varied. Typical anions currently used are rosin and tall-oil fatty acids, saturated and unsaturated naturally occurring long-chain monocarboxylic fatty acids with 7 to 22 carbon atoms, and naphthenic, 2-ethylhexanoic, and the newer synthetic tertiary acids.

Acid soaps contain free acid (positive acid number), whereas neutral (normal) soaps contain no free acid (zero acid number). The basic soap is characterized by a higher metal-to-acid equivalent ratio than the normal metal soap. Particular properties are obtained by adjusting the basicity.

Properties are furthermore determined by the nature of the organic acid, the type of metal and its concentration, the presence of solvents and additives, and the method of manufacture. Generally, metals of low atomic weight form soaps with higher melting points (eg, lithium). Higher melting points are also characteristic of soaps made of high molecular weight, straight-chain, saturated fatty acids. Branched-chain, unsaturated fatty acids form soaps with lower melting points. Light powders are characteristic of precipitated soaps, whereas soaps made by fusion are usually hard solids. Liquids and pastes are mixtures or solutions of soaps in hydrocarbon and/or nonvolatile solvents. Liquid soaps are manufactured as fluid as possible compatible with maximum metal content. Newer techniques are available in which liquid soaps of very high metal content (18–24%) are obtained which still possess excellent fluidity. Both liquid and paste soaps are manufactured to strict viscosity and flow specifications.

The odor of a soap is determined by its organic constituents. Color, another important property of metal soaps, is determined by the type and amount of metal used, the color and quality of the organic raw materials, the method of manufacture, and the care in processing.

Table 1. Applications of Metal Soaps

Application	Metal
stabilizers for plastics	Ba, Cd, Sn, Sr
fungicides	Cu, Hg, Zn
catalysts	Co, Cu, Mo, Mn, Cr, Ni
driers	Bi, Ca, Co, Fe, Pb, Mn, Zn, Zr
fuel additives	Ba, Fe, Mg, Mn, Pb

Manufacture

Metallic soaps are manufactured by three methods: the precipitation or double-decomposition process; the fusion process; and the direct metal reaction (DMR) process. The choice of process and variation depends upon the metal, the desired form of the product, the desired purity, raw material availability and cost, metal content, etc. The desired final composition determines the stoichiometry of the starting materials, ie, whether an acid, neutral, or basic soap is obtained.

In the case of metallic soap solutions, the metal is assumed to be the active part and the acid portion serves as carrier for the metal, conferring oil solubility, water insolubility, and compatibility with other components of the system in which it is to be used. Therefore, it is economically advantageous to incorporate as much metal per unit of acid as possible, providing the resulting soap is oil soluble and fulfills the requirements of the application.

Health and Safety Factors

The hazards encountered in the manufacture, processing, handling, and use of metal soaps are associated for the most part with the inherent toxicity of the metals and solvents present. In general, the acid portion of the metal soap is low in toxicity. However, naphthenic acid is highly irritating to the skin on prolonged contact. High concentrations have a narcotic effect.

Metallic soaps are used in driers, waterproofing agents, fuel additives, rubber, greases, lubricants, and chemical thickeners, plastics, catalysts, fungicides, and cosmetics and pharmaceuticals.

FREDERICK J. BUONO
MARTIN L. FELDMAN
Nuodex Inc.

W.J. Stewart in W.H. Madson, ed., *Paint Driers and Additives*, *Federation Series of Coatings Technology*, Federation of Societies for Paint Technology, Philadelphia, Pa., 1969, Unit 11, pp. 1–26.

DRYCLEANING AND LAUNDERING

DRYCLEANING

Drycleaning can be defined as the cleaning of fabrics in a substantially nonaqueous, liquid medium. This process has evolved during the last 60–70 yr into a highly effective, low cost, safe method of removing soils from all types of textiles. The industry enjoyed steady growth until the introduction of wash-and-wear garments, and after a period of adjustment, again started to expand and diversify. Recent growth has been aided by the development of new equipment, cleaning techniques, and applications.

The important distinction between dry and wet solvents is that the latter (principally water, glycols, and other hydroxylic compounds) swell the hydrophilic textile fibers, but the drycleaning solvents do not. Dimensional changes that fibers undergo as they swell in water are transmitted throughout the textile structure and can cause serious fabric damage. These changes can be local distortions (wrinkles) or more extensive, causing shrinkage. Drycleaning solvents do not swell the textile fibers and thus do not cause wrinkles or shrinkage. This is one of the principal advantages of drycleaning processes over laundering. Another advantage is that drycleaning solvents remove oily soils at low temperatures, whereas a high temperature, colloidal suspension mechanism is needed in wet laundering processes.

The main difference between drycleaning and laundering processes is the solvent used. Since both methods remove soil by the same mechanisms, similar operations are used. Garments are tumbled in the cleaning liquid in horizontal washers to separate soil and fabric as thoroughly as possible without damage. Detergents are used in both cases to emulsify or solubilize the soils insoluble in the cleaning liquid. Special detergent ingredients peptize and suspend insoluble soils.

The Drycleaning Operation

The following sequence of operations is used in the normal drycleaning process: (1) Soiled garments are marked and sorted. (2) Prespotting is performed when required. (3) Garments are rotated in a tumble-type washer containing the drycleaning solvent. (4) Solvent is drained from the tumbler and most of the residual solvent removed by centrifugal extraction. (5) The small amount of remaining solvent is removed in heated dryers. (6) Dirty solvent from the wash cycles is continuously passed through diatomaceous earth and activated carbon or disposable cartridge filters to remove as much of the fugitive dyes and insoluble soils as possible. (7) Soils not removed by filters, diatomaceous earth, or activated carbon can be removed from the solvent by distillation. (8) Dry, solvent-free garments are inspected and spot cleaned a second time, if necessary. (9) A garment may be "wet cleaned" at this point, but this is usually omitted when operating a charged-solvent system. (11) After a final inspection, the garments are assembled according to customer's order. (12) Garments are bagged and placed on racks ready to be picked up. Mechanical conveyors aid in fast recovery of a customer's garments.

Many drycleaning operations provide special services, such as garment repair, dyeing, and moth- and waterproofing treatments. Many operators add optical brighteners, antistatic agents, and sizing to improve the brightness and feel (hand) of a garment (see Brighteners, fluorescent; Antistatic agents; Waterproofing).

Materials

Drycleaning solvents remove oily soils easily by simple solution, and hence detergents are not required. However, detergent addition greatly enhances the removal of the many other soils that are generally present. Detergents probably inhibit redeposition of soil by being adsorbed on the soil particles to cause a lowering of the interfacial energy between them and the solvents, at the same time increasing the interfacial energy between particle and fiber surface.

A modern procedure called the charged system uses a detergent that forms a stable colloidal solution in the solvent. The term colloidal solution, as used here, means a solution containing dispersed aggregates called micelles. The key to the success of the charged system is the reduced effective vapor pressure of water dissolved in detergent micelles. This prevents the solubilized water from being completely removed from solution by the garment (see Emulsions).

Sulfonates of mixed petroleum hydrocarbons, the original charged-system detergents, were used increasingly after the expiration of the Reddish patent in 1950. They have now been replaced by purer synthetic detergents, particularly the amine and sodium salts of alkylarenesulfonic acids. Drycleaning detergents are not formulated products similar to household laundry detergents but principally surfactants or a mixture of surfactants, with concentrations of active ingredients ranging from 40–90%.

Water control. Sensitive and reliable control systems are required to maintain the 70% rh in a drycleaning unit. These controls are designed to measure the moisture, inject water into the system, and close the injection valve at optimum rh. Some water also enters the unit with the garments.

Two types of water-control systems are widely used. The direct type uses an electric hygrometer to measure relative humidity in the atmosphere above the solution. An indirect method based on the electrical conductivity of the solution has proved the most successful.

Solvents. An acceptable drycleaning solvent must be an effective solvent for fats and oils, sufficiently volatile to permit easy drying, easily purified, and of low toxicity. Solvents should not weaken, dissolve, or shrink the ordinary textile fibers or cause bleeding of dyes. They must be noncorrosive to metals commonly used in drycleaning machinery and have a flash point above 37.7°C or be nonflammable. It is estimated that about 29% of the volume of drycleaning in the United States is carried out in Stoddard solvent, a petroleum fraction (see Solvents, industrial).

The amount of drycleaning carried out with chlorinated hydrocarbon solvents in the United States is estimated to be 70% of the business. Their main advantage over the petroleum solvents is their nonflammability, permitting their use in many areas where zoning laws prohibit the use of a flammable solvent. Perchloroethylene (tetrachloroethylene) is the most widely used drycleaning solvent (see Chlorocarbons).

Solvent purification, distillation. The principal method of purifying drycleaning solvents in the United States is distillation, which has displaced the caustic clarification process. There has been an unfortunate tendency in recent years for cleaners to omit distillation and depend entirely upon activated charcoal and fatty acid-adsorbing sweetener powders for solvent purification. Lower quality cleaning results from this cost-cutting practice.

Carbon recovery. Devices called sniffers should be an essential part of any drycleaning operation to protect workers from exposure to drycleaning solvents and increase operating life.

Health and Safety Factors

Environmental effects of all chemicals, including drycleaning solvents, are receiving more attention since the enactment of the Occupational Safety and Health Act of 1970. Many studies of the long-term (chronic) toxicity of solvents have been performed. The IFI (International Fabricare Institute) has been working with various government agencies since 1973 to develop realistic safety standards for the drycleaning industry.

Self-Service Drycleaning

Coin-op drycleaners use the same detergents as commercial cleaners, but always as a one-bath (no rinse) operation. In general, the coin-op machines compare favorably with commercial drycleaning and have excellent filtration and rapid solvent turnover.

Industrial Drycleaning

In the last decade a new service, known as industrial drycleaning, has become significant in the drycleaning industry. With improvement in fabric appearance, color retention, hand, odor, and moisture penetration, customer resistance to fabrics of synthetic fibers has disappeared. Rental and maintenance service of career apparel, industrial garments, and linen supply items became economically feasible. In the most successful processes, high concentrations of water were added to the cleaning solvent or a 100% water cycle incorporated.

DONALD A. REICH
CHARLES L. CORMANY
PPG Industries

E.R. Phillips, *Drycleaning*, National Institute of Drycleaning, 1961.

A.R. Martin and G.P. Fulton, *Drycleaning: Technology and Theory*, Textile Book Publishers, Inc., New York, 1958.

U.S. Pat. 1,911,289 (May 30, 1933), W.T. Reddish (to Emery Industries, Inc., Cincinnati, Ohio).

J.C. Parker and co-workers, *Tetrachloroethylene (Perchloroethylene), Current NIOSH Intelligence Bulletin #20*, March, 1978.

LAUNDERING

Laundering is the process in which soils and stains are removed from textiles in an aqueous medium. Together with drycleaning, which accomplishes a similar result in nonaqueous solvents, commercial laundering constitutes a multi-billion dollar (10^9) industry. However, the largest portion of consumer textiles is laundered in the home.

Commercial Laundering

Commercial laundry practice differs in many ways from home laundering. For example, in the commercial and industrial laundry plant, ion exchange is always used as water softener. Caustic and silicated alkalies are usually employed as detergent builders, and tallow soap is much more widely used than in home laundering. Classification of articles is generally by soil intensity.

Home Laundering

The laundry is first sorted, primarily by color. Further sorting by type of fabric and garment construction is advisable. Pretreatment is often effective, loosening stains that might be firmly set at the higher temperatures employed in the regular washing cycle. The regular washing cycle for white and colored fabrics, from heavily soiled work clothes to lightly soiled lingerie, normally employs a heavy-duty detergent. Water softening agents are added in areas where water supplies are unusually hard, in addition to bleaches such as liquid sodium hypochlorite, dry chlorine bleach, or oxygen bleach. Fabric softeners impart softness and fluffiness to finished fabrics and reduce static cling, especially in synthetic fabrics. Some household detergents include a fabric softener, combining both laundering and softening in a single application.

Materials

Textiles are either from natural or synthetic fibers (see Fibers, chemical; Fibers, vegetable; Textiles). More than 90% of the natural fibers encountered in laundering are cotton (qv); the only other natural fiber laundered of any importance is wool. The chemical constitution of these fibers is quite different, and therefore they must be handled differently in laundering.

Water is the most important material used in laundering, assisted by detergents. Water provides the medium through which mechanical action is transmitted to the fabrics to be cleansed and serves as a wetting agent to penetrate the soil–fiber interface and displace the soil from fibers. It carries chemicals to the soiled area and removes suspended soil. The quality of the water used for washing is crucial to the quality of the finished product (see Water).

Chemicals

Chemicals used in laundering are classified as follows: alkalies, detergents, bleaches, sours, and finishing specialties.

JOSEPH C. SHERRILL
Sherrill Associates

Learning About Laundering, Proctor and Gamble, Cincinnati, Ohio, 1977.

Detergents in Depth, Soap and Detergent Association, New York, 1974, p. 13.

C.L. Riggs and J.C. Sherrill, *Textile Laundering Technology*, Linen Supply Association of America, Hallandale, Fla., 1979.

DRYING

Drying is an operation in which a volatile liquid is separated from a solid or semisolid material by vaporization. In dehydration, vegetable and animal products are dried to less than their natural moisture contents, or water of crystallization is removed from chemical compounds. In freeze drying, the wet material is cooled to freeze the liquid; vaporization then occurs by sublimation of ice. Evaporation differs from drying in that feed and product are both pumpable fluids; different equipment is employed. Gas drying means the separation of a condensable vapor from noncondensable gases by adsorption on solid surfaces or absorption in a solution.

Reasons for drying include customer convenience or preference, reduction of shipping costs, maintenance of product stability, and removal of toxic or noxious liquids. Waste recycling and disposal are growing applications. Most manufacturing operations that produce solid products include drying steps; eg, pigments, paper, polymers, ceramics, leather, wood, and foods. Drying operations involving flammable and toxic liquids are becoming common. For these applications, gas-tight, inert-gas recirculating equipment is manufactured with integral dust collectors, vapor condensers, and recycle gas heaters.

When a material dries, heat is transferred to evaporate the liquid, and mass is transferred in the form of liquid and vapor. In most dryers, heat is transferred first to the surface of the material and then into the interior. The usual heat-transfer mechanisms are convection from a hot gas that is brought in contact with the material; conduction from a hot surface that contacts the material; and radiation from a hot gas or hot surface that either contacts or is in close proximity to the material. In dielectric heating, a fourth but less common method, the energy source is an alternating electric field that generates heat inside the material by molecular friction. High internal vapor pressures may develop, and the temperature inside the material may be greater than at the surface (see Microwave technology). Many dryers employ more than one of these predominant heat-transfer mechanisms. Mass transfer (qv) during drying involves the removal of vapor from the material surface and the movement of internal moisture to the surface. Material structure and the mechanisms of internal liquid and vapor flow usually control the drying rate when internal moisture is removed during drying.

Because all drying operations require the handling of solids, the capability of the equipment to handle solids is of greatest importance. The material must reliably enter one end, leave the other, and not linger too long in passage. If possible, drying should be preceded by a mechanical separation operation to minimize the liquid quantity that has to be vaporized. Liquid separation from a solid without vaporization is always less expensive than drying. Evaporators with low equipment and energy costs are also used to reduce dryer vaporization (see Evaporation).

Drying costs are determined by labor costs, equipment, materials of construction, fuel, and plant size. Generally, the product form desired, the chemical and physical properties of the liquid and solid, and process options dictate drying method and equipment, not cost. However, continuous dryers are less expensive than batch dryers.

Drying Mechanisms

Drying periods. Although it is sometimes possible to select a suitable drying method simply by evaluating variables such as humidities and temperatures, the goal of many operations is not only to separate a volatile liquid, but also to produce a dry solid of specific size, shape, porosity, density, texture, color, or flavor; an understanding of liquid and vapor mass-transfer mechanisms is essential for quality control. Mass-transfer mechanisms are best identified by measuring drying time and drying-rate behavior under controlled conditions. No two materials behave alike and a change in material handling method or any operating variable, such as heating rate, may affect mass transfer.

Constant-rate drying. During the constant rate period, vaporization occurs as from a free liquid surface of constant composition and vapor pressure; material structure has no influence.

Critical moisture content. The critical moisture content, which is the average material moisture content at the end of the constant-rate drying period, is a function of material properties, the constant-drying rate, and particle size.

Equilibrium moisture content. Equilibrium moisture content is a steady-state equilibrium obtained by the gain or loss of moisture when a material is exposed to an environment of specific temperature and humidity. This equilibrium condition is independent of drying rate or method but is a material property. Only hygroscopic materials have equilibrium moisture contents.

Falling-rate drying. The principal internal mass-transfer mechanisms that control falling-rate drying are liquid diffusion in continuous, homogeneous materials; capillarity in porous and fine granular materials; gravity flow in granular materials; vapor diffusion in porous and granular materials; flow caused by shrinkage-induced pressure gradients; and pressure flow of liquid and vapor when a material is heated on one side and vapor escapes from the other.

Dryer Classification

A batch dryer is one into which a charge is placed, then removed after the dryer runs through its cycle. A continuous dryer operates under steady-state conditions, with continuous feed-in and product removal. Table 1 gives a classification of industrial dryers according to their predominant heat-transfer mechanisms. An alternative classification is based on material handling characteristics and specific problems.

In direct-heat dryers, heat is transferred to the material from a hot gas. Direct-heat dryers include batch dryers such as the tray or compart-

Table 1. Industrial Dryers

Heat-transfer mechanism	Continuous dryers	Batch dryers
direct	tray	parallel-flow tray
	tunnel	through-circulation tray
	turbo-tray	fluid-bed
	through circulation	spouted-bed
	suction-drum	rotary
	foam-mat	
	festoon	
	multipass-loop	
	tenter-frame	
	float	
	rotary	
	fluid-bed	
	spouted-bed	
	vibrating-conveyor	
	pneumatic-conveyor	
	spray	
indirect	steam-tube	vacuum
	rotary	shelf
	fluid-bed	freeze
	screw-conveyor	rotating
	high-speed agitator	rotary
	drum	pan
	can	fluid-bed
	belt	rotary
	tunnel	screw-conveyor
radiant	rf[a]	rf[a]
dielectric	microwave	microwave

[a] rf = radio frequency.

ment dryer which is an insulated housing in which particulate solids are placed in tiers of trays, and large objects may be stacked in piles or on shelves. Through-circulation dryers are batch dryers that employ perforated- or screen-bottom trays and suitable baffle arrangements so that gas is forced through the material.

Continuous direct-heat dryers are characterized by continuous material flow without mixing, and thus, dryer residence time is uniform for all material increments. Tray and tunnel dryers consist of long, insulated housings through which material is moved on trucks or trays fastened together. Gas flow is usually parallel to material surfaces; it may be cocurrent, countercurrent, or it may flow from side to side and be recirculated through heaters and fans mounted on each side of the housing. Conveyor movement may be continuous or semicontinuous. Performance is comparable to that of batch compartment dryers. In through-circulation dryers, a bed of permeable material is conveyed through an enclosure on one or more perforated plate or screen conveyors. The enclosure is a series of independent compartments, each having its own fans and heaters. Humid gas is removed below the conveyor and fresh gas enters under the heaters.

The direct-heat through-circulation dryer obtains a relatively high drying rate because drying gas flows through, rather than in parallel flow above and below the material. However, if bed depth and porosity are not uniform, gas channels through thin spots and larger passages and all particle surfaces are not contacted equally. A better arrangement is one in which the particles are completely separated in such a manner that the gas can flow freely among them. The drying rate for all particles of a given size should then approach a constant. The purpose of various forms of mechanically- and fluid-agitated dryers is to obtain optimum conditions of particle separation and surface exposure for products of various drying and material handling characteristics. The latter usually govern the choice of dryer. A disadvantage of agitated compared to tunnel dryers is the loss of the plug-flow property.

Mechanically and fluid-agitated dryers include the rotary dryer: a horizontal, cylindrical rotating shell through which gas is blown to dry material that is tumbled inside; fluid- and spouted-bed dryers used mostly for particulate and pelleted polymers, grain, coal, sand, and mineral ores; the vibrating conveyor which dries by conveying the wet material on a screen-covered, perforated deck; the pneumatic-conveyor dryer which can be adapted for drying by heating the conveying gas and is used principally for materials that are not sticky and can be thoroughly dispersed in the drying gas so that drying is mostly at a constant rate; and the spray dryer: a large, usually vertical, chamber through which a hot gas is blown, and into which a liquid, slurry, or paste is sprayed by a suitable atomizer. Common spray-dryer applications include soluble coffee, detergents, milk, agglomerated instant food products, pigments, dyes, and inorganic chemicals.

Indirect-Heat Dryers

Although heat is transferred mostly by conduction, radiation heat transfer may be effective when the conducting surface temperature exceeds 150°C. Steam is the most common surface-heating medium. Hot water is used for low temperature heating, and heat-transfer oils are used for high temperatures. Indirect-heat dryers include atmospheric-pressure dryers and vacuum dryers.

Atmospheric-pressure dryers include the steam-tube (rotary) dryer: a horizontal, cylindrical rotating shell in which are installed one or more circumferential rows of steam-heated tubes and which can be used for any particulate material that does not stick to metal when dry; a rotary dryer with short turning bars that are used in the place of lifting flights, hence allowing material to roll in contact with the shell, and with a shell enclosed in a furnace to provide indirect heat; fluid-bed dryers which have pipe or plate coils installed in the fluid bed for indirect-heat drying and are ideal for solvent-recovery and inert-gas operations; the screw-conveyor dryer which has a double-wall, steam- or oil-heated screw; high speed agitator dryers which take advantage of the higher heat-transfer rates obtained by intense gas-particle mixing near a heated surface; drum dryers in which a thin film of slurry or solution is applied to the outer surface of a rotating heated drum and dried in a few seconds; and cylinder dryers used to dry plastic, textile, and paper webs that are not self-supporting and cannot be handled in festoon, multipass-loop, or tenter-frame dryers.

Vacuum dryers include the shelf dryer in which wet material is spread on trays that rest on heated shelves in a pressure-tight insulated chamber and which is used for drying small quantities of valuable materials; the freeze dryer, which is a vacuum shelf dryer operated at less than 100 Pa (750 μm Hg) pressure, so that water is sublimed from ice; the rotating vacuum dryer which is formed by providing a double-cone mixer with a jacket and a stationary, internal vapor tube passing through a rotary joint mounted on one trunnion, and is used for materials that do not stick to metal and do not pelletize during drying; rotary dryers which are stationary, horizontal, jacketed cylinders with an internal ribbon agitator, and are used for materials that do not stick to metal when dry; and pan dryers: a stationary, vertical, jacketed cylinder with a jacketed bottom and scraper agitator designed to handle doughlike materials that would overload or break the ribbon agitator in a rotary dryer.

Radiant-Heat Dryers

Heat transfer by radiation is controlled by the radiant-source temperature and the radiation-adsorption properties of the material. For drying, the source may consist of a number of incandescent lamps, reflector-mounted quartz tubes, electrically heated ceramic-enclosed gas burners. Radiant-heat dryers are not suitable for large objects and deep layers of material in which the drying rates are controlled by internal heat and mass-transfer mechanisms. The best applications for radiant-heat dryers are thin coatings and paint films on metal, polymers, wood, and paper during continuous and uniform conveying.

Dielectric Dryers

Dielectric dryers include radio-frequency dryers that operate in the range of 3–150 MHz, and microwave dryers that operate at 915 and 2450 MHz. A radio-frequency dryer may consist of two flat, metal plates between which a wet material is placed or conveyed. Dielectric dryers are used for drying rayon cakes, sand molds, lumber, and food products. Generally, because of high capital and operating costs, a dielectric dryer is best suited for removing a small quantity of moisture that is difficult to reach by conventional, surface-heating methods. At present, microwave dryers are only used for removing small quantities of water from materials in situations in which the energy must be accurately applied.

PAUL Y. McCORMICK
E.I. du Pont de Nemours & Co., Inc.

R.H. Perry and C.H. Chilton, *Chemical Engineers' Handbook*, 6th ed., McGraw-Hill, New York, 1983, Section 20.

R.B. Keey, *Drying: Principles and Practice*, Pergamon Press, New York, 1972.

G. Nonhebel and A.A.H. Moss, *Drying of Solids in the Chemical Industry*, CRC Press, Cleveland, Ohio, 1971.

DRYING AGENTS

There are many substances that take up (sorb) water from their surroundings by one or more of a number of different physical or chemical mechanisms, or both. These substances are widely used for removing water from gases, liquids, and solids. For this reason they are called drying agents or desiccants. Drying agents may be liquids or solids. They may be used repetitively by regenerating the desiccant after use to return it to its active state. In some cases they are used only once, and the spent desiccant is discarded. Drying agents are used either in a static (batchwise) or dynamic (continuous or semicontinuous) mode.

In the last twenty years, the use of drying agents has grown rapidly. For example, dehydration of process streams to extremely low water contents has become much more prevalent partly because of the popularity of cryogenic processing for the separation of gas mixtures such as air and natural gas (see Cryogenics).

Mechanism

The dehydration mechanism of drying agents may be classified as follows: *Type 1*—Chemical reaction: (*a*) formation of a new compound; (*b*) formation of a hydrate. *Type 2*—Physical absorption with constant relative humidity (solid + water = saturated solution). *Type 3*—Physical absorption with variable relative humidity (a solid or liquid + water = dilute solution). *Type 4*—Physical adsorption. These mechanisms are characterized by the relative magnitudes of the heats of reaction, solution, or adsorption.

Static Drying Agents

Many liquids are dried batchwise rather than continuously. The drying agent is added to the liquid, and sufficient time is allowed to dry the product. The liquid is then separated from the drying agent by filtration, decantation, or distillation. Drying agents employing type 1 or 2 mechanisms are generally used for these applications. The most commonly used, from a long list of these materials, are barium oxide, calcium chloride, calcium oxide, calcium sulfate, lithium chloride, perchlorates, phosphorus pentoxide, and sodium potassium hydroxides.

Dynamic Drying Agents

Continuous drying is employed in many operations where it is not practical to dry a volume of gas or liquid in a batchwise fashion. When a solid dynamic drying agent is used, the flowing stream is passed over a fixed bed of drying agent which must have the physical properties to allow the flowing stream to pass readily through. When liquids are used, the drying is usually achieved by countercurrent contact of the gas (flowing up) against the liquid drying agent (flowing down). Most dynamic drying agents are regenerable. They include liquid agents such as sulfuric acid, glycerol, ethylene glycol, and other polyhydric alcohols, as well as solid agents like activated alumina, silica gel, and molecular sieves (qv). The capacities of some selected liquid drying agents are compared in Figure 1.

Design of Adsorption Drying Systems

Adsorbent drying systems are typically operated in a regenerative mode with an adsorption half cycle to remove water from the process stream and a desorption half cycle to remove water from the adsorbent and to prepare it for another adsorption half-cycle. Usually, two beds are employed to allow for continuous processing. In most cases, some residual water remains on the adsorbent after the desorption half cycle because complete removal is not economically practical. The difference between the amount of water removed during the adsorption and desorption half cycle is termed the differential loading. This is the working

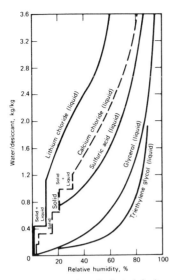

Figure 1. Drying capacity of selected drying agents.

capacity available for dehydration. The two most common types of drying systems operate on either a pressure-swing cycle or a thermal-swing cycle.

Adsorption Plots

Among the methods of plotting adsorption data, isotherm plots are the most common, wherein the water-vapor pressure in equilibrium with the adsorbent is plotted vs the adsorbed water content of the adsorbent with lines of constant temperature being shown. Another commonly used plot is called an isostere plot. In this case, temperature is plotted vs the water-vapor pressure in equilibrium with the adsorbent, with lines of constant adsorbed water content on the adsorbent (see Adsorptive separation).

Cost effectiveness of adsorbents. The cost effectiveness of a dynamic drying agent can be evaluated in three respects: the water capacity; the water-adsorption rate; and the life of the drying agent for the service intended.

JOSEPH P. AUSIKAITIS
Union Carbide Corporation

R.A. Ford and A.J. Gordon, *The Chemist's Companion: A Handbook of Practical Data, Techniques, and References,* John Wiley & Sons, Inc., New York, 1972, pp. 445–447.

O.A. Hougen and F.W. Dodge, *The Drying of Gases,* Edwards Brothers, Inc., Ann Arbor, Mich., 1947.

DRYING OILS

Since 1945, drying oils have gradually been displaced in many of their traditional uses by water-based latex emulsions. One goal of modern coating chemists is to find both vehicles and pigments that bond more strongly with the substrate and thus last for longer periods. How big a part drying oils will play in these new developments remains to be determined. It is, however, clear that drying oils are declining in use in the United States and elsewhere. New technology in the petroleum and coating industry and competition for land from food crops is expected to continue to decrease use of linseed oil (see also Vegetable oils).

Occurrence

Linseed oil is produced from the seeds of the common flax plant, *Linum usitatissimum.* Although flax had been grown in Texas and the Imperial Valley of California, the Dakotas and Minnesota are the main areas of production at present. Flaxseed for oil is also grown extensively in Argentina, Canada, India, and the USSR.

Soybean oil is obtained from the almost spherical seeds of the legume *Glycine max* (L) Merrill. Soybeans are widely cultivated throughout the world but grow best at 30–45° latitude. The United States, Brazil, and China are major producers of soybeans. The oil, which is more saturated than linseed oil, is classified as a semidrying oil. It is seldom the only drying component in coating (see Soybeans; Alkyd resins).

Tung or Chinawood oil comes from the seed kernels of the *Aleurites fordii* tree. Because of its scarcity in the late 1930s and early 1940s, synthetic products were developed to replace tung-oil coatings, and its use has declined.

Certain varieties of fish, such as menhaden, pilchard, and sardine are high in oil content. For coating use, the oils must be cooled to remove saturated glycerides which are present in relatively large amounts. The oils also must be relatively free of fish odor to be acceptable. Because the winterized oil produces brittle films and yellows on aging, its use is declining.

Oiticica and isano oil are somewhat similar to tung oil. Oiticica contains a conjugated keto fatty acid (licanic acid, 4-oxo-9,11,13-octa-decatrienoic acid) and isano oil contains conjugated acetylenic bonds such as found in isanic acid (17-octadecene-9,11-diyonic acid). Both oils are extremely reactive, isano reacting even faster than tung. Oiticica oil is derived from the fruit of *Licania rigidia,* a tree that grows in northeastern Brazil, and isano oil is obtained from the fruits of *Onguekoa gore* and was imported for a short period from Zaire. Oiticica oil continues to be imported on a relatively small scale, but importation of isano oil has apparently ceased.

Castor oil (qv) derived from the plant *Ricinus communis* is not a drying oil but can be converted to dehydrated castor oil which produces superior alkyd resins. The beans contain 48–52% oil of which about 80% is ricinoleic acid (12-hydroxy-9-octadecenoic acid) which can be dehydrated to octadecadienoic acids.

An extensive search for potential new oilseeds launched by the USDA in 1958 has led to much new information on oils containing epoxy groups, hydroxy-conjugated dienes, conjugated trienes, and potential replacements for sperm oil. Some of these may become commercially available, and other-now available may continue to compete successfully with petroleum-based products.

The characteristics of a number of commercially available linseed oils are listed in Table 1 and those of some other drying oils are shown in Table 2.

Table 1. Characteristics of Some Commercial Linseed Oils

Linseed oils	Viscosity, mPa·s (= cP)	Color Gardner, max	Iodine value	Acid value, max	Density, g/mL
nonbreak	50	12	180	7	0.92
alkali-refined	50	6	180	0.5	0.94
acid-refined	50	6	180	6	0.94
polymerized oils	1,760	8	118	7	0.96
polymerized oils	2,700	8	118	9	0.96
polymerized oils	6,340	8	118	9	0.96
polymerized oils	59,000	8	118	9	0.97
boiled	50–65	7	175	5	0.94
blown	3,620–6,340	10	125	11	0.99
treated with					
dicyclopentadiene	4,630–6,340	10	156	4	0.98
styrene	3,620–6,340	10	70	4	0.96

Table 2. Some Chemical and Physical Properties of Drying Oils

Property	Soybean	Safflower (nonbreak)	Tung	Dehydrated castor oil	Oiticica (raw)	Menhaden
specific gravity 25/25°C	0.920	0.924	0.935	0.931	0.970	0.926
viscosity, mPa·s (= cP)	50	50	165–250	165–200	solid	50
iodine value	120–140	142–150	158–166	136	148	170–178
acid value	0.5–1.6	0.2–0.5	0.3–4.0	4	4	<8
$n_D^{25°C}$	1.473	1.475	1.5150	1.482	1.510	1.480
unsaponifiable, %	1.5		0.4		1	1.2
color (Gardner)		9–10	6–10	4	8–10	5–7

Preparation

Drying oils are usually separated from their sources by one or more of the following operations: (a) hydraulic pressing; (b) continuous screw pressing, sometimes called expelling; and (c) solvent extraction. Soybeans and sources containing a low percentage of oil are extracted. Hexane, a mixture of petroleum hydrocarbons, is the solvent of choice except in hot weather when a slightly higher boiling hydrocarbon such as heptane may be used. Fish oils are separated by treatment with steam. Flaxseed is usually reduced to 20–30% oil in a screw press and then extracted with hexane.

Functionality

Oils are generally divided into three classes, drying, semidrying, and nondrying, according to their iodine values. They can be defined as follows: drying oils, iodine value > 140; semidrying, 125–140; and nondrying, < 125. Actually, any spreadable liquid that reacts with the oxygen of the air to form a comparatively dry film would be classified as a drying oil. Although a solvent might be used to improve the spreadability, the film-forming product should itself be liquid before the film sets. This definition includes a wide range of products, but chemists generally define oil more narrowly as a liquid triacyl glycerol or triglyceride. Thus, most oils are the mixed esters of glycerol with fatty acids. Glycerol (qv) is the alcoholic component and a variety of monobasic fatty acids, usually C_{18}, are the acidic part of the esters. They are low melting oils rather than fats when the amount of unsaturation is high enough.

Reactions of Drying Oils

Reactions include alcoholysis, acidolysis, ester–ester exchange, autooxidation to hydroperoxides at reactive (allylic) sites, and polymerization, heat polymerization of conjugated and nonconjugated oils, castor-oil dehydration, and isomerization of linoleic and linolenic acids to conjugated dienoic and trienoic products.

Maleic adducts. In recent years, the maleic anhydride adducts of linseed oils have had considerable use in various vehicles for coatings, as in polymers used for electrodeposition of coatings. The latter have proved useful in coating hard-to-cover hidden areas in automobile and other frames (see Maleic anhydride).

Polyols other than glycerol. Linseed oil has an apparent functionality of six when each double bond has a functionality of one. A highly cross-linked three-dimensional polymer should result on high temperature bodying or oxidation of the film. Linseed oil dries to a characteristic film with oil tackiness. Soybean oil with an apparent functionality of 4.5 dries on oxidation to a soft, tacky film that is inferior in hardness, gloss, alkali resistance, etc. Improvement in drying and film properties can be effected by replacing the glycerol with a polyol of higher functionality, such as pentaerythritol, dipentaerythritol, sorbitol, poly(vinyl alcohol), α-methyl-O-glucoside, and poly(allyl alcohol). By far the most commercially successful synthetic oil has been one made with pentaerythritol containing some dipentaerythritol (see Alcohols, polyhydric).

Uses

Drying oils are used in paints, enamels, and varnishes, to protect concrete, in epoxy plasticizers and stabilizers for poly(vinyl chloride), for epoxy derivatives converted to coatings, in baits to attract ants, in coatings for electrodeposition, and in variations in the preparation of alkyd resins, urethane oils, and carboxy derivatives of fatty acids.

JOHN C. COWAN
Bradley University

Federation Series on Coatings Technology, 25 volumes, Federal Society of Paint Technology, Philadelphia, Pa., 1965–1977.

R.R. Myers and J.S. Long, eds., *Treatise on Coatings*, Vol. 1, *Film-Forming Compositions*, Marcel Dekker, New York, pt 1, 1967; pt 2, 1968; and pt 3, 1972.

DUST, ENGINEERING ASPECTS. See Air pollution control methods.

DUST, HYGIENIC ASPECTS. See Air pollution.

DYE CARRIERS

Dye carriers are used to achieve complete dye penetration of polyester fibers. They loosen the interpolymer bonds and allow the penetration of water-insoluble dyes into the fiber. The polyester polymer does not contain an ionic group and is hydrophobic. Therefore, it cannot be dyed with water-soluble ionic dyes. It can be dyed, however, with certain water-insoluble dyes called disperse dyes (see Fibers, synthetic; Polyester fibers). Deep shades and full fastness properties on polyester can be achieved using disperse dyes and carriers, or temperatures over 100°C with or without carriers.

Many substances show carrier behavior. Of these, some have found more acceptance than others for various reasons among which are availability, cost, toxicity, ease of handling, odor, etc. Most carriers are aromatic compounds and have similar solubility parameters to the poly(ethylene terephthalate) fibers and to some disperse dyes. There are many chemicals theoretically suitable as carriers. The boiling point is one of the major criteria in selection.

o-Phenylphenol was one of the earliest carrier-active compounds of industrial use. Originally it was used as its water-soluble sodium salt. By lowering the pH of the dye bath, the free phenol was precipitated in fine form and made available to the fiber. However, proprietary liquid preparations containing the free phenol are available today that afford a greater ease of handling. Compounds most commonly used as dye carriers include o-phenylphenol, bp 280–284°C; p-phenylphenol, bp 305–308°C; methyl cresotinate, bp 240°C; o-dichlorobenzene, bp 172–178°C; 1,3,5-trichlorobenzene, bp 214–219°C; biphenyl, bp 255.9°C; methylbiphenyl, bp 255.3°C; diphenyl oxide, bp 259.0°C; 1-methylnaphthalene, bp 244.6°C; 2-methylnaphthalene, bp 241°C; methyl benzoate, bp 198–200°C; butyl benzoate, bp 250°C; benzyl benzoate, bp 323–324°C; dimethyl phthalate, bp 298°C; diethyl phthalate, bp 298°C; diallyl phthalate, bp 290°C; and dimethyl terephthalate, bp 284°C.

Because the carrier-active compounds have little or no solubility in water, emulsifiers are needed to disperse these compounds in the dyebath (see Emulsions). Many proprietary carriers are available as solids (flakes or pellets) or in preemulsified form. Currently, the industry prefers clear products easily emulsified by premixing with water at the time of use.

Carrier Formulation

The formulation of a carrier depends on three basic ingredients: (1) the carrier-active chemical compound; (2) the emulsifier; and (3) special additives. Additional parameters to be considered in the formulation of a carrier product with satisfactory and repeatable performance are provided by the equipment in which the dyeing operation is to be carried out.

Basis for Carrier Selection

A carrier is selected by the dyer according to various criteria. The type of equipment and conditions under which it is to be used have been already mentioned. Other considerations are color yield, dye migration, product and emulsion stability, and cost.

Health and Safety Factors

Specific handling information can be obtained from the supplier or manufacturer of the carrier. OSHA and EPA have established exposure limits that must be carefully considered in relation to the waste disposal method available and the environment in which dye carriers are to be used.

ANTHONY DEMARIA
SYBRON Corp.

C.M. Hansen, *J. Paint Technol.* **39**, 104 (1967).

M.C. Keen and R.J. Thomas, "Absorption Properties of Latyl Disperse Dyes on Application to Dacron Polyester Fibers," *Dyes and Chemicals Technical Bulletin*, E.I. du Pont de Nemours & Co., Inc., Organic Chemicals Dept., Wilmington, Del.

DYES AND DYE INTERMEDIATES

Dyes are intensely colored substances used for the coloration of various substrates, including paper, leather, fur, hair, foods, drugs, cosmetics, waxes, greases, petroleum products, plastics, and textile materials. They are retained in these substrates by physical adsorption, salt or metal-complex formation, solution, mechanical retention, or by the formation of covalent chemical bonds. The methods used for the application of dyes to the substrates differ widely, depending upon the substrate and class of dye. It is by application methods, rather than by chemical constitutions, that dyes are differentiated from pigments. During the application process-dyes lose their crystal structures by dissolution or vaporization. The crystal structures may in some cases be regained during a later stage of the dyeing process. Pigments, on the other hand, retain their crystal or particulate form throughout the entire application procedure. They are usually applied in vehicles, such as paint or lacquer films, although in some cases the substrate itself may act as the vehicle, as in the mass coloration of polymeric materials.

The optical properties of dyes are determined by electronic transitions between the various molecular orbitals of the dye molecules that absorb some, but not all, of the incident radiation. These properties are defined by the terms color, intensity, and brightness. The color, also frequently referred to as the shade or hue, of a dye is determined by the energy differences between the molecular orbitals. The intensity, strength, or saturation is determined by the probability of the electronic transitions and the amount of dye present. The brightness or purity depends upon the width of the waveband adsorbed by the dye molecules (see Color).

The energy, probability, and distribution of the electronic transitions are, to a large extent, governed by the chemical constitution of the molecules. The chemical constitution also determines the other properties of a dye, such as the suitability for dyeing a specific substrate and the fastness properties of dyeings produced by the application of the dye to the substrate. A large number of dyes, with widely differing properties, is necessary because of the great variety of materials to be dyed. There are at present some 1200 different commercial dyes manufactured in the United States, and a further 800 are imported. On a worldwide basis, over 8000 chemically different dyes have achieved commercial significance. To assist both the dye users and dye manufacturers, dyes are classified into groups in two ways. The first method of classification is by chemical constitution, in which the dyes are grouped according to the chromophore or color-giving unit of the molecule. The second method is based on the application class or use of the dye. The first, from a chemical standpoint, satisfies the needs of the manufacturer, and the second is used by the dyer.

Table 1. Usage Classification of Dyes

Class	Major substrates	Method of application	Chemical types	Relevant articles in *ECT*
acid	nylon, wool, silk, paper, inks, and leather	usually from neutral to acidic dyebaths	azo, including premetallized dyes, anthraquinone, triphenylmethane, azine, xanthene, nitro, and nitroso	Azo dyes
azoic components and compositions	cotton, rayon, cellulose acetate, and polyester	fiber impregnated with coupling component and treated with a solution of stabilized diazonium salt	azo	Azo dyes
basic	acrylic, modified nylon and polyester, paper, and inks	applied from acidic dyebaths	methine, diphenylmethane, triarylmethane, azo, azine, xanthene, thiazole, acridine, oxazine, and anthraquinone	Azine dyes; Azo dyes
direct	cotton, rayon, paper, leather, and nylon	applied from neutral or slightly alkaline baths containing additional electrolyte	azo, phthalocyanine, stilbene, oxazine, and thiazole	Azine dyes; Azo dyes
disperse	polyester, polyamide, cellulose, acetate, acrylic, and plastics	fine, aqueous dispersions often applied by higher temperature–pressure or lower temperature carrier methods; dye may be padded on cloth and baked on or thermofixed	azo, anthraquinone, nitro, and methine	Azo dyes; Dyes, anthraquinone
fluorescent brighteners	soaps and detergents, all fibers, oils, paints, and plastics	from solution, dispersion or suspension in a mass	stilbene, azoles, coumarin, pyrazine, and naphthalimides	Brighteners, fluorescent
food, drug, and cosmetic	foods, drugs, and cosmetics		azo, anthraquinone, carotenoid, and triarylmethane	Colorants for foods, drugs and cosmetics
mordant	wool, leather, and anodized aluminum	applied in conjunction with chelating Cr salts	azo and anthraquinone	Azo dyes; Dyes, applications and evaluation
natural		applied as mordant, vat, solvent, or direct and acid dyes	anthraquinone, polymethine, ketone imine, flavones, indigoids, quinones, chlorophylls, etc	Dyes, natural
oxidation bases	hair, fur, and cotton	aromatic amines and phenols oxidized on the substrate	aniline black and indeterminate structures	Azine dyes
pigments	paints, inks, plastics, and textiles	printing on the fiber with resin binder or dispersion in the mass	azo, basic, phthalocyanine, quinacridone, oxazine, anthraquinone, and indigoid	Pigments
reactive	cotton, rayon, wool, silk, and nylon	reactive site on dye reacts with functional group on fiber to bind dye covalently under influence of heat and proper pH	azo, anthraquinone and phthalocyanine	Dyes, reactive
solvent	gasoline, varnish, lacquer, stains, inks, fats, oils, and waxes	dissolution in the substrate	azo, triphenylmethane, anthraquinone and phthalocyanine	Azo dyes; Dyes, anthraquinone; Phthalocyanine compounds
sulfur	cotton and rayon	aromatic substrate vatted with sodium sulfide and reoxidized to insoluble sulfur containing products on the fiber	indeterminate structures	Sulfur dyes
vat	cotton, rayon, and wool	H_2O insoluble dyes solubilized by reducing with sodium hydrosulfite, then exhausted on fiber and reoxidized	anthraquinone (including polycyclic quinones) and indigoids	Dyes, anthraquinone

The principal usage or application classes of dyes, accounting for 85% of production in the United States, are as follows: acid dyes, basic dyes, direct dyes, disperse dyes, fluorescent brighteners, reactive dyes, sulfur dyes, and vat dyes. Table 1 (p. 377) is arranged according to the CI application classification.

Dye Intermediates

The precursors of dyes are called dye intermediates. They are obtained from simple raw materials, such as benzene and naphthalene, by a variety of chemical reactions. Usually, the raw materials are cyclic aromatic compounds. They are derived from two principal sources, coal tar and petroleum (qv). Coal tar results from the pyrolysis of coal and is obtained chiefly as a by-product in the manufacture of coke for the steel industry (see Coal, carbonization). The petroleum industry is now the main supplier of benzene, toluene, the xylenes, and naphthalene.

The United States International Trade Commission lists ca 1000 compounds under the heading of cyclic intermediates. They include cycloaliphatic and heterocyclic species among a majority of carbocyclic aromatic compounds.

In addition to cyclic intermediates, many aliphatic reagents and inorganic chemicals are used in the dye industry. These include sulfuric acid and oleum for sulfonation, nitric acid for nitration, chlorine and bromine for halogenation, caustic soda and caustic potash for fusion and neutralization, and sodium nitrite for diazotization, as well as ammonia, hydrochloric acid, chlorosulfonic acid, and sodium carbonate, bicarbonate, and sulfide.

The reactions for the production of intermediates and dyes are carried out in kettles made from cast iron, stainless steel or steel lined with rubber, glass (enamel), brick, or carbon blocks. Jackets or coils serve for heating and cooling by circulating through them high boiling fluids, steam or hot water to raise the temperature, and air, cold water, or chilled brine to lower it. Products are transferred from one piece of equipment to another by gravity flow, pumping, or by blowing with air or inert gas. When possible, the intermediates are taken for the subsequent manufacture of other intermediates or dyes without drying. Where drying is required, air, or vacuum ovens (in which the product is spread on trays), rotary dryers, spray dryers, or less frequently drum dryers (flakers) are used.

Batch processes remain the rule but the progress in computer science and technology, electronic circuitry, and instrumentation has led to a growing use of automatic process control in recent years.

The chemistry of dye intermediates. Organic dyes and their intermediates range in structure from simple substitution products of benzene, naphthalene and anthraquinone to fairly complex ring systems. Their prime building block is the hexagonal skeleton of benzene, itself the prototype of aromatic compounds. Typical for the reactivity of an aromatic compound is the presence of delocalized π-electrons in the ring. These permit facile polarization by certain reactants so that electrons become readily available for electrophilic substitution reactions such as sulfonation, nitration, halogenation and acylation, which are indispensable for the manufacture of synthetic organic dyes and intermediates.

The first step in the manufacture of dye intermediates from aromatic raw materials is usually a direct electrophilic substitution. A variety of reactions then permits conversion of the primary substitution products to dye intermediates. They can be grouped under the following headings: chemical change of a substituent; removal of a substituent; replacement of a substituent with another by nucleophilic substitution; special reactions and rearrangements; condensations and ring closures, usually done to form polycyclic or heterocyclic structures; dimerizations; and oxidations.

Health and Safety Aspects

Biological treatment, either by the individual factory or in conjunction with a regional or municipal system, is the method most widely used for the treatment of effluents from dye and dye intermediate factories.

The toxic nature of some dyes and intermediates has long been recognized. Acute, or short term, effects are generally well known. They are controlled by keeping the concentration of the chemicals in the workplace atmosphere below prescribed limits and avoiding physical contact with the materials. Chronic effects such as benign and malignant tumor growth, on the other hand, frequently do not become apparent until after many years of exposure.

D.W. BANNISTER
A.D. OLIN
H.A. STINGL
Toms River Chemical Corporation

Colour Index, 3rd ed., Society of Dyers and Colourists, Bradford, Eng., and American Association of Textile Chemists and Colorists, Lowell, Mass., 1971–1975.

H.A. Lubs, ed., "The Chemistry of Synthetic Dyes and Pigments," *ACS Monogr.* **127**, (1955).

K. Venkataraman, *The Chemistry of Synthetic Dyes*, Vols. I–VII, Academic Press, New York, 1952–1974.

DYES, ANTHRAQUINONE

Anthraquinones are among the earliest natural products to be used as dyestuffs. The characteristic chromophore of the anthraquinone series consists of one or more carbonyl groups in association with a conjugated system. Amino or hydroxyl groups, as well as their substituted forms, and more complex heterocyclic systems may be present as auxochromes. In a number of cases, the parent carbonyl compound of complex fused-ring systems, such as pyranthrone, violanthrone, and dibenzpyrenequinone, is colored even though no auxochrome is present.

Anthraquinone (**1**), the parent compound for the carbonyl dyes, exhibits a long uv band ($\lambda_{max} = 327$ nm, in CH_2Cl_2) that extends into the visible spectrum producing a faint yellow color; however, it is not a dye. The introduction of relatively simple electron donors to the basic chromophore structure creates compounds that absorb at various regions of the visible spectrum depending upon the strength of the electron donor (OH < NH_2 < NR_2 < NHAr).

The position of substituents and the consequent formation of hydrogen bonds between substituent and carbonyl group influence not only the absorption maximum but also other properties such as the resistance to sublimation of disperse dyes, affinity for the substrate, lightfastness, and vatting properties.

(1)

Anthraquinone derivatives with β-hydroxyl or β-amino groups are capable of forming intermolecular hydrogen bonds and generally exhibit better resistance to sublimation, better solubility, and better affinity for textile substrates than the α-substituted compounds. On the other hand, intramolecular hydrogen bonds reduce the acidity of the carbonyl groups in the peri-position (more positive redox potential) which generally is advantageous with respect to wash- and lightfastness.

Disperse Dyes

The anthraquinone disperse dyes have their primary use in coloring synthetic fabrics. The present day trend toward the use of polyesters (qv) with truly hydrophobic fibers has created new problems in dyeing and has nurtured interest in the disperse dyes. The problem of preparing stable dispersions of insoluble or sparingly soluble dyes, required for such fabrics, has been met with the use of more powerful dispersing agents, with improved methods of dispersion, and with the inclusion of alkanol, carboxyl, amide, and similar functional groups in the dye molecules. Aside from making dyes more readily dispersible such functionality in the molecule has also increased affinity, improved leveling, and rendered minor modifications in shade.

The violet, blue, and green anthraquinone dyes used on cellulose acetate and polyester fibers are commonly synthesized from 1,4-diaminoanthraquinone, employing two different alkyl or aryl groups. These substituted diamino compounds have good affinity and level-dyeing power as well as excellent fastness. In addition, it has been found that in coloring polyester and rayon acetate fibers blue with these dyes, a mixture of two or three such dyes gives brighter and more intense shades than any one of the mixture components alone. It should be noted that these anthraquinones are primarily used for violet, blue, and green shades, and the yellow to red shades are commonly, though not exclusively, obtained by means of azo dyes. With the exception of pure green, however, the anthraquinones for synthetic fibers cover the spectrum from yellow through violet.

The affinity of a dye for cellulose acetate and polyester fibers is proportional to its basicity. The ideal dye for this type of synthetic fiber would demonstrate good dispersibility, substantivity, light and sublimation fastness, as well as resistance to gas-fading and washing. Although these may be the goals for a disperse dye, this in practicality, is not always the case. Many of the anthraquinone dyes also show poor sublimation fastness; this has been an especially significant problem in the yellow and red shades because of the rather simple molecular structure of the dye. This problem is less severe for the darker-colored anthraquinone dyes, since their structures usually include polar functional groups, such as $-CONHCH_2CH_2OH$, $-N(R)CH_2CF_2H$, $-N(R)CH_2CH_2OH$, $-SO_2NHCH_2CH_2OH$, and $-N(R)CH_2CHOHCH_2OH$, that have been found to improve sublimation fastness.

Anthrarufin and chrysazin derivatives. Another large class of disperse dyes other than those based on a 1,4-diaminoanthraquinone is that in which all of the α-positions are occupied, primarily by hydroxy, nitro, and amino groups, and their derivatives. Dinitroanthrarufin (**2**) and dinitrochrysazin (**3**) are two such dyes that have been in use for many years. These two, however, have shown poor affinity for synthetic fibers. To improve the dyeing qualities of dinitroanthrarufin, one nitro group is replaced with an arylamino-bearing hydroxyalkyl, hydroxyalkoxy, etc, group. This produces blue dyes for acetate and polyester with high affinity and outstanding fastness to light and gas. Similar dyes derived from dinitrochrysazines are slightly greener and faster to light than the 1,4-hydroxy isomers (anthrarufin derivatives).

(2)   (3)

As with many of the anthraquinone parent compounds, diaminoanthrarufin and diaminochrysazine can be substituted in the 2- and/or 6 position in order to improve the qualities of the dyes. These are frequently derived from the 2,6-disulfonic acid derivatives through the monosulfonic acid. Fastness properties of this type of dye on polyester have recently been improved by substituting the phenylalkoxy group in position 2. Recently, a wide variety of substituents and substitution patterns on the anthraquinone nucleus have been developed. The manipulation of position and type of substituent has proved a useful tool in improving the qualities of these dyes, eg, to structure dyes that are clear blue on polyester with good sublimation and lightfastness.

Vat Dyes

Despite their relatively high cost and difficult methods of application, anthraquinone vat dyes are one of the most important groups of synthetic dyes because of their all-around superior fastness. The anthraquinoids are rich in blues, greens, browns, khakis, and blacks, but a serious defect of many of the yellow and orange dyes is their property of accelerating the degradative action of light and bleaching agents on cellulose. The vats undergo characteristic coloration in sulfuric acid and nitric acid, as well as in concentrated sulfuric acid upon the addition of small amounts of potassium persulfate, or divanadyl trisulfate. The dyed fiber, treated successively with acidified permanganate and hydrogen peroxide, undergoes color changes that are useful as supplementary tests, particularly for certain groups of dyes such as the halogenated indanthrone and benzanthrone derivatives.

The shade of many of these dyes also tends to change by hot-pressing and water-spotting, but the change is reversible and the original color can be restored. These changes have been attributed to a variety of reactions including hydration and dehydration, changes in the state of aggregation of the pigment, and oxidation. Furthermore, the effects of each process can be reduced by vigorous soaping of the dyed material which removes surface color and assists in crystallization of the dye within the fiber.

An important factor influencing the ease of vatting and affinity for the fabric is the physical form of an anthraquinone vat dye. A simple and effective method of producing vat dyes in a very highly dispersed form is to oxidize a solution of the leuco salt or the leuco sulfuric ester in the presence of a protective colloid. For high tinctorial power, vat dyes may be prepared by adding a surface-active reagent such as an alkylnaphthalenesulfonic acid, diethylene glycol, or various substances (see Surfactants and detersive systems). Such agents improve solubility, assist in the vatting process, increase penetration, and produce higher color value as well as brighter, clearer prints.

The anthraquinone vat dyes are frequently sold in the insoluble, oxidized form, sometimes as dry solids but more often as aqueous pastes. They are applied in the reduced form as vats, the reduction being effected by the use of the sodium hydrosulfite under strongly alkaline conditions. As a consequence of this alkalinity, the use of vat dyes is mainly confined to cellulosic fibers, but under suitable conditions they may also serve as pigments.

The fastness of the dyes to alkali renders them suitable for coloring paper pulp (eg, for soap wrappers and printed wall papers), washable distempers, and cement. As a consequence of their excellent lightfastness and great stability, the anthraquinone vat dyes are of interest as pigments, but their use is limited by the cost as well as the relatively low covering power of many of them. They are useful in coloring plastics (including rubber) since they can withstand the temperatures used in molding and the chemicals used in vulcanization. They may be used for producing photographic prints on textiles by padding them successively with a dispersion of a vat dye and a solution of ferric ammonium citrate. Vat dyes manufactured primarily for dyeing and printing textiles, as well as vat dyes made especially for use as pigments have been employed also in the coloration of paints, varnishes, enamels, and like materials.

Acylaminoanthraquinones. The acylaminoanthraquinones are noted for their simplicity both of structure and method of preparation. As a group they are easily vatted and, in general, give level shades. Furthermore, they have good fastness to light and chemical agents and their substantivity has been found to increase with increases in molecular size.

Although the aliphatic carboxylic acids are unable to impart the necessary intensity of color and substantivity for technically valuable dyes, almost all of the benzoylated α-aminoanthraquinones can be used practically. The most useful dyes of this type, however, have been derived from diamino- and tetraaminoanthraquinones or from 1-aminoanthraquinone condensed with an aromatic dicarboxylic acid.

The benzamido group has found a large role among the anthraquinone vat dyes. In addition to its use with simple aminoanthraquinones, it has been found sufficient in preparing vat dyes from the more complex anthraquinone derivatives such as the pyrimidoanthrones. Furthermore, the introduction of a benzamido group provides a simple and convenient method for modifying the shades and improving the dyeing properties of other types of anthraquinone vat dyes. The range of shades covered is limited mainly to yellows and reds except for two important violet dyes.

Aside from the diamide type of linkages used to couple to aminoanthraquinones, the dyes in this series have been prepared by using an aminoanthraquinonecarboxylic acid, with carboxyl group at positions 2-, 3- or 6-, or a substituent at these positions to which a carboxyl group is attached, in order to acylate a second aminoanthraquinone. Examples of such dianthraquinones may possess a simple amide bridge as in dye (**4**).

Anthrimides. In general, α,α-dianthraquinonylamines exhibit poor affinity for fiber and are not fast to light when employed as vat dyes. α,β-Condensates, on the other hand, yield very satisfactory vat dyes. The anthrimides dye somewhat dull shades of orange, red, bordeaux, and gray and display excellent fastness properties.

The linkage between anthraquinone nuclei may be either a direct —NH— type or one that involves two —NH— groups separated by one of several possible residues, frequently aromatic hydrocarbons. The simplest anthrimide used practically is CI Vat Orange 20. More useful dyes have been obtained by introducing benzamido groups to one or both of the anthraquinone nuclei and by increasing the number of anthraquinone residues in order to form trianthrimides. All of these are used to dye cotton shades of red from cold, weakly alkaline hydrosulfite vats. They have good all-around fastness but, as with other yellow and red anthraquinone vat dyes, possess the disadvantage of causing the dyed cotton to deteriorate, especially on exposure to sunlight (tendering).

(4) Amide-bridged dianthraquinone

Indanthrones. Indanthrone (CI Vat Blue 4) is employed today largely for dyeing cotton but the use of indanthrone blues also includes dyeing and printing on linen and rayon. In addition, they function as pigments in graphic art and the lacquer industry.

As is customary with the anthraquinone vat dyes, the leuco compound is used for actual dyeing. Although indanthrone is composed of two anthraquinone residues, it requires only two hydrogen atoms (or two reducing equivalents, such as hydrosulfite and alkali) to convert it into a leuco compound.

Preparation of the dyeing vat and application of the dye to the fiber necessitates careful control of dye and hydrosulfite–caustic soda concentrations, as well as temperature. The disodium salt dihydro (or normal leuco) indanthrone is sparingly soluble in the vat solution and consequently may separate if its concentration is too high. Temperatures are maintained ca 50–60°C because, if higher, duller shades and lower color values result from over reduction. Excess alkali must also be monitored carefully. If present at the end of the dyeing process after air oxidation has been completed, it must be removed immediately to avoid further oxidation of the dye to the azine with loss of the beautiful reddish-blue shades.

The shades of blue produced by indanthrone are much faster than indigo to light and washing, and in fact, all of the indanthrones possess good lightfastness, as well as fastness to boiling alkali. The imino groups of these dyes are very sensitive to oxidizing agents (eg, hypochlorites and chlorine) since their dihydrazine rings are readily and reversibly oxidized to yellow azine derivatives. Despite the problems associated with indanthrone, however, it is used extensively because of its beautiful shades for which there is no equivalent available.

Flavanthrone. The only significant member of this group is flavanthrone itself (CI Vat Yellow 1). It is a by-product of the potash fusion of 2-aminoanthraquinone during the manufacture of indanthrone, but is usually prepared in an Ullmann reaction by heating two equivalents of 2-amino-1-chloroanthraquinone in the presence of copper powder. The resulting dianthraquinonyldiamine is then condensed with two equivalents of benzaldehyde to yield the dianil derivative, or more frequently, with phthalic anhydride to afford the phthalimido derivative.

Flavanthrone dyes yellow-orange shades from a blue alkaline hydrosulfite vat. If more drastic reduction is carried out, the result is a brown vat that may undergo facile reoxidation. Flavanthrone serves as a useful yellow component of lightfast greens, olives, etc, and is phototropic, ie, undergoes a change in shade by severe exposure to light. This light change is also readily reversible by contact with air or by a mild soap treatment. In general, flavanthrone is an important dye because it is a nontenderer and has good lightfastness. On the other hand, since it exhibits reducibility by cellulose and boiling alkali, it is not recommended for goods that are bleached after dyeing.

Pyranthrone. Pyranthrone is closely related to flavanthrone; it has with two methine groups in place of flavanthrone's nitrogen atoms. It has gained great commercial importance under the designation CI Vat Orange 9. Preparation is accomplished by two possible routes: from 2,2'-dimethyl-1,1'-dianthraquinone, by heating with zinc chloride or alcoholic potassium hydroxide; or condensation of pyrene with benzoyl chloride and aluminum chloride to give dibenzoylpyrene, which upon heating with aluminum chloride provides pyranthrone. This dye produces bright orange shades, which show remarkable fastness to chlorine and alkali washes, from a cherry-red vat.

Anthraquinone carbazoles. 1,1'-Anthrimide is itself of no value as a vat dye, but its treatment with aluminum chloride or caustic yields the useful carbazole derivative CI Vat Yellow 28. As a dye, this particular heterocyclic compound colors bright orange-yellow and is a tenderer.

As a class these compounds include some of the most important anthraquinone vat dyes. They produce orange, olive, khaki, and brown shades with good overall fastness properties when applied by the hot-dyeing process. Anthrimides containing benzamido groups in the 4- or 5-position, with reference to the imino group, are cyclized much more readily than the parent anthrimides, providing brighter dyes with better tinctorial power. CI Vat Orange 15 is one such dye that represents a series of colorants distinguished for their excellent fastness.

Anthraquinone thiazoles. Anthraquinones with condensed thiazole rings have found only limited value as vat dyes. The oldest and most important member of this class is CI Vat Yellow 2. It is prepared by reaction of benzotrichloride and sulfur with 2,6-diaminoanthraquinone, and yields bright greenish-yellow shades that are not especially fast to light.

Anthrapyrazolones. Pyrazolanthrone itself can be converted to a yellow vat dye (CI Dat Dye 70315) by fusion with potassium hydroxide. This dye, possessing two anthraquinone nuclei, dyes from a blue vat, bright yellow shades that exhibit fastness to chlorine, but poor fastness to alkali because of its two acidic hydrogen atoms. Replacement of the two acidic hydrogen atoms with ethyl groups affords the more useful dye CI Vat Red 13.

Anthrapyrimidines. Many yellow anthraquinone vat dyes possess tendering properties; dyes of the anthrapyrimidine groups are an exception; they are nontendering. Beyond the yellow pyrimidanthrones, a series of violet, blue, gray, and black dyes can be prepared. They are characterized by high tinctorial power, excellent fastness to light and wet treatment, and stability for printing.

Dibenzanthrones. Benzanthrone is a yellow compound devoid of dyeing properties. The fusion of two molecules with potassium hydroxide, however, yields a dark blue dyestuff of excellent fastness to light, chlorine, and washing. It is used to dye cotton from a strongly alkaline hydrosulfite vat. From a technical and commercial point of view, the series of dibenzanthrones that are derivatives of violanthrone (CI Vat Blue 20), and its isomer isoviolanthrone are some of the most interesting polycyclic vat dyes available.

Dibenzopyrenequinones. Dibenzopyrenequinone (CI Vat Yellow 4) is a dye closely related to anthanthrone. It is prepared by application of the Scholl cyclization with aluminum chloride to 3-benzoylbenzanthrone or 1,5-dibenzoylnaphthalene.

Anthanthrones. The dichloro and dibromo derivatives of anthanthrones are valuable level dyes—CI Vat Orange 19 and CI Vat Orange 3.

Anthraquionone acridones. The anthraquionone acridones are an important group of dyes with relatively simple molecular structure. They cover a wide range of hues including orange, red, volet, blue, brown, and khaki, with good general fastness especially to light. Dyeing is accomplished from a cold, readily soluble vat with good substantivity, but the dyes themselves demonstrate only moderate affinity for fabrics.

Solubilized vat dyes. Soluble esters of the leuco compounds of the anthraquinone dyes can be made directly from the parent dyes. These soluble vat dyes are generally sold as spray-dried powders and, because of their cost, are used mainly for pale shades and special applications.

Anthraquinonesulfonic Acid Dyes

The introduction of sulfonic groups to anthraquinones produces the anthraquinone acid dyes. The predominant shades in this series of dyes are blues and greens, not available among the azo dyes and unequalled by any other class of dye. They are characterized as a group by their excellent fastness to light, which falls within the range 5–7 (9 being the maximum rating) and a wet-fastness that is moderate to good.

The presence of sulfonic acid groups in the molecule can be achieved by one or a combination of three possible routes: by condensation of bromamine acid with aromatic amines which results in a sulfonic acid group on the same ring of the anthraquinone system as contains the auxochromes; by sulfonation of the anthraquinone nucleus itself with oleum; or when sulfonation is the final stage, using strong sulfuric acid or weak oleum which sulfonates the arylamino group.

With the exception of azo dyes, the anthraquinonesulfonic acids constitute the most important group of dyes for wool and silk. Beyond their additional use in dyeing nylon, they have also demonstrated the advantage of leaving cellulose fibers unaffected and have, therefore, been found useful for cross-dyeing union materials of the cellulose fibers and wool or silk. These dyes are usually sold and used as the sodium salts or as other alkali-metal salts.

Bromamine acid. Bromamine acid (1-amino-4-bromo-2-anthraquinonesulfonic acid) is one of the most important intermediates in the synthesis of acid dyes. The reaction is customarily run in boiling aqueous alkaline solution in the presence of a copper catalyst The simplest dye derived by this procedure is CI Acid Blue 25. These 1-amino-4-arylaminoanthraquinones are valuable as wool dyes because they can dye bright, level shades with very good fastness to light and washing. They are thus superior to earlier dyes even through their milling fastness is only 1–2.

Mordant Dyes

Because of their high cost, mordant dyes are no longer manufactured in large amounts. Nevertheless, they offer certain attractive features as dyes: a wide range of bright shades and excellent fastness properties.

Di- and trihydroxyanthraquinones are the older forms of mordant dyes. Having originally been obtained from natural sources, eg, madder root and morinda root, many of these were later produced synthetically from coal-tar chemicals. Dyes such as these were used extensively in printing and exhibited fastness to light but inferior fastness to washing, wet treatments, and perspiration.

Mordant dyes are capable of combining with various metals to form complexes of different types. Although the salts of such multivalent metals as iron, copper, aluminum, and cobalt have been used for mordanting purposes, the preferred metal is chromium. Colors of the resulting dyes range from yellow through red, orange, brown, green, blue, violet, and black, depending upon the metal used and the structure of the anthraquinone compound.

Alizarin. The simplest and most important constituent of this group is 1,2-dihydroxyanthraquinone, commonly known as alizarin (CI Mordant Red 11). Alizarin crystallizes from alcohol in brilliant brownish-yellow needles and sublimes at 110°C as orange-red needles, mp 290°C, bp 430°C. It is readily soluble in common organic solvents and dissolves in aqueous caustic soda producing a purple color. This dye is produced commercially by heating the sodium salt of anthraquinone-2-sulfonic acid (silver salt) with aqueous soda and sodium nitrate (or chlorate) in an autoclave at 200°C. Alizarin is a polygenetic dyestuff; consequently, the color varies with the metal mordant.

Heat-Transfer Dyes

Transfer printing refers to a process in which a thin film of volatile dyes applied to an inert substrate, such as paper, is transferred by dry heat to a suitable fabric held in contact with the paper. The heating and good contact required for the procedure are achieved with a hot iron, hot press, or continuous calender. The fabrics colored are those that demonstrate an affinity for disperse dyes.

The heat-transfer printing process is expected to encourage the present trend toward the use of polyester as the dominant clothing fabric. Its advantages include the ability to reproduce intricate patterns with excellent clarity. On the other hand, limitations exist in range of colors and the degree of colorfastness exhibited on all fabrics, including polyester.

RACK HUN CHUNG
General Electric Co.

RUSSELL E. FARRIS
Sandoz Colors & Chemicals

H.E. Fierz-David and L. Blangy, *Fundamental Processes of Dye Chemistry*, Interscience Publishers, New York, 1949, pp. 224–230, 314–323, 454–465.

K. Venskataraman, *Chemistry of Synthetic Dyes*, Academic Press, Inc., New York, Vol. 2, 1952, pp. 803–945; Vol. 3, 1970, pp. 385–423; Vol. 5, 1971, pp. 57–240.

E.N. Abahart, *Dyes and Their Intermediates*, Chemical Publishing, New York, 1977, pp. 121–141.

DYES, APPLICATION AND EVALUATION

APPLICATION OF DYES

Myriads of dyes have been synthesized in laboratories around the world, a few hundred of which have found their way to large-scale production and industrial application. However, one can expect that this process will slow down in the near future as development of new dyes becomes more expensive, and as safety, health, and ecological restraints become more stringent.

Dyeing describes the imprintation of a new and often permanent color, especially by impregnating with a dye, and is generally used in connection with textiles, paper, and leather. Printing may be considered as a special dyeing process by which the dye is applied in locally defined areas in the form of a thickened solution and then is fixed.

Generally, dyes are dissolved or dispersed in a liquid medium before being applied to a substrate into which they are fixed by chemical or physical means, or both. Owing to its suitability, its availability, and its economy, water usually is the medium used in dye application; however, nonaqueous solvents have been studied extensively in recent years (see Dye carriers).

Textile substrates can be classified in three groups: cellulosic, protein, and synthetic polymer fibers. Even and economical distribution of a small amount of dye (generally ≤ 3 wt % of goods) throughout the substrate and fixation of the dye are the keys to dyeing, ie, with regard to fastness to washing and to other deteriorating influences.

Production of dyeings of acceptable quality requires the use of many auxiliary products and chemicals. These include chemicals that improve fastness properties such as bleaching agents, wetting and penetrating agents, leveling and retarding agents, and lubricating agents. Other agents speed the dyeing process or are used for dispersion, oxidation, reduction, or removal of dyes from poorly dyed textiles.

Dyes of similar or identical chromophoric class are used for widely differing applications and, therefore, are classified according to their usage rather than their chemical constitution. As all dyes are water soluble at one stage of their application, it is also convenient to classify them according to their solubility groups; these can be permanent or temporary:

Permanent solubilizing group	*Types of dye*
$-SO_3Na$	acid, direct, mordant, fiber reactive
$-\overset{+}{N}H_3Cl^-$, $-\overset{+}{N}R_3Cl^-$	basic
$-OH$, $-NH_2$, $-SO_2NH_2$	disperse
Temporary solubilizing group	*Types of dye*
$-ONa$	vat, insoluble azo
$-OSO_3Na$	solubilized vat

Dyes with identical or similar solubilizing groups generally display similar dyeing behavior even though their main structure may vary substantially. Another important consideration of the use of a given dye for a specific application and fastness properties of commercial dyes is found in the pattern cards issued by their manufacturers. The following classification of colorants for dyeing is used: acid, basic, direct, disperse, insoluble azo, sulfur, vat, fiber-reactive, miscellaneous dyes, and pigments.

Theories of Dyeing

Affinity and diffusion are fundamental aspects of the dyeing process. The former describes the force by which the dye is attracted by the fiber, and the latter describes the speed with which it travels within the fiber from areas of higher concentration to areas of lower concentration. Alone or together with fiber characteristics, affinity and diffusion determine the speed of dyeing, its temperature dependency, the equilibrium exhaustion of the dye, and some of its fastness properties.

Two fundamental dyeing processes have been distinguished: the non-ionic and the ionic dyeing processes. The former refers to the dyeing of synthetic fibers such as polyester, nylon, and triacetate by disperse dyes. The ionic dyeing process implies ionic groups in fibers and in dyes and the dyeing process involves ion-exchange mechanisms.

In recent years, dyeing theory also has concerned itself with optimal dyeing conditions to achieve level dyeing. Two fundamental approaches are possible. In the first, the dye is applied under carefully controlled conditions at a uniform rate of absorption so which level (uniform) dyeing is maintained from the beginning. In the other approach, the dye is exhausted very rapidly, possibly under isothermic conditions at the maximum dyeing temperature, and the initially created unlevelness is then leveled out in the subsequent extended migration period.

Dyeing Machinery

In the application of dyes, there have developed over the years three chief principles of dyeing textiles. (1) The dye liquor is moved as the material is held stationary. An illustration of this is raw-stock or package-machine dyeing (Fig. 1). (2) The textile material is moved without mechanical movement of the liquor. Examples of this principle are jig dyeing and continuous dyeing which involves the padding of fabric (Fig. 2). (3) A combination of (1) and (2) is exemplified by a Klauder-Weldon skein-dye machine in which the dye liquor is pumped as the skeins are mechanically turned. Another example is a jet dyeing machine in which both the goods and the liquor are constantly moving.

Control of dyeing equipment. Today there are many completely automated computer-controlled exhaust dyehouses, especially those

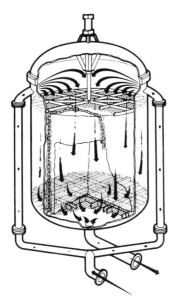

Figure 1. Schematic raw-stock dye machine. Courtesy of Gaston County Dyeing Machine Company.

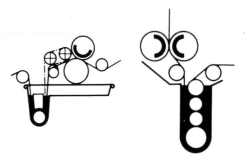

Figure 2. Dye padder. Courtesy of Zima Corporation, USA-Kuesters.

processing 100% synthetics. Some firms have a no-add procedure in the dyehouse by which the dyer loads the fabric or yarn, weighs the dye, punches a button and lets the computer take over the entire process. This procedure ensures a constant dyeing cycle and the only variables are the dye index of the fiber or the quality of the dyestuff.

In continuous dyeing, there is less automation because of the many variables and the rapidity of the dyeing process which require many adjustments during the period in which several thousand meters of textile are dyed. However, instrumental science has continued to advance rapidly so that at present there is at least one continuous range available that is entirely computed-controlled except for the make-up of the dye mix.

Preparation for Dyeing

The proper preparation of textile fibers, yarns, and fabrics is an essential and integral part of dyeing and of finishing. The purposes of the various preparative procedures are removal from the fibers of natural or applied impurities that might impede the penetration and leveling of the dyes on the fibers, conversion of the materials into proper physical condition and form for necessary mechanical operations, and in some cases, application of chemicals to the goods prior to dyeing. Improper finishing techniques, or the failure to recognize the physical and chemical properties of the treated fibers, may lead to a variety of problems during the dyeing cycle. The appearance of resist marks from silicate or lime-soap deposits, chafe marks from mechanical damage, poor absorbency, crease marks, and changes in the surface and dimensional characteristics of certain thermoplastic fabrics result from improper preparation.

The fibers may be classified into two general groups: natural and synthetic. Many of the synthetic fibers have structural and physical properties similar to those of the natural animal and vegetable fibers. Blends of various fibers require handling techniques that will protect all fabric's fibers from damage.

Acid (Anionic) Dyes

The most common fiber types to be dyed with acid dyes are (in sequence of their importance): polyamide, wool, silk, modified acrylic, and polypropylene fibers as well as blends of the aforementioned fibers with other fibers, such as cotton, rayon, polyester, regular acrylic, etc. Approximately 80–85% of all acid dyes sold to the U.S. textile industry are used for dyeing nylon, 10–15% for wool, and the balance for those fibers mentioned above. Acid dyes are organic sulfonic acids; the commercially available forms are usually their sodium salts, which exhibit good water solubility.

According to their structure, acid dyes belong to the following chemical groups: azo (qv); anthraquinone (av); triphenylmethane (qv); pyrazolone; azine (qv); nitro; and quinoline (qv). Azo dyes represent the largest and most important group and are followed by anthraquinone and triarylmethane dyes. Of the other dye groups, very few products are of any commercial value.

Acid dyes can be divided into four groups:

(1) These are the level-dyeing acid dyes with one sulfonic acid group. They offer excellent leveling, migration, and coverage of barré properties. Fastness to light is very good, but the wetfastness properties in heavier shades generally are only marginal. The latter can be improved with an

aftertreatment of either tannic acid/tartar emetic or any other synthetic aftertreating agent.

(2) These are the neutral-dyeing acid dyestuffs. They are also monosulfonated and are very similar in their leveling, migration, and coverage of barré properties to group 1; however, because of their chemical structure (larger molecular size), these dyes exhibit superior wetfastness properties.

(3) These are the milling-type acid dyes. They are disulfonated dyes and provide dyeings with highest wetfastness properties. The leveling and migration properties of these products are much inferior to those of the monosulfonated dyes in groups 1 and 2. It is obvious that there are no acid dyes that combine all desirable properties. Those that offer the excellent dyeing characteristics such as good leveling, migration, and coverage of barré, have only marginal wetfastness properties; those that provide high wetfastness do not level very well. The dyes in group 2 represent the best compromise. When the fabric is aftertreated with a synthetic aftertreating agent, wetfastness—in most instances—will be equal or similar to dyeings obtained from milling dyes.

(4) These are the premetallized dyes that also include mono- and disulfonated types. Premetallized dyes exhibit a rather high dyeing strike rate even at pH values of 7–8 and, therefore, are extremely difficult to dye in a level manner. Their wetfastness properties are either comparable or superior to milling-type dyes.

Dyeing processes. The two major polyamide types commercially available today are nylon 6 and nylon 6,6 (see Polyamides). Both fiber types are very receptive to acid dyes under certain conditions.

A direct relationship exists between the chemical structure of an acid dye and its dyeing and wetfastness properties. The dyeing process is influenced by a number of parameters, such as: dyestuff selection, type and quantity of auxiliaries, pH, temperature, and time.

Dyeing of nylon carpet. A great percentage of acid dyes used in the textile industry is used in dyeing of nylon and wool carpets. Nylon carpet fibers can be dyed in many different ways, eg, in raw stock form, where usually highest wetfastness dyes are applied, in yarn form and in piece form. The latter is divided into piece or beck dyeing and continuous dyeing. For the continuous dyeing process, the most popular machine is the Kuesters carpet-dyeing range, which consists of a wet-out padder, a dye applicator, a loop steamer, and 3–4 wash boxes. Usually a drying oven is used after a dyeing range.

Dyeing of wool. Wool (qv) has lost its former importance and represents, today, only a minor factor in fiber dyeing. Dyeing of wool may be carried out in the various stages of processing. Wool fabrics are dyed primarily by exhaust-dyeing methods. Although continuous dyeing of wool has been discussed for some time, it has not become very important. Only a few units are running today which continuously dye wool top and raw stock. Wool is dyed as: raw stock, yarn, and piece goods.

Basic (Cationic) Dyes

Cationic dyes are currently used in large quantities to dye acrylics (Orlon, Acrilan, Creslan, Zefran) and modified acrylics (Verel). Subsequent developments led to the introduction of acidic groups to polyester and polyamide fibers, further increasing the market for these dyes.

Dyeing of acrylic fibers. Basic dyes are water-soluble and dissociate into anions and colored cations. The cations have a strong affinity for the acidic group (sulfonic or carboxylic) in the acrylic fibers and form salts. Because of these strong bonds, washfastness is usually outstanding, and lightfastness varies considerably, depending on the dyestuff. Basic dyes are usually applied in a batch process with skein, stock, and package dyeing more prevalent than piece dyeing on becks or jet machines.

The fundamental steps in the acrylic dyeing mechanism are adsorption of the basic dyestuff at the fiber surface, which occurs only when the glass-transition point of the fiber is exceeded; diffusion of the dye into the fiber as the fiber molecules acquire enough energy to move; formation of the dye–fiber bond; and migration of dyestuff from one dyesite to another or from within the fiber to the surface. The degree of migration varies considerably, depending on the dyestuff, but generally basic dyes

do not migrate well. Leveling of basic dyestuffs can be a major problem that often can be traced partially to differing exhaustion rates of individual dyestuffs that may occur in combination shades. To overcome the high affinity of some basic dyestuffs and to prevent unlevel dyeing, often a cationic retarder is used. Cationic retarders act as a colorless dyestuff and compete with the basic dyes for available dyesites. The optimum amount of retarder can be calculated.

Dyeing of basic-dyeable polyester. Cationic-dyeable polyester has achieved a moderate degree of success because it lends itself to two- or three-color effects in blends with regular disperse-dyeable polyester, cellulosic fibers, or wool. Typical basic-dyeable polyester fibers are Dacron T-64 and T-92, Fortrel 402, and Trevira 640.

Nonionic products must be present to act as antiprecipitants or to suspend particles resulting from the reaction between the cationic dyes and anionic dyestuffs when blends of basic-dyeable and disperse-dyeable polyester are dyed.

Direct Dyes

Direct dyes are water-soluble dyes that exhaust onto cellulosic fibers, such as cotton, rayon, and linen, from a salt bath. Because of this property, these dyes are called substantive dyes.

Nearly all direct dyes are azo products containing one or more sulfonic radicals which impart water solubility. A few contain carboxyl groups ortho to hydroxyl, a configuration that permits aftertreatment for improvement of fastness. Most direct dyes contain two or three azo groups, although monoazo and higher polyazo products are known. The following structural properties govern direct dyestuffs with substantivity or excellent affinity for cellulose fibers: elongated molecular structure with few side chains; several aromatic rings, which may be arranged in one plane and which contain chemical groups that can form hydrogen bonds relatively easily; and few solubilizing chemical groups.

Direct dyestuffs form anions (negative charges) by dissociation. In aqueous solution, the molecules of direct dyestuffs are linked by hydrogen bonds, forming larger agglomerates or colloidal solutions. This reduces their solubility and promotes disposition on the fibers. The elongated structure of these agglomerates adapts them to the parallel arrangement of the cellulose molecule to which they are firmly linked by multiple hydrogen bonds.

Uses. Direct dyes are used extensively for coloring cotton and rayon textiles, on which the dyes give a full range of color. Also, direct dyes are used on paper and on many other fibers, such as linen, jute, hemp, silk, polyamide, and mixtures of other fibers. Direct dyes, in addition to ease of application, also give excellent penetration, even coverage, level dyeing, and they do not generally impair luster of the material to be colored.

Application. The amount of salt used in dyeing is governed by the liquor-to-fabric ratio, the total amount of dye, and the exhausting properties of the particular dyes. Direct dyes normally do not exhaust completely. Manufacturers' circulars and shade cards indicate the appropriate dye method, the general salt requirements and the optimum dyeing temperatures of specific dyes.

Aftertreatment. The wetfastness of many direct dyeings can be improved in various ways: aftertreatment with cation-active organic fixing agents; aftertreatment with formaldehyde; diazotization of the dyes on the fiber and development with amines, phenols, or naphthols; and aftertreatment with diazonium compounds. The lightfastness of some direct dyeings can be improved by aftertreatment with copper salts. Copper salts are used alone, in combination with chrome, or in combination with cation-active organic condensation products.

Disperse Dyes

Disperse dyes were introduced in the 1920s, and primarily were used for the dyeing of cellulose acetate fibers; however, later many of these dyes were found to be suitable for polyamide fibers. Subsequent synthetic fibers, such as triacetate and polyester, had different molecular structures that required new disperse dyes. The use of disperse dyes for the dyeing of polyacrylonitrile fibers is limited. The dyes possess very limited wash- and lightfastness and are generally used for pastel shades where leveling is the major consideration.

Hydrophobic fibers, such as acetate, triacetate, and polyester, show no affinity for ionic dyestuffs unless chemically modified to accept them. These fibers absorb nonionic organic colorants applied from a dispersion. Disperse dyes are colored organic substances with very low solubility in water and which, during the course of a dyeing, are transferred from dispersion via monomolecular solution to the fiber, ie, the dyestuff is believed to form a solid solution in the fiber, which acts as a solvent.

Fastness properties. The fastness properties of the acetate dyes were not adequate on triacetate and polyester fibers, but newly developed disperse dyes can be dyed in a full range of shades that satisfy the fastness demands of the market. Fastness properties of disperse dyes vary widely, depending on the molecular structures of the dyestuff and of the fiber to which they are applied. When selecting dyestuffs to meet specific fastness requirements, special consideration must be given to wetfastness, sublimation-fastness, gas-fastness, and lightfastness.

Dyeing methods. Typical dyeing methods are atmospheric exhaust, used primarily for dyeing polyester blends when pressure equipment is not available; high temperature exhaust, the most prevalent method since it is faster and more economical than atmospheric dyeing; and continuous dyeing, the most economical method of dyeing because speeds in excess of 90 m/min are possible.

Fiber-Reactive Dyes

Fiber-reactive dyes are the newest primary dyestuff group and have had their significant developments occur within the last 20 yr. Fiber-reactive dyes supplement vat, direct, and naphthol dyes, and are available in a full range of colors of outstanding brilliance and fastness properties. Fiber-reactive dyes can be applied to wool, silk, and polyamide; however, they are used primarily for cellulosic fibers with exhaust dyeing, cold pad-batch dyeing, and continuous dyeing. By proper selection of dyes, one can apply fiber-reactive dyes by almost all known batch processes, as well as semi-continuous and full-continuous dyeing methods.

Sulfur Dyes

The fastness properties of sulfur dyes lie between those of diazotized direct dyes and vat dyes. They have better washfastness than diazotized or aftertreated direct dyes but are somewhat inferior in washfastness to vat dyes. Lightfastness varies from mediocre to good. Chlorine fastness is rather poor in most cases, but there are a few dyes that show fairly good fastness. Sulfur dyes in earlier days gave cotton an inadsorbent harsh hand (texture); however, with the use of chelating agents, ester softeners, and new oxidizing agents, such as bromites, bromates, and iodates, much, but not all, of this problem has been overcome.

Some sulfur dyes in cellulosic fibers, especially blacks, cause loss of tensile strength upon storage under conditions of high humidity and temperature. This problem has been largely minimized by proper oxidation, pH, buffering, and durable-press resin finishing.

Sulfur dyes can be applied by exhaust dyeing on raw stock and on packages or on piece goods on the jig. However, continuous dyeing is the process by which most of the sulfur dyes are applied today. The prereduced dyes are padded and steamed for 30–45 s and then are washed, oxidized, soaped, rewashed, and dried.

Insoluble Azo Dyes (Naphthols)

As a rule of thumb, those naphthols exhibiting less than 25% substantivity are used for continuous dyeing. Since, in exhaust dyeing, the fiber is salt-rinsed to improve crock and washfastness, too much naphthol is lost with low substantivity naphthols. On the other hand, those naphthols with low affinity are highly suitable for continuous dyeing because they offer almost no substantivity in padding. The reverse is true of substantive naphthols with "fast strike" for all types of exhaust dyeing, since they not only give better economy, but also give superior washfastness because they can be salt-rinsed before coupling to remove unfixed naphthol.

At present, the number of naphthols and coupling agents, as well as the number of firms manufacturing these products, have shrunk substantially. However, for extremely bright red and cranberry shades, naphthols are the only dyes that give top all-around fastness, including fastness to chlorine bleaching.

Vat Dyes

Vat dyes are complex organic molecules that are insoluble in water, but when their carbonyl groups are properly reduced in a solution of caustic soda and sodium hydrosulfite to the so-called leuco or soluble state, they exhibit an affinity for cellulosic fibers. Chemical oxidizing agents are usually used to hasten oxidation of the reduced dye within the fiber back to its insoluble form, which is physically trapped. This results in shades usually of excellent wash-, chlorine-, and lightfastness. The dyes are sold as powders or pastes that form dispersions upon dilution with water. Vat dyes are usually applied by either continuous or exhaust procedures, and occasionally by the pad-jig method.

Soluble vats—leuco esters of vat dyes. The procedure for modifying vat dyes was discovered in 1921 by Bader and Sunder and involved the reaction of leucoindigo in pyridine solution with chlorosulfuric acid, then conversion of this sulfonated compound to the sodium salt by addition of the proper amount of caustic soda. This compound is water-soluble and slightly substantive to cellulose.

Miscellaneous Dyes

Miscellaneous dyes include dyes developed on the fiber, oxidation dyes, mordant dyes, natural dyes, and mineral dyes.

Dyeing of Fiber Blends

The differences in dyeability between the many fibers on the market open a wide field of multicolored yarns and fabrics to the stylist. Fiber blends can be dyed into union shades or tone-on-tone. Often multicolor effects are obtained by coloring the individual components in different shades or by reserving one fiber. A complete reserving of a fiber is not possible in all cases.

Pigment Dyeing

Modern pigment application had its practical beginning in 1942 with the introduction of the continuous application of a bath consisting of an oil-in-water emulsion of organic pigments with thermosetting and/or thermoplastic resinous binders. The resultant shades generally have outstanding lightfastness and good washfastness. Fabrics being used range from lightweight poplins and sheetings to corduroy. Pigment dyeing has become of even greater importance with the growing concern for energy and pollution because of the following factors: (a) cold pad bath and no heat requirements beyond curing at 170–175°C; (b) no reducing or oxidizing chemicals required; and (c) less hot-soaping and rinsing required as for fiber-reactive, sulfur, or vat dyes.

Solvent Dyeing

Solvent dyeing generally refers to dyeing in nonaqueous media. In the early 1970s, solvent dyeing was expected to become the dyeing process of the future and was discussed and researched extensively. The primary interest was in the possibility of reduced water consumption and water-effluent treatment costs; secondarily, improved productivity was expected. Although the argument for or against nonaqueous dyeing has not been finally settled by any means, it has at present been decided in favor of aqueous dyeing. For all practical purposes, there is today no commercial solvent dyeing (except for a few very specialized applications).

Textile Printing

The term textile printing is used to describe the production through a combination of various mechanical and chemical means, of colored designs or patterns on textile substrates. In printing on textiles, a localized dyeing process takes place, whereby in general the chemical and physical parameters of dyeing apply.

The process of textile print coloration can be divided into three steps. First, the colorant is applied as pigment dispersion, dye dispersion, or dye solution from a vehicle called print paste or printing ink, containing in addition to the colorant such solutions or dispersions of chemicals as may be required by the colorant or textile substrate to improve and assist in dye solubility, dispersion stability, pH, lubricity, hygroscopicity, rate of dye fixation to the substrate, and colorant-fiber bonding. The second step in processing printed textiles is the fixation process. During the following after-scouring, the third step in the textile printing process, the prints are rinsed and scoured in a detergent solution in order to remove auxiliary chemicals, thickening agents, and portions of unfixed dyes remaining on the surface of the printed fibers.

Dyes used in textile printing include pigments, disperse dyes, acid dyes, premetallized dyes, direct dyes, reactive dyes, basic dyes, vat dyes, azoic dyes, and phthalocyanine dyes.

Paper Coloring

Paper may be colored by dyeing the fibers in a water suspension by batch or continuous methods. Paper may also be colored by surface application of dyestuff solutions after the paper has been formed and dried or partially dried by utilizing size-press addition, calendar staining, or coating operations on the paper machine. In addition, paper may be colored in off-machine processes by dip dyeings or absorption of dyestuff solution and subsequent drying, such as for decorative crepe papers.

Among the colorants that have been and are being used for the dyeing of paper are natural inorganic pigments (ochre, sienna, etc); synthetic inorganic pigments (chromium oxides, iron oxides, carbon blacks, etc); natural organic colorants (indigo, alizarin, etc); and synthetic organic colorants. The last is the largest and most important group (see Paper-making additives).

Leather Dyeing

The dyeing of leather (qv) presents some difficulties not encountered in the dyeing of textiles. Unlike textiles, leather is not a homogeneous product of definite composition whose chemical properties may be closely and accurately defined, but is rather a product derived from protein collagen (skin or hide substance) treated with one or more tanning agents. Not only may the compound used to convert hide substance into leather vary chemically over a wide range, but the quantities used, the method of application, and the physical condition of the hide prior to tanning or dyeing may vary, with each factor in turn affecting the dyeing properties of the resultant leather. Also, leather retains many of the properties originally associated with the parent substance, and these affect profoundly and, in many ways, limit the dyeing properties of the final product. Chief among these properties are sensitivity to extremes of pH, thermolability, and the tendency to combine with acidic or basic compounds. The four main classes of dyes employed in the coloring of leather are the acid/direct, direct, and basic types. The methods used in dyeing leather are quite simple and they obtain their names from the equipment employed, such as drum, wheel, paddle, brush, tray, or spray dyeing.

R.G. KUEHNI
J.C. KING
R.E. PHILLIPS
W. HEISE
F. TWEEDLE
P. BASS
C.B. ANDERSON
H. TEICH
Mobay Chemical Corp.

C.L. Bird and W.S. Boston, eds., *The Theory of Colouration of Textiles*, The Dyers Company Publications Trust, London, 1975, distributed by the Society of Dyers and Colourists, Bradford, Eng.

D.M. Nunn, ed., *The Dyeing of Synthetic-polymer and Acetate Fibres*, The Dyers Company Publications Trust, London, 1979, distributed by the Society of Dyers and Colourists, Bradford, Eng.

1982/83 Technical Manual of the American Association of Textile Chemists and Colorists, Vol. 58, 1982.

E.R. Trotman, *Dyeing and Chemical Technology of Textile Fibers*, C. Griffin & Col, Ltd., London, 1964.

DYES, NATURAL

Although the use of natural colorants for textiles and fibers has declined, the use of nature-produced compounds for coloring foods and ingested products has recently increased. This has been prompted, in part, by government regulations, such as FDA's banning of such synthetic dyes as FD & C Reds 2 and 4. With extensive testing of all synthetic products for carcinogenicity and mutagenicity, the food-dye industry has returned to natural products as the source of its coloring bodies.

Natural dyes comprise those colorants (dyes and pigments) that are obtained from animal or vegetable mater without chemical processing. They are mainly mordant dyes, although some vat, solvent, pigment, direct, and acid types are known. There are no natural sulfur, disperse, azoic, or ingrain types.

Anthraquinones

Some of the most important natural dyes are based on the anthraquinone structure. They include both plant (madder, munjeet, emodin) and animal (lac, kermes, cochineal) types (see Dyes, anthraquinone).

Plant anthraquinone dyes. Madder, or now as it is better known by its common chemical name, alizarin, is one of the most ancient of natural dyes. The coloring material is concentrated in the roots of the madder plant, occurring to a maximum concentration of 4%. These perennial plants are allowed to remain in the ground for 18–28 mo at which time the roots are dug up, washed, dried, and finely ground. The primary use of madder is dyeing cotton Turkey Red by mordanting with alum in the presence of lime.

Animal anthraquionone dyes. With the exception of lac dyes from ancient India, kermes or kermesic acid is the oldest of all insect dyes. It is obtained from an oriental shield louse that lives on the leaves and stems of low, shrubby trees, the holm oak, *Quercus ilex*, and the shrub oak, *Quercus coccifera*. Kermesic acid is slightly soluble in cold water; it is soluble in hot water giving a yellowish-red solution. On dissolving in concentrated sulfuric acid, a violet red is obtained that turns blue with the addition of boric acid. The fame of its color is owing to the brilliant scarlet obtained when dyed on alum mordant.

Chemically, cochineal is similar to kermes. Of current interest, the alumina lake of cochineal, called carmine, has found use in the food-dye industry. It is permanently listed as approved by the FDA for ingested products. The lake is composed of not less than 50% carminic acid. Recent uses have included coloring toothpaste, bakery goods, apple sauce, and pharmaceutical tablets (see Colorants for foods, drugs, and cosmetics).

Lac dye is derived from the gum lac or viscous fluid made by the insects *Coccus laccae* which bore into the bark of trees and become enclosed in an exuding juice, which hardens into a resin (see Shellac; Resins, natural). The dye yields scarlet and crimson shades that exhibit good fastness properties, especially to light and water.

Just as in the fused three-ring system of anthraquinones, in the two-ring naphthalene base system nature has developed enough conjugation to achieve color. Three principal chemical systems yield natural dyeing materials: α-naphthoquinones, flavones, and anthocyanidins.

Naphthoquinone Types

α-Naphthoquinones. The leaves of the tropical shrub henna, *Lawsonia inermis*, cultivated mainly in Egypt and India, upon extraction with hot water yield the yellow pigment lawsone which dyes wool and silk an orange shade. Lawsone has been identified as 2-hydroxy-1,4-naphthoquinone. The isomeric 5-hydroxy compound, commonly called juglone, nucin, or regianin, is found in the shells of unripe walnuts, usually in the colorless leuco form, α-juglone. Lapachol is found in lapachol or taigu wood from South America and in bethabara wood from West Africa. The pigment forms a bright red, water-soluble sodium salt.

Flavones. A majority of the naturally occurring yellows have the flavone basic structure. Flavone (2-phenylbenzopyrone), a colorless material of mp 97°C, is obtained from the primrose.

Closely related in chemical structure to the above flavones are substituted dihydropyrans. The most important naturally occurring members are haematin and its leuco form, haematoxylin. These are the principal coloring bodies of logwood (CI Natural Black 1, CI 75290) historically one of the most important natural dyes for dark shades of silk, wool, cotton, leather, wood, and animal bristles, hair, and fur. It is still used extensively in dyeing and tanning leather.

Anthocyanidins (flavylium salts) are glycosides of hydroxylated 2-phenylbenzopyrylium salts. They are usually isolated as the chlorides after hydrolytic fission of the glycoside with hydrochloric acid. The color developed by the plant is determined by the pH of the cell sap. For example, cyanin is red under acidic conditions, violet at neutral pH, and blue under alkaline conditions.

Natural dyes, particularly red shades, have become increasingly important for use in food coloring. With the FDA's withdrawal of FD & C Reds 2 and 4, new sources for food colors have been required. The initial approach by the food-dye industry has been to return to natural dyes such as those of the anthocyanin class.

Alloxan adduct. A yellow dye currently being used as a food colorant is the phosphate salt of riboflavin, vitamin B_2 (see Vitamins). This is found in varying amounts in all plant and animal cells; particularly good natural sources are milk, eggs, malted barley, yeast, liver, kidney, and heart. Excretion of riboflavin in the urine is responsible for the yellow color. Structurally, riboflavin is an alloxan derivative.

Betanin

Another natural dye that is currently being used in food to take the place of the delisted FD & C Reds 2 and 4 is obtained from red beet extracts, *Beta vulgarus*. This is available as beet-juice concentrate, dehydrated beet root, and spray-dried extracts. Red beet root contains both red and yellow pigments of the class betaines; these are quaternary ammonium amino acids. The red bodies are betacyanine pigments of which the major constituent is betanin.

Indigoid Dyes

Two very important natural dyes have the indigoid structure. Indigo, the main product that gives this dyestuff class its name, is still one of the primary dyes of the world, although now made via a synthetic route. Tyrian or royal purple, although no longer in demand, was once the prize sought by the Caesars. The source of Tyrian purple is the purpura shellfish or *Murex Brandaris* found in shallow waters throughout the Mediterranean. Each mollusk contains a few drops of glandular mucous; this fluid at first appears white, but on exposure to light changes to yellow-green and eventually violet or reddish-purple. Natural indigo is obtained from the leaf of *Indigofera tinctoria*, a leguminose, widely distributed in Asia, Africa, and America. The dye is contained only in the leaf of the plant, unlike other vegetable dyes where the coloring matter is also found in the stalk, pods, and twigs. Natural indigo also contains Indigotin (CI 75780), Indirubin (CI 75790), Indigo Brown, Indigo Gluten, and Indigo Yellow (CI 75640).

Carotenoids

Although most color in the visible spectral range is generated by conjugated aromatic ring systems, another class of dyes found in nature obtains its color owing to the presence of long, conjugated double-bond chains (see Terpenoids). These dyes are the carotenoids, the class name being derived from the orange pigment found in carrots, carotene. They are also known as lipochromes because of their solubility and occurrence in fats.

Carotene (Natural Yellow 26; Natural Brown 5; CI 75130) is widely distributed throughout the vegetable and animal kingdoms. The carotenoids have use as colorants in foods and ingested drugs. β-Carotene is permanently listed by the FDA as an FD & C approved colorant (see Colorants for foods, drugs, and cosmetics). At present it is available commercially as a 30% liquid suspension; a 24% semisolid suspension; a 10% beadlet-water dispersion, and a 3% emulsion.

Another carotenoid, extract of annatto, is also used in food coloring. Currently its main use is in dairy products such as butter, margarine,

and cheese. Annatto is obtained from the pulpy portion of the seeds of the plant *Bixa orellana*, found in India, Central America, and Brazil.

Saffron is obtained from the pistils of the *Crocus sativus*, a plant that flowers in the fall and is quite different from the common spring variety crocus. The principal chemical constituent of saffron is crocin, the digentiobiose ester of crocetin.

Chlorophyll

Unlike the other natural products discussed in this article, one might not consider chlorophyll in the same class as dyestuffs. This is probably because chlorophyll does not have affinity for the natural fibers, especially cotton and wool, which can be made into wearing apparel. However, if one overlooks this most common dye requirement, chlorophyll is indeed a dyestuff (CI Natural Green 3; CI 75810). It is used extensively for coloring soaps, resins, inks, waxes (eg, candles); because it is physiologically harmless, it is used in the coloring of edible fats and oils (eg, chewing gum, confectionery, egg white, gelatin), and for cosmetics, liniments, lotions, mouthwashes, and perfumes. In the usual sense of a dye, it is used to color leather, where it exhibits good penetrating power and is especially light-stable.

Chemically pure chlorophyll is very difficult to prepare; therefore, the commercial product, like that found in nature, is a mixture along with several colored substances of the carotenoid family. The main components of the natural mixture have been designated chlorophyll a and chlorophyll b, in a ratio of approximately 3 to 1, along with yellow, orange, and red bodies. The excellent hiding power of the green chlorophylls usually mask the rest whose presence is unknown until the chlorophyll is destroyed, as in the case of autumn leaves.

RUSSELL E. FARRIS
Sandoz Colors & Chemicals

W.F. Legget, *Ancient and Medieval Dyes*, Chemical Publishing Company, New York, 1944.

A.G. Perkin and A.E. Everest, *The Natural Organic Colouring Matters*, Longmans, Green and Co., New York, 1918.

F. Mayer, *The Chemistry of Natural Coloring Matters*, ACS Monograph No. 89, translated and revised by A.H. Cook, Reinhold Publishing Co., New York, 1943.

DYES, REACTIVE

Reactive dyes are colored compounds that contain functional groups capable of forming covalent bonds with active sites in fibers such as hydroxyl groups in cellulose, amino, thiol, and hydroxyl groups in wool or amino groups in polyamides. This bond formation between the functional group and the substrate results in high wetfastness properties. These dyes differ fundamentally from other types of dyes that owe their wetfastness to physical adsorption or mechanical retention (see Dyes and dye intermediates). The principal commercial applications of reactive dyes are in the dyeing of cellulose, wool, and nylon, either individually or as components of fiber blends. They are also used in dyeing silk, hair, and leather (see Dyes, application and evaluation).

Intensive research on the reaction of soda cellulose with cyanuric chloride (1) led to a useful industrial method for the production of dyeings in which a covalent bond was formed between the dye and the fiber. This development resulted in the introduction of the first range of reactive dyes for cellulose in 1956. These dyes contained either a monochloro- or a dichloro-s-triazine moiety, and were marketed as the Procion H (ICI) or Procion M (ICI) ranges.

(1)

During the 1960s, several reactive dye systems were developed, based upon other heterocyclic aromatic compounds containing labile leaving groups. For example, dyes based upon 2,4,5-trihalopyrimidine, (2), (eg, Reactone (CIBA-GEIGY), Drimarene (Sandoz), and Levafix EA and PA (Bayer)), [2,3-dichloroquinoxaline, (3), (eg, Levafix E (Bayer) and Cavalite (DuP)), and 4,5-dichloro-6-pyridazinone, (4) (eg, Primazin P (BASF)) became commercially available. Hoechst also marketed the Remazol reactive dyes containing vinyl sulfone groups, which combine with cellulose through nucleophilic addition reactions to form cellulose-ether derivatives.

(2)   (3)   (4)

The Procinyl (ICI) group of reactive disperse dyes were designed specifically for application to nylon. The reactive group, however, is not limited to one particular system and may contain dichloro-s-triazine, epoxy, 2-chloro-3-hydroxyalkyl or aziridine groups. In 1963, the Procilan (ICI) dyes were marketed as the first range of reactive dyes, specifically for application to wool. The Lanasol (CIBA-GEIGY) dyes were subsequently introduced for the coloration of wool. These dyes, containing a α-bromo-acrylamide group, are noted for their bright shades, high reactivity and good wetfastness properties. Similar properties are exhibited by the Drimalan F (Sandoz), Reactolan (CIBA-GEIGY), and Verofix (Bayer) dyes containing the difluorochloropyrimidine group. The Hostalan dyes (Hoechst) were developed to overcome the problems of unlevel dyeing, and are the reaction products of the Remazol (vinyl sulfone type) dyestuffs and N-methyltaurine. Since 1978, polyester–cotton fabrics have been printed successfully by using a combination of a reactive dye containing a phosphonic acid group with a disperse dye containing an alkali-labile ester group. These dye combinations are known as the Procilene PC (ICI) dyes.

A reactive dye molecule may, for convenience, be regarded as a combination of the following units: dye—B—Y—X where the dye is the chromophore (usually an azo, anthraquinone, or phthalocyanine residue); B is a bridging atom or group although this, in many cases, is part of the chromophoric system; Y is the unit carrying the reactive group (the activity of the reactive group depends to a large extent upon the nature of Y); and X is the group that reacts with the fiber. An additional water-solubilizing group, which is not part of the chromophoric unit, may be found as part of the reactive group such as the sulfuric acid ester group of β-hydroxyethyl sulfone reactive dyes.

Chromophoric system. In principle, practically any desired chromophoric system can be combined with reactive groups to produce reactive dyes. The properties of the resulting dye, however, are affected by both of these groups. Proper combination is, therefore, needed to obtain dyes with good qualities such as high tinctorial strength, good solubility, good fastness properties, and economy.

Commonly used chromogens include azo, metallized-azo, anthraquinone, phthalocyanine, and metal-complex formazan derivatives.

The bridging group. The nature of the bridging group between the chromophore and the reactive group not only affects the shade, strength, and affinity of the dye, but can significantly affect its reactivity and the stability of the dye–fiber bond. In addition, the length and flexibility of the bridging group has an affect on the degree of dye–fiber fixation. Therefore, amino and alkylamino groups are generaly used as bridging groups in heterocyclic reactive dyes in view of the ease of synthesis, stability to hydrolysis, and minimum interference with solubility.

The reactive system. The combined unit Y—X in the above formula can be regarded as the reactive system. Generally, the reactive systems used in commercially available reactive dyes can be classified into two groups: reactive systems based on nucleophilic substitution reactions, in which the mobile reactive group X is replaced by an attacking base; and reactive systems based on nucleophilic addition reactions, in which a 1,2 trans addition of a nucleophile occurs across a polarized double bond.

Synthesis

The methods of synthesis of reactive dyes depend largely on the nature of the reactive component. In general, the following methods are used: (1) prepare an azo dye containing a nucleophilic group, eg, —NH$_2$, —OH, and then condense the azo dye with a heterocyclic aromatic reactive system, eg, cyanuric chloride, tetrachloropyrimidine, etc; (2) combine a diazotized aromatic amine with a coupling component containing a reactive system, or a diazonium compound containing a reactive system with a coupling component; (3) condense an aromatic compound containing a reactive system to an anthraquinone derivative, in particular, 1-amino-4-bromoanthraquinone-2-sulfonic acid (bromaminic acid); and (4) for phthalocyanine dyes that do not carry a suitable nucleophilic group for the condensation with the reactive system, it is necessary to introduce a bridging group, eg, a sulfonamide, to act as the nucleophile.

Reactive Dyes For Cellulosic Material

Any group that is capable of forming a covalent bond with alcoholic hydroxyl groups is a potential reactive system to be used in a dye molecule for cellulosic fiber. However, certain criteria must be met in order for this reactive system to be of practical use. The most important criterion is that the dye can be applied in an aqueous medium. Another important criterion is that the covalent bond formed between the dye and the fiber must be sufficiently stable to resist subsequent aftertreatments. In addition, the dye must have a suitable shelf life, and be nontoxic and economical to produce.

The reaction mechanisms involved in the combination of reactive dyes with cellulose can be classified generally as follows: reactive systems based on nucleophilic substitution and/or addition; dyes that react under acidic conditions; dyes containing bifunctional groups; and dyes that react with polyfunctional fixing agents.

Reactive Dyes for Wool

The first reactive dyes used on wool were those that were originally designed for application to cellulose. However, difficulties were encountered with unlevel dyeing properties and hydrolysis of the dye during dyeing. These problems were overcome through the development of reactive groups with little or no tendency to hydrolyze during the dyeing process, the use of special auxiliary products that minimize unlevel dyeing properties, and modification of the dyeing process.

The mechanism of the reaction of wool with reactive dyes is more difficult to interpret than the reactive dye–cellulose systems. Nevertheless, reactive dye–wool systems can be classified in the same general manner as the reactive dye–cellulose systems, namely: reactive systems based on nucleophilic substitution and/or addition; formation of asymmetrical disulfide bonds; and reaction of dyes with modified wool to produce level, fast, bright dyeings.

Reactive Dyes for Nylon

The level dyeing of nylon with high washfastness properties was initially unobtainable with conventional acid and disperse dyes, owing to variations in the physical and chemical characteristics of the fiber. However, the introduction of reactive disperse dyes under the Procinyl (ICI) trade name solved these previous problems. These dyes are stable in water and can be applied as conventional disperse dyes from weakly acid dyebaths. At the conclusion of the dyeing process, the dyebath is made alkaline in order to promote reaction between the dye and the fiber, resulting in a level dyeing with high fastness properties. Chemical combination of the Procinyl dyes with nylon has been demonstrated. It is suggested that reactivity is associated not only with the primary amino end groups of the fiber, but also with amide groups in the polymer chain. The possibility that dimerization or polymerization of the dye itself within the fiber, however, has not been eliminated.

JOHN ELLIOTT
CIBA-GEIGY Corporation

PATRICK P. YEUNG
Parke, Davis & Company

W.F. Beech, *Fibre-Reactive Dyes*, Logos Press, London, 1970.

D. Hildebrand, K-H. Schündehütte, and E. Siegel in K. Venkataraman, ed., *The Chemistry of Synthetic Dyes*, Vol. 6, Academic Press, Inc., New York and London, 1972.

P. Rosenthal, *Rev. Prog. Color. Relat. Top.* **7**, 23 (1976).

R.R. Davis, *Rev. Prog. Color, Relat. Top.* **3**, 73 (1972).

DYES, SENSITIZING

Spectral sensitizing dyes extend the wavelengths of light to which inorganic and organic semiconductors (qv), and chemical reactions can be photosensitized. Spectral sensitizers are needed for the blue, green, and red portions of the visible spectrum as well as for the infrared. The sensitizing dyes can be ionic or nonionic, heterocyclic or nonheterocyclic, and in many cases exhibit more than one type of sensitization reaction. Prime considerations for spectral sensitizing include the range of wavelengths needed for sensitization and the absolute efficiency of the spectrally sensitized process. Because both sensitization wavelength and efficiency are important, optimum sensitizers vary considerably in their structures and properties.

Sensitization Wavelength and Efficiency

The wavelengths for which useful spectral sensitization can be achieved are best illustrated by the spectral sensitivity of commercial photographic plates and films. Spectral sensitivity can be extended beyond the natural ultraviolet and blue-light photosensitivity of silver bromide crystals to include green light, red light, and infrared wavelengths.

Spectral Sensitization of Silver Halides

The efficiency of spectral sensitization in photographic film (silver halide/gelatin) is a function of a dye's electrochemical reduction potential for spectral generation of photoelectron carriers and oxidation potential for spectral generation of photohole carriers. The relationships are diagrammed in Figure 1. Dyes with very negative reduction potentials are difficult to reduce and provide efficient spectral sensitizers where the latent image is formed by photoelectron carriers. Dyes with high positive oxidation potentials are difficult to oxidize and provide efficient spectral sensitizers where the latent image is formed by photohole carriers. In Figure 1, the symbols for the typical sensitizing chromophores (BN—blue sensitizers; GN—green sensitizers; RN—red sensitizers; MN—merocyanine sensitizer; and IRN—infrared sensitizers), some of which appear in Figure 2, are included on the abscissa to indicate the usual electrochemical potential ranges exhibited by these various classes of dyes.

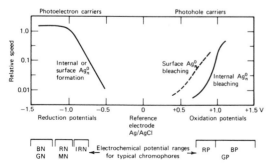

Figure 1. Spectral sensitization of silver bromides.

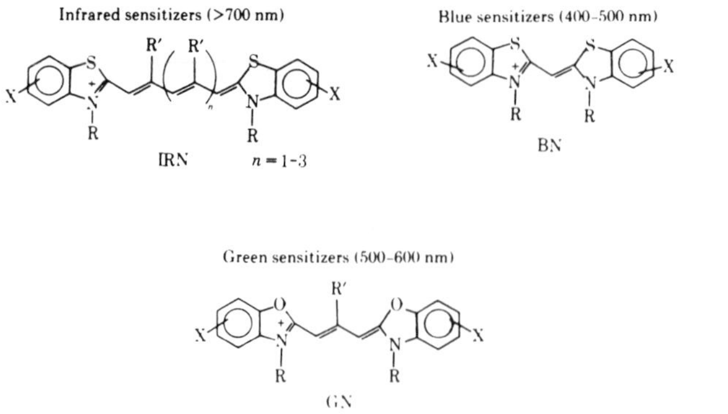

Modification of dye solubility by substituent changes affects migration of spectral sensitizers between layers in a multilayer film and dye removal by photographic processing solutions. These changes are primarily in the substituents attached to heterocyclic nitrogen atoms. Aggregation (self association) of spectral sensitizers into ordered arrays is of prime importance in silver halide films, so that sensitization of a layer in multilayer color films can be selective for one color in the visible region of the spectrum. The tendency of dyes to aggregate in ordered arrays is not only a function of the type of substituent but also the length of the chromophore and the surface of the solid substrate.

Spectral Sensitization of Inorganic and Organic Solids

Many semiconductors other than silver halides exhibit spectrally sensitized photoconductivity. Inorganic semiconductors include selenium, germanium, CdS, HgO, HgI_2, ZnO, PbO, Cu_2O, thallium halides, and TiO_2. Spectrally sensitized photoconduction also has been observed for organic semiconductors: anthracene, poly(N-vinylcarbazole), polyacetylene, copper phenylacetylide, phthalocyanines, and other solid dyes.

Spectral Sensitization of Photoresists, Photopolymers, and Photopolymerization

Photoreactive polymers (qv) may be classed as negative-working or positive-working. Poly(vinyl cinnamate) (PVCN) is a commonly used negative photoresist that is spectrally sensitized to wavelengths longer than its intrinsic absorption (315 nm).

Dye-Sensitized Reactions in Solution: Singlet-Oxygen Generation and Photodynamic Reactions

Metastable singlet oxygen exists in two states $^1\Sigma_g$ and $^1\Delta_g$ which are 159 and 92 kJ (38 and 22 kcal) above the $^3\Sigma_g$ ground state, respectively. Because of the low energies involved, dyes of a wide variety of structures can sensitize the formation of singlet oxygen via the dye triplet state. The xanthene dyes are widely used as efficient sensitizers.

Sensitizing dyes, light, and oxygen can damage and inactivate virtually all classes of organisms through the photooxidation of proteins, polypeptides, individual aminoacids, lipids with allylic hydrogens, tocopherols (see Vitamins), sugars, and cellulosic materials. This general type of photoxidation has been termed photodynamic action.

DAVID STURMER
Eastman Kodak Co.

W. West and P.B. Gilman in T.H. James, ed., *The Theory of the Photographic Process*, 4th ed., Macmillan, New York, 1977, p. 251; P.B. Gilman, *Pure Appl. Chem.* **49**, 357 (1977).

J.L.R. Williams in E. Sélégny, ed., *Polyelectrolytes*, D. Reidel Publishing Co., Dordrecht-Holland, 1974, p. 507.

J.L.R. Williams and co-workers, *Pure Appl. Chem.* **49**, 523 (1977).

D.A. Lightner and C-S. Pak, *J. Org. Chem.* **40**, 2724 (1975); D.A. Lightner, *Photochem. Photobiol.* **26**, 427 (1977).

Infrared sensitizers (>700 nm)

IRN $n = 1–3$

Blue sensitizers (400–500 nm)

BN

Green sensitizers (500–600 nm)

GN

Figure 2. Spectral sensitizers.

ECONOMIC EVALUATION

Economic evaluation is the financial evaluation of business ventures. The increased profit accruing to each venture is estimated and systematically compared to the required capital investment.

Definitions

Profit before taxes equals net sales less production costs and costs of marketing, research, and administration. Net sales equals revenue received from customers less the producer's costs for transportation and terminal use. For chemical production ventures, which involve the chemical conversion of raw materials to products, production cost is the sum of the costs of raw materials, labor, utilities, maintenance, plant overhead, and depreciation. Production costs can be classified into variable costs, which on an annual basis are directly proportional to the quantity of product made; and fixed costs, which on an annual basis are independent of the quantity of product made. Depreciation is determined as a fraction of fixed capital. Return on investment is equal to net profit divided by total fixed and working capital expressed as a percentage. Cash flow in a given year equals net profit plus depreciation minus new fixed and working capital invested in that year.

Fixed capital is that required to pay for all tangible facilities. It is the sum of process and offsites capital. Process capital is the sum of battery limits and ancillaries capital. Battery limits refers to process equipment, instruments, and piping within the immediate production area; ancillaries refers to facilities such as storage tanks that are in direct support of a specific process unit though located elsewhere in the plant. Offsites refers to general support facilities, eg, office buildings, storehouses, maintenance shops, waste-disposal facilities, land, and utilities such as steam, electrical, and cooling-water systems. Working capital refers to the liquid assets required to meet continuing operating costs, eg, payroll and utilities.

Capacities Economics

The first step in an economic evaluation is to estimate the return on investment for the venture at full capacity operation. Return on investment should be greater than some specified value, eg, 15% per year. As an example, consider a proposal to produce 135,000 t/yr of a product with a net sales return of \$0.66/kg. Ventures A, B, and C are proposed, as shown in Table 1. Venture C must be rejected, but A and B merit further consideration.

Year-By-Year Economics

Ventures that show satisfactory return on investment at capacity should be further evaluated. Net sales, production costs, profit, fixed capital, and working capital are calculated for each year of the life of the venture, commencing at the time the initial expenditure of capital is made. Most chemical production ventures will be evaluated over a period of at least ten years. Results are tabulated as in Table 2. It is important

Table 1. Capacity Economics for Three Ventures

Factor	A	B	C
capacity, 1000 t/yr	135	135	135
fixed capital, millions of $	100	50	200
net sales return, ¢/kg	66.0	66.0	66.0
costs, ¢/kg			
utilities	2.9	7.3	8.8
raw materials	17.6	34.1	13.2
other variable costs	1.1	1.1	11.0
depreciation	7.3	3.7	14.7
other fixed costs	5.5	2.6	1.1
Total production cost	34.3	48.8	48.8
plant profit, ¢/kg	31.7	17.2	17.2
sales, administrative, and research, ¢/kg	3.3	3.3	3.3
profit before federal income tax, ¢/kg	28.4	13.9	13.9
profit after 48% federal income tax, ¢/kg	14.7	7.3	7.3
return after taxes on original fixed capital, %/yr	20	20	5

Table 2. Cash Flow Table of Year-by-Year Economic Data, in Thousands of Dollars

Year	Added net sales	Added profit after taxes	Added depreciation	New fixed capital	New working capital	Cash flow	Cumulative cash flow
1		(260)		9,000		(9,260)	(9,260)
2	5,000	310	600	6,000	200	(5,290)	(14,550)
3	10,000	2,050	1,000		400	2,650	(11,900)
4	14,000	3,300	1,000		600	3,700	(8,200)
5	15,000	3,200	1,000		300	3,900	(4,300)
6	15,000	3,200	1,000			4,200	(100)
7	15,000	3,200	1,000			4,200	4,100
8	14,000	2,700	1,000			3,700	7,800
9	14,000	2,700	1,000			3,700	11,500
10	14,000	2,700	1,000			3,700	15,200
11	13,000	2,200	1,000			3,200	18,400
12	13,000	2,200	1,000			3,200	21,600
13	13,000	2,200	1,000			3,200	24,800
14	12,000	1,700	1,000			2,700	27,500
15	12,000	1,700	1,000			2,700	30,200
16	12,000	1,700	1,000			2,700	32,900
Total	191,000	34,800	14,600	15,000	1,500	32,900	

that the net sales from the venture be calculated as the increase in sales of new products less any decrease in sales of existing products.

Profitability

Payout time is the number of years required for cumulative cash flow to equal total fixed and working capital. In Table 2, payout time is six years. This might be acceptable, though five years or less would be better. Return on investment can be calculated for each year of the venture. In the sixth year of the venture described in Table 2, return on investment is 19%/yr. Discounted cash flow (DCF), interest rate and net present worth can be calculated from the last column in Table 2, which correlates cumulative cash flow with time. For the data shown, the DCF interest rate is 15%/yr over the life of the investment. This is satisfactory because it is equal to the assumed minimum acceptable ("hurdle") rate, 15%/yr. Correspondingly, net present worth is zero at an assumed hurdle rate of 15%/yr. This value of net present worth is satisfactory because it is not negative.

Comparison of Ventures

When comparing mutually exclusive ventures with one another, the calendar time span representing the life of the venture should be the same for all ventures being considered. Each should be evaluated separately.

Estimating Working Capital

For large plants, working capital can be approximated as 10–20% of fixed capital. For small plants, the percentage can be considerably higher, in fact up to 100%. Another rapid estimate of working capital is 25–35% of net sales. Working capital can be best estimated as the sum of cash on hand (5% of net sales), inventory of raw materials and products (typically 2–4 weeks), and accounts receivable less accounts payable.

OLIVER AXTELL
J.M. ROBERTSON
Celanese Chemical Company

F.C. Jelen, *Hydrocarbon Process.*, 161 (Sept. 1974 and Oct. 1974).

F.A. Holland and co-workers, *Chem. Eng.*, 72 (Apr. 1, 1974).

O.P. Kharabanda, *Process Plant and Equipment Cost Estimation*, Sevak Publications, Bombay, India, 1977.

M.S. Peters and K.D. Timmerhaus, *Plant Design and Economics for Chemical Engineers*, 2nd ed., McGraw-Hill Book Company, New York, 1968.

EGGS

The (hen's) egg is an important part of the human food supply. It is highly nutritious, contributing proteins, fats, vitamins, and minerals. Because of its unique composition, the egg has extremely useful functional properties when used by itself or as an ingredient in other foods. There are also nonfood uses for eggs and components of eggs.

Eggs are marketed in the shell and also as egg products in three different forms: liquid, frozen, and dried. Eggs are produced and utilized throughout the world. There has been a trend toward reduced per capita consumption, and an increased usage of egg products.

Chemical Properties

The egg can be divided into three basic parts: ca 31% yolk, ca 58% white, and ca 11% shell, which is 94% $CaCO_3$ and 1% each of $MgCO_3$ and $Ca_3(PO_4)_2$. The mixture of white and yolk is called whole egg. Gross chemical composition of the egg contents is shown in Table 1. The pH of egg white is normally about 9.0. However, egg white in freshly laid eggs has a pH of about 7.6 pH increases as carbon dioxide is lost during storage. The pH of yolk is normally about 6.6, and in freshly laid eggs it is 6.0.

Functional Properties

Eggs function in many different ways to give food products in which they are used certain desirable characteristics. Functional properties are as follows: coagulation and thickening ability, foam-forming, emulsifying power, tenderizing, moisture retention, flavor, and color.

Production and Marketing of Shell Eggs

Today, most eggs are produced from flocks of 30,000 birds or more. Eggs are graded and sorted according to size and quality factors, which include both shell quality and interior quality. Eggs are flash-candled on a continuous conveyor within a short time after being laid. Because of their freshness, most of these eggs have uniformly high interior quality with the proportion of thick to thin egg white being relatively high. The USDA has an egg-grading program which is run on a voluntary basis in cooperation with each state. Most states also have their own egg-grading laws. Following are United States size or weight classifications for eggs:

Size or weight class	Minimum net weight per dozen, kg
jumbo	0.85
extra large	0.77
large	0.68
medium	0.60
small	0.51
pee wee	0.43

Eggs are downgraded according to the following conditions of the shell:

Dirty. The shell is unbroken and has adhering dirt or foreign material, prominent stains, or moderate stains covering more than 1/4 of the shell surface.

Check. The individual egg has a broken shell or a crack in the shell, but with its shell membrane intact and its contents do not leak.

Leakers. The shell and membrane are broken so that the contents are leaking. USDA regulations prohibit use of this type of egg for human consumption.

Grade AA and A quality eggs are generally recommended for most household uses, such as frying, poaching, and cooking in the shell. However, lower grades can usually be used for many cooking and baking purposes. Higher-grade eggs have somewhat better functional properties such as foam-forming power.

The present handling and processing methods for shell eggs is now highly mechanized. This includes the steps of collecting, sorting, washing, drying, candling, grading, classifying by weight, and packing. The large end of the egg is generally spray-coated with mineral oil in order to retard escape of carbon dioxide and water, and thus retain quality for a longer period.

Egg Products

There are three basic types of egg products: liquid, frozen, and dried. All of these have many uses, eg, in baked goods, bakery mixes, noodles, mayonnaise and salad dressing, candies, and ice cream. Tremendous changes have taken place in the egg-products industry through the past 20–30 years, including those in government regulations, customer specifications, nutritional labeling, and waste disposal and loss control.

Egg breaking. Egg breaking is the beginning step in the production of all types of egg products. At present, almost all eggs are broken and separated by machine. Eggs are loaded onto the system, washed, rinsed, sanitized, flash-candled, and then fed to an egg breaking and separating machine. This machine cracks and opens the shell and separates yolks from whites. Each system can handle 18,000–25,000 eggs per hour. Three persons are required to operate this type of system: one to load; one to flash-candle eggs coming out of the washer; one to operate the egg-breaking and separating machine. Three streams flow from the egg-breaking and separating machine: egg white, yolk, and mix. Each component is strained or clarified and thoroughly mixed before going to further processing.

Pasteurization. Except for egg white to be dried, it is necessary to pasteurize all liquid eggs whether they are for liquid usage or are to be frozen or dried. These pasteurization conditions are based on a bacterial kill equivalent to that obtained when heating whole egg liquid to 60°C for 3 1/2 min. This has been established as effective for destruction of salmonella.

Liquid egg products. The most common liquid egg products are white, yolk, and whole egg. Liquid eggs are generally consumed by large users, such as bakeries, who have the necessary handling equipment. They are transported in tank trucks holding approximately 20 metric tons. Temperature of egg yolk and whole egg must be held below 5°C and egg white must be below 7°C.

Frozen egg products. The most common frozen egg products are frozen egg white, frozen whole egg, frozen fortified whole, frozen plain egg yolk, frozen fortified whole egg with syrup, frozen sugared egg yolk, frozen salted egg yolk, and frozen salted whole egg.

Dried egg products. Dried egg products are classified as follows: dried egg white, dried plain whole egg and yolk, dried blends of whole egg and yolk with carbohydrate, and special dried egg products (eg, scrambled egg mix).

Other special egg products include simulated egg products with low cholesterol.

Dwight H. Bergquist
Henningsen Foods, Inc.

Table 1. Chemical Composition of Egg Components, wt%

Component	Egg white, %	Egg yolk, %	Whole egg, %
water	88.1	48.8	74.6
solids	11.9	51.2	25.4
protein	10.1[a]	16.4	12.1
lipids	trace	32.9[b]	11.2
carbohydrates (both bound and free)	1.1	1.0	1.1
free glucose	0.4	0.2	0.3
ash	0.6	1.7	1.0

[a] Principal proteins are ovalbumin (54 wt%), conalbumin (13 wt%), and ovomucoid (11 wt%).

[b] Lipids are glycerides (65.5 wt%), phospholipids (29.3 wt%), and cholesterol (5.2 wt%). Phospholipids are lecithin (73 wt%), cephalin (15 wt%), and others (12 wt%).

W.D. Powrie, "Chemistry of Eggs and Egg Products" in W.J. Stadelman and O.J. Cotterill, eds., *Egg Science and Technology*, Avi Publishing Co., Inc., Westport, Conn., 1977, pp. 65–91.

Regulations governing the grading of shell eggs and United States Standards, grades, and weight classes for shell eggs, *USDA 7 CFR*, Part 56, U.S. Government Printing Office, Washington, D.C., July 1, 1971.

H.H. Palmer, "Eggs" in P.C. Paul and H.H. Palmer, eds., *Food Theory and Applications*, John Wiley & Sons, Inc., New York, 1972, pp. 527–561.

Egg Pasteurization Manual, USDA ARS74-48, U.S. Government Printing Office, Washington, D.C., March 1969.

ELASTOMERS, SYNTHETIC

SURVEY

Polymer chemists can, to a considerable degree, now control the average chain length, distribution of chain lengths and isomeric unit structures in the polydiene macromolecule. Furthermore, the chemical and stereochemical structural requirements to provide elastomeric behavior over a specified temperature interval can now be predicted. Many elastomers have now been developed from monomers other than dienes.

A basic requirement for a rubbery material is that it must consist of long flexible chainlike molecules that can undergo rapid rotation as a result of thermal agitation. A further requirement is that the long linear-chain molecules must be cross-linked by a few intermolecular bonds during processing to form an insoluble three-dimensional network.

The theories of rubber elasticity relate the change in molecular dimensions or chain vector lengths resulting from a deformation to the macroscopic strain. The work of deformation per unit volume (W) for the most general type of strain is given below.

$$W = \tfrac{1}{2}NkT\left(\lambda_1^2 + \lambda_2^2 + \lambda_3^2 - 3\right)$$

where λ_1, λ_2, and λ_3 are the three principal extension ratios of a cube of rubber, N is the number of network chains per unit volume, k is Boltzman's constant, and T is absolute temperature.

The larger-volume rubbers are often referred to as the general-purpose (GP) rubbers. Similarly, the lower-volume, higher-priced systems are termed specialty rubbers (eg, silicones, fluoroelastomers, polyacrylates, epichlorhydrin, chlorosulfonated polyethylene, chlorinated polyethylene, ethylene–acrylic, and propylene oxide). The accepted abbreviations and chemical structures for CP rubbers are shown in Table 1.

JAMES F. McGRATH
Virginia Polytechnic Institute and State University

H. Staudinger, *From Organic Chemistry to Macromolecules*, John Wiley & Sons, Inc., New York, 1970.

M. Morton, "Advantages in Synthetic Rubber," in E.B. Mano, ed., *Proceedings of the International Symposium on Macromolecules*, Elsevier, Amsterdam, 1975, p. 287.

W.M. Saltman, ed., *The Stereo Rubbers*, Wiley-Interscience, New York, 1977.

ACRYLIC ELASTOMERS

Acrylic elastomers are rubbery polymers and copolymers in which esters of acrylic acid such as ethyl acrylate and butyl acrylate constitute a large portion. They exhibit heat and oil resistance which generally places them between nitrile rubber and silicone rubber in the spectrum of specialty elastomers. They make up a large part of the market in various automotive applications such as seals and gaskets because of their ability to retain properties in the presence of hot oils and other fluids, and to resist cracking or softening and other effects of exposure to heat and air. A closely related class of elastomers are the ethylene–acrylic elastomers, copolymers of ethylene and methyl acrylate.

Acrylic elastomers, as noted above, are polymers consisting largely of one or more esters of acrylic acid. The resulting structure of a saturated backbone with pendant ester groups is characteristic of all acrylic elastomers and accounts for their two most important properties. These are resistance to aging in air at moderately elevated temperatures up to about 200°C, and resistance to swelling, hardening, and other changes that take place in hot oils. The saturated backbone also makes them completely resistant to ozone, an important factor in weathering. Polymers of ethyl acrylate and copolymers of ethyl, butyl, and methoxyethyl acrylates provide products that remain flexible on cooling to as low as −40°C.

Table 1 gives the typical vulcanizate properties of acrylic elastomers. The most widely used vulcanization system at the present time consists of a metal carboxylate (soap) and sulfur. Inclusion of a low concentration of an active chlorine-containing monomer in the polymerization recipe typically provides the reactive site for cross-linking.

Manufacture

Esters of acrylic acid, the raw material for acrylic elastomers, can be manufactured by a variety of methods (see Acrylic acid and derivatives). Acrylate esters polymerize readily, and for various purposes all the usual polymerization methods, ie, bulk, solution, emulsion, and suspension, have been used. For elastomers, only emulsion and suspension polymerization are important.

Acrylic elastomers are processed in the same manner as the other elastomers. A Banbury mixer is used on a large scale and mill mixing on a smaller scale or for occasional batches. Parts can be shaped by compression, transfer, or injection molding as well as extrusion and calendering. The distinctive compounding feature of acrylic elastomers is

Table 1. Nomenclature and Structure of General Purpose Synthetic Rubbers

Name	Chemical structure
natural rubber (NR); ca 6% nonhydrocarbon content	> 99% *cis*-1,4-polyisoprene $-\!(CH_2C\!=\!CHCH_2)_n$ with CH_3 pendant
isoprene rubber (IR); Ziegler-Natta catalysts	> 97% *cis*-1,4-polyisoprene
butadiene rubber (BR); Ziegler-Natta catalysis, anionic polymerization, emulsion polymerization	(a) > 97% *cis*-1,4-polybutadiene $-\!(CH_2CH\!=\!CHCH_2)_n$ (b) ca 90% 1,4 with mixed cis/trans (c) ca 80% 1,4 with mixed cis/trans ca 20% 1,2 with mixed cis/trans
chloroprene rubber (CR); emulsion polymerization	ca 90% *trans*-1,4 $-\!(CH_2C\!=\!CHCH_2)_n$ with Cl pendant
isobutylene–isoprene (IIR) rubber (butyl); cationic polymerization	$-\!(CH_2C\!)_m(CH_2C\!=\!CHCH_2)_n$ with CH_3, CH_3 pendant; ca 98% ca 2%
nitrile–butadiene rubber (NBR); emulsion random copolymerization	$-\!(CH_2CH\!=\!CHCH_2)_m(CHCH_2)_n$ with $C\!\equiv\!N$ pendant
styrene–butadiene rubber (SBR); emulsion random copolymerization also some anionic random copolymerization	$-\!(CH_2CH\!=\!CHCH_2)_m(CH_2CH)_n$ with phenyl pendant
ethylene–propylene copolymer (EPM); Ziegler-Natta catalysis, random copolymerization	$-\!(CH_2CH_2)_l(CH_2CH)_n$ with CH_3 pendant
ethylene–propylene–diene terpolymer; Ziegler-Natta catalysis, random copolymerization, ethylidene norbornene also used	$-\!(CH_2CH_2)_l[CH_2CH]_m(CH_2CH)_n$ with CH_3 and side chain CH_2–$CH\!=\!CH$–CH_3 pendant

Table 1. Typical Properties of Acrylic Elastomers

Property	Regular	Low temperature
100% modulus, MPa[a]	10.3	8.3
tensile strength, MPa[a]	15.2	10.3
elongation, %	215	150
hardness, durometer A	80	75
compression set, %		
70 h at 150°C	28	35

[a] To convert from MPa to psi, multiply by 145.

Table 2. Typical Properties of an Ethylene–Acrylic Elastomer, Press Cure 20 Min at 177°C

Property	Value
100% modulus, MPa[a]	2.8
200% modulus, MPa[a]	6.4
tensile strength, MPa[a]	14.1
elongation, %	450
hardness, durometer A	63

[a] To convert MPa to psi, multiply by 145.

their vulcanization chemistry. For most applications, they employ carbon black fillers. Mineral fillers (qv) are used to some extent, although acidic fillers are to be avoided because of their retarding effect on the vulcanization.

Although it is true that acrylic elastomers are used in many different applications, the largest consumer is the automobile industry. As a result, the annual sales figures follow new-car sales rather closely.

Ethylene–Acrylic Elastomers

The new ethylene–acrylic elastomer, Vamac, is a copolymer primarily of methyl acrylate and ethylene. Like the traditional acrylic elastomers, the structure provides a fully saturated backbone which gives excellent resistance to aging, especially to ozone attack. Typical properties of a commercial ethylene–acrylic elastomer are shown in Table 2.

T.M. VIAL
American Cyanamid Company

Product bulletins on Cyanacryl acrylic elastomer from American Cyanamid Company, Wayne, N.J.; Hycar acrylic elastomer from B.F. Goodrich Company; Vamac ethylene–acrylic elastomer from E.I. du Pont de Nemours & Co., Inc.

T.M. Vial, *Rubber Chem. Technol.* **44**, 344 (1971).

C.H. Fisher, G.S. Whitby, and E.M. Beavers in G.S. Whitby, ed., *Synthetic Rubber*, John Wiley & Sons, Inc., New York, 1954, pp. 900–910.

BUTYL RUBBER

The physical properties of butyl-type polymers responsible for their commercial success as elastomers are low glass transition temperature, T_g ca $-70°C$, high level of impermeability to common gases including water vapor over a broad temperature range, and high hysteresis over a useful temperature range. No other elastomer displays this combination of characteristics.

The isoprene in butyl rubber is linked predominantly by 1,4 addition at a level from ca 0.5 to 2.5 mol per 100 mol of monomers, depending on grade; the residual olefin is in approximately 50:50 cis:trans ratio. The substantially saturated hydrocarbon nature of the polymer endows it with good chemical resistance, and good resistance to oxidative deterioration and to cracking by ozone. The presence of hydrogen on carbons α to the double bond permits vulcanization by sulfur and accelerators. In halogenated butyl rubber, allylic halogen is present and the double bond initially present in butyl rubber has been shifted.

Virtually all of the world's production of butyl rubber is made by a precipitation (slurry) polymerization process in which isobutylene and a minor amount of isoprene are copolymerized in methyl chloride diluent at temperatures of -100 to $-90°C$ using aluminum chloride as catalyst. The polymer precipitates as it is formed as a finely divided, milky slurry.

Table 1. Exposure Limits[a] for Compounds Used in Butyl Rubber Manufacture

Compound	Exposure limit, ppm (vol)	Details
methyl chloride	TWA[b] = 100	OSHA
	max peak exposure = 300	for 5 min in a 3 h period
hexane	TWA = 500	OSHA
	TWA = 100	ACGIH (1976)
	STEL[c] = 125	ACGIH (1976)
	TWA = 100	with 15-min ceiling conc of 510 ppm (vol), NIOSH, 1977
chlorine	TWA = 1	OSHA
	STEL[c] = 3	ACGIH
bromine	TWA = 0.1	OSHA
	STEL[c] = 0.3	ACGIH

[a] As of April 1, 1978.
[b] Time-weighted average calculated for 8-h time period of exposure. Normal exposure based on 8-h shift, 40-h week.
[c] Short term exposure limit, ceiling that should not be exceeded during a 15 min exposure period with no more than four such exposures per day permitted.

Halogenated butyl rubbers are produced commercially by dissolving butyl rubber in hydrocarbon solvent and contacting the solution with gaseous or liquid elemental halogens.

Eighty-five percent of all butyl-type elastomers is used in the tire industry. The other 15% is used in mechanical goods, electrical applications, and miscellaneous applications.

Table 1 lists the chemicals employed in the production of butyl rubber and halogenated butyl rubber that have personnel exposure limits established by OSHA or recommended by other groups.

Derivatives

There are two other basic modifications of butyl rubber manufactured in low volume. A cross-linked terpolymer of isobutylene, isoprene, and divinylbenzene containing some unreacted substituted vinylbenzene appendages is available in two grades of differing degrees of cross-linking. It is employed primarily in the manufacture of sealant tapes and caulking compounds (see Sealants). The other modification is a low-molecular weight "liquid" butyl rubber manufactured by controlled molecular fragmentation of a standard butyl rubber and used in sealants, caulks, potting compounds (see Embedding), and coatings.

F.P. BALDWIN
R.H. SCHATZ
Exxon Chemical Co.

D.J. Buckley, *Rubber Chem. Technol.* **32**, 1475 (1959).

A. van Tongerloo and R. Vukov, *Proc. Int. Rub. Conf.*, Milan, Italy, 1979, p. 70.

R.A. Labine, *Chem. Eng.* **66**(24), 60 (1959).

CHLOROSULFONATED POLYETHYLENE

Chlorosulfonated polyethylene is the term used to represent a group of synthetic elastomers derived from the reaction of a mixture of chlorine and sulfur dioxide on any of the various plastic polyethylenes (see Olefin polymers). The product of this reaction is a chemically modified form of the original polyethylene, and may contain 20–40% chlorine (mainly as secondary alkyl chlorides, $RR'CHCl$) and ca 1–2% sulfur, present mostly as secondary sulfonyl chloride groups ($RR'CHSO_2Cl$). Chlorosulfonated polyethylene is thus a saturated chlorohydrocarbon elastomer having sulfonyl chloride functions that are used as cross-linking or curing sites.

Chlorosulfonated polyethylenes (suitably compounded) provide elastomeric vulcanizates resistant to ozone and oxygen attack even under sunlight or ultraviolet radiation. They furthermore resist deterioration by heat, chemicals, and solvents. The largest uses are in the areas of wire covering, hose, tubing, and sheet goods.

Chlorosulfonated polyethylene, as sold by its manufacturer, is a raw synthetic rubber. To convert it to useful articles, other manufacturers must mix, or compound the polymer with selected fillers (qv), processing aids, stabilizing chemicals, and cross-linking agents or catalysts. The compound must then be shaped or molded and finally vulcanized. The properties of the finished product depend on the exact type and amount of chlorosulfonated polyethylene used, as well as the kind and amount of other agents.

The dynamic shear modulus of a 28% chlorine, 1.24% S, chlorosulfonated free-radical-based polyethylene ranges from 7 MPa to 2.1 GPa (1,000–300,000 psi).

The working properties of chlorosulfonated polyethylenes are a function of the starting polyethylene and the chlorine content. As chlorine content increases from 0 to 25%, the crystallizing tendency of the parent polymer decreases. At a chlorine content of ca 30% for free-radical-based polyethylene, and ca 35% for linear polyethylene, the product possesses minimum stiffness. As the chlorine content is further increased, stiffness increases again and the polymers become more and more glassy in character, as well as more soluble in common solvents.

The chemical properties of chlorosulfonated polyethylene can be reliably predicted from the types of chemical functions present and the known chemistry of these functions in low molecular weight substances.

The principal commercial producer of chlorosulfonated polyethylenes is the DuPont Company which markets a line of nine products under the name Hypalon synthetic rubber. Commercial chlorosulfonated polyethylenes are available under specifications as to chlorine and sulfur content, Mooney viscosity, and general physical appearance.

Health, Safety, and Environmental Factors

The process for manufacturing chlorosulfonated polyethylene involves the well-recognized hazards of using chlorine, sulfur dioxide, and carbon tetrachloride. Safe handling procedures and toxicity limits for Cl_2 and SO_2 are well documented and their irritation potential provides significant warning if exposure occurs. Carbon tetrachloride exposure is more insidious and constant monitoring is necessary to maintain work-area concentrations below the OSHA promulgated limit of 10 ppm in air.

The manufacturing waste products, ie, excess SO_2 and HCl, are recycled and neutralized, respectively, with minimal environmental effect. Extensive efforts are necessary, however, to prevent significant loss of carbon tetrachloride vapors to the atmosphere during the polymer isolation step.

PAUL R. JOHNSON
E.I. du Pont de Nemours & Co., Inc.

Hypalon, Synthetic Rubber Bulletin HP-210.1 (E-12336), E.I. du Pont de Nemours & Co., Inc., Wilmington, Del., 1976.

R.M. Straub, *General Compounding of Hypalon, Synthetic Rubber Bulletin A62957*, E.I. du Pont de Nemours & Co., Inc., Wilmington, Del. 1969.

ETHYLENE–PROPYLENE RUBBER

Ethylene–propylene rubber was first introduced in the United States in limited commercial quantities in 1962, and is now the fastest-growing elastomer. It is rapidly becoming a general-purpose elastomer. The terpolymer modification, EPDM, includes, in addition to ethylene (E) and propylene (P), a small amount of a third monomer, a diene (D), to permit conventional sulfur vulcanization at pendant sites of unsaturation. Minor amounts of ethylene–propylene copolymer without diene unsaturation, called EPM, are also produced (see Copolymers). Both EPDM and EPM show outstanding resistance to heat, light, oxygen, and ozone because one double bond is lost when the diene enters the polymer and the remaining double bond is not in the polymer backbone but external to it.

The properties of a typical raw EPM are shown in Table 1; they are essentially the same for EPDM.

The two principal raw materials for EPM and EPDM, ethylene (qv) and propylene (qv), both gases, are abundantly available at high purity. The manufacturing processes of EPDM are proprietary. A continuous solution process has been patented.

For curing EPDM, common rubber accelerators are used that are found in recipes for other synthetic rubbers. The EPDM compounds can

Table 1. Properties of Raw Ethylene–Propylene Copolymers

Property	Value
specific gravity	0.86–0.87
appearance	colorless
Mooney viscosity	varied
heat capacity, kJ/(kg·K)[a]	2.18
thermal conductivity, mW/(m·K)	355.2
brittle point, °C	−95
glass-transition temperature, °C	−60

[a] To convert J to cal, divide by 4.184.

be successfully and economically mixed and processed on the machinery commonly found in various rubber-fabricating plants. In general, EPDM compounds extrude readily on all commercial extruders.

EPDM compounds are used in wire applications, automotive profile extrusions, membrane applications, oil viscosity index improving, and in blends with other rubbers for many general purpose uses.

SAMUEL E. HORNE, JR.
Polysar, Inc.

Reviewed by
E.L. BORG
Uniroyal Chemical Co.

U.S. Pat. 3,341,503 (Sept. 12, 1967), J.L. Paige and S.M. DiPalmar (to Uniroyal, Inc.).

R.J. Kelly, *Ind. Eng. Chem. Prod. Res. Dev.* **1**, 210 (1962).

F.P. Baldwin and G. Ver Strate, *Rubber Chem. Technol.* **15**(3), 709 (1972).

FLUORINATED ELASTOMERS

Fluorinated elastomers are a class of synthetic elastomers that are designed for demanding service applications in environments where combinations of extreme temperature ranges, chemicals, fluids, and/or fuels exist. The three basic fluorinated elastomer types are: fluorocarbons, fluorosilicones, and fluoroalkoxyphosphazenes (see Fluorine compounds, organic). Table 1 lists the general physical properties of the main classes of fluoroelastomers. A list of the principal commercial fluorocarbon elastomers is given in Table 2 (p. 394).

Military interest in the development of fuel- and heat-resistant elastomers for low temperature service initiated the development of fluorocarbon elastomers. Independent of the development of fluorocarbon elastomers for military applications, fluorosilicone elastomers were developed in the late 1950s at Dow Corning. They impart improved fluid resistance through the incorporation of fluorine, and they maintain the low temperature flexibility of the silicone elastomers. The polyfluoroalkoxyphosphazenes are a recent commercial addition to the fluorinated elastomer family. These elastomers are vulcanizable (by

Table 1. Fluorine-Containing Elastomers

Type	Temperature use range, °C	Fluorine content, %	Valuable characteristics
fluorocarbon	−46 to 316	53–70	low compression set, high temperature stability, fuel oil, chemical resistance Specific gravity, 1.80–1.86
fluorosilicone[a]	−54 to 232	30–40	low temperature flexibility, softness specific gravity, 1.35–1.65
fluoroalkoxyphosphazenes[b]	−54 to 232	30–40	low temperature flexibility, fuel oil, chemical resistance

[a] Produced by Dow Corning (Silastic) and GE (FSE).
[b] Produced by Firestone (PNF).

Table 2. Commercial Fluorocarbon Elastomers

Copolymer	Trade names	Suppliers
poly(vinylidene fluoride-co-hexafluoropropylene)	Fluorel	3M
	Viton A	DuPont
	Tecnoflon	Montedison
	Dai-El	Daikin
poly(vinylidene fluoride-co-hexafluoropropene-co-tetrafluoroethylene)	Viton B	DuPont
	Dai-El G-501	Daikin
poly(vinylidene fluoride-co-hexafluoropropylene-co-tetrafluoroethylene) plus cure site monomer	Viton G (peroxide curable)	DuPont
poly(vinylidene fluoride-co-tetrafluoroethylene-co-perfluoromethyl vinyl ethyl) plus cure site monomer	Viton GLT (peroxide curable)	DuPont
poly(tetrafluoroethylene-co-perfluoromethyl vinyl ether) plus cure site monomer	Kalrez	DuPont
poly(tetrafluoroethylene-co-propylene)	Aflas 100, 150	Asahi
poly(vinylidene fluoride-co-chlorotrifluoro-ethylene) poly(vinylidene fluoride-co-1-hydropentafluoropropylene)	Kel-F 3700	3M
	Tecnoflon SL	Montedison
poly(vinylidene fluoride-co-1-hydropentafluoropropylene-co-tetrafluoro-ethylene)	Tecnoflon T	Montedison

peroxide via incorporation of a cure-site monomer) and offer a unique set of low temperature properties similar to those of the fluorosilicones; however, they exhibit improved oil and solvent-resistant properties from -55 to $175°C$.

Manufacture and Processing

Fluorocarbon elastomers have been prepared by high pressure, free-radical aqueous emulsion-polymerization techniques. The initiators (qv) can be organic or inorganic peroxy compounds such as ammonium persulfate. The emulsifying agent is usually a fluorinated carboxylic acid soap, and the temperature and pressure of polymerization ranges from $80–125°C$ and 2.2 to 10.4 MPa (300–1500 psig). The molecular weights of the resultant polymers are controlled by the ratios of initiator to monomer or choice of chain-transfer reagents, or both. Typical chain-transfer reagents that can be employed are carbon tetrachloride, methanol, acetone, diethyl malonate, and dodecyl mercaptans.

Like the fluorocarbon elastomers, fluorosilicones and fluoroalkoxy-phosphazene elastomers are available in a range of viscosities, compounded masterbatches, and physical properties.

Processing of fluorocarbon elastomers. When one has a firm idea of use requirements and rubber response to specific additives, a formulation for the elastomer may be selected. Uses generally fall into one of two classes: O-rings or molded goods.

Fluoroelastomers can provide environmental resistance without sacrificing processability. The most workable formulations are compounded with raw gums containing the phenolic or incorporated-cure systems. These rubbers offer the best starting point for maximum processability.

Fluoroelastomer formulations are processible by any standard technique. Open-mill mixing is frequently used since most commercial gums mix well.

Extrusion preforming is easily accomplished if relatively cool barrier temperatures are used with either a screw or piston-type extruder (Barwell). Compression molding is generally used when it is desirable to conserve material and when a molding operation is set up to allow preparation of large numbers of preforms with minimum labor costs (see Plastics processing).

Post-curing at elevated temperatures develops maximum tensile strength and compression-set resistance in fluoroelastomers. General post-cure conditions are 16–24 h at 233–260°C.

Health and Safety Factors

In general, under normal handling conditions, the fluoroelastomers have been found to be low in toxicity and irritation potential. The specific toxicological, health, and safe-handling procedures are provided by the manufacturer of each fluoroelastomer product upon request (see also Fluorine compounds, organic).

ARTHUR C. WEST
ALLEN G. HOLCOMB
3M Company

"Fluoropolymers" in K.J.L. Paciorek and L.A. Wall, eds., *High Polymers*, Vol. XXV, Wiley-Interscience, New York, 1972, pp. 291–313.

"Fluoroelastomers" in D.A. Stivers, ed., *The Vanderbilt Rubber Handbook*, R.T. Vanderbilt Co., New York, 1978, p. 244.

Rubber Technology, M. Morton, ed., Van Nostrand Reinhold Co., New York, 1973, pp. 407–439.

NEOPRENE

The broad range of physical and chemical properties available in the family of chloroprene homo- and copolymers permits neoprenes to fulfill the requirements of many applications. This versatility arises from the chemistry of free-radical emulsion polymerization of chloroprene. All neoprene polymers are now prepared by free-radical emulsion polymerization.

Neoprene, as sold by its manufacturers, is a raw synthetic rubber. To convert it into useful objects it must be mixed or compounded with selected chemicals, fillers, and processing aids. The resulting compound is then shaped, or molded, and finally vulcanized. The properties of the finished product depend on the exact type and amount of neoprene and the compounding ingredients. Hence, the physical and chemical properties are closely related to the compounding recipe (see Rubber compounding).

The basic physical properties of representative raw polymer, gum vulcanizate, and carbon black-filled vulcanizate are given in Table 1.

Analysis of crystalline behavior and infrared absorption spectra showed that polychloroprenes consist of linear sequences of predominantly *trans*-2-chloro-2-butenylene units. Small moieties of the structures formed by 1,2-, 3,4-, and cis 1,4-polymerizations are also present. The amounts of these polymerizations present increase regularly with increasing polymerization temperature from a total of 5% at $-40°C$ to about 30% at 100°C.

Properly compounded polychloroprene vulcanizates have outstanding resistance to natural and accelerated aging if protected by an antioxidant. Volatile degradation products liberated as temperature rises from 100–500°C include ethylene, a trace of chloroprene, and hydrogen chloride equivalent to at least 90% of the chlorine in the polymer. The degradation mechanism involves random, intramolecular evolution of HCl, leaving behind a predominance of triene structures in the polymer. Polychloroprene resists ignition.

Table 1. Physical Properties of Polychloroprene

		Vulcanizates	
Properties	Raw polymer	Gum	Carbon black
density, g/cm³	1.23	1.32	1.42
Thermal properties			
glass transition temp, K	228	228	230
heat capacity, Cp, kJ/(kg·K)[a]	2.2	2.2	1.7–1.8
thermal conductivity, W/(m·K)	0.192	0.192	0.210
Mechanical			
ultimate elongation, %		800–1000	500–600
tensile strength, MPa[b]		25–38	21–30
Young's modulus[c], MPa[b]		1.6	3–5
resilience[d], %		60–65	40–50

[a] To convert J to cal, divide by 4.184.
[b] To convert MPa to psi, multiply by 145.
[c] Initial slope.
[d] Rebound.

Neoprene polymers are mainly manufactured by batch emulsion polymerization, and the polymers are isolated by freezing–drying procedures. The polymerization of chloroprene involves the same steps as the emulsion polymerization of other diene monomers, namely: emulsification; initiation and catalysis; heat transfer; monomer conversion and molecular weight control; short-stopping and stabilization; monomer recovery; and polymer isolation.

Polychloroprene can be compounded in all types of rubber-mixing equipment and shaped by molding, extrusion, and calendering. It can be vulcanized by practically all the methods used by the rubber and the wire and cable industries (see Rubber compounding).

Health, Safety, and Environmental Factors

The fire and health hazards of chloroprene monomer are of serious concern in the manufacture of polychloroprene (see Chlorocarbons and chlorohydrocarbons, chloroprene). However, after steam stripping the monomer from the emulsion, the concentration of chloroprene in the system presents no acute hazards.

The major effluents from a chloroprene polymerization plant are the gaseous vents from monomer transfers, the vapor exhausts from the polymer dryers, and the water washings from the polymer isolation lines. The gaseous exhausts are generally vented to the atmosphere in high level stacks. The main organic components of the wash-belt effluent are salts of acetic acid. This stream is sent to sewage treatment plants where the BOD is reduced to acceptable levels.

Applications of solid neoprene polymers include automobile tires, cable sheating and insulation, cellular rubber goods, conveyer belts, fabric coatings, footwear, hoses, large sealing applications, linings and covers, moldings and extrusions, power-transmission belts, coating rolls, and roofing membranes.

Emulsion polymerization of chloroprene monomer produces a colloidal dispersion (or latex) of the polymer in which the polymer particle size is in the range of 5–200 mm. Neoprene latex has the general properties as natural rubber latex and can be used in the same applications. Because of its smaller particle size, it can also be used in novel applications where the natural rubber latex cannot be used (see Latex technology; Rubber, natural).

All but three of the current commercial polychloroprene latex products are anionic dispersions prepared from soaps formed *in situ* by the interaction of a rosin-derived acid and an alkali-metal hydroxide. The special features of each product are best obtained from the producer's trade literature. Some of the applications of neoprene latexes are adhesives, asphalt modification, binders, cement modification, coatings and proofing, dipped goods, extruded thread, impregnation, latex foam, mastics, saturants, and sealants.

PAUL R. JOHNSON
E.I. du Pont de Nemours & Co., Inc.

P.R. Johnson, "Polychloroprene Rubber," *Rubber Chem. Technol.* **49**, 650 (1976).

NITRILE RUBBER

Nitrile rubbers are classified as specialty rather than general purpose elastomers, as the vulcanized forms are used primarily for their oil, solvent, and chemical resistance, even though they are capable of displaying excellent rubbery properties suitable for many applications at moderate cost. ASTM has recommended the designation NBR to identify the family of poly(acrylonitrile-*co*-1,3-butadiene) or nitrile–butadiene rubbers.

Nitrile rubber grades available in the marketplace have a 20–50% acrylonitrile content. When properly compounded and cured, the fuel and solvent resistance, abrasion resistance and resistance to gas permeation increase with increasing nitrile content; with decreasing nitrile content the low temperature properties and resilience improve.

Perhaps the most valuable property of NBR is its ability to be converted by conventional compounding, forming, and curing practices

into parts that function reliably in service when in contact with fluids that swell or degrade other elastomers. Commercially established products for such service include seals, hoses, belts, wire-and-cable insulation, rolls, sponges and numerous molded or extruded mechanical items.

Vulcanized NBR compounds are serviceable for continuous use up to 120°C in air and to 150°C in complete oil immersion and in the absence of air. For short exposure conditions, flexibility is retained at temperatures over 200°C. Manufacturers, therefore, are able to specify NBR for most automotive under-the-hood applications.

Resistance to oxygen may be improved by compounding with 2 to 3 parts of a commercial amine antioxidant, for example, Aminox (a polymeric amine) for black stocks, or a phenolic or an organic phosphite antioxidant, eg, Polygard (TNTP, Weston), for light-colored parts. To minimize ozone attack, a combination of chemical antiozonant, typically a *p*-phenylenediamine derivative such as one of the Flexzone (Santoflex, Monsanto, Vulkanox, Bayer) series, at the level of 3 parts by wt in combination with about 3 parts of wax, such as Sunproof (Sunolite 240, Witco), is most beneficial (see Antioxidants and antiozonants).

The polar nitrile groups in NBR strongly retard the diffusion of gases and liquids through hoses and diaphragms in direct proportion to the nitrile content (see Barrier polymers). This polymer is often used to contain air, hydrogen, nitrogen, and carbon dioxide and the medium-to-high nitrile types are used widely to contain both liquid and gaseous Freon refrigerants.

Amorphous nitrile rubbers display inferior mechanical properties unless they are reinforced with appropriate carbon blacks or mineral pigments and unless they are vulcanized. In addition to tensile strength and modulus development, the reinforcing filler enhances other desirable properties such as abrasion resistance, tear strength, reduced compression set, controlled resilience and low water absorption.

Manufacture and Compounding

The process of emulsion polymerization, which uses water as the carrier medium, has been developed extensively. Its commercial applications include production of SBR, styrene–butadiene rubber (qv), as well as other rubber, eg, polychloroprene (CR) (see Neoprene) and polyacrylates (see Acrylic elastomers) and NBR.

Conversion of NBR Latex to Dry Polymer

Low-boiling unreacted butadiene and higher boiling unreacted acrylonitrile are removed from short-stopped latex by vacuum-venting and steam distillation, respectively. These stripping operations, either batch or continuous, are significant users of electrical and thermal energy. The recovered monomers are condensed and stored for recycling to subsequent polymerizations in combination with fresh monomers.

Compounding and Curing

In addition to reinforcing fillers, proper preparation for thermal curing also includes vulcanization agents. The most important of these are elemental sulfur and sulfur donors (see Rubber compounding). Organic peroxides are also used as cross-linking agents and they impart good high temperature resistance. Other chemicals that may be required to balance the vulcanization system are activators, such as zinc oxide and stearic acid, and accelerators that reduce the time and temperature requirements for sulfur cure. Any standard vulcanization technique used in the rubber industry may be used with nitrile compounds.

Health and Safety Factors

Environmental restrictions on nitrile-rubber manufacturing operations include OSHA's 8-h TWA of 2 ppm acrylonitrile (Title 29, 1910.1045 in 1978). EPA has no air-pollution effluent guidelines for acrylonitrile. The EPA water-pollution effluent guidelines for synthetic-rubber manufacture (including NBR, SBR, etc) are outlined in 40 CFR, part 428 (1974).

Uses

NBR products are used in footwear, belts and belting, hose, o-rings, packing, gaskets, automotive and other mechanical goods, adhesives, sealants, coatings, rolls, coated fabric, film, sheeting, wire and cable insulation, and blends (NBR/PVC, etc).

H.W. ROBINSON
Uniroyal, Inc.

J.R. Dunn and co-authors, "Advances in Nitrile Rubber Technology," *Rubber Chem. Technol.* **51**, 389 (1978).

H.H. Bertram in *Developments in Rubber Technology*-2: *Synthetic Rubbers*, Applied Science Publishers Ltd., London, 1981.

W. Hofman, "Recent Developments in Nitrile Rubber Technology," *Prog. Rubber Technol.* **46**, 43 (1984).

POLYBUTADIENE

The merits of polybutadiene, aside from cost, became quickly evident as it has applications involving both tire and nontire uses. With its importance to the rubber industry, it ranks second in production only to emulsion SBR among all synthetic rubbers.

In the United States butadiene is used primarily in the production of general purpose rubbers. Tires and other fabricated rubber products account for 81% of butadiene's consumption.

Butadiene is produced commercially in the United States as a co-product in the manufacture of ethylene and propylene feedstocks and by dehydrogenation of butenes from refineries and butanes from natural gas. From the latter process, so-called primary butadiene is obtained. Since domestic consumption exceeds production, additional butadiene (co-product) is imported from Europe and Japan (see Butadiene).

Microstructure

The polymerization of butadiene is an example of an addition polymerization wherein the repeating structural unit within the polymer chain has the same molecular weight as the entering monomer unit. The configurations of polybutadiene are cis, trans, and vinyl (isotactic and syndiotactic). Since either or both of the double bonds in butadiene can be involved in the polymerization, the resulting polymer may have a variety of configurations. These result from the fact that the spatial arrangement of the methylene groups in polybutadiene allow geometric isomerism to occur along the polymer chain.

The four structurally different configurations of polybutadiene give rise to notably different behavior. Since the pure stereoisomers are highly regular, they are likewise highly crystalline. The cis-1,4 isomer crystallizes upon stretching over 200%. The trans-1,4 and vinyl isomers are crystalline without elongation. The high cis-1,4-polybutadiene is a soft, easily solubilized elastomer that exhibits excellent dynamic properties, low hysteresis, and good abrasion resistance. The trans-1,4-polybutadiene, in contrast, is a tough elastomer which, in addition to its high hardness and thermoplasticity, is sparingly soluble in most solvents.

The 1,2-isotactic and 1,2-syndiotactic polybutadienes are rigid, crystalline materials with poor solubility characteristics.

Polymerization Process

Most commercial processes employ solution polymerization. In general these systems are based upon organic lithium compounds or coordination catalysts containing metals in reduced valence states. Polymerizations are carried out using essentially pure dry monomer and solvents, such as aromatic, aliphatic, and alicyclic hydrocarbons.

In cationic polymerizations, the reaction is initiated by complex ions or ion pairs, produced by the interaction of Lewis acids with water, hydrogen halides, and other halogen-containing compounds. When the Lewis acids are halides of aluminum, boron, titanium, or tin, a low solubility polybutadiene is produced, comprised of trans-1,4 and 1,2 units.

Free-radical polymerizations, whether performed in emulsion or solution, are initiated by a free radical (R·) formed by the decomposition of a peroxide, peroxysulfate, or similar free-radical-forming reactions. The free-radical polymerization of butadiene is a function of the reaction temperature. As the polymerization temperature decreases, the cis-1,4

content also decreases; no cis-1,4 content occurs below −15°C. This same observation is made for polybutadiene segments of copolymers with styrene, acrylonitrile, and others.

The most outstanding characteristic of the Ziegler or Ziegler-Natta catalyst (see Olefin polymers) is its ability to produce highly stereoregular polymers. These catalysts are comprised, in general, of an organometallic compound and a transition-metal compound from Groups IV–VIII, and may be homogeneous or heterogeneous. Much academic and industrial effort has centered on synthesizing the four stereoregular isomers of polybutadiene. Mostly it has been directed toward developing anionic catalyst systems for the synthesis of high (> 90%) cis-1,4-polybutadiene, because of its excellent elastomeric properties.

Although a wide variety of possible catalysts for polymerizing butadiene has been reported, only catalysts containing titanium, cobalt, and nickel have been successfully developed commercially.

Processing and Curing

The processing characteristics of polybutadiene are influenced by the polymer microstructure, molecular weight, molecular weight distribution, and degree of branching. The polymer undergoes mastication, mixing, molding, and curing. A Banbury mixer and a roll mill are employed for mastication and mixing, a calender and extruder for molding.

Most polybutadienes are highly resistant to breakdown and have poor millbanding characteristics and rough extrusion appearance compared to SBR elastomers. The solution polybutadienes process satisfactorily when blended with other elastomers such as SBR. Emulsion polybutadiene processes better than solution polymer, but not as well as SBR, and is commonly blended with other elastomers for enhanced processing.

Certain chemical peptizers slightly increase breakdown and improve processing. A lower Mooney viscosity also improves processing, but may lead to cold-flow problems. In addition, a broad molecular weight distribution and branching both improve milling and extrusion behavior as compared to a linear polymer. Excessively high molecular weight polybutadiene tends to crumb on the roll mill.

Very high cis-polybutadiene prepared with a uranium catalyst is reported to have greatly improved millability, calenderability, tack, green strength, and adhesion to fabric. Such processability would eliminate the need for blending with other elastomers.

Curing and compounding recipes are chosen by the individual manufacturer. Polybutadienes are cured employing conventional sulfur recipes or peroxide systems. Normally, the polymer is blended with another elastomer, then mixed with filler (eg, HAF or ISAF blacks) typically an aromatic processing oil, wax, antioxidant, antiozonant, and curing ingredients are added at a later stage. Cure rates are close to SBR using similar loads of sulfur.

Medium vinyl polybutadiene may replace SBR partially or fully and has only recently become commercially important to the rubber industry. It could compensate for styrene, should a shortage of this monomer occur in the petrochemical industry.

Safety Factors and Testing

Polybutadiene rubbers, in dry form, are nontoxic. Compounding ingredients, however, may be slightly toxic or cause dermatitis in frequent contact. Special consideration must be given to ingredients added to polybutadienes used for food-packaging materials.

L.J. KUZMA
W.J. KELLY
The Goodyear Tire & Rubber Co.

E. Tornqvist in J.P. Kennedy and E. Tornqvist, eds., *Polymer Chemistry of Synthetic Elastomers*, John Wiley & Sons, Inc., New York, 1968, Chapt. 2.

W. Cooper in W.M. Saltman, ed., *Stereo Rubbers*, John Wiley & Sons, Inc., New York, 1977, Chapt. 2.

D.D. Babiţskii and V.A. Krol in I.V. Garamonov, ed., *Synthetic Rubber* (in Russian), Khimiya, Leningrad, USSR, 1976, pp. 76–99.

POLYETHERS

This article discusses only polyether elastomers of sufficiently high molecular weight to be processed and fabricated by conventional rubber equipment and which are then cross-linked (ie, vulcanized) in a separate step. The preparation of such polyether elastomers required the development of new catalyst systems, specifically coordination catalysts, for polymerizing epichlorohydrin (ECH), propylene oxide, and other epoxides.

In early work on epoxide polymerization, high molecular weight, largely amorphous ECH homopolymer, ECH–ethylene copolymer and propylene oxide–unsaturated epoxide copolymers were made and their potential value as improved elastomers was recognized. Subsequently, properties of similar propylene oxide–unsaturated epoxide copolymer elastomers were reported. These new types of polyether elastomers became commercially available under the trademarks of Herclor and Hydrin for ECH elastomers and Parel for propylene oxide elastomers. Parel 58 elastomer is a sulfur-curable copolymer of propylene oxide and allyl glycidyl ether (AGE) (see Allyl monomers and polymers).

The epichlorohydrin and propylene oxide elastomers are available only in dry form. At present, the ECH elastomers are manufactured only by Hercules and Goodrich.

The properties and structures of epichlorohydrin and propylene oxide elastomers are listed in Table 1.

Polymer Preparation

Epichlorohydrin self-polymerizes and copolymerizes with ethylene oxide by a coordination polymerization mechanism using aluminum alkyl–water or an aluminum alkyl–water–acetylacetone-type catalyst system, generally in hydrocarbon solvents. Propylene oxide is copolymerized with allyl glycidyl ether in a solution polymerization in an aliphatic, aromatic, or chlorinated hydrocarbon, in the presence of coordination catalysts such as the aluminum alkyl–water–acetylacetone, diethyl zinc–water, and complex cyanide catalysts, as well as others. Both epichlorohydrin elastomers and propylene oxide elastomers can be processed and fabricated exceptionally well, and can be molded, calendered, and extruded.

Compounding

Epichlorohydrin elastomers are cured or vulcanized (cross-linked) by adding reagents that can react difunctionally on the chloromethyl side chains of these polymers via nucleophilic-type substitution reactions. In addition, a suitable acid acceptor is necessary for best vulcanization, generally to accept by-product HCl and sometimes to catalyze other aspects of the curing reaction. Propylene oxide elastomers are readily cured with a variety of sulfur accelerator systems.

Health and Safety

Many epoxides are suspected carcinogens. Thus, appropriate measures must be taken to obtain elastomers that are essentially free of monomers.

Although a recent epidemiological study shows no evidence for ethylenethiourea (curative for ECH elastomers) causing cancer and birth defects in humans, it is recommended that skin contact and breathing of dust of NA-22 be avoided.

Uses

Epichlorohydrin elastomers are employed in many conventional specialty rubber applications where the homopolymer and ECO or combinations improve oil resistance, high and low temperature properties, environmental resistance, and fabricability over lower-cost rubbers. The most important application for propylene oxide elastomer is in motor mounts because of its unusual heat resistance combined with its good rubber properties.

Epichlorohydrin Elastomer Derivatives

Extensive work has been directed toward the synthesis of derivatives from epichlorohydrin elastomers, generally via nucleophilic substitution reactions on the chloromethyl group. Diverse applications have been reported for these modified polyethers, including flame retardants, photosensitive materials, a variety of applications for water systems, and shrinkproofing agents for wool.

E.J. VANDENBERG
Hercules Inc.

E.J. Vandenberg, *J. Polym. Sci.* **47**, 486 (1960).

W.D. Willis and co-workers, *Rubber World* **153** (1), 88(1965).

E.J. Vandenberg and A.E. Robinson in E.J. Vandenberg, ed., *Polyethers*, American Chemical Society, Washington, D.C., 1975, p. 101.

POLYISOPRENE

Natural rubber is *cis*-1,4-polyisoprene. Worldwide production of synthetic *cis*-1,4-polyisoprene as a replacement for natural rubber is the result of a long period of research by many people. The ASTM nomenclature for natural rubber is NR and for synthetic *cis*-1,4-polyisoprene is IR.

In 1983 there was one producer of *cis*-polyisoprene in the United States, Goodyear, and one each in France, Italy, and the Netherlands, three in Japan, one in Romania, and at least five in the USSR. Only Shell in the Netherlands is known to have used the lithium catalyst. All others are believed to use catalyst systems related to the Ziegler system.

Isoprene polymerizes to yield four different basic structures of polyisoprene. The cis-1,4, trans-1,4, and 3,4 polymers can be made in high purity, but the poly-1,2 structure is obtained only in conjunction with the other three structures. Many catalyst systems yield polyisoprenes that are random copolymers of all four of the structural units. Some

Table 1. Properties of Commercial Polyether Elastomers

| | Epichlorohydrin elastomers | | Propylene oxide elastomer |
	ECH homopolymer	ECO	Parel 58
structure	$+\text{CH}_2\text{CHO}\overline{)_n}$ CH_2Cl	$+\text{CH}_2\text{CHOCH}_2\text{CH}_2\text{O}\overline{)_n}$ CH_2Cl	$+\text{CH}_2\text{CHO}\overline{)_n}(\text{CH}_2\text{CHO}\overline{)_{0.025n}}$ CH_3 $\text{CH}_2\text{OCH}_2\text{CH}=\text{CH}_2$
name	chloromethyloxirane homopolymer	chloromethyloxirane copolymer with oxirane	methyloxirane copolymer with [(2-propenyloxy)methyl] oxirane
chlorine, %	38.4	26	
specific gravity	1.36	1.27	1.01
Mooney viscosity of raw polymer at 100°C	48[a]	50–140	75
brittleness, temperature, °C	−18	−40	

[a]Reduced specific viscosity of about 1.4–1.6 (0.1%, α-chloronaphthalene at 100°C), corresponding to a weight average molecular weight of about 500,000 and a Mooney viscosity (ML-4 at 100°C) of ca 50.

Table 1. Raw Polymer Properties

	Hevea[a]	Synthetic *cis*-1,4-polyisoprene	
catalyst		RLi	Ti–Al
cis-1,4 content, %	100	92	98
intrinsic viscosity [η], dL/g		ca 6.5	ca 3.5
Mooney viscosity at 100°C, ML-4			80
volatiles, %	1.0	0.1	0.1
ash, %	0.5–1.5	0.1	0.4
second-order transition, °C	−72	−70	−72
specific gravity	0.92	0.91	0.91
M_w/M_n	> 3	2	2–3

[a]*Hevea* rubber is natural *cis*-1,4-polyisoproprene.

typical values of the raw polymer properties for natural rubber and polyisoprene are shown in Table 1.

The Ziegler coordination catalyst, which consists of an aluminum alkyl, R_3Al, and titanium tetrachloride, $TiCl_4$, polymerizes isoprene to stereoregular polyisoprene with 98% to 99% cis-1,4 structure, compared to natural rubber of 100% cis-1,4 structure. In the R_3Al, R may be an alkyl or an aromatic substituent (see Organometallics). High purity isoprene that is free of polar compounds as well as acetylenes and cyclopentadiene is essential to the polymerization of isoprene.

The polymerization may be carried out in aliphatic or aromatic hydrocarbon solvents and in certain chlorinated aromatic solvents such as chlorobenzene and dichlorobenzene.

Use of high purity isoprene and high purity solvents produces a polymer molecular weight such that the polymer is useful for tires and other elastomer products. A change in the temperature of polymerization is a more effective method than a change in the amount of catalyst to achieve adjustments in molecular weight. The molecular weight changes inversely with temperature. The anionic polymerization of isoprene by organolithium catalyst in a hydrocarbon solvent also leads to a polymer of high cis-1,4 content.

The polymerization of isoprene to the trans-1,4 structure is best carried out using a mixed VCl_3–$TiCl_3$–R_3Al catalyst. The resulting *trans*-1,4-polyisoprene is similar in properties to natural *Balata* rubber.

Since *cis*-polyisoprene has the same basic structure as natural rubber, techniques for the compounding and product application of natural rubber are directly applicable to *cis*-polyisoprene.

About 60% of synthetic *cis*-polyisoprene is used in tires. *trans*-1,4-Polyisoprene is used largely in golf ball covers. Other uses for trans are hot-melt adhesives and orthopedic splints (see Prosthetic and biomedical devices).

A shift in the pattern of tire production with increased production of radial tires in the United States will result in an increase in demand for the combination of natural rubber and *cis*-polyisoprene as a percentage of total tire rubber. The ability of the synthetic *cis*-polyisoprene to participate in this market will be determined, not by the technology of the synthetic-rubber industry, but rather by the availability of cheaper by-product isoprene, and, ultimately the price of natural rubber.

HAROLD TUCKER
S.E. HORNE, JR.
BF Goodrich Co.

Belg. Pat. 543,292 (June 2, 1956), (to Goodrich-Gulf Chemical Co.); S.E. Horne, Jr., and co-workers, *Ind. Eng. Chem.* **48**, 784 (1956).

W.M. Saltman, ed., *The Stereo Rubbers*, John Wiley & Sons, Inc., New York, 1977, Chapt. 2.

C.F. Gibbs and co-workers, *Kautschuk Gummi Kunst.* **13**, WT 336 (1960).

POLYPENTENAMERS

Polypentenamer is obtained by the transition-metal-catalyzed polymerization of cyclopentene, and is one member of the homologous series of linear, unsaturated polymers designated *polyalkenamers*. Ring open-

ing occurs via cleavage of the carbon-carbon double bond in cyclopentene, and thus the polymer may be represented structurally by bisalkylidene repeat units, but other equivalent representations have been employed:

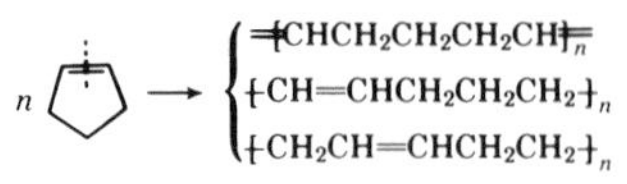

Cycloolefins such as cyclopentene possessing low ring-strain energy homopolymerize very poorly via addition across the double bond. Low molecular weight oligomers are normally obtained, although when ultrahigh pressures are employed (eg, 6,580 MPa or 65,000 atm), cyclopentene can be converted in good yield to high molecular weight, semicrystalline saturated polymers.

Of the polyalkenamers, *trans*-polypentenamer has generated the greatest interest for several reasons. The requisite monomer can be economically obtained from cyclopentadiene, an abundant petrochemical by-product. In addition, polypentenamer is a readily vulcanizable elastomer and possesses an intriguing combination of properties that indicate significant potential for use in tires and a variety of other applications.

This trans polymer has a glass transition temperature of ca −95°C, close to that of *cis*-1,4-polybutadiene, yet has a crystalline melting point just below room temperature. This fortuitous combination of properties results in a stress-crystallizing elastomer that is easily processed in conventional rubber-mixing equipment and exhibits good tensile, resilience, and abrasion characteristics.

Cyclopentene

The most commercially significant process for obtaining cyclopentene would utilize refinery C_5 fractions as starting material. Cyclopentene itself is a minor recoverable component of these streams, but much larger amounts of monomer can be obtained by recovery and hydrogenation of the more abundant component cyclopentadiene (see Cyclopentadiene). Numerous methods have been proposed for the hydrogenation step. Two particularly attractive routes utilize titanium or nickel catalysts. These processes are said to be at least 97% selective for the formation of cyclopentene under conditions where cyclopentadiene is quantitatively consumed.

Polypentenamer from Cyclopentene

Cyclopentene polymerizations have been conducted in bulk, but control of the reaction exotherm of ca 19 kJ/mol (4.5 kcal/mol) is best accomplished by solution polymerization methods if rapid rates of polymerization are anticipated. Suitable inert solvents include benzene, toluene, chlorobenzene, and methylene chloride. Polymerizations appear to be much slower in aliphatic solvents owing to reduced catalyst solubility or activity. Usually reactions are carried out at RT or below, because polymer yields decrease markedly at elevated temperatures. This is a consequence of the rather low ceiling temperature for this polymerization. Rather strict precautions for control of purity of monomer and solvent are required for optimum catalyst activity, as with other Ziegler-Natta polymerizations.

Mechanism of Cyclopentene Polymerization

The olefin metathesis reaction. Cyclopentene polymerization represents one application of a remarkable general reaction of olefins known as the olefin metathesis reaction. Several features of cyclopentene polymerizations are a direct consequence of olefin metathesis chemistry. Of prime importance is the fact that metathesis catalysts continue to react with all double bonds present in a given system throughout the course of polymerization reactions; thus, initially formed polymer vinylene units continually undergo redistribution reactions. These reactions also consume acyclic olefins, which are usually present in traces adventitiously, but may be added intentionally for molecular weight control.

Thermodynamics. For small rings such as cyclopentene, polymerization is an anti-entropic process. Therefore, polymerizability is enthalpy-dependent and is derived from the release of ring strain. Calculated estimates of the low strain energy in cyclopentene have produced values of 20.5 kJ/mol (4.9 kcal/mol) and 28.5 kJ/mol (6.8 kcal/mol). These compare well with the calculated enthalpy for the hypothetical ring-opening polymerization of cyclopentane of -21.8 kJ/mol (-5.2 kcal/mol).

Other Routes to Polypentenamer

Two methathesis-based alternatives to cyclopentene polymerization have been considered. The polymerization of 1,6-cyclodecadiene (a dimer of cyclopentene) has been proposed as an indirect means of converting low cost ethylene and butadiene to polypentenamer. The metathesis-induced conversion of α, ω-dienes to oligomeric polyalkenamers has been proposed as a general method.

Properties of Polypentenamer

The early characterization of polypentenamer was accomplished by ozonolysis. The reccurrence of double bonds every five carbon atoms along the chain was established by the near-quantitive recovery of pentamethylenediol in reductive ozonolyses.

The solid-state properties of polypentenamers are quite dependent on microstructure. Samples varying from ca 98% cis to 70% trans are amorphous under essentially all conditions. cis-Polypentenamer is quite resistant to crystallization, even at low temperatures. However, a sample possessing 99% cis content was induced to crystallize by being annealed at $-75°C$ for one week, and gave a melting point of $-41°C$. Crystallographic data have not been obtained on this material.

Vulcanization. Because of the high level of unsaturation present, polypentenamers respond well to conventional cross-linking systems based on sulfur or sulfur-donor agents and accelerators. Several laboratories have examined the properties of vulcanized polypentenamer formulations in comparisons with those obtained from conventional elastomers, and the effects of variations in vulcanization parameters have been explored in some detail. The cross-linking process is highly efficient in polypentenamer (see Rubber compounding). In a detailed evaluation of the effects of cure parameters on vulcanizate properties, optimum results were obtained by using low levels of zinc oxide, stearic acid, and accelerator, and high cure temperatures of about 170°C were desirable. Cross-links appear to be quite stable, and cure reversion is minimal.

Modification of Polypentenamer

Copolymerization. Metathesis polymerizations provide several pathways for the formation of random, block, or graft copolymers, but the properties of these materials have not been widely examined.

Chemical modification. Most chemical alterations of polypentenamer have been through reactions at the double bond.

The hydrogenation of polypentenamer has been accomplished using toluene-sulfonyl hydrazide and in a series of reactions, ionic derivatives of polypentenamer have been prepared.

EILERT A. OFSTEAD
The Goodyear Tire & Rubber Company

G. Dall'Asta, *Rubber Chem. Technol.* **47**, 511 (1974).

STYRENE-BUTADIENE RUBBER

Butadiene and styrene are the chief raw materials required to manufacture SBR. Others, which are required in smaller amounts, are the various emulsifiers, modifiers (eg, thiols), catalysts, initiators, shortstops, coagulating agents, antioxidants, and antiozonants. Water is a principal ingredient of the emulsion polymerization recipe as well as of the coagulation and product-washing operations.

Two methods of polymerization are in widespread use today and both are used for SBR. The emulsion process is used for standard SBR and the solution process for the other varieties.

It is possible to prepare solution copolymers of styrene and butadiene having a nonrandom, block structure. Block copolymers are molecules in which two or more chemically or structurally dissimilar segments are joined. Each segment or block usually is a long sequence of units of a single monomer, but it may also be a long sequence of randomly copolymerized units. Both types of block copolymers are known (see Copolymers). Physical properties for a typical SBR polymer, containing about 23.5% bound styrene content, are given in Table 1.

Table 1. Poly(butadiene-*co*-styrene), SBR (About 23.5% Bound Styrene Content)

Property	Unvulcanized	Pure-gum vulcanizate	Vulcanizates with ca 50 phr[a] of carbon black
density, g/cm^3	0.933 (0.9325–0.9335)	0.980 (0.940–1.000)	1.150
thermal			
glass transition temperature, K	209–214	221	221
heat capacity, C_p, kJ/(kg·K)[b]	1.89	1.83	1.50
thermal conductivity, W/(m·K)		0.190–0.250	0.300
heat of combustion, MJ/kg CO_2[b]	-56.5		
refractive index, n_D	1.5345 (1.534–1.535)		
dielectric constant (1 kHz)	2.5	2.66	
mechanical			
compressibility, $B = -(1/V_0)(\delta V/\delta P)$, Pa^{-1}[c]	530	510	400
bulk modulus (isothermal), GPa[d]	1.89	1.96	2.50
ultimate elongation, %		400–600	400–600
tensile strength, MPa[e]		1.4–3.0	17–28
initial slope of stress–strain curve, Young's modulus, E (60s), MPa[e]		1.6 (1.0–2.0)	3–6
shear modulus, G (60s), MPa[e]		0.53 (0.3–0.7)	2.0 (2.0–2.5)
creep rate, $(1/J)(\delta J/\delta \log t)$, %/unit $\log t$		7 (3–10)	12
storage modulus, G', MPa[f]	0.66 (0.66–0.71)	0.76 (0.44–1.6)	8.7 (2.5–8.7)

[a] phr = parts per hundred of rubber.
[b] To convert J to cal, divide by 4.184.
[c] To convert Pa to psi, divide by 6895.
[d] To convert GPa to psi, multiply by 145,000.
[e] To convert MPa to psi, multiply by 145.

All types of SBR use compounding recipes, as do other unsaturated hydrocarbon polymers that share the common ingredients of sulfur, accelerators, antioxidants, antiozonants, activators, fillers, and softeners or extenders. SBR requires less sulfur than natural rubber for curing. The usual range is 1.5–2.0 phr of sulfur; however, this range should be based on the rubber hydrocarbon only for oil-extended SBR. Because of their lower unsaturation, all styrene–butadiene rubbers are slower curing than natural rubber and require more acceleration. Processing SBR compounds is similar to that of natural and polybutadiene rubbers. The ingredients are mixed in internal mixers or on mills and may be extruded, calendered, molded and cured in conventional equipment.

Uses

About 65% of all SBR elastomer produced in the United States is used in the manufacture of passenger-car tires. Two expanding markets for SBR are adhesives (qv) and chewing gum. A wide variety of SBRs is available for adhesive applications, and several of the crumb forms were designed specifically for the adhesives industry.

The block styrenic copolymers are intended for applications in adhesives, caulks, sealants, coatings, food packaging, toys, tubing, sheeting, molding equipment, belting, shoe soles and heels, and miscellaneous uses.

R.G. BAUER
The Goodyear Tire & Rubber Company

S.S. Medvedev, *International Symposium on Macromolecular Chemistry*, Pergamon Press, New York, 1959, pp. 174–190.

THERMOPLASTIC ELASTOMERS

Thermoplastic resins are polymeric structures that soften or melt at elevated temperatures, allowing them to be processed into fabricated products that, when cooled, recover the physical and chemical properties of the original resin. Of the three classes of thermoplastic elastomers to be discussed, the styrene–diene block copolymers are the largest volume (> 50,000 metric tons), the thermoplastic polyurethanes are next (> 15,000 t), and copolyester ethers, the newest entry, are now > 2000 t.

Styrene – Diene Thermoplastic Block Copolymers

Preparation of styrene–diene block copolymers is achieved by forming a living polymer, a term coined to describe the product of a polymerization that has no termination or chain-transfer reactions. Shown in Table 1 is a comparison of mechanical properties of SBS block copolymers with vulcanized SBR and natural rubber illustrating the range inherent in the thermoplastic elastomers.

The chemical characteristics of the copolymers are determined by the nature of the components. Alteration of the chemical characteristics is achieved by altering one or more of the blocks.

The styrene–diene thermoplastic elastomers have excellent resistance to water, acids, and bases. Resistance to hydrocarbons, solvents, and oils is poor. The thermoplastic nature limits their utility to temperatures below 65°C depending on the stress. Elastic recovery, compression set, and creep properties are usually inferior to the chemically cross-linked elastomers.

Table 1. Typical Properties of ABA Thermoplastic Elastomers and Conventional Rubbers

	Kraton 1101[a]	Kraton 1107[b]	Natural rubber	SBR rubber
styrene, %	30	14		
tensile strength, MPa[c]	31.8	21.4	20.8	14.5
modulus at 300% ext, MPa[c]	2.8	0.7	3.5	2.1
elongation at break, %	880	1300	600	800
hardness, Shore A	71	37	55	45
specific gravity	0.94	0.92		

[a]SBS (styrene–butadiene–styrene).
[b]SIS (styrene–isoprene–styrene).
[c]To convert MPa to psi, multiply by 145.

Table 2. Typical Properties of Segmented Polyether Esters

4GT hard segment, %	33	58	76
polymer melt temperature (by dsc[a]), °C	176	202	212
specific gravity	1.15	1.20	1.22
tensile strength, MPa (psi)	39.3 (5700)	44.1 (6400)	47.5 (6900)
elongation at break, %	810	760	510
flexural modulus, MPa (psi)	44.8 (6500)	206 (30,000)	496 (72,000)
oil swell (ASTM NO. 3 oil, 7 days at 100°C), % vol increase	22.0	12.2	6.6

[a]Differential scanning calorimetry.

The SBS elastomers may be processed by a wide variety of techniques including solution processing, extrusion, calendering, injection molding, blow molding, and vacuum forming. Standard rubber and plastics equipment is useful for processing the elastomers.

Uses for the thermoplastic elastomers fall into two main sectors: primary raw materials for rubber products without vulcanization, and modifiers to upgrade the qualities of the rubbers and plastics. The largest markets for the styrene–diene block copolymers are footwear, adhesives (qv), and mechanical goods.

Thermoplastic Urethane Elastomers

Thermoplastic polyurethane (TPU) elastomers are a special class of urethanes that can be processed as plastics and as cements for a wide range of applications (see Urethane polymers). Generally, polyester-based materials are selected for high strength, tear, chemical and heat resistance, and polyether-based materials are selected for low temperature flexibility, high humidity conditions, and resistance to attack by fungi and bacteria.

Since urethane elastoplastics incorporate exceptional resistance to abrasion, fuel and oils, and have high tensile, tear- and load-bearing properties, and are available in a broad durometer range, they are candidates for demanding applications in such areas as automotive, sporting, general mechanical goods, fabric coatings, and biomedical applications such as intra-aortic balloons (see Prosthetic and biomedical devices).

Thermoplastic Copolyester – Ether Elastomers

Segmented copolyester–ethers represent a novel family of commercial thermoplastic elastomers derived from terephthalic acid (T), polytetramethylene ether glycol (PTMEG), and 1,4-butanediol. They offer an unusual combination of easy processing and high performance under environmental extremes (see Polyesters; Polyethers). The polyester–ether copolymers are prepared by titanate ester (tetrabutyl titanate)-catalyzed melt transesterification of a mixture of dimethyl terephthalate, polyether glycol, and excess 1,4-butanediol. Some typical physical properties are listed in Table 2.

The thermoplastic copolyester–ether elastomers commercialized as Hytrel by DuPont can be processed by injection, blow, compression, transfer, or rotational molding. Some of the many uses of these elastomers include as a replacement for cured rubber and rubber-metal parts with a one-component elastomer unit.

A.F. FINELLI
R.A. MARSHALL
D.A. CHUNG
The Goodyear Tire and Rubber Co.

A. Noshay and J.E. McGrath, *Block Copolymers: Overview and Critical Survey*, Academic Press, Inc., New York, 1976.

U.S. Pat. 3,265,766 (Aug. 9, 1965), G. Holden and R. Milkovich (to Shell Oil Co.).

U.S. Pat. 2,871,218 (Jan. 27, 1959), C.S. Schollenberger (to B.F. Goodrich Co.).

U.S. Pat. 3,651,014 (Mar. 21, 1972), W.K. Witsiepe (to E.I. du Pont de Nemours and Co., Inc.).

ELECTRICAL CONNECTORS

Electrical connectors are mechanical devices that connect wires, cables, printed circuit boards, and electronic components to each other and to related equipment. Connector designs include miniature units for microelectronic applications, specialized cable, rack, and panel designs for incorporating combinations of a-c, d-c, and r-f conducting contacts, and high current connectors for industrial application and for transmission and distribution of electrical power in overhead and underground networks.

Connector Configurations

Electronic connectors. The complexity and size of many electronic systems necessitate construction from relatively small building blocks which are then assembled with connectors.

Electronic connectors may connect internally or externally. Internal connections may be between a component and a printed circuit board or wire (Fig. 1a); a printed circuit board and a wire or another printed circuit board which is in a chassis; and between chassis in the same cabinet. External connectors join separate pieces of equipment (Fig. 1b). Another type of electronic connector joins coaxial conductors. These are used primarily in radio-frequency circuits.

The contacts of an electronic connector have spring elements that press the mating surfaces together with a predetermined force, usually in the range of 0.25–5 N (0.056–1.12 lbf); this range depends on the connector design and the materials from which the contact is made.

The most widely used techniques for the termination of wires to separate contacts are the soldering (see Solders), welding (qv), crimping, solderless wrapping, and slotted-beam methods. Except for crimping and welding, it is usually possible to replace wires to a contact a limited number of times if repair or wiring changes are necessary.

Splicing connectors. Splicing connectors are used to permanently join wire to wire. Some are simple sleeve barrels that are crimped to bare wire; others are preinsulated with the crimp made by compressing the sleeve with its positioned insulation onto wires that may or may not have insulation.

Terminals. Terminals are connectors with which individual wires are designed to be screwed down at their separable ends, and to which conductors are permanently joined at the back end, usually by crimping.

(a)

(b)

Figure 1. Some types of electronic connectors. (a) Receptacle for dual-in-line package (DIP) semiconductor integrated circuit. (b) Circular connector for cable. Courtesy of Burndy Corporation.

Either ring or open-tongue configurations provide a terminal for the screw connection.

Utility and industrial connectors. Connectors in power distribution systems are nearly always of the permanent type and are usually made for single conductors. For example, sleeve-barrel connectors are crimped to large-diameter cables.

Methods of application. Attachment of separable contacts to conductors may be achieved with automated machinery or with specialized hand tools.

Contact Principles

Metallic contact between surfaces of separable connectors usually is obtained either by using noble metals, which are essentially film-free, or by designing the contact so that any films that are present are broken before or as the surfaces are brought together. A large part of practical connector engineering is devoted to designing metallic contacts that need minimal maintenance, especially for those intended to serve in chemically aggressive atmospheres and in high temperature applications.

The contact resistance of any electrical connector in a circuit must be stable and low for proper functioning of that circuit. Low voltage circuits, which are common in modern electronic systems, have open circuit voltages of not more than a few volts, insufficient levels for the fritting of films that grow on base metals from environmental exposure. Low contact resistance should be achieved with the use of noble metal contacts or of base metals concurrently with methods that mechanically perforate any insulating films that are present. Typical separable electronic connectors have contact resistances that range from several milliohms to tenths of ohms (The contact resistances in utility–industrial connectors are in the microohm range).

Materials and Processing

Contact substrates. The substrate must be able to be terminated readily as well as be a good electrical conductor. In electronic connectors the substrate may serve as a spring element. The most widely used spring materials for connectors are the copper alloys: 98.1 wt% Cu; 1.9 wt% Be (Copper Development Association (CA) 172); 94.8 wt% Cu, 5.0 wt% Sn, 0.19 wt% P (phosphor bronze, CA 510); and 88.2 wt% Cu, 9.5% Ni, 2.3 wt% Sn (CA 725) (see Copper alloys). Connectors for aluminum power cables are usually made of aluminum.

Contact finishes. Oxides, tarnish, or corrosion films developed in humid, polluted atmospheres may prevent adequate metal-to-metal contact when the connector or connector–conductor surfaces are mated. Therefore, coatings of other metals commonly are used to obtain corrosion resistance, to provide conductivity, or to facilitate termination to conductors by soldering, wire wrapping, or by other means. Application of finishes is achieved by electroplating (qv), cladding, and by hot-dipping when low melting metals such as tin are used (see Metal surface treatments). The chief noble contact finishes are gold, palladium, and rhodium. The nonnoble finishes are tin, silver, nickel, and indium. Alloys of most of these metals also are widely used.

Conductive elastomers. Conductive elastomers, rubbers that are made conductive by molding metal or carbon powders in them, have characteristics of both a contact material and a spring. Silicone rubbers, neoprene, polyurethane, and other elastomers have been used; however, silicones are the most popular because they have a low compressive set and operate over a wide temperature range, from ca −65 to 200°C (see Polymers, conductive; Antistatic agents).

Contact lubricants. Debilitating wear of separable connector contacts, which may occur if the metallic coating is thin or if forces normal to the contact surfaces are high, can be minimized by coating the contacts with thin films of organic lubricants. Viscous mineral oils and poly(phenyl ethers), soft microcrystalline waxes, and petrolatum have been used on electrical connectors.

Joint aids. It has been found that finely divided zinc incorporated in grease can be coated on the interior surfaces of an aluminum connector to provide lower initial contact resistance and better long-term resistance stability. Other soft metal coatings are effective, such as indium which is used as a plating on copper alloy connectors intended for permanent joints to aluminum communications cables having slotted-beam connectors.

Insulators

Molded plastics serve as insulators for multi-contact connectors and glass is the insulating material used in hermetic connectors intended for bulk-head mounting. Ceramics (qv) are employed in some high voltage power connectors. Hard rubber shells insulate connectors that serve underground power-distribution cables (see Insulation, electric). Nylon and poly(vinyl chloride) sleeving are used in preinsulated terminals. A wide variety of plastics is employed in electronic connector bodies depending on their size, the strength requirements, the complexity of the design, and the service environment.

Failure rates. Acceptable failure rates range from < 1 in 10^9 operating hours for contacts in air-frame electrical systems and in some telecommunications equipment, to 100–1000 in 10^9 operating hours in instruments, to even larger rates for contacts in many consumer products. A failure is defined as an excess of contact resistance, which can be as little as twice the initial contact resistance, that causes circuit malfunction. The required lifetimes of connectors may be ≥ 40 yr, although most required application times are shorter. The causes of connector contact failure can be of a thermal, chemical or mechanical nature, in addition to misapplication and physical abuse.

MORTON ANTLER
Bell Telephone Labs

R. Holm, *Electric Contacts Handbook*, 4th ed., Springer-Verlag, New York, 1967.

G.L. Ginsberg, ed., *Connectors and Interconnections Handbook*, Basic Technology, Vol. 1, The Electronic Connector Study Group, Inc., Camden, N.J.; 1977, Vol. 3, Fort Washington, Pa., 1981.

Physical Design of Electronic Systems, Integrated Device and Connection Technology, Vol. III, Prentice Hall, Englewood Cliffs, N.J., 1972.

ELECTRIC INSULATORS.   See Insulators, electric.

ELECTROANALYTICAL METHODS.   See Analytical methods.

ELECTROCHEMICAL MACHINING.   See Electrolytic machining methods.

ELECTROCHEMICAL PROCESSING

INTRODUCTION

Electrochemical processes involve the interconversion of electrical and chemical energy by means of a reaction at an electrode. Electrical charge may be fed to an electrolysis cell to induce chemical reactions (synthesis, metal winning (see Extractive metallurgy) or refining, etc), or chemical reactions may be run in a cell to generate electricity (batteries, accumulators, fuel cells) (see Batteries). Since the electrode reaction occurs at a surface, electrochemical techniques may also be used for surface treatment (electroplating (qv), electropolishing, anodizing) (see Metal surface treatments) or machining (see Electrolytic machining methods).

Electrosynthesis was first carried out by Davy in 1807 for the production of sodium and potassium, and in 1833 by Faraday who performed the first known example of the Kolbe reaction. Today, this reaction remains one of the most useful in organic electrosynthesis. Generally speaking, all chemical reactions may be performed electrochemically and there are often great advantages to be obtained in doing so: electrochemical reactions allow control of selectivity and reaction rate through the electrode potential; are inherently pollution-free; have high thermodynamic efficiency; make possible reactions at ambient temperature; can often reduce the number of reaction steps; can often use cheaper starting materials; by means of electrolytic regeneration can use catalytic quanti-

ties of chemical oxidizing or reducing agents; and can often perform a synthesis electrogeneratively.

However, disadvantages are electrochemical engineering and technology are far less developed than chemical engineering; many reaction variables may be involved with complex interdependencies; long-term stability of process components is often poor; and work-up of product is often costly.

The Electrochemical Cell

An electrochemical cell consists of at least two electrodes (anode and cathode) that dip into an electrolyte contained in a cell or reactor housing. The cell may be constructed so that the electrolytes at the anode and cathode are separated (anolyte and catholyte). One of the first cells studied was the Daniell cell and it is used as the basis of many introductions to the thermodynamics of electrochemical processes. It consists of a zinc anode and copper cathode dipped into solutions of their sulfates. A porous separator prevents mixing of the two solutions. Symbolically the cell is written:

$$Cu|Cu^{2+}||Zn^{2+}|Zn$$

The overall cell reaction is the sum of a cathodic and an anodic reaction:

$$Cu^{2+} + Zn \rightleftarrows Cu + Zn^{2+} \qquad (1)$$

When the two electrodes are connected externally through a load (high resistance is assumed so that the reaction proceeds reversibly), nF coulombs of charge per mole flow from the cathode to the anode through the potential difference E_{rev}, which is the reversible potential of the cell. Now ΔG is the free energy of the system exchanged with the surroundings minus the $P\Delta V$ work. Since the system is designed so that no other work is performed in addition electrical work:

$$\Delta G = -nFE_{rev} \qquad (2)$$

Potential is the energy scale in electrochemical processes just as temperature serves for chemical reactions, and wavelength for spectroscopic and photochemical methods.

The reversible cell potential may be calculated from the free energy change ΔG. For electricity sources, the reversible cell potential is the maximum voltage available from the element, whereas for synthesis its significance is the minimum voltage required to bring about the reaction. The voltage efficiency (ξ) for synthesis may thus be defined:

$$\xi = E_{rev}/E_{cell} = -\Delta G/nFE_{cell} \qquad (3)$$

The actual cell potential deviates from the reversible potential owing to potential drop in the electrolyte and the phenomenon of overvoltage, which are important when the cell current is not zero, as is the case in processing applications.

The concentration dependence of the reversible potential is given by the Nernst equation which is derived from equation 2:

$$E_{rev} = E^0 + \frac{RT}{nF} \ln \Pi a_i^{v_i} \qquad (4)$$

where E^0 $(= -\Delta G^0/nF)$ is the standard potential, ie, the cell potential with all species in their standard states, a_i is the activity of the ith species and v_i is its stoichiometric coefficient. The stoichiometric coefficient is positive for the reactants and negative for the products. The product Π is taken for all species involved in the electrode reaction. Thus, for the Daniell cell:

$$E_{Daniell} = -0.7628 - (-0.3402) + \frac{RT}{2F} \ln \frac{a_{Cu^{2+}}}{a_{Zn^{2+}}} \qquad (5)$$

It has been found useful to define a single or standard electrode potential as its potential versus the normal hydrogen electrode (NHE). The potential of the normal hydrogen electrode ($H^+ + e \rightleftarrows 1/2\ H_2$) is by convention taken to be 0.000V. Single electrode potentials may therefore be obtained by measuring the potential of the cell:

$$Pt|H_2|H^+||X|M \quad \text{where} \quad X|M \text{ represents the electrode of interest}$$

For example, to obtain the potential of the copper electrode, the potential of the cell $Pt|H_2|H^+||Cu^{2+}|Cu$ should be measured. The cell has been written here according to the IUPAC convention with the NHE on the left. Tables listing standard electrode potentials are available. By using the Nernst equation, the reversible potential for a particular set of conditions may be calculated. The cell potential is given by the difference of the two separate electrode potentials. For organic processes, which are usually irreversible, it is often useful to calculate a standard (reversible) electrode potential from ΔG using equation 2.

Rate of Electrode Reactions

In 1905, Tafel found empirically that the rate of an electrode reaction could be expressed by an equation of the form

$$E = a - b \ln j \tag{6}$$

The electrode reaction rate, observable through the current density j, is exponentially proportional to potential. The exchange-current density may be obtained from a Tafel plot, ie, a plot of $\ln j$ against E, by extrapolating the linear section to the reversible potential. For irreversible reactions, a pseudo j_0 may be obtained by extrapolating the plot back to the E_{rev} calculated from ΔG (see also Corrosion).

Factors limiting the overall reaction rate. The Tafel plot shows that the current (proportional to reaction rate) obeys an exponential law over only a limited range of overpotential. At high overpotentials, the current takes a limiting value. Thus, only at low overpotentials is the process charge-transfer (or activation) controlled.

In addition, chemical reactions involving electroactive species, eg,

$$A + O \underset{\overleftarrow{k}}{\overset{\overrightarrow{k}}{\rightleftharpoons}} B \tag{7}$$

(where O is an electroactive species) can also influence the rate of transport to or from the electrode. An example is the manufacture of chlorates.

Lastly, adsorption and desorption must not be neglected since they can greatly alter the surface concentration of species and the rate of the electron transfer.

Productivity. In order to maximize the productivity, usually an industrial process is performed with as high a mass-transfer rate to the electrode as possible. This is determined often by the hydrodynamic conditions present. Some important mass-transfer correlations are shown in Table 1.

However, because electrochemical processes are heterogeneous, large electrode areas are typical. It is therefore desirable to select a geometric form that allows a compact structure.

Current and Potential Distribution

An electrosynthesis is particularly easy to control simply through the selection of the electrode potential. The full advantage may be taken only when the potential is uniform over the whole electrode area. Generally, a good current-density distribution is obtained at large Wagner number, $Wa = k(d\eta/di)/l$, where k is the conductivity of the electrolyte, S/cm, η is the overpotential, V, and l is a characteristic length.

Energy Balance

The heat produced by an electrochemical reaction is determined by the enthalpy change accompanying the reaction and the electrical energy input to the cells. The heat balance for the electrolysis system is thus: $h_2 - h_1 + IE - Q = 0$, where h_1 and h_2 are the enthalpy fluxes for reactants and products, respectively (kJ/s), IE is the electrolysis power input and Q is the heat flux to the environment. For isothermal operation, Q is the heat that must be dissipated in a heat exchanger.

Optimization

The optimum operating conditions of an electrochemical process may be equated to the minimum total cost for a given production rate or to the maximum return on invested capital. There are also several different operating modes, eg, batch or continuous and direct or indirect, that should be considered in the optimization, as well as the possibility of different electrochemical routes, eg, oxidation or reduction to the desired product.

Process Trends

Two of the most important areas of development are improving electrode materials through the use of catalytic coatings, eg, DSA (Dimensionally Stable Anodes, Diamond Shamrock) electrodes (see Metal anodes), and the development of compact cells with large electrode areas.

Nomenclature

a	= activity; Tafel coefficient, specific electrode cost
b	= Tafel coefficient; unit cost of electricity
D	= diffusion coefficient
D_e	= equivalent diameter
d_h	= hydraulic diameter
d_p	= particle diameter
E^0	= standard potential (V)
F	= faraday constant (96,485 C/M)
ΔG	= change in Gibbs free energy
$Gr = g(\rho_b - \rho_s)$ $L^3/\nu^2\rho_b$	= Grashof number
h	= enthalpy flux
j	= current density
k	= mass transfer coefficient; rate
L	= electrode length
Δl	= promoter spacing
Q	= heat flux
R	= cylinder diameter; resistance; reductant
$Re = ud_h/\nu$	= Reynolds number
$Re_p = ud_p/\nu$	= Reynolds number based on particle diameter
$Sc = \nu/D$	= Schmidt number
Sh	= Sherwood number
Wa	= Wagner number
κ	= electrolyte conductivity
ξ	= voltage efficiency of reversible cell

PETER M. ROBERTSON
Eidgenossische Technische
Hochschule, Zürich

J.O'M. Bockris and A.K.N. Reddy, *Modern Electrochemistry*, Plenum Press, New York, 1970.

K.J. Vetter, *Electrochemical Kinetics*, Academic Press, Inc., New York, 1967.

D.J. Pickett, *Electrochemical Reactor Design*, Elsevier Scientific Publishing Company, Amsterdam, Neth., 1977.

INORGANIC

Electrochemical processes involve the transfer of electrons between an electrode and a substrate in solution. The energy required is in the range 0–35 eV, and depends on the electrode material and the substrate. This is a moderate energy input in comparison to photochemical-or-radiation activation methods. Table 1 lists the most important processes that are now practiced.

Table 1. Mass Transfer Correlations of Various Hydrodynamic Regimes in Electrolytic Cells

parallel-plate reactor with turbulent flow	$Sh = kd_h/D = 0.023 Re^{0.8} Sc^{0.33}$
parallel-plate reactor with turbulence promoters	$Sh = k \cdot d_h/D = 0.272\, Re^{0.631} Sc^{0.33} (D_e/\Delta l)^{0.357}$
rotating cylinder	$Sh = k \cdot R/D = 0.0642 \nu^{0.3} Sc^{-0.644} (R \cdot u_r)^{0.7}/D$
packed bed	$Sh = k \cdot d_p/D = 1.44 Re_p^{0.58} Sc^{0.33}$
natural convection at vertical electrodes ($10^4 < GrSc < 10^{12}$)	$Sh = k \cdot L/D = 0.66(Gr \cdot Sc)^{0.25}$

Table 1. U.S. Production of Electrolytically Manufactured Elements and Compounds, Listed in Order of Amount Produced

Metals	Chemicals
aluminum[a]	chlorine/caustic
copper	soda ash
zinc	potash/caustic
sodium[a]	sodium chlorate
chromium	lithium carbonate
magnesium[a]	manganese oxide
nickel	permanganates
titanium[a]	fluorine
cadmium	perchlorates
cobalt	hydrogen/oxygen

[a] By fused-salt processes.

Hardware for Electrochemical Processing

The distinguishing feature of an electrochemical process is the electrolysis cell and its power supply. Like many chemical processes, the feedstock must be made up by the addition of solvents/electrolytes and the product must be extracted or worked up.

In an electrochemical cell, feedstock is transported to the electrode/electrolyte interface. The design of an electrolysis cell is therefore based, among other factors, on optimization of transport of electroactive species from the cell volume to the electrode surface, materials and topography of the electrode, and possible need to separate the products or reactants of the anode and cathode reactions. The main design possibilities of electrolysis cells are summarized in Table 2.

Production Conditions

Electrolysis of chloride solutions. Chloride may be oxidized electrochemically to chlorine or hypochlorite and chlorate and perchlorate. The electrolysis of brine is one of the oldest and certainly one of the most important and widespread industrial electrochemical processes. The overall reaction is

$$2\,NaCl + 2\,H_2O \xrightarrow[\text{energy}]{\text{electrical}} 2\,NaOH + Cl_2 + H_2 \qquad (1)$$

There are three main technologies available for carrying out this process: diaphragm cells, mercury cells (diaphragmless), and membrane cells. The latter is a recent development and should make a great contribution to future production. The most frequently used technology is the diaphragm cell such as Solvay Type DBT 26, DeNora-Diamond Type Elincor 46, or Diamond Type DS 85, which operate at ca 1–3 kA/m^2 current density, a current efficiency of > 90%, and an energy consumption of ca 3 (kW·h)/kg.

Chlorates. The manufacture of chlorates is carried out by the electrolysis of brine at pH 6.5–7.4 with graphite or dimensionally stable anodes (DSA) and steel cathodes. The formation of chlorate occurs by a purely chemical reaction between OCl$^-$ and HOCl.

$$2\,HOCl + ClO^- \rightarrow ClO_3^- + 2\,Cl^- + 2\,H^+ \qquad (2)$$

Table 2. Cell Types Used for Inorganic Electrochemical Processing

Cell type	Example of application
One-compartment cells	
inert vessel with immersed electrodes	
monopolar electrodes	chlorates, perchlorates, MnO$_2$, etc
bipolar electrodes	chlorates, metal refining
vessel is one electrode (monopolar electrodes)	chlorates, fluorine
horizontal liquid-metal cathode	aluminum, chlorine/caustic
Two-compartment cells	
diaphragm cells	
monopolar electrodes	chlorine/caustic
bipolar electrodes	
filter press	
bipolar electrodes	hydrogen/oxygen

The main use for sodium chlorate is as a precursor for chlorine dioxide, which is used as a bleach in the pulp and paper industry (see Chlorine oxygen acids and salts).

Perchlorates. Chloride solutions may be electrolyzed directly to perchlorate; however, only low yields are obtained. The normal procedure is to electrolyze a concentrated solution (300–600 g/L) of the chlorate. Perchlorates are used mainly for the production of the rocket fuel, ammonium perchlorate.

Manganese dioxide. Although most battery needs are met with selected MnO$_2$ ores, there are special applications, such as high capacity cells, that require the very pure product that is provided by electrolytically produced manganese dioxide. The production method involves the leaching of reduced manganese ore (MnO) with H$_2$SO$_4$ (or spent electrolyte) and anodic deposition of MnO$_2$ on a graphite or titanium anode at 90°C.

Water electrolysis. Where a source of cheap electricity is available, electrolytic hydrogen may provide the feedstock for an ammonia plant, as is the case at Aswan, Egypt, where 56 Brown-Boveri EBK 385-70 electrolyzers produce hydrogen at the rate of 210 m^3/min.

Fluorine. Fluorine (qv) is the most reactive product of all electrochemical processes. It has been manufactured on a technical scale since the 1940s mainly for use in uranium-enrichment plants. Manufacture of fluorine is now based on medium temperature (65–75°C) electrolysis of KF: 12 HF.

Permanganate. Potassium permanganate is prepared from manganese dioxide or directly from manganese metal or alloys. The synthesis of permanganate from manganese dioxide first involves the preparation of potassium manganate by roasting manganese dioxide with potassium hydroxide in air. The manganate is dissolved, filtered and then oxidized to permanganate at a concentration of 120–180 g/L with 80–120 g/L KOH in an electrolytic cell.

Electrowinning of Metals

The metals that are currently produced by electrolysis are shown in Table 1. Fused-salt processes are used when the reactivity of the metal does not allow deposition from aqueous solutions.

Safety and Environmental Aspects

The electrolysis of aqueous solutions often leads to the formation of gaseous products at both the anode and the cathode which may be hazardous if combined. The main examples are the H$_2$/Cl$_2$ and H$_2$/O$_2$ products of the chlor-alkali and water-electrolysis plants.

Chlorine production can cause environmental problems owing to losses from mercury-cell operation originating from brine treatment, drainage, thick mercury treatment, emissions from the cell-room area, and entrainment with caustic and hydrogen. Mercury may be removed from liquids by either reduction and filtration or oxidation (with hypochlorite) and absorption by an ion-exchange resin. Good results have also been obtained by absorption in beds of active charcoal. Modern mercury-cell plants have an emission of only 1 g Hg/t Cl.

The operation of aluminum electrolysis at 950°C results in considerable evaporation and generation of waste gases containing fluorine in various forms. This emission is particularly undesirable owing to the high toxicity of fluorine to livestock and plants.

PETER M. ROBERTSON
Eidgenossiche Technische
Hochschule, Zürich

J. O'M. Bockris and A.K.N. Reddy, *Modern Electrochemistry*, Plenum Press, New York, 1973.

ORGANIC

Synthesis of organic chemicals by electrolysis is a very old science but a relatively new technology. Electrolytic oxidation, reduction, and coupling of organic molecules have been carried out in the laboratory for more than a century, but only in the past 10–15 yr have organic chemicals been produced on an industrial scale by electrochemical means.

Factors stimulating recent interest in this area are development of stable, noncorroding anodes; availability of durable, ion-permselective membranes; significant improvements in design and construction of electrolytic cells; and a better understanding of and some novel solutions to mass-transfer problems in electrolysis systems.

Electrochemical synthesis of organic chemicals may offer economic advantages over the more conventional chemical processing schemes for one or more of the following reasons: lower-cost raw materials; fewer reaction steps; higher selectivities; avoidance of environmentally troublesome by-products; and simplified product recovery and purification. Against these must be weighed the cost of the required electrical energy and relatively high electrolysis system capital cost. However, in many instances the electrochemical route provides a means of selectively directing the reaction via potentiostatic control to a product not readily obtained by the usual oxidation–reduction agents.

Hardware for Electro-Organic Processing

Electrochemical cells for the production of organic compounds are as diverse in design and construction as are chemical reactors for nonelectrolytic processes. Cell materials and geometry must be tailored to meet specific process and economic considerations and the interrelationship between these factors may differ greatly from one organic system to the next. However, all systems will include electrodes, electrolytes, and diaphragms. The cell types may be classified as follows: plate-and-frame cells; planar or cylindrical electrodes in tanks; and particulate electrode cells, either fixed-bed or fluidized-bed types.

Scale-up considerations. In the scale-up of an electro-organic reaction system from the bench-top experiment to a commercial prototype, many of the engineering considerations are the same as for a conventional chemical process. Assuming an electrode–electrolyte combination has been identified in the laboratory that gives satisfactory product selectivity and current efficiency, the following factors generally need to be addressed during scale-up: electrode and bulk-electrolyte kinetics; mass transfer in the cell; heat removal; cell productivity and cost; current distribution; voltage minimization; and materials of construction. Electro-organic systems often pose unique problems in mass transfer because of the low solubility of the reactants in the electrolyte. Similarly, heat removal and current distribution may require special attention because of the lower conductivity electrolytes frequently employed. The commercialization of an electrochemical process for large-scale production of organic chemicals depends primarily on three economic criteria: product selectivity, electrical power usage, and electrolysis system capital.

Industrial Electro-Organic Processes

Two principal electro-organic processes that are being operated on an industrial scale are the Nalco facility which produces 15,000–18,000 t/yr of tetraalkyllead compounds (Fig. 1), and the Monsanto plants that manufacture 180,000 t/yr of adiponitrile by electroreductive coupling of acrylonitrile (Fig. 2).

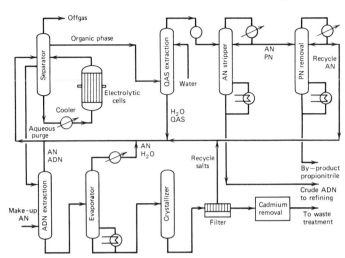

Figure 2. Monsanto adiponitrile process. QAS, quaternary ammonium salt. ADN, adiponitrile; PN, propionitrile; AN, acrylonitrile. Courtesy of Gulf Publishing Company

Although the scale of operation of other known electro-organic processes is considerably below those of the Nalco and Monsanto processes, several are important in providing routes to organic compounds that are difficult to prepare selectively by more conventional means. As an example, dihydrophthalic acids can be produced in yields exceeding 90% by reduction of phthalic acid at lead and mercury cathodes, whereas catalytic hydrogenation leads to a mixture of the di-, tetra-, and hexahydrophthalic acids. Similarly, the electroreductive coupling of acetone to pinacol has received considerable attention over the years as a highly selective route to this intermediate, useful in production of pharmaceuticals and pesticides. An anodic process of commercial importance is the Kolbe synthesis of dimethyl sebacate from monomethyl adipate, which has been conducted on an industrial scale in the USSR. Electrochemical fluorination of organic chemicals has been practiced commercially by the 3M Company for a number of years, and more recently, extensive laboratory studies in this area have been described by Phillips Petroleum company (see Fluorine compounds, organic).

Donald Danly
Monsanto Chemical Intermediates Co.

L.L. Bott, *Hydrocarbon Process.* **44**(1), 115 (1965).

ELECTROCHEMISTRY. See Electrochemical processing.

ELECTRODECANTATION

Electrodecantation (also called electrogravitational separation and electrophoretic convection) is based on a stratification phenomenon that may take place when colloidal dispersions are subjected to an electrical field between vertical membranes permeable to the electrical current and retained on the membrane. Semicolloids and electrolytes also stratify in a similar manner on selectively permeable membranes; however, the basic phenomena differ somewhat in the mechanism of the formation of the stratifying layers on the membranes. With colloids, the charged particles migrate under the influence of the electrical field toward one of the poles (see Electrodialysis) and are retained and sometimes accumulated on the membranes, and freed by reversal of polarity. Under certain conditions, the thin concentrated layer on the membrane surface moves up or down between the surrounding liquid boundary and membrane interface, according to its relative density; at the opposite membrane the movement of the dilute layer takes place in the other direc-

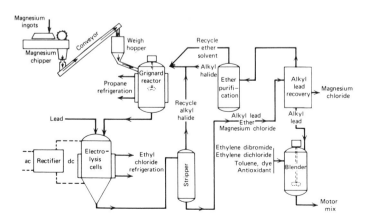

Figure 1. Nalco tetramethyllead process. Courtesy of Gulf Publishing Company.

tion. This phenomenon was first observed by Pauli in experiments on electrodialysis and had been used by him and his co-workers for studying the structure and composition of a great variety of colloidal dispersions including biological liquids and proteins.

The separation into stratified layers was found to be independent of the particle size. The behavior of the material retained on the membrane depends upon the migration velocity of the particles in the electrical field vel_e, and the velocity of gravitational movement of the enriched layer vel_g. If vel_g is greater than vel_e, stratification will take place continuously, but if vel_e is greater, a deposit will accumulate on the membranes.

Research and Development

Electrodecantation of biological systems has been the subject of frequent publications. It has been used for the isolation of immunoglobulin antibody activity in relation to the hepatitis-associated antigen. The active immunoglobulin hepatitis-associated antigen (HAA) was separated by an electrodecantation method. Various isoelectric points were used by means of convectional electrophoresis on modified electrodecantation fractions of human gamma globulin. Electrodecantation is used for enzyme purification and concentration of macromolecules by electrophoresis sedimentation.

The first commercial application of electrodecantation has been in the concentration of rubber latex. Electrodecantation has also been used for concentrating electrolytic coating baths.

PAUL STAMBERGER
Crusader Chemical Co.

W. Pauli, *Anz. Akad. Wiss. Wien. Math Naturwiss. Kl.* **60**, 10 (1923).

W. Pauli, *Trans. Faraday Society* **31**, 19 (1933).

W. Pauli, *Helv. Chim. Acta* **25**, 155 (1942).

ELECTRODIALYSIS

Electrodialysis (ED) is a process for moving ions across a membrane from one solution to another under the influence of a direct electric current (see also Dialysis; Ion exchange).

Membranes

In 1950, ion-selective membranes having high selectivity, low electrical resistance, good mechanical strength, and good chemical stability were described. These were essentially insoluble, synthetic, polymeric organic ion-exchange resins in sheet form. Typical modern membranes of this type are based on cross-linked polystyrene. Those selective to cations usually have sulfonate groups bound to the polymer; those selective to anions usually have quaternary ammonium groups bound to the polymer. Commercial membranes are usually reinforced with woven, synthetic fabrics to improve the mechanical properties. The electrochemical properties of most importance for ED are the electrical resistance per unit area, the ion-transport number (related to current efficiency), the electrical water transport (related to process efficiency), and the back-diffusion (also related to process efficiency). Commercial ion-selective membranes have thicknesses of ca 0.15–0.6 mm and electrical resistances of ca 3–20 $\Omega \cdot cm^2$ at RT when in equilibrium with 0.5 N sodium chloride. They have ion transport numbers in dilute solutions of ca 0.85–0.95, electrical water transport ca 100–200 cm^3 per equivalent of ions transferred (toward the low end of the range for anion-selective membranes and toward the high end for cation-selective membranes), and back-diffusion at sodium chloride solutions of 0–1 N amounting to less than ca 2×10^{-6} meq/(s·cm²).

Apparatus

ED apparatus is fundamentally an array of alternating anion-selective and cation-selective membranes terminated by electrodes. Membranes are separated from each other by gaskets which form fluid compartments. Compartments that have anion-selective membranes on the side facing the (positively charged) anode are electrolyte-depletion compartments. The remaining compartments are electrolyte-enrichment compartments (see Fig. 1). The enrichment and depletion compartments also alternate through the array. Holes in the gaskets and membranes register with each other to provide two pairs of hydraulic manifolds to carry fluid into and out of the compartments, one pair communicating with the depletion compartments and the other with the enrichment compartments.

In the case of concentrated electrolytes, electric current applied to a stack (up to several hundred cell pairs) is limited by economic considerations; the higher the current I, the greater the power consumption W in accordance with: $W = I^2 R_s$, where R_s is the electrical resistance of the stack. In the case of relatively dilute electrolytes, the electric current applied is limited by the ability of ions to diffuse to the membranes.

Polarization. At some applied current density (i_{max}), proportional to the concentration of the effluent from the depletion compartments, the anion membrane (and the apparatus) are said to be concentration polarized or simply polarized. At i_{max}, the fluid in the depletion compartments at the surface of the anion-selective membranes is depleted of electrolyte essentially completely because of diffusion limitations and the electrical resistance of the apparatus increases substantially even though the bulk solution in such compartments still may contain an appreciable concentration of electrolyte. This is one sign of polarization. Another sign is a decrease in the pH of the enrichment compartments and an increase in the depletion compartments. Finally, if i_{max} is exceeded substantially, then electrolytes that are insoluble at high pHs may precipitate at the interface between the anion-selective membrane and the enriched solution. This is the third sign.

Performance. The performance of an ED stack may be estimated by considering the material balance around the stack:

$$\Delta(vc) = v_i c_i - v_o c_o = \frac{\bar{i} E A_p N_p}{F}$$

where v_i and v_o are the flow rates in L/s into and out of the depletion-compartments of the stack, respectively; c_i and c_o are the concentrations in equiv/L into and out of the depletion compartments, respectively; $\bar{i}$ is the average current density in A/cm²; E is the net ion transfer ($\bar{t}_-^A - \bar{t}_+^C$ where t_-^A is the fraction of current carried by anions in the anion selective membranes and t_+^C is the fraction of current carried by anions in the cation-selective membranes; A_p is the current-carrying area per cell pair, cm²; N_p is the number of cell pairs in the stack; and F is Faraday's constant. Typically, a stack is designed so that $v_o c_o$ is ca 50% of $v_i c_i$. This is because $\bar{i}_{max}$ is determined by c_o.

Uses

More than 1000 electrodialysis plants have been installed to produce high purity water and for the production of potable water from brackish water, most such plants of the reversing type. ED also plays a prominent role in the extraction of salt from seawater and in the production of low-ash cheese whey. Other applications include recycling of cooling tower blowdown; and electrowinning of metals.

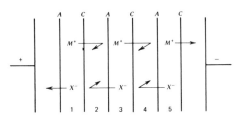

Figure 1.   Principle of multicompartment electrodialysis.

Nomenclature

A_p = current-carrying area per cell pair, cm^2
c = concentration, eq/L
E = net ion transfer or transport, $\bar{t}_-^A - \bar{t}_+^C$
F = Faraday's constant, 96.5 kAs
i = current density, A/cm^2
I = current, A
N_p = number of cell pairs in a stack
R_s = electrical resistance of stack, Ω
$\bar{t}_-^A$ = fraction of current carried by anions in anion selective membrane
$\bar{t}_+^C$ = fraction of current carried by anions in cation selective membrane
v = flow rate, L/s
W = power consumption, W

Wayne A. McRae
Ionics, Inc.

J.R. Wilson, ed., *Demineralization by Electrodialysis*, Butterworths, London, UK, 1960.

W. Juda and W.A. McRae, *J. Am. Chem. Soc.* **72**, 1044 (1950).

E.A. Mason and T.A. Kirkham, *Chem. Eng. Prog. Symp. Ser.* **55**, 173 (1959).

ELECTROLESS PLATING

Electroless plating is the controlled autocatalytic deposition of a continuous film by the interaction in solution of a metal salt and a chemical reducing agent. Electroless deposition can give films of metals, alloys, compounds, and composites on both conductive and nonconductive surfaces. The theory and practice of electroless plating parallels that of electrolytic plating. The main difference is that the electrons used for reduction are supplied by a chemical reducing agent present in solution. This means that electroless solutions are not thermodynamically stable because the reducing agent and the metal salt are always present and ready to react.

Electroless solutions contain a metal salt, a reducing agent, a pH adjuster or buffer, a complexing agent, and one or more additives to control stability, film properties, deposition rates, etc. Only a few metals —nickel, copper, gold, and silver—are used on any significant scale. The ideal electroless solution would deposit metal only on an immersed article, never as a film on the sides of the tank or as a fine powder. Many electroless copper and electroless nickel baths now closely approach this ideal.

A great advantage of electroless solutions is their ability to give conductive metallic films on properly prepared nonconductors, and their ability to uniformly coat any platable object.

Equipment

Plating tanks are preferably plastic or lined with a plastic or rubber coating. The tank linings must be stripped of metal deposits with acid at periodic intervals. Tank and rack design are less important than tank loadings and rack coatings, especially when plating nonconductors. The critical point is not to overload the plating bath. Electroless baths rapidly become overactive and decompose when too much surface area is being plated at once. Most vendors recommend a maximum loading of ≤ 365 cm^2/L (1.5 ft^2/gal).

Safety and Waste Disposal

Most solutions are not particularly hazardous although they may be highly acidic or basic. All reducing agents (solids or liquids) should be stored separately from oxidizing substances such as chromic acid, in case of spills. Waste treatment should always conform to current rules. The vendors' recommendations should be followed for specific cases.

Plating on Metals

The first large-scale process was the Kanigen electroless-nickel process from General American Transportation Co. which uses a hypophosphite

reducing agent. The Ni–P alloy (1–15 wt% P) has good corrosion resistance, lubricity, and especially hardness. It can be heat-treated to a hardness equivalent to electrolytic hard chromium, and its lubricity is comparable. Thus, it is not surprising that many applications for electroless nickel are in replacement of hard chromium.

The advantages over hard chromium include safety of use, ease of waste treatment, plating rates of as much as 40 μm/h, low porosity films, and the ability to uniformly coat any geometric shape without burning or use of special anodes. Electroless nickel has superior corrosion and erosion resistance as compared to electrolytic nickel. It is used extensively on molds, pistons, pump parts, oil-field equipment, dies, compressors, tanks, and piping. Electroless-nickel systems based on dimethylamineborane or sodium borohydride reducing agents have advantages in the plating of titanium for the aircraft industry. They are also used to minimize the use of or completely replace gold in connector applications in the electronics industry. Specialized nonmagnetic electroless nickels are becoming the coating of choice for aluminum memory disks in the computer industry (see Electrical connectors).

Plating on Nonconductors

All nonconductors can be electrolessly plated but only a few can be plated to give good adhesion and appearance.

Plating of plastics. This market began in 1963 on a very small scale using electroless copper. It was soon discovered that ABS engineering thermoplastic was the easiest plastic to plate. Plated ABS has about 90% of the market; most of the remainder is modified polycarbonate and modified poly(phenylene oxide). The principal suppliers of plating-grade resins are Borg-Warner, General Electric, Monsanto, and Dow. Automotive items make up over 60% of the market on a plated-area basis; the remainder is hardware, plumbing, and decorative items. The largest new growth market for plating on plastics is the field of EMI/RFI (Electromagnetic Interference/Radio Frequency Interference) shielding. Pure electroless nickel, pure electroless copper, and combination coatings of copper plus nickel are being used. A typical process cycle for plating on plastics is shown in Table 1.

Printed circuits. This is by far the most diverse field of electroless plating. Numerous variations exist in solution processes, and techniques, both in laboratory and commercial form, and are used to create a great variety of products. All have the basic purpose of producing a layer of highly conductive copper in specified areas. Modern electroless copper films have a ductility and conductivity identical to that of electrolytic copper.

Printed-circuit boards are composed of epoxies, phenolics, and other heat-stable dielectrics; flexible films of polyester and polyimide are also plated. The ratio of rigid to flexible surface areas plated is about 10 : 1. They are used in communications, instruments, control, consumer markets, military, aerospace, and business applications (see also Integrated circuits). Multilayer printed circuit boards are rapidly increasing their market share, especially for computer applications.

Other Applications

Mirror production is the primary application of electroless silver. The other important commercial glass-plating application is for production of

Table 1. Plating on Plastics[a]

Step	Solution	Temperature, °C	Time, min
etchant	CrO$_3$–H$_2$SO$_4$	60	5–10
neutralizer	mild Cr^{6+} reducer	25	$\frac{1}{2}$–2
catalyst	PdCl$_2$–SnCl$_2$–HCl	25	1–5
accelerator	mild acid or alkali	50	1–2
electroless nickel bath	nickel salts, complexer, buffer, hypophosphite; pH 8–10	25	5–10
or			
electroless copper bath	copper salts, complexer, buffer, formaldehyde; pH 12–13	25–40	5–10

[a] As in all plating processes, thorough rinsing is necessary after each step. This is a typical ABS plating cycle; for other plastics plating differs mainly in the pretreatment or etchant and neutralizer steps. The electroless metal deposit is typically 0.15–0.5 μm thick; electrolytic plating follows.

architectural reflective glasses. An electroless-nickel matrix is used commercially to securely bond diamonds to cutting tools.

GERALD A. KRULIK
Borg-Warner Corp.

"Symposium on Electroless Plating," *Am. Soc. Test. Mater. Spec. Tech. Publ.* **265**, (1959).

K.M. Gorbunova and A.A. Nikiforova, *Physiochemical Principles of (Chemical) Nickel Plating*, Academy of Sciences, Moscow, U.S.S.R., 1960 (Engl. transl. available from U.S. Government Office of Technical Services, Washington, D.C., 1963).

The Engineering Properties of Electroless Nickel Deposits, International Nickel Co., New York.

R.G. Wedel, *Int. Metal Rev.* **217**, 97 (1977).

G.A. Krulik, *J. Chem. Ed.* **55**, 361 (1978).

ELECTROLYTIC MACHINING METHODS

One of the promising novel machining methods for extremely hard alloys is the atom-by-atom removal of a metal by anodic corrosion which has been called electrochemical machining (ECM). Because the metal is removed atom-by-atom, ECM affords the opportunity to machine a given workpiece without work-hardening, burring or smearing the metal, without regard to the hardness of the metal being cut, and virtually without tool wear.

In a chemical reaction, electron transfer between the reacting species occurs by the oxidation (loss of electrons) and the reduction (gain of electrons) taking place at the same site, and the energy liberated appears as heat and possibly light. For an electrochemical reaction to occur, the oxidation must take place at a site remote from the reduction. This situation is accomplished by interposing an electrolyte between the conductor (anode) at which oxidation occurs and that (cathode) at which reduction occurs (see Electrochemical processing).

When iron (qv) dissolves, ferrous ions enter the electrolyte by donating electrons to the anode. Eventually, the anode becomes so negative that further dissolution of iron is prevented by the electrostatic attraction between negatively charged anode and the positive ions in solution. If a sink for the electrons on the anode is provided by a battery that is connected in an external circuit between the anode and cathode, continuous dissolution of the iron anode can be obtained. Current is carried in the circuit by electrons (electronic conduction), and internally by ions (electrolytic conduction) through the electrolyte. The dissolution of metal at an anode driven by an external source of current, such as a battery or rectifier, is termed anodic corrosion (see Corrosion; Metal anodes).

Principles of the ECM Process

In the ECM operation, the metal workpiece to be shaped is the anode, and the tool that produces the shaping is the cathode. The electrodes are connected to a low voltage source of direct current (dc). The anode and cathode are held in close proximity (0.5 mm) by a properly designed fixture, and a solution of a strong electrolyte is pumped at high rates (2–6 m/s) between the two electrodes. If there are no side reactions, the passage of each Faraday of electrical charge (96,500 C) results in the dissolution of an equivalent weight of metal. Under these conditions, very high current densities may be passed (50–500 A/cm^2) so that high metal removal rates (2 cm^3/(min·kA)) for 4340 steel) may be obtained with good tolerances, with integrity of the geometry, and with excellent surface finishes. Voltages applied between the anode and cathode are normally 5–25 V.

Types of ECM operations are shown in Figure 1. Innovations in ECM operations include static fixture-finishing and sizing, embossing and broaching.

ECM is commonly employed in the transportation and aerospace industries. Turbine blades, vanes, disks, and transmission parts are made

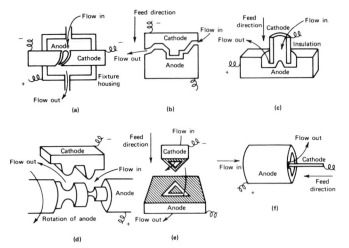

Figure 1. Various ECM operations. (**a**) External shaping; (**b**) cavity-sinking; (**c**) plunge-cutting; (**d**) turning; (**e**) trepanning; (**f**) internal grooving.

by ECM. Electrochemical machining is also used in the manufacture of precision guidance parts for missiles.

JAMES P. HOARE
MITCHELL A. LoBODA
General Motors Research Labs

A.E. DeBarr and D.A. Oliver, *Electrochemical Machining*, Elsevier Scientific Publishing Co., Inc., New York, 1968.

J.F. Wilson, *Practice and Theory of Electrochemical Machining*, John Wiley & Sons, Inc., New York, 1971.

J.P. Hoare and M.A. LaBoda in E. Yeager and A. Salkind, eds., *Techniques of Electrochemistry*, "Electrochemical Machining," Wiley-Interscience, a division of John Wiley & Sons, Inc., New York, 1978, Chapt. 2, p. 48.

J.A. McGeough, *Principles of Electrochemical Machining*, Chapman and Hall Ltd., London, Eng., 1974.

ELECTROMIGRATION

Electromigration or electrotransport refers to the net motion of atoms owing to the passage of an electrical current through a metallic conductor in either the solid or liquid state. Simple metals may be considered to be constituted of atoms stripped of their valence electrons, existing as ions surrounded by a sea of electrons. In the presence of an electric field, the ion cores are subjected to a force directly resulting from the field, as in an electrolytic solution. However, in electromigration there is, in addition to this direct field force, a force resulting from the friction between the numerous and rapidly moving electrons and the ion cores. This force, known as the electron-wind force, is in the opposite direction to the field force in electron conductors, and for many metals is considerably greater than the field force. In metals with complex electronic structures that are not electron conductors but are hole conductors, the concept of electron-wind force is extended to refer to the friction resulting from the motion of the charge carriers regardless of their sign. Whereas in electrolytic solutions the current results entirely from the motion of ions, in usual electromigration phenomena the fraction of the current contributed by ion motion is quite negligible in comparison to that fraction of the current caused by the motion of the charge carriers, either electrons or holes.

Fundamental Aspects

Phenomenologically, electromigration results from a slight bias in the random motion of atoms and can be considered as a special case of diffusion (see Diffusion separation methods). The velocity of the moving atoms v is given by the product of the mobility and the force, according to the Nernst-Einstein relation. This can be written as:

$$v_i = \frac{D_i}{kT} \cdot F_i \qquad (1)$$

where D is the diffusion coefficient, F is the force on the moving atoms, k is the Boltzmann constant, T is the temperature, and the subscript i refers to the atoms of the ith species.

The force on an atom in the presence of a current is usually considered to consist of two parts, one owing to the field F_e and one owing to the friction of the charge carriers F_{wd}. Of course, this division is somewhat arbitrary and ideally one would hope to arrive at a single formulation. However, the consideration of those two types of force is intuitively suggestive. The force in equation 1 is of a statistical kind with a unique value averaged over a large number of atomic jumps.

In general, the electron-wind force is expressed as:

$$Z_{wd}^* = -Z \cdot \gamma \cdot \frac{N \cdot \rho_d}{N_d \cdot \rho} \qquad (2)$$

where Z is the valency of the matrix atoms, γ is an averaging term, ρ_d is the specific resistivity of the mobile defects, N_d/N is the concentration of moving defects, and ρ is the resistivity of the metal.

Common experimental techniques. For most solid materials, the diffusion coefficients and the effective charges are sufficiently small so that experimental measurements of electromigration transport can be made only under circumstances that maximize the effect of both terms. In practice this means that electromigration experiments on bulk samples, as distinguished from thin films, are conducted at the highest current densities that are compatible with the requirement that the samples should not melt as a result of Joule heating. Generally, the current densities used are of the order of 10 kA/cm². Although one finds a variety of experimental techniques, it may be said that most of these conform to one of the three different types: the drift technique, the marker-motion technique, and the steady-state technique.

Purification by Electromigration

The extensive and successful use of electrolysis methods for the purification of metals such as Pb and Cu stimulated attempts to use electromigration in purification of other metals. Many early reports on electromigration used the term solid-state electrolysis although the two processes are basically different.

Purification by electromigration differs from classical chemical separation processes in that redistribution of an impurity occurs within one phase rather than by partition between two different phases. Electromigration purification has so far been a relatively slow process, restricted to small portions of a metal and requiring large amounts of current. It has been used mostly with very reactive metals that have such a high affinity for carbon, nitrogen, and oxygen that common chemical methods have failed. Most of the metals that have been purified were used in research investigations. For this purpose, the small amount of metal processed per run and the high processing cost are not overwhelming considerations.

Formal Basis of Purification

The steady-state concentration profile. The usual form of a specimen to be purified by electromigration is a rod through which a direct electric current is passed longitudinally. This simple geometric shape is both the easiest to model with mathematical relationships and also seems to be so far unsurpassed in achieving the highest purity. The migration of a solute in this specimen, if the solute does not escape, must result in an increase in concentration in one end and a depletion at the other. For a rod of uniform cross-section at constant temperature and with an initial uniform concentration, the concentration profile will change with time. Ultimately, a steady-state condition will be reached such that the flux owing to back diffusion will be equal to the electromigration flux.

The potential of electromigration as the basis of a purification method has been evaluated for a number of metal-impurity systems. For available electromigration and diffusion data, the average purity of the purest half of a rod was calculated by use of realistic values of the length of the specimen and the electric field. In many systems, the ratio of the electric mobility to the diffusion coefficient is large enough that very substantial purification can be achieved if the steady state can be reached.

Most of the study of purification by electromigration has been on interstitial solutes in solid metals. For many reactive metals, it has been established as a simple and effective purification method for small quantities of materials. In solid metals fast-diffusing substitutional impurities can be removed but the time required for solutes with a normal coefficient of diffusion would be excessive. Liquid metals could be purified quite effectively if mixing in the liquid could be suppressed.

Thin Films

The most prevalent manifestations of electromigration have been the deleterious effects discovered in microcircuit technology. It was found that thin-film conductors of aluminum used in planar silicon circuits can become discontinuous after prolonged passage of a direct current. This discovery set the stage for a period of intense activity in the investigation of electromigration phenomena that are almost entirely specific for thin films. The problems considered below are related exclusively to the metallic conductors that lead the electric current to, or away, from transistors; they do not concern the behavior of the semiconducting devices themselves (see Electrical connectors).

Failure formation and failure mechanisms. In Figure 1 can be seen a discontinuity, a crack across the width of an Al thin-film conductor, resulting from the passage of a current with a density of ca 2000 kA/cm² at a conductor temperature of 160°C for ca 1 wk.

Cracks causing electrical discontinuity are certainly the most dramatic failures caused by electromigration, yet hillocks or whiskers may result in short circuits between conducting lines lying side by side, or lines superimposed with a separating insulation layer (eg, sputtered silicon oxide). Both mass depletions (cracks and holes), and mass accumulations (hillocks and whiskers) result from some specific discontinuity or divergence in the atomic flux along the length of a conductor.

Lifetime tests, failure models. Technological necessities require that at a specified time the rate of failure in electronic devices should remain lower than some maximum, which can be chosen to be as small as desired. The selection of acceptable rate and time values varies according to the type of application considered. Practically, on account of time limitations, this means that one must conduct accelerated failure tests at high temperatures and at high current densities as well as with a limited set of samples. The results thus obtained are extrapolated to normal use conditions (low current densities, low temperatures, long times, and very large device populations). Such extrapolations require an understanding of the laws relating failure times to current density and temperature and of the statistical distribution of these failure times.

Methods used to increase electromigration lifetimes. The most obvious way of increasing lifetimes is selection of a material with a low

| (a) | (b) |

Figure 1. (a) A scanning electron microscope image of an electromigration crack across a thin film Al conductor. (b) Random damage formed elsewhere along the same conductor. The conductor had a width of 80 μm and a thickness of 0.3 μm. The crack caused failure after 160 h (ca 1 wk) at 170°C and 2×10^3 kA/cm².

diffusion coefficient, yet this solution cannot be used frequently because of other technological requirements. The resistivity must be low and the metal must make ohmic contact to Si, adhere well to silicon oxide, not diffuse into Si and poison transistor junctions, and preferably be easy to deposit and fabricate, usually by vacuum evaporation and photoetching techniques. Aluminum is almost ideally suited for this purpose. Gold, which is also used extensively, must be used in conjunction with other metals because of its poor adhesion to silicon dioxide. Material selection cannot be justified entirely on the basis of the resistance to electromagnetic damage. Thus, when electromigration failures became a matter of serious concern to the electronic industry, in the late 1960s, the various solutions which became adopted were based on the continued use of aluminum films. Diverse measures, the control of the grain size, the deposition of dielectric overlayers and alloying may be used to enhance the resistance of aluminum (and other metal) films to electromigration damage. Probably the most widely spread method of lifetime improvement has been the addition of a few (1 to 4) at.% of copper to aluminum.

The discovery of electromigration failures led to a flurry of activities and publications about thin-film conductors about 1970. As a result of the methods introduced then, the rate of investigations and publications fell almost to zero during the middle and late 1970s. However, with the ceaseless decrease of the size of silicon devices unaccompanied by a corresponding reduction in the magnitude of the processed electrical signals, a renewed concern arose with electromigration in conductors having a width of $\leq 1 \, \mu m$. As well as the obvious interest in the effect of diminishing line width on lifetimes, the problem of the effect of very short electrical pulses involving complex matters of the rate of hole nucleation and thermal distribution continues to attract attention. Because of the reduced geometrical dimensions, the conductors are not only very narrow, they are also placed very close to one another, resulting in the fear that failure may result not only from electrical discontinuities within a single conductor, but also that metal extrusions from one conductor may cause short circuits to adjoining conductors.

FRANÇOIS M. D'HEURLE
IBM

DAVID T. PETERSON
Iowa State University

J.S. Arzigian, J.M. Towner and E.P. van de Ven, and J.R. Lloyd in *Proceedings of the 21st Reliability Physics Symposium*, IEEE, New York, 1983.

ELECTRONIC-GRADE CHEMICALS.   See Fine chemicals.

ELECTRON SPIN RESONANCE.   See Analytical methods.

ELECTROPHORETIC DEPOSITION.   See Coating processes; Electromigration; and Powder coatings.

ELECTROPHOTOGRAPHY

It has been estimated that by the year 1985 almost 46% of all reprographic prints will be made by one or another electrophotographic process. The other chief contender for this trillion (10^{12}) reproduction-copy market is offset duplication (44%), the remainder being serviced by stencil (5%), spirit (3%), and an array of other reproduction technologies (2%) (see Reprography; Printing processes). Electrophotography can be defined broadly as a process in which photons are captured to create an electrical image analogue of the original. This electrical analogue is, in turn, manipulated through a variety of steps that result in a physical image. In electrophotography the primary quantum yield is generally much higher than that of silver halide photography.

The necessary amplification in the case of projection imaging, focusing an illuminated original through a lens onto the photosensitive element, is obtained by some form of physical development. The free energy required to initiate image gain is supplied in the form of an electrostatic potential, typically applied across a photosensitive dielectric layer before exposure, or to a bias electrode (a closely spaced metal plate to which a potential is applied) during development, eg, by the attraction of relatively large, visible particles in response to only a few surface charges.

Transfer Xerography

The most widely used form of electrophotography today is called transfer xerography. Figure 1 illustrates one cycle of transfer xerography starting at the nine o'clock position.

In transfer xerography, almost all process steps are carried out on the photoreceptor surface. The first step necessary for creating a latent electrostatic image is the application of a uniform charge pattern on the surface of the photoreceptor; at the same time a uniform field from front to back is provided. Charges must be held on the surface. Corona-discharge devices have proven the most reliable means of applying a stable, uniform charge layer to large-area photoreceptors. Typically, these consist of one or more thin wires connected to a positive or negative high voltage supply and backed by a grounded shield that serves to stabilize the discharge.

As in other photographic systems, the image must be brought to the photosensitive surface with efficiency. Light sources are selected to match the spectral response of the photoreceptor and to provide adequate illumination of the original to be imaged. Automatic copying machines usually require continuous motion of the photoreceptor in order for all other process steps to go on at the same time.

Xerographic development. The electrostatic latent image can be made visible by a variety of techniques. Toner particles can be attracted only where electrostatic field lines extend above the surface of the receptor. In the absence of a development electrode, external field lines appear only along the periphery of charged areas, permitting toner attraction only to the outline of the area. Field lines can be extended outward across the entire charged surface by ringing a development electrode (Fig. 2), connected to the back electrode of the photoreceptor, close to its surface. Charged toner particles introduced into this narrow space follow the external field lines and deposit more uniformly to provide solid area development.

For the development process to occur, toner particles must first be charged reproducibly. Conditions in the development zone must be carefully controlled to avoid dusting and contamination elsewhere. The field configuration is crucial to uniform development.

Main dry-development systems include powder cloud development, cascade development, magnetic brush development, and touchdown or

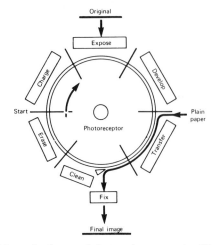

Figure 1.   Schematic of one cycle in transfer xerography. The process steps take place on the surface of the photoreceptor, which is coated on the central imaging drum.

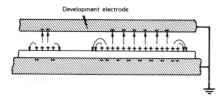

Figure 2. Addition of a closely spaced development electrode forces field lines to extend into the space above the photoreceptor. The field strength becomes everywhere proportional to the local photoreceptor potential, and the entire image area can be developed. Courtesy of Focal Press, London (25).

impression development. Image reversal development can be achieved by dry- or liquid-development techniques.

Transfer of developed image. To be useful, the developed toner image must be transferred to a receiving sheet, usually paper. With the aid of an externally applied field, the toner particles are induced to jump to the paper as it separates from the photoreceptor.

To become permanent, the transferred toner image must be fixed to paper. A thermoplastic resin base in the toner powder permits fixing by fusing onto the paper. In noncontact fusing, a radiant-heat source provides energy to the toner image as the paper passes below it. In contact fusing, heated rollers with good release properties (eg, silicone or fluorocarbon elastomers), backed by a pressure roller, melt the toner and force it into the paper.

Post-imaging process steps. In cyclic-transfer xerography, the photoreceptor must be restored to its original condition after image transfer. Between 10 and 30% of the toned image and residual charge patterns remain on the receptor. Flooding the photoreceptor surface with light and applying neutralizing charges aid in removing toner. Other aids are lubricants that are added to the toner or applied directly to the photoreceptor surface. Both help form low friction surfaces, increasing transfer efficiency and cleanability. Fatty acid metal salts and fluorinated organics have been used for this purpose. Toner is removed by rotating brushes made of fur or of natural or synthetic fibers, and is collected by vacuum suction. The photoreceptor can also be cleaned by a conforming wiper blade.

Liquid development. The most widely used liquid-development scheme relies on electrophoretic particle migration. This development system offers very high resolution, approaching that of the latent image itself; 800 lines and spaces (or line pairs) per millimeter have been achieved. Although liquid development has found its most prominent use in direct electrophotography, successful liquid-transfer systems have been devised and marketed recently.

Materials for Transfer Xerography

Photoreceptors. The photoreceptor is an electronic device that forms the electrostatic image charge pattern in response to light. It typically comprises three layers (Fig. 3): a conductive base layer which can be omitted if the support of the photoreceptor is metallic; a thin dielectric barrier layer which is required only to prevent charge injection into the photoconductive layer; and the photoconductor layer on top. If a separate base layer is required, as on polymeric film supports, it consists of a sputtered or evaporated metal or conductive metal oxide layer well < 100 nm thick. Deliberately applied barriers may consist of inorganic oxides, sulfides, or thin polymeric films.

The optimal thickness of the photoconductor layer is determined by the potential required for development, typically 300–800 V. The exposure required to produce this potential measures the photosensitivity of the device. Practical photoconductors are between 15 and 25 μm thick if they are made up of organic or binder-type layers of low dielectric constant ($K = 3$–5); selenium and its alloy layer ($K = 6$–9.5) are typically between 30 and 100 μm thick.

High photosensitivity requires effective light absorption in the desired spectral region (molar extinction coefficient 10^4–10^5 L/(mol·cm); efficient generation of carriers, ie, separation of the photoexcited electron/hole pair in the applied field; and subsequent complete transport of the mobile charge through the photoconductive layer. Most photoconductor materials exhibit efficient transport for either electron or holes.

Vitreous selenium has long been the basis of the most commonly used photoreceptors. Tellurium may be added to extend red response. Arsenic alloys have been found to enhance both photosensitivity and stability against crystallization. Dispersion of photoconductive pigments such as cadmium sulfide, zinc oxide, and phthalocyanine in dielectric binders may be used.

Organic photoconductors compete today effectively with selenium. Figure 4 shows a new family of double-layer photoreceptors in which a relatively thin, visible-light-sensitive, carrier generator film is adjacent to a much thicker, transparent, polymeric, active-transport layer. The former supplies photosensitivity; the latter, dielectric strength and voltage contrast. This allows greater freedom in designing photoreceptors that have desirable mechanical properties as well as high photosensitivity.

Developer materials. The toner particles used for developing xerographic images generally consist of 8–15 μm particles of a thermoplastic powder colored by a dispersion of 5–10% carbon-black particles of less than 1 μm. Cyan, magenta, or yellow colorings may be substituted for use in color xerography (see Color). Pigment concentration and dispersion must be adjusted to impart a conductivity to the toner mass that is appropriate to the development process. For efficient induction development a conductivity greater than 10^{-4} S/cm is found to be desirable; most other development processes require the toner to retain charge applied by contact electrification for extended time periods. The toner thermoplastic is generally selected on the basis of its fusing characteristics; it must melt sharply at the lowest temperature consistent with stability to storage and to vigorous agitation that occurs in xerographic development chambers.

Direct Electrophotography

Electrofax process. The key innovation in Electrofax was to produce images directly on the final sheet in fewer steps and make the photoreceptor an inexpensive consumable item. Although Electrofax has its widest use for low volume black-and-white copying, at least one color copier has been marketed.

Transparent-film processes. Direct electrophotography on transparent organic photoconductor films has been used to produce transparencies. Since the transfer step is omitted, images resolving 200 line

Figure 3. Cross section through typical xerographic photoreceptor. [1] Photoconductive layer, [2] injection barrier, [3] base electrode (not drawn to scale).

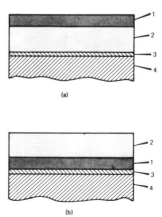

Figure 4. Active matrix photoreceptor with (**a**) generator layer on top, (**b**) at interface. [1] Generator (sensitizing) layer; [2] active transport layer; [3] blocking interface; [4] base electrode. (Layers not drawn to scale.)

pairs/mm and better can be readily produced, particularly with electrophoretic liquid developers.

Transfer of Electrostatic Images (TESI)

An alternative that obviates the toner transfer and cleaning steps in conventional xerography is the transfer of the electrostatic latent image to a suitable receiving sheet prior to development. Transfer of an electrostatic charge pattern can be accomplished either during photoreceptor exposure or after a stable charge pattern is formed on the photoreceptor by conventional charging and exposure.

Persistent Photoconductivity Imaging

A number of attempts have been made to reverse the usual xerographic sequence, by exposing the *neutral* photoconductor first and then charging it to form a latent image. A process known as Magne-Dynamic is one approach to chargeless electrophotography.

Photoconductography. In photoconductography, a conductivity pattern formed in a photoconductive insulator by light exposure is developed by electrochemical deposition. Photographic gain is achieved by drawing multiple charges through the photoconductor-electrolyte series circuit, or by secondary chemical amplification, or both.

Photoactive-Pigment Electrophotography

Photoconductive toner. A number of processes have been developed on the principles of photoconductivity of a pigment and the ability of such pigment particles to move under the influence of a field. A successful approach to the formation of images by moving photoactive particles requires their containment in a liquid or plastic.

Particle-migration imaging. A totally new form of electrophotography uses the migration of microscopic selenium particles embedded in the top surface of a 10 μm plastic layer. The selenium moves through the layer under the influence of a field when the plastic is softened with a solvent.

Photoelectrophoretic imaging processes. Photoconductive pigments such as zinc oxide and phthalocyanine can be caused to print out directly onto paper interposed between a backing electrode and a pigment suspension covered by a transparent imaging electrode. This process, which is called photoelectrophoresis or PEP, has been applied to an experimental printer capable of converting light input to very clean black print on paper.

Color photoelectrophoresis. Perhaps the most challenging use of photoactive pigments and their electrophoretic motion is a single-step, full-color electrophotographic process that uses the spectral response of individual photosensitive particles in a mixture of cyan, magenta, and yellow pigments (see Color photography). Under the influence of an electric field, the particles migrate selectively according to illumination to produce instant full-color images.

Applications for Electrophotography

By far the most widespread application of electrophotography today is plain-paper copying; fast xerographic duplicators can now produce two copies per second, and sort and collate them into finished booklets. Computer printers can print at high speed by means of a deflected and modulated laser beam. These machines can be operated on-line from a computer or off-line from magnetic tape and disk drives. Such machines can handle very large volumes of individualized printing of high quality (500,000 or more copies per month).

NIKOLAUS E. WOLFF
Consultant, Hanover, NH

JOHN WEIGL[‡]
Xerox Corp.

C.F. Carlson in P.A. Spencer, ed., *Progress in Photography: 1955–1958*, "Xerography," Focal Press, London, UK, 1958.

[‡]Deceased.

J.H. Dessauer and H.E. Clark, *Xerography and Related Processes*, Focal Press, London, UK, 1965.
R.M. Schaffert, *Electrophotography*, 2nd ed., Focal Press, London, UK, 1975.
J.W. Weigl, *Angew. Chem. Int. Ed.* **16**, 374 (1977).

ELECTROPLATING

Electroplating is the electrodeposition of an adherent metallic coating on an electrode in order to form a surface with properties or dimensions different from those of the basis metal.

Electroplating is a surface treatment. The material (work) being treated is made the cathode in an electroplating solution, or bath. Such baths are almost always aqueous solutions, so that only those metals that can be reduced from aqueous solutions of their salts can be electrodeposited. The only major exception at present is aluminum, which is plated on a semicommercial scale from organic electrolytes. Some of the refractory metals (eg, niobium, tantalum) can also be deposited from fused salts as coherent plates, but little commercial use has resulted from this development.

The thickness of deposit applied by electroplating varies with the application: from as little as 0.025 μm for decorative gold deposits through 25–50 μm for standard nickel–chromium plate on exterior automotive hardware, to 1 mm or more for electroforms.

The properties conferred by electroplating include improved corrosion resistance, appearance, frictional characteristics, wear resistance and hardness, solderability, specific electrical properties, and many others. Electroforming (including electrotyping) is used to manufacture articles that cannot be made as economically in any other way.

The Substrate in Electroplating

There is much more to electroplating than the final step of laying down a coating of the plating metal, and much more has to be considered than the properties of the plated metal. The final article will consist not of the deposit alone, but of deposit plus substrate. It is the properties of this combination that often determine the right metal to plate and the right solution from which to plate it.

Preparation of the Substrate

Before a useful electroplate can be deposited on a surface, the surface must be in condition to receive. Useful means, among other things, adherent. The preplating treatments necessary to prepare the surface to accept an adherent deposit are generally aspects of cleaning. The ideal surface would be one consisting entirely of atoms of the metal to be plated upon, and having no foreign material at all. This is virtually impossible to attain, even in the laboratory. A practical definition, then, of a satisfactorily clean surface would be a surface containing no foreign material that interferes with the formation of an adherent deposit. In general, this connotes the removal of gross dirt and soil, heavy oxide or tarnish films, and in some cases surface skins of damaged metal produced by prior mechanical operations.

Cleaning. Choice of the proper cleaning or preparative cycle depends primarily upon the nature of the substrate to be prepared and the nature and amount of the soils to be removed. Ferrous and nonferrous metals generally require different types of cleaners. The more active metals such as aluminum require special techniques to prepare them for plating. And obviously, the more contaminated the surface the more cleaning it will require. A typical cleaning cycle includes the following steps: pickling to remove gross scale; any mechanical preparation such as polishing or buffing; cleaning to remove oils, greases, shop dirt, and polishing and buffing compounds; acid dipping to remove oxide films; and rinsing.

Rinsing. Adequate rinsing between all steps in cleaning and plating is of the utmost importance. Rinse waters should be clean and should not be allowed to become contaminated by drag-in of preceding solutions. Countercurrent rinsing is often employed as a means of conserving water while ensuring that the last rinse is relatively pure. Hot water is more efficient than cold for removing contaminants; on the other hand, it entails the risk that the work may dry before entering the next operation. Quality of the available water supply is often of importance;

softened, deionized, or distilled water may be required for final rinses in many instances.

Adequate and efficient rinsing assumes additional importance in relation to waste disposal. To the extent that it cannot be fed back to the plating cycle, the last rinse constitutes effluent from the plant. Almost universally it must be treated to reduce harmful contaminants forbidden by EPA regulations, and the less of these it contains the less costly waste treatment will be.

Acid dipping. Acid dipping, which generally follows cleaning, serves two purposes: it removes slight tarnish or oxide films formed in the cleaning step, and it neutralizes the alkaline film that even good rinsing cannot completely remove from the surface. It is thus particularly important when the plating solution is acid. The acid dip is usually a 10–30 vol% solution of hydrochloric or sulfuric acid, the latter solution being more common.

Drag-out and drag-in. Every solution in a plating cycle is contaminated, to a greater or lesser extent, by the solution that precedes it in the cycle. How serious this situation is depends on many factors, some controllable and some not. The shape of the parts and how they are positioned in the tank is of great importance; drain times vary, and some solutions are more free-rinsing than others. Contamination caused by this drag-in may be serious or inconsequential. The same is true of drag-out which also must be considered in connection with the problem of waste disposal.

Special preparation cycles. For plating on the common substrates such as ferrous metals and copper and its alloys, the preparation cycles briefly outlined above usually suffice. Other substrates require more specialized treatment. Among those requiring special preparation are aluminum and magnesium; zinc-base die castings; refractory metals like titanium, zirconium, tantalum, niobium, molybdenum, and tungsten; and nonmetallics like synthetic plastics, leather, wood, and plaster.

The Electroplating Process

The operations of electroplating, including the cleaning, rinsing, plating, and postplating treatment, can be carried on manually or with almost any desired degree of automation. Parts may be hung in the plating tank on wires or on racks; when many small parts are to be plated, they may be contained in wire baskets or, more commonly, in barrels that rotate in the plating tank. Movement from one operation to another may be by hand or by machine.

The necessary d-c power is derived from motor-generators or, increasingly, by rectifiers. Solid-state rectifiers may be of three types: selenium, germanium, or most important, silicon (see Semiconductors). Power is conveyed to the plating tanks by bus bars; the anodes are hung into the tank from the positive bus bar, usually along the two sides (see Figure 1) and the work to be plated from the negative or cathode bar, usually down the center. Tank voltage is read from a voltmeter, and current from an ammeter: these two instruments should be available for each plating tank. A third instrument, an ampere-hour meter, is often helpful as a means of regulating the thickness of deposit and for general control of the operations.

Some agitation of plating solutions or of the work is usually helpful. The oldest and simplest method consists of an operator merely swishing the work around at intervals, but automatic cathode-rod agitation is preferable.

Temperature control is almost always desirable in plating operations because the characteristics of plating solutions, of the deposit, or of both usually depend to a large extent on the temperature of operation. Heating or cooling coils in the tank itself may be used, or the solution may be circulated through a heat exchanger. Electric immersion heaters may be used. Occasionally, plating tanks are heated by open gas flames beneath the tank.

Figure 1 shows a typical plating tank but there are many variations on this basic design. In addition to the basic equipment—power source, plating, cleaning, and rinsing tanks, and bus bars—most plating installations require one or more of the following: filters, for either continuous or intermittent purification of solutions; drying facilities, which may range from a simple jet of compressed air to large ovens; racks of design appropriate to the part being plated, and a racking station where the work can be conveniently hung on these racks and unracked after plating; one or more stripping tanks for stripping faulty deposits so that parts can be reworked and for stripping the plating racks themselves; reclaim tanks if the drag-out is valuable and worth reclaiming; portable pumps for transferring solutions; and at least one empty tank so that a plating tank can be emptied and worked on.

In addition to the equipment required for electroplating and allied processes, the modern plating shop must include apparatus for waste treatment, metal recovery, or both.

Continuous plating. Electrolytic processes are well-suited to plating continuous coils of strip or wire. The substrate can be uncoiled, pickled, cleaned, plated, given postplating treatments, and recoiled in one continuous operation.

Materials of construction. Plating tanks and auxiliary equipment are constructed of materials resistant to the particular solution involved. This usually means plain steel for alkaline solutions and rubber- or synthetic-lined steel for acid solutions.

Safety. Hazards in plating operations arise from the nature of the materials routinely handled, many of them highly toxic. Thus, certain minimum precautions are absolutely necessary. Most of the normal hazards of a plating room can be adequately handled by a combination of proper ventilation and appropriate protective clothing.

Waste disposal and metal recovery. There are two basic approaches to waste treatment: destruction and recovery. Original emphasis was on the former; the objectionable metals were precipitated as sludges and disposed of in landfills, or similar means. As landfills become both less available and more restrictive in their acceptance of such sludges, emphasis has shifted to recovery of the metals for reuse in the plant or for sale to a refiner.

The disadvantages of precipitation methods have turned attention more and more to recovery of the valuable metals from effluents before their discharge. Among the methods proposed for recovery systems are reverse osmosis (qv) (RO); evaporative recovery; ion exchange (qv), and various combinations of two or more techniques. Most of these methods are technically feasible and the principal snag is their relatively high cost, large capital expense for equipment, energy for evaporation, membrane life for RO, etc.

Plating Solutions

Plating solutions are usually aqueous. Every plating solution contains ingredients to perform at least the first, and usually several of the following functions: provide a source of ions of the metal(s) to be deposited; form complexes with ions of the depositing metal; provide conductivity; stabilize the solution against hydrolysis or other forms of decomposition; buffer the pH of the solution; regulate the physical form of the deposit; aid in anode corrosion (see below under Anodes); and modify other properties peculiar to the solution involved.

Current density range. Current density is the average current in amperes divided by the area through which that current passes; the area is usually nominal area, since the true area for any but extremely smooth

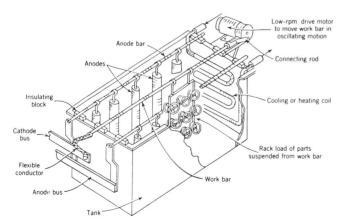

Figure 1. Cut-away view of plating tank.

Low-rpm drive motor to move work bar in oscillating motion

Anode bar

Anodes

Connecting rod

Insulating block

Cooling or heating coil

Cathode bus

Flexible conductor

Rack load of parts suspended from work bar

Anode bus

Work bar

Tank

electrodes is seldom known. Units used in this regard are A/m^2 (92.9 mA/ft^2).

Current densities at the anode and cathode are both important; they may differ considerably although not by so large a factor as to make necessary very great differences between the anode area and the cathode area. Most solutions exhibit a range of current densities within which deposits are satisfactory and outside of which they are not.

Throwing power. Except in the special case of concentric anode and cathode, the current density over the electrodes varies from point to point. In general, the area on the cathode nearest to the anode receive a higher current density than those more remote. Thus, more metal is plated on the projections than in the recesses. Many plating solutions, however, have the ability to moderate this condition to some degree. Throwing power may be defined as the improvement in metal distribution over primary current distribution on a cathode. (Primary current distribution is the distribution of the current that depends only on the geometry of the cell.)

Acidity. Plating baths are either acid, neutral, or alkaline, and for most of them close control of the pH is essential to successful operation.

Anodes. Anodes in a plating bath perform two functions: they act as the positive electrode, introducing current into the solution; and, in most cases, they replenish the metal deposited at the cathode, thus maintaining the balance of the bath.

Temperature. Control of temperature is important in almost all plating processes; each solution is characterized by a range of temperatures within which it gives best results. Temperature affects almost all the variables of the solution: conductivity, current efficiencies, nature of the deposit, and stability.

Current efficiency. Faraday's laws predict the amount of metal that is deposited at the cathode or dissolved at the anode, but these amounts are not always obtained; the deficiency is owing to evolution of hydrogen at the cathode or oxygen at the anode, or to other side reactions. In practical plating operations, current efficiency is not usually of direct concern so long as it is known, but it is often important to equalize the efficiencies at cathode and anode so that the bath remains in balance.

Purity. Plating processes differ in sensitivity to the presence of impurities in the bath. Specific means are available for purifying most plating baths by chemical treatment; filtration through activated carbon is a generally useful method for removing organic contaminants.

Bright plating. Most deposits from simple plating solutions are mat unless the substrate is bright and the plate very thin. If the use is purely functional, this is no drawback, but for decorative purposes a bright appearance is usually desired. This brightness was formerly obtained by mechanical treatment, buffing after plating, but today most metals can be bright as they come from the bath, and require little or no further treatment, at least as regards appearance. This bright appearance is produced by the addition to the plating bath of small amounts of one or more addition agents or brighteners, usually organic compounds.

Maintenance of plating baths. All plating baths require more or less chemical control. They must be analyzed, often or seldom, depending on individual circumstances, and adjustments must be made. For some baths a determination of specific gravity suffices for routine control; other baths require frequent and complete analysis.

Individual Plating Baths

Plating baths include cadmium, which affords good corrosion protection to steel and other substrates in marine atmospheres; chromium, the final finish on the great majority of plated consumer items; cobalt, used for its magnetic properties; copper, used most often as an undercoat for other metals; gold, which has become extremely important in the electronics and computer industries because of the low contact resistance, corrosion resistance, and solderability of the metal; indium, used in bearings; iron, which has minor electroforming and other engineering uses; lead, used in battery parts and chemical construction; nickel, used in a host of engineering uses; platinum-group metals, used for decorative applications as well as for engineering purposes; silver, used for its electrical and mechanical characteristics; tin; zinc, plated on all types of hardware; and various kinds of alloy plating.

Nonelectrolytic Plating Processes

There are many other ways than electroplating to deposit a coating of metal on a substrate. Hot dipping, vacuum evaporation, chemical vapor deposition, and various aqueous processes not requiring current are some of the best developed. Nonelectrolytic aqueous deposition includes immersion plating and chemical, autocatalytic, or what has come to be known as electroless plating (qv).

Postplating Treatments

Postplating treatments include chromate conversion coatings for zinc and cadmium, phosphate treatments for iron, and various passivating and brightening dips.

Applications

Applications for plating are classified according to the principal function of the plate; thus the plating may be applied mainly for appearance, protection, special surface properties, or engineering or mechanical properties.

FREDERICK A. LOWENHEIM[‡]
Consultant

W.J. PIERCE
3M Company

F.A. Lowenheim, ed., *Modern Electroplating*, 3rd ed., John Wiley & Sons, Inc., New York, 1974.

A.K. Graham, ed., *Electroplating Engineering Handbook*, 3rd ed., Van Nostrand Reinhold Co., New York, 1971.

N. Hall, ed., *Metal Finishing Guidebook Directory*, Metal and Plastics Publication, Inc., Hackensack, N.J., annual.

ASTM B322-68 (1973), Recommended Practice for Cleaning Metals Prior to Electroplating.

ELECTROSTATIC PRECIPITATION. See Air pollution control methods.

ELECTROSTATIC SEALING

The sealing technique described in this article is unusual in that the glass typically remains at a viscosity > 1 TPa·s (10^{13} P) throughout the sealing process, and in that no foreign materials such as binders, fluxes, or adhesives are introduced. The name electrostatic sealing (also field-assisted or anodic bonding) is derived from the central feature of the technique: the application of a voltage across the bonding surfaces to establish a high electric field at the interface.

Theory

Electrostatic sealing may be viewed as a two-step process: electrostatic forces at the glass–metal interface bring the parts into intimate physical contact; and bond formation between the glass and the metal. Although the former is readily explained in terms of basic electrostatic theory, the mechanisms in the latter are not well understood at this time.

The Technique

The electrostatic process has been applied to the sealing of glass to metals and semiconductors (qv). Since the methodology is identical, semiconductors for simplicity are included under the metals. Basically, the method consists of the following steps. The parts that are to be sealed are machined so that their faces are smooth and conform to each other as closely as possible. The clean faces are then placed in contact and heated to a temperature at which the viscosity of the glass approaches the annealing range (1–32 TPa·s or (1–32) $\times 10^{13}$ P). If a d-c voltage of 400–1000 V is applied for 1–5 min across the glass–metal

[‡] Deceased.

interface with the metal positive with respect to the glass, many glass–metal systems form strong, hermetic seals. The experimental ranges cited above are typical. Any glass–metal system that can be sealed at all by the electrostatic method invariably has been found to be sealable at temperatures in the annealing range (ie, between the strain point and the annealing point). Seals are evaluated with regard to strength, hermeticity, and resistance to thermal shock.

Uses

Although there are broad applications for the electrostatic sealing method, only a few examples that demonstrate its benefits are given here. Since sealing temperatures are low and macro-distortion of the glass is minimal, the method is useful in the mounting of optical components. The application that has been investigated most widely is concerned with the mounting of silicon pressure transducers.

The electrostatic sealing process is commercially utilized by a number of companies under licensing agreements with Duracell Inc. which holds the applicable patents. As far as is known, one of the most popular applications is in the construction of pressure transducers.

GEORGE WALLIS
Duracell Inc.

G. Wallis and D.I. Pomerantz, *J. Appl. Phys.* **40**, 3946 (1969).

U.S. Pat. 3,397,278 (Aug. 13, 1968), D.I. Pomerantz (to P.R. Mallory & Co.).

EMBEDDING

The embedment of objects, the complete encasement of objects in a medium, practiced for centuries, has today become a science involving large numbers of scientists and engineers in all parts of the industrial world. The objectives of embedding may be either functional or decorative. The most important functional area is the embedding of electrical and electronic circuitry.

Embedding Process Considerations

Often, a given embedded product can be made by two or more of the various embedding processes such as casting, molding, potting, impregnating, or transfer molding. Thus, an analysis of the important comparative advantages of the methods is required. This section presents some of the most important of these considerations.

Primary embedding materials. Although waxes and bitumens are still used occasionally, most materials used for embedding are plastics. Liquid or easily liquefied plastics are most commonly used. As a class plastics are insulators. They therefore provide the necessary insulation required in electronic and electrical applications. Useful plastics fall into two large classes, thermoplastics and thermosets. Two properties to consider when choosing a plastic include viscosity and exothermic properties.

Transfer-molding resins. The most widely used materials for transfer molding are epoxies, although low pressure molding materials have been developed from silicones, diallyl phthalates, phenolics, and alkyds (see Alkyd resins), the first two enjoying wide usage.

Thermosetting Embedding Resins and Filled Compounds

Many chemically distinct embedding resins are available, and there are many variations in each group. They include epoxies, silicones, polyesters, polyurethanes, polysulfides, polybutadienes, low density foams, allylic resins, most often incorporating fillers.

Waxes and Thermoplastic Embedding Materials

Owing to the low temperature stability of waxes and bitumens, these materials have only limited embedding applications. The use of most thermoplastic resins is also restricted by the high molding pressures and temperatures required in their processing.

Decorative Embedding

The vast majority of embedding applications serve electrical or mechanical functional objectives, but decorative embedding also enjoys widespread use. Examples are primarily decorative items embedded in an unfilled resin having some degree of transparency, although nontransparent art objects are also cast using materials similar to those used in functional embedding. Also, scientific specimens are often embedded for cross-sectioning investigations.

Standards and Controls for Embedding

The importance of proper controls cannot be stressed too strongly from the standpoint of higher yield, lower cost, and increased efficiency, and from the standpoint of safety. Frequently, the potential hazard of handling embedding resins is not nearly so well recognized in an electronics plant as it would be in a chemical plant. Quality control should cover both manufacturing and production control. Finally, the key to reliable embedding techniques lies in material selection and maintaining the material quality by extensive testing.

CHARLES A. HARPER
Westinghouse Electric Corp.

C.A. Harper, *Handbook of Plastics and Elastomers*, McGraw-Hill Book Co., Inc., New York, 1975.

C.A. Harper, *Handbook of Materials and Processes for Electronics*, McGraw-Hill Book Co., Inc., New York, 1973.

H. Lee and K. Neville, *Epoxy Resins*, McGraw-Hill Book Co., Inc., New York, 1957.

EMULSIONS

An emulsion is a mixture of two or more immiscible liquids, one being present in the other in the form of droplets. Industrial emulsions frequently contain a solid as a third ingredient. Strictly speaking, a mixture of a solid dispersed in a liquid is a suspension. A variation in which molten waxes are emulsified is called a wax emulsion, although it is in fact a suspension or dispersion at room temperature. In the classic emulsion, the oil may either be dispersed in the water (oil-in-water or o/w emulsion) or the water dispersed in the oil (water-in-oil, w/o, or inverse emulsion). This terminology is important because of the external phase, a key factor in emulsion formulation and design.

Emulsions are found in nature, two of the principal examples being milk and rubber latex. These emulsions are stabilized by natural emulsifying agents. In a like fashion, commercial emulsions require emulsifying agents.

Emulsions are used in a variety of fields such as textiles, leather, and metal treatment; foods, cosmetics, pharmaceuticals, and paints; in agricultural chemicals, polymerizations, cleaning, and polishing; and ore and petroleum recovery.

Emulsions are inherently unstable systems and the risk of deteriorating during storage is greater than with a nonemulsified product. Emulsion technology, though seemingly based on simple interface principles, is highly complex, especially when dynamic and static conditions are considered (see Cosmetics; Food processing; Textiles; Leather; Pharmaceuticals; Latex technology; Paint; Polymers; Polishes, Petroleum).

Properties

Emulsion properties and characteristics are not necessarily related to the properties of the main active or key ingredients. The properties can usually be tailor-made to suit various use/applications requirements. They are built into an emulsion during formulation. The physical properties of an emulsion depend chiefly upon the properties of the external phase; the phase-volume relationship, and the particle size and they are under a surprising degree of control. The properties of emulsions and the related controlling features are shown in Table 1.

Table 1. Properties of Emulsions

Property	Related controlling features
appearance	
clarity	
clear	small particle size; matched refractive indexes
translucent	medium particle size
opaque	large particle size; unmatched refractive indexes
color	
white	large particle size; unmatched refractive indexes
gray	medium–small particle size; unmatched refractive indexes
colors	colors in continuous phase
viscosity	
thick (high)	HIP[a] emulsion; small particle size, thickeners in outside phase
thin	LIP[b] emulsion with no thickener
dispersibility	
in water	o/w
in oil	w/o
ease of preparation	
high, easy	emulsifier, solution level; low viscosity concentrate
low, difficulties	low emulsifier level
re-emulsification	emulsifier selection; emulsifier level
stability	
high, good	emulsifier selection; emulsifier moderately high
low, poor	low emulsifier levels; emulsifier selected for other property
stable to electrolytes	emulsifier selection
on evaporation (o/w)	emulsifier selection; emulsifier level
spoilage	preservative selection; sterile packaging
wetting-spreading	
high	emulsion type; emulsifier selection
low	emulsifier selection
particle size	
small	emulsifier selection; emulsifier level
large	emulsifier level

[a] HIP = high internal phase.
[b] LIP = low internal phase.

Type of emulsion. The choice of the phase that is external or continuous is defined by the type of emulsion: o/w or w/o. In general, o/w emulsions conduct electricity, are dilutable with water, feel like water, dry (lose water by evaporation), are corrosive in comparison with w/o emulsions, and in general, exhibit properties expected of an aqueous liquid. On the other hand, w/o emulsions conduct electricity poorly, if at all, are usually diluted with oils or solvents rather than with water, feel more like oil, resist drying or loss of water although they will lose a volatile solvent readily if present in the continuous phase, are difficult to wash away, are less corrosive than o/w emulsions, and, in general, exhibit the properties of the continuous oil phase. There are also, in very few instances, dual emulsions, in which some water is dispersed in oil which is in turn dispersed in water, or the reverse. They exhibit properties of the external phase.

Volume ratio of external to internal phase. A low internal phase emulsion assumes the overall characteristics of the external phase. On the other hand, high internal phase emulsions exhibit higher apparent viscosities as the internal phase volume ratio increases. This effect is not pronounced until the internal phase volume exceeds 50–60% of the total volume, then the apparent viscosity increases rapidly and can exhibit characteristics of a gel or paste.

Particle size. Table 2 shows the effect of particle size on the appearance of the emulsion. The same range of particle sizes is of interest in formulating for stability; the smaller the particle size, the greater the stability. However, stable emulsions can be formulated with particles as large as 1–2 μm. Since a smaller particle size emulsion is usually more costly to manufacture, formulation is usually a compromise of application needs/cost. The particle size of an emulsion may be reduced by increasing the amount of emulsifier, improving the selection of emulsifier with respect to both HLB (hydrophile–lipophile balance) and chemical type, modification of the method of preparation of the emulsion, ie, by the use of phase inversion to provide an extended internal phase at the time of inversion to the final emulsion type, and improved and/or increased energy input by changing or enhancing the agitation.

Formulation

In emulsion formulation, the goal is to achieve the best combination of emulsion properties to fulfill the application needs and stability requirements. The key or active ingredient(s) may or may not be the major ingredient. The main ingredient is most frequently the continuous phase and this dictates the type of emulsion, o/w or w/o, a pivotal decision. For economic as well as technical reasons, most commercial emulsions are o/w and have low oil (internal) phase levels. If a higher viscosity is desired with an emulsion of this type, a suitable soap (gel) may be incorporated in the emulsifier system or thickeners (gums, starches, proteins, and polymers) may be added. Moderate amounts of emulsifier(s) are usually used, ranging from 5–20 wt% of the oil/wax mixture. The level of energy input during manufacture will influence the level of emulsifier. The emulsifier blend is best if soluble or dispersible in the oil/wax. Coupling with alcohols, glycol polyethers, fatty acids, traces of water, etc., are desirable. The emulsifier should be chosen to give the best balance of emulsification ease and emulsion stability; these characteristics do not always occur at the same HLB or with the same chemical type. If the product is to be an emulsifiable concentrate, it must be formulated to be self-emulsifiable. Some products may require deposition via controlled instability or a chosen particle charge that renders the particles substantive.

Choice of emulsifier. The use of the least amount of emulsifier is possible when a choice is made that most nearly matches the requirements for ionic type, HLB, and emulsifier chemical type. The choice of ionic type—anionic, cationic, amphoteric, or nonionic—will influence many properties of the final emulsion. The HLB is an expression of the relative simultaneous attraction of an emulsifier for water and oil (or for the two phases of the emulsion system being considered). It is determined by the chemical composition and the extent of ionization of the emulsifier. For example, ionic emulsifiers change HLB values radically with change in pH and/or salt content of the formula whereas nonionic emulsifiers exhibit a more constant HLB under these circumstances. The HLB value of commercial emulsifiers is usually available from their suppliers. It can be approximated by observing the dispersibility of the emulsifier in water, Table 3. It is usually best to achieve the desired HLB by blending a low HLB emulsifier with a high HLB emulsifier, both being of the same chemical family. An indication of required HLB values for many petroleum oils and waxes is shown in Table 4. Various oils have differing required HLB values, especially for differing applications, but even with these rough guidelines testing of a selected variety of blends of emulsifiers can cover a range of HLB and chemical types that will allow selection of the best emulsifier. Emulsifier chemical type is a more elusive match, though trial of surfactants, eg, laurate, stearate, oleate, etc, at the same HLB blended value will allow correlation of chemical type with stability and other properties.

Table 2. Effect of Particle Size of Dispersed Phase on Emulsion Appearance

Particle size	Appearance
macroglobules	two phases may be distinguished
greater than 1 μm	milky-white emulsion
1–0.1 μm	blue-white emulsion, especially a thin layer
0.1–0.05 μm	gray semitransparent, dries bright
0.05 μm and smaller	transparent, dries bright

Table 3. HLB of Emulsifiers by Dispersibility in Water

Dispersion characteristics	HLB range
no dispersibility in water	1–4
poor dispersion	3–6
milky dispersion after vigorous agitation	6–8
stable milky dispersion	8–10
translucent to clear dispersion	10–13
clear solution	13 +

Table 4. Emulsion Properties vs HLB or Required HLB Values for Hydrocarbon Oils / Waxes

Required HLB	Type of emulsion	Particle size, μm
4 to 8	w/o	0.5 to 3
9 to 12	o/w	0.1 to 3
13 to 14	o/w, detergency	
15 to 18	o/w, solubilization	less than 0.1

Emulsion testing. During formulation studies, the test emulsions should be evaluated not only for ease of emulsification but also for the important desired application properties and for stability to various temperatures including freeze-thaw conditions, and to agitation as anticipated in shipping. It is important to note that stability for commercial purposes may occur by either creaming or by coalescence of the internal phase. The latter is far more serious with respect to final use unless the product is re-emulsifiable.

Manufacturing procedures. In any formulation study, a tentative manufacturing procedure, including types of equipment and order of addition of ingredients, should be selected based upon equipment costs and availability. Power input cost must be balanced versus emulsifier costs. For example, an emulsion prepared on a colloid mill may require less than a tenth of the amount of emulsifier required if the emulsion were prepared in a tank with propeller agitation. On the contrary, the latter method can provide much smaller particle size. Order and rate of addition of ingredients can have a profound effect on the quality of the emulsion and must be tested in a variety of permutations and combinations. After a suitable choice is made, all further testing should follow the procedure unfailingly. Generally, all the oils and waxes and other oil-soluble ingredients are combined as the oil phase. Polyols are combined with the water. However, salts are best added to o/w emulsions with the last part of the water, after inversion has occurred and a good primary emulsion has been established. In most instances of liquid–liquid emulsification, ambient temperature is preferred. Whenever elevated temperatures are used, the partition of the emulsifier between the oil and water may change, and modify the emulsification. With some equipment, having large heat input, some type of cooling must be effected.

Preservation. Since emulsions are often combinations of fatty materials and water and sometimes contain carbohydrates, they are good substrates for bacterial growth. Rarely are emulsions packaged under sterile conditions, hence preservatives are usually required (see Industrial antimicrobial agents). Some of the more common preservatives employed in emulsions include benzoic, propionic, and acetic acids and their salts and esters, also aldehydes and biguanides.

Safety

Most emulsifiers are of low toxicity. Toxicity is mainly of concern in food and medicinal fields. These emulsions are subject to FDA scrutiny. Emulsions containing toxicants must be carefully formulated since they offer differing paths for entering the environment. Of course, emulsifiers can aid antipollution efforts by helping to clean up oil spills, etc., and aid in tank cleaning.

WILLIAM C. GRIFFIN
ICI Americas Inc.

P. Becher, *Emulsions: Theory and Practice*, Reinhold Publishing Corp., New York, 1965.

P. Becher, ed., *Encyclopedia of Emulsion Technology*, Marcel Dekker, Inc., New York, 1979.

J.C. Johnson, "Emulsifiers and Emulsifying Techniques 1979," *Chem. Technol. Rev.* (125), 16 (1979).

ENAMELS, PORCELAIN OR VITREOUS

In the United States, the term porcelain enamel designates the glassy coating on metal; however, in some other countries, vitreous enamel is the more common term for the same glassy coating. The ASTM defines porcelain enamel as a substantially vitreous or glassy inorganic coating bonded to metal by fusion at a temperature above 425°C. Ceramic coatings, another term used for coatings on metal, connotes emphasis on the protective feature of the coating for the metal (see Refractory coatings). Ceramic coatings are often formulated and designed to contain mainly crystalline rather than glassy material (see also Colorants for ceramics; Glass).

Porcelain enameling protects against corrosion, decorates, and resists the attack of alkalies, acids, and other chemicals. This material is a nonporous sanitary coating imparting no odors or tastes. Since it is entirely inorganic, it does not serve as a feedstock for microorganisms. Because of its sanitary aspects, its protective and strengthening function, and its decorative character, porcelain enamel has been adopted as the most suitable material for bathtubs, laundry appliances, ranges, sinks, and refrigerator liners.

Typical physical properties include density 2.5–3.5 g/cm³; hardness 5–6 (Mohs scale); tensile strength 34–103 MPa (4,930–14,935 psi); compressive strength 1380–2760 MPa (200,100–1400,200 psi); modulus of elasticity 55–83 GPa (7,915,000–12,035,000 psi); and dielectric constant 5–10.

Classifications. Porcelain enamel is commonly classified by function, eg, ground coats, single frit, applied to ground-coated metal; by service, eg, acid resistant, alkali resistant, electrically insulating, thermal-shock resistant; by composition, eg, alkali borosilicate, titania, lead-bearing or leadless enamels; by metal coating, eg, sheet steel, cold-rolled steel; by decorative character, eg, clear, colored, stippled, mat; by opacifying material, eg, titania, zirconia, molybdenum oxide; by method of application, eg, wet process or dry process; by type of product, eg, appliances, jewelry, silos; and by firing temperature, eg, 540°C low temperature, 595–760°C.

The Enameling Process

The porcelain enameling process involves the re-fusing of powdered glass on the metal surface. The powdered glass is prepared by ball-milling a porcelain enamel glass engineered for specific properties. First the glass is smelted from raw batch materials. The enamel smelter is usually a box-shaped tank furnace. Continuous smelters, wherein the thoroughly mixed raw batch is fed in at one end and molten glass is flowing out at the other end, are common in commercial operations. Decomposition, gas evolution, and solution occur during smelting. After the molten glass has been smelted to a homogeneous liquid, it is poured in a thin stream of water or onto cooled metal rollers. This quenched glass, termed frit, is a friable material easily reduced to small particles by a ball-milling operation. Ball-milling the glass frit into small-sized particles can be carried out whether the frit is wet or dry (see Size reduction). Dry powders are used for dry-process cast-iron enameling and for electrostatic application on sheet steel (see Powder coating). Dry powders are also prepared and marketed for the subsequent preparation of slurries and slips used in the wet-process application techniques.

The electrostatic powder processing of porcelain enamel is similar to the electrostatic techniques used in preparing organic coatings (see Coating processes). Compared to the conventional wet-process porcelain enameling of sheet steel, the dry system eliminates the need for the clay suspending agents which represent refractory additions. As a consequence, these dry powders permit firing at a lower temperature as compared to that of the wet powders, thereby saving energy. At the lower firing temperatures, approaching the ferrite–austenite transition (ca 725°C), metal sagging and distortion are minimized, and additional materials savings can be achieved as higher carbon, nonpremium steels may then be used and less structural bracing of the steel article is required.

Electrophoresis also is employed commercially and provides a dense uniform coating. In this process the sheet-metal article is positively charged and the negative electrodes are located in the slip. The metal article, which is immersed in the slip, attracts the charged frit particles which then pack into a tightly adhering coating layer. This process can be automated for uncomplicated shapes.

Preparation of metals. Enameling cannot be successful unless the metal is thoroughly cleaned and kept clean until the final coat is fired. Simply touching the surface may cause defects. Cast-iron and thick steel parts are sandblasted without danger of excessive loss of metal and excessive warping. Sand, silicon carbide, and steel grit are satisfactory abrasives (qv). Products made from thin sheet materials are most satisfactorily and economically cleaned by chemical methods which require alkali and soap solutions to remove grease and dirt, and acid solutions to remove oxidized metal.

The commercial chemical treatment in the metal preparation process may be carried out with continuous cleaning and pickling equipment. The fabricated sheet-metal articles are supported on a special continuously moving rack which is first passed through the cleaning stage. The ware subsequently is subjected to pickling acids, nickel flashing, and neutralizing treatments. Metal articles also are cleaned and pickled in a batch process whereby baskets of metal articles are immersed in large tanks of various solutions and then water rinses.

Fundamental Considerations in Porcelain Enamels

Cobalt-bearing materials incorporated in the frit composition enhance the adherence of the enamel to sheet steel. The mechanism by which glass adheres to metal is still not completely clear and the role of cobalt in promoting adherence is a continuing fundamental research subject. There are additives or ingredients of the frit other than cobalt which seem to add adherence. Molybdenum compounds incorporated in the enamel are considered adherence promoters.

The mechanism of adherence can be considered of two main types: physical adherence, which involves the physical gripping of the glass by a metal surface that has been roughened mechanically prior to enameling or roughened during enameling by the corrosive attack of glass on the metal and by dendritic attachments to the metal formed during the enamel-firing treatment; and a mechanism involving the chemical bonding of the metal, metal oxide, and enamel glass.

Strain in enamels that leads to failure is on the order of 0.002–0.003 cm/cm. Thinner enamels with their high residual compressive stresses are more flexible and can be strained to a greater degree.

Decorative porcelain enamels involve either a one-coat or multiple-coat system. White enamel is the most common cover, or direct-on, enamel. Whiteness or high diffuse reflectance is called opacity, the white, high opacity enamel depending on crystalline opacifying agents, such as antimony oxide and zirconium oxide or titanium oxide, which either remain well-dispersed in the glass during smelting and subsequent firing or are recrystallized from the enamel glass during the firing process. Opacifying pigments have an index of refraction much different from the 1.50–1.55 range of the glass matrix. The most effective opacifiers are (index of refraction): NaF (1.336); CaF_2 (1.434); Sb_2O_3 (2.087–2.35); SnO_2 (1.997–2.093); ZrO_2 (2.13–2.20); TiO_2 (anatase) (2.493–2.554); and TiO_2 (rutile) (2.616–2.903).

A.L. FRIEDBERG
University of Illinois

A.I. Andrews, *Porcelain Enamels*, 2nd ed., Garrard Press, Champaign, Ill. 1961, p. 633.

L.M. Dunning, *Proc. Porcelain Enamel Institute Technical Forum* **38**, 37 (1976).

ENCAPSULATION.   See Embedding; Microencapsulation.

ENDORPHINS.   See Anesthetics; Antibiotics—Peptides: Hormones; Hypnotics, sedatives, and anticonvulsants.

ENERGY MANAGEMENT

Energy management should be considered in the broadest possible sense, not just as a plant-energy audit or an energy conservation program. Energy must be considered an essential element of industry together with land, labor, capital, and raw materials, and managed to optimize progress towards the industry's desired goal. Many of the concepts important to the successful management of other elements in the chemical industry apply equally to the management of energy.

Available Energy Forms

The dominant energy sources in the chemical industry are fuel and purchased electricity, often referred to as basic energy commodities. The use of energy often involves a derived energy commodity. The most significant one for the chemical industry is process steam (qv); others include compressed air, refrigeration (qv), circulated hot oil, and organic heat-transfer vapors (see Heat-exchange technology).

Another basic energy source that is often important is the raw material or feedstocks entering the process. If the dominant reactions are exothermal, the energy liberated by the process must be considered in the same way as energy liberated by burning purchased fuel. Waste for disposal is a possible energy source derived from feedstocks or raw material, if the materials have a positive heating value (see Fuels from waste).

In terms of energy management, the first and second law of thermodynamics imply that all energy used in an industrial plant is ultimately wasted to the atmosphere or to a heat-rejection system. The objective of energy management is to ensure that the minimum amount of energy is expended and that maximum benefit is obtained from every unit of energy throughout its unavoidable path of degradation from the highest practical release level to the lowest available rejection level.

Economics of energy levels. The value of energy is related to both its cost and its use. The problem is simplified somewhat if energy value is always thought of as being related to energy level. This is reasonable since the amount of work or other benefit that can be derived from energy is a function solely of its level. For instance, steam at 10.34 MPa (1500 psia) and 538°C has an enthalpy of 3462.5 kJ/kg (1490.1 Btu/lb). This energy is far more valuable than an equivalent number of heat units in saturated steam at 68.9 kPa (10 psia) because the steam at high temperature and pressure can be expanded in a steam turbine to produce high value prime energy and still leave the exhaust of the turbine with enough heat at a usable level for process or space heating.

The value of the exhaust steam would be the cost of the high pressure steam less the net cost of providing the power developed by the turbine with an electric motor instead. The shaft work developed in this way is often referred to as by-product power. By-product power obtained represents a considerable energy economy. Typically, its cost is 20–40% of the cost of shaft work from an electric motor using purchased electricity.

Recycling of energy by applying relatively small amounts of shaft work to restore the level of the energy, the basis of the heat pump or vapor recompression, is another method of economizing on energy use. An application of the economics of this is the organic heat pump diagrammed in Figure 1. Vapor compression or heat-pump applications are evaluated by the coefficient of performance (COP), the ratio of the amount of heat delivered to the heat use divided by the thermal (joule) equivalent of the work required to raise the energy level. The COP is a function of the difference in energy level between the heat source and the heat use; the amount of heat-transfer surface involved; its effectiveness; and the efficiency of the compression device. To be useful for an application, the COP must usually be substantially greater than the cost

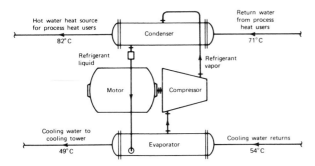

Figure 1. Application of organic heat pump.

of shaft work (purchased electricity) in energy units divided by the cost of thermal energy (fuel or steam) in the same energy units.

Waste heat recovery. The effect of heat recovery is to lower the final energy rejection level so that additional benefit can be obtained before the energy is ultimately wasted. Some examples of heat recovery found in chemical plants include: boiler economizers (heat exchangers that recover some of the heat in boiler flue gases to preheat the boiler feedwater); furnace convection sections (heat exchangers added to industrial process furnaces that use flue gas heat to preheat incoming process material or to provide part of the heat for another process); combustion air preheaters (flue-gas-to-air heat exchangers that recover heat from flue gases to heat the combustion air); and product-to-feed heat exchange, heat exchangers that cool the product of a thermal process by preheating the raw material feed to that process (see Heat exchange technology).

Some of the equipment and systems employed in heat recovery are direct mixing, contact condensing and recycling; conventional heat exchange; regenerative heat exchangers (wherein a thermal mass is first heated by direct contact with the high temperature heat recovery source stream then placed in contact with the use stream where it gives up heat); heat pipe, a pipe sealed on both ends and partially filled with vapor–liquid phase heat-transfer medium; liquid run-around systems; waste-heat boilers which generate steam from hot gases, hot liquids or condensing process vapors. Energy balances are made and used as tools in developing heat-recovery opportunities. These are tabular or diagrammatic summaries of all of the energy sources and all of the energy rejections for a single item of process equipment, for a process unit, or for an entire manufacturing plants. Table 1 is a tabular balance for a propane-fired dryer.

Table 1. Product Dryer Analysis Heat Balance, As Found

	kg/h	Temperature, °C	kJ/kg[a]	Energy, MJ/h[b]
Inputs				
fuel, C_3H_8	130	16	50254	6553
combustion air	6,817	16	15.6	106
secondary air	14,846	16	15.6	232
air in-leakage	4,289	16	15.6	67
water with product	1,731	49	204	354
dry product solids	4,478	49	55.64	249
Totals in	*32,291*			*7561*
Outputs				
water vapor from product	1,445	74	2632	3808
water vapor from combustion of H_2	212	74	2632	560
air and combustion products	25,870	74	74.61	1933
Subtotal to bag house	*27,527[c]*			*6301*
dry product solids	4,478	122	102.1	458
water with product out	286	122	510.4	146
radiation and convection losses				656
Totals out	*32,291*			*7561*

[a] At 0°C; to convert kJ/kg to Btu/lb, divide by 2.32.
[b] To convert MJ/h to Btu/h, divide by 1.054×10^{-3}.
[c] Air to bag house: rh, 24.5%; dew point, −83°C.

Energy Systems and Equipment

The overall energy efficiency of a plant depends heavily on the efficiency of the individual energy systems. Some losses are inherent in all energy delivery systems; in most cases these cannot be completely

eliminated but very substantial and economically attractive savings can be realized by minimizing them. For the steam system, the dominant losses are boiler flue-gas losses, steam leaks, steam vents, hot condensate losses, and thermal losses through uninsulated or inadequately insulated piping and equipment surfaces. All of these can be controlled (see Steam).

Electrical distribution system energy losses are usually less significant than those from the steam system. An electrical system that is poorly maintained or overloaded will, however, cause direct energy losses. Correcting the deficiencies will produce some immediate energy savings, but the main result will be improved reliability and safety.

Some of the most significant items of energy system equipment found in chemical manufacturing plants, where energy may be saved by proper design, are boilers and process furnaces, steam turbines, steam traps, electric motors, compressors, and pumps.

A.F. WATERLAND
Waterland, Viar & Associates, Inc.

W.C. Turner, *Energy Management Handbook*, John Wiley & Sons, Inc., New York, 1982.

T.F. O'Sullivan and co-workers, *Hydrocarbon Process.* **57**(7), 95 (1978).

ENGINEERING AND CHEMICAL DATA CORRELATION

Quantitative measurements are required to establish even the most philosophical theories of chemistry, as well as its practice, and the design of the equipment and plants of the chemical engineer. There are never enough dependable experimental data, hence estimates and predictions of the most probable values are always necessary.

Experimental values may be correlated to minimize or average their deviations from equations which may be developed to represent them. The most probable values may be obtained by interpolation or extrapolation graphically or from the equation. Often the equation depends not on pure theory but requires practical assumptions or empirical corrections to represent the properties of real matter. Or the model may be entirely empirical and require considerable search and calculations to determine the constants, as many as 8 to 20 in some empirical equations representing the variations of even the simplest physical properties.

Helpful in the development of such mathematical models is the interrelation of data for different properties of matter in the quantitative understanding of each, because the properties themselves are so interrelated. Equally useful is the comparison of how the properties of different similar materials vary with the same changes of conditions. Such comparison usually allows the reduction of complicated relations to linear equations.

Because of the simplicity and accuracy in use of all graphical, algebraic, and computer techniques, linear relations are highly desirable for all correlation methods, and much effort is justified to find a means of arranging the expression in a linear form. A simple illustration is a plot of data which is most readily used if linear; it can have the "best" line drawn through the points by moving a piece of black thread held taut over the points until it gives the best representation of the points.

The following presentation shows how to quantify from a minimum of experimental points many chemical and physical properties, most often in linear form, through comparison with the exactly known values of a reference substance.

Fitting Curves to a Variable

Most practical correlations are a compromise with some equation suggested from theoretical considerations and the constants determined from experimental data. Representation of experimental data by mathematical expressions or models is most desirable since it facilitates interpolation and extrapolation and minimizes experimental requirements. Prediction of data is possible with surety if a model of the variation of a property is known. This model allows the development of computer programs, a shorthand method of expression instead of the use of vast tables of data, as well as extensive assistance in calculations.

Developing such a mathematical model involves three steps: establishing the form of the equation; determining the numerical values of the constants in the equation; and appraising quantitatively the reliability of the resulting equation.

Determining a suitable form of the model equation may be based on underlying theoretical principles as well as considerations of simplicity and ease of use. There may be no theoretical basis for the relations of two variables or it may be complex or obscure. Thus the search for a suitable linear form may start along empirical lines. A simple plot of the data on ordinary rectangular graph paper may give an immediate indication of the essential form of the equation (see Fig. 1 and Table 1).

On each of the graphs shown in Figure 1, an equation is suggested by which, having the same general form as the graph, data often may be correlated. For some of these types, the method of reduction to a linear form may be apparent immediately, possibly through a transformation of variables. Often this may be done by a plot of logarithms, reciprocals, power functions, etc, of one or both variables that are calculated—or in some cases, plotted on specially ruled graph paper, which immediately includes these functions. Developed with ingenuity, such a plot often gives a straight line, and the constants of the resulting linear equation can be found directly from the slope and intercept of this straight line, or by means of analytical techniques, some of which are discussed below.

Sometimes the simple plot of experimental values of one variable against those of the other or of some simple function of these variables produces a type of curve, allowing a derived or alternative plot which will produce lines which are nearly straight, and thus represent a linear equation.

Having determined some basis of a linear correlation, the values of the constants in the equation must be determined so that the differences between calculated and observed values are within the range of assumed experimental error in the original data.

The classical evaluation of the linear equation that best represents such a collection of points is by the method of least squares. This minimizes the sum of the squares of the deviations observed from those which are predicted by the equation and the line which may be regarded as the best representation of the experimental data.

Curve fitting is achieved best when the form of the equation used is based on a known theoretical relationship between the variables associated with the data points. When no theoretical relationship exists, polynomials may be used to describe a curve. Polynomials may be evaluated by standard methods; their coefficients occur linearly, and often a better fit is obtained by increasing their degree, ie, the highest power appearing in the equation.

Fitting polynomials to data points involves essentially the same technique as for the classic correlation of an expression for multiple variables. Thus, the single dependent variable may be expressed as the sum of several or more terms, each including a higher power of the dependent variable. There results in the polynomial an expansion of the quadratic of Table 1:

$$y = a + bx + cx^2 + dx^3 + \cdots \Omega x^n \tag{1}$$

Table 1. Some Common Curves and Their Equations

Curve	Equation	Method
linear	$y = a + bx$	plot of y vs x; a = intercept on y axis at $x = 0$ and b = slope
parabolic through origin	$y = cx^n$	plot log y vs log x directly or on log (double) coordinate paper; c = antilog of the y intercept at $x = 1$ and n = slope
general	$y = a + cx^n$	first obtain a as the y intercept on plot of y vs x, then plot log $(y - a)$ vs log x and proceed as in parabolic through origin
polynomial quadratic	$y = a + bx + cx^2$	first obtain a as the y intercept on a plot of y vs x, then plot $y - y_n/x - x_n$ vs x where y_n, x_n are the coordinates of any point on a smooth curve through the experimental points. The slope of this new plot gives c, and the intercept is equal to $b + cx_n$
general	$y = z + bc + cx^2 + dx^3 + \cdots$	graphical procedures are almost impossible, but analytical procedures are easy to use and work well
exponential	$y = ab^x$	plot log y vs x directly or on semilogarithmic coordinate paper with y on the log scale; a = antilog of intercept and b = antilog of slope
	$y = ae^{bx}$	plot ln y vs x directly or on semilog paper as above, a = antilog of intercept and b = slope
hyperbolic	$y = a + b/x$	plot y vs $1/x$; a = intercept and b = slope
logarithmic	$y = a \log x$	plot y vs log x directly or on semilog paper with x on the log scale; a = slope
logistic	$1/y = a + be^x$	plot $1/y$ vs e^x; a = intercept and b = slope; alternatively, plot $1/y$ vs x, a = intercept; then plot $1/y - a$ vs x on double log paper; b = antilog of y intercept

A set of simultaneous equations can be developed by continuing this substitution and then can be solved to generate the coefficients. This process of polynomial regression depends on the equation being linear in the coefficients a through Ω.

A technique that may be used with large computers employs the Chebyshev polynomials in a series up to those that contain as many as 20, or any number of terms. These will determine the coefficients of a correlating equation, and algorithms have been developed for such usage. For example, this entirely mathematical or empirical technique can be used to represent the vapor pressure of water from the triple point to the critical point. The Chebyshev polynomial with 11 terms was found to be the lowest-order equation that would correlate the available vapor pressure data for water within the estimated errors of the observed data points. Obviously, such equations can only be worked on large computers.

Absolute Correlations of Individual Properties

Most of the hundreds of reported theoretic correlations of chemical and engineering data are for individual properties, based on an absolute method. They represent attempts to build up from molecular functions a model of the physical mechanism controlling the property, into which some numerical values are substituted, usually to evaluate constants in the equations. Since the properties of molecules are not known exactly, the solution or correlation usually requires numerous assumptions and empirical constants, unique for each substance. Often there are subsidiary equations and tables of values of different parameters which must be used.

More complex mathematical or empirical approaches and equations have been developed for binary and multicomponent systems. Other methods require large computers and their programming, along with extensive tables of constants for each individual liquid and correction factors for other parameters, etc. Sometimes assumed values and trial-and-error calculations are necessary.

The Reference-Substance Concept

Experimental data are usually obtained as a function of the values of a controlling condition, such as temperature. Any correlation is then made either theoretically or empirically, referring to the performance of one particular material as if its performance was unique for that material.

In general, however, different substances, and their molecules, will react to the same physical changes according to a similar pattern or law. In particular, the absolute values of the property of one substance may be expected to vary similarly to those of another substance at the same values of the controlling condition, eg, temperature.

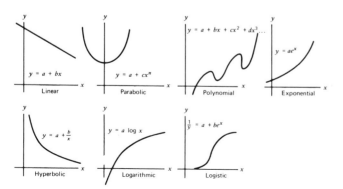

Figure 1. Some common curves and their equations.

To take advantage of this recognized fact that different materials usually have similar variations under a similar change of conditions, the less well-established data for one substance may be related to the corresponding values of data for another material, which are well known, studied, and tabulated in all handbooks. This comparison would always be made at the same condition, eg, temperature, and may be regarded as the relative variation of the property of one substance in relation to the variation of the established and true values of the reference substance. The comparison may be made by using a table of data, an algebraic equation, a geometric plot, or a readily written computer program.

More nearly correlated results, ie, a closer fit to the relevant equation or curve, are always obtained for these ratios of data for the two substances than by a consideration of the absolute values of the less well known one. This is because the variations for both substances follow the same physical mechanisms or laws under changes of the particular condition, eg, temperature.

The reference-substance concept has long been used with various properties. Thus, specific gravity is defined as the ratio of the density of the substance to that of water always taken at the same temperature. For gases, specific gravity is the ratio of density referred to that of dry air always taken at the same temperature and the same pressure.

The reference-substance relation, either as an equation or a graph, is simple in use and exact in expression of the variations of many different physical properties under changing conditions.

Linear relations, with their simplicity of both algebraic and geometric treatment, usually result. Then not more than two terms express—often with theoretical rigor—such linear equations: the multiplying constant, m, expressing the slope of a straight line; and the additive constant C, that of the intercept. Often a family of such lines has a common point, and the second constant becomes zero to give an equation with a single constant, that of the slope m. This slope constant is usually a function of the activation energy, ie, latent heat or some other related term.

Reference-substance and corresponding states. The reference substance is a considerably simpler concept than its outgrowth, the so-called principle of corresponding states, which assumes that different substances exhibit similar thermodynamic behavior when compared at the same reduced conditions. The term corresponding states refers to conditions where pressure, temperature, and volume are at an equal fraction of their critical values. All gases are thus considered at equally reduced and generalized properties. This is less satisfactory for liquids.

Correlating charts based on corresponding states of fractions of critical values of pressure, temperature, and volume of all compounds cannot approach in precision and accuracy the presentation of the properties at the same single condition, eg, temperature of one substance versus the properties of one other reference substance. This is because of the very generality of attempting to encompass all compounds on one graph, compared with the specificity of comparing the behavior of one compound versus that of a single other compound under just one fixed condition rather than many.

Charts of corresponding states convert the P, T, V variables for every substance to an involved system of empirical curves. These curves are too difficult to prepare for any single use, but are available in handbooks. There are very substantial errors of the generalized correlations in the principle of corresponding states, and these must be tolerated when their use is the only way to estimate data. Also, the reference-substance concept can be used in working with many times as many properties as those possible by the use of corresponding states.

Properties of Liquids

The thermodynamic scale of temperatures for gases uses absolute zero, $-273°C$, as the fixed or zero point. This point has substantially no meaning in relation to the molecular structure or properties of a liquid. A more relevant base or starting point for referring properties of liquids to temperature has been shown to be the critical point of the particular liquid, ie, the highest temperature where the substance can be a liquid. Thus any other temperature is measured downwardly from that fixed point, and is usually expressed as $T_c - T$, where $T_c =$ critical temperature. Hereinafter this difference is referred to as the critical difference temperature, T_D.

Surface tension. Surface tension, γ, has been related to the critical difference temperature, T_D:

$$\log \gamma = a \log T_D + b \tag{2}$$

However, if γ' is the surface tension of a reference liquid and if the values of surface tensions of both liquids are always taken at the same value of T_D, then the equation for $\log \gamma'$ may be combined with the above equation by combining constants to give new constants m and C:

$$\log \gamma = m \log \gamma' + C \tag{3}$$

Here m represents the ratio of the surface activation energy of the two liquids. For hundreds of liquids—all of those tried—this equation or its nomogram gives values within the accuracy of experimental data for either pure liquids or solutions of liquids or solids in a liquid.

Density. Similarly, an almost exact linear equation was derived and tested with thousands of points of density data for 100 liquids:

$$\log d = m \log d' + C \tag{4}$$

Here the slope constant, m, is almost unity for most organic liquids, d is density of the material, d' is density of the reference liquid at the same critical difference temperature for each T_D, and C is a constant. This nearly exact equation predicts densities of a liquid or of its solutions, either graphically (usual plot or nomogram) or algebraically, if densities at two points are known.

Also, $\log d$ has a direct linear relation with $\log T_D$; and without use of the reference substance:

$$\log d = s \log T_D + c \tag{5}$$

This linear equation, with s and c as constants, is more convenient than the previous one with the reference substance; however, data deviate from it slightly more at higher temperatures, ie, up to within $25°C$ of the critical temperature.

A generalized equation, good for all liquids (also good to within $25°C$ of the critical temperature), was obtained which requires only one value of the density of a particular liquid, d_1, at a temperature T_{D1}:

$$\log d_1 = s \log T_{D1} + c \tag{6}$$

If this equation is subtracted from the previous equation to eliminate the constant c, and the result is changed to an exponential function, then:

$$d = d_1 (T_D/T_{D1})^s \quad \text{or} \quad \log d = \log d_1 + s \log T_D - s \log T_{D1} \tag{7}$$

The value of s is always close to, and may be averaged to 0.2437 to express the data of 3500 experimental points for a wide variety of compounds. The average deviation of these experimental values from those obtained using this general equation, which is linear when used in its logarithmic form, was 1.15% up to temperatures within $25°C$ of the critical temperature.

Viscosity. Viscosity appears in many multivariable equations and dimensionless groups, eg, of heat transfer and mass transfer. Thus, a simple expression for viscosities and temperature is useful not only by itself but also for insertion algebraically into more complicated equations.

A simple and exact equation may be based on a temperature scale depending not on the usual expansion of a gas, liquid or solid, but on the vapor pressure of a liquid.

The ordinates of Figure 2 (p. 422) are erected on double logarithmic paper at points calibrated on the X axis for temperatures at values corresponding to the vapor pressure of water. On these ordinates, the logarithms of the viscosity are plotted directly. Straight lines, thus linear equations, are obtained except for the break for water at about $30°C$, as often noted for other physical properties of water. (This is equally noticeable, but was not mentioned above, for density or surface tension.)

Of more interest is the theoretical derivation. The horizontal calibration, the vapor pressure of water, is defined by the Clausius-Clapeyron equation:

$$\frac{d \ln P}{dT} = \frac{L}{RT^2} \tag{8}$$

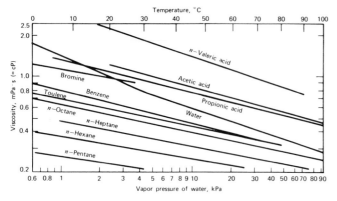

Figure 2. Plot of viscosities of eleven representative liquids against temperatures which are obtained from the corresponding vapor pressures of water. To convert kPa to mm Hg, multiply by 7.5. Courtesy of *Industrial and Engineering Chemistry*.

where T is temperature, L is latent heat, P is vapor pressure of water, and R is the gas constant. A similar expression has been developed for viscosities:

$$\frac{d \ln u}{dT} = -\frac{E}{RT^2} \tag{9}$$

where u is viscosity and E is the activation energy of viscosity. If values are always considered at the same temperatures, the equations may be combined to give:

$$\frac{d \ln u}{d \ln P} = -\frac{E}{L} \tag{10}$$

$$\log u = -\frac{E}{L} \log P + C \quad \text{or} \quad \log u = -m \log P + C \tag{11}$$

This linear equation (eq. 11) gives excellent correlation, and $-E/L$ is known to be constant over wide temperature ranges for a large number of substances. It may be used with assurance for liquid solutions. When these are aqueous, the use of viscosities of solutions vs viscosities of water corrects for the known abnormality of water at about 30°C.

Solubilities in liquids. These most important data may be correlated readily also against a temperature scale based on vapor pressures, eg, of water. A special case of the van't Hoff equation is often cited:

$$d \log x = (\bar{h}_l - h_s)/RT^2 \, dT \tag{12}$$

where x is the solubility as mole fraction of a solid in solution, $\bar{h}_l$ is partial molal heat of the solute in its saturated solution, h_s is the molal heat content of the pure solid, R is the gas constant, and T is the temperature. This equation may be divided by the Clausius-Clapeyron equation (values always at the same temperature). When the resulting differential equation is integrated, assuming constancy of the difference of heats, there results approximately:

$$\log x = \frac{(\bar{h}_l - h_s)}{L'} \cdot \log P' + C \tag{13}$$

where at the same temperature as for the solubility, L' is the latent heat of the reference substance—here water—and P' is its vapor pressure.

For many solutions of solids, this equation has been found to be linear within the experimental error and to give better correlations than the sometimes used $\log x$ vs $1/T$ plot.

Distribution coefficients between two liquids. Distribution coefficients of a solute between two solvent layers can be correlated similarly for use particularly in extraction calculations:

$$\log K_D = \Delta H/L' \log P' + C \tag{14}$$

where K_D = distribution coefficient at the given temperature, ie, concentration of solute in one layer divided by the concentration in the other layer, and ΔH is molal heat of transfer of solute between the two layers. This heat is very difficult to measure, but is readily obtained by this equation from the easily determined distribution data.

Diffusion coefficients. Diffusions of gases, liquids, and solids in liquids, often water, and particularly their rates are of great importance in transport calculations and engineering operations. A differential equation has been derived:

$$d \log D = (dT/RT_2)E_d \tag{15}$$

where D is the diffusion coefficient (cm²/s) and E_d is the energy of activation for diffusion. This may be combined with the Clausius-Clapeyron equation in differential form, values taken at the same temperature. If the result is integrated, there is obtained approximately:

$$\log D = (E_d/L) \log P' + C \tag{16}$$

This linear equation (eq. 16) correlates well diffusion data in any solvent.

Since diffusion is closely related to viscosity, the differential form of this diffusion equation may be divided by the differential form of the viscosity equation above (eq. 11), always at the same temperature. There results by integration, approximately:

$$\log D = (-E_d/E_v) \log u_w + C' \tag{17}$$

where u_w are viscosities of reference substance (eg, water) taken at the same temperatures from tables and E_v is the activation energy of viscosity. This linear equation represents accurately data for all systems for which it has been tried.

However, the absolute values of activation energies of diffusion and of viscosity are almost the same at any given temperature, and their ratio is almost equal to -1. The average value of this ratio for many systems was found to be -1.1, which is the slope of the line. This value gives a general linear equation giving a slope equal to -1.1, good for most substances. There is a single remaining constant, C, and thus requires only one data point to predict diffusion coefficients throughout the entire temperature range.

Moreover, C', in the last equation, is the intercept of every line, where $\log u_w = 0$ or $u_w = 1$ Pa·s (= 1 cP) which for water is very close to 20°C. Fortunately, 20°C happens to be the temperature where most diffusion coefficients are reported, so that when using the value at this single temperature, the equation gives immediately those for all other temperatures.

The volume of the diffusing molecule v may be determined as roughly equal to the sum of the atomic volumes. These are known, and v has been found to be a reciprocal function of the diffusion coefficient D. When combined with the above relation, this gives:

$$D = 0.00014/v^{0.6}u_w 1.1 \tag{18}$$

This has given a lower apparent error than the various other correlations for all available data for solids, liquids and gases diffusing in water.

Vaporization Functions of Pure Liquids

Vapor pressures using reference equation at the same temperatures. Vapor pressures, P, for one substance, and P' for a reference substance, are defined in terms of temperature, T, molar latent heats, L and L', and the gas constant R by the Clausius-Clapeyron equation. Thus, if pressures and latent heats are taken always at the same temperatures,

$$\frac{dP}{LP} = \frac{dT}{RT^2} \quad \text{and} \quad \frac{d \log P}{d \log P'} = \frac{L}{L'} = m_T \tag{19}$$

If m_T, which is the ratio of the latent heats per mole, is assumed to be constant, when values are always taken at the same temperature, integration gives almost exactly:

$$\log P = m_T \log P' + C \tag{20}$$

This linear equation gives a straight line when plotted on double logarithmic paper, more easily if values of the vapor pressures of water as a reference substance are first laid out on the logarithmic X axis; and ordinates are erected at the corresponding values of temperature. The constants m_T and C for about 500 compounds have been averaged and published using for each material the known vapor pressures at about 10 values of temperature to define the function. Excellent correlation is obtained from the equation using tables of logarithms and vapor pressures and latent heats of water to determine from two data points of the material its latent heat (from m) and vapor pressure at any temperature.

Reference equations at same reduced temperatures. An equation with only one term expresses a linear logarithmic relation of reduced vapor pressures, P_R, to that of the reference substance P_R' always taken at the same reduced temperatures. The single constant m_{TR} represents the slope:

$$\log P_R = \frac{LT_c'}{L'T_c}\log P_R' \quad \text{or} \quad \log P_R = m_{TR}\log P_R' \quad (21)$$

This precise linear equation is much simpler and more readily used than other reduced equations for vapor pressures. It allows correlation and prediction of vapor pressures of all materials at all temperatures, and represents a sheaf of lines through the origin. It also allows the calculation of the critical pressure of a compound from its critical temperature. This equation also represents the reduced vapor pressures within 2% of all normal hydrocarbons from ethane up to C_{10} because of the orderly progress of the series, where n is the number of carbon atoms:

$$\log P_R = 0.7025 n^{0.1802}\log P_R' \quad (22)$$

Reference equations, values of temperatures at the same pressures. The first use of the reference substance was the quite inaccurate Duhring relation. The same technique as obtained in equation 20 gives:

$$\log T = m_P\log T' + C \quad \text{where} \quad m_P = \frac{L'\Delta V}{L\Delta V'} \text{ is a constant} \quad (23)$$

This equation does not use the reduced concept and covers the entire range of temperatures for any substance, using corresponding data always at the same pressure.

It is possible to use water as the reference substance at any temperature because water, as one of its many unusual properties, is in the liquid state throughout the greatest range of saturation pressures of any material, as it also has the highest critical pressure. This equation may be the most inclusive and for some purposes, the most convenient nonreduced equation available. It predicts more accurately than even equation 20, one reason being that it does not have the errors inherent to the Clausius assumption. For this same reason, the molar volumes of saturated vapors may be predicted more accurately from the $\Delta V'$ term in this equation than by other much more complicated empirical equations; also, the volume of a gas in a solid-absorption or a liquid-absorption relation.

Latent heats of vaporization, using compressibility factor. The compressibility factor corrects for the use of the gas laws in the Clausius modification of the Clapeyron equation, although the ratio in m has reduced this error substantially. To make this correction, L should be multiplied by the difference of the compressibility factor for the gas

$$Z_G = PV/RT \quad \text{and} \quad Z_L$$

for the liquid, thus by $(Z_g - Z_L)$ or ΔZ. When this is considered for both liquids:

$$m_T = \frac{L}{L'}\cdot\frac{\Delta Z'}{\Delta Z} = \frac{L}{L'}\frac{P'\Delta V'}{P\Delta V} \quad (24)$$

The latent heats of many hundreds of compounds have been tabulated by use of this equation. Also useful in many calculations is the $P\Delta V$ term, the heat equivalent of the external work in the vaporization of a liquid, and entropies of vaporization may also be determined readily from the equation giving temperatures at the same pressures of the reference substance.

Vapor Pressures of Solutions of Nonvolatile Solutes

Molecules of a solvent may be hindered from evaporation by those of the solute. The vapor pressure of the liquid is lowered and the boiling point is elevated. The mechanism and the physical laws expressing vapor pressures are similar for solvents when pure and when containing a solute. Solutions having fixed concentrations have very similar relations which may be correlated immediately and/or predicted by the nonreduced equations and graphs previously mentioned. Boiling-point elevations are immediately shown by either, as are freezing points of aqueous solutions from the known vapor pressures of ice.

Since the slopes of the straight lines, represented by the values of m in the equations, measure the heat of evaporation, for solutions this must include the heat of solution; and this may be obtained readily and accurately by difference compared with calorimetric determination. From these heats of solution at different concentrations, enthalpy charts for solutions are constructed.

Binary Solutions of Volatile Liquids

Vapor pressures and heat quantities. In systems of two or more volatile components, in addition to the variables of temperature and total pressures, the variables of compositions of both the liquid composition x, and of the equilibrium vapor phase y, must be considered along with the related partial pressures of each component. Also important is the heat of mixing of the two liquids.

The linear equations and plots using the reference-substance technique are applied directly to solutions of two or more volatile liquids. These give the familiar linear functions with their m and C constants for total pressures, partial pressures, vapor compositions, activity coefficients, relative volatilities, and equilibrium constants. It is convenient, but not necessary, to use one of the two liquids as the reference material.

An entirely different correlation without the reference concept is that linear logarithmic equations and straight-line plots result when any of these functions are related to the total pressure on the system. All of these linear functions have been found to hold equally well for azeotropic mixtures and their compositions.

Use of the heat quantities, derived from the values of m or the slopes of the lines, allow the construction of enthalpy or Ponchon diagrams, so exact for use in designing distillation equipment.

Properties of Gases

Reference equation of state for gases. The compressibility factor $Z = PV/RT$ is the precise criterion of an individual gas and its equation of state. This may be expressed in a virial form for a pressure series where B is the second virial coefficient. It is the only one that can be calculated from known molecular properties. Fortunately, the higher coefficients may be dropped to give, within the experimental error, by the method following:

$$Z = \frac{PV}{RT} = 1 + \frac{B}{RT}P + \frac{(C - B^2)}{R^2T^2}P^2 + \frac{(D - 3BC + 2B^3)}{R^3T^3}P^3 + \cdots$$

$$(25)$$

Again, the reference concept is helpful. Thus

$$Z = \phi(RT + BP)$$

where

$$\phi = Z'/\left(RT + B'P_R P_c'T_c/T_c'\right) \quad (26)$$

(The values with primes represent properties of the reference substance. T_c is the critical temperature; P_c is the critical pressure.)

Alternatively, the equation expresses V directly:

$$V = \phi RT(RT/P + B) \quad (27)$$

Using nitrogen as a reference, 2000 data points on 21 systems were tested with this equation having only the one constant B. Deviations were always less than 1%. Other equations with up to 21 empirical constants, each to be determined for each gas, do no better for a single system in a narrow range where there are only nine data points per empirical constant to be evaluated, and then represented.

Mixtures of gases. The above, single-constant equation of state correlates and predicts data for gas mixtures, as shown by representing over a thousand data points for almost forty systems with an average deviation of 2.34%. Using critical values for the individual gases, those for a representative pseudo gas are determined, which represent the mixture effectively and accurately.

Gas–liquid relations. Gas–liquid relations are of great industrial importance in absorption and other industrial systems, the design of which depends on solubility; ie, relation of the amount of gas dissolved in a liquid to the temperature and gas pressure. In developing this relation, the familiar linear logarithmic equations and graphs were obtained for the gas pressure exerted at a different temperatures by solutions of different fixed concentration of the gas dissolved in the liquid on a scale of temperatures derived from the vapor pressures of water. Heats of solution of the gas are readily evaluated from the values of m, the slopes of the straight lines. Similarly, Henry's law constants were represented accurately by the same relation.

Gas–solid relations. The same linear logarithmic equations and straight lines express the equilibrium pressures and temperatures of different fixed amounts of gas or vapor adsorbed on a solid adsorbent. (The classic Freundlich isotherms are curves, indicating that temperature is not the variable to fix in such an analysis.) The slopes of the straight lines include the total heat of disengagement of the vapor molecules from the solid or, in the solution case, from the liquid. Thus, the heat of adsorption or of absorption may be determined readily.

Other properties of gases. Other properties also have been correlated and may be predicted by linear relations developed from modifications of the reference-substance techniques: viscosities; diffusion coefficients of gases in gases; permeabilities of membranes; entropies; and enthalpies.

Further Uses of These Correlation Methods

Expansion of the above techniques allow the ready development of charts for: heats of solution and of enthalpy; the prediction of construction of Ponchon diagrams from either vapor pressures or boiling-point composition data; Equilibrium flash vaporization (EFV) data for the design of petroleum distillation towers, the heats of solution, activities, and electromotive forces of electrolytes in cells, psychrometric or humidity charts related to water, its solutions, or other liquids. The same techniques have also been shown to give linear functions of data relative to chemical reactions such as chemical equilibrium constants (over a total range of 10^{40} times and a temperature range of over 1000°C), also reaction rate constants and heats of chemical reaction.

DONALD F. OTHMER
Polytechnic Institute of New York

D.F. Othmer and H.T. Chen, *Ind. Eng. Chem.* **60**, 39 (April, 1968).

Applied Thermodynamics, American Chemical Society, Washington, D.C., 1968.

R.C. Reid, J.M. Prausnitz, and T.K. Sherwood, *Properties of Gases and Liquids*, 3rd ed., McGraw-Hill, New York, 1977.

ENGINEERING PLASTICS

Any discussion of engineering plastics falters at the outset on the matter of definition. A practical definition must include not only property/performance criteria, but also market/pricing criteria that place certain resins in the engineering category to the exclusion of all others, eg, nylon, acetal, thermoplastic polyester molding compounds, poly-(phenylene oxide)-based resin (PPO), and polycarbonate.

Properties

In terms of properties, engineering plastics have a good balance of high tensile properties, stiffness, compressive and shear strength, as well as impact resistance, and they are easily moldable. Their high physical-strength properties are reproducible and predictable, and they retain their physical and electrical properties over a wide range of environmental conditions (heat, cold, chemicals). They can resist mechanical stress for long periods of time. Flame retardance is not an essential requirement, but it has become an important added asset (see Flame retardants).

In terms of market/pricing criteria, the engineering plastics form a distinct group as compared to the high volume/low price commodity plastics and the low volume/high price specialty plastics.

The commodity plastics have sales in the 500–3300 thousand metric ton range, and a large number of suppliers. The engineering plastics are sold in the 20–120 thousand metric ton range, and have few suppliers. The annual consumption of specialty plastics is roughly 10–15 thousand

metric tons. Each of the specialty molding compounds has just one supplier.

In general, engineering plastics are characterized by predictable properties over a wide range of loadbearing conditions, patent protection on composition or process, or polymer and/or monomer, specialty chemical raw-material bases, minimum number of competitors, high development costs and high cost of supplying markets, low volume as compared to commodities, high selling price, and high capital cost/output.

Processing

The melt-processability of thermoplastic resins is a characteristic that distinguishes them from thermosets. This pertains not only to the advantages of injection molding as compared to compression or transfer molding, but also to the variety of processing alternatives that extend the utility of the thermoplastics. There are thermosets that can be injection molded, but only the thermoplastics offer the options of extrusion into sheet, film and profiles, or blow molding (see Plastics technology).

The degree to which each of the engineering plastics is amenable to alternative process methods varies, and the relative potential of each of them depends also on their potential in alternative processes, not just on their utility in injection molding.

In addition to its use as an injection molding resin, nylon can be extruded into filaments, rods, and flexible film, and it can be blow molded. Acetal can also be extruded and blow molded, but its use is almost entirely in injection molding. The thermoplastic polyesters used in engineering application are generally injection molded as well.

PPO-based resin is generally injection molded, but its use for extruded profiles and sheet is increasing. Polycarbonate excels in both injection molding and extrusion, and it also has use in blow molding.

Engineering Plastics

Nylon. Nylons, also known as polyamides, carry numerical designations that refer to the number of carbon atoms in the amide links. Nylon-6,6 is the reaction product of hexamethylenediamine and adipic acid, each of which contains six carbon atoms. Nylon-6 results from the polymerization of caprolactam (see Polyamides). These two types of nylon account for over 90% of nylon-resin consumption.

In the auto industry, many of the applications are replacements for die-cast zinc. In industrial machinery parts, nylon is widely used because of its natural lubricity, wear resistance, and chemical resistance. It is the oldest established resin for mechanical drive components such as bearings, gears, sprockets, pulleys, rollers, races, and chains. In these applications, acetal is nylon's closest competitor.

Because of their hydrocarbon resistance, nylon compounds perform well in lawn mowers, chain saws, and other gasoline-powered tools. In electric/electronic components, nylon is particularly important in wiring devices such as plugs and connector bodies, receptacles, and other parts, sometimes in replacement of phenolic molding compounds or rubbers (see Electrical connectors; Phenolic resins; Rubber, natural).

Acetals. Like nylon, the acetals are excellent for parts that require lubricity and resistance to chemicals and abrasion. Acetal's use in electric/electronic applications is very minor. Because of its flammability, in particular, it is not found in wiring devices, connectors, and other U.L.-regulated parts.

Acetal is used in a wide variety of consumer products, including lighters, zippers, fishing reels, and writing pens, but its greatest strengths are in those consumer products that require chemical and water resistance, such as garden-chemical sprayers, household water softeners, paint-mixing paddles and canisters, and particularly plumbing applications where the copolymer exhibits superior resistance to continuous exposure to hot water.

Thermoplastic polyester molding compounds. The thermoplastic polyesters are used primarily for electric/electronic applications. Because of their high heat resistance and property retention in flame-retardant grades, they have been able to replace phenolic molding compounds and other thermosets.

Apart from electric/electronic usage, they are used in industrial machinery parts, where certain kinds of chemical resistance are required, or where glass-reinforced grades provide necessary heat resistance.

Poly(phenylene oxide)-based resin. The PPO-based engineering thermoplastic is General Electric's Noryl, a blend of poly(phenylene oxide) and impact polystyrene (see Acrylonitrile polymers; Styrene plastics). The principal competitors for Noryl are high heat grades of ABS and polycarbonate–ABS alloys.

Flame retardance is imparted to Noryl by the incorporation of a proprietary flame retardant into the compounding procedure. Among the engineering resins, it has the highest percentage of volume in flame-retardant grades. This feature has been important in appliances, business machines, and other electric/electronic applications.

In the automotive market, Noryl is used for some exterior plated parts and interior parts that have to withstand greenhouse temperatures of 107–110°C. Appliances represent the largest single use category for Noryl.

It is also used for motor housings because of its heat resistance; for electrical enclosures because of its U.L. approval as sole support for current-carrying parts; for mixer housings, TV deflection yokes, and other parts where flame retardance is required; for dishwasher pump covers where hot-temperature moisture resistance is necessary; and for hydromassages, steam curlers, and other units using moist heat.

In the electric/electronic category, Noryl is used for outlet boxes, switch plates, connectors, compressor terminal covers, terminal blocks and strips, and other parts.

Polycarbonate. Polycarbonate is an amorphous polyester of carbonic acid, produced from dihydric or polyhydric phenols through a condensation reaction with a carbonate precursor. Because of its transparency and processability, polycarbonate is used in high-glass volume replacement applications not open to any of the other engineering thermoplastics.

Polycarbonate is very popular for small appliances; it is generally the first choice where heat or impact resistance, or both, is required.

It is also used in electric/electronic applications in a variety of uses that include covers for magnetic storage disks, switches, diode blocks, telephone dials, rings and push buttons, and enclosures for control equipment, and business machines.

Other plastics used in engineering applications. The five engineering plastics discussed in this article compete with other plastics in some markets, and to the extent that these other plastics do compete, they can also be called engineering plastics. These include glass- and/or mineral-filled polypropylene, flame-retardant ABS, and polystyrene.

Thermosets have long been available as insulators in electric/electronic applications, offering a wide range of capabilities in resistance to heat and environmental conditions. Processing advantages have allowed thermoplastics to replace thermosets in many markets, but competition remains in some of the more demanding uses.

The engineering resins do not compete directly with epoxies, but some of the high temperature resins are being used for that purpose (see Epoxy resins).

Fluoropolymers are often categorized along with the engineering plastics, but the two groups seldom compete. As a class, fluoropolymers do not offer the loadbearing capability of the engineering plastics. In nonloadbearing uses, however, fluoropolymers have outstanding and unique properties, including resistance to very high and low temperatures, exceptional electrical properties, and low coefficient of friction (see Fluorine compounds, organic).

Specialty Plastics

Specialty plastics generally have high temperature capability but this capability involves complex, costly synthesis and usually difficulty in processing. In this group, the polyimides (qv) can be used continuously at temperatures in the 260°C range.

There are only two materials in this group that compete directly with the engineering plastics as defined here. These are poly(phenylene sulfide) (PPS) and polysulfone. PPS, available for molding in glass- and/or mineral-reinforced compounds, offers U.L. Thermal Index ratings of 240°C at a relatively low price. PPS also has outstanding heat and chemical resistance, and it is inherently flame retardant.

Polysulfone is a transparent amorphous resin, selected instead of polycarbonate where the higher use temperature is required, and sometimes for better stress-crack resistance at lower temperatures.

Plastics with still higher heat-resistance capabilities include some thermoplastics (ie, other polysulfones), some thermosets (ie, polyimides and aramids), and some that can be melt-processed but need cross-linking for optimal property development (ie, poly(amide-imide) resins).

MARILYN BAKKER
Business Communications Co., Inc.

C.A. Harper, ed., *Handbook of Plastics and Elastomers*, McGraw-Hill Book Co., New York, 1975.

Modern Plastics Encyclopedia, annual, McGraw-Hill Book Co., New York.

ENHANCED RECOVERY. See Petroleum.

ENKEPHALINS. See Hormones.

ENZYME DETERGENTS

Enzymes are a vital part of all living processes and are believed to be as old as life itself. The term enzyme is derived from the Greek word *enzymos* and literally translated means in the cell or ferments. This name was first used by W. Kuhne in 1876 when he defined enzymes as unformed or unorganized ferments whose action may take place in the absence of organisms and outside of organisms. It is now known that enzymes are organic proteinaceous catalysts produced by all living cells. These enzymes are of two types: exoenzymes, which are excreted by the manufacturing cell into the cell's environment, and, once excreted, penetrate and break down organic matter, such as proteins, starches, and fats, into soluble derivatives that can be absorbed or transported through the cell membrane (as a general rule, exoenzymes produce little energy directly for the cell to use); and endoenzymes, which remain with the living cell, and are transformed and/or broken down by the action of coenzymes to produce energy and cell components needed for life. Relatively large amounts of energy accompanies these reactions, which is in turn made available to the host cell. The detergent enzymes described here are exoenzymes, characterized by hydrolytic activity under alkaline conditions (see Enzymes, industrial).

The development of enzyme supplements in laundry products is mainly the result of work by O. Rohm, founder of Rohm & Haas, Darmstadt, Federal Republic of Germany, and E. Jaag, of the Swiss firm, Gebrueder Schnyder, Biel. Rohm obtained a patent that describes application of pancreatic enzymes for washing purposes.

Properties

Serine-active alkaline proteolytic enzymes constitute > 95% (est) of all enzymes sold worldwide for enzyme laundry detergents. The use of effective detergent enzymes from bacterial sources is based primarily on the development of alkaline-stable and active proteases although amylolytic enzymes are also employed for the breakdown of starch-containing soils. With the exception of nonbuilt heavy-duty liquid laundry detergents, the pH of laundry detergents is generally in the 9–10.5 range. Another important property for a detergent proteolytic enzyme is the thermal stability, or thermal versatility; the so-called European laundry wash utilizes temperatures from ambient to 100°C and thus, the more heat-stable the enzyme, the better. A third factor in the suitability of a detergent enzyme is the nature of the active sites for the enzyme.

In addition to stability at relatively high temperatures and pH values, the enzyme should be compatible with other detergent ingredients, such as surfactants, sequestering agents (see Chelating agents), (phosphates,

EDTA, nitrilotriacetic acid (NTA), etc), optical brighteners (see Brighteners), perfumes (qv), etc. This is the case with the serine-active alkaline proteolytic enzymes.

Production

Detergent enzymes are derived from specific microorganisms. Since these properties are inherent in the organisms, they cannot be altered by variations of the growth conditions. Most detergent enzymes used today are excreted by bacteria of the genus *Bacillus*, especially members of the *Bacillus subtilis* group, a group regarded as harmless saprophytes which have been employed in the production of proteases and amylases for the textile and baking industries for many years. The advantages of detergent proteolytic enzymes from bacteria are improved performance under washing conditions and unlimited supply. The vegetable- and animal-derived proteases fall short of both requirements (see also Microbial transformations).

Fermentation requires isolation of the proper strain of bacterium. The most effective nutrient medium and growth conditions are then selected. Culturing is initiated in glass laboratory flasks, followed by transfer to pilot-plant fermenters. The medium functions as a source of protein and carbohydrate, and also contains the necessary nutrients and minerals such as phosphorus. Feed materials, eg, corn, starch, and soybean meal, are especially suitable in this regard. A typical medium for enzyme production is 3.0% potato starch; 2.5% soybean meal; 8.0% ground barley; 0.4% $CaCO_3$; 0.4% soybean oil; and ca 85.7% tap water.

The recovery of the enzyme begins as soon as the fermentation has been terminated. The medium is transported through a cooling stage to a separation process utilizing centrifuges where bacteria and gross insoluble substrate components are removed. The enzyme solution is concentrated by vacuum evaporation or ultrafiltration (qv) and the enzyme is removed from the filtrate by the addition of a protein precipitation agent. In order to prevent the growth of bacteria during the recovery process, the enzyme precipitate is maintained at the lowest temperature possible.

Health and Safety

Repeated inhalation of enzyme dust is associated with a comparatively high risk of respiratory allergy (hay fever, asthma) in susceptible persons. Detergent enzymes are proteinases and will, therefore, irritate moist skin, the eyes, and mucous membranes.

Uses

Alkaline-active enzymes have been used continuously since their introduction in the U.S. in the so-called presoak and "booster" laundry pretreatment products (see Drycleaning and laundering; Surfactants and detersive systems).

C.A. STARACE
Novo Laboratories Inc.

HANS C. BARFOED
Novo Industri A/S

Ger. Pat. 283,923 (1913), O. Röhm.

C. Dambmann, P. Holm, and V. Jensen, *Dev. Ind. Microbiol.* **12**, 11 (1971).

G. Jensen, *paper presented at the meeting in Deutsche Gesellschaft für Fettwissenschaft*, Giessen, FRG, Oct. 8–12, 1972.

ENZYMES, IMMOBILIZED

Enzymes, the natural linear polypeptides that catalyze biochemical reactions, have a mol wt of 10^5–10^6. The single polypeptide chains of enzymes can aggregate in solution to yield molecular weights as high as several million (10^6).

The properties of purified enzyme, can make their use very difficult. Generally purified enzymes can be stored dry and cool for periods of months or even years, but they lose their catalytic activity, often unpredictably, in a matter of minutes or hours when in solution. It is also well known that the folded structure by which catalytic activity is achieved is destroyed > 50–$70°C$ (see Biopolymers). Although many enzymes are sensitive to transition-metal ions that act as inhibitors of enzyme activity, the particular transition-metal ions that inhibit are characteristic of each enzyme. Likewise, an enzyme may be affected by organic inhibitors, activators, or both.

Methods of Immobilization

Many of the characteristics that make enzymes difficult to use *in vitro* can be overcome by immobilizing the enzyme on a solid support. Generally, immobilized enzymes are more stable to changes in pH and temperature than are their soluble counterparts. When suspended in water, these insoluble enzyme derivatives are usually stable for months, providing the opportunity of reusing the catalyst many times. Separation of products and reactants is also quite easy since these compounds are usually soluble and can simply be rinsed from the insoluble catalyst material. Of course, the reuse of expensive immobilized enzymes can make enzymes economical catalysts where the cost of soluble enzymes had traditionally been prohibitive (see Catalysis).

Figure 1 illustrates some of the general methods of immobilization. Several review articles describing the methods in detail are available. Probably more enzymes have been immobilized by activation of polysaccharides with cyanogen bromide than by any other method. Cellulose or a commercial beaded polysaccharide is allowed to react at an alkaline pH with cyanogen bromide; the intermediate is usually not stored but allowed to react immediately with the soluble enzyme. Stability of the soluble enzyme in the alkaline region is usually necessary, although some enzymes have been immobilized at neutral pH. The method can seldom be used in large-scale applications because cyanogen bromide is poisonous and expensive.

Enzyme cross-linking is a method often used in an application requiring a very high enzyme activity per volume of immobilized-enzyme material. Glutaraldehyde is the bi-functional reagent most often used for enzyme cross-linking. The soluble enzyme can be immobilized without a support material by addition of glutaraldehyde to the enzyme solution which usually results in precipitation of a polymerized enzyme; if an enzyme is first absorbed onto a support material, glutaraldehyde addition leads to formation of a sheet of polymerized enzyme on the surface of the support material.

Physical and Chemical Properties

The physical properties of immobilized enzymes are determined by the support material. Many organic support materials offer a wider choice of functional groups than typical inorganic supports, thus higher enzyme loading is possible which is important in applications where the space available to house the catalyst is limited.

The chemical properties of an enzyme often change during immobilization because of changes in the microenvironment or in the tertiary or folded structure of the enzyme molecule. Changes in microenvironment may affect the catalytic activity as a function of pH, temperature, ionic strength, etc. Changes in apparent activity may also result from changes

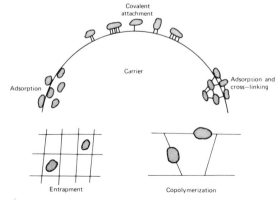

Figure 1. Some general methods of immobilization.

in the solubility of reactant molecules between the bulk solution and the microenvironment. In other cases, apparent changes in specificity of the enzyme and reaction rate are owing to the rates of diffusion of reactants and products. These chemical properties of an immobilized enzyme depend on the method of immobilization employed and the support material used.

Biochemical Processing

The first step in the development of a commercial process is to select the type of reactor that best fits the process being considered. Generally, reactors are classified as either tank or column reactors. In the simplest type of tank reactor, the batch reactor, a soluble or immobilized enzyme is mixed with reactant solution until the product is formed. Then the tank must be cleaned for the next batch. A continuously stirred tank reactor (CSTR) is used to accomplish continuous operation; the initial solution can be pumped into the tank while solution containing product exits from the other side of the tank. A useful modification of the CSTR is the inclusion of a membrane or ultrafilter at the exit port to allow the retention of soluble and insoluble macromolecules. The most commonly used column reactor is the packed-bed type in which the immobilized enzyme is packed in a cylinder through which reactant solution is pumped. In a fluidized bed, the immobilized enzyme is placed loosely in a column and the upward flow adjusted to cause the particles of immobilized enzyme to entrain or boil (see Fluidization; see also Reactor technology). Tables 1 and 2 list some of the immobilized enzymes for the two reactors.

Uses

Commercial applications. The production of high fructose corn syrup (HFCS) is the largest commercial application of immobilized enzymes. The discovery of microorganisms that produce glucose isomerase, which converts glucose to fructose until an equilibrium mixture is obtained, made it possible to prepare sweeter corn syrups than previously available via acid hydrolysis.

Immobilized enzymes are used in analytical testing. There are two types of analytical instruments designed to use immobilized enzymes. The enzyme electrode, in which the immobilized-enzyme derivative is placed on the tip of an electrochemical sensor. For example, to detect glucose the enzyme glucose oxidase (catalase) on a polycarbonate support is utilized with a glutaraldehyde plus collagen immobilization method; penicillinase on a polyacrylamide support via entrapment to detect penicillin, and 5'-AMP deaminase on membrane supports via

Table 1. Immobilized Enzymes for CSTR

Reactant/product	Enzyme(s)	Support	Immobilization method
maltose/glucose	amylo-glucosidase	DEAE-cellulose	covalent bond (2-amino-4,6-dichloro-s-triazine)
glucose/gluconic acid	glucose oxidase (catalase)	polyacrylamide	entrapment
lysis of *Micrococcus lysodeikticus*	lysozyme	polyacrylamide or cellulose	covalent bond (diazonium intermediate)
cornstarch/glucose	glucoamylase	glass	covalent bond (two methods)

Table 2. Immobilized Enzymes for Packed-Bed Reactors

Reactant/product	Enzyme(s)	Support	Immobilization method
starch/maltose	β-amylase	cellulose beads	covalent bond
sucrose/glucose	invertase	glass/cellulose	covalent bond (azo linkage)
N-acetyl-DL-methionine/ methionine	aminoacylase	ceramic	covalent bond (glutaraldehyde)
lactose/glucose	lactase	ceramic	covalent bond (glutaraldehyde)

solution-entrapment immobilization to detect 5'-AMP. The other instrument is the partitioned-enzyme sensor, which separates the immobilized-enzyme component from the detector and connect them by means of a flowing stream. Immobilized enzymes used in this instrument are generally prepared on inorganic supports or on the inside surface of organic tubing, eg, glucose dehydrogenase on a nylon tube using the glutaraldehyde immobilization method to detect glucose, and L-amino acid oxidase on a controlled-pore glass support using the glutaraldehyde immobilization method to detect amino acids.

Immobilization of enzymes has also been useful in biochemical research in biochemical synthesis, and is being researched for use in fuel cells and in the utilization of solar energy (qv).

MELVIN H. KEYES
Owens-Illinois, Inc.

M. Salmona, C. Saronia, and S. Garattini, eds., *Insolubilized Enzymes*, Raven Press, New York, 1974.

L. Goldstein and G. Manecke in L.B. Wingard, Jr., and E. Katchalski-Katzir, eds., *Applied Biochemistry and Bioengineering*, Vol. 1, 1976, p. 23.

Methods Enzymol. **44**, (1976).

ENZYMES, INDUSTRIAL

Industrial enzymes may be defined as enzyme preparations manufactured for use as catalysts either directly or for use in the production of a manufactured commodity (see Antibiotics, β-lactams; Microbial transformations; Beer; Fermentation; Yeasts).

Industrial enzymes vary from very pure crystalline material, such as glucose isomerase, to dried ground tissue, such as pancreas glands dried and ground to form pancreatin, and whole microbial cells, eg, immobilized and fixed with glutaraldehyde.

Nomenclature

Enzymes, particularly industrially important enzymes, are known principally by their trivial (common historical) names. These names were given based on the source of the enzyme by adding the suffix -in or -ain to a root indicated the source (eg, papain from papaya or pancreatin from pancrease), or named after the substrate or action by adding the suffix -ase to a root indicative of the substrate (eg, lactase acts on lactose, glucose oxidase oxidizes glucose).

In 1956, the Third International Congress of the International Union of Biochemistry (IUB) established a Commission on Enzymes to develop a systematic approach to name enzymes. Their system includes six classes: oxidoreductases, active in biological oxidation and reduction and therefore in respiration and fermentation processes, a class also including the peroxidases which use H_2O_2 as the oxidant, the hydroxylases which introduce hydroxyl groups, and the oxygenases, which add molecular O_2 to a double bond in the substrate; transferases, which catalyze the transfer of one-carbon groups (methyl, formyl, carboxyl groups), aldehydic or ketonic residues, acyl, glucosyl, or alkyl groups, nitrogenous groups, and phosphorus- and sulfur-containing groups; hydrolases, which include the esterases, phosphatases, glycosidases, peptidases, and others; lyases, which remove groups from their substrates (not by hydrolysis) and may or may not add groups to double bonds, and which include decarboxylases, aldolases, dehydratases, and others; isomerases, including racemases, epimerases, cis-trans isomerases, intramolecular oxidoreductases, and intramolecular transferases; and ligases, which catalyze the joining of two molecules coupled with the breakdown of a pyrophosphate bond in adenosine triphosphate (ATP) or a similar triphosphate (also known as synthetases).

Characteristics

The enzyme molecule consists of a chain of amino acids (qv) that has a particular geometric configuration specific for that arrangement of amino acids. The twisting and turning forms some locations that are catalytically active and these are referred to as the active sites.

Catalytic action. Most reactions in biological systems proceed almost infinitely slowly in the absence of the specific enzyme catalyst, although at tremendous speed in its presence. This applies even to simple reactions, such as the formation of carbonic acid from carbon dioxide and water, a reaction of great importance in rapidly moving carbon dioxide from the site of formation in the tissues to the lungs for expiration:

$$CO_2 + H_2O \rightarrow H_2CO_3.$$

This reaction is catalyzed by the enzyme carbonic anhydrase. It proceeds ten million (10×10^6) times faster in the presence of the enzyme than in its absence. Carbonic anhydrase has a catalytic-center activity (turnover number) of 100,000. The catalytic-center activity is the number of molecules of reactant that can be converted to product by one catalytic center of an enzyme in one second. In enzyme reactions, the reactant or substance converted by the enzyme is called the substrate. An optimum temperature and an optimum pH are characteristic of every enzyme, and so are the minimum and maximum values of pH and temperature at which the enzyme is active (see Figs. 1 and 2).

Specificity. Most enzymes are highly specific and catalyze only one reaction or act upon only one isomer of a particular compound. The transformation of a compound expressed by a simple chemical equation quite frequently requires the cooperation of a number of enzymes. This explains the presence of hundreds of different enzymes in a single cell.

Pure enzymes form the following groups with regard to the degree of specificity:

(1) Absolute enzymes catalyze the reaction of only one substrate. For example, urease breaks down urea into CO_2 and ammonia, but does not attack any other substrate. (2) Stereospecific enzymes catalyze reactions with one type of optical isomer, but may react with a series of related compounds with the same configuration. For example, many proteolytic enzymes hydrolyze only peptide bonds linking L-amino acids (natural amino acids). (3) General hydrolyzing enzymes react with a specific type of chemical linkage. For example, most lipases hydrolyze a wide range of organic esters and many phosphatases break down phosphate esters into phosphoric acid and alcohol. (4) Enzymes that attack certain specific point of a molecule. For example, some proteolytic enzymes act on points where the adjacent amino acid contains a benzene ring. Some hydrolytic enzymes attack at the center of the molecule and others at the ends. For example, α-amylase attacks the center of the starch molecule and of the long glucose chains deriving from starch, whereas β-amylase attacks the ends, splits off two glucose units, and thus forms maltose.

Composition and chemical nature. All enzymes are proteins, metalloproteins, or conjugated proteins. Metals are often an integral part of the enzyme, where separation can result in irreversible loss of activity (eg, Fe^{3+} in catalase). In other cases, metal ions are required for activity or stability (eg, Ca^{2+} for bacterial α-amylase, Mg^{2+} or CO_3^{2-} for glucose isomerase). The nonprotein portion of an active enzyme separated from the protein portion is called the prosthetic group or coenzyme; the remaining portion is called the apoenzyme and the combined apoenzyme and prosthetic group is called the holoenzyme.

The more general term, cofactor, includes the metallic portion of metalloprotein enzymes, and the coenzyme or prosthetic-group portions of holoenzymes. In many enzymes, the coenzyme portion is the reactive part. Most coenzymes contain a nucleotide, a vitamin, a five-carbon sugar, and possibly other compounds tied together. The vitamins in coenzymes are almost exclusively members of the vitamin B group (see Vitamins).

Production

Collection or production of enzyme-rich material. Animal enzymes are generally derived from specific organs (eg, lipase from throat glands of young animals, pancreatin from pancreas) although urokinase, a proteolytic enzyme used to dissolve blood clots *in vivo*, is extracted from human urine. Plant enzymes from higher plants are obtained from roots (horseradish, peroxidase), seeds (malt), fruit (papain), sap (ficin), or other plant parts. Microbial enzymes are obtained by growing the desired organism either on a semisolid medium or in a submerged culture (agitated tank) virtually identical to the equipment and means developed for antibiotic production (see Antibiotics).

Extraction. Intracellular enzymes, if used as a cell-free preparation, must be extracted. Similarly, intra- or extracellular enzymes produced on a semisolid medium must be extracted from the semisolid mass. Before or during extraction, the cell must be ruptured and the enzyme solubilized. Often, the process is spontaneous, following death of the cell, carried out by autolytic enzymes released within the cell. Cell rupture can be accomplished by enzyme addition, ultrasonics, ball-milling, grinding, freezing and thawing, and homogenization (see Mixing and blending; Ultrasonics). Enzyme solubilization is accomplished spontaneously by indigenous enzymes (autolysis), surfactants, adding enzymes such as lysozyme, pressurization and sudden release, grinding in the frozen state, and other similar operations. Once solubilized, the extraction is merely a washing-out process.

Purification and concentration. For most industrial enzymes, the aqueous extract is concentrated and stabilized. The crude enzyme is precipitated from the aqueous extract by the addition of water-miscible solvents such as acetone, methanol, ethanol, 2-propanol, or mixtures of these. The precipitated enzyme is separated by filtration, as in a plate-and-frame filter with the addition of diatomaceous-earth filter aid, or separated by centrifugation through a high force centrifuge such as a Sharples supercentrifuge. Enzymes precipitated by solvents are rapidly dried at low temperature (20–50°C), often under vacuum. The dry enzyme is then pulverized, assayed, standardized, and packaged. Ultrafiltration (qv) is often used to substantially reduce the volume of aqueous phase, and hence the volume of solvent needed to precipitate the enzyme. Indeed, ultrafiltration has, in conjunction with the trend to liquid enzyme preparations, often eliminated the use of solvents (see also Filtration).

Formulations, stabilization, and standardization. Commercial enzyme preparations must be formulated to contain the necessary activities and made stable to microbial degradation. Added materials are of various types: pH modifiers, eg, phosphates, citrates, caustics, inorganic and organic acids; preservatives, either antimicrobial agents such as benzoate or benzoate ester, sorbate, or hyperosmic agents such as NaCl, glycerol, sorbitol, propylene glycol, sugars and other solutes; sequestrants, eg, citrate, EDTA; activates, eg, calcium salts, sulfite, cobalt salts; other enzymes, eg, pepsin mixed with rennet; and standardizing materials

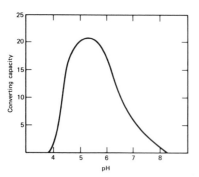

Figure 1. The effect of pH on the activity of malt amylase.

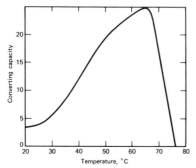

Figure 2. The effect of temperature on the activity of malt amylase.

(diluents), eg, water, diatomaceous earth, whey, lactose, sawdust, flour, mannitol.

Important Enzymes

Lipases. Lipases are esterases hydrolyzing esters of glycerol and fatty acids. Since fatty acids are attached to carbons 1, 2, and 3, the hydrolysis of a triglyceride results in the formation of diglyceride, a monoglyceride, and finally glycerol as one, two and finally three fatty acid ester linkages are hydrolyzed. Thus, in the course of hydrolysis of a fat, at least three potential substrates exist: triglyceride, diglyceride, and monoglyeride.

Animal lipases, the most important of which is pancreatic lipase, are almost always used as a crude dried pancrease powder blended with other materials.

Pectic enzymes. Pectic enzymes, which take part in the hydrolytic degradation of pectic substances, eg, carbohydrate derivatives, are universally present in plant tissue. The pectic enzymes acting directly on pectic substances are comprised of three enzymes, of which polygalacturonase (PG) and pectin methylesterase esterase (EP) are hydrolytic. The third is nonhydrolytic and is known as pectate lyase or pectin transeliminase (PTE). Pectic substances are essentially chains of anhydrogalacturonate residues, with varying amounts of methyl ester linkages. These chains are attacked at random by PTE and are thus reduced to short pieces of six to eight methoxyl anhydrogalacturonate residues which cannot be attacked by PTE. Thus PTE rapidly reduces viscosity caused by pectic substances, but does not break the polymer into its monomeric constituents.

Pectic enzymes are principally used for: production of fruit juices (see Fruit juice) and fruit-juice products; production of wines; fermentation of coffee (qv) and cocoa beans; rehydration of dehydrated foods; production of galacturonic acid and low methoxyl pectin; and recovery and stabilization of citrus oil (see Food processing).

Lysozyme. Lysozyme, a mucopolysaccharidase, is a globulin protein consisting of a single polypeptide chain of 129 amino acid residues cross-linked by four or five disulfide bridges. The enzyme is stable between pH 2.8 and 11.8, and is most soluble at pH 4.5. It is being used, or suggested to be used, in the treatment of cancer, virus diseases, eye infections, blood diseases, infectious postoperative complications, hemorrhagic conditions, varicose ulcers, multiple sclerosis, and as a food preservative; in Europe, lysozyme is sold as an over the counter "cure-all."

Hyaluronidase. Hyaluronidase, a transglucosidase, which also hydrolyzes the mucopolysaccharide hyaluronic acid, is present in snake venom, leeches, in some pathogenic bacteria, and in the testicles of mammals. Commercial hyaluronidase preparations are made from beef seminal vesicles. It is used together with local anesthetic agents in surgery and dentistry, in insulin-shock therapy, and with injections of antibiotics, adrenaline, and heparin.

Carbohydrases. Carbohydrases constitute the largest group of naturally occurring organic materials, with more than 50% of the dry substance of all plant material made up of carbohydrates, including cellulose, hemicelluloses, starches, and complex and simple sugars (see Carbohydrates). Invertase, also called sucrase or saccharase, or β-fructofuranosidase, is one of the simplest commercial carbohydrases. It catalyzes the hydrolysis of the β-D-fructofuranosyl linkage in sucrose, raffinose, gentibiose, methyl- and β-fructofuranoside. By hydrolyzing sucrose into glucose and fructose, it produces invert sugar, which is widely used in confectioneries, candies, syrups, cordials, ice cream, and sweets because it is sweeter than sucrose and begins to crystallize at much higher concentrations than glucose or sucrose syrups.

The most important carbohydrases, produced in larger quantities than any other enzyme, belong to the group of amylase or diastase enzymes which hydrolyze starch and its hydrolytic degradation products. Although all amylases hydrolyze the D-glucosidic linkage, they differ in many respects. Most of them hydrolyze only linkages between carbons 1 and 4; others, in addition to the $(1 \rightarrow 4)$ linkage also hydrolyze bonds between carbons 1 and 6. α-Amylase, present in pancrease, saliva, plants, molds and bacteria, is more active in hydrolyzing larger molecules, cleaving them at random close to the middle of long glucose chains; β-amylase splits maltose molecules at the nonreducing end of a chain; and glucoamylase splits off single glucose units, attacking at nonreducing

ends of both long and short chains, and cleaves both $(1 \rightarrow 4)$ and $(1 \rightarrow 6)$ linkages.

Plant proteases. The plant protease papain, derived from the green fruit of *Carica papaya*, is the principal enzyme for beer chillproofing, meat tenderizing, and for softening wheat gluten for crackers. Commercial papain usually also contains chymopapain, which preferentially cleaves peptide bonds where the carbonyl group is from aspartate or glutamate. It is very active over the pH range 3–9. It is used in digestive aids, wound debridement, protein hydrolysates, liquefying protein waste products, tooth-cleaning powders, and for the recovery of silver from used film.

Glucose oxidase. Glucose oxidase or β-D-glucopyranose aerodehydrogenase, a typical aerobic dehydrogenase enzyme produced by *Aspergillus niger*, *A. oryzae*, and *Pencillium notatum*, oxidizes glucose to gluconic acid in the presence of molecular oxygen. Applications are based on the removal of oxygen from beverages or from the air space in a closed food container, or on the removal of glucose from a food ingredient or produce. For example, glucose oxidase is used when the removal of either oxygen or glucose is desirable, eg, it is added to canned beverages to impede fading of sensitive colors and to retard iron pickup, added to dehydrated foods to prevent oxidative deterioration and extend shelf life, and added to preserve the freshness of salad dressings and mayonnaise.

Health and Safety

Enzymes are consumed in large quantities in raw foods; in general, pure enzymes are inherently safe for industrial use. However, enzymes are proteinaceous in nature and, like all proteins, can cause allergic reactions. Proteolytic enzymes, eg, bromelin or papain, can cause skin irritation after frequent or lengthy exposure.

Uses

Industrial enzyme preparations include pure crystalline enzymes classified into use for production of foods and food ingredients, beverages, and miscellaneous goods.

DON SCOTT
Fermco Biochemics Inc.

T. Godfrey and J. Reichelt, *Industrial Enzymology*, The Nature Press, New York, 1983.

G. Reed, *Enzymes in Food Processing*, 2nd ed., Academic Press, Inc., New York, 1975.

J.R. Whitaker, *Principles of Enzymology for the Food Sciences*, Marcel Dekker Inc., New York, 1972.

ENZYMES, THERAPEUTIC

The manufacture or processing of enzymes for use as drugs is a minor but important facet of today's pharmaceutical industry. This importance derives from several cardinal attributes of enzymes; they are catalysts of remarkable efficiency; the unique configuration of amino acid residues in their catalytic centers accelerates highly specific chemical and pharmacologic reactions; they function effectively under chemically mild conditions, and often optimally under physiologic conditions; and they are eliminated from the body in a reasonable time.

The principal therapeutic enzymes include: L-asparaginase, streptodinase, arvin, urokinase, L-glutaminase, deoxyribonuclease, hyaluronidase, penicillinase, and superoxide dismutase.

Physical and Chemical Properties

All enzymes are at present known as proteins (see Enzymes, industrial). Attached may be lipid, carbohydrate, nucleotide, or metal-containing prosthetic groups. The pH activity optima of theratherapeutic enzymes range from < 2.0 for pepsin to > 8.0 for L-asparaginase.

The types of reactions catalyzed by therapeutic enzymes are generally restricted to those able to operate with cofactors or cosubstrates avail-

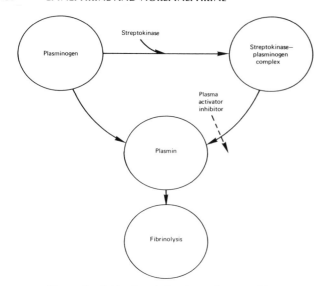

Figure 1. Activation of plasminogen by streptokinase.

able in the extracellular environment; this restriction is dictated by the relative impermeability of the ordinary cell membrane to molecules having the dimensions of proteins. For this reason, the majority of the therapeutic enzymes used to date are hydrolases.

Health and Safety

FDA requires that the safety of any enzyme intended for parenteral use in humans first be established in test animals.

Uses

Therapeutic enzymes have a broad variety of specific uses as oncolytics, as anticoagulants or thrombolytics, and as replacements for metabolic deficiencies. As oncolytic enzymes, there are two primary classes: those that degrade small molecules for which neoplastic tissues have a requirement, eg, L-asparaginase, and those that degrade macromolecules such as membrane polysaccharides, structural and functional proteins, or nucleic acids, eg, ribonuclease (see Chemotherapeutics, antimitotic). They are also administered to tumor-bearing subjects along with a prodrug conjugated to a functionality susceptible to attack by the enzyme (see Microbiological transformations). As anticoagulants, the plasminogen activators (eg, streptokinase and urokinase) produce controlled enzymatic fibrinolysis *in vivo* for the management of thromboembolic vascular disease (see Fig. 1).

Proteolytic enzymes have been widely used as anti-inflammatory agents, reducing soft fibrin and clearing proteinaceous debris found in inflammatory exudates. Enzymes are also used as antidotes; eg, rhodanese (thiosulfate-sulfurtransferase), administered along with its cosubstrate, sodium thiosulfate, can counteract cyanide poisoning.

DAVID A. COONEY
GEORGE STERGIS
H.N. JAYARAM
Division of Cancer Treatment, National Cancer Institute
National Institutes of Health

M.D. Prager in R.L. Clark and co-eds., *Oncology 1970*, Vol. 2, Yearbook Medical Publishers, Inc., Chicago, Ill., 1971, p. 237.

T.M.S. Chang, *Methods Enzymol.* **44**, 676 (1977).

EPINEPHRINE AND NOREPINEPHRINE

Epinephrine and norepinephrine, the hormones of the adrenal medulla, are the final products of the biosynthetic pathway, phenylalanine → tyrisine → dihydroxyphenylalanine or DOPA → dopamine → norepi-

nephrine → epinephrine. Figure 1 shows the structure of the two hormones.

Epinephrine and norepinephrine released from the adrenal medulla are involved in the control of heart rate, blood pressure, and lipid and carbohydrate metabolism. In the sympathetic nervous system, norepinephrine released from postganglionic nerve terminals acts as a transmitter of the nerve impulse across the gap, or synaptic cleft, between the nerve terminal and a postsynaptic site of action. In certain neurones in the central nervous system, a neurotransmitter function is served by dopamine, norepinephrine, and epinephrine.

Physical Properties

D-(−)-Epinephrine, $C_9H_{13}NO_3$, has mp ca 209–210°C (dec). It is soluble in water to the extent of ca 0.1 g/100 mL, and is sensitive to air, light, heat and alkalies. Metals, notably copper, iron, and zinc, destroy its activity. In solution with sulfite or bisulfite, it slowly forms an inactive sulfonate. The red color that forms when neutral or alkaline solutions are exposed to air is caused by adrenochrome. D-(−)-Norepinephrine, $C_8H_{11}NO_3$, has mp ca 212°C (dec). It is slightly soluble in water. The compound is unstable in light and air especially at neutral and alkaline pH. Oxidation to noradrenochrome occurs in the presence of oxygen and such divalent metal ions as copper, manganese, and nickel.

Physiology

In the chromaffin cells of the adrenal medulla, epinephrine and norepinephrine are present in small vesicles. Stimulation of the splanchnic nerve leading to the adrenal medulla causes the secretion of the sympathomimetic amines into the blood for action at receptor sites through the body with the exception of the central nervous system. At physiological concentrations, epinephrine and norepinephrine do not pass the blood–brain barrier to a marked extent. Catechoalmines located in nerve terminals are released when an electrical impulse or action potential travelling along the nerve reaches the terminal. Although the mechanisms regulating release of catecholamines in the adrenal medulla and nerve terminals are different, in both cases the final step is believed to involve exocytosis. The storage vesicles fuse with the external membrane of the cell and the membrane opens, releasing the contents of the vesicles into the extracellular space. The rate at which norepinephrine is released from noradrenergic nerve terminals is subject to control by the central nervous system and by local feedback mechanisms.

Synthesis

The original source for commercial use was bovine adrenal glands from which epinephrine was extracted by Takemine's procedure. Synthetic epinephrine has replaced the natural product as a commercial source. The procedure used is basically that developed by Stolz: Friedel-Crafts acylation of pyrocatechol with chloroacetyl chloride gives the α-chloroacetophenone derivative. Displacement of the chlorine by methylamine gives the methylamino derivative, which on catalytic reduction yields (±)-epinephrine. Substitution of ammonia for methylamine in the sequence gives the amino derivative, which on reduction yields norepinephrine. The isomers are resolved with (+)-tartaric acid to give the physiologically active (−)-isomers.

Health and Safety

Both epinephrine and norepinephrine are highly toxic substances. The most serious effects of an overdose of epinephrine are a rapid large increase of blood pressure, cerebral hemorrhage, and cardiac arrhymias. The effects of norepinephrine are similar, but less severe.

R = H, (−)-Norepinephrine
R = CH$_3$, (−)-Epinephrine

Figure 1. Structures of epinephrine and norepinephrine.

Uses

Epinephrine is used for its vasoconstrictor property in topical application to reduce bleeding from superficial wounds; in conjunction with a local anesthetic to prolong the anesthetics action by reducing absorption rate (see Anesthetics); to counteract bronchospasms in asthma; and in anaphylactic shock for its bronchodilator and cardiac stimulant properties (see Cardiovascular agents; Anti-asthmatic agents).

JOHN R. MCLEAN
Warner-Lambert/Parke-Davis

U.S. Pats. 730,175; 730,176; 730,196, 730,197 (June 22, 1903), J. Takamine.

F. Stolz, *Chem. Ber.* **37**, 4149 (1904).

The U.S. Pharmacopeia, XX, The United States Pharmacopeial Convention, Inc., Rockville, Md., 1980, p. 278.

EPOXIDATION

Epoxidation is the formation of cyclic three-membered ethers (oxiranes) by the reaction of peracids (see Peroxides) and hydrogen peroxide (qv) with olefinic and aromatic double bonds. Oxiranes may also be formed by an internal S_N2 reaction of a chlorohydrin (see Chlorohydrins). The three-membered ethers formed are also designated as 1,2-epoxides.

Epoxidation Processes

The epoxidation processes can be divided into two basic types: either the peracid is preformed or it is formed *in situ*, ie, in the primary reaction vessel.

$$\underset{\text{cat.}}{\overset{\text{RO}_2\text{H}}{\longrightarrow}} \qquad (1)$$

Epoxidation with peracetic acid. Epoxidation with peracetic acid does not necessarily utilize a catalyst at 20–80°C if preformed peracetic acid is used.

$$\text{RCOOH} + \text{H}_2\text{O}_2 \overset{\text{H}^+}{\rightleftharpoons} \text{RCOOH} + \text{H}_2\text{O} \qquad (2)$$

The peroxidation of acetic acid with hydrogen peroxide is not efficient except at high molar ratios of acetic acid/hydrogen peroxide; however, large amounts of acetic acid must be removed if high ratios are involved. In addition, concentrations of peracetic acid above 40–45 wt% in acetic acid are explosive at epoxidation temperatures, and therefore, related epoxidation processes require large-volume production on an essentially continuous basis since the preformed peracid cannot be safely stored (see Health and Safety).

Epoxidation with peracid formed *in situ*, acid catalyzed. Although no reaction that involves peroxides is without hazard, 30 years of experience has shown the *in situ* processes to be safer than the preformed-peracid processes In general, a peroxide solution (35–70% H_2O_2 in H_2O) containing small quantities of a mineral-acid catalyst (such as sulfuric acid or phosphoric acid) is added to a mixture of an epoxidizable substrate and acetic acid or formic acid. As the reactants mix, the hydrogen peroxide and the organic acid react in the presence of the mineral-acid catalyst to form the peracid.

$$\text{CH}_3\text{COOH} + \text{RCH}=\text{CHR} \rightleftharpoons \text{RCH}-\text{CHR} + \text{CH}_3\text{COH} \qquad (3)$$

Oxidation by hydroperoxides. Hydroperoxides have been observed to convert olefins to epoxides in the presence of transition-metal ions such as Mo, W, Cr, or V, as well as Ag for ethylene oxide. Recently, it has been observed that hydrogen peroxide in the presence of catalytic quantities of arsenic compounds reacts with olefins to yield epoxides.

Manufacture

Epoxidized materials are manufactured mainly from peracetic acid (peroxyacetic acid) and performic acid (peroxyformic acid). Peracetic acid can be prepared by the oxidation of acetaldehyde by hydrogen peroxide. Performic acid can be prepared by a similar oxidation of formaldehyde. Although there are several routes to these peracids, preformed peracetic acid (via the air oxidation of acetaldehyde) and *in situ* performic acid are the chief reagents. Epoxy plasticizer production in the U.S. is about equally divided between the two processes. On a worldwide basis, performic-process epoxy plasticizers comprise 60–70% of the market.

The use of the epoxidation reaction in manufacture has involved hydrogen peroxide as the principal source of epoxidizing agents. Peracetic acid derived from acetaldehyde has become increasingly important.

Merchant peracetic acid is produced from hydrogen peroxide via sulfuric acid catalysis. Peracetic acid (40%) may be prepared at the site of epoxidation, however, by mixing 1.6 mol of glacial acetic acid with 1 mol of 90% hydrogen peroxide in the presence of ca 2–3% sulfuric acid. The mixture attains equilibrium if allowed to stand overnight.

Several *in situ* processes for the epoxidation of soybean oil based on the use of acetic acid with sulfuric acid as a catalyst have been developed. In one process the sulfuric acid catalyst is added last admixed with 0.5 mol of glacial acetic acid. In another, an inert solvent, like benzene or hexane, is used to reduce the effect of sulfuric acid in catalyzing epoxy ring opening.

Other epoxidation processes include the repeated-resin process, used for the epoxidation of fatty oils and esters and eliminating unsaturation efficiently while producing the highest epoxy–oxygen values; and the minimal-resin process which requires the use of ca 2% resin (dry wt) based on the weight of the material to be epoxidized.

Health and Safety

Special attention must be given to the construction of storage containers and reaction vessels that hold hydrogen peroxide or peracid compositions; eg, stainless steel is commonly used for mixing and reaction vessels.

Uses

Uses for many of the important epoxides are described in the Ethylene oxide, Propylene oxide, Polyethers, Epoxy resins, and Chlorohydrins articles. Epoxy plasticizers include epoxidized vegetable oils (qv) using substrates such as linseed, safflower, soybean, corn, cottonseed, rapeseed, and peanut oils; and epoxidized alkyl esters such as the alkyl 9,10-epoxystearates which have the lowest oxirane-oxygen content of the commercial epoxy plasticizers.

JOHN T. LUTZ JR.
Rohm and Haas Co.

F.P. Greenspan, *Ind. Eng. Chem.* **39**, 847 (1947).

A.F. Chadwick and co-workers, *J. Am. Oil Chem. Soc.* **35**, 355 (1958).

EPOXIDE POLYMERS. See Glycols; Polyethers.

EPOXIDES. See Ethylene oxide; Propylene oxide; Chlorohydrins; Butylenes; Styrenes.

EPOXY RESINS

Epoxies are monomers or prepolymers that further react with curing agents to yield high performance thermosetting plastics.

Chemical and Physical Properties

Epoxy resins are characterized by the presence of a three-membered cyclic ether group commonly referred to as an epoxy group, 1,2-epoxide, or oxirane.

The most widely used epoxy resins are diglycidyl ethers of bisphenol A (1), derived from bisphenol A and epichlorohydrin. These are most frequently cured with anhydrides, aliphatic amines, or polyamides, depending upon desired properties.

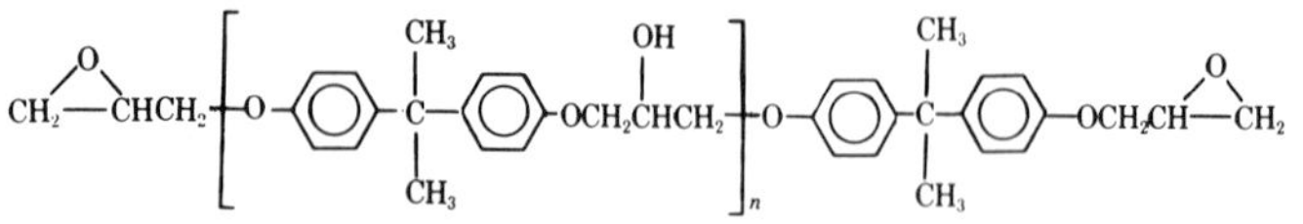

(1)

The outstanding performance characteristics of the resins are conveyed by the bisphenol A moiety (toughness, rigidity, and elevated-temperature performance), the ether linkages (chemical resistance), and the hydroxyl and epoxy groups (adhesive agents) (see also Phenolic resins), In addition to bisphenol A, other starting materials such as aliphatic glycols and both phenol and o-cresol novolacs are used to produce specialty resins (see Table 1). Epoxy resins may also include compounds based on aromatic amine, triazine (see Table 2), and cycloaliphatic backbones. Recent developments have utilized a heterocyclic hydantoin (see Hydantoin and derivatives) nucleus as a building block to produce a family of epoxides with superior electrical properties and outdoor weatherability. Conventional epoxy resins range from low viscosity liquids to solid resins. Curing agents commonly used to convert epoxies to thermosets include anhydrides, amines, polyamides, Lewis acids, and others.

In order to impove high temperature performance and other select properties, various multifunctional resins were synthesized including epoxy cresol novolac resins (ECN) (2) and polynuclear phenol-glycidyl ether-derived resins (3).

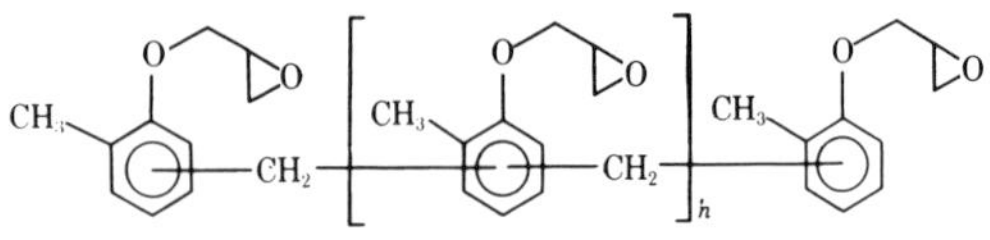

(2)

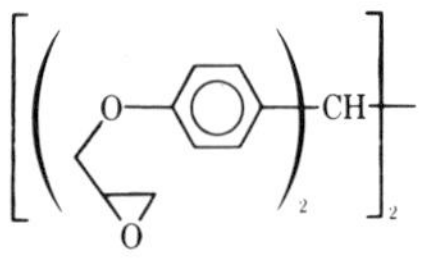

(3)

Table 1. Typical Properties of Epoxy Cresol-Novolac Resins

Property	ECN 1235	ECN 1273	ECN 1280	ECN 1299
mol wt	540	1080	1170	1270
epoxy value, equiv/100 g	0.500	0.445	0.435	0.425
softening point, °C	35	73	80	99
density, g/cm³	1.11	1.16	1.17	1.19

Table 2. Typical Properties of Multifunctional Epoxy Resins

Property	Amine based 0500	MY 720	Other 0163	PT 810
physical form	liquid	semisolid	solid	solid
viscosity at 25°C, Pa·s (= P/10)	1.5–5.0	10–35		
epoxy value, equiv/100 g, minimum	0.86	0.75	0.38	1.05

Resin Synthesis and Manufacture

Epichlorohydrin and bisphenol A-derived resins. Liquid epoxy resins may be synthesized by a two-step reaction of an excess of epichlorohydrin to bisphenol A in the presence of an alkaline catalyst. The reaction consists initially in the formation of the dichlorohydrin of bisphenol A and further reaction via dehydrohalogenation of the intermediate with a stoichiometric quantity of alkali. The use of a large excess of epichlorohydrin minimizes the formation of higher molecular weight species, ie, further reaction of the diglycidyl ether of bisphneol A with bisphenol A results in the formation of polymeric species. Side reactions such as hydrolysis, formation of bound chlorine, incomplete dehydrohalogenation and abnormal addition of epichlorohydrin can reduce the theoretical epoxide functionality. The pure diglycidyl ether of bisphenol A is a crystalline solid (mp 43°C) with a weight per epoxide (WPE) of 170. The typical commercial unmodified liquid resins are viscous liquids with a viscosity of 11–16 Pa·s (110–160 P) at 25°C, and an epoxy equivalent weight of ca 188. Resins of higher molecular weight can be prepared by two routes:

Taffy process. Bisphenol A reacts directly with epichlorohydrin in the presence of stoichiometric amounts of caustic, the molecular weight of the product being governed by the ratio of epichlorohydrin–bisphenol A. In practice, the taffy process is generally employed for only medium molecular weight resins ($n = 1$–4). The polymerization reaction results in a highly viscous product (emulsion of water and resin) and the condensation reaction becomes very dependent on agitation. At the completion of the reaction, the heterogeneous mixture consists of an alkaline brine solution and a water–resin emulsion, and recovery of the product is accomplished by separation of phases, washing of the taffy resin with water, and removal of water under vacuum.

Advancement process. In the advancement process, sometimes referred to as the fusion method, the liquid epoxy resin (crude diglycidyl ether of bisphenol A) is chain-extended with bisphenol A in the presence of a catalyst to yield higher polymerized products. The molecular weight of the resin is a function of the ratio of excess liquid epoxy resin to bisphenol A. The terminal groups are preponderantly epoxy groups. Since no by-products are generated, the advancement process is more convenient and can be used to prepare high molecular weight resins directly. Liquid resins containing bromine (ca 49 wt%) can also be prepared directly from tetrabromobisphenol A and epichlorohydrin and are used for critical applications where a high degree of flame retardancy is required (see Bromine compounds; Flame retardants).

Specialty Resins

Epoxy cresol–novolac resins. These resins (2) are prepared by glycidylation of o-cresol–formaldehyde condensates, which are prepared under acidic conditions with formaldehyde–o-cresol ratios of less than unity.

Tetraglycidylmethylenedianiline-derived resins (4) can be formulated into hot-melt or solution-binder systems with various reinforcements, eg, glass, graphite, boron, and aramid. These systems provide simple cure cycles with excellent mechanical and adhesive properties at room and elevated temperatures.

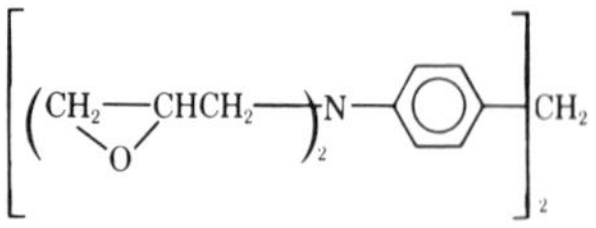

(4)

Curing Reactions

A variety of reagents have been described for converting the liquid and solid epoxy resins to the cured state, which is necessary for the development of the inherent properties of the resins. Liquid epoxy resins contain mainly epoxy groups and solid resins are composed of both epoxy and hydroxyl curing sites. The curing agents or hardeners are categorized as either catalytic or coreactive and the functional groups of the resins are terminal epoxides together with a pendant hydroxyl per repeat unit of the polymer chain. Catalytic curing agents initiate resin homopolymerization, either cationic or anionic, as a consequence of using a Lewis acid or base in the curing process (eq. 1). The Lewis-acid catalysts frequently employed are complexes of boron trifluoride with amines or ethers and boron trichloride complexes. The most important Lewis bases are tertiary amines or polyamines converted into tertiary amines upon reaction with epoxide groups (eq. 2).

$$RCH\overset{}{\underset{\diagdown O\diagup}{-}}CH_2 \xrightarrow{BF_3OH + H^+} HOCH_2CH\overset{}{\underset{R}{|}}\left[OCH_2CH\overset{}{\underset{R}{|}}\right]_n OCH_2\overset{+}{C}H\overset{}{\underset{R}{|}} + \bar{B}F_3OH \qquad (1)$$

$$RNH_2 + CH_2\underset{\diagdown O\diagup}{-}CHCH_2OR' \longrightarrow RNHCH_2CHCH_2OR'\underset{OH}{|}$$

$$RNHCH_2CHCH_2OR'\underset{OH}{|} + CH_2\underset{\diagdown O\diagup}{-}CHCH_2OR' \longrightarrow RN(CH_2CHCH_2OR')_2\underset{OH}{|} \qquad (2)$$

Coreactive curing agents are polyfunctional reagents that are employed in stoichiometric quantities with epoxy resins and possess active hydrogen atoms. The important classes include polyamines, polyaminoamides (formed from polyamines and dimerized fatty acid), polyphenols, polymeric thiols, polycarboxylic acids, and anhydrides. Polyamines constitute a large class of hardeners with aliphatic, aromatic, cycloaliphatic, and heterocyclic groups (eq. 3).

$$CH_2\underset{\diagdown O\diagup}{-}CHR' + R_3N \longrightarrow R_3\overset{+}{N}CH_2CHR'\underset{O^-}{|} + CH_2\underset{\diagdown O\diagup}{-}CHR' \longrightarrow R_3\overset{+}{N}CH_2CH\underset{R'}{|}\left[OCH_2CH\underset{R'}{|}\right]_n O^- \qquad (3)$$

Uses

The largest simple use is in the protective coatings market where high chemical resistance and adhesion is important (see Coatings). Epoxies have gained wide acceptance in protective coatings and electrical and structural applications because of their exceptional combination of properties such as toughness, adhesion, chemical resistance, and superior electrical properties.

STANLEY SHERMAN
JOHN GANNON
GORDON BUCHI
CIBA-GEIGY Corporation

W.R. HOWELL
Dow Chemical USA

W.G. Potter, *Epoxide Resins*, Springer-Verlag, New York, 1970, p. 12.
H. Lee, D. Stoffey, and K. Neville, *New Linear Polymers*, McGraw-Hill Book Co., New York, 1967, pp. 17–60.

ESTERIFICATION

This article describes methods for the production of carboxylic esters. For the properties of these compounds, see Esters, organic. For esters of inorganic acids, see the articles on Nitric acid, Phosphoric acid, Sulfuric acid, etc.

The most usual method for the preparation of esters is the reaction of a carboxylic acid and an alcohol with the elimination of water. Esters are also formed by a number of other reactions, including the use of acid anhydrides, acid chlorides, amides, nitriles, unsaturated hydrocarbons, ethers, aldehydes, ketones, dehydrogenation of alcohols, and ester interchange.

On the basis of bulk production, poly(ethylene terephthalate) manufacture is the most important esterification process today. This polymer is produced by the direct esterification of terephthalic acid and ethylene glycol or by the transesterification of dimethyl terephthalate and ethylene glycol (see Polyesters). As a result of polyester manufacture, the production of dimethyl terephthalate is the second most important esterification process (see Phthalic acids).

Reactions between Organic Acids and Alcohols

In the esterification of organic acids with alcohols, in most cases under acid catalysis, the union is between acyl and alkoxy groups. The rate at which different alcohols and acids are esterified as well as the extent of the reaction depend upon the structure of the molecule and types of radicals present.

With acetic acid at 155°C, the primary alcohols are esterified most rapidly and completely (methanol gives the highest yield and most rapid reaction). Ethyl, n-propyl, and n-butyl alcohols react with about equal velocities and yields. Under the same conditions, the secondary alcohols react much slower and have lower limits of esterification; however, wide variations are observed among the different members of this series. The tertiary alcohols react slowly and yields are generally low (1–10% conversion at equilibrium). Tests with isobutyl alcohol at 155°C and various acids show that those acids containing a straight chain (acetic, propionic, and butyric) and phenylacetic and β-phenylpropionic acids are esterified readily. The introduction of a branched chain in the acid decreases the rate of esterification, and two branches cause a still greater retarding effect. Double bonds also have a retarding influence. However, the limits of esterification of these substituted acids are higher than for the normal straight-chain acids. Aromatic acids, benzoic and p-toluic, react slowly but have high esterification yields.

The nitrile group has a pronounced inhibiting effect on the rate of esterification of aliphatic acids. With the chloroacetic acids, the velocity decreases with increased substitution. From tests on substituted acrylic acids, it is shown that an α,β-unsaturated acid is esterified much less easily than its saturated analogue. A triple bond in the α,β position has about the same effect as a double bond. A β,γ double bond has less of a retarding action. If the double bond is sufficiently removed no effect is noted. Conjugated double bonds, when one is in the α,β position, give a great retarding effect. Cis-substituted acids esterify more slowly than the trans isomers.

Kinetic considerations. Generally, the rate of esterification with acid catalyst is proportional to the acid or hydrogen-ion concentrations as well as the concentration of the alcohols and organic acid. The effect of temperature on the reaction rate is given by the well-known Arrhenius equation. These factors are interrelated and may be used to predict optimum operational conditions for the production of a given ester if the necessary data are available, ie, the order of the reaction under the conditions to be used; a mathematical relation describing the yield with time; and an empirical equation relating the reaction rate constant to temperature, catalyst concentration, and proportions of reactants.

Completion of esterification. Because the esterification of an alcohol and an organic acid involves a reversible equilibrium, these reactions do not go to completion. Conversions approaching 100% are desirable and can often be achieved by the simple method of upsetting the equilibrium by removing one of the products formed, either the ester or the water. In general, esterifications are divided into three broad classes, depending upon the volatility of the esters: (1) Esters of high volatility, eg, methyl formate, have lower boiling points than those of the corresponding alcohols, and therefore, can be readily removed from the reaction mixture; (2) esters of medium volatility are capable of removing the water formed by distillation, eg, propyl, butyl, and amyl formates. In some cases, tertiary mixtures of alcohol, ester and water are formed; (3) esters of low volatility have several modes of esterification. In the case of esters of butyl and amyl alcohols, water is removed as a binary mixture with the alcohol. To produce esters of the lower alcohols (methyl, ethyl, and propyl), it may be necessary to add a hydrocarbon such as benzene or toluene to increase the amount of distilled water. With high boiling alcohols (eg, benzyl, furfuryl, and β-phenylethyl), an accessory liquid is required to eliminate the water by codistillation.

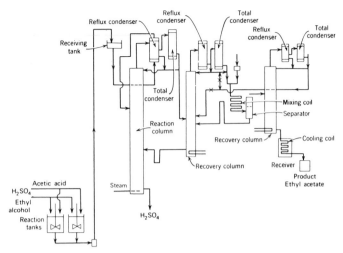

Figure 1. Continuous ethyl acetate process.

Catalysts. The choice of the proper catalyst for an esterification reaction depends upon several factors. Most common catalysts are strong mineral acids but other agents, eg, tin salts, organo-titanates, silica gel, and cation-exchange resins, are often employed.

The use of acid regenerated cation exchangers (see Ion exchange) as catalysts offers distinct advantages over other conventional methods. The nature of the cation exchanger is of relatively minor importance as long as it contains strongly acidic groups.

Production

The production of esters may be carried out by batch or continuous processing. In general, large-volume production favors continuous esterification methods. However, the older batch processing, based on the use of a still-pot reactor and an ordinary fractionating column (bubblecap or packed type) is still used.

The law of mass action, the laws of kinetics, and the laws of distillation all operate simultaneously in continuous esterification processes. Continuous esterification to produce ethyl acetate is an excellent example of the use of azeotropic principles to obtain a high yield of ester (Fig. 1).

The choice of the esterification process to obtain a maximum yield depends upon many factors, with no single process having universal applicability.

Use

The applications of esterification, such as alcoholysis in analysis, research, and manufacturing are too numerous to mention in detail.

E.G. ZEY
Celanese Corporation

E.E. Reid "Esterification," in P. Groggins, *Unit Processes in Organic Synthesis*, 4th ed., McGraw-Hill Book Co., New York, 1952.

S. Patai, *The Chemistry of Carboxylic Acids and Esters*, Wiley-Interscience, New York, 1969.

K.S. Markley in K.S. Markley, ed., *Fatty Acids*, Part 2, Wiley-Interscience, New York, 1961, p. 757.

ESTERS, ORGANIC

Esters are compounds that on hydrolysis yield alcohols or phenols and acids according to the equation:

$$RA + H_2O \rightleftharpoons ROH + HA$$

where R is a hydrocarbon fragment and A is the anionic portion of an organic or inorganic acid. For carboxylic acid esters, the reaction can be represented as:

$$RCOOR' + H_2O \rightleftharpoons R'OH + RCOOH$$

where R and R′ are the same or different hydrocarbon radicals. The reverse reaction constitutes the usual method for preparing esters.

Many molecules contain both carboxy and hydroxy groups, and the specific nature of the esters formed by interaction of these groups depends largely upon the distance between them in the molecule. Aliphatic compounds that contain carboxy and hydroxy groups attached to the same carbon atom usually give a cyclic ester formed from two molecules of the compound.

Physical and Chemical Properties

The lower esters are colorless, volatile liquids, and many of them have pleasant odors (see Flavors and spices). Most of the esters of the higher saturated fatty acids are colorless and odorless crystalline solids; those of the very long-chain acids and alcohols are hard, brittle, and lustrous crystalline solids and are generally referred to as waxes (qv). Most of these waxes are not known in pure form but only as complex mixtures comprising the naturally occurring waxes.

In general, the melting points of the esters of fatty acids are lower than those of the corresponding acids. With increasing chain length of the alcohol, the boiling points increase and ultimately become much higher than those of the corresponding acids.

The data for the methyl and ethyl esters of the saturated fatty acid series are fairly complete and relatively well known, but data for the esters of the higher alcohols are often lacking. Data on boiling point, density, molar volume, viscosity, solubility, heat of combustion, and other physical constants of esters of aliphatic acids have been tabulated.

The esters are generally insoluble in water, but are soluble in various organic liquids.

Reactions. The most familiar reactions of esters involve nucleophilic attack at the electron-deficient carbonyl carbon atom and can be generalized by the equation:

$$
\underset{\text{RCOR}'}{\overset{O}{\|}} + Nu^- \longrightarrow \underset{\underset{OR'}{|}}{\overset{O^-}{\underset{|}{RCNu}}} \longrightarrow \underset{\text{RCNu}}{\overset{O}{\|}} + OR^-
$$

where $Nu = OH^-$, OR''^-, HN_3, NHR''_2, etc. These reactions are frequently acid catalyzed; the acid polarizes the carbonyl group and increases its susceptibility to nucleophilic attack.

Hydrolysis. Although hydrolysis of low molecular weight esters to their corresponding acids and alcohols occurs slowly even in the absence of added catalysts, for practical purposes it is usually carried out at elevated temperatures in the presence of either acidic or basic catalysts. The ease of hydrolysis decreases with increasing molecular weight and with the presence of bulky groups which impede attack of the nucleophile. Basic hydrolysis (saponification), which results in the formation of the carboxylate anion in the last step, is irreversible. Acidic hydrolysis is an equilibrium reaction. The formulas below illustrate the most common base- and acid-catalyzed hydrolysis pathways, respectively.

Ammonolysis and aminolysis. In a reaction that is exactly analogous to their hydrolysis to alcohols and acids, esters react with ammonia to give alcohols and amides. The reaction is usually carried out in aqueous or alcoholic ammonia. Simple esters react with ammonia at a satisfactory rate at room temperature; higher esters may require elevated temperatures and pressures. Other ammonia derivatives, such as primary and secondary amines, react analogously to give N-substituted amides. When esters are passed with ammonia in the vapor phase over contact catalysts at 400–500°C, they are rapidly converted to nitriles.

Reduction. Hydrogenation of esters over copper chromite catalyst at 200–300°C and 10–30 MPa (100–300 atm) reduces them to alcohols:

$$\text{RCOOR}' + 2\,\text{H}_2 \rightarrow \text{RCH}_2\text{OH} + \text{R}'\text{OH}$$

When R is saturated and no halogen or sulfur is present, the reaction is smooth and almost quantitative; but when R is an aromatic nucleus, such as benzene or pyrrole, the reaction proceeds beyond the alcohol step, as in the reduction of ethyl benzoate to toluene. This cleavage can be minimized by carrying out the reaction at low temperatures with a high ratio of catalyst to ester.

Reaction with Grignard reagents. The reaction of esters with Grignard reagents is similar to that of ketones: the nucleophilic alkyl or aryl group of the Grignard reagent attacks the carbonyl carbon (see Grignard reaction).

Acetoacetic ester condensation. In the presence of certain bases (usually sodium alkoxide), an ester having hydrogen on the α-carbon atom reacts with a second molecule of the same ester or with another ester (which may or may not have hydrogen on the α-carbon atom) to form a β-ketoester:

$$\text{R}_2\text{CHCOR}' + \text{R}''_3\text{CCOR}' \xrightarrow{\text{R'O}} \text{R}''_3\text{COCR}_2\text{COR}' + \text{R}'\text{OH}$$

These reactions are special cases of the Claisen reaction which includes ketone–ester condensation to form 1,3-diketones.

Preparation of acyloins. When aliphatic esters are allowed to react with metallic sodium in inert solvents, acyloins (α-hydroxyketones) are formed:

$$2\,\text{RCOR}' + 4\,\text{Na} \xrightarrow[-2\,\text{NaOR}']{} \begin{array}{c}\text{RCONa} \\ | \\ \text{RCONa}\end{array} \xrightarrow[-2\,\text{NaOH}]{\text{H}_2\text{O}} \left[\begin{array}{c}\text{RCOH} \\ \| \\ \text{RCOH}\end{array}\right] \rightarrow \begin{array}{c}\text{RC}{=}\text{O} \\ | \\ \text{RCHOH}\end{array}$$

Pyrolysis. The pyrolysis of simple esters of the formula RCOOCR'R''CHR$_2$''' to form the free acid and an olefin is a general reaction that is used for producing olefins. The pyrolysis is generally carried out at 300–500°C over an inert heat-transfer agent such as Pyrex glass or 96% silica glass chips. Esters of tertiary alcohols are pyrolyzed more readily than esters of secondary alcohols, and esters of primary alcohols are the most difficult to pyrolyze.

Miscellaneous reactions. Ketones can be obtained in substantial yields from fatty acid esters as well as from the fatty acids by heating the esters or acids in the presence of certain metals or their oxides. For example, ethyl laurate (ethyl dodecanoate) in the vapor phase over a thoria-gel catalyst at 300°C gives a 92.5% yield of laurone (12-tricosanone). Esters can be used as acylating agents in the Friedel-Crafts ketone synthesis although, except in special cases, the more reactive acyl halides or anhydrides are usually chosen (see Friedel-Crafts reactions).

Occurrence and Preparation

Currently, most of the simple esters used commercially are of synthetic origin although a number of them occur in nature. Recovery of naturally occurring esters is accomplished by steam distillation, extraction, or pressing, or by a combination of these processes. Synthetic esters are generally prepared by reaction of an alcohol with an organic acid in the presence of a catalyst such as sulfuric acid or p-toluenesulfonic acid. Ion-exchange resins of the sulfonic acid type can also be used, and an azeotroping agent such as benzene or toluene can be used to remove water and force the reaction to completion (see Esterification).

Health and Safety

No general statement can be made about the toxicity of esters; the degree of toxicity covers a wide range. Many are highly volatile, and can act as asphyxiants or narcotics. Skin absorption, as well as inhalation, can be a significant hazard with esters that are volatile and have high solvent action.

Uses

The greatest uses of esters are in the solvent and plasticizer fields with lower esters used in lacquers, paints (qv), and varnishes, and higher esters used chiefly as plasticizers. Many polymeric materials in commercial use are based on esters, eg, vinyl polymers made from unsaturated esters, cross-linked polyesters made from polyhydric alcohols and dibasic acids, polyester resins (qv) and plastics (qv) (see Coatings, industrial; see also Methacrylic acid; Acrylic acid). In the form of natural fats, oils and waxes, esters are used as lubricants. Esters are also used in perfumes, flavors, cosmetics, and soaps, as surface-active agents, and in medicinals.

EDWARD U. ELAM
Tennessee Eastman Company

K.S. Markley, ed., *Fatty Acids, Their Chemistry, Properties, Production, and Uses*, 2nd ed., Interscience Publishers, Inc., New York, Part 1, 1960; Part 2, 1961.

S. Patai, ed., *The Chemistry of Carboxylic Acids and Esters*, Wiley-Interscience, New York, 1969.

ESTROGENS.   See Hormones.

ESTRONE.   See Hormones.

ETHANE.   See Hydrocarbons C$_1$–C$_6$.

ETHANOIC ACID.   See Acetic acid.

ETHANOL

Ethanol, or ethyl alcohol, $\text{CH}_3\text{CH}_2\text{OH}$, has unique combinations of properties as a solvent, a germicide, a beverage, an antifreeze, a fuel, a depressant, and is versatile as a chemical intermediate.

Physical Properties

Ethanol, under ordinary conditions, is a volatile, flammable, clear, colorless liquid. Its odor is pleasant, familiar, and characteristic, as is its taste when suitably diluted with water. Normal boiling point 78.32°C; fp −114.1°C; d_4^{20} 0.7893 g/cm^3; ref in (n_D^{20}) 1.36143; viscosity at 20°C 1.17 mPa·s (= cP); and it is miscible in water.

Chemical Properties

The chemistry of ethanol is largely that of the hydroxyl group, namely, reactions of dehydration, dehydrogenation, oxidation, and esterification.

Other reactions involving the hydrogen atom of the hydroxyl group in ethanol include the opening of epoxide rings to form hydroxy ethers, and the addition to acetylene to form ethyl vinyl ether:

$$\text{CH}_3\text{CH}_2\text{OH} + \text{CH}{\equiv}\text{CH} \rightarrow \text{CH}_3\text{CH}_2\text{OCH}{=}\text{CH}_2$$

The hydroxyl group can be replaced by halogens from inorganic acid halides or phosphorus halides to give two different products, ethyl esters of the acid, and ethyl halide.

The halogen acids also produce alkyl halides:

$$\text{CH}_3\text{CH}_2\text{OH} + \text{HX} \rightarrow \text{CH}_3\text{CH}_2\text{X} + \text{H}_2\text{O}$$

The halogen influences the rate of reaction, and in general, the order of reactivity is HI > HBr > HCl. Important uses of ethyl chloride include the manufacture of tetraethyllead and ethyl cellulose. Ethyl bromide can be used to produce ethyl Grignard reagent and various ethylamines (see Amines).

Esterification. Esters are formed by the reaction of ethanol with inorganic and organic acids, acid anhydrides, and acid halides. Organic esters are formed by the elimination of water between an alcohol and an organic acid. The reaction is reversible and reaches equilibrium slowly. Generally, acidic catalysts are used.

Ethanol also reacts with acid anhydrides or acid halides to give the corresponding esters.

Dehydration. Dehydration of ethanol forms both ethylene or ethyl ether to some extent, but the conditions can be altered to favor one reaction of the other.

Dehydrogenation. Dehydrogenation of ethanol to acetaldehyde can be effected by a vapor-phase reaction over various catalysts.

Haloform reaction. Ethanol reacts with sodium hypochlorite to give chloroform.

$$CH_3CH_2OH + NaOCl \rightarrow CH_3CHO + NaCl + H_2O$$

$$CH_3CHO + 3\ NaOCl \rightarrow CCl_3CHO + 3\ NaOH$$

$$CCl_3CHO + NaOH \rightarrow CHCl_3 + HCOONa$$

Manufacture

Industrial ethanol can be produced either synthetically from ethylene, as a by-product of certain industrial operations, or by the fermentation (qv) of sugar, starch, or cellulose.

There are two main processes for the synthesis of ethanol from ethylene that follow.

Indirect hydration (esterification–hydrolysis) process, variously called the strong sulfuric acid–ethylene process, the ethyl sulfate process, the esterification–hydrolysis process, or the sulfation–hydrolysis process, which was the earliest to be developed. It involves three steps: absorption of ethylene in concentrated sulfuric acid to form mono- and diethyl sulfates; hydrolysis of ethyl sulfates to ethanol; and reconcentration of the dilute sulfuric acid.

Direct hydration of ethylene. Direct hydration has completely supplanted the old sulfuric acid process.

There are two main process categories for the direct hydration of ethylene to ethanol. Vapor-phase processes contact a solid or liquid catalyst with gaseous reactants. Mixed-phase processes contact a solid or liquid catalyst with liquid and gaseous reactants. Generally, ethanol is produced by a vapor-phase process; mixed-phase processes are used for the analogous hydration of propylene to isopropyl alcohol (see Propyl alcohols).

Figure 1 shows a simplified flow diagram for the process employed by Union Carbide to produce ethanol by the direct hydration of ethylene. An ethylene-rich gas is combined with process water, heated to the desired reaction temperatures, and passed through a mixed-bed catalytic reactor to form ethanol. The reactor product is cooled by heat exchange with the reactor feed stream and is separated into liquid and vapor streams.

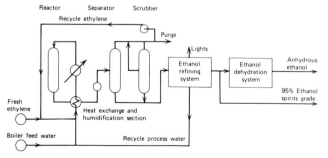

Figure 1. Manufacture of ethanol by direct hydration of ethylene. Courtesy Gulf Publishing Co.

Synthesis gas. The production of ethanol from synthesis gas, a mixture of carbon monoxide and hydrogen, is receiving a great deal of attention. The use of synthesis gas as a base raw material has the same drawback as fermentation technology: low yields limited by stoichiometry.

$$2\ CO + 4\ H_2 \rightarrow CH_3CH_2OH + H_2O$$

Its appeal lies in the fact that synthesis gas can be produced from trash, municipal sewage, scrap wood, sawdust, newsprint, or other waste (see Fuels, synthetic; Fuels from biomass; Fuels from waste).

Ethanol can also be obtained by the reaction of methanol with synthesis gas at 185°C and 6.9–20.7 MPa (68–204 atm) in the presence of a cobalt octacarbonyl catalyst. However, although ethanol is the main product, methyl formate and methyl, propyl, and butyl acetates and propyl and butyl alcohols and methane are all present in the product.

Fermentation ethanol. Ethanol is made from a variety of agricultural products such as grains, molasses, fruit, whey and sulfite waste liquor.

Ethanol can be derived by fermentation processes from any material that contains sugar. Sugars can be converted directly to ethanol (see Sugar); starches must first be hydrolyzed to fermentable sugars by the action of enzymes from malt or molds (see Beer; Yeast). Cellulose (qv) must likewise be converted to sugars, generally by the action of mineral acids. Once simple sugars are formed, enzymes from yeast can readily ferment them to ethanol.

Recovery and purification. The specifications for an industrial alcohol depend upon the intended use. Premium-grade ethanol has low water content, low odor ratings, and high permanganate-time test values.

Purification schemes generally emphasize the following techniques: extractive distillation using water reflux to distill a large share of the impurities and concentrate of the crude alcohol–water mixture (see Azeotropic and extractive distillation); efficient fractionation to produce approximately 190 proof alcohol (see Distillation; Beverage spirits); hydrogenation to convert aldehyde impurities to alcohols, together with the use of chemicals such as inorganic bases and sodium sulfite; and ion-exchange resins or azeotropic distillation to dehydrate 190 proof to 200 proof, or absolute, alcohol.

Chemicals Derived From Ethanol

These include acetaldehyde (qv), glycol ethers (see Glycols), Vinegar (qv), ethylamines (see Amines by reduction; Amines, aliphatic; Amines, cyclic), ethyl acrylate (see Acrylic acid and derivatives), ethyl ether (see Ethers), ethyl vinyl ether (see Acetylene-derived chemicals; Vinyl polymers), ethyl acetate (see Esterification), and ethyl chloride (see Chlorocarbons and chlorohydrocarbons).

Units

In the U.S., the proof is twice the alcohol content by volume, thus 190° proof alcohol contains 95% ethanol by volume. According to Federal statutes, proof spirits shall be held to be that alcoholic liquor which contains one-half its volume of alcohol of a specific gravity of 0.7939 at 15.5°C.

Uses

Industrial ethanol is one of the largest volume organic chemicals used in industrial and consumer products. The main uses for ethanol are as an intermediate in the production of other chemicals and as a solvent. Specially denatured alcohol has many uses: resins and lacquers; proprietary purposes; toilet preparations and pharmaceuticals; industrial processing; and cleaning preparations. As a solvent, ethanol is second only to water, and it is a key raw material in the manufacture of drugs, plastics, lacquers, polishes, plasticizers, perfumes, and cosmetics.

PAUL DWIGHT SHERMAN JR.
PERCY R. KAVASMANECK
Union Carbide Corporation

L.F. Hatch, *Ethyl Alcohol*, Enjay Chemical Co., a division of Humble Oil and Refining Co., New York, 1962.

S.A. Miller, ed., *Ethylene and Its Industrial Derivatives*, Ernest Benn Std., London, Eng. 1969.

ETHANOLAMINES.   See Alkanolamines.

ETHERS

Ethers are compounds of the general formula Ar–O–Ar', Ar–O–R, and R–O–R' where Ar is any aryl group and R is any alkyl group. If the two R or Ar groups are identical, the compound is a symmetrical ether. Examples of symmetrical ethers are methyl ether, CH_3OCH_3, and phenyl ether, $C_6H_5OC_6H_5$; examples of unsymmetrical ethers are methyl ethyl ether, $CH_3OCH_2CH_3$, and methyl *tert*-butyl ether, $CH_3OC(CH_3)_3$ (see also Glycols, glycol ethers). Cyclic ethers are oxygen heterocycles such as tetrahydrofuran, $\overgroup{OCH_2CH_2CH_2CH_2}$, *p*-dioxane, $\overgroup{OCH_2CH_2OCH_2CH_2}$, and 1,2-propylene oxide, $\overgroup{OCH_2CHCH_3}$.

Simple ethers derive their name from the two groups attached to the oxygen followed by the word ether, eg, ethyl ether, $CH_3CH_2OCH_2CH_3$. If one group has no simple name, the compound may be named as an alkoxy derivative, eg, 2-ethoxyethanol, $CH_3CH_2OCH_2CH_2OH$.

Physical and Chemical Properties

In general, ethers are neutral, pleasant-smelling compounds that have little or no solubility in water, but are easily soluble in organic liquids. Table 1 lists properties of some representative ethers.

Ethers are comparatively unreactive compounds because the carbon–oxygen bond is not readily cleaved. For this reason, they are frequently employed as inert solvents in organic synthesis. Ethers do react with exceptionally powerful basic reagents, particularly certain alkali–metal alkyls, to give cleavage products. They are weakly basic and are converted to unstable oxonium salts by strong acids such as sulfuric acid, perchloric acid, and hydrobromic acid; relatively stable complexes are formed between ethers and Lewis acids such as boron trifluoride, aluminum chloride and Grignard reagents (qv).

Preparation

The most versatile method of preparing ethers is the Williamson ether synthesis, particularly in the preparation of unsymmetrical alkyl ethers. The reaction of sodium alcoholates with halogen derivatives of hydrocarbons gives the ethers:

$$RX + NaOR' \rightarrow ROR' + NaX$$

Table 1.   Physical Properties of Some Representative Ethers

Ether	bp, °C	d_4^{20}, g/cm³	n_D^{20}
Aliphatic			
saturated, symmetrical			
methyl	−23.7	1.617 (air)	
2-methoxyethyl (diglyme)	162	0.9451	1.4097
ethyl	34.5	0.7138	1.3526
unsymmetrical			
methyl *n*-propyl	38.9	0.738	1.3579
methyl isopropyl	32.5	0.7237	1.3576
methyl *n*-butyl	70.5	0.7443	1.3736
unsaturated			
vinyl	28–31	0.773	1.3989
allyl	95.0	0.8260	1.4163
Aromatic			
methyl phenyl	153.8	0.9954	1.5179
4-methoxytoluene	176.5	0.9689	1.5124
ethyl phenyl	172	0.9792	1.5076
Cyclic			
furan (oxole)	31.36	0.9514	1.4214
tetrahydropyran (oxane)	88	0.8810	1.4200

Health and Safety

Although ethers are not particularly hazardous, their use involves risks of fire, toxic effects, and several unexpected reactions. Ethers tend to absorb and react with oxygen from the air to form unstable peroxides that may detonate with extreme violence when concentrated by evaporation or distillation, when combined with other compounds that give a detonable mixture, or when disturbed by heat, shock, or friction.

Uses

Alkyl ethers are used for organic reactions and extractions, as plasticizers, as vehicles for other products. The vapors of certain ethers are toxic to insects and are useful as agricultural insecticides (see Insect control technology). Aryl ethers have distinctive, pleasant odors and flavors which make them valuable to perfume (qv) and flavor industries (see Flavors and spices).

Commercially Important Ethers

Diethyl ether is probably the most important member of the ether family. It is a colorless, very volatile, highly flammable liquid with a sweet pungent odor and burning taste. As a commercial product it is available in several grades and is used in chemical manufacture as a solvent, extractant, or reaction medium, as well as a general anesthetic.

Ethyl ether is relatively nontoxic, and its greatest hazards are fire and explosion.

Other ethers include *n*-butyl ether, important as a solvent for the Grignard and other reactions that require an anhydrous, inert medium; isopropyl ether, of moderate importance as an industrial solvent, since its boiling point lies between that of ethyl ether and acetone and it readily forms hazardous peroxides and hydroperoxides (much more so than other ethers); and methyl *tert*-butyl ether (MTBE), which is finding increasing use as a blending component in high octane gasoline as current gasoline additives based on lead, manganese and bromine are phased out (see Gasoline).

DONALD E. KEELEY
General Electric Company

S. Patai, ed., *The Chemistry of the Ether Linkage*, Wiley-Interscience, New York, 1967.

C.A. Buehler and D.E. Pearson, *Survey of Organic Synthesis*, Vol. 2, Wiley-Interscience, New York, 1977.

Chemical Economics Handbook, Stanford Research Institute, Menlo Park, Calif., 1978.

ETHYL ALCOHOL.   See Ethanol.

ETHYL BENZENE.   See Xylenes and ethylbenze; Styrene.

ETHYLENE

Ethylene (ethene), $H_2C{=}CH_2$, is the largest volume organic chemical produced today. It is the most important building block of the petrochemical industry and is converted to a multitude of intermediates and products on a large scale, mainly polymeric materials such as plastics, resins, fibers, and elastomers, as well as solvents, surfactants, coatings, plasticizers, and antifreeze.

Physical and Chemical Properties

Ethylene, the lightest olefinic hydrocarbon, is a colorless, flammable gas with a slightly sweet odor. Bp −103.71°C; d 20.27 mol/L at normal boiling point; sp heat 67.4 J/(mol·K) at normal boiling point.

The chemistry of ethylene centers about its double bond, which reacts chiefly by addition to give saturated hydrocarbons, their derivatives, or polymers. The geometry of ethylene is relatively simple, all six atoms are in one plane. The double bond prevents rotation, except at high temperatures.

The chemical reactions of ethylene may be divided into those of commercial significance, eg, polymerization, oxidation and addition reactions, and those that have mainly academic interest.

Polymerization. Polymerization of ethylene represents the largest segment of the petrochemical industry, polyethylene ranking first as an ethylene consumer. Ethylene (99.9% purity) is polymerized under specific conditions of temperature and pressure in the presence of an initiator or catalyst:

$$n \ CH_2 {=} CH_2 \rightarrow {+} (CH_2CH_2)_n$$

is an exothermic reaction that may involve homogenous initiation (radical or cationic) or heterogenous initiation (solid catalyst). The products range in molecular weight from below 1000 up to 5×10^6 or more (ultrahigh molecular weight polyethylene) (see Olefin polymers).

There are two commercially important systems for polyethylene production: high pressure polymerization by free-radical initiation using oxygen, peroxides, or other strong oxidizers as initiators; and low pressure polymerization by heterogenous catalysis using transition-metal oxides such as molybdenum oxide or chromium oxide supported on inorganic carriers, or Ziegler catalysts such as aluminum alkyls and titanium halides which may also be supported on inorganic carriers.

High pressure processes produce so-called low density polyethylene (LDPE) with a density of 0.910–0.940 g/cm³. Pressures of 60–350 MPa (8,700–50,750 psi) at up to 350°C are applied. LDPE is used primarily in the manufacture of films for injection molding, and as wire, cable and paper coatings. The conventional low pressure processes produce so-called high density polyethylene (HDPE) with a density of 0.941–0.970 g/cm³, used primarily for blow- and injection-molded products. Novel low pressure processes employ advanced catalyst systems for the production of linear low density polyethylene (LLDPE) which differs from conventional LDPE significantly in physical properties. It is replacing gradually most of the LDPE and some of the HDPE in existing markets and is also entering new markets.

Oxidation. Oxidation reactions give ethylene oxide–glycol, ranking second, vinyl acetate ranking seventh, and acetaldehyde ranking eighth, as consumers of ethylene in the United States. Direct oxidation processes are preferred today because they are more economical than older multistage processes. Direct vapor-phase air oxidation over silver oxide catalysts, at 200–300°C and 1.5–3.0 MPa (217.5–435 psi) is one such process (see Ethylene oxide).

About two-thirds of the ethylene oxide produced is converted to ethylene glycol by noncatalytic high pressure, high temperature hydrolysis. Higher glycols such as di- and triethylene glycol are produced as by-products.

A vapor-phase ethylene-based process is now the preferred route to vinyl acetate. Palladium on carbon, alumina, or silica-alumina is employed as catalyst at ca 175–200°C and 0.4–1.0 MPa (58–145 psi).

During direct oxidation of ethylene, the preferred route to acetaldehyde today, palladium chloride is reduced at room temperature to palladium.

Addition. Addition reactions of ethylene have considerable importance: in halogenation–hydrohalogenation for the production of ethylene dichloride, ethylene chloride, ethyl chloride and ethylene dibromide; in alkylation or the production of ethylbenzene, ethyltoluene, and aluminum alkyls; in oligomerization for the production of alpha olefins and linear primary alcohols; in hydration for the production of ethanol; and in hydroformylation for the production of propionaldehyde.

The most important intermediate produced by addition is ethylene dichloride, $ClCH_2CH_2Cl$, which ranks third in the United States ethylene consumption (see Chlorocarbons). By far the leading derivative of ethylene dichloride (EDC) is vinyl chloride monomer (VCM) used to produce poly(vinyl chloride) (PVC) resins and chlorinated hydrocarbons.

Manufacture

Single-component feedstocks, such as ethane or propane, or multicomponent hydrocarbon feedstocks, such as natural-gas liquids (NGL) and naphthas, and gas oils from crude oil, may be used for the production of olefins (see Feedstocks).

Pyrolysis of hydrocarbons in tubular reactor coils. Pyrolysis in coils, which represent a thermal reactor, accounts for almost all the ethylene produced today. The properties of the feedstock and the conditions at which the reactor coils are operated determine reactor effluent product distribution. High selectivity toward the production of desired olefins and diolefins (ie, ethylene, propylene, and butadiene) with minimum methane and coking in the coils leading to longer heater runs are achieved by operating the pyrolysis heaters at high temperatures (750–900°C), short residence times (0.1–0.6 s), and low hydrocarbon partial pressure.

Modern ethylene plants are normally designed for near-maximum cracking severity because of economic considerations. The production of ethylene from liquid feedstocks is accompanied by the production of other olefins which invariably entails coproduction of many industrially important petrochemicals.

Polymer-grade ethylene and propylene are produced from the effluent of an ethylene plant. Other pyrolysis products affect operating economics significantly. With propane as feed, the C_3 fraction of the pyrolysis product may contain 40–60 wt% propylene, and for heavier feedstocks, it may contain 90–95 wt% propylene.

The pyrolysis heaters and their adjacent transfer-line exchangers (also called quench coolers) are the heart of an ethylene plant. They consist of (1) an upper central part with tubes arranged horizontally in the direction of the heater axis for feedstock preheat, feedwater preheat, and steam superheating coils designed for heat transfer from the rising flue gases essentially by convection, and (2) lower heater boxes containing the reactor coils arranged in vertical planes paralleling the heater axis and designed predominantly for radiant heat transfer from heater-box walls and radiating flue gases. Design of the reactor coils is a formidable task which may be facilitated by a computer. Characteristics of the pyrolysis heater are outlined in Table 1.

The gaseous effluent from the pyrolysis section leaves the transfer-line exchangers at temperatures in the range of 300°C (gaseous feedstocks) to 600°C (liquid feedstocks) to enter the recovery and purification section. The effluent is resolved into the desired products by compression in conjunction with condensation and fractionation at gradually lower temperatures. Absorption is not used for recovery because of its higher energy requirements. Separation is carried out by low temperature fractionation. Pyrolysis gases derived from ethane feedstock require the least effort and apparatus for separation, whereas those from heavy naphthas and gas oil producing ethylene, propylene, butadiene, and substantial amounts of liquid products, require more complicated treatment.

Ethanol dehydration. Ethanol dehydration for the production of ethylene cannot compete with pyrolysis of hydrocarbons in large olefins units. The economic feasibility of producing ethylene from ethanol (qv) depends mainly on the availability and price of fermentation ethanol. The process offers, however, several advantages to a country with abundant fermentation materials but limited hydrocarbon resources. Ethanol dehydration goes back to 1797 when ethylene was first being produced by passing ethanol or ether over heated alumina or silica.

Table 1. Pyrolysis Heater Characteristics[a]

Characteristics	Value
number of coils	2–20
coil length, m	50–80
process gas outlet temperature, °C	750–900
tube wall temperature, reactor coil, clean, °C	950–1050
max reactor tube wall temperature, °C	1040–1100
average heat absorption, reactor coil, kW/m²	50–80
residence time, reactor coil, s	0.15–0.6
inside coil diameter, mm	50–200

[a] Per heater unit; for a yearly ethylene production capacity of 20,000–70,000 t.

Later, activated alumina and phosphoric acid on a suitable support became the choices for industrial processes.

In industrial production, the ethylene yield is 94–99% of the theoretical value, depending on the processing scheme. Traces of aldehydes, acids, higher hydrocarbons and carbon dioxide, as well as water, have to be removed.

Ethylene from single carbon feedstocks. Ethylene from single-carbon feedstocks (synthesis gas, methane) is presently produced in South Africa only where coal is mined at very low cost. There are various routes to ethylene from single-carbon feedstocks (derived from such raw materials as coal, lignite, heavy oil fractions, biomass, natural gas) by conventional and by new process technology. These processing routes (which include such processing steps as manufacture of synthesis gas by gasification of various raw materials, Fischer-Tropsch processing, steam reforming, methanol production, homologation and cracking (Mobil catalyst), direct catalytic or thermal conversion (Benson process) of methane to ethylene, and ethanol dehydration) are not economical today but may become increasingly of commercial importance toward the end of this century.

Health and Safety

At room temperature and atmospheric pressure, ethylene is a colorless gas with a mild odor, of a density practically equal to air, non-irritating to eyes and the respiratory system. Released to the atmosphere, it quickly diffuses and a the very low concentrations of 2.7 vol% forms a flammable mixture with air. As it does not give adequate warning of its presence, any accidental leak must be approached with the greatest caution and in strict observance of safety procedures contained in plant operating manuals. Chronic injuries to humans from the temporary inhalation of ethylene have not been reported.

Uses

Almost all ethylene is consumed by the chemical industry as feedstock for a variety of petrochemical products.

LUDWIG KNIEL[‡]
OLAF WINTER
CHUNG-HU TSAI
The Lummus Company

S.A. Miller, *Ethylene and Its Industrial Derivatives*, Ernest Benn, Ltd., London, UK, 1969.

S.B. Zdonik, E.J. Green, and L.P. Hullee, *Manufacturing Ethylene*, The Petroleum Publishing Company, Tulsa, Okla., reprinted from *Oil Gas J.*, (1966–1970).

"Ethylene" in *Chemical Economics Handbook*, Stanford Research Institute, Menlo Park, Calif.

L. Kniel, O. Winter, and K. Stork, *Ethylene, A Keystone to the Petrochemical Industry*, Marcel Dekker, Inc., New York, 1980.

ETHYLENE GLYCOL, CH_2OHCH_2OH.   See Glycols.

ETHYLENEIMINE, $HN(CH_2)_2$.   See Imines, cyclic.

ETHYLENE OXIDE

Ethylene oxide (oxirane) was first prepared by Wurtz in 1859 by the reaction of 2-chloroethanol (ethylene chlorohydrin) with aqueous potassium hydroxide (see Epoxidation).

Physical and Chemical Properties

Ethylene oxide is a colorless gas condensing at low temperatures to a mobile liquid. It is miscible in all proportions with water, alcohol, ether, and most organic solvents. Its vapors are flammable and explosive. Bp 10.4°C· fp −112.5°C; n_D 1.3597.

[‡]Deceased.

Ethylene oxide is a highly reactive molecule; industrially it is primarily a chemical intermediate for a wide variety of compounds (see Derivatives). The three-membered ring is opened in most of its reactions, except in the formation of oxonium salts with strong anhydrous mineral acids.

Reaction of ethylene oxide with many compounds having a labile hydrogen atom introduces the hydroxyethyl group:

$$RH + CH_2{-}CH_2 \ (O) \longrightarrow RCH_2CH_2OH$$

The terminal hydroxyl thus formed can react further with ethylene oxide:

$$RCH_2CH_2OH + n\ CH_2{-}CH_2 \ (O) \longrightarrow RCH_2CH_2O{-}(CH_2CH_2O)_n{-}H$$

These reactions give a series of poly(ethylene glycol) derivatives of increasing chain length and water solubility.

Catalytic or chemical reduction of ethylene oxide produces ethanol (qv).

A continuation of the reaction of ethylene oxide with alcohols or water forms low molecular weight polymers. High polymers of average molecular weights from 90,000 to 4,000,000 are formed by a process of coordinate anionic polymerization. The patent literature describes numerous organometallic compounds, alkaline-earth compounds and mixtures as catalysts. This process is also important as the mechanism of formation of nonvolatile residue (NVR) during ethylene oxide storage. The primary catalyst for NVR formation is rust: no inhibitor has been found.

Ethylene oxide forms cyclic oligomers (crown ethers) in the presence of fluorinated Lewis acids such as boron trifluoride, phosphorus pentafluoride, or antimony pentafluoride (see Chelating agents).

Wurtz was first to obtain ethylene glycol when heating ethylene oxide and water in a sealed tube. Hydration is slow at ambient temperatures, but is speeded by heat or acid or base catalysts.

Reactions with alcohols parallel those of ethylene oxide with water; the primary products are monoethers of ethylene glycol.

With organic acids and anhydrides, the carboxyl group of an organic acid reacts with ethylene oxide to give initially the corresponding ethylene glycol monoester. This may react further with ethylene oxide to give a poly(ethylene glycol) ester, or with another acid to give the glycol diester.

In the presence of aluminum chloride, ethylene oxide undergoes a Friedel-Crafts reaction (qv) with aromatic hydrocarbons. Ethylene oxide and hydrogen chloride passed into a mixture of benzene and aluminum chloride yields dibenzyl ($C_6H_5CH_2CH_2C_6H_5$) and a small amount of 2-phenylethanol ($C_6H_5CH_2CH_2OH$).

Compounds containing active $-CH_2-$ or $-CH\big\langle$ groups, such as malonic and monosubstituted malonic esters, ethyl cyanoacetate, and β-keto esters, condense with ethylene oxide and other alkylene oxides to form a wide variety of compounds including α-carboethoxy-γ-butyrolactone and 2-acetyl-4-butyrolactone.

Manufacture

Ethylene oxide is commercially produced by two routes: the ethylene chlorohydrin process, the older process, and the direct oxidation process, which completely dominates the field today. The first process involves the reaction of ethylene with hypochlorous acid followed by dehydrochlorination of the resulting chlorohydrin, $ClCH_2CH_2OH$, with lime to produce ethylene oxide and calcium chloride.

The fundamental reaction of the direct oxidation process is the catalytic oxidation of ethylene with oxygen over a silver-based catalyst to yield ethylene oxide. Direct oxidation processes are divided into two categories depending on the source of the oxidizing agent: the air-based process, and the oxygen-based process. In the first, air or air enriched with oxygen is fed directly to the system. In the second, a high purity oxygen stream, (> 95 mol%) from an air-separation unit is employed as the source of the oxidizing agent.

Direct oxidation processes. Compared to the chlorohydrin process, direct oxidation eliminates the need for large volumes of chlorine, there are no chlorinated hydrocarbon by-products to be sold, and processing facilities can be made simpler and operating costs are lower. The main disadvantage of the direct oxidation process is the lower yield or selectivity of ethylene oxide per unit of feed ethylene consumed. The major inefficiency in the process results from the loss of ca 25–30% of the ethylene to carbon dioxide and water.

According to the best current understanding of the mechanism, ethylene oxide is formed when ethylene reacts with one atom of diatomic oxygen species adsorbed on a silver surface. A large amount of heat is released by the reaction. At 600 K, each kg of ethylene converted to ethylene oxide releases 3.756 MJ (3564 Btu); each kg of ethylene converted to carbon dioxide and water releases 60.58 MJ (48,083 Btu).

Commercial processes operate under recycle conditions in a packed-bed multitubular reactor. Reaction temperatures of 200–300°C are typical, and operating pressures of 1–3 MPa (10–30 atm) have been reported. The reactor is of the shell-and-tube type, comprised of several thousand mild steel or stainless-steel tubes, 20–50 mm inside diameter. The selectivity or yield (moles of product produced per mole of ethylene consumed) is normally 60–77% depending on catalyst type, per pass conversion, reactor design, and a large number of process operating variables.

A schematic flow diagram of the ethylene oxide processes is shown in Figure 1, which represents the salient concepts involved in the manufacturing process. Published information on the detailed evolution of commercial ethylene oxide process is very scanty, and precise information regarding process technology is proprietary. The process can be conveniently divided into three primary sections: reaction system, oxide recovery, and oxide purification.

Even though the fundamental reaction and the ultimate results are the same, there are substantial differences in detail between air- and oxygen-based processes. Virtually all the differences arise from the change in the oxidizing agent from air (ca 20 mol% O_2) to pure oxygen (> 95 mol% O_2). Figure 2 represents the oxygen-based process.

The relative economics of the air versus the oxygen process are reported. Two main process characteristics dictate the difference in the capital costs; the air process requires additional investment for the purge reactors and their associated absorbers, and for the energy recovery from the vent gas. However, this is offset by the need for an oxygen-production facility and a carbon dioxide removal system for an oxygen-based unit. In a comparison of necessary investments for small- to medium-capacity units (< 50,000 t/yr), oxygen-based plants have a lower capital cost even if the air-separation facility is included. However, for medium- to large-scale plants (75,000–150,000 t/yr), the air-process investment is smaller than that required for the oxygen process and the air-separation unit, unless the oxygen is purchased from a very large air-separation unit serving many customers.

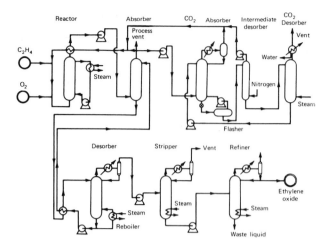

Figure 2.   Oxygen-based-direct oxidation process for ethylene oxide.

Chlorohydrin process. The chlorohydrin process is commercially attractive to a producer only when an adequate supply of captive low cost chlorine and lime or caustic are available. Further, a market for the by-products ($ClCH_2CH_2Cl$, ($ClCH_2CH_2)_2O$, and $CaCl_2$ or $NaCl$) plays an important role in process economics.

New routes for manufacture. New routes which may make a favorable impact on the overall production costs for making ethylene oxide, are arsenic-catalyzed liquid-phase process; thallium-catalyzed epoxidation process; Lummus hypochlorite process; electrochemical process; unsteady-state direct-oxidation process; and fluid-bed oxidation process.

Health and Safety

Ethylene oxide is relatively toxic in liquid and gas. Liquid causes severe eye injury and gas may cause eye irritation. Vapors are irritating to the nose and throat. A TLV of 50 ppm air is recommended as a safe 8-h daily concentration in a 40-h wk. Liquid or dissolved ethylene oxide on exposed skin does not cause immediate irritation but may cause severe delayed skin burns. A 50% aqueous solution of ethylene oxide is the most irritating combination. Ethylene oxide has been found to be a mutagen in at least 13 test species. It has not found to be a carcinogen to date.

Pure ethylene oxide vapor will explode in the presence of common igniters. The commonly accepted autoigition value for the vapor if 560°C.

Uses

Ethylene oxide is an excellent fumigant and sterilizing agent. Because of its high chemical reactivity, most ethylene oxide produced is converted to other compounds, especially ethylene glycol and surfactants (qv).

JAMES N. CAWSE
JOSEPH P. HENRY
MICHAEL W. SWARTZLANDER
PARVEZ H. WADIA
Union Carbide Corporation

Ethylene Oxide, Brochure F-7618E, Union Carbide Corporation, New York, 1973.

A. Rosowsky in A. Weissberger, ed., *Heterocyclic Compounds*, Vol. 19, pt. 1, Wiley-Interscience, New York, 1964.

F.W. Stone and J.J. Stratta in N.M. Bikales, ed., *Encyclopedia of Polymer Science and Technology*, Vol. 6, Wiley-Interscience, New York, 1967, pp. 103–145.

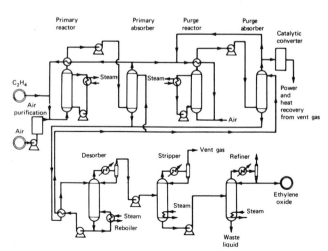

Figure 1.   Air-based direct-oxidation process for ethylene oxide.

ETHYLENIC ALCOHOLS.   See Acetylene-derived chemicals.

ETHYLIDENE DIACETATE, CH$_3$CH(OOCCH$_3$)$_2$.   See Acetic acid.

ETHYL MERCAPTAN (ETHANETHIOL), CH$_3$CH$_2$SH.   See Thiols.

ETHYNYLATION.   See Acetylene-derived chemicals.

EUTROPHICATION.   See Water pollution.

EVAPORATION

Evaporation by its broadest definition is the conversion of a liquid to a vapor. It applies to such widely diverse equipment as boilers, cooling towers, dryers, and humidifiers, and losses from fields, storage tanks, and reservoirs. In the narrower chemical-engineering sense, evaporation is the removal of volatile solvent from a solution or a relatively dilute slurry by vaporizing the solvent.

The highly varied purposes for which evaporators are used industrially include: reducing the volume to economize on packaging, shipping, and storage costs (eg, of salt, sugar, caustic soda); obtaining a product in its most useful form (eg, salt from brine or sugar from cane juice); eliminating minor impurities (eg, salt, sugar); removing major contaminants from a product, eg, diaphragm-cell caustic soda solutions contain more salt than caustic when produced but practically all the salt can be precipitated by concentrating to a 50% NaOH solution; concentrating a process stream for recovery of resources, eg, pulp-mill spent cooking liquor, if concentrated sufficiently in an evaporator, can be used to reconstitute fresh cooking liquor; concentrating wastes for easier disposal, such as nuclear-reactor wastes, dyestuff plant effluents, and cooling-tower blowdown streams; transforming a waste into a valuable product, such as spent distillery slop after alcohol recovery, which can be concentrated to produce an animal feed; and recovering distilled water from impure streams such as seawater and brackish waters.

Steam-Heated Evaporators

The three principal requirements of steam-heated evaporators are transfer to the liquid of large amounts of heat needed to vaporize the solvent; efficient separation of the evolved vapor from the residual liquid; and accomplishing these aims with the least expenditure of energy justifiable by the capital cost involved.

Natural-circulation evaporators.   Natural-circulation evaporators were the first to be developed commercially and still represent the largest number of units in operation. These evaporators utilize the density difference between the liquid and the generated vapor to circulate the liquid past the heating surface and thereby give good heat-transfer performance. The heat-transfer tubes may be either vertical or horizontal, with the liquid either inside or outside the tubes (Fig. 1).

Forced-circulation evaporators.   Forced-circulation evaporators, suitable for the largest variety of applications, are the most expensive type. They usually consist of a shell-and-tube heat exchanger, a vapor-liquid separator (variously called vapor head, vapor body, separator, flash

chamber, or body), and a pump to circulate the liquor from the body through the heater and back to the body.

Film-type evaporators.   Film-type evaporators were developed 50 to 80 years ago. The rising-film, or long-tube vertical (LTV) evaporator is the most widely used in the United States, and consists of a vertical shell-and-tube heat exchanger surmounted by a vapor-liquid separator (Fig. 2). Because of the simplicity of construction, costs per heating surface are the lowest of any type.

The LTV is classified as a film-type evaporator because boiling takes place within the tubes and the vapor-liquid mixture is usually in the annular or film-flow regime for much of the tube length. High vapor velocities are generated and the interfacial shear also causes the liquid to move at high velocity, yielding good heat-transfer coefficients.

The falling-film evaporator is an inverted version of the LTV that greatly reduces the adverse effects of pressure drop on available ΔT exhibited by the rising-film LTV. The hydrostatic head loss is eliminated, acceleration losses are lower because the liquid film is not accelerated substantially by the vapor flow, and the frictional pressure drop is generally only a little more than that of vapor flowing alone in a dry tube.

The principal advantages of the falling-film evaporator are its good heat-transfer performance, even at low temperatures and low temperature differences, its low initial cost, and its excellent vapor-liquid separation characteristics. Principal applications have been for citrus juices, where performance at low temperature and low holdup is important, and applications requiring operation at low temperature differences, such as vapor compression or multiple-effect evaporators needing a large number of effects to be economical.

Energy Conservation

Most of the complexity and cost of an evaporator installation is a result of attempts to reduce energy consumption, which is usually by far the most important element of operating cost. Evaporators are not normally rated directly in efficiency of energy usage since they only have to separate the solvent from the solution, which requires very little theoretical energy in an ideal system. Thus, for separating water from a salt solution having a boiling point rise of 5 K at its atmospheric boiling point, the minimum theoretical energy requirement is only 30.1 kJ/kg of water removed (13.0 Btu of work per lb) whereas it takes 2250 kJ/kg (970 Btu of heat per lb) to vaporize the water. Since heat and work energy are not directly interchangeable in utility or cost, steam-heated evaporators are generally rated in terms of steam economy, ie, kg of water evaporated per kg of steam used, also called gained output ratio or performance ratio, and frequently standardized at pounds of water evaporated per 1000 Btu extracted from the stream (kg evaporated per

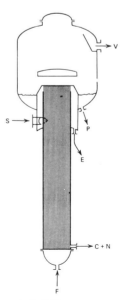

Figure 2.   Long-tube vertical (LTV) evaporator. Symbols: see Figure 1.

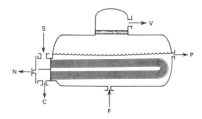

Figure 1.   Horizontal-tube natural-circulation evaporator. Symbols: C = condensate, E = entrainment return, F = feed, N = noncondensables' vent, P = product or concentrate, S = steam, and V = vapor.

2324 kJ). Even evaporators that do use work energy (electrical or mechanical) for their operation are not usually rated in terms of efficiency but instead in such terms as J/m^3 ($= 1.05 \times 10^{-6}$ kW·h/1000 gal) evaporated.

The single-effect evaporator is the simplest arrangement. It uses steam from an outside source and exhausts its vapor to the atmosphere or to an air- or water-cooled condenser. Such an evaporator requires about 1 kg steam per kg of water evaporated and somewhat more if the feed is colder than the product and heat cannot be recovered by preheating the feed with concentrate and condensate.

The single-effect evaporator produces almost as much vapor as the amount of steam used, the only difference being that the vapor is at a lower temperature and hence lower pressure. Compressing the vapor for reuse as the heat source was put into operation in the 19th century. This thermocompression or vapor-recompression operation can be accomplished by either mechanical or steam-jet compressors. Mechanical compressors are by far the more efficient and may be driven by electric motors, gas or diesel engines, or steam or gas turbines. Compressor efficiencies are usually in the range of 70–75% for single-stage centrifugal machines (for compression ratios to about 1.5) and 80–85% for axial-flow machines (for compression ratios of about 1.15 per stage). Steam-jet compressors are only about 20–30% efficient.

If steam instead of power is the source of energy, multiple-effect evaporation is the principal means of energy conservation. In this operation, the vapor from one effect is used to heat another effect boiling at a lower temperature and the vapor from this effect is used to heat yet another effect boiling at a still lower temperature. Steam consumption goes down and capital cost goes up in almost direct proportion to the number of effects used. Minimum total cost is usually achieved with eight effects or more at today's energy prices.

Other Evaporation Methods

Solar evaporation was the first evaporative method developed and still is in widespread use, primarily for production of salt from seawater (see Chemicals from brine). Evaporation rates vary widely with climate and can be predicted from weather data and the properties of the solution being evaporated. Present laws require elimination of seepage to avoid groundwater contamination. The cost of providing the seepage barrier exceeds by far all other costs of solar-pond development. Solar evaporation has also been used for the production of potable water from seawater. A barrier is required between the ponds and the atmosphere to admit the solar energy and to condense and collect the evaporated water. This barrier is even more expensive than a pond liner, making the method uneconomical even at today's energy prices.

FERRIS C. STANDIFORD
W.L. Badger Associates, Inc.

J.G. Knudsen and co-workers, in R.H. Perry and C.H. Chilton, eds., *Chemical Engineers' Handbook*, 5th ed., McGraw-Hill Book Co., New York, 1973, Section 10, pp. 32–35.

F.L. Rubin in R.H. Perry and C.H. Chilton, eds., *Chemical Engineers' Handbook*, 4th ed., McGraw-Hill Book Co., New York, 1963, Section 11, pp. 35–36.

Energy Conservation Tools for the Process Engineer: Upgrading Existing Evaporators to Reduce Energy Consumption, ERDA, Technical Info. Center, Oak Ridge, Tenn., 1977.

EXHAUST CONTROL, AUTOMOTIVE

Atmospheric Chemistry

Automobile exhaust contains carbon monoxide, carbon dioxide, oxides of nitrogen (NO_x) (primarily NO), water, and nitrogen. The exhaust can also contain hundreds of hydrocarbons (HC) and particulates including carbon and oxidized carbon compounds, metal oxides, oil additives, fuel additives, and breakdown products of the exhaust system.

The principal health effects of concern are carbon monoxide, which acts by substituting for oxygen in the blood to form carboxyhemoglobin

(the high concentrations of CO in the urban atmosphere can threaten the health of individuals with angina or other cardiovascular ailments); oxides of nitrogen, which cause irritation of the respiratory system and may reduce the individual's ability to resist infection; and oxidants which include a wide amount of reaction products so that the EPA has proposed that ozone measurement be used as a surrogate. As in the case of oxides of nitrogen, the principal concern is irritation of the respiratory system resulting in stress during exercise and possible infection.

Control Devices

There are three sources of vehicle emission: hydrocarbons from the crankcase ventilation system; evaporative loses including hydrocarbons from the carburetor and the gas tank as well as the various lines leading from these sources to the engine; and hydrocarbons, carbon monoxide, and oxides of nitrogen from the exhaust system. Since both the crankcase and evaporative emissions are ultimately controlled in the engine, controls for the first two are considered before discussing the other engine controls.

Crankcase fumes enter the engine at the intake manifold during most of the vehicle operation at a rate that is controlled by a pressure-control valve (PCV) of the type shown in Figure 1. Since the lowest air flow occurs during engine idling and deceleration, and since these conditions also produce the highest vacuum at the intake manifold, the valve moves to a position where the smallest opening is presented. As air flow increases and the vacuum level decreases, the orifice opens, allowing a greater amount of fumes to pass. The net result is that disturbances of the fuel intake charge are minimized.

The gas-tank cap must provide a positive seal against exhaust of gasoline vapors. In addition, the cap must be able to pop open if the vapor pressure exceeds the maximum design limit of the tank, ca 13.8 kPa (2 psi). During engine operation, it must be vented to the atmosphere to compensate the resulting vacuum as fuel is pulled out of the tank. While the vehicle is at rest, a charcoal canister collects fumes from the gas tank and from the carburetor; during vehicle operation these fumes are pulled into the intake manifold for further processing by the engine.

An exhaust-gas recirculation system (EGR) acts to reduce the formation of NO_x during the combustion process in the engine. Oxides of nitrogen are formed at the very high temperatures encountered in the combustion flame, and NO_x formation increases as the temperature increases (see also Burner technology). EGR reduces the peak combustion temperature (and also the combustion rate) and thereby reduces the amount of NO_x formed. However, EGR also produces increased hydrocarbon emissions, and these must be controlled by other means.

Recirculated exhaust gas is introduced into the intake manifold by way of a control valve. Care is needed in applying EGR to assure optimum distribution among the cylinders and to minimize any adverse effect on the fuel and air distribution so that uniform combustion is maintained in each of the cylinders for efficient engine operation.

In theory, the control of any carbon monoxide and hydrocarbon that manage to elude the combustion chamber should be comparatively simple. In practice, this control is a very difficult task because there is not sufficient time, temperature is too low, and it is difficult to provide the extra air when needed without reducing the temperature. The first systems that were tried for post-engine control of hydrocarbons and CO featured the introduction of air at the exhaust ports which are the

Figure 1. PCV valve.

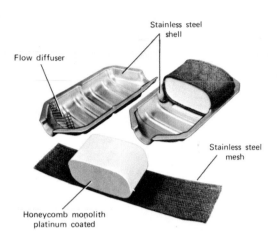

Figure 2. Monolith catalytic converter. Courtesy of the Chrysler Corporation.

hottest points in the exhaust stream. Considerable reduction in hydrocarbons and CO can be brought about by this technique, but the amount of air must be very carefully controlled in order to prevent unnecessary cooling of the exhaust stream.

Catalysts. The first catalysts used were oxidative and required a lean fuel-exhaust mixture. Supplementary air was added by an air pump or by an aspirator preceding the catalyst.

Oxidation catalysts promote further oxidation of hydrocarbons and carbon monoxide in the exhaust gas. Effective operation of this type of catalyst requires temperatures of 315°C or higher as well as an adequate supply of oxygen. Oxidation catalysts in current use normally start oxidizing within two minutes after the start of a cold engine.

Catalytic conversion efficiencies as measured during the official Federal emission-test procedure, vary with catalyst location and availability of supplemental air. Catalysts now in use that are located ca 0.7 m downstream in the exhaust from the engine and that do not have an air pump or aspirator have conversion efficiencies of ca 65% for HC and 45% for CO after being driven for 80,500 km. With the addition of an air pump and a start-up catalytic unit designed to operate on cold exhaust gases, catalyst systems have shown efficies of ca 80% for HC and ca 75% CO after 80,500 km.

Current oxidation catalysts consist of platinum and mixtures of platinum and other noble metals, notably palladium. These metals are deposited on alumina of high surface area (see Figure 2).

The most feasible approach to reducing NO_x emissions to levels at or below 0.93 g/km (1.5 g/mi) uses a reducing-type catalyst based on a material such as rhodium. When this catalyst is combined with a conventional oxidation catalyst (using platinum and rhodium as the active materials), it can provide three-way emission control of HC, CO and NO_x.

CHARLES M. HEINEN
Chrysler Corporation

B.G. Ferris, Jr., *J. Air Pollut. Control Assoc.* **28**, 482 (1978).

O.A. Ludicke and D.L. Dimick, *Diesel Exhaust Particulate Control System Development*, S.A.E. paper, 1983.

EXHAUST CONTROL, INDUSTRIAL

Many commercial operations and industrial processes generate gaseous chemical by-products which may, when disposed of, become environmental pollutants, pollutants which change the ambient air and have a deleterious effect on both living organisms and material objects. A high value is conferred upon engineering technology used to meet the regulatory objectives for improving environmental quality. A principal element

in this technology seeks to minimize the generation of undesirable by-products by modifying specific process materials or operating conditions. However, process economics or product quality may restrict the general applicability of this approach, and thus considerable technical activity is directed to developing specialized equipment that would retrofit (modify) or augment existing processing equipment and could capture the objectional by-products or convert them to harmless effluents.

A major developing technology for control of exhaust-gas pollutants is their catalyzed conversion into innocuous chemical species, such as water and carbon dioxide. This is typically a thermally activated process commonly called catalytic incineration, or catalytic combustion or oxidation, or catalytic fume abatement; the devices for its application are termed either catalytic or catalyst afterburners, abaters, reactors, or filters.

A comparative view of the technical basis of exhaust emission control systems can distinguish between capture devices, which involve several types of physical interaction, and conversion devices which primarily involve chemical reaction. Functions of capture devices include condensation, precipitation, filtration, adsorption, and adsorption emulsification, whereas functions of conversion devices include thermal activation, chemical activation (ozone, peroxide, acid–base, etc), electrical activation, and photoactivation.

Catalytic Incineration

An incinerator is a chemical reactor in which the reaction is activated by heat and is characterized by a specific rate of consumption of reactants. In this redox reaction there are at least two chemical reactants: an oxidizing agent and a reducing agent. The rate of reaction is related to the nature and to the concentration of reactants and to the conditions of activation, that is the temperature at which the reactants meet under conditions otherwise suitable for reaction. Implicit in such a simplified description are practical concepts of temperature (activation), turbulence (proper mixing of reactants), and time (period in which the reaction can accomplish a significant change in the quantities of reactants present).

Thermal incineration relies on a homogeneous gas–phase reaction condition. Proper activation requires establishing the minimum required temperature (650–800°C) for an adequate time (0.1–0.3 s). General design consideration is given to minimizing heat input and reactor size under the constraints of time, turbulence, and temperature.

Catalytic incineration relies on heterogenous gas–solid interface reaction conditions. Activation necessitates establishing an interaction of one or both of the reactants with the solid catalyst to generate more reactive chemical species, and a required temperature at the catalyst surface for an adequate time, 200–400°C and 0.0005 to 0.01 s for supported platinum catalysts.

Catalytic incineration performs the same chemical reactions as thermal incineration but at much lower temperatures, and hence at considerably reduced fuel costs. The time requirement varies inversely with temperature in both systems. Carbon dioxide and water are products of the complete reaction of hydrocarbons with oxygen for both systems.

Typical System

A typical system may utilize the following components (see Figure 1): blower (primary) to move the gas stream; blower (secondary) to supply air, if required; preheater (burner) or heat exchanger (primary) to heat reactants to operating temperatures; catalyst bed and reactor for the chemical reaction of pollutant; heat exchanger (secondary) or recirculation ducting for recovery of heat for other uses; instrumentation and controls (not shown) for maintaining operating conditions and assessing performance variations; and filter/mixer which is the section between preheater and catalyst bed used to assure flow distribution, shield the bed from flame impingement, remove noncombustible particulates, and vaporize entrained liquids and aerosols.

Rate processes. The general phenomenon of heterogeneous catalysis involves specific rate processes (types of physical or chemical transitions) that can be visualized in the following simple steps: gas-phase mass (heat) transfer of reactants to catalyst surface; diffusion which estab-

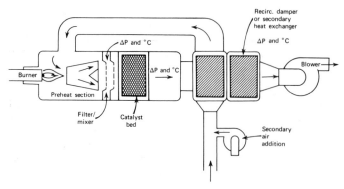

Figure 1. Typical components of a catalytic incinerator. ΔP and °C indicate pressure drop and heated gas, respectively.

lishes distribution of adsorbed reactive species; chemical reaction between oxidizing and reducing species; diffusion and desorption of reaction products; and gas-phase mass (heat) transfer of products away from the catalyst surface.

Catalyst Composition

Precious-metal catalysts were found to be superior to base or transition-metal oxide catalysts for automobile exhaust emission control systems. Of the potentially useful heterogeneous supported catalysts, only the precious metal elements rhodium, palladium, iridium, platinum, and gold maintain a stable metal surface under the typical operating conditions. Platinum is the primary choice for catalytic incineration.

The precious metal is formulated as a supported catalyst consisting of very small metal particles (3 to 20 nm dia) distributed uniformly upon the surfaces of a support or carrier. The word support has been used both in reference to the catalyst carrier upon which the active catalytic ingredient is distributed, and the geometric structure (pellet, honeycomb, etc) upon which a catalyst coating is applied. The nature of the metal dispersion (eg, the particle size range or the ratio of surface atoms/total atoms of metal present, and the particle distribution upon the carrier) has an important bearing on catalytic activity and stability.

The individual matrix elements (eg, nonbonded cylindrical, spherical, random-shaped particles (pellets); fibers; ribbons; wires; rods; monoliths) all form a foraminous substrate when used in catalytic incineration.

Operating conditions. General guidelines are as follows. Temperature at catalyst inlet is a minimum of 250°C, maintained for complex hydrocarbon pollutants to prevent possible organic-char formation, and a minimum of 120°C needed for simple, reactive reducing species to prevent water condensation on the catalyst surfaces. At the catalyst outlet, a maximum temperature of 650°C used with systems that release nominal heat. A linear gas velocity of 3–6 m/s is required for normal performance index and heat release and higher velocities are used for high heat release systems. Reactant concentrations vary with heat of reaction, but should be maximized when concentrations are low, up to 10–15% of lel (lower explosive limit); higher concentrations should be reduced toward this level. Optimal pressure drop is 750–1250 Pa (3 to 5 in. H$_2$O), with a minimum of 500 Pa (2 in. H$_2$O).

Deactivation and Maintenance

The performance index of a catalyst is highest with initial use, and during use, performance decreases at varying rates. Deactivation may be the sole result of: (1) physical changes resulting from thermal effects, such as recrystallization of the carrier, the precious metal, or both, and is referred to as thermal aging or sintering. Thermal aging is accompanied by large changes in surface and by dispersion of the supported metal. For this reason, most supported catalysts are over-designed relative to precious-metal composition and loading; (2) physical changes associated with the accumulation of contaminants that form a barrier or mask between the gas stream and catalyst surface, and referred to as catalyst binding. Masking may result from deposition of noncombustible material entrained in the exhaust emissions. Deactivation may also result from

poisoning, or chemical changes caused by the combined effects of thermal condition and contamination characterized by chemical reaction of a contaminant with the supported catalyst; inhibition, where the extent of chemical change is quantitatively lower than poisoning and is reversible.

Applications

Chemical Processes

Abatement. The principal pollutants associated with nitric acid manufacture are the residual unreacted nitrogen dioxide, NO$_2$, and nitric oxide, NO, from the absorption column. These two oxides constitute the NO$_x$ that must be maintained at or below the regulatory levels, by a reducing process known as abatement. The predominant abatement method is the catalytic reduction of the NO$_x$ to elemental nitrogen with any available hydrocarbon as the reactant; the most commonly used fuel is methane, but other fuels, eg, ammonia-synthesis purge gas, propane, butane, propylene, kerosene, and naphtha, are being employed with increasing frequency. Spurred by possible natural gas shortages, the selective reduction of NO$_x$ in the presence of oxygen with NH$_3$ is also being used for NO$_x$ abatement.

Acrylonitrile manufacture. Off-gases containing from 1–3% of CO plus various hydrocarbons are emitted to the atmosphere. Catalytic beds of platinum-group metals are used to reduce the regulated compounds to acceptable levels. Because of the high thermal energy content of the off-gases, considerable heat recovery is possible in abating acrylonitrile-plant emissions.

Phthalic/maleic anhydride manufacture. Partial oxidation of a suitable hydrocarbon feedstock over a fixed-bed catalyst (usually V$_2$O$_5$) produces phthalic anhydride or maleic anhydride, or both, which are collected as condensed particulates on oil-filled switch-condenser surfaces (see Maleic anhydride; Phthalic acids). The gaseous effluent of the phthalic anhydride process contains up to 0.5% carbon monoxide and smaller quantities of organic acids (phthalic acid, maleic acid, benzoic acid, etc) derived from the feedstock and its partial oxidation products. Catalytic oxidation of the carbon monoxide and the organic compounds in this exhaust stream has been successfully demonstrated.

Vinyl monomer manufacturing. Process vent gases containing small quantities of halogenated hydrocarbons and substantial quantities of nonhalogenated hydrocarbons have been successfully reduced to comply with air-pollution agency objectives in large-scale laboratory pilot catalytic fume abaters, with satisfactory long-term catalyst performance.

Heat Processing and Miscellaneous Operations

The sources of emissions requiring control are numerous and of wide variety, eg, the EPA states that 44% of the 1975 estimated 19,000,000 metric tons of emissions from stationary sources was contributed by evaporation of organic solvents, the principal sources of these emissions being auto and light-truck manufacturing, can manufacturing, coil coating, fabric coating, paper coating, appliance manufacturing, tire manufacturing, printing, pressure-sensitive and magnetic-tape manufacturing, dry cleaning-wire coating, and textile manufacturing. The list does not include other combustible gaseous emissions such as carbon monoxide, organic sulfides, some organic halides, organic resins, and organic acids that remain in the off-gases from many chemical, papermaking, and plywood manufacturing processes. Catalytic afterburners and fume abaters, both exothermic devices, are used to control solvent evaporation pollution.

The selective catalytic reduction of NO$_x$ with ammonia has been used on exhaust gases of gas-fired power equipment such as power boilers and peak-shaving turbines, as well as excess air-fired stationary internal-combustion engines.

The catalytic reduction of NO$_x$ from the exhausts of internal-combustion engines operating at or near stoichiometric mixtures of fuel/air has become an increasingly popular pollution-control method during the past few years.

JAMES H. CONKLIN
DONALD M. SOWARDS
JOHN H. KROEHLING
E.I. du Pont de Nemours & Co., Inc.

Control of Volatile Organic Emissions From Existing Stationary Sources, EPA, Research Triangle Park, N.C., 1977.

R.W. Rolke and co-workers, *Afterburners Systems Study,* EPA Contract EHS-D-71-1, Shell Development Company, Emeryville, Calif., 1972.

EXPECTORANTS, ANTITUSSIVES, AND RELATED AGENTS

Expectorants

Expectorants enhance the production of respiratory-tract fluid and thus facilitate the mobilization and discharge of bronchial secretions. Those that increase respiratory-tract secretion by a direct effect on the bronchial secretory cells are sometimes called stimulant expectorants, whereas those that act by gastric reflex stimulation are sometimes called sedative expectorants.

Guaiacols. Creosote, which is obtained from the pyrolysis of beechwood, and its active principles guaiacol (1) and creosol (2), have long been used in expectorant mixtures. Creosol is obtained by the Clemmensen reduction of vanillin (qv) whereas guaiacol can be prepared by a number of methods including the mercuric oxide oxidation of lignin (qv), the zinc chloride reduction of acetovanillone, and the diazotization and hydrolysis of *o*-anisidine. Because of its bitter taste and water insolubility, guaiacol has been chemically modified to improve is properties. The synthetic guaiacol derivative that has received by far the greatest acceptance is guaifenesin (3) formerly known as glyceryl guaiacolate. This compound is widely used today, both in single-entity cough preparations, and in combination with other active ingredients.

(1) R = H, guaiacol
(2) R = CH$_2$, creosol

(3) guaifenesin

Volatile oils. The use of volatile oils, sometimes called essential oils, has been traced back 2000 yr to Pliny who used turpentine internally to relieve coughing. Evidence suggests that the compounds act by direct stimulation of the bronchial secretory cells. In general, the volatile oils are isolated from plant sources and are terpenoid (qv) in structure. Turpentine is one of the most familiar oils used in expectorant formulations. It is prepared by the process of rectification which consists of steam distillation of crude turpentine from sodium hydroxide. This removes the acidic and resinous components. The rectified oil contains primarily α-pinene (4) along with small amounts of β-pinene. Terpin hydrate (5), one of the most well-known expectorants, is isolated from crude pine rosin left after the distillation of volatile terpene hydrocarbons and alcohols. Menthol, commercially the most important terpene alcohol, is obtained by crystallization from peppermint oil, or prepared synthetically in racemic form by the hydrogenation of thymol. Menthol's local anesthetic activity may contribute to its antitussive properties.

(4) α-pinene

(5) terpin hydrate

Iodides and other inorganic compounds. Iodides and other compounds such as potassium iodide, hydriodic acid, antimony potassium tartrate, and ammonium chloride, are thought to act by gastric reflex stimulation. Of these, only the iodides have been studied to any appreciable extent. A number of toxic reactions have been associated with both antimony potassium tartrate and inorganic iodides.

Mucolytics

These are agents that reduce the viscosity of tenacious and purulent mucus, thus facilitating removal. Steam, sometimes in conjunction with surfactants or volatile oils, has long been used to decrease viscosity by physical hydration. Recently, however, agents that chemically depolymerize certain components of mucus have become available. Trypsin and other proteolytic enzymes have shown good clinical activity, by virtue of their ability to cleave glycoproteins. Pancreatic dornase (deoxyribonuclease), which depolymerizes DNA found in purulent mucus, has also shown clinical utility (see Enzymes). *N*-acetyl-L-cysteine (6) is an important synthetic compound that reduces the viscosity of mucus by cleaving disulfide bonds.

(6) *N*-acetyl-L-cysteine

Antitussives

Through the centuries, cough remedies have remained a popular item and are found today in most medicine cabinets. Over 300 prescriptions and over-the-counter prescriptions are available.

Centrally active antitussives. Centrally active antitussives exert their effect by depressing the medullary cough center, thus raising the threshold for sensory cough impulses. The most well-known compounds in this category are narcotics, eg, codeine (7), one of the most widely used and effective compounds. The majority of the codeine sold today is prepared by methylating the phenolic hydroxy group of morphine. When prescribed for cough, the usual oral dose is 10–20 mg, three-to-four times daily. At these doses, there are few adverse effects, such as nausea, anorexia, and constipation. Most narcotics diminish ciliary activity and produce a drying effect on the respiratory tract mucosa.

Nonnarcotic antitussives, such as dextromethorphan (8), by far the most important, are similar to codeine in terms of potency and mechanism of action, eg, it is a direct depressant of the cough center. It is unique in that even though it is structurally related to codeine it is not addictive, does not depress ciliary activity, secretion of respiratory tract fluid, or respiration.

(7) R = CH$_3$, codeine

(8) dextromethorphan

Levopropoxyphene, the optical antipode of the dextrorotary analgetic propoxyphene, is an antitussive without analgetic activity (see Analgesics). It is widely used in the form of the 2-naphthalenesulfonate salt which has a less unpleasant taste than the hydrochloride salt.

Peripherally active antitussives. In general, these antitussives act by raising the threshold for cough at the sensory nerve ending (eg, local anesthetics), or by facilitating bronchial drainage, mucociliary clearance, or both (eg, expectorants and mucolytics).

Health and Safety

Safety and efficacy data on a number of antitussives and expectorants have been reviewed by the FDA's Advisory Review Panel on Over-The-Counter (OTC) Cold, Cough, Allergy, Bronchodilator, and Antiasthmatic Products in the *Federal Register* 41(176), 38312 (1976).

WILLIAM J. WELSTEAD, JR.
A.H. Robins Company

H. Salem and D.M. Aviado, eds., *International Encyclopedia of Pharmacology and Therapeutics,* Vols. 1–3, Sect. 27, Pergamon Press, Ltd., Oxford, UK, 1970.

EXPLOSIVELY CLAD METALS.   See Metallic coatings.

EXPLOSIVES AND PROPELLANTS

EXPLOSIVES

Propellants and explosives are chemical compounds or their mixtures that rapidly produce large volumes of hot gases when properly initiated. Propellants burn at relatively low rates measured in centimeters per second, whereas explosives detonate at rates of kilometers per second.

The energy liberated by explosives and propellants depends on the thermochemical properties of the reactants. As a rough rule of thumb, these materials yield about 1000 cm^3 of gas and 4.2 kJ (1000 cal)/g. A comparison of the characteristics associated with propellant burning, explosive detonation, and the performance of conventional fuels is shown in Table 1. The greatest difference is the rate at which energy is evolved.

General Characteristics

Exothermic oxidation–reduction reactions provide the energy released in both propellant burning and explosive detonation. The reactions are either internal oxidation–reductions, as in the decomposition of nitroglycerin and pentaerythritol tetranitrate, or reactions between discrete oxidizers and fuels in heterogeneous mixtures such as RDX (see below) and aluminum or ammonium perchlorate and a hydrocarbon fuel.

An activation energy of 125–250 kJ/mol (30–60 kcal/mol) is usually required to initiate the reaction, at which point the heat evolved is sufficient to cause the reaction to continue and become self-sustaining. Most explosives and propellants are organic compounds or mixtures of compounds that contain carbon, hydrogen, oxygen, and nitrogen. Metallic fuels such as aluminum may be added to increase the heat of reaction. Industrial dynamites have traditionally used nitroglycerin, nitrocellulose, and inorganic salts as sources of oxygen but are now increasingly formulated with ammonium nitrate as the oxygen source. Composite propellants commonly use ammonium perchlorate to supply the oxygen required. The most common gaseous products of the oxidation–reduction reactions are hydrogen, water, carbon monoxide, carbon dioxide, and nitrogen.

Table 1.   Characteristics of Burning and Detonation

| Characteristics | Burning | | Explosive detonation |
	Fuel	Propellant	
typical material	coal–air	propellants	explosives
linear reaction rate, m/s	10^{-6}	10^{-2}	$2-9 \times 10^{-3}$
type of reactions	oxidation–reduction	oxidation–reduction	oxidation–reduction
time for reaction completion, s	10^{-1}	10^{-3}	10^{-6}
factor-controlling reaction rate	heat transfer	heat transfer	shock transfer
energy output, J/g[a]	10^4	10^3	10^3
power output, W/cm^2	10	10^3	10^{9}[b]
most common initiation mode	heat	hot particles and gases	high temperature–high pressure shock waves
pressures developed, MPa[c]	0.07–0.7	$0.7-7 \times 10^2$	$7 \times 10^3 - 7 \times 10^4$
uses	source of heat and elecricity	controlled gas pressure, guns, and rockets	brisance, blast, munitions, civil engineering

[a] To convert J to cal, divide by 4.184.
[b] This may be compared with the total United States electric-generating capacity of about 30×10^6 kW.
[c] To convert MPa to psi, multiply by 145.

Explosive detonation.   Detonations proceed as a result of a reaction front moving in a direction normal to the surface of the explosive. However, detonation is a hydrodynamic phenomenon that differs in a fundamental sense from burning. Upon initiation, burning first occurs at an increasing rate for a period of time up to several microseconds. A high pressure shock wave is formed which passes through the explosive at high velocity. As it does so, it causes exothermal decomposition of the explosive. The continued passage of the wave is supported by transfer of energy from the spent explosive to the unreacted explosive by shock compression. The rate of reaction depends on the rapid rate of transmission of a shock wave rather than on the relatively slow rate of heat transfer associated with propellant burning. The detonation rate is stable and constant and is primarily governed by the physical and chemical properties of the explosive, its geometry, degree of confinement, and particularly its density ρ, with which the detonation rate varies in a linear fashion for most explosives:

$$D_i = D_o + M(\rho_i - \rho_o)$$

where D_i = linear detonation rate dx/dt at density ρ_i, D_o = linear detonation rate at density ρ_o, and M = a constant characteristic of the explosive composition. Typical values of D_o at $\rho_o = 1.0$ g/cm^3 are about 5000–6000 m/s. Values of M are about 3000–4000 m/s.

Primary Explosives

Explosives are commonly categorized as primary, secondary or high explosives and propellants in order of decreasing sensitivity to energy output.

Primary, or initiator, explosives are the most sensitive to heat, friction, impact, shock, and electrostatic energy. They have been studied in considerable detail because of their almost unique capability, even when present in small quantities, to transform rapidly a low energy stimulus into a high intensity shock wave. Most recent evidence indicates that there is a minimum thickness or run-up distance before steady state detonation occurs and a gradual transition from an unreacted shock to a stable detonation.

Primary explosives are used to initiate the next element in a series which consists of explosives of increasing mass and decreasing sensitivity. They are arranged in sequence to amplify the input stimulus to an output level of sufficient intensity to maximize the probability of initiating the main charge. Overall energy intensification is about $10^7:1$. Primary explosives are used in military detonators, commercial blasting caps, and in stab and percussion primers.

Mercury fulminate, mercuric cyanate, [$Hg(ONC)_2$], is a gray-white powder, which explodes at 160°C, has a detonation rate 5.4 km/s at 4.2 g/cm^3; is soluble to the extent of 0.1 g/100 H_2O at 20°C, and is no longer used as a primary explosive in the United States. It is prepared by the Chandelen process, essentially a large-scale laboratory process.

Lead azide ($Pb(N_3)_2$), is the primary explosive used in military detonators in the United States. Detonation rate 5.1 km/s at 4.8 g/cm^3. The azides are among the very few useful explosive compounds that do not contain oxygen. Lead azide is very stable at ambient and elevated temperatures, and has good flow characteristics. Since lead azide is less sensitive to ignition than mercury fulminate, a more readily ignitable material such as lead styphnate or NOL 130 is often used as cover charge to ensure initiation. Lead azide is made in small batches buffered by the reaction solutions of **lead nitrate** or lead acetate with sodium azide:

$$Pb(NO_3)_2 + 2\,NaN_3 \rightarrow 2\,NaNO_3 + Pb(N_3)_2$$

All phases of the manufacturing process are conducted by remote control in stainless-steel vessels using either distilled or demineralized water and filtered solutions. The overall precipitation time is about 60 min.

Silver azide.   Silver azide, AgN_3, has received attention as a potential replacement for lead azide because it may be used in smaller quantities as an initiator and, therefore, offers the possibility of miniaturization of fuse components. Silver azide requires somewhat less energy for initiation than lead azide and fires with a shorter time delay; detonation rate 6.8 km/s at 5.1 g/cm³. It is less apt to hydrolyze and is more sensitive to heat than lead azide. It is incompatible with sulfur compounds, with tetrazene, and with some metals, including copper. Silver azide is made in the same manner as lead azide, except that silver nitrate is used in the reaction with sodium azide.

Lead styphnate. Lead styphnate, lead 2,4,6-trinitroresorcinate, is one of a number of compounds used in priming compositions to start the ignition-to-detonation process in the explosive sequence. Detonation rate 5.2 km/s at 2.9 g/cm³. Lead styphnate monohydrate is precipitated as the basic salt from a mixture of solutions of magnesium styphnate and lead acetate followed by conversion to the normal form by acidification with dilute nitric acid.

Diazodinitrophenol. DDNP (**1**) is an orange-yellow compound made by diazotizing picramic acid, (**2**), with sodium nitrite and hydrochloric acid, washing the product with ice water, and recrystallizing it from hot acetone. Mp 157°C and detonation rate 6.9 km/s at 1.60 g/cm³; it is somewhat less stable than lead azide. Since it is about equivalent to TNT in brisance, it is more effective for some purposes than lead azide and is used as an initiator in commercial blasting caps.

(**2**) picramic acid   (**1**) DDNP

Tetrazene. Tetrazene (**3**) is a pale-yellow crystalline explosive (explodes at 140–160°C) made by adding sodium nitrite to a solution of 1-aminoguanidine hydrogen carbonate in dilute acetic acid at 30°C. It is used as a component in priming compositions.

(**3**) tetrazene

Secondary Explosives

Aliphatic nitrate esters. Aliphatic nitrate esters are among the most powerful explosives available. They are generally less stable than aromatic nitro compounds or nitramines because they tend to hydrolyze autocatalytically to form nitric and nitrous acids which further accelerate decomposition.

Nitroglycerin. Nitroglycerin, NG, glyceryl trinitrate, is primarily used as an explosive in dynamites and as a plasticizer for nitrocellulose in double- and multibase propellants. It is very sensitive to shock, impact, and friction and is employed only when desensitized with other liquids or absorbent solids or when compounded with nitrocellulose. It is readily soluble in many organic solvents, acting as a solvent for many explosive ingredients. Unconfined nitroglycerin burns without exploding if present in thin layers and in small quantities but detonates if confined. Aerated nitroglycerin and other liquid explosives containing microbubbles are especially sensitive to shock.

Manufacture. Nitroglycerin was traditionally made by the batch process. However, the hazard of handling large quantities led to the development and wide use of continuous processes, such as the Biazzi continuous process and the Nitro Nobel Injector process which uses an injector to mix the glycerol or other polyol with precooled nitration acid. In general, the methods used to make nitroglycerin are applicable to the manufacture of other liquid aliphatic polyhydroxy alcohol nitrates. Nitroglycerin is made from very pure glycerol (qv) to ensure stability of the product. Mixed acid (90% nitric acid and 25–30% oleum) is used in both batch and continuous processes.

Other glycol nitrates, eg ethylene glycol dinitrate (nitroglycol) and diethylene glycol dinitrate, have been used as explosive plasticizers for nitrocellulose. They are made like nitroglycerin with mixed-acid nitration.

Nitrocellulose. NC, cellulose nitrate, derives its exceptional properties from the polymeric and fibrous structure of the cellulose from which it is made. Nitrocellulose provides mechanical strength as well as readily available energy to gun and rocket propellants. It is manufactured by nitration of cellulose with mixed acid.

Properties. D 1.66 g/cm³; detonation velocity = 7.3 km/s at 1.20 g/cm³; hygroscopicity, heat of formation, heat of combustion and explosion point vary with nitrogen content. Dry nitrocellulose burns rapidly and furiously. It may detonate if present in large quantities or if confined and is dangerous to handle in the dry state since it is sensitive to friction, static electricity, impact and heat. Nitrocellulose is always shipped wet with water or alcohol. The higher the nitrogen content, the more sensitive it tends to be. Even nitrocellulose with 40% water detonates if confined and sufficiently activated.

Manufacture. The batch nitration processes have included the pot process, the centrifugal process, the Thompson displacement process, and the Mechanical Dipper process. In the batch process, the raw materials proceed by gravity through the processing operations. The nitration of cellulose occurs very rapidly at first and the nitrocellulose is separated from spent nitrating acid. Semicontinuous nitration processes have been developed for military and industrial grades. Semicontinuous nitration uses a multiple-cascade system and a continuous wringing operation. The controlling factors in the nitration process are the rates of diffusion of the acid into the fibers, the composition of the mixed acid, and the temperature.

Nitramines

The four most important nitramines are: cyclotrimethylenetrinitramine (RDX) (**4**); cyclotetramethylenetetranitramine (HMX) (**5**); nitroguanidine (NQ) (**6**); and 2,4,6-trinitrophenylmethylnitramine (tetryl) (**7**). Tetryl has been increasingly replaced by RDX; both RDX and HMX are used as high energy explosives and may also be incorporated in high performance rocket propellants. Nitroguanidine is employed almost exclusively in gun propellants.

(**4**) RDX   (**5**) HMX

(**6**) nitroguanidine   (**7**) Tetryl

RDX and HMX. The properties of RDX (**8**) and HMX (**5**) are quite similar: HMX has a higher density and a higher detonation rate, yields more energy per unit volume and has a higher melting point and higher explosion and cook-off temperatures.

Both are white, stable crystalline solids, less toxic than TNT and capable of being handled with no physiological effect if appropriate precautions are taken to assure cleanliness of operations. Both RDX and HMX detonate to form mostly gaseous, low molecular weight products with little intermediate formation of solids. The calculated mole fractions of detonation products of RDX is 3.00 H_2O, 3.00 N_2, 1.49 CO_2, and 0.02 CO. RDX has been stored for as long as ten months at 85°C without perceptible deterioration. At present, HMX is the highest-energy solid explosive produced on a large scale, primarily for military use. It exists in four polymorphic forms of which the beta form is the least sensitive and most stable and the type required for military use. The mole fractions of products of detonation of HMX in a calorimetric bomb are 3.68 N_2, 3.18 H_2, 1.92 CO_2, 1.06 CO, 0.97 C, 0.395 NH_3, and 0.30 H_2.

Both RDX and HMX are substantially desensitized by mixing with TNT to form cyclotols (with RDX) and octols (with HMX) or by coating with waxes, synthetic polymers, and elastomeric binders.

Manufacture.   The two most common processes use hexamethylenetetramine (hexamine) (9) as starting material. The Woolwich or direct nitrolysis process used in the United Kingdom, proceeds according to:

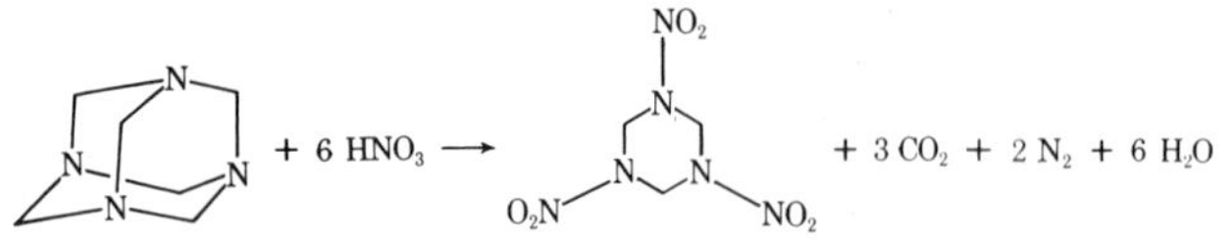

The Bachmann process, now used exclusively in the U.S. is a simplification of a series of complex reactions which may be summarized as follows:

$$C_6H_{12}N_4 + 4\ HNO_3 + 2\ NH_4NO_3 + 6\ (CH_3CO)_2O$$
$$\rightarrow 2\ RDX + 12\ CH_3COOH$$

In the Bachmann process an 80–84% yield is obtained, ca 5–10% of which is cyclotetramethylenetetranitramine (HMX). The Woolwich process gives a 70–75% yield containing only a trace of HMX.

A continuous process for medium-scale production of RDX based on the Woolwich process has also been developed. A modification of the Bachmann process used to make RDX with the same starting materials and in similar equipment is employed for the manufacture of HMX.

Nitroguanidine.   NQ (6) has been used to some extent as an industrial explosive but not as a military explosive because of its relatively low energy content and difficulty of initiation. It is stable and nonhygroscopic. Produced in the alpha crystalline form with a bulk density of about 0.2 g/cm³, the crystals are needlelike and often hollow and are about five μm in diameter and 15 μm long.

Manufacture.   Nitroguanidine may be made by several methods. In the typical Welland process, calcium cyanamide is first made from calcium carbonate and converted to cyanamide by acidification of a water slurry. The dimer dicyandiamide, which is then formed by filtration and evaporation of the filtrate, is fused with ammonium nitrate to form guanidine nitrate. Dehydration with 96% sulfuric acid gives nitroguanidine. In these processes, guanidine nitrate is the intermediate which is then dehydrated with sulfuric acid. In the Marquerol and Lorriette process, nitroguanidine is obtained directly in about 90% yield from dicyandiamide by reaction with sulfuric acid to form guanidine sulfate, followed by direct nitration with nitric acid.

Tetryl.   Tetryl (7) has been used in pressed form mostly as a booster explosive and as a base charge in detonators and blasting caps because of its sensitivity to initiation by primary explosives and its relatively high energy content. It is highly stable, losing virtually no weight on prolonged storage at 80°C. The calculated molar fraction products of detonation at a density of 1.70 g/cm³ are 4.16 $C_{(sol)}$, 2.66 CO_2, 2.50 H_2O, 2.50 N_2, and 0.17 CO. Its specific heat is 0.92 J/(g·K) [0.22 cal/(g·°C)] and thermal conductivity 0.25 W/(m·K) [6 × 10⁻⁴ cal/(cm·s·°C)]. It is manufactured by a batch process in which dimethylanisidine dissolved in concentrated sulfuric acid is nitrated with mixed acid.

Nitroaromatics

The commonly used nitroaromatic explosives contain three NO_2 groups, generally in the 1, 3, and 5 positions. Aromatics are most often nitrated to the trinitro stage with mixed acid. Further nitration is difficult, and aromatics with four or more nitro groups attached to the ring tend to be relatively unstable. The most extensively used explosive is trinitrotoluene (TNT); however, hexanitrostilbene (HNS), hexanitroazobenzene (HNAB), and di- and triaminotrinitrobenzene (DATB and TATB) have found increasing application because of their low sensitivity to impact, shock, and friction and their excellent stability at elevated temperatures. Ammonium picrate (AP) has been used in armor-piercing gun projectiles because of its insensitivity to impact and shock.

Trinitrotoluene.   2,4,6-trinitrotoluene (TNT) is very stable and may be stored indefinitely at temperate conditions without deterioration. TNT is nonhygroscopic and relatively insensitive to impact, friction, shock and electrostatic energy. It has been fired in high acceleration gun projectiles with reported premature rates of less than one in a million

(10⁶). Bombs and projectiles are filled with steam-melted TNT by the casting process. Melted TNT also serves as the liquid carrier for RDX, HMX, aluminum, ammonium nitrate, and other high melting ingredients to form a wide range of castable slurries.

Manufacture.   Trinitrotoluene is made by batch or continuous process. Toluene is nitrated in a three-stage operation by using increasing temperatures and mixed-acid concentrations to successively introduce nitro groups to form mononitrotoluene (MNT), dinitrotoluene (DNT), and trinitrotoluene. The steps used in each stage are similar and include acid mixing, addition of the oil, digesting the reaction to completion, cooling and settling the mix, and separating the oil from the acid. The TNT is purified by treatment with sodium sulfite solution.

Continuous processes used in many countries are based on nitration of toluene with an organic phase flowing countercurrent to the acid phase. A series of eight nitrator–separator reactors are used to conduct the nitration in six stages (see Nitration).

Picric acid and ammonium picrate.   Picric acid (2,4,6-trinitrophenol) (PA) is of historic interest as the first modern high explosive to be used extensively as a burster in gun projectiles. It was first obtained by nitration of indigo, and used primarily as a fast dye for silk and wool. Picric acid has an energy content somewhat greater than that of TNT and a higher detonation rate. A large disadvantage is its tendency to form sensitive salts with calcium lead, zinc, and other metals. Picric acid is no longer used as a military explosive. It can be made by gradually adding a mixture of phenol and sulfuric acid at 90–100°C to a nitration acid containing a small excess of nitric acid. The picric acid crystals are separated by centrifuging, washed, and dried.

Ammonium picrate (explosive D) (AP) is now used only where a high explosive is required that is particularly insensitive to shock. It has been used in pressed form primarily as a burster in naval projectiles for armor penetration.

Safety and Environmental Considerations

Standard safety procedures include attention to intraline distances to minimize explosion propagation, cleanliness, nonsparking equipment and explosion-proof motors, and location in selected areas away from heavy population. Many regulations cover the classification, shipping, and handling of explosives. Rigorous Federal standards exist for permissible quantities of contaminants added to the environment.

Uses

Military: TNT, RDX or HMX, nitrocellulose, and nitroglycerin are the single-component explosives most commonly used for military compositions. Nitrocellulose and nitroglycerin are used exclusively to make propellants (see under Propellants). Other binary mixtures include the octols (HMX + TNT), cyclotols (RDX + TNT), pentolites (PETN + TNT), tetrytols (tetryl + TNT), amatols (ammonium nitrate + TNT), and picratols (ammonium picrate + TNT).

Industrial: the quantity of industrial explosives used in the United States continues to increase, but the type used has changed radically in the last forty years. The use of black powder and liquid oxygen disappeared completely and dynamites declined significantly, whereas the consumption of dry and water-based ammonium nitrate consumption increased greatly.

VICTOR LINDNER
U.S. Army Armament Research
and Development Command (ARRADCOM)

T. Urbanski, *Chemistry and Technology of Explosives*, Vols. 1–3, Macmillan, New York, 1964–1967.

B.T. Federoff and O.E. Sheffield, eds., *The Encyclopedia of Explosives and Related Items*, PATR 2700 CO. 1–10. Picatinny Arsenal, Dover, N.J.

PROPELLANTS

Propellants are mixtures of chemical compounds that produce large volumes of gas at controlled, predetermined rates. Their chief applications are in launching projectiles from guns, rockets, and for missile systems. Propellant-actuated devices are used to drive turbines, move pistons, operate rocket vanes, start aircraft engines, eject pilots, jettison stores from jet aircraft, pump fluids, shear bolts and wire, and act as sources of heat in special devices. They are employed in guns in the form of dense grains or sheets of plasticized nitrocellulose. Those that contain only plasticized nitrocellulose are known as single-base propellants; double-base propellants also contain a liquid explosive plasticizer such as nitroglycerin. Multi- or triple-base propellants incorporate a crystalline explosive such as nitroguanidine in the double-base formulation. Double- and triple-based nitrocellulose propellants are used in rockets as well as guns.

Typical components of nitrocellulose propellants are listed in Table 1.

Polymer-based rocket propellants, referred to as composite propellants, contain a cross-linked polymer which acts as a viscoelastic matrix for holding a crystalline inorganic oxidizer such as ammonium perchlorate and for providing mechanical strength. Many other substances may be added including metallic fuels, plasticizers, extenders, and catalysts. Polymer-based composite propellants are too erosive to be used in guns because of the residues formed after repeated firings. Typical components of composite propellants are listed in Table 2.

Selection

In addition to energy and burning-rate considerations, a propellant must meet other criteria including mechanical characteristics, stability, sensitivity, cost of manufacture, and uniformity of performance.

The Burning Process

The mass rate of propellant burning at a given pressure and temperature depends on the amount of heat evolved during decomposition and the amount of heat transferred to the burning surfaces of the propellant from the hot gases above it. It is also influenced by the tangential velocity of the propellant gases and the radiation from the surrounding gases. Propellants burn in parallel layers so that the surface recedes in all directions normal to the original surface (Piobert's law). The geometry of the grain on completion of burning is similar to its geometry at the start. Propellant burning at high gun pressures proceeds more smoothly and is less subject to erratic behavior than burning at very low pressures because the conditions are appropriate for maximum energy transfer in minimum time. The burning rate at gun pressures usually varies somewhat less than the first power of the pressure. It changes more slowly at rocket pressures of 3.45–10.34 MPa (500–1500 psi), often to less than the square root of the pressure.

Table 1. Typical Components of Nitrocellulose Propellants and Their Function

Component	Application
nitrocellulose	energetic polymeric binder
nitroglycerin, metriol trinitrate, diethylene glycol dinitrate,	plasticizers:
triethylene glycol dinitrate, dinitrotoluene	energetic
dimethyl, diethyl or dibutyl phthalates, triacetin	plasticizers: fuels
diphenylamine, diethyl centralite, 2-nitrodiphenylamine	stabilizers
organic and inorganic salts of lead; eg, lead stannate, lead stearate, lead salicylate	ballistic modifiers
carbon black	opacifiers
lead stearate, graphite, wax	lubricants
potassium sulfate, potassium nitrate, cryolite (potassium aluminum fluoride)	flash reducers
ammonium perchlorate, ammonium nitrate	oxidizers inorganic
RDX, HMX, nitroguanidine and other nitramines	organic
aluminum	metallic fuels cross-linking catalysts
lead carbonate	defouling agents
tin	

Table 2. Typical Components of Composite Rocket Propellants

	Characteristics
Binders	
polysulfides	reactive group, mercaptyl (—SH), is cured by oxidation reactions; low solids loading capacity and relatively low performance; now mostly replaced by other binders
polyurethanes polyethers polyesters	reactive group, hydroxyl (—OH), is cured with isocyanates; intermediate solids loading capacity and performance
polybutadienes copolymer, butadiene and acrylic acid	reactive group, carboxyl (—COOH) or hydroxyl (—OH), is cured with difunctional epoxides or aziridines; intermediate solids loading capacity and better performance than polyurethanes; less than adequate cure stability and mechanical characteristics
terpolymers of butadiene, acrylic acid and acrylonitrile	superior physical properties and storage stability
carboxy-terminated polybutadiene	cured with difunctional epoxides or aziridines; have very good solids loading capacity, high performance and good physical properties
hydroxy-terminated polybutadiene	cured with diisocyanates; have very good solids loading and performance characteristics and good physical properties and storage stability
Oxidizers	
ammonium perchlorate	most commonly used oxidizer; it has a high density, permits a range of burning rates, but produces smoke in cold or humid atmosphere
ammonium nitrate	used in special cases only, it is hygroscopic and undergoes phase changes, has a low burning rate and forms smokeless combustion products
high energy explosives (RDX–HMX)	have high energy and density; produce smokeless products; have a limited range of low burning rates
Fuels	
aluminum	Al most commonly used; has a high density; produces an increase in specific impulse and smoky and erosive products of combustion
metal hydrides	provide very high impulse, but generally inadequate stability, give smoky products, and have a low density
Ballistic modifiers	
metal oxides	iron oxide most commonly used
ferrocene derivatives	permit a significant increase in burning rate
other	coolants for low burning rate and various special types of ballistic modifiers
Modifiers for physical characteristics	
plasticizers	improves physical properties at low temperatures, and processability; may vaporize or migrate; can increase energy if nitrated
bonding agents	improve adhesion of binder to solids

The operating pressure of the system has the predominant effect on the burning rate of propellants. Photographic evidence shows that increasing the pressure decreases the distance between the flame zone in the gas phase and the propellant surface. The rate of heat transfer to the propellant surface increases accordingly.

Burning-rate equations. Burning-rate equations have been developed to describe the performance of solid propellants based on the assumption that all the exposed propellant surfaces are ignited simultaneously and burn at the same linear rate. For example:

$$r = a + bP \tag{1}$$

$$r = cP^n \tag{2}$$

where r is the linear burning rate, P is the pressure, n is the pressure exponent, and a, b, and c are constants that vary with temperature.

Equation 1 is often used for propellants burning at high gun pressures, whereas equation 2 is associated with low pressure rocket systems.

Mechanism of burning of nitrocellulose-based propellants. Much of the information available on the burning process of nitrocellulose propellants is based on the decomposition of nitrate esters and the reaction of oxides of nitrogen with the products of decomposition. The three reaction zones identified are (a) the surface of the solid is a reaction zone ca 0.01 cm in thickness and ranging up to ca 600 K. Molecular bond breakage occurs in this zone, primarily the O—N bond in cellulose nitrate–nitroglycerin-type propellants; (b) above the foam zone is a nonluminous gaseous reaction zone, several thousandths of a centimeter thick. This zone attains a temperature of ca 1500 K as a result of partial reaction among the material ejected from the surface. Aldehydes and alcohols are converted to smaller molecules nitrogen, water, carbon monoxide, carbon dioxide, and nitric oxide. About half the total heat evolved by the propellant is liberated in the fizz zone which is prominent at low pressures and disappears at high pressure; (c) the reaction continues to completion and thermodynamic equilibrium is established in the flame zone. This zone, ca 0.001 cm thick, defines the flame temperatures of the propellant which may range from ca 1500 K for cool propellants to 3500 K for very hot ones. The nitric oxide reacts with the reaction products formed to produce carbon monoxide, carbon dioxide, hydrogen, water, nitrogen, and a small percentage of other molecules.

Mechanism of burning of composite propellants. A number of analytical models have been developed to permit estimations of the burning rates of composite propellants as a function of the factors that affect them most: caloric value, propellant temperature, chamber pressure, fuel–oxidant ratio, particle size distribution of the oxidizer, the presence of decomposition catalysts, and the incorporation of aluminum in powder, wire, or staple form.

The Beckstead-Derr-Price model (Figure 1) considers both the gas-phase and condensed-phase reactions. It assumes heat release from the condensed phase, an oxidizer flame, a primary diffusion flame between the fuel and oxidizer decomposition products, and a final diffusion flame between the fuel decomposition products and the products of the oxidizer flame. If aluminum is present, additional heat is subsequently produced at a comparatively large distance from the surface. Only small aluminum particles ignite and burn close enough to the surface to influence the propellant burn rate. The temperature of the surface is ca 500 to 1000°C compared with ca 300°C for double-base propellants.

Manufacture

Nitrocellulose propellants. Nitrocellulose propellants are made with or without incorporation of a solvent as plasticizer by five processes: solvent extrusion, solvent emulsion, solventless extrusion, solventless rolling, and casting. All require colloiding the nitrocellulose to eliminate its fibrous structure and cause it to burn predictably in parallel layers. Mechanical working of the ingredients contribute to plasticization and uniformity of composition. Although the manufacture of extruded nitrocellulose propellants has been traditionally a batch process, continuous processes are being developed for single-, double-, and multi-base propellants.

Solvent extrusion batch process. Almost all standard gun propellants and small-webbed rocket propellant grains are made by the solvent extrusion process. Purified and blended, nitrocellulose of the required nitrogen content and wet with ca 30% water is received from the nitrocellulose plant. A charge is transferred to a double-acting hydraulic dehydration press and compressed at low pressure to remove some of the water. The remainder is removed by pumping 95% ethyl alcohol through the press. The final blocks, containing ca 18% alcohol, are broken up and screened to remove lumps or oversized particles. Mixing is usually conducted in a water-jacketed bladed mixer and consists essentially of solid–solid and solid–liquid incorporation, the solution of stabilizers and ballistic modifiers, and the absorption of solvents and liquid plasticizers by the nitrocellulose. After mixing, the dough-like composition (single-, double-, or multibased) is transferred to a vertical block press where it is consolidated after first purging the press chamber with an inert atmosphere of carbon dioxide or nitrogen. The blocks of propellant are transferred to a vertical or horizontal graining press and extruded at relatively low pressures of 10.3–17.2 MPa (1500–2500 psi) through dies designed to produce the required dimensions, allowances being made for shrinkage occurring during drying. To ensure safety during pressing, explosive mixtures of air and solvent have to be carefully excluded during the ramming operation. The strands of perforated propellant are fed to a mechanical cutter and sliced to specified lengths.

Continuous solvent extrusion process (for single-base propellants). Continuous processes for making solvent propellant for cannon offer substantial reductions in labor, hazard, and pollutants. The main features in which the continuous process differs from the operation are thermal dehydration, water-wet nitrocellulose vacuum dried on a continuous vacuum belt filter followed by hot-air transfusion (80°C) to reduce the moisture to less than 2%; compounding, in which the alcohol-wet nitrocellulose is transferred from a surge feeder to a compounder by a continuous weigh-belt along with the other ingredients of the composition, which are also weighed and added automatically. The loosely mixed paste is fed by a conveyor to a heavy duty reciprocating screw mixer that is temperature controlled and specially designed to thoroughly mix and work the paste by forcing it past pins in the mixer barrel and out through a die plate. As the paste is extruded, it is cut into small pellets which are fed continuously to water-jacketed screw extruders and forced through multiple dies. The strands are cooled as they are extruded to facilitate cutting by an adjustable roll-type cutter. After the cut grains are screened to remove clusters and odd sizes, the solvent is removed in the solvent recovery–water dry system where the propellant is treated first with hot inert gas and then hot water, then air dried until the moisture content is reduced from 12 to less than 0.8%.

Solventless extrusion process. The solventless process for making double-base propellant has been used in the United States primarily for the manufacture of rocket-propellant grains with web thickness ca 1.35–2.0 cm and for thin-sheet mortar (M8) propellant. It is also used in Europe for making cannon propellant. In the water-slurry process for making solventless propellants, explosive and nonexplosive liquid plasticizers and water-insoluble constituents are incorporated into nitrocellulose suspended as a slurry in a large volume of hot water. After removing the excess water, the resulting wet mass is partially dried and passed through heated rolls to thoroughly mix the colloidal composition into sheets of homogeneous propellant and remove any remaining water. These sheets may then be cut and rolled into scrolls for insertion in the press and extruded as propellant grains or cut up and used as a sheet or flake propellant.

Ball Powder

Ball powder is typically used in small-caliber weapons. It consists of spherically shaped or flattened ellipsoidal grains, ca 0.04–0.09 cm dia. The process permits the recovery and use of nitrocellulose from obsolete granular propellant.

Batch process. Water-wet fresh or extracted nitrocellulose is transferred as a slurry to a graining still. Calcium carbonate is added to neutralize any free acid released by the dissolved nitrocellulose. The

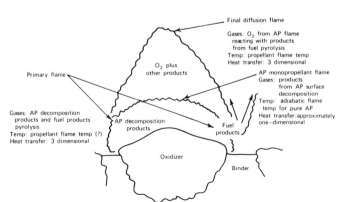

Figure 1. The postulated flame structure for an AP composite propellant, showing the primary flame followed by the final diffusion flame. AP = ammonium perchlorate.

required amount of ethyl acetate is added as well as other soluble components. The contents are heated to ca 70°C and agitated to dissolve the nitrocellulose and to form a lacquer. When of proper viscosity, a protective colloid is added to form an emulsion of nitrocellulose globules which prevents their coalescing. Sodium sulfate is added to form an osmotic-pressure differential between the water-laden nitrocellulose globules and the concentrated salt solution in the still, and water is extracted. Small spheres of dissolved nitrocellulose and the other soluble ingredients are formed, the ethyl acetate is distilled, and the remaining spherical particles are grained. After graining, the slurry is water-washed to remove adherent salt solution and then wet-screened into the required particle sizes.

Nitrocellulose-Based Cast Propellants

Cast nitrocellulose propellant is made by a two-step process. In the first stage, casting powder is produced by techniques that are almost identical to those used for the manufacture of conventional solvent-extruded small-grain gun propellants. The second stage consolidates the casting powder by filling the interstices of the granules with a fluid plasticizer that diffuses into the powder and causes swelling and ultimate coalescence of the granules into a monolithic grain. The plasticizer generally consists of a mixture of an explosive energy-producing liquid and an inert fluid. The process of consolidation is a physical one. No chemical reaction occurs and there is virtually no shrinkage during curing.

Polymer-based cast propellants. These propellants have an extensive range of performance characteristics, have excellent thermal and mechanical stability, and are relatively inexpensive. The manufacturing operations for making composite propellants are very similar, although a variety of polymeric binders may be used. A flow chart of a typical batch process is shown in Figure 2.

Black Powder

Black powder is mainly used as an igniter for nitocellulose gun propellants and to some extent in safety blasting fuses, delay fuses, and in firecrackers. Potassium nitrate black powder is used for military applications. The slower-burning, less costly, and more hygroscopic sodium nitrate black powder is used industrially.

The critical relative humidity of black powder is 60%. It gains ca 2% moisture in 48 h at 90% rh and 25°C. Ignitability decreases rapidly at ca 3 to 4% moisture. The structure of the granules also deteriorates during cycling through high humidity atmospheres. However, it can be stored satisfactorily for many years if dry. The hygroscopicity of black powder is caused by the carbon present and impurities in the potassium nitrate.

Manufacture. Since black powder is a mechanical mixture, its performance is critically dependent on the degree of intimacy of the components in the product. A typical procedure for bringing the ingredients into maximum mutual contact is: dry potassium nitrate is pulverized in a ball mill and sulfur is milled into cellular charcoal to form a uniform mix in a separate ball mill. The nitrate and the sulfur-charcoal mix are screened and then loosely mixed in a tumbling machine. Magnetic separators may be used to ensure the absence of ferrous metals. The preliminary mix is transferred to an edge-runner wheel mill with large, heavy cast iron wheels. The magnitude of the gap between the pan and the wheels, a clearance needed for safety purposes, contributes to the density of the black powder granules obtained. Water is added to minimize dusting and improve incorporation of the nitrate into the charcoal. The milling operation requires ca 3 to 6 h. The moist milled powders is consolidated into cakes in a hydraulic press, with the density of the powder increasing to 1.6 to 1.8 g/cm^3, depending on the pressure applied. The cakes are transferred to a corning mill, crushed, screened and formed into a product close to the required grains, and then polished, dried, mixed with graphite and blended.

Benite. Benite is an extrudable composition consisting of ca 60 parts of black powder in the matrix of ca 40 parts of plasticized nitrocellulose. It is used as a propellant igniter to reduce the residue formed compared to use of black powder alone. Its approximate wt composition is nitrocellulose (13.15% N), 40%; potassium nitrate, 44%; sulfur, 6.5%; carbon, 9.5%; ethyl centralite, 0.5%. It is made by the single-base process, followed by air-drying to remove volatile solvents.

VICTOR LINDNER
U.S. Army Armament Research
and Development Command (ARRADCOM)

C. Boyars and K. Klager, eds., *Proceedings of Symposium on Propellants Manufacture, Hazards, and Testing*, American Chemical Society, Washington, D.C., 1969.

T. Urbanski in *Chemistry and Technology of Explosives*, Vol. 3, Pergamon Press, New York, 1967.

EXT. D & D DYES. See Colorants for food, drugs, and cosmetics.

EXTRACTION, LIQUID – LIQUID

Liquid–liquid extraction is sometimes also referred to as solvent extraction or merely as extraction. It is a separation process that depends on the transfer of the component to be separated (the consolute component) from one liquid phase to a second liquid phase that is immiscible with the first. In fractional extraction, there are two or more consolute components. Applications may be classified broadly as (*1*) purifications of the consolute component(s), eg, the preferential extraction of uranyl nitrate from an aqueous solution of mixed metal ions; and (*2*) purification of a liquid phase by extraction of a contaminant, eg, the extraction of phenols from aqueous industrial effluents.

Whether an extraction process is carried out in the laboratory or industry, it always involves contact of the liquid phases with an approach toward equilibrium, and separation of the contacting liquid phases. On the laboratory scale, this is done by shaking the phases in a separating funnel, leaving them to settle, and then separating the phases. On a larger scale, extractions are nearly always carried out continuously using a variety of equipment designs (see under Equipment). The process is shown schematically in Figure 1 which describes the standard nomenclature used in extraction technology. The feed solution contains the consolute component C dissolved in a component A, and it is brought into contact with a solvent B. Mass transfer of C occurs between the two liquid phases: A and B are chosen to be immiscible or very sparingly miscible with each other (nonconsolute components). After contact and

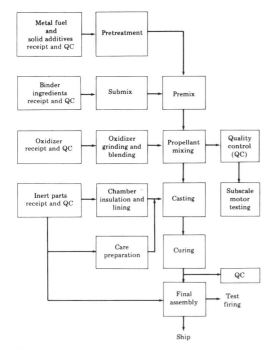

Figure 2. Batch process for cast-composite polymer-based propellants.

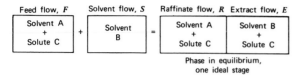

Figure 1. Single contact.

separation, the distribution of C between A and B is represented by a quantity known as the extraction factor of this component of C:

$$\epsilon = m\frac{B}{A} = \frac{\text{quantity of C in phase B}}{\text{quantity of C in phase A}} \qquad (1)$$

where m = distribution coefficient, B = quantity of solvent B, A = quantity of solvent A, and ϵ = extraction factor. The extraction factor plays a most important part in extraction calculations.

The separated extract from a single contact shown in Figure 1 can be subjected to a second liquid–liquid contact whereby the consolute component is extracted back into the raffinate phase under different conditions (eg, temperature). This process is known as stripping and the stripped solvent can then be used again for extraction. Alternatively, the solvent B can be recovered from the extract by a distillation or evaporation step.

Principles

Physical equilibria. Liquid–liquid equilibria are governed by the phase rule:

$$F = C - P + 2$$

where C is the number of components, P is the number of phases (two in this case), and F is the number of degrees of freedom, ie, independent variables permitted by the system.

A binary liquid–liquid system is, therefore, bivariant; eg, if the pressure and the temperature are fixed, the composition of each phase in equilibrium is fixed.

As the temperature of equilibration is increased, the equilibrium phase compositions generally approach each other until the upper critical solution temperature, at which the two phases become identical, is reached. In some cases, there is also a lower critical solution temperature below which two liquid phases cannot coexist in equilibrium.

Ternary liquid–liquid systems are allowed three degrees of freedom according to the phase rule. At a given temperature and pressure, the composition of one phase determines the composition of the other phase at equilibrium. Figure 2 shows a typical equilateral triangular diagram for components A, B, and C. The compositions are read along axes inclined at an angle of 60° so that the sum of mass fractions x_A, x_B, and x_C must always be unity.

Chemical equilibria. In many cases, equilibria in liquid–liquid systems are affected by chemical changes; eg (1) equilibrium with a chemical change occurring in the bulk of one or both of the phases but with components transferring between phases without chemical change; and (2) equilibrium with a chemical change occurring at the interface itself; no molecular species has significant solubility in both phases.

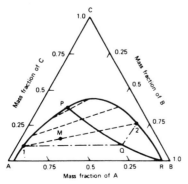

Figure 2. Typical equilibrium curve (ternary system) on triangular coordinates.

Interfacial mass transfer. Although the equilibrium relationships are important in determining the ultimate separation obtainable in a liquid–liquid system, the rate of equilibrium attainment is of equal importance in the design or operation of extraction processes. In a batch process, this rate determines the contact time required; in a continuous process, the required size of the equipment and the residence time (volume of equipment divided by volume flow rate of phases) depend on mass-transfer rates as well as on equilibrium properties (see Mass transfer).

In the absence of any chemical reactions, the rate of transfer of a consolute component from one phase to another is governed by the diffusion laws. The diffusional mass-transfer rate of solute per unit area, relative to a stationary medium, is given by Fick's first law:

$$N = -D\frac{\partial c}{\partial z} \qquad (2)$$

where N refers to transport in the z direction, c is the concentration of the consolute component, and D is its molecular diffusivity with respect to the solvent. It would also be proper to relate the flux to the activity gradient, particularly as activity coefficients in liquid–liquid systems are not usually unity. In general, molecular diffusivities of solutes in liquids are within the range 10^{-6}–10^{-5} cm^2/s.

If the transfer space contains a volume fraction h (the holdup of the dispersed phase in the spherical droplets of diameter d_m), it can be shown that:

$$a = 6h/d_m \qquad (3)$$

where a is the interfacial area available per unit volume. Thus a and the mass-transfer effectiveness can be maximized by increasing the holdup and reducing the drop size.

Calculation of equilibrium stages. In practice, multistage contacting of two immiscible phases can be arranged in a cocurrent, crosscurrent, or countercurrent manner. Multistage, cocurrent operation increases only the residence time and, therefore, does not increase the separation above that obtained in a single stage. Crosscurrent contact in which fresh solvent is added in each stage increases the separation beyond that obtainable in a single stage. However, it can easily be shown that the enhancement is not as great as can be obtained by countercurrent operation with a given amount of solvent, nor is it as economical.

Crosscurrent extraction. For the case of a partially miscible solvent system, the calculations for the equilibrium stages for a particular separation are carried out on a ternary diagram, and the method is well documented. For the case of solvents that are completely immiscible, the calculation can be simplified and carried out graphically. Assuming that a quantity A of a solvent A' containing a solute at a mass ratio (C to A) of X_1 is mixed with a quantity B of solvent B', the mass ratios in the two exit phases may be given by the point (X_1, Y_1) on the equilibrium curve. From the material balance on the two phases, the point can be located by the following equation:

$$Y_1/(X_f - X_1) = A/B \qquad (4)$$

where x_f is the mass ratio (C to A) at the feed.

For the case where the solvents are completely immiscible and the ideal distribution law applies ($Y = mX$), the number of theoretical stages required for a particular crosscurrent separation can be calculated by the following equation:

$$N = \frac{\log(X_f/X_N)}{\log(1 + \epsilon)} \qquad (5)$$

where ϵ is the extraction factor mB/A.

Countercurrent extraction. The best compromise between high extract concentration and high recovery is obtained using multistage countercurrent extraction, as shown in Figure 3. The feed entering stage 1 is brought into contact with a B-rich stream which has already been through the other stages, while the raffinate leaving the last stage has been in contact with fresh solvent. Because of the economic advantages of continuous countercurrent extraction, this type of operation is preferable for commercial-scale operation.

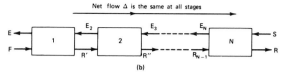

Figure 3. Countercurrent stagewise extraction. E = extract flow, F = feed flow (A-rich), S = solvent flow (B-rich), R = raffinate flow.

The economic optimum solvent–feed ratio is commonly one and one-half to two times the minimum value and is determined for a given case by considering the capital costs (largely dependent on the number of stages) and the operating costs which are strongly influenced by the cost of solvent recovery.

Fractional extraction. Although solvent extraction is considered in its simplest form as the transfer of a consolute component C from a binary feed solution (A-rich) to a solution in another phase (B-rich), it is often used to separate one or more consolute components. This process is known as fractional extraction. In one type of application, a binary feed (C and A) is introduced into a countercurrent cascade through which a solvent B is circulated. The solvent preferentially dissolves one component (eg, C) as it passes down the column. For this type of extraction to be practicable, the solvent should be only partially miscible with either A or C in order that two phases can exist in all parts of the column. This process is closely analogous to distillation of a binary mixture; a stagewise procedure has been designed on this basis.

If no solvent is available that satisfies the requirement of being sparingly miscible with each of the feed components, fractional extraction may be carried out using two solvents. The feed components C and D may be wholly miscible with either solvent as long as the two solvent streams are immiscible. Thus two-solvent fractional extraction involves two phases and four or more components.

Hydrodynamic factors in extractor design. A successful extraction device should create droplets small enough to provide interfacial area for reasonably fast extraction, but also allowing the two liquid phases to separate properly after extraction. In the case of countercurrent contactors, there is a third hydrodynamic requirement, namely, that the flow of the two phases in opposite directions through the equipment should be stable.

Equipment and Processing

Industrial applications of solvent extraction have increased rapidly in the last 25 years. New and improved multistage and differential contactors employing mechanical energy input to achieve a high rate of mass transfer have been developed since the late 1940s and have found wide commercial application. In the past decade, several reviews have appeared that describe the various types of extractors.

Because of the great variety available, the choice of a commercial extractor for a new process can be bewildering, and the following criteria should be considered in the selection of one for a particular application: stability and residence time, settling characteristics of the solvent system, number of stages required, capital costs and maintenance, available space and building height, and throughput.

Commercial Extractors

Contactors can be classified according to the methods applied for interdispersing the phases and producing the countercurrent flow pattern.

Figure 4 summarizes the classification of the main types of commercial extractors. Table 1 summarizes their chief characteristics.

Economics of Extraction Design

These include capital costs, ie, equipment and inventory of material held within plant such as solvents, and the operating (prime) costs, ie, extractor operation, solvent recovery, and solvent losses.

Solvent recovery is usually the dominant factor because of large energy consumption. Many processes would be more economical with a larger number of extractor stages which would reduce the amount of solvent to be recovered and hence the cost.

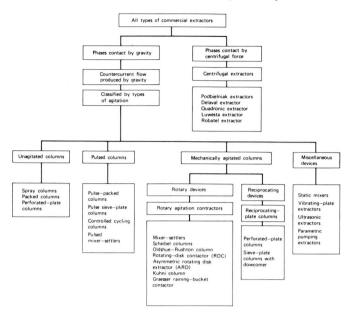

Figure 4. Classification of commercial extractors.

Table 1. Summary of Features and Fields of Industrial Application of Commercial Extractors

Types of extractor	General features	Fields of industrial application
unagitated columns	low capital cost, lower operating and maintenance cost, simplicity in construction, handles corrosive material	petrochemical, chemical
mixer-settlers	high stage efficiency, handles wide solvent ratios, high capcacity, good flexibility, reliable scale-up, handles liquids with high viscosity	petrochemical, nuclear, fertilizer, metallurgical
pulsed columns	Low HETS, no internal moving parts, many stages possible	nuclear, petrochemical, metallurgical
rotary agitation columns	reasonable capacity, reasonable HETS, many stages possible, reasonable construction cost, low operating and maintenance cost	Petrochemical, metallurgical, pharmaceutical, fertilizer
reciprocating-plate columns	high throughput, low HETS, great versatility and flexibility, simplicity in construction, handles liquids containing suspended solids, handles mixtures with emulsifying tendencies	pharmaceutical, petrochemical, metallurgical, chemical
centrifugal extractors	short contacting time for unstable material, limited space required, handles easily emulsified material, handles systems with little liquid-density difference	pharmaceutical, nuclear, petrochemical

Note: HETS = height of an equivalent theoretical stage.

Industrial Applications

Solvent extraction is generally applied where direct separation methods, such as distillation and crystallization, fail and where it provides a less expensive process than a competitive physical or chemical method.

Organic Processes

Petroleum and petrochemical processes.   There is an ever-increasing need for jet-fuel kerosene and lubricating oil which require a low aromatic content obtained by solvent extraction (see Aviation and other gas-turbine fuels). Solvent extraction is used extensively to meet the ever-increasing demand for the high purity aromatics benzene, toluene, and xylene (BTX) as feedstock for the petrochemical industry (see BTX processing). The separation of aromatics from aliphatics is still one of the largest applications of solvent extraction today (see Petroleum, refinery processes).

Pharmaceutical processes.   Many pharmaceutical intermediates and products are heat sensitive and cannot be processed by methods such as distillation.

Food processing.   Industrial refining of fats and oils with propane is known as the Solvexol process.

Inorganic Processes

Many hydrometallurgical uses of liquid–liquid extraction have been developed, as well as a number of other applications involving nonmetallic, inorganic products (see Extractive metallurgy). These include nuclear fuel reprocessing; uranium extraction; recovery of copper from ore leach liquors; and the separation of nickel and cobalt.

TEH C. LO
Hoffmann-La Roche Inc.

MALCOLM H.I. BAIRD
McMaster University

R.E. Treybal, *Liquid Extraction*, 2nd ed., McGraw-Hill Book Co., New York, 1963.

C. Hanson, ed., *Recent Advances in Liquid–Liquid Extraction*, Pergamon Press, New York, 1971.

T.C. Lo, M.H.I. Baird, and C. Hanson, eds., *Handbook of Solvent Extraction*, Wiley-Interscience, New York, 1983.

G.M. Ritcey and A.W. Ashbrook, *Solvent Extraction: Principles and Application to Process Metallurgy*, Elsevier, Amsterdam, Neth., 1978.

EXTRACTION, LIQUID – SOLID

The extraction from a solid material of a soluble component by means of a solvent liquid–solid extraction or leaching is an industrial operation which predates large-scale chemical technology, having been applied in earlier centuries to extract alkaline salts from wood ashes.

The solid can be contacted with the solvent in a number of different ways but traditionally that part of the solvent retained by the solid is always referred to as underflow or holdup whereas the solid-free, solute-laden solvent separated from the solid after extraction is called the overflow. In a typical multistage countercurrent liquid–solid extraction process, based on immersion and decantation, fresh solid enters the first stage and fresh solvent enters the final stage; the latter is gradually enriched in solute until it leaves the extraction battery as overflow from the first stage.

Process Design

Solvent.   Solvent choice is in the first place determined by the chemical structure of the material to be extracted, the rule "like disolves like" providing useful guidance. Thus, vegetable oils (qv) consisting essentially of triglycerides of fatty acids, are normally extracted with hexane whereas for the free fatty acids, which are more polar than the triglycerides, more polar alcoholic solvents are used. Where a choice of solvent (other than water) exists on the grounds of comparable solubility of the solute in question, the following criteria are likely to be considered: selectivity—considering the purity of the recovered extract, maximum selectivity consistent with solvent capacity is desirable; physical properties, low surface tension facilitating wetting of the solids in the first extraction stage whereas low viscosity improves diffusion in the solvent phase, and low solvent density is desirable as it results in a reduced mass of solvent held up in the solid being extracted; thermal stability complete at the processing temperatures in order to avoid expensive solvent purification and losses; hazards such as fire and explosion; and cost.

Extractors

Extractors are devices used in liquid–solid extraction to provide for batch and continuous operation.

Batch extractors.   The pot extractor is a batch extraction plant that offers the small-scale processor the advantage of carrying out extraction and solvent recovery from the exhausted solids in one vessel. These extractors are normally provided with mechanical agitation which may however only be required during solvent recovery. Pot extractors range from 2 to 10 m^3 (500–2500 gal), beyond which the battery system discussed below is more attractive.

The diffuser battery, which had its origins in beet-sugar extraction technology, is a semibatch extractor operating on a cyclical basis (see Sugar). The individual units in the battery are charged sequentially with the solids to be extracted and the extracting-liquor flow designed to provide apparent countercurrent flow. The number of diffusers in the battery depends on extraction conditions and equilibria. An additional diffuser above the estimated number is required to permit cyclical operation. The size of individual units depends on factors related to the mass balance, the contact time required, and the hydrodynamic behavior of the bed of solids.

Continuous extractors.   Although continuous extraction may be operated cocurrent, crosscurrent, or countercurrent, considerations of extraction depth and of solvent consumption generally preclude the first two modes. The countercurrent extraction devices considered here may operate on either the percolation or the immersion principle, with the percolation rate of solvent through the comminuted solid playing an important part in determining the choice.

Percolation.   In the Rotocel extractor, the material to be extracted is fed continuously as a slurry in the extraction solvent or as a dry feed to sector-shaped cells arranged around a horizontal rotor (see Fig. 1). The cells have a perforated base to permit drainage of the solvent into stage basins from which it is pumped to the next cell on the countercurrent principle. In the last cell, where fresh solvent is supplied, an extended drainage period is provided (by allowing a proportionately larger arc of the rotary motion for this cell), and thereafter the extracted solids are dumped.

The endless-belt extractor (Lurgi; De Smet) is, in principle of operation, closely related to the Rotocel. Extraction time and percolation rate, established experimentally for the material as comminuted, determine the belt speed and the amount of drainage area required.

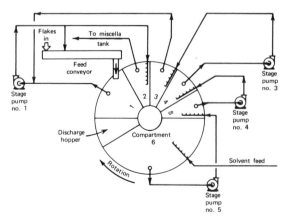

Figure 1.   Downflow Rotocel flow diagram. Courtesy of Dravo Corporation.

Immersion. The main advantage of immersion extraction is the ability to handle finely ground material. Since only the drainage rate must be considered, the operation is simpler. Immersion extraction may also be used where the percolation rate of the material to be extracted is too great for effective diffusion within the bed. The De Danske Sukkerfabriker (DDS) diffusion extractor is an example of an extractor of this type.

The use of immersion extractors is contingent upon the ability to transport solids without excessive backmixing. They need less space than percolation extractors and less power for the band drive and liquor circulation.

Health and Safety

Factors that constitute a hazard in extraction plants are solvent flammability, solvent content of the work atmosphere and of the products in the case of edible material; and the dust content of the work atmosphere. Recommendations published annually by the ACGIH should be consulted for currently acceptable limits of the threshold limit values (TLV) for solvents and dusts. Liquid effluents and gaseous emissions are subject to the usual environmental considerations (see Wastes, industrial).

Applications

The largest application in terms of scale of liquid–solid extraction is the aqueous extraction of sugar from sugar beets, and of vegetable oils from oilseeds using hexane, and more recently, the extraction of sugar from sugarcane. Alcoholic solvents are used to remove residual oil from fish meal or fish waste as a step in the production of fish-protein concentrate. In chemical leaching, hydrometallurgy comprises the most significant applications, a wide range of metals, occurring naturally in the native, oxide, or sulfide state, being recovered from the ore bodies as mined and crushed or concentrated (see Extractive metallurgy).

WOLF HAMM
Unilever PLC

W.L. McCabe and J.C. Smith, *Unit Operations in Chemical Engineering*, 3rd ed., McGraw-Hill Book Co., London, UK, 1976.

R.N. Rickles, *Chem. Eng.*, *(N.Y.)* **72**(6), 157 (Mar. 15, 1965).

A.R. Burkin, ed., *Leaching and Reduction in Hydrometallurgy*, The Institution of Mining and Metallurgy, London, UK, 1975.

H.G. Schwartzberg, *Chem. Eng. Prog.* **76**, 67 (1980).

EXTRACTIVE METALLURGY

Extractive metallurgy deals with the extraction of metals from naturally occurring compounds and their refinement to a purity suitable for commercial use. These operations, known as the winning and refining of metals, follow the mining and beneficiation of the ore, and they precede the fabricating processes.

Extractive metallurgy is usually divided into: pyrometallurgy, consisting of processes that use high temperatures to carry out smelting and refining operations; hydrometallurgy, characterized by the use of aqueous solutions and inorganic solvents to achieve the desired reactions; and electrometallurgy, which applies electrical energy to extract and refine metals by electrolytic processes carried out either at high temperature or in aqueous solution.

Pyrometallurgy

The essential operations of an extractive-metallurgy flow sheet are the decomposition of a metallic compound to yield the metal and the physical separation of the reduced metal from the residue. This is usually achieved by a simple reduction or by controlled oxidation of the nonmetal with simultaneous reduction of the metal. In a simple pyrometallurgical reduction, the reducing agent R combines with the nonmetal X in the metallic compound, MX, according to a substitution reaction of the type:

$$MX + R \rightarrow M + RX \qquad (1)$$

The product RX and the gangue material constitute the residue. In order to achieve a spontaneous process, the free-energy change for equation 1 must be negative.

Preparatory processes. Preparatory processes and calcination are the simplest pyrometallurgical operations: the evaporation of free water and the decomposition of hydrates and carbonates, eg, the decomposition of pure limestone ($CaCO_3$) to calcium oxide and carbon dioxide in a strongly exothermic reaction. Most nonferrous metals occur in nature mainly as sulfides, which cannot be easily reduced directly to the metal. Burning metallic sulfides in air transforms them into oxides or sulfates which are more easily reduced.

Oxide ores and concentrates react at high temperature with chlorine gas to produce volatile chlorides of the metal. This reaction can be used for the common nonferrous metals, but it is particularly useful for refractory metals like titanium and zirconium and for reactive metals like aluminum.

The beneficiation of ores by physical techniques requires the liberation of metal-bearing minerals, and this is usually achieved by reducing the raw material to a very fine size. The high throughput reduction processes, such as the iron blast furnace, however, cannot be used with a finely divided feed. Sintering and pelletizing are techniques for agglomerating finely divided solid particles into a coarse material suitable for charging a blast furnace (see Size enlargement).

Reduction to liquid metal. This type of reduction is the most common of metal-reduction processes and is the preferred method, used for metals of moderate melting point and low vapor pressure. Because most metallic compounds are fairly insoluble in molten metals, the separation of the liquefied metal from a solid residue or from another liquid phase of different density is usually complete and relatively simple. Because the product is in condensed form, the throughput per unit volume of reactor is high, and the number and size of the units is minimized. The common furnaces for production of liquid metals are the blast furnace, the reverberatory furnace, the converter, the flash smelting furnace, and the electric-arc furnace (see Furnaces, electric).

Reduction to gaseous metal. The volatile metals can be reduced to metallic vapors. These are easily and completely separated from the residue before being condensed to a liquid or a solid product in a container physically separated from the reduction reactor. Reduction to gaseous metal is industrially significant for zinc, mercury, magnesium and calcium, and possibly as well for cadmium and the alkali and alkaline-earth metals.

Reduction to solid metal. Metals with very high melting points cannot be reduced in the liquid state. Because the separation of a solid metallic product from a residue is usually difficult, the raw material is purified before reduction. Tungsten and molybdenum, for instance, are prepared by reduction of a purified oxide (WO_3, MoO_3) or a salt ($[NH_4]_2MO_4$) by hydrogen following a reaction such as:

$$WO_3 \text{ (s)} + 3\,H_2 \text{ (g)} \rightarrow 3\,H_2O \text{ (g)} + W \text{ (s)} \qquad (2)$$

The metallic product consists of fine particles which are fabricated into various shapes by the techniques of powder metallurgy (see Tungsten and tungsten alloys; Molybdenum and molybdenum alloys; Powder metallurgy).

Refining processes. All the reduction processes discussed above yield an impure metal that still contains some of the minor elements present in the concentrate (eg, cadmium in zinc) or some elements introduced during the smelting process (eg, carbon in pig iron), and which must be removed from the crude metal in order to meet the specifications for its use. Refining operations may be classified according to the kind of phases involved in the process: separation of a vapor from a liquid or solid, separation of a solid from a liquid, or transfer between two liquid phases. In addition, they may be characterized by whether or not they involve oxidation–reduction reactions.

In volatilization, the simplest separation process, the impurity or the base metal is removed as a gas. Lead containing small amounts of zinc is refined by batch vacuum distillation of the zinc. Most of the zinc produced by the smelting processes described earlier contains lead and cadmium and it must be purified for commercial use. Crude zinc is refined by a two-step fractional distillation. In the first column, zinc and cadmium are volatilized from the lead residue, and in the second column cadmium is removed from the zinc.

In the simplest case of precipitation, the solubility of an impurity in the liquid metal changes with temperature, and it precipitates as a solid phase upon cooling. For instance, the removal of iron from tin and the removal of copper from lead are achieved by precipitation of a solid phase. When the solid is lighter than the liquid, it floats as a dross on the surface of the melt where it is easily removed by scraping, a process called drossing.

Slag refining is effective for the purification of the less reactive nonferrous metals: gold, silver, copper, lead, and others. For instance, in the recovery of precious metals from anode slimes, one of the steps is a fusion, usually in a small reverberatory furnace, where impurities are removed by oxidation with air and dissolution into successive slags: an oxide slag, a sodium carbonate slag, and a sodium nitrate slag. The product of this operation is a silver–gold alloy (ca 90% Ag and 9% Au) called doré metal, which is further refined by electrolysis.

Hydrometallurgy

The two main steps of hydrometallurgical flow sheets are leaching, or dissolution of the metal in a suitable aqueous solvent, and recovery or precipitation of the metal or its compound from solution. In addition, several steps are usually required to achieve the physical separation of the solid phase from the liquid: washing, clarification, thickening, filtering, drying, evaporation, etc.

Leaching chemistry. The purpose of leaching is to dissolve the desired mineral and to separate it from the gangue material. The reaction should be selective and fast, and the solvent should be inexpensive or easily regenerated. Several leaching agents are commonly used, including water, acid solutions, alkaline solutions, complex-forming solutions such as ammonium hydroxide and ammonium carbonate, oxidizing solutions such as ferric sulfate or ferric chloride, and oxidizing and complex-forming solutions (eg, the leaching of gold or silver can be achieved only by oxidation of the metal by air and formation of a stable cyanide complex).

Leaching techniques include *in situ*, heap and dump (crushed ore is placed in large heaps and sprayed with the leaching solution), percolation, agitation, and pressure leaching, depending upon the type of reaction and the characteristics of the ore.

Recovery of metal from solution. Electrowinning, electrolytic deposition, is the most efficient way of recovering a valuable metal from solution (see Electrometallurgy below). It is quite selective and yields a pure product which can be marketed directly as cathodes, or after casting into commercial shapes.

Precipitation of aluminum trihydroxide in the recovery step of the Bayer process is achieved either by lowering the temperature or by diluting the pregnant liquor and reducing its pH.

A metal can be removed from solution by cementation, displacing it with a more active metal. This simple and cheap method has been commonly used to recover copper from dilute solution (1–3 g/L) using scrap iron or tin cans as a cheap reducing agent.

The use of a gaseous reducing agent is attractive because it produces the metal as a powder that can easily be separated from the solution. Carbon dioxide, sulfur dioxide, and hydrogen can be used to precipitate copper, nickel, and cobalt, but only hydrogen reduction is applied on an industrial scale.

Metallic ions can be removed from an aqueous solution by exchange with ions at the surface of an organic resin. The technology for this process is borrowed from the water-treatment industry using mainly packed columns (see Ion exchange; Water).

Solvent extraction, a liquid–liquid extraction process, was first used in extractive metallurgy for the processing of uranium. More recently, the development of selective extractants for copper has made economically feasible the extraction of copper from dilute solutions (3 g Cu/L) and its transfer to a more concentrated sulfuric acid solution (30 g Cu/L. 150 g H_2SO_4/L), from which it is recovered by electrowinning.

Electrometallurgy

Electrowinning from aqueous solution. Electrowinning is the recovery of a metal by electrochemical reduction of one of its compounds dissolved in a suitable electrolyte. Various types of solutions can be used, but sulfuric acid and sulfate solutions are preferred because they are less corrosive than others available and the reagents are fairly cheap. From an electrochemical viewpoint, the high mobility of the hydrogen ion leads to high conductivity and low ohmic losses, and the sulfate ion is electrochemically inert under usual conditions.

The generalized flow sheet of an aqueous electrowinning process consists of at least three main steps: (*1*) The metal is put into solution by leaching of a calcine or by direct leaching of low grade ores containing oxidized minerals or weathered sulfides. (*2*) The pregnant solution is purified to remove any metallic impurities more noble than the metal to be recovered, or any impurities that could reduce the current efficiency. (*3*) The purified solution is fed to the electrolysis tanks where the metal is plated on a cathode and oxygen is evolved in an inert anode, usually made of lead.

Electrolysis of molten salts. Metals more active than zinc and manganese cannot be recovered by electrodeposition from aqueous solutions. Most of them, however, can be obtained from a molten electrolyte. Usually, a compound of the metal to be electrodeposited is dissolved in a mixture of salts of more active metals in order to achieve a low melting point, a suitable viscosity and density, and a high conductivity. Solid electrodeposits obtained from molten salts are mostly dendritic or powdery, and they are difficult to separate from the melt.

Electrorefining. Electrolytic refining is a purification process in which an impure metal anode is dissolved electrochemically in a solution of a salt of the metal to be refined, and then recovered as a pure cathodic deposit. Electrorefining is a more efficient purification process than other chemical methods because of its selectivity (see Electrochemical processing).

PAUL DUBY
Columbia University

T. Rosenqvist, *Principles of Extractive Metallurgy*, McGraw-Hill Book Co., New York, 1974.

J. Newton, *Extractive Metallurgy*, John Wiley & Sons, Inc., New York, 1959.

J.D. Gilchrist, *Extractive Metallurgy*, Pergamon Press Ltd., Oxford, UK, 1967.

F

(a)   (c)

FACE POWDER. See Cosmetics.

FANS AND BLOWERS

Fan is the generic term for a low pressure air- and gas-moving device using rotary motion. Fans are subdivided into centrifugal and axial-flow types, depending on the direction of air flow through the impeller (see Mixing and blending).

Manufacturers' performance ratings are generally based on atmospheric pressure (101.3 kPa) at sea level, 20°C, and 50% rh. Changes in temperature, gas density, and fan speed affect fan performance, as predicted by fan laws. Fan laws can be used to construct performance curves.

Centrifugal Fans

In centrifugal fans, the rotation of the wheel causes air between the blades to be rotated, the resulting centrifugal force causing the air to be compressed and ejected radially from the wheel. The compression results in an increase in static pressure in the fan scroll. The static pressure produced at the blade tips depends on the ratio of the velocity of air leaving the tips to the velocity of air entering at the heel of the blades. Thus, the longer the blades, the greater the static pressure developed by the fan at a constant speed.

When a centrifugal fan is used to force air through a system under positive pressure, it is called a blower. It generally implies a fan developing a reasonably high static pressure of at least 500 Pa (several inches of water). An exhaust fan or exhauster is a fan placed at the end of a system so that most of the system pressure drop is on the suction side. This term also applies to a ventilating fan which primarily exhausts air from a room or an open hood. Centrifugal compressors or turbocompressors are high volume centrifugal devices capable of gas compression of 105–1500 kPa (0.5 to several hundred psig) (see High pressure technology).

Design operating efficiencies of fans under test conditions are in the range of 40–80%. Actual efficiency can be affected appreciably by the arrangement of inlet and outlet duct connections. The air power in W (hp) of a fan is given by the equation:

$$\text{air power(W)} = Q \Delta p \ \left(\text{in units of hp, } 41.21 Q' \Delta p'/10^5\right)$$

where Q is the volume of gas handled in m^3/s (for hp, Q' in ft^3/min), and p is the pressure rise across the fan, Pa (for hp, p' in psi).

Types and Characteristics

Figure 1 shows the four main types of centrifugal fans. They have the following characteristics: (**a**) Forward-curved blades are generally shallow, spaced close together, and have both the heel and tip of the blade curved forward in the direction of the rotation so that air leaves the tip of the wheel at a velocity greater than the wheel-tip speed. Such fans are commonly used for low pressure, high volume ventilating applications. (**b**) Backward-curved blades have single-thickness straight or curved blades inclined backward (opposite to rotation direction) from the point of heel attachment on the wheel (usually 12–16 per wheel). Air leaves the blade at a velocity less than wheel-tip speed since the increasing flow passage through the blade provides for expansion of the air. (**c**) Straight radial blades, the simplest and also the least efficient for a given speed, tend to develop a higher static pressure than other wheel designs and are attractive for high speed, high pressure fans compressing air to 108–120 kPa (1–3 psig). (**d**) Airfoil design has the highest rotational speed of any of the wheel designs.

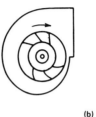 

(b)   (d)

Figure 1. Shape of fan blades and typical performance curves: (**a**) forward-curved blades; (**b**) backward-inclined blades, at left, straight backward blades at top, curbed backward blades at bottom; (**c**) straight radial blades; (**d**) airfoil blades.

Fan Selection

A fan is selected according to its location in the air-flow system, system performance and control characteristics, cost, efficiency, control stability, flexibility, and noise level (see Insulation, acoustic; Noise pollution). For example, a fan operating on the highest-density inlet gas available is smaller and less expensive, and has lower operating costs and requires less maintenance. The wisest fan choice is frequently not the cheapest fan. A small fan operates well on its curve but may not have adequate capacity for maximum flow control, future needs, or process upset conditions. Manufacturers provide tables of operating ranges of quietest operation. There is no set fan-discharge velocity that is applicable to all fans to ensure quiet operation. Fans do not operate as quietly when throttled back as when they are allowed to handle substantial quantities of air. Both inlet and outlet duct connections can influence fan performance significantly, actually changing the shape of the curve. Poor performance can result from fan-inlet eccentric or spinning flow, and discharge ductwork that does not permit development of full fan pressure.

Other selection problems include problems arising when fans must handle solids or gases of low density, or must be operated in parallel or series. A complicated flow system involving several fans in parallel, all of which are in series with a common exhaust fan, can lead to surging and vibration unless selected carefully. Maximum tip speed, bearing types, single- and double-inlet fans, and wheel and shaft natural frequency and rigidity must also be considered.

Likewise, mechanical considerations include bearings, which on fans may be either sleeve or antifriction types, and must be designed to withstand loads due to dead weight, unbalance, and rotor thrust, and be able to operate at the intended maximum speed without excessive heating (see Bearing materials). When natural convection from the bearings is inadequate, some other cooling method must be provided. Lubricating oil may be circulated through an external cooler or the pillow blocks may be cored with passages for forced circulation of air or water. Fans operated at high temperatures increase the bearing cooling problem caused by heat conduction along the shaft. A small external fan wheel on the shaft, called a heat slinger, is frequently provided, or forced-circulation water cooling is used.

Fan and Ductwork System Operating Problems

Vibration results from out-of-balance rotating parts, flow pulsation, or an inadequate foundation. Many fan forces are transmitted to the foundation even by a smoothly operating fan. Pulsation problems resulting from multiple fans have also been mentioned, but the interaction of fan and system can lead to serious flow-pulsation problems resulting from many causes. Fans may surge or give pulsing flow because they are operated in an unstable portion of their performance curve, eg, to the left of the peak discharge pressure. Poor inlet or outlet connections can cause a modification in the shape of the fan curve from nonuniform wheel loading and result in unstable operation when not expected. A resonating duct system can amplify such pulsation. In addition, the flow system can amplify or reinforce acoustic waves generated by compression in the fan.

Axial-Flow Fans

In axial-flow fans, the air continues to move directly forward through the fan along the axis of the shaft. Kinetic energy is imparted to the air by the shape and arrangement of the blades. After discharge through the blades, although the general flow direction is still forward, a spiral component of velocity generally has been added to the air.

The most common of this type is the propeller fan, typically having from two to six blades mounted on a central shaft and revolving within a narrow mounting ring, either driven by belt drive or directly connected. Some type of close-clearance shroud at the blade tips is desirable to prevent air recirculation from the discharge side of the blades back to the suction side. Such fans generally operate from free delivery up to a static head of 125–250 Pa (0.5–1 in. of water). If driven at tip speeds of 61–81 m/s (12,000–16,000 ft/min), they are capable of discharge pressures of up to 375 Pa (1.5 in. of water) but the noise level may be objectionable.

Tube-Axial Fans

Tube-axial fans are a refinement of the propeller fan in both wheel design and mechanical strength, with improved capacity, pressure level, and efficiency. Designs are often capable of operating over a greater range of speeds. The cheapest fans may have an open-type propeller wheel with the motor enclosed in a tube if directly connected. In more refined types, the blades are shorter and of airfoil cross section mounted on a large diameter hub which may approach 50% of the wheel diameter.

Vane-Axial Fans

These fans have short, stubby airfoil blades mounted on a hub which may be as large as 75% of the wheel diameter. The air leaving the axial-flow wheel has an appreciable rotational component, which can be converted to static pressure in a suitably designed set of stationary straightening vanes which are shaped to pick up the air leaving the wheel blades without shock. Some units with mechanical efficiencies above 90% have been developed but many commercial units have efficiencies no higher than 80%. High efficiency vane-axial fans are more efficient than comparable centrifugal fans.

Applications

Fans and blowers are the most widely used mechanical devices for moving air and gases in both large and small volumes in a tremendous range of applications, including ventilation, mechanical draft for combustion (including forced- and induced-draft fans and primary and secondary fans), local exhaust for fume and dust containment, cooling for spray towers, cooling towers, and air-cooled heat exchangers and conveying of solids (see also Heat-exchange technology).

BURTON B. CROCKER
Monsanto Company

Industrial Ventilation, 15th ed., American Conference of Governmental and Industrial Hygienists, Lansing, Michigan, 1978.

W.C. Osborne, *Fans*, 2nd ed., Pergamon Press, New York, 1977.

J.B. Graham in *Fan Application—Testing and Selection, Symposium Papers, San Francisco, Calif., Jan 19–22*, 1970, ASHRAE, New York, 1972, pp. 15–24.

FAST COLOR SALTS. See Azo dyes.

FATS AND FATTY OILS

Fats and fatty oils are made up predominantly of triesters of glycerol (qv) with fatty acids, and commonly are called triglycerides. Customarily fats, also known as lipids, are solids at ambient temperatures and oils are liquids.

Fats are widely distributed in nature. They are derived from vegetable, animal, and marine sources and often are by-products in the production of vegetable proteins or fibers and animal and marine proteins. The chemical structures of fats are very complex owing to the combinations and permutations of fatty acids that can be esterified at the three (enzymatically nonequivalent) hydroxyl groups of glycerol. A generalized triglyceride has structure (1) without regard to optical activity.

When $R = R' = R''$, the trivial name of the triglyceride is derived from the parent acid by means of a termination— in, eg, for stearic acid where $R = R' = R'' = C_{17}H_{35}$, the triglyceride is called tristearin (2).

(1)   (2)   (3) phospholipid $m, n = 10$–16

The vast majority of vegetable and animal fats made up of fatty acid molecules of more than 16 carbon atoms. Marine fats (and some cruciferae fats) are characterized by their content of longer-chain (up to C_{24}) fatty acids.

Composition

Glycerides. The number of triglycerides in a given natural fat is a function of the number of fatty acids present and the specificity of the enzyme systems involved in the particular fat-synthesis reactions.

Fatty acids. Most of the fatty acids in fats are esterified with glycerol to form glycerides. However, in some fats, particularly where abuse of the raw material has occurred leading to enzymatic activity, considerable (> 5%) free fatty acid (FFA) is found. Most producers of fats attempt to prevent the formation of FFA, because if they are present, certain penalties are assessed in the trading of crude and refined fats.

Phospholipids. Compounds (3) occur in most natural fats with differing amounts and compositions depending on the source of the fat. Owing to their complexity, these fat-soluble, biologically important compounds also have presented some intriguing analytical problems to chemists and biochemists. Phospholipids, eg, from soybeans (qv), are composed mainly of lecithin (qv), cephalin (phosphatidyl-ethanolamine—similar to lecithin but with 2-aminoethanol substituted for choline (qv)), or phosphatidylinositol (where inositol (see Vitamins) is linked through its 1-hydroxyl). These products are sold as commercial lecithin used in margarine, confection, and shortening where a fat-soluble emulsifier is required.

Antioxidants. The most commonly occurring natural antioxidants (qv) in vegetable fats are the tocopherols (vitamin E active). These derivatives of 6-chromanol ($2H$-3,4-dihydro-1-benzopyran-6-ol) are not synthesized by mammals and occur in their fats only through ingestion of plant materials and vegetable fats. Antioxidants tend to protect fats by inhibiting autoxidation and subsequent rancidity.

Pigments. The chief pigments of fats are the carotenoids. Palm oil, usually bright reddish-orange, contains as much as 0.2% β-carotene. Many seed oils, particularly if processed from immature seeds, also contain significant amounts of chlorophyll pigments that lend a greenish tinge to the fats (see Pigments).

Vitamins. The principal components in vegetable fats with vitamin activity are the tocopherols. Vitamin A is found in butterfat and in fish oils. The carotenes (pro-vitamin A) are found at significant levels in palm oil, in butterfat and as traces in other fats. Vitamin D is found primarily in some fish oils (see Vitamins).

Sterols. Most of the unsaponifiables in vegetable and animal fats are sterols. The animal fats predominantly contain cholesterol and most vegetable fats contain only traces of this sterol. Plant sterols—collectively called phytosterols—are made up mainly of sitosterols and stigmasterol but individual vegetable fats contain additional phytosterols. The pattern of typical sterols has been suggested for use in detecting adulteration of one oil with another.

Minor constituents. In addition to the materials listed above, hydrocarbons, ketones, aldehydes, and mono- and diglycerides are found in fats and oils at varying but low levels.

Physical Properties

Densities (specific gravity) of fats in the liquid state do not differ much for most of the common fats, and are 0.914–0.964 at 15°C. An equation developed for specific gravity of liquid oils is specific gravity = 0.8475 + 0.00030(saponification number) − 0.00014(IV, iodine value).

Melting and freezing points. The melting and freezing points of most commercial fats and oils are, at best, crude indicators of the product under examination. Melting points of various polymorphs of simple triglycerides range from −44.6 to 73.1°C, depending on the fatty acids and polymorphs present. More complex mixed triglycerides cover a greater mp range. Freezing points are not often used. The congeal-point (set point) method provides some control in the hydrogenation of fats where it principally is used.

Solubility and miscibility. At temperatures above their melting points, fats and oils are freely miscible with most organic solvents except alcohols. Castor oil exhibits the peculiarity of free miscibility with alcohols and limited miscibility with hydrocarbons at usual ambient temperatures. Ordinary refined liquid oils dissolve about 0.07% of their own weight of water at −1°C and ca 0.14% at 32°C. Liquid oils dissolve increasing amounts of common gases and air with increasing temperatures. In all cases, there is a linear relationship between solubility and temperature.

Refractive index. Refractive index, n, is easily determined on small samples. Impurities (fatty acids, mono- and diglycerides, oxidation products, and conjugated olefinic bonds) cause positive deviations in n. However, n is widely used in quality control to check purity of materials and to follow and control hydrogenation and isomerization procedures. Roughly, n_D^{60} ranges from 1.4468 (0 IV) to 1.4568 (100 IV) to 1.4687 (200 IV) for many neutral oils.

Chemical Properties

The chemical reactions of fats are principally those of esters, olefins and hydrocarbons. Triglyceride esters usually are hydrolyzed with acid or base catalysts to yield soaps (fatty acids salts) (see Soaps; Surfactants and detersive systems) or fatty acids (see Carboxylic acids), and glycerol (qv) is recovered. Triglycerides also undergo interesterification or alcoholysis with added alcohols (usually with basic catalysts, eg, sodium methoxide). Amines and other nuclephilic reagents react readily with triglycerides to yield amides and other derivatives. The principal reactions of the long-chain olefins are oxidation and cis-trans isomerization.

Manufacture

Commercially important sources include oilseeds, fruit pulps, animals, and fish. In oilseeds, the fat is concentrated in the kernel and varies widely as to amount, eg, soybeans (18–22%), sunflower (42–63%), peanut (46–50%), castor seed (65–70%) and safflower (46–54%). The two principal fruit pulps (fat contents) are palm (30–55%) and olive (38–58%). Land-animal fatty tissue ranges from 60–90% and whole fish contain 10–20% fat.

Oilseed processing. As most oilseeds are dried to a narrow moisture range (7–13% H_2O depending on the seed type) before storage, further drying is normally not required. Depending upon the seed type, oil content, hull content, etc, one of a variety of processing systems is employed. High hull, high fat-content oilseeds are often dehulled (decorticated) using various mechanical schemes (cf cottonseed). This reduces the fiber content of the meal produced which often has a significant value in the marketplace as food for both humans and animals.

The decorticated kernels are cooked (conditioned) with steam and prepressed in screw presses (expellers) where the oil content is reduced to 10–15%. The press cake is flaked through large steel rollers, often with moisture adjustment, and is conveyed to a solvent-extraction plant (see Extraction). The flakes are countercurrently contacted with solvent: hexane is almost universally employed except for castor seed where the higher boiling heptane is preferred; then taken to a solvent-stripper which removes all traces of solvent. The oil–solvent mixture (called miscella) can be processed per se (see Refining) after filtration, but the vast majority of plants remove the solvent in a vacuum stripper and store the dried, crude oil.

Recovery of oils from fruit pulps. Some of the outstanding accomplishments of the last decade in the fat industry are the improvements in the palm-oil industry leading to a high-quality fat with less than 2% FFA.

Rendering of animal fats. See Meat products.

Refining. The refining of crude fats or crude fats combined with degummed oils removes FFA, gums (phosphatides) and some pigments. Refining processes include alkali refining, steam refining and miscella refining, which is very similar to alkali refining except that the fat is dissolved in a solvent. Usually the process is carried out at the solvent-extraction plant where the oil extracted from the seed by hexane is not stripped of solvent but is taken directly to the refinery where it is treated with gum conditioners and alkali; heated; mixed; and centrifuged before bleaching and removal of the solvent.

Bleaching and decolorization. Adsorptive bleaching is the most widely used process, at 108–110°C in open bleaching, or preferably, *in vacuo*.

Hydrogenation. Hydrogenation is one of the most important processes in the fat industry because it allows the production of many different functional edible fats (eg, margarines, shortening, and confectionary fats) from a wide variety of liquid or partially solid fats. Depending on the particular use, hydrogenation, actually partial hydrogenation, yields products that are interchangeable for many applications.

Commercial hydrogenation is mainly carried out in batch reactors equipped with gas-dispersing agitators, heating and cooling coils, gas-handling systems, catalyst introduction and removal equipment, and attendant safety and operation controls.

The well-refined and bleached fat is introduced, vacuum-deaerated as it is heated, and the catalyst is introduced. Normally, wet or dry reduced nickel on a support is added at a predetermined concentration dependent upon the type of product to be produced. For most edible applications 0.02–0.15 wt% Ni is employed. High purity hydrogen gas is introduced through a sparge ring at the bottom of the converter and the agitator is started at high speed. As hydrogenation commences, the temperature rises. This is an exothermic reaction and converters often are equipped with automatic cooling systems for temperature control. Normal operating pressures are 200–700 kPa (1–6 atm) gauge and most pressure vessels are rated to only 1 MPa (10 atm).

Deodorization. Three types of deodorization equipment are employed: batch, continuous and semicontinuous. The unit operation is steam stripping.

Health and Safety

Fats have no particular technical problems with respect to health and safety. If extremely overheated, traces of glycerol in fats can decompose to form acrolein (qv) which is a toxic and lacrimatory substance. Polymeric substances can be developed by overheating unsaturated fats either in the presence or absence of air and, if isolated from such fats, are toxic to animals when included in diets at significant levels.

Uses

Principal edible oil uses are in salad and cooking oils, shortening, margarines, filled dairy products (see Milk products), and prepared foods. Primary industrial uses are soaps and detergents (see Surfactants and detersive systems), paints (qv), plastic additives (see Alkyd resins), and lubricants (qv) (see also Carboxylic acids).

THOMAS H. APPLEWHITE
Kraft, Inc.

D. Swern, ed., *Bailey's Industrial Oil and Fat Products*, 4th ed., John Wiley & Sons, Inc., New York, vol. 1, 1979; Vol. 2, 1982.

A.R. Baldwin, *J. Am. Oil Chem. Soc.* **53**, 221–461 (1976); *J. Am. Oil Chem. Soc.* **60**, 137A–430A (1983).

J. Baltes, *Gewinnung und Verabeitung von Nahrungsfetten*, P. Parey, Berlin, Ger., 1975.

FATTY ACIDS. See Carboxylic acids.

FATTY ACIDS FROM TALL OIL. See Carboxylic acids.

FEEDSTOCKS

Petrochemical feedstocks are derived almost exclusively from natural gas or petroleum. Therefore, the use of these materials as feedstocks is competitive with their use in the so-called energy sector, ie, with the raw materials for space heating, transportation fuels, industrial fuels, or generation of electricity.

For the petrochemical producer, the selection of the proper feedstock is largely an economic matter, involving as well a knowledge of supply logistics.

Throughout the world in recent years, forces other than those based on free-market economics have been applied to direct the flow of natural gas and petroleum in the market place. In the U.S., government regulatory control on natural gas has existed since 1954, when the U.S. Supreme Court ruled that the 1938 Natural Gas Act established price-control for gas in interstate commerce (see Gas, natural). More recently, Federal regulatory pricing policies were established by the Natural Gas Policy Act (NGPA) of 1978. Although the U.S. now seems to be moving towards deregulation, many olefins plants were designed for a feedstock of natural gas liquids recovered from the regulated, low cost interstate natural gas. Similarly, the national energy policy in Canada, which kept natural gas under price and export control, recently triggered a major round of world-scale olefin plant building in Alberta; much of the product flows into export markets despite relatively high shipping costs from Alberta.

The application of price setting and allocating regulations to crude petroleum and several related products also tends to complicate, if not distort, the economics at the interface of the energy/feedstock alternative. The selection of a chemical feedstock almost invariably involves its diversion from the alternative fuel use. Political and other forces frequently tend to modify idealized economic forces.

Definitions

Petrochemical feedstocks are mainly hydrocarbons that are transferred from the energy sector to the chemical sector, from operations commonly performed in the preparation of usable energy, to operations in which they are first used as raw materials in processes involving chemical change or physical separation before eventual use. An example involving chemical change is the steam pyrolysis of naphtha to make ethylene; an example involving physical separations is the extraction of an aromatic concentrate from a refinery-produced reformate with subsequent distillation of the extract to high purity benzene.

Selection of Feedstocks

The choice of a feedstock is an important decision in planning for petrochemicals because large capital investments and the kind and volume of by-products generated in the plant depend on it. Petrochemical manufacturers have built more plants with so-called heavier feedstocks (ie, higher molecular weight feeds), with boiling ranges that characterize them as naphthas, light gas oil (ie, the kerosene boiling range), or heavy gas oils having boiling points as high as 450°C at their 90% point.

Form value. There are a number of commonly used empirical relations that aid in the selection of preferred feedstocks for desired products. Many of these relate to a property of the feed called form value, a

Table 1. Demand for Petrochemical Feedstocks, 10^6 t coe[a]

Feedstock	1965	1975	1981
natural gas	9.5	17.8	17.0
ethane	3.1	8.8	10.5
propane, butanes	7.2	8.9	12.2
refinery of gases	3.8	5.3	4.2
naphthas, including coal tar	12.8	19.2	25.4
gas oils and heavier	3.4	8.6	14.7
Total gross	*39.8*	*68.6*	*84.0*
less by-products returned to energy sector	3.4	10.9	17.8
Total net	*36.4*	*57.7*	*66.2*
as 10^6 m³/d	*0.11*	*0.18*	*0.21*
(10^6 b/d[b] coe)	*(0.71)*	*(1.12)*	*(1.28)*

[a] Crude oil equivalent.
[b] b/d = barrels per day.

useful concept that is usually intuitively applied by experts in the petrochemical industry. The form value of a feedstock is not an exact property, but in a qualitative sense it refers to its molecular structure and whether the feed is a good one or a poor one for the desired petrochemical product. Feedstock naphthas can be ranked according to their suitability in making certain primary petrochemicals, using the paraffin, isoparaffin, and cycloparaffin content of the feed as parameter.

Characterization factor. A relationship that has been correlated with form value for several processes is characterization factor (CF)

$$CF = \frac{1.216(bp, K)^{1/3}}{sp \; gr_{15.6}^{15.6}}$$

Feedstocks are often compared at roughly similar boiling points because they are normally produced in gas plant or refinery distilling units where cuts between fractions are made under boiling point control. Under these conditions, high CF will in general be associated with low specific gravity and high paraffin/isoparaffin content. Feedstocks with relatively high CF tend to be good for ammonia and light olefin manufacture and poor for BTX manufacture (benzene, toluene, xylene) (see BTX processing).

Demand pattern. The feedstock and energy demands of the petrochemical industry are only a small percentage of the total consumption of natural gas and petroleum (see Table 1). In addition to the net feedstock consumption of 66×10^6 t crude oil equivalent (coe) in 1981 there is 59×10^6 t/yr coe total energy consumed by the industry. Total energy is all fuels plus electricity consumed. The combined feedstock plus energy use of 125×10^6 t/yr coe is equal to 5.3×10^{12} GJ/yr (5.0×10^{18} Btu/yr). This is about 6.5% of the estimated total U.S. energy consumption in 1981 of 83×10^{12} GJ (79×10^{18} Btu).

Note that no implication is being made here that some feedstocks in the future will not be obtained from sources called unconventional today. Indeed, energy experts almost unanimously predict that new source material such as coal liquids, shale oil, biomass, etc, will be commercially developed in the United States before the year 2000. Any need based on traditional natural gas and petroleum can probably be satisfied by alternative feedstocks. However, the economics of feedstock choice logically should determine whether the new source material is consumed in the energy sector or is used as a petrochemical feedstock.

F.A.M. BUCK
King, Buck & Associates

MERRITT G. MARBACH
Shell Chemical Company

"The Petrochemical Industry and the U.S. Economy," *Report to the Petrochemical Energy Group*, C74903, A.D. Little, Inc., Boston, Mass., Dec., 1978.

Proceedings Conference on Chemical Feedstock Alternatives, Houston, Oct. 2–5, 1977, Amer. Inst. Chem. Engrs., New York, 1978.

S.B. Zdonik, E.J. Green, and L.P. Hallee, *Manufacturing Ethylene*, the Petroleum Publishing Co., Tulsa, Okla., 1978.

FELTS

Felt is a homogeneous fibrous structure created by interlocking fibers using heat, moisture, and pressure. Wool or wool combined with other fibers may be meshed using mechanical work and chemical action while the fibrous mass is kept warm and moist; no weaving is used to make this type of felt, which is often called a pressed felt. These felts may be combined with resins or chemicals or laminated with other materials, and many industrial applications result from the cutting, molding, or shaping of such felts.

Needled felts are strong, mechanically interlocked structures created by barbed needles penetrating and compressing synthetic fibers. Such felts may incorporate a woven base (foundation). They are used for products such as ink rollers or filters, or where chemical inertness or thermal stability may be required. With or without bases, they are also used on paper machines for pressing or drying operations.

Woven felts include those made with a fabric of a special weave and usually incorporate wool fibers. Wool enables the use of heat, moisture, and mechanical action to create a compact, interlaced structure that may have a deep nap or pile of surface fibers. These felts are used similarly to needled felts.

Felts have many useful properties, including: thermal and chemical stability, depending upon the fiber content; high permeability and porosity; vibration and shock absorption; and wear resistance.

Function of Felts on the Paper Machine

Papermaking generally follows the same principles for all machine types and most paper and board: a web of paper is formed from an aqueous suspension of fibers (furnish) by gravity and suction through the screen or fabric; the web is transferred to the pressing section where more water is removed by pressure and vacuum; the sheet enters the dryer section, where steam-heated dryers and hot air complete the drying process; and if desired, the sheet may be finished by coating or calendering before it is wound onto the reel.

Manufacture of Felts for Papermaking

Finish, drainage (for pressing), and durability are factors considered in the design of a custom-made felt.

Materials. Depending on the type of felt, the designer may vary: the type of material; the yarn form, weight and count; the weave; the amount and fineness of batt (fibrous web); the seam, if any; and chemical treatment, if any. Both natural and synthetic fibers can be used, with no one fiber best for all applications. Basic fiber types and forms can be combined by blending and twisting operations during manufacturing. One fiber type is frequently used for a machine-direction yarn, and another is used in the cross-machine direction. In press felts, for example, polyester's excellent resistance to stretching is used along with nylon's superior resistance to friction.

Manufacturing processes. The five general manufacturing categories are

(1) Yarn making, with synthetic multifilaments and monofilaments prepared by melting the polymer, forcing it through fine extrusion dies, and stretching the filaments to ensure orientation in molecular chains to build high strengths. Spun yarns are prepared by blending, carding, and spinning;

(2) Weaving, for different patterns chosen for dimensional stability, strength, finish characteristics, drainage ability, wear, and bulk. Felts can be woven either as flat material to be joined later or in endless form without a seam;

(3) Burling and joining, most felts being inspected and burled, during which minor imperfections are corrected;

(4) Needling, which creates mechanical bonding and replaces fulling in conventional felts by creating controlled contraction and stabilization of the felt. A needled felt is composed of the base fabric, which determines the strength and stability, and the batt, formed by carding fibers, which determines the finish, openness, and other characteristics of the felts; and

(5) Finishing and treating, with conventional press felts not needled but wet with a soap solution and then fulled or felted on a rotary fulling

Table 1. Felt Application Chart, Typical Felts on Their Representative Machines[a]

Paper grade/Machine	Position	Machine speed, m/s (ft/min)	Press load, kN/m (pli[b])	Weight per unit area, kg/m^2 (oz/ft^2)
fine paper/three-roll inclined press	suction pickup	6.1–14 (1200–2800)	35–61 (200–350)	0.92–1.1 (3.0–3.7)
	second press	6.1–14 (1200–2800)	39.4–65.7 (225–375)	0.98–1.2 (3.2–3.8)
	third press	6.1–14 (1200–2800)	44–79 (250–450)	0.98–1.2 (3.2–4.0)
tissue/Yankee Fourdrinier (two-felt)	Yankee pickup	7.6–23 (1500–4500)	58–96 (330–550)	1.0–1.2 (3.4–3.9)
	Yankee bottom	7.6–23 (1500–4500)	30.6–44 (175–250)	0.79–0.88 (2.6–2.9)
board/multicylinder	top	0.76–3.3 (150–650)	26–44 (150–250)	0.61–0.82 (2.0–2.7)
	bottom	0.76–3.3 (150–650)	26–44 (150–250)	0.76–0.92 (2.5–3.0)
	press	0.76–3.3 (150–650)	35–149 (200–850)	0.76–1.4 (2.5–4.5)
pulp/three-press Fourdrinier	third press (grooved)	1.3–2.5 (250–500)	175–263 (1000–1500)	1.2–1.3 (3.9–4.3)
asbestos/two-cylinder cement	bottom	0.56–0.81 (110–160)	18–44 (100–250)	1.4–1.5 (4.5–5.0)

[a] From *Paper Machine Felts and Fabrics*, Albany International Corp., Albany, New York, 1976.
[b] pli = pounds-force per lineal inch.

mill to crowd and squeeze the felt to shrink. This process determines final size and tension of the finished felt by mechanically bonding the individual wool fibers For both conventional and needled felts, the trademark line is applied next, and the felt is washed, extracted, and possibly treated with proprietary chemicals.

Application of Press Felts

Table 1 lists some felts used on typical machines; the examples given are represenative and have large ranges in basis weights.

Other industrial applications include use as a mounting under precision instruments; as sound absorbers; as felt seals; and for wet and dry filtration (qv).

MARY K. PORTER
Consultant, Troy, N.Y.
and the Albany International Corp.

J.B. Casey, *Pulp and Paper; Chemistry and Chemical Technology*, 2nd ed., 3 Vols, Interscience Publishers, Inc., New York, 1961.

J.F. Atkins, ed., *Paper Machine Wet Press Manual*, TAPPI Monograph Series, no. 34, Technical Association of the Pulp and Paper Industry, Atlanta, 1972.

Papermachine Clothing, *Monographs in Paper and Board Making*, Vol. 5, Ernest Benn, Ltd., for the Technical Section, The British Paper and Board Industry Federation, London, UK, 1974.

FERMENTATION

Although the word fermentation referred originally to the anaerobic metabolism of organic compounds by microorganisms or their enzymes to produce products simpler than the starting material, the modern definition is that of any microbial action controlled by man to make useful products. Some of the substances produced from carbohydrates on a commercial scale by anaerobic microbial metabolism include ethanol (qv) and lactic acid (see Hydroxy carboxylic acids) and products containing these materials such as beer (qv), wine (qv), vinegar, pickles, and sauerkraut (see also Food processing). Products from aerobic metabolism include antibiotics (qv), organic acids (see Carboxylic acids), enzymes (qv), and vitamins (qv).

Molds, yeasts (qv), bacteria, and streptomycetes are among the microorganisms used in fermentation processes today (see Microbial transformations). Most of the successful fermentation operations have some degree of simplicity, a rather uniform series of techniques, and the possibility of using the same manufacturing facilities to make many chemically unrelated substances.

Microbiological Aspects

The most important consideration in industrial fermentations is the selection of the proper microorganism. This choice must provide for production of product at the highest possible concentration and yield. Stock cultures of microorganisms useful in industrial fermentations are usually maintained by the manufacturer, and less frequently, in commercial collections or collections at academic or government laboratories.

For antibiotic-producing fermentations, the careful selection of a microorganism to produce high titers of drugs appears to be the most important method for increasing productivity of the process. Other important factors are formulation of the medium used for growth of the microorganism, and changes in bioengineering technology. New developments in recombinant DNA techniques and microbial genetics may lead to technology that can be generally applied to all microorganisms of interest for industrial processes.

Fermentation Operations

The microorganisms used in industrial fermentations require various nutrients for growth, including carbohydrates, nitrogen-containing compounds (or ammonium salts), growth factors, vitamins, and minerals. In most of the fermentations, these nutritional requirements are met by including in the medium peptones, yeast products, agricultural materials, eg, cornsteep liquor (steepwater), soybean meal, cottonseed flour, and buffer salts.

Specific precursors are sometimes added to fermentations to increase the rate of formation and amount of product formed. These include substances such as cobalt salts for the vitamin B_{12} fermentation, and phenylacetic acid and phenoxyacetic acid for fermentations producing penicillin G (benzylpenicillin) and penicillin V (phenoxymethyl penicillin) respectively.

In most fermentation processes, the nutrient medium is first sterilized to inactivate or kill extraneous microorganisms which may be present. This sterilization is usually accomplished by heating at high temperatures (eg, 150°C) for a few minutes or at lower temperatures (eg, 126°C) for longer periods. Any material that enters the fermenter during the process must be sterilized. Air may be sterilized by filtration through hydrophobic sterilizable filters, and nutrients are often sterilized by continuous or batch sterilization at 126°C.

Experience accumulated in the past 35 years suggests that the desired metabolic changes taking place in the fermentation frequently have a rather narrow temperature optimum. Product yields can often be increased through control of the pH of the growing culture or the level of dissolved oxygen and this can be programmed and controlled by using computerized process control to optimize product yield.

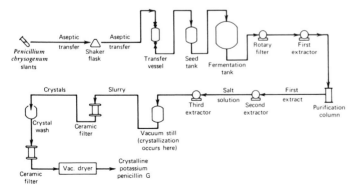

Figure 1. Flow sheet for penicillin G process.

Vessels for fermentation operations range in size from 380-L (100-gal) for production of specialized products, including enzymes and vaccines, to 900 m³ (237,000 gal) (see also Vaccine technology). A flow sheet of a typical fermentation process for production of penicillin G is shown in Figure 1.

The production of antibiotics and the use of microorganisms as replacements for chemical operations in complicated chemical syntheses are becoming increasingly common. Uses of microorganisms are being considered for preparation of peptide hormones, eg, somatostatin, insulin (qv), and growth hormone, substances that are quite foreign to the microbiologists's domain. This potential application of genetics to fermentations, is so large that there may be a complete revolution in the fermentation industries in the next 10 years in that previously scarce human proteins will be produced at a low cost by large-scale fermentation procedures (see Genetic engineering).

D. PERLMAN‡
University of Wisconsin

Reviewed by
MICHAEL C. FLICKINGER
National Cancer Institute
Frederick Cancer Research Facility

H.J. Peppler and D. Perlman, eds., *Microbial Technology*, 2nd ed., Vol. I and II, Academic Press, Inc., New York, 1979.

D.I.C. Wang and co-authors, *Fermentation and Enzyme Technology*, John Wiley & Sons, Inc., New York, 1979.

A.H. Rose, ed., *Economic Microbiology*, Vols. 1–5, Academic Press, Ltd., London, UK, 1977–1981.

B. Atkinson and F. Mavituna, eds., *Biochemical Engineering and Biotechnology Handbook*, The Nature Press, New York, 1983.

FERRITES

The term ferrite has been misused but is commonly a generic term describing a class of magnetic oxide compounds that contain iron oxide as a primary component as distinct from the metallurgical term that applies to the compound Fe_3C (see Iron compounds; Magnetic materials). More specifically, there are several crystal-structure classes of compounds loosely defined as ferrites, such as spinel, magnetoplumbite, garnet and perovskite structures. All ferrites share the property of spontaneous magnetization, a magnetic induction in the absence of an external magnetic field.

Structure and Chemistry

The magnetic properties of ferrites derive directly from the electron configuration of the ion and their interactions with each other. Although the specific structures differ, they can all be considered to be composed of two sublattices; a rigid anion lattice composed of the relatively large (r = 0.132 nm) oxygen anions and the cation sublattice formed by the filling of holes (interstitial sites) with the smaller cations.

Spinel ferrites. This important class has the general composition AB_2X_4 and is isostructural with the mineral spinel $MgAl_2O_4$. The structure is a cubic close packing of the anions with a variety of A and B cations capable of filling the interstitial sites. The smallest crystallographic unit cell that has the required cubic symmetry contains eight formula units of AB_2X_4. Each unit cell has two types of interstitial sites that can be occupied by the A and B cations; thus, there are 64 tetrahedral sites (8 of which are occupied) and 32 octahedral sites (16 of which are occupied).

The spinel structure may be formed from various anions including sulfur (thiospinels), chlorine (halospinels), and oxygen (see Fig. 1). The cations in the structure must satisfy both size and neutrality charge considerations. For a simple spinel $A^m + B_2^n + O_4$, this would mean that $m + 2n = 8$. The materials of greatest technical and commercial interest

‡ Deceased.

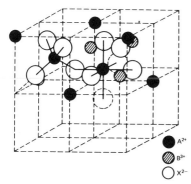

Figure 1. Spinel structure.

are the oxide spinels based on oxides of iron (ferrospinels). Within this group, materials such as lithium ferrite $Li_{0.5}^{1+}Fe_{2.5}^{3+}O_4$ and the whole family of $M^{2+}Fe_2^{3+}O_4$ spinels are found (M^{2+} = Mg., Mn, Fe, Co, Ni, Zn, Cu, etc).

The spontaneous magnetization (M_s) arising in the spinel ferrites is a result of the interactions of transition metal cations and the oxygen anions. Consequently, the magnetic properties can be considered to be the sum of the magnetic interactions of the A-site sublattice and B-site sublattice. In the spinel ferrites, these two sublattices have magnetic moments that are oppositely aligned and the magnetic moment per molecule can be calculated from the distribution of the cations between the A and B sites.

Because the magnetic saturation depends on the site location and the d-electron structure of the transition-metal cations, it is possible to systematically alter the net magnetic moment by chemical substitutions. The classic example is the increase of the magnetic moment in the mixed zinc ferrites. In this case, the addition of nonmagnetic zinc ions, which prefer to occupy the tetrahedral sites, reduces the magnetic moment on the A-site sublattice and thus increases the net magnetic moment with increasing zinc substitution.

Hexagonal ferrites. Hexagonal ferrites are a group of ferromagnetic oxides in which the principal component is Fe_2O_3 in combination with a divalent oxide (BaO, SrO, or PbO) and a divalent transition-metal oxide. Phases are designated as W, Y, Z, and M hexagonal ferrites. Most hexagonal ferrite materials in use today as permanent-magnet materials are of the M hexagonal type which is isostructural with the mineral magnetoplumbite, $PbFe_{7.5}Al_{0.5}Ti_{0.5}O_{19}$. This structure can be considered as being built of oxygen (anion) layers that have hexagonal cubic packing. The fifth layer contains a barium or other divalent cation in place of one fourth of the oxygen ions. The resultant structure contains interstitial sites having a fourfold (tetragonal), fivefold (trigonal bipyramid), and sixfold (octahedral) coordination. The octahedral sites are of three different types. Consequently, the cations have five different types of environments.

In contrast to the spinel ferrites, where the object is to produce a material with the lowest possible value of the magnetocrystalline anisotropy (typically $0–10^{-11}$ J/cm^3 at room temperature) in order to maximize permeability and reduce hysteresis losses, the M-type hexagonal ferrites are useful because of their high anisotropic value (typically 3×10^{-1} J/cm^3). This large anisotropy, caused by the crystal structure, makes it difficult to turn the magnetization away from the easy direction of magnetization, and furthermore, if the size of the magnetic particles is kept below a certain value (1.6 μm for barium ferrite), the particles are single-domain and very difficult to demagnetize.

Garnets. Garnets represent a series of compounds having the general structure $M_3Fe_5O_{12}$ and are isostructural with the ideal mineral $M_3Al_2Si_3O_{12}$. Although cubic, the garnet structure is considerably more complicated than the spinel. The unit cell contains eight formula units or 160 ions. Within the structure there are 24 tetrahedral and 16 octahedral sites which can accommodate the small Fe cation (r = 0.055 nm) and other cations of similar size. Additionally, there are 24 dodecahedral sites (eightfold coordination) that can accommodate Y, La, Ca, the rare earths, and other large cations. As was the case with both the

hexagonal and spinel ferrites, there are two magnetic sublattices opposed to each other.

Preparation and Processing

Most ferrites are prepared as ceramic materials by what is commonly known as standard ceramic processing. In this technology, the constituent raw materials, oxides, hydroxides, or carbonates are weighed and first milled in a steel mill using steel balls as the milling media and water as the carrier. The raw materials are mixed to yield a homogeneous mixture on a macroscopic scale. In the calcining (called presintering) reaction, the raw materials are heated to 1073–1573 K and form the ferrite compound. The carbonates decompose and react by solid-state diffusion to form the final compound. Following the calcining reaction, the material is milled again to further homogenize the material and reduce the particle size to permit subsequent pressing and sintering. The material is then granulated so that it will be free flowing and can be dry pressed into the desired shape. In the sintering process, the ceramic material is densified and the final magnetic properties are developed. Some materials such as the iron-deficient nickel–zinc ferrites and the M-type hexagonal ferrites may be fired in air because all the cations exist at their highest valence state. However, this is not the case with the manganese–zinc ferrites.

Applications

Main applications of spinel ferrites are linear or low signal applications; telecommunications circuit elements switched-mode power supplies; and digital recording heads for computer memory applications. Hexagonal ferrites are low cost, have low density and high coercive force. They would be best applied where a high coercive force is needed, eg, the ceramic magnet is having increasing application in d-c permanent-magnet motors, especially in automotive applications (eg, window-lift, blower, windshield-wiper motors).

Thomas G. Reynolds, III
Ferroxcube Corporation

A. Broese van Groenov, P.F. Bongers, and A.L. Stuifts, *Mater. Sci. Eng.* **3**, 317 (1968–1969).

J. Smit and H.P.J. Wijn, *Ferrites*, John Wiley & Sons, Inc., New York, 1959, p. 143.

T.G.W. Stijntjes and co-workers in Y. Hoshino, S. Iida, and M. Sugimoto, eds., *Ferrites: Proceedings of the International Conference*, University Park Press, Baltimore, Md., 1971, p. 195.

FERROCYANIDES. See Iron compounds.

FERROELECTRICS

The term ferroelectric was first used to emphasize the analogy between the nonlinear hysteretic dielectric properties of Rochelle salt (sodium potassium tartrate tetrahydrate, $NaKC_4H_4O_6 \cdot 4H_2O$) and the magnetic behavior of ferromagnetic iron. More recently, the concept has been generalized and the term ferroic has been used to describe all materials that exhibit one or more phases showing a twin or domain structure in which the individual domain states may be reoriented by applied magnetic, electric, or elastic stress fields or combinations of such fields (see Magnetic materials).

It is evident that in the ferroelectric, the domain states differ in orientation of spontaneous electric polarization, and that the ferroelectric character is established when it is evident that the states can be transformed one into another by suitably directed electric fields. Spontaneous polarization (P_s), remanent polarization (P_r), and coercive field (E_c) are defined by analogy with corresponding magnetic quantities. Clearly, it is the reorientability of the domain-state polarizations that distinguishes ferroelectrics as a subgroup from the larger class of pyroelectric crystals in the ten polar point symmetries.

At present there are more than 38 different structural families of ferroelectrics. As an example, within the perovskite group, in addition to barium titanate, there are more than 60 ferroelectric compounds, and crystalline solid solutions between more than 150 of these different member compositions have been explored.

Properties

In many (not all) ferroelectrics, the domain-state spontaneous polarization is a decreasing function of temperature, going to zero at a phase-transition temperature T_c which is called the Curie point. Above T_c a number of ferroelectrics exhibit very high, temperature-dependent dielectric susceptibility (η) such that one component of the susceptibility tensor follows a Curie-Weiss law:

$$\eta_{ii} = \frac{C}{T - \theta}$$

where C is called the Curie constant and θ the Curie-Weiss temperature. In practice, this means that very high useful dielectric susceptibilities (permittivities) persist for a wide range of temperature above T_c in the paraelectric phase. It is this high intrinsic "softness" in the dielectric response that is the phenomenon most used in the practical application of ferroelectrics, and polycrystalline ceramic ferroelectric dielectrics with relative permittivities ranging up to 10,000 are very widely used in compact disk, tubular, and multilayer capacitors (see Dielectrics below).

In ferroelectrics, both P_s and P_r are strong functions of temperature, particularly at temperatures close to T_c thus, these crystals and poled ceramics have very high pyroelectric coefficients and may be applied in thermometry and in bolometric sensing devices (point detectors) of infrared radiation.

Definition

A ferroelectric crystal is a material that exhibits one or more ferroelectric phases in a realizable range of temperature and pressure. In the ferroelectric phase, the crystal is spontaneously electrically polarized and the polarization has more than one possible equilibrium orientation. To establish ferroelectricity, it must be demonstrated that the polarization can be reoriented between orientation states by a realizable electric field.

Classification schemes. A comprehensive classification of all possible ferroelectrics has been given on symmetry grounds. An earlier classification—hard–soft ferroelectrics—was based on the qualitative elastic properties and is no longer widely used.

Antiferroelectrics. Antiferroelectric is a term reserved for crystals with anomalously high dielectric permitivity that exhibit phase changes into ferroelastic forms in which the volume of the primitive unit cell is an integral multiple of that in the prototype structure and appears to be well described as a superlattice of two interpenetrating but compensating ferroelectric-like displacement systems.

Ferroelectric Materials

Perovskites. The simple cubic perovskite structure, which is the high temperature form for many mixed oxides of the ABO_3 type, was one of the first simple structures to exhibit compounds with ferroelectric properties and is still probably the most important ferroelectric prototype.

The very simple cubic structure (point symmetry $m3m$) (Fig. 1) is made up of a regular array of corner-sharing oxygen octahedra with smaller highly charged cations like Ti, Sn, Zr, Nb, Ta, W, etc, occupying the central octahedral B site, and lower-charged larger cations like Na, K, Rb, Ca, Sr, Ba, Pb, etc filling the interstices between octahedra in the larger 12-coordinated A sites.

Many simple ABO_3 compounds have interesting high temperature ferroelectric or antiferroelectric phases. The perovskite structure is, however, tolerant of a very wide range of multiple cation substitution on both A and B sites, so that many more complex compounds such as $(K_{1/2}Bi_{1/2})TiO_3$, $Pb(Fe_{1/2}Ta_{1/2})O_3$, $Pb(Co_{1/4}Mn_{1/4}W_{1/2})O_3$ can be prepared.

Perovskite-related octahedral structures. Lithium niobate, $LiNbO_3$, and $LiTaO_3$ have a structure related to the perovskites, but of lower symmetry ($3m$), due to the fact that the oxygen octahedra share edges. (Fig. 2). In sequence along the trigonal c axis in the low temperature

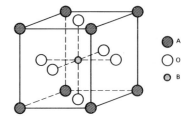

Figure 1. Cubic ($m3m$) prototype structure of perovskite-type ABO_3 compounds. For $BaTiO_3$, $A = Ba^{2+}$, $B = Ti^{4+}$, and $O = O^{2-}$.

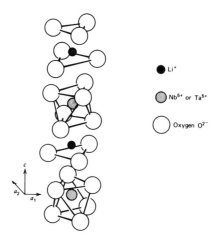

Figure 2. Structure of ferroelectric $LiNbO_3$ and $LiTaO_3$.

ferroelectric phase, the octahedra are occupied in the sequence Nb(Ta), vacancy, Li, Nb(Ta), vacancy, and Li. Lithium niobate is widely used in the single-domain form for both piezoelectric and electrooptic devices. The tantalate has a wide application in simple pyroelectric point detectors, and more recently, for acoustic wave devices.

KDP family. Potassium dihydrogen phosphate and the isomorphous rubidium and cesium phosphates, RbH_2PO_4, CsH_2PO_4, and arsenates (RbH_2AsO_4, CsH_5AsO_4 were the first important family of ferroelectrics to be discovered. In the single-domain ferroelectric phase, the protons order along the bonds in such a manner that they are all near upper or lower oxygens of the tetrahedral phosphate groups, and it is this coupling that pushes the phosphorus and potassium from the high-symmetry positions giving rise to a switchable moment along the c axis.

Principal applications of KDP family materials use properties in the paraelectric prototypic form. Frequently, the properties are far from ideal, but the tractable water-solution growth of KDP has permitted large crystals of excellent quality, which are difficult if not impossible to duplicate in the more desirable refractory oxides.

Preparation of Ferroelectric Ceramics

Powder preparation. Conventional (mixed oxide) methods use mostly oxides, eg, TiO_2, ZrO_2, PbO, or carbonates, eg, $BaCO_3$, $SrCO_3$, $CaCO_3$, as raw materials. These have to meet certain application-dependent specifications with respect to purity and grain-size distribution. The final powder product with perovskite structure is formed by a solid-state reaction (calcination) of these starting raw materials. A typical example is

$$BaCO_3 + TiO_2 \rightarrow BaTiO_3 + CO_2$$

Usually, the required product is a fully homogeneous compound having perovskite structure with small particle size ($< 1 \ \mu m$). The solid-state reaction occurs by solid-state diffusion with a relatively slow diffusion rate.

Currently, several methods for wet-chemical preparation of ferroelectric ceramics are used (mostly on a laboratory scale): thermolysis of compounds consisting of the critical elements in an exactly defined stoichiometric ratio in a closely packed arrangement on the atomic scale; homogeneous distribution of the different ions in a liquid solution which can be transferred into the solid state by coprecipitation of the different ions from aqueous solutions; precipitation from metal alcoholates by hydrolysis; precipitation from aqueous solutions by spraying into a liquid in which there is no solubility; and reaction of an aqueous citrate solution with a polyhydric alcohol to form an organic glass (polyester resin).

Shaping. Various amounts of binders (mostly organic compounds) are added to the powders for better rheological properties, eg, thin sheets are made by pouring a controlled thickness of special slip onto a polished metal belt, then the slip is dried and stripped.

Sintering. The green ferroelectric ceramics are sintered at high temperatures for 1–10 h to achieve a high density, 95–100% of the theoretical. Proper sintering is most important for the quality of the products and is generally carried out in electric furnaces for better control of temperature and atmosphere.

Uses

Applications of ferroelectrics include piezoelectric devices, eg, spark generators, bandpass filters, and transducers; dielectrics, eg, in capacitor applications; semiconducting, positive temperature coefficient (PTC) materials used as current limiters, degaussing units in color-television sets, and self-controlling heating elements.

L.E. CROSS
The Pennsylvania State University

K.H. HARDTL
Philips Forschungslaboratorium Aachen GmbH

A. Glass and M.E. Lines, *Principles and Applications of Ferroelectrics and Related Materials*, Clarendon Press, Oxford, UK, 1977.
Landolt-Bornstein, *Ferroelectric and Antiferroelectric Substances*, Vols. 3, 9, New Series Group III, Springer-Verlag, Berlin, Ger., 1969, 1975.

FERTILIZERS

Like all living things, vegetation requires certain foods for its survival and growth. Fertilizers are materials added to the soil, and sometimes to foliage, to supply nutrients to sustain plants and promote their abundant and fruitful growth. The elements that constitute these plant foods are divided into three classes: primary—nitrogen (N), phosphorus (usually expressed as P_2O_5) and potassium (expressed as K_2O); secondary—calcium (Ca), magnesium (Mg), and sulfur (S); and minor or so-called micronutrients—iron (Fe), manganese (Mn), copper (Cu), zinc (Zn), boron (B), and molybdenum (Mo) (see Mineral nutrients). In addition to their role as nutrients, calcium and magnesium are important in adjusting the pH and tilth of the soil.

All the nutrient elements are present to some extent in soil or other natural materials in forms useful for plants but, with the frequent exception of micronutrients, such supply is generally not adequate for sustained and economic production of crops. Fertilizer is, therefore, one of the most important products of the chemical industry.

Nomenclature

Discussing the fertilizer industry requires a number of special terms peculiar to fertilizers, some of which are traditional. Analysis denotes the concentration of plant nutrients; high analysis denotes fertilizers of relatively high nutrient concentration. Grade defines the nutrient content expressed in weight percentage, eg, 10-10-10 usually denotes a fertilizer containing 10 wt% N, 10 wt% P_2O_5, and 10 wt% K_2O. Unit refers to 1% of a ton of a nutrient. In the metric system, 1 unit is 10 kg and a 10-10-10 fertilizer contains 10 units or 100 kg each of N, P_2O_5, and

K_2O. When the units are in short tons, as in the United States, a 10-10-10 fertilizer contains 91 kg each of N, P_2O_5, and K_2O. BPL (bone phosphate of lime) is the phosphate content of phosphate ore expressed as tricalcium phosphate, $Ca_3(PO_4)_2$. APA (available phosphoric acid) is the phosphate content (as P_2O_5) available to plants as determined by an empirical solubility test. Muriate denotes potassium chloride. Straight fertilizer is usually one containing only one nutrient that is applied to the soil uncombined with other materials. Mixed fertilizer is one containing two or more nutrients. Complex fertilizer is a multinutrient fertilizer such as ammonium phosphate or nitric phosphate, made by processes based mainly on neutralization of an acid or other chemical interaction of ingredients. Compound fertilizers contain two or more nutrients (not used in U.S.). These terms are often used interchangeably so they are vague and often confusing. Materials denotes nutrient carriers used in making mixed fertilizers. Formulation refers to the amounts of materials used for a given weight of finished multinutrient fertilizer of specific grade. Direct application is the application to the soil of a primary fertilizer, such as diammonium phosphate, superphosphate, or ammonia, without first combining it with other fertilizer materials. Conditioning is the treatment of fertilizer to reduce hygroscopicity and caking in storage.

The following definitions of terms are widely used in the phosphate rock industry. Phosphate rock designates either an apatite-bearing rock containing enough P_2O_5 to be processed into fertilizer or phosphorus, or a beneficiated apatite concentrate. Phosphorite is a sedimentary rock in which a phosphate mineral is a principal constituent. Ore or matrix is material that can be mined at a profit. Grade of ore is expressed in terms of phosphate content, either as P_2O_5, BPL, or TPL (triphosphate of lime).

Raw material resources. The main raw materials required for manufacturing fertilizers are hydrogen and nitrogen for ammonia synthesis, phosphate rock, and potassium minerals for phosphorus and potassium, and sulfur. Sulfur, in addition to being a secondary macronutrient, is a key raw material for the fertilizer industry because of its use in production of sulfuric acid, the material most often used to solubilize phosphate rock (85% of fertilizer phosphate being derived from phosphoric acid is made by use of sulfuric acid), although nitric acid can be used (nitrophosphate, about 12% of fertilizer phosphate, is made by use of nitric acid).

The other secondary macronutrients, calcium and magnesium, are widespread in agricultural soils. Nutrient deficiencies of these are unusual, although serious magnesium deficiency occasionally occurs.

Nitrogen. Virtually all commercial nitrogen fertilizer is made from ammonia and more than 80% of all industrial nitrogen is used in fertilizer production. Of this at least 85% of the world supply of ammonia for fertilizers is manufactured by fixation of atmospheric nitrogen with process hydrogen (see Ammonia; Ammonium compounds).

Hydrogen. Hydrogen, like nitrogen, in absolute amount is virtually limitless, principally in the forms of hydrocarbons and water. Natural gas is by far the preferred source of hydrogen; it has been cheap and its use is more energy efficient than that of other hydrocarbons (see Gas, natural). Since 1960, about 95% of the synthetic ammonia in the U.S. has been made from natural gas; worldwide the proportion is about 85%.

Phosphate rock. Natural mineral deposits are the source of phosphorus for all manufactured phosphate fertilizer. Phosphate deposits are numerous throughout the world but their size and quality vary widely. Those minerals that contain enough phosphorus to be potential sources for industrial use are called phosphate rock and phosphate ore (see Phosphoric acids and phosphates).

Potash. Potassium is the seventh most abundant element in the earth's crust. The raw materials from which potash fertilizer is derived are principally bedded marine evaporite deposits but other sources include surface and subsurface brines and, to a limited extent, by-products of the sugar and cement industries. Both underground and solution mining are used to recover evaporite deposits and fractional crystallization is used for the brines. The potassium salts of marine evaporite deposits occur in beds in intervals of halite (NaCl) which also contains bedded anhydrite ($CaSO_4$) and clay or shale. The K_2O content of such deposits varies widely (see Potassium compounds). The British Sulphur

Corporation recently reported the world recoverable K_2O at ca 90×10^9 t. In addition, there are extensive insoluble minerals and ores, by-product K_2O, and the oceans which contain 3.9×10^5 t K_2O/km^3 of seawater.

Sulfur. About 83% of manufactured fertilizer is based on the use of sulfuric acid. The weight ratio of sulfur to phosphorus in fertilizer manufacture has increased steadily from 1.32 in 1950 to 1.8 in 1977. World reserves are estimated at 2.2×10^9 t, excluding coal and oil shales. U.S. reserves are 254×10^6 t.

Secondary Macronutrients

Calcium, the fifth most abundant element in the earth's crust, is widely distributed in nature as calcium carbonate, chalk, marble, gypsum, fluorspar, phosphate rock, and also other rocks and minerals. Magnesium content of soils is usually less than calcium because Mg is more soluble. Important sources are dolomitic limestone, epsom salt, calcined kieserite, magnesia, potassium–magnesium sulfate, and a few organic sources (see Magnesium compounds).

Micronutrients

Boron production in the U.S. and in about three fifths of the world comes from bedded deposits and lake brines in California. Copper resources are estimated at 408×10^6 t Cu worldwide, one-fifth in the U.S. Iron world reserves are placed at 236×10^9 t of ore containing 90×10^9 t iron (qv). Manganese world reserves are ca 1.8×10^9 t Mn U.S. reserves of manganese ore containing 35% or more of Mn virtually nonexistent (U.S. resources are ca 67×10^6 t containing Mn). Molybdenum world reserves and resources are ca 6 and 23×10^6 t, respectively; U.S. reserves and resources are ca 3 and 13×10^6 t, respectively (see Molybdenum). Zinc (qv) deposits in the U.S. are extensive and reserves are estimated at ca 27×10^6 t.

Nitrogen

Nitrogen is the nutrient used in fertilizer in the largest amount, being present in amounts almost double that of each of the other primary nutrients.

Ammonia. Ammonia is the basis for nearly all commercial nitrogen fertilizer, and about 85% of all industrial ammonia worldwide is used in fertilizer.

Ammonium nitrate. Broadly defined, fertilizer ammonium nitrates include high grade or straight ammonium nitrate, AN (33–34% N), ammonium sulfate–nitrate, ASN (26% N), and calcium–ammonium nitrate, CAN (20.5–28% N). The two lower grades are used primarily because of their greater safety and better storage and handling qualities. Additional ammonium nitrate is a component of nitrophosphate but this complex fertilizer is treated as a separate material later in the article (see Ammonium compounds).

Urea. About 75% of all urea is used as fertilizer, recently becoming the largest-volume nitrogen fertilizer. Other uses are as animal feed and industrial chemicals (see Urea). The growth of urea granulation results from its use in bulk blends, as forest and rice fertilizer, and for coated fertilizer (lower surface per unit weight) and its generally simpler storage and handling.

Properties particularly important to urea use as fertilizer include: N content (pure, 46.6%), 45–46%; solubility at 25°C, 119 g/100 g H_2O; melting point, 132.7°C; hygroscopic point (critical relative humidity at 30°C), 75.2%; specific gravity d_4^{20}, 1.335; and heat of solution, -60 J/kg (-14.34 cal/kg).

The raw materials for urea synthesis are ammonia and the by-product of ammonia production, carbon dioxide. The ammonia and CO_2 are fed to a high pressure reactor (up to 30 MPa or 300 atm) and at temperatures of about 200°C where they react to form ammonium carbamate, urea, and water according to the following reactions:

$$2 NH_3 + CO_2 \rightleftharpoons NH_4COONH_2 + \text{heat} \qquad (1)$$

$$NH_4COONH_2 \rightleftharpoons NH_2CONH_2 + H_2O - \text{heat} \qquad (2)$$

Prilling has been the most widely used method of producing solid urea, but the use of granulation is increasing rapidly. In prilling, anhydrous molten urea is forced through spray heads at the top of a tower (21–52

m in height) to produce droplets which fall through a countercurrent stream of air where they solidify to form prills. The temperature and rate of flow of air also affect the size and quality of prills.

Urea granules of preferred size are produced in a drum granulator (spherodizer) process developed by Cominco Inc, in conjunction with C & I Girdler, by pan-granulation methods developed by TVA and by Norsk Hydro, and by a falling-curtain drum-granulation method recently developed by TVA.

Nitrogen solutions. Nitrogen solutions are an important class of fertilizer in the U.S. and they are increasing in popularity abroad. The high analysis solutions are both nonpressure and low pressure, 170–212 kPa (10–16 psig) at 40°C and containing free ammonia. The most popular is the nonpressure material made from urea and ammonium nitrate, UAN.

The largest use for nitrogen solutions is in direct application, but those solutions containing free ammonia are also very convenient sources of supplemental nitrogen in formulating complete liquid fertilizers and in granulation processes.

The basis for nitrogen solutions is the far greater solubility of mixtures of ammonium nitrate and urea as compared to that of either material alone. At the optimum ratio of ammonium nitrate and urea, 9 kg of water will dissolve 36 kg of the mixture. A system of nomenclature has been devised, making simple identification possible. The three-digit number preceding, parenthesis shows the nitrogen content to one decimal place but omitting the decimal, eg, 320 signifying 32% N. Inside the parenthesis, the percentages of ammonia, ammonium nitrate, and urea are given in that order; 320 (0–45–35) signifying a 32% N solution containing no ammonia, 45% ammonium nitrate, and 35% urea; the balance is water.

The production of nitrogen solutions is simple. The ingredient solutions are blended in a mixing tank in either a batch or continuous process. The usual feed solutions used contain about 84% ammonium nitrate and 72–85% urea.

Process economics is attractive in comparison to production of solids because evaporation is decreased and granulation, drying, and conditioning are avoided, greatly lowering investment cost.

Ammonium sulfate. Ammonium sulfate, $(NH_4)_2SO_4$ (21.2% N, 24.3% S) is used almost exclusively as fertilizer.

Other nitrogen fertilizers. Other nitrogen fertilizers are used to a small extent. Individually, the usage of several water-soluble forms (calcium nitrate, sodium nitrate, calcium cyanamide, and ammonium chloride) ranges from 0.4% to 0.6% of total world N use; collectively, they amount to about 2% of world N. Their low analysis, poor physical quality and poor compatibility with other materials in mixtures are leading reasons for their status.

Controlled-release materials. Most nitrogen fertilizers are highly water-soluble and tend to be leached, producing cycles of luxury and deficient feeding. Altogether they are inefficient; on average only about 50% of applied N is recovered by crops. To use multiple applications to feed crops more efficiently is labor-intensive, thus costly. Another means for improving the efficiency of N use is through use of fertilizer whose release of N is in some way regulated or slowed to lengthen the period of feeding and decrease leaching and other losses. Such materials are known as controlled-release fertilizers.

Some properties and mechanisms that provide a measure of controlled-release nutrient are low solubility; resistance to attack by soil bacteria (urea–aldehyde reaction products); coating soluble materials to provide a barrier to rapid dissolution, and application with the fertilizer of other materials that inhibit nitrification, denitrification, or urease activity.

Use of controlled-release fertilizers in the U.S. in 1975 was ca 180,000–220,000 t, mostly on nonfarm crops, a use that has continued to grow. Such fertilizers include urea–formaldehydes (ureaforms), isobutylidene diurea, crotonylidene diurea, and sulfur-coated urea.

Phosphate

The principal routes for the production of phosphate fertilizers are sketched in Figure 1.

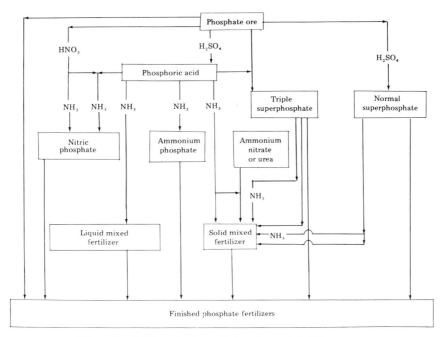

Figure 1. Main routes for producing phosphate fertilizers.

Wet-process phosphoric acid. This type of phosphoric acid is the key intermediate for the phosphate fertilizer industry; over 90% of all the acid produced is used in fertilizer. In the U.S., some 85% of all fertilizer phosphate is based on phosphoric acid.

The main chemical reaction in the acidulation of phosphate rock with sulfuric acid is

$$Ca_{10}F_2(PO_4)_6 + 10 H_2SO_4 + 10x H_2O \rightarrow$$
$$6 H_3PO_4 + 10 CaSO_4 \cdot x H_2O + 2 HF \quad (3)$$

where $x = 0.5$–0.7, or 2.0. This acidulation process is strongly exothermic but with temperature control it can be operated to produce either the calcium sulfate dihydrate (gypsum), the hemihydrate ($CaSO_4 \cdot 1/2H_2O$), or anhydrite ($CaSO_4$).

Wet-process superphosphoric acid. SPA (ca 95%) is used in liquid fertilizers. SPA fertilizer phosphate production, ca 10% of the U.S. fertilizer phosphate consumption, is still growing.

SPA is produced by two methods: vacuum evaporation and submerged-combustion direct heating. About 70% of U.S. production is by vacuum evaporation. Both types of evaporators are fed with 50–54% P_2O_5 acid, which is usually desludged. When the feed acid is made from calcined rock, the product acid is green, but otherwise it is black. Liquid fertilizers are made from SPA because of its capacity for sequestering metallic impurities.

Normal superphosphate. NSP, also called ordinary and single superphosphate, from its beginning as the first commercial phosphate fertilizer has continued among the top fertilizers of the world.

NSP is produced by the reaction of phosphate rock with sulfuric acid. The method quickly yields a solid mass containing monocalcium phosphate monohydrate and gypsum, according to the following simplified equation:

$$Ca_{10}F_2(PO_4)_6 + 7 H_2SO_4 + 3 H_2O \rightarrow$$
$$3 CaH_4(PO_4)_2 \cdot H_2O + 7 CaSO_4 + 2 HF$$

Important properties of NSP are free acid content, as H_2SO_4, 1–2%; moisture content, 5–8%; citrate-soluble P_2O_5, in neutral citrate solution, 20–21%; hygroscopicity at 30°C, 94% relative humidity; and bulk density, nongranular, 0.8 g/cm^3, and granular, 0.97 g/cm^3.

A wide variety of mixers and curing dens has been used over the long history of NSP production but the more advanced technology is illustrated in Figure 2.

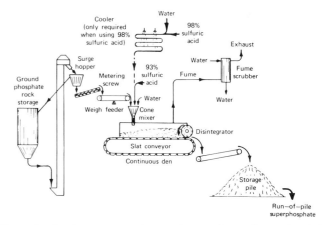

Figure 2. Continuous process for manufacture of normal superphosphate. Courtesy of TVA.

Triple superphosphate. TSP was the first high analysis phosphate fertilizer, containing 45–47% P_2O_5 compared to 20% P_2O_5 in NSP. Simplicity of production, high analysis, and excellent agronomic quality are reasons for the sustained high production and consumption of TSP.

TSP is essentially impure monocalcium phosphate monohydrate ($CaH_4(PO_4)_2 \cdot H_2O$) made by acidulating phosphate rock with phosphoric acid according to the following reaction:

$$Ca_{10}F_2 \cdot (PO_4)_6 + 14 H_3PO_4 + 10 H_2O \rightarrow 10 CaH_4(PO_4)_2 \cdot H_2O + 2 HF$$

Phosphate rock and wet-process phosphoric acid are the raw materials for manufacturing TSP. The grade of rock can be a little lower than that needed for NSP production. Over the years, a large number of process modifications, both batch and continuous, have been used.

Potash

World fertilizer potash production equals about 95% of all commercial potash. This industry is essentially a mining and beneficiation industry: the two main fertilizer materials, KCl and K_2SO_4, are produced by beneficiating ores at the mine site. The upgraded salts are then shipped to distributors and manufacturers of mixed goods.

Potassium chloride. The three beneficiation methods for producing fertilizer grades of KCl are thermal dissolution, flotation, and electrostatic refining.

Potassium sulfate. K_2SO_4 is the preferred form for crops that have a low tolerance for chloride. It is produced from langbeinite by the metathetical reaction in aqueous solution:

$$K_2SO_4 \cdot 2MgSO_4 + 4 KCl \rightarrow 3 K_2SO_4 + 2 MgCl_2 \qquad (4)$$

Potassium nitrate. KNO_3, known but little used as a fertilizer for many years, may be reclaimed as a by-product of the production of sodium nitrate from natural deposits of caliche in Chile, and by the double decomposition reaction between sodium nitrate and potassium chloride.

Potassium phosphates. Potassium phosphates, because of their very high analysis, freedom from chloride, and high solubility in liquid fertilizers, have long been the subject of a search for methods of economical production, not yet attained on a commercial scale.

Secondary Nutrients

Calcium. Soil minerals are a main source of calcium for plants, making nutrient deficiency of this element in plants rare. As a soil amendment for pH adjustment and other soil conditioning, calcium is frequently underused in many acidic soils.

Magnesium. Magnesium, like calcium, is present in soil minerals, as well as being widespread in massive mineral forms. Magnesium minerals, however, are more soluble than those of calcium and its content in soils is lower than that of calcium. Dolomite, calcium–magnesium carbonate, is often used as a liming agent and to provide nutrient magnesium.

Sulfur. Sulfur occurs in some plants at amounts as high as that of phosphorus. It is absorbed by plants as the sulfate anion (SO_4^{2-}), an important constituent of many fertilizers. Normal superphosphate contains much calcium sulfate. Wet-process phosphoric acid often contains a few percent of sulfur as sulfuric acid and calcium sulfate, and this sulfur appears in fertilizers made from the acid.

Micronutrients

The generally recognized micronutrient elements are boron, copper, iron, manganese, molybdenum, and zinc (and sometimes cobalt). These are required only in micro amounts and some can be harmful to plants when used in excess. Prescription formulation is being used increasingly as deficiencies are better identified but also used is the shotgun approach (adding a range of micronutrients to fertilizer at amounts believed low enough to avoid harmful affects but adequate to prevent deficiencies).

Generally, soluble materials are more effective than insoluble ones, especially insoluble oxides. For this reason, many soil minerals containing micronutrient compounds are ineffective for plants (see Mineral nutrients).

The surest way to achieve even distribution of micronutrients in fertilizer granules, and to achieve even distribution with solid fertilizers, is to incorporate the micronutrient material as one of the feeds in the granulation process. Granular grades that have been produced in this way without causing problems are 6-12-12-10 S–2 Mg–0.10 B–0.10 Fe–0.45 Mn–0.35 Zn and 4-12-24-6 S–2 Mg–0.10 B–0.10 Fe–1.0 Zn. Prescription formulation of micronutrient fertilizers is more practical in the production of bulk blends and fluid fertilizers, as these processes are more adaptable to rapid changes in grades and, thus, to custom formulation.

Fluid fertilizers are probably the best micronutrient carriers in terms of homogeneity and even distribution. These are usually nitrogen solutions, clear liquid mixtures, or suspensions. Some micronutrients, however, are applied as simple water solutions or suspensions. Foliar micronutrient sprays often contain only the micronutrient material.

Multinutrient Fertilizers

These materials, generally called compound fertilizers in Europe and other places (but not in U.S.), represent a very important class of fertilizers both in terms of amount and in the large and varied technologies for their production. They contain more than one primary nutrient,

Table 1. Properties of Pure MAP and DAP

Property	MAP	DAP
N, %	12.2	21.2
P_2O_5, %	61.7	53.8
specific gravity	1.803	1.619
ammonia vapor pressure, kPa[a]		
73°C	insignificant	107.6
100°C	insignificant	140.0
125°C	0.004	300.0
solubility at 20°C, g/L	299.0	569.0
critical humidity at 30°C, %	92.0	83.0

[a] To convert kPa to psi, multiply by 0.145.

in whatever form—liquid, solid, granular, or nongranular—combined in a single chemical compound or as several compounds.

Four principal types of mixtures are discussed: those made generally by ammoniation–granulation (includes ammonium phosphates, ammonium phosphate-based materials, and superphosphate-based materials and some granular materials made without ammoniation); nitrophosphates; blends of granular materials; and fluid mixed fertilizers (clear liquids and suspensions).

Ammonium phosphates. In addition to DAP and MAP, ammonium polyphosphate (APP) is used extensively in liquid fertilizers, and recent developments of economical methods for producing it in granular form open the way for its increasing use in solids applications.

Because of the economy of their production and distribution, excellent physical properties, and excellent agronomic performance, these materials are applied in all phases of fertilizer production and use. The properties of monoammonium phosphate (MAP), $NH_4H_2PO_4$, and diammonium phosphate (DAP), $(NH_4)_2HPO_4$, are shown in Table 1.

Nitrophosphates. There are 80–85 nitrophosphate plants in the world, two thirds of which are in Europe, having a total annual capacity of about 3.18×10^6 t P_2O_5 (ca 12% of the currently estimated world capacity for chemically produced phosphate fertilizer).

Nitrophosphates are made by acidulating phosphate rock with nitric acid followed by ammoniation, addition of potash as desired, and granulation or prilling of the slurry. The acidulate contains calcium nitrate and phosphoric acid or monocalcium phosphate according to the equations below:

$$Ca_{10}F_2(PO_4)_6 + 20 HNO_3 \rightleftharpoons 10 Ca(NO_3)_2 + 6 H_3PO_4 + 2 HF \qquad (5)$$

$$Ca_{10}F_2(PO_4)_6 + 14 HNO_3 \rightleftharpoons 3 CaH_4(PO_4)_2 + 7 Ca(NO_3)_2 + 2 HF$$
$$(6)$$

Bulk Blends

A modern bulk-blend fertilizer is a mechanical mixture of granular fertilizer materials. The bulk blender often mixes grades to the farmer's order or to recommendations based upon soil analysis. The mixtures, which may contain micronutrients and pesticides, is commonly marketed through bulk spreading services, usually within a radius of about 40 km.

Bulk-blend plants consist of bulk storage areas for raw materials, facilities for weighing and transferring materials, a simple mixer (usually a rotary drum) and a discharge and loading system.

Liquid Mixed Fertilizers

Those discussed here are multinutrient fluids, both clear solutions and suspensions. Like bulk-blend plants, liquid-mix plants purchase the basic ingredients used, and many of them market other materials as well as liquid mixes. Most plants are small, directly serving farmers in limited areas with prescription-blended liquids, spreading services, soil-needs advice, and other services. A recent survey of the U.S. industry showed an average annual materials throughput per plant to be 2400 t represented by 1050 t of liquid mixtures, 650 t of anhydrous ammonia, 630 t of nitrogen solution (UAN) and 570 t of liquid direct-application materials such as 10-34-0 and 8-20-0.

The main advantages of clear liquid fertilizers are ease of handling, homogeneity, accurate application, water solubility, freedom from dust, and the ability of many blends to carry micronutrients. Their chief

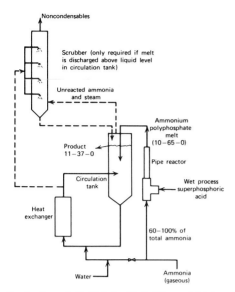

Figure 3. Production of 11-37-0 liquid fertilizer of high polyphosphate content from low conversion, wet-process superphosphoric acid.

disadvantages are comparatively low analysis and generally higher cost of their raw materials.

Suspensions have most of the advantages of clear liquids and, in addition, can compete in nutrient content and cost with high analysis granular fertilizers and bulk blends. They are compatible with most pesticides, and usually micronutrients can be added (dissolved and suspended) in the amounts desired. High purity, high cost raw materials are not required for their successful production. Suspensions represent the fastest-growing segment of the industry in the early 1980's.

Production methods. Many small liquid-fertilizer plants in the U.S. use cold-mix batch processes. A base liquid such as 8-24-0, 10-34-0, or 11-37-0 is blended with other raw materials such as UAN solution, potash, and water to make the desired product such as 7-21-7. Clear liquids of this kind are used mostly as starter fertilizers.

A hot-mix plant is used for the reaction of phosphoric acid and ammonia to furnish the ammonium phosphate of the NPK fertilizer. The widely used NP base-solution (11-37-0) process is based on a pipe-reactor method. By feeding low conversion wet-process superacid (20–30% of the P_2O_5 as polyphosphate) and ammonia vapor through the T ports of a pipe reactor, the heat generated in the reactor (315–400°C) converts some 90% or more of the phosphate to polyphosphate. See Figure 3.

Environmental Aspects of Fertilizer Production and Use

Of the six primary industrial air pollutants deemed injurious to human health, three are directly involved in fertilizer manufacture: sulfur oxides (from sulfuric acid production), nitrogen oxides (from nitric acid and ammonium nitrate production), and particulates.

In addition to ambient air-quality standards and emission standards, fertilizer manufacturers are required to control the environment in the working area in accordance with OSHA regulations. Table 2 gives OSHA standards for air contaminant associated with fertilizer manufacture in 1979.

The future. The continuing upward trend in U.S. natural-gas prices has been a large factor in depressing ammonia production in the United States. Natural-gas prices are substantially lower in Mexico, Venezuela, parts of Europe, and the Middle East, providing advantages for ammonia production in such areas. As a result, imports of ammonia into the United States have increased, causing shutdown of numerous ammonia plants. Importation of ammonia may continue to increase, at least over the foreseeable short range. Along with this trend, exports of U.S. fertilizers have fallen.

The U.S. phosphate industry is also in some difficulty owing to excessive installed capacity at this time (see Phosphoric acids and phosphates).

Table 2. Industrial Hygiene Threshold Limit Values for Fertilizer Manufacture

Fertilizer material	Threshold limit values[a], mg/m³ (ppm)
ammonia	35[b] (50[b])
fluoride (as F)	2.5
HF	2 (3)
H_3PO_4	1
CO	55 (50)
H_2SO_4	1
SO_2	13 (5)
HNO_3	5 (2)
NO_2	9 (5)
dusts	
silica	10
nuisance or inert, total	15
respirable	0.5
noise (8-h)	90 dB[c]

[a] 8-h time-weighted average value.

[b] Values probably will be lowered to 18 mg/m³ and 25 ppm, respectively.

[c] Higher noise levels are permitted when exposure times are less than 8 h daily.

These factors, together with the unknown rate of expected general economic recovery here and abroad, leave the question of rate of recovery of the U.S. fertilizer industry rather speculative.

Fertilizer need worldwide has not diminished. It should be noted, however, that many developing nations are striving to produce as much as possible of their needs in order to reduce outflow of scarce funds. Thus, the pattern of future world fertilizer production may be changed substantially from that of the recent past.

In order to continue in a dominant position in world fertilizer production, the United States will need to overcome the problem of relative excessive cost for ammonia-synthesis feedstock.

E.O. HUFFMAN
Consultant

Worldwide Study of the Fertilizer Industry: 1975–2000, International Centre for Industrial Studies, UNIDO/ICIS, United Nations Industrial Development Organization, Vienna, Austria, 1976.

Fertilizer Manual, International Fertilizer Development Center, United Nations Industrial Development Organization, Reference Manual IFDC-R-1, December, 1979, P.O. Box 2040, Muscle Shoals, Ala.

FIBERBOARD. See Laminated and reinforced wood.

FIBER OPTICS

Fiber-optic lightguides are being developed to improve optical-fiber communication. This article is principally a description of the processes that have allowed lightwave communications to become practical and efficient.

Electromagnetic radiation in the visible or infrared region has frequencies ten thousand times those used for radio communications. Since information-carrying capacity increases with frequency, the potential for lightwave communications is enormous.

A lightwave communication system is composed of three parts: light source, transmission medium, and detector. The earliest light source was the sun in a photophone patented in 1880 by Alexander Graham Bell. The transmission medium in that case was the atmosphere. Then, as now, the loss encountered along the transmission path together with the brightness of the light source and the amplification of the detector, determined the transmission distance limit of the system. In the case of the photophone, disruptions due to fog, rain and snow severely limited the system.

The concept of optical communication lay dormant until 1960 when the laser was first demonstrated to be a bright light source which could be modulated at high bit rates to be used in communication. Again, it was clear that the atmosphere was an unsuitable transmission medium. Light pipes of various types were studied but proved to be inefficient and costly. Glass and plastic waveguides had been used in the past for short distance applications as illuminators and medical scopes. For long distance transmission, higher quality glasses were required.

Light is guided in an optical fiber by total internal reflection according to Snell's law. This is illustrated in Figure 1, and the critical angle is controlled by the difference in refractive index between the core and cladding glasses. There are many designs of optical fiber that fall into two main categories; single mode and multimode.

A critical characteristic of optical fibers is the attenuation or loss that the structure manifests. Early work on glass for optical-fiber transmission concentrated on the conventional glass systems: soda-lime silica, and sodium borosilicate. These glasses contained unacceptable levels of transition-metal impurities when prepared from naturally occurring materials. This led to high absorption losses. Purification methods were developed for the starting materials, and combined with the conventional glass-melting techniques of melting, fining, and drawing into fiber. Fibers of acceptable quality could be prepared in this way. Multimode characteristics were achieved by diffusion between the core and cladding glasses to obtain acceptable mode dispersion. This technique was used successfully by researchers at the U.K. Post Office and AT&T Bell Laboratories.

The performance of multicomponent glass fibers, though acceptable, was eclipsed in the early 1970s by the emergence of high silica fibers prepared by vapor-phase techniques. These fibers are mainly silica (SiO_2) with additions of 1–25 wt% germania (GeO_2). The loss of these fibers was greatly improved owing to the purity of the starting materials: silicon tetrachloride and germanium tetrachloride. The four processes that have emerged based on vapor-phase reactions of these materials are outside vapor-phase oxidation (OVPO), modified chemical-vapor deposition (MCVD), plasma (activated) chemical-vapor deposition (PCVD), and vapor-axial deposition (VAD). Processing differs greatly in form and chemistry. However, each is capable of producing fibers with excellent optical and physical properties.

OVPO. OVPO, outside vapor-phase oxidation, is a process that has been developed at Corning Glass Works and has remained largely proprietary to that company. The process makes use of flame hydrolysis to generate submicrometer-sized amorphous particles. These are deposited on a rotating mandrel to produce a "soot" preform. This preform is subsequently dehydrated, consolidated and finally drawn into fiber. These steps are shown in Figure 2. In the first part of the figure (2**a**), soot is being deposited from a torch which is translated across the growing "soot" boule as it rotates in a lathe. Halide vapors are transported in an oxygen or inert-gas stream through the central port of the torch. Surrounding the central port is a ring of additional ports to provide a shield flow which prevents soot from building up on the torch face. Further out along the diameter of the torch face is a ring of burner ports which provide both fuel and oxygen to support the flame and allow reaction of the halides with oxygen and/or water vapor to form oxides. The composition of the gas stream is changed with time to control the refractive index profile as a function of radius.

The mandrel is removed once the soot boule has been formed. Sintering of the porous soot preform is accomplished by passing it vertically through the hot zone of a furnace at ca 1500°C. The atmosphere consists

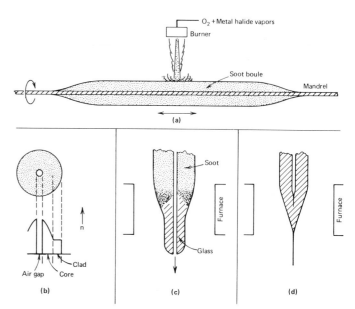

Figure 2. OVPO process: (**a**), soot deposition; (**b**), soot preform cross section; (**c**), preform sintering; (**d**), fiber drawing.

of He with Cl_2 in some form. The chemistry is complex owing to the reactions of Cl_2 with GeO_2. However, chlorine is necessary for dehydration of the preform according to the reaction: $H_2O + Cl_2 \rightarrow 2\ HCl + 1/2\ O_2$. The central hole in the preform may be collapsed during the drawing process.

MCVD. MCVD, the modified chemical-vapor deposition process invented at AT&T Bell Laboratories, involves depositing high optical quality glass on the inner surface of a tube, collapsing this composite to form a preform rod comprising the core/clad structure necessary for light guidance, then drawing this preform into controlled diameter light-guide fiber (Fig. 3). A commercial silica tube is first mounted in a glass-working lathe where it is aligned and straightened. The entrance end of the tube is connected to a chemical delivery system via a rotating seal or joint. The exit end is typically flared and connected to a large tube which serves as a collector of unincorporated particulate material, which in turn is coupled to a chemical scrubbing system. This configuration provides a closed chemical environment eliminating the need for clean-room processing conditions. Halide vapors are carried in an oxygen and/or inert gas stream and flow through a rotating seal into the rotating tube. The tube is heated by a traversing oxyhydrogen torch which provides heat for reaction of the halides with oxygen to form submicrometer glassy particles which deposit downstream of the hot zone. The same hot zone sinters the deposited powder to a high optical quality glass layer typically 1–5 μm thick. The dimensions and refractive index profile are easily controlled by depositing successive layers of controlled chemical composition, then collapsing the composite to a solid rod preform. This preform is then drawn to fiber of controlled diameter.

PCVD. PCVD, the plasma-activated chemical-vapor-deposition process, was first reported in 1975 by researchers from Philips Research

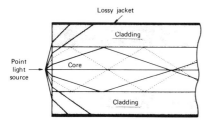

Figure 1. Light from a point source collected and guided by a glass-fiber lightguide. Courtesy of Academic Press, Inc.

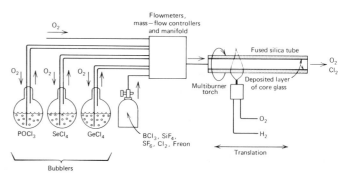

Figure 3. MCVD process.

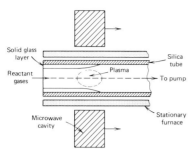

Solid glass layer
Silica tube
Reactant gases
Plasma
To pump
Microwave cavity
Stationary furnace

Figure 4. PCVD process.

S.R. Nagel, J.B. MacChesney, and K.L. Walker, *IEEE J. Quantum Elec.* **QE-18**, 459 (1982).

K. Inada, *IEEE Trans. Microwave Theory and Tech.* **MTT-30**, 1412 (1982).

P.C. Schultz, *Proc. IEEE* **68**, 1187 (1980).

Laboratory. It is an inside process like MCVD using the same precursor chemicals; however, the reaction is not thermally initiated but occurs as a result of a nonisothermal plasma. In this process, a vitreous deposit is formed directly rather than soot particles produced in the other processes. The setup is shown in Figure 4. The silica tube is positioned inside of a furnace. The microwave cavity operating at 2.45 GHz traverses inside the furnace and creates a plasma in the interior of the tube which is maintained at 0.1–2.5 kPa (1–25 millibar). Chemical flows are controlled and the temperature inside the tube is 1000–1250°C. At these temperatures, a thermally activated reaction is kinetically inhibited. Reaction occurs only where the plasma is present due to electronic excitation and the formation of highly reactive free radicals. The process produces very thin layers so that a preform is comprised of ca 1000 layers contrasted to less than 100 in MCVD. Because of the perceived low deposition rate coupled with high water-related absorption losses, PCVD had been regarded with skepticism as a mass-production process; however, recent results have altered this assessment and it is now regarded as competitive with the other three processes.

VAD. VAD, the development of vertical-axial deposition, was the last of the four processes to be announced. It is a flame hydrolysis process like OVPO; however, it differs in that deposition occurs on the end of a vertically growing cylindrical boule. This process has the advantage of producing a soot-form without a center hole. Such a body can be consolidated without as great a risk of breakage due to thermal expansion mismatch between the core and cladding. Conceptually, VAD has the further advantage of being a continuous process. As commercially practiced, however, it is a five-step process. The deposition step consists of a rotating mandrel that is withdrawn from the deposition zone. Amorphous oxide particles formed by flame hydrolysis are deposited to form a porous boule of controlled geometry and composition. The soot boule is then treated at low temperature (ca 1200°C) with halide vapors to remove hydroxyl. At higher temperatures (ca 1500°C), the boule is sintered to a vitreous body. The resulting preform consists of the core and a cladding layer. The cladding thickness is insufficient at this point, and it is necessary to elongate the preform and overclad it with a commercial silica tube. The preform is then drawn to fiber as in the other processes.

All of the above processes produce high quality optical fiber. Initially, fibers were optimized for transmission at 0.82 μm owing to the good performance and reliability of light sources and detectors at that wavelength, specifically GaAlAs devices. Recent advances in sources and detectors of a quaternary composition (InGaAsP) have led to good reliability and performance at high bit rate and long wavelength. The availability of long-wavelength sources and detectors has allowed manufacturers and researchers to take advantage of the low loss window in silica fibers of 1.3–1.6 μm. At the latter wavelength, fibers with losses of approximately 0.2 dB/km and bandwidths of tens of GHz·km are common and multimode fibers with similar losses and bandwidths in the vicinity of 1 GHz·km are available commercially. New fiber structures have emerged using single-mode designs having minimum dispersion at the operating wavelength.

S.G. Kosinski
John B. MacChesney
AT & T Laboratories

FIBERS, CHEMICAL

The generally used term man-made, referred to here as chemical, as applied to fibers differentiates between the natural fibers and those produced by chemical and physical means in manufacturing processes. The advent of nylon, and the subsequent proliferation of fibers from synthetic polymers, has brought about the subdivision of chemical fibers into two broad classes: those made from natural polymers such as the cellulosics (see Cellulose acetate and triacetate fibers; Rayon; Fibers, vegetable), and those derived from synthesized polymers. The latter, which are called synthetics (or sometimes noncellulosics), now represent the majority of chemical fibers. In this category fall the nylons, the polyesters, the acrylics, the polyolefins, and many others (see Acrylic and modacrylic fibers; Olefin fibers; Polyamides; Polyester fibers).

Chemical Fibers Based on Natural Polymers

Rayon. The predominant method of formation is the viscose process based on the finding of Cross, Bevan, and Beadle that cellulose is soluble in a mixture of caustic soda and carbon disulfide. Relatively purified wood-pulp cellulose, called dissolving pulp or chemical cellulose, is treated first with caustic soda, then with carbon disulfide, to form a xanthate which is, in turn, soluble in aqueous alkali solution. Control of conditions in the spinning solution, called viscose, and of the bath into which the viscose is extruded, importantly affect final fiber properties.

Cellulose acetate. Cellulose acetate is dry-spun, predominantly in continuous filament form. The fiber-forming base polymer is made by acetylating dissolving pulp with acetic anhydride usually in acetic acid solvent with sulfuric acid as catalyst. The fully acetylated cellulose initially produced is partially hydrolyzed to produce secondary acetate. Following catalyst removal, precipitation of the acetate flake, washing, and drying, the flake is dissolved in acetone containing a small amount of water. This spinning "dope" is subjected to extensive filtration and then dry-spun to evaporate the acetone solvent.

Cellulose triacetate. Cellulose triacetate production is similar to that of acetate, differing in that the initial reaction product is not hydrolyzed to a product with less than 92% of the hydroxyl groups esterified and the spinning solution for the triacetate flake is made with a mixture of methylene chloride and a small amount of methanol or ethanol.

Other fibers. Other fibers based on natural polymers include: alginate fibers, based on alginic acid from seaweed, and protein fibers, several of which have been based on casein from milk, the fibers being produced by wet spinning (see Milk products).

Synthetic Fibers

Nylon-6,6. Nylon-6,6 is so designated because both the adipic acid and the diamine from which the polymer is synthesized contain six carbon atoms. The fiber-forming polymer is formed by heating nylon-6,6 salt under pressure to remove water as the adipic acid and hexamethylene diamine condense to form a long-chain linear polymer in what is called a melt polymerization. When the desired average molecular weight is reached, pressure is reduced, and the polymer is usually extruded in spaghetti-like strands, cooled, and cut into chips. In some cases, the molten polymer is conveyed directly to melt-spinning equipment.

Nylon-6. Nylon-6 manufacture is based on the use of caprolactam, the internal cyclic amide of aminocaproic acid, derived petrochemically (see Polyamides). Nylon-6 polymer is also melt-spun, and the manufacture, fiber properties, and uses are very similar to those of nylon-6,6.

Acrylic fibers. The fiber-forming acrylic polymers are high in molecular weight and are produced primarily in aqueous medium by free-radical-initiated addition polymerization. The regular acrylics are usually copolymers having minor amounts of one or more comonomers such as methyl acrylate, having to provide accessibility or sites for dyestuffs.

Table 1. Relevant F.T.C. Generic Names for Manufactured Textile Fibers

Generic name	Definition of fiber-forming substance[a,b]
acetate	cellulose acetate; triacetate where not less than 92% of the cellulose is acetylated
acrylic	at least 85% acrylonitrile units
aramid	polyamide in which at least 85% of the amide linkages are directly attached to two aromatic rings
azlon	regenerated naturally occurring proteins
glass	glass
modacrylic	less than 85% but at least 35% acrylonitrile units
novoloid	at least 85% cross-linked novolac
nylon	polyamide in which less than 85% of the amide linkages are directly attached to two aromatic rings
nytril	at least 85% long chain polymer of vinylidene dinitrile where the latter represents not less than every other unit in the chain
olefin	at least 85% ethylene, propylene, or other olefin units
polyester	at least 85% ester of a substituted aromatic carboxylic acid, including but not restricted to substituted terephthalate units and para-substituted hydroxybenzoate units
rayon	regenerated cellulose with less than 15% chemically combined substituents
saran	at least 80% vinylidene chloride
spandex	elastomer of at least 85% of a segmented polyurethane
vinal	at least 50% vinyl alcohol units and at least 85% total vinyl alcohol and acetal units
vinyon	at least 85% vinyl chloride units

[a] All percentages are by weight.
[b] Except for acetate, azlon, glass, novoloid, and rayon, the fiber-forming substance is described as a long chain synthetic polymer of the composition noted.

Like other synthetics, acrylic fibers are stretched to develop orientation and fiber strength.

Polypropylene. Polypropylene constitutes the majority of commercial polyolefin fibers. In stereoregular form, polypropylene has a crystalline melting point of about 165°C, compared to ca 125°C for low pressure (stereoregular) polyethylene and ca 110–120°C for high pressure polyethylene (see Olefin polymers).

These fibers are melt-spun and produced in monofilament, multifilament, and staple forms. For uses where color is requisite, either the base polymer is modified to provide dyeability or the fiber is spun-colored that is, produced with pigments incorporated in the polymer and fiber during melting.

Other synthetic fibers are summarized in Table 1.

WILLIAM J. ROBERTS
Consultant

G. Clayton and co-workers, *Text. Prog.* **8**, 1, 27 (1976).

R.W. Moncrieff, *Man-Made Fibers*, 6th ed., John Wiley & Sons, Inc., New York, 1975, pp. 522–532.

E.M. Hicks and co-workers, *Text. Prog.* **8**, 1 (1976).

FIBERS, ELASTOMERIC

An elastomeric fiber can be made from any natural or synthetic polymeric material that has high elongation and good recovery properties. At present, only natural rubber (qv) and urethane polymers (qv) are raw materials for commercially successful elastomeric fibers.

The term spandex has come into common trade usage throughout the world in referring to fibers based on elastomeric urethane polymers.

Properties

In both rubber thread and spandex, mechanical properties may be varied over a relatively broad range. In rubber, variations are made by changes in the degree of cross-linking or vulcanization which is accomplished by changing the amount of vulcanizing agent, generally sulfur, and accelerants used. In spandex, many more possibilities for variation are available. Again by definition, however, any polymer containing urethane linkages in the repeat structure

$$\cdots R-OCNH-R'-NHCO\cdots_n$$

may be classified as a urethane polymer. Most urethane polymers in current use for the manufacture of spandex are made by the reaction of a ca 2000 molecular weight polyester or polyether glycol with a diisocyanate at a molar ratio of ca 1:2 followed by reaction of the resulting isocyanate-terminated prepolymer with a diamine to produce a high molecular weight urethane polymer.

The long-chain urethane polymer molecules in spandex fibers are substantially linear block copolymers comprising relatively long blocks in which molecular interactions are weak, interconnected by shorter blocks in which interactions are strong (see Copolymers). The weakly interacting blocks, commonly referred to as soft segments, are from the polyester or polyether glycol component, whereas the blocks having strong interactions result from the diisocyanate and diamine and are referred to as hard segments. Any deformation of the network, either extension (stretching) or compression results in an increase in the degree of order (ie, more uniform molecular alignment) and is opposed by entropy forces within the soft segments.

Manufacture

Cut rubber. In producing cut-rubber thread, smoked rubber sheet or crepe rubber is milled with vulcanizing agents along with stabilizers and pigments. The milled stock is calendered into sheets of thicknesses of 0.3–1.3 mm, depending on the final size of rubber thread desired. Multiple sheets (usually 20 to 40) are layered, heat-treated to vulcanize, then slit to form ribbons containing the same number of threads as layered sheets.

Extruded latex thread. In the manufacture of extruded latex thread, a concentrated natural-rubber latex is blended with aqueous dispersions of vulcanizing agents, pigments, and stabilizing agents. The compounded latex is held under carefully controlled temperature conditions for 8–36 h in order to allow some initial vulcanization to occur. The matured latex is extruded at constant pressure through precision-bore glass capillaries into an acetic acid bath where coagulation into thread form occurs. Threads are removed from the coagulation bath, washed, dried, then a finish is applied and the threads are wound on individual bobbins or formed into multi-end ribbons and wound on ribbon spools or boxed in the free-ribbon form. The thread may be finally vulcanized by heating the ribbon spools or bobbins in vulcanizing ovens for up to 18 h; however, most modern manufacturers vulcanize continuously in ribbon form before winding up on boxing.

Spandex. Spandex is produced commercially by four very different processes—melt extrusion, reaction spinning, solution dry spinning, and solution wet spinning. These are illustrated in Figure 1.

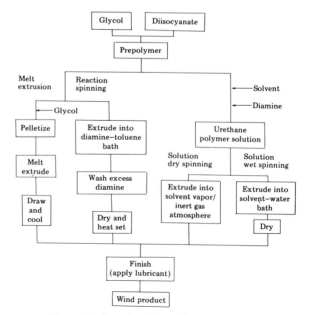

Figure 1. Spandex fiber production methods.

Use

Both spandex and natural-rubber fibers may be used in any area where extendability with good recovery is required. Spandex has replaced most rubber in foundation garments, swimwear, and other critical performance-use areas, owing to better aging properties, higher modulus, availability of finer deniers, and superior durability.

TIMOTHY V. PETERS
Consultant

H. Oertel, *Chem. Ztg.* **98**, 344 (1974); *Chemiefasern Text. Anwendungstech. Text. Ind.* **27**, 1090, 1095 (1977); *ibid.*, **28**, 44 (1978).

E. Hicks, Jr., A. Ultee, and J. Drougas, *Science* **147**, 373 (1965).

FIBERS, MAN-MADE. See Fibers, chemical.

FIBERS, MULTICOMPONENT

The designation for fibers composed of two polymers is bicomponent fibers (bico fibers). Since there are few examples of three or more fiber-forming polymer components, bicomponent tends to be used as a general term. It is important to distinguish between composite fibers (where every individual fiber or filament has two or more components) and composite yarns, also called heteroyarns.

Bicomponent fibers are classified on the basis of the cross-sectional and longitudinal arrangement of the components. The simplest class is known as side–side fibers irrespective of the cross-sectional shape, which is not necessarily circular. This class includes fibers with a single interface and with both components possessing an external boundary. The second class is composed of fibers with a single interface but in which only one component has an external boundary. These fibers are known as sheath–core fibers.

Materials and Processes

The manufacture of bicomponent fibers differs little from that of single-component fibers (see Acrylic and modacrylic fibers; Olefin fibers; Polyamides; Polyester fibers). Many pairs of fiber-forming polymers may be employed, and it is not necessary for both components to be capable of forming fibers if spun alone. The choice of components depends on the fiber application and they must be compatible in various respects as well as in spinning.

Frequently, component pairs do not adhere sufficiently to behave as single fibers when passing through various textile processes or in the final application. Splitting and breakage may occur on rolls, guides, or texturing devices resulting in heavy, sticky, or dusty deposits, particularly with matrix-filament fibers. Typical properties of an acrylic bicomponent fiber and a single-component fiber are given in Table 1.

Table 1. Properties of Cashmilon Acrylic Fibers

Property	Cashmilon GW, bicomponent	Cashmilon FW, monocomponent
strength, N/tex[a]		
dry	0.25	0.25–0.32
wet	0.21	0.23–0.28
elongation, %		
dry	40–45	28–40
wet	45–50	30–45
loop strength, N/tex[a]	0.31	0.18–0.27
loop elongation, %	20–25	5–20
modulus of tensile elasticity, %	99 ± 2	(90–92) ± 3
specific gravity	1.18	1.188
moisture regain at 20°C, %	1.8–1.6	1.6
shrinkage in hot water, %	4	2–4

[a] To convert N/tex to gf/den, multiply by 11.33.

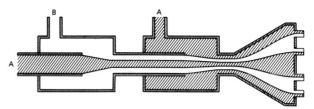

Figure 1. Pipe-in-pipe method for mixed-stream formation.

The processes by which multicomponent (mainly bicomponent) fibers are produced are largely independent of the polymers used, except for the formation of self-skinning fiber in the wet-spinning process. Production processes rely on the stability of the interfaces between the components to maintain the filament cross-sectional geometry as the cross-section area is reduced, perhaps several hundred-fold, during extrusion and drawdown.

A form of multicomponent fibers is produced by spinning a blend of immiscible polymers, eg, nylon and polyester. A matrix–fibril fiber is produced in which the minor component forms short individual fibrils dispersed in a fiber composed of the principal component. True multicomponent fibers are prepared by the controlled combination of liquid polymer streams after the individual streams have entered the spinning pack. If only side–side fibers are required and a degree of uncertainty regarding the position of the interface can be tolerated, a conventional spinnerette may be fed with a polymer stream containing alternating layers of the two components.

Other methods of preparing multicomponent fibers require the controlled flow of the components to the spinnerette holes or beyond. Most of these methods require the components to be combined and formed into the required cross-sectional configuration just above the spinnerette holes.

In another method, both components must be of the same material but with different orientation. Each filament is formed by the combination of two polymer streams below the spinnerette hole. Side–side bicomponent fibers are also produced by feeding a mixed polymer stream to a simple spinnerette with the holes aligned with the stream interfaces. The mixed polymer streams are generated as concentric rings or parallel laminae.

The pipe-in-pipe method is the simplest for the formation of mixed streams (see Fig. 1). When it is employed for the production of matrix–filament fibers, a number of core–sheath filaments are combined and narrowed in a conical chamber below the spinnerette holes before extrusion (combining-chamber method). One form of the septum method is similar to the pipe-in-pipe technique. The core stream is inserted directly into the sheath stream via a fine tube. The wall of this tube may be regarded as a circular septum.

The need to control the sheath polymer has led to the castellation method, in which a flow of polymer sheath is controlled by passage through a fine gap between the upper (distribution plate) and lower spinnerettes. The gap is ca 100 μm and extends evenly over the spinnerette, resulting in the production of uniform fibers.

Matrix–filament fibers are also manufactured by a side-injection technique with a septum device in which a tube with fine side holes passes through a chamber with one component to connect another chamber containing the second component to the spinnerette hole.

Uses

Where the components are inert, nontoxic, and nonallergenic, multicomponent fibers behave similarly. By far the most important application is the creation of desirable textile properties or aesthetic effects. For example, although the initial aim was to achieve a steric crimp similar to that of wool, much more sophisticated effects are now obtained.

Microdenier (< 0.1 tex per filament) fibers formed from matrix–filament fibers are used in the production of knitted or woven fabrics with appearance different from those made of fiber in the normal filament range (0.1–0.5 tex per filament). In other applications, one component serves as an adhesive. Core-sheath fibers are generally used in which the sheath melts ca 20°C lower than the core. Multicomponent fibers are used to modify physical properties of textiles, eg, for antistatic applications.

The inclusion of more than one component in a fiber may confer various optical properties. For example, a bichromic shot effect is obtained with a fiber of flat cross section and a second component located at the rounded-end portions. Multicomponent fibers have been proposed as containers or diffusion barriers.

K. PORTER
ICI Fibres

P.A. Koch, *Ciba-Geigy Rev.* **1**, 1 (1974).

R. Jeffries, *Ciba-Geigy Rev.* **1**, 12 (1974).

R. Jeffries, *Biocomponent Fibres*, Merrow Publishing Co., Ltd., UK, 1971 pp. 2–4.

FIBERS, VEGETABLE

Natural fibers of vegetable origin are constituted of cellulose (qv), a polymeric substance made from glucose molecules, bound to lignin (qv) and associated with varying amounts of other natural materials. A small number of a vast array of these vegetable fibers have industrial importance for use in textiles (qv), cordage, brushes and mats, and paper (qv) products. Cotton fiber makes up about three-quarters of the world vegetable-fiber tonnage (see Cotton).

Vegetable fibers are classified according to the part of the plant where they occur and from which they are extracted, as shown in Table 1.

Table 1.   Selected Vegetable Fibers of Commercial Interest

Commercial name	Botanical name	Geographical source	Use
Leaf (hard) fibers			
abaca	*Musa textilis*	Borneo, Philippines, Sumatra	cordage
cantala	*Agave cantala*	Philippines, Indonesia	cordage
caroa	*Neoglaziovia variegata*	Brazil	cordage, coarse textiles
henequen	*Agave fourcroydes*	Australia, Cuba, Mexico	cordage, coarse textiles
istle (generic)	*Agave* (various species)	Mexico	cordage, coarse textiles
Mauritius	*Furcraea gigantea*	Brazil, Mauritius, Venezuela, tropics	cordage, coarse textiles
phormium	*Phormium tenax*	Argentina, Chile, New Zealand	cordage
bowstring hemp	*Sansevieria* (entire genus)	Africa, Asia, South America	cordage
sisal	*Agave sisalana*	Haiti, Java, Mexico, South Africa	cordage
Bast (soft) fibers			
China jute	*Abutilon theophrasti*	China	cordage, coarse textiles
flax	*Linum usitatissimum*	north and south temperate zones	textiles, threads
hemp	*Cannabis sativa*	all temperate zones	cordage, oakum
jute	*Corchorus capsularis; C. olitorius*	India	cordage, coarse textiles
kenaf	*Hibiscus cannabinus*	India, Iran, USSR, South America	coarse textiles
ramie	*Boehmeria nivea*	China, Japan, United States	textiles
roselle	*Hibiscus sabdarifa*	Brazil, Indonesia (Java)	cordage, coarse textiles
sunn	*Crotalaria juncea*	India	cordage, coarse textiles
cadillo	*Urena lobata*	Zaire, Brazil	cordage, coarse textiles
Seed-hair fibers			
cotton	*Gossypium* sp.	United States, Asia, Africa	all grades of textiles, cordage
kapok	*Ceiba pentranda*	tropics	stuffing
Miscellaneous fibers			
broom root (roots)	*Muhlenbergia macroura*	Mexico	brooms, brushes
coir (coconut husk fiber)	*Cocos nucifera*	tropics	cordage, brushes
crin vegetal (palm leaf segments)	*Chamaerops humilis*	North Africa	stuffing
piassava (palm leaf base fiber)	*Attalea funifera*	Brazil	cordage, brushes

a Ref. 2.

Table 2.   Chemical Composition of Various Vegetable Fibers, wt%

Type of fiber	Cellulose	Hemi-cellulose	Pectins	Lignin	Water-soluble compounds	Fats[a] and waxes[b]
cotton	91.8		6.4		1.1	0.7
flax (retted)	71.2	18.6	2.0	2.2	4.3	1.7
flax (nonretted)	62.8	17.1	4.2	2.8	11.6	1.5
Italian hemp	74.4	17.9	0.9	3.7	2.3	0.8
jute	71.5	13.4	0.2	13.1	1.2	0.6
Manila	70.2	21.8	0.6	5.7	1.5	0.2
New Zealand cotton	50.1	33.4	0.8	12.4	2.4	0.1
ramie	76.2	14.6	2.1	0.7	6.1	0.3
sisal	73.1	13.3	0.9	11.0	1.4	0.3

[a]See Fats and fatty oils.
[b]See Waxes.

Physical Properties

The vegetable fibers are stronger but less extensible than cotton, ie, they have a higher breaking length and elasticity modulus with a lower strain (extensibility) and work modulus. They approach glass in stiffness (resistance to deformation) and are considerably stiffer than man-made fibers, but have lower toughness (ability to absorb work) (see Fibers, chemical). Kapok and other seed-hair fibers are relatively low in strength but have great buoyancy.

Chemical Composition

The chemical compositions (Table 2) vary greatly between plants and within specific fibers depending on genetic characteristics, part of the plant, and growth harvesting and preparation conditions.

Economic data. Vegetable fibers, excluding cotton and flax, have been relatively unimportant in the total fiber supply and have become less important in relation to converted cellulosic and noncellulosic fibers.

JOHN N. McGOVERN
The University of Wisconsin

J.E. Atchison, *Science*, **191**, 768 (1976).

D. Lapedes, ed., *Encyclopedia of Science and Technology*, Vol. 5, McGraw-Hill Book Co., New York, 1977.

G.E. Linton, *The Modern Textile and Apparel Dictionary*, 4th ed., Textile Book Service, Plainfield, N.J., 1973.

FIBRINOGEN.   See Blood fractionation.

FIELD-ASSISTED SEALING.   See Electrostatic sealing.

FILLERS

A filler is a finely divided solid that is added to a liquid, semisolid, or solid composition to modify the composition's properties and reduce its cost. A filler can constitute either a major or a minor part of a composition. The structure of filler particles can range from irregular masses to precise geometrical forms such as spheres, polyhedra, or short fibers.

Fillers are usually classified according to their source, function, composition, or morphology (see Table 1). However, none of these classifica-

Table 1.   Classification of Fillers

Method of classification	Examples
source	natural (mineral); synthetic
function	opacification; reinforcement
composition	carbonate; silicate; lignin
morphology	hollow sphere; fiber; platelet

tion schemes is entirely adequate owing to overlap and ambiguity of their categories. An understanding of the physical properties of fillers and the associated functions is most important in applying filler technology. Mineral fillers represent over 80% of the filler market primarily because of their lower cost.

Properties

Particle morphology. Filler-particle morphology affects rheology and loading of filled compositions. Particles with fibrous, acicular (needlelike), or irregular morphology yield compositions more resistant to flow than compositions filled with spheres, polygons, or other fillers without regular morphologies.

Surface area and energy. Surface area is the available area of fillers, including the surfaces of cracks, crevices, and pores. It helps determine the ease of dispersion, the rheology, and the optimum loading of filled compositions. The free surface energy of fillers (wettability) is a function of surface area and compositions. Many commercial fillers are surface-coated or treated to modify their free surface energy.

Functional properties. Functional properties of fillers are those that fillers impart to compositions to enhance their performance or economic utility. Although the properties that are required by compositions vary from one application to another, a given physical or chemical property of the filler may or may not be functional and depends on the requirements of the application in which it is used. A quantification of functional properties per unit cost, therefore, provides a valid criterion for filler comparison and selection.

Among the determining functional properties are specific gravity; bulk density; optical reflectance; refractive index; free moisture; thermal stability; and thermal expansion.

Uses

Fillers such as carbon black and pyrogenic silica are used in the rubber industry as reinforcing agents. They impart abrasion resistance, tear resistance, tensile strength, and stiffness to an elastomer after vulcanization. Fillers are used in paper to impart superior optical and mechanical properties, eg, clay, titanium dioxide, and calcium carbonate. In plastics, the term filler refers to particulate materials added to plastic resins in relatively large-volume loadings; plastic compounders tend to compound with the objective of optimizing properties at minimum cost rather than maximizing properties at optimum cost (see Laminated and reinforced plastics). Fillers such as barium sulfate, calcium carbonate, clay, mica, silica, precipitated diatomite, talc and zinc oxide are used in paint (they are referred to as inerts, extender pigments, and supplemental pigments) to contribute to application, durability, protection, and decoration and to impart optical and mechanical properties.

Health and Safety

Inhalation of airborne particles in the respirable size range ($< 10 \ \mu m$ aerodynamic diameter) represents the largest hazard involved in the handling and use of fillers. Filler dusts are classified into carcinogens, fibrogens, and nuisance particulates. Suppliers usually have information on the safe handling of their product.

JOHN G. BLUMBERG
JAMES S. FALCONE, JR.
LEONARD H. SMILEY
PQ Corporation

DAVID I. NETTING
ARCO Chemical Company

J.V. Milewski in H.S. Katz and J.V. Milewski, eds., *Handbook of Fillers and Reinforcements for Plastics*, Van Nostrand Reinhold, New York, 1978, p. 66.

D.H. Solomon and D.G. Hawthorne, eds., *Chemistry of Pigments and Fillers*, John Wiley & Sons, Inc., New York, 1983.

T.C. Patton, ed., *Pigment Handbook*, Vols. 1–3, John Wiley & Sons, Inc., New York, 1973.

FILM AND SHEETING MATERIALS

A film, as defined by the *Modern Plastics Encyclopedia* and as dealt with for the purposes of this article, is a flat section of a thermoplastic resin or a regenerated cellulosic material that is very thin in relation to its length and breadth and has a nominal thickness not greater than 0.25 mm (see also Film-deposition techniques). The same materials in similar configuration but greater thicknesses are classified as sheets.

Numerous film- and sheet-forming materials are currently available in various densities, melt indexes, copolymers, and blends, having additives for plasticizing, coloring, impact modification, slip antifogging, ultraviolet stabilization, fire retardance, and biodegradability. The principal basic materials for film and sheet manufacture consist of about 49 thermoplastic resins and regenerated cellulose.

Properties

Tear strength is the force required to propagate an initiated tear in a film under controlled conditions. Impact strength measures the force required to rupture a film or sheet and can be measured in various ways.

Table 1.   Physical Properties of Some Film and Sheeting Materials

Material	Specific gravity	Tensile strength, MPa[a]	Elongation, %	Tear strength, gf/mm
acrylonitrile–butadiene–styrene	1.04	35.2–70.3	10–50	
acrylonitrile–methyl acrylate copolymer, rubber modified	1.15	63.3–77.3	50–200	
cellophane (regenerated cellulose)	1.40–1.50	49.2–126.5	10–50	0.8–7.9
cellulosics, ethyl cellulose	1.15	56.2–70.3	20–30	2.8–14.2
cellulose acetate	1.28–1.31	49.2–115.3	15–55	1.6–3.9
cellulose acetate butyrate	1.19–1.20	35.2–63.3	50–100	2.0–3.9
cellulose propionate	1.20	28.1–35.2	60–80	39.4
cellulose triacetate	1.28–1.31	63.3–112.5	10–40	1.6–3.9
Fluoroplastics, ETFE	1.7	49.2–56.2	300	600–900
FEP fluoroplastic	2.15	17.6–21.1	300	125
polytrifluorochloroethylene copolymers	2.08–2.15	35.2–70.3	50–150	2.5–40
polytetrafluoroethylene	2.1–2.2	10.6–31.6	100–350	10–100
poly(vinyl fluoride)	1.38–1.57	49.2–126.5	115–250	12–100
ionomer	0.94	35.2	250–450	30
polyamides, nylon-6	1.13	63.3–126.5	250–550	50–90
nylon-6,6	1.14	63.3–84.4	200	35–40
nylon-11	1.03	63.3–77.3	250–400	400–500
nylon-12	1.01	49.2–84.4	290–330	
polybutylene	0.908–0.917	26.7–30.9	300–380	
polycarbonate	1.20	59.1–61.9	85–105	7.9–9.8
polyester (PE terephthalate)	1.38–1.41	281[b]	50[b]	4.7–10.6
polyethylene, LDPE	0.91–0.925	10.6–21.1	100–700	19.7–118
MDPE	0.926–0.940	14.1–24.6	50–650	19.7–118
HDPE	0.941–0.965	16.9–42.9	10–650	5.9–118
UHMWPE[c]	0.940	21.1–38.7	300	
ethylene copolymers, vinyl acetate	0.924–0.940	7.0–21.1	400–800	19.7–118
methylacrylate	0.93	7.0–21	400–700	120
polyimide	1.42	17.6	70	2.5
poly(methyl methacrylate), standard	1.18–1.19	57.7–61.9	4–12	
type A (Korad acrylic)	1.14	35.9	75	
polypropylene, extrusion cast	0.885–0.905	31.6–49.2	550–1000	236TD–10MD
biaxially oriented	0.902–0.907	52.7–281	35–475	1.2–3.9
polystyrene, oriented	1.05–1.06	56.2–84.4	3–40	2
foam		2.1–9.1	2–8	
sulfone polymers, polysulfone	1.24–1.25	59.1–74.5	64–110	2.5–4.7
polyethersulfone	1.37	70.3–84.4	20–150	2.8–6.3
polyurethane elastomer	1.11–1.24	35.2–84.4	200–700	87–279
poly(vinyl alcohol)	1.26	27.6–69	180–600	250–800
poly(vinylidene chloride)	1.65	256–113	30–80	10–90
poly(vinyl chloride), nonplasticized	1.2–1.5	49.2–70.3	25–50	10–700
plasticized	1.2–1.8	98.4–112.5	100–500	60–1400
vinyl chloride-acetate copolymer nonplasticized	1.3–1.4	38.7–56.2	3–100	10–30
plasticized	1.2–1.35	17.6–35.2	100–500	30–1400

[a] To convert MPa to psi, multiply by 145.
[b] Tensilized.
[c] Ultrahigh mol wt polyethylene.

Tensile impact measures the energy extracted from a pendulum-type hammer rupturing the film. Stiffness of film and sheeting can be reported as the tensile modulus of elasticity as measured (as are the tensile strength and elongation) on the Instron tensile tester. The moisture-vapor transmission rate (MVTR) and the gas permeability are important properties in packaging applications, since moisture exclusion or retention is necessary for many packaged products (see Barrier polymers). The permeability of a film can be reduced by increasing its thickness, by coating or coextruding it with another material, or by laminating it to another film, paper, or foil. Rockwell hardness is a measure of resistance to indentation from a round steel ball, and is reported as a whole number with a scale symbol. Among the optical properties, transparency refers to the percentage of light transmitted by the sample, and haze is the percentage of transmitted light scattered. Gloss is the amount of light reflected from the film surface toward the viewer.

Table 1 lists important physical properties of some film and sheeting materials.

Chemical properties may be found under individual articles in the *Encyclopedia*.

Manufacture

Of the six fundamental processes for forming a basic polymeric composition into a film, four are commercially significant. Chemical regeneration from a solution and melt forming are by far the most important. The four largest-volume film materials currently in use are made by chemical regeneration (cellophane) and melt forming (polyethylene, vinyls, and polypropylene). Acrylic, cellulosic, and vinyl films are made by solution forming. Emulsion forming (together with plastisol and organosol) is used mainly for coating substrates. A modest volume is made by fiber forming. Powder forming is at present commercially not important.

Chemical regeneration. In this process, wood cellulose is first steeped in caustic soda solution to convert it to alkali cellulose. Then it is pressed to remove excess caustic, and shredded into light, fluffy crumb form. After aging (from a few hours to several days, depending on the viscose viscosity desired), the alkali cellulose is converted by reaction with carbon disulfide to sodium cellulose xanthate, which is dissolved in dilute aqueous caustic solution forming viscose. Cellophane is made by extruding viscose through a long narrow slot into a coagulating regenerating bath consisting of sulfuric acid and sodium sulfate dissolved in water. The film is washed to remove acids, salts, and other products, desulfurized, washed again, and bleached if desired. After washing in cold water, the film goes through a softening bath, where plasticizer is added, and is then dried to the desired moisture level. Dried film is wound onto rolls for commercial applications.

Melt forming. Melt-forming processes encompass calendering and three methods of extrusion: blown bubble, slot-die casting, and coating on a substrate. Calendering forms a continuous film by squeezing a thermoplastic material between two or more horizontal metal rolls. The polymers and other ingredients are compounded into a plastic mass and fed to the top rolls of the calender.

In the blown-bubble extrusion process, a tube is extruded from an upright annular die, inflated with air to a size dictated by the film properties desired and limited by feasibility, cooled with a refrigerated blown air, collapsed to a flat tube, and wound into rolls of either slit or tubular film.

In slot-die extrusion, a thin section of polymer melt is extruded from a slot die onto a chill or quenching roll that rapidly cools the melt. The quenching roll is overdriven relative to the linear rate of melt flow from the die to effect a drawdown, ie, a reduction in thickness from ca 0.51 mm to the range of 0.013–0.028 mm required for most applications.

In the third extrusion method, coating on a substrate, a high temperature melt is extruded from the lips of a flat die into the nip of two rolls, where it is bonded to the substrate under pressure. The roll that is in contact with the melt is chilled to solidify the melt; the roll that is in contact with the substrate is a rubber-surface pressure roll.

Solution forming (solvent casting). This method of film and sheet formation is not used extensively today for commercial production since it presents serious problems of solvent recovery and safe handling of

flammable solvents. However, it is widely used in polymer research because enough film for testing can be made from a very small sample of polymer.

Biaxial orientation of film and sheet. Biaxially oriented film and sheeting is produced by tentering (stretching a formed flat sheet first in the machine direction over heated rolls and then in the transverse direction with a tenter-frame within an oven) or bubble blowing (stretching the film simultaneously in machine and transverse directions). In both processes, the film is subject to a postorienting heating cycle for annealing (heat set) purposes.

Postfinishing operations. Film and sheet properties can be altered by treating for printability, cross-linking the polymer by ionizing radiation, and depositing a thin metal coating using vacuum metallizing (see Film-deposition techniques).

Uses

Polyethylene is the largest-volume film material, and more than 56% of the low density polyethylene (LDPE) production is consumed as film. Nonpackaging applications account for ca 45% of the usage of all film materials. Among industrial applications, agriculture and construction consume about half the volume of film materials. The relatively small-volume industrial application of film substrates is a high value–high performance market consisting of magnetic recording tape, electrical insulation, photographic film, microfilm, and pressure-sensitive tape. Among these, the highest-volume materials are made of biaxially oriented polyester and polypropylene (see Magnetic tape; Insulation, electric).

E. Lea Crump
Gulf Oil Chemicals Company

Reviewed by
Frank J. Barborek
Gulf Oil Company

J.H. Briston and L.L. Katan, *Plastics Films*, John Wiley & Sons, Inc., New York, 1974.

Modern Plastics Encyclopedia, Vol. 60, McGraw-Hill Book Co., New York, 1983–1984.

Films, Sheets, and Laminates, The International Plastics Selector, Inc., San Diego, Calif., 1979.

FILM-DEPOSITION TECHNIQUES

Thin films have been used for decorative purposes for well over a thousand years (eg, one of the earliest uses was "gold leaf" formed by controlled beating of gold to a thickness of 0.1 μm). Thin films include electroplated films used for decorative and protective purposes as well as the conductive, resistive, magnetic, insulating, semiconducting, and even superconducting films used in microelectronics (see Electroplating; Magnetic materials; Semiconductors).

Properties

The ideal thin-film coating adheres well and is chemically inert to the substrate and atmosphere. However, the film is not a separate entity and must also be considered together with the attached substrate. Among the properties of interest are its crystallographic orientation, the density of dislocations or imperfections, and the distribution of adsorbed gases.

Other properties are related to the intended use of a film. For optical films, the main physical properties of interest are refractive index and the transmission value at a particular wavelength. Laser components require a high transmittance, necessitating low impurity concentrations of metals that absorb in the desired wavelength. Thin polymeric coatings considered for modern microelectronic applications must be pure and homogeneous solid layers of uniform thickness. Good electrical properties, including high intrinsic breakdown voltage, low dielectric constant, low electrical conductivity, and small dissipation factor, can be enhanced or otherwise affected in use. For example, the electric field in a polymeric film may be large because of the relatively small thickness of insulation and the measured and predicted geometric capacitances may be different. Thin-film insulators can also function as conductors, a phenomenon that has been ascribed to field emission, electron tunneling, and impurity, and space-charge limited conductions.

The adherence of thin films is governed by the substrate surface, crystallographic orientation (if any), density of dislocations on imperfections, gases adsorbed, and by the bonding type and spatial distribution of surface atoms.

There are two types of bonding forces: chemical and physical. Divided further, chemisorption processes usually lead to maximum bonding a chemical bond being formed if there is a reaction between the arriving film material and the substrate, but physiosorption alone occurs in some systems. By comparison, these bonding forces—van der Waal's and electrostatic—are weak. Film–substrate systems that involve only physical bonding (eg, gold on glass) have few practical uses, though they are of some laboratory significance.

The most significant film parameter is film thickness. Thickness can be continually monitored during the deposition process or after deposition is complete. The choice of technique is dictated by the film thickness and accuracy required, and by the film use which is related to the uniformity of film thickness, hardness, transparency, substrate size, and physical properties. Film-thickness measurement techniques utilize light wavelength, mechanical, electrical and magnetic approaches.

Deposition of Films From Solution

Electrolytic deposition in cathodic films. This process involves the movement of metallic ions in solution to the electrodes under the influence of the applied electric field. At the cathode, the ions accept electrons, are neutralized, and are incorporated into the metal lattice (see Electroplating).

Electrolytic deposition in anodic films. Anodized aluminum coatings have been known for ca 150 years. The first practical coating, that of a "duraluminum" seaplane, employed chromic acid which was subsequently replaced with sulfuric acid as the anodizing electrolyte. On becoming the anodes of electrolytic cells, many metals form a protective oxide film, frequently an extremely hard, compact, well-adhering coating.

Metals on which anodic films can be formed are known as valve metals because of the rectifying characteristic of their anodic oxides. These include: aluminum, antimony, beryllium, bismuth, germanium, hafnium, magnesium, niobium, silicon, tantalum, tin, titanium, tungsten, uranium, and zirconium.

The rate of increase of film mass is proportional to the current under conditions where the electronic leakage current is negligible, so that dl/dt is constant where l = film thickness. The voltage/time relation is linear if the structure and composition of the film are assumed to be constant during formation and a constant differential field strength dV/dl is required (V = applied voltage) to maintain a constant ionic current.

$$\frac{dV}{dt} = \frac{dV}{dl} \cdot \frac{dl}{dt}$$

Anodic oxidized products are used in aluminum protective coatings, aluminum and tantalum electrolytic capacitors, and electronic tunnel devices.

Chromate conversion coatings. Chromate conversion coatings, which are applied in thicknesses of 2.5 to 5 μm, are used for protection of aluminum and its alloys. The coating is applied by immersion or is rolled on rather than being electrodeposited. Although these coatings offer good chemical resistance to weather, they are not wear-resistant. They may be colorless or dyed. Some proprietary chromate coatings such as Alodine and Iridite employ a special coating and rinsing procedure.

Electroless plating. Electroless plating or chemical reduction plating, (also known as autocatalytic plating and chemical plating) is a process that requires neither electrodes nor any external source of electricity (see Electroless plating). The plating process is based on a chemical process involving the catalytic reduction of metal salts. The metal salt and chemical reducer, most often sodium hypophosphite, formaldehyde,

sodium borohydride, and amine boranes, react in presence of a catalyst. The base material can itself be catalytic or be activated by use of a gold, platinum, or palladium compound. The most common electroless coating is electroless nickel plating which can be applied virtually to any substrate material including metals, plastics, ceramics, and other insulators. Electroless plating permits plating in selected areas and inside holes, making it very popular in the electronics industry for printed circuit board patterns, connector contacts, semiconductor devices, ceramic–metal integrated circuit packages, aluminum connector shells, and lead frames.

Polymeric coatings. The simplest method of forming a polymer film is to dissolve the polymeric material in a solvent, paint or spread the solution over the surface to be coated, and allow the solvent to evaporate. "Lacquer film" capacitors have been successfully manufactured by this technique.

Vacuum Deposition of Films

The five principal vacuum-deposition techniques of sputtering, glow discharge, evaporation, vapor plating, and ion plating are summarized diagrammatically in Figure 1. The starting materials (in square boxes) may be either solid or gas, and the type of energy applied either plasma or thermal (in the diamonds). Sputtering, for example, employs a solid target and a plasma energy source; glow discharge, a gas feedstock and a plasma energy source.

Evaporation of inorganic materials. The vacuum evaporative process is a direct thin film deposition technique in which material is deposited by vaporization in a vacuum. Advantages of vacuum deposition are its simplicity, economy, and efficiency. The heat required to evaporate metals and metal oxides can be introduced by resistance heating, electron bombardment, or audio-frequency induction. The rate of deposition by vacuum evaporation is generally higher than by sputtering. This may lead to structural disorder in the deposited thin film. To obtain evaporative films having well-defined chemical and physical properties, certain guidelines must be observed: the greater the vacuum, the better chance that the film will resemble the starting material; substrates should be clean, free of any grease or dust, and preferably subjected to a final glow-discharge cleaning; the entire vacuum chamber and substrate should be baked at ca 400°C to remove adsorbed material such a water that would prevent good adhesion of the film.

Vacuum-deposition techniques are used extensively for the thin-film metallization of glass and ceramic substrates. Nickel–chrome alloys, rhenium, nickel, titanium, and other resistive materials have been vacuum-deposited for thin-film resistors. For dielectric coatings, silicon monoxide is very common. Magnesium fluoride has been used. Thin-film conductor coatings employ aluminum, silver, copper and other good metallic conductors. Typical substrates include glass, alumina, and beryllia.

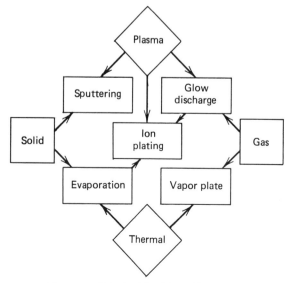

Figure 1. Relationship of deposition process.

Evaporative coating with polymers. The wide variety of plastics available has led to considerable interest in the possibility of coating materials by simple evaporation. The evaporation of a polymer appears at first to be a process of depolymerization on heating, followed by repolymerization of the fragmented polymers and oligomers on the substrate surface.

Successful vacuum deposition has been reported using Teflon (PTFE), polyethylene, and nylon. Coatings resulting from these polymers have been deposited by electrically heating the powders to a full red heat. In a typical example, Teflon heated for only a few minutes in a crucible or tungsten boat deposited films of the order of 1 μm.

Vapor-phase polymerization. Vacuum-deposited polymers have certain capabilities that suit them uniquely for thin polymeric coatings. One such polymer is Parylene, which is produced by vapor-phase polymerization and deposition of p-xylylene (or its substituted derivatives). Parylene has been successfully applied in the vapor phase via a thermal decomposition free-radical mechanism in the absence of either high temperature or high vacuum. The vapor is pervasive, but it coats without bridging so that holes can be jacketed evenly. Moreover, with Parylene, the object to be coated remains at or near room temperature, eliminating the risk of thermal damage. Coating thickness is controlled easily and very accurately simply by regulating the amount of dimer vaporized.

In current commercial applications, Parylene is deposited in thicknesses ranging from 6 to 38 μm, in a single tumble-coating operation for use as a capacitor dielectric material in extended foil Flat Kap capacitors (Union Carbide), as a circuit-board conformal coating for high reliability applications, as a semiconductor device, discrete-component protective coating, in thermistors, thermocouples, sensing probes, and photocells, and for ferrite toroids.

Ion plating. The techniques used for the evaporation of a coating material in the positive-glow region of a gas discharge is known as ion plating. The ionized evaporant is accelerated to the negative electrode surface under the influence of electrical fields in a low pressure d-c glow discharge (usually of argon). In the process the deposition of coating on a substrate is maintained at a faster rate than the sputtering removal of material from the substrate surface. As a result, the film grows in thickness with time (see also Ion implantation).

Sputtering. Sputtering is a process in which material is removed from a source or target (the cathode), carried by a plasma and deposited on the substrate (the anode). It is the main competition to the vacuum evaporative process and yields more consistent sheet-resistivity values and an improved temperature coefficient of resistance. For the deposition of complex metal oxides and alloys, sputtering is generally preferred to vacuum evaporation. The disadvantages of sputtering as compared with vacuum evaporation are the higher cost of equipment and operation, more complex equipment, and slower deposition rates. In addition, when bias sputtering is not employed or when gas pressure is excessive, gas entrapment becomes a factor.

One improved sputtering technique is planar magnetron sputtering, first patented in 1939, which uses a closed magnetic-field loop to confine and compress the plasma, causing the ionized gas to sputter more efficiently. Under the influence of the two fields, the electrons travel in a cycloidal path rather than in straight lines. This extended path, and its formation into an endless loop, greatly increases the probability of ionizing collisions and produces a highly dense plasma near the target surface. The net result is to create a higher ion flux with deposition rates of 5 to 20 times those of conventional d-c or radiofrequency (r-f) diode methods. Diode or conventional deposition employing r-f equipment is the sputtering technique used for most applications involving conductive and insulating materials.

Deposition of Films in Gaseous Discharge

R-f glow-discharge of inorganic materials. At high temperatures or under reduced pressures, gases will conduct electricity when subjected to a strong electrical field. At this time, the gas becomes partially ionized and contains free electrons and ions that carry current through the gas. In a glow or gaseous discharge, the free electrons in the gas acquired from the electrical field collide with gas molecules, and transfer energy to them. Electron–molecule collisions result in molecular excitation and in dissociation and ionization of the gas molecules. Thin inorganic films of silicon and of silicon nitride have been prepared in an r-f plasma by allowing silane to react with water or ammonia, respectively.

Other gas-discharge techniques are gaseous anodization, and glow-discharge polymerization.

Deposition of Films at Atmospheric Pressure

Metallo-organic deposition. Metallo-organic deposition is a simple process compared to sputtering or vacuum evaporation and requires simple equipment. The chemical compounds used are organo–metallic halides, hydrides, or oxides that are readily dissociated or reduced below the temperature of the substrate material. They are mixed in liquid or paste form in a homogeneous single-phase solution called a formulated resinate, applied by brushes, dipping, etc, and then dried and fired. This treatment decomposes or oxidizes the organic portion, depositing the adherent thin film of metal oxide on the substrate. The resultant film thicknesses range between 0.1–0.2 μm, similar to the thickness obtained by vacuum techniques.

Chemical vapor deposition (CVD). CVD, also called vapor plating or vapor forming, is a process in which a surface is coated with vapors of volatile stable chemical compounds at a temperature below the melting point of the surface. The compound is reduced or dissociated, resulting in an adherent coating of material.

STUART M. LEE
Ford Aerospace and Communications Corp.

W.F. Gorham, *J. Polym. Sci.* **4**, 3027 (1966).

J.L. Vossen and W. Kern, eds., *Thin Film Processes*, Academic Press, Inc., New York, 1978.

G. Hass and R.E. Thun, eds., *Physics of Thin Films*, Vols. 1–4, Academic Press, Inc., New York, 1964–1967.

FILM THEORY OF FLUIDS. See Mass transfer.

FILTERS, OPTICAL. See Optical filters.

FILTRATION

Filtration is the separation of particles of solids from fluids (liquid or gas) by use of a porous medium (see also Air-pollution control methods; Gas cleaning; Ultrafiltration). This article discusses only separation of solids from liquids.

The two main mechanisms of filtration are cake and depth filtration. In cake filtration, solids form a filter cake on the surface of the filter medium. In depth filtration, solids are trapped within the medium using either cartridges or granular media such as sand or anthracite coal. Cartridges are usually disposable, have media of various kinds of fibers or porous structures, and are generally mounted in pressure enclosures. In depth filtration, also called bed filtration, gravity flow as well as pressure operation are used (see Water, industrial water treatment; municipal water treatment; water pollution; water reuse).

Filtration can also be accomplished by centrifugal force (see Centrifugal separation), or by ultrafiltration, which involves a cross-flow filtration that is actually a classification based on molecular size. In straining and screening, closely related to filtration, the objective is to separate only a proportion of the solids (see also Size classification).

There is no sharp dividing line between the need for depth and cake, or surface filtration. However, cartridge depth filtration is seldom used when solids recovery is required. The finer the solids and the more dilute the suspension (conc < 0.01%), the more likely that depth filtration will be economically preferable.

Cake Filtration

In the last 25 years, several developments have shaped the course of evolution of filtration. First, polyelectrolyte flocculants permit the nature of feeds to be modified, increasing equipment possibilities. Second, synthetic fibers and membranes vastly enlarged the choices of filter media and consequently a suitable filter material is available for most applications. Third, endless-belt filters with a variety of configurations are of increasing importance their key feature being the inclusion of continuous medium washing. Fourth, the filter press, after steadily losing to continuous filters since their introduction, is receiving renewed interest because of automated discharge, grooved-rim recessed plates, and the inherent capability of the filter press to dewater to the lower moisture content necessary for landfill, or conserving fuels in cake drying and incineration. Membrane development further extends its performance capability.

Theoretical aspects. Cake filtration is unique in that filter testing provides directly the rate, cake moistures, washing effectiveness, etc, without application of theory. Theoretical considerations, however, contribute to the understanding of the filtration mechanism, estimating effects of variables, and analyzing data. The four filtration operations shown in Figure 1 are all used with compressible materials and all but expression are used on incompressible materials.

Cake formation (Fig. 1a) is the process of removing liquid from a suspension by maintaining a hydraulic pressure difference between the feed slurry supplied from one chamber and the filtrate received by another chamber, the two chambers being separated by a filter medium. For constant pressure filtration, the basic equation is:

$$v^2 \left(\frac{\eta \bar{a} C^*}{2} \right) + v (\eta R_m) = \Delta p\, t$$

where v is volume of filtrate per unit area of filtration surface, η the viscosity of the liquid phase, $\bar{a}$ the average specific resistance of the filter cake, C^* the mass of dry solids in the cake per unit volume of filtrate, R_m the hydraulic resistance of the medium, Δp the hydraulic pressure differential across the cake-medium system, and t the filtration time.

Cake washing (Fig. 1b) is the process of removing a dissolved solute from the pores of a filter cake by forcing a wash liquid to flow through the cake. The objective may be either to increase the purity of a solid product or to improve the recovery of the liquid product, or even to reduce the liquid viscosity to aid in dewatering. The washing process involves an initial hydraulic displacement stage, a transition stage, and a final diffusion.

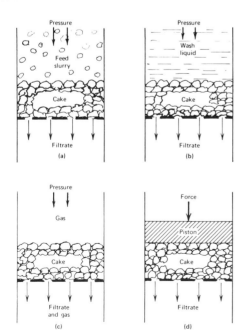

Figure 1. Basic filtration processes: (**a**) cake formation; (**b**) cake washing; (**c**) cake dewatering by gas displacement; (**d**) cake dewatering by expression.

In gas-displacement dewatering (Fig. 1c), a gas (usually air or steam) is forced through the cake and the liquid phase is substantially displaced by the gas. However, the flow channels in the cake may be of such a small diameter that capillary forces prevent the gas at the applied pressure differential from displacing the liquid significantly. Factors affecting cake moisture are best correlated by an approach or correlation factor:

$$F_c = \frac{v_{air} \Delta p_d\, \theta_d}{\mu w_0}$$

where v_{air} is the air velocity measured at downstream pressure, Δp_d the pressure drop, θ_d the dewatering time, η the viscosity of the liquid, and w_0 the cake weight of dry solids per unit area.

Dewatering by expression (Fig. 1d) is the process of applying mechanical pressure to compress a cake and squeeze out the liquid. The basic mechanism is operable during dewatering in equipment such as the filter-belt press and filter presses equipped with movable diaphragms.

Practical aspects. For cake formation, increased pressure differential (Δp) on incompressible cakes is productive, its benefit on rate calculable. However, on compressible cakes it increases cake resistance to an extent determinable only empirically.

High initial Δp tends to drive fine particles into and through the filter medium, increasing the medium resistance and the solids content of the filtrate. The same force can add to the resistance of initial cake layers, favoring low pressure start of cake formation. However, the main reason for variable rate and pressure operation in batch pressure filtration is the practicality and economics of centrifugal-pump selection, dictated by their performance curves. Some filter cakes formed from flocculated solids have critical pressures at which the cakes densify significantly, decreasing rates substantially. It is then desirable to operate below such pressures or modify the feed with filter aids to maintain porosity.

Formation time: Increased length of cake-formation time at constant pressure usually affects the rate by the square-root function. Thus, any doubling of cake thickness or liquor throughput per cycle takes about four times as long.

Liquor viscosity is changed by temperature and its effect on rate is often unexpectedly large. Doubling viscosity doubles the time to form the same cake thickness, thus cutting the rate in half.

Cake specific resistance is mostly the result of basic feed nature. Particle shape and size distribution are the key elements. Even a 1% shift in concentration of extreme fines can make a critical change in filtration performance. Any shift in natural flocculating tendencies or in conditions of flocculant addition may create a large shift in effective particle size. Slurry aging and pH can affect particle size or flocculation.

As feed-solids concentration approaches the cake concentration, its effect on rate and cake moisture are magnified.

Cake washing is extremely efficient during the displacement stage and becomes quite inefficient in late diffusion stages. Usually it is most economical to terminate washing during the transition stage, after only two to four displacement volumes of wash.

With moderately porous cakes of low–moderate compressibility where dewatering must be by gas displacement, the approach factor correlates well with cake moisture and often simplifies to cubic meters of air per kilogram of cake solids.

Decreasing liquor viscosity by feed heating or by a hot wash will effectively reduce moisture. Steam is highly efficient in reducing moisture because of the high heat release into the cake.

Surface-tension reduction often reduces the residual moisture, but does not change the factor.

Filter media. Selection of a filter medium has to be based on many factors, overall economics being the final criterion. The most permeable medium consistent with process requirements permits the highest output.

The important properties of media (excluding precoats) are particle retention on surface vs depth; minimum flow resistance; good cake release; resistance to blinding and ease of cleaning vs disposable media; flexibility for the intended use, eg, caulking or sewing bags; strength and durability for the particular equipment design and process conditions; dimensional stability under tension or pressure; and suitability as a gasket between filter-press plates.

Woven fabrics. Wool has been all but eliminated from the filter medium market and cotton is limited to only a small share, being replaced by synthetics with better cake release, nonblinding characteristics, long life, and resistance to attack from chemicals, bacteria, or fungi. Polypropylene is moderately priced and the most versatile and popular woven synthetic fabric, nylon is a close second. Other synthetics such as polyethylene, polyesters, Dynel, Kynar, Orlon, Saran and Teflon are also available. Metal wire cloths perform similarly to monofilament synthetics.

The filtering characteristics of woven fabrics depend mostly on the type of yarn and weave. Yarns can be monofilaments, multifilament, spun from staple fiber, or a combination of the latter two. A high twist can make a multifilament behave more like a monofilament.

Nonwoven media. Felts, specifically needle felts, comprise about 60% of the total nonwoven filter market and 80–90% of the laid nonwoven filters (including cartridge-filtration applications). Felts are available in most of the same fibers as woven cloths, polypropylene predominating.

Precoats. In conjunction with filter cloths, precoats provide a renewable filter-medium system that can accomplish optimum clarification with a negligible risk of blinding. Only diatomaceous earth, perlite, and cellulose fibers are marketed extensively.

Feed pretreatment. Filterability may be affected by changes in the following feed characteristics or feed treatments: average particle size of solids; ratio of slimes to coarser particles; agitation and pumping (causing solids degradation); age, pH and temperature of slurry; flocculation or dispersion; heat treating or freeze-thawing; viscosity and vapor pressure of liquid; dissolved materials; aeration; surface tension of liquid; filterable solids concentration; filter thickening; and admix of filter aid.

Cake-Filtration Equipment

A guide to matching filterability with suitable types of equipment is presented in Table 1.

The distinction between continuous and batch filters is not clear-cut because many automated cyclic batch operations become almost as continuous as sectionalized rotary-drum filters. Units with a continuous output of both filtrate and cake are treated here as continuous.

ROBERT M. TALCOTT
CHARLES WILLUS
M.P. FREEMAN[†]
Dorr-Oliver Incorporated

R.J. Wakeman, *Filtration Post Treatment Processes*, Elsevier Scientific Publishing Co., Amsterdam, Neth., 1975.

D.B. Purchas, ed., *Solid-Liquid Separation Equipment Scale-Up*, Uplands Press Ltd., Croydon, UK, 1977.

C. Orr, *Filtration Principles and Practices*, Pt. 1, Marcel Dekker, Inc., New York, 1977.

FINE ART EXAMINATION AND CONSERVATION

Scientific Examination

The explosive development in analytical techniques has given birth to many instrumental methods which, through high sensitivities and small sample requirements, are extremely appropriate for the study of art objects. For example, metal-alloy analysis has a long history of application in numismatics, where the evaluation of relative values of contemporary coinages depends on the knowledge of their intrinsic values. With the traditional wet-chemical analysis, this often meant the loss of half a coin. Although the analytical results are certainly very accurate and precise, and hence allow meaningful and valuable conclusions, the reluctance of a curator to sacrifice an irreplaceable part of humanity's cultural heritage for this purpose is not only understandable, but justified. Because of the extreme importance of the proper preservation and prevention of damage, it is absolutely necessary that any study involving the removal of even the tiniest sample be discussed by the curator and scientist in order to determine the goals of the proposed project, the value of the results, and the extent of the damage which will be inflicted. In cases where questions of authenticity are raised, the scientist will identify the materials, analyze their chemical composition, and investigate whether this corresponds to what has been found in comparable objects of unquestioned provenance. Dating techniques may allow the establishment of the date of manufacture.

Most fruitful and satisfying are those truly interdisciplinary projects in which an object or group of objects of unquestionable provenance is studied jointly by the historian and the scientist, in order to better evaluate the historical context.

Methodologies

The most important tool in a museum laboratory is the low power stereomicroscope. This instrument, usually used at magnification of 3–50 ×, has enough depth of field to be useful for the study of surface phenomena on many types of objects without the need for sample preparation (see Fig. 1). The information thus obtained can relate to toolmarks and manufacturing techniques, wear patterns, the structure of corrosion and paint layers, artificial patination techniques, or previous restorations. At higher powers, reflected-light microscopy may be useful in the study of a polished metal specimen. Transmitted-light microscopy is the preferred tool in the study of thin sections from stone or ceramic objects.

The very high powers of magnification afforded by the electron microscope, either scanning electron microscopy (SEM) or scanning transmission electron microscopy (STEM), are in great demand for identifications, such as of wood species, studies of ancient metallurgical techniques, and ceramics, as well as in the study of deterioration processes. Ultraviolet light has been in use for a long time in the examination of paintings and other objects, especially for the detection of repairs and restorations. Because of the variations in fluorescent behavior of different materials with otherwise similar optical properties,

Table 1. Slurry Filterability vs Filter Capability[a, b]

Slurry characteristics	Fast filtering	Medium filtering	Slow filtering	Dilute	Very dilute
Filterability aspects					
cake formation	cm/s	cm/min	1–5 mm/min	<1 mm/min	no cake
normal solids concentration	>20%	10–20%	1–10%	<5%	<0.1%
dry solids rate, kg/(h·m²)[c]	>2500	250–2500	25–250	<25	
filtrate rate, m³/(h·m²)[d]	>10	1–10	0.1–1	0.02–5	0.02–5
solids settling rate	difficult to suspend	fast	slow	slow	
Filter application					
continuous vacuum					
horizontal table, scroll discharge	++	++			
tilting pan	++	++			
horizontal belt	++	++	+		
top feed drum	++				
internal drum	++	++	+		
single-compartment drum	++				
multicompartment drums					
most dischargers	+	++	++		
roll discharge		+	++		
rotary precoat				++	++
disc		++	++		
filter belt press			++		
batch nutsche	++	++	++	++	++
batch vacuum leaf		+	++	++	
batch pressure					
filter press		++	++	++	++
horizontal leaf	++	++	++	++	++
tubular				++	++
vertical leaf		++	++	++	++
tube press		++	++		
automated diaphragm press		++	++		
cartridge, disc, edge					++
filter thickeners					
high shear or cross flow		++	++		
cycling backflush				++	++

[a] Excluding pulp and paper which is fast filtering but dilute giving high filtrate loadings.
[b] Crosses indicate degree of efficiency.
[c] To convert kg/(h·m²) to lb/(h·ft²), multiply by 0.205.
[d] To convert m³/(h·m²) to gal/(h·ft²), multiply by 24.5.

[†] Deceased.

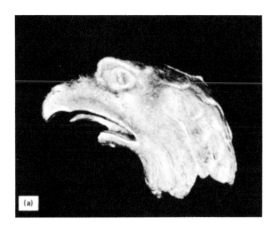

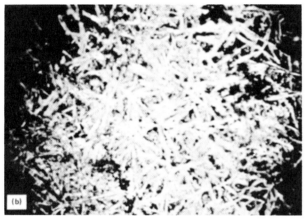

Figure 1. (a) A bronze hawk's head, forgery of a Roman finial. (b) The patina, studied through a stereomicroscope, shows a structure typical for an artificially induced corrosion, probably effected by pickling. Courtesy of the Research Laboratory, Museum of Fine Arts.

areas of repaint or replacements of losses with color-matched filling materials often can be observed easily. Infrared irradiation may enable the detection of changes in composition, called pentimenti, restorations, or underdrawings.

The best known structural-examination technique is probably x-radiography. The x rays are primarily absorbed by the heavy metals present in some pigments. X radiographs of paintings yield information regarding the development of the composition, eg, the blocking out of certain compositional elements and changes therein, as well as the technique of the artist. X radiography is also used extensively in the examination of other types of art objects, in order to obtain information regarding manufacturing techniques or repairs. Information relating to shaping or joining techniques, can be obtained. β-radiography is used on paper.

In neutron-activation autoradiography, a painting is exposed to a flux of thermal neutrons produced in a nuclear reactor. These neutrons, which penetrate the painting completely, interact with a small fraction of the nuclei of the various chemical elements present in the painting. Radioactive isotopes are produced, each with its own characteristic half-life. A so-called autoradiograph produced by exposure of a photographic emulsion placed in close contact with the painting, shows the distribution of the predominant activities within the painting at the time of the exposure of the film. Information obtained relates to a number of pigments other than lead white, and complements rather than duplicates that obtained from x radiography.

The damage resulting from the removal of a sample for analytical study has to be weighed carefully against the expected benefits. Often, the sample size is dictated not so much by the requirements of analytical techniques as by the nature of the material under study.

Dating. Radiocarbon dating depends upon the equilibrium between the formation of the radioactive isotope ^{14}C and its decay at a half-life of 5730 yr. After its formation in the upper stratosphere, the ^{14}C is rapidly mixed into the carbon-exchange reservoir, ie, oceans, atmosphere, and biosphere. If, however, materials from dead plants or animals are used in the manufacture of an artifact, no exchange takes place, and the radioactive-decay process results in a continual decrease of the ^{14}C concentration in the material. The time elapsed since the death of the plant or animal can be determined by comparison of the measured radioactive concentration of the artifact with that of the exchange reservoir (see Radioactive tracers).

The concentration of ^{14}C is determined by measurement of the specific β activity. The application of this technique in dating of art objects has been limited severely by the large samples needed. However, the combination of specially designed miniature gas counters and complex electronic equipment has resulted in a reduction of the required sample.

Dendrochronology depends on the variations in annual growth rates for trees with the climatic conditions during the growing season. Wet summers result in thicker growth rings than dry ones. Dendrochronological sequences have been established for many climatic regions. This is important not only for the corrections thus obtained for radiocarbon dates, but also for the direct dating of the wood used in the manufacture of art objects.

In thermoluminescence dating, a sample of the material is heated, and the light emitted as a result of de-excitations of the electrons or holes freed from the traps at luminescence centers, is measured to indicate the trap-population density. The signal is compared with that obtained from the same sample after a laboratory irradiation of known dose. The annual dose rate is calculated from determined concentrations of radioisotopes in the material and assumed or measured environmental radiation intensities. This technique is used for ceramics and in geological studies. The accuracy is ca 5–10%, and is less important in authenticity questions than in dating.

In amino acid racemization dating, the degree to which racemization has progressed since the death of an organism serves as a measure of the time elapsed after that event. The technique is applied to the dating of shell, bone, or ivory objects.

Another technique for fossil materials is known as nitrogen–fluorine dating. Bone or ivory, under burial conditions, loses nitrogen because of groundwater leaching of the amino acids resulting from hydrolysis of proteins. At the same time, fluorine is absorbed from the groundwater by the hydroxyapatite in the bone, forming fluoroapatite. The ratio of nitrogen and fluorine concentrations in the bone decreases with time.

In indirect dating, or dating by inference, certain conclusions are reached regarding the date of manufacture of an object from other information.

Paintings. A close visual inspection under normal illumination reveals many indications of the condition of the painting and previous repairs. The information obtained by examination with the stereomicroscope at low magnification relates to the characteristics of the craquelure, the pigments, buildup of paint layers, technique of the artist, and condition.

The ir photograph, or the image from an ir-sensitive vidicon system, gives evidence of restorations as well as pentimenti and reveals underdrawings. The radiograph reveals evidence of damage and losses in the paint layers and the support.

Identification of pigments present on paintings of unquestioned attribution provides reference information regarding the use of certain pigments in given periods or schools.

Microscopic examination of cross sections through the paint layers gives definite information regarding the paint-layer sequence in the area from which the sample was taken.

Identification of the binding media includes solubility tests, ir absorption spectrophotometry, specific staining techniques, gc and ms.

Metal objects. Examinations of metal objects generally include the characterization of the metal, determination of the techniques involved in the manufacture, and study of aging phenomena. Of the latter, the state of corrosion is especially important, both in the examination of an object for the purpose of determining its authenticity and an assessment of its state of conservation. The layer of corrosion products covering the surface of the metal, the so-called patina, can be studied for its composition and structure. Identification of corrosion products is often performed by means of x-ray diffraction analysis.

For x radiography, high energy x rays are needed, obtained from either an industrial x-ray unit or radioisotopes (γ radiography). The radiograph may give direct evidence of small cracks, casting defects, failing joints, etc. It also provides information relating to the type of joining and forming.

Nondestructive, energy-dispersive x-ray fluorescence spectrometry only allows a surface analysis. For precise quantitative analyses, samples have to be taken from below the surface, to be analyzed by atomic absorption or spectrographic techniques.

Trace-element analysis of metals can give indications of the geographic provenance. Both emission spectroscopy and activation analysis have been used.

Ceramics. Clays differ in working behavior and firing properties. Pots can be made by working a slab of clay into the desired form, by building up with coils, and by throwing on the wheel. Clay can also be cast into porous molds, which became the most important manufacturing technique in the porcelain industry (slip casting).

Glazes are actually glasses. The first glazes were ash glazes. The Egyptians used alkaline glazes, which were already colored through the addition of metal oxides. In the Roman period, glassmaking reached unprecedented heights. During this time, glassblowing was invented.

Natron and ash of seaweeds provided the sodium which served as the principal flux in glasses until the medieval period. Wood ashes then came into use in Europe, which changed the glass formulation to such a degree that potassium salts became the principal fluxing alkalies.

Examinations of ceramic objects involve a variety of techniques, depending on the type of information sought. If an assessment of condition and state of repair is made, the most important tools are the low power stereomicroscope, x radiography, and examination under uv light. For the identification of glass degradation products, a number of chemical analytical techniques can be used, especially x-ray diffraction.

Elemental chemical analysis provides information regarding the formulation and coloring oxides of glazes and glasses. Energy-dispersive x-ray fluorescence spectrometry is very convenient. The electron-beam microprobe is an extremely useful tool for the analysis of elements with low atomic numbers.

Trace-element analysis, using emission spectroscopy and activation analysis has been applied successfully in provenance studies on archaeological ceramics. The attribution of a certain geographic origin for the clay of an object excavated elsewhere has a direct implication on past trade and exchange relationships. Microscopic examination of ceramic paste, both at low magnification and at high power with prepared cross sections, can be used for petrographic study of the mineral composition and for the determination of techniques involved in the manufacturing process. For ware decorated with slips and glazes, microscopic examination of a cross section is a means of studying the decoration technique.

Stone objects. The technical examination of stone objects begins with the low power stereomicroscope, which yields information regarding toolmarks, cutting techniques, and wear patterns. X-ray diffraction can be used for identification.

The polarizing microscope is another important tool. Petrographic study of a thin section can be extremely helpful and often yields information with regard to geographic origin. Trace-element analysis provides another approach, and activation analysis has been applied successfully in provenance studies of, eg, limestone sculpture.

For marble-provenance studies, the most successful technique seems to be the measurement, through mass spectrometry, of the abundance ratios of the stable isotopes of carbon and oxygen.

Other materials. Some specific questions come up in the study of textiles. The first one is with regard to the identification of the fiber, a second relates to the dyes used. Elemental chemical analysis, eg, emission spectroscopy, is used for identification of the mordant, and dye identification is most often done spectrophotometrically or with tlc.

Inks, watercolor pigments, and media, etc, are analyzed similarly to the pigments and media for oil paintings.

Conservation

One of the principal problems with restorations in the past was that the effort was directed entirely toward the restoration of the appearance of the object. To avoid recurrence of deterioration, it is, however, necessary to remove its cause.

The conservator possesses the necessary skills, a good grasp of scientific methodology, and a sufficient knowledge of chemistry and physics to be able to understand the basic causes of deterioration. A number of ethical standards have been formulated; no action is allowed that could damage the object. Repairs, restorations, and other treatments should be reversible. The arrest of the deterioration and the prevention of its recurrence has higher priority than restoration. Overpainting of original surfaces is not considered to be an ethical practice. Extensive documentation, including photographs during all stages of the work, must be kept.

In the United States, a few universities have established graduate conservation programs. Most conservators are members of the International Institute for the Conservation of Historic and Artistic Works (IIC) in London. The International Committee of Museums (ICOM) has its own subcommittee on conservation.

Deterioration Processes

Corrosion processes for various metals are vastly different. In addition to the metal, the reactive species involved also determine the nature of the corrosion process. Water is involved in all corrosive processes. Other factors, eg, environment, climate and pollutants, influence the reaction rates and the nature of the products (see Fig. 2).

Chemical damage is dependent on the nature of the stone. Limestones and marbles, for example, are very susceptible to acidic attack. On the other hand, granites and basalts have survived millennia without much effect. A mechanical effect is caused by water-soluble salts.

Low fired ceramics can suffer through rehydration. The deterioration of glasses, including glazes, involves devitrification.

Glazed ceramics can be subject to an additional problem: if the thermal-expansion coefficient of the deteriorated glaze does not match that of the body, changes in temperature can result in loss of adhesion.

Figure 2. Influence of air pollution. The surface of this statue has been almost completely lost through pollution-induced corrosion; the light-colored areas correspond to losses of up to 2-mm thickness. Courtesy of the Research Laboratory, Museum of Fine Arts.

Wooden objects are subject to warping when exposed to high humidity, which also enhances the growth of molds; extreme dryness results in cracking.

Organic substances have survived under the dry conditions of a desert climate. Sudden changes are often disastrous.

Many organic materials are light-sensitive. Chemical degradation is especially harmful to materials that contain cellulose. Composite objects are highly vulnerable. A special problem is caused when one of the components promotes chemical decay in another. For example, copper pigments promote the deterioration of silk.

No cure can undo the damage that often could have been prevented. Of the various environmental factors, temperature is probably the easiest to control. A temperature of ca 20°C is fairly safe. More difficulties are encountered with humidity. Many institutions solve this problem through the installation of extensive climate-control systems or the creation of microclimates, where the objects are placed within smaller spaces, such as cases, in which an ideal environment is maintained.

In many museums, natural daylight is preferred on aesthetic grounds. In such cases, the ultraviolet part of the spectrum should be removed with appropriate filters. Artificial-light sources that produce a significant amount of uv radiation should be provided with a filter. Storage areas should be dark.

One method of reducing exposure of extremely sensitive objects is to exhibit them only for limited periods.

The problems posed by air pollutants have become very serious in recent times. Within the museum, measures can be taken to remove harmful substances. In the preservation of outdoor objects, the problems caused by air pollution are overwhelming. The ravages inflicted on acid-sensitive stones have already resulted in the virtual destruction of innumerable sculptures.

Objects should be guarded against vandalism or damage inflicted by ill-informed admirers. Preventive conservation should include a training program for all who handle art objects.

All natural-resin varnishes exhibit, with age, problems related to chemical deterioration. The most common problem in the paint layers is loss of adhesion. Flaking paint is treated by infusion of an adhesive. Failure of a canvas support occurs when the fabric becomes brittle and the fibers break. The common solution is lining or relining.

Wooden panels have a tendency to warp and crack. Wall paintings become endangered when the wall decays. A technique now routinely used for the removal of fresco paintings from the wall is the so-called strappo technique. The paint layer is removed and adhered to a stable support; the assembly can be remounted.

The degradation of paper has been recognized to be the result of hydrolytic degradation and oxidation. Another problem is so-called foxing or the formation of small brown spots.

Metal and stone objects. On highly polished surfaces, the slightest tarnish constitutes a disfiguring effect and must be removed. One way that recurrence can be prevented is by applying lacquer. Disfiguring corrosion crusts are removed by mechanical or chemical cleaning. Stones are cleaned by removing grime and stain with solvents and detergents. Consolidation involves the introduction of a supporting material to increase mechanical strength.

Ceramics and glass. To mend broken ceramics, most conservators use adhesives based on poly(vinyl acetate), acrylic resins, and epoxies. So-called weeping glass is treated by removing the free alkali with acid, following by rinsing and drying.

When a textile has lost its mechanical strength, it must be backed with a lining fabric. Consolidation by application of a polymer may be necessary. Dyes fade through photodegradation, and little can be done to restore them. Other causes of textile deterioration are mold and insect damage.

Wooden objects recovered from underwater sites may suffer irreparable damage. Treatment with ethylene glycol or impregnation with a solution of rosin in acetone are possibilities.

L. van Zelst
Museum of Fine Arts, Boston

W.J. Young, ed., *Applications of Science in Examination of Works of Art*, Museum of Fine Arts, Boston, Mass., 1973.

W.J. Young, ed., *Application of Science to Dating Works of Art*, Museum of Fine Arts, Boston, Mass., 1976.

J.C. Williams, ed., *Preservation of Paper and Textiles of Historic and Artistic Value, II, Advances in Chemistry Series No. 193*, American Chemical Society, Washington, D.C., 1981.

FINE CHEMICALS

Fine chemicals generally are considered to be chemicals that are manufactured to high and well-defined standards of purity, as opposed to heavy chemicals which are considered as being made in large amounts to technical levels of purity. Although fine chemicals usually are thought of as being produced on a small scale, this is by no means always the case. The production of some fine chemicals is in tens or hundreds of kilograms per year, others are in thousands of metric tons (especially fine chemicals used as drugs or food additives). For example, the United States annually produces ca 14,000 t of aspirin.

Table 1 shows standards for sodium hydroxide, ranging from technical grade through the drug, food, and reagent grade, to the electronic grade. The advances in purity represented by these various grades of chemicals are based upon the special uses of the chemicals.

USPC. The legal standards for drugs in the United States are established by the United States Pharmacopeial Convention (USPC), and have been published in 20 revisions of the *United States Pharmacopeia* (*USP*). In the past, standards for many drugs that were not in the USP were established by the American Pharmaceutical Association, and were published in the *National Formulary* (*NF*). In 1974, the USPC acquired the NF. *USP XX* and *NF XV* have been published in a combined volume. In this compendium, drug substances and dosage forms of drug substances are designated USP, and pharmaceutic ingredients, used to make the active ingredients into a suitable dosage form for use by the patient, are designated NF.

Food-additive chemicals. Standards recognized as defining food-grade chemicals in the United States are set by the Committee on Food Chemical Codex of the NAS which publishes them in the *Food Chemicals Codex* (FCC) (see Regulatory agencies). Standards for laboratory reagents are set by the ACS Committee on Analytical Reagents, and are published in *Reagent Chemicals—ACS Specifications*. The *SEMI Book* standards for electronic-grade chemicals are quite new and a number have recently been published in the *1983 Book of SEMI Standards*.

National Bureau of Standards (NBS). The NBS is the source of many of the standards used in chemical and physical analysis. The standards prepared and distributed by the NBS are used to calibrate measurement systems and to provide a central basis for uniformity and accuracy of measurement. At present, nearly 900 Standard Reference Materials are available and are described in *NBS Special Publication 260*. Included are many steels, nonferrous alloys, high purity metals, primary standards for use in volumetric analysis, microchemical standards, clinical laboratory standards, biological material certified for trace elements, environmental standards, trace-element standards, ion-activity stan-

Table 1. Typical Specification for Various Grades of Sodium Hydroxide, wt%

	Technical	NF	FCC	Reagent	Electronic
assay (NaOH)	94.0 min	95.0 min	95.0 min	97.0 min	98.0 min
Na_2CO_3		3.0 max	3.0 max	1.0 max	0.4 max
heavy metals (as Pb)		0.003 max	0.003 max		
heavy metals (as Ag)				0.002 max	0.001 max
lead (Pb)		0.001 max			
mercury (Hg)		0.00001 max	0.00001 max	0.00001 max	0.00001 max
arsenic (As)		0.0003 max			
potassium (K)	to pass test			0.02 max	0.01 max
iron (Fe)				0.001 max	0.0003 max
nickel (Ni)				0.001 max	0.0005 max
ammonium hydroxide ppt				0.02 max	0.02 max
chloride (Cl)				0.005 max	0.001 max
sulfate (SO4)				0.003 max	0.0005 max
phosphate (PO4)				0.001 max	0.0002 max
nitrogen compounds (N)				0.001 max	0.0003 max

dards (for pH and ion-selective electrodes), freezing and melting-point standards, colorimetry standards, optical standards, radioactivity standards; particle-size standards, and density standards.

SAMUEL M. TUTHILL
JOHN A. CAUGHLAN
Mallinckrodt, Inc.

The United States Pharmacopeia XX, (USP XX-NF IV), The United States Pharmacopeial Convention, Inc., Rockville, Md., 1980.

Food Chemicals Codex, 3rd ed., National Academy of Sciences—National Research Council, Washington, D.C., 1981.

Reagent Chemicals—American Chemical Society Specifications, 6th ed., American Chemical Society, Washington, D.C., 1981.

1983 Book of SEMI Standards, Vol. 1, Semiconductor Equipment and Materials, Inc., Mt. View, Calif., 1983.

FIREBRICK. See Refractories.

FIRECLAY. See Refractories.

FIRE-EXTINGUISHING AGENTS

Fire is a rapid oxidation that is self-sustaining and accompanied by the evolution of heat and light. The prerequisites for combustion are heat, fuel, an oxidizing agent, and a suitable chemical reaction path. Any method for inhibiting combustion must involve one or more of the following: removal of heat faster than it is released; separation of fuel and oxidizing agent; dilution of the vapor-phase concentration of fuel and oxidizing agent below that which is necessary for combustion; and termination of the chain reaction.

Classification of Fires

The following classifications are used in the United States and Canada: Class A fires are fires in ordinary combustible materials, eg, wood, cloth, paper, rubber, and many plastics; Class B fires are fires in flammable and combustible liquids, gases, and greases; Class C fires are fires that involve energized electrical equipment; and Class D fires are fires in combustible metals such as magnesium, titanium, zirconium, sodium, and potassium. Different classifications are used in Europe.

Mechanism of Extinguishment

The cooling effect of water and carbon dioxide is well known as is the dilution of oxygen by the latter. Separation of fuel from air (oxygen) is usually achieved by foams. Halon and CO_2, being heavier than air, can also form a barrier between the fuel and air. However, the ability of the Halons and dry chemicals to extinguish fires cannot be completely explained by the above mechanisms. Their action may be due to the interruption of the chain reaction. Other factors contributing to the extinguishment of fires, including radiation shields, shock waves, critical vibrations, and ion separation owing to magnetic fields, require more research.

Specifications

In the United States, Underwriters' Laboratories, Inc. (ULI) and Factory Mutual (FM) evaluate the performance of fire-extinguishing equipment. In Canada, the Canadian Government Specification Board (CGSB) and the Underwriters' Laboratories of Canada (ULC) perform a similar function. Portable fire extinguishers are shown in Figure 1.

Underwriters' Laboratories developed the following rating system: Class A rating is obtained by extinguishing crib, panel, and excelsior fires. A Class B rating is obtained by extinguishing the fire of a certain area of heptane. The Class C rating is obtained by an electrical-conductivity test. The Class D rating is used for molten metals, castings, or chips; specific fires are noted for a given agent, ie, Met-L-X is an agent for sodium, potassium, sodium-potassium alloys, and magnesium, whereas G-1 powder is rated only for magnesium.

Water

Water is the most widely used agent because of its Class A effectiveness, thermal properties, cost, and availability. When applied in spray form, water can be an effective agent on flammable liquids with a flash point $> 66°C$. The disadvantages of water are its freezing point at $0°C$, high surface tension (78 mN/m or dyn/cm), and low viscosity (1 mm^2/s or cSt). The two latter properties limit the ability to penetrate a burning mass. Foaming increases the effectiveness of water. Because of its versatility, water can be applied from portable extinguishers, hose lines, sprinkler systems, and monitor nozzles. Addition of a surfactant reduces the surface tension, and so-called wetted water penetrates cotton bales, paper rolls, mattresses, and other solid objects.

Water thickeners, such as organic gelling agents or bentonite in combination with ammonium sulfate or phosphate, have been used successfully against forest fires.

Alkali-metal salts, eg, potassium carbonate or acetate are used to lower the freezing point of water. Calcium chloride, a good freezing-point depressant, is corrosive to some metals. The potassium salts have become known as loaded-stream agents. They are used in stainless steel "water" extinguishers found in public places.

Dilute solutions of poly(ethylene oxide), so-called slippery or rapid water, reduce water to nonturbulent flow in pipes, fire hoses, etc.

Foams

Foam-producing chemicals are used to modify water. Chemical foams are almost obsolete because the cheaper liquid-foam-forming concentrates are easier to handle (see Foams).

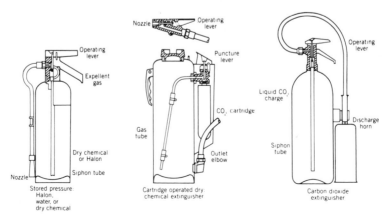

Figure 1. Portable extinguishers.

Protein-foam. Protein-foam concentrate is produced by hydrolyzing natural protein materials, eg, fish meal, feather meal, and horn-and-hoof meal. The meal is cooked in an alkaline solution, neutralized, and filtered. Stabilizers and other additives are blended in. Foam solutions contain 3 or 6% of the protein concentrate. Protein foam is applied to runways to assist disabled aircraft in landing. The foam acts as lubricant and as barrier against fires. A fluoroprotein is made in the same manner; a fluorocarbon surfactant is added to the blend.

Aqueous-film-forming foam (AFF). A fluorinated synthetic foam has been developed for the control of certain liquid fires, in particular those caused by petroleum products. The foam concentrate can be diluted with fresh or seawater. It releases a film which retards vapor formation and excludes air; it is also used for underground fires.

Synthetic foams (Syndets). These foams are blends of surfactants. They are used in concentrations of 1–6%.

High expansion foam (Hi Ex). Hi Ex foam is an aggregate of bubbles generated by the passage of air or other gases through a net, screen, or porous material which is wetted with an aqueous solution of surfactant foaming agents. Hi Ex foam is particularly suited for fires in confined spaces.

Foams resistant to alcohol or other polar solvents include a polysaccharide which forms an insoluble film over water-miscible liquids. It is available in 3 and 6% concentrates.

The principal use of foams is in the extinguishment of fires caused by liquids lighter than water. Fire may be prevented by spreading foam on the area in danger. However, foams cannot be used on fires involving gases, liquefied gases, or cryogenic liquids. Foam is a conductor and cannot be used on electrical fires. Certain wetting agents and dry chemicals are incompatible with foams.

Liquefied Gases

Carbon dioxide is a clean agent which leaves no residue; it is cheap and readily available; it can be used on electrical equipment; and it discharges under its own pressure.

Halogenated hydrocarbons (Halons). Halons are used in portable fire extinguishers, mobile equipment (trucks), and total-flooding systems. Halons leave no corrosive or abrasive residue. As nonconductors, they are used on fires in electrical and electronic equipment.

Halon 1301 (bromotrifluoromethane) and Halon 1211 (bromochlorodifluoromethane) are the most widely used halogenated agents; both are gases. Halon 1301 has a vapor pressure of 1.5 MPa (199 psi), and Halon 1211 has a vapor pressure of 253 kPa (22 psig) at 21°C. The latter is generally used in portable extinguishers.

Exposure to Halon concentrations up to ca 7 vol% and Halon 1211 concentration of 2–3 vol% have little toxic effect. At higher concentrations, dizziness may be observed, indicating a mild anesthetic effect. High concentration may lead to a loss of consciousness. The degradation products of halogenated agents are more toxic than the agents.

Halogenated agents are ineffective on flammable materials which contain an oxidizing agent, eg, gunpowder, rocket propellants, and some metals and hydrides.

Dry Chemicals

Dry chemicals are the most effective fire-extinguishing agents, which in the United States include sodium bicarbonate, potassium bicarbonate, monoammonium phosphate, and potassium chloride. In Europe, potassium carbamate and potassium sulfate are also used.

Sodium bicarbonate is frequently mixed with magnesium (or zinc or barium) stearate to promote water repellency. Recently, silicones have replaced the stearates. Both stearate and siliconized sodium bicarbonate are ULI and ULC rated for Class B:C fires.

Products based on potassium bicarbonate are usually referred to as Purple K, because of the purple color they impart to a flame. Purple K is highly effective on high pressure gas fires. It is classified as a Class B:C agent by ULI and ULC. Potassium sulfate is not used in Purple K. It has been used in Europe in conjunction with potassium bicarbonate; however, the product is not called Purple K.

Agents based on monoammonium phosphate have Class A capabilities. They are usually referred to as multipurpose or ABC. In the U.S. and Canada, they have to be tinted yellow (to prevent mixing with other agents) and are classified A:B:C by ULI and ULC.

Potassium chloride was developed as an agent with Class B capabilities compatible with protein foam.

Dry chemicals do not present a health hazard, since they do not degrade to toxic materials.

Agents against metal fires are based on sodium chloride or graphite compounds. Met-L-X is based on sodium chloride and contains a vinylidene dichloride copolymer to seal off the oxygen from the burning surface. G-1 powder is based on graphite and contains an organophosphate plasticizer that aids in excluding oxygen. Lith-X is a free-flowing graphite-based agent specifically developed for lithium fires. Na-X contains sodium carbonate for use around stainless steel. Foundry flux protects molten magnesium from air; it consists of a mixture of inorganic chlorides. Natrix is based on sodium carbonate. Pyromet is a mixture of sodium chloride, phosphates, proteins, clay, and a waterproofing agent. It is effective on sodium or calcium fires. TEC is a mixture of inorganic chlorides which has been successfully used on small uranium and plutonium fires.

E. EUGENE STAUFFER
The Ansul Company

R.L. Tuve, *Principles of Fire Protection Chemistry*, National Fire Protection Association, Boston, Mass, 1974, p. 143.

W.M. Haessler, *The Extinguishment of Fire*, National Fire Protection Association, Boston, Mass., 1974, p. 2.

Fire Protection Handbook, 15th ed., National Fire Protection Association, Quincy, Mass. 1981, Sect. 18, Chapt. 2.

FIRE PREVENTION AND EXTINCTION. See Fire-extinguishing agents; Plant safety.

FIRE-RESISTANT TEXTILES. See Flame-retardants for textiles.

FIREWORKS. See Pyrotechnics.

FISCHER-TROPSCH PROCESS. See Fuels, synthetic.

FISH-LIVER OILS. See Aquaculture.

FISH OILS. See Fats and fatty oils.

FLAME PHOTOMETRY. See Analytical methods.

FLAMEPROOFING. See Flame retardants.

FLAME RETARDANTS

AN OVERVIEW

Terminology

The application of such terms as fireproof, flameproof, self-extinguishing, nonburning, noncombustible, etc, has often led to ambiguities about the relative flammabilities of different materials. The preceding terms were challenged by the Federal Trade Commission as misleading to users of cellular plastics. The terms fire or flame retardant are preferred.

Fire retardant (flame retardant) is used to describe polymers whose basic flammability has been reduced by some modification as measured by one of the accepted test methods. A fire-retardant chemical denotes a compound or mixture of compounds that when added to, or incorporated chemically, into a polymer serves to slow up or hinder the ignition or growth of fire. A flame-retardant chemical technically is one that has the foregoing effect primarily in the vapor phase.

Means of Achieving Fire Resistance

The materials to be made resistant to fire are invariably polymers—natural and synthetic. Some are inherently harder to ignite and burn than others. Low flammability may be achieved by making a polymer rich in ring structures and low in readily oxidizable side groups. The aromatic polyimides (qv) have excellent fire resistance, but they are expensive and therefore limited to specialty uses such as aerospace applications. Some of the many candidate aromatic polymers will enter markets now held by more flammable polymers treated with flame-retardant chemicals (see also Aramid fibers; Novoloid fibers).

The more common approach to obtain fire resistance is to add one or more of the commonly used fire-retardant elements to a flammable polymer system. This may be done either with an inert additive (eg, surface treatments of fibers and films, or unreactive admixtures to molding or extrusion mixes) or by copolymerizing a moderate amount of a monomer containing the requisite element. Occasionally, a polymer can be treated reactively after polymerization as in chlorinating polyethylene (see Olefin polymers).

Requirements for Fire Resistance

Nearly all markets for fire-retardant chemicals are created by—and are as varied as—fire regulations. Textile products are covered by Federal standards for apparel, carpets, and mattresses, as well as by local building codes for textile coverings used in places of public assembly (see Textiles). The building codes control flammability of wall and ceiling coverings for the public parts of most buildings.

A survey in the mid-1970s estimating the market for chemicals used as fire retardants in plastics and related products showed about two thirds of these chemicals are used in plastics (see Flame retardants articles); 1986 estimated volumes in thousands of metric tons were: phosphate esters 32; halogenated phosphate esters 20; chlorinated hydrocarbons 32; brominated hydrocarbons 20; antimony trioxide (Sb_4O_6) 22; borates 7; polyols containing phosphorus 12; brominated bisphenol A 7; others 7.

JOHN W. LYONS
National Bureau of Standards

J.W. Lyons, *The Chemistry and Uses of Fire Retardants*, Wiley-Interscience, New York, 1970, Chapt. 5.

ANTIMONY AND OTHER INORGANIC COMPOUNDS

In many polymers, the high concentration of halogenated organic compounds needed to impart flame retardancy adversely affects their physical properties. In practice, halogen-containing flame retardants are formulated with inorganic compounds that behave synergistically with the halogen. This enables formulators to use fewer additives without diminishing flame retardance. Indeed, in many instances, flame retardancy is improved when inorganic halogen synergists are used.

Antimony Compounds

Antimony trioxide. Antimony trioxide is commonly referred to as antimony oxide. Antimony trioxide is manufactured by oxidizing molten antimony sulfide ore and/or antimony metal in air at 600–800°C. Typical properties are listed in Table 1.

Antimony oxides and antimonates are not, *per se*, fire retardants. They must be converted to volatile species. This is usually accomplished by having halogenated organics release halogen acids in the presence of fire temperatures. The halogen acids react with the antimony-containing materials to form antimony trihalide and/or antimony halide oxide.

Table 1. Properties of Antimony Trioxide

Physical properties	High pigmenting	Low pigmenting
specific gravity	5.3–5.8	5.3–5.8
average particle size (Fisher subsieve method), μm	0.8–1.8	0.9–2.5
oil absorption, g/100 g Sb_2O_3 (ASTM 332-64)	9–12	9–12
refractive index, n_D^{20}	2.087	

Table 2. Some Properties of Sodium Antimonate

Property	Value
physical form	white powder
antimony content	61.7 wt %
bulk density, kg/m^3	39.4–46.4
specific gravity	4.8
fineness	99.75% through 74 μm (200 mesh) sieve

These materials act both in the substrate (condensed phase) and in the flame to suppress flame propagation. In the condensed phase, they promote char formation which acts as a physical barrier to flame and inhibits the volatilization of flammable materials. In the flame, the antimony halides and halide oxides, generated in sufficient volume, provide an inert gas blanket over the substrate, thus excluding oxygen and preventing flame spread. These compounds alter the chemical reactions occurring at fire temperatures in the flame, thus reducing the ease with which oxygen can combine with the volatile products.

Other antimony compounds. Other compounds include antimony pentoxide, primarily available as a stable colloid or as a redispersible powder. Significantly more expensive than antimony trioxide, it is designed primarily for highly specialized applications, although manufacturers suggest it has potential use in fiber and fabric treatment.

Sodium antimonate, $Na_2OSb_2O_5.1/2H_2O$, made by the oxidation of antimony trioxide in base, is recommended for formulations in which deep tone colors are required (see Table 2).

Recent developments in inorganic flame-retardant synergists have centered on mixed products that contain antimony and other metals which reportedly give excellent performance at reduced cost, eg, a series of antimony–silico complexes (trade name, Oncor), which are less opacifying than either high or low tint antimony oxide, and less expensive, although they are less effective flame retardants. Another example of a low cost replacement for antimony oxide is Thermoguard CPA, which is of equal efficiency as a flame retardant.

Boron Compounds

In this class of compounds, by far the most widely used is zinc borate (see Boron compounds), and a variety is available that vary in zinc, boron, and water content. Zinc borate is rarely used alone; it acts synergistically with antimony oxide, enabling compounders to extend antimony trioxide in some formulations. It is also used with high levels of alumina trihydrate in some halogenated unsaturated polyester resins.

Boric acid and sodium borate (borax). These are two of the oldest known flame retardants, used primarily with cellulosics, eg, cotton and paper. Both products are inexpensive and fairly effective, but their use is limited to products for which nondurable flame retardancy is acceptable since both are very water-soluble.

Alumina Hydrates

Alumina trihydrate (ALTH) is made either from bauxite by the Bayer process (see Aluminum) or from recovered aluminum by the sinter process (see Aluminum compounds). Alumina trihydrate is the only aluminum compound of commercial significance as a flame retardant, and functions in both the condensed and flame phases. When exposed to temperatures over 250°C, it forms water and alumina, with the evolution of water absorbing heat by cooling the flame and diluting the flammable gases and oxidant in the flame.

Molybdenum Oxides

Molybdenum compounds have been used as flame retardants of cellulosics for many years (see Molybdenum compounds) and have recently been used with other polymers. They appear to function as condensed-phase flame retarders.

JOHN M. AVENTO
IRVING TOUVAL
M & T Chemicals, Inc.

I. Touval and H.H. Waddell, *Handbook of Fillers and Reinforcements for Plastics*, Van Nostrand Reinhold, New York, 1978, 13, p. 219.

P.V. Bonsignore and H.P. Hsieh, "Alumina Trihydrate and Other Inorganic Additives for Polyester Resins," *Flame-Retardant Chemicals Association (FRCA) Meeting*, Houston, Tex., March, 1978.

F.W. Moore, *Molybdenum Compounds as Smoke Suppressant for Poly(vinyl chloride) Formulations*, Society of Plastics Engineers (SPEANTEC), Montreal, Can., Apr., 1977.

HALOGENATED FLAME RETARDANTS

Current commercial halogenated products used as flame retardants for plastics are mainly compounds containing high (50–85 wt%) levels of either chlorine or bromine, ie, decabromodiphenyl oxide, chlorendic acid, tetrabromophthalic anhydride, etc. These materials fall into two distinct types: additives and reactives. The advantage of the additive is that it is readily added to a polymer by mechanical means with a minimum of reformulation required. The reactives, on the other hand, require the development of essentially new polymer systems.

Only massive polymer forms are considered, though the materials and concepts discussed are almost similarly applicable to fibers, fabrics, coatings, and elastomers. Halogenated phosphorus compounds are included under Flame retardants—phosphorus compounds.

Principles of Developing Flame-Retardant Polymers

There are five fundamental methods used to fire-retard both synthetic and natural polymer systems: raise the decomposition temperature of the polymer, generally, by increasing the cross-linking density of the polymer (eg, ladder polymers); reduce the fuel content of the system, eg, by halogenating the polymer backbone, adding inert fillers, or resorting to inorganic systems; induce polymer flow by selective chain scission, usually applied to thermoplastic polymer systems where interrupting the polymer backbone reduces the viscosity of the polymer and promotes dripping; induce selective decomposition pathways, a method most applicable to cellulosics where introducing phosphorus compounds generates phosphorus acids which catalyze the loss of water and the retention of the carbon as char; use mechanical means, including bonding a nonflammable skin on the polymer, covering the polymer with an intumescent coating, designing the system, and using sprinklers.

Polymer Classes

Broadly classed, synthetic polymers are either thermoplastics (generally linear polymers that melt readily and are molded with heat and/or pressure), or thermosets (generally handled as liquids, set by cross-linking the polymer chains, and do not melt after they are cured). Flexible polyurethane foams are an intermediate class of polymers that are lightly cross-linked and flow, or shrink, upon the application of sufficient heat.

Thermoplastics such as polystyrene are frequently flame-retarded by adding a small amount (1–4%) of an aliphatic bromine compound having a low (< 250°C) decomposition temperature. Upon heating, the bromide decomposes, promotes chain scission, and allows the polymer to melt and drip away from the ignition source.

Thermoset polymers generally require ca 20–30 wt% chlorine or 12–15 wt% bromine in order to be significantly flame retardant in moderately severe tests, such as the HLT-15 test.

The number of chlorinated compounds used as fire retardants has decreased in recent years, and most of the newer agents are based on bromine. This shift has occurred at least partially as a result of the increasingly severe flammability standards imposed on plastic products, standards normally requiring levels of chlorine that are difficult to obtain without destroying the desirable properties of the polymer. In addition, the decreased use of lead in gasoline has freed large quantities of bromine that were previously used in the production of ethylene dibromide. The anticipated large excess in bromine capacity prompted new uses for bromine, among them a number of flame-retardant chemicals.

It is desirable for the agent to decompose with the liberation of halogen at a somewhat lower temperature, ca 60°C, than the decomposition temperature of the polymer. Bromine-containing compounds generally decompose in the range of 200–300°C because of the relatively low carbon–bromine bond energy and consequently have decomposition temperatures that mesh quite well with the decomposition temperatures of most common plastics.

Additive Flame Retardants

Nonreactive additives have commonly been used for flame-retarding polymers. The ideal additive is inexpensive, colorless, easily blended, compatible, heat and light stable, efficient, permanent, and has no deleterious effects upon the properties of the base polymer. The most important limitations are incompatibilities that affect the physical properties of the polymers and the tendency for additives to be fugitive.

The principal commercially available flame retardants are listed individually below.

Polychlorinated paraffins. Polychlorinated paraffins are produced by the controlled chlorination of normal paraffinic hydrocarbons, chlorination being controlled to yield mainly liquid products containing 20–70 wt% chlorine. Several solid resinous products containing 70 wt% chlorine are also available (see Chlorocarbons), in a variety of formulations ranging from neat oils and solids to water-based emulsions (qv). A number of products are marketed with heat stabilizers (qv) and/or antimony oxide present.

Decabromodiphenyl oxide (DBDPO). DBDPO is a free flowing white powder that usually contains more than 85 wt% of decabromodiphenyl oxide, and less than 15 wt% of other brominated diphenyl oxides, principally nanobromodiphenyl oxide. It is manufactured by the bromination of diphenyl oxide in the presence of a Friedel-Crafts catalyst (see Friedel-Crafts reactions). It is recommended for use in impact polystyrene, polycarbonates, polyester resins, polyolefins, ABS, polyamides, PVC and rubber, although its acceptance in many of these systems is limited to special applications. DBDPO has limited utility in polymers that have relatively low decomposition temperatures since the high thermal stability of the aromatic bromine tends to make the agent inefficient.

Other additive flame retardants include: pentabromochlorocyclohexane (PBCCH), hexabromocyclododecane (HBCD), 2,3,4,5,6-pentabromoethylbenzene (PBEB), and 1,2-bis(2,4,5-tribromophenoxy)ethane (BTBPE).

Reactive flame retardants. Reactive halogenated flame-retardant agents usually consist of compounds that contain, in addition to the halogen, a chemically reactive moiety that can enter into the polymerization reaction. The halogen-containing monomer can be a vinyl group, for agents intended for use in vinyl polymers, eg, polystyrene, polyacrylates, etc; or an anhydride, carboxyl, or hydroxyl group, for agents intended for use on condensation polymers, eg, polesters, polyurethanes, epoxies, etc.

Reactive agents generally lose their chemical identity and become a permanent part of the polymer backbone. As a consequence, many of the problems associated with additive agents do not apply to reactive agents. The development of polymer systems based on reactive agents is more time-consuming and more expensive because in using these agents, the manufacturer is developing new copolymer systems.

Principal reactive flame retardants. Tetrachlorophthalic anhydride (TCPA) is manufactured by the ferric chloride-catalyzed chlorination of phthalic anhydride.

Chlorendic acid (hexachlorobicyclo[2,2,1]-hept-5-ene-2,3-dicarboxylic acid) and its corresponding anhydride are among the most important reactive flame retardants and precursors in use today (see Cyclopentadiene). Their principal use has been in the manufacture of unsaturated polyester resins. Because the esterification rate of chlorendic anhydride is similar to that of phthalic anhydride, it can be used in place of phthalic anhydride in commercial polyester formulations. The double bond in chlorendic acid is not reactive as a cross-linking site, hence reactive monomers, such as maleic anhydride, must be included in the polyester backbone to achieve cross-linking.

Tetrabromobisphenol-A (TBBP-A), is a versatile brominated bisphenol which is the primary raw material for several other widely used fire-retardant agents. It has a high melting point (179–181°C), and is a dense white solid containing 58.8 wt% of bromine. Although TBBBP-A is commercially important as a flame retardant for polycarbonates, its primary use is as an intermediate in the manufacture of epoxy resins where it reacts with epichlorohydrin and caustic to give the diglycidyl ether of TBBP-A (see Epoxy resins).

Tetrabromophthalic anhydride (TBPA), a widely used reactive flame retardant in unsaturated polyesters as well as the precursor to a number of other flame retardants designed for use in other polymer systems, eg, polyurethane foams.

Vinyl bromide (VBR), is a colorless gas at ambient temperature and pressure, which is a reactive flame-retardant monomer. Through its vinyl functionality, this monomer is capable of homopolymerization or copolymerization with other reactive monomers. The homopolymer, because of its thermal and photochemical instability, is of very little interest *per se*; however, various copolymers are suitable for manufacture of films, fibers, and extruded or molded objects.

Vinyl bromide is used as a flame retardant in the manufacture of modacrylic fibers (see Acrylic and modacrylic fibers). It is copolymerized with vinyl acetate and maleic acid or anhydride to produce finely divided granular products, or with vinyl chloride to yield relatively low molecular weight polymers.

E.R. LARSEN
Dow Chemical USA

J.W. Lyons, ed., *The Chemistry and Uses of Fire Retardants*, Wiley-Interscience, New York, 1970.

W.C. Kuryla and A.J. Papa, eds., *Flame Retardancy of Polymeric Materials*, Vol. 1–3, Marcel Dekker, Inc., New York, 1973–1975.

E.R. Larsen and R.B. Ludwig, *J. Fire Flammability* **10**, 69 (1979).

PHOSPHORUS COMPOUNDS

The main fire retardants currently used in plastics and textiles fall into several distinct classes: alumina trihydrate; halogenated compounds, usually used in combination with antimony oxide; borax and boric acids; and the phosphorus, phosphorus–nitrogen, and phosphorus–halogen compounds.

Mechanism of Action of Phosphorus Flame Retardants

Condensed-phase mechanisms. The flame-retardant action of phosphorus compounds in cellulose is believed to proceed by way of initial phosphorylation of the cellulose, probably by initially formed phosphoric or polyphosphoric acid. The phosphorylated cellulose then breaks down to water, phosphoric acid, and an unsaturated cellulose analogue, and eventually to char by repetition of these steps. Certain nitrogenous compounds such as melamines, guanidines, ureas, and other amides appear to catalyze the cellulose phosphate-forming steps and are found to enhance or synergize the flame-retardant action of phosphorus on cellulose.

In poly(ethylene terephthalate) and poly(methyl methacrylate), the mechanism of action of phosphorus-based flame retardants has been shown to involve both a similar decrease in the amount of combustible volatiles and a similar increase in the amount of residue (aromatic residues and char). The char thus formed also acts as a physical barrier to heat and gases. In rigid polyurethane foams, the action of phosphorus flame retardants also appears to involve char enhancement. In flexible foam, the mechanism is less understood.

Vapor-phase mechanisms. In addition to the condensed-phase mechanisms discussed above, phosphorus flame retardants can exert vapor-phase flame-retardant action. It has been demonstrated that trimethyl phosphate retards the velocity of a methane–oxygen flame with about the same molar efficiency as $SbCl_3$.

Phosphorus-Based Flame Retardants in Commercial Use

Inorganic phosphorus compounds. Ammonium phosphates were recommended for treating theater curtains as early as 1821. Their use in forest-fire control is well established. Monoammonium phosphate and diammonium phosphate, or mixtures of the two, which are more water-soluble and nearly neutral, are still used in large amounts for nondurable flame-retarding of paper, textiles, disposable nonwoven cellulosic fabrics, and wood products. Their advantage is high efficacy and low cost. Ammonium phosphate finishes are not resistant to laundering or even to leaching by water, but they are resistant to organic solvents eg, dry-cleaning solvents. A water-insoluble ammonium polyphosphate is used mainly in flame-retardant paints and mastics. Melamine phosphates are also used in these same applications.

Triethyl phosphate is a colorless liquid boiling at 209–218°C and containing 17 wt% phosphorus. It is manufactured from diethyl ether and phosphorus pentoxide via a metaphosphate intermediate. Triethyl phosphate has been used commercially as an additive for polyester laminates and in cellulosics. In polyester resins it functions as a viscosity depressant, (eg, in polyester resin it permits high loadings of alumina trihydrate, a fire-retardant smoke-suppressant filler) and as a flame retardant.

2-Chloroethanol phosphate (3:1), (tris(2-chloroethyl) phosphate), is a liquid containing 10.8 wt% P and 36.7 wt% Cl, made from ethylene oxide and phosphorus oxychloride. This ester is widely used as a flame retardant because of low color, low odor, favorable economics, high percent phosphorus, and compatibility with most polymers except polyolefins and polystyrene. It is used commercially in polyurethane and polyisocyanurate foams, carpet backing, flame-laminated polyurethane foam, flame-retardant paints and lacquers, polyester resins, and wood–resin composites such as particle boards.

Tris(1,3-dichloro-2-propyl) phosphate (1) is a liquid made from epichlorohydrin and phosphorus oxychloride and contains 49 wt% chlorine and 7.2 wt% phosphorus. The resultant dichloropropyl groups are mainly 1,3-dichloro-2-propyl groups. Compared to tris(2-chloroethyl) phosphate (3:1), (1) has substantially reduced volatility, and much lower water solubility. It has an extremely low rate of hydrolysis, and it resists attack by bases. Because of its good stability, this phosphate is a very useful additive flame retardant for flexible urethane foams. It does not interfere with the catalyst system, cause discoloration, nor adversely affect the tensile strength of flexible urethane foams.

$$O{=}P{\left(\!\begin{array}{c} CH_2Cl \\ | \\ OCH \\ | \\ CH_2Cl \end{array}\!\right)}_3$$

(1) (Fyrol FR-2, Stauffer)

Tris(1-chloro-2-propyl) phosphate (Fyrol PCF, Stauffer) is made from propylene oxide and phosphorus oxychloride, and contains 33 wt% chlorine and 9.5 wt% phosphorus. Its physical properties are rather close to those of the chloroethyl analogue, but the chloropropyl compound has still lower reactivity to water and bases.

Related chloroalkyl diphosphates and oligomeric phosphates have low volatility, low water solubility and good-to fair-thermal stability. In products such as open-cell foams, they show resistance to dry and humid aging. Three 2-chloroethyl bisphosphates with a similar pattern of utility as low volatility flame retardant additives are sold commercially. The first introduced was Monsanto's Phosgard 2XC20, and it is recommended as a nonvolatile flame-retardant additive for flexible urethane foam, acrylics, and epoxies. More recently introduced was Olin's Thermolin 101. Stauffer's Fyrol 99, an oligomeric 2-chloroethyl phosphate, is used in similar applications. Mobil's Antiblaze 19 is an oligomeric halogen-free methylphosphonate with good thermal stability and is also useful in urethane foams.

Triaryl phosphates. Isopropylphenyl diphenyl phosphate and *tert*-butylphenyl diphenyl phosphate are thermally stable liquids which are used as flame-retardant additives in certain engineering thermoplastics. These compounds are also used as flame-retardant vinyl plasticizers and as flame-retardant functional fluids.

Organic phosphorus compounds — reactive types. Bis(2-chloroethyl) vinylphosphonate is a commercially available phosphorus-containing monomer made from ethylene oxide and phosphorus trichloride. Several applications have been found for bis(2-chloroethyl) vinylphosphonate. Emulsion copolymers with other vinyl monomers have been introduced as flame-retardant latexes for textiles and paper applications. The vinylphosphonates have been found to copolymerize particularly well with vinyl halides and vinylidene halides.

The commercial development of several phosphorus-containing diols occurred in response to the need to flame-retard rigid urethane foams used for insulation in the transportation and construction industry. Important commercial usage has been attained by a phosphonate diol containing a tertiary amine structure which may exert a synergistic effect on the flame-retardant action of the phosphorus. This diol, called Fyrol 6 (diethyl N,N-bis(2-hydroxyethylamino)methylphosphonate), is used in rigid urethane-foam insulation.

Health and Safety

The toxicology of the compounds themselves, and the effect of the flame retardants on combustion-product toxicology must both be considered. No valid generalization concerning the toxicology of phosphorus compounds as a broad class can be made. Some are essential components of all living tissues. Phosphoric acid and its various salts as well as salts of pyrophosphoric acid are commercial food additives. In general, the acute toxicity of commercial phosphorus flame retardants is slight. In the case of tris(2,3-dibromopropyl)phosphate, possible chronic toxicity (mutagenicity) was determined by use of the Ames test and this compound has been removed from the market.

EDWARD D. WEIL
Stauffer Chemical Company

M. Lewin, S.M. Atlas, and E.M. Pearce, eds., *Flame-Retardant Polymeric Materials*, Plenum Press, New York, 1978.

W.C. Kuryla and A.J. Papa, eds., *Flame Retardancy of Polymeric Materials*, Marcel Dekker, Inc., New York, 1975–1979.

A. Granzow, *Accounts Chem. Res.* **11**(5), 177 (1978).

J.W. Lyons, *The Chemistry and Uses of Flame Retardants*, Wiley-Interscience, New York, 1970.

Mod. Plast. **59**(9), 57 (1982).

FLAME RETARDANTS FOR TEXTILES

The terms used in connection with flame-resistant fabrics are sometimes confusing. Fire resistance and flame resistance are often used in the same context as the terms fireproof or flameproof. A textile that is flame resistant or fire resistant does not continue to burn or glow once the source of ignition has been removed, although there is some change in the physical and chemical characteristics. On the other hand, fireproof or flameproof refer to material that is totally resistant to fire or flame, eg, asbestos. No appreciable change in the physical or chemical properties is noted.

Most organic fibers undergo a flowing action after the flame has been extinguished, and flame resistant fabrics should also be glow resistant. Afterglow may cause as much damage as the flaming itself since it can completely consume the fabric. The burning (decomposition) temperature of cellulose is about 230°C, whereas afterglow temperature is approximately 345°C.

Chemical modification of cellulose with fire retardants gives products whose resistance to laundering and weathering is superior to that of finishes based on the physical deposition of the flame retardant within the fabric, yarn, or fiber. The reactions involved are either esterification or etherification. The latter is preferred because ether linkages are more stable to hydrolysis.

Flame Resistance

The flame resistance of a textile fiber is affected by the chemical nature of the fiber, the ease of combustion, the fabric weight and construction, the efficiency of the flame retardant, the environment, and laundering conditions.

The weight and construction of the fabric affect its burning rate and ease of ignition. Lightweight, loose-weave fabrics usually burn much faster than heavier-weight fabrics; therefore, a higher weight add-on of fire retardant is needed to impart adequate flame resistance.

Phosphorus-containing materials are by far the most important class of compounds used to impart durable flame resistance to cellulose (see also Flame retardants, phosphorus compounds). They usually contain either nitrogen or bromine and sometimes both. A combination of urea and phosphoric acid imparts flame resistance to cotton fabrics at a lower add-on than when the acid or urea is used alone. Other nitrogenous compounds, such as guanidine, or guanylurea, could be used instead of urea. Amide and amine nitrogen generally increase flame resistance, whereas nitrile nitrogen can detract from the flame resistance contributed by phosphorus. The most efficient flame-retardant systems contain two retardants, one acting in the solid and the other in the vapor phase.

Mechanisms. Imparting flame resistance to cellulose has been explained by the following theories:

Coating theory suggests that fire resistance is due to the formation of a layer of fusible material which melted and formed a coating, thereby excluding the air necessary for the propagation of a flame. This theory, suggested by Gay-Lussac in 1821, was based on the efficiency of some easily fusible salts as flame retardants. Carbonates, borates, and ammonium salts are good examples of coating materials that produce a foam on the fiber by liberation of gases such as carbon dioxide, water vapor, ammonia, etc.

Gas theory theorizes that the flame retardant produces noncombustible gases at the burning temperature, which dilutes the flammable gases produced by decomposition of the cellulose to a concentration below the flaming limit.

Thermal theory suggests that heat input from a source is dissipated by an endothermic change in the retardant and the heat supplied from the source is conducted away from the fibers so rapidly that the fabric never reaches temperature of combustion.

Chemical theory says that strong acids, bases, metal oxides, and oxidants that tend to degrade cellulose, especially under the influence of heat, usually impart some degree of flame resistance. Flame retardants for cotton may possibly act through a dehydration process by Lewis acid or base formation through a carbonium ion or carbanion mechanism.

Durability

Nondurable finishes. Flame-retardant finishes that are not durable to laundering and bleaching are, in general, relatively inexpensive and efficient. In some cases, mixtures of two or more salts are much more effective than any one of the components alone. For example, an add-on of 60% of borax $Na_2B_4O_7 \cdot 10H_2O$ is required to prevent fabric from burning; boric acid, H_3BO_3, by itself, is ineffective as a flame retardant even in amounts that equal the weight of the fabric. However, a mixture of seven parts borax and three parts boric acid imparts flame resistance to a fabric with as little as $6\frac{1}{2}\%$ add-on.

Semidurable finishes. Semidurable finishes are those that resist removal for one to about 15 launderings. Such retardants are adequate for applications such as drapes, upholstery, and mattress ticking. If they are sufficiently resistant to sunlight or can be easily protected from actinic degradation, they can also be applied to outdoor textile products. The principle advantage of water-soluble flame retardants—their lack of durability—can be overcome by precipitating inorganic oxides on the fabric; for example, $WO_3 \cdot xH_2O$ and $SnO_2 \cdot yH_2O : 2\ Na_2WO_4 + SnCl_4 + (2x + y)H_2O \rightarrow 4\ NaCl + 2\ WO_3 \cdot xH_2O + SnO_2 \cdot yH_2O$. These codeposits add flame- and glow-resistance properties to textile fabrics though some insoluble deposits may also degrade the fabrics.

Durable finishes. Earlier studies to produce durable finishes for cellulose were based on treatment with inorganic compounds containing antimony and titanium. Numerous patents were issued based on these types of treatments, eg, DuPont's erifon process and the Titanox FR process of the Titanium Pigment Corporation.

The abbreviation, FWWMR, for fire, water, weather, and mildew resistance has frequently been used to describe treatment with a chlorinated organic metal oxide. A plasticizer, coloring pigments, fillers, stabilizers, or fungicides are usually added. However, hand, drape, flexibility, and color of the fabric are more affected by this type of finish than by other flame retardants.

Types of Retardants

The types of retardants include mesylated ($CH_3SO_2—$) and tosylated ($CH_3C_6H_4SO_2—$) celluloses, phosphonomethylated ethers (prepared by the reaction of cotton cellulose with chloromethyl phosphonic acid in the presence of sodium hydroxide by the pad-dry-cure technique), and amide-based systems such as Pyroset CP.

Another retardant, tetrakis(hydroxymethyl) phosphonium chloride (THPC), $[HOCH_2]_4P^+Cl^-$, is used alone or as part of a retardant system. More recently the chloride has been replaced by the sulfate compound (THPS). THPOH is made by neutralization of THPS when the fabric is wet with the water solution, then dried and cured with anhydrous ammonia followed by oxidation and washing. This flame-retardant system is used more than any other system for imparting flame resistance to cotton. It is sold under the trade name Firestop. The THPOH-NH₃ flame-retardant system was developed USDA scientists at the Southern Regional Laboratory.

Flame retardants suitable for cotton are also suitable for rayon. A much better product is obtained by incorporating flame retardants in the viscose dope before fiber formation. The main classes of flame retardants used in the dope are: phosphorus (eg, alkyl and aryl phosphates, phosphonates and phosphites, polyphosphonates, Sandoflam 5060) which can be mixed with halogenated materials for synergistic effects; phosphorus–nitrogen (eg, phosphazenes, phosphoryl or thionophosphonyl amides or ester amides, spirocyclotriphosphazenes, THPC-amines of condensates) which are very efficient and effective and, in general, have no unusual toxicological effects; phosphorus–halogen (eg, halogenated alkyl or aryl phosphonates or polyphosphonates, halogenated alkyl or aryl phosphates, phosphites, or phosphazenes), probably the largest class but whose useful members have severe toxicological problems: halogen (eg, poly(vinyl or vinylidene halides) as latexes, poly (halogenated acrylate) latexes, emulsion or dispersions of alkyl or aryl halides, halogenated paraffins), which are not effective by themselves are excellent synergists when combined with phosphorus compounds or colloidal antimony pentoxide; and antimony oxide (eg, colloidal antimony perntoxide), which is not compatible with viscose but must be mixed with suitable halogenated synergist.

Thermoplastic fibers, eg, nylon and polyester, are considered less flammable than natural fibers because they possess a relatively low melting point and the melt drips rather than remaining to propagate the flame when the source of ignition is removed. Most common synthetic fibers have low melting points, with reported values for polyester and nylon of 225–290°C and 210–260°C, respectively.

Flame retardancy of the synthetic fibers is obtained by either mechanically building the retardant with the polymer before it is drawn into a fiber, or chemically modifying the polymer itself. Incorporation of chemicals in the dope before spinning the fiber has not been very successful.

Mobil developed a flame retardant, Antiblaze 19, for polyester fibers. A nontoxic mixture of cyclic phosphonate esters, it is 100% active whereas Antiblaze 19T is a 93% active, low-viscosity formulation for textile use. Both are miscible with water and are compatible with wetting agents, thickeners, buffers, and most disperse dyes formulations. Another flame retardant finding use in imparting flame resistance to cotton and cotton polyester blends of any ratio is White Chemical

Company's Caliban P-44 which is based on decabromodiphenyl oxide–antimony oxide in a latex emulsion.

GEORGE L. DRAKE, JR.
U.S. Department of Agriculture

W.C. Kuryla and A.J. Papa, eds., *Flame Retardancy of Polymeric Materials*, Vol. 5., Marcel Dekker, Inc., New York, 1979.

J.W. Lyons, *The Chemistry and Uses of Fire Retardants*, Wiley-Interscience, New York, 1970.

W.A. Reeves, G.L. Drake, Jr., and R.M. Perkins, *Fire Resistant Textiles Handbook*, Technomic Publishing Co., Inc., Westport, Conn., 1974.

FLAMETHROWERS.   See Chemicals in war.

FLARES.   See Pyrotechnics.

FLAVIANIC ACID.   See Naphthalene derivatives.

FLAVOR CHARACTERIZATION

Flavor characterization is concerned with the similarities in human flavor perception using methods that are designed to average out the differences. A collection of people (a panel) tastes or smells the same material and reports the perceptions according to previously explained guidelines. Using appropriate statistical methods, the similarities, if any, in the panelists' perceptions can be isolated. Such a process, previously called organoleptic testing but now called sensory analysis, requires a large amount of time to design, execute, and analyze, and is therefore expensive. Consequently, manufacturers concerned with flavor are motivated to find less expensive instrumental procedures to predict human flavor perception. However, such methods are indirect and their accuracy can only be determined by direct sensory analysis.

Flavor is the combined perception of three components: odor, taste, and texture. Odor is a result of stimuli interacting with specialized receptors in the nose, and taste results from the interactions between stimuli and receptor organs on the tongue. There are, however, no single identifiable organs involved with the perception of texture. The presence of pain, the sense of touch, and the detection of sound all contribute to the sense of texture. Thus the texture of a food includes all perceptions of flavor detected in the mouth that are owing to neither odor nor taste (see Flavors and Spices).

Sensory perception is both qualitative and quantitative. Sweet, bitter, salty, fruity, floral, etc, are quite different flavor qualities, each produced by different chemical compounds, and the intensity of a particular sensory quality is determined by the amount of the stimulus present. The saltiness of a sodium chloride solution becomes more intense if more of the salt is added, but its quality does not change. However, if hydrochloric acid is substituted for sodium chloride then the flavor quality changes from salty to sour. For this reason, quality is substitutive and quantity, intensity, or magnitude is additive.

Descriptive Sensory Analysis

In a descriptive sensory analysis, persons are asked to associate a name or number with their sensory perceptions. The set from which these names or numbers are chosen is called a scale (nominal, ordinal, interval, and ratio) each with different properties and allowable statistics.

A nominal scale is always used to determine the quality or qualities of a flavor. It consists of an unordered array of names that people are asked to associate with their perceptions of flavor quality, eg, sweet, floral, crisp, etc. The measurement of flavor intensity, unlike the evaluation of quality, requires an ordered scale, the simplest of which is an ordinal scale. A scale consisting of the words weak, moderate, strong, or the numbers 1, 2, 3 is an example of an ordinal scale with three points.

Discriminant Sensory Analysis

Discriminant sensory analysis is used to determine whether a quantitative or a qualitative difference (or both) can be detected in the flavor of two or more samples. The nature and the magnitude of the difference are not revealed; discriminant analyses merely detect differences. The most popular type of discriminant analysis is called the triangle test. In this test panelists are asked to identify the odd sample in a set of three samples, two of which are the same. This procedure is used when a manufacturer wants to substitute one component of a food product with another safer or less expensive one without changing the flavor in any way.

TERRY E. ACREE
Cornell University

I.D. Morton and A.J. Macleod, *Food Flavors: Part A, Introduction*, Elsevier Scientific Publishing Co., Amsterdam, Neth., 1982.

J.J. Powers and H.R. Moskowitz, *Am. Soc. Test. Mater. Spec. Tech. Publ.* 594 (1976).

R.A. Scanlan, ed., *Flavor Quality: Objective Measurement*, ACS Symposium Series No. 51, 1977.

FLAVORS AND SPICES

In this article, flavors and spices are presented separately, although they are employed for similar purposes in the food and flavor industries.

Flavors

Flavor is that property of a substance (commonly a food or one used in food) that causes a simultaneous reaction or sensation of taste on the tongue and odor in the olfactory center in the nose. Thus defined, taste and odor are descriptive of a sensation, whereas taste in its colloquial meaning is practically synonymous with flavor.

The preferred definition is proposed by the society of Flavor Chemists: "Flavor is the sum total of those characteristics of any material taken in the mouth, perceived principally by the sense of taste and smell and also the general senses of pain and tactile receptors in the mouth as perceived by the brain."

Food acceptance. There are four significant factors of food which when present in proper qualitative proportions, make a food generally acceptable. These are appearance, mouthfeel (a texture or kinesthetic factor evaluated by the skin or muscles in the mouth which includes smoothness, roughness, stickiness, slickness, and related surface characteristics, as well as brittleness and viscosity which are a food's physical characteristics), flavor, and nutritive value. When all four are interdependent, the sense of sight (appearance) takes precedence over the others and nutritive value is obviously not an important consideration.

Taste. Certain basic principles are involved in the physiology of flavor perception. It is generally accepted that there are four tastes (sweet, salty, bitter, and sour) although at various times three more have been proposed (alkaline, meaty and metallic) (see Flavor characterization).

There are several factors that affect the extent and character of the sensation. A hereditary or genetic factor may cause a variation between individual reactions. Thus phenylthiourea, which can cause a bitter taste sensation, may not be perceptible to certain individuals whose general ability to distinguish other tastes is not noticeably impaired. The variation of pH in saliva (which acts as a buffer) may influence the perception of acidity differently in individuals. Enzymes in saliva can cause rapid chemical changes in basic good ingredients, such as protein and carbohydrates, with variable effects on the individuals. Although the values cannot be considered absolute, approximate magnitudes or taste sensitivity has been measured (Table 1).

Table 1. Approximate Taste Thresholds

Basic taste	wt%	Test standard
sour	0.007	hydrochloric acid
salty	0.25	sodium chloride
sweet	0.50	sucrose
bitter	0.00005	quinine

Several generalizations can be made regarding taste: there are four, or perhaps five, taste qualities (salty, sour, bitter, and sweet, and possibly, metallic); the material must be in water solution (eg, saliva) to have taste; only acids are sour with sourness not identical to chemical acidity or pH (which is a function of the hydrogen ion concentration), but which appears to be a function of the entire acid molecule (a combination of pH and acid concentration determines the actual degree of sour taste); only salts are salty, however not all salts are salty, some are sweet, bitter, or tasteless, with the salty taste exhibited by ionized salts and the anions contributing the greatest to their salty taste; and organic molecules are usually sweet or bitter, or a combination of these, or tasteless (probably owing to lack of water solubility).

Odor. Olfactory response is only observed when the substance contacts the olfactory membrane, also called the olfactory mucosa or olfactory epithelium, which occupies an area of ca 2.5 cm² in each nostril. Above the nasal passages, the two olfactory clefts are separated by the nasal septum. For a substance to have an odor, it must be capable of reaching the olfactory epithelium high up in the nose, ie, it must possess sufficient volatility and low molecular weight so that, by inspiration, it may make contact in the nasal passages.

Flavor materials. Materials for flavoring may be divided into several groups, most commonly labelled artificial and natural. Under natural materials are spices and herbs; essential oils and their extracts, concentrates, and isolates; fruit and fruit juices (qv); animal and vegetable materials and their extracts; and aromatic chemicals isolated by physical means from natural products, eg, citral from oil of lemongrass, and linalool from bois de rose. Under artificial materials are found aliphatic, aromatic and terpene compounds that are made synthetically as opposed to those isolated from natural sources. Examples of each grouping are benzaldehyde made synthetically (artificial) or isolated from oil of bitter almond (natural); and 1-menthol made synthetically (artificial) or isolated from oil of Mentha arvensis (natural).

Compounding. In the compounding technique, constituents are selected or rejected because of their odor, and the results of flavor tests in water, syrup, milk, etc. After the compound is considered to be characteristic, it is tested as a flavor in the final product by an applications laboratory. Flavoring proportions used in high-fat foods must be greater than in low-fat food because flavors are absorbed by fat thereby lowering flavor impact. Specifications for many of these essential oils and artificial flavorings may be found in monographs compiled by the Essential Oil Association (EOA) of the U.S.

The flavor chemist is responsible for the basic knowledge of sensory and application properties of each of the large number of raw materials. The tremendous number of possible combinations of these items to produce specifically flavored finished compounds is readily apparent. It is not uncommon to develop a flavor that combines essential oil, plant extractives, fruit juices, and synthetics. As mentioned previously, the choice of materials depends upon the type of product, conditions of manufacture, and intended use.

Health and safety. The Pure Food and Drug Act of 1906 introduced Federal regulations "to combat the atrocities of food and drug alteration and fraud." The law was superseded by the Food, Drug and Cosmetic Act of 1938 and the Amendments of 1965, 1958, and 1962 (see Food additives).

Spices

A spice may be any of various aromatic plant materials used to flavor foods, or a variety of dried plant products (sometimes used in the form of extracts) that exhibit a pronounced aroma and flavor, but these descriptions are incomplete. Spices are derived from vegetable substances that have flavor and aroma from which no volatile or flavoring principles have been removed.

Spices can be divided into 4 categories: the so-called true or tropical spices, such as black and white pepper, cloves, cinammon, ginger, nutmeg, and mace; these may be buds, fruit, bark, roots, or other parts of tropical plants; herbs, such as sage, rosemary, majoram and oregano; these are usually the leafy parts of plants that grow in the temperate zones; spice seeds, such as mustard, celery, caraway, dill, fennel, and anise which grow in both tropical and temperate zones; and dehydrated aromatic vegetables such as onion, garlic, parsley, leek and sweet pepper, many of which are now produced in the United States.

Condiments are sometimes classified as spices although they are usually employed in a different manner. Spices are used during the cooking or processing of food, condiments are added to food after it has been served at table (chutney, piccallili, catsup, and salad dressings might be mentioned in this category). Black pepper may be considered a spice or a condiment, depending on its use.

Health and safety. The USDA considers most spices GRAS (generally recognized as safe) and there are not standards of identity or legal definitions of spices. Those used in drugs must meet the official standards of the USP.

Use. The essential oils carry the aroma of the spice (or aromatic plant) in concentrated form (see Oils, essential). The essential oils are volatile and do not contain the very high boiling or nonvolatile constituents that are responsible for the characteristic taste of certain spices. For example, piperine and zingerone, the biting principles of black pepper and ginger root, respectively, are completely lacking in the essential oils of these two spices. To isolate these compounds the spice has to be extracted with a volatile solvent (ethanol, acetone, hexane, etc). By removing the solvent by careful evaporation of the solution *in vacuo*, a viscous, semisolid or sometimes solid residue is obtained (oleoresin). These products have lately achieved great importance in the food industry; they offer many advantages over straight spices, eg, cleanliness and uniformity of flavor from lot to lot. This is especially important when foods are produced by an automated process; the use of oleoresins permits better distribution of flavor and a standardization of the flavor level that is difficult to obtain with straight spices. Moreover, oleoresins, when properly prepared, are almost free of bacteria, molds, spores, etc, which is a major advantage over crude spices and aromatic herbs, especially in prepared foods. Because of their high concentration, a few kilograms of oleoresins (or in many cases of essential oils, where the nonvolatile portion is of minimal significance to the total flavor) can replace many kilograms of natural spices. Oleoresins (see also Resins, natural) are often added to the food product in a diluted form; the most common method is distribution on a dry edible carrier, such as table salt or dextrose. These products, commonly known as dry solubles, are usually equivalent to the corresponding natural spice.

J.A. ROGERS, JR.
F.FISCHETTI, JR.
Fritzsche Dodge & Olcott, Inc.

"Spices," *Foreign Agricultural Circular FTEA 1-83*, U.S. Department of Agriculture, Washington, D.C., April, 1983.

FLAX. See Fibers, vegetable.

FLAXSEED OIL. See Drying oils (linseed oil); Fats and fatty oils.

FLOCCULATING AGENTS

The flocculants currently in commercial use are conveniently classified as inorganic, synthetic organic, and those derived from natural products

(naturally occurring organic polymers which may have been modified chemically). In 1978 the category with largest sales value was synthetic organic, although inorganic flocculants had the largest sales volume on a weight basis (see Water; Sedimentation; Dispersants; Extractive metallurgy).

Flocculation

Consider a suspension of solid particles in water, with a distribution in sizes of particles. This suspension is called a substrate. In the usual situation, if the suspension is allowed to stand quietly, the large particles settle out first. There is also a portion of very fine material that settles only very slowly or not at all. This resistance to settling by the very fine particles relates to the size and density of the particles, to their surface properties, and to the composition of the suspending medium. For a suspension of chemically homogenous particles, there is no fundamental distinction between the particles that remain in suspension and those that settle out. Any difference is in degree rather than in kind, and reflects the much larger surface-area-to-volume ratio of the smaller particles.

In water suspensions, the solid surfaces tend to have a net electrical charge, balanced by ions of opposite charge in solution in the vicinity of the particle surface. The primary sources of this surface charge are ionization of surface groups (such as OH groups for mineral oxides or carboxyl groups for latex particles), isomorphic substitution in the solid lattice, and preferential adsorption of ions or ionizable species from the suspending medium. This situation of a charge localized on the particle surface, with a diffuse distribution of counterions extending into the liquid medium, is referred to as an ionic double layer.

The surface charge generates a replusive force between particles, tending to keep them apart. For the large particles, gravitational forces are dominant and make them unstable with respect to suspension, but for individual particles of colloidal dimension the electrostatic and microhydrodynamic forces dominate and make their suspensions relatively stable. If these colloidal particles aggregate into larger entities for which gravitational forces dominate, suspension stability is lost. This formation of aggregates is termed flocculation; an aggregate thus formed is termed a floc or floccule, and any chemical agent that enhances the process is termed a flocculating agent or a flocculant.

Destabilization. There appear to be five general modes by which flocculating agents bring about destabilization of aqueous suspensions: double-layer compression, specific-ion adsorption, sweep flocculation-enmeshment, polymer charge patch, and polymer bridging. More than one mode of action may be involved at the same time in any real flocculating system.

Double-layer compressions. Increasing the ionic strength of the suspending medium reduces the thickness of the double layer thereby reducing the range of interparticle repulsion. Ions of similar charge to the surface are repelled by the surface, and counterions are attracted to its vicinity. As the counterion concentration increases, the volume of the diffuse ionic layer necessary for electroneutrality decreases. Simple electrolytes whose effect is purely electrostatic in nature and that have no specific chemical interaction with the surface are termed indifferent electrolytes, and their effectiveness as flocculants increases with charge on the counterion (Schulze-Hardy rule).

Specific-ion adsorption. Some counterions can participate in a chemical reaction with the particle surface, losing at least a portion of their hydration sphere and complexing with other atoms in the surface structure. Counterion adsorption may be sufficient to reduce the net surface charge to zero, or even to reverse the sign of the net surface charge (charge reversal). When the net surface charge is reduced to zero by this mechanism, electrical repulsion between particles is minimized and the rate of flocculation should be high. Actually the surface charge need not be reduced to zero to effect flocculation at a measurable rate by this mechanism; it need only be lowered enough to allow close approach of particles.

Sweep flocculation (enmeshment). Hydrolyzing metal salts usually are applied as flocculants in a concentration/pH region that is oversaturated with respect to the neutral metal hydroxide. The rate of precipitation of the hydroxide, which is a function of temperature, other ions present, and nucleation sites, may be slow unless a certain degree of supersaturation is exceeded. Under appropriate conditions, however, a fluffy amorphous hydroxide forms rapidly, trapping and enmeshing colloidal particles as it settles, and giving rise to the term sweep flocculation.

Polymer charge patch. The particles in many common substrates have relatively fixed sites of negative charge on their surfaces. When organic polymer molecules containing positively charged sites are added to such a substrate, they are adsorbed onto the particle surface via electrostatic bonds. In most respects this is specific-ion adsorption, with the effects described previously, but for some situations there is a significant additional feature that justifies designation as a separate mode. One general type of cationic polymer has a high charge density along the chain (such as a cationic charge every few carbon atoms) providing a greater density of charge in the polymer molecule than is present on the surface of a particle. When this type of polymer molecule is adsorbed onto the surface, it not only neutralizes the negative charge within the geometric area where it is attached but it also provides excess cationic charge to compensate for other negative charge sites on the surface.

Polymer bridging. Water soluble polymers can bond to a particle surface by a variety of mechanism, of which the electrostatic bond is only one. Other types include the hydrophobic bond, van der Waals bond, and a group of rather specific physiochemical interactions ranging from hydrogen bonding to covalent bonding. Hydrogen bonding probably is the most common of this group but other chemical interactions can be crucial at times.

When very long polymer molecules (high molecular weight linear polymers) are adsorbed ono the surface of particles, they tend to form loops that extend some distance from the surface into the aqueous phase, and the ends may also dangle. These loops and ends may come into contact with, and attach to, another particle, forming a bridge between the two particles. This is the bridging mode of flocculation.

Chemical precipitation. Flocculating agents participate in several commercially important precipitation reactions including not only those that remove a component from solution, but those that flocculate the resulting suspended solids. These include precipitation of soluble phosphates from waste streams with Al^{3+}, Fe^{3+}, or Ca^{2+} ions, removal of fulvic and humic acids from water supplies with hydrolyzing cations (Al^{3+}, Fe^{3+}), or with high charge density cationic polymers, and removal

of soluble anionic color components generated in certain types of wood pulping operations (see Pulp). Frequently, the flocs formed under these conditions are small and settle slowly. If more rapid clarification is desired, a high molecular weight anionic polymer is added to form larger flocs for faster settling but this does not effect intrinsic removal of the soluble component.

Inorganic Materials

Many soluble inorganic salts can function as indifferent electrolytes following the Schulze-Hardy rule in general effectiveness, but few are significant commercially. More important are the inorganic salts that have hydrolyzing ions and specific ion interactions with the substrate.

Aluminum derivatives. The hydrolysis behavior of the aluminum ion in aqueous solution has been the subject of many investigations. Although differences in detail are reported, there seems to be a consensus on general features. Kinetic data show that mononuclear products form rapidly and reversibly. Although formation of polynuclear species is relatively fast, breakdown is a slow process. For most effective use as a flocculant, aluminum solutions are prepared in concentrated form and then diluted at time of use.

Alum. In the literature on flocculants, the term alum refers to a commercial aluminum sulfate hydrate, $Al_2(SO_4)_3 \cdot x H_2O$ where x is about 14. It also is called papermakers' alum or filter alum, and is available either in the dry form or in solution. Dry alum is available in several grades, with a minimum aluminum content of 17% expressed as Al_2O_3. Liquid alum is about 49 wt% solution of $Al_2(SO_4)_3 \cdot 14H_2O$, or about 8.3 wt% aluminum as Al_2O_3. It can be stored indefinitely without deterioration (see Aluminum compounds).

Iron derivatives. Compared with aluminum, hydrated ferric ion is more acidic, it forms stronger complexes, with simple anions, and its amorphous hydroxide is less acidic but the two show a gross similarity in hydrolysis reactions. Aging characteristics of the polynuclear products of the ferric ion are more dependent on the anions.

Liquid ferric chloride, a dark brown oily-appearing solution which is 35–45 wt% $FeCl_3$, is the customary form for flocculant use, although it is also available in solid form.

Lime. The term lime refers to a number of chemicals that are principally calcium and oxygen, but which may contain a considerable amount of magnesium (dolomitic lime). Hydrolysis reactions of calcium do not produce polymeric species, nor is there significant specific ion adsorption (by itself) under the usual conditions of commercial use. Lime

Table 1. Principal Commercial Synthetic Organic Flocculant Structures

Polymer	Molecular formula	Charge	Comments
poly(ethyleneamine)	$\left(CH_2CH_2N\right)_{\overline{n}}$ ⏐ H (idealized structure)	cationic	homopolymer; charge density function of pH
poly(2-hydroxypropyl-1-N-methylammonium chloride)	OH CH₃ ⏐ ⏐ $\left(CH_2CHCH_2N\right)_{\overline{n}}$ Cl⁻ ⏐⁺ H (idealized structure)	cationic	homopolymer; charge density function of pH
poly(2-hydroxypropyl-1,1-N-dimethylammonium chloride)	OH CH₃ ⏐ ⏐ $\left(CH_2CHCH_2N\right)_{\overline{n}}$ ⁺⏐ Cl⁻ CH₃	cationic	homopolymer; strongly cationic, pH-insensitive, chlorine resistant; major representative
poly[N-(dimethylaminomethyl)-acrylamide]	$\left(CH_2CH\right)_{\overline{n}}$ ⏐ C—NHCH₂N⟨CH₃ / CH₃ ‖ O	cationic	solution equilibria; charge density function of pH; major commercial importance
poly(2-vinylimidazolinum bisulfate)	$\left(CHCH_2\right)_{\overline{n}}$ HN⟨⟩NH H_2SO_4	cationic	homopolymer; charge density function of pH (amidine structure)
poly(diallyldimethylammonium chloride)	$\left(CH_2\quad CH_2\right)_{\overline{n}}$ N⁺ Cl⁻ CH₃ CH₃	cationic	homopolymer; strongly cationic, pH-insensitive; chlorine resistant; major representative; copolymers with acrylamide also of significance

Table 1.   Continued

Polymer	Molecular formula	Charge	Comments
poly(N,N-dimethylaminoethyl methacrylate), neutralized or quaternized	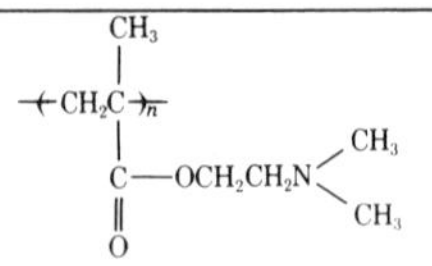 $\left(CH_2C\right)_{\overline{n}}$ with CH_3, $C-OCH_2CH_2N(CH_3)_2$, O	cationic	homopolymer; copolymer with acrylamide of more importance
poly[N-(dimethylaminopropyl)-methacrylamide]	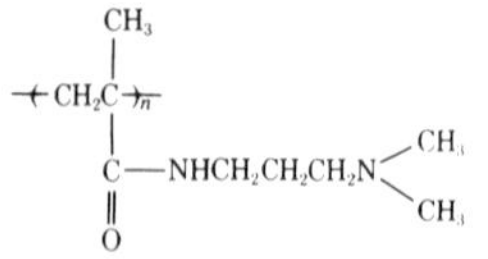 $\left(CH_2C\right)_{\overline{n}}$ with CH_3, $C-NHCH_2CH_2CH_2N(CH_3)_2$, O	cationic	hydrolytically stable cationic acrylamide derivative
poly(sodium or ammonium acrylate)	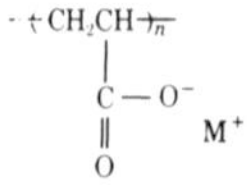 $\left(CH_2CH\right)_{\overline{n}}$ with $C-O^-$ M^+, O	anionic	homopolymer and major anionic constituent with acrylamide
poly(sodium styrenesulfonate) (PSS)	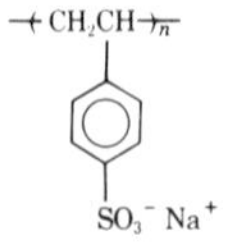 $\left(CH_2CH\right)_{\overline{n}}$ with C_6H_4, $SO_3^-\ Na^+$	anionic	homopolymer of some significance as strongly anionic, pH-insensitive flocculant
polyacrylamide (PAM)	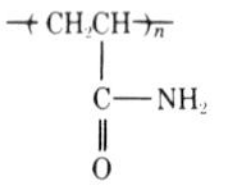 $\left(CH_2CH\right)_{\overline{n}}$ with $C-NH_2$, O	nonionic	principal nonionic flocculant and a major constituent in copolymers
poly(ethylene oxide) (PED)	$\left(CH_2CH_2O\right)_{\overline{n}}$	nonionic	effective nonionic flocculant for special applications
poly(vinylpyrrolidinone)	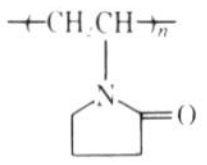 $\left(CH_2CH\right)_{\overline{n}}$ with N-pyrrolidinone	nonionic	flocculant for special applications

is used primarily for pH control of chemical precipitation (including water softening) in water and wastewater treatment, or both. At the same time it assists flocculation, frequently functioning as a coflocculant. As the largest tonnage chemical involved (ca 2×10^6 t during 1978) it represents an important flocculating agent.

Synthetic Organic Materials

Polyamine quaternies are cationic polymers shown in Table 1.

Guar gum. Guar gum is a water soluble high molecular polysaccharide, and has been of interest since its commercialization in the early 1950s as a hydrocolloid. Although it is a legume seed extract, mechanical seeding and harvesting has made it available on a large scale. Chemically, guar gum is a galactomannan, more specifically it is a poly-(1 → 4)-β-D-mannose substituted on alternative mannose units with D-galactose through α-(16) links. It is nonionic and, hence, an effective flocculant over a wide range of pH and ionic strengths.

Protein colloid. Protein colloid, such as animal glue or gelatin, is derived from fibrous collagen, the major protein of skin, bones, and connective tissues. Collagen has a multistranded structure of helically coiled polypeptide chains, varying somewhat with origin of the collagen (see Biopolymers). Animal glue is used as a flocculant in mineral processing but is being supplanted by synthetic products. Edible gelatin is used for clarification of wine, beer, vinegar, and nonalcoholic beverages.

Health and Safety

Most flocculants are considered practically nontoxic; follow manufacturers' recommended procedures.

Uses

Representative applications are wastewater treatment; water supply clarification; pH adjustment in wide range of substrates; thickening agents; sludge dewatering; mineral processing.

F. HALVERSON
H.P. PANZER
American Cyanamid Company

W.J. Weber, Jr., *Physiochemical Processes for Water Quality Control*, Wiley-Interscience, New York, 1972.

J. Vostrcil and F. Juracka, *Commercial Organic Flocculants*, Noyes Data Corporation, Park Ridge, N.J., 1976.

W.K. Schwoyer, ed., *Polyelectrolytes for Water and Wastewater Treatment*, CRC Press, Boca Raton, Fla., 1981.

FLOTATION

Froth flotation is a process for separating finely ground valuable minerals from their associated gangue. The process is based on the affinity of properly prepared surfaces for air bubbles. A froth is formed by introducing air into a pulp of finely divided ore in water containing a frothing or foaming agent. Minerals with a specific affinity for air bubbles rise to the surface in the froth and are thus separated from those wetted by water. Prior to flotation, the ore must first be ground to liberate the intergrown valuable mineral constituent from its worthless gangue matrix. The size reduction, usually to about 208 μm (65 mesh) reduces the minerals to such a particle size that they may be easily levitated by the bubbles (see Extractive metallurgy; Flocculating agents; Gravity concentration; Sedimentation; Size classification; Size reduction).

Froth flotation is usually used to separate one solid from another, but may also be used for solid–liquid separations, as in dissolved air flotation, or for liquid–liquid separation, as in foam fractionation. The process also has the potential to make a particle size separation since fine particles are more readily flocculated and floated than are coarse ones.

Fundamentals

The flotation separation of one mineral species from another depends on the relative wettability of surfaces. Typically, the surface free energy is lowered by the adsorption of heteropolar surface-active agents. The hydrophobic coating then acts as a bridge so that the particle may be attached to an air bubble.

Contact angles. An air bubble when brought into contact with a clean mineral surface usually does not adhere to the surface. However, if

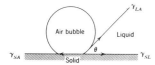

Figure 1. The contact angle formed by an air bubble on a solid submerged in water; γ_{SA}, γ_{SL}, and γ_{LA} are the interfacial tensions at the solid–air, solid–liquid, and liquid–air interfaces, respectively.

a suitable reagent is added, the mineral acquires a hydrophobic coating and an air bubble may be attached quite readily. This is illustrated in Figure 1.

The electrical double layer. Particulate matter in aqueous solution invariably has an electrical charge because an excess of cations or anions exists at the solid surface. This may be the result of, for example, the dissociation of surface groups, unequal dissolution of lattice ions, atomic defects occurring in the natural solid, or broken bonds between atoms resulting from comminution.

Heteropolar surfactants of the colloidal electrolyte class are known to adsorb onto oppositely charged minerals through electrostatic attraction and change the sign of the charge of a mineral in solution. These ions probably adsorb singly at dilute concentrations, but at greater concentrations (ca 10^{-4}–10^{-5} mol/L) they may tend to form in patches called hemimicelles. At this concentration they may cause the charge on the mineral to be reversed.

Phases. In flotation, gas, liquid, and solid phases have to be considered.

Gas phase. The gas universally used is air, though often any gas will serve. Oxygen plays a special role in the flotation of the sulfide minerals. Theoretical analysis indicates that the equilibrium adsorption density of the collector at the solid–gas interface exceeds its density at the solid–liquid interface. Experimental work with the adsorption of 1-hexanethiol by gold and of sulfonate by mercury appears to substantiate this thesis.

Liquid phase. The universal liquid for froth flotation is water, though in a few instances saturated brine or seawater is used. To this phase various reagents are added for selective control of the wettability of various mineral surfaces and to achieve the desired frothing. There are three general classes of reagents: collectors (promoters); modifiers; and frothers.

The collector may be adsorbed on a mineral by chemical or physical forces. Attachment by chemisorption to the surface is the preferred method in practice because it generally leads to greater selectivity in separation of the mineral species, and the flotation can be achieved at a lower reagent concentration and, hence, at a lower cost. Physical adsorption of the collector may involve van der Waals, hydrogen, hydrophobic, bonding, etc, but the usual adsorption mechanism is electrostatic attraction between an ionic collector and a mineral of opposite charge.

Modifying agents are classed as pH-regulating agents, activators, depressants, dispersants, and flocculants.

pH may indirectly affect potential-determining ions through chemical equilibria that alter the concentrations of potential-determining ions. In addition, pH controls the ionization of the fatty acid and amine collectors from molecular to ionic species, which, in turn, influences the collector adsorption and, hence, flotation.

Activators are added to a flotation system to permit better collector attachment to the mineral to be floated. The classic example is the addition of copper sulfate to a flotation pulp containing sphalerite, ZnS. The cupric ion attaches to the sphalerite, allowing the mineral to be floated with a xanthate.

The electrostatic model is of some help in understanding the function of depressants in achieving selectivity. Where H^+ and OH^- ions determine the potential, their use as a depressant assures that the mineral has the opposite charge to that of the collector. Sulfate ion depresses corundum when using an anionic collector. Barite depression by Ba^{2+} when using a cationic amine collector is another example. Many depressant phenomena should more appropriately be called deactivation.

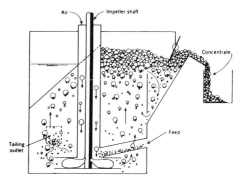

Figure 2. Froth-flotation cell.

The activating ion may be incorporated into the mineral lattice by minor elemental substitution, or may be on the mineral surface through chance adsorption. The deactivating type of depressant action is mostly accomplished by use of a complexing agent to sequester the offending ion.

In sulfide mineral flotation, separation is ordinarily made in the presence of slimes, or very fine particles. For nonsulfide flotation, where selectivity between species must be accomplished by subtle means, slime control is almost always necessary. A dispersant can be added to a flotation system to ensure that coflocculation of the desired mineral and gangue does not occur.

Flocculating agents are normally added after flotation to aid in the filtration of the products or in thickening finely ground pulps. Lime and alum are commonly used as are natural and synthetic polymeric substances such as starches, gums, polyethylene oxides, polyacrylamides, and copolymers of polyacrylamide with acrylates or quaternaries (see Flocculating agents).

Solid phase. Crystal chemistry plays an extremely important role in flotation, and the nature and structure of the solid phase are critically important. In addition to the direct influence of crystal structure, minor elemental substitution or chance adsorption of ions is sometimes the dominant feature. The solid phase may be roughly categorized into five groups: sulfides, slightly soluble salts (eg, $BaSO_4$), insoluble oxides, soluble salts, and naturally floatable compounds (eg, MoS_2 and S_8).

Industrial Technology

Successful flotation requires a rather delicate balance of many operating variables, only some of which are under the direct control of the mineral engineer, notably, ore, water, reagents, and machinery. The texture or physical associations of the minerals comprising an ore is one of the most important considerations, determining whether the valuable constituents may be economically removed, and the degree of treatment complexity required for the separation. As a general rule, in the most economical treatment, separation takes place at as coarse a size as possible to secure liberation of the minerals. It is impossible to state exactly the conditions for any one flotation separation because of the numerous variables involved and the constantly improving techniques.

The most commonly used flotation machines are the Agitair (Galigher Co.), Denver D-R (Denver Equipment Co.), and Fagergren (Envirotech Corp.) machines. The basic elements of such machines are shown in Figure 2.

Uses

Froth flotation is the principal means of concentrating copper, lead, molybdenum, zinc, phosphate, and potash ores, and a host of others. Separations by flotation also include widely divergent applications such as the separation of ink from repulped paper stock, peas from pea pods, oils from industrial wastes, and metal ions, bacteria, proteins, and colloidal particles from water.

FRANK F. APLAN
The Pennsylvania State University

M.C. Fuerstenau, ed., *Flotation, A.M. Gaudin Memorial Volume*, 2 vols., AIME, New York, 1976.

D.W. Fuerstenau, ed., *Flotation—50th Anniversary Volume*, AIME, New York, 1962.

J. Leja, *Surface Chemistry of Froth Flotation*, Plenum Press, New York, 1982.

FLOUR.   See Bakery processes and leavening agents.

FLOW MEASUREMENT

When selecting a flow-measurement method, fluid properties, ambient environment, measurement requirements, and economics must be considered. A great variety of equipment is available for measuring clean, low viscosity, single-phase fluids at moderate temperatures and pressures. Any extreme fluid characteristic, such as a high operating temperature, greatly reduces the range of equipment available and should be given first consideration.

Main fluid-related factors are operating pressure, temperature, viscosity, density, corrosive and erosive characteristics, flashing and cavitational tendencies, and fluid compressibility.

The fluid velocity profile can be evaluated by calculating the dimensionless Reynolds number, Re, which represents the ratio of inertial to viscous forces within the fluid.

$$Re = \frac{\rho V D}{\mu}$$

where ρ = fluid density, μ = fluid viscosity, V = average velocity, and D = diameter of pipe or meter inlet. A low Reynolds number indicates laminar flow. If the operating Reynolds number is increased > ca 2000, the flow becomes turbulent. Most flowmeters are designed and calibrated for turbulent flow and may be in error if used under laminar conditions.

Measurement requirements include accuracy, repeatability, and the range needed for the application.

The principal economic consideration is total installed system costs, including the initial cost of the flow primary, flow secondary, and related ancillary equipment as well as the material and labor required for installation.

Calibration Methods

Calibrations are generally performed with water, air, or hydrocarbon fuels using weigh tanks, volumetric tanks, pipe provers, and master flowmeters as standards. These standards can be used statically (flow rate is quickly started and stopped), dynamically (readings are taken at the beginning and end of the test), or in a dynamic start-and-stop, static reading mode which provides a more accurate liquid calibration. The key to the accuracy of this method is the operation of the flow-diverter valve, which switches the flow in and out of the standard.

In liquid displacement gas calibration, a known volume of liquid is displaced with gas. The gas enters an inverted bell, the ball rises, and a volume increment can be timed. In pipe-proving systems, a piston or elastic sphere is driven along a controlled pipe section using the fluid energy and timed. Prover systems can be very accurate as well as compact and rugged. They are widely used for field testing of oil-field meters.

The most common calibration method is to compare the output of the meter tested with one or more meters of high resolution and proven accuracy, so-called master flowmeters which have been calibrated against one of the above standards. The experimentally determined discharge coefficient, K, is described in equation 1:

$$q = K\beta^2 A Y F_a \sqrt{2g\frac{(P_1 - P_2)}{(\rho)}} \qquad (1)$$

where q equals the volumetric rate of flow; β equals the restriction : pipe diameter ratio ($d : D$); A equals the inside cross-sectional area of pipe; Y equals the gas expansion factor (for an adiobatic change from P_1 to P_2;

F_a equals the thermal expansion factor for the restriction; P_1 equals the upstream pressure; P_2 equals the restriction pressure; and ρ equals the fluid density.

Self-Generating Flowmeters

Positive-displacement meters separate the incoming fluid into chambers of known volume. The total quantity of fluid passing through the meter is the product of the internal-meter volume and the number of fillings. Included in this category are reciprocating piston meters, bellows or diaphragm meters, mutating-disk meters, and rotary-impeller-vane and gear meters. Positive-displacement meters are utilized in the distribution of natural gas because of their mechanical nature and ability to maintain accuracy over a long time. Wear tends to increase leakage, and errors tend to be in the direction of underregistration. Meters are periodically recalibrated and adjusted to read within 1% of the true flow. Filters minimize seal wear.

Differential-pressure flowmeters.   These flowmeters are based on the Bernoulli principle that, in a flowing stream, the total energy remains constant. Extensive tables of discharge coefficients (resulting from live constrictions) in terms of the β ratio (the ratio of restriction to pipe diameter as shown in equation 1) and the Reynolds number are accurate within 1–5%. Because of the square relationship between differential pressure and flow, the practical range of most differential meters is about 4 : 1.

The most common form of differential pressure meter is the thin, flat plate orifice. Orifice plates of various types are inserted perpendicular to the flow; they have a clean, sharp-edged opening and are simple, hydraulically predictable, readily interchangeable, and reliable. They are widely used in both liquid and gas services where moderate accuracy and limited range suffice (see Fig. 1).

Figure 1.   Quadrant-edge orifice plate.

A venturi tube consists of two hollow truncated cones, the smaller diameters of which are connected by a short circular section known as the throat (see Fig. 2). The pressure differential is measured between the upstream and throat sections and can be related to flow by equations or tables. The downstream cone recovers part of the differential pressure. Venturi tubes have a low permanent pressure drop and reduced sensitivity to hydraulic conditions.

A flow nozzle is a constriction with an elliptical inlet section. Pressure differential is measured between taps located on pipe diameter upstream and one-half diameter downstream of the inlet. Flow nozzles are used for steam and other high velocity fluids where erosion can occur.

In elbow meters, fluid passing through a common pipe elbow generates a differential centrifugal pressure between the inside and outside of the elbow. Measurement of this differential provides a measurement of flow.

In pitot tubes the opening into the flow stream measures the total or stagnation pressure of the stream; a wall tap senses the static pressure. Laminar flowmeters use a series of capillary tubes, rolled metal, or sintered elements to divide the flow conduit into innumerable passages,

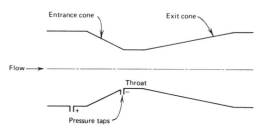

Figure 2.   Venturi meter.

small enough that the Reynolds number in each is kept below 2000 for all operating conditions. Under these conditions, the pressure drop is a measure of the viscous drag and is linear with flow rate. Laminar flowmeters can be used only with clean fluids.

Variable-area flowmeters.   In variable-area meters, the differential pressure is maintained constant, and the restriction area is allowed to change in proportion to the flow rate. These are simple and inexpensive meters, which provide readings accurate to within several percent of full-scale flow.

Head-area meters (weirs and flumes).   The Bernoulli principle, the basis of closed-pipe differential-pressure flow measurement, can also be applied to open-channel liquid flows. When an obstruction is placed in an open channel, the flowing liquid backs up, and with the help of the Bernoulli equation, the flow rate can be shown to be proportional to the head.

Weirs are dams or obstructions across open channels that have along their top edge an opening of fixed dimensions and shape through which the stream can flow. Flumes are flow channels with gradual rather than sharp restrictions. Flumes for open channels are closely analogous to venturi meters for closed pipes, and weirs are analogous to orifice plates.

Velocity flowmeters.   A number of flowmeters designs use a rotating element that is kept in motion by the kinetic energy of the flowing stream and whose speed is thus a measure of fluid velocity. Anemometers are used to measure wind velocity, turbine flowmeters are used to measure closed pipes, and current meters are used for open-channel liquids. Turbine meters are used in a broad range of industrial environments, including liquids, gases, and steam. Their share of the flowmeters market is exceeded only by differential-pressure devices.

Target flowmeters.   Target flowmeters use a drag-producing body in the flow stream to generate a force proportional to velocity. This force is sensed with strain gauges or a force-balance system. Target meters are used in the measurement of hot, viscous, or sediment-bearing fluids.

Vortex-shedding flowmeters.   Vortex-shedding flowmeters provide excellent accuracy and range on liquids, gases, and steam. They offer the advantages of digital output and no moving parts and are the fastest-growing type of flow-measuring devices. Current vortex-sensing techniques include differential-pressure-sensing diaphrams (with copassitive or inductive pic-off), strain gauges, piezoelectric elements, or velocity-sensing thermistors.

Electromagnetic flowmeters.   In electromagnetic flowmeters, a pair of coils produces an electromagnetic field through an insulating tube that carries the liquid. Electrodes at a right angle to both the flow and the field sense an induced voltage E, whose magnitude, by Faraday's Law, is,

$$E = CBdV$$

where C = calibration factor, B = average magnetic-flux density, d = distance between electrodes, and V = average fluid velocity. Electromagnetic flow meters have the additional advantage of sensing flows with entrained gas or solids on a flowing volume basis provided the flow is well mixed and travelling at a common overall velocity. Electromagnetic flowmeters are commonly used on drinking water, raw sewage and wastewater flows, paper-pulp slurries, viscous polymer solutions, milk, and pharmaceutical products.

Ultrasonic flowmeters.   Doppler ultrasonic flowmeters sense the Doppler shift in apparent frequency of an ultrasonic beam as it is reflected from air bubbles or other acoustically reflective particles that are moving in a liquid flow. Both wetted-sensor and clamp-on Doppler meters are available. In pulse or transit time flowmeters, a pair of ultrasonic transducers are mounted diagonally on opposite sides of the pipe section and may be wetted. The transit time for the pulse moving in the direction of the flow is less than for the pulse moving with the flow. They are generally used in large pipes where flow profile and fluid properties are relatively constant. Both single- and multiple-path designs are sensitive to swirl.

Laser Doppler flowmeters.   These measure liquid and gas velocities in both open and closed conduits. Velocity is measured by detecting the frequency shift in the light scattered by contaminant particles in the flow. Laser Doppler meters can be applied to very low flows and have the

advantage of sensing at a distance without mechanical contact or interaction.

Correlation flowmeters.   In correlation flowmeters, a quantity of a foreign substance is injected into the stream and the time interval for this substance to reach a detection point or pass between detection points is measured (tracer type).

In cross-correlation meters, some characteristic pattern in the flow is computer-identified at some point or plane in the flow. It is detected again at a measurable time later at a position slightly downstream. The correlation signal can be electrical, optical, or acoustic.

Thermal flowmeters.   Hot-wire devices depend on the removal of heat from a heated wire or film sensor exposed to the fluid velocity (see Temperature measurement). Hot-wire signals are dependent on the heat transfer from the sensor and thus on both the fluid velocity and density. Hot-wire and hot-film anemometers are primarily used in clean liquids and gases where they can be calibrated to the exact conditions.

Differential-temperature thermal flowmeters inject heat into the fluid and measure the resulting temperature rise or the amount of power required to maintain a constant temperature differential. The thermal meter measures the mass-flow rate of a particular gas independent of pressure, provided the specific heat is constant.

Nuclear magnetic resonance flowmeters use a field coil to generate a strong magnetic field. The liquid nuclei align in parallel, spinning like tops. A modulation frequency is produced which is linear with liquid velocity. The NMR meter has a straight-through configuration with the measurement system external to the flow conduit.

THOMAS H. BURGESS
Fischer & Porter Company

T.H. Burgess and co-workers, *Flow* **2**, 777 (1981).

Fluid Meters—Their Theory and Application, 6th ed., American Society of Mechanical Engineers, New York, 1971.

R.W. Miller, *Flow Measurement Engineering Handbook*, McGraw-Hill Book Co., New York, 1983.

FLUIDIZATION

The term fluidization is used to designate the gas–solid contacting process in which a bed of finely divided solid particles is lifted and agitated by a rising stream of process gas. At low velocities, the amount of lifting is slight and the bed behaves as a boiling liquid, hence the term boiling bed. At high velocities, the particles are fully suspended in the gas stream and are carried along with it and the terms suspension, suspensoid, and entrainment contact have been used to describe this effect.

Since 1942, the application of the fluidization technique has spread rapidly to metallurgical ore roasting, limestone calcination, synthetic gasoline, petrochemicals, and even to the design of nuclear reactors (qv). Fluidization as a unit operation has touched almost every process industry.

In most fluidized-solids processes, the solid material is handled in one or more steps and transferred from step to step through pipelines. To raise the material to a higher level, it is carried as a suspension in a gas stream; to take it to a lower level or to a region of higher pressure, the settled material is allowed to flow down a pipeline. This is illustrated in Figure 1. Process gas enters a plenum below a bed-supporting grid and bubbles up through the bed of catalytic or reacting solids thus fluidizing the bed. The bubbles burst at the surface and entrain some particles into the disengaging height. The exiting gases pass through a cyclone which returns the entrained particles to the bed and exhausts clean product gases to downstream equipment. If bed particles require regeneration, they can be withdrawn through a standpipe and returned to the fluidized bed via a pneumatic lift line and separator. From the separator they again flow through a standpipe seal leg into the reactor. The host of variations in solids flow between processing vessels can be visualized best by reference to the so-called fluids-solids phase diagram.

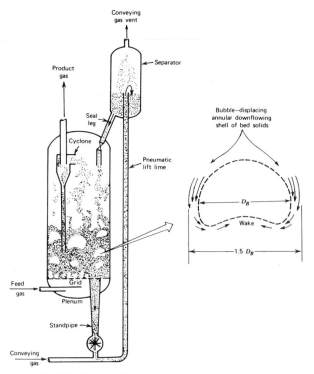

Figure 1. Bubble rise via displacement by inflow of a surrounding down-flowing shell of bed solids; D_B = bubble diameter.

Advantages and Disadvantages of the Fluidization Technique

Advantages: Temperature Control

The ability of the fluidized-solid bed to approach isothermal conditions is the outstanding advantage of this method over other methods of carrying out reactions. Of the several reaction variables, temperature is one of the most important, for reaction rates change exponentially with temperature (often doubling for a 10°C change). Commonly, there are several competing reactions in which a temperature change of a few degrees may shift the balance of the several rates.

Three factors, in order of their importance, leading to the close control of temperature in a fluidized solid bed are

(1) Turbulent agitation within the fluidized mass, which breaks and disperses any hot or cold spots throughout the bed before they grow to significant size. The catalytic activity differs somewhat for each particle, and those with greater activity accelerate the reaction in their neighborhood to a greater extent; consequently, their temperature is different than that of the surrounding particles of lower activity;

(2) High heat capacity of the bed relative to the gas within it. This factor stabilizes the temperature of the bed, permitting it to absorb relatively large heat surges with only small temperature changes.

(3) A high heat-transfer rate, which is possible because of the large amount of transfer surface per unit volume of the fluidized bed. This permits rapid leveling of any temperature surges either from the incoming gas or from reactions within the bed. Although the heat-transfer coefficients are not unusually high, the amount of surface per unit volume is very large.

Disadvantages

Solids that do not flow freely or tend to agglomerate cannot be processed in a fluidized-solid reactor; instead, rotary kilns and tray-type reactors are used. As a reaction proceeds, fine solid particles may be formed that become entrained in the gas leaving the fluidized bed, and recovery methods must usually be included in the design. The pressure drop in the gas system of a fluidized-solid boiling-bed reactor is larger than in kilns or tray reactors because the gas supports and fluidizes the solid. This pressure drop may sometimes be a serious objection to the fluidized-solid reactor because of the larger compressors required.

Operating Characteristics

Fluidization occurs in a bed of particulates when an upward flow of fluid through the interstices of the bed attains a frictional resistance equal to the weight of the bed. At this point, an infinitesimal increase in the fluid rate lifts or supports the particles. Hence, the particles are envisioned as barely touching, or as floating on a film of fluid.

Bubbles form at the grid ports where fluidizing gas enters the bed because the velocity at the interface of the bed just above the hole represents a gas input rate in excess of what can pass through the interstices. The gas input has a frictional resistance less than the bed weight, hence the layers of solids above the holes are pushed aside until they represent a void through whose porous surface the gas can enter at the incipient fluidization velocity. As the void grows, the interface velocity becomes insufficient to hold back the walls of the void, and hence, the bubbles cave in from the sides cutting off the void and presenting a new interface to the incoming gas. This sequence is illustrated in Figure 2.

From grid design, operating superficial velocity, and fluid particle properties, it is possible to calculate the initial bubble size at the grid, the maximum stable size, and the bed depth over which the bubbles may grow from their initial to their stable diameter. Once having reached their maximum stable diameter, any further unlikely mergers would also lead to collapse, so that bubble diameter may be considered constant once having reached the stable size. The bubbles represent a flow superimposed on the superficial incipient bubbling velocity passing up through the bed; they are purged continuously as they rise. Their local size, velocity, and residence times are calculable from grid to bed surface, and it is possible to calculate the degree to which they are purged before bursting at the surface, ie, to assure that no bubble gas by-passes contact with the bed solids. It may be assumed that the minimum bed depth required to avoid any feed-gas breakthrough (ie, 100% bubble purging) represents the minimum bed depth required for the desired reaction.

The superficial gas velocity minus the incipient bubbling velocity is approximately equal to the volumetric bulk-solids movement across any unit-bed cross section per unit time. This amounts to a relatively substantial mass movement, hence a fluidized bed exhibits reasonably uniform particle size distribution and bed temperature throughout its volume. Reasonable quantitative estimates of such local solids-mixing rates are important principally to determine allowable solids' feed rates.

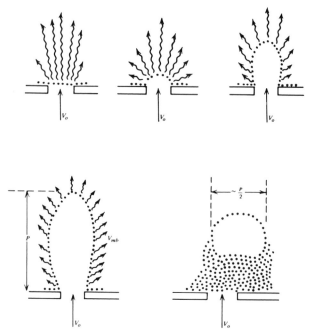

Figure 2. Bubble formation from bed-penetrating gas jets at the grid ports; V_o = fluidizing medium velocity, V_{mb} = superficial velocity at point of incipient bubbling through grid hold, and P = jet penetration depth.

In order to avoid accumulation at any feed-pipe location, the bed mixing rate must involve removing the fed material as rapidly as it enters. Such solids' mixing rates have been correlated.

The substantial heat capacity of the bed solids relative to the gas inventory represents an enormous flywheel which, coupled with the high solids mixing rate, leads to a uniform temperature throughout the bed. The heat transfer between gas and solids is nearly instantaneous owing primarily to the high particle surface area per unit of bed volume; however, the transfer of heat between the bulk bed and the vessel walls, or any other heat-transfer surface, represents a composite of mechanisms such that the average heat-transfer coefficient is 56.7–567 W/(m$^2 \cdot$K) [10–100 Btu/(h$\cdot$ft$^2 \cdot$°F)] depending on particle size, bubble size, fluid properties, and superficial fluidizing velocity. The homogeneity and relative uniformity of bed temperature make a fluidized bed an attractive vehicle in which to conduct exothermic as well as endothermic reactions controlled by immersed boiler tubes, exchangers, platecoils, or other heat-transfer surfaces when the bed walls do not offer sufficient area for cooling or heating via a fluid circulated through a surrounding jacket.

FREDERICK A. ZENZ
Consultant

F.A. Zenz and D.F. Othmer, *Fluidization and Fluid–Particle Systems*, Reinhold Publishing Corp., New York, 1983.

Report Nos. DOE/MC/14141-1158 and DOE/MC/14141-1304, N.T.I.S., Springfield, Va.

F.A. Zenz, *Chem. Eng.*, 81–91 (Dec. 1977).

FLUID MECHANICS

Fluid mechanics is both a descriptive science of the phenomena that occur when fluids flow and a quantitative science showing how these phenomena may be described in mathematical terms. To the technologist, fluid mechanics is an entire body of knowledge, theoretical and empirical, qualitative and quantitative, allowing analysis of the performance of complex plant equipment handling moving fluids. Calculation of the details of the flow is secondary to understanding the phenomena well enough to accomplish the process task. At times, the technologist's needs are best satisfied by an empirical correlation; at other times the necessary skills consist largely of knowing how to fit the idealized solutions of the mathematician into a practical situation.

It differs from conventional mechanics in that it treats mainly bodies that are capable of unlimited deformation. Flow is regarded as primary and elasticity as secondary, in contrast to solid mechanics where elasticity is primary and flow is regarded as secondary. Instabilities in flow, eg, turbulence, play a large role in fluid mechanics whereas they are of minor concern in solid mechanics. However, there is no essential difference between the two disciplines. In some instances, the two converge, as in the treatment of extrusion of very viscous or plastic materials in which fluid-like and solid-like behavior are of equal importance.

Flow Phenomena

Flow past bodies. A fluid moving past a surface of a solid exerts a drag force on it, usually manifested as a drop in pressure in the fluid. Locally, at the surface, the pressure loss stems from the stresses exerted by the fluid on the surface and the equal and opposite stresses exerted by the surface on the fluid. Both shear stresses and normal stresses can contribute; their relative importance depends on the shape of the body and the relationship of fluid inertia to the viscous stresses, commonly expressed as a dimensionless number called the Reynolds number, $LV\rho/\mu$. The same phenomena that affect drag also affect heat and mass transfer to the surface (see Heat exchange technology; Mass transfer). Because of the importance of flows around bodies, relatively detailed description of such flows and their associated pressure changes is given.

Flow along smooth surfaces. When the flow is entirely parallel to a smooth surface, eg, in a pipe far from the entrance, only the shear stresses contribute to the drag; the normal stresses are directed perpendicular to the flow. The shear stress is usually expressed in terms of a dimensionless friction factor:

$$f = \frac{2\tau_w g_c}{\rho \overline{V}^2} \tag{1}$$

Two distinct regimes of flow are observed in a pipe. At low velocities the flow is laminar, so-called because the fluid flows in concentric cylindrical sheets or laminae that do not mix with each other. At some high Reynolds number ($D\overline{V}\rho/\mu$), typically about 2100 in commercial pipe, an abrupt change occurs. Random instabilities grow and the laminar flow pattern is replaced by a chaotic, turbulent one with rapid lateral mixing of the fluid.

Flow past bluff bodies. When a fluid passes around a bluff body, both the shear and normal stresses can contribute to the drag force exerted on the body. The force produced by the shear stresses is commonly referred to as skin drag to distinguish it from the form drag arising from an imbalance in normal stresses. Skin drag is dominant at low velocities, and form drag is dominant at high velocities. The combined drag force is usually expressed in terms of a dimensionless drag coefficient:

$$C_D = \frac{2Fg_c}{A_p \rho V_\infty^2} \tag{2}$$

Flow around a cylinder with axis normal to the flow provides a good example of the many unusual features of such flows, as illustrated in Figure 1.

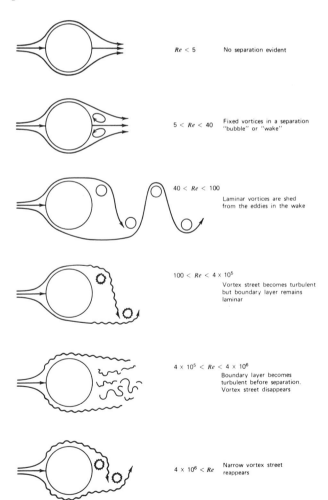

Figure 1. Flow past a circular cylinder. Courtesy of J.H. Liendhard, *Bulletin 300*, College of Engineering, Washington State Univ., Pullman, Wash., 1966.

Table 1. Boundary Layer Properties for 100 m / s Flow of Air

Property	Value
distance to laminar turbulent transition, cm	6.0
thickness of laminar boundary layer at transition, cm	0.05
distance to complete transition, cm	60.0
thickness of turbulent boundary layer, cm	1.1
laminar sublayer, cm	0.002
buffer layer, cm	0.010
velocity at edge of laminar sublayer, m/s	19
buffer layer, m/s	53

Courtesy of: H. Schlichting, *Boundary Layer Flow*, McGraw-Hill Book Company, New York, 1960.

The boundary layer. This is a concept developed principally by Prandtl to facilitate the solution of equations of motion in the presence of large velocity gradients perpendicular to the flow. The boundary layer is the fluid flowing close to the surface. It expresses the observation that the flow field can often be divided into two reasonably well-defined regions: a thin region close to the surface, the boundary layer, in which the gradient of tangential velocity is large and shear stresses are important; and a region outside the boundary in which velocity gradients are small and the potential flow solutions of inviscid flow theory eg, Bernoulli's equation, are very nearly valid. The transition between the two regions is gradual, not abrupt, with the edge of the boundary layer taken to be the point at which the velocity differs from that for inviscid flow by some small amount, eg, one percent. Table 1 illustrates the boundary-layer properties for a 100 m/s flow of air past a thin wing.

Flow past deformable bodies. The flow of fluids past deformable surfaces is often important, eg, contact of liquids with gas bubbles or with drops of another liquid. Proper description of the flow must allow both for the deformation of these bodies from their shapes in the absence of flow and for the internal circulations that may set up within the drops or bubbles in response to the external flow. Deformability is related to the interfacial tension and density difference between the phases, and internal circulation is related to the drop viscosity. A proper description of the flow involves not only the Reynolds number ($dV\rho/\mu$) but also other dimensionless groups, eg, the viscosity ratio (μ_d/μ_c), the Eötvos number ($g\Delta\rho\, d^2/\sigma$), and the Morton number ($g\mu_c^4\,\Delta\rho/\rho_c^2\sigma^3$).

Where surface-active agents are present, the notion of surface tension and the description of the phenomena become more complex (see Surfactants and detersive systems). As fluid flows past a circulating drop (bubble), fresh surface is created continuously at the nose of the drop. This fresh surface can have a different concentration of agent (hence a different surface tension) than a surface further downstream that was created earlier. Neither of these values need equal the surface tension developed in a static, equilibrium situation. A proper description of the flow under these circumstances involves additional dimensionless groups related to the concentrations and diffusivities of the surface-active agents.

Gas-liquid flow. When two or more fluids flow together, a much greater range of phenomena occurs as compared with flow of a single phase. In a conduit, many of the technically significant phenomena have to do with the positions assumed by the phase boundaries, and these are governed by the flow conditions rather than by the walls of the conduit. Figure 2 shows typical results for flow in small pipes.

Flow Measurement

There are dozens of flowmeters available for the measurement of fluid flow. The discussion in this section is limited to typical devices grouped according to the primary measurement used to determine flow. These primary measurements include: differential pressure, variable area, liquid level, and electromagnetic effects. Most of the devices discussed are those used commonly in the process industries.

Measurement by differential pressure. The most widely used devices for the flow of fluids are those that utilize a fixed constriction in the path of flow to produce a difference in pressure between the upstream and

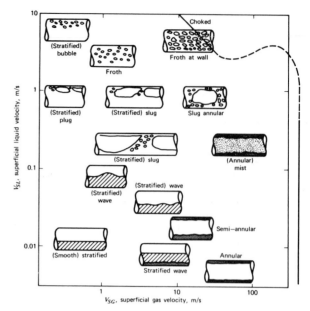

Figure 2. Flow regimes for air-water in a 2.5-cm horizontal pipe. Courtesy of Shell Development Company.

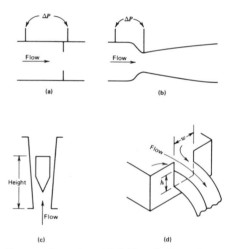

Figure 3. Common flowmeters. (**a**) Orifice. (**b**) Venturi. (**c**) Rotameter. (**d**) Weir.

downstream measuring points. The devices discussed here are orifice meters, venturi meters, and target meters. The metering orifice (Fig. 3**a**) has been standardized in most countries as a centered, circular, square-edged opening in a thin pipe. It is simple and convenient to use, but it has the disadvantage that a large fraction of the pressure drop generated for the measurement of flow is dissipated as frictional energy.

The venturi meter (Fig. 3**b**) consists of a standardized conical nozzle followed by a gradually expanding downstream cone. The pressure change is measured between the upstream fluid and the narrow throat. The gradual expansion downstream of the throat allows the fluid to recover as much as 90% of the pressure difference that would otherwise be lost with a sudden expansion. The venturi meter is a rugged precision meter but suffers from the disadvantages of high cost as well as large space requirements.

Measurement by variable area. Meters that operate on the principle of variable area incorporate an adjustable constriction in the path of the flow that may be varied so as to maintain a constant differential pressure. The most commonly used device of this type is the rotameter, a tapered vertical tube containing a free plummet or float (Fig. 3**c**).

Measurement by liquid level. The flow rate of liquids flowing in open channels is often measured by the use of weirs, the most common type being the rectangular weir shown in Fig. 3**d**.

Measurement by electromagnetic effects. The magnetic flow meter is a device that measures the potential developed when an electrically conductive flow moves through an imposed magnetic field. The voltage developed is proportional to the volumetric flow rate of the fluid and the magnetic field strength. The process fluid sees only an empty pipe so that the device has a very low pressure drop.

Nomenclature

A_p	= area of a body projected in a plane perpendicular to the flow
C_D	= drag coefficient
d	= diameter of drop or bubble
D	= diameter, eg, of pipe, nozzle, cylinder, strand
$E\ddot{o}$	= Eötvos number = $g\,\Delta\rho\,d^2/\sigma$
F	= force exerted on a body by a fluid
g_c	= constant in Newton's second law = 9.81 m/s^2 = 32.17 ft/s^2
L	= characteristic length in a system; length to achieve a specified approach to fully developed
Mo	= Morton number $g\mu_c^4\,\Delta\rho/\rho_c^2\sigma^3$
Re	= Reynolds number, $DV\rho/\mu$, $dV\rho/\mu$, etc
V	= velocity, characteristic system velocity
$\overline{V}$	= average velocity, eg, in a pipe
V_x	= uniform velocity at a great distance from a bluff body
μ, μ_c, μ_d	= shear viscosity; viscosity of continuous, dispersed phase
ρ, ρ_i	= density; density of fluid i
ρ_c, ρ_d	= density of continuous phase, dispersed phase
$\Delta\rho$	= density difference
σ	= interfacial tension
τ_w	= shear stress at the wall

A.M. Benson
G.Q. Martin
J.S. Son
C.V. Sternling
Shell Development Company

R.B. Bird, W.E. Stewart, and E.H. Lightfoot, *Transport Phenomena*, John Wiley & Sons, Inc., New York, 1960.

R.S. Brodkey, *The Phenomena of Fluid Motions*, Addison-Wesley Publishing Co., Inc., Reading, Mass., 1967.

FLUORAPATITE, $Ca_{10}F_2(PO_4)_6$. See Fertilizers.

FLUORESCEIN. See Xanthene dyes.

FLUORESCENT PIGMENTS (DAYLIGHT). See Brighteners, fluorescent.

FLUORINE

Fluorine, F_2, is a diatomic molecule that exists as a pale yellow gas at ordinary temperatures. It is formed by combination of atoms of the element F, atomic 9, the first member of the Group VII halogens of the periodic table. The name is derived from the Latin word fluere, meaning to flow, alluding to the well-known fluxing power of the mineral fluorite, the most abundant naturally occurring compound of the element. The single naturally occurring isotope has an atomic weight of 18.9984; other radioactive isotopes (qv) between atomic weight 17 and 22 have been artificially prepared, and have half-lives between 4 s for ^{22}F and 110 min for ^{18}F. Fluorine is the most electronegative element, and the most reactive nonmetal. The electron configuration of the atom is $1s^2 2s^2 2p^5$. The tendency to complete the outer shell with eight electrons explains its extreme chemical reactivity.

Fluorine does not occur free in nature, except for trace amounts in radioactive materials, but is widely found in combination with other elements and accounts for ca 0.065 wt% of the earth's crust. The most important natural source for industrial purposes is the mineral fluorspar, CaF_2, containing about 49% fluorine. Among the elements, fluorine is about thirteenth in abundance.

The only feasible method for preparing elemental fluorine is electrolysis. Fluorine was first isolated in 1886 by the French chemist Moissan, and first produced commercially almost 50 years after its discovery.

Physical Properties

Fluorine is a pale yellow gas that condenses to a yellowish-orange liquid at $-188°$C. It solidifies to a yellow solid at $-220°$C, and turns white in a phase transition at $-228°$C. Mp $-219.61°$C; critical temperature $-129.1°$C; refractive index of liquid at bp 1.2. Fluorine has a strong odor which is easily detectable at concentrations as low as 20 ppb. The odor resembles that of the other halogens and is comparable to strong ozone.

Chemical Properties

Fluorine is the most reactive element, combining with most organic and inorganic materials at or below room temperature. Organic and hydrogen-containing compounds especially can burn or explode when exposed to fluorine. Fluorine forms compounds, in which it shows valence minus one, with all the elements except helium, neon, and argon. Nitrogen and oxygen form fluorides but do not react directly with fluorine except in the presence of an electric discharge.

Fluorine is the most electronegative element and has a normal potential of -2.85 V. Its extreme electronegativity is responsible for its ability to oxidize many elements to their highest oxidation state. The small size of the atom permits the arrangement of a large number of fluorine atoms around an atom of another element. These properties of high oxidation potential and small size allow the formation of many simple and complex fluorides in which elements show their highest oxidation states. Another unique property of the nonionic metal fluorides is their volatility. Volatile compounds such as tungsten hexafluoride, WF_6, and molybdenum hexafluoride, MoF_6, are produced by the reaction of the metal with elemental fluorine.

Fluorination reactions occur generally with the liberation of intense heat and frequently occur in situations where other halogenations do not.

Reactions

Metal. At ordinary temperatures, fluorine reacts vigorously with most metals to form fluorides. A number of metals, including aluminum, copper, iron, and nickel, form a protective surface film of fluoride, so that the temperature must be increased to obtain further reaction. Susceptibility to attack depends, to a great extent, on the physical state of the metal, eg, powdered iron 0.84-mm size (20 mesh) is not attacked by liquid fluorine whereas in the 0.14-mm size (100-mesh) it ignites and burns violently.

Nonmetals. Sulfur reacts with fluorine to yield the remarkably stable sulfur hexafluoride, SF_6. A mixture of the lower fluorides such as disulfur difluoride, S_2F_2, disulfur decafluoride, S_2F_{10}, and sulfur tetrafluoride, SF_4, is also obtained as by products; operating conditions must be controlled to minimize their formation.

Silicon and boron burn in fluorine forming silicon tetrafluoride, SiF_4, and boron trifluoride, BF_3, respectively. Selenium and tellurium form hexafluorides, whereas phosphorus forms tri- or pentafluorides. Fluorine reacts with the other halogens to form eight interhalogen compounds (see Fluorine compounds, inorganic—halogens).

Water. Fluorine forms hydrofluoric acid, HF, and oxygen difluoride, OF_2, with water. The overall reaction under controlled conditions provides a method for the disposal of fluorine by conversion to a soluble salt:

$$2\,F_2 + 4\,NaOH \rightarrow 4\,NaF + O_2 + 2\,H_2O$$

Oxygen. Oxygen does not react directly with fluorine under ordinary conditions. In addition to oxygen difluoride, three oxygen fluorides are known. Dioxygen difluoride, O_2F_2, trioxygen difluoride, O_3F_2, and tetra-oxygen difuloride, O_4F_2, are produced in an electric discharge at cryogenic temperatures by controlling the ratio of fluorine to oxygen.

Noble gases. Fluorine has the unique ability to react with the heavier noble gases (krypton, xenon, and radon) to form binary fluorides.

Hydrogen. The reaction between fluorine and hydrogen is self-igniting and extremely energetic. It occurs spontaneously at ambient temperatures as evidenced by minor explosions which sometimes occur in fluorine-generating cells from the mixing of the H_2 and F_2 streams. The controlled high temperature reaction of fluorine atoms (generated thermally or photolytically from fluorine gas) with hydrogen or deuterium, is one of the most significant energy sources for the high power chemical laser. Currently, the HF–DF laser is the most promising chemical laser system under development (see Lasers; Hydrogen energy).

Organic compounds. Generally, the reaction of fluorine with organic compounds is accompanied by ignition or violent explosion. Since the heats evolved are always high, heat removal is the main problem in direct fluorination. Reactions can be moderated by dilution with large amounts of an inert gas, such as nitrogen, by reducing reaction temperatures ($-78°C$), or by the presence of finely divided metal packing.

Polymers. The dilution of fluorine with an inert gas such as nitrogen or helium significantly reduces its reactivity so that even reactions with hydrocarbon polymers at elevated temperatures can be controlled. High density polyethylene containers can be blow-molded with 1–10% fluorine-in-nitrogen mixtures to produce a fluorocarbon barrier layer on the inside of the container.

Manufacture

Fluorine is produced by the electrolysis of anhydrous potassium bifluoride, KHF_2, or KF.HF, containing various concentrations of free HF. The fluoride ion is oxidized at the anode to liberate fluorine gas, and the hydrogen ion is reduced at the cathode to liberate hydrogen. Anhydrous HF cannot be used directly as the electrolyte because of its low conductivity.

Fluorine-generating cells are classified into three distinct types, depending on their operating temperature: low temperature cells ($-80–20°C$), medium temperature cells ($60–100°C$), and high temperature cells ($220–300°C$).

Commercial cells. Today, all commercial fluorine installations employ medium temperature cells with operating currents of $\geq 6000A$. Carbon anodes overcome the excessive corrosion encountered with metal. The medium temperature cell offers the following advantages: the vapor pressure of HF over the electrolyte is less; the composition of the electrolyte can vary over a relatively wide range for only a small variation in the operation of the cell; reduced corrosion; tempered water can be used as cell coolant; the formation of a highly resistant film on the anode surface is considerably reduced compared to the high temperature cell. The C and E types of the Atomic Energy Commission (AEC) (now Department of Energy) cell design are used primarily in the United States and Canada. The other cell type used in the United States is a proprietary design developd by Allied Chemical Corporation, illustrated in Figure 1.

Raw material. The principal raw material for fluorine production is high purity anhydrous hydrofluoric acid. Each kilogram of fluorine generated requires ca 1.1 kg HF. The hydrofluoric acid consumed in fluorine production represents only ca 2% of the total United States hydrogen fluoride consumption.

Process. The generation of fluorine on an industrial scale is a complex operation. The basic raw material, anhydrous hydrogen fluoride, is stored in bulk and charged to a holding tank from which it is fed continuously to the cells. Electrolyte for the cells is prepared by mixing KF.HF with HF to form KF.2HF. The newly charged cells are depolarized by starting up at a lowcurrent, which is gradually increased at a conditioning station separate from the cell operating position until full current is obtained at normal voltages.

The total commercial production capacity of fluorine in the United States and Canada is > 5000 t/yr (est.), of which 70–80% is devoted to uranium hexafluoride production. The primary raw material, HF, used in F_2 production, is in ample supply and readily available.

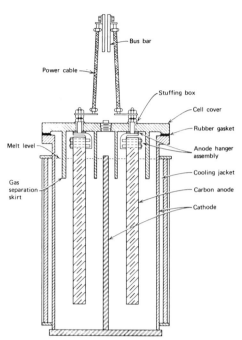

Figure 1. Fluorine generator. Courtesy of Allied Chemical Corp.

Health and Safety

Fluorine is a dangerous material that can be handled safely under proper precautions. For even short intervals of contact with low pressure fluorine, the operator should wear safety glasses, a neoprene coat and boots and clean neoprene gloves to afford overall body protection effective against both fluorine and hydrofluoric acid which may form by reaction with moisture in the air. Fluorine is extremely corrosive and irritating to the skin. Inhalation even at low concentrations irritates the respiratory tract and at high concentrations results in severe lung congestion. ACGIH has established a TLV of 1 ppm or 2 mg fluorine per cubic meter of air. Fluorine has a sharp penetrating odor detectable at levels well below the TLV.

Uses

Elemental fluorine is used captively by most manufacturers for the production of various inorganic fluorides. Fluorine's main use is in the manufacture of uranium hexafluoride, UF_6, by the following reaction:

$$UF_4 + F_2 \rightarrow UF_6$$

Uranium hexafluoride is used in the gaseous-diffusion process for the separation and enrichment of uranium 235 which exists in low concentrations in natural uranium. The other important use for elemental fluorine is in the production of sulfur hexafluoride, SF_6, a gaseous dielectric for electrical and electronic equipment (see Fluorine compounds, inorganic —sulfur). Fluorine reacts with halogens to produce several commercially important interhalogen compounds, eg, chlorine trifluoride, bromine trifluoride, and iodine pentafluoride.

Andrew J. Woytek
Air Products & Chemicals, Inc.

J.F. Tompkins and co-workers, *The Properties and Handling of Fluorine*, *Technical Report No. ASD-TDR-62-273*, Air Products & Chemicals, Inc., Allentown, Pa., 1963.

A.T. Kuhn, ed., *Industrial Electrochemical Processes*, Elsevier Publishing Company, Ltd., Amsterdam, Neth., 1971.

A.J. Rudge, *The Manufacture and Use of Fluorine and Its Compounds*, Oxford University Press, Inc., New York, 1962.

FLUORINE COMPOUNDS, INORGANIC

Fluorine is the most electronegative element and is much more reactive than the next most electronegative elements oxygen and chlorine. Fluorine reacts with virtually every element in the periodic table. Indeed, the Group VIII elements were commonly called inert gases until 1962 when xenon, radon and krypton were shown to react with fluorine (see Helium-group gases). The unique properties of fluorine have caused the element to be called a superhalogen, and several of its compounds to be called superacids. (The term superacid is used for systems with higher acidities than anhydrous sulfuric or fluorosulfuric acid.) A number of fluorine species are superacids in HSO_3F or HSO_3F–SO_3 solutions; most attention is given to SbF_5–HSO_3F.

The basic fluorine raw materials are fluorspar ore, CaF_2, and by-product fluorosilicic acid from phosphate production. Calcium fluoride reacts with sulfuric acid to produce hydrogen fluoride, the most common reagent for production of fluorine compounds. Elemental fluorine is produced by electrolysis of anhydrous potassium fluoride–hydrogen fluoride melts.

The fluoride ion is a small anion, 0.136 nm (0.045 nm smaller than the chloride ion) and is the least polarizable anion. The relatively small size of F^- allows higher coordination numbers and leads to different crystal forms, different solubilities, and higher bond energies than do the other halides. Bonds between fluorine and other elements are very strong, whereas the fluorine–fluorine bond is much weaker than the chlorine–chlorine bond. A number of elements exhibit their highest oxidation state only as fluorides and oxidation states of 6 and 7 are not uncommon.

Fluorine's special properties lead to many applications. Its complexing properties account for its use as a flux in steelmaking and as an intermediate in aluminum manufacture. The reaction of fluorides with hydroxyapatite $(Ca_5(PO_4)_3OH)$, which is found in tooth enamel, to form less soluble and/or more acid-resistant compounds, led to the incorporation of fluorides in drinking water and dentifrices (qv) to reduce dental caries. Because many fluorine compounds are very volatile, they are useful in applications such as the enrichment of uranium (UF_6) and the vapor deposition of metals (WF_6, MoF_6, ReF_6) and as unreactive dielectrics (SF_6). Because fluorine forms very stable bonds, its compounds can be either extremely reactive $(F_2, HF,$ ionic $MF_x)$ or extremely stable $(CF,$ covalent $MF_x)$. Properties and uses of selected inorganic fluoride compounds are given in Table 1.

Table 1. Properties of Inorganic Fluoride Compounds[a, b]

Compound	Formula	Crystal (parameters)	Mp, °C	Density, g/cm³	Solubility, g/L H₂O	Safety	Uses
Aluminum							
aluminum trifluoride	AlF_3	white, trigonal rhombohedrons (a = 0.504 nm alpha 58°31′)	1278 (sublimes)	3.1	4.1	TLV 88, mg/m³	electrolyte component, aluminum manufacture
cryolite	Na_3AlF_6	white, monoclinic (a = 0.546, b = 0.561, c = 0.780)	1012	2.97	0.042	LD$_{50}$, 200 mg/kg	aluminum manufacture
Ammonium							
ammonium bifluoride	NH_4HF_2	colorless, orthorhombic	126.1	na	41.5 (wt%)	LD$_{50}$, 150 mg/kg (guinea pig)	pickling metals, treating glass
Antimony							
antimony pentafluoride	SbF_5	colorless liquid	7	3.14	hydrolyzes	corrosive	fluorinating reagent, oxidizer
antimony trifluoride	SbF_3	white to tan solid	292	4.38	384.7 g/100 g H₂O	corrosive	fluorinating reagent
Arsenic							
arsenic pentafluoride	AsF_5	gas	(−79.8)	2.33	hydrolyzes	carcinogen	fluorinating reagent
arsenic trifluoride	AsF_3	colorless liquid	−6	2.67	hydrolyzes	carcinogen	fluorinating reagent
Barium							
barium fluoride	BaF_2	colorless solid	1290–1355	4.9	1.6	LD$_{50}$ 350 mg/kg (guinea pig)	fluxing agent
Boron							
boron trifluoride	BF_3	colorless gas	(−128.37)	0.591	369.4/100 g H₂O (6°C)	TLV 0.3 ppm	Lewis acid catalyst
sodium fluoroborate	$NaBF_4$	orthorhombic (a = 0.68358, b = 0.62619, c = 0.67916)	406 dec	2.47	1080 (26°C)	na	catalyst, flame retardant
fluroboric acid	HBF_4	colorless liquid	na	1.37 (48% soln)	infinite	corrosive	Electroplating baths, catalyst
Calcium							
calcium fluoride	CaF_2	colorless (when pure), fcc	1402	3.181	0.146	LD$_{50}$ 4250 mg/kg (rats), may irritate skin	HF manufacture, steelmaking
Cobalt							
cobalt trifluoride	CoF_3	light brown powder	dissociates	3.88	hygroscopic	carcinogenic	"hard" fluorinating reagent
cobalt difluoride	CoF_2	pink powder	1200	4.46	13.6	carcinogenic	
Copper							
copper(II) fluoride	CuF_2	white solid	785 ± 10	4.85	47 (20°C)	na	catalyst
Halogens							
bromine trifluoride	BrF_3	colorless liquid	8.8	2.803	na	highly toxic	well-drilling oxidizing agent
chlorine trifluoride	ClF_3	pale yellow liquid	−76.3	1.62	na	highly toxic	converts uranium to UF_6
iodine pentafluoride	IF_5	straw-colored liquid	8.5	3.252	na	highly toxic	fluorinating reagent

Table 1. Continued

Compound	Formula	Crystal (parameters)	Mp, °C	Density, g/cm³	Solubility, g/L H₂O	Safety	Uses
Hydrogen							
hydrogen fluoride	HF	colorless liquid	−83.55	0.9576	very sol	toxic, skin irritant	aluminum production, manufacture of fluorocarbons
Iron							
iron(III) fluoride	FeF₃	lime green rhombic	1000 (sublimes)	3.52	slightly sol	na	fluorinating reagent, catalyst
Lead							
lead(II) fluoride	PbF₂	white crystalline powder	824	8.24	0.66	TWA 0.1 mμ (Pb)/m³	fluorinating agent
Lithium							
lithium fluoride	LiF	white cubic (a₀ = 0.401736 nm)	842	2.635	0.133 (25.4°C)	LD₅₀, 200 mg/kg (guinea pig)	ceramics
Magnesium							
magnesium fluoride	MgF₂	white powder, tetragonal (a = 0.4623 nm, c = 0.3052 nm)	1263	3.127	0.013	LD₅₀, 1000 mg/kg (guinea pig)	flux
Mercury							
mercury(II) fluoride	HgF₂	white solid	645	8.95	reacts	extremely toxic	fluorinating agent
Nickel							
nickel difluoride	NiF₂	pale yellow solid	1100	4.63	slightly sol	LD₅₀, 130 mg/kg	catalyst
Nitrogen							
nitrogen trifluoride	NF₃	colorless gas	−206.8	1.533 (bp)	very slightly sol	TLV 29, mg/m³, toxic	fluorinating source for lasers
Oxygen							
oxygen difluoride	OF₂	colorless gas	−224	2.190 −0.00523 × K (from −145 to 153°C)	sol	highly toxic	experimental oxidizer
Phosphorus							
phosphorus pentafluoride	PF₅	colorless gas	−91.6	na	hydrolyzes	toxic	catalyst
phosphorus trifluoride	PF₃	gas	−151.5	1.6	reacts slowly	toxic	forms metal complexes
Potassium							
potassium fluoride	KF	white hygroscopic salt	857	2.48	923 (18°C)	LD₅₀, 245 mg/kg (rats)	flux
Rhenium							
rhenium hexafluoride	ReF₆	yellow liquid	18.5	3.58	hydrolyzes	corrosive	chemical vapor deposit of rhenium
Silver							
silver fluoride	AgF	golden yellow solid	435	5.85	1.82 kg	toxic, irritant	cathode material in batteries
Sodium							
sodium fluoride	NaF	white powder	992	2.56	42 (10°C)	LD₅₀, 188 mg/kg (rats)	fluorination of water supplies
Sulfur							
sulfur hexafluoride	SF₆	colorless gas	−63.9 (sublimes)	2.836 (−195.2°C)	na	nontoxic	gaseous dielectric
fluorosulfuric acid	HSO₃F	colorless liquid	−88.98	1.726	reacts violently	extremely toxic acid	catalyst, reagent organic reactions
Tantalum							
tantalum pentafluoride	TaF₅	white solid	96.8	4.74	soluble	LD₅₀, 110 mg/kg	catalyst
Tin							
stannous fluoride	SnF₂	white crystalline salt	215	na	709	LD₅₀, 188 mg/kg (rats)	dental products
Tungsten							
tungsten hexafluoride	WF₆	white octahedron	2	3.441 (15°C)	hydrolyzes	toxic, irritant	chemical vapor deposit of tungsten
Zinc							
zinc fluoride	ZnF₂	white solid	872–910	na	16	LD₅₀, 280 mg/kg (frogs)	fluorinating agent
Zirconium							
zirconium tetrafluoride	ZrF₄	white solid	640 (sublimes)	1.388	sparingly sol	toxic	catalyst, optical glass

[a] Properties are listed at 25°C, unless otherwise noted.

[b] Table information was extracted from articles by the following authors: Richard E. Eibeck (Allied Chemical Corporation), John F. Gall (Philadelphia College of Textiles and Science), H.S. Halbedel (The Harshaw Chemical Company), Charles B. Lindahl (Ozark-Mahoning Company, a subsidiary of the Pennwalt Corp.), Donald R. Martin (University of Texas at Arlington), Whitney Mears (Allied Chemical Corporation), Dayal T. Meshri (Ozark-Mahoning Company, a subsidiary of the Pennwalt Corp.), Thomas E. Nappier (The Harshaw Chemical Company), Irvine J. Solomon (IIT Research Institute), Kenneth Wachter (Olin Corporation), and Andrew J. Woytek (Air Products and Chemicals, Inc.).

Sources

The earth's crust consists of 0.09% fluorine. Among the elements it ranks about 13th in terrestial abundance. The most important ores are fluorospar, CaF_2; fluoroapatite, $Ca_5(PO_4)_3F$, and cryolite, Na_3AlF_6. Fluorspar is the primary commercial source of fluorine. The steel industry continues to be the largest user of fluorspar and accounts for about 50% of the total U.S. consumption. Most of the balance of the CaF_2 is treated with sulfuric acid to produce hydrogen fluoride. About 30% of all hydrogen fluoride is used to produce aluminum fluoride and synthetic cryolite for the Hall aluminum process.

Manufacture

Most inorganic fluorides are prepared by the reaction of hydrofluoric acid with oxides, carbonates, hydroxides, chlorides or metals. Routes starting with carbonate or hydroxide or oxide (if reactive) are the most common and the choice is determined by the most economical starting material. In many cases, the water produced by the reaction cannot be removed without at least partial hydrolysis of the metal fluoride. This hydrolysis frequently can be reduced by dehydrating in a stream of hydrogen fluoride. If hydrolysis is unavoidable, reaction of anhydrous HF with the metal or the metal chloride is required.

Typical manufacturing processes of fluorine compounds are: the reaction of fluorspar with sulfuric acid to produce HF (hydrogen fluoride) and $CaSO_4$

$$CaF_2(s) + H_2SO_4 \rightarrow CaSO_4 + 2\,HF(g)$$

and the preparation of aluminum fluoride by reaction of alumina trihydrate with hydrogen fluoride

$$Al_2O_3.xH_2O + 6\,HF \rightarrow 2\,AlF_3 + (x+3)H_2O$$

and the preparation of uranium hexafluoride from the oxide by successive reactions with hydrogen fluoride and fluorine.

$$UO_2 + 4\,HF \rightarrow UF_4 + 2\,H_2O$$

$$UF_4 + F_2 \rightarrow UF_6$$

Health and Safety

Hazards associated with fluorides are several. Anhydrous or aqueous hydrogen fluoride is extremely corrosive to skin, eyes, mucous membranes and lungs and can cause permanent damage and even death. Fluorides susceptible to hydrolysis can generate aqueous hydrogen fluoride. Ingestion of excess fluorides may cause poisoning or damage to bones and/or teeth. Fluorine containing oxidizers can react with the body in addition to causing burns.

CHARLES B. LINDAHL
DAYAL T. MESHRI
Ozark-Mahoning Company, a Pennwalt Subsidiary

P. Tarrant, ed., *Fluorine Chemistry Reviews*, Marcel Dekker, Inc., New York, Vol. 1, 1967; Vol. 2, 1968; Vol. 3, 1969; Vol. 4, 1969; Vol. 5, 1971; Vol. 6, 1973; Vol. 7, 1974; Vol. 8, 1977.

J.H. Simons, ed., *Fluorine Chemistry*, Academic Press Inc., New York, Vol. 1, 1950; Vol. 2, 1954; Vol. 3, 1963; Vol. 4, 1965; Vol. 5, 1964.

M. Stacey and co-eds., *Advances in Fluorine Chemistry*, Butterworth Inc., Washington, D.C., Vol. 1, 1960; Vol. 2, 1961; Vol. 3, 1963; Vol. 4, 1965; Vol. 5, 1965, Vol. 6, 1970; Vol. 7, 1973.

FLUORINE COMPOUNDS, ORGANIC

INTRODUCTION

Physical Properties

Substitution for hydrogen in an organic molecule has a profound influence on the molecule's chemical and physical properties. Several factors that are characteristic of fluorine and that underlie these effects are the large electronegativity of fluorine, its small size (slightly larger than hydrogen), the low degree of polarizability of the carbon–fluorine bond, and their weak intermolecular forces.

The replacement of chlorine by fluorine results in a nearly constant boiling point drop of ca 50°C for every chlorine atom that is replaced. This effect is the result of low dispersion forces of internal pressure.

Preparation

Halogen exchange. The exchange of a nonfluorine halogen atom in an organic compound for a fluorine atom is most widely used method of fluorination. The relative ease of replacement follows the general order I > Br > Cl. The most commonly used fluorinating agents are the fluorides of alkali metals (especially KF), antimony, and mercury. Two other widely used exchange reagents are hydrogen fluoride and sulfur tetrafluoride; the latter also is effective in replacing carbonyl oxygen and hydroxyl groups with fluorine.

Replacement of hydrogen. Three methods of substitution of a hydrogen atom by fluorine are reaction of a C—H bond with elemental fluorine (direct fluorination); reaction of a C—H bond with a metal fluoride, eg, AgF_2 or CoF_3; and electrochemical fluorination in which the reaction occurs at the anode of a cell containing fluoride.

Direct fluorination. The principal disadvantage of the use of elemental fluorine as a fluorination agent is the exothermicity of the reaction. A considerable degree of carbon–carbon bond scission can occur as well as polymer formation. In order to prevent these complications, the fluorine is diluted with nitrogen and the reaction zone is constructed so that good heat conductivity is possible.

Reaction with a metal fluoride. A second technique for hydrogen substitution is the reaction of a higher metal fluoride with a carbon–hydrogen bond:

$$2\,CoF_2 + F_2 \rightarrow 2\,CoF_3$$

$$RH + 2\,CoF_3 \rightarrow RF + 2\,CoF_2 + HF$$

The principal advantage to this approach is that the heat evolved for each carbon–fluorine bond formed, 192.5 kJ/mol (46 kcal/mol), is much less than that obtained in direct fluorination, 435.3 kJ/mol (104 kcal/mol), and the reaction yields are much higher. The only two metal fluorides that are of practical use are AgF_2 and CoF_3.

Electrochemical fluorination. The electrochemical synthesis of fluorinated organic compounds involves the electrolysis of an organic substrate in liquid anhydrous hydrogen fluoride at a voltage below that for the liberation of fluorine. The reaction is limited by temperature (usually 0°C) and by the solubility of the substrate in HF. This method is used to fluorinate aryl halides, sulfonyl halides, ethers, carboxylic acids, and amines. The product is a fluorocarbon having no residual hydrogen; olefins and carbocyclics, as well as heterocyclic compounds, are saturated.

Chemical Properties and Applications

Substitution of fluorine in an organic molecule has two principal effects on the molecule's chemical properties: the chemical stability, eg, resistance to hydrolysis, is greatly increased; and the reactivity of functional groups is drastically altered resulting from the large inductive effect of fluorine. These effects become more pronounced as the degree of fluorine substitution is increased, and reach their maxima in the fluorocarbons and their derivatives.

Fluorinated alkanes. As the fluorine content increases, the chemical reactivity decreases until complete fluorination, after which they are inert to most chemical agents including the highly reactive element, fluorine.

Fluorinated olefins. In electrophilic addition reactions, the reactivity of an unsaturated linkage is reduced by the inductive effect of fluorine. Nucleophilic additions are enhanced by this same effect. Fluoroolefins undergo free-radical polymerization producing a wide range of polymers derived from tetrafluoroethylene, vinylidene fluoride, vinyl fluoride, and hexafluoropropylene.

Fluorinated drugs. The preparation of fluorinated compounds for use in medicine has increased rapidly based on the following considerations:

fluorine most closely resembles bioactive hydrogen analogues with respect to steric requirements at receptor sites; fluorine alters electronic effects, owing to its high electronegativity; fluorine imparts improved oxidative and thermal stability to the parent molecule; and fluorine improves lipid solubility in membranes, thereby increasing rates of absorption and transport, *in vivo*.

OGDEN R. PIERCE‡
Dow Corning Corporation

Reviewed by
YUNG K. KIM
Dow Corning Corporation

J.H. Simons, *J. Electrochem. Soc.* **95**, 47 (1949).

R. Filler "Fluorine-Containing Drugs," in R.E. Banks, ed., *Organofluorine Chemicals and Their Industrial Applications*, Ellis Horwood Litd., Chichester, UK, 1979.

AMA Drug Evaluations, 3rd ed., Publishing Sciences Group, Inc., Littleton, Mass., 1977.

DIRECT FLUORINATION

Although early literature references to direct fluorination of organic compounds allude to degradation, cross-linking, and even combustion, there are now well-established techniques for controlling the reaction and deriving useful products. A carefully programmed exposure to the proper concentration of elemental fluorine at selected temperatures for limited periods of time can result in protective films on pre-formed objects or high yields of reasonably complex perfluoro/partially fluorinated molecules. Fluorination of various industrial grades of acetylene black, activated carbons and graphitic materials occurs at 200–600°C to yield various perfluorocarbon gases, liquids and solids including C_4F, C_2F and CF_x where $x = 0.6$ to 1.2.

In the early work (before 1970), the concentrations of fluorine in the reactor were usually kept at a constant level, often 10% or greater, by premixing an inert gas (N_2, He, etc) with fluorine in the desired proportion and then introducing this mixture into the reactor. The rates of reaction between hydrocarbons and such a 10% fluorine mixture are relatively high and the very exothermic reactions can lead to fragmentation, degradation and even to combustion. Much lower concentrations of fluorine (ca 1%) are more appropriate for nondestructive fluorinations.

Various relaxation processes, including vibrational or rotational relaxations and thermal conduction, make it possible to dissipate the energy released during controlled fluorinations. In the initial stages of fluorination, it is necessary to reduce the probability of simultaneous reactions at adjacent sites occurring in the same molecule, or in adjacent molecules in a crystal, by diluting the reactants and relying on relaxation processes to distribute the energy over the entire system and thus avoid fragmentation. Ideally, one chooses experimental parameters so that the rate of fluorination will be slow enough to allow relaxation processes to occur and provide a heat sink to remove excess reaction heat and keep temperatures down.

Most direct fluorination reactions with organic compounds are performed at or near room temperature unless: reaction rates are so slow that even a fluorine pressure of 101.3 kPa (1 atm) does not lead to appreciable reaction; a vapor-phase reaction is desired and it is necessary to heat the sample being treated to get enough vapor; or reaction rates are so fast that excessive fragmentation, charring or decomposition occurs and a much lower temperature is desirable. Examples of all types are known. Thus, in the production of partially fluorinated lubricating oils, fluorine gas is bubbled at 101 kPa through medium-weight oils at 50, 75, and 100°C. For production of fluorocarbon coatings on natural or synthetic rubber or polypropylene, temperatures of 25–50°C are optimum. Although fluorination of hot hydrocarbon vapors like gaseous naphthalene or gaseous anthracene is not too satisfactory because fragmentation and charring lead to low yields, it is possible to obtain good

yields of the perfluoro-alicyclic derivatives eg, perfluorodecalin, by room temperature fluorination of solid naphthalene.

Low temperature fluorination techniques (−78°C) are promising for the preparation of complex fluorinated molecules, especially where functional groups are present; eg, fluorination of hexamethylethane to perfluorohexamethylethane, of norbornane to perfluoro- and 1-hydroundecafluoronorbornane, of adamantane to 1-hydropentadecafluoroadamantane, of dioxane to yield perfluorodioxane and of polymethylene oxides to yield polyperfluoroethers.

Uses

Many unique simple and complex fluoro-organic molecules can be produced. For example, direct fluorination of CS_2 yields several derivatives including SF_5CF_3, $(SF_3)_2CF_2$, $SF_5CF_2SF_3$, SF_3CF_3, and $F_3SCF_2SF_3$; and the treatment of $Si(CH_3)_4$ or $Ge(CH_3)_4$ with F_2 can make most of the Si–C–H–F or Ge–C–H–F species possible. The direct fluorination of uracil by bubbling F_2 through a Freon slurry has been used for commercial production of 5-fluorouracil which is effective in cancer chemotherapy. Direct fluorination can be used for the synthesis of fluorocarbon polymers, otherwise difficult to produce, and for producing fluorocarbon coatings on hydrocarbon polymers either by fluorination of preformed objects or by blow-molding with a F_2/N_2 gaseous mixture. Fluorination of natural or synthetic rubber creates a fluorocarbon coating which is smooth, slippery and water-repellent (see Waterproofing).

JOHN L. MARGRAVE
R.H. HAUGE
R.B. BADACHHAPE
Rice University

R.J. LAGOW
University of Texas

R.E. Banks, ed., *Preparation, Properties, and Industrial Application of Organofluorine Compounds*, John Wiley & Sons, Inc., New York, 1983.

M. Hudlicky, *Chemistry of Organic Fluorine Compounds*, 2nd ed., Halsted Press, a division of John Wiley & Sons, Inc., New York, 1976. U.S. Pat.; 3,758,540 (Sept. 11, 1973), R.J. Lagow and J.L. Margrave; U.S. Pat. 3,775,489 (Nov. 27, 1973), R.J. Lagow and J.L. Margrave; and U.S. Pat. 3,992, 221 (Nov. 16, 1976), R.B. Badachhape, C. Homsy, and J.L. Margrave.

FLUORINATED ALIPHATIC COMPOUNDS

Fluorocarbons and Fluorohydrocarbons

Properties. The physical properties of fluorocarbons appear abnormal when compared with those of their hydrocarbon counterparts; many of the differences owing to the low forces of attraction between fluorocarbon molecules. They are much more volatile than hydrocarbons of similar molecular weight. Table 1 lists some of the physical properties.

Manufacture. The direct fluorination (qv) of hydrocarbons with elemental fluorine is extremely exothermic and difficult to control. The disadvantages of direct fluorination have been overcome by the use of fluorine carriers, in particular, high valence metal fluorides such as cobalt trifluoride, CoF_3.

Chlorofluorocarbons

Properties. The physical properties of aliphatic fluorine compounds containing chlorine are similar to those of the fluorocarbons. They usually have high densities and low boiling points, viscosities, and surface tensions. The irregularity in the boiling points of the fluorinated methanes and ethanes, however, does not appear in the chlorofluorocarbons. Their boiling points consistently increase with the number of chlorines present.

Manufacture

The most important commercial method for manufacturing chlorofluorocarbons is the successive replacement of chlorine by fluorine using hydrogen fluoride. The traditional liquid-phase process uses antimony

‡Deceased.

Table 1. Physical Properties of Some Fluorinated Aliphatic Compounds

Compound	Solubility in water at 101.3 kPa^a and 25°C, wt%	Bp, °C	Mp, °C	Liquid density, g/cm³ at °C	Refractive index, n_D at °C
Fluorocarbons and fluorohydrocarbons					
CF_4	0.0015	−128.06	−183.6	1.613_{-130}	1.151_{-73}
CF_3CF_3		−78.2	−100.6	1.600_{-80}	1.206_{-73}
CHF_3	0.10	−82.16	−155.2	1.442_{-80}	1.215_{-73}
C_4F_8 (cyclic)		−5.85	−41.4	1.500_{25}	1.217_{25}
Chlorofluorohydrocarbons					
CCl_3F	0.11	23.82	−111	1.476_{25}	1.374_{25}
CCl_2F_2	0.028	−29.79	−158	1.311_{25}	1.287_{25}
$CHClF_2$	0.30	−40.75	−160	1.194_{25}	1.256_{25}
CCl_2FCCl_2F	0.012^b	92.8	26	1.634_{30}	1.413_{25}
Bromofluorohydrocarbons					
$CBrF_3$	0.03	−57.8	−168	1.538_{25}	1.238_{25}
$CBrClF_2$		−4	−161	1.850_{15}	
$CHBrClCF_3$		50.2		1.860_{20}	1.3700_{20}
$CBrF_2CBrF_2$		47.3	−110	2.163_{25}	1.367_{25}

^{a}To convert kPa to mm Hg, multiply by 7.5.
^{b}At the vapor pressure.

pentafluoride or a mixture of antimony trifluoride and chlorine as catalysts.

Fluorocarbons Containing Other Halogens

Properties. The physical and chemical properties of bromo- and iodofluorocarbons are similar to those of the chlorofluorocarbons except for higher densities and generally decreased stability. The stability of these compounds decreases as the ratio of bromine or iodine to fluorine increases.

Manufacture. Brominated fluoromethanes are prepared industrially by the halogen exchange of tetrabromomethane or by the bromination of CH_2F_2 or CHF_3 at elevated temperatures. Other bromo- or iodofluorocarbons can be prepared by halogenating suitable fluorocarbons, including fluoroolefins.

Health and safety. Completely fluorinated alkanes are essentially nontoxic; however, some fluorochemicals, eg, functionalized derivatives and fluoroolefins, can be lethal. Monofluoroacetic acid and perfluoroisobutylene are notoriously toxic. The toxicity of aliphatic halofluorocarbons generally decreases as the number of fluorine atoms increases. Fluorocarbons containing bromine or iodine are more toxic than the corresponding chloro compounds.

Uses. Carbon tetrafluoride is used as a degreasing agent for the plasma etching of semiconductor devices (see Ion implantation), among other uses. Experimental applications of liquid fluorocarbons, particularly perfluorodecalin, are for oxygen and carbon dioxide transport in artificial blood (see Blood-replacement preparations). Chlorofluorocarbons are mainly used in the aerosol field as propellants; air conditioning and refrigeration account for over 90% of the $CHClF_2$, ca 32% of the CCl_2F_2, and 8% of the CCl_3F manufactured (see Refrigeration). Of the other fluorocarbons, $CBrClF_2$ and $CBrF_3$ are commercially the most important as fire-extinguishing agents. Halothane, $CF_3CHClBr$, is widely used as an anesthetic.

BRUCE E. SMART
E.I. du Pont de Nemours & Co., Inc.

R.E. Banks, ed., *Organofluorine Chemicals and Their Industrial Applications*, Ellis Horwood Ltd., Chichester, UK, 1979.

R.E. Banks, ed., *Preparation, Properties, and Industrial Applications of Organofluorine Compounds*, Ellis Horwood Ltd., Chichester, UK, 1982.

FLUOROETHANOLS

2-Fluoroethanol

2-Fluoroethanol (ethylene fluorohydrin, β-fluoroethyl alcohol), FCH_2CH_2OH, is a colorless liquid with an alcohol-like odor; mp

−26.45°C; bp 103.55°C; d_4, 1.1297 g/cm³; and n_D^{18}, 1.13647. Like the other fluoroethanols, it is miscible with water, stable to distillation, and low in flammability. It is the least acidic of the fluoroethanols, although more acidic than ordinary alcohols. Its most notable difference from the other fluoroethanols is its extreme toxicity; in mice, an LD_{50} of 10 mg/kg has been measured.

In its chemical reactions, 2-fluoroethanol behaves like a typical alcohol. Oxidation yields fluoroacetaldehyde or fluoroacetic acid; reaction with phosphorus tribromide gives 1-bromo-2-fluoroethane; addition to olefins results in ethers; and additions to isocyanates gives carbamates.

The alcohol has been used to control rodent populations (see Poisons, economic), or when labelled with ^{18}F as a radiodiagnostic agent (see Radioactive drugs).

2,2,2-Trifluoroethanol

2,2,2-Trifluoroethanol, CF_3CH_2OH, is a colorless liquid with an ethanol-like odor; mp −45°C; bp 73.6°C; d_4^{25}, 1.3823 g/cm³; and n_D^{20}, 1.2907. It is stable to distillation and miscible with water and many organic solvents. Because of its excellent combination of physical and thermodynamic properties, 2,2,2-trifluoroethanol–water mixtures have application as working fluids in Rankine-cycle engines being developed for recovering energy from waste heat sources (see Power generation).

Trifluoroethanol was first prepared by the catalytic reduction of trifluoroacetic anhydride. More recently, hydrogenation of 2,2,2-trifluoroethyl trifluoroacetate over a copper(II) oxide catalyst has been reported to give the alcohol in 95% yield.

GARY ASTROLOGES
Halocarbon Products Corporation

F.L.M. Pattison, *Toxic Aliphatic Fluorine Compounds*, Elsevier Publishing Co., New York, 1959, p. 65.

Trifluoroethanol Brochure, Halocarbon Products Corp., Hackensack, N.J., 1979.

M. Hudlicky, *Chemistry of Organic Fluorine Compounds*, 2nd ed., John Wiley & Sons, New York, 1976.

FLUORO ETHERS AND AMINES

The perfluoroaliphatic ethers and perfluoroalkyl tertiary amines, together with the perfluoroalkanes, comprise a class of materials known as perfluorinated inert fluids. They are colorless, odorless, dense, virtually nontoxic, and nonflammable compounds with an extremely nonpolar character.

Physical Properties

The perfluorinated inert liquids boil lower than their hydrocarbon analogues even though they have much higher molecular weights. The low boiling points of the perfluoroaliphatic ethers and perfluoroalkyl tertiary amines are especially remarkable; they boil very close to the perfluoroalkanes having the same number of carbon atoms. The hetero atoms apparently contribute little polarity to the molecules, but do have a marked effect on the freezing points, eg, C_{12}-perfluoro ether freezes at −90°C, and C_{12}-perfluoro tertiary amine freezes at −50°C, vs the perfluoroalkane $C_{12}F_{26}$, which is a solid at room temperature.

Many of the unusual properties of the perfluorinated inert fluids are the result of their low intermolecular forces. An example is their low surface tensions, which are 9–18 mN/m (= dyn/cm) at 25°C, enabling these liquids to wet any surface, including polytetrafluoroethylene. Their refractive indexes are lower than any other organic liquids, as are their acoustic velocities. They have isothermal compressibilities almost twice as high as water.

Chemical Properties

The inert character of the perfluoro ethers and tertiary amines is demonstrated by their lack of basicity or reactivity normally associated with hydrocarbon ethers or amines. They do not form salts with acids or complexes with boron trifluoride. They are not attacked by most oxidizing or reducing agents or strong acids or bases. Like the perfluoroalkanes, the perfluoro ethers and tertiary amines may react violently with fused alkali metals. Unlike the alkanes, they also react with aluminum chloride at ca 150–200°C with some substitution of the fluorine by chlorine.

Solvent properties. Water and alcohols are almost completely insoluble. Most hydrocarbons and nonfluorine-containing aliphatic halogen compounds are only slightly soluble. However, lower aliphatic hydrocarbons, such as pentanes and hexanes, and fluorine-containing liquids may be completely miscible with the perfluoro compounds at room temperature.

Preparation

In the electrochemical fluorination process, organic reactants are dissolved in anhydrous hydrogen fluoride to form conductive solutions. The solutions are electrolyzed with direct current at 4–8 V in a single-compartment cell, usually with nickel electrodes. Essentially complete replacement of organic hydrogen atoms by fluorine atoms occurs at the anode and hydrogen gas is released at the cathode. A low temperature condenser strips HF and liquid products from the hydrogen stream. Liquid products, higher in density and insoluble in HF, are recovered by phase separation.

The products of electrochemical fluorination contain appreciable amounts of the perfluorinated analogues of the starting materials. In addition, because of extensive molecular rearrangement, many isomers and perfluorinated cleavage products are also present. The products are usually purified by treatment with base and distillation.

Health and Safety

The perfluorinated inert fluids are nonirritating to the eyes and skin, and are practically nontoxic by oral ingestion, inhalation, or intraperitoneal injection.

Uses

These compounds are used extensively in electronics industry as test baths because of their compatibility with semiconductor devices and their ability to evaporate completely without leaving residues.

RICHARD D. DANIELSON
Minnesota Mining and Manufacturing Co.

J.H. Simons, ed., *Fluorine Chemistry*, Vol. 1, Academic Press, Inc., New York, 1950, pp. 486–491; Vol. II, 1954, pp. 242, 391, 396; Vol. V, 1964, pp. 311–314, 325, 332.

A.M. Lovelace and co-workers, *Aliphatic Fluorine Compounds*, Reinhold Publishing Corp., New York, 1958, pp. 155, 285.

M. Stacey, J.D. Tatlow, and A.G. Sharpe, eds., *Advances in Fluorine Chemistry*, Vol. 1, Butterworths Scientific Publications, London, UK, 1960, pp. 139–145; Vol. 2, p. 134; Vol. 3, p. 32.

HEXAFLUOROACETONE AND DERIVATIVES

Hexafluoroacetone

Physical properties. $CF_3C(O)CF_3$ is a reactive, nonflammable, gaseous perfluoro ketone. Bp at 101.3 kPa 27.28°C; mp −125.45, −129°C; liquid density at 25°C 1.318 g/cm^3.

Chemical Properties

The chemical properties of hexafluoroacetone differ markedly from those of halogen-free ketones such as acetone. The inductive effect of the electronegative fluorine substituents causes the carbonyl group to be exceptionally electron deficient. As a consequence, hexafluoroacetone is very resistant to attack by electrophilic reagents. It is not protonated by strong acids, and can even be distilled from concentrated sulfuric acid. As would be expected, the electron-deficient carbonyl group is very susceptible to nucleophilic reagents, and this is the basis for the exceptional reactivity of hexafluoroacetone.

Reactions with nucleophiles. Hexafluoroacetone reacts exothermically with water to form an acidic (pK_a 6.58) solid (mp = 49°C) gem-diol, $(CF_3)_2C(OH)_2$. With more water, a liquid sesquihydrate is formed which is a useful polymer solvent. These hydrates are potent donors for hydrogen bonding and form stable complexes with acceptors such as ethers and amines.

Aliphatic alcohols form stable hemiketals.

Compounds that contain hydroperoxide groups add across the carbonyl group of hexafluoroacetone to give peroxides that are oxidizing agents and polymerization catalysts. Thiol compounds add similarly.

Hexafluoroacetone oxime, a colorless liquid (bp 75°C) and an excellent solvent for polyamide and polyester polymers, or hexafluoroacetone hydrazone, a colorless liquid (bp 96°C), are prepared via the imine.

Hexafluoroacetone reacts with Grignard reagents, organolithium compounds, and other organometallics to give tertiary alcohols, eg, perfluoro 2-methyl-3-buten-2-ol, $CF_2=CFC(CF_3)_2OH$, is prepared by reaction with trifluorovinyllithium, $CF_2=CFLi$.

Preparation

Laboratory syntheses for hexafluoroacetone include: oxidation of perfluoroisobutylene with potassium permanganate; isomerization of hexafluoropropylene oxide with Friedel-Crafts catalysts; oxidation of the dimer of hexafluorothioacetone with potassium iodate; and direct oxidation of hexafluoropropene with oxygen in the presence of a fluorinated Al_2O_3. Hexafluoroacetone is manufactured by the reaction of hexachloroacetone and hydrogen fluoride in the presence of a chromium catalyst.

Health and Safety

Hexafluoroacetone is moderately toxic; it has a TLV established at 0.1 ppm.

WILLIAM J. MIDDLETON
E.I. du Pont de Nemours & Co., Inc.

C.G. Krespan and W.J. Middleton in P. Tarrant, ed., *Fluorine Chemistry Reviews*, Vol. 1, Marcel Dekker, Inc., New York, 1967, pp. 145–196.

FLUORINATED ACETIC ACIDS

Fluoroacetic Acid

FCH_2COOH, has mp 33°C; bp 165°C, and d$_{36}$ 1.369 g/cm^3. Chemically, fluoroacetic acid behaves like a typical carboxylic acid, although its acidity is higher than average ($K_a = 2.2 \times 10^{-3}$). It can be prepared from the commercially available sodium salt by distillation from sulfuric acid.

Difluoroacetic Acid

$F_2CHCOOH$ is a colorless liquid with a sharp odor; mp 35°C; bp 134°C; d$_4$ 1.539 g/cm^3; and n_D^{20} 1.3428. It is a moderately strong acid, and undergoes reactions typical of a carboxylic acid such as forming an ester when heated with an alcohol and sulfuric acid.

Trifluoroacetic Acid

CF_3COOH is a colorless liquid with a sharp odor resembling that of acetic acid; mp −15.36°C; bp 72.4°C; and d$_{25}$ 1.489 g/cm^3.

Trifluoroacetic acid undergoes reactions typical of a carboxylic acid, the trifluoromethyl group being inert to most common reagents including lithium aluminum hydride. However, there are important differences; for example, the amides and esters are more easily hydrolyzed than is typical for carboxylic acids and this has led to the use of the acid and its anhydride to make derivatives of carbohydrates, amino acids, and peptides from which the trifluoroacetyl protective group can be removed with relative ease. Its peracid, formed from the reaction of trifluoroacetic anhydride and hydrogen peroxide, is a stronger oxidizing agent than other peroxycarboxylic acids and gives better yields of epoxides from alkenes, esters from ketones, and nitrobenzenes from anilines (see Peroxides and peroxy compounds).

Because of its stability to further oxidation, trifluoroacetic acid can be prepared by the oxidation of compounds containing a trifluoromethyl group bonded to carbon.

Health and Safety

Fluoroacetic acid is noted for its high toxicity to animals, including humans. Sodium fluoroacetate is nonvolatile, chemically stable, and not toxic or irritating to the unbroken skin of workers. Difluoroacetic acid is much less toxic than fluoroacetic acid. Trifluoroacetic acid presents no unusual toxicity problems, but owing to its strong acidity, its vapors can be irritating to tissue, and the liquid can cause deep burns if allowed to contact the skin.

Uses

Sodium fluoroacetate is one of the most effective all-purpose rodenticides known, usable in either a water solution or in bait preparations (see Poisons, economic).

GARY ASTROLOGES
Halocarbon Products Corporation

M. Hudlicky, *Chemistry of Organic Fluorine Compounds*, 2nd ed., John Wiley & Sons, Inc., New York, 1976.

FLUORINATED HIGHER CARBOXYLIC ACIDS

Properties

The boiling points (°C) and densities (d_4^{20}, g/cm^3) of some straight chain acids are perfluoroacetic acid, CF_3COOH, bp 72.4, d 1.489; perfluorobutyric acid, C_3F_7COOH, bp 120, d 1.561; perfluorohexanoic acid, $C_5F_{11}COOH$, bp 157, d 1.762. The boiling points are about 50°C lower than those of the corresponding nonfluorinated acids. The density range of these acids is 1.5–1.8 g/cm^3. The acids with fewer than eight carbon atoms are liquids at room temperature. Because of the extremely electronegative character of fluorine, these compounds are strong acids, completely ionized in aqueous solution. They exhibit the usual stability associated with fully fluorinated materials. They can be treated with strong alkali without hydrolysis of fluorine, and are not attacked by aqueous solutions of oxidizing or reducing agents.

Reactions

The reactions of perfluoro acids are similar to those of hydrocarbon acids (see Carboxylic acids; Dicarboxylic acids). Salts are formed with the ease expected of strong acids; the heavy metal salts are very hygroscopic. Esterification takes place readily with primary and secondary alcohols. Acid anhydrides can be prepared by distillation of the acids from phosphorus pentoxide and halides can be prepared by standard procedures. The amides are readily prepared by ammonolysis of the acid halides, anhydrides, or esters, and can be dehydrated to the corresponding nitriles.

Preparation

Many of the perfluoro carboxylic acids have been prepared by the electrochemical fluorination of the corresponding carboxylic acid halides which involves the electrolysis of a solution of the acid halide in anhydrous hydrogen fluoride (see Electrochemical processing).

Uses

The acids and their salts are very surface-active and have a variety of surfactant uses.

RICHARD GUENTHNER
3M Company

E.A. Kauck and A.R. Diesslin, *Ind. Eng. Chem.* **43**, 2332 (1952).

R.A. Guenthner and M.L. Vietor, *Ind. Eng. Chem. Prod. Res. Dev.* **1**, 165 (1962).

A.M. Lovleace, W. Postelnek, and D.A. Rauch, *Aliphatic Fluorine Compounds*, ACS Monograph 138, Reinhold Publishing Co., 1958.

FLUORINATED AROMATIC COMPOUNDS

This article covers three types of fluorinated aromatic compounds: ring-fluorinated aromatics, eg, fluorobenzenes, -biphenyls, -naphthalenes, and fused-ring systems; side-chain fluorinated aromatics, eg, benzotrifluorides and other poly(fluoroalkyl)benzenes; fluorinated nitrogen heterocyclics, eg, fluoropyridines, -quinolines, -pyrimidines, -triazines, and miscellaneous nitrogen systems (see Table 1).

Preparative Methods

Ring-fluorinated aromatics and heterocyclics. Fluoroaromatics are produced on an industrial scale by diazotization of substituted anilines with sodium nitrite in anhydrous hydrogen fluoride, followed by *in situ* decomposition of the aryldiazonium fluoride.

Fluoroaromatics (ortho and para) that were activated by nitro or cyano groups can be prepared by exchange fluorination of the corresponding chloro compounds in dipolar aprotic solvents, eg, dimethyl sulfoxide (DMSO). The role of aprotic solvents permits less solvation of fluoride ion (as compared with protic solvents), a kinetically significant amount of fluoride ion in solution, and greater insolubility of potassium chloride which, in turn, provides a further reaction driving force. The degree of fluorination can be limited by the thermal stability of the solvent or by its reaction with basic potassium fluoride through proton abstraction.

The commercial route to perfluorinated aromatics is based on a multistage saturation–rearomatization process. In the first stage, benzene is fluorinated by a high valency oxidative metal fluoride (cobalt trifluoride) to give a mixture of polyfluorocyclohexanes which is then subjected to a combination of dehydrofluorination (with alkali) and/or defluorination (with heated iron, iron oxide or nickel packing) to give hexa- penta-, and tetrafluorobenzenes.

Side-chain-fluorinated aromatics and heterocyclics. Benzotrifluorides are generally prepared from trichloromethylaromatics with metal fluorides or hydrogen fluoride. Benzotrifluorides are also prepared from aromatic carboxylic acids (and their derivatives) with sulfur tetrafluoride (SF_4).

Ring-Fluorinated Benzenes

Fluorobenzene. C_6H_5F (monofluorobenzene), is a colorless mobile liquid with a pleasant aromatic odor. It has a high order of thermal stability and undergoes no detectable decomposition when kept at 350°C for 24 h at pressures of up to 40.5 MPa; mp −42.22°C, bp 84.73°C, density 1.0183 g/cm^3 at 25°C, refractive index (n_D^{25}) 1.4629.

Reactions

Application of the Friedel-Crafts ketone synthesis to fluorobenzene has commercial importance, eg, the reaction of 4-chlorobutyryl chloride with fluorobenzene gives 4-chloro-4′-fluorobutyrophenone, which is the key intermediate in the synthesis of fluorobutyrophenone tranquilizers such as Haloperidol.

It is difficult to replace the fluorine atom in fluorobenzene by nucleophilic attack. Only a very slow reaction occurs with potassium amide at 80°C or with aqueous ammonia (with a cuprous chloride catalyst) at 250°C. The presence of activating groups, eg, nitro groups, makes aromatic fluorine very reactive in nucleophilic displacement reactions.

Table 1. Physical Properties

Property	Fluorobenzene	Hexafluorobenzene	Benzotrifluoride
melting point, °C	−42.22	5.10	−29.02
boiling point, °C	84.73	80.26	102.1
density, 25°C, g/mL	1.018	1.607	1.181
coefficient of expansion, (°C)$^{-1}$	0.00116		
refractive index, n_D^{25}	1.4629	1.3761	1.4114
solubility in water, g/100 g^a	0.154		0.045

^{a}At 30°C.

Numerous applications have been based on the lability of fluoronitroaromatics; 4-fluoro-3-nitroaniline, a commercial intermediate; 2,4-dinitrofluorobenzene (Sanger's reagent) for amino acid characterization; 4-fluoro-3-nitrophenylazide, an antibody tagging reagent.

Less activated substrates such as fluorohalobenzenes also undergo nucleophilic displacement and thereby permit entry to other useful compounds, eg, bromine is preferentially displaced in *p*-bromofluorobenzene by hydroxyl ion under certain conditions.

Manufacture. Fluorobenzene is produced by diazotization of aniline in anhydrous hydrogen fluoride at 0°C followed by *in situ* decomposition of benzenediazonium fluoride at 20°C. The spent hydrogen fluoride layer (which contains water and sodium bifluoride) from this process is treated with sulfur trioxide, and hydrogen fluoride is distilled for recycle to the next batch.

Hexafluorobenzene. C_6F_6 (perfluorobenzene), is a colorless liquid with a sweet odor, miscible with many organic solvents and immiscible with water. Table 1 lists some physical properties. Hexafluorobenzene has good thermal stability; it forms a 1:1 molecular complex with benzene, mp 23.7°C.

Reactions. Hexafluorobenzene is very susceptible to attack by nucleophilic agents to given pentafluorophenyl compounds of the general formula, C_6F_5X, where X is OCH_3, NH_2, OH, SH, $NHNH_2$, H, C_6H_5, CH_3, etc. Radicals attack hexafluorobenzene and cause displacement of a fluorine atom.

Manufacture. The commercial process features a three-stage saturation–rearomatization technique using benzene and fluorine gas as raw materials. (*1*) Benzene is fluorinated with cobalt trifluoride at about 150°C giving a mixture of fluorinated cyclohexanes containing one to four hydrogen atoms. (*2*) The mixture of polyfluorocyclohexanes is heated with an alkali-metal hydroxide. During this process dehydrofluorination occurs and a mixture of polyfluorocyclohexenes and -hexadienes results. (*3*) The next stage is the defluorination of the product produced in stage (*2*), accomplished by passing the product of stage (*2*), in vapor form, through a reactor packed with a defluorinating agent at 400–600°C. The product from stage (*3*) consists of hexafluorobenzene, pentafluorobenzene, and the isomeric tetrafluorobenzenes which are purified by fractional distillation and chemical methods.

The development of commercial routes to hexafluorobenzene included an intensive study of the chemistry of its derivatives. Particularly noteworthy was the development of high temperature lubricants, heat-transfer fluids, and radiation-resistant polymers.

Derivatives. Numerous derivatives of hexafluorobenzene have utility in analytical and other applications.

Pentafluorophenol has been prepared from the reaction of hexafluorobenzene with potassium hydroxide in *t*-butyl alcohol. Fluorophenyl esters prepared from pentafluorophenol illustrate the key features of a rapid stepwise peptide-synthesis technique. Commercial high performance elastomers based on copolymerization of tetrafluoroethylene, perfluoro(methyl vinyl ether), and a third monomer incorporating a pentafluorophenoxy group as a cure site, give vulcanizates with good chemical and fluid resistance and high temperature oxidative resistance (see Elastomers, synthetic).

Pentafluorotoluene has been prepared from the reaction of methyllithium with hexafluorobenzene or from pentafluorophenylmagnesium bromide with dimethyl sulfate. Products based on pentafluorotoluene are used as derivatizing agents for gas-chromatographic analysis of biologically active compounds via electron-capture detection.

Fluorobiphenyls

Recent pharmaceutical applications for fluorobiphenyls include analgesics and anti-inflammatory agents such as Diflunisal (**1**) and Flurbiprofen (**2**).

(**1**) Diflunisal   (**2**) Flurbiprofen

Mono- and difluorobiphenyls can be prepared by the Balz-Schiemann reaction (or modification in HF); eg, 4,4'-difluorobiphenyl was formed in 80% yield from 4,4'-diaminobiphenyl by the Balz-Schiemann reaction. 2,4-Difluorobiphenyl, a key precursor to the analgesic/anti inflammatory agent Flurbiprofen is prepared by diazotization of 2,4-difluoroaniline and subsequent coupling with benzene.

Side-Chain-Fluorinated Aromatics

Benzotrifluoride. $C_6H_5CF_3$ is a colorless liquid. Physical properties are listed in Table 1.

Reactions. Benzotrifluoride undergoes electrophilic substitution reactions, eg, halogenation, nitration, typical of an aromatic containing a strong electron-withdrawing group. The trifluoromethyl group is meta directing.

Manufacture. Benzotrifluoride is produced by the high pressure reaction of benzotrichloride with anhydrous hydrogen fluoride (AHF).

Other important side-chain aromatics. These include 4-chlorobenzotrifluoride, whose derivatives are widely used as herbicides, eg, Trifluralin, Dinitramine, and Fluorodifen. Other benzotrifluoride derivatives are also widely used as drugs, eg, Trifluromazine, Flumethiazide, and Trifluperidol; and as germicides, eg, Flurosalan and Cloflucarban.

Fluorinated Nitrogen Heterocycles

Examples of this class include monofluoropyridines, eg, 2-, 3- and 4-fluoropyridine; difluoropyridines, eg, 2,6-difluoropyridine alkaline hydrolysis of which forms 2-fluoro-6-hydroxypyridine, a precursor to Dowco 275, an insecticide–nematocide; tri-, tetra-, and pentafluoropyridines; and perfluoroalkylpyridines.

Fluoroquinolines

Some ring- and trifluoromethylquinolines exhibit biological properties, eg, Mefloquine is an antimalarial.

Fluoropyrimidines

5-Fluoropyrimidine derivatives are of tremendous importance in cancer chemotherapy, eg, 5-fluorouracil (5-FU); as antifungal agents, eg, 5-fluorocytosine and 2'-deoxy-5-fluorouridine; as antiviral and antineoplastic agents (see Chemotherapeutics, antibacterial and antimycotic; Chemotherapeutics, antimitotic; Chemotherapeutics, antiviral).

Ring-fluorinated triazines. Triazines such as 2,4,6-trifluoro-1,3,5-triazine (cyanuric fluoride) are the basis for fiber-reactive dyes.

Uses

The ring-fluorinated aromatics are broadly applied in pharmaceutical areas, eg, in tranquilizers (eg, haloperidol and fluspirilene), hypnotics (eg, flurazepam hydrochloride), sedatives, antibacterial agents (eg, floxacillin) and anti-inflammatory agents (eg, flazalone).

Trifluoromethyl aromatics are widely used in the production of drugs, crop-protection chemicals, germicides, dyes, etc.

Commercial uses for fluorinated nitrogen heterocyclics include drugs, and reactive dyestuffs. Numerous patents have been issued for trifluoromethylpyridine derivatives as crop-protection chemicals.

MAX M. BOUDAKIAN
Olin Chemicals

A.E. Pavlath and A.L. Leffler, *Aromatic Fluorine Compounds*, ACS Monograph No. 155, Reinhold Publishing Co., New York, 1962.

G. Schiemann and B. Cornils, *Chemie und Technologie Cyclischer Fluorverbindungen*, F. Enke Verlag, Stuttgart, 1969.

R.E. Banks, ed., *Organofluorine Chemicals and Their Industrial Applications*, Society of Chemical Industry, Ellis Horwood Ltd., Chichester, UK, 1979.

FLUOROALKOXYPHOSPHAZENES

The phosphazenes are compounds containing the —P=N— moiety as the basic unit in either cyclic or linear structures. In fluoroalkoxyphosphazenes, fluoroalkoxy groups are attached to the phosphorus atom of the basic unit. Commercial importance was established with a poly(fluoroalkoxyphosphazene) elastomer introduced by the Firestone Tire and Rubber Co. under the trademark PNF (phosphonitrilic fluoro-elastomer).

PNF elastomers are serviceable at a broad range of temperatures and have excellent solvent resistance and good mechanical properties. At present, fluorosilicone is the only other oil-resistant elastomer that functions below $-40°C$. However, its flex and tear resistance are inferior. PNF is tougher than fluorosilicone and complements the currently available solvent-resistant elastomers.

Preparation

The phosphazene system is obtained by the reaction of phosphorus pentachloride with ammonium chloride.

$$n \text{ PCl}_5 + n \text{ NH}_4\text{Cl} \longrightarrow \underset{\substack{| \\ \text{P}=\text{N}}}{\overset{\text{Cl} \quad \text{Cl}}{}}_n + 4n \text{ HCl}$$

A mixture of oligomers is obtained whose composition can be varied widely by control of reaction conditions.

High molecular weight, linear phosphazene polymers form at 250°C from the cyclic trimer.

(1) hexachlorocyclotriphosphazene   (2) poly(dichlorophosphazene)

This ring-opening polymerization has a classical ring–chain type of equilibrium.

Alkoxyphosphazenes are generally derived from the reaction of poly-(dichlorophosphazene) with sodium alkoxides:

$$\underset{\substack{| \\ \text{P}=\text{N}}}{\overset{\text{Cl} \quad \text{Cl}}{}}_n + 2n \text{ NaOR} \longrightarrow \underset{\substack{| \\ \text{P}=\text{N}}}{\overset{\text{RO} \quad \text{OR}}{}}_n + 2n \text{ NaCl}$$

Inert Fluids

Fluid fluoroalkoxyphosphazenes are readily obtained from reaction of sodium fluoroalkoxides with either the cyclic trimeric or tetrameric dichlorophosphazene. The oily products possess remarkable chemical and thermal stability.

Homopolymers

Fluoroalkoxyphosphazene homopolymers are obtained by substitution of poly(dichlorophosphazene) with a single sodium fluoroalkoxide. They generally resist solvents, chemicals, and water. These films are unaffected by moisture, glacial acetic acid, alcohols, pyridine, concentrated caustic solution, and high intensity uv radiation.

Copolymers

Reaction of two fluoroalkoxides of different chain lengths with poly(dichlorophosphazene) inhibits crystallization and, therefore, produces an elastomer. Poly(trifluoroethoxyheptafluorobutoxyphosphazene) has good high and low temperature properties. In addition, prolonged contact with boiling water, common organic solvents, concentrated sulfuric acid, or concentrated potassium hydroxide has no apparent effect.

Manufacture

The polymeric $-\overset{|}{\underset{|}{\text{P}}}=\text{N}-$ backbone is attained by thermal polymerization of hexachlorocyclotriphosphazene. Subsequently, the chlorines are replaced nearly quantitatively by refluxing poly(dichlorophosphazene) in benzene with the sodium fluoroalkoxides in THF.

$$\underset{\substack{| \\ \text{P}=\text{N}}}{\overset{\text{Cl} \quad \text{Cl}}{}}_n + n \text{ NaOCH}_2\text{CF}_3 + n \text{ NaOCH}_2\text{CF}_2\text{CF}_2\text{CF}_3 \longrightarrow \underset{\substack{| \\ \text{P}=\text{N}}}{\overset{\text{CF}_3\text{CH}_2\text{O} \quad \text{OCH}_2\text{CF}_2\text{CF}_2\text{CF}_3}{}}_n + 2n \text{ NaCl}$$

poly(trifluoroethoxyheptafluorobutoxyphosphazene)
(an elastomer, $T_g = -77°C$)

Uses

PNF is used in the aerospace, aircraft, industrial, automotive, and oil-exploration industries. Products utilizing PNF include oil seals, gaskets, diaphragms, hoses, shock mounts, foams and coatings. No commercial applications exist for the inert fluids.

DAVID P. TATE
T.A. ANTKOWIAK
Firestone Tire and Rubber Co.

H.R. Allcock, *Phosphorus–Nitrogen Compounds*, Academic Press, Inc., New York, 1972, p. 356.

D.P. Tate, *J. Polym. Sci. Symp.* **48**, 33 (1974).

G.S. Kyker and T.A. Antkowiak, *Rubber Chem. Technol.* **47**, 32 (1974).

PERFLUOROALKYLENETRIAZINES

Perfluoroalkylene triazines consist of *sym*-triazine rings with per-fluorinated substituents in the 2, 4, and 6 positions.

Fluids

Properties. Mp -100 to 30°C; bp 98 to 407°C and above; density 1.5 to 11.8 g/cm³; viscosity 0.3 to >1000 m²/s $\times 10^6$. The perfluoroalkylene triazine fluids are inert toward oxidizing agents and strong acids such as H_2SO_4, P_2O_5, and HCl, but are attacked by HNO_3, amines, and others strong alkalies, especially in the presence of water. Compounds of silver, tin, antimony, arsenic, and copper cause rearrangements of the substituents of the triazine ring at high temperatures.

Elastomers

Properties. High polymers in the range of 150,000–200,000 $\overline{M}_n$ (number-average molecular weight) are water-clear tough gums that can be processed on conventional equipment associated with the rubber industry. Molded products have 1000 h life at 260°C and shorter times in the 350 to 400°C range.

Health and Safety

Perfluoroalkylenetriazine fluids and elastomers are inert and no health or safety problems have been reported, however, the high vapor pressure of stable liquids is reason for cautious handling.

Uses

The tris-trifluoromethyl-*sym*-triazine as well as the tris-perfluoroethyl-, propyl-, heptyl-, and nonyl-*sym*-triazines are utilized as mass spectroscopic standards. The perfluoroalkylene(perfluoroalkyl)-*sym*-triazines and perfluoroalkyleneoxy(perfluoroalkyoxy)-*sym*-triazines are suitable for nonflammable hydraulic fluids at -25 to $+300°C$. Elastomeric polymers, with their excellent rubberlike properties, are potentially useful as electrical wire insulation and seals for hydraulic, lubricating, and fuel systems of aircraft (see Elastomers, synthetic).

The fluids can be made available for specific applications that would justify the relatively high cost. The elastomers are attacked by high temperature, high humidity environments. They have no specific uses.

WARREN R. GRIFFIN
Air Force Materials Laboratory

C.E. Snyder, *ASLE Trans.* **14**(3), 237 (1971).

C.E. Snyder, Jr., and C.A. Svisco, *Lub Engr.* **35**, 451 (1979).

W.R. Griffin, *Rubber Chem. Technol.* **39**(Pt. 2), 1175 (1966).

PERFLUOROALKANESULFONIC ACIDS

Trifluoromethanesulfonic Acid

Properties. This acid is a stable, hygroscopic liquid which fumes in air. An equimolar amount of water with the acid results in a stable, distillable monohydrate, mp 34°C, bp 96°C at 0.13 kPa (1 mm Hg). Measurement of conductivity of strong acids in acetic acid has shown the acid to be one of the strongest simple protonic acids known, similar to fluorosulfonic and perchloric acid. Trifluoromethanesulfonic acid is miscible in all proportions in water, and is soluble in many polar organic solvents such as dimethylformamide, dimethyl sulfoxide, and acetonitrile. In addition, it is soluble in alcohols, ketones, ether, and esters, but these generally are not suitably inert solvents.

Derivatives. Alkyl esters of trifluoromethanesulfonic acid, commonly called triflates, have been prepared primarily from the silver salt and an alkyl iodide, or by reaction of the anhydride with alcohol.

Higher Perfluoroalkanesulfonic acids

The longer-chain perfluoroalkanesulfonic acids are very hygroscopic oily liquids. Distillation of the acid from a mixture of its salt and H_2SO_4 gives hydrated mixtures with melting points above 100°C. Many of the higher perfluoroalkanesulfonic acids have been prepared by electrochemical fluorination. The longer-chain acids, particularly $C_8F_{17}SO_3H$ and higher, are surface-active agents in aqueous media.

Health and Safety

Extreme caution is needed when handling perfluoroalkanesulfonic acids. Use normal procedures for treating strong acid burns.

Uses

Derivatives of the acids have unique surface-active properties and have formed the basis for a number of commercial fluorochemical surfactants.

RICHARD A. GUENTHNER
3M Company

R.N. Hazeldine and J.M. Kidd, *J. Chem. Soc.*, 4228 (1954).

R.D. Howells and J.D. McCown, *Chem. Rev.* **77**, 69 (1977).

R.A. Guenthner and M.L. Vietor, *Ind. Eng. Chem. Prod. Res. Dev.* **1**, 165 (1962).

PERFLUOROEPOXIDES

Almost all work on perfluoroepoxides has been with only three compounds, tetrafluoroethylene oxide (TFEO); hexafluoropropylene oxide (HFPO); and perfluoroisobutylene oxide (PIBO). The main use of these epoxides is as intermediates in the preparation of other fluorinated monomers.

Physical Properties

In general, the perfluoroepoxides have boiling points that are quite similar to those of the corresponding fluoroalkenes. They can be distinguished easily from the olefins by ir spectroscopy, specifically by the lack of olefinic absorption and the presence of a characteristic epoxide band in the 1550 cm^{-1} region.

Chemical Properties

There are three general reactions of perfluoroepoxides: pyrolyses (thermal reactions), electrophilic reactions, and reaction with nucleophiles and bases.

Nucleophilic reactions are by far the most important. The strong electronegativity of fluorine permits the facile reaction of perfluoroepoxides with nucleophiles. Nucleophilic attack on the epoxide ring takes place at the more highly substituted carbon atom to give ring-opened products.

Preparation

The reaction of perfluoroalkenes with alkaline hydrogen peroxide is a good general method for the preparation of perfluoroepoxides with the exception of the most reactive of the series, TFEO. Usually, the perfluoroalkene is allowed to react at low temperatures with a mixture of aqueous hydrogen peroxide, base, and a cosolvent to give a low conversion of the alkene. The direct oxidation of fluoroalkenes is also an excellent general synthesis procedure for the preparation of perfluoroepoxides. This method makes use of the low reactivity of the epoxide products to both organic and inorganic free radicals. The oxidation may be carried out with an inert solvent thermally, with a sensitizer such as bromine, with ultraviolet radiation, or over a suitable catalyst.

Recently, internal and cyclic fluoroalkenes have been epoxidized in high yield by reaction with hypochlorites.

Hexafluoropropylene Oxide

HFPO is the most important of the perfluoroepoxides and has been synthesized by almost all of the methods noted above. HFPO reacts with a large number of acyl fluorides in a general reaction to give 2-alkoxytetrafluoropropionyl fluorides which in turn may be converted to trifluorovinyl ethers. These ethers readily copolymerize with tetrafluoroethylene and other fluoroalkenes to give commercially significant plastics, elastomers, and ion-exchange resins, eg, Teflon PFA, Kalrex, and Nafion (see Perfluorinated ionomer membranes).

PAUL R. RESNICK
E.I. du Pont de Nemours & Co., Inc.

D. Sianesi and co-workers, *Gazz. Chem. Ital.* **98**, 265, 277, 290 (1968).

P. Tarrant and co-workers, *Fluorine Chem. Rev.* **5**, 77 (1971).

N. Ishikawa, *Yuki Gosei Kagaku Kyokai Shi* **35**, 131 (1977).

P.L. Coe, A.W. Mott, and J.C. Tatlow, *J. Fl. Chem.* **20**, 243 (1982).

POLYTETRAFLUOROETHYLENE

Polytetrafluoroethylene (PTFE), a perfluorinated straight-chain high polymer having the chemical formula $(CF_2CF_2)_n$, is made by polymerizing the tetrafluoroethylene (TFE) monomer. The white-to-translucent solid polymer has an extremely high molecular weight—in the 10^6–10^7 range—and, consequently, has a viscosity of about 10 GPa·s (10^{11} P) at 380°C. Its high thermal stability results from the strong carbon–fluorine bond and characterizes PTFE as a very useful high temperature polymer. Its heat resistance, chemical inertness, electrical insulation properties, and its low coefficient of friction in a very wide temperature range make PTFE the most outstanding plastic in the industry.

Commercially available PTFE is manufactured by two entirely different polymerization techniques that result in two completely different types of chemically identical polymers. Suspension polymerization produces a granular resin PTFE and emulsion polymerization produces the coagulated dispersion that is often referred to as a fine powder, and PTFE dispersion.

Because of its chemical inertness and high molecular weight, PTFE is a nonflowing material that does not undergo conventional polymer-fabrication techniques. Therefore, an extensive processing technology developed.

The suspension-polymerized PTFE polymer (referred to as granular PTFE) generally is fabricated using modified powder metallurgy techniques. Emulsion-polymerized PTFE behaves entirely differently from granular PTFE and coagulated dispersions of PTFE are processed by a cold extrusion process (like lead). Stabilized PTFE dispersions, made by emulsion polymerization, usually are processed according to latex processing techniques (see Latex technology).

Manufacturers of PTFE include Allied Chemical (Halon), Daikin Kogyo (Polyflon), DuPont (Teflon), Hoechst (Hostaflon), ICI (Fluon), Montecatini (Algoflon), Ugine Kuhlman (Soreflon), and the USSR (Fluoroplast). The People's Republic of China recently introduced some PTFE products.

Monomer

The tetrafluoroethylene manufacturing process involves the following steps: $CaF_2 + H_2SO_4 \rightarrow CaSO_4 + 2\ HF$; $CH_4 + 3\ Cl_2 \rightarrow CHCl_3 + 3\ HCl$; $CHCl_3 + 2\ HF \rightarrow CHClF_2 + 2\ HCl$; $2\ CHClF_2 \rightarrow CF_2{=}CF_2 + 2\ HCl$. TFE is a colorless, tasteless, odorless, and relatively nontoxic gas at RT. Although TFE is relatively nontoxic, other fluorocarbon compounds that may be extremely toxic (such as perfluoroisobutylene) may contaminate it at high temperatures.

Basic polymer properties. The properties discussed in this section are those of polymers prepared by suspension polymerization, and those prepared by emulsion polymerization, since the latter relate to the basic structure of PTFE. The two valence forces binding the PTFE structure are the carbon–carbon bonds, which form the backbone of the polymer chain, and the carbon–fluorine bonds. Both of these chemical bonds are extremely strong and are the key contributors in imparting the outstanding combination of properties of PTFE. The fluorine atoms form a protective sheath over the chain of carbon atoms. If the atoms attached to the carbon-chain backbone were either smaller or larger than fluorine, the sheath would not form a regular, uniform cover. This sheath shields the carbon chain from attack by various chemicals and gives PTFE its chemical inertness and stability. It also lowers the surface energy giving PTFE a low coefficient of friction and nonstick properties.

Extensive work has been reported on transitions. Effects of structural changes on properties, eg, specific heat, specific volume, and/or dynamic mechanical and electrical properties, are observed at various temperatures. PTFE transitions occur at specific combinations of temperature and at the frequency of mechanical or electrical vibrations. Transitions, sometimes called dielectric relaxations, can cause wide fluctuations in the value of the dissipation factor.

Mechanical properties of PTFE are highly influenced by many processing variables, eg, preforming pressure, sintering temperature and time, cooling rate, void content, and crystallinity level. Some of the more frequently used average values for molded and sintered samples of PTFE have been listed in Table 1.

Filled resin properties. Besides retaining the desirable properties of uncompounded resins, filled compositions meet the requirements of an increased variety of mechanical, electrical, and chemical applications. Table 2 lists a number of physical properties of a variety of filled granular compounds.

Chemical properties. Vacuum thermal degradation of PTFE results in monomer formation. Mass spectroscopic analysis of PTFE thermal degradation shows degradation begins at ca 440°C, peaks at 540°C, and continues until 590°C. Radiation degrades PTFE. In the absence of oxygen, stable secondary radicals are produced. An increase in stiffness

Table 1. Typical Mechanical Properties of PTFE Resins

Property	Granular resin	Fine powder
tensile strength at 23°C, MPa[a]	7–28	17.5–24.5
elongation at 23°C, %	100–200	300–600
flexural strength at 23°C, MPa[a]	does not break	
flexural modulus at 23°C, MPa[a]	350–630	280–630
impact strength, J/m[b], at 21°C	106.7	
hardness durometer, D	50–65	50–65
compressive stress, MPa[a]		
at 1% deformation at 23°C	4.2	
at 1% offset at 23°C	7.0	
coefficient of linear thermal expansion per °C, 23–60°C	12×10^{-5}	
thermal conductivity, 4.6 mm, W/(m·K)	0.24	
deformation under load, %		
26°C, 6.86 MPa[a], 24 h		2.4
26°C, 13.72 MPa[a], 24 h	15	
water absorption	< 0.01	< 0.01
static coefficient of friction against polished steel	0.05–0.08	

[a] To convert MPa to psi, multiply by 145.
[b] To convert J/m to (lbf·ft)/in., divide by 53.38.

Table 2. Filled PTFE Compounds and Their Properties

Property	Unfilled PTFE	15 wt% glass fiber	25 wt% glass fiber	15 wt% graphite	60 wt% bronze
specific gravity	2.18	2.21	2.24	2.16	3.74
tensile strength, MPa[a]	28	25	17.5	21	14
elongation, %	350	300	250	250	150
stress at 10% elongation, MPa[a]	11	8.5	8.5	11	14
thermal conductivity, mW/(m·K)	0.244	0.37	0.45	0.45	0.46
creep modulus, kN/m[b]	2	2.21	2.1	3.4	6.2
hardness, Shore durometer, D	51	54	57	61	70
Izod impact, J/m[c]	152	146	119		
wear factor, 1/PPa[d]	5013	28.0	26.2	102	12
coeff. of friction static, 3.4 MPa[a] load	0.08	0.13	0.13	0.10	0.10

[a] To convert MPa to psi, multiply by 145.
[b] To convert kN/m to lbf/in., divide by 0.175.
[c] To convert J/m to (lbf·ft)/in., divide by 53.38.
[d] To convert 1/PPa to (cm·s)/(bar·m·min), multiply by 6×10^{23}.

in the vacuum-irradiated material indicates cross-linking. In air or oxygen, reactions take place and accelerated scission and rapid degradation occurs.

Mechanical strength and chemical and thermal stability in combination with excellent electrical properties characterize PTFE as a very versatile electrical insulator. It does not absorb water; therefore, volume resistivity remains unchanged even after prolonged soaking in water.

Fabrication Techniques

Granular resins are sold in different forms. Generally, an optimum balance between resin handleability and product properties is sought. A resin that can flow freely is used in small moldings and automatic moldings. A finely divided resin that is more difficult to handle provides a good leveling in large moldings and develops superior properties in sintered articles; it is used for large billet and sheet-molding operations. Reduced crystallinity and superior handleability make a presintered resin an ideal product for ram-extrusion applications.

Many manufacturers of PTFE give detailed descriptions of molding equipment and molding procedures which include automatic molding, isostatic molding, sintering and ram extrusion. The most frequently used molds are round piston molds for the production of solid or hollow cylinders.

Fine powder resins are extremely sensitive to shear. As a result, the manufacturer and the processor must take great care to handle the powder gently. Since fine powder is suitable for the manufacture of tubing and wire insulation and since compression molding is not suitable for the preparation of thin-walled continuous lengths, a post-extrusion process is used.

PTFE dispersions generally contain 30–60 wt% polymer particles and some surfactant in the aqueous medium. The type of surfactant and the particle characteristics depend on the dispersion application. These dispersions are applied to various substrates by spraying, flow-coating, dipping, coagulating, or electrodepositing.

Effects of fabrication on physical properties. The basic factors that affect physical properties are molecular weight, which can be decreased by degradation but cannot be increased during processing; void content; and crystallinity. These factors can be controlled during molding through choice of resin and fabricating conditions.

Safety and Health

Exposure to PTFE can arise from the ingestion, skin contact, or inhalation of the unheated material, which is physiologically inert, and from the inhalation of the heated polymer. Only the heated polymer is a source of a possible health hazard.

Applications

Consumption of PTFE continuously grows as new applications are developed. The chief applications of these resins are electrical, mechanical, and chemical. Electrical applications consume half of the PTFE

Table 3.   Polytetrafluoroethylene Resins and Their Uses

Type	Resin grade	Method of processing	Description	Main uses
granular	agglomerates	molding, preforming-sintering, ram extrusion	free-flowing powder	gaskets, packing, seals, electronic components, bearings, sheet, rod, heavy-wall tubing; tape and molded shapes for chemical, electrical, mechanical, and nonadhesive (adherent) applications
	coarse	molding, preforming-sintering	granulated powder	tape, molded shapes for chemical mechanical, electrical, and nonadhesive applications
	finely divided	molding, preforming-sintering	powder for use where highest quality, void-free moldings are required	molded sheets, tape, wire wrapping, tubing, gaseting
	presintered	ram extrusion	granular, free flowing powder	rods and tubes
fine powder	high reduction-ratio resin	paste extrusion	agglomerated powder	wire coating, thin-walled tubing
	medium reduction-ratio resin	paste extrusion	agglomerated powder	tubing, pipe, overbraided hose, spaghetti tubing
	low reduction-ratio resin	paste extrusion	agglomerated powder	thread-sealant tape, pipe-liners, tubing, porous structures
dispersion	general purpose	dip coating	aqueous dispersion	impregnation, coating, packing
	coating	dip coating	aqueous dispersion	film, coating
	stabilized	cocoagulation	aqueous dispersion	bearings

produced, and mechanical and chemical applications share equally the other half (see Table 3).

S.V. GANGAL
E.I. du Pont de Nemours & Co., Inc.

D.I. McCane in N.M. Bikales, ed., *Encyclopedia of Polymer Science and Technology*, Vol. 13, Wiley-Interscience, New York, 1970, pp. 623–670.

C.A. Sperati in J. Brandrup and E.H. Immergut, eds., *Polymer Handbook*, 2nd ed., John Wiley & Sons, Inc., New York, 1975, p. V-29–36.

FLUORINATED ETHYLENE-PROPYLENE COPOLYMERS

Low melting perfluorinated copolymers of tetrafluoroethylene (TFE) and hexafluoropropylene (HFP) are produced as alternatives to polyte-trafluroethylene (PTFE) (qv), which has a high melt viscosity. The copolymers retain most of the desired properties of PTFE and, unlike the latter compounds, can be processed by conventional melt-processing techniques. They also have excellent thermal stability and chemical inertness, and their dielectric constants and dissipation factors are low, and remain unchanged when subjected to wide ranges of temperature and frequency. The copolymers retain mechanical properties up to 200°C, even in continuous service, better than most plastics. They react with fluorine, fluorinating agents, and molten alkali metals at high temperatures. Fluorinated ethylene–propylene copolymers are commercially available under the DuPont trademark Teflon FEP fluorocarbon resin. A product similar to Teflon FEP is manufactured by Daikin Kogyo of Japan and sold under the trademark Neoflon.

Monomers

HFP, bp $-29.4°C$, does not homopolymerize easily, so it can be stored as a liquid. HFP undergoes many addition reactions typical of an olefin. Some reactions that HFP undergoes include preparation of linear dimers and trimers, and cyclic dimers; decomposition at 600°C with subsequent formation of octafluoro-2-butene and octafluoroisobutylene; oxidation with formation of the epoxide (1), an intermediate for a number of perfluoroalkyl perfluorovinyl ethers; and homopolymerization to low molecular weight liquids and high molecular weight solids.

$$CF_3CF{-}CF_2$$
$$O$$
(1)

Table 1.   Typical Properties of FEP Copolymer Resins

Property	Low MV resin	Extrusion trade	High MV resin
specific gravity	2.12–2.17	2.12–2.17	2.12–2.17
melting point, °C	253–272	262–282	262–282
tensile strength at 23°C, MPa[a]	20.68	20.68	20.68
elongation at 23°C, %	300	300	300
flexural modulus 23°C, MPa[a]	655	655	754
dielectric constant at 60 Hz to 1 GHz	2.1	2.1	2.1
dielectric strength for a short time at 0.254 mm, kV/mm	82.7	82.7	82.7
water absorption, %	< 0.01	< 0.01	< 0.01

[a] To convert MPa to psi, multiply by 145.

Copolymers

Aqueous and nonaqueous dispersion polymerizations appear to be the most convenient routes to commercial production of HFP–TFE copolymers. The polymerization conditions used to produce these polymers are very similar to those used in TFE homopolymer dispersion polymerization. The copolymer of HFP–TFE is a true copolymer, and HFP units add to the growing chains at random intervals. The optimization composition of the copolymer requires that the retention of mechanical properties is in the usable range and the melt viscosity of the product is low enough for easy melt processing.

Fluorinated Ethylene–Propylene Copolymers

FEP copolymers are available in four forms: the low melt viscosity, extrusion grade, high melt viscosity resins, and the FEP copolymer dispersion. Typical properties of these products are listed in Table 1.

FEP copolymer dispersion is available as a 55 wt% aqueous dispersion containing 6% nonionic surfactant (on a solids basis) and a small amount of anionic dispersing agent. The average particle size of this dispersion is ca 0.2 μm.

Fabrication.   Standard thermoplastic processing techniques can be used to fabricate useful products from FEP. Factors of primary importance in FEP fabrication include avoiding thermal degradation of the polymer, and maintaining a homogeneous structure and good surface quality. Techniques include injection molding, extrusion, and dispersion processing.

The principal electrical applications of FEP include hook-up wire, interconnecting wire, coaxial cable, computer wire, thermocouple wire, and molded electrical parts. Principal chemical applications are lined pipes and fittings, overbraided hose, heat exchangers, and laboratory ware. The primary mechanical uses are antistick applications, such as conveyor belts and roll covers (see Abherents).

S.V. GANGAL
E.I. du Pont de Nemours & Co., Inc.

D.I. McCane in N.M. Biklaes, ed., *Encyclopedia of Polymer Science and Technology*, Vol. 13, John Wiley & Sons, Inc., New York, pp. 654–670.

Teflon Fluorocarbon Resins, Mechanical Design Data, 2nd ed., E.I. du Pont de Nemours & Co., Inc., Wilmington, Del., 1965.

Electrical/Electronic Design Data for Teflon, E.I. du Pont de Nemours & Co., Inc., Wilmington, Del.

TETRAFLUOROETHYLENE COPOLYMERS WITH ETHYLENE

Ethylene–tetrafluoroethylene copolymer is isomeric with poly(vinylidene fluoride) and is available commercially as du Pont's Tefzel, Asahi's Aflon COP, and Hoechst's Hostaflon ET. The tetrafluoroethylene segments of the polymer molecules account for about 75 wt% of a 1:1 copolymer:

$${+}CH_2CH_2CF_2CF_2{+}_n \qquad {+}CH_2CF_2{+}_n$$
poly(ethylene-*co*-tetrafluoroethylene)   poly(vinylidene fluoride)

The properties of ETFE copolymers vary considerably with composition; those containing 40–90 wt% tetrafluoroethylene soften at 200–300°C, depending on their composition. These copolymers exhibit good dimensional stability below their softening points and have tensile strengths of 41–70 MPa (6,000–10,200 psi), ie, between two and three times the values of either polytetrafluoroethylene or polyethylene. Resistance to oxidation increases with increasing tetrafluoroethylene content. For an article of commerce, a copolymer with a mole ratio of almost 1 : 1 of tetrafluoroethylene to ethylene gives the best balanced composition with respect to chemical resistance and physical characteristics. Several commercial polymers also contain termonomers to enhance performance properties.

Monomer

Tetrafluoroethylene, perfluoroethylene, $CF_2{=}CF_2$, is prepared on a commercial scale by the pyrolysis of chlorodifluoromethane ($CHClF_2$). Material that is of purity suitable for granular or dispersion polymerizations is acceptable for copolymerization with ethylene. Polymerization-grade ethylene is suitable for copolymerization with tetrafluoroethylene. Ethylene and tetrafluoroethylene are copolymerized by suspension or emulsion techniques, generally under such conditions that tetrafluoroethylene may homopolymerize but not ethylene.

Polymer

The unit cell of an alternating copolymer of ethylene and tetrafluoroethylene has been determined by x-ray diffraction. The unit cell is probably either orthorhombic or monoclinic and has the following parameters: $a = 0.96$ nm, $b = 0.925$ nm, $c = 0.50$; ($\gamma = 96°C$).

Ethylene–tetrafluoroethylene resins are less dense, tougher, stiffer, and exhibit a higher tensile strength and creep resistance than polytetrafluoroethylene and fluorinated ethylene–propylene copolymer. They are, however, similarly ductile. Ethylene–tetrafluoroethylene copolymer displays the relatively nonlinear stress–strain relationships characteristic of nearly all ductile materials.

Thermal stability, chemical inertness, strength, toughness, nonflammability, clarity, excellent electrical properties, weatherability, and ease of processing are useful properties of these polymers.

Ethylene–tetrafluoroethylene copolymers are high temperature thermoplastics that can be readily processed by conventional methods including extrusion and injection molding.

Health and Safety Factors

Initial tests in exposing rats to the pyrolysis decomposition products of Tefzel 200 at 400°C suggest a close comparison with Teflon FEP. Accordingly, all safety precautions recommended for processing of Teflon FEP should be followed.

Uses

Modified ETFE copolymers have gained rapid acceptance as a high performance insulation for wire and cable (see Insulation, electric).

RICHARD L. JOHNSON
E.I. du Pont de Nemours & Co., Inc.

F.C. Wilson and H.W. Starkweather, *J. Polym. Sci. Poly. Phys. Ed.* 11, 919 (1973).

Handling and Use of Teflon Fluorocarbon Resins at High Temperatures, E.I. du Pont de Nemours & Co., Inc., Wilmington, Del., 1961.

J.C. Reed and J.R. Perkins, *paper presented at 21st International Wire and Cable Symposium, Atlantic City, N.J., Dec. 6, 1972.*

TETRAFLUOROETHYLENE COPOLYMERS WITH PERFLUOROVINYL ETHERS

Teflon PFA fluorocarbon resins are thermoplastic polymers that combine the ease of melt fabrication with properties rivaling those of polytetrafluoroethylene (TFE). Introduced by DuPont in 1972, these polymers combine the carbon–fluorine backbone ($-CF_2-$) of polytetrafluoroethylene resins with a perfluoroalkoxy (PFA) side chain:

$$-CF_2CF_2\overset{\displaystyle F}{\underset{\displaystyle OR_f}{C}}CF_2CF_2-$$

portion of a PFA molecule
$R_f = CF_3, C_2F_5, C_3F_7$, etc

The pendant perfluoroalkoxy groups exert their greatest effect on the crystallinity of the polymer. Unlike polytetrafluoroethylene homopolymers that have virtually no melt flow at or above their melting point, Teflon PFA resins are readily melt-processible by conventional extrusion or injection and transfer molding.

Monomer

Tetrafluoroethylene and perfluoroalkyl vinyl ethers copolymerize in either aqueous or nonaqueous media. Soluble initiators are preferred, such as inorganic persulfates for aqueous media or fluorinated acyl peroxides for nonaqueous media.

Polymer

Table 1 summarizes the physical properties of Teflon PFA resin. The dielectric constant is slightly higher than that of Teflon TFE, and the dielectric losses are somewhat higher. The high temperature properties enhance the attractive electrical characteristics of the resin. Weather-Ometer tests indicate virtually unlimited outdoor life. All the perfluorinated resins are severely degraded at 5×10^5 Gy (5×10^7 rad). The friction coefficient of Teflon PFA is 0.236 compared to a control of Teflon FEP at 0.333. Teflon PFA may be pigmented to a wide variety of colors, but pigments with good heat stability must be selected.

Teflon PFA fluorocarbon resin can be fabricated by conventional melt-processing techniques (see Plastics processing). Equipment should be constructed of corrosion-resistant materials and be capable of operating between 315 and 425°C.

Safety Practices

The usual safety practices applied to fluorocarbon resins are adequate for Teflon PFA. Adequate ventilation is essential in processing areas and contamination of smoking tobacco or cigarettes should be avoided.

Uses

Teflon 340 is the general-purpose grade of Teflon PFA designed for extended service in environments that demand high levels of chemical, thermal, and mechanical performance. It is widely used in extrusion applications such as extruded tubing, shapes, primary insulation, and

Table 1. Physical Properties of Teflon PFA Fluorocarbon Resins

Property	Teflon 340	Teflon 350
nominal melting point, °C	302–306	302–306
specific gravity, g/cm³	2.13–2.16	2.13–2.16
melt flow rate, g/10 min	10.6	1.8
continuous use temperature, °C	260	260
tensile strength, MPa[a]		
23°C	27.6	31.0
tensile yield, MPa[a]		
23°C	13.8	15.2
ultimate elongation, %		
23°C	300	300
flexural modulus, MPa[a]		
23°C	655	690
creep resistance tensile modulus, MPa[a]		
23°C	270	270
water absorption, %	0.03	0.03
coefficient of linear thermal expansion, cm/(cm·°C)		
at 20–100°C	12×10^{-5}	12×10^{-5}

[a] To convert MPa to psi, multiply by 145.

jacketing for wire and cable, injection- and blow-molded components, and compression-molded articles (see Insulation, electric).

RICHARD L. JOHNSON
E.I. du Pont de Nemours & Co., Inc.

Teflon PFA Fluorocarbon Resin—Melt Processing of Teflon PFA TE-9705, PIB #1, bulletin, E.I. du Pont de Nemours & Co., Inc., Wilmington, Del., 1973.

E.W. Fasig, D.I. McCane, and J.R. Perkins, *paper presented at the 22nd International Wire and Cable Symposium, Atlantic City, N.J., Dec. 1973.*

Teflon PFA Fluorocarbon Resin—Injection Molding of Teflon PFA TE-9704, PIB #4, bulletin, E.I. du Pont de Nemours & Co., Inc., Wilmington, Del., 1973.

POLYCHLOROTRIFLUOROETHYLENE

PCTFE properties are governed by the molecular weight and percent crystallinity. The high molecular weight thermoplastic has a melt temperature (T_m) of 211–216°C and glass-transition temperature (T_g) of 71–99°C, and is thermally stable up to 250°C. Because of the loss of strength above the T_m, the useful range of operation is from −240 to 205°C.

The specific gravity of PCTFE for the amorphous and crystalline polymers has been calculated to be 2.075 and 2.185, respectively.

The combination of mechanical properties that makes PCTFE a unique engineering thermoplastic is shown in Table 1.

The high fluorine content contributes to resistance to attack by essentially all chemicals and oxidizing agents. This plastic is compatible with liquid oxygen, remains flexible at cryogenic temperatures, and retains its properties when exposed to either uv or gamma radiation.

At present, there are several commercial suppliers of PCTFE and vinylidene fluoride-modified copolymers including Daikin Koygo Yodogawa, 3M Company, Halocarbon Products Corporation, Ugine Kuhlmann, and Allied Chemical.

In general, the PCTFE resins have been found to be low in toxicity and irritation potential under normal handling conditions.

The chief uses for PCTFE plastics today have been in the areas of electrical/electronics, cryogenic, chemical, and medical instrumentation industries.

A.C. WEST
3M Company

R.P. Bringer, "Influence of Unusual Environmental Conditions on Fluorocarbon Plastics," SAMPE (Society of Aerospace Material and Process Engineers) Symposium, St. Louis, May 7–9, 1962.

R.P. Bringer, and C.C. Solvia, *Chem. Eng. Prog.* **56**(10), 37 (1960).

R.E. Schamm, A.F. Clark, and R.P. Reed, *A Compilation and Evaluation of Mechanical, Thermal and Electrical Properties of Selected Polymers*, NBS Report, AEC SAN-70-113, SANL 807 Task 7, SSANL Task 6, Sept. 1973, pp. 335–443.

Table 1.   Mechanical Properties of Polychlorotrifluoroethylene

Property	Value
tensile strength, MPa[a]	32–39
compressive strength, MPa[a]	380
modulus of elasticity, MPa[a]	1423
hardness, Shore D	76
deformation under load, 0.3% at 24 h, MPa[a]	6.86
heat deflection temperature, °C at 1.78 MPa[a]	126

[a] To convert MPa to psi, multiply by 145.

BROMOTRIFLUOROETHYLENE

The monomer, bromotrifluoroethylene, $CF_2{=}CFBr$, is a colorless gas; bp −3.0°C at 101 kPa (754 mm Hg) and d_4^{25} 1.86 g/cm³. Since it is spontaneously flammable in air, its odor is that of its oxidation products, mixed carbonyl halides.

Bromotrifluoroethylene is manufactured and sold in commercial quantities with a purity of 99.9% by the Halocarbon Products Corporation.

Polymers

The olefin can be polymerized in trichlorofluoromethane solution at −5°C for 7 d with a halogenated acetyl peroxide, or it may be polymerized in an aqueous suspension with 2 parts by weight distilled water, 0.01 part ammonium persulfate, 0.004 part sodium bisulfite, and 0.001 part hydrated ferrous sulfate present for each part of the monomer. Prepared either way, the homopolymer is a white powder that is soluble in acetone and useful as a hard, chemically resistant coating for metal or fabric surfaces.

Telomers of bromotrifluoroethylene have been prepared using chain-transfer agents such as CF_3SSCF_3, C_2F_5I, CBr_4, or CBr_3F. Commercially available bromotrifluoroethylene telomers have densities of 2.14–2.65 g/cm³ and viscosities of 2–4000 mm²/s (= cSt). These fluids are expensive but are made in small volume for the aerospace industry. They are nontoxic and noncorrosive to many common metals, alloys, and other materials.

GARY ASTROLOGES
Halocarbon Products

R.E.A. Dear and E.E. Gilbert, *J. Fluorine Chem.* **4**, 107 (1974).

Ger. Pat. 2,235,885, (Feb. 7, 1974), J. Kuhls, H. Fitz, and P. Haasemann (to Faberwerke Hoechst A.-G.).

M. Hudlicky, *Chemistry of Organic Fluorine Compounds*, 2nd ed., John Wiley & Sons, New York, 1976, pp. 389, 409, 430, 717.

POLY(VINYL FLUORIDE)

Monomer

Vinyl fluoride, VF, bp −72.2°C, is a colorless gas with an ethereal odor. The unreactivity of vinyl fluoride, relative to other monomers, is attributed to the electron-withdrawing effect of the fluorine atoms. High pressure (30–100 MPa or 4350–14,500 psi), moderate temperature (60–140°C) polymerizations are carried out with free-radical catalysts.

Polymer

Poly(vinyl fluoride) (PVF) is tough, transparent, high melting, inert, and highly resistant to weathering. This combination of properties accounts for most of its uses. Physical, thermal, and electrical properties are listed in Table 1 (see also Barrier polymers).

PVF is converted into thin film by a plasticized melt extrusion. It is available from E.I. du Pont de Nemours & Co., Inc., both as transparent and pigmented films under the trademark Tedlar PVF film.

Health and Safety

Exposure to vinyl fluoride monomer had temporary or no effects on rats. The monomer is flammable in air between limits of 2.6–22.0% by volume. Minimum ignition temperature for mixtures of air and VF is 400°C.

The self-ignition temperature of PVF film is 390°C. The limiting oxygen index (LOI), which is 22.6% for PVF (60), is raised to 30% for antimony-modified film (see Flame retardants). The relative toxicity hazards of thermally degraded poly(vinyl fluoride) and other polymers have been reported.

Uses

Pigmented films, available in white and in a variety of colors, are used as surfacing for aluminum and galvanized steel for exterior residential

Table 1. Physical and Thermal Properties of Poly(Vinyl Fluoride) Film

Property	Value
bursting strength, kPa[a]	131–482
coefficient of friction (film/metal)	0.15–0.30
density, g/cm³	1.38–1.57
impact strength, kJ/m[b]	10–22
refractive index, n_D	1.46
tear strength, kJ/m[b]	
initial	174–239
tensile modulus, MPa[c]	1700–2600
ultimate tensile strength, MPa[c]	48–120
ultimate elongation, %	115–250
ultimate yield strength, MPa[c]	33–41
linear coefficient of expansion, cm/(cm·°C)	0.00005
useful temperature range, °C	
continuous use	−72–107
zero strength, °C	260–300
thermal conductivity, W/(m·K)	
60°C	17
self-ignition temperature, °C	390
dielectric constant at 1 kHz at 23°C	8.5–9.9
dielectric strength, kV/µm	0.13
dissipation factor, %	
1 MHz at 23°C	1.4–1.6

[a] To convert kPa to psi, multiply by 0.145.
[b] To convert kJ/m to (ft·lbf)/in., divide by 0.0534 (see ASTM D 256).
[c] To convert MPa to psi, multiply by 145.

and industrial building siding. Nonadhering film is used as release sheets in plastic processing, particularly in high temperature pressing of epoxy resins (qv) for circuit boards in aerospace parts (see Abherents). PVF provides resistance to weather, chemicals, staining, and abrasion.

DONALD E. BRASURE
E.I. du Pont de Nemours & Co., Inc.

G.H. Kalb and co-workers, *J. Appl. Polym. Sci.* **4**, 55 (1960).

M.L. Wallach and M.A. Kabayanna, *J. Polym. Sci. A-1* **4**, 2667 (1966).

R.F. Boyer, *J. Polym. Sci. Poly. Symp.* **50**, 189 (1975).

POLY(VINYLIDENE FLUORIDE)

Poly(vinylidene fluoride) is the polymer of 1,1-difluoroethylene (vinylidene fluoride), $+CH_2-CF_2\frac{}{n}+$. Poly(vinylidene fluoride), PVF$_2$, or (preferably) PVDF, is a semicrystalline polymer containing 59.4% fluorine. The symmetrical arrangement of the hydrogen and fluorine atoms in the chain contributes to the unique polarity which influences the polymer's dielectric properties and solubility. The polymer is readily melt-fabricated by standard methods of molding and extrusion or solvated by polar solvents to form coatings. PVDF displays the stability characteristic of fluorocarbon polymers when exposed to harsh thermal, chemical, and uv environments. PVDF has many applications wherever high performance polymers are required, eg, construction or coating materials.

Monomer

Vinylidene fluoride, or 1,1-difluoroethylene, VF$_2$ or (preferably) VDF, bp −85.7°C, is a colorless and nearly odorless gas. Acute exposure to the monomer has not produced liver damage in rats. Commercially significant polymerizations are those that involve radical-initiated systems in emulsion or suspension.

Polymer

PVDF was introduced in 1961 and has been commercially available since 1965. Producers of PVDF and their trademarks are listed in Table 1. Physical property data of PVDF are listed in Table 2.

Table 1. Poly(vinylidene Fluoride) Producers

Manufacturer	Trade name
Kureha Chemical Co., Ltd. (Tokyo, Japan)	KF
Pennwalt Corporation (Philadelphia, Pa.)	Kynar
Produits Chimiques Ugine Kuhlmann (Paris, Fr.)	Foraflon
Solvay & Cie (Brussels, Belg.)	Solef
Suddeutsche Kalkstickstoff-Werke, A.G. (Trostberg, FRG)	Vidar

Table 2. Typical Properties of Poly(vinylidene Fluoride)

Property	Value
specific gravity	1.75–1.80
refractive index, n_D^{25}	1.42
melting point (T_m), °C	154–184
glass transition (T_g), °C	−40
water absorption, %	0.04–0.06
tensile strength at break, MPa[a]	36–56
elongation at break, %	25–500
tension modulus of elasticity, MPa[a]	1340–2000
Izod impact (notched), kJ/m[b]	150–530
hardness, Shore D	70–82
thermal conductivity, W/(m·K)	0.1–0.13
specific heat, J/(g·K)[c]	1.26–1.42
coefficient of linear expansion, K⁻¹	$7.9–15.7 \times 10^{-5}$
limiting oxygen index, %	43.6
deflection temperature (1.82 MPa[a]), °C	112–150
low temperature embrittlement, °C	−62 to −64

[a] To convert MPa to psi, multiply by 145.
[b] To convert kJ/m to ft·lbf/in., divide by 0.0534.
[c] To convert J to cal, divide by 4.184.

The properties of the polymer depend on molecular weight, molecular weight distribution, extent of irregularities along the polymer chain, and crystalline form; the first three factors are controlled by the polymerization process. Although vinylidene fluoride predominantly adds to the growing chain in head-to-tail sequence, 4–6% of monomer units add head-to-head and tail-to-tail, depending on the polymerization process. The polymer crystallizes in three crystalline forms: form I (β), form II (α), and form III (γ).

Polymorphism of PVDF and the two distinct dipole groups, (CF$_2$) and (CH$_2$), contribute to the polymer's exceptional dielectric properties. It has a very high dielectric constant which is a function of frequency, temperature, thermal history, and orientation.

PVDF is available in a wide range of melt viscosities as required by the fabricator. The polymer is readily extruded into film sheet, bar stock, pipes, and filament and molded in conventional compression, transfer, and injection equipment.

Health and Safety

Based upon toxicity studies (including acute oral, subacute contact, and inhalation tests with rats), PVDF is considered nontoxic.

Uses

PVDF is the basic raw material used in many diverse industrial products that require high mechanical strength and resistance to severe environmental stresses.

JULIUS E. DOHANY
LESTER E. ROBB
Pennwalt Corp.

I.W. Kuleshov, A.A. Remisova, and G.M. Gartenev, *Plaste Kautsch.* **24**, 635 (1977).

M. Goerlitz and co-workers, *Angew. Makromol. Chem.* **29–30**, 137 (1973).

C.W. Wilson and E.R. Santee, Jr., *J. Polym. Sci. C* **8**, 97 (1965).

POLY(FLUOROSILICONES)

The role of fluorine and the carbon–fluorine bond in achieving a high degree of solvent resistance and stability is well known in organic polymer systems. About 25 years ago, the two properties were combined in silicone polymers by incorporation of fluorine into polyalkylsiloxane systems.

Table 1 lists properties at RT both before and after exposure to elevated temperatures. The stability of the fluorosilicone elastomer is evident in the small changes in hardness resulting from aging at elevated temperatures. Low temperature properties of fluorosilicone elastomers are outstanding. The TR-10 (ASTM D 1329) of these elastomers is typically $-59°C$. No other commercially available fluorocarbon elastomers have this low TR-10 value. Like silicone polymers, fluorosilicone elastomers have inherently good electrical insulating properties. Bondability of fluorosilicone elastomers to metals and to other polymeric composites can be significantly improved with silastic A-4040 primer. In general, other properties of fluorosilicone elastomers are similar to general-purpose dimethylsilicone elastomers.

Manufacture

The dichlorosilane, $CF_3CH_2CH_2Si(CH_3)Cl_2$, is the principal starting material for cyclotrisiloxane (1) which is purified by distillation and serves as the monomer for polymer synthesis. Preparation of fluorosilicone polymers from the cyclic trimer can be accomplished by base catalysis at elevated temperatures in the absence of solvent.

(1)

Fluorosilicone elastomers can be formulated to provide specific properties. Materials that are designed to resist tearing, exhibit a tear strength of 44 kN/m (250 ppi). Moduli may range from 1.0–6.0 MPa (150–900 psi) at 100% elongation. Compression sets as low as 10% (22 h/177°C) can be achieved.

In the production of various rubber products, fluorosilicone gums are compounded, generally with fumed and/or precipitated silica fillers, hydroxy-containing low viscosity silicone oils, and readily available peroxides. Fluorosilicone elastomers can be molded, extruded, or calendered by any of the conventional methods employed in the industry. Compression molding is the most widely used method and is ideal for a great many fabrications at 115–170°C and 5.5–10.3 MPa (800–1500 psi).

Fluorosilicones are used most often as elastomers and fluids (oils), mainly in the automotive and aerospace industries.

Y.K. KIM
Dow Corning Corp.

O.R. Pierce and co-workers, *Ind. and Chem.* **52**, 783 (1960).

Table 1. Physical Properties of Fluorosilicone Elastomers Heat-Aged at 204°C

Test properties and conditions	Silastic LS-2311[a]	Silastic LS-2332[b]	Silastic LS-2380
original properties			
Durometer (Shore A-2)	75	52	70
tensile strength, MPa[c]	6.2	8.6	6.9
elongation, %	100	500	150
100% modulus, MPa[c]	6.1	1.0	4.8
tear strength kN/m[d], die B	8.8	46.4	
aged 70 h, 204°C, change			
Durometer, points	+4	+2	nil
tensile strength, %	nil	−9	
elongation, %	−20	−20	+6
wt loss, %	+2		
tear, %		−30	

[a] Molded 10 min, 171°C; postcured 8 h, 204°C (with benzoyl peroxide).
[b] Molded 5 min, 127°C; postcured 8 h, 204°C (with Varox).
[c] To convert MPa to psi, multiply by 145.
[d] To convert kN/m to ppi, divide by 0.175.

FLUORINE COMPOUNDS, ORGANIC, SURFACE CHEMISTRY.
See Surfactants and detersive systems.

FLUORINE COMPOUNDS, SURFACE CHEMISTRY OF FLUORO-CHEMICALS. See Surfactants and detersive systems.

FLUORITE, FLUOROSPAR, CaF₂.
See Fluorine compounds, inorganic—Calcium fluoride.

FLUOROALUMINATES, FLUOROBERYLLATES, FLUOROPHOSPHATES, FLUOROSILICATES, AND SIMILAR ENTRIES. See Fluorine compounds, inorganic.

FLUOROCHEMICALS. See Fluorine compounds, organic.

FLUOROMETRY. See Analytical methods; Color; Dyes, application and evaluation.

FOAMED PLASTICS

Foamed plastics, otherwise known as cellular plastics or plastic foams, have been important to human life since primitive man began to use wood (qv), a cellular form of polymer cellulose. Cellulose (qv) is the most abundant of all naturally occurring organic compounds, comprising approximately one third of all vegetable matter in the world.

Cellular polymers have been commercially accepted in a wide variety of applications since the 1940s. The total usage of foamed plastics in the United States is projected to rise to about 6.6×10^6 metric tons in 1990.

A cellular plastic has been defined as a plastic the apparent density of which is decreased substantially by the presence of numerous cells disposed throughout its mass. The gas phase in a cellular polymer is usually distributed in voids or pockets called cells. If these cells are interconnected in such a manner that gas can pass from one to another, the material is termed open-celled. If the cells are discrete and the gas phase of each is independent of that of the other cells, the material is termed closed-celled.

Theory of the Expansion Process

Foamed plastics can be prepared by a variety of methods. The most important process consists of expanding a fluid polymer phase to a low density cellular state and then preserving this state. This is the foaming, or expanding, process. Other methods of producing the cellular state include leaching out solid or liquid materials that have been dispersed in a polymer, sintering small particles, and dispersing small cellular particles in a polymer. The latter processes, however, are relatively straightforward processing techniques but are of minor importance.

The expansion process consists of three steps: creating small discontinuities or cells in a fluid or plastic phase; causing these cells to grow to a desired volume; and stabilizing this cellular structure by physical or chemical means.

Manufacturing Processes for Cellular Polymers

Cellular plastics and polymers have been prepared by a wide variety of processes involving many methods of cell initiation, cell growth, and cell stabilization. The most convenient method of classifying these methods appears to be based on the cell growth and stabilization processes. The growth of the cell depends on the pressure difference between the inside of the cell and the surrounding medium. Such pressure differences may be generated by lowering the external pressure (decompression) or by increasing the internal pressure in the cells (pressure generation). Other methods of generating the cellular structure are by dispersing gas (or solid) in the fluid state and stabilizing this cellular state, or by sintering polymer particles in a structure that contains a gas phase.

Foamable compositions in which the pressure within the cells is increased relative to that of the surroundings have generally been called expandable formulations. Both chemical and physical processes are used to stabilize plastic foams from expandable formulations. There is no single name for the group of cellular plastics produced by decompression processes. The various operations used to make cellular plastics by this principle are extrusion, injection molding, and compression molding. Either physical or chemical methods may be used to stabilize products of the decompression process. See Table 1 for some representative methods for production of cellular polymers.

The properties of commercial rigid foamed plastics, commercial flexible foamed plastics, and commercial structural foams (density > 0.3 g/cm^3) are shown in Tables 2, 3, and 4, respectively.

Environmental aging of cellular polymers is of primary importance to most applications of plastic foams. The response of cellular materials to the action of light and oxygen is governed almost entirely by the composition and state of the polymer phase. Expansion of a polymer into a cellular state increases the surface area; reactions of the foam with vapors and liquids are correspondingly faster than those of solid polymers.

Applications

Concern over energy conservation and consumer safety has provided a rapid growth in applications for insulation and cushioning in transport.

A healthy and affluent economy is also expected to increase the consumer demand for comfort cushioning in furniture, bedding, and flooring, as well as for packaging such as drinking cups, and packaging inserts for consumer products (molded and loose-fill materials). The cost of finished wood articles has forced the industry to develop structural foams that have important applications as replacements for wood, metal, or solid plastics (see Engineering plastics).

Table 1. Methods for Production of Cellular Polymers

Type of polymer	Extrusion	Expandable formulation	Spray	Froth foam	Compression mold	Injection mold	Sintering	Leaching
cellulose acetate	X							
epoxy resin		X	X	X				X
phenolic resin		X						
polyethylene	X	X			X	X	X	X
polystyrene	X	X				X	X	
silicones		X						
urea–formaldehyde resin				X				
urethane polymers		X	X	X		X		
latex foam rubber				X				
natural rubber	X	X			X			
synthetic elastomers	X	X			X			
poly(vinyl chloride)	X	X		X	X	X		X
ebonite					X			
polytetrafluoroethylene							X	

Table 2. Physical Properties of Commercial Rigid Foamed Plastics

Property	Cellulose acetate	Epoxy		Phenolic		Extruded plank	
density, kg/m^{3a}	96–128	32–48	80–128	32–64	112–160	35	53
mechanical properties							
compressive strength, kPa^b at 10%	862	138–172	414–620	138–620		310	862
tensile strength, kPa^b	1,172		345–1,240	138–379		517	
compression modulus, MPa^c	38–90	3.9	14.5–44.8			10.3	
thermal properties							
thermal conductivity, W/(m·K)^c	0.045–0.046	0.016–0.022	0.035–0.040	0.029–0.032	0.035–0.040	0.030	
max service temperature, °C	177	205–260	205–260	132	205	74	
specific heat, kJ/(kg·K)^d						1.1	
electrical properties							
dielectric constant	1.12			1.19–1.20	1.19–1.20	< 1.05	< 1.05
moisture resistance							
water absorption, vol%	4.5			13–51	10–15	0.02	0.05

^{a}To convert kg/m^3 to lb/ft^3, multiply by 0.0624.
^{b}To convert kPa to psi, divide by 6.895.
^{c}To convert MPa to psi, multiply by 145.
^{d}To convert kJ/(kg·K) to Btu/(lb·°F), divide by 4.184.

Table 3. Physical Properties of Commercial Flexible Foamed Plastics

Property	Expanded acrylonitrile–butadiene rubber	Expanded butyl rubber		Expanded natural rubber		Expanded neoprene		Expanded SBR	Latex	Foam	Rubber
density, kg/m^{3a}	160–400	128–144	224–304	56	320	112	192	72	80	130	160
cell structure	closed	closed	closed	closed	closed	closed	closed	closed	open	open	open
tensile strength, kPa^b	275				206		758	551	103		
rebound resilience, %		39–36	30–16						73		
max service temperature, °C	100			70	70	105	70	70			
thermal conductivity W/(m·K)	0.036–0.043			0.036	0.043	0.040	0.065	0.030		0.050	

^{a}To convert kg/m^3 to lb/ft^3, multiply by 0.0624.
^{b}To convert kPa to psi, divide by 6.895.

Table 4.   Typical Physical Properties of Commercial Structural Foams

Property	ABS		Noryl	Nylon-6,6 glass-reinforced	Poly-carbon-ate	Poly-ethylene, high density	Poly-propylene		High impact polystyrene		Polyurethane		
glass reinforced	no	yes	no	yes	no	no	no	20%	no	20%	no	no	no
density, kg/m³	0.80	0.85	0.80	0.97	0.80	0.60	0.60	0.73	0.70	0.84	0.40	0.50	0.60
tensile strength, kPa[a]	18,600	48,000	22,700	101,000	37,900	8,900	13,800	20,700	12,400	34,500	11,000	17,200	23,400
compression strength, kPa[a] at 10% compression	6,900		34,500		51,700	8,900					5,500	12,400	19,300
max use temperature, °C	82		96	203	132	110	115						

[a] To convert kPa to psi, divide by 6,895.

Health and Safety

Plastic foams are organic in nature and, therefore, are combustible when exposed to open flames. They do vary in their response to small sources of ignition because of composition and/or additives. All plastic foams should be handled, transported, and used according to manufacturers' recommendations as well as applicable local and national codes and regulations.

The presence of additives or unreacted monomers in certain plastic foams can limit their use where food or human contact is anticipated. The manufacturers' recommendations or existing regulations again should be followed for such applications.

K.W. SUH
R.E. SKOCHDOPOLE
Dow Chemical USA

K.C. Frisch and J.H. Saunders, *Plastic Foams*, Vol. 1, Pts. 1–2, Marcel Dekker, Inc., New York, 1972 and 1973.

C.J. Benning, *Plastic Foams*, Vols. 1–2, Wiley-Interscience, New York, 1969.

J.L. Throne and B.C. Wendle, eds., *Engineering Guide to Structural Foams*, Technomic Publishing Co., Westport, Conn., 1976.

FOAMS

Foams are coarse dispersions of gas in a relatively small amount of liquid. Bubbles vary in size from ca 50 μm (microbubbles) to several mm. Foam properties depend primarily on the chemical composition and properties of the adsorbed films and are influenced by numerous factors such as the extent of adsorption from solution to the liquid–gas surface, the rheology of the adsorbed layer, gaseous diffusion out of and into bubbles, size distribution of the bubbles, surface tension of the liquid, and external pressure and temperature (see also Sprays; Surfactants).

Foams that can remain stable for an appreciable time cannot be created from pure liquids. The third component is usually a solute in the liquid, although it may also be a finely divided solid, a liquid-crystal phase, or an insoluble monolayer.

The two laws of bubble geometry (Plateau's laws), which hold for all assemblies of bubbles and for the morphology of foams, are based on the minimizing of surface area of liquid films, which is the direct result of the tension of liquid surfaces. (1) Along an edge, three, and only three, liquid lamellae meet; the three lamellae are equally inclined to one another all along the edge; hence, their mutual or dihedral angles of inclination equal 120°; (2) At a point, four, and only four, of those edges meet; the four edges are equally inclined to one another in space; hence, the angle at which they meet is the tetrahedral angle (109°28′16″). These two statements are not independent; either follows from the other.

Whatever the mechanism of their stability, the shape of the bubbles or their size distribution, the foam at any stage of its existence can be described by an equation of state that relates the external pressure P, the volume of gas in the foam V, the moles of gas in the foam n, the area of the liquid films A, the surface tension of the liquid σ, and the temperature T. The contained gas is taken to be an ideal gas:

$$PV + \frac{2}{3}\sigma A = nRT \qquad (1)$$

Evanescent Foams

Evanescent foams, eg, champagne, are stabilized by a mechanism of film elasticity that depends on the possibility of creating local differences of surface tension of a solution by expansion or contraction of its surface. If an adsorbed layer of solute is present on the surface, expansion dilutes the layer and the surface tension rises toward that of the pure solvent; contraction of the surface compresses and concentrates the layer and the surface tension declines. A flow of the surface, along with underlying solution carried by viscous drag, then occurs toward the region of higher surface tension. This flow, known as Marangoni flow, offsets the hydrodynamic and the capillary drainage, and so restores the thickness of the lamella and acts to stabilize it.

Homogeneous solutions. A diagram for the two-component system, 2,6-dimethyl-4-heptanol and ethylene glycol, is shown in Figure 1; interpolated lines of equal foam stability (isaphroic lines, from *aphros* the Greek word for foam) are superimposed. The isaphroic lines center about a point close to the critical point as a maximum and decrease in value the farther they are from it.

Measurements of foam stability, reported by superimposition of isaphroic contours on a phase diagram, markedly resemble contours of surface activity, named cosorption lines, obtained by plotting the Gibbs excess surface concentration Γ_2 superimposed on the same phase diagram.

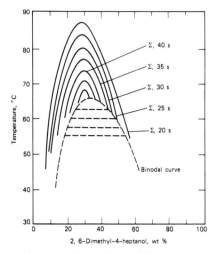

Figure 1. Phase diagram and, superimposed thereon, isaphroic contours of the two-component system 2,6-dimethyl-4-heptanol and ethylene glycol showing maximum foaminess near the critical-solution point.

A useful rule of thumb can be obtained from the phase diagram; foaming of a solution reaches its maximum at the temperature and concentration where a transition into two separate liquid phases is imminent.

Heterogeneous mixtures. The heterogeneous portions of phase diagrams of polycomponent systems may include two conjugate solutions in equilibrium at a given temperature. Foam inhibition occurs when the dispersed liquid phase can spread spontaneously on a foamable solution.

Slow-Draining Lamellae

Another mechanism, leading to a different foam morphology than evanescent foams, comes into operation when the lamellae can drain to extremely thin films before rupture. This effect is a surface phenomenon of solutions of polymers such as proteins, or of long-chain solutes or mixtures of solutes in which adsorption at the surface produces coherent monolayers. The observed low drainage rates, as well as the observed surface viscosity, at least in some cases, are the result of a positive, electrostatic disjoining pressure caused by adsorbed ions. The presence of a macroscopic liquid-crystalline phase at the surface is another cause of greatly enhanced viscosity and non-Newtonian flow (see also Rheological measurements; Liquid crystals).

When a slow-draining film is formed, a sufficient increase in temperature may cause a change to a fluid type of draining. The transition occurs over so slight a range of temperature, ie, within 0.2°C, that it resembles a phase transition, like the melting of a condensed monolayer. Thus, foams with slow-draining lamellae are stable below the transition temperature and suddenly become evanescent foams when the temperature is raised above that point.

Multiphase Stabilization Factors

Finely divided solids that are poorly wetted by the liquid can stabilize foam. For instance, soot shaken up in tap water gives a persistent foam. The contact angle of the liquid on the surface of the particle is the determining property that enables particles to adhere to the liquid–air surface in seeming defiance of gravity. The work of adhesion is given by W (adhesion) $= \sigma_L(1 + \cos\theta)$. Contact angles between 40° and 70° are optimum.

Liquid crystals as stabilizers. The stability of foams containing a liquid-crystalline phase in equilibrium with an isotropic liquid phase has been investigated in a few cases. To obtain a stable foam, a combination of the isotropic solution and a liquid crystal is necessary.

Desorption of adsorbed solute. The reduction of surface area, consequent to rupture of the lamellae, requires that adsorbed solute be returned to the bulk solution. As the adsorption is spontaneous, desorption requires work, which is an increase of the free energy. The more strongly adsorbed the solute, the larger is this contribution to the stability of the lamella. This purely thermodynamic generalization provides no explicit mechanism but supplements the idea of film stabilization caused by positive or negative adsorption.

Drainage of liquid from lamellae. The thinning of a liquid lamella by drainage is a destabilizing factor. Thick lamellae are elastic but become brittle when thinned.

Marangoni flow. The draining of a liquid, including also any motion of a surface that carries an adsorbed layer, creates local differences of surface tension, a nonequilibrium temporary condition. This local difference is removed by the motion of the surface from regions of lower to that of higher tension. This flow (Marangoni flow) opposes the movement of drainage and acts as a stabilizing factor to restore the thickness of the lamella.

Negative pressure from the electric double layer. Adsorbed ions establish electrostatic repulsion between the two surfaces of the lamella. The presence of excess counterions in the interstitial solution creates a hydrostatic pressure forcing the surfaces apart and thus maintains the lamella thickness.

Surface viscosity and slow draining. The draining rate is reduced by the overlapping of electrical double layers; or by the presence of coherent, tightly packed, adsorbed monolayers of solute.

Stability Measurement

The stability of short-lived or evanescent foams is measured by the dynamic foam meter. For foams of high stability, rate of loss of interfacial area is a direct indicator of foam stability. Other parameters related to the interfacial area are given by the equation of state of foam in its various forms.

Uses

Foam applications are based on the following properties: difference in composition between the bulk liquid and the collapsed foam (froth flotation); low density (fire-extinguishing foams); large surface (foam fractionation); mechanical properties (replacement of drilling mud as a circulating medium in oil-well completion); immobilization of gas and liquid (retention of airborne particulate matter); and thermal properties (heat insulation).

SYDNEY ROSS
Rensselaer Polytechnic Institute

J.J. Bikerman, *Foams*, Springer-Verlag, New York, 1973.
J.A. Kitchener and C.F. Cooper, *Q. Rev. London* **13**, 71 (1959).

FOOD ADDITIVES

Defining food additives is not a straightforward task since there are no fewer than two important factors that can affect description. In the technical sector, the term food additives has been used to describe broadly any chemical substance used to augment or modify the characteristics of a food product. The legal definition of food additives is much more complex and subject to controversy, interpretation, and opinion. In 1959, the Food Protection Committee of the National Research Council (NRC) defined a food additive as a substance or mixture of substances, other than a basic foodstuff, that is present in food as a result of any aspect of production, processing, storage, and packaging. The term does not include chance contaminants.

In most nations, direct food-additive legislation limits the use of a substance to maximum doses in specified foods and are subject to premarketing approvals.

In the United States, a new substance gains approval for use in food through the successful submission of a Food Additive Petition (FAP) in which the following information is documented: safety, including negative carcinogenic studies; declaration of intended use(s); efficacy data at specific levels in specified food systems; manufacturing details including intermediates and final product specifications backed by analytical methods; methods for analyzing the substance in foods; and an environmental impact statement.

It is important to note that, in the United States, colorants are handled in a special way. A new colorant intended for use in foods is the subject of a Color Additive Petition (CAP). This follows the basic format of an FAP but does not require specific proof of efficacy, other than imparting color, and is not stringently limited to specified use levels. Moreover, colorants are classified as either certified or uncertified (see Colorants for foods, drugs, and cosmetics). Substances that are not intentionally added to foods but nonetheless find their way into foods from packaging materials, agriculture chemicals residues, etc, are also regulated by FDA and the agencies of other nations. In the United States, these are termed indirect food additives and are subject to other petitioning procedures. For the most part, the issue in this instance centers on proof of safety.

Categories

Depending on the nation, food ingredients and additives are described in a variety of categories. An industrial publication describes the following 52 food additive categories: acidulants, amino acids, anticaking agents (drying agents), antimicrobial agents (fumigants), antioxidants, baking aids (yeast foods), binders (plasticizers, fillers), bleaching agents

Table 1. Selected Chemical Additives and Estimated Level of Use (1975) in the U.S. Food Supply

Type of additive	Use per year (est), 1000 metric tons	Examples
antioxidants	7–9	butylated hydroxyanisole, erythorbic acid, nitrogen gas
colorants	40	caramel, certified colors, beet powder
emulsifiers and surface-active agents	105	mono- and diglycerides propylene glycol sorbitol, glycerol monostearate
enzymes	7–10	papain, rennet, glucose-isomerase, pectinmethylesterase
flavor enhancers	25	monosodium glutamate
flavors and spices	115	mustard, pepper, ethyl acetate, carbon dioxide, vanillin
leavening agents and baking aids	55–60	benzoyl peroxide, sodium acid pyrophosphate, various carbonates
nonnutritive sweeteners	1–2	sodium saccharin
nutrient supplements	10–15	ascorbic acid, niacin, ferrous sulfate, thiamin hydrochloride, zinc sulfate
pH control agents	215	acetic acid, calcium carbonate, calcium phosphates, citric acid, phosphoric acid, sodium bicarbonate, malic acid
preservatives	20–25	sodium benzoate, sulfur dioxide, sorbic acid, propionic acid
stabilizers, thickeners	200	modified food starch, various gums, carboxymethyl cellulose, cellulose
anticaking agents	4–5	calcium silicate, magnesium stearate
stimulants	1	caffeine
protein hydrolysate	20	
sequestrants	0.5	citric acid, disodium ethylenediaminetetraacetate

(**1**) Anoxomer

(oxidizing agents), buffering agents, bulking agents, caramelization aids, carriers (disintegrating agents, dispersing agents), clarifying agents, clouding agents (density adjusters for essential oils), colorants (pigments, lakes), color fixatives (color retention aids), cooking media, defoamers, dough conditioners, emulsifiers (foaming agents, whipping agents), enzymes, filter aids, firming agents, flavor enhancers (flavor intensifiers), flavors, foam stabilizers, food starch modifiers (starch modifiers), gums, humectants (moisturizing agents), hydrocolloids, leavening agents, lubricants (antistick agents, release agents), masticatory substances (chewing gum base), maturing agents, microorganisms (molds, yeasts), neutralizers (acids, bases, alkalies), nonnutritive sweeteners, nutrients and dietary supplements, preservatives, propellants, protein sources, refrigerants, salt substitutes, sequestrants (chelating agents), solvents (solubilizers, vehicles, diluents, extractants), stabilizers (suspending agents), surface-active agents (wetting agents), surface-finishing agents (glazes, coating

agents, film forms, microencapsulating agents), sweeteners (carbohydrates), texturizers (texture modifiers), thickeners (gelling agents), and vitamins.

The technology, specific uses, and effects of food additives in various food systems have been described in detail and kept current by updates.

Table 1 is a list of a few chemical additives and their estimated levels of use.

On April 26, 1983, the FDA approved the first new direct food additive in over a decade. The compound (Anoxomer (**1**)) is an antioxidant and represents the first in a class of non absorbable, polymeric food additives.

THOMAS FURIA
Intechmark Corp.

T.E. Furia, *Handbook of Food Additives*, 2nd ed., Vol. 1 (1972) and Vol. 2 (1980), CRC Press, Inc., West Palm Beach, Fla.

FOOD PROCESSING

Food processing, regardless of the type of food, can be divided into three classes: separation (eg, rendering, skimming, boning, coring, defeathering, husking, peeling, shelling, etc); assembly (eg, coating, enrobing, baking, homogenization, roasting, pelleting, stuffing, etc); and preservation. These can occur at harvest, at the food-processing plant, or even at the point of retail sales.

Foods can be categorized as living-tissue or raw foods, and as nonliving tissue. Living-tissue foods include fresh fruits, vegetables, meats, and grains.

The genetically controlled taste, color, nutritive value, cell structure, and conformation of fresh foods usually are best at harvest. Protection of these qualities is achieved by retarding or inhibiting the detrimental action of microbes and enzymes, insects, bruising, and chemical degradation, while maintaining the integrity of the membranes, enzyme systems, and gross structure of the food.

The intensity of processing operations is limited by the tissue itself in terms of temperature range, water activity, respiratory-gas composition, mechanical stresses, and concentrations of chemicals. Except for certain fruits and vegetables, storage should be to 0°C without ice formation.

Most processed foods, eg, canned, frozen, and dried products, are marketed as nonliving tissue or manufactured foods. Separation, assembly, and preservation operations determine the final quality of the product. Nonliving tissue products must be defined in terms of composition, absence of defects, nutritional value, fill of container, etc, rather than of genetic quality.

Table 1 lists modes of preservation used to treat various spoilage vectors by the food industry. Typical processing temperatures, pressures, and pH values are shown in Table 2.

Toxicological Implications of Food Deterioration and Spoilage

A primary objective of food processing is to ensure a safe food supply. Certain harvested foods can contain naturally occurring poisonous substances, compounds with pharmacological effects, or compounds that can interfere with the utilization of nutrients present in other foods when they are consumed together. Foods just harvested also may contain or become contaminated with spore or vegetative forms of microbes, parasites, or the wastes of higher life forms, as well as pesticides and other chemicals used for crop protection. A third source of potential health hazards in food supplies is toxic compounds formed during the chemical degradation of processed and stored foods.

Food-Processing Facilities

Factors important in the selection of sites for food-processing operations include: availability of raw materials; abundant potable water supply; low cost waste-disposal facilities; adequate, low cost energy supply; adequate seasonal and nonseasonal labor supply; ease of access to rail and truck transportation; proximity to consumers; and adequate storage areas.

Table 1. Preservation Unit Operations of the Food Industry Arranged by Mode of Preservation and Spoilage Vectors

Spoilage vector	Mode of preservation				
	Physical			Chemical	Mechanical
	Heat	Cold	Drying		
microbes	frying, boiling, pasteurizing, retorting, pasteurization	cooling, freezing	dehydration, desiccation, evaporation, lyophilization	acidification, brining, fermentation, fumigating, irradiation, isomerization, pickling, smoking	cleaning, centrifugation, compressing, washing
enzymes	blanching, boiling, scalding,	hydrocooling		acidification	
chemical	exhausting, deodorizing	chilling		hydrogenation, esterification, smoking	cleaning, degassing, washing
mechanical damage		freezing			aspiration, cleaning, centrifugation, sorting, washing

Table 2. Typical Temperatures, Pressures, and pH Values in Food-Processing Operations

Temperature	Example of use	Limits of application
Heat-transfer media		
cryogenic freezing		
liquid nitrogen, −196°C	rapid freezing to minimize cell damage, moisture loss	cost, stress cracking during freezing
solid carbon dioxide, −78.5°C	rapid freezing	cost
air, plate, and aqueous base freezants, −40 to −5°C	commercial freezing; parasite and insect destruction, commercial freezing, storage; freeze drying	foods not suitable for freezing
refrigerated storage, 0–10°C	commercial refrigerated storage	microbial growth possible
room temperature storage, air, 20–40°C	canned and dry food storage	suitable only for preserved and packaged foods
water; air; atmospheric steam, 50–100°C	pasteurization, milk, eggs, blanching, sterilization of acid foods (pH 4.5); air drying (product temperature); "cooking"	excessive times will cause poor color, flavor, and structure
steam, 110–130°C	thermal sterilization of nonacid foods; destruction of antinutritive factors	rapid thermal degradation of nutrients, pigments, structure, flavors
oil, steam, infrared radiation, 180°C	frying, roasting, baking, generation of browning reaction products	short duration surface treatments
various infrared radiation sources, 180°C	surface heating for flash drying, peeling	charring, pyrolysis without precision control
Pressure[a]		
<0.5 kPa	freeze-drying	cost
0.5–4 kPa	hypobaric storage, deaeration, vacuum concentration, vacuum cooling	cost
101 kPa	most processing operations	
0.1–1 MPa	steam sterilization, over pressure for glass and pouch packs; carbonated and aerosol packaged foods	cost
>1 MPa	extruders, hydraulic pressing, homogenization, potential microbial and enzyme inactivation at 1 GPa (1000 atm) and higher	cost, specialized products
Hydrogen-ion concentration		
pH 1	acid hydrolysis	neutralize to pH range of 7
pH 2.5	lemons, limes, vinegar, organic acids	taste
pH 3–4.5	fruits, acidified foods	noncompatability of foods; protein denaturation
pH 7–3.5	normal pH of most foods	
pH 8–9	solubilization of certain proteins for extraction, alkali process cocoa	neutralize to normal pH for use
pH 12	limed corn	color, flavor, nutrient loss
33% sodium hydroxide	peeling of fruits and vegetables	surface treatment only, neutralize to normal pH

[a] To convert kPa to mm Hg, multiply by 7.5; to convert MPa to atm, divide by 0.101.

Equipment Standards

Standards for food-processing equipment are specified by the USDA in the dairy-, meat-, poultry-, and egg-processing industries. Equipment and processing methods used in the manufacturing of other foods also are covered.

Food Packaging

Food packaging includes rigid glass or metal containers, easily molded paper, plastics, and aluminum, and pouches formed by heat-sealing paper–foil–plastic laminates.

Because of the potential migration of packaging materials into foods, current food and drug regulations should be consulted prior to marketing foods packaged in nonstandard materials.

D.F. FARKAS
University of Delaware

D.F. Farkas, *Chem. Technol.* **7**, 428 (1977).

Code of Federal Regulations (CFR), Title 21, Subchapt. B, pt. 110.

Food Chemicals Codex, 3rd ed., Committee on Specifications, Committee on Food Protection, National Research Council, National Academy of Sciences, Washington, D.C., 1980.

FOODS, DIET. See Sweeteners.

FOODS, NONCONVENTIONAL

Nonconventional foods differ from the usual materials of plant and animal origin that are used for human food or animal feed in that they can be produced from chemicals such as carbohydrates, hydrocarbons, or industrial organic chemicals by processes such as microbiological, enzymatic, or chemical synthesis, or from existing natural products containing carbohydrates, proteins, and fats and by physical, chemical, microbiological, or enzymatic modification.

Single-Cell Protein (SCP)

Two broad classes of microorganisms are of interest for single-cell protein (SCP) production: photosynthetic organisms including the algae and certain bacteria; and nonphotosynthetic organisms, including bacteria, actinomycetes, yeasts, molds, and higher fungi. In addition, two different uses of SCP are distinguished: food for humans and feed for animals.

Photosynthetic organisms. Mass cultivation of algae in ponds or tanks under photosynthetic conditions, using incident sunlight as the energy sources and CO_2 as the carbon source, has been investigated in Japan, Mexico, Algeria, and India, and in California in the United

States. Artificial-illumination systems have been used for experimental mass cultivation of algae in Japan and Czechoslovakia and in bioregenerative systems for converting CO_2 and human wastes into breathable oxygen and food as part of life-support systems for long duration, space-exploration missions.

Key factors influencing growth include temperatures, pH, availability of CO_2, nitrogen, phosphorus, and other inorganic nutrients, and availability of sunlight as influenced by latitude, cloud cover, and depth of the culture pond or tank. Slow, erratic growth results from wide fluctuations between day and night temperatures in outdoor ponds, and from season to season.

Table 1 lists proximate analyses of selected algae as determined by animal feeding studies. It is important to note that the N × 6.25 factor used to calculate the crude protein contents of the algae does not reflect accurately the true protein content because algal cells may contain nonprotein nitrogen substances.

Algae tend to have lower contents of methionine than is desirable in human and animal nutrition and supplementation with this amino acid is necessary with many species. *Spirulina maxima* apparently has a higher protein quality than other species since its protein efficiency ration (PER) is nearly equivalent to that of casein (2.5%).

Dried algae are not accepted readily by livestock. Palatability of dried algae for swine was improved by pelletizing with steam-rolled barley. In general, many species of algae have cell walls that are resistant to digestive enzymes, have dark colors, and bitter flavors; all of which must be altered to make an acceptable food or feed product.

Nonphotosynthetic organisms. Nonphotosynthetic microorganisms of interest in SCP production include bacteria, actinomycetes, yeasts, molds, and higher fungi. Carbon and energy sources that have been considered for growing these organisms include carbohydrates such as simple sugars, starches, cellulose, and agricultural, forestry, pulp, and paper, or food processing wastes containing these carbohydrates, hydrocarbons, and chemicals derived from them including alcohols and organic acids. Commercial-scale operations are conducted in batch-semicontinuous or continuous culture systems.

The product quality considerations for nonphotosynthetic microorganisms are similar to those mentioned previously for algae. Table 2 presents typical proximate analyses for selected bacteria, yeasts, molds, and higher fungi that have been produced on a large pilot-plant or commercial scale.

It is apparent that most of the bacteria, yeasts, molds, and higher fungi of interest for SCP production are deficient in methionine and must be supplemented with this amino acid to be suitable for animal feeding or human food applications. Also, lysine–arginine ratios should be adjusted in poultry rations in which yeast SCP is used. Human feeding studies have shown that only limited quantities of yeast such as *Candida utilis* can be added to food products without adverse effects on flavor.

Derived Plant and Animal Products

Plant protein products—leaf protein concentrates. Leaf protein concentrates (LPC) are prepared by crushing plant material, extracting the juices, using the juice *per se* or recovering protein from the juice by heating or chemical precipitation.

Table 3 gives proximate analyses of several LPC products. In general, the composition of LPCs varies widely with the raw materials and processes used. Typical protein-efficiency-ratio values for LPCs derived from alfalfa range from 1.87 without supplementation to 2.57 when 0.4% methionine was added as compared with casein adjusted to a PER of 2.50.

It is apparent that there are definite limits to the use of LPC products in human diets and that raw materials used for LPC production should be evaluated for possible allergenic problems.

The present market for LPC-type products in the United States is for dried alfalfa meal for animal feed.

Plant protein products—seed-meal concentrates and isolates. Seed-meal protein concentrates and isolates, particularly soy protein products, can be used as extenders for meat, seafood, poultry, eggs, or cheese. In the United States, FDA regulations cite the requirements for using vegetable protein products as substitutes in whole or in part for conventional animal proteins (see Soybeans and other seed proteins).

Animal proteins—fish-protein concentrates. Fish-protein concentrates (FPC) are distinguished from fish meal and fish solubles in that the former are produced for human food use from whole edible species of fish using sanitary processing methods rather than for animal feed (see Aquaculture).

FDA regulations for FPC require a minimum protein content of 75% and maximum moisture and fat contents of 10 and 0.5% respectively. Hake and hakelike fish, herring of the genera *Clupea*, menhaden, and anchovy of the genus *Engraulis mordax*, are permitted. Residues of isopropyl alcohol or ethylene dichloride cannot exceed 250 and 5 ppm, respectively, and fluorides cannot exceed 100 ppm as F. The limit for total bacterial count is 10,000/g and *Escherichia coli*, *Salmonella*, and pathogenic organisms must be absent.

Synthetic Protein Products

Plastein synthesis. Plasteins are mixtures of high molecular weight proteinaceous peptides. They are synthesized by enzyme-catalyzed growth of peptide chains from lower molecular weight peptides. The

Table 1. Proximate Analyses of Selected Algae Grown Photosynthetically

Organism	Content, % (dry wt basis)			
	Nitrogen	Crude Protein N × 6.25	Fat	Ash
Chlorella pyrenoidosa 71105 (*Sorokiniana*)	9.6	60	8.1	8.9
Chlorella regularis	9.3	58	16	6.7
Chlorella ellipsoidea and *Scenedesmus obliquus*	9.7	61	19.5	13.4
Scenedesmus acutus 276-3A	8–8.8	50–55	12–19	6–8
Spirulina maxima	9.9	62	2–3	

Table 2. Proximate Analyses of Selected Nonphotosynthetic Microorganisms Grown on Various Substrates

Organism	Substrate	Composition, g/100 g dry wt				
		Nitrogen	Protein	Fat	Crude fiber	Ash
Bacteria						
Acinetobacter cerificans	hexadecane	11	72			
Cellulomonas sp	bagasse	14	87	8		7
Methylophilus methylotrophus	methanol	13	83	7	< 0.05	8.6
Yeast						
Candida lipolytica (Toprina)	*n*-alkanes	10	65	8.1		6
Candida lipolytica (Toprina)	gas oil	11	69	1.5		8
Candida utilis	ethanol	8.3	52	7	5	8
	sulfite liquor	9	55	5		8
Kluyveromyces fragilis	cheese whey	9	54	1		9
Saccharomyces cerevisae	molasses	8.4	53	6.3		7.3
Molds and higher fungi						
Morchella hortensis	glucose	5.4	34	1.4		
Paecilomyces varioti (Pekilo)	sulfite waste liquor	9.1–10.11	57–63			

Table 3. Proximate Analysis of Leaf Protein Products

Leaf protein	Composition, % (dry wt)			
	Protein N × 6.25	Fat	Fiber	Ash
Pro-Xan II (alfalfa)	47	14	4	16
cauliflower (pressed leaves)	26	2.9	12.5	10.7
pea vine (spray-dried juice)	18.5	2.5	1.4	12.8

process by which plasteins are formed is called the plastein reaction and is the reverse of the proteolytic enzyme hydrolysis of peptide bonds of proteins. At present, plasteins are in the experimental stage of development.

Synthetic proteins. Apparently, these protenoids are digestible by mammalian proteinases and can serve as sources of nutrients for *Lactobacillus arabinosus* and *Proteus vulgaris*. The possibility exists that products could be nutritionally imbalanced, have mammalian toxicity and undesirable tastes, odors and stability. These problems must be investigated further before any assessment of the utility of polyamino acids or protenoids can be made.

Textured and Formulated Foods

The availability of protein concentrates and isolates prepared from soy bean protein has made it possible to produce a wide range of textured protein analogue foods in which the soy protein is substituted for animal protein. The Protein Advisory Group, *ad hoc*, is the working group of the WHO United Nations system involving WHO, FAO, and UNICEF. It has developed guidelines for the evaluation of novel sources of protein.

JOHN H. LITCHFIELD
Battelle, Columbus Labs.

M. Milner, N.S. Scrimshaw, and D.I.C. Wang, eds., *Protein Resources and Technology: Status and Research Needs*, Avi Publishing Co., Westport, Conn., 1978.

L.P. Hanson, *Vegetable Protein Processing*, Noyes Data Corp., Park Ridge, N.J., 1974.

J.H. Litchfield in H.J. Peppler and D. Perlman, eds., *Microbial Technology*, 2nd ed., Vol. 2, Academic Press, Inc., New York, 1979, pp. 93–155.

FOOD STANDARDS. See Food processing; Regulatory agencies.

FOOD TOXICANTS, NATURALLY OCCURRING

Toxicants are substances that, upon ingestion, produce changes in homeostasis that are threatening to the normal function of the organism. Some naturally occurring food toxicants are shown in Table 1.

Toxic Proteins, Peptides, Amides, and Amino Acids

Nitrogenous compounds are present in every living cell, and are the most frequently implicated natural toxicants in foods. Their mode of action is often obscure but they may be loosely grouped according to either gross manifestations or specific structural characteristics. Accordingly, vitamin-destroying enzymes, hemaglutenins, enzyme inhibitors, and many hepatotoxins (many of which are carcinogens) are of protein, peptide, or amino acid composition.

Thiocyanates and Related S-Containing Compounds

Goitrogens are compounds that produce goiter by interfering with thyroxine synthesis in the thyroid gland. Food-borne goitrogens are often characterized by the presence of sulfur and most are thiocyanates or closely related compounds. Because of their widespread occurrence in *Cruciferae*, eg, cabbage, kale, onions, cress, broccoli, cauliflower, rutabaga, turnip and radish, goitrogens are among the most common and longest-recognized substances of toxic nature in the human food supply.

Oxalates, Phytates, and Other Chelates

Of nutrient chelates in the human diet, oxalates and phytates are the most common. Oxalic acid, found principally in spinach, rhubarb leaves, beet leaves, and mushrooms, is a primary chelator of calcium. Phytic acid, found mainly in grain products, complexes a broader spectrum of minerals than does oxalic acid. Decreased availability of P is probably the most wiely recognized result of excessive intakes of phytic acid, yet Ca, Cu, Zn, Fe, and Mn are also complexed and rendered unavailable by this compound.

Vasoactive and Psychoactive Amines and Alkaloids

Most compounds producing hypertensive eposides are classified as amines and are found in greatest concentration in banana, plantain, tomato, avocado, pineapple, broad beans, and various cheeses. Amines that are vasoactive include dopamine, tyramine, histamine, tryptamine, noradrenaline, and dihydroxyphenylalanine (DOPA).

Caffeine, a xanthine derivative, is perhaps the best known of the natural stimulants and is found in coffee (qv) beans, tea leaves, and cola nuts.

Depressant symptoms, which include burning abdominal pain, decreased excitability, convulsions, nausea, and coma, become the general syndrome for all oral alkaloid poisoning. Discorine, alkaloid derivatives of lysergic acid produced by a parasitic fungus, and myristicin are some common alkaloids.

Antinutrients

Any substance that destroys, inactivates, or in other ways renders unavailable an essential dietary constituent can be termed an antinutrient. The most widely studied are antivitamins. Some antivitamins are enzyme thiaminase, niacin inhibitors, biotin antagonist, avidin, linatine, a pyridoxine antagonist, and pantothenic acid antagonists.

Vitamin Toxicity

Because fat-soluble vitamins (eg, A and D) tend to accumulate in the body with a relatively inactive mechanism for excretion, they cause greater toxicological difficulties than do water-soluble vitamins (see Vitamins, survey).

Of the water-soluble vitamins, overdoses of nicotinic acid and folic acid may cause toxic effects as may ascorbic acid in doses over 9 g/d.

Essential Minerals and Heavy Trace Elements

Toxic ingestion of essential minerals from naturally occurring foods is almost beyond comprehension. Cases involving human toxicity from heavy trace elements, such as Pb, Hg, As, and Cd, are not uncommon but almost exclusively traced to accidental contamination rather than true natural occurrences.

Cyanogenic Glycosides

Complex glycosides, which upon hydrolysis yield hydrocyanic acid, are found commonly among plant materials (see Cyanohydrins). The toxicity of this class of compounds is directly related to their liberation of HCN upon digestive hydrolysis. They are found in the bitter almond, pits of stone fruits, sorghum, and lima beans.

Table 1.   Some Common Naturally Occurring Food Toxicants

Compound	Toxin classification	Typical food sources
aflatoxin B_1	mycotoxin	moldy grains, nuts, oilseeds
amygdalin	cyanogenic glycoside	apricot pits, peach pits
avidin (chicken), mol wt ca 60,000	forms insoluble complex with biotin	raw egg white
caffeic acid	destroys thiamine	bracken fern
caffeine	alkaloid, stimulant	coffee, tea, cola nuts
capsaicin (5)	amide	*Capsicum* peppers
goitrin	goitrogen	cabbage, kale, onions, cress, cauliflower, broccoli, turnips
myristicin	alkaloid, psychoactive	nutmeg, mace, carrots
oxalic acid	reacts with calcium to reduce availability	rhubarb
phytic acid	reacts with calcium to reduce availability	oats
solanine	glycoalkaloid, antiacetylcholinesterase	potatoes, tomatoes, apples, eggplant
thiaminase, mol wt ca $(75–100) \times 10^3$	enzyme, inactivates thiamine	raw fish
tyramine	vasoactive amine	cheeses, bananas, plantains, tomatoes, pineapple

Nitrates, Nitrites, and Nitrosamines

The carcinogenicity of nitrosamines has created widespread concern over the safety of food products that are significant sources of nitrates and nitrites. Nitrates are found in fairly high concentrations in beets, spinach, kale, collards, eggplant, celery, and lettuce. Additionally, nitrates and nitrites are commonly used in the curing solutions of bacon, ham, and other cured meats.

Sodium Chloride

The fact that excessive intake of NaCl contributes to increased fluid retention has led to the conclusion that there may be a relationship between NaCl intake and hypertension. The Senate Select Committee on Nutrition and Human Needs has recommended the reduction of sodium chloride intake in average American diets.

Mycotoxins

The condition produced by the consumption of moldy foods containing toxic material is referred to as mycotoxicosis. Molds and fungi fall into this category, and several derive their toxicity from the production of oxalic acid, although the majority of mycotoxins are much more complex (see also Fungicides.)

Seafood Toxins

Several species of the moray eel have caused toxic reactions. The toxic principle appears to be proteinaceous, and is found predominately in the blood but it may occur also in the flesh as well. Pufferfish toxin has been identified as having the empirical formula $C_{11}H_{17}N_3O_8$. The liver of sharks, and other oily fishes sometime accumulate toxic levels of vitamin A.

FRED H. HOSKINS
Louisiana State University

Food Protection Committee, *Toxicants Occurring Naturally from Foods*, 2nd ed., NAS/NRC, Washington, D.C., 1973.

C. Moreau, *Moulds, Toxins, and Food*, 2nd ed., John Wiley & Sons, Inc., New York, 1979.

W.H. Lewis and M.P.F. Elvin-Lewis, *Medical Botany*, John Wiley & Sons, Inc., New York, 1977.

FORENSIC CHEMISTRY

Forensic chemistry is the field of knowledge that involves the application of the principles of chemistry and related sciences to the examination of physical evidence collected at the scene of a crime, and the presentation of the results of that examination in a court of law or public forum by an expert witness, a forensic chemist, who has either performed or supervised the tests (see also Analytical methods).

Arson

Several clues suggest arson. Fires that spread with exceptional rapidity for the type of structure in which they occur imply that an accelerant was used. A liquid accelerant such as gasoline, kerosene, or paint thinner may simply be poured, causing sites of contact to burn more rapidly than the adjacent areas. Surprising as it may seem, the charred residues contain detectable amounts of the liquid that can be identified by various analytical methods. Gas–liquid chromatography is by far the method preferred by forensic chemists. Another valuable method is ir spectroscopy (see Analytical methods).

Crimes of Violence

Hair samples, fiber samples, body fluids, soils, tire prints, shoe prints, glass fragments, paint fragments—virtually any physical specimen can yield valuable information to the crime-scene investigator. Gunshot residues deposited on a hand that discharges a pistol consist primarily of traces of barium, antimony, and lead. The amounts, on the order of micrograms, can be detected only by very sensitive analytical techniques such as neutron-activation analysis and atomic-absorption spectroscopy.

Strands of hair examined microscopically can reveal a number of significant characteristics, and permits distinguishing human from nonhuman hairs, as well as the race, color, body site and condition of human hair.

In a case of suspected homicide, a forensic pathologist may be required to make an autopsy to ascertain the cause of death. If there is evidence of poisoning, the forensic chemist or toxicologist analyzes blood and tissue samples for toxic substances.

The precipitin test is the most widely used to distinguish between human and animal blood and is based on serological principles. Today the analysis of blood stains has gone far beyond the classification into the four blood groups. Electrophoretic analysis yielding information as to enzymatic composition has become more and more valuable in establishing the origin of blood samples.

Driving While Under the Influence of Intoxicating Beverages (DWI)

In 1962 the National Safety Council in conjunction with the American Medical Association defined the limit of 0.10% blood alcohol (10 mg/100 mL) as presumptive evidence of being under the influence of alcohol. Analysis of blood, urine, or breath determines the alcohol content, with blood analysis being the most useful. It measures the alcohol in the blood stream that directly affects the central nervous system.

Drug Offenses

Confiscated drug samples are often adulterated. The components may be separated by the use of appropriate solvents. Generally, the steps taken for analysis proceed from microscopic examinations to color tests followed by any one or a combination of the tests shown in Table 1. The last two methods listed, although effective and reliable, are too expensive for routine examinations.

After automobile or other accidents in which the possible presence of drugs or alcohol creates a legal issue, blood analyses are performed. Although tlc, glc, and uv spectrophotometry are the primary means of identifying drugs, gc-ms, and fluorometric analyses are likely to be used increasingly in the future.

Explosive Devices

The type of explosive used in an incident can frequently be identified. Traces of explosive material may be present on the hands or clothes of a suspect or in a vehicle that was used, and may be identified. The residues are dissolved with acetone, and the components separated by tlc.

A commercial device called the Vapor-Phase Analyzer (VPA, Hydronautics, Inc., Lauren, Md.) can be employed to identify explosion vapors and even residues on a suspect's hand. The device is based on gas chromatography. In another method, the bomb residue is first leached with hot acetone and then with hot water followed by a series of instrumental tests.

Offenses Against Property

A burglar is likely to leave traces of his presence and to take with him evidence of the site. Paint fragments are first examined microscopically; then organic components are identified by pyrolysis gas chromatography, metals by optical emission spectroscopy. A variety of other techniques are also employed (see Analytical methods).

Table 1.   Analytical Methods for Drugs of Abuse

Method	Abbreviation
chromatography	
thin-layer	tlc
gas–liquid	glc
high-pressure liquid	hplc
spectroscopy	
ultraviolet	uv
visible	vis
infrared	ir
nuclear magnetic resonance	nmr
mass spectrometry	ms

Soil designates mineral substances found at a crime site, generally outdoors. It is identified by microscopic examinations followed by other analytical techniques discussed here.

Questioned Documents

The basic issue relating to questioned documents is their integrity. The forensic chemist may investigate possible alteration of date or contents by examination of the ink using tlc, hplc, uv and visible spectrophotometry.

Sexual Offenses

In virtually all sexual offenses, the identification of seminal fluid is vital. Microscopic examinations of vaginal swabs to identify human spermatozoa are routine procedures. Also, a sensitive color test for acid phosphatase, a major constituent of seminal fluid, is used.

THEODORE P. PERROS
George Washington University

P.M. Cain, *J. Forensic Sci. Soc.* **15**, 301 (1975).

E.T. Blake and G.E. Sensabaugh, *J. Forensic Sci.* **21**, 784 (1976).

FORMALDEHYDE

Formaldehyde, $H_2C=O$, is the first of the series of aliphatic aldehydes. Annual worldwide production capacity now exceeds 12×10^6 metric tons (calculated as 37% solution). Because of its relatively low cost, high purity, and variety of chemical reactions, formaldehyde has become one of the world's most important industrial and research chemicals.

At ordinary temperatures, pure formaldehyde is a colorless gas with a pungent, suffocating odor. Physical properties are summarized in Table 1.

Formaldehyde is produced and sold as water solutions containing variable amounts of methanol. These solutions are complex equilibrium mixtures of methylene glycol ($CH_2(OH)_2$) poly(oxymethylene glycols), and hemiformals of these glycols. Commercial formaldehyde–alcohol solutions are clear and remain stable above 16–21°C. They are readily

obtained by dissolving a high concentration formaldehyde in the desired alcohol.

Chemical Properties

Formaldehyde is surprisingly stable. Uncatalyzed decomposition is very slow below 300°C; extrapolation of kinetic data to 400°C indicates that the rate of decomposition is ca 0.44% min at 101 kPa (1 atm). The main products are CO and H_2. Metals such as platinum, copper, chromia, and alumina also catalyze the formation of methanol, methyl formate, formic acid, carbon dioxide, and methane.

At ordinary temperatures, formaldehyde gas is readily soluble in water, alcohols, and other polar solvents. Its heat of solution in water and the lower alipihatic alcohols is approximately 63 kJ/mol (15 kcal/mol). The reaction of unhydrated formaldehyde with water is very fast; the first-order rate constant at 22°C is 9.8 s^{-1}.

Formaldehyde is readily reduced to methanol by hydrogen over many metal and metal oxide catalysts. It is oxidized to formic acid or carbon dioxide and water. Formaldehyde condenses with itself in an aldol-type reaction to yield lower hydroxy aldehydes, hydroxy ketones, and other hydroxy compounds.

Since 1974, the cost of petrochemicals has increased rapidly. Thus, there has been renewed interest in condensation products of formaldehyde with synthesis gas (CO, H_2), such as ethylene glycol:

$$HCHO + CO + H_2O \xrightarrow[\substack{>30\,MPa \\ (300\,atm)}]{H_2SO_4} \underset{\text{glycolic acid}}{HOCH_2COH} \longrightarrow (CH_2OH)_2$$

A variety of condensation catalysts, including cobalt carbonyl, rhodium oxide, hydrogen fluoride, and fluoroboric acid, are reported to be effective (see Glycols).

Today, all of the world's commercial formaldehyde is manufactured from methanol and air by an older process using a metal catalyst and a newer one using a metal oxide catalyst. Reactor feed to the former is on the methanol-rich side of a flammable mixture and virtually complete reaction of oxygen is obtained; conversely, feed to the metal oxide catalyst is lean in methanol and almost complete conversion of methanol is achieved.

Silver-catalyst process. Methanol is oxidized over a silver catalyst. Domestic licensors include Borden Chemical Company and Davy Powergas, Inc. (Imperial Chemical Industries Ltd., process).

The silver-catalyzed reactions occur at essentially atmospheric pressure, at 600 to 650°C and can be represented by two simultaneous reactions:

$$CH_3OH + \tfrac{1}{2}O_2 \rightarrow HCHO + H_2O \ (\Delta H = -156 \text{ kJ or } -37.28 \text{ kcal})$$

$$CH_3OH \rightarrow HCHO + H_2 \ (\Delta H = +85 \text{ kJ or } 20.31 \text{ kcal})$$

Figure 1 is a flow diagram of a typical formaldehyde plant employing silver catalyst.

Table 1. Properties of Monomeric Formaldehyde

Property	Value
density, g/cm³	
at −20°C	0.8153
boiling point at 101.3 kPa^a, °C	−19
melting point, °C	−118
vapor pressure, Antoine constants, Pa^b	
A	6.32074
B	970.6
C	244.1
heat of vaporization, ΔH_v	
at 19°C, kJ/mol^c	23.3
heat of formation, $\Delta H_f°$ at 25°C, kJ/mol^c	−115.9
std free energy, $\Delta G_f°$ at 25°C, kj/mol^c	−109.9
heat capacity, $Cp°$, J/(mol·K)^c	35.4
entropy, $S°$, J/(mol·K)^c	218.8
heat of combustion, kJ/mol^c	561–571
critical constants	
temperature, °C	137.2–141.2
pressure, MPa^d	6.784–6.637
flammability in air	
lower/upper limits, mol %	7.0/73
ignition temperature °C	430

^{a}To convert kPa to mm Hg, multiply by 7.5.

^{b}Log$_{10}P_{Pa} = A - B/(C + t)$; $t = $ °C. To convert log$_{10}P_{Pa}$ to log$_{10\,mm\,Hg}$, add 0.87506 to A.

^{c}To convert J to cal, divide by 4.184.

^{d}To convert MPa to atm, divide by 0.101.

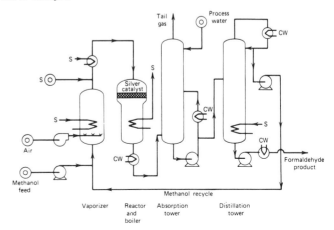

Figure 1. Flow scheme of a typical silver-catalyst process. S = steam; CW = cooling water.

Table 2. Health and Safety Factors for Formaldehyde Concentrations

	Value
exposure limits	
8-h time-weighted average, ppm	3
ceiling	5
max peak above ceiling, ppm for 30 min	10^a
minimum levels for sensory detection	
airborne, odor detectable at ppm	1
in water, odor detectable at mg/L	20–50
profuse lachrymation, ppm	20
eye irritation[b], ppm	0.05–0.5

[a]ACGIH: Threshold limit value (TLV): 2 ppm, ceiling.
[b]Experienced by some persons.

Metal oxide-catalyst process. An iron oxide–molybdenum oxide catalyst is the choice today. Catalysts are improved by modification with small amounts of other metal oxides, support on inert carriers, and methods of preparation and activation.

Storage and Handling

Solutions of formaldehyde are unstable. Both formic acid (acidity) and paraformaldehyde concentrations increase with time and depend on temperature. Materials of construction preferred for storage vessels are 304, 316, and 347-type stainless steels or lined carbon steels.

Health and Safety Factors

Table 2 shows health and safety data.

Uses

As a basic chemical building unit, formaldehyde is an intermediate in a large variety of organic compounds ranging from amino and phenolic resins (qv) to slow-release fertilizers (see Fertilizers; amino resins). Formaldehyde also is used in the synthesis of chelating agents, textile finishes, and acetal resins.

Derivatives of formaldehyde include 1,4-butanediol, hexamethylenetetraamine, paraformaldehyde, and trioxane and tetraoxane.

H.R. Gerberich
A.K. Stautzenberger
W.C. Hopkins
Celanese Chemical Company, Inc.

J.F. Walker, *Formaldehyde*, 3rd ed., Reinhold Publishing Corp., New York, 1974.

H. Diem and A. Hilt "Formaldehyde," in *Ullman's Encyclopädie der technischen Chemie*, Verlag Chemie, Weinheim/Bergstr., Ger., 1976, Vol. 2, pp. 687–702.

FORMALS. See Formaldehyde.

FORMAMIDE, HCONH₂. See Formic acid and derivatives.

FORMIC ACID AND DERIVATIVES

FORMIC ACID

Formic acid (methanoic acid) is a colorless, odorous acid, the first and by far the strongest of the unsubstituted series of carboxylic acids (qv). It is important in textile souring, leather preparation, cattle-fodder preservation, and manufacture of chemicals. It can be degraded biologically or chemically to innocuous substances in most environments. The anhydrous acid must be handled with the same care as concentrated sulfuric acid. It is a powerful dehydrating agent and serious burns can result from its action on the skin. It burns when ignited.

Table 1. Physical Properties of Formic Acid

Property	Value
melting point, °C	8.4
boiling point, °C	100.7
density at 20°C, g/cm³	1.220
refractive index, n_D^{25}	1.369
surface tension at 20°C, mN/m (= dyn/cm)	37.58
viscosity at 20°C, mPa·s (= cP)	1.784
heat of vaporization at 25°C, kJ/mol[a]	20.10
heat capacity at 17°C, C_p, J/(mol·K)[a]	98.78
molar entropy at 25°C, J/(mol·K)[a]	129.0
specific conductance, Ω·cm	6.08×10^{-5}
dielectric constant, 25°C	56.1
self-ionization constant	2.20×10^{-7}
dissociation constant in H₂O	3.739 ± 0.001

[a]To convert J to cal, divide by 4.184.

Physical properties of formic acid are given in Table 1.

Formic acid may react both as an acid and as an aldehyde because the carboxyl is bound to a hydrogen rather than an alkyl group. It is not a ready electron donor and its ion has far less tendency than acetate to enter the coordination spheres of various transition-metal ions (see Coordination compounds). It decomposes readily by dehydration, dehydrogenation, or through a bimolecular redox reaction.

Formic acid is produced mainly as a by-product of liquid-phase oxidation of hydrocarbons to acetic acid. It is shipped in tank cars, tank trucks, drums and carboys; a DOT white label is mandatory. Stainless steel is the construction material of choice for handling formic acid; glass and ceramics are also satisfactory. Polyethylene and rubber furnish satisfactory liners for equipment exposed to it.

When one is in contact with formic acid, its hazardous properties should be considered and safety precautions should be observed. Mask, gloves, aprons, boots, and goggles are needed whenever the concentrated acid is handled. Safety shower, eye-washing facilities, and good ventilation of the area should be provided. The OSHA standard is 5 ppm per 8 h TWA.

Frank S. Wagner, Jr.
Celanese Chemical Co.

H.W. Gibson, *Chem. Rev.* **69**, 673 (1969).

P. Mars, J.J.F. Scholten, and P. Zwieteing, *Adv. Catal.* **14**, 35 (1963).

N.M. Emanuel, E.T. Denisov, and Z.K. Maizus, *Liquid Phase Oxidation of Hydrocarbons* (Engl. translation by B.J. Hazzard), Plenum Press, New York, 1967.

FORMAMIDE

Formamide (methanamide) is a clear, hygroscopic liquid with viscosity similar to that of glycerol and a faint odor of ammonia. It is soluble in water, lower alcohols, and glycols, and insoluble in hydrocarbons, chlorinated solvents, and ethers. Formamide is a good solvent for proteins and salts owing to its high dielectric constant. Main applications are as solvent in the chemical industry, as softener for paper, as intermediate for the manufacture of formic acid, its esters, and hydrocyanic acid, and as a reaction medium.

Formamide has been detected in space and is the only molecule found in interstellar clouds containing four different elements.

Table 1 lists the important physical properties of formamide.

Today, formamide is produced by one of two continuous processes: in the direct synthesis, NH₃ and CO react at 10–30 MPa (100–300 atm) and 80–100°C in methanolic sodium methoxide. The two-stage synthesis, developed by BASF, presents fewer problems:

$$CO + CH_3OH \rightarrow HCOOCH_3$$
$$HCOOCH_3 + NH_3 \rightarrow HCONH_2 + CH_3OH$$

Table 1. Physical Properties of Formamide

Property	Value
boiling point, °C	210
density at 20°C, g/cm³	1.133
dielectric constant	109 ± 1.5
freezing point, °C	2.55
heat of vaporization, kJ/mol[a]	75.4
refractive index, n_D^{25}	1.4468
specific heat at 19°C, kJ/(kg·K)[a]	2.30
vapor pressure, kPa[b]	
at 70.5°C	0.13
flash point, °C	175

[a] To convert J to cal, divide by 4.184.
[b] To convert kPa to mm Hg, multiply by 7.5.

Recently, a process using ammonia, hydrogen, carbon dioxide, and a special metal–organophosphorus catalyst has been proposed.

Health and Safety Factors

Formamide has a low acute toxicity. Precautions that should be observed when handling formamide include avoidance of prolonged inhalation of vapors or contact of the liquid with the skin or eyes. Women of child-bearing age should not be employed in the production and processing of formamide.

CLAUDIO L. EBERLING
BASF Aktiengesellschaft

Formamide, Technical Leaflet, BASF Aktiengesellschaft, 67 Ludwigshafen, FRG, 1973.

H. Kiefer and H. Lang, Ullmann's Encyklopadie der Technischen Chemie, Vol. 11, Verlag Chemie GmbH, Weinheim, FRG, 1976, pp. 703–705.

DIMETHYLFORMAMIDE

Dimethylformamide, $HCON(CH_3)_2$, DMF, mol wt 73.09, is a colorless, high-boiling, polar liquid with a faint but typical odor. It can be distilled without decomposition and is completely miscible with water and many organic compounds but not with aliphatic hydrocarbons. As an aprotic liquid with a high dielectric constant, DMF is an excellent solvent. It was first synthesized in 1893; however, it gained technical importance only in the 1950s as solvent for the spinning of acrylic fibers. Since then, many new applications have been found; for example, as solvent for polyurethane and polyamide coatings and films; selective medium for extraction of aromatics from crude oil; reaction medium; selective solvent of gases; solvent for dyes; solvent for electrolytes in galvanization baths; and paint remover and cleaner.

The physical properties of DMF are listed in Table 1.

Table 1. Physical Properties of DMF

Property	Value
boiling point, °C	153
density at 25°C, g/cm³	0.9445
dielectric constant at 25°C	36.7
dipole moment, C·m (D)	12.7×10^{-30} (3.82)
freezing point, °C	−61
heat of vaporization at 153°C and 101 kPa (1 atm), kJ/mol[a]	42.3
refractive index, n_D^{20}	1.4301
specific heat, kJ/(kg·K)[a]	
20°C	2.03
vapor pressure at 20°C, kPa[b]	0.35
flash point, °C	58

[a] To convert J to cal, divide by 4.184.
[b] To convert kPa to mm Hg, multiply by 7.5.

The most commonly used processes are the one-stage reaction of carbon monoxide with dimethylamine and the two-stage process with methyl formate and dimethylamine. Manufacturers of DMF in the United States include Air Products and E.I. du Pont de Nemours & Co., Inc.

Health and Safety Factors

Single exposures to DMF by inhalation, skin contact, or ingestion are not considered hazardous.

CLAUDIO L. EBERLING
BASF Aktiengesellschaft

Dimethylformamide Technical Leaflet, BASF Aktiengesellschaft, 67 Ludwigshafen, FRG, 1980.

Properties and Uses of Dimethylformamide, E.I. du Pont de Nemours & Co., Inc., Wilmington, Del., 1972.

FRICTION MATERIAL.   See Brake linings and clutch facings.

FRIEDEL-CRAFTS REACTIONS

A number of reactions bear the general name Friedel-Crafts and a large number of other reactions are related to this type. Today, Friedel-Crafts-type reactions are considered to be any substitution, isomerization, elimination, cracking, polymerization, or addition reaction that takes place under the catalytic effect of Lewis acid-type acidic halides (with or without co-catalysts) or related proton acids.

The important industrial processes for production of high octane gasoline, ethylbenzene, synthetic rubber, plastics, and detergent alkylates are all based on the Friedel-Crafts principle.

Despite the seemingly wide variety of types, Friedel-Crafts reactions can be divided into two general categories: alkylations and acylations. Within these two broad areas there is considerable diversity.

Alkylation

Alkylation of aromatic compounds.   In the usual Friedel-Crafts alkylation of aromatics, a hydrogen atom of the aromatic nucleus is replaced by an alkyl group through the interaction of an alkylating agent in the presence of Friedel-Crafts catalyst. The most frequently used alkylating agents are alkyl halides, alkenes, alkynes, alcohols, esters (of carboxylic and inorganic acids, ethers, aldehydes and ketones, paraffins and cylcoparaffins, thiols (mercaptans) sulfides, amines (via diazotization), and thiocyanates.

The overall reaction using alkyl halides, alcohols, and alkenes as alkylating agents in the presence of aluminum chloride may be written as follows:

$$C_6H_6 + RX \xrightarrow{AlCl_3} C_6H_5R + HX$$
$$C_6H_6 + ROH \xrightarrow{AlCl_3} C_6H_5R + H_2O$$
$$C_6H_6 + RCH{=}CH_2 \xrightarrow{AlCl_3} \underset{\underset{C_6H_5}{|}}{RCHCH_3}$$

The action of aluminum chloride is not restricted to the introduction of alkyl groups into aromatics. It also removes alkyl groups from alkylbenzenes (dealkylation, transalkylation, disproportionation). One of the most important commercial alkylations, is the preparation of ethylbenzene from benzene and ethylene.

Alkylation frequently occurs in a heterogeneous reaction system, specifically in the acidic-catalyst layers. The tendency toward polysubstitution during Friedel-Crafts alkylation is owing to the extraction of the more nucleophilic alkylation reaction products in this catalyst layer. The tendency toward polysubstitution may be minimized by the use of an excess of the aromatic compound to be alkylated and a mutual solvent for the hydrocarbon and catalyst layer, or by operating in the vapor phase over heterogeneous catalysts.

An additional consideration is the orientation involved in the introduction of more than one alkyl group. Alkylbenzenes generally direct into the ortho and para positions in kinetically controlled reactions. Under the more vigorous conditions (reaction time and temperature, relative amount of catalyst, absence of solvent, etc), the tendency is toward the formation of the thermodynamically favored meta derivatives.

A characteristic of the Friedel-Crafts alkylations is the tendency of alkyl groups to rearrange during the alkylation process. As isomerization and disproportionation are always possible during alkylations, these side reactions should be considered wherever Friedel-Crafts conditions are employed.

Alkylation of benzene with *tert*-butyl chloride at RT in the presence of $AlCl_3$ and an isoparaffin as the solvent, produces not only *tert*-butylbenzene but also alkylbenzene in which the alkyl group is formed from the isoparaffin.

Stereoselective alkylation of benzene with optically active 4-valerolactone catalyzed by $AlCl_3$ with 50% net inversion of configuration was observed.

Haloalkyl groups can be introduced directly by processes similar to Friedel-Crafts alkylations into aromatic and, to a certain degree, into aliphatic compounds.

As a rule, Friedel-Crafts alkylation of aromatics with difunctional compounds leads partly to open-chain alkylates but, if conditions are suitable, mainly to cyclialkylation products.

Aliphatic Compounds

Alkylation of olefins by branched-chain paraffins, first carried out over aluminum chloride, made it possible to alkylate olefins with paraffins of normal and branched structures.

$$\text{isobutane} \xrightarrow{\text{ethylene}} \text{hexanes} \xrightarrow{\text{ethylene}} \text{octanes} \xrightarrow{\text{ethylene}} \text{decanes}$$

The first industrial aliphatic alkylation process, the sulfuric acid catalyzed alkylation of branched paraffins with olefins, was developed in 1938. Because of its simplicity, low cost, the wide variety of raw materials, and the high quality alkylate, this process is still widely used. In the presence of 96–97% sulfuric acid, isobutane $(CH_3)_2CHCH_3$ and other isoalkanes are easily alkylated by olefins (with the exception of ethylene), producing alkylates with high octane numbers. The suggested reaction pathway is shown below:

$$R_2C{=}CH_2 \xrightarrow{H^+} R_2\overset{+}{C}CH_3 \xrightarrow{R_3CH} R_2CHCH_3 + R_3'C^+$$

$$R_3'C^+ + R_2C{=}CH_2 \rightarrow R_2\overset{+}{C}{-}CH_2CR_3' \xrightarrow{H^-} R_2CHCH_2CR_3'$$

Arylation of Aromatic Compounds

In contrast to the Friedel-Crafts alkylations, arylations of aromatics are by no means as well known or as easy to carry out. They include dehydrogenating condensations (Scholl reactions) and arylation with diazonium halides.

The elimination of the two ring hydrogens accompanied by the formation of an aryl–aryl bond under Friedel-Crafts conditions is known as the Scholl reaction. The dehydrogenating condensations can take place in either an inter- or an intramolecular way. Intermolecular Scholl reactions are numerous and include such reactions as formation of biphenyl from benzene:

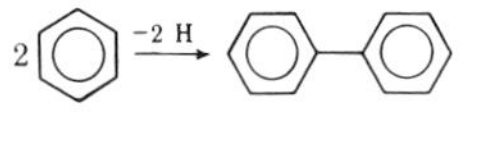

These reactions generally involve radical ions formed under oxidative conditions. In the decomposition reaction of aryldiazonium tetrafluoroborates in aromatics (eg, nitrobenzene), it was found that in addition to fluorobenzene, arylated aromatics are formed. Ionic or radicalcation type arylation reactions seem to take place.

Isomerization of Saturated Hydrocarbons

The ability of Friedel-Crafts-type superacids to protonate not only π-bonds but also the weaker bonds of hydrocarbons at relatively low temperatures allows the isomerization of saturated hydrocarbons which is of substantial practical importance. Isomerization of straight-chain hydrocarbons is of particular importance in obtaining so-called lead-free gasoline (qv) of suitable octane number Addition of higher octane aromatic hydrocarbons or olefins is less advisable because of pollution problems and their carcinogenic nature. Efficient additives are methyl or *tert*-butyl ethers.

Polymerization

The Friedel-Crafts polymerization of olefins did not attract much attention until the 1930s (see Polymerization). Since then a considerable amount of interest has been focused on olefin polymerizations catalyzed by $AlCl_3$ and BF_3 and a variety of other Lewis-acid-type halides and protic acids; three main types are recognized: conversion of low molecular weight olefins to gasoline-range olefins; conversion to intermediate molecular weight polymers for use as synthetic lubricant oils; and conversion to high molecular weight polymers having 5,000–100,000 monomer units (eg, butyl rubber).

Acylation

Acylation of Aromatic Compounds

In the Friedel-Crafts ketone synthesis, an acyl group is introduced into the aromatic nucleus by an acylating agent such as an acyl halide, acid anhydride, ester, or the acid itself:

$$C_6H_6 + RCOX \xrightarrow{AlCl_3} C_6H_5COR + HX$$

$$X = \text{halide, —OH, —OR, —OCOR}$$

Ketenes, amides, and nitriles also may be used; aluminum chloride and boron trifluoride are the most common catalysts (see Ketones).

Friedel-Crafts acylation of aromatics is of considerable practical value owing to the importance of aryl ketones and aldehydes.

Cyclic ketones can be prepared by intramolecular Freidel-Crafts acylation of an aromatic ring that has an acyl halide group in an attached side chain. The method is used in the preparation of hydrindones, tetralones, chromones, xanthones, etc.

In the presence of aluminum chloride and a small amount of cuprous halide, a mixture of hydrogen chloride and carbon monoxide serves as a formylating agent of aromatics:

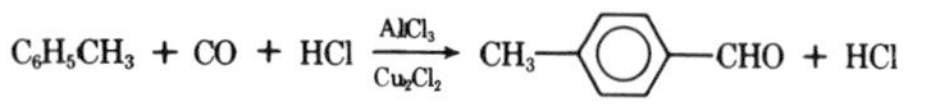

$$C_6H_5CH_3 + CO + HCl \xrightarrow[Cu_2Cl_2]{AlCl_3} CH_3{-}\!\!\bigcirc\!\!{-}CHO + HCl$$

p-tolualdehyde

$HF{-}BF_3$ is also a particularly effective catalyst for the reactions that gained importance in the manufacture of terephthalic acid from toluene by subsequent oxidation of *p*-tolualdehyde.

The Gattermann-Koch synthesis is suitable for the preparation of simple aromatic aldehydes from benzene and its substituted derivatives, as well as from polycyclic aromatics.

The stability of formyl fluoride, HCOF, allows a direct, simple Friedel-Crafts formylation reaction:

$$C_6H_6 + FCHO \rightarrow C_6H_5CHO + HF$$

The Gattermann-Koch reaction cannot generally be applied to formylation of phenolic compounds. To overcome this difficulty, a mixture of hydrogen cyanide and hydrogen chloride is allowed to react with phenols in the presence of zinc or aluminum chloride.

Aromatic and heterocyclic compounds are formylated by reaction with dialkyl- or alkylarylformamides in the presence of phosphorus oxychloride or phosgene.

$$C_6H_6 \;+\; \underset{R'}{\overset{R}{\diagdown}}N-\overset{\overset{O}{\|}}{\underset{H}{C}} \xrightarrow{POCl_3} C_6H_5CHO \;+\; HN\underset{R}{\overset{R'}{\diagup}}$$

Mercuric fulminate reacts with benzene and aluminum chloride containing some hydrated salt to give benzaldoxime.

The general Friedel-Crafts acylation principle can be successfully applied to the preparation of aromatic carboxylic acids. Carbonyl halides are diacyl halides of carbonic acid. Phosgene or oxalyl chloride react with aromatic hydrocarbons in the presence of a Friedel-Crafts catalyst to give aroyl chlorides which yield acids on hydrolysis.

Under Friedel-Crafts reaction conditions, sulfonyl halides and sulfonic acid anhydrides sulfonylate aromatics, a reaction that can be considered as the analogue of the related acylation with acyl halides and anhydrides. The products are sulfones (instead of ketones obtained in the acylation reaction). Sulfonyl chlorides are the most frequently used reagents, although the bromides and fluorides also react:

$$C_6H_6 + RSO_2Cl \xrightarrow{AlCl_3} C_6H_5SO_2R + HCl$$

The general Friedel-Crafts acylation principle can be applied in the case of sulfonation to the halides and anhydride of sulfuric acid (halosulfonic acids, sulfur trioxide).

Benzene and its homologues react with SO_2 in the presence of $AlCl_3$ and HCl to form sulfinic acids.

$$C_6H_6 + SO_2 \xrightarrow[HCl]{AlCl_3} C_6H_5SO_2H$$

The general Friedel-Crafts acylation principle can also be applied to nitrations involving nitryl halides and dinitrogen pentoxide (the halides and the anhydride of nitric acid).

The direct introduction of the amino group into aromatic hydrocarbons was first reported by Graebe in 1901. Hydroxylamine hydrochloride in the presence of aluminum or ferric chlorides reacts with benzene or toluene to give a low yield of aniline or toluidine.

The stability of perchloryl fluoride, $FClO_3$, made possible the development of a new type of aromatic substitution reaction, perchlorylation, a reaction closely related to Friedel-Crafts acylation.

The halogenation of a wide variety of aromatic compounds, a generalized acylation reaction, proceeds readily in the presence of ferric chloride, aluminum chloride, and related Friedel-Crafts catalysts. Halogenating agents used are elementary chlorine, bromine, or iodine, and interhalogens (such as iodine monochloride and bromine monochloride):

$$C_6H_6 + X_2 \xrightarrow{FeCl_3} C_6H_5X + HX \; (X = Cl, Br, I)$$

The direct synthesis of phenol from benzene is of special interest since phenol is widely used in industry and benzene is a relatively cheap starting material for phenol. The electrophilic hydroxylation of benzene with H_2O_2 or peracids, catalyzed by $AlCl_3$, BF_3, etc, however frequently results in a mixture also containing polyhydroxybenzenes and oxidized products.

Acylation of Aliphatic and Cycloaliphatic Compounds

Similar to alkylation, not only aromatic but also aliphatic and cycloaliphatic compounds undergo Friedel-Crafts acylation reactions.

Olefins and cycloolefins give unsaturated ketones with acylhalides. Saturated chloroketones are formed as intermediates, usually with elimination of HCl. Similar products may be obtained by using acid anhydrides. Acylation of acetylenic compounds produces β-chlorovinyl ketones. Even saturated hydrocarbons may give ketones with acyl chlorides.

Friedel-Crafts-type aliphatic aldehyde syntheses are considerably rarer than the corresponding aromatic syntheses. However, the hydroformylation reaction of olefins and the related oxo synthesis are effected by a catalyst that is a strong acid. (see Oxo reaction).

Alkenes are carbonylated in the presence of acid catalysts at 75–100°C and under pressures of 60–90 MPa (600–900 atm) to give carboxylic acids. Olefins are carbonylated in concentrated sulfuric acid at moderate temperatures (0–40°C) and low pressures with formic acid, which serves as the source of carbon monoxide.

Alkanes also undergo electrophilic nitration with nitronium salts such as $(NO_2)^+(PF_6)^-$ in a protic solvent such as CH_2Cl_2 or sulfolane.

Catalysts

Friedel-Crafts-type catalysts are electron acceptors, and fall into the general class of acids as defined by Lewis (see Catalysis). The most frequently used are aluminum chloride and bromide.

Acidic halide catalysts effect Friedel-Crafts reactions with non-bonded electron-donor reagents (halides, oxygen, sulfur, etc, compounds) through their Lewis acid activity. Consequently, Lewis acidic metal alkyls and metal alkoxides also possess catalytic activity in Friedel-Crafts reactions.

Friedel-Crafts reactions of hydrocarbons carried out with metal halide catalysts are, indeed, proton-catalyzed as shown by the very close relationships of these reactions to those carried out with proton acid (Brønsted acid) catalysis. Sulfuric acid and hydrogen fluoride are among the most widely used.

Chalcogenide catalysts include a great variety of solid oxides and sulfides. The most widely used comprise alumina, silica, and mixtures of alumina and silica (either natural or synthetic), in which other oxides such as chromia, magnesia, molybdena, thoria, tungstic oxide, and zirconia may also be present, as well as certain sulfides such as sulfides of molybdenum.

Brønsted acid catalytic activity has resulted in the successful use of so-called solid acids (acidic chalcogenides) in a variety of reactions.

In the 1960s a class of acids millions (10^6) of times stronger than mineral acids was discovered; acids stronger than 100% sulfuric acid are called superacids. They include protic superacids, Brønsted-Lewis superacid mixtures, and solid superacids.

A number of effective Friedel-Crafts systems are capable of forming cations in methathetic reactions. Thus, their activity is not catalytic because they are generally consumed in the cation-forming reaction. Anhydrous silver salts in conjunction with halide reagents are representatives of this class of compounds.

Solvents

Generally, in Friedel-Crafts processes, an excess of the hydrocarbon that is undergoing substitution is used as solvent. Therefore, in alkylations or acylations of benzene, the latter compound is used as solvent. Other frequently used solvents include carbon disulfide, petroleum ether, ethylene chloride, methylene chloride, carbon tetrachloride, nitromethane, nitrobenzene, etc.

GEORGE A. OLAH
DAVID MEIDAR
University of Southern California

G.A. Olah, ed., *Friedel-Craft and Related Reactions*, Vols. 1–4, Interscience Publishers, a division of John Wiley & Sons, Inc., New York, 1963–1965.

G.A. Olah, ed., *Friedel-Crafts Chemistry*, John Wiley & Sons, Inc., New York, 1973.

C.A. Thomas, *Anhydrous Aluminum Chloride in Organic Chemistry*, Reinhold Publishing Corp., New York, 1961.

FRUIT JUICES

The composition of a number of fruit juices and concentrates is shown in Table 1.

Variety and maturity are important factors affecting suitability for juice production. For some juices such as citrus or grape, only a few varieties are used for a distinctive flavor or because of freedom from bitterness.

Table 1. Composition of Fruit Juices, per 100 g

Fruit juice	Water, %	Food energy, kJ[a]	Proteins, g	Fat, g	Carbohydrates, g	Ash, g	Ca, mg	P, mg	Fe, mg	Na, mg	K, mg	Vitamin, A, RE[b]	Thiamin, mg	Riboflavin, mg	Niacin, mg	Ascorbic acid, mg
acerola	94.3	96	0.4	0.3	4.8	0.2	10	9	0.5	3			0.02	0.06	0.4	1600
apple	87.8	197	0.1	tr[d]	11.9	0.2	6	9	0.6	1	101		0.01	0.02	0.1	1
grapefruit, S-S[c]	89.2	172	0.5	0.1	9.8	0.4	8	14	0.4	1	162	33	0.03	0.02	0.2	34
sweetened	86.4	222	0.5	0.1	12.8	0.4	8	14	0.4	1	162	33	0.03	0.02	0.2	31
conc frozen	62	607	1.9	0.4	34.6	1.1	34	60	0.4	4	604	100	0.14	0.06	0.7	138
grapefruit orange blend, S-S[c]	88.7	180	0.6	0.2	10.1	0.4	10	15	0.3	1	184	333	0.05	0.02	0.2	34
sweetened	86.9	209	0.5	0.1	12.2	0.4	10	15	0.3	1	184	333	0.05	0.02	0.2	34
conc[e]	59.1	657	2.1	0.5	37.1	1.2	29	47	0.4	2	623	1276	0.23	0.03	1.1	144
grape, S-S[c]	82.9	276	0.2	tr[d]	16.6	0.3	11	12	0.3	2	116		0.04	0.02	0.2	tr[d]
conc[e] sweetened	52.8	766	0.6	tr[d]	46.3	0.3	10	15	0.4	3	118	67	0.06	0.10	0.7	15
grapejuice drink	86.0	226	0.1	tr[d]	13.8	0.1	3	4	0.1	1	35		0.01	0.01	0.1	16
lemon, S-S[c]	91.6	96	0.4	0.1	7.6	0.3	7	10	0.2	1	141	67	0.03	0.01	0.1	42
conc[f]	58.0	485	2.3	0.9	37.4	1.4	33	47	0.9	5	658	266	0.14	0.06	0.3	230
lemonade, conc[e]	48.5	816	0.2	0.1	51.1	0.1	4	6	0.2	2	70	67	0.02	0.03	0.3	30
lime	90.3	109	0.3	0.1	9.0	0.3	9	11	0.2	1	104	33	0.02	0.01	0.1	21
limeade, conc[d]	50.0	783	0.2	0.1	49.5	0.2	5	6	0.1	tr[d]	59	tr[d]	0.01	0.01	0.1	12
orange, S-S[c]	87.4	201	0.8	0.2	11.2	0.4	10	18	0.4	1	199	666	0.07	0.02	0.3	40
sweetened	86.5	218	0.7	0.2	12.2	0.4	(10)[c]	18	0.4	1	(199)[g]	666	0.07	0.02	0.3	40
conc frozen[e]	58.2	661	2.3	0.2	38.0	1.3	33[c]	55	0.4	2	657	2360	0.30	0.05	1.2	158
dehydrated	1.0	1590	5.0	1.7	88.9	3.4	84	134	1.7	8	1728	5590	0.67	0.21	2.9	359
orange apricot juice drink	86.7	209	0.3	0.1	12.7	0.2	5	8	0.1	tr[d]	94	1930	0.02	0.01	0.2	16
pear nectar	86.2	218	0.3	0.2	13.2	0.1	3	5	0.1	1	39	tr[d]	tr[d]	0.02	tr[d]	tr[d]
pineapple, S-S[a]	85.6	230	0.4	0.1	13.5	0.4	15	9	0.3	1	149	167	0.05	0.02	0.2	9
conc frozen[e]	53.1	749	1.3	0.1	44.3	1.2	39	29	0.9	3	472	167	0.23	0.06	0.9	42
pineapple-grapefruit juice drink	86.0	226	0.2	tr[d]	13.6	0.2	5	5	0.2	tr[d]	62	33	0.02	0.01	0.1	16
prune	80	322	0.4	0.1	19.0	0.5	14	20	4.1	2	235		0.01	0.01	0.4	2
tangerine, S-S[a]	88.8	180	0.5	0.2	10.2	0.3	18	14	0.2	1	178	1400	(0.06)[g]	(0.02)[g]	(0.01)[g]	22
conc frozen[e]	58	678	1.7	(0.7)[g]	38.2	1.3	62	48	0.7	2	613	4860	0.20	0.06	0.4	96
tomato	93.6	79	0.9	0.1	4.3	1.1	7	18	0.9	200	227	2660	0.05	0.03	0.8	16

[a] To convert kJ to kcal, divide by 4.184.
[b] RE = 33.3 IU; RE = Retinal Equivalent; IU = International Units.
[c] Single-strength.
[d] Trace.
[e] Concentrated, must be diluted with water before use.
[f] Lemon juice concentrate is a product in itself.
[g] Imputed value.

Juice factories frequently employ field persons to monitor the application of sprays to the growing crops so that residues on harvested fruit are within prescribed limits (see Trace and residue analysis). They also may sample the crop before harvest for analysis and coordinate harvesting with factory-production schedules.

Manufacture

On delivery to the plant, the fruit is passed over a roller conveyor or between revolving brushes to remove leaves or other debris before entering a tank containing a chlorine solution or detergent. Strong currents of wash solution direct the floating fruit toward a roller elevator over which sprays of fresh water remove the wash solution.

The fruit receives final inspection on roller conveyors to ensure freedom from imperfections before it is segregated into different size grades.

A variety of methods is used to break the fruit for juice release depending on the structure of raw material, the clarity desired in the final juice, and enzymatic discoloration and destruction of pectin. The most common disintegrator is the hammermill or a variation of it where fixed or free-swinging hammers force the fruit particles through a screen.

When mechanical disintegration is not feasible because of pit breakage, whole peaches or plums are heated to $\geq 93°C$ in an enclosed screw or ribbon conveyor by direct injection of culinary-grade steam. Retention time may be 10–20 min until oxidizing enzymes are inactivated and the flesh separates easily from the whole pits in a pulper.

Stemmers are used to separate grape berries and other fruit harvested as clusters from the stems, vine, or other debris and reduce the berries to a pulp.

Many fruit pulps have a mucilaginous character which makes juice extraction difficult, especially with continuous presses. The nature of this mucilage is not known but it is commonly believed to be due to pectin (see Gums). The slipperiness of the pulp can be overcome somewhat by the addition of press aid. Several types of press aid are used: diatomaceous earth (see Diatomite), coarse wood flour, rice hulls, and bleached and unbleached wood pulp.

Owing to composition and structural differences in fruits, a variety of procedures is used to extract juice. They are divided into two classes: those used for citrus and those used for deciduous fruits.

The clarities of juices vary widely; citrus and tomato juice have high insoluble-solids contents, whereas cranberry, grape, and many apple juices have little or no suspended solids and are nearly translucent. Thus, clarification processes are sometimes used in the production of clear juices. The preference for cloud in some juices probably results from the adherence of the flavoring constituents to fruit solids.

Preservation

Fruit juices often are pasteurized to prevent spoilage. Chemical preservatives such as potassium and sodium salts of sorbic, benzoic, and sulfurous acids also are used (see Sorbic acid). Some juices may be preserved simply by freezing.

Concentration and Dehydration

Many juices are sold to the consumer in concentrated form. During concentration by evaporation (the most common method), aromatic flavoring materials may be lost and must be recovered.

Fruit juices may also be dehydrated and sold in powdered form, although drying pure fruit juices is difficult because of the low melting points of the solids and their hygroscopicity.

JAMES C. MOYER
New York State Agricultural Experiment Station

R.M. Smock and A.M. Neubert in *Apples and Apple Products*, Interscience Publishers, New York, 1950, pp. 313–376.

L.R. Mattick and J.C. Moyer, *J. Assoc. Off. Anal. Chem.* **66**(5), 1251 (1983).

FUEL CELLS. See Batteries and electric cells, primary.

FUELS, SURVEY

Until a few years ago, fuel generally referred only to those substances that could be ignited in air and continued to burn and to release heat that could be used to meet the needs of society. The principal fuels include the so-called noncommercial fuels (wood, wastes, agricultural products, and animal dung), and commercial fuels (peat, lignites, coal, oil, and natural gas). On reacting with oxygen in the air, these fuels produce heat and residual products that consist mainly of carbon dioxide and water vapor.

Recently, with advances in science and technology, the term fuels has been expanded to include other means of supplying heat and energy; these include reactions of a variety of types, eg, rocket fuels, and various forms of nuclear energy, eg, uranium and, possibly at some later date, thorium. The renewable energy resources include hydroelectric, solar, wind, geothermal, and tidal energy sources, which can also be used to provide heat and power.

Fuel Consumption and Productivity

The historical relationship between economic growth and energy use suggests that to increase economic growth, additional amounts of energy are required; however, there may be substantial flexibility in the amount of additional energy required per unit of increased gross domestic product (GDP); the aggregate value of domestically produced goods and services.

After the oil embargo of 1973–1974, which was accompanied by an abrupt increase in the price of oil and of other energy forms, a concern emerged regarding the availability of energy, and a fear that an energy scarcity could develop. This was intensified in 1979 with the political conditions in Iran and the sharp price increases promulgated by the Organization of Petroleum Exporting Countries (OPEC). These new concerns are in sharp contrast to the situation over the past 100 yr when energy was abundant; its real value has declined steadily.

Structure of energy demand. Both world energy consumption and population increased rapidly after the turn of the century. By 1980, average annual per capita consumption was ca 66.0 GJ (62.4×10^6 Btu).

The Organization for Economic Cooperation and Development (OECD) countries, following the 1973–1974 embargo, embarked on a conservation program and have projected a reduction of primary energy consumption from 1.44 metric tons of crude oil equivalent (coe) per thousand U.S. dollars to 1.35 t of coe in 1985. A metric ton of coe is equal to 41.9 GJ (39.8×10^6 Btu). However, even with vigorous conservation efforts, the International Energy Agency (IEA) countries project increases in annual energy consumption of 3.7% from 1976–1985 and 3.5% from 1976–1990. Although use is still increasing, this is a significantly lower projection of energy consumption growth rates than was made in earlier years.

Energy utilization, and health and the environment. The type and magnitude of environmental impacts are different for each fuel form during its utilization. The principal concern regarding the air pollutant emitted from the combustion of fossil fuels, for instance, is the long-term chronic health effects that appear to be related to large concentrations of air pollutants. In addition, the sulfur or nitrogen oxides (or both) are believed to be the cause of acid rain which has harmful effects on foliage and soils and on aquatic flora and fauna.

These environmental considerations affect the level of fuel utilization, their cost, and the share that each fuel form contributes to the energy supply. These effects occur because of the costs involved in reducing emission. Methods for pollution reduction include switching to other higher quality fuels, upgrading the fuels before use, finding new technology to burn the fuel cleanly, or applying control technology to remove pollutants from the combustion products (see Air pollution control methods).

World fossil-fuel resources and reserves. A summary of the estimates of world conventional fossil-fuel resources and reserves is shown in Table 1. Not included in the table are the large quantities of unconventional resources not being produced.

Unconventional gas resources. In the United States, where production of natural gas from conventional sources appears to have peaked in the early 1970s, rough estimates have been made of the amount of natural gas that might be recoverable from unconventional gas deposits. These are listed in Table 2.

Nuclear energy resources. Uranium (qv) is an energy resource that is being used to produce heat and to generate steam and electricity by nuclear fission (see Nuclear reactors). The chief deposits of uranium resources are found in the United States, Canada, Africa, and Western Europe.

Projections of Energy Demand

The original estimate for energy use by IEA countries in 1985 (made in 1976) was 203 EJ (193×10^{15} Btu); by 1977, this was reduced to 196 EJ (186×10^{15} Btu), and later to 193 EJ (183×10^{15} Btu). The 1982 IEA projection of 1985 energy use is 170 EJ (161×10^{15} Btu) and represents an expected reduction in energy use from the 1977 projection of 16%.

HARRY PERRY
Resources for the Future

World Energy Resources, 1985–2000, World Energy Conference, IPC Science and Technology Press, Guildford, UK, 1978.

S.H. Schurr, J. Darmstadter, H. Perry, W. Ramsey, and M. Russell, *Energy in America's Future, The Choices Before Us*, Johns Hopkins University Press for Resources for the Future, Baltimore, Md., 1979.

Table 1. World Conventional Fossil Fuel Reserves and Resources[a]

	Reserves		Resources	
	10^3 EJ	Percent of total	10^3 EJ	Percent of total
coal	19.7	59.4	314	94.0
peat	6.7	20.4		
oil	4.0	12.1	8.2–11.9	3.0
natural gas	2.7	8.1	10.0–10.3	3.0
Total	*33.1*	*100.0*	*332–336*	*100.0*

[a] To convert EJ to 10^{15} Btu (quad), divide by 1.054.

Table 2. Estimated Unconventional Gas Resources

	10^{12} m³ [a]
coal-bed degasification	8.7–23.3
devonian shale	14.6–17.4
tight formations	17.4
geopressured gas	85–1458

[a] To convert m³ to ft³, multiply by 35.31.

FUELS FROM BIOMASS

Renewable carbon resources capable of conversion to synthetic fuels that are not depletable must ultimately be developed if the economy of the world continues to rely on organic fuels as a primary energy source. Technology for manufacturing large supplies of organic fuels and chemicals from nonfossil carbon must be commercialized. Land- and water-based vegetation and photosynthetic organisms, all of which are categorized as biomass, represent a practical source of this carbon.

The capture of solar energy as fixed carbon in biomass via photosynthesis is the key initial step in the growth of biomass:

$$CO_2 + H_2O + light \xrightarrow{chlorophyll} (CH_2O) + O_2$$

The main features of biomass-to-energy technology are illustrated in Figure 1. Conventionally, biomass is harvested for feed, food, and materials-of-construction applications or is left in the growth areas where natural decomposition occurs. The decomposing biomass or the waste products from the harvesting and processing of biomass, if disposed of on or in land, can in theory be partially recovered after a long period of time as fossil fuels. This is indicated by the dashed lines in Figure 1. Alternatively, the biomass and any wastes that result from its processing or consumption could be directly converted into synthetic organic fuels if suitable conversion processes were available. Another route to energy products is to grow certain species of hydrocarbon-producing biomass, such as the rubber tree (*Hevea braziliensis*), in which high energy hydrocarbons are formed within the tree by natural biochemical mechanisms.

Another approach to the development of fixed-carbon supplies from renewable carbon sources is to convert carbon dioxide outside the biomass species into synthetic fuels and organic intermediates. The ambient air, which contains an average of about 320 ppm of carbon dioxide, the dissolved carbon dioxide and carbonates in the oceans, and the earth's large terrestrial carbonate deposits could serve as renewable carbon sources. But because carbon dioxide is the final oxidation state of fixed

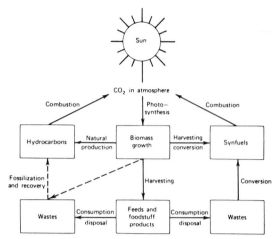

Figure 1. Biomass-to-energy technology.

carbon, it contains no chemical energy. Energy must be supplied in a reduction step. A convenient method of supplying the required energy and simultaneously reducing the oxidation state is by reduction of carbon dioxide with elemental hydrogen. The product can be methane, CH_4, the dominant component of natural gas:

$$CO_2 + 4 H_2 \rightarrow CH_4 + 2 H_2O$$

Distribution of Carbon

The estimation of the amounts of biomass carbon on the earth's surface is the ultimate problem in global statistical analysis. The results of one study are summarized in Table 1. Biomass carbon is a very small fraction of the total carbon inventory on earth, but it is a very important fraction that helps to maintain the delicate balance in support of all life forms. About 0.7% of the total standing biomass carbon on earth, or ca

Table 1. Estimates of Net Photosynthetic Production of Dry Biomass Carbon and Standing Biomass Carbon for World Biosphere[a]

Ecosystem	Area, 10^6 km^2[b]	Mean net carbon production t/(hm$^2 \cdot$ yr)[b]	Mean net carbon production 10^9 t/yr[b]	Standing biomass carbon t/hm^2	Standing biomass carbon 10^9 t
tropical rain forest	17.0	9.90	16.83	202.5	344
boreal forest	12.0	3.60	4.32	90.0	108
tropical seasonal forest	7.5	7.20	5.40	157.5	118
temperate deciduous forest	7.0	5.40	3.78	135.0	95
temperate evergreen forest	5.0	5.85	2.93	157.5	79
Total	*48.5*		*33.26*		*744*
extreme desert-rock, sand, ice	24.0	0.01	0.02	0.1	0.2
desert and semidesert scrub	18.0	0.41	0.74	3.2	5.8
savanna	15.0	4.05	6.08	18.0	27.0
cultivated land	14.0	2.93	4.10	4.5	6.3
temperature grassland	9.0	2.70	2.43	7.2	6.5
woodland and shrubland	8.5	3.15	2.68	27.0	23.0
tundra and alpine	8.0	0.63	0.50	2.7	2.2
swamp and marsh	2.0	13.50	2.70	67.5	14.0
lake and stream	2.0	1.80	0.36	0.1	0.02
Total	*100.5*		*19.61*		*85*
Total continental	*149.0*		*52.87*		*829*
open ocean	332.0	0.56	18.59	0.1	3.3
continental shelf	36.6	1.62	4.31	0.004	0.1
estuaries excluding marsh	1.4	6.75	0.95	4.5	0.6
algae beds and reefs	0.6	11.25	0.68	9.0	0.5
upwelling zones	0.4	2.25	0.09	0.9	0.04
Total marine	*361.0*		*24.62*		*4.5*
Grand total	*510.0*		*77.49*		*833.5*

[a] Dry biomass is assumed to contain 45% carbon.
[b] 1 km^2 = 1 $\times$ 10^6 m^2 (0.3861 sq mi); to convert t/(hm$^2 \cdot$ yr) to short ton/(acre$\cdot$yr), divide by 2.24.

7% of one year's net fixed-carbon production, would contain approximately the same energy content as all the gas and oil consumed in the world in 1980.

Energy Impact

By examining the potential amounts of synfuels that might be produced from renewable carbon sources and comparing these amounts with fossil-fuel demands, the percentage of demand that might be satisfied by particular nonfossil energy resources can be estimated.

Presuming the existence of suitable conversion technology, Table 2 summarizes the quantities of synthetic crude oil and natural gas that might be manufactured at a thermal efficiency of conversion of 60% from the organic wastes estimated to be produced in the United States in 1980. However, the gas and oil potentials indicated in Table 2 cannot actually be realized in practice because most of this waste is not collected and is thus not available for conversion to synfuels.

It should be emphasized, however, that even if all the wastes and residues generated were converted to substitute organic fuel, only a small fraction of the total fossil-fuels needs could be satisfied. Less than 10% of the United States fossil-fuel demand could be met by wastes if a conversion efficiency of 60% to energy products is assumed. Nevertheless, organic wastes can be a valuable supplemental source of synfuels (see Fuels from waste).

Virgin biomass, in contrast, represents an excellent source of synfuels. As already pointed out, extremely large quantities of carbon are fixed each year in the form of land- and water-based biomass. A realistic assessment of biomass as an energy resource can be made by calculating the average surface areas needed to produce sufficient biomass at different annual yields to meet certain percentages of fuel demand for a particular country, and then to compare these required areas with those that might be made available (see Table 3). Relatively large areas are required but not so much as to make the use of land- or freshwater-based biomass for energy applications impractical.

The third source of renewable carbon consists of the deposits and reservoirs of essentially nonenergy carbon forms, ie, carbon dioxide and the carbonates. The availability of such raw materials cannot be questioned, although low cost separation and energy-efficient recovery of very small concentrations of carbon dioxide present technological challenges. Another basic problem resides in the fact that all the energy must be supplied by a second raw material such as elemental hydrogen. Hydrogen (qv) would have to be made available in large quantities and from a nonfossil source or the purpose of the synfuel system would be defeated. Conceptually, there is no difficulty in developing such hydrogen sources. Hydrogen can be produced by water electrolysis and thermochemical and photolytic splitting of water. Electrical power and thermal energy can be supplied by nonfossil-powered nuclear reactors, and by means of wind, ocean thermal gradients, wave action, and solar-actuated devices. Hydrogen can also be manufactured from biomass by reforming and by direct action of solar energy (qv) on certain catalytic surfaces.

Chemical Characteristics

In Table 4, typical analyses and the energy contents of land- and water-based biomass (wood, grass, kelp) and wastes (manure, urban refuse, primary sewage sludge) are compared with those of cellulose, peat, and bituminous coal.

Conversion

Various methods can be used to produce the energy shown in Table 4. Chemicals can be produced by a wide range of processing techniques. Table 5 summarizes the important feed, process, and product variables considered for the development of a synfuel-from-biomass process.

Biomass conversion processes can be divided conveniently into four groups: physical, biological-biochemical, thermal, and chemical. Physical processes include particle size reduction, separation, drying, and fabrication. Biological-biochemical processes include anaerobic digestion for methane production, alcoholic fermentation for ethanol production, biophotolysis for hydrogen production, and natural processes for lipid and hydrocarbon production. Thermal processes include direct combustion for heat, direct gasification for the production of fuel and synthesis gas, and direct liquefaction for the production of liquid fuels and chemicals.

Production

The manufacture of synfuels and energy products from biomass requires that suitable quantities of biomass be grown, harvested, and transported to the conversion plant site. In the ideal case, biomass chosen for energy applications should be high yield, low cash-value

Table 2. Estimates of Synfuels by Conversion of Organic Wastes Generated in the United States in 1980, 10^6 metric tons / yr[a]

Source	1980
agricultural crop and food wastes[b]	354
manure	241
urban refuse	201
logging and wood manufacturing residues	54
miscellaneous organic wastes	54
industrial wastes	45
municipal sewage solids	13
Total	962
net oil potential, L/yr[c]	227×10^9
net gas potential, m^3/yr[d]	219×10^9

[a] Unless otherwise stated.

[b] Assumes 70% dry organic solids in major crop-waste solids.

[c] Based on an oil yield of 236 L/dry t of organic wastes (1.35 petroleum bbl/short ton) and 60% thermal efficiency. To convert bbl to L, multiply by 159.0.

[d] Based on a gas yield of 228 m^3/dry t of organic waste (3.65 ft^3/lb) and 60% thermal efficiency.

Table 3. Potential Substitute Natural Gas in the United States from Biomass at Different Crop Yields

1980 Demand, %[a]	Average area required, km^2[b]	
	25 t/(hm²·yr)[c]	100 t/(hm²·yr)[c]
1.58	20,400	5,100
10	129,000	32,300
50	645,500	161,000
100	1,291,000	323,000

[a] U.S. demand in 1980 estimated to be 244×10^8 GJ or 653×10^9 m^3 (231×10^{11} ft^3).

[b] 1 km^2 = 1×10^6 m^2 (0.3861 sq mi).

[c] Yields expressed as dry metric ton; to convert t/(hm²·yr) to short ton/(acre·yr) divide by 2.24.

Table 4. Typical Compositions and Heating Values of Wastes, Biomass, Peat, and Coal

	High heating value MJ/kg[c] (dry)
pure cellulose	17.51
Pine wood	21.24
Kentucky blue grass	18.73
giant brown kelp[a]	10.01
feedlot manure	13.37
urban refuse[b]	12.67[d]
primary sewage sludge	19.86
reed sedge peat	20.79
Illinois bituminous coal	28.28

[a] *Macrocystis pyrifera*.

[b] Combustible fraction.

[c] To convert MJ/kg to Btu/lb, multiply by 430.

[d] As received with metals.

Table 5.   Feeds, Processes, and Products

Biomass feed	Primary conversion process		Primary energy products	
land-based	separation			thermal
trees	combustion	energy		steam
plants	pyrolysis			electric
grasses	hydrogenation			
	anaerobic fermentation	solids fuels		char
water-based	aerobic fermentation			combustibles
single-cell algae	biophotolysis			
multicell algae	partial oxidation			methane (SNG)
water plants	steam reforming			hydrogen
	chemical hydrolysis	gaseous fuels		low-thermal-value gas
	enzyme hydrolysis			medium-thermal-value gas
	other chemical conversions			light hydrocarbons
	natural processes			
				methanol
		liquid fuels		ethanol
				higher hydrocarbons
				oils
		chemicals		

species that have short growth cycles, that grow well in the area and climate chosen for the biomass energy system, and that have no competitive markets. Fertilization requirements should be low and possibly nil if the species selected fix ambient nitrogen, thereby minimizing the amount of external nutrients that have to be supplied to the growth areas. In areas having low annual rainfall, the species grown should have low water needs and be able to utilize efficiently available precipitation. For land-based biomass, the requirements should be such that the crops can grow well on low grade soils and do not need the best classes of agricultural land. After harvesting, growth should commence again without the need for replanting. Surprisingly, several biomass species meet many of these idealized characteristics and appear to be quite suitable for energy applications.

Photosynthesis. Yield is plotted against solar-energy capture efficiency in Figure 2 for insolation values of 150 and 250 W/m² [1142 and 1904 Btu/(ft² · d)] which span the range commonly encountered in the United States, and for biomass energy values of 12 and 19 MJ/dry kg (5160 and 8170 Btu/lb). The higher the efficiency of photosynthesis, the higher the biomass yield. However, for a given solar-energy capture efficiency and incident solar radiation, the yield is projected to be lower at the higher biomass energy values (curves A and B, curves C and D). From an energy production standpoint, this simply means that a higher-energy-content biomass could be harvested at lower yield levels and still compete with higher-yielding but lower-energy-content biomass species. It is also apparent that for a given solar-energy capture efficiency, yields similar to those obtained with higher-energy content species should be possible with a lower-energy-content species even when it is grown at a lower insolation (curves B and C). Finally, at the solar-energy-capture efficiency usually encountered in the field, about 1% or less, the spread in yields is much less than at the higher-energy capture efficiencies.

Because of the many uncontrollable factors in the field, such as changes in climate and seasonal changes in biomass composition, departures from the norm given here can be expected.

The biochemical pathways involved in the conversion of carbon dioxide to carbohydrates play an important role in our understanding of the molecular events of biomass growth. Carbon dioxide fixation proceeds via three different biochemical energy-transfer pathways. The first is called the Calvin three-carbon cycle and involves an initial three-carbon intermediate of 2-phosphoglyceric acid (PGA). The second is called the C_4 cycle because the carbon dioxide is initially fixed as the four-carbon dicarboxylic acids, malic or aspartic acids, and the third pathway is called crassulacean acid metabolism (CAM). It refers to the capacity of chloroplast-containing tissues to fix carbon dioxide in the dark via phosphoenolpyruvate carboxylase leading to the synthesis of free malic acid.

Appreciable differences in net photosynthetic assimilation of carbon dioxide are apparent between C_3, C_4, and CAM biomass species. The generally lower yields of C_3 biomass are caused by its higher photorespiration rate. The specific carbon dioxide-fixing mechanism used by plants affects the efficiency of photosynthesis. Therefore, it is desirable from an energy utilization viewpoint to choose plants that exhibit high photo-

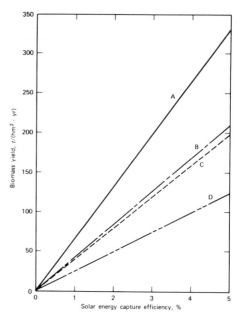

Figure 2. Effect of solar energy capture efficiency on biomass yield. To convert W/m² to Btu/(ft² · d), multiply by 7.616; to convert MJ/kg to Btu/lb, multiply by 430. To convert t/(hm² · yr) to short ton/(acre · yr), divide by 2.24.

	I, W/m^2	F, MJ/kg (dry)
A	250	12
B	250	19
C	150	12
D	150	19

I = average insolation; F = energy content of dry biomass.

synthesis rates to maximize the yields of biomass in the shortest possible time.

Climate and environment. The biomass species selected for energy applications must be compatible with the climate to facilitate operation of fuel farms. Insolation, rainfall, and temperature have the most influence on the productivity of indigenous or transplanted species. Atmospheric carbon dioxide concentration and the availability of nutrients are also important.

Land availability. The availability of land suitable for production of land-based biomass is estimated by the land-capabilities classification scheme developed by the USDA in which land is divided into eight classes varying from few limitations that restrict use largely to recreation, wildlife habitat, water supply, or to esthetic purposes. A survey of actual usage of this land suggests that there is ample opportunity to produce biomass for energy applications on non-Federal land that is not used for foodstuffs production. Large areas of land in Classes V–VIII not suited for cultivation would appear to be available, and sizable areas in Classes I–IV seem to be available as well.

Land-based biomass. Most land-based biomass plantations operated for energy production or synfuel manufacture will probably also yield products for nonenergy markets; production of single-energy products will be the exception rather than the rule.

Water-based biomass. With the exception of phytoplankton, aquatic biomass seems to exhibit higher net organic yields than land biomass. Water-based biomass considered to be the most suitable for energy applications include the unicellular and multicellular algae and water plants.

Systems Analysis

In the ideal case, the synfuel production plants would be equivalent to an independent system with inputs of solar radiation, air, carbon dioxide, and minimal water, and one output, synfuel. The nutrients are kept within the ideal system eliminating addition of external fertilizers and chemicals. Environmental and disposal problems are minimized.

Economics. With few exceptions, comparison of the projected synfuel costs in 1985 with the projected costs of conventional fuels on the same basis shows that synfuels from biomass are more expensive than conventional fuels. As the price of crude oil and natural gas continues to increase, it is expected that biomass energy products will become more competitive with fossil fuels.

Energetics. The net energy production efficiency of an integrated biomass energy system is extremely important to its development since the goal is to design and operate environmentally acceptable systems to produce new supplies of salable energy from biomass at the lowest possible cost and with the minimum consumption of energy inputs. It is necessary to quantify how much energy is expended and how much salable energy is produced in each fully integrated system.

Extensive research programs have continued on biomass production for energy applications and on the gasification and liquefaction of biomass and wastes for fuels, energy, and chemicals. Commercialization of this technology appears to be increasing at a higher rate, particularly for combustion of wood, wood wastes, and municipal solid wastes for heat, steam, and electric production; anaerobic digestion of industrial wastes for combined waste disposal and methane production; and use of fermentation ethanol as a motor fuel. Ethanol fuel usage more than doubled in 1982 in the United States as compared to 1981, and plant capacity is expanding rapidly. Methanol has not yet begun to compete with ethanol because of Federal limitations on the concentration of methanol in blends with gasoline (see Alcohol fuels).

Donald L. Klass
Institute of Gas Technology

D.L. Klass in D.L. Klass, ed., *Symposium Papers: Energy from Biomass and Waste VII*, Institute of Gas Technology, Chicago, Ill., Jan. 24–28, 1983, pp. 1–65.

D.L. Klass and G.H. Emert, eds., *Fuels from Biomass and Waste*, Ann Arbor Science Publishers, Ann Arbor, Mich., 1981.

D.L. Klass, ed., *Biomass as a Nonfossil Fuel Source*, ACS Symposium Series 144, ACS, Washington, D.C., 1981.

FUELS FROM WASTE

Wastes from which energy can be recovered are many and diverse; the following types are most important: wood wastes from pulp mills, sawmills, and plywood mills; municipal solid waste (MSW); crop residues such as bagasse, corn cobs, and straw; and animal manure, principally from cattle feedlots (see Wastes, industrial and municipal).

The annual generation and utilization rates for the various waste-based fuels given in Table 1 show that wood wastes are the dominant energy feedstock.

Characteristics of Wastes For Fuels

Waste fuels share the following common characteristics: they are solid, hygroscopic, modest in carbon content compared to coal or oil, and they are low in bulk density. Thus, they are fuels for which markets are local, and the economics of their use is based upon fossil-fuel replacement rather than on established market pricing mechanisms. The driving characteristic is that the fuel value of each type of waste is determined by its carbon content on a dry weight basis. Thus the following regression holds with an $r = 0.982$:

$$HHV_D = 0.475C - 2.38 \qquad (1)$$

where HHV_D = higher heating value (HHV), MJ/kg dry matter; C = % carbon.

From equation 1 the net heat content of all waste fuels can be estimated by the following equation:

$$NHV_G = HHV_D(1 - MC) \qquad (2)$$

where NHV_G = net heating value (green) and MC = moisture content, decimal basis.

Table 1. Waste Generation and Utilization

Waste type	Dry waste generated annually, $t \times 10^6$	Dry waste used annually, $t \times 10^6$	Waste used annually, %
primary manufactured wood residue[a]	57	45	79
logging residue	82	negl.	negl.
crop residue[b]	363	2	0.5
feedlot manure	190	negl.	negl.
municipal solid waste	150	6	4.0

[a] Does not include coarse wood residues used for manufacturing pulp or particleboard.
[b] Mostly bagasse, cotton gin waste, and corn seed burned in one Indiana power plant.

Wood-waste characteristics. Wood wastes exhibit the most potential among wastes for an increase in their energy contribution and are already frequently used. Their present and potential sources include sawmills, plywood mills, logging operations, and unmerchantable stand conversions. The problem of heterogeneity in higher heating value (HHV), moisture content (MC), net heating value (NHV), particle size distribution, and ash content is inherent in these fuels. On a net heating-value basis, cubic volumes of wood residues are quite modest, eg, 3.3 GJ/m^3 (90×10^3 Btu/ft^3).

Municipal waste characteristics. Municipal solid waste (MSW) varies in composition by city, by season of year, and by weather conditions. Paper waste ranges 21.5–53.3% about a mean of 37.4%. Wood-leather, and rubber waste ranges 0.3–2.1% about a mean of 1.2%. The U.S. Bureau of Mines analyzed ten MSW samples. The as-received mean HHV was 11.2 MJ/kg (4.83×10^3 Btu/lb) with a standard deviation (σ) of 0.9/kg (385 Btu/lb).

Characteristics of agricultural residues. Agricultural and crop processing residues include bagasse (qv), rice straw, rice hulls, peanut shells, cotton-gin trash, corn cobs, grasses, and a host of grain stalks. Variations in plant material come from type of plant and type of processing. Similar variations in manures are a function of animal type (eg, cattle, hogs, chicken) and age of manure.

Wastes as Fuel

Most wastes are burned directly with little, if any, upgrading; hence, the primary consideration must be the use of wastes as fuel.

The basic combustion equations are stated as

$$C + O_2 \rightarrow CO_2 \quad \Delta H = 33.9 \text{ MJ/kg (14.6 Btu/lb)} \qquad (3)$$

$$H_2 + 2 O_2 \rightarrow 2 H_2O \quad \Delta H = 142.9 \text{ MJ/kg (61.5 Btu/lb)} \qquad (4)$$

However, the pathways of combustion are critical to the conversion and energy recovery process. The process follows the following sequential steps for any given fuel particle: initial heat up, drying, pyrolysis to volatiles and char, and then combustion. Pyrolysis is the most critical step. Temperatures of > 420 K are required for drying; 550 K is a minimum for effective pyrolysis. The volatiles produced burn in flaming combustion with rapid rates of heat release. The oxidation of char occurs as glowing combustion, with slower heat release.

Wood-Waste Combustion Systems

Wood-combustion systems include pile-burning and modified pile-burning systems (such as Dutch ovens and cell furnaces), semisuspension or spreader-stoker boilers, and suspension burners for dry material. Of these, spreader-stokers are by far the most commonly employed as they can handle varying fuel quality and changes in steam load.

Over 90% of all wood waste burned is used to raise steam for process purposes, to generate electricity, or both. Process steam generated is typically 345 kPa to 1.72 MPa (50–250 psi) and 475–560 K. Such wood-fueled systems have high capital and operating costs.

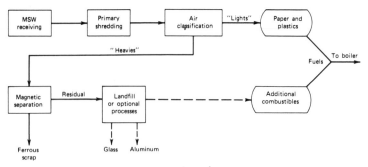

Figure 1.   Simplified schematic of municipal solid waste (MSW) processing.

Municipal-Waste Combustion

For municipal-waste processing ranges from virtually nothing (in water-wall incineration) to systems for MSW shredding, ferrous metals recovery, glass and nonferrous metals removal, and possibly reshredding the light MSW fraction to < 255-mm size. A typical simplified flow sheet for refuse-derived fuel (RDF) is shown in Figure 1.

Combustion of municipal wastes as fuel must be considered in three segments: combustion of unprocessed refuse; combustion of RDF as the sole fuel; and cocombustion of RDF with coal or oil.

Incineration systems are used to raise industrial steam as in Saugus, Massachusetts, where the incinerator supplies steam to the Lynn Works of General Electric; and to raise district heat and circulate cooled water, as in Nashville, Tennessee. In the Saugus system, cogeneration is employed.

For the RDF fuels, the overwhelming market is utilities—hence the generation of electricity.

Conversion Systems Applied to Wastes

Numerous pyrolysis, gasification, and liquefaction systems developed for coal conversion have been adapted to wastes. These systems are in various stages of research, development, and demonstration. Additional research is applying anaerobic digestion, acid hydrolysis, and enzymatic hydrolysis of wastes to production of methane-rich gases and alcohol fuels. Among these systems, gasification is considered most promising since it capitalizes on the high ratio of volatile-to-fixed carbon associated with wastes, the higher atomic hydrogen-to-carbon ratios, and the virtual absence of sulfur. Gasification also permits continued use of many existing natural-gas fired boilers threatened by uncertainties in existing fuel supplies (see Fuels, synthetic).

DAVID A. TILLMAN
Consultant

L.L. Anderson in L.L. Anderson and D.A. Tillman, eds., *Fuels From Waste*, Academic Press, Inc., New York, 1977, p. 2–15.

F. Shafizadeh and W.F. DeGroot in F. Shafizadeh, K.V. Sarkanen, and D.A. Tillman, eds., *Thermal Uses and Properties of Carbohydrates and Lignins*, Academic Press, Inc., New York, 1976, pp. 1–18.

J.S. Bethel and co-workers, *Renewable Resources for Industrial Materials*, National Research Council, National Academy of Sciences, Washington, D.C., 1976.

FUELS, SYNTHETIC

GASEOUS FUELS

Only a trivial amount of fossil fuels has been used for gasification since natural gas became a dominant energy source. Plentiful supplies of coal (qv), oil shale (qv), peat, and biomass (see Fuels from biomass), however, have led to extensive research and development of gasification processes (see also Lignite and brown coal). The following processes could become

economically competitive with natural gas because of changes in the price structure and may be needed in future on short notice in view of U.S. dependence on overseas oil and gas supplies (see Fuels, survey).

Coal

Coal is gasified either by partial combustion or by hydrogasification. Very often a process is a combination of both modes, depending on the product desired: low, medium, or high heat-value gas.

Low heat-value gas. Low heat-value (low-Btu) gas usually has a heating value of 3.3–5.6 MJ/m^3 (90–150 Btu/SCF), and is formed by partial combustion of coal with air, usually in the presence of steam. The primary reactions are

$$C + 1/2\ O_2 \rightarrow CO \tag{1}$$

$$C + H_2O \rightarrow CO + H_2 \text{ (steam)} \tag{2}$$

$$CO + H_2O \rightarrow CO_2 + H_2 \tag{3}$$

It is now being proposed for use in electric power and for small requirements of manufacturing concerns. A popular choice for these "backyard" gas producers is the atmospheric pressure, air-blown Wellman-Galusha gasifier because of its long history of successful operation. One kilogram of good bituminous coal will make 4.0 m^3 (64.1 ft^3/lb) of the gas shown in Table 1.

Medium heat-value gas. Medium heat-value gas has a heating value between 9.3 and 20.5 MJ/m^3 (250–550 Btu/SCF). At the lower end of this range, the gas is produced like low heat-value gas, except that an air separation plant is added and relatively pure oxygen is used instead of air to partially oxidize the coal. This eliminates the nitrogen and doubles the heating value from 5.2 to 10.6 MJ/m^3 (140–285 Btu/SCF). The Winkler gasifier (Fig. 1) is a commercial example of medium heat-value gas processes as is the Koppers-Totzek (K-T) process.

High heat-value gas. High heat-value (high-Btu) gas is generally referred to as SNG (substitute natural gas) or pipeline-quality gas. Its composition is as close to pure methane as possible (see Hydrocarbons, C_1–C_6). Any of the medium heat-value gases that consist of carbon monoxide and hydrogen (often called synthesis gas) can be converted to SNG by methanation, a low temperature catalytic process that combines carbon monoxide and hydrogen to form methane and water.

$$CO + 3\ H_2 \rightarrow CH_4 + H_2O \tag{4}$$

Table 1.   Typical Gas Compositions, %

Components	Bituminous	Anthracite
CO	25	27.1
CH_4	2.3[b]	0.5[a]
illuminants	0.6	
H_2	14.5	16.6
CO_2	4.7	5.0
N_2	52.8	50.8
MJ/m^3 [b] HHV (higher heating value)	6.2	5.4

[a] Methane increases with increasing volatile content.
[b] To convert MJ/m^3 to Btu/ft^3, multiply by 26.87.

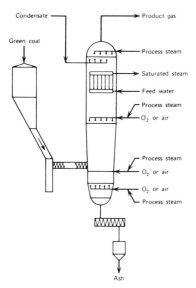

Figure 1. Winkler generator.

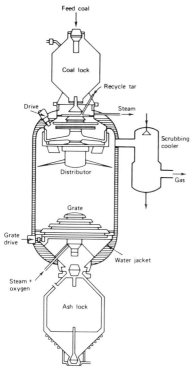

Figure 2. Lurgi pressure gasifier.

Upon drying, the methanated gas is nearly pure methane. The Lurgi gasifier (Fig. 2), similar in principle to the Wellman-Galusha unit, is one commercial SNG process.

Peat

The United States, as well as the rest of the world, contains immense resources of peat, a cellulosic plant residue (see Lignite and brown coal.) The peat is easily obtained because it lies above ground. However, it contains 90–95% water. In the United States, it has only been used in small quantities for agricultural purposes. The PEATGAS process is a two-stage gasification process which is now ready for large pilot-plant operation (Fig. 3).

Oil Shale

Oil shale (qv) is another immense natural resource that is not being exploited for a variety of reasons. Much work has been done on con-

verting it to oil. Essentially, the only efforts to produce SNG from it have been done by IGT under American Gas Association sponsorship.

Hydrocarbon Liquids

Extensive work has been done in the United States, the United Kingdom, and Japan on the production of SNG from hydrocarbon liquid feedstocks, and many naphtha-reforming plants have been constructed (13 U.S. plants with a total capacity of 37.55×10^6 m^3 of SNG/d (1326×10^6 ft^3/d)). Most plans for future construction, however, have been shelved, and little interest has been shown in oil gasification since the Middle East war of 1973 and the subsequent Arab oil embargo.

Processes used in the production of high heat-value oil gas included thermal cracking (eg, the Petrogras process (Fig. 4), and the Hall process), cyclic catalytic (eg, UGI Cyclic Catalytic Reforming, Segas process, and O.N.I.A.-G.E.G.I. process), continuous thermal cracking (eg, Fluid-Bed Thermal Cracking (FTC) process, and Thermofor Pyrolytic Cracking (TPC) process), partial oxidation, and steam-reforming (eg, British Gas Corporation's CRG process, (Fig. 5)).

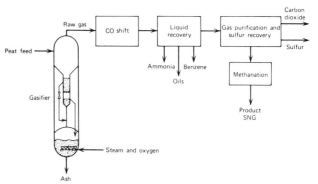

Figure 3. PEATGAS process.

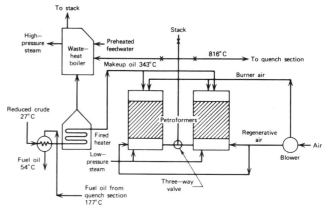

Figure 4. Improved heat recovery system for the Petrogas process.

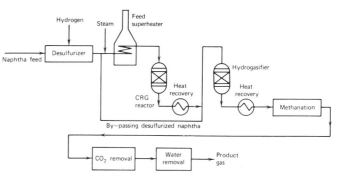

Figure 5. British Gas Council CRG process.

Economic Aspects

The economics of gaseous-fuel production by gasification varies significantly by type of process, type of feedstock, and the scale of the project.

JACK HUEBLER[‡]
JOHN C. JANKA
Institute of Gas Technology

Report of the Committee on Production of Manufactured Gases, Rep. No. IGU/B-76, International Gas Union, London, Eng., 1976.

"Produce Your Own Gas for Your Own Business," *Brochure No. 3M 177 CA 4104*, Applied Technology Corp., Houston, Tex., 1977.

LIQUID FUELS

The term synthetic liquid fuels has come to refer to prepared liquids with characteristics approaching those of fuel liquids in current commerce, specifically gasolines, kerosenes, jet fuels, and fuel oils. Synthesis may include liquefaction of coals, destructive distillation of kerogen or coal, and extraction or hydrogenation of organic matter in coke liquids, coal tars, tar sands, or bitumen deposits, in addition to processes that fit the classical interpretation of the organic chemist.

Carbonaceous fuel substances, other than natural petroleum, are widely distributed throughout the world (see also Coal; Oil shale, Asphalt; Lignite; Tar sands).

In the Western hemisphere, Canada has large tar sand, bitumen (or very heavy crude oil), and coal deposits. The United States has very large reserves of coal and shale.

The total resource base is immense by any estimate. Unfortunately, estimates of the quantities of in-place reserves that may be recoverable with current technology range no higher than ca 10%. Hence, considerable pressure for new recovery technology will develop in the next century, even if fuel usage is obviated by other developments.

Most carbonaceous substances may be combusted more or less directly in suitable equipment to produce heat or power. Direct combustion may represent the means for deriving maximum benefit of the heating or energy value of particular fuel substances, although the necessity to limit emissions of noxious compounds, especially of sulfur or nitrogen, and to control disposition of noncombustible residues, may significantly reduce the overall thermal efficiency of the combustion process and add to the cost of work so produced.

Liquid fuels possess inherent advantageous characteristics. They are much more readily stored, transported, and metered than solids or tars, and are much easier to process or clean by chemical means. Noncombustibles may be reduced to very low levels by treatment, and sulfur and nitrogen compounds may be removed. The energy densities of clean hydrocarbon liquids may be very high relative to solid or semisolid fuel substances, which may contain mineral matter and moisture.

In general, liquid hydrocarbon fuels produced from natural petroleum, as well as the crude petroleum itself, are chemically distinguished from natural solid-fuel substances (coal, peat, lignite, bitumen, kerogen) by the higher hydrogen-to-carbon content of the liquids.

If paraffinic liquids are extracted from solid fuel substances, as by the destructive distillation of bitumen or oil shale, the hydrogen content of the residual material is reduced even further, and the residues become more refractory. The yields of liquids so derivable are generally low, even when a significant fraction of the hydrogen is extractable. Thus, production of fuel liquids from nonliquid carbonaceous fuel substances may be enhanced only by the introduction of additional hydrogen in a synthesis process. Coal liquefaction can proceed either directly or indirectly. In the direct coal-liquefaction process, coal reacts with gaseous hydrogen and a solvent at high temperature and pressure.

Coal Pyrolysis

Coal pyrolysis has been studied at both reduced and elevated pressures, and in the presence of a variety of agents and atmospheres (see Coal).

[‡]Deceased.

Table 1. Typical COED Syncrude Properties[a]

Property	Utah A-seam	Illinois no. 6 seam
specific gravity, (° API)	0.934 (20)	0.929 (22)
pour point, °C	16	−18
flash point, closed cup, °C	24	16
viscosity, mm^2/s (= cSt) at 38°C	8	5
ash, wt%	0.01	0.01
moisture, wt%	0.1	0.1
metals, ppm	10	10
hydrocarbon type analysis, liquid vol%		
paraffins	23.7	10.4
olefins	0	0
cycloparaffins (naphthenes)	42.2	41.4
aromatics	34.1	48.2

[a]An elemental analysis (wt%): C, 87.2; N, 11.0; N, O.2; O, 1.4; S, 0.1.

Although important to the study of coal structure and reactions, coal pyrolysis, as a means to generate liquids, has thus far proved to have limited commercial value. Several modern coal-conversion processes, however, enhance the recovery of potential fuel liquids. These include the COED process, the Occidental Petroleum Coal Conversion process (Garett Coal Pyrolysis, Oxy Coal Conversion), and the TOSCOAL process.

Properties of typical COED syncrude are listed in Table 1. Heating values of TOSCOAL process gas on a dry, acid gas-free basis are in the natural-gas range if butanes and heavier components are included. The production of gas undiluted with nitrogen or excess carbon dioxide is a considerable advantage.

Coal Carbonization

In the by-product recovery of a modern coke oven, coal tar is removed first by cooling the emitted gases, and light oil is removed last by scrubbing the gas with solvents (see Coal chemicals and feedstocks; Coal-conversion processes). The precise compositions of the light oil and coal tar recovered from coke-oven gas is a distinct function of the design of the recovery system, as well as of the properties of the starting coal. In general, 12.5–16.7 m^3 light oil/metric ton (3–4 gal/short ton) of coal carbonized is recovered from high temperature coke-oven operations. Light oil may contain 55–70% benzene, 12–20% toluene, and 4–7% xylene. Unrecovered light oil appearing in the effluent coal gas may comprise ca 1 vol% and contribute ca 5% of the gas's heating value. Refining of light oil consists mainly of sulfuric acid washing, followed by fractional distillation. The total United States production of BTX from coke-oven operations is barely significant compared to current petroleum product consumptions (see BTX processing).

Coal and Coal-Tar Hydrogenation

Experimental plants for hydrogenating coal or coal tar were operated in Japan, France, Canada, and in the United States before, or during, World War II. In general, coal-in-oil slurries containing iodine or stannous oxalate catalyst were subjected to liquid-phase hydrogenation at pressures of 25–70 MPa (250–700 atm). Liquids produced were fractionated, and the middle oils were then subjected to vapor-phase hydrogenation over molybdenum–cobalt-, or tungsten sulfide-on-alumina catalysts. About one metric ton of liquid fuel was recovered from 4.5 t of coal, from which all necessary hydrogen and power requirements for the production were also obtained.

In the 1960s, the Office of Coal Research (OCR) funded the development of a number of coal-liquefaction process variants incorporating hydrogenation, including the H-Coal process, the Solvent-Refined Coal processes (SRC) I and II, the Consol Synthetic Fuel process (CSF), and the EDS (Exxon Donor Solvent) Coal Liquefaction process (see Coal chemicals and feedstocks, hydrogenation).

Fischer-Tropsch Synthesis

Fischer-Tropsch Synthesis, or indirect coal liquefaction, involves five steps: synthesis-gas manufacture; gas purification by removal of water and dust, and hydrogen sulfide and organic sulfur compounds; synthesis of hydrocarbons; condensation of liquid products and recovery of gasoline from product gas; and fractionation of synthetic products.

Fischer-Tropsch liquid obtained with cobalt catalysts is roughly equivalent to a very paraffinic natural petroleum oil (see Petroleum, petroleum composition) but is not so complex a mixture. Straight-chain, saturated aliphatic molecules predominate, but monoolefins may be present in an appreciable concentration. Alcohol, fatty acids, and other oxygenated compounds may represent less than 1% of the total liquid product. The normal pressure synthesis yields ca 60% gasoline, 30% gas oil, and 10% paraffin (mp 20–100°C). The medium pressure synthesis yields approximately 35% gasoline, 35% gas oil, and 30% paraffin. The octane rating of the gasoline is too low for direct use as motor fuel.

Extensive research continues to interpret the mechanism of the Fischer-Tropsch synthesis. The most significant recent achievement in a related field is the methanol conversion process announced by Mobil in 1976. A wide variety of heteroorganic compounds, including methanol and dimethyl ether, may be efficiently transformed to C_2–C_{10} hydrocarbons using a new class of shape-selective zeolites (see Molecular sieves). Gasoline so produced is chemically similar to petroleum-derived gasoline, and has unleaded research octane numbers of 90–95. This Mobil gasoline process (MTG) is superior in both yield and quality to that produced by the traditional Fischer-Tropsch processes.

Shale Oil

In the United States, shale oil, or oil derivable from oil shale, represents a large potential source of liquid hydrocarbons that can be readily processed to fuel liquids similar to those derived from natural petroleum (see Oil Shale).

A number of foreign countries produce liquid fuels from oil shale, and the pioneer plant in the U.S. is that of Union Oil Company.

Heavy Oil

Heavy crude oil is widely distributed, and it is difficult to estimate reserves separate from normal crude-oil reserves or from tar-sands deposits. Current estimates of petroleum reserves frequently include a large heavy-oil component, which can only be produced at significantly higher cost than light oil. Most heavy-oil production is currently concentrated in California, Canada, and Venezuela.

Tar Sands

Tar sands are considered to be sedimentary rocks with natural porosity, whose pore volume is occupied by viscous, petroleum-like hydrocarbons. Similar semantic problems may be incurred with the terms oil sands, rock asphalts, asphaltic sandstones, and malthas, or malthites, all of which have been applied to the same resource. The hydrocarbon component of tar sands is properly bitumen (see Tar sands).

Alcohol Fuels

The lower alcohols, especially methanol and ethanol, have been widely used separately, and in blends with gasolines and other hydrocarbons, to fuel internal-combustion engines. Since these compounds, as a class, may be considered to be already partially oxidized, incorporating a mole of oxidized hydrogen in each molecule, they differ from the hydrocarbons that make up gasoline principally in their lower heating values and in their higher vaporization heat requirements. Their solubilities in gasolines and their effects on gasoline vapor pressure are of special importance in the formulation of fuel blends, eg, when alcohols are used to extend volumes of gasolines (see Alcohol fuels; Gasoline).

Claims relating to improvements in fuel economy, exhaust emission, and driveability for alcohol–gasoline blends must be examined critically. The fuel economy, in km/L, must decrease proportionally to increases in the alcohol content of the blend. Exhaust emissions should be changed to the same extent as with gasolines alone, as carburetting is modified to accommodate the blended fuel. Driveability, starting, vapor lock, and corrosion problems must also increase with increasing alcohol content when blends are introduced to current systems designed to operate with straight gasolines.

A fuel consisting essentially of methanol has the potential to be a significant factor in the liquid-fuel market. The technology for its production is already available, it can be produced from a variety of feedstocks, and it offers use efficiency advantages compared to other liquid fuels. The ICI and Lurgi methanol synthesis processes, normally integrated with natural-gas reforming, are currently in use in the chemicals industry. Possible feedstocks include any substance that can be converted to synthesis gas (CO and H_2), for example: coal, natural gas, and biomass. In an internal-combustion engine specifically designed for methanol, the thermal efficiency is projected to be 10–50% higher than a conventional internal-combustion gasoline engine. These efficiency improvements have yet to be demonstrated in other than prototype models. The actual efficiency improvement depends on the extent to which advantage is taken of methanol's high octane and low lean flammability limit. These properties facilitate a higher compression ratio, lean-burn operation, and use in a stratified-charge engine. When considered from an overall systems viewpoint (resource through ultimate use), methanol in optimized vehicles is the most attractive liquid synthetic fuel. Although personal transportation appears to be the largest potential market, other applications with efficiency-improvement potential include highway and nonhighway freight, electric-power generation, and residential/commercial heating.

Ethanol is produced both from ethylene derived from the cracking of petroleum fractions (35%) and by the fermentation of sugars derived from grains and other biomass. Many of its relevant properties are similar to those of methanol. Although it may be a more desirable fuel or fuel component than methanol, its significantly higher cost (volume basis) outweighs these advantages.

In the United States, and especially in the Midwest, gasohol is associated with the production of ethanol from grain. In addition, a number of Federally sponsored research programs seek to effect the microbiological degradation of cellulosic biomasses, and the direct conversion to ethanol, or to soluble sugars (see Fuels from biomass).

CHARLES D. KALFADELIS
HENRY SHAW
Exxon Research and Engineering Company

Assessment of Technologies for the Liquefaction of Coal: Summary, prepared by the Ad Hoc Panel on Liquefaction of Coal of the Committee on Processing and Utilization of Fossil Fuels, Commission on Sociotechnical Systems, National Academy of Sciences, 1977.

H. Shaw, C.D. Kalfadelis, and C.E. Jahnig, *Evaluation of Methods to Produce Aviation Turbine Fuels From Synthetic Crude Oils—Phase I*, Technical Report AFAPL-TR-75-10, Vol. 1, Air Force Aero Propulsion Laboratory, Wright-Patterson Air Force Base, Ohio, March, 1975.

FULLER'S EARTH. See Clays (uses).

FULMINATES. See Explosives.

FUMARIC ACID, *trans*-HOOCCH=CHCOOH. See Maleic acid, fumaric acid, and maleic anhydride.

FUMIGANTS. See Insect control technology.

FUNCTIONAL FLUIDS. See Hydraulic fluids.

FUNGICIDES, AGRICULTURAL

The use of fungicides in agriculture is necessitated by the great losses caused by a wide variety of plant-pathogenic fungi. To be economic, the cost of controlling plant diseases by the application of fungicides must be offset by potential gains of several fold. The largest tonnages of fungicides are required by apples, pears, bananas, cereals, cocoa, coffee, cotton, potatoes, tobacco, and grapes. Fungicides usually are applied in water suspension with hydraulic sprayers but are sometimes used as dusts, granules, or fumigants. They are applied not only to foliage but also to seeds, to soil, and to harvested plant parts for control of fungus diseases.

Early fungicides included sulfur and polysulfides, and heavy-metal fungicides, all of which have been supplemented by more recently developed fungicides.

Organic Fungicides

Quinones. Hundreds of quinones have been evaluated, but only chloranil (1) and dichlone (2) have been developed commercially.

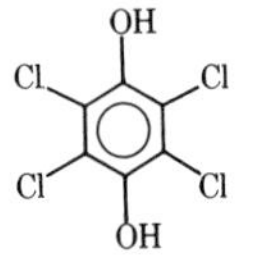

(1) 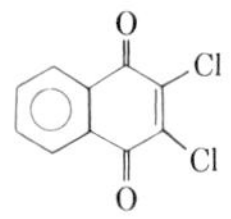   (2)

Organic sulfur compounds. Dithiocarbamates were developed as fungicides in the early 1940s. The first derivative developed was tetramethylthiuram disulfide, still used widely as a vegetable and corn-seed treatment. Later, a new group of carbamates was introduced which included disodium ethylenebisdithiocarbamate, $NaSSCNH(CH_2)_2$-$NHCSSNa$, or nabam. The manganese-stabilized material (maneb) was later found to be even more effective and is the dominant form used today.

Imidazolines and guanidines. Heptadecyl-2-imidazolinium acetate is effective for the control of apple scab and cherry leaf spot. This compound, known as glyodin, is synergistic with another fungicide, dodecylguanidinium acetate, known as dodine.

Trichloromethylthiocarboximides. Captan, (N-(trichloromethylthio)-4-cyclohexene-1,2-dicarboximide) is one of the least phytotoxic agricultural fungicides and is widely used as a foliage fungicide on fruit, as well as a seed treatment and soil fungicide. Folpet, (N-(trichloromethylthio)-phthalimide), controls powdery mildew, against which captan is ineffective.

Chlorinated and nitrated benzenes. Among the chlorinated and nitrated benzenes, 2,3,4,6-tetrachloronitrobenzene, pentachloronitrobenzene (PCNB), 1,2,4-trichloro-3,5-dinitrobenzene, 1,3,5-trichloro-2,4,6-trinitrobenzene, and hexachlorobenzene have been used as soil or seed fungicides (see Nitrobenzenes). Also, a new, broad-spectrum foliar fungicide, tetrachloroisophthalonitrile, was introduced. This safe, effective material, called chlorothalonil, has become one of the most important protectant agricultural fungicides.

Systemic Fungicides

Oxathiins. The oxathiins permitted systemic protection of new growth. This systemic activity marked the beginning of the third era of agricultural fungicides. One of the more effective of the series is 2,3-dihydro-5-carboxanilido-6-methyl-1,4-oxathiin, known as carboxin (3). Carboxin is used as a seed treatment for smuts of cereals, and for Rhizoctonia diseases of seedlings.

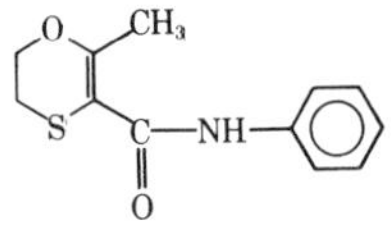

(3)

Benzimidazoles. The first of this series of systemic fungicides was 2-(4-thiazoyl)benzimidazole. Methyl (1-butylcarbamoyl)-2-benzimidazolylcarbamate (benomyl) (4) is an excellent systemic fungicide with protectant and eradicant activity. The benzimidazoles are probably the most successful of the systemic fungicides currently in use.

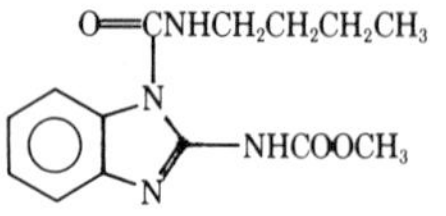

(4)

Pyrimidines. The key to development of systemic fungicidal activity for the hydroxypyrimidines was the oxygen function placed at position 4. Both 5-butyl-2-dimethylamino-6-methyl-4(1H)-pyrimidinone (dimethirimol) (5) and the 2-ethylamino analogue (ethirmol) (6) have achieved commercial use as systemic agents against powdery mildew.

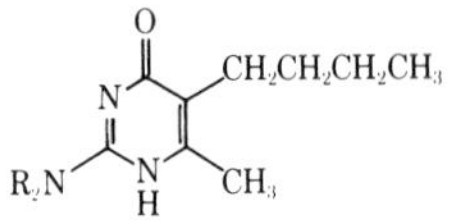

(5) R = CH₃
(6) R = C₂H₅

Fungicides other than those described above include Dinocap, Fenaminosulf, and the antifungal antibiotic cycloheximide (Actidone).

Mode of Action

In general, the broad-spectrum, protectant-type materials apparently have several sites of action that cause deleterious effects in many enzyme and membrane systems. In contrast, the site of action of the highly specific, systemic compounds seems to be confined to one system. These fungicides exert their action though one of three different methods: inhibition of energy production; interference with biosynthesis; or disruption of cell structure.

E.J. BUTTERFIELD
D.C. TORGESON
Boyce Thompson Institute for Plant Research, Inc.

M.R. Siegel and H.D. Sisler, eds., *Antifungal Compounds*, 2 vols., Marcel Dekker, Inc., New York, 1977.

R.W. Marsh, ed., *Systemic Fungicides*, John Wiley & Sons, Inc., New York, 1972.

D.C. Torgeson, ed., *Fungicides, An Advanced Treatise*, 2 vols., Academic Press, Inc., New York, 1967, 1969.

FUNGICIDES, INDUSTRIAL. See Industrial antimicrobial agents.

FURAN DERIVATIVES

Furan derivatives may be simple furans, ie, when the nucleus occurs as a free monocycle, or condensed furans, when the nucleus is fused to another ring.

The furan nucleus is a cyclic, dienic ether with some aromatic character. A stabilization energy of 96 ± 4 kJ/mol (23 + 1kcal/mol) is observed for the furan ring.

Reactivity of the furan nucleus is directly related to the electron density at particular ring atoms. Canonical forms (1) through (4) represent the overall electron distribution.

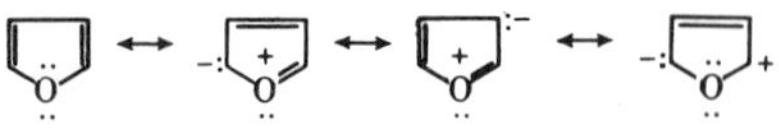

(1)   (2)   (3)   (4)

The balance between aromatic and aliphatic character is markedly affected by substituents. Furan and its homologues, for example, function as dienes in the Diels-Alder reaction. On the other hand, furans with electron-withdrawing substituents (eg, furfural, 2-furoic acid, and nitrofurans) fail as dienes even with the strongest dienophiles.

Furan and its homologues are readily cleaved on acid hydrolysis to open-chain dicarbonyl compounds. In view of the susceptibility of the furan nucleus to cleavage and/or polymerization, considerable care is required in the selection of reagents and reaction conditions to achieve desired products in reasonable yield.

There are two classes of partially saturated monocyclic furans, the 2,3-(5) and the 2,5-dihydrofurans (6).

(5)   (6)

Tetrahydrofuran is a fully saturated cyclic ether. It is manufactured by cyclodehydration of 1,4-butanediol and by catalytic hydrogenation of furan (see Ethers).

The chemical properties of the industrially significant furan derivatives are characteristic of the functional groups affixed to the furan nucleus. In many instances, however, those properties are modified because of the highly reactive furan ring.

Furfural

Furfural (2-furancarboxaldehyde), a colorless liquid aldehyde with a pungent, aromatic odor reminiscent of almonds when freshly distilled, darkens appreciably on exposure to air. It is miscible with most of the common organic solvents but only slightly miscible with saturated aliphatic hydrocarbons. Inorganic compounds, generally, are quite insoluble in furfural.

In Table 1 are collected the most important physical properties of furfural.

The chemical properties of furfural are characteristic of aromatic aldehydes but with unique variances attributable to the furan ring.

Furfural is produced from annually renewable agricultural sources such as nonfood residues of food crops and wood wastes. Of the various components of these vegetable materials, the pentosan polysaccharides (xylan, arabinan) are the primary precursors of furfural and are almost as widely distributed in nature as cellulose (see Alcohols, polyhydric).

Furfural is produced commercially in batch or continuous digesters where the pentosans are hydrolyzed to pentoses and the pentoses subsequently cyclodehydrated to furfural.

Furan derivatives including furfural, furfuryl alcohol, and tetrahydrofurfuryl alcohol are used as solvents in a variety of industrial applications. Furfural and furfuryl alcohol are specialty solvents.

Furfural, the key member of the industrial furan chemicals, is known as a moderately toxic substance. Specific toxicological data are presented in Table 1.

Furfuryl Alcohol

Furfuryl alcohol (2-furanmethanol) is a colorless, liquid, primary alcohol with mild odor. On exposure to air, it gradually darkens in color. It is completely miscible with water, alcohol, ether, acetone, and ethyl acetate, and is soluble in most organic solvents with the exception of the paraffinic hydrocarbons. The physical constants of furfuryl alcohol are listed in Table 1.

The chemical properties of furfuryl alcohol are characteristic of the hydroxymethyl group; however, the reactivity of the furan ring, intensified by carbonium-ion formation under acid catalysis, introduces complexities. Furfuryl alcohol undergoes the reactions expected of a primary alcohol such as oxidation, esterification, and etherification. Although stable to strong alkali, furfuryl alcohol is sensitive to acid; thus it imposes limitations on the conditions used in many of the typical alcohol reactions. Ethoxylation and propoxylation of furfuryl alcohol provide ether alcohols useful in coatings, printing ink, and paint-stripper formulations.

Acid-catalyzed resinification is the most important industrial reaction of furfuryl alcohol and manufactured resins have important applications in bonding sand for cores and molds in foundry metal casting.

Furfuryl alcohol has been manufactured on an industrial scale by employing both liquid-phase and vapor-phase hydrogenation of furfural. Copper catalysts are preferred because they are selective and do not promote hydrogenation of the ring.

Furfuryl alcohol, classed as moderately toxic, is less toxic than furfural. Like furfural, this compound shows a tendency to penetrate intact skin (rabbit test). See Table 2 for specific toxicological information.

Furan

Furan, a colorless liquid with strong ethereal odor, is both low boiling and highly flammable. It is miscible with most common organic solvents but only slightly soluble in water. The physical properties of furan are listed in Table 1.

Furan is a heat-stable compound although, at 670°C in the absence of catalyst, or at 360°C in the presence of nickel, it decomposes to form a mixture consisting mainly of carbon monoxide, hydrogen, and hydrocarbons. Substitution and addition reactions can be effected under controlled conditions, reaction occurring first in the 2 and 5 positions.

Furan is produced commercially by decarbonylation of furfural. It is used as a chemical building block in the production of other industrial chemicals for use as pharmaceuticals, herbicides, stabilizers, and fine chemicals.

Furan is much more toxic than the other furan chemicals, and with its high volatility, constitutes a more serious hazard. Human exposures have been controlled routinely to less than 10 ppm. Because furan is one of the volatile constituents in tobacco smoke, its toxicity (and metabolism) in the lung is now being studied.

Tetrahydrofurfuryl Alcohol

Tetrahydrofurfuryl alcohol (2-tetrahydrofuranmethanol) is a colorless, high boiling liquid with a mild, pleasant odor. It is completely miscible

Table 1. Physical Properties of Furan Derivatives

	Furfural	FA[a]	Furan	THFA[b]
molecular weight	96.09	98.10	68.08	102.13
boiling point at 101.3 kPa (1 atm), °C	161.7	170	31.36	178
freezing point, °C	−36.5		−85.6	< −80
refractive index, n_D^{20}	1.5261	1.4868	1.4214	1.4250
density, d_4, at 20°C, g/cm³	1.1598	1.1285	0.9378	1.0511
solubility, wt%, in water				
20°C	8.3	∞		∞
25°C			1	
alcohol; ether	∞	∞	∞	∞
viscosity, mPa·s (= cP)				
20°C			0.38	6.24
25°C	1.49	4.62		
explosion limits (in air), vol%	2.1–19.3	1.8–16.3	2.3–14.3	1.5–9.7
flash point, °C				
Tag closed cup	61.7	65	−35.5	
Tag open cup				83.9

[a] Furfuryl alcohol.
[b] Tetrahydrofurfuryl alcohol.

Table 2. Toxicology of Furan Derivatives

	Furfural	FA[a]	THFA[b]
LD₅₀, oral, mg/kg, rat	149	149	5250
occupational exposure limit, 8 h-TWA, ppm	5	50	na[c]

[a] Furfuryl alcohol.
[b] Tetrahydrofurfuryl alcohol.
[c] Not available.

with water and the common organic solvents. Tetrahydrofurfuryl alcohol is a moderately hydrogen-bonded solvent, essentially nontoxic, biodegradable, and with a low photochemical oxidation potential. The more important physical properties of tetrahydrofurfuryl alcohol are listed in Table 1.

The reactions of tetrahydrofurfuryl alcohol are characteristic of its bifunctional structure involving the carbinol and/or ether moieties. As a primary alcohol, it undergoes normal displacement or condensation reactions affording new functional groups (eg, halides, esters, alkoxylates, ethers, glycidyl ethers, cyanoethyl ethers, amines, etc). As a cyclic ether, it is susceptible to hydrolytic or hydrogenolytic cleavage to a variety of open-chain compounds, some of which may be recyclized to different heterocyclic species.

Tetrahydrofurfuryl alcohol is produced commercially by the hydrogenation of furfuryl alcohol. It is of interest in chemical and related industries where low toxicity and minimal environmental impact are important. For many years THFA has been used as a specialty organic solvent. The fastest-growing applications are in formulations for crop sprays, cleaners, paint strippers, water-based paints, and the dyeing and finishing of textiles and leathers.

Since it is completely saturated, tetrahydrofurfuryl alcohol shows low toxicity. Other toxicological data are given in Table 2.

2,5-Bis(hydroxymethyl)furan

2,5-Bis(hydroxymethyl)furan (BHMF) is a colorless, crystalline monomer, molecular weight, 128.13, bp 162°C, mp 76.5–77°C, that is soluble in water, ethanol, acetone, pyridine, and tetrahydrofuran.

BHMF monomer behaves as a diol and undergoes esterification (base-catalyzed transesterification), alkoxylation, glydicyl ether formation, cyanoethylation, etherification, urethane formation, carbonate formation, etc.

Monomeric 2,5-bis(hydroxymethyl)furan and its oligomeric resins are produced from furfuryl alcohol and formaldehyde. A linear diol based on these oligomeric resins has commercial application as a polyol component in Class I urethane foam. 2,5-Bis(hydroxymethyl)furan may be used to prepare linear polymers that can subsequently be cross-linked under a variety of conditions, eg, acid catalysis, heat, Diels-Alder condensation, etc.

2,5-Bis(hydroxymethyl)furan exhibits low toxicity. In recent tests, no mutagenic effects were detected.

W.J. McKILLIP
E. SHERMAN
The Quaker Oats Co.

A.P. Dunlop and F.N. Peters, *The Furans*, ACS Monograph 119, Reinhold Publishing Corp., New York, 1953.

A.P. Dunlop, W.J. McKillip, and S. Winderl, "Furan und Derivate," Vol. 15, *Ullmanns Encyklopadie der Technischen Chemie*, 1976.

W.J. Pentz, W.R. Dunlop, and R.H. Leitheiser, *J. Consumer Prod. Flammability* **9**, 149 (Dec. 1982).

FURFURAL AND OTHER FURAN COMPOUNDS. See Furan derivatives.

FURNACES, ELECTRIC

INTRODUCTION

The term electric furnace applies to all furnaces that use electrical energy as their sole source of heat, as distinguished from fuel-fired furnaces (see Furnaces, fuel-fired) in which heat is produced directly by combustion. Electric furnaces are used mainly for heating solid materials to desired temperatures below their melting points for subsequent processing, or melting materials for subsequent casting into desired shapes, (ie, electric heating furnaces, electric melting furnaces). According to the manner in which the electrical energy is converted into heat,

three distinct types of widely used industrial furnaces can be distinguished; they are: electric-resistance furnaces (qv), electric-arc furnaces (qv), and electric-induction furnaces (qv).

The power supply is normally ac (dc is rarely used) at standard power line frequency (60 Hz in North America) and it usually includes a transformer so that the most suitable furnace voltage is obtained. However, certain induction furnaces require a higher frequency, obtained from the power line through solid-state power-frequency converters.

MARIO TAMA
Ajax Magnethermic Corp.

W. Trinks, *Industrial Furnaces, Fuels, Furnace Types and Furnace Equipment—Their Selection and Influence Upon Furnace Operation*, 4th ed., Vol. 2, John Wiley & Sons, Inc., New York, 1967.

A.G.E. Robiette, *Electric Melting Practice*, John Wiley & Sons, Inc., New York, 1972.

ARC FURNACES

Arc furnaces used in electric melting, smelting, and electrochemical operations are of two basic designs: the indirect and the direct arc. The arc of the indirect-arc furnace is maintained between two electrodes and radiates heat to the charge. The arcs of the direct-arc furnace are maintained between the charge and the electrodes, making the charge a part of the electric power circuits. Not only is heat radiated to the charge, but the charge is heated directly by the arc and the current passing through the charge.

Indirect-Arc Furnaces

Indirect-arc furnaces are used primarily in foundries for melting copper, copper alloys, and other nonferrous metals having a low melting point (see Copper). They have also been used for producing molten iron and, occasionally, molten steel (see Iron; Steel). The typical indirect-arc furnace is a single-phase furnace utilizing two horizontally mounted graphite electrodes, each of which project into an end of a refractory lined horizontally mounted cylindrical steel shell.

Direct-Arc Furnaces

Open-arc furnaces. Most of the open-arc three-phase furnaces are used in melting and refining operations for steel and iron. They are also used in melting operations for copper, slag, refractories, and other materials having melting points above 1000°C. Today's standard furnace consists of a refractory-lined steel shell to contain the charge and/or molten metal; a set of graphite electrodes; electrical equipment, bus bars, and flexible conductors to energize the electrode equipment to regulate the position of the electrodes; a means to slag the furnace; and inspect the refractory; a method to tilt the furnace to empty it; and a means to allow the furnace to be recharged.

Figure 1 shows a typical arc-furnace power curve for a given voltage input.

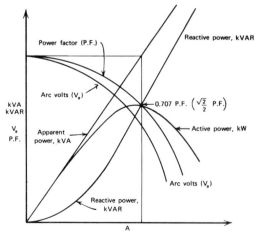

Figure 1. Typical arc-furnace power curve.

Other forms of open-arc furnaces include vacuum-arc and plasma-arc furnaces.

Submerged-arc furnaces. Furnaces used for smelting and for certain electrochemical operations are similar in general design to the open-arc furnace in that they are usually three phase and have three vertical electrode columns and a shell to contain the charge. They are used in the production of ferroalloys, pig iron, phosphorus, calcium carbide, copper, other metals and compounds, and some types of high temperature refractories.

Arc-resistance furnace. The arc-resistance furnace is very similar to the submerged-arc furnace except the electrodes are immersed in the molten material—usually slag or a nonmetallic material. Most of the submerged-arc and arc-resistance furnaces do not tilt and usually do not have roofs. Where the volatilized materials and gases are toxic or cannot be exposed to air, the shell is covered with a sealed monolithic roof so that the gases and vapors can be ducted away from the furnace for subsequent collection.

DAN OAKLAND
Whiting Corp.

Electrode Arc Furnace Electrode Digest, Union Carbide Corp., Carbon Products Division, 1975.

The Electric Arc Furnace, International Iron and Steel Institute, Committee on Technology, Brussels, Belgium, 1981.

C.J. Szymkowski, "Uprating of Hanna's 12 MVA Silicon Metal Furnaces," *AIME Electric Furnace Proceedings*, Houston, Tex., 1981, Vol. 39, pp. 319–323.

INDUCTION FURNACES

Induction furnaces are transformers in which the electrically conducting furnace charge constitutes the secondary and load resistance, wherein electric energy is converted into heat. They are suitable for heating and melting all metals; principally iron, copper, aluminum, zinc, and their alloys. Furnace capacities range from a few kilograms to several hundred metric tons, and power ratings range from 20–20,000 kW. There are three basic types: channel induction-melting furnaces, coreless induction-melting furnaces, and induction-heating furnaces.

Channel Induction-Melting Furnaces

A channel induction-melting furnace operates at line frequency. It has the same electrical elements as a power transformer; however, the secondary coil and load resistance are formed by one or two single closed turns of molten metal. The magnetic iron core and primary coil assembly thread a steel-encased refractory block which contains the annular secondary cavity or channel for the melt. The assembly of steel casing, refractory block, transformer core, and primary coils is the inductor which is then attached to the part of a channel induction furnace. Electromagnetic molten-metal circulation distributes energy from the channel throughout the melt in the hearth. Several inductors often are attached to the same heart to form a powerful multiple-inductor furnace (Figure 1).

Channel induction-melting furnaces are very efficient, and preferred for steady production (at least two shifts per day) where large alloy changes are not required.

Coreless Induction-Melting Furnaces

Coreless furnaces lack a magnetic iron core to link the primary and secondary.

The charge is placed in a crucible-shaped refractory container surrounded by a cylindrical primary coil. Coreless induction melting is made possible by the electromagnetic phenomenon known as the skin effect, whereby the current is concentrated in a surface layer defined by the "depth of penetration".

High currents in a molten charge create electromagnetic pressures resulting in a convex shape of the melt surface, and also cause vigorous stirring of the melt. Higher frequency decreases the depth of penetration and thereby increases the secondary resistance. A lower secondary current then is required to attain a desired power level, and stirring is reduced.

In practice, high frequency (180–3000 Hz) is preferred for smaller furnaces (50–5000 kg), whereas larger coreless furnaces (10–60 t) are predominantly powered at line frequency. High frequency coreless furnaces are preferred for operations with frequent alloy changes and those with less than two shifts per day.

Induction-heating furnaces. Induction heating furnaces are generally known as induction heaters or induction coils. They also rely on the skin effect for the efficient transfer of energy to the charge. Most induction heaters are for steel, copper, or aluminum (qv). Three types are used extensively: static induction heaters, continuous induction heaters, and induction heaters for surface hardening. The first two are used for through-heating. The latter apply localized heat only to specific surface areas of steel products.

MARIO TAMA
Ajax Magnethermic Corp.

M. Tama, *J. Metals* **26**(1), 18 (1974).

Electric Melting and Holding Furnaces in the Ironfounding Industry (Conference Record), University of Technology, Loughborough, Sept. 20–22, 1967, British Cast Iron Research Assoc., Birmingham, UK, 1968.

C.A. Tudbury, *Basics of Induction Heating*, 2 Vols., J.F. Rider, New York, 1960.

RESISTANCE FURNACES

Resistance furnaces are categorized by a combination of four factors: batch or continuous; protective atmosphere or air atmosphere; method of heat transfer (radiation, convection, and conduction); and operating temperature. The primary method of heat transfer in an electric furnace is usually a function of the operating temperature range. The three methods of heat transfer are radiation, convection, and conduction. Of the three, radiation and convection apply to all of the furnaces described above. Conductive heat transfer is limited to special types of furnaces covered later. Operating temperature ranges are classed as low (< 760°C), medium (760–1150°C), and high (> 1150°C).

Batch furnaces. Batch furnaces carry out the desired time–temperature cycle for the work to be processed by subjecting the entire furnace

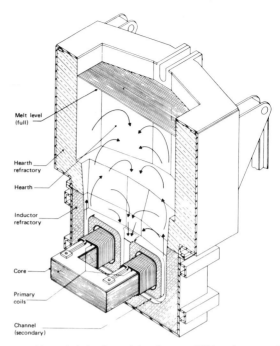

Figure 1. Channel induction-melting furnace. Tilting furnace with one high-powered twin-coil inductor. Courtesy of Ajax Magnethermic Corporation.

Melt level (full)

Hearth refractory

Hearth

Inductor refractory

Core

Primary coils

Channel (secondary)

and change of work to the particular cycle. Batch furnaces are most often used for one or more of the following applications: very large and/or heavy charges; low production rates; infrequent operation, variable time-temperature cycle; and processing material that must be in batches because of previous or subsequent operations.

Continuous furnaces. Continuous furnaces are applicable for: uniform charges of work that arrive at the furnace continuously; moderate to high production rates; constant time–temperature cycle; and continuous operation over at least one and preferably two or three shifts per day.

Air-atmosphere furnaces. These furnaces are applied to processes where the work load can tolerate the oxidation that occurs at elevated temperatures in air.

Protective-atmosphere furnaces. Protective-atmosphere furnaces are used where the work cannot tolerate oxidation or where the atmosphere must provide a chemical or metallurgical reaction with the work. Protective-atmosphere furnaces are of two general types. In one type, the work is inside a muffle (retort) and the protective atmosphere is inside the muffle. The other type is gas-tight and the atmosphere is introduced directly into the furnace, obviating the expensive and expendable retort or muffle.

Low Temperature Convection Furnaces

Low temperature convection furnaces are designed to transfer the heat from the heating elements by forced convection. Convection is normally used in furnaces operating below 760°C because it is the most effective means of heat transfer that can maintain good uniformity of temperature on various configurations of work load. Convection furnaces also are used in this range of temperatures where it is important that no part of the work load exceed the controlled temperature.

Low Temperature Radiation Furnaces

These furnaces are of the infrared-heater type. Heat transfer is by direct radiation from a high temperature heating element. This type of furnace is normally used for such applications as drying of paint films.

Medium Temperature Radiation Furnaces

Medium temperature radiation furnaces generally are used at 760–1150°C. The primary part of the heat is transferred directly to the work by radiation from the heating elements and by radiation to the furnace refractory which reradiates the energy to the work.

High Temperature Radiation Furnaces

High temperature radiation furnaces operating above 1150°C are similar in construction to medium temperature radiation furnaces. The insulation system must be designed to withstand the high temperatures, and internal structural parts become critical.

Vacuum Radiation Furnaces

Vacuum radiation furnaces are used where the work can be satisfactorily processed only in a vacuum or in a protective atmosphere. Most vacuum furnaces use molybdenum heating elements. Because all heat transfer is by radiation, metal radiation shields are used to reduce heat transfer to the furnace casing. The casing is water-cooled and a sufficient number of radiation shields between the inner cavity and the casing reduce the heat flow to the casing to a reasonable level.

Conduction Furnaces

Conduction furnaces utilize a liquid at the operating temperature to transfer the heat from the heating elements to the work being processed. Conduction furnaces are of two general types. One has a pot or crucible with suitable exterior insulation. The salt-bath furnace is the other type of conduction furnace where the molten salt provides the heat transfer medium and is the heating resistor.

Direct-Heat Electric-Resistance Furnaces

These furnaces use the material to be heated as the resistor, and the furnace consists of an insulated enclosure to return the heat, a power source of suitable voltage, and means of attaching the power leads to the

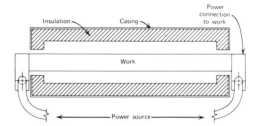

Figure 1. Direct-heat resistance furnace.

work (see Fig. 1). Since the work is the resistor, it must have a uniform cross section between power connection points, and the material must be homogeneous.

There are large-scale operations mainly in melting bulk materials where the liquid material serves as a uniform resistor. The material is contained in a crucible of fixed dimensions which, coupled with a given resistivity of the material, fixes the total resistance within reasonable limits.

RICHARD M. MILLER
Wellman Thermal Systems Corporation

Reviewed by
O.C. SHEESE
Wellman Furnaces, Inc.

W. Trinks and M. Mawhinney, *Industrial Furnaces, Principles of Design and Operation*, 5th ed., Vol. 1, John Wiley & Sons, Inc., New York, 1961.

V. Paschkis and J. Persson, *Industrial Electric Furnaces and Appliances*, 2nd ed., Interscience Publishers, New York, 1960.

C. Cone, *Energy Management for Industrial Furnaces*, Wiley-Interscience, New York, 1980.

FURNACES, FUEL-FIRED

A furnace is a device (enclosure) for generating controlled heat with the objective of performing work. In fossil-fuel furnaces, the work application may be direct (eg, rotary kilns) or indirect (eg, plants for electric-power generation). The furnace chamber is either cooled (waterwall enclosure) or not cooled (refractory lining). All modern power-plant furnaces of any significant megawatt rating are of waterwall construction suspended from a frame structure.

Fuels

Oil and coal are currently the most important fuels in industrial use. Although the last few decades witnessed an upsurge in oil- and gas-fired power-plant furnaces, the increasing worldwide shortage of these fuels in the 1970s has resulted in new power-plant furnaces being almost exclusively coal-fired. Furnaces that do not supply electric power to a regional grid (eg, rotary power kilns, taper-dash mill recovery furnaces, package boilers, etc.) are referred to under the collective designation of industrial furnaces. These are mainly gas and oil fired.

Power-Plant Furnaces

In 1977, 13% of the electric power consumed in the United States was generated by nuclear energy (see Nuclear reactors). It is expected that before the year 2000 this will have increased to 25%, the remainder supplied by fossil-fuel and hydroelectric power. The bulk of electric-power generation will probably be supplied by coal-fired power-plant furnaces supplying steam to turbogenerators. In terms of megawatts supplied, the coal-fired power plant is, therefore, the foremost component of importance in the energy supply system. As stated previously, power-plant furnaces are of waterwall type and are generally designed for steam pressures in the range of 12.4–24.1 MPa (1800–3500 psi) turbine throttle. Examples of power-plant furnaces include the lignite-fuel furnace (Fig. 1) and the cyclone furnace.

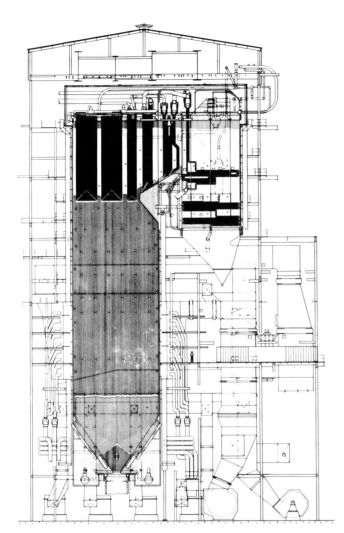

Figure 1. Martin Lake unit of Texas Utilities Services, Inc., 750 MW; a representative of modern supercritical design. Courtesy of Combustion Engineering, Inc.

Industrial Furnaces

Generally speaking, industrial furnaces are an order of magnitude smaller than power-plant furnaces since the applications are usually on an individual basis (hospital complex, chemical plant, paper mill, etc.) rather than feeding power to a regional electric grid. Like the power-plant furnace, the function of the industrial furnace usually is to generate steam, generally for a chemical process, mechanical power, or heating application, rather than electric-power generation. Industrial furnaces include package boilers, paper-mill (recovery) furnaces, large industrial furnaces (waterwall), and refractory-wall furnaces.

New Developments

New furnace concepts in an evolutionary stage include fluidized-bed furnaces, coal-gasification furnaces, and MHD furnaces. The primary attraction for the first two is the intrinsically low pollution potential.

KEES A. BUETERS
Combustion Engineering, Inc.

ROBERT H. ESSENHIGH
The Ohio State University

R.H. Essenhigh, *Future Fuel Supplies for Industry: Bases for Choice*, ASHRAE Transactions, Vol. 84, American Society of Heating, Refrigerating, and Air-Conditioning Engineers, New York, 1978, Pt. 1.

J.G. Singer, ed., *Combustion, Fossil Power Systems*, 3rd ed., Combustion Engineering, Inc., Windsor, Conn., 1981.

FURS, SYNTHETIC

Simulating natural furs in the broadest sense can encompass a wide spectrum of constructions from the simplest fleeces and pile fabrics in natural fur colorations to complex synthetic furs, duplicating to the degree possible the ombré-contoured multilayered appearance of a coat made from natural mink pelts.

Of the twelve to fifteen main categories of chemical fiber commercially available today, only three (acrylic, polyester, and modacrylic categories) are important for the manufacture of synthetic fur, although individual fibers from other categories (and even from some natural-fiber categories) continue to be incorporated.

Historically, the two methods used to produce artificial fur have been based upon the weaving and the knitting processes. In each instance, insertion of an extra set of yarns into the basic woven or knit structure results in a fabric face made up of yarn loops, which with cutting or shearing and unraveling of the individual fibers making up the face-yarn tufts remaining, produces the effect of a pile or furlike surface on one side of the fabric.

Direct material costs commonly constitute more than 50% of the total costs associated with such a fabric at the manufacturing stage.

From the standpoint of value added, the fibrous materials commonly used to manufacture synthetic fur fabrics can be categorized in ascending order as fiber tow, staple, flock, sliver, slit-film yarn, textured-filament yarn, and spun yarn. Fabric substrate structures in ascending order of value added are nonwovens, knit, and woven fabrics (see Nonwoven textile fabrics). The 13 methods most commonly used to produce synthetic furs, roughly ranked in their order of cost competitiveness based upon the value added to the materials employed are needle punch, sliver knitting, flocking, stitch bonding (Malipol bonding), spring-needle knitting, loop knit (single-jersey plush knitting), pile-warp knitting, modified double knitting, weaving, surface-pile raising (napping, teaseling, etc), fine-gauge tufting, I-tuft bonding, and U-tuft bonding.

Synthetic-Fur Finishing

In all cases where a surface loop of yarn is first produced, the loop must be cut either on the forming equipment, by napping, or by a later shearing process. The formed fabric produced in practically all cases also must be sheared to evenly crop the pile prior to further processing.

In most cases, particularly where a knit structure is involved, the fabric itself must be stabilized, usually by tentering it and either heat-setting or applying a polymer-based backcoating to the nonpile side of the backing structure. Where a free-flowing pile with a high luster is desired, synthetic-fur fabrics are commonly subjected to alternating electrifications and shearings in final finishings.

Finishing agents can be introduced during face-fiber or yarn-preparation stages, such as picking, carding, or spinning, at napping or immediately prior to an electrification. These chemicals can include various waxes, resins, silicones, and fluorochemical-based topical finishes designed to enhance the appearance performance characteristics (see Textiles).

Energy Requirements

Despite the recent large increases in the price of all forms of energy, the current cost of energy used in the actual manufacture of synthetic furs is relatively small, normally falling below 3% of the total cost of sales.

Health and Safety Factors

Toxicology problems relating to the manufacture of synthetic furs appear to be minor. Of all the problems facing the synthetic-pile industry, product liability claims based upon the potential flammability of its product is probably receiving the greatest industry attention. In the cases of fabrics made predominantly from the acrylic family, a certain portion of a flame-resistant fiber, most commonly a modacrylic, may have to be blended into the fabric face in order to achieve an acceptable flammability rating.

GUY KIECKHEFER
NORMAN C. ABLER
Borg Textile Corp.

M. Bachrach, *Fur—A Practical Treatise: Geography of the Fur World*, Gordon and Breach Science Publisher, Inc., New York, 1977.

B.P. Corbman, *Textiles, Fiber to Fabric*, 5th ed., McGraw-Hill Book Co., New York, 1975.

N. Hollen, J. Saddler, and A. Langford, *Textiles*, 5th ed., Macmillan Co., New York, 1979.

FUSION ENERGY

Several impressive advances over the past ten years have led to a well-founded feeling of optimism that fusion will become practical during the 21st century. The long-range role for fusion in the energy economy is in central-station electric-power generation.

In order to effect a fusion reaction between two atomic nuclei, it is necessary that they be brought together closely enough to experience each other's nuclear forces. Because all nuclei are positively charged, they repel one another via the electrostatic law of the repulsion of like charges. This electrostatic barrier can be surmounted by imparting sufficient kinetic energy to the reacting species. These repulsive forces increase rapidly with the magnitude of the nuclear charge; therefore, nuclear fusion is always limited to elements with low atomic numbers.

Deuterium – Tritium Fusion

The so-called D–T reaction, $D + T \rightarrow {}^4He + n$, is especially attractive because of its relative ease of ignition. D and T are the heavy isotopes of hydrogen (see Deuterium and tritium). The products of this reaction are an alpha particle (the helium nucleus, ${}^4He$) and a free neutron (n) carrying kinetic energies of 3.5 and 14.1 MeV, respectively. In an electric-power-generating facility, the neutrons would be absorbed in a blanket surrounding the fusion region, and their kinetic energy would be converted to heat. Conventional power-conversion systems would be used to transform this heat into electric energy. Because of its relatively high reactivity, the D–T fusion fuel will be employed in the first generation of fusion reactors. Deuterium is easily extracted from seawater, whereas tritium must be bred from lithium, a natural element which could supply fusion fuel for several hundred years.

Plasma Condition Required For Fusion Reaction

The most promising approach to attaining a significant reaction rate is to heat the reacting species to a high temperature, thereby imparting sufficient kinetic energy to the nuclei in the form of thermal motions. By doing so, the particles (eg, deuterons and tritons) may scatter amongst each other several times, keeping their energy within the system, before undergoing fusion reactions.

At these temperatures all matter exists in the plasma state, ie, it consists of totally ionized, positively charged atomic nuclei (ions) moving energetically throughout a cloud of negatively charged electrons, the system as a whole being electrically neutral. Generally speaking, the reactivity, or the fusion-power output rate, increases with increasing temperature. However, as the temperature of the plasma is raised, the radiation losses are also increased, primarily because of bremsstrahlung, or continuum radiation from the electrons. For any fusion-fuel system there exists a unique temperature at which the fusion-power production is precisely balanced by the radiation losses. This temperature is called the ideal ignition temperature, and equals ca 46×10^6 K for a D–T plasma (see Plasma technology).

Besides having to satisfy a minimum temperature requirement, a plasma must be sufficiently dense and must be contained for a long enough time to yield net power. If the plasma burns above the ideal ignition temperature for some time period τ, the fusion energy released must at least equal the energy required to heat the plasma to that temperature plus the energy radiated during that period. It can be shown that this condition is met providing that the product of plasma density N and confinement time exceeds a characteristic value that depends only on temperature. The minimum value of the product $N\tau$ represents the least stringent condition for the plasma to be a net

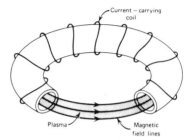

Figure 1. Cutaway view of toroidal field configuration.

producer of fusion energy. The minimum $N\tau$ product is called the Lawson criterion for net power production and can be generally defined for any fusion-fuel system or confinement scheme.

Magnetic Confinement

The two diverse approaches, magnetic and inertial confinement, are being pursued and form the basis of a large number of fusion research programs throughout the world. Magnetic confinement is based upon the fact that charged particles, such as electrons and ions (eg, deuterons, tritons), tend to be fixed to lines of magnetic force. Thus, the essence of the magnetic-confinement approach to fusion is to heat and trap a plasma in a suitably chosen magnetic-field configuration in the hope of exceeding the Lawson criterion.

A possible approach is to construct a configuration hereby the field lines close on themselves so that a particle following such a line remains within the system. The simplest way of producing such a configuration is to employ a torus, or doughnut-shaped container, with current-carrying coils wrapped around the minor diameter as shown in Figure 1.

Better confinement can be achieved with more complex field configurations as used in devices such as tokamaks, magnetic mirrors, the Elmo Bumpy torus (EBT) and the ZT-40 pinch.

Inertial Confinement

A totally different approach to controlled fusion attempts to create a much denser reacting plasma that, therefore, needs to be confined for a correspondingly shorter time. This concept is the basis of a relatively new research area known as inertial confinement. In this approach, small pellets containing fusion fuel are compressed to extremely high densities by intense, focused beams of photons or energetic charged particles. Two approaches to inertial confinement are laser-driven fusion and particle-beam fusion.

Environmental Aspects

For D–T fusion, the principal concerns in fusion energy relate to the tritium inventory in the power plant, ca 10 kg or 3.7×10^{18} Bq (10^8 Ci), and the activation of the structural materials of the reactor by the 14 MeV neutrons. Safety systems must be engineered to deal with unexpected tritium releases, and remotely operated mechanisms will be required to perform routine periodic replacement of the activated structures.

In the long term, advanced fuel cycles will avoid the use of tritium. Although more difficult to ignite, these fuels will produce less radiation and can be extracted from natural resources of virtually unlimited supply.

Gerald M. Halpern
Exxon Research and Engineering Company

J. Dawson, *Bull. Am. Phys. Soc.* **24**, 42 (1979).

H.A. Bethe, *Nucl. News*, **41** (May, 1978).

L. Booth, *Dept. of Energy Report CONF-770593*, U.S. Government Printing Office, Washington, D.C., 1977.

G

G ACID, (OH)C₁₂H₅(SO₃H)₂. See Dyes and dye intermediates.

GADOLINITE, Be₂FeY₂Si₂O₁₁, GADOLINIUM. See Rare-earth metals.

GALENA, GALENITE. See Lead.

GALLIUM AND GALLIUM COMPOUNDS

Gallium is a scarce element and is found most commonly in association with its immediate neighbors in the periodic table: zinc and aluminum. Gallium is not a rare element: its concentration in the earth's crust, 10–20 g/t (10d–20 ppm) is comparable to that of lead an arsenic.

Because of the abundance of aluminum (qv) and alumina plants (and less importantly zinc ores and plants), these two are the main sources of gallium.

Gallium has two stable isotopes, ^{69}Ga (60.4%) and ^{71}Ga (39.6%). Twelve unstable isotopes (mass 63–76) are known. The radius of the atom is 0.138 nm, and the ionic radius of Ga^{3+} is 0.062 nm and that of Ga^+ is 0.133 nm. Solid gallium has a metallic, slightly bluish appearance; the liquid is more white than silver.

The physical properties of gallium—especially its thermal properties—are exceptional. It has the longest liquid interval of the elements. Principal physical properties of normal gallium are listed in Table 1.

In accordance with its normal potential, gallium is chemically similar to zinc and is somewhat less reactive than aluminum. Just as is aluminum, it is protected from air oxidation at ambient temperature by a fine film of oxide. It is entirely oxidized by air and by pure oxygen at ca 1000°C.

Table 1. Physical Properties of Normal (I or α) Gallium

Property	Value
melting point, °C	29.77–29.78
boiling point, °C	ca 2200
density at melting point, g/cm³	
solid	5.904
liquid	6.095
heat of fusion, J/g^a	79.8
heat capacity, J/(kg·K)^a	
30°C	381.5
thermal conductivity, W/(m·K)	
solid at 20°C axes	a 88.4 / b 16.0 / c 40.8
liquid at 77°C	28.7
cubic coefficient of expansion, °C^{-1}	
solid, at 0–20°C	5.95×10^{-5}
liquid, at 103°C	1.20×10^{-4}
vapor pressure at 1198 K, Pa^b	0.14
crystallographic properties	orthorhombic Cmca space group
axes	a = 0.45186 nm / b = 0.76602 nm / c = 0.45258 nm
electrical resistivity at 20°C, μΩ·cm	polycrystal 15.05
liquid at 30°C	25.795
viscosity, dynamic at 32°C, mPa·s (= cP)	1.810
surface tension at 30°C, mN/m (= dyn/cm)	709

^{a} To convert J to cal, divide by 4.184.
^{b} To convert Pa to mm Hg, divide by 133.3.

Normally, gallium is trivalent; however, it also may be monovalent. Gallium forms some compounds of these two valences. It cannot have a valence of +2.

Extraction

The sodium aluminate liquor in the Bayer-process plants that produce alumina is the best source of gallium. There is a considerable quantity of alumina produced in the world. The high concentration of gallium that is readily available from this solution obviates the preliminary extraction. Gallium can be extracted by fractional carbonation, by direct electrolysis of the aluminate solution, or by reaction with Lewis-base organic complexes containing oxygen or, preferably, nitrogen.

The gallium content of sphalerites generally is concentrated in the residues of zinc distillation and in the iron mud resulting from purification of zinc sulfate solutions. Gallium is extracted from these media with acidic solutions and $GaCl_3$ is extracted with an organic solvent.

Alloys and Intermetallic Compounds

Gallium has complete miscibility in the liquid state with aluminum, indium, tin, and zinc, without formation of compounds. The systems obtained with gallium and bismuth, cadmium, germanium, mercury, lead, silicon, and thallium present miscibility gaps, and they give no intermetallic compounds.

Numerous compounds formed with the transition elements have been reported.

Lanthanides (and yttrium). The following compounds are known: M = lanthanide: Dy, La, Nd, Sm, and Tb, MGa_2, hexagonal; MGa, orthorhombic; M_5Ga_3, tetragonal; Eu and Yb, MGa_2, hexagonal; Lu, Tm, MGa_3, cubic; MGa, orthorhombic; M_5Ga_3, hexagonal; Er_5Ga_3, hexagonal; $GdGa_2$, hexagonal; GdGa, orthorhombic; Gd_3Ga_2, hexagonal; GdGa, orthorhombic; Gd_3Ga_2, tetragonal; Gd_5Ga_3, tetragonal; $HoGa_3$, cubic; $HoGa_2$, hexagonal; HoGa, orthorhombic; Ho_5Ga_3, hexagonal; $PrGa_2$, hexagonal; PrGa, orthorhombic; Pr_5Ga_3, orthorhombic; YGa_2, hexagonal; YGa, orthorhombic; and Y^5Ga_3, hexagonal.

Actinides. The following compounds are known: U_2Ga_3, orthorhombic; UGa_2, hexagonal; UGa_3, cubic; $PuGa_6$, tetragonal; $PuGa_4$, orthorhombic; $PuGa_3$, hexagonal; $PuGa_2$, hexagonal; PuGa, tetragonal; Pu_5Ga_3, tetragonal; and Pu_3Ga, tetragonal.

Compounds Other than Intermetallic

Compounds other than intermetallic include hydrides (GaH_3, GaH_3, $MGaH_4$ [M = Li, Na, K, Rb, and Cs]); halides of gallium (I, II, and III) (Ga_2X_4 [X = Cl, Br, or I], GaF_3, $GaCl_3$, $GaCl_2OH$, $GaBr_3$, $GaBr_2OH$, GaI_3); gallium oxyhalides (GaOCl, GaOBr), gallium halogenates (Ga-$(ClO_4)_3$); sulfohalides and sulfohalogenates (GaSF, Ga_9S_8Cl, $Ga_9S_8Br_{11}$, $Ga(SO_3Cl)_3$); compounds with ammonia ($GaF_3 \cdot NH_3$, $GaF_3 \cdot 3\ NH_3$, $GaCl_3 \cdot NH_3$, $GaBr_3 \cdot NH_3$, $GaI_3 \cdot NH_3$); mixed halides (eg fluorogallates (III) (M_3GaF_6 where M = Cs, K, Li, Na, Rb, Tl, and NH_4), chlorogallates (III) ($MGaCl_4$ where M = Cs, K, or Rb), bromogallates (III) ($MGaBr_4$ where M = Cs, K, or Rb), and iodogallates (III) ($MGaI_4$ where M = Cs or Rb); gallium oxides (I (Ga_2O), II (GaO), and III (Ga_2O_3)); gallates (III) (Li_5GaO_4, $LiGaO_2$, and $LiGa_5O_8$), gallium chalcogenides (Ga_2Se, Ga_2Se_3, and Ga_2Te_3), mixed chalcogenides ($MGaS_2$ (or Se_2 or Te_2) where M = Ag, Cu, In, or Tl), sulfates and selenates ($Ga_2(SO_4)_3$), gallium compounds with nitrogen (GaN), gallium compounds with phosphorus, arsenic and antimony (GaP, GaAs, and GaSb), and carbon compounds of gallium (Mn_3GaC, Nd_3GaC, and Mo_2GaC).

Toxicology

The toxicity of metallic gallium or gallium salts is very low. The corrosive, poisonous or irritating nature of some of its compounds is attributable to the anions or radicals with which it is associated.

PIERRE DE LA BRETÈQUE
Alusuisse France S.A.

K. Wade and A.J. Banister, *The Chemistry of Al, Ga, In, and Tl*, Pergamon Press, New York, 1974.

GAS CLEANING

Gas cleaning comprises those steps that are necessary for removal of impurities from a gas stream for process reasons. In order to design and select gas-cleaning equipment, the flow rate, composition, temperature, and pressure of the gas to be cleaned must be known, in addition to the amount of the contaminant and its chemical species and physical state. All equipment used for air-pollution effluent abatement can generally be employed for gas cleaning. When present as particulates, the particle size distribution and apparent density of the particle are important.

Five methods applicable for removing gaseous impurities are absorption, adsorption, chemical reaction, condensation, and incineration. The last item is valid only for combustible impurities. Absorption may be water-based or employ an organic liquid. Typically, packed and plate-type columns are used (see Absorption). Carbon adsorption can remove organic compounds almost completely from large gas volumes.

Particulates may be removed by inertial impaction and direct interception in centrifugal and impingement devices, diffusional deposition, and flux-force interaction. Selection depends on the particle size and the required removal efficiency. Common treatment devices include cyclones, fabric and granular-bed filters, impaction devices, electrostatic precipitators, and both low and high energy wet scrubbers (see Air-pollution control methods). Except for fabric and granular-bed filters, these devices can be used for removal of both liquid and solid particulates.

Dust is removed from gas or air with filters referred to as air cleaners or air filters (see also Filtration). They are designed for lower concentrations of particulate matter in the carrier gas such as that encountered in outdoor air or air recirculated within a building. Although they are typically employed in ventilation, heating, and air conditioning, they are very important for processes requiring clean air.

B.B. CROCKER
Monsanto Company

W. Strauss, *Industrial Gas Cleaning*, 2nd ed., Pergamon Press, Oxford, UK, 1975.

R.H. Perry and C.H. Chilton, eds., *Chemical Engineers' Handbook*, 6th ed., McGraw–Hill, New York, 1983.

GAS HYDRATES. See Hydrocarbons, C_1–C_6; Water, supply and desalination.

GASKETS. See Packing materials.

GAS, MANUFACTURED. See Fuels, synthetic.

GAS, NATURAL

Natural gas is defined as a naturally occurring mixture of hydrocarbon and nonhydrocarbon gases found in the porous geologic formation beneath the earth's surface, often in association with petroleum. To obtain a marketable product, the raw natural gas flowing from gas or oil wells must be processed to remove water vapor, inert or poisonous constituents, and condensable hydrocarbons. The processed gas is principally methane, with small amounts of ethane, propane, butane, pentane, carbon dioxide, and nitrogen. This gas can easily be transported from the producing areas to the market in underground pipelines under pressure or liquefied at low temperatures and transported in specially designed ocean-going tankers (see Pipelines; Transportation).

Natural gas is used principally as a source of heat in residential, commercial, and industrial service. Because of its clean-burning quality, the convenience of utilization, low cost, and abundance, natural gas supplies approximately one third of the total energy requirements of the United States.

Reserves

Natural-gas reserves in the United States are classified into two categories based on the reservoir occurrence: nonassociated gas is defined as free natural gas not in contact with crude oil in the reservoir, and associated, dissolved gas is the combined natural gas that occurs in crude-oil reservoirs, either as free gas or as gas in solution with crude oil. Total proved gas-reserve estimates are the sum of all nonassociated gas, associated, dissolved gas, plus all of the recoverable gas volumes available in underground storage reservoirs. Proved gas reserves in the United States increased steadily each year until 1967, when they peaked at ca 8.3×10^{12} m^3 (293×10^{12} ft^3) and have declined since that time.

The world reserves continue to increase, spurred by improved price and technological advances in the developing countries. Large gas deposits are now being discovered in Mexico, South America, and in the Far East (see also Fuels, survey; Petroleum, resources). Studies indicate that the total remaining recoverable reserves of natural gas from conventional sources in the world amount to ca $(2.6–2.7) \times 10^{14}$ m^3 $[(9.18–9.53) \times 10^{15}$ ft$^3]$.

Properties and Composition

Methane, the principal constituent of natural gas, is colorless and odorless. It is a simple, nonpoisonous asphyxiant. When properly mixed with air, it burns with a faint, luminous flame; however, when mixed with air in concentrations ranging between the lower and higher limits of flammability, natural gas explodes. Table 1 shows some of the more important physical properties of the common hydrocarbon compounds and impurities found in gas streams (see Hydrocarbons, C_1–C_6).

Natural gas is classified in several broad categories based on the chemical composition: wet gas contains condensable hydrocarbons such as propane, butane, and pentane; lean gas denotes an absence of condensable hydrocarbons; dry gas is a gas whose water content has been reduced by a dehydration process; sour gas contains hydrogen sulfide and other sulfur compounds; and sweet gas denotes an absence of hydrogen sulfide and other sulfur compounds. Natural gas sold to the public is described as lean, dry, and sweet.

The composition of natural gas at the wellhead varies widely from field to field. Many undesirable components may be present that must be removed by processing before delivery to the pipeline. Natural-gas liquids must be extracted from the wet gas in sufficient quantity to eliminate the possibility of condensation in the pipeline. Accumulations of these liquids are a safety hazard and restrict the gas flow. Natural-gas liquids are usually propane and higher hydrocarbons; ethane may also be included. Propane and butane are known as liquefied petroleum gas (LPG), whereas pentane and higher hydrocarbons are known as natural gasoline (see Gasoline; Liquefied petroleum gas). There is a ready market for all of these natural-gas liquids that helps offset the cost of processing.

Table 1.   Physical Constants^a of Common Natural-Gas Constituents

Component	Molecular weight	Boiling point, °C	Liquid at 16°C/ 16°C	Gas^a (air = 1.00)	Gross heating value^a, MJ/m$^{3\,b}$
methane	16.043	−161.5	0.3^c	0.5539	37.56
ethane	30.070	−88.6	0.3564^d	1.0382	65.80
propane	44.097	−42.0	0.5077^d	1.5225	93.65
n-butane	58.124	−0.5	0.5844^d	2.0068	121.36
isobutane	58.124	−11.7	0.5631^d	2.0068	121.01
n-pentane	72.151	36.1	0.6310	2.4911	149.16
isopentane	72.151	27.8	0.6247	2.4911	148.82
carbon dioxide	44.010	−78.5	0.827^d	1.5195	
hydrogen sulfide	34.076	−60.3	0.79^d	1.1765	23.70
nitrogen	28.013	−195.8	0.808^e	0.9672	
water	18.015	100.0	1.000	0.6220	

Specific gravity column spans "Liquid at 16°C/16°C" and "Gas^a (air = 1.00)".

^{a}At 101.35 kPa (1 atm) and 16°C.
^{b}To convert MJ/m^3 to Btu/ft^3, multiply by 26.88.
^{c}Apparent value at 16°C.
^{d}At saturation pressure and 16°C.
^{e}Density of liquid, g/mL, at normal boiling point.

Hydrogen sulfide must be removed from the natural gas since it is extremely dangerous and poisonous, and in the presence of air or water, corrosive. Natural gas also is odorized before distribution to provide a distinctive odor which alerts customers to possible gas leaks in their equipment. Commercial odorant, containing mercaptans, aliphatic sulfides, or cyclic sulfur compounds, is added at rates varying from 4 to 24 mg/m^3 (0.25–1.5 lb/10^6 ft^3).

Gas Conditioning

Gas conditioning is required to alter the composition of the wet natural gas flowing from the wellhead to provide uniform lean, dry, sweet gas for consumer use. Processing facilities are usually located in the fields and vary in size from small wellhead installations to large, centrally located plants. The conditioning of a natural gas involves a large variety of processes, each designed to remove a specific impurity.

Water vapor, generally present in gas produced from underground reservoirs is removed either by absorption by hygroscopic liquids or adsorption on activated solid desiccants (see Drying agents). To remove acid gases such as carbon dioxide, hydrogen sulfide, and other sulfur compounds from natural gas, many processes are available, such as chemical solvent, physical solvent, and dry-bed processes.

The EPA has issued stringent sulfur-emission standards which limit the amount of sulfur dioxide that may be released into the atmosphere. Hence, practically all hydrogen sulfide removed from natural gas must be converted to elemental sulfur (see also Sulfur recovery).

Transportation and Storage

Production wells in the United States are connected to the consumers by a complex network of buried pipelines, consisting of 13.9×10^4 km of field-gathering lines, 43.5×10^4 km of transmission lines, and 114.2×10^4 km of distribution mains. Underground storage near market areas provides the most economical means of balancing variations in market demand with delivery capability of transmission lines.

Recent technological advancement in the use of liquefied natural gas (LNG) provided the gas industry with new methods to solve the problems of storage and transportation. Natural gas can be reduced to 1/600 of the volume occupied in the gaseous state by cryogenic processing, safely stored or transported in double-walled insulated metal containers at near atmospheric pressure and $-162°C$, and when required, can be regasified (see Cryogenics).

Outlook

The future prospects for the United States gas industry depend on maintaining production from conventional sources in the 48 contiguous states and development of several higher-cost supplemental gas sources shown in Table 2.

Table 2. Estimated Total Future United States Gas Supply

Sources	Annual gas volumes, 10^{10} m^{3a}	
	1990	2000
contiguous 48 states production		
from conventional sources	42.5–58.2	34.0–39.7
supplemental sources		
Canadian imports	4.8	5.7
Mexican imports	2.8	5.7
LGN imports	2.0	2.0
Alaskan gas	3.7	7.9
coal gasification development	2.0	7.6
SNG^b from petroleum	0.8	0.8
tight gas formation development	1.1	3.4
nonconventional gas source	0.6	2.8
Total	*60.3–66.0*	*69.9–75.6*

^{a}To convert m^3 to ft^3, multiply by 35.3.
^{b}Substitute natural gas.

JOHN H. HILLARD
Mountain Fuel Supply Co.

Gas Facts, American Gas Association, Arlington, Va., 1981.

J.D. Parnet and H.R. Linden, *A Survey of United States and Total World Production, Proved Reserves and Remaining Recoverable Resources of Fossil Fuels and Uranium as of December 31, 1975*, Institute of Gas Technology, Chicago, Ill., Jan., 1977.

D.L. Katz and co-workers, *Handbook of Natural Gas Engineering*, McGraw-Hill Book Co., Inc., New York, 1959.

The Gas Energy Suply Outlook 1980–2000, a report of the American Gas Association Gas Supply Committee, Arlington, Va., May, 1982.

GASOHOL. See Alcohol fuels; Ethanol; Fuels, synthetic, liquid fuels; Gasoline and other motor fuels.

GASOLINE AND OTHER MOTOR FUELS

To satisfy modern high performance automotive engines, today's gasolines must meet exacting specifications. Moreover, some of their characteristics must be tailored to fit prevailing local conditions.

Taken together, the Reid vapor pressure and the ASTM distillation curve tell much about a gasoline's volatility, ie, whether it has the proper vaporizing characteristics for the climate and altitude where it will be used. This is essential; otherwise, the gasoline may not provide proper warmup characteristics on the one hand, or, on the other hand, may cause vapor lock. Table 1 shows average volatility data for U.S. regular, unleaded, and premium-grade gasolines, both in summer and winter.

To prevent annoying, fuel-wasting, potentially damaging engine knock at all engine speeds and loads, gasoline must have high antiknock quality (octane number) throughout its entire distillation range.

Fairly early in the study of knock, it was recognized that the chemical structure of fuel hydrocarbons largely determines their tendency to cause knock and that straight-chain paraffins are more prone to knocking than branched-chain paraffins, olefins, or cyclic hydrocarbons. Likewise cycloparaffins (naphthenes) are more knock-prone than aromatics. This knowledge has led to development of several octane-improvement processes. Knock also is combated by negative catalysts, the antiknock compounds used to head off or to slow precombustion reactions that might otherwise convert fuel hydrocarbons to autoigniting, detonating compounds.

To date, extensive research has failed to turn up any other antiknock with the cost-effectiveness of the alkylleads. Soon after refiners began adding tetraethyllead to gasoline, it became evident that some yardstick was needed for measuring the antiknock quality of motor fuels. In 1926, Graham Edgar of Ethyl Corporation developed the octane scale, which has become the worldwide standard for that purpose. In order to rate fuels of more than 100 octane number, a method has since been devised to match the antiknock quality of the test fuel with that of 2,2,4-trimethylpentane (isooctane) plus a known amount of tetraethyllead. A curve similar to Figure 1 is used convert such knock values to octane numbers above 100.

The CFR (Cooperative Fuel Research) knock-test engine has been adopted as the standard for determining octane number The antiknock quality of today's gasolines is rated by two laboratory knock-test procedures, the Motor method and the Research method.

Table 1. Average Volatility Data for U.S. Motor Gasolines

Property	Unleaded		Regular		Premium	
	Winter	Summer	Winter	Summer	Winter	Summer
initial boiling point, °C	28	32	28	32	28	32
10% point, °C	42	49	42	48	42	48
50% point, °C	100	104	94	99	99	102
90% point, °C	165	167	168	172	164	166
end point, °C	207	209	210	213	208	209
Reid vapor pressure, kPa^a	86	67	85	67	85	68

^{a}To convert kPa to psi, multiply by 0.145.

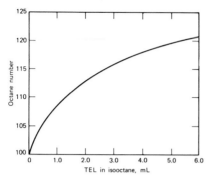

Figure 1. Relationship between "knock value" and octane numbers above 100. Courtesy of Ethyl Corporation.

The importance of octane number in evaluating gasoline stems from its role in permitting increases in engine compression ratio. The power obtainable from combustion of a gasoline–air mixture in an engine can be increased by more highly compressing the mixture before igniting it. In recent years, the compression ratios of U.S. cars have been considerably reduced to permit their operation on the unleaded gasoline required by their catalytic emission-control systems (see Exhaust control, automotive).

A good gasoline must be substantially free of gum-forming materials in order to avoid clogging of engine fuel systems, carburetor malfunction, sticking of engine intake valves, and manifold deposits that would cause exhaust smoke and loss of engine power. Modern catalytic refining processes produce very little or none of the worst gum-forming unsaturates once present in thermally cracked and thermally reformed gasoline fractions. And certain processes in the manufacture of today's gasolines are aimed at converting any potential gum formers into unreactive forms prior to final blending of the finished gasoline. Most motor gasolines contain less than 3 mg of such gum per 100 mL.

Gasolines containing hydrogen sulfide or the thiols (mercaptans) naturally present in crude oil would be extremely offensive to the motorist. These and other sulfur compounds are also corrosive to metals and could cause serious damage both to refinery equipment and to the engine. Moreover, many of the sulfur compounds present in crude oils poison refining catalysts and are antagonists toward antiknock compounds. Thus, a number of refinery processes have been developed either to remove sulfur compounds from petroleum products or to convert them to inoffensive forms. Actual removal is preferred.

Petroleum refining begins with the distillation of crude oil into fractions of different boiling ranges (see Distillation; Petroleum). The crude oil is heated to 370–430°C, pumped into a fractionating tower, and separated into the following principal fractions: light naphtha; heavy naphtha; kerosene; light gas oil; heavy gas oil; and reduced crude.

Most modern plants cover the gasoline-range hydrocarbons present in "wet" natural gas by absorbing in petroleum fractions somewhat heavier than straight-run gasoline. The wet gas in passed through a column in contact with the absorption oil, and the absorbed "natural gasoline" is recovered from the oil by a form of distillation, using steam to increase the separation of lighter from heavier hydrocarbons.

Petroleum fractions boiling above the gasoline range can be broken down (cracked) to gasoline by heating to high temperatures. Modern catalytic-cracking units employ catalysts in the form of small beads that are finely divided, and fluidized powder to obtain higher cracked-product yield and higher octane gasoline than thermal treatment alone can provide. The carbon deposits formed on the surface of these catalysts during cracking are burned off (thereby regenerating the catalyst) by subjecting the hot catalyst to a stream of air in a separate catalyst-regeneration step.

Gasoline-range fractions produced by catalytic cracking make highly desirable blending components for high octane gasoline. The lower-octane paraffinic and cycloparaffinic hydrocarbons in the feedstock are largely converted to higher-octane olefins and aromatics. Thus, in addition to its prime function of being a gasoline-production process, cracking also serves, in a sense, as an octane-improvement process.

Hydrocracking, which consists of cracking in the presence of added hydrogen, has recently been gaining favor as a highly flexible refining process because it permits wide variations in yields of gasoline and furnace oils to meet seasonal demand changes (see Petroleum).

All reforming processes have the same general purpose: to convert low octane gasoline-range hydrocarbons into higher-octane ones. Reforming is largely limited to the upgrading of heavier gasoline fractions, such as straight-run stocks boiling from 90–200°C, because lighter fractions do not contain substantial amounts of hydrocarbons suitable for reforming.

Unlike cracking, alkylation makes larger hydrocarbons from smaller ones; it produces gasoline-range liquids from refinery gases. Because of its cost, however, its chief value stems from the exceptionally high antiknock quality of its product, rather than from the fact that it offers another way to make gasoline.

Like alkylation, polymerization is a way of making gasoline from refinery gases. But in polymerization only the olefinic gases in the feed react, linking to form olefinic liquids. Any paraffinic gases in the feed pass through the process unchanged. A typical polymerization reaction is that of two molecules of isobutylene (C_4H_8) combining into one molecule of a branched-chain octylene (C_8H_{16}).

To date, petroleum-refinery use of isomerization to convert straight-chain hydrocarbons into branched-chain hydrocarbons of the same molecular weight has been confined mainly to the conversion of normal butane to isobutane needed for alkylation feed.

Sulfur removal usually is accomplished by hydrogen treating, wherein hydrogen reacts with the sulfur in objectionable compounds, forming easily removable hydrogen sulfide.

Additives to gasoline include antiknock compounds, blending agents, antioxidants (qv), metal deactivators, antirust agents, anti-icing agents, detergents (qv), multipurpose additives, upper-cylinder lubricants, dyes, and proprietary additives.

Blending of components.   To save on tank requirements, vapor losses during storage and transfer, and blending manpower, many refiners have now adopted in-line blending systems that automatically and continuously combine a multiplicity of gasoline components in a blending manifold, from which finished gasoline exits in a constant stream. Owners of older cars designed for premium-grade leaded gasoline can mix the best unleaded gasoline available with a good leaded regular-grade gasoline. The result is a surprisingly good low lead gasoline, in some cases one that is considerably higher in octane number than either of its components.

Gasoline Substitutes

Processes developed to produce gasoline substitutes include the Fischer-Tropsch process, Mobil's methanol-to-gasoline process, and coal liquefaction (see Fuels, synthetic). Exploitation of oil shale (qv) and tar sands (qv), and gasohol and other alcohol fuels also may provide gasoline substitutes in the future.

Alcohol vs hydrocarbons as fuel.   Comparing the properties of ethyl alcohol with those of gasoline-range hydrocarbons (Table 2) suggests some of the advantages and disadvantages of alcohol.

Diesel Fuel

The procedure used in rating the ignition quality of diesel fuel is similar to that used to measure gasoline octane number. Cetane (*n*-

Table 2.   Ethyl Alcohol vs Hydrocarbons

	Ethyl alcohol	Isooctane
heating value, J/g[a]	29,800	36,500
heating value, MJ/cm^3[b]	28,430	33,445
specific gravity	0.794	0.702
stoichiometric A/F ratio[c]	9	15
latent heat of evaporation, J/g[c]	840	328
Research octane number	106	100
Motor octane number	89	100

[a] To convert J/g to Btu/lb, multiply by 0.430.
[b] To convert J to cal, divide by 4.184.
[c] A/F = air–fuel ratio.

hexadecane), which has very high ignition quality (ie, low ignition delay), represents 100 on the cetane-number scale. Blends of cetane with a heptamethylnonane isomer (HMN) represent intermediate ignition qualities; their cetane numbers are calculated to the nearest whole number from the percentages of cetane and HMN in the blend. The equation is as follows: cetane number = % cetane + 0.15(% HMN). Thus, a reference blend containing 35% cetane and 65% HMN has a cetane number (CN) of 45. Like laboratory octane-number determinations, cetane-number ratings are made in a single-cylinder engine having a variable compression ratio.

Desirable diesel-fuel volatility characteristics are obtained by careful balancing of light and heavy components available from distillation and cracking at the refinery.

The temperature at which a fuel ceases to flow is called its pour point. Most diesel-engine manufacturers recommend that fuels have a pour point 6°C below the lowest atmospheric temperature at which the engine will be operated.

When cooled to a temperature somewhat above the pour point, diesel fuel becomes cloudy, forming wax crystals and other solid materials that can clog fuel filters and supply lines. The temperature at which these solids begin to form is the cloud point. In general, engine manufacturers recommend that cloud point be below the temperature of use and not more than 6°C above the pour point.

For optimum performance of diesel-engine injector pumps, the fuel should have the proper "body" or viscosity (see Rheological measurements). Too low viscosity may make more frequent maintenance of the injection system necessary; too high viscosity may cause excessively high pressures in the injection system.

The greater the density of a diesel fuel, the greater is its heat content per unit volume. Because diesel fuel is purchased on a volume basis, density is stipulated in purchasing specifications and often is measured during delivery inspections. In the United States, the common measurement of density is API gravity, (°API at 15.6°C = (141.5/sp gr) − 131.5).

The flash point of a fuel indicates the temperature below which it can be handled without danger of fire. This is the temperature to which the fuel must be heated to create sufficient fuel vapors above the surface of the liquid fuel for ignition to occur in the presence of an open flame. Specifications on flash point vary with the grade of diesel fuel, 38°C being the lowest ever specified.

The tendency of a diesel fuel to form carbon deposits in an engine can be roughly predicted by one or the other of two carbon-residue tests: the Ramsbottom coking method or the Conradson carbon test.

During intermittent engine operation or operation at low temperatures, when condensation of moisture is appreciable, sulfur compounds in the fuel may cause cold corrosion and increased engine wear. Specifications on the sulfur content of diesel fuel usually are based on a bomb test (ASTM D 129).

Because the fuel injectors of diesel engines are designed to very close tolerances, they are sensitive to any abrasive material in the fuel. The amount of unburned residue or ash should not exceed 0.01–0.02% of the weight of the sample in high speed engines, but it may be as high as 0.10% for low speed and medium speed engines operated under sustained loads at nearly constant speed.

Grades of diesel fuel include the following: Grade No. 1-D, which includes volatile fuel oils from kerosene to the intermediate distillates; Grade No. 2-D, which includes distillate gas oils of lower volatility; and Grade No. 4-D, which includes the more viscous distillates and their blends with residual fuel oils.

Diesel fuel manufacture. High quality diesel fuels usually are made by blending a high proportion of straightrun distillates with lesser amounts of cracked fractions. Additives include diesel ignition improver, stability improvers, corrosion inhibitors, multipurpose additives, conductivity improvers, broad-spectrum distillate additives, and ethyl alcohol.

Liquefied Petroleum Gases

Liquefied petroleum gases (LPG), which consist primarily of propane, butanes, and their mixtures, have long been of interest as engine fuels because of their relatively low cost and their high octane number (see Liquefied petroleum gas).

Natural gas, the main source of propane, varies in composition, but a typical analysis might show 80–90% by volume of methane, 5–8% ethane, 4–6% propane, 1–3% butanes, and 1–5% pentanes and heavier hydrocarbons (see Gas, natural). Generally, propane and heavier hydrocarbons are separated from the methane and ethane by passing the natural gas at a pressure of about 3.45 MPa (500 psi) upward through an absorber tower, countercurrent to a flow of oil.

The antiknock performance of commercial liquefied petroleum gases can be made to equal or exceed that of pure propane by addition of tetramethyllead (TML) antiknock compounds.

Because LPG is considerably cheaper than diesel fuel in certain sections of the United States, some U.S. railroads have considered converting diesel locomotives to operate on propane-rich LPG. For maximum savings in fuel cost, a locomotive should be able to alternate between fuels at will, depending on which is cheaper in the area where the locomotive happens to be at the time. Indications are that dual-fuel operation (burning two fuels simultaneously) may prove to be the key to successful use of LPG in locomotives.

Natural Gas

In many ways, natural gas is a highly desirable fuel for internal-combustion piston engines. On the basis of cost per joule of heat energy, it is often the least expensive engine fuel available in a given area, and its high antiknock quality permits design of highly efficient, high compression engines. Moreover, in the dry, sulfur-free form readily available from pipelines, it is an extremely clean fuel that burns cleanly in engines, producing very little engine deposits and little or no corrosion of engine parts. However, because natural gas cannot be stored in a vehicle's fuel tank, its use as engine fuel is limited to stationary installations, such as engines that power heavy-duty compressors or drive generators in central power stations (see Gas, natural).

A number of large stationary diesel engines are operated on the dual-fuel principle, using natural gas as the primary fuel and a small pilot charge of diesel fuel for compression ignition.

The Total-Energy Concept

As central air conditioning has become almost as prevalent as central heating in public, commercial, and industrial buildings, interest has grown in systems that can supply all the energy requirements of a large building or complex of buildings, ie, total-energy systems. By 1990, total-energy systems are predicted to be a 3×10^9/yr business, and home installations will account for almost two-thirds of this. Total-energy systems vary in components employed, but all have one thing in common: an engine to provide the power required for heating, cooling, lighting, etc.

JOHN C. LANE
Ethyl Corporation

Hydrocarbon Processing, Refining Handbook Issue, **9**, 91 (Sept. 1982).

W.A. Gruse, *Motor Fuels*; *Performance and Testing*, Reinhold Publishing Corp., New York, 1967.

H.S. Bell, *American Petroleum Refining*, D. Van Nostrand, New York, 1959.

GASTROINTESTINAL AGENTS

The compounds currently used as gastrointestinal agents may be divided into several classes according to their general activity: agents for peptic-ulcer therapy, laxatives, antidiarrheals, and antiemetics. Most of the compounds described in the article are discussed in the *United States Pharmacopeia* (USP-NF) and approved for use by the FDA.

Antipeptic-Ulcer Therapy

The market for antiulcer agents is very large and is comprised of both prescription-only products, such as cimetidine and anticholinergics, and over-the-counter (OTC) products that are mainly antacids.

Cimetidine. Cimetidine (**1**) (Tagamet, SKF 92334), has a mol wt of 252.33 and an mp of 141–143°C. This compound has not shown significant toxicity in humans on chronic dosing. It is the first and only approved histamine (H-2) antagonist currently used widely for therapy of duodenal ulcers (see Histamines and histamine antagonists).

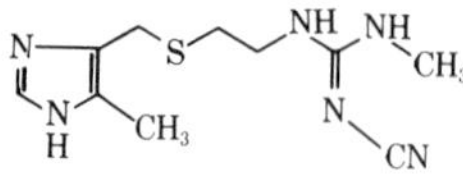

(1)

Antacids. Antacids include aluminum hydroxide gel, precipitated calcium carbonate, magnesia and alumina oral suspension, magnesium oxide, magnesium trisilicate, magaldrate, and sodium bicarbonate.

Laxatives

Laxatives are agents that facilitate the passage and elimination of feces. These agents are used most commonly as self-treatment and are rarely prescribed by physicians. Laxatives have traditionally been classified into four categories: irritant (stimulant) cathartics; saline; bulking agents (the agent of choice in most instances); emollients (lubricants); and fecal softeners.

Cathartics. Cathartics include danthron (1,8-dihydroxyanthraquinone), bisacodyl (4,4'-(2-pyridiylmethylene)diphenol diacetate), phenolphthalein (**2**) (3,3-bis(p-hydroxyphenyl) phthalide), Cascara sagrada, castor oil, and mineral oil.

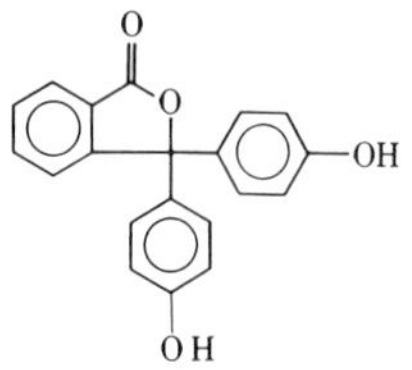

(2)

Saline cathartics. Saline cathartics include magnesium citrate solution ($C_{12}H_{10}Mg_3O_{14}$), magnesium sulfate ($MgSO_4.7H_2O$), and sodium phosphate ($Na_2HPO_4.7H_2O$).

Bulking agents. Bulking agents include plantago seed and psyllium hydrophilic mucciloid.

Emollients and fecal softeners. These include dioctyl calcium sulfosuccinate (calcium salt of 1,4-bis(2-ethylhexyl)sulfosuccinate), and dioctyl sodium sulfosuccinate (sodium 1,4-bis(2-ethylhexyl)sulfosuccinate).

Antidiarrheals

Diarrhea is a commonly occurring problem. It is usually self-limiting and of short duration. Commonly used antidiarrheals work by one of two mechanisms. Narcotic analgesics are constipating and are antidiarrheal owing to their effect on intestinal propulsion (see Analgesics). This is apparently the mechanism of such commonly used antidiarrheals as codeine, morphine, diphenoxylate hydrochloride (ethyl 1-(2-cyano-3,3-diphenylpropyl)-4-phenylpiperidinecarboxylate hydrochloride), and loperamide (4-(4-chlorophenyl)-4-hydroxy-N,N-dimethyl-α,α-diphenyl-1-piperidine butanamide. The other types of agents, such as polycarbophil (1,5-hexadiene-3,4-diol) are much less effective and appear to act by binding water in the intestine or by adsorption of toxins. Two other antidiarrheals are the systemic ambecidal iodochlorhydroxyquin (5-chloro-7-indo-8-quinolinol), and sulfasalazine, 5-[[p-(2-pyridylsulfanoyl)-phenyl]azo]salicyclic acid, which is used in the treatment of inflammatory bowel diseases.

Antiemetics

Nausea and vomiting are frequent symptoms of disease and can occur from a variety of causes. These drugs all appear to act at the chemoreceptor trigger zone or on the vestibular apparatus (see also Histamine and antihistamine antagonists). They include cyclizine hydrochloride (1-(diphenylmethyl)-4-methylpiperazine hydrochloride), dimenhydrinate,

8-chlorotheophylline compounds with 2-(diphenylmethoxy)-N,N-dimethylethylamine), meclizine hydrochloride (**3**) (1-(p-chloro-α-phenylbenzyl)-4-(m-methylbenzyl)piperazine dihydrochloride), and procloperazine maleate (**4**) (2-chloro-10-[3-(4-methyl-1-piperazinyl)propyl]phenothiazine maleate.

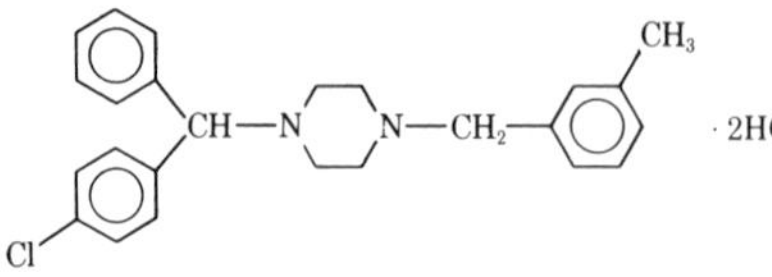

(3)

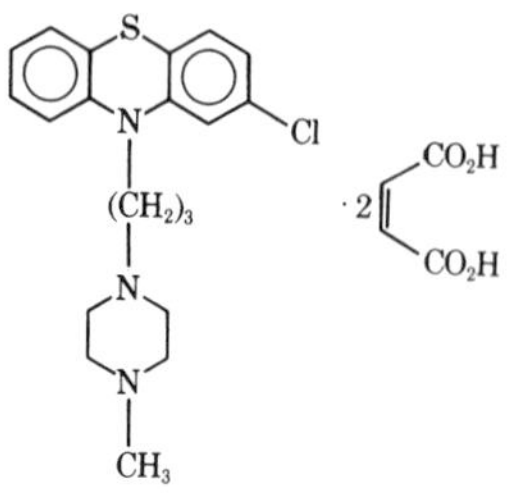

(4)

HENRY I. JACOBY
McNeil Pharmaceuticals

The United States Pharmacopeia XX (USP XX-NF XV), The United States Pharmacopeial Convention, Inc., Rockville, Md., 1980.

F.P. Brooks, *Gastrointestinal Pathophysiology*, 2nd ed., Oxford University Press, New York, 1978.

A. Osol and J.E. Hoover, *Remington's Pharmaceutical Sciences*, 15th ed., Mack Publishing Co., Easton, Pa., 1975.

GELATIN

Gelatin, Type A and Type B, is obtained by the partial hydrolysis of collagen, the chief protein component in skins, bones, hides, and white connective tissues of the animal body (see also Glue; Proteins). Type A is produced by acid processing of collagenous raw materials and has an isoelectric point between pH 7 and 9, whereas Type B is produced by alkaline or lime processing and has an isoelectric point between pH 4.6 and 5.2. Mixtures of Types A and B as well as gelatins produced by modifications of these processes may exhibit isoelectric points outside of the stated ranges. Gelatin is a hydrolysis product obtained by hot water extraction and does not exist in nature.

Gelatin is used in the food, pharmaceutical, and photographic industries which take advantage of its unique properties such as the reversible gel-to-sol transition of aqueous solution; viscosity of warm aqueous solutions; capability to act as protective colloid; water permeability; and insolubility in cold water but complete solubility in hot water.

Physical and Chemical Properties

Commercial gelatin produced in the United States ranges from coarse granules to fine powder. Dry commercial gelatin contains about 9–12% moisture and is an essentially tasteless, odorless, brittle solid with specific gravity of 1.3–1.4. The physical and chemical properties of gelatin, such as solubility, stability, and swelling, are measured on aqueous gelatin solutions and are functions of the source of collagen, the method of manufacture, conditions during extraction and concentration, thermal history, pH, and the chemical nature of impurities or additives. Gelatin is classified as derived protein since it is obtained from collagen by a controlled partial hydrolysis. This makes gelatin a heterogeneous mixture of polypeptides; properties desired are obtained by blending products from selected extracts.

Gelatin is not a single chemical substance. The main constituents of gelatin are large and complex polypeptide molecules of the same amino acid composition as the parent collagen, covering a broad molecular weight distribution range. Most commercial gelatin contains molecular species from 15,000 to ca 250,000 mol wt, the average being between 50,000 and 70,000 mol wt.

The amphoteric character of gelatin is due to the functional groups of the amino acids and the terminal amino and carboxyl groups created during hydrolysis. The distribution of these groups and the resulting isoionic point is determined by the manufacturing process.

A most useful and unique property of gelatin solution is its capability to form reversible gel–sols. A gelatin solution gels at temperatures below 35°C. This conversion temperature is determined as the setting point (sol to gel) or melting point (gel to sol). Commercial gelatins melt between 23 and 30°C, the setting point being lower by 2–5°C.

Gelatin is used mainly by the food, pharmaceutical, and photographic industries.

Derivatized Gelatins

Commercially successful derivatized gelatins are made mostly for the photographic gelatin and microencapsulation markets. In both instances, the amino groups are acylated.

FELIX VIRO
Kind & Knox, a division of Knox Gelatine, Inc.

A.G. Ward and A. Courts, eds., *The Science and Technology of Gelatin*, Academic Press, Inc., New York, 1977.

Gelatin, Gelatin Manufacturers Institute of America, Inc., New York, 1982.

R.J. Croome and F.G. Clegg, *Photographic Gelatin*, Focal Press, New York, 1965.

R.J. Cox, ed., *Photographic Gelatin—II*, Academic Press, Inc., New York, 1976.

GEMS, SYNTHETIC

Today, many traditional gemstones can be produced in the laboratory, including diamond, sapphire, ruby, and emerald. Such gems are called synthetic gemstones since they are of exactly the same chemical composition and structure as the stones occurring in the earth's crust.

At the present time, commercial sizes and quantities of synthetic ruby, sapphire, emerald, and opal are readily available on the market; diamonds, however, although routinely produced commercially for use as abrasives, are not generally available in sizes suitable for gemstones (see Carbon, diamond, synthetic).

The properties desired for good gemstones are hardness, high refractive index, high dispersion, color and clarity.

Synthetic gemstones can be separated into the classical synthetics, which are identical to those found in nature, and gemstone substitutes, especially diamond. The latter are synthetic crystals which may or may not be found in nature but that appear somewhat like diamond.

Classical Synthetic Gems

One of the classical synthetic gems is sapphire (corundum), Al_2O_3, which when pure is known as white sapphire. Addition of iron and titanium produces a blue sapphire, whereas cobalt yields green sapphire. Addition of other dopants produces colored stars such as star ruby (Ti, Cr), or the black star sapphire (Ti, V). Sapphires are made by the Verneuil process.

Spinel, $MgAl_2O_4$, is a colorless crystal of cubic structure and is usually made by the Verneuil process. It also is nearly always doped with cations to produce colored stones.

Beryls (emerald), $Be_3Al_2Si_6O_{18}$, range in color from very pale aquamarine to green. Synthetic emeralds are produced by flux growth and the hydrothermal technique.

Diamonds are produced through a process that uses tapered pistons, a hydraulic press, and the famous belt of tungsten carbide.

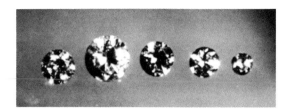

Figure 1. Comparison of simulated diamonds with genuine diamond. Left to right: strontium titanate, YAG, GGG, cubic zirconia, diamond. Photograph by V.F.S. Yip.

Diamond Substitutes

Diamond substitutes include strontium titanate, $SrTiO_3$, rutile, TiO_2, yttrium aluminum garnet (YAG), $Y_3Al_5O_{12}$, (until recently the best known diamond substitute), gadolinium gallium garnet (GGG), $Ga_3Gd_5O_{12}$, and cubic zirconia, ZrO_2, (now the best of the diamond imitations, Fig. 1).

Manufacture

The Verneuil process is the most widely used. In this process, a seed crystal is placed upright in a ceramic, insulated furnace and an oxyhydrogen flame is directed from above on the seed. When the seed is slightly molten on its upper surface, the raw material in very finely divided powder form is sprinkled downward onto the molten cap. The powder melts and forms part of the liquid adhering to the upper tip. As the process continues, the seed is very slowly lowered and more powder is added. As the liquid reaches a cooler zone, it solidifies and a crystal is built onto the seed.

Two other processes commonly used include the hydrothermal technique and the cold-crucible process.

L.R. ROTHROCK
Union Carbide Corporation

D. Elwell, *Man-Made Gemstones*, Ellis Horwood Ltd., Chichester, UK, 1979, p. 26.

K. Nassau, *Gems Made by Man*, Chilton, Radnor, Pa., 1980.

GENETIC ENGINEERING

Genetic engineering usually implies deliberate manipulation of genes of various organisms, procaryotic or eucaryotic, in order to achieve useful products of metabolism or to cause a permanent hereditary change in the organism. Although the techniques for engineering the genetic make-up of higher eucaryotes have not been developed very well, those for lower eucaryotes (eg, yeast (qv)) and procaryotes have been developed to an extent by which foreign DNA from any source can be introduced and stably maintained in bacteria such as *Escherichia coli*.

Two techniques usually are employed for manipulating the genes of microorganisms: *in vivo* genetic engineering in which the changes in genetic constitution are brought about in cells by processes analogous to those occurring in nature; and *in vitro* recombinant-DNA techniques in which foreign genes from entirely different sources can be ligated with stably replicating plasmid (or phage) DNA and introduced within the cells. The *in vitro* recombinant-DNA technique may involve production of entirely new substances (eg, substances of animal origin) in microorganisms and, therefore, may involve both qualitative and quantitative changes. Because recombinant-DNA technology allows the incorporation and replication of DNA from various animal sources in bacteria—and this type of exchange is considered extremely rare in nature—this technology has generated a considerable amount of controversy regarding the safety and well-being of mankind. There are strict guidelines in various countries of the world regulating the type of experiments, the nature of the host bacterium and foreign DNA inserts, and the volume of experimental materials that can be handled in laboratories involved in recombinant-DNA work (see also Microbial transformations).

In Vivo Genetic Engineering

In vivo genetic engineering may involve simple mutational alteration of the genetic material leading to an enhanced yield of the product or an improvement in the quality of the product. Such techniques have led to the isolation of mutant *Actinomyces* or bacterial strains capable of producing antibiotics (qv), vitamins (qv), or amino acids (qv) in high yield. Another widely used technique employs plasmid transfer between different bacterial species or genera. Thus, entirely new genetic functions can be transferred from the chromosome of one bacterial genus to different genera in the form of plasmids.

In addition to plasmid-mediated transfer of chromosomal genes from one bacterium to another, plasmids themselves may specify functions that can be used for construction of novel strains. Wide host-range plasmids, harboring chromosomal genes encoding useful functions, eg, nitrogen-fixation (qv), have been constructed that allow the transfer of nitrogen fixation genes to root nodules and other bacteria. Considerable progress also is being made to introduce nitrogen fixation and other desirable characteristics, such as resistance against harmful pests or plant viruses, to the plants themselves.

In Vitro Genetic Engineering (*in vitro* Recombinant-DNA Technology)

One technique that has made a tremendous contribution to the development of genetic engineering research is the *in vitro* recombinant-DNA technology, also known as molecular cloning or gene cloning. In its most widely used and simplest form, the technique allows the incorporation of any segment of a foreign DNA, procaryotic or eucaryotic, into a piece of phage or bacterial plasmid DNA and the recombinant-DNA segment then is reintroduced into the bacterial cell by transfection or transformation. Since the vector is the phage or bacterial plasmid DNA, it is not restricted within the cell and the foreign DNA segment replicates stably as part of the bacterial vector DNA. The simple technique of joining segments of DNA derived from procaryotes or eucaryotes with the vector DNA and the subsequent introduction into the bacterium may mark the beginning of a revolutionary way of manufacturing biologically functional proteins such as insulin (qv) and growth hormones (qv), antibodies, interferons, blood-clotting factors, and a host of other pharmacologically important compounds, by fermentation (qv).

The procedures used in combining a recombinant-DNA molecule *in vitro*, using a plasmid as vector, are shown in Figure 1. The first step is to isolate and purify the desired phage or plasmid DNA to be used as a vector. Once the vector and foreign DNA are available, the DNA segments are cut in areas that should not interfere with the biological activities of the gene(s) to be cloned or of the replication, maintenance, or selectable characteristics of the vector. The specifically cut DNA fragments are mixed and joined using the enzyme DNA ligase. The recombinant molecules thus generated are introduced into the bacterial host cells.

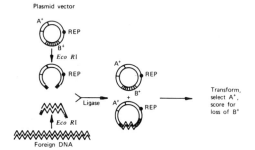

Figure 1.   The procedure for cloning foreign DNA in *E. coli*. Both A and B genes represent selectable traits, so that introduction of foreign DNA in B gene leads to the loss of an identifiable function. REP represents replication and maintenance genes. Courtesy of CRC Press.

Identification of clones with specific genes.   Once a segment of foreign DNA has been ligated with the vector DNA and introduced into *E. coli* cells by transformation, often a significant problem is identifying the transformant colony that harbors the recombinant-DNA molecule. Several methods have been developed to facilitate scoring of transformants with a foreign DNA insert: insertional inactivation; direct scoring of transformants for the traits coded by the foreign DNA; indirect selection of recombinant molecules based on size or density difference of the insert DNA; radioactive RNA probes; and *in situ* immunoprecipitation reactions.

Industrial applications of *in vitro* genetic engineering include expression of eucaryotic genes and production of eucaryotic proteins in bacteria; bacterial production of somatostatin; bacterial production of rat and human insulin; and bacterial production of interferon, human growth hormone, and vaccines against viruses such as hepatitis B and foot and mouth disease virus (see Supplement).

A.M. CHAKRABARTY*
University of Illinois at the Medical Center Chicago

R.F. Beers, Jr., and E.G. Bassett, eds., *Recombinant Molecules: Impact on Science and Society*, Raven Press, New York, 1977.

Research with Recombinant DNA, National Academy of Sciences, Washington, D.C., 1977.

A.M. Chakrabarty, ed., *Genetic Engineering*, CRC Press, West Palm Beach, Fla., 1978.

GEOTHERMAL ENERGY

In the broadest sense, geothermal energy comprises all the thermal energy stored in the earth's crust. The source and transport mechanisms of geothermal heat are unique to this energy source. Heat flows through the crust of the earth at a rate of almost 5.9 cW/m^2 [1.4 × 10^{-6} cal/(cm^2·s)] on the average. The intrusion of large masses of molten rock can increase this normal heat flow locally, but for most of the continental crust, the heat flow is due to the conduction of heat from the mantle and heat generated by the decay of radioactive elements in the crust, particularly isotopes of potassium, uranium, and thorium.

Certain conditions must be met before geothermal energy can be utilized. The first requirement is accessibility, which is created by natural transport processes such as convective heat transfer in porous and/or fractured regions of rock and conduction through the rock itself. If the physical extent and heat content of a deposit are large enough and the deposit is near the surface, a heat-extraction scheme must be designed based upon the hydrologic and geologic situation including the *in situ* rock properties. After such a scheme is successfully implemented, the desired raw material (hot water or steam) can be produced. Techniques to convert the steam or hot water into a marketable product (electricity, process heat, or space heat) and to dispose adequately of the waste products complete the process. Many aspects of geothermal heat extraction are similar to those found in the oil, gas, coal, and mining industries. Because of these similarities, equipment, techniques, and terminology have been borrowed or adapted for use in geothermal development.

Commercial utilization of the resource requires that the process be economically competitive. Consequently, commercial geothermal systems developed to date have been limited to a relatively few, accessible high grade deposits scattered throughout the world. Fossil-fuel shortages and price increases will make a significant fraction of lower-grade geothermal resources commercially feasible in the coming decades.

There are basically two types of geothermal systems: those that spontaneously produce hot fluid (steam or water) from a reservoir and those that do not. Systems that spontaneously produce hot fluids are hydrothermal or convection-dominated systems (Fig. 1). Large reservoirs

*Supported in part by a grant from the National Science Foundation (PCM81-13558).

in sedimentary rock containing pore fluids under confining pressures much greater than the hydrostatic head are called geopressured basins. Conduction-dominated systems also known as hot dry rock (HDR) and magma make up most of the world's geothermal resources (Fig. 2).

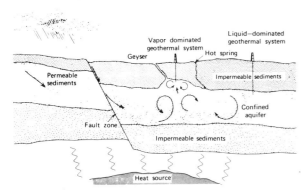

Figure 1. Schematic of hydrothermal system geometry.

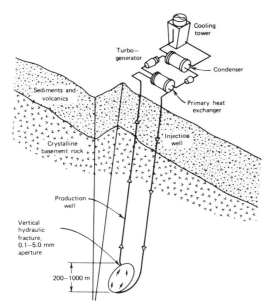

Figure 2. Hot dry rock reservoir concept for low permeability formations.

Table 1. Comparison of Estimates of Accessible United States Resource Base in EJ[a] from 0 to 10 km Referenced to 15°C

Resource category	USG[b]	EPRI[c]	LASL[d]
fluid			
convection	10,100	669,000	
geopressured	895,900		
Subtotal	*906,000*	*669,000*	
rock			
conduction	34,800,000	36,500,000	23,900,000
igneous	1,100,000		
Subtotal	*> 35,900,000*	*36,500,000*	*23,900,000*
Total	*> 36,000,000*	*37,170,000*	*23,900,000*

[a]1 EJ = 0.949 Quad (10^{15} Btu).
[b]United States Geological Survey.
[c]Electric Power Research Institute; The difference between the EPRI fluid resource base estimate and the total accessible resource base (recalculated to 10 km) is assumed to be the rock resource base. The depth of circulation of fluid is taken to be 3 km.
[d]Los Alamos Scientific (National) Laboratory; Recalculated from 0 to 10 km and assuming an average gradient of 25°C/km.

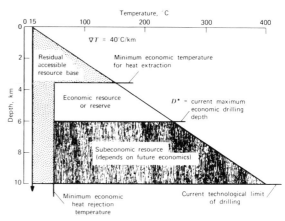

Figure 3. Factors limiting the fraction of the accessible resource base that can be economically extracted.

Comparisons of accessible resource base estimates for the United States from several sources all referenced to a mean annual temperature of 15°C and a maximum depth of 10 km are given in Table 1. Two important conclusions can be drawn from these estimates. First, the amount of energy is enormous. Second, probably 97% of the energy contained in the upper 10 km of the earth's surface is contained in the rock itself as a conduction-dominated resource and, therefore, requires a hot dry rock extraction scheme.

Several factors limit the amount of energy that can be extracted from the earth (Fig. 3). Some of these are now and always will be fixed; others depend upon future technological and economic developments. The first serious limitation is the effective geothermal temperature gradient. This largely determines the potential for a given area. If the gradient is small, the depth to the minimum acceptable initial rock temperature is large. Another large limitation is the maximum economic drilling depth D^*. The final limitation is imposed by heat-rejection conditions for the geothermal fluid in the power plant (taken here to be 50°C).

Reservoir Productivity and Fluid Characteristics

To justify the cost of developing a geothermal field, estimates of the total amount of energy that can be extracted and the production rate must be made. Typically, computer models are used to simulate performance for a given set of reservoir properties. Two general categories of reservoir models, one involving matrix and one fracture-dominated flow, are usually encountered. The first type is typically applied to high matrix permeability hydrothermal formations, and the second to flow in fractured, low matrix permeability media.

Extraction of heat from many natural hydrothermal systems now in operation is quite simple—because the reservoirs are frequently pressurized, the fluid passes directly to the surface under artesian flow when the reservoir is penetrated. Efficient production of a pressurized field is largely a matter of proper well spacing and flow management for a specified set of *in situ* properties.

Because they are artificially stimulated in a conduction-dominated geologic setting, HDR reservoirs can be designed to meet lifetime production capacity and temperature requirements of the user. The efficiency of electric-power production over the 20- to 40-yr lifetime of a HDR power plant depends primarily on the effective heat-transfer area and rock volume in contact with the circulating fluid. This determines the fluid temperature and its drawdown in time as well as its composition.

More than 158 wells including some deep ones (over 2400 m) have been drilled over the past 50 yr with basically no hydrothermal utilization; because of a very rough cost figure of $250,000 per well, private industry has invested ca 40×10^6 with virtually no return. Scaling, corrosion, and waste disposal are major problems.

Environmental Impact

The effect of a geothermal development on its surroundings is important. Occasionally, there are significant costs for the restoration of an area to its original state. These costs can be minimized somewhat by careful evaluations of the environmental impact and installation of facilities to minimize the impact. Environmental considerations should include pollution, uses of resources such as water and land, seismic hazard and subsidence.

Energy Conversion and Utilization

Electric-power generation and space and process heating have been applied to commercial geothermal operations throughout the world. Engineering components and thermodynamic issues associated with geothermal power and heating systems include the following: A Rankine cycle using dry- or saturated-steam expansion in a low pressure condensing turbine with direct contact condensation is now used to generate power for vapor-dominated resources such as those at The Geysers, California, and Lardarello, Italy. The multistage flashing cycle depicted in Figure 4 is used for liquid-dominated resources such as those at Wairakei, New Zealand, and Cerro Prieto, Mexico. In this case, saturated steam is created by flashing the geothermal fluid at the surface to a lower pressure, followed by expansion in a condensing steam turbine being the unflashed liquid fraction either reinjected or discarded.

If noncondensable gas concentrations are too high, indirect binary or two-phase flow conversion cycles might be the only feasible alternatives. If the geothermal fluid contains large amounts of dissolved material that may corrode or deposit on heat-exchange surfaces, a combined multistage flash and organic binary cycle may offer a reasonable solution.

Cycles proposed to deal with the deposition or scaling problem are less efficient than the multistage, direct-steam flashing systems that may be practical when dissolved gas contents are low. Sperry Research is investigating an unusual binary system that involves a downhole heat exchanger and organic turbine-driven pump in a gravity head cycle for resources in the 120–200°C range.

Scientists at the Lawrence Livermore Laboratory have evaluated total-flow, two-phase expansion systems for geothermal applications. Total-flow expanders may be particularly useful for high temperature, high salinity brines such as those found in the Niland-Salton Sea area of Imperial Valley, California.

With the possible exception of a few small binary-fluid power plants operating in the U.S., Japan, and the USSR, direct-steam injection and steam-flashing cycles are the only ones in actual commercial service today for geothermal resources.

Nonelectric systems can be described very simply as a primary heat-exchange loop and a cascaded utilization system. Although electric transmission costs are generally lower, commercial nonelectric applications for space heating can be economical for transportation of hot fluids for several kilometers as demonstrated in Iceland, USSR, Japan, and Boise, Idaho. Extensive use is presently being made of geothermal fluids ranging from 40–200°C for process heat in mining and pulp-and-paper industries, and for district space heating, therapeutic baths, absorption air conditioning, and agricultural and aquacultural purposes.

The thermodynamic utilization of both fossil and geothermal energy for electric-power production can be improved by a hybrid arrangement.

In this concept, geothermally supplied heat is used for feedwater preheating. A fossil-fired unit supplies additional heat before the steam enters the turbogenerator.

Accurate data on the thermodynamic properties of proposed working fluids are required to calculate electric-power cycle performance. Heat capacity at constant pressure in the ideal states, vapor pressure, pressure–volume–temperature behavior, enthalpy and entropy changes, and liquid density at saturation can be expressed in semiempirical equations.

Factors other than desirable thermodynamic properties frequently determine practical working fluids. These include fluid thermal and chemical stability, flammability, toxicity, material compatibility (corrosion), and cost.

Any real process for generating electricity or heat has inefficiencies or nonreversible steps. Efficient use of the resource may be necessary for commercial feasibility if reservoir development costs are high. Efficient utilization means that: most of the heat is extracted from the geothermal fluid before disposal or reinjection; temperature differentials across heat-transfer surfaces are maintained at minimum practical levels; turbines and feed pumps are carefully designed for optimum efficiency; and heat is rejected from the thermodynamic cycle at a temperature near the minimum ambient temperature, and the temperature of the coolant (water or air) to which the power plant rejects waste heat.

In principle, utilization efficiency η_μ can assume any value between zero (P, power extracted = 0) and unity ($P = \eta_\mu \dot{m}_{gf} W_{net}^{max}$) where $\dot{m}_{gf}$ = geothermal fluid flow rate and W_{net}^{max} = maximum net work; in practice, its value is determined from economic considerations by balancing the cost of obtaining the heat (drilling and piping costs) against the cost of processing it (heat exchangers, turbines, pumps) to generate electricity in the power station.

In a typical analysis of cycle performance, a set of equations is developed to describe the work and heat-flow rates to and from the chief plant components. After selecting a working fluid and a geothermal-fluid inlet temperature, the principal independent design variables are the maximum cycle operating pressure at the turbine inlet, the condensing temperature and the heat-exchanger approach temperature.

Worldwide Developments: Present and Future

The implication of significant growth for the geothermal industry in the coming years is clear. This is perhaps best illustrated by recent activities in the United States. As of 1983 at least nine specific sites are under development. These include: The Geysers-Calistoga area in California; Imperial Valley, California; Desert Peak, Nevada; Valles Caldera, New Mexico; Roosevelt Hot Springs, Utah; Puna, Hawaii; Raft River, Idaho; Klamath Falls, Oregon; Gulf Coast region of Texas and Louisiana; and Fenton Hill, New Mexico.

If one grants that the technical reservoir-engineering problems associated with hydrothermal, geopressured, magma, and hot dry rock resources are solved, perhaps the most critical factors affecting the growth of geothermal energy include drilling and plant capital-cost escalation relative to alternatives, the price escalation of alternative fossil and nuclear fuels, tax treatment and royalties, and finally, the assessment of risk and of the rate of return required to stimulate private investment.

JEFFERSON TESTER
Massachusetts Institute of Technology

CHARLES O. GRIGSBY
Los Alamos National Laboratory

J. Kestin, ed., *Sourcebook on the Production of Electricity from Geothermal Energy*, U.S. Government Printing Office, Washington, D.C., 1980.

L.M. Edwards and co-editors, *Handbook of Geothermal Energy*, Gulf Publishing, Houston, Tex., 1982.

Proceedings of 2nd UN Symposium on the Development and Utilization of Geothermal Resources, Lawrence Berkeley Laboratory, San Francisco, Calif., 1975.

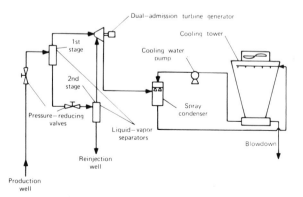

Figure 4. Dual-stage flashed-steam process for liquid-dominated systems.

GERANIOL. See Terpenoids.

GERANIUM OIL. See Oils, essential; Perfumes.

GERMANIUM AND GERMANIUM COMPOUNDS

Germanium (Ge) is a semiconducting metalloid element found in Group IVA and period 4 of the periodic table. Although it looks like a metal, it is fragile like glass. Its electrical resistivity is about midway between that of metallic conductors and that of good electric insulators. The entire modern field of semiconductors owes its development to the early successful use of germanium. After ca 40 yr of commercial use, this is still a significant application for germanium in the world today. However, its market picture has changed significantly. In the United States, its use in the field of ir optics surpassed its use in electronics in 1978, and this trend became worldwide during the early1980s. Germanium has also found wide use as a detector in gamma-ray spectroscopy and some use as a catalyst. Other applications with considerable potential demand are in the fields of fiber optics (qv), medicine, superconductivity, and solar energy (qv) (see Semiconductors; Integrated circuits; Superconducting materials).

The crust of the earth is estimated to contain 1.5–7 g Ge/t (1.5–7 ppm). Germanium usually occurs widely dispersed in minerals such as sphalerite and rarely in concentrated form. The physical, thermal, and electronic properties of germanium metal are shown in Table 1.

Germanium is quite stable in air up to 400°C where slow oxidation begins. The metal resists concentrated hydrochloric acid, concentrated hydrofluoric acid, and concentrated sodium hydroxide solutions, even at their boiling points. Germanium reacts readily with mixtures of nitric and hydrofluoric acids, with molten alkalies, and more slowly with aqua regia. It also reacts readily with the halogens to form the respective tetrahalides.

Germanium halides. Halides include germanium tetrachloride ($GeCl_4$), germanium tetrabromide ($GeBr_4$), and tetraiodide (GeI_4).

Germanium oxides. Oxides include germanium dioxide (GeO_2) and germanium monoxide (GeO).

Germanates. Germanates include sodium heptagermanate ($Na_3H-Ge_7O_{16}$).

Germanides. Germanides include magnesium germanide (Mg_2Ge).

Organogermanium compounds. Organogermanium compounds are generally characterized as having low chemical reactivity and relatively high thermal stability.

Manufacturing and Processing

Almost all of the germanium recovered worldwide is a by-product of other metals, mostly zinc and copper. Regardless of the source of the germanium, all germanium concentrates are purified by similar techniques. The ease with which concentrated germanium oxides and germanates react with concentrated hydrochloric acid and the convenient boiling point of the resulting $GeCl_4$ (83.1°C) make chlorination a standard refining step. The purified $GeCl_4$ is hydrolyzed with deionized

water to produce GeO_2 which is removed by filtration and dried. The dried GeO_2 is reduced with hydrogen at about 760°C to germanium metal powder which is subsequently melted and cast into so-called first-reduction or as-reduced bars. These bars are then subjected to zone refining to produce intrinsic or electronic-grade germanium metal (see Fine chemicals).

J.H. ADAMS
Eagle-Picher Industries, Inc.

F. Glocking, *The Chemistry of Germanium*, Academic Press, Inc., London, UK, 1969.

M. Lesbre, P. Mazerolles, and J. Satgé, *The Organic Compounds of Germanium*, John Wiley & Sons, Ltd., London, UK, 1971.

V.A. Nazarenko, *Analytical Chemistry of Germanium*, translated by N. Mandel, John Wiley & Sons, Inc., New York, 1974.

GETTERING. See Vacuum technology.

GETTERS. See Vacuum technology.

GILSONITE

Gilsonite (uintaite) is a natural hydrocarbon substance of the class known as asphaltites (see Asphalt), occurring as a coal-like solid that is mined much like other minerals and sold essentially in its native state. The only commercially important deposits of gilsonite in the world are located in the Uinta Basin, in the northeast corner of Utah.

Gilsonite is classed as one of the asphaltites, which are natural asphaltlike substances, characterized by their high softening points (above 110°C). Glance pitch and grahamite are other members of this group.

The tests applied to gilsonite are in many cases the same as those used for asphalt (qv). Typical properties of commercial-grade gilsonite (American Gilsonite Selects) are shown in Table 1.

Table 1. Properties of Germanium

Property	Value
crystal structure	diamond cubic
density at 25°C, g/cm^3	5.323
lattice constant at 25°C, nm	0.565754
surface tension, liquid at mp, mN/m (= dyn/cm)	650
Mohs hardness	6.3
Poisson's ratio at 125–375 K	0.278
melting point, °C	937.4
boiling point, °C	2830
heat capacity at 25°C, J/(kg·K)^a	322
coefficient of linear expansion at 300 K, 10^{-6}/K	6.0
thermal conductivity at 300 K, W/(m·K)	59.9
intrinsic resistivity at 25°C, Ω·cm	53

^{a}To convert J to cal, divide by 4.189.

Table 1. Properties of Gilsonite^a

Property	Value
color in mass	black
color of streak or powder	brown
fracture	conchoidal
specific gravity, 15/15°C	1.04
softening point (ring and ball method), °C	160
flash point, COC^b, °C	315
heating value, kJ/g^c	41.8
specific heat, J/kg^c at 260°C	2550
thermal coefficient of expansion (volumetric), per °C	0.0005
viscosity (Brookfield viscometer), mPa·s (= cP) at 260°C	500
volatility (5 h), wt% at 260°C	< 6
hardness (Mohs scale)	2
penetration, 100 g, 5 s, mm/10 at 25°C	0
resistivity, Ω·m	1.9×10^{10}
acid value^d	2.3

^{a}American Gilsonite Selects.
^{b}Cleveland open cup.
^{c}To convert J to cal, divide by 4.184.
^{d}mg KOH/g substance to give pH 7.

Uses

Large quantities of gilsonite now are used by the oil-drilling, foundry, building board, explosive and nuclear-graphite industries, in addition to the continuing use by the automotive and ink industries.

KENNETH R. NEEL
American Gilsonite Co.

E.T. Hodge, *Bull. Am. Assoc. Pet. Geol.* **11**, 395 (1927).

J.M. Sugihara and D.F. Sorenson, *J. Am. Chem. Soc.* **77**, 963 (1955).

T.F. McCullough, *Hydrocarbons and Other Compounds Obtained from Gilsonite*, Ph.D. thesis, University of Utah, Salt Lake City, Utah, 1955.

GIN.   See Beverage spirits; distilled; Oils, essential.

GLASS

Glass was formed naturally from common elements in the earth's crust long before anyone ever thought of experimenting with its composition, molding its shape, or putting it to the myriad of uses that it enjoys in the world today.

Glass technology has evolved for six thousand years, and some of today's principles date back to early times. This includes what is today known about the structure of glass, its composition, properties, method of manufacture, and uses (see also Glass-ceramics, Glassy metals).

Common usage of the term glass follows the definition of Morey: "Glass is an inorganic substance in a condition that is continuous with, and analogous to, the liquid state of that substance, but which, as the result of a reversible change in viscosity during cooling, has attained so high a degree of viscosity as to be, for all practical purposes, rigid." Both organic and inorganic materials may form glasses if their structure is noncrystalline, ie, if they lack long-range order. This includes some plastics, metals, and organic liquids. In principle, rapid cooling could prevent crystallization of any substance if the final temperature is sufficiently low to prevent structural rearrangement.

Glass is not merely a supercooled liquid. This distinction is illustrated by the volume–temperature diagram shown in Figure 1. When a liquid that normally does not form a glass is cooled, it crystallizes at or slightly below the melting point (path A). If there are insufficient crystal nuclei or if the viscosity is too high to allow sufficient crystallization rates, undercooling of the liquid can occur. However, the viscosity of the liquid rapidly increases with decreasing temperatures, and atomic rearrangement slows down more than would be typical for the supercooled liquid.

This results in the deviation from the metastable equilibrium curve that is depicted by paths B and C in Figure 1. This change in slope with temperature is characteristic of glass. Figure 1 shows that the point of intersection of the two slopes defines a transformation point (glass-transition temperature) T_g, for a given cooling rate. Practical limitations on cooling rate define the transformation range $T_g \rightarrow T_g'$ as the temperature range in which the cooling rate can affect the structure-sensitive properties such as density, refractive index, and volume resistivity. The structure, which is frozen in during the glass transformation, persists at all lower temperature. Thus, a glass has a configurational or a "fictive" temperature that may differ from its actual temperature. The fictive temperature is the temperature at which the glass structure would have been the equilibrium structure. It describes the structure of a glass as it relates to the cooling rate. A fast-quenched glass would have a higher fictive temperature than a slowly cooled glass. Glasses can be prepared by methods other than cooling from a liquid state, including solution evaporation, reactive sputtering, vapor deposition, neutron bombardment, and shock-wave vitrification.

Structure

The basic structural unit of silicate glasses is the silicon–oxygen tetrahedron in which a silicon atom is tetrahedrally coordinated to four surrounding oxygen atoms. Oxygens shared between two tetrahedra are called bridging oxygens. The similarity between crystalline and vitreous structure on the basis of silicon–oxygen–silicon bond angles and, hence,

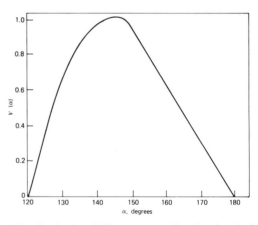

Figure 2.   The distribution of silicon–oxygen–silicon bond angles in vitreous silica. The function $V(\alpha)$ is the fraction of bonds with angles normalized to the most probable angle, 144°. This distribution gives quite a regular structure on the short range, with gradual distorting over a distance of 3 or 4 rings (2–3 nm). Crystalline silica such as quartz or cristobalite would have a narrower distribution around specific bond angles.

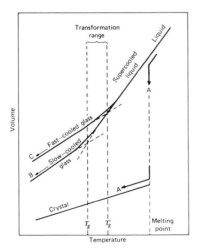

Figure 1.   Volume–temperature relationships for glasses, liquids, supercooled liquids, and crystals. Courtesy of Corning Glass Works.

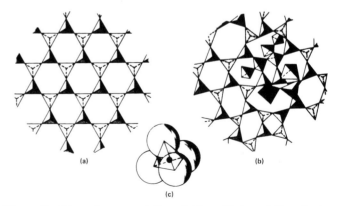

Figure 3.   Schematic representation of (**a**) an ideal crystalline structure (Si–O–Si bond angles = 180°) and (**b**) a simple glass (Si–O–Si bond angles = 144° ± according to Fig. 2). The tetrahedra in the schematics represent four oxygens clustered around a silicon as shown (**c**). Courtesy of Corning Glass Works.

the relative orientations of the silicon tetrahedra is pointed out in Figures 2 and 3.

Composition

Glass formers generally have cation–oxygen bond strengths greater than 335 kJ/mol (80 kcal/mol). In multiple-component systems, oxides with lower bond strengths do not become part of the network and are called modifiers. Oxides with energies of ca 335 kJ/mol may or may not become part of the network and are referred to as intermediates.

Glass formation of individual oxides can be predicted from the melting point and individual bond energies. A low melting point and high bond energy favors glass formation.

Glass-composition work starts with the application of structural and bonding rules of glass formation. Numerous ternary systems and their glass-forming regions have been investigated. There are three types of ternaries: single-former and two modifiers; two formers and one modifier; and three glass formers.

Single-Phase Glasses

Vitreous silica is the most important single-component glass. Highly cross-linked vitreous silica is viscous and has a thermal expansion coefficient within the 0–300°C range of about $5.5 \times 10^{-7}/°C$. It is an excellent dielectric and resists attack by most chemicals, except fluorides or strong alkali (see Silica).

Most glasses fall into the category of silicates containing modifiers and intermediates. Addition of a modifier such as sodium oxide, Na_2O, to the silica network alters the structure by cleaving the Si–O–Si bonds to form Si–O·Na linkages.

Mixtures of alkalies and alkaline earths give glasses of higher durability than the alkali silicates. Soda-lime glass accounts for nearly 90% of all the glass produced. It is used for containers, flat glass, pressed and blown ware, and lighting products where exceptional chemical durability and heat resistance are not required.

Replacement of alkali by boric oxide in a glass network give a lower-expansion glass. Borosilicate glass is applied as ovenware, laboratory equipment, piping, and sealed-beam headlights.

Aluminosilicate glasses are used commercially because they can be chemically strengthened and withstand high temperatures. Thus, applications include airplane windows, frangible containers, lamp envelopes, and top-of-stove uses.

Lead glasses may be easily melted and have a long working range and a high refractive index which makes them useful for lead crystal, optical glass, and hand-formed art ware.

Borates, including vitreous B_2O_3, have been studied more than any other glass-forming system with the exception of silicates. The very low durability of borate glasses precludes their use in all except the most special applications. Low molecular weight Lindemann glasses (Li_2O. $BeO.B_2O_3$) were developed as x-ray-transmitting glasses.

Phosphate glasses, like borates, tend to have low durability. Important commercial applications of phosphate glasses do exist, however. Almost-clear, heat-absorbing glasses with several percent iron oxide are possible. Phosphate-based glasses also are more resistant to fluoride than silicate glasses. Some of the optical glasses produced by Schott, Hoya, Owens-Illinois, and Corning-Sovirel use phosphate as the primary glass former.

Germanium, arsenic, and antimony oxides all form stable glasses and their structures have been predicted.

Glasses based upon sulfur, selenium, or tellurium rather than oxygen are well known. These glasses, although often opaque to visible light, transmit ir radiation of a much longer wavelength than oxide systems, and many are also semiconductors (qv).

Although zinc chloride glasses are known, BeF_2-containing glasses are the more common of the halide glasses.

Under highly specialized conditions, the crystalline structure of some metals and alloys can be suppressed and they form glasses. These amorphous metals can be made from transition-metal alloys, eg, nickel–zirconium, or transition or noble metals in combination with metalloid elements, eg, alloys of palladium and silicon or alloys of iron, phosphorus, and carbon (see Amorphous magnetic materials).

Glass-Ceramics and Phase-Separated Glasses

Glass-ceramics are melted and formed by conventional glass manufacturing techniques and then given a subsequent heat treatment (ceram) to transform then into fine-grained crystalline materials. By definition, glass-ceramics are more than 50% crystalline after heat treatment; frequently, the product is more than 95% crystalline. The operation is generally accompanied by an increase in viscosity which increases the product's use temperature.

Aluminosilicate glass-ceramics are the most useful commercial products. The presence of alkali or alkaline earth oxides affects the type of crystal structure. The morphology is regulated by a nucleating agent, usually TiO_2 or $TiO_2 + ZrO_2$, and a heat treatment. Many other components are added that further optimize the crystalline phases and the glass-ceramic properties.

The formation of a glass-ceramic is extremely complex, but generally follows a four-step sequence: (1) a dispersed amorphous phase, structurally incompatible with the host glass, usually highly unstable, and enriched in one or two key oxides (eg, TiO_2 or ZrO_2), forms on either cooling or reheating the glass; (2) primary crystalline nuclei, which are often titanates or zirconates, form either heterogeneously at the phase boundaries or homogeneously within the second phase; (3) a metastable crystalline phase heterogeneously nucleates on the primary crystallites and grows, generally at the expense of the glassy second phase. This produces a typically fine-grained, metastable, solid-solution material; and (4) the metastable material breaks down to the final, stable, fine-grained crystalline structure by means of isochemical phase transformations, reactions between metastable phases, or exsolution.

Devitrification is the uncontrolled formation of crystals in glass during melting, forming, or secondary processing. It can adversely affect the optical properties, mechanical strength, and sometimes the chemical durability of the glass. Devitrification is most likely to occur in glasses in which the optimum temperatures for maximum nucleation rate and for maximum growth rate nearly coincide. If these glasses are held too long in this critical temperature range or are cooled too slowly through it, the glass starts to crystallize.

Glasses that derive their color, optical transparency, or chemical durability from a small amount of a finely dispersed second phase are termed phase-separated glasses. They are distinguished from glass-ceramics by virtue of their predominantly glassy character. Therefore, the control of the nucleation and morphology of the second phase is not so critical.

Photosensitive glasses employ structural transformations, ie, precipitation and crystallization, which are sensitized by exposure to electromagnetic radiation, usually ultraviolet, but sometimes visible or x-ray radiation. The transformation is generally initiated by the photoreduction of a metal ion. Subsequent or simultaneous heat treatment morphologically transforms the glass in the regions exposed to the radiation leading, in general, to a change in optical or physical properties (see Chromogenic materials, photochromic).

Properties

The viscosity of a glass determines its melting, forming, and annealing procedures as well as the limitations of its use at high temperature. Viscosity is ordinarily measured between 10^{13} and 10 Pa·s (10^{14} and 100 P); at room temperature, it is greater than 10^{19} Pa·s (10^{20} P). The rapid but smooth change of viscosity with temperature is shown for several glasses in Figure 4. Reference points on the viscosity–temperature curve in Figure 4 have been chosen to characterize properties of an individual glass and to facilitate comparisons of similar glasses. Those used most frequently are the working, softening, annealing, and strain points. The working point of a glass is the temperature at which its viscosity is exactly 1 kPa·s (10^4 P). At this viscosity, glass is sufficiently fluid for forming by most common methods. Glass generally deforms under its own weight at the softening point. Residual stress can be annealed out of the glass in a matter of minutes at the annealing point and in a matter of hours at the strain point.

The thermal expansion of a glass determines the range of materials to which it can be sealed safely. It also affects the glass's ability to survive

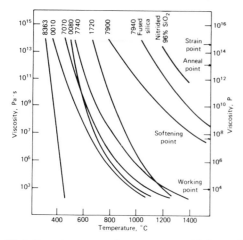

Figure 4. Viscosity vs temperature for some commercial glasses. Courtesy of Corning Glass Works.

thermal shock or cycling. A glass's usefulness as a heat exchanger or as a thermal barrier and its ease of melting and forming depend upon its heat-transfer properties and emissivity. The upper use temperature is a function of all of these properties.

Stresses caused by steady-state thermal gradients may or may not cause failure, depending on the degree of constraint imposed by some parts of the item upon others or by the external mounting. Thus, under minimum constraint and maximum uniformity of gradient through the thickness, very large temperature differences can be tolerated.

At room temperature, the thermal conductivity of glasses is 0.67–1.21 W/(m·K), the most common composition being near the upper end of the range. Thermal conductivity of glass-ceramics is 1.7–3.8 W/(m·K). At a mean temperatures of 200°C, the values are greater by approximately 25%.

The liquid's temperature determines the susceptibility of a glass to devitrification and, therefore, influences its forming limitations and often its heat-treating requirements.

Gas permeation though glass is of crucial importance to any high-vacuum system. To permit the advent of high intensity tungsten–halogen lamps, the lamp envelopes must contain gas under pressure for a long time at high temperatures. The permeation rate varies with temperature, glass type and composition, and gas.

Because of its amorphous structure, glass is brittle, reasonably abrasion-resistant, and almost perfectly elastic as long as its temperature is low enough to prevent viscous flow. Glasses do not contain crystallographic planes that slip relative to each other, and therefore, permit the material to deform plastically when stress is applied. Slip tends to relieve stress and allow broken bonds to form new bonds before breakage occurs. Since no such phenomena take place in glass, a bond that is broken because of excessive stress quickly forms a crack and eventually the glass fractures.

Glasses are used in the electrical and electronic industries as insulators, lamp envelopes, cathode-ray tubes, and encapsulators and protectors for microcircuit components, etc. Besides their ability to seal to metals and other glasses and to hold a vacuum and resist chemical attack, their electrical properties can be tailored to meet a wide range of needs. Generally, a glass has a high electrical resistivity, a high resistance to dielectric breakdown, and a low power factor and dielectric loss.

First developed for military applications, light-amplification devices have now been adapted as special microchannel-plate eyeglasses for people who are almost totally blind. Briefly, an acid-soluble glass rod is first selected and placed inside an insoluble glass tube, and the combination is drawn into a coaxial rod. The rods (ca 10 μm dia) are then fused in a bundle, sliced into thin plates, polished, and leached in acids so that the core glass is removed, and thus formed into a microchannel glass plate. The cladding glass is then rendered electronically conducting on the surface, and electrodes are deposited and assembled into the device. In combination with a photocathode located in the front end of the tube

and a phosphor screen located at the exit end, a highly efficient light-amplification device can be made.

The properties of some glasses and glass-ceramics make them more suitable for microwave application than others. The relative suitability of a glass or glass-ceramic material is expressed by a factor of merit that is proportional to the material's strength and inversely proportional to its Young's modulus, expansion coefficient, and rate of microwave-energy absorption.

The resistance of glass to chemical corrosion is frequently the reason for its use. However, the durability of a glass varies from highly soluble to highly durable, depending upon its composition and the solvent considered. Glasses low in silica are usually not durable to acid. Hydrofluoric acid, however, attacks even high silica glasses. Comparisons are usually based upon measurements of weight loss, changes in surface quality, or the analysis of solutions that were in contact with a glass.

Optical glasses are usually described in terms of their refractive index at the sodium D line (589.3 nm) and their v value (or Abbé number) which is a measure of the dispersion or variation of index with wavelength. These data are incorporated by Schott into a six-digit numbering system used to identify optical glasses. Glasses with index $n_D < 1.60$ and a v value of ≥ 55 are defined as crown glasses; those with a v value below ca 50 are defined as flint glasses. Glasses with $n_D > 1.60$ are defined as crown glasses if v is ≥ 50. Crowns are usually alkali silicate glasses and flints are lead alkali silicates.

The spectral transmission of glass is determined by reflection at the glass surfaces and the optical absorption within the glass. Overall transmission of a flat sample at a particular wavelength is equal to $(1 - R)^2 e^{-\beta t}$, where β is the absorption coefficient, t the thickness of glass, and R the air–glass reflection coefficient. Transmission is a function of wavelength. It may be controlled by the type of glass used or by the control or addition of coloring additives, melting atmosphere, melting temperatures, and cooling schedules.

When strained, glass becomes doubly refracting, or birefringent. Stress-optical effects are presumed to be elastic; therefore, strain is proportional to stress. The stress-optical coefficient, B, or Brewster's constant, is a measure of this proportionality, and varies greatly with glass composition.

Interaction of glass with low-energy visible and ultraviolet radiation may result in the alteration of the electronic states. These changes may cause coloration or luminescence effects, eg, photochromic or photosensitive glass. When changes in color are produced by sunlight, the effect is frequently referred to as solarization. Effects produced by ions, gamma rays, and x-rays are relevant to glasses used as electronic tube envelopes, radiation-shielding windows, and dosimeters.

Manufacture and Processing

Most glass articles are manufactured by a process in which raw materials are converted at high temperatures to a homogeneous melt that is then formed into the articles. The flow diagram in Figure 5 summarizes the details of conventional glass manufacturing. The vapor deposition of SiO_2 from a flame fed with $SiCl_4$ and oxygen is the basis for manufacturing high-purity glass used for blanks that are redrawn into optical-waveguide fibers (see Fiber optics). Fused silica items that cannot be formed from viscous melts of SiO_2 or quartz are prepared by vapor deposition. Raw materials are selected according to purity, supply, pollution potential, ease of melting, and cost. Sand is the most common ingredient. Limestone is the source of calcium and magnesium. Powdered anthracite coal is a common reducing agent, and common colorants for glass include iron, chromium, cerium, cobalt, and nickel.

Melting and fining depend on the batch materials interacting with each other at the proper time and in the proper order. Thus, extreme care must be taken to obtain materials of optimum grain size, to weigh them carefully, and mix them intimately. The efficiency of the melting operation and the uniformity and quality of the glass product are very often determined in the mix house. Batch handling systems vary widely through the industry, from manual to fully automatic. Melting units range from small pot furnaces for manual production to large, continuous tanks for rapid machine forming. The two common feeding designs used today are the screw feeder and the reciprocating pusher. Control

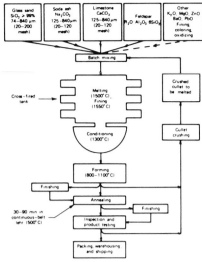

Figure 5. Glass manufacture. Temperatures are for common soda-lime glass. Other glasses may require appreciably different temperatures.

devices have advanced from manual control to sophisticated computer-assisted operation. Radiation pyrometers in conjunction with thermocouples monitor furnace temperatures at several locations. Natural gas, oil, and electricity are the primary sources of energy; propane is used as a backup reserve in emergencies.

Molten glass is either molded, drawn, rolled, or quenched, depending on desired shape and use. Bottles, dishes, optical lenses, television picture tubes, etc, are formed by blowing, pressing, casting, and/or spinning the glass against a mold to cool it and to set its final shape. Window glass, tubing, rods, and fiber are formed by freely drawing the glass in air (or in a bath of molten tin as in the float process) until it sets up and can be cut to length. Art glass is usually hand-formed by freely blowing and shaping it while soft. Glass that is intended to be crushed into powder, called frit, is quenched between water-cooled rollers or ladled or poured directly into water and then dried (dri-gauged).

Simultaneous forming of two glasses to produce a single article is possible if the two have similar viscosities in the appropriate temperature range (see Laminated materials, glass).

Glass articles formed at high temperatures must be cooled in order to reduce the strain and associated stress caused by temperature gradients in the glass to a low level to prevent damage during finishing and subsequent use. The continuous solidification of glass on cooling and the methods of reducing strain are unique to glasses and do not apply to crystalline materials that undergo phase transitions. Lehring processes include annealing, tempering, densifying or compacting, and postheat treatments. The term annealing generally refers to removal of stress, and terms such as fine annealing or trimming of optical glass imply structural changes associated with lehring.

In secondary forming, a piece of preformed glass is reheated and reworked into the finished product. Secondary operations include mechanical finishing, chemical finishing, and cleaning.

Economic aspects. Glass manufacture is classified according to the product into flat, container, fiber, or specialty glass. The flat and container companies produce over 14×10^6 t soda-lime glass per year. The specialty glass manufacturers of pressed and blown ware, television bulbs, and lighting and optical glasses melt hundreds of glass composition to fulfill the need for a large variety of products.

DAVID C. BOYD
DAVID A. THOMPSON
Corning Glass Works

R.H. Doremus, *Glass Science*, John Wiley & Sons, Inc., New York, 1973.

W. Geigerich and W. Trier, *Glass Machines Construction and Operation of Machines for the Forming of Hot Glass*, Springer-Verlag, Berlin, 1969 (translated by N.J. Kreidl).

E.B. Shand, *Glass Engineering Handbook*, 2nd ed., McGraw-Hill Book Co., New York, 1958.

GLASS-CERAMICS

Glass-ceramics are polycrystalline solids produced by the controlled crystallization of glasses. They are primarily silicate-based materials that can be formed by highly automated glass-forming processes and converted to a ceramic-like product by the proper heat treatment.

The Glass-Ceramic Process

Figure 1 is a flow chart of a glass-ceramic manufacturing process. By definition, a glass-ceramic initially must be a glass. Thus, the first stages of the manufacturing process are similar to the processing of most commercial glasses; however, there are significant differences in the postforming stages. The environmentally related effects resemble those encountered in conventional glass processing. Pollution problems are generally minimal except for emisions during the melting of some compositions.

Postforming. The shaped body may be annealed (cooled slowly to avoid stresses) and inspected prior to heat treatment (Fig. 1), or the shaped object may be put directly into a heat-treatment furnace. The heat-treatment process may be continuous or intermittent.

Nucleation. One of the main factors that permits the formation of a conventional glass without crystallization is the very slow nucleation rate. Although some glass compositions are self-nucleating, most of the commercial glass-ceramic compositions rely on the presence of a nucleating agent, usually a noble metal or a transition-metal oxide, to initiate internal crystallization.

Crystallization. After nucleation, the material must generally be heated to higher temperatures in order for crystal growth to proceed; these temperatures are 750–1150°C, depending on the desired crystalline phase.

After heat treatment, the material may undergo additional finishing steps such as cutting, grinding, polishing or decorating, or it may be packaged directly for shipment.

Properties. The properties of a glass-ceramic are dependent on such microstructural parameters as: the amount and properties of the crystalline phase(s); the crystal sizes, shapes, and orientations; and the amount, composition, and distribution of the glassy phase.

Thermal properties. Glass-ceramic materials have been reported with thermal-expansion coefficients ranging from ca $(-60$ to $200) \times 10^{-7}/°C$. The thermal-expansion properties, particularly the low expansions, are the primary properties that have been exploited commercially. The thermal-conductivity range is ca $1.7–5.4 \text{W}/(\text{m}\cdot\text{K})$.

Optical properties. Transparent and opaque glass-ceramics can be made. The degree of transparency is a function of the optical properties of the crystalline species, the grain size, and the difference of the refractive indexes of the glass and the crystals.

Chemical properties. The chemical durability of the common lithia aluminosilicate glass-ceramics is generally on a par with borosilicate glasses.

Mechanical properties. Glass-ceramics are brittle materials like ordinary glasses and ceramics and exhibit many of the same properties. The abraded modulus of rupture for most glass-ceramic materials is 55–250 MPa (8,000–36,000 psi) compared with 40–70 MPa (6,000–10,000 psi) for common glasses under identical abrasion and test conditions.

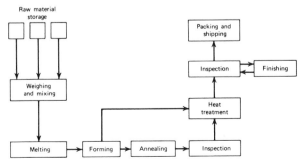

Figure 1. The glass-ceramic manufacturing process.

The high strengths are generally attributed to the formation of a surface-compressive layer, either through differential crystallization of a low expansion crystal phase, or by an external chemical ion exchange. The elastic properties are an addition function of the concentration and elastic properties of the individual phases.

Electrical properties. Most glass-ceramics are insulators.

Magnetic properties. Magnetic phases can be crystallized from a glass, but magnetic glass-ceramics are not commercially available.

Commercial Applications

Lithium silicate systems have been used for the manufacture of a photosensitive glass that can be chemically machined. The commercial lithia aluminosilicate glass-ceramics are among the best glass-ceramic materials in terms of formability. The combination of the magnesia aluminosilicates properties and formability has led to their use as radomes (see Microwave technology). In addition to machinability, mica glass-ceramics are strong, thermally shock-resistant and have good dielectric properties.

Glass-ceramic materials called slagcerams have been produced by the crystallization of glasses made from metallurgical slags. Products from these slags such as floor and wall coverings, are attractive because of the low cost and the availability of the raw materials, and have been manufactured in the USSR and several other countries for many years.

DANIEL R. STEWART
Owens-Illinois

P.W. McMillan, *Glass-Ceramics*, Academic Press, Inc., New York, 1979.

GLASSY METALS

The most familiar glasses are oxide glasses such as silicate or borate glasses. However, they are not the only members of the family of glasses. A recent addition is the group of glassy metals, or metallic glasses, also called amorphous alloys (see Amorphous magnetic alloys). They offer excellent possibilities in the electric, electronic, and structural industries. Several compositions of glassy metals, particularly those containing about 80 atom% transition metals and about 20 atom% metalloids, such as B, Si, P, C, are now available in commercial quantities.

Unlike crystalline solids, glasses do not have a long-range ordered atomic structure. They are in essence frozen liquids. Most glassy metals cannot be obtained by simply cooling liquid metals at rates commonly employed in metals industries because the samples would crystallize. However, when liquid metals are quenched at a very high rate, often exceeding 10^5 K/s, it is possible to avoid crystallization and thus obtain glassy metals that largely preserve the atomic structure of liquid metals.

The properties of glassy metals are sometimes similar to those of crystalline metals. It is, therefore, not possible to judge unambiguously whether a solid is crystalline or amorphous by observing its macroscopic properties alone. Glassiness can be determined either directly by studying the atomic structure with diffraction methods, or indirectly by carefully studying the thermodynamic properties.

Properties

The mechanical properties of a glassy metal are characterized by a tensile strength that exceeds 1% of Young's modulus, the formation of shear bands, ductile flow on the fracture surface, and in many cases, embrittlement after an annealing that does not cause crystallization. Most glassy metals show an appreciable amount of mechanical creep at temperatures as low as 200°C. Because of their ductility, glassy metals can be processed mechanically by cutting or punching. Many glassy metals containing transition metals or rare earths, or both, are ferromagnetic or ferrimagnetic (see Magnetic materials; Ferroelectrics) (see Table 1).

Glassy metals usually have high electrical resistivities. The temperature dependence of the resistivity is often small, and can be either positive or negative or show a weak minimum. Various metal–metal

Table 1. Representative Alloy Composition of Glassy Metals

Composition	Remarks
$Fe_{80}B_{20}$	Metglas 2605[a]
$Fe_{40}Ni_{40}P_{14}B_6$	Metglas 2826[a]
$Fe_{82}B_{10}Si_8$	Metglas 2605S[a]
$Pd_{80}Si_{20}$	
$Pd_{77.5}Si_{16.5}Cu_6$	very stable alloy, easy to vitrify
$Pt_{60}Ni_{15}P_{25}$	very stable alloy, easy to vitrify
$Cu_{60}Zr_{40}$	
$Mg_{70}Zn_{30}$	
$Gd_{17}Co_{83}$	sputtered sample can support magnetic bubbles
$Mo_{64}Re_{16}P_{10}B_{10}$	superconducting below 8.7 K
$W_{60}Ir_{20}B_{20}$	crystallization temperature above 1200 K

[a]Allied Chemical Corp.

glasses such as $La_{80}Au_{20}$ and metal–metalloid glasses such as $(Mo_xRu_{1-x})_{80}P_{20}$, show superconductivity.

Glassy metals containing Cr, such as chromium–iron alloys, are extremely corrosion resistant because of the formation of a dense chromium oxo–hydroxide layer on the surface of the metal.

Thermal Behavior

Glassy metals are thermodynamically unstable and crystallize when they are heated. Crystallization is usually accompanied by drastic property changes. Furthermore, the glassy metals as produced are not in a stable state within the amorphous phase because of rapid quenching. An annealing treatment near the glass-transition temperature (T_g) can also result in substantial property changes even if it does not lead to crystallization. This stabilization phenomenon is called structural relaxation.

Formation

Any liquid can be made into glass if the quenching rate is high enough to avoid the minimum time of the crystallization curve shown in Figure 1. In practice, the quenching rate that may be achieved is restricted by the limited heat conductivity and heat capacity of the sample. All the glassy metals stable at RT are alloys with two, three, or more constituent elements.

Preparation

A number of methods are used to prepare glassy metals from the corresponding crystalline alloys but they generally belong to one of three categories. The first is the liquid-quenching method mentioned earlier. The second is vapor deposition including the ion-sputtering technique (see Film deposition techniques). The third is the wet chemical technique of electrodeposition or electroless chemical deposition (see Electroless plating).

The liquid-quenching methods are, in general, simple and suited to large-scale mass production but limited as to the composition ranges.

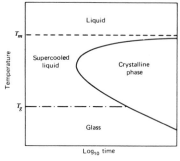

Figure 1. A schematic time–temperature–transformation (TTT) diagram for solidification of liquid; T_m and T_g denote the melting and the glass-transition temperatures, respectively.

The vapor-deposition technique affords a wider range, particularly the ion-sputtering method. The equipment for vapor deposition is complex and expensive and the deposition rate is significantly lower (by a factor of 10^6 to 10^{10}) than the production rate in liquid quenching.

Liquid-quenching techniques include splat quenching, the roller quenching technique, the Pond-Maddin technique, and the melt-spinning technique. Vapor-quenching techniques include the thermal vapor-deposition method and the ion-sputtering technique.

Applications

Applications of glassy metals include those as a permalloy replacement, in power devices, bubble memory devices, and delay lines. Mechanical applications include their use as strengthening materials, cutting tools, and brazing materials.

TAKESHI EGAMI
Department of Materials Science and Engineering,
University of Pennsylvania

P. Duwez, *Ann. Rev. Mat. Sci.* **6**, (1976).

H.J. Guntherodt and H. Beck, eds., *Glassy Metals I*, Springer-Verlag, Berlin, 1981.

F.E. Luborsky, ed., *Amorphous Metallic Alloys*, Butterworths, London, UK, 1983.

GLOBULINS.   See Blood fractionation.

GLUCOSE, $CH_2OH(CHOH)_4CHO$.   See Syrups; Sugar; Carbohydrates.

GLUE

The origins of animal glue in antiquity and its wide range of adhesive applications have contributed to the persistent use of the term glue to indicate any adhesive (see Adhesives). Animal glue refers to that material produced by the hydrolysis of the protein collagen, which is found in specific animal tissues such as skin, hides, bones, sinews, and tendons. Animal glue and gelatin (qv) are almost identical as they are produced by the same methods and from the same raw materials. Gelatin requires clean and edible raw materials, more purification, and more stringent processing conditions and control. Animal glue, however, can be processed from raw materials that otherwise would be wasted. The color and purity of animal glues, unlike that of gelatin, do not necessarily indicate its effectiveness as an adhesive.

Commercial animal glues usually are named according to the raw materials from which the product is made; thus, there is a bone glue and a hide glue.

Properties

Animal glue is marketed as a dry solid, the color of which may range from light yellow through tan, brown, and sometimes almost black. Commercial glues usually are sold ground to a size of 2–2.4 mm (8–10 mesh). Glue, the hydrolysis product of collagen, is an amorphous composition of protein fragments ranging in size from simple dipeptides to large multichain polypeptides of colloidal size. Glue is a polydisperse system and, as such, gives molecular weights that are statistical averages; molecular weights have been reported from 20,000 to 90,000 and some fractions are as high as 250,000. Many of the properties of glue can be attributed to its amino acid structural constituents.

The color, odor, and clarity of animal glues is related to the quality of the raw materials and to the care and cleanliness of the manufacturing process; contamination by blood, metals, and sugars darken colors, and fat and dirt contribute opacity. The fats and their oxidation products, as well as some of the amine by-products, contribute to objectional odors.

The most characteristic property of animal glue is its ability to form reversible gels in water solution. Numerous agents exert a pronounced effect on the gelling properties of glue. Salts are used to control the gelation and setting times of adhesive formulations that contain animal glue. Citrates, tartrates, and maleates tend to increase the rate of gelation but not the rigidity of the gel.

Manufacture

Preparation of animal glue is essentially a treatment of a collagen source with heat and water in order to hydrolyze it to a soluble product as rapidly and as efficiently as possible. The resulting solution is filtered, centrifuged to remove fat, concentrated to a suitable solids content, chilled to gel the concentrate, extruded or cut into particles, dried, ground, and analyzed.

Economic aspects.   The number of United States producers of animal glue has declined steadily as raw material, energy, and wage costs have increased. The versatility of synthetic adhesives and the modern developments in vegetable adhesives have also contributed to the decline in animal-glue production (see Adhesives).

Since 1980, the amount of glue used in gummed tape has declined rapidly to about 10–15% of the market and its use in cork, and paper has just about disappeared. At the same time, the amounts used in fiber glass, paints, caulks, and plasters has grown. The use of sterile glue and glue hydrolysates for cosmetics has become a considerable market. To meet the effect of rising costs, the industry has innovated the application of such techniques as reverse osmosis (qv), ultrafiltration (qv), and waterless high pressure hydrolysis.

Specialty Glues

Although the market for specialty glues has undergone a great deal of change over the past decade, a significant portion of both domestic and imported animal glue is used in this market. Liquid glues are almost any animal glue that can be converted into a liquid formulation. Opaque glues are formulations made by the addition of pigments to concentrated glue solutions and are used widely where a pigmented glue is used to correct a fault. Flexible glues are glue mixtures that are formulated using humectants and solubilizing agents which allow them to be applied with heat or to flow at ambient temperatures.

THOMAS F. MITCHELL
Darling and Company

A. Veis, *The Macromolecular Chemistry of Gelatin*, Academic Press, Inc., New York, 1964.

A.G. Ward and A. Courts, *The Science and Technology of Gelatin*, Academic Press, Inc., New York, 1977.

I. Skeist, *Handbook of Adhesives*, Reinhold Publishing Corp., New York, 1962.

GLUTAMIC ACID, $HOOCCH_2CH_2CH_2CH(NH_2)COOH$.   See Amino acids.

GLUTARIC ACID.   See Dicarboxylic acids.

GLUTEN.   See Bakery processes and leavening agents.

GLYCERIDES.   See Drying oils; Fats and fatty oils.

GLYCEROL

Glycerol, propane-1,2,3-triol, glycerin (USP), $CH_2OHCHOHCH_2OH$, a trihydric alcohol having a sweet taste, is a clear, water-white viscous, hygroscopic liquid at ordinary room temperatures, which are above its melting point. Glycerol occurs naturally in combined form as glycerides in all animal and vegetable fats and oils, and is recovered as a by-product when these oils are saponified in the process of manufacturing soap, or during the direct splitting of fats in the production of fatty acids. Since 1949 it also has been produced commercially by synthesis from propylene. At one time synthetic glycerol accounted for about 50% of U.S. production. However, since then the production of synthetic glycerol in the U.S. has ceased.

The uses of glycerol number in the thousands but the largest amounts go into the manufacture of synthetic resins and ester gums, drugs, cosmetics, and toothpastes. Tobacco processing and foods also consume large amounts either as glycerol or glycerides.

Properties

Physical properties of glycerol are shown in Table 1. Glycerol is completely soluble in water and alcohol, lightly soluble in diethyl ether, ethyl acetate and dioxane, and insoluble in hydrocarbons.

Glycerol, the simplest trihydric alcohol, forms esters, ethers, halides, amines, aldehydes, and such unsaturated compounds as acrolein (qv). As an alcohol, glycerol also has the ability to form salts such as sodium glyceroxide (see also Alcohols, polyhydric).

Grades. Two grades of crude glycerol are marketed: soap-lye crude glycerol obtained by concentration of lyes from the soap kettle contains ca 80% glycerol; and crude saponification glycerol resulting from hydrolysis of fats contains ca 88–91% glycerol and a small amount of organic salts.

Economic aspects. Commercial production and consumption of glycerol has generally been considered a fair barometer of industrial activity as it enters into such a large number of industrial processes. It generally tends to rise in periods of prosperity and fall in recession times.

Glycerol production in the United States has risen from 19,800 metric tons in 1920 to its peak of 166,100 t in 1967. World production of glycerol then was 450,000–600,000 t/yr. Today only natural glycerol is produced in the U.S. as domestic synthetic producers have ceased their operations. Much of the U.S. demand is now met through imports.

Health and Safety Factors

Glycerol, since 1959, has been generally recognized as safe (GRAS) as a miscellaneous or general-purpose food additive (qv) under the CFR, and it is permitted in certain food-packaging materials.

Derivatives

Glycerol derivatives include acetals, amines, esters, and ethers. Of these the esters are the most widely employed. Commercially, the most

Table 1. Physical Properties of Glycerol

Property	Value
mp, °C	18.17
bp, °C	
at 101.3 kPa[a]	290
sp gr, 25/25°C	
100% glycerol in air	1.262
n_D^{20}	1.47399
vapor pressure, Pa[b]	
at 50°C	0.33
surface tension at 20°C, mN/m (= dyn/cm)	63.4
viscosity at 20°C, mPa·s (= cP)	1499
specific heat at 26°C, J/g[c, d]	2.425
heat of vaporization, J/mol[d]	
at 55°C	88.12
heat of solution to infinite dilution, kJ/mol[d]	5.778
flash point, °C	
Cleveland open cup	177

[a] To convert kPa to mm Hg, multiply by 7.5.
[b] To convert Pa to mm Hg, divide by 133.3.
[c] 99.94% glycerol.
[d] To convert J to cal, divide by 4.184.

important are the alkyd resins (qv), which are esters of glycerol and phthalic anhydride. The next most important is glyceryl nitrate (nitroglycerin), used in explosives (qv) and as a heart stimulant (see Cardiovascular agents).

JOYCE C. KERN
Glycerine Producers' Association

Physical Properties of Glycerine and Its Solutions, Glycerine Producers' Association, New York, 1975.
The United States Pharmacopeia XX (USP XX-NFXV), The United States Pharmacopeial Convention, Inc., Rockville, Md., 1980.

GLYCOLIC ACIDS.   See Hydroxy carboxylic acids.

GLYCOLS

ETHYLENE GLYCOL AND PROPYLENE GLYCOL

Glycols, also called diols, are compounds containing two hydroxyl groups attached to separate carbon atoms in an aliphatic chain. Although glycols may contain heteroatoms, those discussed here, with the exception of thiodiglycol, are composed solely of carbon, hydrogen, and oxygen.

Simple glycols are those in which both hydroxyl groups are attached to an otherwise unsubstituted hydrocarbon chain as represented by the general formula, $C_nH_{2n}(OH)_2$.

Polyglycols are adducts of simple glycols and are distinguished by intervening ether linkages in the hydrocarbon chain, as represented by the general formula $C_nH_{2n}O_x(OH)_2$. Many commercially important polyglycols have physical, chemical, and toxicological properties similar to those of the simple glycols (see also Polyethers; Alcohols, polyhydric; Glycerol).

Physical properties of the ethylene and propylene glycols are listed in Table 1. Glycols undergo reactions common to monohydric alcohols forming esters, acetals, ethers, and similar products.

Glycols as a class exhibit a low order of toxicity. They have low vapor pressure at normal temperatures and are, therefore, not an inhalation hazard. With the notable exception of propylene glycol, glycols should, however, not be used for internal consumption.

Ethylene Glycol

Ethylene glycol and its lower polyglycols are colorless, odorless, high boiling, hygroscopic liquids completely miscible with water and many organic liquids. Ethylene (and propylene) glycols markedly reduce the freezing point of water. Ethylene glycol undergoes the reactions typical of monohydric alcohols. Its unique chemical behavior is attributed to the presence of hydroxyl groups on adjacent carbon atoms. Hydrolysis of ethylene oxide remains the main commercial source of ethylene glycol (see Ethylene oxide).

Ethylene glycol is one of the more toxic simple glycols by oral administration, and serious injury or death may result from swallowing about 60 mL.

About 40% of the ethylene glycol produced domestically is used as a nonvolatile antifreeze for liquid-cooled motor vehicles. Approximately 35% of total production is consumed in the manufacture of polyester fiber and film (see Polyester fibers; Polyesters, thermoplastic).

There are several important classes of derivatives: monoethers, diethers, esters, ether–esters, dialdehydes, acetals, and ketals. Glyoxal, CHOCHO, is a highly reactive dialdehyde with many important commercial applications including in paper manufacture, textiles, medicine, bacteriology and pest control (see also Esters; Ethers).

Diethylene Glycol

Diethylene glycol is similar in many respects to ethylene glycol but contains an ether group. It is the main co-product of ethylene glycol

Table 1. Properties of Glycols

Property	Ethylene glycol	Diethylene glycol	Triethylene glycol	Tetraethylene glycol	Propylene glycol	Dipropylene glycol
formula	$HOCH_2CH_2OH$	$HO(CH_2CH_2O)_2H$	$HO(CH_2CH_2O)_3H$	$HO(CH_2CH_2O)_4H$	CH_3CHCH_2OH | OH	isomers (see text)
mol wt	62.07	106.12	150.17	194.23	76.10	134.18
sp gr 20/20°C	1.116	1.119	1.126	1.125	1.038	1.023
bp at 101.3 kPa[a], °C	197.6	245.8	288	dec	187.3	232
vapor pressure at 20°C, Pa[b]	< 1.3	< 13	< 1.3	< 1.3	< 13	< 1.3
mp, °C	−13.0	−6.5	−4.3	−4.1	−60[c]	−40[d]
viscosity at 20°C, mPa · s (= cP)	19.83	36	49	61.9	60.5	107
refractive index, n_D^{20}	1.4318	1.4475	1.4561	1.4598	1.4326	1.4407
heat of vaporization at 101.3 kPa[a], kJ/mol[e]	52.24	52.26	61.04	62.63	52.30	53.64
flash point of commercial material, °C	116[f]	138[g]	172[g]	191[g]	101[f]	118[f]

[a] To convert kPa to mm Hg, multiply by 7.5.

[b] To convert Pa to mm Hg, divide by 133.3.

[c] Sets to glass below this temperature.

[d] Pour point.

[e] To convert J to cal, divided by 4.184.

[f] Determined by ASTM method D 56, using the Tag closed cup.

[g] Determined by ASTM method D 92, using the Pensky-Martens closed cup.

manufacture. About 35% of diethylene glycol is used as an intermediate in the manufacture of unsaturated polyester resin and polyols for polyurethanes. Approximately 15% is converted to triethylene glycol by reaction with ethylene oxide, 10% is consumed as a softening agent and lubricant for textiles (see Textiles), and 10% for natural-gas dehydration. The UDEX process extraction solvent consumes ca 5% (see Adsorptive separation) and the remaining 25% is distributed among other industrial applications including plasticizer and surfactant manufacture and antifreeze mixtures.

Important classes of diethylene glycol derivatives include monoethers, esters, thioethers, and morpholines (see Amines, cyclic).

Triethylene Glycol

Triethylene glycol has chemical and physical properties essentially identical to those of diethylene glycol. Triethylene glycol is a co-product of ethylene glycol produced via ethylene oxide hydrolysis. It is an efficient hygroscopicity agent and as such 35% of triethylene glycol produced is used as a liquid drying agent for natural gas (see Drying agents). Another 45% is used in roughly equal amounts in the manufacture of vinyl plasticizers, as a humectant, and as a solvent.

Ethers and esters, prepared like the monoethylene glycol derivatives, are the most important triethylene glycol derivatives.

Tetraethylene Glycol

Tetraethylene glycol has properties similar to diethylene and triethylene glycols, and may be used preferentially in applications requiring a higher boiling point and lower hygroscopicity. Tetraethylene glycol is used to separate aromatic from nonaromatic hydrocarbons by selective extraction.

Propylene glycol

1,2-Propylene glycol is a clear, viscous, colorless liquid that is practically odorless and has a slight characteristic taste. It is produced by the hydrolysis of propylene oxide under pressure and at high temperature without catalyst.

Propylene glycol is used in polyester resins, pet food, tobacco humectants, cellophane, foods and pharmaceuticals.

Dipropylene Glycol

Dipropylene glycol is a co-product of propylene glycol in the hydrolysis of propylene oxide. The approximate isomer distribution is as follows:

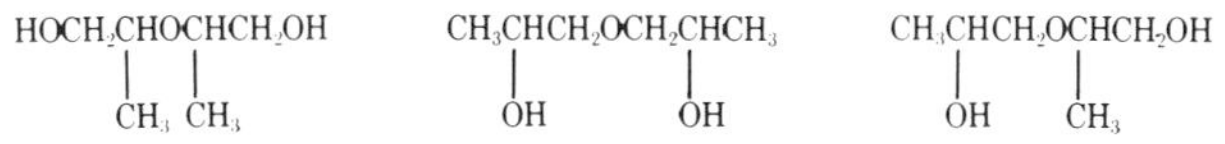

HOCH₂CHOHCH₂OH (with CH₃, CH₃)   CH₃CHCH₂OCH₂CHCH₃ (with OH, OH)   CH₃CHCH₂OCHCH₂OH (with OH, CH₃)

4%   43%   53%

The distribution of dipropylene glycol uses in polyester resin manufacture is polyester resins, 60%; plasticizers, 30%; and alkyd resins, hydrocarbon extraction, urethane polyols, and other, 10%.

E.S. Brown
C.F. Hauser
R.V. Berthold
B.C. Ream
Union Carbide Corp.

Glycols, Brochure F-41515B, Union Carbide Corporation, New York, 1978.

G.O. Curme and F. Johnston, Glycols, ACS Monograph No. 114, Reinhold Publishing Corp., New York, 1952, Chapt. 2.

1,3-BUTYLENE GLYCOL

1,3-Butylene glycol, butane-1,3-diol, $CH_3CHOHCH_2CH_2OH$, is a colorless, mildly bittersweet liquid. Some of its esters occur in nature. It is soluble in water, lower alcohols, ketones, and esters, and it is not soluble in the aliphatic hydrocarbons or most of the common chlorinated solvents. Nylon is quite soluble in hot 1,3-butylene glycol; shellac and rosin are partially soluble in it. It has a low toxicity and is nonirritating to normal skin. When rubbed on, it penetrates and gives a marked warming sensation. It is now manufactured by catalytic hydrogenation of acetaldol.

$$2\ CH_3CHO \xrightarrow{OH^-} CH_3CHOHCH_2CHO \xrightarrow[H_2]{Ni} CH_3CHOHCH_2CH_2OH$$

Celanese Chemical Company is the sole United States manufacturer of 1,3-butanediol.

Some of the common physical properties are listed in Table 1.

Uses

Simple esters of 1,3-butylene glycol with monocarboxylic acids have been used as plasticizers (qv) for cellulosics and poly(vinyl chloride) resins. 1,3-Butylene glycol is used as a terminal group in oil-free alkyd

Table 1. Physical Properties of 1,3-Butanediol

Property	Value
bp, °C	207.5
mp, °C	−77
density at 20°C, g/cm³	1.004
viscosity at 25°C, kPa·s (= 10^4 P)	103.9
refractive index, n_D^{20}	1.441
surface tension at 25°C, Pa[a]	378
heat of vaporization, J/g[b] at bp	585
heat capacity of liquid, J/(kg·K)[b] at 20°C	2302

[a] To convert Pa to mm Hg, divide by 133.3.
[b] To convert J to cal, divide by 4.184.

resins (qv). It is used also for deicing aircraft (see Antifreezes and deicing fluids).

FRANK S. WAGNER, JR.
Celanese Chemical Co.

1,3-Butylene Glycol, Celanese Chemical Co., Dallas, Tex., 1979.

OTHER GLYCOLS

Glycols such as neopentyl glycol, 2,2,4-trimethyl-1,3-pentanediol, and 1,4-cyclohexanedimethanol are used in the synthesis of polyesters (qv) and urethane foams (see Foamed plastics). Their physical properties are shown in Table 1.

Neopentyl glycol, or 2,2-dimethyl-1,3-propanediol, $HOCH_2C(CH_3)_2$ CH_2OH, is a white crystalline solid at room temperature, soluble in water, alcohols, ethers, ketones, and benzene but relatively insoluble in alkanes. Two primary hydroxyl groups are provided by the 1,3-diol structure, making this glycol highly reactive as a chemical intermediate. Neopentyl glycol can undergo typical glycol reactions such as esterification (qv), etherification, condensation, and oxidation.

2,2,4-Trimethyl-1,3-pentanediol, trimethylpentanediol, $CH_3CH(CH_3)$ $CH(OH)C(CH_3)_2CH_2OH$, is a white, crystalline solid. It is soluble in

most alcohols, other glycols, aromatic hydrocarbons, and ketones, but it has only negligible solubility in water and aliphatic hydrocarbons.

1,4-Cyclohexanedimethanol, 1,4-dimethylolcyclohexane or 1,4-bis-(hydroxymethyl) cyclohexane, is a white waxy solid.

$$HOCH_2 - \langle \rangle - CH_2OH$$

The commercial product consists of a mixture of cis and trans isomers. 1,4-Cyclohexanedimethanol is miscible with water and low molecular weight alcohols and appreciably soluble in acetone. It has only negligible solubility in hydrocarbons and diethyl ether.

PAUL VON BRAMER
JOHN H. DAVIS
Eastman Chemical Products, Inc.

Publications, Nos. N-153, N-154, and N-199, Eastman Chemical Products, Inc., Kingsport, Tenn., 1982–1983.

GLYCOLS — 1,4-BUTYLENE GLYCOL AND BUTYROLACTONE.
See Acetylene-derived chemicals.

GLYOXAL. See Glycols, ethylene glycol.

GOLD AND GOLD COMPOUNDS

Gold, atomic number 79, is a third row transition metal in Group IB of the periodic table. It occurs naturally as a single stable isotope of mass 197. Its electronic configuration is $[Xe] 4f^{14}5d^{10}6s^1$. Common oxidation states are 0, 1, and 3. Selected properties of gold are shown in Table 1. Gold is characterized by high density, high electrical and thermal conductivity, and high ductility.

Gold is the most noble of the noble metals. Other than in the atomic state, the metal does not react with either oxygen, sulfur, or selenium at any temperature. It does react, however, with tellurium at elevated

Table 1. Physical Properties of Several Glycols

	Neopentyl glycol	2,2,4-Trimethyl-1,3-pentanediol	1,4-Cyclohexanedimethanol
melting range, °C	124–130	46–55	45–50
boiling point at 101.3 kPa (1 atm), °C	210		286
density, g/cm³			
at 20°C			1.150
at 21°C	1.06	0.897	
crystallization point, °C			35
viscosity at 50°C, mPa·s (= cP)			675
flammability, flash point, COC[a], °C	129	113	167
heat of fusion (estd), kJ/mol[b]	21.77	8.63	
heat of vaporization, kJ/mol[b]			
at 32 kPa[c] 170°C	67.1		
at 101.3 kPa (1 atm)			95.6
hygroscopicity[d], wt% H_2O			
at 50% rh		0.1–0.2	
at 51% rh	0.3		

[a] Cleveland Open Cup.
[b] To conver kJ to kcal, divide by 4.184.
[c] To convert kPa to mm Hg, multiply by 7.5.
[d] At equilibrium; neopentyl glycol at 25–38°C, and 2,2,4-trimethyl-1,3-pentanediol at 25°C.

Table 1. Gold Properties

Property	Value
melting point, K	1337.6
boiling point, K	3081
crystal structure	fcc.4 atoms/unit cell
lattice constant at ambient temperature, nm	0.407
density at 273 K, g/cm³	19.32
Brinell hardness (10/500/90) (annealed at 1013 K), kgf/mm²	25
modulus of elasticity at 293 K (annealed at 1173 K), MPa[a]	7.747×10^4
tensile strength (annealed at 573 K), MPa[a]	123.6–137.3
elongation (annealed at 573 K), %	39–45
heat of fusion, J/mol[b]	1.268×10^4
vapor pressure, Pa[c] at 1000 K	5.5×10^{-8}
specific heat at 298 K, J/(g·K)[b]	1.288×10^{-1}
thermal conductivity at 273 K, W/(m·K)	311.4
electrical resistivity at 273 K, Ω·cm	2.05×10^{-6}

[a] To convert MPa to psi, multiply by 145.
[b] To convert J to cal, divide by 4.184.
[c] To convert Pa to mm Hg, divide by 133.3.

temperatures (ca 475°C) to produce $AuTe_2$ (also found as the naturally occurring mineral, calaverite). It also reacts with the halogens, particularly in the presence of moisture. At low temperatures ($\leq 200°C$), chlorine is adsorbed on the gold surface with formation of surface chlorides. The rate of further chlorination is limited by the rate of diffusion of Cl through the surface chloride layer. At higher temperatures (700–1000°C), the reaction is kinetically controlled as gold chlorides sublime and fresh surface is continually exposed. In this temperature region, adsorbed Cl_2 dissociates into Cl atoms that then react with the surface.

At high temperatures, attack by concentrated sulfuric and nitric acids is slow and is negligible for phosphoric acid. Gold is very resistant to fused alkalies and to most fused salts except peroxides. Gold readily amalgamates with mercury. Gold is very corrosion- and tarnish-resistant and imparts corrosion resistance to most of the commonly used gold alloys, especially to alloys containing ≥ 50 atom% of gold. Although gold is resistant to organic acids, some of the alloys used in jewelry may become tarnished by perspiration.

Extraction and Refining

At present, most gold is obtained by deep mining, notably in South Africa. Mined gold ore is milled sufficiently to allow separation of the gold; recoveries usually are 92–96% of the ore's gold content. After milling, various recovery processes may be used such as amalgamation, cyanidation, gravity concentration (qv), flotation (qv), and roasting, or a combination of these processes.

Production

Gold is widely distributed and the average content in the earth's crust is estimated to be 3.5 ppb. Gold is produced in many countries, eg, the Republic of South Africa, the USSR, and Canada. In the United States, about 60% of the gold originates from ores and the remainder primarily from the refining of base metals, eg, copper, zinc, lead, and mercury. The largest United States gold producer is Homestake Mine at Lead, S. Dakota, which supplies ca 30% of the U.S. production. The chief gold-producing states are S. Dakota, Nevada, Utah, and Arizona. Demand for fabricated gold in the United States is estimated to grow ca 2.5% annually which requires increasing importation.

Health and Safety Factors (Toxicology)

Disodium gold (I) thiomalate (Myocrysin) and gold (I) thioglucose (Solganal) are two of the few effective treatments for rheumatoid arthritis. Excretion of gold during chrysotherapy is relatively slow (ca 20% within two weeks and less thereafter) and toxic side effects (such as kidney and liver damage, dermatitis, stomatits, thrombocytopenia, bone marrow depression, and other haematopoietic disorders) can be long lasting and sufficiently severe to require cessations of treatment in about 35% of the patients. Toxicity and the requirement of parenteral administration has led to further efforts to synthesize less toxic gold drugs that can be administered orally.

Radioactive ^{198}Au, prepared by irradiating natural gold in a nuclear reactor, has been used in radiotherapy either as grains that can be implanted in cancerous tissue or infused in colloidal form as in the treatment of bladder cancer. Colloidal ^{198}Au also has diagnostic applications such as bone-marrow scanning and visualization of organs such as the lungs and liver (see Radioactive drugs).

Uses

In addition to its use for monetary reserves, gold is used in the private sector principally for investment and fabrication (eg, carat jewelry; electronics; dentistry; medals, medallions and unofficial coins; official coins; and decorative uses).

Derivatives

Gold compounds. The chemistry of nonmetallic gold predominantly is that of Au(I) and Au(III) compounds and complexes. For the most part, the chemistry of gold is more closely related to that of its horizontal neighbors in the periodic table, platinum and mercury, than to the other members of its subgroup, copper and silver.

Halides include gold(III)chloride, Au_2Cl_6, tetrachloroauride $[AuCl_4]^-$, the corresponding bromides, and AuI. Cyanides include $[Au(CN)_2]^-$. Sulfides include gold sulfide, Au_2S.

Both alkyl and aryl complexes of Au(I) and Au(III) as well as olefin and acetylene complexes have been prepared and studied.

J.G. COHN
ERIC W. STERN
Englehard Corp.

R.J. Puddephatt, *The Chemistry of Gold*, Elsevier Scientific Publishing Co., Amsterdam, The Netherlands, 1978.

W.S. Rapson and T. Groenewald, *Gold Usage*, Academic Press, Inc., New York, 1978.

E.M. Wise, ed., *Gold: Recovery, Properties and Applications*, D. Van Nostrand Co., Inc., Princeton, N.J., 1964.

GONADOTROPIC HORMONES.   See Hormones—Anterior-pituitary hormones.

GOSSYPOL.   See Vegetable oils.

GRAMICIDIN.   See Antibiotics, peptides.

GRAPHITE.   See Carbon and artificial graphite; Carbon, natural graphite.

GRAVITY CONCENTRATION

Gravity concentration is the general term used to describe those concentration processes where particles of mixed sizes, shapes, and specific gravities are separated one from another by use of the force of gravity or centrifugal force. It is employed chiefly in the processing of ores and coal, although similar processes are used for preparation of cereal grains, degritting of paper pulp, and the recycling of municipal solid waste (see Centrifugal separation; Coal; Flotation; Fluidization; Paper; Size separation; Wastes, industrial).

Principles

Under the influence of gravity, particles settle differentially in a fluid, usually water, although pneumatic devices using air as the fluid medium are not uncommon. Gravity concentration depends on the fact that light, small, and flat particles settle more slowly in a fluid than heavy, large, and spherical particles.

Density. The basis of gravity concentration processes is the difference in density between two minerals. In general, the greater the difference between two minerals, the more effective the separation.

Free settling. Particles settle in a still fluid at rates depending on their density, size, and shape, according to the equations of Newton and Stokes. For the settling of coarse, spherical particles, Newton derived the following equation:

$$v_m = \sqrt{\frac{4}{3Q} \frac{\rho - \rho'}{\rho'} d \cdot g} \tag{1}$$

where v_m = the terminal settling velocity of the particle, cm/s; Q = the coefficient of resistance; for spherical particles, $Q = 0.4$; ρ = density of the particle, g/cm^3; ρ' = density of the fluid, g/cm^3; d = particle diameter, cm; and g = gravitational acceleration, cm/s^2.

For the settling of fine spherical particles, the Stokes equation applies:

$$v_m = \frac{1}{18} \frac{\rho - \rho'}{\mu} d^2 g \tag{2}$$

where μ = fluid viscosity in $mPa \cdot s$ (cP).

Hindered settling. This process may be described as that condition "where particles of mixed sizes, shapes, and densities in a crowded mass, yet free to move among themselves, are sorted in a rising [fluid] current." Collision between particles is continuous and the assemblage settles at a much slower rate than the individual particles. The hindered settling phenomenon can be induced by crowding, a constriction, or inducing turbulence. It can be used to stabilize a suspension against rapid settling of the particle mass. Such suspension stability is favored by fine particles, light particles, and a high concentration of particulate solids.

Particle shape. The settling rate of those particles with other than spherical shape is lower than that of a sphere of equivalent volume. The practical settling rate of coarse particles settling under Newtonian conditions may thus be substantially different than that calculated for spheres. For fine particle, however, equation 2 is a good approximation of reality. Even for those fine particles of pronounced tabular form, eg, clays, the particles are found to settle at roughly one-half the value calculated for spheres.

Equal-settling particles. A dense particle (H) may be so selected as to settle in a fluid at the same rate as a larger particle of lower density (L). If both are relatively coarse particles, whose individual settling rates are described by equation 1, then since their settling rates are equal, the free settling (F.S.) ratio of the two particles, ie, the ratio of their diameters, settling under Newtonian conditions may be given by:

$$\left.\frac{d_L}{d_H}\right]_{\text{F.S,Newt.}} = \frac{(\rho_H - \rho')^1}{(\rho_L - \rho')^1} \tag{3}$$

For particles settling under Stokesian conditions, the equation is identical except the exponent, n, is 0.5. For particles in the intermediate or Allen range, the value of n is between 0.5 and 1. Furthermore, since diametrical ratios between the settling particles are usually relatively small, both particles may be assumed to settle under approximately the same condition, eg, Newtonian or Stokesian.

Processes and Devices

Range of applicability. The range of applicability of the various devices to be described here is given in Figure 1 and compared to other concentration techniques. Gravity concentration devices are available for particles ranging from 10 μm (1250 mesh) to 10 cm. Because they can recover fine particles, flotation or magnetic-separation (qv) techniques are preferred where applicable (see Fig. 1). Gravity concentration processes are evaluated by the metallurgical balance or the partition-curve method.

Heavy-medium separation. The ideal fluid to separate two particles of differing density would be a liquid with a density intermediate between the density of the two particles.

Heavy-liquid separation processes are used in laboratory evaluations and have been tested industrially for a few materials. Currently, the Otisca process, which uses a proprietary halogenated hydrocarbon, is undergoing pilot-plant testing for the cleaning of bituminous coal.

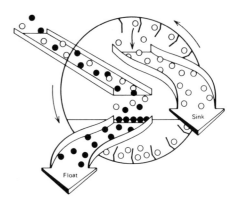

Figure 2. Drum separator. Courtesy of Wemco Division, Envirotech Corp.

Table 1. The Concentration Criterion

Concentration criterion	Separation
< 2.5	effective down to ca 74 μm (200 mesh)
2.5–1.75	effective to 150 μm (100 mesh)
1.75–1.50	possible to 2030 μm (10 mesh) but difficult
1.50–1.25	possible to 635 μm but difficult
< 1.25	gravity concentration processes inapplicable except for heavy-liquid and heavy-media separations

Suspensions of finely divided heavy solids in water are commonly used as separating "fluids" of high density. Today, quartz and especially magnetite are used for coal separation, whereas ferrosilicon is nearly always used for ore separations. Heavy-medium separation is the most rapidly growing gravity concentration method and offers great promise for the future.

To serve as an effective heavy-medium solid, a substance must produce a relatively stable suspension of the required specific gravity. Suspension stability can be achieved by (1) using a high concentration of solids, (2) using finely divided solids, (3) introducing a suspending agent such as clay, or (4) introducing turbulence or rapidly rising currents in the separating vessel.

Devices for Gravity Concentration

Separatory vessels include the Chance cone separator, the rotary-drum separator (Fig. 2), the water-only cyclone, sluices, hindered-settling classifiers, jigs, flowing-film concentrators, the shaking table, and pneumatic concentrators.

Device Selection

The selection of the appropriate device to accomplish a given separation is based on the size range of feed (see Fig. 1), the mineral density, and on the concentration criterion. The applicability of a gravity concentration device to effect a given separation may thus be estimated from Table 1.

FRANK F. APLAN
The Pennsylvania State University

R.H. Richards, C.E. Locke, and R. Schuhmann, *Textbook of Ore Dressing*, 3rd ed., McGraw Hill Book Co. Inc., New York, 1940.

J.W. Leonard, ed., *Coal Preparation*, 4th ed., AIME, New York, 1979.

A.F. Taggart, *Handbook of Mineral Dressing*, John Wiley & Sons, Inc., New York, 1945.

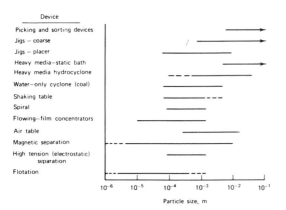

Figure 1. Approximate range of applicability of various concentration devices.

GREAT SALT LAKE CHEMICALS. See Chemicals from brine.

GRIGNARD REACTION

The Grignard reaction may be regarded as the formation of an organomagnesium halide (Grignard reagent) and its reactions with organic, inorganic, or elemental substrates. Few, if any, general reactions have had as great an impact on the development of organic chemistry as the Grignard reaction. It has probably been the most widely used general method for the synthesis of organic molecules.

In 1898 Barbier reported that 2,6-dimethylhept-5-en-2-ol was obtained by the reaction of methyl iodide and 6-methylhept-5-en-2-one in the presence of magnesium. He assigned the task of investigating the general utility of the method to a student, Victor Grignard. In the course of his initial studies, Grignard realized that Barbier's method consisted of three general steps:

Grignard reagent preparation:

$$CH_3I + Mg \longrightarrow CH_3MgI$$

Grignard reaction: CH_3MgI +

Hydrolysis:

The great importance of Grignard's studies was that the components of Barbier's original observations were separated and characterized; this was the key that permitted the exploitation of the general method.

Grignard reagents are generally prepared in ether solvents. As in the original report, an organohalogen compound reacts with magnesium to form the organomagnesium halide which is usually soluble in the ether medium. This basic methodology is employed in the industrial manufacture and use of Grignard reagents.

The Grignard reaction remains a widely used synthetic reaction and is frequently employed in the production of pharmaceuticals and their intermediates, flavor and fragrance chemicals (see Flavors and spices; Perfume), polymerization catalysts, and other organometallics.

Nature of the Reagent

The question of the Grignard-reagent structure has continued to receive serious attention. The entire history of this question is summarized in a comprehensive review in which it is concluded that all of the reported compositions of the reagent may be found but that the degree of their presence is highly dependent on solvent and concentration characteristics. For almost all practical purposes, however, the generalized structure RMgX is sufficient.

Preparation of Grignard Reagents

Grignard reagents are generally prepared as in the original report: an organohalogen compound is allowed to react with magnesium in an anhydrous ether solvent to yield the organomagnesium halide. The main difficulty encountered in the preparation of Grignard reagents is initiation. Numerous methods and catalysts for the initiation of reaction have been reported. In the batch manufacture of Grignard reagents, the most satisfactory procedure is to retain an excess of magnesium from the preceding batch (activated heel) and to add a small amount of the Grignard solution. In this fashion, little difficulty is usually encountered in initiating the reaction.

The structure of the organohalogen component, RX, has a pronounced effect on both the ease of Grignard-reagent formation and its subsequent reactivity. In general, the halides may be expected to react in the order: X = I > Br > Cl ≫ F. Fluorides are inert to the usual conditions.

For most purposes, chlorides are the only economically attractive halides for commercial use. Difficulties in the preparation of Grignard reagents from chlorides in ethyl ether are common but the use of cyclic or higher alkyl ethers, eg, tetrahydrofuran or di-n-butyl ether greatly facilitates reagent formation.

The constitution of the carbon-based portion of the organohalogen compound has a pronounced effect on the ease of Grignard-reagent formation. For a series of homologous alkyl iodides, the relative order of reactivity is reported as: $CH_3I > C_2H_5I > n\text{-}C_3H_7I > n\text{-}C_4H_9I > n\text{-}C_6H_{13}I$. A series of isomeric butyl bromides gives the order: $sec\text{-}C_4H_9Br > i\text{-}C_4H_9Br > n\text{-}C_4H_9Br$. A further generalization may be made regarding the relative reactivities of carbon structural types: allylic, alkyl > cycloalkyl ≥ monocyclic aromatic > polycyclic aromatic ≥ heterocyclic ≥ vinylic.

Indirect methods of Grignard-reagent formation. A variety of indirect methods permit the preparation of Grignard reagents inaccessible by the usual method.

Organic derivatives of Group IA metals react with magnesium halides to generate the Grignard reagent, eg:

$$RNa + MgCl_2 \rightarrow RMgCl + NaCl$$
$$RLi + MgBr_2 \rightarrow RMgBr + LiBr$$

In certain cases, hydrocarbons may be converted to Grignard reagents by an exchange process:

$$RH + R'MgX \rightarrow R'H + RMgX$$

Acetylene is capable of yielding a di-Grignard reagent (see also Acetylene-derived chemicals):

$$HC{\equiv}CH + 2\,C_2H_5MgBr \rightarrow BrMgC{\equiv}CMgBr + 2\,C_2H_6$$

Cyclopentadiene (qv) affords the cyclopentadienylmagnesium halide on reaction with a lower-alkyl Grignard reagent.

Limitations on substituents within the organohalogen compound. In general, only those functionalities that are inert to the action of the reagents can be tolerated. In some cases, the functionality can be masked by pretreatment with another reagent.

Manufacture

The processes involved in batch manufacture of Grignard reagents are formation of the Grignard reagent, reaction with a carbonyl or other reactive compound, and hydrolysis of the resulting complex.

Most of the continuous processes use a column of magnesium particles through which a solution of the organohalogen compound in an ether-class solvent is passed. In this fashion, a solution of the Grignard reagent is generated continuously. The principal advantage in using such continuous processes appears to lie in the reduction of side reactions, chiefly Wurtz-type coupling, which renders production of many Grignard reagents impractical.

Economic Aspects

Main producers of Grignard reagents in the U.S. are Laramie Chemical Co., PCR Research Chemicals, Inc., Alfa Products, M & T Chemicals, Inc., and Vineland Chemical Co., Inc.

Reactions of Grignard Reagents

Reactions with inorganic halides. The reactions of Grignard reagents with inorganic halides have been used widely for the commercial preparation of a variety of organometallic compounds. In general, halides of the Group IIB, IIIA, IVA, and VA elements react with Grignard reagents to give the corresponding organometallic compounds:

$$RMgX + MX_n \rightarrow RMX_{n-1} + R_2MX_{n-2} + \cdots$$

Reactions with organic halides. The reaction of Grignard reagents with organic halides has been studied extensively and often provides a means of forming carbon-carbon single bonds from readily accessible materials.

Reactions with carbonyl groups. Grignard reagents most frequently are used in reactions with carbonyl groups. Reaction with aldehydes and ketones leads to the formation of secondary and tertiary alcohols, respectively (see Alcohols). Primary alcohols can be prepared by the reactions of formaldehyde and ethylene oxide with Grignard reagents which adds one and two carbon atoms, respectively.

Addition to carbon–carbon double bonds. Grignard reagents add to carbon–carbon double bonds that are activated by electron-attracting substituents such as aldehyde, ketone, or ester functionalities.

Reactions with carbon–nitrogen multiple bonds. Grignard reagents add to nitriles (qv) and imines much as they add to carbonyl groups:

$$R'C \equiv N + RMgX \longrightarrow R'C = N - MgX$$
$$\qquad\qquad\qquad\qquad\quad |$$
$$\qquad\qquad\qquad\qquad\quad R$$

$$R'CH = NR'' + RMgX \longrightarrow R'CHN - MgX$$
$$\qquad\qquad\qquad\qquad\qquad\quad |\quad\;\, |$$
$$\qquad\qquad\qquad\qquad\qquad\quad R\;\; R''$$

Reactions with oxygen, sulfur, and halogens. Grignard reagents react with oxygen, sulfur, and the other elements of Group VIA, as well as the elements in the halogen series to form products with C—O, C—S, etc, and C—X bonds, respectively.

Reactions with active hydrogen compounds. The reactions of Grignard reagents with active hydrogen compounds are used on a commercial scale for the indirect preparation of Grignard reagents (see Preparation of Grignard Reagents). In this manner, Grignard reagents derived from acetylenes, cyclopentadienes, indoles, pyrroles, and other materials are prepared.

THOMAS E. MCENTEE
Syntex Chemicals, Boulder Division

G.E. Coates, M.L.H. Green, and K. Wade, eds., *Organometallic Compounds*, Vol. I, Methuen and Co., Ltd., London, UK, 1967.

G. Wilkinson, F.G.A. Stone, and E.W. Abel, eds., *Comprehensive Organometallic Chemistry*, Vol. I, Pergamon Press, Ltd., Oxford, UK, 1982.

Methoden der Organischen Chemie (Houben-Wayl), Band XIII/2a, George Thieme Verlag, Stuttgart, 1973.

GRINDING. See Abrasives; Size reduction.

GUAIAC. See Resins, natural.

GUAIACOL, o-CH$_3$OC$_6$H$_4$OH. See Ethers.

GUANIDINE AND GUANIDINE SALTS

Guanidine (**1**) is a crystalline, strong organic base present in combined form in guano, egg albumen, nucleic acids, and other natural products. It may be regarded as the imide of urea (**2**) or the amidine of carbamic acid (**3**).

$$\underset{(1)}{\overset{\overset{\textstyle NH}{\|}}{H_2NCNH_2}} \qquad \underset{(2)}{\overset{\overset{\textstyle O}{\|}}{H_2NCNH_2}} \qquad \underset{(3)}{\overset{\overset{\textstyle O}{\|}}{H_2NCOH}}$$

Guanidine is a monoacid base with a strength equivalent to that of sodium hydroxide. This marked basicity may be due to the ability of the guanidinium ion to resonate among three equivalent structures.

Properties

Free guanidine forms colorless, hygroscopic crystals melting at ca 50°C. It absorbs carbon dioxide rapidly from the air. In aqueous solutions, it hydrolyzes slowly to urea and then to carbon dioxide and ammonia.

Guanidine is liberated from its salts only by special procedures, such as the treatment of guanidinium carbonate with sodium alkoxide in a solution of absolute alcohol. A variety of salts is formed by double decomposition. Containing two amino groups, guanidine can be used as the starting material for imidazoles, triazines, and pyrimidines.

Basic solutions of guanidine or its salts react readily with aqueous formaldehyde to give guanidine–formaldehyde resins (see Amino resins). A reaction of special importance is the dehydration of guanidinium nitrate by sulfuric acid or other dehydrating agents to nitroguanidine.

Guanidinium sulfate and an equimolar amount of aluminum sulfate in aqueous solution form guanidinium aluminum sulfate, $(CH_6N_3)Al(SO_4)_2 \cdot 6H_2O$, which exhibits ferroelectric properties.

Analytical Methods

Guanidine is determined quantitatively by potentiometric titration with perchloric acid, or it can be precipitated as the picrate.

Health and Safety Factors

Guanidine salts do not have any detectable toxic effects on humans. Precautions should be observed against skin and mucous, membrane contact when handling solid guanidine hydrochloride or concentrated solutions.

Manufacture

Guanidine is made from dicyandiamide and ammonium nitrate in the presence of anhydrous ammonia (see Cyanamides).

$$\underset{\text{dicyandiamide}}{\overset{\overset{\textstyle NH}{\|}}{NH_2CNHCN}} \xrightarrow{NH_4NO_3} \underset{\text{biguanide nitrate}}{\overset{\overset{\textstyle NH}{\|}\;\;\overset{\textstyle NH}{\|}}{NH_2CNHCNH_2 \cdot HNO_3}} \xrightarrow{NH_4NO_3} \underset{\text{guanidinium nitrate}}{\overset{\overset{\textstyle NH}{\|}}{2\; NH_2CNH_2 \cdot HNO_3}}$$

The mixture is heated at 160°C for 1 h, producing pressures up to 13.8 MPa (2000 psi); the ammonia is flashed off and yields of 91–93% are obtained.

Guanidine salts can also be produced by treating cyanamide with ammonia. The nitrate can be produced directly from ammonium nitrate and urea. Ammonolysis of ammonium thiocyanate and urea gives guanidine.

Uses

Guanidine is marketed in the form of its salts, which are named as guanidine salts (or guanidinium salts). Guanidine hydrochloride is an intermediate in the synthesis of pharmaceuticals, eg, sulfaguanidine, sulfadiazine, and sulfamerazine (see Antibacterial agents, synthetic). Nitroguanidine is an exceptionally cool and flashless explosive and is used as a propellant (see Explosives and propellants). Aminoguanidines are useful as dye intermediates. Guanidine soaps are effective dry-cleaning agents. Dodecylguanidine salts (Cyprex fungicides) are used in agricultural applications. N-Substituted guanidine salts are used as antimicrobials.

PAUL PATTERSON
Cyanamid Canada Inc.

"The Chemistry of Guanidine," *Nitrogen Chemicals Digest*, Vol. IV, American Cyanamid Company, New York, 1950.

Guanidine Products Study, SKW-Trostberg, A.G. Trostberg, FRG, 1976.

GUMS

The term gums denotes a group of industrially useful polysaccharides or their derivatives that hydrate in hot or cold water to form viscous solutions or dispersions.

Gums are classified as natural and modified gums. Natural gums include seaweed extracts, plant exudates, gums from seeds or roots, and gums obtained by microbial fermentation (see Fermentation). Modified (semisynthetic) gums include cellulose and starch derivatives and certain synthetic gums such as low methoxyl pectin, propylene glycol alginate,

Table 1.   Comparative Viscosities of Botanical Gums, mPa·s (= cP)

Concen-tration, %	Natural plant exudates					Seed gums	
	Gum tragacanth	Gum karaya	Gum ghatti	Gum arabic	Larch arabino-galactan	Guar	Locust bean
1	54	3,000	2			4,500	100 (300)ᵃ
1.5						15,000	
2	906	8,500	35				1,600 (3,400)ᵃ
3	10,605	20,000					
5	111,000	45,000	288	7			
10			1,012	17	2		
20			2,240	41	10		
30				200	21		
40				936	47		
50				4,163			

ᵃViscosities measured in cold water; hot water viscosities indicated within parentheses.

Table 2.   Physical Properties of Refined Sodium Alginate

Property	Value
moisture content, %	13
ash, %	23
powder color	ivory
specific gravity	1.59
bulk density, kg/m³	87.39
browning temperature, °C	150
charring temperature, °C	340, 460
ashing temperature, °C	480
heat of combustion, J/gᵃ	10.46

ᵃ To convert J to cal, divide by 4.184.

triethanolamine alginate, carboxymethyl locust bean gum, and carboxymethyl guar gum (see Cellulose derivatives; Resins, water-soluble; Starch).

Gums are used in industry because their aqueous solutions or dispersions possess suspending and stabilizing properties. In addition, gums may produce gels, act as emulsifiers, adhesives, flocculants, binders, film formers, lubricants, or friction reducers, depending on the shape and chemical nature of the particular gum. Gum uses are divided about evenly between food and industrial applications. Gum arabic has the largest annual world production—between 24,000 and 68,000 metric tons (av). Locust bean gums are second at 15,000–18,000 t (av). Others range between 1,000 and 10,000 t.

The properties of a botanical gum are determined by its source, the climate, season of harvest, and extraction and purification procedures. Table 1 illustrates one of the important basic properties of all gums, ie, the relationship between concentration and solution viscosity. The considerable viscosity variation observed among gums from different sources determines, in part, their uses.

Gums From Algal Sources (Seaweed)

Agar.   Agar is extracted from certain marine algae belonging to the class Rhodophyceae, red seaweed. Today, it is produced in Japan, Spain, Taiwan, Republic of Korea, Morocco, Chile, and Portugal, and by one manufacturer in the United States. The uses of agar, which result from its ability to form gels, include the preparation of dessert gels and of solid culture media for culturing microorganisms.

Agarose, the gelling portion of agar, has a double helical structure. Double helices aggregate to form a three-dimensional framework that holds the water molecules within the interstices of the framework. Thus, thermoreversible gels are produced.

Agar is insoluble in cold water but is soluble in boiling water. On cooling to about 35°C, a firm gel forms that does not melt or liquefy below about 85°C.

Algin

Algin occurs in all members of the Phaeophyceae, brown seaweed, as a structural component of the cell walls in the form of the insoluble mixed calcium, magnesium, sodium, and potassium salt of alginic acid. Alginic acid comprises D-mannuronic acid and L-guluronic acid residues.

The many industrial uses of algin result from its thickening, stabilizing, emulsifying, suspending, film-forming, and gel-producing properties.

Alginic acid contains three kinds of polymer segments: one consisting essentially of D-mannuronic acid units, a second of L-guluronic acid units, and the third of alternating D-mannuronic acid and L-guluronic acid residues. Table 2 gives typical physical properties of a refined sodium alginate.

Carrageenan

The term carrageenan is the generic description for a complex mixture of sulfated polysaccharides that are extracted from certain genera and species of the class Rhodophyceae, red seaweed. Commercial carrageenan has a molecular weight in the range of 100,000 to 1,000,000. It is sold as a powder ranging from white to beige, depending on the grade. Carrageenan produces high viscosity solutions and gels in water that react with proteins, especially casein. This allows the preparation of high strength milk gels. Most of the uses for carrageenan are in the food industry. A small amount of carrageenan is used in cosmetics and pharmaceuticals.

Carrageenan is a mixture of galactans that carry varying proportions of half-ester sulfate groups linked to one or more of the hydroxyl groups of the galactose units.

Gums from Botanical Sources

Plant Exudates

Most plant families include species that exude gums, and those that produce copious quantities represent a ready supply of gums. These exudates were the first gums to be used commercially and still represent a significant but diminishing segment of the natural-gum market. The plants are usually shrubs or low-growing trees. Collection is by hand and labor costs represent a large proportion of the cost of these gums.

Natural gums are exuded in a variety of shapes characteristic of the species of origin. These shapes include the globular shape of gum arabic and the flakes or thread-like ribbons of gum tragacanth.

The quality of individual gums is mainly determined by color and taste or odor. Many gums are colorless when secreted but darken on aging. Most gums are usually tasteless unless contaminated by the bitter flavors of tannins which precludes their use in foods.

Although many plant gum exudates are known, only gum arabic, ghatti, karaya, and tragacanth have wide industrial use. Gum arabic is used mainly as a stabilizer and thickener in foods. Gum karaya is used as a thickening and suspending agent and as a stabilizer in foods, as a bulk laxative, and as a binder for pulp in the paper industry. Gum tragacanth also is used in food products as a stabilizer and emulsifier. Also it is used for the preparation of pharmaceutical emulsions and jellies and in the cosmetic industry. Gum ghatti is used particularly as an emulsifier for oil and water emulsions in foods and pharmaceuticals, as well as in many of the applications described for gum arabic.

Seed Gums

Although most seeds contain starch as the principal food reserve, many contain other polysaccharides and some have industrial utility. Today, only guar and locust bean gum are used, particularly in food applications; quince, pectin and psyllium gums are used only in specialized applications.

Harvesting of these gums is expensive; however, harvesting from annual plants costs less than from perennial plants or trees. This is clearly demonstrated by the tremendous increase in the use of guar, a gum that is extracted from an annual leguminous plant. Today, more guar gum is consumed in the food, paper and mining industries than all other gums combined.

Microbial Gums

The longest-known microbial gum is dextran. This polysaccharide is produced from sucrose by certain species of *Leuconostoc* (see also

Fermentation; Microbial transformations). To date, however, only xanthan gum produced by *Xanthomonas campestris* has achieved commercial significance.

Xanthan gum is a cream-colored powder that dissolves in either hot or cold water to produce solutions with high viscosity at low concentration. These solutions have unique rheological properties. They exhibit pseudoplasticity, ie, the viscosity decreases as the shear rate increases. The most unusual property of xanthan gum is the reactivity with galactomannans, such as guar gum and locust bean gum. Xanthan gum/locust bean gum combinations form elastic gels whereas xanthan gum/guar gum combinations provide higher than expected viscosities but do not form a gel.

The unique properties of xanthan gum make it suitable for many applications in the food, pharmaceutical, and agricultural industries.

I.W. COTTRELL
J.K. BAIRD
Kelco Division of Merck & Co., Inc.

R.L. Whistler, *Industrial Gums*, Academic Press, Inc., New York, 1973, p. 6–7.

M. Glicksman, *Gum Technology in the Food Industry*, Academic Press, New York, 1969, pp. 210–213.

E. Percival and R.H. McDowell, *Chemistry and Enzymology of Marine Algal Polysaccharides*, Academic Press, London, UK, 1967.

"Extracellular Microbial Polysaccharides," P.A. Sandford and A. Larkin, eds., *ACS Symposium Series No. 45*, American Chemical Society, Washington, D.C., 1977, p. 301.

GYPSITE.   See Calcium compounds.

GYPSUM, CaSO$_4$.   See Calcium compounds.

H

HAFNIUM AND HAFNIUM COMPOUNDS

Hafnium, Hf, is in the same subgroup as zirconium with which it is always associated in nature. The two elements have more similar properties than any other pair in the periodic table. This similarity in chemical behavior is related to the electron configuration of the valence electrons, $4d^2, 5s^2$ and $5d^2, 6s^2$ for zirconium and hafnium, respectively; and the similarity in ionic radii of the M^{4+} ions, Zr^{4+}, 0.074 nm, and Hf^{4+}, 0.075 nm (a consequence of the lanthanide contraction). The principal valence state of hafnium is +4. The aqueous chemistry of hafnium compounds is characterized by the high degree of hydrolysis exhibited, the formation of polymeric species, and the multitude of complex ions that can be formed.

Hafnium is used as a neutron-absorber material, usually in the form of control rods, in nuclear reactors (qv) because of its high thermal-neutron absorption cross section, 1.05×10^{-26} m² (105 barns), its good ductility, and excellent corrosion resistance in pressurized hot water. Hafnium is used as a high temperature alloy strengthening agent, it is burned in flash bulbs, and it has been substituted for tantalum in cemented carbide tool bits when tantalum becomes scarce and more expensive (see Carbides).

Hafnium is a hard, shiny, ductile metal with a color very similar to that of stainless steel, although the color of hafnium sponge metal is a dull powder gray. Selected physical properties of hafnium are given in Table 1.

The average amount of hafnium in the earth's crust has been estimated at 4.5 ppm, which makes it slightly more plentiful than uranium, boron, tantalum, or bromine.

The commercially important hafnium mineral is zircon, zirconium orthosilicate, which usually contains 1.2% hafnium oxide. Zircon is found in large quantities in the beach sands of Australia, the United States, Africa, and India, and in the alluvial sands in the tin placers of Malaya and Nigeria (see also Gems, synthetic).

Manufacture. Zircon sand is obtained as a by-product of strip mining or dredging operations that recover ilmenite and rutile from alluvial deposits containing heavy minerals. Zircon sand is treated chemically to remove the silica, which is ca 35% of the zircon, and to convert the zirconium and hafnium contents into water-soluble compounds. Commercially, this is done by one of two methods in the United States: caustic fusion, or direct chlorination.

The separation of hafnium from zirconium is the most difficult step in the preparation of hafnium. The commercial method used in the United States is a countercurrent liquid–liquid extraction process first proposed

as an analytical procedure and developed into full-scale production at the Oak Ridge National Laboratory (see Extraction, liquid–liquid).

Health and safety. Most hafnium compounds require no special safety precautions because hafnium is nontoxic.

Hafnium Compounds

Most hafnium compounds have been of slight commercial interest aside from intermediates in the production of hafnium metal. They include hafnium boride, HfB_2, mp 3250°C; hafnium borohydride, $Hf(BH_4)_4$, mp 29°C; hafnium carbide, HfC, mp 3950°C; hafnium tetrafluoride, HfF_4, mp > 968°C; hafnium tetrachloride, $HfCl_4$, mp 432°C; hafnium tetrabromide, $HfBr_4$, mp 424°C; hafnium tetraiodide, HfI_4, mp 449°C; hafnium hydride, HfH_2; hafnium nitride, HfN, mp 2982°C mp; hafnium dioxide, HfO_2, mp 2900°C; and hafnium disulfide, HfS_2.

RALPH H. NIELSEN
Teledyne Wah Chang, Albany

D.E. Thomas and E.T. Hayes, eds., *The Metallurgy of Hafnium*, U.S. Government Printing Office, Washington, D.C., 1960, p. 331.

E.M. Larsen, "Zirconium and Hafnium Chemistry" in H.J. Emeleus and A.F. Sharpe, eds., *Advances in Inorganic Chemistry and Radiochemistry*, Academic Press, New York, 1970, pp. 1–133.

C.T. Lynch, "Hafnium Oxide" in A.M. Alper, ed., *High Temperature Oxides*, Academic Press, New York, 1970.

HAHNIUM. See Actinides and transactinides.

HAIR PREPARATIONS

Shampoos

Cleanliness of hair and scalp are among the most important personal grooming considerations today. The emphasis placed on cleanliness is a relatively new cultural phenomenon and, as mass markets have developed, shampoos have increased in importance beginning with soap-based products and expanding as a growing number of synthetic surfactants (qv) became available.

The introduction of synthetic surfactants marked the advent of a new era in the field of shampoo chemistry. Greater formulating flexibility and control, more efficient cleansing under hardwater conditions, and the development of surfactants with low irritation are all hallmarks of this era. The product must clean, but it must do so gently, leaving the hair in a soft, shiny, full-bodied, manageable state.

Cleansing of hair and scalp requires an effective shampoo that, with some degree of hand and finger manipulation, safely and efficiently removes the soil and yet leaves hair and scalp in relatively good condition. Furthermore, the wide variations in physical and cosmetic effects produced by absorbed surfactants, various cosmetic treatments, exposure to sunlight and atmospheric conditions, and the normal variation in hair diameter must be considered.

In addition, the formulator must deal with a number of essential and desirable shampoo properties if a satisfactory product is to be evolved. During application, the shampoo must spread easily over the head and into the hair. It should foam quickly and copiously in both soft and hard water and then rinse out thoroughly, leaving no detectable residues. Neither hair nor scalp should feel drawn or dry and they should convey a fresh clean scent. The hair should be left in a softer, lustrous, full-bodied, manageable state. The product must be nontoxic, nonirritating to the hand and scalp, and properly preserved against microbial and fungal contamination. In the event of accidental introduction into the eyes, no permanent damage should result. Finally, the particular blend of product color, clarity, pH, viscosity, and fragrance, including the appropriate stability, may often affect success or failure of the product.

Almost without exception, shampoos consist of an aqueous solution, emulsion, or dispersion of one or more cleansing agents, together with additives employed for the purposes of modifying and stabilizing the various functional and aesthetic properties of the finished product. As cleansing agents, soaps now have very limited use and have, with few

Table 1. Physical Properties of Hafnium

Property	Value
atomic weight ($^{12}C = 12$)	178.49
density, g/cm³	13.28
melting point, K	2467 ± 15
heat of fusion, J/kg[a]	1.35×10^5
specific heat at 298 K, J/(kg·K)[a]	144
thermal conductivity at 323 K, W/(m·K)	22.3
thermal neutron capture cross section, m²[b]	
absorption	$(1.05 \pm 0.05) \times 10^{-26}$
scattering	$(8 \pm 2) \times 10^{-28}$
linear thermal expansion,	
10^{-6}/K from 293 to 700 K	5.9
electrical resistivity at 273 K, Ω·m	3.57×10^{-7}

[a] To convert J to cal, divide by 4.184.
[b] To convert m² to barn, multiply by 10^{28}.

Table 1. Typical Shampoo Formulations

Ingredient	Wt%	Function
clear liquid		
sodium lauryl sulfate, 30% active	40.0	cleansing agent
lauramide DEA[a]	4.0	foam stabilizer
disodium EDTA[b]	0.1	sequestering agent
formaldehyde	0.04	preservative
fragrance	0.5	fragrance
FD & C Blue No. 1[c]	0.001	color
FD & C Yellow No. 5[c]	0.004	color
deionized or distilled water	55.36	
opaque, pearlescent		
TEA[d] lauryl sulfate, 40% active	20.0	cleanser
sodium lauryl sulfate, 28% active	20.0	cleanser
cocoamide DEA[a]	5.0	foam stabilizer
glycol stearate	1.0	opacifier, pearlescent agent
disodium EDTA[b]	0.1	sequestering agent
methylparaben	0.1	preservative
propylparaben	0.01	preservative
fragrance	0.5	fragrance
deionized or distilled water	53.29	

[a] Diethanolamine.
[b] Ethylenediaminetetraacetic acid
[c] See Colorants for food, drugs, and cosmetics.
[d] Triethanolamine.

exceptions, been supplanted by their synthetic counterparts. Additives are employed to provide fragrance and color, to thicken, opacify, and convey specific tactile attributes to the product. Others additives include solubilizers, foam modifiers, lime-soap dispersants, conditioning agents, medicaments, pH controls, preservative systems, and the like. Almost all shampoos are in clear liquid and opaque lotion form.

Typical shampoo formulations are given in Table 1.

Synthetic surfactants. The rapidly expanding cosmetics and toiletries industry makes widespread use of synthetic surfactants in a large assortment of product applications.

Surfactants are classified generally according to the electrical properties of their hydrophilic groups in aqueous solution into anionic, nonionic, amphoteric, and cationic surfactants. Anionic surfactants yield negatively charged hydrophiles in aqueous solutions; they are used extensively in shampoos because of their excellent foaming and cleansing features. Nonionic surfactants, although polar in nature, exhibit zero formal charge in water; in general they are poor foamers but excellent thickeners, solubilizers, and foam modifiers. Amphoteric surfactants form both positive and negative ions in aqueous solution; they are generally regarded as mild, low foaming, and somewhat inefficient cleansing agents. Cationic hydrophiles are positively charged in aqueous solution; they are also poor foamers but are substantive to hair and are used mostly as conditioning additives.

Additives include viscosity-control agents, opacifiers, conditioners, preservatives, fragrance oils, sequesterng agents, and ethyl alcohol, isopropyl alcohol, propylene glycol, glycerol, and sodium xylene sulfonate to maintain shampoo clarity.

Specially formulated shampoos include baby and medicated dandruff shampoos.

Rinses and Conditioners

Almost every product in the hair-preparation category has some conditioning effect. Conditioning is an all-encompassing term involving any or all of the desired qualities related to the finished hair, ie, ease of combing, detangling, body, shine, texture, split-end mending, prevention of static buildup, and manageability (see Antistatic agents). Rinses and conditioners are specifically designed for this purpose and are used separately or in conjunction with other hair products.

"Creme" rinses. Immediately after shampooing, the wet hair often is tangled and difficult to comb. Creme rinses are highly effective in reducing this resistance. In addition, they may improve feel, reduce static charge, and improve dry combing. These beneficial effects are due

to the cationic substances sorbed onto hair. The active ingredients in most creme rinses are quaternary ammonium compounds (qv) such as stearyldimethylbenzylammonium chloride, cetyltrimethylammonium bromide and dimethylbis(hydrogenated tallow)ammonium chloride.

Conditioners and creme rinse – conditioner combinations. The combination creme rinse–conditioner offers additional benefits, including manageability, body, and split-end mending and generally improves the finish of the hair. Protein hydrolysate, fatty alcohols, silicones, lanolin and its derivatives, and cationic polymers are among the many agents that have been incorporated into these products.

Hairdressings

Hairdressings is a broad term for the products applied for final grooming. They may be oil- or water-based preparations that give the hair a natural healthy appearance while holding it in place and improving luster and condition. Hairdressings are applied by rubbing a small amount on the palm of the hands, and then massaging it into the hair. This application results in a uniform distribution. Some hydroalcoholic preparations are applied directly onto the hair from a shaker bottle and then combed through. Hairdressings include brilliantines, alcohol-based lotions, emulsion creams and clear gels.

Fixatives

Hair sprays and wave sets provide a temporary setting effect or curl to hair. These products are generally composed of film-forming additives; components that modify the film to provide specific properties; a solvent, usually alcohol or a mixture of alcohol and water; fragrance; and in the case of aerosol products, a propellant. The setting action is achieved by a deposition of the product on hair by spraying, finger application, or dipping a comb into the product and combing through the hair. When dried, the coating on the hair causes it to retain its set or styling configuration. The resulting effect depends on the nature of the film-forming additive and its adhesive and cohesive properties.

Hair sprays and wave sets have similar compositions, ie, the principal component is a film-forming resin, modified with plasticizers and other additives, solvents, and perfumes to promote performance and aesthetics. Aerosol hair sprays also include propellants.

Hair sprays are generally finishing sprays, ie, they are applied and allowed to dry on hair that already has been set in a styled configuration. The application of a product to hair as it is being styled or set and then allowed to dry classifies it as a wave set.

The following are the main characteristics for an acceptable hair fixative: (1) hair-holding properties, ie, extra-hold, normal, or soft; (2) curl-retention properties should not be affected appreciably by the degree of humidity; (3) as little flake or powder on combing as possible; (4) rapid drying; (5) nonstickiness; (6) lustrous effect without an oily feel; (7) the sprayed film should be easily removed by shampooing; (8) nontoxic and nonirritating; and (9) adequately preserved against microbial contamination. In addition, these hair-care products should function simply, apply uniformly and have limited flammability and good stability.

Health and safety factors. All aerosol hair-spray products, whether containing hydrocarbon or carbon dioxide as propellants, are classified as flammable by virtue of their flame-propagation properties according to established flammability tests and the flammable nature of their contents (see Aerosols).

Deliberate inhalation of aerosols poses a potential health hazard to the consumer. Their misuse by inhalation of contents can be harmful or fatal. As a result, appropriate label warnings and precautionary statements are found on all aerosols, including hair sprays, to inform consumers of this fact.

Wave sets. Today's wave sets are far superior to early versions. They not only offer improved holding characteristics but also provide a variety of effects such as good feel, nonflaking, ease of combing, as well as ranges of physical consistencies from gels to liquids, as a result of the many types of synthetic resins that are now available for formulation. Among the recommended resins are poly(vinylpyrrolidinone), poly(vinyl acetate), carboxyvinyl polymers, quaternary ammonium polymers, vinyl acetate-crotonic acid copolymer, cationic starch, acrylic polymers, and others.

Colorant Preparations

Among the desired properties of a good hair dye, toxicological safety is of primary importance. The coloration should be achievable in 10–30 min at ambient temperature and from a limited dye bath. These requirements mean that only relatively small molecules can penetrate into hair keratin and for this reason oxidative and nitro dyes have received considerable attention. In addition, the dye should impart natural appearance to the hair under a variety of lighting conditions. It should produce a minimum of scalp staining and be convenient to use. Sunlight fastness must be good, and though fading depends on the type of product, the various components of the dye mixture must fade uniformly. Though hair tends to be nonuniform in diameter, shade, and history of abuse, the color should be fairly uniform or level from root to tip.

Modern hair colorants can be divided conveniently into permanent, semipermanent, and temporary systems. These categories are characterized by the durability of the color imparted to the hair, the type of dye employed, and the method of application (see Colorants for foods, drugs, and cosmetics). Coloring preparations include plant extracts (henna), metallic dyes, and the more modern synthetic organic compounds.

Health and safety factors; government regulations. *p*-Phenylenediamine, the main component of oxidation hair dyes, is known to be a sensitizer and capable of producing contact dermatitis. Because of this property, hair colorants in the United States carry on the label as a legal requirement a warning and instructions for a 24-h patch test with the intermediates and hydrogen peroxide mixed in the same manner as in use.

Bleaches

The main coloring component of hair is the brown-black pigment, melanin, that occurs as granules (1 μm by 0.3 μm) embedded in the cortex. Natural red hair is reported to owe its color to an iron-containing pigment called trichosiderin. The aim of bleaching is to decolorize selectively the natural pigments with minimal damage to the hair matrix. Today, hair bleaches rely almost exclusively on hydrogen peroxide (see Bleaching agents; Hydrogen peroxide).

Dye removal. Removal of permanent hair dye is a difficult problem that at present has not been solved. A reducing agent used for this purpose is zinc formaldehyde sulfoxylate at a pH of 3 to 4. Urea or oxidizing agents accelerate this process.

Permanent-Waving Preparations

Wavy hair not only surpasses straight hair in opportunities for more diverse styling, but owing to its geometry, it is more luxurious and full of body and shine. Basic precepts of modern permanent waving are based on work in the 1930s on the chemical and physical properties of wool.

The chemistry of hair waving. Each hair has a geometry that is the result of processes of keratinization and follicular extrusion which transform a viscous mixture of polypeptide chains into strong, resilient, and rigid keratin fiber. In principle, waving can be viewed as a stepwise restaging of these processes as it entails softening of keratin, molding it to a desired shape, and annealing the newly imparted configuration. The underlying mechanism of waving is thus essentially molecular and involves manipulation of physico-chemical interactions that stabilize the keratin structure. Reductive fission of hair disulfides by mercaptans has become the preferred technique. Under relatively mild conditions, adequate cleavage levels can be achieved and the ensuing formation of cysteine residues provides an opportunity to reform the severed linkages by simple oxidative treatment to complete the process.

The reagent most frequently used for the reduction is thioglycolic acid although alkali sulfites are also frequently employed. The principal active ingredient of cold-wave neutralizers is an oxidizing agent, usually 1–2 wt% hydrogen peroxide.

Health and safety factors. The dermal toxicology of alkaline solutions of thioglycolic acid has been extensively reviewed. The reagent has been found harmless to normal skin when used under conditions adopted for cold waving. Hand protection is recommended for the professional hairdressers who routinely handle these products.

Hair-Straightening Preparations

Temporary hair straightening. In hot combing, the most frequently used technique, an oily substance (pressing oil) is applied to hair that is then combed under tension with a heated comb. The straightening effect is immediate but is lost quickly on exposure of the hair to moisture.

Permanent hair straightening. The basic technical premise underlying permanent hair straightening is similar to that in waving. Hair is softened, maintained straight under tension for a period of time (by means of the high viscosity of the product and repeated combing), and after rinsing, rehardened by application of the neutralizer. It is thus not surprising that many hair-straightening compositions are just thickened versions of permanent-waving products. Alkaline thioglycolate (6–8%) is formulated into a thick oil–water emulsion or cream using generous concentrations of cetyl and stearyl alcohols and high molecular weight polyethylene glycol together with a fatty alcohol sulfate as emulsifier.

An important class of permanent straighteners still in frequent use is based on alkali as an active ingredient. Great care must be exercised in the use of the alkaline relaxers as even a short contact with skin can cause blistering.

RAYMOND FEINLAND
FRANK E. PLATKO
LUCIEN WHITE
RICHARD DeMARCO
JOSEPH J. VARCO
LESZEK J. WOLFRAM
Clairol, Inc.

M.S. Balsam and E. Sagarin, eds., *Cosmetics: Science and Technology*, 2nd ed., Wiley-Interscience, New York, 1972.

N.F. Estrin, ed., *CTFA Cosmetic Ingredients Dictionary*, 3rd ed., Cosmetic, Toiletry, and Fragrance Association, Inc., Washington, D.C., 1982.

R.G. Harry, ed., *Harry's Cosmeticology*, 7th ed., edited by J.B. Wilkinson and R.J. Moore, Chemical Publishing Co., New York, 1982.

HALOGENATED FIRE RETARDANTS. See Flame retardants, halogenated.

HANSA RED. See Azo dyes.

HARDNESS

Hardness is the resistance of one body to local deformation (indentation, scratching, cutting, abrasion, or wear) by another body (see Abrasives; Corrosion and corrosion inhibitors). It usually is considered to be a relative rather than an absolute property. The relative shape, deformation resistance, and other conditions strongly influence the significance of the comparison or conflict between the indenting and the indented bodies. Hardness normally is measured by either static penetration of the specimen with a standard indenter at a known force, or dynamic rebound of a standard indenter of known mass dropped from a standard height, or scratching with a standard pointed tool under a known force. Hardness as a property is a complex combination of three absolute properties that are expressed in units of stress (force per unit area) and in strain (change in length per unit length): elastic modulus (stress/recoverable strain), yield strength (stress to cause a small permanent strain), and strain-hardening capacity (gain in stress per unit permanent strain). These properties are illustrated for different materials in Figure 1. The complexity of the interrelationships between the hardness test and the material under test dictates that the type of test, indenter, and force must be stated when the measured hardness is reported. Hardness testing is used mainly to identify, categorize, inspect, or investigate materials.

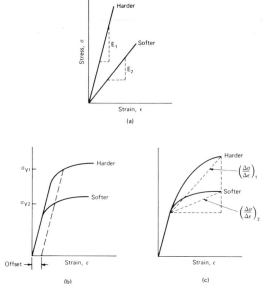

Figure 1. Stress–strain relationships for materials with different: (**a**) elastic moduli (Young's modulus), the slope of elastic stress–strain relationship; (**b**) yield stresses (at a prescribed permanent strain or offset), but same strain-hardening and elastic moduli; and (**c**) strain-hardening capacities, but same elastic moduli and elastic limits.

Indentation Hardness Tests

A hard indenter (usually hardened steel, tungsten carbide, or diamond) is brought into static mechanical equilibrium with, and at normal incidence to, the test specimen using a known force between the indenter and specimen. Tests on the width of indentation include the Brinell hardness test and the Vickers hardness test. Tests based on the depth of indentation include the Rockwell hardness test. Tests based on Dynamic rebound include the Scleroscope hardness test. Microhardness testers using loads as small as a few grams are used for evaluation of heterogeneous specimens.

Scratch Hardness Tests

The two scratch hardness tests are the Microcharacter hardness test and the Mohs scale of hardness.

General Considerations

General considerations for the use of hardness tests include accuracy and reproducibility of the test, conversion of hardness numbers, suitability of the specific test, the physical meaning of hardness, and the structure sensitivity of hardness.

Hardness Tests Developed for Specific Materials

The Pfund test is used for organic coatings such as paint. The durometer is used for rubber. The Barcol Impressor is used for rigid plastics, and the Sward Rocker is used for quantifying the softening of an organic coating caused by contact with another polymer.

GEORGE LANGFORD
Drexel University

"Hardness Testing" in H.E. Boyer, ed., *ASM Metals Handbook*, 8th ed., Vol. 11, American Society for Metals, Metals Park, Ohio, 1976, pp. 1–20 and 425–427.

1978 Annual Book of STM Standards, Part 10, American Society for Testing and Materials, Philadelphia, Pa.

G.L. Kehl, *The Principles of Metallographic Laboratory Practice*, McGraw-Hill, New York, 1949, pp. 212–265.

HEAT-EXCHANGE TECHNOLOGY

HEAT TRANSFER

Heat is transferred by conduction, convection, and radiation. Because chemical processing involves heating and cooling, a thorough understanding of the basic principles of heat transfer is essential for the chemical-process engineer. In many instances, the efficiency and economics of a chemical process may depend directly on how effectively fundamental heat-transfer principles are applied. Furthermore, the evolution of heat transfer as a unit operation represents an interesting example of the expanded scope of engineering, and of the increasing diversity of engineering disciplines (see also Evaporation; Fluid mechanics; Simultaneous heat and mass transfer).

Basic Design Equation

Most steady-state heat-transfer applications involve two fluids at different temperatures, separated by a solid barrier, such as the wall of a tube or pipe. The rate of heat flow q from the hot fluid to the cold equals the product of a proportionality factor commonly referred to as the overall heat-transfer rate U, the amount of barrier surface through which the heat flows A, and the temperature difference between the two fluids Δt.

$$q = UA \, \Delta t \qquad (1)$$

In the design of heat-transfer equipment, equation 1 applies for local conditions only, unless the three factors on the right side remain constant with respect to the flow paths of the fluids. With changing conditions, equation 1 must be expressed in differential form.

$$dq = U \Delta t \, dA \qquad (2)$$

Equations 1 and 2 are directly applicable to systems involving conductive and convective heat transfer and may be adapted easily for use in systems involving radiant heat transfer.

Rate of heat flow. The rate of heat flow q often called duty or heat exchanged, is the rate of transfer of heat energy with respect to time through the heat-transfer surface.

Overall heat-transfer rate. The overall heat-transfer rate U accounts for the multiple resistances r to heat flow that are present in a system. Such resistances include the fluids themselves, fouling deposits that may be present on the hot and cold surfaces of the barrier separating the fluids, and the barrier wall resistance itself. The overall heat-transfer rate is equal to the inverse of the sum of all resistances in series as expressed by equation 3. This equation applies directly to heat transfer through barriers with constant area normal to the direction of heat flow, eg, flat plates.

$$U = 1/(r_h + r_{hf} + r_w + r_{cf} + r_c) \qquad (3)$$

The inverse of each of these resistances is referred to as conductance, $h \, (h = 1/r)$. The higher the conductance, or lower the resistance, the greater is the ability of the fluid or material involved to transmit heat.

In the case of the hot and cold fluids, the conductances, h_h and h_c are generally referred to as film coefficients. These coefficients depend on the mode of heat transfer, ie, conduction, convection, radiation, or a combination thereof. Numerous other factors influence fluid-film coefficients, including but not limited to system geometry, fluid-flow velocities, fluid physical properties, difference between bulk fluid and wall temperature, etc.

The steady-state rate of heat flow, q, is equal through all series resistances.

Temperature difference. Equation 1 is strictly valid only under constant conditions of overall heat-transfer rate, temperature difference, and heat-transfer area along the flow path of the fluids; otherwise, equation 2 must be used.

Assuming all above criteria are met except that of constant temperature difference, an expression for overall effective temperature difference for fluids in countercurrent flow, in terms of the hot- and cold-fluid inlet and outlet temperatures, may be derived. Equation 5 assumes each fluid exhibits a straight-line relationship between heat release (or gain) and temperature over its entire cooling (or heating) range. Sensible cooling (or heating) with constant values of fluid specific heat capacity, or isothermal condensation or boiling, would result in such linear relationships. From equation 2,

$$\int_o^q dq = U \int_o^A (t_h - t_c)\, dA \tag{4}$$

With the appropriate substitution of heat-balance relationships, integration of both sides, and rearrangement, the following relationship results:

$$q = UA \left[\frac{(t_{hi} - t_{co}) - (t_{ho} - t_{ci})}{\ln \dfrac{(t_{hi} - t_{co})}{(t_{ho} - t_{ci})}} \right] \tag{5}$$

Therefore, the effective overall temperature difference under the above conditions, commonly referred to as the log-mean temperature difference is

$$\Delta t_{lm} = \frac{(t_{hi} - t_{co}) - (t_{ho} - t_{ci})}{\ln \dfrac{(t_{hi} - t_{co})}{(t_{ho} - t_{ci})}} \tag{6}$$

Heat-transfer surface area. The heat-transfer surface area A is provided by the barrier wall between the hot and cold fluids through which the flow of heat occurs. The direction of heat flow through the barrier wall is normal to the heat-transfer surface. For tubular equipment, the heat-transfer area is generally based on the tube outside surface. The overall rate for tubular heat-transfer equipment with resistances referred to the tube outside surface is

$$U = 1 \left/ \left[\left(\frac{A_o}{A_i} \right) r_i + \left(\frac{A_o}{A_i} \right) r_{if} + \left(\frac{A_o}{A_{lm}} \right) p_w + p_{of} + p_o \right] \right. \tag{7}$$

Conduction

Conduction refers to the molecular transmission of heat through a body from point to point within the body, or from one body to another in direct contact with it. Conduction is generally limited to the transfer of heat through a solid such as a metal tube wall, or a series of solids in direct contact with each other such as in a multicomponent refractory wall. Heat transfer by conduction may, however, occur in stagnant liquids or gases when no mixing occurs.

Fourier equation. The basic theory of heat transfer by conduction was established over a century ago by Fourier who expressed it as follows:

$$q = dQ/d\theta = -kA(dt/dx) \tag{8}$$

It states that the rate of heat flow is equal to the thermal conductivity k multiplied by the area of heat transfer and the temperature gradient. A is the surface area normal to the direction of heat flow and dt/dx is the rate of change of temperature with distance in the direction of heat flow.

Convection

In convection, heat transfer is effected by movement of material. Obviously, such heat transfer occurs in liquids and gases only.

Movement of fluids in connection with convective heat transfer occurs because of density differences between portions of the fluid at different temperatures or an external driving force such as a pump, fan, or agitator mixing the fluid. In the case of a pump or fan, mixing occurs as the result of downstream resistance to fluid flow at the heat-transfer surface, rather than at the impeller or fan blade itself. The former condition is commonly referred to as natural or free convection, the latter as forced convection.

Condensation. The change of phase of a saturated vapor to its liquid state is termed condensation. Transfer of heat associated with condensation is generally classed as convective; however, conduction is often a significant contributor to the heat-transfer process. Heat transfer from a condensing vapor is widely experienced in chemical processing applications. Steam heaters, overhead condensers in distillation systems, and vent condensers are but a few examples (see also Distillation).

Boiling. The change of phase of a saturated liquid to its vapor state is commonly termed boiling or vaporization. Transfer of heat during

Figure 1. Continuous circular cross fin. Courtesy of Escoa Fintube Corporation.

boiling is appropriately classed as convective, since substantial fluid turbulence is generally developed. Boiling is commonly found in chemical process applications and is often the mechanism by which the heat to supply basic process needs is transferred. For instance, steam boilers generate a most versatile heat source and column reboilers provide the heat required for distillation processes.

Temperature Difference

The temperature difference existing between two fluids is the driving force by which heat is transferred from the higher temperature fluid to the one at lower temperature. Without a temperature difference, there can be no transfer of heat. Because of its importance, the effective mean temperature difference or driving force has to be established before the amount of heat-transfer surface required can be determined for a particular application.

By use of equations 4–6, the logarithmic mean temperature difference Δt_{lm} is presented as the applicable overall temperature difference provided the following assumptions apply: (1) constant overall heat-transfer rate U over the entire heat-transfer surface area; (2) linear heating and cooling enthalpy–temperature profiles over the entire heating and cooling ranges of the fluids; and (3) true countercurrent flow direction of the fluids over the entire heat-transfer surface area.

Finned heat-transfer surface. When the heat-transfer film coefficient on one side of a surface is much lower than on the other wide, it is often economical to use fins on the low coefficient side to increase the surface exposed to the low film coefficient fluid. In cylindrical tubing, the fins are normally installed or formed on the outer surface; however, internally finned tubes also are available. Figure 1 shows the continuous circular cross fin. This type of fin is widely used in applications involving the recovery of heat from fuel gases, waste gas streams, gas-turbine exhaust, and even low pressure process streams.

Equipment for Convective Heat Transfer

A large number of equipment types are available that transfer heat by convection. Selection depends on the particular application and an economic evaluation. Equipment includes the double-pipe exchanger, the shell-and-tube exchanger, the fire-tube boiler, the falling-film exchanger, the fixed-bed catalytic reactor, the fluidizied-bed reactor, the compact exchanger, the spiral-tube exchanger, the plate exchanger, the spiral-plate exchanger, jacketed vessels, the air-cooled exchanger, and the trombone cooler.

Radiation

Radiant heat transfer does not depend on an intervening medium and can, in fact, transfer heat across an absolute vacuum.

Stefan-Boltzmann Law. The rate at which radiant energy is released by a body is directly proportional to the amplitude of oscillation of its molecules. Since molecular movement or oscillation increases with a rise in temperature, radiation from a body increases with the body temperature.

The rate W_B at which radiant heat energy is emitted by a perfect black body may be determined from the Stefan-Boltzmann law as follows,

$$W_B = \sigma T^4 \qquad (9)$$

where σ is the Stefan-Boltzmann constant, 5.73×10^{-8} W/(m$^2 \cdot$K^4), and T is the surface temperature of the body.

Emissivity. Most materials radiate less heat than a perfect black body at the same temperature. Emissivity or emittance ϵ expresses the ratio of the heat emitted by a body to the heat that would be emitted by a perfect black body at the same temperature.

Absorptivity. A perfect black body absorbs all of the incident thermal radiation to which it is exposed, whereas a nonperfect black body absorbs only a fraction of such incident thermal radiation. The fraction of the total incident radiant heat that is absorbed by any body is defined as the absorptivity or absorptance α which varies not only with the temperature and type of surface of the body, but also with the wavelength distribution as determined by the effective temperature of the incident radiation.

Radiation heat-transfer coefficient. Consider a gray body at temperature T_1 with surface A_1 of emissivity and absorptivity ϵ_1 completely enclosed by black-body surroundings at temperature T_2. The radiation emitted by the gray body is given by $\sigma \epsilon_1 T_1^4 A_1$. The heat absorbed from the black-body surroundings is given by $\sigma \epsilon T_2^4 A_1$. Hence, the net heat released from the gray body is given by

$$q = \sigma \epsilon_1 \left(T_1^4 - T_2^4 \right) A_1 \qquad (10)$$

Industrial boilers and furnaces. Steam boilers and process furnaces transfer a large percentage of the total heat by direct radiation to the heat-absorbing surfaces from a flame. The receiver must be able to "see" the emitting source since radiation travels essentially in straight lines.

Combined heat-transfer mechanisms. In industrial boilers and furnaces, all three basic heat-transfer mechanisms occur in series: radiation emitted from the flue gases and absorbed by the outside tube surface, conduction through the tube metal and through any fouling deposits on the surfaces, and convection within the fluid being heated inside the tube. No one mechanism is more or less significant than another.

Nomenclature

A = heat-transfer surface area, m^2
h = heat-transfer conductance, film coefficient, W/(m$^2 \cdot$K)
k = thermal conductivity, W/(m$\cdot$K)
q = rate of heat flow with respect to time, W
Q = amount of heat energy, W$\cdot$s
r = resistance to heat flow, (m$^2 \cdot$K)/W
t = temperature, °C
T = absolute temperature, K
U = overall heat-transfer rate, W/(m$^2 \cdot$K)
ϵ = emissivity
θ = time, s
W_B = radiant heat flux, W/m^2
x = distance in direction of heat flow, m
α = absorptivity
Subscripts
σ = Stefan-Boltzmann constant, W/(m$^2 \cdot$K^4)
c = cold side, cold end, condensate, critical
cf = cold-side fouling
ci = cold-side inlet
co = cold-side outlet
h = hot side, hot end
hf = hot-side fouling
hi = hot-side inlet
i = inside
if = inside fouling
lm = logarithmic mean
o = outside
of = outside fouling
w = wall

J.P. FANARITIS
J.A. KWAS
A.T. CHASE
Struthers Wells Corp.

W.M. Rohsenhow and J.P. Harnett, *Handbook of Heat Transfer*, McGraw-Hill Book Co., New York, 1973.

W.H. McAdams, *Heat Transmission*, 3rd ed., McGraw-Hill Book Co., 1954.

R.H. Perry and C.H. Chilton, *Chemical Engineers' Handbook*, 6th ed., McGraw-Hill Book Co., New York, 1983.

HEAT-TRANSFER MEDIA OTHER THAN WATER

The economics of using water as a heat-transfer medium must be evaluated for each installation; at temperatures below 0°C or above 200°C, it is desirable to consider heat-transfer media other than water.

High Level Heat-Transfer Media

The ideal high level heat-transfer medium would be low in cost, noncorrosive to common materials of construction, nonflammable, nontoxic, ecologically safe, and thermally stable. It also would remain liquid at winter ambient temperatures and would afford high rates of heat transfer. In practice, the value of a heat-transfer medium depends upon several factors: its efficiency, its thermal stability at the service temperature, its adaptability to various systems, and certain of its physical properties, eg, vapor pressure, freezing point, flash point, fire point, and autoignition temperature.

Vapor-Phase and Liquid-Phase Operations

When establishing whether liquid-phase or vapor-phase systems are better, it is necessary to consider the overall process and economics, the thermal tolerance of the process, and the required equipment. In many cases, the costs for the two types of systems do not differ significantly. With vapor-phase systems, heat is transferred at the saturation temperature of the vapor, which affords uniform and precisely controlled temperatures. With liquid-phase systems, the temperature of the fluid necessarily changes as heat is transferred; therefore, temperatures are not uniform even if large circulation rates are employed for the heat-transfer fluid. In systems with multiple heat users, a combination of both vapor and liquid phases may be preferred. For small, compact systems, vapor-phase systems generally are preferred. For large systems, heat losses from fluid piping may be greater for vapor-phase systems. Vapor-phase systems frequently require forced-circulation condensate return when there are several heat users at different temperature levels.

Principal advantages of liquid-phase systems over vapor-phase systems are (1) no condensate-return equipment is required; (2) simpler and more easily operable systems result when heating and cooling must be alternated; (3) there is no temperature gradient as a result of pressure drop in the supply piping; (4) liquid systems afford a positive flow through equipment and minimize problems associated with improper venting; and (5) liquid phase eliminates the problems associated with condensate removal from complex geometries.

Principal advantages of vapor-phase systems over liquid-phase systems are (1) vapor-phase systems provide much more heat per kilogram of heat-transfer fluid; (2) condensing or boiling affords more uniform heat removal or addition and more precise temperature control; (3) vapor-phase heat transfer has an advantage when using equipment that does not permit easy control of liquid flow pattern and velocity; (4) natural-circulation systems can be employed, thereby obviating pumps; (5) vapor systems require lower working inventories of the heat-transfer fluid; and (6) vapor systems frequently permit higher rates of heat transfer.

Petroleum oils. The most widely used heat-transfer medium at temperatures above those obtained with moderate-pressure steam is a high-

Table 1. Physical Properties of Commonly Used Petroleum Oils

	Mobiltherm Light	Mobiltherm 603	Caloria HT 43	Thermia Oil C
pour point, °C	−29	17	−9	−12
flash point, °C	116	193	204	235
autoignition temperature, °C	410	349		
recommended maximum bulk temperature, °C	204	288	316	288
recommended maximum fluid film temperature, °C	218	304	360	316
vapor pressure at the recommended maximum bulk temperature, kPa[a]	21.3	21.3	80.0	4.7

[a] To convert kPa to mm Hg, multiply by 7.5.

Table 2. Physical Properties of Synthetic Heat-Transfer Fluids

	Dowtherm A	Dowtherm G-20	Marlotherm L	Therminol 60	Therminol 66	KSK-330
molecular weight	166	215	182	250	240	250
initial freezing (crystal) point, °C		−6.5				
freezing point, °C	12					
pour point, °C			−70	−68	−28	−30
flash point, °C	116	152	120	154	180	165
fire point, °C	135	157		160	194	
autoignition temperature, °C	621	555	500	445	374	
recommended maximum bulk temperature, °C	400	370	350	315	340	320
recommended maximum film temperature, °C				370	335	375	350
vapor pressure at maximum bulk temperature, kPa[a]	1051	310	320	101.4	86.7	99.4

[a] To convert kPa to mm Hg, multiply by 7.5.

boiling petroleum fraction; several oils are used. In general they are safe, essentially nontoxic, relatively low cost, noncorrosive fluids that have been refined to standard physical-property specifications for heat-transfer service. Table 1 summarizes the physical characteristics of these oils.

Synthetic fluids. Synthetic fluids are safe, noncorrosive, essentially nontoxic, and thermally stable when operated under recommended conditions. Generally, they are more expensive than petroleum oils, but usually they can be reprocessed to remove degradation products. There are several classes of chemicals offered that permit a wide temperature range of application. Examples of synthetic fluids include Tetralin, Dowtherm heat-transfer fluids, Therminol heat-transfer fluids, Marlotherm heat-transfer fluids, KSK-oils, Ucon HTF-500, Ucan Thermofulid 17, and Hitec heat-transfer salt (see Table 2).

Gases and liquid metals also are used as heat-transfer media. Commonly used gases include air, flue gases, nitrogen, carbon dioxide, hydrogen, helium, and argon as well as superheated steam. The most commonly used liquid metal is sodium–potassium eutectic.

Comparison of heat-transfer fluids. A large number of heat-transfer fluids are available for use in moderately high temperatures (100–300°C). Several of these are utilized at temperatures up to, and sometimes exceeding, 400°C. A dozen fluids may fulfill the operating requirements of a specific application. Final fluid selection should be based on heat-transfer rate, operating pressure drop, and system cost.

Low Level Heat-Transfer Media

Refrigeration is required to cool to temperatures lower than those attainable with cooling water or ambient air. Low temperature processing also requires a suitable low temperature fluid. Several fluids are used for both of these types of services. There are several types of refrigeration systems, each of which requires a suitable working fluid or refrigerant [eg, gas-cycle systems (air); steam-jet systems (water); absorption systems (ammonia–water and water–lithium bromide); and mechanical compression systems (ammonia, halocarbon compounds, hydrocarbons, carbon dioxide, sulfur dioxide and cryogenic fluids)]. Refrigerants absorb heat not wanted or needed and reject it elsewhere.

Secondary coolants (brines). In many refrigeration applications, heat is transferred to a secondary coolant that is in turn cooled by the refrigerant. The most commonly used secondary coolant is ethylene glycol.

Heat pumps involve the application of external power to pump heat from a lower temperature to a higher temperature.

Thermal engine cycles operating with organic refrigerants are employed to recover energy from waste-heat streams at temperatures below 150°C. The most frequently used refrigerants are halocarbons and hydrocarbons (see Refrigeration).

PAUL E. MINTON
CHESTER A. PLANTS
Union Carbide Corp.

ASHRAE Handbook of Fundamentals, American Society of Heating, Refrigerating, and Air-Conditioning Engineers, New York.

R.C. Lord, P.E. Minton, and R.P. Slusser, "Design of Heat Exchangers," *Chem. Eng.*, 96 (Jan. 26, 1970).

P.L. Geiringer, *Handbook of Heat Transfer Media*, Reinhold Publishing Corp., New York, 1962.

HEAT PIPE

The heat pipe is a heat-transfer device that is receiving increasing attention for application in the chemical industry. The current use is primarily in regenerative systems where heat losses from combustion processes are recovered by preheating inlet air (see Burner technology). In its simplest form, the heat pipe possesses the property of extremely high thermal conductance, often several hundred times that of the metals. It makes an almost ideal heat-transfer element. In another form, it can provide positive, rapid, and precise control of temperature under conditions that vary with respect to time.

The heat pipe is self-contained, has no mechanical moving parts, and requires no external power other than the heat that flows through it. A typical heat pipe may require as little as one thousandth the temperature differential needed by a copper rod to transfer a given amount of power between two points. The heat pipe has been called a thermal superconductor.

Principles of Operation

The heat pipe achieves its high performance through the process of vapor-state heat transfer. A volatile liquid employed as the heat-transfer medium absorbs its latent heat of vaporization in the evaporator (input) area. The vapor thus formed moves to the heat output area, where condensation takes place. Energy is stored in the vapor at the input and released at the condenser.

The heat pipe consists of the following components: a closed, evacuated chamber, a wick structure of appropriate design, and a thermodynamic working fluid with a substantial vapor pressure at the desired operating

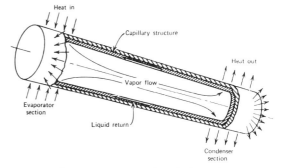

Figure 1. Cutaway view of a heat pipe.

temperature. A schematic drawing of an elemental heat pipe is shown in Figure 1 (p. 581). The following basic condition must be satisfied for proper operation:

$$\Delta P_c > \Delta P_l + \Delta P_v + \Delta P_g$$

That is, for liquid return, the pressure difference owing to capillarity ΔP_c must exceed the sum of the opposing evaporator-to-condenser pressure differential in the vapor ΔP_v, plus the pressure differential in the liquid caused by gravity ΔP_g, plus that caused by frictional losses ΔP_l.

Selection of Materials

Working fluid. Qualitatively, for high power throughout under typical operating conditions, it is advantageous to have a high latent heat of vaporization, high surface tension, high liquid and vapor densities and low liquid and vapor viscosities.

Operating lifetime. The operating lifetime of a given heat pipe is usually determined by corrosion mechanisms (see Corrosion and corrosion inhibitors). Life of tens of thousands of hours can generally be expected.

Wick. The selection of a suitable wick for a given application involves consideration of its form factor or geometry. The basic material generally is chosen on the basis of the wetting angle and compatibility considerations.

Vessel. The vessel in which a heat pipe is enclosed must be impermeable to assure against loss of the working fluid or leakage into the heat pipe of air or other undesired materials from the external environment. It should conform with established pressure-vessel codes, considering both rupture and creep strengths. The vessel, as well as the wick, must be compatible with the working fluid. High temperature heat pipes for operation in hydrogen-containing environments, including flames, must be protected by a nonmetallic (eg, glass or ceramic) material.

G. YALE EASTMAN
DONALD M. ERNST
Thermacore, Inc.

G.Y. Eastman, "The Heat Pipe," *Sci. Am.* **218**(5), 38 (1968).
P.D. Dunn and D.A. Reay, *Heat Pipes*, Pergamon Press, New York, 1976.

NETWORK SYNTHESIS

The relatively new discipline of process synthesis has spawned a technology that can be used to solve the problem of efficient energy utilization in process-plant design. The process of developing a good heat-exchange network is most easily viewed as a multiple-tier optimization problem. Minimum cost is the objective. Efficient network-synthesis techniques relieve the process engineer of the burden of accepting designs based on art that cannot be shown to be superior. The opportunities for process improvement are best before the structure of the network is determined. Recent developments in network analysis have progressed to the point that energy use in turbines can also be considered along with heat recovery.

Problem Specification

The impetus for heat-exchange-network design was the development of an adequate means to represent the problem. The temperature and heat (enthalpy) relationship for any process stream can be represented on a temperature–enthalpy diagram. Figure 1 shows four different streams. Stream A is a pure component that is condensing (eg, steam). Streams B and C are streams with constant heat capacity (C_p) that are to be heated (B) or cooled (C). Stream D represents a multicomponent mixture that changes phase as it is cooled. Individual streams can be lumped together in an ensemble of single hot and single cold composite streams.

Alternative representations include the heat-content diagram, which represents each stream as an area on a graph.

The flow sheet is a means of representing the matching of streams for heat transfer that is traditionally used by process engineers.

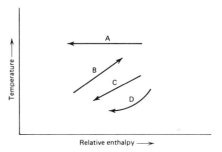

Figure 1.   Temperature–enthalpy representation of streams.

Idealization. The final rating or design of a heat exchanger requires detailed knowledge of the fluids to be heated and cooled as well as information on the specific type of exchanger. For the purpose of network synthesis, the overall heat-transfer coefficient is usually idealized as a constant value. This independence makes possible the iterations necessary to solve the network problem.

Limiting network conditions. Heat-exchange-network design has been made easier by the development of various limiting conditions. The technique of limiting conditions is common in many areas of chemical engineering and can provide great insight into a problem.

A network must be in heat balance, but more important is the establishment of the utility requirements, which can be determined with the help of the concept of composite streams.

The limiting utility rates for any given temperature of approach at the pinch (ΔT_p) can be found from the temperature-enthalpy diagram. A low temperature of approach for the network reduces utilities but raises heat-transfer area requirements. For most of the published problems, utility costs are normally more important than annualized capital costs.

If idealized double-pipe exchangers are used, a heat-exchange network with minimum area can quickly be developed for any ΔT_p. In the limiting case, where all heat-transfer coefficients are assumed to be equal, the area for this network can easily be obtained from the composite streams by the following integration:

$$A_{\Delta T_p} = [1/U]\int_{T_{\text{lowest}}}^{T_{\text{highest}}}\left[dQ/(T_j - T_i)\right]$$

where A is the minimum possible network area at the given ΔT_p, U is the heat-transfer coefficient, and Q is the heat transferred. Methods have been developed that are able to generate a minimum-area network for any given energy recovery (ΔT_p)

The smallest number of matches or exchangers that are required in a network can be developed as a limit. The number needed (E_{min}) is generally one less than the total number of streams (process and utility) involved in the network:

$$E_{\text{min}} = S_{\text{process}} + S_{\text{utilities}} - 1$$

Unfortunately, the minimum number of exchangers is not the same as the number of shells required if conventional shell-and-tube heat exchangers are used. If ΔT_p is large, the difference is not significant. However, as more energy is recovered, ΔT_p must be reduced.

Synthesis algorithms. The synthesis of heat-exchange networks takes place after the feasibility of developing a network has been established for the selected ΔT_p and other constraints.

Synthesis methods can be classified in many different ways: combinatorial methods express the synthesis problem as a traditional optimization problem. Heuristic methods rely on the application of sets of rules to lead to a specific objective. Inventive synthesis methods develop networks by exploiting selected characteristics of heat exchange among groups of hot and cold streams. Evolutionary methods develop new heat-exchange networks from those generated by other means.

Network Optimization

Process calculations for traditional unit-operations equipment can be divided into two types: design and performance. Sometimes the performance calculation is called a simulation (see Simulation and process

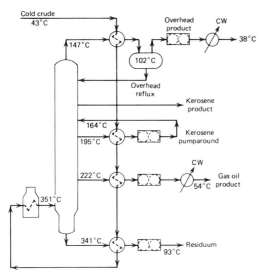

Figure 2. Schematic diagram of a crude distillation unit.

design). Detailed network calculations are time-consuming and computer-aided methods are required to evaluate the alternatives in the optimization process.

However, many of the calculations necessary for heat-exchange-network design can be done by hand. A clear understanding of the methodology involved can be most easily accomplished by developing curves and networks on a sheet of graph paper.

The earliest use of heat-exchange-network synthesis was in the analysis of crude distillation units. A crude-oil feed stream and five hot streams leaving the crude column are shown in Figure 2.

Heat-exchange networks are concerned with just a portion of the total energy in a process plant. As indicated earlier, significant steps have been made toward integrating other parts of the energy system. Ideally, a process plant should be examined for its total energy consumption.

Most separations are conducted by distillation, and there is considerable interest in the process of designing sequences of distillation columns.

EDWARD HOHMANN
California State Polytechnic University

E.C. Hohmann, *Optimum Networks for Heat Exchange*, PhD thesis, University of Southern California, Los Angeles, Calif., 1971.

B. Linnhoff and J.A. Turner, *Chem. Eng.* **88**(22), 56 (1981).

B. Linnhoff and co-workers, *A User Guide on Process Integration for Efficient Use of Energy*, Institute of Chemical Engineers, Rugby, UK, 1982.

HEAT-RESISTANT POLYMERS

The study of heat-resistant polymers as a scientific discipline had its start in the late 1950s and several promising materials were discovered rather quickly. Two experimental materials—HT-1, aromatic polyamide fiber, and H-film, aromatic polyimide film—were introduced by the DuPont Co. and later commercialized as, respectively, Nomex fiber and Kapton film. By the late 1960s, it clearly was evident that the limits of stability of the carbon–carbon bond in particular and the poor resistance to oxidation of carbon compounds in general would not permit polymers having long-term stability much in excess of 400°C (300°C in air) and even short-term stability much in excess of 600°C.

Perhaps ironically, the most important commercial uses of the heat-resistant polymers have little or nothing to do with their useful thermal properties. Thus, fibers of aromatic polyamides are being used in flame-resistant fabrics (see Flame-retardants for textiles), in reinforced composites of the flexible (eg, tires, V-belts) and rigid (eg, panels) types, and in separation of both gaseous and liquid mixtures (see also Membrane technology; Tire cords).

Thermal Stability

The most important criterion of thermal stability is the retention of physical properties in a given environment at an elevated temperature for a specified time. In general, physical properties of heat-resistant polymers that are highly desirable are a high glass-transition temperature ($\geq 250°C$); a high melting point ($\geq 400°C$); a high decomposition temperature ($\geq 400°C$) as determined by differential thermal analysis (dta); and a high resistance to weight loss as measured by thermogravimetric analysis (tga). As a rule of thumb, a material in its ultimate application may be said to be thermally stable if it can retain its properties at 260°C for ca 1000 h, at 538°C for 1 h, or at 815°C for 5 min.

Synthesis and Properties

Historically, the use of ring structures has led the way to higher melting polymers. Thus, incorporation of terephthalic acid in polyesters transformed the low melting compositions pioneered by Carothers into materials having useful melting points and greatly improved hydrolytic stability (see Polyesters, thermoplastic). It is for similar reasons that polymers of maximum heat resistance contain phenylene rings.

Heat-resistant polymers having reasonable tractability are prepared by use of phenylene rings joined by means of condensation linkages. Among the phenylene rings, the para-oriented isomer usually provides the optimum heat resistance because of its rigidity, contribution to polymer crystallinity, and ability to participate in extended resonance conjugation.

In general, three broad types of entirely aromatic polymers have been prepared (Table 1). For the sake of simplicity, these polymers are simple condensation polymers, heterocyclic polymers, and combination condensation–heterocyclic copolymers such as poly(amide-imide)s and poly(ester-imide)s. More complex forms of the heterocyclic polymers are called ladder polymers because of their double strands of linkages; in view of the large amount of literature associated with them, these are treated as a separate class.

Table 1. Heat-Resistant Polymers Grouped According to Broad Types

Structure	Names
Simple condensation polymers	
⎡Ar—X⎤	polyamides
	polyazomethines
	polyesters
	polyhydrazides

Ar = ⬡ , ⟨◯⟩ , ⟨◯—◯⟩ , ⟨◯◯⟩ , etc , ◯—Z—◯

Z = —O— , —S— , —C(=O)— or absence of functional group

X = —C(=O)—NH— , —C(=O)—O— , —C(=O)—NH—NH—C(=O)— , —CH=N—

Heterocyclic polymers	
⎡Ar—Z⎤	polybenzimidazoles
	polybenzoxazoles
	polyimides
	polyoxadiazoles
	polyquinoxalines
	ordered heterocyclic copolymers

Ar = ⬡ , ◯—◯ , ◯—◯ , ◯◯ , ◯—O—◯ , etc

X = —O— , —S— , —N(R)—

X' = —O— , —SO₂— , —C(=O)— or absence of functional group

R = H, CH₃

Ladder polymers	pyrrones
	polyquinoxalines

Inorganic polymers	
	organic radicals + B, P, Si, etc

Health and Safety Factors

The heat-resistant polymers probably represent little in the way of a health risk. However, some of the monomers employed, eg, aromatic diamines and tetraamines, must be handled with great care because of their potential risk as carcinogens. In particular, solutions of low molecular weight polymers, precursors, or the various monomers should be used with care. Furthermore, the solvents for the various polymers or their precursors pose some risk because of their toxicity and care must be used to ensure adequate ventilation when processing the polymers from solution.

J. PRESTON
Monsanto Triangle Park
Development Center, Inc.

J.E. Mulvaney, "Heat-Resistant Polymers" in N.M. Bikales, ed., *Encyclopedia of Polymer Science and Technology*, Vol. 7, Wiley-Interscience, New York, 1967, pp. 478–506.

A.H. Frazer, *High Temperature Resistant Polymers*, Wiley-Interscience, New York, 1968.

C. Arnold, Jr., "Stability of High-Temperature Polymers," *J. Polym. Sci. Macromol. Reviews* **14**, 265 (1979).

HEAT STABILIZERS

Heat stabilizers are chemical additives used to protect certain polymers, eg, halogenated thermoplastic resins, particularly those that contain regular, repeating units of chlorine or bromine, against the effects of heat or exposure to uv radiation.

By virtue of their chemical compositions and modes of function, many heat stabilizers also act as light-stabilizing agents, often to the extent that additional, specific organic light-stabilizing agents (ie, light screeners, uv-radiation absorbers, triplet-state energy quenchers, etc) are rendered unnecessary for many commercial applications (see Uv stabilizers). Some heat stabilizers interact synergistically with a variety of organic light-stabilizing agents, and others act in an antagonistic manner.

Deterioration of color, rheological characteristics, mechanical properties, electrical properties, chemical resistance, optical properties and, indirectly, resistance to long-term aging and weathering are all aspects of heat-induced deterioration. The basic cause underlying this deterioration is chemical structure. End groups containing allylic chlorine structures frequently result from some of the free-radical termination reactions that occur as a consequence of polymerization. Tertiary chlorines and internal allylic chlorines also may be present and these are suspected of being initiating sites for eventual degradation.

Function of Stabilizers

Polymer deterioration proceeds by free-radical processes and ionic processes. Arrestive stabilization likewise proceeds by both routes although, in PVC, ionic mechanisms appear to play the principal role in arresting sequential changes that are thought to be autocatalytically induced. Stabilizing agents, depending on their structure, may function in any one or combination of the following ways. It should be understood that not all of the reactions that may occur are necessarily helpful reactions in a stabilization sense and, further, that some reactions may work adversely to the objective, occasionally depleting the active concentration of useful stabilizing reagent, and thus reducing the effectiveness of that reagent (see also Vinyl polymers, vinyl chloride, poly(vinyl chloride)).

Reactions

Reactions include hydrogen chloride neutralization which is common to all PVC-stabilizing agents or coagents. The preponderance of PVC-stabilizer systems contain one or more metallic salts or soaps (ie, salts of weakly acidic fatty acids) that react with HCl:

$$M(X)_n + n\ HCl \rightarrow M(Cl)_n + n\ HX \qquad (1)$$

where M = metal, and X = acid anion.

Other reactions include displacement of labile chlorine atoms from the resin chain, addition to unsaturated sites on the resin, antioxidation, and complex formation (chelation).

Stabilizer interaction (synergism). The mode of action of a mixed heavy metal-alkaline earth metal stabilizer system can be described by the following reaction scheme. Cadmium is used to represent the heavy metal and barium the alkaline-earth metal component, although alternatives for both are widely used

$$(2)$$

$$(3)$$

Also (coincidentally with eq. 3):

$$Ba(OOCR)_2 + 2\ HCl \rightarrow 2\ RCOOH + BaCl_2 \qquad (4)$$

Equation 4 is a specific example of the acid neutralization discussed and illustrated in general form as equation 1.

Knowledge of these interactions has been used to interpret the synergism that one observes when such substances are used in combination to stabilize PVC. Optimum stabilization is achievable only when the additives are used in low concentrations and within fairly closely regulated ratios (usually between 2:1 and 1:2 of barium to cadmium, eg perhaps slightly different when alternative metals are used). Higher concentrations of stabilizer lead to a diminishing-returns situation and may be disadvantageous if the heavy-metal concentration in the resin becomes too high, regardless of the alkali or alkaline-earth carboxylate concentration.

Economic and Production Aspects

Table 1 lists the main United States producers and distributors in each stabilizer category.

Health and Safety Aspects

Toxicological considerations in stabilizer selection and use are of great concern to food packagers. The acceptability of specific foodstuff materials is decided by the FDA. Also, the Meat Inspection Division of USDA judges the suitability of meat-wrap film and thermoformed meat packages and containers. Other regulatory agencies include the Bureau of Alcohol, Tobacco, and Firearms, and EPA, the Office of Toxic Substances, the OSHA, the NIOSH, and the CPSC. The National Sanitation Foundation indirectly exerts control over what stabilizers and other compounding ingredients are used in PVC water pipes (see Regulatory agencies).

Commercial Stabilizers for Poly(vinyl Chloride)

Commercial stabilizers include: (*1*) lead stabilizers (eg, basic lead carbonate, tribasic lead sulfate, basic lead silicate/sulfate (24 wt% SiO_2 and 60 wt% SO_3), dibasic lead phthalate, dibasic lead stearate, and dibasic lead phosphite); (*2*) mixed-metal combinations, eg, barium bis(4-nonylphenoxide) 23.0 wt%, cadmium 2-ethylhexanoate 6.5 wt%, zinc 2-ethylhexanoate 3.5 wt%, isodecyl alcohol 10.0 wt%, mineral spirits 33.0 wt%, and diphenyl decyl phosphite 24.0 wt%, also coprecipitated barium–cadmium soap of C_{10}–C_{14} mixed fatty acids (1.5:1 metal ratio) 80 wt%, pentaerythritol (finely ground) 17 wt%, and phenolic antioxidant (ie, bisphenol-A) 3 wt%; also calcium stearate 29 wt%, zinc stearate 19

Table 1. U.S. Stabilizer Producers, According to Type

Lead	Mixed-metal	Organotin	Antimony	Organic
American Cyanamid	Ferro	Cardinal	Interstab	various, depending
Associated Lead	Interstab	Carstab	M & T	on specific types
Eagle-Picher	Vanderbilt	Interstab	Synthetic	involved;
Hammond	Synthetic	M & T	Witco	ordinarily
	Tenneco	Witco		supplied under
	Witco			chemical name

wt%, 2,6-di-*tert*-butyl-4-methylphenol 2 wt%, and epoxidized soybean oil 50 wt%—this combination for FDA-sanctioned applications); (*3*) organotin stabilizers (all derivatives of tetravalent (stannic) tin); and (*4*) antimony stabilizers.

LEONARD I. NASS
Consultant

L.I. Nass in L.I. Nass, ed., *Encyclopedia of PVC*, Vol. 1, Marcel Dekker, Inc., New York, 1976, Chapts. 8–9.

HELIUM-GROUP GASES

The helium-group gases are helium (He), neon (Ne), argon (Ar), krypton (Kr), xenon (Xe), and radon (Rn), members of group VIIIA (sometimes called group 0) of the periodic table, and they are characterized by completely filled valence electrons shells. Historically, they have been called the rare gases and the noble or inert gases. But although comparatively rare, krypton, xenon, and radon are not completely inert; all three form stable molecules with highly electronegative materials (see Helium-group gas compounds). And although inert enough, helium and argon are not rare; both are bulk items of commerce.

However, when helium-bearing natural gases are eventually depleted and the atmosphere becomes its only source, helium will return to being a truly rare gas.

Occurrence

In the cosmos, helium is plentiful; it is a principal product of the nuclear fusion reactions that are the prime source of stellar energy. The other members of the helium-group gases are thought to have been created like other heavier elements by further nuclear condensation reactions occurring at the extreme temperatures and densities found deep within stars and in supernovas. On earth, the only practical sources of the stable helium-group gases are the atmosphere and certain helium-bearing natural gases.

Physical Properties

All of the helium-group elements are colorless, odorless, and tasteless gases at room temperature and pressure. They are monatomic and are considered to have perfect spherical symmetry. Some of the physical properties of the helium-group elements are summarized in Table 1.

Quantum-mechanical effects. The very light gases show significant deviations from the classical law of corresponding states, especially at cryogenic temperatures. This anomalous behavior is caused by quantum-mechanical effects that become increasingly significant with decreasing

Table 1. Physical Properties of the Helium-Group Elements[a]

Property	^{3}He	^{4}He	Ne	Ar	Kr	Xe	Rn
atomic weight, ^{12}C	3.0160	4.0026	20.183	39.948	83.80	131.30	222
critical point							
temperature, K	3.324	5.2014	44.40	150.86	209.4	289.74	378
pressure, kPa[b]	116.4	227.5	2654	4898	5502	5840	6280
density, kg/m^3	41.3	69.64	483	535.7	908	1100	
normal bp, K	3.1905	4.224	27.102	87.28	119.79	165.02	211
triple point (tp)							
temperature, K	no tp	no tp	24.562	83.80	115.76	161.37	202
pressure, kPa[b]	no tp	no tp	43.37	68.90	73.15	81.66	[70]
density							
gas at 101.3 kPa[b] and 0°C, kg/m^3	0.1347	0.17850	0.9000	1.7838	3.7493	5.8971	9.73
gas : liquid volume ratio,[c]	437.4	700	1340	781	644	518	452
heat of vaporization, normal bp, J/mol[d]	25.48	81.70	1741	6469	9012	12640	18100
heat of fusion, tp, J/mol[d]	no tp	no tp	335	1183	1640	2313	3247
heat capacity							
c_p, gas at constant pressure, 101.32 kPa, 25°C, J/(mol · K)[d]	20.78	20.78	20.79	20.85	20.95	21.01	[21]
sonic velocity, gas,							
101.32 kPa, 0°C, m/s	[1122]	973	433	307.8	213	168	
thermal conductivity							
gas, 101.32 kPa, 0°C, mW/(m · K)	[163.6]	141.84	46.07	16.94	8.74	5.06	
viscosity							
gas, 101.32 kPa, 25°C, Pa · s	[17.2]	19.85	31.73	22.64	25.3	23.1	23.3
solubility in water, 20°C, β[e]		8.61	10.5	33.6	59.4	108.1	230
1st ionization potential, eV		24.586	21.563	15.759	13.999	12.129	10.747
min excitation energy, eV		19.818	16.618	11.548	9.915	8.315	6.772
stable isotopic abundance, at%	1.3×10^{-4} (atm)	100	^{20}Ne = 90.5	^{36}Ar = 0.337	^{78}Kr = 0.35	^{124}Xe = 0.096	no
			^{21}Ne = 0.27	^{38}Ar = 0.063	^{80}Kr = 2.27	^{126}Xe = 0.090	stable
	1.7×10^{-5} (wells)		^{22}Ne = 9.23	^{40}Ar = 99.60	^{82}Kr = 11.56	^{128}Xe = 1.92	isotopes
					^{83}Kr = 11.55	^{129}Xe = 26.44	
					^{84}Kr = 56.90	^{130}Xe = 4.08	
					^{86}Kr = 17.37	^{131}Xe = 21.18	
						^{132}Xe = 26.89	
						^{134}Xe = 10.44	
						^{136}Xe = 8.87	

[a] Numbers in brackets are estimated.

[b] To convert kPa to psia, divide by 6.896.

[c] Volume of gas at 101.32 kPa and 0°C equivalent to unit volume of liquid at nbp.

[d] To convert J to cal, divide by 4.184.

[e] mL (101.32 kPa, 0°C) dissolved per kg water with partial pressure of pertinent gas of 101.32 kPa.

molecular weight. Only small quantitative effects can be observed in neon's properties; the effects in hydrogen are somewhat more pronounced; but these quantum-mechanical effects have profound qualitative effects on the behavior of helium at low temperature.

The liquid and solid phases of the two helium isotopes exhibit physical characteristics found in no other substances.

Solid phases. Quantum-mechanical descriptions of solids do not allow the complete cessation of molecular motion even at absolute zero. Instead, they require a certain zero-point motion about the molecule's average position in the crystal lattice. The helium isotopes have the largest zero-point motion of any substance. This is manifested in several unique characteristics of condensed helium. The heliums are the only known substances that will not freeze under their own vapor pressure.

Even at the lowest temperatures, a substantial pressure is required to solidify helium, and then the solid formed is one of the softest, most compressible ones known.

It was predicted in 1950 by Pomeranchuk from quantum-mechanical arguments that at very low temperatures the entropy of liquid helium-3 should be less than that of the solid; hence, the melting-curve slope should be negative, and the heat of solidification should also be negative. This is indeed the case, and adiabatic compression refrigerators based on the Pomeranchuk effect are used to produce temperatures in the millikelvin (mK or 0.001 K) range (see also Refrigeration; Cryogenics).

Liquid helium-4. Quantum mechanics admits of two fundamentally different types of particles: bosons, which have no unpaired quantum spins, and fermions, which do have unpaired spins. Because the helium-4 atom contains an even number of fermions, it is a boson. When saturated liquid helium is cooled below 2.172 K, it undergoes what is generally recognized to be the manifestation of a Bose-Einstein condensation. The liquid displays a striking and unique change of properties—it becomes a superfluid.

Superfluid helium (helium II) can pass easily through openings so small that they cannot be detected by conventional leak-detection methods. Another unique phenomenon exhibited by liquid helium II is the Rollin film. All surfaces below the lambda-point temperature that are connected to a helium II bath are covered with a very thin (several hundredths μm), mobile film of helium II.

Liquid helium-3. The helium-3 atom contains an odd number of fermions; thus it is itself a fermion. For many years, it was expected that liquid helium-3 might have relatively normal properties at even the lowest temperatures. However, in 1971, a helium-3 adiabatic compression refrigerator was used to discover helium-3's superfluid transition at temperatures below 0.003 K. It was soon found that helium-3 has not one but three superfluid phases, and that they have properties quite different from superfluid helium-4. They are magnetic, and many of their physical properties are anisotropic.

Production

Helium is separated from helium-bearing natural gases usually, but not always, in the process of improving the fuel value of the gas. Thus, in a sense, the greatest part of commercial helium is a by-product. Argon, neon, krypton, and xenon, as well as small quantities of helium, are obtained from by-product streams in which they concentrate during the separation of air to produce oxygen and nitrogen. Radon is collected as a daughter product of the fission of radium, and helium-3 is the daughter product of tritium.

Helium-group gases from the atmosphere. Atmospheric air contains 0.93% argon, lesser amounts of neon, helium, krypton, and less of xenon. Since air is the only practical present-day source of four of these elements, it is fortunate that industry has constructed very large plants for the production of oxygen and nitrogen by the distillation of air. In the distillation–liquefaction process, the helium-group gases concentrate as noncondensable gases (helium, neon, plus hydrogen) at liquid nitrogen temperatures; as a primary inert contaminant (argon) which could reduce oxygen purity up to 5% if not removed; and as a trace inert material, Kr–Xe (up to 10 ppm), also in the oxygen product, which may be ignored or removed for its own value (see Cryogenics; Oxygen; Nitrogen).

Krypton and xenon from nuclear power plants. Both xenon and krypton are products of the fission of uranium and plutonium. They are present in the spent fuel rods of nuclear power plants in the ratio 1 Kr : 4 Xe. Recovered krypton contains ca 6% of the radioactive isotope Kr-85, having a 10.7 year half-life, but all xenon radioactive isotopes have short half lives.

Radon separation. Owing to its short half-life, radon is normally prepared close to the point of use in laboratory-scale apparatus. Radium salts are dissolved in water and the evolved gases periodically collected.

Uses

Helium uses (1981) in the United States by category are cryogenics, 33.0%; welding and industrial atmospheres, 19.4%; pressurizing and purging, 18.0%; medical and breathing mixtures, 8.4%; chromatography, 4.4%; leak detection, 4.2%; lifing gas, 3.7%; heat transfer 3.5%; research and other 2.9%; controlled atmospheres, 2.5%; total use 1981, estimated, 2.45×10^6 m^3.

The main uses for argon are in metallurgical applications and in electric lamps. Neon, krypton, and xenon, because of their high costs, are limited to specialized uses in research, instrumentation, and electric lamps. There are now no significant technical uses for radon.

BRADLEY S. KIRK
ALFRED H. TAYLOR
AIRCO, Inc.

P.C. Tully, "Helium," *U.S. Bureau of Mines Yearbook*, USBM, Washington, D.C., 1981.

C.L. Davis, "Helium" in *Mineral Facts and Problems*, *U.S. Bureau of Mines Bulletin 667*, U.S. Government Printing Office, Washington, D.C., 1980.

R.E. Stanley and A.A. Moghissi, eds., *Noble Gases*, CONF 730915 *(Library of Congress 75-27055)*, ERDA TIC, 1973, 680 pp.

K.D. Timmerhaus and co-eds., *Advances in Cryogenic Engineering*, Vols. 1–24, Plenum Press, New York, 1960–1978.

HELIUM-GROUP GASES, COMPOUNDS

The helium-group gases are characterized by extreme chemical inertness and for many years were believed to be incapable of combining with other elements to form binary or ternary compounds. In recent years, however, it has been shown that the heavy gases, krypton, xenon, and radon, can be combined with fluorine and other powerful oxidants to form a number of stable products. Some physical properties of krypton and xenon compounds are given in Table 1.

Radon Compounds

Radon fluoride. Experiments with trace amounts of radon-222 (half-life, 3.82 d) have shown that radon forms a stable, nonvolatile fluoride when it is heated to approximately 400°C with fluorine gas.

Complex salts. Radon reacts at room temperature with solid oxidants, such as $O_2^+ SbF_6^-$, $O_2^+ Sb_2F_{11}^-$, $N_2F^+ SbF_6^-$, and $BrF_2^+ BiF_6^-$, to form nonvolatile complex salts.

Uses

Unstable noble gas halides are widely used as light-emitting species in lasers (see Light-emitting diodes and semiconductor lasers). Stable compounds of the helium-group gases have no industrial uses at present but are frequently utilized in laboratories as fluorinating or oxidizing reagents.

LAWRENCE STEIN
Argonne National Laboratory

"Edelgasverbindungen," in *Gmelins Handbuch der anorganischen Chemie*, 8th ed, Main Supplement, Vol. 1, Verlag Chemie, Weinheim, 1970.

Table 1. Physical Properties of Some Krypton and Xenon Compounds

Compound	Melting point, °C	Color	Symmetry	a	b	c	Density, g/cm³
				Lattice parameters, nm			
KrF_2	dec ca 25	colorless	tetragonal	0.6533		0.5831	3.24
$KrF^+SbF_6^-$	45, dec	white					
$KrF^+Sb_2F_{11}^-$	50, dec	white					
$Kr_2F_3^+SbF_6^-$	dec ca 25	white					
$KrF^+Ta_2F_{11}^-$	dec ca -20	white					
XeF_2	129.03[a]	colorless	tetragonal	0.4315		0.6990	4.32
XeF_4	117.10[a]	colorless	monoclinic	0.5050	0.5922 $\beta = 99.6°$	0.5771	4.04
XeF_6	49.48	colorless[b] or yellow-green,[c,d]	monoclinic[e]	0.933	1.096 $\beta = 91.9°$	0.895	3.56
			orthorhombic[f]	1.701	1.204	0.857	3.71
			monoclinic[g]	1.680	2.393 $\beta = 90°40'$	1.695	3.82
			cubic[h]	2.506			3.73
XeO_3	dec ca 25 (explosive)	white	orthorhombic	0.6163	0.8115	0.5234	4.55
XeO_4	dec < 0 (explosive)	yellow					
$XeOF_2$	dec ca 0 (explosive)	yellow					
$XeOF_4$	-46.2	colorless					
XeO_2F_2	30.8 (explosive)	colorless	orthorhombic	0.6443	0.6288	0.8312	4.10
$Na_4XeO_6.6H_2O$	loses water at 100; dec 360	colorless	orthorhombic	1.844	1.0103	0.5873	2.59
$XeF^+RuF_6^-$	110–111	yellow-green	monoclinic	0.7991	1.1086 $\beta = 90.68°$	0.7250	3.78
$XeF^+Sb_2F_{11}^-$	63	yellow	monoclinic	0.807	0.955 $\beta = 105.8°C$	0.733	3.69
$Xe_2F_3^+AsF_6^-$	99	yellow-green	monoclinic	1.5443	0.8678 $\beta = 90.13°$	2.0888	3.62
$XeF_3^+SbF_6^-$	109–113	yellow-green	monoclinic	0.5394	1.5559 $\beta = 103.10°$	0.8782	3.92
$XeF_3^+Sb_2F_{11}^-$	81–83	yellow-green	triclinic	0.8237 $\alpha = 72.54°$	0.9984 $\beta = 112.59°$	0.8004 $\gamma = 117.05°$	3.98
$XeF_5^+RuF_6^-$	152	green	orthorhombic	1.6771	0.8206	0.5617	3.79
$XeF_5^+AsF_6^-$	130.5	white	monoclinic	0.5886	1.6564 $\beta = 91.57°$	0.8051	3.51
$Xe_2F_{11}^+AuF_6^-$	145–150	yellow-green	orthorhombic	0.9115	0.8542	1.5726	4.24

[a] Triple point.
[b] Solid.
[c] Liquid.
[d] Vapor.

[e] Phase I.
[f] Phase II.
[g] Phase III.
[h] Phase IV.

N. Bartlett and F.O. Sladky, "The Chemistry of Krypton, Xenon, and Radon" in A.F. Trotman-Dickensen, ed., *Comprehensive Inorganic Chemistry*, Vol. 1, Pergamon Press, Oxford, 1973, pp. 213–330.

J.G. Malm and E.H. Appelman, *At. Energy Rev.* **7**(3), 3 (1969).

HELMHOLTZ (YOUNG-HELMHOLTZ). See Color.

HERBICIDES

The term herbicides refers to agents designed to provide weed control. Design of chemical agents has traditionally been based on structure–activity relationships. This approach, although successful in the past, is slowly giving way to bioresearch of the new chemicals made possible by increased understanding of the biochemistry of herbicidal action and herbicide metabolism in plants.

Modes of Action and Metabolism

Two important processes in plant life are photosynthesis and respiration. Because photosynthesis does not occur in mammals, herbicides that act as photosynthetic inhibitors usually demonstrate low acute toxicity toward mammals. Other processes affected may include inhibition of normal growth.

Respiration inhibitors. The respiration process comprises several stages. The first is glycolysis, during which carbohydrates, fats, or proteins are degraded by acetyl coenzyme-A in the presence of oxygen. In the second stage, acetyl coenzyme-A is utilized in the Krebs tricarboxylic acid cycle which generates the reduced form of nicotinamide adenine dinucleotide ($NADH_2$) and the flavoprotein portion of the enzyme succinate dehydrogenase. Finally, in the coupled processes of electron transport and oxidative phosphorylation, ADP is converted to ATP by release of the energy from the transfer of electrons to oxygen, which is converted to water. Herbicides are known to interfere with several of these stages in plants. For example, arsenical compounds in the form of arsenite ion may react with thiol groups in enzyme systems, whereas as arsenates they may uncouple oxidative phosphorylation or may mimic the role of the phosphate ion, thus interfering with the normal function of ATP. Dinitrophenol herbicides affect the coupling of electron transport and oxidative phosphorylation. Bromoxynil and other hydroxybenzonitriles may possess a similar mode of action. Respiratory inhibitors such as benzoic acids, picolinic acids, and α-haloacetamides reduce both oxygen uptake and oxidative phosphorylation in mitochondria.

Photosynthesis inhibitors. There are two light reactions in photosynthesis linked by the photosynthetic electron-transport pathway which is coupled to the phosphorylation of ADP to ATP (photophosphorylation). In light reaction I, NADP is reduced to NADPH and, during light reaction II, electrons are removed from water and oxygen is produced. In the presence of an electron acceptor, cell-free preparations of chloroplasts catalyze the light-dependent reduction of oxygen to water, known as the Hill reaction. Light reaction II is probably blocked by herbicides that inhibit the Hill reaction (eg, diuron, herbicidal ureas, triazines, anilides, hydroxybenzonitriles, uracils, phenyl ethers, pyridazinones, dinitrophenols, N-phenylcarbamates, and pyrimidinones).

Bipyridylium herbicides interfere with photosynthesis because their positive ions readily accept electrons to form relatively stable free radicals. The original ion is then regenerated by reaction with oxygen. The other product of the reaction is hydrogen peroxide, which rapidly destroys plant tissue.

Plant-growth regulators. Physiological processes in plants are controlled by hormones (qv). Several herbicides act by mimicking the activity of natural plant-growth regulators or by interfering with their synthesis or action. Such a herbicide may release natural growth hormone, combine with the hormone, prevent hormone synthesis, or affect transport or the fate of the hormone in the plant. Examples include phenoxyalkanoic acids, benzoic acids, picloram, endothall, α-chloroacetamides, and amitrole.

Safety Factors and Regulations; Disposal

Weed killers are generally less toxic to mammals than are other classes of pesticides, except certain highly toxic compounds such as acrolein, acrylonitrile, arsenical compounds, paraquat, and nitrophenol derivatives. The user of pesticides should observe appropriate precautions, including the use of protective clothing and respiratory devices. Safe handling and disposal procedures are the responsibilities of the user.

The labeling and marketing of pesticides in interstate commerce are regulated by the *Federal Insecticide, Fungicide and Rodenticide Act* as amended, which is administered by the EPA. Most states have similar laws. Therefore, herbicides may require both Federal and state registrations.

Herbicide Classes

Chlorinated phenoxyalkanoic acids. The formulas, melting points, and solubilities of some chlorinated aryloxyalkanoic acids are given in Table 1. These compounds are used as the free acids, metal salts, amine salts, ammonium salts, or esters. The problem of possible harmful effects of the phenoxyalkanoic acids is linked to the presence of the chlorinated dioxins that may arise during the manufacture of chlorophenols and may be present as contaminants.

s-Triazines. The triazines, especially the 2-chloro derivatives, are extremely effective herbicides for weed control in corn. Properties of some s-triazines are given in Table 2.

Urea herbicides. Initially, the phenylureas were developed as industrial herbicides. Subsequently, they have also found uses in selective applications in agriculture. The properties of important phenylureas are shown in Table 3 (see also Urea and urea derivatives).

Carbamates. The biological activity of the carbamates varies greatly. Suitable substitution gives a range of compounds used as herbicides, insecticides, medicinals, nematicides, miticides, or molluscicides (see Poisons, economic). Two widely used herbicidal carbamates are propham (isopropyl phenylcarbamate, mp 87–88°C, solubility in water, 250 ppm) and chloropropham (isopropyl 3-chlorophenylcarbamate, mp 38–40°C, solubility in water 88 ppm).

Table 1. Properties of Chlorinated Aryloxyalkanoic Acids,

Name	R	X	Y	Z	n	Melting point, °C	Solubility, ppm, water °C
2,4-D[a]	H	Cl	Cl	H	0	140.5	0.07[25]
2,4-DB	H	Cl	Cl	H	2	117–119	46[25]
dichlorprop	CH$_3$	Cl	Cl	H	0	118	350[20]
MCPA	H	CH$_3$	Cl	H	0	119	825
MCPB	H	CH$_3$	Cl	H	2	100	44
mecoprop	CH$_3$	CH$_3$	Cl	H	0	94–95	600

[a]Solubility in acetone is 45 ppm at 33°C.

Table 2. Properties of s-Triazine Herbicides,

Name	R	R′	R″	Melting point, °C	Solubility, ppm, water °C
atrazine	C$_2$H$_5$	i-C$_3$H$_7$	Cl	173–175	70[27]
simazine	C$_2$H$_5$	C$_2$H$_5$	Cl	225–227	5[20]
	i-C$_3$H$_7$	C$_2$H$_5$	OCH$_3$	91–92	730[20]
prometryn	i-C$_3$H$_7$	i-C$_3$H$_7$	SCH$_3$	118–120	48[20]
cyanazine	C$_2$H$_5$	C(CH$_3$)$_2$CN	Cl	166–167	171[20-25]
terbutryn	C$_2$H$_5$	C(CH$_3$)$_3$	SCH$_3$	104–105	58[20-25]

Table 3. Properties of Urea Herbicides,

Name	X	Y	R	R′	Melting point, °C	Solubility, ppm, water °C
monuron	Cl	Cl	CH$_3$	CH$_3$	174–175	230[25]
diuron	Cl	Cl	CH$_3$	CH$_3$	158–159	42
					180–190 (dec)	
fluometuran	H	CF$_3$	CH$_3$	CH$_3$	163–164	90[25]
linuron	Cl	Cl	CH$_3$	OCH$_3$	93–94	75[25]
siduron	H	H		H	133–138	18
chloroxuron	H	C$_6$H$_4$Cl	CH$_3$	CH$_3$	151–152	3.7[20]

Table 4. Properties of Thiocarbamate Herbicides

Name	Boiling point, °C	Solubility, ppm, water °C
EPTC	127$_{2.67}$	370[25]
vernolate	140$_{2.67}$	109[24]
triallate	148$_{1.2}$	4
butylate	137.5$_{2.8}$	45[25]
molinate	202$_{1.33}$	80

Thiocarbamates. *S*-Ethyl dipropylthiocarbamate (EPTC) was introduced as a herbicide in 1959 (see Table 4).

Amides. The chloroacetamides are used in preemergence treatments to control many weeds in corn and numerous other crops. Alachlor and metolachlor inhibit germination of grassy weed seedlings and certain broadleaf weed seedlings. Properties of some herbicidal amides are shown in Table 5.

Table 5.   Properties of Amide Herbicides

Name	Melting, point[a], °C	Solubility, ppm, water °C
propanil	93–94	500[b]
diphenamid	134–135	260[27]
alachlor	110$_{0.03}$[c]	242[20]
naptalam (NPA)	185	200
propachlor	77	580[20]

[a] Unless otherwise stated.
[b] Grams per 100 g water.
[c] Boiling point.

Chlorinated aliphatic acids. The chlorinated aliphatic acids are strongly acidic, eg, TCA (trichloroacetic acid; mp 59°C, bp 197.5°C, solubility 1036 g/100 g water) and fenac (2,3,6-trichlorophenylacetic acid).

Chlorinated benzoic acids. The chlorinated benzoic acid or nitrile herbicides include a number of polychlorinated aromatic compounds with a variety of ring substituents. Table 6 lists the properties of some of the members of this group.

Table 6.   Properties of Chlorinated Benzoic Acids and a Related Compound

Name	Melting point, °C	Solubility, ppm, water °C
chloramben	210	700[25]
dicamba	114–116	0.45[b]
dichlobenil	145–146	18[20]

[a] Unless otherwise stated.
[b] Grams per 100 g water.

Dichlobenil. 2,6-Dichlorobenzonitrile inhibits germination of weed seeds and has been used with fruit trees, alfalfa, and dormant cranberries.

Phenols. Bromoxynil (2,6-dibromo-4-cyanophenol, mp 194–195°C, solubility in water 5 ppm at 20°C) has been used for postemergence control of broadleaf weeds in corn.

Substituted dinitroanilines. The development of dinitroaniline herbicides has continued rapidly during the past decade. Trifluralin (*N*,*N*-dipropyl-2,6-dinitro-4-trifluoromethylaniline, mp 48.5–49°C, solubility in water 1 ppm at 27°C) is one of the most successful commercial products.

Bipyridinium herbicides. The bipyridinium compounds (diquat and paraquat salts) are marked by their very strong adsorption to clay minerals. Consequently they may be sprayed between rows of crops to eliminate weeds. The herbicide is rapidly adsorbed and inactivated by the soil and does not damage the growing crop.

Miscellaneous Herbicides

Picloram, 4-amino-3,5,6-trichloropicolinic acid, a herbicide and growth regulator, is rapidly translocated within plants and is resistant to metabolic breakdown. It is classified as an auxin-type herbicide and causes growth responses similar to those of 2,4-D.

Amitrole, 3-amino-*s*-triazole, interferes with synthesis of chlorophyll in plants; it is readily translocated.

Glyphosat, *N*-(phosphonomethyl)glycine, is a herbicide with a very broad spectrum of activity. It is usually applied as the isopropylamine salt and is readily absorbed through foliage and translocated throughout the plant. It is believed to be inactivated in soils by chelation with iron. Glyphosine is closely related to glyphosate and has similar chemical but quite different phytotoxic properties. It speeds maturity and maintains high levels of sucrose, and is used as a growth regulator in sugarcane production.

Uracils. Substituted pyrimidines exhibit herbicidal activity: bromacil (**1**), isocil (**2**), and terbacil (**3**) are crystalline solids soluble in aqueous bases:

(1)   (2)   (3)

The uracils inhibit photosynthesis in plants.

Diphenyl ethers. The diphenyl ethers are applied to rice paddies for weed control because their toxicity to shellfish is low and they are not injurious to rice.

Inorganic Herbicides

A number of inorganic chemicals have been used as herbicides, eg, sulfamic acid and ammonium sulfamate, various boron compounds, arsenical compounds, and sodium chlorate.

Herbicide Antidotes or Safeners

To protect corn against damage by thiocarbamate herbicides, herbicide antidotes or safeners, which do not affect weed control, were developed in 1972. A safener is a chemical used in a pesticide spray to prevent damage to plants by the other spray ingredients. It seems that the selective antidote *N*,*N*-diallyl-2,2-dichloroacetamide (R-25788) increases the glutathione (GSH) content of corn. Several amides exert a similar protective effect, but R-25788 is most effective when the compounds are applied to soils. Seed-corn treatment with naphthalic anhydride effectively protects corn plants from herbicidal injury by EPTC, butylate, ethiolate, and vernolate.

Plant Growth Regulators

Many herbicides exert their phytotoxic effect by affecting plant growth. Although a lethal effect is easily observed, more subtle changes in plant growth may be advantageous to the cultivator who seeks to maximize crop yield and reduce production costs. Thus chemicals may be used to promote or retard growth, assist setting or abscission of fruit, improve quality, cause defoliation, or prevent flowering. Yield may be improved by changed plant growth, morphology, or metabolism. Chemical treatment may increase resistance to drought, heat, cold, weather, or other environmental factors.

J.R. PLIMMER
United States Department of Agriculture

Reviewed by
MAY INSCOE
United States Department of Agriculture

Herbicide Handbook, 4th ed., Weed Science Society of America, Champaign, Ill., 1979.

F.M. Ashton and A.S. Crafts, *Mode of Action of Herbicides*, Wiley-Interscience, New York, 1973; L.J. Audus, ed., *Herbicides: Physiology, Biochemistry, Ecology*, 2nd ed., Vols. 1–2, Academic Press, Inc., New York, 1976; P.C. Kearney and D.D. Kaufman, eds., *Herbicides: Chemistry, Degradation, and Mode of Action*, Vols. 1–2, Marcel Dekker, Inc., New York, 1975.

HEXAMETHYL PHOSPHORIC TRIAMIDE. See Phosphorus compounds, organic.

HEXANES. See Hydrocarbons, C_1–C_6.

HIGH PRESSURE TECHNOLOGY

A wide range of chemicals and materials is manufactured using high pressure processes, and many types of equipment depend on the action of high pressure fluids for their operation. In addition, the pressure variable is used extensively in scientific research.

Typical process pressures are illustrated in Table 1. In a few processes, such as production of cubic boron nitride and diamond, the nature of the process imposes a lower limit on the pressure range (see Boron compounds; Carbon). In other cases, the pressure chosen is a compromise between the improvement of yield or reaction rate at higher pressure, and the higher costs. In some gaseous reactions, however, equipment size and costs can be decreased by operating at increased pressure.

The internationally accepted unit of pressure is the pascal Pa. For most industrial processes, the megapascal, MPa (9.87 atm) is a convenient unit.

There is no sharp dividing line between high and low pressure. Often a pressure value such as 5 MPa (50 atm) is chosen arbitrarily. However, the basic design of high pressure equipment is similar up to 600 MPa (ca 6000 atm) where limitations of construction materials become critical. Special designs are required above this pressure.

Pressure Vessels and Closures

The following factors must be considered when designing a high pressure vessel and selecting its materials of construction: (*1*) maximum pressure to which the vessel will be subjected; (*2*) volume; (*3*) shape of the vessel based on its use (eg, cylindrical or spherical, single- or double-end closures); (*4*) temperature, including maximum or minimum, control, rate of change, and allowance for removal of heat generated within the vessel by chemical reaction, etc; (*5*) fatigue induced by pressure cycling; (*6*) corrosion; and (*7*) miscellaneous factors such as fast opening and closing of the end closure.

A vessel must usually meet the codes adopted by various regulatory bodies such as the ASME in the United States.

Thin-walled vessels. The body, end cap, and closure section are welded together in these vessels. If the vessel is long, the body may be constructed of several welded sections. Thin-walled vessels include forged cylinders; solid-wall, rolled-plate vessels, multiwall, rolled-plated vessels; multilayer vessels; coil-layer or strip-wound pressure vessels; and wrapped, interlocking-band vessels.

Thin-walled vessels are used extensively in the chemical industry for pressures ranging up to 30 MPa (ca 300 atm). They are usually constructed of inexpensive, low alloy steels, although liners are sometimes of stainless steel or glass, particularly for service with corrosive materials. Fiber-reinforced composites are also used occasionally.

The tangential stress σ_t in a long, thin-walled vessel can be calculated approximately by assuming that the stress is uniform throughout the wall thickness. For a vessel diameter d of thickness t, with internal pressure P, Barlow's equation applies

$$\sigma_t = P\frac{d}{2t} \tag{1}$$

Thick-walled vessels. Useful design criteria for both thick- and thin-walled vessels can be derived for a cylindrical vessel of great length. The radial and tangential stresses in the walls are then only functions of the radius coordinate, and the internal pressure. Maximum stresses are developed at the inner wall. The longitudinal stress component is relatively small and is sometimes neglected.

The curve shown in Figure 1 shows the increased values of pressure that can be obtained with double-walled cylinders operating within elastic limits compared to single-walled cylinders.

An alternative strengthening mechanism is called autofrettage. If a cylinder is intentionally overpressurized, a plastic-elastic interface is produced that moves outward as pressure increases. When pressure is released, the residual stresses left as a result of radial expansion allow the vessel to be used up to the pressure at which it was subjected to autofrettage without exceeding the yield stress of the material. The calculation for autofrettage assumes that the material has a perfectly elastic-plastic stress-strain curve. Burst pressure for an autofrettaged cylinder is given in Figure 1, D.

Seals. A large variety of seals is available for pressure-vessel closures: confined-gasket and compressed gasket seals (Poulter seals), self-energizing ring seals, and pressure-energized seals (Bridgman seals, Fig. 2). Bridgman seals are used for critical seals at the highest pressure (eg, > 200 MPa).

Cyclic loading fatigue. The fatigue life can be crucial when considering safe operating conditions of a pressure vessel or tubing, and particularly the critical components of a reciprocating compressor. Design usually starts with consideration of the stress-to-failure vs the number-of-cycles-to-failure (S-N) curve obtained for the particular material.

Temperature. Temperature factors are particularly important in autoclaves (pressure vessels heated to high temperature) and vessels subjected to low temperature. The latter are subject to possible embrittlement. In high temperature operation, a decrease of material strength

Table 1.   Typical Pressures for Industrial Processes

Process	Pressure, MPa[a]
liquefaction of gases	2–20
pneumatic equipment and gas storage	2–200
hydraulic equipment	1–100
typical gas- and liquid-phase chemical reactions	0.1–50
synthesis of low density polyethylene	100–500
synthesis of single crystal materials	10–200
cold- and hot-isostatic pressing	50–200
metal working	100–500
liquid-jet cutting	100–1000
autofrettage of pressure vessels and gun barrels	1000–2000
diamond and cubic boron nitride synthesis	5000–7000

[a] To convert MPa to atm, divide by 0.101.

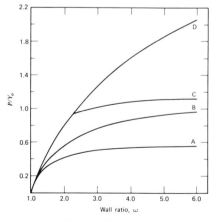

Figure 1. Pressure obtainable with thick-walled vessels within elastic limits. A, monobloc; B, two-piece shrink fit; C, autofrettaged cylinder; D, burst pressure of a single-walled cylinder before autofrettage. Y_o = yield point.

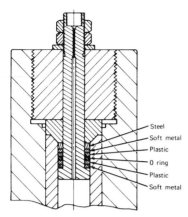

Figure 2. Bridgman seal with mushroom plug.

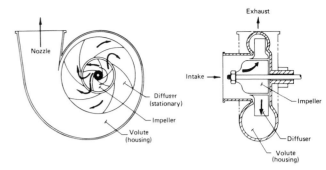

Figure 4. Schematic of a centrifugal compressor (single stage) with vaned diffuser.

may occur, particularly if the temperature rises above the tempering temperature of a heat-treated steel. If the vessel is held at high pressure for an extended period of time, creep can occur.

Materials for cryogenic service are usually fcc metals, such as austenitic stainless-steels, copper alloys, aluminum alloys, and superalloys. Low-alloy steel, manganese or austenitic stainless-steel autoclaves are available for use to 450°C. For higher-temperature operation, special alloys must be used, such as Hastalloy (see Cryogenics; Sterilization techniques).

Environmental effects — corrosion. The effect of corrosive substances is often increased at high pressure. The synergistic effect of corrosive and high stress environments may lead to failure of materials well below the normal yield strength. This type of failure is called stress corrosion.

A severe problem for alloys with bcc structure is hydrogen embrittlement, such as hydrogen blistering, decarburization, and attack by diffusion of atomic hydrogen into the region near the crack tip. Failure can occur rapidly after a stressed body is subjected to a hydrogen-containing vapor.

Pressure vessels lined with a protective sleeve cost less than those made entirely of corrosion-resistant metal. In addition, seal designs can minimize corrosive effects.

Voids, cracks, and areas of high stress concentration. Plastic flow and tearing initiate at regions of high stress concentration, such as voids and cracks in the material and surface flaws, or in certain areas such as side-entry ports, notches, regions where the bore diameter changes, thermal discontinuities, etc. Therefore, the design should avoid regions of high stress concentration, and include specifications for surface finish and maximum flaw size.

Pumps and Compressors

A pump is a unit for pressurizing or circulating a liquid. A compressor is a unit for compressing a gas or compressible fluid. If the temperature is above the critical temperature, the fluid can be compressed continuously into the liquidlike state and, therefore, the distinction between pump and compressor may not always be clear cut.

Pumps and compressors are either rotary or reciprocating (Fig. 3). In reciprocating units, fluid is drawn in through an inlet check valve, and expelled through the outlet check valve. In this action, they may be

considered as positive displacement-type pumps since a discrete volume of fluid is trapped, then expelled. Some rotary units are also of this type, eg, gear, vane, and lobe pumps. Other rotary designs impart kinetic energy to the fluid, which may then be converted to potential energy (compressional energy). The most important units of this type are the turbine and centrifugal compressor (Fig. 4).

The most important design factors are the operating pressure and the flow rate. The inlet pressure required to reach the desired outlet pressure depends on the fluid being compressed. Another important design criterion is the temperature reached by the fluid during compression, which varies with fluid type.

Pumps for hydraulic equipment are usually of the positive-displacement, rotary type. Large-volume industrial compressors are mostly of the centrifugal type for pressures up to 40 MPa (ca 400 atm) and of the axial-turbine type for large-volume, lower-pressure applications (eg, 1 MPa, 10 atm). However, reciprocating compressors are still preferred for lower capacity (eg, 0.5 m^3/s at standard inlet conditions) and higher pressures (eg, 40 MPa, ca 400 atm). Contamination-free compression can be achieved with diaphragm units up to 200 MPa (ca 2000 atm). They are used for charging storage bottles, diving operations, and critical nuclear applications. Small intensifiers are used in laboratory and pilot-plant equipment. The principle of intensification is also used in reciprocating pumps and compressors, such as hydraulically driven compressors for polyethylene plants.

High Pressure Components

Valves. The importance of valves can be gauged from the fact that their cost may represent 8–10% of new plant expenditures and 10% of the maintenance budget.

When choosing a valve, the following factors have to be considered: (1) pressure range; (2) temperature; (3) leak-tightness when closed; (4) restriction to flow when open; (5) flow-rate adjustment capability; (6) maintenance costs and frequency vary from one type to another; and (7) size range.

The needle valve (Fig. 5) is preferred for high pressure operation (up to 1.5 GPa, 14,800 atm). The needle tip is usually made of steel, and of tungsten carbide for highest pressure applications.

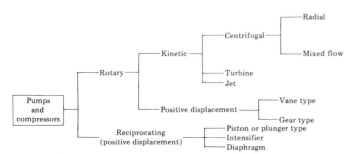

Figure 3. Classification of pumps and compressors.

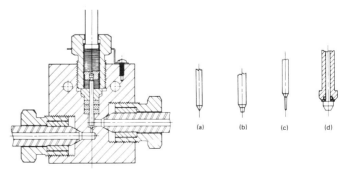

Figure 5. A needle valve. Designs of valve stems are also shown (**a–d**). Courtesy of American Instrument Company.

Pipes, tubes, and hoses.   Tubes and pipes can be seamless, welded, or welded and drawn. Seamless tubing is always preferred for the highest pressure service (eg, above 100 MPa or ca 1000 atm). Flexible hose is available for pressures up to 80 MPa (ca 800 atm). It can be useful for situations where fatigue is important, and can handle surge pressure very well.

The effects of bends in seamless tubing or piping would also be considered. The bending radius should always be at least four times the outer diameter of the tube, and preferably greater than six times. In addition, tubing or piping should always be anchored to prevent whipping if failure of a connection occurs.

Fittings.   Several types of fittings are used for tubing connections, but cone or lens-ring connections are preferred for pressures of 1.5 GPa (ca 15,000 atm). A wide range of auxiliary fittings is available, eg, crosses tees, elbows, end plugs, and adapters.

Properties of Matter at High Pressure

Pressure is an important variable in the control of equilibrium and transport properties of fluid and solid phases. Gaseous phases become liquidlike at sufficiently high pressure even if the temperature is higher than the critical temperature T_c. At even higher pressure, further condensation occurs to a solid phase.

Below the critical temperature, the transformation from gaseous to liquid phases occurs with a discontinuity in volume and entropy (first-order transformation), but at the critical point these changes tend to zero. Solid–liquid and solid–gas phase changes are also first order. The triple point defines the simultaneous equilibrium between solid-liquid-vapor. Other triple points are possible if there are distinct solid phases (solid–solid–liquid, or solid–solid–solid).

The equation of state $V(P,T)$ of a substance allows thermodynamic properties to be derived over a region of pressure and temperature:

$$\Delta H = \int C_p \, dT + \int V(1 - \alpha T) \, dP \qquad (2)$$

$$\Delta S = \int \frac{C_p}{T} \, dT - \int V\alpha \, dp \qquad (3)$$

where H = enthalpy, S = entropy, and α = thermal expansion coefficient $\{(1/V)[\partial V \cdot \partial T]_p\}$.

Hence, thermodynamic properties of a substance can be computed over a region of temperature and pressure if the specific heat is known as a function of temperature at a certain pressure (traditionally 101.3 kPa or 1 atm), the absolute entropy at one point, and the equation of state over the required range of temperature and pressure.

Fluids.   Gaseous fluids approximately obey the perfect gas equation at sufficiently low density:

$$PV = nRT \qquad (4)$$

where n is the number of moles and R is the universal gas constant, 8.314 J/(mol·K) [1.987 cal/(mol·K)]. This equation becomes inadequate at higher pressure where interactions between molecules become more frequent. The equation of state and other thermodynamic properties can be reduced to universal curves for classes of fluids using reduced coordinates $P_r = P/P_c$, $T_r = T/T_c$, $V_r = V/V_c$, where P_c, T_c, V_c, are values at the critical point. This is a convenient method of estimating properties of fluids in the absence of more accurate data.

The most complete measurements of the equation of state combine conventional and derivative (ultrasonic velocity) measurements.

A useful concept for obtaining thermodynamic properties is the fugacity, which acts as a pseudopressure in the equation for the molar Gibbs free energy (μ). For a perfect gas:

$$\mu = \mu°(T) + RT \ln P/P° \qquad (5)$$

which is replaced by

$$\mu = \mu^\varnothing + RT \ln f \qquad (6)$$

for an imperfect gas, where $\mu°$ is the molar Gibbs free energy at the reference pressure $P°$ (101.3 kPa or 1 atm), and $\mu^\varnothing$ at unit fugacity, f.

The fugacity, which plays the role of a pseudopressure, may be obtained from equation of state data:

$$\ln f(P')/P' = \int_O^P \left\{ \frac{V_m}{RT} - \frac{1}{P} \right\} dP$$

$$= \int_O^P \frac{Z - 1}{P} \, dP \qquad (7)$$

Solids.   Since solid, crystalline materials may be anisotropic, length changes induced by hydrostatic pressure may depend on direction. The isothermal bulk modulus ($B_T - V(dp/dV)_T$) can be derived from elastic stiffness constants, C_{ij}.

For isotropic solids of general engineering interest, the following two relationships are useful:

$$B_T = \frac{E}{2(1 - 2\sigma)} \qquad (8)$$

$$G = \frac{E}{2(1 + \sigma)} \qquad (9)$$

where E = Young's modulus (GPa), σ = Poisson's ratio (dimensionless) and G = shear modulus (GPa).

A useful parameter for the equation of state is the reduced pressure P/B_o where B_o is the isothermal bulk modulus at $P = 101.3$ kPa (1 atm).

Phase changes.   Vapor–liquid–solid equilibria and solid–solid transformations are strongly affected by pressure. If two phases (α, β) of a pure component are in equilibrium, transfer of molecules between the two phases at constant temperature and pressure is characterized by constancy of the Gibbs function ($G = H - TS$). Accordingly, the molar Gibbs function ($g = G/n$) is equal for each phase

$$g_{(\alpha)} = g_{(\beta)} \qquad (10)$$

Reactions at High Pressure

Reactions are carried out at high pressure to increase yield and reaction rate and in gaseous reactions to reduce the size of the reaction vessel.

Equilibrium yield.   A general reaction of the form:

$$aA + bB + \cdots \rightarrow lL + mM + \cdots \qquad (11)$$

can be described by the equation

$$O = \sum_{i=1}^{n} \nu_i M_i \qquad (12)$$

where there are n chemical components M_i and the stoichiometry coefficients ν_i are positive for products and negative for reactants.

At equilibrium, the Gibbs function takes a minimum value for a system in which pressure and temperature are held constant. If the chemical potential for each constituent is written

$$\mu_i = \mu_i^\varnothing + RT \ln a_i \qquad (13)$$

where $\mu_i^\varnothing$ is the value in the standard state at unit pressure and a_i is the activity (a_i is equated to the fugacity for fluid states), then the equilibrium state is characterized by

$$\sum_i \nu_i \mu_i^\varnothing = -RT \sum_i \nu_i \ln a_i \qquad (14)$$

$$\equiv -RT \ln K_\alpha \qquad (15)$$

where

$$K_a = \prod_i a_i^{\nu_i} \qquad (16)$$

K is the equilibrium constant for the reaction and the symbol $\prod_i$ represents a product summation, eg, for equation 16.

$$K_a = \frac{a_L^l \cdot a_M^m \cdots}{a_n^a \cdot a_B^b \cdots} \qquad (17)$$

The left side of equation 14 is the standard free energy change for the reaction (Δg_T^o) at the temperature of reaction.

$$\text{if} \begin{cases} \Delta g_T^o > 0 \text{ reaction is not favored} \\ \Delta g_T^o < 0 \text{ reaction is favored} \end{cases} \qquad (18)$$

Values of Δg_T^o can be found for many compounds in standard compilations, and can be derived for other compounds using the additivity law. For solid and liquid reagents, the activities are not strongly dependent on the total pressure. However, for gaseous reactions, the activities increase strongly with pressure, eg, for a perfect gas mixture the activities equal the partial pressures. Thus, as a total pressure is increased, the reaction is moved in the direction in which volume is decreased (Le Chatelier's principle).

Kinetics. A simple molecular view of reaction kinetics is obtained by considering a bimolecular reaction of the form

$$aA + bB \rightarrow X^* \rightarrow cC + dD \qquad (19)$$

where X^* represents the activated complex of highest energy through which the reaction must proceed.

Generally, catalysts decrease the energy of the activated complex, and thus increase the reaction rate (see Catalysis). However, pressure increases the surface density of adsorbed molecules at all pressures (see Adsorptive separation). Diffusion processes are more complex and depend on the predominant scattering mechanism (molecule–molecule and molecule–surface) (see Diffusion separation methods). In certain pressure ranges, increase of pressure may increase or decrease molecular flux rates.

Typical High Pressure Processes

A large number of gas-phase, catalytic reactions are carried out in the pressure range of 20 MPa (ca 200 atm) and even higher. Higher pressures may be used in the future because of the increase in fuel costs. Reactions of this type are ammonia and methanol syntheses; reactions involving carbon monoxide, hydrogen, or carbon dioxide; the Fischer-Tropsch synthesis of liquid fats, oils, and fuels (see Fuels, synthetic); oxidation of ethylene to ethylene oxide (qv); carbonylation of alcohols to give acids (see Acetic acid) and hydrocarbonylation of olefins to ketones, aldehydes, and alcohols (see Oxo process); hydrocarbon cracking and hydrocracking in petroleum processing (see Petroleum).

Particular problems related to the operation of high pressure reactors are hydrogen attack, particularly decarburization, if walls are above 400°C, and decay of catalyst supports. Moderate pressures are used in liquid-phase reactions, such as hydrogenation of oils and fats, and syntheses of organic compounds.

Pressure Measurement

Pressure is measured by primary systems that rely on the measurement of force (F) and area (A) [$P = F/A$], and secondary systems that utilize the change of a physical property with pressure and are calibrated against primary standards. Among a large number of physical properties, elastic strain and electrical resistance are the most commonly used.

Safety Factors

High pressure equipment poses a safety hazard because of the high values of stored energy that can be released and result in a catastrophe under certain conditions. In addition to the resulting blast wave, parts of the equipment can be ejected with very high speed. A simple calculation can determine the large values of forces and equivalent explosive content of typical systems. Other hazards include the effects of leaking gas (toxicity, asphyxiation, flammability, secondary explosions, etc), noise, and vibrations from pumps and compressors.

Applications

Mechanical applications include machinery in which hydraulic pressure is used to supply power, whereas metallurgical applications include processing of materials. High pressure circulating systems are used in nuclear plants for gas decontamination (see Nuclear reactors). High pressure jets are used for mining minerals and cutting metal sheets and fabrics. Materials, chemicals, and components that are synthesized, refined, processed, or formed using high pressure processes are numerous. High pressure technology also is used in single-crystal synthesis, synthesis of diamond and superhard materials, isostatic compaction and hot isostatic compaction, hydrostatic extrusion of metals, and shock applications.

IAN L. SPAIN
Colorado State University

J.F. Harvey, *Theory and Design of Modern Pressure Vessels*, Van Nostrand Reinhold Company, New York, 1974.

W.R.D. Manning and S. Labrow, *High Pressure Engineering*, CRC Press, Cleveland, Ohio, 1971.

I.L. Spain and J. Paauwe eds., *Equipment, Design, Materials and Properties*, Vol. I, and *Applications and Processes*, Vol. II of *High Pressure Technology*, Marcel Dekker, New York, 1977.

HIGH TEMPERATURE ALLOYS

High temperature alloys are those combinations of metals that are used specifically for their heat-resisting properties. Physical properties such as melting temperature, elastic moduli, densities, and thermal conductivities of the elemental metals that serve as the basis for most high temperature alloys are listed in Table 1.

Mechanical Behavior

Creep rupture. Metals and their alloys lose appreciable strength at elevated temperatures. Creep-rupture tests must be conducted in which the time-dependent deformation and fracture are determined from periodic measurements under a fixed stress or load.

Relaxation. If a specimen is stretched or compressed and is held over a period of time at a high temperature with its ends in fixed positions, the stresses within the specimen gradually diminish. After enough time has elapsed, the tensile or compressive stresses may relax to a fraction of their original values.

Fatigue. Engineering components often experience repeated cycles of load or deflection during their service lives; under repetitive loading, most metallic materials fracture at stresses well below their ultimate tensile strengths by a process known as fatigue. The actual lifetime of the part depends upon both service conditions (eg, magnitude of stress or strain, temperature, environment, surface condition of the part) and the microstructure.

Thermal fatigue. An important source of failure in high temperature components is the change in stress produced in a constrained part by temperature cycling. This thermal fatigue is particularly pronounced in materials displaying low thermal conductivity, in which high thermal gradients are likely to occur.

Strengthening Mechanisms

Solid-solution strengthening. Commercial high temperature alloys generally contain substantial alloying additions in solid solution to provide strength, creep resistance, or resistance to surface degradation.

Table 1. Physical Properties for Selected High Temperature Metals

	Crystal structure at 22°C	at 1000°C	Melting point, °C	Density, g/cm³	Coefficient of expansion at RT, 10⁶/°C	Thermal conductivity at RT, W/(m·K)	Young's Modulus GPa[a]
Co	hcp	fcc	1495	8.85	13.8	69.0*	206.8
Ni	fcc	fcc	1453	8.90	13.3	92.0	206.8
Fe	bcc	fcc	1537	7.87	11.76	75.3	196.5
Cr	bcc	bcc	1890	7.2	6.2	66.9	248.2
Nb	bcc	bcc	2468	8.6	7.1	52.3	103.4
W	bcc	bcc	3410	19.3	4.5	20.1	344.7
Ta	bcc	bcc	2996	16.6	6.6	54.4	186.2
V	bcc	bcc	1900	6.1	9.7	31.0	125
Mo	bcc	bcc	2610	10.22	5.4	146	324.1

[a] To convert GPa to psi, multiply by 1.45×10^5.

In addition, the stronger alloys may contain elements that after suitable heat treatment or thermal-mechanical processing help form small coherent particles of an intermetallic compound or carbide.

Short-range order. Concentrated solid solutions are likely to show appreciable short-range order, with a preponderance of unlike nearest-neighbor atoms surrounding each lattice. Short-range order strengthening should provide an athermal increment to flow stress. However, because the degree of short-range order increases with decreasing annealing temperature, the short-range-order component of flow stress is sensitive to thermal history.

Grain-boundary strengthening. Grain boundaries are strong barriers to plastic flow at low and intermediate temperatures, provided that there is a significant lattice misorientation between the adjacent grains.

Cold working. Cold working is an important means of imparting strength to metals for low temperature service. When a cold-worked part is used at elevated temperature, the heavily dislocated structure anneals, leading eventually to recovery or recrystallization. Since the lattice vacancies produced by cold working tend to accelerate atomic rearrangements during annealing, cold working can actually be detrimental to high temperature strength. As a practical matter, cold working is applied principally to the group VIa refractory metals to improve strength and ductility; and to zirconium alloys, which tend to operate at low temperatures.

Precipitation hardening. Except for ferritic steels, which can be hardened either by the martensitic transformation or by eutectoid decomposition, most heat-treatable alloys are of the precipitation-hardening type. During heat treatment of these alloys, a controlled dispersion of submicroscopic particles is formed in the microstructure. The final properties of the alloy depend upon how the particles are dispersed, and upon their size and stability. Because precipitation-hardening alloys can retain their strength at temperatures above those at which martensitic steels become unstable, they are considered the most important class of high temperature materials.

Dispersion strengthening. Dispersion-hardened alloys are strengthened by particles of a second phase dispersed in their microstructures by methods other than heat treatment. Although similar in some respects to precipitation-hardened alloys, dispersion-hardened alloys differ in the types of secondary phases, the means for dispersing them, and the mechanism of strengthening. The temperature at which precipitation-hardened alloys retain useful strength is limited, because in order to be capable of heat treatment, the second phase must dissolve into the matrix at some temperature below the melting point of the alloy. No such limitation exists in the case of dispersion-strengthened alloys.

Composite strengthening. An alternative strengthening method that holds great promise for producing advanced high temperature alloys involves incorporation of fibers or lamellae of a strong, often brittle phase, in a relatively weak, ductile, metallic matrix. This technique has been commercially exploited for polymer-matrix, glass-reinforced materials such as fiber glass but has not yet been applied commercially to high temperature systems with a metal matrix. The principal benefit of composite strengthening is the high strength achievable at extremely high fractions of the melting point. In this respect, composite and dispersion strengthening offer substantially identical advantages for tensile or creep strength and fatigue resistance (see Laminated and reinforced metals).

Surface Stability

Oxidation. The terms metal oxidation and scaling in this context describe the attack of a metal or an alloy by oxidizing gases such as oxygen, sulfur, the halogens, or water vapor.

Oxide-scale morphology. The scale morphology is dependent on the condition of reaction, the time of oxidation, on the composition of the corrosive medium, and on the type and composition of the alloy involved. The most desirable scale morphology is the nonporous (compact) oxide. In order to maintain good oxidation resistance, at least one of the layers must be a compact and preferably slow-growing oxide.

Oxidation kinetics. The process of compact scale formation can be thought of as a series of partial processes occurring in the solid phase and at phase boundaries. The kinetics of scale formation can be further subdivided into individual heterogeneous surface reactions and the processes of transport of matter through the scale. Design of corrosion-resistant materials is based upon maintaining a slow-growing compact protective scale.

Hot corrosion. Hot corrosion is an accelerated form of oxidation that arises from the presence not only of an oxidizing gas, but also of a molten salt on the component surface. Most commonly, hot corrosion is associated with condensation of a thin molten film of Na_2SO_4 on superalloys used in first-stage gas blades, vanes, and other components of gas turbines. It is generally conceded that chromium content is the most important factor in hot-corrosion resistance.

Coatings. Some type of protective coating is often applied to extend the surface stability of superalloy components. These coatings provide an aluminum reservoir for growth of protective Al_2O_3 scales and inhibit further oxidation. Addition of chromium enhances hot-corrosion resistance and improves coating ductility, giving superior protection in a hot-corrosion environment.

Specific Alloy Systems

Plain-carbon and low-alloy steels. In this discussion, plain-carbon and low-alloy steels are those alloys containing up to 10 wt% chromium and 1.5 wt% molybdenum, plus small amounts of other alloying elements. These steels are generally cheaper and easier to fabricate than the more highly alloyed steels, and they are the most widely used class of alloys within their serviceable temperature range.

Plain-carbon steel is used where corrosion and oxidation resistance are not important and temperatures are below 425°C, although temperatures up to perhaps 535°C can be withstood for short periods. Although creep-rupture strength increases with the carbon content, carbon is generally kept below 0.20% in order to simplify welding. Plain-carbon steel is generally used in the hot-rolled or normalized condition (see Steel).

Of the common alloying elements in steel, molybdenum is the most effective in increasing creep-rupture strength, and the carbon–molybdenum steels generally have more than twice the creep-rupture strength of plain-carbon steel at the same temperature. Carbon–molybdenum steels are about equivalent to plain-carbon steels in metallurgical stability and resistance to corrosion and oxidation. The straight carbon–molybdenum steels should not be used continuously above 470°C because of graphitization.

Chromium is the most effective addition to improve resistance to corrosion and oxidation at elevated temperatures, and the chromium–molybdenum steels are an important class of alloys for use in steam power plants, petroleum refineries, and chemical-process equipment.

Stainless steels. The prime characteristics of stainless steels, those containing $\geq 11\%$ chromium, are corrosion and oxidation resistance, which increase as the chromium content is increased.

Although the stainless steels usually are specified in the wrought condition, a number of iron–chromium, iron–chromium–nickel, and nickel or cobalt-base alloys are produced as castings. Castings are classified as heat resistant when utilized in applications at 650°C or higher.

Modified stainless steels. The 12%-chromium ferritic superalloys are a group of proprietary steels that are essentially modifications of AISI 403 stainless steel. Examples are Crucible 422, Lapelloy (AISI 619), and Jessop-H46. The modifications include addition of up to several percent of molybdenum and/or tungsten to stiffen the matrix and of up to 0.5% of niobium and vanadium to improve the dispersion and stability of the carbides. Up to 2% nickel, copper, and aluminum also may be present in these steels. The modified steels have a substantially greater creep-rupture strength than the standard AISI 403 stainless steel, and about the same corrosion and oxidation resistance. Typical applications include high temperature bolts, blades for jet-engine compressors and for high temperature steam turbines, compressor and turbine disks for jet engines, boilers, superheater and reheater tubes, and valve parts. These steels are available in most wrought forms.

The highly alloyed austenitic stainless steels are proprietary modifications of the standard AISI 316 stainless steel. They have higher creep-rupture strengths than the standard steels, yet they retain the good corrosion resistance and forming characteristics of the standard austenitic stainless steels. The austenitic stainless steels play a large role in nuclear applications.

Nickel-base superalloys. The nickel-base superalloys are the most complex in composition and microstructures and in many respects the most successful high temperature alloys in current use. Among the most notable of these very strong alloys developed in the 1960s were Inconel 713 and a low-carbon version, 713 LC, as well as IN-100, B-1900, and MarM-200. Alloys developed in the United States and United Kingdom are widely used in French aircraft engines, although there are now efforts to develop alloys independently in France.

Apart from γ' and solid-solution strengthening, many alloys benefit from the presence of carbides, carbonitrides, and borides. An undesirable feature of the most highly alloyed superalloys is their tendency to develop unwanted phases such as sigma and mu phases.

The temperature capability of nickel-base alloys has been improved markedly by new processing techniques. Vacuum-arc melting was introduced to avoid loss of titanium and aluminum by oxidation and to keep gaseous-element contaminants down to acceptable levels. The vacuum-induction furnace, with capacity to 45.5 metric tons, provides optimum capabilities for close compositional control and removal of deleterious gases and other elements. Air-induction melting is still used for some high temperature alloys, but there is little opportunity for refining during melting. Vacuum-arc remelting is often used to develop optimum solidification structures (see Furnaces, electric).

The coarse grains developed by casting processes are usually deleterious to fatigue life. For parts such as turbine disks that are life-limited by fatigue rather than creep, fine grains are produced by powder metallurgical techniques (see Powder metallurgy).

The main applications of nickel-base alloys are as blades, disks, and sheet-metal parts of gas turbines.

Oxide dispersion-strengthened alloys. The latest class of ODS alloys to be developed, based upon Ni, Cr, and Al, relies upon an Al_2O_3 protective scale for dynamic oxidation resistance; either ThO_2 or Y_2O_3 is the dispersoid. The critical parameter in such alloys is the Cr : Al ratio, which determines the composition of the protective scale. The oxidation kinetics may depend on the alloy conditions, particularly Nb, Mo, and W content, some of which have an adverse effect on the oxidation resistance of the nickel-base superalloys.

Oxidation and hot corrosion. The oxidation behavior of multicomponent γ'-strengthened alloys can be estimated by considering the Ni–Cr–Al content of the alloy. The critical parameter in such alloys is the Cr–Al ratio which determines the composition of the protective scale. The oxidation kinetics of such alloys are generally quite complex and may depend on the alloy additions, particularly the Nb, Mo, and W content, some of which have an adverse effect on the oxidation resistance of the nickel-base superalloys. In general, good cyclic oxidation resistance is associated with Al_2O_3 and/or $NiAl_2O_4$ formation.

Iron–nickel-base superalloys. Iron–nickel-base superalloys have been developed primarily from the austenitic stainless steels. In the United States, these alloys included 19-9 DL, 16-25-6, and A-286. Later, higher nickel contents were employed to take advantage of the superior oxidation resistance of nickel and the beneficial effects of γ'-forming elements. All iron–nickel-base superalloys rely somewhat upon solid-solution hardening.

Since iron–nickel alloys tend to contain large amounts of ferrite stabilizers such as chromium and molybdenum, the minimum nickel content required to maintain a fcc matrix is about 25 wt%. High iron contents lower cost, increase fabricability, and tend to raise the melting point at the expense of poorer oxidation resistance than nickel-base alloys. Chromium is added for surface protection and solid-solution strengthening of γ. Molybdenum, too, is added for solid-solution strengthening, but is present also in carbides and γ'. Small quantities of boron or zirconium are added to improve workability and stress-rupture properties, and carbon is useful as a deoxidant and to provide MC carbides (eg, NbC, TaC, TiC, VC, and ZrC) to help refine grain size during hot working. Finally, ductility can be improved by small additions of magnesium, calcium, and certain rare-earth elements. Iron–nickel alloys are utilized in high temperature gas-cooled reactor (HTGR) steam plants, and are candidate materials for the liquid-metal-cooled, fast-breeder reactor (LMFBR).

Cobalt-base superalloys. The cobalt-base superalloys are used principally at 650–1000°C and relatively low stresses. They are strengthened primarily by carbide precipitation and solid–solution effects. They are now widely used as forgings and castings for nozzle vanes in gas-turbine engines because of good thermal shock and hot corrosion resistance, and in sheet-metal assemblies, such as combustion-chamber liners, tail pipes, and afterburners. Some of the industrial uses of cobalt-base alloys include grates for heat-treating furnaces, quenching baskets and pouring funnels for molten copper.

Refractory metals and their alloys. Many elements that could be called refractory are present in the periodic table, but those that have received the most attention for potential structural applications are the bcc metals, tantalum, molybdenum, niobium, and tungsten, all of which melt above 2000°C.

The technology of molybdenum is well advanced, and niobium too has received considerable attention because of its good fabricability. Tungsten, of course, has highly developed metallurgical technology because of its use as lamp and electron-tube filaments. The extreme brittleness of chromium has limited its use in the past. The low temperature ductility and corrosion resistance of tantalum make it attractive for cryogenic applications and for chemical equipment, but it has received only limited use at elevated temperatures.

Extrusion, forging, and sheet-rolling technologies have advanced rapidly as a result of demands by the aerospace industries, so that many of the refractory-metal alloys are now available in various mill forms. Electron-beam melting, plasma-arc spraying, fused-salt electroplating, and vapor deposition are among the specialized techniques used to produce and fabricate the refractory metals.

In many of their high temperature applications in the electrical and electronics industry, the refractory metals are protected by a vacuum or an inert gas, so that oxidation is not a problem. However, for most other high temperature applications, poor oxidation resistance limits their use.

New materials and processes. Metal–matrix composites produced by directional solidification of eutectic or near-eutectic compositions are considered prime candidates for application in advanced gas-turbine engines (see Laminated and reinforced metals). Several eutectic composites promise significant improvement over conventional nickel and cobalt-base superalloys in terms of tensile strength, impact strength, creep-rupture properties, and fatigue resistance.

Oxide and fiber-reinforced superalloys. Apart from the aligned eutectics, two other approaches promise to improve the maximum service temperature of current turbine-blade materials: oxide dispersion-strengthened superalloys (ODSS) and oxidation-resistant alloys reinforced with refractory-metal fibers (FRS).

Intermetallic compounds and other ordered phases. Intermetallic compounds and other ordered phases offer several very attractive features for high temperature applications. Within a long-range ordered lattice, all diffusion-controlled processes are slowed, as reflected by unusually high activation energies. Consequently, such processes as creep, recrystallization, and oxidation are markedly hindered by long-range order.

Potential applications include fusion reactors (inert gas, lithium, or vacuum environments) or hardware for use in outer space, eg, in small power sources.

Nomenclature
γ = fcc
γ' = Ni_3Al or Ni_3Al, Ti (long-range ordered and fcc)

N.S. STOLOFF
S.R. SHATYNSKI
Rensselaer Polytechnic Institute

C.T. Sims and W.C. Hagel, eds., *The Superalloys*, John Wiley & Sons, Inc., New York, 1972.

E. F. Bradley, ed., *Source Book on Materials for Elevated Temperature Applications*, American Society for Metals, 1979.

G.W. Meetham, ed., *The Development of Gas Turbine Materials*, Halsted Press, a division of John Wiley & Sons, Inc., New York, 1981.

HIGH TEMPERATURE COMPOSITES

High temperature composites are a special class of composites (see Composite materials) in which the purpose is to produce a more desirable balance of properties over a range of elevated temperatures or at a given required maximum temperature than can be readily achieved with a homogeneous material. The motivation is often the reduction of weight as in the case of aircraft, space vehicles, or rotating machinery; the potential of raising the allowable temperature in a thermally limited application such as a gas turbine; and the need for better thermal protection.

In addition to satisfying the principles applicable to composites in general, high temperature composites are subject to further restrictions. Fibers, matrices, surface treatments, and fabrication procedures must be selected with great care because of stresses that can result from thermal expansion mismatch, and because chemical reaction, dissolution, or microstructural changes may degrade the materials. Environmentally caused degradation of the constituent materials during service must be considered also.

Polymer Matrix

Polymers are particularly attractive as matrix materials because of their relatively easy processibility, low density, good mechanical properties, and often good dielectric properties (see Laminated and reinforced plastics). High temperature resins for use in composites are of particular interest to the high speed aircraft, rocket, space, and electronic fields. There has been a substantial search for resin structures capable of sustaining temperatures of 300°C or even $\geq$ 500°C. The most successful approaches have been to build thermally stable molecules based on aromatic and on heterocyclic ring structures. Among such polymer families that have been explored are those based on phenylene or on N-, O-, and S-containing aryl-heterocyclic repeat groups, such as quinoxaline, triazole, and imide structures (see Heat-resistant polymers; Polyimides).

A representative list of commercially available high temperature matrix-resin systems includes NR-150 (DuPont), Torlon (Amoco), PPQ 401 (Whittaker), Kapton (DuPont), Skybond (Monsanto), P13N (TRW), Kerimid-Kinel (Rhodia). These can be used to bond any inorganic reinforcing phase. However, they have been studied mainly in combination with carbon or glass fibers. Broadly speaking, these resins yield polymer-bonded composites that increase their upper-use temperature relative to epoxy resin systems by 100–150°C. These resins can also serve as useful high temperature adhesives.

Carbon Matrix

Carbon–carbon (C—C) composites (see Ablative materials) are especially desirable where extreme temperatures may be encountered, such as in rocket nozzles, ablative materials for reentry vehicles and disk brakes for aircraft. Other uses include bearing materials (qv) and hot-press die components.

The carbon matrix is produced by pyrolysis of resin or pitch infiltrants, or by chemical vapor deposition of carbon within the pores by the pyrolysis of methane, acetylene, benzene, etc, at reduced pressure (see Film deposition techniques). Representative properties are given in Table 1.

Metal Matrix

Metal–matrix composites offer less pronounced anisotropy and greater temperature capability in oxidizing environments than do the polymeric and carbonaceous counterparts. Although most metals or alloys could serve as matrices, in practice the choices are sharply limited. The low density metals aluminum, magnesium, and titanium are particularly favorable for aircraft applications. Chemical and physical stability are particularly important and pervasive concerns in metal–matrix composites. Composite systems are classified according to their degree of reaction as follows: Class I, filament and matrix do not react and are mutually insoluble; Class II, filament and matrix do not react but exhibit some solubility; and Class III, filament and matrix react to form a surface coating. Examples of each class are given in Table 2.

Aluminum and magnesium matrices. The most widely studied and best developed metal–matrix composite system is that made using boron or Borsic SiC-coated boron fibers in combination with aluminum alloys. Spun silicon carbide fibers as well as SiC whiskers incorporated into aluminum impart improved high temperature strength, stiffness, and dimensional stability. In part because carbon fibers are less expensive, there has been continuing interest in the use of carbon-fiber reinforcement (see Carbon and artificial graphite). The use of refractory oxide fibers to reinforce aluminum for automobile engine parts has been reported by Toyota. Magnesium composites have also been explored based on these fibers. Other reinforcements that have been used include steel, Be, W, Al_2O_3, and vitreous silica.

Titanium matrix. Because of their high temperature potential, there has been a considerable interest in titanium–matrix composites similar to the case for aluminum and magnesium. Although more dense than the latter metals, titanium is nevertheless a relatively low density material having a high potential use temperature and good mechanical properties that diminish with increasing temperature. It is a very reactive metal often used as a getter and does not form a protective oxide. The most important reinforcements have been boron, silicon carbide-coated boron, and silicon carbide.

Superalloy matrices. For even higher temperature usage, iron-, cobalt-, and nickel-based superalloys are commonly used as structural materials. Simultaneous unrelated or antagonistic requirements are often demanded of materials for use as gas-turbine blades, combustion systems, or rocket components, such as high strength and stiffness, combined with oxidation resistance, creep resistance, and hot-corrosion resistance. The use of high strength filaments as reinforcements can reduce the problems introduced by these constraints.

Table 1. Properties of Carbon–Carbon Composites Made with High Strength Carbon Fibers[a]

| Fiber and orientation | Composition, vol% | | | Density, g/cm$^{3\,b}$ | Flexural strength, MPa[c] | Flexural modulus, GPa[d] |
	Fiber	Matrix carbon	Voids			
high modulus carbon fiber						
Modmor type I; unidirectional lay-up	55–65	15–27	10–25	1.63–1.69	345–524	138–172
high strength carbon fiber						
Grafil type II; unidirectional lay-up	62–65	20–24	13–18	1.47–1.49	1034–1241	152–172

[a]Courtesy of the Propellants Explosives and Rocket Motor Establishment Procurement, Ministry of Defence, UK.
[b]To convert g/cm^3 to lb/in^3, divide by 27.68.
[c]To convert MPa to psi, multiply by 145.
[d]To convert GPa to psi, multiply by 145,000.

Table 2. Classification of Composite Systems[a]

Class I	Class II	Class III
copper–tungsten	copper(chromium)–tungsten	copper(titanium)–tungsten
copper–alumina	eutectics	aluminum–carbon (> 700°C)
silver–alumina	columbium–tungsten	titanium–alumina
aluminum–BN-coated B	nickel–carbon	titanium–boron
magnesium–boron	nickel–tungsten[b]	titanium–silicon carbide
aluminum–boron[c]		aluminum–silica
aluminum–stainless steel[c]		
aluminum–SiC[c]		

[a]Courtesy of A.G. Metcalfe and Academic Press, Inc.
[b]Becomes reactive at lower temperatures with formation of Ni$_4$W.
[c]Pseudo-Class I system.

Glass Matrix

Glass (qv) is a convenient matrix candidate since it is an inorganic thermoplastic material. It is a relatively inert material and lends itself to some of the composite-processing methods applicable to polymers such as melt infiltration and compression molding. However, glass has an elastic modulus that is comparable to that of the reinforcing fibers in most cases and a failure strain that is less than that of the fibers.

Carbon-fiber reinforcement. Glass has been used as a matrix material in combination with carbon, typically by infiltrating the fibers with a slurry of finely ground glass. Work of fracture of glass–carbon composites as great as 25 kJ/m^2 (11.9 ft·lbf/in^2) has been reported.

Oxide and silicon carbide-fiber reinforcement. High strength, high stiffness, and oxidation-resistant fibers, such as sapphire and silicon carbide, are potentially useful and are readily wetted by molten glass.

Ceramic matrix

These composites offer potential for overcoming some of the inherent flaw sensitivity of monolithic ceramics (qv), for improving the thermal shock resistance of otherwise very serviceable high temperature materials, and for facilitating fabrication in some cases (see also Glass-ceramics).

Nonfiber Composites

Particulate and layer reinforcements or second-phase additions are also often used in high temperature materials. Laminated-metal structures, such as titanium-clad steel, are used for chemical-process reaction vessels. Other metal combinations are used as thermostat-sensor elements. Mixtures of ceramics or nonductile intermetallics are particularly interesting because in some cases they offer extreme temperature capability, eg, as rocket-nozzle materials.

WILLIAM B. HILLIG
General Electric Co.

R.A. Signorelli in E. Scala and co-eds., *Proceedings of the 1975 International Conference on Composite Materials*, Vol. 1, Metallurgical Society of AIME, New York, 1976, p. 411.

J.N. Fleck, *Bibliography on Fibers and Composite Materials—1969–1972*, *MCIC Report 72-09*, National Technical Information Service, Springfield, Va., 1972 (AD 746214).

HISTAMINE AND HISTAMINE ANTAGONISTS

Histamine

Histamine (1), 1*H*-imidazole-4-ethylamine, β-imidazoylethylamine, 4-(β-aminoethyl)glyoxaline, C$_5$H$_9$N$_3$, derives its name from the amino acid histidine (2), from which it may be formed by decarboxylation.

(2) histidine   (1) histamine

As a biogenic amine responsible for a wide variety of physiological and pathological effects, histamine has been extensively studied as one of the fundamental chemical mediators of neural, secretory, and musculotropic action (inflammation).

Chemical properties. The free base histamine, mp 86°C, bp at 2.4 kPa (18 mm Hg) 209–210°C, crystallizes as deliquescent plates from chloroform. The best-known salts are the diphosphate, C$_5$H$_9$N$_3$·2H$_3$PO$_4$, mp about 140°C, and the dihydrochloride, C$_5$H$_9$N$_3$·2HCl, mp 244–246°C.

Pharmacology. Histamine is liberated in the tissues by the antigen–antibody reaction. Its pharmacology is consistent with the symptoms seen in anaphylaxis and the allergic state. The numerous manifestations of such reactions include asthma, hay fever, urticaria, vasomotor rhinitis, and angioneurotic edema. The effects of histamine are considered to be mediated by two sets of receptors termed H-1 and H-2.

Histamine Antagonists

Antihistamines diminish or abolish the actions of histamine on the body. They do so by a mechanism that involves the occupation of receptor sites in the effector cells to the exclusion of histamine. They do not prevent the production or release of histamine.

The conventional antihistamines—those that were found first and those that are given the generic label—are histamine H-1 receptor antagonists. They block most of the actions of histamine on the body but do not have an effect on the gastric acid secreting action of histamine. The secretion of gastric acid is inhibited by a class of antihistamines termed H-2 receptor antagonists.

H-1 receptor antagonists. Most of the histamine H-1 receptor antagonists belong to one of the following five chemical classes: ethanolamines (eg, diphenhydramine (3)); ethylenediamines (eg, tripelennamine alkylamines (eg, chlorpheniramine (4)); cyclizines (eg, hydroxyzine (5)) and phenothiazines (eg, promethazine (6)).

(3) diphenhydramine, X = H, R = OCH$_2$CH$_2$N(CH$_3$)$_2$

(4) chlorpheniramine, X = Cl, R = CH$_2$CH$_2$N(CH$_3$)$_2$

(5) hydroxyzine, X = Cl, R = —N⌒N—CH$_2$CH$_2$OCH$_2$CH$_2$OH

Antihistamines have wide therapeutic application, chiefly for the symptomatic control of allergic disease. Some antihistamines are used for their powerful antiemetic action, and still others are used chiefly for their sedative properties (see Hypnotics, sedatives, and anticonvulsants). Side effects of antihistamines vary in severity and incidence with each patient as well as with each drug. The most common side effect is sedation.

H-2 receptor antagonists. The history of the discovery of H-2 receptors and of H-2 receptor antagonists is very recent. The first drug with H-2 antagonist action, cimetidine (7), was introduced into medical practice in 1976–1977 as a gastric acid antisecretory agent. It is used widely

(6) promethazine

(7) cimetidine

and with considerable therapeutic success in treatment of peptic ulcer and related diseases (see Gastrointestinal agents).

R.W. FLEMING
Warner-Lambert Co.

J.M. GRISAR
Merrell-National Laboratories

"Histamine and Antihistamines" in M. Rocha e Silva, ed., *Handbook of Experimental Pharmacology*, Vol. 18, Pt. 1, Springer-Verlag, New York, 1966.

"Histamine II and Antihistamines, Chemistry Metabolism, and Physiological and Pharmacological Action" in M. Rocha e Silva, ed., *Handbook of Experimental Pharmacology*, Vol. 18, Pt. 2, Springer-Verlag, New York, 1978.

B.J. Hirschowitz, *Ann. Rev. Pharmacol. Toxicol.* **19**, 203 (1979).

HOLLOW-FIBER MEMBRANES

The development of hollow-fiber membrane technology has been greatly inspired by intensive research and the development of reverse-osmosis membranes during the 1960s. The excellent mass-transfer properties conferred by the hollow-fiber configuration soon led to numerous applications. Commercial applications have been established in the medical field (see Blood fractionation) and in water reclamation (purification and desalination) (see Water, supply and desalination), and others are in various stages of development. A hollow-fiber membrane is a capillary having a diameter of < 1 mm, and whose wall functions as a semipermeable membrane. The fibers can be employed singly or grouped into a bundle which may contain tens of thousands of fibers. In most cases, hollow fibers are used as cylindrical membranes that permit selective exchange of materials across their walls. However, they can also be used as "containers" to effect the controlled release of a specific material or as reactors to chemically modify a permeant as it diffuses through a chemically activated hollow-fiber wall (eg, loaded with immobilized enzymes) (see Enzymes, immobilized).

Hollow-fiber membranes, therefore, may be divided into two categories: "open" hollow fibers where a gas or liquid permeates across the fiber wall, while flow of the lumen medium—gas or liquid—is not restricted; and "loaded" fibers where the lumen is filled with an immobilized solid, liquid, or gas. Potential applications for the two types of hollow-fiber membranes are presented in Figure 1.

Hollow fibers offer two primary advantages over flat-sheet or tubular membranes. First, hollow fibers exhibit higher productivity per unit volume; second, they are self-supporting. The primary disadvantage of the hollow-fiber unit as compared to the other membrane configurations is its vulnerability to fouling and plugging by particulate matter.

Hollow fibers can be prepared from almost any spinnable material. The fiber can be spun directly as a membrane or as a substrate that is post treated to achieve desired membrane characteristics. Analogous fibers have been spun in the textile industry and are being investigated for the potential production of high bulk, low density fabrics. The technology employed in the fabrication of synthetic fibers applies also to the spinning of hollow-fiber membranes from natural and synthetic polymers.

Properties

Basic morphology. The desired fiber-wall morphology frequently dictates the spinning method. The basic morphologies are isotropic, dense or porous; and anisotropic, having a tight surface (interior or exterior) extending from a highly porous wall structure. Membrane-separation technology is achieved by use of these basic morphologies.

The anisotropic configuration is of special value. In the early 1960s, the development of the anisotropic (asymmetric) membranes exhibiting a dense, ultrathin skin on a porous structure provided momentum to the progress of membrane separation technology. Membranes with this structure permit high transport rates, yet can yield excellent separation. In addition, mechanical integrity problems associated with isotropic ultrathin membranes are obviated by use of anisotropic morphologies.

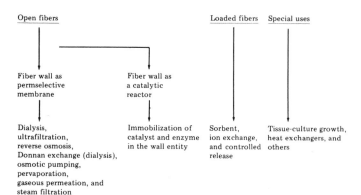

Figure 1. Hollow-fiber applications.

Figure 2. Anisotropic composite hollow-fiber membrane for separation of gas mixtures. The macrovoids-free walls support an ultrathin (0.1 μm) selective silicone polymer barrier. (The fiber was developed at State University of New York-C.E.S.F., Syracuse, N.Y.)

Anisotropic composite hollow-fiber membranes consisting of ultrathin dense silicone polymer resting on porous glassy polymer (below the glass-transition temperature) is shown in Figure 2. Such membranes are used for gas separations (eg, oxygen from nitrogen in air). These membranes are the frontrunners in industrial technology in the early 1980s. A market of 250×10^6 and a research and development effort of half this sum was expended for the first five years of the decade for gas separation.

Mechanical consideration and fiber dimensions. Because the hollow fiber is self-supporting, it is more inclined to mechanical failure than a supported flat-sheet membrane. A hollow fiber that is exposed to external pressure would exhibit a collapse pressure P_c that depends on the inner and outer fiber radii (IR, OR) and the Young's modulus E and Poisson ratio v of the material. The approximate relationship is given by the expression:

$$P_c = \frac{2E}{(1 - v^2)} \left[(OR - IR)/(OR + IR) \right]^3$$

Processing

Spinning of hollow fibers. There are three conventional synthetic-fiber spinning methods that can be applied to the production of hollow-fiber membranes: (1) melt spinning, in which a polymer melt is extruded into a cooler atmosphere which induces phase transition and controlled solidification of the nascent fiber; (2) dry spinning, in which the spinning dope, consisting of the polymer(s) predissolved in a volatile solvent mixture, is spun into an evaporative column; and (3) wet spinning, in which the polymer solution is spun into a liquid coagulating bath. A combination of the last two methods is applied for hollow-fiber fabrication in the dry-jet wet-spinning technique, in which the spinneret is positioned above a coagulation bath. In this process, all three mechanisms of formation (temperature gradient, solvent evaporation, and solvent-nonsolvent exchange) can be combined.

Macrovoids. Hollow-fiber membranes that are solution-spun by the above methods frequently exhibit large voids in conical, droplet, or lobe configurations. These voids may extend through the entire fiber cross section. The voids, in general, result from fast coagulation of a spinning solution that is relatively low in either polymer concentration or viscosity. Anisotropic fibers that are prepared by employing a strong coagulant often display large macrovoids and cavities. The use of a less severe quenching medium, on the other hand, yields a macrovoid-free hollow fiber.

The presence of macrovoids in hollow-fiber membranes is a serious drawback since it increases the fragility of the fiber and limits its ability to withstand hydraulic pressure. Such fibers have lower elongation and tensile strength.

Materials

Materials used in the manufacture of hollow-fiber membranes include cellulose, cellulose esters, polysulfones, polyacrylonitrile (PAN), Poly-

(methyl methacrylate) (PMMA), polyamides, polybenzimidazole (PBI), and glass hollow-fiber membranes.

Sorbent Fibers

Filled fibers. A growing interest in the encapsulation of specific active materials (eg, activated charcoal, enzymes, drugs) led to the development of encapsulation spinning, usually employing a wet- or dry-jet wet-spinning process. In the encapsulation process, the filling ingredient is suspended or dissolved in the core liquid medium (usually a coagulant) which is injected through the internal orifice during the spinning process. Subsequent quenching and washing in a water bath are employed.

The rationale for the development of such fibers is demonstrated by their application in the medical field, notably hemoperfusion.

Hollow fiber with sorbent walls. A cellulose sorbent and dialyzing membrane hollow fiber was reported in 1977 by Enak Glanzstoff AG. This hollow fiber, having an inside diameter of about 300 μm, has a double-layer wall. The advantage of such a fiber is that it combines the principles of hemodialysis with those of hemoperfusion.

ISRAEL CABASSO

State University of New York, Syracuse, NY

I. Cabasso, K.Q. Robert, J.K. Smith, and E. Klein, *J. Appl. Polym. Sci.* **21**, 1883 (1977).

I. Cabasso and A.P. Tamvakis, *J. Appl. Polym. Sci.* **23**, 1509 (1979).

A.F. Allegrezza, Jr., R.D. Burchesky, G. Gotz, R.B. Davis, and M.J. Coplan, *Desalination* **20**, 87 (1977).

HOLMIUM. See Rare-earth elements.

HOLOGRAPHY

A hologram is a recording of a wave, whereas a photograph is a recording of an image (see Photography). Rather than using a lens and recording an image of an object, as in a camera, the light distribution at an arbitrary distance from the object is preserved. The recording of the light distribution is called a hologram. When the hologram is illuminated, the same wave can be reconstructed and appears at the location at which it was recorded. The wave then continues to propagate through space (see also Reprography).

Fundamentals

Recording the wave. If a piece of photographic film was placed a distance from the object, the silver in the emulsion would respond to the energy in the wave. However, the parameters of interest are its amplitude and phase. To record these parameters, a second wave is needed to serve as a reference. Only a few waves of different wavelength can be recorded at one time. Assuming only one wavelength, the wave from the object can be described in the xy plane:

$$U(x, y)e^{i\Psi(x, y)} \equiv U(x, y) \tag{1}$$

where $U(x, y)$ gives the amplitude, $\Psi(x, y)$ describes the deviation of the phase of the reference wave from the plane, and $i = \sqrt{-1}$.

The wave of equation 1 propagates in the z-direction according to the expression:

$$U(x, y)e^{i(\gamma z - \omega t)}$$

where τ = wavelength

$$\gamma = 2\pi/\lambda.$$

Under the assumption that all waves of interest propagate in this fashion, $e^{i(\gamma z - \omega t)}$ can be dropped.

A second wave is now introduced to serve as a reference wave. For simplicity, a plane wave is assumed to be propagating at an angle ϕ with respect to the z-axis represented by:

$$V_o e^{i\zeta x}$$

and

$$\zeta = (2\pi/\lambda) \sin \phi$$

When the two waves superimpose on the photographic emulsion, the response is to the energy content of the sum of the two waves, ie, the exposure is proportional to the square of the sum of the complex amplitudes of the two waves:

$$E \sim |U(x, y)e^{i\Psi(x, y)} + V_o e^{i\zeta x}|^2$$
$$= U^2(x, y) + V_o^2 + 2V_o U(x, y) \cos [\zeta x - \Psi(x, y)] \tag{2}$$

If a hologram is recorded on a photographic emulsion, the transmittance of the emulsion can be assumed to be proportional to the exposure. If the hologram is illuminated by a single-frequency uniform plane wave moving in the same direction as the original reference wave (read-out wave):

$$W_o e^{i\zeta x}$$

the resulting distribution just beyond the hologram is proportional to the product of equation 2 and the read-out wave. If equation 2 is written in exponential form, the product can be written as:

$$W_o [U^2(x, y) + V_o^2] e^{i\zeta x} + W_o V_o U(x, y) e^{i2\zeta x} e^{-i\Psi(x, y)}$$
$$+ W_o V_o U(x, y) e^{i\Psi(x, y)}$$

The first term describes a wave moving in the direction of the hologram illuminating wave. The second term describes a wave moving at twice the angle of the illuminating wave (the sum of the angles of the reference and illuminating waves). This wave is proportional to the complex conjugate of the object wave. The third term describes a wave that is, aside from multiplying constants, exactly the original object wave. Thus, the recorded wave is reconstructed.

Phase holograms. A phase hologram is obtained if the recording produces a transparency that changes the refractive index or thickness to modify the phase of the illuminating wave.

Thick holograms. Still greater efficiencies can be achieved if the volume of the hologram-recording medium is utilized. The reference and object waves interfere throughout the medium. In a photographic emulsion, this causes surfaces of deposited silver to be formed. The interference fringes, caused by two interfering waves, lie along the bisector of the angle between the directions of propagation of the two waves.

Efficiency. The efficiency of a hologram is determined by the percentage of the illuminating energy that appears in the image. The efficiency depends upon whether the illuminating wave is modulated by absorption or by changes in the refractive index. In addition, the hologram recording material has to be thick enough for the Bragg effect to contribute to the image brightness. If not, the hologram acts like a thin, modulated, diffraction grating.

Recording materials. Photographic emulsions are the most commonly used materials for recording holograms. However, for fine interference fringes, high resolution emulsions are needed. Recording materials include dichromated gelatin, thermoplastic embossing, photoresists, photopolymers, and organic recording media.

Types of Holograms

The principle of recording a wave front by use of a reference wave is relatively straightforward. However, the configuration of the holographic setup dramatically influences the resolution, field of view, depth of field, and location of the image. In addition, special recording techniques provide various desirable characteristics.

The recording medium and manner in which the reference wave is introduced determine whether a surface or volume recording is obtained. Either amplitude or phase-modulation holograms is possible.

The location of the object is another distinguishing feature. An image of the object made on the hologram gives an image-plane hologram. If a lens is used to form a two-dimensional Fourier transform of the object, a Fourier-transform hologram results. Many other types have been described including Fresnel-zone holograms, far-field holograms, image-plane holograms, rainbow holograms, and pulsed-laser holograms.

Effect of the reference wave. Modification of the reference wave has vastly differing effects upon the image. In the case of a spherical wave, coming from a point source, it can be shown that when the point source is at infinity, thereby producing a uniform plane wave, the effect of the resolution of the holographic recording material is to limit the resolution in the image in a manner similar to placing a mask over the hologram to reduce its size. A smaller hologram, like a small lens, gives an image of lower resolution. However, when the point reference source is in the plane of the object, a reduction in the resolution of the recording material reduces the field of view over which the image can be seen as if a mask had been placed in the object plane and centered about the point reference. If the point reference is somewhere between, the effect is that of placing the mask somewhere between the object and the hologram; both the resolution and field of view are affected.

Another effect of the reference source is that with a nonpoint source, the resolution of the image can be degraded. Alternatively, the reference wave can be coded in special cases.

Recording and Illuminating Wavelength

The same wavelength need not be used in recording and illuminating a hologram. In fact, the goal of the original work by Gabor was to record a hologram in an electron imaging system and to use it in an optical system to correct for aberrations in the electron lenses. Holograms can be recorded with microwaves, acoustic waves, or infrared and illuminated in the visible region. The effect of a change in wavelength is to introduce aberrations and possible magnification.

Holographic Displays

Display holograms are either image-plane reflection holograms or transmission holograms; the latter may also be formed in a hybrid fashion. Reflective display holograms are the most common display holograms. Other holographic displays include rainbow holograms and multiplex holograms.

Holographic Microscopy

In holographic microscopy, the entire wave is recorded for future use, ie, all the information is stored and any plane can be viewed later.

Holographic Interferometry

In conventional interferometry, a high quality optical system is needed to provide the reference beam and the object must be either transparent or have a highly polished reflecting surface. In holographic interferometry, a rough or diffusely reflecting object can be used and high quality optical components are not needed. With this procedure, so-called live fringes are obtained, ie, changes in the object, and consequently, the fringes can be observed as they occur.

Data Storage

Holography can be used for high density data storage in a distributed manner. If a portion of the recording is damaged, the information is not lost and the signal-to-noise ratio simply decreases. Alignment is not critical. The chief problems relate to the noise, the means of recording the information, and how to get the information into the recording system (see also Deformation recording media).

WADE THOMAS CATHEY
University of Colorado

W.T. Cathey, *Optical Information Processing and Holography*, John Wiley & Sons, Inc., New York, 1974, Chapt. 6.

R.J. Collier, C.B. Burckhardt, and L.H. Lin, *Optical Holography*, Academic Press, Inc., New York, 1971, Chapt. 12.

J.W. Goodman, *Introduction to Fourier Optics*, McGraw-Hill, New York, 1968, Chapt. 8.

HORMONES

SURVEY

Vertebrate Hormones

A hormone is a chemical compound produced in specialized cells, usually in a ductless gland called an *endocrine gland*. The hormone is delivered directly to the blood stream and it exerts a physiological effect at a site remote from its origin. Hormones are chemical messengers that control and coordinate numerous chemical reactions in the living body; such reactions are under humoral control. Two other classes of compounds that act at extremely low levels to sustain life are enzymes and vitamins (see Enzymes, therapeutic; Vitamins). A fourth group of substances that is classified among the hormones is the prostaglandins (qv). These arahcidonic acid metabolites are present in most cells of the human body and exert a wide variety of effects.

Hormones are secreted at varying rates that follow established patterns (eg, diurnal rhythms for ACTH, reproductive cycles for the gonadotropins, carbohydrate content of the diet for insulin). Analysis by radioimmunoassay has greatly helped in the understanding of the mechanisms that control these fluctuations (see Radioactive tracers). These competitive binding assays allow measurements of the low hormonal concentrations in the venous circulation.

It is believed that hormones initiate their biological effects in a cell by first combining with a stereospecific receptor site. The mediators of the secretory responses in endocrine cells appear to represent energy-requiring processes, including changes in membrane permeability and in the rate of transport of substrates (eg, glucose), modifications of ion fluxes (eg, calcium), and changes in the concentration of cyclic nucleotides.

The pioneering accomplishments in structure analysis of hormones were the establishment of the primary structures of insulin by Sanger and of vasopressin and oxytocin by du Vigneaud. The determination of the complete amino acid sequence of 191 residues of human growth hormone also has been reported. The most recent development in the nature and role of hormones is the discovery of small peptides in the brain and in parts of the gastrointestinal tract that exert numerous actions on the brain, the pituitary, the pancreas, and on various vegetative functions (eg, blood pressure, etc) (see Opioids, endogenous).

Insect hormones. Hormones also are found among invertebrates; the best known are those from insects and crustaceans. Insects secrete minute amounts of two hormones: ecdysone from the prothoracic gland and juvenile hormone (allatum hormone) from the corpora allata.

Plant hormones. β-Indolylacetic acid (auxin) is a phytohormone that effects elongation of cells. It is formed in the apex of the plant, especially in the germinal shoot, and promotes stretching of the basally located cells (see Plant-growth substances). The gibberellins, originally isolated as fungal metabolites, also effect cell elongation and cause gigantism, and they also may stimulate cell division (see also Herbicides).

Pheromones. Substances that evoke humoral responses from different individuals of a given specie are pheromones. An example would be the sex attractants of insects (see Insect-control technology).

CATHERINE RIVIER
JEAN RIVIER
Salk Institute

W. Vale, C. Rivier, and M. Brown, *Ann. Rev. Physuol.* **39**, 473 (1977).

A.J. Prange, Jr., C.B. Nemeroff, M.A. Lipton, G.R. Breese, and I.C. Wilson in L.L. Iversen, S.D. Iversen, and S.H. Snyder, eds., *Handbook of Psychopharmacology*, Vol. 13, Plenum Press, New York, 1978, pp. 1–107.

R. Collu, A. Barbeau, J.R. Ducharme, and J.-G. Rochefort, eds., *Central Nervous System Effects of Hypothalmic Hormones and Other Peptides*, Raven Press, New York, 1979.

ANTERIOR-PITUITARY HORMONES

The anterior pituitary directly controls the adrenal cortex, the thyroid, and the gonads, and influences general body growth and carbohydrate metabolism. Hence, some of the hormones secreted by the anterior pituitary are called tropic hormones, because each stimulates a specific target organ—an endocrine gland—whose activity thereby is either initiated or enhanced. All eleven anterior pituitary hormones have been purified and characterized. Their primary structures are known, and seven have been either partially or totally synthesized. All have been grouped according to common structural features (see Table 1) (see also Opiods, endogenous).

Table 1. The Eleven Hormones of the Adenohypophysis

Group	Hormone	No. of amino acids[a]	Principal function
simple peptides	corticotropin (ACTH)	39	stimulates the adrenal cortex to produce cortical hormones
	β-endorphin	31	opiatelike activities
	lipotropin (LPH)		fat-mobilizing activity;
	β-LPH	91	β-LPH is the
	γ-LPH	58	prohormone for endorphins
	melanotropin (MSH)		darkening of skin
	α-MSH	human 13	(pigmentation)
	β-MSH	bovine 18	
	β-MSH	human 22	
simple proteins	prolactin (lactogenic hormone)	199	development and lactation of the mammary gland
	somatotropin (growth hormone, GH)	191	general body growth
glycoproteins	follitropin (FSH)	ovine 196 human 210	affects reproduction
	lutropin (LH, ICSH)	ovine 215 human 204	affects reproduction
	thyrotropin (TSH)	bovine 209 human 211	stimulates thyroid gland to produce thyroid hormones

[a] Values are for human extracts, unless indicated otherwise.

CHOH HAO LI
University of California, San Francisco

C.H. Li, *Proc. Am. Philos. Soc.* **116**, 365 (1972).

J.G. Pierce and T.F. Parson, *Ann. Rev. Biochem.* **20**, 465 (1981).

ANTERIOR-PITUITARYLIKE HORMONES

There are three extensively studied protein and glycoprotein hormones of nonpituitary origin whose chemistry and biological actions closely resemble their pituitary hormone counterparts. These hormones, the anterior-pituitarylike hormones, are products of the placenta (chorionic tissue).

Human Chorionic Gonadotropin (hCG)

The most commonly employed pregnancy tests are based on the immunological detection of hCG, usually in the urine. A radio-receptor assay for hCG enables detection of pregnancy at the day of the missed period and is useful in screening for twins and for predicting abortions.

Chemical analyses show hCG to be a glycoprotein containing 29–34 wt% sugar. The high sialic acid content imparts an acidic character to hCG and is involved in maintaining the presence of the hormone in the circulation for long periods of time. hCG is characterized by a high content of crystine and proline, relatively low numbers of histidine and methionine, and an absence of tryptophan. Amino terminal-group analyses indicate the presence of alanine and serine, which supports the presence of two polypeptide chains.

Equine Chorionic Gonadotropin (eCG; Pregnant-Mare Serum Gonadotropin, PMSG)

This hormone, like hCG, is present in high concentrations in the blood during the first trimester of a mare's pregnancy (40–160 d). Biologically, eCG has actions similar to both luteinizing hormone (LH), and follicle-stimulating hormone (FSH).

Physicochemical studies show eCG to be a large molecule of ca 55,000–60,000 mol wt. It is a very acidic material with an isoelectric point (pI) of ca 1.8. Chemical analyses show eCG to be a glycoprotein containing 41–45 wt% carbohydrate; thus it is similar to hCG. The high carbohydrate content of eCG is unique among the glycoprotein hormones, and the long half-life of this hormone in the venous circulation when injected into experimental rats is probably a reflection of the high sialic acid content. As with hCG and other gonadotropins, eCG possesses a high content of cystine and proline and a low content of histidine and methionine. Tryptophan appears to be absent. Amino-terminal-group analysis shows the presence of phenylalanine and serine, which is indicative of two polypeptide subunit chains.

Human Choriosomatotropin (hCS)

During pregnancy, hCS is detected rapidly in the blood and urine. Its precise physiological role in pregnancy is uncertain. The mol wt of hCS is 21,500—a value almost identical to that of human growth hormone (hGH); chemically, hCS and hGH are very similar molecules. Both are single polypeptide chains consisting of 191 amino-acid residues.

HAROLD PAPKOFF
University of California, San Francisco

S. Birken and R.E. Canfield, *Structure and Function of the Gonadotropins*, Plenum Press, New York, 1979, p. 47.

T.A. Bewley in C.H. Li, ed., *Hormonal Proteins and Peptides*, Vol. 4, Academic Press, New York, p. 61.

H. Papkoff, *Theriogenology* **15**, 1 (1981).

POSTERIOR-PITUITARY HORMONES

The hormones of the human posterior pituitary include the two nonapeptides, oxytocin and arginine vasopressin (ADH or antidiuretic hormone). The structures of oxytocin and the vasopressins and the other natural neurohypophyseal principles are depicted in Table 1. Each compound contains nine amino acids and each has a disulfide bridge between cysteine residues at 1 and 6. Variations among naturally occurring nonapeptides are found only in positions 3, 4, and 8. The relative activities of these compounds, when contrasted with their phyletic

Table 1. Structures of the Major Neurohypophyseal Principles

	Peptide hormone[a]
oxytocin	Cys–Tyr–Ile–Gln–Asn–Cys–Pro–Leu–GlyNH₂
arginine vasotocin	Cys–Tyr–Ile–Gln–Asn–Cys–Pro–*Arg*–GlyNH₂
mesotocin	Cys–Tyr–Ile–Gln–Asn–Cys–Pro–*Ile*–GlyNH₂
valitocin	Cys–Tyr–Ile–Gln–Asn–Cys–Pro–*Val*–GlyNH₂
aspartocin	Cys–Tyr–Ile–*Asn*–Asn–Cys–Pro–Leu–GlyNH₂
glumitocin	Cys–Tyr–Ile–*Ser*–Asn–Cys–Pro–*Gln*–GlyNH₂
isotocin	Cys–Tyr–Ile–*Ser*–Asn–Cys–Pro–*Ile*–GlyNH₂
arginine vasopressin	Cys–Tyr–*Phe*–Gln–Asn–Cys–Pro–*Arg*–GlyNH₂
lysine vasopressin	Cys–Tyr–*Phe*–Gln–Asn–Cys–Pro–*Lys*–GlyNH₂

[a] Structural changes relative to oxytocin are italicized.

distributions and structural variations, suggest an evolutionary relationship among the various peptides.

Medicinal Uses

Oxytocin has been used widely to induce labor and facilitate nursing. Vasopressin has been used in the management of gastrointestinal hemorrhage and as an antidiuretic agent.

ARNO F. SPATOLA
University of Louisville

B. Berde, *Handb. Exp. Pharmacol.* **23**, 1 (1968).

R. Walter, *Ann. N.Y. Acad. Sci.*, 1248 (1975).

J. Meienhofer and C.W. Smith in C.W. Smith and J. Meienhofer, eds., *The Peptides*, Vol. 8, "Chemistry and Biology of Neurophysical Hormones," Academic Press, Inc., New York, 1984.

ADRENAL-CORTICAL HORMONES

All natural adrenocorticoids are derivatives of the planar ring system 5α-pregnane (**1**). Substituents lying above the plane of the rings are assigned a β-stereochemical configuration which is indicated by a solid line, and substituents lying below the plane of the rings are assigned an α-stereochemical configuration which is indicated by a dotted line. Angular methyl groups at C-10 and C-13 have the β-configuration and are often shown simply by solid bonds. Tertiary hydrogen atoms at C-8, C-9, C-14, and C-17 are usually omitted unless their stereochemistry differs from that shown in (**1**).

(**1**) 5-α-pregnane

The antiarthritic effects of cortisone were discovered in 1949. Today, corticosteroids are commonplace drugs with a world market of approximately 10^9. They are used mainly in topical preparation for the treatment of inflammatory and allergic conditions of the skin (see also Analgesics; Anti-asthmatic agents; Histamine and histamine antagonists). The biological effects of the corticoids include those on carbohydrate metabolism and inflammation (the glucocorticoids) and on electrolyte balance (the mineralocorticoids). Biosynthesis of the adrenocorticosteroids takes place in specialized zones within the adrenal glands.

Classification of the adrenocortical steroids generally is based upon their main biological activities. Thus corticoids, which act mainly in the carbohydrate/thymus/anti-inflammatory group of assays, are termed glucocorticoids, whereas those that act mainly upon electrolyte balance are termed mineralocorticoids. However, the principal human glucocorticoids, cortisol (**2**), has significant mineralocorticoid activity so that its classification as a glucocorticoid is open to criticism.

(2) R =
(3) R = O

The mineralcorticoids, the most important of which is the hemiketal aldosterone (**4**) ⇌ (**5**), have only weak anti-inflammatory activity. They control the balance of Na^+, K^+, and H_2O in the body, thus ensuring Na^+ and H_2O retention without a buildup of K^+, which in high concentration is a tissue poison. The market for mineralocorticoids is extremely modest. Their main value is in the treatment of Addison's disease.

(4)   aldosterone   (5)

Glucocorticoids

Hydrocortisone and prednisolone. Following the discovery of the anti-inflammatory actions of cortisone (**3**) and cortisol (**2**), there was a need not only to develop highly efficient routes to the corticoids, but to discover novel structures with fewer side effects than those of the corticoids eg, sodium and water retention, reduced carbohydrate tolerance (steroid diabetes), osteoporosis, and depressed host defense.

A major difficulty in the manufacture of corticosteroids was the lack of an abundant raw material containing an 11-oxygenated function. Microorganisms capable of introducing an 11-hydroxyl group (both α and β) into a steroid were discovered to form the basis for most industrial manufacture procedures. *Corynebacterium simplex* converts cortisone (**3**) and cortisol (**2**) into their 1-dehydro derivatives, prednisone (**6**) and prednisolone (**7**), respectively. These steroids surpass their parent hormones in antirheumatic and antiallergic activity and produce lower mineralocorticoid activity and other side effects. Nearly all corticoids on the market—other than cortisone—are 1-dehydro steroids. The Upjohn Company is the major producer of hydrocortisone and prednisolone in the United States.

(6) R = OH prednisone
(7) R = O prednisolone

Other glucocorticoids include 9-fluoroderivatives of corticoids, 16α-hydroxy derivatives of corticoids and their acetonides, methylated glucorticoids, 6-fluorocorticoids, 17β-acylated corticoids, 20-ketopregnan-21-oic acids, the 17-carboxy androstanes, and the D-homocorticoids.

Biological activity. In addition to endowing an organism with a capacity to resist stress, noxious stimuli (eg, bacterial endotoxins), and environmental change, corticoids affect the central nervous system and can induce euphoria. They have an osteoporotic (catabolic) effect on bone. They also lower the immune response and thus are particularly valuable in treating hypersensitivity diseases. Some of the anti-inflammatory actions of the corticoids result from their inhibition of the release of prostaglandins (qv). Corticoids are used widely in the treatment of eczematoid and allergic skin disease and pruritis, inflammatory eye diseases, rheumatoid arthritis when direct injection into the joints may be employed, bronchial asthma, lymphocytic leukemia, chronic ulcerative colitis, congenital adrenal hyperplasia, nephrotic syndrome when attributable to lupus erythematosus or to primary renal disease, systemic lupus erythematosus, cerebral edema and increased intracranial pressure, and substitution therapy.

Glucorticoid antagonists. Certain steroids, eg, progesterone, 11-deoxycortisol and Δ¹-11-oxa-11-deoxocortisol, antagonize glucocorticoid activity. 17α-Carboxamido steroids antagonize glucocorticoid action at the target cell.

Transport Proteins and Receptors

Transport mechanisms. Corticosteroids are transported in the blood as free steroids, as loosely bound complexes with specific proteins, and as conjugates with sulfuric and glucuronic acids. The primary high affinity binding protein for transport of C-21 steroids in human blood is called transcortin or corticosteroid-binding globulin (CBG).

Intracellular recognition mechanism. All steroid-responsive cells have a recognition mechanism involving specific macromolecules in both the cytoplasm and nucleus that bind covalently to, and with high affinity to, the hormone. These receptors appear to be asymmetric proteins with molecular weights between 50,000 and 150,000. In conformity with their function, the receptors recognize subtle differences in steroid structure that are able to influence steroid specifity or selectivity in the binding process. Such selectivity and binding affinity bears directly upon the biological activity.

VLADIMIR PETROW
Consultant

R.H. Silber, *Ann. N.Y. Acad. Sci.* **82**, 821 (1959); S. Tolksdorf, *Ann. N.Y. Acad. Sci.* **87**, 829 (1959); I. Ringler in R.I. Dorfman, ed., *Methods in Hormone Research*, Vol. 2, Academic Press, Inc., New York, 1964, p. 227. O.J. Lorenzetti, *Curr. Therap. Res.* **25**, 92 (1979); T.L. Popper and A.S. Watnick in *Medicinal Chemistry, a Series of Monographs*, Vol. 13, R.A. Scheerer and M.W. Whitehouse, eds., Academic Press, Inc., New York, 1974, p. 245.

"Adrenocortical Steroids" in L.S. Goodman and A. Gilman, *The Pharmacological Basis of Therapeutics*, Macmillan Co., New York, Chapt. 72.

J.R. Stockigt in M.H. Briggs and G.A. Christie, eds., *Advances in Steroid Biochemistry and Pharmacology*, Vol. 5, Academic Press, Inc., New York, 1976, p. 161.

BRAIN OLIGOPEPTIDES

The numerous oligopeptides that have been identified in the central nervous system exert gastrointestinal and behavioral effects as well as modify pituitary function. The oligopeptides have emerged as a class of new putative extracellular messenger substances that may play an important role in the transmission of information and in the regulation of physiological mechanisms (see Neuroregulators). Primary structures of some of these brain peptides are given in Table 1.

Thyroliberin (TRF)

TRF's greatest hypothalamic concentration is found in the mammalian median eminence. Its ability to stimulate thyroid-stimulating hormone (TSH) secretion in the rat and mouse justified its purification from the hypothalamic extracts. It was subsequently found to release prolactin (PRL) and growth hormone (GH) under specific conditions. TRF has been reported to alleviate depressive symptoms, to reverse the duration of anesthesia and hypothermia induced by a number of substances, and to increase spontaneous motor activity. As a neurotransmitter candidate, it modifies the rate of discharge of neurons and the secretion of monoamines (see Analgesics; Hypnotics, sedatives, and anticonvulsants).

Gonadoliberin (LRF)

The highest concentration of LRF is found in the hypothalamus. LRF acutely stimulates LH and FSH secretion. Additionally, there is evidence that it can act on the central nervous system to modulate sexual behavior. Paradoxically, however, its long-term administration is associated with antigonadal effects that include termination of pregnancy, decreased gonadal weights, and lowered steroid secretion, possibly through desensitization at pituitary and gonadal levels and through alterations in steroidogenesis.

Somatostatin (SS)

SS is distributed throughout the brain and the gastrointestinal tract. Apart from its inhibitory effect on growth hormone, TSH, and on glucagon and insulin secretion, it exerts some neurotropic actions, eg, as a tranquilizer and as a spontaneous motor-activity depressor. It also lengthens barbiturate anesthesia time and induces sedation and hypothermia.

Table 1.   Primary Structure of Some Characterized Brain Oligopeptides

Hormone	Primary structure
angiotensin II (Ang II)	Asp–Arg–Val–Tyr–Ile–His–Pro–Phe–OH
bombesin (frog) (BN)	pGlu–Gln–Arg–Leu–Gly–Asn–Gln–Trp–Ala–Val–Gly–His–Leu–Met–NH₂
corticoliberin (ovine) (CRF)	Ser–Gln–Glu–Pro–Pro–Ile–Ser–Leu–Asp–Leu–Thr–Phe–His–Leu–Leu–Arg–Glu–Val–Leu–Glu–Met–Thr–Lys–Ala–Asp–Gln–Leu–Ala–Gln–Gln–Ala–His–Ser–Asn–Arg–Lys–Leu–Leu–Asp–Ile–Ala–NH₂
α-endorphin (End)	Tyr–Gly–Gly–Phe–Met–Thr–Ser–Glu–Lys–Ser–Gln–Thr–Pro–Leu–Val–Thr–OH
β-endorphin (ovine) (End)	Tyr–Gly–Gly–Phe–Met–Thr–Ser–Glu–Lys–Ser–Gln–Thr–Pro–Leu–Val–Thr–Leu–Phe–Lys–Asn–Ala–Ile–Ile–Lys–Asn–Ala–His–Lys–Lys–Gly–Gln–OH
γ-endorphin (End)	Tyr–Gly–Gly–Phe–Met–Thr–Ser–Glu–Lys–Ser–Gln–Thr–Pro–Leu–Val–Thr–Leu–OH
[Leu⁵]-enkephalin (Enk)	Tyr–Gly–Gly–Phe–Leu–OH
[Met⁵]-enkephalin (Enk)	Tyr–Gly–Gly–Phe–Met–OH
gonadoliberin (LRF)	pGlu–His–Trp–Ser–Tyr–Gly–Leu–Arg–Pro–Gly–NH₂
hypothalamic somatoliberin (human) (GRF)	Tyr–Ala–Asp–Ala–Ile–Phe–Thr–Asn–Ser–Tyr–Arg–Tys–Val–Leu–Gly–Gln–Leu–Ser–Ala–Arg–Lys–Leu–Leu–Gln–Asp–Ile–Met–Ser–Arg–Gln–Gln–Gly–Glu–Ser–Asn–Gln–Glu–Arg–Gly–Ala–Arg–Ala–Arg–Leu–NH₂
neurotensin (NT)	pGlu–Leu–Tyr–Glu–Asn–Lys–Pro–Arg–Arg–Pro–Tyr–Ile–Leu–OH
somatostatin (SS)	Ala–Gly–Cys–Lys–Asn–Phe–Phe–Trp–Lys–Thr–Phe–Thr–Ser–Cys–OH
substance P (SP)	Arg–Pro–Lys–Pro–Gln–Gln–Phe–Phe–Gly–Leu–Met–NH₂
thyroliberin (TRF)	pGlu–His–Pro–NH₂

Neurotensin (NT)

The highest concentration of NT is in the hypothalamus. The many pharmacological actions of neurotensin include induction of hypotension, increased vascular permeability, hyperglycemia, increased intestinal motility, and inhibition of gastric secretion. Its effects on insulin, glucagon, and SS secretion appear to depend on glucose concentration. Its effects on prolactin, growth hormone, and gonadotropin secretion seem to depend on whether it is administered intravenously or intracerebrally.

Substance P (SP)

Substance P was originally detected in the acid-alcohol extracts of equine brain and intestine, and subsequently was isolated from bovine hypothalami. It was first isolated for its hypotensive action and stimulation of rabbit jejunum contraction and later by following its ability to produce salivation in rats. Substance P also stimulates glucagon secretion and produces hyperglycemia in the rat, stimulates smooth-muscle contraction in the guinea pig vas deferens and ileum, and elevates GH and PRL secretion.

Bombesin (BN)

Bombesin has been found throughout the mammalian intestine, lung, brain, plasma and gut. It releases gastrin, gastric acid, and cholecystokinin; increases pancreatic secretion; produces contraction of the gall bladder; and increases blood pressure. It also produces hypothermia in rats exposed to a cold atmosphere. Its intravenous administration stimulates PRL and GH secretion, but has been reported to inhibit basal PRL secretion after central administration.

Endorphins and Enkephalins

These peptides specifically displace bound ^{3}H-naloxone from brain opiate receptors. β-endorphin is more potent than the short peptides.

Additionally, the peptides elicit a number of morphinominetic activities following intracerebroventricular injection, eg, analgesia and catatonia, and exerts behavioral effects (see Opioids, endogenous).

Corticotropin-releasing factor (CRF), originally isolated from the ovine hypothalamus, stimulates the secretion of ACTH, β-endorphin and glucocorticoids. When injected into the brain, CRF elevates mean arterial pressure, heart rate, and plasma–catecholamine levels. It also differentially modifies behavioral responses to familiar and novel environments.

Somatoliberin (GRF), originally characterized from human pancreatic tumors and later shown to be identical to human hypothalamic GRF, stimulates the secretion of GH when injected into humans, fish, chickens, the calf, pig, etc. It may have great therapeutic impact to treat dwarfism, and may be used commercially to promote growth of domestic animals.

Miscellaneous Peptides

Miscellaneous peptides include gastrin amino acids, vasoactive intestinal polypeptide, cholecystokinin, renin, angiotensin-forming enzyme and bradykinin.

JEAN RIVIER
CATHERINE RIVIER
Salk Institute

W. Vale, C. Rivier, and M. Brown, *Ann. Rev. Physiol.* **39**, 473 (1977).

R.L. Moss, *Ann. Rev. Physiol.* **41**, 617 (1979).

W. Vale, C. Rivier, and M. Brown in *Physiology of the Hypothalamus* Vol. 2, 1980, Chapt. 3.

W. Vale, C. Rivier, J. Spiess, M.R. Brown, and J. Rivier in D. Krieger, M. Brownstein, and J. Martin, eds., *Brain Peptides*, John Wiley & Sons, Inc., New York, 1983.

SEX HORMONES

The natural sex hormones are derivatives of the planar tetracyclic structures 5α-androstane (**1**), estrane (**2**), gonane (**3**), estra-1,3,5(10)-trien-3-ol (**4**), and 5α-pregnane (**5**).

(1)
5α-androstane

(2)
estrane

(3)
gonane

(4)
estra-1,3,5(10)-trien-3-ol

(5)
5-α-pregnane

The androgens are derived from 5α-androstane (**1**) and, with few exceptions, have hydroxyl or oxo groups at *C*-3 and *C*-17. Removal of the *C*-19 methyl group gives the tetracyclic structure, estrane (**2**), also known as 19-norandrostane. This ring system is important because many anabolic agents and contraceptive gestagens are derived from it. The fully demethylated ring system (**3**) is represented by a number of totally synthetic structures derived from 13β-ethylgonane, also known as 18-methylestrane or 18-methylnorandrostane. Naturally occurring estrogens are derivatives of (**4**) and have oxo or hydroxy substituents at *C*-17, *C*-16, etc. The natural gestagens are derivatives of (**5**) and most have hydroxy or oxo substituents at *C*-3 and *C*-20.

The sex hormones are a group of steroids (qv) secreted by the gonads (ovaries or testes), the placenta, and the adrenals. In addition, many tissues have the capacity to biosynthesize the sex hormones from pre-formed precursors that contain 19 and 21 carbon atoms. Generally, the biological properties of the androgens are associated with maleness, those of the estrogens with femaleness, and those of the gestagens with pregnancy. Both the male and the female secrete androgens and estrogens and the difference between the sexes is the amounts of secreted hormones—not the absence of one or the other group.

Steroid hormones are chemical messengers. On entry into the target cell, they are recognized by it and set in motion the specific intracellular events that constitute their biological responses. The intracellular-recognition mechanism involves specific macromolecules–receptors in the cytoplasm and nucleus that bind covalently and with high affinity to the hormone. The receptors appear to be asymmetric proteins having molecular weights of 50,000–150,000. Such receptors recognize subtle steroid structural differences that can influence steroid specificity or selectivity in the binding process. Such selectivity and binding affinity for the receptor bear directly upon the biological activity.

Androgens

The androgens are biosynthesized in the body from cholesterol. Listed in decreasing order of quantity, the most important androgens that are secreted are dehydroepiandrosterone (**7**), androstenedione (**6**), testosterone (**8**), and 5α-dihydrotestosterone (**9**); their biological activities follow the reverse order. In the female, similar quantitative relationships exist, except that, in addition, the female secretes progesterone in cyclical fashion from the ovaries. The estrogens are formed from androstenedione (**6**) and testosterone (**8**) and, again, their release in the female is cyclical.

(6) R = O, X = H
androstene-3,17-dione

(7)
dehydroepiandrosterone
(prasterone; DHA)

(8)
testosterone

(9)
5α-dihydrotestosterone

In addition to the above androgens, cortisol and cortisone (see Hormones, adrenal-cortical) are degraded into 11β-hydroxyandrost-4-ene-3,17-dione and androst-4-ene-3,11,17-trione (adrenosterone), respectively, which have weak androgenic activity.

The androgens stimulate the growth and development of male sexual characteristics, eg, facial hair, certain muscles and bones, kidney weight, genitalia, prostate, and seminal vesicles. Significant nitrogen, phosphorus, calcium, and potassium retention follows treatment with testosterone in men and women. These growth effects, or anabolic effects, are not related strictly to sexual function.

Testosterone. Testosterone (**8**) is the most important androgen secreted by the testes. It is prepared readily from dehydroepiandrosterone (prosterone; androstenolone; DHA) (**7**), which is a preferred intermediate for the manufacture of many androgenic and anabolic agents.

Testosterone has a limited market in androgen-replacement therapy. Its value in the treatment of male impotence is highly controversial unless androgen deficiency is present. Its use as an anabolic or body-building agent is limited strictly by androgenic side effects.

Anabolic agents. There has been considerable research effort to develop testosterone analogues with improved anabolic-androgenic indexes. Testosterone derivatives include 17β-hydroxy-17α-methylandrosta-1,4-dien-3-one (methandrostenolone), 9α-fluoro-11β,17β-dihydroxy-17α-

methylandrost-4-en-3-one(fluoxymesterone), and 7β,17α-dimethyl-testosterone (calusterone). Dihydrotestosterone derivatives include 2-hydroxymethylene-17α-methyl-17β-hydroxy-5α-androstan-3-one (oxymetholone), 2α,17α-dimethyl-17β-hydroxy-5α-androstan-3-one, propionate (dromostanolone propionate) and 17β-hydroxy-5-α-androst-1-en-3-one. 19-Nortestosterone derivatives include 19-nortestosterone decanoate (nandrolone decanoate injection) and 17β-acetoxyestra-4,9,11-trien-3-one (trienolone acetate). Heterosteroid derivatives include 17β-hydroxy-17α-methylandrostane-[3,2-c]pyrazole (stanozolol) and 17β-hydroxy-17α-methyl-2-oxa-5α-androstan-3-one (oxandrolone).

Anabolic steroids are used for senile debility, anorexia, asthenia, and convalescence. They also are widely used by athletes in training. They can and often do improve athletic performance, but their side effects on prolonged use represent a health hazard. Most 17-alkylated anabolic agents produce varying degrees of BSP (bromosulfophthalein) retention in the liver.

Production of 19-norsteroids. Some of the main industrial groups that produce 19-norsteroids include Roussel-Uclaf, Hoffmann-LaRoche Inc., Schering A.G., G.D. Searle & Co., and Wyeth Laboratories.

Estrogens

The estrogens are characterized by their ability to induce estrus in the female mammal. The three most important estrogens in humans are estrone (10), estradiol (11), and estriol (12).

(10) R = H
estrone

(11) R = H
estradiol

(12)
estriol

Quantitatively, androstenedione (6) is the main substrate for the production of estrogens in the body, with smaller amounts of estrogens derived from testosterone (8). In addition, the female secretes significant quantities of estriol (12). Small quantities of estrogens are secreted by the adrenal glands in both sexes and by the gonads in the male. In addition to the low ground level of estrogen production in the ovaries, the ovulatory menstrual cycle is characterized by increasing estrogen production by an ovarian follicle to a peak level just prior to ovulation on day 14. This is followed by a second estrogen peak on ca day 22 as a result of estrogen biosynthesis by the corpus luteum. Unless pregnancy occurs, the corpus luteum degenerates with a consequent drop in estrogen production to ground levels. Should pregnancy occur, estrogen levels continue to rise through estrogen biosynthesis by the developing placenta. At menopause, ovarian estrogen production terminates.

In addition to the adrenals and gonads (and placenta in the pregnant woman), many tissues contain the enzyme aromatase, and thus are able to convert circulating androgens (mainly androstenedione (6)) into estrogens. These include the liver, hair, skin, brain, and (abdominal) fat. The latter, in particular, serves as an important extragonadal source of estrogen in the postmenopausal female.

Estrogen-replacement therapy. Estrogens are used primarily in estrogen replacement therapy in the hypogonadal female and, in particular, in the menopausal and postmenopausal female. Their use is indicated in the treatment of menopausal flushing and in such conditions as senile vaginitis. Estrogens used in replacement therapy include estrone sulfate, estradiol and its esters, ethinyl estradiol and mestranol, and catechol estrogens.

Several manufacturers synthesize estrogens including Roussel-Uclaf, Hoffmann-LaRoche Inc., and Schering A.G.

Gestagens

The human corpus luteum, or yellow body, secretes estrogen and progesterone which prepare the uterine lining for implantation of a fertilized ovum and maintain pregnancy. Maintenance of pregnancy represents the historical use for progesterone, which is only weakly active when taken orally. Development of potent, orally active progestational agents, the gestagens, has expanded their uses to replacement therapy in gynecological disorders, as the progestational component in oral contraceptives, the treatment of certain endocrine-dependent tumors, and in the veterinary field.

Progesterone. Progesterone is a naturally occurring hormone. It is formed in the adrenals, in the ovaries (the testes do not produce significant quantities of progesterone), and in the placenta from cholesterol. Cholesterol is converted enzymically into (20R,22R)-20,22-dihydroxycholesterol which is degraded to pregnenolone and then to progesterone.

Most of the progesterone used in the United States is manufactured by the Upjohn Company from stigmasterol. All gestagens used in the United States in oral contraceptives may be regarded as derived structurally from ethisterone. All other gestagens in use are derived structurally from either progesterone or from ethisterone.

Gestagens derived from progesterone. These gestagens include hydroxyprogesterone hexanoate and acetate, medroxyprogesterone acetate, megestrol acetate, melengestrol acetate, medrogestone and dydrogesterone.

Gestagens derived from ethisterone. These include dimethisterone, norethindrone, norethynodrel, ethynodiol diacetate, norethindrone 17-acetate, quingestanol acetate, lynestrenol, norgestrel and norgestrienone.

Hormone Antagonists

Androgen antagonists. It seems likely that dihydrotestosterone (9) is the true somatic androgen in man and that testosterone is a pro-hormone insofar as hormonal support of the secondary sexual characteristics is concerned. Androgens therefore can be antagonized by inhibiting the enzyme, 5α-reductase, which converts (8) to (9), or by means of an enzyme that is effective at the receptor level. Both types of compounds are known.

$$\underset{\text{testosterone}}{(8)} \xrightarrow{\text{5α-reductase}} \underset{\text{dihydrotestosterone}}{(9)}$$

5α-Reductase inhibitors include 3-oxoandrost-4-en-17β-carboxylic acid, 17-N,N-diethylcarbamoyl-4-methyl-4-azo-5α-androstan-3-one, and 6-methyleneprogesterone. The last compound is an irreversible inhibitor of the k_{cat} type. Androgen antagonists at the receptor level include cyproterone acetate and 17β-hydroxy-2,2,17α-trimethylestra-4,9,11-trien-3-one. Many compounds of both types have been reported. Though still experimental, their main uses will probably be in the treatment of acne, hirsutism, androgenic alopecia, virilism, prostatic hypertrophy, and prostatic cancer. Cyproterone acetate has use in hirsutism and prostatic cancer.

Estrogen antagonists. Nonsteroidal antiestrogens have limited utility, eg, in the treatment of certain forms of breast cancer. Many steroids show antiestrogenic activity; one of the new compounds is 11α-methoxyethynylestradiol (see Hormones, nonsteroidal estrogens).

Gestagen antagonists. An effective antagonist of progesterone is expected to find wide utility as a contraceptive anti-implantation agent. Such an antagonist [11,β-(dimethylaminophenyl)-17β-hydroxy-17α-propynylestra-4,9(11)-dien-3-one; RV-48b] has recently been reported and appears to be effective in preliminary clinical trials.

VLADIMIR PETROW
Consultant

R.J.B. King and W.I.P. Mainwaring, *Steroid-Cell Interactions*, Butterworths, London, UK, 1974.

R.I. Dorfman in R.I. Dorfman, ed., *Methods in Hormone Research*, Vol. 2, Academic Press, Inc., New York, 1962, p. 304; F.A. Kincl, Vol. 4, 1965, p. 21.

M.K. Agarwal, ed., *Antihormones*, Elsevier, Amsterdam, Neth., 1979.

NONSTEROIDAL ESTROGENS

Only DES (**1**), dienestrol (**2**), and chlorotrianisene (**3**) are used in human medicine, and only zeranol (**4**) and DES can be used in animals intended for human consumption. The carcinogenicity of DES in humans has stigmatized estrogen use, and although carcinogenicity has not been determined for each estrogen, estrogen use has been limited.

(**1**) DES

(**2**) dienestrol

(**3**) chlorotrianisene

(**4**) zeranol

Estrogens

Diethylstilbestrol. Diethylstilbestrol (DES) (**1**) can exist in a cis or trans form. X-ray crystallography shows that DES has a nonplanar structure and at the planes of the aromatic rings parallel having a 62° angle to the plane of the ethylenic double bond. The similarities in the distances between the hydroxyl groups of DES and of estradiol may contribute to the estrogenicity of DES.

The original six-step synthesis of DES by way of 4,4'-dimethoxydeoxybenzoin has been superseded by a route that stems from *p*-hydroxypropiophenone, and that is characterized by a pinacol–pinacolone rearrangement.

Although DES is used most commonly without modification, its esters —particularly the dipropionate—have been investigated because of their long duration of action. The diphosphate has been used as a water-soluble derivative. Medical applications include estrogen-replacement therapy, suppression of lactation, postcoital contraception, treatment of breast and prostate carcinoma, and treatment of acne.

Acute side effects of DES and estrogen therapy in general are breakthrough bleeding, endometrial hyperplasia, withdrawal bleeding, nausea, vomiting, fluid retention, weight gain, breast enlargement and tenderness, reduced carbohydrate tolerance, and hypertension. Estrogens are contraindicated for patients with thromboembolic disease, estrogen-dependent neoplasia, liver and gall bladder disease, endometriosis, and uterine fibroids. Changes in coagulation factors and an increased incidence of liver tumors have also been attributed to estrogen therapy.

DES and other estrogens cause mammary carcinoma in susceptible strains of female mice and pituitary and testicular tumors in male mice. DES also causes abnormal changes in the reproductive tracts of mice thirteen months after dosing during the neonatal stage.

The FDA has ordered that no estrogen may be prescribed during pregnancy and that if DES is administered during pregnancy, the patient should be advised of the risks to the fetus and that abortion should be considered. The use of DES as a postcoital contraceptive is lawful only following rape or incest. The FDA has ordered that a lay-language brochure accompany estrogen prescriptions and that it include a warning of the risks of endometrial cancer.

Dienestrol. The 3,4-disubstituted hexa-2,4-diene structure of dienestrol can exist in three geometrical isomeric forms; the most potent estrogen of the three is α-dienestrol, mp 227–228°C. The other isomers are β-dienestrol, mp 184–185°C, and γ-dienestrol, mp 121–122°C. Based on uv, nmr, and x-ray diffraction data, the structure of dienestrol is established as (*E,E*) (*trans,trans*). The original synthesis of dienestrol involved dehydration of 3,4-bis(*p*-hydroxyphenyl)-3,4-hexanediol using acetyl chloride. Dienestrol is effective in treating menopausal symptoms.

Hextestrol. The meso and racemic forms of 3,4-bis(*p*-hydroxyphenyl)hexane have been separated and the racemic form has been resolved. The meso form (mp 185°C) is the parent estrogen, hexestrol, which is one thousand times more potent than the racemic form (mp 128°C). The oxygen–oxygen distance is 0.863 nm, which is considerably less than the corresponding distance in DES (**1**) or estradiol, and may account for the marked reduction in estrogenic potency. The shortest and most common route to hexestrol makes use of a Wurtz-type coupling of *p*-(1-halopropyl)anisole for which many coupling conditions have been reported.

Hexestrol is an orally active estrogen, but is less potent than DES, and its clinical usefulness is limited. Weight gain increases and feed efficiency improvements have been demonstrated in lambs and beef steers with an enhancement in the amount of lean meat at the expense of fat.

Benzestrol. Modification of the aliphatic portion of the hexestrol molecule led to diphenylpropane analogues, including benzestrol, which is the most active of the analogues and has an estrogenic potency between estradiol and estrone.

Chlorotrianisene. Triphenylethylenes possess estrogenic activity when given orally and are characterized by a longer duration of action than the stilbenes. Chlorotrianisene resembles other nonsteroidal and natural estrogens in its effects on the vagina, uterus, and breast. It differs from them, however, in that at oral doses in excess of the minimum required to produce an estrogenic response, it shows a prolonged duration of effect; it is stored in body fat and may be considered an oral implant. Today the risk of endometrial cancer must be considered, particularly with a drug having a long residence time. Chlorotrianisene is also effective in relieving painful breast engorgement and suppressing the onset of lactation. Chlorotrianisene is effective in treating prostatic carcinoma, but comparisons have not been made with other estrogens.

Carbestrol. Carbestrol may be considered a seco-analogue of doisynolic acid. An all-cis configuration has been assigned to the ring substituents of carbestrol and to the isomer based on nmr correlations. Carbestrol is not approved for clinical use.

Fenestrel. Fenestrel, like carbestrol, is a *seco*-analogue of doisynolic acid. It is a potent postcoital antifertility agent in the rat and interrupts pregnancy in the rabbit and monkey.

Zearalenone. The discovery of zearalenone arose from the observation that pigs feeding on moldy corn developed estrogenic changes of their sex organs. A mycological survey was made of moldy corn producing vulvular enlargement of swine. One of the isolates, *Fusarium graminearum* (the imperfect stage of *Gibberella zeae*), when grown on ground corn, provided a partially purified extract that induced uterine enlargement in mice and increased the rate of weight gain and improved feed efficiency in sheep. The active principle, zearalenone, was isolated and its structure was determined. It is a weak estrogen which has 0.1% of the oral activity of DES as determined by a uterotrophic assay in rats.

Fusarium. Fusarium contamination effects include hyperestrogenism, infertility (lack of production), and abortion. Swine are the most sensitive of domestic animals to these effects, and poultry and dairy cattle are affected to a lesser extent.

Natural zearalenone is optically active, $[\alpha]_D - 190°$ (chloroform), and has an *S* configuration. The macrocyclic ring of zearalenone is resistant to acid hydrolysis but opens with base with concomitant decarboxylation.

Zearalenone is manufactured by large-scale fermentation. It is an intermediate for the production of zeranol and its 7-hydroxy epimer.

Zeranol and taleranol. The absolute stereochemistries and optical rotations in methanol of zeranol and taleranol are $(3S,7R)$ $[\alpha]_D^{25} + 46°$ and $(3S,7S)$ $[\alpha]_D^{25} + 39°$, respectively. Both isomers are produced by sodium borohydride reduction or catalytic hydrogenation of zearalenone. Both zeranol and taleranol are under clinical investigation for treatment of the menopausal syndrome.

Zeranol was approved for use as a growth promotor in cattle and sheep by the FDA in 1969. It is a weak estrogen.

Other estrogens. Other estrogens include cyclofenil, coumestrol, genistein, miroestrol, and containing 15–20% *o,p*–DDT.

Antiestrogens

Antiestrogens counteract the biological effects of estrogens; the following discussion is limited to those compounds that do so at the receptor level. The clinical applications of antiestrogens have been reviewed and in particular, clamiphene citrate and tamoxifen citrate are now drugs of great importance for the treatment of female infertility and advanced breast cancer, respectively.

Medical applications of antiestrogens include the treatment of infertility and breast cancer, and the suppression of postpartum lactation.

Biochemistry of Estrogens and Antiestrogens

The triphenylethylenic antiestrogens combine with the cytoplasmic estrogen receptor to form a complex that translocates to the nucleus. The antiestrogen ER complex interacts with nuclear sites and elicits many of the estrogen responses, eg, stimulation of RNA polymerase and uterine growth. In contrast to estrogens, however, the complex is retained in the nucleus for periods well in excess of 24 h, nafoxidine is retained over 19 d in the complex and does not stimulate replenishment of the cytoplasmic receptor. Nuclear retention of the ER complex by antiestrogen does not cause continuous stimulation of uterine growth. Although uses of nafoxidine (5) doubles uterine weight, it is incapable of stimulating the five-fold increase of which estradiol is capable. Tamoxifen also causes nuclear retention of the ER for long periods but a recent study shows that, despite this retention, replenishment of cytoplasmic receptor still occurs.

(5) nafoxidine

Although the replenished receptor is capable of translocation, the cells remain intransigent to further uterine growth; this implies that the inability to clear the receptor from the nucleus rather than its failure to return to the cytoplasm is important for antiestrogenicity.

GRAHAM C. CRAWLEY
Imperial Chemical Industries Limited

U.V. Solmssen, *Chem. Rev.* **37**, 504 (1945); J. Grundy, *Chem. Rev.* **57**, 300 (1957).

C.B. Lunan and A. Klopper, *Clin. Endocrinol.* **4**, 551 (1975).

R.F. Harrision and J. Bonar, *Pharmac. Ther.* **11**, 451 (1980).

HORMONES—PEPTIDE HORMONES. See Hormones, anterior-pituitary; —anterior-pituitarylike; —posterior-pituitary; —brain oligopeptides; Peptides.

HUMIDIFICATION. See Air conditioning; Simultaneous heat and mass transfer.

HYDANTOIN AND DERIVATIVES

Hydantoin (1), (2,4-imidazolidinedione, glycolyleurea) was discovered by Baeyer in 1861 as a hydrogenation product of allantoin (2), itself a degradation product of uric acid.

Allantoin, one of several naturally occurring hydantoins, is a product of purine metabolism and is thus found in the urine of most animals (but not humans). It also occurs in plant and animal embryos and in the leaves of some plants. Hydantoin itself has been isolated from such diverse sources as the buds of the oriental plane tree and the white shoots of sugar beets. Methylated hydantoins have been isolated from a testicular extract containing pressor substances and from decomposed whale meat.

Hydantoins are important intermediates in the synthesis of several amino acids (qv). Hydantoin derivatives are used as anticonvulsants in the treatment of epilepsy, Saint Vitus's dance (chorea), and heart arrhythmia (see Hypnotics, sedatives, and anticonvulsants). Derivatives of 1-aminohydantoin are used in the treatment of urinary-tract infections, as muscle relaxants, and as bactericides. In the chemical industry, various 5,5-substituted hydantoins are the basis of a new generation of weatherproof high-temperature-stable epoxy resins (qv). Other hydantoin derivatives are used in numerous consumer products, such as hair sprays, cosmetics, acne medications, and photographic film. 1,3-Dichloro-5,5-dimethylhydantoin is an important germicide and slow-releasing bleach, but its use has declined since the introduction of the chloroisocyanurates (see Bleaching agents).

Physical Properties

Hydantoins, when unsubstituted in the *N*-1 and *N*-3 positions, are high melting crystalline solids and have been used as solid derivatives of liquid aldehydes and ketones in qualitative analysis. Hydantoin is a colorless, platelike solid. Tables 1 and 2 list some physical properties of hydantoin.

Chemical Properties

Hydrolysis. Hydrolysis with concentrated aqueous or alcoholic alkali, an important synthetic method, gives α-amino acids.

Substitution at nitrogen. The hydantoin ring shows nucleophilic activity at the nitrogen atoms. In most cases, *N*-3 substitution is preferred over *N*-1 because of the polarizability of the N—H bond at *N*-3 and the ready formation of the *N*-3 anion, which is more nucleophilic than the unionized *N*-1 nitrogen.

Substitution at the C-5 position. Hydantoin reacts with aromatic or heterocyclic aldehydes at the C-5 position to give unsaturated hydantoins.

Reactions at the C-2 and C-4 positions. The only type of reaction involving the C-2 or C-4 carbon atoms of hydantoins is the reduction of 5,5-disubstituted hydantoins with metal hydride reducing agents to give good yields of 2-imidazolidinones.

Table 1. Melting Points of Some Common Hydantoins

Hydantoin	Melting point, °C
hydantoin	217–218
5,5-dimethylhydantoin	177–178
5-ethyl-5-methylhydantoin	145–146
5,5-diphenylhydantoin	295–298
5,5-pentamethylenehydantoin	214–215
allantoin	228–235 (dec)[a]

[a] Depending on method of determination.

Table 2. Physical Properties of Hydantoin

Property	Value
pK_a, 25°C	9.12
heat of combustion, MJ/mol[a]	1.31
dielectric coefficient, dE/dc[b]	−6.4
partial molar volume (H$_2$O, 25°C), cm^3/mol	65.0

[a] To convert J to cal, divide by 4.184.
[b] E = dielectric constant; c = concentration.

Hydantoins as reagents in organic synthesis. Because of their accessibility and reactivity, hydantoin derivatives are useful intermediates in organic synthesis. For example, the unstable nature of 3-acetylhydantoins led to a method of selectively acylating phenols in the presence of alcohols.

Synthesis

Hydantoins can be synthesized from aldehydes and ketones, from amino acid derivatives (eg, organic isocyanates, potassium cyanate, urea, cyanamide, and ethyl chloroformate, diphenyl carbonate, or phosgene), and from urea with α-dicarbonyl compounds, α-hydroxy acids or nitriles, or unsaturated carboxylic acids.

Derivatives of hydantoin include allantoin (2,5-dioxo-4-imidazolidinylurea, 5-ureidohydantoin), 1,3-bis(2-hydroxyethyl)-5,5-dimethylhydantoin (mp 62–65°C), 1,3-dibromo-5,5-dimethylhydantoin (mp 185–190°C), 1,3-dichloro-5,5-dimethylhydantoin, 5,5-dimethylhydantoin, 5,5-dimethylhydantoin–formaldehyde adducts, 5,5-diphenylhydantoin, hydantoin epoxy resins, nitrofurantoin, and 1-[[5-(4-nitrophenyl)-2-furanyl]methylene]aminohydantoin.

JOHN H. BATEMAN
CIBA-GEIGY Corp.

E. Ware, *Chem. Rev.*, 403 (1950).

D.E. Cadwallader and H.W. Jun, *Anal. Profiles Drug Subst.* **5**, 345 (1976).

HYDRAULIC FLUIDS

The moving parts of many industrial machines are actuated by oil that is under pressure. A system used to apply the oil can consist of a reservoir, a motor-driven pump, control valves, a fluid motor, and piping to connect these units, eg, a hydraulic system. Generally, lubricating petroleum oils and sometimes water are used as the pressure-transmitting or hydraulic fluids. Lubricating oil not only is suitable for pressure transmission and controlled flow, but it also minimizes friction and wear of moving parts (see Lubrication and lubricants) and protects ferrous surfaces from rusting (see Corrosion and corrosion inhibitors).

Hydraulic actuation is based on Pascal's discovery that pressure that has developed in a fluid acts equally and in all directions throughout the fluid and behaves as an hydraulic lever or force multiplier (see Pressure measurement).

Types

Antiwear premium hydraulic fluids were formulated during the early 1960s, and are essentially the same today. The largest volume of hydraulic fluids in use is mineral oil-based with additives to meet specific requirements. These fluids comprise 98% of the world demand [ca 1.9 GL (500×10^6 gal)] for hydraulic fluids. The world demand for fire-resistant fluids is 136 ML (36×10^6 gal) or ca 6.5% of the total industrial-fluid market. Fire-resistant fluids are classified as high-water-base fluids, water-in-oil emulsions, glycols, and phosphate esters.

Mineral oil-based fluids. Premium mineral oils are ideally suited for use in most hydraulic systems and are, by themselves, excellent hydraulic fluids. They are high viscosity index (VI) oils and are available in a wide range of viscosity grades. Unusually high VI products are especially suitable for use under low temperature conditions. All of the oils contain additives, eg, rust and oxidation inhibitors. In the event that the additives are consumed or removed in service, these oils would continue to serve effectively for long periods (see Additives). These oils are carefully processed to have good water-separating ability and resistance to foaming. Because of high oxidation resistance, these qualities are maintained over long service periods.

Synthetic Fluids

The starting materials for synthetic lubricants are synthetic base stocks which often are manufactured from petroleum. The base fluids are made by synthesizing low molecular-weight compounds that have ade-

Table 1. Advantages and Limiting Properties of Synthetic Base Stocks

Stock	Advantages over mineral oil	Limiting properties
synthetic hydrocarbon fluid	high temperature stability; long life; low temperature fluidity; high viscosity index; improved wear protection; low volatility, oil economy; compatibility with mineral oils and paints; no wax	solvency/detergency[a]; seal compatibility[a]
organic esters	high temperature stability; long life low temperature fluidity; solvency/detergency	seal compatibility[a]; mineral oil compatibility[a]; antirust[a]; antiwear; and extreme pressure[a]; hydrolytic stability; paint compatibility
phosphate esters	fire resistant; lubricating ability	seal compatibility; low viscosity index; paint compatibility; metal corrosion[a]; hydrolytic stability
polyglycols	water versatility; high viscosity index; low temperature fluidity; antirust; no wax	mineral-oil compatibility; paint compatibility; oxidation stability[a]

[a] Limiting properties of synthetic base fluids that can be overcome by formulation chemistry.

quate viscosity for use as lubricants. The process of combining individual units can be controlled so that a large proportion of the finished base fluid is comprised of one or only a few compounds. Depending on the starting materials and the combining process that is used, the compound (or compounds) can have the properties of the most effective compounds in a mineral-base oil. It can also have unique properties, eg, miscibility with water or complete nonflammability, that are not found in any mineral oil.

The primary performance features of synthetic lubricants are their outstanding flow characteristics at extremely low temperatures and their stability at extremely high temperatures. Advantages, as well as limiting properties, are outlined in Table 1.

Synthesized hydrocarbons (eg, olefin oligomers, alkylated aromatics, polybutenes, and cycloaliphatics), organic esters (eg, dibasic acid esters and polyol esters), polyglycols, and phosphate esters account for > 90 vol% of synthetic lubricant bases in use. Other synthetic lubricating fluids include a number of materials that generally are used in low volumes, such as silicones, silicate esters, and halogenated fluids.

Additives

Practically all lubricating oils contain at least one additive, and some oils contain several. The amount of additive that is used varies from < 0.01 to ≥ 30%. Additives can have detrimental effects, especially if the dosage is excessive or if interactions with other additives occur. Some additives are multifunctional, eg, certain VI improvers also function as pour-point depressants or dispersants (qv). The additives most commonly used in hydraulic fluids include pour-point depressants, viscosity-index improvers, defoamers (qv), oxidation inhibitors, rust and corrosion inhibitors, and antiwear compounds.

Properties

Hydraulic fluid functions include transmitting pressure and energy; sealing close-clearance parts against leakage; minimizing wear and friction in bearings and between sliding surfaces in pumps, valves, cylinders, etc; removing heat; flushing away dirt, wear particles, etc; and protecting surfaces against rusting. The hydraulic-fluid properties that are used to characterize a suitable product are listed in Table 2.

Availability

The main suppliers of petroleum-based hydraulic fluids in the United States are Amoco, Arco, Chevron, Citgo, Exxon, Gulf, Mobil, Shell, Sun, Texaco, and Union.

Uses

Hydraulic actuation is applied to machine tools, presses, draw benches, jacks, and elevators as well as to die-casting, plastic-molding, welding,

Table 2. Properties of Antiwear Hydraulic Oils and Lubricants[a]

Property	Light	Medium	Heavy-medium
specific gravity, 16°C	0.860–0.876	0.868–0.887	0.871–0.881
(°API)	(30–33)	(28–31.5)	(29–31)
pour point, max, °C	ca −15	ca −15	ca −15
flash point, min, °C	ca 200	ca 200	ca 200
viscosity system			
SUs, 38°C	149–182		317–389
40°C		194–236	
mm²/s (= cSt), 40°C	28.8–35.2	41.4–50.6	61.2–74.8
viscosity index (VI), min	90	90	90
color, max	2.0	3	3.0
neutralization no., mg KOH/g oil [should not contain any mineral acids]	1.5 max	1.5 max	1.5
rust	pass	pass	pass

[a] For low temperature mineral oils, there also are tests for emulsion (ASTM D 1401) and foam (ASTM D 892).

coal-mining, and tube-reducing machines. Hydraulic loading is used for pressure-, sugar-mill-, and paper-machine-press rolls, and calendar stacks. Numerous other applications of hydraulics include mechanisms for tilting ladles and operating clamps, brakes, valves, furnace doors, and loading platforms. There are also many hydraulic applications in aircraft, automobiles, trucks, contractor, and farm equipment.

The most recent use of hydraulic power has been in the hydrostatic transmissions that are used in many self-propelled harvesting machines and garden tractors and in large tractors and construction machines.

Fire-resistant hydraulic fluids are used where the fluid could spray or drip from a break or leak onto a source of ignition.

J. George Wills[‡]
Mobil Oil Corporation

Reviewed by
J. M. Allen
Mobil Oil Corporation

Hydraulic Systems for Industrial Machines, Mobil Oil Corporation, New York, 1970.

1978 Report on U.S. Lubricating Oil Sales, National Petroleum Refiners Association, Washington, D.C., 1979.

R.H. Schmitt, R.J. Poole, and J. Shim, *From Water to "Super-Stabilized" Antiwear Hydraulic Oils*, Automotive Engineering Congress, SAE, Detroit, Mich., 1978.

HYDRAULIC SEPARATION. See Gravity concentration.

HYDRAZINE AND ITS DERIVATIVES

Hydrazine, NH_2NH_2, the simplest diamine and parent of innumerable derivatives, was first prepared in 1887 by Curtius as the sulfate salt from diazoacetic ester. Thiele (1893) suggested the oxidation of ammonia with hypochlorite should yield hydrazine. F. Raschig (1906) first demonstrated this process, variations of which constitute the chief commercial methods of manufacture today.

Hydrazine and its simple methyl and dimethyl derivatives have endothermic heats of formation and high heats of combustion—hence their use as rocket fuels. Hydrazine is a base, slightly weaker than ammonia, and forms a series of useful salts. As a strong reducing agent, hydrazine is used for corrosion control in boilers and hot-water heating systems; also for metal plating, and for reducing noble-metal catalysts and un-

[‡] Deceased.

saturated bonds in organic compounds. It is also an oxidizing agent under suitable conditions. With two active nucleophilic nitrogens and four replaceable hydrogens, hydrazine is the starting material for many derivatives, among them foaming agents for plastics, antioxidants, polymers, polymer cross-linking and chain-extending agents, as well as biologically active pesticides, herbicides, plant-growth regulators, and pharmaceuticals. As expected, hydrazine is a good complexing ligand so that numerous complexes have been studied. Many heterocyclics are based on hydrazine, some rings containing from one to five nitrogen atoms as well as other heteroatoms.

Physical Properties

Pure hydrazine is a colorless, mobile, fuming liquid with an ammoniacal odor. It has a relatively high melting point (2°C) and boiling point (113.5°C) as well as an abnormally high heat of vaporization. These data indicate considerable hydrogen bonding in the solid and liquid phase. Predictably, therefore, hydrazine is miscible with water and soluble in other polar solvents such as alcohols, amines, ammonia, etc. Ethyl ether serves as an aprotic solvent, but solubility is limited. Anhydrous hydrazine undergoes some self-ionization, to a greater extent than anhydrous ammonia but less than water.

Hydrazinium salts, $N_2H_5^+ X^-$, are acids in anhydrous hydrazine and metallic hydrazides, $M^+ N_2H_3^-$, are bases. The hydrazine molecule represents an intermediate valence state for nitrogen suggesting that hydrazine can function both as an oxidizing agent and as a reducing agent.

Chemical Properties

Although hydrazine is an endothermic compound with a high heat of formation (see Table 1), elevated temperatures are needed (250°C) before any appreciable decomposition occurs. However, this decomposition temperature is lowered significantly by many catalysts, particularly copper, cobalt, molybdenum, and iridium, and their oxides. Iron oxides (rust)

Table 1. Physical Properties of Commercial Hydrazines

Property	Hydrazine (AH)	Monomethylhydrazine (MMH)	Unsymmetrical dimethylhydrazine (UDMH)
formula	N_2H_4	CH_3NHNH_2	$(CH_3)_2NNH_2$
freezing point, °C	2.0	−52.4	−57.2
boiling point, °C	113.5	87.5	63
vapor pressure, 25°C, kPa[a]	1.92	6.62	20.93
critical constants			
P_c, MPa[b]	14.69	8.24	5.42
T_c, °C	380	312	250
d_c, g/cm³	0.231	0.290	0.23
liquid density, 25°C, g/mL	1.0045	0.874	0.786
surface tension, 25°C, mN/m (= dyn/cm)	66.67	34.3 (20°C)	24.0
liquid viscosity, 25°C, mPa·s (= cP)	0.90	0.771	0.509
dielectric constant, 25°C	51.7		
refractive index, n_D^{25}	1.4644		1.4508
heat of vaporization, kJ/mol[c]	45.27	40.37 (25°C)	35.02
heat of fusion, kJ/mol[c]	12.66	10.42	10.08
heat capacity at 25°C, J/(mol·°C)[c]	98.87	134.93	164.05
heat of combustion, kJ/mol[c]	−622.1	−1304.2	−1979
heat of formation, kJ/mol[c]	50.63	53.97	49.37
free energy of formation, kJ/mol[c]	149.24	179.9	206.69
entropy of formation, J/(mol·°C)[c]	121.21	165.9	197.99
flash point, COC[d], °C	52	1.0	−15

[a] To convert kPa to mm Hg, multiply by 7.5.
[b] To convert MPa to atm, divide by 0.101.
[c] To convert J to cal, divide by 4.184.
[d] COC = Cleveland open cup.

also catalyze decomposition. Hydrazine, especially in high concentrations, should be handled with care using scrupulously clean systems.

Acid-base reactions.   As an Arrhenius base, hydrazine is somewhat weaker than ammonia. In principle, it can form two series of salts with monobasic acids, one having the hydrazinium (1 +) cation, $N_2H_5^+$, and the other, the hydrazinium (2 +) cation, $N_2H_6^{2+}$.

Reductions.   Hydrazine is a very strong reducing agent. In aqueous solution, it yields primarily nitrogen and water on reaction with ammonia and hydrazoic acid.

The techniques available for reducing organic compounds with hydrazine and its derivatives have been reviewed. These procedures have some advantages over conventional pressure hydrogenations in being more selective in their attack, sometimes stereospecifically yielding only cis isomers, and in not requiring the use of high pressure. They include carbonyl reductions, catalytic reductions, diazene reductions, aldehyde syntheses, and olefin syntheses.

Alkyl hydrazines.   Since it has four replaceable hydrogens and two unbonded electron pairs, hydrazine can form many alkyl (and aryl) derivatives, including mono-, di-, tri-, tetra-, and pentasubstituted derivatives and their isomers.

Aromatic hydrazines.   Generally, aromatic hydrazines are not made from haloaromatics and hydrazine unless the halo substituent is activated by neighboring electronegative groups.

Hydroxyalkyl hydrazines.   Epoxides react with hydrazine to give hydroxyalkylhydrazines (hydrazinoalkanols). Thus, ethylene oxide (qv) yields hydroxyethylhydrazine. This plant-growth regulator has been extensively tested for control of pineapple flowering; it is also used to make a coccidiostat, Furazolidone (see Chemotherapeutics, antiprotozoal), and has been proposed as a stabilizer in the polymerization of acrylonitrile.

Hydrazides and related compounds.   Substitution of the hydroxy group in carboxylic acids with a hydrazino moiety gives carboxylic acid hydrazides. In this formal sense, a number of related compounds fall within this product class although they are not necessarily prepared this way. Some of the more common of these compounds are carboxylic acid

hydrazides, thiohydrazides, sulfonylhydrazides, semicarbazide, thiosemicarbazide, carbohydrazide, thiocarbohydrazide, phosphoric acid hydrazide, phosphoric acid dihydrazide, and thiophosphoryltrihydrazide. Hydrazides are important as pharmaceuticals, photographic chemicals, and polyolefin stabilizers. They also are precursors for many heterocyclics.

Hydrazones and azines.   Depending on reaction conditions, hydrazines react with aldehydes and ketones to give hydrazones (**1**), azines (**2**), and diaziridines (**3**), the latter formerly known as isohydrazones.

Many of these compounds are highly colored and have found use as dyes and photographic chemicals. Several pharmaceuticals and pesticides are members of this class.

Heterocycles.   One of the most characteristic and useful properties of hydrazine and its derivatives is the ability to form heterocyclic compounds. Many pharmaceuticals, pesticides, and dyes, to name a few, are based on these rings.

Manufacture

Many methods for the preparation of hydrazine have been proposed. At present, however, the only commercially feasible processes involve partial oxidation of ammonia (or urea) with chlorine, hypochlorite, or hydrogen peroxide. Most hydrazine is now produced by some variation of the Raschig process, which is based on the oxidation of ammonia with alkaline hypochlorite according to the following overall reaction:

$$2\,NH_3 + NaOCl \rightarrow N_2H_4 + NaCl + H_2O$$

One adaptation of the basic Raschig process is used by the Olin Corp. in its Lake Charles, Louisiana plant (Fig. 1).

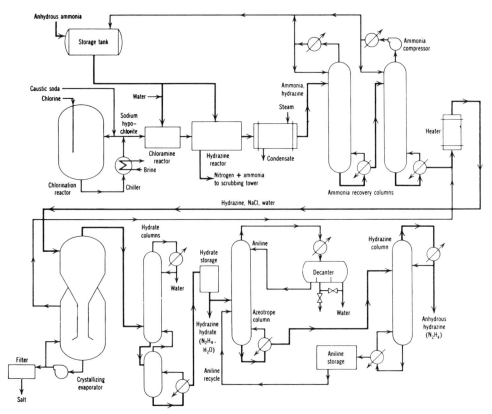

Figure 1.   Olin Raschig process flow sheet.

Other processes include the ketazine process, the urea process, the PCUK peroxide process, and nitrogen fixation (qv).

Health and Safety Factors (Toxicology)

Hydrazine is highly toxic and readily absorbed by oral, dermal, or inhalation exposure. It is irritating to the skin, eyes, and respiratory tract. Permanent damage to the cornea may occur if the liquid is splashed into the eyes. At high doses, it is a strong convulsant, but even low doses may cause depression of the central nervous system. Lowering of the hydrazine TLV to 0.1 ppm to conform to the European standard has been proposed. This is far below the olfactory limit of 3–5 ppm, so that the working-space environment must be monitored by suitable means. There is no specific antidote; treatment is based on the symptoms.

Uses

Once used chiefly as a rocket propellant, hydrazine and its derivatives are now used mainly as blowing agents for foamed plastics (33%), as agricultural pesticides (40%), and for water treatment (ca 15%) (see Explosives and propellants).

HENRY W. SCHIESSL
Olin Corporation

P.A.S. Smith, *The Chemistry of Open-Chain Nitrogen Compounds*, Vols. 1–2, W.A. Benjamin, Inc., Menlo Park, Calif., 1965–1966.

Houben-Weyl, *Methoden der Organischen Chemie*, 4th ed., Band X/2, Stcikstoff Verbindungen 1, Teil 2, Georg Theime, Verlag, Stuttgart, FRG, 1967.

C.C. Clark, *Hydrazine*, Mathieson Chemical Corp., Baltimore, Md., 1953.

L.F. Audrieth and B.A. Ogg, *The Chemistry of Hydrazine*, John Wiley & Sons, Inc., New York, 1951.

HYDRIDES

Hydrides are compounds that contain the element hydrogen in a reduced or electron-rich state. They are either simple binary or complex hydrides. In the former, negative hydrogen is bonded ionically or covalently to a metal, or is present as a solid solution in the metal lattice. The latter comprise a large group of chemical compounds containing complex hydridic anions, such as BH_4^-, AlH_4^-, and derivatives of these. Hydrides are important industrial chemicals manufactured and used on a large scale.

Simple (Binary) Hydrides

Ionic Hydrides

The ionic or saline hydrides contain metal cations and negatively charged hydrogen ions. They crystallize in the cubic lattice similar to the corresponding metal halide, and when pure, are white solids. When dissolved in molten salts or hydroxides and electrolyzed, hydrogen gas is liberated at the anode. Their densities are greater than those of the parent metal, and their formation is exothermic. All are strong bases.

Alkali metal hydrides. Physical properties of the alkali metal hydrides are given in Table 1.

Table 1. Physical Properties of Alkali Metal Hydrides

Hydride	Mp, °C	$S°$, J/(mol·K)^a	Density, g/cm^3
LiH	688	25	0.77
NaH	420 dec	48	1.36
KH	dec	61	1.43
RbH	300 dec		2.60
CsH	dec		3.4

^{a}To convert J to cal, divide by 4.184.

Alkaline earth metal hydrides. Table 2 gives some thermochemical data of alkaline earth metal hydrides; all form orthorhombic crystals.

Covalent Hydrides

In all hydrides, hydrogen is bound to an atom of lower electronegativity than itself (H = 2.1). In covalent hydrides, the hydrogen–metal bond is completed through a common electron pair. Beryllium and magnesium hydrides are included in this group and are polymeric materials, as is aluminum hydride. The simple hydrides of silicon, germanium, tin, and arsenic are gaseous or easily volatile compounds. Table 3 gives some properties of the main group metal hydrides.

Transition metal hydrides. Transition metal hydrides, or interstitial metal hydrides, have metallic properties, conduct electricity, and are less dense than the parent metal. Metal valence electrons are involved in both the hydrogen and metal bonds. Compositions can vary within limits and stoichiometry may not always be a simple numerical proportion. These hydrides are much harder and more brittle than the parent metal, and most have catalytic activity. They include titanium hydride (TiH_2) and zirconium hydride (ZrH_2).

Rare earth hydrides. Activated rare earth metals react directly with hydrogen even at room temperature. They are activated by heating to 300°C in H_2, followed by cooling under H_2. Rare earth hydrides include lanthanum dihydride (LaH_2) and trihydride, and cerium hydride (CeH_2) and hydrides of high hydrogen content ($CeH_{<3}$). The most important uses of cerium and lanthanum hydrides are in the hydrogen-storage alloys $LaNi_5$ and $CeMg_2$ (see Hydrogen energy).

Group VB hydrides. These hydrides are formed from the metals, preferably by heating the metals in powder form in a hydrogen atmosphere up to 1000°C. Trace impurities (oxides, nitrides) in the metal prevent complete hydriding. These hydrides are brittle powders that can be handled in air. They are manufactured in small amounts, mainly for research and development work in powder metallurgy. Group VB hydrides include vanadium hydride (VH), niobium hydride (NbH), niobium dihydride (NbH_2), and tantalum hydride (TaH).

Hydrogen-storage alloys. A number of metal alloys are extremely useful for safely storing large volumes of hydrogen because of their easy solubilization of hydrogen at relatively low temperatures and pressures, presumably to form hydrides. Many metals and binary and ternary alloys have been thoroughly studied for this application; the most

Table 2. Physical Properties of Alkaline-Earth Metal Hydrides

Hydride	$\Delta H°_{(298)}$, kJ/mol^a	$S°$, J/(mol·K)^a	Density, g/cm^3
CaH$_2$	−186.3	42	1.90
SrH$_2$	−180.5	54	3.27
BaH$_2$	−171.2	67	4.16

^{a}To convert J to cal, divide by 4.184.

Table 3. Properties of Covalent Hydrides

Hydride	Formula	Mp, °C	Bp, °C	Density, g/L
beryllium hydride	BeH$_2$	125 dec	220^b	
magnesium hydride	MgH$_2$	280 dec		1.45 g/cm^3
aluminum hydride	AlH$_3$			
silane	SiH$_4$	−185^c	−119.9	1.44^d at 20°C
germane	GeH$_4$	−165	−90	3.43^d at 0°C
stannane	SnH$_4$	−150	52	
arsine	AsH$_3$	−116.9^c	−62	2.695^d

^{a}To convert J to cal, divide by 4.184.
^{b}Begins to dissociate.
^{c}Freezing point.
^{d}Gas at atmospheric pressure.

important appear to be $FeTiH_{1-2}$ (also modified with rare earth metals, Ni, or Mn) and AB_5 alloys where A is a rare earth metal, Ca or Th, and B is Co or Ni ($LaNi_5$ has shown special promise).

Complex Hydrides

The complex hydrides are a large group of compounds in which hydrogen is combined in fixed proportions with two other constituents, generally metallic elements. These compounds have the general formula $M(M'H_4)_n$, where n is the valence of M, and M' is a trivalent Group IIIA element such as boron, aluminum, or gallium. Sodium borohydride in particular and lithium aluminum hydride are by far the most important commercially.

Borohydrides

The alkali metal borohydrides are the most important complex hydrides. They are ionic, white, crystalline, high melting solids that are sensitive to moisture, but not to oxygen. Group IIIA and transition metal borohydrides, on the other hand, are covalently bonded and are either liquids or sublimable solids. The alkaline earth borohydrides are intermediate between these two extremes, and display some covalent character (see also Boron compounds, boron hydrides). Borohydrides include lithium borohydride ($LiBH_4$) and sodium borohydride ($NaBH_4$). $NaBH_4$ and KBH_4 are unique among complex hydrides for their stability in alkaline aqueous solution. Decomposition by hydrolysis is accelerated by increasing acidity or temperature.

The principal uses of $NaBH_4$ are in synthesis of pharmaceuticals and fine organic chemicals; removal of trace impurities from bulk organic chemicals; wood-pulp bleaching, clay leaching, and vat-dye reductions; and removal and recovery of trace metals from plant effluents.

Borohydride Derivatives

Modification of the BH_4^- anion has provided derivatives of widely differing reducing properties. Unusual reducing properties can be obtained with borohydride derivatives formed *in situ*. A variety of reductions have been reported, including hydrogenolysis of carbonyls and alkylation of amines with sodium borohydride in carboxylic acids such as acetic and trifluoroacetic, in which the acyloxyborohydride is the reducing agent. Thiol-activated borohydrides reduce nitro, amide, and ester groups; the anilidoborohydride has been used for selective ester reductions. Trialkylborohydrides, eg, alkali metal tri-*sec*-butylborohydride, show outstanding stereoselectivity in ketone reductions. Borohydride derivatives include sodium cyanoborohydride ($NaBH_3CN$), which is uniquely stable in aqueous acid to ca pH 3.

Aluminohydrides

In general, the aluminohydrides are more active and powerful reducing agents than the corresponding borohydrides. They decompose vigorously with water. Reaction also occurs with alcohols, although more moderately, providing a route to substituted derivatives. Aluminohydrides include lithium aluminum hydride ($LiAlH_4$), sodium aluminum hydride ($NaAlH_4$), and aluminohydride derivatives [$LiAlH(OCH_3)_3$, $LiAlH(OC_2H_5)_3$), and $LiAlH(O-t-C_4H_9)_3$].

Health and Safety Factors

Generalizations on the toxicity of hydrides as a class of compounds are difficult to make because of variations in their chemical reactivity. Furthermore, there is little published toxicological information available.

In general, hydrides react exothermically with water, resulting in the generation of hydrogen. Since the flammable gas hydrogen is formed, a potential fire hazard may result unless adequate ventilation is provided. Ingestion of hydrides must be avoided because hydrolysis to form hydrogen could result in gas embolism. Another aspect of the hydrolysis is the alkalinity which can cause chemical burns in skin and other tissues.

EDWARD A. SULLIVAN
ROBERT C. WADE
Thiokol/Ventron Division

E.R.H. Walker, *Chem. Soc. Rev.* **5**, 23 (1976).

R.M. Adams and A.R. Siedle, *Boron, Metallo-Boron Compounds and Boranes*, Interscience Publishers, New York, 1964, Chapt. 6, pp. 373–506.

W.M. Mueller, J.P. Blackledge, and G.G. Libovitz, *Metal Hydrides*, Academic Press, Inc., New York, 1968, Chapt. 12, pp. 546–674.

B.D. James and M.G.H. Wallbridge, *Progr. Inorg. Chem.*, Vol. 11, Wiley-Interscience, New York, 1970, pp. 99–231.

HYDROBORATION

By 1960, the hydroboration reaction, ie, reaction of an alkene with borane (BH_3), usually in an ether solvent, was already established for the *in situ* laboratory synthesis of organoboranes. However, apart from their direct oxidation to give alcohols, organoboranes had few applications as synthetic reagents. Since then, hydroboration–oxidation as a means of effecting the anti-Markovnikov hydration of alkenes (eq. 1) has become one of the best known and most widely used organic laboratory synthetic methods.

$$RR'C{=}CHR'' \xrightarrow[OH^-]{BH_3 \quad H_2O_2} RR'CHCHOH \atop R'' \qquad (1)$$

Furthermore, in these few years, a vast array of new and synthetically useful reactions of organoboranes has been discovered, such that now organoboranes must be considered among the most useful synthetic reagents. They tolerate many functional groups, are often formed and react in a stereospecific manner, and undergo a number of unique transformations.

The Hydroboration Reaction

With borane–tetrahydrofuran. Many features of the hydroboration reaction are common to all hydroborating agents, and borane–THF can thus be used as a standard against which to compare other reagents. It is commercially available as a $1M$ solution from Aldrich or Cambrian Chemicals.

Of alkenes. The hydroboration of an alkene involves addition of a B—H bond across the carbon–carbon double bond. The reagent adds in a cis manner, and for alkenes that do not have powerfully polarizing substituents, addition is preferentially from the less-hindered face, placing boron primarily at the less-substituted end of the double bond. All of these features are admirably illustrated by the hydroboration of α-pinene (eq. 2).

The reactions of alkenes with borane–THF occur in sequential stages to give first monoalkylboranes, then dialkylboranes, and finally, trialkylboranes (eq. 3).

$$RCH{=}CH_2 + BH_3 \to RCH_2CH_2BH_2 \xrightarrow{RCH{=}CH_2} (RCH_2CH_2)_2BH$$
$$\xrightarrow{RCH{=}CH_2} (RCH_2CH_2)_3B \qquad (3)$$

Mono- and dialkylboranes usually exist as bridged dimers, *sym*-dialkyldiboranes and *sym*-tetraalkyldiboranes, respectively, but for most purposes they may be considered as the simple monomers.

Of polyunsaturated compounds. The hydroboration of dienes and polyenes with borane–THF opens up possibilities for formation of a variety of polymeric and cyclic organoboranes. In some cases, however, the reactions can be controlled to allow the high yield synthesis of specific cyclic organoboranes. Examples include the synthesis of 3,5-dimethylborinane, 3,5-dimethylborepane, and 9-borabicyclo[3.3.1]nonane (9-BBN).

Hydroboration with other reagents. Borane–methyl sulfide (BMS). The complex of borane with dimethyl sulfide is more stable than that with THF. Consequently, it can be prepared and handled as a pure liquid. It is commercially available from Aldrich and other sources. In its reactions with alkenes, BMS appears to behave in a way very similar to borane–THF and for most purposes can be considered as an interchangeable hydroborating agent.

Dialkylboranes. Dialkylboranes hydroborate alkenes, alkynes, and allenes provided that the unsaturated compound is not too hindered. The degree of hindrance that may be tolerated varies greatly. Thus, 9-BBN, which is relatively unhindered, hydroborates even hindered alkenes, such as 2,3-dimethyl-2-butene, whereas diisopinocampheylborane $[(IPC)_2 BH]$ hydroborates simple trans alkenes only with concomitant elimination of α-pinene. Diisopinocampheylborane is valuable for asymmetric hydroborations (eq. 4).

(4)

$[\alpha]$ −136°
92% optical purity

Monoalkylboranes. Only a very limited number of monoalkylboranes can be synthesized directly by simple hydroboration of an alkene, and of these, only thexylborane has been studied extensively.

Unhindered alkenes, such as 1-hexene and styrene, undergo clean reactions with thexylborane in 2:1 mol ratio to give the corresponding thexyldialkylboranes (eq. 5), but reactions cannot be controlled to give thexylmonoalkylboranes.

(5)

Monohydroboration of thexylborane can be achieved with more hindered alkenes, such as 2-methyl-1-butene and cyclopentene. These intermediates are valuable for many synthetic applications.

Halogenoboranes. The reagents of interest for hydroboration are the mono- and dichloroboranes, both as their etherates and their complexes with dimethyl sulfide, and mono- and dibromoborane dimethyl sulfide complexes.

The chloroborane etherates are prepared by stoichiometric reactions of lithium tetrahydroborate with trichloroborane in diethyl ether (eqs. 6–7). They have limited stability and, therefore, must be used within a short time of their preparation.

$$LiBH_4 + BCl_3 \xrightarrow{(C_2H_5)_2O} LiCl + 2 BH_2Cl.O(C_2H_5)_2 \qquad (6)$$

$$LiBH_4 + 3 BCl_3 \xrightarrow{(C_2H_5)_2O} LiCl + 4 BHCl_2.O(C_2H_5)_2 \qquad (7)$$

Catecholborane and other reagents. At elevated temperatures, 1,3,2-benzodioxaborole (catecholborane) hydroborates alkenes and alkynes, giving good yields of the corresponding *B*-alkyl- and *B*-alkenyl-catecholboranes (eg, eq. 8). These compounds are converted into mixed trialkylboranes by reduction–hydroboration or by reaction with Grignard reagents.

(8)

Reactions of Organoboranes

Replacement of boron by hydrogen or a hetero atom. Under appropriate conditions, the alkyl groups of a trialkylborane can be transferred to H, N, O, S, Se, halogen, or metal atoms. The combination of hydroboration and functionalization therefore corresponds overall to the anti-Markovnikov addition of the elements of HX to a double bond (eq. 9):

(9)

Protonolysis. Trialkylboranes are not readily protonolyzed by water, alcohols, amines, or aqueous or anhydrous mineral acids, and with each successive dealkylation the reaction is inhibited further. Alkenyl-, alkynyl-, benzyl-, and allylboranes are more readily hydrolyzed than the saturated compounds, but total hydrolysis may still require vigorous conditions. By contrast, even trialkylboranes are readily protonolyzed by carboxylic acids.

Oxidation. Organoboranes are oxidized by a variety of oxidizing agents, of which the most widely used is alkaline hydrogen peroxide, which gives alcohols in high yield (eq. 10).

$$R_3B + 3 H_2O_2 + NaOH \rightarrow 3 ROH + NaB(OH)_4 \qquad (10)$$

Synthesis of halogenoalkanes. Trialkylboranes are readily converted into alkyl bromides (eq. 11) or iodides, by treatment with the appropriate halogen and sodium methoxide. Only two alkyl groups are utilized in the case of iodides. The reactions are accompanied by inversion of configuration at the functionalized carbon atom.

$$R_3B + 4 NaOCH_3 + 3 Br_2 \rightarrow 3 RBr + NaB(OCH_3)_4 + 3 NaBr \qquad (11)$$

Synthesis of nitrogen-containing compounds. Trialkylboranes react with chloramine or hydroxylamine-*O*-sulfonic acid to give primary amines (eq. 12; X = Cl, OSO_3H). Only two alkyl groups are utilized, and the transfer occurs with retention of configuration.

$$R_3B + 2 NH_2X \rightarrow RB(NHR)_2 + 2 X^- \xrightarrow{NaOH} 2 RNH_2 \qquad (12)$$

Secondary amines are produced in the reactions of trialkylboranes with alkyl azides, but dialkylchloroboranes and alkyldichloroboranes (eq. 13) allow milder reaction conditions and better utilization of alkyl groups.

$$RBCl_2 + R'N_3 \rightarrow Cl_2BNRR' \xrightarrow{NaOH} RR'NH \qquad (13)$$

Synthesis of sulfur and selenium compounds. Thiocyanates, RSCN, or selenocyanates, RSeCN, can be obtained by treatment of trialkylboranes with potassium thiocyanate or selenocyanate in the presence of iron compounds. Mixed thioethers are obtained from the air-catalyzed, radical reactions of trialkylboranes with disulfides; two alkyl groups may be utilized (eq. 14; R' = CH_3, C_6H_5).

$$R_3B + 2 R'SSR' \rightarrow 2 RSR' + RB(SR')_2 \qquad (14)$$

Synthesis of organometallic compounds. Aqueous mercury(II) chloride reacts with aryldihydroxyboranes and triarylboranes to give arylmercury chlorides, utilizing all aryl groups. Mercury(II) acetate in THF or mercury alkoxides in alcohols are more suitable for preparation of alkylmercury compounds (eq. 15). All three groups of tri-*n*-alkylboranes, or two of tri-*sec*-alkylboranes are utilized. There is some loss of stereochemical integrity during reaction.

(15)

Contrathermodynamic isomerization of alkenes. Under action of heat, organoboranes isomerize to place boron predominantly at the least-hindered site. This makes possible the synthesis of organoboranes not available by direct hydroboration. For example, hydroboration–oxidation of β-pinene gives *cis*-myrtanol. Thermal treatment of the borane, prior to oxidation, results in *trans*-myrtanol.

Coupling of organic groups attached to boron. Treatment of alkylboron compounds with alkaline silver nitrate solution brings coupling of the alkyl groups (eq. 16) in good yields for primary groups; rather less for secondary. Anodic oxidation has also been used for coupling.

$$2\,R\!-\!B\!\!\diagup \xrightarrow{\text{AgNO}_3} R\!-\!R \qquad (16)$$

Synthesis of cyclopropanes. Treatment of (3-chloroalkyl)-9-BBN derivatives with base produces cyclopropanes (eq. 17).

$$H_2C\!\!=\!\!CHCHCH_2Cl \xrightarrow{\text{9-BBN}} BCH_2CH_2CHCH_2Cl \xrightarrow{\text{OH}^-} \triangle\!\!-\!\!CH_2Cl \qquad (17)$$

Single-carbon insertion reactions can be used for synthesis of aldehydes, ketones, and alcohols. With single-carbon insertion units, three types of process occur (eqs. 18–20).

$$RB\!\!\diagup \longrightarrow RCH_2OH \text{ or } RCHO \qquad (18)$$

$$-BR_2 \longrightarrow R_2CHOH \text{ or } R\overset{O}{\overset{\|}{C}}R \qquad (19)$$

$$BR_3 \longrightarrow R_3COH \qquad (20)$$

Carbonylation of organoboranes. At elevated temperatures (ca 150°C), trialkylboranes undergo reaction with carbon monoxide to give products in which all three alkyl groups are transferred from boron to carbon. Oxidation gives the corresponding tertiary alcohol (eq. 21).

$$R_3B \xrightarrow[\text{ca 150°C}]{\text{CO,HOCH}_2\text{CH}_2\text{OH}} R_3C\!-\!B\overset{O}{\underset{O}{\diagdown}} \xrightarrow{[O]} R_3COH \qquad (21)$$

The cyanoborate reaction. Alkali metal cyanides dissolve in trialkylborane solutions to give trialkylcyanoborate salts. Treatment of these with electrophilic reagents, particularly trifluoroacetic anhydride, induces rearrangement to give cyclic intermediates that are oxidized to ketones (eq. 22).

$$R_3B + NaCN \longrightarrow Na^+[R_3BCN]^- \xrightarrow{(CF_3CO)_2O} \underset{\substack{RB\diagdown O}}{\overset{R\diagup R}{\diagup N}}\!\!CF_3 \xrightarrow{[O]} R\overset{O}{\overset{\|}{C}}R \qquad (22)$$

The DCME reaction. In the presence of a strong, sterically hindered base, trialkylboranes react with α,α dichloromethyl methyl ether (DCME) to transfer all three alkyl groups from boron to carbon under remarkably mild conditions. Oxidation provides the corresponding tertiary alcohol, and the reaction can even be applied to synthesis of triarylmethanols and highly hindered trialkylmethanols (eq. 23):

$$(23)$$

Reactions with acyl carbanion equivalents; synthesis of aldehydes, ketones, and alcohols. Acyl carbanion equivalents (represented by the general symbol, −CRXY, where X and Y are potential leaving groups) react with organoboranes to give one or two rearrangements. Oxidation leads to ketones (aldehydes in the special case of R=H) or tertiary (secondary) alcohols, respectively (see below). The products differ from those obtained in one-carbon insertion reactions in that one of the alkyl groups originates in the acyl carbanion equivalent (eq. 24).

$$(24)$$

In these reactions, three types of acyl carbanion equivalents have been employed: lithiated thioacetals, aldimines, and enol ethers.

Reactions with simple carbonyl compounds; synthesis of unsaturated alcohols and aryl ketones. In general, trialkylboranes do not add in Grignard fashion to carbonyl compounds. Formaldehyde is exceptional in undergoing such a simple alkylation reaction under catalysis by air. This provides an approach to primary alcohols.

By contrast, allylboron compounds add readily to aldehydes and ketones. Use of dialkyl(allyl)boranes allows quantitative utilization of allyl groups under mild conditions (eq. 25). Rearrangement of the allyl group accompanies reaction, which allows the synthesis of hindered tertiary homoallylic alcohols.

$$(25)$$

α-Alkylation of carbonyl compounds. Organoboranes react with α-diazocarbonyl compounds or α-halogenoenolates to give products that hydrolyze to α-alkylcarbonyl compounds (eq. 26; X = halogen, N_2^+; Y = OC_2H_5, CH_3, H, etc.). Nitriles can be alkylated in a similar manner.

$$R_3B + \overset{}{\underset{X}{\bar{C}}}HCOY \longrightarrow R_2BO\overset{}{\underset{Y}{C}}\!\!=\!\!CHR \xrightarrow{H_2O} RCH_2COY \qquad (26)$$

α-Halogenocarbonyl compounds. The reaction between ethyl bromoacetate and a trialkylborane is initiated by addition of a *tert*-butoxide, but for more sensitive cases (eg, bromoacetone, chloroacetonitrile), the more hindered base, 2,6-di-*tert*-butylphenoxide, is recommended. Utilization of only one alkyl group of the trialkylborane is circumvented by use of B-alkyl-9-BBN derivatives from which the alkyl group is specifically transferred. B-Aryl-9-BBN derivatives allow transfer of aryl groups. This method has greater generality than alkylation of enolates with alkyl halides.

Diazocarbonyl compounds. Diazoacetone, ethyl diazoacetate, diazoacetonitrile, and diazoacetaldehyde react with trialkyl- or triarylboranes to give the corresponding α-alkylated (or arylated) carbonyl compounds. Trialkynylboranes and triallyboranes also react with the ester, and the ketone reaction has been extended to cover mono- and dialkylation of cyclic ketones (eqs. 27–28).

$$(27)$$

$$(28)$$

β-Alkylation of carbonyl compounds and related conjugate additions. Trialkylboranes undergo air-catalyzed 1,4-addition to α,β-unsaturated alkdehydes and ketones. Suitable substrates include acrolein, methyl vinyl ketone, 2-bromoacrolein, cycloalkenones, and α-methylenecycloalkanones prepared *in situ* from appropriate Mannich bases.

Reactions of alkynyl- and alkenylborates with electrophiles; synthesis of alkynes, alkenes, ketones, and alcohols. Addition of an alkynyllithium or alkenyllithium to an organoborane gives the corresponding lithium trialkylalkynyl borate or trialkylalkenyl borate.

Iodine; Synthesis of Alkynes, Alkenes, Enynes, Diynes

Treatment of trialkylalkynyl- or trialkylalkenylborates with iodine leads to alkynes (including terminal alkynes if the alkynyllithium used is ethynyllithium–ethylenediamine complex) or alkenes, respectively. Conjugated enynes of known stereochemistry are obtained by iodination of alkenylalkynyldisiamyl borates. Di-*sec*-alkyldialkynylborates react with iodine to give conjugated diynes.

Acids and alkylating agents. Protic acids and alkylating agents, such as alkyl halides, react with trialkylalkynylborates to give mixtures of the two geometrical isomers of the corresponding dialkylalkenylborane (E = H, alkyl).

Alkynylborates have reacted with a variety of other electrophiles, including many possessing heteroatoms such as B, Si, Sn, N, P, or O.

Trialkylalkynylborates of the type, $Li^+[R_3BC{\equiv}CCOX]^-$ (X = Ar, OCH_2CH_3) react with iodine to give α,β-unsaturated ketones and esters in good yields.

1-Lithioalkynes possessing a leaving group (halide, acetoxy) in the propargylic position give unstable trialkylalkynylborates that rearrange with displacement of the leaving group.

Ethoxyethynyltrialkylborates give homologated ethoxyalkynes on reaction with iodine at low temperature. Lithium 4-tosyloxy-1-alkynes (where tosyloxy = TsO) react with trialkylboranes to give cyclopropyl derivatives which may be converted into acylcyclopropanes or homopropargyl alcohols.

Rearrangement of α-halogenoorganoboranes; synthesis of tertiary alcohols, ketones, alkenes, or 1,2,3-trienes. The free-radical reaction of organoboranes with bromine can be controlled to allow isolation of intermediate α-bromoorganoboranes. Alternatively, such compounds can be obtained using *N*-bromosuccinimide. On treatment with water, rearrangement occurs. Indeed, the reaction is often conveniently carried out in one step by light-induced catalyzed bromination in the presence of water. In the case of tri-*sec*-alkylboranes, the reaction gives possible access to two different types of tertiary alcohols.

Cyclic dialkyl(bromoalkenyl)boranes undergo ring expansion on treatment with base and subsequent reaction of the intermediate with iodine and base. Modifications of the reaction allow synthesis of (*E,E*)-dienes or α,β-unsaturated ketones, or of 1,2,3-trienes.

Miscellaneous syntheses of unsaturated compounds. Alkenylborane reactions with iodine and base: synthesis of (*Z*)-alkenes and (*E,Z*)-dienes. (*Z*)-Alkenes and (*E,Z*)-dienes may be obtained by treatment of appropriate alkenylboranes with iodine and base.

Allylborane reactions with alkynes or ethoxyethene; synthesis of 1,4-dienes. Allylboron compounds undergo a series of interesting reactions with reactive unsaturated compounds, eg, alkynes. Careful control of the reaction with alkynes allows the initial intermediate to be hydrolyzed, giving a useful synthesis of substituted 1,4-dienes. Reactions with enol ethers also give 1,4-dienes, and in the case of cyclic enol ethers, the products contain a hydroxyalkyl unit.

Coupling of allylborates with allyl halides; synthesis of 1,4-dienes. Addition of an allyllithium to tri-*n*-butylborane, followed by addition of an allyl halide, gives a 1,4-diene resulting from head to tail coupling. This overcomes the lack of regioselectivity normally observed in the direct reaction of allyllithium with allyl halides.

Reactions of certain arylborates with electrophiles. Whereas lithium trialkylphenylborates react with acyl chlorides to give simple phenyl ketones, the magnesium analogues react with rearrangement, giving rise to (2-alkylphenyl) ketones.

Treatment of lithium trialkyl-1-naphthylborates with methyl fluorosulfonate involves a similar rearrangement, resulting in 1,2-dialkylnaphthalenes. Finally, treatment of borates obtained from 2-lithioindole derivatives with iodine gives the corresponding 2-alkylindoles.

HERBERT C. BROWN
Purdue University

KEITH SMITH
University College of Swansea, Wales

H.C. Brown, *Organic Synthesis via Boranes*, Wiley-Interscience, a division of John Wiley & Sons, Inc., New York, 1975.

A. Pelter and K. Smith, in D.H.R. Barton and W.D. Ollis, eds., *Comprehensive Organic Chemistry*, Vol. 3, Pergamon Press, Oxford, UK, 1979, pp. 689–940.

H.C. Brown, M. Zaidlewicz, and E. Negishi in G. Wilkinson, F.G.A. Stone, and E.W. Abil, eds., *Comprehensive Organometallic Chemistry*, Vol. 7, Pergamon Press, Oxford, UK, 1982, pp. 111–363.

HYDROBROMIC ACID.   See Bromine compounds.

HYDROCARBON OXIDATION

Hydrocarbon oxidation has been studied for over 100 years but the mechanisms began to be understood less than 50 years ago (see also Hydrocarbons). Radical reactions in the gas phase, and in not strongly polar solvents, do not differ appreciably in their kinetic parameters. Differences that appear to exist can be traced back to fairly well understood kinetic phenomena.

Vapor-Phase Oxidation

Above 250°C, the vapor-phase oxidation (VPO) of many organic substances becomes self-sustaining. Vapor-phase oxidation is characterized by a lengthy induction period. The initial radical-forming reaction may be:

$$RH + O_2 \rightarrow R\cdot + HO_2\cdot \qquad (1)$$
$$\text{hydroperoxy radical}$$

Because of the high activation energy of reaction 1, a termolecular alternative with lower activation energy is sometimes proposed:

$$2\,RH + O_2 \rightarrow 2\,R\cdot + H_2O_2 \qquad (2)$$

In any event, the initiating reactions are of great importance only during the induction period. When sufficient chain-branching intermediates have accumulated, the reaction rate increases rapidly.

Under self-sustaining conditions, reactions may be conveniently divided into high and low temperature regions. In an overlapping region, fundamental changes of mechanisms appear to be occurring. When the temperature of a typical low temperature reaction (eg, the oxidation of propane at 300°C) conducted at atmospheric pressure in a flow reactor is increased, the rate initially rises. However, a maximum is soon reached (at 350°C) and the rate begins to fall. This is known as the negative temperature-coefficient region. The rate subsequently goes through a minimum (at ca 375°C) and thereafter increases again.

In the low temperature region, mainly oxygenated compounds, such as aldehydes, alcohols, ketones, and carbon oxides, are produced. As the region of negative temperature coefficient is approached, an olefin with the same skeleton as the parent hydrocarbon is obtained, reaching a maximum in the region of the minimum rate. At still higher temperatures, lower molecular weight olefins and usually some methane are the products.

The apparent change in mechanism may be a function of the reversibility of the addition of oxygen to alkyl radicals:

$$R\cdot + O_2 \rightleftharpoons RO_2\cdot \qquad (3)$$

When the necessary conditions are met, a cool flame seems to arise when heat generated during the low temperature oxidation exceeds heat losses. This leads to increasing temperature and increasing rates because of a higher radical generation rate by the low temperature chain-branching agent. At a critical temperature, a cool flame appears. This causes the temperature to rise into the negative temperature coefficient region. Provided the temperature does not rise enough to permit the high temperature chain-branching agent to become effective in providing new radicals, the flame can be quenched and the temperature drops. When the temperature returns to the low temperature reaction region, if sufficient amounts of the reactants remain, the whole process can proceed through another cycle.

Although disagreements are numerous, the concept of a low temperature chain-branching agent, a (different) high temperature chain-branching agent, and a transition region where one is replacing the other, is widely accepted.

The effect of pressure in VPO has not been extensively studied but is rather informative. The negative temperature coefficient region and cool-flame phenomena are associated with low pressures, usually at atmospheric pressure. As pressure is increased, the production of olefins is suppressed and the negative temperature coefficient region disappears. The reaction rate also increases significantly and, therefore, essentially complete oxygen conversion can be attained at lower temperatures.

Molecular structure strongly influences the reaction rate in low temperature VPO but has less effect in the high temperature region.

The similarity of oxidation rates of different hydrocarbons in the high temperature region is probably related to the predominance of alkyl radical cracking reactions under these conditions. The products of such reactions would be similar for most common hydrocarbons.

Simple paraffins.   Methane is reported not to oxidize by a low temperature mechanism. The oxidation rate is rather low below 475°C and not high below 500°C. Direct oxygen attack (eq. 4) has been reported to be of dominant importance in initiation up to 2000 K, but direct dissociation (eq. 5) is also reported to be the main initiation step at high temperature:

$$CH_4 + O_2 \rightarrow CH_3\cdot + HO_2\cdot \tag{4}$$

$$CH_4 \rightarrow CH_3\cdot + H\cdot \tag{5}$$

Formaldehyde appears at a very early point in the reaction and plays a critical role in the chain-branching mechanism.

Ethane oxidation occurs at lower temperatures than methane oxidation but requires higher temperatures than the higher hydrocarbons. This can be regarded as a transition case, demonstrating mixed characteristics. Low temperature oxidation, cool flames, and a negative temperature coefficient region do occur. The main low temperature products are formaldehyde, acetaldehyde ($[HCHO]:[CH_3CHO]$ = ca 5), and carbon monoxide. These products probably arise largely through ethoxy radicals.

As temperature is raised, ethylene and hydrogen peroxide become important products. In the high temperature (600–630°C) region, hydrogen peroxide is a significant product but it is isolated in much smaller quantities than ethylene.

The VPO of propane could perhaps be called the classic case. The low temperature oxidation (beginning at ca 300°C) readily produces oxygenated products. A prominent negative temperature region is encountered on raising the temperature and cool flames are extensively reported as complicated functions of composition, pressure, and temperature.

The VPO of butane is, in most respects, quite similar to the VPO of propane. There is intriguing evidence, however, that a β-dicarbonyl component may be a significant intermediate and that it may arise by intramolecular hydrogen abstraction. This intermediate could be the source of the acetone found in the product.

The oxidation of isobutane is in some ways similar to that of other low molecular weight paraffins. However, among the significant differences is that isobutylene can be a main product even in the low temperature region.

Higher hydrocarbons.   The VPO of the higher hydrocarbons is similar to that of the lower members of the series. However, the analogous mechanisms and product distributions are more complicated, and some mechanisms and products arise that are not characteristic of the lower hydrocarbons. The fundamental change that occurs appears to be the greater opportunity for intramolecular hydrogen abstraction by alkylperoxyradicals, particularly in the β- and γ-positions.

Liquid-Phase Oxidation

There are many similarities between vapor- and liquid-phase oxidations (LPO); however, there are important differences. The latter is generally conducted at lower temperatures; it may take place in the presence of variable-valent metal-ion catalysts; it may exhibit a limiting chemical rate; and it is usually much less subject to oxygen starvation, although there are mass-transfer rate limitation effects in the transfer of oxygen to the liquid phase. There does not appear to be a negative temperature coefficient region, although the reaction may stop with increasing temperature if vapor dilutes the oxygen too much. Little conjugate olefin appears to be produced.

The transfer of oxygen from the vapor to the liquid phase is a critical part of any LPO. A simplified view of this process can provide a rationalization of the observed system behavior. The typical air-sparged LPO system can be considered to consist of two zones. The zone near the sparger is chemically rate-limited since the liquid contains enough dissolved oxygen to scavenge alkyl radicals by reaction 3. Under these conditions, the reaction is zero order with respect to oxygen; the kinetics are thus determined by the initiating and chain-branching, chain-terminating, and chain-propagating reactions.

The two zones can interact in such a way that the overall activation energy for the oxidation process can be very low as long as both zones are present. This lends stability to the operation. If the chemically rate-limited zone becomes so large that the overall energy of activation becomes significant, the operation becomes unstable because a small temperature fluctuation (reduction) can result in a significantly reduced reaction rate (because of the high activation energy). This causes a reduction in the heat evolution rate and a further reduction of temperature ensues.

Stable operation of LPO reactors thus requires the presence of a mass-transfer rate-limited zone. It is usually desirable to limit this zone to minimize oxygen starvation problems.

These views lead to the surprising conclusion that, within limits, pressure has relatively little effect in air-sparged LPO reactors.

The horizontal cross-sectional area is a critical parameter since it determines the degree of interaction of the two types of zones. It is acceptable, eg, for a reaction with a high intrinsic rate (which is largely mass-transfer rate-limited) to be conducted in a reactor with a small horizontal cross section. Slower reactions should be conducted in reactors with larger horizontal cross sections.

The lower temperature and reduced degree of oxygen starvation in LPO (vs VPO) generally significantly lower the carbon monoxide production. Also, acids, from further oxidation of aldehydes, are usually the main products.

Butane.   The main product of butane LPO is acetic acid. Most of the intermediates can generate acetic acid on further oxidation and acetic acid is remarkably resistant to further attack by even highly reactive oxy radicals. Ethyl methyl ketone, acetaldehyde, and ethanol are significant intermediates in butane LPO. Understanding the sequential relationships and relative oxidation rates of these components is key to understanding the mechanisms of the process.

Esters are also highly resistant to oxidation. Thus, esterification of alcohols during LPO considerably complicates the intermediate relationships.

Acetone is a minor coproduct of LPO of butane by a variety of mechanisms. It is also produced in part from isobutane, an impurity present in all commercial butane.

Other commercial LPO processes include the following: production of terephthalic acid (TPA) and dimethyl terephthalate (DMT) based on the LPO of *p*-xylene in an acetic acid solvent using various metal-ion catalysts;

$$CH_3 - \bigcirc - CH_3 \xrightarrow{[O]} CH_3 - \bigcirc - COOH \xrightarrow{[O]} HOOC - \bigcirc - COOH \tag{6}$$

production of synthetic phenol (qv) based on the cumene process (Cumene (qv) is produced by alkylating benzene with propylene, and is oxidized to cumyl hydroperoxide which is subsequently decomposed by acid catalysis to given phenol and acetone.); production of phenol from *sec*-butylbenzene with ethyl methyl ketone as a coproduct; production of synthetic phenol from toluene (toluene is oxidized to benzoic acid in a conventional LPO process. Oxidative decarboxylation with copper-containing catalyst gives phenol.); liquid phase oxidation of cyclohexane to produce raw materials needed for nylon, including cyclohexanol, cyclohexanone, adipic acid (qv), and 6-hydroxyhexanoic acid; and a propylene oxide process based on noncatalytic LPO of isobutane to *tert*-butyl hydroperoxide developed by Oxirane.

CHARLES C. HOBBS
Celanese Chemical Company, Inc.

R.R. Gould and F.R. Mayo, eds., "Oxidation of Organic Compounds, I and II," *Advances in Chemistry Series 75 and 76*, American Chemical Society, Washington, D.C., 1968.

F.R. Mayo, *Acc. Chem. Res.* **1**(7), 193 (1968).

S.W. Benson and P.S. Nangia, *Acc. Chem. Res.* **12**(7), 223 (1979).

HYDROCARBON RESINS

The products commonly referred to as hydrocarbon resins are low molecular weight, thermoplastic polymers derived from cracked petroleum distillates, turpentine fractions, coal tar, and a variety of pure olefinic monomers. They are used extensively in adhesives, rubber, hot-melt coatings, printing inks, paint, flooring, and other applications. They generally modify other materials and are only rarely used alone. The average molecular weight usually is below 2000. Hydrocarbon resins range from viscous liquids to hard, brittle solids. The hard resins are sold in flake or solid form, or in solution, according to customer demand. The colors range from water-white in the case of resins from pure vinyl aromatic monomers polymerized with boron trifluoride, to pale yellow and amber, and to very dark brown in the case of certain inexpensive petroleum by-product resins. Some resins are also available in anionic, cationic, or nonionic types of aqueous emulsions.

Properties

Certain properties such as the softening point, molecular weight distribution, color, color stability, and solubility (solubility contours) determine the characteristics of the resin. Manufacturers frequently aim to tailor-make a resin for a specific purpose by selection of comonomers and polymerization conditions.

Coumarone–Indene or Coal-Tar Resins

The name coumarone–indene or coal-tar resins is derived from coumarone (**1**) and indene (**2**). It is, however, not particularly accurate because coumarone is only a small component.

(**1**)

(**2**)

The commercial importance of these resins has diminished over the past decade, and they are being replaced by petroleum resins. Coumarone–indene resins with a softening point of 100°C and higher are used in coatings together with film-forming materials such as drying oils.

Petroleum Resins

Petroleum resins are low molecular weight, thermoplastic hydrocarbon resins derived from cracked petroleum fractions. They are to be distinguished from high molecular weight polymers such as polyethylene, polypropylene, or polystyrene which, are made from essentially pure starting materials (see also Petroleum; Olefin polymers). The current large-volume uses for petroleum resins are in the rubber industry, and in

Table 1. Properties of Hydrocarbon Resins

Property	A Typical C$_5$ Aliphatic Olefin– Diolefin Resin	A Typical High Melting, Aromatic Petroleum Resin	A Typical Cyclic Diolefin Resin
color, Gardner scale (50% in toluene)	11	11	11
softening point, ring and ball, 100°C	100	140	100
specific gravity	0.98[16]	1.08[25]	1.11[16]
refractive index, n_D^{20}	1.53		1.58
acid number	< 1	< 1	< 1
bromine number	45	21	55
iodine number			180
melt viscosity, at 245°C, Pa·s[a]		0.1	

[a] To convert Pa·s to poise, multiply by 10.

printing inks, adhesives, and coatings. Resin colors vary from pale yellow to dark brown.

Properties of a typical C$_5$ aliphatic olefin–diolefin resin, of a typical high melting aromatic petroleum resin (C$_q$ type) and of a typical cyclic diolefin (dicyclopentadiene) resin are given in Table 1.

Terpene Resins

Terpene resins are obtained by polymerizing various unsaturated terpene hydrocarbons, most of which occur naturally (see Resins, natural; Terpenoids). The most important terpene resins are those made from limonene or dipentene (**3**), β-pinene (**4**), and α-pinene (**5**) via cationic polymerization.

(**3**)   (**4**)   (**5**)

Resins from Pure Monomers

Thermoplastic resins from high purity monomers such as styrene, substituted styrenes, isobutylene, and the like are generally lighter in color than Gardner 1 (water-white). Among the earliest resins in this category were those made from styrene and sold as Piccolastic.

J.F. HOLOHAN, JR.
J.Y. PENN
W.A. VREDENBURGH
Hercules Incorporated

P.O. Powers in N.M. Bikales, ed., *Encyclopedia of Polymer Science and Technology*, Vol. 4, John Wiley & Sons, Inc., New York, 1966, p. 272.

J. Findlay in N.M. Bikales, ed., *Encyclopedia of Polymer Science and Technology*, Vol. 9, John Wiley & Sons, Inc., New York, 1968, p. 853.

C.T. Gonzenbach, M.A. Jordan, and R.P. Yunick in N.M. Bikales, ed., *Encyclopedia of Polymer Science and Technology*, Vol. 13, John Wiley & Sons, Inc., New York, 1970, p. 575.

HYDROCARBONS

SURVEY

Hydrocarbons are compounds of carbon and hydrogen. Hydrocarbon compounds are structurally classified as aromatic and aliphatic; the latter includes alkanes (paraffins), alkenes (olefins), alkynes (acetylenes), and cycloparaffins. An example of a low molecular weight paraffin is methane; of an olefin, ethylene; of a cycloparaffin, cyclopentane; and of an aromatic, benzene. Crude petroleum oils, which span a range of molecular weights of these compounds, excluding the very reactive olefins, have been classified according to their content as paraffinic, cycloparaffinic (naphthenic), or aromatic.

In the paraffin series, methane (CH_4) to butane (C_4H_{10}) are gases at ambient conditions. Propane (C_3H_8) and butanes are sometimes considered in a special category since they can be liquefied at reasonable pressures. These compounds are commonly referred to as liquefied petroleum gases (qv) (LPG). The pentanes (C_5H_{12}) to n-pentadecane ($C_{15}H_{32}$) are liquids, commonly called distillates, that include gasoline, kerosene, and diesel fuels (see Gasoline and other motor fuels). n-Hexadecane ($C_{16}H_{34}$) and higher molecular weight paraffins are solids at ambient conditions and are referred to as waxes (see Wax). All classes of hydrocarbons are used as energy sources and feedstocks for petrochemicals.

Hydrocarbons are important sources for energy and chemicals and are directly related to the gross national product. The United States has led the world in developing refining and petrochemical processes for hydrocarbons from crude oil and natural gas.

The earliest oil marketed in the United States came from springs at Oil Creek, Pennsylvania, and near Cuba, New York. It was used for medicinal purposes and was an article of trade among the Seneca Indians. At that time, the term Seneca Oil applied to all oil obtained from the earth.

Hydrocarbon resources can be classified as organic materials that are either mobile such as crude oil or natural gas, or immobile materials including coal, lignite, oil shales, and tar sands. Worldwide estimates of carbon and hydrogen reserves are given below.

Reserve	Carbon, 10^{12} t	Hydrogen, 10^{12} t	Molar H:C ratio
organic mobile	1.47	0.33	2.7
organic immobile	11.33	0.76	0.8
Total	*12.80*	*1.09*	

Most hydrocarbon resources occur as immobile organic materials that have a low hydrogen:carbon ratio. However, most hydrocarbon products now in demand have a H:C higher than 1.0.

Immobile hydrocarbon sources require refining processes involving hydrogenation. Additional hydrogen is also required to eliminate sources of sulfur and nitrogen oxides that would be emitted to the environment.

Resources can be classified as mostly consumed, proven but still in the ground, and yet to be discovered. A reasonable maximum for the ultimate reserves for crude oil is estimated at 250×10^9 metric tons (1.8×10^{12} bbl).

Another factor to be considered is the recovery of crude oil that can be obtained from a reservoir. At present, the average recovery from a reservoir is about 34% of the crude oil in place. Technology for secondary and tertiary recovery now under development might, given the proper economic incentive, recover 34–75% of the oil still remaining in place (see Petroleum, enhanced oil recovery).

To conserve hydrocarbons, certain reclaiming technologies are being developed, involving the re-refining of lubricating oils and the reclaiming of rubber. New technology for re-refining oils without creating acid-sludge-disposal problems is being marketed by Phillips Petroleum Company. Other commercial re-refining processes have been developed by Kinetics Technology International, B.V., Motor Oils Refining Company, Division of Estech, Inc., and Eko Tek Lube Inc. (see Recycling).

Uses. Hydrocarbons from petroleum are still the principal energy source for the world. About 74% of the world's energy is supplied by gas and oil and about 18% from coal. Petroleum lubricants continue to be the mainstay for automotive, industrial, and process lubricants.

Hydrocarbons in the Chemical Industry

At present, petroleum is the main source of raw materials for chemical conversion, ie, in 1978 about 98% of the feedstocks (qv) used by United States petrochemical producers came from oil and gas hydrocarbons.

Petrochemicals are classified into inorganic compounds; compounds derived from synthesis gas (hydrogen and carbon monoxide); compounds derived from aliphatic compounds; and cyclic compounds (aromatic and cycloparaffinic). The most important chemical from synthesis gas is methanol (qv). The aliphatic compounds include acetic acid, acetone, ethanol (qv), ethylene glycol (see Glycols), polyethylene, polypropylene (see Olefin polymers), and butadiene (qv), generally synthesized from C_2–C_4 olefins. The cyclic compounds include cyclohexane, benzene, toluene, and the xylenes which are the starting materials for caprolactam (see Polyamides), ethylbenzene, isocyanates, and polyesters (qv), respectively.

Consumer Products Based on Hydrocarbons

Many of the high volume products consumed in the world today are derived from hydrocarbons. In the food chain, beginning with fertilizers and chemicals for the treatment of seeds, hydrocarbons play an important role. To an even greater degree, materials for clothing and shelter are produced from hydrocarbons. Cleaning agents, surfactants, and solvents are derived largely from hydrocarbons. In addition to the durable plastic and rubber articles produced from hydrocarbons, great quantities of hydrocarbons are converted to wrappings and other single-use (throwaway) products. Furthermore, many of today's paints are hydrocarbon based.

Future Sources

Today, hydrocarbons are almost completely produced from petroleum and natural gas (see Fuels, survey; Petroleum, resources). Many conversion processes for alternative sources are under investigation and development. Most require large capital investment and are not competitive with petroleum recovery processes. Alternative sources include coal, oil shale (qv), tar sands (qv) and heavy oils, and biomass (see Fuels from biomass).

Toxicity

Most hydrocarbons are nontoxic or of low toxicity.

GLENN H. DALE
D.P. MONTGOMERY
Phillips Petroleum Company

J.G. Erdman and D.G. Petty, *Rev. Inst. Fr. Pet.* **35**(2), 199 (1980).

E.M. Goodger, *Prog. Combust. Sci. Technol.* **8**(3), 233 (1982).

NOMENCLATURE

Hydrocarbons, organic compounds containing only carbon and hydrogen, are classified as aliphatic, alicyclic, and aromatic.

Aliphatic or open-chain hydrocarbons include saturated and unsaturated structures and are named by IUPAC rule or, where these rules are indefinite or ambiguous, by the current index practices used in *Chemical Abstracts* (*CA*).

Saturated Hydrocarbons

Saturated aliphatic compounds, known as alkanes or paraffins, $C_nH_{(2n+2)}$, include branched and unbranched structures, and are characterized by the ending -*ane*. Names of straight-chain paraffins are derived from the Greek numeral corresponding to the number of carbon atoms with the exception of the first four members of the series (methane, ethane, propane, and butane), the nine-carbon nonane and the eleven-carbon undecane, eg, pentane, hexane, and heptane.

Unsaturated Hydrocarbons

Unsaturated aliphatic compounds, which are subdivided into a number of homologous series according to the number of double or triple bonds, have names based on the stem names of the corresponding paraffins. Olefins contain one or more double bonds. Alkenes, or mono-olefins, C_nH_{2n}, have only one double bond, and are characterized by the ending -*ene*, (Alkynes, or acetylenes, $C_nH_{(2n-2)}$, have a triple bond and are characterized by the ending -*yne*).

Alicyclic hydrocarbons. Alicyclic hydrocarbons are cyclic compounds with aliphatic properties, and are named as outlined in the IUPAC rules and the *Parent Compound Handbook* and its supplements. Saturated monocyclic compounds are cycloalkanes, or cycloparaffins, C$_n$H$_{2n}$, and are named by prefixing cyclo to the names of corresponding aliphatic structures, eg, cyclopropane.

Polycyclic hydrocarbons. In the literature, trivial names for certain sets of polycyclic hydrocarbons such as paddlanes, propellanes, peristylanes, asteranes, buttaflanes, and polymantanes have sprung up. Polycyclic compounds include bicycloalkanes, tricycloalkenes, etc.

Aromatic Hydrocarbons

Arenes. Aromatic hydrocarbons, or arenes, contain at least one benzene ring, and can be subdivided into four groups. The first is benzene and its derivatives, which are named by the rules of IUPAC and *CA*.

Polyphenyls. The second group is the polyphenyl series which includes ring assemblies in which two or more benzene rings are linked together, but which have no atoms common to more than one ring, and are named biphenyl (*CA*: 1,1'-biphenyl), terphenyls, quaterphenyls, etc.

Fused polynuclear aromatics. The third subclassification includes polynuclear aromatics that contain two or more fused rings, at least one of which must be a benzene ring. They are named according to the IUPAC rules and the *Parent Compound Handbook* and its supplements.

Phanes. The fourth group includes phanes which are bridged aromatic compounds. Phanes containing a carbocyclic nucleus are called carbophanes. Phanes containing a benzene nucleus are called cyclophanes.

KURT L. LOENING
Chemical Abstracts Service

J. Rigaudy and S.P. Klesney, eds., *Nomenclature of Organic Chemistry: Sections A, B, C, D, E, F, and H*, International Union of Pure and Applied Chemistry, Commission on Nomenclature of Organic Chemistry, Pergamon Press, New York, 1979.

"Selection of Index Names for Chemical Substances," Appendix IV of the *1984 Index Guide*, Chemical Abstracts Service, Columbus, Ohio, 1984.

Parent Compound Handbook, Chemical Abstracts Service, Columbus, Ohio, 1979.

HYDROCARBONS C$_1$ – C$_6$

METHANE, ETHANE, AND PROPANE

Methane, ethane, and propane are the first three members of the alkane hydrocarbon series having the composition, C$_n$H$_{2n+2}$. Selected properties of these alkanes are summarized in Table 1.

The main commercial source of methane, ethane, and propane is natural gas (see Gas, natural).

The largest chemical use for methane is in the production of synthesis gas for conversion to ammonia (qv) and methanol (qv). Ethane and propane are used most often in the production of ethylene (qv) by way of high temperature (ca 1000 K) thermal cracking.

D.S. MAISEL
Exxon Chemical Company

R.D. Goodwin, *NBS Technical Note 653*, U.S. Department of Commerce, NBS, Cryogenics Div., Boulder, Col., Apr. 1974; R.D. Goodwin, R.M. Roder, and G.C. Straty, *NBS Technical Note 684*, Aug. 1976; R.D. Goodwin, NBS PB272-355, July, 1977.

Oil Gas J., 65 (June 25, 1979).

Hydrocarbon Process., 192 (May 1977); 79 (Mar. 1979).

BUTANES

Butanes are naturally occurring alkane hydrocarbons that are produced primarily in association with natural-gas processing and certain refinery operations such as catalytic cracking and catalytic reforming. The term butanes includes the two structural isomers n-butane, CH$_3$CH$_2$CH$_2$CH$_3$, and isobutane, (CH$_3$)$_2$CHCH$_3$ (2-methylpropane).

Properties

The properties of butane and isobutane are summarized in Table 1.

The alkanes have low reactivities as compared to other hydrocarbons. Much alkane chemistry involves free-radical chain reactions that occur under vigorous conditions, eg, combustion and pyrolysis. Isobutane ex-

Table 1. Selected Properties of Methane, Ethane, and Propane

	Methane CH$_4$	Ethane C$_2$H$_6$	Propane C$_3$H$_8$
molecular weight	16.04	30.07	44.09
mp, K	90.7	90.4	85.5
bp, K	111	185	231
explosivity limits, vol%	5.3–14.0	3.0–12.5	2.3–9.5
autoignition temperature, K	811	788	741
flash point, K	85	138	169
heat of combustion, kJ/mol[a]	882.0	1541.4	2202.0
heat of formation, kJ/mol[a]	84.9	106.7	127.2
heat of vaporization, kJ/mol[a]	8.22	14.68	18.83
vapor pressure at 273 K, MPa[b]		2.379	0.475
specific heat at 293 K, J/(mol·K)[a]	37.53	54.13	73.63
density at 293 K, kg/m^3 [c]	0.722	1.353	1.984
critical point			
pressure, MPa[b]	4.60	4.87	4.24
temperature, K	190.6	305.3	369.8
density, kg/m^3 [c]	160.4	204.5	220.5
triple point			
pressure, MPa[b]	0.012	1.1×10^{-6}	3.0×10^{-10}
temperature, K	90.7	90.3	85.5
liquid density, kg/m^3 [c]	450.7	652.5	731.9
vapor density, kg/m^3 [c]	0.257	4.51×10^{-5}	1.85×10^{-8}

[a] To convert J to cal, divide by 4.184.
[b] To convert MPa to atm, divide by 0.101.
[c] To convert kg/m^3 to lb/ft^3, divide by 16.0.

Table 1. Properties of Butanes

Property	n-Butane C$_4$H$_{10}$	Isobutane CH(CH$_3$)$_3$
molecular weight	58.124	58.124
normal bp at 101.3 kPa (1 atm), K	272.65	261.43
flash point, K	199	190
heat of combustion, gross, kJ/mol[a]		
real gas at 101.3 kPa (1 atm) and 288.7 K	2880	2866
ΔG_f° at 101.3 kPa (1 atm) and 298.15 K, kJ/mol[a]	−17.15	−20.88
heat of fusion, kJ/mol[a]	4.660	4.540
heat of vaporization at normal bp, kJ/mol[a]	22.39	21.30
vapor pressure at 310.93 K, kPa[b]	356	498
thermal conductivity at 101.3 kPa (1 atm), and 273.15 K, W/(m·K)	0.0136	0.0140
density, kg/m^3 [c]		
gas, at 101.3 kPa (1 atm) at 288.7 K	2.5379	2.5285
critical point		
pressure, MPa[d]	3.797	3.648
temperature, K	425.16	408.13
density, kg/m^3 [c]	228.0	221.0
C_p°, ideal gas, J/mol·K[a] at 288.7 K	95.04	94.16
refractive index, n_D^{25}		
gas at 101.3 kPa (1 atm)	1.001286	
stoichiometric combustion flame temperature, K		
in air	2243	2246
maximum flame speed, m/s		
in air	0.37	0.36

[a] To convert J to cal, divide by 4.184.
[b] To convert kPa to atm, divide by 101.3.
[c] To convert kg/m^3 to lb/ft^3, divide by 16.0.
[d] To convert MPa to atm, divide by 0.1013.

hibits a different chemical behavior than *n*-butane, in part because of the presence of a tertiary carbon atom and to the stability of the associated free radical.

Reactions. The most important industrial reactions of *n*-butane are thermal cracking to produce ethylene (qv), dehydrogenation to produce butadiene (qv), liquid-phase oxidation to produce acetic acid (qv) and oxygenated by-products, isomerization to form isobutane, and vapor-phase oxidation to form maleic anhydride (qv). Various C$_1$–C$_6$ alkenes are alkylated with isobutane to produce a high octane-rated gasoline blending stock.

Health and Safety

n-Butane and isobutane are colorless, flammable, and nontoxic gases. They are simple asphyxiants, irritants, and anesthetics at high concentrations.

Uses

Butanes are used primarily as gasoline blending components (isobutane is first alkylated) and less so as liquefied gas fuel and in the manufacture of chemicals.

D.D. Boesiger
R.H. Nielsen
M.A. Albright
Phillips Petroleum Co.

Gas Processors Suppliers Association Engineering Data Book, 9th ed., Gas Processors Suppliers Association, Tulsa, Okla., 1972, 1977.

L'Air Liquide Division Scientifique, Encyclopedie Des Gaz, Engl. transl. by N. Marshall, Elsevier Scientific Publishing Co., Amsterdam, Neth., 1976.

J.R. Beatty and R.E. Cannon, *1980 Proceedings—American Petroleum Institute*, Refining Dept. (45th Midyear Meeting, May 12–15, 1980, preprint 10-80) **59**, 52 (1980).

PENTANES

Pentanes, C$_5$H$_{12}$, are members of a class of saturated aliphatic hydrocarbons called alkanes and each pentane contains five carbon atoms. The pentanes exist in three isomeric forms, normal, iso-, and neo- which are represented by the following structures:

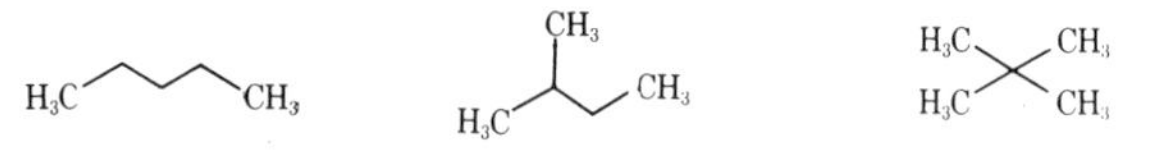

n-pentane   isopentane   neopentane

Properties

Each isomer has its individual set of physical and chemical properties; however, these properties are similar (see Table 1). The fundamental chemical reactions for pentanes are sulfonation to form sulfonic acids, chlorination to form chlorides, nitration to form nitropentanes, oxidation to form various compounds, and cracking to form free radicals.

Health and Safety

Pentanes are only slightly toxic. Because of their high volatilities, and consequently, their low flash points, they are highly flammable.

The threshold limit value for the time-weighted average (8-h) exposure to pentanes is 600 ppm or 1800 mg/m^3 (51 mg/SCF). Pentanes are classified as simple asphyxiants and anesthetics.

Uses

The main use for pentanes is in gasoline which is intended for internal combustion engines, eg, in automobiles, trucks, tractors, and light planes.

T.S. Schmidt
D.M. Haskell
Phillips Petroleum Co.

Table 1. Properties of Pentanes

Property	*n*-Pentane	Isopentane	Neopentane
molecular weight	72.151	72.151	72.151
normal freezing point, K	143.429	113.250	256.57
normal bp, K	309.224	301.002	282.653
water solubility at 25°C, g C$_5$H$_{12}$/100 kg H$_2$O	9.9	13.2	
spontaneous ignition temperature in air, K	557.0	700.0	729.0
flash point, K	233.0	213.0	198.0
critical point			
pressure, MPa^a	3.369	3.381	3.199
temperature, K	469.7	460.39	433.75
density, kg/m$^{3\ b}$	231.9	234.0	237.7
heat of combustion, kJ/mol^c (at 298 K)			
gas	3272	3264	3253
heat of fusion, kJ/mol^c	8.39	5.15	3.15
heat of vaporization, kJ/mol^c	25.77	24.69	22.75
dipole moment, C·m^d	0.0	3.336 × 10^{-31}	0.0
surface tension at 20°C, mN/m (= dyn/cm)	16.00	15.00	12.05
refractive index, n_D^{25}			
liquid	1.35472	1.35088	1.339
gas	1.001585		

^{a}To convert MPa to atm, divide by 0.101.
^{b}To convert kg/m^3 to lb/ft^3, divide by 16.0.
^{c}To convert J to cal, divide by 4.184.
^{d}To convert C·m to debye, divide by 3.336 × 10^{-30}.

H.H. Szmant, *Organic Chemistry*, Prentice-Hall, Englewood Cliffs, N.J., 1957, p. 69.

R.C. Reid, J.M. Prausnitz, and T.K. Sherwood, *The Properties of Gases and Liquids*, 3rd ed., McGraw-Hill, New York, 1977.

R.R. Dreisbach, "Physical Properties of Chemical Compounds, II," *Advances in Chemistry Series 22*, American Chemical Society, Washington, D.C., 1959.

HEXANES

Hexane refers to the straight-chain hydrocarbon, C$_6$H$_{14}$; branched hydrocarbons of the same formula are isohexanes. Hexanes include the branched compounds, 2-methylpentane, 3-methylpentane, 2,2-dimethylbutane, 2,3-dimethylbutane, and the straight-chain compound *n*-hexane. Commercial hexane is a narrow-boiling mixture of these compounds and methylcyclopentane, cyclohexane, and benzene (qv); minor amounts of C$_5$ and C$_7$ hydrocarbons also may be present. Hydrocarbons in commercial hexane are found chiefly in straight-run gasoline, which is produced from crude oil and natural-gas liquids (see Gasoline and other motor fuels; Gas, natural). Smaller volumes occur in certain petroleum refinery streams.

Properties

The flash point of *n*-hexane is −21.7°C and the autoignition temperature is 225°C. The explosive limits of hexane vapor in air are 1.1–7.5%. Above 2°C, the equilibrium mixture of hexane and air above the liquid is too rich to fall within these limits. A few other properties are described in Table 1.

Manufacture

Commercial hexanes are manufactured by two-tower distillation of a suitable charge stock, eg, straight-run gasolines that have been distilled from crude oil or natural-gas liquids.

Health and Safety

Hexane is classified as a flammable liquid by the ICC, and normal handling precautions for this type of material should be observed. According to the ACGIH, the maximum concentration of hexane vapor in air to which a worker may be exposed without danger of adverse

Table 1. Properties of Hydrocarbons Found in Commercial Hexanes

Hydrocarbon	Freezing point, °C	Bp, °C	Liquid density $kg/m^{3\ a}$ (at 20°C)	Liquid refractive index, n_D^{20}
2-methylbutane	−159.90	27.85	619.7	1.35373
n-pentane	−129.73	36.07	626.2	1.35748
cyclopentane	−93.87	49.26	745.4	1.40645
2,2-dimethylbutane	−99.87	49.74	649.2	1.36876
2,3-dimethylbutane	−128.54	57.99	661.6	1.37495
2-methylpentane	−153.66	60.27	653.2	1.37145
3-methylpentane		63.28	664.3	1.37652
n-hexane	−95.32	68.74	659.3	1.37486
methylcyclopentane	−142.46	71.81	748.6	1.40970
benzene	5.53	80.10	879.0	1.50112
cyclohexane	6.55	80.74	778.6	1.42623
2,2-dimethylpentane	−123.81	79.20	673.9	1.38215
2,4-dimethylpentane	−119.24	80.50	672.7	1.38145
1,1-dimethylcyclopentane	−69.80	87.85	754.5	1.41356

health effects is 125 ppm; benzene is rated at 10 ppm. n-Hexane can be grouped with the general anesthetics (qv) in the class of central nervous system depressants.

Uses

Most commercial hexane is used in motor fuel; its primary commercial uses are as a solvent for oil-seed extraction (see Extraction, liquid–solid) and as a reaction medium for various polymerization reactions (see Polymerization mechanisms and processes). It also is used in conjunction with more polar solvents, eg, furfural, in the separation of fatty acids.

G.H. DALE
L.E. DREHMAN
Phillips Petroleum Co.

C.E. Miller and F.D. Rossini, eds., *Physical Constants of Hydrocarbons C_1 to C_{10}*, American Petroleum Institute, New York, 1961.

Industrial Hygiene Monitoring Manual for Petroleum Refineries and Selected Petrochemical Operations, prepared for American Petroleum Institute, contract no. LER-40-73, by Clayton Environmental Consultants, Inc., p. 69.

The Solvent Extraction of Oil Seed—An Informational Survey, Circular Vol. 12, No. 5, State Engineering Experiment Station, Georgia Institute of Technology, June, 1950, pp. 78–80.

CYCLOHEXANE

Cyclohexane, C_6H_{12}, is a clear, essentially water-insoluble, noncorrosive liquid. It is easily vaporized, readily flammable, and less toxic than benzene. Structurally, it is a cycloparaffin. Essentially all high purity cyclohexane is made by hydrogenation of benzene (qv). Cyclohexane is present in all crude oils in concentrations of 0.1–1.0%.

Properties

Properties of cyclohexane are given in Table 1.

Stereochemistry. Cyclohexane can exist in two molecular conformations: the chair and boat forms. Conversion from one conformation to the other involves rotations about carbon–carbon single bonds.

Reactions. The most important commercial reaction of cyclohexane is its oxidation (liquid phase) with air in the presence of a soluble cobalt catalyst to produce cyclohexanol and cyclohexanone (see Hydrocarbon oxidation; Cyclohexanol and cyclohexanone).

Health and Safety

The threshold limit value (TLV) for cyclohexane is 300 ppm (1050 mg/m^3). At high concentrations, it is an anesthetic and narcosis may occur.

Table 1. Properties of Cyclohexane

Property	Value
fp, °C	6.554
molal fp lowering, °C	20.3
bp, °C	80.738
flammability limits (in air), vol%	1.3–8.4
flash point (closed cup), °C	−17
heat of transition, kJ/kg^a	80.08
heat of fusion, kJ/kg^a	31.807
heat of vaporization at 25°C, kJ/kg^a	392.50
vapor pressure at 30°C, kPa^b	16.212^c
transition point, °C	−87.05
surface tension at 20°C, N/m^c	0.0253 ± 0.3
refractive index, $n_D^{20\ d}$	1.4623
density, $d_4^{20\ d}$, g/m^3	0.77855
dynamic viscosity at 20°C, mPa·s ($=$ cP)	0.980
specific heat relative to water at 25.9°C	0.440

[a] To convert J to cal, divide by 4.184.
[b] To convert kPa to atm, divide by 101.3.
[c] To convert N/m to dyn/cm, multiply by 10^3.
[d] For air-saturated liquid at 101.3 kPa.

Uses

Almost all of the cyclohexane that is produced in concentrated form is used as a raw material in the first step of nylon-6 and nylon-6,6 manufacture (see Polyamides).

M.L. CAMPBELL
Exxon Chemical Company

R.C. Reid, J.M. Prausnitz, and T.K. Sherwood, *The Properties of Gases and Liquids*, 3rd ed., McGraw-Hill Book Co., New York, 1977, pp. 61, 151, 184, 188, 226, 443.

Selected Values of Hydrocarbons and Related Compounds, American Petroleum Institute Research Project 44, Carnegie Press, Pittsburgh, Pa., 1960, Table 23-2 (3.1110)-m, p. 1.

HYDROCARDANOL, m-PENTADECYLPHENOL, m-$C_{15}H_{31}$-C_6H_4OH. See Alkylphenols.

HYDROCHLORIC ACID. See Hydrogen chloride.

HYDROCOLLOIDS. See Colloids; Gums.

HYDROGEL. See Contact lenses; Polyelectrolytes.

HYDROGEN

Hydrogen, the lightest and most abundant element, has an atomic number of one and an atomic weight of 1.00782519 on the ^{12}C scale. The single proton comprising the hydrogen nucleus has a single orbiting electron filling one half of the $1s$ electron shell. Hydrogen atoms, designated H, can exist in a pure state under certain circumstances. Normally, however, pure hydrogen exists as H_2, the hydrogen molecule, and the lightest of all gases.

Hydrogen (H_2) is a product of many reactions, but it is present at only extremely low levels in the earth's atmosphere (0.1 ppm levels). The H_2 molecule exists in two forms, depending on the nuclear spins of the atoms. The forms are designated ortho hydrogen (nuclear spins parallel) and para hydrogen (nuclear spins antiparallel). Many physical and thermodynamic properties of H_2 depend upon the nuclear-spin orientation. In particular, those properties that involve heat, such as enthalpy, entropy, and thermal conductivity, can show definite differences for ortho and para hydrogen; other properties show little difference.

Hydrogen is a very stable molecule because of its high bond strength (435 kJ/mol or 104 kcal/mol). It is not particularly reactive under ordinary conditions, but at elevated temperatures and with the aid of catalysts, it undergoes many reactions. Hydrogen forms compounds with almost every other element, often by direct reaction of the elements. The explanation of its ability to form compounds with such chemically dissimilar elements as alkali metals, halogens, transition metals, and carbon lies in the intermediate electronegativity of the hydrogen atom.

Hydrogen has two isotopes; deuterium D of atomic weight 2.01410222 and tritium T of atomic weight 3.0160497. Deuterium and tritium occur naturally on earth, but at very low concentrations. Deuterium, tritium, and hydrogen have different physical properties, but nearly the same chemical properties (see Deuterium and tritium).

Hydrogen is an important industrial commodity. It is used in the reduction of metal oxides, such as iron ore, in other metalworking operations, and in welding. Some of the industrial syntheses using hydrogen include the production of ammonia, hydrochloric acid, aluminum alkyls, methanol, higher alcohols, and aldehydes. Hydrogen is used to hydrogenate various petroleum and edible oils, coal, and shale oil. Liquid hydrogen is a very important cryogenic fluid (see Cryogenics). Table 1 outlines many of the physical and thermodynamic properties of gaseous para- and normal hydrogen.

Hydrogen gas diffuses rapidly through many materials, including metals. This property is used in separating hydrogen from other gases and in purifying hydrogen on an industrial scale. Hydrogen diffusion through metals is also used as an analytical technique for hydrogen determination in gas chromatography. Hydrogen is only slightly soluble in water, but is somewhat more soluble in organic compounds.

Solid hydrogen usually exists in the hexagonal close-packed form, with $a_o = 378$ pm and $c_o = 616$ pm. Solid deuterium also exists in the hexagonal close-packed configuration, with unit cell dimensions $a_o = 354$ pm, $c_o = 591$ pm.

In addition to H_2, D_2, and T_2, the following isotopic mixtures exist: HD, HT, and DT.

A mixture of solid and liquid para hydrogen has been termed slush hydrogen. Slush hydrogen is thought to be better for fuel purposes than liquid hydrogen because of the greater density and high heat capacity of the solid–liquid mixture.

Solid hydrogen is known to undergo phase transitions with increased pressure. One phase of solid hydrogen that is postulated to exist under conditions of extreme pressure is metallic hydrogen. The great interest in this material lies in the fact that metallic hydrogen may be a superconductor of electricity, although solid hydrogen under usual conditions is an insulator.

Hydrogen gas chemisorbs on the surface of many metals in an important step for many catalytic reactions.

Reactions

Hydrogen-producing reactions. Hydrogen is produced by a number of reactions. On a laboratory scale, it can be made from the action of an aqueous acid on a metal or from the reaction of an alkali metal in water:

$$Zn + 2\ HCl \rightarrow H_2 + ZnCl_2$$

$$2\ Na + 2\ H_2O \rightarrow H_2 + 2\ NaOH$$

These reactions can be carried out at room temperature. Hydrogen gas can also be produced on a laboratory scale by the electrolysis of an acidic solution. This involves the following reaction at the cathode of the electrochemical cell:

$$H^+ (aq) + e \rightarrow \tfrac{1}{2}\ H_2$$

Production of hydrogen through electrolysis is also used industrially. Electrolytic production of hydrogen may have a central role in any future hydrogen-based energy system. Other industrial methods of producing hydrogen involve the use of hydrocarbons such as natural gas, petroleum, and coal. One method of hydrogen production is known as steam reforming of hydrocarbons (see Petroleum):

$$CH_4 + H_2O \xrightarrow[\substack{3.0\ \text{MPa (30 atm)}\\ \text{Ni catalyst}}]{800°C} CO + 3\ H_2$$

On an industrial scale, hydrogen is also produced in significant quantities during the uncatalyzed steam pyrolysis of aliphatic hydrocarbons:

$$\text{alkane} \xrightarrow[\text{heat}]{\text{steam}} H_2,\ \text{olefins, aromatics, other paraffins}$$

The carbon–steam reaction has been used to produce hydrogen gas.

$$C + H_2O \rightarrow CO + H_2$$

Reactions leading to hydrogen often have carbon monoxide as a coproduct. The water-gas shift reaction is used industrially to reduce the CO levels in a gas stream, and to produce additional H_2:

$$CO + H_2O \xrightarrow[\text{catalyst}]{400°C} CO_2 + H_2$$

This reaction is extremely important in ammonia production, in detoxifying town gas, and in other processes (see Fuels, synthetic).

Bonding of hydrogen to other atoms. The hydrogen atom can either lose the $1s$ valence electron when bonding to other atoms to form the H^+ ion, or conversely, it can gain an electron in the valence shell to form the hydride ion, H^-. The formation of the H^+ ion is a very endothermic process:

$$H(g) \rightarrow H^+ (g) + 1\ e \qquad \Delta H = 1310\ kJ/mol\ (313.1\ kcal/mol)$$

The formation of the hydride ion is also endothermic:

$$\tfrac{1}{2}\ H_2(g) + e \rightarrow H^- (g) \qquad \Delta H = 151\ kJ/mol\ (36.1\ kcal/mol)$$

Because the hydrogen molecule has a high bond strength, it is not particularly reactive under normal conditions. For this reason, high temperatures and catalysts are often used in hydrogen reactions.

Reactions of hydrogen and other elements. Hydrogen forms compounds with almost every other element with direct reaction of the elements possible in many cases. Hydrogen combines directly with halogens, as follows:

$$H_2 + X_2 \rightarrow 2\ HX$$

$$X = F,\ Cl,\ Br,\ or\ I$$

Table 1. Physical and Thermodynamic Properties of Gaseous Normal and Gaseous para-Hydrogen[a]

Property	Para	Normal
density at 0°C, (mol/cm³) × 10³	0.05459	0.04460
compressibility factor $Z = \dfrac{PV}{RT}$ at 0°C	1.0005	1.00042
adiabatic compressibility $\left(-\dfrac{1}{V}\dfrac{\partial V}{\partial P}\right)_s$ at 300 K, MPa^{-1}[b]	7.12	7.03
coefficient of volume expansion $\left(\dfrac{1}{V}\dfrac{\partial V}{\partial T}\right)_p$ at 300 K, K^{-1}	0.00333	0.00333
C_p at 0°C, J/(mol·K)[c]	30.35	28.59
C_v at 0°C, J/(mol·K)[c]	21.87	20.30
enthalpy at 0°C, J/mol[c]	7656.6	7749.2
internal energy at 0°C, J/mol[c]	5384.5	5477.1
entropy at 0°C, J/(mol·K)[c]	127.77	139.59
velocity of sound at 0°C, m/s	1246	1246
viscosity at 0°C, mPa·s (= cP)	0.00834	0.00834
thermal conductivity at 0°C, mW/(cm·K)	1.826	1.739
dielectric constant at 0°C, ε	1.00027	1.000271
isothermal compressibility $\dfrac{1}{V}\left(\dfrac{\partial V}{\partial P}\right)_T$ at 300 K, MPa^{-1}[b]	−9.86	−9.86
heat of dissociation at 298.16 K, kJ/mol[c]	435.935	435.881

[a] All values at 101.3 kPa (1 atm).
[b] To convert MPa to atm, divide by 0.101.
[c] To convert J to cal, divide by 4.184.

Hydrogen combines directly with oxygen, either thermally or with the aid of a catalyst:

$$2 H_2 + O_2 \rightarrow 2 H_2O$$

Hydrogen reacts with carbon to form methane:

$$2 H_2 + C \text{ (graphite)} \rightarrow CH_4$$

One of the most important reactions of hydrogen industrially is in the production of ammonia (qv), as follows:

$$3 H_2 + N_2 \rightarrow 2 NH_3$$

Hydrogen reacts directly with a number of metallic elements to form hydrides (qv). The ionic hydrides, or saline hydrides, are formed from the reaction of hydrogen with the alkali metals and with some of the alkaline earth metals. Examples of these reactions are

$$Li + \tfrac{1}{2} H_2 \rightarrow LiH$$

$$Ca + H_2 \rightarrow CaH_2$$

As a reducing agent. Hydrogen reacts with a number of metal oxides at elevated temperatures to produce the metal. Examples of these reactions are

$$FeO + H_2 \rightarrow Fe + H_2O$$

$$Cr_2O_3 + 3 H_2 \rightarrow 2 Cr + 3 H_2O$$

Under certain conditions, hydrogen reacts with nitric oxide, an atmospheric pollutant and contributor to photochemical smog, to produce N_2:

$$2 NO + 2 H_2 \rightarrow N_2 + 2 H_2O$$

Sulfur, nitrogen, and oxygen heteroatoms, which are abundant in many fuel sources such as petroleum, coal, and shale oil, are considered pollutants and detriments to the refining process. Hydrogen is used to reduce the amounts of these contaminants in fuels.

Hydrogenation reactions. Hydrogenation reactions, which involve the addition of hydrogen to various compounds, are important in the production of fuels, in fuel conversions and upgrading, and in the synthesis of various pure chemicals. Almost all reactions of this type involve catalysts.

The catalyzed hydrogenation of carbon monoxide and carbon dioxide has been studied for many years; it is now receiving renewed attention for the production of various fuels and chemicals. Some of these reactions are as follows:

$$CO + 3 H_2 \rightarrow CH_4 + H_2O$$

$$CO_2 + 4 H_2 \rightarrow CH_4 + 2 H_2O$$

$$CO + 2 H_2 \rightarrow CH_3OH$$

$$2 CO + 5 H_2 \rightarrow C_2H_6 + 2 H_2O$$

$$m\, CO + (2m + 1) H_2 \rightarrow C_mH_{2m+2} + m\, H_2O$$

$$CO + H_2 \rightarrow \text{glycols}$$

A reaction of industrial importance involving an olefin reacting with hydrogen is the oxo process (qv) that uses a homogeneous cobalt catalyst in the production of aldehydes and alcohols:

$$RCH{=}CH_2 + CO + H_2 \xrightarrow[\substack{20.3 \text{ MPa} \\ (200 \text{ atm})}]{150°C} RCH_2CH_2CHO$$

Another reaction of industrial importance involving an olefin and hydrogen is the direct synthesis of aluminum alkyls (see Organometallics):

$$Al + 1.5 H_2 + 3 \text{ (1-olefin)} \rightarrow \text{(alkyl)}_3Al$$

Reactions of atomic hydrogen. Atomic hydrogen is a very strong reducing agent and a highly reactive radical which can be produced by various means, among them subjecting H_2 to high temperatures or radiation, and decomposition of hydrocarbon radicals.

Absorption of hydrogen in metals. Many metals and alloys absorb hydrogen in large amounts. A striking example is a palladium electrode which, during electrolysis, can absorb several hundred times its volume of hydrogen.

Manufacture

The principal commercial processes for hydrogen manufacture are catalytic steam reforming, partial oxidation, coal gasification, and water electrolysis. In the United States, the bulk of the industrial hydrogen is manufactured by steam reforming of natural gas. Relatively small quantities of hydrogen are produced by steam reforming of naphtha and by partial oxidation of oil. Methane reforming:

$$CH_4 + 2 H_2O \rightarrow CO_2 + 4 H_2$$

Resid partial oxidation:

$$CH_{1.8} + 0.98 H_2O + 0.51 O_2 \rightarrow CO_2 + 1.88 H_2$$

Coal gasification:

$$CH_{0.8} + 0.6 H_2O + 0.7 O_2 \rightarrow CO_2 + H_2$$

Water electrolysis:

$$2 H_2O \rightarrow 2 H_2 + O_2$$

Other processes using hydrocarbon feeds. Other processes include: the internal combustion-engine process using oxidation of methane by oxygen; thermal decomposition of hydrocarbons or natural gas on a brick checkerwork heated to 1100°C by hot combustion gases; the use of coke-oven gas (typical analysis 58% H_2, 4% CH_4, 8% CO, 2.5% higher hydrocarbon, 1% O_2, CO_2, 2.5% N_2); and the oxidation of hydrocarbons with steam and oxygen or with steam only at 425–730°C.

H_2 purification. The impurities usually found in raw hydrogen are CO_2, CO, O_2, N_2, H_2O, CH_4, and higher hydrocarbons. Removal of these impurities can be accomplished by shift catalysis, H_2S and CO_2 removal, PSA (pressure-saving adsorption) process, and nitrogen wash. Very pure hydrogen can be produced by the PSA process or by cryogenic methods.

Health and Safety Factors

Hydrogen gas is not considered toxic, but it can cause suffocation by the exclusion of air. The main danger in the use of liquid and gaseous hydrogen lies in its extreme flammability in oxygen or air.

B.G. MANDELIK
DAVID NEWSOME
Pullman Kellogg

R.D. McCarty, *Hydrogen Technological Survey—Thermophysical Properties (NASA SP-3089)*, GPO, Washington, D.C., 1975, pp. 518–519.

H.P. Leftin, D.S. Newsome, T.J. Wolff, and J.C. Yarze, *Industrial and Laboratory Pyrolysis*, ACS Symposium Series 32, ACS, Washington, D.C., 1976, p. 363.

H.G. Corneil, F.J. Heinezelman, and U.S. Nicholson, *Hydrogen in Oil Refinery Operations—Hydrogen for Energy Distribution*, IGT, Chicago, Ill., 1978.

HYDROGEN BROMIDE.   See Bromine compounds.

HYDROGEN CHLORIDE

Hydrochloric acid is a solution of hydrogen chloride in water. The relationship of the various density units to the concentration of the three standard commercial grades of hydrochloric acid is shown in Table 1. High purity hydrogen chloride is made preferably by direct combination of the purified elements.

Hydrochloric acid is present in the digestive system of most mammals. Hydrogen chloride was present in the original atmosphere of the earth and in the oceans. Considerable quantities of hydrochloric acid occur in the gases evolved from many volcanoes.

Physical and Thermodynamic Properties

Anhydrous hydrogen chloride. Anhydrous hydrogen chloride is a colorless gas that condenses to a colorless liquid and freezes to a white

Table 1. Density and Concentration of Commercial Grades of Hydrochloric Acid

| Specific gravity | Density unit[a] | | | % HCl |
	°Baumé	°Twaddell		
1.1417	18	28.34		27.92
1.1600	20	32.00		31.45
1.1789	22	35.78		35.21

^{a}sp gr $= \dfrac{145}{145 - °Bé} = 0.005 °Tw + 1.$

crystalline solid. The quantitative physical and thermodynamic properties are shown in Table 2 for selected temperatures and pressures.

Hydrogen chloride and water form four hydrates: dihydrate (mp $-17.7°C$ in a sealed tube), monohydrate (mp $-15.35°C$), trihydrate (mp $-24.9°C$), and hexahydrate (mp $-70°C$).

Hydrogen chloride and water also form constant-boiling mixtures.

HCl–H$_2$O–inorganic compound systems. Salts. The salting out of metal chlorides from aqueous solutions by the common ion effect as the HCl concentration is increased has been studied extensively and put to practical use in many cases. The general form of the solubility–temperature curves is not altered by the lowering of solubility.

The properties of the system $FeCl_2$–HCl–H_2O are important in the steel-pickling industry, both for development of pickling procedures and for regeneration of spent pickling liquors (see Metal surface treatments). Many metal chlorides can be almost completely salted out by addition of hydrogen chloride to their aqueous solutions. Important salts in this group are the chlorides of sodium, potassium, magnesium, strontium, and barium. Ferric chloride solutions consist of complex mixtures of chloro and hydroxy groups, which give a wide range of compounds with hydrogen chloride.

Certain metal chlorides are not readily salted out by hydrochloric acid, and require high concentrations for precipitation. This property is put to use in recovery of hydrogen chloride from azeotropic mixtures. Calcium chloride has been found particularly useful for this application.

Chlorine. The solubility of chlorine in hydrochloric acid is significant in the purification of by-product hydrochloric acid. The concentration of chlorine in solution is proportional to the partial pressure of chlorine in the gas phase, and its slope is a function of the HCl concentration and pressure. The chlorine is present as a combination of Cl_2 and Cl_3^- in relatively concentrated HCl solution ($> 2N$). As a result of formation of the trichloride ion, the solubility of chlorine increases in relatively concentrated HCl solutions.

HCl–organic compound systems. Although the solubility of hydrogen chloride is usually considered to deviate more from ideality in water than in most other solvents, in fact this is not the case. In terms of Raoult's law, the solubility of hydrogen chloride in many solvents deviates more widely from ideality than does its solubility in water.

The solubility of hydrogen chloride in many solvents shows a fairly good correlation with Henry's law. Notable exceptions are polyhydroxy compounds, such as ethylene glycol, which have characteristics similar to those of water.

Chemical Properties

Reactions of anhydrous hydrogen chloride. Reactions with inorganic compounds. Hydrogen chloride can react by either heterolytic or homolytic fission of the H—Cl bond. Because the kinetic barrier to either type of fission is high for the anhydrous material, it is relatively inert. The following reactions have been studied: protonation of Group V hydrides, MH_3 (M = N, P, As); reaction with nitrides, borides, silicides, germanides, carbides and sulfides; reaction with silicon, germanium, and boron hydrides; reaction with metal oxides, metals, and oxidizing agents; and reaction with oxyacids and their salts.

Reactions with organic compounds. These reactions include addition to olefins and acetylenes and replacement of aliphatic hydroxyl with chloride.

Table 2. Physical and Thermodynamic Properties of Anhydrous Hydrogen Chloride

Property	Value
melting point, °C	-114.2
boiling point, °C	-85.05
heat of fusion at $-114.22°C$, kJ/mol^a	1.992
heat of vaporization at $-85.05°C$, kJ/mol^a	16.142
entropy of vaporization, J/(mol·K)^a	85.85 (Trouton constant)
triple point, °C	-114.3
critical temperature (T_c), °C	51.54
critical pressure (P_c), MPa^b	8.316
critical volume (V_c), L/mol	0.069
critical density, g/L	424
critical compressibility factor (Z_c)	0.117
$\Delta H_f°$ at 298 K, kJ/mol^a	-92.31 (measured)
	-100.4 (calculated)
$\Delta G_f°$ at 298 K, kJ/mol^a	-95.303
$S°$ at 298 K, J/(K·mol)^a	186.8
dissociation energy at 298 K, kJ^a	431.62 (measured)
	427.19 (calculated)
liquid vapor pressure (P (kPa))^c (160–260 K)	$\log_{10} P$ (kPa) $= -905.53\,T^{-1}$ $+ 1.75 \log T - 0.0050077\,T$ $+ 3.78229\,(T$ in K)
compressibility coefficient, λ	0.00787
internuclear separation, nm	0.1251
dipole moment, C·m^d	3.716×10^{-30} (calculated)
	3.74×10^{-30} (measured)
ionization potential, J^a	20.51 (measured)
	20.45 (calculated)
heat capacity (C_p), J/(mol·K)^a	
vapor (constant pressure) at/ 273.16 K, J/(mol·K)^a	29.162
liquid at 163.16 K, J/(mol·K)^a	60.378
solid at 147.16 K, J/(mol·K)^a	48.98
surface tension at 118.16 K, mN/cm	
(= dyn/cm)	23
viscosity, mPa·s (= cP)	
liquid at 118.16 K	0.405
vapor at 273.06 K	0.0131
thermal conductivity, mW/(m·K)	
liquid at 118.16 K	335
vapor at 273.16 K	13.4
density, g/cm^3	
liquid at 118.16 K	1.045
solid (rhombic form) at 81 K	1.507
solid (cubic form) at 98.36 K	1.48
refractive index	
liquid at 283.16 K	1.254
gas at 273.16 K	1.0004456
dielectric constant	
liquid at 158.94 K	14.2
gas at 298.16 K	1.0046
electrical conductivity, s/m	
at 158.94 K	1.7×10^{-7}
at 185.56 K	3.5×10^{-7}

^{a} To convert J to cal, divide by 4.184.
^{b} To convert MPa to atm, divide by 0.101.
^{c} To convert kPa to mm Hg, multiply by 7.5.
^{d} To convert C·m to debye, divide by 3.336×10^{-30}.

Chloromethylation reactions. The chloromethyl group can be introduced in other aliphatic and aromatic compounds by reaction with paraformaldehyde and hydrogen chloride. Methyl chloromethyl ether and benzyl chloride can be made by this reaction. A highly carcinogenic by-product, bischloromethyl ether, restricts the use of this reaction.

Hydrochloric acid. Reactions with inorganic compounds. These include reactions with metals, oxides, and hydroxides. Elemental chlorine can be obtained with oxidizing agents and by various electrolytic processes.

Reactions with organic compounds. Hydrochloric acid catalyzes many organic reactions, but most of these reactions are not specific to hydrochloric acid.

Manufacture: Synthesis, Recovery, and Purification

The chloride component of hydrogen chloride is always obtained from the chloride salts formed initially when the crustal deposits of the earth reacted with primal hydrochloric acid. Hydrochloric acid can be obtained in a reversal of this process by pyrohydrolysis of such salts as calcium and magnesium chlorides. Synthetic metal chlorides such as $TiCl_4$ can also be hydrolyzed. Hydrochloric acid can also be made by thermal dissociation of various chloride hydrates made synthetically, such as $AlCl_3 \cdot 6H_2O$. In most of these processes, the primary product is the resultant metal hydrolysate, and the hydrogen chloride is a secondary by-product. The various commercially significant processes for making hydrogen chloride include the decomposition of alkali-metal chlorides by acids (eg, the salt–sulfuric acid process and the Hargreaves process), direct synthesis from hydrogen and chlorine, by-product hydrogen chloride from chemical processes (eg, recovery and purification of anhydrous gas; recovery and purification of aqueous solutions; production of HCl gas from hydrochloric acid solutions), and hydrogen chloride produced from waste organics.

Materials of Construction for Handling, Storage, and Process Equipment

Gaseous hydrogen chloride. Pure hydrogen chloride gas does not react with standard metallic and nonmetallic materials of construction at ambient conditions. The reaction with most metals is slow at all but elevated temperatures. Because HCl gas reacts with metal oxides to form chlorides, oxychlorides, and water, steel equipment should be pickled to remove mill scale before it is put in service.

Aqueous hydrochloric acid. Aqueous hydrochloric acid attacks most metals in accordance with the equation

$$M + n\, H_3O^+ \rightarrow M^{n+} + n\, H_2O + n/2\, H_2$$

The two metals with the highest resistance to hydrochloric acid are tantalum and zirconium. Plastics and elastomers, carbon and graphite, and glass and ceramics also are resistant to HCl.

Health and Safety Factors

Hydrogen chloride in air is a respiratory irritant. The maximum allowable concentration under normal working conditions has been set at 5 ppm. There is no evidence of chronic systemic effects. Hydrogen chloride in the lungs can cause pulmonary edema, a life-threatening condition. Concentrated hydrochloric acid in contact with the skin can cause chemical burns or dermatitis.

In air, hydrogen chloride can also be a phytotoxicant.

Uses

The uses of hydrogen chloride and hydrochloric acid derive from their special properties and the properties of the reaction products. The uses of anhydrous hydrogen chloride include reactions that form $MH_4^+ Cl^-$ (M = N, P, As), reaction with carbides and nitrides, reactions associated with metals, reaction with metal oxides, reaction with oxidizing agents to produce chlorine, reaction of oxyacids and salts, addition to olefinic and acetylenic compounds, replacement of organic hydroxyl with chloride, and chloromethylation reactions.

Uses of aqueous hydrochloric acid include metal cleaning (eg, steel pickling), hydrometallurgical processes (eg, production of alumina and titanium dioxide), chlorine dioxide synthesis, and hydrogen production.

Applications of hydrochloric acid to organic processing are based primarily on the catalyzing properties common to other strong acids. The availability of high purity hydrochloric acid combined with its distinctive physical properties makes it the preferred acid for many such applications. Carbohydrate reactions promoted by hydrochloric acid are often analogous to those in the digestive tracts of mammals. Hydrochloric acid is used to make high fructose corn syrup. Many reactions in organic synthesis employ hydrochloric acid. Among them are esterification of aromatic acids, transformation of acetaminochlorobenzene to chloroanilides, and inversion of menthone.

DAVID S. ROSENBERG
Hooker Chemical Co.

F.D. Rossini, D.D. Wagman, W.H. Evans, S. Levine, and I. Jaffe, "Selected Values of Chemical Thermodynamic Properties," *Nat. Bur. Stnd. U.S. Circ.* **500** (1952).
"Chlor," Sect. B in *Gmelins Handbuch der Anorganische Chemie*, 8th ed., Frankfurt, Main, FRG 1968.
JANAF Thermochemical Tables, The Dow Chemical Company, Midland, Michigan, 1965.

HYDROGEN ENERGY

A change to hydrogen as the universal fuel would make possible more extensive use of coal because the fuel produced from it would be clean burning and the hydrogen could be made from coal in coastal plants where the effluents, including carbon dioxide, could be disposed of in the ocean. Also, the use of hydrogen fuel would allow the use of atomic reactors not only for providing electricity, but for providing hydrogen fuel for transportation. Thus, the alternative of small-range and low-speed electric cars would be obviated.

Sources

The increasingly utilized sources of energy will be coal, fission power, and solar energy; the latter will be used only if hydrogen is used with it. If second-law efficiencies and environmental factors are taken into account, the three energy sources in order of increasing cost are the solar-thermal production of hydrogen, the photovoltaic production of hydrogen (using concentrators), and the use of coal to produce hydrogen with carbon dioxide rejection into the oceans. Solar energy (qv) may become an economically acceptable solution for the health hazards of atomic energy and coal if hydrogen is used as an intermediate. Water splitting also may be a source using electrolysis of water to give pure hydrogen (which may, at first, be used in NH_3 synthesis, hence for fertilizer manufacture).

Transmission and Storage

If the principal source of energy is coal, it will be necessary to transmit the energy in a gaseous form over long distances from seashore plants (because of the necessity of CO_2 injection into the sea) to city centers. With breeder reactors, the problem is environmental, as it will be necessary to transmit energy from rural plants to cities. Hydrogen transportation is less expensive than electricity if the distance is more than ca 700 km. However, with regard to transportation across water, it would not be economically acceptable to transfer hydrogen for more than 48–80 km in an underwater pipe. In one energy-transmission method, the power is transmitted to a satellite and then is beamed down to energy receptors. The necessary relay satellite is less than one-tenth the size and weight of a platform receiving solar energy from the sun. Hydrogen would be the storage medium.

Underground storage of hydrogen involves pressures of ca 10 MPa (100 atm). The rate of outflow from a natural-gas field that has been filled with hydrogen would be about half that of natural gases that have been stored at the same pressure. The compressors for hydrogen storage would have to be larger and natural-gas impurities would be present; however, the cost would be four times greater than that for the same energy quantity of natural gas. Nevertheless, such a method would be the best way to store hydrogen in large amounts. Small quantities of hydrogen could be stored in iron–titanium alloys. The initial cost is high, but several hundred charges and discharges are possible without difficulty.

Liquid hydrogen storage is expensive and liquefaction costs ca \$4/GJ (\$4/10^6 Btu) because ca 35% of the hydrogen energy is used for the liquefaction process. Liquid hydrogen has the highest energy density of any liquid. Storage for 30 days is possible.

Uses

Transportation. Hydrogen can be used as the only really clean chemical fuel for transportation. Several hundred hydrogen-fueled automobiles exist. It is possible to burn hydrogen efficiently in the internal combustion engine with an increased efficiency of 25–50% over gasoline. The difficulty in using hydrogen, apart from the fact that it is not yet produced from coal on a large scale, is storage. However, hydrogen could be stored at intermediate pressures of 10 MPa (100 atm) within the tubular structure of a vehicle.

Hydrogen from fuel cells and in combination with batteries can maximize the efficiency of hydrogen for mechanical and electrical power. All aircraft would be able to carry greater payloads if run on liquid hydrogen. Hypersonic aircraft will have to use hydrogen to obtain an acceptable range. Supersonic aircraft would have a far better range if liquid hydrogen were their fuel. Cooling the airframe with liquid hydrogen-based refrigerant would allow the use of lightweight alloys and consequently permit a payload increase (see Refrigeration).

Metallurgy. When abundant hydrogen is available from coal and, eventually, from breeder reactors and solar power, direct reduction of materials by hydrogen will take place with low pollution consequences. The reduction of alumina and iron oxide will be feasible. Hydrogen-using reductions also are applied in many chemical and industrial processes (see Hydrogen).

Food. The synthetic production of food on a large scale using hydrogen-eating bacteria has been suggested. Hydrogen is produced electrolytically in a bath that contains hydrogen-eating bacteria and through which carbon dioxide is bubbled. The bacteria reproduce and, in the presence of atmospheric CO_2 and N_2, form protein (see Proteins; Foods, nonconventional).

Pollution control. Water treatment and other antipollution systems on a massive scale would be possible if massive amounts of oxygen were available. This would be the case if hydrogen was produced from water.

Household uses. Hydrogen can be utilized for lighting by using phosphors activated in the presence of hydrogen, and for heating and cooking by using catalytic nonluminous heaters. Utilization of a hydrogen economy is favorable environmentally (see Fig. 1), and it could lower the cost of energy to the consumer, if the costs could take into account

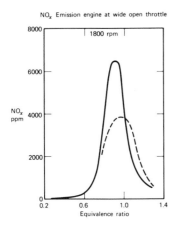

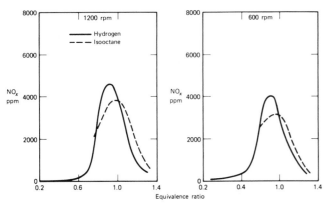

Figure 1. The superiority of hydrogen in lessening pollution.

both the greater efficiency of conversion of hydrogen to mechanical energy compared with gasoline and with the diminished environmental damage costs (eg, from acid rain) associated with the use of hydrogen.

J.O'M. BOCKRIS
Texas A&M University

J.O'M. Bockris, *Energy: The Solar–Hydrogen Alternative*, Australia and New Zealand Book Co., Sydney, Australia, 1975, p. 25.

J. O'M. Bockris, *Energy: The Solar–Hydrogen Alternative*, 2nd ed., Australia and New Zealand Book Co., Sydney, Australia, and Halsted Press, a division of John Wiley & Sons, Inc., New York, 1980, Chapt. 8.

J. O'M. Bockris and T.N. Veziroglu, *Int. J. Hydrogen Energy* **8**, 323 (1983).

HYDROGEN FLUORIDE, HF.   See Fluorine compounds, inorganic.

HYDROGEN IODIDE, HI.   See Iodine compounds.

HYDROGEN-ION ACTIVITY

The effective concentration of hydrogen ion in solution is expressed in terms of pH, which is the negative logarithm of the hydrogen activity:

$$pH = -\log_{10} a_{H^+} \tag{1}$$

The relationship between activity and concentration is

$$a = \gamma c \tag{2}$$

where the activity coefficient γ is a function of the ionic strength of the solution and approaches unity as the ionic strength decreases; ie, the difference between the activity and the concentration of hydrogen ion diminishes as the solution becomes more dilute. The pH of a solution may have little relationship to the titratable acidity of a solution that contains weak acids or buffering substances; the pH of a solution indicates only the free hydrogen-ion activity. If total acid concentration is to be determined, an acid–base titration must be performed.

Thermodynamically, the activity of a single ionic species (and, eg, the pH) is an inexact quantity, and a conventional pH scale has been adopted which is defined by reference solutions with assigned pH values. These references solutions, in conjunction with equation 3, define the pH.

$$pH(X) = pH(S) - \frac{(E_X - E_S)F}{2.303RT} \tag{3}$$

E_S is the electromotive force (emf) of the cell: reference electrode KCl ($\geq 3.5\ M$) solution S $H_2(g)$, Pt, and E_X is the emf of the same cell when the reference buffer solution S is replaced by the sample solution X. The quantities R, T, and F are the gas constant, the thermodynamic temperature, and the Faraday constant, respectively.

pH Determination

There are two methods used to measure pH. The most common utilizes the commercial pH meter with a glass electrode. This procedure is a determination of the difference between the pH of an unknown or test solution and that of a standard solution. The instrument measures the emf developed between the glass electrode and a reference electrode of constant potential. The difference in emf when the electrodes are removed from the standard solution and placed in the test solution is converted to a difference in pH. The second method is the indicator method, but has more limited applications. The success of this procedure depends upon matching the color that is produced by the addition of a suitable indicator dye to a portion of the unknown solution with the color produced by adding the same quantity of the same dye to a series of standard solutions of known pH.

pH Measurement System Electrodes

Glass electrodes. The glass electrode is the hydrogen-ion sensor in most pH-measurement systems. The pH-responsive surface of the glass electrode consists of a thin membrane formed from a special glass which, after suitable conditioning, develops a surface potential that is an accurate index of the acidity of the solution in which the electrode is immersed. To permit changes in the potential of the active surface of the glass membrane to be measured, an inner reference cell of constant potential is placed on the opposite side of the glass membrane. The inner reference cell consists of a solution that has a stable hydrogen-ion concentration and that contains counterions to which the inner electrode is reversible. The ionic concentrations are sufficient to guarantee constancy of the inner cell over extended periods of time. The choice of an inner cell has a bearing on the temperature coefficient of the emf of the pH assembly. The inner cell commonly consists of a silver–silver chloride electrode or calomel electrode in a buffered chloride solution.

Immersion electrodes are the most common glass electrodes. These are roughly cylindrical and consist of a barrel or stem of inert glass which is sealed at the lower end to a tip, which is often hemispherical, of special pH-responsive glass. The tip is completely immersed in the solution during measurements.

Reference electrodes and liquid junctions. The electric circuit of the pH cell is completed through a salt bridge which usually consists of a concentrated solution of potassium chloride. The solution makes contact at one end with the test solution and at the other with a reference electrode of constant potential. The liquid junction is formed in the area of contact between the salt bridge and the test solution. The mercury–mercurous chloride electrode—the calomel electrode—provides a highly reproducible potential in the potassium chloride bridge solution and is the most widely used reference electrode. However, because mercurous chloride disproportionates above 80°C, producing an unstable potential, the silver–silver chloride electrode and the thallium amalgam–thallous chloride electrode often are used for measurements at elevated temperatures.

RICHARD A. DURST
National Bureau of Standards

ROGER G. BATES
University of Florida

R.G. Bates, *Determination of pH*, 2nd ed., Wiley-Interscience, New York, 1973.

R.A. Durst, *Standardization of pH Measurements*, NBS Spec. Publ. 260-53, U.S. Govt. Printing Office, Washington, D.C., 1975.

G. Eisenman, ed., *Glass Electrodes for Hydrogen and Other Cations*, Marcel Dekker, Inc., New York, 1967.

HYDROGEN PEROXIDE

Hydrogen peroxide, (H_2O_2), mol wt 34.014, is a weakly acidic, clear colorless liquid, miscible with water in all proportions. The four atoms are covalently bound in a nonpolar H—O—O—H structure. It is commercially available in aqueous solution over a wide concentration range, and its scale of manufacture and use has increased markedly since 1925 when electrolytic processes were introduced to the United States and industrial bleach applications were developed. Now prepared primarily by anthraquinone autoxidation processes, hydrogen peroxide is used widely to prepare other peroxygen compounds and as a nonpolluting oxidizing agent.

Physical Properties

Mp, $-0.41°C$; bp, 150.2°C; density at 25°C, 1.4425 g/cm³; viscosity at 20°C, 1.245 mPa·s ($= cP$); specific heat at 25°C, 2.628 J/g (0.628 cal/g); dissociation constant at 20°C, 1.78×10^{-12} at zero ionic strength.

Chemical Properties

The reactions of hydrogen peroxide are

$$decomposition, 2 H_2O_2 \rightarrow 2 H_2O + O_2 \tag{1}$$
$$molecular\ additions, H_2O_2 + Y \rightarrow Y.H_2O_2 \tag{2}$$
$$substitutions, H_2O_2 + RX \rightarrow ROOH + HX \tag{3}$$
$$H_2O_2 + 2 RX \rightarrow ROOR + 2 HX \tag{4}$$
$$oxidations, H_2O_2 + W \rightarrow WO + H_2O \tag{5}$$
$$and\ reductions\ H_2O_2 + Z \rightarrow ZH_2 + O_2 \tag{6}$$

Hydrogen peroxide may react directly or after it has first ionized or dissociated into free radicals. In many cases, the reaction mechanism is extremely complex and may involve catalysis or be dependent upon the reaction environment.

Manufacture

Hydrogen peroxide can be formed from compounds that contain the peroxy group, from hydrogen–oxygen mixtures, or water and oxygen, by thermal, photochemical, electrochemical, or similar processes, and by the uncatalyzed reaction of molecular oxygen with an appropriate hydrogen-containing species. It has been manufactured commercially by processes based on the reaction of barium or sodium peroxide with an acid, the electrolysis of sulfuric acid solutions, and the autoxidation of anthraquinones, isopropyl alcohol, and hydrazobenzene. All manufacturing facilities constructed in the United States since 1957 have been based on the autoxidation of an anthraquinone.

Autoxidation Methods

Anthraquinone autoxidation. A 2-alkylanthraquinone dissolved in a suitable solvent or solvent mixture is reduced catalytically to the corresponding anthraquinol (or anthrahydroquinone) (equation 7). The anthraquinone is commonly called the reaction carrier or working material, whereas the anthraquinone–solvent mixture is called the working solution. The working solution containing the anthrahydroquinone is separated from the hydrogenation catalyst and aerated with an oxygen-containing gas, usually air, to reform the anthraquinone and simultaneously produce hydrogen peroxide (eq. 8). The hydrogen peroxide is extracted with water and the aqueous solution purified and concentrated to the degree required. The extracted solution is recycled.

$$\tag{7}$$

$$\tag{8}$$

Health and Safety

In high concentrations, hydrogen peroxide is a high energy material and a strong oxidant which can be handled safely provided proper precautions are taken. It is irritating to the skin, eyes, and mucous membranes. Decomposition hazards increase with increasing concentration. Contact with some inorganic reagents should be avoided, eg, mercurous oxide and hydrogen peroxide react with explosive violence.

Uses

Used mainly in chemicals (30%), textiles (17%), pulp and paper (17%), pollution control and metallurgy (26%), and miscellaneous (10%). The largest single use is in the bleaching of cotton textiles (see Bleaching agents).

J.R. KIRCHNER
E.I. du Pont de Nemours & Co., Inc.

W.C. Schumb, C.N. Satterfield, and R.L. Wentworth, *Hydrogen Peroxide*, Reinhold Publishing Cop., New York, 1955.

P.A. Giguere "Peroxyde d'Hydrogne et Polyoxydes d'Hydrogen," in P. Pascal, ed., *Complements au Nouveau Traite de Chemie*, Vol. 4, Masson et Cie, Paris, France, 1975.

H. Pistor "Die Peroxoverbindungen," in K. Winnacker and L. Kuchler, eds., *Chemische Technologie*, Band 1 Anorganische Technologie 1, Hanser, Munich, FRG., 1969.

HYDROGEN SULFIDE, H₂S.   See Sulfur compounds.

HYDROMETALLURGY.   See Extractive metallurgy.

HYDROPROCESSES.   See Petroleum refinery processes.

HYDROQUINONE, RESORCINOL, AND CATECHOL

Structures and Physical Properties

Hydroquinone. Hydroquinone (**1**), (1,4-benzenediol, 1-4-dihydroxy-benzene, *p*-dihydroxybenzene), is a white crystalline compound that first was obtained in 1820 by dry distillation of quinic acid. Mp, 172°C; bp, 287°C at 101.3 kPa (1 atm); d_4^{15}, 1.332 g/cm³; solubility (at 60°C g/100 g solution) is 26.0 in water, 45.7 in ethyl alcohol, 12.1 in ethyl ether, 45.9 in acetone, and 0.8 in benzene. Hydroquinone has three crystalline modifications. By sublimation, a labile form is obtained as monoclinic prisms, whereas the labile forms separates as trigonal prisms on crystallization from methanol. The stable form is obtained as white trigonal needles or prisms by crystallization from water. Its crystals are triboluminescent. Its solutions are oxidized in air.

Resorcinol. Resorcinol (**2**), (1,3-benzenediol, 1,3,-dihydroxybenzene, *m*-dihydroxybenzene), is a crystalline compound with a faint aromatic odor, and a sweet and bitter taste. Mp, 110°C; bp, 277°C at 101 kPa; d_4^{15}, 1.272 g/cm³; n_D^{20}, 1.620; solubility (at 60°C g/100 g solution) is 83.3 in water, 73.0 in ethyl alcohol. 75.1 in acetone, 14.1 in benzene. Resorcinol crystallizes in the orthorhombic hemimorphic system, and its crystals are colorless, triboluminescent, and piezoelectric. X-ray diffraction measurements indicate that resorcinol crystallizes in the α form which is converted to the β form at 74°C; the latter is more dense (d = 1.33 g/cm³) as a result of hydrogen bonding between oriented molecules in the crystal lattice. It acquires a pink tint when exposed to light and air.

Catechol. Catechol (**3**), (1,2-benzenediol, 1,2-dihydroxybenzene, *o*-dihydroxybenzene), is a crystalline compound with a phenolic odor and a sweet and bitter taste. Mp, 105°C; bp at 101.3 kPa, 245°C; d_4^{15}, 1.344 g/cm³; n_D^{20}, 1.615; solubility (at 60°C g/100 g solution) is 804 in water, 74.0 in ethyl alcohol, 65.3 in ethyl ether, 75.2 in acetone, and 7.3 in benzene. Catechol crystallizes in the monoclinic system, and the crystals are colorless. It sublimes and is volatile with steam. Air and light cause discoloration.

(1)   (2)   (3)

Chemical Properties

Dihydroxybenzenes (DHBs) are weak acids with two dissociation constants. They form mono- and di-salts in solutions of alkali hydroxides or carbonates. With its two adjacent hydroxyl groups, catchol is able to complex most metallic salts. Many of the complexes of catechol and catechol derivatives are useful analytical reagents (see Analytical methods; Chelating agents). Dihydroxybenzenes are more easily oxidized than

phenol. However, only catechol and hydroquinone are converted by most oxidizing agents to the corresponding *o*- and *p*-benzoquinones.

Resorcinol behaves similarly to monophenols; following one-electron oxidation with potassium ferricyanide or ceric sulfate, it affords an unstable meta-semibenzoquinone which produces a mixture of C—C- and C—O-coupled dimeric and polymeric compounds. Resorcinol and its derivatives are only slowly attacked by periodate, but reaction with hydrogen peroxide in presence of tungstic oxide yields maleic acid.

DHBs undergo all of the typical reactions of phenols.

DHBs react with phthalic anhydride to form multi-ring systems according to the reaction conditions. Hydroquinone yields quinazarin (1,4-dihydroxy-9,10-anthracenedione) (**4**), resorcinol yields fluorescein (**5**), and catechol gives hystazarin (2,3-dihydroxy-9,10-anthracenedione) (**6**) and alizarin (1,2-dihydroxy-9,10-anthracenedione) (**7**).

(4)   (5)

(6)   (7)

Both mono- and diethers of DHBs can be prepared readily by the usual methods (see Ethers). However, it is difficult to confine etherification to the monoether stage.

Catechol with its two adjacent hydroxyl groups undergoes a cyclization reaction with methylene chloride to give methylenedioxybenzene (1,3-benzodioxole) (**8**) and with di(2-chloroethyl) ether yields *s*-dibenzo-18-crown-6-polyether (**9**).

(8)   (9)

Acylation of DHBs with carboxylic halides or esters gives corresponding mono- and diesters which can undergo Fries rearrangement (see Esters, organic).

Hydroquinone, resorcinol, and catechol react with ammonia to give the corresponding *p*-, *m*-, and *o*-aminophenols; alkyl- and arylamines react similarly. They react with formaldehyde under both acidic and basic conditions to give ethylol derivatives which undergo condensation to yield high molecular weight condensation products; resorcinol has the highest reactivity of the three and has its most important use in this reaction. DHBs readily couple with aryldiazonium salts to give azo and biazo compounds which are used in dyes.

DHBs undergo the Reimer-Tiemann reaction. Catechol reacts with chloroform under alkaline conditions to give a mixture of proto- and catechuic aldehydes. DHBs undergo Kolbe-Schmitt carboxylation by reaction of carbon dioxide with their alkali-metal salts. Hydroquinone gives gentisic acid (2,5-dihydroxybenzoic acid) and 2,5-dihydroxy-terephthalic acid. Resorcinol gives β-resorcylic acid (2,4-dihydroxybenzoic acid) and γ-resorcylic acid (2,6-dihydroxybenzoic acid).

Manufacture

Manufacture is based on the processes listed in Table 1.

Hydroxylation of phenol. Although phenol is less easily oxidized than dihydroxybenzenes, it can be hydroxylated directly to catechol and

Table 1. Manufacturing Processes for Dihydroxybenzenes

Dihydroxybenzene	Process	Location
hydroquinone	aniline oxidation	U.S.
		FRG
		Japan
		U.K.
		COMECON[a]
		People's Republic of China
	phenol hydroxylation	France
		Italy
		Japan
	p-diisopropylbenzene oxidation	U.S.
		Japan
resorcinol	benzene disulfonation	U.S.
		Italy
		FRG
		U.K.
		Puerto Rico
		Japan
	m-diisopropylbenzene hydroperoxidation	Japan
catechol	o-chlorophenol hydrolysis	none
	phenol hydroxylation	France
		Italy
		Japan
	coal-tar distillation	U.K.
		COMECON[a]

[a]COMECON = Council for Mutual Economic Assistance (Communist-bloc nations).

hydroquinone. The hydroxylation agent is hydrogen peroxide and the reaction occurs in the presence of catalytic amounts of strong mineral acids or ferrous or cobaltous salts.

Alkaline fusion of m-benzenedisulfonic acid. The manufacture of resorcinol was first described in 1878 as disulfonation of benzene followed by alkaline fusion of the disodium salt of m-benzenedisulfonic acid; this process is still used.

Hydroperoxidation of diisopropylbenzene. Analogous to the cumene process for phenol production, the preparation of resorcinol from m-diisopropylbenzene (m-DIPB) has been studied and reviewed (see Phenol; Cumene). From p-diisopropylbenzene (p-DIPB) and by the same process, hydroquinone is produced according to the dihydroperoxide reaction (eq. 1). Hydroquinone and resorcinol can be produced alternatively in the same plant. o-Diisopropylbenzene (o-DIPB) is an inert compound in radical oxidation because its two adjacent isopropyl groups are sterically hindered.

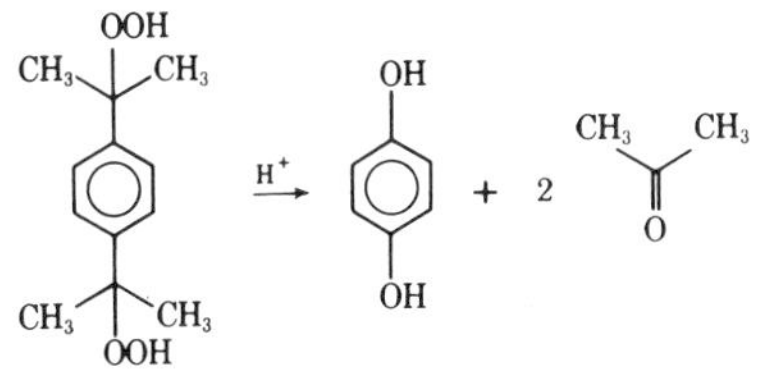

dihydroperoxide (DHP)

Aniline oxidation. Aniline oxidation to benzoquinone was the first process used in the United States by DuPont after World War I and originated at Usines du Rhone in France. Sodium bichromate was used as the oxidizing agent and the quinone was reduced to hydroquinone with sulfur dioxide. Oxidation of aniline to quinone is still one of the most widely used processes for hydroquinone production; a more current oxidizing agent being manganese dioxide.

Health and Safety

Except for resorcinol, dihydroxybenzenes are more toxic than phenol. Experimental studies on humans and animals have demonstrated low chronic toxicity and rapid excretion of DHBs. Systemic effects from industrial exposure have not been observed. Threshold limit value (mg/m^3) for hydroquinone 370; resorcinol 45; and catechol 3890.

Uses

Dihydroxybenzenes are high priced chemical specialties. In decreasing order of importance, the main outlets are photographic developers, tire adhesives, rubber antioxidants and antizonants and monomer inhibitors, wood adhesives, ultraviolet absorbers and optical brighteners, dyestuffs, and miscellaneous derivatives (see Photography; Antioxidants and antizonants; UV absorbers; Brighteners, fluorescent).

J. VARAGNAT
Rhone-Poulenc

C.J. Pedersen, *Org. Synth.* **52**, 66 (1972).

Tecquinol, Hydroquinone Technical Grade, D-104 B, Eastman Chemical Products, Inc., Kingsport, Tenn.

Resorcinol, Technical Bulletin CD-2-424, Koppers Co., Inc., Pittsburgh, Pa.

Catechol, Technical Bulletin CD-0-527, Koppers Co., Inc., Pittsburgh, Pa.

HYDROXYBENZALDEHYDES

Hydroxybenzaldehydes are organic compounds of the general formula:

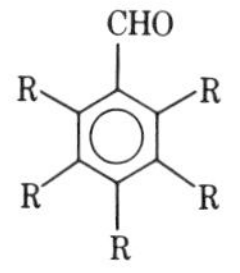

where the R groups are H or OH; and at least one R group is OH. All of the nonidentical isomeric mono-, di- and trihydroxybenzaldehydes are known. Of the higher polyhydroxybenzaldehydes, only 2,3,4,5-tetrahydroxybenzaldehyde has been reported.

There are two commercially important hydroxybenzaldehydes, the o- and p-hydroxy isomers, the ortho isomer, commonly known as salicyaldehyde (salicylic aldehyde, o-hydroxybenzaldehyde, salicylal); being the more important of the two. Salicyaldehyde is a clear, straw-colored liquid with a strong almondlike odor, and occurs naturally in oils of spirea plants, bird cherries, and cassia oil (see Benzaldehyde). p-Hydroxybenzaldehyde (4-formylphenol) is a colorless-tan solid with a slight, agreeable, aromatic odor. It occurs naturally in some plants in small amounts. This article deals primarily with salicylaldehyde and p-hydroxybenzaldehyde, which represent more than 99% of the hydroxybenzaldehydes market.

Physical Properties

With the exception of salicylaldehyde, all of the hydroxybenzaldehydes are solids at room temperature. The location of the hydroxyl and aldehyde groups ortho to one another in salicylaldehyde results in intramolecular hydrogen bonding, and is the reason for the lower melting point. Solubility in water and other polar solvents generally increases as the number of hydroxyl groups increases.

Chemical Properties

The effect of the aldehyde group on the phenolic hydroxyl is primarily an increase in its acidity; all three monohydroxybenzaldehydes are stronger acids than phenol. The aldehyde group, however, has little affect on the reactions of the hydroxyl group. The hydroxybenzaldehydes undergo most of the normal aldehyde reaction such as bisulfite addition and oxime and hydrazone formation, and oxidation and reduction. Of interest also is the reaction of p-hydroxybenzaldehyde with sodium cyanide and ammonoum chloride, a Strecker synthesis, which yields p-hydroxyphenylglycine, a key intermediate in the manufacture of semisynthetic penicillins and cephalosporins.

Manufacture

The two main processes for the manufacture of hydroxybenzaldehydes are both based on phenol. The most widely used process is the saligenin process. Saligenin (o-hydroxybenzyl alcohol) and p-hydroxybenzyl alcohol are produced from based-catalyzed reaction of formaldehyde with phenol. Air oxidation of saligenin over a suitable catalyst such as platinum or palladium produces salicylaladehyde. In a refinement of the process, the reaction of phenyl metaborate with formaldehyde followed by catalytic oxidation under atmospheric pressure has been reported to give salicyaldehyde directly from phenol without isolation of any intermediate products.

Health and Safety

Salicyaldehyde has a moderate acute oral toxicity; the LD_{50} for rats is 0.3–2.0 g/kg of body weight. o-Hydroxybenzaldehyde has a low acute oral toxicity; the LD_{50} for rates is 4.0 g/kg of body weight.

Uses

The hydroxybenzaldehydes are primarily used as chemical intermediates to make a variety of products. Major uses are as ingredients in agricultural chemicals, electroplating, perfumes, petroleum chemicals, polymers and fibers. The largest single use of salicylaldehyde is in the manufacture of coumarin (qv).

RICHARD M. MULLINS
The Dow Chemical Company

H.D. Dakin, *Organic Synthesis Col. Vol. I*, Wiley-Interscience, New York, 1941, pp. 149–153.

H. Wynberg, *Chem. Rev.* **60**, 169 (1960).

P. Polss, *Hydrocarbon Process.* 61 (Feb. 1973).

HYDROXY CARBOXYLIC ACIDS

Lactic Acid

Lactic acid (2-hydroxypropanoic acid, 2-hydroxypropionic acid), $CH_3CHOHCOOH$, is a naturally occurring organic acid. It is present in many foodstuffs, eg, as a primary acid component in sour milk and a constituent in animal blood and muscle tissues. It is the simplest hydroxy acid and is optically active. The active forms are water-soluble, colorless liquids at ordinary temperatures or are low melting solids in the pure state.

Physical properties. Lactic acid in aqueous solutions readily forms intermolecular esters which complicate the preparation of crystalline lactic acid. Vacuum distillation can remove the excess water at low temperatures and the lactic acid then is distilled from the intermolecular esters. Lactic acid can be crystallized only when it is substantially free of lactoyllactic acid. Both optical isomers of lactic acid occur in nature, but the commercial acid is the optically inactive form. Lactic acid has an asymmetric carbon. The following formulas were proposed for the stereo isomers in 1891:

L(+) lactic acid   D(−) lactic acid

Chemical properties. Because lactic acid has both hydroxyl and carboxyl functional groups, it undergoes self-esterification (when it is concentrated by evaporation) and forms linear polyesters (qv). The first esterification product is the dimer lactoyllactic acid.

lactoyllactic acid

However, as the concentration increases, higher linear esters, ie, poly-lactic acids (anhydrides) or trimeric, tetrameric and polymeric lactic acids are formed. The following formula shows the general structure of the polyacids:

$$CH_3CHCOOH$$
$$(CH_3CHCOO)_n$$
$$CH_3CHOHCOO$$

Under mild conditions and particularly in the presence of a mineral acid catalyst, the lactic esters undergo self-alcoholysis to produce esters of polylactic acid:

$$n\ CH_3CHOHCOOR \overset{H^+}{\rightleftharpoons} HO[CH(CH_3)COO]_nR + (n-1)ROH$$

The diesters of lactic acid, however, show a completely different behavior when subjected to pyrolysis; acrylic esters and acetic acid are formed from acetylated lactic esters:

$$CH_3CH(OOCCH_3)COOR \overset{550°C}{\longrightarrow} CH_2{=}CHCOOR + CH_3COOH$$

Lactic acid shows the typical reactions of organic acids; many salts have been reported and generally they are water-soluble. Lactic acid readily undergoes esterification with many alcohols, and numerous lactic esters have been prepared by acid-catalyzed esterification.

Lactamide, $CH_3CHOHCONH_2$, and substituted lactamides can be prepared readily by ammonolysis or aminolysis of methyl lactate or by the dehydration of amine or ammonia salts of lactic acid.

Lactic acid also shows many of the reactions characteristic of alcohols, eg, it can be esterified with organic acids, anhydrides, and acid chlorides; it can be alkylated with alkylating agents such as diazomethane and dimethyl sulfate. Lactic acid and its esters can be converted into chloroformates with phosgene (qv), into carbamates with cyanic acid, and into allophanates and urethanes (see Urethane polymers) with isocyanates (see Isocyanates, organic).

Manufacture. Lactic acid can be manufactured either by fermentation or by synthesis, both methods being used commercially. In the United States, more than 85% of the lactic acid that is sold is made by a synthetic route, and all of the lactic acid that is produced in Japan is synthetic. Lactic acid from European manufacturers, whose combined capacity is approximately half of the estimated world capacity, is produced by fermentation processes. The world capacity is estimated at 28,000–30,000 metric tons per year.

Fermentation. Two types of microorganisms produce lactic acid; heterolactic fermentation organisms produce some lactic acid as well as other fermentation products, ie, carbon dioxde, ethyl alcohol, and acetic acetic, and are of little use for industrial production. Commercial production is by homolactic fermentation organisms, bacteria such as *Lactobacillus delbrueckii*, *L. bulgarcius*, and *L. leichmanii*, which form lactic acid exclusively or predominantly from carbohydrates (qv).

Synthesis. Synthesis of lactic acid is based on lactonitrile, which is a by-product from the acrylonitrile (qv) synthesis. The lactic acid synthesis reaction was discovered in 1863 by Wislicenus who prepared lactonitrile from acetaldehyde and hydrogen cyanide and hydrolyzed it to lactic acid.

$$CH_3CHO + HCN \rightarrow CH_3CHOHCN$$
$$CH_3CHOHCN + 2\ H_2O + HCl \rightarrow CH_3CHOHCOOH + NH_4Cl$$

The same reactions are used today. The lactic acid is isolated and purified by esterification with methyl alcohol, and the resulting methyl lactate is purified by distillation. Once the methyl ester has been produced and distilled, it can be hydrolyzed with a strong acid catalyst to produce a semirefined lactic acid. Purification is achieved by a combination of steaming, carbon treatment, and ion exchange. The synthetic acid is produced in three grades: technical, food, and USP, and in two concentrations (50 and 88%) (see Fine chemicals). Synthetic lactic acid is water-white and has excellent heat stability.

Uses. Primary uses are in foodstuffs and pharmaceutical products; a relatively small volume is used in industrial applications (see Food additives; Pharmaceuticals). The main use for food-grade synthetic lactic acid is in the manufacture of calcium and sodium stearoyl-2-lactylates for the baking industry. As a food acidulant, lactic acid has a mild acidic taste in contrast with the sharp taste of some of the other food acids. Lactic acid occurs naturally in many food ingredients. Its salts are soluble and can partly replace the acid. Lactic acid and its sodium and calcium salts are completely nontoxic and are classified as GRAS (generally recognized as safe) for general-purpose food additives by the FDA.

Hydroxyacetic Acid

Glycolic acid, $HOCH_2COOH$, is the first and simplest member of the series of hydroxy carboxylic acids. It is a colorless, translucent, and odorless solid. Hydroxyacetic acid is used in textile and leather (qv) processing, detergents, metal cleaning and plating, and dairy sanitations (see Metal surface treatment, cleaning, pickling, and related processes).

Properties. Mp, 79–80°C; bp, 100°C (dec); d_4^{25} 1.49 g/cm³; K_a at 25°C, 1.5×10^{-4}; pH of 1 M solution, 2.4; heat of combustion, 697.1 kJ/mol (166.6 kcal/mol); heat of solution (infinite dilution), -154 J/g (-36.8 cal/g). Glycolic acid is slightly volatile in steam but cannot be distilled because it readily loses water by self-esterification to form poly(hydroxyacetic acid)[poly(glycolic acid)]. Glycolic acid is very soluble in water, methanol, ethanol, acetone, and ethyl acetate; slightly soluble in ethyl ether; and only sparingly soluble in hydrocarbon solvents.

Manufacture. Hydroxyacetic acid occurs in nature in sugar beets, unripe grapes, and spent sulfite liquor from pulp processing. However, it has never been produced commercially from these sources. Commercial production in the United States is as an intermediate in the manufacture of ethylene glycol (see Glycols). Under high pressure, 30.4–91.2 MPa (300–900 atm) and from 160–200°C, formaldehyde reacts with carbon monoxide and water in the presence of an acidic catalyst, ie, sulfuric, hydrochloric, or phosphoric acids, to form hydroxyacetic acid. In the presence of HF catalyst, this reaction rapidly produces hydroxyacetic acid even at 20–60°C; moderate pressures are required for the reaction.

Uses. Hydroxyacetic acid is produced in large volume. Nevertheless, it has found uses in a number of areas, eg, adhesives, metal cleaning, electroplating (qv), dairy cleaning, biodegradable polymers, dyeing, water-well cleaning, masonry, textiles, and detergents.

Other Hydroxy Acids

Preparation. α-Hydroxy acids may be prepared by the hydrolysis of an α-halo acid or by the acid hydrolysis of the cyanohydrins (qv) of an aldehyde or ketone:

$$RCHO \xrightarrow{HCN} \underset{\underset{OH}{|}}{R}CHCN \xrightarrow[HCl]{H_2O} \underset{\underset{OH}{|}}{R}CHCOOH + NH_4Cl$$

Aliphatic α-hydroxy acids that do not have side chains can be prepared in good yields by the hydrolysis of α-nitrato acids with aqueous sulfite solutions. The α-nitrato acids are obtained by the reaction of olefins (qv) and N_2O_4 in the presence of oxygen. The α-hydroxy acids that are obtained can be esterified or acylated directly to yield anhydro ester acids which, in turn, give the α-hydroxy acid on saponification.

β-Hydroxy acids may be made by catalytic reduction of β-keto esters followed by hydrolysis, or obtained by the Reformatsky reaction.

Uses. Almost all of the normal hydroxy aliphatic acids occur in nature. The literature is filled with biochemical studies of these acids, particularly 3-hydroxy- and 4-hydroxybutyric acids. γ-Butyrolactone and its hydrolysis product, γ-hydroxybutyric acid, exert a variety of pharmacological effects when administered systematically to humans and animals (see Psychopharmacological agents). Bacteria appear to use γ-hydroxybutyric acid as a carbon source after converting it to succinic acid. In vitro, γ-hydroxybutyric acid is utilized as a mitchondrial substrate and for investigating the properties of the enzymes of body metabolism. γ-Hydroxybutyric acid (GHB) and its derivatives, particularly its sodium salt, have been studied as anesthetics (qv), tranquillizers,

sedatives, and hypnotics (see Hypnotics, sedatives, anticonvulsants). GHB is used as an anesthetic for surgery and in general obstetrics.

J.H. VAN NESS
Monsanto Industrial Chemicals Company

Symp. N.Y. Acad. Sci. Ann. N.Y. Acad. Sci. **119**, 851 (1965).

HYDROXY DICARBOXYLIC ACIDS

Many natural and synthetic organic compounds are hydroxy dicarboxylic acids (see also Hydroxy carboxylic acids). This article deals with malic, thiomalic, tartaric, tartronic, and phloionic acids; thiomalic acid is included because of its structural similarity to malic acid.

Malic Acid

Malic acid (hydroxysuccinic acid, hydroxybutanedioic acid, 1-hydroxy-1,2-ethanedicarboxylic acid), $C_4H_6O_5$, is a white crystalline material. The levorotatory isomer, $S(-)$-malic acid (L-malic acid) is a natural constituent and common metabolite of plants and animals. It is the principal acid found in apples. The racemic compound, R,S-malic acid, (DL-malic acid) is a widely used organic food acidulant. This material also is used in some industrial applications to sequester ions, neutralize bases, and is used as a buffer for pH control. $R(+)$-Malic acid (D-malic acid) is available as a laboratory chemical (see Chelating agents; Food additives).

Physical properties. Malic acid crystallizes from aqueous solutions as white, translucent, anhydrous crystals. The $S(-)$ isomer melts at 100–103°C and the $R(+)$ isomer at 98–99°C. On heating, malic acid decomposes at ca 180°C. Under normal conditions, malic acid is stable; under conditions of high humidity, it is hygroscopic. Malic acid is a relatively strong organic acid; $K_1 = 4 \times 10^{-4}$. Mp of R,S-Malic acid is ca 130°C and d_4^{20} 1.601 g/cm³. It is soluble to the extent of 58 g/100 g solution in water (25°C) and 39.15 g/100 g solution in ethanol. The density of 58 g/L aqueous malic acid solution (15°C) is 1.212 g/cm³.

Chemical properties. Configuration: malic acid is optically active because of its chiral center. The levorotatory enantiomer was confirmed as having the spatial configuration (1) when tartaric acid was first reduced to malic acid. The other enantiomer (2) is assigned the R configuration. The optically inactive compound has the R,S symbol.

(1) $S(-)$-malic acid   (2) $R(+)$-malic acid

Reactions. Malic acid undergoes many of the characteristic reactions of dibasic acids, monohydric alcohols, and α-hydroxy carboxylic acids. When heated to 170–180°C, it decomposes to fumaric acid and maleic anhydride which sublimes on further heating (see Maleic anhydride, maleic acid, and fumaric acid). Malic acid forms two types of condensation products: linear malomalic acids and the cyclic dilactone or malide; it does not form an anhydride.

As a dibasic acid, malic acid forms the usual salts, esters, amides and acyl chlorides (see Esters, organic). Thus, malic acid yields the usual diesters with an alcohol in the presence of an esterification catalyst (see Esterification). With polyhydric alcohols and polycarboxylic aromatic acids, malic acid yields alkyd polyester resins (see Alcohols, polyhydric; Alkyd resins).

Manufacture. In the United States, Canada, and Europe, only the synthetic R,S-malic acid is produced commercially, whereas both the S and the R,S forms are produced in Japan.

Biosynthesis of $S(-)$-malic acid. Aqueous fumaric acid is converted to levorotatory malic acid by the intracellular enzyme, fumarase, which is produced by various microorganisms. A Japanese process for continu-

Table 1.   Properties of Thiomalic Acids

Acid	Mp, °C	Solubility		$[\alpha]_D^{17}$ (5% in ethanol)	Dissociation at 25°C	
		Water	Ethanol		pK_{a1}	pK_{a2}
R,S-	151	very sol	very sol		3.30	4.94
R	154	sol	sol	$+64.4°$		
S	152–153	sol	slightly sol	$-64.8°$		

ous commercial production of $S(-)$-malic acid from fumaric acid is based on the use of immobilized *Brevibacterium ammoniagenes* cells. The yield of pyrogen-free $S(-)$-malic acid that is suitable for pharmaceutical use is ca 70% of the theoretical.

Commercial synthesis of R,S-malic acid. Synthesis involves hydration of maleic or fumaric acid at elevated temperature and pressure. A Japanese patent, describing a manufacturing procedure for malic acid, claims the direct hydration of maleic acid at 180°C and 1.03–1.21 MPa (150–175 psi). These workers suggest that in the hydration of maleic acid, fumaric acid is formed as a by-product and is hydrated slowly under the conditions of the reaction. If the amount of fumaric acid equivalent to that formed at equilibrium is charged with the maleic acid, it is possible to hydrate maleic acid to malic acid in a short time without the formation of additional fumaric acid.

The conventional commercial processes are commonly carried out in aqueous solution at elevated temperatures above 150°C. The resulting mixture contains, primarily, malic acid and fumaric acid in equilibrium with a small percentage of maleic acid.

Thiomalic acid

Thiomalic acid (mercaptosuccinic acid), $C_4H_6O_4S$, is a sulfur analogue of malic acid. The properties of the crystalline, solid thiomalic acids are given in Table 1.

R,S-Thiomalic acid can be prepared from bromosuccinic acid by reaction with K_2S. The enantiomers can be obtained from the corresponding optically active potassium bromosuccinates.

Tartaric Acid

Tartaric acid (2,3-dihydroxybutanedioic acid, 2,3,-dihydroxysuccinic acid), $C_4H_6O_6$, is a dihydroxy dicarboxylic acid with two chiral centers. It exists as the dextro- and levorotatory acid, the meso-form (which is inactive owing to internal compensation) and the racemic mixture (which commonly is known as racemic acid). The commercial product in the U.S. is the natural, dextrorotatory form, $(R-R^*,R^*)$-tartaric acid, $(L(+)$-tartaric acid). This enantiomer occurs in grapes as its acid potassium salt (cream of tartar).

Physical properties. When crystallized from aqueous solutions above 5°C, natural $(R-R^*,R^*)$-tartaric acid is obtained in the anhydrous form. Below 5°C, tartaric acid forms a monohydrate which is unstable at room temperature. Mp 169–170°C; d_4^{20} 1.76 g/cm³; at 25°C $pK_1 = 1.04 \times 10^{-3}$; $[\alpha]_D^{20}$ (for concentration, c, from 20–50%) $15.050-0.1535c$. $(R-R^*,R^*)$-tartaric acid is soluble in H_2O to the extent of 115 g/100 g H_2O at 0°C. One hundred grams of absolute ethanol dissolves 20.4 g of tartaric acid at 18°C, and 100 g of ethyl ether dissolves 0.3 g at 18°C. Densities of $(R-R^*,R^*)$-tartaric acid solutions are listed in Table 2.

Manufacture. The raw materials for the manufacture of $(R-R^*,R^*)$-tartaric acid and its salts are by-products of wine making. Crude tartars

Table 2.   Density of $(R-R^*,R^*)$-Tartaric Acid Solutions at 15°C

wt% (at 15°C)	d_4^{15} g/cm³
1	1.0045
10	1.0469
20	1.0969
30	1.1505
40	1.2078
50	1.2696

are recoverable from the following sources: (1) the press cakes from grape juice are boiled with water, and alcohol, if present, is distilled, then the mash is settled, decanted, and the clear liquor is cooled to crystallize (the recovered high-test crude cream of tartar has an 85–90% cream of tartar content); (2) lees, which are the dried slimy sediments in the wine fermentation vats (total tartaric acid equivalent ranges from 16–40%); and (3) the crystalline crusts that form in the vats in the secondary fermentation period (argols). It is usually advantageous to combine the manufacture of tartaric acid, cream of tartar, and Rochelle salt in one plant. The chemical reactions involved are: formation of calcium tartrate from crude potassium acid;

$$2\ KHC_4H_4O_6 + Ca(OH)_2 + CaSO_4 \rightarrow 2\ CaC_4H_4O_6 + K_2SO_4 + 2\ H_2O$$

formation of tartaric acid from calcium tartrate;

$$CaC_4H_4O_6 + H_2SO_4 \rightarrow H_2C_4H_4O_6 + CaSO_4$$

formation of Rochelle salt from argols;

$$2\ KHC_4H_4O_6 + Na_2CO_3 \rightarrow 2\ KNaC_4H_4O_6 + CO_2 + H_2O$$

and formation of cream of tartar from tartaric acid and Rochelle salt liquors;

$$2\ H_6C_6O_6 + 2\ KNaC_4H_4O_6 + K_2SO_4 \rightarrow 4\ KHC_4H_4O_6 + Na_2SO_4$$

Health and Safety

Malic acid. The FDA has affirmed R,S- and $S(-)$-malic acids as substances that are generally recognized as safe (GRAS) as flavor enhancers, flavoring agents, and adjuvants, and pH control agents in amounts varying from 6.9% for hard candy to 0.7% for miscellaneous food uses. R,S- and $S(-)$-malic acid may not be used in baby foods.

Tartaric acid. In the past, the FDA considered $(R-R^*,R^*)$-tartaric acid as a GRAS food substance.

Uses

Malic acid. R,S-Malic acid is used in food and nonfood applications because of its pleasant tartness, flavor-retention characteristics, high water solubility, chelating and buffering properties, and frequently, lower effective cost; it is also a reactive intermediate in chemical synthesis. Thiomalic acid is an antidote in heavy-metal poisoning; a component of cold permanent hair-waving solutions (see Hair preparations) and rust-removing and corrosion-inhibiting compositions (see Corrosion and corrosion inhibitors); the well known insecticide malathion is the thiomalate S-ester of O,O-dimethylphosphonodithioic acid (see Insect control technology).

Tartaric acid. $(R-R^*,R^*)$-Tartaric acid is used as an acidulant in carbonated and still beverages, including beverage powders, as well as in a number of other acidulated food products; Rochelle salt is well known as an important bath component in electroplating of many metals and alloys; cream of tartar is used in baking powder and prepared baking mixes (see Bakery processes and leavening agents).

S. EDMUND BERGER
Allied Chemical Corporation

J.F. Stoddart ed., *Comprehensive Organic Chemistry*, Vol. 1, Pergamon Press, Oxford, 1979, pp. 3–33.

Jpn. Pat. 4360 (Dec. 16, 1950), K. Saito, Y. Ono, and Y. Mikawa.

HYGROMETRY. See Air conditioning; Drying.

HYPNOTICS, SEDATIVES, ANTICONVULSANTS

Hypnotics are central nervous system (CNS) depressants that induce sleep when given in appropriate doses (see Neuroregulators). At lower doses, these substances frequently exhibit a calming or sedative action and at higher doses they may produce anesthesia, coma, or possibly death. Anticonvulsants are agents that are used primarily to prevent epileptic seizures and to control convulsions produced by seizures. Because their main use is in epilepsy, they frequently are referred to as antiepileptics.

Hypnotics and Sedatives

Insomnia is the inability to sleep properly. The cause for the sleep disturbance frequently is related to chronic anxiety, psychiatric dis-

orders, stress, noises, and overwork (see Noise pollution). The ideal hypnotic agent should alleviate these problems without affecting the ability of the patient to function normally after sleep.

Ethyl alcohol in the form of wine or beer has been and still is used since the earliest days of history to induce sleep. Progress in the development of hypnotics and sedatives proceeded slowly until 1960 when the first 1,4-benzodiazepine was introduced (see Psychopharmacological agents). This class of compounds has led to the development of a number of safe and effective hypnotics, and sedatives, eg, diazepam and flurazepam, that dominate the field. The properties of the hypnotics and

Table 1. Properties of Hypnotics and Sedatives

Class	Example structures	Compound	Appearance	Mp, °C	Trade name, supplier	Common form dosage, mg
alcohols					Placidyl, Abbott	capsule, 500
		ethchlorvynol	liquid		many proprietary and generic names	capsule, 500
		chloral hydrate	white solid	55		
cyclic amides	(1)	methyprylon	white solid	74–77	Noludar, Roche	capsule, 300
		glutethimide	white crystalline	86–89	Doriden, USV	tablet, 500
	(2)	methaqualone	white solid	114–117	Quaalude, Lemmon	tablet, 150
acylureas and urethanes		meprobamate	white powder	103–107	Miltown, Wallace; many proprietary and generic names	tablet, 400
benzodiazepines	(3)	diazepam	off-white crystals	125–126	Valium, Roche	tablet, 5
		flurazepam hydrochloride	pale yellow crystals	215.5–217.5	Dalmane, Roche	capsule, 30
phenothiazines	(4)	mesoridazine besylate	white powder	178 (dec)	Serentil, Boehringer–Ingelheim	tablet, 25
		1-methotrimetrazine maleate	(light-sensitive) solid	190	Levoprome, Lederle	ampul, 20 mg/cm³
		promethazine hydrochloride	white (air-sensitive) powder	230–232	Phenergan, Wyeth; many proprietary and generic names	tablet, 25

(1)   (2)   (3)   (4)

Table 2. Properties of Some Anticonvulsants

Class	Compound	Appearance	Mp, °C	Trade name, supplier	Common form, dosage, mg
hydantoins					
	phenytoin	white solid	295–298	Dilantin, Parke-Davis	capsule, 100
	mephenytoin	white crystalline powder	136–137	Mesantoin, Sandoz	tablet, ca 100
oxazolidinediones					
	trimethadione	white crystalline solid	45–47	Tridione, Abbott	tablet, 150
	paramethadione	colorless liquid		Paradione, Abbott	capsule, 300
succinimides					
	phensuximide	crystalline powder	68–74	Milontin, Parke-Davis	tablet, 500
	methsuximide	white crysatalline powder	50–56	Celontin, Parke-Davis	tablet, 300
	ethosuximide	white solid	47–52	Zarontin, Parke-Davis	capsule, 250
other					
	primidone	white powder	279–284	Mysoline, Ayerst	tablet, 250
	clonazepam	white crystalline solid	36.5–238.5	Clonopin, Roche	tablet, 1
	carbamazipine	white or yellowish white solid	189–193	Tegretol, CIBA-GEIGY	tablet, 200
	phenacemide	white powder	212–216	Phenurone, Abbott	tablet, 500
	valproic acid	colorless liquid		Depakene, Abbott	capsule, 250
	diazepam (see Table 1)				

sedatives, that are discussed below, are listed according to their chemical classes in Table 1 (p. 633).

Anticonvulsants

The main application of anticonvulsant drugs is in the control and prevention of seizures associated with epilepsy. A recent international classification of epileptic seizures recognizes partial, generalized, unilateral, and unclassified seizures.

Most antiepileptic drugs in use are weak acids that presumably exert their action on neurons or glial cells, or both, of the CNS. The majority of these compounds is characterized by the presence of at least one amide unit and one or more benzene rings that are present as a phenyl group or part of a cyclic system. A notable exception to this is valproic acid (2-propylpentanoic acid) which lacks both of these features. Most antiepileptic drugs are inactivated by the mixed-function oxidase system (cytochrome P-450 system) located in hepatic microsomes that produce an oxidized metabolite which is excreted free or in conjugated form. Some physical properties of anticonvulsants are listed in Table 2 (p. 633).

The hydantoins are useful in the control of generalized convulsive seizures and all forms of partial seizures. The oxazolidinediones are useful in the control of generalized nonconvulsive seizures. The succinimidis, like their structural analogues, the oxazolidinediones, are useful in generalized nonconvulsive seizures.

WILLIAM J. HOULIHAN
GREGORY B. BENNETT
Sandoz Inc.

S. Garattini, E. Mussini, and L.O. Randall, eds., *The Benzodiazepines*, Raven Press, New York, 1973.

Physicians Desk Reference, 38th ed., Medical Economics Company, Oradell, N.J., 1984.

Redbook 1984, Medical Economics Company, Oradell, N.J., 1983.

M.E. Wolff, ed., *Burger's Medicinal Chemistry*, 4th ed., Pt. 3, John Wiley & Sons, Inc., New York, 1981.

HYPOGLYCEMIC AGENTS. See Insulin and other anti-diabetic agents.

I

ICE.   See Refrigeration; Water.

ICE COLORS.   See Azo dyes.

ICE CREAM.   See Milk products.

IDITOL, $CH_2OH(CHOH)_4CH_2OH$.   See Alcohols, polyhydric.

IDOSE, $CH_2OH(CHOH)_4CH_2O$.   See Carbohydrates; Sugars.

ILANG-ILANG OIL.   See Oils, essential.

ILMENITE, $FeTiO_3$.   See Titanium.

IMINES, CYCLIC

Imines are those compounds containing the bivalent —NH— (imine) group. This term, when used to name cyclic imines such as ethyleneimine, denotes a combination of the cyclic group (ethylene, —CH_2CH_2—) and the imine group. The term is also used as a class name for compounds containing the imine group attached with a double bond to a single carbon atom, such as acetaldimine ($CH_3CH=NH$).

Cyclic imines are classified according to the number of atoms in the alkyleneimine ring. Each ring size is individually named as a different heterocyclic series. Thus, derivatives of ethyleneimine are named as substituted aziridines, trimethyleneimines as azetidines, tetramethyleneimines as pyrrolidines, and pentamethyleneimines as piperidines. This article is concerned only with the simplest series of cyclic imines, the aziridines. Aziridine (ethyleneimine) is the most important member of this group, followed by 2-methylaziridine (propyleneimine) and various aziridine derivatives.

Physical Properties

The low molecular weight aziridines are colorless mobile liquids. Aziridine, 2-methylaziridine, and 1-(2-hydroxyethyl)aziridine are miscible in all proportions with water and most organic solvents. Aziridine has mp $-73.96°C$; bp $56.72°C$; n_D^{20} 1.4123; d^{25} 0.831 g/cm^3. 2-Methylaziridine has mp $-65.00°C$; bp $66.0°C$; n_D^{20} 1.4084; d^{25} 0.802 g/cm^3.

Reactions.   The reactions discussed here refer to aziridines unsubstituted on the nitrogen atom, unless otherwise stated.

Aziridines react with protonic acids to form salts which are usually unstable if the anion of the acid is nucleophilic and there are no substituents on the aziridine.

Aziridines react readily with acyl (or other acid) halides at room temperature or below. Acylaziridines are obtained in good yields in an inert solvent in the presence of a base, such as triethylamine. Polyfunctional derivatives also are prepared readily. If the base in these reactions is omitted, the ring is opened and N-2-chloroethyl amides are obtained in good yield.

Aziridines may be converted to 1-alkyl- or 1-arylaziridines by treatment with the alkyl or aryl halides. The reaction with reactive chlorides or bromides is usually performed at room temperature or below in the presence of a base in an inert solvent.

Aziridines form 1:1 adducts with aldehydes and ketones that are relatively stable when compared to such adducts formed from other secondary amines.

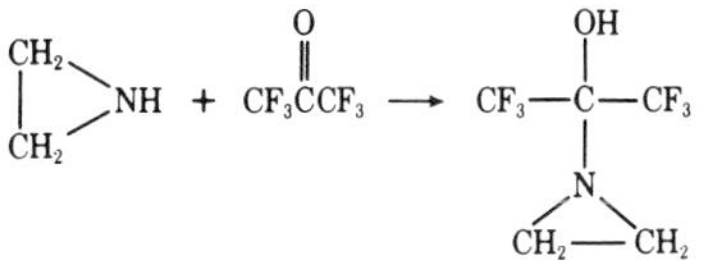

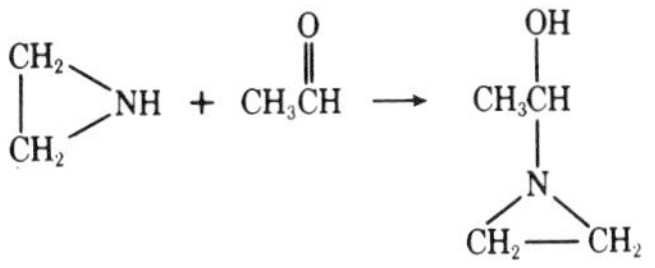

Aziridines or 1-alkylaziridines do not react with most anionic reagents. If, however, the anion is highly nucleophilic such as a carbanion, ring opening occurs at elevated temperatures. The Grignard reagent 3-indolemagnesium bromide reacts similarly with aziridine to give tryptamine.

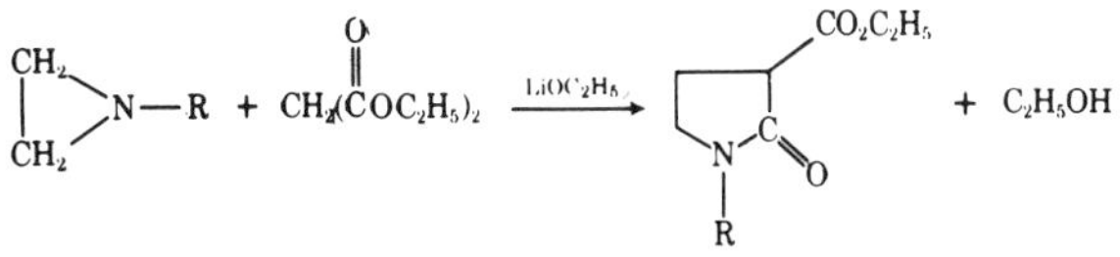

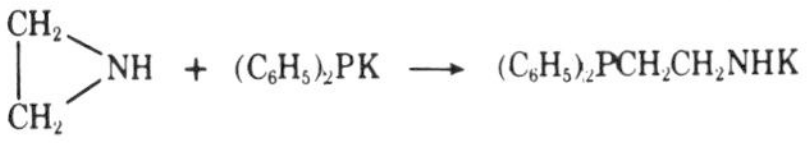

Aziridines polymerize at elevated temperatures in the presence of catalytic amounts of acid. The reaction is highly exothermic and can be violent with concentrated solutions. After polymerization has been initiated with acid and heat, cooling is frequently necessary.

Manufacture

The first commercial production of aziridine was based on the following reactions:

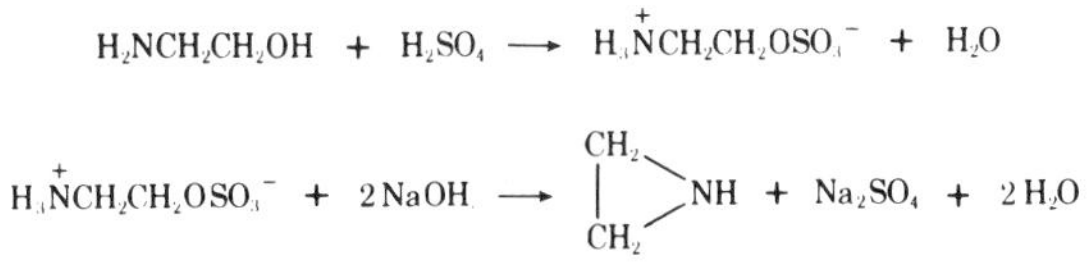

$$H_2NCH_2CH_2OH + H_2SO_4 \longrightarrow H_3\overset{+}{N}CH_2CH_2OSO_3^- + H_2O$$

2-Chloroethylamine hydrochloride can be substituted for 2-aminoethyl hydrogen sulfate. A number of modifications have improved yields by more rapid removal of the product from the reaction mixture.

A process based on ethylene dichloride and excess ammonia or an inorganic acid acceptor, such as calcium oxide, was patented in 1967:

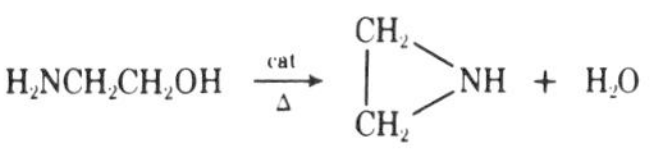

2-Methylaziridine may be produced by either of the above processes. The process based on sulfuric acid ester gives better yields.

The catalytic dehydration of monoethanolamine is particularly suitable for commercial production of aziridine, although no such process is known at this time.

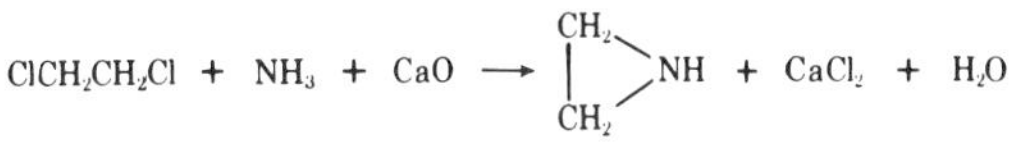

Health and Safety

Aziridines are hazardous chemicals because they are generally very toxic, can polymerize violently if undiluted, and the lower molecular weight analogues, such as aziridine itself, are extremely flammable.

Vapors may cause severe irritation of the eyes and throat and can produce inflammation of the upper and lower respiratory tract. A TLV of 0.5 ppm in air for normal exposure has been established by the ACGIH. Liquid aziridine is readily absorbed through most living tissue (eye, skin) and causes serious burns from contacts of even a few seconds; the appearance of burns may be delayed. Precautions must be taken to prevent all contact when handling aziridines.

Uses

Aziridines are used widely in industrial application. The single most important derivative is polyaziridine which is cationic and substantive to many naturally occurring anionic materials, such as cellulose; it can be used in small amounts to promote binding effects such as adhesion, improvement of paper strength, or flocculation of colloids (see Flocculating agents). The biological effects of the aziridines make these compounds useful agents for producing mutations in plant breeding and for controlling insect pests by sexual sterilization (see Insect control technology). Other uses are in the paper industry to impart wet strength to paper; as an adhesive; in textiles to attach various modifying chemicals to fabrics or fibers; in wastewater treatment and flocculation; in coatings and plastics; in ion-exchange and metal-ion complexing; and in biological applications.

G.E. HAM

Dow Chemical U.S.A.

O.C. Dermer and G.E. Ham, *Ethylenimine and Other Aziridines*, Academic Press, Inc., New York, 1969, pp. 87–105.

G.E. Ham in K.C. Rrisch, ed., *Cyclic Monomers*, Vol. 26 of *High Polymers*, John Wiley & Sons, Inc., New York, 1972, pp. 313–339.

G.E. Ham in N.M. Bikales, ed., *Encyclopedia of Polymer Science and Technology*, Suppl. Vol. 1, John Wiley & Sons, Inc., New York, 1976, pp. 25–51.

G.E. Ham in E.J. Goethals, ed., *Polymeric Amines and Ammonium Salts*, Pergamon Press, Inc., New York, 1980, pp. 1–19.

IMMUNOSUPPRESSANTS. See Analgesics, antipyretics, and anti-inflammatory agents; Chemotherapeutics, antimitotic; Immunotherapeutic agents.

IMMUNOTHERAPEUTIC AGENTS

Immunotherapeutic agents are substances that, because of their effects on the immune system, can be used in the treatment of various diseases or conditions involving immunological disorders.

The Immune System

The immune system is constantly exposed to antigens from many sources, including, exogenously, multicellular parasites and organ transplants. At the unicellular level, the body often is infested with protozoa and a host of microorganisms (see Chemotherapeutics, antiprotozoal). Red and white blood cells containing antigens are introduced during transfusions. Viruses, food, and blood substitutes are subcellular sources of antigens.

Endogenously, physically, or chemically induced tissue injury releases antigenic material. Cell turnover, which may be accelerated under various conditions, leads to increased antigen load. Many of the human biopolymers (qv), eg, proteins (qv), nucleic acids, polysaccharides, and immunoglobulins, can be antigenic, especially when denatured or partially metabolized. As the body's anabolic and catabolic processes increase or decrease, the availability of these antigens fluctuate with the state of health or disease. During aging, active diabetes, starvation, or a drastic crash diet program, deterioration of tissue components may present a stream of antigens. The ability of the immune system to cope with the sum total of these exogenous and endogenous antigenic materials determines the difference between normality and health, or pathology and disease.

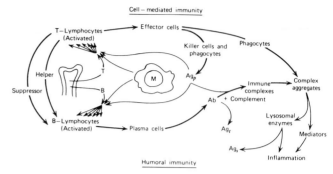

Figure 1.   Mechanisms involved in immune regulation.

The immune system is composed of two chief components, the two limbs: the humoral immune system, primarily the domain of the B-lymphocytes; and the cell-mediated immune system, primarily the domain of the T-lymphocytes. In a normal healthy state, regulatory mechanisms control the two limbs so that they function in proper balance. The humoral immune system produces antibodies that react specifically with the antigen, and the cell-mediated immune system mobilizes the phagocytic leucocytes to ingest and destroy the invading organisms that contain the antigen. In the healthy state, the two limbs are regulated by complex feedbacks through numerous mediators and by cell-to-cell cooperation. The mechanisms involved in immune regulation are depicted in Figure 1.

Immunotherapeutic Agents

Immunosuppressants. Immunosuppressive agents are used to alter normal immune function to prevent destruction of transplanted tissues. However, known agents often have undesirable side effects and narrow safety margins of a number of drugs with immunosuppressive activity have been used in the treatment of a variety of chronic inflammatory diseases with immunological features. In general, immunosuppressants may have a mild favorable effect on the clinical course of the disease, but their undesirable side effects can be fatal or chronically disabling.

Chemotherapeutic agents. Many chemical compounds have been synthesized and tested in experimental animals, but only a few were found to be useful as chemotherapeutic agents with acceptable risks of toxicity. Those that generally are available may be classified as alkylating agents, antimetabolites, and natural products.

The alkylating agents are subdivided into five types according to their basic chemical structures. These are nitrogen mustards, aziridines (ethyleneimines), alkylsulfonates, nitrosoureas, and triazines. The antimetabolites are subdivided into three types: folic acid analogues, eg, methotrexate (Amethopterin); pyrimidine analogues, eg, azauridine (Triazure); and purine analogues, eg, azathioprine. The natural products are *vinca* alkaloids (qv), antibiotics (qv), and enzymes (qv). These chemotherapeutic agents have a marked cytotoxic effect on lymphoid tissue by disturbing the fundamental mechanisms involved in cell growth, mitotic activity, differentiation, and function. The ability of these drugs to interfere with mitosis, precancer and cell division in rapidly proliferating tissues provides the basis for their therapeutic use in oncology as well as their toxic and immunosuppressive activity.

Steroids and nonsteroids. The adrenocorticosteroids have numerous and diversified physiologic functions and pharmacologic effects. The adrenal cortex is essential for the maintenance of homeostasis, which is mediated through mineral-corticoid and fluorocorticoid activity. Since the discovery and identification of the naturally occurring glucocorticosteroids, hydrocortisone and corticosterone, a number of synthetic analogues have been synthesized. Synthetically induced changes in structure have resulted in greater separation between anti-inflammatory activity and sodium retention, but many share the same side effects and differ only with respect to potency. There is no member of this group of compounds that is significantly safer than another (see Steroids). Nonsteroidal anti-inflammatory agents include acetaminophen, aspirin, phenylbutazone, indomethacin, ibuprofen, and naproxen.

Immunostimulants. Immunostimulants include a spectrum of substances that enhance or augment immunologic function. Immunomodulators, agents than enhance one immunologic component or function and elicit a decrease in a corresponding but opposite component, are included as well as immunoadjuvants which have little or no effect without the presence of an antigen. Therefore, the term immunostimulant describes the direct activity of the agent.

Both humoral and cell-mediated immune responses to an antigen can be enhanced by combination with Bacillus-Calmette-Guerin (BCG). Strains of other mycobacteria and certain fractions of BCG have similar activity. BCG has been used as an immunotherapeutic agent for the treatment of neoplasia during the past decade. Prolongation of chemotherapeutic remissions were reported for acute myeloblastic leukemia, metastatic myeloma, and metastatic breast cancer. Postoperative remission of malignant melanoma and colon cancer was delayed by the administration of BCG.

Polysaccharides have been extracted from a number of natural sources and have potent immunostimulant and anti-tumor activity. Lentinan, an adjuvant derived from mushrooms, enhances helper T-cell response and stimulates antibody-dependent cell-mediated cytotoxic activity with no apparent effect on other T-cell responses.

The anthelmintic agent, levamisole, modulates the immune system in animals and humans. Its immunopharmacologic profile suggests that levamisole may be an important prototype for the research and development of future immunologic agents to be used in chronic inflammatory diseases and neoplasia. It has been shown to influence virtually all functions involved in cell-mediated immune reactions and to restore these functions in compromised hosts, but has little or no effect in the normal animals. Responses to levamisole depend not only on the initial immune status of the host, but also on the concentration of antigen, dose of levamisole, timing, and duration of treatment. Antigen-stimulated phagocytosis and intracellular killing by macrophages also are significantly stimulated by levamisole.

Since 1971, levamisole has been used in human disease where immune balance has been either known or postulated; for example, in rheumatoid arthritis, preliminary results indicated significant improvement in 50–70% of the patients. It had been assumed that inflammation and hyperactive immunological events were responsible for autoimmunity and joint destruction in rheumatoid arthritis. Treatment, initially, was aimed at suppressing these processes; however, some investigators have demonstrated that autoimmunity primarily results from immune deficiency and not immune hyperfunction.

Health and Safety

Oral toxicity: The LD_{50}s of a few immunotherapeutic agents and anti-inflammatory agents are: azathioprine 2500 mg/kg; levamisole [20 mg/(kg·d) for 2]; ibuprofen 1600 mg/kg.

STEWART WONG
Boehringer-Ingelheim Ltd.

E.S. Golub, *The Cellular Basis of the Immune Response*, 2nd ed., Sinauer Associates, Inc., Sunderland, Mass., 1980.

S. Cohen, P.A. Ward, and R.T. McClusky, eds., *Mechanisms of Immunopathology*, John Wiley & Sons, Inc., New York, 1979.

"Mechanisms of Tissue Injury," *Ann. N.Y. Acad. Sci.* **256**, (1975).

INCENDIARIES. See Chemicals in warfare; Pyrotechnics.

INCINERATORS

The concept of a furnace operated specifically to burn refuse originated in England about 100 years ago. These furnaces were originally called destructors, the term incinerator coming into usage only in the early twentieth century.

Without recovery of energy, incineration of municipal refuse is not the cheapest method for the disposal of municipal waste. Landfill is generally cheaper, but the growing shortage of disposal sites near population centers, the increasing cost of transportation, and the growing reluctance of smaller communities and rural areas to accept waste from other localities eliminates it as a future disposal method.

Combining recovery and utilization of heat with incineration in a well-designed and operated plant may be justified. The economics of installing heat-recovery equipment is based on the income from the sale of steam or power to offset the additional cost of utilizing the excess heat. The two basic designs used for heat recovery are (*1*) a combustor combined with a waste-heat boiler to produce steam; and (*2*) a waterwall combustion chamber where heat is removed from the combustion chamber to produce steam.

Refractory-lined chambers require 150–200% excess air and waterwall chambers 50–100%. Low excess air reduces the volume of flue gas, increases the recoverable heat and reduces the size of the necessary air-pollution control equipment. Because of the variations in energy available from the solid waste, production ranges from 1 to 3.5 kg steam/kg solid waste burned.

Different classes of wastes require incinerators of different design. In general, industrial wastes can be handled by providing a special incinerator for each type of waste product. This unit may be more closely designed to meet the particular wastes, eg, fume incinerators for paint solvents and sewage-sludge incinerators.

Incinerator Pollutants

Atmospheric emissions are generally controlled by careful incinerator design, and, where necessary, the installation of control equipment. Particulate material may be removed by cyclones and electrostatic precipitators. Scrubbers are generally required for gaseous pollutants and can remove particulates as well. Baffle towers with overflow weirs and impingement baffle screens with flushing sprays are frequently used. Where fine particulates must be caught, high energy venturi scrubbers may be used. Scrubbers are generally avoided where possible because they create corrosion problems and problems of treating the effluent water (see Air-Pollution control methods).

Solid-Waste Incineration

Figure 1 is a diagram of a typical modern municipal-waste incinerator with heat recovery in the form of steam. The most recent facility in the United States uses many of the features shown in the figure. The system is composed of three main areas: the front-end system is composed of units 1 to 4 and provides a feed for the thermal part of the system composed of units 5 to 9. Units 10 to 13 are associated with the proper disposal of the incinerator discharge, both solid and gaseous.

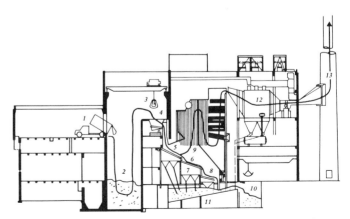

Figure 1. Typical waterwall furnace incinerator. RESCO Facility, Saugus, Mass. *1.* Unloading shed; *2.* refuse pit; *3.* loading crane; *4.* vibrating feeder; *5.* drying grate; *6.* combustion grate; *7.* grate movement; *8.* burnout grate; *9.* boiler section; *10.* residue bunker; *11.* residue channel; *12.* electrostatic precipitator; *13.* stack.

Moving-grate-incinerators. In a grate furnace, the refuse fed to the furnace is first dried and preheated by radiation from the hot combustion gases and refractory furnace lining. The refuse, as it is heated further, first pyrolyzes and then ignites. Combustion takes place both in the solid to burn out the residue and in the gas space to burn out the pyrolysis products. Overfire air jets greatly assist the mixing and combustion in the overfire air space.

The grate must provide support for the refuse; admit the underfire air through openings; transport the refuse from the feed chute to the ash quench; and agitate the bed to bring fresh charge to the surface.

Industrial Waste Incineration

In many medium-sized or large industrial plants, the costs of waste collection and disposal may warrant an on-site operation. Industrial waste is usually more homogeneous waste than municipal waste. It includes packaging, office, and process wastes and scrap lumber. For the chemical industry, process wastes are often classified as hazardous, and a specially designed system may be required. An industrial system is usually smaller than a municipal incinerator. Any system below 150 kg/h is uneconomical except for disposal of extremely hazardous wastes. The moving grate system for municipal waste may be employed.

Rotary kiln. The ability to accept diverse feeds is the outstanding feature of the rotary kiln and, therefore, this type is often selected by the chemical industry. Indeed, any liquid capable of being air-atomized can be incinerated, as well as heavy tars, sludges, pallets and filter cakes.

Multi-hearth furnaces. These furnaces are often used for the incineration of municipal and industrial sludge and for the generation and reactivation of activated char (see Furnaces).

Fluidized-bed incinerators. Fluidized-bed incinerators have been employed in the paper and petroleum industries, in the processing of nuclear wastes, and the disposal of sewage sludge. They are versatile and can be used for the disposal of solids, liquids, and gaseous combustible wastes. Several hundred fluid-bed installations are operating throughout the world for a variety of thermal processes. The world's largest sludge incineration of chemical waste is in the FRG and consists of three fluid-bed reactors (see Fluidization).

The basic fluid-bed unit consists of: a refractory-lined vessel; a perforated plate that supports a bed of granular material and distributes air; a section above the distributor containing granular solids referred to as the fluid bed; a space above the fluid bed referred to as the freeboard; an air blower to move the air through the unit; an air preheater for thermal economy; an auxillary heater for start-up; and a system to move and distribute the feed in the bed.

Thermal Conversion Processes

An alternative method of energy recovery from organic wastes is the conversion of organic waste into fuel forms that are acceptable in conventional combustion equipment. Gas of medium heating value may be substituted for natural gas in package boilers and pyrolytic liquids for fuel oil in utility boilers. A wide variety of thermal process systems is being developed that convert the waste to a more useful fuel form. These systems involve pyrolysis, gasification, or liquefaction, and have the advantages of: production of storable fuel as opposed to steam; recovery of char which may be used as fuel, converted to activated carbon or synthetic gas; cost reduction of air-pollution control because of lower gas volumes compared to incinerators; and utilization of existing boiler facilities by providing a substitute fuel.

BURTON B. CROCKER
Monsanto Co.

RICHARD C. BAILIE
West Virginia University

R.C. Corey, ed., *Principles and Practices of Incineration*, Wiley-Interscience, New York, 1969.

W.R. Niessen, *Combustion and Incineration Processes*, Marcel Dekker, New York, 1978.

B. Baum and co-workers, *Solid Waste Disposal*, Ann Arbor Science, Ann Arbor, Michigan, 1973.

INCLUSION COMPOUNDS. See Clathration.

INDANTHRENE, INDANTHRENE DYES. See Anthraquinone dyes.

INDENE. See Coumarone–indene resins under Hydrocarbon resins.

INDICATORS. See Hydrogen-ion concentration.

INDIGOID DYES. See Dyes, natural.

INDIUM AND INDIUM COMPOUNDS

Indium

Indium, In, is an element of Group IIIA in the periodic table between gallium and thallium. It is a soft, lustrous, silver-white metal, highly malleable and ductile having a face-centered tetragonal crystalline structure ($a = 0.4583$ nm, $c = 0.4936$ nm). The abundance of indium in the earth's crust is probably ca 0.1 ppm. It is found in trace amounts in many minerals, particularly in the sulfide ores of zinc and to a lesser extent in association with sulfides of copper and iron. Its content in these ores is often related to that of tin. Indium follows zinc through flotation (qv) concentration, and commercial recovery of the metal is achieved by treating residues, flue dusts, slags and metallic intermediates in zinc smelting and associated lead smelting (see Extractive metallurgy).

Properties

Mp 156.6°C; bp 2080°C; Brinell hardness number 0.9 HB; tensile strength 2.645 MPa (383.5 psi); elongation 22%; modulus of elasticity 10.8 GPa (1.5×10^6 psi). The highly plastic properties which are indium's most notable feature are due to deformation from mechanical twinning. It retains these plastic properties at cryogenic temperatures. Indium does not work-harden, can endure considerable deformation through compression, cold-welds easily, and has a distinctive "cry" like tin on bending.

Indium metal is not oxidized by air at ordinary temperature, but at red heat it burns to form the trioxide, In_2O_3. On heating, indium also reacts directly with metalloids (arsenic, antimony, selenium, tellurium) and with halogens, sulfur, and phosphorus. It dissolves in mineral acids and mercury but is not affected by alkalies, boiling water, and most organic acids. In general, the chemistry of indium in its trivalent compounds stems from nonionic or covalent bonding characteristics.

Health and Safety

Physiologically, indium is a nonessential element (see Mineral nutrients). It is classified as toxic but there have been no reported cases of systemic effects in human exposure to indium. Threshold limit value of the ACGIH is 0.1 mg/m³. The primary toxic effect of ionic indium is on the kidneys.

Uses

The first important use of indium was in the production of bearings for heavy duty and high speed service (see Bearing materials). The solder and alloy market, including the low-melting point or fusible alloys, is the largest single use (see Solders and brazing alloys). Indium generally increases the strength, corrosion resistance, and hardness of a system to which it is added. Low melting point alloys of indium with lead, tin, bismuth, and cadmium having melting points as low as 57°C are used in surgical casts, patternmaking, lens blocking, and electrical fuses.

E.F. MILNER
Cominco Ltd.

C.E.T. WHITE
Indium Corporation of America

J.M. Ramaradhya, R.C. Bell, S. Brownlow, and C.J. Mitchell in S.D. Snell and L.S. Etltre, eds., *Encyclopedia of Industrial Chemical Analysis*, Vol. 19, Wiley-Interscience, New York, 1971, p. 518.

M.T. Ludwick, *Indium*, 2nd ed., Indium Corporation of America, Utica, N.Y., 1959.

"Indium" in C.A. Hampel, ed., *Encyclopedia of the Chemical Elements*, Reinhold Book Corp., New York, 1968.

G.F. Smitt in A.R. Katritzky, ed., *Advances in Heterocyclic Chemistry*, Vol. 2, Academic Press, Inc., New York, 1953, p. 300.

C.W. Rees and C.E. Smithen in A.R. Katritzky, ed., *Advances in Heterocyclic Chemistry*, Vol. 3, Academic Press, Inc., New York, 1964, p. 57.

W.J. Houlihan in A. Weissburger and E.C. Taylor, eds., *The Chemistry of Heterocylic Compounds*, Vol. 25, Pt. 1, Wiley-Interscience, New York, 1972, p. 232.

INDOLE

Indole (1) (1*H*-indole) is a benzopyrrole in which the benzene and pyrrole rings are united through the 2,3-positions of the pyrrole. The indole nucleus is found in a large number of naturally occurring compounds; the compound constitutes about 2.5% of jasmine oil and 0.1% of orange-blossom oil, and in both cases, it contributes to their fragrances (see Oils, essential).

Isoindole (2) (2*H*-isoindole), the isomer in which the benzene and pyrrole rings are fused through the 3- and 4-positions of the pyrrole, is not stable. A few of its derivatives are known, the simplest being *N*-methylisoindole.

Properties

Indole is a colorless crystalline solid; mp 52–54°C; bp 254°C (dec); and heat of combustion at constant volume 4.268 MJ/mol (1020 kcal/mol). The molecule is planar and has only moderate polarity. Indole has good solubility in a wide range of solvents including petroleum ether, benzene, chloroform, and hot water. The solubility in cold water is only 1 : 540 at 25°C; thus, water is a good solvent for purification by recrystallization.

Reactions

Indole undergoes halogenation, nitration, sulfonation, diazo coupling, alkylation, acylation, Mannich reaction, Michael reactions, Vilsmeier-Haack formylation, Gatterman and Houben-Hoesch reactions, oxidation, and reduction. The high resonance energy and abundance of electrons makes indoles resistant to reduction under neutral conditions. Catalytic reduction at 8.6 MPa (85 atm) and 90–100°C with Raney nickel gives indoline, whereas at 25.3 MPa (250 atm) and 250°C, the product is octahydroindole.

Preparation

Although synthetic methods have been described for the manufacture of indole, extraction from the 240–260°C coal-tar distillate fraction is the only commercial source.

The methods used for the synthesis of indole derivatives involve the formation of the pyrrole ring by the cyclization of an appropriately substituted aniline derivative. In the Fischer indole synthesis, the phenylhydrazone of an aldehyde, ketone, or ketonic acid is converted to a substituted indole by heating with an acid catalyst, such as zinc chloride, sulfuric acid, formic acid, acetic acid, or boron trifluoride. This synthesis is widely applicable to the preparation of indole derivatives that are substituted on the benzene or pyrrole rings, but fails for indole itself.

Uses

Indole itself is commercially important as a component of perfumes (qv). It is also the parent substance of a large number of natural compounds including alkaloids (qv), hormones (qv), and psychopharmacological agents (qv).

D.W. BANNISTER
CIBA-GEIGY Corp.

INDOPHENOL.   See Sulfur dyes.

INDULINES.   See Azine dyes.

INDUSTRIAL ANTIMICROBIAL AGENTS

Industrial antimicrobial agents are chemical compositions that are used to prevent microbiological contamination and deterioration of commercial products, materials and systems. The following chemical classes of antimicrobial agents are recognized in industry: phenolics; halogen compounds; quaternary ammonium compounds (qv); metal derivatives; amines (qv); alkanolamines (qv) and nitro derivatives; anilides; organosulfur and sulfur–nitrogen compounds; and miscellaneous compounds (see Table 1, p. 640) (see Antibacterial agents; Antibiotics; Fungicides).

A given antimicrobial agent may either destroy all the microbes present or just prevent their further proliferation to numbers that would be significantly destructive to the substrate or system being protected. The terms, microbes and microorganisms, refer primarily to bacteria and fungi. Each of these groups is subdivided into two general subclasses: gram-positive and gram-negative bacteria, and among the fungi, molds and yeasts (qv). A number of use-oriented terms include antibiotic (which refers to a medicinal or therapeutic agent), preservative, disinfectant, antiseptic, antifoulant, slimicide, and mildewcide. A preservative is used for a variety of products, eg, food, cosmetics, paint, etc, and a disinfectant is employed for household needs, dairy machinery, etc.

Choosing an Antimicrobial Agent

The agent is chosen that is claimed to combat the microorganism(s) of concern to the degree desired; appears to be physically and chemically compatible with the system, ie, will not upset desirable physical and chemical properties of the system and (conversely) will not be inactivated by the ingredients of the system; maintains stability under use and storage conditions (pH, temperature, light, etc), for the required length of time; is safe and nontoxic in handling, formulation, and use; is environmentally acceptable; and is economically acceptable and cost effective.

Health and Safety

In general, one should expect the supplier to provide the common group of acute tests, ie, one-time dose responses; acute oral LD_{50}; acute dermal LD_{50}; skin irritation; eye irritation; and in some cases, acute inhalation and TLV. Generally, antimicrobials are toxic and irritating and should be handled with care.

Uses

Areas of application include cosmetics (qv), disinfectants and sanitizers, wood (qv) preservation, food and animal feeds, paint (qv), cooling water, metalworking fluids, hospital and medical uses, plastics and resins, petroleum, pulp (qv) and paper (qv), textiles (qv), latex, adhesives (qv), leather (qv) and hides, and paint slurries (see Disinfectants and antiseptics; Food additives; Latex technology; Pet and other livestock feeds; Petroleum; Plastics, Processing; Resins, natural; Water, industrial water treatment; and treatment of swimming pools).

SAMUEL I. TROTZ
JAMES J. PITTS
Olin Corporation

Table 1. Selected Industrial Antimicrobial Agents

Common name	Structure	Trade names	Producers	Applications
Phenolic Antimicrobial Agents				
pentachlorophenol (PCP)		Dowicide 7	Dow Reichhold Vulcan	wood and paper construction materials leather paint textiles
o-phenylphenol		Dowicide 1	Dow	hard-surface cleaners metalworking fluids automotive gaskets leather paint polish formulating
Halogen-Containing Antimicrobial Agents *Chlorine compounds*				
trichloroisocyanurate		ACL-85 CDB-90 Pace	Monsanto FMC Olin	swimming-pool sanitizer disinfectant
Iodine compounds iodine–poly(vinylpyr-rolidinone) complex		PVP–Iodine	GAF	germicide
Bromine compounds 2,2-dibromo-2-cyanoacet-amide		X-7287L	Dow	cooling-water slimicide secondary oil recovery metalworking fluids
2-bromo-2-nitropropane-1,3-diol		Onyxide 500	Onyx	cosmetic preservative
Quaternary Antimicrobial Agents				
benzalkonium chloride	R = C$_{12-18}$ R = C$_{12-16}$ R = C$_{12-16}$ R = C$_{16}$	BTC-100 Maquat MC-1416 Bio-Quat 50-24 Hyamine-3500 Roccal II Barquat MB 50 Dibactol	Onyx Mason Bio-Lab Rohm and Haas Hilton-Davis Lonza Hexcel	germicide disinfectant
cetylpyridinium chloride (CPC)			Hexcel	disinfectant
Industrial Metal-Containing Antimicrobial Agents *Mercurials*				
phenylmercuric acetate		PMA-100 Troysan PMA	Cosan Troy	paint preservative paint mildewcide leather
Arsenicals 10,10'-oxybisphenoxy-arsine		Vinyzene BP-5 Durotex	Ventron Ventron	plastics textiles
Organotins tributyltin oxide	$(n\text{-}C_4H_9)_3\text{-Sn}\text{-O-Sn-}(C_4H_9\text{-}n)_3$	Cotin 300 Keycide X-10 Intercide 340A	Cosan Ferro Interstab	antifoulant paint latex paint plastics
Copper compounds chromated copper arsenate	$Cu_3(AsO_4)_2 \cdot NaCr_2O_7$		Koppers	wood
Anilide Antimicrobial Agents trichlorocarbanilide (TCC)			Monsanto	soap germicide

Table 1. Continued

Common name	Structure	Trade names	Producers	Applications
Amine, Alkanolamine, and Nitro-Containing Antimicrobial Agents				
Cyclic				
hexahydro-1,3,5-tris (2-hydroxyethyl)-s-triazine		Onyxide 200	Millmaster-Onyx	metalworking fluids
		Grotan	Lehn & Fink	
Bisthiocyanates				
methylenebisthiocyanate (MBT)	NCSCH₂SCN	Cytox 3522 Biocide N-948	American Cyanamid Stauffer	paper-mill slimicide
Dithiocarbamates				
sodium dimethyldithiocarbamate (Sodam)		Thiostat Thiostop N	Uniroyal	paint mildewcide slimicide
Sulfones				
sodium pyridinethione		Sodium Omadine	Olin	metalworking fluids cosmetics
Organo-Sulfur and Sulfur–Nitrogen Antimicrobial Agents				
Nitrogen–sulfur heterocyclics				
2-*n*-octyl-4-isothiazolin-3-one		Skane M-8	Rohm and Haas	latex paint mildewcide
		Kathon LP	Rohm and Haas	leather hides
Miscellaneous Antimicrobial Agents				
tetrachloroisophthalonitrile		Nopcocide N-96	Diamond Shamrock	latex paints
1-(3-chloroallyl)-3,5,7-triaza-1-azoniaadamantane chloride		Dowicil 100	Dow	adhesives floor waxes latex emulsions metalworking fluids paint

R.A. Payne and J.P. Stekla, *Biocides U.S.A. 1979*, C.H. Kline & Co., Inc., Fairfield, N.J., 1979.

S.S. Block, ed., *Disinfection, Sterilization, and Preservation*, 2nd ed., Lea and Febiger, Philadelphia, Pa., 1977.

F.J. Buono and G.A. Trautenberg in N.M. Bikales, ed., *Encyclopedia of Polymer Science and Technology*, Suppl. 1, Wiley-Interscience, New York, 1976, pp. 95–115.

INDUSTRIAL HYGIENE AND TOXICOLOGY

Industrial Hygiene

Industrial hygiene began in the fourth century, BC, with Hippocrates' account of lead toxicity in the mining industry. Some 500 years later, Pliny the Elder recognized the dangers encountered in handling zinc and sulfur, whereas Galen, a Greek physician residing in Rome in the second century, described the dangers of acid mists to copper miners.

The profession of industrial hygiene embraces the total realm of control, including recognition and evaluation of those factors of the environment originating in the place of work which may cause illness, discomfort, or lack of well-being, either among workers or among the community as a whole.

Industrial hygiene begins with the recognition of industrial health problems such as biological problems caused by insects and mites, molds, yeasts and fungi, bacteria and viruses; chemical problems caused by liquids, dusts, fumes, mists, vapors, or gases; energy problems caused by electromagnetic and ionizing radiations, noise, vibration and temperature extremes, and ergonomic problems caused by body position in relation to task, monotony, repetitive motion, boredom, work pressure, anxiety, and fatigue.

After recognition of the hazard, the work atmosphere is evaluated for long- and short-range effects on health. Finally, corrective measures are developed such as the replacement of harmful or toxic materials; changing work processes; new ventilation procedures; increasing distance and time between exposures to radiation; introduction of water to reduce dust emission in certain fields, eg, mining; maintenance of "good housekeeping" and adequate methods of waste disposal; and use of proper protective working clothes, including respirators and ear protectors when necessary (see also Air pollution; Noise pollution).

Evaluation. Recognition of potential hazards includes an inventory of physical and chemical agents encountered in a specific process, a periodic review of different activities in a work area, and a study of existing control measures. In the preparation of a field study, the proper instruments are selected and checked, and the appropriate analytical methods are developed. When conducting a field study, the manner of sampling is extremely important. Finally, the results of the survey are interpreted. Comparison should be made with health standards and previously obtained data.

Control Procedures

Administrative Control. By limiting the amount of time a worker is exposed to a particular substance of group of substances, the 8-h average exposure is kept below the permissible limit. For example, the airborne contaminant limit for oil is 5 mg oil/m³ air based on an 8-h exposure. Therefore, a worker exposed to a level of 10 mg mg/m³ is limited to 4 h/d.

Engineering control. These fall into four categories:

(1) Process change: Many processes used today were developed when water was inexpensive, air pollution was disregarded, and employee protection a minor concern. Today, the emphasis has shifted and control of air and water contaminants has greatly increased costs. As a result of Federal legislation and union efforts, employee health and welfare are of primary concern. In process redesign, pollution and industrial hygiene problems are minimized, eg, the reduction of air contamination by a process change may mean replacing spray painting with dip painting.

(2) Substitution of a product with low toxicity for a more toxic product to bring an excessive exposure into acceptable limits, eg, replacement of carbon tetrachloride (exposure limit 10 ppm) by methylchloroform (exposure limit 350 ppm).

(3) Isolation, or containment, usually reserved for highly toxic materials. OSHA defines an isolated system as a fully enclosed structure, other than the vessel of containment, which is impervious to the contaminant and prevents its loss to the environment in case of leakage or spillage, eg, a glove box.

(4) Ventilation controls contaminants by dilution with large volumes of air which must be exhausted and replaced (see also Exhaust control, industrial; Air pollution control, methods). The operation can be expensive in cold climates and should not be used as a primary control method. It is more economic to control worker exposure by local ventilation, using general ventilation as a secondary exhaust system.

Industrial Toxicology

Industrial toxicology is largely concerned with the development and use of toxicity information as a means of predicting safe or harmful amounts of materials which may be encountered in the workplace, home, or the general environment. Knowledge of the toxicity of industrial chemicals is extensive.

Definitions. LD_{50}, average lethal dose, is the dosage that, when administered to animals, kills half (50%) of them. It is usually expressed in mg toxicant per kg body weight, and the test route (oral, intravenous, or subcutaneous) is usually given.

TD_{Lo}, toxic dose low, is the lowest known dose of a substance that has produced any toxic, carcinogenic, mutagenic, teratogenic, or neoplastic (tumorous) effects.

TC_{Lo}, toxic concentration low, is the lowest concentration of a substance that is reported to have any toxic effect.

TLV, threshold limit value, is an estimate of the average safe toxicant concentration that can be tolerated on a repetitive basis, usually an 8-h period on a day-to-day basis.

TWA, time weighted average, is a mathematical expression summing the products of toxicant concentrations and durations of exposure to those toxicants, and dividing by the total exposure time. In simple terms, the concentrations of the various toxicants are multiplied by the duration of exposure to each individual toxicant, the results are added, and then divided by the time of exposure.

Inhalation. Methods of evaluating the hazard of inhaled environmental substances have been studied more intensively than percutaneous hazards.

Skin contact. The skin has ca 1.86 m^2 (20 ft^2) of surface area for possible contact with harmful substances. It retains a large portion of the total available body water and has, in addition to its protective and excretory functions, a principal role in controlling heat exchange with the environment.

The most commonly employed methods for determining acute dermal toxicity generally use rabbits or guinea pigs. Substances are applied to the clipped skin in varying quantities and held in place for 24 h by a sleeve of impervious plastic sheeting or rubber dam. Observations are made for at least two weeks, much as in the case of the acute oral LD_{50}. Such tests allow an estimate of the hazard of serious systemic effects by dermal contact, and they may give an idea of the rapidity of absorption. Generally they do not provide a quantitative measure of the percentage of the applied dose that has penetrated the skin. Some substances may be absorbed into the skin but not penetrate it. Quantitative data can be obtained in a variety of ways: by tracer substances, by measuring concentration in blood or excretion in urine, and by *in vitro* methods.

Ingestion. In the industrial environment, ingestion is the least important of the three modes of entry to the body. Few substances can be ingested, and the frequency and degree of contact are very limited. Furthermore, and most important, toxicity by mouth is generally of a lower order than that by inhalation.

Exposure limit values. Although it is the responsibility of OSHA to establish legal standards for the workplace, many organizations in the United States determine TLV data or their equivalents, the most prominent probably being the ACGIH which publishes a list every year. These TLVs represent concentrations of airborne substances and physical agents below which workers may be exposed eight hours per day, forty hours per week, without adverse effect. Medical surveillance is recommended to detect workers who are oversensitive to specific chemicals or physical agents. These workers should be exposed to these substances only under special protection. Ceiling values represent exposure limits related to substances that are fast acting and whose threshold limits are more appropriately based on a particular biologic response. In instances where the cutaneous route is an important source of absorption, a substance is so labeled since the TLV refers only to inhalation as the source of entry of the agents into the body.

GEORGE CLAYTON
Clayton Environmental Consultants, Inc.

G.D. Clayton, *The Industrial Environment—Its Evaluation and Control*, U.S. Dept. of Health, Education, and Welfare, PHS, Superintendent of Documents, Washington, D.C., 1973, Chapt. 1.

G.D. Clayton and F.E. Clayton, eds., *Patty's Industrial Hygiene and Toxicology*, Vols. 1, 2A, 2B, and 2C, John Wiley & Sons, Inc., New York, 1980.

INFORMATION RETRIEVAL

The information explosion is more than a catch phrase to describe the proliferation of technical information in recent years. For many chemists, it marks a destruction of the traditional ways of finding chemical facts, with a resultant information fallout—the unexpected, often bewildering products of this sudden surge. Information is more abundant and more specialized than ever before; but chemists, amid this vast wealth of information, are often unable to obtain easily the particular facts they need.

Developments in Information-Retrieval Technology

To discuss the advances in information retrieval requires familiarity with the terms used for the sources of searchable information. Primary sources represent either new information, new interpretations of old knowledge, or new compilations of known information. They include books, journals, and patents. Secondary sources serve as locators for primary information—they are the organizers and the condensers of primary literature. These include encyclopedias, dictionaries, and handbooks. Chemical Abstracts Service (CAS) is an example of an abstracting and indexing service that organizes the primary periodical literature. CAS has emerged as the single most important secondary source of information for chemists and, through international agreements, has indeed become responsive to an international community. Tertiary sources direct the user to both the primary and secondary sources; among tertiary sources are guides and directories.

The computer has become indispensable in information retrieval. There is, however, a wealth of material that predates computer-generated information. In the 1960s, computers generated indexes to the literature and, in specialized systems, identified structures. These indexes were produced as paper products for manual searching. Extensive use of the computer to actually store the indexes for manipulation started about 1970. The computer offers the ability to retrieve current information in ways never before possible, but until technology develops ways of providing access to the accumulated knowledge of the past, the traditional retrieval techniques are still needed.

In the late 1960s, the computer-readable tapes that were used in the production of printed publications of the main abstracting and indexing services began to be used for alerting purposes. This was the beginning of a significant change in information retrieval. These tapes were manipulated further to produce a variety of products. They became searchable in batch-mode; ie, they could be queried, and then scanned by the computer in one operation. Then, on-line systems were developed. In on-line systems, data are transmitted directly between a computer and remote terminals, and are immediately processed by the computer. These systems provided a way to search the tapes interactively. The person at the terminal, in direct communication with the computer, could refine the query during the scanning process. The development of these systems led to a proliferation of machine-readable files called data bases.

Because purchasing and maintaining these data bases for in-house use was expensive, companies and search services merged their collected data bases and sold access to them either for a fee based on use or by subscription. Once connected to the retrieval system, the user can look at the contents of that search service's data bases. This process of examining data bases from remote locations using a computer terminal is called on-line searching.

Several types of files are searchable on-line. Bibliographic data bases provide references to source material. Nonbibliographic or numerical data bases allow the user to retrieve specific data directly and offer the additional capabilities of adding and manipulating data for analysis. On-line dictionary and vocabulary files guide the user to formulas, synonyms, particular spellings of words, and particular terms for optimum retrieval of information from other on-line data bases.

Automated systems offer two modes for information retrieval: retrospective searching, by which the user identifies material on a given subject; and current awareness, by which the user identifies the most recent material on a given subject to keep abreast of the latest developments. This later activity, called selective dissemination of information (SDI), is a continual searching service.

Retrieving chemical information requires manipulation of textual material and chemical structures. The bibliographic data elements by which a document is stored and retrieved include its title, author(s), subject, journal sources, date, and reference number. Keywords and descriptors identify the contents of the document. Because structural diagrams are extremely difficult to convey in written text, systematic names have been developed. Chemists revert to structural diagrams to define the chemical structures of compounds unequivocally. The most commonly used systems for the retrieval of structures are topological systems and linear-notation systems.

Chemical Abstracts Service Registry Numbers serve as a useful bridge between textual and structural information. A Registry Number is a unique number assigned to a chemical compound by CAS in sequential order when the compound initially enters the registry system. This system, which originated in 1965, has registered over 8×10^6 substances.

Registry Numbers for a particular substance can be found in the CAS *Registry Handbook*, or in the CAS *Chemical Substance Index*, *Formula Index*, or *General Index*. They may be found also in chemical dictionary files. Registry Numbers have appeared regularly in *Chemical Abstracts* volume indexes since 1972 and are an integral part of the CA SEARCH data base. The Registry Number appears as a data element in an ever increasing number on on-line chemical files and technical journals, and it is even required in reporting to the EPA.

Bibliography of Chemical Literature Sources

In addition to the full version of this article (*ECT*, 3rd ed., **13**, 278–336 (1981)), the following bibliography lists some current literature guides:

T.P. Peck, ed., *Chemical Industries Information Sources (Management Information Guide Series 29)*, Gale Research Company, Detroit, Mich., 1979.

S.H. Wilen, *Use of the Chemical Literature, An Introduction to Chemical Information Retrieval*, ACS Audio Course, American Chemical Society, Washington, D.C., 1978.

M.G. Mellon, *Chemical Publications, Their Nature and Use*, 5th ed., McGraw-Hill, New York, 1982.

H. Skolnik, *The Literature Matrix of Chemistry*, Wiley-Interscience, New York, 1982.

R.H. Powell, *Handbooks and Tables in Science and Technology*, Oryx Press, 1980.

J.T. Maynard, *Understanding Chemical Patents*, American Chemical Society, Washington, D.C., 1978.

Information in the Laboratory

When working at the bench, chemists are looking for facts, experimental procedures or techniques. Until now, the information resources for these needs were in traditional book form. New systems deliver factual, handbook information in electronic form via the laboratory's computer terminal, already in use for automated instrument control and data analysis. For the chemical industry, delivery of handbook information is workable if the data, equations, and the like are accompanied by design modules that allow retrieved information to be manipulated. In contrast, simple electronic reference systems are not as useful.

Handbooks. Handbooks are the most convenient reference tools for the laboratory. They contain just the kind of information required in the middle of things—physical constants and mathematical tables. The data may or may not be evaluated but they are usually reliable—the hard core of chemistry.

Bibliography

J.A. Dean, *Lange's Handbook of Chemistry*, 12th ed., McGraw-Hill Book Co., New York, 1979.

D.W. Green, *Perry's Chemical Engineers' Handbook*, 6th ed., McGraw-Hill Book Co., New York, 1984.

R.H. Powell, *Handbooks and Tables in Science and Technology*, 2nd ed., Oryx Press, 1983.

G.J. Shugar, *Chemical Technicians' Ready Reference Handbook*, 2nd ed., McGraw-Hill Book Co., New York, 1981.

R.C. Weast, ed., *CRC Handbook of Chemistry and Physics*, 64th ed., CRC Press, Inc., Boca Raton, Florida, 1984.

Data compilations. Data compilations can save chemists' time if they provide quality information. The institutions that produce these collection spend considerable effort in evaluating data. The complex organizational schemes designed to accommodate updates require the user to read the introductions carefully.

An example of a classic resource is:

E.W. Washburn, ed., *International Critical Tables of Numerical Data—Physics, Chemistry, and Technology*, published for the National Research Council by McGraw-Hill Book Co., New York, 1926–1933.

Encyclopedias and treatises. Encyclopedias provide background information. The treatises, by dealing in-depth, can help the nonspecialist become more knowledgeable. Treatises generally are descriptive, contain evaluated data, discuss methodologies, and are likely to have better indexes than handbooks or monographs.

C.H. Bamford and C.F.H. Tipper, *Comprehensive Chemical Kinetics*, 18 vols., American Elsevier Publishing Company, New York, 1969–1977.

D.H.R. Barton and W.D. Ollis, eds., *Comprehensive Organic Chemistry: The Synthesis and Reactions of Organic Compounds*, 6 vols., Pergamon Press, New York, 1979.

S. Coffey, ed., *Rodd's Chemistry of Carbon Compounds, A Modern Comprehensive Treatise*, 2nd ed., Elsevier Publishing Company, New York, 1964.

M. Grayson and D. Eckroth, eds., *Kirk-Othmer Encyclopedia of Chemical Technology*, 3rd ed., Wiley-Interscience, New York, 26 vols, 1978–1984.

J.W. Mellor, *A Comprehensive Treatise on Inorganic and Theoretical Chemistry*, Longman's, Green & Co., Ltd.; John Wiley & Sons, Inc., New York, 16 vols. (main series), 1922–1937, supplements 1956–1972.

J.C. Bailar, Jr., H.J. Emeleus, R. Nyholm, and A.F. Trotman-Dickenson, eds., *Comprehensive Inorganic Chemistry*, 5 vols., Pergamon Press, New York, 1973.

Beilstein and Gmelin. The Beilstein and Gmelin chemical treatises hold a special place in the minds of chemists because they present evaluated information on a comprehensive scale. Both are compound-based but have different organizational schemes. The value of these handbooks is immense, and retrieval is rapid; instructional information is available for each.

Beilstein's Handbuch der Organischen Chemie, simply known as *Beilstein*, is a comprehensive reference source of carbon-containing compounds. The chemical literature is evaluated, and data are checked to ensure that misleading or erroneous results are not documented. *Beilstein* is a key source to factual chemical information, free from the inaccuracies and redundancies of the open chemical literature.

Beilstein gives specific data on and references to nearly 4×10^6 compounds of known constitution. The description of such compounds includes their composition, configuration, natural occurrence and isolation from natural products, preparation and purification, structural and energy parameters, physical properties, chemical properties, their characterization and analysis, and their salts and addition compounds. A booklet entitled *How to Use Beilstein* is available free of charge from the publisher, Springer-Verlag, Berlin, Heidelberg, and New York.

The Gmelin Handbuch der Anorganischen Chemie, or simply *Gmelin*, is the most authoritative and comprehensive treatise on inorganic compounds. Each volume and its supplement discuss an element and its compounds, their formation and occurrence, methods of preparation, and physical and chemical properties. The Gmelin Institute is progressively translating its entire collection from German to English; the complete handbook eventually will be available in English. The value of the *Gmelin Handbook* must not be underestimated. The current edition contains revaluated material from the older editions and thus represents a comprehensive documentation of the entire body of inorganic chemical literature. The handbook is described in detail in the Institute's brochure *Was ist der Gmelin?*, available in either English or German from the Gmelin Institute, Frankfurt.

Experimental methods. Chemists have a long tradition of publishing methodological compilations. These make it easy to locate synthetic preparations or discussions of laboratory techniques at a more practical level than found in encyclopedias. Two examples are:

Weissberger's *Techniques of Chemistry* series, a successor to his *Technique of Organic Chemistry*, 3rd., John Wiley & Sons, Inc., New York, is organized into 16 multipart volumes, 1971–.

Organic Syntheses, John Wiley & Sons, Inc., New York, 1921–; annual volumes published since 1921.

Technical dictionaries, reviews series and review journals are also valuable information resources.

Computer Information Resources

Chemical information is available through many on-line data base vendors, Lockheed DIALOG, System Development Corporation ORBIT, Chemical Abstracts Service CAS ONLINE, BRS SEARCH, NIH/EPA Chemical Information System, and the National Library of Medicine MEDLARS are key U.S. providers. European systems are DATA STAR, Telesystems DARC/QUESTEL, ESA/IRS, Pergamon INFOLINE, INDA and DIMDI. Useful data bases for chemists are available on the Canadian systems CAN/OLE and QL Systems.

Three of these vendors provide the ability to work with chemical substances using structure graphics as well as with registry numbers and chemical nomenclature.

CAS ONLINE is a comprehensive system for using all aspects of the *Chemical Abstracts* data base since 1967—structures, registry numbers, chemical names, keywords, controlled vocabulary index terms, bibliographic information, and the full abstract. The CAS Registry file, with Registry Numbers only, began in 1965. CAS has announced plans to build a worldwide network and to add other data bases.

Bibliography

Chemical Abstracts Service Using CAS Online, 3 volumes, Columbus, Ohio, 1983.

P.G. Dittmar, and co-workers, *Journal of Chemical Information and Computer Sciences*, **23**(3), 93–102 (1983). "The CAS Online Search System. 1. General System Design and Selection, Generation and Use of Search Screens."

N.A. Farmer, and M.P. O'Hara, *Database*, **3**, 10–25 (1980); "CAS Online: A New Source of Substance Information from Chemical Abstracts Service."

DARC/QUESTEL uses the CAS Registry file in conjunction with graphic or textual structure input including bond information, to retrieve structures, bibliographic information, and spectral data.

The Chemical Information System (CIS) of the National Institutes of Health and the Environmental Protection Agency (NIH–EPA) contributes data analysis programs for use with various numeric data bases in the system. Structure or nomenclature techniques are used to retrieve substances. The CAS Registry Number, or pseudo-Registry Number for undefined substances, is used to retrieve biological, chemical, and regulatory information.

The Structure and Nomenclature Search System (SANSS) is the heart of CIS. SANSS works on a unified data base, built from many files that identify chemicals of commercial importance or known biological activity (ie, drugs, pesticides, commodity chemicals, food additives), and therefore relevant to these regulatory agencies. The CAS Registry Number is used as the unifier for the various files. Reported or regulated substances that are structurally undefined get pseudo-Registry Numbers. Ultimately, because of the overlap in files, and because of the limited number of chemicals actually used commercially, the file is expected to total 175,000–200,000 substances.

Among the chemical-data components currently available in CIS are the Mass Spectral Search System, X-Ray Crystallographic Search System, and Carbon-13 Nuclear Magnetic Resonance Spectral Search System.

Bibliography

G.W.A. Milne and S.R. Heller, *Am. Lab.* **8**(9), 43–54 (1976).

S.R. Heller, G.W.A. Milne, and R.J. Feldman, *J. Chem. Inf. Comput. Sci.* **16**(4), 232–233 (1976).

R.J. Feldman, G.W.A. Milne, S.R. Heller, A. Fein, J.A. Miller, and B. Koch, *J. Chem. Inf. Comput. Sci.* **17**(3), 157–163 (1977).

Chemical Abstracts Service. *Chemical Abstracts* (CA) is the most comprehensive chemical data base. Chemical Abstracts Service's mission is to abstract and index the world's chemical literature. CAS adopted three principles to manage the rapidly growing literature, the technological advances, and the changing environment—international cooperation, a unified data base concept, and the use of Registry Numbers for substance identification.

Chemical Abstracts has indexes of varying depth and type. In 1963, CA added a permuted keyword-subject index and patent concordance to its author and numerical patent indexes for the weekly issues.

The volume, semiannual indexes, and the collective indexes, now issued every five years, are much more extensive. These indexes are Author Index, Formula Index, Index of Ring Systems, Numerical Patent Index, Patent Concordance, General Subject Index, and Chemical Substance Index.

CA indexes can also be searched by on-line, interactive computer techniques from several services as mentioned earlier.

CAS has a number of current-awareness services. When individual issues became unwieldy, *CA* established five section groups, which could be subscribed to separately. These divide the 80 sections into biochemistry, organic chemistry, macromolecular chemistry, applied chemistry and chemical engineering, and physical and analytical chemistry.

Personalized SDI profiles based on *CA*, a computer technique discussed earlier, are available in many organizations.

CA Selects are topical current-awareness bulletins that draw across *CA* sections, include the abstracts, and are of typeset-printout quality.

CAS Bibliography

A.J. Beach, H.F. Dabek, Jr., and N.L. Hosansky, *J. Chem. Inf. Comput. Sci.* **19**, 149–155 (1979).

J.E. Blake, V.J. Mathias, and J. Patton, *J. Chem. Inf. Comput. Sci.* **18**, 187–190 (1978).

P.G. Dittmar, R.E. Stobaugh, and C.E. Watson, *J. Chem. Inf. Comput. Sci.* **16**, 111–121 (1976).

R.G. Dunn, W. Fisanick, and A. Zamora, *J. Chem. Inf. Comput. Sci.* **17**, 212–218 (1977).

M.D. Huffenberger and R.L. Wigington, *J. Chem. Inf. Comput. Sci.* **15**, 43–47 (1975).

I.R. McKinley and A.K. Kent, *Chem. Br.* **12**(1), 4–6 (1976).

R.E. O'Dette, *J. Chem. Inf. Comput. Sci.* **15**, 165–169 (1975).

O.B. Ramsay, "Chemical Abstracts: An Introduction to its Effective Use," *Tape/Slide Audio Course*, American Chemical Society, Washington, D.C., 1979.

Institute for Scientific Information. The Institute for Scientific Information (ISI) offers a number of information tools for the chemist. The level of complication varies to serve a variety of needs. Most chemists are already familiar with the current-awareness series, *Current Contents*. These reproduce, by permission, the contents pages of core journals in particular disciplines, in weekly, pocket-sized publications for personal scanning. *Current Contents, Physical and Chemical Sciences* is the primary publication in the series for chemists.

Current Abstracts of Chemistry (*CAC*), published weekly, provides chemists with a guide to new organic chemistry literature. It contains abstracts of articles from 107 core journals reporting the synthesis, isolation, and identification of new compounds. Extensive use of flow diagrams and use-profile and technique-data symbols alert the user to chemical activity of a compound and to analytical procedures used by the investigator. The technique data also highlight articles containing new synthetic methods. Details on these methods appear in the monthly publication, *Current Chemical Reactions* (*CCR*). CAC includes reaction schemes, experimental data, bibliographic information, authors' abstracts and an index section containing journal, author, permuted subject terms, and corporate addresses.

Index Chemicus (*IC*), the companion index to *CAC*, and now on-line, is incorporated into the weekly issues. It contains the following sections: molecular formula, author, subject, bibliographic information, biological activity, analytical techniques, and new synthetic methods.

All new compounds are encoded using the Wiswesser line notation (WLN). These are then cumulated and permuted monthly and annually and are available on microfilm as the *Chemical Substructure Index* (*CSI*).

A personalized current-awareness service that is a product of this family of ISI products is the *Automatic New Structure Alert* (*ANSA*).

A third family of ISI products is based on their *Science Citation Index* (*SCI*). The SCI, by providing the unique ability to identify papers citing an earlier paper, allows users to search forward in time.

Patents. In general, patents are an underutilized source of technical information. The immense volume of patent literature may be intimidating. There are 100,000–150,000 basic (first disclosed) chemical patents issued annually throughout the world and 150,000 equivalent chemical patents. But patents are technical publications: once the reader is comfortable with their format and jargon, patents have more experimental detail than most journal articles.

Patent literature affects research and development in three areas as: a fruitful source of information regarding the economic potential of existing or near future art; a teacher of art; and a guide to application areas for marketing activities. An extension of this commentary is the ability to mine the richness of patent information data bases as intelligence tools. For instance, patent information can be used to predict trends or to analyze competitors' research and development in a given area.

Individual national patent offices, and the European Patent Office (EPO), issue bulletins disclosing the title, assignee, and abstract of approved applications. In the United States, the *Official Gazette*, published weekly on Tuesdays, is organized by classification. About 30% of the patents reported are chemical.

The American Petroleum Institute Patent data base (APIPAT), dates from 1964. Although its scope is relatively narrow, the indexing is done from a controlled vocabulary, which provides a simple chemical fragment system, and uses role indicators to search for a chemical as a starting material or as a product.

Bibliography

E.J. Saxl, *Am. Lab.* **10**(9), 31–35 (1979).

H. Skolnik, *J. Chem. Inf. Comput. Sci.* **17**, 114–121 (1977).

J.T. Maynard, *Understanding Chemical Patents: A Guide for the Inventor*, American Chemical Society, Washington, D.C., 1978.

J.W. Lotz, "Patents (Literature)," in M. Grayson and D. Eckroth, eds., *Kirk-Othmer Encyclopedia of Chemical Technology*, 3rd ed., Vol. 16, Wiley-Interscience, New York, 1981, pp. 889–945.

Derwent Publications, Ltd. covers patents issued in 24 countries, plus EPO patents and international Patent Cooperation Treaty (PCT) patents. Some of these are from slow-issue countries, which examine applications, and some are from fast-publishing ones. Derwent's special value is the comprehensiveness and timeliness of its service.

For keeping abreast of new inventions, Derwent targets two bulletins for chemists. The *Alerting Bulletins—Classified* provides very rapid notice of disclosures with brief abstracts of all basic patents and all examined equivalents. The *Profile Booklets* are a few weeks behind the *Alerting Bulletins* but include longer abstracts and cover more defined subject areas.

Chemical Abstracts does not treat patents as legal documents, so their coverage is selective. To be covered, a patent must disclose new chemical information. In some countries, *CA* covers all basics; in others, only patents issued to nationals. However, CA's chemical indexing is not restricted to products. It indexes all substances, uses, and methods of manufacture included in the claims and additional uses or substances if actually prepared.

IFI-Plenum has several levels of patent-retrieval services. The most readily accessible is the CLAIMS family of data bases. Together, these cover an extended time period, but the indexing is shallow, except for CLAIMS/UNITERM. UNITERM adds deep uniterm indexing and chemical fragmentation coding to the data base (starting 1950).

Other bibliographic data bases. There are many data bases of interest to chemists beyond those directed to chemistry. The number of data bases is overwhelming; however, the quality and variety should encourage chemists to use these resources.

On-line chemical dictionary files. On-line chemical dictionary data bases, companions to both bibliographic and numerical data bases, are not traditional dictionaries at all. They do not define chemicals, but are designed to assist chemists in identifying CAS Registry Numbers and all the names and synonyms of chemicals. They are primarily identifier or locator tools with some substructure-searching capability. They thus enhance comprehensive searching in other data bases.

Bibliography

R.E. Buntrock, *Database* **2**(1), 33–34 (1979).

S.R. Heller and G.W.A. Milne, *Database* **2**(3), 71 (1979).

J. Kasperko, *Database*, **2**(3), 24–35 (1979).

Nonbibliographic data bases. Another aspect of the information explosion is the increasing number of data bases that contain numerical information such as physical properties, product consumption, or social statistical data. These are not only fact-finding systems but also tools for manipulating facts for analyses or evaluation. Users can even add their own data and merge or compare them with that in the data base.

Very rapid developments are in progress to make the retrieval systems easier to use. This will enable the chemists to have direct access to many data bases, and will reserve the assistance of an information specialist for complex searches and training.

Business Information

To conduct and manage its business, the chemical industry requires vast amounts of primary information to be selected, analyzed, organized, and transformed for use by management. The principal areas of important nontechnical information are the political, social, legal, and economic environments; past, present, and future markets; competition; and finances, prices, taxes, and tariffs.

Bibliography

Published Data on European Industrial Markets, Industrial Aids, Ltd., London, Eng., 1975.

L. Daniells, *Business Information Sources*, Center for Business Information, University of California Press, Berkeley, Calif., 1976.

B. Lawrence, *Chemtech* **5**, 678–681 (1975).

K.D. Vernon, ed., *Use of Management and Business Literature*, Butterworths Pub., Inc., Woburn, Mass., 1976.

F.C. Pieper, *SISCIS, Subject Index to Sources of Comparative International Statistics*, CBD Research, Ltd., Beckenham, UK, 1978.

Market and product information. Many resources are available for obtaining information on markets and products. The individual countries' foreign-trade offices are a source of market information abroad. Chambers of commerce, special agencies, and offices established in most countries are another source of market information. Names and activities of these organizations appear in publications such as the *Encyclopedia of Business Information Sources*, 5th ed., P. Wasserman and co-eds., Gale Research Company, Detroit, Mich., 1983.

Bibliography

D. Degen and T.E. Miller, eds., *Findex: The Directory of Market Research Reports, Studies and Surveys*, annual, Information Clearing House, Inc., New York.

L.J. Wheeler, ed., *International Business and Foreign Trade Information Sources*, Gale Research Co., Detroit, Mich., 1968.

T. Landau, ed., *European Directory of Market Research Surveys*, Gower Press, distributed by Teakfield, Ltd., Hampshire, UK; Unipub, New York, 1976.

E. Tarnell, "Market and Marketing Research," in M. Grayson and D. Eckroth, ed., *Kirk-Othmer Encyclopedia of Chemical Technology*, 3rd ed., Vol. 14, Wiley-Interscience, New York, 1981, pp. 895–910.

SRI International offers diverse services to its subscribers from a wide range of industries. SRI developed a number of information resources specifically for the chemical, petroleum, and engineering sectors. These include the following:

Process Economics Program (PEP) report series, SRI International; evaluates the processes for the production of chemicals and petroleum products.

Chemical Economics Handbook (CEH), SRI International; provides data on production, sales, exports, imports, consumption, stocks and shipments of raw materials, primary and intermediate chemicals or end products such as fertilizers or polymers.

The *World Hydrocarbons Program*, 1979, developed by SRI International; series of reports on approximately 60 main chemicals of interest to the international petrochemical industry.

Directory of Chemical Producers—USA and *Directory of Chemical Producers—Western Europe*, 1979, SRI International, provide information on companies and products, list chemicals by manufacturer and plants by location.

Company information. Although available information on companies is diverse, the most common needs are for company names, addresses, and products. Information requested next is usually for a company's financial performance.

Company names, addresses, ownership, and products appear in directories and handbooks that have international or national coverage. Some of these are listed below.

Who Owns Whom, Dun & Bradstreet, Ltd., London, UK, lists the parent and associated companies in a series of volumes covering the world.

The Worldwide Chemical Directory, prepared by ECN Chemical Data Services, IPC Business Press Ltd., New York, annual.

Chem Sources—USA, annual, Directories Publishing Co., Inc., Flemington, N.J., and *Chem Sources—Europe*, Chemical Sources Europe, Mountain Lakes, N.J.

OPD Chemical Buyers Directory, annual, Schnell Publishing Co., Inc., New York.

Chemical Week Buyers' Guide Issue, McGraw-Hill, Inc., New York.

Directory of Chemical Producers—United States and *Directory of Chemical Producers—Western Europe*, SRI International, Menlo Park, Calif., 1979; provide company names, addresses, ownerships, subsidiaries, plants, and products and list the chemicals by manufacturing company and plants by location.

Business-information services and data bases are available through on-line services, just like the scientific data bases. Some of these are bibliographic, many have time series data. The data bases are important to chemical business, management, and corporate policy.

Data Resources, Inc. (DRI), in Boston, Mass., makes available information on the economies of several countries from data bases that contain time series from United States, foreign governmental and nongovernmental sources and organizations. DRI's services allow the user to relate external economic developments to the corporate planning of manufacturing companies in terms of future sales, cash flow, capital budgeting and investment, inventories, production, pricing, business cycles, or the implications of governmental decisions. Furthermore, DRI has developed economic models and forecasts for individual regions, ie, the United States, Canada, the European Economic Communities, and Japan.

Chase Econometrics provides access to several data bases and to econometric models with full simulation capability. These models allow the user to link the economies of the following countries: the United States, FRG, Belgium, France, Italy, the Netherlands, the United Kingdom, Spain, Sweden, Japan, Canada, Mexico, and Brazil. The system also comprises historical and forecast data bases. The historical data bases include detailed financial series and wholesale and consumer price series. The forecast data bases contain Chase Manhattan's estimates as derived from their models.

Automatic Data Pioneering, Inc. (ADP), through ADP Network Services International and Information Services, offers files including the EXSTAT data base from Exter, COMPUSTAT, which provides financial information on the United States, COPPER DEVELOPMENT DATA BASE, McGraw-Hill's METALS WEEK, and the IPC CHEMICAL DATA BASE, which covers plant and international product information on chemicals.

Stanford Research Institute's (SRI) WORLD PETROCHEMICAL DATA BASE (WP Data base) covers approximately 80 petrochemicals and derivatives is keyed to individual products and subdivided according to country. Within each country, product information is expressed in terms of manufacturing company, plants in operations during the period from 1974 to the current year, and forecast figures, captive use of product, and raw materials. The supply-and-demand section of the WP Data base provides the year-end capacities, production years, imports and exports, and consumption figures. Ownership relations and subsidiary names are searchable too.

A few files are of particular interest to chemical business and marketing. References to published articles, and often full abstracts, on the chemical industry's activities and economic outlook may be found in the American Petroleum Institute's APILIT and P/E NEWS, which provides technical-economic information for the petrochemical sector. The chemical industries are also covered by files such as Predicasts' PTS PROMPT and PTS F & S INDEXES which supplement each other. In addition, Predicasts offers the EIS INDUSTRIAL PLANTS file which is an on-line directory of manufacturing industries providing for each plant the name and address of the operating company and its parent company, the principal product, the market share, total sales, the employment-size class, and size of shipments.

MARGARET H. GRAHAM
Exxon Research and Engineering Company

ALEXIS B. LAMY
Essochem Europe, Inc.

BARBARA LAWRENCE
Exxon Corporation

LORRAINE Y. STROUMTSOS
Exxon Research and Engineering Company

INFRARED DETECTORS. See Infrared technology; Photodetectors.

INFRARED TECHNOLOGY

The electromagnetic spectrum is divided arbitrarily into a number of wavelength regions, called bands, distinguished by the methods utilized to produce and detect the radiation. There is no fundamental difference between radiation in the different bands of the electromagnetic spectrum; they are all governed by the same laws and the only differences are those due to the differences in the wavelength.

Thermography makes use of the infrared spectral band. At the short-wavelength end, the boundary lies at the limit of visual perception, in the deep red. At the long-wavelength end, it merges with the microwave radio wavelengths, in the millimeter range.

The infrared bands commonly are subdivided further into four lesser bands, the boundaries of which also are chosen arbitrarily. They include: the near infrared (0.75–3 μm), the short infrared (3–6 μm), the long infrared (6–15 μm), and the far infrared (15–1000 μm).

Blackbody Radiation

A blackbody is defined as an object that absorbs all radiation that impinges upon it at any wavelength. The apparent misnomer "black" related to an object with the ability of absorbing and emitting radiation is explained by Kirchoff's law, which states that a body capable of absorbing all radiation at any wavelength is equally capable of emitting all radiation.

The construction of a blackbody source is very simple, in principle. The radiative characteristics of an aperture in an isothermal cavity made of an opaque absorbing material represents almost exactly the property of a blackbody.

An isothermal cavity with a suitable heater is termed a cavity radiator. Such a cavity heated to a uniform temperature generates blackbody radiation, the characteristics of which are Plankian, ie, determined solely by the temperature of the cavity. Such cavity radiators are commonly utilized as sources of radiation in temperature reference standards in the laboratory for calibrating thermal-measurement instruments (see Temperature measurement).

Ir Detectors

An ir detector is a converter that absorbs ir energy and converts it to a signal, usually an electrical voltage or current. There are two principal types: thermal detectors and photon detectors. Thermal detectors have been conceived on the notion of the temperature rise produced in an absorbing receiver, such as in the thermopile, pneumatic (Golay) cell, the thermocouple, the bolometer, and the new pyroelectric (capacitor) detector. The most important thermal detector today is the thermistor bolometer, which utilizes the change in resistance (or conductivity) of a semiconductor (qv) film when it is heated by the radiation. Typical of thermal detectors is "flat" spectral response. If they have been properly blackened, the output signal remains practically constant over a very wide range of wavelengths. The main drawback of most thermal detectors is the comparatively slow response to radiation variations, due to the thermal processes involved. The pyroelectric thermal detector, however, has relatively fast response owing to its use of the ferroelectric effect of certain crystals (see Ferroelectrics).

Photodetectors are sensitive to changes in the absorbed number of infrared photons and are not actually heated by the incident radiation as are thermal detectors. These detectors respond with changes in voltage (photovoltaic) or in current (photoconductive) depending on the type of element and circuitry used. The most common types are indium antimonide, lead selenide and mercury cadmium–telluride. These detectors generally are much more sensitive and have much faster response than thermal detectors.

A widely accepted figure of merit for expressing sensitivity of ir detectors is a quantity called detectivity, represented by the symbol D^*. D^* is a normalized figure of merit that is particularly convenient for comparing the performance of detectors having different electric bandwidths. D^* assumes higher values as detector sensitivity improves.

Measurement Instruments

Some confusion has existed concerning the term infrared photography, as contrasted with thermography (see Color photography). The distinction is one of wavelength. Conventional infrared film emulsions are sensitive to wavelengths no longer than 1.2 μm. For this reason, the wavelength span 0.75–1.2 μm is the photographic infrared spectrum. Beyond the 2 μm wavelength lies the so-called thermal infrared. Thus, infrared photography exploits the differences in the absorptive and emissive properties of surfaces, and depends upon the reflection of very short infrared wavelengths generated by outside sources such as the sun, or lamps, that are much hotter than the object.

One of the most common instruments used for measuring thermal energy is the point radiometer, a relatively simple device which most often uses a thermal detector, such as a thermopile, or a pyroelectric detector. Generally, there are two primary types of point radiometers, a hand-held device which is battery powered and normally in a pistol shape (see Fig. 1), and a second type intended as a fixed unit which is often permanently mounted in some type of process or manufacturing facility.

A somewhat more complex system involves scanning the object with a refractory or reflective electromechanical scanning system and displaying the resulting graph or image on an oscilloscope screen. This information may also be fed into a computer. Of the two types of line scanning, traversing radiometer and electro-optical scanner, the first is by far the most common. These systems are used in diverse manufacturing fields, but are very commonly found in monitoring paper machines, cement kilns, rubber processing and steel mills, and plastic-film extruders.

Although only recently introduced, nonmeasurement imaging systems have found many uses in the processing industry. These systems are most often very simple and less expensive than measuring systems and offer the user a clearly portable, rugged, and practical alternative for thermal imaging.

The number of types of simple imaging systems available fall into two categories: The first, still considered by many to be in the development stage, is based on the use of the pyroelectric vidicon tube (PEV). Similar to a television vidicon, the PEV is sensitive to infrared radiation in the 8–14 μm region. These systems offer high scanning rates with full TV compatibility. Their disadvantage is that the detector must receive changing thermal information in order to generate an electric signal. Thus, the unit must be panned continuously in order to maintain an image. An alternative means of operating such systems is to chop the field of view; however, this complicates the system and reduces its reliability, which already is normally limited to less than 2000 h.

The second type is based on an electro-optical-mechanical scanning system using multielement detectors. Such systems often offer a highly

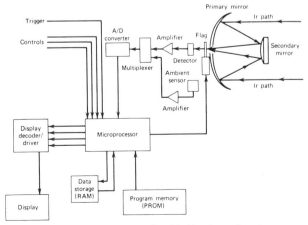

Figure 1. Schematic of a hand-held point radiometer.

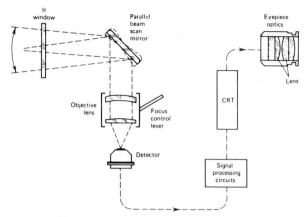

Figure 2. A nonmeasuring imaging system.

reliable package, often without the need for liquid nitrogen cooling. The most modern of these devices utilize thermoelectric cooling which, when properly constructed, allows use of the system at any time and any place as long as the batteries are properly charged (see Fig. 2).

Uses

Electrical power supply systems. The use of thermal measurement instruments for electrical inspection is well established as a relatively simple, fast, and direct application. As electrical components corrode or deteriorate (age) the electrical properties also change, mainly in electrical resistance. These changes are usually evident as increased surface temperature. The component's temperatures, differentiated with respect to ambient temperature, can indicate a partial failure. Most components deteriorate gradually before complete failure. Consequently, electrical inspection can be performed at infrequent intervals, allowing for planned and regular maintenance schedules.

Refractory and insulation conditions. Inspection here is based on the theory that a uniform internal temperature exists within a vessel and that the resulting exterior-surface temperature is a direct function of the heat conduction through the insulating medium and the external wall (see Refractories; Insulation, thermal).

Product flow. In most product-flow inspections, thermal energy is used as an indicator of thermally related defective conditions, eg, leakages, blockages, thinning conditions, corrosion and erosion. In each case, however, an analysis must be made to determine the correlation between the thermal information and the defect.

Clifton Warren
AGA Infrared Systems AB

T.A. Smith, F.E. Jones, and R.P. Chasmar, *The Detection and Measurement of Infrared Radiation*, Oxford University Press, London, 1968.

W.L. Wolf and G. Zissis, *The Infrared Handbook*, Office of Naval Research, Washington, D.C., and Environmental Research Institute of Michigan, Ann Arbor, Mich., 1978.

INITIATORS

Initiators are chemical substances or energy sources that are used to initiate a chemical reaction and that are consumed during the initiation process. This is in contrast to catalysts, which are used in small quantities to either start or speed up a reaction but which are not consumed in the process, thereby allowing the catalyst to be reused (see Catalysis).

The amount of chemical initiator required for reaction depends upon the type of chemical reaction and can vary from stoichiometric or greater quantities in certain nonchain-propagating reactions to very small quantities in chain-propagating reactions (eg, polymerizations) (see Polymerization mechanisms and processes). A chain propagating reaction generally involves three steps: initiation, propagation, and termina-

tion. For example, in a vinyl monomer polymerization, these steps can be represented by the following equations:

$$\textit{Initiation}: \text{I}^* + \text{M} \rightarrow \text{I}{-}\text{M}^*$$
$$\textit{Propagation}: \text{I}{-}\text{M}^* + n\,\text{M} \rightarrow \text{I}{-}\text{M}_n\text{M}^*$$
$$\textit{Termination}: \text{I}{-}\text{M}_n\text{M}^* \rightarrow \text{dead polymer}\left(\text{eg, I}{-}\text{M}_{(n+1)}\text{X}\right)$$

where I* is the active-initiator species, M is the vinyl monomer, M* is the propagating active-monomer species, and X is some termination function that results from the reaction of M* with a terminating species.

Although initiators are used widely in the polymer industry, they are also employed for initiating a variety of other chemical reactions, eg oxidation and autoxidation, chlorination, bromination, and nonpolymeric addition to double bonds. The same types of initiators generally are used for initiating polymerizations and nonpolymer chemical reactions.

Free-Radical Initiators

Since active initiator free-radical species ($\text{I}^* = \text{I}\cdot$) are reactive intermediates having short lifetimes [half-lives ($t_{1/2}$) of less than 10^{-3} s], they are generated during the initiator process from a precursor. (Free-radical initiator is used synonymously with the precursor or free-radical generator rather than with the free-radical species). There are three general ways in which radicals can be produced: thermal bond homolysis, one-electron redox reactions, and irradiation processes. Commercial free-radical generators (or initiators) are, primarily, organic peroxides and aliphatic azo compounds.

Organic peroxides. Organic peroxides have the general structure, ROOR′ or ROOH, and decompose thermally by the initial cleavage of the oxygen–oxygen bond to produce two free radicals:

$$\text{ROOR}' \xrightarrow{\Delta} \text{RO}\cdot + \cdot\text{OR}'$$

Certain organic peroxides also can be decomposed by specific promoters or activators. Such decompositions occur well below the peroxides' normal thermal decomposition temperatures and usually generate one free radical instead of two in a redox reaction:

$$\text{ROOH} + \text{Fe}^{2+} \rightarrow \text{RO}\cdot + \text{OH}^- + \text{Fe}^{3+}$$

There are more than 40 different organic peroxides in more than 80 formulation that are produced as free-radical initiators primarily for the polymer and resin industries (see Peroxides and peroxy compounds, organic). Certain organic peroxides are susceptible to radical-induced decomposition:

$$\text{R}'\cdot + \text{ROOR} \rightarrow \text{ROR}' + \text{RO}\cdot$$

Radical-induced decompositions of peroxides result in inefficiency in radical production since the peroxide decomposes without adding more radicals to the system. Such decompositions generally do not occur in vinyl monomer polymerizations because the monomer quickly scavenges the initially generated radicals.

Azo compounds. Azo compounds commercially available as initiators are symmetric (**1**) and unsymmetrical (**2**) azonitriles:

R′   R′   CH₃   R

$$\underset{\text{CN}}{\text{RC}}{-}\text{N}{=\!=}\text{N}{-}\underset{\text{CN}}{\overset{\text{R}'}{\text{CR}}} \qquad \underset{\text{CH}_3}{\text{CH}_3\text{C}}{-}\text{N}{=\!=}\text{N}{-}\underset{\text{CN}}{\overset{\text{R}}{\text{CR}'}}$$

(**1**)   (**2**)

The symmetric azonitriles (**1**) are solids with somewhat limited solubility and the unsymmetrical *tert*-butylazonitriles (**2**) are liquids or low melting solids that are completely miscible in all common organic solvents, including petroleum hydrocarbons. Azonitriles are not susceptible to radical-induced decompositions and their decomposition rates are little affected by the environment. In contrast to most peroxides, azonitrile decomposition rates show only minor solvent effects and are not affected by transition metals, acids, bases and other conaminants. Thus azonitrile decomposition rates are predictable. They are ideally suited as thermal initiators for curing resins that contain a variety of extraneous materials without affecting cure rates and displaying exceptional shelf (or pot) lives at ambient storage temperatures. The symmetrical azonitriles are less stable than the corresponding unsymmetrical azonitriles, eg, 2,2′-azobis(2-methylpropionitrile) has a 10-h half-life at 64°C, whereas the analogous 2-(*tert*-butylazo)-2-methylpropionitrile has a 10-h half-life at 79°C. Azo initiators decompose thermally by cleavage of two carbon–nitrogen bonds to form two alkyl radicals and nitrogen:

$$RN{=}NR' \rightarrow R\cdot + N_2 + R\cdot$$

The alkyl radicals generated from commercial azo initiators are all tertiary alkyl radicals which are relatively more stable than most radicals generated from peroxide initiators. Thus, when these azonitriles are used as initiators for vinyl monomer polymerizations, the primary initiator radicals generally do not attack growing polymer backbones; such attack can occur with peroxide initiators. This results in more linear polymers with less long-chain branching.

Initiation through radiation. Free-radical reactions and polymerizations (primarily curing systems) have been initiated by high energy radiation, ie, x-rays, gamma rays, and beta rays. These radiation processes fragment chemical bonds into ions as well as free radicals. Ionizing radiation is often used in these processes (see Radiation, curing). Initiating free-radical reactions with uv radiation and photoinitiators is widely used in industrial processes (see Photochemical technology). Photoinitiators are chemical compounds that generate initiating free radicals when subjected to uv radiation.

Ionic Initiators

Ionic initiators are compounds that are capable of initiating ionic polymerization reactions.

Anionic. Alkali metals and aromatic complexes of alkali metals have been used as initiators in anionic polymerization (see Elastomers, synthetic). The mechanism of initiation involves an electron-transfer step (as depicted below), which results in the formation of a radical anion.

$$Li + CH_2{=}CHCH{=}CH_2 \rightarrow \dot{C}H_2CH{=}CH\bar{C}H_2\ Li^+$$

$$2\ \dot{C}H_2CH{=}CH\bar{C}H_2\ Li^+ \rightarrow Li^+\bar{C}H_2CH{=}CHCH_2CH_2CH{=}CH\bar{C}H_2\ Li^+$$

This can subsequently couple to form a dianion, which is capable of propagating the chain. The radical anion also can participate in a second electron-transfer step (as shown below) and still form a dianion species:

$$Li + \dot{C}H_2CH{=}CH\bar{C}H_2\ Li^+ \rightarrow Li^+\bar{C}H_2CH{=}CH\bar{C}H_2\ Li^+$$

The polymerization system, in general, and the alkali metal, in particular, influence the magnitude of these individual reactions.

Various organolithium compounds are commonly used as initiators in anionic polymerization. Their solubility in a variety of solvents has contributed to their versatility.

Cationic. Cationic polymerization can be initiated by using a variety of compounds. It is preferred to refer to them as catalysts, since a co-catalyst generally is necessary to initiate the polymerization reaction. The initiation mechanism is quite complex and is influenced strongly by the individual polymerization system. Catalysts that are used in cationic polymerization can be classified as follows: protonic or Brønsted acids; aprotonic acids, eg, the various Lewis acids and Friedel-Crafts halides; stable carbonium ion salts, eg, antimony and aminium salts; and substances capable of generating cations—iodine is by far the most well-known compound in this category.

Health and Safety

Azo and organic peroxide initiators in pure and highly concentrated states are thermally sensitive and can decompose extremely rapidly when overheated.

CHESTER S. SHEPPARD
VASANTH KAMATH
Pennwalt Corporation

D. Swern, ed., *Organic Peroxides*, Vol. I, Wiley-Interscience, New York, 1970.
C.S. Sheppard and V.R. Kamath, *Polym. Eng. Sci.* **19**, 597 (1979).

R. Zand in N.M. Bikales, ed., *Encyclopedia of Polymer Science and Technology*, Vol. 2, Interscience Publishers, a division of John Wiley & Sons, Inc., New York, 1965, pp. 278–295.
S.P. Pappas, ed., *UV Curing: Science and Technology*, Technology Marketing Corporation, Stamford, Conn., 1978, Chapt. 1.

INKS

Printing Inks — General Considerations

Printing ink is a mixture of coloring matter dispersed or dissolved in a vehicle or carrier, which forms a fluid or paste that can be printed on a substrate and dried. The colorants used are generally pigments, toners, and dyes, or combination of these materials, which are selected to provide color contrast with the background on which the ink is printed. The vehicle used acts as a carrier for the colorant during the printing operation, and, in most cases, serves to bind the colorant to the substrate (see also Reprography; Printing processes).

Printing inks are applied in thin films on many substrates, such as paper, paperboard, metal sheets and metallic foil, plastic films, and molded plastic articles, textiles, and glass (see Film and sheeting materials; Paper). Printing inks can be designed to have decorative, protective, or communicative functions or combinations of these.

There are four classes of printing ink (letterpress, lithographic, flexographic, and rotagravure inks) which vary considerably in physical appearance, composition, method of application, and drying mechanism. Four properties of inks are of cardinal importance—drying, printability (which is largely a function of the rheology of the ink), color, and use properties (those considerations that determine how printed substrates function throughout all processing and usage from the time of printing throughout the useful life of the printed product).

Drying. Drying is accomplished by one or more of the following physical or chemical mechanisms: absorption, evaporation, precipitation, oxidation, polymerization, cold setting, gelation, and radiation curing.

Rheology. The rheology of inks is of the greatest importance. In a Newtonian liquid, any stress produces a flow, and the rate of flow is proportional to the stress. But inks are generally non-Newtonian, and the most common terms used by the ink maker to describe rheology are viscosity (resistance to flow); yield value (point at which a liquid starts to flow under stress); thixotropy (decreasing viscosity with increasing agitation); and dilatancy (opposite of thixotropy). Mayonnaise is an example of a material having a high yield value but a low viscosity.

The ink-distribution systems of flexo and gravure printing presses are very simple and do not provide the means to distribute and level highly viscous inks. Therefore, the viscosity must be low, on the order of 50–100 mPa·s (= cP). Yield value must be low also to permit pumping of ink from reservoir to fountain. Thixotropy should also be avoided for the same reason.

Letterpress and offset inks can vary in viscosity from under 500 mPa·s for a letterpress-type news ink to over 500 Pa·s (5000 P) for special litho-ink formulation. The viscous nature of letterpress and litho inks has forced press designers to use a multitude of rollers in the ink distribution unit to ensure uniform and adequate transfer of ink to the printing plate.

Color. Color is the third common denominator for all types of inks formulations and may very well be the most important one because it has such tremendous psychological impact on the ultimate consumer. Color appears in three different dimensions referred to as hue, saturation or chroma, and value (see Color).

Pigments are not only used for color but also for other physical properties such as bulk, opacity, specific gravity, viscosity, yield value, and printing qualities. Different pigments of the same color have different permanency to light, heat, chemicals, and bleed or staining in water, oil, alcohol, fat, grease, acid, or alkali. Five dye families of interest to ink manufacturers are azo, triphenylmethane, anthraquinone, vat, and phthalocyanine (see Dyes and dye intermediates; Dyes—application).

Other Ingredients. Other ingredients are (*1*) driers (qv), in the form of soaps of cobalt, manganese and lead formed with organic acids such as linolenic, naphthenic and octanoic acids, which catalyze oxidation of

Table 1. Letterpress, Lithographic, and Dry-Offset Inks[a]

Ink type and use	Ink class, printing method	Application, stock feed	Ink consistency	Approx printing speed
news (papers and comics)	LP, litho	web-fed	liquid medium paste	300–450 m/min
publications (magazines, periodicals)	LP, litho	web-fed, sheet-fed	soft–medium pastes, medium pastes	300–370 m/min, 3500–7500 sph
commercial (display, document, pamphlet, label)	LP, litho	sheet-fed	medium pastes	3500–7500 sph
business documents and forms (duplicator, narrow web)	litho	sheet-fed, web-fed	medium pastes	2000–4000 sph, 10–180 m/min
folding cartons (food, soap, drug, liquid, parts, detergent)	LP, litho, DO	sheet-fed	medium–heavy pastes	3500–8000 sph
containers and boxes (corrugated, linear)	LP	sheet-fed	medium–heavy pastes	2500–5000 sph
books (school, directories, religious, library)	LP, litho, DO	sheet-fed, web-fed	soft–medium pastes	2000–7000 sph, 120–240 m/min
bags (paper, cloth, plastic)	LP, DO	web-fed	medium pastes	80–180 m/min
wrappers (bread, box, produce)	LP	web-fed	medium pastes	120–240 m/min
metal containers (beverage and food cans, pails, drums, displays)	litho, DO	sheet-fed, pr. can	medium–heavy pastes	2000–5000 sph, 100–500 cpm
plastic (cups, bottles, displays, book covers)	DO, litho	pr. can, sheet-fed	medium–heavy pastes	100–200 cpm, 2000–4000 sph

[a]Abbreviations: LP, letterpress; DO, dry offset; pen, penetration; pr can, preformed can; sph, sheets per hour; cpm, containers per minute.

drying oils, and are used in inks that dry by oxidation; (2) waxes (qv), mainly dispersions of polyethylenes, hydrocarbons, or vegetable and animal waxes in a vehicle system used to impart slip and scruff-resistance to ink films (Lately, polyolefinic waxes are also used directly in micropulverized form.); (3) antioxidants (qv), such as the universally-accepted eugenol and ionol, which retard premature oxidation of inks on the press at 1% or lower concentrations; and (4) miscellaneous additives such as lubricants, surface-energy reducing agents, thickeners, gellants, defoamers, wetting agents and shorteners.

Letterpress, Dry Offset, and Lithographic (Paste) Inks

Table 1 summarizes the properties of inks of these three classes, divided among eleven use classifications.

Radiation-Cure Inks

In response to air-pollution regulations, the shortage or high cost of natural gas, and because of a general desire to conserve energy, new ink formulations/drying systems have emerged that are based on radiant energy: ultraviolet, electron-beam, and infrared radiation. Lithographic and letterpress inks are now available for web-fed and sheet-fed presses that will cure–dry instantly when exposed to radiators of the required intensity.

Flexographic and Rotogravure Inks

Although these inks have many elements in common with inks used for other printing processes, they constitute a distinct class because of the specific characteristics of their printing processes, their application and ingredients, and methods of manufacture. The colorants, pigments, and dyes, except for their surface treatment, are the same as those used for letterpress, offset, and other methods. However, the flexographic and rotogravure inks differ in that they are of very low viscosity, they almost always dry by evaporation of highly volatile solvents (eg, low boiling point alcohols, esters, aliphatic and aromatic hydrocarbons, ketones, and water), and generally they are produced by different manufacturing processes from those used for other inks.

Manufacture. Ink manufacture is related to two operations—vehicle preparation and colorant dispersion. Vehicle preparation can be as basic as polymerization of resins or as simple as cold-dissolving of vehicle solids in appropriate solvents. The equipment includes autoclaves for polymerization reactions and high speed mixers for simple dissolving.

Pigment dispersion usually is done in shot or ball mills which lend themselves to volatile fluid formulations. Pebble mills using porcelain balls and linings are used for white and light colors because steel ball mills, although more effective, cause discoloration. Color concentrates and pigment resin dispersions called "chips" are made in dough mixers and two-roll rubber mills. These high shear methods often result in better dispersion and consequently higher gloss than is achieved with shot or ball mills.

Flexographic inks. Flexographic inks can be classified into five general composition categories: alcohol-dilutable inks containing nitrocellulose as a main constituent of the vehicle; polyamide inks; dye inks; acrylic inks; and water-based inks.

Rotogravure inks. Since there is no rubber contact with solvents in gravure-ink formulations, it is permissible to us solvents such as ketones and aromatic hydrocarbons which cannot be tolerated in flexo inks. This provides the gravure-ink formulator with much greater latitude with regard to binder selection and constitutes the principal difference between gravure and flexo inks. In other respects, the compositions used are similar.

Health and Safety

OSHA has enforced strict, and in some cases, prohibitive control of materials such as lead and chromium salts. The FDA, although not controlling inks that are not in direct contact with food, influences formulations used on meat and poultry packages. Listing of all non-vehicle ingredients under Title 21, Code of Federal Regulations, is a specific requirement. Solvent recovery via carbon absorption and incineration equipment is currently used to satisfy state and Federal emission requirements.

BOHDAN V. BURACHINSKY
HUGH DUNN
JAMES K. ELY
United Technologies Inmont

E.A. Apps, ed., *Ink Technology for Printers and Students*, 3 vols., Chemical Publishing Co., Inc., New York, 1964.

H.J. Wolfe, *Printing and Litho Inks*, 6th ed., MacNair-Dorland Co., New York, 1967.

F.A. Askew, ed., *Printing Ink Manual*, 2nd ed., W. Heffer & Sons, Ltd., Cambridge, Eng. 1969.

T.C. Patton, ed., *Pigment Handbook*, 3 vols., John Wiley & Sons, Inc., New York, 1973.

INORGANIC HIGH POLYMERS

Inorganic polymers include solid metals, ionic crystals, ceramics, silica, silicates, etc (see Ceramics; Silica, etc). In addition to these inorganic polymeric materials, there are inorganic polymers that exhibit elastomeric and plastic properties. The polymer backbones consist essentially of inorganic elements and occasionally have organic side groups to modify properties. Polysiloxanes are prime examples of inorganic high polymers. They possess a unique combination of high temperature stability and excellent low temperature elastomeric properties. Siloxanes are commercially important inorganic high polymers (see Silicon compounds, silicones).

Polyphosphazenes

Polyphosphazenes are polymers having alternating phosphorus and nitrogen atoms in the polymer backbone:

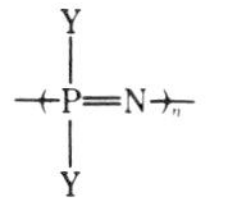

$$\begin{matrix} & Y & \\ & | & \\ \left(\!\!\!\!\!\!\!\!\!\!\!\!+ \right. & \!\!P\!=\!N & \left.\!\! +\!\!\!\!\!\!\!\right)_n \\ & | & \\ & Y & \end{matrix}$$

where Y can be halogen, pseudo halogen, or alkyl, aryl, alkoxy, aryloxy, arylamino, or alkylamino groups (see Phosphorus compounds; Fluorine compounds, organic, fluoroalkoxyphosphazenes).

Physical properties. The two main transitions in polymers are the glass–rubber transition (T_g), and the crystalline melting point (T_m). The T_g is the most important parameter of an amorphous polymer because it determines whether the material will be a hard solid or an elastomer at specific use-temperature ranges and at what temperature its behavior pattern changes. The dichloro- and difluorophosphazene polymers exhibit low glass-transition temperatures. Copolymers of trifluoroethoxy and heptafluorobutoxy (50:50) groups are elastomeric a glass transition of $-77°C$.

Thermogravimetric analysis (tga) of several polyphosphazenes exhibit the onset of weight loss at temperatures above 300°C.

Manufacture. Poly(dichlorophosphazene) is prepared by the thermal polymerization of hexachlorocyclotriphosphazene:

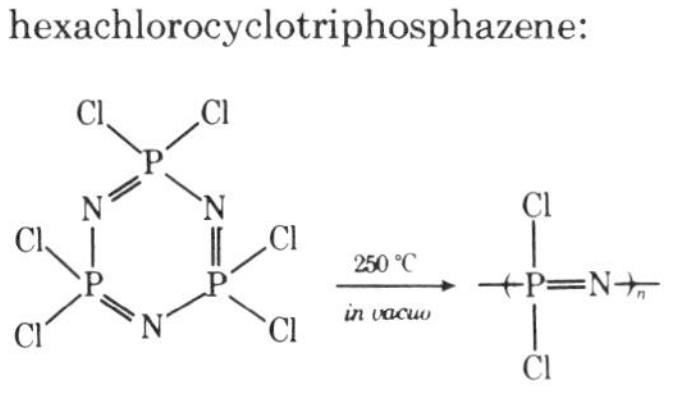

Generally a cross-linked rubbery material is obtained. However, under carefully controlled reaction conditions and purification of trimer, the linear polymer can be prepared. It is important to terminate the polymerization reaction at an intermediate stage (250°C for 24–48 h) in order to prevent extensive branching and cross-linking.

Alkoxyphosphazene and fluoroalkoxyphosphazene polymers are prepared by treating poly(dichlorophosphazene) with a slight excess of sodium alkoxide in refluxing benzene–tetrahydrofuran (60–80°C).

Properties of vulcanizates. The polyphosphazene copolymers and terpolymers containing equal amounts of trifluoroethoxy and octafluoropentoxy substituents are amorphous a low T_g ($-67°C$) and have been studied in some detail. The terpolymer contains a small amount of a substituent that facilitates vulcanization.

Typical elastomer formulations consist of three primary components: reinforcing and extending fillers, a high activity magnesium oxide for efficient cure and enhanced thermal stability, and a peroxide curing agent.

Uses. The unusual resistance to oils, gasoline, jet fuel, and hydraulic fluids (qv) as well as the low temperature flexibility of fluorophosphazene elastomers (see Elastomers, synthetic, fluorinated) have led to utility in the preparation of O-rings, gaskets, and hydrocarbon-fuel hoses.

Poly(carborane–siloxanes)

Carborane–siloxanes are a family of polymers that have a linear structure:

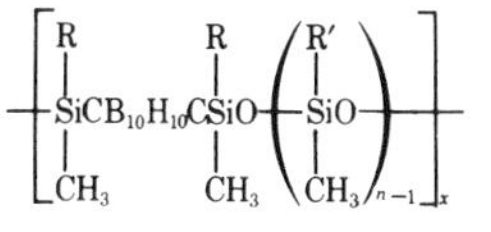

where R and R′ can be methyl or fluoroalkyl, and in addition R′ can be phenyl. The term carborane is commonly used in a generic sense to describe compounds composed of boron, hydrogen, and carbon whose molecular geometrics are polyhedra or polyhedral fragments (see Boron compounds, boron hydrides, carboranes, and their metalloderivatives.) Several families of carboranes exist with the general formulas $CB_nH_{(n+2)}$, $CB_nH_{(n+4)}$, $C_2B_nH_{(n+2)}$, etc. Of these families, the neutral, closed, polyhedral $C_2B_nH_{(n+2)}$ ($n = 5$ and 10) species have been used in the preparation of carborane–siloxane polymers.

Physical properties. Transition temperatures: for the homologous series of carborane–dimethylsiloxanes [(**1**) R = R′ = CH_3, $n = 1$ through 5] the glass-transition temperature decreases with increasing siloxane content (ie, greater n). The D_1 and D_2 polymers have crystalline melting points. This crystalline phase adversely affects elastomeric properties. In the case of the D_2 polymer, this crystallinity can be eliminated by the incorporation of phenyl groups on the polymer backbone (R′ = C_6H_5).

Thermogravimetric analyses (tga) in an inert atmosphere reveal that a rapid weight loss does not occur until heating above 400°C for carborane–siloxanes. For the homologous series of carborane–dimethylsiloxanes, the stability of the polymers increases with decreasing siloxane content per repeat unit.

Manufacture. Different structures of carborane–siloxane polymers are obtainable via several synthetic routes. The D_3 and D_5 polymers can be prepared by hydrolysis condensation of a carborane-based siloxane (eq. 1.):

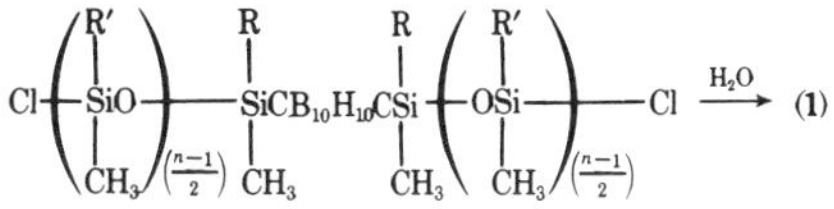

$$(1)$$

The D_4, D_5, and D_6 polymers are prepared by the cohydrolysis condensation shown in equation 2:

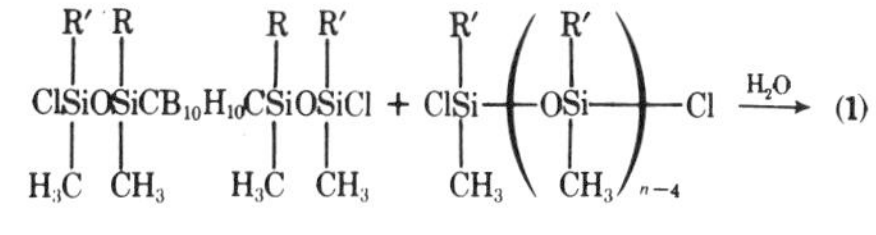

$$(2)$$

The D_2 polymer can be prepared directly from 1,7-carborane in a facile, one-pot synthesis. Compound (**2**) is first converted to its dilithio salt (**3**) by reaction with *n*-butyllithium (eq. 3):

$$HCB_{10}H_{10}CH + 2\,CH_3(CH_2)_3Li \rightarrow LiCB_{10}H_{10}CLi + CH_3(CH_2)_2CH_3 \qquad (3)$$
$$\quad\ (\mathbf{2}) \qquad\qquad\qquad\qquad\qquad\quad (\mathbf{3})$$

Properties of vulcanizates. The amorphous polymers have useful elastomeric properties and can be formulated with fillers and other additives and vulcanized using standard silicone technology.

Uses. The high temperature capabilities of these polymers have utility as liquid phases in gas chromatography. Carborane–siloxanes have been fabricated into O-ring gaskets, and wire coatings that are capable of performing at temperatures exceeding 300°C (see Heat-resistant polymers).

Poly(Sulfur Nitride)

Poly(sulfur nitride), or polythiazyl, $(SN)_x$, is composed of nonmetallic elements. However, it exhibits metallic electronic conducting properties. At low temperatures, it becomes superconducting (see Polymers, conductive; Superconducting materials).

Properties. Polymeric sulfur nitride is a crystalline, fibrous material that is soft and malleable and can be flattened readily by mild pressure. Its density is 2.30 g/cm³.

Electrical properties. The room temperature conductivity of $(SN)_x$ crystals parallel to the fibers can be as high as ca $3700/\Omega \cdot cm$. The conductivity increases over 200 times upon lowering the temperature to 4.2 K. At 0.26 K, the crystals become superconducting.

Manufacture. Poly(sulfur nitride) has been prepared by thermal decomposition of S_4N_4 into S_2N_2 which is transformed to $(SN)_x$ by solid-state polymerization.

Uses. Polymeric sulfur nitride has been used as an electrode material in aqueous solution.

Polysilanes

Polysilanes or polysilylenes are polymers composed of disubstituted silicon atoms:

$$\left(\begin{array}{c} R \\ | \\ Si \\ | \\ R \end{array} \right)_n$$

Physical properties. Compared to polysiloxanes, polysilanes are less stable and more rigid. The relatively low bond strength of the silicon–silicon bond results in lower thermal stability. Hydrogen substitution on silicon results in increased oxidative sensitivity. For example, $(SiH_2)_n$ is spontaneously flammable at room temperature in air.

Manufacture. The parent polysilane, $Si_nH_{(2n+2)}$, has been prepared from the reaction of acidic reagents with metallic silicides in an inert atmosphere, usually in poor yields.

Uses. Polysilanes can be pyrolyzed to form microcrystalline beta-silicone carbide fibers (see Carbides).

EDWARD N. PETERS
General Electric Company

H.R. Allcock, *Angew. Chem. Int. Ed. Engl.* **16**, 147 (1977).

E.N. Peters, *J. Macromol. Sci. Rev. Macromol. Chem.* **C17**, 173 (1979).

M. Akhtar, C.K. Chiang, M.J. Cohen, A.J. Heeger, J. Kleppinger, A.G. MacDiarmid, J. Milliken, M.J. Moran, and D.L. Peebles in C.E. Carraher, J.E. Sheats, and C.U. Pittman, eds., *Organometallic Polymers*, Academic Press, New York, 1978, p. 301.

J.P. Wesson and T.C. Williams, *J. Polym. Sci. Polym. Chem. Ed.* **17**, 2833 (1979).

INORGANIC REFRACTORY FIBERS.   See Refractory fibers.

INOSITOL.   See Vitamins.

INSECT-CONTROL TECHNOLOGY

Insects are the most numerous of living organisms and nearly one million (10^6) described species constitute approximately 70% of all animal species. Of these, about 1% are considered significant pests; they attack humans and/or their domestic animals; transmit human, animal, and plant diseases; destroy structures; and compete for available supplies of food and fibers. In the United States, at least 600 species of insects are important pests. Estimates suggest that the total annual loss to agriculture in the U.S. is ca 10% of production, and worldwide agricultural losses are about 14% of production.

Integrated pest management (IPM). The widespread use of chemical insecticides since 1946 has resulted in increasing difficulties in practical pest control. The difficulties include hereditary selection of races of more than 400 insect pests that are resistant to one or more classes of insecticides and some to every available material; resurgences of pests and outbreaks of secondary pests that result from elimination of natural enemies by the use of broad-spectrum biocides; adverse human-health effects from injudicious use of highly toxic insecticides; pollution of virtually every segment of the environment by persistent, lipophilic organochlorines; and exponentially increasing costs of new insecticides. Therefore, an ecologically based insect-control strategy relying on multiple control interventions with minimal disturbances to the ecosystem provides the only effective way in which to deal with serious insect control problems in agricultural and public health. Such a general strategy is under worldwide development as integrated pest management (IPM), which relies heavily on protection and conservation of the natural enemies, parasites, predators, and diseases that regulate or balance populations of insect pests. Thus, IPM rejects the regular or preventive use of broad-spectrum insecticides and the general philosophy of eradication of insect pest species which has proved unworkable. IPM programs are based upon the concept of the economic threshold, or that level of increasing insect-pest population at which control measures should be applied to prevent economic injury, eg, where loss caused by the pest equals the cost of control. Under this concept, many conventional insecticide applications become unnecessary and often damaging because of their needless destruction of natural enemies.

IPM represents a highly skilled technology that almost always reduces the need for chemical insecticides by $50- \geq 90\%$ over conventional spray programs. By encouraging natural enemies—parasites, predators, and diseases, IPM practices greatly decrease the rigor of natural selection by pesticides which is responsible for resistance. These natural enemies also prevent the great fluctuations and surges in insect-pest populations observed after injudicious use of broad-spectrum insecticides. Under the IPM concept, insecticides generally are to be used when other practices are inadequate and the pest population reaches the economic threshold. In order to make the concept effective, insecticides must be used in as selective a manner as possible, with minimal disturbances to all other elements of the ecosystem. In IPM systems, insecticides with resistant host plants, biological control, cultural control, attractants, and repellents are the hardware for the systems approach to regulation of pest densities.

Insecticides

Insecticides are chemicals that are used to control damage or annoyance from insects. Generally, control is achieved by poisoning the insects by oral ingestion of stomach poisons, contact poisons that penetrate through the cuticle, or fumigants that penetrate through the respiratory system. Ancillary chemicals are also employed in insect control and include attractants and repellents, which influence insect behavior, and chemosterilants, which influence reproduction.

Insecticide formulation. The successful employment of any insecticide depends upon its proper formulation in a preparation that can be applied for insect control with safety to the applicator, animals, and plants. Insecticides are commonly formulated as dusts, water dispersions, emulsions, and solutions. The preparation and use of such formulations involves accessory agents such as dust carriers, solvents, emulsifiers, wetting and dispersing agents, stickers, and deodorants or masking agents.

Application. The usefulness of any insecticide is substantially dependent upon its proper application, and this is determined by the properties of the insecticide, the habits of the pest to be controlled, and the site of the application to be made. The three general methods of applying insecticides are spraying (with water or oil as the principal carrier), dusting (with a fine dry powder as the carrier), and fumigation (where the insecticide is applied as a gas).

Registration and regulation. The registration and sale of pesticides in the United States is controlled by the Federal Insecticide, Fungicide, and Rodenticide ACT (FIFRA), Public Law 92-516 (Oct. 21, 1972) and its amendments, Public Law 94-140 (Nov. 28, 1975) and Public Law 95-396 (Sept. 30, 1978). This law requires registration of all pesticides with the EPA.

Types of Insecticides

Inorganic stomach poisons. Various arsenicals have been widely used as stomach poisons for insects, arsenous oxide, As_2O_3, secured from flue dust after the roasting of various metallic ores, being the principal source.

The insecticidal activity of an arsenical usually is directly related to the percentage of metallic arsenic it contains; although other metals in combination, eg, lead or copper, also may add to the toxicity. Lead arsenate, $PbHAsO_4$, and $Ca_3(AsO_4)_2$, were the most widely used but are now obsolete.

The properties of fluoride compounds that have had insecticidal usage are salts of hydrofluoric acid, HF, fluorosilicic acid, H_2SiF_6, and fluoroaluminic acid, H_3AlF_6. Cryolite or sodium fluoroaluminate, Na_3AlF_6, and sodium fluoride are still used to some extent.

Contact poisons of plant origin. Nicotine from tobacco was one of the earliest insecticides and was recommended for use in 1763 as a tea for the destruction of aphids. Nicotine, α-1-methyl-2-(3'-pyridyl)pyrrolidine (bp 247°C, d 1.009 g/cm³), is found in the leaves of *Nicotiana tobacum* and *N. rustica* in amounts ranging from 2 to 14%, and also is found in *Duboisia hopwoodii* and in *Aesclepias syriaca*. Anabasine is the chief alkaloid of *Anabasis aphylla*, where it occurs from 1–2% in the shoots and is found to 1% in *Nicotiana glauca*.

Nicotine and anabasine affect the ganglia of the insect central nervous system, facilitating transsynaptic conduction at low concentrations and blocking conduction at higher levels. The extent of ionization of the nicotinoids plays an important role in both their penetration through the ionic barrier of the nerve sheath to the site of action and in their interaction with the site of action, which is believed to be the acetylcholine post-synaptic receptor protein.

Nicotine is used as a contact insecticide for aphids attacking fruits, vegetables, and ornamentals, and as a fumigant for greenhouse plants and poultry mites. Nicotine sulfate is safer and more convenient to handle and the free alkaloid is rapidly liberated by the addition of soap, hydrated lime, or ammonium hydroxide to the spray solution.

The insecticidal properties of pyrethrum from the ground flowers of *Chrysanthemum cinerariaefolium* and *C. coccineum* result from six esters, the pyrethrins I and II, the cinerins I and II, and the jasmolins I and II (1), which are present in the flowers, mostly in the achenes, from 0.7–3% in selected strains. The structure of the pyrethrin esters is as follows:

(1)

	R	R'
pyrethrin I	CH_3	$-CH_2CH=CHCH=CH_2$
pyrethrin II	$-COOOCH_3$	$-CH_2CH=CHCH=CH_2$
cinerin I	CH_3	$-CH_2CH=CHCH_3$
cinerin II	$-COOCH_3$	$-CH_2CH=CHCH_3$
jasmolin I	CH_3	$-CH_2CH=CHCH_2CH_3$
jasmolin II	$-COOCH_3$	$-CH_2CH=CHCH_2CH_3$

The pyrethrins readily penetrate the insect cuticle as shown by the LD_{50} values to the cockroach, *Periplaneta*: topical, 6.5 mg/kg; injected, 6.0 mg/kg. A characteristic of pyrethrin action on insects is rapid knockdown followed by substantial recovery. Thus, an approximately threefold increase in dosage is required to produce a 24-h mortality equivalent to the 25-min knockdown of houseflies. This recovery from paralysis is the result of rapid enzymatic detoxification in the insect, apparently by microsomal oxidases. Because of their very rapid knockdown, the pyrethroids have been favored for use in household, fly, and cattle sprays at ca 0.03–0.1%.

The use of rotenone-bearing roots as insecticides in the United States was developed as a result of Federal laws against residues of lead, arsenic, and fluorine upon edible produce. Rotenone is harmless to plants, highly toxic to many insects, and relatively innocuous to mammals. Sixty-eight species of plants, including twenty-one species of *Tephrosia*, twelve of *Derris*, twelve of *Lonchocarps*, ten of *Milletia*, and several of *Mundulea*, have been reported to contain rotenone or rotenoids (2).

(2) rotenone

Insects poisoned with rotenone exhibit a steady decline in oxygen consumption, and the insecticide has been shown to have a specific action in interfering with the electron transport involved in the oxidation of reduced nicotinamide adenine dinucleotide (NADH) to nicotinamide adenine dinucleotide (NAD) by cytochrome *b*. Poisoning, therefore, inhibits the mitochondrial oxidation of Krebs-cycle intermediates which is catalyzed by NAD.

Synthetic organic insecticides. DDT (3), 1,1,1,-trichloro-2,2-bis(*p*-chlorophenyl)ethane, was first synthesized in 1874 and its insecticidal properties were discovered in 1939. Chemically pure *p,p'*-DDT consists of white needles (mp 109°C; density 1.6; and vapor pressure ca 20 μPa (0.15 nm Hg) at 20°C. In alkaline solution, DDT is readily dehydrochlorinated to form a noninsecticidal product, 1,1-dichloro 2,2-bis(*p*-chlorophenyl)ethylene, DDE (mp 85°C). This compound may be oxidized to *p,p'*-dichlorobenzophenone by ultraviolet radiation catalysis. These two reactions apparently account for the decomposition of DDT residues.

(3) DDT

DDT has a highly specific insecticidal action affecting peripheral sensory organs of insects to produce violent trains of afferent impulses that cause hyperactivity and convulsions. The paralysis and death that ensue are thought to occur from metabolic exhaustion or from elaboration of a naturally occurring neurotoxin. The insecticidal action of DDT is facilitated by the ease of absorption through the insect cuticle, eg, the topical LD_{50} to the cockroach *Periplaneta americana*, is 10 mg/kg compared to the injected LD_{50} of 8 mg/kg.

DDT is the most permanent and durable of the commonly used contact insecticides because of its insolubility in water, very low vapor pressure, and resistance to destruction by light and oxidation. The unusual stability of DDT and its high lipid/H_2O partitioning (> 100,000) has resulted in many environmental problems, eg, soil persistence with a half-life of 2.5–10 yr, bioaccumulation from water to fish at factors > 1,000,000, transport through food chains, and ubiquitous tissue storage in humans and animals. These properties are shared by DDE, which is even more environmentally recalcitrant.

DDT is slowly converted *in vivo* by reductive dechlorination to DDD and by further dechlorinations to 4,4'-dichlorodiphenylacetic acid (DDA), the predominant excretory metabolite. DDT is highly toxic to fish (LC_{50} for trout and blue gill: 0.0002–0.008 ppm) and is only moderately toxic to birds (oral LD_{50}: mallard 1300 and pheasant > 2240 mg/kg). However, widespread bird kills have resulted from bioconcentration of DDT through food chains, ie, from fish or earthworms.

DDT has been employed for control of hundreds of species of insect pests of orchard, garden, field, and forest. It was generally applied as a dust or as water sprays suspensions or emulsions of DDT at 0.1–5%. At the peak of its usage in the United States, DDT was registered for use on 334 agricultural commodities, but these registrations were cancelled in

1973 because of widespread environmental contamination, and the use of DDT is restricted to essential public-health usage. Similar restrictions have been invoked in most countries of Europe and in Japan.

Cyclodienes. The cyclodienes are polychlorinated cyclic hydrocarbons with endomethylene-bridged structures, prepared by the Diels-Alder diene reaction. The development of these materials resulted from the discovery in 1945 of chlordane, the chlorinated adduct of hexachlorocyclopentadiene and cyclopentadiene (qv). Technical chlordane (4) is an amber liquid (bp 175°C at 267 Pa (2 mm Hg), d 1.61 g/cm³) which contains about 60% of isomers of chlordane plus variable amounts of heptachlor (5), or 1,4,5,6,7,8,8-heptachloro-3a,4,7,7a-tetrahydro-4,7-methannindene (mp 95°C). Heptachlor (5), which is about 3–5 times as active insecticidally as chlordane, is more efficiently produced by chlorinating chlordene (hexachlorodicyclopentadiene) with SO_2Cl_2.

(4) β-chlordane   (5) heptachlor

Many variations of the Diels-Alder reaction with hexachlorocyclopentadiene produce useful insecticides. The adduct with bicycloheptadiene is aldrin, or 1,2,3,4,10,10-hexachloro-1,4,4a,5,8,8a-hexahydro-1,4-endo-exo-5,8-dimethanonaphthalene (mp 104°C). Endrin, 1,2,3,4,10,10-hexachloro-6,7-epoxy-1,4,4a,5,6,7,8,8a-octahydro-1,4,endo,endo-5,8-dimethanonaphthalene (mp 245°C dec) is the endo, endo isomer of dieldrin and differs in the spatial arrangement of the two rings.

Compounds, eg, aldrin and heptachlor (5), that have unsubstituted double bonds, readily add oxygen to form epoxy derivatives. This is not only of importance in the synthesis of the insecticide dieldrin but also markedly affects biological behavior. These epoxides form in the tissues of animals that ingest or absorb the compounds and are preferentially concentrated and stored in fats.

The cyclodienes, like most other good contact insecticides, are efficiently absorbed by insect cuticle and, for aldrin, the LD_{50} values with the cockroach, *Periplaneta*, are 1.9 mg/kg topically and 1.5 mg/kg injected. Symptoms of poisoning include hypersensitivity, hyperactivity, convulsions, prostration, and death. The site of these disturbances is the ganglia of the central nervous system.

The high lipophilicity of the cyclodienes and the prolonged persistence of dieldrin and heptachlor epoxide (soil half-lives 2–10 yr) have resulted in severe environmental contamination.

Aldrin and heptachlor have been widely used as soil insecticides, for cotton insect control, for tree-fruit pests, and against grasshoppers. Chlordane is used for control of cockroaches, ants, termites, and certain pests of vegetable and field crops. Dieldrin is more residual and has been used in the treatment of houses to control the mosquito vectors or malaria and the triatominae vectors of American trypanosomiasis, and for mothproofing. In 1974, uses of aldrin and dieldrin were cancelled in the United States and most of those for heptachlor and chlordane were cancelled in 1978.

Organophosphorus insecticides. It is estimated that more than 100,000 different organophosphorus compounds have been synthesized and evaluated as pesticides.

Organophosphorus toxicants are available with very short residual action eg, TEPP and mevinphos, or with prolonged activity, eg, diazinon (6) and azinophosmethyl (7). There are broad spectrum insecticides, eg, parathion (8) and malathion (9), and materials with highly specific action, eg, schradan (10). The unique properties of demeton and dimethoate have resulted in successful plant systemic insecticides and this activity has been further refined in seed and soil treatments with phorate (11), and disulfoton which will protect newly developed seedlings from insect attack. Compounds, eg, cruformate, can be fed to cattle to kill grubs living in the animals bodies, whereas others, eg, trichlorfon, have pronounced activity as stomach poisons but almost no contact action. By taking advantage of differences in the processes of detoxifica-

tion in the Insecta and Mammalia, compounds, eg, malathion (9) and fenitrothion, incorporate a high degree of insecticidal action and safety to the human user and domestic animals.

(6) diazinon   (7) azinphosmethyl

(8) parathion   (9) malathion

(10) schradan   (11) phorate

Systemic insecticides for plants. Systemic insecticides for plants, when applied to seeds, roots, or leaves of plants, are absorbed and translocated to the various plant parts in amounts lethal to feeding insects. This method of plant protection has the advantages of minimizing to some extent the inequalities of spray coverage, increasing the length of residual control by protection of the spray residue from attrition from weathering, protecting new plant growth formed subsequent to application, and having less damaging effects on beneficial predatory and pollinating insects. Systemic compounds may be applied as direct sprays to the foliage by drenching the soil (as in the irrigatio., water); by direct application to, or injection of, concentrates into the trunk or stem; by side-dressing about the root with granular or encapsulated material; or by treatment of seeds before planting. Examples include schradan (10), or octamethyl pyrophosphoramide, which provides long-term protection from aphids and mites; Dimethoate, O,O-dimethyl S-(N-methylcarbamoyl)methyl phosphorodithioate, used as a persistent systemic for fruit-fly larvae and for side-dressing of soil about plants; and phosphamidon, dimethyl 2-chloro-2-diethylcarbamoyl-1-methylvinyl phosphate, which is absorbed rapidly by plant surfaces and is decomposed quickly in the plant to provide a short-lived systemic.

Systemic insecticides for animals. When fed or topically applied to domestic animals, systemic insecticides will move through the body tissues in quantities lethal to such internal parasites as cattle grubs, screwworm larvae, and helminths and, over a shorter period, to external parasites, eg, horn and stable flies, mites, lice and ticks, all without causing injury to the host animal. The insecticides are slowly destroyed by enzymic action in the animal body, so that after a safe period of 60 d the animal can be used for milk production or slaughter for meat (see Chemotherapeutics, antiprotozoal).

The organosphosphorus compounds owe their biological activity to the capacity of the central P atom to phosphorylate the esterified portion of the enzyme, acetyl cholinesterase (AChE), which is an essential constituent of the nervous system not only of Insecta but also of all higher animals. The phosphorylated enzyme is irreversibly inhibited and, therefore, is no longer able to carry out its normal function of the rapid removal and destruction of the neurohormone, acetylcholine (ACh), from the nerve synapse. As a result, ACh accumulates and disrupts the normal functioning of the nervous system, giving rise to the typical cholinergic symptoms associated in insects with organophosphorus poisoning, ie, hyperactivity, tremors, convulsions, paralysis, and death.

The most widely used organophosphorus insecticides are methyl parathion [(O,O-dimethyl O-(p-nitrophenyl) phosphorothionate]; diazinon (6) [O,O-diethyl O-2-isopropyl-4-methyl pyrimidyl-6-phosphorothionate]; azinphosmethyl (7) [O,O-dimethyl S-4-oxo-1,2,3-benzotriazin-3-(4-H)-ylmethyl phosphorodithioate]: malathion (9) [O,O-dimethyl S-(1,2-dicarbethoxyethyl)phosphorodithioate]: and phorate (11) [O,O-diethyl S-(ethylthio)-methyl phosphorodithioate].

Organophosphorus insecticides are intrinsically reactive and readily degrade by oxidation and hydrolysis and in living organisms. Therefore, their use does not present serious problems of biomagnification and food-chain transfer.

Carbamates. The carbamate insecticides are synthetic relatives of the alkaloid physostigmine from *Physostigma venenosum* (see Alkaloids). Activity is associated with considerable variation in structure, and the individual compounds show a remarkable degree of insecticidal selectivity. The most widely used compounds include: carbaryl (12) (1-naphthyl-N-methylcarbamate); carbofuran (13) (2,3-dihydro-2,2-dimethyl-7-benzofuranyl N-methylcarbamate); propoxur (14) (2-isopropoxyphenyl N-methylcarbamate); and the oxime carbamates methomyl (15) [S-methyl-N-(methylcarbamoyloxy)thioacetimidate]; and aldicarb (16) [2-methyl-2-(methylthio)propionaldehyde O-methylcarbamoyl oxine].

(12) carbaryl   (13) carbofuran   (14) propoxur

(15) methomyl   (16) aldicarb

All of the insecticidal carbamates are cholinergic, and poisoned insects and higher animals exhibit violent convulsions and other neuromuscular disturbances. The compounds are strong carbamylating inhibitors of acetylcholinesterase and may have a direct action on the post-synoptic acetylcholine receptors because of their structural resemblance to acetylcholine.

Insect growth regulators. The great majority of insecticides are neurotoxins and produce biochemical lesions at target sites, eg, AChE, that are common to both insects and vertebrates. For this reason, these insecticides often are lacking in selectivity, and thus man, domestic animals, and wildlife frequently are the unintended victims of insecticide use. The insect growth regulators that interfere with biochemical and physiological processes that are unique to the arthropods, eg, molting, ecdysis, and formation of the chitinous exoskeleton, are much more truly selective insecticides and there is an urgent need for their exploitation. These regulators include: juvenile hormone analogues such as methoprene (17) and chitin-synthesis inhibitors such as diflubenzuron (18), (1-(4-chlorophenyl)-3-(2,6-difluorobenzoyl) urea (mp 239°C), which inhibits the action of the enzyme chitin synthetase which polymerizes uridine-diphospho-acetylglucosamine to give chitin.

(17) methoprene   (18) diflubenzuron

Synthetic Pyrethroids. Structural optimization of the naturally occurring pyrethrum insecticides has produced a variety of more active and more persistent insecticides that represent the newest group of insect-control agents to be widely exploited. They are often effective against insects resistant to the older groups of synthetic insecticides. Representative compounds include allethrin (19) and resmethrin (20), widely used in household insecticides, and permethrin (21) and fenvalerate (22), used for insects attacking field and vegetable crops. The synthetic pyrethroids have the marked advantages of moderate toxicity

to humans and domestic animals and effectiveness at application rates one-tenth or less of those required for organophosphorus and carbamate insecticides.

(19) R = —CH₂CH=CH allethrin

(20) R = —CH₂ ... CH₂ ... resmethrin

(21) R = —CH₂ ... O ... permethrin

Microbial insecticides. Insects are attacked by a multitude of pathogens and ca 450 viruses, 80 bacteria, 460 fungi, 250 protozoa, and 20 rickettsial diseases are effective natural enemies. A number of these are adaptable for mechanical dissemination as microbial insecticides for the innoculation of insect populations, soils, fields, orchards, or forests with spores, microbial toxins, or virus suspensions. Microbial insecticides are highly specific in action against a few closely related pest species and generally are harmless to other animals.

The classic utilization of a microbial insecticide is that of *Bacillus popillae* spores which are produced from infected larvae of the Japanese beetle and are diluted with talc to a standardized powder containing 10^8 spores/g. The powder is applied to soil (2.24–22.4 kg/ha or 2–20 lb/acre) and results in the very slow spread of the disease and in satisfactory control of the beetle larvae in turf within three years. *Bacillus thuringiensis* toxin (Bt) is a standardized microbial insecticide widely used to control caterpillars attacking vegetable crops, ornamentals, and forests.

Genetic control. Manipulation of the mechanisms of inheritance of the insect pest populations has occurred most successfully through the mass release of sterilized males, but a variety of other techniques have been studied, including the environmental use of chemosterilants and the mass introduction of deleterious mutation, eg, conditional lethals and chromosomal translocations (see Genetic engineering).

fenvalerate
(22)

Petroleum oils. Petroleum oil sprays are used as insecticides for dormant sprays for the control of scale insects, mites, and insect eggs; summer foilage sprays for aphids, mealybugs, mites, thrips, psyllids, whiteflies, and scale insects; livestock sprays for the control of lice, fleas, and mites; and mosquito larvicides. They are also used as carriers for contact insecticides to increase their effectiveness.

Fumigants. Fumigants are chemicals that are distributed through space as gases, and therefore, at a given temperature and pressure, must exist in the gaseous state in sufficient concentration to be lethal to the insect pest. This physical requirement greatly limits the number of insecticides that may be usefully employed as fumigants. Compounds boiling at about room temperature, eg, hydrogen cyanide, methyl bromide, and ethylene oxide, are the most useful general fumigants. For soil fumigation, however, the slower release of vapors from substances, eg, ethylene dibromide and β,β'-dichlorodiethyl ether, that boil as high as 180°C, has proved effective. Other organic toxicants of relatively high vapor pressure, eg, naphthalene and p-dichlorobenzene, sublime readily enough to have special uses as fumigants; and contact insecticides, eg, azobenzene, lindane, dichlorvos, and mevinphos, may kill insects by vapor action under certain circumstances.

Several types of chemosterilants are known to produce adequate sterility in insects by preventing the production of ova or sperm, causing death of sperm or ova, or producing severe injury to the genetic material of sperm or ova so that the zygotes that are produced do not develop into mature progeny. These include alkylating agents such as apholate (**23**) and triethylphosphoramide and the antimetabolites 5-fluorouracil and amethopterin, a folic acid antagonist, which produce sterility in female flies when fed at 0.01–0.05% in the diet.

(**23**) apholate

Repellents and attractants. Repellents (qv) are substances that protect animals, plants, or products from insect attack by making food or living condition unattractive or offensive. These substances, which may not be poisonous or only mildly toxic, are rarely, if ever, effective against all kinds of insects. Such chemicals can sometimes be employed to advantage where it is impossible to use an insecticide and may afford a greater or lesser degree of protection to manufactured products, growing plants, or the bodies of animals and humans. Among the many examples are trichlorobenzene and other chemicals used to protect buildings from termites; the application of sulfur to the body to keep chiggers from attacking; and the use of pine-tar oil and diphenylamine to keep screwworm flies from laying eggs about the wounds of animals.

A number of chemicals have been identified that influence insect behavior as they search for food, oviposition sites, or mates. Many of these chemicals are used to attract insect pests into traps or to poison baits for both population control and for measurement of population densities, eg, for timing spray applications.

ROBERT L METCALF
University of Illinois at Urbana-Champaign

R.L. Metcalf and J.J. McKelvey, Jr., eds., *The Future for Insecticides*, John Wiley & Sons, Inc., New York, 1976.

R.L. Metcalf and W.H. Luckmann, eds., *Introduction to Insect Pest Management*, 2nd ed., John Wiley & Sons, Inc., New York, 1982.

C.F. Wilkinson, ed., *Insecticide Biochemistry and Physiology*, Plenum Press, Inc., New York, 1976.

A.W.A. Brown, *Ecology of Pesticides*, John Wiley & Sons, Inc., New York, 1977.

M.S. Quraishi, *Biochemical Insect Control*, John Wiley & Sons, Inc., New York, 1977.

INSTRUMENTATION AND CONTROL

Automatic control in its simplest form is self-correcting or feedback control, whereby a control instrument continuously monitors an output variable of the controlled process and compares this output with its established desired value. The instrument then uses any resulting error obtained from the comparison to compute the required correction to the setting of a fluid-flow valve, or other basic activating element of the piece of equipment being controlled. This so-called servomechanism type of control is discussed in this article.

The design and use of a servomechanism control system requires a knowledge of each element of the control loop. For example, in the method illustrated in Figure 1, the distillate composition is regulated by controlling the distillate take-off rate and, thus, the reflux ratio of the column. The engineer must know the dynamic response or complete operating characteristics of each pictured device.

In the automatic process, the control instrument continuously monitors some output variable of the controlled process, eg, a temperature,

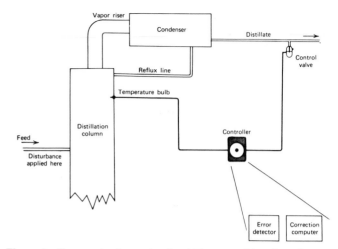

Figure 1. Servomechanism system in which a control loop is used to regulate the reflux ratio of the distillation column.

pressure, composition, etc. It then compares this output with some established desired value or set point of the controlled variable. The error resulting from this comparison is used by the instrument to compute a correction to the setting of the process-control valve or other final control element in order to return and sustain the value of the output variable to its desired level. If the set point is altered, the response of the control system in bringing the process to the new operating level is known as that of a servomechanism or self-controlling device. The action of holding the process at a previously established level when the process is opposed by external disturbances is that of a regulator.

In considering the possible responses of a system to various inputs or upsets, the control engineer usually confines the study to two types of response: the transient response, which is the action of the system when it is subjected to a step function, or other sudden change in operating point, which is applied as a forcing function or input to the system. The second or frequency response is the equivalent action of the system when subjected to a sinusoidally varying forcing function. A transient response is obtained when solving the differential equations of the system analytically or by means of electronic computers (qv). The frequency response is especially valuable for those systems that can be described by linear ordinary differential equations.

Instrumentation of an Automatic Control System

The large number of variables of a typical industrial plant comprises a wide variety of flows, levels, temperatures, compositions, positions, and other parameters that are to be measured by the sensor elements of the control system. Such devices sense some physical, electrical, or chemical property of the variable under consideration and use this property value to develop an electrical, mechanical, or pneumatic signal that is representative of the magnitude of the variable in question. The signals, as continuous representations of the variables involved, are analogue signals in contrast to single on-off signals or digital signals.

The size of most industrial plants necessitates connection of the transducer by long electrical cables to the controller unit. Quite often, the connecting cable must run close to a source of high voltage, resulting in noise pickup on the leads, which manifests itself as a voltage added to the transducer output or series-mode voltage. The series-mode voltage often is of a much higher frequency than the information frequency of the signal, and therefore can be attenuated sufficiently by a low-pass R–C (resistor–capacitor) filter between the connecting cable and the input selection switch. This and other techniques, when properly applied, can usually nullify the effects of such noise pickup. In other circumstances, there may be considerable external noise pickup at the sensor, eg, common-mode voltage noise or voltage. Again, several grounding and shielding techniques can counteract the effects of such pickups.

Computer control. Digital computers have removed the limitations on complexity imposed by earlier analogue devices and have become popular elements of industrial-plant control systems. The three classes of computer application to industrial-plant control problems are supervisory or optimizing control; direct control; and hierarchy control, which is a combination of the other two affecting all levels of decision making in the plant.

Supervisory or optimizing control places the computer in an external or secondary control loop to the primary plant control system which remains as the conventional plant instruments and individual electronic or pneumatic analogue controllers. Direct digital control uses the computer to replace a group of single-loop analogue controllers. Hierarchy control, the most recent and most ambitious of the concepts, attempts to apply computers to all plant control situations at the same time. As such, it requires the best of computer capabilities and automatic-control potentialities to carry out its function of integrating the plant operation completely from management decision to final valve movement.

Advanced Control Methods

A short listing of some of the advanced control topics that have vastly extended the capabilities of industrial control systems follows. As indicated, a computer is almost always necessary for their implementation.

Adaptive control is the capability of the control system to modify its own operation so as to achieve the best possible mode of operation.

Steady-state optimization allows the computer to determine for the process a new best operating level if exterior conditions require such changes in order to maintain the process operation at some optimum (usually economic) criterion.

Dynamic optimizing control is the requirement that the control system itself operates so that a specific performance criterion is satisfied. This criterion usually is formulated so that the controlled system must move from the original position to a new position in the minimum possible time or at the minimum total cost.

Learning control implies that the control system contains sufficient computational ability to develop representations of the mathematical model of the system being controlled and to modify its operation in order to compensate for the newly developed knowledge.

Multivariable, noninteracting control concerns large systems, the size of whose internal variables is dependent upon the values of other interrelated variables of the process.

Design methods. The information obtained from transient and frequency-response experiments or from mathematical modelling is used to design the best control system for a process. The Nyquist method and the Bode method are two methods most commonly used for plotting open-loop frequency-response data in order to determine which controller to use on a process and to check the response and stability of the resulting process-controller complex.

Overall Plant and Company Control Systems

As the overall requirements, both energy- and productivity-based, become more complex, more complex and capable control systems are necessary. To achieve these, the field gravitated more and more toward the digital computer-based systems to carry out the needed work. In order to obtain highly advanced control responses, an overall system with the following capabilities is needed:

A tight control of each operating unit of the plant to assure that it is operating at its maximum efficiency of energy utilization and/or production capability based upon the production level set by the scheduling and coordination functions listed below. This control reacts directly to any emergencies that may occur in its own unit.

A coordination system that determines and sets the production level of all units working together between inventory locations. This system assures that no unit exceeds the general area level and thus does not use excess energy or raw materials. The coordination system responds to the existence of emergencies or upsets in any of the units under its control by shutting down or systematically reducing the output in these and related units.

A system capable of carrying out the scheduling for the plant from customer orders or management decision, so as to produce the required

products for these orders at the optimum combination of time, energy, and raw materials, ie, cost function.

THEODORE J. WILLIAMS
Purdue University

D.M. Considine, ed., *Process Instruments and Controls Handbook*, 2nd. ed., McGraw-Hill Book Company, New York, 1974.

F.H. Raven, *Automatic Control Engineering*, 2nd ed., McGraw-Hill Book Company, New York, 1968.

T.J. Williams and F.M. Ryan, *Progress in Direct Digital Control*, Instrument Society of America, Pittsburgh, Pa., 1969.

J.G. Truxal, *Automatic Feedback Control Synthesis*, McGraw-Hill Book Company, New York, 1955.

T.J. Harrison, *Handbook of Industrial Control Computers*, John Wiley & Sons, Inc., New York, 1972.

INSULATION, ACOUSTIC

Acoustic insulation materials control sound and are divided into four categories: sound absorbent, sound containment, damping, and vibration isolation materials.

Sound-Absorbent Materials

Porous materials are the most common materials used to provide sound absorption. As sound passes into the intercommunicating pores, air moves back and forth rapidly within the material, and sound energy is converted into heat by frictional and viscous forces at the air–material surface. Most porous sound-absorbing materials are applied to walls or ceilings in living or office spaces (architectural treatment) and in unoccupied structures, eg, machinery housing (acoustic linings) to control reflected sound. In both cases, the amount of acoustic energy dissipated by a sound-absorbent material is a function of the physical properties and positioning of nearby reflecting surfaces. Special standard tests are used to measure the performance of sound-absorbent materials or products intended as architectural treatments or machine linings.

Absorption coefficients. The fundamental unit for describing sound-absorbing performance in both architectural and acoustic-lining treatments is the absorption coefficient which indicates the percentage of incident sound energy dissipated upon reflection during specific test conditions. An absorption coefficient of one implies that no sound energy is reflected under the test conditions (all absorbed), whereas a coefficient of zero implies that incident sound energy is reflected completely.

Acoustic performance is dependent on the frequency of the sounds involved. Therefore, a series of absorption coefficients at various frequencies is needed to portray performance adequately. Normally, for a given material or product, six coefficients are measured—one for each of the six octave-frequency bands in the range from 125 to 4000 Hz. The ANSI recognizes two standard tests: the reverberation-room method (ASTM C 423) and the impedance-tube method (ASTM C 384).

With the reverberation-room method, which is preferred for use in design work, measurements are made of the rate that sound decays in a reverberant room. The room absorption value, A, is established through the following relationship:

$$A = 0.9210Vd/c$$

where A is the amount of room absorption, m^2; V is the room volume, m^3; d is the logarithmic rate of decay, dB/s; and c is the speed of sound, m/s. By subtracting the amount of absorption occurring in the empty room from the amount of absorption with the test specimen present, a value for the absorption provided by the test specimen is determined:

$$\alpha_R = A_{\text{test material}}/S$$

where S is the surface area of the test specimen, m^2. The subscript R indicates that the procedure provides a random incidence value for absorption.

The impedance-tube method requires an apparatus consisting of a tube with a test specimen at one end and a loudspeaker at the other. A probe microphone is placed inside the tube and between the loudspeaker and the specimen. Sound is emitted from the loudspeaker, propagates toward the specimen and is reflected; the intensity of the reflection depends on how well the test material absorbs sound. A standing-wave pattern develops inside the tube, and the probe microphone establishes the nature of the pattern. The absorption coefficient is related directly to the standing-wave pattern through:

$$\alpha_N = 1 - \left[\frac{\log_{10}^{-1}(L/20) - 1}{\log_{10}^{-1}(L/20) + 1} \right]^2$$

where the quantity L is the standing-wave ratio or difference, in dB, between the maximum sound pressure of the standing wave. The subscript α_N indicates that the absorption coefficient is determined only for sound incident in one direction: perpendicular (normal) to the absorbing material.

Materials. The characteristics that determine sound absorption are possessed by thousands of materials, including heavy woven fabrics, felts (qv), and fibrous thermal-insulation materials, particularly fiber glass (see Textiles; Insulation, thermal). Thus, many materials can serve as acoustic insulation; however, a small portion of these materials is marketed intentionally for acoustical purposes.

Sound-Containment Materials

These are materials that contain or redirect sound. When used in sound-containment systems, the physical properties of the intervening structure that isolates the noise source from the receiver usually are the most critical for preventing sound from being transmitted across the construction. Redirection of sound is provided by partial height, impervious, structural elements (barriers and shields). The performance of such structures usually depends on how much sound diffracts (bends) around their edges or on the effect of sound reflections from nearby surfaces.

The performance measure of interest for structural elements that contain sound is the sound-transmission loss TL. By definition, $TL = 10 \log_{10} W_i/W_t$ in dB, where W_i is the incident sound power (which induces structural deformation of the intervening structure), and W_t is the transmitted sound power (which results from the radiation of sound on the opposite side of a wall because of induced structural deformations). Structural stiffness, internal damping, and mass are attributes affecting the performance of a sound-reflective material. The TL values must be known in considerable detail to ensure that the transmission loss of a particular construction or material is characterized adequately. A standard set of TL data consists of 16 data points, one for each contiguous one-third octave band from 125 to 4000 Hz.

In the TL test for building partitions (ASTM E 90), the specimen is mounted as a partition between two reverberation rooms. One room is a source room and the other is a receiver room. Transmission loss is found from the following:

$$TL = NR + 10 \log_{10}(S/A_2)$$

where S is the area of sound-transmitting surface of the specimen, m²; A_2 is the sound absorption in the receiving room, metric sabins; and NR is the difference in space-time average sound-pressure levels in the two rooms.

Damping Materials

Damping is any process that reduces vibration by converting the ordered mechanical energy of vibration into thermal energy. Interface friction, fluid viscosity, turbulence, and magnetic hysteresis are examples of the many ways of damping vibration. Only one damping method depends on material properties—mechanical hysteresis, otherwise known as material damping. The stresses in the material caused by vibratory motions are relieved partly as material strains develop, and a portion of the mechanical energy available for vibration is converted into heat in the process. Because no material is perfectly elastic, a molecular-level energy transformation occurs in all materials, including common structural materials. Damping materials, however, dissipate more strain en-

ergy than ordinary materials because, when under stress, they tend to flow like a very viscous liquid. This behavior is characteristic of viscoelastic materials used extensively to dampen vibration. When properly joined to the surface of a vibrating element, the damping material participates in the vibration of the structure, is stressed, and dissipates vibrational energy.

Loss factor. Material damping is characterized by the loss factor, a dimensionless quantity:

$$\eta = D/2\pi W_0$$

where D is the energy dissipated per cycle of vibration, and W_0 is the average total energy of the vibrating system.

Extensional damping. Two forms of material-damping applications are used: extensional and constrained layer. In extensional damping, the material is applied directly as a surface treatment. As the damping material is flexed and extended, energy is dissipated. Generally, a damping material is applied in a continuous surface to ensure that the application participates in the surface motions.

Constrained-Layer Damping

In constrained-layer damping, the damping material is attached to a base material and is sandwiched by an outer constraining layer. When such a system is subjected to flexure, the damping material is put into shear as well as bending. Energy losses associated with this form of damping are related primarily to the damping material which relieves the shear strains. The composite system-loss factor depends on the geometry of the structure, the moduli of elasticity of the system components, and the shear modulus of the damping material.

Vibration-Isolation Materials

Structural vibrations can be inhibited by introducing a vibration isolator in the path of the disturbance. A vibration isolator is an elastic material or structure that, through its dynamic properties, tends to prevent the transmission of vibration. The performance of an isolator is characterized by its transmissibility—the ratio of the force transmitted to the quiet side of the isolator compared with the driving force introduced by the motion of the object to be isolated:

$$\text{transmissibility} = \text{output force/input force}$$

Transmissibility varies according to frequency and is a function of the natural frequency of an isolator and its internal damping, assuming the isolator rests on a foundation that is considerably stiffer than the isolator. For most materials, the natural frequency of the isolator, f_0, is related directly to the static deflection of the isolator through the relationship:

$$f_0 = 5/\sqrt{x}$$

where x is the static deflection, cm.

Vibration isolators are used extensively at support points of vibrating equipment and components, such as mechanical equipment, noisy pipes, and industrial machinery (see Noise pollution). When so used, they prevent vibrational energy from being transmitted as noise or to vibration-sensitive areas. Isolators also are used to insulate a noise- or vibration-sensitive machine or area from external vibration.

CHARLES R. JOKEL
Bolt, Beranek & Newman

Reviewed by
JACK D. VERSCHOOR
Manville Corp.

Compendium of Materials for Noise Control, HEW (NIOSH) Publication 75-165, HEW, Washington, D.C., June 1975.

L.L. Beranek, ed., *Noise and Vibration Control,* McGraw-Hill Book Co., New York, 1971.

R.D. Berendt and E.L.R. Corliss, *Quieting: A Practical Guide to Noise Control NBS Handbook 119,* NBS, Washington, D.C., July 1976.

Industrial Noise Manual, 3rd ed., American Industrial Hygiene Association, Akron, Ohio, 1975.

INSULATION, ELECTRIC

PROPERTIES AND MATERIALS

Electrical devices generally consist of four separate components: conductors, a magnetic circuit, mechanical supports, and electrical insulation. Insulation confines, restrains, and directs the flow of electric currents. Gases, liquids, and solids are used to insulate electrical devices, depending on the specific needs of the system. Although the function of an insulator is to restrain current flow from conductor to ground or to a lower potential, insulations often must provide mechanical support, protect the conductor from environmental degradation, and transfer heat to the surrounding environment. The insulating system influences the reliability of the device, and the type and quality of the insulation affects its cost, weight, size, performance, and life.

Insulator Properties

Dielectric constant. The dielectric constant (ϵ') is a fundamental property of every insulating material. The dielectric constant of a polymer is the ratio of the capacitance of a capacitor with the polymer as the dielectric to a similar capacitor with a vacuum as the dielectric. The dielectric constant is determined by the contribution of the energy stored in the material. Typical values of ϵ' for polymers vary from 2–12 as compared with clean dry air or vacuum as 1. Inorganic compounds, eg, titanates, have dielectric constants of over 1000; when blended with polymers, these compounds permit insulations with high values of ϵ'.

The dielectric constant for a material depends upon molecular polarity, field frequency, and temperature. Nonpolar materials do not vary greatly in ϵ' with frequency, but polar materials have a great dependency on frequency.

Dissipation factor. When an a-c voltage is imposed on a dielectric material, a current flows and leads the voltage in time by an amount δ as shown in Figure 1. This current results in a power loss within the dielectric of $W = 2\pi f C_p \tan\delta E^2$ where f = frequency, C_p is the equivalent parallel capacitor, E is the voltage, and δ is the dissipation factor. Defined vectorially, the dissipation factor is the ratio of parallel reactance to parallel resistance in a capacitor having the dielectric material. The angle δ is the loss angle, θ is the phase angle, $\tan\delta$ is the dissipation factor, and $\cos\theta$ is the power factor.

Note that for small values, the dissipation factor closely approximates the power factor. It is desirable to use dielectric materials with low values in an insulation system (if consistent with other system requirements) in order to limit the thermal losses in the dielectric. Thermal runaway can occur with poorly selected insulation systems wherein conduction losses in the dielectric material increase its temperature. The increasing temperature increases the dissipation factor which increases the losses, resulting in ever-increasing dissipation factor and thermal buildup, until dielectric failure occurs. Dissipation factor is affected by temperature, frequency, humidity, voltage, and partial discharges that occur in or on the insulation. Standard test methods are defined by ASTM D 150.

Loss factor. The loss factor ϵ'' is the product of the dissipation factor and the dielectric constant or the ratio of the unrecoverable energy to the total energy that is introduced by an electrical stress E applied to an insulator.

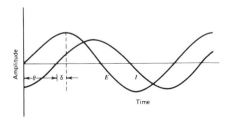

Figure 1. Relationship between the voltage (E) and the current (I) for a dielectric material in an a-c field, θ = phase angle and δ = loss angle.

Resistivity. Resistivity is the inverse value of conductivity. It is commonly expressed by: volume resistivity, surface resistivity, and insulation resistance as defined in ASTM D 257.

Dielectric breakdown. Every insulator possesses a few free electrons. Under the stress of an electric field, these electrons tend to move between the electrodes. A voltage level is reached wherein the electrons impact with atoms and molecules, therein dislodging other electrons and causing an electron avalanche or breakdown. Three theories regarding the basis of dielectric breakdown have been postulated: the thermal theory, based on a transport of current in certain portions of the insulation that have high conductivity resulting in ever-increasing current flow, temperature, and conductivity until a failure occurs; the theory of ionic-species generation within the insulator which postulates ionic movement which uses energy and generates increasingly more ions until failure; and physical failure of the insulation opening direct paths for current flow and causing massive electrical failure.

In practice (at least in solids), probably all three conditions exist and often are initiated by the physical failure of the insulation. All of these conditions describe dielectric strength which is the voltage at failure divided by the thickness of the insulator (MV/m). Values of dielectric strength are highly dependent on the density of the insulator, the electron-absorbing quality, the applied voltage and the rate of its application, thickness, the moisture content of the insulator, frequency, time, and temperature. In general, the electric strength of solids is greater than that of liquids which is greater than that of gases.

Thermal conductivity. Thermal conductivity controls heat transfer from the conductor to the ambient environment. Heat is generated in the conductor, the magnetic circuit, and the insulator. The thermal conductivity of insulation systems is much lower than that of metals.

Thermal expansion. Many industrial applications depend on the adhesion of insulation to conductors, the magnetic core, or the structural components (see Adhesives). Electrical components are enameled magnet wire, resin encapsulations, and cast structures. Large differences in the coefficient of thermal expansion can cause debonding which introduces problems in heat transfer, corona, and partial-discharge formation. Coefficients can vary over several orders of magnitude from $2.13 \times 10^{-4}/°C$ for polyethylene to $13 \times 10^{-6}/°C$ for mica.

Physical properties. Electrical insulations must have good physical properties to function satisfactorily since their role in the system often is mechanical as well as electrical. In addition to withstanding the mechanical forces of the manufacturing process, the insulators are designed for specific mechanical performance. For example, high tensile strengths are required of filament-wound rings that are used to support and restrain the end turns of turbine generators during short-circuit conditions. Slot wedges, which are used in motors to hold the coils in place against centrifugal forces, require laminates with high flexural strength. Many insulations must have great flexibility to permit their application to conductors; and tear strength and burst strength are required of film insulation.

Thermal life. Where insulation systems are used in air environments, oxidation reactions become important and must be considered in determining the life of the insulation system. Most polymers oxidize, and the aliphatic resin systems are particularly oxidation sensitive. Oxidation causes several responses: production of volatile by-products which lead to partial discharges and to electrical breakdown; generation of corrosive by-products, mostly acidic chemicals, which attack the insulation system; and chemical changes to the molecular structure, which make the polymer more rigid and brittle, increasing its propensity to shrink, crack, or break prematurely.

Pyrolysis is the decomposition of organic polymers caused by the effects of heat exclusive of oxidation. Pyrolysis usually results in decreased molecular weight or even a reversion to the monomer and, even in air, occurs within the bulk of the material.

Chemical resistance. Any insulation-system design must provide resistance to chemical environments to which the system may be exposed during manufacture or use. For example, many are exposed to cleaning solutions, fluxes, electroplating solutions, and paint and varnish solvents during manufacture (see Paint and varnish removers). Organic solvents can be particularly harsh on many polymers. The thermoplastic resins often are dissolved by many organic solvents and, even when they do not dissolve, they can be locally crazed and cracked, particularly at areas of high mechanical stress.

Flammability. Most organic polymers burn readily in air. However, the presence of the halogens (chlorine, fluorine, and bromine) and of nitrogen and phosphorus add to the flame resistance. Some polymers, eg, PVC, can be ignited but extinguish quickly when the ignition source is removed (see Flame retardants).

Design of Insulation Systems

Economic aspects. A total system cost includes the cost of the materials, the amount needed, the application process, and the life expectancy. Material costs are related to the polymers used, with polyolefins and phenolics (see Olefin polymers; Phenolic resins) being low in cost and fluorocarbons and polyimides (qv) being extremely high (see Fluorine compounds, organic). Insulation with high electric strength can be used in thinner sections than other materials. Other cost factors include the availability of the material, the safety factors to be used, and maximum temperatures being considered.

Health and safety. When fully polymerized, organic polymers are generally inert and harmless. However, certain polymers or their monomers exhibit positive responses in the Ames test (see Industrial hygiene and toxicology). Notable among these are certain bisphenol-A epoxies. Many of the amines that are used to cure epoxies are harsh chemicals that, when in contact with skin, can cause dermatitis (see Amines; Epoxy resins).

Gas Insulation

Gases are used as electrical insulation. Highly compressible gases have low conductivity and dielectric constants close to one. Air is a common dielectric gas (ASTM D 3283). Nitrogen (qv) was widely used in cables in the past; however, its use is decreasing. In transformers, dry nitrogen is often used to blanket the oil, thereby decreasing oxidation of the oil and other insulating components. Hydrogen is used in turbine generators although it contributes only minor insulation functions. Its high thermal conductivity, high specific heat, and light weight make it desirable as a coolant. Sulfur hexafluoride (SF_6) is used in power circuit breakers, switchgear, and compressed gas-insulated transmission systems as an arc-interrupting and insulating gas (see Fluorine compounds, inorganic).

Liquid Insulation

The most common insulating liquids are largely aliphatic mineral oils consisting of $C_nH_{(2n+2)}$ and C_nH_{2n}. The oils usually are complex mixtures of straight and branched-chain compounds and cyclic structures. They may have sizable aromatic contents consisting mostly of benzene and naphthalene derivatives with a wide molecular weight range.

Solid Insulation

Solid insulations are the most common form of insulating materials and, even in those devices utilizing gases or liquids, solid materials are invariably used in series. Although many insulators are supplied by the manufacturer in their solid form, a large number of insulations are supplied as solutions, eg, *B*-stage (partly polymerized) or monomeric systems, which are applied to the device and polymerized to the final form. The latter case applies to impregnates and varnishes. Most solid insulations play a mechanical role as well as an electrical one. Insulations that separate high voltage members from earth are ground insulations. Insulating materials applied to wires, cables, and conductors comprise turn insulation or conductor insulation. Sheet materials used to separate discrete layers of windings are layer insulation.

Solid insulation includes: enameled magnet wire; cable insulation such as polyolefins; varnishes; paper; mica tapes; polymer films; vacuum/ pressure-impregnating resins (VPI); casting and potting resins (see Embedding); laminates; molded products such as thermoset and thermoplastic polymers (qv); elastomers (see Elastomers, synthetic); and glasses and ceramics.

R.N. SAMPSON
Westinghouse Electric Corporation

G.L. Moses, *Insulation Engineering Fundamentals*, Lake Publishing Co., Lake Forest, Ill., 1958, p. 8.

D.K. Fink, *Standard Handbook for Electrical Engineers*, McGraw-Hill Book Co., Inc., New York, 1978, Sect. 260, pp. 4–140.

WIRE AND CABLE

Cross-Linked (Thermoset) Insulations

The approximate composition of cross-linked insulation compounds is shown in Table 1, and the properties of these insulations are listed in Table 2.

Thermoplastic Insulations

Poly(vinyl chloride). Poly(vinyl chloride) (PVC) is a dry hard material which in powder form is mixed with modifiers, and is usually converted into feedstock granules (see Vinyl polymers, vinyl chloride, and poly(vinyl chloride)). PVC provides a general-purpose, low voltage insulation at reasonable cost.

Polyethylene. The dielectric constant of 2.3 for low density polyethylene is close to the theoretical minimum that can be obtained for a solid substance that is mechanically suitable for insulation use, and the dissipation factor is about as low as that of any available material. These properties in combination with high dielectric strength are ideal for high voltage insulation.

Polypropylene. Some manufacturers prefer polypropylene copolymer instead of high density polyethylene for telephone cable insulation. The choice is based on mass volume cost (cost per kg × specific gravity) (see Elastomers, synthetic, Ethylene–propylene rubber).

Fluorinated plastics. Poly(tetrafluoroethylene) (PTFE) and fluorinated ethylene–propylene copolymer (FEP) are thermoplastics with outstanding heat and chemical resistance and are suitable for use over a range of −90 to 250°C (see Fluorine compounds, organic).

Taped Insulations

Impregnated paper. Many cables are covered with insulation applied in the form of tape. Until ca 1960, most cable that was rated ≥ 15 kV was insulated with oil-impregnated paper. Aside from a few prototype installations of extruded solid dielectric-insulated cables that were rated 115–138 kV, it is the standard insulation used at present for 115–550 kV circuits.

Table 1.   Composition of Typical Cross-Linked Insulation Compounds

Function in compound	Approximate content, %	Typical ingredients
elastomer	25–40	styrene–butadiene (SBR), Buna S
		isobutylene–isoprene (IIR), butyl
		ethylene–propylene (EPM)
		ethylene–propylene–diene (EPDM)
		poly(dimethyl siloxane) (silicone)
		polyethylene
filler and reinforcing agent	40–70	clay
		silica
		calcium carbonate
		barium sulfate
		talc
		vulcanized natural asphalts
		solid bitumen
		carbon black
vulcanization system	<5	sulfur
		selenium diethyl dithiocarbamate (Selenac)
		dibenzoylquinone dioxime (Dibenzo GMF)
		dicumyl peroxide (Dicup)
		mercaptobenzothiazole (MBT)
		zinc dimethyldithiocarbamate
		tetramethylthiuram monosulfide (Monex)
		zinc oxide
		litharge
		stearic acid
aging inhibitors	<10	zinc salt of mercaptobenzimidazole
		alkylated diphenylamine
		polymerized trimethyldihydroquinoline
processing aids	<2	resins, waxes, petroleum oils, plasticizers

Table 2. Properties of Various Cross-Linked Insulations

				Elastomer		
	SBR or NR	Butyl	Ethylene–propylene	Polyethylene	Silicone	Polyethylene
				Filler		
Property	Oil base	Mineral	Mineral	Unfilled	Mineral	Carbon
electrical at 20°C						
dielectric constant (max)	5.0	4.5	3.5	3.0	3.5	6.0
100 tan δ (max)	5.0	3.5	1.0	2.0	1.0	2.0
resistivity (min), ohm–cm	10^{14}	10^{15}	10^{15}	10^{15}	10^{14}	10^{14}
dielectric withstand reel test (min avg stress), kV/mm (V/mil)	3.9 (100)	3.9 (100)	5.9 (150)	7.9 (200)	3.9 (100)	2.9 (75)
sample dielectric strength, rapid rise, kV/mm (V/mil) avg stress	23.6 (600)	23.6 (600)	23.6 (600)	27.6 (700)	19.7 (500)	19.7 (500)
impulse, 1.5×40 μs wave (max stress), kV/mm (V/mil)	47.2 (1200)	47.2 (1200)	47.2 (1200)	59.1 (1500)	39.4 (1000)	31.5 (800)
mechanical						
tensile strength (min), MPa (psi)	3.1 (450)	4.1 (600)	4.8 (700)	12.4 (1800)	5.5 (800)	12.4 (1800)
elongation at rupture (min), %	250	350	300	300	250	300
cont. high temp limit, °C	75	85	90	90	125	90
resistance to						
oxidation	good	excellent	excellent	excellent	excellent	excellent
moisture	good	good	excellent	excellent	fair	excellent
corona discharge	good	good	excellent	fair	excellent	fair
flame	poor	poor	poor	poor	fair	fair
solvents	poor	poor	poor	excellent	fair	excellent
brittleness at low temp	poor	excellent	excellent	excellent	excellent	excellent
radiation	fair	fair	excellent	good	good	good

Coated fabrics. Cotton (qv) sheeting in thicknesses of 0.13–0.30 mm and in widths of 91–122 cm that is impregnated by multiple dips into a bath of asphalt base, and tung- and linseed-oil varnish, and subsequently baked is called varnished cambric insulation. The varnish impregnation provides the electrical insulation; the cotton sheeting supports the varnish film.

Splicing tape. Poly(vinyl chloride) is calendered into a film 0.18–0.25 cm thick, and a thin coating of rubber-based adhesive is applied on one side, the film-slit into 1.9-cm-wide strips and is wound into rolls for splicing applications.

Component tapes. Polyester and polypropylene film (see Film and sheeting materials) that is slit into tape is used commonly as a component of thermoplastic-insulated and -jacketed cables. Because of their high strength, a thickness of 0.02–0.05 mm is adequate, resulting in reduced diameter. Such tapes are not particularly suited for thermoset-insulated and -jacketed cables because thermoset materials do not bond to them during vulcanization as they do to fabric tapes that are filled with rubberlike compounds.

Coated Insulations

Enamels. Resin-insulated wires as distinguished from heavier conductors are insulated by passing clean, smooth wire through a bath of liquid enamel and a baking chamber and repeating the process as often as necessary to build the desired thickness of enamel, usually less than 0.05 mm on wire sizes AWG (American Wire Gauge) 40-8. Insulated wires of this type possess an excellent space factor (ratio of conductor volume to total volume) and are used wherever the space requirements are severe.

JOHN E. HOGAN
The Okonite Co.

R.O. Babbit, *The Vanderbilt Rubber Handbook*, R.T. Vanderbilt Co., Norwalk, Conn., 1978.

A.R. Von Hippel, *Dielectric Materials and Applications*, Technology Press of Massachusetts Institute of Technology, Cambridge, Mass., and John Wiley & Sons, Inc., New York, 1954.

F.M. Clark, *Insulating Materials for Design and Engineering Practice*, John Wiley & Sons, Inc., New York, 1962.

INSULATION, THERMAL

Thermal insulation is defined by ASTM C 168 as a material or assembly of materials that is used primarily to provide resistance to heat flow. Thus, its distinguishing feature is high thermal resistance, which is measured by the ratio of the difference between the average temperatures of two surfaces to the steady-state heat flux that is in common through them (time rate of heat flow per unit area of one surface). Thermal resistance is expressed as $(K \cdot m^2)/W$ and is directly related to thickness, indirectly to thermal conductivity (k). Thermal properties vary with temperature and must be quoted at a specific mean temperature.

A low thermal conductivity (high thermal resistivity) is a relative concept depending upon the application since other physical properties are also important in a practical insulation. Thermal insulation may be specified for a number of purposes: to conserve energy and, therefore, costs; to increase the comfort of living spaces; to facilitate control of the temperature of a process; to reduce the temperature of the shell of a pressure vessel; to control the external temperature of the insulated space in order to avoid danger to personnel; to protect structural members from damage by high temperature; to reduce the temperature of working spaces; and, in the case of equipment that is operated below ambient temperature, to prevent condensation or icing within the structure.

Principles of Heat Transfer

Heat is transferred through a thermal insulation under the pressure of a thermal gradient across the insulation by the mechanisms of radiation, gas conduction, gas convection, and solid conduction (see Heat-exchange technology, heat transfer). Each mechanism obeys its own laws and each, in a gas-filled insulation, is operative at all temperatures. The relative importance of each of the mechanisms depends on the physical properties of the insulation as well as on environmental conditions. The most favorable design is adjusted so as to give minimum total heat transfer.

Materials

Thermal insulations are used at temperatures from $\leq -200°C$ to 2200°C. Table 1 presents a listing of block, board, and pipe insulation products for use from -268 to 1040°C (see also Refractory materials).

Table 1. Thermal Conductivities of Block, Board, and Pipe Insulation

Material	Temp range or max, °C	Thermal conductivity, W/(m·K)						ASTM specification
		− 20°C	24°C	100°C	200°C	400°C	540°C	
polyurethane	− 73 to 110	0.026	0.025					C 591
polystyrene			0.035					C 578
cellular elastomeric		0.040	0.043					C 534
cork pipe insulation			0.048					C 640
cellulose fiber board			0.055					C 208
mineral fiber								
blanket	204		0.037	0.051				C 553
block	204		0.041	0.052	0.071			C 612
board	982				0.083	0.116		C 612
pipe insulation	650		0.045	0.058	0.075			C 547
cellular glass	− 268 to 427		0.060	0.074	0.100			C 552
calcium silicate								
block	649				0.066	0.079	0.106	C 533
board (577 kg/m³ or 36 pcf)	650				0.123	0.126		C 656
diatomaceous earth	870				0.100	0.109	0.117	C 517
diatomaceous earth	1040				0.108	0.121	0.130	C 517
expanded perlite	816				0.079	0.105		C 610

Health and Safety

Since 1978, the sale of PCB has been greatly restricted (see Chlorocarbons). Asbestos-containing insulation should be removed with great care. When urea-based foams are generated, formaldehyde gas may be released.

R.H. Neisel
J.D. Verschoor
Manville Corporation

R.P. Tye, *Thermal Conductivity*, Vol. 1, Academic Press, Inc., New York, 1969.

J.R. Welty, *Engineering Heat Transfer*, John Wiley & Sons, Inc., New York, 1974.

ASHRAE Handbook, 1981 Fundamentals, American Society of Heating, Refrigerating and Air Conditioning Engineers, Inc., Atlanta, Ga., 1981.

INSULIN AND OTHER ANTIDIABETIC AGENTS

Since the causes of diabetes are not understood, no totally satisfactory definition of the disease is possible. It is certainly a generalized chronic metabolic condition which, fully developed, may be characterized by hyperglycemia (elevated blood glucose), glycosuria (elevated urine glucose), increase protein breakdown, and ketosis and acidosis (elevated blood ketones: β-hydroxybutyric acid and acetoacetic acid).

Although the disease is very heterogeneous with respect to age onset, severity, treatment, associated endocrine, autoimmune, and metabolic factors, etc, diabetic patients may be classified as either Type I, ketoacidosis-prone (juvenile, insulin-requiring) or Type II, ketoacidosis-resistant (adult, maturity-onset, noninsulin-requiring). The Type I patient generally experiences the onset of the disease prior to age 20 and is dependent upon exogenous insulin to prevent ketoacidosis. Type II patients generally experience the onset of diabetes after age 40, tend to be overweight, and do not require insulin.

Insulin

Insulin is the hormone that promotes the processes by which the various tissues in all parts of the body may use glucose, either as a fuel for liberation of energy, or as storage forms of energy such as glycogen, or fat. In the diabetic subject, exogenous insulin restores, temporarily, the ability to utilize carbohydrates and fats in a comparatively satisfactory manner; the concentration of sugar in the blood may be confined within normal limits; the urine becomes free of sugar and ketone bodies; and diabetic acidosis and coma are prevented.

Chemical properties. Bovine insulin consists of two polypeptide chains with specific amino acid sequences. The A chain has 21 amino acids, and the basic B chain has 30 amino acids. These chains are linked by two disulfide (—S—S—) bonds of cystine, with an additional cystine bridge within the A chain. In aqueous solution, the insulin monomer polymerizes to form macromolecules of 12,000 or 36,000 mol wt, depending on pH, temperature, and concentration. The isoelectric point of insulin is 5.3. It has been shown that about 20% of intact insulin is right-handed α-helix and like other proteins is levorotatory. Preparations of crystalline insulin often contain about 0.5% zinc, the physiological function of which is unknown.

Preparation. Commercial insulin is obtained by a combination of acid–alcohol extractions, isoelectric precipitations, and separation of the hormone as an insoluble salt. In one preparation, frozen bovine or porcine pancreas glands in 45–68 kg lots are ground and placed in large extraction tanks filled with cold acidified 50–85% ethanol (pH 2.5–4.0) to extract the insulin. The mixtures are centrifuged, extracted, and the pH adjusted to 5.5–8.5. The solution is then filtered and its pH is readjusted to 3 with sulfuric acid. By a subsequent evaporation and heat treatment, the fats are separated, the crude insulin solution is filtered through diatomaceous earth, and the clean filtrate is purified by the addition of zinc to produce crystals in an amount of ca 3100–4000 units of insulin per kilogram of glands processed.

Adverse reactions. Insulin overdosage causes weakness, headache, and fatigue, nervousness, tremor, pallor or flushing, and profuse sweating —a characteristic sign. If not treated immediately, the condition may lead to insulin reaction or shock. Allergic reactions to insulin are fairly common, especially local reaction at the site of injection which occurs ten times more frequently than systemic generalized reactions.

Individual evaluations. The seven forms of insulin currently marketed in the United States differ with respect to the time of onset and duration of action. They may be divided into rapid-, intermediate-, and long-acting forms. Globin zinc isophane (NPH), and protamine zinc insulins are conjugated with large protein molecules. As a result, their absorption from subcutaneous sites is delayed and their duration of action is prolonged. Absorption is also delayed by the large particle size and crystalline form of extended insulin zinc suspension (Ultralente insulin). Amorphous Semilente insulin, having a smaller particle size, is more rapidly absorbed, and shorter-acting Lente, which is intermediate acting, is a combination of 70% Ultralente and 30% Semilente insulins.

Oral Antidiabetic Agents

Two groups of drugs form the basis of the current oral therapy of diabetes mellitus: the arylsulfonylureas, known simply as the sulfonylureas (see Table 1), which are the only oral agents used in the United

Table 1. Sulfonylureas Available in the United States

	Acetohexamide (**1**) (Dymelor)	Chlorpropamide (**2**) (Diabenese)	Tolazamide (**3**) (Tolinase)	Tolbutamide (**4**) (Orinase)
daily effective dose[a]	0.25–1.5 g	125–750 mg	0.1–1 g	0.5–3 g
plasma half-life, h[b]	6–8	30–36	7	4–6
tablet sizes, mg	250, 500	100, 250	100, 250, 500	500, 1000
duration, h	12–24	60	≤ 24	6–12

[a] With all compounds except chlorpropamide, divided doses are used when the larger dosage is required.

[b] Biological half-life regarding hypoglycemic potential may be longer.

$$R \!-\!\langle\bigcirc\rangle\!-\! SO_2NHCONH\!-\!R'$$

(**1**) R = CH₃CO—, R' = —$\langle\hexagon\rangle$

(**2**) R = Cl, R'=CH₂CH₂CH₃

(**3**) R = CH₃, R' = —N$\langle\hexagon\rangle$

(**4**) R = CH₃, R' = CH₂CH₂CH₂CH₃

States, and the biguanides. The only commercially available U.S. preparation in the biguanide series of hypoglycemic agents was phenformin until its removal from the U.S. market in 1978. Its action in reducing hyperglycemia occurs without promoting insulin release, and may include delayed gastrointestinal absorption of nutrients as well as direct promotion of glucose uptake by peripheral tissues. The drug did not lower the fasting blood glucose of nondiabetics.

G.F. TUTWILER
McNeil Pharmaceuticals

L. S. Goodman and A. Gilman, eds., *The Pharmacological Basis of Therapeutics*, 6th ed., MacMillan Publishing Co., New York, 1980.

A.J. Lewis, ed., *Modern Drug Encyclopedia and Therapeutic Index*, York Medical Books, New York, 1977.

INTEGRATED CIRCUITS

Silicon integrated circuits are pervasive in all modern electronic systems, most of which would not exist without the availability of inexpensive high performance integrated circuits (ICs).

The production of a modern, complex IC requires a coordinated effort covering three technical areas: the system design determines the overall system architecture, partitioning the system into sub-systems that can be put together with one or more ICs; the circuit design specifies the components with available IC elements; and the technology provides the capability of fabricating the IC. This review is concerned with the latter area and only with silicon IC technology.

An IC is comprised of passive and active electronic devices that are interconnected to provide a desired electronic function. The devices are formed by selectively introducing impurities into the silicon, so as to effect local change in its electrical properties. Two active devices are the metal-oxide semiconductor (MOS) transistor and the bipolar junction transistor; cross sections of these are shown in Figure 1. These devices divide silicon IC technology into two branches, MOS and bipolar; usually both devices are not used on the same IC because of the fabrication complications that would result. The different impurity (electrical) regions, n (excess electrons) and p (excess holes), are introduced into the silicon from the same side of the wafer; silicon IC technology is concerned with the processes used to form the required impurity regions and to interconnect them. The interconnection processes involve the formation and patterning of appropriate conducting and insulating layers (thin films) on the silicon wafer.

Bipolar devices (see Figs. 1b and 1c) require a separate isolation region to prevent interaction between devices, whereas MOS devices are self-

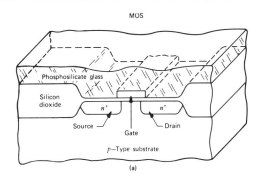

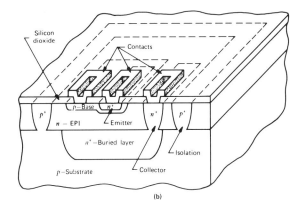

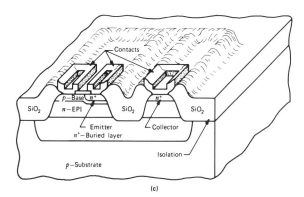

Figure 1. Cross-sections of electronic devices used in ICs. (**a**) MOS transistor, (**b**) bipolar transistor, junction isolated, (**c**) bipolar transistor, oxide isolated.

isolating. Until recently, all bipolar ICs were junction-isolated with the p^+ isolation region being reverse-biased with respect to the n-type collector.

A typical MOS device is shown in Figure 1a. It is an n-channel transistor and, by applying an appropriate voltage to the gate electrode, an n-type conducting region can be induced in the silicon which connects the n^+ source and drain regions. Self-isolation is achieved in normal operation by not allowing the n^+ source and drain to become forward-biased with respect to the p-type substrate. This self-isolation has meant that, in general, MOS ICs can be made with higher packing densities than bipolar ICs.

Manufacture and Processing

Planar process. The majority of ICs are manufactured using the planar process, as illustrated in Figure 2. The patterning and reoxidation steps are repeated as many times as necessary to form the various impurity regions required. The key to the planar process is the layer of silicon dioxide (which can be thermally grown with excellent uniformity and reproducibility for thicknesses from 0.01–2 μm). The thermally grown silicon dioxide is an insulator, and the diffusion coefficient of impurities of interest for selectively doping silicon are significantly lower for silicon dioxide than for silicon. Thus, silicon dioxide can be used as a diffusion mask for forming localized impurity regions in silicon. To serve as a diffusion mask, the silicon dioxide must be patterned by photolithography (see Printing process).

Silicon materials. The substrates that are used for silicon ICs are single-crystal silicon wafers which are precisely aligned to crystallographic axes. An extremely high purity and degree of crystalline perfection are required for optimum performance.

Silicon oxidation. Various techniques exist for the formation of thin films of silicon dioxide upon silicon substrates and include anodization, chemical vapor deposition (CVD), and sputter deposition (see Film deposition techniques); but the majority of thin silicon dioxide layers are formed by thermal oxidation which yields thin silicon dioxide films of excellent uniformity, high passivation quality, and low pinhole density. Furthermore, the films are amorphous and are characterized by an open random lattice of the Si–O tetrahedra.

Prior to any silicon substrate oxidation process, the surface of the wafers must be carefully cleaned, which involves organic and particulate removal, strong oxidizing treatment, chemical oxide removal, and drying.

Photolithography. Photolithography is used to transfer patterns from a mask containing circuit-design information to thin films on the surface of a silicon wafer. The pattern transfer is accomplished with a photoresist, an ultraviolet light-sensitive organic polymer (see Photoreactive polymers). A wafer that is coated with photoresist is illuminated through a mask and the mask pattern is transferred to the photoresist by chemical developers. Further pattern transfer is accomplished by appropriate liquid or gaseous etchants. The photoresist must be resistant to the etches used for patterning thin films of thermally grown silicon dioxide. The pattern in the silicon dioxide can be etched chemically by high purity (electronic-grade) hydrofluoric acid that is buffered with ammonium bifluoride using the exposed and developed photoresist as a mask. Various gaseous plasmas are also used to perform etching. Plasma etching provides better control and minimizes undercutting of the photoresist mask. The masking process is usually repeated 5–10 times in the fabrication of an IC.

Metallization. Metallization provides contact to the silicon and interconnects devices. Silicon contact is made at windows that are etched in a final glass layer which is formed over the entire wafer, either by thermal oxidation or vapor deposition, after all impurity regions have been defined. The metal is deposited uniformly and then patterned, the pattern being determined by the circuit function.

Aluminum is the most widely used metal for metallization and, for MOS ICs, it is predominant. For bipolar devices, two systems are used: one is based on aluminum but is modified to meet unique bipolar requirements, and the other is based on gold (see Electrical connectors).

E.F. LABUDA
J.T. CLEMENS
Bell Telephone Laboratories, Inc.

A.B. Glaser and G.E. Subak-Sharpe, *Integrated Circuit Engineering*, Addison-Wesley Publishing Co., Reading, Mass., 1977.

P.E. Gise and R.E. Glanehard, *Semiconductor and Integrated Circuit Fabrication Techniques*, Reston Publishing Co., Inc., Reston, Va., 1979.

A.S. Grove, *Physics and Technology of Semiconductor Devices*, John Wiley & Sons, Inc., New York, 1967.

IODINE AND IODINE COMPOUNDS

Iodine

Iodine is a nonmetallic element belonging to the halogen family in Group VIIA of the periodic table. It is the heaviest common member of this family and the only one that is solid at ordinary temperatures. The principal valence numbers are -1, $+1$, $+3$, $+5$, and $+7$. An oxide of composition IO_2 with an average valence number of $+4$ is also known. Compounds are known in all these oxidation states (with the single exception of $+4$), that are thermodynamically stable with respect to their constituent elements and can be formed from them. Examples are KI, ICl, ICl_3, IF_5, and Na_5IO_6. Iodine is a simple element, and only one stable atomic species with a mass number of 127 is known, but 10 or more radioisotopes have been prepared artificially (see Isotopes).

Iodine is the forty-seventh most abundant element in the earth's crust, counting the rare earths as a single element. It is widely distributed in nature, occurring in rocks, soils, and underground brines in small quantities; seawater contains about 0.05 ppm. Iodine has been identified in less than a dozen minerals; lautarite (anhydrous calcium iodate), the form in which iodine occurs in the Chilean nitrate deposits, is probably the most important of these.

Physical properties. In massive form, iodine is a soft bluish-black solid, whereas the familiar resublimed material consists of almost opaque, doubly refracting orthorhombic crystals with a nearly metallic luster, a high index of refraction, and pronounced pleochrism. Both forms yield

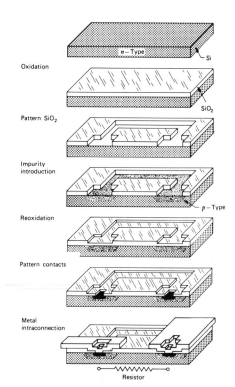

Figure 2. Pictorial view of the planar process used almost exclusively for silicon IC fabrication.

the characteristic violet-colored vapor when heated, hence the name iodine from the Greek, ioeides, violet-colored.

Mp 113.6°C; bp 184°C; d_4^{20} 4.940 g/cm^3 at 101.3 kPa (1 atm) (solid), and n_D 3.34 (solid). Iodine is slightly soluble in water, the solubility increasing with the temperature; no hydrate is formed. Iodine dissolves in many organic solvents.

Chemical properties. Iodine, like the other halogens, is very active chemically, but less violent in its action. As the heaviest of the common halogens, it has a lower electronegativity than the others. As a result, the iodides are less stable than the other halides and the iodine can be readily replaced by the other halogens. On the other hand, its oxides and other compounds in which iodine is in a positive valence state are much more stable than those of the other halogens, and iodine can replace chlorine and bromine in such compounds.

Iodine forms binary compounds with all other elements except sulfur, selenium, and the noble gases. It does not react directly with carbon, nitrogen, or oxygen, and only at high temperature with platinum, but it does react with most other elements.

Production and processing. In the production of iodine from brine, the first step is the clarification of the brine to remove oil and other suspended material. In one process, a silver nitrate solution is added to precipitate silver iodide, which is filtered and treated with scrap iron to form metallic silver and a solution of ferrous iodide. The silver is redissolved in nitric acid for recycle, and the solution is treated with chlorine to liberate the iodine.

In another process, chlorine is added after clarification to liberate the iodine as the free element in solution. This solution is then passed over bales of copper wire, and insoluble cuprous iodide precipitates. At intervals, the bales are agitated with water to separate the adhering iodide; the bales are then recycled. The cuprous iodide suspended in the water is filtered, dried, and shipped as such.

Much of the U.S. iodine production employs a chlorine–oxidation, air-blowout method for the recovery of iodine as the free element from natural, subsurface brines in Oklahoma and Michigan. The brines, containing in excess of 100 ppm iodine in Oklahoma and 30–40 ppm iodine in Michigan, are acidified with sulfuric acid and treated with a slight excess of chlorine to liberate the iodine in a denuding tower, where the brine gives up its iodine to a countercurrent stream of air. The iodine-enriched air then passes to a second tower where the iodine is absorbed by a solution of hydriodic and sulfuric acids. This solution is treated with sulfur dioxide to reduce the iodine to hydriodic acid; part is drawn off to a reactor for the recovery of iodine, and the remainder is recirculated to the absorption tower. The liquor in the reactor is treated with chlorine and the liberated iodine is settled, filtered, and melted in a kettle under concentrated sulfuric acid. It is then either poured onto a drum flaker or cast into pigs. Brine stripped of iodine is returned to its source. This process is similar to that for the recovery of bromine from seawater (see Bromine).

Health and safety. Iodine is much safer to handle at ordinary temperatures than the other halogens, because it is a solid and its vapor pressure is only 1 kPa (7.5 mm Hg) at 25°C, compared to ca 28.7 kPa (215 mm Hg) for bromine and 700 kPa (6.91 atm) for chlorine. However, the vapor is irritating to the lungs and eyes and prolonged exposure to a high concentration should be avoided.

Uses. At present well over half of the iodine produced is utilized in industrial applications, but significant amounts are used in nutrition, sanitation, and medicine.

Inorganic Iodine Compounds

Iodides. In general, the iodides are very soluble in water and many are hygroscopic. However, a few are insoluble, eg, cuprous, lead, silver, and mercurous iodides. With the exception of cuprous iodide, they dissolve in concentrated alkali iodide solutions. Some iodides are more stable in contact with water than the chlorides and bromides. A number of iodides are also soluble in nonaqueous solvents, eg, mercuric iodide is more soluble in ethanol and acetone than in water. Potassium iodide is much more soluble in acetone than the chloride, and also dissolves in liquid sulfur dioxide and ammonia.

The iodides have less tendency to form complexes than the other halides. They vary widely in their stability to heat. The iodides of alkali metals and those of heavier alkaline earths are resistant to oxygen on heating, but most others can be roasted to oxide in air or oxygen.

Chlorine and bromine readily displace iodine from the iodides, converting them to the corresponding chlorides and bromides.

Alkali iodides, dissolved in acetone and other organic solvents, react with aliphatic chloro and bromo compounds to give the corresponding iodo compounds, a convenient method of preparation for many of these substances.

Organic Iodine Compounds

Organic iodine compounds differ markedly from their chlorine and bromine analogues. They have a higher density, lower vapor pressure, greater reactivity, and a correspondingly lower stability. Because of a tendency to lose hydriodic acid and the relatively high cost of iodine compared with that of bromine or chlorine, their use is limited to specialty applications such as pharmaceutical and organic intermediates for special syntheses. Aliphatic iodine compounds are formed by the reaction of an alcohol with phosphorus triiodide or hydriodic acid; the addition of iodine monochloride or monobromide and, in a few cases, iodine itself to an olefin; replacement reactions in which the organic compound containing chlorine or bromine is heated with an alkali iodide in a suitable solvent; and the reaction of triphenyl phosphite with methyl iodide and an alcohol. For the preparation of aromatic iodine compounds, oxidizing agents such as nitric acid, fuming sulfuric acid, or mercuric oxide react with elemental iodine and the aromatic system. Organic iodine compounds are used in relatively small quantities in industry.

Charles J. Mazac
PPG Industries

F.G. Sawyer, F.G. Ohman, and F.E. Lush, *Ind. Eng. Chem.* **41**, 1547 (1949); Industrial and Engineering Chemistry Staff, eds., *Modern Chemical Processes*, Vol. 1, Reinhold Publishing Corp., New York, 1950, p. 26.

Minerals Yearbook, U.S. Dept. of Interior, Bureau of Mines, U.S. Govt. Printing Office, Washington, D.C., 1981.

Mineral Industry Surveys, Annual Advanced Summary, Iodine in 1978, U.S. Dept. of Interior, Bureau of Mines, Washington, D.C., Aug. 31, 1979, pp. 1–5.

IODINE VALUE. See Fats and fatty oils; Carboxylic acids.

IODOACETIC ACID. See Acetic acid.

IODOFLUOROHYDROCARBONS. See Fluorine compounds, organic.

ION EXCHANGE

Ion exchange is the reversible interchange of ions between a solid and a liquid in which there is no permanent change in the structure of the solid that is the ion-exchange material. The utility of ion exchange rests with the ability to use and reuse the ion-exchange materials.

Physical Properties of Resins

Conventional ion-exchange materials contain ion-active sites throughout their structure with a uniform distribution of activity, as a first approximation. A cation-exchange resin with a negatively charged matrix and exchangeable positive ions (cations) is shown in Figure 1.

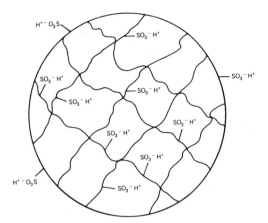

Figure 1. Cation-exchange resin schematic showing negatively charged matrix and exchangeable positive ions.

Chemical Properties of Resins

Capacity. Total ion-exchange capacity, ie, the total number of sites available for exchange, normally is determined after converting the resin by chemical regeneration techniques to a given ionic form. The ion then may be chemically removed from a measured quantity of the resin and quantitatively determined in solution by any one of several conventional analytical methods.

Operating capacity is a measure of the useful performance obtained with the ion-exchange material when it is operating in a column under a prescribed set of conditions. It is determined by the inherent (total) capacity of the resin, the level of regeneration (the extent to which the resin has been converted to the proper ionic form), the composition of the solution treated, resin specificity and, for one ion with respect to another, the flow rates through the column, temperature, particle size, and numerous other factors.

Swelling equilibria. The solvent-retention capacity of an ion exchanger is a reproducible, equilibrium quantity that is dependent upon ion-exchange capacity, ionic form, the kind of solvent used, the composition of the solution, the extent of cross-linkage, relative humidity, and temperature. Water swelling of an ion exchanger is primarily a hydration of the fixed ionic groups.

Ion equilibria. Ion-exchange reactions are reversible. By washing a resin with an excess of electrolyte, the resin can be converted entirely to the desired salt form:

$$R^- A^+ + B^+ \rightarrow R^- B^+ + A^+$$

However, with a limited quantity of solution B^+ in batch contact, a reproducible equilibrium is established which is dependent on the proportions of A^+ and B^+ and on the selectivity of the resin.

Kinetics. Overall exchange rates may be influenced by a change in solvent nature and content, particle size, temperature, and the functional group (kind and concentration). Exchange rates are most rapid in water systems, and become increasingly slow with less polar solvents.

Manufacture

The manufacture of ion-exchange resins is a combination of processes including the preparation of a cross-linked bead copolymer by suspension polymerization, sulfonation, or chloromethylation and the amination of the copolymer and an ionic conversion. A schematic layout of an ion-exchange production facility is given in Figure 2.

The copolymerization of styrene with varying amounts of divinylbenzene (DVB) is carried out in an agitated vessel, in suspension in an aqueous phase. Suspending agents and reaction agitation control the particle size and particle size distribution of the resulting spherical copolymer beads. It is at this point that the properties of the resin can be modified by changes in DVB content or by the addition of a diluent. Following the copolymerization under controlled temperature, catalyst, and agitation conditions, the copolymer is transported to a vessel where it is washed with water of controlled quality to remove the suspending agents. It is dried by centrifuging and/or thermal drying and then is

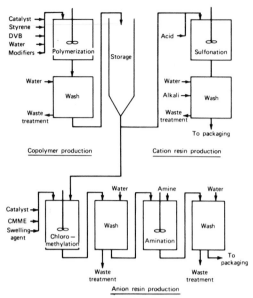

Figure 2. Ion-exchange production processes. DVB = divinylbenzene; and CMME = chloromethyl methyl ether.

sent to a screening operation to obtain the desired particle size distribution. Sulfonation of the copolymer using sulfuric acid yields a cation-exchange resin, whereas chloromethylation and amination of the copolymer yields an anion-exchange resin.

Health and Safety

As manufactured, the only resin ionic forms that present any significant chemical safety hazards are the hydrogen forms of the cation-exchange resins and the hydroxide forms of the anion-exchange resins; these resins produce acid or base, respectively, when contacted with salt solution. Such toxic ions as chromate, cyanide, or heavy metals may be concentrated by exchange from operating systems.

Uses

Ion exchange is used in water softening and deionization. An example of a demineralization reaction–cation exchange is

$$R^- H^+ + Na^+ Cl^- \rightleftarrows R^- Na^+ + H^+ Cl^-$$

anion exchange which, in series, completes the demineralization:

$$R^+ OH^- + H^+ Cl^- \rightarrow R^+ Cl^- + HOH$$

(Water softening accounts for the major tonnage of resin sales.) It also provides a method of separation that is useful in many chemical processes and in analyses. It has special utility in chemical synthesis, medical research, food processing (qv), mining, agriculture, and a variety of other areas. Radiation waste systems in nuclear power plants include mixed bed ion-exchange systems for the removal of trace quantities of radioactive nuclides from water that will be released to the environment.

R.M. WHEATON
L.J. LEFEVRE
Dow Chemical U.S.A.

F. Helfferich, *Ion Exchange*, McGraw-Hill, New York, 1962.

S.B. Applebaum, *Demineralization by Ion Exchange*, Academic Press Inc., New York, 1968.

W. Rieman, III, and H.F. Walton, *Ion Exchange in Analytical Chemistry*, Pergamon Press, New York, 1970.

ION IMPLANTATION

Ion implantation is a process for electrically injecting atoms of any element into any solid material to selected depths and concentrations in

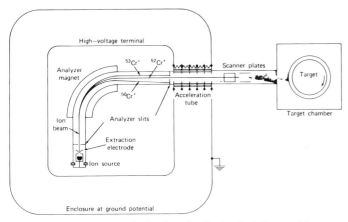

Figure 1. A schematic drawing of an ion-implantation machine.

order to produce an alloy or other solid mixture that has a different composition from the original solid and that, therefore, exhibits different and sometimes highly desirable chemical and physical properties. An ion-implantation accelerator, illustrated schematically in Figure 1, is similar to an isotope separator but has an added final acceleration stage (100–1000 keV) for giving the selected ions enough energy to penetrate significantly the lattice of the target (or workpiece) material. Atoms of the selected chemical element are ionized by collisions with electrons in an electrical discharge in a gas at low pressure. These ions pass through an orifice into a high vacuum region where they are accelerated by a negative potential on the extraction electrode to a moderate energy (10–30 keV) and then separated on the basis of mass by a magnetic field. Figure 1 illustrates the passage of chromium isotopes through the analyzer system, the acceleration to final energy, and the scanning of the beam over the exposed portion of the workpiece.

Ion – Target Interactions

Atomic collisions. When the ions strike the workpiece, they collide with the lattice atoms, lose energy, and after a sufficient number of collisions come to rest. For calculation purposes in predicting the depth distributions of various species of ions and energies, implanted in various solid materials, two types of energy transfer are considered: elastic transfer (in which the two partially screened nuclei experience each other's Coulomb field, thereby transferring momentum to the lattice atoms as in a billiard-ball collision) and inelastic transfer (in which energy is transferred to the electrons of the lattice atoms).

Ion range. Variations in the range (the projected length of the ion path in the solid along to the beam's initial velocity vector) from ion to ion arise because of statistical fluctuations in the number of elastic collisions and in the directions of the ion after each such collision. The calculated (and measured) shape of the distribution of ion ranges is approximately the same as that of the normal error function. Typically the mean range for implantation of metals for modification of surface properties is a few tens of nanometers.

Metastable alloys. Because the ions are individually deposited at selected depths, the equilibrium laws of thermodynamics governing thermal diffusivity and solubility limits do not prevail; hence, the required concentrations of ions can be achieved quickly at room temperature, and supersaturated solid solutions can be prepared. Such capabilities enable the preparation of alloys that are predicted to have desirable properties but are not producible by conventional means.

Radiation damage. To displace an atom from its lattice site requires the expenditure of a minimum amount of energy (20–40 eV) for most materials; the average displacement energy E_d is ca 25 eV. When a collision between an incident ion and a target atom causes an energy transfer that is much greater than E_d, the struck atom collides with other atoms, knocking some of them out of their lattice sites. The process creates a cascade of collisions, producing secondary and tertiary knock-on atoms. Thus an ion leaves a path of severe lattice disarrangement in its wake. But most of these disarrangements are promptly

corrected by spontaneous movement of defects (interstitials and vacancies being attracted to each other). The residual lattice disarrangement is called radiation damage.

Although radiation damage is almost always deleterious in semiconductor devices, it may be beneficial in metal surfaces, because the residual interstitial atoms have the effect of wedges driven into a surface. Since the much thicker substrate does not permit the thin surface layer to expand under the action of the wedges, the surface layer is under compressive stress, which tends to prevent cracks from forming and propagating, as happens in the fatigue failure of a part.

Sputtering. Whenever the cascade of knock-on atoms intersects the surface, an atom may sputter, or leave the surface, if its energy exceeds the surface binding energy (usually 2–5 eV). Sputtering, which is another form of radiation damage, is a significant factor for heavy implantation doses, which are common for metal targets, because it affects the depth concentration profile of implanted ions and limits the maximum achievable concentration. The sputtering yield (defined to be the number of sputtered atoms per incident ion) depends on ion energy, species, flux, angle of incidence, target material, dose, crystal state, and surface binding energy. For incident ions from the lower half of the periodic table, sputtering yields generally are 1–10 (see also Film deposition techniques).

Power dissipation. The power density of the beam at the target can be sufficient to cause thermal damage if precautions are not taken. For example, a 5-mA beam of 200-keV ions has a power of 1 kW. If the beam has a cross section of 2 cm^2, then the energy-transport density is 5000 kW/m^2. Such power densities are not readily dissipated by a surface in a vacuum. For beam currents below about 2 mA, the beam can be electromagnetically scanned over the front surface of the workpiece (in much the same manner as the raster of a television picture tube) to prevent unwanted temperature excursions. For larger beam currents, electromagnetic manipulation of the beam is ineffective because of neutralizing electron space-charge effects; hence, it is necessary to rotate (or otherwise manipulate) the workpieces to control temperature excursions. Such techniques can keep the workpiece temperature excursions within tolerable limits (eg, 100°C).

Applications

For more than a decade, ion implantation has been a chief step in the industrial processing of semiconductor devices; in particular, large-scale integrated circuits. Also, a large amount of research effort in a number of laboratories worldwide has shown that the implantation of appropriate doses of selected ions into metals can produce major changes in the surface properties, eg, greatly reduced vulnerability to corrosion, sliding wear, or fatigue. Treatments for the reduction of wear have been reduced to commercial practice in the United Kingdom. Examples are mill rolls for metal-rolling operations, injection molds for plastic parts, paper cutters, and wire-drawing dies. Increases in lifetimes for these items as a result of implantation treatments range from a factor of two to a factor of ten. Typical treatments involve the implantation of 100-keV nitrogen ions to a fluence of about 2×10^{17} ions/cm^2 into the critical wear surface of each item.

An example of the reduction of corrosion is the implantation of chromium ions and phosphorus ions with energies of about 100-keV and fluences of ca 10^{17} ions/cm^2 into the surfaces of high performance-aircraft-engine bearings. Laboratory tests have shown a remarkable increase in resistance to corrosion under accelerated conditions, and in-service flight tests are being conducted (1982–1984).

Although most experimental effort on the implantation of metals has involved wear and corrosion properties, there is a substantial amount of evidence that indicates that the fatigue resistance of metals can be substantially improved by the implantation of nitrogen or certain other ions. Other experiments have shown that the index of refraction of optical-grade silicon can be modified as a function of depth by the implantation of nitrogen ions.

JAMES W. BUTLER
Naval Research Laboratory

J.K. Hirvonen, ed., *Ion Implantation*, Academic Press, Inc., New York, 1980.

V. Ashworth, W.A. Grant, and R.P.M. Procter, eds., *Ion Implantation Into Metals*, Pergamon Press, Oxford, 1982.

S.T. Picraux and W.J. Choyke, eds., *Metastable Materials Formation by Ion Implantation*, Elsevier, Amsterdam, Neth., 1982.

IONOMERS

Ionomers contain a certain number of inorganic salt groups attached to a polymer chain. This article describes ionomers with a maximum of ca 10 mol% ionic groups pendent to a hydrocarbon or perfluorinated polymer chain. The pendent ionic groups interact to form ion-rich aggregates contained in the nonpolar polymer matrix. A typical ionomer structure can be depicted as follows:

$$-[CH_2CH_2]_{\overline{m}}[CH_2\underset{\underset{O}{\overset{|}{\underset{}{CO^-M^-}}}}{\overset{CH_3}{\overset{|}{C}}}]-$$

Typically, the ratio of m/n is on the order of 10–100, reflecting a low overall content of ionic groups.

The ionic interactions and resultant properties are dependent on the type of polymer backbone (plastic or elastomeric), ionic functionality (ionic content, generally 0–10%), type of ionic moiety (carboxylate, sulfonate, or phosphonate), degree of neutralization (0–100%), and type of cation (amine, metal, monovalent, or multivalent).

Structure

Analytical methods applied to the structure elucidation include small-angle x-ray scattering, small-angle neutron scattering, transmission electron microscopy, ir spectroscopy, Raman spectroscopy, Mossbauer spectroscopy, and direct measurements of mechanical and dielectric properties.

Above a critical ion concentration, two types of ion aggregates can be defined: multiplets and clusters.

At higher ion-pair contents, the aggregation of multiplets forms clusters. This aggregation is favored by electrostatic interaction of the ion pairs within the multiplets, but opposed by the forces caused by the elastic nature of the hydrocarbon polymer chains. Clusters, therefore, contain ion-pair aggregates interspersed with polymer backbone chains. As the temperature of an ionomer system is increased, the clusters contained in the polymer become unstable, and the multiplets redistribute.

Various models for ionic aggregation have been proposed, but no single model can adequately describe the wide range of ionomer structures. Most systems are intermediate between the homogeneous aggregate and the phase-separated cluster, depending on backbone polarity and ionic functionality.

Properties

The poly(ethylene-*co*-methacrylic acid)s, introduced by DuPont in 1964, were the first ionomers whose physical properties were recognized as a consequence of phase-separated, ionic-rich domains (see Table 1).

Clearly, tensile strength is enhanced with increasing neutralization, and the melt flow is markedly reduced. Thus, the presence of even monovalent cations neutralizing the acrylic acid moiety gives rise to an ionic association which increases the viscosity markedly and leads to an enhancement of physical properties as progressively more of the acid species is ionized.

The number of ionic groups existing in clusters is strongly dependent on polarity of the polymer matrix, ionic functionality, and temperature.

The rheological behavior of ionomers offers convincing evidence that cluster formation or microphase separation has a significant impact on properties. With poly(styrene-*co*-methacrylic acid) ionomers, for example, stress–relaxation data at low ionic content (< 6 mol%) give satisfactory time–temperature curves. A significant difference in water absorption is observed for polymers of different ion contents.

The dynamic mechanical properties of ionomer systems provide definitive evidence that salt forms of these systems are dramatically different from either the acid form or the parent polymers. The neutralized forms of these materials generally display a rubbery plateau in the modulus–temperature curves that is not present in the parent materials.

Preparation and Manufacture

Ionomers are prepared by copolymerization of a functionalized monomer with an olefinic unsaturated monomer or direct functionalization of a preformed polymer.

Poly(ethylene-*co*-methacrylic acid) ionomers. The copolymerization of ethylene and acrylic or methacrylic acid is effected at high pressures using a free-radical initiator. Typically, 3–6 mol% acid is incorporated in the commercial polymers. Neutralization can be effectively achieved with bulk polymer on two-roll rubber mills at 150–200°C by addition of sodium hydroxide or other bases in water to the fluxed copolymer. As the water evaporates, the melt viscosity increases markedly, and at sufficiently high neutralization, the ionomer can be stripped off the mill to yield a tough, flexible sheet.

Carboxylated elastomers. Acrylic and methacrylic acid groups are incorporated into synthetic elastomers via free-radical copolymerization.

Table 1.   Properties of Poly(Ethylene-*co*-Acrylic Acid) Salts

	Percent neutralized	Melt index at 298 kPa, 10 g/min[a]	Melt index at ca 3 MPa, 10 g/min[b]	Secant modulus 1% extension, MPa[a]	Ultimate tensile strength, MPa[a]	Elongation at break, %
control material, 14.8 wt% acrylic acid	0	67	2.57	48.26	14.8	470
sodium salt	12.0	12.2	256	230	21.7	420
	30.0	3.9	92	540	27.6	330
	47.5	1.0	30	293	31.7	310
	66.0	0.3	7.6	273	33.0	280
potassium salt	8.0	16.3	360	202	21.0	470
	25.0	4.5	110	362	25.5	410
	51.0	2.7	49	341	30.6	370
	63.0	0.57	15	308	34.4	390
lithium salt	12.0	18.7	442	181	21.7	410
	28.5	5.2	116	337	26.5	350
	52.5	1.4	38	334	28.2	260
	67.5	0.2	5.4	254	31.7	250

[a]To convert kPa to psi, multiply by 0.145.

[b]To convert MPa to psi, multiply by 145.

In telechelic polymers, the carboxyl functionality terminates both ends of the polymer chain.

Most commercial carboxylated elastomers are prepared by emulsion polymerization. In acidic emulsion formulations, the free acid copolymerizes much more readily than the neutral salt.

Sulfonated ethylene–propylene–diene terpolymers. Sulfonate groups are introduced into ethylene–propylene–diene monomer (EPDM) systems via electrophilic attack of the sulfonation reagent on the polymer unsaturation. A number of different diene termonomers, amenable to sulfonation, can be copolymerized with ethylene and propylene. A preferred starting material is 5-ethylidene-2-norbornene (ENB). Polymer sulfonic acids are not highly associated, and are thermally less stable than metal-neutralized sulfo-EPDM. When treated with metallic bases, bulk and solution properties change markedly (see Elastomers, synthetic).

XR resin is a high molecular-weight polymer which can be melt-fabricated and processed into sheet or tubes by standard techniques. Hydrolysis gives the perfluorosulfonate ionomer Nafion, which has an equivalent weight of ca 1000–1500 per sulfonate group.

Perfluorocarboxylate ionomers (Flemion) are based on the copolymerization of the appropriate functional monomers with tetrafluoroethylene and subsequent membrane fabrication (see Fluorine compounds, organic).

$$-[CF_2CF_2]_n CF_2CF- \qquad \qquad O$$
$$\bullet \quad [OCF_2CF]_m O-[CF_2]-CO^- \quad NA^+$$
$$| \qquad \qquad$$
$$CF_3$$

(*m* = 0 or 1, *n* = 1–5)
Flemion polymer

Plasticization

In ionomers, plasticization offers an advantage, since either the polymer backbone or the inorganic salt groups respond to fugitive and nonfugitive additives in a selective manner. For example, the addition of 5 wt% glycerol as a plasticizer in sulfonated polystyrene (S-PS) reduces the melt viscosity by a factor of 1000 at elevated temperatures. To obtain the same degree of reduction, 40% dioctyl phthalate is required.

Sulfo-EPDM can be plasticized with metal carboxylates to enhance tensile properties and lower melt viscosity. A variety of thermoplastic elastomers can be created by the addition of plasticizers.

The behavior of ionomers in dilute solution can be interpreted as a special case of plasticization phenomena where solvents interact to different degrees with the polymer backbone or the ionic groups.

Sulfo-EPDM and sulfonated polystyrene of low functionality are effective thickeners for hydrocarbon solvents. At sufficiently high metal sulfonate content, sulfonated ionomers are insoluble in hydrocarbon diluents.

Elastomers

The copolymerization of methacrylic acid with 1,3-butadiene leads to a tougher, less elastic elastomer than is obtained with polybutadiene.

Sulfonated ethylene–propylene–diene terpolymers. The sulfonation of EPDM can be conducted in hydrocarbon diluents at low concentration without substantial change in the molecular weight. The polymer sulfonic acid is neutralized, preferably with a metal acetate, to yield a fully neutralized metal-sulfonated EPDM. The metal cation affects the melt viscosities and other physical properties (see Table 2).

Thermoplastics

Ionomers based on the polyethylene backbone exhibit a number of tensile properties, good clarity, and high melt viscosities, but a marked decrease in melt flow and in haze.

In a study of the behavior of poly(styrene-*co*-sodium methacrylate), at low ionic contents, the glass transition increases, consistent with multiplet formation. At ionic contents > 6 mol%, a pronounced rubbery plateau appears.

Lightly sulfonated polystyrene (S-PS), generally as the sodium salt, is obtained by sulfonation of polystyrene. The melt viscosity of fully neutralized S-PS is extremely high compared to that of polystyrene, the

Table 2. Effect of Cation on Flow and Physical Properties of Sulfo-EPDM^a

Metal	Apparent viscosity,^{b} μPa·s^c	Melt fracture at shear rate, Hz	Melt index, 190°C, 3.3 MPa,^{d} 10 g/min	Room temperature	
				Tensile strength, MPa^d	Elongation, %
Mg	55.0	< 0.88	0	2.2	70
Co	52.3	< 0.88	0	8.1	290
Li	51.5	< 0.88	0	5.2	320
Na	50.6	< 0.88	0	6.6	350
Zn	12.0	147	0.75	10.2	400

^{a}Sulfonate content: 31 meq/100 EPDM.
^{b}At 200°C and 0.88 s^{-1}.
^{c}To convert μPa·s to centipoise, divide by 1000.
^{d}To convert MPa to psi, multiply by 145.

unneutralized polymer sulfonic acid, or even to an analogous sodium carboxylated polystyrene.

Health and Safety Factors

Ethylene–methacrylic acid-based ionomers (Surlyn) produce no hazardous by-products upon heating to normal processing temperatures (160–300°C). Sulfonic acid polymers are of very low toxicity.

Uses

Surlyn is used in packaging films, including vacuum packaging for processed meats, and skin packaging for electronic and hardware items.

Sulfo-EPDM is used in rubber applications, eg, adhesives, impact modifiers, footwear, calendered sheets, and garden hoses.

Nafion products exhibit outstanding chemical and thermal stability. They are used in membrane applications as films and tubing-fuel cells, electrodialysis, spent acid regeneration, and selectively permeable separations in chemical processing (see Membrane technology).

ROBERT D. LUNDBERG
Exxon Research and Engineering Company

W.J. MacKnight and T.R. Earnest, *J. Macromol. Res.* **16**, 41 (1981).

C.G. Bazuin and A. Eisenberg, *Ind. Eng. Chem. Prod. Res. Dev.* **20**, 271 (1981).

A. Eisenberg and M. King, *Ion-Containing Polymers*, Academic Press, Inc., New York, 1977.

IONOPHORES. See Antibiotics, polyethers; Antibiotics, polypeptides; Chelating agents.

ION-SELECTIVE ELECTRODES

Ion-selective electrodes (ISEs) are membrane electrodes as compared to the older and better known metal/metal-salt redox electrodes. ISEs are electrochemical devices whose voltage output at virtually zero current is directly related to the concentration of some species which, generally, is in solution (see also Membrane technology).

Modern solid-state ISEs, which are based on crystalline solids rather than glass, date back to the work of Kolthoff in 1937, when it was shown that silver halide membranes could function as the sensing element in electrodes. However, such membranes were light-sensitive, and it was not until the discovery in 1966 that the addition of Ag_2S to the halide eliminated the light sensitivity that such electrodes became feasible commercially.

Types of Electrodes

The properties and performance of the electrode depend on the composition and choice of membrane material.

Glass. Commercially available glass electrodes include pH and sodium electrodes. Glass electrodes consist of a three-dimensional silicate network containing occasional SiO$^-$ sites. Because of the relatively open network structure, it is possible for monovalent cations to move from site to site. Although this provides a convenient pathway for monovalent cations, anions cannot enter, and glass electrodes usually have a response that is very close to theoretical. Because polyvalent cations are held at more than one site simultaneously, their mobility in glass is very low, and no commercially useful divalent glass electrode has been found.

Since monovalent cations are mobile in the silicate network, the selectivity of glass electrodes depends upon ion-exchange sites at the surface. In the case of pH electrodes, there is general agreement that a hydrated silica layer occurs at the surface and is responsible for the selectivity. For sodium electrodes, the substitution of some aluminum for silica results in more acidic sites and less tendency to bind hydrogen ion.

Liquid-membrane. Liquid-membrane electrodes are similar, both in mechanism and in response to interferences, to the glass electrodes. The difference is that in liquid-membrane electrodes, the ion-exchange sites are dissolved in a high molecular weight, water-insoluble organic solvent, and are themselves mobile.

Commercially available liquid-membrane electrodes include calcium, fluoroborate, nitrate, potassium, and water-hardness electrodes.

Solid-state. With the exception of the fluoride electrode, all of the commercial solid-state electrodes are silver-based (eg, chloride, cupric, cyanide, iodide, lead, and silver/sulfide). In a number of silver compounds (eg, the silver halides and silver sulfide), there are more lattice sites available than silver ions which are necessary for charge neutralization. Since a number of silver salts are quite water-insoluble, they are almost ideal membrane materials. Any of them can be used as specific electrodes for the silver ion.

The fluoride electrode, which has a sensing membrane made of a rare earth fluoride (eg, LaF$_3$), with or without doping to increase conductivity, is the only based membrane material to achieve commercial success. Hexagonally crystallizing rare-earth fluorides (LaF$_3$, CeF$_3$, NdF$_3$, and SmF$_3$) have a lattice structure in which there are alternate layers of LaF^{2+} and F$^-$ ions; the latter require only a small activation energy for mobility.

Because of the small size of the fluoride ion, the membrane does not appear to be permeable to any other species.

The fluoride electrode is made from disks that are sliced from a rod of single-crystal LaF$_3$, which is then sealed to an epoxy tube.

Gas sensing. In gas-sensing electrodes, the sending electrode is placed in a small amount of solution which is separated from the sample by a water-impermeable, gas-permeable membrane. In the case of the CO$_2$ electrode, a pH electrode and a reference electrode are in an extremely small amount of solution on one side of a silicone rubber membrane, and the sample solution is on the other side. CO$_2$ diffuses through the silicone rubber and changes the pH of the solution on the inside.

$$CO_2 + H_2O \rightarrow HCO_3^- + H^+$$

If its level is fixed by placing a relatively large amount of bicarbonate ion in the inside solution, the pH of that solution will vary directly with the CO$_2$ vapor pressure in the external sample.

Ammonia and carbon dioxide electrodes are commercially available forms of this type.

Uses

The nitrate electrode is used for soil-nitrate measurements, to indicate soil-fertilizer requirements (see Fertilizers). Biomedical applications are measurements of Na$^+$ and K$^+$ in whole blood and the measurement of ionized calcium in blood serum (see Blood). In electroplating, the cyanide electrode is used to measure cyanide in plating baths (after dilution to 10^{-3} M or less) and in waste streams. In steam and power uses, the purity of feedwater and steam (qv) condensate in the ppb range is monitored by sodium electrodes. In petroleum refining, chloride is measured at low levels in water that is used to desalt crude oil, and NH$_3$, H$_2$S, HCN, and HCl trapped by sour-water scrubbers and measured by the appropriate electrode.

MARTIN S. FRANT
Foxboro Analytical,
a division of The Foxboro Company

G.J. Moody and J.D.R. Thomas, *Sensitive Ion-Selective Electrodes*, Merrow Publishing Co., Watford, UK, 1971.

P.L. Bailey, *Analysis with Ion-Selective Electrodes*, Heyden, London, UK, 1976.

Handbook of Electrode Technology, Orion Research, Inc., Cambridge, Mass., 1982.

IRIDIUM.   See Platinum-group metals.

IRON

Iron, Fe, is in Group VIII of the periodic table and is between manganese and cobalt in period 4. It has a valence of +3 or +2 and combines readily with other elements. Iron has four stable isotopes, of mass 54 (6.04%), 56 (91.57%), 57 (2.11%), and 58 (0.28%); it is the fourth most abundant element in the earth's crust, outranked only by aluminum, silicon, and oxygen. Although gold, silver, copper, brass, and bronze were in common use before iron, it was not until humans discovered how to extract iron from its ores that civilization developed rapidly.

Pure iron is a silvery white, relatively soft metal that has a bcc crystal lattice at normal temperatures. Native metallic iron is rarely found in nature because iron combines readily with other elements, eg, oxygen and sulfur. Iron oxides are the most prevalent natural form of iron. Generally, these iron oxides (iron ores) are reduced to iron in a blast furnace. The iron from the blast furnace is refined in steelmaking furnaces to make steel (qv).

Physical and Chemical Properties

Properties of high purity iron are given in Table 1.

Beneficiation

As it is applied to iron ores, beneficiation means to improve the ore's chemical and/or physical properties. When the iron content of the ore is increased, the process generally is referred to as concentration; when the physical structure is improved by making larger particles out of smaller particles, the process is referred to as agglomeration. Some ores can be concentrated simply by crushing, screening, and washing. Other ores must be ground to very small particles before the iron oxides can be separated from the rest of the material. This separation normally is accomplished by magnetic drums (see Magnetic separation) or by flotation (qv). Agglomeration is one of the principal methods of physical beneficiation of iron ores. Fine particles, whether in natural ores or in concentrates, are undesirable as blast-furnace feed; the most desirable size for blast-furnace feed is 6–25 μm. Of the numerous methods available, sintering and pelletizing are the most important (see Size enlargement).

Blast furnace. By far the most important method for the manufacture of iron is the blast-furnace process. The blast furnace essentially is a large chemical reactor into which are charged iron ore, coke, and limestone. As the iron ore descends through the furnace, it is reduced to iron which is melted by a countercurrent flow of hot, reducing gas. The gas is produced by burning coke with preheated air near the bottom of the furnace. The molten iron and slag (formed from the gangue in the ore,

Table 1.   Properties of Relatively Pure Iron

Property	Value
mp, °C	$1532 \pm 5°$
bp, °C	$3000 \pm 150°$
density (hot rolled), g/cm^3	7.87
thermal conductivity, at 0°C, W/(m·K)	79
electrical resistivity, at 20°C, $\mu\Omega \cdot$cm	9.71
tensile strength, MPa^a	245–280
Brinell hardness	82–100

^{a}To convert MPa to psi, multiply by 145.

Table 2. Principal Iron-Bearing Minerals

Class and mineralogical name	Chemical composition of pure mineral	Common designation
oxide		
magnetite	Fe_3O_4	ferrous–ferric oxide
hematite	Fe_2O_3	ferric oxide
ilmenite	$FeTiO_3$	iron–titanium oxide
limonite	$HFeO_2{}^a$	
	$FeO(OH)^b$	hydrous iron oxides
carbonate		
siderite	$FeCO_3$	iron carbonate
silicate		
chamosite	various	
silomelane	and	iron silicates
greenalite	sometimes	
minnesotaite	complex	
grunerite		
sulfide		
pyrite (iron pyrites)	FeS_2	
marcasite (white iron pyrites)	FeS_2	iron sulfides
pyrrhotite (magnetic iron pyrites)	FeS_2	

^{a} Goethite.
^{b} Lepidocrocite.

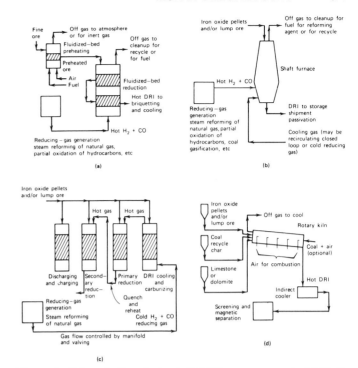

Figure 1. Schematic flow sheets of chief DR process concepts. (**a**) Fluidized-bed processes. (**b**) Moving-bed shaft processes. (**c**) Fixed-bed retort processes. (**d**) Rotary-kiln processes. DRI = direct-reduced iron.

the ash in the coke, and the limestone) are removed about every two to four hours from the hearth of the furnace. The raw materials are charged continually into the top of the furnace, thus making the blast furnace a continuous process. The blast furnace is one of the most efficient countercurrent processes. The main product from the blast furnace is hot metal or pig iron, generally refined to make steel or used to make iron castings.

Direct reduction. The next most important process for the reduction of iron ore is direct-reduction processes. Although these processes do not account for a large percentage of the world's production of iron, they are alternatives to the blast-furnace process and production from direct reduction is increasing every year.

Minerals. Iron-bearing minerals are numerous and are present in most soils and rocks. However, only a few minerals are important sources of iron (see Table 2). Commercially, the oxides are the most important class of minerals, followed by carbonates, silicates, and sulfides.

W.A. KNEPPER
United States Steel Corporation

H.E. McGannon, *The Making, Shaping, and Treating of Steel*, 8th ed., U.S. Steel Corp., Pittsburgh, Pa., 1964.

W.A. Knepper, *Agglomeration*, Interscience Publishers, a division of John Wiley & Sons, Inc., New York, 1963.

J.H. Strassburger, *Blast Furnace Theory and Practice*, 2 vols., published by AIME, Gordon and Breach Science Publishers, New York, 1969.

IRON BY DIRECT REDUCTION

Direct-reduction (DR) processes produce metallic iron from its ores by removing the associated oxygen at temperatures below the melting temperature of any of the materials involved in the process. Products from the DR processes are referred to as direct-reduced iron (DRI).

Chemistry

The reduction of iron ore in any DR process is accomplished by the same reactions that occur in the blast-furnace stack, including reduction by CO and H_2 and, in some cases, solid carbon, through successive oxidation states to metallic iron, ie, $Fe_2O_3 \rightarrow Fe_3O_4 \rightarrow FeO \rightarrow Fe$-(metallic iron). Where the reduction reactions are carried out below 1000°C, the reducing agents are usually restricted to CO and H_2, and the DRI that is produced is porous and has essentially the same size and

shape as the original iron ore particle or agglomerate. Above about 1000°C, the reduced material begins to sinter and, at ca 1200°C, a pasty, porous mass forms. This latter temperature represents close to an upper limit for the DR processes. Above 1200°C, the metallic iron that has been formed absorbs any carbon that is present, which results in freezing-point depression (from 1530°C) and subsequent fusing or melting of the solid.

The effectiveness of the commercial processes, eg, fluidized beds, moving-bed shafts, fixed-bed retorts, and rotary kilns, depends on sound process design and engineering to ensure that good contact is obtained between reacting gases and solids. Schematic flow sheets of the process concepts are presented in Figure 1.

Production

As of the end of 1983, there were 64 DR plants for reducing steelmaking-grade DRI with an annual capacity of 19.5×10^6 metric tons. At that time, 19 of these plants having an annual capacity of 4.5×10^6 metric tons were shut down. The chief fuel in 43 of these plants is natural gas. Eighteen plants are based on coal and two plants are based on liquid hydrocarbon fuel.

Uses

DRI is mainly used as prime metallic iron units for electric furnace steelmaking and foundry operations; as a coolant in BOF (BASIC-oxygen furnace) steelmaking; as a replacement for scrap in the open hearth; and for increasing productivity and decreasing coke ratio in blast furnaces and other smelting processes.

J. FEINMAN
United States Steel Corporation

L. von Bogdandy and H.J. Engell, *The Reduction of Iron Ores, Scientific Basis and Technology*, Verlag Stahleisen GmbH, Düsseldorf, Springer-Verlag, Berlin, Ger., 1971.

R.L. Stephenson, ed., *Direct Reduced Iron—Technology and Economics of Production and Use*, ISS/AIME, Warrendale, Pa., 1980.

J.K. Tien and J.F. Elliott, eds., *Metallurgical Treatises*, The Metallurgical Society of AIME, Warrendale, Pa., 1981, pp. 211–228.

IRON COMPOUNDS

The usual oxidation states of iron are $+2(d^6)$ and $+3(d^5)$, the ferrous and ferric states, respectively. The preferred nomenclature indicates oxidation states as iron(II) and iron(III), and this convention is used here. Oxidation states from -4 (as in the phthalocyanine complexes) to $+6$ (as in the ferrates) are known, but the $+2$ and $+3$ oxidation states are of overwhelming importance.

Iron metal dissolves in dilute mineral acids and yields, in the absence of other oxidizing agents (including air), the iron(II) ion. Salts with a tremendous variety of anions may be formed by evaporation of the corresponding aqueous solution. Such salts are typically green or yellow hydrated substances and are subject to air oxidation and either hydration or efflorescence.

Iron(III) salts generally are white or pale-colored hydrates. Aqueous solutions of iron(III) salts quickly hydrolyze to form aquo species, eg, $[(H_2O)_4Fe(OH)_2Fe(H_2O)_4]^{4+}$, and a pH of about zero must be maintained to prevent hydrolysis. At a pH of two, more polynuclear species are formed and colloidal gels appear, and at a pH above two, red-brown hydrated ferric oxide precipitates. Iron(III) complexes are typically octahedral, but a number of tetrahedral complexes are also known.

The Russel-Saunders ground-state term for the free iron(II) ion is ^5D. An octahedral or tetrahedral field splits the configuration into a $^5T_2$ state and a ^5E state. All tetrahedral complexes have magnetic moments of $(4.6–4.8) \times 10^{-23}$ J/T $(5–5.2\ \mu\beta)$. Octahedral complexes may be high spin $(4.8 \times 10^{-23}$ J/TR or $5.2\ \mu\beta)$. Low-spin octahedral complexes result in strong ligand fields and a fair number of these are known. In a few cases where strong ligand fields are accompanied by large tetragonal distortion (ie, square planar coordination), the orbital energies increase as $d_{yz}, d_{xz} < d_{z^2} < d_{xy} \ll d_{x^2-y^2}$, and the orbital occupation is $(d_{yz})^2(d_{xz})^2(d_{z^2})^1(d_{xy})^1$; a triplet ground state results.

Acetates

If scrap iron is treated with acetic acid, the iron may be solubilized as acetate salts. The initial black liquor solution is concentrated to a 12% solution from which iron acetate may be obtained. This salt appears to be a mixture of the iron(II) and iron(III) oxidation states of indefinite proportions. A pure iron(II) acetate, $Fe(C_2H_3O_2)_2$, also may be prepared. It is a colorless compound which may be recrystallized from water to yield hydrated species. Iron(II) acetate is used widely in the preparation of dark shades of inks (qv) and dyes and is used as a mordant in dyeing (see Dyes and dye intermediates).

Iron(III) acetate, $Fe(C_2H_3O_2)_3$, is prepared industrially by treatment of scrap iron with acetic acid, followed by oxidation of the resulting solution with air. Iron(III) acetate is used as a mordant, as a catalyst in organic oxidation reactions, and as a convenient source of iron in the preparation of other compounds.

Carbonates

Iron(II) carbonate, $FeCO_3$, appears as a white precipitate when solutions of alkali carbonate salts are added to solutions of iron(II) salts. The compound darkens when subjected to air oxidation. Iron(II) carbonate is used as a flame retardant and as an iron supplement in animal diets (see Flame retardants; Food additives; Pet and livestock feeds).

Citrates

Iron citrate is a compound of indefinite ratio of citric acid (qv) and iron in mixed oxidation state.

Cyanides

The variety, complexity, and utility of iron cyanide compounds is astounding. All of the iron cyanide compounds are complex coordination compounds and, therefore, the names hexakiscyanoferrate $(4-)$ and hexakiscyanoferrate $(3-)$ are preferred to iron(II) cyanide or iron(III) cyanide or ferrocyanide or ferricyanide (see Coordination compounds).

Hexakiscyanoferrate $(4-)$, $[Fe(CN)_6]^{4-}$, may be formed by the displacement of water from simple iron(II) salts by aqueous cyanide ion, accompanied by the evolution of a large amount of heat. Salts of hexakiscyanoferrates $(3-)$ may be considered salts of ferricyanic acid,

trihydrogen hexakiscyanoferrate, $H_3[Fe(CN)_6]$, which may be prepared by adding sulfuric acid to trilead or tribarium bis(hexakiscyanoferrate). Red-brown needles are obtained by evaporation of the aqueous solution.

Formates

Iron(II) formate dihydrate, $Fe(HO_2)_2.2H_2O$, is a green salt which is only slightly soluble in water. It may be prepared from iron(II) sulfate and sodium formate in an inert atmosphere. It is fairly stable to air oxidation and thus is a useful reagent. The anhydrous salt also is known.

Iron(III) formate, $Fe(HCO_2)_3$, exists as red crystals or powder which are soluble in water, but only very slightly soluble in alcohol. Aqueous solutions are subject to hydrolysis, which results in the formation of basic formates and eventually results in precipitation of iron hydroxide and liberation of formate. The compound is conveniently prepared from iron(III) nitrate and formic acid in alcohol solution.

Fumarates

Iron(II) fumarate, $Fe(C_4H_2O_4)$, is the iron salt of *trans*-2-butenedioic acid. This red-orange to red-brown powder is favored as a human and veterinary hematinic because it produces few cases of gastrointestinal upset (see Veterinary drugs). Its solubility in water is 0.0014 g/mL at 25°C, but it is more soluble in dilute acid solutions. It is prepared by adding a hot aqueous solution of sodium fumarate to a hot aqueous solution of iron(II) sulfate, followed by filtration of the slightly soluble iron(II) fumarate.

Halides

All of the binary halides of iron(II) and iron(III) are known except for the iodide of iron(III), which is stable only in the vapor phase. Complex iron(II) and iron(III) halides have been characterized, again with the exception of complex iron(III) iodides. The number and variety of these are large.

Nitrates

Iron(II) nitrate hexahydrate, $Fe(NO_3)_2.6H_2O$, forms green rhombs which melt at 60.5°C. Its solubility in water is 0.835 g/mL at 20°C or 1.67 g/mL at 61°C. It is prepared by the action of cold nitric acid having a density of less than 1.034 g/cm³ of iron. Iron(II) nitrate is used as a catalyst for reduction reactions and as a convenient reagent in the synthesis of other iron compounds.

Iron(III) nitrate hexahydrate, $Fe(NO_3)_3.6H_2O$, forms colorless, cubic crystals which melt at 35°C. Its solubility in water at 20°C is 0.835 g/cm³, and it is infinitely soluble in hot water. Iron(III) nitrate is used as a mordant, in tanning, as a catalyst of oxidation reactions, and as a convenient source of iron in the preparation of other compounds.

Oxides and Hydroxides

Iron(II) oxide, FeO, is a black substance and forms cubic crystals. Its density is 5.7 g/cm³ and its mp is 1369 ± 1°C. It is insoluble in water, alcohol, or alkali, but it reacts with acids. It occurs naturally as the mineral wüstite, but may be prepared synthetically by thermal decomposition of iron(II) oxalate in a vacuum.

Iron(III) oxide, Fe_2O_3, forms red-brown to black trigonal crystals of density 5.24 g/cm³. It is insoluble in water, but is soluble in hydrochloric or sulfuric acid. It melts and begins to decompose at ca 1565°C. Iron(III) oxide occurs naturally as the mineral hematite, the principal ore of iron, and it may be prepared synthetically by heating brown iron hydroxide oxide, hydrous, FeO(OH), at 200°C. Iron(III) oxide is used in large quantity as a red pigment for paint, rubber, ceramics, and paper, as a coating for steel and other metals, in magnetic recording material, and as a catalyst of oxidation reactions (see Ceramics; Magnetic tape; Pigments, inorganic).

Sulfates

Iron(II) sulfate heptahydrate, $FeSO_4.7H_2O$, forms blue-green monoclinic crystals having a density of 1.898 g/cm³ and melting at 64°C. It is very soluble in water, soluble in absolute methanol, but only slightly soluble in ethanol. In dry air, the compound is efflorescent and, in moist air, the compound oxidizes to basic iron(III) sulfate. Aqueous solutions also are subject to oxidation, and the rate of oxidation increases with an increase in alkali, temperature, and light. Upon warming to 56°C, the compound loses three waters of hydration to form iron(II) sulfate tetrahydrate, $FeSO_4 \cdot 4H_2O$, and further warming to 65°C forms iron sulfate monohydrate, $FeSO_4 \cdot H_2O$, which is stable to 300°C.

Iron(III) sulfate, $Fe_2(SO_4)_3$, is a yellow substance which is slightly soluble in cold water, but which decomposes in hot water. A series of hydrates is known, including the monohydrate, the hexahydrate, the heptahydrate, and the nonahydrate. Iron(III) sulfate may be obtained by oxidation of iron(II) sulfate or by treating iron(III) oxide with sulfuric acid. It is used as a pigment, as a coagulant in water and sewage treatment, and as a mordant (see Water, industrial water treatment; Water, sewage).

Chelates

A chelate ligand is one that binds at more than one site to a metal atom (see Chelating agents). Chelating ligands may bind at two sites (a bidentate ligand), or they may act as tridentates, tetradentates, etc. A closed-ring structure is formed and, in general, five-membered rings are more stable than six-membered rings. Examples of iron chelates are the iron complexes of bipyridines, diketones, and ethylenediamine tetraacetic acid.

Phenanthrolines

Complexes of 1,10-phenanthroline with iron(II) are intensely colored and 1,10-phenanthroline serves as a reagent for the qualitative and quantitative determination of iron. A large variety of complexes may be synthesized from the many substituted phenanthrolines.

JAMES V. McARDLE
Smith Kline Beckman Corporation

F.A. Cotton and G. Wilkinson, *Advanced Inorganic Chemistry*, 4th ed., John Wiley & Sons, Inc., New York, 1980.

J.E. Huheey, *Inorganic Chemistry*, 2nd ed., Harper and Row, New York, 1978.

J.W. Mellor, *A Comprehensive Treatise on Inorganic and Theoretical Chemistry*, Vol. 13, Longmans, Green, and Co., London, UK, 1934.

ISOCYANATES, ORGANIC

Organic isocyanates are compounds in which the isocyanate group —NCO is attached to an organic group. They are frequently classified as esters of isocyanic acid, HNCO. The reactivity of the highly unsaturated isocyanate group has led to its study and use in a great variety of reactions. Polyfunctional isocyanates, usually with two or three isocyanate groups in the molecule, have been particularly useful for the systematic buildup of polymer molecules having tailored properties; as a result, they have become the cornerstone of a large new branch of the plastics industry. The methods of preparation and the reactions of the isocyanates have been reviewed, as have the applications of these compounds in the polymer field.

Properties

Some of the physical properties of a number of representative mono- and diisocyanates are summarized in Table 1. Physical and chemical properties have been compiled and published. The isocyanates react readily with a great variety of organic compounds and also may react with themselves.

Reactions with active-hydrogen compounds. The normal isocyanate reaction ultimately provides addition to the carbon–nitrogen double bond. In reactions involving compounds with an active hydrogen, ie, one that can be replaced by sodium, the hydrogen becomes attached to the nitrogen of the isocyanate, and the remainder of the active-hydrogen compound becomes attached to the carbonyl carbon:

$$RN{=}C{=}O + HA \longrightarrow RNHC{\overset{\displaystyle O}{\|}}A$$

In most reactions, especially those involving active hydrogen compounds, the aromatic isocyanates are more reactive than the aliphatic isocyanates. Substitution of electronegative groups on the aromatic ring enhances the reactivity, whereas electropositive groups reduce the reactivity of the isocyanates. All of the reactions are catalyzed by acids and, usually, more strongly by bases.

The most important reaction of isocyanates is with alcohols. Primary alcohols react at room temperature, and secondary and tertiary alcohols react much more slowly. The usual reaction of an alcohol and an isocyanate leads to a urethane (see Urethane polymers). When this reaction occurs between a diisocyanate and a dihydric (or higher) alcohol, straight-chain (or cross-linked) polymers arise that are used in foams, elastomers (see Elastomers, synthetic), and coatings (qv) (see Foamed plastics).

$$RNCO + R'OH \longrightarrow \underset{\text{urethane}}{RNHC{\overset{\displaystyle O}{\|}}OR'}$$

Reaction of isocyanates with water is very significant for two reasons: (1) rapid hydrolysis gives the corresponding disubstituted urea and CO_2; and (2) the evolution of CO_2 may lead to rupture in metal and glass containers. Polyethylene or polypropylene containers of isocyanates may harden and rupture, presumably because of diffusion of moisture through the plastic with deposition of urea in the plastic and development of CO_2 pressure in the container.

$$RNCO + H_2O \rightarrow RNHCOOH \rightarrow RNH_2 + CO_2$$
$$\underset{\text{disubstituted urea}}{RNH_2 + RNCO \rightarrow RNHCONHR}$$

Essentially all compounds containing a hydrogen attached to a nitrogen are reactive. In addition to the usual reactions with active hydrogen, Friedel-Crafts and Grignard reactions may be performed, both yielding amides (see Friedel-Crafts reactions; Grignard reaction). Because isocyanates are highly reactive toward such a large number of active hydrogen compounds, more than one reaction may occur in a system at a given time.

Isocyanates are capable of reaction with themselves to form dimers, trimers, and polymers, with ureas to form biurets, and with urethanes to form allophanates. Extended heating of isocyanates results in formation of carbodiimides with elimination of CO_2.

Table 1. Physical Properties of Some Isocyanates

Compound	mp, °C	bp, °C$_{kPa}$[a]	Density, g/cm^3	Refractive index, n_D^t	Flash point open cup, °C
methyl isocyanate		38$_{101}$	0.96$_4^{20}$	1.36^{20}	−7
n-butyl isocyanate	< −70	115$_{101}$	0.89$_4^{20}$	1.4064^{20}	ca 20
cyclohexyl isocyanate	< −80	171$_{101}$	0.96$_4^{20}$	1.4557^{20}	54
octadecyl isocyanate	21	170$_{0.27}$	0.86$_{15}^{25}$	1.4468^{30}	184
hexamethylene diisocyanate	−67	127$_{1.3}$	1.05$_4^{20}$	1.4530^{20}	140
phenyl isocyanate	−30	165$_{101}$	1.10$_4^{20}$	1.5362^{20}	60
p-chlorophenyl isocyanate	28	88$_{1.3}$	1.25$_4^{40}$	1.5543^{40}	110
3,4-dichlorophenyl isocyanate	42	113$_{1.3}$	1.39$_4^{50}$	1.5710^{50}	142
m-phenylene diisocyanate	51	110$_{1.5}$	1.21$_4^{60}$	1.5573^{50}	107
2,4-toluene diisocyanate	22	120$_{1.3}$	1.22$_{15}^{25}$	1.5654^{25}	132
80/20-toluene diisocyanate	11–14[b]	120$_{1.3}$	1.22$_{15}^{25}$	1.5666^{25}	132
65/35-toluene diisocyanate	3–5[b]	120$_{1.3}$	1.22$_{15}^{25}$	1.5663^5	132
4,4′-diphenylmethane diisocyanate	38	196$_{0.7}$	1.19$_{50}$	1.5906^{50}	201

[a] To convert kPa to mm Hg, multiply by 7.5.
[b] Freezing point.

Manufacture and Processing

Phosgenation. The reaction of amines with phosgene (phosgenation) has, for economic reasons, been used almost exclusively for the manufacture of isocyanates. The details of processing may vary with the specific isocyanate and, in particular, for aromatic and aliphatic isocyanates, but the general approach is the same. The primary reactions involved in the phosgenation of a simple amine are indicated below (see also Amines, aromatic; Phosgene).

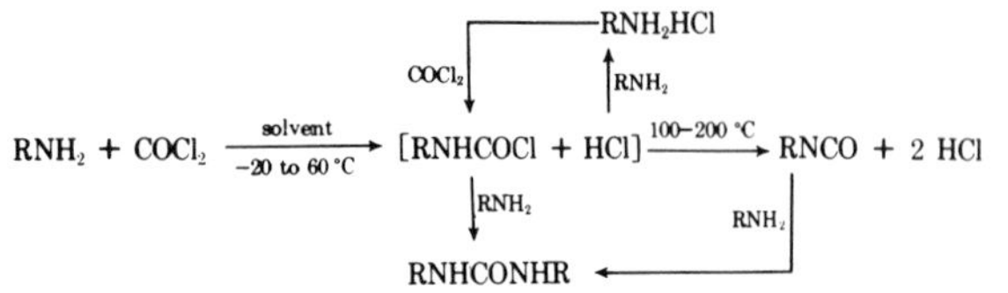

The overall chemistry of an integrated facility for the manufacture of 80/20 toluene diisocyanate (TDI) is illustrated by the following equations:

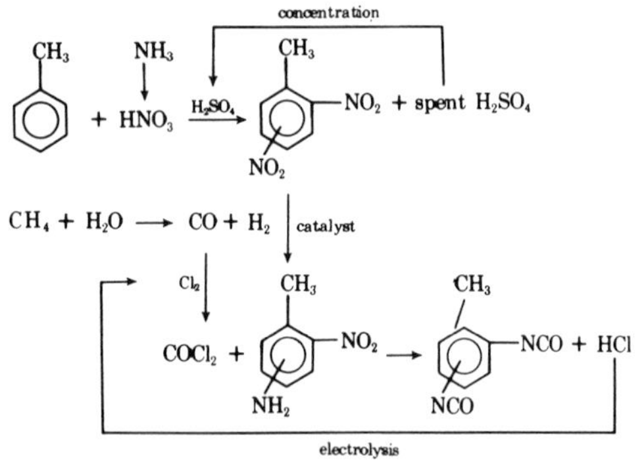

80/20 TDI (toluene diisocyanate)

Health and Safety

NIOSH has recommended a maximum exposure for all diiosocyanates of a 10-min ceiling of 20 ppb and a time-weighted average for a 10-h day/40-h week of 5 ppb.

Uses

For a detailed discussion of polyurethane products and chemistry, see Urethane polymers and Foamed plastics. In the last decade, the most significant development in the flexible-foam area has been the introduction of molding formulations that require little, if any, oven curing and produce more resilient and higher load-bearing foams. Known as cold cure, ambient cure, or high resiliency, among other names, these products are based on highly reactive polyether polyols containing relatively high fractions of primary hydroxyl groups. With rigid and semirigid (elastomeric) foams, the reaction-injection molding RIM technique has led to higher-density products and applications similar to those of the usual injection-molded plastics.

D.H. CHADWICK
T.H. CLEVELAND
Mobay Chemical Corporation

J.H. Saunders and K.C. Frisch, *Polyurethanes: Chemistry and Technology, Part I: Chemistry*, Interscience Publishers, a division of John Wiley & Sons, Inc., New York, 1964; *Part II: Technology*, 1964.

Title 29, Code of Federal Regulations, 1910.1000; *Criteria for a Recommended Standard, Occupational Exposure to Diisocanates*, National Institute for Occupational Safety and Health, U.S. Department of Health, Education, and Welfare, Sept. 1978.

ISOCYANURIC COMPOUNDS. See Cyanuric and isocyanuric acids.

ISOPHORONE (3,5,5-TRIMETHYL-2-CYCLOHEXENE-1-ONE). See Ketones.

ISOPRENE

Isoprene, 2-methyl-1,3-butadiene, 2-methyldivinyl or 2-methylerythrene, is a colorless, volatile liquid that is soluble in most common hydrocarbons, but is practically insoluble in water. Isoprene forms binary azeotropes with methanol, *n*-pentane, methylamine, methyl formate, ethyl bromide, methyl sulfide, acetone, propylene oxide, isopropyl nitrite, methylal, formaldehyde, ethyl ether, and perfluorotriethylamine. Ternary azeotropes with water–acetone, water–acetonitrile, methyl formate–ethyl bromide are known.

Mol wt, 68.11; density (of liquid) 0.668 g/cm³ at 25°C; fp, −146°C; bp, 34.1°C (at 101.3 kPa or 1 atm); n_D^{30}, 1.41524; flash point, −48°C.

Isoprene may be considered as an equilibrium of two conformations, namely, a cisoid (*s-cis*) conformation in which both vinyl groups are located on the same side of the C—C bond, and a transoid (*s-trans*) one with the vinyl groups located on the opposite sides of the bond. The predominance of the trans-planar or nonplanar configuration has been supported by experimental data.

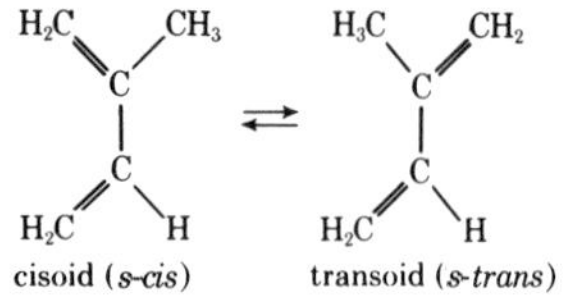

cisoid (*s-cis*)   transoid (*s-trans*)

Reactions. Isoprene is highly reactive both as a diene and through its allylic hydrogens, and its reaction processes are similar to those of butadiene (qv).

Apart from polymerization, the most widely investigated isoprene reactions are the formation of six-membered rings by the Diels-Alder reaction:

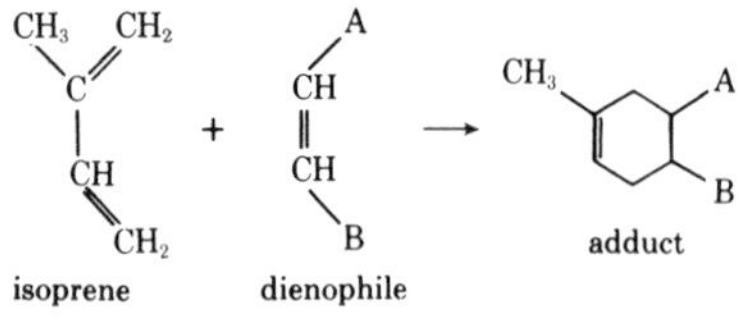

isoprene   dienophile   adduct

The reaction proceeds readily, depending on the nature of the dienophile and, normally, no catalyst is effective.

Free radicals attack isoprene and two competing mechanisms are postulated:

$$R\cdot + R'H \xrightarrow{\text{solvent}} RH + R'\cdot$$

Polymerization of isoprene can be either by 1,4 or vinyl addition (see Elastomers, synthetic–polyisoprene). 1,4 Addition leads to two possible structures which differ in the configuration of the remaining double bond:

cis-1,4   *trans*-1,4

Synthesis

Besides the synthetic routes, economical recovery of isoprene as a superfractionated by-product from naphtha cracking recently has gained paramount commercial importance.

Dehydrogenation of tertiary amylenes. The starting material is a C_5 fraction which is cut from the catalytic cracking of petroleum. Two of the tertiary amylene isomers, 2-methyl-1-butene and 2-methyl-2-butene, are extracted from the C_5 stream by cold, aqueous sulfuric acid. The amylenes are mixed with steam and are dehydrogenated over a catalyst:

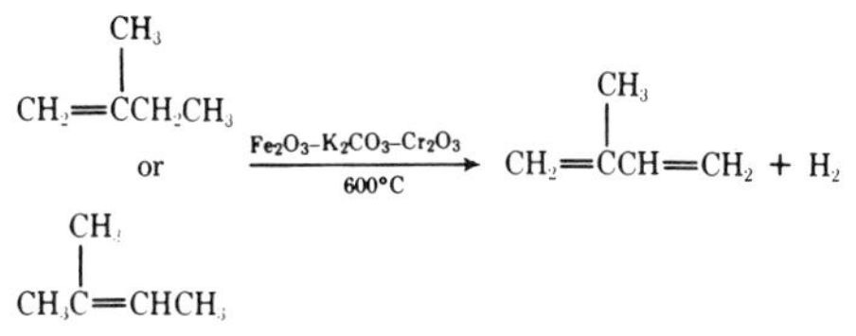

Isobutylene–formaldehyde. Isobutylene is condensed with formaldehyde at 95°C to give the principal product, 4,4-dimethyl-m-dioxane. In the second step, the dioxane is decomposed in the presence of an acid catalyst to isoprene, formaldehyde, and water.

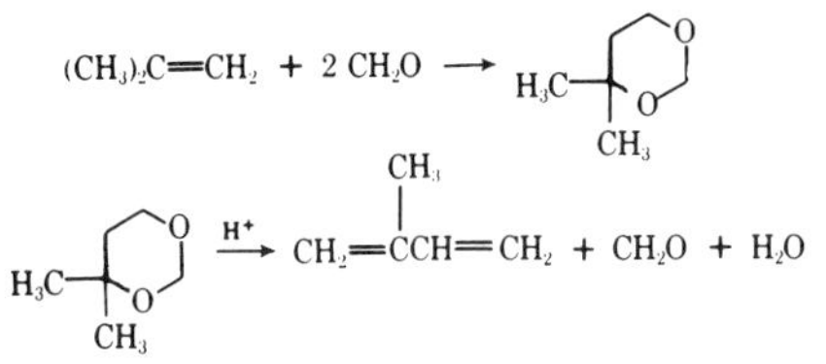

Isopentane dehydrogenation. A dehydrogenation process used industrially in the USSR process dehydrogenates isopentane or a C_5 fraction from a catalytic cracker to yield isoprene:

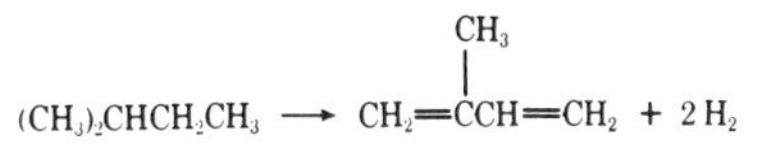

Petroleum cracking. Petroleum cracking has gained commercial importance in recent years. Isoprene is recovered from C_5 streams that are obtained in the thermal cracking of naphtha and gas oil. When naphtha or gas oil is cracked to produce ethylene (qv), a host of by-products is created. Values differ depending on the cracking process, but isoprene yields may be from 2–5% of the ethylene yield. Because of the enormous demand for ethylene-based products, this by-product isoprene has become a commercial source.

Health and Safety

Isoprene is not known to present serious toxicological hazards in handling. A concentration of 2% isoprene in air does not narcotize mice, but produces bronchial irritation; 5%, concentrations are fatal to mice. Isoprene is classified by the ICC as a flammable liquid requiring a red label. It forms dangerous peroxides on exposure to air in the absence of inhibitors.

Uses

Almost all isoprene that is produced is used for the preparation of cis-1,4-polyisoprene or in small proportions in copolymers with isobutylene for butyl rubber and in thermoplastic elastomeric, SIS block polymers. cis-Polyisopene is by far the largest (> 95%) produce of isoprene.

WILLIAM M. SALTMAN
The Goodyear Tire & Rubber Company

Reviewed by:
S.E. HORNE, JR.
Polysar Inc.

W.J. Bailey in E.C. Leonard, ed., *Vinyl and Diene Monomers*, Part II, John Wiley & Sons, Inc., New York, 1971, Chapt. 5.

E. Schoenberg, H.A. Marsh, S.J. Walters, and W.M. Saltman, *Rubber Chem. Tech.* **52**, 526 (1979).

ISOPROPANOLAMINES. See Alkanolamines.

ISOTOPES

The term isotope was coined to describe members of the natural radioactive series that have the same atomic number, but different atomic weight. The modern definition of an isotope, based on the nuclear theory of the atom, derives from the hypothesis that the atomic nucleus is composed of protons and neutrons. The number of protons (or atomic number Z, which is equal to the number of electrons surrounding the nucleus) is associated with the chemical properties of the element, whereas the nuclear properties are dependent upon both Z and N (the latter being the neutron number). The mass number A ($= N + Z$) is nearly equal to the atomic mass; the difference (ca < 0.1 mass unit) arises mainly from variations in the average nuclear binding energy. (The atomic mass scale is defined by $M(^{12}C) \equiv 12$). Many of the elements are polyisotopic. Terrestrial materials, moon rocks, and meteorites display a remarkably uniform isotopic composition for all but a few elements; however, the small variations that can be measured provide a wealth of information about the geochemical and biological history of the earth.

Separation

Although stable isotopes can be produced by transmutation, economically feasible processes involve the separation of the isotopes of natural elements. The most important classes of processes that have been used or proposed for isotope separation are (1) processes used for small-scale commercial production of many isotopes: electromagnetic [large ($\gg 1$) separation factors], thermal diffusion, distillation (C, N, O), chemical exchange (Li, B), and photochemical or photophysical (laser) processes [large ($\gg 1$) separation factors, pilot plants planned or under construction]; (2) processes used for large-scale commercial production: chemical exchange (H), distillation (H), electrolysis (H), gaseous diffusion (U), gravitational (gas centrifuge) (U, commercial plants planned), aerodynamic (separation nozzle) (U, commercial plants planned), and photochemical or photophysical (laser) processes (U, H) [large ($\gg 1$) separation factors, pilot plants planned or under construction]; and (3) other processes: chromatographic [ion exchange (qv)], mass and sweep diffusion, adsorption, electromigration, and biological processes. Because most of these processes have small intrinsic separation capabilities, many separations in series are required in order to achieve a product of high isotopic purity. The use of repeated separations requires the application of cascade theory in most separation processes (see Diffusion separation methods).

The effectiveness of the separation is measured by the enhancement of the desired isotope and is defined by the separation factor:

$$\alpha = \frac{y/(1-y)}{x/(1-x)}$$

where y = the composition of the heads stream and x = the composition of the tails stream. α is nearly independent of the compositions for many separation processes, whereas the heads separation factor,

$$\beta = \frac{y/(1-y)}{z/(1-z)}$$

where z is the composition of the feed stream, depends on the cut or ratio of the heads flow rate to the feed flow rate (x, y, and z are expressed as mole fractions of the desired isotope). Separation factors (αs) for some processes are listed in Table 1.

Table 1. Typical Separation Factors for Some Isotope Separation Processes

Process	α
electromagnetic isotope separator,	
medium-weight element,	
eg, Z = 40 (zirconium)	≈ 100
countercurrent thermal diffusion column, ^{84}Kr–^{86}Kr	2.6
distillation: ortho H_2–HD at $-252.9°C$	1.81
H_2O–D_2O at 80°C	1.04
chemical exchange (at 25°C):	
$H_2O(l) + HDS \rightleftarrows HDO(l) + H_2S$	2.37
$^6LiOH(aq) + {}^7Li(Hg) \rightleftarrows {}^7Li(OH) + {}^6Li(Hg)$	1.07
gaseous diffusion of $^{235}UF_6$–$^{238}UF_6$	
through a porous barrier	1.004
countercurrent gas centrifuge, $^{235}UF_6$–$^{238}UF_6$	1.4
laser isotope separation, ^{235}U–^{238}U	≈ 100

Applications of Stable Isotopes

Nuclear electric-power generation involves a number of isotope applications and accounts for most of the separated isotope production (see Nuclear reactors). The use of isotopes as tracers is a powerful method for following the path of an element through a chemical, biological, or physical system. Radioisotopes are most commonly employed as tracers; however, stable isotopes are often used because no suitable radioisotope exists, or because the emitted radiation would cause damage to the system under study, to a patient or experimenter, or to the environment. Separated stable isotopes are also used in the production of radioisotopes, in nuclear physics research, and in nmr studies. Isotope effects (changes in the equilibrium or rate of chemical reactions upon isotopic substitution) provide information about the reaction mechanism. Isotope substitution in biological systems provides a means for producing labeled organic compounds and a tool for the production of isotopically labeled organic compounds (see Radioactive drugs; Radioactive tracers).

C. MICHAEL LEDERER
Lawrence Berkeley Laboratory
University of California

C.M. Lederer and V.S. Shirley, eds., *Table of Isotopes*, 7th ed., John Wiley & Sons, Inc., New York, 1978.

M. Benedict, T.H. Pigford, and H.W. Levi, *Nuclear Chemical Engineering*, 2nd ed., McGraw-Hill Book Co., New York, 1980.

J. Hoefs, *Stable Isotope Geochemistry*, Springer-Verlag, Berlin, FRG, 1973.

ISOTOPES, RADIOACTIVE.   See Radioisotopes.

ITACONIC ACID AND DERIVATIVES

Itaconic acid (methylenebutanedioic acid, methylenesuccinic acid), $CH_2=C(COOH)CH_2COOH$, is a crystalline, high melting acid produced commercially by fermentation. Isolated from the pyrolysis of products of citric acid in 1836, this α-substituted acrylic acid received its name by rearrangement of aconitic, the acid from which it is formed by decarboxylation.

Physical Properties

Mp, 167–168°C; crystal density, 1.49 ± 0.01 g/cm³; dissociation constant K_1, 1.40×10^{-4} in H_2O at 25°C; soluble to the extent of 9.5 g/100 mL water; 5.8 g/100 mL acetic acid; and 35.8 g/100 mL methanol.

Chemical Properties

Itaconic acid (**1**) is isomeric with citraconic (**2**) and mesaconic (**3**) acids. Under acidic, neutral, or mildly basic conditions and at moderate temperatures, itaconic acid is stable. At elevated temperatures or under strongly basic conditions, the isomers are interconvertible, probably via the species (**A**) where $Y = H^+$ or a cation.

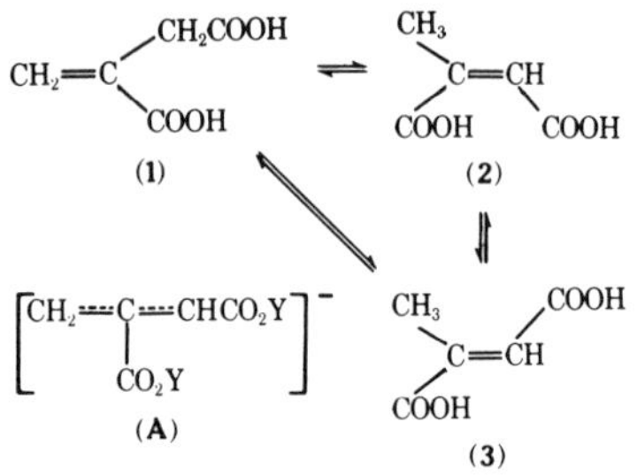

Itaconic anhydride is readily formed by treatment of the acid with acetic anhydride or acetyl chloride. Both the acid and anhydride are converted to itaconyl chloride by phosphorus pentachloride. Standard esterification methods provide dialkyl esters in good yields. Esterification proceeds almost entirely through the 4-alkyl ester (**4**), which is also the main product obtained from itaconic anhydride.

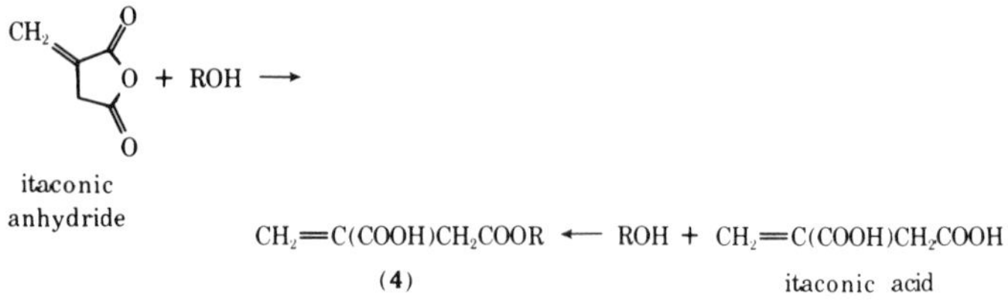

Nucleophilic reagents such as alkoxides, thiols, and amines add to the itaconic double bond.

Manufacture

Itaconic acid is manufactured by fermentation of carbohydrates. Large-scale production utilizes molasses as the raw material because of low cost, but glucose, cane, or corn sugar are acceptable substrates. Initial carbohydrate concentration in the fermenter charge may be on the order of 15–25%. Smaller amounts of other ingredients, such as ammonium salts to supply nitrogen, and essential metals, eg, magnesium and zinc, are also introduced. After the mixture is sterilized and inoculated with a selected strain of *Aspergillus terreus*, the broth is agitated and aerated at 35–40°C for 3–7 d. Itaconic acid is produced in the broth from citric acid generated in the tricarboxylic acid cycle.

Health and Safety

Itaconic acid is a high melting solid exhibiting essentially no volatility at room temperatures. Acute toxicities in the rat are oral LD_{50}, 4000 mg/kg; intraperitoneal LD_{50}, 450 mg/kg.

Uses

Itaconic acid is a specialty monomer that affords performance advantages to certain polymeric coatings when relatively low levels of the acid are incorporated. Styrene–butadiene latices containing under 10%, typically 1–5%, of itaconic acid, are widely used in carpet backings and paper coatings (see Coatings; Paper). Acrylic latices containing itaconic acid have applications as nonwoven fabric binders (see Nonwoven textile fabrics).

BRYCE E. TATE
Pfizer Inc.

L.S. Luskin in R.H. Yocum and E.B. Nyquist, eds., *Functional Monomers*, Vol. 2, Marcel Dekker, Inc., New York, 1974, pp. 465–501.

B.E. Tate in E.C. Leonard, ed., *Vinyl and Diene Monomers*, Wiley-Interscience, New York, 1970, Pt. 1, pp. 205–261.

B.E. Tate, *Adv. Polym. Sci.* **5**, 214 (1976).

J

J ACID, $H_2N(OH)C_{10}H_5SO_3H$. See Dyes and dye intermediates.

JASMIN, JASMINE. See Oils, essential; Perfumes.

JUNIPER. See Oils, essential.

JUTE. See Fibers, vegetable.

K

KETENES AND RELATED SUBSTANCES

Ketenes and analogous compounds are characterized by a hetero-cumulene structure, $R_2C=C=X$, where R substituents may be any combination of hydrogen, alkyl, aryl, acyl, halogen, and various functional groups. If X is O, S, or NR, the compounds are ketenes, thioketenes, or keteniminies, respectively. The parent compound ketene, $CH_2=C=O$, is the only compound of this type that is manufactured commercially, but ketenes play an important role in many organic reactions and processes. In general, ketenes and their analogues are highly reactive; however, the electrophilic center (the central carbon on the heterocumulene triad) is strongly affected by the nature of the substituents. Although a few ketenes are moderately stable, most of these compounds are transient species, not isolable as such.

Monomeric Ketenes

Physical properties. Ketenes range in properties from colorless gases such as ketene and carbon suboxide to highly colored liquids such as diphenylketene and carbon subsulfide. Many ketenes are unstable and cannot be isolated. The ketenes listed in Table 1 are the best known examples in the chemical literature.

Chemical properties. The chemistry of ketenes is dominated by addition reactions. Reagents with labile hydrogen atoms (HX where X is hydroxyl, amino, halogen, acyl, and the like) form the corresponding carboxylic acid derivatives. The most important application of this reaction is the manufacture of acetic anhydride from ketene and acetic acid (see also Acetic acid and derivatives).

$$CH_2=C=O + CH_3CO_2H \rightarrow (CH_3CO)_2O$$

Many other reagents add to ketenes, eg, halogens and nitrosyl chloride. A second general pathway in ketene chemistry is cycloaddition to reactive olefins, enol ethers, and enamines to give cyclobutanones as primary products.

In the absence of active ketenophiles, most ketenes undergo self-addition to form dimers, trimers, and polymers. Exceptions are very electrophilic or sterically hindered ketenes, such as bis(trifluoromethyl)ketene, diphenylketene, and di-*tert*-butylketene. Ketenes dimerize either symmetrically to form 1,3-cyclobutanediones, or unsymmetrically to give β-lactones.

Preparation. Ketenes are the product of abstraction of one mole of water per mole of corresponding carboxylic acid, but few ketenes are prepared directly. The parent compound ketene is manufactured by

Table 1. Monomeric Ketenes

Name	Properties	
	Physical state	mp or bp,[a] °C
ketene	colorless gas	bp −49.8
methylketene	colorless gas	
dimethylketene	yellow liq	bp 34
diphenylketene	orange liq	bp 118–120 (0.1 kPa)
pentamethyleneketene	yellow liq	bp 40–41 (0.4 kPa)
dichloroketene	not isolable	
bis(trifluoromethyl)ketene	colorless gas	bp 5
tert-butylcyanoketene	not insolable	
trimethylsilylketene	colorless liq	bp 81–82
triphenylphosphoranylidene-ketene	white cryst	mp 172–174

[a] At 101.3 kPa, unless otherwise stated; to convert kPa to mm Hg, multiply by 7.5.

pyrolysis of acetic acid at 700–800°C under reduced pressure (10–50 kPa, or 0.1–0.5 atm). A phosphate ester is injected to provide an acidic catalyst. After removal of water and unconverted acetic acid, the gaseous ketene is absorbed immediately in an appropriate reaction medium, eg, acetic anhydride is prepared by passage into a mixture of acetic acid and anhydride.

Health and Safety

Ketenes are sensitive to moisture and many are sensitive to air. The lower dialkylketenes react instantly with oxygen to form dangerously explosive polymeric peroxides. In the case of dimethylketene, peroxides can be avoided by handling the material above 70°C. Diarylketenes, although very sensitive to oxygen, do not form hazardous peroxides.

ROBERT H. HASEK
Tennessee Eastman Company

Reviewed by
JOHN A. HYATT
Tennessee Eastman Company

D. Borrmann in E. Müller, ed., *Methoden der Organischen Chemie* (*Houben-Weyl*), Vol. 7, (Sauerstoff Verbindungen II), Georg Thieme Verlag, Stuttgart, FRG, 1968, Pt. 4, pp. 53–339.

R.N. Lacey in R.A. Raphael, E.C. Taylor, and H. Wynberg, eds., *Advances in Organic Chemistry: Methods and Results*, Vol. 2, Interscience Publishers, Inc., New York, 1960, pp. 213–263.

S. Patai, ed., *The Chemistry of Ketenes, Allenes and Related Compounds*, Wiley-Interscience, New York, 1980, Pts. 1 and 2.

KETONES

Ketones are organic compounds containing one or more carbonyl groups bound to two carbon atoms and are represented by the general formula:

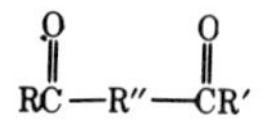

$$\underset{RCR'}{\overset{\overset{\textstyle O}{\|}}{}}$$

$$\underset{RC-R''-CR'}{\overset{\overset{\textstyle O}{\|}\qquad\overset{\textstyle O}{\|}}{}}$$

Depending on the nature of the hydrocarbon groups (R, R', and R'') attached to the carbonyl group, ketones are classified as symmetric or simple aliphatic, such as acetone, CH_3COCH_3, symmetric aromatic, such as benzophenone, $C_6H_5COC_6H_5$, unsymmetric or mixed aliphatic–aromatic, such as methyl ethyl ketone, $CH_3COC_2H_5$, and acetophenone, $CH_3COC_6H_5$, alicyclic, such as cyclopentanone, or heterocyclic, such as 3-pyrrolidinone.

Physical Properties

Simple and mixed low molecular-weight aliphatic and cycloaliphatic ketones are stable, colorless liquids and generally have a pleasant, slightly aromatic odor. They are relatively volatile with boiling points slightly above those of corresponding paraffins. The members of the series up to C_5 are soluble in water and are excellent solvents for nitrocellulose, vinyl resin lacquers, cellulose ethers and esters, and various natural and synthetic gums and resins. In contrast, even the simplest aromatic ketone ($CH_3COC_6H_5$) is a high boiling colorless liquid with a

Table 1. Physical Properties of Ketones

CAS name	Common name	Mol wt	Freezing point, °C	Melting point, °C	Boiling point at 101.3 kPa,[a] °C	Refractive index, n_D^{20}	Apparent specific gravity, 20/20°C	Viscosity at 20°C, mPa·s (= cP)	Solubility at 20°C, wt% In water	Water in
Simple and mixed acyclic ketones										
acetone, 2-propanone	acetone, dimethyl ketone	58.08	−94.7		56.1	1.3590	0.7905	0.33	complete	complete
acetophenone	acetophenone, phenyl methyl ketone	120.15	19.7	19–20	201.7	1.5342	1.0296	0.93	0.55	1.65
benzophenone	benzophenone, diphenyl ketone	182.22		48–49.5	305					
2-heptanone	methyl *n*-amyl ketone	114.19	−35		151.5	1.4087	0.8166	0.77	0.43	1.45
3-hexanone	ethyl propyl ketone	100.16			123.2	1.4003	0.8174		1.57	
5-methyl-2-hexanone	methyl isoamyl ketone	114.12	−73.9		144.9	1.4069	0.8127	0.77	0.54	1.28
Unsaturated ketones										
3-butene-2-one	methyl vinyl ketone	70.09	−6		79–80 at 98.1 kPa[a]	1.4130				
Substituted ketones										
2,4-dihydroxy-benzophenone		214.2	[b]	142	194 at 0.13 kPa[a]					
Cyclic ketones										
cyclohexanone pimelic ketone		98.15	−31.2	−40.5	155.8	1.4502	0.9482	2.2	2.5	8.0

[a] Unless otherwise stated; to convert kPa to mm Hg, multiply by 7.5.
[b] Sets to glass below −80°C.

Table 2. Azeotropes of Ketones

Components	Component bp at 101.3 kPa,[a] °C	Azeotrope bp at 101.3 kPa,[a] °C	Azeotrope composition at 20°C, wt%
acetone	56.1		59
hexane	68.7	49.8	41
acetone	56.1		56.5
isopropyl ether	68.5	53.3	43.5
acetone	56.1		48
methyl acetate	57	55.6	52
acetone	84[b]		98.7
water	126[b]	81.4[b]	1.3
acetophenone	201.7		18.5
water	100	99.1	81.5
diacetone alcohol	169.2		13.0
water	100	99.6	87.0
diisobutyl ketone	169.4		48.1
water	100	97.0	51.9
isobutyl heptyl ketone	218.2		16
water	100	99	84
isophorone	215.3		16.1
water	100	99.5	83.9
mesityl oxide	129.8		65.3
water	100	91.8	34.7
methyl ethyl ketone	79.6		37.5
benzene	80.1	78.4	62.5
methyl ethyl ketone	79.6		66
ethanol	78.3	74.8	34
methyl ethyl ketone	79.6		70
isopropanol	82.3	77.3	30
methyl ethyl ketone	79.6		88
water	100	73.4	12
methyl isobutyl ketone	116.2		76
water	100	87.9	24
2,4-pentanedione	140.4		59
water	100	94.4	41

[a] To convert kPa to mm Hg, multiply by 7.5.
[b] At 137.9 kPa.

fragrant odor. Aromatic ketones are useful as intermediates in chemical manufacture. Physical properties of some common ketones are listed in Table 1. Pure ketones are not associated liquids and tend to form azeotropes (see Table 2) with water and/or a variety of organic compounds. This property is used in industrial extractive distillation processes (see Azeotropic and extractive distillation).

Chemical Properties

Ketones undergo addition, redox, and condensation reactions forming alcohols, ketals, acids, and amines. The reactivity of ketones is a function of the polarity and electrophilic nature of the carbonyl group and its influence on the activity of nearby functional groups (eg, hydrogens alpha to the carbonyl group). In diketones, cyclic and unsaturated ketones, such as 2,4-pentanedione, cyclohexanone, and mesityl oxide, respectively, these chemical properties are enhanced, increasing their usefulness as chemical intermediates.

Reduction. Most ketones are reduced readily to the corresponding alcohols by a variety of hydrogenation processes. For example, 4-methyl-2-pentanol (methyl isobutyl carbinol) is commercially produced by catalytic reduction of methyl isobutyl ketone in the vapor phase with a nickel catalyst:

$$CH_3COCH_2CH(CH_3)_2 \xrightarrow[Ni-Cr]{H_2} CH_3CH(OH)CH_2CH(CH_3)_2$$

Oxidation. Ketones are oxidized readily with air or oxygen to produce peroxides; oxidation at about 100°C with powerful oxidizing agents such as chromic or nitric acids, causes carbon–carbon bond cleavage and gives carboxylic acids. Ketone peroxides may explode in concentrated form (> 30%) but dilute solutions are useful in curing unsaturated polyester resin mixtures (see Peroxides, organic; Initiators).

Condensations. *Base catalyzed*: Depending on the nature of the hydrocarbon groups attached to the carbonyl, ketones can under either undergo self-condensation or condense with other activated reagents in the presence of base. Name reactions that describe these condensations include the Aldol reaction, the Darzens-Claisen condensation, the Claisen-Schmidt condensation, and the Michael reaction. Production of diacetone alcohol and 4-ethoxy-4-methyl-2-pentanone are Aldol and Michael reactions, respectively.

Acid catalyzed: Although ketonic carbonyl groups are less reactive than aldehydic carbonyls in the presence of basic catalysts, this is not the case with acid catalysts. Thus, acetone undergoes aldol condensation in the presence of sulfuric acid to give mesityl oxide, which then condenses with a third molecule of acetone to give a mixture of phorone (2,6-dimethyl-2,6-heptadien-4-one) and mesitylene. This reaction undoubtedly proceeds through the enol form of acetone.

Ketones also condense with activated aromatic compounds in the presence of sulfuric acid to give coupled aromatic products. For example, acetone and phenol condense to 4,4′-isopropylidenediphenol, bisphenol A, which is used to prepare epoxy resins, modified phenolic resins, polycarbonates, aromatic polyesters, and polysulfones.

Thermal stability. The C_4–C_{12} ketones are thermally stable up to pyrolysis temperatures (500–700°C). At these high temperatures, decomposition of ketones can be controlled to produce useful ketene derivatives. Ketene (CH_2=C=O) itself is produced commercially by pyrolysis of acetone at a temperature just below 550°C.

Health and Safety

As a general class of chemicals, low molecular-weight (C_3–C_{12}) saturated aliphatic ketones present a low toxicity hazard. However, the toxicity of diketones and α,β-unsaturated ketones is significantly greater. The primary health hazard of ketones is owing to inhalation of vapors, especially those of low molecular-weight derivatives, which cause central nervous system depression. For example, rats died after a 4-h exposure to 2,000–40,000 ppm of ketone vapors. Exposure at lower concentrations (about 550 ppm) for a few hours typically produces irritation of the mucous membranes of the eyes, nose, and respiratory tract, usually providing adequate warning of overexposure.

Examples of Manufacture of Aliphatic and Unsaturated Ketones

Aliphatic Ketones

Acetone, cyclohexanone, methyl ethyl ketone, and methyl isobutyl ketone are among the 100 largest-volume organic chemicals (in order of decreasing volume). The three aliphatic ketones are used as solvents, whereas cyclohexanone is primarily used in polymer manufacture. Next in importance are aliphatic ketones such as diacetone alcohol, mesityl oxide, and diisobutyl ketone, which are primarily used as solvents.

Methyl ethyl ketone. Methyl ethyl ketone is usually made by the vapor-phase dehydrogenation of 2-butanol, like the production of acetone from 2-propanol. A two-step process from butenes, which are first hydrated to give 2-butanol, is involved. The dehydrogenation is catalyzed by zinc- or copper-based catalysts at high temperatures and low pressures. The process gives high conversion of 2-butanol and high selectivity to methyl ethyl ketone, about 90–95 mol% for each:

Unsaturated Ketones

Mesityl oxide is produced by liquid-phase dehydration of diacetone alcohol in the presence of acidic catalysts at 100–120°C and 101 kPa (1 atm). Preferred catalysts are dilute aqueous phosphoric acid and sulfuric acid. Yields of about 95% with conversions of greater than 90% are obtained.

Table 3. Relationship of Cyclic Ketone Ring Size to Odors

Ring size	Odor
5	bitter almonds
6	peppermint
7–9	transition to camphorlike
10–12	camphorlike
13	cedarwoodlike
14–18	musklike

[a] Ref. 200.

Table 4. Properties of Cyclic Ketones, $(CH_2)_xC{=}O$

CAS name	Formula, $x =$	Bp, °C (kPa[a])	Melting point, °C	Refractive index, n_D^{20}	Density at 20°C,[b] g/cm³	Enol, %
cyclopropanone	2	[c]				
cyclobutanone	3	100–102		1.4195	0.938	0.55
cyclopentanone	4	130		1.4359	0.951	0.09
cycloheptanone	6	179–181		1.4611	0.951	0.56
cyclooctanone	7	74 (1.6)	42		0.958	9.3
cyclononanone	8	93–95 (1.6)	34	1.4770	0.959	4.0
cyclodecanone	9	107 (1.7)	29	1.4820	0.958	6.1
cycloundecanone	10	108 (1.6)	10	1.4804		
cyclododecanone	11	125 (1.6)	61		0.906	
cyclotridecanone	12	138 (1.6)	32	1.4790	0.927	
cyclotetradecanone	13	155 (1.6)	53			
cyclopentadecanone	14	120 (0.04)	63		0.897	
cyclohexadecanone	15	138 (0.04)	60			
cycloheptadecanone	16	141 (0.13)	63			
cyclooctadecanone	17	158 (0.04)	72			
cyclononadecanone	18	160 (0.04)	72			
cyclocosanone	19	171 (0.04)	59			

[a] To convert kPa to mm Hg, multiply by 7.5.
[b] For solids, the density is given for the liquid at melting point temperature.
[c] Rapidly polymerizes at room temperature. A stable hydrate, mp 71–72°C, is formed in water.

Dehydration is generally conducted batch-wise in the base of a distillation column. Dry acetone is removed overhead and mesityl oxide recovered from a sidestream as an azeotropic mixture with water; mesityl oxide so produced is saturated with water (3.4 wt%). Distillation gives an anhydrous product with 98–99.5% purity.

Cyclic Ketones

Cycloaliphatic ketones are colorless liquids with boiling points that increase regularly with increasing molecular weight. Virtually all members of the series have characteristic odors, depending on ring size. Neither ring substitution nor nature of the ring atoms (eg, lactones) have much effect on odor. Relationship of ring size to odor is given in Table 3 and physical properties of some compounds are listed in Table 4.

Cyclohexanone. Most cyclohexanone is produced as a mixture with cyclohexanol (KA oil) by air oxidation of cyclohexane. Some is also produced by hydrogenation of phenol, followed by dehydrogenation of the resulting cyclohexanol. The principal use of pure cyclohexanone is in manufacture of γ-caprolactam (by rearrangement of its oxime) for nylon-6. Crude cyclohexanone–cyclohexanol is used to make adipic acid for nylon-6,6 (see Cyclohexanol and cyclohexanone; Polyamides).

ANTHONY J. PAPA
PAUL D. SHERMAN, JR.
Union Carbide Corporation

S. Patai, ed., *The Chemistry of Functional Groups*, Vol. 2, Interscience Publishers, a division of John Wiley & Sons, Inc., New York, 1966.

K. Schmitt, *Chem. Ind. Düsseldorf* **18**(4), 204 (1966).

C.R. Noller, *Chemistry of Organic Compounds*, 3rd ed., W.B. Saunders Company, Philadelphia, Pa., 1965.

KRYPTON. See Helium-group gases.

L

LAC.   See Shellac.

LACQUERS.   See Coatings, industrial; Paint.

LACRIMATORS.   See Chemicals in war.

LACTIC ACID.   See Hydroxy carboxylic acids.

LACTONITRILE, $CH_3CHOHCN$.   See Cyanohydrins.

LACTOSE, $C_{11}H_{22}O_{11}$.   See Carbohydrates; Milk and milk products; Sugars.

LAMINATED AND REINFORCED METALS

Metallic laminates and fiber-reinforced metals are serious contenders for structural applications. In this article, these laminates are called metallic matrix laminates (MMLs). MMLs are relatively expensive because fibers, eg, boron fibers, are expensive and the manufacturing processes are time-consuming and costly. They have been used mainly in aerospace structures (see Ablative materials).

Definitions of Selected Laminates

The types of MMLs that are reviewed herein include those made from fiber-reinforced metals, superhybrids, and those made from layers of different metals. Fiber-reinforced metals consist of unidirectional fiber composite (UFC) laminates, as depicted schematically in Figure 1(a), and angleplied (APL) laminates, Figure 1(b). In both UFC and APL laminates, metallic foils may be used between plies to enhance certain mechanical

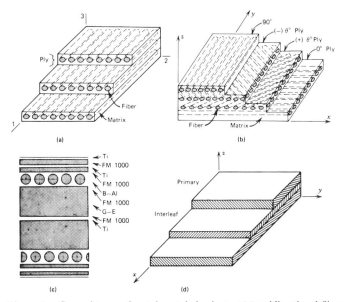

Figure 1.   General types of metal–matrix laminates: (a) unidirectional fiber composite (UFC); (b) angleplied laminate (APL); (c) superhybrid composite (SCH); (d) metal–metal laminate (MML).

Table 1.   Summary of Constituent Materials for Metallic Laminates

Fiber-reinforced metal laminates		Superhybrids — Metal-matrix composite		Metal-metal laminates	
Fiber	Metal	Fiber	Matrix	Primary	Interleaf
alumina (FP)	aluminum	boron	aluminum	aluminum	aluminum
	lead			beryllium	titanium
	magnesium	*Resin-matrix composite*			
beryllium	titanium	Fiber	Matrix	steel	steel
boron	aluminum	graphite	epoxy		
	magnesium		polyimide	titanium	aluminum
	titanium				titanium
borsic	aluminum	kevlar	epoxy		
	titanium			tungsten	copper
graphite	aluminum	S-glass	epoxy		superalloy
	niobium				
	copper			tungsten	tantalum
	lead	*Metal foil*			
	magnesium	titanium			
	nickel				
	tin				
	zinc				
molybdenum	superalloy				
silicon carbide	aluminum				
	superalloy				
	titanium				
steel	aluminum				
	nickel				
tantalum	superalloy				
tungsten	niobium				
	superalloy				

properties, as discussed later. Superhybrid composites (SHC) consist of outer metallic foils, boron–aluminum plies (B–Al), graphite-fiber-resin (UFC) inner or core plies, and adhesive film between these as shown in the photomicrograph in Figure 1(c). Metallic laminates consist usually of alternate layers from two or more metals as depicted schematically in Figure 1(d). The constituents that have been used for metallic laminates are summarized also in Table 1.

Hybrids

A general definition for a hybrid composite is a composite that combines two or more different types of fibers in the same matrix, or one fiber type in two different matrices or combinations of these (ICCM, International Conference on Composite Materials). Superhybrids (Fig. 1(c)) are a generic class of composites that combine appropriate properties of fiber–metal–matrix composites, fiber–resin–matrix composites and/or metallic plies in a predetermined manner in order to meet competing and diverse design requirements. Tiber hybrids are trivial names for titanium–beryllium adhesively bonded metallic laminates.

Fabrication Procedures

Laminates or composites from MMLs are fabricated using diffusion bonding, roll bonding, coextrusion, explosive bonding, and brazing in general.

In diffusion bonding, the filament or the interleaf layer (plies) are hot-pressed between layers of the matrix material. The pressure is usually 6.9–20.7 MPa (1000–3000 psi) and the temperature is 455–540°C. In roll bonding, the layers of metal–metal laminates, the ply (monolayer) is formed by diffusion bonding, or any of the other methods, then the laminate of the specified number of plies is made by roll bonding. In coextrusion, the constituents are assembled as in a billet and are extruded through a given die at specified temperatures and pressures depending on the constituents used. The primary bonding mechanism in the coextrusion process is diffusion bonding. Coextrusion is particularly suited for round and rectangular bar stocks. In explosive bonding, the constituent metal plies are bonded into a laminate by the high pressure generated through explosive means (see Metallic coatings, explosively clad metals). The amount of charge used is determined by the metallurgical bond required between the plies. This method is especially suitable for fabricating MMLs from metal plies with widely different melting temperatures. In brazing, bonding of the constituent metal plies into a laminate is accomplished by a third metal (brazing foil) which acts as a wetting liquid–metal phase and which has a lower melting temperature than either of the constituent metals (see Solders and brazing alloys). The plies to be bonded are stacked into a laminate with brazing foils between them. Then the temperature is raised between the melting temperature of the brazing foils and the constituent plies and appropriate pressure is applied. Upon solidification, the brazing foil bonds the adjacent constituent plies together into a laminate. Boron–aluminum plies are fabricated at about 600°C and < 1.4 MPa (200 psi).

C.C. CHAMIS
National Aeronautics and Space Administration

M.F. Smith, *Metal Matrix Composites*, NTIS/PS-78/0684, N79-10155, Vol. 2, National Technical Information Service, Springfield, Va., 1978.

K.G. Kgeider, ed., Composite Materials, Vol. 4 of L. Broutman and R. Krock, eds., *Metallic-Matric Composites*, Academic Press, Inc., New York, 1974.

W.J. Renton, ed., *Hybrid and Select Metal-Matrix Composites: A State-of-the-Art Review*, American Institute of Aeronautics and Astronautics, New York, 1977.

LAMINATED AND REINFORCED PLASTICS

Reinforced plastics are combinations of fibers and polymeric binders or matrices that form composite materials (qv). The strongest geometry in which any solid can exist is as a wire, a fiber, or a whisker, ie, the strength of any solid is determined by the defects it contains—voids, cracks, discontinuities, etc—and the magnitude of their weakening influence depends upon their absolute size. Thus, if a fine-diameter fiber can be drawn or grown from a material, any defect it contains must be very small; the material will be stronger than in bulk form. But fibers, even if they are strong when pulled, have very limited structural utility by themselves. They bend easily, especially if they are small in diameter, and when pushed axially, buckle under very low forces. To remedy these inadequacies, a supporting medium is required, surrounding each fiber, separating it from its neighbors, and stabilizing it against bending and buckling. These are the functions of the matrix and they are best fulfilled when good adhesion exists between the two.

Properties

The specific gravity of reinforced plastics is low, ie, in the range of 1.5–2.25, compared to 3 for aluminum, 7.9 for steel, 2.5 for concrete, and 2.7 for natural granites and marbles. Only wood, at 0.5, is lower among the structural materials. This low density and the ease of forming into intricately curved or corrugated forms make possible very high stiffness-to-weight structural elements and shapes. These two factors, the high specific stiffness and specific strength, as they are called, are the reasons why both commercial and military aircraft use fiber-reinforced plastics so widely (see Composites, high performance).

The same general characteristics make reinforced plastics attractive to the automobile designer, now under such pressure to reduce vehicle weight. Reinforced plastics offer another advantage: complex subassemblies involving welds, rivets, screws, and bolts executed in metals often can be molded in one piece, eliminating substantial labor costs, even though the plastics may be more expensive in terms of material costs. A classic example of this is the front-end grille assembly which also supports the many lights used in current automobiles. In metals, several dozen different parts require manufacture and assembly; in reinforced plastics, the whole unit is made in one molding operation, and it weighs 50% less than earlier metal designs.

Fibers

Table 1 presents data about most of the fibers used in composites and laminates. The two predominant fibers are glass and cellulose.

Polymeric Binders

In principle, any polymeric resin that can be liquefied and thereby used to wet the reinforcing fibers can be used as a matrix for a composite material. In practice, there are examples of almost any resin in some kind of composite formulation. Realistically, however, the bulk of reinforced plastics produced is based on polyester, epoxy, or a few thermo-

Table 1. Fiber Properties

Material	Specific gravity	Tensile strength, MPa[a]	Tensile modulus, GPa[a]
E-glass	2.6	3570	85
carbon[b]	1.6	2035	357
carbon[b]	1.9	1790	430
Kevlar 29[c,d]	1.44	2860	64
Kevlar 49[c,d]	1.44	3750	135
bulk spruce wood	0.46	104	10

[a] To convert MPa to psi, multiply by 145; to convert GPa to psi, multiply by 145,000.
[b] Properties depend upon carbon: graphite ratio.
[c] E.I. du Pont de Nemours & Co., Inc.
[d] See Aramid fibers.

plastic matrix materials (see Elastomers, synthetic—thermoplastic; Epoxy resins; Polyesters).

Fillers

Fillers (qv) differ from fibers in that they are small particles of very low cost materials; they are used extensively in reinforced plastics and laminates. Typical fillers include clay, silica, calcium carbonate, diatomaceous earth, alumina, calcium silicate, carbon black, and titanium dioxide. Sometimes they comprise as much as half the volume of a composite. They provide bulk at low cost, and they confer other valuable properties such as hardness, stiffness, abrasion resistance, color, reduced molding shrinkage, reduced thermal expansion, flame resistance, chemical resistance, and a sink for heat evolved during curing. If coupling agents are used on their surfaces, many of them improve the impact strength of the composite.

Products and Processes

Statistics in this area are imprecise, but a reasonable estimate of use categories is marine (boats, decks, shields, etc)—20%; transportation (autos, trucks, and trailer body components)—24%; construction (corrugated sheet, space dividers, showers, tubs, and light-control panels)—21%; chemical (pipes, ducts, hoods, and tanks)—12%; electrical (printed circuits, insulation panels, and switchgear)—9%; appliances, aircraft, recreational (housing, partitions, luggage racks, and floor panels)—13%.

FREDERICK J. MCGARRY
Massachusetts Institute of Technology

G. Lubin, ed., *Handbook of Composites*, Van Nostrand Reinhold Co., New York, 1982.

LAMINATED MATERIALS, GLASS

A laminate is an orderly layering and bonding of relatively thin materials. A commonly laminated material is glass. Usually, two pieces of float or sheet glass are bonded with poly(vinyl butyral) (PVB) (see Vinyl polymers, poly(vinyl acetal)s) to provide a highly transparent safety glass, eg, automotive windshields. This combining of transparent abrasion-resistant glass and resilient plastic achieves the durability and safety demanded of such products. Other materials that may be incorporated in laminated glass are colorants, electrically conducting films or wires, and rigid plastics.

Laminated glass is not a true composite material (see Composite materials). The glass needs the safety-net effect of the interlayer if impacted, and the interlayer needs the durability and rigidity of the glass for useful service other than during impacts.

Properties

Laminated materials frequently have limits on properties below those found in one of the components. Laminated glass, with a PVB interlayer, has a service temperature not exceeding 70°C (conservative), far below that of solid glass. Laminated glass becomes more rigid with a decrease in temperature and, below −7°C, it approaches the performance of solid glass. A temperatures above 38°C, it responds more nearly like two glass plies separated by a fluid. Some applications utilize heat-strengthened or tempered glass for additional strength.

The optical properties of laminated glass are required to be equal to solid glass, since most applications are in vision areas. Light scattering by the interlayer essentially is nonexistent if PVB is used. Clean-room practices can reduce the dust and lint that is attracted to the surfaces (see Sterilization techniques). Visible-light transmittance of a typical automotive laminate (2.1 mm glass–0.76 mm PVB–2.1 mm glass) is nearly equal to solid glass and noticeable color usually is absent. Visible-light transmittance is about 88% for clear glass laminates and 78% for tinted laminates.

Manufacture

Practically all laminated glass utilizes plasticized poly(vinyl butyral) (PVB) as the interlayer. Curved, laminated windshields are by far the principal product; silicone and cast-in-place urethane resins are seldom used. Laminators purchase PVB in rolls up to 500 m long, up to 270 cm wide, and from 0.38–1.52 mm thick. There are several plasticizers that are used and at different ratios of plasticizer-to-resin content, depending on the product being manufactured. Flexol 3 GH (bis(2-ethylbutanoic acid), triethyleneglycol ester), used for over 40 years, is being replaced by di-n-hexyl adipate and tetraethylene glycol di-n-heptanoate.

The glass for laminating may be annealed, heat-strengthened, tempered, flat or curved, clear or colored. Thicknesses of 1.5–≥ 12 mm are used. For flat laminates, the glass is cut to size, edged, and treated, if specified, washed, and delivered to the clean room by conveyor (see Conveying).

In order to manufacture curved laminates, the glass is preshaped before laminating. This is achieved by simultaneously bending a pair of glass templates which are usually cut to the shape of the finished windshield and are separated by an inert powder to prevent fusing of the plies. Glass temperatures of 600°C are required to achieve the shape, and the shaping is followed by annealing. Banded windshields usually are constructed with one or more pieces of tinted, heat-absorbing glass to enhance occupant comfort and to reduce air-conditioning load.

The interlayer is placed on one piece of glass with the gradient band, if present, carefully positioned above the designated eye position. The adjacent piece is superimposed, excess interlayer is trimmed, and the sandwich is conveyed from the room through a series of heaters and rolls that press the assembly together while expelling air. Temperature is increased stepwise to 90°C and pressures of 170–480 kPa (25–70 psi) are applied. Solid rubber rolls usually are used with flat laminates and curved glass requires segmented rolls on a swivel frame. For more complex shapes, peripheral gaskets may be applied on the assembly, which is then placed in a bag, and evacuated.

Uses

Laminated glass is used in windshields, and in architectural products where conditions such as safety, sound attenuation, solar control, and security are factors.

ROBERT M. SOWERS
Ford Motor Company

R.C. Cunningham, *U.S. Glass Metal and Glazing*, U.S. Glass Publications, Memphis, Tenn., Jan. 1979, p. 28.

Ward's Automotive Yearbook, 39th and 41st eds., Ward's Communications, Inc., Detroit, Mich., 1977 and 1979.

U.S. Pat. 2,983,635 (May 9, 1961), R.E. Richardson (to Pittsburgh Plate Glass Company).

U.S. Pat. 2,994,629 (Aug. 1, 1961), R.E. Richardson (to Pittsburgh Plate Glass Company).

LAMINATED WOOD-BASED COMPOSITES

During the last 35 years, the plywood industry, by far the largest user of wood adhesives, has become an extremely important factor in the construction field. The structural-laminating industry also has shown healthy growth as improved glues and design information have become available. Adhesives for furniture have shifted more and more from those based on the naturally occurring materials to synthetics. Bonding of wood with adhesives, which generally is far more efficient than with mechanical fasteners, has made possible a wide range of products and uses for which wood was not considered a few decades ago.

Structural Panel Products

There are many wood-based composite panel products on the market and new products that are designed for specific purposes are being developed at an expanding rate. Some of the panel products provide strength and stiffness needed for a particular use; others provide required finish, sound reduction, insulation, covering, or other characteristics. These products are partly or entirely of wood-based material, ie, plywood, insulation board, hardboard, laminated paperboard, and particle board (see Insulation, thermal). Manufacturing and finishing methods vary greatly to provide materials with specific desirable properties.

Although the resulting products vary widely in appearance and properties, all are manufactured in panel form and can cover large areas quickly and easily. The most easily recognizable and widely used form is plywood, which is constructed from veneers that are unrolled from the tree (a log is rotated against a knife in a lathe) (Fig. 1). Adjacent layers of veneers are placed together with the grain at right angles and then they are glued to form the product. Production of most other panel products involves breaking the wood into small chips, flakes, or particles and then reassembling them into boards or panels.

Plywood. Plywood is a glued-wood panel that is composed of relatively thin layers, or plies, with the grain of adjacent layers at an angle to each other (usually 90°). The usual constructions have an odd number of plies to provide a balanced construction. If thick layers of wood are used as plies, often two corresponding layers with the grain directions parallel to each other are used; plywood that is so constructed often is called four ply or six ply. The outer pieces are faces or face and back plies, the inner plies are cores or centers, and the plies between the inner

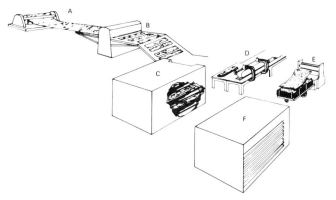

Figure 1. Manufacture of hardwood plywood. A, Veneer (80–90%) is cut by the rotary lathe method. As the lathe spindles move, the log is rotated against a knife. Speed with which knife and knife carriage move toward center of log regulates thickness of veneer. Before cutting, logs are steam-heated to assure smooth texture and easier cutting. Another cutting method, slicing, is used primarily to cut face veneers from walnut, mahogany, cherry, and oak. Flitch is attached to log bed which moves up and down, cutting slice of veneer on each downward stroke. B, The clipper cuts veneer sheets into various widths. C, Dryers then remove moisture content to a level compatible with gluing. D, Veneer sheets of various sizes are clipped and spliced to make full-size sheets. E, Veneers then are coated with liquid glue, front and back, with a glue spreader. F, Heat and pressure that are applied in the hot press bonds the veneer into plywood. Panels are trimmed, sanded, and stacked for conditioning and inspection, after which they are ready for grading, strapping, and shipping. Courtesy of Hardwood Plywood Manufacturers Association.

Table 1. Classification of Wood Species Used for Plywood

Group 1	Group 2	Group 3	Group 4	Group 5
apitong[a,b]	cedar, Port	maple, black	alder, red	aspen	basswood
beech,	Orford	mengkulang[a]	birch, paper	bigtooth	fir, balsam
American	Douglas-fir[c]	meranti, red[a,d]	cedar, Alaska	quaking	poplar,
birch	fir	mersawa[a]	fir, subalpine	cativo	balsam
sweet	California red	pine	hemlock,	cedar	
yellow	grand	pond	eastern	incense	
Douglas fir[c]	noble	red	maple, bigleaf	western red	
kapur[a]	pacific silver	Virginia	pine	cottonwood	
keruing[a,b]	white	western white	jack	eastern	
larch, western	hemlock,	poplar, yellow	lodgepole	black (western	
maple, sugar	western	spruce	ponderosa	poplar)	
pine	lauan	red	spruce	pine	
Caribbean	almon	sitka	redwood	eastern white	
ocote	bagtikan	sweetgum	spruce	sugar	
pine, southern	mayapis	tamarack	black		
loblolly	red lauan		engelmann		
longleaf	tangile		white		
shortleaf	white lauan				
slash					
tanoak					

[a] Each name represents a trade group of woods consisting of a number of closely related species.
[b] Species from the genus *Dipterocarpus* are marketed collectively: apitong if originating in the Philippines; keruing if originating in Malaysia or Indonesia.
[c] Douglas fir from trees grown in Washington, Oregon, California, Idaho, Montana, Wyoming, and the Canadian provinces of Alberta and British Columbia are classed Douglas fir no. 1. Douglas fir from trees grown in Nevada, Utah, Colorado, Arizona, and New Mexico are classed Douglas fir no. 2.
[d] Red meranti are limited here to species having a specific gravity of ≥ 0.41 based on green volume and oven dry weight.

and outer plies are crossbands. The core may be veneer, lumber, or particle board, the total panel thickness typically being not less than 1.6 mm or more than 7.6 cm.

Plywood has several advantages compared to solid wood, eg, it is relatively isotropic, has greater resistance to splitting, and has a form permitting many useful applications where large sheets are desirable. Because in some applications it is permissible to use plywood that is thinner than that normally available as sawed lumber, large areas may be covered with a minimum amount of wood fiber in the form of plywood.

The bulk of softwood plywood is used where strength, stiffness, and construction convenience are more important than appearance. Some grades of softwood plywood are made with faces that are selected primarily for appearance and are used either with clear natural finishes or with pigmented finishes (see Table 1).

Hardwood plywood is used normally where appearance is more important than strength. Most of the production is intended for interior or protected uses. Typical uses of unfinished hardwood plywood include containers, curved or molded products, door skins, die boards, marine-grade materials, pin blocks, kitchen cabinets, furniture components, and exterior siding. Examples of prefinished hardwood-plywood uses include wall paneling or interior uses as surface-covering materials, laminated hardwood block flooring, furniture components, door skins, kitchen and bathroom cabinets, and laminated door sides.

The insulation board-hardboard-paperboard group is known as building fiberboard and includes such proprietary products as Celotex, Insulite, Masonite, Beaverboard, and Homasote.

Insulation board includes two categories: (1) the semirigid type consists of low density products (d < 0.5 g/cm³) that are used as insulation and cushioning, and the rigid type includes both interior board which is used for walls and ceilings and exterior board which is used for wall sheathing; and (2) hardboard (d 0.53–1.44 g/cm³) is a grainless, smooth, hard product that is used as prefinished wall paneling and exterior siding. Laminated paperboard (d 0.51–0.53 g/cm³) serves as sheathing as well as in other covering applications but is not used as much as the other building fiberboards.

Glued Structural Members

Glued-laminated or parallel-grain construction, as distinguished from plywood or other crossbanded construction, refers to two or more layers of wood that are joined with an adhesive so that the grain of all layers or laminations is approximately parallel. The size, shape, number, thickness of the laminations, and uses, may vary greatly.

Although the properties of glued-laminated products are similar to those of solid wood of similar quality, manufacture by gluing permits production of long, wide, and thick items from small and inexpensive material and often with less waste of wood than if solid wood were used alone. Curved members may be fabricated by simultaneously bending and gluing thin laminations to shapes that would be very difficult or impossible to produce from solid wood. The essentially parallel direction of grain of the wood to the longitudinal axis of these laminated products gives them strength that often is far superior to solid wood that is cut to the same size and shape.

The advantages of glued-laminated or parallel-grain construction include: ease of manufacture of large structural elements from standard commercial sizes of lumber or veneer; minimization of checking or other drying defects associated with large, one-piece wood members, in that the laminations are thin enough to dry readily before manufacture of members; the opportunity of designing on the basis of the strength of dry wood, for dry service conditions, inasmuch as the individual laminations can be dried to provide members that are thoroughly seasoned throughout; the opportunity to design structural elements that vary in cross section along their length in accordance with strength requirements; the possible use of lower-grade material for less highly stressed laminations without adversely affecting the structural integrity of the member; and the manufacture of large laminated structural members from smaller pieces is increasingly adaptable to future timber economy, as more lumber will be produced in smaller sizes and lower grades than at present. Generally, laminating provides the opportunity to produce engineered wood components for specific use applications.

Glued-laminated timbers (glulam). Structural glued-laminated timber (glulam) is defined as three or more layers of sawed lumber that are glued together with approximately parallel grain direction. The layers may vary as to species, number, size, shape, and thickness. Laminated wood was first used in the United States for furniture parts, cores of veneered panels, and sporting goods but now is widely used for structural timbers in building.

Glued structural timbers may be straight or curved. Curved arches have been used to span more than 91.4 m. Straight members spanning up to 30.5 m are not uncommon, and some span as much as 39.6 m. Sections deeper than 2.1 m have been used. Straight beams can be designed and manufactured with horizontal laminations (lamination parallel to the neutral plane) or vertical laminations (laminations perpendicular to the neutral plane). The horizontally laminated timbers are the most widely used. Curved members are horizontally laminated to permit bending of laminations during gluing.

Quality of glue joints. Glue-joint quality is one of the most critical manufacturing considerations. Using proven adhesives, it is possible to produce joints that are durable and essentially as strong as the wood itself. *U.S. Product Standard PS 56-73* and *CSA Standard 0122-1969* describe minimum production requirements for glulam.

Uses. Ninety percent of glulam is used in making straight or slightly cambered members for use in roof-support structures. Curved members for arch-type roof supports account for the other 10% of the production.

Wood–plywood glued structural members. Highly efficient structural components or members can be produced by combining wood and plywood through gluing. The plywood is utilized in load-carrying capacity and in filling large opening spaces. These components include box beams, I-beams, stressed-skin panels, and folded plate roofs.

Laminated-veneer structural members. Parallel-laminated veneer in thick sheets of any width or length is being examined as an alternative to solid sawed timber or glulam for structural-sized or specialty-type members.

Laminated-veneer structural members appear to have the greatest market potential when used for specialty items (eg, truck decking and scaffold planking) or where high tensile strength is primary and stiffness is secondary (eg, in truss chords and flanges of built-up beams).

JOHN A. YOUNGQUIST
United States Department of Agriculture

Reviewed by
ALAN D. FREAS
Professional Engineer

Wood Handbook: Wood as an Engineering Material, U.S. Dept. Agric., Agric. Hand. 72, Rev. U.S. Forest Products Laboratory, Madison, Wisc., 1974.

T.M. Maloney, *Modern Particleboard and Dry-Process Fiberboard Manufacturing*, Miller-Freeman Publications, San Francisco, Calif., 1977.

A.D. Freas and M.L. Selbo, *U.S. Dept. Agric. Tech. Bull.* 1069, U.S. Government Printing Office, Washington, D.C., 1954.

LAMPBLACK.   See Carbon.

LANOLIN.   See Wool.

LANTHANUM.   See Rare-earth elements.

LARD.   See Fats and fatty oils; Meat products.

LASERS

Lasers are sources of light. The many different types of lasers produce light at a variety of different wavelengths in the visible, infrared, and ultraviolet spectrum. Today, lasers have become familiar tools for applications such as alignment, metrology, and industrial materials processing. There are still many exciting applications for future development, including laser-assisted thermonuclear fusion, and laser-assisted separation of isotopes. Among the most important chemical applications are high speed transient monitoring of the kinetics of chemical reactions, extremely high resolution spectroscopic techniques, and control of the course of a chemical reaction by selectively exciting certain molecular states.

Laser radiation can be used for a variety of material processing functions such as welding, cutting, shaping and drilling. It is employed for many such operations in industry. Lasers also have been used for measurement of many different physical parameters, including length and distance, velocity of fluid flow and of solid surfaces, dimensions of manufactured goods, and the quality of surfaces, including flow detection and determination of surface finish. Holography (qv) is a photographic process that yields three-dimensional images and is dependent on lasers for a light source. A variety of medical procedures also depend on lasers, including the repair of tears and holes in the retina, removal of tattoos and colored birthmarks, and simultaneous cutting and cauterization of tissue. Laser-based optical-fiber telecommunication systems are becoming increasingly important for local and long-distance communication (see Fiber optics).

Fundamentals of Lasers

Laser light is produced in transitions between energy levels of atoms or molecules. The relevant energy levels may be those of atoms or molecules in a gas, as in the case of the helium–neon laser, or of ions embedded in a solid host material, as in a ruby laser, or they may be energy levels that belong to the crystalline lattice as a whole, as in the case of the aluminum gallium arsenide laser. The important requirement is that there are two energy levels that can be connected by a transition that emits light.

The interaction between the light and the energy levels to produce laser operation relies on the phenomenon of stimulated emission. For stimulated emission, an atom or molecule must be in one of its excited levels and have a vacant energy level of lower energy. Then incoming light of the proper frequency can trigger a transition from the upper level. To do so, the photon energy of the incident light must nearly equal the energy difference between the two levels. The light can then stimulate the atomic or molecular system to make the transition. At the same time, the energy stored in the atomic or molecular system is given up as light. The light that is emitted by the atomic or molecular system travels in the same direction as the original light and remains in phase with it.

This latter property gives rise to many of the important properties of lasers, including coherence and directionality.

There are three requirements to produce a laser: first, a material that possesses an appropriate set of energy levels (the active medium); second, some means for excitation or pumping the atoms or molecules to excited upper levels, and at the same time leaving empty, lower-lying energy levels (a condition called a population inversion); third, some means of resonant feedback to allow the light to pass back and forth through the active medium. During these passes, the light is amplified and builds in intensity. The resonant feedback is provided by a pair of mirrors, one on each side of the active medium.

Properties of Laser Light

The light emitted by a laser has a number of unusual properties that distinguish it from light emitted by conventional light sources. Among the unusual properties are a high degree of collimation, a narrow spectral linewidth, good coherence, and the ability to focus to an extremely small spot. Because of these unusual properties, there are many possible applications for which lasers are better suited than conventional light sources. For example, alignment applications utilize the collimation of laser light. Spectroscopic and photochemical applications depend on the narrow spectral linewidth of tunable laser sources. Communications applications require a high degree of coherence.

Spatial profiles.   The cross sections of laser beams have certain well-defined spatial profiles called transverse modes. The word mode in this sense should not be confused with the same word as used to discuss the spectral linewidth of lasers. Transverse modes represent configurations of the electromagnetic field determined by the boundary conditions in the laser cavity. The modes are denoted by the nomenclature TEM_{mn}, where TEM stands for transverse electromagnetic, and where m and n are small integers. The notation can be interpreted as meaning that m and n specify the number of low intensity points (nulls) in the spatial pattern in each of two perpendicular directions, transverse to the direction of beam propagation. The TEM_{00} mode is an especially desirable case. This transverse mode, the Gaussian mode, is symmetric and has no nulls. It is preferred for many applications.

Temporal Characteristics

Laser operations may be characterized as either pulsed or continuous. There are a number of distinctive types of pulsed laser operations, having widely different pulse durations.

In order to produce pulses with smooth temporal characteristics and also to increase the peak power, methods of pulse control were developed. The most common method is called Q-switching. In a Q-switched laser, a switchable shutter is inserted between the laser material and one of the mirrors. When the laser is first pumped, the shutter is switched to its closed position. No light can reach the mirror and the process of amplification through stimulated emission cannot build up. The laser medium is excited to the point that the population inversion considerably exceeds the threshold for laser operation. Then the shutter is rapidly switched from its closed position to its open position. Because of the large population inversion, the gain can be very high, and the laser pulse develops rapidly. The energy stored in the population inversion can be swept out in a single pulse of high peak power. The pulse duration is much shortened, from the millisecond region to the region of perhaps 30 nanoseconds. The total energy release in the laser pulse is reduced by the Q-switching operation. Because the pulse duration is decreased by many orders of magnitude, the peak power can be much higher than for a normal pulse. Such pulse control is accomplished using switchable shutters.

In many cases, there is substructure in the Q-switched pulse. The output of the Q-switched laser can consist of a train of much shorter pulses. The individual pulses in the pulse train are separated by the round-trip transit time of the cavity, $2L/c$, where L is the distance between the laser mirrors and c is the velocity of light (3×10^{10} cm/s). The widths of the individual pulses are very short, on the order of tens of picoseconds. This train of pulses is commonly called a mode-locked train, and the individual pulses are called mode-locked pulses or picosecond pulses.

It is possible to switch a single picosecond pulse out of the train of mode-locked pulses with an electrooptic switch. Thus, one can obtain a single pulse with duration in the picosecond regime.

Laser Types

There are many different types of laser showing a wide variety of methods of construction, and based on different classes of materials. Table 1 summarizes properties of some lasers that are available commercially. They include gas lasers (Fig. 1), solid-state lasers (Fig. 2), organic-dye lasers, and semiconductor lasers (see Light-emitting diodes and semiconductor lasers). There are a number of other lasers that are under development that may become important. These include chemical lasers, excimer lasers, and other tunable lasers.

Table 1.   Common Commercial Lasers

Laser	Wavelength, μm	Operation	Typical output	Typical applications
helium–neon	0.6328	continuous	mW	alignment, metrology, holography
carbon dioxide	10.6	continuous or pulsed	to 10 kW CW (or more) or 1 J/pulse	welding, hole drilling, heat treating
neodymium–yttrium–aluminum–garnet (YAG) and neodymium–glass	1.06	continuous or pulsed	to hundreds of watts CW or 100 J/pulse	welding, hole drilling, trimming
argon	0.4880, 0.5145 and other lines	continuous	to 18 W	velocimetry, holography
ruby	0.6943	pulsed	few J/pulse	spot welding
aluminum–gallium–arsenide	0.85	usually continuous	mW	communications, ranging
dye	tunable 0.4–1	continuous	1–2 W	spectroscopy, photochemistry
helium–cadmium	0.4416 or 0.3250	continuous	mW	photoresist exposure
krypton	0.6471	continuous	to 6 W	displays

[a]CW = continuous wave
[b]To convert J to Btu, divide by 1.054×10^3

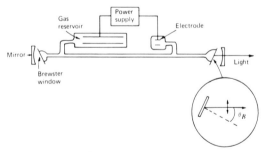

Figure 1.   Gas laser.

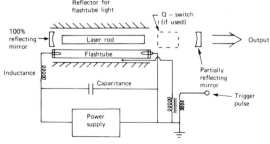

Figure 2.   Solid-state laser.

Laser Safety

Effects of laser radiation on the body. The structure of the body most easily damaged by laser light is the retina, the photosensitive surface at the back of the eyeball, because the lens and cornea are transparent to light in the wavelength region 0.4–1.4 μm, laser light in this wavelength can reach the retina. High power lasers also can cause serious skin burns.

Most discussions of laser safety emphasize the hazards produced by the optical beam, but there are many other possible hazards associated with the use of lasers. The most serious is the hazard of electrical shock associated with the high voltage electrical power supplies commonly used with lasers. In addition, the poisonous or corrosive substances used either in the laser itself (eg, dye materials and solvents) or in equipment used in association with lasers, such as modulators, present serious hazards.

The basic practice for protection from the laser beam is not to allow laser radiation to strike the human body at a level higher than the maximum permissible exposure level. Maximum permissible exposure levels for both eyes and skin have been defined in various laser-safety standards. One of the most common safety measures is to use protective eyewear to reduce the exposure of the eye, but it is important to realize that such eyewear offers limited protection.

Nonlinear Optics

The electromagnetic field of a light beam produces an electrical polarization vector in the material through which it passes. In ordinary optics, which may be termed linear optics, the polarization vector is proportional to the electric field vector $\vec{E}$. However, the polarization can be expanded in an infinite series:

$$P = \chi E\left(1 + a_2 E + a_3 E^2 + a_4 E^3 + \cdots\right)$$

In this equation P is the polarization, E is the electric field, and χ is the linear polarizability. This equation is a simplified scalar representation of a tensor equation. The nonlinear coefficients a_i are very small compared to unity. For reasonably small values of electric field, only the first term in the equation will be important. Thus, in the prelaser era, one could use the approximation that polarization is linearly proportional to electric field. When high power lasers became available, the electric fields became much higher. The product $a_i E^{i-1}$ could become large enough to be observable. The second term in the equation is of the form $\chi a_2 E^2$. If E has a sinusoidal variation of the form $E_0 \cos \omega t$, the components of polarization owing to it are of the form:

$$a_2 \chi E_0^2 \cos^2 \omega t = \tfrac{1}{2} a_2 \chi E_0^2 \left(1 + \cos 2\ \omega t\right)$$

The second term on the right-hand side, a component oscillating at frequency 2ω, represents the second harmonic of the incident beam. This component of the polarization vector can radiate light at the frequency 2ω.

Only certain types of crystalline materials can exhibit second harmonic generation. Because of symmetry considerations, the coefficient a_2 must be identically equal to zero in any material that has a center of symmetry. Thus, the only candidates for second harmonic generation are materials that lack a center of symmetry. Some common materials that are used in nonlinear optics include barium sodium niobate ($Ba_2NaNb_5O_{15}$), lithium niobate ($LiNbO_3$), and iodic acid (HIO_3). The coefficient a_3 may be nonzero even in centrosymmetric crystals, so that effects that depend on mixing of three waves ($a_3 E^3$) may be observed in such crystals.

A variety of other nonlinear optical effects, including parametric amplification, stimulated Raman scattering, and optical-phase conjugation, have been observed.

JOHN F. READY
Honeywell Corporate Technology Center

J.F. Rady, *Industrial Applications of Lasers*, Academic Press, New York, 1978.

D.C. O'Shea, W.R. Callen, and W.T. Rhodes, *An Introduction to Lasers and Their Applications*, Addison-Wesley, Reading, Mass., 1977.

J.T. Knudtson and E.M. Eyring, *Ann. Rev. Phys. Chem.* **25**, 255 (1974).

LATEX TECHNOLOGY

Since World War II, latex or emulsion-polymerization technology has expanded rapidly to include vinyl acetate, vinyl chloride, acrylics, acrylonitrile, and ethylene, in addition to the established monomers of styrene, butadiene, and isoprene. Latex describes all the different emulsion polymers made (see Emulsions; Polymerization mechanisms and processes). There are two main uses for latices: surface coatings and thermoplastics. The increasing number of new latex applications in surface coatings (taking over a share of the decreasing solvent-based coatings), and a new family of modifiers for thermoplastics promise future growth.

Emulsion Polymerization

Mechanisms. Emulsion polymerization is a heterogeneous noncatalytic reaction. The mechanistic reaction model is based on the mixing of a monomer and a surfactant in water to yield both monomer droplets and solubilized monomer in micelles. The free radical is generated in the aqueous phase and can enter either the monomer droplet or monomer-swollen micelle.

Generally, the course of batch-emulsion polymerization is subdivided into three intervals. Particle formation takes place during interval I and, depending on the water solubility of the monomer, normally ends at ca 1–5% conversion. Interval II is characterized by the growth of the primary particles and ends when the separate monomer phase disappears. In interval III, the particle volume decreases, ie, the monomer concentration in the particle decreases with a corresponding increase in polymer viscosity.

Colloid and Surface-Chemical Aspects

Solubilization and emulsification of monomers. Solubilization is the spontaneous passage of molecules of an insoluble substance in water into the interior of the micelle of a dilute surfactant solution. Solubilization involves the process of diffusion from a separate monomer phase to the micelle. Emulsification occurs beyond the monomer saturation point of micelles with further addition of monomer. There is a strong temperature effect on the solubilization of an oil in the aqueous solution of a nonionic surfactant that results from a change in the hydrophilic–lipophilic balance (HLB) of the surfactant (see Emulsions).

Surface stability. Emulsion polymers generally are considered to be metastable systems. The large surface represented by the particles is thermodynamically unstable and any perturbation affecting the balancing forces results in a change of the kinetics of particle agglomeration. The chief forces that affect stability are a repulsive force, which results from the interaction of similarly charged particle surfaces; entropic repulsive forces, which are generated by steric hindrance of solvation of the adsorbed layers; and attractive forces of the London–van der Waals dispersion type. The balance of these forces determines the stability of the polymer colloid. The parameters that can perturb these forces are electrolyte addition, mechanical agitation, thermal effects (freeze-thaw), and organic solvent addition.

Addition of electrolyte. An interface, eg, the latex polymer particle-aqueous-phase interface, can acquire an electrostatic charge by several mechanisms. The most important mechanisms are those in which the organic interface is controlled by the ionization of surface chemical groupings or the adsorption of ionized, surface-active material.

Mechanical agitations. Bulk coagulation of latex in a shear field results from the increased energy of collisions between particles and, as a result, larger numbers of collisions have sufficient energy to surmount the repulsive barrier.

Thermal effects. Freeze–thaw stability is important to shipping and storing of latices. Generally, the freezing of drums that contain latex is a slow, equilibrium process. Other than the incorporation of ionogenic comonomers, perhaps the most straightforward attempt to incorporate freeze–thaw stability is the prevention of ice formation. This can be achieved easily by the addition of water-soluble solvents (eg, ethylene glycol, propylene glycol, etc) which depress the freezing point (see Glycols; Antifreezes).

Organic solvent addition. With ionogenically stabilized latices, the addition of a water-soluble organic solvent swells the latex particles but does not dissolve them. The latex particles are stabilized by charge and hydration of polar surface groups. The degree of charge stabilization is determined by the balance between the surface-active agent concentration on the particle surface and particle size.

Particle size. The stability and the performance properties of latices depend in part on the particle size and distribution. The standard of reference and most reliable methods are microscopic. The most recent and promising development for particle-size distribution determination is hydrodynamic chromatography (see Size measurement of particles).

Viscosity. Latex rheology is important in manufacturing and handling and in coating applications; measurements are carried out with rotational viscometers. At low solids content, latices behave according to the Einstein model. At higher concentrations, particle–particle interactions become significant and there is a steep rise in viscosity with increasing polymer–volume fraction (see Flow measurement; Rheological measurements).

Film Properties

When dried above their respective glass-transition temperatures, emulsion polymers form a continuous polymer phase. When a substrate is coated with a latex and the latex is dried, the resultant structure has a polymer-coated surface. The latex can be filled with pigments or extenders, or it may be kept as the pure polymer. The properties of the film determine the latex's surface-coating applications.

Bulk Properties

Bulk polymer properties are dependent on monomer types and modifying comonomers, eg, chain-transfer and cross-linking agents and surface-modifying polar monomers.

Initiation and Process Conditions

Initiator systems. The salts of peroxydisulfate are used predominantly in thermal initiation. The combination of oxidizing and reducing agents, which generate free radicals at low temperatures, are particularly useful in obtaining high molecular-weight polymers with low branching (see Initiators; Peroxide).

Batch process. The performance properties of latex largely depend on the way it is made. Only the isothermal batch process was initially considered. Industrially, in addition to the isothermal batch, the adiabatic batch, the semibatch, and continuous processes are important.

Artificial latices. Several kinds of polymers, eg, polyurethanes, polyesters (qv), polypropylene, epoxy resins (qv), and stereoregular butadiene (qv) and isoprene (qv), cannot be polymerized by emulsion polymerization techniques; however, they can be prepared in emulsion form (see Urethane polymers; Olefin polymers). The methods for the preparation of these artificial latices are either by emulsification of a polymer solution with subsequent stripping of the solvent or by phase inversion.

Inverse emulsion polymerization is utilized for water-miscible monomers, eg, sodium *p*-vinylbenzenesulfonate, 2-sulfoethyl methacrylate, acrylamide, etc.

Cationic emulsion polymerization. Cationic emulsion polymers can be prepared by using several techniques, eg, artificial latex technology, copolymerization of a functional monomer which can be converted at the particle surface after polymerization, or copolymerization of a cationic monomer.

A. KLEIN
Lehigh University

J.W. Vanderhoff, "The Mechanism of Emulsion Polymerization" in G.E. Ham, ed., *Vinyl Polymerization*, Vol. I, Part II, Marcel Dekker, Inc., New York, 1969, pp. 1–138.

D.C. Blackley, *High Polymer Latices*, Vols. I and II, Palmerton Publishing Co., Inc., New York, 1966.

P.C. Becher and M.N. Yudenfreund, eds., *Emulsions, Latices and Dispersions*, Marcel Dekker, Inc,. New York, 1977, pp. 23–195.

LAUNDERING.   See Drycleaning and laundering.

LEAD

Lead is an essential commodity in the modern industrial world, ranking fifth in tonnage consumed after iron, copper, aluminum, and zinc. Its outstanding properties are low melting point, ease of casting, high density, low strength, ease of fabrication, acid resistance, electrochemical reaction with sulfuric acid, chemical stability in air, water, and earth, and the unique ability of lead tetraethyl and tetramethyl to suppress knocking in gasoline engines. The principal uses of lead and its compounds, in descending order, are storage batteries (about half of total U.S. consumption), tetraethyllead (see Lead compounds, organolead), pigments, ammunition, solders, plumbing, cable covering, bearings, and caulking. In addition, the use of lead to attenuate sound waves, atomic radiation, and mechanical vibration shows promise for growth. Because of its softness and high density, lead is used as an alloy in these applications (see Lead alloys).

Lead and its compounds are cumulative poisons and should be handled with recommended precautions. They should not be used in contact with food and other substances that may be ingested.

The occurrence of lead ore deposits is unexpectedly high and they are widely distributed throughout the world. The most important mineral is galena (PbS, 87 wt% Pb), followed by anglesite (PbSO$_4$, 68 wt% Pb), and cerussite (PbCO$_3$, 77 wt% Pb). The two latter minerals result from the natural weathering of galena.

Most lead mined in the United States comes from seven mines in Missouri (88%), whereas the rest comes from eight mines in Colorado, Idaho, and Utah. The recovery of lead from scrap is of prime importance in supplying U.S. demands.

Physical Properties

Lead is a member of Group IVA of the periodic table. The crystal structure of lead is face-centered cubic, and the length of the edge of the cell is 4.9389 nm. The number of atoms per unit cell is 4. Other properties are listed in Table 1.

Chemical Properties

Lead forms two series of compounds corresponding to the oxidation states of +2 and +4, the most common is +2. Compounds of Pb(IV) are regarded as covalent and those of Pb(II) primarily ionic. Lead is amphoteric and forms plumbous and plumbic salts as well as plumbites and plumbates.

Although lead readily tarnishes in the atmosphere, it is one of the most stable fabricated materials because of its excellent corrosion resistance to air, water, and soil. Initial reaction with these environments takes place, but protective coatings of insoluble lead compounds form.

Lead should replace hydrogen from acids because of its position relative to hydrogen in the electropotential series, but the difference is small, and the high hydrogen overvoltage prevents replacement. Reaction with oxidizing acids releases oxidants that combine with hydrogen and depress the overvoltage, resulting in replacement.

Processing

Ore dressing.   The lead mineral in most crude ores is separated from the gangue and other valuable minerals. Occasionally, the ores are sufficiently rich in lead and low in impurities to be smelted directly. The

Table 1.   Physical Properties of Lead

Property	Value
atomic weight	207.2
melting point, °C	327.4
boiling point, °C	1770
specific gravity, g/cm^3	
20°C	11.35
327°C (solid)	11.00
327°C (liquid)	10.67
specific heat, J/(kg·K)^a	130
latent heat of fusion, J/g^a	25
latent heat of vaporization, J/g^a	860
vapor pressure at 980°C, kPa^b	0.133
thermal conductivity at 28°C, W/(m·K)	34.7
thermal conductivity	8.2
(relative to Ag = 100)	
coefficient of linear expansion, at 20°C per °C	29.1×10^{-6}
electrical resistivity at 20°C, μΩ/cm	20.65
specific conductance at 0°C, S/cm	5.05×10^4
electrical conductivity	7.8
(relative to Cu = 100)	
normal electrode potential,	0.22
V, standard hydrogen electrode = 0	
electrochemical equivalent of Pb^{2+}, g/(A·h)	3.8651
velocity of sound in lead, cm/s	122,700
surface tension a 360°C, mN/m (= dyn/cm)	442
viscosity at 440°C, mPa·s (= cP)	2.12
magnetic susceptibility at 20°C, m^3/kg (emu/g)	-0.29×10^{-6} (-23×10^{-6})
hardness, Mohs	1.5
Brinell	
common lead	3.2 to 4.5
chemical lead	4.5 to 6
Young's modulus, GPa^c	16.5
tensile strength of common lead, kPa^d	
−100°C	42,000
20°C	14,000
150°C	5,000
elongation (% in 5-cm gauge length)	50–60

^{a} To convert J to cal, divide by 4.184.
^{b} To convert kPa to mm Hg, multiply by 7.5.
^{c} To convert GPa to psi, multiply by 145,000.
^{d} To convert kPa to psi, divide by 6.895.

primary operations of ore dressing are crushing, grinding, and concentration (beneficiation).

Blast-furnace smelting.   Although the lead sulfide concentrate could be oxidized and reduced to metallic lead in one operation in a blast furnace, sintering the concentrate prior to charging lowers the operating temperature and greatly increases the capacity and efficiency of the furnace. In addition to the formation of lead oxide, sintering drives off volatile oxides such as SO$_2$, SO$_3$, As$_2$O$_3$, and Sb$_2$O$_3$ and agglomerates (conditions) the charge to produce a suitably hard but porous feed.

In the blast furnace, lead and other metal oxides, not reduced during sintering, are reduced to metals, the molten lead is coalesced in the hearth, and the gangue material is separated into a molten slag. The molten lead serves as a solvent for the valuable metallic impurities that are recovered during the refining, separated, and ultimately purified for marketing. Typical blast-furnace feed and products are shown in Table 2.

Hearth smelting.   Because of limitations on the type of feed material and the superior economics of blast-furnace operations, this method is used no longer in the United States.

Drossing.   Certain compounds and elements are entrained and dissolved in the separated molten lead from the blast furnace. Among these are portions of slag, matte, speiss, lead oxide, and copper which must be removed before the bullion is refined. The operation is referred to as copper drossing, and is performed in one or two 250-t cast-iron kettles. The process consists of skimming off the dross, stirring the lead, and reskimming.

Table 2. Typical Blast-Furnace Analysis of Materials and Products, %

Feed material		Slag		Bullion	
Ag	0.05		0.0003		0.2
				Au	0.0003
Cu	0.6–1.5		0.10		1.0–2.5
Pb	35–50		1.5–3.5		
S	0.75–1.6				
Fe	12–15.5	FeO	25–33	Fe	0.6–0.8
SiO₂	13.5–15.6				
CaO	9.0–10.5		10–17		
Zn	9.5–12.5		13–17		
		insoluble matter	22–27		
		MnO	2.4–4.5		
		As	0.10		0.7–1.1
		Sb	0.10		1.0–1.75
				Bi	0.01–0.03

ᵃ As FeO.

Refining

Pyrometallurgical methods. To prepare blast-furnace bullion for commercial sale, certain standards must be met either by the purity of the ores and concentrates smelted or by a series of refining procedures. Since the separated impurities have market value, the refining operations serve not only to purify the lead, but also to recover valuable by-products.

The pyrometallurgical processes (furnace-kettle refining) are based on the higher oxidation potentials of the impurities of antimony, arsenic, and tin in comparison to lead, and the formation of insoluble intermetallic compounds by reaction of metallic reagents such as zinc with the impurities, gold, silver, and copper, and calcium and magnesium with bismuth.

The oxidizing process used to remove antimony, arsenic, and tin has been termed softening because lowering these impurities results in a readily detectable softening of the lead. Softening processes include the reverberatory process, continuous softening process, and the Harris process.

After softening, the impurities that may still remain in the lead are silver (ca 0.2 wt%), gold (ca 0.006 wt%), copper, tellurium, platinum metals, and bismuth (up to 0.1 wt%). Although these concentrations may be tolerable for some lead applications, their market value encourages separation and recovery.

Electrolytic refining (Betts process) in one step is capable of removing all these impurities, but in general, with several exceptions outside the U.S., pyrometallurgical techniques are preferred, namely the Parkes process for removing noble metals and any residual copper, and the Kroll-Betterton process for bismuthizing.

Electrolytic refining. Electrolytic refining is used by Cominco Ltd. in Trail, B.C., Canada, and Cerro de Pasco Corp., La Oroya, Peru, as well as several refineries in Europe and Japan.

Electrolytic refining of lead is conducted in electrolytic tanks or cells. The electrolyte is a solution of lead fluorosilicate, PbSiF₆, and free fluorosilicic acid. Cathode starting sheets are made from pure electrolytic lead. The concrete electrolytic cell boxes and tanks are usually 2.7-m long, approximately 80-cm wide, and ca 1.2-m deep. They are lined with asphalt or a plastic material such as polyethylene.

After the corroded anodes are washed, and the adhering slimes are scraped off, filtered, and dried, approximately 8 wt% moisture is left to prevent dusting. The general practice is to smelt the slimes in a small reverberatory furnace, which produces a slag that is 10–12 wt% of the slimes. This slag is taken to a second small reverberatory furnace, in which it is partially reduced to remove precious metals; the remainder is transferred to the smelting department for production of antimonial lead. The reduced portion, containing the precious metals, is returned to the original slimes-smelting furnace.

Slag and litharge formed during cupellation are segregated and reduced to a metal containing ca ≥ 20–25 wt% bismuth, depending upon the bismuth content of the original bullion, and transferred to a bismuth-recovery plant.

Imperial Smelting Process

The Imperial smelting process was originally designed as a method of producing zinc in a blast furnace. This smelting process is applied particularly for treating mixed zinc–lead ores from mines in which the mineralization is such that separation of the lead and zinc portions is difficult.

New Developments

The mainstay of the lead-smelting industry is the sinter-blast-furnace process. Improvements in sintering have resulted from the installation of updraft sintering machines in new smelters or replacements for downdraft machines in old plants. Much greater capacity per machine, more economical production of sinters, better blast-furnace performance, and improved hygienic conditions have been reported.

Progress in improving the blast-furnace operation has been sought through enriching the air blast with oxygen, perheating the blast, replacing expensive coke with liquid or gaseous fuels, and adopting better methods of measurement and control.

However, problems of air pollution and waste disposal and rising fuel costs have stimulated the development of new approaches to lead smelting. A new process for extracting lead should take advantage of the autogenous nature of the pyrometallurgical reduction of sulfide ores, and pyrometallurgical methods should be developed that consume less energy.

Recycled Lead

The recovery of lead from scrap is an important source for the lead demands of the United States and the rest of the world. In the United States, 50% of the lead requirements are satisfied by recycled lead products (see Recycling). The principal types of scrap are battery plates, drosses, skimming, and industrial scrap such as sodlers, babbitts, cable sheathing, etc.

Health and Safety Factors

Measures to control exposure include provision of proper ventilation, application of proper work practices, following rules of good hygiene, and wearing respirators and protective clothing. Detailed regulations by OSHA and other Federal agencies governing occupational exposure to lead may be found in Part 1910 of Title 29 of the Code of Federal Regulations.

H.E. Howe[‡]
Consultant

Reviewed by
William P. Roe
ASARCO

P. McIllroy, *Availability of U.S. Primary Lead Resources*, Bureau of Mines IC 8646, 1974.

"Properties and Selection of Nonferrous and Pure Metals: Lead" in J.F. Smith and R.R. Kubalak, eds., *Metals Handbook*, 9th ed., Vol. II, Metals Park, Ohio, 1979.

LEAD ALLOYS

In order to strengthen it, lead is used mostly in the form of alloys. The principal alloying elements are copper, antimony, calcium, tin, arsenic, tellurium, silver, and strontium. Among the less important elements are bismuth, cadmium, and indium.

Lead and its alloys are generally melted, handled, and refined in cast-iron, cast-steel, welded-steel, or spun-steel melting kettles without fear of contamination by the iron. Normal melting procedures require no flux cover for lead. Special reactive metal alloys require special alloying elements, fluxes, or covers to prevent dross formation and loss of alloying elements.

Lead and lead alloys are commonly joined by a welding process called lead burning that requires great skill. It differs considerably from conventional welding because no flux is used, no inert gas cover is required, and the welding rod used is the same as the base metal (see Welding).

[‡] Deceased.

Lead – Copper Alloys

The lead–copper system has a eutectic point at 0.06 wt% copper at 326°C; disregarding the high copper alloys, only lead alloys with copper contents below 0.10 wt% have practical applications such as use in sheet lead, lead pipes, cable sheathings, wire, and other fabricated lead products.

The alloys with small copper contents produce considerable grain refinement and give resistance to grain growth, particularly at elevated temperatures. The addition of copper to lead in the range 0.04–0.08 wt%, with or without the addition of silver, produces the ASTM B 29–79 grades: chemical and copper-bearing (see Lead). These grades are specified for use in the chemical industry because of higher strength, superior creep resistance, and greater resistance to recrystallization than pure lead. In addition, the copper–lead alloys show extremely good resistance to H_2SO_4 corrosion at elevated temperatures.

Properties. The mechanical properties of the copper-bearing lead are compared to common lead in Table 1.

Lead – Antimony Alloys

The most important lead alloys are the lead–antimony alloys. Virtually all lead products utilize antimony alloys of some composition.

Properties. The lead–antimony system has a eutectic point at 252°C at 11.1 wt% antimony. Most lead–antimony alloys of commercial importance are produced with ≤ 11 wt% antimony. Table 2 gives the physical and mechanical properties of 1 wt%, 4 wt%, and 6 wt% lead–antimony alloys in both cast and wrought form.

Lead–antimony alloys, in a manner similar to lead–copper alloys, develop a protective coating of lead oxide and lead carbonate that renders them practically inert to further atmospheric attack. Therefore, they are used for flashing and roofing in 1.2-mm thicknesses.

Uses. The principal use for lead–antimony alloys is in the manufacture of grids, straps, and terminals of lead-acid batteries.

Lead – Calcium Alloys

Lead-low-calcium alloys are rapidly replacing the lead–antimony alloys in many traditional applications such as batteries, anodes, sleeving, and other castings.

Properties. Calcium, a reactive metal, combines readily with lead. Because of its reactive nature, calcium oxidizes preferentially from lead–calcium alloys, and care must be taken to maintain the desired composition. Aluminum has been added to calcium alloys to reduce calcium oxidation and maintain the calcium content of the alloys. The mechanical properties of common lead, lead–antimony, lead–copper, and lead–calcium alloys are compared in Table 3.

Uses. The main use of binary lead–calcium alloys has been as a grid alloy for large stationary standby power batteries.

Lead – Calcium – Tin Alloys

Lead-calcium-tin alloys are significantly stronger than the binary alloys and have become the principal alloys for maintenance-free batteries in both cast and wrought form. These alloys exhibit great stability, extremely high mechanical properties, and creep resistance. Because of their superior mechanical properties, the cast and wrought lead-calcium-tin alloys are increasingly used as anodes in electroplating applications, and as shielding, where the high strength permits the lead to stand alone without reinforcement.

Lead – Silver Alloys

The lead–silver system is eutectic having the eutectic point at 2.5 wt% silver and a eutectic temperature of 304°C. Only low-silver-lead alloys are used in industry. Lead-silver alloys show high resistance to recrystallization and grain growth.

Uses. Lead-silver alloys are used mainly as anodes and solders (qv).

Lead – Strontium Alloys

The lead-strontium system is a peritectic much like the lead-calcium system. Lead-strontium-tin alloys are being used as anodes in electrowinning. Cast-lead-strontium-tin alloys are stronger than lead-calcium-tin alloys and age harden much more rapidly.

Lead – Tellurium Alloys

Low tellurium (0.04–0.10 wt%) alloys are used for pipe and sheet in chemical installations where high strength with minimal alloying is desired. They also are used for nuclear shielding (see Nuclear reactors).

Lead – Tin Alloys

The main use of lead–tin alloys is as solders (qv) for the sealing and joining of metals.

Lead – Antimony – Tin Alloys

The lead–antimony–tin alloy system constitutes a series of alloys that are used widely in industry. These alloys are used for printing, lead-base sleeve bearings, special casting alloys, slush castings, and decorative casting alloys. These alloys generally are characterized by low melting points, excellent replication of detail in casting, extremely high strength for lead alloys, and excellent retention of strength at elevated temperatures.

Table 1. Comparison of Mechanical Properties of a Lead–Copper Alloy and Common Lead

Property	Copper-bearing	Common or corroding
ultimate tensile strength, MPa[a]	16.7–19.6	11.8–13.2
elongation, %	50–55	55
creep at RT, %/h		
at 1.4 MPa[a]	0.4×10^{-5}	5×10^{-5}
resistance to bending, bends	100	55
fatigue load limit at 50 MHz, MPa[a]	4.2	1.8–2.7

[a] To convert MPa to psi, multiply by 145.

Table 2. Properties of Lead–Antimony Alloys

Property	1 wt% Sb	4 wt% Sb	6 wt% Sb
sp gr at 20°C	11.27	11.04	10.88
solidification shrinkage, %	3.72	3.36	3.11
liquidus temperature, °C	320	299	285
solidus temperature, °C	313	252	252
coeff. of linear thermal expansion $\times 10^{-6}$ per °C between 20 and 100°C	28.8	27.8	27.2
sp heat between 20 and 100°C, J/g[a]	0.131	0.133	0.135
thermal conductivity at 0°C, W/(m·K)	33.5	30.5	28.9
electrical resistivity at 20°C, $\mu\Omega \cdot$ cm	22.0	24.0	25.3
tensile strength, MPa[b]	20.9	28.1	31.4
elongation, %	50	50	50
Brinell hardness	7	8.1	10.7
modulus of elasticity, GPa[b]	13.9		
minimum creep strength[c], MPa[b]			
at 0°C	4.51	3.63	2.75
casting temperature range, °C	400–500	400–500	400–500

[a] To convert J to cal, divide by 4.184.
[b] To convert MPa to psi, multiply by 145; to convert GPa to psi, multiply by 145,000.
[c] To produce 1% elongation in 10,000.

Table 3. Mechanical Properties of Pure Lead and Lead Alloys

	Pure lead	Lead–0.06 wt % Cu	Lead–1 wt % antimony	Lead–0.04 wt % calcium
tensile strength, MPa[a]	12.2	17.4	20.9	27.9
elongation, %	55	55	35	35
fatigue strength, MPa[a]	2.75	4.90	6.28	9.02

[a] To convert MPa to psi, multiply by 145.

Fusible Alloys and Special Alloys

Fusible alloys consist of Pb, Cd, Bi, and Sn in varying compositions. Special alloys include lead–indium, lead–lithium, and lead–lithium–tin alloys.

Health and Safety Factors

Because of its toxicity, special care and precautions must be observed when working with lead, its compounds, and alloys (see Lead compounds, industrial toxicology).

Uses

Lead alloys are used for their antifriction properties, for sound attenuation, radiation shielding, and corrosion resistance.

R. DAVID PRENGAMAN
RSR Corporation

M. Hansen, *Constitution of Binary Alloys*, McGraw-Hill Book Co., New York, 1958.
Lead in Modern Industry, Lead Industries Association, New York, 1952.
W. Hofmann, *Lead and Lead Alloys Properties and Technology*, Springer-Verlag, New York, 1970.

LEAD COMPOUNDS

LEAD SALTS

Lead has four electrons in its outer, or valence, shell. However, the usual valence of lead is $+2$, rather than $+4$, because of the reluctance of its two s electrons to ionize. As a result, tetravalent lead exists as a free, positive ion only in minimal concentrations. Furthermore, the bivalent or plumbous ion differs from the other Group IVA bivalent ions, such as the stannous ion of tin, because it does not have reducing properties. In general, the chemistry of inorganic lead compounds is similar to that of the alkaline-earth elements.

The ease with which lead monoxide combines with silicon dioxide to form a low melting silicate has been utilized in the ceramics industry in the preparation of glazes and in the manufacture of certain types of glasses (see Glass-ceramics; Silicon compounds).

The lead storage battery, the largest single user of lead and its compounds, is made possible by the high degree of reversibility, both chemical and physical, of the reactants in the fundamental chemical reaction:

$$Pb + PbO_2 + 2 H_2SO_4 \rightleftharpoons 2 PbSO_4 + 2 H_2O$$

All lead-containing compounds are produced from pig lead through a series of suitable steps, except for the small amount of lead in leaded zinc oxide, for which high grade lead ore is used. Most lead compounds are prepared directly or indirectly from the monoxide (PbO), commonly known as litharge, which is made from pig lead. In general, lead compounds may be formed by one or more of the following three methods, or variations thereof: reaction between a slurry of litharge and the desired acid or soluble salt of that acid; reaction between the solution of a lead salt and the desired acid; and fusion or calcination of litharge and the desired oxide.

Table 1. Physical Properties of Lead Halides

Property	PbBr$_2$	PbCl$_2$	PbF$_2$	PbI$_2$
mol wt	376.04	278.1	245.21	461.05
mp, °C	373	501	855	402
bp, °C	916	950	1290	954
d, g/cm^3	6.66	5.85	8.24	6.16
soly, g per 100 mL H$_2$O				
at 0°C	0.455	0.673		0.044

Halides

Properties of some halides are listed in Table 1. Some examples of the application of the halides include the following. Lead fluoride is used in low power fuses. Lead chloride is used in asbestos clutch or brake linings (qv). Lead bromide is a photopolymerization catalyst for acrylamide monomer. Lead iodide is used in the manufacture of thin lead iodide films.

Oxides

Lead oxides are used most often in lead–acid storage batteries. Physical properties for lead monoxide, lead dioxide, lead sesquioxide, and lead tetroxide are listed in Table 2. Lead hydroxide (Pb(OH)$_2$), mol wt 241.23, is used in sealed nickel–cadmium battery electrolytes.

Table 2. Physical Properties of Lead Oxides

Property	PbO	PbO$_2$	Pb$_2$O$_3$	Pb$_3$O$_4$
mol wt	223.21	239.21	462.42	685.63
mp, °C	897[a]			830[c]
dec, °C	1472[b]	290	370	500
d, g/cm^3	9.53 (α)	9.375		9.1
	9.6 (β)			

[a] Begins to sublime before melting.
[b] Boiling point.
[c] When decomposition is prevented by oxygen pressure.

Sulfide and Telluride

Lead sulfide (galena, lead glance, PbS), mol wt 239.25, mp 114°C, is used, among other applications, in photoconductive cells. Lead telluride (PbTe), mol wt 334.79, mp 917°C, is used in pyrometry.

Sulfates

Basic and normal lead sulfates are fundamental components in the operation of lead–sulfuric acid storage batteries. Basic lead sulfates are also used as pigments and heat stabilizers in vinyl and certain other plastics.

Sulfates include lead sulfate and monobasic lead sulfate, whose physical properties are listed in Table 3. Other sulfates are dibasic lead sulfate (2PbO.PbSO$_4$), tribasic lead sulfate (3PbO.PbSO$_4$.H$_2$O), and tetrabasic lead sulfate (4PbO.PbSO$_4$).

Table 3. Physical Properties of Lead Sulfates

Property	PbSO$_4$	PbO.PbSO$_4$
mol wt	303.25	526.44
mp, °C	1170[a]	977
d, g/cm^3	6.2	6.92
soly, g/100 mL H$_2$O	4.25×10^{-3} at 25°C	4.4×10^{-3} at 0°C
	5.6×10^{-3} at 40°C	

[a] Decomposes above 900°C.

Lead Nitrate

Lead nitrate (Pb(NO$_3$)$_2$), mol wt 331.23, is used in many industrial processes ranging from ore processing to pyrotechnics (qv) to photothermography (see Reprography).

Phosphite

In commercial applications of poly(vinyl chloride) polymers where weathering resistance, thermal stability, and electrical insulating properties are required, a stabilizer system based upon dibasic lead phosphite (2PbO.PbHPO$_3$.1/2H$_2$O), mol wt 742.63, provides a unique balance of properties.

Lead Azide

Lead azide ($Pb(N_3)_2$), mol wt 291.23, is used to prepare electrophotographic layers and for information storage on styrene–butadiene resins.

Lead Antimonate

Lead antimonate ($Pb_3(SbO_4)_2$), mol wt 993.07, is used as a modifier for ferroelectric lead titanates, as a pigment in oil-base paints, and as a colorant for glasses and glazes (see Ferroelectrics; Colorants for ceramics).

Acetates

Some physical properties of lead acetates are given in Table 4. Lead acetate is often used for the preparation of other lead salts by the wet method. Lead acetate trihydrate is used in the preparation of basic lead carbonate and lead chromate. Oxidation with lead tetraacetate is often used in organic syntheses.

Lead Benzoate

Lead benzoate monohydrate, $Pb(C_6H_5CO_2)_2.H_2O$, mol wt 467.43, is used as an antioxidant in organolead engine lubricants.

Carbonates

Lead carbonate, $PbCO_3$, mol wt 267.22, catalyzes the polymerization of formaldehyde to high molecular-weight crystalline poly(oxymethylene) products, among other applications. Basic lead carbonate, $2PbCO_3.Pb(OH)_2$, mol wt 775.67, is used as a catalyst for the preparation of polyesters from terephthalic acid and diols among other applications.

Phthalates

Two commercial forms of lead phthalates, both dibasic, are widely used as heat stabilizers in poly(vinyl chloride) (PVC) polymers and copolymers. Dibasic lead phthalate, $2PbO.Pb.C_6H_4(CO_2)_2 \frac{1}{2}H_2O$, has a mol wt of 826.87, and dibasic lead phthalate (coated) has a sp gr of 3.5–3.9.

Silicates

Lead forms acid, basic and normal, or metasilicates. Commercial lead silicates are used for the glass, ceramics, paint, rubber, and plastics industries. Some properties of lead silicates are listed in Table 5.

Borate

Lead borate monohydrate (lead metaborate), $Pb(BO_2)_2.H_2O$, mol wt 310.82, mp 500°C, is used in glazes.

Lead Titanate

Lead titanate (lead metatitanate), $PbTiO_3$, mol wt 302.09, is used in surface coatings as a pigment in outdoor paints, among other applications.

Table 4. Physical Properties of Lead Acetates

Property	Anhydrous	Basic	Trihydrate	Tetraacetate
mol wt	325.28	807.69	379.33	443.77
mp, °C	280	75 (200 dec)	75 (200 dec)	175
d, g/cm^3	3.25		2.55	2.228
refractive index, n_D			1.567^a	

^{a}Along the β-axis.

Table 5. Physical Properties of Lead Silicates

Property	Monosilicate	Bisilicate	Tribasic silicate
mol wt	294.85	343.37	729.63
mp, °C	700–784	788–816	705–733
d, g/cm^3	6.50–6.65	4.60–4.65	7.52
refractive index, n_D	2.00–2.02	1.72–1.74	2.20–2.24

Lead Zirconate

Lead zirconate, $PbZrO_3$, mol wt 346.41, mixed with lead titanate has particularly high piezoelectric properties. Mixed lead titanate–zirconates are used in high power acoustic-radiating transducers, hydrophones, and specialty instruments.

Health and Safety Factors

Lead is poisonous in all forms, but to different degrees, depending upon the chemical nature and solubility of the lead compound. The U.S. Center for Disease Control recommends a blood level of less than 30 micrograms per 100 mL for children. OSHA regulations limit exposure to inorganic lead compounds of an employee without a respirator to 50 $\mu g/m^3$ air as a time weighted average (TWA) in an 8-h shift.

DODD S. CARR
International Lead Zinc
Research Organization, Inc.

D. Greninger, V. Kollonitsch, and C.H. Kline, *Lead Chemicals*, International Lead Zinc Research Organization, Inc., New York, 1975.

J.W. Mellor, *Inorganic and Theoretical Chemistry*, Vol. III, Longmans, Green & Co., New York, 1930, pp. 636–888.

A.T. Wells, *Structural Inorganic Chemistry*, 3rd ed., Clarendon Press, Oxford, UK, 1962, particularly pp. 475–479, 902–903.

ORGANOLEAD COMPOUNDS

Organolead compounds comprise the broad class of organometallic structures that are characterized by at least one carbon atom bonded directly to a lead atom. Because of the widespread use of the tetraethyl and tetramethyl derivatives of lead as gasoline motor-fuel antiknock additives, organolead compounds constitute the largest single industrial application of organometallic chemistry (see Gasoline and other motor fuels, Organometallics).

Since an organolead compound has at least one lead–carbon bond, then the type formulas are R_nPbG_{4-n} and R_2Pb. R is any of a wide variety of aliphatic, unsaturated, aromatic, heterocyclic, or cyclopentadienyl groups, and G is from a wide variety of metallic and nonmetallic substituents. The lead atom carries the equivalent of positive charge, so that a G that is noncarbon usually is anionic, although apparently covalently bonded compounds exist. The R_2Pb and R_1PbG_3 series have limited thermal stability and only a few examples are known for each. The most common form for the organolead series is tetravalent lead, in contrast to the divalency of most inorganic lead compounds.

The tetraorganolead compounds, R_4Pb, appear to be typical organic compounds in physical properties, ranging from the liquid, volatile $(CH_3)_4Pb$ to the crystalline, low volatility tetraarylleads. The R_3PbG and R_2PbG_2 compounds act more as ionic salts. G is derived from alkali and alkaline earth metals, halide and other nonmetalic anions (eg, of oxygen, sulfur, and nitrogen), and metals in various forms. If a chelating anion (as from acetylacetone) is used, products are formed that behave as if intramolecular complexing were occurring (see Chelating agents).

Physical Properties

Most R_4Pb compounds resemble typical nonpolar organic compounds. Boiling points parallel molecular weights, and melting points reflect the usual symmetry and physical force factors. Solubilities are similar to those of R—R compounds. The R_nPbG_{4-n} compounds usually are polar, ionic solids, and often are crystalline and have anions from the nonmetallic elements and the alkali metal compounds.

For R_4Pb, physical properties, such as dipole moment, heat of vaporization, heat capacity, and viscosity are comparable to those of hydrocarbons of analogous structure. Most R_4Pb compounds are colorless to yellow. Some typical physical properties are given in Table 1.

Chemical Properties

The chemistry of the carbon–lead bond is similar to that of a relatively unreactive organometallic. Although covalent, the bond chemistry

Table 1. Physical Properties of Tetramethyllead and Tetraethyllead

Properties	Tetramethyllead	Tetraethyllead
mp, °C	−30.2	−137.6 to −130.3
bp, °C (kPa[a])	110 (101); 6 (1.3)	78 (1.3)
d_4^{20}, g/cm³	1.995	1.653
viscosity, mPa·s (= cP)	0.572	0.864
vapor pressure, kPa[a]	$\log p = 6.06257 - \dfrac{1335.317}{t°C + 219.084}$	$\log p = 8.551 - \dfrac{2960.0}{t°C + 273.1}$
heat of fusion, kJ/mol[b]	10.86	8.80
heat of combustion, kJ/mol[b]	3507–3644	6473

[a] To convert kPa to mm Hg, multiply by 7.5.
[b] To convert J to cal, divide by 4.184.

corresponds to that of a polar species, ie, it is subject to a wide range of free-radical and ionic reactions. Although R_4Pb compounds are reasonably stable photochemically, they can be decomposed by light, especially near uv, to give products that are comparable to those from pyrolysis.

Chemistry involving electron-transfer or redox processes is common. Reagents are diverse, eg, reactive metals, oxidizing agents, and halogens.

The toxicity of tetraalkyllead compounds requires that efficient decontamination systems be available where work is done. Most involve oxidant use.

Synthesis

Methods that both break and form a Pb—C bond are widespread. Formation of Pb—C bonds involves the use of organometallic techniques. Lead metal reacts with ethyl or methyl chloride if the lead is combined with sodium, preferably in 1:1 alloy. Catalysts, eg, acetone or higher alcohols, normally are used with C_2H_5Cl.

Manufacture and Processing

Raw materials and commercial processes. The basic raw materials for the manufacture of tetraalkyllead compounds are lead, sodium, and alkyl chlorides. These react according to the following equation:

$$4\,NaPb + 4\,RCl \rightarrow R_4Pb + 3\,Pb + 4\,NaCl$$

Tetraethyllead (TEL) is produced by two different processes based on the ethylation of sodium–lead alloy (NaPb). Most tetramethyllead (TML) is manufactured from methyl chloride in a manner similar to the batch TEL process, except that higher temperatures and pressures are used in conjunction with different catalysts.

TML is also manufactured employing an electrolytic Grignard reaction:

$$2\,CH_3MgCl + 2\,CH_3Cl + Pb \rightarrow (CH_3)_4Pb + 2\,MgCl_2$$

A third type of commercial lead antiknock mix contains a chemically redistributed mixture of alkylleads. In the presence of certain Lewis acid catalysts, a mixture of TEL and TML undergoes a redistribution reaction to produce an equilibrium mixture of five possible tetraalkyllead compounds.

Pollution control. In the modern plant, atmospheric emissions from all process stages and storage tanks are controlled by a variety of scrubbers, absorbers, conservation vents, and bag filters and by incineration. Aqueous discharges have been subjected to considerable abatement and much effort has been spent in the areas of solids separation, chemical precipitation, and adsorption.

Production and Shipment

Commercial manufactures of tetraethyllead and lead alkyl antiknock additives include Nalco Chemical Co., Inc., E.I. du Pont de Nemours & Co., Inc., the Ethyl Corp., and the Associated Octel Co., Ltd.

Health and Safety Factors

All lead compounds are considered toxic, especially the organoleads. The current OSHA permissible exposure limit for organolead is 75 µg Pb/m³ (or ca 9 ppb) at ambient conditions.

W.B. McCormack
Ralph Moore
Charles A. Sandy
E.I. du Pont de Nemours & Co., Inc.

H. Shapiro and F.W. Frey, *The Organic Compounds of Lead*, John Wiley & Sons, Inc., New York, 1968.

L.C. Willemsens, *Organolead Chemistry*, International Lead Zinc Research Organization, Inc., New York, 1964.

R.W. Leeper, L. Summers, and H. Gilman, *Chem. Rev.* **54**, 101 (1954).

INDUSTRIAL TOXICOLOGY

Today there is a very low incidence of overt lead poisoning from industrial exposure.

Sources of lead in the environment include soils, paint, cars and point sources, and water and food.

The two main routes of entry for lead compounds are the gastrointestinal tract and the lungs. About 90% of the ingested lead passes through the gastrointestinal tract unabsorbed. About 10% of the ingested lead is excreted in urine with lesser amounts in sweat, hair, and nails. Under conditions approximating steady state, more than 90% of the lead in the body is in the skeleton, where it remains in a relatively inert state. In addition to bone, lead in the bloodstream can be deposited in many other tissues.

High levels of lead in the body produce toxic effects. The most common signs of plumbism are gastrointestinal. Symptoms include anorexia, nausea, vomiting, diarrhea, and constipation followed by colic. Lead can also affect hemoglobin synthesis and red blood cell survival time as well as the central and peripheral nervous systems.

One of the earliest effects of lead on the kidney is the development of intranuclear inclusion bodies in the renal tubular lining. With continued exposure, swelling and mitochondrial changes occur in proximal tubular lining cells and subsequent functional impairment resulting in aminoaciduria, glycosuria, and hyperphosphaturia (Fanconi Syndrome).

Reproductive effects have been associated with lead. Decreased fertility, increased abortion, and neonatal morbidity have reportedly occurred in the past in women exposed to massive lead concentrations.

The diagnosis of lead poisoning should be based on history of exposure, clinical symptoms and signs, evidence of increased lead absorption, and significant biochemical changes. The symptoms of lead overexposure are rather nonspecific and all relevant factors need to be considered. The use of blood lead as a biological monitoring tool to determine current absorption and serve as a guideline for prevention of undue lead absorption is appropriate.

The current OSHA regulations require action at blood lead levels about 40 µg Pb/100 mL.

Gary Ter Haar
Ethyl Corporation

Committee on Biologic Effects of Atmospheric Pollutants, *Airborne Lead in Perspective*, National Academy of Sciences, National Research Council, Division of Medical Sciences, Washington, D.C., 1972, p. 74.

S.S. Brown, *Clinical Chemistry and Chemical Toxicity of Metals*, Elsevier North-Holland by Biomedical Press, 1977.

LEAD COMPOUNDS, CONTROLS OF INDUSTRIAL HAZARDS. See Lead compounds, industrial toxicology.

LEAD COMPOUNDS, HEALTH AND SAFETY FACTORS. See Lead compounds, industrial toxicology.

LEAD POISONING.   See Lead compounds, industrial toxicology.

LEAD STORAGE BATTERIES.   See Batteries, secondary.

LEATHER

Despite efforts, particularly during the last 30 years, to produce leatherlike materials (qv) from other fibers, sheet materials, and petrochemicals and to substitute them in the marketplace for leather products, leather still is the product of choice by the consumer. The demand for leather far exceeds the supply, and the principal function of the substitutes has been to fill those needs for which the supply of leather is inadequate.

Physical Properties

Leather has unique properties that make it ideally suited for use in the manufacture of a variety of products, the most notable being footwear. Even if leather is considered for a single use, eg, shoe uppers, a large variation in properties is advisable to allow for different shoe types, eg, casual, military, work, or dress shoes. A range of values for various physical properties of cattlehide shoe-upper leather and the representative value are listed in Table 1.

One of the most important properties of leather, and one for which there is a paucity of quantitative data, is the ability of leather to conform to the shape of the foot. Plastic flow imparts the necessary give to provide foot comfort; elastic flow is responsible for the maintenance of shape by ensuring proper recovery during flexing action. Leather substitutes have not achieved this combination of properties. Nor are they able to remove moisture from the foot as leather does. The mechanism of the transmission of water vapor through leather has been demonstrated to be a function of the material rather than a linear diffusion process. The evidence indicates that water is sorbed at polar groups of the protein molecules and is conducted by an activated diffusion process along the fibers, even against an air-pressure head.

Manufacture

Raw material.   Fresh cattlehides, like most biological materials, contain 65–70% water, 30–35% dry substance, and less than 1% ash. The dry

substance is largely made up of the fibrous proteins (qv), collagen, keratin, elastin, and reticulin. The main components of the ash, listed in decreasing concentration, are phosphorus, potassium, sodium, arsenic, magnesium, and calcium.

Defects in leather include those that result from poor flaying technique in the slaughterhouse, inadequate preservation and handling methods, or improper use of chemicals and heat in the tannery. The scar produced by heat-branding, those left by parasitic skin diseases, and biological defects also mar the surface appearance and impair the physical properties of leather.

Chrome tanning.   The primary function for a tanning agent is to stabilize the collagen fibers so that they are no longer biodegradable. Over 95% of all the leather manufactured in the United States is chrome-tanned. The manufacture of leather can be divided into three separate phases. Historically, the first took place in the beamhouse where the skin was prepared for tanning. The next phase was done in the tanyard. The last phase is finishing. Although methods have changed greatly, the beamhouse and tanyard still exist in modern tanneries.

The hides generally are received by the tanner in a salt- or brine-cured condition. The first step taken by the tanner is to remove the salt and to rehydrate the fibers by soaking. Next, the hair must be removed.

In many tanneries, the relatively brief unhairing step is followed by a longer (4–16 h) liming step. The action of lime not only loosens the hair but opens up the collagen fiber structure. This swelling leads to subsequent fiber separation and allows rapid penetration of tanning chemicals.

The liming step is followed by deliming and bating during which the pH is lowered and extraneous proteins are removed enzymatically. Immediately after the bating, the hides are pickled with sulfuric acid to lower the pH to less than 3. Once the hide is in the acid condition, it is prepared for the tanning operation.

At a pH of 2.8, chrome sulfate (usually ca 33% basicity) is soluble. After the tanning solution has been allowed to fully penetrate the hide, the pH is raised slowly with sodium bicarbonate. By the time a pH of 3.4–3.6 is obtained, the chrome has reacted with the collagen to produce a fully preserved, tanned hide.

The tanned hides generally are stacked overnight and the chrome further fixes onto the collagen. They are then put through a hide wringer so that they are almost dry to the touch and sorted for quality and thickness.

Chrome tanning generally is carried out by adding the acidified hide to an aqueous solution of trivalent chromium sulfate of 30–50% basicity. The reactions of these basic chromium sulfate tanning solutions with hide collagen have been studied. It is firmly established that cross-linking is accomplished by bonding of the various chromium species with free carboxyl groups in the collagen side chains. The liming of hides is effective in providing additional carboxyl groups by chemical hydrolysis of amide side chains.

Vegetable tanning.   Vegetable tanning has decreased considerably in importance in the United States. Leathers made with a full vegetable tannage are used for shoe soles, belts, saddles, upholstery, lining, and luggage. Vegetable tanning produces a fullness and resiliency characteristic of only this type of tannage. It has certain molding characteristics so that, in sole leather, a shoe is produced that adapts to the shape of the individual foot. Its hydrophilic character is a great aid in shoe linings for the removal of perspiration from the foot. Vegetable-tanned leather also has good strength and dimensional stability and, thus, finds use in power-transmission belts.

Vegetable tannins are the water-soluble extracts of various parts of plant materials, including the wood, bark, leaves, fruits, pods, and roots. In general, the same steps are carried out in preparing the hide for vegetable tannage as are carried out in preparing the hide for chrome tannage, eg, the hair is removed by the same chemicals. However, a much slower tanning process is used which frequently involves as much as 15 days of soaking in pits containing solutions of the tannins.

Recently, a more rapid, minimum-effluent vegetable-tanning system, known as the Liritan process, has been developed. The limed and bated hides are treated for 24 h in a pit with 5% sodium hexametaphosphate (Calgon) solution and sufficient sulfuric acid to achieve a pH of 2.8, and can be tanned in a shorter period of time. First introduced in 1960, the

Table 1.   Physical Properties of Shoe Upper Leather

Property	Range	Representative value
tensile strength, MPa[a]	15.26–37.48	27.6
elongation at break, %	29.5–73.0	40
stitch tear strength, N/cm[b]	1280–2275	1751
tongue tear strength, N/cm[b]	226–961	525
thickness, mm	1.5–2.4	1.8
bursting strength, kN/cm[b]	1.10–24.5	17.5
grain-cracking, N/cm[b]	525–1489	1051
wet shrinkage temperature,[c] °C	96–120	100
apparent density, g/cm^3	0.6–0.9	0.75
real density, g/cm^3	1.4–1.6	1.5
flexibility (Flexometer)		
bending length, cm	6–9	7
flexural rigidity, mg/cm	10,000–50,000	20,000
bending modulus, MPa[a]	19.7–68.9	34.5
compression modulus, MPa[a]		0.345
cold resistance,[d] K		
(without finish or fat liquor)		
workable without cracking at		92
heat resistance	shrinks depending on moisture content, anhydrous decomposition at 160–165°C	

[a] To convert MPa to psi, multiply by 145.
[b] To convert N/cm to lbf/ft, divide by 14.6.
[c] Chrome-tanned leather values.
[d] Ordinarily limited by finish-crack or lubricant hardening properties.

Liritan process has spread through the world and is used by the principal vegetable-leather tanners in the U.S.

The tanned hides are further processed in order to clean the surface of the hides of excessive amounts of tannins (usually unbound tannin) and then are wrung free of excess water and are oiled. The grain surface then is sponged with a dilute oil preparation and is rolled repeatedly under considerable pressure with a highly polished metal (usually brass) cylinder on a large pendulum-type machine. The operation packs the fibers of the leather and imparts a characteristic gloss to the grain. The leather is allowed to dry, dip-washed in a solution containing a small amount of wax, redried, and given a final dry rolling.

Functionally, the vegetable tannins are polyphenolic compounds. They are empirically divided into two groups: the hydrolyzable tannins and the condensed tannins. It is highly likely that no cross-linking of the protein, other than by the formation of hydrogen bonds, takes place as a result of vegetable tanning.

Other tannages. Most of these tannages are pretannages or post-tannages for chrome or vegetable processes and are employed to give such qualities as filling, lighter shade for dyeing, and reduction of tannery effluent. Only occasionally are they used alone. They include mineral tannages, polyphenolic syntans, resin tannages, oil tannage, sulfonyl chloride tannage, and aldehyde tannage.

Post-tanning. During the early aging period, when the leather is stacked, there is a gradual decline in the pH of the stock as increased chrome fixation, olation, and oxolation occur. Consequently, after transfer to a drum, the stock is neutralized to a desired pH, frequently is retanned to develop certain properties for different end uses, and then is dyed and fatliquored in rapid succession in the same drum. Fatliquoring is the application of oil-in-water emulsions to the leather.

Once all of the wet operations have been completed, the leather can be dried, which not only involves removal of excess water, bringing the moisture content close to that of the finished leather, but completes the reactions of some of the materials (eg, tanning agents, fatliquors, and dyes) with which the hide has been treated (see Drying).

It is the finishing process with its application of natural or synthetic polymers and of colorants within and on the surface of the leather that produces the uniformity, appearance characteristics, and resistance to scuffing and abrasion that are required for a commercial product. Coatings (qv) for leather function as decoration and protection.

Energy consumption. On the basis of quantity consumed per unit of product produced, the leather-manufacturing industry would be categorized with aluminum, paper, steel, cement, and petroleum-manufacturing industries as a gross consumer of energy. An energy wastage in the amount of 70% of energy input offers considerable opportunity for improvements.

Health and Safety Factors

The effluent streams that must be treated can be separated into two areas of the tanning operation, the beamhouse and the tanyard. The main pollutants found in beamhouse effluent are sulfide and BOD, and the pH is ca 12. The second, and even more difficult, stream to treat is that from the tanning operation. This effluent contains chromium as well as conventional pollutants, eg, BOD and suspended solids.

Uses

The principal domestic markets for leather are leather footwear, luggage, handbags, small personal leather goods, leather apparel, and gloves. Hide collagen, when it can be separated from the hide in a native state, has use in food products, cosmetics (qv), pet treats, and biomedical products.

DAVID G. BAILEY
PETER R. BUECHLER
ALFRED L. EVERETT
STEPHEN H. FEAIRHELLER
United States Department of Agriculture

F. O'Flaherty, W.T. Roddy, and R.M. Lollar, eds., *The Chemistry and Technology of Leather*, Vols. I–IV, Reinhold Publishing Corp., New York, 1956–1965.

T.C. Thorstensen, *Practical Leather Technology*, Robert E. Krieger Publishing Co., Inc., Huntington, New York, 1975.

LEATHERLIKE MATERIALS

Leatherlike materials normally refers to synthetic materials that are used as substitutes for leather in those applications where the latter traditionally has been used. Periodic shortages caused by the cyclical nature of leather supplies cause continued inroads by new products into traditional leather markets. These include the use of urethane-coated fabrics in shoes and apparel and the development of poromerics for men's shoes (see Urethane polymers). Each of these products is designed to simulate certain properties of leather that are relevant to a specific application (see Leather). The suitability of synthetics as leather substitutes depends upon their ability to duplicate certain desirable characteristics of leather. These characteristics arise from the physicochemical structure of the leather. The high permeability for air and moisture vapor is one of leather's most desirable properties in its main use area, ie, footwear.

Types

Leatherlike materials have one common characteristic: they look like leather in the goods in which they are used. For purposes of a technological description, however, they may be divided into two general categories: coated fabrics (qv) and poromerics. Poromerics are manufactured so as to resemble leather closely in both its barrier and transition properties, eg, for heat or moisture, and its workability and machinability. The barrier or permeability properties normally are obtained by manufacturing a controlled microporous structure.

The modulus characteristics of synthetics should match those of leather at 80°C rather than at room temperature. Although the latter temperature commonly is used as the standard in the industry, the former more closely matches the lasting conditions.

Coated fabrics. All commercial nonporomerics are coated fabrics. Several types of fabrics have been used, and the coating may be either vinyl or urethane, depending upon the desired balance between physical properties and economic considerations.

Poromerics. Poromerics have been developed in order to match the moisture-vapor permeability, skivability, and nonfraying characteristics of leather; the products have microscopic, open-celled structures. The structure of a poromeric is more homogeneous than that of a coated fabric, since it is either all polymer or polymer reinforced with nonwoven fibers. The size and configuration of the cells is such that the passage of moisture vapor but not liquid water can occur. Although the vapor transmission rates of commercial poromerics vary considerably, the better ones can equal that of leather.

Manufacture and Processing

Coating of fabrics. The type of fabric used as a coating substrate is determined by the intended application of the product. Woven fabrics are used for most footwear products, whereas knits are used for upholstery, apparel, or accessories where greater extensibility is required.

The weight of solid vinyl coatings varies from 200–270 g/m^2 (6–8 oz/yd^2) and the thickness of the vinyl from 0.30–0.75 mm, depending upon the specific use. The vinyl usually is a plasticized vinyl chloride homopolymer that is compounded with various stabilizers and pigments as desired. The formulated vinyl is applied to the continuously moving textile web by coating, calendering, or extrusion (see Vinyl polymers).

Expanded vinyl coatings are prepared by either the transfer-coating or calender-coating technique. Foaming or blowing the interlayers is accomplished by including a blowing agent in the formulation of the vinyl. Upon being heated, the blowing agent decomposes to yield a gas, eg, N_2 or CO_2. The gas, in turn, produces a closed-cell foam structure (see Foamed plastics).

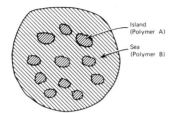

Figure 1. Cross-sectional view of islands-in-the-sea fiber.

Urethanes have an advantage over vinyl in that their molecular structure renders them inherently flexible. The commercial products are block, or segmented, ester–urethane or ether–urethane copolymers. Urethanes also have tensile strengths that are as much as ten times greater than those of vinyls. Urethanes do not require the stabilization against thermal degradation that vinyl does, but they do require stabilization against hydrolytic decomposition. One of the main disadvantages of urethanes, with respect to vinyls, is that urethane solutions require strong solvents, eg, dimethylformamide (DMF) or tetrahydrofuran, both of which are expensive and toxic.

Formation of poromeric structures. The manufacture of poromeric structures, particularly those with sufficient permeability for footwear applications, has presented the largest technological problem in the leather-substitute field.

The simplest poromeric products are those that contain no reinforcing fiber, eg, Porvair. The most direct method of forming such a homogeneous microporous web is by polymer coagulation, which involves extruding a web of urethane solution onto a suitable support. The web is immersed in a nonsolvent (for the polymer) which is miscible with the solvent used. The usual solvent/nonsolvent combination is DMF/water. When the urethane web is immersed in the water, DMF migrates to the aqueous phase and water begins to migrate into the web; simultaneously, the polymer precipitates from solution. The net effect of this precipitation/migration is the formation of a solid urethane that contains microscopic channels or pores.

The various techniques for producing fiber-reinforced poromerics result in a large variety of structures. The main variations include the type of fiber structure used and the degree to which it is impregnated with polymer. Although some poromerics, eg, Corfam, include woven-fabric reinforcement, recent practice has tended toward using only nonwoven materials; the nonwoven-based products tend to match leather more closely in modulus (see Nonwoven textile fabrics).

The Japanese have developed an elegant nonwoven approach in the production of Clarino by producing fibers that are close in properties to the collagen fibers in leather; the method that was developed yields an islands-in-the-sea fiber. Two immiscible molten polymers are extruded to give a fiber with a cross section, as shown in Figure 1. By subsequent selective solvent extraction of either the island or the sea phase, a hollow supple, compressible fiber or a bundle of very fine fibers is obtained.

The best overall properties in laminated structures have been obtained with a combination woven/nonwoven structure. The structure is prepared by heavily napping a sateen fabric, shearing the nap, and then impregnating the fabric with a reinforcing resin. The process results in a backing that is skivable and that has a leatherlike hand.

Finishing. In addition to having suitable physical properties, it is necessary for a synthetic leather to have a leatherlike appearance. The appearance of a piece of leather results principally from two factors: color and texture. The finished surface of the material must have some resistance to abrasion and appropriate tactile properties. Three types of surfaces are commercially important: grained, patent, and suede, and each requires different finishing techniques.

F.P. Civardi
G. Frederick Hutter
Inmont Corp.

T. Hayasaki, *Chem. Tech.*, 28 (Jan. 1975).
W. Blume, *J. Ind. Fabr.* 1(1), 29 (1982).

LECITHIN

Phosphatides were first observed in egg yolk in 1864, and twenty years later the choline component in lecithin was identified (see Eggs).

Structural formulas (where R and R′ are fatty acids) for the main phosphoglycerides that are present in commercial lecithins are

lecithin (phosphatidylcholine, 1,2-diacylglycero-3-phosphorylcholine)

phosphatidylethanolamine (cephalin)

phosphatidylinositol

Commercial lecithin is derived almost entirely from soybeans because of the relatively large amount of lecithin in crude soybean oil, ie, usually 2.5–3.25% in terms of commercial lecithin, and because of the enormous quantities of soybeans grown and processed. The material consists roughly of 64% mixed phosphatides and 36% crude soybean oil. The term lecithin is used below only to denote phosphatidylcholine, ie, pure lecithin.

All commercial lecithin products are not the same but can vary according to their source, component phosphatides and other constituents, and with the processing to which they are subjected. Resulting differences in properties have distinct biological significance, eg, a saturated lecithin, dipalmitoyl phosphatidylcholine in lung tissue, is involved in oxygen uptake, whereas dilinoleyl phosphatidylcholine is involved in lipoprotein formation and in cholesterol mobilization.

Lecithin and phosphatides generally are of universal occurrence in living organisms. They are constituents of biological membranes and are involved in permeability, oxidative phosphorylation, phagocytosis, and chemical and electrical excitation. Lecithin has been identified in many animal tissues and organs, especially in the brain, nervous system, liver, heart, lungs, kidneys, blood, milk, and sperm; in microorganisms of all kinds; and throughout the vegetable kingdom, eg, in seeds.

In animal and vegetable lecithins, both saturated and unsaturated fatty acids are present but their proportions vary over a wide range in relation to the source and, in particular, to the composition of the triglyceride with which the lecithin is associated in nature.

Commercial soybean lecithin and commercial vegetable lecithin from other sources, in contrast to animal lecithins, are characterized by a relatively high percentage of free and bound carbohydrates that essentially are plant sugars, ie, dextrose, raffinose, galactose, and stachyose. Cholesterol is absent but substantial amounts of phytosterols and sterol glycosides are present.

The phosphatides are characterized by their solubility in ethyl ether and petroleum naphtha but, unlike the glycerides, they are insoluble in acetone. This difference in solubility affords a convenient method of separating the phosphatides from triglycerides.

In addition to differences in the kind and location of the fatty acids, all of these classes of phosphatides possess many possibilities for stereoisomerism, especially the phosphoinositides, in which all six carbon atoms of inositol are asymmetric. The preponderance of lecithin molecules are in the internally neutralized zwitterionic form at about pH 7.

Unlike animal phosphatides, soybean phosphatides contain no cholesterol. However, like some of the animal phosphatides, they contain D-galactose in various combinations and in the adsorbed state. In addition, soybean phosphatides can be associated with compounds, eg, stigmasterol, dihydrositosterol, the sitosterols, and compounds of inositol.

Whereas animal phosphatides are likely to be in labile combination with protein, eg, the so-called lecithoprotein or vitellin of egg yolk, vegetable phosphatides more generally are in less labile combination with carbohydrates, eg, dextrose, D-galactose, and even pentoses.

Physical Properties

Commercial lecithin is brown to light yellow in color, depending on whether it is unbleached or bleached. When properly refined, it has practically no odor and has a bland taste. In consistency, it may vary from plastic to fluid. It is soluble in aliphatic and aromatic hydrocarbons, including the halogenated hydrocarbons; however, it is only partially soluble in aliphatic alcohols. Pure lecithin, phosphatidyl choline, is soluble in ethanol. In water, a particle of lecithin exhibits myelin growths. Like other antipolar, surface-active agents, the phosphatides are insoluble in polar solvents, eg, ketones and, particularly, acetone.

Commercial lecithin is soluble in mineral oils and fatty acids but is practically insoluble in cold vegetable and animal oils. It is insoluble but infinitely dispersible in water. When commercial lecithin is mixed with water, it readily hydrates (depending on concentration) to a thick yellow emulsion.

Commercial lecithin is a wetting and emulsifying agent inasmuch as its constituents, ie, fatty acid-containing phosphatides, are amphipathic in chemical structure, having strongly lipophilic, fat-forming acid nuclei at one end of the molecule and a strongly hydrophilic amino or phosphoric acid group at the opposite end.

Chemical Properties

In general, the presence of fatty acid groups in the phosphatide molecule permits reactions, eg, saponification, hydrolysis, hydrogenation, hydroxylation, halogenation, sulfonation, phosphorylation, elaidinization, and ozonization.

Manufacture and Processing

Commercial lecithin is produced in conjunction with the solvent processing of soybeans (qv) by which most of it is made. It may be recovered from other seed oils where the expeller process is used (see also Vegetable oils).

In the United States, the solvent that has been adopted is n-hexane. n-Hexane permits extraction of about half the total phosphatides of the soybean, but the lecithin that is produced contains smaller amounts of carbohydrates, bitter principles, and color. There are about 95 processors of soybeans, and virtually all use the solvent-extraction method (see also Extraction).

Commercial grades. There are six commercial grades of lecithin: the unbleached, single-bleached, and double-bleached lecithins of plastic consistency and of fluid consistency. Users prefer fluid lecithin because it is easier to handle and blends more readily. Table 1 shows the general properties (approximate) of the six common types of commercial lecithin.

Health and Safety Factors

Environmental considerations encourage degumming of crude vegetable oils. There are no known health hazards involved in the production of commercial lecithin from crude vegetable oils because the phos-

phatides are nonvolatile and are a nonirritating food material. Lecithin is GRAS for use in foods as required.

Uses

Lecithin is used in animal feeds, baking products and mixes, candy, chewing gum, chocolate, cosmetics and soaps, dehydrated foods, dyes, edible oils and fats, food, ice cream, instant foods, insecticides, inks, leather, macaroni and noodles, margarine, paints, petroleum products, pharmaceuticals, plastics, release agents, rubber, sealing and caulking compounds, textiles, and whipped toppings.

Derivatives

Derivatives of commercial lecithin include fractionated lecithins and modified lecithins; pure compounds may be synthesized.

JOSEPH EICHBERG
American Lecithin Co.

H. Wittcoff, *The Phosphatides*, Reinhold Publishing, New York, 1951.

G.B. Ansell and J.N. Hawthorne, *Phospholipids—Chemistry, Metabolism and Function*, Elsevier, Amsterdam, 1964.

R.M.C. Dawson, in *Form and Function of Phospholipids*, Elsevier, Amsterdam, 1973, p. 11.

W. Van Nieuwenhuyzen, *J. Am. Oil Chem. Soc.* **53**, 425 (1976).

LEUCO BASES.   See Vat dyes under Dyes.

LEVULOSE (FRUCTOSE).   See Carbohydrates; Sugar.

LIGHT-EMITTING DIODES AND SEMICONDUCTOR LASERS

The p–n Junction Diode

The invention of the transistor by Shockley, Brattain, and Bardeen in 1948 can be considered the genesis of the development of an electronics technology that has vastly modified the handling of electronic signals. This technology primarily has utilized silicon as the semiconductor material (see Integrated circuits; Semiconductors). Light-emitting diodes (LEDs) and semiconductor (injection) lasers convert electricity into optical radiation, whereas the solar cell and a variety of other light detectors convert radiation into electricity (see Photodetectors; Photovoltaic cells). Although some light detectors are made of silicon, all light-emitting semiconductor devices utilize other semiconductor materials, the most important of which, particularly for the visible and near-infrared portion of the spectrum, are the zinc-blended compounds that are composed of elements from Groups IIIA and VA of the periodic table. Compounds composed of elements from Groups IVA and VIA yield LEDs and lasers (qv) that emit further into the infrared at wavelengths up to about 10 μm. However, the III–V compounds are the more technologically important.

The significant energy bands for semiconductors are the conduction and valence bands which are separated from each other by an energy gap or band gap of width E_g (radiation energy of combined electrons). Ideally, in the pure semiconductor, there are no energy states within the energy gap and, at absolute zero (0 K), all of the valence-band states are occupied by electrons that do not contribute to the conductivity. Increasing the temperature causes some electrons to be thermally excited from the valence to the conduction band and, at any temperature above absolute zero, there is a dynamic equilibrium between electrons that are excited upward in energy across the band gap and electrons returning. Thus, there is at any time a net excited-electron concentration as is illustrated in Figure 1. The excited electrons are in conduction-band energy states and there are a corresponding number of unoccupied energy states in the valence band. The latter are called holes and behave like positively charged electrons. Both the electrons in the conduction

Table 1. General Properties (Approximate) of Commercial Lecithin

Property	Value	Property	Value
iodine value	95	phosphorus, %	1.9
saponification value	196	unbound carbohydrates, %	5.0
sp gr at 25°C	1.03	choline, %	2.2
pH	6.6	inositol, %	1.4
isoelectric point, pH	3.5	tocopherols, %	0.1
ash, %	7.0	sterols and sterolglycosides, %	3.5
total nitrogen, %	0.8	moisture, %	0.75
amino nitrogen, %	0.4	other mineral matter (mostly potassium), %	1.75

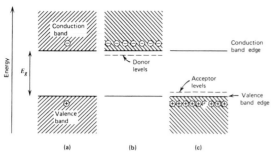

Figure 1. Band-edge energy diagram at $T > 0$: (a) undoped semiconductor with a few thermally excited carriers; (b) an n-type semiconductor with shallow donors; (c) a p-type semiconductor with shallow acceptors.

band and the holes in the valence band are not localized in space, are mobile, and are capable of carrying an electric current.

For semiconductors of interest as LEDs and as semiconductor lasers, the band-gap energy is large compared to the thermal energy at room temperature, and the conductivity resulting from thermally excited carriers is small; however, the addition of very small amounts of impurity elements that are incorporated into the host crystal structure can vastly change that conductivity. Such elements may contain more or fewer valence electrons than are required for bonding of the impurity atom into the crystal structure. If they are deficient in electrons, there is a tendency for them to attract an electron and, thus, become ionized. If the impurity atom has more valence electrons than are needed, there is a tendency for the atom to lose an electron and become ionized. These two kinds of impurities are referred to as acceptors or donors, respectively, and the n-type regions of the semiconductor have more ionized donor than acceptor impurities and the p-type regions have more ionized acceptor than donor impurities. The common boundary of the regions is the $p-n$ junction and the simplest device incorporating such a junction is called a junction diode. All technologically important light-emitting diodes and injection lasers (see Semiconductor Lasers) are $p-n$ junction diodes.

Radiative and Nonradiative Transitions

Several processes occur when majority carrier holes and electrons reach the $p-n$ junction. In the processes that are of interest for light-emitting devices, the majority carriers are injected across the $p-n$ junction and, thereby, become minority carriers in the region of opposite type. They rapidly come into thermal equilibrium with the host lattice.

The forward biasing of the junction diode causes current to flow and causes the injection of carriers across the $p-n$ junction. The injected carrier must recombine and, in the recombination process, must give up energy of approximately the band gap of the semiconductor. If the recombination occurs by a radiative transition, light of approximately the band-gap energy is generated. If the device structure is such that a useful fraction of this light can escape the surroundings, the device is a light-emitting diode.

The internal quantum efficiency of an LED largely depends upon the competition between radiative and nonradiative processes. Rather than producing electromagnetic radiation, the recombining electron can give up its energy to another conduction-band electron which, thereby, is raised to a higher energy in the conduction band. That electron then thermalizes through the continuum of states in the conduction band. This process is called Auger recombination. In other nonradiative processes, the electron may lose all or part of its energy by generating quanta of lattice vibrations (phonons), thereby simply heating the crystal. In indirect-gap semiconductors, except in special cases, transitions must involve a phonon.

Light-Emitting Diodes

Homostructures. It is necessary to optimize the amount of the generated radiation that can be emitted from a light-emitting diode. The configuration of the device depends upon its ultimate use, ie, whether as an illuminator, eg, for telephone numeral illumination; a section of an alphanumeric device (see Digital displays); an indicator light; or an optical source for a fiber-optic communication system (see Fiber optics). Except for the latter application, LEDs usually are fabricated from a single semiconductor or, in the most complicated case, from a composite structure formed by epitaxial growth of a ternary compound semiconductor on a binary compound semiconductor substrate. In that case, there may be a gradual grading of the composition in the epitaxial layer to achieve the proper band gap. Structures that are made of a single semiconductor, or where there is not an abrupt compositional change near the $p-n$ junction, are homostructures. Such LEDs usually are intended to operate at low current densities, eg, 1–10 A/cm² of junction area.

Heterostructures. Several semiconductor systems permit the growth of high quality, thin-layered structures that have large and usually abrupt changes in composition within a single semiconductor crystal. The compositional changes correspond to changes in both the band gap and refractive index that are very important for high radiance LEDs and for injection lasers. The most important of the heterostructures, the double heterostructure, is a sandwich of a narrow band-gap material between two layers of wider band-gap material.

Semiconductor Lasers

The operation of a laser is based on the fact that, in addition to being absorbed in the process of causing a quantum transition, a photon can stimulate such a transition. In many conventional lasers, the photon interacts with an electron and, if that electron is in a properly excited state, the photon stimulates the electron to make the transition to a lower energy state such that an identical photon is emitted. A laser is then formed by providing feedback of some of this emission.

A semiconductor laser, like any other laser, requires a medium to provide the transitions, a means of excitation, and means for feedback so that a resonant cavity may be formed. The semiconductor is the medium, the conduction and valence bands providing the necessary electronic structure for electronic transitions. Feedback is provided by crystal surfaces that usually are two opposite cleaved ends of the crystal chip. The crystal surfaces that result from cleaving two ends of the crystal are perfectly parallel to each other and are partial mirrors, so that some of the generated light is reflected back into the crystal.

There are several ways of providing excitation in semiconductors. The most common are electron bombardment or optical pumping to excite electrons from the valence to the conduction band, or injection of carriers across a $p-n$ junction. Electron bombardment and optical pumping provide means of studying stimulated emission in semiconductor structures not characterized by $p-n$ junctions. The injection laser is, however, by far the most important semiconductor laser. In injection lasers, the injected minority carriers are excited, there being a thermodynamic driving force for them to recombine with the emission of radiation at an energy corresponding to the near band-gap transition.

Semiconductors for LEDs and Injection Lasers

The main uses for light-emitting diodes result from the need for visible indicator lights and alphanumeric devices, for which the spectral range of interest coincides with the visible spectrum. An important use of semiconductor (injection) lasers and LEDs is as light sources for fiber-guide, optical communications. Such light sources are most useful in the near-infrared spectral range at wavelengths of 0.9–1.5 µm, which are dictated largely by the absorption and dispersion properties of the fibers.

The choice of the semiconductor of an LED or laser depends upon the wavelength of interest, the most important factor being the band-gap energy. However, other considerations seriously limit the usefulness of some classes of semiconductors. In particular, Si and Ge, which are indirect-gap semiconductors, are not useful in LEDs because of their low radiative recombination efficiency. The II–VI compounds, eg, ZnS, although they have potential for LEDs that emit in the blue region of the spectrum, have not become commercially important because of difficulty in $p-n$ junction formation, and SiC LEDs are very difficult to manufacture because of the extremely refractory properties of SiC.

MORTON B. PANISH
Bell Laboratories

W. Shockley, *Electrons and Holes in Semiconductors*, Van Nostrand, New York, 1950.

A.A. Bergh and P.J. Dean, *IEEE Proc.* **60**, 156 (1972).

H.C. Casey and M.B. Panish, *Heterostructure Lasers*, Academic Press, New York, 1978.

LIGNIN

Lignin is, after cellulose (qv), the principal constituent of the woody structure of higher plants. It is now well-established that about one quarter of dry wood consists of lignin, in part deposited in the xylem cell walls and in part located in the intercellular spaces where it may constitute as much as 70%, or even more, of the solid materials present. Lignin acts as a cementing agent to bind the matrix of cellulose fibers into a rigid woody structure. Woody plants are so widespread that lignin ranks second as the most abundant organic material. Trees are the most abundant sources of lignin. Coniferous woods (softwoods) usually contain more lignin (ca 28%) than deciduous woods (hardwoods, ca 24%). Always closely associated with cellulose, most, if not all, lignin is bound to plant polysaccharides, especially the hemicelluloses.

Lignin from coniferous trees is almost surely a polymeric substance resulting from random combination of the products of an enzymatically induced oxidation (dehydrogenation) of coniferyl alcohol. In deciduous trees, structures from the polymerization of both coniferyl alcohol and sinapyl alcohol are present in the lignin. Functional groupings such as hydroxyl, methoxyl, and carbonyl have been identified in the polymer. There are probably many lignins, the properties and composition of each depending on the source and method of isolation.

Spent liquors from the wood-pulp industry are the principal commercial sources of lignins available in the free or acid form, as sodium lignates, and as various metal salts of ligninsulfonic acid. These products are available in a wide range of grades as brown powders or aqueous solution (see Pulp).

Lignin in spent liquors is usually called waste lignin. Such a designation may have been valid during the period when fuel costs were low, but with the very rapid rise in energy cost associated with petroleum price increases, lignin in spent liquors has assumed greater importance as fuel.

Unmodified lignins are used as binders, fillers, and resin extenders. The uses as dispersants and sequestering agents are more specifically dependent on the nature of the lignin. Chemical modification of isolated lignin can improve its utility in adhesives and structural polymers, coating, and dispersants. Lignin also can be converted to pure, low molecular-weight chemicals; it may serve as an important feedstock for the chemical industry. At present, only vanillin (qv) and dimethyl sulfide are produced commercially, but economically competitive processes for conversion to phenolic and aromatic chemicals are real possibilities.

General Properties

The lignins isolated experimentally or available commercially are friable solids, usually powders, without any observable crystalline form under the electron microscope. They have densities of 1.3–1.4 g/cm^3 and a refractive index of 1.6. They are usually brown, although specimens of very light colors have been made. Free lignins are insoluble in water, strong mineral acids, and hydrocarbons. Their insolubility in 72% sulfuric acid is the basis for a quantitative test to differentiate them from cellulose and other carbohydrates in woods and wood pulps. Commercial lignins are soluble in aqueous alkaline solutions and some are also soluble in many oxygenated organic compounds and in amines. Lignins do not melt, but some soften and all char on heating. X-ray diffraction studies indicate lignins are amorphous polymers. The low specific viscosity of

Table 1. Average Elemental Analysis of Wood Lignin, %

Species	C	H	O	OCH$_3$
coniferous	63.8	6.3	29.9	15.8
deciduous	59.8	6.4	33.7	21.4

lignin solutions suggests that lignin is a branched-chain polymer with a lower degree of polymerization than cellulose. The heat of combustion of lignin is about 29.5 MJ/kg (12,700 Btu/lb) and, thus, it comprises nearly 40% of the fuel value of softwood, whereas it constitutes only some 28% of its dry mass.

Composition. A typical elemental analysis for wood lignin is shown in Table 1.

Commercial Lignins

Commercial lignins are by-products of the wood and cellulose industries; the largest source of lignin products is the paper industry. Spent pulping liquors from wood-pulping processes contain lignin compounds which are a source of chemicals and fuel. The sulfite process results in a spent liquor that contains 40–55% of the original wood. The principal components are ligninsulfonates. The commonly used term for these and often for the total solids in the spent liquor is lignosulfonate. In most uses, the whole solids are processed into the products. Some 14–17% of lignosulfonates are sold and the remainder burned or disposed of. About 48% of the sulfite pulp is obtained from the Magnefite process for magnesium lignosulfonate. Such liquors are burned in recovery furnaces to provide chemicals for the subsequent pulping cycle.

The kraft- and soda-pulping processes result in spent liquors called black liquors containing alkali lignins also called kraft and sulfate lignin. The hydrolysis of wood is another source of lignin, which at present is not commercially available in the United States. It is, however, a growing source in other countries, particularly the USSR.

Processes are under development for the conversion of whole wood into hydrocarbons; a pilot plant for this process is located in Albany, Oregon.

A large quantity of wood lignin is altered by high temperature steaming in the production of particle-board pulp. The finished product is called Presdwood.

Ligninlike materials. Many residues from processes utilizing wood or cereals contain considerable lignin or ligninlike materials. Such products are sometimes classed as lignins, though for the most part their manufacturers offer them under trade names. They include Furafil, products from bark, and hydrolysis lignin.

D.W. GOHEEN
C.H. HOYT
Crown Zellerbach Corp.

A. Payen, *Compt. Rend.* **7**, 1052 (1838).

K. Freudenberg, *Science* **148**, 595 (1965).

E. Adler, *Wood Sci. Technol.* **11**, 169 (1977).

T.E. Lindemuth, *Biomass Liquefaction at Albany, Oregon*, report published by DOE, Solar Div., Washington, D.C., 1978.

K.V. Sarkanen and C.H. Ludwig, *Lignins*, Wiley-Interscience, New York, 1971, p. 800.

LIGNITE AND BROWN COAL

The common names lignite and brown coal are given to coals whose properties are intermediate between peat and bituminous coal as a result of a limited degree of coalification. In general, the term brown coal designates a geologically younger or less coalified material than lignite. In this article, lignite or lignitic coal is used as a comprehensive term.

The principal use of lignite is for combustion in steam generation of electric power (see Power generation). Lesser amounts are used (generally in the form of briquettes) for industrial and domestic heating. The briquettes are pressed and often carbonized at low temperatures to provide a smokeless fuel (see Size enlargement). The by-product tars obtained from briquette production have been used for liquid fuels and chemical manufacture. Lignite has been converted by gasification to synthesis gas for motor fuels, chemicals, and ammonia-based fertilizers in large integrated plants.

Lignite is less valuable than coals of higher rank, primarily because of its much higher water content (up to 70% as mined). The higher water content and the high combined oxygen content produce a low heating value. The expense of shipping a fuel with a very high moisture content has limited the market largely to the vicinity of the mine. The increasing worldwide demand for all forms of energy and the desire of nations for self-sufficiency has increased the importance of these coals. The pace of scientific research and development of all technologies for mining, utilization, and improvement of properties has accelerated. Production and utilization have grown rapidly. In the United States, development of the lignite resources had been inhibited by the large supplies of alternative fuels. Recently, however, the lower sulfur content of many lignite deposits has led to a rapid growth in the rate of production and utilization of this fuel.

Geology

Lignite was deposited relatively recently. Most of the reserves were produced during the Tertiary era (ca $(2.5-60) \times 10^6$ yr ago). These include the deposits of the Dakotas, Montana and Wyoming, Saskatchewan, the FRG, Asiatic USSR, Pakistan, northern India, Borneo, Sumatra, Manchuria, Alaska, and northwest Canada.

Composition, Properties and Analysis

Macroscopic appearance. Lignitic coals vary from brown to dull black when moist. The color may appear considerably lighter when the coal is dried. Because of its weak structure and tendency to shrink and crack on drying, brown coal disintegrates through all stages of use including handling, transportation, and storage.

Physicochemical structure. Water-filled pores and capillaries of differing diameters permeate the organic gel material that makes up as-mined lignitic coal. The structure of the pores and capillaries permit significant retention of moisture upon air drying. Mineral matter is nonuniformly distributed through these coals. The alkaline content is usually owing to salts of the humic acids in these coals.

Properties. The apparent density of lignitic coals is $1.05-1.25$ g/cm^3. Density values for lignitic coals tend to be lower than those for higher rank coals.

More mature coals generally have greater elasticity and lower plasticity. The plasticity index or ratio of elastic energy to plastic energy involved in compressing coals was used to indicate the ease of briquette formation. Briquetting without a binder is possible only with the softer, less mature coals. The tar yield is usually higher than that for more mature coals.

The high reactivity of lignites with oxygen requires special care during mining, transportation, and storage to avoid spontaneous combustion from the oxidation.

Analysis. The results of proximate and ultimate analyses, and heat values of a number of lignitic coals from North Dakota are given in Table 1.

Resources and Production

The eastern European reserves of lignitic coals provide the primary solid fuel for the GDR, Czechoslovakia, Hungary, Yugoslavia, and Bulgaria. In the United States, lignite deposits are located in the northern Great Plains and the Gulf states. Australia is the only country outside Europe and the USSR that is among the largest ten producers.

Most lignitic coal is now mined by strip-mining (open-cast or open-cut) methods.

Storage. Concern about spontaneous ignition has led some operators to try to match the mining rate to the consumption rate, with little if any reserve, as in a mine-mouth power-generation station. When the coal must be stockpiled, careful stacking minimizes oxygen reaction and overheating. To limit drying, spraying with cold water is useful. Underwater storage is sometimes used in drier climates such as in Australia.

Transportation. For short distances from the mine, transportation is by truck or conveyor belt.

Integrated projects. The need to efficiently use available resources, coupled with the ability to produce a range of useful products from lignitic coals has led to the development of large integrated complexes. At one site, a mine produces large tonnages of lignitic coal that is processed to provide a variety of fuels and chemicals. Examples of these can be seen in South Africa, the GDR, and India (see Fuels, synthetic).

Health and Safety Factors

The chief hazard involves the tendency to spontaneous combustion of the coal as it dries, especially at the exposed seam. This danger requires careful planning and continued reclamation efforts to cover these faces. Adequate water for revegetation has been of some concern in the arid areas of the northwestern U.S. Vegetative growth is slow and reclamation is expected to take many years.

KARL S. VORRES
Institute of Gas Technology

Symposia on the Technology and Use of Lignite have been held at the University of North Dakota, Grand Forks, North Dakota, in conjunction with the Grand Forks Energy Technology Center.

M.A. Elliot, ed., *The Chemistry of Coal Utilization*, Second Supplementary Volume, Wiley-Interscience, New York, 1981.

LIGNOSULFONIC ACID. See Lignin.

LIME AND LIMESTONE

The elements calcium and magnesium, which are distributed very widely in the earth's crust, most commonly occur in carbonate forms of rock, generally classified as limestone. Although vast strata of this ubiquitous rock are buried so deeply as to be inaccessible, great tonnages of this stone are extracted for commercial use. Since limestone occurs in varying degrees in nearly every country, annual world production is virtually incalculable but has been estimated at 2×10^9 metric tons in the 1970s. Literally, it is one of the most basic raw materials of industry and construction and has been compared to one leg of a six-legged stool

Table 1. Analyses of Lignitic Coals

Coal	Proximate analysis, wt %					Ultimate analysis, %					Heat value, gross dry, MJ/kg
	Moisture	Ash	Volatile matter	Fixed carbon	Volatile matter, daf	C	H	S	N	O	
North Dakota (range for coals from five areas)	33.9–41.2	3.5–8.5	25.4–27.6	26.9–31.7	45.2–48.8	71.1–74.4	4.8–5.3	0.3–2.3	1.0–1.1	16.9–22.7	27.8–29.6

upon which modern industry rests, the other legs being coal, oil, iron ore, sulfur and salt.

Limestone may be classified as to origin, chemical composition, texture of stone, and geological formation. Chemically, it is composed primarily of calcium carbonate, $CaCO_3$, and secondarily of magnesium carbonate, $MgCO_3$, and varying percentages of impurities. Although these carbonates occur in many other rocks, ores, and soils, in its broadest definition, limestone is distinguished by a content of more than 50% total carbonate. In a cursory manner, limestone can be distinguished from most other rock by applying a dilute hydrochloric acid solution to it. If the stone effervesces, it is a basic carbonate rock with a definite alkaline reaction. Limestone's most important chemical characteristic is that when subjected to high temperature it decomposes chemically into lime, calcium oxide, CaO, decarbonation occurring through the expulsion of carbon dioxide gas. This primary product, known as quicklime, can be hydrated, or slaked, into hydrated lime, calcium hydroxide, $Ca(OH)_2$, the water being chemically combined with the calcium oxide in an equimolecular ratio (see Calcium compounds).

Definition. In addition to showing varying degrees of chemical purity, limestone assumes a bewildering number of widely divergent physical forms, including marble, travertine, chalk, calcareous marl, coral, shell, oolites, stalagmites, and stalactites. All these materials are essentially carbonate rocks of the same approximate chemical composition as conventional limestone.

Geology

Limestone, as a constituent of the earth's crust, is a rock of sedimentary origin from material precipitated by chemical and organic action on drainage waters. Calcium is a common element, estimated to comprise 3–4% of the earth's crust, and the calcium constituent of limestone must have come originally from igneous rocks.

Geographical occurrence. Limestone is widely distributed through the world in deposits of varying sizes and degrees of purity. Limestone is present in the majority of geological formations. Significant deposits in the UK are chiefly confined to the Devonian, lower Carboniferous, Jurassic, and cretaceous systems. In the U.S., the geological distribution is broader, quantities of stone occurring in the older Cambrian and Silurian systems, as well as in the previously mentioned groups.

Impurities. The chemical composition and properties of lime and limestone depend upon the nature of the impurities and the degree of contamination of the original stone. The contaminating materials either were deposited simultaneously with the $CaCO_3$ or during some later stage. Impurities include alumina in combination with silica, iron compounds, sodium and potassium compounds, carbonaceous matter, and sulfur and phosphorus compounds.

Properties of Limes and Limestones

The chemical and physical properties of limestone vary tremendously, owing to the nature and quantity of impurities present and the texture (crystallinity and density). These same factors exert a marked effect on the properties of the limes derived from the diverse stone types. In addition, calcination and hydration practices can profoundly influence the properties of lime. Therefore, many of these characteristics assume a range of values.

Color. The purest forms of calcite and magnesite are white, often with an opaque cast, but most conventional limestone, even relatively pure types, are gray or tan.

Quicklime is usually white of varying intensity, depending upon chemical purity; some species possess a slight ash-gray, buff, or yellowish cast. Invariably quicklime is lighter in color than the limestone from which it is derived. Hydrated limes, except for hydraulic and impure hydrates, are extremely white in color, invariably whiter than their quicklimes.

Odor. Except for highly carbonaceous species, most limestones are odorless. Quick and hydrated limes possess a mild characteristic odor that is difficult to describe except that it is faintly musty or earthy but not offensive.

Texture. All limestones are crystalline, but there is tremendous variance in the size, uniformity, and arrangement of their crystal lattices.

Hydrated lime is invariably a white, fluffy powder of micrometer and submicrometer particle size. Commercial quicklime is used in lump, pebble, ground, and pulverized forms.

Hardness. Pure calcite is standardized on the Mohs scale at 3. Most limestone is soft enough to be readily scratched with a knife.

Strength. The compressive strength of limestone varies tremendously, 8.3–196 MPa (1,200–28,400 psi). Marble generally has the highest value and chalk and calcareous marl the lowest.

Melting and boiling points. Since all limestone is converted to an oxide before fusion or melting occurs, the only melting point is applicable to quicklime. These values are 2570°C for CaO and 2800°C for MgO. The boiling points for CaO are 2850°C, and for MgO, 3600°C.

Heat of hydration. In the hydration of quicklime, the considerable heat that is generated has been measured as follows: $Ca(OH)_2$, 63.6 kJ/mol (488 Btu/lb) of quicklime; $Mg(OH)_2$, 32.2–41.8 kJ/mol (247–368 Btu/lb).

Plasticity. An innate characteristic of a lime putty of pastelike consistency is its plasticity or its ability to be molded under pressure and to retain its altered shape in masonry mortar and plaster.

Stability. All calcitic and dolomitic limestones are extremely stable compounds, decomposing only in fairly concentrated acids or at calcining temperatures of 898°C for high calcium and about 725°C for dolomitic stones at 101.3 kPa (1 atm).

Quicklime and hydrated lime are reasonably stable compounds but not nearly as stable as their limestone antecedents.

Chemical Reactions

Neutralization. In water, lime ionizes readily to Ca^{2+}, Mg^{2+}, and OH^- ions, forming a strong base or alkali. Both $Ca(OH)_2$ and $Mg(OH)_2$ are diacid bases, and only one molecule of either is necessary to neutralize such strong monobasic acids as HCl and HNO_3, yielding neutral salts and heat.

pH. Lime solutions develop a high pH of over 12.5 and approach 13 at maximum solubility at 0°C.

Causticization. Lime, particularly the high calcium type, reacts with carbonates such as Na_2CO_3 and Li_2CO_3 to form other hydroxides and carbonates through double decomposition as follows:

$$Na_2CO_3 + Ca(OH)_2 \rightarrow 2\ NaOH + CaCO_3 \downarrow$$

Silica and alumina. The manufacture of portland cement is predicated on the reaction of lime with silica and alumina to form tricalcium silicate and aluminate. Limestone does not react with silica and alumina under any circumstances, unless it is first calcined to lime, as in the case of hydraulic lime or cement manufacture (see Cement).

Other reactions. Dry hydrated lime adsorbs halogen gases, such as Cl_2 and F_2, to form hypochlorites and fluorides. It reacts with hydrogen peroxide to form calcium peroxide, a rather unstable compound. At sintering temperatures, quicklime combines with iron to form dicalcium ferrite.

Limestone Production

More than 99% of U.S. limestone production is sold or used as crushed and broken stone, rather than dimension-stone. Most stone is obtained by open-pit quarrying methods. There is, however, a slight trend toward increased mining which is expected to continue.

Stone processing. Next to blasting, primary crushing is the most effective method of reducing stone size. Primary crushers are of two basic types: compression or impact. For fine pulverization, both dry and wet processes are utilized, but increasingly the dry process is more popular since wet grinding ultimately requires drying and is much more energy-intensive.

To comply with stringent specifications, some plants beneficiate the semiprocessed stone by removing clay and soil clinging to limestone. Several wet methods are used: washing, scrubbing, flotation, and heavy-media separations are used for removing silica (flint and quartz). Optical mineral sorters use compressed air to deflect stone particles deviating from a preset color standard; and hand-picking from conveyor belts is still practiced by some plants.

Table 1. Typical Analyses of Commercial Quicklimes

Component	High calcium range,[a] %	Dolomitic, range,[a] %	Component	High calcium, range,[a] %	Dolomitic, range,[a] %
CaO	93.25–98.00	55.50–57.50	Al_2O_3	0.10–0.50	0.05–0.50
MgO	0.30–2.50	37.60–40.80	H_2O	0.10–0.90	0.10–0.90
SiO_2	0.20–1.50	0.10–1.50	CO_2	0.40–1.50	0.40–1.50
Fe_2O_3	0.10–0.40	0.05–0.40			

[a]The values given in this range do not necessarily represent minimum and maximum percentages.

The main environmental problem is dust control, which requires collection of particulate emissions from point sources and suppression of fugitive dust from a multitude of areas.

Lime Manufacture

Most lime plants worldwide produce their own kiln feed from a contiguous quarry or mine, making them integrated lime producers. However, ten unintegrated lime plants located on the Great Lakes obtain their kiln feed from large commercial quarries in northern Michigan. Most of these plants, situated in the Chicago and Detroit areas and in northern Ohio, are among the largest U.S. lime-producing plants.

Theory of calcination. Although the reversible reaction involved in the calcination and recarbonation of lime-limestone is one of the simplest, most fundamental of all chemical reactions, as given above, in practice, lime burning can be quite complex and many empirical modifications are often necessary for efficient performance.

There are three essential factors in the thermal decomposition of limestone: (1) The stone must be heated to the dissociation temperature of the carbonates; (2) this minimum temperature (but in practice a higher temperature) must be maintained for a certain duration; and (3) the carbon dioxide evolved must be removed rapidly.

Kilns. In the United States, nearly 90% of commercial lime capacity and 50% of captive lime is calcined in rotary kilns. Outside the United States, the vertical kiln is the most commonly used kiln.

Calcination products. Table 1 summarizes the chemical analyses of commercial quicklime in the United States.

Hydrated Lime Manufacture

Although most lime is sold as quicklime, production of hydrated lime is also substantial. This product is made by the lime manufacturer in the form of a fluffy, dry white powder, and its use obviates the necessity of slaking.

The manufacture of hydrated lime proceeds by the slow addition of water to crushed or ground quicklime in a premixing chamber or a vessel known as a hydrator, both of which mix and agitate the lime and water.

Uses

Lime and limestone are used in construction applications (concrete aggregate, roadstone, railroad ballast, asphalt filler, mortar and limestone sand), building materials (portland cement manufacture, concrete products, and insulation), fillers (micro-calcium carbonate products), building lime, lime soil stabilization, agricultural applications (liming soils, fertilizer filler, and mineral animal feed), chemical and industrial applications (iron and steel metallurgy and nonferrous metallurgy), environmental uses (water treatment, wastewater treatment, air pollution control, solid-waste disposal and filter beds), chemicals manufacture (lime manufacture, alkalies, calcium carbide, and plastics), and other industrial uses (pulp and paper, glass manufacture, industrial fillers, coal-mine dusting, sugar, and petroleum).

R.S. BOYNTON
Consultant

R.S. Boynton, *Chemistry and Technology of Lime and Limestone*, John Wiley & Sons, Inc., New York, 1966.

J. Gilson and co-workers, "Carbonate Rocks," in *Industrial Minerals and Rocks*, AIME, New York, 1960, pp. 123–201.

D.D. Carr and L.F. Rooney, "Limestone and Dolomite," in *Industrial Minerals and Rocks*, 4th ed., AIME, 1975, pp. 757–789.

LINCOSAMINIDES. See Antibiotics, lincosaminides.

LINEN. See Fibers, vegetable.

LINOLEIC ACID, LINOLENIC ACID. See Carboxylic acids.

LINSEED. See Fibers, vegetable.

LINT. See Cotton.

LINTERS. See Cellulose; Vegetable oils.

LIPASES. See Enzymes.

LIPIDS. See Fats and fatty oils; Vegetable oils.

LIPSTICK. See Cosmetics.

LIQUEFIED PETROLEUM GAS

Liquefied petroleum gas (LPG) is a class of versatile petroleum products that is produced with natural gas or are derived from the refining of crude oil (see Gas, natural; Petroleum). Its main uses are as fuels and as feedstocks (qv) for the production of a wide variety of chemicals. It is commercially available as propane, butane, and butane–propane mixtures (see Hydrocarbons, C_1–C_6). Ethane is rapidly becoming a principal product and, in most cases, is produced in the same plants as LPG.

Properties

In general, the specifications for LPG involve limits for physical properties; consequently, the composition of the commercial-grade products varies between wide limits. Physical properties of the principal components of LPG are summarized in Table 1.

Table 1. Physical Properties of LPG Components

	bp (at 101.3 kPa[a]), °C	Vapor pressure (at 37.8°C), kPa[a]	Liquid density, $d_{15.6}^{15.6}$, g/cm^3 (at saturation pressure)
ethane	−88.6		354.9
propane	−42.1	1310	506.0
propylene	−47.7	1561	520.4
n-butane	−0.5	356	583
isobutane	−11.8	498	561.5
1-butene	−6.3	435	599.6
cis-2-butene	3.7	314	625.4
trans-2-butene	0.9	343	608.2
n-pentane	36.0	107	629.21

[a]To convert kPa to psi, multiply by 0.145; to convert kPa to mm Hg, multiply by 1.5.

Manufacture and Processing

LPG that is recovered from natural-gas processing is essentially free of unsaturated hydrocarbons, eg, propylene and butylenes. LPG is recovered from natural gas principally by three recovery methods: absorption, compression, and adsorption. Selection of the process is dependent upon the conditions and the degree of recovery desired.

Purification. LPG generally requires treatment for removal of H_2S, organic sulfur compounds, and water in order to meet specifications. Several methods are used, including amine treatment, caustic treatment, coalescing, solid-bed dehydration, molecular-sieve treatment, solid-bed caustic treatment, and fractionation (see Adsorptive separation; Carbon dioxide; Molecular saves).

Uses

Residential and commercial demands represent nearly half of the LPG sales, and home heating is by far the largest use of LPG. The second largest use is in the manufacture of petrochemical and polymer intermediates.

F.E. SELIM
Phillips Petroleum Co.

R.E. Cannon, *Oil Gas J.*, (July 18, 1983).
Petroleum Products Handbook, McGraw-Hill Book Co., Inc., New York, 1960.

LIQUID CRYSTALS

Liquid crystals are highly anisotropic fluids that exist between the boundaries of the solid and conventional, isotropic liquid phase. The phase is a result of long-range orientational ordering among constituent molecules that occurs within certain ranges of temperature in melts and solutions of many organic compounds. The ordering is sufficient to impart some solidlike properties to the fluid, but the forces of attraction usually are not strong enough to prevent flow. In those cases where a liquid–crystalline substance is substantially solid in terms of flow, there are other fluid aspects to its physical state. This dualism of physical properties is expressed in the term liquid crystal. Liquid crystallinity also is referred to as mesomorphism. Liquid-crystals are in thermodynamic equilibrium over wide temperature ranges and undergo well defined phase changes.

Many thousand organic substances and some polymers exhibit liquid crystallinity. The general, common molecular feature is an elongated, narrow molecular framework, which usually is depicted as a rod- or cigar-shaped entity. Some disk-shaped molecules also adopt liquid-crystal structures. The geometric anisotropy of individual molecules translates throughout the entire fluid medium. The orientational association of the molecules is only partial and, as the nature of intermolecular forces is delicate, liquid crystals are extraordinarily sensitive to external perturbations, eg, temperature, pressure, electric or magnetic fields, or foreign vapors. Liquid crystals may be used as practical devices to

Table 1. Some Thermotropic Liquid Crystalline Compounds

Formula	Name	Liquid crystalline range, °C
Nematic liquid crystals		
	p-methoxybenzylidene-*p'*-*n*-butylaniline (MBBA)	21–47
	p-azoxyanisole (PAA)	117–137
	p-*n*-hexyl-*p'*-cyanobiphenyl	14–28
Cholesteric (spontaneously twisted nematic) liquid crystals cholesteric esters		
	cholesteryl nonanoate	78–90
noncholesteryl, chiral compound	(−)-2-methylbutyl-*p*-(*p'*-methoxybenzylidene-amino)cinnamate	53–97
Smectic liquid crystals smectic A	ethyl *p*-(*p'*-phenyl-benzalamino)benzoate benzalamino)benzoate	121–131
smectic B	ethyl *p*-ethoxybenzal-*p'*-aminocinnamate	77–116
smectic C	*p*-*n*-octyloxybenzoic acid	108–147

monitor ambient changes or to transduce an environmental fluctuation into a useful output.

Liquid crystals are used widely in electric display devices, eg, digital watches and calculators. Oscillographic and television displays using liquid-crystal screens also are being developed. Other applications include radiation and pressure sensors, optical switches and shutters, and thermography. Polymers that form the intermediate phase are important in the fabrication of lightweight, ultrahigh strength, and temperature-resistant fibers (see Aramid fibers; Heat-resistant polymers). Liquid crystals also appear to play an important role in the structure and biochemical function of living tissue, where their characteristic combination of order and flow mobility is particularly well-suited to life processes.

Categories, Classes, and Structures

Liquid crystals may be divided into two broad categories according to the principal means of breaking down the complete order of the solid state. Lyotropic liquid crystals result from the action of a solvent and, hence, are multicomponent mixtures. Thermotropic liquid crystals, which also may be mixtures of compounds, result from the melting of mesogenic solids and, hence, are thermally activated mesophases.

Within each category, three distinctive structural classes of liquid crystals have been identified: the smectic, nematic, and cholesteric structures. These structures are related to the dimensionality and packing aspects of the residual molecular order. A few representative examples of thermotropic liquid crystals of each structure and their mesomorphic thermal ranges are listed in Table 1 (p. 703).

Polymorphism

A liquid-crystal compound can take on more than one type of mesomorphic structure as the conditions of temperature or solvent are changed. In thermotropic liquid crystals, transitions between various structural subclasses occur at definite temperatures and are accompanied by definite changes in the latent heat. The order of appearance of different mesomorphic structures as the temperature is raised usually is consistent with gradual breakdown upon heating of long-range molecular order.

An exception to these hypotheses that involves the association of the symmetry of the liquid-crystal structure to temperature occurs in some cyano compounds where a smectic phase that has formed by cooling from the nematic structure reverts to the nematic phase at a still lower temperature (re-entrant nematic phase). The re-entrant phase also can be formed under the influence of pressure.

Electric or magnetic fields also may induce mesomorphic phase transitions. If a cholesteric liquid crystal is composed of molecules with a positive dielectric or diamagnetic susceptibility, an applied field tends to align all the molecules or molecular clusters along the field direction. At sufficiently high field strengths, a transition to nematic order may occur as the helical organization is unwound.

Synthesis

It is not true that because a molecule is elongated, it will engage in a mesomorphic structure; eg, n-paraffins and homologues of acetic acid do not display any properties that are characteristic of liquid crystals, although these molecules are long, narrow, and meet the requirements of geometric anisotropy. The forces of attraction between these molecules are not sufficiently strong for an ordered, parallel arrangement to be retained after the melting of the solid. The particular mesomorphic structure that occurs, ie, smectic, nematic, or cholesteric, not only depends on molecular shape but is intimately connected with the strength and position of polar groups within the molecule, the molecule's overall polarizability, and the presence of chiral centers.

Molecular interactions that lead to attraction are dipole–dipole interactions, dipole–induced dipole interactions, dispersion forces, and hydrogen bonding.

In order for dipole–dipole and dipole–induced dipole interactions to be effective, the molecule must contain polar groups and/or be highly polarizable. Ease of electronic distortion is favored by the presence of aromatic groups and double or triple bonds. These groups frequently are found in the molecular structure of liquid-crystal compounds.

Table 2. Some Central Linkages Found in Mesomorphic Series

X	Series name
—CH=N—	Schiff bases
—N=N—	diazo compounds
—N=N— ↓ O	azoxy compounds
—CH=N— ↓ O	nitrones
—CH=CH—	stilbenes
—C≡C—	tolans
—O C— ‖ O	esters
— (nothing)	biphenyls

The most common nematogenic and smectogenic molecules are of the type:

$$R \!-\!\bigcirc\!-\!X\!-\!\bigcirc\!-\!R$$

where R and R′ are alkyl, alkoxy, or acyl groups, and X linkages are shown in Table 2.

Although it is difficult to predict exactly which type of mesomorphic structure formed by a molecule meets the general requirements of liquid crystallinity, rough trends can be recognized. The presence of functional groups that lead to strong lateral interactions, eg, dipoles operating across the molecule's long axis, favor the layered, smectic structure. When these structural elements are not present, but the molecule is otherwise suitable for mesomorphism, ie, is long and narrow, the nematic modification is likely. An asymmetric center is necessary for the cholesteric, ie, spontaneously twisted nematic, and chiral smectic mesophases.

Goals in liquid-crystal synthesis include the design of room temperature thermotropics with enhanced thermal, chemical, and photochemical stabilities for use in practical devices (eg, in watch and calculator displays, and in television screens). Extended mesomorphic temperature ranges also are sought. Ferroelectric liquid crystals, mesomorphic free radicals for epr studies, and colorless, large-pitch cholesterics also are being developed.

Polymer Liquid Crystals

A number of high molecular weight polymeric substances naturally exist in or can form liquid-crystal states. Liquid crystallinity has important implications in biological systems which, typically, are composed of large molecules. Technological advances in the fabrication of ultrahigh strength fiber materials also depend upon a liquid-crystal precursor organization of the macromolecules.

Polymers would seem predisposed to liquid crystallinity since they may adopt an extremely elongated shape. In many cases, however, the incorporation of rigid polarizable segments into the polymer chain results in increased thermal stability and decreased solubility. The polymers decompose at the high temperatures that are needed to produce a fluid state and/or prove intractable in conventional solvents. It is possible, however, in some cases by design of the monomer, to optimize polymer rigidity and to retain useful thermal or solubility properties.

Thermotropic phases. In the quiescent state, certain polymer melts also exhibit anisotropic, liquid-crystal properties that can be understood in terms of extended chain rigidity. Firm evidence for fluid melts that are optically and rheologically anisotropic exists for macromolecules, eg, petroleum pitches, polyesters (qv), polyethers (qv), polyphosphazines (see Inorganic high polymers), α-poly-p-xylylene, and polysiloxanes (see Silicon compounds). Synthesis goals include the incorporation of a mesogenic entity into the main chain of the polymer to increase the strength and thermal stability of the materials that are formed from the liquid-crystal precursor, the locking in of liquid-crystal properties of the fluid into the solid phase, and the production of extended-chain polymers that are soluble in weaker solvents than high strength sulfuric acid.

Liquid Crystals in Biological Systems

Many biological systems exhibit the properties of liquid crystals. Considerable concentrations of mesomorphic compounds have been found in many parts of the body, often as sterol or lipid derivatives. A liquid-crystal phase has been implicated in at least two degenerative diseases, eg, atherosclerosis and sickle-cell anemia. Living tissue, such as muscle, tendon, ovary, adrenal cortex, and nerve, show the optical-birefringence properties that are characteristic of liquid crystals.

There are two important classes of fibers that are characterized by nematiclike organization and are in the cytoplasm of many plant and animal cells. These are microfilaments and microtubules that play a central part in the determination of cell shape, either as the dynamic element in the contractile mechanism or as the basic cytoskeleton.

There is also a correlation between the type of subsurface organizations of these fibers and gross cell shape in tissue other than muscle.

DONALD B. DuPRÉ
University of Louisville

P.G. deGennes, *The Physics of Liquid Crystals*, Oxford University Press, (Clarendon), London, UK, 1974, pp. 24, 30, 45.

E.B. Priestly and co-eds., *Introduction to Liquid Crystals*, Plenum Press, New York, 1974, Chapts. 12 and 14.

S. Chandraeskhar, *Liquid Crystals*, Cambridge University Press, London, 1977.

LIQUID-LEVEL MEASUREMENT

Liquid-level-measuring devices have been used throughout the centuries and rank among the oldest measuring instruments. For most chemical processes, measurement of the liquid level is a basic requirement. This measurement may perform a primary function in a control system when it is used to maintain an optimum level in a distillation column or in a reactor that is an essential part of a continuous chemical process. On the other hand, the liquid-level device may provide a record of the amount of a particular liquid in a storage tank or the amount added to a batch process or of the quantity disbursed to a customer.

Liquid-level measurements may be either direct or inferential. In direct measurement, the actual height of a liquid surface or interface is gauged by means of a sight glass or a mechanical float gauge, or by ultrasonic, capacitance, radiation, or microwave techniques (see Microwave technology; Ultrasonics). The measurement is inferential when the hydrostatic head exerted by the liquid is determined by any pressure-measuring device mounted on the bottom of an open tank or by a differential-pressure-measuring device if the tank is closed. The head can be measured also by a bubbler system or by the buoyant force exerted on a displacer. Knowing the geometry of the tank, the static-head measurement can be converted to the total mass of the liquid in the tank, or—knowing the density of the fluid—to the liquid-level height. Recently introduced techniques utilize fiber optics (qv), solid-state devices, and fluidics. Furthermore, the use of microprocessors in the controller sections of liquid-level systems is under investigation.

Mass Versus Actual Level

If the quantity of fluid in a storage tank is handled on a mass basis, the temperature must be determined accurately, in addition to the exact level-surface measurement, since the density of the liquid is temperature dependent. The mass can then be calculated according to the formula:

$$M = A \cdot H \cdot D$$

where M = total mass of liquid in kilograms; A = average area of the tank in square meters for the height H; H = height of liquid in meters; and D = density of liquid in kilograms per cubic meter.

Measurement of the pressure of the hydrostatic head at the bottom of a tank permits the direct determination of the mass without measuring temperature since:

$$M = (P \cdot A)/g$$

where P = pressure due to the hydrostatic head in Pascals (or psi $\times$ 6895); A = area in m^2; and g = gravitational constant (9.80665 m/s^2).

Direct Level Measurements

Three of the more popular direct level-measurement systems are the capacitance, ultrasonic, and radiation types.

Capacitance method. If the liquid to be measured has a dielectric constant between 2 and 100, a simple capacitance system can be constructed. A bare metal rod is mounted vertically in the vessel and serves as one plate of a capacitor. The metal wall of the vessel acts as the other capacitor plate, and the liquid is the dielectric. The probe is connected to an oscillator circuit which is adjusted for a dielectric value of unity. As the liquid rises and covers the probe, the oscillator frequency changes, producing an electrical output signal that is proportional to the height of the liquid.

Ultrasonic type. In liquid-phase sonic-type level detectors for continuous level measurement, pulses of ultrasonic energy are transmitted from a transducer at the tank bottom to the liquid surface. From there they are reflected to a receiver where they are converted to electrical impulses. The elapsed time between the sonic-pulse transmission and the receipt of its echo is a measure of the liquid level.

Radiation type. Nuclear radiation is utilized in a variety of liquid-level measuring devices. Figure 1 illustrates a type in which a gas-ionization cell is used. A group of radioactive cells, commonly ^{60}Co, is stacked vertically on the outside of the tank. A stack of detecting cells, such as the Ohmart cell, is placed at another position on the outside of the tank in such a way that a vertical section of liquid is interposed between the transmitting and receiving cells as the level rises. The gamma rays emitted from the source are partially absorbed by the interposed liquid. The radiation received by the detector decreases proportionally to the rise in liquid level.

Inferential Level Measurements

One of the most widely used level-measuring devices of the inferential type is the differential-pressure transmitter. Today, most differential-pressure transmitters designed for liquid-level measurements use solid-state electronics and have a two-wire 4–20 mA d-c output. For open tanks, the high pressure side of the transmitter is connected to a tap in the vessel at the minimum level, and the low pressure side of the transmitter is left open to atmospheric pressure.

Figure 2 shows a typical installation of a differential-pressure instrument for closed tanks. Connections from the instrument are made to taps in the vessel at minimum and maximum levels. Between the instrument and the maximum-level tap is a constant-reference leg. This leg is filled with liquid until its head is equivalent to the head of the

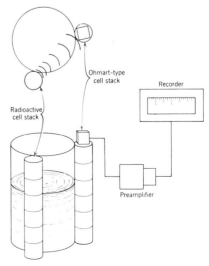

Figure 1. Nuclear radiation level gauge, using an Ohmart-type cell stack.

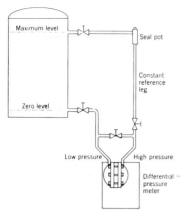

Figure 2. Differential-pressure-gauge system.

liquid in the vessel at maximum level. The reference leg must remain constant, with no formation of vapor under varying ambient conditions.

ROBERT C. WHITEHEAD, JR.
Honeywell, Inc.

P.A. Elfers and C.M. Johnson, "Liquid-Level Measurement" in D.M. Considine, ed., *Process Instruments and Controls Handbook*, 2nd ed., McGraw-Hill Book Co., Inc., New York, 1974.
C.F. Cusick, "Liquid Level Measurement," *Combustion* **41**(11), 23 (1969).
J. Williams, *Instrum. Control Syst.* **52**(1), 47 (1979).

LITHIUM AND LITHIUM COMPOUNDS

The applications of lithium metal are mainly in metallurgy and batteries and in the manufacture of lithium compounds, such as the amide, hydride, and nitride, and organolithium compounds. Commercial production of lithium in the United States has become important since 1930.

Many of the properties of lithium are similar to those of magnesium and of the alkaline-earth metals. The resemblance to magnesium includes the high solubility of the halides (except the fluoride) in both water and polar organic solvents and the high solubility of the alkyls in hydrocarbons; the low aqueous solubility of the carbonate, phosphate, fluoride, and oxalate; the thermal instability of the carbonate and nitrate; the formation of the carbide and nitride by direct combination; and the reaction with oxygen to form the normal oxide.

Sources

Geochemistry. Lithium is widely distributed in nature; trace amounts are present in many minerals, in most rocks and soils, and in many natural waters. The lithium content of the earth's crust is estimated to be about 20 ppm.

Only spodumene, petalite, and lepidolite are important lithium sources from minerals. They occur mainly in granitic pegmatites, which are coarse-grained igneous rocks composed largely of quartz, feldspar, and mica (qv).

Production

Recovery from ores and brines. Spodumene is the chief lithium source for the two domestic producers of lithium products. The preferred method of extraction of lithium from spodumene is the sulfuric acid process. Methods suitable for extraction from spodumene also can be used for petalite, since the latter mineral converts to a β-spodumene-SiO_2 solid solution on heating to a high temperature.

Other recovery processes from silicate ores. Most of the numerous processes that have been described for the extraction of lithium can be classified as either alkaline methods or ion-exchange methods.

Recovery from brines. Natural predominately chloride brines vary widely in composition, and the economical recovery of lithium from such

sources depends not only on the lithium content but on the concentration of interfering ions, especially calcium and magnesium. The location and the availability of solar evaporation capability also are important factors (see Chemicals from brine).

Lithium Metal

Properties. Lithium is an alkali metal with a silvery luster and an atomic weight of 6.941. It is the first member of Group IA in the periodic system. Two stable isotopes are present in natural lithium (^7Li has an abundance of 92.6 at% and ^6Li 7.4 at%). Lithium has a density of 0.531 g/cm^3 at 20°C, and is the lightest of all solid elements. In general, the properties of lithium are similar to those of the other alkali metals. Physical properties of lithium are listed in Table 1.

The reaction of hydrogen and lithium readily gives a hydride, LiH, which is stable at temperatures from the melting point up to 800°C. Lithium reacts with nitrogen, even at ordinary temperatures, to form the reddish-brown nitride, Li_3N. Lithium burns when heated in oxygen to form the white oxide, Li_2O.

Manufacture

An electrolytic process was devised in 1893 that resembles the one used today consisting of molten-salt electrolysis from a lithium chloride–potassium chloride mixture using a graphite rod as an anode. Modern U.S. industrial installations employ a 55 wt% LiCl–45 wt% KCl electrolyte at about 460°C using high purity lithium chloride as a feed. The current efficiency is about 80% and the lithium recovery is better than 98%, based on the charged chloride. The purity of the metal is 99.8% or better and the metallic impurities are less than 0.1%.

Inorganic Lithium Compounds

Lithium acetate. Lithium acetate, $CH_3COOLi.2H_2O$, is used as an alcoholysis catalyst for alkyd resin manufacture (see Alkyd resins).

Lithium amide. Lithium amide, $LiNH_2$, is used in the pharmaceutical industry in the synthesis of antihistamines (see Histamines and antihistamine antagonists) and analgesics (qv).

Lithium borates. Lithium metaborate, $LiBO_2.2H_2O$, mp 849°C, is used in special glass and enamel formulations. Lithium tetraborate, $Li_2B_4O_7$, mp 917°C, is used in the ceramic industry and as a flux in emission x-ray spectroscopy.

Lithium carbonate. Lithium carbonate, Li_2CO_3, mp 726°C, is one of the most important lithium salts because it may be prepared from most of its water-soluble compounds and it is the starting material for many

Table 1. Physical Properties of Lithium

Property	Value
at wt	6.941
at vol, cm^3	13.0
mp, °C	180.5
bp, °C	1336
electronic configuration	$1s^2\,2s^1$
first ionization potential, kJ/mol^a	519
electron affinity, kJ/mol^a	52.3
crystal structure	bcc
lattice constant, pm	350
metallic radius, pm	122.5
ionic radius, pm	60
d_{20}, g/cm^3	0.531
specific heat at 25°C, J/g^a	3.55
specific heat of liquid at mp, J/g^a	4.39
heat of fusion, J/g^a	431.8
heat of vaporization, kJ/g^a	ca 21.3
electrical resistivity at 20°C, $\mu\Omega\cdot cm$	9.446
characteristic spectrum lines, nm	
red	670.8
orange	610.4
vapor pressure at 702°C, kPa^b	0.065

^{a} To convert J to cal, divide by 4.184.

^{b} To convert kPa to mm Hg, multiply by 7.5.

other lithium salts as well as for the hydroxide. The ceramic and glass industries use substantial amounts of lithium carbonate for glasses, enamels, specialty glasses, and special ceramic ware having low thermal-expansion coefficients. A special grade of high purity lithium carbonate is used in the treatment of manic-depressive psychoses in acute mania and in schizo-affective illnesses (see Psychopharmacological agents).

Lithium formate. Lithium formate, $HCOOLi.H_2O$, is obtained in the form of its monohydrate by the interaction of aqueous formic acid and either lithium hydroxide or lithium carbonate.

Lithium halides. Lithium fluoride, LiF, mp 848°C, bp 1681°C, is used as a flux in enamels, glasses, and glazes; in brazing and welding fluxes; and in molten-salt chemistry. Lithium chloride, LiCl, is the raw material in molten-salt electrolysis for the production of lithium metal. It also is used, generally in its eutectic-melting composition with potassium chloride, in molten-salt chemistry and metallurgy. Lithium bromide, LiBr, is used on a laboratory scale as a catalyst and as a dehydrohalogenating agent in organic chemistry. It is a swelling agent for organic fibers such as wool (qv), hair, etc. Lithium bromide also is a hypnotic and a sedative (see Hypnotics, sedatives, and anticonvulsants).

Lithium iodide, LiI, mp 469°C, bp 1142°C, commercially is the least important halide of lithium.

Lithium hydride. Lithium hydride, LiH, melts at 686.4°C. Silicon halides react with lithium hydride that is dissolved in the lithium chloride–potassium chloride eutectic to yield high purity silane, which can be decomposed pyrolytically for semiconducting silicon substrates of high purity (see Silicon; Semiconductors). It also reacts with $AlCl_3$ in ether solutions to form lithium aluminum hydride ($LiAlH_4$), an important reducing agent.

Lithium hydroxide. Lithium hydroxide anhydrous, LiOH, is used in the manufacture of lithium stearate or other lithium soaps (see Soap), used in lubricant compositions.

Lithium hypochlorite. Lithium hypochlorite, LiOCl, in dry form has 35% available chlorine, is available commercially for use as a sanitizing agent, eg, in swimming pools, and as a laundry bleach (see Water, treatment of swimming pools; Bleaching agents).

Lithium nitrate. Lithium nitrate, $LiNO_3$, mp 251°C, forms very low melting eutectics with other alkali nitrates; mixtures are used as heat-transfer media (see Heat-exchange technology, heat transfer).

Lithium nitride. Lithium nitride, Li_3N, is the only stable nitride in the alkali-metal group. It melts at 813°C and can be obtained by direct combination of the elements. Lithium nitride is used as a catalyst in the manufacture of cubic boron nitride, which is close to diamond in its hardness (see Boron compounds; Nitrides).

Lithium oxide. Lithium oxide, Li_2O, can be prepared by heating pure lithium hydroxide to about 800°C in vacuum; however, a more convenient method is thermal decomposition of lithium peroxide.

Lithium perchlorate. Lithium perchlorate, $LiClO_4$, mp 247°C, has been used as a rocket-fuel oxidizer (see Propellants).

Lithium peroxide. Lithium peroxide, Li_2O_2, is obtained by the interaction of hydrogen peroxide with a lithium hydroxide solution in boiling ethyl alcohol or by the reaction of hydrogen peroxide in aqueous slurries of $LiOH.H_2O$.

Lithium metasilicate. Lithium metasilicate, Li_2SiO_3, is used in the ceramic industry for the production of enamels and glazes.

Lithium sulfate. Lithium sulfate, Li_2SO_4, is used in the production of special high strength glass.

Organolithium Compounds

Organolithium compounds are polar organometallic compounds in which the lithium is bonded directly to carbon. Because of the considerable amount of covalent character in these bonds, many of the compounds exist as liquids or as low melting solids and are soluble in organic solvents. They are reactive to oxygen and moisture and can ignite spontaneously in the pure state or in solution on exposure to air. They are useful in many Grignard-type syntheses (see Grignard reaction).

***n*-Butyllithium.** *n*-Butyllithium, $CH_3CH_2CH_2CH_2Li$, is the most important organolithium compound. Uses for *n*-butyllithium are mainly as an initiator in solution polymerization to produce elastomers, predominantly of the styrene–butadiene type, although butadiene alone and a

small amount of isoprene rubber also is made with *n*-butyllithium (see Elastomers, synthetic; Initiators).

***sec*-Butyllithium.** *sec*-Butyllithium, $CH_3CH_2CH(Li)CH_3$, is a clear, colorless-to-pale-yellow, pyrophoric liquid, d_{25} 0.783 g/cm³, viscosity$_{25}$ 20.1 mPa.s (= cP). Its uses generally are the same as those for *n*-butyllithium.

***tert*-Butyllithium.** *tert*-Butyllithium, $(CH_3)_3CLi$, is a colorless, crystalline solid that can be sublimed at 70–80°C and 13.3 Pa (0.1 mm Hg). It is a useful reagent in syntheses where the high reactivity of the carbon–lithium bond and the small size of the lithium atom promote the synthesis of sterically hindered compounds.

Methyllithium. Methyllithium, CH_3Li, crystallizes from benzene or hexane solution giving cubic crystals that have a saltlike constitution. It is a useful reagent in the synthesis of vitamins A and D and in the synthesis of various analgesics (qv).

Lithium acetylide-ethylenediamine complex. Lithium acetylide-ethylenediamine complex, $LiC{\equiv}CH.H_2NCH_2CH_2NH_2$, is obtained as colorless-to-light-tan, free-flowing crystals from the reaction of *n*-lithioethylenediamine with acetylene in an appropriate solvent. Its principal uses are ethynylation of halogenated hydrocarbons to give long-chain acetylenes and ethynylation of ketosteroids and other ketones in the pharmaceutical field to yield the respective ethynyl alcohols.

Phenyllithium. Phenyllithium, C_6H_5Li, forms colorless, monoclinic, pyrophoric crystals that do not melt before decomposition at 150°C. Phenyllithium can be used in Grignard-type reactions involving attachment of a phenyl group.

Organo-dilithium compounds. Organo-dilithium compounds have utility in anionic polymerizations. Living chains grow from both lithium sites.

RICARDO BACH
R.B. ELLESTAD
C.W. KAMIENSKI
J.R. WASSON
Lithium Corporation of America

"Lithium," Syst.-Nr. in *Gmelin's Handbuch der Anorganischem Chemie*, 8th ed., Verlag Chemie, Berlin, 1920; Suppl. Vol. Syst.-Nr 20 Weinheim, FRG, 1960.

W.A. Hart and O.F. Beumal, Jr., "Lithium and Its Compounds," in J.C. Bailar, Jr., and co-eds., *Comprehensive Inorganic Chemistry*, Vol. I, Pergamon Press, Oxford, UK, 1978.

LUBRICATION AND LUBRICANTS

The primary purpose of lubrication is separation of moving surfaces to minimize friction and wear. The use of additives in lubricants has progressed rapidly since ca 1930; synthetic lubricants have developed largely since World War II. Use of synthetics in automobiles is growing and expected to expand to 38×10^6 L (10^7 gal) by 1985 (see also Hydraulic fluids).

Lubrication Principles

Several distinct regimes are commonly described to systematize the fundamental principles of lubrication. These range from complete separation of moving surfaces by a fluid lubricant, through partial separation in boundary lubrication, to dry sliding where solid material properties and surface chemistry dominate. Frequently, a change from sliding motion to rolling contact with elastohydrodynamic lubrication also is possible.

Petroleum Lubricants

Lubricating oils from petroleum consist essentially of complex mixtures of hydrocarbon molecules (see Petroleum). These generally range from low viscosity oils with molecular weights as low as 250 to very viscous lubricants with molecular weights as high as about 1000. Physical properties, such as viscosity, viscosity–temperature–pressure characteristics, and performance, depend largely on the relative distribution

Table 1. Representative Petroleum Lubricating Oils

Type	Viscosity, mm²/s (= cSt) at 100°C	Flash point, °C	Pour point, °C	Sp gr (at 15°C)	Viscosity index	Common additives[a]	Uses
automobile (SAE)							
10W	4.9	204	−28	0.878	106	R, O, D, VI,	automobile, truck, and
10W-30	10.3	208	−36	0.880	155	P, W, F	marine reciprocating engines
gear (SAE)							
80W-90	14.0	192	−22	0.900	93	EP, O, R, P,	automotive and industrial
85W-140	27.5	210	−14	0.907	91	F	gear units
automatic transmission	6.9	196	−44	0.870	160	R, O, W, F, VI, P	automotive hydraulic systems
turbine							
light	5.4	206	−10	0.863	107	R, O	steam turbines, electric
heavy	9.9	230	−6	0.879	103		motors, industrial circulating systems
hydraulic fluids							
medium	6.5	210	−23	0.871	98	R, O, W	machine tool hydraulic systems
extra low temp	5.1	96	−62	0.859	370	R, O, W, VI, P	aircraft hydraulic systems
aviation							
grade 80	14.7	232	−23	0.887	105	D, P, F	reciprocating aircraft
grade 120	23.2	244	−18	0.893	95		engines
refrigeration							
medium	5.9	198	−37	0.892	73	none	refrigerator compressors and low temperatures

[a] R, rust inhibitor; O, oxidation inhibitor; D, detergent–dispersant; VI, viscosity-index improver; P, pour-point depressant; W, antiwear; EP, extreme pressure; F, antifoam.

of paraffinic, aromatic, and alicyclic (naphthenic) components. Typical properties of commercial petroleum oils are listed in Table 1.

Viscosity. The viscosity of an oil is its stiffness or internal fraction. The wide range of viscosities in commercial petroleum oils is illustrated by some representative types listed in Table 1. Despite this range, by far the largest proportion of oils is in the kinematic viscosity range of 25–75 mm²/s (= cSt) at 40°C (see Rheological measurements).

Viscosity-temperature. The viscosity of oils decreases with increasing temperature. Viscosity index (VI), although empirical, is the most common measure that is applied to the decrease in viscosity of petroleum oils with increasing temperature. Oils having a VI above 80 or 90 generally are desirable. These oils are composed primarily of saturated hydrocarbons of the paraffinic and alicyclic types that give long life, freedom from sludge and varnish, and generally satisfactory operation when they are compounded with proper additives for a given application.

Viscosity-pressure. Lubricating oils appear to drop slightly in viscosity as they are exposed to higher pressures in equilibrium with nitrogen. The thinning effect of the dissolved gas apparently more than offsets the increase in viscosity that normally occurs with pressure increase.

Additives

Lubricant-additive production has grown to a 10^9 segment of the chemical industry. Although the bulk of additives is used in automotive lubricants, they also are used in oils for turbines, diesel and aircraft engines, two-cycle engines, hydraulic equipment, gears, and metal working. The common types of additives in the approximate order of their frequency of application are oxidation inhibitors, rust inhibitors, antiwear agents, detergents-dispersants, pour-point depressants, viscosity-index (VI) improvers and foam inhibitors.

Synthetic Oils

The greatest utility of synthetic oils has been for extreme temperatures. Above about 100–125°C, petroleum oils oxidize rapidly; high viscosity and wax separation generally set a low temperature limit of −20 to −30°C. Outside this range, synthetics are almost a necessity; the same types of additives as those discussed for petroleum oils usually are used. Fire resistance, low viscosity-temperature coefficient, and water

solubility are among the unique properties of synthetic oils. Properties and uses of representative synthetics are listed in Table 2. Selection of appropriate paints, seals, gaskets, hoses, plastics, and electrical insulation is necessary so as to prevent the pronounced solvency and plasticizing action of many of the synthetic oils. Synthetic oils contain organic esters, polyglycols, synthetic hydrocarbons, phosphates, silicones and fluorochemicals.

Greases

A grease is a lubricating oil that is thickened with a gelling agent, eg, a soap (qv). Because of design simplicity, decreased sealing requirements, and less need for maintenance, greases almost universally are given first consideration for ball and roller bearings in electric motors, household appliances, automotive wheel bearings, machine tools, aircraft accessories, and railroad apparatus. They also are used for the lubrication of small gear drives and for many slow-speed sliding applications.

Oils in greases. Essentially the same type of oil is employed in compounding a grease as would normally be selected for oil lubrication. Petroleum oils are used in over 99% of the greases produced and commonly range from SAE 20 to 30.

Although oils derived from many crudes and refined by widely different processes are used in making greases, less highly refined oils and the alicyclic types are the most widely used and are more adaptable to grease compounding.

Thickeners. Common gelling agents are the fatty-acid soaps of lithium, calcium, sodium, aluminum, and barium in concentrations of 8–25 wt%.

Additives. Chemical additives similar to those used in lubricating oils also are added to grease to improve oxidation resistance, rust protection, and extreme-pressure (EP) properties.

Synthetic greases. Although all of the synthetic oils mentioned previously have been used in the formulation of lubricating greases, synthetic production is less than 1% of the total grease market; this reflects the ability of petroleum greases to meet most of the operating requirements of ball and roller bearings.

Mechanical properties. Greases vary in consistency from soap-thickened oils that are fluid at room temperature to brick-type greases that

Table 2. Properties of Representative Synthetic Oils

Type	Viscosity, mm²/s (= cSt)			Pour point, °C	Flash point, °C	Typical uses
	at 100°C	at 40°C	at −54°C			
silicones						
SF-96 (50)[a]	16	37	460	−54	316	hydraulic and damping fluids
organic esters						
MIL-L-7808	3.2	13	12,700	−62	232	jet engines
phosphates						
tricresyl phosphate	4.3	31		−26	240	fire-resistant fluids
Fyrquel 150[b]	4.3	29		−24	236	for die casting, air
Fyrquel 220[b]	5.0	44		−18	236	compressors and hydraulic systems
synthetic hydrocarbons						
Mobil 1[c]	7.3	42		−54	236	auto engines
polyglycols						
LB-300-X[d]	11	60		−40	254	rubber seals
polyphenyl ether						
OS-124[e]	13	373		4	288	radiation resistance and high temperatures
silicate						
Coolanol 45[e]	3.9	12	2,400	−68	188	aircraft hydraulics and cooling
fluorochemical						
11-21[f]	3.7	30		−18	none	oxygen compressors, liquid-oxygen systems

[a] General Electric Co.
[b] Stauffer Chemical Co.
[c] Mobil Oil Corp.
[d] Union Carbide Chemicals Co.
[e] Monsanto Co.
[f] Halocarbon Products Corp.

are so hard they must be cut with a knife. Grade 2 greases are the most commonly used. They generally are sufficiently stiff to provide the mechanical stability necessary to avoid mechanical churning, which would break down the soap fibers, and are adequately soft and oily to supply the lubrication needs of most bearings.

Solid-Film Lubricants

A wide variety of solid lubricants recently has been coming into much more general use for high temperature applications and in vacuum, nuclear radiation, aerospace devices, and other environments that prohibit the use of oils and greases. In such applications, the solid material is interposed as a film between two moving surfaces to prevent metal-to-metal contact.

To minimize friction and wear, an ideal solid-film lubricant should provide low shear strength, strong adhesion to the substrate material, good malleability, complete surface coverage, and freedom from abrasive impurities. Many solid-film lubricants are deficient in two respects. They often have poor wear resistance, and any breaks in the film may not be self-healing. In general, the wide variety of materials in use at one time or another as solid lubricants can be classified as inorganic compounds, chemical conversion coatings, solid organic compounds, or metal films. The most common solid lubricants are indicated in Table 3.

Inorganic compounds. The most important inorganic materials are layer-lattice solids in which the bonding between atoms in an individual layer is by strong covalent or ionic forces and those between layers are relatively weak van der Waals forces. Because of their high melting points, high thermal stabilities in vacuum, low evaporation rates, good radiation resistance, and effective friction-lowering ability, molybdenum disulfide (MoS_2) and graphite are the preferred choices in this group.

Bonded solid-film lubricants. Although a thin film of solid lubricant that is burnished onto wearing surfaces often is useful for break-in operation, use of adhesive binders generally is desirable as their use permits application of coatings 5–20 μm thick by spraying, dipping, or brushing as dispersions in a volatile solvent. Some commonly used bonded lubricant films are listed in Table 4.

Table 3. Common Solid Lubricants

Material	Acceptable usage temperature, °C		Av friction coefficient, f	Remarks
	Min	Max		
molybdenum disulfide, MoS_2	−240	370	0.10–0.25	low f, carries high load, good overall lubricant, can promote metal corrosion
polytetrafluoroethylene (PTFE)	−70	290	0.02–0.15	lowest f of solid lubricants, load capacity moderate and decreases at elevated temp
fluoroethylene–propylene copolymer (FEP)	−70	200	0.02–0.15	low f, lower load capacity than PTFE
graphite	−240	540	0.10–0.30	low f and high load capacity in air, high f and wear in vacuum, conducts electricity
niobium diselenide, $NbSe_2$		370	0.12–0.40	low f, high load capacity, conducts electricity (in air or vacuum)
tungsten disulfide, WS_2	−240	430	0.10–0.20	f not as low as MoS_2, temp capability in air a little higher

Table 4. Commonly Used Bonded Solid-Film Lubricants

Trade name	Generic description		Cure temp, °C	Max use temp, °C	Advantages	Limitations
	Lubricant, pigment	Binder system				
MIL-L-23398 (air drying)	MoS_2, graphite		RT		easy to apply, cure oven not necessary	short wear life, poor adhesion to base metal, and can promote corrosion
Vitrolube 1220 (ceramic bonded)	MoS_2, graphite, Ag	ceramic	540	320	very reliable, best wear life of all SFL at temps to 320°C	must be used on corrosion-resistant metals capable of withstanding cure cycle
Proprietary 425 (polyimide bonded)	MoS_2, Sb_2O_3	polyimide		260	higher temp capability than MIL-L-8937, can be applied to low temp alloys except aluminum	processing techniques must be carefully controlled

Substrate properties. High hardness of the substrate lowers friction. Wear rate of the film also is generally lower with higher substrate hardness. Phosphate undercoatings on steel considerably improve wear life of bonded coatings by providing a porous surface that holds reserve lubricant.

Chemical conversion coatings. Chemical conversion coatings are inorganic compounds that are developed on the surface by chemical or electrochemical action (see Metal surface treatments). One of the best known treatments for steel is phosphating by which the surface is coated with a layer of mixed zinc, iron, and manganese phosphates.

Metal films. In many respects, soft metals are ideal solid lubricants. They have low shear strength, can be bonded strongly to the substrate metal as continuous films, have good lubricity, and have high thermal conductivity. Some of the more promising metals are gallium, indium, thallium, lead, tin, gold, and silver. Application of metal films can be carried out either by electroplating (qv) or vacuum processes, eg, evaporation, sputtering, and ion plating.

Bolt lubricants. Numerous severe-duty, bolting applications, eg, in large, high speed rotating machinery (eg, steam and gas turbines) and nuclear reactors, require that torquing stresses on the bolt assemblies must be accurately determined. Bolting materials and lubricants are selected on the basis of their ability to meet bolting design load requirements while resisting galling under the extreme contact stresses on the threads.

Metalworking Lubrication

Metalworking commonly involves one of two processes: cutting or machining, or deformation to change shape without melting or cutting.

In general, water-base fluids containing $\geq 5\%$ of additives for improved lubrication, rust protection, and better wetting are used at high speeds where cooling is a principal requirement. Oils containing sulfur and chlorine EP additives are more common for slower cutting speeds where galling is a main limitation. For rolling, drawing, and other forming operations, a large variety of proprietary oil-water emulsions and mineral oils that are compounded with both fatty oils and some solid lubricants are used (see Fats and fatty oils). Solid lubricant films are being applied increasingly on cutting and forming tools and dies for longer tool life and to enable higher processing speeds.

Extreme Ambient Conditions

Gas lubrication. Despite its severe limitations, gas lubrication of bearings has received intensive consideration for its resistance to radiation, attainment of high speeds, applicability to temperature extremes, low friction at high speeds, and use of the working fluid (gas) in a machine as its lubricant. A main limitation, however, is the very low viscosities of gases which commonly are less than one-thousandth of that of a lubricating oil.

Gases that have been used for lubrication of bearings are hydrogen, helium, nitrogen, oxygen, uranium hexafluoride, air, carbon dioxide, and argon.

Liquid metals. If operating temperatures rise above 250–300°C (where many organic fluids decompose and water exerts a high vapor pressure), liquid metals may be used, eg, mercury for limited application in turbines; sodium, especially its low melting eutectic with 23 wt% potassium, as a hydraulic fluid and as a coolant in nuclear reactors; and potassium, rubidium, cesium, and gallium in some special uses.

Cryogenic bearing lubrication. Cryogenic fluids, such as liquid oxygen, hydrogen, or nitrogen, are used as lubricants in liquid rocket-propulsion systems, turbine expanders used in liquefaction and refrigeration processes, and pumps to transfer large quantities of liquefied gases.

Lubricants for missile systems. Use of liquid fuels and oxidizers as the propellant system for missiles imposes a special lubrication requirement beyond those of cryogenic applications; the lubricant must be inert, ie, insoluble and unreactive, to both the fuels and oxidizers. Solid-film lubricants on linkages, PTFE liners for gimbals, and ball bearings and self-lubricating films commonly are used (see Fluorine compounds, organic).

Nuclear radiation effects. Radiation damage to petroleum oils and greases in a nuclear plant results primarily from ionization as a secondary effect from bombardment of lubricant molecules by fast neutrons, slow (or thermal) neutrons, and rays emanating either directly or indirectly from the nuclear reactor. Some greases with radiation-resistant oils and stable thickeners, eg, polyurea, copper phthalocyanine, sodium amate, calcium complex soaps, indanthrene dyes, or clay particles appear capable of withstanding doses of 10^7 Gy (10^9 rad) while maintaining satisfactory consistency for lubrication purposes.

Lubrication with glass. Softening glass (qv) is used as a lubricant for extrusion, forming, and other hot-working processes with steel and nickel-base alloys at about 1000°C, for extrusion and forging titanium, and less frequently for extruding copper alloys. Principal types of glasses are pure fused silica, 96% silica, soda-lime glass, lead alkali silicates, borosilicates, and aluminosilicates.

Environmental and Health Factors (Toxicology)

Considerable pressure has developed in recent years from the EPA and similar governmental organizations to discontinue disposal of lubricants in streams, chemical dumps, and other environmental channels. Consequently, changes being made in practices have resulted in 43% of waste oil being burned, 18% used for dust control and road oils, and about 10% being recycled (see Recycling).

Toxic and hazardous constituents. Questionable constituents of lubricating oils are polynuclear aromatics (some of which are carcinogens) in the oil plus many additives.

E.R. BOOSER
General Electric Co.

D.F. Wilcock and E.R. Booser, *Bearing Design and Application*, McGraw-Hill Book Co., New York, 1957.

E. Rabinowicz, *Friction Wear of Materials*, John Wiley & Sons, Inc., New York, 1965.

International Conference on Solid Lubrication, American Society Lubricating Engineers, Park Ridge, Ill., 1978.

LUMINESCENT MATERIALS

PHOSPHORS

Luminescence is broadly defined as the emission of electromagnetic radiation in excess of thermal radiation. Luminescent materials are called phosphors. Phosphors are generally prepared as powders and contain one or more impurity ions or activators present in 1–5 mol%. Perhaps fewer than one hundred phosphors are commercially manufactured on a large scale.

Most phosphors are excited by high energy photons or electrons. In addition, there are phosphors that are excited by infrared radiation (ir to visible conversion) (see Infrared technology), electric fields (electroluminescence), chemical reactions (chemiluminescence) (qv), and mechanical stress (triboluminescence).

Although phosphors are used in a variety of applications, including fluorescent lamps, cathode-ray and television screens, display devices (see Digital displays), x-ray-intensifying screens, stamp marks, fabric brighteners (see Brighteners, fluorescent) etc, in this article only phosphors primarily used for lighting are discussed.

Luminescence theory. Absorption, relaxation, and emission processes occurring in a phosphor are often schematically illustrated in a configuration coordinate diagram, such as shown in Figure 1.

After absorbing a photon, the impurity center undergoes severe anharmonic vibrational motion. The weak electromagnetic radiation that occurs during this relaxation is called hot luminescence. Ordinary luminescence takes place after the excited state has relaxed and established thermal equilibrium, emission occurring from the lowest vibrational levels of the excited state. Examples of broad-band-absorbing and -emitting phosphors are those than undergo interconfigurational elec-

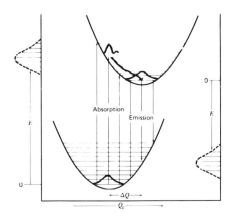

Figure 1. General configuration-coordinate diagram for band absorbers and emitters. The ordinate represents the total energy of the activator center and the abscissa Q is a displacement of the atoms surrounding the defect; ΔQ is the difference between the ground- and excited-state minima. On the left side is the absorption spectrum and on the right side is the emission spectrum.

tronic transitions in absorption or emission. Important broad-band activators include Sb(III), Pb(II), Bi(III), and Sn(II) which undergo $(ss) \rightleftharpoons (sp)$ transitions; Eu(II) and Ce(III) which have $f \rightleftharpoons d$ transitions; and Cu(II) which undergoes $s \rightleftharpoons d$ transitions.

In sharp contrast to the broad-band adsorbers are the intraconfiguration f-f transitions among the rare-earth and actinide activators and certain spin-flip intraconfiguration transitions in transition-metal activators. For these transitions $\Delta Q \approx 0$, that is, there is little difference in the size or shape of the defect in its ground and excited states. The absorption and emission occur mainly to and from the no-phonon line. The width of the narrow band or line is determined mainly by inhomogeneous broadening, ie, different activator-ion environments. Weak vibrational structures may accompany a narrow-band absorption or emission.

Nonradiative decay. To have any technical importance, luminescent material should have a high quantum efficiency, the ratio of the number of quanta emitted to the number of quanta absorbed. Phosphors for use in fluorescent lamps usually have quantum efficiencies of at least 0.75. All the quanta absorbed would be reemitted if there were no nonradiative losses. Nonradiative processes are illustrated in Figure 2.

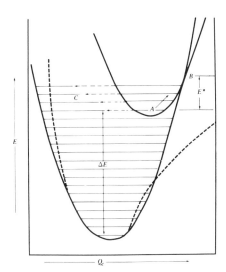

Figure 2. A configuration–coordinate diagram showing mechanisms of radiationless decay to the ground state. Nonradiative decay to the ground-state vibrational manifold can occur via the paths $A \to B$ or $A \to C$. The ordinate represents the total energy of the activator center and the abscissa Q_c is a displacement of the atom surrounding the defect. The dashed line represents a Morse curve ground-state potential.

Energy transfer. In addition to either emitting a photon or decaying nonradiatively to the ground state, an excited defect center may also transfer energy to another center either radiatively or nonradiatively. The ion that transfers energy is called a sensitizer (see also Dyes, sensitizing).

Specific Phosphors

Halophosphated phosphors. Since their discovery in the 1940s, the calcium halophosphate phosphors, $Ca_5(PO_4)_3(F, Cl):Sb, Mn$ remain the most extensively used phosphors in fluorescent lamps. An analogous structural system exists with strontium replacing calcium in the apatite structure, $Sr_5(PO_4)_3(F, Cl)$. Commercial phosphors in the strontium system are usually fluorophosphates.

New Phosphor Blends

Because of an increasing awareness of the need for efficient, ie, energy-saving lamps and the development of rare-earth-activated phosphors, new multicomponent systems have been introduced and are currently receiving an increasing emphasis in the lighting industry. For instance, a two-component phosphor mixture, ie, a calcium fluorophosphate phosphor with a high manganese concentration plus strontium chloroapatite: Eu(II), is designed to approximate more closely an optimum spectral-power distribution (SPD). Although the SPD gives the same color as the cool-white calcium halophosphate phosphor visually, the phosphor blend increases the light output by 7%. Similarly, a blend of three narrow-line-emitting phosphors with emissions in the blue (ca 450 nm), green (ca 550 nm) and red (610 nm) gives very good color rendition (overcoming the objection that fluorescent lamps badly distort colors) at efficiencies approaching 50% higher than previous "deluxe" blends.

Phosphors and phosphor systems also are designed to optimize the color and output of high pressure mercury-vapor lamps which have significant radiation in the near uv and visible mercury lines. These phosphors must respond to the near-ultraviolet region and must be efficient at the high operating temperatures of the bulb wall of these lamps. Currently, the most common phosphor for improving the color of high pressure mercury lamps is the red-emitting yttrium vanadate phosphate: Eu(III).

Rare-Earth Phosphors for Lighting

The largest area of advancement in phosphor materials has been the use of rare earths as activators, which has permitted the use of highly specific spectral-power distribution with the efficacy required for commercial fluorescent lamps. Because of the demand for rare-earth phosphors for television, x-ray-excited materials, and fluorescent lamps, the chemical technology for preparation and purification of these compounds has been improved greatly, providing consistent and reliable materials in purity levels required for efficient phosphor absorption and emission.

The most important rare-earth activators are, Ce(III), Eu(II), Eu(III), and Tb(III). A recently developed family of structure types composed of aluminates has provided the largest and most versatile group of phosphors utilizing rare earths as activators. These are hexagonal aluminates which can provide hosts for efficient emission of Tb(III), Ce(III), Eu(II), and Eu(II) with Mn(II) for activator–sensitizer combinations.

Other Eu(II)-activated phosphors. The Eu(II) ion has been useful in providing efficient absorption and emission in several other structures, mostly based on strontium compounds. Commercially important phosphors include $SrB_4O_7:Eu(II)$, $Sr_5(PO_4)_3(Cl):Eu(II)$, $Sr_2P_2O_7:Eu(II)$, $SrMgP_2O_7:Eu(II)$, $Sr_3(PO_4)_2:Eu(II)$, and $Ba_2MgSi_2O_7:Eu(II)$.

Phosphors for Special Application

Infrared-to-visible phosphors. With some combinations of rare earths, infrared energy can be absorbed in a multiple-step excitation process and emitted in a single step in the visible region. The most efficient of these contain Yb as an absorber and Er for green emission or Tm for blue emission, usually in a host lattice of YF_3. Typical compositions are $Y_{0.79}Yb_{0.20}Er_{0.01}F_3$, $La_{0.86}Yb_{0.12}Er_{0.02}F_3$, and $Y_{0.639}Yb_{0.35}Tm_{0.001}F_3$. Fluorides or oxysulfides have also been used to provide a suitable host lattice for the other emitters.

X-ray-excited phosphors.   Another application of rare-earth-activated phosphors is in x-ray screens that convert x-ray radiation, such as used in medical units, to ultraviolet or visible light which then sensitizes a photographic film. The following combinations have been used for this purpose: $GdOS : Tb(III)$, $LaOS : Tb(III)$, $LaOBr : Tb(III)$, $LaOBr : Tm(III)$, and $Ba(F,Cl)_2 : Eu(II)$.

Phosphors excited by rare-gas discharges.   Phosphors are also used in combination with very small rare-gas discharge lamps that are currently used primarily for display devices. A suitable green-emitting phosphor is $YPO_4 : Tb(III)$ (see Helium-group gases).

Health and Safety Factors

During the manufacture of phosphors, controls on particulate emissions are used on furnace ventilation and on grinding equipment. The most hazardous material present in phosphors is cadmium.

THOMAS F. SOULES
MARY V. HOFFMAN
General Electric Company

P. Goldberg, ed., *Luminescence of Inorganic Solids*, Academic Press, Inc., New York, 1963.

B. DiBartolo, *Optical Interactions in Solids*, John Wiley & Sons, Inc., New York, 1968.

FLUORESCENT DAYLIGHT

There are many types of luminescent materials, some of which require a special source of excitation such as an electric discharge or ultraviolet radiation.

Daylight-fluorescent pigments, in contrast, require no artificially generated energy. Daylight, or an equivalent white light, can excite these unique materials not only to reflect colored light selectively, but to give off an extra glow of fluorescent light, often with high efficiency and surprising brilliance (see also Dyes, sensitizing; Pigments).

Daylight-fluorescent pigments, with a few exceptions, consist of particles of colorless resins containing dyestuffs that not only have color but are capable of intense fluorescence in solution.

A fluorescent substance is one that absorbs radiant energy of certain wavelengths and, after a fleeting instant, gives off part of the absorbed energy as quanta of longer wavelengths. In contrast to ordinary colors in which the absorbed energy degrades entirely to heat, light emitted from a fluorescent pigment adds to the light returned by simple reflection to give the extra glow characteristic of daylight-fluorescent materials.

Important dyestuffs used for daylight-fluorescent pigments include Rhodamine B, Rhodamine F5G (BASF), Xylene Red B (Sandoz Chemical), Fluorescent Yellow Y (L.B. Holliday), Maxilon Brilliant Flavine 10GFF (CIBA-GEIGY), Alberta Yellow, Potomac Yellow (Day-Glo Color), and Macrolex Fluorescent Yellow 10GN (Bayer).

Structure

Virtually all important dyes contain aromatic rings in their structures. According to the theory of chromophores and auxochromes, a number of groups such as

$$-N{=}N-, \quad {>}C{=}C{<}, \quad {>}C{=}O, \quad -N{=}O, \text{ and } -NO_2,$$

so-called chromophores, have to be present on benzenoid rings in order for compounds to have appreciable light absorption or color. Certain basic groups, so-called auxochromes, chiefly $-NH_2$, $-NHR$, $-NR_2$, where R is alkyl or the phenolate oxygen, are necessary in addition to bring out or intensify the color.

The electronic theory of atoms and molecules, wave mechanics, the theories of valence-bond resonance and molecular orbitals have increased the understanding of colored and fluorescent substances and their interaction with light. The benzene ring with its six π electrons can act in conjunction with electron-donating groups (auxochromes), and only some of the electron-accepting (chromophoric) groups, to produce strong ab-

sorption in the ultraviolet or visible regions to give rise to fluorescence. Such a system of atoms, responsible for significant absorption of photons in the uv or visible regions, is referred to as a chromogen. A chromogen that absorbs in the uv as a rule can be modified chemically to absorb visible light, thus becoming colored. This is often accomplished by adding benzene rings to the molecule or introducing an unsaturated chain of atoms.

Chromogens.   Organic dyes can be divided into four classes, depending on the type of chromogen or unsaturated system present: (1) $n \to \pi^*$ chromogens, (2) donor-acceptor chromogens, (3) cyanine-type chromogens (see Cyanine dyes), and (4) acyclic and cyclic polyene chromogens. Almost all strongly fluorescent dyes fall into classes (2) and (3), whereas only a few have cyclic polyene chromogens of groups (4). The chromogens of class (1) are detrimental to fluorescence. Nitro and azo compounds generally have little or no fluorescence.

Electronic States

Dye molecules in the ground state.   At RT and in the absence of exciting light, the electrons of a dye molecule are in their lowest energy states, referred to as ground singlet states. Electrons are in constant motion in the neighborhood of the atomic nuclei, primarily in certain regions referred to as orbitals. In the process of combining atoms to form an organic compound, the orbitals blend and modify their shapes and electron densities, attaining the lowest energy patterns for the interrelated system of atoms of the molecule.

Energy levels and light absorption.   Figure 1 shows the typical transition between various energy states that the electrons of a dye molecule can undergo. The singlet ground state of the electrons in the molecule is designated S_0 and represents the lowest electronic energy level possible for the molecule. The molecule can be excited to higher electronic states such as S_1 or S_2 with an associated set of vibrational energy levels represented by a series of lines above the particular electronic level.

In Figure 1, after absorption A and vibrational deactivation VD occur, the lowest or nearly lowest level of the singlet excited state S_1 is reached.

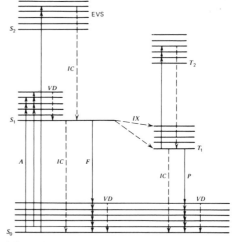

Figure 1.   Schematic energy-level diagram for a dye molecule.

Electronic states	*Transitions*
S_0 = ground singlet state	A = absorption to excited states
S_1 = first excited singlet state	VD = vibrational deactivation
S_2 = second excited singlet state	IC = internal conversion
T_1 = first excited triplet state	F = fluorescence
T_2 = second excited triplet state	IX = intersystem crossing
EVS = excited vibrational states	P = phosphorescence

Lifetime, s
10^{-15}
10^{-13} to 10^{-11}
10^{-13} to 10^{-11}
10^{-8}
10^{-8} to 10^{-7}
10^{-4} to several

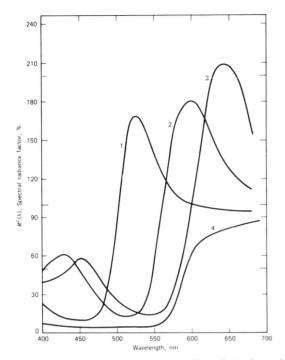

Figure 2. Curves 1, 2, and 3 show the spectral radiance factor for equivalent coatings of separate toluenesulfonamide–melamine–formaldehyde Day-Glo pigments containing 0.5% of a dye, either Alberta Yellow, Rhodamine F5G, or Rhodamine B Extra. Curve 4 is for a bright nonfluorescent red-orange printing ink. The illuminant was Source C. A magnesium oxide-coated block was used as a comparison white.

If the molecule is fluorescent with a high quantum efficiency, fluorescent emission of a quantum of light generally occurs, indicated by F.

Color Formation

Spectral-energy-ratio curves. Figure 2 shows the spectral-energy-ratio curves of three daylight-fluorescent dyes in pigment drawdowns and a curve for a nonfluorescent ink. The lower left part of each of the first three curves is essentially the same as the transmittance or reflectance spectrum of the dye. With a strongly fluorescent substance, most of the absorbed energy is stored in the S_1 excited state and is largely given off as fluorescent light of longer wavelengths covering a considerable range.

Effect of two or more dyes. A most remarkable effect in daylight fluorescence, the transfer of energy from one fluorescent dye to another, both dissolved in the same polymer, can be made use of to produce colors more brilliant for their particular spectral regions than one dye alone could exhibit.

Manufacture

The majority of daylight-fluorescent pigments are produced by bulk polymerization of the dye-carrying polymer. Today, most fluorescent pigment media are of the toluenesulfonamide–melamine–formaldehyde resin matrix type (see Amino resins). Special ester and amide polymers are also utilized.

Polymerization of a dyed polar prepolymer as the internal phase in an oil-based external phase gives a fluorescent ink-base in which spherical fluorescent particles are dispersed. This base is suitable for litho and letterpress inks (see Inks). In recent years, several pigment manufacturers have also developed fluorescent polymers intended to be used in flexo-solutions for application to various substrates. The following companies are the principal producers of fluorescent pigments Dane and Company, Day-Glo Color Corp., Lawter Chemical Corp., Nippon KK, Radiant Color Division of Magruder Color Co., Inc., Sinloihi Division of Dainippon Paint, and Sterling Color Division of ICI.

Health and Safety Factors

Good safety practices are recommended when handling fluorescent pigments, including a respirator and dust-collection equipment. Daylight-fluorescent pigments are considered to be nontoxic. Because they are combinations of polymers and dyestuffs, the combined effect of the ingredients must be taken into account when considering any net toxic effect of these materials.

Uses

Fluorescent colors are remarkable for their extremely high visibility and their ability to attract attention. It is in applications utilizing these properties where fluorescent color has gained the greatest acceptance, eg, plastics, paint, gravure inks, flexographic inks, silk-screen inks, lithographic and letterpress printing inks, and vinyl products.

RICHARD A. WARD
EDWARD L. KIMMEL
Day-Glo Color Corp.

U.S. Pat. 2,938, 873 (May 31, 1960), Z. Kazenas (to Switzer Brothers, Inc.).
U.S. Pat. 3,412, 036 (Nov. 19, 1968), M.D. McIntosh (to Switzer Brothers, Inc.).

M

MAGNESIUM AND MAGNESIUM ALLOYS

Magnesium, Mg, at no. 12, is in Group IIA of the periodic table of the elements between beryllium and calcium. It has a valence of 2 and occurs as three isotopes with mass numbers 24, 25, and 26 existing in the relative frequencies of 77, 11.5, and 11.1%, respectively. The electron arrangement in free atoms is $1s^2\, 2s^2\, 2p^6\, 3s^2$ (2) (8) 2.

Magnesium occurs widespread in nature in the ores dolomite, magnesite, and carnallite, and as the chloride in seawater, underground natural brines, and salt deposits. Metallic magnesium is produced by electrolysis of molten magnesium chloride or thermal reduction of the oxide.

Substantial quantities of magnesium-base scrap are recycled annually in the United States. The main sources of scrap are automotive components, primarily Volkswagen crankcase and transmission housings, etched photoengraving sheets, chain-saw housings, scrap from wrought-product manufacture and fabrication, general die-cast scrap, magnesium turnings, and sludges from various magnesium-melting operations. Products obtained from secondary magnesium include magnesium alloy ingot for die-cast and foundry use, cast anodes for cathodic protection, and ingot for aluminum- and zinc-base alloys.

Metallic magnesium is silvery white. With a specific gravity of 1.74, it is the lightest structural metal. For engineering applications, it is alloyed with one or more elements of a group that includes aluminum, manganese, rare-earth metals, lithium, silver, thorium, zinc, and zirconium, to produce alloys having very high strength-to-weight ratios and the highest strength-to-weight ratio of the ultralight metals at elevated temperatures.

Magnesium alloys are available in practically all the usual metal forms, including cast ingots, slabs, and billets; sand, permanent-mold, dies, and investment castings; forgings; extruded bars, rods, tubes, structural and special hollow and solid shapes; and rolled sheet and plate. Magnesium alloys are used widely in a great variety of applications.

Properties

Table 1 gives the physical properties of 99.9% pure magnesium.

Reactions. Although magnesium is high in the electromotive series, at an electronegative standard potential of -2.4 V, it is resistant to atmospheric attack and certain chemical media because of its ability to acquire a protective film (oxide, sulfate, chromate, fluoride, etc). It is relatively easy to ignite fine powders or thin foils of magnesium with the production of a dazzling, strongly actinic light. Metallic magnesium reacts slowly with water. Under ordinary atmospheric conditions or in pure or salt-free water of high pH, the reaction is terminated because of the formation of an insoluble hydroxide film:

$$Mg + 2\,H_2O \rightarrow Mg(OH)_2 + H_2$$

The high chemical activity of magnesium is indicated by its position in the electromotive series. It can displace zinc from solutions of zinc salts; similar reactions take place with Fe^{3+}, H^+, and Hg^{2+}. In neutral or basic solutions these reactions do not proceed at any significant rate. However, in relatively concentrated acid solutions, or even in weak solutions of some acids such as hydrochloric acid, the reactions occur freely because the acid prevents the formation of hydroxide coatings. Magnesium is attacked readily by all acids except hydrofluoric acid.

If ignited, an intimate mixture of magnesium and iron oxide reacts vigorously with the production of heat:

$$3\,Mg + Fe_2O_3 \rightarrow 3\,MgO + 2\,Fe$$

Metallic magnesium combines directly with free halogens. The reaction takes place in certain organic solvents or at elevated temperatures in the absence of oxygen or nitrogen, which would form the oxide or the nitride.

Table 1. Physical Properties of Magnesium, 99.9% Pure

Property	Value
at vol, cm^3/mol	14.0
crystal structure	close-packed hexagonal
lattice parameters, nm	$a_0 = 0.3203$, $c_0 = 0.5199$
axial ratio, $c : a$	1.623
modulus, GPa^a	
of elasticity	44.8
of rigidity	16.6
Poisson's ratio	0.35
density at 20°C, g/cm^3	1.74
volume contraction, 650°C (liq) to 650°C (solid), %	4.2
linear contraction, 650°C (solid) to 20°C (solid), %	1.8
electrical resistivity at 20°C, $\mu\Omega\cdot cm$	4.46
temperature coefficient at 20°C, $\mu\Omega\cdot cm/°C$	0.017
electrical conductivity at 20°C^b	
mass %	198.0
vol %	38.6
mp, °C	650
bp, °C	1110 ± 10
critical temperature^c, °C	1867
flame temperature^d, air, °C	2820
coefficient of expansion at 20–100°C,	
coefficient of thermal expansion/°C	0.0000261
specific heat at 20°C, $J/(g\cdot°C)^e$	1.02
thermal conductivity, $W/(cm\cdot K)$, at 20°C	1.55
thermal diffusivity, cm^2/s, at 20°C	0.87
heat of combustion, $kJ/(mol)^e$	606
latent heat of fusion, J/g^e	368
latent heat of evaporation, J/g^e	5.27
heat of reaction with oxygen, at 2900 K, 101.3 kPa^f magnesium	
kJ/kg^e metal	19,027
kJ/kg^e O_2	28,912
thermal neutron absorption cross section $m^2/atom^g$ (cm^2/cm^3)	0.059×10^{-28} (0.00254)

^{a}To convert GPa to psi, multiply by 145,000.
^{b}Annealed copper standard.
^{c}Calculated.
^{d}Theoretical.
^{e}To convert J to cal, divide by 4.184.
^{f}To convert kPa to atm, divide by 101.3.
^{g}To convert m^2 to barn, divide by 10^{-28}.

Grignard reagents are an important class of compounds made from magnesium and an organic halide, usually in an ether under complete exclusion of water. These reagents are used extensively in organic synthesis (see Grignard reaction).

Corrosion characteristics. Magnesium exhibits good stability in the presence of a variety of dry and liquid chemical substances. In general, the chemical behavior of magnesium is considered to be the opposite of that of aluminum.

The rate of corrosion of magnesium in aqueous solutions is strongly influenced by the hydrogen-ion concentration or pH. Magnesium is resistant to alkalies but is attacked by acids that do not promote the formation of insoluble films. In regard to salts, neutral or alkaline fluorides form insoluble magnesium fluoride, and consequently, magnesium alloys are resistant to them. Chlorides are usually corrosive even in solutions having pH values above that required to form magnesium hydroxide. Acid salts are generally destructive but chromates, vanadates, and phosphates form films that usually retard corrosion except at elevated temperatures. Most mineral acids attack magnesium rapidly. Hydrofluoric acid, except at low concentration and elevated temperature, is an exception to the rule. Chromic acid has a low rate of attack except when chlorides or sulfates are present. Most organic acids attack magnesium alloys readily. Organic compounds normally cause little or no corrosion of magnesium. Magnesium is not attacked seriously by dry chlorine, iodine, bromine, or fluorine gas. Some tests indicate that magnesium alloys are resistant to loam soil.

Resistance to tarnishing outdoors is increased by the addition of up to 9% aluminum. More than 5% zinc decreases tarnishing resistance. With only 9.5% humidity, magnesium alloys have remained untarnished for eighteen months.

The working electrode potential of primary magnesium is of the order of only 1.4 V vs the standard hydrogen electrode (SHE) because of voltage (IR) losses due to the protective magnesium hydroxide film.

The rate of oxidation of magnesium alloys increases with time and temperature.

Producers

Currently, there are five magnesium producers in the United States: Titanium Metals Corporation, Oregon Metallurgical Corporation, Northwest Alloys, The Dow Chemical Company and Amax Magnesium Corporation.

Health and Safety Factors

A magnesium powder or dust ignites readily. If ignited while in suspension in air in concentrations above the lower explosive limit of 0.04 g/L, powder will explode violently. Magnesium is essential to most plant and animal life; dietary deficiency rather than toxicity has been the main problem (see Mineral nutrients).

Alloys

The ASTM designations for alloys are based on chemical composition, and consist of two letters representing the two alloying elements specified in the greatest amount, arranged either in decreasing percentages or alphabetically, if of equal percentage. The letters are followed by the respective percentages rounded off to whole numbers, with a serial letter at the end. The serial letter indicates some variation in composition. Experimental alloys have the letter X between the alloy and serial numbers.

Table 2 shows the chemical compositions and physical properties of the magnesium alloys used most commonly in cast form. The chemical compositions and physical properties of selected wrought commercial magnesium alloys are given in Table 3.

Most commercial magnesium alloys are either of the magnesium-rich, solid-solution, or hypoeutectic type with intermediary phases as second

Table 2. Nominal Chemical Composition and Physical Properties of Magnesium Casting Alloys

Form	Alloy ASTM	Alloy UNS	Temper[a]	Al	RE[b]	Ag	Th	Zn	Zr	Other	Density at 20°C, g/cm³	Melting point, °C	TC[d], W/(m·K)	ER[e], μΩ·cm
sand and permanent-mold castings	AZ63A	M11630	−T4	6.0				3.0			1.82	610	50.3	14.0
			−T6										62.8	11.8
	AZ91C	M11914	−T4	8.7				0.7			1.80	596	46.1	17.0
			−T6										54.6	14.5
	AZ92A	M11920	−T4	9.0				2.0			1.82	593	46.1	16.8
			−T6										58.6	12.4
permanent-mold castings	AM100A	M10100	−T4	10.0						0.2 Mn	1.80	593	41.9	17.2
			−T6										58.6	12.4
die casting	AZ91B	M11912	−F	9.0				0.6			1.80	596	54.6	12.8

[a] Temper designations: −T4, solution heat treated and naturally aged to a substantially stable condition; −T6, solution heat treated and then artificially aged; and −F, as fabricated.
[b] Rare earths.
[c] Balance Mg.
[d] Thermal conductivity, at 20°C.
[e] Electrical resistivity, at 20°C.

Table 3. Nominal Chemical Compositions and Physical Properties of Wrought Magnesium Alloys

Form	Alloy ASTM	Alloy UNS	Temper[a]	Al	Mn	Th	Zn	Zr	Density at 20°C, g/cm³	Melting point, °C	TC[c], W/(m·K)	ER[d] μΩ·cm
sheet and plate	AZ31B	M11311	−H24	3.0			1.0		1.77	627	75.4	9.2
			−H26	3.0			1.0		1.77	627	75.4	9.2
			−O	3.0			1.0		1.77	627	75.4	9.2
	HK31A	M13310	−H24			3.0		0.7	1.79	649	113.1	6.1
			−O			3.0		0.7	1.79	649	104.7	6.6
extruded bars, rods, and solid shapes	AZ31B	M11311	−F	3.0			1.0		1.77	627	75.4	9.2
	HM31A	M13312	−T5		0.1	3.3		0.7	1.81	605	104.7	6.6
	ZK60A	M16600	−F				5.7	0.55	1.83	635	117.3	6.0
			−T5				5.7	0.55	1.83	635	121.4	5.7
extruded hollow shapes and tubes	AZ31B	M11311	−F	3.0			1.0		1.77	627	75.4	9.2
	ZK60A	M16600	−F				5.7	0.55	1.83	635	117.3	6.0
			−T5				5.7	0.55	1.83	635	121.4	5.7

[a] Temper designations: −H24, strain hardened and then partially annealed to $\frac{1}{2}$ hardness; −H26, strain hardened and then partially annealed to $\frac{3}{4}$ hardness; −O, annealed recrystallized, −F, as fabricated; and −T5, artificially aged only.
[b] Balance Mg.
[c] Thermal conductivity, at 20°C.
[d] Electrical resistivity at 20°C.

constituents. Magnesium alloys are fabricated by common methods, including melting followed by casting, rolling, extrusion, and forging. Further fabrication includes forming, joining, and machining after which stan·lard assembly methods are used.

Uses

Nonstructural.   Magnesium is an important constituent in aluminum alloys and improves the properties and stability of zinc die-casting alloys. It acts as a reducing agent in the manufacture of titanium and zirconium, and as a desulfurization agent of iron and steel. In the manufacture of ductile cast iron, magnesium spherodizes the graphite to provide improved ductility and strength.

Owing to its high potential energy, magnesium is utilized in explosives and many pyrotechnic devices, including flares and various incendiary devices (see Pyotechnics; Chemicals in war).

Galvanic anodes made of magnesium increase the life of water heaters, underground pipelines, tank bottoms, and tower footings. Magnesium alloys are also used in dry and reserve-cell type batteries (see Batteries).

Structural.   Structural uses of magnesium generally take advantage of the weight savings over other metals and alloys, especially in aircraft and aerospace applications, and military and electronic equipment.

L.F. LOCKWOOD
G. ANSEL
P.O. HADDAD
Dow Chemical U.S.A.

Metals Handbook—Properties and Selection: Non-ferrous Alloys and Pure Metals, 9th ed., Vol. 2, American Society for Metals, Metals Park, Ohio, 1979, pp. 525–609.

C.S. Roberts, *Magnesium and Its Alloys*, John Wiley & Sons, Inc., New York, 1960.

Magnesium: Light, Strong, Versatile, International Magnesium Association, Dayton, Ohio, June 1978.

MAGNESIUM COMPOUNDS

Magnesium is the eighth most abundant element in the earth's crust and the third most abundant element in seawater. More than sixty magnesium-containing minerals are known, the most important of which can be divided into three classes according to their commercial importance: the carbonates; the salts and double salts; and the silicates. Magnesium also occurs as the hydroxide brucite, $Mg(OH)_2$, and in combination with aluminum as spinel, $MgAl_2O_4$. Magnesium readily forms salts with mineral acids. Anhydrous and hydrated salts are common. In addition, a whole range of organometallic compounds is available, including salts of carboxylic acids, alkyls, Grignard reagents, and alkoxides. Magnesium displays a great propensity for the formation of double salts.

Magnesium Acetate

Anhydrous magnesium acetate occurs in two forms: α-$Mg(C_2H_3O_2)_2$, and β-$Mg(C_2H_3O_2)_2$. The commercial product is the tetrahydrate, $Mg(C_2H_3O_2)_2.4H_2O$. Physical properties of magnesium acetate and its hydrates are given in Table 1.

Table 1.   Physical Properties of Magnesium Acetate and Its Hydrates

Property	α-Mg-$(C_2H_3O_2)_2$	β-Mg-$(C_2H_3O_2)_2$	$Mg(C_2H_3O_2)_2.$ H_2O	$Mg(C_2H_3O_2)_2.$ $4H_2O$	β-$Mg(C_2H_3O_2)_2.$ $4H_2O$
mol wt	142.36	142.36	160.38	214.42	214.42
crystal system	orthorhombic	triclinic	orthorhombic	monoclinic	monoclinic
density, g/cm³					
calculated	1.524		1.553	1.453	1.545
observed	1.507	1.502		1.454	
mp, °C	323 dec			80	
color	white	white			

Uses

By far the largest use for magnesium acetate is in the production of rayon (qv) fiber, which is used for cigarette filter tow (see Cellulose acetate and triacetate fibers).

Magnesium Alkyls

Magnesium alkyls are white, crystalline, pyrophoric solids that react vigorously with water, alcohols, and other active hydrogen compounds.

Uses.   Magnesium alkyls are used as polymerization catalysts for olefins and dienes, especially butadiene, often in combination with aluminum alkyls and the transition-metal halides.

Magnesium Carbonate

Magnesite is widely distributed throughout the earth's crust and is used for the preparation of magnesia and magnesium compounds. The physical properties of the normal carbonates are given in Table 2, those of the basic carbonates in Table 3.

Table 2.   Physical Properties of Magnesium Carbonate and Its Hydrates

Property	$MgCO_3$, magnesite	$MgCO_3.2H_2O$, barringtonite	$MgCO_3.3H_2O$, nesquehonite	$MgCO_3.5H_2O$, lansfordite
mol wt	84.32	120.35	138.37	174.40
crystal system	trigonal	triclinic	monoclinic	monoclinic
calculated density, g/cm³	3.009	2.825	1.837	1.73
mp, °C	350 dec			
index of refraction	1.510, 1.700	1.458, 1.473, 1.501	1.412, 1.501, 1.526	1.456, 1.476, 1.502
color	white	colorless	colorless to white	white
heat of formation, ΔH°_{298}, kJ/mol[a]	−1095.8			

[a] To convert J to cal, divide by 4.184.

Table 3.   Physical Properties of Basic Magnesium Carbonates

Property	$MgCO_3.$ $Mg(OH)_2.$ $3H_2O$, artinite	$4MgCO_3.$ $Mg(OH)_2.$ $4H_2O$, hydromagnesite	$4MgCO_3.$ $Mg(OH)_2.$ $5H_2O$, dypingite	$4MgCO_3.$ $Mg(OH)_2.$ $8H_2O$,
mol wt	196.70	467.67	485.69	539.74
crystal system	monoclinic	monoclinic	monoclinic	
calculated density, g/cm³	2.039	2.254		
index of refraction	1.488, 1.534, 1.556	1.523, 1.527, 1.545	1.508, 1.510, 1.516	1.515, 1.521, 1.522
color	white	white	white	white–grey

Magnesium carbonate forms many double salts, including $MgCO_3.$ $MgCl_2.7H_2O$, $2MgCO_3.MgBr_2.8H_2O$, $MgCO_3.MgBr_2.7H_2O$, $MgCO_3.$ $NH_4CO_3.4H_2O$, $MgCO_3.K_2CO_3.8H_2O$, $MgCO_3.KHCO_3.4H_2O$ (Engel's salt), $MgCO_3.Rb_2CO_3.4H_2O$, and $MgCO_3.Na_2CO_3$.

Uses.   Calcined magnesite and dolomite are used in the manufacture of basic refractory bricks and fettling materials for the steel industry. Precipitated magnesium compounds in the paint and printing inks industries as well as in the manufacture of fireproofing, fire-extinguishing, flooring, and polishing compounds and as fillers (qv) and smoke suppressants in the plastics and rubber industries. Magnesium carbonate is used as an additive to table salt to keep it free flowing, a bulking compound in powder formulations, and an antacid (see Gastrointestinal agents).

Magnesium Chloride

Magnesium chloride, one of the most important magnesium compounds, forms hydrates with 2, 4, 6, 8, and 12 molecules of water, but only the hexahydrate is of commercial importance. The physical properties of anhydrous $MgCl_2$ and the hexahydrate bischofite are given in Table 4.

Table 4. Physical Properties of Magnesium Chloride and Its Hexahydrate

Property	MgCl$_2$	MgCl$_2$.6H$_2$O, bischofite
mol wt	95.22	203.31
crystal system	hexagonal	monoclinic
density, g/cm^3		
calculated	2.333	1.585
observed	2.325	1.56
mp, °C	708	116–118 dec
index of refraction	—, 1.675, 1.59	1.498, 1.505, 1.525
color	white lustrous	colorless
heat of formation, ΔH_{298}°, kJ/mol^a	−641.3	−2499.0

^{a}To convert J to cal, divide by 4.184.

Magnesium Bromide

Magnesium bromide, MgBr$_2$, occurs in seawater, brines, inland seas and lakes, and salt deposits. The physical properties of MgBr$_2$ and MgBr$_2$.6H$_2$O are given in Table 5.

Table 5. Physical Properties of Magnesium Bromide and Magnesium Bromide Hexahydrate

Property	MgBr$_2$	MgBr$_2$.6H$_2$O
mol wt	184.13	292.22
crystal system	hexagonal	monoclinic
density, g/cm^3		
calculated	3.855	2.076
observed	3.722	
mp, °C	711	174.2
heat of formation, ΔH_{298}°, kJ/mol^a	−524.3	−2410.0

^{a}To convert J to cal, divide by 4.184.

Magnesium Iodide

The physical properties of anhydrous magnesium iodide, MgI$_2$, and of the hexa- and octahydrates are shown in Table 6.

Table 6. Physical Properties of Magnesium Iodide and Its Hexa- and Octahydrates

	MgI$_2$	MgI$_2$.6H$_2$O	MgI$_2$.8H$_2$O
mol wt	278.12	386.21	422.24
crystal system	hexagonal	monoclinic	orthorhombic
density, g/cm^3			
calculated	4.496	2.353	2.098
observed	4.43		
mp, °C	637 dec		41 dec
color	white	white	white
heat of formation, ΔH_{298}°, kJ/mol^a	−364.0		

^{a}To convert J to cal, divide by 4.184.

Magnesium Nitrate

Anhydrous magnesium nitrate, Mg(NO$_3$)$_2$, is difficult to isolate and the commercial product is the deliquescent hexahydrate Mg(NO$_3$)$_2$.6H$_2$. Physical properties are given in Table 7.

Magnesium Oxide and Hydroxide

Magnesium oxide (magnesia), MgO, is the principal product of the magnesium compounds industry. The principal sources of magnesia for commerce are magnesite, dolomite, and magnesium hydroxide. The physical properties of magnesium oxide are given in Table 8.

Table 7. Physical Properties of Anhydrous Magnesium Nitrate and Its Hexahydrate

Property	Mg(NO$_3$)$_2$	Mg(NO$_3$)$_2$.6H$_2$O
mol wt	148.32	256.38
crystal system		monoclinic
mp, °C		89
heat of formation, ΔH_{298}°, kJ/mol^a	−790.7	−2613.3

^{a}To convert J to cal, divide by 4.184.

Table 8. Physical Properties of Magnesium Oxide, Periclase

Property	Value
mol wt	40.31
crystal system	cubic
density (x-ray), g/cm^3	3.581
color	colorless, transparent
mp, °C	2852
electrical resistivity at 27°C, $\Omega\cdot$cm	1.3×10^{15}
specific heat at 27°C, kJ/(kg·K)^a	0.92885
heat of formation, ΔH_{298}°, kJ/mol^a	−601.70

^{a}To convert J to cal, divide by 4.184.

Uses. Dead-burned magnesia is used almost exclusively for refractory applications in the form of basic granular refractories and brick (see Refractories). Crushed fused magnesia is used as an electrical insulating material in Calrod heating elements for industrial electrical furnaces and domestic appliances.

The uses for caustic-calcined and specified magnesias range from mineral supplements in animal feeds to catalysts in carbonate leach systems to recover uranium oxide from uranium ore.

Magnesium Sulfate

Magnesim sulfate, MgSO$_4$, is widespread in nature, occurring either as a double salt or as a hydrate. Physical properties of the anhydrous salt and kieserite and epsomite are given in Table 9.

Uses. Magnesium sulfate is used as a fertilizer, a dietary supplement in animal feedstuffs, and in medicine as a cathartic and analgesic. It is used also in the building industry and textile industry.

Magnesium Sulfite

The system Mg(OH)$_2$–SO$_2$–H$_2$O is used in flue-gas desulfurization processes where magnesium hydroxide is the alkaline scrubbing medium. The properties of the two hydrates are given in Table 10.

Magnesium Sulfonate

The main use of overbased magnesium sulfonates is as an acid acceptor and sludge dispersant in crankcase lubricating oils with a smaller market in fuel additives (see Lubrication and lubricants).

Magnesium Vanadates

Magnesium forms several vanadates; Table 11 gives their physical properties. Vanadium compounds cause severe slagging and corrosion

Table 9. Physical Properties of Magnesium Sulfate and the Mono- and Heptahydrates

Property	MgSO$_4$	MgSO$_4$.H$_2$O, kieserite	MgSO$_4$.7H$_2$O, epsomite
mol wt	120.37	138.38	246.48
crystal system	orthorhombic	monoclinic	orthorhombic
density, g/cm^3			
calculated	2.908	2.571	1.678
observed	2.93		1.677
index of refraction	1.557, 1.582	1.520, 1.533, 1.584	1.4325, 1.4554, 1.4609
color	white	colorless	colorless
heat of formation, ΔH_{298}°, kJ/mol^a	−1284.9	−1602.1	−3388.6

^{a}To convert J to cal, divide by 4.184.

**Table 10.   Physical Properties of Magnesium Sulfite Tri-
and Hexahydrates**

Property	$MgSO_3 \cdot 3H_2O$	$MgSO_3 \cdot 6H_2O$
mol wt	158.42	212.47
crystal system	orthorhombic	hexagonal
calculated density, g/cm^3	2.117	1.723
mp, °C		200 dec
index of refraction	1.552, 1.555, 1.595	1.464, 1.511
color	colorless	white
heat of formation, ΔH°_{298}, kJ/mol[a]	−1931.8	−2817.5

[a] To convert J to cal, divide by 4.184.

Table 11.   Physical Properties of Some Magnesium Vanadates

Property	$Mg_{1.9}V_3O_8$	MgV_3O_8	$Mg_2V_2O_7$	$Mg_3V_2O_8$
mol wt	327.01	305.12	262.50	302.79
crystal system	monoclinic	orthorhombic	triclinic	orthorhombic
density, g/cm^3				
calculated	3.41	3.42	3.26	3.473
observed	3.37	3.39	3.1	
heat of formation, ΔH°_{298}, kJ/mol[a]			−2835.9	

[a] To convert J to cal, divide by 4.184.

problems in utility boilers. These problems are alleviated by treating the fuel oil with an oil dispersion of magnesium oxide or by separately injecting a magnesium oxide dispersion into the flame zone.

A.N. COPP
R. WARDLE
CE Basic

B. Petkov, *Mineral Industries Surveys—Magnesium Compounds 1983*, U.S. Department of the Interior, Bureau of Mines, U.S. Government Printing Office, Washington, D.C., 1983.

Powder Diffraction File, Sets 1–33, JCPDS (Joint Committee Powder Diffraction Standards) International Center for Diffraction Data, Swarthmore, Pa., 1983.

MAGNETIC MATERIALS, BULK

All materials that are magnetized by a magnetic field are magnetic materials. Magnetism is classified according to the nature of the magnetic response, ie, diamagnetism, paramagnetism, ferromagnetism, antiferromagnetism, ferrimagnetism, metamagnetism, parasitic ferromagnetism and mictomagnetism (spin glass). Most commercially important magnetic materials are comprised of ferromagnets and ferrimagnets (see also Amorphous magnetic materials, Ferrites, Ferroelectrics; Magnetic tape; Recording disks).

Theory

Types of magnetism.   The two atomic origins of magnetism are the spin and orbital motions of electrons. Diamagnetism occurs when the orbital rotation of the electrons is induced electromagnetically by an applied field. This weak magnetism is characterized by magnetization that is directed opposite to the applied field. The susceptibility $\kappa = M/H$ (where M is the magnetization and H is the magnetic-field strength) is ca 10^{-5} and, with few exceptions, is independent of temperature. Many of the metals and most of the nonmetals are diamagnetic.

In paramagnetism, the magnetization is aligned parallel to the applied field and the susceptibility is 10^{-3} to 10^{-5}. The susceptibility is positive for the alkali metals and negative for Cu, Ag, and Au.

Where the permanent magnetic moments are aligned as a result of a strong positive interaction among neighboring atoms or ions, the material exhibits a spontaneous magnetization and ferromagnetism results. Examples include iron, nickel, cobalt, and their alloys as well as many of the rare-earth elements.

In the case where the permanent magnetic moments are aligned antiparallel as a result of a strong negative interaction, the complete cancellation of the neighboring atomic moments results in zero net magnetization and antiferromagnetism. Examples include chromium, manganese, MnO, and NiO. Where the cancellation is incomplete, as in Fe_3O_4, a net magnetization remains. This is the case of ferrimagnetism.

Metamagnetism refers to the appearance of a net magnetization resulting from a transition from antiferromagnetism to ferromagnetism by the application of a strong field or by a change of temperature. $MnAu_2$ and $FeCl_2$ undergo the transition by field application, and heavy rare-earth metals, eg, terbium, dysprosium, and holmium, undergo the transition by temperature change.

Parasitic ferromagnetism is a weak ferromagnetism that accompanies antiferromagnetism, eg, in $\alpha\text{-}Fe_2O_3$.

Mictomagnetism, or spin glass, refers to the onset of short-range magnetic order in alloys that have spin orientations that are frozen at a critical low temperature but, in contrast to ferromagnetism and antiferromagnetism, that do not have long-range magnetic order. Spin glass is associated with a cusp in the susceptibility at a critical temperature and generally occurs in alloys containing dilute magnetic atoms embedded in a nonmagnetic matrix, eg, CuMn, or AuFe.

Magnetic domains.   Magnetic domains are associated with ferromagnetic, antiferromagnetic, or ferrimagnetic solids. In the demagnetized condition, these materials do not possess a net magnetization in the bulk because there are domains, ie, local regions, within which the magnetic moments of all atoms are aligned. The direction of these moments, however, changes from one domain to another such that the net magnetization is zero for the solid. The transition region between domains is the domain wall or boundary and is a region of high energy. Typical domain sizes are 10^{-4}–10^{-7} m and domain-wall thicknesses are ca 10^{-7} m for Fe. Domains arise as a result of total energy minimized from three principal types of competing energy.

Magnetic anisotropy energy.   There are several kinds of magnetic anisotropoy energy and perhaps the most well known is the magnetocrystalline anisotropy. This is a property only of a crystalline solid since the energy is dictated by the symmetry of the crystal lattice. For example, in bcc Fe, the easy axis is in a $\langle 100 \rangle$ direction and in fcc Ni, it is in a $\langle 111 \rangle$ direction. Other kinds of magnetic anisotropy include magnetostrictive anisotropy, shape anisotropy, thermomagnetic anisotropy, and slip-induced anisotropy.

Technical magnetic behavior.   When a magnetic-field strength H is applied to a ferromagnetic or ferrimagnetic material, the latter develops a flux density or induction B as a result of orientation of the magnetic domains. The relation between B and H is

$$B = \mu_o(H + M) = \mu_o H + J$$

where B, the number of lines of magnetic flux per unit of cross-sectional area, is in T ($= 1.0 \times 10^{-4}$ G) or Wb/m^2; H is an A/m ($= Oe/79.58$); M, the magnetization, is in A/m; and μ_o, the permeability of free space, is a constant equal to $4\pi \times 10^{-7}$ (T · m)/A (1.0 G/Oe). The product $\mu_o M$ is the magnetic polarization and is given in T, and generally is denoted by J.

Soft Magnetic Materials

Soft magnetic materials are characterized by high permeability and low coercivity. There are six major groups of commercially important soft magnetic materials in use: iron and low carbon steels, iron–silicon alloys, iron–aluminum and iron–aluminum–slicon alloys, nickel–iron alloys, iron–cobalt alloys, and ferrites. In addition, some amorphous soft magnetic alloys appear promising. Typical magnetic properties of some soft magnetic materials are in Table 1. Table 2 lists some characteristics of amorphous soft magnetic alloys.

Uses.   Because of their low resistivity, iron and low carbon steels tend to be used in static applications, eg, pole pieces of electromagnets and

Table 1. Typical Magnetic Properties[a]

	B_s, T[a]	Resistivity, $\mu\Omega\cdot cm$	H_c ($B_m = 1$ T)[b], A/cm[c]
cast magnetic ingot iron	2.15	10.7	0.68
magnetic ingot iron, 0.2-cm sheet	2.15	10.7	0.88
electromagnet iron, 0.2-cm sheet	2.15	12.0	0.81
hydrogen-annealed iron	2.15	10.1	0.04
low carbon steel, decarburized	2.14	12.5	0.70
M36 cold-rolled Si–Fe	2.04	41.0	0.36
M22 cold-rolled Si–Fe	1.98	49.0	0.31
M6 (110)[001] 3.2% Si–Fe	2.03	48.0	0.06

[a] Fully annealed.
[b] To convert T to G, multiply by 10^4.
[c] To convert A/cm to Oe, divide by 0.7958.

Table 2. Typical Characteristics of Some Amorphous Soft Magnetic Alloys

Material	J_s (at 20°C), T[a]	T_c, °C	Resistivity, $\mu\Omega\cdot cm$
$Fe_{80}B_{20}$	1.60	374	140
$Fe_{80}P_{16}C_3B_1$	1.49	292	150
$Fe_{80}P_{14}B_6$	1.36	344	
$Fe_{40}Ni_{40}B_{20}$	1.03	396	
$Fe_{40}Ni_{40}P_{14}B_6$	0.82	247	180
$Fe_3Co_{72}P_{16}B_6Al_3$	0.63	260	

[a] To convert T to G, multiply by 10^4.

cores of d-c magnets or relays. Low carbon steels and the lower-grade Fe–Si alloys are used in small motors and generators, whereas the high grade Fe–Si alloys tend to be used in power and distribution transformers and large rotating machinery. By weight, the Fe–Si alloys are used the most of all magnetic materials. Fe–Al and Fe–Al–Si alloys are used sparingly and primarily as recording-head materials because of their high hardness and resistivity. Ni–Fe alloys are used widely in high quality relays, transformers, converters, and inverters in the electronics industry; they have much higher permeability and lower loss compared with Si–Fe alloys. The Co–Fe alloys compete with the Ni–Fe alloys in applications, having the advantages of higher saturation polarization and Curie temperature but the disadvantages of poorer workability and higher cost. Because of their exceptionally higher resistivities, ferrites are particularly suitable for high frequency applications.

Hard Magnetic Materials

Hard or permanent magnetic materials are characterized by high coercivity and high energy product. The most important commercial hard magnetic materials in use today are Alnico alloys which are characterized by high energy product ($BH)_{max} = 40–70$ kJ/m^3 (5–9 × 10^6 G·Oe)), high remanent induction ($b_r = 0.7–1.35$ T (7–13.5 kG)), and moderate coercivity ($H_c = 40–160$ kA/m (500–2010 Oe)); and ferrites which are characterized by moderate energy product ($BH)_{max} = 8–30$ kJ/m^3 (1–4 × 10^6 G·Oe)), low remanent induction ($B_r = 0.2–0.39$ T (2–3.9 kG)), and high coercivity ($H_c = 150–270$ kA/m (1.9–3.4 kOe), $H_{cJ} = 200–320$ kA/m (2.5–4.0 kOe)). Rare-earth cobalt magnets are produced on a large commercial scale. They exhibit values of intrinsic coercivity to 3200 kA/m (40,200 Oe) and ($BH)_{max}$ to 240 kJ/m^3 (30 × 10^6 G·Oe), thus far surpassing any other commercial material. Commercially, new Cr–Co–Fe alloys have magnetic properties similar to the Alnicos but have the advantage of being cold-formable. Thus, these alloys are classed with Cunife (Cu–Ni–Fe) and Vicalloy (V–Co–Fe) in the family of ductile hard magnets. Some physical and magnetic properties of selected hard magnetic materials are given in Table 3.

Table 3. Properties of Permanent (Hard) Magnet Materials

Magnet material	Chemical composition	B_r, T[a]	H_c, kA/m[b]	$(BH)_{max}$, kJ/m^{3c}
$3\frac{1}{2}$% Cr steel	3.5Cr, 1C, bal Fe	1.03	5	2.4
3% Co steel	3.25Co, 4Cr, 1C, bal Fe	0.97	6	3.0
17% Co steel	18.5Co, 3.75Cr, 5W, 0.75C, bal Fe	1.07	13	5.5
Alnico 2	10Al, 19Ni, 13Co, 3Cu, bal Fe	0.75	45	13.5
Alnico 5[d]	8Al, 14Ni, 24Co, 3Cu, bal Fe	1.28	51	44.0
Alnico 5 DG[d]	8Al, 14Ni, 24Co, 3Cu, bal Fe	1.33	53	52.0
Alnico 6[d]	8Al, 16Ni, 24Co, 3Cu, 1Ti, bal Fe	1.05	62	31.0
Alnico 8[d]	7Al, 15Ni, 35Co, 4Cu, 5Ti, bal Fe	0.82	130	42.0
Alnico 9[d]	7Al, 15Ni, 35Co, 4Cu, 5Ti, bal Fe	1.05	120	72.0
sintered Alnico 5[d]	8Al, 14Ni, 24Co, 3Cu, bal Fe	1.09	49	31.0
sintered Alnico 6[d]	8Al, 16Ni, 24Co, 3Cu, 1Ti, bal Fe	0.94	63	23.0
sintered Alnico 8[d]	7Al, 15Ni, 35Co, 4Cu, 5Ti, bal Fe	0.74	120	32.0
Ceramic 1	$MO·6Fe_2O_3$ ⎫ M represents one or more of the metals chosen from the group barium, strontium, lead	0.23	150/260[e]	8.4
Ceramic 5	$MO·6Fe_2O_3$	0.38	190	27.0
Ceramic 7	$MO·6Fe_2O_3$ ⎭	0.34	260/320[e]	22.0
bonded ceramic[f]	plastic-molded ferrite	0.16	110/240[e]	4.4
bonded ceramic[d,g]	flexible anisotropic ferrite	0.24	170/215[e]	11.0
ESD 32[a]	18.3Fe, 10.3Co, 72.4Pb	0.68	76	24.0
ESD 42	18.3Fe, 10.3Co, 72.4Pb	0.48	66	10.0
Cunife 1[d]	60Cu, 20Ni, 20Fe	0.55	42	11.0
Vicalloy 1	10V, 52Co, bal Fe	0.75	20	6.4
R-Co 16z[d]	R represents one or more of the metals chosen from the rare-earth group	0.83	600/1440[e]	127.0
R-Co 18		0.87	640/1600[e]	143.0
rare-earth cobalt[d,h]	25.5Sm, 8Cu, 15Fe, 1.4Zr, 50Co	1.10	510/520[e]	240.0
Cr–Co–Fe[d,i]	10Co, 30Cr, 1Si, bal Fe	1.17	46	34.0
Cr–Co–Fe[d,j]	23Co, 31Cr, 1Si, bal Fe	1.25	52	40.0
Cr–Co–Fe[d]	11.5Co, 33Cr, bal Fe	1.20	60	42.0
Cr–Co–Fe[d]	5Co, 30Cr, bal Fe	1.34	42	42.0
Mn–Al–C[d]	70Mn, 29.5Al, 0.5C	0.56	180	44.0

[a] To convert T to G, multiply by 10^4.
[b] To convert kA/m to Oe, divide by 7.958×10^{-2}.
[c] To convert kJ/m^3 to G·Oe, multiply by 12.57×10^4.
[d] Anisotropic.
[e] Intrinsic coercive force, H_{cJ}.
[f] TDK FB Plastic Magnet.
[g] BQ A14 Rubber Magnet.
[h] TDK REC-30.
[i] Hitachi YHJ-MA.
[j] Sumitomo CKS500.

Uses. Hard ferrites are used widely in electromechanical devices, electronic applications, holding devices, and in toy designs. Loudspeakers are the largest consumer of permanent magnets (ca 50%).

The exceptionally large values of maximum energy product and coercivity of the rare-earth magnets permit their use in devices where small size and superior performance are desired. Magnets for electronic wristwatches and for traveling-wave tubes are largely made of rare-earth alloys.

Because of their cold ductility, Cunife magnets are used in speedometer and timing motors where parts are precision stamped at high speed.

Vicalloy is used widely in antitheft labels in department-store articles and library books.

ESD Magnets

ESD (elongated single-domain) magnets represent the synthesis of a hard magnet based on a permanent-magnet theory. These magnets often are referred to as Lodex, a trademark of General Electric and Hitachi Magnetics. Because of dimensional precision and magnetic uniformity, Lodex often is used in timing motors, meters, thermostats, and relays.

Magnet Steels

Magnet steels are carbon steels containing ca 1% C and various percentages of Co, W, and Cr. They are among the first steels made specifically for permanent magnets.

Manganese – Aluminum – Carbon Alloys

Anisotropic Mn–Al–C permanent magnet alloys have been developed using warm working. The Mn–Al–C magnets have good mechanical properties and can be machined readily. Their use could expand since their manufacture does not require expensive raw materials. However, manufacture is restricted to warm extrusion, a relatively expensive process.

Semihard Alloys

Coercivities of semihard magnets are from 10–100 A/cm (12–126 Oe). A good number of them are used in hysteresis motors; in such applications, the magnet steels, Cunife and Vicalloy, are commonly used. More recent development involves the use of semihard magnets in self-latching remanent-reed electrical contacts in the telecommunications industry.

G.Y. CHIN
J.H. WERNICK
Bell Laboratories

M. McCaig, *Permanent Magnets in Theory and Practice*, John Wiley & Sons, Inc., New York, 1977.

E.P. Wohlfarth, ed., *Ferromagnetic Materials—A Handbook on the Properties of Magnetically Ordered Substances*, North Holland, New York, Vols. 1 and 2, 1980; Vol. 3, 1982.

C.W. Chen, *Magnetism and Metallurgy of Soft Magnetic Materials*, North-Holland, New York, 1977.

MAGNETIC MATERIALS, THIN FILM

The largest use of magnetic films and particles is in memory and storage technologies; a great deal of progress has been made as a result of the explosion in computer applications (see Computers). However, of the myriads of magnetic materials studied for potential device applications, only a few are technologically important. Price per bit of information and performance, as denoted by access time, generally are used to characterize the various memory technologies, but power modular capacity, reliability, nonvolatility, etc, also are factors describing the efficacy of memories. Memories that are based on magnetic films, eg, bubble, fixed-head disk/drum, moving-head disk, and tape, have low cost-access-time ratios compared to the fast memories based on semiconductors, eg, metal oxide semiconductor (MOS) and bipolar random-access memories (RAM) (see Semiconductors).

Magnetic Properties and Structure

The static or low frequency magnetic properties pertinent to thin-film materials generally are utilized by materials scientists to characterize magnetic materials and serve to suggest, as a first approximation, their utility for device applications. Saturation magnetization M_s and Curie temperature T_{Ci} are intrinsic (structure insensitive) and are equal to the bulk values when the films are made properly. The extrinsic or structure-sensitive properties depend on size, shape, and surface topography of films; size, shape, and orientation of crystallites in polycrystalline films; concentration and distribution of imperfections, impurities, and alloying elements; and state of residual stress. The extrinsic properties can be classified further as static or dynamic depending on whether the property displays a frequency dependence. The importance of the deposition process and techniques on magnetic properties cannot be overemphasized in view of the above and of the large number of process variables associated with thin-film preparation techniques.

Shape anisotropy generally causes the magnetization in thin films to be in the plane of the film. In addition to shape anisotropy, an induced anisotropy (constant K_u) can be present in films as a result of deposition in a magnetic field. These anisotropies result from short-range directional order or an anisotropic distribution of atom pairs. The induced anisotropy constant is related to the anisotropy field H_k, where $H_k = 2K_u/M_s$. H_k is the field required to rotate the magnetization from the easy axis into the hard directions.

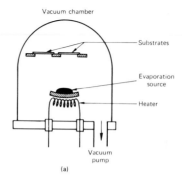

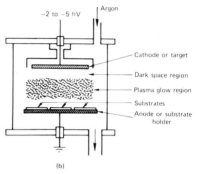

Figure 1. Simple schematic representation of (**a**) vacuum evaporation, (**b**) cathodic sputtering.

Permalloys, eg, 81.5% Ni–18.5% Fe, exhibit very low magnetocrystalline anisotropy and magnetostriction. Very low or zero magnetostriction is necessary for storage elements because dimensional changes, which can lead to stresses, are absent when the magnetization is switched. Substrate temperature and deposition rate influence the kinetics of film growth, the degree of impurity incorporation, and the residual stress distribution.

Fabrication

Thermal evaporation. Thermal evaporation (qv) in vacuum, the oldest and most economical method of thin-film preparation (Fig. 1a) consists of heating the material that is to be deposited to a temperature at which appreciable vapor pressure is developed. The vapor condenses onto an appropriately placed substrate that may be maintained at any temperature.

Sputtering. Cathodic sputtering processes (Fig. 1b) have come into widespread production use recently. Material is sputtered from a source target (cathode) by inert energetic ions (argon) and deposits on a substrate (anode).

Other fabrication methods include chemical vapor deposition (CVD), electrolytic and electroless deposition, and growth from solution (see Film depositon techniques).

Materials

Magnetic thin-film materials. Some of the magnetic materials and materials that have been or continue to be studied for memory or storage devices, switching elements for logic manipulation, thin-film recording heads, thermomagnetic writing (Curie-point writing) are iron, iron–nickel alloys, cobalt–nickel, cobalt–phosphorus, cobalt–nickel–phosphorus, iron–nickel–chromium, Vicalloy II (vanadium–iron–cobalt), Cunife I (copper–nickel–cobalt), Cunico II, MnBi, MnAlGe, MnGaGe, Sendust alloy, RCo(Fe) amorphous alloys, and Co–Fe–Cr–P–C–B amorphous alloys.

Bubble-domain devices. Magnetic bubbles are cylindrical magnetic domains, formed by an external field in thin, single-crystal films of garnets, and memory and logic devices are based on them. Materials based on the ferrimagnetic rare-earth iron garnets ($R_3Fe_5O_{12}$) are preferred for practical devices.

Tapes, drums, and disks. Most modern magnetic tape consists of a dispersion of ferromagnetic or ferrimagnetic oxide particles on a mechanically strong, flexible, plastic substrate (see Magnetic tape). The oxides for these applications can be considered to be hard magnetic materials exhibiting coercivities of several thousand amperes per meter. $\gamma\text{-Fe}_2\text{O}_3$ is the most popular tape material. $\gamma\text{-Fe}_2\text{O}_3$ or Fe_3O_4 generally are used for drum storage.

Recording head. Materials that are suitable for read–write recording heads for tapes and disks should be characterized by high saturation flux density, low remanent induction to avoid erasure of information when the writing current ceases, and low hysteresis loss and low eddy-current loss, particularly for high data rates or high frequency operation. In addition, because of the small air gap between the head and recording medium, the head material should be abrasion resistant because dust particles and the magnetic attraction between head and tape could lead to abrasion. Most of the recently manufactured recording heads are based on bulk Ni–Fe (50–50) Permalloys and manganese–zinc ferrite, which are soft magnetic materials (see Ferrites).

Magnetooptics

The application of magnetooptic effects to optical memory systems (laser-beam writing, magnetooptic read) has been the subject of much research during the past fifteen years. However, commercial memory systems based on laser writing and reading through the interaction of electromagnetic radiation, either through reflection (utilizing the Kerr effect) or by transmission (utilizing the Faraday effect), have not been produced.

Magnetic fluids. Magnetic fluids are stable colloidal suspensions of ferromagnetic or ferrimagnetic particles, such as Fe_3O_4 of subdomain size (ca 10 nm) in aqueous or organic bases. The fluid behaves as a homogeneous Newtonian liquid and reacts to a magnetic field. These materials are used in bearings, rotary-shaft seals, and feedthroughs. Other applications, eg, as jet inks and for float separation, are being studied (see Magnetic separation).

J.H. WERNICK
G.Y. CHIN
Bell Laboratories

R. Glang in L.I. Maissel and R. Glang, eds., *Handbook of Thin Film Technology*, McGraw-Hill, New York, 1970.

A.H. Bobeck and E. Della Torre in E.P. Wohlfarth, ed., *Magnetic Bubbles*, Vol. XIV, North-Holland, New York, 1975.

C. Heck, *Magnetic Materials and Their Applications*, Crane, Russak and Co, New York, 1974.

F.N. Bradley, *Materials for Magnetic Functions*, Hayden Book Co., Inc., New York, 1971.

S.W. Charles and J. Popplewell in E.P. Wohlfarth, ed., *Ferromagnetic Materials*, North-Holland, New York, 1980.

MAGNETIC SEPARATION

Magnetic separation involves the use of magnetic forces to separate particles of varying composition on the basis of their magnetic properties. Recently, advances in magnetic-separator technology have expanded old applications and made possible entirely new uses for this venerable industrial process.

Magnetic separation involves transport processes for the feed and the products of the separation, the less magnetic (nonmags or tailings), the more magnetic material (mags), and in some cases, an intermediate product (middlings). In magnetic separators, the separation occurs by virtue of a competition between the magnetic forces that act on the magnetic particles, and inertial, frictional and gravitational forces that act on all the particles transported through the separator. The existence of interparticle forces tends to reduce the degree of separation between the nonmagnetic and the magnetic particles and in some magnetic separators an intermediate product is produced.

Table 1. Separators Using Magnetic Fields

magnetostatic	magnetic attractive
magnetohydrostatic	magnetic buoyant
magnetohydrodynamic	magnetic force on a moving charge
eddy current	magnetic force on a moving charge
differential coercivity	magnetic repulsive

The principal determinant of the character of a magnetic separation process is the magnetic response of the particles to be separated. Materials vary in their magnetic response from strongly magnetic materials like iron to very weakly magnetic materials like paramagnetic cupric oxide or diamagnetic organic compounds. The magnetic response of a material is the measure of the degree to which it becomes magnetized in the presence of a background magnetic field. The magnetic separation devices considered here are magnetostatic magnetic separators. As indicated in Table 1, at least five types of separators may be considered to utilize magnetic forces in separating materials. Magnetic separators are distinguished from each other by the nature of the magnetic response they utilize in the particles to be separated.

The main uses of magnetic separation, until recently, have been for the concentration of large particles of strongly magnetic iron ores and the removal of unwanted or tramp iron from a variety of industrial feeds. Now the development of filamentary matrix-type, high gradient magnetic separators that optimize the magnetic force on a fine particle of even weakly magnetic material has made possible a variety of new applications in mineral processing, water purification and other applications.

Basic Principles

Interaction of magnetic fields and matter. The magnetic response of a substance is an intrinsic property of its atomic structure and may be roughly characterized as strongly magnetic, modestly magnetic, and weakly magnetic. The magnetization M of a substance may be expressed as follows: $M = M_0 + M(H)$ where M_0 is the spontaneous magnetization of the substance in the absence of background magnetic field and $M(H)$ is the portion of the magnetization induced by the applied magnetic field H.

Dipolar nature of magnetic forces. Magnetic forces are based on dipolar interactions between magnetic substances and magnetic fields. All magnetic particles have both north and south poles and all magnetic field lines are closed loops; therefore, magnetic forces always involve the interaction of both the north and south poles of a magnetized body or particle with the external magnetic field. If the magnetic field is uniform, it may orient a magnetized particle along the direction of the field but will not exert an attractive force on it. If the magnetic field is nonuniform, then a magnetized particle will experience an attractive or repulsive force in the direction of increasing field intensity, depending on the direction of the particle's magnetization. Thus, it is the gradient of the magnetic field that interacts with the magnetization of the particle to produce an attractive magnetic force.

The production of magnetic fields and gradients. The creation of magnetic fields also involves the creation of magnetic-field gradients that depend on the geometry of the magnet. Magnetic-field lines tend to fan out in space away from the source of the magnetic field which may be an electric current or magnetized body. The magnitude of the magnetic-field gradient depends on a variety of factors, including the magnetization of the substance, the detailed geometry, and the distance from the magnetized body, but the gradient is generally proportional to $1/r$ where r is the overall radius curvature of the magnetized structure. The range of the magnetic force is therefore indirectly proportional to r. This relationship between radius of curvature in the magnetic-field gradient may be used to estimate the magnitude of the magnetic forces and the range of those forces available in different types of magnetic separators.

In magnetic separators, magnetic fields may be produced in several ways. A great variety of magnetic separators make use of electromagnetic coil/iron-pole combinations.

Magnetic separator designs. Two basic magnet designs for magnetic separators exist. Single-surface-type magnetic separators are those in which the magnetic-field-producing elements also directly produce the magnetic-field gradients. Examples of these single-surface separators include grates, suspended magnets, pulleys, and drums. In the matrix-type magnetic separators, an electromagnetic coil or coil/iron-pole combination is utilized to produce a relatively uniform magnetic field in an open working volume. The working volume contains a filamentary ferromagnetic structure that is magnetized to produce strong magnetic-field gradients on the sharp edges or filaments in the structure.

In single-surface magnetic separators, the active separation volume is that directly adjacent to the surface and is defined by the area of the surface and the range of the magnetic forces. These separators generally have relatively low magnetic-field gradients with ranges of several centimeters and lend themselves to the separation of larger, more strongly magnetic particles. Matrix-type separators allow much higher magnetic forces to be produced by much larger field gradients on fine magnetized structures. In order to achieve sufficient active trapping volume adjacent to these fine structures, a relatively large working magnetized volume must be produced by the external magnet. Advances both in the development of structures with large surface areas and high field gradients as well as magnetic circuits for large volumes have led to the development of new, effective matrix-type, high gradient magnetic separators for fine, weakly magnetic particles.

There are two main design factors in magnetostatic magnetic separators: the creation of the magnetic field and gradients; and the specification of the competing or transport forces for the feed and separator products through the magnetic separator.

Magnetic Separators

A wide variety of magnetic separators whose designs and operating characteristics are appropriate for widely differing applications are commercially available. Both the single-surface and matrix-type devices include designs in which the separation surface or matrix may be stationary or moving. In stationary separators, the trapped magnetics are held on the surface until removed by a separate operation at a later time. In fully continuous magnetic separators, the collection surface moves and carries the trapped particles out of the separation.

Single-surface devices — stationary collection-surface plates. Plate magnetic separators are simple, low cost devices, utilizing permanent magnetic materials arranged to form a flat surface similar to that for a drum surface. The plate is generally hinge-mounted above the conveyor belt or in the side of a narrow duct where the feed material is transported by gravitational forces. These devices include the suspended magnet, and the grate.

Single-surface devices — moving collection surfaces. In these devices the surface against which magnetic particles are attracted moves away

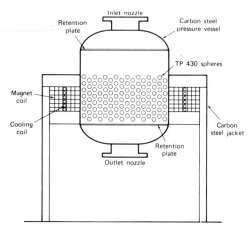

Figure 2. Ball matrix separator.

from the feed stream in such a way that magnetic material can be continuously removed from the separation zone and fresh unloaded surface is available for trapping. These separators are completely continuous and a periodic, separate operation for cleaning is unnecessary. Devices of this type include pulleys, belt separators, and a variety of drum separators (an example is shown in Fig. 1).

Matrix-type magnetic separators. The basic distinction between matrix-type magnetic separators (Fig. 2) and single-surface separators is that magnetic forces are created throughout the volume of a matrix bed that has been magnetized by an external background magnetic field. Separators of this type permit the optimization of strong, short-range magnetic forces, and provide sufficient capacity for effective operation. The design differences among matrix magnetic separators relate to the matrices and the magnetic circuits used. As in the case of single-surface separators, both stationary and moving-surface devices exist. Stationary matrix devices include a grid, a ball matrix, and a high gradient magnetic separator (HGMS). Moving matrix separators include ball-selenoid separators, grids, grooved plates, and high gradient devices.

JOHN OBERTEUFFER
IONEL WECHSLER
Sala Magnetics, Inc.

J.A. Oberteuffer, *IEEE Trans. Magnetics* **Mag-10**, 223 (1974).

A.H. Morrish, *The Physical Principles of Magnetism*, John Wiley & Sons, Inc., New York, 1965.

J.E. Lawver and D.M. Hopstock, *Min. Sci. Eng.* **6**(3), 154 (1974).

MAGNETIC TAPE

Magnetic tape has become a principal means of storing information because it has high capacity, is permanent and versatile, and yet may be easily erased and used repeatedly. Its design and manufacture have improved markedly in the past 15–20 yrs, and applications have been expanded to a multiplicity of formats in reels and in packaged systems, eg, cassettes and disks of diminishing size (see Recording disks).

Properties

Magnetic tape consists of two basic components: base film or support, and a magnetic coating, as shown in Figure 1.

Formulations of the magnetic coating vary with the use of the tape. However, a coating normally contains the following ingredients (by volume): magnetic material (40–55%), binder (30–35%), conductive agent (0–10%), dispersant (2–10%), lubricants (2–8%), miscellaneous additives (1–3%) (see Magnetic materials, bulk; Dispersants; Lubrication and lubricants).

For many tape applications, a backcoating is not used, but in some precision tapes, a conductive, nonmagnetic coating is employed on the

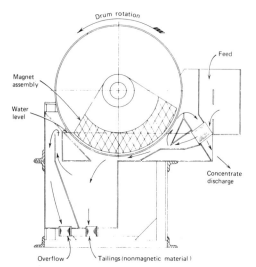

Figure 1. Countercurrent separator used in roughing stage.

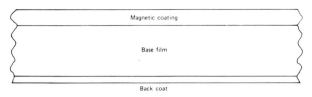

Figure 1. Magnetic tape: base film, 6.4–38.1 μm; magnetic coating, 5.1–15.2 μm; back coat, 1.0–1.5 μm.

back side of the tape. This coating improves the following: winding and handling characteristics of the tape, abrasion resistance, and control of static buildup.

Base support. Poly(ethylene terephthalate), commonly referred to as polyester, has an almost ideal combination of features for a magnetic-tape support, including mechanical properties, eg, tensile strength and modulus, tear resistance, chemical stability, availability, and cost (see Polyesters; Film and sheeting materials). The main suppliers of base film to the industry are DuPont, ICI, Hoechst, Mitsubishi, Toray, Sunkyong, Teijin, and Rhodia (Rhone-Poulenc).

Magnetic coating. Magnetism and magnetic-recording theory. Magnetic recording involves the conversion of electrical signals to a corresponding magnetization on a magnetic tape. Thus, signals obtained from audio, video, or various other kinds of data are fed to a transducer, or head, which converts the signals to magnetic fields of varying strength and polarity and induces an analogous magnetization in the tape which passes across it. In the reproduction of the stored information, the tape is moved across a playback head, which may be the same as the record head, and the magnetic flux in the tape links through the core of the head and induces a voltage in its windings.

Magnetic particles. Four chemical types of magnetic particles are used in magnetic tape: γ-ferric oxides, doped iron oxides, chromium dioxide, and metallic particles that usually consist of elemental iron, cobalt, and/or nickel.

Metallic particles have been prepared in several ways. Practical processes include reduction with hydrogen of iron oxides and oxalates, and reduction of aqueous ferrous salt solutions with borohydrides or hypophosphites. Several suppliers to magnetic tape manufacturers produce metallic powders from which audio cassette tapes are being made and sold in a limited market and vigorous study is in progress on video tape application.

Vacuum metallization technology is now being exported very widely to prepare magnetic tapes with still higher signal: noise ratios and very high data-packing densities. Cobalt or metal alloys thereof are evaporated by electron-beam energy and deposited into special polyester film wrapped around a revolving drum under vacuum.

The metal film is normally 120–150 nm thick; coercivity and retentivity are 44 kA/m (550 Oe) and 1.2 T (1.2×10^4 G), respectively. The recording density, when used for dictation purposes in the form of a microcassette, is about ten times as high as conventional tape. Application of vacuum deposition of thin magnetic layers can be extended to flexible disks and to audio and video tapes (see Film deposition techniques).

Binders. Binders are required in all magnetic tapes containing particulate magnetic materials. The binder is the organic polymer that holds the magnetic material together in an immobile yet flexible matrix that adheres to the substrate. It should be capable of holding the maximum amount of magnetic material without shedding, and it should have a high degree of toughness, hardness, abrasion resistance, and chemical stability. Also, it must be soluble in suitable solvents in order to permit its use in a coating process.

In most modern tapes, thermoset binders are preferred because of their superior overall performance, especially in recorders operating at high head-tape speeds in a wide range of recorder configurations and environments. Polyurethanes are one of the most important polymers because of their unusual toughness and abrasion resistance (see Urethane polymers).

Vacuum-deposited metal tapes do not incorporate binders.

Dispersants. Dispersants (qv) aid in the deagglomeration of magnetic particles. Individual particles are separated to the maximum possible degree since tape performance critically depends on dispersion. Lecithin (qv) is one of the most common dispersants.

Conductive pigment. The electrical resistivity of tapes made with γ-ferric oxide is relatively high, eg, 10^{10}–10^{14} Ω/sq. In some applications, especially in video systems and instrumentation in which high head-tape speeds are common, static-charge buildup can be a problem. Addition of a conductive pigment to the magnetic coating reduces the resistivity to an acceptable level of 10^6–10^8 Ω/sq. Conductive carbon blacks normally are used for this purpose.

Lubricants. Lubricants are incorporated in most tape formulations to minimize head-tape friction which, if not controlled, causes excessive tape wear, headwear, and audible squeal. A wide variety of substances has been used as lubricants including silicones; fatty acids and esters; silanes; glyceryl esters; fatty acid amides; fluorinated hydrocarbons; hydrocarbons (eg, squalane), and silica. Lubricant is sometimes added as a tape coating in a separate step instead of part of the main coating formulation.

Solvents. Solvents are used to dissolve the raw binder polymers and to provide a fluid medium for the pigment dispersion in the coating mix. Examples of common solvents are tetrahydrofuran, cyclohexanone, methyl ethyl ketone, methyl isobutyl ketone, and toluene.

Other additives. Sometimes mild abrasives (qv), eg, silica, alumina, or chromia, are added so that the tapes clean the heads during use and prevent loss of output resulting from buildup of traces of undesirable materials on the heads.

Manufacture and Processing

Tape performance depends on the materials used in its construction and how they are processed. The sequence of processes involves formulating the coating mix, dispersing the magnetic pigment, preparing the base film, coating, orienting the particles, drying, surface finishing, slitting, rewind/assembly, testing, and packaging. Each of these steps must be conducted in a manner depending on the use of the tape.

Future trends. Processes designed to conserve energy and raw materials, especially solvents, had begun to receive attention by principal tape manufacturers in the late 1970s. Waterborne coatings for magnetic tape have been demonstrated at the research level but have not been extended to commercial products.

Radiation curing (qv) in magnetic-tape processing is a rapidly developing alternative to chemical curing processes. A wide variety of radiation-curable monomers, oligomers, and prepolymers is becoming available for magnetic-tape applications, and many new raw materials will be commercialized in the next several years.

Toxicity

The ingredients used in the manufacture of magnetic tape exhibit relatively low levels of toxicity.

Uses

Magnetic tape uses may be grouped in the following classifications: audio, video, instrumentation, and computer products.

R.H. PERRY
A.A. NISHIMURA
Ampex Corp.

F. Jorgensen, *The Complete Handbook of Magnetic Recording*, Tab Books, Blue Ridge Summit, Pa., 1980.

G. Bate, *J. Appl. Phys.* **52**, 2447 (1981).

G. Bate, *paper presented at 4th International Conference on Video and Data Recording*, University of Southhampton, UK, April 20, 1982.

F. Kalil, ed., *Magnetic Tape Recording for the Eighties*, NASA Reference Publication 1075, April, 1982.

MAINTENANCE

The concern of plant managers, production supervisors, plant engineers, and maintenance managers is that a facility and its equipment are maintained in a way that enables a company to produce its product at the lowest possible cost. The goal of a good maintenance program is to operate a facility and its equipment in such a manner that production can continue with the least possible interruption and that the reasonable cost of such a program contributes to the profitability of the company. Maintenance of facilities and equipment can result in reduction of operating costs and increased productivity.

Maintenance is as much an art as it is a science. The problems of cost, downtime, and technical knowledge increase with more complicated automation, safety requirements, environmental controls, and energy conservation. With the cost of labor and material spiraling upwards, ways of making the workforce more productive and of securing longer operating times for equipment and material must be found. This can be accomplished only by better organization, management, and control of maintenance.

Organization of maintenance should not start after the first breakdown or catastrophic failure; it must start with the planning of a new facility and with the choice of equipment. A complete systems approach to organize, manage, and control the plant maintenance function may include the following steps: designing, planning, and constructing a facility with maintenance in mind; providing reliable utilities and their proper usage; developing a maintenance organization and program while the facility is under construction; developing a simple, practical preventive-maintenance (PM) system; supplementing an in-house program with contract maintenance; motivating the maintenance worker; implementing safety measures; developing an energy conservation program; cooperating with other service groups of the facility; and controlling costs and providing budget responsibility.

ERIC M. BERGTRAUN
National Semiconductor Corp.

L.R. Higgins and L.C. Morrow, *Maintenance Engineering Handbook*, 3rd ed., McGraw-Hill Book Co., Inc., New York, 1977.

B.T. Lewis, *Management Handbook for Plant Engineers*, McGraw-Hill Book Co., Inc., New York, 1977.

S.E. Fuchs, *Complete Building Equipment Maintenance Desk Book*, Prentice-Hall, Inc., Englewood Cliffs, N.J., 1983.

MALEIC ANHYDRIDE, MALEIC ACID, AND FUMARIC ACID

Maleic anhydride, maleic acid, and fumaric acid are polyfunctional chemicals of significant commercial interest worldwide. They are chemically related in that each has α,β unsaturation associated with acid carbonyl functions. Maleic anhydride and maleic acid are industrially important raw materials in the manufacture of alkyd and polyester resins, surface coatings, lubricant additives, plasticizers (qv), copolymers (qv), and agricultural chemicals (see Alkyd resins; Polyesters; Lubrication and lubricants).

Fumaric acid is found naturally in many plants and is named after the genus *fumaria*. It is used as a food acidulant and in the manufacture of unsaturated polyester resins, quick-setting inks, furniture lacquers, and paper sizes.

Physical Properties

Physical constants, including thermal and solution properties, are listed in Table 1.

Chemical Properties

Both the ethylenic double bond and the carbonyl groups are susceptible to attack. Because the C=C bond is conjugated with two C=O bonds, both groups may participate in certain reactions.

Table 1. Physical Properties of Maleic Anhydride, Maleic Acid, and Fumaric Acid

Property	Maleic anhydride	Maleic acid	Fumaric acid
structural formula			
formula weight	98.06	116.07	116.07
mp, °C	52.85	138–139 (from water) 130–130.5 (from alcohol)[a]	287
bp, °C	202	ca 138 (dec)	290
sp gr (at 20/20°C, solid)	1.48	1.590	1.635
sp gr (at 70/70°C, molten)	1.3		
heat of formation, kJ/mol[b]	−470.41	−790.57	−811.03
free energy of formation, kJ/mol[b]		−625.09	−655.63
heat of combustion, MJ/mol[b]	−1.390	−1.358	−1.335
flash point, °C			
open cup	110		
autoignition temperature, °C	147		
crystal structure	orthorhombic	monoclinic	monoclinic, prismatic, needles, or leaflets
solubility, g/100 g soln			
water, at 25°C	[c]	44.1	0.70
benzene, at 25°C	50[d]	0.024	0.003
chloroform, at 25°C	52.5[d]	0.11	0.02
carbon tetrachloride, at 25°C	0.6[d]	0.002	0.027
vapor pressure, kPa[e] at 44.0°C	0.13		
viscosity at 60°C, mPa·s (= cP)	1.61		

[a]Sublimes at 165°C at 0.23 kPa.
[b]To convert J to cal, divide by 4.184.
[c]Hydrolyzes slowly.
[d]In g/100 g solvent.
[e]To convert kPa to mm Hg, multiply by 7.5.

Acid chloride formation. The monoacid chlorides of maleic and fumaric acids are not known. Fumaric acid can be converted to fumaroyl chloride in the reaction with phosgene (qv), phthaloyl chloride, phosphorus pentachloride, or thionyl chloride in the presence of iron. Under similar conditions, maleic acid or anhydride yields an isomeric dichloroisocrotonic lactone, maleyl chloride.

Acylation. Maleic anhydride reacts with aromatic hydrocarbons in the presence of aluminum chloride to form β-aroylacrylic acids. Chlorinated hydrocarbons generally are used as solvents.

Alkylation. Addition of an alkyl group is effected by the reaction of maleic anhydride with an olefin or aromatic compound in which a C—H bond is activated by α-β unsaturation or by adjacent aromatic resonance.

Amidation. Maleic anhydride and the isomeric acids react with ammonia (qv), primary amines, and secondary amines to give mono- or diamides (see Amines, lower aliphatic).

Concerted nonpolar reactions. The C=C double bond of maleic anhydride reacts with other unsaturated species in a number of ways. The Diels-Alder reaction of maleic anhydride and its derivatives has been studied extensively.

These 1,4-cycloaddition reactions usually are reversible and often afford interesting stereochemistry.

[2 + 2] Cycloaddition. Acetylene and its derivatives photolytically add to the double bond of maleic anhydride in a [2 + 2] manner to yield cyclobutene adducts.

Ene reaction. Maleic anhydride serves as an efficient enophile in the ene reaction. The ene must have an allylic hydrogen; the product is an alkenylsuccinic anhydride.

Decomposition and decarboxylation. Maleic anhydride is decomposed to carbon oxides over catalysts containing transition-metal oxides. It also can be exothermally decomposed in the liquid phase, affording carbon dioxide, carbon monoxide, ethylene, and/or acetylene, depending on the reaction conditions. The decarboxylation of maleic acid or its monoalkyl esters is accomplished readily by heating with catalysts, eg, copper oxide. Diaryl fumarates decarboxylate at high temperatures in a stepwise manner, producing stilbenes.

Electrophilic addition. The electron-deficient double bond of maleic anhydride and its derivatives is susceptible to attack by various electrophilic reagents.

Esterification. Mono- and dialkyl maleates and fumarates are prepared by heating the alcohol or alkoxide with maleic anhydride, maleic acid, or fumaric acid. The esterification (qv) usually is acid-catalyzed, eg, by sulfuric acid or toluenesulfonic acid.

Linear unsaturated polyesters are obtained from glycols or epoxides. Ethylene glycol and maleic anhydride react to give the repeating unit

$$-\!\left(CH_2CH_2OCCH\!=\!CHCO\right)_{\overline{n}}$$
$$\quad\quad\quad\underset{O}{\|}\qquad\quad\underset{O}{\|}$$

This reaction is industrially important as it is the first step in the production of polyester resins (see Polyesters).

Free-radical reactions. Nucleophilic radicals add to maleic anhydride to form the corresponding alkylsuccinic anhydride.

Grignard-type reactions. Grignard reactions involve the nucleophilic addition of various organometallic reagents to the maleyl carbonyl groups (see Grignard reaction). Both 1,2- and 1,4-additions are possible.

Halogenation. Addition of a halogen to the ethylenic function of maleic or fumaric acid forms a dihalosuccinic acid. Chlorine and bromine also add in this manner.

Hydration/dehydration. Maleic anhydride is hydrolyzed easily to maleic acid in water at room temperature. Malic acid results when fumaric acid is hydrated at elevated temperatures and pressures. Maleic acid can be dehydrated thermally or by azeotropic distillation to maleic anhydride.

Isomerization. Maleic acid is isomerized to fumaric acid by applying heat or using chemical catalysts. Derivatives of maleic acid also can be isomerized (see Manufacture, Fumaric acid).

Nucleophilic addition. Because of the electron-withdrawing capacity of the carbonyl groups in the conjugated maleic and fumaric systems, the β-carbon atom is highly susceptible to attack by nucleophilic reagents. This type of reaction usually is carried out under basic conditions according to the following general equation:

$$\underset{\displaystyle ROCCH=CHCOR}{\overset{\displaystyle O\quad\quad O}{\|\quad\quad\|}} + \underset{\displaystyle HOOCCH_3}{\;} \xrightarrow{\text{base}} \underset{\displaystyle\underset{OCOCH_3}{|}}{\underset{\displaystyle ROCCH_2CHCOR}{\overset{\displaystyle O\quad\quad O}{\|\quad\quad\|}}}$$

Oxidation. Ozonolysis gives glyoxalates, tartaric acid, or oxalic acid, depending on the reaction conditions and solvent.

Polymerization. Maleic anhydride homopolymerizes with some difficulty, but copolymerizes with ease (see Polymerization mechanisms and processes). Monomers that are used with maleic anhydride to make commercial copolymers include ethylene, styrene, methyl vinyl ether, and vinyl chloride.

Unsaturated polyester resin manufacture comprises the largest single use of maleic anhydride. It is esterified with a glycol to form a linear polyester, and cross-linked with a vinyl monomer, eg, styrene, to form a rigid, insoluble, and strong structure. The resins commonly are reinforced with glass fiber.

Reduction. The heterogeneous catalytic reduction of maleic anhydride and its derivatives gives succinates, γ-butyrolactone, tetrahydrofuran, or 1,4-butanediol, depending on the catalyst used.

Sulfonation. Sulfomaleic anhydride, an intermediate in the production of sulfosuccinate derivatives, results from the reaction of maleic anhydride and sulfur trioxide.

Manufacture

Maleic anhydride and maleic acid. The predominant commercial route to maleic anhydride is the vapor-phase oxidation of hydrocarbons in air over a solid catalyst. Benzene, *n*-butane, and *n*-butylenes are used.

Fumaric acid. Essentially all of the fumaric acid in commerce is derived from maleic acid by catalytic isomerization.

Health and Safety Factors, Toxicology

Maleic anhydride is a strong irritant. The ACGIH threshold-limit value in air is 0.25 ppm (1 mg/m³), and the OSHA permissible exposure level (PEL) also is 0.25 ppm.

Although not a severe fire hazard, maleic anhydride can burn. Dry chemicals are not recommended for extinguishing fires, nor is water or foam. Violent reactions can occur when maleic anhydride contacts alkali or alkaline-earth metals, carbonates, hydroxides or amines. Hot water, not alkaline solutions, should be used to wash equipment that may contain maleic anhydride.

Although maleic acid does not present the hazard of hydrolyzing exothermally that the anhydride does, it is a strong irritant. Fumaric acid, however, is practically nontoxic.

W.D. ROBINSON
R.A. MOUNT
Monsanto Co.

L.H. Flett and W.H. Gardner, *Maleic Anhydride Derivatives*, John Wiley & Sons, Inc., New York, 1952.

"Maleic Anhydride," *Chemical Economics Handbook*, Report No. 46B, Stanford Research Institute, Menlo Park, Calif., April 1983.

B.C. Trivedi and B.M. Culbertson, *Maleic Anhydride*, Plenum Press, New York, 1982.

MALONIC ACID AND DERIVATIVES

Malonic acid (propanedioic acid, methanedicarboxylic acid), $HOOCCH_2COOH$, was discovered and isolated as a product of malic acid, $HOOCCH_2CHOHCOOH$, oxidation. Malonic acid derivatives have found particular use in the preparation of pharmaceutical chemicals (see Fine chemicals).

Malonic Acid

Physical properties. Malonic acid is trimorphic. One of the two triclinic forms is stable up to 94°C and the other is unstable. A monoclinic form exists and is stable above 94°C and melts at 135.6°C; above 135.6°C, decomposition occurs. Its solubilities in 100 mL water are 61.1 g at 0°C, 73.5 g at 20°C, and 92.6 g at 50°C. It also is very soluble in alcohols, although it is 5–10% soluble in pyridine and ethyl ether. The alkali salts also are very soluble in water. The barium and lead salts, however, are insoluble and can be used for the determination of the acid.

The ionization constants (pK_a) for malonic acid in water are 2.83 and 5.69; in methanol, 7.5 and 8.8; in ethanol, 7.91 and 11.95; in 2-propanol, 8.3 and 9.9; in *t*-butanol, 9.7 and 12.5; in acetonitrile, 15.3 and 20.0; and in dimethyl sulfoxide, 7.2 and 10.26.

Reactions. Malonic acid undergoes most of the typical reactions of dicarboxylic acids (qv). It forms acid and neutral salts with metals and amines; double and complex salts of malonic acid also are known. It forms most of the functional derivatives formed by other dicarboxylic acids, eg, esters, amides, nitriles, and mixed-function derivatives. The most outstanding exception to this occurs when an attempt is made to dehydrate malonic acid to malonic anhydride; in this case, carbon suboxide is obtained. An important reaction of malonic acid is the loss of carbon dioxide upon heating. Malonic acid is oxidized in acidic solution by potassium permanganate to carbon dioxide and formic acid. Sulfuryl chloride converts malonic acid into chloromalonic acid, whereas thionyl chloride converts the acid, depending upon the conditions, into the mono- or dichloride. Bromine substitutes in the methylene group, yielding monobromomalonic acid and dibromomalonic acid.

Preparation. Malonic acid often is an intermediate in the preparations of its diesters and generally can be obtained by omitting the esterification step. However, the malonic acid thus obtained often is in admixture with sodium chloride. Purification is achieved through the calcium salt of malonic acid.

Health and safety factors (toxicology). Malonic acid is a strong acid; it can damage skin and mucous membranes, although not as severely as oxalic acid.

Uses. Malonic acid can be used to inhibit the catalytic effect of copper in the degradation of some materials. It also has been used as a blowing agent in making foamed plastics (qv). When hides are treated with malonic acid and formaldehyde, a Mannich reaction occurs, producing a surface that binds mineral tanning agents in a superior fashion. Malonic acid also has been used in the electrolyte for silver plating in order to increase the reflecting power of silver electroplates. Malonic acid has been used in the preparation of malonato(1,2-diamminocyclohexane)platinum II which is effective against mouse leukemia and exhibits lower renal toxicity than DDP, dichlorodiammineplatinum (II), a dichloroplatinate which is effective in humans.

Diethyl Malonate and Dimethyl Malonate

Physical properties. Selected physical constants of diethyl malonate and dimethyl malonate are listed in Table 1.

Reactions. Malonic esters are reactive both at the ester site and at the intermediate methylene position. The latter site is reactive because of the β-dicarbonyl structure—a major reason for the use of malonic esters.

Preparation. The classic method of cyanation of monochloroacetic acid to cyanoacetic acid, hydrolysis to malonic acid, and esterification is the important commercial process for the synthesis of diethyl malonate and dimethyl malonate.

Health and safety factors (toxicology). No specific hazards are posed by either malonates if handled under normal conditions.

Uses. The bulk of the production of malonate esters is consumed by the pharmaceutical industry. The following are some of the more important applications: the manufacture of amino acids, barbiturates, phenyl butazone, additives (antioxidants, light stabilizers), and dyes.

Malononitrile

Physical properties. Malononitrile is a colorless crystalline solid (mp 30–31°C and bp 218–219°C). The heat of formation is 265.9 kJ/mol (63.6 kcal/mol) and the specific heat is 100 J/(mol·K) (23.9 cal/(mol·K)) at 0°C.

Reactions. As with malonic esters, malononitrile has two potential sites for reaction: the reactive methylene and the cyano groups. The chemistry of the reactive methylene of malononitrile very closely parallels that of diethyl malonate.

Preparation. The only commercially important process for the preparation of malononitrile is the Lonza process. This continuous process, based on the pyrolysis of cyanogen chloride and acetonitrile, provides extremely pure material.

Health and safety factors (toxicology). Malononitrile is highly toxic and can be absorbed through the skin, by inhalation, or by swallowing; acute oral toxicity of LD_{50} 18.6 mg/kg has been reported.

Uses. Malononitrile can replace sodium cyanide in leaching gold from refractory carbonaceous ores and yields 10–20% higher recovery. Many uses take advantage of its particularly good ability to form heterocyclic compounds. Perhaps the most important use of malononitrile is in the synthesis of Vitamin B_1.

DAVID W. HUGHES
Dow Chemical U.S.A.

Two useful technical brochures are available from Lonza, Inc., 22-10 Route 28, Fair Lawn, N.J. 07410 or Lonza Ltd., P.O. Box CH-4002 Basel, Switzerland; *Malonic and Cyanoacetic Esters*, and *Malononitrile*, 2nd ed.
Beilsteins Handbuch der Organischer Chemie, Springer-Verlag, Heidelberg.

MALTS AND MALTING

Malting is essentially the same process as occurs when seeds fall to the ground or are planted, moistened by water, and germinate. During germination, rootlets (sprouts) and a nascent stem (acrospire) emerge; simultaneously, enzymes are produced or activated and the cellular structure and composition are modified, resulting in a product that can be used as a substrate for fermented beverages and as a food adjunct. The terms malts and malting can apply to any germinated grain; however, nearly all commercial malting involves barley. Because the brewing process and finished beer characteristics are a function of malt properties, malting is considered to be part of the brewing process. Brewers malt is designed to provide fermentable carbohydrates, assimilable nitrogen, as well as precursors for beer flavor (see Beer).

Most of the malt not produced for beer is used to make distilled spirits and a small amount is used in foods (see Beverages spirits, distilled).

Manufacturing and Processing

Two chief types of malting-grade barley are in use, ie, six-row and two-row. The main growing areas for barley are North Dakota, eastern South Dakota, and western Minnesota; six-row barley is predominant.

The malting process consists of three steps: steeping, germination, and kilning (see Fig. 1). Prior to steeping, barley is cleaned and then sized according to kernel width. After kilning, the malt is cleaned, stored, and blended with other malt to meet customer specifications.

Table 1. Physical Properties of Diethyl Malonate and Dimethyl Malonate

	Diethyl malonate	Dimethyl malonate
mp, °C	−49.8	−61.9
bp, °C		
101.3 kPa[a]	198.9	181.4
2.4 kPa[a]	92.0	
n_D^{20}	1.41428	1.4138
d_4^{25}, g/cm³	1.055	1.153
flash point, °C	75	90
solubility (at 20°C), g/100 mL H_2O	2.8	
organic solvents	very soluble	very soluble

[a]To convert kPa to mm Hg, multiply by 7.5.

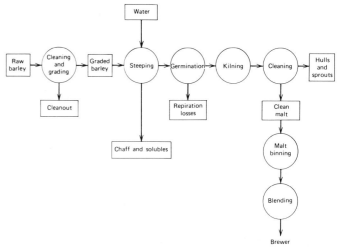

Figure 1. Malting process.

New technology. Barley breeding. The barley-breeding programs at state experimental stations involve conventional crossbreeding techniques and have resulted in barley varieties of better yield, disease resistance, and malt quality (National Barley and Malt Laboratory, Madison, Wisc.; Brewing and Malting Research Institute, Winnipeg, Canada).

A substantial breakthrough in improving malting quality through the use of mutagenic techniques is the development of a proanthocyanidin-free variety that yields a beer that is colloidally stable and, thus, does not require stabilization in the brewery.

Growth regulators. Perhaps the most significant scientific contribution to malting technology has been the use of gibberellic acid and an increased understanding of its role in the malt modification process. When small amounts are added, eg, 0.01–1 ppm, dry barley basis, germination time is reduced by at least 24 h.

Nitrosamines and kilning. In order to avoid the formation of nitrosamines, which are attributed to the reaction of nitrogen oxides and amines that are preset in barley, either indirect heating of kiln air or using low $(NO)_x$ burners (Maxon Corporation, Muncie, Ind.) on direct-fired kilns are being applied (see *N*-Nitrosamines).

Health and Safety Factors

Dust-control systems and good housekeeping are employed in the barley and malt elevators, steephouse, and other processing areas to eliminate or minimize the potential of dust explosions and inhalation. Although low levels of sulfur dioxide are employed (0.1–2.0 kg S/(h·t malt), the potential of toxic sulfur dioxide concentrations resulting from process or operator error does exist. Fumigants and various cleaning agents are used routinely in the malting industry in accordance with safe operating practices.

By-Products

The main by-products from the malting industry are malt sprouts, clean-out material, and small-kernel barley. Since the protein is readily available, malt sprouts are used in various animal-feed blends. Occasionally, malt hulls and barley chaff are blended with malt sprouts. The remainder of the clean-out material and small-kernel barley is sold as feed (see Pet and livestock feeds).

MICHAEL R. SFAT
JAMES A. DONCHECK
Bio-Technical Resources, Inc.

E.B. Adamic, *The Practical Brewer*, Master Brewers Association of the Americas, Madison, Wisc., 1977, pp. 21–39.

A.H. Cook, *Barley and Malt*, Academic Press, Inc., New York, 1962.

J.S. Hough, D.E. Briggs, and R. Stevens, *Malting and Brewing Science*, Chapman and Hall, London, UK, 1971.

MANGANESE AND MANGANESE ALLOYS

Manganese, atomic number 25, atomic weight 54.95, belongs to Group VIIB in the periodic table. Its isotopes are ^{51}Mn, ^{52}Mn, ^{54}Mn, ^{55}Mn, and ^{56}Mn, but ^{55}Mn is the only stable one. A gray metal resembling iron, manganese is hard and brittle and of little use alone. Its principal use in a metallic form is as an alloying element and cleansing agent for steel, cast iron, and nonferrous metals. Manganese is essential to the steel industry where it is used normally in the form of a ferroalloy (see Steel).

Minerals and Ores

Manganese occurs in many minerals that are widely distributed in the earth's crust. It constitutes about 0.1% of the earth's crust and is the twelfth most abundant element. Since 1960, the principal sources of commercial grades of manganese ore for the noncommunist world have been Australia, Brazil, Gabon, India, and the Republic of South Africa. The chief minerals of manganese are pyrolusite, psilomelane, manganite, and hausmannite (see also Ocean raw materials).

Table 1. Properties of Manganese

Property	Value
melting point, °C	1244
boiling point, °C	2060
density at 20°C, g/cm^3	7.4
specific heat at 25.2°C, J/g^a	0.48
latent heat of fusion, J/g^a	244
linear coefficient of expansion (0–100°C), per °C	22.8×10^{-6}
hardness, Mohs scale	5.0
compressibility	8.4×10^{-7}
solidification shrinkage, %	1.7
latent heat of vaporization at bp, J/g^a	4020
standard electrode potential, V	1.134
magnetic susceptibility, m^3/kg	1.21×10^{-7}

^{a}To convert J to cal, divide by 4.184.

Properties

The properties of manganese metal are given in Table 1.

Alloy Processing

Smelting. The largest demand for manganese is in ferrous metallurgy. Manganese suitable for addition to steel and cast iron can be made by reduction of manganese ore with carbon.

The largest-tonnage manganese addition alloy used in the steel industry is high carbon ferromanganese, commonly known in the United States as standard ferromanganese. Since 1978, the production of high carbon ferromanganese by the blast-furnace method has been discontinued in the United States and Japan, and is rapidly declining in Europe.

To take up the shortfall and satisfy the increased demand for ferromanganese resulting from the large increase in world production of steel, many new large high-carbon ferromanganese electric furnaces have been built, ranging from 20 to 40 MW in size.

Other alloys include silicomanganese, low carbon silicomanganese, and refined ferromanganese.

Electrolytic Processes

Electrolysis of aqueous solutions. The electrolytic process for manganese, which was pioneered by the U.S. Bureau of Mines, is used by Elkem Metals Company, Foote Mineral Company, and Kerr-McGee Chemical Corporation.

Manganese ore is roasted to reduce the higher oxides to manganese(II) oxide which is acid soluble, or as practiced by Elkem Metals, manganese(II) oxide is supplied from slag produced in the high carbon ferromanganese smelting operation. The reduced ore or slag is then leached with sulfuric acid at pH 3 to give manganese(II) sulfate. The solution is neutralized with ammonia to pH 6–7 to precipitate iron and aluminum which are later removed by filtration. Arsenic, copper, zinc, lead, cobalt, and molybdenum are removed as sulfides after introducing hydrogen sulfide gas. Ferrous sulfide or ammonium sulfide plus air is then added to remove colloidal sulfur, colloidal metallic sulfides, and organic matter. The purified liquid is electrolyzed in a diaphragm cell.

Fused-salt electrolysis. This process, developed and used by the Chemetals Corporation, is in many ways similar to the Hall process for producing aluminum (see Aluminum). The feed is manganese ore that has been reduced to the manganese(II) state. It is charged to the electrolytic cell which contains molten calcium fluoride and lime.

Health and Safety Factors

Manganese in trace amounts is an essential element for both plants and animals and is among the trace elements least toxic to mammals including humans; however, exposure to abnormally high concentrations of manganese, particularly in the form of dust and fumes, is known to have resulted in adverse effects to humans (see Mineral nutrients).

Uses

In steel (qv) manganese serves first as a control of hot shortness in the finishing process caused by sulfur, by combining with the sulfur to given manganese sulfide; second, as a dioxidizer; and third, most importantly, as an alloying agent to improve strength, toughness, hardness, hardenability, and abrasion resistance. In case iron, manganese neutralizes the sulfur and adds strength. The use of manganese in aluminum has increased substantially with the growing use of aluminum containers for beverages, where manganese imparts strength, hardness, and stiffness.

L.R. MATRICARDI
J.H. DOWNING
Elkem Metals Co.

G. Volker and co-workers, "Ferromangan and Mangan" in *Metallurgie der Ferrolegierungen*, Springer, New York, 1972.

P.L. Dancoisne, *Mater. Tech.*, special issue 15 (Dec. 1977).

G.L. DeHuff and T.S. Jones, *Manganese, Mineral Commodity Profiles*, U.S. Bureau of Mines, Washington, D.C., July 1979.

MANGANESE COMPOUNDS

Manganese is contained so ubiquitously in the earth's crust and its water bodies that it can be regarded as being everywhere. At present, the United States depends completely on imports for its manganese needs even though sizable stockpiles are maintained by the government for emergencies. World reserves of manganese ore of all grades are estimated to exceed 2×10^9 metric tons. The largest known deposits are in South Africa, followed by USSR, Gabon, India, Australia, and Brazil in that order.

Physical Properties

Table 1 lists some properties of the more important manganese compounds, many of which are commercial products.

Chemical Properties

Hydrogen dissolves to some extent in manganese metal, but does not react with it. Exposure to oxygen leads to the ready formation of manganese oxides, especially at high temperatures. Nitrogen above $740°C$ forms solid solutions as well as several nitrides such as MnN, Mn_6N_5, Mn_3N_2, Mn_2N, and Mn_4N. Manganese nitrides are used in steelmaking as nitrogen-containing intermediate alloys. Carbon reacts with molten manganese with formation of various carbides including $Mn_{23}C_6$, Mn_3C, Mn_7C_3, Mn_2C_7, and $Mn_{15}C_4$. Manganese tricarbide reacts with water to yield about 75% H_2, 12–15% CH_4, and 6–8% ethylene. It is an important factor in a fuel-alloy process designed to produce liquid hydrocarbons. In steel and other ferrous alloys, manganese carbides achieve the desired mechanical properties. With silicon, manganese forms a series of silicides (Mn_3Si, Mn_5Si_3, $MnSi$, $MnSi_{1.7}$) that have excellent heat-resisting properties. Manganese forms compounds only with a limited number of metals, ie, Au, Be, Zn, Al, In, Ti, Ge, Sn, As, Sb, Bi, Ni, and Pd. In the commercially important iron–manganese system, no compounds are formed.

In the first series of transition metals, manganese is the element with the most numerous oxidation states. In its highest oxidation state, manganese resembles chlorine, ie, the pairs Mn_2O_7–Cl_2O_7 and $HMnO_4$–$HClO_4$ have similar characteristics. Otherwise, manganese behaves more like iron with regard to both the physical and chemical properties of its compounds.

Although manganese compounds of all valence states from -3 to $+7$ are known, the most important ones have valences of $+2$, $+4$, and $+7$.

The basicity of manganese declines from MnO to Mn_2O_7 (ie, with increasing state of oxidation) and this accounts for the fact that oxo anions are more readily formed in the higher valence states. Another characteristic of manganese chemistry of the higher valence states is the abundance of disproportionation and reverse disproportionation (synproportionation) reactions.

Lower oxidation states (-3 to $+1$). A new family of highly reduced metal carbonyls, based on the anion $Mn(CO)_4^{3-}$, has recently been described. The oxidation state -2 is represented by an anionic complex of Mn with phthalocyanine. Manganese pentacarbonyl anion $Mn(CO)_5^-$ serves as an illustration of -1-valent Mn. A manganese compound with the oxidation state of zero is manganese carbonyl (dimanganese decacarbonyl), $Mn_2(CO)_{10}$.

Divalent manganese. Most soluble forms of manganese occurring in nature are of the divalent state which is stable under acid or neutral condition. In alkaline media, however, a marked susceptibility to oxidation is noted. Divalent manganese forms coordination compounds with both inorganic and organic ligands, such as manganese(II) cetylacetonate and cationic (eg, $[Mn(NH_3)_6]^{2+}$) and anionic (eg, $[Mn(CN)_6^{4-}]$ complexes.

Manganese(II) compounds are characteristically pink or colorless, with the exceptions of MnO (green), $Mn(OH)_2$ (white), and a MnS (green). Some complexes are also green to greenish-yellow. Manganese(II) salts act as oxidation catalysts.

Trivalent manganese. The most stable forms of Mn(III) are the dark-colored manganese sesquioxide, Mn_2O_3 and $Mn_2O_3 \cdot nH_2O$, and manganese oxide, Mn_3O_4. Trivalent manganese also forms a number of anionic complexes, such as the pink to dark-red fluoro- and chloromanganates(III) of the types $M(I)MnF_4$ and $[M_2(I)]MnF_5$, and alkali-metal salts of Mn(III) disulfato and diorthophosphato complexes.

Although the color of Mn(III) compounds in the solid state can vary from red to green, their aqueous solutions mostly have a reddish-purple, almost permanganatelike appearance.

Tetravalent manganese. The simple cation Mn^{4+} is so unstable in aqueous solution that it decomposes instantly. In view of this fact, the great stability of manganese(IV) oxide in contact with water is explained by its extreme insolubility. Only a few stable Mn(IV) compounds are known, such as the blue manganese tetrafluoride. The most important manganese(IV) compound is manganese dioxide in its various forms, which occur in many manganese minerals.

Alkali-metal manganate(IV) is the starting material for hydrate-type manganese dioxide for dry-cell battery applications (see Batteries).

Pentavalent manganese. Manganese(V) appears to exist only as the oxyanion MnO_4^{3-}, generally referred to as manganate(V), but sometimes also as hypomanganate. The most important manganese(V) compound is K_3MnO_4 which is a key intermediate in the manufacture of potassium permanganate.

Hexavalent manganese. The hexavalent state of manganese is represented by a few alkali-metal and alkaline-earth metal salts of the hypothetical manganic acid, H_2MnO_4, namely, sodium, potassium, rubidium, cesium, and barium manganates(VI) and strontium manganate(VI). Alkali manganates(VI) are used as oxidants in synthetic organic chemistry.

Heptavalent manganese. Numerous heptavalent manganese salts are known, including their acids and anhydrides. These compounds are relatively stable and readily accessible. Permanganic acid, $HMnO_4$, is an example.

Many salts of permanganic acid have been prepared, usually from commercial potassium permanganate. Most stable salts of permanganic acid can be made by a metathetic reaction between $KMnO_4$ solutions and the particular hexafluorosilicate.

Permanganate is one of the most versatile oxidizing agents known. Depending on the requirements of a given substrate, MnO_4^- oxidizes inorganic or organic compounds under neutral, acidic, or alkaline conditions in aqueous or nonaqueous media. It exhibits this remarkable flexibility because of its ability to use different reaction paths, and the low enthalpies of activation on the order of 21–42 kJ/mol (5–10 kcal/mol) in conjunction with high redox potentials in the various stages of permanganate ion reduction. An important new development is the adaptation of phase-transfer catalysis to permanganate oxidation.

Commercial Manganese Compounds and Their Manufacture

Manganese carbonate. Synthetic manganese carbonate is made from a water-soluble Mn(II) salt, usually the sulfate, by precipitation with an alkali carbonate.

Table 1. Physical Properties of Selected Manganese Compounds

Compound	Formula	Oxidation state	Appearance	Density, g/cm³	mp, °C	bp, °C	Solubility
methylcyclopentadienylmanganese tricarbonyl	$C_9H_7Mn(CO)_3$	+1	light amber liquid	1.39_{20}	1.5	233	insol H_2O, sol most org solvents
manganese acetate tetrahydrate	$Mn(C_2H_3O_2)_2 \cdot 4H_2O$	+2	pale red crystals	1.589			sl sol H_2O, sol ethanol, methanol
manganese borate	$MnB_4O_7 \cdot 8H_2O$	+2	white to pale red solid				insol H_2O, ethanol, sol dil acids
manganese carbonate (rhodochrosite)	$MnCO_3$	+2	pink solid	3.125	dec > 200		sol prod H_2O: 8.8 × 10^{-11}, sol in dil acids
manganese chloride	$MnCl_2$	+2	pink crystal solid	2.977_{25}	652	1190	v sol H_2O, sol pyridine, ethanol, insol ether
manganese hydroxide (pyrochroite)	$Mn(OH)_2$	+2	white to pink	3.26_{25}	dec 140		sol acid, sol base at higher temp
manganese nitrate hexahydrate	$Mn(NO_3)_2 \cdot 6H_2O$	+2	colorless to slightly pink crystals	1.81	25.8	dec	v sol, H_2O, sol ethanol
manganese(II) oxide (manganosite)	MnO	+2	green	5.37_{23}	1945		insol H_2O
manganese sulfate	$MnSO_4$	+2	almost white cryst solid	3.25	dec 850		sol 52 g/100 g H_2O, sl sol methanol, insol ether
trimanganese tetraoxide alpha phase (Hausmannite)	Mn_3O_4	+2, +3	black crystals with metallic sheen	4.84	1560		insol H_2O
manganese(III) acetylacetonate	$Mn(C_5H_7O_2)_3$	+3	brown to black crystal solid		172		insol H_2O, sol org
manganese(III) fluoride	MnF_3	+3	red crystals	3.54	dec (stable to 600)		dec H_2O
manganese(III) oxide α	Mn_2O_3	+3	black to brown solid	4.89_{25}	871–887 dec		insol H_2O
manganese(III) oxide γ, hydrated	$MnO(OH)$	+3	black solid	4.2–4.4	250 dec to gamma Mn_2O_3		insol H_2O, disproportionates in dilute acids
pentamanganoctoxide	Mn_5O_8	+2, +4	black solid	4.85_{20}	550 dec to alpha Mn_2O_3		insol H_2O
manganese(IV) oxide β, pyrolusite	MnO_2	+4	black to gray crystal solid	5.118_{25}	500–600 dec		insol H_2O
potassium manganate(IV)	K_2MnO_3	+4	black microscopic crystals	3.071_{25}	1100		dec H_2O (disproportionates)
barium manganate(V)	$Ba_3(MnO_4)_2$	+5	emerald green crystals	5.25	dec 960		insol H_2O
potassium manganate(V)	K_3MnO_4	+5	turquoise blue microscopic crystals	2.78	dec 800–1100		v sol H_2O dec, hygroscopic, sol 40% KOH at −15°C
potassium manganate(VI)	K_2MnO_4	+6	dark green to black needles	2.80_{23}	dec 600		sol H_2O dec
manganese heptoxide	Mn_2O_7	+7	dark red oil	2.396	5.9 dec 55		v sol H_2O, hygroscopic
ammonium permanganate	NH_4MnO_4	+7	dark purple rhombic bipyramidal needles	2.22_{25}	dec > 70		8 g/100 g H_2O at 15°C (86 g/L at 25°C)
calcium permanganate tetrahydrate	$Ca(MnO_4)_2 \cdot H_2O$	+7	black crystals, solutions look purple	Ca 2.49	dec 130–140		388 g/100 g H_2O at 25°C, deliquescent
potassium permanganate	$KMnO_4$	+7	dark purple bipyramidal rhombic prisms	2.703_{20}	dec 200–300		sol H_2O, acetic acid, trifluoroacetic acid, acetic anhydride, acetone, pyridine, benzonitrile, sulfolane

Manganese chloride. Manganese chloride is made by reaction of manganese metal, oxide, hydroxide, or carbonate with hydrochloric acid.

Manganese nitrate. Manganese nitrate is prepared from manganese(II) oxide or carbonate with dilute nitric acid, or from MnO_2 and a mixture of nitrous and nitric acids.

Manganese(II) oxide. Manganese(II) oxide is an important precursor of many commercial manganese compounds as well as an ingredient in fertilizer and feedstuff formulations. It is made from manganese dioxide ores by reductive roasting.

Dimanganese trioxide, Mn_2O_3, **trimanganese tetroxide,** Mn_3O_4. These oxides are usually manufactured from manganese dioxide by means of thermal deoxidation.

Chemical manganese dioxide (CMD). Chemical manganese dioxide is made either by chemical reduction of permanganate (Type 1) or by thermally decomposing manganese salts, such as $MnCO_3$ or $Mn(NO_3)_2$ under oxidizing conditions (Type 2), followed if necessary, by oxidation in the liquid phase.

Electrolytic manganese dioxide (EMD).   The first anodic oxidation of an Mn^{2+} salt to manganese dioxide goes back to 1830 but the usefulness of electrolyically prepared manganese dioxide for battery purposes was not recognized until 1918. The remarkable electrochemical characteristic of EMD is its ability to function superbly as a solid-state oxygen electrode in dry-cell batteries over a wide range of discharge conditions.

Potassium permanganate.   The only one-step method of producing potassium permanganate is based on electrolytic conversion of ferro-manganese to permanganate. The others start from MnO_2 ore and need two steps, ie, thermal synthesis of potassium manganate(VI) followed by the electrolytic oxidation of MnO_4^{2-} to MnO_4^-.

Manganese sulfate.   Manganese sulfate, an important intermediate as well as product, is made by dissolving manganese carbonate ore (rhodochrosite) or manganese(II) oxide in sulfuric acid. It is also obtained as a by-product in the manufacture of hydroquinone from aniline sulfate and manganese dioxide, usually a manganese ore with 78–85% MnO_2.

Other Mn salts and compounds.   Most water-soluble manganese(II) salts are made from the corresponding acids and MnO, $MnCO_3$, or Mn metal.

Health and Safety Factors; Environmental Considerations

As an essential trace element, manganese plays an important role in plant and animal life but it can, if present in excessive amounts, have distinctly toxic effects.

Because of the widespread industrial uses of manganese, a considerable number of people in the United States are exposed occupationally to airborne Mn. To prevent excessive exposure, maximum limits for the presence of manganese in air have been set in the United States and other industrialized countries.

Environmental control.   In view of the toxic effects of manganese and its compounds, it is necessary to limit emissions from industrial and other synthetic sources. In addition, airborne manganese is known to catalyze the oxidation of sulfur dioxide to sulfur troxide, thereby increasing the pollution effects of atmospheric sulfur. For wastewater to be discharged into a storm sewer, the current limit is 1 ppm. Industrially, manganese-bearing particulate matter is usually removed from air with dust-collecting devices including electrostatic precipitators, fabric filter systems (baghouse), cyclones, and wet scrubbers. In case of liquid effluents, soluble manganese is precipitated as a hydrous higher oxide before separation.

The final disposal of manganese dioxide-containing sludges depends on local conditions and regulations; usually they are deposited in landfills.

Uses

Among the manganese oxides, the dioxide is by far the most important. Some of the uses of manganese dioxide are based on its available or active oxygen. Thus, manganese dioxide is an essential component in dry-cell batteries, and carbon–zinc, zinc chloride, and alkaline cells.

Manganese dioxide is an oxidizing agent for the production of various organic compounds and in the curing of polysulfide rubber sealants (qv).

Other applications of MnO_2 are based primarily on its manganese content and include brick, tile, glass, and frit manufacture, and the production of ceramic magnets (ferrites) and welding rods and fluxes.

Manganese(II) oxide is used in agriculture as a fertilizer and feed additive.

The uses of potassium permanganate are based on the active oxygen content of the $KMnO_4$, which effects controlled or degradative oxidation of organic molecules. Potassium permanganate also is used in the purification of potable and wastewater.

ARNO H. REIDIES
Carus Chemical Company

S.A. Weiss, *Manganese—The Other Uses, A Study of the Non-Steelmaking Applications of Manganese*, Metal Bulletin Books, London, UK, 1977.

"Mangan" in *Gmelin's Handbuch der Anorganischen Chemie*, 8th ed., System-Nummer 56, Verlag Chemie, Weinheim, FRG, 1973.

R.D.W. Kemmit, "Manganese" in J.C. Bailar, H.J. Emeleus, R. Nyholm, and A.F. Trotman-Dickenson, eds., *Comprehensive Inorganic Chemistry*, Vol. 3, Pergamon Press Ltd., Oxford, UK, 1973, pp. 771–876.

MANNITOL.   See Alcohols, polyhydric.

MANNITOL HEXANITRATE.   See Explosives.

MANNOSE.   See Carbohydrates; Sweeteners.

MANUFACTURED GAS.   See Fuels, synthetic, gaseous fuels.

MARCASITE.   See Iron; Iron compounds.

MARGARINE.   See Vegetable oils.

MARKET AND MARKETING RESEARCH

Market research is a well-established tool for management decisions in the chemical-process industries. Marketing research is an extension of the former procedure to identify opportunities for sales and profits more specifically.

Market research may be short or long term. Some market analysts use these time frames: short term—up to 18 mo; intermediate term—19 mo to 5 yr; long term—5 to 15 yr; and futuristic—more than 15 yr. In general, short-term market research is synonymous with sales analysis and is used to assist the sales manager in setting goals, measuring performance, and giving the production department operating targets.

Intermediate or long-term market research has as its objectives the quantifying of markets for a particular chemical in terms of tonnages, growth potentials, general location of markets, competitive factors, and the impact of existing or potential government regulations on the market.

Marketing research, as compared with market research, is more directly concerned with identifying existing or potential users of a product, their present sources of supply, the nature and duration of any contracts that exist between seller and buyer, competitors' strategies in product development and pricing, requirements for facilities and personnel to compete successfully, and increasingly, the status of competition from plants in other countries. Also, particularly in the United States, government regulations involving production of chemicals, their transportation, and disposal of wastes and by-products have a marked influence on the profitability of most chemical-process industry products.

Market-research studies usually originate in the sales or marketing groups of a company. As a general rule, the sales analysis or short-term-type study is done by in-house personnel, often on a continuous basis. Field sales personnel are often used to assist the market-research group in securing data. Long-term market-research studies often originate in sales or marketing groups if the company already produces the product. If a new product is involved, the study may originate in the development (R&D) group or at the corporate planning level. Marketing-research studies usually originate in the higher levels of management (eg, general manager or vice president). This is especially true if the proposed study is for a product new to the company.

Use of consultants to conduct chemical-market- and marketing-research studies began in the early 1950s in the United States and grew rapidly during the mid-1960s. Today, there are at least one hundred well-known and capable consultants or consulting firms in the United States performing this function for individual clients or on a multiple-client basis.

Purchasing Research

A relatively new but growing responsibility for some market-research practitioners is purchasing research. The objective of this kind of research is to ensure availability of raw materials at competitive prices for three or more years ahead (see also Feedstocks).

EDWARD TARNELL
Roger Williams Technical & Economic Services, Inc.

D.D. Lee, *Industrial Marketing Research, Techniques and Practices*, Technomic Publishing Co., Westport, Conn., 1978.

R.L. Carlsen, *Chem. Purchas.* **13**, 71 (Sept. 1977).

MARTENSITE. See Metal treatments; Steel.

MASS SPECTROMETRY. See Analytical methods.

MASS TRANSFER

Diffusional mass transfer involves the migration of one substance through another under the influence of a concentration gradient. Chemical engineering applications in separation processes involve diffusional transport of some component within a single phase or between two immiscible phases that have been brought into contact to enable transfer of the component from one phase to the other. Components may migrate from the bulk of one phase to the interface between phases and remain there, as in adsorption or crystallization. Alternatively, penetration of the interface may occur, followed by diffusion into the bulk of the other phase, as in the operations of distillation, gas absorption, and liquid–liquid extraction.

PART 1: TRANSFER BY MOLECULAR DIFFUSION

Flux definitions. If the moles and the mass of component A per unit volume of mixture are c_A and ρ_A, respectively, then the mole fraction of A is c_A/c or x_A, and the mass fraction is ρ_A/ρ or w_A.

In a nonuniform fluid mixture with n components that is experiencing bulk motion, the statistical mean velocity of component i in the x direction with respect to stationary coordinates is u_i. The molal and mass-average velocities of the mixture in the x direction are then defined as

$$U = \frac{1}{c} \sum_{i=1}^{n} c_i u_i, \qquad u = \frac{1}{\rho} \sum_{i=1}^{n} \rho_i u_i \qquad (1)$$

The corresponding definition of molal fluxes in the x direction for component i are as follows:

relative to stationary coordinates: $N_{ix} = c_i u_i$ (2)

relative to the mass average velocity: $I_{ix} = c_i(u_i - u)$ (3)

relative to the molal average velocity: $J_{ix} = c_i(u_i - U)$ (4)

To relate molal fluxes J_{ix} and N_{ix}, consider equations, 1, 3, and 4:

$$\begin{aligned} J_{ix} &= c_i u_i - c_i U \\ &= N_{ix} - \frac{c_i}{c} \sum_{i=1}^{n} c_i u_i = N_{ix} - x_i \sum_{i=1}^{n} N_{ix} \end{aligned} \qquad (5)$$

and for a binary mixture,

$$J_{Ax} = N_{Ax} - x_A(N_{Ax} + N_{Bx}) \qquad (6)$$

For nonreacting systems of two components A and B, Fick's first law of molecular diffusion for steady one-dimensional transfer with constant c is

$$J_{Az} = -D_{AB}\frac{dc_A}{dz}, \qquad J_{Bz} = -D_{BA}\frac{dc_B}{dz} \qquad (7)$$

and for a perfect gas:

$$J_{Az} = -\frac{D_{AB}}{RT}\frac{dp_A}{dz}, \qquad J_{Bz} = -\frac{D_{BA}}{RT}\frac{dp_B}{dz} \qquad (8)$$

In the general case in which a steady total or bulk flow is imposed upon the fluid mixture in the direction in which component A is diffusing, equation 6 gives

$$N_{Az} = (N_{Az} + N_{Bz})\frac{p_A}{P} - \frac{D_{AB}}{RT}\frac{dp_A}{dz} \qquad (9)$$

Integrating for constant N_{Az}, N_{Bz}, and D_{AB},

$$N_{Az} = \frac{D_{AB}P}{RTz}\left(\frac{1}{1+\gamma}\right)\ln\left[\frac{1-(1+\gamma)\dfrac{p_{A2}}{P}}{1-(1+\gamma)\dfrac{p_{A1}}{P}}\right], \qquad \gamma = N_{Bz}/N_{Az} \tag{10}$$

This reduces to two special cases of molecular diffusion: equimolal counterdiffusion, often approximated in distillation, and unimolal unidirectional diffusion, where only one molecular species, component A, diffuses through component B, which is motionless relative to stationary coordinates. This is frequently approximated in gas absorption, liquid–liquid extraction, and adsorption.

Steady-state equimolal counterdiffusion in gases. In this case $N_{Az} = -N_{Bz}$. Applying l'Hopital's rule to equation 10 for $\gamma = -1$,

$$N_{Az} = \frac{D}{RTz}(p_{A1} - p_{A2}) \qquad (11)$$

Steady-state unimolal unidirectional diffusion in gases. In this case N_{Bz} equals zero, so equation 10 reduces to

$$N_{Az} = \frac{DP}{RTz}\ln\frac{p_{B2}}{p_{B1}} = \frac{DP}{RTz}\left(\frac{p_{B2}-p_{B1}}{p_{BLM}}\right) = \frac{D}{RTz}\left(\frac{P}{p_{BLM}}\right)(p_{A1}-p_{A2})$$

$$p_{BLM} = \frac{p_{B2}-p_{B1}}{\ln(p_{B2}/p_{B1})}, \qquad z = z_2 - z_1 \qquad (12)$$

Molecular diffusion in liquids. In the absence of a fully developed kinetic theory for liquids, the relationships for molecular diffusion are usually assumed to parallel those for gases. Thus, for equimolal counterdiffusion,

$$N_{Az} = \frac{D}{z}(c_{A1} - c_{A2}) = \frac{Dc}{z}(x_{A1} - x_{A2}) \qquad (13)$$

For unimolal unidirectional diffusion, the liquid-phase analogue of equation 12 is

$$N_{Az} = \frac{D}{z}\left(\frac{c}{c_{BLM}}\right)(c_{A1} - c_{A2}) = \frac{Dc}{z}\frac{(x_{A1}-x_{A2})}{x_{BLM}} \qquad (14)$$

Steady-state diffusion from a sphere. The equation for steady-state diffusion from the surface of a single sphere into an infinite, stagnant, surrounding fluid is

$$N_{Sh} = \frac{k_c d_s}{D} = 2 \qquad (15)$$

where the mass-transfer coefficient k_c is defined as

$$4\pi r^2 N_{Ar} = k_c \pi d_s^2 (c_{As} - c_{A\infty}) \qquad (16)$$

Unsteady-State Diffusion

In unsteady-state diffusion processes, the concentration at a given point varies with time. Consideration of the rate of accumulation of A within the volume element $dx\,dy\,dz$ in an isotropic medium that is free from convection leads to Fick's second law of molecular diffusion as follows:

$$\frac{\partial c_A}{\partial t} = D\frac{\partial^2 c_A}{\partial z^2} \qquad (17)$$

Unsteady-state diffusion in a slab. For analytical purposes, it is assumed that the edges of the slab are sealed against mass transfer, with diffusion occurring through the two large, opposite surfaces. The origin of coordinates is at one large surface of the slab, with the other large surface at $z = 2a$. Equation 17 is then solved by separation of variables for the following boundary conditions:

$$c_A(z, 0) = c_{A0}, \qquad c_A(z, \infty) = c_A^*$$
$$c_A(0, t) = c_A^*, \qquad c_A(2a, t) = c_A^*$$

The result is

$$c_A = c_A^* + \frac{4}{\pi}\left(c_{A0} - c_A^*\right)\sum_{n=0}^{\infty} \frac{1}{2n+1}\sin\left[\frac{(2n+1)\pi z}{2a}\right]$$
$$\times \exp\left[\frac{-D(2n+1)^2\pi^2 t}{4a^2}\right] \tag{18}$$

This gives c_A at any (z, t). The average concentration throughout the slab at time t is given by $\bar{c}_A$ as

$$\frac{c_{A0} - \bar{c}_A}{c_{A0} - c_A^*} = 1 - \frac{\bar{c}_A - c_A^*}{c_{A0} - c_A^*}$$
$$= 1 - \frac{8}{\pi^2}\sum_{n=0}^{\infty}\frac{1}{(2n+1)^2}\exp\left[\frac{-D(2n+1)^2\pi^2 t}{4a^2}\right] \tag{19}$$

Numerical application of equation 19 is facilitated by Newman's plot (1) in the form of $1 - (c_{A0} - \bar{c}_A)/(c_{A0} - c_A^*)$ vs Dt/a^2. Analogous results are available for diffusion in spheres, cylinders, etc (1).

Diffusivities in gases. An empirical estimation for the binary system $A + B$, based on the form predicted by kinetic theory, is

$$D_{AB} = \frac{0.0150 T^{1.81}}{P(T_{cA}T_{cB})^{0.1405}\left(V_{cA}^{0.4} + V_{cB}^{0.4}\right)^2}\sqrt{\frac{1}{M_A} + \frac{1}{M_B}} \tag{20}$$

in which critical temperature T_c is in K, critical volume V_c is in cm^3/mol, P is in atm (101.3 kPa), and D is in cm^2/s.

Diffusivities in liquids. For dilute diffusion of nonelectrolytes in organic solvents, the Lusis and Ratcliff equation is

$$\frac{D_{AB}\mu_B}{T} = 8.52(10^{-8})V_{bB}^{-1/3}\left[1.40\left(\frac{V_{bB}}{V_{bA}}\right)^{1/3} + \frac{V_{bB}}{V_{bA}}\right] \tag{21}$$

where A is solute, B is solvent; D_{AB} is in cm^2/s; μ_B is the viscosity of the solvent, $\text{mPa}\cdot\text{s}$ (= cP); T is in K; V_b is the molal volume at normal boiling temperature in cm^3/mol. For dilute aqueous solutions, a recent relation is

$$D_{AB} = \frac{13.26 \times 10^{-5}}{\mu_{wT}^{1.4} V_{bA}^{0.589}} \tag{22}$$

where μ_{wT} is the viscosity of water at T in $\text{mPa}\cdot\text{s}$ (= cP). Average errors of ca 8%, 29%, and 11% have been reported for equations 20, 21, and 22, respectively.

Mass-transfer coefficients. A solute such as component A is distributed between two immiscible fluid phases in contact with each other, where one phase may be a gas and the other a liquid.

It is assumed that local equilibrium prevails at the interface between phases, where the compositions are Y_A^* and X_A^*. If transfer takes place from the gas to the liquid, the individual coefficients k_Y and k_X for the gas and liquid phases, respectively, are defined as follows:

$$N_A A = k_Y A\left(Y_A - Y_A^*\right) = k_X A\left(X_A^* - X_A\right) \tag{23}$$

A is the area of the interface, where the flux is N_A, and Y_A and X_A are concentrations of component A in the bulk of the gas and liquid phases. Interfacial concentrations (X_A^*, Y_A^*) are often unknown; it is then convenient to use overall coefficients K_Y and K_X, defined in terms of overall concentration differences or "driving forces," as shown below:

$$N_A A = K_Y A\left(Y_A - Y_{AL}\right) = K_X A\left(X_{AG} - X_A\right) \tag{24}$$

It can be shown that

$$\frac{1}{K_Y} = \frac{1}{k_Y} + \frac{m}{k_X}, \qquad \frac{1}{K_X} = \frac{1}{k_X} + \frac{1}{m'k_Y}, \qquad \frac{1}{K_X} = \frac{1}{m''K_Y} \tag{25}$$

where

$$m = \frac{Y_A^* - Y_{AL}}{X_A^* - X_A}, \qquad m' = \frac{Y_A - Y_A^*}{X_{AG} - X_A^*}, \qquad m'' = \frac{Y_A - Y_{AL}}{X_{AG} - X_A} \tag{26}$$

When the equilibrium curve is linear on an $X_A - Y_A$ plot, its slope is $m = m' = m''$. When the interfacial area A is unknown, the rate equations become

$$N_A A = k_Y a\left(Y_A - Y_A^*\right)_m V_0 = k_X a\left(X_A^* - X_A\right)_m V_0$$
$$= K_Y a\left(Y_A - Y_{AL}\right)_m V_0 = K_X a\left(X_{AG} - X_A\right)_m V_0 \tag{27}$$

where subscript m denotes a suitable mean, V_0 is the contacting volume of the equipment, a is the interfacial area per unit volume, and the combined quantities $k_X a$, $k_Y a$, $K_X a$, and $K_Y a$ are called volumetric or capacity coefficients.

The two-film theory. This theory supposes that motion in the two phases dies out near the interface, and the entire resistance to transfer is considered as being contained in two fictitious films on either side of the interface, in which transfer occurs by purely molecular diffusion. It is postulated that local equilibrium prevails at the interface and that the concentration gradients are established so rapidly in the films compared to the total time of contact that steady-state diffusion may be assumed. The theory shows the mass-transfer coefficients to be independent of solute concentration for equimolal counterdiffusion but not for unimolal unidirectional diffusion. In both mechanisms, the two-film theory predicts that the mass-transfer coefficient is directly proportional to the molecular diffusivity to the power unity.

The penetration theory. The penetration theory supposes that turbulence transports eddies from the bulk of the phase to the interface, where they remain for a short but constant time before being displaced back into the interior of the phase to be mixed with the bulk fluid. Solute is assumed to penetrate into a given eddy during its stay at the interface by a process of unsteady-state molecular diffusion, in accordance with Fick's second law (eq. 17). The penetration theory predicts that the mass-transfer coefficient is directly proportional to the square root of the molecular diffusivity, in contrast to the two-film theory. Many other theories are reviewed elsewhere (1).

PART 2: TRANSFER BY MOLECULAR DIFFUSION PLUS CONVECTION WITH KNOWN VELOCITY FIELDS

Mass Transfer in Laminar Flow

The laminar-boundary layer on a flat plate. Consider a fluid, designated component B, which is in laminar flow over a flat plate oriented parallel to the undisturbed stream. Suppose that component A is diffusing from the surface of the plate into the fluid stream. The thickness of the concentration-boundary layer δ_c at a given distance x from the leading edge of the plate is defined as the normal distance from the plate surface at which $(\rho_{A0} - \rho_A) = 0.99(\rho_{A0} - \rho_{A\infty})$, where $\rho_{A\infty}$ is concentration in the undisturbed stream. The thickness of the momentum-boundary layer δ is defined as the normal distance from the late surface at which the velocity u is 99% of its value in the undisturbed stream u. In the general case, δ and δ_c are not identical at a given x. Assuming a cubic polynomial form for the velocity and concentration distributions in the respective boundary layers, the mean mass-transfer coefficient over the range $0 \le x \le L$ is derived (1) as

$$(N_{Sh})_m - \frac{k_{\rho m}L}{D} = 0.646(N_{ReL})^{1/2}(N_{Sc})^{1/3} \tag{28}$$

Transition to a turbulent boundary layer normally begins in the range

$$3 \times 10^5 \le x_c u_\infty \rho/\mu \le 3 \times 10^6$$

Laminar natural convection on a vertical plate. Variations in concentration with position in a fluid may be associated with corresponding changes in density. This results in natural convection flows whose influence on mass transfer may be significant, particularly in the absence of forced convection (1).

Laminar flow through a tube. Applications of mass transfer in flow through a tube include hemodialysis in the artificial kidney, desalination of seawater by reverse osmosis, transpiration and film cooling in jet engines and rocket motors, and absorption in wetted-wall columns, reactors, and the like. The appropriate differential equation in fully developed flow is

$$\frac{\partial \rho_A}{\partial x} = \frac{D}{u}\left[\frac{\partial^2 \rho_A}{\partial r^2} + \frac{1}{r}\frac{\partial \rho_A}{\partial r}\right] \tag{29}$$

Solutions for plug flow and for a parabolic velocity distribution, respectively, appear in graphical form in Figure 1, where $\rho_{AW} \neq f(x)$. The coefficient in the ordinate is for use in the equation

$$\frac{\pi d_t^2}{4} V(\rho_{AB0} - \rho_{Ai}) = k_{\rho a}(\pi d_t L)\frac{(\rho_{AW} - \rho_{Ai}) + (\rho_{AW} - \rho_{AB0})}{2} \tag{30}$$

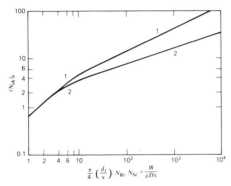

Figure 1. Average Sherwood number for mass transfer in laminar flow through a tube with uniform wall concentration. Curve 1: plug flow. Curve 2: fully developed parabolic velocity distribution.

Transfer with a high mass flux. Distortion of the velocity field caused by mass transfer may occur if the solute flux is sufficiently high. Correction procedures are available (1).

Mass Transfer in Turbulent Flow

The turbulent-boundary layer on a flat plate. Prandtl's "one-seventh power law" forms of concentration and velocity distributions are used in the respective boundary layers (for $xu_\infty\rho/\mu$) up to about 10^7 to obtain the integrated mean $k_{\rho m}$ over the range $0 \leq x \leq L$

$$(N_{Sh})_m = \frac{k_{\rho m}L}{D} = 0.0365(N_{ReL})^{0.8} \tag{31}$$

The derivation assumes $\delta = \delta_c$, for which $N_{Sc} = 1.0$.

Turbulent natural convection on a vertical plate. Results analogous to those for laminar flow are available (1).

Analogies between momentum and mass transfer. Osborne Reynolds was the first to recognize an analogy between the convective transfer of heat and momentum, and his expressions are extended readily to mass transfer. Improvements in the theory were made by Prandtl and Taylor and subsequently in 1939 by von Kármán (1).

The refinement by von Kármán allows for three regions of flow, ie, the laminar sublayer, the buffer zone, and the turbulent core. The result for tube flow is

$$N_{Sh} = \frac{(f/2)N_{Re}N_{Sc}}{1 + 5\sqrt{f/2}\left\{N_{Sc} - 1 + \ln\left[1 + \frac{5}{6}(N_{Sc} - 1)\right]\right\}} \tag{32}$$

A balance on component A in tube flow is as follows:

$$(\pi d_t^2/4)Vd\rho_{AB} = k_\rho \pi d_t(\rho_{AW} - \rho_{AB})\,dx \tag{33}$$

Integrating,

$$\ln\frac{\rho_{AW} - \rho_{AB1}}{\rho_{AW} - \rho_{AB2}} = \frac{4L}{d_t} \cdot \frac{N_{Sh}}{N_{Re}N_{Sc}} \tag{34}$$

where $N_{Sh}/N_{Re}N_{Sc}$ is given by equation 32, to evaluate ρ_{AB2}. Further refinements in the analogy between heat, mass, and momentum transfer were subsequently made (1, 2).

An alternative approach to the analogy between momentum, heat, and mass transfer in forced convection is through the j factors, defined in general terms by Chilton and Colburn as (1)

$$j_H = \frac{N_{Nu}}{N_{Re}N_{Pr}^{1/3}}, \qquad j_D = \frac{N_{Sh}}{N_{Re}N_{Sc}^{1/3}} \tag{35}$$

For any fully developed flow at a given N_{Re} (1):

$$j_H = j_D = f/2 \quad \text{or} \quad C_D/2 \text{ (no form drag present)} \tag{36}$$

$$j_H = j_D \neq f/2 \quad \text{or} \quad C_D/2 \text{ (form drag present)} \tag{37}$$

Interfacial turbulence. A spontaneous agitation of the interface between two immiscible liquids has been observed when a solute is passing from one phase to the other. The effect appears to depend upon the direction of transfer of solute and is apparently associated with local variations in interfacial concentration, and therefore, in interfacial tension, often referred to as the Marangoni effect.

Transfer with a high mass flux. Distortion in the velocity field may occur when the mass flux is high, in a way that is qualitatively similar to that noted earlier for laminar flow conditions. Correction procedures are available (1).

PART 3: TRANSFER BY MOLECULAR DIFFUSION PLUS CONVECTION WITH UNKNOWN VELOCITY FIELDS

In mass transfer between two phases in countercurrent flow through packed and plate columns and in agitated vessels, the details of velocity distribution in the two phases are unknown, so that the convective contribution to transfer cannot be quantitatively formulated. One must therefore resort to empirical correlations of mass-transfer rates with operating conditions, physical properties, and system geometry in the manner described below.

Design of Continuous Columns from Rate Equations

Many types of mass-transfer operations are carried out in either continuous or stagewise columns. Continuous columns are usually packed with Raschig rings, Pall rings, Berl saddles, Lessing rings, or other types of packing to promote intimate contact between the two phases. Continuous contact is therefore maintained between the two countercurrent streams throughout the equipment, necessitating a differential type of treatment (see Fig. 2).

The operating line. The relationship giving the composition of the two phases at a given section of the column is obtained by material balance and is

$$y_A = \frac{L}{G}x_A + \frac{G_1 y_{A1} - L_1 x_{A1}}{G} \tag{38}$$

Transfer units in equimolal counterdiffusion (ecd). This mechanism prevails in distillation, provided that the heats of vaporization per mole

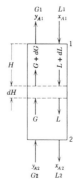

Figure 2. Terminology for a continuous column: A, total interfacial area; G, L, flow rates of phases G and L, mol/(time·area of empty column cross section); H, column height; S, cross-sectional area of empty column; x_A, y_A, concentration of component A in phases L and G, mol fraction.

are equal for all components and that sensible heat exchanges throughout the column, heats of mixing, and heat losses to the surroundings are all negligible.

The rate equations 27 in differential form are written in terms of x_A and y_A as

$$\frac{d(N_A A)}{S} = k'_y a (y_A - y_A^*)\, dH = k'_x a (x_A^* - x_A)\, dH$$

$$= K'_y a (y_A - y_{AL})\, dH = K'_x a (x_{AG} - x_A)\, dH \quad (39)$$

where $dV_0 = S\, dH$ and primes denote the mechanism of equimolal counterdiffusion. Integrating with the aid of the two-film theory (1),

$$H = \frac{G}{k'_y a} \int_{y_{A1}}^{y_{A2}} \frac{dy_A}{(y_A - y_A^*)} = \frac{L}{k'_x a} \int_{x_{A1}}^{x_{A2}} \frac{dx_A}{(x_A^* - x_A)}$$

$$= \frac{G}{K'_y a} \int_{y_{A1}}^{y_{A2}} \frac{dy_A}{(y_A - y_{AL})} = \frac{L}{K'_x a} \int_{x_{A1}}^{x_{A2}} \frac{dx_A}{(x_{AG} - x_A)} \quad (40)$$

Each integral is a measure of the difficulty of separation and has been defined as the number of transfer units (NTU). Clearly, the ratio $H : NTU$ may be called the height of a transfer unit (HTU) and is given by the quantity outside each integral in equation 40. More briefly,

$$H = (HTU)_G (NTU)_G = (HTU)_L (NTU)_L$$

$$= (HTU)_{OG} (NTU)_{OG} = (HTU)_{OL} (NTU)_{OL} \quad (41)$$

By setting the NTU equal to unity, equation 40 shows that the HTU is the column height necessary to effect a change in phase composition equal to the average driving force in the region under consideration.

Transfer units in unimolal–unidirectional diffusion (uud). This mechanism is approximated in such operations as gas absorption, liquid–liquid extraction, and adsorption. These are the rate equations for this case:

$$\frac{d(N_A A)}{S} = k_y a (y_A - y_A^*)\, dH = k_x a (x_A^* - x_A)\, dH$$

$$= K_y a (y_A - y_{AL})\, dH = K_x a (x_{AG} - x_A)\, dH \quad (42)$$

The absence of primes on the coefficients compared with equation 39 signifies the difference in mechanism. Integrating with the aid of the two-film theory (1),

$$H = \left[\frac{G}{k_y a (1 - y_A)_{iLM}}\right]_{av} \int_{y_{A1}}^{y_{A2}} \frac{(1 - y_A)_{iLM}\, dy_A}{(1 - y_A)(y_A - y_A^*)}$$

$$= \left[\frac{L}{k_x a (1 - x_A)_{iLM}}\right]_{av} \int_{x_{A1}}^{x_{A2}} \frac{(1 - x_A)_{iLM}\, dx_A}{(1 - x_A)(x_A^* - x_A)}$$

$$= \left[\frac{G}{K_y a (1 - y_A)_{0LM}}\right]_{av} \int_{y_{A1}}^{y_{A2}} \frac{(1 - y_A)_{0LM}\, dy_A}{(1 - y_A)(y_A - y_{AL})}$$

$$= \left[\frac{L}{K_x a (1 - x_A)_{0LM}}\right]_{av} \int_{x_{A1}}^{x_{A2}} \frac{(1 - x_A)_{0LM}\, dx_A}{(1 - x_A)(x_{AG} - x_A)} \quad (43)$$

where

$$k_y (1 - y_A)_{iLM} = k_y (y_A - y_A^*)/\ln\left[(1 - y_A^*)/(1 - y_A)\right] \quad (44)$$

$$k_x (1 - x_A)_{iLM} = k_x (x_A^* - x_A)/\ln\left[(1 - x_A)/(1 - x_A^*)\right] \quad (45)$$

$$K_y (1 - y_A)_{0LM} = K_y (y_A - y_{AL})/\ln\left[(1 - y_{AL})/(1 - y_A)\right] \quad (46)$$

$$K_x (1 - x_A)_{0LM} = K_x (x_{AG} - x_A)/\ln\left[(1 - x_A)/(1 - x_{AG})\right] \quad (47)$$

Each integral again defines an NTU expression, whereas the quantities in brackets outside each integral constitute the corresponding HTU. Individual and overall expressions for each of the L and G phases are identified by associating equation 41, term by term, with equation 43.

Evaluation of the NTU. The integrals in equations 40 and 43 are usually evaluated numerically by graphical integration. Information for this procedure is obtained from the equilibrium curve–operating line

plot on (x_A, y_A) coordinates, as sketched in Figure 3 for equimolal counterdiffusion (distillation) and in Figure 4 for unimolal unidirectional diffusion (gas absorption) (1).

The relationship between overall and individual HTU. Equation 25 may be used to show that for equimolal counterdiffusion,

$$(HTU)_{OG} = (HTU)_G + \frac{mG}{L}(HTU)_L \quad (48)$$

$$(HTU)_{OL} = (HTU)_L + \frac{L}{m'G}(HTU)_G \quad (49)$$

In the case of unimolal unidirectional diffusion,

$$(HTU)_{OG} = (HTU)_G \frac{(1 - y_A)_{iLM}}{(1 - y_A)_{0LM}} + \frac{mG}{L}(HTU)_L \frac{(1 - x_A)_{iLM}}{(1 - y_A)_{0LM}} \quad (50)$$

$$(HTU)_{OL} = (HTU)_L \frac{(1 - x_A)_{iLM}}{(1 - x_A)_{0LM}} + \frac{L}{m'G}(HTU)_G \frac{(1 - y_A)_{iLM}}{(1 - x_A)_{0LM}} \quad (51)$$

Empirical correlations of experimental $(HTU)_L$ and $(HTU)_G$ values have been presented for the operations of distillation, gas absorption, and stripping (1). Application of equation 41 for the appropriate mechanism then gives the required column height for the prescribed separation in Figure 2.

Design of perforated-plate extraction columns from rate equations. A procedure has been developed for the design of perforated-plate extraction columns which essentially consists of using rate equations for mass transfer during droplet formation (preferably under jetting conditions), free rise (or fall), and coalescence on each plate, to locate a pseudo-equilibrium curve. This curve is used in place of the true equilibrium relationship when stepping off the desired number of actual stages between the pseudo-equilibrium and operating curves on the $x_A - y_A$ diagram.

The design procedure is written in Fortran IV computer language and includes drop formation under jetting conditions. The computer print-out

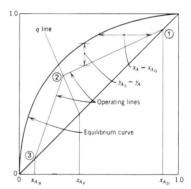

Figure 3. Evaluation of components in expressions for NTU in distillation.

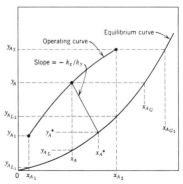

Figure 4. Evaluation of components in expressions for NTU in gas absorption.

gives the number of real plates required for a prescribed separation, the number of perforations per plate, the column diameter, and the cross-sectional area of the downcomers. Predictions from the method agreed well with published experimental data (3).

Mass Transfer in Agitated Vessels

In liquid–liquid systems correlations are available for the minimum impeller speed to achieve complete dispersion of one immiscible liquid in another, for drop size, and for the continuous phase coefficient of mass transfer (4). Optimized design and scaleup are also treated. Reviews of mass transfer in agitated gas–liquid and solid–liquid systems are given (2).

Nomenclature

A	= area; total interfacial area
a	= half-thickness of a slab or radius of a cylinder; interfacial area per unit volume
C_D	= drag coefficient, $2(\tau_0)_{av}/\rho u_\infty^2$
c	= total concentration, total mol/vol
c_A, c_B	= molar concentrations of components A and B, mol/vol
c_A^*	= equilibrium concentration at the surface or interface, mol/vol
$\overline{c_A}$	= average concentration throughout the phase, mol/vol
c_{A0}	= uniform concentration of solute throughout at the start of diffusion ($t = 0$), mol/vol
c_{As}	= concentration at surface of a sphere, mol/vol
$c_{A\infty}$	= concentration at infinity, mol/vol
c_{BLM}	= logarithmic mean concentration of component B between points 1 and 2, distance z apart, mol/vol
c_i	= molar concentration of component i, mol/vol
c_p	= specific heat, energy/(mass·temp)
D, D_{AB}, D_{BA}	= (volumetric) molecular diffusivity: of A in B: of B in A, length^2/time
d_s	= diameter of a sphere
d_t	= tube diameter
f	= Fanning's friction factor
G	= flow rate of phase G, mol/(time·area of empty-column cross section)
g	= acceleration due to gravity, length/time^2
g_c	= conversion factor, (mass·acceleration)/force
H	= column height
$(HTU)_G, (HTU)_L$	= heights of individual G- and L-phase transfer units, length
$(HTU)_{OG}, (HTU)_{OL}$	= heights of overall G- and L-phase transfer units, length
h	= individual heat-transfer coefficient, energy/(time·area·temp)
I_{Ax}, I_{ix}	= molal flux of components A and i relative to the mass average velocity, all in the x direction, mol/(area·time)
J_{Ax}, J_{ix}	= molal flux of components A and i relative to the molal average velocity, all in the x direction, mol/(area·time)
J_{Az}, J_{Bz}	= molal flux of components A and B relative to the molal average velocity, all in the z direction, mol/(area·time)
j_D, j_H	= j factors for mass and heat transfer
K_X, K_Y, k_X, k_Y	= overall and individual mass-transfer coefficients based on ΔX and ΔY, mol/(area·time)(unit ΔX or ΔY)
K_x, K_y, k_x, k_y	= overall and individual mass-transfer coefficients based on Δx_A and Δy_A, for a mechanism involving bulk flow, eg, as in unimolal unidirectional diffusion, mol/(area·time)
K_x', K_y', k_x', k_y'	= overall and individual mass-transfer coefficients based on Δx_A and Δy_A for equimolal counter diffusion, mol/(area·time)

k_c, k_c'	= individual mass-transfer coefficients based on Δc_A, for mechanisms with and without bulk flow (equimolal counterdiffusion), respectively, length/time
$k_\rho, k_{\rho a}, k_{\rho m}$	= mass-transfer coefficient based on $\Delta \rho_A$; for use with an arithmetic mean driving force, mean value, length/time
L	= length; plate or tube length; flow rate of phase L, mol/(time·area of empty column cross section)
M, M_A, M_B	= molecular weight; of components A, B
m	= slope of equilibrium curve
N_A	= molal flux of component A relative to stationary coordinates, mol/(area·time)
N_{Ax}, N_{Bx}, N_{ix}	= molal flux of components A, B, and i relative to stationary coordinates, all in the x direction, mol/(area·time)
N_{Az}, N_{Bz}	= molal flux of components A and B relative to stationary coordinates, both in the z direction, mol/(area·time)
N_{Ar}	= molal flux of A in the r direction relative to stationary coordinates, mol/(area·time)
N_{Nu}	= Nusselt number, h(length)/(thermal conductivity)
N_{Pr}	= Prandtl number, $c_p\mu$/thermal conductivity
N_{Re}	= $d_t V\rho/\mu$
N_{ReL}	= $Lu_\infty\rho/\mu$
N_{Sc}	= Schmidt number, $\mu/\rho D$
$N_{Sh}, (N_{Sh})_a, (N_{Sh})_m$	= Sherwood number, $k_\rho L/D$ or $k_\rho d_t/D$; using $k_{\rho a}$; using $k_{\rho m}$
$(NTU)_G, (NTU)_L$	= number of individual G- and L-phase transfer units
$(NTU)_{OG}, (NTU)_{OL}$	= number of overall G- and L-phase transfer units
n	= an integer
P	= total pressure, force/area, or atm (101.3 kPa)
p_A, p_B	= partial pressures of components A and B
p_{BLM}	= defined below equation 12
R	= gas constant, (length·force)/(mol·temp)
r	= radius
S	= cross-sectional area of empty column
T	= absolute temperature
T_{cA}, T_{cB}	= critical temperatures of A and B
t	= time
U	= molal average velocity, equation 1
u	= velocity in the x direction, length/time
u_i	= statistical mean velocity of component i in the x direction with respect to stationary coordinates, length/time
u_∞	= velocity outside the boundary layer, length/time
V	= mean velocity in the x direction, length/time
V_{bA}, V_{bB}	= molal volumes of components A and B at their normal boiling temperatures, vol/mol
V_{cA}, V_{cB}	= critical volumes of A and B, vol/mol
V_0	= contacting volume
W	= mass-flow rate, mass/time
w_A	= mass fraction of component A
X_A	= composition of L phase, any convenient units
X_A^*	= local equilibrium concentration in the L phase at the interface, any convenient units
X_{AG}, x_{AG}	= L-phase concentration that would be in equilibrium with existing G-phase concentration, any convenient units, mol fraction
x	= distance along surface or tube in direction of flow
x_A, x_B	= mol fraction of component A, B
x_A^*	= local equilibrium concentration in the L phase at the interface, mol fraction
x_{BLM}	= as for c_{BLM}, in x_B units
x_c	= critical value of x at which laminar flow ends in the boundary layer

Y_A	= composition of G phase, any convenient units
Y_A^*	= local equilibrium concentration in the G phase at the interface, any convenient units
Y_{AL}, y_{AL}	= G-phase concentration that would be in equilibrium with existing L-phase concentration, any convenient units, mol fraction
y	= normal distance from surface or tube wall
y_A	= mol fraction of component A
y_A^*	= local equilibrium concentration in the G phase at the interface, mol fraction
z	= distance in the direction of diffusion
γ	= N_{Bz}/N_{Az}
δ, δ_c	= thicknesses of momentum and concentration boundary layers
μ, μ_B, μ_{wT}	= viscosity; of component B; of water at T, mass/(length·time)
ξ	= an association factor for the solvent (B)
$\rho, \rho_0, \rho_\infty$	= density; at $y = 0$; at $y = \infty$, mass/vol
$\rho_A, \rho_{AB}, \rho_{AB0}, \rho_{Ai},$ ρ_{A0}, ρ_{AW}	= mass concentration of component A; bulk average or mixing-cup value of ρ_A; ρ_{AB} at the outlet; mass concentration of A at the inlet; at the surface $y = 0$; at the wall, mass/vol
$\rho_{A\infty}$	= mass concentration of A in the bulk or outside the concentration boundary layer, mass/vol
ρ_i	= mass concentration of component i, mass/vol
$\tau_0 = (\tau_{yx})_{y=0}$	= shear stress at the surface $y = 0$, force/area

A.H.P. Skelland
Georgia Institute of Technology

1. A.H.P. Skelland, *Diffusional Mass Transfer*, Wiley-Interscience, New York, 1974.

2. T.K. Sherwood, R.L. Pigford, and C.R. Wilke, *Mass Transfer*, McGraw-Hill Book Co., Inc., New York, 1975.

3. A.H.P. Skelland and Y.F. Huang, *AIChE J.* **25**(1), 80 (1979).

4. A.H.P. Skelland and J.M. Lee, *AIChE J.* **27**, 99 (1981); **29**, 174 (1983).

General Reference

E.L. Cussler, *Multicomponent Diffusion*, Elsevier Scientific Publishing Co., Amsterdam, Neth., 1976.

MATCHES

In common parlance, a match is a short, slender, elongated piece of wood or cardboard suitably impregnated and tipped to permit, through pyrochemical action between dry solids with a binder, the creation of a small transient flame. The word match is also used for fuse lines which after ignition on one end serve as fire-transfer agents in fireworks and for explosives (qv).

Manufacture

The low price of book matches is mainly the result of high speed, mechanized production methods. Book matches are punched from 1-mm thick, lined chipboard in strips of one hundred splints of ca 3.2 mm width each. In an eight-hour shift, a single machine can produce about 20×10^6 match splints and deliver them half an hour later as completed, strikable matches, ready for cutting and stapling into books. In this half hour, the tips of the punched-out splints are first immersed in molten paraffin wax. Immediately following wax application, the tip composition is affixed by dipping the ends of the strips into a thick but smoothly fluid suspension carried on a cylinder rotating in a relatively small tank at the same speed as the match strips move over it in the clamps of an endless chain. Then, matches enter a dryer where the main object is not so much the speedy removal of the water in the match

composition as the prior congealing of the match head, which takes place at a temperature of about 24°C and a rh of 45–55%.

Formulations

The following formulation with some adjustments permits the preparation of workable matches: Animal (hide) glue, 9–11%; starch, 2–3%; sulfur, 3–5%; potassium chlorate, 45–55%; neutralizer (ZnO, $CaCO_3$), 3%; diatomaceous earth, 5–6%; other siliceous fillers (powdered glass; "fine" silica), 15–32%; burning-rate catalyst ($K_2Cr_2O_7$ or PbS_2O_3), to suit; water-soluble dye, to suit.

The striking strip contains the nontoxic red forms of phosphorus.

Health and Safety

Potassium chlorate is the only active material that can be extracted in more than traces from a match head and only milligrams are contained in one head. This, even multiplied by the content of a whole book, is far below any toxic amount, even for a small child.

HERBERT ELLERN
Consultant

I. Kowarsky, "Matches," in F.D. Snell and L.S. Ettre, eds., *Encyclopedia of Industrial Chemical Analysis*, Vol. 15, John Wiley & Sons, Inc., New York, 1972.

MATERIALS RELIABILITY

Reliability is a parameter of design like a system's performance or load ratings and is concerned with the length of failure-free operation. It is difficult to conceptualize reliability as part of the usual design calculations; thus, it frequently is unquantified.

Design Reliability

The total picture of product reliability from the customer's viewpoint is rather complex. Since reliability and the related measures are essentially design parameters, improvements are most easily and economically accomplished early in the design cycle.

Design review. A design review is a formalized, documented, and systematic audit of a design by senior company technical personnel. It addresses the complex design tradeoffs and assures early design maturity. It should be multiphased and performed at various stages of the product development cycle.

Failure-mode analysis. The product-design activity usually emphasizes the attainment of performance objectives in a timely and cost-efficient fashion. Failure-mode analysis (FMA) determines how the product might fail. It begins with the selection of a subsystem or component and then documents all potential failure modes. Their effect is traced up to the system level. A documented worksheet is used on which the following elements are recorded: function (a concise definition of the functions that the component must perform); failure mode (a particular way in which the component can fail to perform its function); failure mechanism (a physical process or hardware deficiency causing the failure mode); failure cause (the agent activating the failure mechanism, eg, saltwater seepage owing to an inadequate seal might cause corrosion as a failure mechanism); identification of effects on higher level systems (this determines whether the failure mode is localized, causes higher level damage, or creates an unsafe condition); criticality rating (a measure of severity and probability of failure occurrence used to assign priority-design actions).

Life-cycle cost. The total cost of ownership of a system during its operational life can be accounted for, including not only the initial design and acquisition cost but also cost of personnel training, spare-parts inventories, repair and operations, etc. A complete projection of system costs might point out the wisdom of investing more initially in order to forego high maintenance costs owing to poor reliability and serviceability.

System Reliability Models

Static reliability models are used in preliminary analyses to determine necessary reliability levels for subsystems and components. A subsystem is a particular low level grouping of components. Some trial and error is usually necessary to obtain reasonable groupings for any particular system. Early identification of potential system weaknesses facilitates corrective action.

Series systems. The series configuration is the most commonly encountered in practice. In a series system, all subsystems must operate successfully for the system to be successful. The reliability block diagram is given in Figure 1.

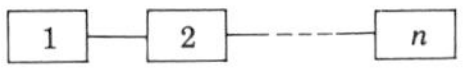

Figure 1. Series block diagram.

The system reliability is

$$R_s = \prod_{i=1}^{n} R_i$$

where R_i = the reliability for the ith subsystem; and R_s = system reliability.

In a series system, if each subsystem had an exponential time to failure given by:

$$f(t) = \lambda_i e^{-t\lambda_i}, \qquad t \geq 0$$

where λ is the failure rate for the ith subsystem, the system failure rate is

$$\lambda_s = \sum_{i=1}^{n} \lambda_i$$

Reliability Measures

The reliability function $R(t)$ is defined as

$$R(t) = P(\mathbf{t} > t) = 1 - F(t)$$

where **t** is the time to failure random variable and $F(t)$ is the cumulative distribution. In terms of the probability density function $f(t)$, the reliability function is given by

$$R(t) = \int_t^\infty f(u)\, du$$

For example, if the time to failure is given as an exponential distribution, then

$$f(t) = \lambda e^{-\lambda t}, \qquad t \geq 0, \quad \lambda > 0$$

And the reliability function is found as follows:

$$R(t) = \int_t^\infty \lambda e^{-\lambda u}\, du = e^{-\lambda t}, \qquad t \geq 0$$

Failure rate and hazard function. The failure rate is defined as the rate at which failures occur in a given time interval. Considering the time interval $[t_1, t_2]$, the failure rate is given by

$$\frac{R(t_1) - R(t_2)}{(t_2 - t_1) R(t_1)}$$

The hazard function is defined as the limit of the failure rate as the interval of time approaches zero. The resulting hazard function $h(t)$ is defined by

$$h(t) = \frac{f(t)}{R(t)}$$

The hazard function can be interpreted as the instantaneous failure rate.

The Weibull Distribution

The Weibull distribution is a versatile failure model widely used to estimate product reliability because it can be analyzed graphically with Weilbull probability paper. Although the graphical form of analysis is presented here, other procedures are available.

Basic statistical properties. The reliability function for the three-parameter Weibull distribution is given by

$$R(t) = \exp\left[-\left(\frac{t - \delta}{\theta - \delta} \right)^\beta \right], \qquad t \geq \delta \geq 0, \quad \beta > 0, \quad \theta > \delta$$

where δ = minimum life; θ = characteristic life; and β = Weibull slope.

The two-parameter Weilbull has a minimum life of zero and the reliability function is

$$R(t) = e^{-(t/\theta)^\beta}, \qquad t \geq 0$$

The hazard function for the two-parameter Weibull is

$$h(t) = \frac{\beta}{\theta^\beta} t^{\beta - 1}, \qquad t \geq 0$$

This hazard function decreased with $\beta < 1$, increases with $\beta > 1$, and remains constant for $\beta = 1$. The value of β can give some indication of wearout or infant mortality.

The expected life for the two-parameter Weibull distribution is

$$\mu = \theta \Gamma(1 + 1/\beta)$$

where $\Gamma(\cdot)$ is a gamma function and can be found in gamma tables. The variance for the Weibull is

$$\sigma^2 = \theta^2 \left[\Gamma\left(1 + \frac{2}{\beta}\right) - \Gamma^2\left(1 + \frac{1}{\beta}\right) \right]$$

The characteristic life parameter θ has a constant reliability associated with it. Evaluating the reliability function at $t = \theta$ gives

$$R(\theta) = e^{-1} = 0.368$$

And this is the same for any parameter values. Thus, it is a constant for any Weibull distribution.

Parameter estimation. Weibull parameters can be estimated using the usual statistical procedures; however, a computer is needed to readily solve the equations. Graphical estimation can be done on Weibull paper without the aid of a computer; however, the results cannot be expected to be as accurate and consistent.

The two-parameter cumulative Weibull distribution is

$$F(t) = 1 - e^{-(t/\theta)^\beta}$$

which, after rearranging and taking logarithms twice, becomes

$$\ln\left(\ln \frac{1}{1 - F(t)} \right) = \beta \ln t - \beta \ln \theta$$

This would give a straight-line plot on rectangular graph paper. Weibull graph paper plots $[F(t), t]$ as a straight line.

Binomial Distribution

To determine in the laboratory if a component will survive in use, a test bogey is frequently established based on past experience. The test bogey is correlated with the particular test used to duplicate (or simulate) field conditions. The bogey can be stated in cycles, hours, revolutions, stress reversals, etc. A number of components are placed on test and each component either survives or fails. The reliability for this situation is estimated.

The failure model is the binomial distribution given by

$$p(y) = \binom{n}{y} R^y (1 - R)^{n - y}, \qquad y = 0, 1, 2 \ldots n$$

where R = the product reliability; n = the total number of products placed on the test; and y = the number of products surviving the test. The lower $100(1 - \alpha)\%$ confidence limit on the reliability given by

$$R_L = \frac{y}{y + (n - y + 1) F_{\alpha, 2(n - y + 1), 2y}}$$

where

$$F_{\alpha, 2(n - y + 1), 2y}$$

is obtained from tables for values of F.

Probabilistic Engineering Design

The probabilistic approach to design is an attempt to quantify the design variables from a reliability standpoint.

The basic approach for probabilistic engineering design is to realize that a component has a certain strength which, if exceeded, results in failure. Broadly speaking, strength indicates the ability to resist failure, whereas stress indicates the agents that tend to induce failure. The factors determining strength or stress of a component are random variables.

LEONARD LAMBERSON
Wayne State University

R.E. Barlow and F. Proschan, *Mathematical Theory of Reliability*, John Wiley & Sons, Inc., New York, 1965.

J.H. Bompas-Smith in R.H.W. Brook, ed., *Mechanical Survival: The Use of Reliability Data*, McGraw-Hill, New York, 1973.

W. Nelson, *Applied Life Data Analysis*, John Wiley & Sons, Inc., New York, 1982.

MATERIALS STANDARDS AND SPECIFICATIONS

Standards have been a part of technology since building began, either at a scale that exceeded the capabilities of an individual, or for a market other than the immediate family. Standardization minimizes disadvantageous diversity, assures acceptability of products, and facilitates technical communication. There are many attributes of materials that are subject to standardization, eg, composition, physical properties, dimensions, finish, and processing. Implicit to the realization of standards is the availability of test methods and appropriate calibration techniques. Apart from physical or artifactual standards, written or paper standards also must be examined, ie, their generation, promulgation, and interrelationships.

A standard is a document, definition, or reference artifact intended for general use by as large a body as possible; whereas a specification, although involving similar technical content and similar format, usually is limited in its intended applicability and its users.

Standards

Objectives and types. The objectives of standardization are economy of production by way of economies of scale in output and optimization of varieties in input material; improved managerial control; assurance of quality; improvement of interchangeability; facilitation of technical communication; enhancement of innovation and technological progress; and promotion of the safety of persons, goods, and the environment.

Physical or artifactual standards are used for comparison, calibration, etc, eg, the national standards of mass, length, and time maintained by the NBS or the Standard Reference Materials collected and distributed by NBS.

Paper or documentary standards are written articulations of the goals, quality levels, dimensions, or other parameter levels that the standards-setting body seeks to establish.

Regulatory standards most frequently derive from value standards but also may arise on an *ad hoc* consensus basis. They include industry regulations or codes that are self-imposed; consensus regulatory standards that are produced by voluntary organizations in response to an expressed governmental need, especially where well-defined engineering practices or highly technical issues are involved; and mandatory regulatory standards that are developed entirely by government agencies.

Voluntary or consensus standards are especially prevalent in the U.S. and are generated by various consortia of government and industry, producers and consumers, technical societies and trade associations, general interest groups, academia, and individuals. These standards are voluntary; however, some standards of voluntary origin have been adopted by governmental bodies and made mandatory in certain contexts.

Generation, administration, and implementation. The development of a good standard is a lengthy and reiteratively involved process, whether it be for a private organization, a nation, or an international body. The generic aspects of this process are shown in Figure 1.

Standard reference materials. An important development in the United States, relative to standardization in the chemical field, is the establishment by the NBS of standard reference materials (SRMs), originally called standard samples. The objective of this program is to provide materials that may be used to calibrate measurement systems and to provide a central basis for uniformity and accuracy of measurement. SRMs are well-characterized, homogeneous, stable materials or simple artifacts with specific properties that have been measured and certified by NBS.

Standards for nondestructive evaluation (NDE). Nondestructive evaluation standards are important in materials engineering in evaluating the structure, properties, and integrity of materials and fabricated products. Such standards apply to test methods, artifactual standards for test calibration, and comparative graphical or pictorial references. In addition to standard reference materials, material scientists and engineers also frequently require access to standard reference data. Such information may enable them to identify unknown materials, describe structures, test theories or draft new standard specifications. Standard reference data refers to a set or collection of data that has passed some screening and evaluation by a competent body (see Nondestructive testing).

Basic standards for chemical technology. There are many numerical values that are standards in chemical technology; a brief review of a few basic and general ones is given below. Numerical data and definitions quoted in this section are those established by the cognizant international bodies (see Units) and are expressed in the International System of Units (SI).

Atomic weight. The present definition of atomic weights (1961) is based on ^{12}C, which is the most abundant isotope of carbon and whose atomic weight is defined as exactly 12 (see Isotopes).

Temperature. This is the measurement of average kinetic energy, resulting from heat agitation, of the molecules of a body. The most widely used scale, ie, Celsius, uses the freezing and boiling points of water as defining points (see Temperature measurement).

Pressure. Standard atmospheric pressure is defined to be the force exerted by a column of mercury 760 mm high at 0°C. This corresponds to 0.1.1325 MPa or 14.695 psi. Reference or fixed points for pressure

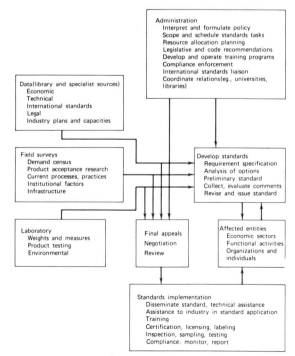

Figure 1. Flow chart of the standardization process.

calibration exist and are analogous to the temperature standards cited above. These are based on phase changes or resistance jumps in selected materials. For the highest pressures, the most reliable technique is the correlation of the wavelength shift, $\Delta\lambda$, with pressure of the ruby R_1 fluorescence line and is determined by simultaneous specific-volume measurements on cubic metals correlated with isothermal equations of state which are derived from shock-wave measurements (see Pressure measurement).

Length. One meter is defined as exactly 1,650,763.73 wavelengths of the radiation in vacuum corresponding to the unperturbed transition between the levels $2p_{10}$ and $5d_5$ of krypton-86 (the orange-red line).

Mass. The unit of mass is the kilogram, and is the mass of a particular cylinder of Pt–Ir alloy which is preserved in France by the International Bureau of Weights and Measures.

Time. In the International System of Units, the unit of time is the second: "The second is the duration of 9,197,631,770 periods of the radiation corresponding to the transition between the two hyperfine levels of the fundamental state of the atom of cesium-133."

Standard cell potential. A very large class of chemical reactions is characterized by the transfer of protons or electrons. Substances losing electrons in a reaction are said to be oxidized, those gaining electrons are said to be reduced. Many such reactions can be carried out in a galvanic cell which forms a natural basis for the concept of the half cell, ie, the overall cell is conceptually the sum of two half cells, one corresponding to each electrode. The half-cell potential measures the tendency of one reaction, eg, oxidation, to proceed at its electrode; the other half-cell of the pair measures the corresponding tendency for reduction to proceed at the other electrode. Measurable cell potentials are the sum of the two half-cell potentials. Standard cell potentials refer to the tendency of reactants in their standard state to form products in their standard states. The standard conditions are 1 M concentration for solutions, 101.325 kPa (1 atm) for gases, and for solids, their most stable form at 25°C. Since half-cell potentials cannot be measured directly, numerical values are obtained by assigning the hydrogen gas–hydrogen ion half reaction the half-cell potential of zero V.

Standard cell potentials are meaningful only when they are calibrated against an emf scale. To achieve an absolute value of emf, electrical quantities must be referred to the basic metric system of mechanical units. If the current unit, the ampere, and the resistance unit, the ohm, can be defined, then the volt may be defined by Ohm's law as the voltage drop across a resistor of one standard ohm when passing one standard ampere of current (see Electrochemical processing).

Concentration. The basic unit of concentration in chemistry is the mole, which is the amount of substance that contains as many entities, eg, atoms, molecules, ions, electrons, protons, etc, as there are atoms in 12 g of ^{12}C, ie, Avogadro's number $N_A = 6.022045 \times 10^{23}$. Solution concentrations are expressed on either a weight or volume basis. Molality is the concentration of a solution in terms of the number of moles of solute per kilogram of solvent. Molarity is the concentration of a solution in terms of the number of moles of solute per liter of solution.

A particular concentration measure is that of the acidity of aqueous solutions, the pH value, which usually is regarded as the common logarithm of the reciprocal of the hydrogen-ion concentration. More precisely, the potential difference of the hydrogen electrode in normal acid and in normal alkali solution (-0.828 V at 25°C) is divided into 14 equal parts or pH units; each pH unit is thus 0.0591 V. Operationally, pH is defined by pH = pH (soln) + E/K, where E is the emf of the cell:

$$H_2 \,|\, \text{solution of unknown pH} \,\|\, \text{saturated KCl} \,\|\, \text{solution of known pH} \,|\, H_2$$

and $K = 2.303RT/F$, where R is the gas constant, 8.314 J/(mol·K) or [1.987 cal/(mol·K)], T is the absolute temperature, and F is the value of the faraday, 9.64845×10^4 C/mol (see Hydrogen-ion activity).

Energy. The SI unit of energy is the joule which is the work done when the point of application of a force of one newton is displaced a distance of one meter in the direction of the force. The newton is that force which, when applied to a body having a mass of one kilogram, accelerates the body one meter per second squared. The calorie is the quantity of heat absorption of water per gram per degree Celsius at 15°C and it is equal to 4.184 J (see Units).

Specifications

Objectives and types. A specification establishes a basis for assurance of the fitness of a material, product, process, or service for use. Such fitness usually encompasses safety and efficiency in use as well as technical performance. Material specifications may be classified as to whether they are applied to the material, the process by which it is made, or the performance or use that is expected of it. Product or design specifications are not relevant to materials.

Content. Although formats of materials specifications may vary according to the need, the principal elements are title, statement of scope, requirements, quality assurance provisions, applicable reference documents, preparations for delivery, notes, and definitions.

Great reliance was formerly placed on competitional specifications for materials, and improvements in materials control were sought by increasing the number of elements specified and decreasing the allowable latitude (maximum, minimum, or range) in their concentration. Increasingly, it is now realized that property requirements alone or in combination with a less exacting compositional specification, are usually a more effective solution. The most effective specification is that which accomplishes the desired result with the fewest requirements.

Requirements are not included if they cannot be assessed by a prescribed method of tests or quantitative inspection technique. Wherever possible, tests should be called for that are both easy to perform and highly correlative with service performance, especially those having predictive capability with respect to service life.

Economics of Standards and Specifications

The costs and benefits associated with standardization are determined by direct and indirect effects. A proper assessment depends on having suitable base-line data with which to make a comparison. Several surveys have shown typical dollar returns for the investment in standardization in the range of 5 : 1–8 : 1 with occasional claims made for a ratio as high as 50 : 1.

Savings include reduced costs of materials and parts procurement; savings in production and drafting practice; reduction in engineering time, eg, design, testing, quality control, and documentation, and reduction in maintenance, field service, and in warranty repairs.

The DOD estimates conservatively that materials and process specifications represent almost 1% of total hardware-acquisition costs. The total U.S. cost for material and process specifications is greater than 3×10^8.

Savings can usually be realized by adopting a standard already established by an organization at a higher level, eg, national or international trade associations.

The ideal specification elicits only those properties required to ensure satisfactory performance in the intended application, properties that are quantitative and measurable in a defined test. Excessively stringent requirements not only carry with them their own direct costs for compliance and test verification, but also constitute indirect costs by restriction of sources of the material. Analogously, there is an optimal level of standardization. Increased standardization lowers costs but restricts choice and can effectively stall new technological innovation.

Sources

There are many hundreds of standards-making bodies in the United States. These comprise branches of state and Federal government, trade associations, professional and technical societies, consumer groups, and institutions in the safety and insurance fields. Individuals seeking access to standards and specifications are referred to the directories listed in the general references.

J.H. WESTBROOK
General Electric Company

S.J. Chumas, "Directory of United States Standardization Activities," NBS SP 417, Washington, D.C., 1975.

E.J. Sturglia, *Standards and Specifications—Information Sources*, Gale Research, Detroit, Michigan, 1965.

N.E. Promisel and co-workers, *Materials and Process Specifications and Standards*, NMAB Report-33, Washington, D.C., 1977.

R.E. Englehardt, "Bibliography of Standards, Specifications, and Recommended Practices" in *Nondestructive Testing Information Analysis Center Handbook*, March, 1979.

MEAT PRODUCTS

Meat, an excellent source of protein, iron, and B vitamins, was processed as early as prehistoric times, probably by drying under the sun and later by smoking and drying over wood fires. The purpose of meat processing was to prepare products that could be stored for considerable time periods at ambient temperatures. Today, meat is processed with salt, color-fixing ingredients, and seasonings in order to impart desired palatability traits to intact meat products (eg, bacon, corned beef, ham, smoked butt, and pork hocks) and comminuted meat products (eg, all types of sausage items). Products intermediate to these categories are sectioned, or chunked and formed meats (see Food processing).

Curing

Meat-curing agents include sodium chloride, nitrite, ascorbate or erythorbate, and possibly sodium phosphate, sucrose, dextrose, or corn syrup and seasonings. The salt content of processed meats varies from 1–12% according to the type of product. Salt is used for flavor, preservation, and extraction of myofibrillar protein, whereas nitrite promotes color development, flavor, and preservation by inhibiting the growth of microorganisms and fat oxidation, and ascorbate or erythorbate inhibits undesirable nitrite reactions. Phosphates facilitate myofibrillar protein extraction, inhibit fat oxidation, and improve color development. Sugars and seasonings are used principally for flavor (see Food additives).

Comminution

Comminuted meat may be cured or a fresh product. The degree of comminution varies considerably from one product to another. Sectioned or chunked and formed products may be composed of particles that weigh more than 450 g each, whereas finely comminuted meats are chopped to a pastelike texture of very small particles. Comminution equipment includes grinders, bowl choppers, emulsion mills, and flaking machines. In addition to comminution, the meat is blended with other ingredients. Blenders, mixers, tumblers, and massagers are used to subject the meat protein to mechanical action in the presence of salt. This causes the salt to extract the principal myofibrillar protein, myosin, from the muscle. The extracted myosin gels when the comminuted meat is heated to form a matrix which entraps water and fat and binds the meat particles to each other.

Smoking

Many intact and comminuted, cured meat products are smoked to impart a desirable smoked flavor and color. The smoking process may also include a drying or cooking cycle.

Canning

Canned meats may be processed to be commercially sterile or semipreserved. The objective of commercial sterilization is the destruction of all harmful bacteria, or bacteria that may cause spoilage of the product under normal unrefrigerated storage. However, the process does not kill the spores of all heat-resistant bacteria. Therefore, it is essential to cool the cans rapidly after processing and avoid storage above 35°C. The most persistent type of bacteria are sporeforming organisms. The common vegetative and nonsporeforming pathogenic bacteria are killed with adequate processing and are of little or no importance in spoilage.

Dehydration

Freeze-drying. Freeze-drying of meat results in a product with a spongelike appearance, practically devoid of moisture but resembling the original meat. The composition and type of meat influence the acceptability and stability of freeze-dried meat. Freeze-drying extends shelf life

and reduces weight. The meat is rapidly defrosted by immersing in water before cooking. Under optimal processing and storage conditions, reconstituted meats have acceptable flavor, color, texture, and nutrient retention.

Air-drying. Precooked comminuted lean meat is dried in a forced-air rotary or tunnel dryer to less than 10% moisture for use as an ingredient in dried soups and stews. The dried product is often compressed to reduce its volume to about one third of the fresh meat volume for shipment.

By-Products

By-products in the meat-packing industry represent a substantial part of the sales value of the production derived from the slaughter of animals. By-products include variety meats, edible and inedible fats, and hides and other inedibles. The value of meat or by-products depends upon the species and age of the animal, degree of finish, and price.

Approved portions of the carcasses of clean, sound meat animals yield edible fats including lard (from swine), edible tallow (from cattle and sheep), and related products such as oleo stock. These may be consumed in the form of shortening for baking and similar food applications, or as frying fats.

Inedible fats are used widely in animal feeds and for other industrial uses. The principal products of this type are inedible tallow and grease (see Fats and fatty oils).

Glenn R. Schmidt
R.F. Mawson
Colorado State University

J.C. de Holl, *Encyclopedia of Labeling Meat and Poultry Products*, *Meat Plant Magazine*, St. Louis, Mo., 1978.

J.E. Price and B.S. Schweigert, *The Science of Meat and Meat Products*, W.H. Freeman and Co., San Francisco, Calif., 1971.

MECHANICAL TESTING. See Materials reliability.

MEDICAL DIAGNOSTIC REAGENTS

Generally, a medical diagnostic reagent is a product used to measure an analyte (material undergoing analysis), either in concentration or by activity in a biological matrix, and thereby to help assess health or disease in humans or animals.

Tests

Measurements of clinical body-fluid enzymes. One of the important constituents of life is the protein in particular enzymes, the biochemical catalysts without which all life would cease. Under conditions of health and absence of pathology, a quantity of cellular enzyme leaks into the body-fluid compartments, eg, the blood, urine, and cerebrospinal fluid. Reference values for these enzymatic activities have been established by method. In many disease and other abnormal states, eg, myocardial infarction, hepatitis, pancreatitis, and prostatic carcinoma, or in pregnancy, rises in particular enzyme activities are noted and measured in serum and other body fluids.

Cardiac enzymes. The heart is heavily dependent on oxidative metabolism and as a result is rich in vascularities and mitochondria containing large quantities of cytochromes. Ultimately, the blood vessels of the heart separate into very small branches of the coronaries. Occlusion of these blood vessels deprives the heart of oxygen, and in its early phases, this deprivation is manifested by angina. Later phases of anoxia result in cell death and coronary or myocardial infarction (see Cardiovascular agents). When cells of the heart die, enzymes that are within them enter the blood. Among the more frequently measured enzymes that are measured by medical diagnostic reagents are serum oxaloacetic transaminase (SGOT, aspartate transferase or AST); creatine phosphokinase isozymes; lactic dehydrogenase (LDH) and its isozymes; and α-hydroxybutyric acid dehydrogenase (α-HBD). An isoenzyme is an enzyme that exists in more than one form with differing pH optima, temperature optima, and substrate specificities.

There are three basic ways of measuring enzyme activity, eg, kinetic, static, and immunological coupled with kinetic. Another method which soon will be available is the measurement of enzyme concentration rather than activity. Kinetic enzyme methods involve rate measurements of enzyme activity, static enzyme methods are end-point analyses, and the immunological-coupled-to-kinetic enzyme measurement is a special example of immunological methods (see Immunologic measurements of hormones and drugs). The latter involve immunoprecipitation of one or more isozymes by specific antibodies and the measurement of activity by kinetic methods of the remaining isozymes.

Liver enzymes. Serum enzyme elevations indicate liver damage. The three principal enzymes that are measured are alkaline phosphatase, alanine transferase (SGPT), and γ-glutamyl transpeptidase.

Pancreatic enzymes. Acute pancreatitis is manifested by a rise in serum and urine pancreatic enzymes. The main enzyme activities that are measured are amylase and lipase.

Measurements of abnormal levels of chemical constituents. Many medical diagnostic reagent systems are designed for measurement of levels of chemical constituents as well as enzymatic activity. Abnormalities of these constituents have been associated with pathology. Among the substances measured by reagents in kit form are blood urea nitrogen (BUN), albumin, total protein, cholesterol, triglycerides, bilirubin, calcium, glucose, creatinine, and phosphorus.

Measurement of hormones. Although many of the hormones of clinical importance are measured by other techniques, eg, glc, hplc, and RIA, many laboratories still use colorimetric analytical techniques (see Hormones, survey). These techniques can give clinically useful results but, in general, they are less specific than the techniques employing chromatography or immunoisotopic principles. Hormones, eg, 17-hydroxycorticosteroids (see Hormones, adrenocortical), 17-ketosteroids, vanilmandelic acid, and 5-hydroxyindolacetic acid, can be indexed using colorimetric analysis. In most instances, urine is the body fluid used for analytical purposes. The analysis of 17-ketosteroids is illustrative of the test principles used in the measurement of all the hormones mentioned above.

17-Ketosteroids. The glucuronide conjugates of the 17-ketosteroids are cleaved by mild acid hydrolysis to yield free 17-ketosteroids. The free 17-ketosteroids are extracted into ethyl ether and the extract is washed with alkali to remove interfering compounds. The extract is evaporated to dryness and the 17-ketosteroids react with m-dinitrobenzene in alkaline solution to form a reddish-purple complex.

Immunologic measurements of hormones and drugs. With development of radioimmunoassay (RIA) as an analytical technique, the field of analytical endocrinology and immunopharmacology has burgeoned (see Radioactive tracers). Radioimmunoassay permits measurement of picogram quantities of hormones or drugs. The three prominent permutations of this technology are RIA, EMIT (enzyme-multiplied immunotechnique), and ELISA (enzyme-linked immunosorbent analysis). Competition occurs between an antigen without a label, ie, the analyte, and a labeled analyte for a fixed amount of a specific antibody for that antigen:

$$Ab + Ag \rightarrow Ab\text{-}Ag$$
$$Ab - Ag + Ag^* \rightarrow Ab - Ag^* + Ag$$

where Ab = antibody, Ag = unlabeled antigen (free unlabeled), Ag^* = labeled antigen (free labeled), Ab–Ag = antibody–antigen complex unlabeled (bound unlabeled, zeolite type), and Ab–Ag^* = antibody–antigen complex labeled (bound labeled, zeolite type).

HERBERT E. SPIEGEL
Hoffmann-La Roche Inc.

J.H. Wilkinson, *An Introduction to Diagnostic Enzymology*, Ed. Arnold Ltd., London, UK, 1962.

J.E. Logan in *Clinical Biochemistry*, Academic Press, Inc., New York, 1981, pp. 43–83.

MELTING AND FREEZING TEMPERATURES. See Temperature measurement; Calorimetry.

MEMBRANE TECHNOLOGY

This article deals with the mechanism of transport processes in membranes, membrane structure, and fabrication, how membranes can be formed into functional modules, and important applications of this technology. Because separation processes represent a large use of membranes, many of the current and potential concepts for separation and purification via membrane technology are presented. Membrane technology also is used in a variety of other contexts: a sampling of these is presented to suggest possibilities.

Mechanisms of Transport

Transport of fluids or solutes through membranes can occur by any of several different mechanisms, depending on the structure and nature of the membrane. In all cases, transport of any species through the membrane is driven by a difference in free energy or chemical potential of that species across the membrane. These driving forces may result from differences in pressure, concentration, electrical potential, or combinations of these factors between the fluid phases on the upstream and downstream sides of the membrane. Species fluxes can be related to the driving forces, through generalized abstract relationships from irreversible thermodynamics; however, various specific relationships applicable to individual cases are introduced here.

Most often, the transport of a permeant through a membrane can be specified in terms of a permeability coefficient P defined as

$$P = \frac{Jl}{\Delta\phi} \tag{1}$$

where J = permeant flux in appropriate units, l = membrane thickness, and $\Delta\phi$ is the difference in hydrostatic pressure, partial pressure, concentration, or other potential between the upstream and downstream fluid phases.

Solution-diffusion model. A very important and fundamental means by which a species can be transported through a membrane involves dissolving of the permeate molecules into the membrane at its upstream surface followed by molecular diffusion down its concentration gradient to the downstream face of the membrane. There it is evaporated or dissolved into the adjacent fluid phase. This solution-diffusion mechanism is applicable when the membrane does not contain pores and may be regarded for thermodynamic purposes as a fluid phase. This mechanism operates only in dense membranes or in the dense skin of asymmetric membranes. In most cases, thermodynamic equilibrium partitioning of the penetrant species between the membrane surface and the adjacent fluid phase takes place. Thus, the permeability defined by equation 1 is the product of a thermodynamic partition coefficient and a molecular diffusivity for the permeate through the membrane.

Gases and vapors. Gas sorption and transport in rubbery amorphous polymers represent the simplest examples of the solution-diffusion mechanism. The equilibrium partitioning of gas molecules between the membrane and the external gas phase follows Henry's law

$$C = Sp$$

where C = gas concentration in the polymer, p = partial pressure of the particular gas in the gas phase, and S = a solubility coefficient (see Fig. 1).

There are extensive tabulations of the permeability coefficient P for various gas-polymer systems, but information on D (diffusion coefficient) and S separately are available less readily (see Barrier polymers). For a given polymer, S increases as the condensibility of the gas (boiling points or critical temperatures provide a convenient measure) increases and D decreases as the molecular diameter of the gas increases. For a

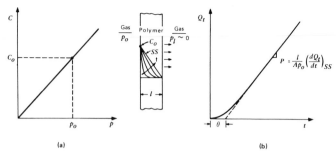

Figure 1.   Gas sorption in polymer membranes.

given gas, the main factor is the segmental mobility of the polymer. Both P and D increase rapidly with temperature whereas S is less temperature-sensitive and may increase or decrease depending on the system.

Hydraulic permeation of liquids.   Hydraulically driven permeation of liquids through dense membranes can occur by a solution-diffusion mechanism. The rate of this process strongly depends on the extent that the permeating liquid swells the polymer membrane. The upper part of Figure 2 shows a membrane adjacent to a porous support plate which freely transports liquid component l. Application of a pressure differential, $p_o - p_l$, causes a flux of liquid l through the membrane that is related by the definition of the hydraulic permeability

$$J_1 = K(p_o - p_l)l \qquad (2)$$

Mechanical arguments can be used to show that there is no pressure gradient within the membrane provided that it is nonporous, but that the pressure is uniformly p_o as shown in Figure 1. However, the pressure differential does induce a concentration gradient of component l in the membrane. A thermodynamic analysis shows that the volume fraction of 1 in the membrane at the upstream face, v_{1o} remains at its equilibrium value v_1^*; the volume fraction of 1 at the downstream face is reduced to v_{1l}. This reduction occurs because the pressure difference between the membrane and fluid phases at $x = l$ results in a reduced activity in the membrane phase for component 1, a_{1l}^M, compared to its activity in the fluid phase a_{1l}^L.

$$a_{1l}^m = a_{1l}^L \exp\left[-(p_o - p_l)V_1/RT\right] \qquad (3)$$

Combining knowledge of the sorption isotherm, ie, $v_1 = f(a_1)$, with equation 3 permits calculation of v_{1l}.

Electrochemical phenomena.   When the solute or membrane is ionized, electrical potential can play a role in transport just as do concentration and pressure. An extension of Fick's law for an ionic

species i in an electrical field $d\phi/dx$ gives

$$J_i = D_i\left[\frac{dC_i}{dx} + \frac{Z_i C_i F}{RT}\frac{d\phi}{dx}\right] \qquad (4)$$

where Z_i = ionic valence (including sign) and F = Faraday constant. This is the Nernst-Planck equation in a form that assumes thermodynamic ideality of the solution. Application of an electrical field on such a system may induce a transport flux of ions which, in general, is opposed by molecular diffusion from the concentration gradient induced. Conversely, molecular diffusion by an imposed concentration gradient of ions induces an electrical field, or streaming potential, that opposes this transport and affects all other ions in the system.

Other mechanisms.   Membrane phenomena such as osmosis, reverse osmosis, and electrodialysis can be understood and analyzed in terms of mechanisms and models presented earlier. Temperature gradients may lead to species transport that can be explained in terms of irreversible thermodynamics. Mechanistic models of thermal diffusion can be developed in certain instances, eg, in the Knudsen regime.

Physical Structure of Membranes

The most important synthetic membranes are formed from organic polymers. They perform functions that also could be performed by metals, carbon, inorganic glasses, and other materials, but because of their predominant importance in current membrane technology, the focus here is on organic polymers. Early artificial membranes were based on natural polymers such as cellulose (qv) and these still are used. Because of the demand for more versatile and highly tailored membranes, membrane technology currently employs a wide range of other polymeric materials, some synthesized especially for this purpose. The chemical structures of these polymers range from simple hydrocarbons (like polyethylene or polypropylene) to polar structures (like polyamides) or ionic structures in which cations or anions are attached to the backbone (see Olefin polymers; Polyamides). Performance, therefore, may depend upon physiochemical interactions between the permeating species with the membrane material including strong ionic interactions, weaker dipolar interactions, and quite weak van der Waal forces. However, in all cases, both the physical microstructure and the macro-form of the membrane are important considerations.

Basically, there are two microstructural forms—porous and nonporous. Composite membranes may be formed in which these two types are arranged in series or other combinations. Such composite membranes potentially can be highly tailored to meet specific demands, including both transport and mechanical properties.

Dense films.   In its usage here, the term dense implies that there are no pores of microscopic dimensions and that all unoccupied volume is simply free space between the segments of the macromolecular chains. In an amorphous polymer above its glass-transition temperature, the chain segments undergo motion similar to the thermal motions of molecules in liquids. In fact, the local structure may be regarded as identical to that of a simple liquid. Therefore, transport through any membrane of this type must be by a solution-diffusion mechanism. The penetrating molecules may interact strongly or weakly with the polymer segments depending on the structure of each; however, the rate of transport is strongly affected by the cooperative molecular motions of the polymer segments. Temperature has a large affect on these motions, and transport parameters frequently follow an Arrhenius relationship. As a result, this is frequently referred to as activated diffusion.

An important consideration in dense membranes is the extent to which the adjacent fluid phase is imbibed into it or acts as a solvent of the polymer. The extent of sorption is determined by the interaction of the solvent with the polymer segments—a well known problem in considerations of polymer solubility and chemical resistance. Sorption of solvent may reduce the melting point of the polymer and its glass-transition temperature. Sorption of significant quantities of solvents, vapors, or gases plasticizes the polymer by increasing the mobility of its segments and its diffusion coefficients.

Dense membranes can be formed by a wide variety of techniques including solution techniques, melt processing, or direct polymerization. Solution methods generally involve casting a film and then completely

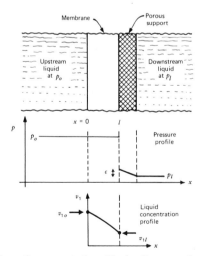

Figure 2.   Schematic representation of hydraulic permeation through membranes. When the transport is by a solution-diffusion mechanism, the middle part shows the pressure profile and the lower part shows the liquid or solvent concentration profile in the membrane. Courtesy of Marcel Dekker, Inc.

evaporating the solvent. Melt-extrusion techniques follow the same strategy as in plastics processing. If the membrane must be cross-linked, it is formed directly by polymerization with a cross-linking agent. In some cases the membrane can be made by melt or solution methods followed by cross-linking as a secondary reaction.

Porous membranes. Membranes containing voids that are large in comparison with molecular dimensions are considered porous. In these membranes, the pores are interconnected, and the polymer may comprise only a few percent of the total volume. Transport, whether driven by pressure, concentration, or electrical potential, occurs within these pores. The essential transport characteristics are determined by the pore structure with selectivity being governed primarily by the relative size of solute molecules or particles compared to the membrane pores. The mechanical properties and chemical resistance of the membrane are greatly affected by the nature of the polymer.

Asymmetric and composite membranes. The ability of dense membranes to transport species selectively makes possible molecular separation processes such as desalination of water or gas purification, but with normal thicknesses these rates are extremely slow (see Water, supply and desalination). In principle, the membranes could be made thin enough that the rates would be attractive, but such thin membranes would be very difficult to form and to handle, and they would have difficulty supporting the stresses imposed by the application. Conversely, microporous membranes have high transport rates but very poor selectivity for small molecules. One of the greatest advances in the field of membrane technology was the resolution of the dilemma through use of asymmetric membranes in which a thin, dense membrane is placed in series with a porous substructure.

Hollow fibers. The economics of using membranes for separation processes dictates the development of a high membrane surface area per unit of volume of container. An ideal geometry for this purpose is fine hollow fibers. Moreover, hollow fibers may be self-supporting and thus eliminate the need for expensive support hardware. Hollow fibers with inside diameters as small as 10 μm can be formed using spinning technology adapted from the synthetic fiber industry. Basically, the polymer is extruded through an annular hole, and an appropriate fluid is injected into the bore to prevent collapse. Early work on hollow fibers was directed toward reverse-osmosis (qv) applications but now these geometrics are used in a variety of other membrane applications (see Hollow-fiber membrane).

Membrane Module Configurations

The first requirement in any membrane process is a membrane capable of the function needed, but successful implementation requires packaging the membrane in a module whose configuration is engineered for the specific application. Membranes may be formed as flat sheets, tubes of relatively large diameter, or fine hollow fibers. Modules have been developed to accommodate each of these. Important economic considerations in their design and operation include the cost of the supporting and containing vessels (which is largely determined by the ratio of membrane area per module volume that can be achieved), power consumption in fluid pumping, and how much of the module hardware can be reused when the membrane is replaced.

Uses

Membranes are used in separation processes where they offer potential for energy-efficient separations, particularly water purification. Other filtration processes, such as dialysis (qv), are also important. Analytical uses of membranes include the measurement of osmotic pressure of dilute polymer solutions to determine the molecular weight of polymers. Membranes are used in controlled-release technology, where uniform concentration over time is needed (see Microencapsulation; Pharmaceuticals, controlled-release).

DONALD R. PAUL
University of Texas

G. MOREL
Université de Paris-Nord

J. Crank and G.S. Park, eds., *Diffusion in Polymers*, Academic Press, Ltd., London, UK, 1968.

S.T. Hwang and K. Kammermeyer, *Membranes in Separations*, Vol. VII in A. Weissberger, ed., *Techniques of Chemistry*, John Wiley & Sons, Inc, New York, 1975.

U. Merten, ed., *Desalination by Reverse Osmosis*, M.I.T. Press, Cambridge, Mass., 1966.

V.T. Stannett, W.J. Doros, D.R. Paul, H.K. Lonsdale, and R.W. Baker, *Adv. Polym. Sci.*, **32**, 69 (1979).

MEMORY-ENHANCING AGENTS AND ANTIAGING DRUGS

Memory-Enhancing Agents

The processes underlying recent or short-term memory probably are different from those underlying long-term memory. Although short-term processes alter the synaptic conductance within the brain, protein and macromolecular synthesis is related intimately to longer-term processes of memory retention, particularly consolidation. There are several ways in which macromolecules might be involved in memory storage, including the synthesis of new molecules or a change in the composition or conformation of preexisting macromolecules. Neural coding also has been related to memory storage as a result of the chemical identification in animals of a behavior-inducing peptide, scotophobin.

Neuropeptides. A number of peptide hormones act directly on the brain to affect motivation, learning, and memory processes (see Hormones, brain oligopeptides). The pituitary hormones adrenocorticotrophic hormone (ACTH), and melanocyte-stimulating hormones (MSH) appear to function as modulators of neuronal activity and facilitate several types of conditional behaviors in animals (see Hormones, anterior-pituitary). Vasopressin, which is synthesized in the hypothalamus and is stored and released by the pituitary, appears to have a long-term effect on learned behavior (see Hormones, posterior-pituitary). The ACTH-related peptides have a short-term behavioral effect, whereas those related to vasopressin have a long-term one. Studies in human subjects using ACTH and vasopressin fragments have shown beneficial effects in attention and memory tests, although not all the results have been conclusive.

Cholinergic mechanisms. Central cholinergic mechanisms have long been implicated in learning and memory consolidation. Changes in free acetylcholine after stress or learning have been observed in different brain regions of mice (see Choline; Cholinesterase inhibitors). Differences in ACh metabolism in the temporal lobes of good- and poor-performing mice have been reported. Deanol, an ACh precursor capable of penetrating the blood-brain barrier more effectively than choline, was reported to improve performance of trained rats in a maze but only when administered during the tasks. Cholinesterase inhibitors, eg, physostigmine and diisopropyl fluorophosphate (DFP), which prevent the destruction of ACh at the synapse, facilitate maze learning and enhance weak or nearly forgotten memory, but they have a blocking effect upon well-established memory. Human subjects receiving physostigmine, arecoline, or choline show improvement of long-term memory processes and serial learning. Thus, therapies to boost functioning of the cholinergic system may be valuable in reversing memory deficits resulting from aging.

Catecholaminergic mechanisms. Drugs that deplete the biogenic amines of the central nervous system have been found to impair the consolidation of memory, suggesting that normal levels of catecholamines during a critical period of memory storage may be essential for retention. Post-training administration of exogenous norepinephrine (NE) or of amphetamine, which release brain catecholamine (CA), improves memory in several learning tests.

CNS stimulants and analeptics. Numerous studies using subconvulsive doses of CNS stimulants on learning facilitation have been reported. Experiments using strychnine, as well as other stimulants such as picrotoxin, bemegride, pentylenetetrazole, and xanthines (see Alkaloids) have been interpreted as indicating that these agents affect learning by increasing attention and acquisition or enhancing storage processes.

Miscellaneous agents. Pemoline (2-amino-5-phenyl-4-oxazolidinone) and piracetam (nootropyl) are two agents reported to give beneficial effects on memory and performance. However, clinical trials have not yet fully demonstrated efficacy for these agents.

Antiaging Drugs

Antioxidants and free-radical scavengers. Vitamin E and several chemically unrelated antioxidants and free-radical scaveners, eg, cysteine, 2-mercaptoethylamine (2-MEA), 2,2-diaminoethyl disulfide, vitamin C, glutathione, butylated hydroxytoluene (BHT), ethoxyquin, and seleno amino acids suppress lipid peroxidation of biological membranes in vitro and have been tested for antiaging effects. BHT and 2-MEA are among the more effective agents in increasing the mean life span in animal colonies but have no significant effect on maximum life span.

Agents protecting against senility. *Lipofuscin Inhibitors*: meclofenoxate (centrophenoxine), the *p*-chlorophenoxyacetyl ester of dimethylaminoethanol, is a geriatric therapeutic agent believed useful for treatment of patients suffering from confusional states, Parkinsonian, and other senile mental disorders. Meclofenoxate reverses the accumulation of lipofuscin pigments in neurons of senile guinea pigs, and it has been suggested that reduction of lipofuscin may be one of the ways by which the drugs exerts its beneficial effects on the CNS.

Agents protecting against membrane damage. Labilization of the lysosomal membrane and rampant lysosomal action have been suggested as causes of cellular death during aging. Leakage of lysosomal enzymes into the cytoplasm could, in principle, be responsible for several processes suggested as primary aging mechanisms, eg, reduced fidelity of protein synthesis. Agents, eg, dimethylaminoethanol (DMAE) which stabilize lysosomal membranes, produce significant increases in the life span of senile mice. FDA-approved clinical studies with DMAE are believed to be underway.

Uses

On the basis of current knowledge, it is not possible to formulate a rational chemotherapeutic approach to aging, and the available rejuvenating drugs such as procaine hydrochloride (Gerovital H$_3$) are of dubious utility. A clearer understanding of the fundamental aging process is needed to provide at least a rational chemical approach to retarding human aging.

JASJIT S. BINDRA
Pfizer Inc.

P.E. Gold. *Ann. Rep. Med. Chem.* **12**, 30 (1977).
D. de Wied, *Life Sci.* **20**, 195 (1977).
S. Kent, *Geriatrics* (7), 77 (1979).

MENDELEVIUM.   See Actinides and transactinides.

MERCURY

Mercury, Hg, atomic number 80, also called quicksilver, is a heavy, silvery-white metal which is liquid at room temperature. Below its melting point, mercury is a white solid and above its boiling point a colorless vapor. The symbol Hg is taken from the Latin word hydrargyrum, meaning liquid silver. In nature, mercury occurs mainly in combination with sulfur to form more than a dozen different minerals. Commercially, the most important one is the red sulfide, cinnabar, HgS (86.2 wt% mercury and 13.8 wt% sulfur). Mercury metal produced from mining operations is called prime virgin mercury and is usually more than 99.9% pure. It has a clean, bright appearance, and contains less than 1 ppm of any base metals. Prime virgin mercury is acceptable for most industrial uses. Higher-purity mercury necessary for some applications is obtained by multiple distillation or electrolytic refining.

Properties

Mercury has a uniform volume expansion over its entire liquid range which, in conjunction with its high surface tension and therefore an inability to wet and cling to glass, makes it extremely useful for barometers, manometers, and thermometers, as well as many other measuring devices. This ability is enhanced by the liquidity of mercury at room temperature. Mercury also has a propensity to form alloys (amalgams) with almost all other metals except iron, and at high temperatures even with iron. Mercury is rated as one of the best electrical conductors among the metals. It has a high thermal-neutron-capture cross section enabling it to absorb neutrons and act as a shield for atomic devices; its high thermal conductivity also permits it to act as a coolant.

Mp $-38.37°C$; bp $356.9°C$; density at $20°C$ 13.546 g/cm^3; emf at $100°C$ relative to Pt-cold junction at $0°C$ -0.60 mV; latent heat of fusion 11.80 J/g (2.82 cal/g) refractive index at $20°C$ $1.6–1.9$; soluble to the extent of $20–30$ g/L water; viscosity at $20°C$ 1.55 mPa·s ($=$ cP).

Production

Mercury ore is mined by both surface and underground methods; the latter furnish about 90% of the world's production. Mercury is recovered also as a by-product in the mining and processing of precious and base metals. In past years, small quantities of mercury have been produced by processing soils under and adjacent to the sites of less efficient ore-burning furnaces used in early-day mercury recovery operations. Mercury is also produced by working mine dumps and tailing piles, particularly those accumulated during turn-of-the-century mining operations. The average grade of mercury ore mined from large mines throughout the world in recent years has ranged from 4 to 20 kg/t, total recovery approaching 95%. The average grade of ore generally has declined over the years partly as a result of the practice of mining the richest parts of ore bodies to realize a higher profit and partly because prices generally have increased over the years, thus allowing lower-grade ores to be exploited.

Processing

Mercury metal is primarily produced from its ores by standard methods throughout the world. The ore is heated in retorts or furnaces to liberate the metal as vapor which is cooled in a condensing system to form mercury metal. Retorts are inexpensive installations for batch-treating concentrates and soot, and require only simple firing and condensing equipment. For large operations, either continuous rotary kilns or multiple-hearth furnaces with mechanical feeding and discharging devices are preferred. With careful control at properly designed plants, 95 percent or more of the mercury in the ore can be recovered as commercial grade, 99.9 percent purity, mercury.

Health and Safety

Mercury metal, its vapors, and most of its organic and inorganic compounds are protoplasmic poisons that can be fatal to humans, animals, and plants. The most toxic are the organic mercury compounds, such as the alkyl types. Factors that determine the effect of mercury poisoning on humans are the amount and rate of absorption, the physicochemical properties of the compounds, and individual susceptibility. Mercury may enter the body through the skin, gastrointestinal tract, and respiratory tract. Chronic poisoning may develop gradually without conspicuous warning signs.

The immediate causes of industrial mercury poisoning are usually the absorption and retention of small quantities of mercury metal vapor, or compounds over a long period of time. Recommended safety measures include the use of efficient respirators, adequate ventilation and air-exhaust systems, employee warning signs and messages, training in accident emergency procedures, immediate, thorough, and safe cleanup of spills, air-tight storage of mercury wastes, and soiled clothing, and frequent monitoring of mercury levels in the work area.

Uses

Mercury consumed in the United States is utilized in agriculture (fungicides, pesticides, bactericides, and disinfectants) (see Mercury compounds); as catalysts (for the production of vinyl chloride monomers urethane foams (see Urethane polymers); anthroquinone derivatives in electrical applications such as batteries and electrical lamps which employ an electric discharge tube usually made of fused silica that contains varying volumes of mercury vapor; and in industrial control instruments such as thermometers, manometers (flowmeters), barometers and other pressure-sensing devices, gauges, valves, seals, and navigational devices (see Instrumentation and control).

HAROLD J. DRAKE
U.S. Bureau of Mines

E.H. Bailey, A.L. Clark, and R.M. Smith, *U.S. Geol. Surv. Pap.* **820**, 401 (1973).

J.W. Pennington, *U.S. Bur. Mines. Inform. Circ.* **7941**, 29 (1959).

Compilation of Air Pollution Factors, U.S EPA, AP-42, 1972.

MERCURY COMPOUNDS

Mercury salts exist in two oxidation states: mercurous (valence +1) and mercuric (valence +2). They exist as double salts, eg, mercuric chloride is Hg_2Cl_2 in both solution and the solid states, as shown by conductance studies and x-ray analysis. Many mercury compounds are volatile and often may be purified by sublimation. They are labile and are easily decomposed by light, heat, and reducing agents. Organic compounds of weak reducing activity, such as amines, aldehydes, and ketones, often break them down to compounds of lower oxidation state and mercury metal. This lack of stability makes it relatively easy to recover the mercury values from the various wastes that accumulate with the production of compounds of economic and commercial importance (see Recycling).

The covalent character of mercury compounds and their ability to complex with various organic compounds explains their unusually wide solubility characteristics, including alcohols, ethyl ether, benzene, etc. Small amounts of chemicals such as amines, ammonia, ammonium acetate, can have a profound solubilizing effect (see Coordination compounds).

Examples of Mercury Salts

Mercuric acetate. Mercuric acetate, $Hg(C_2H_3O_2)_2$, is a white water-soluble, crystalline powder, also soluble in many organic solvents. It is prepared by dissolving mercuric oxide in warm 20% acetic acid. A slight excess of acetic acid is helpful in reducing hydrolysis. Glass-lined equipment is preferred, although stainless steel may be used.

Mercuric fulminate. Mercuric fulminate, $Hg(ONC_2)$, is used as a catalyst in the oxynitration of benzene to nitrophenol. Its most common use is as a detonator for explosives.

Mercuric chloride. Mercuric chloride, $HgCl_2$, is the corrosive sublimate of mercury or mercury bichloride. It is extremely poisonous and is particularly dangerous because of its water solubility, 71.5 g/L at 25°C, and high vapor pressure. It sublimes without decomposition at 300°C, and has a vapor pressure of 13 Pa (0.1 mm Hg) at 100°C and 400 Pa (3 mm Hg) at 150°C. The vapor density is high (9.8 g/cm³) and, therefore, mercuric chloride vapor dissipates slowly.

Mercuric oxide. Mercuric oxide, HgO, is a red or yellow water-insoluble powder, rhombic in shape when viewed microscopically. The color and shade depends on the particle size, the finer particles (under 5 μm) appearing yellow and the coarser particles (over 8 μm) redder. The product is soluble in most acids, organic and inorganic, but the yellow form, which has greater surface area, is more reactive and dissolves more readily. It decomposes at 332°C and has a high specific gravity, 11.1.

Although the compound has some medicinal value in ointments and other such preparations, its primary use is as a starting raw material for other mercury compounds, eg, Millon's base, Hg_2NOH, is formed by the reaction of aqueous ammonia and yellow mercuric oxide.

Red mercuric oxide is chemically identical to the yellow form but is somewhat less reactive and more expensive to produce. An important use is in the Ruben-Mallory dry cell, where it is mixed with graphite to act as a depolarizer (see Batteries). The overall cell reaction is

$$Zn + HgO \rightarrow ZnO + Hg$$

Organomercury Compounds

Phenylmercuric acetate. PMA, $C_6H_5HgOC(O)CH_3$, melts at 149°C, is slightly soluble in water, soluble in various organic solvents, but much more soluble in solutions of ammonium acetate in aqueous ammonia. It is primarily used in latex paint; at low levels it is a preservative preventing putrefaction of the liquid paint and at higher levels it protects the dry film from fungal attack or mildew.

Alkyl mercuric compounds. Alkyl mercuric compounds, $RHgX$, are no longer manufactured in most of the world because of the long-lasting toxic hazards they present, and their destructive effect on the brain and central nervous system, where they tend to accumulate. They were, until recent years, widely used as seed disinfectants. They have some utility in organic synthesis and in the preparation of other organometallics (qv).

Health and Safety

Alkyl mercury compounds were used widely as seed disinfectants until their use in the United States was prohibited by the EPA in 1970. Subsequently, in 1972, the EPA prohibited the use of all mercury compounds in agriculture. At present, only mercuric chloride and mercurous chloride are permitted for use on turf to control specific fungi.

Inorganic mercury compounds, aryl mercury compounds, and alkoxy mercurials are generally considered to be quite similar in their toxicity. Alkyl mercury compounds are considered to be substantially more toxic and hazardous.

WILLIAM SINGER
MILTON NOWAK
Troy Chemical Corporation

L. Clendening, *Source Book of Medical History*, Dover Publications, New York, 1960; W. Singer and E.A. Underwood, *A Short History of Medicine*, 2nd ed., Oxford University Press, New York, 1962.

H.L. Friedman, *Ann. N.Y. Acad. Sci.* **65**, 461 (1957).

MERCURY, RECOVERY BY ELECTROOXIDATION. See Mercury.

MESITYLENE, 1,3,5-$(CH_3)_3C_6H_3$. See (Polymethyl)benzenes.

MESITYL OXIDE, $CH_3COCH=C(CH_3)_2$. See Ketones.

METAL ALKYLS. See Organometallics.

METAL ANODES

In any electrolytic process, the anode is the positive terminal through which electrons pass from the electrolyte. The selection of the materials of construction depends on reaction requirements and the specifics of the process. In practice, the choice of available materials is limited. Soluble, or insoluble metal anodes are used in a variety of processes, such as electroplating, electrorefining, cathodic protection, etc, whereas in other processes, such as batteries, fuel cells, etc, the dimensional stability of the anode is retained (see Batteries, primary; Electroplating).

Performance Characteristics

The operating behavior of metal anodes, indeed of any electrode, depends to the greatest extent on the electrolytic conditions under which they operate. Factors such as cell design, electrolyte flow, concentration of electrolyte and its pH, operating temperature, current density, and the presence or absence of impurities affect the voltage and operating lifetime of the anode. Only a laboratory comparison with standard testing conditions can evaluate fairly the differences among various anodes, and even then there is no guarantee that equivalent differences will occur under the conditions of commercial electrolytic processes. However, after over a decade of commercial use, several generalized features of each type of metal anode have been identified.

Platinum–iridium. With respect to chloride electrolysis, the principal characteristic of standard 70/30 wt% platinum–iridium coating is a low overpotential for chlorine evolution, together with a relatively high overpotential for oxygen evolution. The result is an anode coating that exhibits high current efficiency for production of chlorine.

Ruthenium–titanium oxides. Mixed oxide coating of the dimensionally stable anode (DSA) type exhibit stable operation for long periods of time at low operating voltages. In the production of chlorine, this results in energy savings of 20% over graphite (the previous anode material of choice), a reduction in labor required to maintain cell operation, and a reduction in downtime for cell cleaning and adjusting.

Manufacture

Generally, the details of preparation of commercial metal-anode coatings are considered proprietary by the manufacturer. Coatings are prepared by applying the solution to a treated titanium surface, followed by heat treatment. Chemically etching the titanium surface prior to application improves the adhesion of the coating. Methods of application include roller coating, dip coating, brushing, and spraying (see Coating processes). The coating usually is applied in many thin layers, each followed individually by heat treatment. A final heat treatment, or annealing step, sometimes is applied after the final coating thickness is obtained.

Precious-metal oxide coatings used in chlorine–caustic production are based primarily on ruthenium oxide. In the usual procedure, the solvent is evaporated, and the substrate baked at elevated temperatures (max 400–500°C) in air.

Health and Safety

Metal anodes do not have adverse health or safety effects.

Uses

The two largest electrochemical processes, the production of aluminum and of chlorine and caustic, used graphite as anode material (see Aluminum; Alkali and chlorine products). In both processes, the graphite anode is consumed and, in the case of aluminum production, considerable research efforts to date have yielded no substitute (metal anode) material; however, practically all of the Cl_2 in North America is now manufactured with metal anodes. Metal anodes have been rapidly accepted by the chlorine–caustic industry, where ruthenium–titanium oxide DSA coatings are used. The ever-increasing use of low strength hypochlorite solutions to sterilize drinking water and prevent fouling in cooling systems has resulted in the development of specialized electrolytic cells capable of producing the hypochlorite at point of use from low strength brine solutions or from seawater. Such cells are also used in odor-control installations, secondary oil recovery, and for the destruction of cyanides.

H. Stuart Holden
Diamond Shamrock Technologies, S.A.

James A. Kolb
Diamond Shamrock Corporation

Reviewed by
Thomas A. Liederbach
Eltech Systems Inc.

A.T. Kuhn and P.M. Wright in A.T. Kuhn, ed., *Industrial Electrochemical Processes*, Elsevier Publishing Co., Amsterdam, Neth., 1971, pp. 525–574.

R. Baboian, "Clad Metal Anodes," *paper presented at the Chlorine Bicent. Symp.*, San Francisco, Calif., May, 1974.

P.C.S. Hayfield and W.R. Jacob, "Platinum–Iridium-Coated Titanium Anodes in Brine Electrolysis," *paper presented at Advances in Chlor-Alkali Technology*, London, UK, 1979.

METAL-CONTAINING POLYMERS

Transition-Metal Polymers

Cyclopentadienyl and arene metal π-complexes act as electron-rich aromatic systems that undergo many reactions typical of benzene and other aromatic compounds (see Organometallic compounds). At the same time, they possess unusual properties introduced by the metal atom such as the possibility of variable oxidation states, ligand exchange on the metal atom, enhanced absorption of ultraviolet and visible radiation, electrical conductivity, and the ability to liberate finely divided particles of metal or metal oxide upon pyrolysis. Therefore, many attempts have been made to incorporate them into polymers.

Vinylic polymers. *Properties*: The compounds discussed are shown in Table 1. Vinyl metallocenes undergo either cationic or radical polymerization but not anionic polymerization to form soluble polymers with molecular weights of 2,000–20,000. The high ratios of weight-average molecular weight $\overline{M}_w$, to number-average molecular weight $\overline{M}_n$ indicate highly branched structures, probably the result of extensive chain transfer. A few monomers such as (**11**) and (**14**) do not homopolymerize. Polymers of compounds (**1–10**) are soluble in aromatic and chlorinated solvents but decompose in chlorinated solvents upon standing, particularly in the presence of light. Solutions of the polymers, when evaporated, produce transparent, highly brittle films.

More flexible polymers with higher molecular weight may be obtained if the organometallic group is moved away from the growing chain. For instance, polymers of (**13**) have been prepared with mol wt up to 7×10^5. The glass-transition temperature, T_g, is 140–145°C over the range $\overline{M}_n = 1.3 \times 10^4$ to 1.5×10^5 but can be lowered to 90°C by addition of a plasticizer.

The electrical conductivity of ferrocene-containing polymers has been extensively investigated. Polyvinylferrocene is an insulator (10^{-15} S/cm) but can be oxidized by mild oxidizing agents, such as silver salts, benzoquinone, or dichlorodicyanobenzoquinone (DDQ) to a mixed-valence ferrocene–ferrocenium polymer in which electrons probably move from one ferrocenyl (Fc) group to the next.

Preparation. Vinylic polymers may be prepared from monomers (**1**)–(**8**) by cationic or radical polymerization. Cationic polymerization may be initiated by Lewis acids such as $BF_3 \cdot O(C_2H_5)_2$, $(C_2H_5)_2$-$AlCl/M(CH_3COCHCOCH_3)_2$ (M = Ni, Cu, VO), or H_2SO_4. Low molecular weight, highly branched polymers are obtained along with varying amounts of insoluble cross-linked material. Because of the great ability of transition metal π-complexes to stabilize carbonium ions in the α-position, attack occurs preferentially upon the vinyl group. Attack of the carbonium ion upon the metallocene rings would lead to branching and cross-linking. Internal hydride transfer also could lead to branching.

Applications. The vinylic polymers of compounds (**1**)–(**14**) have limited use because of their poor thermomechanical properties. Ferrocene- and ruthenocene-containing polymers have been of interest because of their ability to absorb uv radiation without being degraded. Use as uv or radiation-resistant coatings has been proposed. Species such as cyclopentadienyldicarbonylcobalt have been used extensively as catalysts for olefin trimerization and for stereospecific synthesis of complex molecules (see Oxoprocess).

Polymers Containing Trialkyltin Esters

Polymers containing trialkyltin esters have been investigated extensively by several groups. The monomers listed in Table 2 (**18–25**) and cross-linking agents in Table 3 (**26–32**) are employed in preparing these polymers.

(33)

Table 1. Transition Metal Vinyl Monomers

Structure	Name	Formula
Vinyl metallocenes		
(1)	vinylferrocene	$C_5H_5FeC_5H_4CH=CH_2$
(2)	vinylruthenocene	$C_5H_5RuC_5H_4CH=CH_2$
(3)	ethynylferrocene	$C_5H_5FeC_5H_4C≡CH$
(4)	1,1'-divinylferrocene	$Fe(C_5H_4CH=CH_2)_2$
(5)	diisopropenylferrocene	$Fe[C_5H_4C(CH_3)=CH_2]_2$
(6)	1,3-butadienylferrocene	$C_5H_5FeC_5H_4CH=CHCH=CH_2$
(7)	(1-methyleneallyl)ferrocene	$C_5H_5FeC_5H_4CCH=CH_2$ $\overset{\|}{CH_2}$

(8)	poly(3-vinyl-1,1″:1′,1‴-bisferrocene)	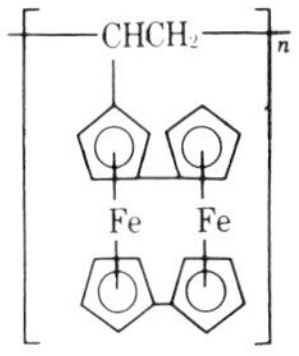
(9)	tricarbonyl[(1,2,3,4,5-η)-1-vinyl-2,4-cyclopentadiene-1-yl]-manganese	$(CO)_3MnC_5H_4CH=CH_2$
(10)	dicarbonyl[(1,2,3,4,5-η)-1-vinyl-2,4-cyclopentadien-1-yl]-nitrosylchromium	$(CO)_2NOCrC_5H_4CH=CH_2$
(11)	tricarbonyl[(1,2,3,4,5,6,-η) – vinylbenzene]chromium	$(CO)_3CrC_6H_5CH=CH_2$
(12)	dicarbonyl[(1,2,3,4,5-η)-1-vinyl-2,4-cyclopentadien-1-yl]cobalt	$(CO)_2CoC_5H_4CH=CH_2$
(13)	tricarbonylmethyl[(1,2,3,4,5-η)-1-vinyl-2,4-cyclopentadien-1-yl]-tungsten	$(CO)_3CH_3WC_5H_4CH=CH_2$
(14)	tricarbonyl[(1,2,3,4-η)-1,3,5-hexatriene]-iron, stereoisomer	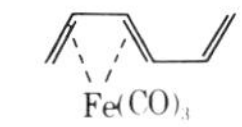
(15)	chloro(4-vinylphenyl)bis(tributylphosphine)palladium	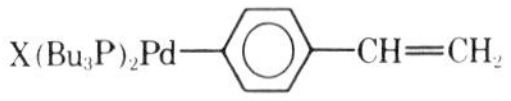
(16)	chloro(4-vinylphenyl)bis(triphenylphosphine)platinum	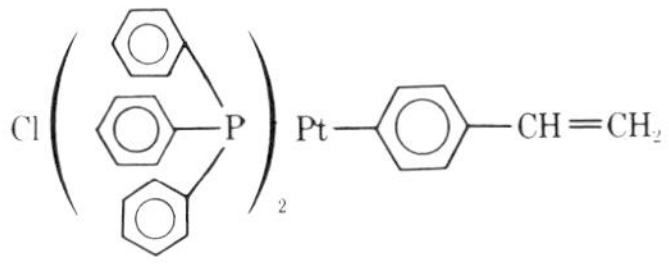
(17)	chloro(4-vinylphenyl)bis(tributylphosphine)platinum	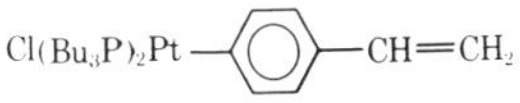

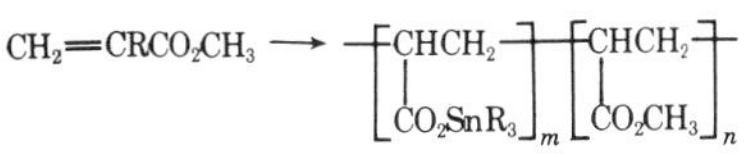

$X = Cl, Br, CN, C_6H_5$

$CH_2=CRCO_2CH_3 \longrightarrow$ [structure]

Properties. Physical properties of polymers with structures (33)–(35) are summarized in Table 4.

Applications. Polymers containing trialkyltin esters have been used primarily in antifouling paint to prevent growth of fungi and barnacles on ship bottoms and shore installations (see Coatings, marine).

Table 2. Monomers Containing Trialkyltin Ester Groups

Structure	Structure
(18) $(Bu_3Sn)_2O$	(22) $Bu_3SnO_2CCH_2NH_2$
(19) $Pr_3SnO_2CC(CH_3)=CH_2$	(23) $Bu_3SnO_2C(CH_2)_3NH_2$
(20) $Bu_3SnO_2CCH=CH_2$	(24) $Bu_3SnO_2C(CH_2)_5NH_2$
(21) $Bu_3SnO_2CC(CH_3)=CH_2$	(25) $Bu_3SnO_2C(CH_2)_{10}NH_2$

Condensation Polymers

These can be prepared by copolymerizing a difunctional metal halide with a difunctional Lewis base that may contain a metallocene. The structures of typical polymers are shown below:

$R_2MCl_2 + HO_2CZCO_2H \rightarrow \text{+}MR_2O_2CZCO_2\text{+}_n$

$R_2MCl_2 + HSZSH \rightarrow \text{+}MR_2SZS\text{+}_n$

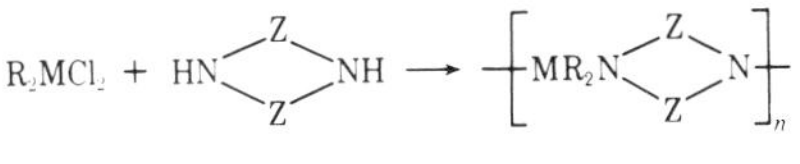

$R_2MCl_2 + HN\overset{Z}{\underset{Z}{\diagdown\diagup}}NH \rightarrow \text{[structure]}_n$

Table 3. Cross-linking Agents Employed in Preparing Polymers Containing Trialkyltin Esters

Structure	Structure
(26) 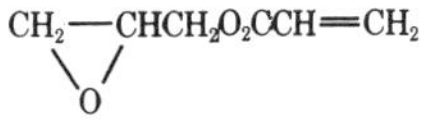 $CH_2{-}CHCH_2O_2CCH=CH_2$ (O)	(29) $H_2NCH_2CH_2NHCH_2CH_2NH_2$
(27) $CH_2{-}CHCH_2O_2CC(CH_3)=CH_2$ (O)	(30) $H_2NCH_2CH_2NHCH_2Ch_2{-}NHCH_2CH_2NH_2$
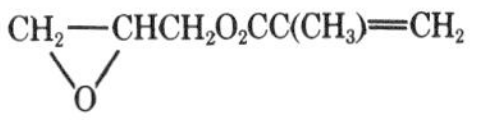 (28) $(CH_2{-}CHCH_2O{-}\bigcirc)C(CH_3)_2$	(31) $1,3{-}(NH_2)_2C_6H_4$
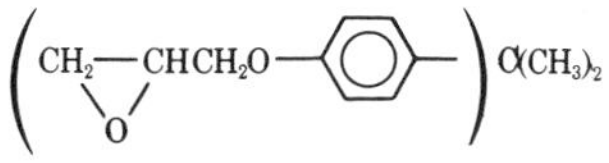	(32) $(4{-}H_2NC_6H_4)_2CH_2$

Table 4. Physical Properties of Polymers

Structure	Polymer	$\overline{M}_n$	$\overline{M}_w$	$\overline{M}_w/\overline{M}_n$	Properties
(33)		1,430,000			soluble only in ketone solvents, forms poor films, remains tacky upon drying (rejected)
(34)	1,1,1-copolymer of (19), (21), and methyl methacrylate	134,000	131,600	0.98	good solubility in organic solvents, good film-former[a]
(35)	copolymer of (21) and methyl methacrylate	109,000	160,800	1.46	good film-former[a]

[a]Suitable for incorporation into paint.

Properties. Because of their limited solubility, the Group IVA and VA polymers do not form good films when cast from solution. Plasticizers such as long chain esters or alkyl phosphates enhance the solubility of the polymers and greatly improve the quality of the films produced.

Thermal stability of the Group IVA and VA polymers is similar to that of the monomers. Thermogravimetric analysis in either nitrogen or air shows degradation commencing below 200°C and rising to 75–80% weight loss at 300°C. The polymers are particularly susceptible to hydrolysis.

JOHN E. SHEATS
Rider College

C.E. Carraher, Jr., J.E. Sheats, and C.U. Pittman, Jr., eds., *Organometallic Polymers*, Academic Press, Inc., New York, 1978.

C.U. Pittman, Jr. in E. Becker, and M. Tsutsui, eds., *Organometallic Reaction*, Vol. 6, Plenum Press, New York, 1977, pp. 1–62.

J.E. Sheats, *J. Macromol. Sci. Chem.* **A15**(6), 1173 (1981).

C.E. Carraher, Jr., J.E. Sheats, and C.U. Pittman, Jr., eds., *Advances in Organometallic and Inorganic Polymer Science*, Marcel Dekker, New York, 1982.

METALDEHYDE, $(C_2H_4O)_n$. See Acetaldehyde.

METAL FIBERS

Properties

Fiber. Fiber properties result from a combination of the material properties, the effect of processing the material into fiber form, and in some cases, the geometry of the final fiber. The mechanical, physical, and chemical characteristics of metals and alloys are high modulus and high strength, high density, high hardness, good electrical and thermal conductivity, can be magnetic, high temperature stability, good oxidation resistance, and corrosion resistance to varied chemical groups.

Most of the metal-fiber industry is directed to markets where one or more of the following properties predominate: strength-stiffness, corrosion resistance, high temperature oxidation resistance, and electrical conduction. These markets generally are satisfied by the following materials: carbon and low alloy steels, stainless steels, iron, nickel, and cobalt-based superalloys (see High temperature alloys).

Manufacture and Processing

Often there are two distinct elements in the fabrication of fiber products, ie, the formation of the fiber and the assembly of the fibers into a useful structure or form. The majority of commercial applications involve a large degree of secondary processing by various fiber-manipulation techniques.

Certain fiber-forming processes yield free fiber and others tend to produce a primitive fiber assembly, eg, a tow or mechanically interlocked bundles comparable to the bale of natural fiber. Thus, the commercially available form depends on the type of forming process employed. It also may depend on the business strategy of the manufacturer who may limit the availability of the primitive form in order to attain the benefits of the value added by further in-house processing.

Fiber dimensions, as defined by the natural- and synthetic-fiber industry, tend to be restricted to diameters or equivalent diameters for noncircular cross sections of less than 250 μm. This limitation also applies to metal fibers.

Fiber forming. The principal methods that have been developed for metal-fiber forming relate to the basic starting material form. Mechanical processing incorporates processes that rely on plastic deformation to produce a fiber from a solid precursor. In liquid–metal processing or casting, the fiber is formed directly from the liquid phase.

Mechanical processing. This involves material attenuation by gross deformation or the parting of material from a source, eg, a strip or rod. The first group encompasses wire-drawing techniques and solid-state extrusion, and the second group consists of cutting or scraping-type operations, ie, slitting, broaching, shaving, and grinding.

The fiber characteristics and relative economics of the commercially significant fiber-forming techniques by mechanical processing are summarized in Table 1.

Liquid–metal or casting processes. Melt spinning of glass and certain polymers is an established technique for mass production of fine filaments or fibers. The liquid material is forced through a carefully designed orifice or spinerette and solidifies in a cooled environment, usually after considerable attenuation, before being wound on a spool. The development of liquid–metal fiber-forming processes has revolved around overcoming the inherently low viscosity of molten metals. The low viscosity and the high surface tension of liquid metals make it extremely difficult to establish free-liquid jet stability over a length sufficient to allow freezing of the metal into a fiber before the jet separates into droplets. The problem has been solved with varying degrees of success by one of the following approaches: altering the surface of the liquid jet by chemical reaction; promoting jet stabilization by indirect means, eg, an electrostatic field; or accelerating the removal of heat from the jet to promote solidification before breakup occurs.

Melt spinning involving a free-liquid jet is used for a variety of low-melting-point metals or alloys including Pb, Sn, Zn, or Al. Various techniques are involved in the cooling and quenching process including cocurrent gas flow, mists, and liquid media. The processes permit production of fiber of continuous length and 25–250 μm diameter.

Another approach to fiber forming from the melt is accelerated heat transfer. The processes essentially are based on the quenching of the molten-metal jet on a chill plate (usually a revolving disk or wheel) positioned close to the orifice or at a point prior to the onset of jet instability and breakup.

The crucible melt extraction (CME) process is an extension of the chill-plate processes. There is no spin orifice and, thus, no need for forming a liquid stream. The chill surface, in this case, the shaped rim of a revolving disk, is dipped into the crucible so that the shaped edge just

Table 1. Mechanical Fiber-Forming Processes and Related Fiber Characteristics

Process	Typical fiber diameter, μm	Typical length	Materials	Cross-section shape
conventional wire drawing	≥ 12	continuous	all ductile metals and alloys	round (other sections possible)
bundle drawing	≥ 4; typically 8 or 12	continuous	ductile metals and alloys	rough surface
broaching or shaving, eg, wire rod, and billet	≥ 8	short to continuous	most ductile metals and alloys	generally triangular
slitting and shaving, eg, foil and sheet	ca 25 and greater	short (0.0004–4 cm) or continuous	most ductile metals and alloys	ductile square or rectangle

contacts the liquid–metal surface. The cool disk edge immediately causes solidification of a small volume of liquid metal, which is carried out of the crucible and is ejected from the wheel by centrifugal action. The shape of the fiber cross section is dependent on the wheel-edge geometry and the depth of immersion, but it can be made circular for small diameter filaments (25–75 μm); larger filaments tend to be crescent-shaped in cross section. A further evolution in the chill-plate concept is the pendant-drop melt-extraction process (PDME). In this process, the orifice for jet forming is eliminated and the crucible is replaced by a suspended drop held by surface tension to the feedstock which is a wire or wire rod.

The PDME and CME processes have been varied to provide discrete fiber lengths as opposed to continuous filament. The introduction of discontinuities on the chill-wheel rim at fixed intervals results in the casting of short fibers with the discontinuity spacing defining the fiber length.

Uses

Applications for metal fibers are of two principal types. One is the substitution for other fiber types in textile product configuration, or for metal powder in the case of porous metals, to improve performance or to provide a cost benefit, eg, high temperature oxidation resistance or improved permeability in porous structures. The other application is in the development of new products that are based on the unique fiber properties or property combinations of metals, eg, in high gradient magnetic separation (see Magnetic separation) or fiber reinforcement in composite materials (see Composite materials).

JOHN A. ROBERTS
Atlantic Richfield Co.

C.Z. Carroll-Porczynski, *Advanced Materials*, Chemical Publishing Co., Inc., New York, 1969.

R.E. Maringer and C.E. Mobley, *J. Vac. Sci. Technol.* **11**, 1067 (1974).

J.L. Broutman and R.H. Krock, eds., *Modern Composite Materials*, Addison-Wesley Publishing Co., Reading, Pa., 1967.

METALLIC COATINGS

SURVEY

Metallic coatings provide a basis material having the surface properties of the metal being applied as coating. The functional composite so produced has an appearance or utility not achieved by either component singly and becomes a new material (see Composite materials). The base material, most often another metal or a ceramic, paper, or a synthetic fiber, almost always provides the load-bearing function, and the coating metal serves as a corrosion- or wear-resistant protective layer (see Corrosion and corrosion inhibitors). The bulk of all metallic coatings provide a protective function in one of five principal ways: they are anodic to iron and can protect it by cathodic protection, eg, Al, Mg, Zn, and Cd; they form highly protective, passive films in aqueous media, eg, Cr, Ni, Ti, Ta, and Zr; their oxides are slow growing and adherent, and therefore, protect at high temperatures, eg, Al, Cr, and Si; they are noble metals and corrode little or not at all and function as barriers to corrosive agents, eg, Au, Ag, Cu Pt, Rh, Pd, etc (see Metal anodes); and the compounds, which are formed by uniting with the basis metal (alloy) or one of its constituents, are very hard and provide wear resistance, eg, B_4C, SiC, TiC, WC, Cr_2O_3, etc. In certain coating processes, eg, chemical vapor deposition (CVD), the hard-phase compounds are grown directly on the substrate.

Diffusion

In the various processes for diffusion coating, the basis metal is contacted with the coating metal, which is in the liquid or solid state or is brought to the surface by vapor transport. The two materials are held at an elevated temperature for a sufficient time to allow lattice interdif-

fusion of the two materials. The growth of the coating often is limited by diffusion of one species through one of the intermetallic layers, resulting in a parabolic rate of coating thickness increase with time. Whether the coating metal is brought into contact as a liquid, or is carried by another solvent, or is transported by a vapor-phase mechanism, surface contamination and oxide films must be removed by a suitable fluxing process either prior to or during the contact period and oxidation-free conditions must be maintained during diffusion. Three basic techniques are used in diffusion coating: hot dipping, cementation, and use of a liquid carrier. Examples of hot-dipped coatings include aluminum, lead, zinc, and aluminum–zinc. The diffusion-coating methods are generally considered low in cost, and are extensively used commercially.

Cementation coatings. The cementation process is conducted in a mixture of inert diluent particles, eg, alumina or sand; the coating metal in powder form; and a halide activator, which is poured or packed into a metal container with the part to be coated. The container usually is sealed against air entrainment but is arranged so that volatilization of the activator can drive the air from the pack mix. The pack is heated to 800–1100°C and held for ca 1–24 h depending on the thickness of the desired coating. The coating metal is transferred to the basis material by the formation of a volatile metal halide which is transferred through the pack mix by volume diffusion. Decomposition of the halide at the part surface provides a coating metal which diffuses into the basis metal, thus forming compounds as dictated by the equilibrium phase diagram but limited by the activity of the coating metal. This activity often is controlled by using alloys or intermetallic compounds of the coating metal. The size of the part that can be coated in this manner is somewhat limited by the time required to heat a large pack vessel and the ability to heat the contents uniformly. Small parts, eg, turbine blades, screws, and nuts, are handled commercially; however, pipe and tubing as long as 14 m and up to 2 m dia that are intended for refinery and chemical-plant service have been coated by pack aluminizing. The pack cementation process has been used primarily to apply Al, Cr, B, and Zn (Sherardizing) to other materials, eg, Si, Ti, and Mo, and many other elements on an experimental basis.

Liquid-carrier diffusion coating. Diffusion coatings can also be made by immersion of the basis metal in a liquid bath containing the dissolved coating metal. The bath can be composed of fused-salt mixtures or liquid metals, eg, Ca or Pb, which can dissolve small amounts of the coating metal but in which the basis metal is not dissolved.

Spraying

Sprayed coatings generally are applied to structures or parts which either are not conveniently coated by other means because of their size and shape or are susceptible to damage by the heating requirements of other coating techniques.

Slurry coatings and electrostatic powder coatings (qv) require heating to the fusion temperature either by massive heating of the part or by localized heating, eg, by induction, electron-beam, or laser techniques and generally in a protective atmosphere.

Flame-spraying and arc-spraying techniques are used in a large variety of industrial applications, eg, in both shop and field situations because equipment usually is portable and can be taken to the work site. However, laser, electrostatic, and slurry coatings must be formed in the shop, and generally on small-sized components.

Flame spraying. *Oxyacetylene*: Flame spraying is the simplest of the thermal spray techniques; it is used when heating the substrate above ca 315°C would cause undesirable tempering, recrystallization, oxidation, or warping. Flame spraying uses oxyacetylene or oxypropane flames with flame temperatures of ca 2750°C which is adequate to spray most ferrous and nonferrous coatings and oxides, eg, alumina and zirconia, but only to densities from 85–95%. Either wire or powder is fed into the flame. The heat of the flame melts the coating material and accelerates it toward the workpiece where particles fuse as interlocking laminates, each layer fused to the previous one. The flame is suitably oscillated over the part surface, giving a uniform coating.

Arc spraying. *Wire-Arc Spraying*: In wire arc spraying, two wires are fed to a gun through two electrical conduits which bring the wires together at a 30° angle. On contact, an arc is struck and melts the wire ends. Compressed air or nitrogen drives the liquid metal forward to the work. As the arc is broken, the wires are advanced to repeat the process. The arc temperature (ca 3800°C) causes deposition of molten droplet ca 3–8 times faster and with more fluidity than oxyacetylene flame-spray units. Because all of the heat is used to melt metal, this method is most energy-efficient.

Plasma: The plasma spraying process utilizes the available energy in a controlled electric arc to heat gases to ≥ 8000°C. The low voltage arc is ignited between a water-cooled tungsten cathode and a cylindrical water-cooled copper anode (see Plasma technology).

Although expensive as a spray method, the plasma spraying process is suitable for specialty coating and has wide usage commercially, particularly for thermal-barrier coatings on jet-engine blades and diesel-engine parts.

Laser coating. Laser power is applied to produce sprays of powdered material for coating purposes (see Lasers). Powder particles can be accelerated in the laser beam and melted before striking the substrate material where rapid solidification takes place. Power from a 25-kW CO_2 laser is directed into a suitable chamber and focused on the substrate surface. A carrier gas, eg, helium, is used to transport powdered material through a gold-plated water-cooled nozzle which projects powder into the laser beam. The accelerated molten particles impact on the substrate and, depending on the power density at the substrate, either coat the substrate with a quenched structure or are incorporated into a region of the substrate that has been melted by the beam. Solid particles, eg, carbides, in the latter process can be incorporated into the molten matrix with little dissolution, thereby producing a modified surface region that is up to 1 mm thick and impregnated with hard particles for wear resistance (see Lasers).

Electrostatic powder coating. A relatively new application of an old technique is the electrostatic deposition of powders, eg, Al, Cr, Ni, and Cu. Cleaned strip or sheet is electrostatically coated with metal powder, cold-rolled to compact the coating, and sintered by heating to develop a bond to the steel. The principal advantage is the protection provided at 500°C in service, eg, heating appliances, automotive exhaust systems, and heat exchanger (see Powder coatings).

Slurry coatings. Metallic coatings can be applied simply by applying powders of the desired metal or alloy in a paint medium and by brushing, dipping or spraying it onto the basis material. The coating is cured and then fired, during which time the organic portion vaporizes and the metallic particles fuse to form a dense, metallurgically bonded coating. Slurry coatings, because of the multiple handling steps and the extensive heat treatments required, tend to be expensive, and uses are limited to specialty coatings.

Mechanical and Liquid–Metal Cladding

Metallic coatings also can be applied by mechanical methods where the coating material is forced into intimate contact with the basis metal such that the forces at the interface disrupt and disperse the boundary oxide films existing on each of the constituent metals. Formation of a metallurgical bond may be augmented by mechanical attachment and thermal interdiffusion. Metallic coatings also can be melted into place by a weld-surfacing or casting operation. These techniques generally are applied to large, heavy basis metals in the form of plates or large forgings and usually produce quite thick coatings in comparison to other types of processes. Weld surfacing is a cost-effective method.

The selection criterion for a given method of producing a composite plate is the required thickness of the desired protective material and the thickness of the basis material to which it is applied. If 1 cm is adequate, the solid coating (cladded) alloy often is used alone. For a basis material > 1-cm and up to 6-cm thick, roll-bonding techniques are used. Explosion bonding is most often used when the basis metal thickness is ca 6–8 cm (see Metallic coatings, explosively clad metals). Beyond a 10-cm basis-metal thickness, only weld-overlay techniques or electroslag-casting techniques are practical.

Electroslag cladding. A technique similar to the submerged-arc process, but which works with more massive equipment and produces much thicker coatings, is the electroslag weld-overlay coating process. The heat energy necessary to melt the basis metal and the filler alloy is generated by electrical current in a molten, electrically conducting slag which also purifies and protects the filler metal as it advances through the slag layer into the molten pool. The molten slag and metal pool are held against the basis metal by water-cooled copper retainers, which must conform to the contour of the object to be clad. One application for this process is in the refurbishing of mill rolls which can be performed on ESR (electroslag remelted) ingot-production units with only minor change. Electroslag is ca 25% more efficient than submerged arc, largely because of the lower labor requirements of the former process. It is applicable only to large objects using equipment repetitively.

Chemical Coatings

Chemical vapor deposition (CVD). Chemical vapor deposition is the gas-phase analogue of electroless plating (qv): CVD is catalytic, occurs on surfaces, and involves a chemical reduction of a species to a metallic or compound material which forms the coating; the reactions are temperature dependent but occur at much higher temperatures than in plating.

Chemical reactions utilized in CVD are reduction reactions, displacement reactions, and disproportionation reactions. These generally require temperatures from 500 to 1200°C and often as high as 1500°C. For example, a CVD process may involve a metal carrier compound, ie, $SiCl_4$, which is reduced by a gaseous reducing agent such as H_2 to deposit a metallic coating, or on the other hand, may occur by thermal decomposition of an unstable compound such as $Ni(CO)_4$, nickel carbonyl, into its parts, one of which is a metal, which subsequently forms the deposit as pure metal, or in some cases as a compound by reaction with another gas species (see Film deposition techniques).

Chemical vapor deposition coatings tend to be purer than non-CVD-produced coatings. Control of chemical composition is a matter of controlling the gaseous reactants entering the reactor; graded coatings and mixed (sequential) coatings are possible through selection of appropriate gases. Control of the nucleation and growth of coating-metal crystallites is a matter of great importance because it affects strongly chemical properties of the composite. Recently, plasma-assisted CVD has resulted in effective processing of some semiconductor materials at much lower temperatures, eg, ca 300°C, by using energy from the plasma to promote the desired reaction.

Vacuum Coatings

In vacuum deposition, the desired coating metal is transferred to the vapor state by a thermal or ballistic process (sputtering) at low pressure. The vapor is expanded into the vacuum toward the surface of the precleaned basis metal. Diffusion-limited transport and gas-phase prenucleation of the coating material is avoided by processing entirely in a vacuum that is sufficiently low to ensure that most of the evaporated atoms arrive at the basis metal without significant collision with background gas. This usually requires a background pressure of 0.665–66.5 mPa (0.005–0.5 μm Hg). At the basis metal, the arriving atoms of coating metal are condensed to a solid phase. The condensation process involves surface migration, nucleation of crystals, growth of crystals to impingement, and often renucleation. Thermal sources based on resistance (I^2R) heating, induction heating, electron-beam heating, and laser irradiation have been used to vaporize the coating material. These processes are physical vapor-deposition techniques, ie, thermal energy raises the material to its melting point or above, whereupon it is vaporized by evaporation and adiabatic expansion. If the evaporated coating metal is made to intercept atoms or ions of a special background gas, with which it may react to form a compound, and then strikes the basis metal, the process is reactive evaporation. Other vapor-generating methods such as sputtering and ion implantation (qv) may also be used. The use of r-f sputtering allows oxides and other compounds to be vaporized. Because of the specialized equipment involved, vapor-deposited coatings tend to be expensive, but for some uses this is not unattractive because of the properties of the resulting composite. Ion implantation, though quite expensive, is attractive for use in mitigating wear in a number of industrial applications. Recent work has shown ion implantation produces a reduction of 400–1000 times in wear of metallic prosthetic hip joints against Teflon synthetic cartilage; a situation where the anaerobic *in vivo* environment does not allow the naturally protective passive films to form on the alloy (see Prosthetic and biomedical devices).

Although the vapor techniques are capital-intensive to initiate, they have become economically attractive processes for thin-film coating applications, as shown by the continually descending cost of modern integrated circuits (qv) for computers.

Health and Safety

All coating techniques are based on technologies that have inherent hazards, eg, high temperature, the use of liquid–metal or molten salt baths, high voltages, and often the use of toxic chemicals, that require methods and precautions consistent with their use.

R.C. KRUTENAT
Exxon Research and Engineering Company

R.D. Gabe, *Principles of Metal Surface Treatment and Protection*, 2nd ed., Pergamon Press, Inc., Elmsford, N.Y., 1978.

V.E. Carter, *Metallic Coatings for Corrosion Control (Proc. Conf.)*, Newnes-Butterworths, London, UK, 1977.

Materials and Coatings to Resist High Temperature Corrosion (Proc. Conf.), Verein Deutscher Eisenguttenleute Dusseldorf, FRG, New York, May 1977, Applied Science Publishers, Ltd., 1978.

B. Chapman and J.C. Anderson, *Science and Technology of Surface Coatings*, Academic Press, Inc., New York, 1974.

EXPLOSIVELY CLAD METALS

Explosives (qv) were used increasingly in the 1950s in metal-working operations because the explosives provided an inexpensive source of energy and precluded the need for expensive capital equipment. Explosive cladding, explosion bonding, and explosion welding, are methods wherein the controlled energy of a detonating explosive is used to create a metallurgical bond between two or more similar or dissimilar metals. No intermediate filler metal, eg, a brazing compound or soldering alloy, is needed to promote bonding and no external heat is applied. Diffusion does not occur during bonding.

Advantages and Limitations

The explosive-cladding process provides the following advantages over other metal-bonding processes:

A metallurgical, high quality bond can be formed between similar metals and between dissimilar metals that are incompatible for fusion or diffusion joining. Brittle intermetallic compounds, which form in an undesirable continuous layer at the interface during bonding by conventional methods, are minimized, isolated and surrounded by ductile metal in explosion cladding. Examples of these systems are titanium–steel, tantalum–steel, aluminum–steel, titanium–aluminum, and copper–aluminum. Immiscible metal combinations, eg, tantalum–copper, also can be clad.

Explosive cladding can be achieved over areas that are limited only by the size of the available cladding plate and by the magnitude of the explosion that can be tolerated.

Metals with tenacious surface films that make roll bonding difficult, eg, stainless steel/Cr–Mo steels, can be explosion clad.

Metals having widely differing melting points, eg, aluminum (660°C) and tantalum (2996°C), can be clad.

Metals with widely different properties, eg, copper/maraging steel, can be bonded readily.

Multiple-layered composite sheets and plates can be bonded in a single explosion and cladding of both sides of a backing metal can be achieved simultaneously. When two sides are clad, the two primer or clad metals need not be of the same thickness nor of the same metal or alloy.

Limitations of the explosive bonding process are

There are inherent hazards in storing and handling explosives and undesirable noise and blast effects from the explosion.

Metals to be explosively bonded must be somewhat ductile and resistant to impact. Alloys with as little as 5% tensile elongation in a 5.1-cm gauge length, and backing steels with as little as 13.6 J (10 ft-lbf) Charpy V-notch impact resistance can be bonded. Brittle metals and metal alloys fracture during bonding.

The preparation and assembly of clads is not amenable to automated production techniques, and each assembly requires considerable manual labor.

Theory and Principles

Parallel and angle cladding. The arrangements shown in Figures 1 and 2 illustrate the operating principles of explosion cladding. Figure 1 illustrates angle cladding which is limited to cladding for relatively small pieces. Clad plates with large areas cannot be made using this arrangement because the collision of long plates at high stand-offs, ie, the distance between the plates, on long runs is so violent that metal cracking, spalling, and fracture occur. The arrangement shown in Figure 2 is by far the simplest and most widely used.

Processing

Explosives. The types of high velocity explosives (4500–7600 m/s) that have been used include trinitrotoluene (TNT), cyclotrimethylenetrinitramine (RDX), pentaerythritol tetranitrate (PETN), and composition C_4. Low–medium velocity (1500–4500 m/s) explosives include ammonium nitrate, ammonium perchlorate, and amatol (see Explosives).

Health and Safety

All explosive materials should be handled and used following approved safety procedures either by or under the direction of competent, experienced persons and in accordance with all applicable federal, state, and

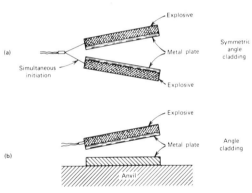

Figure 1. Angle arrangements to produce explosion clads.

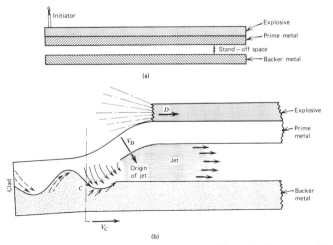

Figure 2. Parallel arrangement for explosion cladding and subsequent collision between the prime and backer metals that leads to jetting and formation of wavy bond zone.

local laws, regulations, and ordinances. The Institute of Makers of Explosives (IME), New York, provides education publications to promote the safe handling, storage, and use of explosives.

Uses

Cladding and backing metals are purchased in the appropriately heat-treated condition because corrosion resistance is retained through bonding. It is customary to supply the composites in the as-bonded condition because hardening usually does not affect the engineering properties. Occasionally, a postbonding heat treatment is used to achieve properties required for specific combinations. Applications such as chemical-process vessels and transition joints represent ca 90% of the industrial use of explosion cladding.

ANDREW POCALYKO
E.I. du Pont de Nemours & Co., Inc.

B. Crossland, *Explosive Welding of Metals and Its Applications*, Clarendon Press, Oxford, UK, 1982.

"Explosion Welding," *Metals Handbook*, 9th ed., Vol. 6, American Society for Metals, Metals Park, Ohio, 1983, p. 705.

A.A. Ezra, *Principles and Practices of Explosives Metal Working*, Industrial Newspapers, Ltd., London, UK, 1973.

A.H. Holtzman and G.R. Cowan, *Weld. Res. Counc. Bul. No. 104*, Engineering Foundation, New York, April, 1965.

METALLIC SOAPS.   See Driers and metallic soaps.

METALLOCENES.   See Organometallics.

METAL PLATING.   See Electroplating; Metallic coatings.

METAL SURFACE TREATMENTS

CLEANING, PICKLING, AND RELATED PROCESSES

Cleaning

Cleaning usually is necessary before painting, bonding, plating or other surface treatments in the manufacture of metal products. It involves the removal of unwanted surface materials by a chemical or physical process or by a combination of these methods. The primary characteristics of important cleaning methods are given in Table 1.

Pickling

Pickling is the removal of oxides by converting them to soluble compounds with acids or alkaline solutions or molten alkali salts.

Acid.   Traditional acid-pickling solutions are based upon inhibited sulfuric, hydrochloric, or phosphoric acids. The process parameters are concentration, temperature, time, and agitation. An increase in any of these variables increases the rate of oxide removal.

Alkaline.   Light scale and light-to-medium rust can be removed by aqueous, strongly caustic solutions with sequestrants, eg, sodium gluconate or ethylenediaminetetraacetic acid, usually in proprietary formulations. The solutions are used at concentration of ca 121–363 g/L, at 71–91°C for 15 min to several hours. Action is considerably slower than in acidic pickling. Agitation is desirable and the use of electric current, either with the work as the anode or with the periodic reversal of polarity, is very effective. The same baths may be used for cleaning and paint stripping.

Related Processes

Aluminum etching is a widely practiced commercial procedure utilizing modified, aqueous caustic soda baths that are often proprietary. Etching of metals other than aluminum uses acids and often baths containing chlorides, eg, hydrochloric acid, ferric chloride, etc. Blasting involves the mechanical hurling of hard materials, eg, sand, metal shot, abrasive grit, or ground nut shells, against the surface to be cleaned (see Abrasives). Electropolishing and brightening involve removal of the metal at the high spots of irregularities with little or no dissolution of metal in the low spots or valleys (see Electrolytic machining methods).

G.L. SCHNEBERGER
GMI Engineering and Management Institute

"Heat Treating and Finishing" in *Metals Handbook*, Vol. II, 8th ed., American Society for Metals, Metals Park, Ohio, 1964.

Metal Finishing Guidebook and Directory 79, Metals and Plastics Incorporated, Hackensack, N.J., 1979.

R.C. Snogren, *Handbook of Surface Preparation*, Palmerton Press, New York, 1974.

CHEMICAL AND ELECTROCHEMICAL CONVERSION TREATMENTS

PHOSPHATING

Phosphating is the treatment of a metal surface to provide a coating of insoluble metal phosphate crystals which strongly adhere to the base material. Such coatings affect the appearance, surface hardness, and electrical properties of the metal. They do provide some corrosion resistance, but they are not sufficiently protective to be used by themselves in most corrosive atmospheres. Phosphating is of primary industrial importance in the production of iron and steel surfaces, eg, in the automotive and appliance industries.

Coating formation.   Commercial phosphating products are complex and proprietary, and thus it often is difficult to classify precisely the reactions that take place during phosphate-coating deposition. Simplified equations, however, may be used to illustrate the basic chemistry. When a ferrous surface is treated with a phosphating solution, it is attacked by the free phosphoric acid:

$$Fe + 2 H_3PO_4 \rightarrow Fe(H_2PO_4)_2 + H_2$$

Hydrogen is liberated and iron is introduced into the solution as soluble, primary ferrous phosphate. The primary zinc or manganese iron phosphates, which are present as bath constituents, hydrolyze readily in aqueous solutions and produce the less soluble secondary and tertiary

Table 1. Metal-Cleaning Methods

Metal	Common cleaning methods[a]	Avoid	Comments
Ferrous alloys			
low carbon steel	1, 3, 4, 5, 6	2, temperature over 77°C	use 4, 5, or 6 to clean greasy surfaces
cast iron	1, 3, 4, 5, 6	2, temperature over 49°C	remove scale and tarnish with 1
other alloys	1, 3, 4, 5, 6	2	
Stainless steel			
12% or less Cr	1, 3, 4, 5, 6	2, check temperature limits	watch for different stainless alloys used in the same assembly
400 series	1, 3, 4, 5, 6	2	
300 series	2, 3, 4, 5, 6	water containing chloride and moist atmospheres if chlorinated solvents are used	nonoxidizing acids except HCl may be used on 300 stainless
Copper alloys			
copper	1, 3, 4, 5, 6	2	
brass	1, 3, 4, 5, 6	2	
bronze	1, 3, 4, 5, 6	2	
aluminum	5, 6	1, except H₃PO₄ with used chromate	some formulations of 2, 3, and 4 may be used; check with supplier
ternplate	1, 3, 4, 5, 6	2	
zinc, galvanized steel	5, 6	1, 2, 3, 4; sometimes very mild alkalies may be used	zinc is a rather reactive chemical

[a] Cleaning-method code: 1 = inhibited, nonoxidizing acids, such as HCl, H₃PO₄, some H₂SO₄; 2 = HNO₃; 3 = organic acids; 4 = alkaline cleaners; 5 = chlorinated degreasing solvents; 6 = other organic solvents.

metal phosphates according to the following equations:

$$M(H_2PO_4)_2 \rightleftharpoons M(HPO_4) + H_3PO_4$$
$$3\,M(H_2PO_4)_2 \rightleftharpoons M_3(PO_4)_2 + 4\,H_3PO_4$$
$$3\,M(HPO)_4 \rightleftharpoons M_3(PO_4)_2 + H_3PO_4$$

where M = zinc, manganese or iron.

The phosphoric acid that is produced in these reactions is consumed by dissolving iron from the treated part. The equilibrium shifts from left to right because of the precipitation of the sparingly soluble secondary and tertiary phosphates. The insoluble metal phosphates precipitate from solution and onto the surface of the iron to form a tightly adherent, highly interlocked crystalline layer.

Process parameters. The complete phosphating process cycle, as conducted on a commercial scale, generally consists of the following steps: preparation of the surface to be processed, ie, cleaning, rinsing to remove cleaning agents, and special pretreatments; application of phosphate-coating process; rinsing of the coated surface, ie, water and posttreatment rinsing, and drying. The characteristics of the coating depends to a large extent on the conditioning of the surface prior to coating.

Anodizing

Anodizing involves the formation of an oxide surface on nonferrous metal by electrochemical means. These surface oxide films supplement the natural oxide which occurs in very thin layers on such metals and results in a significant increase in their corrosion resistance. Aluminum, in particular, forms a thin, tenaciously adhering oxide film which provides an excellent barrier against corrosion.

Anodizing of aluminum involves electrochemical conversion of the surface to aluminum oxide; the aluminum serves as the anode and the oxygen is provided by the electrolytic dissociation of water. Chromic acid, sulfuric acid, and oxalic acid electrolytes have been widely used. Other electrolytes, eg, borates, citrates, carbonates, sulfamic acid, and phosphoric acid, have been used in specific applications. The structure of the coating normally is amorphous, although in certain electrolytes, eg, boric acid, a crystalline structure sometimes is observed. Treatment of the coating in boiling water causes partial hydration of the anodically formed oxide to a crystalline mono- or trihydrate. As a result, the porosity of the coating is progressively reduced resulting in improved corrosion resistance.

Sulfuric acid, chromic acid, and oxalic acid are the most common electrolytes used in producing corrosion-resistant anodic coatings. Film color, porosity, flexibility, and other characteristics differ depending on the electrolyte used.

G.L. SCHNEBERGER
GMI Engineering and Management Institute

Metal Finishing Guidebook and Directory—1979, Metals and Plastics Publication Inc., Hackensack, N.J., 1979.
Finishing Industry Yellow Pages, Special Technical Publications Inc., Oxnard, Calif.

CASE HARDENING

Case hardening is a metal-treatment process that produces a hard surface (the case) on a metal (the core) which remains relatively soft. The product is a hard, wear-resistant case backed by a strong, ductile, and tough core. Because of increased energy scarcity and the need for cost control, case depths should be minimized without sacrificing operational quality. A well-documented approach for determining case-depth requirements with an optimum carburizing cycle has been reported. Wear characteristics are achieved by a hard but not necessarily deep case. However, deeper cases with strong and resilient cores are needed when supporting extremely heavy loads.

In general, a case-hardening process is not established on the basis of cost, but upon the metallurgical requirements of the surface for the function it is to perform. However, the greatest cost of any process is bringing the furnace and the parts to be processed up to the operating temperature and maintaining that temperature. The bulk of all the heat treating costs can be attributed to fuel, ie, electricity or natural gas, and labor.

Processes

Carburizing. Carburizing, which is the most common of the case-hardening processes, diffuses nascent carbon into a steel surface (see Steel). Subsequent hardening by quenching at > 790°C produces a hard case in the high carbon areas. Carburizing generally is limited to low carbon steels, ie, below 0.30 wt% C. It is performed between 845 and 955°C. At these temperatures, nascent carbon is most soluble in the austenitic phase and the diffusion rate is sufficient for economical use. Case depth is controlled primarily by controlling carburizing temperature and time. The origin of nascent carbon depends upon the particular carburizing process used, ie, gas, liquid, or pack (solid) carburizing.

Carbonitriding. In cabonitriding, carbon and alloy steels are held at an elevated temperature in a gaseous atmosphere from which they absorb both carbon and nitrogen simultaneously. This process is used primarily to produce a hard, wear-resistant case, generally 76–760 μm deep. A carbonitrided case has better hardenability than a carburized case; consequently, the former can be produced with less energy within the same case-depth range.

Cyaniding. Cyaniding is the liquid-bath form of carbonitriding. The most common production cyanide-bath composition is 30% sodium cyanide, 40% sodium carbonate, and 30% sodium chloride. Oxygen from the air oxidizes the sodium cyanide to sodium cyanate which, at high temperatures, decomposes to form carbon monoxide and nascent nitrogen. The carbon monoxide that reacts to form nascent carbon produces carbon dioxide which reacts with the cyanide to produce more carbon monoxide. Both the nascent carbon and CO products of these reactions are utilized in the carburizing action. Nascent nitrogen is absorbed by the steel and increases the metal's surface hardness.

Nitriding. Nitriding case-hardens steel by addition and diffusion of nascent nitrogen into the surface of the steel where it reacts to form nitrides. The process temperatures are 495–595°C, which are less than the temperature at which the transformation to austenite occurs. A case hardness is produced directly and, therefore, quenching is unnecessary. Nitriding is accomplished by using either a gas atmosphere or a liquid bath.

Although some growth occurs in nitriding, there generally is very little distortion. Before nitriding, machined parts should be stress-relieved at least 10°C over the nitriding temperature to minimize distortion.

Microcasing. Microcasing processes are, in most cases, highly proprietary. They usually are used for surface or slightly subsurface treatments. The processes yield excellent wear resistance, gliding properties, ductility, and fatigue strength in a superficial layer.

Applied-energy hardening. Applied-energy hardening is a selective hardening process that produces a case by locally heating and quenching an area. The very rapid application of heat results in the surface being heated to the hardening temperature but very little heat being conducted inward. Since no carbon or nitrogen is added during the process, the carbon content of the ferrous metal determines the hardness response.

Induction hardening is a selective hardening process in which the localized heat is produced by electromagnetic induction.

The rate of heating by the induction coil depends upon the strength of the magnetic field at the area being heated. Heating is confined to the surface because of the skin effect resulting from the high frequency. The higher the frequency, the more shallow the heating effect. Shallow case depths up to 1.5 mm require frequencies from 10 to 2000 kHz and greater case depths require frequencies from 1 to 10 kHz. The depth of heating is determined by the duration of heating, the frequency used, and the power density.

Both the coil design and the quenching arrangement are important to the success of the process. Automatic timing is necessary for control because of the short cycles involved.

Health and Safety

Heat-treatment equipment purchased since 1970 must comply with OSHA regulations. Any chemical used in the processes must meet the control standards and the waste-disposal requirements of the Toxic Substance Control Act (TSCA) of 1976.

LESTER E. ALBAN
Fairfield Manufacturing Company

Occupational Safety and Health Act, Public Law 91-596, Title 29, Chapt. XVII, Dec. 1970.

Toxic Substance Control Act, Public Law 94-469, Oct. 11, 1976.

"Heat Treating" in *Metal Handbook*, Vol. 4, American Society for Metals, Metals Park, Ohio, 1981.

METAL TREATMENTS

Metal treatments are operations performed on consolidated metals and alloys. Most of these are mechanical and/or thermal. Mechanical treatments involve shape changes by forming or machining. Forming entails plastic deformation which changes the microstructure, and therefore, properties. In thermal treatments, heat is applied to alter structures and properties. Metal treatments such as joining and coating of metals are not discussed here (see Welding; Metallic coatings).

Mechanical Treatments

Forming processes and techniques available for a particular alloy depend upon workability, which is the ability to be plastically deformed.

Hot working. Hot working involves plastic deformation at temperatures sufficiently high that strain hardening does not result. The temperature range for successful hot working depends on composition and other factors such as grain size, previous cold working, reduction, and strain rate. The lack of strain hardening is due to sufficient thermal energy for recrystallization, which refers to the formation of new grains.

Cold working. Cold working involves plastic deformation well below the recrystallization temperature Required stresses for cold working are greater than for hot working and the amount of strain without heat treatment is limited. Advantages are close dimensional control, good surface finish, and increased low temperature strength because of strain hardening. Grain refinement can be achieved by annealing, which entails heating after cold working to temperatures where recrystallization occurs.

Primary forming processes. These operations are usually hot-working operations directed toward converting case ingots into wrought blooms, billet, bars, or slabs. In primary working operations, the large grains typical of cast structures are refined, porosity is reduced, segregation is reduced, inclusions are more favorably distributed, and a shape desirable for subsequent operations is produced. The principal operations used for ingot breakdown are forging, extruding, and rolling. Extrusion differs from forging and rolling in that more deformation occurs in one pass. Forging and rolling include many passes and some reheating.

Secondary forming processes. The objective of these processes —either cold- or hot-working—is to form a shape. Such processes include rolling, open and closed-die forging, upset forging, extruding, roll forging, ring rolling, deep drawing, spinning, bending, stretching, stamping, drawing and high velocity forming.

Thermal Treatments

Annealing. In annealing, a cold-worked material is heated to soften it and improve its ductility. The three stages of annealing are recovery, recrystallization, and grain growth (see Fig. 1). Recovery occurs at relatively low temperature and may result in some softening caused mainly by the arrangement of dislocations into a more favorable distribution. Recrystallization is the formation of new grains with a relatively low dislocation density and little internal strain which replaces strained grains with high dislocation densities. At increasing temperature, the

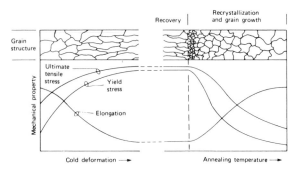

Figure 1. Variation of tensile properties and grain structure with cold working and annealing.

newly formed grains exhibit grain growth. Prolonged exposure at a given temperature also tends to promote grain growth.

Precipitation hardening. Also called age hardening, this process involves fine particles precipitated from a suprsaturated solid solution. These particles impede the movement of dislocations, thereby making the alloy stronger and less ductile. In order for an alloy to exhibit precipitation hardening, it must exhibit partial solid solubility and decreasing solid solubility with decreasing temperatures.

An example of the many alloy systems satisfying these requirements is the aluminum–copper system. At about 500–600°C, an alloy with 4.5 wt% Cu consists only of alpha, a solid solution of Cu in Al. Below 500°C, the phase $CuAl_2$ exists in addition to alpha. The objective of precipitation hardening is to distribute the second phase ($CuAl_2$) as fine particles which are effective in blocking dislocation motion.

Precipitation hardening consists of dissolving "solutioning" (which entails heating above the solvus temperature in order to form a homogenous solid solution), quenching (rapidly to room temperature to retain in solid solution a maximum amount of the alloying elements (Cu)), and aging (heating the alloy below the solvus to permit precipitation of fine particles of a second phase ($CuAl_2$)). The solvus represents the boundary on a phase diagram between the solid-solution region and a region consisting of a second phase in addition to the solid solution).

LARRY A. JACKMAN
Special Metals Corporation

L.E. Doyle, *Manufacturing Processes and Materials for Engineers*, Prentice-Hall, Inc., Englewood Cliffs, N.J., 1969.

A.G. Guy, *Introduction to Materials Science*, McGraw-Hill, Inc., New York, 1972.

L.H. Van Vlack, *Materials Science for Engineers*, Addison-Wesley Publishing Co., Reading, Pa., 1970.

METANILIC ACID, m-$NH_2C_6H_4SO_3H$. See Amines, aromatic, aniline and its derivatives.

METHACROLEIN, CH_2=$C(CH_3)CHO$. See Acrolein and derivatives.

METHACRYLIC ACID AND DERIVATIVES

Methacrylic acid and its most important derivative, methyl methacrylate, are commercially important monomers for the manufacture of specialty polymers. Methyl methacrylate polymers are used in applications that take advantage of their excellent optical and weathering properties. Plastic sheet is used in glazing, signs, and building materials. Molding powder is used largely for automobile lights and other parts. Methacrylates are also used as components of monomer mixtures to prepare emulsion and solution polymers.

Table 1. Physical Properties of Commercially Available Methacrylates, CH_2=$C(CH_3)COOR$

Compound	mp, °C	bp, °C	Refractive index, n_D^{25}	Density, d_5^{25} g/cm³	Flash point, °C COC[a]	TOC[b]	Typical inhibitor level[c], ppm
methacrylic acid	14	159–163[d]	1.4288	1.015	77		100 MEHQ
methyl methacrylate	−48	100–101[d]	1.4120	0.939		13	10 MEHQ
ethyl methacrylate		118–119[d]	1.4116	0.909	35	21	15 MEHQ
n-butyl methacrylate		163.5–170.5[d]	1.4220	0.889	66		10 MEHQ
isobutyl methacrylate		155[e]	1.4172	0.882		49	10 MEHQ
isodecyl methacrylate		120[f]	1.4410	0.878	121		10 HQ + MEHQ
lauryl methacrylate[f]	−22	272–343[d]	1.444	0.868	132		100 HQ
stearyl methacrylate[g]	15	310–370[d]	1.4502	0.864	> 149		100 HQ
2-hydroxyethyl methacrylate	−12	95[h]	1.4498	1.064	108		250 MEHQ
2-hydroxypropyl methacrylate	< −70	96[h]	1.4447	1.027	121		250 MEHQ
2-dimethylaminoethyl methacrylate	ca −30	68.5[h]	1.4376	0.933		74	200 MEHQ
2-t-butylaminoethyl methacrylate	< −70	93[h]	1.4400	0.914	11		1000 MEHQ
glycidyl methacrylate		75[h]	1.4482	1.073		84	25 MEHQ
ethyl dimethacrylate[i]		96–98[j]	1.4520	1.048	113		60 MEHQ
1,3-butylene dimethacrylate[h]	226	110[e]	1.4502	1.011	124		200 MEHQ
trimethylolpropane trimethacrylate[k]	−14	155[l]	1.471	1.06	> 149		90 HQ

[a] Cleveland open cup.
[b] Tag open cup.
[c] MEHQ-monomethyl ether of hydroquinone; HQ = hydroquinone.
[d] At 101 kPa (1 atm).
[e] At 0.4 kPa (3 mm Hg).
[f] Prepared from a mixture of higher alcohols, predominantly lauryl alcohol.
[g] Prepared from a mixture of higher alcohols, predominantly stearyl alcohol.
[h] At 1.3 kPa (9.8 mm Hg).
[i] Diester.
[j] At 0.53 kPa (4.0 mm Hg).
[k] Triester.
[l] At 0.13 kPa (1.0 mm Hg).

Physical Properties

The physical properties of the commercially available methacrylates are summarized in Table 1.

Chemical Properties

Methacrylic acid and its esters undergo reactions typical of their constituent functional groups. Polymerization of these monomers is avoided by conducting the desired reaction in the presence of inhibitors and under mild conditions. The acid is converted to salts, methacrylic anhydride, methacryloyl chloride and esters using standard reaction conditions for carboxylic acids.

Methyl methacrylate, the chief methacrylate product, undergoes typical ester reactions—saponification, transesterification, amidation. Transesterification is useful for the manufacture of higher alkyl methacrylates. Active methylene compounds, halogens, halogen halides and hydrogen cyanide add to the unsaturated group to form α-methyl and β-substituted propionates. Methacrylates also participate with dienes in Diels-Alder reactions.

Manufacture

The commercial method and proposed routes for the manufacture of methyl methacrylate and methacrylic acid are summarized in Figure 1.

Acetone cyanohydrin (ACN) process. The commercial process for the manufacture of methyl acrylate is based on well-established technology and on readily available raw materials (acetone, hydrogen cyanide, methanol, and sulfuric acid). Acetone cyanohydrin, formed by the base-catalyzed addition of hydrogen cyanide to acetone, is treated with excess sulfuric acid to form methacrylamide sulfate.

The preparation of methacrylamide sulfate is performed in a series of reactors at 80–110°C with a residence time of about one hour. The reactor effluent is heated in the range of 125 to 145°C to complete the reaction and it is then transferred to the esterification section.

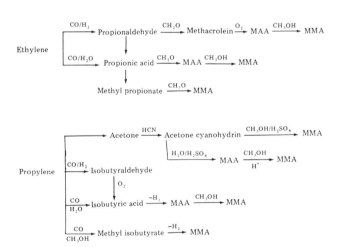
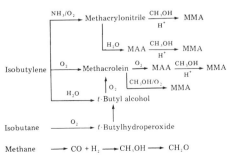

Figure 1. Routes to methyl methacrylate (MMA) and methacrylic acid (MAA).

There are several modifications of the esterification step, In one version, the methacrylamide sulfate stream, excess methanol and recycle streams react at 80–110°C with 2–4 h residence time. The reactor effluent is distilled in a series of columns to give methyl methacrylate and recycle streams. The waste acid is reconverted to sulfuric acid for recycle or treated with ammonia to give fertilizer-grade ammonium sulfate.

Methacrylic acid is manufactured by hydrolysis of the methacrylamide stream using facilities and conditions similar to those used in the esterification step.

C$_4$-Oxidation.   The proposed C$_4$-oxidation process is based on the two-stage oxidation of isobutylene or t-butyl alcohol to methacrylic acid. It is similar to the highly successful process for the manufacture of acrylic acid from the catalytic oxidation of propylene (see Acrylic acid and derivatives). A basic hydrocarbon raw material is used to produce methacrylic acid and the only other required reactant is air. Isobutylene is oxidized in the first stage to methacrolein and, in the second stage, the methacrolein is oxidized to methacrylic acid. The methacrylic acid is separated from the by-products and is esterified with methanol to methyl methacrylate. For this route to be attractive, as compared with the ACN process, both catalysts must be selective and have a life of more than one year and, low cost, ie, relative to other organic feedstocks, isobutylene must be available (see also Hydrocarbon oxidation).

Isobutyrate dehydrogenation.   In this proposed route, isobutyric acid or methyl isobutyrate is catalytically dehydrogenated to methacrylic acid or methyl methacrylate, respectively. The acid-catalyzed addition of carbon monoxide to propylene in the presence of water or methanol is employed for the formation of the isobutyric intermediate. Improved vapor-phase catalysts are claimed for the oxidative-dehydrogenation step. This may result in the development of an attractive process based on propylene (see also Oxo process).

Storage and Handling

Methacrylates must be stabilized to be handled safely. The inhibited materials can be handled as flammable liquids. However, the monomers should not be stored for longer than one year and contamination must be avoided.

Health and Safety

The lower methacrylate esters have low flash points and can form explosive mixtures with air. Methacrylates exhibit low to moderate acute toxicity, produce slight skin and eye irritation and generally are considered to be sensitizers. The workplace exposure limit established by ACGIH for methyl methacrylate is 100 ppm TWA.

JOSEPH W. NEMEC
LAWRENCE S. KIRCH
Rohm and Haas Company

Storage and Handling of Acrylic and Methacrylic Esters and Acids, Bulletin CM-17, Rohm and Haas Company, Philadelphia, Pa., 1975.

Acrylic and Methacrylic Monomers, Properties and Specifications, Bulletin CM-16, Rohm and Haas Company, Philadelphia, Pa., 1978.

L.S. Luskin in E.C. Leonard, ed., *High Polymers, Vinyl and Diene Monomers*, Vol. 24, Pt. I, "Acrylic Acid, Methacrylic Acid, and the Related Esters," Wiley-Interscience, New York, 1970.

L.S. Luskin in R.H. Yocum and E.B. Nyquist, eds., *Functional Monomers, Their Preparation, Polymerization and Application*, Vol. 2, "Basic Monomers: Vinyl Pyridines and Aminoalkyl Acrylates and Methacrylates," Marcel Dekker, New York, 1974.

METHACRYLIC POLYMERS

Methacrylic ester monomers have the generic formula CH$_2$= C(CH$_3$)COOR, and it is the nature of the R group that generally determines the properties of the corresponding polymers. Methacrylates differ from acrylates in that the hydrogen of the acrylate is replaced by a methyl group (see Acrylic ester polymers). It is the α-methyl group of the polymethacrylate that imparts the stability, hardness, and stiffness to methacrylic polymers. The methacrylate monomers are unusually versatile building blocks since they are moderate-to-high boiling liquids that readily polymerize or copolymerize with a variety of other monomers. All of the methacrylates readily copolymerize with each other and with the acrylate series; thus, extreme ranges of properties can be built into the polymer.

The uniqueness of methyl methacrylate as a plastic component accounts for the large production volume of methyl methacrylate compared to the combined volumes of all of the other methacrylates.

Hard methacrylates easily copolymerize with soft acrylates to form polymers having a wide range of hardness; thus, polymers that are designed to fit specific application requirements can be tailored readily from these versatile monomers. The properties of the polymers can be varied to form extremely tacky adhesives (qv), rubbers, tough plastics, and hard powders. Although higher in cost than many other common monomers, the methacrylates' unique stability characteristics, ease of use, efficiency, and the associated high quality products more than compensate for their expense.

Physical Properties

The nature of the alcohol group of the methacrylate monomer unit within the polymer chain and the molecular weight of the polymer largely determines the physical and chemical properties of methacrylate ester polymers. Typically, the mechanical properties of methacrylate polymers improve with increasing molecular weight; however, beyond a critical molecular weight, eg, ca 100,000–200,000 for amorphous polymers, the increase in properties is slight and tends to level off asymptotically. The atactic (random) configuration is the result of polymerization unless there are special circumstances favoring sterospecific addition to form isotactic (cis) or syndiotactic (trans) chains (see Olefin polymers; Polymerization mechanisms and processes). Physical and chemical properties of the ordered or stereospecific polymer can differ significantly from the atactic type.

Mechanical and thermal.   Substitution on the main chain of the methyl group in the methacrylates for the α-hydrogen of the acrylates results in restricted freedom of rotation and motion of the polymer backbone yielding harder polymers of higher tensile strength and lower elongation that the acrylate counterparts.

Electromagnetic spectrum.   Poly(methyl methacrylate) transmits light almost perfectly, ie, 92% compared to 92.3% theoretical, at 360–1000 nm (The wavelengths of visible light are a 400–700 nm). At thicknesses of > 2.5 cm, poly(methyl methacrylate) absorbs virtually no visible light. Beyond 2800 nm, essentially all ir radiation is absorbed. Commercial grades of poly(methyl methacrylate) often contain uv radiation absorbers which block light in the 290–350-nm range (see Uv stabilizers). The absorber, acting as a sunscreen, protects the user from sunburn and the polymer against long-term degradation from light. Poly(methyl methacrylates)'s transparency to x-ray radiation is about the same as that of human flesh or water. Sheets are opaque to α particles, and above 6.35 mm, the polymer is essentially opaque to β-ray radiation; poly(methyl methacrylate) also is used as a transparent neutron barrier.

Electrical.   The surface resistivity of poly(methyl methacrylate) is higher than that of most plastic materials. Weathering and moisture affect it to a minor degree. High resistance and nontracking characteristics have resulted in its use in high voltage applications.

Solution.   As with the acrylates, the solubility of a methacrylate polymer is affected by the nature of the alcohol-derived side group. Methacrylate polymers that contain short ester side chains are relatively polar and are soluble in polar solvents, eg, ketones, esters, ethers, and alcohols. With increasing length of an alkyl side chain, the polymer becomes less polar and dissolves in relatively nonpolar solvents, eg, aromatic or aliphatic hydrocarbons.

Chemical Properties

Methacrylate polymers have a greater hydrolytic resistance to both acidic and alkaline hydrolysis than do acrylate polymers; both are far

Table 1. Chemical Resistance of Poly(Methyl Methacrylate)

Not affected by	Attacked by
most inorganic solutions	lower esters, eg, ethyl acetate, isopropyl acetate
mineral oils	aromatic hydrocarbons, eg, benzene, toluene, xylene
animal oils	phenols, eg, cresol, carbolic acid
low concentrations of alcohols	aryl halides, eg, chlorobenzene, bromobenzene
paraffins	aliphatic acids, eg, butyric acid, acetic acid
olefins	alkyl polyhalides, eg, ethylene dichloride, methylene chloride
amines	high concentrations of alcohols, eg, methanol, ethanol,
alkyl monohalides	isopropanol
aliphatic hydrocarbons	high concentrations of alkalies and oxidizing agents
higher esters, ie, >10 carbon atoms	

more stable than poly(vinyl acetate) and vinyl acetate copolymers. There is a marked difference in the chemical reactivity among the noncrystallizable and crystallizable forms of poly(methyl methacrylate) relative to alkaline and acidic hydrolysis. Conventional, ie, free-radical, bulk-polymerized, and syndiotactic polymers hydrolyze relatively slowly compared with the isotactic type. Polymer configuration is unchanged by hydrolysis. The chemical resistance is summarized in Table 1.

Manufacture

The preparation and properties of methacrylic esters are discussed in detail in the article Methacrylic acid and derivatives.

Methacrylate polymers are produced commercially in the form of sheets, rods, tubes, and blocks as well as pellets, solutions, latices, and beads. The physical characteristics of the neat polymers range from soft and flexible to hard and rigid, depending upon the monomer composition and the method of polymerization. Most of the commercially available polymers are prepared by free-radical processes involving initiation, propagation, chain transfer, and termination. The type and level of initiators and chain-transfer agents depends upon the polymerization method (see Initiators).

Methacrylate polymerizations are accompanied by liberation of heat and a decrease in volume. In general, the percent shrinkage decreases as the size of the alcohol substituent increases; on a molar basis, the shrinkage is constant.

Methacrylate polymerizations are markedly inhibited by oxygen; therefore, considerable care is taken to exclude air during polymerization.

Bulk polymerization. The bulk polymerization of monomeric methacrylic esters is used principally to manufacture sheets, rods, tubes, and molding powder. In terms of volume, sheet casting is the most important process and probably accounts for half of the methacrylate monomer produced. The monomer that is used is usually methyl methacrylate, often with minor portions of other monomers, uv absorbers, pigments or dyes, and other additives. The polymerization is a free-radical process which is initiated by heat and radical initiators, eg, peroxides and azo compounds.

Solution polymerization. The solution polymerization of methacrylic monomers to form soluble methacrylic polymers or copolymers is an important commercial process for the preparation of polymers for use as coatings, adhesives, impregnates, and laminates. Typically, the polymerization is accomplished batchwise by adding the monomer to an organic solvent in the presence of a soluble initiator which usually is a peroxide or an azo compound.

Emulsion polymerization. The emulsion polymerization of methacrylic esters to form aqueous dispersion polymers is used for the preparation of polymers suitable for applications in the paint, paper, textile, floor polish, and leather industries where they are used principally as coatings or binders. Copolymers of methyl methacrylate with either ethyl acrylate or butyl acrylate are the most common (see Latex technology).

Suspension polymerization. Suspension polymerization yields polymethacrylates in the form of tiny beads which are used primarily as molding powders and ion-exchange resins (see Ion exchange). Suspension polymers that are prepared for molding powders generally are poly(methyl methacrylate) copolymers containing up to 20 wt% acrylate for reduced brittleness and improved processibility.

Health and Safety

In general, methacrylate polymers are nontoxic. Various methacrylate polymers are used in the packaging and handling of food, in dentures, and dental fillings, and as medicine dispensers and contact lenses (qv). Most of the health and safety aspects are involved with their manufacture and fabrication. Considerable care is exercised to reduce the potential for violent polymerizations and to reduce exposure to flammable and potentially toxic monomers and solvents.

Uses

Methacrylate polymers are used in lubricating-oil additives, surface coatings, impregnates, adhesives, binders, sealers, and floor polishes. The largest use for the polymethacrylates is as glazing, lighting, or decorative material; they are also used in medicine and light-focusing plastic fibers in optics.

BENJAMIN B. KINE
R.W. NOVAK
Rohm and Haas Company

E.H. Riddle, *Monomeric Acrylic Esters*, Reinhold Publishing Corp., New York, 1954.

L.S. Luskin, J.A. Sawyer, and E.H. Riddle, "Hamufacture of Acrylic Polymers," in W.M. Smith, ed., *Polymer Manufacturing and Processing*, Reinhold Publishing Co., New York, 1964.

Emulsion Polymerization of Acrylic Monomers, CM-104, Rohm and Haas Co., Philadelphia, Pa.

METHANE. See Hydrocarbons, C_1–C_6.

METHANOL

Methanol (methyl alcohol), CH_3OH, is a clear water-white liquid with a mild odor at ambient temperatures. It has been called wood alcohol, or wood spirit, because it was obtained commercially from the destructive distillation of wood for over a century. However, the true wood alcohol contained more contaminants (primarily acetone, acetic acid, and allyl alcohol) than the chemical-grade methanol available today.

Physical Properties

Selected physical properites are shown in Table 1.

Chemical Reactions

Methanol undergoes reactions that are typical of alcohols as a chemical class. Those of particular importance from an industrial standpoint

Table 1. Physical Properties of Methanol

Property	Value
freezing point, °C	−97.68
boiling point, °C	64.70
heat of formation (liquid) at 25°C, kJ/mol[a]	−239.03
heat of vaporization at boiling point, J/g[b]	1129
flammable limits in air	
lower, vol%	6.0
upper, vol%	36
autoignition temperature, °C	470
surface tension, mN/m (= dyn/cm)	22.6
solubility in water	miscible
density at 25°C, g/cm³	0.78663
refractive index, n_D^{20}	1.3284
viscosity of liquid at 25°C, mPa·s (= cP)	0.541

[a] To convert kPa to mm Hg, multiply by 7.5.
[b] To convert J to cal, divide by 4.184.

are dehydrogenation and oxidative dehydrogenation to formaldehyde (see Formaldehyde) over silver or molybdenum–iron oxide catalysts and carbonylation to acetic acid catalyzed by cobalt or rhodium (see Acetic acid). Dimethyl ether can be formed by the acid-catalyzed elimination of water. The acid-catalyzed reaction of isobutylene and methanol to form methyl *tert*-butyl ether (MBTE), an important gasoline-octane improver, has increasing application. Methyl esters of carboxylic acids can be prepared by acid-catalyzed reaction with azeotropic removal of water to force the reaction to completion. Methyl hydrogen sulfate, methyl nitrate, and methyl halides are formed by reaction with the appropriate inorganic acids. Mono-, di-, and trimethylamines result from the direct reaction of methanol with ammonia (see Amines).

Manufacture

The oldest industrially significant method of methanol manufacture was the destructive distillation of wood. Practiced from the mid-19th century to the early 1900s, it is no longer used in the U.S. The technology became obsolete with the development of a synthetic route from hydrogen and carbon oxides in the mid-1920s. Methanol also was produced as one of the products of the noncatalytic oxidation of hydrocarbons, a practice discontinued in the U.S. since 1973 (see Hydrocarbon oxidation). Methanol also can be obtained as a by-product of Fisher-Tropsch synthesis (see Fuels, synthetic).

Modern industrial-scale methanol production is based exclusively on synthesis from pressurized mixtures of hydrogen, carbon monoxide, and carbon dioxide gases in the presence of metallic heterogeneous catalysts. The required synthesis pressure depends on the activity of the particular catalyst. By convention, technology is generally distinguished by pressure as follows: low pressure processes, 5–10 MPa (50–100 atm); medium pressure processes 10–25 MPa (100–250 atm); and high pressure processes, 25–35 MPA (250–350 atm).

Raw materials. Common feedstocks used in producing synthesis gas for methanol manufacture in the U.S. are natural gas and petroleum residues. Other suitable feedstocks are naphtha and coal. Combined, natural gas, petroleum residues, and naphtha account for 90% of worldwide methanol capacity, miscellaneous off-gas sources making up the difference.

Natural-gas feedstock accounts for most of worldwide methanol capacity. A typical schematic of the process steps is shown in Figure 1.

Health and Safety

The most generally known health hazard associated with methanol is blindness, usually as a result of ingestion. The ingestion of methanol has resulted in a wide range of responses, probably owing to the concurrent intake of varying amounts of ethanol. Ethanol is selectively metabolized

by the body, allowing detoxification by respiration to occur to some extent. By mouth, 25–100 mL of methanol is reported to be fatal. The recommended maximum exposure level to vapor is 200 ppm TWA per 40-h work week. It is not clear that blindness has resulted from inhalation.

Uses

Historically, almost half of all methanol produced has been used to produce formaldehyde. Acetic acid manufacture uses ca 8% of the methanol produced. Fuel use, either directly or as MTBE, consumed about 10% of 1983 production and may continue to grow if the methanol price remains competitive with hydrocarbons (see Alcohol fuels).

L.E. WADE
R.B. GENGELBACH
J.L. TRUMBLEY
W.L. HALLBAUER
Celanese Chemical Company, Inc.

J.A. Monick, *Alcohols, Their Chemistry, Properties and Manufacture*, Reinhold Book Corp., New York, 1968, pp. 93–101.

Acceptable Concentrations of Methanol, American National Standards Institute, New York, 1971.

J.K. Paul, *Methanol Technology and Application in Motor Fuels*, Noyes Data Corporation, Park Ridge, N.J., 1978.

METHIONINE, $CH_3SCH_2CH_2CH_2CH(NH_2)COOH$. See Amino acids.

METHYL ACETATE, CH_3COOCH_3. See Acetic acid; Esters, organic.

METHYLACETYLENE. See Acetylene.

METHYL ALCOHOL, CH_3OH. See Methanol.

METHYLAMINES. See Amines, lower aliphatic.

METHYL SULFATE, $(CH_3)_2SO_4$. See Sulfuric and sulfurous esters.

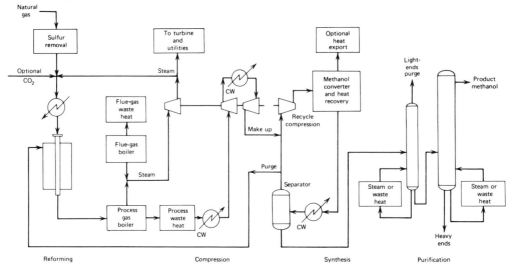

Figure 1. Methanol from natural gas. CW = cooling water.

MICAS, NATURAL AND SYNTHETIC

Mica is the name for a group of complex hydrous aluminum silicate minerals constructed of extremely thin cleavage flakes and characterized by near perfect basal cleavage, and a high degree of flexibility, elasticity, and toughness. Laminae as thin as 15 μm are obtained readily and utilized. The various micas, although structurally similar, may vary widely in chemical composition, particularly the rarer types; however, within any one variety the composition does not vary greatly.

The properties of mica derive from the periodicity of weak chemical bonding alternating with strong bonding. Each ultimate single sheet of mica has one weakly bonded layer (potassium) and three strongly bonded layers (two silicon and one aluminum or magnesium). The ultimate sheets are about 1 nm in thickness.

The principal minerals of the mica group are: muscovite $(K_2Al_4(Al_2Si_6O_{20})(OH)_4$; phlogopite, $K_2Mg_6(Al_2Si_6O_{20})(OH, F)_4$; biotite, $K_2(Mg, Fe)_6(Al_2Si_6O_{20})(OH)_4$; and lepidolite, $K_2Li_3Al_3$-$(Al_2Si_6O_{20})(OH, F)_4$. Other minerals include: roscoelite, K_2V_4-$(Al_2Si_6O_{20})(OH)_4$; fuchsite, $K_2Cr_4(Al_2Si_6O_{20})(OH)_4$; fluorophlogopite, $K_2Mg_6(Al_2Si_6O_{20})F_4$; and paragonite, $Na_2Al_4(Al_2Si_6O_{20})(OH)_4$. In general, the Si:Al ratio is about 3:1. The general formula gives the number of ions in the unit cell and is the repeat unit of the inorganic mica structure.

Selected properties of muscovite, phlogopite, and fluorophlogopite are shown in Table 1.

Natural Micas

Mica is widely distributed as small flakes in many igneous, metamorphic, and sedimentary rocks. Some schists and kaolin deposits contain sufficient muscovite to be recovered for use as scrap mica (see also Clays).

Sheet muscovite is obtained from coarse-grained igneous rocks called pegmatites, which, when well-zoned, are the best source of commercial sheet muscovite. Generally, zones are successive concentric shells around a core, complete or incomplete, that often reflect the shape or structure of a whole pegmatite body. From the outermost inward is the border zone, the wall zone (containing the best quality and greatest concentration of sheet muscovite), the intermediate zones, and the core margin. When the pegmatite is poorly zoned, muscovite is scattered throughout the rock.

Phlogopite deposits are found in areas of metamorphosed sedimentary rocks intruded by masses of pegmatite-rich granitic rocks. The phlogopite is found as veins or pockets in pyroxenite interlayered with or intersecting marble or gneiss.

Mining, preparation, and processing. The unique properties of mica and the erratic character in its natural occurrence are such that methods employed in mining and processing other types of minerals are applicable only to a limited extent in the development of mica deposits.

Mining methods for sheet-mica mines tend to be rather primitive, small-scale, often underground operations. Open-pit mining is used when feasible. Care must be taken to avoid drilling through good mica crystals. A low velocity explosive may be needed to blast around a pocket of mica. Several holes are shot at one time to avoid destruction of available mica books.

Grades, qualities, and specifications of sheet mica. Sheet mica is classified according to: preparation, including crude, hand-cobbed, thumb-trimmed, half-trimmed, three-quarter trimmed, and full-trimmed; thickness, ie, block (min 0.18 mm thick with min. usable area of 6.5 cm²), film (thin sheet or sheets split form better-quality block mica with thickness range from 0.032 to 0.010 mm and tolerance of 0.0062 mm) and splittings (thin pieces of irregularly shaped mica with max thickness of 0.03 mm); size; appearance, including color, flatness, stain, crack, air inclusion, waviness, herringbone, sandblast and other; electrical quality; thermal stability (for phlogopite)—(high heat phlogopite must withstand 750°C for 30 min with less than 25% increase in thickness after cooling and with no evident transformations).

Synthetic Fluorine Micas

Compared to the natural-mica industry, the synthetic-mica industry is very small and, because of costs and substitution for mica products in general, will most likely remain so.

In the synthetic fluorine micas, fluoride replaces the hydroxyl ion. They are true micas, and the potassium varieties have the characteristics of natural mica and can be split as easily. The fluorine micas are formed at atmospheric pressure by melting or solid-state reaction of the proper raw materials. Owing to the absence of hydroxyl ion, fluoromicas do not decompose as readily as natural micas and have a much higher thermal stability. By ionic substitution, a wide variety of fluoromicas may be synthesized, some of which have different or very unusual properties.

Properties. The properties of fluorophlogopite are given in Table 1. In general, the physical, electrical, and chemical properties of fluorophlogopite are similar to those of the better grades of natural muscovite or phlogopite. Single crystals of fluorophlogopite are clear and transparent, and readily cleave parallel to the (001) crystallographic plane. The main differences are the higher heat stability (mp = 1387°C) and the high purity of fluorophlogopite which contains < 0.04% Fe_2O_3.

Phase reactions / batch compositions. Fluorophlogopite is prepared easily in small or large amounts from potassium fluorosilicate, feldspar, magnesia, alumina, and quartz sand. The < 150 μm (−100 mesh), or finer, raw materials are mixed and fired to temperatures above melting (1387°C). Crystallization in crucibles may be accomplished by stopping heating and letting the furnace cool at its natural rate or by programmed reduction of temperature. Larger crystals are obtained by slow cooling (below 5° and preferably 1°C/h) through the crystallization range. Fluorophlogopite has been grown in sheets up to 7.5 × 12.5 cm.

Special Products

Built-up mica. This is sheet or board made by alternate layers of overlapping splittings and a suitable binder, consolidated by heat and pressure at ≥ 150°C and 0.34–6.9 MPa (3.4–68 atm).

Glass-bonded mica. Glass-bonded mica is a composite material made from natural or synthetic mica or combinations of both which combines the molding ability of glass with the superior electrical properties of mica. It is machinable, and smooth surfaces with close dimensional tolerances are obtained. By using preheated precision dies, the product may be made by compression or injection molding in a one-shot rapid process. Glass-bonded mica does not change with age, does not char or track with an arc, and it is essentially unaffected by oxygen, ozone, or contaminants in the atmosphere. Metallic inserts may be molded in place, and by appropriate techniques, the product may be plated with conducting films. The high temperature stability of the fluoromicas allows the use of higher-temperature glasses that possess better electrical qualities. Its manufacturing process is essentially similar to the molding of plastics (see Plastics processing).

Health and Safety

At present no definite health hazard caused by mica has been established. Its crystal lattice does not produce fibers although some deposits

Table 1. Properties of Natural and Synthetic Micas

Property	Natural		Synthetic fluorophlogopite
	Muscovite	Phlogopite	
density, g/cm³	2.6–3.2	2.6–3.2	2.8
hardness, Mohs	2.0–3.2	2.5–3.0	3.4
orientation of optic plane to plane of symmetry	perpendicular	parallel	parallel
refractive index, n_D			
α	1.552–1.570	1.54–1.63	1.522
tensile strength, MPa[a]	225–296	255–296	310–358
melting point, °C	decomposes	decomposes	1387 ± 3
power factor at 25°C (highest qualities), %			
at 60 Hz	0.08–0.09		
at 1 MHz	0.01–0.02	0.3	0.02

[a] To convert MPa to psi, multiply by 145.

and products may contain fibers of other minerals as impurities. Mica produces bladed or platy particles when ground or milled.

Uses

Block and film mica are used in the electronic and electrical industries and in the fabrication of vacuum-tube spacers. Splittings and built-up mica are the largest use of sheet mica and indispensable as an insulating medium in the electrical industry. High quality scrap mica is used to produce mica paper. Ground mica is used primarily in gypsum plasterboard cement as a filler, and in the production of rolled roofing and asphalt shingles (see Fillers).

ALVIN B. ZLOBIK
U.S. Bureau of Mines

L.G. Berry and B. Mason, *Mineralogy—Concepts, Descriptions, Determinations*, W.H. Freeman and Company, San Francisco, Calif., 1959.

Standards for Manufactured Electrical Mica, Pub. FI-1977, National Electrical Manufacturers Association, New York, 1977.

A.B. Zlobik, *General Yearbook, 1978–1979*, U.S. Dept. of Interior, Bureau of Mines, U.S. Government Printing Office, Washington, D.C., Oct. 1979, 18 pp.

MICELLULAR FLOODING.   See Petroleum, enhanced oil recovery.

MICROBIAL POLYSACCHARIDES

Polysaccharides of microorganisms occur as intracellular-storage amylosaccharides, lipid-associated substances in conjunction with cytoplasmic membranes, structural glycans that impart rigidity to cell walls, both discrete and diffuse capsular slimes that remain attached to the cells, and extracellular products in the media. Only the capsular and extracellular (exo-) polysaccharides can be produced in sufficiently high yields to merit commercial interest. Because physical, chemical, and enzymatic means are required to free the capsular types from cells, these types generally have not been considered for industrial production. Exopolysaccharides that either have been produced or are being considered for industrial production are described in this article.

Dextran

Dextrans are α-D-glucans in which $(1 \rightarrow 6)$ linkages predominate, ie, 50% or more of the α-D-glucopyranosyl residues are linked as such. Dextrans are produced from sucrose by bacteria belonging to the genera *Leuconostoc*, *Streptococcus*, and *Lactobacillus*, all of which are in the family *Lactobacillaceae*. The majority of known dextrans is formed by strains of *Leuconostoc mesenteroides*. Because they interfere in the production of sucrose, dextrans were the first extracellular microbial polysaccharides to be investigated. Aside from impeding the filtration and handling of cane and beet-sugar juices, dextran causes sucrose to crystallize in the form of impure, elongated needles (see Sugar).

Biosynthesis of dextran from sucrose was the first direct enzymic polymerization demonstrated for a disaccharide donor substrate. The reaction is catalyzed by an inducible enzyme, dextransucrase (sucrose: 1,6-α-D-glucan 6-α-D-glucosyltransferase, E.C.2.4.1.5), which may be either cellbound or extracellular.

Sucrose is the only donor substrate known and it is the initial acceptor; repetitive α-D-glucopyranosyl transfer occurs so rapidly that high molecular weight products are formed without detectable oligosaccharide intermediates. At high concentrations, (> 10–70%) sucrose competes as the acceptor with growing dextran chains, and low molecular weight products are formed.

Enzymatic biosynthesis.   In industrial production of dextran, the extracellular enzyme is used. Initially, dextransucrase is produced. After adjustment of culture pH and removal of cells, the culture fluid is distributed into vessels containing sugar solutions (and, perhaps, low molecular weight dextran primers) where the polymerization reaction takes place. Advantages of this approach are more efficient use of equipment, virtually complete conversion of substrate into product; ease of product recovery and purification; control of reaction conditions; ie, pH, temperature, and primer addition; and the possibility of recovering the coproduct D-fructose, which otherwise would be consumed by the cells metabolically with concomitant production of lactic acid.

Maximum elaboration of dextransucrase by *L. mesenteroides* NRRL B-512 (F) occurs in cultures that are maintained at pH 6.7. The enzyme is unstable at that pH but has maximum stability and activity at pH 5.0–5.2. Because the activity of dextransucrase is highly sensitive to temperatures above 25°C, incubation temperatures do not exceed 30°C. A sucrose level of 2% is optimal for enzyme production, because higher levels lead to formation of an amount of dextran that interferes with removal of cells.

Uses.   Unlike most other microbial polysaccharides, the utility of B-512 (F) dextran depends much less on its ability to impart high viscosity to aqueous solutions than on its inherent structural features. For example, a 2% soln (polymer wt/soln wt) of the dextran at 25°C gives a viscosity of ca 100 mPa·s (= cP); the same concentration of xanthan gum displays a viscosity of ca 7 Pa·s (70P) (see Gums). The useful characteristics of B-512 (F) dextran derive from its primary structural features, whereas the properties of xanthan gum arise from secondary and tertiary macromolecular structural effects.

Pharmaceutical uses probably are the main outlets for dextran. Clinical use of dextran as fractions having specific molecular size ranges is based on its compatibility with human tissues and complete metabolic utilization, whether it is ingested or administered parenterally. Material of $\overline{M}_w$ 75,000 $\pm$ = 25,000 is specified for parenteral infusion as a blood-volume expander in treatment of shock, eg, from hemorrhage or burns; the lower limit was set because lower molecular-weight material clears too rapidly through excretion from the kidneys.

Extracellular Polysaccharides As Metabolic Products

Xanthan gum.   The extracellular polysaccharide of *Xanthomonas campestris* B-1459 was the first biopolymer product of a fermentation based on corn sugar that attained commercial status. The associated rapid fermentation gives high conversions of carbohydrate into a highly viscous product of unusual stability and rheological properties of the gum. The FDA has approved xanthan gum for use as a food additive where such use is not precluded by Standards of Identity Regulations. By the end of 1983, the world capacity for biopolymer production (mainly xanthan gum, but excluding dextran) exceeded 30,000 metric tons. The structure of xanthan gum consists of a β-$(1 \rightarrow 4)$-linked D-glucopyranosyl backbone chain, eg, as in cellulose:

sucrose   +   (1,6-α-D-glucopyranosyl)$_n$ $\rightarrow$

D-fructose   +   (1,6-α-D-glucopyranosyl)$_{n+1}$

xanthan gum

To the chain are appended trisaccharide side chains composed of D-mannopyranosyl (Man) and D-glucopyranosyluronic acid (GlcA) residues. The β-(1 → 2)-linked mannosyl residues have 6-O-acetyl substituents. An average of about half of the β-D-mannosyl end groups bear 4,6,-O-(1-carboxyethylidene) substituents, ie, 4,6-acetal-linked pyruvic acid.

Physical properties. Aqueous dispersions of xanthan gum exhibit several novel and remarkable rheological properties. Low concentrations of the gum have relatively high viscosities which permit its economical use in applications, eg, emulsion stabilization and mobility control of water-flooding fluids in petroleum reservoirs. Xanthan gum solutions are characterized by high pseudoplasticity; ie, over a wide range of gum concentrations, rapid shear-thinning occurs that is instantaneously reversible. If a salt is present, the viscosity is independent of pH from pH 1.5–13. Viscosity also is stable to heat over a wide temperature range, and this stability is enhanced by salts of mono- and divalent cations.

Structure and Conformation. The mol wt of xanthan gum is (2–15) × 10^6. Many of the properties of xanthan gum dispersions indicate that the molecule assumes a rodlike, ordered secondary structure. A sharp increase in the viscosity of 0.5–1% aqueous dispersions has been noted at 50–60°C and has been confirmed by optical rotation, circular dichroism, and nmr measurements. Increasing the ionic strength raises the temperature at which the transition occurs; such melting out or denaturation is a helix-to-coil conformational transition which is similar to that observed for double-stranded nucleic acids, triple-stranded collagen, and certain polysaccharides. X-ray diffraction studies with xanthan gum fibers indicate a single-stranded helix of fivefold symmetry with side chains folded along the backbone in a hydrogen-bonded structure (see Biopolymers).

Production. The industrial process for xanthan gum has been outlined (see Gums). Inoculum build-up from a stock culture proceeds through stages to a seed tank, 5 vol% of the contents is added to the production fermentor directly. An aerated, 2-d fermentation is carried out at pH 6.0–7.5 and at 28–31°C. After fermentation, the culture is pasteurized and then treated with isopropyl alcohol to recover the polysaccharide. The precipitate is dried, milled, tested, and packaged.

Uses. Over half of microbial polysaccharides sales are to the food industry. Xanthan gum interacts synergistically with galactomannans, eg, locust bean and guar gums, to produce combined viscosities greater than would be expected from the individual polysaccharides. The interaction with locust bean gum is much stronger, and appropriate combinations can form a heat-reversible gel.

Others. These include gellan gum, an agarlike polysaccharide; bacterial alginic acid; succinoglucan; and neutral glucans such as curdlan, scleroglucan, and pollulan.

MOREY E. SLODKI
Northern Regional Research Center
U.S. Department of Agriculture

M. E. Slodki and M.C. Cadmus, *Adv. Appl. Microbiol.* **23**, 19 (1978).

K.S. Kang and I. Cottrell in H.J. Peppler and D. Pelman, eds., *Microbial Technology*, 2nd ed., Vol. 1, Academic Press, Inc., New York, 1979, pp. 417–481.

MICROBIAL TRANSFORMATIONS

Microorganisms are of considerable economic importance in the manufacture of antibiotics (qv), alkaloids (qv), vitamins (qv), amino acids (qv), industrial solvents, organic acids, nucleosides, nucleotides, fermented beverages, and fermented foods (see Solvents, industrial; Carboxylic acids; Beer; Beverage spirits, distilled; Wine; Food processing; Foods, nonconventional; Fermentation; Malts and malting). They are also principal to simple and chemically well-defined reactions involving compounds that are not related to these products. Since some of such reactions can be carried out more economically by microbial means than by a strictly chemical manipulation, the reactions have been included in the processes that yield a number of important products, eg, L-ascorbic acid, steroid hormones, 6-aminopenicillanic acid, various L-amino acids,

L-ephedrine, D-fructose, vinegar, and malt (see Hormones, adrenal-cortical; Sugar).

Reactions

Oxidations. Both mono- and polynuclear aromatic hydrocarbons can be oxidized by different microorganisms. For example, p-cymene is converted to cumic acid and p-xylene to p-toluic acid; a high-yielding (98%) process has been developed in Japan for the production of salicylic acid (**2**) from naphthalene (**1**):

Hydroxylation, dehydrogenation, and β-oxidation. When cortisone and hydrocortisone were identified in 1949 as potent anti-inflammatory agents and no adequate synthesis existed to meet the sharply increased demand for these compounds, hydroxylation of steroid intermediates at C-11 became crucial for their large-scale production. The problem of introducing functionality at that site was solved when it was discovered that progesterone (**3**) is oxidized to 11α-hydroxyprogesterone (**4**) by *Rhizopus arrhizus*.

Reduction. Although the reduction of ketonic substrates usually is only partially asymmetric, more stereospecific microbial reduction depends on the size and nature of the substituents in the substrates; eg, the reduction of racemic decalone and hexahydroindanone derivatives and of related di- and tricyclic ketones by *Curvularia falcata* is highly stereospecific. Stereoselective reduction occurs if the ketone is flanked by a large L and small s groups, and yields an alcohol of the S-configuration:

This model may help to predict if the substrate will be reduced to an optically active alcohol which may be difficult to obtain by purely chemical means.

Hydrolysis. Microbial hydrolysis of a large number of esters, glycosides, epoxides, lactones, β-lactams, and amides has been described. Acetylated steroids have been hydrolyzed with varying degrees of selectivity by numerous organisms; (−)-14-acetoxycodeine has been converted to (−)-1-hydroxycodeine; and atropine to tropine; the sugar moiety has been removed from heart glycosides and saponins; and L-amino acids have been produced from their optically inactive (DL) forms.

Condensation. Asymmetric microbial condensation was discovered in 1921 and utilized in 1934 in the synthesis of the natural (1R, 2S)-ephedrine. In this process, benzaldehyde is added to the fermenting yeast and reacts with acetaldehyde which is generated from glucose by the organism, and yields (R)-1-phenyl-1-hydroxy-2-propanone. The latter undergoes reductive chemical condensation with methylamine and yields the desired (1R, 2S)-ephedrine (see Epinephrine).

Amination and hydration. Optically active products, which correspond to the naturally occurring L isomers, have been obtained by the asymmetric addition of water or ammonia to fumaric acid. Thus, aspartase-producing bacteria have been used in the manufacture of L-aspartic acid.

Other reactions and microbial transformations. Other reactions include deamination, dehydration, N- and O-demethylations; decarboxylation, N-acetylation, O-phosphorylation, O-adenylylation, transglycosylation, and isomerization.

Methodology

The selection of the organisms that carry out the desired transformation is of paramount importance. In general, fungi and streptomycetes hydrolyze various esters and carry out specific hydroxylations, yeasts reduce carbonyl groups, and bacteria oxidize alcohols and aldehydes.

For the transformation, the selected organism is usually grown in aerated flasks until a sufficient amount of cells has been generated. The substrate is then added to the cells, the incubation is continued, and the progress of the transformation is monitored by suitable chromatographic, spectroscopic, or biological methods. When the maximum transformation has been obtained, the reaction is terminated and the product isolated and identified (see also Genetic engineering).

OLDRICH K. SEBEK
The Upjohn Company

H. Iizuka and A. Naito, *Microbial Conversion of Steroids and Alkaloids*, University of Tokyo Press, Tokyo, Japan, and Springer-Verlag, Berlin, FRG, 1981.

A. Wiseman, ed., *Principles of Biotechnology*, Chapman and Hall, New York, 1983.

H.-J. Rehm, *Industrielle Mikrobiologie*, 2nd ed. (in German), Springer-Verlag, Berlin, FRG, 1980.

D. Perlman in J.B. Jones, C.J. Sih, and D. Perlman, eds., *Applications of Biochemical Systems in Organic Chemistry*, John Wiley & Sons, Inc., New York, 1976, Pt. 1, p. 47.

MICROCHEMISTRY.   See Analytical methods.

MICROENCAPSULATION

"Small is better" would be an appropriate motto for the many people studying microencapsulation, a process in which tiny particles or droplets are surrounded by a coating to give small capsules with many useful properties. In its simplest form, a microcapsule is a small sphere with a uniform wall around it. The material inside the microcapsule is referred to as the core, internal phase, or fill, whereas the wall is sometimes called the shell, coating, or membrane. Most microcapsules have diameters between a few micrometers and a few millimeters, as illustrated in Figure 1. However, many microcapsules bear little resemblance to these simple spheres. The core may be a crystal, a jagged absorbent particle, an emulsion, a suspension of solids, or a suspension of smaller microcapsules. The microcapsule even may have multiple walls.

Properties Important in Choice of Process

In choosing among processes for a particular application, the following physical properties must be carefully considered:

Figure 1.   Microcapsules come in a wide size range. Courtesy of Eurand America, Inc.

Core wettability.   Coacervation is the formation of a second polymer-rich liquid phase from a polymer solution, eg, by addition of a non-solvent. In coacervation coating, the crucial property is the wettability of the core by the coacervate. As long as solid particles are properly wetted, they are frequently easier to coat than liquid cores.

Core solubility.   In a coacervation system, it is critical that the core not be soluble in the polymer solvent and that the polymer not partition strongly into a liquid core. In interfacial polymerization systems, determination of the solubilities of the reactants in the phase permits a choice of solvents and polymers.

Wall permeability and elasticity.   The polymer permeability indicates whether a core can be isolated or a drug released at the required rate (see Membrane technology). The microcapsules must be able to tolerate handling, but may be required to break above a threshold pressure. The wall polymer, capsule size, and wall thickness determine elasticity and friability.

Techniques

Formation.   The pan coating process, widely used in the pharmaceutical industry, is among the oldest industrial procedures for forming small coating particles or tablets. The particles are tumbled in a pan or other device while the coating material is applied slowly.

Air-suspension coating.   Air-suspension coating of particles by solutions or melts gives better control and flexibility. The particles are coated as they are suspended in an upward-moving air stream.

Centrifugal extrusion.   Several processes have been patented by the Southwest Research Institute in which liquids are encapsulated using a rotating extrusion head containing concentric nozzles. In this process, a jet of core liquid is surrounded by a sheath of wall solution or melt. As the jet moves through the air, it breaks, owing to Rayleigh instability, into droplets of core, each coated with the wall solution. While the droplets are in flight, a molten wall may be hardened or a solvent may be evaporated from the wall solution. Since most of the droplets are within $\pm 10\%$ of the mean diameter, they land in a narrow ring around the spray nozzle. Hence, if needed, the capsules can be hardened after formation by catching them in a ring-shaped hardening bath. This process is excellent for forming particles 400–2000 μm dia.

Vacuum metallizing.   The National Research Corporation has developed vacuum-deposition techniques to coat particles with a wide variety of metals and some nonmetals. The particles can be as small as 10 μm. The entire operation is carried out under vacuum, the particles being conveyed slowly down a refrigerated vibrating table as they are exposed to a metal-vapor beam from a heated furnace. The beam condenses on the cool small particles, eventually coating them evenly. The process can handle only solids and usually forms a coating that is not completely impervious to gases or liquids.

Liquid-wall microencapsulation.   A liquid as the wall of a microcapsule offers several advantages. Since the wall can be broken at will, the wall material can be recovered and reused. The core also can be recovered if desired. In addition, the permeabilities of many molecules through a liquid wall are better controlled and predicted than through a solid wall where morphology can vary greatly.

The liquid core can be a single droplet, an emulsion, or a suspension. It can consist of a variety of chemical reactants, which suggests a wide spectrum of possible uses. The liquid wall is stabilized by addition of a surfactant, such as saponin or Igepal, and through the addition of strengthening agents. It is possible to form thin, aqueous membranes around organic droplets, such as toluene, or oil-based membranes around aqueous solutions.

Spray drying.   Spray drying serves as a microencapsulation technique when an active material is dissolved or suspended in a melt or polymer solution and becomes trapped in the dried particle. The main advantage is the ability to handle labile materials because of the short contact time in the dryer; in addition, the operation is economical (see Drying).

Hardened emulsions.   Since the aim of miroencapsulation is the formation of many tiny particles, the first step in a number of processes is the formation of an emulsion or suspension of the core material in a solution of the matrix material (see Emulsions). This emulsion can be emulsified in another liquid, and the droplets hardened.

Liposomes and surfactant vesicles. The incorporation of drugs into liposomes (phospholipid vesicles) and surfactant vesicles has attracted recent attention. Liposomes, such as lecithin (qv) and phosphatidylinositol, are smectic mesophases of phospholipids organized into bilayers (see Liquid crystals). Liposomes can be prepared in the laboratory by rotary evaporation of a chloroform solution of the phospholipid and cholesterol, and subsequent removal of the thin lipid film from the wall of the flask by shaking with an aqueous buffer. In order to incorporate a hydrophobic drug into these liposomes, the drug is dissolved in the chloroform solution before evaporation. A water-soluble drug can be incorporated by dissolving it in the buffer solution used to form the liposome. The untrapped drug is removed by gel filtration.

Phase separation. In several microencapsulation processes, the core material is first suspended in a solution of the wall material. The wall polymer then is induced to separate as a liquid phase, eg, by adding a nonsolvent for the polymer, decreasing the temperature, or adding a phase inducer, another polymer that has higher solubility in the solvent. In the last case, incompatibility between the two polymers causes the first polymer to separate as another phase. When the wall polymer separates as polymer-rich liquid phase, this phase is called a coacervate and the process is called coacervation.

Complex coacervation. The first commercially valuable microencapsulation process, developed by the National Cash Register (NCR) Company in Dayton, Ohio, was based on coacervation. The coacervate was formed from the reaction product complex between gelatin and gum arabic. The first application was in No-Carbon Required carbonless copy paper. Carbonless copy systems are still by far the largest market for microencapsulated products.

Chemical Methods

Interfacial polymerization. Two reactants in a polycondensation meet at an interface and react rapidly. The basis of this method is the classical Schotten-Baumann reaction between an acid chloride and a compound containing an active hydrogen atom, such as an amine or alcohol. Polyesters, polyureas, polyurethanes, or polycarbonates may be obtained. Under the right conditions, thin flexible walls form rapidly at the interface.

In-Situ Polymerization. In a few microencapsulation processes, the direct polymerization of a single monomer is carried out on the particle surface. In one process, eg, cellulose fibers are encapsulated in polyethylene as they are immersed in dry toluene. In the first step, a Ziegler type catalyst is deposited on the surface of the fiber at 20–30°C by the addition of $TiCl_4$ followed by triethylaluminum. When the catalyst is formed, the surface appears dark brown. The addition of ethylene, propylene, or styrene results in immediate polymerization directly on the surface. This process also has been applied to coating glass. The coating, typically translucent or opaque, can be clarified by melting after formation.

Characterization and Evaluation

Determination of the size distribution of the capsules as produced is an important measurement. The amounts of polymer coating and core can be measured directly by isolating the microcapsules, dissolving or crushing the wall, and making a mass balance on the core and polymer. In many applications, such as carbonless copy paper, microcapsules are ruptured by the pressure. Rupture pressure is measured under controlled conditions similar to those of the application. Most companies have developed their own test apparatus for purposes of quality control and process development.

Health and Safety

In any application of microencapsulation in foods, pharmaceuticals, or veterinary products, only materials approved by the FDA should be used. The GRAS (Generally Recognized As Safe) list can be consulted.

Uses

The most significant application is in carbonless copy paper. Other applications include preserving flavors and essences; in pesticides and herbicides to decrease toxicity upon contact and control release rate (see Herbicides; Insect control technology); and in pharmaceuticals to control the release of a drug (see Pharmaceuticals, controlled-release).

ROBERT E. SPARKS
Washington University

J.R. Nixon, ed., *Microencapsulation*, Marcel Dekker, Inc., New York, 1976.

J.E. Vandegaer, *Microencapsulation: Processes and Applications*, Plenum Press, New York, 1974.

A.C. Tanquary and R.E. Lacey, eds., *Controlled Release of Biologically Active Agents*, Plenum Press, New York, 1974.

MICROPLANTS. See Pilot plants and microplants.

MICROSCOPY, CHEMICAL. See Analytical methods.

MICROWAVE TECHNOLOGY

The application of electrical or electromagnetic (EM) energy to materials as part of some chemical process is a broad subject and of long history in chemical technology. Microwaves are used to ionize gases with sufficient applied power but only through the intermediate process of classical acceleration of plasma electrons to energy values exceeding the ionization potential of molecules in the gas (see also Plasma technology). The term nonionizing radiation is used to distinguish the spectrum below visible light in frequency from the ionizing region well above visible light which exhibits more biological-effect potential whatever the power-flux levels. The distinction between these two parts of the spectrum is so sharp that for years the traditional literature in health physics used the term radiation hazards without qualification, the relation to the ionizing spectral region being obvious.

The distinctive meaning for microwave power applications derives from the fact that for most materials, in particular, biological tissue, maximum penetration of the electromagnetic energy irradiating objects of macroscopic size in human commerce occurs in the microwave range.

Principles of Microwave Power in Processing Materials

In most practical applications of microwave power, the material to be processed is adequately specified in terms of its dielectric permittivity and conductivity. The permittivity is in general taken as complex to reflect loss mechanisms of the dielectric-polarization process, whereas the conductivity may be specified separately to designate free carriers. For simplicity, it is common to lump all loss or absorption processes under one constitutive parameter which can be alternatively labeled a conductivity or an imaginary part of the complex dielectric constant, as expressed in the following equation for complex permittivity:

$$\epsilon = \epsilon_o(\epsilon_r + j\epsilon_i) = \epsilon_o\left(\epsilon_r + j\frac{\sigma}{\omega\epsilon_o}\right) \quad (1)$$

where ϵ is the complex dielectric permittivity in F/m, $\epsilon_o = 8.86 \times 10^{-12}$ F/m, the permittivity of free space ϵ_r is the real part of the relative dielectric constant, ϵ_i is the imaginary part of the relative dielectric constant, and σ is the conductivity in S/m which is equivalent to:

$$\epsilon_i = \frac{\sigma}{\omega\epsilon_o} \quad (2)$$

where ω is the assumed radiant frequency of the fields. It is convenient to define auxiliary terms like the loss tangent σ:

$$\tan\delta = \frac{\epsilon_i}{\epsilon_r} = \frac{\sigma}{\omega\epsilon_r\epsilon_o} \quad (3)$$

From Maxwell's equation, the current density $\vec{J}$ in A/m² is related to the internal electric field:

$$\vec{J} = (\sigma - j\omega\epsilon_r\epsilon_o)\vec{E}_i \quad (4)$$

Thus, the rate of internal density of absorbed energy or power is given by:

$$P = r.p.(\vec{J} \cdot \vec{E}^*) = \sigma |E_i|^2 \qquad (5)$$

or simply

$$P = \omega \epsilon_r \epsilon_0 \tan \delta |E_i|^2 \qquad (6)$$

This is the practical equation for computing power dissipation in materials and objects of uniform composition adequately described by the simple dielectric parameters.

The internal field is that microwave field which is generally obtained when Maxwell's equations are applied to an object of arbitrary geometry and placed in a certain electromagnetic environment. This is to be distinguished from the local field seen by a single molecule which is not necessarily the same.

In a typical application of microwave power, the engineering task is to solve for the internal fields of an object for the given system, compute its heating distribution vs time and resulting change of state, and similar problems. Clearly, the dielectric parameters are key data for such a calculation, particularly their dependence on temperature.

Power Sources

The development of electron tubes, including those for the microwave range, has been a mature field for some time. Today, it is feasible to generate almost any desired power level for most microwave frequencies of practical interest, limited only by costs.

Power sources in the millimeter-wave range are mostly in the category of extended-interaction klystrons or narrow-band backward-wave oscillators. They are quite expensive and suffer from low life and efficiency. Thus, commercial applications in the millimeter-wave range have been hindered, though they are slowly developing including the ISM (industrial, scientific, and medical) bands above 50 GHz. Significant power is generated only at frequencies below 300 GHz.

Above this frequency, the expectation has always been that useful laser sources would eventually be developed, albeit with power limits decreasing with decreasing frequency because of fundamental principles.

The most dramatic evolution of a microwave power source is that of the cooker magnetron for microwave ovens. Such tubes for ca 700-W capability were generally bulky (> 4.5 kg) and water-cooled 25 years ago. Today, cooker magnetrons are air-cooled and weigh < 1.4 kg. These tubes generate well over 700 W at 2450 MHz into a matched load and exhibit a tube efficiency on the order of 70%. Their application is enhanced by the low cost of microwave oven hardware.

Applicators and Instrumentation

The basic elements of a microwave power system for material processing are indicated schematically in Figure 1.

Health and Safety

In addition to the usual mechanical, chemical, thermal, and electrical hazards of power equipment, there are some unique safety considerations in microwave systems. Microwave voltage breakdown can occur in microwave systems and waveguides at power levels far below theoretical values for ideal systems, by a factor of at least 100 below theoretical breakdown. The presence of sharp metal objects, accidental small gaps,

and other situations often can induce localized arcing or coronas which may or may not lead to a basic system breakdown. In this case, the plasma region of the breakdown travels down the feed waveguide toward the source and may cause failure of the tube through cracking of the output window. Therefore, flammable materials should not be processed in microwave systems except with precautions, the more so the greater the degree of flammability.

The most serious hazard is that from incidental interference with other systems. This could be caused by out-of-band radiation, ie, a violation of rfi (radio-frequency indicator) regulations, or by so-called high power effects where the offending radiation is out of the band of the affected system but is still effectively interfering because of its intense level. An example of this would be the incidental interference with cardiac pacemakers. This problem is partly caused by insufficient protection from rfi in the pacemaker unit, ie, susceptibility. In the last ten years, this susceptibility has been reduced and because of government supervision, the problem appears to be under control.

Uses

Literature and patents on microwave power applications are extensive, most of which have been limited to research or small-scale production efforts. Applications are usually limited by economic rather than technical considerations.

The most successful application is that of food processing, cooking, and heating (see Food processing). The domestic microwave oven has been the economic surprise success, surpassing by far all other microwave power applications. Essentially all microwave ovens operate at 2450 MHz except for a few U.S. combination-range models that operate at 915 MHz. The success of this appliance was due to the successful development of low cost magnetrons producing over 700 W for oven powers of 500–700 W.

The most successful industrial application is that of tempering of meat products. Close to 200 units operating at 915 MHz, with output power levels of 25–150 kW are installed throughout the meat-processing industry in the U.S. Another successful application is that of pasta drying.

For at least the last three decades, diathermy at both 27.33 and 2450 MHz has been used extensively in physical therapy. Recent studies suggest that 915 MHz would be more effective than 2450 MHz because of deeper penetration. New diathermy applicators are designed for a more quantitative prediction of tissue-heating patterns and reduced leakage.

An extension of diathermy-heating techniques has been investigated in the last decade with regard to its use in hyperthermia as an adjunct of cancer therapy. The basis for preferential destruction of tumor cells is being studied by hyperthermia alone, or in conjunction with ionizing radiation or chemotherapy. A variety of applicators have been designed, including multiple-focused antennae, injected-probe antennae, and contact applicators.

In textiles, microwave curing imparts crease resistance to fabrics impregnated with amino-resin mixtures (see Textiles).

Microwaves are used in the curing and drying of foundry cores; and for preheating, curing, and drying rubber products.

JOHN M. OSEPCHUK
Raytheon Company

E.C. Okress, *Microwave Power Engineering*, Vols. 1–2, Academic Press, New York, 1968.

M.A. Stuchly and S.S. Stuchly, *IEEE Proceedings*, **130A**(8), 467 (Nov. 1983).

W.A.G. Voss, ed., Special issue on ISM applications of microwaves, *Proc. IEEE* **62** (Jan. 1974).

S.A. Goldblith and R.V. Decareau, *An Annotated Bibliography on Microwaves: Their Properties, Production and Applications to Food Processing*, MIT Press, Cambridge, Mass., 1973.

D.A. Copson, *Microwave Heating*, 2nd ed., Avi Publishing Co., Westport, Conn., 1975.

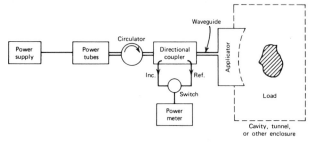

Figure 1. Basic elements of a microwave power system for processing of materials.

MILK AND MILK PRODUCTS

Composition and Properties

Milk consists of 85–89 wt% water and 11–15 wt% total solids (Table 1). The latter comprises solids-not-fat (SNF) and fat. Milk with a higher fat content also has higher SNF content with an increase of SNF of 0.4% for each 1% fat increase. The principal components of SNF are protein, lactose, and minerals (ash). The fat content and other constituents of the milk vary with the species, and for the milk cow, with the breed. Likewise, the composition of milk varies with feed, stage of lactation, health of animals, udder position of withdrawal, and seasonal and environmental conditions.

The nonfat solids, fat solids, and moisture relationships are well-established and can be used as a basis for detecting adulteration with water.

The physical properties of milk are given in Table 2.

Nutritional content. Vitamin D is now added directly to the milk to provide 400 USP units per liter. Vitamin A may be added to low fat or skim milk to provide 1000 RE (retinol equivalents) per liter.

Milk fat is a mixture of triglycerides and diglycerides (see Fats and fatty oils). The triglycerides are short chain, C_{24}–C_{46}; medium chain, C_{34}–C_{54}; and long chain, C_{40}–C_{60}. Milk fat contains more fatty acids than vegetable fats. In addition to being classified according to the number of carbon atoms, the fatty acids in milk may be classified as saturated and unsaturated as well as soluble and insoluble. The fat carries numerous lipids (see Table 3), and vitamins A, D, E, and K, which are fat-soluble (see Vitamins).

Table 1. Constituents of Milk from Various Mammals, Average, wt%

Species	Water	Fat	Protein	Lactose	Ash	Nonfat solids (SNF)	Total solids
human	87.4	3.75	1.63	6.98	0.21	8.82	12.57
cows							
Holstein	88.1	3.44	3.11	4.61	0.71	8.43	11.87
Ayrshire	87.4	3.93	3.47	4.48	0.73	8.68	12.61
Brown Swiss	87.3	3.97	3.37	4.63	0.72	8.72	12.69
Guernsey	86.4	4.5	3.6	4.79	0.75	9.14	13.64
Jersey	85.6	5.15	3.7	4.75	0.74	9.19	14.34
goat	87.0	4.25	3.52	4.27	0.86	8.65	12.90
buffalo (India)	82.76	7.38	3.6	5.48	0.78	9.86	17.24
camel	87.61	5.38	2.98	3.26	0.70	6.94	12.32
mare	89.04	1.59	2.69	6.14	0.51	9.34	10.93
ass	89.03	2.53	2.01	6.07	0.41	8.49	11.02
reindeer	63.3	22.46	10.3	2.50	1.44	14.24	36.70

Table 2. Physical Properties of Milk

Property	Value
density at 20°C of milk with 3–5% fat, average, g/cm³	1.032
weight at 20°C, kg/L[a]	1.03
density at 20°C of milk serum, 0.025% fat, g/cm³	1.035
weight at 20°C of milk serum, 0.025% fat, kg/L[a]	1.03
freezing point, °C	−0.540
boiling point, °C	100.17
maximum density at °C	−5.2
electrical conductivity, S (= Ω⁻¹)	(45–48) × 10⁻⁸
specific heat at 15°C, kJ/(kg·K)[b]	
skim milk	3.94
whole milk	3.92
40% cream	3.22
fat	1.95
relative volume 4% milk at 20°C = 1, volume at 25°C	1.002
40% cream 20°C = 1.0010 at 25°C	1.0065
viscosity at 20°C, mPa·s (= cP)	
skim milk	1.5
whole milk	2.0
whey	1.2
surface tension of whole milk at 20°C, mN/m (= dyn/cm)	50
acidity, pH	6.3 to 6.9
titratable acid, %	0.12 to 0.15
refractive index at 20°C	1.3440–1.3485

[a] To convert kg/L to lb/gal, multiply by 8.34.
[b] To convert kJ/(kg·K) to Btu/(lb·°F), divide by 4.183.

Table 3. Composition of Lipids in Cow Milk

Class of lipid	Range of occurrence
triglycerides of fatty acids, %	97.0–98.0
diglycerides, %	0.25–0.48
monoglycerides, %	0.016–0.038
keto acid glycerides, %	0.85–1.28
aldehydrogenic glycerides, %	0.011–0.015
glyceryl ethers, %	0.011–0.023
free fatty acids, %	0.10–0.44
phospholipids, %	0.2–1.0
cerebrosides, %	0.013–0.066
sterols, %	0.22–0.41
free neutral carbonyls, ppm	0.1–0.8
squalene, ppm	70
carotenoids, ppm	7–9
vitamin A, ppm	6–9
vitamin D, ppm	0.0085–0.021
vitamin E, ppm	24
vitamin K, ppm	1

Milk is an emulsion of fat in water (serum) stabilized by phospholipids which are absorbed on the fat globules. In treatments such as homogenization and churning, the emulsion is broken.

Processing

The processing operations for fluid milk or manufactured milk products include centrifugal sediment removal and cream separation, pasteurization and sterilization, homogenization, and packaging, handling, and storing.

Cooling. After removal from the cow by a mechanical milking machine, usually at 35°C, the milk should be cooled as rapidly as possible to 4.4°C or below to maintain quality. At this low temperature, enzyme activity and growth of microorganisms are minimized. Commercial dairy operations usually consist of a milking machine, a pipeline to convey the milk directly to the tank, and a refrigerated bulk milk tank in which the milk is cooled and stored. A meter may be in the line to measure the quantity of milk from each cow. Development of rancidity must be avoided by preventing air from passing through the warm milk as a result of excessive air flow, air leaks, and long risers in the pipeline. The pipelines, made of glass or stainless steel, are usually cleaned by a CIP (cleaning-in-place) process, as described later.

Centrifugation. Centrifugal devices include clarifiers for removal of sediment and extraneous particulates, and separators for removal of fat from milk (see Centrifugal separation). Modifications include a standardizing clarifier that removes fat to provide a certain fat content of the product as sediment is removed; a clarifixator that partially homogenizes while separating the fat; and a high speed clarifier that removes bacterial cells in a bactofuge process. Bactofugation is a specialized process of clarification in which two high velocity centrifugal devices (bactofuges) operate at 20,000 rpm in series. The first device removes 90% of the bacteria. The second removes 90% of the remaining bacteria, providing a 99% bacteria-free product.

Homogenization. Homogenization is the process by which a mixture of components is treated mechanically to give a uniform product that does not separate. In milk, the fat globules are broken up into small particles that form a more stable emulsion in the milk. In homogenized milk, the fat globules do not rise by gravity to form a creamline. The fat globules in raw milk are 1–15 μm in diameter; they are reduced to 1–2 μm by homogenization. The U.S. Public Health Service defines homogenized milk as "milk that has been treated to insure the breakup of fat globules to such an extent that, after 48 h or quiescent storage at 7.2°C, no visible cream separation occurs in the milk..." Today, most fluid milk is homogenized.

Milk is homogenized in a homogenizer or viscolizer. The milk is forced at high pressure through small openings of a homogenizing valve formed by a valve, or a seat, or a disposable compressed stainless-steel conical valve in the flowstream (see Figure 1). The globules are broken up as a result of shearing, impingement on the wall adjacent to the valve, and perhaps to some extent by the effects of cavitation and explosion after the product passes through the valve.

Pasteurization. Pasteurization is the process of heating milk to kill yeasts, molds, and pathogenic bacteria, and most other bacteria, and to

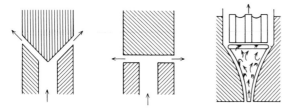

Figure 1. Types of homogenizer valves.

inactivate certain enzymes, without greatly altering the flavor. The principles were developed by and named after Louis Pasteur and his work on wine in 1860–1864 in France. Since then, stringent codes have been developed. The basic regulations are included in the U.S. Public Health milk ordinance which has been adopted by most local and state jurisdictions.

Pasteurization may be carried out by batch or continuous-flow processes. In the batch process, each particle of milk must be heated to at least 62.8°C and held continuously at or above this temperature for at least 30 min. In the continuous process, the milk is heated to at least 71.7°C for at least 15 s. This is known as the HTST (high temperature short-time) pasteurization. For milk products with a fat content above that of milk or with added sweeteners, 65.5°C is required for the batch process and 74.4°C for the HTST process. For either method, following pasteurization, the product should be cooled quickly to 7.2°C or less. Other time–temperature relationships have been established for other products. Ice-cream mix is pasteurized at 79–81°C and held for 25 s.

Another continuous pasteurization process is known as ultrahigh temperature (UHT), in which shorter time, 1–2 s, and higher temperatures, 87–132°C, are employed.

Batch holding. The milk in the batch holding tanks is heated by a hot-water spray on the tank liner, a large-diameter coil which circulates in the milk through which the hot water is pumped, a flooded tank around which hot water or steam is circulated, or by coils surrounding the liner through which the heating medium is pumped at a high velocity.

High temperature short-time pasteurizers. The principal continuous-flow process is the high temperature short-time method. The product is heated to at least 72°C and held at that temperature for not less than 15 s. Other features are nearly the same as in the holding method. The following equipment is needed: balance tank, regenerative heating unit, positive pump, plates for heating to pasteurization temperature, tube or plates for holding for the specified time, flow-diversion valve (FDV), regenerative cooling unit, and cooling unit (see Fig. 2). Other devices often incorporated in the HTST circuit are clarifier, standardizing-clarifier, homogenizer, and booster pump.

Cleaning systems. Both manual and automatic methods of cleaning food-processing equipment are used. Even in a plant with advanced cleaning equipment, some manual cleaning is involved. Today, in most plants, the equipment surfaces are cleaned in place, generally at least once every 24 h. Continuously operated equipment may be cleaned every few days, depending upon the product and the potential build-up of residue on the surfaces.

Cleaning in-place (CIP) systems. CIP systems evolved from recirculating of cleaning solutions in pipelines and equipment of a highly automatic system with valves, controls, and timers. In the early circulation systems, considerable manual operation was required in the assembly and disassembly of units. Homogenizers and heat exchangers were cleaned manually or by a system of circulating solutions.

In the CIP procedure, a cold or tempered aqueous prerinse is followed by the circulation of a cleaning solution for 10 min to 1 h at 54–82°C. Hot water rinses may harden the food product on the surface to be cleaned. The temperature of the cleaning solutions should be as low as possible but high enough to avoid an excess of cleaning chemicals. A wide variety of cleaning solution may be used, depending upon the food product, the hardness of the water, and the equipment. Alkali or chlorinated acid cleaners are preferred. A chlorinated alkaline cleaner may be used separately or in combination with an acid detergent.

Manufactured Products

In the United States, 55% of the fluid milk production is used for manufactured products, mainly cheese, evaporated and sweetened condensed milk, nonfat dry milk, and ice cream. Evaporated and condensed milk and dry milk are made from milk only; other ingredients are added to ice cream and sweetened condensed milk.

Evaporated and condensed milk. Evaporated and condensed milk are produced by removing moisture from milk under a vacuum followed by

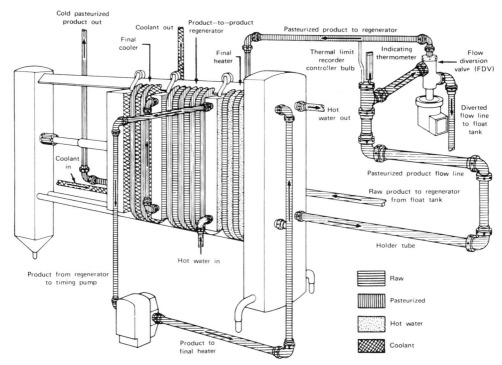

Figure 2. Flow through typical HTST plate pasteurizer. Courtesy of St. Regis Crepaco.

packaging and sterilizing in cans. The milk is condensed to half its volume in single- or multiple-effect evaporators. The product has a fat to SNF ratio of 1 : 2.2785, and is standardized before and after evaporation. It must have at least 7.9% fat and 25.9% of total milk solids, including fat.

Sweetened condensed milk. Sweetened condensed milk, unlike evaporated milk, is not sterilized. Sugar is added as a preservative which replaces sterilization as a means of maintaining keeping quality. The equipment is similar to that used for evaporated milk, except that sugar is added in a hot well before condensing (evaporating) the liquid. According to standards, sweetened condensed milk must contain 8.5% fat min and 28.0% min total milk solids, including fat (fat to SNF ratio = 1 : 2.294). The product contains 43–45% sugar.

Cream. Cream is a high fat product which is secured by gravity or mechanical separation through differential density of the fat and the serum. Fat content may range from 10 to 40% depending on use and state and Federal laws. Whipping cream has a fat content of 34–40%, and table coffee or light cream has a fat content of 20–25%. Half-and-half, suggesting a mixture of cream and milk, has not less than 10.5% milk fat, and in some states, up to 12%.

In the United States, about 10 wt% of the edible fat used is butter, which is defined as a product that contains 80% milk fat with no more than 16% moisture. It is made of cream with 25–40% milk fat.

Cheese. Cheese production is based on the coagulation of casein from milk, or to a minor extent, of the proteins of whey. The casein is precipitated by acidification, which can be accomplished by natural souring of milk. The procedures for making cheese vary greatly and cheese products are countless.

Yogurt. Yogurt is a fermented milk product that is rapidly increasing in consumption in the United States. Milk is fermented with *Lactobacilus bulgaricus* and *Streptococcus thermophilus* organisms producing lactic acid. Usually some cream or nonfat dried milk is added to the milk in order to obtain a heavy-bodied product.

Nondairy Products

Many so-called nondairy products such as coffee cream, topping, and icings utilize caseinates (see Synthetic dairy products). In addition to fulfilling a nutritional role, the caseinates impart creaminess, firmness, smoothness, and consistency of products.

CARL W. HALL
Washington State University

B.H. Webb, A.H. Johnson, and J.A. Alford, *Fundamentals of Dairy Chemistry*, Avi Publishing Co., Westport, Conn., 1974, p. 396.

W.J. Harper and C.W. Hall, *Dairy Technology and Engineering*, Avi Publishing Co., Westport, Conn., 1976, p. 413.

A.W. Farrall, *Engineering for Dairy and Food Products*, John Wiley & Sons, Inc., New York, 1963.

MINERAL NUTRIENTS

Mineral nutrients are involved in the most fundamental processes of life. The oxygen that humans breathe is utilized with the aid of two metal complexes, ie, iron-containing hemoglobin and zinc-containing carbonic anhydrase. With the evolution of life from a reducing to an oxidizing atmosphere, mechanisms involving enzymes were developed by organisms in order to protect the cells from high levels of oxygen. One such class of protective enzymes is the superoxide dismutases which contain metals, eg, manganese, copper, zinc, and iron.

As with other biological substances, a state of dynamic equilibrium exists for the mineral nutrients, and mechanisms exist whereby the system can adjust to varying amounts of minerals in the diet. In the form usually found in foods, and under circumstances of normal human metabolism, most nutrient minerals are not toxic when ingested orally, even in amounts considerably greater than the Recommended Dietary Allowances (RDAs) of the National Academy of Science (see Table 1).

Table 1. Characteristics of the Mineral Nutrients

Element	Body content, mg/kg body wt	Daily requirement, mg/d
Principal elements		
sodium	1,500–1,600	1,100–3,300[a]
potassium	2,000–3,500	1,875–5,625[a]
magnesium	270–500	300[b,c,d]; 350[c,e]
calcium	14,000–20,000	800[c,d]
phosphorus	11,000–12,000	800[c,d]
sulfur	1,600–2,500	[f]
chlorine	1,200–1,500	1,700–5,100[a]
Trace elements		
copper	1.0–2.5	2.0–3.0[a]
zinc	33–50	15[c,d]
selenium	0.2–0.3	0.05–0.2[a]
chromium	0.06–0.2	0.05–0.2[a]
molybdenum	0.1–0.5	0.15–0.5[a]
fluorine	37	1.5–4.0[a]
iodine	0.2–0.4	0.15[c,d]
manganese	0.2–4.0	2.5–5.0[a]
iron	60–66	10[c,e], 18[b,c,d]
cobalt	0.02	0.003[g]

[a] Estimated safe and adequate daily intake, adults.
[b] Female.
[c] RDA, adults.
[d] Increased amounts required during pregnancy and lactation.
[e] Male.
[f] Adequate intake with adequate intake of protein.
[g] As vitamin B_{12}.

Some elements that are found in body tissues have no apparent physiological role and have not been shown to be toxic, eg, rubidium, strontium, titanium, niobium, boron, germanium, and lanthanum. Other elements are toxic when found in greater than trace amounts, and sometimes in trace amounts, eg, arsenic, mercury, lead, cadmium, silver, zirconium, beryllium, and thallium. Numerous elements that have been used in medicine in nonnutrient roles include lithium, bismuth, antimony, bromine, platinum, and gold.

The Principal Elements

Sodium and Potassium

Sodium ion is the most abundant cation in the extracellular fluid; most of the potassium ion is present in the intracellular fluid, although a small amount is required in the extracellular fluid to maintain normal muscle activity.

Metabolic functions. Sodium ion acts in concert with other electrolytes, in particular K^+, to regulate the osmotic pressure and to maintain the appropriate water balance of the body and the acid–base balance (pH); homeostatic control of these functions is accomplished by the lungs and kidneys interacting by way of the blood. Sodium is essential for glucose absorption and transport of other substances across cell membranes, and is involved, as is K^+, in transmitting nerve impulses and in muscle relaxation. Potassium ion acts as a catalyst in the intracellular fluid in energy metabolism, and is required for carbohydrate and protein metabolism.

Active transport: The Na^+/K^+ Pump. Maintenance of the appropriate concentrations of K^+ and Na^+ in the intra- and extracellular fluids involves active transport, ie, a process requiring energy. Sodium ion in the extracellular fluid (0.136–0.145 M Na^+) diffuses passively and continuously into the intracellular fluid and must be removed. Sodium ion is pumped from the intracellular to the extracellular fluid (< 0.14 M K^+). The energy for these processes is provided by hydrolysis of adenosine triphosphate (ATP) which requires the enzyme Na^+/K^+ ATPase, a membrane-bound enzyme which is widely distributed in the body. In some cells, eg, brain and kidney, 60–70 wt% of the ATP is used to maintain the required Na^+–K^+ distribution.

Some disorders associated with mineral nutrients are shown in Table 2.

**Table 2.   Some Disorders Associated with Deficiency, Excess[a],
and / or Faulty Utilization of the Mineral Nutrients**

Nutrient	Deficiency	Excess	Faulty utilization
Principal elements			
sodium	muscle weakness; nausea	hypertension	Addison's disease; Cushing's disease
potassium	muscle weakness		Addison's disease; Cushing's disease
magnesium	neuromuscular irritability; convulsions; muscle tremors; mental changes (confusion, disorientation, hallucinations); heart disease; kidney stones	Mg intoxication (drowsiness, stupor, coma)	kidney stones
calcium	hypocalcemia; tremor; rickets; osteomalacia; osteoporosis; muscle spasm (tetany); possibly heart disease		rickets; osteomalacia; osteoporosis; muscle spasm (tetany); possibly heart disease; Paget's disease
phosphorus	rickets; osteomalacia; osteoporosis		rickets; osteomalacia; osteoporosis, Paget's disease; renal rickets (vitamin D-resistant rickets)
sulfur			homocystinuria
chlorine[b]	impaired growth in infants		
Trace elements			
copper	impaired elastin formation; impaired hemopoiesis		Menkes' kinky-hair syndrome; Wilson's disease
zinc	mental retardation; delayed sexual maturity; dwarfism; sterility; slow wound healing; hypogeusia		acrodermatitis enteropathica
selenium[c]	heart disease; increased cancer		
chromium	impaired glucose tolerance; possibly atherosclerosis		
fluorine[d]		fluorosis; mottled teeth	
iodine	hypothyroidism; cretinism; myxedema; goiter		hyperthyroidism; Grave's disease
manganese	impaired mucopolysaccharide synthesis		
iron	anemia	hemochromatosis; Bantu siderosis	hemochromatosis
cobalt	pernicious anemia	polycythemia	pernicious anemia

[a]Excess of a dietary nature is considered; industrial toxicities are not included in the table.
[b]Abnormalities of chlorine metabolism usually accompany those of sodium metabolism.
[c]For toxicity, see text.
[d]No known deficiency disease.

Magnesium

In the adult human, 50–70% of the magnesium is in the bones associated with calcium and phosphorus. The rest is widely distributed in the soft tissues and body fluids. In these, most of the Mg^{2+}, like K^+, is located in the intracellular fluid in which it is the most abundant divalent cation.

Metabolic functions. Magnesium is essential in numerous metabolic processes because it is the activator of many enzymes, eg, alkaline phosphatases and phosphokinases, pyrophosphates, and thiokinases. Because the phosphokinases are required for the hydrolysis and transfer of phosphate groups, magnesium is essential in glycolysis and in oxidative phosphorylation (see Phosphorus). The thiokinases are required for the initiation of fatty acid degradation. Magnesium is also required in systems in which thiamine pyrophosphate is a coenzyme.

Regulation of serum Mg^{2+} appears to result from a balance among intestinal absorption, renal reabsorption, and excretion.

Calcium

Calcium is the most abundant mineral element in mammals, comprising 1.5–2.0 wt% of the adult human body, over 99 wt% of the calcium

that is present occurring in bones and teeth. The normal calcium content of blood is 9–11 mg Ca/100 mL. About 48% of the serum calcium is ionic; ca 46% is bound to blood proteins; the rest is present as diffusible complexes, eg, citrates. The calcium-ion level must be maintained within definite limits.

Metabolic functions. Bones act as a reservoir of certain ions, in particular Ca^{2+} and PO_4^{3-}, which readily exchange between bones and blood. Bone structure comprises a strong organic matrix combined with an inorganic phase which is principally hydroxyapatite, $3Ca_3(PO_4)_2 \cdot Ca(OH)_2$. Bones contain two forms of hydroxyapatite. The less soluble crystalline form contributes to the rigidity of the structure. The crystals are quite stable but, because of their small size, they present a very large surface area available for rapid exchange of ions and molecules with other tissues. There is also a more soluble intercrystalline fraction. Bone salts also contain small amounts of magnesium, sodium, carbonate, citrate, chloride, and fluoride.

Calcium is necessary for blood-clot formation; Ca^{2+} stimulates release of blood-clotting factors from platelets (see Blood, coagulants and anticoagulants). Neuromuscular excitability depends on the relative concentrations of Na^+, K^+, Ca^{2+}, Mg^{2+}, and H^+. With a decrease in Ca^{2+} concentration (hypocalcemia), excitability increases; if this condition is not corrected, the symptoms of tetany appear (muscular spasm, tremor, even convulsions). Too great an increase in Ca^{2+} concentration (hypercalcemia) may impair muscle function to such an extent that respiratory or cardiac failure may occur.

Blood Ca^{+2} level. In the normal adult, the blood Ca^{2+} level is established by the equilibrium between the blood Ca^{2+} and the more soluble intercrystalline calcium salts of the bone, and by a subtle and intricate feedback mechanism (responsive to the Ca^{2+} concentration of the blood) that involves the less soluble crystalline hydroxyapatite. Also participating in Ca^{2+} control are the thyroid and parathyroid glands, the liver, kidney, and intestine.

Phosphorus

Eighty-five percent of the phosphorus in the body occurs in bones and teeth. There is a constant exchange of calcium and phosphorus between bones and blood but very little turnover of ions in teeth (see Calcium). Phosphorus is the second most abundant element in the human body; the ratio of Ca : P in bones is constant at ca 2 : 1. Every tissue and cell contains phosphorus, generally as a salt or ester of mono-, di-, or tribasic phosphoric acid. Phosphorus is involved in a large number and wide variety of metabolic functions, eg, carbohydrate metabolism, ATP from fatty acid metabolism, and oxidative phosphorylation.

Sulfur

Sulfur is present in every cell in the body, primarily in proteins in the amino acids methionine, cystine, and cysteine. Inorganic sulfates and sulfides occur in small amounts relative to the total body sulfur, but the compounds that contain them are very important to metabolism. Sulfur intake is thought to be adequate if protein intake is adequate and sulfur deficiency has not been reported.

Sulfur is part of several vitamins and cofactors, eg, thiamine, pantothenic acid, biotin, and lipoic acid. Mucopolysaccharides, eg, heparin and chondroitin sulfate, contain a monoester of sulfuric acid with an HSO_3^- group. Sulfur-containing lipids isolated from brain and other tissues usually are sulfate esters of glycolipids. The sulfur-containing amino acid, taurine, is conjugated to bile acids. Labile sulfur is attached to the nonheme iron in stoichiometric amounts in the respiratory chain where it is associated with the flavoproteins and cytochrome b (see Iron).

Chlorine

The chlorides are essential in the homeostatic processes maintaining fluid volume, osmotic pressure, and acid-base equilibrium. Most of the chloride is in the body fluids and a little is in bone salts. It is the principal anion accompanying Na^+ in the extracellular fluid. Less than 15 wt% of the Cl^- is associated with K^+ in the intracellular fluid in which the primary inorganic anions are HPO_4^{2-} and HCO_3^-. Chloride passively and freely diffuses between the intra- and extracellular fluid, through the cell membrane.

Some of the blood Cl^- is used for formation in the gastric glands of HCl, which is required for digestion. Hydrochloric acid is secreted into the stomach where it acts with the gastric enzymes in the digestive processes. The chloride is then reabsorbed with other nutrients into the blood stream. Chloride is actively transported in the gastric and intestinal mucosa. In the kidney, it is passively reabsorbed in the thin ascending loop of Henle and actively reabsorbed in the thick segment of the ascending loop, ie, the distal tubule.

Trace Elements

Copper. All human tissues contain copper, the highest amounts being found in the liver, brain, heart, and kidney. In blood, plasma and erythrocytes contain almost equal amounts of copper, ie, ca 110 and 115 μg per 100 mL, respectively. In plasma, ca 90 wt% of the copper is in the metalloprotein ceruloplasmin (a_2-globulin; mol wt 151,000), which contains eight atoms of copper per molecule. Ceruloplasmin has been identified as a ferroxidase(I) that catalyses the oxidation of aromatic amines and of Fe^{2+} to Fe^{3+}. The ferric ion is then incorporated into transferrin, which is necessary for the transport of iron to tissues involved in the synthesis of iron-containing compounds, eg, hemoglobin (see Iron compounds). Lowered levels of ceruloplasmin interfere with hemoglobin synthesis.

Zinc. The 2–3 g of zinc in the human body is widely distributed in every tissue and tissue fluid. About 90 wt% is in muscle and bone; unusually high concentrations are in the choroid of the eye and in the prostate gland. Almost all of the zinc in the blood is associated with carbonic anhydrase in the erythrocytes. Zinc is concentrated in nucleic acids, but its function there is not clear.

Selenium. Selenium is thought to be widely distributed throughout the body tissues, and animal experiments suggest that greater concentrations are in the kidney, liver, and pancreas, and lesser amounts are in the lungs, heart, spleen, skin, brain, and carcass. The most clearly documented role for selenium is as a necessary component of gluthatione peroxidase, which reduces hydrogen peroxide formed by free-radical oxidant-stressor reaction to H_2O and reduces organic peroxides, eg, those formed by the peroxidation of unsaturated fatty acids, to alcohols and H_2O.

Chromium. Chromium(III) potentiates the action of insulin; it may be considered a cofactor for insulin. Experimental results indicate Cr(III) has an effect in translocating sugars into cells, ie, at the first step of sugar metabolism.

Molybdenum. Molybdenum is a component of the metalloenzymes xanthine oxidase, aldehyde oxidase, and sulfite oxidase in mammals. Two other molybdenum metalloenzymes present in nitrifying bacteria have been characterized: nitrogenase and nitrate reductase. The molybdenum in xanthine oxidase, aldehyde oxidase, and sulfite oxidase is involved in redox reactions; the heme iron in sulfite oxidase also is involved in electron transfer.

Fluorine. Fluorine is present in the bones and teeth in very small quantities. Human ingestion is 0.7–3.4 mg/d from food and water. In the opinion of many authorities, fluoridation of the public water supply is an effective means of significantly reducing the incidence of dental caries (see Water, municipal water treatment). The view is not universally accepted and some have expressed concern regarding the narrow range of safety between effective and toxic concentrations (see also Dentifrices).

Iodine. Of the 10–20 mg of iodine in the adult body, 70–80 wt% is in the thyroid gland; it is present in all tissues. The essentiality of iodine depends solely on its utilization by the thyroid gland to produce thyroxin and related compounds.

Manganese. The adult human body contains ca 10–20 mg of manganese, widely distributed but the largest concentration in the mitochondria of the soft tissues, particularly the liver, pancreas, and kidneys. Manganese concentration in bone varies widely with dietary intake.

Iron. The total body content of iron, ie, 3–5 g, is recycled more efficiently than other metals. There is no mechanism for excretion of iron and what little iron is lost daily, ie, ca 1 mg in the male and 1.5 mg in the menstruating female, is lost mainly through exfoliated mucosal skin or hair cells, and menstrual blood.

A large percentage of the iron in the human body is in hemoglobin: 85 wt% in the adult female, 60 wt% in the adult male. The remainder is present in other iron-containing compounds that are involved in basic metabolic functions or in iron transport or storage compounds. Myoglobin, the cytochromes, catalase, sulfite oxidase, and peroxidase are heme iron enzymes. NADH-dehydrogenase, succinate dehydrogenase, α-glycerophosphate dehydrogenase, monoamine oxidase, xanthine oxidase, and alcohol dehydrogenase are nonheme metalloflavoproteins that contain iron. Aconitase and microsomal lipid peroxidase do not contain iron but do require it as a cofactor. The transport and storage proteins for iron are transferrin, ferritin, and hemosiderin. The iron in transferrin in the blood is a combination of freshly absorbed iron and recycled iron.

Cobalt. Cobalt is nutritionally available only as vitamin B_{12}. Although Co^{2+} can function as a replacement *in vitro* for other divalent cations, in particular Zn, no *in vivo* function for inorganic cobalt is known for humans. In ruminant animals, B_{12} is synthesized by bacteria in the rumen.

CARL L. ROLLINSON
MARY G. ENIG
University of Maryland

H.A. Harper, *Review of Physiological Chemistry*, Lange Medical Publications, Los Altos, Calif., 1975.

Present Knowledge in Nutrition, 4th ed., The Nutrition Foundation, Inc., Washington, D.C., 1976.

E.J. Underwood, *Trace Elements in Human and Animal Nutrition*, 4th ed., Academic Press, Inc., New York, 1977.

MINERAL WOOL. See Refractory fibers.

MINIMUM, Pb₃O. See Lead compounds.

MISCH METAL. See Cerium.

MIXING AND BLENDING

Mixing, an important operation in the chemical process industries, can be divided into five areas: liquid–solid dispersion, gas–liquid dispersion, liquid–liquid dispersion, the blending of miscible liquids, and the production of fluid motion. Mixing performance is evaluated by two criteria: the physical uniformity, described by specifications (ie, a physical relationship is required in terms of samples of uniformity in various parts of the mixing vessel), and a criterion based on mass transfer or chemical reaction. The elements of mixer design are process design, eg, fluid mechanics of impellers, fluid regime required by the process, scale-up, and hydraulic similarity; impeller power characteristics, eg, related impeller power, speed and diameter; and mechanical design, eg, impellers, shafts, and drive assembly.

Fluid Mechanics

Mixer power P produces a pumping capacity Q expressed in kg/s, and a specific velocity work term H expressed in J/kg.

$$P = QH$$

The term H is related to the square of the velocity and, therefore, to fluid shear rates. In low and medium viscosity, the pumping capacity is related to the speed and diameter of the impeller:

$$Q \propto ND^3$$

The power drawn by the impeller is proportional to N^3D^5:

$$\left(P \propto \rho N^3 D^5 \right)$$

These relationships can be combined to show that at constant power input, the flow-to-velocity work ratio is related to the impeller diameter:

$$(Q/H)_P \propto D^{8/3}$$

This is sometimes expressed as the flow-to-fluid shear ratio which is correct conceptually, but is not in terms of the mathematical equation above. This equation also does not hold as a constant on scale-up, and therefore, is used primarily to evaluate the effect of geometric variables. It does not have a ready evaluation in terms of the actual numbers when comparing large to small tanks.

Most mixing applications are sensitive primarily to fluid-pumping capacity. Thus, if the pumping capacity of different impellers is compared, or the overall flow pattern in the tank considered, process results are proportional in the same fashion to the actual circulating capacity of the impeller and in the tank.

If a process is dependent primarily upon pumping capacity, the fluid velocities and the individual shearing rates, both on a macro- and a microscale, are above a certain minimum level to allow other process requirements to proceed unhindered. If the pumping capacity is increased and some of the other velocity and shear-rate values are decreased below some minimum, then fluid shear stress enters into the overall design.

Other process applications sensitive to fluid shear rates include fermentation (qv), crystallization (qv), solids dispersion, polymerization,

Figure 1.   Radial-flow flat-blade disk turbine.

and others. Shear rates should be considered if they are part of the overall process requirement or mechanism.

Impellers

Impellers are either radial flow or axial flow. Figure 1 illustrates a radial-flow disk turbine.

The flat-blade turbine is normally placed 0.5–1 impeller diameter off bottom and has a coverage of 1–2 impeller diameters. The spacing between multiple impellers is somewhere between 1.5 and 3 impeller diameters.

Axial-flow impellers include the fluidfoil square-pitch marine-type propeller shown in Figure 2 and the fluidfoil impeller shown in Figure 3. The former has a variable angle and, therefore, an approximately constant pitch across the impeller face. This gives a relatively uniform flow pattern across the impeller periphery, and the propeller tends to have a high pumping capacity per unit power. The fluidfoil impeller has even greater pumping capacity per unit power.

The axial-flow turbine has a constant blade angle and, therefore, a variable pitch across the surface. It has an effective flow pattern, but is not quite as efficient as the propeller. Axial-flow turbines are used in large-size equipment primarily because of their low cost. The performance of axial-flow impellers in baffled tanks is illustrated in Figure 4. In

Figure 2.   Square-pitch marine-type impeller.

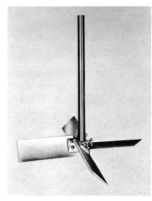

Figure 3.   A fluidfoil impeller.

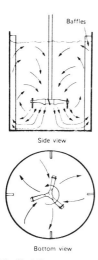

Figure 4. Typical baffled-flow pattern of axial-flow impeller.

general, for applications primarily requiring pumping capacity, such as blending and solid suspension, axial-flow turbines are chosen. For applications requiring gas–liquid or liquid–liquid mass transfer, or in multistage columns, radial–flow turbines are preferred.

Mixing

The mixing requirement may be expressed as pumping or circulating capacity. The impeller flow is defined through the impeller peripheral-discharge area, or total flow in the tank. In draft-tube circulations, shown in Figure 5, the impeller is enclosed in a draft tube. Head, flow, and pumping efficiency are measured as in a pump.

Solid–liquid contacting. The settling velocity of solid particles is a critical factor in the process design of many mixing applications. It is desirable to obtain the settling velocity experimentally by dropping individual particles of various sizes into a graduated cylinder and timing the settling velocities.

In a free-settling system, the settling velocity is above 300 mm/min and is a main factor in the design of the equipment. If the settling velocity is less than 300 mm/min, there is a fair degree of uniformity in the tank once there is motion.

There are three types of solid suspension: complete motion on the tank bottom; all the particles have an upward motion (off-bottom suspension); and complete uniformity. The last is a relative term since the particles have a vertical settling velocity and the top of the tank has a horizontal velocity across the surface. It is, therefore, not possible to obtain complete uniformity in the upper layers of the tank. The relationship between these three types depends upon the settling velocity of the solids.

Gas–liquid operations. Gas–liquid processes are affected by changes in power, speed, pumping capacity, and shear rate. Specifications should give the process conditions required, such as mass-transfer and reaction

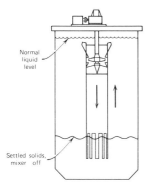

Figure 5. Draft-tube circulator.

rates (see Mass transfer). The dispersion specification alone, eg, 10,000 m^3/min, is not sufficient.

For any gas–liquid process selection, the superficial gas velocity is needed. This is defined as the average volumetric gas-flow rate in and out of the vessel divided by its cross-sectional area at the temperature and pressure at the midpoint of the tank expressed in meters per second. Normally, a value of about 0.1 m/s is the boundary between normal mixer applications and those that must be carefully designed for process considerations such as liquid interface foaming and splashing, and fluid forces on mechanical equipment.

Intimate dispersions are controlled by the mixer flow patterns accompanied by power levels on the order of three times greater in energy in the mixer than in the expanding gas stream.

Minimum dispersion occurs at a point where the energy content of the mixer and gas stream are nearly equal. As a general rule, the flow pattern from axial-flow impellers is governed by the upward velocity of the gas stream. Thus, axial-flow impellers do not operate satisfactorily unless their energy input is five to ten times higher than that of the gas stream. Radial-flow and disk-type impellers are most commonly used.

Liquid–liquid dispersion. Stable emulsions include household products, cosmetics, pharmaceuticals, and a wide variety of other combinations (see Emulsions). A minimum fluid shear rate is usually required for the production of a uniform, stable emulsion. When this fluid shear rate is produced by the mixer, the composition can either maintain its dispersion or the particles coalesce. If, on the other hand, the mixer is not capable of producing the shear rate required, the product will not be satisfactory.

The mixing time depends on the pumping capacity of the unit and on the number of times the particles pass through the high shear zone of the impellers and the length of time each particle spends in that shear zone. Thus, a variable known as the shear work, which is the product of shear rates and the time of contact, gives a measure of the total amount of energy expended into the dispersion.

Blending

Miscible liquids. When blending miscible liquids, two distinct mechanisms are involved. In general, one material is run into the vessel with the mixer, whereas the second is injected into the tank. The uniformity of some particle or a physical or chemical property is then measured. A sampling point has to be chosen and the uniformity required for blending must to be defined.

Large impellers at slow speeds reduce blend times. For a propeller, the blending time is proportional to D^{-2}, whereas for a turbine, the blending time is proportional to D^{-1}. To obtain the same circulating time on scale-up, the ratio P/V increases with the square of the tank diameter. However, this relationship cannot be used for design. Therefore, an increase in blending time is incorporated into the overall plant design.

In another blending method, the tank is initially stratified. After mixing is started, the time required to eliminate the stratified condition is measured. This is typical of large petroleum-storage tanks. Using hot and cold water and the associated temperature differences, the blending equation is

$$\theta \propto P^{-1}(D/T)^{-2.3}\left(\frac{\Delta\rho}{\rho}\right)^{0.9}$$

Viscous fluids. There are several areas of viscous blending. In large industrial tanks, low viscosity is defined as 5 Pa·s (50 P). Either open-axial or radial-flow turbines may be used for the medium-viscosity region at 5–50 Pa·s (50–500 P). The area above 50 Pa·s is defined as high viscosity mixing, in which typically an anchor or helical impeller should be used.

Anchor impellers do not have any tendency for top-to-bottom flow and, therefore, do not provide effective blend time or achieve effective temperature uniformity in heat-transfer applications. However, for many processes anchor impellers are used.

The helical impeller has a strong axial-flow component and can give effective blending and circulation times in both Newtonian and non-Newtonian fluids. Typical are a pitch of 0.5 and a blade width of 1/12 to

1/6 T. The inner flight is only effective on non-Newtonian fluids but gives superior performance over noninner flight.

Nomenclature

D = impeller diameter, m
H = velocity head, J/kg
N = impeller rotational speed, rps
P = power, W
Q = impeller flow; pumping capacity, kg/s
T = tank diameter, m
θ = blend time
ρ = fluid density

JAMES Y. OLDSHUE
Mixing Equipment Co., Inc.

DAVID B. TODD
Baker Perkins Inc.

J.H. Rushon and J.Y. Oldshue, *Chem. Eng. Prog.* **49**(4), 161 (1953); **49**(5), 267.

R.R. Corpstein, R.A. Dore, and D.S. Dickey, *Chem. Eng. Prog.* **75**(2), 66 (1979).

J.Y. Oldshue, *Chem. Eng.*, 82 (June 13, 1983).

J.Y. Oldshue, *Fluid Mixing Technology*, McGraw-Hill Book Co., New York, 1983.

MOLASSES. See Syrups.

MOLD-RELEASE AGENTS.   See Abherents.

MOLECULAR SIEVES

Molecular-sieve zeolites are crystalline aluminosilicates of Group IA and Group IIA elements such as sodium, potassium, magnesium, and calcium. Chemically, they are represented by the empirical formula:

$$M_{2/n}O.Al_2O_3.ySiO_2.wH_2O$$

where y is 2 or greater, n is the cation valence, and w represents the water contained in the voids of the zeolite. Structurally, zeolites are complex, crystalline inorganic polymers based on an infinitely extending framework of AlO_4 and SiO_4 tetrahedra linked to each other by the sharing of oxygen ions. This framework structure contains channels or interconnected voids that are occupied by the cations and water molecules. The cations are mobile and ordinarily undergo ion exchange. The water may be removed reversibly, generally by the application of heat, which leaves intact a crystalline host structure permeated by micropores which may amount to 50% of the crystals by volume. In some zeolites, dehydration may produce some perturbation of the structure such as cation movement and some degree of framework distortion.

The structural formula of a zeolite is based on the crystal unit cell, the smallest unit of structure represented by

$$M_{x/n}\left[(AlO_2)_x(SiO_2)_y\right].wH_2O$$

where n is the valence of the cation M, w is the number of water molecules per unit cell, x and y are the total number of tetrahedra per unit cell, and y/x usually has values of 1–5. However, high silica zeolites have recently been prepared in which y/x is 10–100 or even higher, and in one case, a molecular-sieve silica has been prepared.

Mineral Zeolites

Zeolite minerals are formed over much of the earth's surface, including the sea bottom. Until about twenty years ago, zeolite minerals were considered as typically occurring in cavities of basaltic and volcanic rocks. During the last 20–25 yr, however, the use of x-ray diffraction for the examination of very fine-grained sedimentary rocks has led to the identification of several zeolite minerals which were formed by the natural alteration of volcanic ash in alkaline environments. Of 40 known zeolite minerals, chabazite, erionite, mordenite, and clinoptilolite occur in quantity and reasonably high purity (see Table 1).

Table 1.   Zeolite Compositions

Zeolite	Typical formula
Natural	
chabazite	$Ca_2[(AlO_2)_4(SiO_2)_8].13H_2O$
mordenite	$Na_8[(AlO_2)_8(SiO_2)_{40}].24H_2O$
erionite	$(Ca, Mg, Na_2, K_2)_{4.5}[(AlO_2)_9(SiO_2)_{27}].27H_2O$
faujasite	$(Ca, Mg, Na_2, K_2)_{29.5}[(AlO_2)_{59}(SiO_2)_{133}].235H_2O$
clinoptilolite	$Na_6[(AlO_2)_6(SiO_2)_{30}].24H_2O$
Synthetic	
zeolite A	$Na_{12}[(AlO_2)_{12}(SiO_2)_{12}].27H_2O$
zeolite X	$Na_{86}[(AlO_2)_{86}(SiO_2)_{106}].264H_2O$
zeolite Y	$Na_{56}[(AlO_2)_{56}(SiO_2)_{136}].250H_2O$
zeolite L	$K_9[(AlO_2)_9(SiO_2)_{27}].22H_2O$
zeolite omega	$Na_{6.8}TMA_{1.6}[(AlO_2)_8(SiO_2)_{28}].21H_2O$ [a]
ZSM-5	$(Na, TPA)_3[(AlO_2)_3(SiO_2)_{93}].16H_2O$ [b]

[a] TMA = tetramethylammonium.
[b] TPA = tetrapropylammonium.

Structure

Of the many synthetic and mineral zeolites, 34 structure types are known of which ten are synthetic. There are three structural aspects: the basic arrangement of the individual structural units in space, which defines the framework topology; the location of charge-balancing metal cations; and the channel-filling material, which is water as the zeolite is formed. After the water is removed, the void space can be used for adsorption of gases, liquids, salts, elements, and many other substances. The current concepts of zeolite structures were developed by Pauling in 1930. Modern tools such as x-ray crystallography have provided a very detailed description of many structures.

There are two types of structures. One provides an internal pore system comprised of interconnected cagelike voids; the second provides a system of uniform channels which, in some instances, are one-dimensional and in others intersect with similar channels to provide two- or three-dimensional channel systems. The preferred type has two- or three-dimensional channels to provide rapid intracrystalline diffusion in adsorption and catalytic applications.

In most zeolite structures, the primary structural units—the tetrahedra—are assembled into secondary building units which may be simple polyhedra such as cubes, hexagonal prisms, or octahedra. The final structure framework consists of assemblages of the secondary units.

Structure modification.   Several types of structural defects or variants can occur which figure in adsorption and catalysis:

Surface defects due to termination of the crystal surface and hydrolysis of surface cations.

Structural defects due to imperfect stacking of the secondary units, which may result in blocked channels.

Ionic species may be left stranded in the structure during synthesis (eg, OH^-, AlO_2^-, Na^+, SiO_3^{2-}).

The cation form, acting as the salt of a weak acid, hydrolyzes in water suspension to produce free hydroxide and cations in solution.

Hydroxyl groups in place of metal cations may be introduced by ammonium ion exchange followed by thermal deammoniation. These impart acidity to the zeolite, which is important in hydrocarbon-conversion reactions.

Tetrahedral aluminum atoms can be removed from the framework by internal hydrolysis to produce $Al(OH)_3$ when heated in steam. Chemical treatment with acids or chelating agents may also be used to carry out dealumination, but this may cause severe structural damage.

Properties

Adsorption. Although several types of microporous solids are used as adsorbents for the separation of vapor or liquid mixtures (see Adsorptive separation), the distribution of pore diameters does not enable separations based on the molecular-sieve effect; that is, separations caused by differences in the molecular size of the materials to be separated. The most important molecular-sieve effects are shown by dehydrated crystalline zeolites. Zeolites selectively adsorb or reject molecules based upon differences in molecular size, shape, and other properties, such as polarity. The sieve effect may be total or partial.

Zeolites are high capacity, selective adsorbents because they separate molecules based upon the size and configuration of the molecule relative to the size and geometry of the main apertures of the structures; and they adsorb molecules, in particular those with a permanent dipole moment which show other interaction effects, with a selectivity that is not found in other solid adsorbents.

Separation may be based upon the molecular-sieve effect or may involve the preferential or selective adsorption of one molecular species over another. These separations are governed by several factors discussed below.

The basic framework structure, or topology, of zeolite determines the pore size and the void volume.

The exchange cations, in terms of their specific location in the structure, their population or density, their charge and size, affect the molecular-sieve behavior and adsorption selectivity of the zeolite. By changing the cation types, and number, one can tailor or modify within certain limits the selectivity of the zeolite in a given separation.

The cations, depending upon their locations, contribute electric-field effects that interact with the adsorbate molecules.

The effect of the temperature of the adsorbent is pronounced in cases involving activated diffusion.

Sieving by dehydrated zeolite crystals is based on the size and shape differences between the crystal apertures and the adsorbate molecule. The aperture size and shape in a zeolite may change during dehydration and adsorption because of framework distortion or cation movement.

In some instances, the aperture is circular, such as in zeolite A. In others, it may take the form of an ellipse, such as in dehydrated chabazite. In this case, subtle differences in the adsorption of various molecules result from a shape factor.

Catalytic properties. In zeolites, catalysis takes place within the intracrystalline voids (see Catalysis). The aperture size and channel system affect catalytic reactions caused by diffusion of reactants and products. Activity and selectivity are achieved or altered by modifying the zeolite in several ways. In hydrocarbon reactions in particular, the zeolites with the largest pore sizes are preferred. These include mordenite and zeolites Y, L, and omega. Modification techniques include ion exchange, composition in terms of Si : Al ratio, hydrothermal dealumination or stabilization which produces Lewis acidity, introduction of acidic groups such as OH which impart Brønsted acidity, and introducing dispersed metal phases such as the noble metals. In addition, the zeolite framework structure determines shape-selective effects. Several types have been demonstrated, including reactant selectivity, product selectivity, and restricted transition-state selectivity.

Acidic zeolites have outstanding catalytic activity. Acidity is introduced by the decomposition of the NH_4^+ ion-exchanged form, by hydrogen-ion exchange, or by hydrolysis of a zeolite containing multivalent cations during dehydration.

Dispersed metals. Bifunctional zeolite catalysts, primarily zeolite Y, are used in commercial processes such as hydrocracking. These are acidic zeolites containing dispersed metals such as platinum or palladium. The metals are introduced by cation exchange of the ammine complexes, followed by a reductive decomposition (see Petroleum).

Stabilized zeolites. Thermal and hydrothermal stability of certain zeolites, in particular zeolite Y, is necessary in many catalytic applications. The stability increases with Si : Al ratio and by exchange with polyvalent cations such as rare earths. Increased stability is achieved by hydrothermal treatment of the ammonium- or rare-earth-exchanged form.

Ion exchange. The exchange behavior of nonframework cations in zeolites (selectivity, degree of exchange) depends upon the nature of the cation (the size and charge of the hydrated cation), the temperature, the concentration, and to some degree, the anion species. Cation exchange may produce considerable changes in other properties such as thermal stability, adsorption behavior, and catalytic activity.

The ion-exchange process is represented by

$$z_A B^{z_B^+}(z) + z_B A^{z_A^+}(s) \rightleftharpoons z_A B^{z_B^+}(s) + z_B A^{z_A^+}(z)$$

where z_A and z_B are the ionic charges of cations A and B and (z) and (s) represent the zeolite and solution.

Manufacture

Zeolites are formed under hydrothermal conditions, defined here in a broad sense to include zeolite crystallization from aqueous systems containing various types of reactants. Most synthetic zeolites are produced under nonequilibrium conditions, and must be considered as metastable phases in a thermodynamic sense.

Although more than 150 synthetic zeolites have been reported, many important types have no natural mineral counterpart. Conversely, synthetic counterparts of many zeolite minerals are not yet known. The condition generally used in synthesis are reactive starting materials such as freshly coprecipitated gels, or amorphous solids; relatively high pH introduced in the form of an alkali metal hydroxide or other strong base, including tetraalkylammonium hydroxides; low temperature hydrothermal conditions with concurrent low autogenous pressure at saturated water vapor pressure; and a high degree of supersaturation of the gel components leading to the nucleation of a large number of crystals.

Typical gels are prepared from aqueous solutions of reactants such as sodium aluminate, NaOH, and sodium silicate; other reactants include alumina trihydrate $(Al_2O_3.3H_2O)$, colloidal silica, and silicic acid.

Processes. Processes for the manufacture of commercial molecular-sieve products are classified into three groups: the preparation of molecular-sieve zeolites as high purity crystalline powders or as preformed pellets from reactive aluminosilicate gels or hydrogels; the conversion of clay minerals into zeolites, either in the form of high purity powders or as binderless high purity preformed pellets; and processes based on the use of other naturally occurring raw materials.

The hydrogel and clay-conversion processes may also be used to manufacture products that contain the zeolite as a major or a minor component in a gel matrix, a clay matrix, or a clay-derived matrix. Powdered products are often bonded with inorganic oxides or minerals into agglomerated particles for ease in handling and use. Hydrogel processes are based either on homogeneous gels, that is, hydrogels prepared from solutions of soluble reactants, or on heterogeneous hydrogels which are prepared from reactive alumina or silica in a solid form, for example, solid amorphous silica powder.

In gel-preform processes, the reactive aluminosilicate gel is first formed into a pellet which reacts with sodium aluminate solution and caustic solution. The zeolite crystallizes *in situ* within an essentially self-bonded pellet, or as a component in an unconverted amorphous matrix.

The starting material for clay conversion is kaolin, which usually must be hydroxylated to meta kaolin by air calcination (see Clays). At 500–600°C, meta kaolin forms, followed by a mullitized kaolin at 1000–1050°C.

$$Al_2Si_2O_5(OH)_4 \xrightarrow{550°C} Al_2Si_2O_7 + 2\,H_2O$$
$$\text{kaolin} \qquad\qquad \text{meta-kaolin}$$

$$3\,Al_2Si_2O_7 \xrightarrow{1050°C} Si_2Al_6O_{13} + 4\,SiO_2$$
$$\text{meta-kaolin} \qquad \text{mullite} \quad \text{cristobalite}$$

The zeolites are prepared as essentially binderless preformed particles. The kaolin is shaped in the desired form of the finished product and is converted *in situ* in the pellet by treatment with suitable alkali hydroxide solution. Preformed pellets of zeolite A are prepared by this method.

Health and Safety

Zeolites have applications in foods, drugs, cosmetic products, and detergents. Thus, extensive toxicological and environmental studies have been carried out.

Uses

In addition to a broad spectrum of catalytic applications, molecular-sieve adsorbents are used to perform difficult separations, eg, gases from gases, liquids from liquids, and solute from solutions. Commercial uses for molecular-sieve ion exchangers are based on their unique combinations of selectivity, capacity, and stability not found in the more common cation exchangers (see Ion Exchange).

D.W. BRECK
R.A. ANDERSON
Union Carbide Corporation

D.W. Breck, *Zeolite Molecular Sieves: Structure, Chemistry, and Use*, John Wiley & Sons, Inc., New York, 1974.

J.A. Rabo, ed., *Zeolite Chemistry and Catalysis*, ACS Monograph 171, American Chemical Society, Washington, D.C., 1976.

MOLLUSCICIDES. See Poisons, economic.

MOLYBDENUM AND MOLYBDENUM ALLOYS

Molybdenum was first identified as a discrete element in 1778. Over one-half of the world supply of molybdenum comes from mines where its recovery is the primary objective of the operation. The balance is recovered as a by-product of copper mining. The most abundant mineral, and the only one of commercial significance, is molybdenite, MoS_2. Powellite, $Ca(MoW)O_4$, and wulfenite, $PbMoO_4$, are also known. Primary ore bodies in the Western Hemisphere contain ca 0.12–0.4% molybdenum and give a recovery of 2–4 kg per metric ton of ore.

Molybdenite is concentrated by first crushing and grinding the ore, and passing the finely ground material (called pulp) through a series of flotation cells (see Flotation). Operations recovering molybdenum as a by-product of copper mining produce a concentrate containing both metals. Molybdenite is separated from the copper minerals by differential flotation.

Molybdenite concentrate contains about 90% MoS_2. The remainder is primarily silica. The concentrate is roasted to remove the sulfur and convert the sulfide to oxide. Molybdenum is added to steel in the form of this oxide, known as technical molybdic oxide. In modern molybdenum-conversion plants, the sulfur formed by roasting MoS_2 is converted to sulfuric acid instead of being discharged as SO_2 into the atmosphere.

Physical Properties

Molybdenum is in Group VIB between chromium and tungsten vertically and niobium and technetium horizontally in the periodic table. It has a silvery gray appearance. The most stable valence state is +6; lower, less stable valence states are +5, +4, +3, +2, and 0.

There are many similarities between molybdenum and its horizontal and vertical neighbors in the periodic system. It is a typical transition element having the maximum number of five unpaired $4d$ electrons, which account for its high melting point ($2626 \pm 9°C$), strength, and modulus of elasticity. Coefficient of linear expansion (%) at 200°C 0.23; at 0–800°C 0.46; at 0–1200°C 0.72. Thermal conductivity [W/(m·K)] at 500°C 122, at 1000°C 101, at 1500°C 82. Electrical resistivity (nΩm) at 0°C 50; at 1000°C 320; at 2000°C 61. Density 10.22 g/cm³ at 20°C. Bp 5560°C; heat of vaporization 491 KJ/mol (117 kcal/mol) surface tension of melting point 2240 mNm (= dyn/cm).

Chemical Properties

Resistance to corrosion is one of the most valuable properties of molybdenum metal. It has particularly good resistance to corrosion by mineral acids, provided oxidizing agents are not present. It also resists attack by some liquid metals.

In a reducing atmosphere, molybdenum is resistant at elevated temperatures to hydrogen sulfide, which forms a thin adherent sulfide coating. In an oxidizing atmosphere, however, molybdenum is rapidly corroded by sulfur-containing gases. It has excellent resistance to iodine vapor up to ca 800°C; to bromine up to ca 450°C; and to chlorine up to ca 200°C. Fluorine, the most reactive of the halogens, attacks molybdenum at room temperatures.

Manufacture

The technical-grade MoO_3 formed by roasting molybdenite concentrate can be sublimed to produce high purity molybdenum trioxide or it can be leached with dilute ammonia to give ammonium molybdate. Ammonium molybdate or molybdenum trioxide is reduced to molybdenum powder by hydrogen at 500–1150°C. Boat- and tube-type furnaces are usually selected for this purpose.

In the conventional powder-metallurgy process, molybdenum powder is compacted hydraulically in dies at a pressure of about 207–276 MPa (30,000–40,000 psi) to bars which are sintered electrically at 2200–2300°C in hydrogen-atmosphere bells. Sintering currents supply about 90% of the current required to fuse the bar. In an alternative process, adaptable to larger sections, the bars are sintered in a hydrogen-atmosphere muffle furnace for about 16 h at 1600–1700°C. The longer time at the lower temperatures in the muffle furnace accomplishes the same densification produced by bell sintering. Ingots weighing 2000 kg have been produced by these methods. Rods and bars of molybdenum are made by rolling, forging, or swaging the sintered ingot or extruded billet at 1200–1400°C.

Molybdenum metal can be mechanically worked by almost any commercial process with suitable practices such as forging, extrusion, rolling, forming, bending, punching, stamping, deep drawing, spinning, and power-roll forming. Except for fine wire and sheet, at least a moderate amount of heating is recommended for all working operations.

Health and Safety

Industrial poisoning by molybdenum has not been reported. Lethal doses for rats are 114–117 mg/kg.

Uses

Approximately 85% of all molybdenum is used as an alloying additive to steels and irons.

ROBERT Q. BARR
Climax Molybdenum Company

Molybdenum Metal, Climax Molybdenum Company, Ann Arbor, Michigan, 1960.

J.W. Goth, "Molybdenum in the Eighties," *Second International Ferro-Alloys Conference*, Copenhagen, Denmark, Oct. 1979.

MOLYBDENUM COMPOUNDS

Halides and Oxyhalides

Molybdenum forms halogen compounds of widely different degrees of stability. The hexafluoride and pentachloride deposit molybdenum metal in the vapor phase as result of reaction with hydrogen.

The highest member of each series (MoF_6, $MoCl_5$, $MoBr_4$, and MoI_3) can be made by direct halogenation of molybdenum metal. The hexafluoride, pentafluoride, pentachloride, oxytetrafluoride, oxytetrachloride, and dioxydichloride are volatile at moderate temperatures. The entire series exhibits a wide range of stability, color, and volatility. Few of the compounds are monomeric in their normal state. Generally, the lower halides are prepared by reduction of the highest member of the series with molybdenum metal, hydrogen, or a hydrocarbon.

Oxides and Hydroxy Compounds

Molybdenum trioxide, MoO_3, is a white crystalline powder from which most molybdenum compounds are prepared directly or indirectly. Technical grade MoO_3 is prepared commercially by oxidizing molybdenite, MoS_2, in a multiple-hearth roaster. Air flow and temperature are carefully controlled to provide a low sulfur product with minimum MoO_2 content. High purity chemical grades are prepared either by sublimation of the technical grade or calcination of crystallized ammonium dimolybdate (ADM). Halogens and hydrohalides react with MoO_3 to form oxyhalides. Reaction with thionyl chloride, $SOCl_2$, gives molybdenum oxytetrachloride or pentachloride, depending on the reaction time.

Hydroxides. The structures of molybdenum blues and the series of hydrated molybdenum oxides are intimately related to those of the oxides and are characterized as genotypic. Hydrated oxides are prepared by the carefully controlled reduction of MoO_3 with either atomic hydrogen, zinc and hydrochloric acid, molybdenum metal powder and water, or lithium aluminum hydride.

Molybdates. Isopolymolybdates, a large class of inorganic compounds, consist of a cation and a condensed molybdate anion. The naming of isopolymolybdates is straightforward: they are called simply di, tri, etc, molybdates depending upon the degree of condensation.

Normal or orthomolybdates, $M_2O.MoO_3.xH_2O$ or $M_2MoO_4.xH_2O$, are prepared by direct combination of the oxides, neutralization of slurries of MoO_3 with MOH or M_2CO_3, or by precipitation from the molybdate solution by salts of the desired metals.

Acidification of molybdate ions in alkaline solution in which they exist as $(MoO_4)^{2-}$ results in the formation of polynuclear species.

Heteropolymolybdates. These form a large family of salts and free acids with each member containing a complex and high molecular weight anion.

Sulfides, Selenides, and Tellurides

Molybdenum forms a series of homologous compounds with sulfur, selenium, and tellurium. The disulfide, MoS_2, diselenide, $MoSe_2$, and ditelluride, $MoTe_2$, are isomorphous. These chalcogenides occur as shiny gray plates and may be prepared by the direct combination of the elements or by heating MoO_3 and the appropriate element in potassium carbonate to a high temperature.

Cyanides, Thiocyanates, and Carbonyls

Cyanides and thiocyanates form complex anions with molybdenum in a lower valence state. In most of these, molybdenum atoms have a coordination number of eight. These compounds are highly colored. The cyanides and the thiocyanates are light sensitive, but are generally more stable in air than the complex halides.

The red thiocyanate complex is formed by adding a solution of a soluble thiocyanate to a reduced acid solution of molybdenum. It is utilized in the colorimetric determination of molybdenum. The compound contains a thiocyanate to molybdenum ratio of three; the Mo is pentavalent.

Organomolybdenum Compounds

In addition to the substituted molybdenum hexacarbonyl compounds, organomolybdenum sulfur compounds containing phosphorus or nitrogen have been prepared. Many of these compounds are oil-soluble and, thus, the low friction benefits of the molybdenum–sulfur interaction can be imparted to engine and gear oils. Typical structures for these com-

pounds can be assigned as:

For oil solubility, the R groups are generally C_3 or larger alkyls or aryls.

Health and Safety

Not toxic in humans. In some areas where soil molybdenum is high, and consequently the molybdenum content of forage crops is high, animals develop a toxic reaction called molybdenosis, which can be overcome by injections or feed implementations of copper salts.

Uses

The chemical applications of molybdenum have increased substantially over the past twenty years. New applications have been developed in catalysis, lubrication, corrosion inhibition, protective coatings, inhibitive pigments, and flame and smoke retardants. Molybdenum compounds are used in the transportation and petroleum and petrochemical industries; molybdenum catalysts are prime candidates in the conversion of coal to liquid fuels (see Fuels, synthetic).

H.F. BARRY
Climax Molybdenum Company of Michigan

Reviewed by
I. BOB BLUMENTHAL
Noah Industrial Corp.

R.J.P. Williams, *The Biological Role of Molybdenum*, Climax Molybdenum Company, London, UK, 1978.

Proceedings of the Climax International Conference on the Chemistry and Uses of Molybdenum, Vol. 1, Reading, UK, 1973; Vol. 2, Oxford, UK, 1976; Vol. 3, Ann Arbor, Mich., 1979.

E.J. Stiefel, "The Coordination and Bioinorganic Chemistry of Molybdenum" in S.J. Lippard, ed., *Progress in Inorganic Chemistry*, Vol. 22, John Wiley & Sons, Inc., New York, 1977.

MONAZITE, (Th, Ce, La, Di, Si)PO₄. See Cerium; Rare-earth metals; Thorium.

MONOSODIUM GLUTAMATE, NaOOCCH₂CH₂CH-(NH₂)COOH. See Amino acids, monosodium glutamate.

MORPHOLINE. See Amines—cyclic.

MOUTHWASHES. See Dentifrices.

MUCILAGES. See Gums.

MUNTZ METAL. See Copper alloys, wrought.

MUSCLE RELAXANTS. See Neuroregulators; Psychopharmacological agents.

N

NAPHTHALENE

This article deals mainly with naphthalene. The hydrogenated naphthalenes, the alkylnaphthalenes (particularly methyl- and isopropylnaphthalenes), and acenaphthalene are also discussed (see also Naphthalene derivatives).

Properties

The accepted configuration of naphthalene, ie, two fused benzene rings sharing two common carbon atoms in the ortho position, was established in 1869 and was based on its oxidation product, phthalic acid. Based on its fused-ring configuration, naphthalene is the first member in a class of aromatic compounds with condensed nuclei. Its numbering is shown in structure (1). Naphthalene is a resonance hybrid having three contributing forms:

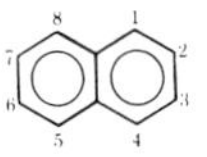

In chemical reactions, naphthalene usually acts as though the bonds were fixed in the position shown in structure (2). For most purposes, the conventional formula in structure (1) is adequate; the numbers represent the carbon atoms with attached hydrogen atoms. The two carbons that bear no numbers are common to both rings and carry no hydrogen atoms. From the symmetrical configuration of the naphthalene molecule, it should be possible for only two isomers to exist when one hydrogen atom is replaced by another atom or group. Therefore, positions 1, 4, 5, and 8 are identical and are designated as α positions.

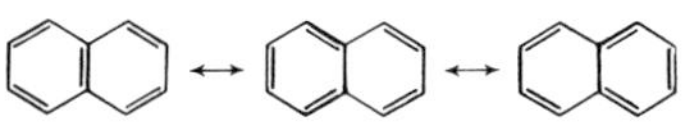

Naphthalene has mp 80.290°C; bp 217.993°C at 101.3 kPa (1 atm); triple point 80.28°C; ignition temperature 79°C; and density 1.175 g/cm³ at 25°C. It is very slightly soluble in water but appreciably soluble in many organic solvents, eg, 1,2,3,4-tetrahydronaphthalene, phenols, ethers, carbon disulfide, chloroform, benzene, coal-tar naphtha, carbon tetrachloride, acetone, and decahydronaphthalene.

Reactions

Substitution. Substitution products retain the same nuclear configuration as naphthalene. They are formed by the substitution of one or more hydrogen atoms with other functional groups. Substituted naphthalenes that are important commercially have been obtained through the use of sulfonation, sulfonation and alkali fusion, alkylation, nitration and reduction, and chlorination.

The hydrogen atoms in the one (4, 5, and 8) positions of naphthalene can be substituted somewhat more easily than hydrogen atoms in benzene, and they tend to do so under mild conditions.

Addition. Addition reactions are characterized by the introduction of an element onto two or more of the adjacent carbon atoms of the naphthalene nucleus. The most important addition products of naphthalene are the hydrogenated compounds; of less commercial significance are those made by the addition of chlorine.

Hydrogen is added to the naphthalene nucleus by reagents that do not affect benzene. It is possible to obtain products into which two, four, six, eight, or ten hydrogen atoms have been added. Of these, only the tetra- and decahydronaphthalenes are commercially significant.

Chlorine addition and some chlorine substitution occurs at normal or slightly elevated temperatures in the absence of catalysts. The chlorination of molten naphthalene under such conditions yields a mixture of naphthalene tetrachlorides, a monochloronaphthalene tetrachloride, and a dichloronaphthalene tetrachloride, as well as mono- and dichloronaphthalenes.

Oxidation. Naphthalene may be oxidized directly to 1-naphthalenol (1-naphthol) and 1,4-naphthaquinone, but yields are not good. Further oxidation beyond 1,4-naphthaquinone results in the formation of o-phthalic acid, which can be dehydrated to form phthalic anhydride. The manufacture of phthalic anhydride is the largest single use for naphthalene. Catalytic vapor-phase oxidation of naphthalene is preferred over liquid-phase oxidation. All vapor-phase oxidation processes involve the use of a catalyst based on vanadium pentoxide (see Pthalic acid and other benzenepolycarboxylic acids).

Manufacture

Two sources of naphthalene exist in the United States: coal tar and petroleum. Since 1970, naphthalene production has decreased at an average rate of 3.3%/yr. Coal-tar naphthalene production decreased 6.5% and petroleum naphthalene dropped 1.5%. As a result of this decline, in the United States in early 1984 there are only two petroleum-naphthalene producers, and three coal-tar naphthalene producers with a capacity distribution of ca 62% coal tar and 38% petroleum.

Coal tar. Coal tar is condensed and separated from the coke-oven gases formed during the high temperature carbonization of bituminous coal in coke plants (see Coal, coal-conversion processes; Coal chemicals and feedstocks). Although some naphthalene is present in the oven gases after tar separation and is removed in subsequent water cooling and scrubbing steps, the amounts are of minor importance. The largest quantities of naphthalene are obtained from the coal tar that is separated from the coke-oven gases. A typical dry coal tar obtained in the United States contains ca 10 wt% naphthalene.

The coal tar first is processed through a tar-distillation step where ca the first 20 wt% of distillate, ie, chemical oil, is removed. The chemical oil, which contains practically all the naphthalene present in the tar, is reserved for further processing and the remainder of the tar is distilled further to remove additional creosote oil fractions until a coal-tar pitch of desirable consistency and properties is obtained.

Principal U.S. producers obtain their crude naphthalene product by batch or continuous fractional distillation of the tar acid-free chemical oil.

Petroleum. The production of naphthalene from petroleum involves two principal steps: production of an aromatic oil in the naphthalene–alkylnaphthalene boiling range by hydroaromatization or cyclization; and dealkylation of such oils either thermally or catalytically. The naphthalene that is produced is recovered as a high quality product and usually by fractional distillation.

Health and Safety

Naphthalene is generally transported in molten form in tank trucks or tank cars that are equipped with steam coils. Depending upon the transportation distance and the insulation on the car or truck, the naphthalene may solidify and require reheating before unloading.

The acute toxicity of naphthalene is low, LD_{50} values for rats of 1780–2500 mg/kg. The inhalation of naphthalene vapors may cause headache, nausea, confusion, and profuse perspiration and, if exposure is severe, vomiting, optic neuritis, and hematuria may occur. Naphthalene can be irritating to the skin and hypersensitivity does occur.

Uses

The largest single use is in the manufacture of phthalic anhydride. The second largest use is as a raw material for the manufacture of 1-naphthyl N-methylcarbamate (carbaryl, Sevin), a chemical used extensively as a replacement for DDT and other products that have become environmentally unacceptable (see Insect control technology).

R.M. GAYDOS
Koppers Company, Inc.

J.P. McCullough and co-workers, *J. Phys. Chem.* **61**, 1105 (1957).

E.E. Sandmeyer in G.D. Clayton and F.E. Clayton, *Patty's Industrial Hygiene and Toxicology*, 3rd rev. ed., Vol. II, Wiley-Interscience, New York, 1981, Chapt. 46.

R. Gerry, "Product Review on Naphthalene," *Chemical Economics Handbook*, Stanford Research Institute, Menlo Park, Calif., Dec. 1978.

NAPHTHALENECARBOXYLIC ACIDS. See Naphthalene derivatives.

NAPHTHALENE DERIVATIVES

Several systems of nomenclature have been used for naphthalene and many trivial and trade names are well established. In this article, the *Chemical Abstracts Index Guide* is employed.

The number of naphthalene derivatives is very large since the number of positional isomers is large: 2 for monosubstitution, 10 for disubstitution/same substituent, 14 for disubstitution/different substituents, 42 for trisubstitution/two different substituents, 84 for trisubstitution/three different substituents, and so on with multiplying complexity. The commercially important compounds are described below.

Naphthalenesulfonic Acids

Naphthalenesulfonic acids are important chemical precursors for dye intermediates, wetting agents and dispersants, naphthols, and air-entrainment agents used for concrete (see Azo dyes; Cement). Generally, the sulfonation of naphthalene leads to a mixture of products. Naphthalene sulfonation at less than ca 100°C is kinetically controlled and produces predominantly 1-naphthalenesulfonic acid. Sulfonation of naphthalene at above ca 150°C provides thermodynamic control of the reaction and 2-naphthalenesulfonic acid as the main product.

Nitronaphthalenes and Nitronaphthalenesulfonic Acids

The nitro group does not undergo migration on the naphthalene ring during the usual nitration procedures. Therefore, mono- and polynitration of naphthalene is similar to low temperature sulfonation. The first nitro group enters the 1-position and a second nitro group tends to enter the 8 or 5 positions to give 1,8- and 1,5-dinitronaphthalene in a 60 : 40 ratio. Many of these compounds are not accessible by direct nitration of naphthalene but are made by indirect methods, eg, nitrite displacement of diazonium halide groups in the presence of a copper catalyst, decarboxylation of nitronaphthalenecarboxylic acids, or deamination of nitronaphthalene amines.

Naphthaleneamines and Naphthalenediamines

Selected physical properties are listed in Table 1.

1-Naphthaleneamine. 1-Naphthaleneamine(1-naphthylamine; α-naphthylamine) can be made from 1-nitroaphthalene by reduction with iron–dilute HCl or by catalytic hydrogenation. 1-Naphthaleneamine is toxic, LD_{50} (dogs) = 400 mg/kg, and a suspected human carcinogen, which mandates appropriate precautions in manufacture and use. 1-Naphthaleneamine is a dye intermediate and is used as the starter material in the manufacture of the rodenticide, Antu, 1-naphthalenethiourea; the rubber antioxidant, *N*-phenyl-1-naphthaleneamine made by the condensation of 1-naphthaleneamine or 1-naphthalenol with aniline; the insecticide and miticide, Nissol; and the herbicide, Naptalam, Alanap or Dyanap, or *N*-1-naphthylphthalamic acid.

Aminonaphthalenesulfonic Acids

Many aminonaphthalenesulfonic acids are important in the manufacture of azo dyes (qv) or are used to make intermediates for azo acid dyes, direct, and fiber-reactive dyes. Usually, the aminonaphthalenesulfonic acids are made by either the sulfonation of naphthalenamines, the nitration–reduction of naphthalenesulfonic acids, the Bucherer-type amination of naphtholsulfonic acids, or the desulfonation of an aminonaphthalenedi- or trisulfonic acid. Most of these processes produce by-products or mixtures which often are separated in subsequent purification steps.

Naphthalenols and Naphthalenediols

Naphthalenols, naphthalenediols, and their sulfonated derivatives are important intermediates for dyes, agricultural chemicals (eg, carbaryl), drugs (eg, propranolol), perfumes (qv), and surfactants. The methods of manufacture include sulfonation–caustic fusion of naphthalene, hydrolysis of 1-naphthaleneamine, oxidation–aromatization of tetralin, and hydroperoxidation of 2-isopropylnaphthalene. As the toxic hazard of 1-naphthaleneamine was recognized, its commercial use was minimized. The sulfonation–caustic fusion process is more difficult to operate than in the past because of increasing difficulties posed by product purity requirements, high investment and replacement costs, and by-product-effluent-handling problems. In the United States, the naphthalenols are made by hydrocarbon oxidation routes.

The chemical properties of the naphthalenols are similar to those of phenol and resorcinol, with added reactivity and complexity of substitution because of the condensed ring system (see Hydroquinone, resorcinol, and catechol). The naphthols and naphthalenediols are listed with some of their physical properties in Table 2.

Hydroxynaphthalenesulfonic Acids

Hydroxynaphthalenesulfonic acids are important as intermediates either for coupling components for azo dyes or azo components and for synthetic tanning agents. These acids can be manufactured either by sulfonation of naphthols or hydroxynaphthalenesulfonic acids, by acid hydrolysis of aminonaphthalenesulfonic acids, by fusion of sodium naphthalenepolysulfonates with sodium hydroxide, or by desulfonation or rearrangement of hydroxynaphthalenesulfonic acids.

Table 1. Physical Properties of Naphthalenamines and Naphthalenediamines

Compound	Mp, °C	Density, g/cm³	Other
1-naphthaleneamine	50	1.13_4^{14}	flash pt, 157°C; sol 0.496 g/L H_2O; vol with steam; bp 301°C (160°C at 1.6 kPa^a)
2-naphthaleneamine	111–113	1.061_4^{98}	sol hot water; vol with steam; bp 306°C (175.8°C at 2.7 kPa^a)
1,2-naphthalenediamine	96–98		sol hot water, alc, either; bp at 0.01 kPa^a 150–151°C
1,4-naphthalenediamine	120		sl sol hot water
1,5-naphthalenediamine	189.5		sol hot water, alc
1,6-naphthalenediamine	78	$1.147_4^{99.4}$	sol hot water, alc
1,7-naphthalenediamine	117.5		sol alc
1,8-naphthalenediamine	66.5	$1.127_4^{99.4}$	sol alc, ether; bp at 1.6 kPa^a 205°C
2,3-naphthalenediamine	191		sol alc, ether
2,6-naphthalenediamine	216–218		sparingly sol alc, ether
2,7-naphthalenediamine	159		

^{a} To convert kPa to mm Hg, multiply by 7.5.

Table 2. Properties of Naphthalenols and Naphthalenediols

Compound	Mp, °C	Density, g/cm³	Other
1-naphthalenol	95.8–96.0	1.224_4^4 1.099_4^{99}	sublimes; sol 0.03 g/100 mL H_2O at 25°C; readily sol alc, ether, benzene; bp 280°C (158°C at 2.6 kPa^a)
2-naphthalenol	122	1.078_4^{130} 1.22_4^{25}	sublimes; sol 0.075 g/100 mL H_2O at 25°C; readily sol alc, ether, benzene; flash pt 161°C; bp 295°C (161.8°C at 2.6 kPa^a)
1,2-naphthalenediol	103–104		
1,3-naphthalenediol	124		
1,4-naphthalenediol	195		heat of combustion 4.77 MJ^b
1,5-naphthalenediol	258		sublimes; sparingly sol water; readily sol ether, acetone
1,6-naphthalenediol	137–138		
1,7-naphthalenediol	181		
1,8-naphthalenediol	144		
2,3-naphthalenediol	159		
2,6-naphthalenediol	222		
2,7-naphthalenediol	194		sol boiling water

^{a} To convert kPa to mm Hg, multiply by 7.5.
^{b} To convert J to cal, divide by 4.184.

Naphthalenecarboxylic Acids

Most of the naphthalenemono-, di-, or polycarboxylic acids have been made by simple routes, eg, the oxidation of the appropriate alkylnaphthalene, or di- or polymethylnaphthalenes, or by complex routes, eg, the Sandmeyer reaction of the selected aminonaphthalenesulfonic acid to give a cyanonaphthalenesulfonic acid followed by fusion of the latter with an alkali cyanide and simultaneous or subsequent hydrolysis of the nitrile groups.

Hydroxynaphthalenecarboxylic and Aminonaphthalenecarboxylic Acids

3-Hydroxy-2-naphthalenecarboxylic acid is commercially important as a dye intermediate. The compounds of technical interest are prepared by the Kolbe-Schmitt reaction, ie, the carboxylation of alkali naphthoxides with CO_2. Less direct syntheses have to be resorted to for the other isomers, eg, the oxidation of hydroxyaldehydes or acylated naphthols, the Sandmeyer reaction of appropriately substituted naphthalene-amines, alkali fusion of sulfonated naphthalenecarboxylic acids, etc. 8-Amino-1-naphthalenecarboxylic acid or its lactam, naphthostyryl, is a valuable intermediate for dyes as well.

Hans Dressler
Koppers Company, Inc.

F. Radt, ed., *Elsevier's Encyclopedia of Organic Chemistry*, Sect. III, Vol. 12B, Elsevier Publishing Co., Amsterdam, Neth., 1949–1955.

N. Donaldson, *The Chemistry and Technology of Naphthalene Compounds*, E. Arnold Ltd., London, UK, 1958.

Ullmanns Encyklopadie Der Technischen Chemie, 4th ed., Verlag Chemie, Weinheim/Bergstr., FRG, 1972–1984.

NAPHTHENIC ACIDS

Naphthenic acids are the carboxylic acids that are derived from petroleum during the refining of the various distilled fractions and are predominantly monocarboxylic acids (see Carboxylic acids). The main distinguishing structural feature of naphthenic acids is a hydrocarbon chain consisting of single or fused cyclopentane rings, which are alkylated in various positions with short aliphatic groups.

The commercial grades of naphthenic acids vary widely in properties and impurities, depending on their source and the refining method. All contain 5–25 wt% hydrocarbons whose composition is the same as the petroleum fraction from which the naphthenic acids are derived. The average molecular weight is higher for acids that are extracted from higher boiling fractions. All contain acidic impurities, eg, phenols, mercaptans, and thiophenols, in small quantities.

Properties

Naphthenic acids are viscous liquids with a characteristic odor resulting from the phenols and sulfur compounds that are extracted with the acids; these impurities are very difficult to remove. Various chemical treatments, such as mild oxidation or reaction with aldehydes followed by distillation, only partially reduce the odor and no economically feasible general treatment has been devised.

Naphthenic acids undergo the same chemical reactions as other saturated carboxylic acids. The most important commercial reactions are amidation, esterifcation (qv), and the formation of metal soaps (see Driers and metallic soaps).

Manufacture

Naphthenic acids are recovered as by-products from the refining of straight-run petroleum distillates. Most of the commercial acids are derived from the kerosene and gas-oil fractions and some higher molecular weight acids are derived from the light lubricating oil cuts (see Petroleum).

Uses

Naphthenates are used primarily in the manufacture of paint driers, corrosion inhibitors, lubricants, catalysts, and preservatives.

W.E. Sisco
W.E. Bastian
E.G. Weierich
CPS Chemical Company

H.L. Lochte and E.R. Littman, *Petroleum Acids and Bases*, Chemical Publishing Co., New York, 1955, pp. 9–279.

U.S. Pat. 1,694,461 (Dec. 11, 1928), G. Alleman (to Sun Oil Co.).

NAVAL STORES.   See Terpenoids.

NEODYMIUM.   See Rare-earth elements.

NEPTUNIUM.   See Actinides and transactinides.

NEUROREGULATORS

The number of endogenous substances that have been isolated and shown to have a neurochemical, neurophysiological, or behavioral effect in animals or humans has increased enormously (see Table 1). Neuroregulator has been suggested as a generic name for these endogenous compounds, and the term includes the neurotransmitter and neuromodulator groups. A neurotransmitter is a substance that conveys a transient and unilateral signal across a specialized synapse. A neuromodulator alters neuronal activity by mechanisms that may or may not involve a synapse. Neuromodulators may be classified as hormonal neuromodulators, which provide direct, short- or long-lasting modulation of neurons in areas far removed from the site of release, and as synaptic neuromodulators, which act indirectly by modulating neurotransmitters function.

Various criteria for establishing the identity of neurotransmitters and neuromodulators are listed below:

Neurotransmitter

(1) The substance must be present in presynaptic elements of neuronal tissue, possibly in an uneven distribution throughout the brain.

(2) Precursors and synthetic enzymes must be present in the neuron, usually in proximity to the site of presumed action.

(3) Stimulation of afferents should cause release of the substance in physiologically significant amounts.

(4) Direct application of the substance to the synapse should produce responses that are identical to those of stimulating afferents.

(5) Specific receptors should interact with the substance and should be in close proximity to presynaptic structures.

(6) Interaction of the substance with its receptor should induce changes in postsynaptic membrane permeability leading to excitatory or inhibitory postsynaptic potentials.

(7) Specific inactivating mechanisms, which stop interactions of the substance with its receptor in a physiologically reasonable time frame, should exist.

(8) Stimulation of afferents and direct application of the substance should be equally responsive to and similarly affected by interventions involving postsynaptic sites or inactivating mechanisms.

Neuromodulator

(1) The substance does not act as a neurotransmitter, in that it does not act transsynaptically.

(2) The substance must be present in physiological fluids and have access to the site of potential modulation in physiologically significant concentrations.

(3) Alterations in endogenous concentrations of the substance should affect neuronal activity consistently and predictably.

Table 1. Neuroregulators

Name	Properties	Formula or structure	Pharmacology
acetylcholine	hygroscopic crystalline powder; sol in H_2O, alcohol; LD_{50} (rats) 250 mg/kg	$(CH_3)\overset{+}{N}CH_2CH_2O\overset{\overset{\displaystyle O}{\|}}{C}CH_3 + CoA$	neurotransmitter at neuromuscular junction in autonomic ganglia, and at postganglionic parasympathetic nerve endings
adenosine	mp 234–235°C; $[\alpha]_D^{11}$ −61.7°; sol in H_2O, insol in alcohol; uv absorb(max) 260 pm		depresses central neurons following iontophoretic application; sedative effect in cats
epinephrine (−)adrenaline	(±)HCl salt mp 157°C; sol in H_2O; (−)mp 211–212°C; $[\alpha]_D^{25}$ −50 to −53.5 (USP)		hormone of adrenal medulla (see Hormones); elevates blood glucose, free fatty acid conc; inhibits insulin secretion *Medical uses:* sympathomimetic vasoconstrictor, cardiac stimulant bronchodilator
ACTH	white powder; sol in H_2O	H–Ser–Tyr–Ser–Met–Glu–His–Phe– 5 Arg–Trp–Gly–Lys–Pro–Val–Gly– 10 Lys–Lys–Arg–Arg–Pro–Val–Lys– 15 20 Val–Tyr–Pro–Asn–Gly–Ala–Glu– 25 Asp–Glu–Ser–Ala–Glu–Ala–Phe– 30 35 Pro–Leu–Glu–Phe–OH. 39	regulates secretory activity of adrenalcortex cells; stimulates steroid synthesis in adrenal glands
androgens angiotensins	(see Hormones, sex hormones) (see Hormones, brain oligopeptides)		
L-aspartic acid	mp 270–271°C; $d^{12.5}$ 1.661 g/cm³; $[\alpha]_D^{20}$ +25°C; pK_1 1.88; pK_2 3.65; sol in H_2O 1 g/57.4 mL at 60°C	$\underset{\displaystyle HOOCHCH_2CHCOOH}{\overset{\displaystyle NH_2}{\|}}$	nonessential amino acid produces excitatory response from CNS neurons
corticosteroids	*corticosterone* mp 180–182°C; $[\alpha]_D^{15}$ +223°C; absorb max 240 nm *cortisone* mp 220–224°C; $[\alpha]_D^{25}$ +209°; absorb max 237 nm *hydrocortisone* mp 217–220°C; $[\alpha]_D^{22}$ +167°; absorb max 242 nm	corticosterone: R = OH, R' = H cortisone: R = O, R' = OH hydrocortisone: R = OH, R' = OH	corticosteroids produce increased gluconeogenesis reduce and inhibit inflamation of tissues; used in treatment of skin and eye disorders, asthma, etc; may act as neuromodulators
dopamine	(HCl) dec 241°C; sol in water, methanol; insol in ether		neurotransmitter in CNS
endorphins, enkephalins	β-endorphin $[\alpha]_D^{24}$ −76.6°; methionine-enkephalin mp 196–198°C $[\alpha]_{589}^{22}$ −21.9°	(see Opoids, endogenous)	produce analgesic effects having opiate-like qualities

Table 1. Continued

Name	Properties	Formula or structure	Pharmacology
estrogens	(see Hormones, sex hormones)	$H_2N(CH_2)_3COOH$	may be an inhibitory neurotransmitter in the CNS
GABA (γ-aminobutyric acid)	sol in H_2O; mp 202°C; K_a 3.7×10^{-11}; K_b 1.7×10^{-10} at 25°C		
L-glutamic acid	dec 247–249°C; $[\alpha]_D^{22.4}$ +31.4; pK_1 2.19; pK_2 4.25; sol in H_2O 8.64 g/L (25°C)	$HOOC(CH_2)_2CHCOOH$ $\quad\quad\quad\quad\quad NH_2$	used as antiepileptic; known to be excitatory amino acid when applied iontophoretically to nerve cells
glycine	dec 233–290°C; d 1.1607 g/cm³; pK_1 2.34; pK_2 9.60; sol in H_2O 0.25 g/mL	NH_2CH_2COOH	may be an inhibitory neurotransmitter in the CNS
histamine	mp 83–84°C; bp (2.4 kPa) 209–210°C; sol in H_2O; sparingly sol in ether; LD_{50} (mice) 13.0 g/kg		intraventricular histamine changes the cortical EEG and produces sedation
serotonin (5-hydroxytryptamine)	mp 167–168°C; sol in H_2O		neurotransmitter in CNS produces inhibitory effects upon iontophoretic adminis
luteinizing hormone, release hormone	(see Hormones, brain oligopeptides)		
melatonin	mp 116–118°C; absorb max 223 nm		skin lightening hormone; may cause sedation in humans
noradrenaline (norepinephrine)	(see Epinephrine)		
phenylethylamine	bp 194.5–195°C; d_4^{25} 0.9640 g/cm³; strong base; sol in H_2O; LD_{50} 366 mg/kg (rats)		increases spontaneous motor activity in animals
piperidine	mp 7°C; bp 106°C; pK (25°C) 2.80; sol in alcohol; misc in H_2O; LD_{50} (rabbits) 500 mg/kg		produces synaptic stimulation, then depression
prostaglandins	(see Prostaglandins)		
somatostatin	(see Hormones, brain oligopeptides)	H–Ala–Gly–Cys–Lys–Asn–Phe–Phe– Trp–Lys–Thr–Phe–Thr–Ser– Cys–OH	inhibits the release of growth hormone (GH) from pituitary and release of thyrotropin and prolactin; used to treat acromegaly, diabetes, ulcers and pancreatis
substance P	sol in H_2O; loses biological activity in aqueous sol in few minutes	H–Arg–Pro–Lys–Pro–Gln–Gln– Phe–Phe–Gly–Leu–Met–NH_2	may be an excitatory neurotransmitter in the CNS
tryptamine	mp 118°C; sol in ethanol; insol in H_2O; absorb max 222, 282, 290 nm		may act as excitatory neurotransmitter in CNS
tyramine	mp 164–165°C; bp (3.33 kPa) 205–207°C; sol in H_2O 1 g/95 mL; sol in boiling alcohol 1 g/10 mL		releases catecholamines from storage sites

(4) Direct application of the substance should mimic the effect of increasing its endogeneous concentrations.

(5) The substance should have one or more specific sites of action through which it can alter neuronal activity.

(6) Inactivating mechanisms, which account for the time course of effects of endogenously or exogenously induced changes in concentrations of the substance, should exist.

(7) Interventions, which alter the effects on neuronal activity of increasing endogenous concentrations of the substance, should act identically when concentrations are increased by exogenous administration.

ALAN S. HORN
University of Gröningen

J.D. Barchas, H. Akil, G.R. Elliott, R.B. Holman, and S.J. Watson, *Science* **200**, 964 (1978).

J.R. Cooper, F.E. Bloom, and R.H. Roth, *The Biochemical Basis of Neuropharmacology*, 4th ed., Oxford University Press, New York, 1983.

M.A. Lipton, A. DiMascio, and K.F. Killam, eds., *Psychopharmacology—A Generation of Progress*, Raven Press, New York, 1978.

NEUTRON ACTIVATION. See Radioactive tracers; Analytical methods.

NICKEL AND NICKEL ALLOYS

Nickel

Physical Properties. Nickel, iron, and cobalt occur in (transition) group VIII of the periodic table. Physical constants of nickel are mp 1453°C; bp (by extrapolation) 2732°C; density at 20°C 8.908 g/cm^3; specific heat at 20°C 0.44 kJ/(kg·K 105 cal (kg·K)); thermal conductivity at 100°C 82.8 W/(m·K); coefficient of thermal expansion at 20–100°C 13.3 × 10^{-6} per °C; electrical resistivity at 20°C 6.97 $\mu\Omega$·cm.

Nickel metal is available in many wrought forms and usually is designated as Nickel 200 or Nickel 201, 205, and 270. Nickel 200 is the general-purpose nickel used in ambient-temperature applications in food-processing equipment, chemical containers, caustic-handling equipment and plumbing, electromagnetic parts, and aerospace and missile components. Nickel 201 has much lower trace carbon content than 200 and is more suitable for elevated-temperature applications where the lower carbon content prevents elevated-temperature stress-corrosion cracking. Nickel 205 is low in carbon but contains trace amounts of magnesium and Nickel 270 is one of the purest, ie, 99.98 wt%, commercial nickels. Duranickel alloy 301, which contains about 4.5 wt% aluminum and 0.5 wt% titanium, can be aged to form very fine γ'-(Ni$_3$Al) precipitates. This type of alloy combines high strength and hardness with the excellent corrosion resistance that is characteristic of Nickel 200. Various of these nickel metals also are used as welding electrodes for joining ferritic or austenitic steels to high nickel-containing alloys and for welding the clad side of nickel-clad steels (see Welding).

Table 1. Typical Properties of Some Nickel Alloys

Composition	A[a,b]	B	C	D	E	F	G	H	J	K
Ni	99.5	66.5	77.0	76.0	32.5	65.4	52.5	60.0	58.0	54.0
Fe	0.15	1.25	0.5	8.0	46.0	2.0	18.5			
Cr			20.0	15.5	21.0	1.0	19.0	9.0	19.5	18.0
Cu	0.05	31.5								
Mo						28.0	3.0		4.3	4.0
Mn	0.25	1.0	1.0	0.5	0.8	1.0	0.2			
Si	0.05	0.25	1.0	0.2	0.5	0.1	0.2			
C	0.06	0.15	0.06	0.08	0.05	0.02	0.04	0.15	0.08	0.08
Al					0.4		0.5	5.0	1.3	2.9
Ti					0.4		0.9	2.0	3.0	2.9
other						2.5 Co	5.1 Cb	10 Co 12 W 1.0 Cb 0.015 B 0.05 Zr	13.5 Co 0.006 B	18.5 Co 0.006 B
melting range, °C	1435–1446	1299–1349	1399	1355–1415	1355–1385	1320–1350	1260–1335	1315–1370	1330–1355	1300–1395
yield strength[c], MPa[d]										
at 20°C	103–931	172–1173	345–1311	285	250	412	1125	840	795	840
at 538°C	139 at 316°C	139 at 316°C		220	180		1020	880	725	795
at 760°C				180	150		800	840	675	730
at 982°C				41				470	140	230
100-h rupture strength, MPa[d]										
at 649°C				160	240		724		760	930
at 812°C				55	63			495	275	305
at 982°C				19	21			179	45	83

[a]A = Nickel 200; B = Monel alloy 400; C = Nichrome; D = Inconel alloy 600; E = Incoloy 800; F = Hastelloy alloy B-2; G = Inconel alloy 718; H = Mar-M200; J = Waspaloy; and K = Udimet 500.

[b]Trademarks—Monel, Inconel, and Incoloy: INCO companies; Hastelloy: Cabot Corporation; Mar-M: Martin Marietta Corporation; Udimet: Special Metals Corporation; and Waspaloy: United Technologies Corporation.

[c]Where two numbers appear, the first refers to the annealed condition, and the second to the condition when maximum strength is achieved by cold working and/or aging. Otherwise, the number refers to the alloy being heat-treated for optimum strength.

[d]To convert MPa to psi, multiply by 145.

Nickel Alloys

Composition and selected properties of nickel alloys are given in Table 1 (p. 781). Nickel-base alloys provide excellent mechanical properties from cryogenic temperatures to temperatures in excess of 1000°C.

Nickel–copper. In the solid state, nickel and copper form a continuous solid solution. The nickel-rich, nickel–copper alloys are characterized by a good compromise of strength and ductility, and are resistant to corrosion and stress corrosion in many environments, in particular, water and seawater, nonoxidizing acids, neutral and alkaline salts, and alkalies. These alloys are weldable and are characterized by elevated and high temperature mechanical properties for certain applications. The copper content in these also ensures improved thermal conductivity for heat exchange (see Heat exchange technology). Monel 400 is a typical nickel-rich, nickel–copper alloy. Coinage nickel is an alloy of 75 wt% Cu and 25 wt% Ni.

Nickel–chromium. Nickel–chromium alloys form a solid solution up to 30 wt% chromium. Chromium is added to nickel to enhance strength, corrosion resistance, oxidation, hot corrosion resistance, and electrical resistance. In combination, these properties result in the nichrome-type alloys used as electrical furnace heating elements.

Nickel–molybdenum. Molybdenum in solid solution with nickel strengthens the nickel and improves its corrosion resistance, eg, the Hastelloy alloys. For example, Hastelloy alloy B-2 is noted for its superior resistance to corrosion by hydrochloric acid at all concentrations up to the boiling point, by other nonoxidizing acids, such as sulfuric and phosphoric, and by hot hydrogen chloride gas.

Nickel–iron–chromium. A large number of industrially important materials are derived from nickel–iron–chromium alloys which are within the broad austenitic, gamma-phase field of the ternary Ni–Fe–Cr phase diagram and are noted for good resistance to corrosion and oxidation and good elevated-temperature strength, eg, Inconel and Incoloy alloys.

Nickel–base superalloys. Superalloys, which are critical to gas-turbine engines because of their high temperature strength and superior creep and stress rupture resistance, are basically nickel–chromium that is alloyed with a host of other elements (see High temperature alloys). The alloying elements include the refractory metals tungsten, molybdenum, or niobium for additional solid-solution strengthening, especially at higher temperatures, and aluminum in appropriate amounts for the precipitation of γ' for coherent particle strengthening. Titanium is added to provide stronger γ', and niobium reacts with nickel in the solid state to precipitate the γ'' phase.

Nickel–iron. A large amount of nickel is used in alloy and stainless steels and in cast irons (see Steel; Iron). Nickel is added to ferritic alloy steels to increase the hardenability and to modify ferrite and cementite properties and morphologies and, thus, to improve the strength, toughness, and ductility of the steel. In austenitic stainless steels, the nickel content is 7–35 wt%. There are also many nickel–iron alloys that have useful magnetic characteristics and are used in a wide range of devices in the electronics and telecommunications fields (see Magnetic materials).

Reserves and Resources

Nickel comprises ca 3% of the earth's composition and is exceeded in abundance by iron, oxygen, silicon, and magnesium. However, although nickel comprises ca 7% of the earth's core, it ranks 24th in order of abundance in the earth's crust, of which it comprises only about 0.009%. Fortunately, ore forms amenable to economic mining exist.

The largest reserves are in Canada (8,600,000 metric tons) and New Caledonia (15,000,000 t), and there are sizable reserves in Indonesia, the Philippines, Australia, and the Dominican Republic (the U.S. has less than 0.4% of the world's estimated reserves).

Nickel Ores

There are two types of nickel ore that can be mined economically and which are classified as sulfide and lateritic. The sulfide deposits currently account for most of the nickel that is produced in the world. The most common nickel sulfide is pentlandite, $Fe_9Ni_9S_{16}$, which is almost always found in association with chalcopyrite, $CuFeS_2$, and large amounts of pyrrhotite, Fe_7S_8. Nickel sulfides were formed far below the earth's

surface by the reaction of sulfur with nickel-bearing rocks. The lateritic ores were formed over long periods of time as a result of weathering of exposed nickel-containing rocks. The lateritic weathering process resulted in nickel solutions that were redeposited elsewhere in the form of oxides or silicates. One type is limonitic or nickeliferous iron laterite which consists primarily of hydrated iron oxide within which the nickel is dispersed in solid solution.

Extraction and Refining

Sulfide ores are initially concentrated by mechanical means, and then are treated by a series of pyrometallurgical processes consisting of roasting, smelting, and converting to produce a copper–nickel matte. The matte is treated to produce a copper–nickel alloy and nickel sulfide which are refined to nickel by electrolysis or the carbonyl process. Lateritic ores may be treated by pyrometallurgical processes followed by reduction or electrolysis, or hydrometallurgical processes involving leaching with ammonia or sulfuric acid may be used.

JOHN K. TIEN
TIMOTHY E. HOWSON
Columbia University

Metals Handbook, 9th ed., Vol. 3, American Society for Metals, Metals Park, Ohio, 1980.

W. Betteridge, *Nickel and Its Alloys*, *Industrial Metals Series*, MacDonald and Evans, Ltd., London, Eng., 1977.

J.R. Boldt, Jr., and P. Queneau, *The Winning of Nickel*, D. Van Nostrand Co., Inc., New York, 1967.

NICKEL COMPOUNDS

Inorganic Compounds

Nickel has a $3d^84s^2$ electronic configuration and forms compounds in which the nickel atom has oxidation states of -1, 0, $+1$, $+2$, $+3$, and $+4$. Ni(II) represents the bulk of all known nickel compounds. Examples of stable crystalline derivatives of Ni(III) and Ni(IV) are the fluoride anions $(NiF_6)^{3-}$ and $(NiF_6)^{2-}$. Examples of the binuclear and diamagnetic species of Ni(I) are the cyanonickelates $K_4[Ni_2(CN)_6]$ and $K_6[Ni_2(CN)_8]$. Nickel(I) dihydrohexacarbonyl, $H_2Ni_2(CO)_6$ is an example of nickel in the -1 oxidation state. Nickel(O) carbonyl, $Ni(CO)_4$, is an example of a large class of Ni(O) compounds.

Nickel Oxides

Properties. Nickel oxide, NiO, is a green cubic crystalline compound, mp 2090°C, density 7.45 g/cm³. The properties of nickel oxide are related to its method of preparation. Green nickel oxide is prepared by firing a mixture of water and high purity nickel powder in air at 1000°C, or by firing a mixture of high purity nickel powder, nickel oxide, and water in air. The latter provides a more rapid reaction than the former method. Single whiskers of green nickel oxide have been made by the closed-tube transport method from oxide powder formed by the decomposition of nickel sulfate using HCl as the transport gas. Green nickel oxide also is formed by thermal decomposition of nickel carbonate or nickel nitrate at 1000°C. Green nickel oxide is an inert and refractory material. Black nickel oxide NiO, is a microcrystalline form resulting from the calcination of nickel carbonate at 600°C. This product typically has more oxygen than its formula indicates, ie, 76–77 wt% nickel compared to green nickel oxide which has 78.5 wt% nickel. The black oxide is chemically reactive and is the chief nickel form used to make simple nickel salts. Black nickel oxide is converted to green nickel oxide at 1000°C.

Manufacture. Several nickel oxides are manufactured commercially. A sintered form of green nickel oxide is made by smelting a purified nickel matte at 1000°C. A powder form of green nickel oxide is made by the desulfurization of nickel matte. Black nickel oxide is made by the calcination of nickel carbonate at 600°C.

Uses. The sintered oxide form is used as charge nickel in the manufacture of alloy steels and stainless steels (see Steel). The oxide furnishes oxygen to the melt for decarburization and slagging.

Nickel Sulfate

Properties. Nickel sulfate hexahydrate, $NiSO_4.6H_2O$, is a monoclinic emerald-green crystalline salt that dissolves easily in water and ethanol. When heated, it loses water and, above 800°C, it decomposes into nickel oxide and SO_3. Its density is 2.03 g/cm^3.

Manufacture. Much nickel sulfate is made commercially by adding nickel powder to hot dilute sulfuric acid. Adding sulfuric acid to nickel powder in hot water enhances a competing reaction whereby the acid is reduced to hydrogen sulfide. Nickel sulfate is also made by the reaction of black nickel oxide with hot dilute sulfuric acid.

Use. The principal use for nickel sulfate is as an electrolyte for the metal-finishing application of nickel in electrorefining and in electroless nickel plating (see Electroless plating).

Nickel Nitrate

Properties. Nickel nitrate hexahydrate, $Ni(NO_3)_2.6H_2O$, is a green monoclinic deliquescent crystal. Mp = 56°C, density = 2.05 g/cm^3. It is extremely soluble in water. Nickel nitrate hexahydrate loses water on heating and eventually decomposes, forming nickel oxide.

Manufacture. Nickel nitrate hexahydrate is made commercially by several methods. Nickel metal reacts vigorously with nitric acid and, if the reaction is not closely controlled, excess heating occurs and causes breakdown of the nitric acid. Nickel powder is added slowly to a stirred mixture of nitric acid and water to yield nickel nitrate. Adding nitric acid to nickel powder results in the formation of nickel ammonium nitrate.

Uses. Nickel nitrate is used as an intermediate in the manufacture of nickel catalysts, and in loading active mass in nickel–cadmium batteries of the sintered-plate type (see Batteries, secondary).

Nickel Halides

Nickel forms anhydrous as well as hydrated halides. Nickel chloride hexahydrate is formed by the reaction of nickel powder or nickel oxide with a hot mixture of water and HCl. Nickel fluoride, $NiF_2.4H_2O$, is prepared by the reaction of hydrofluoric acid on nickel carbonate. Nickel bromide, $NiBr_2.6H_2O$, is made by the reaction of black nickel oxide and HBr. The reaction of HI with nickel carbonate yields nickel iodide, $NiI_2.6H_2O$.

Nickel chloride hexahydrate is an important material in nickel electroplating. It is used with nickel sulfate in the conventional Watts plating bath. Nickel chloride is an intermediate in the manufacture of certain nickel catalysts, and it is used to absorb ammonia in industrial gas masks.

Nickel Carbonate

Nickel carbonate, $NiCO_3$, is a light-green, rhombic crystalline salt that is very slightly soluble in water; density = 2.6 g/cm. The addition of sodium carbonate to a solution of a nickel salt precipitates an impure basic nickel carbonate. The commercial material is the basic salt, $2NiCO_3.3Ni(OH)_2.4H_2O$. It is best prepared by the oxidation of nickel powder in ammonia and CO_2. Boiling away the ammonia causes precipitation of pure nickel carbonate.

Nickel carbonate is used in the manufacture of catalysts, in the preparation of colored glass, in the manufacture of certain nickel pigments, and as a neutralizing compound in nickel electroplating solutions.

Other nickel compounds. Other nickel compounds include nickel hydroxide, nickel fluoroborate, nickel cyanide, nickel sulfamate, nickel sulfide, nickel arsenate, and nickel phosphate.

Organic Compounds

Nickel carbonyl. Nickel carbonyl, $Ni(CO)_4$ is a colorless liquid; bp 42.6°C; crystallization −25°C; density at 17°C 1.3186 g/cm^3; vapor pressure at 0°C 19.2, at 10°C 28.7, and at 20°C 44.0 kPa (144, 215, and 330 mm Hg, respectively); critical t = 200°C. The thermodynamic properties of nickel carbonyl are documented. Vapor density is ca four times that of air. It is miscible in all proportions with most organic solvents and is practically insoluble in water.

Nickel carbonyl can be prepared by the direct combination of carbon monoxide and metallic nickel. The presence of sulfur, the surface area, and the surface activity of the nickel affect the formation of nickel carbonyl.

Uses. Nickel refining (involving nickel carbonyl as an intermediate) capacity at Inco Metals Division of Inco Ltd. is > 1,000,000 t/yr. High purity nickel pellets for melting and dissolving are a product of the carbonyl-refining process. Nickel powders useful in nickel chemical synthesis, and for making nickel alkaline-battery electrodes and powder-metallurgical parts are derived from the carbonyl-refining process.

Health and Safety

Some aqueous solutions of nickel are irritating on contact with the eye, eg, nickel sulfate and nickel chloride. The permissible exposure level in the United States for all forms of nickel and its inorganic compounds is 1 mg/m^3 air.

Uses

Nickel is an important hydrogenation catalyst because of its ability to chemisorb hydrogen, eg, Raney nickel catalyst used widely in laboratory and industrial hydrogenation processes is the most active and least specific of the nickel catalysts. Second largest application is as electrolytes in nickel electroplating (qv). Nickel compounds, eg, black or green nickel oxide, are used extensively in the ceramics industry; to frit compositions used for porcelain enameling of steel; to enhance the adhesion of glass to steel; and in the manufacture of nickel–zinc ferrites (qv) used in electric motors, antennas, and television yokes. Nickel titanate, a chalking yellow pigment, is used extensively in exterior house paint and in vinyl house siding.

D.H. Antonsen
The International Nickel Company, Inc.

J.R. Boldt, Jr. and P. Queneau, *The Winning of Nickel*, Longmans Canada Ltd., Toronto, 1967.

P.W. Jolly and G. Wilke, *The Organic Chemistry of Nickel*, Vols. 1–2, Academic Press, Inc., New York, 1974–1975.

NICKEL SILVER. See Copper alloys.

NICOTINAMIDE, $C_5H_4NCONH_2$. See Vitamins.

NICOTINE, $C_{10}H_{14}N_2$. See Alkaloids; Insect control technology.

NIOBIUM AND NIOBIUM COMPOUNDS

Niobium

Properties. Elemental niobium, Nb, which is also called columbium, Cb, in the United States, has a cosmic abundance of 0.9 relative to silicon $\equiv 10^6$, an average value of 24 ppm in the earth's crust, and a comparable value for the lunar surface. Niobium is a monoisotopic element, although a search for residual radionuclides from the formation of the solar system has established the natural abundance of ^{92}Nb ($t_{1/2}$ 1.7×10^8 yr) to be $1.2 \times 10^{-10}\%$ (see Isotopes). In addition, minute amounts of ^{94}Nb ($t_{1/2}$ 2.03×10^4 yr) and ^{95}Nb ($t_{1/2}$ 35 d) occur in nature: ^{94}Nb from neutron capture by the stable isotope, and ^{95}Nb as the daughter of ^{95}Zr in the fission products of ^{235}U. Niobium-93 has a nuclear spin of 9/2 and a thermal-neutron capture cross section of $1.1 \pm 0.1 \times 10^{-28}$ m^2 (1.1 barns), which makes it of much interest to the nuclear industry.

Niobium is a steel-gray, ductile, refractory metal with a slightly lower melting point than molybdenum, and a lower electron work function than tantalum, tungsten, or molybdenum. Niobium closely resembles tantalum in its properties; the former is only slightly more chemically reactive. The metal is unattacked by most gases below 200°C, but is air oxidized at 350°C with the development of an oxide film of increasing thickness which changes from pale yellow to blue to black at 400°C. Absorption of hydrogen at 250°C and nitrogen at 300°C occurs to form interstitial solid solutions which greatly affect the mechanical properties. Niobium is attacked by fluoride and gaseous hydrogen fluoride and is embrittled by nascent hydrogen at room temperature. It is unaffected by aqua regia and mineral acids at ordinary temperatures, except by hydrofluoric acid, which dissolves it. Niobium is attacked by hot concentrated hydrochloric and sulfuric acids, dissolving at 170°C in concentrated sulfuric acid, and by hot alkali carbonates and hydroxides, which cause embrittling.

The most common oxidation state of niobium is +5, although many anhydrous compounds have been made with lower oxidation states, notably, +4 and +3, and niobium +5 can be reduced in aqueous solution to +4 by zinc. The aqueous chemistry primarily involves halo- and organic acid anionic complexes, but virtually no cationic chemistry because of the irreversible hydrolysis of the cation in dilute solutions. Metal–metal bonding is common with extensive formation of polymeric anions. Niobium resembles tantalum and titanium in its chemistry; thus, its separation from these elements is difficult. In the solid state, niobium has the same atomic radius as tantalum and essentially the same ionic, ie +5, radius as Ta^{5+} (68 pm), Ti^{4+} (68 pm) and Li^{+} (69 pm).

Atomic weight 92.906; mp 2468 ± 10°C; bp 5127°C; density 8.66 g/cm^3 at 20°C; thermal conductivity 52.3 W/m·k at 298 K; coefficient of linear thermal expansion 7.1×10^{-6} per °C (291–373 K); volume electrical conductivity 13.3% IACS (International Annealed Copper Standard); electrical resistivity 13–16×10^{-6} Ω·cm.

Niobium Compounds

Some properties of niobium compounds are listed in Table 1.

Occurrence. Niobium and tantalum usually occur together. Niobium never occurs in the free state; sometimes it occurs as a hydroxide, silicate, or borate; and most often it is combined with oxygen and another metal, forming a niobate or tantalate in which the niobium and tantalum isomorphously replace each other with little change in physical

Table 1. Properties of Some Niobium Compounds

Compound	Lattice	Lattice constant, pm	Density, g/cm^3	Mp, °C	Bp, °C
niobium boride, NbB	orthorhombic	a = 329.8	7.5	2000	
		b = 872.4			
		c = 316.6			
niobium diboride, NbB$_2$	hexagonal	a = 308.9			
		c = 330.3	6.9	3050	
diniobium carbide, Nb$_2$C	hcp	a = 312.7	7.80	3090	
		c = 497.2			
niobium carbide, NbC	fcc	a = 447.1	7.788	3600	4300
niobium pentafluoride, NbF$_5$	monoclinic	a = 963	3.54	79	234
		b = 1443			
		c = 512			
		β = 96.1°			
niobium fluorodioxide, NbO$_2$F	cubic	a = 390.2			
niobium pentachloride, NbCl$_5$	monoclinic	a = 183.0	2.74	208.3	248.2
		b = 1798			
		c = 588.8			
		β = 90.6°			
niobium trichloromonoxide, NbOCl$_3$	tetragonal	a = 1087	3.72	vacuum sublimes at ca 200	
		c = 396			
niobium pentabromide, NbBr$_5$	orthorhombic	a = 612.7	4.36	254	365
		b = 1219.8			
		c = 1855			
niobium tribromomonoxide, NbOBr$_3$				vacuum sublimes at 180	ca 320 dec
niobium pentaiodide, NbI$_5$	monoclinic	a = 1058		ca 200 dec	
		b = 658			
		c = 1388			
		β = 109.1°			
niobium hydride, NbH	bcc		6–6.6		
diniobium nitride, Nb$_2$N	hcp	a = 305.6–304.8	8.08	2050	
		c = 495.6			
niobium nitride, NbN	fcc	a = 438.2–439.2	8.4		
niobium oxide, NbO	cubic	a = 421.08	7.30		
niobium dioxide, NbO$_2$	tetragonal	a = 1371	5.90		
		c = 598.5			
α-niobium pentoxide, α-Nb$_2$O$_5$	monoclinic	a = 2116	4.55	1491 ± 2	
		b = 382.2			
		c = 1935			

properties except density (see Tantalum and tantalum compounds). Ore concentrations of niobium usually occur as carbonatites; they are associated with tantalum in pegmatites and alluvial deposits.

Extraction and refining. The process of extracting and refining niobium consists of a series of consecutive operations, frequently with several steps combined as one; the upgrading of ores by preconcentration; an ore-opening procedure to disrupt the niobium-containing matrix; preparation of a pure niobium compound; reduction to metallic niobium; and refining, consolidation, and fabrication of the metal.

Health and Safety

Toxicity data on niobium and its compounds are sparse. The most common materials, eg, niobium concentrates, ferroniobium, niobium metal, and niobium alloys, appear to be relatively inert biologically. The hydride has moderate fibrogenic and general toxic action with recommended maximum allowable concentrations of 6 mg/m^3.

Uses

Niobium, as ferroniobium, is used extensively in the steel industry as an additive in the manufacture of high strength, low alloy (HSLA) and carbon steels. It acts as a grain refiner to increase yield and tensile strength with additions as low as 0.02 wt%; although normal usage is 0.03–0.1 wt%.

PATRICK H. PAYTON
TRW

G.L. Miller, *Metallurgy of the Rarer Metals–6 Tantalum and Niobium*, Butterworth Scientific Publications, London, UK, 1959.

R.J.H. Clark and D. Brown, *The Chemistry of Vanadium, Niobium and Tantalum*, Pergamon Press, Elmsford, N.Y., 1975.

C.T. Wang, "Composition, Properties and Applications of Niobium and Its Alloys," in *Metals Handbook*, 9th ed., American Society for Metals, 1980.

H. Stuart, ed., "Niobium," *Proceedings of the International Symposium in San Francisco, Calif., Nov. 8–11, 1981*, The Metallurgical Society of AIME, Warrendale, Pa., 1982.

NITRATION

Nitration involves the reaction between an organic compound and a nitrating agent, eg, nitric acid (qv), either to introduce a nitro group into the hydrocarbon or to produce a nitrate. In *C*-nitration, a nitro group, —NO$_2$, is attached to a carbon atom:

$$\diagdown\!\!\diagup\!\!CH + HNO_3 \longrightarrow \diagdown\!\!\diagup\!\!CNO_2 + H_2O$$

O-nitration results in the formation of a nitrate and is regarded as an esterification (qv):

$$\diagdown\!\!\diagup\!\!COH + HNO_3 \longrightarrow \diagdown\!\!\diagup\!\!CONO_2 + H_2O$$

N-nitration is the attachment of a nitro group to a nitrogen atom, eg, in the production of nitramines:

$$\diagdown\!\!NH + HNO_3 \longrightarrow \diagdown\!\!NNO_2 + H_2O$$

Nitrations are highly exothermic, ie, ca 126 kJ/mol (30 kcal/mol). However, the heat of reaction varies with the hydrocarbon that is nitrated. The mechanism of a nitration depends on the reactants and the operating conditions. The reactions usually are either ionic or free radical. Ionic nitrations are used commonly for aromatics; many heterocyclics; hydroxyl compounds, ie, simple alcohols, glycols, glycerol, cellulose, etc; and amines. Nitration of paraffins, cycloparaffins, and olefins frequently involve a free-radical reaction.

Ionic Reactions

Most ionic nitrations involve mixed acids, which nitrate the hydrocarbon. The mixture usually contains nitric acid plus a strong acid, eg, sulfuric acid, perchloric acid, selenic acid, hydrofluoric acid, boron trifluoride, and ion-exchange resins containing sulfonic acid groups. Usually, water also is present in the acid mixture, and some is formed during nitration. Industrially, sulfuric acid is the most frequently used acid because it is highly effective and the least expensive.

Processes for ionic nitrations. Nitrations of aromatic hydrocarbons, glycerol, and many other organic materials are hazardous and require careful control of the kinetics and temperature. If the temperature becomes too high, the rates of reaction and the degree of exothermicity for the reaction may become too high to maintain adequate temperature control. In such a case, runaway reactions, excessive side reactions, or explosions may occur.

Ideally, it is desirable to contact the aromatic hydrocarbon feed with dilute or used mixed acids and to contact the partially nitrated hydrocarbon with the stronger feed or acids. If a continuous flow process is used, then the organic and acid phases should flow countercurrent to each other. In batch systems, nitrations often are conducted in stages. When nitration is partially completed, the phases are separated and a stronger acid is used for further nitrations.

Free-Radical Reactions

The chemistry of free-radical nitrations is complicated since both nitration and oxidation reactions are involved. A chain mechanism frequently occurs in nitrations with nitric acid.

Vapor phase. Vapor-phase nitrations of paraffins (and especially propane) are performed at sufficiently high temperatures and low pressures to obtain gaseous mixtures of hydrocarbons, nitrating agents, products, and inert vapors. When nitric acid is used as the nitrating agent, temperatures vary from ca 350 to 450°C. In addition to the replacement of an H atom with a nitro group, C—C bonds are broken so that nitroparaffins with fewer carbons than the original paraffin are formed.

Considerable amounts of by-products are produced in addition to the nitration products. Nitric acid is an oxidizing agent and produces aldehydes, particularly formaldehyde, carbon monoxide, carbon dioxide, water, lighter paraffins, olefins, and small amounts of alcohols and ketones.

Liquid-phase nitrations of paraffins. Liquid-phase nitrations of paraffins occur predominantly by free-radical reactions. Highly ionized mixed acids are ineffective with relatively nonpolar hydrocarbons; however, nitric acid that is unmixed with any other acid is an effective nitrating agent. Temperatures required for the reaction are ca 100–200°C, and sufficient pressure is required to maintain most, if not all, of the mixture of reactants and products of each of the two phases which normally are present in the liquid state.

Health and Safety

Nitrohydrocarbons generally are hazardous; some are explosive, most are highly flammable, and many are toxic. Dangers of explosion usually increase with the degree of nitration, eg, trinitroaromatics are much more hazardous than mononitroaromatics. Some of the most explosive compounds are TNT, picric acid, glycerol trinitrate (nitroglycerin), and highly nitrated cellulose, which often is used as gunpowder.

LYLE F. ALBRIGHT
Purdue University

J.F. Haggett, R.B. Moodie, J.R. Penton, and K. Schofield, *Nitration and Aromatic Reactivity*, Cambridge University Press, 1971.

G.A. Olah, *ACS Symposium Series #22*, Washington, D.C., 1976, Chapt. 1, pp. 1–47.

K.L. Nelson and H.C. Brown in B.T. Brooks and co-eds., *The Chemistry of Petroleum Hydrocarbons*, Vol. 3, Reinhold Publishing Corporation, New York, 1955, p. 465.

NITRIC ACID

Nitric acid, HNO_3, became an important industrial chemical with the development of the explosives and dyestuffs industries at the end of the nineteenth century.

Physical Properties

Nitric acid is extremely difficult to prepare as a pure liquid because of its tendency to decompose and, thereupon, release nitrogen oxides. When produced by vacuum distillation of a mixture of sodium nitrate and concentrated sulfuric acid with condensation of the liquid at just above its freezing point, a colorless liquid (fp = $-41.59°C$) can be collected. Crystals of the pure acid are quite stable, but the liquid degenerates to a limited extent at any temperature above the melting point, and turns yellow within an hour at room temperature.

Nitric acid is completely miscible with water and generally is known and used as an aqueous solution and sometimes with the addition of dissolved nitrogen oxides at high concentrations. Two hydrates may be crystallized from acid solutions, ie, a monohydrate ($HNO_3.H_2O$), corresponding to 77.77 wt% acid (mp $-37.62°C$) and a trihydrate $HNO_3.3H_2O$, corresponding to 53.83 wt% (mp $-18.47°C$). 100% nitric acid has a density of 1.5129 g/cm^3 at 20°C; fp $-42°C$; bp 86.0°C; specific heat is 1.76 J/(g·K) 0.42 cal(g·K) at 20°C; viscosity, 0.9 mPa·s (= cP) at 20°C; and thermal conductivity, 0.28 W/(m·K) at 20°C.

Fuming nitric acid is concentrated nitric acid that contains dissolved nitrogen dioxide. The density and vapor pressure of such solutions increase with the percentage of NO_2 present. Acid containing ca 45 wt% NO_2 and 55% HNO_3 has a vapor pressure of 101 kPa (760 mm Hg) at 25°C and a density of 1.64 g/cm^3.

Chemical Properties

Nitric acid is a strong, monobasic acid. It reacts readily with alkalies, oxides, and basic materials with the formation of salts. The reaction with ammonia, forming ammonium nitrate for use as a fertilizer, is by far the largest single industrial outlet for nitric acid.

Nitric acid is a strong oxidant. Organic material, eg, turpentine, charcoal, and charred sawdust, are violently oxidized and alcohol may react explosively with concentrated nitric acid. Furfuryl alcohol, aniline, and other organic chemicals have been used with nitric acid in rocket fuels (see Explosives and propellants). Most metals, except platinum metals and gold, are attacked by nitric acid; some are converted into oxides, eg, arsenic, antimony, and tin, but most others are converted into nitrates.

The activity of nitric acid as an oxidizing agent apparently is dependent upon the presence of free nitrogen oxides. Pure nitric acid does not attack copper, but when oxides of nitrogen are introduced, the reaction is at first slow and then proceeds rapidly and violently. Nitric acid also undergoes reactions with organic compounds where the acid serves neither as an oxidizing agent nor as a source of hydrogen ions. The formation of organic nitrates by esterifications, ie, O-nitration, involves reaction with the hydroxyl group:

$$ROH + HONO_2 \rightarrow RONO_2 + H_2O$$

Manufacture and Processing

Virtually all of the nitric acid that is commercially manufactured is obtained by the ammonia oxidation process. Despite the many variations in operating details among the plants producing nitric acid, three essential steps are common: oxidation of ammonia to nitric oxide, NO; oxidation of the nitric oxide to the dioxide, NO_2; and absorption of nitrogen oxides in water to produce nitric acid with the release of additional nitric oxide.

Oxidation of ammonia. Ammonia is oxidized with an excess of oxygen over a catalyst to form nitric oxide and water:

$$NH_3 + 1.25 O_2 \rightarrow NO + 1.5 H_2O$$

$$\Delta H_{298 K} = -226 \text{ kJ/mol} (-54.1 \text{ kcal/mol})$$

The reaction is extremely rapid and goes almost to completion. The principal competing reaction yields nitrogen.

Oxidation of nitric oxide. Nitric oxide undergoes a slow homogeneous reaction with oxygen to yield nitrogen dioxide:

$$2 NO + O_2 \rightleftharpoons 2 NO_2 \qquad \Delta H_{298 K} = -114 \text{ kJ/mol} (-27.3 \text{ kcal/mol})$$

The equilibrium constant for the reaction strongly favors the production of NO_2 at lower temperatures, so that below 150°C almost all nitric oxide combines with any oxygen that is present if sufficient residence time is allowed. Nitrogen dioxide dimerizes almost instantaneously to an equilibrium mixture with dinitrogen tetroxide:

$$2 NO_2 \rightleftharpoons N_2O_4 \qquad \Delta H_{298 K} = -57.4 \text{ kJ/mol} (-13.7 \text{ kcal/mol})$$

Lower temperatures and increasing pressures shift the reaction to the production of tetroxide.

Absorption of nitrogen oxide. In the absorption of nitrogen oxides in water, there are uncertainties about the reaction mechanism, and complexities resulting from mass diffusion in a vapor and in a liquid phase are involved. The overall reaction usually is shown as if only the nitrogen dioxide that is present in the gas reacts with liquid water:

$$3 NO_2(g) + H_2O(l) \rightleftharpoons 2 HNO_3(aq) + NO(g)$$

$$\Delta H_{298 K} = 135.6 \text{ kJ/mol} (-32.4 \text{ kcal/mol})$$

Latest process developments. Changes that have taken place in nitric acid technology in recent years have resulted from governmental restrictions on gaseous emissions and from the skyrocketing increase in the cost of fuels in the 1970s. In 1971, New Source Performance Standards for nitric plants were promulgated by the EPA and resulted in limiting emissions to 1.5 kg of equivalent nitrogen dioxide per metric ton of equivalent 100 wt% nitric acid produced. This quantity corresponds to a concentration of about 230 ppmv of nitrogen oxides in the vented tail gas or 5–10% of the amount typically discharged in earlier facilities.

Nitric acid concentration. Nitric acid is produced by the standard ammonia-oxidation processes as an aqueous solution at a concentration of 50–70 wt%. Such concentrations are suitable for the production of ammonium nitrate but, for use in organic nitrations, anhydrous nitric acid is required. Since nitric acid forms an azeotrope with water at 68.8 wt%, the water cannot be separated from the acid by simple distillation. Two industrial methods for accomplishing the concentration are extractive distillation and reaction with additional nitrogen oxides. The latter are the direct strong nitric (DSN) processes.

Health and Safety

Nitric acid vapors and nitrous oxide fumes or oxides of nitrogen (nitric oxide and nitrogen dioxide) are highly toxic and capable of producing severe injury or death if improperly handled. The extent of injury, signs and symptoms of poisoning, and the nature of the treatment required depend on the concentration of several toxic substances, the time of exposure, and the susceptibility of the individual. The liquid form of the acid is very corrosive and can destroy the skin, respiratory mucosa, and gastrointestinal tissue.

Uses

The largest use of nitric acid has continued to be in the production of ammonium nitrate for use as fertilizer, at first in the form of solid ammonium nitrate granules but increasingly mixed with excess ammonia and/or urea and shipped as aqueous nitrogen solutions for direct application (see Ammonium compounds; see Fertilizers). In the mid-1950s, the use of ammonium nitrate prills mixed with fuel oil was accepted for direct use as an explosive (see Explosives). Nitric acid is used in the manufacture of cyclohexanone (see Cyclohexanol and cyclohexanone), the raw material for adipic acid (qv) and caprolactam, which are monomers used in producing nylon (see Polyamides).

DANIEL J. NEWMAN
Barnard and Burk, Inc.

T.H. Chilton, *Chem. Eng. Prog. Monogr. Ser. #3*, AIChE, New York, 1960.
T.H. Chilton, *Strong Water*, The M.I.T. Press, Cambridge, Mass., 1968.
D.J. Newman and L.A. Klein, *Chem. Eng. Prog.* **68**, 62 (Apr. 1972).

NITRIDES

At elevated temperatures and pressures, nitrogen combines with most of the elements to form nitrogen compounds; with metals and semimetals, it forms nitrides. Atomic nitrogen reacts much more readily with the elements than molecular nitrogen.

Properties

Saltlike nitrides. The nitrides of the electropositive metals of Groups IA, IIA, and IIIB form saltlike nitrides with predominantly heteropolar (ionic) bonding and are regarded as derivatives of ammonia. The composition of these nitrides is determined by the valency of the metal, eg, Li_3N, Ca_3N_2, and ScN. The thermodynamic stability of the saltlike nitrides increases with increasing group number; for example, the nitrides of the alkali metals are only marginally or not at all stable, whereas the rare-earth metals are very effective nitrogen scavengers in metals and alloys. The saltlike nitrides generally are electrical insulators or ionic conductors, eg, Li_3N. The nitrides of the group IIIB metals are metallic conductors or at least semiconductors and, thus, represent a transition to the metallic nitrides. The saltlike nitrides are characterized by their sensitivity to hydrolysis; they react readily with water or moisture and give ammonia and the metal oxides or hydroxides.

Metallic nitrides. Properties of metallic nitrides are listed in Table 1. The nitrides of the transition metals of groups IVB–VIB generally are termed metallic nitrides because of their metallic conductivity and luster and their general metallic behavior. They are characterized by a wide range of homogeneity, high hardness, high melting points, and good corrosion resistance. They are grouped with the carbides (qv), borides (see Boron compounds), and silicides (see Silicon and silicides) as refractory hard metals. They crystallize in highly symmetrical, metal-like lattices; the small nitrogen atoms occupy the interstitial voids within the metallic host lattice forming interstitial alloys similar to the generally isotypic carbides (qv). Metallic nitrides can be alloyed with other nitrides and carbides of the transition metals to give solid solutions. Complete solubility has been demonstrated for a great number of combinations. Properties of metallic nitrides are listed in Table 1.

Metallic nitrides are wetted and dissolved by many liquid metals and can be precipitated from metal baths. The stoichiometry is determined not by the valency of the metal, but by the number of interstitial voids per host atom. The metallic nitrides are very stable against water and all nonoxidizing acids except hydrofluoric acid. The thermodynamic stability decreases with increasing group number from the nitrides of the IVB metals. The nitrides of Mo and W can be prepared only by the action of

nitrogen under high pressure or by reaction with atomic nitrogen or dissociating ammonia.

Nonmetallic (diamondlike) nitrides. Some properties of nonmetallic nitrides are listed in Table 2. The nitrides of some elements of Groups IIIA and IVA eg, BN, Si_3N_4, AlN, GaN, and InN, are characterized by predominantly covalent bonding. They are very stable chemically, have high degrees of hardness, eg, cubic BN, high melting points, and are nonconductive or semiconductive. The structural elements of diamondlike nitrides are tetrahedral, consisting of four metal atoms surrounding one N atom, M_4N, which are structurally related to diamond. Although the most common graphitelike form of BN does not contain these structural elements, boron nitride is considered a diamondlike nitride because of the existence of a diamondlike form at high pressures and its chemical and physical behavior (see Carbon, diamond).

Volatile nitrides. The nitrogen compounds of the nonmetallic elements generally are not very stable; they decompose at elevated temperatures. Some are explosive and decompose upon shock. They form distinct molecules similar to organic compounds, and they behave like organic molecules at low temperatures; they are gaseous, liquid, or easily volatile solids. $(SN)_x$ is polymeric and chemically stable with semimetallic properties. $(PNCl_2)_x$ has attracted some scientific interest as inorganic rubber (see Inorganic high polymers).

Manufacture

Nitride layers. Carbide tips coated with titanium nitride or titanium carbonitride usually are manufactured by the CVD process in equipment like that illustrated in Figure 1.

Silicon nitride. Silicon nitride is manufactured either as a powder as a premateral for the production of hot-pressed parts or as self-bonded, reaction-sintered silicon nitride parts. α-Silicon nitride, which is used in the manufacture of Si_3N_4 intended for hot-pressing, can be obtained by nitriding Si powder in an atmosphere of N_2, H_2, and NH_3. Reaction conditions, eg, temperature, time, and atmosphere, have to be controlled closely.

Health and Safety

The saltlike nitrides decompose with water or moisture with resulting formation of ammonia, which can irritate the respiratory organs. The metallic nitrides are very stable chemically. Very fine powder or dust of the nitrides of the transition metals might be pyrophoric, especially the nitrides of the actinide metals, UN, ThN, and PuN, which ignite in air upon shock and during comminutive operations. The nitrides of the actinide metals are carcinogenic.

Table 1. Properties of Metallic Nitrides

Nitride	Color	Structure	Lattice parameter, RT, nm	Density, g/cm³	Hardness[a]	Mp°C,	Heat conductivity, W/(m·K)	Coefficient of thermal expansion, × 10⁻⁶	Electrical resistivity, μΩ·cm	Transition temperature, K
TiN	golden yellow	fcc NaCl type	0.4246	5.43	HM 2000	2950	29.1	9.35	25	4.8
ZrN	pale yellow	fcc NaCl type	0.4577	7.3	HM 1520	2980	10.9	7.24	21	9
HfN	greenish yellow	fcc NaCl type	0.4518	14.0	HM 1640	3330	11.1	6.9	33	
VN	brown	fcc NaCl type	0.4139	6.10	HM 1500	2350	11.3	8.1	85	7.5
NbN	dark gray	fcc NaCl type	0.4388	8.47	HM 1400	2630 dec	3.8	10.1	78	15.2
ϵ-TaN	dark gray	hexagonal B 35	a: 0.5191	14.3	HM 1100	2950 dec	9.54		128	1.8
			c: 0.2906							
δ-TaN	yellowish gray	fcc NaCl type	0.4336	15.6	HM 3200	2950 dec				17.8
CrN	gray	fcc NaCl type	0.4150	6.14	HM 1090	1080 dec[b]	11.7		640	not superconductive
Mo_2N	gray	fcc	0.416	9.46	HM 1700	790 dec[c]		6.7		5.0
W_2N	gray	fcc	0.412	17.7		dec				
ThN	gray	fcc NaCl type	0.5159	11.9	HM 600	2820			20	
UN	dark gray	fcc NaCl type	0.4890	14.4	HK 580	2800	15.5	8.0	176	
PuN	dark gray	fcc NaCl type	0.4907	14.4		2550				

[a] HM = microhardness; HK = Knoop hardness.
[b] At 0.1 MPa (1 bar).
[c] At 0.7 MPa (7 bar).

Table 2.   Properties of Nonmetallic (Diamondlike) Nitrides

Nitride	Structure	Lattice parameter, RT, nm	Density, g/cm³	Hardness	Temperature stability, up to °C	Heat conductivity, W/(m·K)	Coefficient of thermal expansion, $\beta \times 10^{-6}$
BN	hex	a: 0.2504 c: 0.6661	2.3	like graphite	3000	15	7.51
	fcc Zn blende	a: 0.3615	3.4	approaching diamond			
AlN	hex wurtzite	a: 0.311 c: 0.4975	3.05	HM 1230[a]	2200	30	4.03
GaN	hex wurtzite	a: 0.319 c: 0.518	5.0		600[b]		
Si₃N₄	hex α	a: 0.7748 c: 0.5618	3.2	HM 3340[a]	1900	17	2.75
	hex β	a: 0.7608 c: 0.2911					

[a] HM = microhardness.
[b] *In vacuo.*

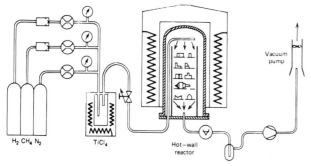

Figure 1.   Typical device for coating of cemented carbides with TiN or Ti(C,N) via a CVD-process. Courtesy of Berna-Bernex AG, Olten, Switzerland.

Uses

Metallic nitrides and carbonitrides were recently fashioned as sintered compositions especially effective in sealing rings for extremely difficult chemical environments. Of the saltlike nitrides, Li₃N has attracted technical interest for its uncommonly high ionic conductivity in the solid state making it a possible suitable electrolyte for lithium–sulfur batteries (see Batteries and electric cells). Hexagonal boron nitride is used in the refractories industry as a mold-facing and release agent. Silicon nitride has found application for runners in the cast iron industry and for cutting tools particularly suited for cutting superalloys.

F. BENESOVSKY‡
Metallwerk Plansee, Reutte, Austria

R. KIEFFER‡
P. ETTMAYER
Technical University, Vienna

P. Schwarzkopf and R. Kieffer, *Refractory Hard Materials*, Macmillan, New York, 1953.

H. Goldschmidt, *Interstitial Alloys*, Butterworths, London, UK, 1967.

L.E. Toth, *Transition Metal Carbides and Nitrides*, Academic Press, Inc., New York, 1971.

NITRIDING.   See Metal-surface treatments.

‡Deceased.

NITRILE RUBBER.   See Elastomers, synthetic.

NITRILES

Nitriles are cyano-substituted organic compounds, ie, organic derivatives of hydrogen cyanide.

Reactions

Nitriles are exceedingly versatile reactants that can be used to prepare amines, amides, amidines, carboxylic acids and esters, aldehydes, ketones, large-ring cyclic ketones, imines, heterocycles, or thioesters, and other compounds. Amides and acids are produced by hydrolysis under either acidic or basic conditions:

$$RCN + H_2O \rightarrow RCONH_2$$
$$RCN + 2\,H_2O \rightarrow RCOOH + NH_3$$

If the water is replaced with alcohols, either esters or substituted amides are obtained. Anhydrides, carboxylic acids, and olefins also react with nitriles to produce substituted amides.

Acetone Cyanohydrin

Acetone cyanohydrin 2-hydroxy-2-methylpropanenitrile, 2-methyllactonitrile, $(CH_3)_2C(OH)CN$, is a colorless liquid, completely miscible in water and most organic solvents but having low solubility in aliphatic hydrocarbons and carbon disulfide. Mol wt 85.10; bp 81.7°C; mp 19°C; density 0.932 g/cm³ at 19°C; n_D^{20} 1.3996. Commercial acetone cyanohydrin is stabilized with sulfuric acid because the commercial product decomposes to hydrogen cyanide and acetone under basic conditions.

Acetone cyanohydrin is produced by a fast, exothermic reaction of acetone with hydrogen cyanide. Cooling is required to maintain the desired reaction temperature of ≤ 25°C.

Uses.   Acetone cyanohydrin is mainly used as an intermediate in the manufacture of methyl methacrylate, but is also used for the manufacture of insecticides, pharmaceuticals, foaming agents, and polymerization initiators. It can be used as a source of hydrogen cyanide for producing cyanohydrins of other ketones, ie, in transcyanohydrination.

Health and safety.   Oral LD_{50} (rats) 17 mg/kg; skin LD_{50} (rabbits) 17 mg/kg; and inhalation LC_{Lo} (rats) 63 ppm/4 h. Effects of acetone cyanohydrin overexposure in humans are similar to those for hydrogen cyanide; that is, it is highly toxic by inhalation, skin contact, and ingestion. Exposure to high concentrations of the vapor may result in instantaneous loss of consciousness and death.

Acetonitrile

Acetonitrile (ethanenitrile), CH_3CN, is a colorless liquid with a sweet, ethereal odor. Mol wt 41.05; bp 81.6°C; fp −45.7°C; density 0.786 g/cm³ at 20°C; n_D^{20} 0.35; coefficient of expansion at 20°C per °C 1.37×10^{-3}; viscosity 0.35 mPa·s (= cP) at 20°C. It is completely miscible with water and its high dielectric strength (38.8 at 20°C) and dipole moment (10.675×10^{-30} C·m) make it an excellent solvent for both inorganic and organic compounds, including polymers.

Although acetonitrile is one of the more stable nitriles, it undergoes typical nitrile reactions and is used to produce many types of nitrogen-containing compounds, eg, amides; amines; higher molecular weight mono- and dinitriles; halogenated nitriles; ketones; isocyanates; and heterocycles, eg, pyridines and imidazolines.

Uses. Because of its good solvency and relatively low boiling point, acetonitrile is widely used as a recoverable reaction medium, particularly for the preparation of pharmaceuticals. Its largest use is for the separation of butadiene from C_4 hydrocarbons by extractive distillation (see Azeotropic and extractive distillation).

Health and safety. Oral LD_{50} (rats) 3030–6500 mg/kg; skin LD_{50} (rabbits) 3884–7850 mg/kg; inhalation LC_{50} (rats) 7,500–17,000 ppm.

Adiponitrile

Adiponitrile (hexanedinitrile), $NC(CH_2)_4CN$, is manufactured principally for use as an intermediate for hexamethylenediamine (1,6-diaminohexane) which is a principal ingredient for nylon-6,6.

Mol wt 108.14; bp 295°C at 101.7 kPa; fp 2.49°C; density 0.965 g/cm³ at 20°C; n_D^{20} 1.4343.

Adiponitrile undergoes the typical nitrile reactions, eg, hydrolysis to adipamide and adipic acid and alcoholysis to substituted amides and esters. The most important industrial reaction is catalytic hydrogenation

$$NC(CH_2)_4CN \xrightarrow{(H)} H_2N(CH_2)_6NH_2$$

A variety of catalysts is used for this reduction including cobalt–nickel, cobalt–manganese, cobalt boride, copper–cobalt, and iron oxide.

Adiponitrile is made commercially by three different processes: dehydration of cyclohexane-based adipic acid; electrodimerization of acrylonitrile obtained by the ammoxidation of propylene; and the addition of HCN to both double bonds of butadiene.

Health and Safety. The following toxicities for adiponitrile have been reported: oral LD_{50} (rats), 300 mg/kg; oral approx lethal dose, 450 mg/kg; skin approx lethal dose (rats) 2000 mg/kg; and inhalation approx lethal concentration (rats) 3.0 mg/kg/0.25 h. Inhalation of adiponitrile may cause nausea, vomiting, irritation of mucosal membranes, and dizziness.

Use. The principal use is for hydrogenation to hexamethylenediamine.

Benzonitrile

Benzonitrile, C_6H_5CN, is a colorless liquid with a characteristic almondlike odor. Mol wt 103.12; bp 191.1°C; fp −12.8°C; density 1.3441 g/cm³ at 20°C; n_D^{20} 1.5289; viscosity 1.24 mPa·s (= cP). It is miscible with acetone, benzene, chloroform, ethyl acetate, ethylene chloride, and other common organic solvents, but immiscible with water at ambient temperatures and soluble to ca 1 wt% at 100°C.

It is produced commercially and in high yield by the vapor-phase catalytic ammoxidation of toluenes:

$$C_6H_5CH_3 + NH_3 + 1.50\,O_2 \xrightarrow{cat} C_6H_5CN + 3\,H_2O$$

Health and safety. Benzonitrile is a very toxic organic cyanide. Toxicity symptoms resulting from industrial accidents include shortness of breath, unconsciousness, and toxic spasms of face and arm muscles.

Uses. Benzonitrile's most important commercial use is in the synthesis of benzoguanamine, which is a derivative of melamine and is used in protective coating and molding resins (see Amino resins, cyanamides).

ROBERT A. SMILEY
E.I. du Pont de Nemours & Co., Inc.

D.H.R. Barton and W.D. Ollis, *Comprehensive Organic Chemistry*, Vol. 2, Pergamon Press, Oxford, UK, 1979, pp. 528–562.

Z. Rappaport, ed., *The Chemistry of the Cyano Group*, Interscience Publishers, London, UK, 1970.

NITRO ALCOHOLS

A nitro alcohol is formed when an aliphatic nitro compound with a hydrogen atom on the carbon that bears the nitro group reacts with an aldehyde in the presence of a base (see Alcohols). Many such components have been synthesized, but only those formed by the condensation of formaldehyde and the lower nitroparaffins are marketed commercially (see Nitroparaffins).

Physical Properties

Tris(hydroxymethyl)nitromethane (Tris Nitro) has mol wt 1151.12; mp 172°C; and solubility 220 g/mL of water at 20°C. 2-Nitro-2-methyl 1-propanol (NMP) has mol wt 119.12; mp 90°C; and solubility 350 g/mL water at 20°C. 2-Nitro-1-butanol (NB) has mol wt 119.12; mp −47 to −49°C; bp 105°C; soluble to the extent of 54 g/100 mL in water at 20°C. 2-Methyl-2-nitro-1,3-propanediol (NMPD) has mol wt 135.12; mp ca 160°C (dec); and is soluble to the extent of 80 g/100 mL water at 20°C. 2-Ethyl-2-nitro-1,3-propanediol (NEPD) has mol wt 149.15; mp 56°C (dec); and is soluble to the extent of 400 g/mL water at 20°C. When pure, these nitro alcohols are white crystalline solids except for nitrobutanol. They are thermally unstable about 100°C and purification by distillation is a hazardous procedure.

Chemical Properties

The nitro alcohols can be reduced to the corresponding alkanolamines (qv). Commercially, reduction is accomplished by hydrogenation of the nitro alcohol in methanol in the presence of Raney nickel. Convenient operating conditions are 30°C and 6900 kPa (1000 psi). Production of alkanolamines constitutes the single largest use of nitro alcohols.

Nitro alcohols form salts upon mild treatment with alkalies. Acidification causes separation of the nitro group as N_2O from the parent compound and results in the formation of carbonyl alcohols, ie, hydroxy aldehydes from primary nitro alcohols, and α-ketols from secondary nitro alcohols.

The hydroxyl groups in polyhydric nitro alcohols readily react with aldehydes to form cyclic acetals.

Chloromethyl ethers may be obtained by the reaction of nitro alcohols with formaldehyde and hydrogen chloride.

Nitro alcohols react with amines to form nitroamines:

$$CH_3\underset{\underset{NO_2}{|}}{\overset{\overset{CH_3}{|}}{C}}CH_2OH + HNRR' \longrightarrow CH_3\underset{\underset{NO_2}{|}}{\overset{\overset{CH_3}{|}}{C}}CH_2NRR' + H_2O$$

Such a reaction can be carried out with a wide variety of primary and secondary amines, both aliphatic and aromatic; a basic catalyst is required if aromatic amines are involved. The products of reactions between dihydric nitro alcohols and amines are nitro diamines, many of which are good fungicides.

The nitrate esters of the nitro alcohols are obtained easily by treatment with nitric acid. The resulting products have explosive properties but are not used commercially.

On dehydration, nitro alcohols yield nitroolefins.

Manufacture

The nitro alcohols that are available in commercial quantities are manufactured by the condensation of nitroparaffins with formaldehyde. These condensations are equilibrium reactions, and potential exists for the formation of polymeric materials. Therefore, reaction conditions such as reaction time, temperature, mol ratio of the reactants, catalyst level, and catalyst removal must be carefully controlled in order to obtain the desired nitro alcohol in good yield. After completion of the reaction, the reaction mixture must be made acidic, either by addition of mineral acid or by removal of base by an ion-exchange resin in order to prevent reversal of the reaction during the isolation of the nitro alcohol (see Ion exchange).

2-Nitro-1-butanol is available as a developmental chemical. Hydroxymethyl-2-nitro-1,3-propanediol is available as the Tris Nitro brand of tris(hydroxymethyl)nitromethane either as a crystalline material or as a 50 wt% aqueous concentrate.

Health and Safety

Because of their low volatility, the nitro alcohols present no vapor inhalation hazard.

Uses

Nitro alcohols are used as intermediates for chemical synthesis, eg, to introduce a nitro functionality and, by reduction of the resultant intermediates, as an amino functionality. Antimicrobial uses of Tris Nitro and NMP in polymers (as sources of formaldehyde for cross-linking) are the uses of greatest importance.

Robert H. Dewey
International Minerals and Chemical Corporation

Allen F. Bollmeier, Jr.
ANGUS Chemical Company

H.B. Hass and E.F. Riley, *Chem. Rev.* **32**, 373 (1943).

ANGUS Technical Data Sheet No. 15, ANGUS Chemical Company, Northbrook, Ill., 1982.

W.E. Noland in H.E. Baumgarten, ed., *Organic Syntheses*, Coll. Vol. 5, John Wiley & Sons, Inc., New York, 1973, pp. 833–838.

NITROBENZENE AND NITROTOLUENES

Nitrobenzene

Nitrobenzene (oil of mirbane), $C_6H_5NO_2$, is a pale yellow liquid with an odor that resembles bitter almonds. Depending upon the compound's purity, its color varies from pale yellow to yellowish brown.

Physical properties. Nitrobenzene is soluble readily in most organic solvents and is miscible in all proportions with diethyl ether and benzene. It is only slightly soluble in water with a solubility of 0.19 parts per 100 parts water at 20°C and 0.8 pph at 80°C. Nitrobenzene is a good organic solvent and, because aluminum chloride dissolves in it, it is used as a solvent in Friedel-Crafts reactions.

Mp 5.85°C; bp 210.9°C; density 1.199 g/cm³; refractive index n_D^{20} 1.55296; viscosity at 15°C 2.17×10^{-2} mPa·s (= cP); dielectric constant at 25°C 34.89.

Chemical properties. Nitrobenzene reactions involve substitution in the aromatic ring and those involving the nitro group. Under electrophilic conditions, the substitution occurs at a slower rate than for benzene, and the nitro group promotes meta substitution. Nitrobenzene can undergo halogenation, sulfonation, and nitration, but it does not undergo Friedel-Crafts reactions. Under nucleophilic conditions, the nitro group promotes ortho and para substitution. The reduction of the nitro group to yield aniline is the most commercially important reaction of nitrobenzene. Usually, the reaction is carried out by the catalytic hydrogenation of nitrobenzene, either in the gas phase or in solution, or by using iron borings and dilute hydrochloric acid, eg, in the Béchamp process. Depending on the conditions, the reduction of nitrobenzene can lead to a variety of products. The series of reduction products is shown in Figure 1 (see Amines by reduction).

Manufacture. Nitrobenzene is manufactured commercially by the direct nitration of benzene using a mixture of nitric and sulfuric acids, which commonly is referred to as mixed acid or nitrating acid. Since two phases are formed in the reaction mixture, and the reactants are distributed between them, the rate of nitration is controlled by mass transfer between the phases as well as by chemical kinetics.

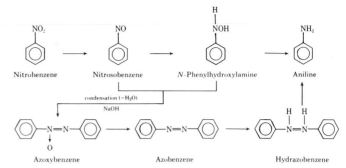

Figure 1. Reduction products of nitrobenzene.

Nitrobenzene can be produced by either a batch or continuous process. With the typical batch process, the reactor is charged with benzene, and then the nitrating acid (56–60 wt% H_2SO_4, 27–32 wt% HNO_3, and 8–17 wt% H_2O) is added slowly below the surface of the benzene. The temperature of the mixture is maintained at 50–55°C by adjusting the feed rate of the mixed acid and the amount of cooling. The temperature can be raised to ca 90°C toward the end of the reaction to promote completion of reaction.

Since a continuous nitration process generally offers lower capital costs and more efficient labor usage than a batch process, it is used by most commercial nitrobenzene producers. The basic sequence of operation is the same as for the batch process, but for a given rate of production the size of the nitrators is much smaller in a continuous process. A 0.144-m³ (30-gal) continuous nitrator has roughly the same production capacity as a 5.68-m³ (1500-gal) batch reactor. In contrast to the batch process, a continuous process typically utilizes a lower nitric acid concentration and, because of the rapid and efficient mixing in the smaller reactors, higher reaction rates are observed.

Health and safety. Nitrobenzene is a very toxic substance; the maximum allowable concentration is 1 ppm or 5 mg/m³. It is readily absorbed by contact with the skin and by inhalation of the vapor. The primary effect of nitrobenzene is the conversion of hemoglobin to methemoglobin, thus eliminating the hemoglobin from the oxygen-transport cycle. Exposure may also irritate the eyes and skin, produce fatigue, headache, vertigo, vomiting, general weakness, and in some cases unconsciousness and coma. There generally is a latent period of 1–4 h before signs or symptoms appear.

Uses.   The most significant use of nitrobenzene is in the manufacture of aniline (see Amines, aromatic). Approximately 97–98 wt% of nitrobenzene produced in the United States is converted to aniline.

Nitrotoluenes

Mononitrotoluenes

The mononitration of toluene results in the formation of a mixture of the ortho, meta, and para isomers of nitrotoluene, $O_2NC_6H_4CH_3$. The presence of the methyl group on the aromatic ring facilitates the nitration, but it also increases the ease of oxidation.

Properties of *o*-nitrotoluene.   *o*-Nitrotoluene is a clear yellow liquid; the solid is dimorphous and the melting points of the α and β forms are −9.55°C and −3.85°C, respectively. *o*-Nitrotoluene is infinitely soluble in benzene, diethyl ether, and ethanol. It is soluble in most organic solvents and only slightly soluble in water, 0.65 g/100 g of water at 30°C. Mp −9.55°C [α form]; −3.85°C [β (stable) form]; d_4^{20} 1.163 g/cm³; refractive index n_D^{20} 1.574. The strong electron-acceptor action of the nitro group in *o*-nitrotoluene confers increased reactivity on the methyl groups; thus, the methyl group is easily oxidized. Oxidation with potassium permanganate or potassium dichromate causes the formation of *o*-nitrobenzoic acid. When boiled with a sodium hydroxide solution, *o*-nitrotoluene exhibits the phenomena of autooxidation and reduction and yields anthranilic acid. When the oxidation is carried out with manganese dioxide and sulfuric acid, *o*-nitrobenzoic acid or *o*-nitrobenzaldehyde is formed, depending on the reaction conditions.

Properties of *m*-nitrotoluene. *m*-Nitrotoluene is a clear yellow liquid that freezes at 16.1°C. It is readily soluble in ethanol, benzene, and diethyl ether. It is soluble in most organic solvents and is only sparingly soluble in water, 0.05 g/100 g of water at 30°C. Mp 16.1°C; bp at 101 kPa 231.9°C; density 1.1571 g/cm^3; refractive index 1.5470. *m*-Nitrotoluene does not have an active methyl group as do the ortho and para isomers. It is oxidized readily to *m*-nitrobenzoic acid by chromic acid and more slowly by potassium hexacyanoferrate(III) in alkaline solution. *m*-Nitrobenzaldehyde is the chief product of the electrolytic oxidation of *m*-nitrotoluene. Acid, neutral, or catalytic reduction of *m*-nitrotoluene yields *m*-toluidine.

Properties of *p*-nitrotoluene. *p*-Nitrotoluene crystallizes in colorless rhombic crystals. It is only slightly soluble in water, 0.044 g/100 g of water at 30°C; moderately soluble in methanol and ethanol; and readily soluble in acetone, diethyl ether, and benzene. Mp 51.7°C; bp 238.5°C; density 1.286 g/cm^3; refractive index $n_D^{62.5}$ 1.5346. The methyl group of *p*-nitrotoluene is activated by the para-nitro group. It is oxidized to *p*-nitrobenzoic acid by potassium hexacyanoferrate(III) in alkaline solution, potassium permanganate, or potassium dichromate. *p*-Nitrotoluene is converted to *p*-nitrobenzaldehye by electrolytic oxidation in an acetic acid/sulfuric acid mixture or by treatment with lead(IV) oxide in concentrated sulfuric acid. *p*-Nitrotoluene is reduced by iron and hydrochloric acid to *p*-toluidine. Alkaline reduction with iron leads to the formation of a mixture of azoxy, azo, and hydrazo compounds, depending upon the reaction conditions.

Manufacture. Mononitrotoluenes are produced by the nitration of toluene in a manner similar to that described for nitrobenzene. The presence of the methyl group on the aromatic ring facilitates the nitration of toluene, as compared to that of benzene, and increases the ease of oxidation which results in undesirable by-products. Thus, the nitration of toluene generally is carried out at lower temperatures than the nitration of benzene to minimize oxidative side reactions.

Mononitrotoluenes can be produced by either a batch or continuous process. With a typical batch process, the toluene is fed into the nitrator and cooled to ca 25°C. The nitrating acid (52–56 wt% H_2SO_4, 28–32 wt% HNO_3, and 12–20 wt% H_2O) is added slowly to the surface of the toluene and the temperature of the reaction mixture is maintained at 25°C. After all of the acid is added, the temperature is raised slowly to 35–40°C. After completion of the reaction, the reaction mixture is put into a separator where the spent acid is withdrawn from the bottom and is reconcentrated. The crude product is washed in several steps with dilute caustic and then water. The product is steam distilled to remove excess toluene and then dried by distilling the remaining traces of water. The resulting product contains 55–60 wt% *o*-nitrotoluene, 3–4 wt% *m*-nitrotoluene, and 35–40 wt% *p*-nitrotoluene. The yield of mononitrotoluenes is ca 96%.

Health and safety. The toxic effects of the mononitrotoluenes are similar to but less pronounced than those described for nitrobenzene. The maximum allowable concentration for the mononitrotoluenes is 5 ppm (30 mg/m^3).

Uses. *o*-Nitrotoluene is used in the synthesis of intermediates for azo dyes, sulfur dyes, rubber chemicals, and agricultural chemicals. *p*-Nitrotoluene is used principally in the production of intermediates for azo and sulfur dyes.

Dinitrotoluenes

Dinitration of toluene results in the formation of a number of isomeric products, eg, 2,4-dinitrotoluene, 2,6-dinitrotoluene, 3,4-dinitrotoluene, 2,3-dinitrotoluene and 2,5-dinitrotoluene. The dinitrotoluenes are moderate fire and explosion hazards when exposed to heat or flame. They are used as intermediates for the production of toluene diisocyanate and dyestuffs, as well as explosives (qv).

K.L. DUNLAP
Mobay Chemical Corporation

L.F. Albright and C. Hanson, eds., *Industrial and Laboratory Nitrations*, American Chemical Society, Washington, D.C., 1976.

W.L. Faith, D.B. Keyes, and R.L. Clark, *Industrial Chemicals*, 3rd ed., John Wiley & Sons, Inc., New York, 1975.

Chemical Economics Handbook, Stanford Research Institute, Menlo Park, Calif., Jan. 1979.

NITROFURANS. See Antibacterial agents, synthetic; Furan derivatives.

NITROGEN

Nitrogen, N, at no. 7, is a nonmetallic element and exists as the colorless, odorless, diatomic gas N_2. There are two stable isotopes ^{14}N and ^{15}N that occur, respectively, at 99.62% and 0.38%, and give an average atomic weight of 14.008. Nitrogen is the lightest element in Group VA of the periodic table. Although nitrogen in compounds can assume a number of valences, its chief valences are +5 and −3. Nitrogen comprises almost 80 wt% of the air and occurs in the protein matter of all living things, in many organic compounds, in ammonia and ammonium salts, and in the nitrate mineral deposits in Chile and Bolivia. Nitrogen in organic compounds may be determined by the Kjeldahl or Dumas methods.

Physical Properties

Gaseous nitrogen condenses to a colorless liquid at −195.8°C and to a white solid at −209.9°C. The solid exists in the α or cubic form, which is stable below −237.5°C, and in the β or hexagonal form, which is stable from −237.5°C to its melting point.

Mp 63.2 K; bp 77.35 K; density of gas at 272.15 K and 101.3 kPa 1.25 kg/m^3; of solid at 63.2 K 1.028 g/m^3; of liquid at bp 808.6 kg/m^3.

Chemical Properties

Nitrogen undergoes a variety of reactions at high temperatures (see Ammonia; Cyanamides; Nitric acid). Nitrogen reacts with ozone (qv) in a hot tube, which gives nitrogen dioxide and some nitrous oxide,

$$3\,N_2 + 4\,O_3 \rightarrow 6\,NO_2$$

A mixture of the nitrogen sulfides is formed from the reaction of nitrogen with elementary sulfur in an electric discharge at 100°C and 102.9 kPa (772 mm Hg) (see Sulfur compounds). Nitrogen reacts with a mixture of oxygen and chlorine gas at 400°C, resulting in the formation of nitrosyl chloride, NOCl.

Nitrogen reacts with a number of the metals, eg, the alkali metals, the alkaline earth metals, manganese and chromium, tantalum, tungsten (wolfram), and titanium, forming nitrides (qv).

Nitrogen reacts with calcium silicide, $CaSi_2$, at 1000°C.

Manufacture

Nitrogen is produced commercially by separating it from air via three important methods: cryogenic distillation; combustion of natural gas or propane and air; and pressure-swing absorption (PSA) (see Adsorptive separation). The choice of method depends primarily on the desired production capacity and nitrogen purity requirements, cryogenic distillation being the most extensively used as the economic choice for large-scale production and high purity.

Health and Safety

Safe handling of nitrogen gas or liquid requires knowledge of its properties and following safe practices; safety bulletins are available from suppliers. The potential hazards result from the same characteristics for which nitrogen is used. It is nontoxic, but its use in an enclosed, inadequately ventilated area could result in oxygen depletion and asphyxiation. Proper protective clothing, glasses, and gloves should be used when working with liquid nitrogen to avoid injuries from contact with its extremely cold temperature.

Uses

Nitrogen is used extensively in the metallurgical, chemical, and food industries as a blanket or purge gas to preclude oxidation during processing, storage, and packing. Nitrogen pressure also is used for pneumatic instruments, hydraulic accumulators, stirring, and pressure transfer. Significant use is also made of liquefied nitrogen for freezing of foods and processing materials at cryogenic temperatures (see Cryogenics).

RONALD W. SCHROEDER
Union Carbide Corporation

D. Mann, ed., *LNG Materials and Fluids*, National Bureau of Standards, Boulder, Colorado, 1977, pp. 1–2.

Cryog. Ind. Gases, 39 (July 1976).

Precautions and Safe Practices for Handling Liquefied Atmospheric Gases (*F-9888*), Union Carbide Corporation, New York.

NITROGEN FIXATION

To appreciate the importance of nitrogen fixation (dinitrogen fixation) fully, it is necessary to understand both the global distribution of nitrogen and its movement within the nitrogen cycle. Global inventories show that more than 99.9% of the nitrogen on earth is present as the dinitrogen molecule (N_2) of which somewhat more than 97% is trapped in primary rocks (2×10^{17} metric tons) and sedimentary rocks (4×10^{14} t) and ca 2% (4×10^{15} t) is free in the atmosphere. In comparison, all plants and animals together contain only 1×10^{10} t of nitrogen. In addition, about 1.2×10^{12} t is distributed in the form of dead organic matter about equally between land and sea. The latter contains 6×10^{11} t of soluble inorganic forms of nitrogen (excluding N_2). Thus, only a very small proportion of the nitrogen present on earth is involved at any one time in the cycle between its usable fixed form and its inert molecular form. Nitrogen fixation is involved with the atmosphere-to-terrestial (land or sea) direction. Nitrification and dentrification convert ammonia to nitrate and then to dinitrogen (via nitrogen oxides) which is lost to the atmosphere; leaching and erosion of soils result in the movement of fixed nitrogen between land and sea. The biological world stays ahead of a nitrogen deficiency because the fixation rate is just above the denitirification rate, but the margin is estimated to be slim. The relationships of the above processes are shown in Figure 1.

Industrial Haber-Bosch Process

A modern ammonia plant performs two distinct functions. The more energy-demanding and complex function is the preparation and purification of the synthesis gas, containing N_2 and H_2 in a 3:1 ratio, from a variety of feedstocks. The second function is the catalytic conversion of synthesis gas to ammonia (see Figure 2). In the years since its commercial introduction in 1913, many process changes have been made, particularly with respect to synthesis-gas production, to lower costs and give greater efficiencies.

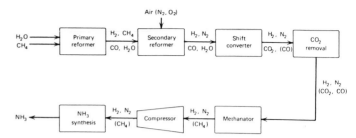

Figure 2. The Haber-Bosch process. Gases in parentheses are minor constituents of the mixture.

In the catalytic steam reforming of natural gas (see Fig. 2), the hydrocarbon stream consisting principally of methane is desulfurized first in order to avoid catalyst poisoning (see Catalysis). Together with superheated steam, it then contacts a nickel catalyst in the primary reformer at ca 3.04 MPa (30 atm) pressure and 800°C to convert methane to dihydrogen.

Reforming is completed in a secondary reformer where air is added to elevate the temperature by partial combustion of the gas stream. The amount of air added is so adjusted that a 3:1 H_2 to N_2 ratio results downstream of the shift converter as required for ammonia synthesis. The water–gas shift converter then produces more H_2 from carbon monoxide and water. The resulting carbon dioxide is removed by various techniques at this stage. A low temperature shift process, using a zinc–chromium–copper oxide catalyst, has replaced the earlier iron oxide-catalyzed high temperature system. Any remaining CO and/or CO_2 at this stage is converted to methane over a nickel catalyst. These gases are then compressed to ca 20.3 MPa (200 atm) for processing in the catalytic ammonia converter over a metallic iron catalyst, promoted with alumina and potassium oxide.

A breakthrough in this area was the development of centrifugal turbine compressors which bring significant savings when compared with reciprocating compressors.

Biological Systems

Biological nitrogen fixation is confined to microorganisms. Only prokaryotes, ie, living things without an organized nucleus (bacteria, blue-green algae, and actinomycetes), can reduce dinitrogen to ammonia. Such bacteria can be either free-livers, such as *Azotobacter* and *Clostridium*, or can form symbiotic associations with higher plants, like the *Rhizobium*-legume system. The latter group is much more important agriculturally. In exchange for the fixed nitrogen supplied by the bacterium, the legume supplies energy in the form of carbohydrates obtained by photosynthesis. Thus, renewable solar energy powers this fertilizer production system, in contrast to the nonrenewable energy sources used commercially. Ammonia from the Haber process involves an additional energy cost in transportation to the user. Furthermore, since commercial operation must be year-round to be economical whereas fertilizer application is seasonal, storage costs are incurred. Thus, as food demands increase and fossil fuel must be conserved, the exploitation of biological nitrogen fixation may well be the answer to the problem.

Plant – bacterial associations. *Rhizobium*–Legume Associations. The legumes (family *Leguminoseae*) include temperature and tropical flowering plants, ranging from small plants, such as clover, to bushes and large trees, such as acacia. Most plants can be infected by bacteria of the genus *Rhizobium*, which colonize their roots and fix dinitrogen within nodules. The best known associations occur with important crops, such as the pulses (peas and beans, including soybeans), the clovers, and alfalfa. These associations show some specificity since certain bacterial species only infect specific plants.

Algal associations. Not only are the blue-green algae (more properly called cyanobacteria) important dinitrogen fixers as free-livers, but they also form effective dinitrogen-fixing associations with a variety of plants ranging from lichens, fungi, through liverworts and ferns to gymnosperms and an angiosperm.

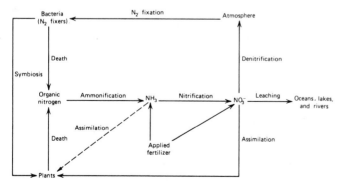

Figure 1. Biological pathways and processes involved in the nitrogen cycle.

Free-living microorganisms. Except for the cyanobacteria, the free-living bacteria are generally not agriculturally important. They contribute to the soil only about 0.1% of the fixed nitrogen of a leguminous association. The free-living cyanobacteria, by comparison, contribute about 2–5% as much as a leguminous association. The difference probably lies in the cyanobacteria's property of photosynthesis, which relieves their dependence on the often limiting carbon (and therefore, energy-yielding) substrates in the soil. Free-living bacteria are used for most of the research on the isolation and characterization of the enzyme nitrogenase, responsible for nitrogen fixation. Nitrogen fixation occurs as a property of aerobes, anaerobes, facultative anaerobes, photosynthetic bacteria, and cyanobacteria.

Nitrogenase. The enzyme system responsible for nitrogen fixation is called nitrogenase. It was first prepared as a cell-free extract from the anaerobe *Clostridium pasteurianum* in 1960. Since then, more than 20 species have yielded partially purified nitrogenase; five have been extensively purified. These extracts are highly oxygen sensitive. Nitrogenase consists of two proteins and needs a source of both reducing potential and energy (in the form of adenosine triphosphate, ATP) to operate. In general, the enzymes from all N_2-fixing bacteria are very similar. Nitrogenase catalyzes many ATP-dependent reactions, the most important of which is reduction with nitrogen to form ammonia plus ADP (adenosine diphosphate) and phosphate. If no substrate is available, hydrogen gas is evolved by reduction of protons, a potentially wasteful reaction.

Genetics and regulation. Nitrogen fixation genetics was not investigated until 1971, when it was shown that the nitrogen fixation (*nif*) genes were clustered closely together on the single circular chromosome of *Klebsiella pneumoniae.* The genetic techniques of transduction and conjugation have both been used since then to transfer the *nif* genes to initially nonfixing bacteria. In transduction, the N_2-fixing cells are infected with a temperate phage, which carries with it the *nif* genes and which can be thought of as a nonlethal bacterial virus. Conjugation involves sexual activity, during which the bacteria are joined and the *nif* genetic material is passed from one to another. In this way, *Escherichia coli,* a common enteric bacterium, and *Salmonella typhimurium,* both of which do not naturally fix dinitrogen, were modified to become dinitrogen fixers (when grown anaerobically).

WILLIAM E. NEWTON
Charles F. Kettering Research Laboratory

W.D.P. Stewart, ed., *Nitrogen Fixation by Free-Living Organisms,* Cambridge University Press, London, UK, 1975.

A.H. Gibson and W.E. Newton, eds., *Current Perspectives in Nitrogen Fixation,* Elsevier-North Holland, New York, 1981.

C. Veeges and W.E. Newton, eds., *Advances in Nitrogen Fixation Research,* Martinus Nijhoff, The Hague, Neth., 1984.

NITROGEN TRICHLORIDE, NCl₃.   See Chloramines.

NITROPARAFFINS

Nitroparaffins (or nitroalkanes) are derivatives of the alkanes in which one hydrogen or more is replaced by the electronegative nitro group ($-NO_2$), which is attached to carbon through nitrogen. The nitroparaffins are isomeric with alkyl nitrites, RONO, which are esters of nitrous acid. The nitro group in a nitroparaffin has been shown to be symmetrical about the R—N axis, and may be represented as a resonance hybrid of the following structure:

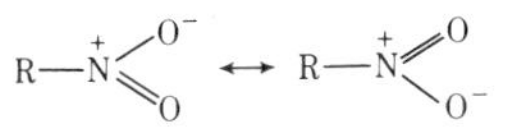

$$R-\overset{+}{N}\overset{O^-}{\underset{O}{\diagdown}} \longleftrightarrow R-\overset{+}{N}\overset{O}{\underset{O^-}{\diagdown}}$$

Nitroparaffins are classed as primary, RCH_2NO_2, secondary, R_2CH-NO_2, and tertiary, R_3CNO_2, by the same convention used for alcohols.

Primary and secondary nitroparaffins exist in tautomeric equilibrium with the enolic or *aci-* forms.

The nitroparaffins are named as derivatives of the corresponding hydrocarbons by using the prefix nitro to designate the $-NO_2$ group; eg, CH_3NO_2, nitromethane; $CH_3CH(NO_2)CH_3$, 2-nitropropane; and $CH_3CH(NO_2)_2$, 1,1-dinitroethane.

The salts obtained from nitroparaffins and the so-called nitronic acids are identical and may be named as derivatives of either, eg, sodium salt of *aci-*nitromethane, or sodium methylenenitronate.

Nitromethane, nitroethane, 1-nitropropane, and 2-nitropropane are produced by a vapor-phase process developed in the 1930s.

Physical Properties

Most polynitroparaffins are colorless crystalline or waxlike solids at or near room temperature. They are insoluble in water and alkanes but soluble in most other organic solvents. The lower mononitroparaffins are colorless, dense liquids with mild odors. The boiling points of the mononitroparaffins are much higher than those of the isomeric nitrites; eg, the normal boiling point of nitromethane is 101.2°C, whereas that of methyl nitrite is -12°C. This phenomenon may be attributed in large part to intermolecular hydrogen bonding. Accurate vapor-pressure determinations for the lower nitroparaffins have been made and adapted to an Antoine equation. A nomograph was constructed from these data.

Most organic compounds, including aromatic hydrocarbons, alcohols, esters, ketones, ethers, and carboxylic acids, are miscible with nitroparaffins, whereas alkanes and cycloalkanes have limited solubility. The lower nitroparaffins are excellent solvents for coating materials, waxes, resins, gums, and dyes.

Chemical Properties

Tautomerism. Primary and secondary mononitroparaffins are acidic substances which exist in tautometric equilibria with their nitronic acids. The nitro isomer is weakly acidic; the nitronic acid isomer (aci form) is much more acidic.

An equilibrium mixture of the isomers usually contains a much higher proportion of the true nitro compound. The equilibrium for each isomeric system is influenced by the dielectric strength and the hydrogen-acceptor characteristics of the solvent medium. The aci- form is dissolved and neutralized rapidly by strong bases, and gives characteristic color reactions with ferric chloride.

Polynitroparaffins are stronger acids than the corresponding mononitroparaffins. Thus, 1,1-dinitroethane has an ionization constant of 5.6×10^{-6} in water at 20°C; trinitromethane is a typical strong acid with an ionization constant in the range of 10^{-2}–10^{-3}. Neutralization of these substances occurs rapidly, and they may be titrated readily.

Salts. Nitroparaffin salts dissociate to form ambidentate anions, which are capable of alkylation at either the carbon or oxygen atom.

Acid hydrolysis. With hot concentrated mineral acids, primary nitroparaffins yield a fatty acid and a hydroxylamine salt.

Halogenation. In the presence of alkali, chlorine replaces the hydrogen atoms on the carbon atom holding the nitro group. If more than one hydrogen atom is present, the hydrogen atoms can be replaced in stages; exhaustive chlorination of nitromethane yields chloropicrin (trichloronitromethane). The chlorination can be stopped at intermediate stages. Bromination or iodination takes a similar course, but bromopicrin and iodopicrin tend to be less stable.

Condensation with carbonyl compounds. Primary and secondary nitroparaffins undergo aldol-type condensation with a variety of aldehydes and ketones to give nitro alcohols. Those derived from the lower nitroparaffins and formaldehyde are available commercially (see Nitro alcohols).

Mannich-type reactions. Nitroparaffins, formaldehyde, and primary or secondary amines can react in one step to yield Mannich bases.

Reduction. The lower nitroparaffins are reduced readily to the corresponding primary amines with a number of reducing agents. Partial reduction yields aldoximes, ketoximes, or N-substituted hydroxylamines. Suitable reduction methods range from iron and hydrochloric acid to high pressure hydrogenation over Raney nickel or noble-metal catalysts.

Oxidation. Nitroparaffins are resistant to oxidation.

Addition to multiple bonds. Mono- or polynitroparaffins with a hydrogen on the carbon atom carrying the nitro group add to activated double bonds under the influence of basic catalysts. Thus, nitromethane forms tris(β-cyanoethyl)nitromethane with acrylonitrile, and 2-nitropropane yields 4-methyl-4-nitrovaleronitrile.

Manufacture. The method currently utilized commercially is vapor-phase nitration of propane, although ethane and butane also can be nitrated readily.

Health and Safety

Commercial nitroparaffins can, in general, be handled similarly to other common flammable solvents, provided certain precautionary measures are observed. Nitroparaffins are toxic by inhalation. The TLVs recommended by the ACGIH should be observed for all operations in which they are used.

Uses

The nitroparaffins are useful intermediates for the synthesis of a variety of chemical compounds, eg, nitro alcohols (qv), alkanolamines (qv), hydroxylamine, and chloropicrin. Large volumes of nitroparaffins are consumed for use as solvents in coatings and inks, for extractions, for crystallizations, and as a reaction medium (see Solvents, industrial).

PHILIP J. BAKER, JR.
ALLEN F. BOLLMEIER, JR.
ANGUS Chemical Company

H.H. Baer and L. Urbas in H. Feuer, ed., *The Chemistry of the Nitro and Nitroso Groups*, Pt. 2, Interscience Publishers, a division of John Wiley & Sons, Inc., New York, 1970, pp. 75–200.

D. Seebach, E.W. Colvin, F. Lehr, and T. Weller, *Chimia* **33**, 1 (1979).

T. Urbanski, ed., *Nitro Compounds: Proc. Int. Symp. Warsaw, Poland, 1963*, Pergamon Press Ltd., Oxford, UK, 1964.

Documentation of Threshold Limit Values, rev. ed., Committee on Threshold Limit Values, American Conference of Governmental Industrial Hygienists, Cincinnati, Ohio, 1980, pp. 305–309.

N-NITROSAMINES

N-Nitrosodialkylamines (*N*-nitrosamines) are typical organic compounds that have been useful as synthetic intermediates or as solvents in addition to possessing interesting structural and spectroscopic properties.

Properties

The single feature common to all of them is the NNO functionality. *N*-Nitrosamines are extremely numerous since there are few restrictions on the groups that can be attached to the remaining two valences on the amine nitrogen, eg:

$$R-\overset{\overset{\displaystyle R'}{|}}{N}-N{=}O$$

(1)

where R, R′ can be alkyl, aryl, or both; for R and R′ = XCH$_2$, X can be H, alkyl, aryl, halogen, alkoxy, etc; for R or R′ = aryl, various substituents may be on the ring(s); and when R or R′ = H, or when X = OH, the resulting primary *N*-nitrosamines or α-hydroxy-*N*-nitrosamines generally are unstable.

N-Nitrosamines typically are volatile solids or oils and are yellow because of absorption of visible light by the NNO group. The electron delocalization in the functionality confers sufficient double-bond character on the N—N bond, so that the E and Z isomers which result from unsymmetrical substitution, eg, structures (**2**) and (**3**), often can be separated.

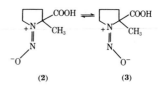

(2)   **(3)**

Synthesis. The classic *N*-nitrosamine synthesis is the reaction of a secondary amine with nitrite ion at ca pH 3, eg, equation 1:

$$\text{>}NH + NO_2^- \xrightarrow{H^+} \text{>}N-N{=}O \tag{1}$$

Reactions. The chemistry of the *N*-nitrosamines is extensive. Most of the reactions, with respect to their biological or environmental behavior, involve one of two main reactive centers, ie, the nitroso group and the C—H bonds adjacent (α) to the amine nitrogen. The nitroso group can be removed readily in a reaction that is essentially the reverse of the nitrosation reaction (eq. 2), or it can oxidized (eq. 3) or reduced (eq. 4).

$$\text{>}N-N{=}O \longrightarrow \text{>}N-H \tag{2}$$

$$\text{>}N-N{=}O + CF_3CO_3H \longrightarrow \text{>}N-NO_2 \tag{3}$$

$$\text{>}N-N{=}O + Zn/CH_3CO_2H \longrightarrow \text{>}N-NH_2 \tag{4}$$

Absorption of uv radiation by the NNO group produces a set of photochemical reactions. Under neutral conditions, and at moderate to high concentrations, these compounds often are chemically stable; although the $E \rightleftharpoons Z$ equilibrium, with respect to rotation around the N—N bond, can be affected.

Health and Safety

Most of the more than 100 nitrosamines tested are carcinogenic. Carcinogenic effects have been observed both with single, relatively large doses and with long-term chronic exposure to lower doses. The *N*-nitrosamines are, in general, mutagenic toward standard bacterial-tester strains.

JOHN S. WISHNOK
Massachusetts Institute of Technology

IARC Working Group on the Evaluation of Carcinogenic Risk of Chemicals to Humans, *Some N-Nitroso Compounds*, *IARC Monograph No. 17*, International Agency for Research on Cancer, Lyons, France, 1978.

J.-P. Anselme, ed., *N-Nitrosamines*, *ACS Symposium Series 101*, American Chemical Society, Washington, D.C., 1979.

R.A. Scanlan and S.R. Tannenbaum, eds., *N-Nitroso Compounds*, *ACS Symposium Series 174*, American Chemical Society, Washington, D.C., 1981.

P.N. Magee, R. Montesano, and R. Preussmann, "*N*-Nitroso Compounds and Related Carcinogens," in C.S. Searle, ed., *Chemical Carcinogens*, *ACS Monograph 173*, American Chemical Society, Washington, D.C., 1976.

NOBELIUM. See Actinides and transactinides.

NOISE POLLUTION

Noise is unwanted sound. Most people today are exposed to some sounds that they describe as noise. These sounds exist in their homes, in their communities, and at work. The effects of exposure to noise depend on the noise level, the length of exposure to noise, and the type of noise. The known effects of noise are permanent irreversible hearing loss, speech interference and telephone communication interference, sleep interference, task interference, and annoyance. Although the causes of industrial hearing loss have been known for many years, many workers are still exposed to excessive levels of noise and are likely to incur permanent hearing loss as a result of noise exposure. This loss is referred to as noise-induced permanent threshold shift.

Industry and government are especially aware of the need to control noise. Federal agencies such as the Bureau of Mines (BOM), the National Institute for Occupational Safety and Health (NIOSH), the Federal Aviation Agency (FAA), the Department of Housing and Urban Development (HUD), the Mine Safety and Health Administration and the Occupational Safety and Health Administration (OSHA) have programs relating to noise.

Terminology

Noise and other sounds can be described by their magnitude, frequency content, and temporal pattern. The magnitude of a noise is expressed in decibels (dB) and is termed the sound-pressure level, L_p. Mathematically, the sound-pressure level is expressed as

$$L_p = 10 \log_{10} (p/p_0)^2 \text{ in dB}$$

where p is the instantaneous sound pressure and p_0 is the reference sound pressure of 20 μPa (0.15 mm Hg). This reference sound pressure is representative of the faintest sound audible to a young person. The frequency of a pure tone indicates the number of times per second that the instantaneous sound pressure oscillates between positive and negative values. The unit of frequency in cycles per second is the Hertz (Hz).

Noise is transmitted from a source along a path to a receiver. Often, the noise being measured consists of individual sound contributions from many different sources. In addition, the noise level may vary during the measurement period.

Measurement

Noise can be measured using a variety of measurement equipment. The essential components are the calibrator, microphone, amplifiers, filter, and readout.

Individual Noise Exposure

A person's noise exposure is a function of occupation, mode of transportation to work, lifestyle, and possibly, socioeconomic status. A single-number measure of a person's noise exposure for a day is the 24-h equivalent sound level, L_{eq}.

The L_{eq} of an individual can be estimated by determining how long is spent in each of the activities during the average day and then measuring the noise level associated with the activity and calculating the equivalent noise exposure. The L_{eq} can be calculated from the following expression:

$$L_{eq}(24 \text{ h}) = 10 \log \left[\frac{t_1 10^{(L_{eq1}/10)} + t_2 10^{(L_{eq2}/10)} + \cdots}{t_1 + t_2 + \cdots} \right]$$

Community Noise Sources

Noise in the community is composed of contributions from many different noise sources. The transportation systems in the United States make significant contributions to the total community noise. Commercial aircraft, general aviation aircraft, highway vehicles, recreation vehicles, and rail systems contribute to the community noise level. In addition, construction equipment and operations, industrial plants, and home-operated internal combustion engines contribute to the community noise problem. Sometimes mechanical equipment in buildings also contributes to the problem.

Noise Control

Noise can be controlled at the source, along the path, and at the receiver. Noise control at the source includes modification of the source itself. Such modification can be accomplished by reducing the motion of radiating surfaces of the source (damping), by vibration isolation of vibrating parts of the source from parts that can radiate the noise, by modification of the operation of the source, by replacement of the source

Table 1. Noise-Control Treatments for Existing Sources

Source	Replace with quieter one	Replace part	Mufflers	Barriers	Lagging	Enclosures
electric motors	√		√	√		√
gears		√		√		√
gas turbines			√			√
steam turbines					√	
furnaces–heaters			√			√
flares		√				
air-cooled heat exchangers	√					
cooling towers			√	√		
compressors			√	√	√	√
pumps					√	
valves	√				√	
blowdowns			√			
fans			√		√	√
granulators			√			√
dicers			√			√
vaporization and air–gas mixing systems			√			
pelletizers			√			
fiber aspirators	√		√			√

with another source that accomplishes the same function but at lower noise emissions, and by better maintenance of equipment.

Noise control along the path can be accomplished by the following means: source location, absorption, mufflers, barriers, and enclosures.

Industrial Noise Regulations

OSHA was charged by the U.S. Congress in 1970 with protecting the health and safety of workers. OSHA issued a noise regulation that requires employers to limit the noise exposure of employees to the following levels, using the A-weighting network.

Noise level, dB(A)	Exposure per day, h
90	8
95	4
100	2
105	1
110	$\frac{1}{2}$
115	$\frac{1}{4}$

If the exposure is to different levels of noise instead of a single noise level, the daily noise dose must be less than or equal to 1; the daily noise dose is defined as

$$\text{daily noise dose} = C_{1/T_1} + C_{2/T_2} + C_{n/T_n} + \cdots$$

where C_n is the actual time exposed to a noise, and T_n is the allowable exposure time.

Noise and Its Control in the Chemical Industry

Table 1 summarizes the types of noise controls available for quieting an existing source (see Insulation, acoustic).

ROBERT D. BRUCE
Hoover Keith & Bruce Inc.

A.P.G. Peterson and E.E. Gross, Jr., *Handbook of Noise Measurement*, General Radio, West Concord, Mass., 1963.

L.L. Beranek, ed., *Noise and Vibration Control*, McGraw-Hill, New York, 1971.

C.M. Harris, ed., *Handbook of Noise Control*, McGraw-Hill, New York, 1979.

NOMENCLATURE

Interest in chemical nomenclature has developed markedly in recent years (particularly in the United States and the United Kingdom) as the importance of good nomenclature has been increasingly recognized. For example, when one compound is known by more than one name, and when one name may refer to more than one compound, confusion inevitably results. Since confusion owing to bad and inconsistent naming practices is a serious problem, especially in indexes and compilations, it is natural that much of the work in systematizing names of chemical compounds has been done in connection with such works. Many committees, both national and international, are now working toward consistent, systematic nomenclature. Especially noteworthy is the greatly increased interest of industrial chemists and government agencies in nomenclature.

Important as good names are, they cannot serve all necessary purposes. The same is true for the other means of identification of chemical compounds, ie, structural formulas, notation systems, and registry numbers. Although structural formulas are often easier to recognize than names, they are space consuming, they may be troublesome to reproduce, and they are difficult to arrange in order. In recent years, systems of representing chemical structures by fragmentation coding, linear notations, and topological coding have been developed, particularly for mechanical recording and sorting. Of the many systems published, the one by Dyson was adopted by IUPAC. Nevertheless, the one by Wiswesser is most widely applied today. Registry numbers also provide an unambiguous identification for chemical compounds, but in general these are assigned arbitrarily and therefore do not have any information built into them. The main generator and user of registry numbers in chemistry is the *Chemical Registry System* of Chemical Abstracts Service.

Inorganic Nomenclature

Berzelius extended and amplified the nomenclature introduced by Guyton de Morveau and Lavoisier. It was he who divided the elements into metalloids (nonmetals) and metals according to their electrochemical character, and the compounds of oxygen with positive elements (metals) into suboxides, oxides, and peroxides. His division of the acids according to degree of oxidation has been little altered. He introduced the terms anhydride and amphoteric. He designated the chlorides in a manner similar to that used for the oxides. This system of nomenclature has withstood the impact of experimental discoveries and theoretical developments that altered greatly the character of chemical thought.

Established practice in the English language.   The nearly literal translation of the French terms into English and other languages gave the system that has become standard practice in English-speaking and other countries. Binary compounds are systematically designated by two words, the first referring to the more electropositive constituent and the second, ending in *-ide*, referring to the more electronegative constituent, eg, sodium chlor*ide*. In case the metal exhibits two oxidation states, the lower is indicated by the termination *-ous* and the higher by *-ic*, as in cupr*ous* oxide and cupr*ic* oxide. Some ternary compounds that contain well-established radicals are named as though they were binary compounds: sodium hydroxide, calcium cyanide, potassium amide. Ternary compounds are also named by giving the more electropositive constituent first.

Modified forms in common use.   There are numerous situations in which the above system does not meet all the requirements. In the formation of binary compounds, several elements exhibit more than two states of oxidation. One method, approved by the IUPAC, of handling these situations is the use of Greek numerical prefixes indicating stoichiometric composition, eg, $TiCl_2$ (titanium dichloride).

Other approved methods of indicating proportions of constituents are the Stock system (parenthetical Roman numeral indicates oxidation state) and the Ewens-Bassett system (parenthetical Arabic numeral indicates ionic change).

Occasionally, an element forms more than four acids and other combinations of prefixes and suffixes have been resorted to: $H_4P_2O_6$, intermediate between H_3PO_3 and H_3PO_4, is known as *hypo*phosphor*ic* acid; and the salts M_2FeO_3, intermediate between M_2FeO_4 and $MFeO_2$, are sometimes known as *perferrites*. Here again the Stock or the Ewens-Bassett system offers definite advantages.

International agreement.   The first report of the Commission for the Reform of the Nomenclature of Inorganic Chemistry was written in 1926 by Delépine. As the work of the commission progressed under Jorissen as chairman, there were several reports, followed in 1940 by a comprehensive set of rules. Remy and Meyer deserve a great deal of credit for the

development of these rules, which were well received and followed widely. These 1940 rules led to the 1957 rules, which in turn evolved into the 1970 *Nomenclature of Inorganic Chemistry*, which is the current basis of naming inorganic compounds. These rules provide for the retention of most of the older well-established names for binary and pseudobinary compounds and for the oxo acids of the nonmetals and their derivatives.

Organic Nomenclature

The International Union and *The Definitive Report*. The *Definitive Report of the Commission on the Reform of the Nomenclature of Organic Chemistry*, adopted by the Commission and by the Council of the International Union of Chemistry in 1930 in Liège, used the rules of the Geneva Congress (1892) as a basis but modified them in certain respects, thus setting up a modified Geneva system. In addition, many of the 68 rules deal with topics not touched in the original Geneva report.

One provision of the *Definitive Report* is that only one kind of function (the principal function) is expressed by the ending of the name, ie, by a suffix. The others, if there are any, are designated by prefixes (Rule 51). Thus, $CH_3CHOHCOCH_3$ is called 3-hydroxy-2-butanone. An important modification of the Geneva system is that the fundamental chain used as a basis in an aliphatic compound will not be necessarily the longest chain in the molecule but will be the longest chain containing the maximum occurrences of the (principal) functional group (Rule 18).

The Commission on the Nomenclature of Organic Chemistry has continued to revise and expand the *Definitive Report*. This has resulted so far in the following IUPAC rules and revisions published in book form: Section A, hydrocarbons; Section B, fundamental heterocyclic systems; Section C, characteristic groups containing carbon, hydrogen, oxygen, nitrogen, halogen, sulfur, selenium, and/or tellurium; Section D, organic compounds containing elements that are not exclusively carbon, hydrogen, oxygen, nitrogen, halogen, sulfur, selenium, and tellurium; Section E, stereochemistry; Section F, general principles for the naming of natural products and related compounds; and Section H, isotopically modified compounds. Section D makes use of the nomenclature rules developed earlier for organosilicon compounds by IUPAC and for organophosphorus compounds by the ACS Committee on Nomenclature, Spelling, and Pronunciation. Section E applies the system of designating absolute configuration devised by Cahn, Ingold, and Prelog. The Organic Commission is presently continuing its work in revising and systematizing the present rules.

The ending *-yl* (or *-oyl*) is the standard one for the names of univalent groups (it also is used for some others, such as oxalyl). It may be combined with a sign of unsaturation, as in prop*enyl*, $CH_3CH{=}CH{-}$, or eth*ynyl*, $CH{\equiv}C{-}$. The ending *-ylene* denotes a bivalent group in which (with the exception of methylene, $H_2C{=}$) the two free valences are on different atoms; as trimethylene, $-CH_2CH_2CH_2-$. When the two free valences of a bivalent group are on the same atom, the ending used (except for methylene) is *-ylidene*, as ethylidene, $CH_3CH{=}$. For three valences on the same atom the ending is *-ylidyne*, as ethylidyne, $CH_3C{\equiv}$.

For indicating a number of groups of the same kind, the prefixes *di-*, *tri-*, *tetra-*, etc, are used when the expressions are simple, and *bis-*, *tris-*, *tetrakis-*, etc, when they are complex; eg, dichloro-, triethyl-, tetracarboxylic, bis(dimethylamino)-. The prefix *bi-* is used to denote the doubling of a group, as in *bi*phenyl, $C_6H_5.C_6H_5$, or the doubling of a compound, as in *bi*arsine, $H_2As.AsH_2$.

Biochemical Nomenclature

The nomenclature of biochemical compounds is in very large measure a part of organic nomenclature. Yet, it has its peculiar problems, arising partly from the fact that most biochemical compounds must be named before their chemical structure has been determined, and partly from the need for grouping them according to their biological function rather than their chemical class.

Since its establishment in 1964, the joint IUPAC/IUB Commission has published many recommendations dealing with nomenclature of natural products: steroids (qv), carbohydrates (qv), cyclitols, tetrapyrroles, corrinoids (see Vitamins, vitamin B_{12}), vitamins (qv), amino acids (qv) and peptides (see Polypeptides), lipids (see Fats and fatty oils), etc.

The IUB Commission on Nomenclature has issued a number of recommendations dealing with nomenclature topics of a more biochemical nature: peptide hormones (see Hormones), conformation of polypeptide chains, abbreviations of nucleic acids and polynucleotides, iron–sulfur proteins, enzyme units (see Enzymes), etc.

Macromolecular Nomenclature

At present, there does not exist any universal systematic nomenclature for polymers. The Commission on Macromolecular Nomenclature has issued recommendations on regular, single-strand organic polymers. Recommendations on naming inorganic polymers and copolymers are expected to be issued in the near future.

Nomenclature in Other Areas of Chemistry

IUPAC commissions have issued nomenclature recommendations in the fields of analytical chemistry (see Analytical methods), physicochemical symbols and terminology, colloid and surface chemistry (see Colloids), ion exchange (qv), spectroscopy, etc.

KURT L. LOENING
Chemical Abstracts Service

International Union of Pure and Applied Chemistry, *Nomenclature of Inorganic Chemistry*, 2nd ed., Butterworths, London, UK, 1970 (the Red book); *Nomenclature of Organic Chemistry*, Sections A, B, C, D, E, F, and H, Pergamon Press, Oxford, UK, 1979 (the Blue book); *Compendium of Analytical Nomenclature*, Pergamon Press, Oxford, UK, 1977 (the Orange book); *Manual of Symbols and Terminology for Physicochemical Quantities and Units*, Pergamon Press, Oxford, UK, 1979 (the Green book).

International Union of Biochemistry, *Biochemical Nomenclature and Related Documents*, Biochemical Society, London, UK, 1978.

"Selective Bibliography of Nomenclature of Chemical Substance," *Chemical Abstracts Index Guide*, Appendix IV, Chemical Abstracts Service, Columbus, Ohio, 1984.

NONDESTRUCTIVE TESTING

The technology of nondestructive testing (evaluation of materials) includes all nondamaging methods for determining material identity; for evaluating material properties, composition, structure, or serviceability; and for detecting discontinuities and defects in materials. Nondestructive tests are commonly used for quality control, process control, and reliability assurance of materials, protective coatings, components, welds, assemblies, structures, and operating systems. Their use helps prevent premature failure of materials during processing, manufacturing, or assembly, and during service under anticipated operating stresses and environments. Their proper use can lower manufacturing and operating costs, minimize insurance risks and help prevent interruptions of service and potential disasters that might cause loss of life or usefulness of costly facilities (see also Analytical methods).

Testing Materials' Serviceability

Nondestructive tests can be performed on materials, components, and structures or systems that actually are to be used. Tests can be made prior to service or during service upon airplanes, pipelines, or plant systems which are used continuously or in repeated operations, or where deterioration or damage could occur as a result of operator error, excessive loading, or damaging environmental or operating conditions.

Large corporations and regulatory agencies have a vital responsibility for specifying and managing the use of nondestructive testing during erection and service of costly facilities and systems such as large ships, bridges, off-shore drilling platforms, and thermal or nuclear power plants where failures during service are potentially disastrous.

Nondestructive tests also can be applied to consumer products to ensure safety and proper operation. Tests might be used periodically

during service to determine whether deterioration or conditions that might lead to premature failure have developed as a result of improper handling or storage, or through misuse or abuse by operators. Of particular concern is evidence of environmental damage such as excessive corrosion and of loading effects such as fatigue cracks, stress corrosion, or corrosion fatigue which can provide nuclei for crack propagation.

Inspector training and qualifications. Since nondestructive tests do not measure material performance properties or defects directly, test operators or inspectors must select test procedures, and evaluate their use in manufacturing, response to environmental conditions, and performance in the intended service. Government, technical societies, and industry codes and specifications often control the certification and qualification of nondestructive-test personnel and their supervision.

Computer automation of tests. In the 1980s, the art of nondestructive testing is in a transition state in which detailed inspection using human manipulation of equipment and searching of all accessible test surfaces is being replaced by automated tasks, and their results are analyzed digitally by more complex electronic systems.

Principles of Nondestructive Tests

Nondestructive tests may be based upon any principles of physics or chemistry that can reveal the identity, geometry, characteristics, or integrity of materials without damaging surface or interior regions. Most tests require use of suitable probing media to cause test objects to emit, transmit, or reflect signals that can be detected and interpreted in terms of material properties or defects. Typical of the many different probes used in nondestructive tests are penetrating electromagnetic radiations such as x rays or gamma rays (see X-ray technology); visible, infrared, or ultraviolet radiation; gaseous or liquid penetrants and leak tracers; sonic or ultrasonic waves (see Ultrasonics); static magnetic or electric fields; dynamic electromagnetic fields of various suitable frequencies; gravitational fields; heat conduction and diffusion phenomena; neutron, electron, or positron radiation; natural-frequency mechanical vibrations; and many other appropriate physical effects for which sources of probing media and transducers for detection of test signals are readily available. The resultant test signals may be converted to visible or audible indications, analogue electrical signals, meter or digital-display readings, computer data, or images of many different types. These output signals, images, or data must be evaluated by the human test operator or by automated test systems.

Probing Media Based on Motion of Matter

Some types of probing media are best suited to inspecting external surfaces or walls for cracks and leaks, or to gauging the dimensions and surface geometry of test objects. Motion-of-matter probing media include movement of solid probes (such as calipers, mechanical gauges, or micrometers), as well as that of fluid media such as gases, liquids, or vapors, or of individual particles of detectable substances. Motion may be imparted to such probing media by mechanical, hydraulic, or electromagnetic forces, or it may result from gas or vapor pressure, diffusion, permeation, osmosis, solubility, evacuation, pumping, or other factors. The primary usefulness of motion-of-matter probing media is based on the general inability of the medium to penetrate solid matter in the absence of discontinuities open to surfaces accessible to the probing medium. Examples of motion-of-matter nondestructive tests are liquid-penetrant inspection, filtered-particle inspection, chemical-reaction tests, radioactive krypton-85 gas-penetrant inspection, electron surface-transit inspection with solid probes, and leak testing and leakage measurements.

Probing Media Based on Transmission of Energy

Energy transmission probing media used in nondestructive testing include those in which energy is transmitted without any significant motion of matter. Typical probing media in this classification include static electric, magnetic, and gravitational fields, dynamic electromagnetic fields, and photon beams. Depending upon their wavelengths relative to the typical spacings between atoms in the test object, such probing media may be limited to testing exposed surfaces (as with light irradiation of opaque materials), or they may be capable of penetrating to great depths, as with high energy x rays or gamma rays. Among the advantages of transmission-of-energy probing media is the possibility of visualizing test object surfaces or interior volumes.

Some tests based on transmission-of-energy probing media include automatic check-weighing of packages and coatings, electrified-particle tests, high voltage probe and corona tests, magnetic particle and magnetographic tests, electric-current conduction tests, magnetic induction and eddy-current tests, microwave tests, infrared and thermal-conduction tests, optical holography tests, and visual inspection and optical tests.

Penetrating X-Ray and Gamma-Ray Tests

X-ray and gamma-ray imaging tests are among the methods most widely used to examine interior regions of metal castings, fusion weldments, and composite and brazed-honeycomb structures. Radiographic tests are made on bridges and structures, ships and barges, pipeline welds, pressure vessels, nuclear fuel rods, and other critical materials and components that may contain three-dimensional voids, inclusions, gaps, or cracks aligned so that portions are parallel to the radiation beam. Since penetrating-radiation tests measure the mass of material per unit area in the beam path, they can respond to changes in thickness or material density, and to inclusions of density different from that of the material in which they are imbedded.

Probing Media Combining Motion of Matter and Transmission of Energy

Uniquely powerful among nondestructive test-probing media are the phenomena in which motion of matter is inseparably combined with energy transmission. In these probing media, energy cannot be transmitted or transferred without an accompanying motion or transfer of matter, and matter cannot be involved in motion without simultaneously transferring energy. Typical of these probes are high energy particles, such as electrons, protons, neutrons, positrons, or other fundamental particles of physics. However, many other useful common phenomena also fall into this class of probing media, including wave motions in solids, liquids, and gases (as with sound waves or ultrasonic beams used in nondestructive tests). Tests include neutron imaging tests, electron imaging tests, ultrasonic tests, acoustic-emission tests, nmr and esr tests, and numerous three-dimensional scanning systems used in medical diagnoses, fundamental scientific measurements, and industrial inspection and materials evaluation.

ROBERT C. MCMASTER
The Ohio State University

R.C. McMaster, ed., *Nondestructive Testing Handbook*, The Ronald Press Co., New York, 1959, 1963, and The American Society for Nondestructive Testing, Columbus, Ohio, 1977, 1979.

H.E. Boyer, ed., *Nondestructive Inspection and Quality Control, Metals Handbook*, 8th ed., Vol. 11, American Society for Metals, Metals Park, Ohio, 1976.

R.C. McMaster, ed., "Liquid Penetrant Inspection and Leak Testing," in *Nondestructive Testing Handbook*, American Society for Nondestructive Testing, Columbus, Ohio, and The American Society for Metals, Cleveland, Ohio, 1982.

NONWOVEN TEXTILE FABRICS

SPUNBONDED

Spunbondeds are a significant and growing area of the nonwovens industry. Spunbondeds result from the preparation of synthetic-fiber nonwovens to achieve chemical-to-fabric routes. The introduction of fabrics prepared from continuous filaments by such routes dates from the mid-1960s in Europe and North America.

Spunbonded lines have a high unit cost but can produce vast quantities of product. Spunbondeds are no longer identified as textile substitutes but as materials in their own right. However, although limited penetration of the domestic textile field has been achieved, spunbondeds

are deficient in esthetic properties, eg, drape, conformity, and textile appeal. These deficiencies result at least in part from their bonded structure which is of rigid fiber-to-fiber links. It is likely that, in the short term, spunbondeds will grow in areas where they have a foothold, eg, in civil engineering and in the carpet, automotive, furniture, durable-paper, disposable-apparel, and coated-fabrics (qv) industries.

Properties

Appearance. Spunbonded nonwoven textiles generally are manufactured as roll goods in widths up to ca 5.5 m at basis weights of 5–> 500 g/m^2 (10^{-3}–> 10^{-1} lb/ft^2). In some cases, shaped articles such as filters may be produced as unit items but these are the exception. The basis weight of the majority of fabrics is 50–180 g/m^2. Generally, they are white, although gray and colored versions are produced.

The fabrics generally have random fibrous texture but they can have orderly arrangements of fibers or be filmlike or netlike. Thicknesses generally are 0.1–5 mm; the majority of fabrics is 0.1–2 mm thick. The fibers comprising the fabrics generally are continuous and circular but, depending on the particular fabric, they may be in short lengths and they may have relatively uniform, noncircular cross sections or nonuniform, circular or noncircular cross sections. Spunbonded nonwovens have a characteristic stiff or feltlike handle that is a result of their rigid structure in which the filaments lack shear and, therefore, have a restricted relative movement. The fabrics may have a glazed, plasticlike finish or a fibrous look.

Physical and chemical. Physical and chemical properties, eg, specific gravity, moisture absorption, electrical behavior, chemical and solvent resistance, temperature resistance, dyeability, resistance to microorganisms and insects, and light stability of spunbonded fabrics, are determined by the material composition. Reference to the equivalent textile or polymer composition generally gives a satisfactory indication of these properties.

Mechanical, hydraulic, and textile properties. Properties of spunbonded textiles are related to the mechanical properties and physical form of their fiber constituents and their geometrical construction and binding. Some important properties are listed in Table 1.

Properties that are common to conventional textiles and nonwovens often are compared when spunbondeds are used in textile applications. These include dimensional stability to washing or to dry heating.

Properties for objective comparison with conventional textiles include shear, drape, and nonrecoverable extension; numerous tests have been devised to assess fabric softness and hand. In some instances, shear, or the resistance of the fabric to in-plane shearing stress, may be the most important single factor determining fabric performance.

Manufacture and Processing

A variety of raw materials and routes is used in the manufacture of spunbonded nonwovens. Virtually all of the spunbonded fabrics that are marketed are prepared from thermoplastic polymers. The basic polymers that are used are those that are common in fiber- and film-forming operations, ie, polyamide (nylon-6 and nylon-6,6), polyester, isotactic polypropylene, and polyethylene.

Spunbonded fabrics are produced by many routes that are combinations of alternative process steps. From these combinations, a small number of chief processes can be identified by which the majority of spunbonded fabrics are produced. The predominant technology consists of continuous filament extrusion, followed by drawing, web formation by the use of some type of ejector, and bonding of the web.

Although continuous filament extrusion followed by needle punching or thermal bonding accounts for most spunbondeds, other processes include melt-blown textiles, flash-spun textiles, foam spinning, integral fibrillated nets, direct extrusion, integral extruded nets, and tack-spun textiles.

Health and Safety Factors

Considerations relevant to textiles of similar composition, which generally are inert, nontoxic, and nonallergenic, probably are broadly applicable. Flammability, allergic reactivity, food-contact considerations, lint-

Table 1. Typical Properties of Spunbonded Materials

Property	Cerex[a]	Corovin PP-S[a]	Fibertex 200[a]	Lillionette 506[a]	Lutradur H7210[a]	Net 909 P520[a]	Novaweb AB-17[a]	Terram 1000[a]	Trevira 30/150[a]	Typar 3351[a]	Tyvek 1073-B[a]
composition	nylon-6	polypropylene	polypropylene	nylon	polyester	high density polyethylene, polypropylene, or blend	polyethylene/polypropylene	low density polyethylene/polypropylene	polyester	polypropylene	high density Polyethylene
bonding	autogenous area bonded	thermally point-bonded	needled	area resin bonded	area thermally bonded	fibrillated film	foamed film	area thermally bonded bicomponent	area resin bonded	area thermally using undrawn segments	area thermally bonded (calendered)
basis weight, g/m^2	10–68	20–100		30–100	50–250	9–54	10–30	70–300	85–170	54–142	39–110
typical basis weight, g/m^2	34[b]	75[c]		60[b]	100[c]	27[d]	16–18[d]	101/170	142[c]	119[d]	75[d]
thickness, mm	0.12[e]		1.27[e]		0.38[f]	0.12[d]		0.7[d]	0.35[f]	0.33[d]	0.2[d]
breaking strength, N[l]	182/116[g]	130[h]	578[g]	103/98[g]	200/180[h]	9.8/7.8[d, i]	15/1.4[d, j]	850[g, k]	62/55[h]	512[g]	79/93[d, i]
breaking elongation, %		45[h]	125[g]	89/77[h]	40/40[h]		30/80[d]	80[g, k]	40/40[h]		26/33[d]
tear strength, N[l]	36/30[m]	15[n]	200[m]	34/32[o]	50/50[p]			250[q]		255[r]	4.5/4.5[s]
burst strength, kPa[t]	276[u]		1724[u]	245[v]				1100[w]			1180[x]
opacity	intermediate	high	high	intermediate	intermediate	intermediate	high	high			88%[y]
flex life, cycles		106[z]									> 10[aa]
air permeability, L/(dm^2·min)	1094[u]	70[bb]	3040[cc]	287[cc, dd]						304[ee]	
water permeability, L/(dm^2·min)			30[ff]					24[gg]			641[hh]
pore size, µm			80–100[ii]		300[d]			100[jj]			

[a] Manufacturers: Cerex, Monsanto Textiles Co., U.S.; Corovin, J.H. Benecke GmbH, FRG; Fibertex, Crown Zellerbach Corporation, U.S.; Lillionette, Snia Viscosa SpA, Italy; Lutradur, Lutravil Spinnvlies, FRG; Net 909, Smith and Nephew Ltd., UK; Novaweb, Bonded Fibre Fabrics, Ltd., UK; Terram, ICI Fibres, UK; Trevira-Spunbond, Hoechst AG, FRG and U.S.; Typar, E.I. du Pont de Nemours & Co., Inc., U.S. and Luxembourg; and Tyvek, E.I. du Pont de Nemours & Co., Inc., U.S.
[b] ASTM D 1910.
[c] DIN 53854.
[d] Test method not given.
[e] ASTM D 1777.
[f] DIN 53855.
[g] ASTM D 1117-1682 (grab test).
[h] DIN 53857 (strip test).
[i] N/10 mm.
[j] N/50 mm.
[k] Sample width 200 mm.
[l] To convert Newton to dyne, multiply by 10^5.
[m] ASTM D 2263.
[n] DIN 53356.
[o] ASTM D 2261.
[p] DIN 53859.
[q] BS (British Standard test) 4303-1968.
[r] Trapezoid test.
[s] Elmendorf test.

[t] To convert kPa to psi, multiply by 0.145.
[u] ASTM D 231.
[v] ASTM D 774-67.
[w] BS 4768-1972.
[x] Mullen test.
[y] Eddy opacity test.
[z] BS 3424; Method 11B.
[aa] MIT Flex test.
[bb] DIN 53887.
[cc] ASTM D 737.
[dd] L/min.
[ee] Frazier Air Porosity.
[ff] U.S. Corp of Engineers, unit cm/s.
[gg] Manufacturer's test method.
[hh] Water vapor (g/m^2/24 h).
[ii] Equivalent opening sieve.
[jj] Pore size 0$_{90}$ µm.

ing in surgical applications, etc, should be considered for health and safety. Specialized information usually is available from manufacturers and from the literature.

Outlook

There is no clear agreement on the detailed direction of the growth of the nonwovens industry. Disposables may outgrow durables from 1978–1988, although durables may lead disposables in short-term growth. All authorities agree that nonwovens will grow at a significant rate in the next few years and that a real-term sales-value growth figure of ca 8% will apply to which will be added any price inflation.

K. PORTER
ICI Fibres

M.S. Burnip and A. Newton, *Text. Prog.* **2**(3), 1 (1970).

A. Newton and J.E. Ford, *Text. Prog.* **5**(3), 1 (1973).

A.T. Purdy, *Text. Prog.* **12**(4), 1 (1983).

STAPLE FIBERS

Nonwoven fabrics comprise a diverse class of textiles that are characteristically two-dimensional and are based upon webs of fibers. In addition to the nonwoven fabrics made from staple-length fibers there are two other primary categories which are composed of (1) endless filaments (spunbond), and (2) very short fibers, primarily wood pulp (see Pulp).

In conventional textile fabrics, the basic elements are yarns or, in special cases, monofilaments. Yarns are composed of fibers that have been parallelized and twisted by a process called spinning to form cohesive and strong one-dimensional elements. In making textile fabrics, the yarns (or the monofilaments) are interlaced, looped, or knotted together in a highly regular repetitive design in any of many well-known ways to form a fabric (see Fig. 1). The fabric strength and other physical properties are derived from friction of individual fibers against each other in each yarn, and friction between adjacent yarns. In nonwoven fabrics, the basic elements are individual fibers, and tensile properties of these fabrics are derived from chemical/adhesive bonding, or frictional forces between individual fibers.

Staple fibers originally were defined as the approximately longest or functionally most important length of natural fibers, eg, cotton (qv) or wool (qv). In the present context, as applied to regenerated cellulose and synthetic fibers, staple fibers are of relatively uniform length, ca 1.3–10.2 cm, and can be processed on conventional textile machinery. Since regenerated and other extruded fibers are endless as formed, they are cut

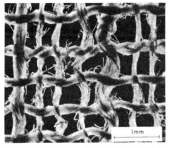

Figure 1. An ancient Egyptian woven linen fabric which closely resembles a modern woven gauze, except for the improved uniformity of weight and twist in modern yarns.

during the manufacturing process to a specific length to meet a processing or market need. Extruded fibers also are produced as continuous filaments. Woven or nonwoven fabrics that are made from staple fibers appear fuzzier and feel softer and fuller than fabrics that are made from endless filaments. The processes for forming webs from staple fibers are entirely different than those in use for continuous filaments. Staple and filament fiber webs lead to products that differ substantially in their properties.

Components and Processes

Fibers. The fibers, as defined by their chemical composition and as a result of their physical–mechanical properties, determine the ultimate fabric properties. Other determinants, such as web structure and bonding, maximize inherent fiber characteristics, eg, strength, resilience, abrasion resistance, chemical properties, and absorbency or hydrophobicity.

Any of the natural fibers that are available in commercial quantities can be used. In practice, wood pulp, which is far shorter than staple-length fibers, is the only one used in large amounts because of its high water absorbency, bulk, and low cost. Regenerated cellulose fibers of viscose rayon are of great commercial importance.

All synthetic fibers are potentially available for use. However, only polyester poly(ethylene terephthalate), nylon (types 6,6 and 6), vinyon, polypropylene (see Olefin polymers), and poly(vinyl alcohol) (see Vinyl polymers) are significant commercially; polyester fibers represent, by far, the largest volume usage in this category, although polypropylene fibers are also used in substantial amounts.

In the early stages of the nonwoven industry, the staple fibers that were used were mainly those available for the conventional textile industry. The exceptions were the small amounts of thermoplastic fibers made primarily as nonwoven binders in the 1940s and 1950s. In recent years, the fiber industry has provided polypropylene fibers and undrawn, only slightly drawn, or low melting copolymer polyester fibers as more effective thermoplastic-fiber binders, and bicomponent fibers, typically comprising a core of high melting polymer and a concentric skin of low melting polymer. The latter are growing rapidly in importance as heat-activated binders.

In addition to the substantial numbers and variety of natural, regenerated, and synthetic fibers that are readily available, mechanical differences are introduced intentionally in extruded fibers to affect processing conditions, and web and product properties. The variations that may be produced include fiber length, diameter, crimp, cross-section shape, spin finish, draw ratio, and inclusion of delustering agent.

Fiber webs. A web is the common constituent of all nonwoven fabrics. The characteristic properties of the base web are determined by fiber geometry, largely as determined by mode of web formation; fiber characteristics, both chemical and mechanical; web weight; and further processing, including compression, fiber rearrangement, and fiber entanglement. Among the important aspects of fiber geometry are the average directional fiber orientation, either isotropic or anisotropic; the longitudinal shape of the fibers, whether predominantly stretched-out, hooked, or curled; interfiber engagement or entanglement; and residual crimp. Of these, fiber orientation properties are the most important.

There are only a few ways to form fiber webs and most are derived from classical textile or papermaking processes. The methods are dry formation, including carding and air-laying, and wet-laying in large volumes of water by modified papermaking methods. Modifications in all these processes are currently under intensive development to enhance their suitability for this industry.

Bonding. Bonding implies use of an adhesive ingredient added overall or intermittently, or heat applied overall or in an embossed pattern, but bonding also can be effected by interfiber friction from mechanical entanglement. The most common commercial adhesive bonding system is based on synthetic vinyl latices. However, bonding by thermal fusion methods is rapidly growing in importance. Bonding by fiber entanglement using high pressure fine jets of water (spunbonded) shows great future promise.

Reinforcements. Fiber webs can be reinforced by woven fabrics, plastic nettings, cross-laid yarn scrims, foams, and polymer films.

Manufacture

The manufacture of nonwoven fabric roll goods is achieved by combining the raw materials, ie, fibers, reinforcing structures, binders, or fibers alone, by the processes of web forming, followed by adhesive bonding or fiber entanglement. Although there are few basic processes and components, each exists in so many variations that there are numerous possible methods.

ARTHUR DRELICH
Chicopee Division, Johnson & Johnson

A.T. Purdy, "Developments in Nonwoven Fabrics," *Text. Prog.* **12**(4), (1983).
Annual Symposium Papers, Clemson Nonwoven Fabrics Forum, Clemson University, Clemson, S.C., 1969 to present.
J.R. Starr, *The Outlook for the World Nonwoven Products Business—1982–1987*, John R. Starr, Boston, Mass., 1982.

NOREPINEPHRINE. See Epinephrine and norepinephrine.

NOVOLOID FIBERS

Novoloid fibers are three-dimensionally cross-linked phenolic–formaldehyde fibers typically prepared by acid-catalyzed cross-linking of a melt-spun novolac resin with formaldehyde (see Phenolic resins). They are infusible and insoluble, and possess physical and chemical properties that clearly distinguish them from all other types of synthetic and natural fibers (see Fibers, chemical). Novoloid fibers are used in flame- and chemical-resistant textiles and papers, in heat- and chemical-resistant composite materials, and as precursors for carbon and activated-carbon fibers, textiles, and composites.

The generic term novoloid has been recognized by the United States Federal Trade Commission as designating a manufactured fiber containing at least 85 wt% of a cross-linked novolac. Novoloid fibers were invented at the Carborundum Company, Niagara Falls, New York; they are commercially manufactured under license by Gun'ei Chemical Industry Co., Ltd. (Japan) and sold through Nippon Kynol, Inc. (Japan) and American Kynol, Inc., under the trademark Kynol.

Fiber Properties

The properties of novoloid fibers are summarized in Table 1. The fibers are light gold in color and darken gradually to deeper shades with age and exposure to heat and light, there is no significant concomitant change in other properties.

Thermal properties. Novoloid fibers are highly flame resistant and have a limiting oxygen index (LOI) of 30–34. When exposed to flame, novoloid fibers do not melt but gradually char until completely carbonized without losing their initial fiber form and configuration. Thus, a 290 g/m^2 novoloid woven fabric withstands an oxyacetylene flame at 2500°C

Table 1. Textile Properties of Novoloid Fibers

Property	Value
diameter, μm (tex[a])	14–33 (0.22–1.1)
specific gravity	1.27
tensile strength, GPa[b] (N/tex[c])	0.16–0.20 (0.12–0.16)
elongation, %	30–60
modulus, GPa[b] (N/tex[c])	3.4–4.5 (2.6–3.5)
loop strength, GPa[b] (N/tex[c])	0.24–0.35 (0.19–0.27)
knot strength, GPa[b] (N/tex[c])	0.12–0.17 (0.10–0.13)
elastic recovery[d], %	92–96
work-to-break, J/g (= mN/tex[e])	26–53
moisture regain at 20°C, 65% rh, %	6

[a] To convert tex to den, multiply by 9.
[b] To convert GPa to psi, multiply by 145,000.
[c] To convert N/tex to gf/den, multiply by 11.33; N/tex = GPa ÷ density.
[d] At 3% elongation.
[e] To convert mN/tex to gf/den, multiply by 0.01133.

for $\geq$ 12 s without breakthrough. This behavior is attributable to the cross-linked, infusible structure of the fibers and to their high (76%) carbon content.

Moreover, since novoloid fibers are composed only of carbon, hydrogen, and oxygen, the products of combustion are principally water vapor, carbon dioxide, and carbon char. Moderate amounts of carbon monoxide may be produced under certain conditions, but the HCN, HCl, and other toxic by-products of combustion of other flame-resistant organic fibers are absent. The toxicity of combustion products is thus very low or negligible; smoke emission is also minimal, less than that of virtually any other organic fiber.

Despite their high resistance to severe attack by flame, however, novoloid fibers are not high temperature fibers in the usual sense of the term. The practical limits for long-term application are about 150°C in air and 200–250°C in the absence of oxygen; above these limits the fibers are subject to gradual oxidative degradation and/or carbonization, and slowly lose weight and strength.

Chemical resistance. Novoloid fibers display excellent chemical and solvent resistance. They are attacked by concentrated or hot oxidizing acids (sulfuric, nitric) and strong bases, but are virtually unaffected by nonoxidizing acids (including hydrofluoric and phosphoric acids), dilute bases, and organic solvents.

Other properties. Novoloid-fiber materials display excellent thermal and electrical insulating properties. Retention of properties at very low temperatures is excellent. Efficiency of sound absorption is high. Ultraviolet radiation, although leading to deepening of color, has minimal effect on fiber properties; resistance to γ-radiation is also high.

Novoloid fibers typically contain 5–6 wt% of methylol groups. These groups are available for further cross-linking with resin and elastomer matrix materials and play a key role in the formation of heat- and chemical-resistant novoloid fiber composites.

Carbon fibers from novoloid. Novoloid fibers and textiles are readily converted by heating in an inert atmosphere to carbon, at high yields of 55–57 wt%. Novoloid-based carbon fibers are amorphous in structure. They have low modulus and moderate strength, in comparison to the high-modulus reinforcing fibers based on polyacrylonitrile (PAN) and pitch. Novoloid-based carbon fiber is soft and pliable, produces little fly or dust on processing, and has good lubricity. Its heat, chemical, and electrical characteristics are similar to those of other carbon fibers.

Activated-carbon fiber from novoloid. Novoloid fibers, felts, and fabrics are excellent precursors for activated-carbon-fiber materials, having effective surface areas (BET) of 2000 m²/g and greater. Pore configuration and the high surface-to-volume ratio of the fibers, compared to granular activated carbon, permit extremely rapid adsorption and desorption. In comparison with activated-carbon fibers produced from other precursors, novoloid-based fibers have significantly higher surface area, strength, and flexibility.

Health and Safety Factors

Toxicity evaluation in accordance with regulations under the Federal Hazardous Substances Act demonstrate that the fiber is neither toxic nor highly toxic under the oral and skin contact categories, is not an eye or primary skin irritant, and has a low order of acute inhalation hazard. Fiber diameters are significantly larger than the range associated with asbestos hazards, and the fiber does not fibrillate. Toxicity of combustion products and smoke evolution are minimal.

Under certain conditions, large, hot masses of tightly packed novoloid fibers may become subject to a combustive degradation process known as punking. This is attributed to exothermic decomposition of peroxides formed at the methylene linkages of the phenolic polymer, and may become self-sustaining under conditions where more heat is generated than is removed by ventilation or other means.

Punking may easily be prevented by appropriate control of storage and use. Thus, novoloid materials should not be stored in bulk at high temperatures or subjected to lengthy heat treatment over 120°C without adequate ventilation, and should be cooled below 60°C after heat treatment or high temperature use. Applications involving prolonged exposure to high temperatures with limited opportunity for heat

escape—such as steam-pipe insulation—should be avoided. Punking does not occur in the absence of oxygen.

JOSEPH S. HAYES, JR.
American Kynol, Inc.

J. Economy in M. Lewin, S.M. Atlas, and E.M. Pearce, eds., *Flame-Retardant Polymer Materials*, Vol. 2, Plenum Press, New York, 1978, pp. 210–219.

J. Economy, L.C. Wohrer, and F.J. Frechette, *J. Fire Flammability* **3**, 114 (April 1972).

NUCLEAR MAGNETIC RESONANCE.   See Analytical methods.

NUCLEAR REACTORS

INTRODUCTION

Engineered nuclear fission reactors supply an appreciable fraction of the electrical energy used in many countries. Their development and use is in a controversial phase such as experienced in the history of industrial development with the use of coal, steam power for railways and ships, the internal-combustion engine, and even electricity itself.

One remarkable characteristic (established early in the history of nuclear reactors operating at high power density) is that, after a shutdown, it may not be possible to restart the chain reaction for 15–40 h. This is owing to transient poisoning of the fission-chain reaction by one of the fission products, ^{135}Xe.

Safety Aspects

Much attention has been paid to safety in establishing the technology of nuclear reactors. The training of operators derived from accumulated experience with these systems must not be underestimated. Although a trained mind is essential in a complex situation, certainty in action requires accurate and adequate information from the necessary monitoring instruments. Containment of the fission products is the basic requirement for minimizing exposure to radiation.

Heat Removal

The rate of the chain reaction can increase to levels at which the safe removal of the heat from fission and, later, from residual fission products becomes the principal engineering concern. Thus, many coolants or caloporteurs have been used, including gases, especially air, CO_2, He, and dry steam; liquids, especially water, single-phase pressurized water, two-phase boiling water or fog, heavy water, organic liquids, especially terphenyl and hydrogenated terphenyl, which is liquid at room temperature; the alkali metals, sodium, NaK alloy, potassium, and lithium; and mercury, molten bismuth, and molten salts, especially fluorides.

Fuels

Depending on the purpose of the reactor, many fuels have been used, although UO_2, ThO_2, PuO_2, and mixtures of these as sintered, high density pellets have been used extensively.

Special Materials

Tritium is produced in reactors not only as a direct fission product, but also by the action of fast neutrons on carbon, nitrogen, and other light elements (see Deuterium and Tritium). Helium is troublesome because it promotes the nucleation of bubbles, especially in nickel and nickel alloys. Zirconium hydride is used in TRIGA (Training Reactor, Isotopes General Atomic) as a solid moderator.

W. BENNETT LEWIS
Queen's University

Reviewed by
G.P.L. WILLIAMS
Atomic Energy of Canada Ltd.

Proceedings of the Four United Nations International Conferences on the Peaceful Uses of Atomic Energy (UNICPUAE), Geneva, Switz., 1955, 1958, 1964, and 1971.

The International Conference on Nuclear Power and its Fuel Cycle, International Atomic Energy Agency (IAEA) Salzburg, Austria, May 2–13, 1977.

International Conference on Nuclear Power Experience, International Atomic Energy Agency (IAEA), Vienna, Austria, Sept. 13–17, 1982.

NUCLEAR FUEL RESERVES

Light-water reactors are the chief type of nuclear power reactor in operation. They are fueled with uranium that has been enriched to ca 3% of ^{235}U from the naturally occurring concentration of 0.71%. Some experimental thorium-fueled reactors are in operation, but they are not expected to be in significant use until well after 2000.

Uranium

The earth's crust contains 2–3 ppm uranium, alkalic igneous rocks tending to be more uraniferous than basic and ferromagnesian igneous rocks. Uranium combines readily with oxygen. Its solubility and distribution in rocks and ore deposits depend largely on its valence state: uranium is highly soluble in the six-valent state and relatively insoluble in the four-valent state. Uraninite, the most common mineral in uranium deposits, contains the four-valent ion.

Resources. WOCA (world outside centrally planned economies) reasonably assumed uranium resources are ca 2×10^6 metric tons. Study of depositry and regional geology suggests that total resources are several times larger.

Availability. A principal factor in considering fuel-resources availability is the time and effort needed to discover and develop the resources and attainable production levels. In 1985, annual planned (WOCA) production capability could reach ca 46,000 t U, production coming chiefly from resources costing $80/kg U. WOCA production capability in 1995 could be 75,000 t U.

Demand. Annual uranium requirements for WOCA were ca 33,000 t in 1982. The demand might be 65,000–90,000 t U in 2000 and 65,000–320,000 t U in 2025. The wide range in the later period reflects the variation in projected nuclear growth and the types of reactors in use. The greater demand would be predominantly for light-water reactors without fuel recycle; the lower-demand cases would involve substantial employment of liquid-metal fast-breeder reactors.

Low grade uranium resources. Low grade resources include shales, seawater, mill tailings, and enrichment-plant tails.

Thorium

Resources. The crustal concentration of thorium is 10–20 ppm. The element is dispersed widely in variable amounts in a number of different rock types. It is commonly associated with rare earths, niobium, titanium, and zirconium.

Demand and supply. The demand for thorium in recent years has been limited to a small amount used for nonnuclear purposes, eg, gas-lamp mantles, catalysts, refractories, and in high temperature alloys. No significant market for thorium as a nuclear fuel has developed; although, if the advanced reactors using thorium were employed in large numbers toward the end of the century, cumulative WOCA demand through 2025 might reach 150,000 t Th.

JOHN A. PATTERSON
International Atomic Energy Agency

Fuel and Heavy Water Availability, Report of Working Group 1, International Nuclear Fuel Cycle Evaluation, Vienna, Austria, International Atomic Energy Agency STI/PUB/534, UNIPUB, Inc., New York, 1980, pp. 222–223.

Uranium Resources, Production, and Demand, Joint Report of OECD Nuclear Energy Agency and the International Atomic Energy Agency, OECD Publications Center, Washington, D.C., 1983.

Nuclear Energy and Its Fuel Cycle, Report of the OECD Nuclear Energy Agency, OECD Publications Center, Washington, D.C., 1982.

WATER CHEMISTRY OF LIGHT-WATER REACTORS

As of January 1980, there were 63 operating commercial nuclear power stations in the United States. Forty-one were pressurized-water reactors (PWRs), and 22 were boiling-water reactors (BWRs). By the end of 1983, there were 204 operating plants.

In a PWR, a closed circuit of high pressure, high temperature water transfers heat from the reactor core to once-through (OTSG) or recirculating U-tube steam generators, where steam is produced. Current secondary-cycle-chemistry control practices parallel those of fossil plants; however, impurity levels are maintained much lower. Initially, secondary water chemistry specified for U-tube generators was based on sodium phosphates. Subsequent to observations of corrosion attack, all volatile treatment (AVT) was adopted at most units. Several new types of corrosion have been observed subsequently. Corrosion of OTSGs has not been as extensive or as serious a problem as with U-tube steam generators. A summary of initial AVT specifications is given in Tables 1 and 2.

Boric acid is added to the PWR primary system to compensate for fuel consumption and to control reactor power. A hydrogen overpressure is maintained, and small amounts of base are added to reduce corrosion rates not only to address component-integrity concerns but also to minimize shutdown radiation levels.

In a BWR, steam is generated on the surface of the nuclear fuel and is used to drive the turbine generator. Since chemical additions are not made, materials corrosion is dependent primarily on coolant oxygen concentrations, which are governed by radiolytic decomposition of water and the steam–liquid oxygen equilibrium.

BWR chemistry specifications are given in Table 3. Because of the importance of maintaining fuel integrity in the direct cycle and of minimizing radiation levels on out-of-core surfaces, transport of corrosion products to the reactor by the feedwater is of considerable importance.

Table 1.   Initial PWR Feedwater Specifications for All-Volatile-Treatment (AVT) Operation[a]

Property	Babcock & Wilcox	Combustion engineering	Westinghouse
pH	9.3–9.5[b]	8.8–9.2[c]	8.8–9.2[c]
	8.5–9.3[d]	9.2–9.5[e]	up to 9.6[e]
O_2, ppb	< 7	< 10	< 5
shutdown		100	
N_2H_4, ppb	20–100	10–50	$[O_2] + 5$
acid conductivity, μS/cm	< 0.5	< 0.5	
abnormal[f]	> 0.5 to ≤ 1 (24 h)	> 1.5 (4 h)	
	> 1 to ≤ 2 (12 h)		
shutdown	> 2		
iron, ppb	< 10	< 10	< 10
copper, ppb	< 2	< 10	< 5
total solids, ppb	< 50		
total silica (as SiO_2), ppb	< 20	< 10	
ammonia, ppm		< 1	< 0.5

[a] Tabulated values are for normal operation unless noted otherwise.
[b] Carbon-steel feedwater heater tubes or combinations of carbon steel and stainless-steel feedwater and/or reheater tubes.
[c] With copper alloys in feedwater heaters, reheaters, or condenser.
[d] With stainless-steel feedwater heater tubes and stainless-steel or copper–nickel reheater tubes.
[e] With no copper alloys in feedwater heaters, reheaters, or condenser.
[f] Corrective action or shutdown recommended within indicated time.

Recently, incidences of intergranular stress-corrosion cracking (IGSCC) in stainless-steel piping have reduced the availability of BWR systems significantly. Numerous remedies for the problem are being evaluated and applied.

Table 2. Initial PWR Recirculating Steam Generator Blowdown Specifications for AVT Operation

| | Combustion engineering | Westinghouse | |
		Freshwater	Seawater or brackish water
pH	8.2–9.2	8.5–9.0	8.5–9.0
abnormal[a]	< 7.5 or > 9.5 (4 h)	8.5–9.2 (2 wk)	8.0–9.2 (2 wk)
shutdown	10.5	< 8.5 or > 9.4	< 8.0 or > 9.4
specific conductivity, μS/cm	< 7		
abnormal[a]	> 15 (4 h)		
suspended solids, ppm	< 1		
abnormal[a]	> 10		
free hydroxide, ppm		< 0.05	< 0.05
abnormal[a]		> 0.05 to ≤ 0.34 (24 h)	> 0.05 to ≤ 0.34 (24 h)
shutdown	5	> 0.34	> 0.34
silica, ppm	< 1		
abnormal[a]	> 10 (4 h)		
cation conductivity, μS/cm		< 2.0	< 2.0
abnormal[a]		> 2 but ≤ 7 (2 wk)	> 2 but ≤ 120 (2 wk)
shutdown		> 7	> 120

[a] Corrective action or shutdown recommended within indicated time.

Table 3. BWR Water-Quality Specifications for Normal and Abnormal Conditions

	General Electric
Reactor water	
conductivity, μS/cm	≤ 1
abnormal (2 wk/yr allowed)	> 1 but ≤ 10
shutdown	> 10
chloride, ppm	≤ 0.2
abnormal (2 wk/yr allowed)	> 0.2 but ≤ 0.5
shutdown	> 0.5
pH	5.6 to 8.6[a]
Feedwater	
conductivity, μS/cm	≤ 0.1
pH	6.5 to 7.5
metals, ppb	< 15 total
	< 2 Cu
oxygen, ppb	> 20 but < 200

[a] Inferred from conductivity limits.

S.G. SAWOCHKA
W.L. PEARL
NWT Corporation

R.B. DeWitt and J.G. Lewis, "Steam Generator Operation and Water Chemistry Experiences at Palisades Plant," *paper presented at the American Nuclear Society Conference*, San Francisco, Calif., Nov.–Dec. 1977.

S.G. Sawochka, N.P. Jacob, and W.L. Pearl, *Primary System Shutdown Radiation Levels at Nuclear Power Generating Stations*, Research Project 404-2 Final Report, Electric Power Research Institute, Palo Alto, Calif., Dec. 1975.

J. Blok, S.G. Sawochka, and D.T. Snyder, *Corrosion Product Deposits on Fuel at the Nine Mile Point Boiling Water Reactor*, General Electric, San Jose, Calif., Aug. 1973 (NEDO-13322R).

ISOTOPE SEPARATION

The difficulty and high cost of isotope separation has limited the use of separated isotopes in nuclear reactors to specific cases where no nonisotopic substitutes are available. The most important example is ^{235}U, the most abundant, naturally occurring, fissionable material. Other isotopes that are separated for nuclear use are ^{2}H or deuterium, D, which is used as D_2O as a neutron moderator in nuclear reactors and is a probable reactant in thermonuclear reactors; ^{3}H or tritium, T, which is produced in nuclear reactors and is a reactant in thermonuclear reactors under development; boron, ^{10}B, whose high neutron cross section is used in control rods and safety devices for reactors; and lithium, ^{7}Li, which is used in reactor cooling water systems because of its low thermal neutron cross section (see Deuterium and tritium; Fusion energy; Isotopes; Radioisotopes).

GEORGE M. BEGUN
Oak Ridge National Laboratory

H. London, ed., *Separation of Isotopes*, George Newnes, Ltd., London, UK, 1961.

S. Villani, *Isotope Separation, American Nuclear Society Monograph*, American Nuclear Society, Hinsdale, Ill., 1976.

H.K. Rae, ed., "Separation of Hydrogen Isotopes," *ACS Symp. Ser.* **68**, (1978).

FUEL-ELEMENT FABRICATION

The fuel assembly forms the basic unit of nuclear-reactor core. Fuel assemblies are fabricated by processing fertile and fissile materials into suitable chemical and physical forms, encapsulating the materials in a protective metallic sheath to produce a fuel rod, and assembling the fuel rods into the required configuration. Different reactor types, eg, water-cooled, gas-cooled, and sodium-cooled breeder reactors, require different fuel-assembly designs. The most widely used fuel assemblies are the metal-clad multirod bundles that are used in light-water reactors (LWR) (Fig. 1) for electric power generation. Similar fuel assemblies are used in breeder reactors that produce more fuel than they consume.

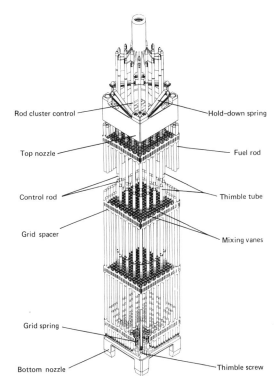

Rod cluster control — Hold-down spring
Top nozzle — Fuel rod
Control rod — Thimble tube
Grid spacer — Mixing vanes
Grid spring
Bottom nozzle — Thimble screw

Figure 1. Light-water reactor fuel assembly.

W.L. LYON
Westinghouse Nuclear Fuels Division

J.M. YATABE
L.H. RICE
Hanford Engineering Development Laboratory

Quality Assurance and Control in the Manufacture of Metal-Clad UO₂ Reactor Fuels, International Atomic Energy Agency, Vienna, Austria, 1976.

Y.S. Tang and co-workers, *Thermal Analysis of Liquid Metal Fast Breeder Reactors*, American Nuclear Society, La Grange Park, Ill., 1978.

H.W. Graves, Jr., *Nuclear Fuel Management*, John Wiley & Sons, Inc., New York, 1979.

CHEMICAL REPROCESSING

Fuel elements that are discharged from light-water nuclear-power reactors (LWRs) typically contain uranium, 0.79 wt% ^{235}U and 0.8 wt% ^{239}Pu, which can be recovered by chemical reprocessing. However, the economic incentive for reprocessing fuel for recycle into light-water reactors in the United States is marginal. Recycle of breeder-reactor fuel is, of course, essential to the breeder concept. Chemical reprocessing must provide for the high recovery of uranium and plutonium, and their separation from each other and from hazardous radioactive contaminants. Reprocessing implies the efficient separation of fertile materials, eg, ^{238}U or ^{232}Th; from one another, eg, ^{239}Pu from ^{235}U and/or ^{233}U; from the highly radioactive fission products; and from materials that are undesirable neutron poisons, including certain fission-product isotopes (see Nuclear reactors—isotope separation).

Purex process. The process used in the United States and abroad is almost exclusively some form of the Purex aqueous solvent extraction process involving tributyl phosphate. Although other solvents have been employed and other processes, ie, precipitation, ion exchange (qv), and volatility processes, have been used, the Purex process and a variant for thorium-based materials, ie, the Thorex process, are almost universally accepted.

The fuel is initially dissolved in nitric acid after an economically optimized period to allow radioactive decay of most of the short-lived isotopes that account for most of the heat and the high intensity gamma radiation associated with spent fuel elements. Preprocessing decay time, which largely eliminates troublesome fission products, eg ^{131}I, generally is one to two years for thermal-reactor fuels and three to nine months for breeder fuels.

Dissolution of the solid fuel is accompanied by complete release of the rare-gas fission products, krypton and xenon, into the off-gas system. Since the dissolver overheads also ordinarily contain fractions of some fission products, eg, iodine, tritium, ruthenium, carbon-14 as CO_2, and tellurium, the off-gas usually is treated to remove or minimize the discharge of these gaseous fractions, as well as entrained particles or aerosols.

The multicycle Purex process depends primarily on countercurrent liquid–liquid extraction techniques that involve the transfers of a large number and variety of solutes between aqueous nitrate dissolver solution and the immiscible, less dense, organic solution (see Extraction, liquid–liquid extraction). The extraction is repeated several times to achieve the desired degree of purification.

Typically, recoveries of > 99.8% of plutonium and uranium and decontamination factors of > 10^6 are achieved.

W.E. Unger
Oak Ridge National Laboratory

R.H. Rainey, W.D. Burch, M.J. Haire, and W.E. Unger, "Fuel Cycle Costs for Alternative Fuels," *Fuel Cycle for the 80's Conference*, CONF-800943, Gatlinburg, Tenn., 1980.

Final Safety Analysis Report—Barnwell Nuclear Fuel Plant, Separations Facility, Docket 50-332, Allied-Gulf Nuclear Services, Barnwell, S.C., 1973.

W.P. Bebbington, "Reprocessing of Nuclear Fuels," *Scientific American*, Dec. 1976, pp. 30–41.

FAST BREEDER REACTORS

A breeder reactor is a nuclear reactor that produces more nuclear fuel or fissile material than it consumes. Although the emphasis for the past 30 yr has been on the liquid–metal fast breeder reactor (LMFBR) and the development of the uranium–plutonium fuel cycle, use of the

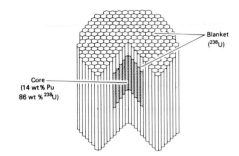

Figure 1.   Core and blanket arrangement for a typical LMFBR.

thorium–uranium fuel cycle for breeder reactors also has been studied, primarily for thermal reactor systems. Consideration has been given to the use of this cycle in fast breeder reactors as a result of the extensive studies arising from considerations of the United States nonproliferation policy.

Design

A fast breeder reactor can be designed in core and blanket configurations to produce more fissionable material than it consumes. This is achieved by converting fertile material, eg, ^{238}U or ^{232}Th, in the blanket into ^{239}Pu or ^{233}U, respectively. Thus, in a typical LMFBR, the central region of the reactor (see Fig. 1), ie, the reactor core, contains the plutonium-bearing assemblies which are comprised of a mixture of plutonium fuel and fertile uranium oxides. The blanket region surrounds the core and holds the uranium-bearing assemblies. Most of the fission occurs in the core, and most of the breeding occurs in the blanket. At intervals, the reactor is shut down to remove selected core and blanket assemblies and to replace them with fresh ones. The spent assemblies then can be reprocessed chemically to reclaim the fuel and the fertile materials for eventual recycling into new and existing reactors.

Breeder Reactor Types

The ability of a reactor to sustain a sufficient conversion or breeding ratio depends on a number of engineering and physical factors that influence the reactor's neutron environment. Average neutron energy is influenced by reactor core geometry, the composition of the fuel, and the presence of other materials, eg, the coolant and construction materials. Design decisions can reflect the different options available; the various factors often are interrelated, and a number of permutations and combinations are possible. The liquid–metal fast breeder reactor (LMFBR), the gas-cooled fast reactor (GCFR), and the light-water breeder reactor (LWBR), have received the most attention. Primary emphasis has been on the LMFBR, which is the most highly developed of these reactor types, and on the uranium/plutonium fuel cycle; eleven experimental and demonstration LMFBRs are now in operation worldwide.

Peter Murray
Westinghouse Electric Corporation

International Nuclear Fuel Cycle Evaluation, INFCE Summary Volume, INFCE/PC/2/9, IAEA, Vienna, Austria, Jan. 1980.

The Development of Breeder Reactors in the United States, H. Kouts, ed., *Annual Review of Energy 1983*, Vol. 8. pp. 385–413.

WASTE MANAGEMENT

On the basis of their packaging, shipping, and possible disposal requirements, fuel-cycle wastes can be grouped into four principal categories: spent fuel, transuranic (TRU) wastes, non-TRU wastes, and ore tailings. The volumes, radioactivity levels, thermal powers, and estimated numbers of shipments to possible disposal sites are given in Table 1 for each waste type based on the production of 1000 MW·yr of electricity.

J.O. Blomeke
Oak Ridge National Laboratory

Table 1. Fuel Cycle Wastes from the Production of 1000 MW·yr of Electricity[a]

Category of waste	Volume, m³	Activity[b], GBq[c]	Power[b], kW	Number of shipments	Possible disposition
spent fuel	14[d]	6.36×10^9	970	11	reprocessed
transuranic (TRU) wastes					
high level solidified	3.1	6.29×10^9	960	2	geologic disposal
cladding hulls	2.7[e]	7.4×10^7	10	2	geologic disposal
intermediate level	140[f]	4.4×10^5	0.057	66	geologic disposal
low level	480[f]	1.9×10^6	0.04	17	geologic disposal
non-TRU wastes					
noble gases	0.1[g]	9.3×10^6	0.37	0.4	near-surface storage
iodine	0.05[h]	37		0.2	geologic or ocean disposal
carbon-14	0.4[i]	740		2	geologic or ocean disposal
LWR tritium	140	2×10^4		9	deep wells
fission-product (FP) tritium	0.35[j]	7×10^5	7×10^{-4}	1	near-surface storage
low level	2,400[f]	7×10^4	0.007	180	burial grounds
ore tailings	42,000	1.9×10^4	0.011		surface storage

[a] Mixed (U–Pu) oxide-fueled light-water reactors (LWRs).
[b] Radioactivity and thermal power at the time of waste generation, which is 160 d of fuel discharge.
[c] To convert Bq to Ci, divide by 3.7×10^{10}.
[d] Based on nominal outside dimensions of the fuel assemblies.
[e] Compacted to 70% of theoretical density.
[f] Volume reduced by a factor of 10 by incineration and compaction.
[g] Pressurized at 15 MPa (150 atm) in standard 50-L steel cylinders.
[h] As $BaIO_3$ incorporated in cement.
[i] As $BaCO_3$ incorporated in cement.
[j] As tritiated $Ca(OH)_2$.

U.S. Department of Energy, *Management of Commercially Generated Radioactive Waste Environmental Impact Statement*, DOE/EIS-0046, Vols. 1–2, National Technical Information Service, Springfield, Va., Nov. 1980.

U.S. Department of Energy, *Technology for Commercial Radioactive Waste Management*, DOE/ET-0028, Vols. 1–5, National Technical Information Service, Springfield, Va., May 1979.

Proc. Symp. Management of Wastes from the LWR Fuel Cycle, Denver, Colorado, July 11–16, 1976, CONF-76-0701, National Technical Information Service, Springfield, Va.

SAFETY IN NUCLEAR FACILITIES

In many countries, including the United States, concern for the safety of nuclear power plants is curtailing their growth rate. The Nuclear Regulatory Commission licenses and regulates nuclear facilities such that there is no undue restraint to the health and safety of the public. One of many requirements is that each nuclear plant must limit occupational exposures to 5 rem/yr and exposures in unrestricted areas to 0.5 rem/yr. In practice, the public exposure at nuclear plant sites generally is some orders of magnitude less. However, it generally is assumed that, for the purpose of evaluating the risks of radiation, the probability of inducing cancer is proportional to dose, no matter how small the dose. If the preceding hypothesis is true (and it may not be at very low dose), the benefits of the activity giving the exposure must be weighed against the risks and the costs of the low level exposure. Furthermore, the average public exposure from nuclear power plants is 0.1 mrem/yr, which is very small in comparison to the routine exposure of ca 200 mrem/yr for the average U.S. citizen.

The Nuclear Fuel Cycle

The nuclear power plant that produces electricity is only one element among many in the nuclear fuel cycle. The other elements include mining, milling, transportation, conversion, enrichment, fuel fabrication, reprocessing, and waste disposal, and each has associated risks that must be considered when assessing the total risk associated with generating electricity by nuclear power. Such risks have been reported in numerous studies and are being investigated by a number of government and nongovernment agencies, eg, DOE, NRC, EPA, and EPRI (Electrical Power Research Institute).

In an EPRI report, only the radiological risks associated with the various elements of the nuclear fuel cycle are addressed, rather than the

Table 1. Summary of the Nuclear Fuel-Cycle Radiological Risks (Consequences Times Probabilities) Involved in the Production of One Gigawatt-Year of Electric Power

Cycle step	Dose (whole body) person-rem	Health effects, number of latent cancers
nuclear power plant operation	257	0.02
mining and milling		
accident	not addressed	not addressed
routine	0.2	2×10^{-5a}
reprocessing	2×10^{-4}	3×10^{-8}
mixed-oxide fuel fabrication	4×10^{-2}	3×10^{-6}
transportation	3×10^{-2}	3×10^{-6}
waste repository		
preclosure	4×10^{-5}	2×10^{-10}
long-term (10^6 yr)	5×10^{-11}	5×10^{-15b}
natural background radiation	7×10^{4c}	

[a] Based on 100 cancer deaths per million person-rems.
[b] Based on 30-yr individual dose rate integrated over 10^6 yr and a population of 10^6.
[c] 3×10^8 persons × 150 mrems/685 GW$_e$ in 2005.

totality of all risks associated with nuclear power production. The results of the EPRI study are summarized in Table 1 and show that the radiological risks of the nuclear fuel cycle contribute ca 1% of the risks of generating nuclear electric power; thus, the radiation risk of nuclear power plants closely approximates the full risk of the entire fuel cycle.

Comparative Risks

The American Nuclear Society stated that the risks associated with the production of electric power utilizing nuclear energy are small when compared with either the risks associated with other methods of electricity production or the societal risks to which people are exposed in their daily lives. In a recent report on serious accidents in all types of power reactors, not a single death or serious injury resulting from radiation exposure from commercial nuclear power plants in the United States could be identified; although there have been some injuries and deaths resulting from industrial-type accidents in nuclear plants. Furthermore, the risk from nuclear plants is shown to be much less than that from most other accepted risks such as fires, plane crashes, dam failures, etc.

Nuclear Reactors

Since it is probable that only light-water-cooled reactors (LWRs) will be built for commercial power production in the United States for the remainder of this century, the following discussion is directed toward LWRs.

Safety philosophy. The safety philosophy has been codified in the regulations and regulatory standards that govern the licensing of nuclear facilities. Implicit in these regulations is the concept of defense in depth, with which three levels of safety are identified.

The first level of safety is achieved by designing a plant for maximum safety in normal operation and maximum tolerance for system malfunctions. The second level of safety is based on the assumption that incidents occur despite care in design, construction, and operation. Safety systems should be provided to protect operators and the public and to prevent or minimize damage when such incidents occur. The third level of safety is the provision of additional safety systems as appropriate, based on the evaluation of the effects of hypothetical accidents, where some protective systems are assumed to fail simultaneously during the accident that they are intended to control.

Other safety considerations in design include quality assurance, surveillance testing, redundancy and coincidence, and independence and diversity.

Safety features. A nuclear power plant abounds with safety features that are installed either to fulfill operational requirements (including safety) of the licensee, or to meet the preceding safety philosophy as

Table 2. PWR Safety Features, Functions, and Purposes

System	Function	Purpose
Reactivity control		
safety rods	shut down reactor	stops fission-heat generation
boron injection	maintain reactor shutdown	prevents fission-heat generation from reoccurring
Pressure control		
pressurizer	maintain system pressure	maintains coolant in liquid state
relief valves	prevent system overpressure	prevents equipment leaks and/or rupture
Emergency core cooling		
high pressure injection	naintain coolant at pressure	maintains cooling of core at pressure
accumulators	provide cooling during large-pipe failure	maintains cooling during transient blowdown event
low pressure injection	maintain cooling of core after blowdown	prevents core from overheating
Emergency heat removal		
auxiliary feedwater system	provide heat sink for reactor coolant	ensures core coolability
decay-heat removal system	maintain coolness of core	prevents core from overheating
Containment		
isolation system	enclosure of radiation	minimizes release of radioactivity to the environment
spray system	control containment temperature and pressure	minimizes containment leakage
recirculation system	cool and cleanup containment atmosphere	minimizes release of radioactivity to the environment
hydrogen recombination	recombine hydrogen with oxygen	minimizes explosion potential
Vital auxiliaries		
emergency power system	supply power for emergency needs	permits plant safety systems to function in the event of loss of normal power
component cooling water	supply cooling water to equipment needed in an emergency	permits plant safety systems to function in the event of an emergency
instrument air	supply instrument air for emergency needs	permits plant safety systems to function in the event of loss of normal instrument air

interpreted by the NRC. Some of the more significant safety features are listed in Table 2.

Accidents. The occurrence and consequences of accidents in nuclear power plants is the ultimate measure of the effective application of the previously described safety concepts to reactor technology. Definitions in the United States of a reactor accident vary widely. Thus, the annual U.S. nuclear accident rate may range from ca 1 to over 4500 annually. The former number is derived from a recent study in which the criteria were death or significant injury, significant activity release, core damage, inadvertent criticality, a precursor to a serious accident, or consequent significant recovery cost. At the other extreme are licensee event reports to the NRC which include all violations of the plant's technical specifications, ie, operating limits and conditions specified in the plant's operating license.

Three accidents in the history of commercial nuclear power in the United States are exceptional because of their significance in terms of cost: fuel melting at Fermi, fire at Browns Ferry Unit 1, and core damage at Three Mile Island Unit 2. It should be noted that there were no deaths or public injuries as consequences of any of these accidents.

Other Nuclear Fuel-Cycle Facilities

Because the nuclear power plant constitutes over 99% of the radiation risk resulting from the nuclear fuel cycle, the radiation risk from the balance of the fuel cycle is small and is distributed among many facilities. There are three potential risk mechanisms: inadvertent criticality, inhalation exposure, and direct radiation. The last mechanism is accommodated readily by adequate radiation shielding, which is well documented.

W.B. COTTRELL
Oak Ridge National Laboratory

H.W. Bertini, *Descriptions of Selected Accidents that Have Occurred at Nuclear Reactor Facilities*, Report ORNL/NSIC-176, Oak Ridge National Laboratory, Oak Ridge, Tenn., Apr. 1980.

Three Mile Island: A Report to the Commissioners and to the Public (The Rogovin Report), Vol. 1, available from the NRC and NTIS, Jan. 1980.

American National Standard for Nuclear Criticality in Operations with Fissionable Materials Outside Reactors, ANS Standard N16.1-1975, American Nuclear Society, LaGrange Park, Ill., 1975.

SPECIAL ENGINEERING FOR RADIOCHEMICAL PLANTS

Concerns that are unique to the nuclear industry result from the handling of radioactive and potentially fissionable materials. The special engineering considerations are categorized in terms of the design for radiation resistance, the nuclear safety design, the mitigation of occupational radiation exposure, and the minimization of environmental impact.

Radiation Resistance

Although a spectrum of radiation types, eg, alpha particles, beta particles, x rays, gamma rays, and neutrons, is emitted from spent nuclear fuel, the primary equipment-design consideration is the effect of gamma radiation, especially on organic materials.

Specially developed organic materials for use in gamma radiation fields have lifetimes that are improved by a factor of 10–100 over the standard materials intended to function for ca 100–10,000 h. They generally are used when no design replacement is available. Local shielding can be provided around the organic material as a means of extending its useful life. Inorganic lubricants, eg, molybdenum disulfide, can also be substituted for organic lubricants.

Safety Design

Fissionable material, ie, the reusable fuels for nuclear reactors, including ^{233}U from the thorium fuel cycle or ^{235}Pu from the uranium fuel cycle, are capable of spontaneous criticality if accumulated in adequate concentration, volume, and geometry. A safe configuration is an infinitely long cylinder, eg, 11.5 cm dia of solution or 4.4 cm dia of solid uranium or plutonium. Other means of preventing nuclear criticality are by limiting the mass of fissile material present; limiting the concentration of fissile material in solution; eliminating or controlling materials that moderate the neutrons and thus reduce the number required; or using liquid or solid poisons or neutron absorbers that greatly increase the number of neutrons required to initiate or maintain the chain reaction or nuclear criticality.

Occupational Radiation Exposure

The massive shielding surrounding the process equipment usually is made of concrete and is 1–2 m thick, depending on the intensity of the radiation being shielded and the density of the concrete. Since the shielding thickness required for alpha, beta, and gamma radiation is a density function, lead and steel are used where the wall thickness must be limited. Visual observation is provided with windows of leaded glass, which are available in densities that are nearly that of steel, with periscopes that penetrate the shielding, or with television cameras. The shielding is provided to protect the plant workers.

Since radioactive gases and fine airborne radioactive particles are generated during reprocessing, the ventilation system for the plant is an important design factor. Conventional ventilation systems maintain a flow of air from the worker-occupied areas to the areas of high contamination to the clean-up systems. In the clean-up systems, particulates are removed and the radioactive gas content is reduced to a safe level before being removed from the plant through the stack.

Environmental Impact

Plant design, as related to environmental impact, refers to the special design requirements that are imposed to ensure that the general population and environment are properly protected. Two key considerations are control and reduction of contaminants from planned or routine releases, and prevention or containment and mitigation of unplanned releases from accident situations.

Remote Operation and Maintenance

The equipment and facilities for radiochemical use are designed to have high efficiencies, high reliability, and the capabilities for remote operation and maintenance. The fundamental keys in the analysis of a remote design are a concept of equipment that, for its stated use, produces the required product from a defined input; selection of construction materials that are compatible with the internal environment (initial form, intermediate products, and final product) and external environment (radiation, atmosphere, and temperature) in which the machine operates; the capability to monitor critical functions so that indication of operational status is available constantly; a modular machine design to provide for adjustment and replacement of key machine segments; and space and structural strength to accommodate remote manipulation.

M.J. FELDMAN
Oak Ridge National Laboratory

R.C. Goertz, *Nucleonics* **10**(11), 36 (1952).

R.C. Goertz and W.M. Thompson, *Nucleonics* **12**(11), 46 (1954).

J. Vertut and co-workers, "Contribution to Define a Dexterity Factor for Manipulators," *Proc. 21st Conf. Remote Syst. Technol.*, San Francisco, Calif., 1973.

NUCLEOSIDE ANTIBIOTICS.   See Antibiotics, nucleosides.

NUTS

Nuts, as generally defined, are hard-shelled seeds enclosing a single edible oily kernel. Most of the common nuts fall within this classification; the many species, however, differ greatly in size, structure, shape, composition, and flavor.

With the exception of peanuts, most of the important nuts of the world are borne on trees, and many are from native seedlings. Among the latter are the beechnut, Brazil nut, butternut, chestnut, filbert, hickory nut, pecan, pine nut, and black walnut. The pecan, English walnut, filbert, and almond are the four chief edible tree nuts produced in the United States.

Chemical Composition

Proximate composition. Most nuts for commercial use are characterized by high oil and protein contents and a low percentage of carbohydrates. The proximate compositions of a number of nuts and of some nut products are given in Table 1.

Chemical Changes During Nut Development

Several investigations have been made of the chemical changes taking place during the development of various nuts. In the macadamia, pecan, almond, Persian walnut, eastern black walnut, and tung, the sugars and other carbohydrates decrease rapidly during oil synthesis, so that when the kernel is mature most of the carbohydrates have disappeared.

Chemical Changes During Storage

Considerable research has been done on the storage of nuts, especially of pecans, which have a high free fatty acid content. Most nuts cannot be held for more than three or four months at ordinary temperatures, especially during the summer, without developing rancidity; however, storage experiments have shown that nuts and nut products keep perfectly from one year to the next under refrigeration at 0–5°C. Nut kernels in tins sealed by the vacuum process also keep most satisfactorily.

An edible and nutritive coating of zein protects tree nuts and peanuts from developing rancidity, staleness, and sogginess during storage.

Nuts are very susceptible to infestation by weevils in storage, especially in warm weather. Infestation is usually prevented or controlled by keeping the nuts in cold storage or by fumigating them either in the open with gas, such as methyl bromide or hydrogen cyanide gas, or under vacuum with a mixture of carbon disulfide and carbon dioxide.

The storage of peanut butter is influenced markedly by temperature, but the extent of roasting, hydrogenation, and added salt have little effect on product stability if packages are sealed properly. Oxygen in the headspace, however, is a main factor in reducing stability, and for that reason peanut butter for the retail trade is packed in air-tight containers or under vacuum.

Aflatoxins

At least eight compounds have been shown to belong to the aflatoxin group. The toxins have similar structures and form a unique group of highly oxygenated, naturally occurring heterocyclic compounds that fluoresce when exposed to ultraviolet light. Although other mycotoxins have been and are being discovered, the aflatoxins retain a position of great importance because of their high toxicity and their common natural occurrence in such foods as cereal grains, oilseeds, and oilseed meals stored under adverse conditions. Although the effects of aflatoxins on humans are not clear at present, the mycotoxins are highly potent carcinogens to a number of animals and should be regarded as potentially hazardous to humans. Practically all agricultural commodities and foods support the development of aflatoxin if conditions are favorable for the growth of *Aspergillus flavus*.

Table 1. Composition of Representative Nuts, 100-g Portions

Nut	Refuse, %	Water, %	Protein, g	Fat, g	Carbohydrates, g Total	Carbohydrates, g Fiber	Ash, g	Calcium, mg	Phosphorus, mg	Iron, mg	Sodium, mg	Potassium, mg	Magnesium, mg	Vitamin A value, IU	Thiamine, mg	Riboflavin, mg	Niacin, mg	Ascorbic acid, mg	Fuel value, MJ (kcal)[a]
almond, dried	49	4.7	18.6	54.2	19.5	2.6	3.0	234	504	4.7	4	773	270	0–580	0.24	0.92	3.5	trace	2.50 (598)
beechnut	39	6.6	19.4	50.0	20.3	3.7	3.7												2.38 (568)
Brazil nut	50	4.6	14.3	66.9	10.9	3.1	3.3	186	693	3.4	1	715	225	10	0.96	0.12	1.6		2.74 (654)
cashew nut		5.2	17.2	45.7	29.3	1.4	2.6	38	373	3.8	15	464		100	0.43	0.25	1.8		2.35 (561)
chestnut, fresh	19	52.5	2.9	1.5	42.1	1.1	1.0	27	88	1.7	6	454	41		0.22	0.22	0.6		0.812 (194)
coconut, fresh meat		50.9	3.5	35.3	9.4	4.0	0.9	13	95	1.7	23	256		0	0.05	0.02	0.5	3	1.45 (346)
hazelnut (filberts)	53	75.8	2.0	0.7	20.8	2.1	0.8						184	440	0.42			2.4	0.364 (87)
hickory nut	80	3.3	13.2	68.7	12.8	1.9	2.0	trace	360	2.4			160		0.6				2.82 (673)
macadamia nut	69	3.0	7.8	71.6	15.9	2.5	1.7	48	161	2.0		264		0	0.34	0.11	1.3	0	2.89 (691)
peanut, raw, with skins		5.9	26.0	47.5	18.6	2.4	2.3	69	401	2.1	5	674	142	200–350	1.14	0.13	17.2	0	2.36 (564)
pecans	56	3.4	9.2	71.2	14.6	2.3	1.6	73	289	2.4	trace	603		130	0.86	0.13	0.9	2	2.87 (687)
pinenut	42																		2.31 (552)
pignolia		5.6	31.1	47.4	11.6	0.9	4.3								0.62				2.31 (552)
pinon		3.1	13.0	60.5	20.5	1.1	2.9		604	5.2				30	1.28	0.23	4.5	trace	2.66 (635)
pistachio nuts	70	5.3	19.3	53.7	19.0	1.9	2.7	131	500	7.3		972	158	230	0.67		1.4	0	2.49 (594)
walnuts, black	78	3.1	20.5	59.3	14.8	1.7	2.3	trace	570	6.0	3	460	190	300	0.22	0.11	0.7		2.63 (628)
English	55	3.5	14.8	64.0	15.8	2.1	1.9	99	380	3.1	2	450	131	30	0.33	0.13	0.9	2	2.72 (651)

[a] kcal = food caloric (cal).

Generally, most aflatoxin production occurs during the harvest period after the nuts have begun to dry but before they attain the moisture level best suited for storage. Therefore, the nuts must pass through this critical moisture zone quickly. Storage of the nuts under proper temperature and humidity conditions will prevent further contamination. Thus, it is recognized that one very effective step to ensure a wholesome product is to divert from edible use any contaminated lots of nuts as early in the food-processing chain as possible.

Processing

Processing of nuts includes shelling, cracking, bleaching, blanching, salting, and roasting.

Peanut butter. By Federal regulation, commercial peanut butter is at least 90% shelled roasted peanuts that are ground and blended with salt, sweeteners, and emulsifiers. No artificial flavors, artificial sweeteners, chemical preservatives, natural or artificial color, purified vitamins, or minerals are allowed.

World Production

Although nuts have been used as a staple food in many countries for generations, their culture in the United States as a chief food crop is relatively recent; until the past few decades, the main supply of Persian walnuts, filberts, and almonds was European. However, the pecan and black walnut are indigenous to North America, and the United States is the principal producer of pecans. Other native nuts, such as beech, butternut, hazel, hickory, pinon, and northern California black walnut, are utilized mainly for local consumption.

Uses

Nuts and nut products. Nuts are used mainly as edible products and are marketed either in the shell or shelled, as the demand requires. The most popular nuts in the shell are the improved, or paper-shell, pecan, Persian walnut, filbert, almond, Brazil nut, peanut, and pistachio; the most popular salted- and roasted-nut kernels include these nuts and the cashew, macadamia, and pignolia. Tung and oiticica are sources of quick-drying oils for the paint and varnish industry (see Drying oils). Coconut, babassu, and palm oils are used chiefly for the manufacture of margarine, soap, shaving cream, cosmetics (qv), and other domestic products. Walnut oil, a fine specialty oil having drying qualities, is used in the preparation of artists' colors. Peanut oil is used as a lubricant and in shaving creams, shampoos, and cosmetics; it is also a good source of edible oil in the manufacture of shortening and margarine. Sweet almond oil is used as a laxative, in the treatment of bronchitis and colds, and in fine soaps and cosmetics. The meal for press cake from oil extraction of pecan, walnut, almond, and other nuts is usually rather bitter, but when refined it can be used in the baking industry or, more commonly, in animal feed. Nutshell waste from shelling and processing plants is used as soft grit in blasting metals.

Waste material from the pecan-shelling industry has shown commercial possibilities for the recovery and production of oil, and tannin, as well as shell flour, and activated charcoal. The peanut has been studied extensively for the development of products ranging from dyes and ink to artificial wool.

CLYDE T. YOUNG
North Carolina State University

J.G. Woodroof, *Peanuts—Production, Processing, and Products*, Avi Publishing Co., Westport, Conn., 1973.

J.G. Woodroof, *Tree Nuts—Production, Processing and Products*, Avi Publishing Co., Westport, Conn., 1979.

H.E. Pattie and C.T. Young, eds., *Peanut Science and Technology*, American Peanut Research and Education Society (APRES), Yoakum, Texas, 1982.

NYLON. See Polyamides.

OCEAN RAW MATERIALS

Ocean raw materials used by the chemical industry (excluding products of living origin such as fish, marine mammals, and plants) are produced from four sources: seawater (magnesium metal and compounds, bromine, salt, water, potassium compounds); beach sands (rutile, ilmenite, monazite, zircon); the submerged surface of the continental shelf (aragonite, marine shells, diamonds, tin ores, sands, and gravel); and beneath the surface of the continental shelf (petroleum, sulfur).

In addition, other raw materials occur in relative abundance on the continental shelves and in the deep-ocean basins. Some potential raw materials such as ferromanganese nodules and certain hydrothermal precipitates occur only in the ocean. Knowledge of these raw materials, their locations, amounts and possible production methods, lags when compared with terrestrial sources (see Chemicals from brine).

Dissolved Materials

More than 70 elements are found in a dissolved state in seawater or have been detected in marine organisms. The technical chemistry of seawater, despite the complexity of the solution, is determined by nine principal components from which industrial products are recovered. Metal cations include sodium, magnesium, calcium, potassium, and strontium. Anions include chloride, sulfate, bicarbonate, and bromide. The other components together constitute less than 1% of the dissolved materials. In order of decreasing value, salt, magnesium metal, magnesium compounds, water, and bromine currently are recovered on an industrial scale.

The feasibility and economic utility of a recovery process depend upon the economic attractiveness of seawater as a source as compared to alternative sources rather than upon the existence of a suitable technology. In general, raw-material recovery from the dissolved state involves an initial separation from solution followed by a second processing step to prepare a salable product sufficiently pure and in a suitable physical form.

Continental Shelves

All the minerals presently being mined from the ocean floor are taken from the continental shelves. In general, the shelves are geologically identical with the continents and because of the variation in the level of the ocean and the forces on the crustal blocks, much of the shelf area has at one time or another been above sea level.

Tinstone, diamonds, iron sands, marine shells, oölitic aragonite, sand, and gravel are mined industrially from the surface of the continental shelves. Sulfur, gas, petroleum, coal, and iron from beneath the shelves, and the beaches at the ocean edges are sources of monazite, zircon, rutile, and ilmenite.

Marine Beaches

Heavy-mineral beach sand usually is commercially processed for the titanium content of the rutile and the ilmenite and the thorium from monazite and zircon. It may be placer-mined with land-based equipment or, after construction of a seawall, with floating cutterhead or bucket-line dredges. Large operations, such as the Australian recovery of titanium ores, employ a suction dredge operating in ponds on the beach dug ahead and back-filled.

Marine beaches are mined for heavy minerals. About 93% of the world's rutile and 77% of the world's zircon plus substantial percentages of ilmenite, monazite, and other sand minerals are produced by this industry.

Submerged beaches are commonly found throughout the world, as are buried river valleys, which were formed during the glacial periods of the Pleistocene epoch, when the level of the ocean was considerably lower than at present. The successful production of diamonds and cassiterite adjacent to proven land deposits has spurred interest in locating submerged heavy-mineral deposits and in locating the source of the heavy minerals found on black-sand beaches.

Deep-Sea Deposits

Deep-sea deposits include metalliferous brines and sediments and ferromanganese nodules.

Ownership of Ocean Materials

The ownership of minerals beyond the continental shelves, on or below the ocean floor, is governed by international law.

In April, 1982, the United Nations Conference on the Law of the Sea adopted a convention dealing with all matters relating to the law of the oceans, including the mineral resources of the deep-sea bed. With regard to the continental shelf, a suggested legal definition sets the depth at 200 m or "to where the depth of the superadjacent water admits the exploitation of the natural resources." This last statement extends the control of the resources to depend upon the ability to exploit the bottom. In 1964, the United States acquired sovereign rights to the continental shelf under the terms of this last statement.

Inorganic Materials from Sea Plants

The production of iodine, bromine, potassium salts, sodium carbonate, and sodium nitrate from sea plants was among the very early chemical ventures on the European continent, and for many years they were the only available sources.

Ocean-Mining Methods

Exploitation of undersea minerals requires location of the deposit, evaluation for economic analysis, extraction, beneficiation, and transportation. The deposits are, in general, surface sediments exposed on the sea floor or *in situ* deposits contained in the sediment or hard rock. Current undersea mining is limited to shallow alluvial deposits worked by dredging and recovery of liquids by solution mining beneath the ocean floor.

W.F. McIlhenny
Dow Chemical U.S.A.

W.F. McIlhenny, "Extraction of Inorganic Materials from Seawater" in J.P. Riley and G. Skirrow, eds., *Chemical Oceanography*, 2nd ed., Vol. 4, Academic Press, London, UK, 1975.

M.J. Cruickshank and co-workers in *SME Mining Engineering Handbook*, "Marine Mining," Society of Mining Engineers, New York, 1973.

J. Mero, *The Mineral Resources of the Sea*, Elsevier, New York, 1965.

OCHER. See Pigments, organic.

OCTANE NUMBER. See Gasoline and other motor fuels.

OCTANOATES. See Carboxylic acids; Driers and metallic soaps.

ODOR CONTROL. See Air-pollution control methods; Odor modification.

ODOR MODIFICATION

Odor modification is the intentional change of one odor by the addition of another. The importance of the process is its usefulness as a method of odor control. Air fresheners, perfumes, and industrial deodorants are examples of odor modifiers. Perfumers employ the principles of odor modification in creating fragrances. Functional fragrances are odor modifiers; thus, odor modification refers specifically to the use of fragrance materials for odor control (see Perfumes).

Modification is a perceptual phenomenon. Although the chemical mixtures that are used to change odor perception usually do not react outside of the olfactory organ, modern theory does postulate that chemical complexes involving enzymes are active within the organ. Changes in the way these complexes form may explain why mixtures of odorants vary from individual chemicals in the way they are perceived.

Odors and Their Mixtures

Odor is that property of a substance that makes it perceptible to the sense of smell. Although it is certain that odor is caused by molecular structure, there remains little predictability about this correlation of odor with structures. It is known from experience that certain chemical classes have certain types of odor (see also Flavor characterization), but even an expert cannot predict what an unfamiliar molecule or mixture will smell like.

Intensity perception. The measurable properties of odor perception are its intensity and its character. The measurement and expression of odor mixture intensity can be achieved in a number of ways, but they are almost all subjective. The simplest but the most subjective method is the use of a hedonic scale. The format possibilities are many, eg, 0 = no odor, 1 = barely perceptible, 2 = distinct, 3 = strong, 4 = very strong, and 5 = overpowering.

One of the most objective measures of an odorant's intensity is its threshold which, however, reflects the intensity of only one specific odorant concentration, ie, the weakest that can be detected.

All the senses, including smell, relate to the psychophysical law (Stevens' law) as expressed by the power equation:

$$\psi = k\phi^{\beta}$$

where ψ is the perceived intensity of the sensation, ie, the odor; k is a constant; and ϕ is the physical intensity of the stimulus, ie, the odorant. Generally, the sensation varies as a power function of the stimulus. In olfaction, the power β is less than one.

Character perception. Odor character is purely subjective. Several chemicals have been experimentally characterized by scaling the degrees to which each chemical possesses subjective reference qualities, eg, burnt, fruity, and spicy. But no method exists that can reliably characterize odors.

Mixture perception (modification). Many odorous and nonodorous chemicals are used to control odors, but only those that work essentially by altering the way the nose perceives the character and intensity are true odor modifiers.

Products

Although the original intent of deodorizers was to reduce malodor, use for decorative purposes is becoming an important factor. The active ingredient of masks and counteractants are oils: essential oils, which are fragrance extracts from flowers, herbs, fruits, and trees; animal extractives; and aroma chemicals, which may be either natural or synthetic.

ROBERT H. BUCKENMAYER
Airwick Industries, Inc.

Ann. N.Y. Acad. Sci. **237**, 1 (1974); H.R. Moskowitz and C.L. Gerbers, p. 1.
Final Report on Room Deodorizers and Air Fresheners: Markets, Manufacturers, and Chemical Compositions, Battelle Memorial Institute, Columbus Division, Columbus, Ohio, Feb. 28, 1979.
H.R. Moskowitz, *Perfum. Flavor.* **2**, 21 (April–May 1977).

OIL SANDS. See Tar sands.

OILS, ESSENTIAL

A simple, though incomplete, definition of essential oils is the following: an essential oil is the predominantly volatile material isolated by some physical process from an odorous single-species botanical. Over 3000 oils have been identified from the vast number of plant species and several hundred have been commercialized. Of these, some are extremely rare and produced in only kilogram quantities, eg, violet oil, concretes (flower extracts), and angelica root oil.

Essential oils are now generally manufactured close to the growing area. Product quality has suffered, however. Essential oils are easily adulterated and this technology has not been overlooked by isolated local producers.

Essential oils are isolated from various plant parts, such as leaves (patchouli), fruit (mandarin), bark (cinnamon), root (ginger), grass (citronella), wood (anyris), heartwood (cedar), gum (myrrh), balsam (tolu balsam), berries (pimenta), seed (caraway), flowers (rose), twigs (clove stems), and buds (cloves). These plant parts are processed to yield their quintessences or essential oils, which are mostly devoid of cellulose, glycerides, starches, sugars, tannins, salts, and minerals which also occur in these botanicals.

The common physical method for isolating essential oils from the botanical is steam distillation, wherein small amounts of nonvolatiles may be carried over. Some oils are expressed. Many flower oils are extracted with a purified petroleum solvent.

Essential oils are generally liquid at room temperature; however, some are semisolid, such as *Mentha arvensis* (Brazilian mint), and several are solid, eg, oil of guaiac wood.

The cultivation of essential oil-bearing plants has kept pace with modern agricultural methods. Hybrids are grown to yield oils of specific odor, flavor, or properties.

Essential oils are used as such for flavors and fragrances, but products derived from, or based on essential oils have large volume usage for specific applications. Essential oils are concentrated, rectified, extracted, or chemically treated to further isolate vital components, purify, adjust properties, or increase the concentration of significant flavor or fragrance components (see Flavors and spices; Perfumes).

Composition

The volatile components of essential oils usually contain fifteen carbon atoms or less. However, seed oils contain long-chain fatty acids or esters and even glycerides that are carried over in distillation; the amounts of these components are very low.

Essential oils basically are made up of carbon, hydrogen, and oxygen, and occasionally nitrogen and sulfur. The largest class of components is the terpenes which have ten carbon atoms and are head-to-tail condensation products of two isoprene molecules (see Terpenoids). The terpenes may be aliphatic, alicyclic, or bi- and tricyclic of varying degrees of unsaturation up to three double bonds.

It is not uncommon for an essential oil to contain over two hundred components and often the trace substances (in ppm) are essential to the odor and flavor. The absence of even one component may change the aroma. The same species of botanical, grown in different parts of the world, usually has the same components; however, climatic and topographical conditions affect plants and can alter the essential oil quantitatively, but rarely qualitatively.

Production

In the botanical, the essential oil is present in oil sacs. It is isolated by comminution, and the action of heat, water, and solvents.

Steam distillation. Steam or hydrodistillation is the preferred method for producing essential oils, employing either water, wet steam, or steam.

Extraction. An essential oil that is sensitive to heat, eg, jasmine or tuberose, or contains an essential nonvolatile constituent, eg, piperine in black pepper, is extracted with a solvent (see Extraction).

Synthetic Substitutes for Essential Oils

Fluctuations in the cost and availability of natural oils and the high cost of some oils have induced users to seek substitutes. Several former large-volume oils have been replaced by synthetics because of large volume demands. The expensive variations in cost are especially evident in essences such as rose, jasmine, violet, lilac, neroli, etc. Duplication of these items is economically worthwhile. Modern techniques and instrumentation offer the possibility for total analysis of these oils. This route has been undertaken by the primary fragrance and flavor companies throughout the world in the attempt to find economic and available substitutes. Nonetheless, there is a trend away from synthetic oils because complete duplications are in most cases not technically, aesthetically, or economically possible.

Health and Safety Factors

Most essential oils, since they are natural and have a history of use, are considered GRAS (generally recognized as safe) by the FDA. Safety and toxicity testing and evaluations and regulations are different for food additives than for fragrance oils and cosmetics (qv). Some oils can be used for both purposes, including celery, rose, and black pepper. Some oils, such as cinnamon oil, can be used in flavors, but not in fragrances because of skin irritation.

Commercial Essential Oils

The commercial essential oils include allspice (pimenta berry), bitter almond, amyris, anise, star anise, sweet basil, bay (myrcia), bergamot, sweet birch, bois de rose (rosewood), camphor, cananga, caraway, cardamom, cassia, cedarwood, cinnamon, citronella, clove, coriander, eucalyptus, geranium, ginger, grapefruit, jasmine, juniper, labdanum, lavandin, lavender, lemon, distilled lime, japanese mint, neroli, nutmeg, ocotea, bitter orange, sweet orange, origanum, orris root, palmarosa, patchouli, black pepper, peppermint, petitgrain bigarade, pine, pinus pumilio, rose, rosemary, dalmatian sage, sage clary (sage muscatel), East Indian sandalwood (santal), spearmint, spike lavender (spike), thuja (cedarleaf), thyme, turpentine, vetiver, wintergreen, and ylang ylang.

JAMES A. ROGERS, JR.
Fritzsche, Dodge & Olcott, Inc.

W.E. Dorland and J.A. Rogers, Jr., *The Fragrance and Flavor Industry*, Dorland Publishing Co., Mendham, N.J., 1977.

Food Chemicals Codex, 3rd ed., National Academy of Sciences, Washington, D.C., 1980.

Circular FTEA 2-83, Foreign Agriculture Service, Horticultural and Tropical Foods Division, U.S.D.A., Washington, D.C., May 1983.

OIL SHALE

Oil shale consists of a marlstone-type sedimentary inorganic material that contains complex organic polymers that are high molecular weight solids. The organic kerogen is a three-dimensional polymer, insoluble in conventional organic solvents, and associated with small amounts of a benzene-soluble organic material, bitumen. Oil-shale deposits were formed in ancient lakes and seas by the slow deposition of organic and inorganic remains from the bodies of water. As the waters stagnated and dried, the deposits compacted. The geology and the composition of inorganic and organic components of oil shale varies with deposit location.

Reserves

Oil-shale deposits occur widely throughout the world; estimates of the resources by continent are given in Table 1. Characteristics of many of the world's best-known oil shales are summarized in Table 2. Oil-shale deposits in the United States occur over a wide area. The most extensive deposits, which cover ca 647,000 km² (250,000 mi²), are the Devonian-Mississippian shales of the eastern United States. The richest U.S. oil shales are in the Green River formation of Colorado, Utah, and Wyoming.

Retorting

Thermal decomposition of oil shale. The thermal decomposition of oil shale, ie, pyrolysis or retorting, yields liquid, gaseous, and solid products. The liquid, which is produced by pyrolysis, is in the form of a vapor or mist. The remaining organic carbon remains on the retorted shale, the mineral matter, as a cokelike deposit. The amounts of oil, gas,

Table 1. Shale-Oil Resources of the Populous Land Areas, 10⁹ m³ᵃ

Shale-oil yield range, L/tᶜ	Total resourceᵇ			Known resources: marginal or submarginal recovery			Recoverable known resources
	21–42	42–104	104–417	21–42	42–104	104–417	42–417
Africa	71,500	12,700	636	small	small	14	1.6
Asia	93,800	17,500	874	na	2	11	3.2
Australia and New Zealand	15,900	3,200	159	na	small	small	small
Europe	22,260	4,100	223	na	1	6	4.8
North America	41,400	8,000	477	350	254	99	13
South America	33,400	6,400	318	na	119	small	8
Total	*278,260*	*51,900*	*2,687*	*350*	*376*	*130*	*30.6*

ᵃ To convert m³ to bbl, divide by 0.159.
ᵇ Includes oil shale in known resources, in extensions of known resources, and in undiscovered but anticipated resources.
ᶜ To convert L/t to gal/short ton, multiply by 0.2397.

Table 2. Properties of Oil Shales

	Australia (Glen Davis)ᵃ	Brazil (Tremembe-Taubate)ᵃ	Canada (Nova Scotia)ᵇ	Manchuria (Fushun)ᵇ	New Zealand (Orepuki)ᵇ	South Africa (Ermelo)ᵃ	Spain (Puertolleno)ᵃ	United States (Colorado)ᵃ
Modified Fischer assay								
oil, L/tᶜ	414	156	257	38	331	228	234	122
oil, wt%	30.0	11.5	18.8	3.0	24.8	17.6	17.6	9.3
water, wt%	0.7	6.2	0.8	4.9	8.3	3.0	1.8	1.0
spent shale, wt%	64.1	78.4	77.7	90.3	57.6	75.6	78.4	87.5
gas and loss, wt%	4.3	3.9	2.7	1.8	9.3	3.8	2.2	1.6
conversion of organic material to oil,ᵈ wt%	66	59	60ᵉ	33	45	34	57ᵉ	70
Rock characteristics								
sp gr (at 16°C)	1.60	1.70		2.29	1.46	1.58	1.80	2.23
heating value, MJ/kgᶠ	18.8	8.2	12.6	3.4	21.3	19.1	12.5	5.1
ash, wt%	51.6	71.4	62.4	82.7	32.7	42.5	62.8	66.9
organic carbon, wt%	39.8	16.5	26.3	7.9	45.7	43.8	26.0	11.3
Assay oil								
sp gr (at 16°C)	0.89	0.88	0.88	0.92	0.90	0.90		0.91
carbon, wt%	85.4	84.3		85.7	83.4	84.8		84.6
hydrogen, wt%	12.0	12.0		10.7	11.8	11.1		11.6
nitrogen, wt%	0.5	1.1			0.6		0.9	1.8
sulfur, wt%	0.4	0.2		0.6	0.6	0.6	0.3	0.5
Ash analysis, wt%								
SiO₂	81.5	55.8	61.1	62.3	44.2	61.3	56.6	43.6
Al₂O₃	10.1	26.7	30.1	26.7	28.1	30.5	27.6	11.1
Fe₂O	3.0	8.5	5.0	6.1	20.5	2.9	9.1	4.6
CaO	0.8	2.8	1.1	0.1	4.6	1.6	2.6	22.7
MgO	0.8	3.7	1.6	1.8	1.4	1.7	2.2	10.0
other oxides	3.8	2.5	1.1	3.0	1.2	2.1	1.9	8.0

ᵃ Average sample.
ᵇ Selected sample.
ᶜ To convert L/t to gal/short ton, multiply by 0.2397.
ᵈ Based on recovery of carbon in oil from organic carbon in shale.
ᵉ Carbon content of oil ca 84 wt%.
ᶠ To convert MJ/kg to Btu/lb, multiply by 430.4.

and coke which ultimately are formed depend on the heating rate of the oil shale and the temperature–time history of the liberated oil.

Numerous kinetic mechanisms have been proposed for oil-shale pyrolysis reactions. The kinetics appear to be adequately represented by first-order rate mechanisms, eg,

$$\text{kerogen} \xrightarrow{k_1} \text{bitumen} + \text{gas} + \text{coke}$$

$$\text{bitumen} \xrightarrow{k_2} \text{oil} + \text{gas} + \text{coke}$$

Most oil-shale retorting processes are carried out at ca 480°C to maximize liquid-product yield.

Aboveground retorting. The first aboveground oil-shale processes were batch or semibatch, and modern commercial contenders are continuous in both feed and product removal. Room-and-pillar mining is used for commercial aboveground retorts. Open-pit mining has the potential for greater resource recovery but requires off-site spent-shale disposal.

The heat required for retorting is transferred to the oil shale in four main types of retorts. In Type-1 retorts, the heat must be transferred through the vessel wall to the oil shale. A combustion zone within the oil-shale bed is characteristic of Type-2 retorts. Heat for Type-3 retorts enters the retorts as externally heated gases. In Type-4 retorts, externally heated solids are used to heat the oil shale.

Multimineral processing.   In the northern portions of the Piceance Creek Basin of Colorado, large deposits of nahcolite and dawsonite occur intermingled with the oil-shale deposits. They appear to be the only significant deposits of the two minerals in the world. The dawsonite resources represent a 600-yr supply of aluminum based on current consumption levels.

Crude Shale Oil

Properties.   The composition of shale oil depends on the shale from which it was obtained as well as on the retorting method by which it was produced. Properties of shale oils from various locations are given in Table 3. The chief difference in shale oils that are produced by different processing methods is in boiling-point distribution.

Upgrading shale oil.   Crude shale oil has a high content of organic nitrogen, ca 2 wt%, which acts as a catalyst poison; contains a large atmospheric residuum fraction, 20–50 wt%; and has a high pour point, generally > 5°C. Prerefining crude shale oil to produce a synthetic crude that is compatible with typical refineries generally is necessary. Upgraded shale oil is a desirable refinery feedstock; it is paraffinic and is characterized by low residuum, nitrogen, and sulfur.

Gasification.   For significant conversion of shale oil or oil shale to gaseous products, considerable hydrogen must be used. Hydrogasification is the main process under consideration for gasification of oil shale.

Health and Safety Factors

Environmental: Green River oil-shale development.   Potential development of the oil-shale reserves of the Green River Formation has caused concern regarding possible environmental degradation and local socioeconomic problems. The population in this region is sparse, the country is rugged, and the towns are small. Oil-shale development will affect the land and water resources, and probably the residents and wildlife as well.

Biological.   Shale oils and certain waste streams from shale-oil processing may contain hazardous substances. The potential exposure of people to carcinogens associated with shale-oil products and whether the risk is greater than for petroleum oils is being studied. Known and suspected carcinogens of the polycyclic aromatic hydrocarbon types have been found in crude shale oil.

Petrochemicals from Shale Oil

A principal difficulty associated with shale-oil production is the high cost of crude shale oil; in particular, the initial hydrogenation to reduce the nitrogen content.

Steam pyrolysis is not a catalytic process; therefore, relatively high nitrogen contents are not necessarily detrimental. Composition of feedstocks strongly affect pyrolysis yields. The yields of light olefins are a primary function of feedstock molecular weight, eg, ethane yields ca 75 wt% ethylene, whereas a heavy gas oil yields 20–25 wt% ethylene. On pyrolysis, the gas oil also yields ca 50% heavier liquids, mostly aromatics, which must be hydrotreated if they are to be used as fuel oil. Lower yields of petrochemicals result with heavier feedstock (see Petroleum, petroleum resources).

Petrochemical utilization of oil shale was proposed in the 1950s, although the removal of specific compounds, eg, phenols, tar bases, and linear olefins, appear to have little impact on the shale-oil industry. However, there is sufficient demand for light olefins and aromatic petrochemical intermediates that are produced by pyrolysis to affect the liquid hydrocarbons industry.

Philip F. Dickson
Colorado School of Mines

Reviewed by
Mack Horton
Union Oil Company of California

S. Siggia and P. Uden, eds., *Analytical Chemistry Pertaining to Oil Shale and Shale Oil*, National Science Foundation Grant Number GP 43807, June 24–25, 1974, pp. 11–13.

G.L. Baughman, ed., *Synthetic Fuels Data Book*, 2nd ed., Vol. 4, Cameron Engineers (Division of The Pace Co.), 1978, pp. 67–104.

Shale Oil, Occidental Petroleum Corp., Los Angeles, Calif., 1980.

OLEFIN FIBERS

Olefin fibers, which also are called polyolefin fibers, are manufactured fibers in which the fiber-forming substance is any long-chain, synthetic polymer of at least 85 wt% ethylene, propylene, or other olefin units. The olefin fibers of commercial importance are polypropylene and, to a lesser extent, polyethylene. Fibers have been produced on a laboratory scale from several other polyolefins, eg, poly(1-butene), poly(3-methylbutene), and poly(4-methyl-pentene) (see Fibers, chemical).

Properties

Physical.   Some physical properties of commercial polyethylene and polypropylene fibers are listed in Table 1. Polyethylene and polypropylene fibers differ from other synthetic fibers in two important respects: their nearly total lack of water absorption ensures that wet properties

Table 3.   Properties of Oils Produced from Shales from Various Sources

Country or company	Retort	Sp gr (°API)	N, wt%	S, wt%	Analysis of distillate (boiling to 315°C), wt%		
					Saturates	Olefins	Aromatics
Australia, Glen Davis	Pumpherston	0.828 (27.9)	0.52	0.56	42	39	19
Brazil, Tremembe	Gas Combustion	0.919 (22.5)	1.06	0.68	23	41	36
South Africa, Ermelo	Salermo	0.906 (24.7)	0.85	0.64	35	44	21
Spain, Puertollano	Pumpherston	0.901 (25.6)	0.68	0.40	51	27	22
United States							
Colorado	Gas Combustion	0.943 (18.6)	2.13	0.69	27	44	29
Colorado	Pumpherston	0.900 (25.7)	1.57	0.77	30	38	32
Superior Shale Oil							
(ibp[b] to 204°C)		0.630 (0.93)	2.0	0.8	25	25	50
Rundle Shale Oil							
(whole oil)		0.636 (0.91)	0.99	0.41	48	2	50
Israeli Shale Oil[c]		0.623 (0.955)	1.2	7.1			

[a]ibp = initial boiling point.
[b]Also contains 79.8 C, 9.7 H, and 2.2 wt% O.

Table 1. Synthetic Fiber Properties

| Polymer | Fiber type | Tenacity, MPa (N/tex[a]) | | Modulus, GPa (N/tex)[a] | Breaking elongation, % | | Elastic recovery, %[b] | Density, g/cm^3 |
		Standard	Wet		Standard	Wet		
low density polyethylene	monofilament	92–280 (0.1–0.3)	92–280 (0.1–0.3)	0.18–1.0 (0.2–1.1)	20–80	20–80	95 (at 5%)	0.92
high density polyethylene	monofilament	290–570 (0.3–0.6)	290–570 (0.3–0.6)	1.7–4.2 (1.8–4.4)	10–45	10–45	100 (at 5%)	0.95–0.96
polypropylene	staple and tow	280–560 (0.3–0.6)	280–560 (0.3–0.6)	0.28–3.3 (0.3–3.5)	20–120	20–120	70–100 (at 5%)	0.90–0.96
	monofilament	270–540 (0.3–0.6)	270–540 (0.3–0.6)	1.6–4.8 (1.8–5.3)	14–30	14–30	98 (at 5%)	0.90–0.91
	multifilament	180–630 (0.2–0.7)	180–630 (0.2–0.7)	1.2–3.2 (1.3–3.5)	20–100	20–100	94–98 (at 5%)	0.90–0.91
polyester	staple and tow	280–830 (0.2–0.6)	280–830 (0.2–0.6)	1.2–2.1 (0.9–1.5)	12–55	12–55	81 (at 3%)	1.38
	multifilament	280–690 (0.2–0.5)	280–690 (0.2–0.5)	1.2–3.6 (0.9–2.6)	24–42	24–42	76 (at 3%)	1.38
nylon-6,6	staple and tow	230–680 (0.2–0.6)	230–570 (0.2–0.5)	1.0–4.5 (1.0–4.5)	16–75	18–28	82 (at 3%)	1.13–1.14
	multifilament	230–570 (0.2–0.5)	230–570 (0.2–0.5)	0.45–2.4 (0.4–2.1)	26–65	30–70	88 (at 3%)	1.13–1.14
acrylic	staple and tow	230 (0.2)	120–130 (0.1–0.2)	1.0 (0.9)	20–28	26–34	73 (at 3%)	1.16
rayon	staple, regular tenacity	150–450 (0.1–0.3)	150–300 (0.1–0.2)	0.75–2.3 (0.5–0.15)	15–30	20–40	82 (at 2%)	1.46–1.54
	staple, high tenacity	450–750 (0.3–0.5)	300–600 (0.2–0.4)	1.7–6.6 (1.1–4.4)	9–26	14–34	70–100 (at 2%)	1.46–1.54
cellulose acetate	staple, multifilament	130 (0.1)	130 (0.1)	0.53–0.70 (0.4–0.53)	25–45	35–50	48–65 (at 4%)	1.32

[a] To convert N/tex to gf/den, multiply by 11.33; N/tex = kJ/g; N/tex × density (g/cm^3) = GPa. To convert GPa to psi, multiply by 145,000.

[b] Parenthetical number denotes the amount of initial extension.

are identical with their properties at standard conditions (65% rh, 21°C) and their low specific gravity leads to a higher covering power, ie, one kilogram of polypropylene can produce a fabric, carpet, etc, with as much as 45% more fiber per unit area than one kilogram of polyester.

Chemical. The general chemical characteristics of olefin fibers are extreme hydrophobicity and inertness to a wide variety of inorganic acids and bases and to organic solvents at room temperature. These properties derive from the hydrocarbon character of the fundamental unit and from the great molecular weights of commercial polymers.

Thermal. Comparatively, polyethylene and polypropylene undergo all thermal transitions at substantially lower temperatures than either polyesters or nylons. The thermal degradation of olefin fibers is oxygen-sensitive, and may be problematic in fiber formation and in use. In general, the absorption of oxygen by the polymer results in chain scission and, therefore, molecular weight degradation as a result of high temperature formation of hydroperoxides.

Ultraviolet degradation. Sunlight affects olefin fibers, and linear polyethylene is somewhat less susceptible than other polyolefins. The effect is similar to that for thermal oxidation, ie, chain scission and molecular weight degradation, although the mechanism of photooxidation is different (see also Photochemical technology; Plastics, environmentally degradable).

Flammability. One fundamental measure of the flammability of polymers is the oxygen index which defines the minimum oxygen concentration necessary to support combustion. Polyolefins are more combustible on the basis of this index than many other common polymers (see Flame retardants).

Dyeing. Olefin fibers are inherently difficult to dye, because there are no sites for the specific attraction of dye molecules, ie, no hydrogen-bonding or ionic groups, and dyeing only can take place by virtue of weak dye–fiber van der Waals forces. As a result, the most common method of coloring olefin fibers is to add pigments to the melt before extrusion.

Stress–strain relationships. As is common for all polymeric fibers, the stress–strain curves for olefin fibers depend on strain rate, temperature, molecular weight, and fiber morphology, especially molecular orientation and crystallinity.

Creep and stress relaxation. Olefin fibers exhibit creep, ie, time-dependent deformation, of which only a portion is recoverable under load; and the inverse phenomenon, ie, stress relaxation, which is the spontaneous relief of internal stress and occasionally is accompanied by small spontaneous length changes. Both of these behaviors depend on fiber molecular weight, molecular orientation, and crystallinity.

Dynamic mechanical properties. The response of olefin fibers to cyclic deformation is determined by the temperature and frequency of the deformation and by the structural characteristics of the molecule and fiber morphology.

Manufacture and Processing

Olefin fibers are fabricated commercially by one of several modifications of the melt-extrusion technique, of which the fundamental elements are the continuous expulsion of molten polymer through a die, ie, spinneret, the solidification of the extrudate by heat transfer to the surrounding fluid medium, and the winding of the solid extrudate onto packages. Further processing may include drawing the fiber to as much as six times its original length and a variety of heat treatments to relieve thermal stresses within the fiber. Texturizing processes, which are combinations of deformations and heat treatments, also may be applied.

Novel olefin fibers. Novel olefin fibers include hard elastic fibers, high modulus fibers, and bicomponent and biconstituent fibers.

Economic Aspects

In the United States, olefin-fiber consumption has increased steadily since about 1966. Most of the increase has resulted from increased demand for continuous-filament yarn and monofilaments rather than from staple products. Polypropylene fiber production involves relatively lower consumption of energy than other synthetic fibers because of the former's low density, low melting temperature, and melt spinnability.

Uses

Olefin-fiber use patterns are concentrated in home furnishing and industrial areas, where good fiber-mechanical properties, relative chemical inertness, low moisture absorption and, sometimes, low density contribute to desirable product properties. Olefin fiber use in apparel has been restricted by low melting temperatures which make ironing of polyethylene and polypropylene fabrics impossible.

D.R. BUCHANAN
North Carolina State University

H. Ahmed, *Polypropylene Fibers—Science and Technology*, Elsevier Scientific Publishing Co., Inc., New York, 1982.

J.G. Cook, *Handbook of Polyolefin Fibers*, Merrow Publishing Co., Watford, UK, 1967.

OLEFIN POLYMERS

LOW PRESSURE LINEAR (LOW DENSITY) POLYETHYLENE

In 1968, commercial production of low pressure linear low density polyethyene (LLDPE) was pioneered by Phillips Petroleum Company with the introduction of two grades of LLDPE at densities of 0.925 and 0.935 g/cm³. LLDPE has grown rapidly in importance during the 1980s owing to the superior physical properties that it displays in a number of applications, and because of economic factors that favor low pressure technology.

Properties

The differences in the nature and amount of short-chain and long-chain branching of LLDPE and LDPE affect physical properties and melt rheology. The magnitude of these differences is illustrated in Table 1, where LLDPE is compared to high pressure LDPE of approximately the same melt index and density.

Structure. In contrast to LDPE molecules, which contain a variety of types of short-chain as well as long-chain branches, LLDPE molecules consist of long sequences of methylene units with periodic uniform short side chains, as illustrated in Figure 1. As would be expected, short-chain branching interferes with crystallization of the main chains, and since the crystalline regions display a higher density than noncrystalline regions, an increase in branching lowers the density.

Linear LDPE

Figure 1. Schematic representation of the LLDPE molecular structure.

Manufacture

In the years since Phillips Petroleum company introduced LLDPE made by the particle-form process, a number of other process technologies have been developed, such as slurry polymerization in a light hydrocarbon, slurry polymerization in hexane, solution polymerization, and gas-phase polymerization. Many different companies have developed processes for the production of LLDPE resins, some in commercial operation and others at an advanced pilot state of development. A summary of various types of processes and their developers is shown in

Table 1. Properties of LLDPE and High Pressure LDPE

Property	LLDPE		LDPE	
	0.922 g/cm³	0.926 g/cm³	0.918 g/cm³	0.927 g/cm³
melt index, 190°C, g/10 min	1.2	0.13	3.5	0.15
tensile strength, MPa[a]	13	20	12	16
elongation, %	800	600	550	600
flexural modulus, MPa[a]	234	510	138	414
hardness, shore D		58		58
environmental stress-crack resistance (ESCR)[b]		1000		65
melting point, dta, °C	122		102	

[a] To convert MPa to psi, multiply by 145.
[b] ASTM D 1693, condition A.

Table 2. LLDPE Processes

Process	Developer	Status	Pressure, MPa[a]	Density, g/cm³	Molecular weight range	Resin form
Gas-phase						
fluid bed	Union Carbide	commercial	2	0.918–0.94	narrow	powder
	Naphtachimie	pilot	2.5	0.918–0.94	full	powder
stirred bed	Amoco	pilot	2	0.920–0.94	full	powder
	Cities Service	pilot	2	0.920–0.94	full	powder
Liquid-phase						
slurry	Phillips	commercial	3.5	0.924–0.94	full	pellets
	Solvay	pilot	3	0.913–0.94	full	pellets
	Montedison	pilot	3	0.918–0.94	narrow	powder
solution	DuPont of Canada	commercial	8	0.917–0.94	narrow	pellets
	Dow	commercial	8	0.917–0.94	narrow	pellets
	Mitsui	commercial	4	0.920–0.94	full	pellets
	DSM	pilot	8		full	pellets

[a] To convert MPa to psi, multiply by 145.

Table 2. The manufacture of LLDPE involves four steps: feed preparation, polymerization, recovery, and finishing.

Processing

The operations associated with conversion of the dry product recovered from the reactor into a form suitable for customer use are termed the finishing process, and comprise stabilization, blending, extrusion, and pelletizing. Various antioxidants, stabilizers, and additives are added to prevent oxidation, discoloration, and thermal degradation during melt processing.

Uses

LLDPE is used in pipe, film, rotational molding, injection molding, and wire and cable resins.

JAMES N. SHORT
Phillips Petroleum

C.T. Levett, J.E. Pritchard, and R.J. Marinovich, *Soc. Plast. Eng. J.* **26**(6), 40 (1970).

U.S. Pat. 4,011,382 (Mar. 8, 1977), I.J. Levine and F.J. Karol (to Union Carbide Corporation).

R.R. Turley, *Soc. Plast. Eng. 37th Ann. Tech. Conf. XXV*, 499 (May 1979).

HIGH PRESSURE (LOW AND INTERMEDIATE DENSITY) POLYETHYLENE

Polyethylene is a clear-to-whitish translucent material that is often fabricated into clear, thin films which are excellent packaging materials (qv). Thick sections are translucent and exhibit a waxlike appearance; the translucency increases with density. With proper colorants, a wide variety of colored products is obtained.

Polyethylene is a polymer of high molecular weight of the formula $\text{+CH}_2\text{CH}_2\text{+}_n$. In commercial polymers, n ranges from ca 500–50,000. Polyethylenes are partially crystalline and partially amorphous. Low and medium density polyethylenes are defined here as those polyethylenes with densities less than 0.94 g/cm³.

Low density polyethylene (LDPE) is produced by three processes: low pressure, high pressure tubular, and high pressure stirred-autoclave polymerization.

In Table 1, physical properties are given in ranges because of their high dependence on the specific resin density although other factors have an influence on these properties. Some attempt has been made to indicate this density dependence by arbitrarily defining LDPE and medium density polyethylene as 0.912–0.925 g/cm³ and 0.925–0.940 g/cm³, respectively.

Table 1. Mechanical and Physical Properties of Low Density High Pressure Polyethylene and Medium Density Polyethylene

Property	ASTM Method	LDPE	MDPE
tensile yield stress, kPa[a]	D 638	80–120	100–180
yield elongation, %	D 638	20–40	10–20
tensile ultimate stress, kPa[a]	D 638	100–170	110–175
ultimate elongation, %	D 638	400–700	100–500
secant modulus of elasticity at 1% strain, kPa[a]	D 638	900–2700	2600–5000
shear strength, kPa[a]	D 732	70–260	210–300
Shore hardness, durometer	D 676	C35–D60	D45–D65
dart drop 38 µm thickness, at F50[b], g/25 µm	D 1709	50–150	60–300
low temperature brittleness at F50[b], °C	D 746	< −76°C	< −76°C
density range, g/cm³		0.912–0.925	0.926–0.940
refractive index, n_D^{25}	D 542	1.50–1.54	1.50–1.54
thermal coefficient of linear expansion, 20–60°C, per °C	D 696	$15-30 \times 10^{-5}$	$15-30 \times 10^{-5}$
heat distortion, 455 kPa (66 psi), °C	D 648	40–50°C	50–65°C
water absorption 24 h, %	D 570	< 0.02	< 0.02

[a] To convert kPa to psi, divide by 6.895.
[b] F50 = number of hours at which 50% fail.

Structure. A chemical analysis of any polyethylene homopolymer yields the empirical formula CH_2. The principal determinants of properties are molecular weight, molecular weight distribution, chain-branching type and size, and branching distribution. A typical polyethylene also contains other chemical groups derived from initiators used in manufacture, or impurities in feedstocks. These generally amount to less than 0.1% of the total polymer and have little effect on the physical properties, although reactivity and electrical properties may be significantly affected.

Molecular weight, molecular weight distribution, and type and amount of branching vary widely with manufacturing and processing conditions, including initiators and chain-transfer agents. Low and medium polyethylene produced by high pressure processes contain significant amounts of long-chain branching, whereas high density polyethylene contains few or no branches. Low pressure linear LDPE contains numerous short-chain branches. Long-chain branches can contain thousands of C atoms, in contrast to short-chain branches that are up to five carbon atoms long.

Crystallinity. Low density polyethylene is partially crystalline. This crystallinity is a function of the polymer density. The density of a pure polyethylene crystal is 1.00 g/cm^3, whereas the density of purely amorphous polyethylene is generally accepted as 0.855 g/cm^3.

Rheology. Although at room temperature low density polyethylene is almost an elastic solid, it exhibits some viscous characteristics, such as low tensile-stress creep, stress relaxation, and the ability to be cold-drawn. At melt-processing temperatures of ca 200°C, polyethylene exhibits chiefly viscous properties. However, significant elastic effects, such as die swell, measurable tensile and elongation properties, and complex die-entrance effects are also evident. Thus, LDPE exhibits viscoelastic behavior over a broad range of temperatures.

Electrical properties. The outstanding electrical properties of LDPE were the basis for its initial commercialization and use as radar-cable insulation during World War II. Today, polyethylene is the principal material used for power- and communications-cable insulation. Its insulation resistance cannot be measured, but is certainly in excess of 10^{16} Ω-cm in the pure state (see Insulation, electric).

Chemical properties. Polyethylene is sensitive to combined chemical and mechanical stress, normally referred to as environmental stress cracking. LDPE, when stressed in contact with certain materials (particularly polar organic liquids such as certain alcohols and esters, metallic soaps and polar liquid hydrocarbons), cracks much more rapidly than expected. Both stress and chemical exposure must be present for this effect to occur.

As would be expected from its substantially paraffinic nature, polyethylene is highly stable and inert. However, it is affected by certain chemicals. The resistance to water is excellent except for possible oxidation at high temperatures.

Liquid and gaseous halogens attack polyethylene slowly at room temperature, but more rapidly at higher temperatures, and may cause discoloration and some change in properties.

Above 50°C, the solubility of polyethylene in hydrocarbons and halogenated hydrocarbons rises sharply although it remains only sparingly soluble in more polar liquids.

Below 40°C, solvents for polyethylene are absorbed slowly and effect swelling without solution. The swelling is accompanied by loss of strength and failure can occur if the polymer is stressed in any way.

Animal, vegetable, and mineral oils, which may be regarded as solvents, are absorbed by polyethylene; the degree and rate of absorption increase markedly with increasing temperature.

Sulfur and certain sulfur-containing organic compounds are absorbed slowly by polyethylene and may affect its electrical properties adversely.

Manufacture

Polymerization. High pressure low and medium density polyethylenes are manufactured by either the stirred-autoclave or the tubular process. Reactor outputs of more than 10 metric tons/h are obtained today.

Quality control. The quality of the reactor output is usually controlled by measuring the melt index and density, and judging the appearance of the final product.

Processing

Compounding of additives. The compounding operation produces specific properties for specific applications. It is carried out by the resin producer, the fabricator, or a producer of polyethylene compounds. Additives can also be introduced by the fabricator during the processing. Almost all additives are mixed into the melt above 110°C.

Fabrication. Virtually all LDPE is processed with a plasticating extruder; less than 2% is processed by rotational molding and calendering (see Plastics processing).

Copolymers

The most important low density copolymers of ethylene are poly(ethylene-*co*-vinyl acetate) (EVA) (1) and poly(ethylene-*co*-ethyl acrylate) (EEA) (2).

Both are random copolymers and have the advantages of greater flexibility, somewhat reduced melting point, greater acceptance of fillers (qv), and in general, physical characteristics approaching those of elastomers (qv). They are available up to ca 30 wt% copolymer.

Poly(ethylene-*co*-acrylic acid) (EAA) polymers have the unique advantage of high adhesion to metal and other substrates for coatings.

ROBERT L. BOYSEN
Union Carbide Corporation

L. Mandelkern and J. Maxfield, *J. Polym. Sci. Poly. Phys. Ed.* **17**, 1913 (1979).

A. Renfrew and P. Morgan, *Polyethylene: The Technology and Uses of Ethylene Polymers*, Interscience Publishers, Inc., New York, 1960.

Plastics Planning Guide, 5th ed., Plastics Publishing, Stamford, Conn., 1979–1980.

LINEAR (HIGH DENSITY) POLYETHYLENE

Properties

Structure. The molecular structure of HDPE (high density polyethylene) homopolymer made by the Phillips process is that of polymethylene, or a very high molecular weight α-olefin, $CH_2{=}CH{-}CH_2CH_2{+}_n H$; as molecular weight is increased, n increases from the hundreds to the hundred thousands.

The nonbranched structure of HDPE permits it to become highly crystalline when cooled to below its crystalline freezing point, provided the molecular weight is not so high that chain entanglements interfere. The crystals in linear polyethylene form in ordered agglomerates called crystallites, giving the resin extremely tough and shatter-resistant characteristics.

Molecular weight and molecular weight distribution. The weight-average molecular weight of HDPE produced by the Phillips process ranges from 40,000 to about 10^6. In a given sample, polymer molecules ranging in molecular weight from as low as 1000 to well over 10^6 are generally present. The molecular weight distribution (MWD) is an important property of HDPE and can be varied to obtain desired properties.

Properties dependent on molecular weight. In the molten state, the viscosity is a function of the molecular weight. Since it is easier to measure the melt flow of polyethylene than to determine molecular weight, polyethylenes are usually characterized by melt index (MI), which is a standard in the industry (ASTM D 1238-65T). The physical properties of linear polyethylene that are dependent on melt index (or molecular weight) are given at different melt indexes in Table 1.

Properties dependent upon MWD. The most important property that is a function of MWD is a melt property termed shear sensitivity or

Table 1. Melt-Index-Dependent Properties of Compression-Molded HDPE

Properties	Melt index[a]				
	0.2	0.9	1.5	3.5	5.0
tensile impact, kJ/m² [b]	210	135	124	86	63
Izod impact, J/m of notch[c]	748	214	107	80	64
elongation at break (50 cm/min), %	30	25	20	15	12
ESCR (Bell test), 50 failures, h	60	14	10	2	1
brittleness temperature, °C	<−118	<−118	<−118	−101	−73

[a] MI = g/10 min.
[b] To convert kJ/m² to ft·lbf/in.², divide by 2.1 (see ASTM D 1822).
[c] To convert J/m to ft·lbf/in. of notch, divide by 53.38 (see ASTM D 256).

shear response. Shear response may be defined as the response of shear rate to changes in shear stress.

Density-dependent properties. A sharp and approximately linear decrease in flexure stiffness with decrease in density is not surprising because of the linear relationship between crystallinity and density. Tensile strength, softening temperature, and hardness decrease as the amorphous fraction of the polymer increases and density is lowered by branching.

In general, the biggest gains to be made by reduction in density of linear polyethylenes are in greater stress-cracking resistance and load-bearing properties.

Chemical properties. In general, HDPE has the chemical properties of saturated hydrocarbon waxes (qv). Polyethylene is not attacked by most inorganic chemicals and is insoluble in most organic solvents at room temperature. All polyethylenes require protection from uv irradiation in outdoor applications involving exposure of more than a few months.

Manufacture

The catalyst permits the polymerization of ethylene at low pressures and moderate temperatures. Phillips catalysts contain chromium in the Cr(VI) state and a powdered substrate of high surface area, usually porous silica. Other supported transition-metal catalysts are used to produce certain polymer grades. Polyethylene is manufactured by the Phillips process in a slurry system (particle-form process). Other methods include solution polymerization with cyclohexane as diluent or gas-phase polymerization in which no hydrocarbon diluent is used and the polymer particles are agitated and suspended by fluidization or by mechanical agitation.

Uses

Linear HDPE is used in wire and cable insulation, containers, pipe, housewares, toys, institutional seating, filament, film, and linear low density polyethylene (LLDPE) (see Piping systems; Film and sheeting materials).

J. PAUL HOGAN
Phillips Petroleum Co.

J.P. Hogan in B.E. Leach, ed., *Applied Industrial Catalysis*, Vol. 1, Academic Press, Inc., New York, 1983, pp. 149–176.
J.P. Hogan, *J. Appl. Poly. Sci. Appl. Poly. Symp.* **36**, 49 (1981).

ZIEGLER PROCESS POLYETHYLENE

In general, Ziegler polymerizations and copolymerizations of ethylene are performed continuously or discontinuously in suspension in a hydrocarbon diluent. This catalytic system can be applied to varied systems to produce high density polyethylene (HDPE) grades for all modern requirements, including waxes of molecular weight ranging from 10,000 to ultrahigh molecular weight HDPE (UMHW–HDPE) of several millions (10^6). Wide variation of branching and density are also possible. Owing to this versatility, Ziegler polyethylene has acquired a leading commercial position throughout the industrialized world.

So-called second- and third-generation Ziegler polymerization technology with highly efficient catalyst systems increases the versatility of HDPE grades and reduces manufacturing cost.

The main usages of HDPE are in blow moldings of small medicine bottles, milk and detergent bottles, jerry cans, drums, and containers up to a volume of 10,000 L. Other items such as bottle crates, buckets, bowls, and thin-walled containers are made by injection molding. Pipes, sheets, monofilaments, cables, raffialike and paperlike film are extruded products.

Properties and Characterization

The properties of finished goods made from HDPE are strongly influenced by the polymerization and processing conditions and the properties of the polymer and the resin.

Physical and mechanical properties of Ziegler HDPE can be summarized as follows: low to medium stiffness and hardness; medium to extremely high toughness; no reduction in toughness to −40°C; unrestricted usage in contact with food; no restriction for disposal, burning, or recycling; high resistance against solvents and chemicals at ambient temperature; and easy and safe processing.

HDPE powder. Normally, HDPE powder is converted into pellets with a compounding extruder, but for blow molding, flake or powder may be processed directly. For reasons of safety (prevention of dust formation) and improved extrudability, powder grades are sold with bulk densities of 400–450 g/L and particle sizes of 100–2000 μm. Fine particles below 50–100 μm cause dust formation and consequently pose an explosion hazard. Usually, HDPE powder exhibits bulk densities of 100–400 g/L and particle sizes of 10–500 μm. Explosion risk is avoided by operating under nitrogen.

Manufacture

Large-scale production of Ziegler HDPE in suspension includes the following steps: (1) The catalyst is usually prepared separately; however, one manufacturer adds the various components directly to the polymerization mixture. Ziegler catalyst is prepared by the reaction of a transition-metal compound and a metal alkyl, eg, titanium tetrachloride and triethylaluminum. (2) In the presence of the catalyst, modifier, and sometimes comonomer, ethylene is polymerized to HDPE in a low boiling hydrocarbon diluent. (3) HDPE powder or flake is separated by centrifugation. (4) Diluent recovery. (5) Drying of the product under nitrogen or after steam stripping in hot air. (6) Compounding the product with additives (colorants) by melting and pelletizing.

EBERHARD PASCHKE
Hoechst AG

D.E. Axelson, G.C. Levy, and L. Mandelkern, *Macromolecules* **12**, 41 (1979).
L.S. Ryder, *Plast. Eng.*, 23 (Jan. 1980).
E. Paschke, *Chem. Eng. Prog.* **76**, 74 (1980).

POLYPROPYLENE

The invention of isotactic polypropylene (Fig. 1) by the Natta group in 1954 opened a new era in macromolecular stereochemistry and created a thermoplastic resin of such excellent properties that it has become one of the fastest-growing in the field. Natta, at Montedison, discovered that propylene could be polymerized with Ziegler catalyst to a new crystalline head-to-tail polymer from which he isolated a highly crystalline and stereoregular fraction (see also Olefin polymers, Ziegler process polyethylene).

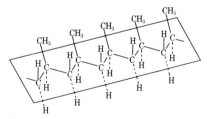

Figure 1. Isotactic polypropylene structure.

Table 1. Typical Properties of Some Propylene Homopolymer Grades[a]

		Grade			
Property	Method[b]	Pipes and sheets	Films	Injection molding	Cast and tubular film
melt flow rate, g/10 min	D 1238 L	0.25–0.35	1.5–2.0	5–7	7–10
intrinsic viscosity at 230°C, dL/g	ME 15071	2.6–3.0	2.0–2.4	1.8–2.1	1.5–1.9
density, g/cm^3	D 1505	0.900–0.905	0.900–0.905		0.900–0.905
crystallinity by x-ray, %	ME 149/67	58–59	59–60	61–62	62.5–64.0
tensile yield strength at 5 cm/mm, MPa[c]	D 638	29–31	32–34	33–35	33–35
tensile yield elongation, %	D 638	10–12	10–12	8–10	8–10
flexural modulus, GPa[d]	D 790	1.1–1.3	1.3–1.4	1.5–1.6	1.6–1.7
impact, Izod at 23°C, J/m[e]	D 256	1.5–2.0	0.5–0.7	0.3–0.5	0.2–0.3
hardness, Rockwell, C scale	D 785 B	55–60	55–60	60–65	60–65
coefficient of linear thermal expansion, × 10^{-5} from 30 to 60°C	D 696	14.0	14.0	14.0	14.0
environmental stress cracking[f]	D 1693	does not break	does not break	does not break	does not break
dielectric constant at 10 kHz	D 1531	2.2–2.6	2.2–2.6	2.2–2.6	2.2–2.6

[a] Determined on Montedison Moplen polypropylene.
[b] D = ASTM; ME = Montedison.
[c] To convert MPa to psi, multiply by 145.
[d] To convert GPa to psi, multiply by 145,000.
[e] To convert J/m to ft·lbf/in., divide by 53.38.
[f] Igepal = 100%.

Commercial polypropylene contains essentially the isotactic structure; depending on the catalyst employed and the polymerization conditions, minor amounts of other structures may be present.

Properties

Table 1 gives the properties of various homopolymer grades. The properties of injection-molded articles depend heavily on morphology, orientation, and the stresses introduced during molding.

Manufacture

Production processes. The choice of the process for polypropylene is principally determined by the performance of the catalysts. Different polymerization techniques are used commercially. Production processes include solvent (slurry) polymerization, gas-phase polymerization, liquid-monomer process, and the Montedison-Mitsui high yield process.

Block and random copolymers. Propylene block copolymers are prepared either by homopolymerization followed by ethylene–propylene copolymerization or by blending polypropylene with ethylene–propylene rubber or polyethylene modified with propylene. Block copolymers are employed specifically in low temperature applications.

Stabilization. Heat, uv radiation, and oxygen attack polypropylene, resulting in loss of physical and chemical properties. Antioxidants (qv), up to 1%, can be added during compounding after polymerization.

Processing

The basic properties of the molten polymer as well as the processing parameters, eg, melt temperature, pressure orientation, and cooling media, determine the morphology and crystalline structure in the manufactured articles. Processing includes molding (injection molding, thermoforming), and extrusion (film extrusion, stretch extrusion, and pipe and plate extrusion).

Uses

When compared to other large-volume molding and extrusion resins, isotactic polypropylene offers an outstanding combination of heat resistance, tensile strength, abrasion resistance, optical gloss and clarity, and low specific density.

Owing to their versatility, injection-molded articles can be used for a wide range of applications, including food containers, car bumpers (reinforced polypropylene all-soft bumpers), large boxes, and tanks.

Polypropylene film is used for the packaging of food, flowers, textile articles, drugs, and in automated packing of food and books (see Film and sheeting materials; Packaging materials). Stretch-extruded products are used for textile yarns and yarns for cords; monofilaments for spinning, cords, and brushes; straps for automatic and nonautomatic packing; decorative ribbons; multifilament for continuous filament; and other products. Extruded polypropylene pipes are used for industrial waste systems (homopolymers), water systems, including irrigation (block copolymers), and for floor-heating systems (random and block-copolymers) (see Pipelines; Piping systems).

GIOVANNI CRESPI
LUCIANO LUCIANI
Montedison SpA

G. Natta and co-workers, *J. Am. Chem. Soc.* **77**, 1708 (1955).

J. Boor, Jr., *Ziegler-Natta Catalyst and Polymerization*, Academic Press, Inc., New York, 1979.

A.V. Galanti and C.L. Mantell, *Polypropylene Fibers and Films*, Plenum Press, New York, 1965.

POLYMERS OF HIGHER OLEFINS

Most olefins with vinyl double bonds can be polymerized with heterogeneous Ziegler-Natta catalysts to isotactic polymers, highly or moderately crystalline materials exhibiting many beneficial physical and mechanical properties. Although one of these polymers, isotactic polypropylene, is produced worldwide in very large volume (see Olefin polymers, polypropylene), few other isotactic polymers of higher olefins (α-olefins with more than three carbons) are produced commercially.

Typically, isotactic polyolefins exhibit high mechanical strength and high dielectric characteristics. They are attacked by strong oxidants, eg, conc HNO_3, but not by other chemical agents. These products are processed easily with the application of standard commercial equipment (the same as used for processing of polyethylene and polypropylene) to form such articles as fibers, films (including easily oriented ones), injection-molded articles, etc. The main drawbacks of these materials are low adhesion ability to metals and other plastics, low dyeability owing to the absence of polar groups in their structure, relatively low rigidity in the nonoriented state, and decreased stability to oxidation in the case of polymers with highly branched side groups. In addition, some of these materials have a relatively high melting point, poor viscosity characteristics, and low rates of crystallization. Table 1 contains some data on structure and properties of two isotactic polymers of higher olefins.

The higher polyolefins are saturated, branched hydrocarbons that exhibit very low chemical reactivity, although they can be halogenated and oxidized at higher temperatures. Substantial susceptibility to oxidative destruction combined with high melting points provide the principal obstacles to commercial production of many higher polyolefins despite their good mechanical, dielectric and processing characteristics. Poly-(4-methylpentene) is highly transparent and is utilized as a high temperature-resistant organic glass. These polymers represent no health hazard.

Poly(1-butene)

Manufacture. Shell Chemical Company, at present the only worldwide producer, manufactures poly(1-butene) with technology developed by Mobil. Recently, however, Mitsui Petrochemical Industries, Ltd., announced its intention of launching into full-scale commercial production using technology imported from Hüls AG of the Federal Republic of Germany.

Catalysts. Catalysts employed in the polymerization of 1-butene are similar to those used in the manufacture of isotactic polypropylene. They belong to the Ziegler-Natta class and typically consist of aluminum-reduced $TiCl_3$ with diethylaluminum chloride as the cocatalyst. Also effective are $MgCl_2$-supported $TiCl_3$ catalysts in conjunction

Table 1.   Properties of Isotactic Polyolefins

Polymer	Structural unit of polymer chain	Melting point, °C[a]	Chain conformation in crystalline state	Crystal type	Density, g/cm³		Glass point, °C
					Crystal	Polymer	
poly(1-butene)[b]		125–130 (136)	helix 3_1	rhombohedral	0.96	0.915–0.91	−20 to −25
poly(4-methylpentene)		235–240 (250)	helix 7_2	tetragonal	0.83	0.83	20–50

[a] Highest observed melting temperature is given in parentheses.
[b] Form I.

with aluminum alkyls and aromatic esters such as ethyl benzoate or ethyl anisate.

Polymerization. The Mobil-Witco-Shell process uses a Ziegler catalyst in excess monomer under enough pressure to keep it liquefied.

Poly(4-methylpentene)

Manufacture. 4-Methylpentene is prepared by dimerization of propylene at 150–200°C in the presence of catalysts, eg, sodium or potassium.

Polymerization. Polymerization of 4-methylpentene may be carried out in hydrocarbon diluents or in excess monomer. The catalysts employed, as in the case of 1-butene polymers, belong to the Ziegler-Natta class.

R.K. KOCHHAR
Gulf Oil Chemicals Co.

YURI V. KISSIN
DAVID L. BEACH
Gulf Research & Development Co.

H.Y. Boenig, *Polyolefins: Structure and Properties*, Elsevier, New York, 1966.

G. Natta and F. Danusso, eds., *Stereoregular Polymers and Stereospecific Polymerizations*, 2 vols., Pergamon Press, New York, 1967.

I.D. Rubin, *Poly(1-Butene)—Its Preparation and Properties*, Gordon & Breach, New York, 1968.

OLEFINS, HIGHER

The higher olefins are important building blocks in the synthesis of certain intermediates that are used in the manufacture of consumer products. Although the demand for the branched olefins, which are produced from propylene and butylenes, has remained relatively constant during the past decade, there has been a significant growth rate in the demand for the lighter linear alpha olefins, particularly in the manufacture of synthetic lubricating oils and both linear low and high density polyethylene, as well as in the established use areas of special detergents and poly(vinyl chloride) (PVC) plasticizer-alcohols manufacture (see Alcohols, higher aliphatic; Lubrication and lubricants; Olefin polymers; Vinyl polymers).

Recent additions to manufacturing facilities for the linear alpha olefins have been based on ethylene oligomerization, using either Ziegler or non-Ziegler catalyst systems (see Ziegler-Natta catalysts). No significant growth in either wax cracking or paraffin dehydrogenation facilities is anticipated in the near future because of feedstock uncertainties, increasing energy costs, and inferior product quality.

Physical Properties

Certain physical properties of the C_6–C_{20} linear alpha olefins are listed in Table 1. The important branched olefins, which are produced by propylene or propylene–butylene polymerization, are mixtures of several isomers and are characterized by the following boiling ranges:

Olefins	Boiling range, °C
heptenes	67–97
nonenes	126–149
dodecenes	182–204

Chemical Properties

The chemistry of the higher olefins is based on the double bond which, in most reactions reacts with an electrophilic reagent.

Rearrangements. Isomerization reactions of the olefins include cis–trans shifts, double-bond migration, and structural arrangements.

Reactivity. The reactivity of the higher olefins is similar to that of the lower olefins, although reaction conditions cannot be extrapolated for the former because of solubility relationships, which are related to the lengths of the hydrocarbon chains.

Electrophilic addition. The addition of acids, eg, hydrogen chloride, is typical of an important class of reactions.

$$RCH{=}CH_2 + H^+ \rightarrow R\dot{C}HCH_3$$

$$R\dot{C}HCH_3 + Cl^- \rightarrow RCHClCH_3$$

The other hydrogen halides, sulfuric acid, organic acids, and halogens add to the olefins in the same way.

Free-radical addition. Radical additions that are initiated by peroxides and by radiation may attack the double bond, as demonstrated in the following equations:

$$HBr + R'\cdot \rightarrow R'H + Br\cdot$$

$$RCH{=}CH_2 + Br\cdot \rightarrow R\dot{C}HCH_2Br$$

$$R\dot{C}HCH_2Br + HBr \rightarrow RCH_2CH_2Br + Br\cdot$$

Of commercial interest is the addition of sodium and ammonium bisulfites to C_{10}–C_{20} α-olefins, particularly C_{15}–C_{18}, to form the primary alkylsulfonates.

Allylic reactions. When an olefin is attacked by a strongly electronegative radical, eg, t-butoxy which is derived from di-t-butyl peroxide, an allyl hydrogen is abstracted, as shown in the following equation:

$$RCH_2CH{=}CH_2 + (CH_3)_3CO\cdot \rightarrow R\dot{C}HCH{=}CH_2 + HOC(CH_3)_3$$

Nucleophilic reactions. Epoxides formed from the reaction of α-olefins with alkaline hydroxide peroxide undergo ring opening to hydroperoxides which are nucleophiles.

$$RCH{=}CH_2 \xrightarrow{H_2O_2/NaOH} RCH\overset{\displaystyle O}{\overset{\diagup\diagdown}{}}CH_2$$

These epoxides react with a variety of other reagents to form consumer products, eg, amine oxides (see Epoxidation; Polyethers).

Transition-metal complexes. The complexes formed between the various transition metals and olefins have been topics for intensive research and numerous surveys in recent years. Olefin reactions involving

Table 1. Properties of C_6–C_{20} Linear α-Olefins

Compound	Freezing point (at 101 kPa[a]), in air, °C	Bp (at 101 kPa[a]), °C	dt/dp (at 101 kPa[a]), °C	Refractive index[b], n_D^{20}	Density[b] (at 20°C), g/cm³	Kinematic viscosity, mm²/s[b, c] 20°C	100°C	Free energy of formation $\Delta F°$ (at 25°C), kJ/mol[d]
1-heptene	−119.03	93.64	0.04447	1.39980	0.69698	0.50		96.02
1-nonene	−81.37	146.87	0.04944	1.41572	0.72922	0.851	0.427	112.8
1-dodecene	−35.23	213.36	0.05522	1.43002	0.75836	1.72	0.678	138.0

[a] 101 kPa = 1 atm.

[b] Values are given for the air-saturated hydrocarbon in the liquid state at 101 kPa[a]. Values of the refractive index are for the sodium D line for which the wavelength is 58.926 μm, which is the intensity-weighted mean of the wavelengths of the D_1 and D_2 lines.

[c] mm²/s = cSt.

[d] To convert J to cal, divide by 4.184.

such catalysts include hydrogenation, dimerization, isomerization, oxidation, and hydroformylation (see Organometallics).

Disproportionation. Linear alpha olefins, eg, 1-octene and 1-decene, disproportionate to produce 7-tetradecene and 9-octadecene at 100–400°C over selected metal catalysts. Also, high molecular weight internal olefins react with ethylene to yield lower molecular weight alpha olefins. Of commercial interest is the disproportionation of mixtures of C_{10} and C_{16} or higher linear alpha olefins to form C_{11}–C_{14} linear internal olefins.

Maleic anhydride. Olefins react with maleic anhydride in a semi-Diels-Alder reaction at 180–200°C. A commercial application is the production of dodecenylsuccinic anhydride from propylene tetramer; the anhydride is used as a corrosion inhibitor (see Corrosion and corrosion inhibitors).

Carboxylation. Olefins react with carbon monoxide in the presence of concentrated sulfuric acid at 20°C to form a carbonyl-containing sulfuric acid ester which, on hydrolysis, yields a carboxylic acid (see Oxo process).

Polymerization. Linear α-olefins polymerize in the presence of Ziegler-Natta catalysts and under conditions used for propylene polymerization to high molecular weight, solid polymers.

Alkylation. The alkylation of benzene with branched olefins, especially propylene tetramer, has been the largest use for the higher olefins, ie, for ultimate conversion to alkylbenzenesulfonates for detergents. Phenol also is alkylated under similar conditions with either linear or branched olefins. The resultant product is used in the preparation of antioxidants (qv), lube-oil additives, plasticizers (qv), and surface-active agents (see Lubrication and lubricants; Surfactants and detersive systems).

Transalkylation. Transalkylation is important in the commercial manufacture of the detergent-range linear α-olefins, ie, C_{12}–C_{16}, by the Ziegler two-step ethylene oligomerization process.

Health and Safety Factors

The toxicity of the higher olefins is considered to be essentially the same as for the corresponding paraffins. Animal toxicity data for hexene, decene, and hexadecene have shown little or no toxic effect except under severe inhalation conditions.

D.G. DEMIANIW
Gulf Oil Chemicals

A.H. Turner, *J. Am. Oil. Chem. Soc.* **60**(3), 623 (1983).

S. Patai, ed., *The Chemistry of the Alkenes*, Wiley-Interscience, New York, 1965.

E. Clippinger, *Ind. Eng. Chem. Prod. Res. Dev.* **3**, 3 (1964).

OLIGOSACCHARIDES. See Antibiotics, oligosaccharides.

OPERATIONS PLANNING

The rapid growth of industry, in terms of size and complexity, has become a hallmark of the twentieth century. Although growth has resulted in many economies of scale, it also has made executive decision-making extremely complicated. This situation has fostered the emergence of scientific methods of decision-making that help organizations utilize their resources in the most effective manner. The problems to which these methods have been applied range from strategic decisions, eg, the sizing and location of important new plants (see Plant layout; Plant location), to tactical problems, eg, the management of inventories or the setting of daily production schedules. This problem-solving approach is referred to as industrial engineering, operations management, systems analysis, operations research, and management science. The art and science of mathematical modeling are fundamental to the process.

In operations planning, mathematical models of decision-making situations are constructed. Quantifiable objectives are prerequisite to building a mathematical model, so that alternative decisions can be compared. Next, the important variables of the problem must be isolated, their interrelationships must be specified, and any constraints or restrictions on their values must be understood. The model is used to choose values of the variables that yield the best outcome for the objectives.

In formulating a mathematical model for decision-making, a balance must be struck between the tractability of simple models and the accuracy of complex systems. Newton's laws of motion, for example, provide a mathematical model of the physical world in terms of only a few basic variables. The three laws are adequate for understanding the properties of many real systems. Other settings require the introduction of additional variables, eg, the effects of friction, electricity, magnetism, and sometimes relativity. Each additional variable increases the accuracy of the model but at the expense of mathematical tractability and clarity of exposition.

General operations-planning methods include linear-optimization models and linear programming, related optimization models, sequential decision processes and dynamic programming, and system simulation using digital computers (see Simulation and process design).

Operations planning can be used for distribution in multi-location systems, equipment replacement, resource allocation, inventory management and production planning, program evaluation and review technique and critical path scheduling, control of chemical processes, and waiting-line models.

RICHARD EHRHARDT
The University of North Carolina at Greensboro

F.S. Hillier and G.F. Lieberman, *Operations Research*, 2nd ed., Holden-Day, Inc., San Francisco, Calif., 1974.

H.M. Wagner, *Principles of Operations Research*, 2nd ed., Prentice-Hall, Inc., Englewood Cliffs, N.J., 1975.

R. Peterson and E.A. Silver, *Decision Systems for Inventory Management and Production Planning*, John Wiley & Sons, Inc., New York, 1979.

E.H. Bowman and R.B. Fetter, *Analysis for Production and Operations Management*, R.D. Irwin, Homewood, Ill., 1967.

OPIATES. See Alkaloids; Analgesics; Hypnotics, sedatives, anticonvulsants.

OPIOIDS, ENDOGENOUS

The concept of the opioid receptor remained hypothetical until 1973 when stereospecific opioid binding in rat brain was independently identified in three separate laboratories. The presence of opioid receptors in vertebrates and the obvious absence of a phylogenetic relationship between vertebrates and the poppy suggested an important physiological role for the opiate receptor. It also suggested the presence of an as-yet-unidentified endogenous ligand. Evidence emerged showing that endogenous opioids were peptides rather than simple morphinelike molecules. The term endorphin (endo + morphine) as a general term for endogenous opioids was adopted. The term enkephalin designates a specific pentapeptide subset.

The precursor relationship of the various opioid peptides has been clarified by cloned (c)-DNA techniques; three distinct precursors, each with a molecular weight of ca 28,000, have been characterized (see Table 1).

Opioid Receptors

The concept of multiple receptors for hormones and neurotransmitters is well documented. For example, the classical neurotransmitter acetylcholine is believed to interact at two distinct cholinergic receptors, the nicotinic and muscarinic sites, producing very different pharmacological effects (see Neuroregulators). Three distinct opioid receptors were originally proposed, termed μ, κ, and σ. A fourth type, the δ receptor, was postulated later.

The true physiological function of each opioid receptor is not known. A hypothesized model for the interactions of the μ- and δ receptors in analgesic processes includes the following: distinct μ and δ receptors coexist in an opioid-receptor complex; Leu-enkephalin is a δ-agonist; β-endorphoin is both a μ- and δ-agonist; the interaction between the two receptors is not considered to be a thermodynamic equilibrium or two interconverting forms of the same receptor; a coupling mechanism in analogy to the adenyl cyclase model; and a coupling mechanism representing the ability of the δ receptor to modulate coupling of the μ-agonists to the effector. Recent advances have led to further characterization of the κ receptor and its potential role in analgetic processes.

Biological Function of Endorphins

The endogenous role of endorphins quite naturally focused on pain modulation because of the early relationship between opioid peptides and pain. The enkephalins are weak analgetics in animal tests because of their short (2–3 min) biological half-life. The finding of analgetic activity of the endorphins created a short-lived hope that these or related peptides might lead at last to an analgesic devoid of dependence liability.

The endorphins influence a wide range of behavior. For example, intraventricular administration of Leu- and Met-enkephalins suppresses operant behavior maintained by food reinforcement, which may indicate that endorphins are involved in regulating biological mechanisms. Studies in rodents have shown that β-endorphin and an enkephalin analogue suppress copulatory behavior.

Metabolic Inactivation

The difference in analgesic effect between the enkephalins and β-endorphins can be explained by their relative stabilities. The half-life of the enkephalins in the presence of synaptic membranes is 2–3 min compared to 2–3 h for β-endorphin.

Intensive research efforts have focused on the discovery of effective inhibitors of enkephalin-degrading enzymes as analgesic agents. A potent (K_I of 4.7 nM) and specific enkephalinase A inhibitor, thiorphan, has been reported to potentiate [D-Ala^2]-Met-enkephalin analgesia. Compared to enkephalin metabolism, little is known about the metabolic inactivation of β-endorphin.

Thiorphan

Structure-Activity Relationships

As with morphine, the identification and structural elucidation of Met- and Leu-enkephalins gave rise to the synthesis of a large number of analogues. Some of the early work was directed toward the similarity between the tyrosine of position 1 of Met-enkephalin and the phenethylamine moiety of morphine; most analogues were considerably less active than Met-enkephalin. An *N*-methyl group increases the potency; removal of the amino group of tyrosine and masking of the phenolic group decreases it. Met-kephamid possesses analgesic potency comparable to that of meperidine in rodents, and clinical data indicate analgesic activity in humans.

Biologically Related Endogenous Substances

The endogenous opioids are members of a rapidly growing family of peptides that exhibit neurotransmitter or hormonal properties. The best candidate for a neurotransmitter of sensory pain information in the spinal cord is substance P (SP), an undecapeptide isolated in the intestines which is linked to endogenous opioids.

Arg–Pro–Lys–Pro–Gln–Gln–Phe–Phe–Gly–Leu–Met–NH$_2$

substance P

Neurotensin is a 13-amino acid peptide that may function as neurotransmitter. Somatostatin is a cyclic tetradecapeptide that inhibits secretion of pituitary growth hormone; it may be a principal transmitter of pain.

*p*Glu–Leu–Tyr–Glu–Asn–Lys–Pro–Arg–Arg–Pro–Tyr–Ile–Leu–OH

neurotensin

Ala–Gly–Cys–Lys–Asn–Phe–Phe–Trp–Lys–Thr–Phe–Thr–Ser–Cys–OH

somatostatin

M. Ross Johnson
David A. Clark
Pfizer, Inc.

Table 1. Structure of Opioid Peptides

Group I. Pro-opiomelanocortin-derived

α-endorphin	H–TyrGlyGlyPheMetThrSerGluLysSerGlnThrProLeuValThr–OH
γ-endorphin	H–TyrGlyGlyPheMetThrSerGluLysSerGlnThrProLeuValThrLeu–OH
β-endorphin	H–TyrGlyGlyPheMet ThrSerGluLysSerGlnThrProLeuValThr LeuPheLysAsnAlaIleValLysAsnAlaHisLysLysGlyGln–OH

Group II. Pro-enkephalin-derived

Leu-enkephalin	H–TryGlyGlyPheLeu–OH
Met-enkephalin	H–TyrGlyGlyPheMet–OH
octapeptide	H–TyrGlyGlyPheMetArgGlyLeu–OH
heptapeptide [73024-95-0]	H–TyrGlyGlyPheMetArgPhe–OH

Group III. Pro-dynorphin-derived

dynorphin (1–17)	H–TyrGlyGlyPheLeuArgArgIleArgProLysLeuLysTrpAspAsnGln–OH
dynorphin (1–8)	H–TyrGlyGlyPheLeuArgArgIle–OH
α-neo-endorphin	H–TyrGlyGlyPheLeuArgLysTyrProLys–OH
β-neo-endorphin	H–TyrGlyGlyPheLeuArgLysTyrPro–OH

R.M. Post, P. Gold, D.R. Rubinow, J.C. Ballenger, W.E. Bunney, Jr., and F.K. Goodwin, *Life Sci.* **31**, 1 (1982).

J.B. Malick and R.M.S. Bell, eds., *Endorphins: Chemistry, Physiology, Pharmacology, and Clinical Relevance*, Marcel Dekker, Inc., New York, 1982.

M.R. Johnson and M. Milne in M.E. Wolff, ed., *Burger's Medicinal Chemistry*, 4th ed., Pt. 3, John Wiley & Sons, Inc., New York, 1981, pp. 699–758.

OPTICAL BLEACHES.   See Brighteners, fluorescent.

OPTICAL FILTERS

Optical filters are used widely in engineering, research, and photography when spectral radiant energy must be altered, modulated, or controlled precisely. These filters may be composed of any number of natural or synthetic materials. Their design and use are governed quite accurately by well-known physical laws (see Color; also Color photography; Photochemical technology; Photography).

Terminology and Definitions

The ratios of the fluxes to the incident flux depicted in Figure 1 are defined as ANSI standard symbols:

$$\text{reflectance } \rho = \frac{\Phi_r}{\Phi_i} \tag{1}$$

$$\text{absorptance } \alpha = \frac{\Phi_a}{\Phi_i} \tag{2}$$

$$\text{transmittance } \tau = \frac{\Phi_t}{\Phi_i} \tag{3}$$

The sum of the three ratios is unity:

$$\rho + \alpha + \tau = 1 \tag{4}$$

τ_{max} is defined as the maximum transmittance of a filter. $\lambda_{1/2}$ is the wavelength where a short-pass or long-pass filter has one half of the τ_{max} value. $\Delta\lambda_{1/2}$, or the half-pass of a band-pass filter, is the wavelength interval between the two points representing one half of the τ_{max} value.

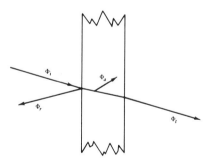

Figure 1.   Incident flux Φ_i; reflected flux Φ_r; absorbed flux Φ_a; transmitted flux Φ_t.

The reflectance from each surface of a filter can be determined using Fresnel's equation:

$$\rho = \left(\frac{n_1 - n_2}{n_1 + n_2}\right)^2 \tag{5}$$

Absorptance is determined by the Bouguer-Lambert law:

$$\Phi_a = \Phi_i e^{k_\lambda x} \tag{6}$$

where Φ_i is the incident flux, e is the Napierian base 2.71828), k_λ is the absorption coefficient for unit concentration, and x is the thickness of the filter. Furthermore,

$$\Phi_a = \Phi_i e^{-\beta_\lambda C x} \tag{7}$$

where β_λ is the absorption coefficient, C is the concentration of the dye or soluble salt, and x is the filter thickness. Some dyes in high concentration do not follow Beer's law, although many do.

Optical spectral density is defined as:

$$D_\lambda = \log_{10} \frac{1}{\tau_\lambda} \tag{8}$$

where τ_λ is the spectral transmittance of the sample.

Filter types.   *Neutral*: a filter that transmits essentially equally at all pertinent wavelengths. *Polarizer*: neutral or colored filters that polarize transmitted light. *Color*: filters that transmit selectively, absorbing or reflecting (nonabsorbing) some wavelengths more than others.

Filter classes.   Absorption filters consist of dyes, soluble salts, or pigments suspended or dissolved in some transparent medium (see Table 1).

Table 1.   Absorption-Filter Types

Type	Mode of dye
liquid	in solution
glass	colorant in glass
Photar	in cement between clear glass
gelatin	in gelatin
plastic	in cast or extruded plastic
imbibed	plastic with dye imbibed into the surfaces

Interference filters consist of stacks of thin layers of essentially colorless materials of differing refractive index arranged for the destruction of Fresnel reflectance (anti-reflection (AR) filters) or the enhancement of this reflectance (most other types). Interference filters reflect nontransmitted light, absorbing very little, and can be used in high intensity beams without fading.

Filter measurement.   The radiant flux is measured with a radiometer. The ratio of the radiant fluxes of two beams is measured with a ratio radiometer. A spectrophotometer is a ratio radiometer that operates with a narrow band of wavelengths divided into two equal-intensity beams; one is a reference beam and the other includes the sample to be measured.

Antireflection Interference Coatings

Equation 5 states that the first surface reflectance in air depends on the refractive index of the filter material. For borosilicate crown glass, the refractive index is 1.52 and the reflectance ρ is 4.26%. If a single layer of MgF_2 is deposited with an optical thickness (ot) of 138 nm or 1/4 of 550 nm (since the refractive index of MgF_2 is 1.38, a physical thickness of 100 nm is used), destructive interference occurs at the air–MgF_2 surface and the reflectance ρ at 550 nm is decreased to 1.4%. The transmission increase is much greater when AR coatings are used on high index infrared-transmitting substrates (see Infrared technology).

Companies offering AR coatings for ultraviolet, visible, or infrared include Balzers Corp., Broomer Research Corp., Continental Optical Corp., Corion Corp., Denton Vacuum, Inc., Eastman Kodak Co., Evaporated Coatings, Inc., Exotic Materials, Inc., Herron Optical Co., Metavac, Inc., Optical Coating Laboratory, Inc., Optical Filter Corp., Schott Optical Glass, Inc., and Valtec Corp.

Neutral Filters

Neutral filters attenuate by reflection, absorbance, scattering, polarization, or combinations of these. Many neutral filters are heterogeneous (scatter light) and cannot be used in image-forming beams. Some filters may be neutral when measured with diffuse geometry but have more scattering at shorter wavelengths and thus are not neutral when used in specular beams. Neutral filters include glass filters, photar filters, carbon dispersions in gelatin, wedges and steps, metal-coated filters, and photographic silver.

Polarizers.   The Nicol prism, invented in 1828, was the first practical polarizer. It is made of calcite, $CaCO_3$, which is a birefringent material, ie, it has refractive indexes of 1.6585 and 1.4862. An incident beam of unpolarized light is divided into two beams that are polarized at right angles; one is reflected from an internal cemented surface and escapes through the side of the prism whereas the other passes through this layer to emerge as strongly polarized light. A second Nicol prism interposed in this beam varies that intensity transmitted by the pair. If the two Nicol prisms are parallel, the intensity is at a maximum. If the second prism is rotated 90° on the optical axis, almost no light is transmitted.

Absorbent Color Filters

Color filters are classified into short-pass, long-pass, and band-pass filters, shaped filters, and other filters. Short-pass filters have higher transmittance at short than at long wavelengths. Long-pass filters have higher transmittance at long than at short wavelengths. Band-pass filters transmit a band of wavelengths with lower transmittance on both sides of the band. Color filters come in a variety of styles including glass filters, gelatin filters, plastic filters, blocking filters, and shaped filters, eg, color-temperature-conversion filters, color-compensating and color-printing filters, response-correcting filters, and filters for multiple-contrast photographic papers.

Nonabsorbent Filters

Nonabsorbent filters include interference filters and Fabry-Perot metallic filters, eg, hot and cold mirrors, all-dielectric band-pass filters, and special interference filters.

Other Filters

Other filters include safelight filters, Kodak Wratten photomechanical filters, and didymium and holmium filter glasses.

Edward K. Letzer
Eastman Kodak Co.

Special Filters from Kodak for Technical Applications, Pamphlet U-73, Eastman Kodak Company, Rochester, N.Y., 1973.

Corning Glass Works Filters, Corning Glass Works, Corning, N.Y.

ORGANOLEPTIC TESTING.   See Flavor characterization; Odor modification; Perfumes.

ORGANOMETALLICS — σ-BONDED ALKYLS AND ARYLS

Organometallics are compounds that contain carbon–metal bonds. Only those compounds with carbon–metal σ bonds, with emphasis on those that are industrially important, are discussed here (see Organometallics—metal π complexes).

Organometallics exhibit a diversity of physical and chemical properties, ranging from highly volatile gases, eg, trimethylborane, to infusible solids that cannot be distilled, eg, di-*n*-butylmagnesium. Many organometallics ignite spontaneously when exposed to air, whereas others are unreactive with air. Most of Group I–III organometallics are highly reactive and, in some cases, are explosive when in contact with water. In contrast, trialkylboranes, tetraalkyltin, and tetraalkyllead compounds are unreactive with water. Sigma-bonded organometallics are comprised primarily of the elements in Groups I–IV.

General synthetic methods for preparing organometallics have evolved steadily since the middle of the nineteenth century. Preparative schemes may be classified as primary or secondary methods. Primary methods involve the direct synthesis of organometallics starting with elemental metal and other reactants, but exclude organometallic reactants. Secondary methods involve the use of organometallics to synthesize other organometallics. In some secondary methods, the organometallic compound that is synthesized may contain the same metal as the reactant except that different substituents are bonded to the metal.

Group IA

The chemistry of Group IA organometallics is dominated by organolithium compounds and, primarily, *n*-butyllithium. Hundreds of metric tons per year are produced and used in the manufacture of synthetic rubber for tires, hoses, footwear, and many other consumer items (see Elastomers, synthetic).

Lithium.   Commercially available organolithium compounds, eg, *n*-butyllithium and *sec*-butyllithium, exist as viscous liquids in neat form, and are usually supplied in aliphatic hydrocarbons. Lithium alkyls are covalently bonded and exist in associated form as tetramers or hexamers, depending on the nature of the alkyl group and the solvent. Lithium alkyls ignite spontaneously on contact with air and react violently with water. Organolithium compounds exhibit very low vapor pressures and are distilled or sublimed with great difficulty even in high vacuum.

Lithium alkyls are manufactured by reaction of the appropriate alkyl halide with a lithium dispersion, usually in a hydrocarbon solvent.

Sodium.   Industrial use of organosodium compounds is significant, although the volume is diminishing because of the increasing use of other chemicals. Organosodium compounds are chemical intermediates in the synthesis of other organometallics, eg, organotin and organomanganese compounds.

In contrast to organolithium reagents, organometallic compounds of sodium, where the R in RNa is a simple alkyl group, are mostly ionic. Generally, simple sodium alkyls are nonvolatile, white, pyrophoric solids which react explosively with water and other protic materials and are insoluble in hydrocarbons. Sodium alkyls are more reactive than organolithium compounds with oxygenated compounds. Sodium alkyls are not stable in long-term storage even at reduced temperatures. On heating, sodium alkyls decompose and liberate alkenes and alkanes.

Simple organosodium compounds are prepared most easily by the reaction of alkyl halides with metallic sodium.

Group IIA

Dialkylmagnesium compounds, R_2Mg, recently have become commercially significant (see Grignard reaction). Organoberyllium compounds are expensive and toxic materials and are primarily of theoretical interest. Little is known about the organometallic chemistry of calcium and heavier Group IIA metals and these compounds are not commercially significant.

Magnesium.   The C_1–C_4 dialkylmagnesium compounds are pyrophoric and react violently with water and other protic materials. The C_1–C_4 straight-chain dialkylmagnesium compounds are white solids that are insoluble in hydrocarbons but are freely soluble in donor solvents, eg, ethers and amines, with which the R_2Mg compounds form complexes. Dialkylmagnesium compounds cannot be distilled and the C_1–C_4 R_2Mg compounds decompose at > 160°C without melting. Commercially available dialkylmagnesium compounds are stable indefinitely at room temperature when stored under an inert atmosphere of dry nitrogen or argon. The most important use of dialkylmagnesium compounds is in the production of modified or second-generation Ziegler-Natta catalysts for olefin polymerization. The use of magnesium alkyls in the polymerization of propylene also may become industrially significant in the near future (see Olefin polymers). Commercial methods of preparing R_2Mg compounds involve reaction of alkyl halides with magnesium in hydrocarbon solvents.

Group IIIA

The organometallic chemistry of gallium, indium, and thallium has not been studied in depth and the organometallic compounds of these elements are not important industrially. However, gallium and indium alkyls in ultrahigh purity, ie, > 99.999 wt% metallic purity, are potentially useful in the production of semiconductor materials for electronic devices (see Semiconductors). Certain high purity gallium compounds, eg, trimethylgallium, may be used in solar-cell technology. Organothallium compounds have been used in laboratory-scale organic synthesis.

Boron.   The volume of organoboron compounds manufactured commercially is small relative to the production volumes of other organometallics, eg, aluminum, lead, and tin alkyls (see Boron compounds, Hydroboration).

Organoboron compounds are very reactive with oxygen. Low molecular weight organoboron compounds, eg, C_1–C_3 trialkylboranes, are spontaneously flammable in air. Unlike other Group IA, IIA, and IIIA organometallics, however, trialkylborane compounds are unreactive with water. Trialkylboranes are nonpolar and are insoluble in water but are freely soluble in and unreactive with saturated aliphatic and aromatic hydrocarbons. Unlike trialkylaluminum compounds, trialkylboranes are monomeric and do not form strong donor-acceptor complexes with ethers.

Methods of preparing trialkylborane compounds include alkylation of borate esters or, preferably, boron halides using zinc alkyls, Grignard reagents, or aluminum alkyls.

Organoboranes can be used in a diversity of organic synthetic applications, although none has been carried out on a large commercial scale. Among the more important recent developments is the use of R_2BH compounds and lithium hydride complexes of R_3B compounds as stereoselective reducing agents for carbonyl compounds.

Aluminum. Because of their utility in Ziegler-Natta polymerization of olefins and dienes, organoaluminum compounds are the most important organometallic derivatives of the Group IIIA metals.

Most development and commercial aluminum alkyls, particularly those with high metal content, are pyrophoric and react violently with water. Aluminum alkyls are freely soluble in and unreactive with aromatic and saturated aliphatic hydrocarbons. Because such solutions are less likely to ignite spontaneously on contact with air, large quantities are sold in hydrocarbon solution. Most of the aluminum alkyls that are sold commercially are used in the Ziegler-Natta polymerization of olefins and dienes to produce thermoplastics and synthetic rubber. The largest-volume application of aluminum alkyls is in the production of linear α-alcohols and α-olefins for use in the manufacture of biodegradable detergents (see Alcohols, higher aliphatic; Surfactants).

Trialkylaluminum compounds are manufactured on a large scale by both direct (Al, hydrogen, and olefin reaction) and displacement (R_3Al and olefin-exchange reaction) processes.

Group IVA

Members of Group IVA elements account for the largest volume of industrially important organometallic compounds. Organosilicon compounds are manufactured on a large scale as intermediates for silicones (see Silicon compounds). Organometallics of tin and lead are more important industrially and are representative of Group IVA organometallics.

Organometallics of Group IVA are characterized by lower chemical reactivity relative to that of neighboring groups. Although most organometallics hydrolyze easily and many ignite on exposure to air, the R_4M compounds of Group IVA are nonreactive with air, water, or in some cases, dilute acid. The relative reactivities increase with increasing atomic weight within the group.

Organometallics of Group IVA elements possess almost exclusively the property of metal–metal bonding, thereby forming catenated structures. Stability of these metal–metal bonds decreases with increasing atomic weight in the group; therefore, no structures are known in which more than three lead atoms are linked in a chain. Association through alkyl bridging is unknown for the tetraalkyls of this group; thus, these compounds exist in monomeric form.

Tin. Generally, the organotin compounds are stable solids or high boiling liquids of low to moderate viscosity. In the pure form, the compounds are colorless to light yellow. With few exceptions, eg, dimethyltin dichloride, the R_4Sn compounds and the various inorganically substituted compounds are insoluble in water but are soluble in many organic solvents. Certain substituted organotin compounds, eg, tricyclohexyltin hydroxide and triphenyltin hydroxide, are nearly insoluble in all solvents; thus, they typically are used as colloidal suspensions or emulsions in water. Organotins are used as stabilizers, biocides, and catalysts. Organotins are manufactured by alkylation of tin tetrachloride or by direct reaction of tin with alkyl halides (see Heat stabilizers; Coatings, marine).

Generally, the inorganic tin compounds are innocuous, and some have been used internally for medicinal purposes. However, the organotin compounds exhibit a broad spectrum of toxicities, which are determined primarily by the number and length of the hydrocarbyl groups rather than the anion groups, provided that the anion groups are not toxic. Of the various classes, R_3SnX compounds are the most toxic to mammals. Although organotins vary widely in toxicity, a TLV of 0.1 mg Sn/m^3 has been established for organotins in the United States.

Lead. The lower tetraalkyllead compounds are colorless, dense liquids with a faintly sweet odor. Tetraarylleads and higher tetraalkylleads, ie, alkyls $> C_9$, generally are colorless solids. All are insoluble in water but are soluble in many organic solvents. Although the least stable of Group IVA organometallics, they are not readily reactive with air and are stable to water and bases at storage conditions. In general, organoleads are thermally unstable above 100°C, but some can be distilled under vacuum. As with organotins, almost all of the stable organoleads are derivatives of tetravalent lead. Virtually all of the lead alkyls that are produced are used for antiknock purposes (see Gasoline).

Basically, three industrial routes to organoleads are utilized. The oldest and most generally used method is the alkylation of sodium–lead alloy by an alkyl halide. A newer process, which was developed by Nalco Chemical Co., involves the electrolysis of a Grignard solution using lead as the anode. More recently, the aluminum alkyl route has been introduced.

The tetraalkyllead compounds are more hazardous than inorganic lead salts because of the former's volatility and ability to be absorbed through the skin. However, significant exposure to these compounds is limited to organolead manufacturing sites and to gasoline-blending facilities. In these areas, engineering control, proper ventilation, and treatment of spills with a strong oxidizer, eg, permanganate, have been effective in minimizing hazards (see Lead compounds, industrial toxicology).

Transition Metals

Only a few σ-bonded alkyl or aryl transition-metal compounds are well known, and most of these are unstable in air and water, and to heat. Strong ligands, eg, ethers and amines, and mixed σ- and π-bonding in the same compound tend to stabilize the C–metal σ bond, but few of the stabilized compounds can be retained at room temperature for an extended time. Although these highly reactive compounds are short-lived at normal reaction temperatures, they serve effectively as catalysts and in organic synthesis. It is the labile nature of these compounds that makes them function so well as catalysts under mild conditions.

The catalytic applications of organotransition metals is diverse and widespread. Undoubtedly, the following reactions, which are catalyzed by transition-metal compounds, involve *in situ* formation of, or an intermediate form of, σ-bonded organotransition metals: isomerization, dimerization, oligomerization, and polymerization of unsaturated hydrocarbons; carbonylation; hydrogenation of unsaturated hydrocarbons; and oxidation of olefins to various organic compounds. Another application of organotransition-metal compounds is in biological methylation by naturally occurring vitamin B_{12} (qv), which is an organoderivative of cobalt (see Catalysis; Oxo process).

Group IIB

The decrease in reactivity of the Group IIB organometallics parallels the increasing electronegativity of the metals. This is shown strikingly by the lower dialkylzincs which hydrolyze with explosive violence, whereas dialkylcadmiums hydrolyze slowly and dialkylmercurials do not hydrolyze. Organometals of Group IIB form much weaker complexes with donor agents, eg, amines and ethers, than do the organometals of Group IIA. The only significant commercial application of organic cadmium compounds is in the use of carboxylate salts as PVC stabilizers (see Cadmium compounds; Heat stabilizers).

Zinc. Organozinc compounds are the most reactive of Group IIB organometallics. The linear structure of dialkylzincs results in nonpolar, monomeric compounds which can be distilled easily. The atmospheric boiling points for dimethylzinc, diethylzinc, and di-*n*-propylzinc are 44°C, 117°C, and 139°C, respectively.

The high cost of organozincs has limited their commercial application. Organozincs are used in syntheses and as catalysts or initiators (qv) for many types of polymerizations. A potential application of diethylzinc is for preservation of books (see Fine art examination and conservation).

Organozinc compounds can be made commercially by alkylation of $ZnCl_2$ with other organometallics or by reaction of alkyl halides with zinc in the presence of aluminum alkyls or copper alloy.

Mercury. The lower dialkylmercurials are volatile liquids that generally are stable when refrigerated. Dimethylmercury (bp 92.5°C) is stable at room temperature, but diethylmercury (bp 159°C) slowly decomposes to the metal. Diarylmercury and arylmercury salts are high melting solids. Alkylmercury salts are solids and are characterized by high vapor pressures and objectionable odors. Organomercurials also are characterized by low reactivity to water. Most applications of organomercury compounds are based on their toxic properties. Organomercurials have been used for therapeutic applications, eg, merbromin and merthiolate as general antiseptics (see Disinfectants and antiseptics; Mercury compounds).

Phenylmercuric acetate is manufactured by the reaction of mercuric acetate and benzene in the presence of glacial acetic acid.

In general, the alkylmercurials are much more toxic than the phenylmercurials, and the therapeutic organomercurials are the least toxic. The simple alkyls and aryls generally are skin vesicants. Organomercury compounds are absorbed readily through the skin. Because C–Hg bonds are very stable, the organomercury compounds are eliminated very slowly from the body and, therefore, are cumulative poisons.

D.B. MALPASS
L.W. FANNIN
J.J. LIGI
Texas Alkyls, Inc.

G. Wilkinson, F.G.A. Stone, and E.W. Abel, *Comprehensive Organometallic Chemistry*, Pergamon Press, New York, 1982, Vols. 1–9.

E.I. Becker and M. Tsutsui, eds., *Organometallic Reactions*, Vols. 1–5, Wiley-Interscience, New York, 1970–1975.

E.I. Negishi, *Organometallics in Organic Synthesis*, Vol. 1, Wiley-Interscience, New York, 1980.

ORGANOMETALLICS — METAL π COMPLEXES

Metal π complexes are characterized by a type of direct carbon-to-metal bonding that is not a typical ionic or covalent σ bond, but rather an unusual σ–π bond.

Numerous molecules and ions, eg, mono- and diolefins, polyenes, arenes, cyclopentadienyl ions, tropylium ions, and π-allylic ion, can form metal π complexes with transition-metal atoms or ions. These are classified as organometallic complexes, because of their direct carbon–metal bond, and as coordination complexes, because the nature and characteristics of the π ligands are similar to those in coordination complexes (see Coordination compounds).

Generally, metal π complexes can be classified into three main groups: metal π complexes of olefins, cyclopentadienyls, and arenes. Mixed complexes are categorized according to structural or chemical analogies within these groups. Allyl π complexes are designated as olefin π complexes in this review. An understanding of metal π complexes has contributed to the elucidation of the mechanisms of Ziegler-Natta polymerization, the oxo reaction, and catalytic hydrogenation, and to the development of the Wacker process, which is used for the oxidation of olefins (see Olefin polymers; Oxo process; Aldehydes).

Properties

The carbon–metal σ,π-bond. Metal complexes are among those that are least satisfactorily described by crystal-field theory (CFT) or valence-bond theory (VBT). The nature of the bonding can be treated more adequately and quantitatively by molecular-orbital theory (MOT) or ligand-field theory (LFT).

Noble-gas formalism. The noble-gas formalism or the effective atomic number (EAN) rule is a useful guideline for predicting the stoichiome-

tries of transition-metal complexes. According to this rule, a transition-metal atom tends to share electrons with its ligands, so as to occupy all low-lying metal orbitals available in the next higher noble gas. Thus, transition metals of the first transition series (Sc–Cu) can accept 36 electrons, as in krypton.

Spectroscopic properties. Infrared spectroscopy. Olefin π complexes. The ir band most generally considered as the criterion of π bonding in olefin π complexes is the C=C stretching frequency. Upon complexation, the band shifts from the band characteristic of free olefins to lower wave numbers.

Acetylenic π complexes. Unlike olefins, some alkynes do not react with transition-metal complexes to give simple addition products. The identity of the alkyne can then be lost through a polymerization process and, in the case of carbonyl complexes, CO insertion reactions are common and unsaturated cyclic ketones are among the reaction products.

Allyl π complexes. The most noticeable features in the ir spectra of a number of allyl compounds of Pd(II) and Ni(III) are the presence of a medium intensity C=C antisymmetric stretching frequency near 1458 cm^{-1}, which is the position expected for a conjugated double-bond system, and a symmetric C=C stretching vibration at 1021 cm^{-1}, which does not occur in ethylene complexes.

η^5-π-Cyclopentadienyl complexes. The C_5H_5 group can be attached to a metal in four different bonding modes: ionically bonded; σ-bonded (monohapto); π-bonded (trihapto); and π-bonded (pentahapto).

Arene π complexes. The characteristic intense ir absorption bands of arene π complexes can be divided ino five frequency ranges: a C—H stretching frequency at 3010–3060 cm^{-1}; a C—C stretching frequency at 1410–1430 cm^{-1}; a C—C stretching frequency at 1120–1140 cm^{-1}; two or three C—H deformation frequencies at 955–1000 cm^{-1}; and one or two C—H deformation frequencies at 740–790 cm^{-1}. These values are also valid for mixed complexes, eg, benzenechromium tricarbonyl (**1**) and π-cyclopentadienylbenzenechromium (**2**).

(1)   (2)

Nuclear magnetic resonance. In addition to its capability of locating magnetically active nuclei by the measurement of chemical-shift and spin-spin coupling parameters, nmr spectroscopy is used to study molecular motions and chemical exchange processes.

Variable-temperature spectral measurements and spin-decoupling experiments have been applied to the study of rearrangements and stereochemically nonrigid molecules.

Olefin complexes. Olefinic protons are shifted from a few tenths to ca 3.5 ppm to higher fields on complexation.

η^3-Allyl : 3 π system. The π allyl radical (**3**) produces three main resonance peaks corresponding to the three types of protons which are designated H_a, H_b, and H_c. The relative intensity ratio is 1 : 2 : 2. H_a ($4 \ll \delta \ll 7$) is the least shielded and, under high resolution, gives a pattern of overlapping triplets because of the splitting by the two equivalent pairs of protons H_b and H_c. The resonance peaks for H_b (3.4–$4.4\ \delta$) and H_c (22–$38\ \delta$) are well separated and generally appear at higher fields.

$$H_b \diagdown \!\!\!\!\!\overset{H_c}{\underset{H_a}{\diagup}}\!\!\!\!\!\!\!\diagdown\!\!\!\overset{}{\underset{H_b}{}} H_c \quad \longleftrightarrow \quad CH_2=CHCH_2\cdot \quad \longleftrightarrow \quad \cdot CH_2CH=CH_2$$

(3)

η^5-Cyclopentadienyl. The protons in a complexed π-cyclopentadienyl ring system appear to be equivalent. The resonance signal usually is observed at ca 5 δ.

Reactions

Metal π complexes display a wide variety of chemical reactions. However, the reaction of the π-olefin-, π-cyclopentadienyl-, and π-arene-metal complexes are distinctly characteristic of each group. π-Cyclopentadienyl complexes, ie, metallocenes, exhibit a high degree of ring stability and undergo many typical aromatic substitution reactions. However, the π-arene complexes generally exhibit such aromatic substitutional chemistry to a lesser degree. Although most physical properties, particularly the structure of metal π complexes, are interpreted by use of the basic principles of coordination chemistry, these principles do not adequately explain some reaction anomalies of the different groups of metal π complexes.

Olefin complexes. Similarly, reactions involving olefin complexes are characteristic both of the uncomplexed and the complexed olefinic functions. Generally, reactions involving the former are not very different from those observed for free olefins. However, reactions of the latter are altered significantly by π-complex formation. Among the reactions of interest are addition (eg, hydrogenation and protonation), elimination and substitution.

Cyclopentadienyl π complexes. The most significant feature of the reactions of π-cyclopentadienyl complexes in general, and ferrocene in particular involves their aromatic nature. The resonance stabilization energy for ferrocene is 210 kJ/mol (50 kcal/mol). Ferrocene undergoes a large number of typical ionic aromatic substitution reactions, eg, Friedel-Crafts acylation and alkylation, metalation, sulfonation, and aminomethylation (see Friedel-Crafts reaction).

Arene-metal π complexes. Generally, arene π complexes do not undergo the reactions that are characteristic of benzene and its derivatives. However, arene π complexes do undergo a limited number of substitution, addition, ring expansion, and condensation reactions.

General Reactions

Ligand and metal exchange. Coordination complexes undergo single- and multiple-ligand exchanges or replacement reactions. Metal π complexes also undergo analogous exchange reactions (see Catalysis).

Arene ligand exchange. Both benzene rings in dibenzenechromium (4) can be exchanged for toluene and carbon monoxide under pressure to form toluenechromium tricarbonyl (5).

Olefin and acetylene ligand exchange. Olefin-exchange reactions are related to catalytic reactions involving metal π complexes. One of the first steps in the hydrogenation, isomerization, and polymerization of olefins is the exchange of a ligand or solvent molecule to form the corresponding metal-olefin or -acetylene π-complex intermediate.

π-Cyclopentadienyl exchange. Several exchange reactions involve the transfer of a π-cyclopentadienyl group from iron to palladium, nickel, cobalt, and titanium; the exchange with titanium is reversible. In some cases, high yields can be attained.

Valence tautomerism and valence isomerism. In solution, some olefin π complexes exist in several interconverting forms, eg, cyclooctatetraeneiron tricarbonyl. The $Fe(CO)_3$ group in this complex shifts about the cyclooctatetraene ring. The structures possess different valence bonds and are isomers of each other. The shifting of a complex group about the ring is a valence tautomerism ("ringwhizzing"). Valence tautomerism can occur if the energy barrier between two forms is small. However, if the energy barrier is high, the individual isomers may be isolated.

Metal exchange. When either iron pentacarbonyl or nickel tetracarbonyl reacts with tetraphenylcyclobutadienepalladium dibromide, the corresponding metal exchanges with the release of palladium. The reac-

tion proceeds only in aromatic solvents and generally is applicable to a variety of transition-metal carbonyls.

Uses

Uses of metal π complexes include catalysis involving metal π-complex intermediates (eg, polymerization of olefins, hydrogenation, stereoregular polymerization of propylene, oxidation of olefins, and addition of carbon monoxide).

Preparation

The most frequently used preparation methods for organometallic π complexes are substitution, elimination, cyclization, ligand or metal exchange, rearrangements, and redistribution reaction.

Minoru Tsutsui[‡]
Texas A & M University

Reviewed by
John J. Eisch
SUNY, Binghampton

G. Wilkinson, F.G.A. Stone, and E.W. Abel, *Comprehensive Organometallic Chemistry*, 9 vols., Pergamon Press, Oxford, UK, 1982.

M.L.H. Green, *Organometallic Compounds*, Vol. 2, Methuen & Co. Ltd., London, UK, 1968.

E.O. Fischer and H. Werner, *Metal π-Complexes*, Vol. 1, Elsevier Publishing Co., Amsterdam, Neth., 1966; M. Herberhold, Vol. 2, 1972.

OSMIUM. See Platinum-group metals.

OSMOSIS, OSMOTIC PRESSURE, AND REVERSE OSMOSIS. See Hollow-fiber membranes; Membrane technology; Reverse osmosis.

OURICURY WAX. See Waxes.

OXALIC ACID

Properties

Oxalic acid, HO_2CCO_2H, or ethanedioic acid, mol wt 90.04, is the simplest of the dicarboxylic acids (qv). The anhydrous form is odorless, hygroscopic, and white to colorless. It exists in two polymorphic forms, ie, the rhombic or α form and the monoclinic or β form. Sublimation of the monoclinic dihydrate gives the monoclinic crystal. Crystallization from a solvent, eg, acetic acid, gives the rhombic structure. The rhombic or pyramidal crystal is thermodynamically stable at room temperature, but the monoclinic form is metastable or slightly stable. The rhombic crystal exhibits a slightly higher melting point and density than the monoclinic form, as shown in Table 1.

Oxalic acid is available commercially as a solid dihydrate, $C_2H_2O_4 \cdot 2H_2O$, mol wt 126.07. The commercial product is comprised of white-to-colorless monoclinic prisms or granules containing 71.42 wt% anhydrous oxalic acid and 28.58 wt% water. The dihydrate is packed in polyethylene-lined paper bags and fiber drums. Oxalic acid dihydrate is stored in cool areas at 50–70% rh to prevent caking.

Oxalic acid is distributed widely in the plant kingdom as the potassium and calcium salts in leaves, roots, and rhizomes of various plants. It also occurs in human and animal urine and the calcium salt is a major constituent of kidney stones.

Manufacture

Of the four general technologies that have been employed for the commercial synthesis of oxalic acid, ie, alkali fusion of cellulose (qv),

[‡]Deceased.

Table 1. Physical and Thermochemical Properties of Oxalic Acid and its Dihydrate

Property	Value
Oxalic acid, anhydrous, $C_2H_2O_4$	
melting point, °C	
α	189.5
β	182
density d_4^{17}, g/cm^3	
α	1.900
β	1.895
refractive index, β, n_4^{20}	1.540
vapor pressure (solid, 57–107°C), kPa[a]	$\log_{10} P = -(4726.95/T)$
	$+ 11.3478$
specific heat (solid, −200 to 50°C), J/g[b]	$Cp = 1.084 + 0.0318\,t$
heat of combustion, ΔE_c° (at 25°C), kJ/mol[c]	−245.61
standard heat of formation, ΔH_f° (at 25°C), kJ/mol[c]	−826.78
standard free energy of formation, ΔG_f° (at 25°C), kJ/mol[c]	−697.91
heat of solution (in water), kJ/mol[c]	−9.58
heat of sublimation, kJ/mol[c]	90.58
heat of decomposition, kJ/mol[c]	826.78
specific entropy, S° (at 25°C), J/(mol·K)[c]	120.08
logarithm of equilibrium constant, $\log_{10} K_f$	122.28
thermal conductivity (at 0°C), W/(m·K)[d]	0.9
ionization constant	
K_1	6.5×10^{-2}
K_2	6.0×10^{-5}
coefficient of expansion (at 25°C), nL/(g·K)	178.4
Oxalic acid dihydrate, $C_2H_2O_4{\cdot}2H_2O$	
mp, °C	101.5
density d_4^{20}, g/cm^3	1.653
refractive index, n_4^{20}	1.475
standard heat of formation, ΔH_f° (at 18°C), kJ/mol[c]	−1422
heat of solution (in water), KJ/mol[c]	−35.5
pH (0.1 M soln)	1.3

[a] To convert $\log_{10} P_{kPa}$ to $\log_{10} P_{mm\ Hg}$, add 0.875097 to the constant, $T = K$ (kelvins).
[b] To convert Cp, J/g, to Cp, cal/g, divide both terms by 4.184.
[c] To convert J to cal, divide by 4.184.
[d] To convert W/(m·K) to (Btu·in)/(h·ft²·°F), divide by 0.1441.

nitric acid oxidation of carbohydrates (qv), by-product formation during cabohydrate or sugar fermentation, and synthesis from sodium formate, only oxidation by nitric acid currently is used.

In the United States, oxalic acid is no longer produced commercially. Prior processes primarily utilized corn starch and, in lesser amounts, ethylene glycol to produce oxalic acid by an atmospheric, one-step, nitric acid oxidation process in the presence of iron/vanadium catalyst and sulfuric acid.

Health and Safety Factors, Toxicology

Oxalic acid and its solutions are corrosive and poisonous. Oxalic acid dust and mist are irritating, especially under prolonged contact. Personnel who handle oxalic acid crystals or solutions should wear rubber gloves, aprons, boots, and goggles. Adequate ventilation should also be provided in areas in which oxalic acid dust fumes are present. NIOSH-approved respirators should be worn when the concentration of oxalic acid in the air exceeds the permissible air concentration of 1 mg/m³.

Waste Disposal

As oxalic acid is toxic and corrosive, neither its crystals nor its solutions should be discarded to the environment without proper treatment. Whenever possible, it is preferable to reuse the material than to treat it. The common treatment methods are acidification, neutralization and incineration.

Uses

Many industrial applications for oxalic acid are based primarily on its calcium ion-removal property and on its chelating and reducing properties. Oxalic acid is used in metal treatment (oxalate coatings, anodizing and metal cleaning), textiles (dyeing, permanent press and flameproofing), plastics, chemicals, powders, catalysts, photography, and wood bleaching.

CICERO A. BERNALES
STEVEN E. BUSHMAN
JOHN KRALJIC
Allied Corporation

Oxalic Acid, Chemical Products Synopsis—June 1978, Mannsville Chemical Products, Mannsville, N.Y., 1978.

OXAZINE DYES.   See Azine dyes.

OXETHANE POLYMERS.   See Polyethers.

OXIRANE.   See Ethylene oxide.

OXO PROCESS

The oxo process is the commercial application of the hydroformylation reaction for making aldehydes and alcohols from olefins. In the oxo process, an olefin reacts with carbon monoxide and hydrogen at elevated temperature and pressure in the presence of a catalyst to produce predominantly two isomeric aldehydes (qv), eg,

$$RCH{=}CH_2 + CO + H_2 \longrightarrow x\ RCH_2CH_2CHO + 1{\cdot}x\ \underset{\underset{CH_3}{|}}{RCHCHO}$$

A variety of Group VIII metal compounds can be used to catalyze the hydroformylation reaction, but only cobalt and rhodium are of commercial importance. The reaction is highly exothermic; the heat release is ca 126 kJ/mol (30 kcal/mol). The proportion of each isomer in the aldehyde product depends upon the olefin type, the catalyst, and the reaction conditions.

In commercial operation, the aldehyde usually is an intermediate which is converted by hydrogenation or by aldolization and hydrogenation to alcohols (qv):

$$RCH_2CHO \xrightarrow{H_2} RCH_2CH_2OH$$

$$2\ RCH_2CHO \xrightarrow{base} \underset{\underset{R}{|}}{RCH_2}\overset{\overset{OH}{|}}{CH}CHCHO \xrightarrow{-H_2O} \underset{\underset{R}{|}}{RCH_2CH{=}}CCHO$$

$$\Big\downarrow H_2$$

$$\underset{\underset{R}{|}}{RCH_2CH_2CH}CH_2OH$$

The aldol-hydrogenation route is used for the manufacture of 2-ethylhexanol from *n*-butyraldehyde.

In the United States, the principal oxo products are 1-butanol and 2-ethylhexanol from propylene and various C_6–C_{13} alcohols from the appropriate olefins. The C_6 and higher alcohols are used in the manufacture of plasticizers (qv), lubricating oil additives, detergents, and defoamers (qv), etc (see Lubrication and lubricants; Surfactants). Most oxo alcohols are produced commercially from mixed olefin streams and, accordingly, are mixtures of many alcohol isomers.

Catalysts

Cobalt. Cobalt was the first of the Group VIII metals to be recognized as having good hydroformylation activity. Dicobalt octacarbonyl, $Co_2(CO)_8$, which either is introduced directly or formed *in situ*, is the primary conventional oxo-catalyst precursor, but the actual active catalyst species is $HCo(CO)_4$.

Rhodium. Rhodium carbonyl complexes are very reactive hydroformylation catalysts. Catalytic activity increases rapidly in the order Ir < Co < Rh. A simple rhodium carbonyl catalyst tends to favor branched-chain isomers in the product, but linear products predominate with rhodium carbonyl-phosphine complex catalysts. Rhodium catalysts are

Table 1. Comparison of Propylene Oxo Processes

Characteristic	Cobalt carbonyl	Rhodium–phosphine complex
pressure, MPa[a]	20–30	< 2
temperature, °C	140–180	ca 100
linear–branched aldehyde ratio	3–4 : 1	8–16 : 1

[a] To convert MPa to psi, multiply by 145.

particularly useful in the continuous hydroformylation of propylene to *n*-butyraldehyde, an intermediate used in the manufacture of 1-butanol and 2-ethylhexanol. The rhodium-catalyzed hydroformylation of propylene has been used commercially by Union Carbide since 1976. A comparison of propylene oxo processes using cobalt and rhodium-phosphine catalysts is shown in Table 1. It is evident from these data that the rhodium-phosphine catalyst system strongly favors the straight-chain isomer in this process.

Kinetics and Mechanism

Hydroformylation is a complex process that proceeds stepwise through a series of consecutive and, for the most part, reversible reactions with more than one possible rate-limiting step. Under conventional oxo conditions, practically any form of cobalt is converted to the active catalyst $HCo(CO)_4$. Various studies have shown the following rate equation to be a good approximation to the kinetics of the cobalt-catalyzed hydroformylation reaction under conventional oxo conditions at elevated pressures:

$$\frac{d[\text{aldehyde}]}{dt} = \frac{k[\text{olefin}][M](P_{H_2})}{(P_{CO})}$$

where M = cobalt and P = partial pressure.

It is generally accepted that the following reaction sequence with, in certain cases, some modification can be used to explain the observed kinetics and product distribution:

$$HCo(CO)_4 \rightleftharpoons HCo(CO)_3 + CO$$

$$RCH{=}CH_2 + HCo(CO)_3 \rightleftharpoons \begin{array}{c} RCH{=}CH_2 \\ | \\ HCo(CO)_3 \end{array}$$

$$\begin{array}{c} RCH{=}CH_2 \\ | \\ HCo(CO)_3 \end{array} \rightleftharpoons \begin{array}{c} R'Co(CO)_3 \\ | \\ HCo(CO)_3 \end{array}$$

where R'— is RCH_2CH_2— and $\begin{array}{c} CH_3 \\ | \\ RCH— \end{array}$

$$R'Co(CO)_3 + CO \rightleftharpoons R'Co(CO)_4 \rightleftharpoons R'COCo(CO)_3$$

$$R'COCo(CO)_3 \xrightarrow{HCo(CO)_4} R'CHO + Co_2(CO)_7$$

$$R'COCo(CO)_3 \xrightarrow{H_2} R'CHO + HCo(CO)_3$$

With some modifications, the basic steps of the hydroformylation mechanism with the rhodium–phosphine complex catalysts are the same as those with cobalt catalysts. However, more than one active catalyst species has been postulated for rhodium. The complex $HRh(CO)_2[P(C_6H_5)_3]_2$ is considered a principal intermediate.

Conventional Cobalt-Catalyzed Process

Raw materials and products. Cobalt, synthesis gas, ie, hydrogen and carbon monoxide in a (1–2) : 1 ratio; and one or more olefins are the raw materials for most of the commercial oxo processes.

Carbon monoxide and hydrogen partial pressures and temperatures are the most important reaction variables that influence isomer distribution. Hydroformylation of linear olefins produces increasing linear-to-

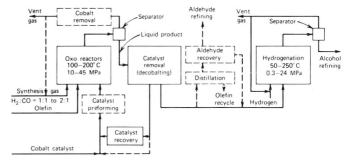

Figure 1. The commercial oxo process. To convert MPa to atm, divide by 0.101.

branched aldehyde ratios as carbon monoxide pressure is increased to ca 5 MPa (50 atm), but there is little further effect if the reaction mixture is saturated with carbon monoxide. Increasing partial pressure of hydrogen also increases the linear-to-branched aldehyde ratio, but the effect is small and apparently is limited to hydrogen pressures of < 10 MPa (100 atm). As reaction temperatures are increased, linear-to-branched aldehyde ratios decrease, but the magnitude of the effect may be influenced by double-bond shifts and possible carbon monoxide diffusional limitations at high reaction rates.

Commercial Process

A commercial oxo process involving the conventional cobalt catalyst consists of at least the following steps: hydroformylation, removal and recovery of catalyst, aldehyde refining, hydrogenation, and alcohol refining. In addition, some plants are equipped for aldolization of *n*-aldehydes, eg, *n*-butyraldehyde to produce 2-ethylhexanol. A schematic diagram of the typical conventional oxo process is shown in Figure 1; the dotted lines indicate optional processing steps. Reaction conditions generally are 100–180°C and 20–35 MPa (200–350 atm).

In the propylene oxo process involving the phosphine-modified rhodium catalyst, the preferred operating conditions are ca 100°C and carbon monoxide and hydrogen pressures of < 0.3 MPa (3 atm) and < 1.4 MPa (14 atm), respectively. During the process, butyraldehydes and by-products are volatilized at their rates of formation, and this effluent is further treated to separate the liquid product and a propylene stream for recycle. The catalyst remains in the reactor, thus avoiding the need for a catalyst recycle and recovery system.

Health and Safety Factors, Toxicology

Metal carbonyls, volatile phosphines, and carbon monoxide can be highly dangerous (see Carbonyls; Carbon monoxide; Phosphorus compounds). All aldehydes are toxic but only those of low molecular weight are considered dangerous (see Aldehydes; Butyraldehyde). The toxicities of many of the oxo-process alcohols have been published (see Alcohols).

I. KIRSHENBAUM
E.J. INCHALIK
Exxon Research and Engineering Company

R.L. Pruett, *Adv. Organomet. Chem.* **17**, 1 (1979).

P. Pino, F. Piacenti, and M. Bianchi in I. Wender and P. Pino, eds., *Organic Synthesis via Metal Carbonyls*, Vol. 2, John Wiley & Sons, Inc., New York, 1977, p. 43.

OXYGEN

Oxygen, the gaseous element that constitutes 20.946% of the earth's atmosphere, is essential to respiration and life in all animals and to most forms of vegetation. Oxygen supports the combustion of fuels that supply mankind with heat, light, and power, and it enters into oxidative combination with many materials. The speed of reaction and effective-

Table 1. Physical Properties of Oxygen

Property	Value
Triple point	
temperature, K	54.359 ± 0.002
pressure, Pa (mm Hg)	146.4 (1.098)
density, g/L (lb/ft^3)	
gas	0.0108 (0.000674)
liquid	1306.5 (81.56)
solid	1300 (81.2)
Boiling point	
temperature, K	90.188
pressure, kPa (psi)	101.325 (14.696)
density, g/L (lb/ft^3)	
gas	4.470 (0.27886)
liquid	1141.1 (71.212)
Critical point	
temperature, K	154.581
pressure, MPa (atm)	5.043 (50.14 atm)
density, g/L (bl/ft^3)	436.1 (26.63)
Gas, at 101.3 kPa (1 atm)	
density at 21°C, g/L (lb/ft^3)	1.327 (0.08281)
heat capacity (Cp) at 25°C, J/(mol·K) (Btu/(mol·°F))	29.40 (0.1549)
dielectric constant at 20°C	1.0004947
n_D^0	1.0002639
viscosity at 25°C, µPa·s (lb/(ft·s))	$20.639 (13.869 \times 10^{-6})$
thermal conductivity at 0°C, mW/(m·K) (Btu/(h·ft·°F))	$2.448 (14.2 \times 10^{-3})$
sound velocity at 0°C, m/s (ft/s)	317.3 (1041)
Liquid	
heat capacity, sat liq, J/(mol·K) (Btu/(mol·°F))	$54.317 (2.862 \times 10^{-2})$
heat of vaporization, J/mol (Btu/(mol))	6820 (6.468)
viscosity, µPa·s (lb/(ft·s))	$189.5 (127.3 \times 10^{-6})$
thermal conductivity, mW/(m·W) (Btu/(h·ft·°F))	$149.87 (86.65 \times 10^{-3})$
sound velocity at 87 K, m/s (ft/s)	904.6 (2968)
surface tension at 87 K, N/m (lbf/ft)	$13.85 \times 10^{-7} (9.490 \times 10^{-4})$
Volume ratio, gas at 21°C to liq at bp	859.9
Solid	
heat of sublimation, J/(g·mol) (Btu/mol)	8204.1 (7.7838)
heat capacity, J/(g·mol·K) (Btu/(mol·°F))	$46.40 (2.4458 \times 10^{-2})$
heat of fusion, J/(g·mol) (Btu/mol)	444.5 (0.4217)
Solubility, mL O$_2$ at STP per mL H$_2$O	
0°C	0.0489
10°C	0.0380
20°C	0.0310
30°C	0.0261
50°C	0.0209
70°C	0.0183
90°C	0.0172

molecule is paramagnetic and has a moment in accord with two unpaired electrons.

Chemical Properties

Oxygen reacts with all other elements except the light rare gases helium, neon, and argon. The reactants usually must be activated by heat before the reaction proceeds, and if the final union releases more than enough energy to activate subsequent portions of both reactants, the overall process may be self-sustaining. The process is known as combustion when light and heat are evolved.

Oxygen usually exhibits an electronegative valence of -2 in combination with other chemical elements, eg, oxides. Most elements combine with oxygen in more than one ratio because of variable valences in the other element, or the existence of complicated molecular structures.

Manufacture

Commercial oxygen is produced by the fractionation of air in units where air is cleaned, dried, compressed, and refrigerated until it partially liquefies (see Cryogenics). It is then distilled into its components.

Production, Pipelines, and Shipping

Oxygen production facilities may be captive plants that are owned and operated by the oxygen user.

On-site plants are located on or immediately adjacent to the premises of the user, but are treated as a utility; they are owned and operated by a second party.

Pipeline plants are usually somewhat larger and more distant from large-volume users. They may supply a number of customers through one or more pipelines radiating from the plant and, in addition, liquefy sizeable amounts for use as merchant product and oxygen for sister plants that may be experiencing insufficient production. Increasingly, pipeline plants are part of a network of producers, pipelines, and users. In general, the pipelines and the producing facilities are owned by one management.

Over 80% of all oxygen currently produced is transferred by pipeline to the location of use. The balance, known as merchant oxygen, is shipped as liquid or gas.

Uses

There is no substitute for oxygen in any of its uses; it cannot be reclaimed or recycled except via the atmosphere. The spectacular increase of oxygen production since World War II is in a very large measure owing to its new availability and its usefulness in steelmaking (see Steel). Other uses are in nonferrous metallurgy, as an oxidizing agent, and in medical and life-support applications.

ALFRED H. TAYLOR
Airco, Inc.

"Key Chemicals: Oxygen," *Chem. & Eng. News*, 10 (September 19, 1983).

Industrial Gases, Current Industrial Reports, Series MA-28C, Annual Production & Shipment, Oxygen Code 28136-00, U.S. Department of Commerce, Bureau of the Census, 1950–1978.

W. Stowasser, "Oxygen" in *Mineral Facts and Problems*, U.S. Bureau of Mines Bulletin 667, U.S. Government Printing Office, Washington, D.C., 1980.

ness of combination increase with oxygen concentrations greater than that of air. Industry has established 99.5% purity for the bulk commercial product.

Oxygen in combination with hydrogen forms the waters of the earth's surface (89% O$_2$). In combination with metals and nonmetals, oxygen is contained in well over 98% of the rocks; it enters into a very large number of known minerals as well as a vast array of organic compounds.

In nature, oxygen occurs in three stable isotopic species: oxygen-16, 99.76%; oxygen-17, 0.038%; and oxygen-18, 0.20%. Commercial fractional distillation of water produces concentrations of ^{18}O as high as 99.98%; ^{17}O concentrations up to 55% are also offered. The ^{18}O isotope has been used to trace mechanisms of organic reactions (see Isotopes).

Physical Properties

Oxygen is colorless, odorless, and tasteless. Oxygen is moderately soluble in water. Physical properties are given in Table 1. The oxygen

OXYGEN-GENERATION SYSTEMS

Oxygen-generation devices are used for the generation of oxygen from chemicals for respiratory support, eg, in submarines, aircraft, spacecraft, bomb shelters, and breathing apparatus. Convenience, long-term storage, and reliability, rather than low cost, are stressed.

Chlorates and Perchlorates

The chlorates and perchlorates of lithium, sodium, and potassium evolve oxygen when heated. These salts are compounded with a fuel to

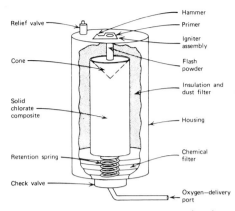

Figure 1. Cutaway view of generator housing.

form a chlorate candle, which produces oxygen by a continuous reaction. The generalized reactions are as follows:

$$2\,NaClO_3 \rightarrow 2\,NaCl + 3\,O_2 \text{ (endothermic, major)}$$

$$2x\,Fe + y\,O_2 \rightarrow 2\,Fe_xO_y \text{ (exothermic, major)}$$

$$4\,NaClO_3 \rightarrow 2\,Na_2O + 2\,Cl_2 + 5\,O_2 \text{ (endothermic, minor)}$$

Components of the composition include the oxygen-producing material, the fuel, a material that absorbs traces of chlorine, and usually an inert binder. Once the reaction begins, oxygen is released from the hot salt by thermal decomposition. A portion of the oxygen reacts with the fuel to produce more heat which produces more oxygen, and so on.

Chlorate candles are quite stable; they have been stored uncontained for as long as 20 yr, and then have been operated successfully with no loss of oxygen output. Thus, they are well-suited as an emergency oxygen-generation system. Chlorate candles also produce oxygen under pressure and, therefore, can be stored in or operated from pressurized cylinders to provide pressurized oxygen.

Materials and reactions. Candle systems vary in mechanical design and shape but contain the same generic components (see Fig. 1). The candle mass contains a cone of material that is high in iron and that initiates the reaction. The reaction of the cone material, in turn, is started by a flash-powder train, which is fired by a spring-actuated hammer, or an electric wire. An outer jacket with a relief valve and a gas-exit port is furnished to collect and deliver the oxygen. The candle is wrapped in insulation and placed within the jacket. A gas-conditioning filter can be made and arrangements provided to prevent vibration and shock damage to the candle.

Oxygen purity. The impurities that form are Cl_2, CO, CO_2, H_2O, and simple organics. All can be minimized by using high purity ingredients, by control of moisture, or by gas conditioning. Candles generally are made from chlorate and BaO_2. All other materials are degreased by burning at 427°C. Iron powders usually are reduced with hydrogen.

Candle fabrication. All ingredients must be grease-free and the candle materials, eg, $NaClO_3$ and BaO_2, should be dry. The oxygen-generating mass is made by mixing and then pressing or casting the ingredients. Care must be taken to assure homogeneous mixing or reaction rates will vary throughout the candle. Shape can be varied as desired, especially if casting is used. With pressing, the shape is limited to some extent, although hydrostatic pressing provides freedom in candle form. All other factors being equal, the rate of oxygen evolution is directly proportional to the cross-sectional area of the unit.

Operational characteristics. Operational characteristics are given in Table 1. Alternative formulations use either Na_2O or Co/Co_3O_4 as the fuel. These candles operate at about 200°C less than iron-fueled candles.

Peroxides and Superoxides

Chlorate candles produce oxygen; however, peroxides and superoxides produce oxygen and absorb carbon dioxide. For every volume of oxygen that a person inhales, 0.82 volumes of carbon dioxide is exhaled. The respiratory coefficient RQ in this example is 0.82. It is desirable that an

Table 1. Chlorate Oxygen-Generator Parameters

Parameter	Value
average composition	80–85 wt % $NaClO_3$, 3–10 wt % Fe, 4 wt % BaO_2, remainder binder
O_2 available (max), wt %	40
sp gr of mixture	2.3–2.5
average reaction rate through shape, cm/min	0.64
heat evolution, J/g^a	837
reaction-zone temperature, °C	538
shape of chlorate unit	unlimited
starting method	hot wire, Bouchon cap
cone material	30 wt % Fe, 60 wt % $NaClO_3$, remainder binder and BaO_2
time before O_2 evolved (max), s	1
gas purity, %	>99.99

[a] To convert J to cal, divide by 4.184.

oxygen-generation device, which is used in a closed system, mirrors this RQ, ie, absorbs 0.82 volume of carbon dioxide and liberates 1.0 volume of oxygen.

Uses. The peroxides and the superoxides must be hermetically sealed for storage. The superoxides especially are strong oxidizing agents and should be kept away from grease, oil, and organic materials. In general, sodium peroxide is used more widely than lithium peroxide. Breathing apparatus based on peroxides often is supplied with bottled oxygen, since the peroxides are not very oxygen-weight efficient. Superoxides are used in breathing applications requiring no auxiliary source of oxygen.

Other Chemical Systems

Regenerative systems that dissociate carbon dioxide to recover the oxygen are of interest to the U.S. space program and in long-duration habitat support. The Bosch process utilizes an iron catalyst for the single-step reaction of hydrogen, carbon dioxide, water, and carbon at 700°C. Oxygen is produced by water electrolysis, and the hydrogen is recycled to the Bosch reactor. Carbon is removed as a solid.

$$CO_2 + 2\,H_2 \rightarrow 2\,H_2O + C$$

$$2\,H_2O \rightarrow 2\,H_2 + O_2$$

Hydrogen peroxide can be dissociated over a catalyst to produce oxygen, water, and heat. It is an energetic reaction, and contaminants can spontaneously decompose the hydrogen peroxide. Water-electrolysis units that produce oxygen for life support are used in submarines.

Health and Safety

Peroxides, superoxides, and chlorates are oxidizing compounds and should not contact organic materials, eg, oils, greases, etc. Caustic residues that may remain after use of peroxides and superoxides require disposal appropriate to Group IA hydroxides. Chlorate candles have no harmful residues (see Peroxides; Chlorine oxygen acids).

Dusts associated with these oxidizing compounds produce caustic irritation of skin, eyes, and nasal membranes. Toxicity is low to moderate and is the same as for the hydroxides; the chlorate materal can also cause local irritation.

J.W. MAUSTELLER
MSA Research Corp.

A.J. Adduci, *Chemtech*, 575 (Sept. 1976).

H. Rind, *Science and Technology*, 40 (March 1969).

OZONE

Ozone has been used since 1903 for the treatment of drinking water. This use, which was developed in Europe, is the largest application for ozone and, as of 1983, there were more than 1200 such water-treatment installations. The treatment of swimming-pool water also was developed in Europe during the 1950s (see Water, treatment of swimming pools).

The second-largest application for ozone is for the treatment of odors from industrial processes and municipal-wastewater treatment plants. Ozone also is used on a large scale for the treatment of municipal secondary effluents (see Water, municipal water treatment). Industrial high quality water supplies also are treated with ozone (see Water, industrial water treatment). The manufacture of specific chemical products and pharmaceutical intermediates involves the use of ozone for carbon-double-bond oxidations. Ozone is used to bleach inorganic products such as clays. Ozone is the second most powerful oxidant, exceeded in its oxidation potential only by fluorine. It is also a sterilant for microorganisms (see Sterilzation techniques).

Chemical Properties

The triangularly shaped ozone molecule has a bond angle of 116°49′ between the three oxygen atoms, according to microwave studies, or 127° according to electron-diffraction studies. The ozone molecule has these resonance structures:

(1)   (2)   (3)   (4)

The actual structure of ozone is a resonance hybrid of (1–4).

The strong electrophilic nature of ozone imparts to it the ability to react with a wide variety of organic and organometallic functional groups. The vast majority of ozone reactions is based on the oxidation of the carbon–carbon double bond which acts as a nucleophile or species having excess electrons.

The oxidation of a double bond proceeds through a variety of intermediate stages. The first stage is a semi-stable intermediate known as a molozonide or a 1,2,3,-trioxolane. This ring opens to form an aldehyde or ketone and a Criegee zwitterion (5). The zwitterion is a very reactive species that can undergo a variety of additional reactions of which the classical reaction is the addition to a carbonyl compound to an ozonide (a 1,2,3,-trioxolane) or a 1,2,4-trioxolane (6).

(5)   (6)

Commercially, ozone is used as a catalyst at $< 15°C$ in the production of peracetic acid from acetaldehyde and ozone. The peracids react with olefins to produce epoxides which are hydrolyzed to diols that then are used in automotive-antifreeze mixtures (see Antifreezes). The pharmaceutical industry employs ozone in organic reactions to produce peroxides as germicides in skin lotions, for the oxidation of intermediates for bacteriostats, and in the synthesis of steroids (qv), eg, cortisone (see Disinfectants and antiseptics; Hormones, adrenal-cortical).

The inorganic chemistry of ozone involves nearly all members of the periodic chart, which is not surprising since ozone has an oxidation potential of 2.09 V. Ozone does not react with metal ions, eg, calcium and sodium, that exhibit only one oxidation state, but it may react with such metals to form ozonides and oxides, or both.

Transition metals are oxidized by ozone to their highest oxidation states, which generally are less water-soluble than the lower oxidation levels. This method of metal-ion removal is employed commercially in many water-treatment plants where iron and manganese must be removed to very low levels. Other heavy metals that can be removed by ozone oxidation and filtration of polluted water include cerium, lead, silver, cadmium, mercury, and nickel.

The ozone oxidation of cyanide is used commercially to treat wastewaters produced from metal-plating, gold-mining, and paint-stripping industries.

Ozone Generation

Ozone can be generated in a variety of ways; most require that the stable bonds of the oxygen molecule first be cleaved into two short-lived oxygen atoms. These reactive oxygen atoms react almost immediately ($k = 1.9 \times 10^{-11}$) with the oxygen molecule to form ozone. Commercially important methods of producing ozone include uv irradiation of air or oxygen and corona discharge.

Atmospheric Ozone

Ozone occurs naturally in the upper layers of the atmosphere where it acts as a shield for large quantities of uv solar radiation. Were it not for this filtering effect of ozone, life as it exists on earth would not be possible. A decrease in the stratospheric ozone concentration and simultaneous increase in the uv radiation would have adverse effects upon climate, plants, and animals. The naturally occurring concentration of ozone at the earth's surface is very low, but this distribution has been altered by the emission of pollutants that lead to the formation of ozone. At higher levels, the naturally occurring ozone concentration changes with jet air streams, the season of the year, and changes in a cyclic manner over a 24-h period.

Ozone can be toxic to plants and animals. The toxicity of ozone is largely related to its being a very powerful oxidant. Small, naturally occurring concentrations may aid in the transformation of refractory compounds into biodegradable compounds.

The odor threshold of ozone varies among individuals but most people can detect 0.01 ppm in air. This is well below the limit for general comfort. The maximum allowable exposure for an 8-h period, as proposed by the OSHA, is 0.10 ppm.

The presence of naturally occurring ozone in the lower stratosphere has created a problem for passengers and crew members of high flying aircraft. The installation of a platinum catalyst decomposes the ozone to oxygen.

Low but potentially harmful concentrations, of ozone also are formed at very low altitudes by atmospheric photolytic reactions.

The strategy for controlling ambient ozone concentrations is based upon the control of hydrocarbons, eg, catalytic converters for the oxidation of hydrocarbons from automobile emissions (see Exhaust control, automotive).

Transfer of ozone into water. When ozone is used to treat water or wastewater, it must be transferred from the gas phase, in which it is generated to the liquid phase. Pure ozone is 12.5 times more soluble in water than oxygen. However, the optimum economic concentrations for generating ozone are 2 and 3 wt% when air and oxygen are used as the feed gases, respectively.

CARL NEBEL
PCI Ozone Corp.

P. Bailey, *Ozonation in Organic Chemistry*, Academic Press, Inc., New York, 1978.

Ozone Chemistry and Technology, Advances in Chemistry Series, Vol. 21, American Chemical Society, Washington, D.C., 1959.

P

PACKAGING MATERIALS, INDUSTRIAL

In any operation involving the production, conversion, and use of chemical substances, it is essential that consideration be given to packaging at an early stage in the manufacturing process. This consideration has become particularly important recently because of increasing government regulation of transportation and increasing civil and criminal liabilities associated with the selection and use of packages in terms of packaging safety and health aspects (see Transportation).

Virtually any chemical can be transported and stored safely and effectively by one of various package types. The choice of the type of containment is generally dictated by operational and economic considerations; however, for a given type, the choice of materials often is influenced primarily by safety and chemical compatibility factors. Both aspects can affect the cost of the product that is being packaged.

Regulations

The number of regulations at all levels of government has proliferated in the 1970s and their application has become increasingly specific and comprehensive. The most important document, which must be consulted for interstate transportation and which can be considered as a guide for intrastate transportation, is the Code of Federal Regulations (CFR). Partial revisions appear from time to time in the *Federal Register*.

Transportation and storage. Three principal considerations apply to both transportation and storage, ie, compliance with legal requirements; compatibility with the manufacturing and physical-distribution procedures including, in particular, safety aspects; and selection of a minimum-cost packaging system that is consistent with the preceding considerations.

Industrial Packaging Materials

Industrial packaging materials include bulk containers (tank cars, bulk cars and semibulk containers), steel drums and pails, plastic drums, wooden barrels, fiber drums, bags (textile, laminated textile, multiwall paper, and plastic), carboys and bottles, and boxes and cartons.

STEPHEN J. FRAENKEL
Technology Services, Inc.

Code of Federal Regulations, Title 49, Pts. 100–109, issued under authority of the Hazardous Materials Transportation Act (Public Law No. 93-633).

J.R. Hanlon, *Handbook of Package Engineering*, McGraw-Hill, Inc., New York, 1971.

S. Sacharow, *Handbook of Packaging Materials*, Avi Publishing Company, Westport, Conn., 1976.

PACKING MATERIALS

Packing materials are used to make a complement of devices known as seals. Packings function in two conventional operating modes, ie, static and dynamic. For the static mode, gaskets and thread sealants are the two principal packings that effectively seal joints and threads. Packings for the dynamic mode, which involves reciprocating, rotating, and helical motions, are mechanical packings. The latter applications are far more demanding of the seal materials and, therefore, require greater expertise in product design and material selection.

Static Seals

Gasket joints. A static seal is formed by the placement of a gasket between two joint faces and the application of greater pressure to the seal than that which is exerted by the contained liquid or gas that tends

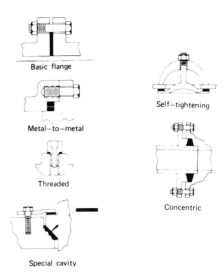

Figure 1. Most common types of gasketed joints. Courtesy of *Machine Design*, Penton Publications, Cleveland, Ohio, 1967.

to leak past the joints. Some of the most common gasketed joint designs are shown in Figure 1.

Gasket design. The ideal gasket is a resilient, predictably compressible composition of one or several materials that conform easily to joint-face irregularities and can compensate for joint-face distortion during operation and thermal fluctuations.

Gasket materials include rubber, asbestos, inorganic fibers, graphite, metals, cork, felt, plant fibers, and plastics.

Sealant materials. Sealants generally are used at lower temperatures and pressures than gaskets. Sealants function in a wide range of applications and are relatively inexpensive compared to gaskets. They are used to exclude dust, dirt, and moisture or to contain a gas or liquid. Their common consistency is pastelike. Their formulations include a variety of fine filler particles that characterize the primary properties of the sealant. Other than good sealing qualities, added features, eg, antistick, antiseize, and anticorrosion properties, are claimed. Some formulations have thermosetting properties, some have adhesive properties, and others remain plastic and never harden (see Sealants).

Paste-type thread-sealing compounds are outmoded in many commercial applications. The development of unsintered TFE (polytetrafluoroethylene) resin tape, which is available in thicknesses of 0.08–0.1 mm and cut to various widths, has made thread sealing more convenient. More recently, Grafoil self-adhesive tape shows promise as a thread sealant with outstanding corrosion resistance and a temperature range from cryogenic to 815°C.

Dynamic Seals

Mechanical packing. Mechanical packings are used to seal rotary, reciprocating, and helical motions. They are seals that restrict fluid or gas leakage between one moving and one stationary surface. Their main use is in pumps, valves, compressors, mixers, swing joints, and all kinds of hydraulic cylinders. The selection of mechanical-packing design and the materials of construction is considerably more critical than with static gaskets. Careful consideration must be given to the effects of motion, eg, heat, friction, wear, and vibration, which result in fatigue, and to fluid compatibility. Mechanical packings are classified as compression or jam-type, automatic or lip- or squeeze-type, and floating or spring-energized segmented-type packings.

Compression packings. Some of the most popular compression packings are shown in Figure 2.

Materials. The essential raw materials required to manufacture mechanical packings include fiber yarns, metals, dry lubricants, and wet lubricants.

Automatic packings and materials. Automatic packings react automatically to system pressure. They also are referred to as rubber packings, because they are made primarily of rubber; lip seals, because of their lip shape; and hydraulic packings, because they are almost exclu-

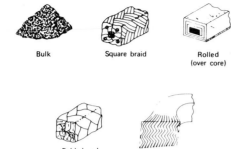

Bulk   Square braid   Rolled (over core)

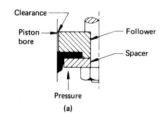

Folded and twisted   W formed

Figure 2. Common types of compression packing.

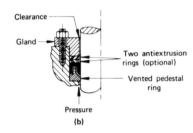

(a)

(b)

Figure 3. Automatic packings and their typical installations. (a) Piston cup. (b) U ring.

sively used in hydraulic applications. Some of the most common designs are illustrated in Figure 3.

The materials of construction for automatic packings include leather, rubber, fabric-reinforced rubber, and plastic.

Floating packings and materials. Floating packings, when in service, are not held in place but float within the walls of their grooves. Because they usually are segmented, they also are known as split-ring seals. As piston rings, they are used widely in compressors, pumps, and internal-combustion engines. Their designs range from simple cross sections to sets of two or more rings made up of three segmented precision parts that are fitted to move radially into a sealing relationship with the rod or cylinder.

Construction materials for floating packings are fairly limited. For years, only solid metals, eg, cast iron, bronze, and stainless steel, were used, but plastics and carbon–graphite materials exhibit improved dry lubricating properties and greater energy savings.

Other. Two sealing devices that greatly extend the sealing range of dynamic packings are oil seals and mechanical end-face seals. The oil seal, which also is known as a lip seal or shaft seal, is suited for fast, continuous, rotary motion that automatic packings cannot tolerate. The mechanical end-face seal extends compression-packing capabilities to high temperatures and pressures, high shaft velocities, and long life with low leakage and maintenance.

I. FREIMANIS
Crane Packing Co.

Engineered Fluid Sealing, Crane Packing Company, Morton Grove, Ill., 1979, p. 7.

H.H. Buchter, *Industrial Sealing Technology*, John Wiley & Sons, Inc., New York, 1979, p. 260.

R.H. Warring, *Seals and Packings*, Trade and Technical Press Ltd., Morden, Surrey, UK, 1967.

PAINT

The paints that are discussed below are commonly called trade-sales paints. They are air-drying and are used primarily for exterior and interior coatings on houses and buildings, in contrast to those paints that are used for factory application to industrial products (see Coatings, industrial). The two main objectives in the use of paint are appearance and protection against weather. There are many other objectives, eg, identification, safety, insulation, vapor barrier, nonskid surface, temperature control, light control, dust control, etc.

Manufacture and Processing

The manufacture of paint involves mixing, grinding, thinning adjustments, filling, and labelings; grinding is the most important and critical step. It consists of breaking up agglomerates into single particles or smaller agglomerates in the liquid.

High speed dispersers. During the 1960s, high speed dispersers were adapted for the grinding step. The large (10^3–10^4-L or 250–2500-gal) tanks are equipped with high speed rotating disks of various shapes. They were first used to dissolve large chips of pigment that were dispersed in solid binders. As pigment production and wetting characteristics improved, pigments dispersed satisfactorily in the high speed dissolvers. The use of high speed dissolvers reduced the cost of manufacturing paint from about \$0.26/L to ca \$0.13/L (\$1.00–0.50/gal) by eliminating the premix step, requiring less labor, and speeding the process (see Mixing and blending).

Bulk pigments. The popularity of water-based paints has led to the use of water slurries of pigments in the manufacture of paint. The use of slurries allows bulk shipping, storage, and easier handling through pumps and pipes as well as simplified dispersion that further reduces manufacturing costs to ca \$0.12/L (\$0.45/gal). Modern paint plants usually utilize slurries of titanium dioxide and common extender pigments.

Formulation principles. Pigment-volume concentration. The percent of the volume of the dry film that is occupied by the pigment is the pigment volume concentration (PVC) and the transition area to minimum gloss is the critical pigment volume concentration (CPVC).

A number of properties of the paint film change rapidly at the CPVC, eg, a decrease in gloss, which reaches a minimum; a rapid increase in permeability; and a rapid decrease in scrubbability, cleanability, and flexibility. Hiding increases above the CPVC. Thus, the volume of the pigment is an important parameter in formulating a paint, particularly near the CPVC.

Parameters. Parameters to be considered in formulations include hiding application, appearance, and cost.

Interior Building Paints

Most interior building paints are based on a latex vehicle. Their popularity with the user compared to alkyd paints results from their easy application, ability to be cleaned with soap and water, much lower odor, and good appearance and appearance retention.

Formulation. There are many latex-polymer types that can be used. Surfactants or protective colloids are used to stabilize the latices; however, they can adversely affect the performance of latices in a paint formula. Pigments are then selected for proper hiding, color, and gloss. They are dispersed in water before the latex is added and surfactants are chosen that complement the pigmentation. Coalescing solvents must be chosen to plasticize the latex temporarily during film formation so that the latex particles coalesce. Because the continuous phase, ie, water, is not very viscous, the paint must be thickened with an additive to promote suspension of the pigment during storage, proper rheology for application, and flow without sagging. Other additives must be added for protection during storage, eg, against bacterial attack, and for prevention of agglomeration by freezing, minimizing foaming, and adjusting pH.

Types. Paints that are used on architectural structures are comprised of primers or undercoats for walls, and woodwork, and flat, semigloss, or gloss finishing coats. The primers and finishing coats differ primarily in function rather than in ingredients. Varying proportions of ingredients can emphasize certain functions such as sealing or sandability for primers or gloss for finishing coats.

Exterior Building Paints

There are several requirements for exterior architectural paints that are not required for interior paints. The principal one is that the paint must be more protective of the substrate in a more hostile atmosphere, eg, rain, dew, temperature extremes and rapid temperature change, uv radiation, etc. In addition, there are substrate problems, eg, weathered wood surfaces, chalky paint surfaces, substrates that are alternately wet and dry because of structural defects, moisture that is forced out of a structure through the paint film by the greater partial vapor pressure of the moisture in the interior of a heated structure, and pollution, particularly sulfur dioxide.

Alkyd vs latex. Because of easy application, cleanability with soap and water, and good service, the latex paints comprise most of the exterior paint market. Alkyd paints are satisfactory and often are preferred for use on a problem-free structure, ie, one that is without defects that would allow moisture entrance, that includes adequate vapor barriers in the walls, and that has not been painted previously with latex paints (see Alkyd resins).

The volume of solids in an alkyd paint is greater than that in a latex paint; therefore, at equal spreading rates, application of the former produces a thicker, longer lasting, better hiding, and more protective paint film.

Because latexes dry by evaporation of the water and not by oxidation and polymerization, they are less subject to shrinkage, and adhesion is less strained when the paint is used on weathered wood.

Although it is recommended that wet substrates should not be painted, one cannot always wait for substrates to dry completely. Since latex is waterborne, it is more tolerant of wet surfaces than alkyds. On the other hand, a rain shower would not damage wet alkyds but may remove wet latex films.

Both alkyd and latex paints require the use of rheology modifiers for adequate pigment suspension, application, and flow and leveling without sagging. Suitable modifiers are available for alkyds. They are also available for latexes but not quite equivalent for flow and leveling. These factors plus the thickness of alkyd films characterize the easier formulation of a one-coat-hiding alkyd paint compared to latex paints.

Latex coatings, particularly in gloss formulation, are advantageous in terms of color fading and gloss retention.

It seems that alkyds are slightly better paints, but most customers prefer latexes.

A latex paint can be applied over an alkyd paint without adverse consequences; however, alkyds should not be applied over latexes. Latex film is considerably weaker, ie, has a lower modulus of elongation, than the alkyd. If the alkyd is used over latex, the latex film may be so weak as to allow shrinkage in the parallel direction, resulting in cracking and peeling of the alkyd.

Formulation. An alkyd resin is the reaction product of a polybasic acid, eg, phthalic anhydride, and a polyhydric alcohol, eg, glycerol. In paints, alkyd resins usually contain a monobasic fatty acid, eg, linseed or soya fatty acids. Polymerization proceeds by way of condensation.

The formulation of exterior and interior latex paints differs primarily in the type of latex used and the range of PVCs that is permissible. In addition, mildewcides are used in the former.

Special-Purpose Coatings

Industrial maintenance. Most of the coating types used for retail and professional paint trades also are used for maintenance in industrial plants. There also are many other coatings designed especially for industrial use, particularly for metal protection. Painting is one of the

important means of combating the corrosion of iron. A coat of practically any kind of paint with a PVC:CPVC ratio ≤ 0.95 protects against rust reasonably well, but for maximum protection, a corrosion-inhibitive primer should be used (see Corrosion and corrosion inhibitors). The additional corrosion inhibition usually is provided by the pigment. Because of the restrictions on the use of lead and chromates, the pigments favored in industrial maintenance coatings are mainly zinc metal, zinc oxide, molybdates, and phosphates.

Often industrial maintenance paints must be chemically resistant to the atmospheres in which they are used. In such cases, the chemical resistance is provided by the polymer (see Coatings, resistant).

Fire retardancy. High PVC paints which are used over flammable substrates tend to retard flame spread. Special paints, which are intended for use over flammable substrates, protect the substrate to such an extent that the damaged paint can be scraped off and the structure can be repainted without affecting appearance. Such intumescent paints contain an ingredient, usually polyammonium phosphate, that emits a gas at elevated temperatures but at lower-than-charring temperatures.

Floor. Paint is an inexpensive floor covering. Floor paints can be formulated to provide antiskid properties even when wet, and they can resist the alkalinity of cement layed by slab construction.

Health and Safety Factors

Regulations. Formulation and applications of architectural coatings are limited by Federal regulations that are intended to curb air and water pollution and to protect consumers from toxic or hazardous materials. Regulations include the Clean Air Act of 1977 and the Lead Poison Prevention Act.

G.G. SCHURR
The Sherwin-Williams Co.

Federation Series on Coating Technology, Federation of Societies for Coating Technology, Philadelphia, Pa., Pamphlet 1 (Oct. 1964), Pamphlet 27 (May 1978).

R. Myers and J.S. Long, *Treatise on Coatings, Formulations*, Pt. 1, Vol. 4, Marcel Dekker, Inc., New York, 1975.

G.G. Sward and H.A. Gardner, *Paint Testing Manual*, American Society for Testing and Materials, Philadelphia, Pa., STP 500, 1973.

PAINT AND VARNISH REMOVERS

Paint and varnish removers are of two main types according to the method of use, ie, application removers and immersion removers.

Organic

Methylene chloride. The most widely used removers are based on methylene chloride (dichloromethane), which is the most versatile stripping agent commonly available (see Chlorocarbons and chlorohydrocarbons). Methylene chloride is the least toxic of chlorinated solvents and the fastest in lifting paint films.

Mechanism of removal. The effectiveness of methylene chloride results from the small molecular size. Its low molecular weight enables it to penetrate rapidly into a coating, and its intermediate solvency enables the coating not to be dissolved so that redeposition on the substrate is avoided. When the methylene chloride has reached the substrate, it swells the film to several times its original volume.

Functions of components. An application remover that is based on methylene chloride has several components, including solvents, eg, methylene chloride or other chlorinated hydrocarbons; cosolvents; activators and corrosion inhibitors; evaporation retarders; thickeners; emulsifiers; and wetting agents.

Modifications. Methylene chloride removers are modified to increase stripping power for special purposes. The modifying chemicals include amines, alkalies, and organic acids.

Manufacture. Most formulas are made by simple mixing of components. Order of addition is important and care must be taken to incorporate the thickener properly.

Other organic paint and varnish removers. Some finishes are relatively easy to strip and can be dissolved by inexpensive solvents and blends of solvents which may be formulated with thickeners and waxes. Shellac is stripped by alcohols, and lacquers are removed by blends of alcohol and acetates, eg, butyl acetate. One remover for exterior paints is a slowly evaporating oil-in-water (o/w) emulsion that is based on xylene and dimethylformamide (see Emulsions).

Inorganic

Caustic soda. Caustic soda is one of the most common industrial strippers. It is used to clean paint-making and paint-application equipment, including jigs, hangers, and conveyors. Caustic soda baths also are used to salvage ferrous metal parts with defective finishes. Some alkali-resistant coatings are not removed in caustic systems. Caustic baths usually must be heated to ca 93°C to be effective.

Others. Paint can be removed by other alkali systems, inorganic acids, fused alkali, and molten-salt baths.

Choosing a Paint and Varnish Remover

Selection of the best compound or method for industrial or commercial paint and varnish removal can be complicated by many factors. Often the problems are best solved by companies who specialize in paint-stripper supplies and services. Selection of a paint stripper depends on the substrate, type of coating to be stripped, how much attack, if any, is allowed on the substrate, number of coats and primer to be stripped, available equipment, time and temperature limitations, odor and flammability restrictions, and disposal of spent stripper.

Health and safety factors. Toxicology. Many potent removers used in the past are no longer used because of carcinogenic and other toxicological effects. These include benzene, benzene derivatives, phenol, cresols, and some chlorinated and fluorinated hydrocarbons. The safe use of methylene chloride has been questioned by Federal agencies, including the CPSC and the FDA.

Handling and safety. Any chemical or formulation that blisters an organic coating film is likely to blister human skin. When working with ordinary methylene chloride removers, one should wear protective clothing and safety glasses. Good ventilation is necessary; established inhalation standards are associated with many of the chemicals and solvents.

Environment. Fume emission must be considered in industrial stripping operations. Emission standards have been established, for many of the solvents used in paint removers, by state and local governments. Disposal of liquid waste and sludge is a serious problem. Minimizing quantities by reusing filtered or decanted stripper solutions is a partial solution.

WALTER R. MALLARNEE
The Sherwin-Williams Company

J. Mazia, *Met. Finish.* **77**(7), 57 (July 1979); **77**(8), 39 (Aug. 1979).

G. Handley, *Prod. Finish. (London)* **28**(12), 9, 12, 14 (1975).

PALLADIUM.   See Platinum group metals.

PAPER

Paper consists of sheet materials that are comprised of bonded, small, discrete fibers. The fibers usually are cellulosic in nature and are held together by secondary bonds which, most probably, are hydrogen bonds (see Cellulose). The fibers are formed into a sheet on a fine screen from a dilute water suspension. The word paper is derived from papyrus, a sheet made in ancient times by pressing together very thin strips of an Egyptian reed (*Cyperus papyrus*).

Table 1.   Main Components of Paper and Paperboard (in 1980), wt%

	World	North America
mechanical/semichemical wood pulp	20.8	21.5
unbleached kraft chemical wood pulp	18.5	25.2
white chemical wood pulp[a]	26.6	28.6
waste fiber	25.3	20.3
nonwood fibers	4.2	0.9
fillers/pigments	4.6	3.5

[a] Includes unbleached sulfite.

Paper is made in a wide variety of types and grades to serve many functions. Writing and printing papers constitute ca 30% of the total production. The balance, except for tissue and toweling, is used primarily for packaging. Paperboard differs from paper in that it is generally thicker, heavier, and less flexible than conventional paper.

Fibrous Raw Materials

The main components used in the manufacture of paper products are listed in Table 1. More than 95% of the base material is fibrous and more than 90% originates from wood (qv). Many varieties of wood, eg, hardwood and softwood, are used to produce pulp. In addition to the large number of wood types, there are many different manufacturing processes involved in the conversion of wood to pulp. These range from mechanical processes, by which only mechanical energy is used to separate the fiber from the wood matrix, to chemical processes, by which the bonding material, ie, lignin (qv), is removed chemically. Many combinations of mechanical and chemical methods also are employed (see Pulp). Pulp properties are determined by the raw material and manufacturing process, and must be matched to the needs of the final paper product.

Physical properties. Most properties of paper depend upon direction. For example, tensile strength is greater if measured in the machine direction, ie, the direction of manufacture, than in the cross-machine direction. For paper made on a Fourdrinier paper machine, the ratio of the two values varies ca 1.5–2.5. An even greater anisotropy is observed if either of the in-plane values is compared to the out-of-plane strength. Paper is quite weak in the thickness direction.

Paper may be considered an orthotropic material, ie, one possessing three mutually perpendicular symmetry planes. The three principal directions are the machine, cross-machine, and thickness directions.

Because the fibers generally are flat or ribbonlike, they tend to be deposited on the wire in layers. There is very little tendency for fibers to be oriented in an out-of-plane direction, except for small undulations where one fiber crosses or passes beneath another. The layered structure results in the different properties measured in the thickness direction as compared to those measured in the in-plane direction. The orthotropic behavior of paper is observed in most paper properties and especially in the electrical and mechanical properties.

The basis weight, W, as commonly expressed in the United States and determined in accordance with T 410 of the Technical Association of the Pulp and Paper Industry (TAPPI), is the mass in grams per square meter. It can also be expressed as pounds of a ream of 500 sheets of a given size, but the sheet sizes are not the same for all kinds of paper. Typical sizes are 43.2×55.9 cm for fine paper, 61.0×91.4 cm for newsprint, and 63.5×96.5 cm for several bookpapers. The most common designation is pounds per 3000 square feet (1.62 g/m^2) for paper. The basis weight of board usually is expressed as pounds per thousand square feet (205 kg/m)^2.

The thickness or caliper is the thickness of a single sheet measured under specified conditions (see TAPPI T 411). It usually is expressed in micrometers.

The tensile strength is the force per unit width parallel to the plane of the sheet that is required to produce failure in a specimen of specified width and length under specified conditions of loading (see TAPPI T 404). The strength of paper also is expressed in terms of a breaking length, ie, the length of paper that can be supported by one end without breaking. Breaking lengths for typical papers are from ca 2 km for newsprint to 12 km for linerboards.

Chemical Properties

The chemical composition of paper is determined by the types of fibers used and by any nonfibrous substances incorporated in or applied to the paper during the papermaking or subsequent converting operations. Paper usually is made from cellulose fibers obtained from the pulping of wood. Occasionally, synthetic fibers and cellulose fibers from other plant sources are used. Paper properties that are affected directly by the fibers' chemical composition include color, opacity, strength, permanence, and electrical properties. Development of interfiber bonding during papermaking also is strongly influenced by the composition of the fibers. Because residual lignin in the fibers inhibits bonding, groundwood pulp is used in newsprint and in some book and absorbent papers which do not require a highly bonded structure. Hemicelluloses in chemical pulps contribute to bonding; therefore, pulps containing hemicelluloses are used for wrapping papers and other grades which require bonding for strength, and in glassine, which requires bonding for transparency. In most papers, the chemical composition largely reflects those nonfibrous materials that were added to the paper to achieve the desired physical, optical, or electrical properties.

Manufacture and Processing

Stock preparation. Stock preparation denotes the several operations that must be undertaken in order to prepare the furnish from which paper is made. During the stock-preparation steps, papermaking pulps are most conveniently handled as aqueous slurries, so that they can be conveyed, measured, subjected to desired mechanical treatments, and mixed with nonfibrous additives before being delivered to the paper machine. In the case of adjacent pulping and papermaking operations, pulps usually are delivered to the paper mill in slush form directly from the pulping operation. Purchased pulps and wastepaper are received as dry sheets or laps and must be slushed before use. The objective of slushing is to separate the fibers and to disperse them in water with a minimum of mechanical work so as not to alter the fiber properties. Slushing is accomplished in several types of apparatus, eg, the Hydrapulper which is illustrated in Figure 1.

Figure 1. The Hydrapulper. Courtesy of Black Clawson Co.

Beating and refining. Virtually all pulps are subjected to certain mechanical actions before being formed into a paper sheet. Such treatments are used to improve the strength and other physical properties of the finished sheet and to influence the behavior of the system during the sheet-forming and drying steps. Beating and refining may be considered synonymous.

Filling and loading. Materials, eg, mineral pigments for filling and loading, are added to the pulp slurry to make the papermaking furnish

(see Papermaking additives). Pigments are used in varying amounts, depending upon the grade of paper, and may comprise 2–50% of the final sheet. Fillers (qv) can improve brightness, opacity, softness, smoothness, and ink receptivity. They almost invariably reduce the degree of sizing and the strength of the sheet. The brightness, particle size, and refractive index of fillers influence the optical properties of the finished sheet, and the particle size and specific gravity are important in regard to the filler retention during sheet formation. All commercial fillers are essentially insoluble in water under the conditions of use.

Sizing. Sizing is the process of adding materials to the paper in order to render the sheet more resistant to penetration by liquids, particularly water. Unsized or waterleaf paper freely absorbs liquids. Writing and wrapping papers are typical sized sheets, as contrasted with blotting paper and facial tissue which usually are unsized. Rosin, various hydrocarbon and natural waxes (qv), starches, glues, casein, asphalt emulsions, synthetic resins, and cellulose derivatives are some of the materials that are used as sizing agents. The agents may be added directly to the stock as beater additives to produce internal or engine sizing, or the dry sheet may be passed through a size solution or over a roll that has been wetted with size solution; such sheets are said to be tub-sized or surface-sized.

Coloring. The color of most papers and paperboards that are made from bleached pulps is achieved by the addition of dyes and other colored chemicals. White papers frequently are treated with small amounts of blue materials to achieve a whiter visual appearance. By far the largest proportion of dyes is added to the stock during stock preparation, although a limited amount of dry paper is colored by dip dyeing or by applying a dye solution during calendering. Water-soluble synthetic organic dyestuffs are the principal paper-coloring materials.

Other beater additives. Beater adhesives are employed widely to enhance fiber-to-fiber bonding. Starches probably are used in the greatest tonnage. Urea–formaldehyde and melamine–formaldehyde polymers provide wet strength to the finished paper sheet (see Amino resins). Other natural and synthetic materials are used to alter the paper properties and to influence the behavior of the system during sheet forming and drying.

Sheet Forming, Pressing, and Drying

Sheet forming. Continuous sheet forming and drying came into use in ca 1800. The equipment was of two types: the cylinder machine and the Fourdrinier machine. Continuous paper machines have undergone extensive mechanical developments in the past half century, although the principles employed have changed little. Cylinder machines still are operated and involve multiples of five to seven cylinders; they are used to produce heavy multi-ply boards. Fourdriniers (Fig. 2) are standard in the industry and are used to produce all grades of paper and paperboard. They vary 1–10 m in width and, including the press and dryer sections, may be > 200 m long.

Sheet pressing. The sheet leaving the wet end contains approximately four parts of water per part of fiber; however, it is possible to remove additional water mechanically without adversely affecting sheet properties. This is achieved in rotary presses, of which there may be one or several on a given paper machine.

Sheet drying. At a water content of ca 1.2–1.9 parts of water per part of fiber, additional water removal by mechanical means is not feasible and evaporative drying must be employed. This is at best an efficient but costly process and often is the production bottleneck of papermaking.

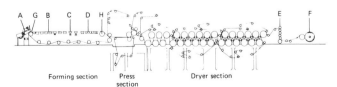

Forming section   Press section   Dryer section

Figure 2. Fourdrinier paper machine with (A) headbox, (B) Fourdrinier wet end with foil boxes, (C) wet and (D) dry suction boxes, pickup and closed transfer of web through the press section, dryer section, (E) calender, and (F) reel. G and H are the breast roll and couch roll, respectively. Courtesy of Beloit Corporation.

Converting. Almost all paper is converted by undergoing further treatment after manufacture. Among the many converting operations are embossing, impregnating, saturating, laminating, and the forming of special shapes and sizes, eg, bags and boxes (see Uses).

Pigment coatings. Pigment coatings are compositions of pigments and adhesives with small amounts of additives and are applied to one or both sides of a paper sheet. They generally are designed to mask or change the appearance of the base stock, improve opacity, impart a smooth and receptive surface for printing, or provide special properties for particular purposes.

Barrier coatings. In packaging applications, a barrier may be needed against water, water vapor, oxygen, carbon dioxide, hydrogen sulfide, greases, fats and oils, odors, or miscellaneous chemicals. A water barrier can be formed by changing the wettability of the paper surface with sizing agents. A grease or oil barrier can be provided by hydrating the cellulose fibers to form a pinhole-free sheet or by coating the paper with a continuous film of a material which is resistant to the particular grease. Gas or vapor barriers are formed by coating the paper with a continuous film of a suitable material (see Barrier polymers).

Environmental Issues and Plant Efficiency

Some Federal laws that affect the pulp and paper industry are the Clean Air Act, 1970, Clean Water Act, 1974, Resource and Recovery Act, 1976, Toxic Substances Control Act (TSCA), 1977, Occupational Safety and Health Act (OSHA), 1970 Federal Hazardous Substances Act, and Federal Insecticide, Fungicide and Rodenticide Act, 1972. The new requirements have resulted in increased lead time for expansion of basic production facilities and have affected the nature of industrial growth.

Pollution control. Methods of pollution control in pulp and paper mills include in-mill control measures, end-of-the-pipe control measures, and monitoring and assessing environmental quality. Control devices for gases may include stripping, scrubbing, condensation, incineration, and adsorption.

Control devices for particulate matter include cyclones, scrubbers, and electrostatic precipitators. Application of these devices individually or in combination has been quite effective in meeting the air-quality criteria.

The pulp and paper industry uses large amounts of water. Since 1972 water-quality requirements have forced mills to install end-of-pipe secondary treatments, which principally are biological. Means other than installation of expensive end-of-the-pipe treatments are being evaluated in terms of recent requirements. Mills are recycling and reusing more process water, which results in higher water temperatures and increased solids in the process water, thereby enhancing problems, eg, corrosion, slime and other deposits. However, chemicals are used to control these problems. Dissolved solids may impart biochemical oxygen demand (BOD), color, and toxicity to mill effluents. The suspended solids impart turbidity and long-term BOD. Directly or indirectly, all of these may affect aquatic life.

Sludge handling and disposal. Most waste-treatment processes generate solid wastes that must be disposed of (see Wastes, industrial; Water, industrial water treatment). Two kinds of sludges are generated by pulp and paper mills: primary sludges that contain fibers, clay filler materials, and other chemical additives, and secondary sludges that are largely biological in nature and are harder to handle and dewater (see Dewatering). The disposal of sludges in landfills is being reevaluated and alternative disposal approaches are being developed.

Uses

Paper is used in paper and paperboard such as folding cartons, fiber cans and tubes, corrugated and solid fiber boxes, paper bags, and multiwall shipping sacks.

> G.A. BAUM
> E.W. MALCOLM
> D. WAHREN
> J.W. SWANSON
> D.B. EASTY
> J.D. LITVAY
> H.S. DUGAL
> The Institute of Paper Chemistry

J.P. Casey, *Pulp and Paper Chemistry and Chemical Technology*, 3rd ed., Vol. 1, Wiley-Interscience, New York, 1980.

J. d'A. Clark, *Pulp Technology and Treatment for Paper*, Miller Freeman Publications, Inc., San Francisco, Calif., 1978.

J.E. Huber, *Kline Guide to the Paper Industry*, 4th ed., Charles H. Kline & Co., Fairfield, N.J., 1980.

B.L. Browning, *Analysis of Paper*, 2nd ed., Marcel Dekker, Inc., New York, 1977.

PAPERMAKING ADDITIVES

In papermaking, chemicals can be added either to the pulp slurry prior to sheet formation, ie, internal addition, or to the resulting sheet after complete or partial drying. The method used depends primarily on the desired effects. For example, strength additives usually are added internally if uniform strength throughout the sheet is wanted, but they are applied to the surface if increased surface strength is needed. If an additive cannot be retained efficiently from a dilute pulp slurry, then it is better to apply it to the surface of the sheet.

Functional internal additives improve specific properties of the paper, eg, sizing, dry strength, wet strength, and optical properties, eg, those that are affected by the presence of light-scattering particulate additives such as clay and titanium dioxide. Color-modifying additives have been reviewed (see Dyes and dye intermediates; Dyes, application and evaluation).

Functional surface treatments involve many of the same chemicals that are used internally. However, since retention of the chemical by pulp fibers is not important to surface treatment, many other materials may be used that are ineffective or, at best, inefficient when added internally.

Environmental constraints on the paper industry have resulted in drastic processing changes in recent years, primarily because very large amounts of water are used to produce paper. If chemical additives are not efficiently retained in the sheet, treatment of the large volumes of effluent is aggravated. Efficiency of various additives used in papermaking has increased and has permitted greater closure of white-water systems, ie, water drained and pressed from the newly formed sheet. In alkaline sizing systems and some alkaline strengthening systems, the additives are retained so completely that effluents from these paper machines contain only very small amounts of dissolved materials.

Because many grades of paper and paperboard are used in direct contact with foods, most mills require paper chemicals that have been cleared for use with food by the FDA. Many mills require that all of the chemicals that they use comply with FDA regulations, so that it is not necessary to segregate machine broke and white water. Most of the chemicals discussed below are approved by the FDA for use in paper and paperboard that are intended for applications in food packaging and processing. However, there are various restrictions and limitations on the amounts of chemicals that can be used and restrictions on some chemical additives in terms of specific functional uses. It is recommended that the FDA status of any paper-chemical additives be determined with the supplier prior to use.

Processing Aids

Processing aids improve the operation of the paper machine, whereas functional chemicals improve the properties of the paper. Processing aids include retention aids, drainage aids, formation aids, flocculants, defoamers (qv), wet-web-strength additives, pitch-control agents, creping aids, and slimicides (see Flocculating agents; Industrial antimicrobial agents). Retention aids, flocculants, and drainage aids are phenomenologically related; all act by agglomerating filler particles, fines, or fibers with themselves or with each other.

Functional Internal Additives

Functional internal additives include sizing agents (rosin-based, cellulose-reactive, wax emulsions, and fluorochemical); dry-strength additives (natural gums, starches, cellulosics, and acrylamide polymers);

wet-strength additives (urea–formaldehyde resins, melamine–formaldehyde resins, aminopolyamide–epichlorohydrin resins, polymeric amine–epichlorohydrin resins, aldehyde-modified resins, and polymeric amines); and fillers, most commonly mineral pigments, eg, clay, calcium carbonate, silica, hydrated alumina, and talc. Kaolin clay is the least expensive and most used filler pigment in the United States.

In recent years, synthetic polymeric pigments have been promoted as fillers for paper. Pigments that are based on polystyrene latices and on highly cross-linked urea–formaldehyde resins have been evaluated for this application. These synthetic pigments are less dense than mineral fillers and, thus, can be used to produce lightweight grades of paper.

Functional Surface Treatments

Although many functional chemicals can be added to the wet end of the paper machine, some grades of paper require special properties that cannot be provided by the low levels of additives that are retained at the wet end. To achieve the properties required for these grades of paper, it is necessary to apply the chemicals to a preformed paper web.

Processes. The most common method for the application of chemicals to the surface of a paper web is by a size press. In the size press, dry paper, which usually is sized to prevent excess water penetration, is passed through a flooded nip and a solution or dispersion of the functional chemical contacts both sides of the paper. Excess liquid is squeezed out in the press and the paper is redried.

A process that is being developed involves the application of chemicals as a foam to the surface of a paper web. The chemical foams contain less water than their liquid equivalent; thus, savings in drying costs can be realized. Also, the foams can be applied to a wet web, eg, in the press section, without breaking the sheet.

Sizing. The most commonly used materials for surface sizing are starches and modified starches, including oxidized, enzyme-converted, hydroxyethylated, and cationic starches. They are used not only for sizing but also to improve strength, especially surface strength, and to impart smoothness. Starches may be applied to the finished sheet by any of the previously discussed methods. Often starch is used with other surface-sizing agents, eg, rosin-based emulsion sizes, wax and wax-rosin emulsions, alkylketene dimer emulsions, and various synthetic polymeric sizing agents. These combinations permit improved sizing against liquid penetrants and increased surface strength and better finish.

Fluorochemical emulsion sizing agents can be applied to the surface of paper or paperboard to provide good oil and grease repellency. If they are used with other sizing agents, eg, alkylketene dimer emulsion, both oil and grease repellency and water repellency are obtained.

Other special sizing agents are the chromium complexes of long-chain fatty acids; these provide water-repellent surfaces and good release properties (see Waterproofing).

Application of dry-strength additives. The various water-soluble natural and synthetic polymers that are used for strength enhancement by internal addition can be applied to paper surfaces. This type of application usually is indicated when surface-strength properties are more important than increased internal strength. Starches and modified starches, especially cationic starches, are used in large quantities to improve the surface strength of paper; they also improve the printing quality of the paper as a result of increased surface strength and reduced linting.

Application of wet-strength resins. Wet-strength resins are seldom applied to the surface of paper for enhancing wet strength because the commercially available, cationic wet-strength resins are retained so effectively internally. However, wet-strength resins are applied frequently to the surface of towels and tissues as creping aids.

Creping. The products that are used as internally added creping aids can be applied to the surface of paper, usually by spraying an aqueous solution or emulsion in front of the Yankee dryer. One of the most commonly used resins for this purpose is an aminopolyamide–epichlorohydrin resin. Such resins yield coatings on the dryer with the required degree of adhesion for optimum creping.

Curl control. Many grades of paper tend to curl, especially as humidity varies, because of the stresses produced during the drying process.

Judicious application of water to the opposite side of the dry sheet followed by redrying may correct the curling.

Pigmented coatings. Paper coatings are applied as coating colors which are aqueous slurries containing 35–65 wt% solids. There are three main components of the solids; pigments, binders, and minor additives. The pigment is the chief component of a paper coating and consists of small, white, particulate materials. Pigments usually are minerals, eg, clay, calcium carbonate, or titanium dioxide.

Binders. Paper-coating binders are either polymers derived from natural sources or synthetic polymers. The largest-volume, naturally derived binder is starch (qv).

Synthetic Fibers

A variety of wet-laid felts and nonwoven fabrics are produced on Fourdrinier-type papermachines (see Nonwoven textile fabrics; Pulp, synthetic). Noncellulosic materials may be included as part or all of the fiber furnish; latexes, water-soluble polymers, or other adhesives are used as bonding agents. Synthetic fibers can make paper highly resistant to wetting; chemical attack; mechanical wear, eg, folding; weathering; and biological degradation. Synthetic fiber-containing papers are used as backings for carpets and vinyl floor coverings, industrial filters, disposable bed linens and hospital garments, heavy-duty wiping materials, tea bags, tissues, labels, and embossable wallpapers.

STEARNS T. PUTNAM
HERBERT H. ESPY
GAVIN G. SPENCE
Hercules Incorporated

J.P. Casey, ed., *Pulp and Paper: Chemistry and Chemical Technology*, 3rd ed., John Wiley & Sons, Inc., New York, 1980.

K.W. Britt, ed., *Handbook for Pulp and Paper Technology*, 2nd ed., Van Nostrand Reinhold Company, New York, 1970.

J. d'A. Clark, *Pulp Technology and Treatment for Paper*, Miller Freeman Publications, Inc., San Francisco, Calif., 1978.

PARAFFIN WAX.   See Waxes.

PARALDEHYDE, $C_6H_{12}O_3$.   See Acetaldehyde.

PARASITIC INFECTIONS, CHEMOTHERAPY.   See Chemotherapeutics.

PARATHION.   See Insect-control technology.

PARTICLE-SIZE MEASUREMENT.   See Size measurement of particles.

PARTICLE-TRACK ETCHING

Particle-track etching is a chemical process that preferentially removes material along the paths of energetically charged particles in certain solids. Invisible until developed by the etching process, the latent particle track is a linear, highly localized region of altered physical and chemical structure compared with the bulk solid. The development process yields an etch pit along the particle path with dimensions that depend upon development conditions but that may range up to optical size.

Latent tracks cannot be formed in conductors because charge neutralization by the conduction electrons prohibits formation of a space-charge region about the particle path. In crystalline nonconductors, however,

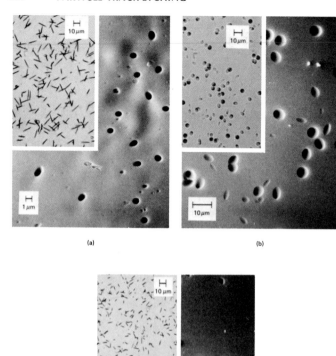

Figure 1. Tracks of ^{252}Cf fission fragments etched (**a**) in polycarbonate plastic for 45 min in 6 N NaOH at 60°C; (**b**) in soda–lime glass for 5 s in 48% HF at 23°C; and (**c**) in mineral, mica, for 15 min in 48% HF at 23°C. Courtesy of Nucleopore Corporation.

the space-charge region is hypothesized as the source of a stress zone identified as the particle track.

Particle-track etching requires a heightened susceptibility to chemical attack along the particle path relative to the bulk solid. The etching is achieved by immersing the solid containing latent particle tracks in an etchant, which after a time yields an etch pit. The etchant used depends on the nature of the material to be etched. Typically, strong base solutions are used as etchants for polymeric materials, and strong acids are used with glasses and minerals. The bulk etch rate in polymers and glasses is isotropic in most cases, yielding symmetrically shaped etch pits. In minerals, however, anisotropic bulk etching often produces irregularly shaped etch pits. Etch pits in the three classes of track-etch materials, polymers, glasses, and minerals, are illustrated in Figure 1.

The term track-etch detector is used here to designate materials that can be etched to reveal the damage track of a charged particle, and the ability of a material to yield etch pits of particle tracks upon etching is known as the track-registration sensitivity.

One of the unique features of particle-track etching is its application to determine the charge, mass, and energy of the particle that creates the latent track. The method of identification is based on the finding that track-etch rate is proportional to the damage caused by the particle per unit path length and, therefore, it is a unique function of the particle residual range in a solid. The relationship between track-etch rate and particle residual range is usually established by a calibration procedure using accelerator ions of known charge, mass, and energy.

Track-etch detectors are inert and require no power or special handling. They are lightweight, inexpensive, and of high mechanical integrity. After exposure, track-etch detectors are capable of storing the damaged trail almost indefinitely under normal conditions.

Particle-track etching as a technique has been used in a number of fields, including astrophysics, nuclear physics, geochronology, radiation dosimetry, cancer-tumor diagnosis, filtration, chemical analysis and microchemical mapping, oil exploration, and uranium prospecting.

Principles

Energetically charged particles penetrate matter along approximately straight paths, slowly dissipating their energy through collisions with the electrons and atoms of the medium. After losing all its kinetic energy, the charged particle may combine chemically with atoms of the medium.

A fast particle deposits energy in discrete collision volumes along its trajectory. This phenomenon is manifest in the grains along particle tracks in emulsions, and in individual bubbles along particle paths in bubble chambers. The increase in stopping power as the particle slows results in an increasing number of discrete collision volumes per unit path length. At some point along the particle trajectory, the volumes come close together, yielding a nearly continuous path of damage. Track etchability requires at least such a continuous damage trail.

The particle track, independent of the type of material, can be envisioned as a highly localized, linearly distributed, secondary phase contained in the primary bulk phase of the material.

The critical damage level that yields etchable tracks is called the track-registration threshold. Track-registration thresholds in a wide range of materials have been explained satisfactorily by the ion-explosion-spike model of track formation.

The etching process. Particle-track etching can be separated into two different etching processes: bulk etching of the solid where no radiation damage has occurred, and core etching along the particle path. Experimental results suggest that core etching occurs in a cylindrical volume that extends to a diameter of ca 5–10 nm. Beyond the core, enlargement of the track diameter occurs by bulk etching. The bulk-etching rate can be determined by a variety of methods—thickness change, weight loss, or diameter of etch pits.

Track etching. The damage core can be thought of as a highly localized secondary phase within the primary bulk phase. It is contained within a cylindrical volume of roughly 5–10 nm diameter and must have a higher chemical reactivity than the primary bulk phase in order to be etched. The track-etching rate is the speed of advancement of degradation front along the particle track. Investigations show that in a given material the track-etch rate depends primarily on the damage density created by the particle, but also on most of the same factors as bulk etching, eg, etchant type, concentration, temperature, and etch products. In addition, marked effects owing to oxygen, including photooxidation and thermooxidation, are observed.

Etch-pit geometry. Certain aspects of the etch-pit geometry affect the analysis of etch tracks, measurement methods, and particle-identification schemes that use track-etch detectors.

Etch-pit geometry, in the simplest case, is governed by the simultaneous action of two etch rates: chemical attack along the particle path at a rate V_t and general attack at the liquid–solid interface of the undamaged bulk solid at a bulk etch rate V_g, which is constant and isotropic.

Applications

The process of particle-track etching for commercial purposes is restricted in its use by patents that cover the particle-irradiation and track-etch development phases of the process. Currently, two companies operate under these patents: Nucleopore Corporation and Terradex Corporation. Particle-track etching is used in cosmic-ray studies, fission-track dating, space-radiation dosimetry, radon dosimetry, and membrane filtration technology.

D.D. PETERSON
M.C. PORTER
Nucleopore Corporation

D.A. Young, *Nature* **182**, 375 (1958).

R.L. Fleischer, P.B. Price, and R.M. Walker, *Nuclear Tracks in Solids*, University of California Press, Berkeley, Calif., 1975.

R.L. Fleischer and co-workers, *Phys. Rev.* **156**, 353 (1967).

PARTING AGENTS. See Abherents.

PATENTS

PRACTICE AND MANAGEMENT

The law of patents is designed to encourage the creation of inventions by granting limited exclusive rights to technological arts which are new, useful, and nonobvious. Patents are often granted for inventions on the leading edge of technical advance, although they also protect relatively small improvements in existing apparatus, products, and processes. As the law of patents has evolved, it has had to address new problems and provide legal solutions for technology not before addressed by those who enacted the patent statutes. Indeed, the change in the U.S. from a largely agrarian society to a highly industrialized one has compelled the law to remain flexible.

Nature of Intellectual Property

A patent is one of a series of intangible, designated rights known collectively as intellectual property. State or Federal laws, or some combination of them, provide means for protecting various types of intellectual creations such as patents for inventions, trade-secret protection for confidential or secret information, trademarks and service marks for the goodwill of a business or individual, and copyrights for the intellectual creations of authors (see Trademarks and copyrights).

Although the Federal and state trademark statutes protect the rights of the first user of a mark in commerce, copyright law protects original authors from the copying of their works, and the law of trade secrets protects the owner of confidential information against the wrongful taking of it, patent law protects the first inventor of new, useful, and nonobvious inventions. The patent grant is a right granted by the U.S. Government to the inventor to exclude all others from making, using, or selling the subject matter claimed as the invention. To be patentable, the subject matter of the application must be new (or novel), useful, and nonobvious. The patent laws are designed to benefit the public by providing an incentive for the creation of inventions and the innovation required to translate a patentable invention into a commercial product. Patent rights are the exclusive domain of Federal law and are only enforceable by suits for infringement in Federal courts.

The patent statutes provide for several types of patents. The most prominent is the utility patent which covers the utilitarian features of electrical, mechanical, or chemical inventions and has a term of 17 yr from its issue date. Design patents are available on new, original, and ornamental designs for articles of manufacture and have a term of 3–14 yr.

Requirements for patentability. An invention involving proper subject matter is patentable if it meets three requirements. The subject matter must be new (eg, novel), nonobvious to a person of ordinary skill in the art, and useful.

Characteristics of the patent right. The right granted by a patent is the right to exclude others from making, using, or selling the invention described and claimed in the patent. Since the patent is an exclusionary right, the scope of a patent is critical so that others can determine what it is they are allowed and not allowed to do. The scope of a patent is determined by its claims.

A second characteristic of a patent as property is that an alleged infringer, upon being charged with infringement or actually sued, may defend by showing that the patent is either invalid, not infringed, or unenforceable on various grounds. Challenges to the validity of a patent cover a wide spectrum but usually focus upon the existence of prior art that was not considered by the United States Patent and Trademark Office (PTO) when the patent in suit was granted.

In addition to challenging the validity of a patent or showing it has been misused, the alleged infringer can also defend on grounds that he or she does not infringe the claims of the patent in suit. This requires the court to consider in detail the patent claims which are word descriptions of the patent owner's invention. Where every element of the claims is found in the defendant's structure, infringement is present unless the claims are invalid.

Benefits of patent ownership. Even though the property rights granted by a patent are subject to some uncertainties, they often constitute an important business asset. There are many examples of inventors of new technology using their patents to found new businesses which develop strong, if not dominant, market positions. A heavy-capital industry usually needs patent protection to protect new technology and to make research expenditures cost-effective.

Patent Procurement

The first step in the patenting process begins with the inventor's mental conception of the invention. The date when the inventor had a mental conception of the complete and operative device or process is often of critical importance in the subsequent stages of patent procurement and/or enforcement. Many companies have wisely instituted programs requiring members of their research and development groups to document their ideas in laboratory notebooks or on invention-disclosure forms. Detailed records should be kept recording each mental and physical step in the conception, development, and reduction to practice of an idea. Such records should be reviewed frequently and witnessed by persons other than the inventor who can attest to having read and understood the concepts on the date of review.

When the inventor has developed an idea to the point where it appears that patent protection is desirable or appropriate, a patent attorney or agent is (or should be) consulted. After the invention is explained fully and understood by the patent attorney, it usually is good practice to have a prior-art or novelty search conducted unless the inventor is well-informed concerning the latest developments in the art to which the invention pertains.

Statutory requirements for patentability. Compliance with the statutory requirements of the patent law entitles the inventor to a patent. To be patentable, the claimed subject matter must be within a subject-matter class for which patents are available, it must be novel, useful, and nonobvious, and it must be described adequately in a written application. A legally complete application must also name the true inventor(s) who must execute an oath (or declaration) that the person believes himself or herself to be the original and first inventor of the subject matter claimed. A statutory filing fee must accompany the application.

Statutory subject matter. The classes of statutory or patentable subject matter are set forth in 35 USC §§101, 161, and 171 and summarized in Table 1.

Requirements of the application. To be entitled to a patent, an applicant must file an application which discloses and claims the invention in accordance with the statutory requirements. The formal application comprises a specification, including one or more claims, drawings if the nature of the invention admits of illustration by drawing, an oath (or declaration) by the inventors, and of course, the prescribed filing fee.

Format of specification. There is no statutory format for the specification of a patent application. However, the PTO suggests a format for all applicants to follow. The Manual of Patent Examining Procedures (MPEP) recommends a form as does the Rule of Practice in order to focus the applicant's attention on the important part of the written description, etc. Under the format suggested by the PTO, the specification is broken down into various parts, each of which is preceded by the following headings: (*1*) Title of the Invention; (*2*) Abstract of the Disclosure; (*3*) Cross-References to Related Applications (if any); (*4*) Background of the Invention (Field of the Invention, Description of the Prior Art); (*5*) Summary of the Invention; (*6*) Brief Description of the Drawings; (*7*) Description of the Preferred Embodiment(s); (*8*) Claim(s).

Prosecution of the application. Since the PTO does not have the facilities necessary to discover all prior devices in use or sale that may bear upon the question of patentability, nor the research facilities to

Table 1. Classes of Statutory or Patentable Subject Matter

Class	Examples
(1) process (§101)	pollution control processes, manufacturing methods, chemical reactions, new uses of old materials, etc
(2) machine (§101)	gas chromatographs, nmr spectrophotometers, photocopying machines, etc
(3) article of manufacture (§101)	ball-point pens, key chains, bobby pins, shingles, etc
(4) composition of matter (§101)	chemical compounds, mixtures, pharmaceuticals, nylon, polyethylene, etc
(5) improvements on subject matter in classes (1–4) (§101)	(see above)
(6) ornamental designs (§171)	cabinets, sculptures, neckties, silverware patterns, etc
(7) asexually reproduced plants (§161)	roses, asparagus plants, apple trees, etc[a]

[a] Sexually reproduced plants are protected by the Plant Variety Protection Act.

verify or disprove points of technological distinctions urged by an applicant, the law imposes upon each applicant a duty to deal with the PTO with the utmost of candor and honesty. To the extent that an applicant may be aware of information, especially prior art, which a reasonable patent examiner would consider important in deciding whether to allow application claims, the applicant has a duty to disclose such information to the PTO. Willful failure to comply with this duty of candor may result in an application being stricken from the files by the PTO or, if a patent was issued, it may be held invalid or unenforceable by a court.

Corrections of patents. Minor errors in issued patents, such as typographical errors, may be corrected by application for and issuance of a certificate of correction. Errors in naming the proper inventors, if made without deceptive intention, may also be corrected by application to the Commissioner.

More serious errors, which are made without deceptive intention, particularly with regard to the manner in which the invention is claimed or which make the patent wholly or partly inoperative or invalid, may be corrected by application for reissue of the patent.

Reexamination. In 1981 a new procedure became available for obtaining a decision from the PTO upon the validity of an issued patent in view of prior art not previously considered in the PTO's initial decision to grant the patent. Under this new procedure, anyone may inform the PTO of prior-art patents and printed publications that are believed to have a bearing upon patent validity and, upon payment of a $1500 fee, request a PTO order that the patent be reexamined for validity. If the PTO considers the newly cited art to raise a substantial new question of patentability, reexamination is ordered. The procedures followed during reexamination are similar to those applied to original application. The patent owner is entitled to present arguments and/or claim amendments during reexamination in an effort to convince the PTO of patentability over the newly cited art. At the conclusion of the reexamination process the PTO issues a certificate of reexamination which cancels unpatentable claims from the patent and/or which reaffirms the patentability of claims as originally allowed or as amended during reexamination. In cases wherein the request for reexamination is not regarded by the PTO to raise a substantial new question of patentability, the PTO will not order reexamination, and a portion of the request fee ($1200) is refunded to the requestor.

It is hoped that reexamination will tend to reduce uncertainties concerning the validity of the patent over prior-art references not before considered by the examiner. It also is hoped that reexamination will do this in a relatively economical way and, hence, reduce the number of cases requiring litigation.

Enforcement of Patent Rights

Patent suits are among the most complex and expensive of all types of litigation. Many of the legal doctrines are unique, relatively complex, and subject to dispute. The essential facts necessary to a resolution of patent litigation may relate to technological issues and scientific principles which may be relatively incomprehensible to a layman. Indeed, the technological facts may have to be resolved in the light of a division of opinion in the scientific community.

Factors to consider. Patents are a business asset and should be treated as such. The decision to litigate should be made within the context of a business judgment on how best to protect that asset and/or obtain an economic return on the investment.

In the decision on whether to litigate, the patent owner should balance probability of success, including the amount of probable damage recovery, against the estimated cost of pursuing the litigation. When it appears that litigation is a possibility, the patent owner should have a thorough validity search made to be certain the patent is as strong as the owner thinks it is.

Of equal importance to the thorough search of the prior art is the thorough search of the patent-owner's files. It is important that the available documents and witness recollection be developed and evaluated before suit is instituted since the suit will be no stronger than its weakest link.

The accused infringer's product should also be studied closely in light of the patent claims and their history of prosecution and amendment to determine whether a good case for infringement exists. Additionally, the infringer's activities in the market should be investigated to arrive at some estimate of the amount of damages which the patentee may expect to recover if successful in the suit.

The above considerations should be balanced against the anticipated cost of the suit. A major cost of litigation is, of course, the attorney and court fees that a patent owner incurs in preparing for a trial. In addition to attorney fees, fees for expert consultants and witnesses who are knowledgeable in the technology involved will have to be incurred. Out-of-pocket expenses may also be substantial. Generally, the parties in patent litigation are located in different states, business headquarters where documents are kept may be in still other states, and the court where the hearings are held may be in still another state. All of this may require extensive travel for the purpose of pretrial discovery and attendance at court hearings, the expense of which must be borne by the patent owner. Further, but to a lesser degree, the patent owner should consider the expense of lost productivity from its officers and other employees who participate in pretrial discovery or trial of the suit. Although the latter is not usually a direct charge, the cost to the company can be considerable.

Prefiling activity. Even though a final decision to litigate has not been made, it is quite common for patent owners to send notice letters to parties they believe are committing acts of infringement. Such letters should advise the addressee of the existence of the patent and the owner's belief that infringement may be occurring.

Litigation issues. The primary issues in litigation involve the questions of the validity of the patent in suit, the enforceability of the patent, and whether it is infringed by the defendant's activities.

Remedies for infringement. Upon proof of infringement of a valid and enforceable patent, the patent owner is entitled to an award of damages adequate to compensate for the infringement. In no event should the sum be less than a reasonable royalty together with interest and costs as fixed by the court.

In addition to general damages, the patent statute authorizes a court to increase the damage award up to three times the amount found or assessed for general damages in proper cases. Generally, a court will award increased damages for infringement only in those cases where the infringement has been found to be willful and wanton.

Where a court finds the case to be exceptional, it may also order an award of reasonable attorney fees to the prevailing party.

In addition to a monetary award, a court is authorized to issue an injunction in a patent case in accordance with the principles of equity governing the grant of injunctions.

Management Policies

Any company that is actively engaged in product development and marketing incurs patent-related problems from time to time. Decisions may be required on how to protect research and development expenditures with patents, how best to preserve a market against those who would be drawn into competition by a new development, or when contemplating the marketing of a new product, whether the new product infringes any existing patent. Some of the policies relating to these types of patent problems and management are the following: procurement policies (maintaining a prior-art file, secrecy of research and development work, agreement for the use of outside ideas, evaluation for possible patent protection, and invention-documentation policy); transfer of patent rights (assignment and license); and marketing new products (conducting an infringement study, undertaking a validity search on an existing patent, and requesting a license under the patent or seeking to purchase the patent).

JAMES B. GAMBRELL
CHARLES M. COX
PAUL E. KRIEGER
Pravel, Gambrell, Hewitt, Kirk,
Kimball & Dodge

D.S. Chisum, *Patent Law*, Matthew Bender & Co., Inc., New York, 1978.

D.R. Dunner, J.B. Gambrell, and M.J. Adelman, *Patent Law Perspectives*, Matthew Bender & Co., Inc., New York, 1970 to present.

I. Kayton, *Patent Preparation & Prosecution Practice*, 6 vols., Patent Resources Institute Inc., Washington, D.C., 1983.

R.A. White, R.K. Caldwell, and J.F. Lynch, *Patent Litigation: Procedure & Tactics*, Matthew Bender & Co., Inc., New York, 1971.

LITERATURE

Patent literature is a key source of technical and scientific knowledge. The information contained in patent documents is by definition new and often can be found nowhere else. In this era of rapid technological change, the chemist or engineer has to maintain a close watch on this massive body of information with regard to scientific development and innovation (see Information retrieval).

Standardization. Fortunately, procedures encouraged under the Patent Cooperation Treaty and administered through the International Bureau of the World Intellectual Property Organization (WIPO) have brought a level of order and standardization to patent documents. The Paris Union Committee for International Cooperation in Information Retrieval Among Patent Offices (ICIREPAT) has developed four standards of significance for the handling and use of patent information. First, they have provided a two-letter standard code system for identifying states and organizations, which facilitates the consistent entry of national information in computer storage and retrieval systems.

The second ICIREPAT standard of importance to the information scientist is the standard code for identification of different kinds of patent documents. This code provides groups of single letters to distinguish the various types of patent documents and the various levels of publication. In addition, any patent office desiring to identify subdivisions unique to its documents may provide a numerical suffix. These codes are used primarily in machine processing of patent information. It also has been recommended that they appear on the first page of patent documents and as identification for references in patent gazettes. Group 1, or the main series, refers to regular patents of invention. Group 2, or the secondary series, refers to patent documents of a secondary nature such as the U.S. reissue and French patent of addition. Group 3 is reserved for special patent documents not classifiable under Groups 1 or 2. U.S. defensive publications are an example of this category. Group 4 refers to a main-series patent that covers a special subject area. French medicament patents and U.S. plant patents are examples. The fifth group is for utility models and the sixth for nonpatent and restricted documents.

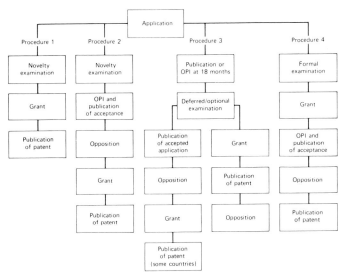

Figure 1. Four generalized examining procedures.

The third standard, to which most Paris Union members adhere, provides that each patent document have a single front page containing a title, bibliographic and filing information, an abstract, and, if appropriate, a drawing.

Dates. Several dates are important in the life of a patent: the filing or application date; the date when the document is first opened to public inspection (often abbreviated OPI); the date when it is published (printed copies available); the date when the application is accepted (U.S., allowed) by the patent office as among all statutory requirements; the date the patent is granted (U.S., issued; UK, sealed); and, of course, the dates the patent term comes into force and expires.

Document numbers. All patent offices assign an application or filing number (U.S. serial number) to applications as received. After the patent is granted, it bears a different patent number. When patents are published before grant (eg, to invite opposition), they usually bear either their application number or a new number that will be the patent number if the potential patent is actually granted. A third alternative is to use a separate number, especially for specifications accepted and published for opposition, as in the Netherlands and Japan.

Information Access

Primary-source access to patent information is through the more than 130 national and regional patent offices throughout the world. All Paris Union members must publish a periodic journal (frequently called a gazette) listing the patent owner and providing an abstract for each patent granted. Most countries provide printed copies of applications or granted patents at prescribed procedural stages during the prosecution of an application. The principal offices also publish annual indexes, usually including a listing of patents by applicant or assignee and by inventor, as well as other summaries and notices.

All patent offices inspect new applications to see that at least the formal requirements have been met (fees, sworn statements, matters of format, subject matter in a patentable category, and often utility of invention). This is all the examination that some countries such as Italy and Spain require. They are the nonexamining or registration countries.

Although no two countries seem to operate in exactly the same manner, most can be grouped into one of the four generalized examining procedures shown in Figure 1. Most procedure-4 countries allow access to the specifications shortly after filing. All procedure-3 countries publish or allow access to specifications 18 months after filing. These are considered the fast-issuing countries. The countries following procedures 1 and 2 require complete examination for novelty before publication and accordingly are the slow-publishing countries.

The United States Patent and Trademark Office. The organization responsible for administering the provisions of U.S. patent law is the

United States Patent and Trademark Office (PTO), an agency of the U.S. Department of Commerce. As its name implies, its mission is to provide patent protection for inventions and a registration system for trademarks. Copyrights are now registered by the copyright office at the Library of Congress and are no longer a responsibility of the PTO.

The PTO grants four kinds of patents: patents (of invention), reissue patents, plant patents, and design patents. It also registers and publishes a patentlike document called a defensive publication. Some 60,000 patents are issued each year; numbered consecutively, they are currently in the four-million (4×10^6) range.

Printed patent documents may be obtained from the PTO on an individual basis by number or through an annual subscription by U.S. Patent Office Classification. They also may be purchased from Rapid Patent International (2221 Jefferson Davis Highway, Arlington, VA, 22202) individually by number or on a subscription basis by class or assignee.

The PTO publishes *The Official Gazette of the United States Patent and Trademark Office* (OG), *Manual of Classification of Patents*, and technical bulletins and class definitions that supplement the *Manual of Classification*. The PTO also maintains microfilm listings, annual indexes of inventors and assignees, and subclass lists.

The following pamphlets and booklets also are available from the Superintendent of Documents, U.S. Government Printing Office, Washington, DC 20402: *General Information Concerning Patents*; *Patents and Inventions, an Information Aid for Inventors*; *Patent Attorneys and Agents Registered to Practice Before the U.S. Patent Office*.

The PTO maintains a public-search center at Crystal Plaza, 2021 Jefferson Davis Highway, Arlington, VA, 22202. In order to accommodate those who cannot visit the public-search facilities, the PTO provides copies of newly issued patent documents to a number of libraries through the country. These are known as patent-depository libraries.

Secondary Sources

Because of the worldwide nature of patent literature and the accompanying language and procurement problems, secondary abstracting and indexing services and custom searching play an important role in the dissemination of patent information. Most secondary services use computer-assisted systems in order to handle the increasing volumes of data now emanating from the world's patent offices (see Information retrieval). Chemical Abstracts Services, INPADOC, Derwent, and IFI (International Federation of Inventors Association)/Plenum are the principal producers of on-line patent data bases (see Table 1). These on-line files, together with the printed services offered by these and other companies and by national patent offices and societies, as well as the custom searching offered by countless commercial and government-sponsored organizations, provide a tremendous store of readily available technological information.

Chemical Abstracts Service. *Chemical Abstracts* (CA), published weekly through the Chemical Abstracts Service (CAS) of the American Chemical Society, provides an excellent abstracting and indexing service for new chemical information. Patent coverage, however, is not as broad as that found in the Derwent and IFI/Plenum files. Patents that are borderline with respect to providing new chemistry and particularly patents on chemical manufactures and applications are frequently not abstracted. Today, coverage has been expanded to all chemical patents from 15 countries and two regional offices plus selected patents issued to nationals in 11 other countries.

The Chemical Abstracts Service maintains its various general literature and patent indexes in computer-readable form and makes them available for on-line searching through System Development's ORBIT, Lockheed's DIALOG, and other on-line systems.

Derwent Publications, Ltd. Currently, the various Derwent products include about 200,000 basic patents (new to the Derwent data base) each

Table 1. On-Line Data Bases Containing Patent Information

File identification	Starting date	Subject coverage	Country coverage[a] data base supplier	Retrieval system[b] LMSC	SDC	BRS
Primary patent files						
APIPAT	1964	petroleum refining, petrochemicals	9 countries, EPO and WPO		×	
Chemical Abstracts	1967	chemical	6–17 countries, partial coverage of 11 more	×	×	×
Derwent CPI	1963	pharmaceuticals	12–15 countries, partial coverage in 11 more		×	
	1965	agriculturals				
	1966	polymers				
	1970	remaining chemical				
Derwent WPI	1974	nonchemical	as in Derwent CPI but lacking Japan		×	
Derwent EPI	1980	electrical/electronics	includes Japan		×	
IFI CLAIMS						
bibliographic	1950	chemical	United States, plus equivalents in 5 others	×		
	1963	nonchemical	United States	×		
Uniterm Index	1950	chemical	United States, plus equivalents in five others	×		
Class		U.S.P.O. Classification		×		
Citation	1947	all	United States	×		
INPADOC IPG	[c]	all	47 countries, EPO and WPO	×		
Pergamon PATSEARCH	1971	all	United States			×
Pergamon PATCLASS	1836	all (class index only)	United States			×

[a] Country coverage for Primary Files and Database supplier for Other On-line Files. EPO-European Patent Office; WPO = World Patent Organization.

[b] LMSC = Lockheed Missiles & Space Company, Inc. SDC = System Development Corporation. BRS = Bibliographic Retrieval Services, Inc.

[c] Most recent six weeks.

year, along with references to over 200,000 equivalents. A full range of patent-information services is offered, including a number of printed bibliographic and abstract publications, associated printed indexes, on-line searchable files, and a search-bureau operation. The printed publications include the *World Patents Index* (WPI), the *World Patent Abstracts* (WPA), the *Central Patents Index* (CPI), and the new *Electrical Patents Index* (EPI). They are organized on the basis of country of origin and/or subject content.

It is generally agreed that Derwent services offer excellent general-purpose alerting and retrospective-search capabilities.

Derwent publishes the *WPI Gazette* and provides alerting abstract services through their *World Patent Abstracts* (WPA) and Country-Order Abstract Booklets. Derwent also prepared cumulative indexes for its CPI, WPI, and EPI services. For retrospective retrieval purposes, Derwent has devised a number of detailed coding schemes including manual codes, punch codes, Ring Index codes, and PLASDOC key terms.

In addition to the alerting abstracts mentioned above, Derwent provides comprehensive, documentation-type abstracts for its chemical patent services, eg, the *Basic Abstract Journal Profile Booklets*, and *Manual-Code Cards*. They cover the basic patents entered into the CPI system. They are available two weeks after the country-alerting abstracts, or about two months after publication of the patent document.

Essentially all Derwent files are searchable on line through facilities of the System Development Corporation. Derwent also makes available microfilm copies of the complete specifications of all basic documents covered in the CPI system except Japanese.

IFI / Plenum Data Company. IFI/Plenum Data Company has provided patent-information services since 1952. During the 1950s, IFI's sole product was the *Uniterm Index to United States Chemical Patents*, comprising a dual-dictionary-type keyword index in book form and a companion volume containing OG claims or abstracts of the accessioned patents. Today IFI produces, in addition the Uniterm book index, two variations of a magnetic-tape index, a number of printed publications including the *IFI Assignee Index* and the *Patent Intelligence and Technology Report*, and several data bases for on-line searching. The latter covers all U.S. patents back to 1963 and U.S. chemical patents to 1950. In addition, IFI provides weekly customized profiles on U.S. patents and a number of other search services.

The International Patent Documentation Center (INPADOC). The International Patent Documentation Center (INPADOC) was founded in 1972 following consultations between the World Intellectual Property Organization and the Austrian Government. An agreement was signed on May 2, 1972, and entered into force on June 22, 1973, after ratification by the Austrian Parliament.

The industrial property offices of the world collectively issue nearly one million (10^6) patent documents a year. Each document contains bibliographic-data items that can be used to identify, classify, and retrieve the document. The general task of INPADOC is to record, in a computer-readable form, the significant bibliographic-data items of patent documents as soon as they are published and then to manipulate the recorded information to provide information services.

Other secondary patent-information sources. Secondary sources include *Research Disclosure*, *Patent Abstracts of Japan*, Rapid Patent International, Search Check, Inc., National Technical Information Service (NTIS), TNO, Nippon Gijutsu Boeki Co., Ltd. (NGB), Polyresearch Service, American Petroleum Institute (API), the Rubber and Plastics Research Association of Great Britain (RAPRA), Research Publications, Inc., and J. Gevers et Cie.

International Aspects

The beginnings of international cooperation in matters of inventions and other intellectual property go back to 1883, and the adoption of the Paris Convention for the Protection of Industrial Property. Since then, numerous meetings have been held, and a number of treaties and agreements have been signed in relation to specific aspects of intellectual-property protection. In 1967, the World Intellectual Property Organization (WIPO) was established. This body was created in 1970 for the purpose of uniting the separate intellectual-property unions that had

been established previously or that might be established in the future. More recent developments of importance to patent literature were the adoption of the Patent Cooperation Treaty (PCT) in 1970 and the International Patent Classification Agreement in 1971. In December of 1974, WIPO became a specialized agency of the United Nations, and its secretariat is known as the International Bureau. This bureau is the administrative arm of WIPO and has direct responsibility for the activities among the various intellectual property unions. With respect to patents, the International Bureau acts as the administrative coordinator for the Paris, PCT, and IPC Unions.

JOHN W. LOTZ
IFI/Plenum Data Company

International Table to Patents, Designs, and Trade Marks, 69th ed., H. Scheer-Verlag, Hürth-Efferen/Köln, FRG, 1979 (wall chart, 2 parts).

R.E. Maizell, *How to Find Chemical Information*, John Wiley & Sons, Inc., New York, 1979, p. 127.

World Intellectual Property Organization, General Information, WIPO Publication No. 400 (E), Geneva, Switz., 1979.

Manual for the Handling of Applications for Patents, Designs and Trademarks Throughout the World, Octrooibureau Los en Stigter, Amsterdam, Neth., 1978 (looseleaf).

International Patent Classification (manual), 3rd ed., WIPO, Carl Heymanns Verlag KG, Munich, FRG, 1979.

PEAT.   See Lignite and brown coal.

PECTIC SUBSTANCES.   See Gums.

PELLETING AND BRIQUETTING.   See Size enlargement.

PENICILLINS.   See Antibiotics, β-lactams.

PENTAERYTHRITOL, C(CH$_2$OH)$_4$.   See Alcohols, polyhydric.

PENTANES.   See Hydrocarbons, C$_1$–C$_6$.

PEPTIDE ANTIBIOTICS.   See Antibiotics, peptides.

PEPTIDES.   See Biopolymers; Hormones; Polypeptides; Proteins.

PERCHLORIC ACID, PERCHLORATES.   See Chlorine oxygen acids and salts.

PERCHLORO COMPOUNDS.   See Chlorocarbons and chlorohydrocarbons.

PERFLUORINATED IONOMER MEMBRANES

The term perfluorinated ionomer membranes generally refers to ion-exchange membranes composed of perfluorinated polymeric backbones (see Membrane technology). Nafion membranes, which are made of a perfluorinated sulfonic acid ionomer (XR resin), opened the way to a new electrolytic process for chlor-alkali production using perfluorinated ion-exchange membranes (see Alkali and chlorine products). Flemion is a carboxylic acid type. The different ion-exchange groups greatly affect membrane properties.

$$+CF_2CF_2 +_x CF_2CF-$$
$$[OCF_2CF +_3 OCF_2CF_2SO_2F$$
$$CF_3$$

Nafion polymer

$$+CF_2CF_2 +_x CF_2CF-$$
$$[OCF_2CF]_m O (CF_2)_n CO_2CH_3$$
$$CF_3$$

(m = 0 or 1, n = 1–5)

Flemion polymer

Both polymers are melt-processable and can be fabricated into films by extrusion-molding. These films can be easily converted to the corresponding ion-exchange membrane by alkaline hydrolysis.

Preparation

The general procedure includes synthesis of a perfluorovinyl ether moiety with a functional group, its copolymerization with tetrafluoroethylene in the presence of a radical initiator in an aqueous or inert organic medium, and the formation of a membrane.

Fabrication. The crystallinity of the copolymer depends upon the content of the functional conomonomer. Amorphous or partly crystalline copolymers are fabricated into films (100–250 μm thick) with conventional extrusion techniques. The films are usually reinforced with Teflon cloth and converted to sulfonic- or carboxylic-acid-type ion-exchange membranes by alkaline hydrolysis.

A sulfonic acid group can be converted to a carboxylic acid group:

$$\cdots -O(CF_2)_n SO_2 X \rightarrow \rightarrow \cdots -O(CF_2)_{n-1}CO_2 H$$
$$(X = Cl \text{ or } F)$$

The sulfonyl halide group is converted to sulfinic acid by reduction and then the carboxylic acid group, having one CF$_2$ less than the original chain of sulfonic acid, is formed through a desulfonylation reaction.

Applications

In the electrolysis of brine, a cation-exchange membrane is used. DuPont has developed a variety of Nafion series. The Nafion 300 series produces 10–20% caustic soda. For the production of 20–28% caustic soda, the Nafion 200 series was developed. The Nafion 900 series membranes are carboxylate–sulfonate two-layer membranes with ca 95% current efficiency at 33% caustic soda.

Asahi Glass has developed the Flemion series. For the production of 35% caustic soda, a standard Flemion 230 is used advantageously with a current efficiency of 94%. With the Flemion 700 series, gas bubbles can be removed easily from the membrane surfaces.

A new electrolytic process with a zero-gap cell, called the AZEC system, combined with Flemion 723 or 753 and a new electrode system, has resulted in drastic reductions in energy consumption.

Asahi Chemical Tokuyama Soda improved the electrolytic performance of Nafion-type membranes by chemical modification of the cathode-side surface of the carboxylic acid-type membrane.

MASAAKI YAMABE
Asahi Glass Company, Ltd.

A. Eisenberg and H.L. Yeager, eds., *Perfluorinated Ionomer Membranes*, *ACS Symposium Series 180*, American Chemical Society, Washington, D.C., 1982.

H. Ukihashi, *Chemtech*, 118 (Feb. 1980).

M. Nagamura, H. Ukihashi, and O. Shiragami, *paper presented at the Symposium on Electrochemical Membrane Technology in 1982 AIChE Winter Meeting*, Orlando, Fla., 1982.

PERFLUORO COMPOUNDS. See Fluorine compounds, organic.

PERFUMES

Perfumery is the art of producing fragrances through the combination of odoriferous substances. The word perfume is derived from the Latin meaning "through smoke". Throughout history, perfumes have played an important role in human lives, and have been associated with notions of happiness, beauty, and satisfaction. Until this century, fragrance materials have been derived from natural sources, which has placed limitations on odor types and markets (see also Cosmetics; Odor modification). The increased use of perfumes in the last thirty years would have been impossible without the development of the chemistry that allowed the invention of totally new odoriferous molecules as well as the synthesis of natural ones. Fragrances are no longer a luxury for the rich but today are incorporated routinely in a great number of products that are in daily use.

Fragrance Raw Materials

Natural products. Essential oils are volatile materials produced from odorous plant material, generally by water or steam distillation or by expressing (see Oils, essential).

A concrete is an extraction almost exclusively from vegetable origin such as leaves, bark, flowers, and fruit. This is normally obtained by extraction with hydrocarbon solvents.

Absolutes are the alcohol-soluble portion of concretes, obtained by extracting the concretes with alcohol. Resinoids are perfume materials obtained by extraction of plant resinous substances with hydrocarbon solvents.

Tinctures are alcoholic solutions. In perfumery, these are generally the solutions obtained by maceration of various odorous materials with alcohol.

Natural products used in perfume include ambergris, benzoin, castoreum, civet, clove leaf oil, galbanum, jasmine absolute, labdanum, maté, melilot, mimosa, musk tonquin, myrrh, oakmoss or mousse de chêne, olibanum, opopanax, orris, patchouli, rosemary oil, sandalwood oil, vetivert oil, and violet leaves absolute.

Aroma Chemicals

During the last 20 years, there has been a rapid advance in the capabilities of instrumental techniques for the separation and identification of volatile organic substances. Of particular importance to the perfumery industry was the development of capillary gas chromatography columns and the ability to use them directly in tandem with a mass spectrometer. Computer technology is used to interpret the vast amount of data generated by such a combination of instruments. These developments along with Fourier transform nmr spectroscopy have allowed discovery and identification of extremely minute odoriferous samples and have revolutionized not only the analysis of essential oils and extractives but also the direction of the synthesis of aroma chemicals.

Research in aroma chemicals can be divided into three general categories: (1) duplication of naturally occurring chemicals, for example, phenethyl alcohol, which occurs in rose oil; (2) chemical modification of abundant, naturally occurring materials, eg, acetylated vetivert oil ("vetivert acetate") from vetivert oil, and vanillin (qv) from lignin (qv); and (3) synthesis based on industrial organic feedstocks, eg, nitro musks.

Aroma chemicals are usually cheap and available in any needed quantity (see also Alcohols, higher aliphatic; Aldehydes; Benzaldehyde; Benzoic acid; Cinnamic acid; Cinnamaldehyde; Cinnamyl alcohol; Coumarin; Esters, organic; Indole; Ketones; Salicyclic acid and related compounds; Terpenoids; Vanillin).

Odor Vocabulary

The descriptions and groups of fragrance raw materials are helpful in evaluating existing aroma chemicals or newly developed materials. To illustrate the use of the odor vocabulary, two well-known materials are described below:

Citronellol   Lyral

Classification of Perfumes

Nature offers the perfumer an unlimited number of models. The impressions, however eccentric and exclusive, always have a recognizable theme and category and are easily identified and described: straight florals/Muguet des Bois (Coty 1936); floral bouquet/Joy (Paou 1930); aldehydic floral/Chanel No. 5 (Chanel 1921), oriental/Shalimar (Guerlain 1925); chypre/Ma Griffe (Carven 1947); woody/Chamade (Guerlain 1970); green/Bill Blass (Blass 1978); citrus/Quartz (Molyneux 1977); fougere/Maja (Myrurgia 1921); Canoe/Ambush (Dana 1969); musk/Musk Oil (Caswell-Massey 1950); animal/Civet (Jovan 1973); and leather/Tabac Blond (Caron 1919).

Men's perfumes can be classified in the following categories: green/Old Spice Herbal (Shulton 1974); citrus/Aqua Velva (Williams 1917); fougere/Fougere Royal (Houbigant 1822); Canoe/Canoe (Dana 1935); spice/Old Spice (Shulton 1937); woody/Deep Woods (Avon 1971); herbal/Polo (Lauren 1978); musk/Musk for Men (Yardley 1971); leather/Ted (Lapidus 1978); oriental/Skin Bracer (Mennen 1931); and chypre/Paco (Paco Rabanne 1973).

Manufacture

Compounding of fragrances. A compounding department is a scaled-up version of the perfumers' laboratory. All equipment in a modern compounding facility is made of stainless steel. The formula is usually rewritten into a working formula which means adding crystals first, more viscous materials second, and then part of the liquids until the mixture is homogeneous. The rest of the liquids are added, and then the valuable, delicate, and often more volatile materials are added. The time required for mixing usually is several hours, because crystals must dissolve without heating.

Quality control in perfumery. All fragrance raw materials exist in many different qualities. The name or description of any essential oil, natural product, isolate, or fragrance chemical has little meaning except to aid in identification. The odor quality of a given product becomes the most important criterion in determining its acceptability according to an established quality. Although there are other physical, chemical, and analytical tests, they have little significance unless the primary requirement of odor quality is satisfied.

The olfactory evaluation of fragrance materials is done on separate odorless slips of blotting paper that are dipped into the standard and the sample simultaneously and to the same depth so that the same amount of material is on both blotters.

The creation of fragrances. There are more than 6000 raw materials at the disposal of the perfumer. It is important to start with a common idea inspired by a new synthetic, a new natural material, or a newly discovered odor facet of an existing material. The perfumer creates an accord based on the inspiring note. The accord or theme can be used in an existing floral or woody floral fragrance, or the perfumer can create a floral composition around the new original theme. A fine fragrance must work on the skin because it has to blend with body odor. It also has to last, and be pleasant and diffusive. It must have the quality of genuine beauty and a signature that distinguishes it from any other fragrance.

A fragrance odor can be analyzed as a blend with a top note continuing into a middle note to a final or end note. The materials with the greater vapor pressures evaporate faster. What the perfumer must do is to smooth the odor profile so that there are no discontinuities in odor impact as the different components volatilize.

Industrial Perfumery

Fragrances play an important role in detergents, soaps, liquid fabric softeners, tumble-dryer softeners, bleach products, shampoos, deodorants and antiperspirants, and talcs and powders.

Fragrance Safety

The fragrance industry has a long and clean record of safety, largely because of the nature, source, and methods of use of its products. However, with the wider-spread application and daily exposure to fragrances in all manner of products, companies have, in recent years, devoted much attention to a reappraisal of their safety criteria. Apart from some national regulations, the companies operate under guidelines issued over the last ten years by the International Fragrance Association (IFRA). This association bases its guidelines on the results of tests carried out by the Research Institute for Fragrance Materials (RIFM) which is supported by 97% of the perfume industry in the United States, Europe, and Japan. As a minimum, the following tests are carried out: acute oral toxicity, skin irritation and sensitization potential, and, where significant, eye irritation, phototoxicity and photosensitization tests.

WILLIAM I. TAYLOR
BERNARD CHANT
GUS VAN LOVEREN
International Flavors & Fragrances

P.Z. Bedoukian, *Perfumery and Flavoring Synthetics*, D. Van Nostrand Co., Inc., Princeton, N.J., 1967.

W.A. Poucher, *Perfumes, Cosmetics and Soaps*, Vols. 1–3, Chapman & Hall, London, UK, 1974.

W. Triebs, *Die Atherischen Ole*, Vols. I–VII, Akademie Verlag, Berlin, Ger., 1956–1961.

PERIODIC ACID, PERIODATES.   See Iodine and iodine compounds.

PERMANGANATES.   See Manganese compounds.

PEROXIDES AND PEROXY COMPOUNDS, INORGANIC

A peroxide or peroxide compound contains at least one pair of oxygen atoms bonded by a single covalent bond. Each oxygen atom has an oxidation number of -1. Peroxides form hydrogen peroxide upon solution in water, eg,

$$Na_2O_2 + 2 H_2O \rightarrow 2 NaOH + H_2O_2$$

All peroxides are strong oxidizing agents; most are classified as oxidizers by the DOT and shipping containers must be so labeled.

Group IA Peroxides

All alkali metals form peroxides and all Group IA peroxides, except lithium peroxide, can be synthesized by reaction of the metal with oxygen at atmospheric pressure. Only sodium and lithium peroxides are available commercially.

Lithium peroxide, a colorless, thermally stable solid, is made by treating lithium oxide hydrate with hydrogen peroxide.

Sodium peroxide, a pale-yellow solid, mp ca 600°C, reacts with water liberating 142 kJ/mol (34 kcal/mol) of heat; hydrogen peroxide and sodium hydroxide are formed. Hydrogen peroxide addition compounds, eg, $H_2O_2 . x Na_2O_2$, have been reported.

Sodium peroxide is produced commercially by oxidizing molten sodium with oxygen-enriched air at ca 300–400°C. Because of its corrosiveness to metals, it is produced in a nickel-alloy reactor coated with graphite and equipped with a zirconium stirrer. Zirconium crucibles are recommended for analytical work requiring the use of sodium peroxide.

Sodium peroxide is irritating to the skin, eyes, and mucous membranes. Carbon dioxide or dry-chemical extinguishers are recommended for sodium peroxide fires. The principal use of sodium peroxide is as a bleaching agent (qv).

Potassium, rubidium, and cesium peroxides are not prepared by direct reaction of the elements because of superoxide formation. They are obtained by oxidizing the metal in liquid ammonia with oxygen or air or by controlled decomposition of the superoxide.

Group IIA Peroxides

All elements of the Group IIA elements, except beryllium and radium, form peroxides. They cannot be synthesized by direct oxidation of the metals. Barium and strontium peroxides are made by direct oxidation of the oxides; the other peroxides are obtained by reaction of the hydroxides with hydrogen peroxide. Group IIA peroxides are more stable in the presence of moisture than the Group IA peroxides, because they are insoluble in water.

Magnesium peroxide is an off-white solid similar to magnesium oxide in appearance and physical properties. Calcium peroxide is a yellow-white powder with good thermal stability. It is made by treating calcium hydroxide with dilute hydrogen peroxide in stainless steel, glass, or plastic equipment. It is used as a dough conditioner and disinfectant for seed and grain.

Both strontium and barium peroxides are light-colored stable solids, made by oxidation of the oxides. Both are irritants and should be handled with care. Their principal use is in pyrotechnic and tracer-bullet formulations (see Pyrotechnics; Explosives and propellants).

Group IIIA Peroxides

Boron is the only element in this group from which well-defined peroxides have been prepared. X-ray studies have shown the following structure of the peroxyborate anion:

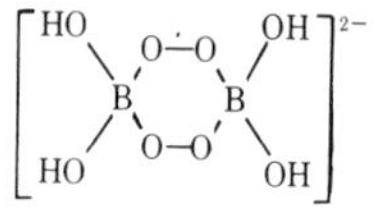

$$\begin{bmatrix} \text{HO} & \text{O}-\text{O} & \text{OH} \\ & \text{B} \quad \text{B} & \\ \text{HO} & \text{O}-\text{O} & \text{OH} \end{bmatrix}^{2-}$$

The alkaline and alkaline-earth metals all form peroxyborates with varying degrees of hydration. The mono-, tri-, and tetrahydrates of sodium peroxyborate are all colorless, crystalline solids. The tetrahydrate is produced commercially by the reaction of a borax solution with sodium hydroxide and hydrogen peroxide in stainless-steel equipment. It is irritating to eyes and mucous membranes. It is used as a carrier of dry hydrogen peroxide and as a bleaching agent; the monohydrate is used in denture and hard-surface cleansers.

Group IVA Peroxides

The carbon peroxides, eg, sodium carbonate peroxyhydrate and sodium and potassium peroxymonocarbonates, are the only well-defined peroxides of Group IVA elements, although peroxides of silicon, germanium, and tin have been reported and the chemistry of organometallic and organometalloid peroxides has been described.

Group VA Peroxides

Only phosphorus peroxides have been isolateed; nitrogen peroxides, such as NO_3 and N_2O_7 have been proposed as have been peroxynitric acid, HNO_4, and peroxynitrous acid, HNO_3.

Peroxymonophosphoric acid (active-oxygen content 14.0 wt%) is a viscous, colorless liquid; its structure probably resembles that of peroxymonosulfuric acid:

$$\begin{matrix} \text{O} \\ \| \\ \text{HOPOOH} \\ | \\ \text{HO} \end{matrix} \qquad \begin{matrix} \text{O} \\ \| \\ \text{HOSOOH} \\ \| \\ \text{O} \end{matrix}$$

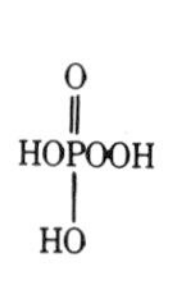

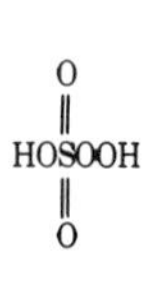

Tetrapotassium peroxydiphosphate and sodium pyrophosphate peroxyhydrate are both colorless crystalline solids. A large number of alkaline or neutral ortho-, pyro-, tripoly-, and trimetaphosphate peroxyhydrates have been prepared.

Group VIA Peroxides

In Group VIA peroxides, the hydrogen atoms of hydrogen peroxide are replaced by a nonmetallic group.

Peroxymonosulfuric acid, Caro's acid, H_2SO_5 (14% active-oxygen content), is a strong oxidizing agent:

$$H_2SO_5 + 2\,H^+ + 2\,e \rightarrow H_2SO_4 + H_2O \qquad E° = -1.44\,V$$

A 30–40% solution is prepared by adding concentrated hydrogen peroxide to concentrated sulfuric acid. Glass, Teflon, and stainless steel are recommended as equipment material for production and handling. Caro's acid is sold in Europe as a 15–20% solution for wastewater treatment.

A number of alkali salts have been reported; only the triple salt, $2KHSO_5 \cdot KHSO_4 \cdot K_2SO_4$ (5.21% active-oxygen content), is available commercially.

Oxone peroxymonosulfate is a white, granular, water-soluble powder. It is used as an oxidizing agent and initiator of free-radical polymerization of vinyl monomers, as well as a bleaching agent and disinfectant (see Initiators; Disinfectants).

Peroxydisulfuric acid, also referred to as persulfuric acid or Marshall's acid, is a colorless crystalline solid, mp 65°C (dec). Solutions are stable at low temperatures, but hydrolyze rapidly when heated at low pH. The peroxysulfate ion decomposes into free radicals, which are initiators (qv) for numerous chain reaction:

$$S_2O_8^{2-} \xrightarrow{\text{heat}} 2\,SO_4^-$$

thus, peroxydisulfate salts are used as initiators for emulsion-polymerization reactions. Peroxydisulfuric acid is prepared by the electrolysis of sulfuric acid. Numerous salts have been prepared, but only the disodium, diammonium, and dipotassium salts are made commercially. They are all crystalline colorless solids that irritate the skin.

Large quantities of diammonium peroxydisulfate are used in metal-cleaning and etching solutions, especially for dissolving copper in the manufacture of electronic circuit boards.

Group IIB peroxides. The Group IIB transition-metal oxides are similar to the alkaline-earth peroxides. They are produced by treating the metal oxide with hydrogen peroxide. Cadmium and zinc peroxides have been prepared, the latter commercially.

Zinc peroxide is a yellow solid, similar to magnesium peroxide. It is stable at room temperature but decomposes above 150°C; at 190–212°C, the decomposition is rapid enough to cause an explosion.

Group IIIB peroxides. Scandium, yttrium, and lanthanide ions form unstable, poorly characterized 1:1 metal–peroxide adducts ($M_2O_3 \cdot H_2O_2$) from the metal hydroxide and hydrogen peroxide.

Group IVB peroxides. Several solid peroxy derivatives of titanium have been isolated, eg, from neutral solution a yellow 1:1 adduct in the form of $TiO_3 \cdot 2H_2O$. Similar compounds can be prepared from cerium, zirconium, hafnium, and thorium.

Group VB peroxides. The number of peroxy groups per number of vanadium atoms increases with alkalinity. A red cation, probably $V(O)(O_2)^+$, forms in dilute acid at low hydrogen peroxide concentration; with excess peroxide, the yellow diperoxy anion, $V(O)(O_2)_2^-$, forms. Peroxy complexes are presumed to be the active agents in vanadium catalysts.

The compounds M_3NbO_8 and M_3TaO_8 are formed from niobium and tantalum oxides, respectively, in basic solutions containing excess peroxide.

Group VIB peroxides. In alkaline solution, chromate ions react with hydrogen peroxide, yielding M_3CrO_8, where M is a monovalent cation.

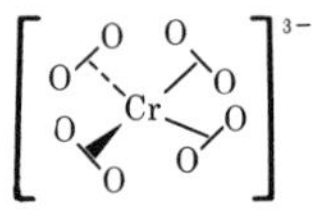

$$\begin{bmatrix} & \text{O} \quad \text{O} & \\ \text{O} & \diagup \quad \diagdown & \text{O} \\ & \text{Cr} & \\ \text{O} & \diagdown \quad \diagup & \text{O} \\ & \text{O} & \end{bmatrix}^{3-}$$

Other metals forming this type of compound are tetravalent titanium and zirconium, pentavalent vanadium, niobium, and tantalum, and hexavalent molybdenum and tungsten.

Table 1. Dissociation Pressure of Potassium Superoxide (KO₂)

temperature, °C	198	279	461	539	589
O₂ pressure, kPa^a	4.0	14.8	43.1	58.8	73.3

^{a} To convert kPa to mm Hg, multiply by 7.5.

Group VIII peroxides. There are two types of 1:1 bonding of the dioxygen ligand to the transition-metal complex:

Such peroxy or dioxygen compounds are formed by the reaction of the metal complex with molecular oxygen. They oxidize various substrates, eg, sulfur dioxide, carbon disulfide, nitrogen dioxide, ketones, and aldehydes under mild conditions. The compounds $[(C_6H_5)_3P]_2PdO_2$ and $[(C_6H_5)_3P]_2PtO_2$ epoxidize olefins. The dioxygen compounds of iron are involved in the transport of oxygen in the blood.

Actinides. Peroxides of most actinide elements are known. In most cases they are formed in solution or as solids of indefinite composition; an exception is crystalline $Th(O_2)SO_4.3H_2O$.

Uranium in the +6 oxidation state forms a variety of peroxy compounds, eg, $UO_4.4H_2O$, $U_2O_7.xH_2O$, and $M_4UO_8.xH_2O$. Other peroxy uranates in solution and in the solid state have been postulated.

Superoxides

The superoxides are characterized by the O_2^- ion. The metal oxides are yellow-to-orange solids. They are strong oxidizing agents, that react vigorously with most organic matter and reducing agents. Oxygen and hydroperoxide are evolved in the reaction with water:

$$2\,KO_2 + H_2O \rightarrow 2\,K^+ + OOH^- + OH^- + O_2$$

Sodium superoxide is thermally unstable at ambient conditions. It is prepared in reasonably high purity from oxygen and sodium peroxide at 490°C and 30.2 MPa (298 atm).

Potassium superoxide, mp 450–500°C, is paramagnetic $[(1.9 \times 10^{-23}$ J/T) (2.04 μ_β)]. Oxygen-dissociation pressures are given in Table 1. It is used in self-contained breathing equipment (see Oxygen generation).

Ozonides

The ozonides are characterized by the presence of the O_3^- ion. They are generally produced by the reaction of the inorganic oxide with ozone. The sodium, potassium, rubidium, and cesium ozonides have been reported. The ammonium and tetramethylammonium ozonides have been prepared at low temperatures. Inorganic ozonides, although not produced commercially, are of potential importance as solid-oxygen carriers in breathing apparatus (see Ozone).

RICHARD E. HALL
FMC Corporation

I.I. Vol'nov, *Peroxides, Superoxides, and Ozonides of Alkali and Alkaline Earth Metals*, Plenum Press, New York, 1966.

W.C. Schumb, C.N. Satterfield, and R.L. Wentworth, *Hydrogen Peroxide*, Reinhold Publishing Corporation, New York, 1955.

D.A. House, *Chem. Rev.* **62**, 185 (1962).

J.A. Connor and E.A.V. Ebsworth in H.J. Emeleus and A.G. Scharpe, eds., *Advances in Inorganic Chemistry and Radio-Chemistry*, Vol. 6, Academic Press, Inc., New York, 1964, p. 279.

PEROXIDES AND PEROXY COMPOUNDS, ORGANIC

Organic peroxides are derivatives of hydrogen peroxide, HOOH, wherein one or both hydrogens are replaced by an organic group, ie, ROOH or ROOR. The source of the O—O linkage is oxygen. Organic peroxides are prepared by direct air oxidation or by reactions of organic compounds with peroxide materials derived from oxygen, eg, hydrogen peroxide, alkali-metal peroxides, ozone (qv), or other organic peroxides. Organic peroxides are intermediates in the air oxidation of many synthetic and naturally occurring organic compounds. They are involved in the development of rancidity in fats, loss of activity in certain vitamin products, gum formation in lubricating oils, prepolymerization of some vinyl monomers, and biological processes. Air oxidation of certain solvents, especially of cyclic and noncyclic ethers, results in the formation of organic peroxides, which when concentrated, can result in highly explosive residues. Oxidation inhibitors are used to prevent formation of undesirable peroxides.

The initially generated species, ie, the free radicals, from the oxygen–oxygen bond homolysis are reactive intermediates with very short lifetimes, eg, $t_{1/2} < 10^{-3}$ s. They quickly undergo a variety of reactions and form stable products and are used as initiators for many free-radical reactions, especially for polymerizations. Organic peroxides also undergo nonradical-forming reactions, eg, heterolysis, hydrolysis, and rearrangement. The classification according to structure is given in Table 1; properties of various peroxides are given in Table 2.

Hydroperoxides

Hydroperoxides are classified into alkyl hydroperoxides, ROOH, and organomineral hydroperoxides, $R_mQ(OOH)_n$, where Q is silicon, germanium, tin, or antimony. The alkyl hydroperoxides are liquids or low melting solids; the lower molecular weight compounds are soluble in water and are explosive. The alkyl hydroperoxides are stronger acids than the corresponding alcohols, and are isolated and purified as the alkali-metal salts.

Table 1. Classification of Organic Peroxides

Class	Structures or characteristic group
hydroperoxides	ROOH $R_mQ(OOH)_n$ (Q = metal or metalloid)
α-oxy- and α-peroxy-hydroperoxides and peroxides	contain the grouping: C(OO⁻)(O⁻)
peroxides	ROOR′ $R_mQ(OOR)_n$ R_mQOOQR_m
peroxyacids	$R(CO_3H)_n$ RSO_2OOH
diacyl peroxides	RCIICR′, ROCOOCOR, RSOCR, RSOOSR, RCOOCOR′ (with O's)
peroxyesters	$R(CO_3R')_n$, $R'(O_3CR)_2$, ROCOOR′, ROOCOOR, NCOOR, RSOOR′

Table 2.   Properties of Organic Peroxides

Hydroperoxide	Structure	Bp, °C$_{kPa}$[a]	Mp, °C	n_D^{20}
methyl hydroperoxide	CH_3OOH	45.5–46.5$_{24.53}$		1.3654^l
ethyl hydroperoxide	C_2H_5OOH	43–44$_{6.67}$		
isopropyl hydroperoxide	i-C_3H_7OOH	38–38.5$_{2.67}$		
n-butyl hydroperoxide	n-C_4H_9OOH	40–42$_{1.07}$		1.4057
sec-butyl hydroperoxide	sec-C_4H_9OOH	41–42$_{1.47}$		1.4050
t-butyl hydroperoxide	t-C_4H_9OOH	33–34$_{2.27}$	4.0–4.5	1.3983^c
cumene hydroperoxide	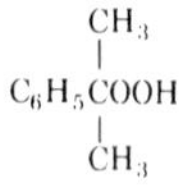 $C_6H_5\overset{\underset{\mid}{CH_3}}{\underset{\underset{\mid}{CH_3}}{C}}OOH$	60$_{0.027}$		1.5242
1,1-dioxybiscyclohexanol	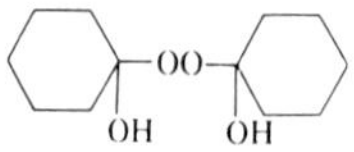		69–71	
1,2,4-trioxolane	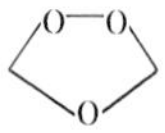	18$_{2.13}$		
3,3-dimethyl-1,2,4-trioxolane	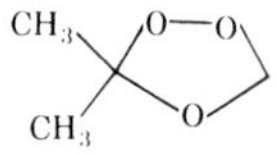	42–42.5$_{18.67}$		
diethyl peroxide	$(C_2H_5O)_2$	62–63$_{101.32}$		
di-t-butyl peroxide	$(t$-$C_4H_9O)_2$	109$_{101.32}$	−18	
9,10-dihydro-9,10-epidoxy anthracene	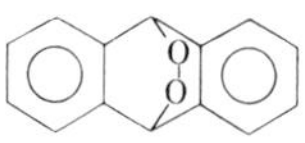		120^b	
2,5-dimethyl-2,5-di(t-butylperoxy)hexane	$(t$-$C_4H_9OOC(CH_3)_2CH_2)_2$	42$_{0.008}$	8	
dicumyl peroxide	$(C_6H_5C(CH_3)_2O)_2$		40–41	
[1,4-phenylenebis(isopropylidene)]bis[(t-butyl)]peroxide	$1,4$-$(t$-$C_4H_9OOC(CH_3)_2)_2C_6H_4$		79	
performic acid	HCO_3H	50$_{13.33}$	−18^b	
peracetic acid	CH_3CO_3H	25$_{1.60}$	0	
perbenzoic acid	$C_6H_5CO_3H$		41–42	
m-chloroperbenzoic acid	3-$ClC_6H_4CO_3H$		112	

^{a} To convert kPa to mm Hg, multiply by 7.5.
^{b} At 21°C.
^{c} At 25°C.

Alkyl hydroperoxides are readily reduced to the corresponding alcohols; many such reductions are quantitative and applicable for analytical use. Alkyl hydroperoxides are used as oxidizing or hydroxylating agents in organic synthesis.

Hydroperoxides react with or without cleavage of the O—O bond. They are thermally sensitive and photosensitive and undergo initial O—O bond homolysis on heating. They are, however, attacked by free radicals:

$$ROOH + R'O\cdot \rightarrow ROO\cdot + R'OH \text{ (H abstraction)}$$

$$ROOH + R'\cdot \rightarrow R'OH + RO\cdot \text{ (displacement on O—O)}$$

Hydroperoxides are readily decomposed by multivalent metal ions, eg, Cu, Co, Fe, V, Mn, Sn, Pb, etc, by an oxidation-reduction or electron-transfer process. The ultimate fate of the oxyradicals that are generated from alkyl hydroperoxides depends upon the decomposition environment. Decomposition-rate studies indicate that alkyl hydroperoxides have 10-h half-life temperatures of 133–172°C. They are among the most stable of the organic peroxides.

Hydroperoxides are prepared from hydrogen or sodium peroxide, molecular oxygen, ozone, or other organic peroxides. Hydrogen peroxide, a powerful nucleophilic reagent, reacts with suitable alkyl compounds, RX, where X is electronegative or R is electropositive, at the carbon joined to X:

$$H_2O_2 + RX \rightarrow ROOH + HX$$

Many hydroperoxides are prepared by autoxidation of suitable substrates with molecular oxygen. These reactions are free-radical chain or nonchain processes, depending on whether triplet or singlet oxygen is involved. The free-radical process consists of three stages:

initiation:	$RH \rightarrow R\cdot + (H\cdot)$
propagation:	$R\cdot + O_2 \rightarrow ROO\cdot$
	$ROO\cdot + RH \rightarrow ROOH + R\cdot$
termination:	$2\,R\cdot \rightarrow RR$ or $RH + R$ (olefin)
	$ROO\cdot + R\cdot \rightarrow ROOR$
	$2\,ROO\cdot \rightarrow O_2 + ROOR$ or alcohol and carbonyl compound
	$ROO\cdot + AH \rightarrow ROOH + A\cdot$ (nonpropagating)

α-Oxy- and α-Peroxyhydroperoxides and Peroxides

The α-oxy- and α-peroxyhydroperoxides and peroxides are represented by the structures shown in Figure 1.

α-Hydroxyhydroperoxides are moderately stable and many can be distilled without decomposition. Short-chain alkyls, especially hydroxymethyl hydroperoxide, can decompose violently. α-Hydroxyhydroperoxides tend to lose hydrogen peroxide and form α,α′-dihydroxyperoxides (5), especially in the presence of water:

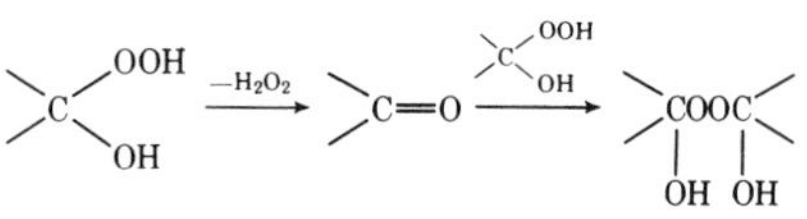

The α-oxy- and α-peroxyhydroperoxides and peroxides are composed of the following structures:

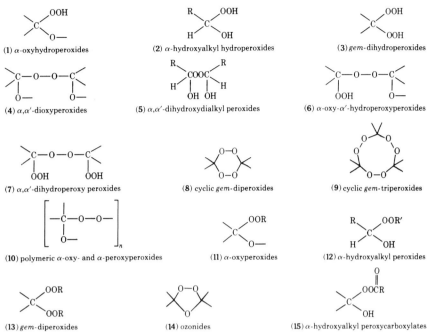

Figure 1. α-Oxy- and α-peroxyhydroperoxides and peroxides.

The short-chain alkyl *gem*-dihydroperoxides (**3**) are soluble in water and are explosive when pure. Thermal decomposition results in initial homolytic cleavage of the O—O bond followed by cleavage of the C—O and C—C bonds yielding a mixture of products, eg, ketones, esters, carboxylic acids, hydrocarbons, and hydrogen peroxide.

The lowest carbon member dihydroxy peroxide, $HOCH_2OOCH_2OH$, is a dangerously explosive solid. The higher members are liquids and, with increasing molecular weight, are solids of decreasing explosive nature and water solubility.

The short-chain alkyl α,α'-dihydroperoxy peroxides (**7**) are highly explosive when pure. Cyclic *gem*-diperoxides (**8**) and *gem*-triperoxides (**9**) are solids and the short-chain alkyl members are shock sensitive and explosive. Acid hydrolysis regenerates the carbonyl compound and hydrogen peroxide and it seems that the cyclic *gem*-diperoxide is formed from the cyclic *gem*-triperoxide. Thermal and photochemical decompositions of cyclic *gem*-diperoxides and *gem*-triperoxides derived from ketones yield macrocyclic hydrocarbons and lactones.

α-Hydroxyalkyl hydroperoxides (**2**) are stable and can be distilled in vacuum; the short-chain alkyl members, eg, hydroxydimethyl peroxide, are unstable and explosive.

gem-Diperoxides (**13**) (diperoxy ketals) are colorless liquids, soluble in organic solvents, but insoluble in water; those derived from *t*-alkyl hydroperoxides are excellent free-radical initiators (qv).

In general, pure ozonides (**14**) are stable in storage and some can be distilled in vacuum. Impure ozonides usually contain heat- and shock-sensitive peroxide contaminants and are highly explosive.

Polymeric α-oxy- and α-peroxyperoxides (**10**) are viscous liquids or amorphous solids which are very prone to explosive decomposition. They are prepared by the addition of hydrogen peroxide to the carbonyl groups of aldehydes and ketones, by the autoxidation of alcohols, ethers, and aldehydes, and by the ozonization of unsaturated compounds.

Peroxides

Dialkyl peroxides are characterized by the formula ROOR', where R and R' are the same or different primary, secondary, or tertiary alkyl, cycloalkyl, and aralkyl hydrocarbon or hetero-substituted hydrocarbon radicals. Organomineral peroxides are characterized by the formulas $R_mQ(OOR)_n$ and R_mQOOQR_m, where at least one peroxy oxygen is bonded directly to the organo-substituted metal or metalloid Q.

Although di-*t*-alkyl peroxides are among the most thermally stable organic peroxides, eg, 10-h $t_{1/2}$, 110–135°C for acyclic products, the short-chain primary dialkyl peroxides are shock-sensitive and explosive. Primary and secondary dialkyl peroxides are more susceptible to radical-induced decomposition than tertiary alkyl peroxides. Acyclic di-*t*-alkyl peroxides are efficient generators of free radicals by thermal or photolytic homolysis.

The polymeric peroxides, $-(OOCH_2CHX)_n$, are viscous liquids or amorphous solids that are characterized by as many as ten repeating units; they explode on heating. The epiperoxides or endoperoxides of polynuclear aromatic compounds are crystalline solids that undergo thermal reversion with chemiluminescence (qv) to singlet oxygen and the aromatic hydrocarbon.

The metalloid peroxides behave as covalent organic compounds; they are mostly insensitive to friction and impact, but decompose violently when heated rapidly. Most solid metalloid peroxides have well-defined melting points; the more stable liquids can be distilled.

Dialkyl peroxides are prepared by the reaction of substrates with hydrogen peroxides or oxygen or from other peroxides. Organomineral peroxides are prepared by the reaction of certain organometallic or organometalloid compounds, R_mQX_n, with hydrogen peroxide or alkyl hydroperoxides. Diaralkyl peroxides have been prepared by autoxidation. Olefins that polymerize readily in the presence of free radicals form copolymers with oxygen and result in the formation of polymeric peroxides.

Peroxyacids

Peroxycarboxylic acids, $R(CO_3H)_n$, where R is an alkyl, aralkyl, cycloalkyl, aryl, or heterocyclic group, and $n = 1$ or 2, are named by prefixing the parent acid with peroxy, eg, peroxypropionic acid. Peroxysulfonic acids are named by prefixing the parent R compound to peroxysulfonic acid, eg, cyclohexaneperoxysulfonic acid.

The aliphatic peroxycarboxylic acids of low carbon number are liquids, whereas the aromatic and diperoxy aliphatic members are solids; ir and x-ray studies show that peroxycarboxylic acids are dimeric in the solid state but monomeric in the vapor and pure liquid state as well as in solution. They are more soluble in water than the parent carboxylic acids. The long-chain aliphatic peroxyacids are insoluble in water but soluble in ether and alcohols. Aliphatic peroxyacids are characterized by sharp, unpleasant odors, the intensity of which decreases with increasing chain length; they are irritating to the skin and mucous membranes.

Organic peroxycarboxylic acids are not very stable and many lose active oxygen on standing; stability increases with increasing molecular weight. Thermal decomposition proceeds by free-radical and nonradical paths, by photodecomposition or radical-induced decomposition, or by the action of metal ions or complexes. Peroxycarboxylic acids are not shock-sensitive but can explode upon heating. They are the most powerful oxidizing agents of all organic peroxides.

Peroxyacids are made by the direct, acid-catalyzed equilibrium reaction of 30–98 wt% hydrogen peroxide with the carboxylic acid, by autoxidation of aldehydes, or from acid chlorides, anhydrides, or boric-carboxylic anhydrides with hydrogen or sodium peroxide.

Diacyl Peroxides

Diacyl peroxides are characterized by the —C(O)OOCO— peroxy moiety with the two ends connected to the same or different alkyl, aryl, heterocyclic, alkoxy, aryloxy, imino, or amino groups. These peroxides may also contain the —SO_2OOSO_2— and —SO_2OOCO— groupings, in which the sulfur is bonded to an alkyl or aryl moiety. Decomposition is induced by radicals or catalyzed by transition-metal ions. The latter generate radicals, and the reaction has been utilized in synthetic applications.

Diacyl peroxides are prepared by the reaction of acyl chlorides with sodium or hydrogen peroxide or percarboxylic acids and a base. The liquid diacyl peroxides and concentrated solutions of the solid compound are unstable at ambient temperature and many must be stored well below 0°C; the solids are stable at ca 20°C, but many are shock-sensitive.

Dialkyl peroxydicarbonates are produced from the reaction of alkyl chloroformates with sodium peroxide. The liquids are unstable at 20°C and require cold storage; the solids are stable. Stability increases with increasing melting point. However, solutions of solid compounds are unstable. Dialkyl peroxydicarbonates are used primarily as free-radical initiators for vinyl monomer polymerizations. Their decomposition is accelerated by certain metals, sulfuric acid, and amines. Neat or concentrated peroxydicarbonates decompose violently.

Symmetrical and unsymmetrical organosulfonyl peroxides are prepared from organosulfonyl chlorides and sodium or hydrogen peroxide or a metal salt of a peroxycarboxylic acid. Organosulfonyl peroxides are liquids or low melting solids.

Peroxyesters

Peroxyesters include the alkyl esters of peroxycarboxylic acids, $R'C(O)O_2R$; monoperoxydicarboxylic acids, $RO_2CR'C(O)O_2R''$; and dipercarboxylic acids, $RO_2(O)CR'C(O)O_2R''$. They are liquids or low-melting solids that are hydrolyzed more readily than the analogous nonperoxide esters. Upon hydrolysis they yield the hydroperoxides from which they were prepared rather than alcohols and peroxyacids.

Primary and secondary alkyl peroxycarboxylates decompose, giving near-quantitative yields of carboxylic acids and carbonyl compound. Peroxyacids derived from benzoic acids and non-α-branched carboxylic acids are more stable than those derived from di-α-branched acids. The instability of *t*-butyl areneperoxysulfonates is increased by the presence of electron-withdrawing substituents on the aromatic ring and decreased by electron-donating substituents; even the most stable members decompose violently on warming.

Peroxyesters are prepared by the reaction of alkyl hydroperoxides with acylating agents. Their main industrial use is as initiators of free-radical reactions, primarily for vinyl monomer polymerization.

Analytical and Test Methods

The most common analytical methods for organic peroxides involve the reduction of the peroxide group followed by the determination of the excess reducing agent or of the oxidized form of the reducing agent. The approximate order of organic peroxides in terms of decreasing ease of reduction, based on polarographic studies and reduction with iodide ion, is peroxyacids > diacyl peroxides > hydroperoxides > ozonides > peroxyesters > dialkyl peroxides.

Health and Safety Factors

Organic peroxides are characterized by low toxicity. Most are oxidants, and most are irritating. The U.S. Dept. of Health, Education, and Welfare (now Health and Human Services) includes nine commercial organic peroxides in its suspected-carcinogenic list. Hydroperoxide solutions can be extremely irritating to the skin, eg, cyclohexanone and methyl ethyl ketone peroxides. The latter are injurious to the eyes and can cause blindness.

Heat or shock may cause violent decomposition, and commercial products are often formulated with desensitizing additives. The self-accelerating decomposition temperature (SADT) test is used to define the maximum allowable temperature of peroxide storage and shipment. Organic peroxides are classified by the DOT as Class-5 hazardous materials.

Uses

More than 65 organic peroxides are commercially available in over 100 formulations, eg, liquids, solids, pastes, powders, solutions, and dispersions. Uses are related mainly to the polymer industry, eg, vinyl monomer polymerization, curing of unsaturated polyester, cross-linking, and as a thermal source of free radicals. The first-order decomposition kinetics largely determine the application of a particular initiator. Other important factors include melting point, solubility, safety, efficiency, cost, and need for refrigerated storage and shipment. Traditionally, some peroxides were used as bleaching agents (qv).

CHESTER S. SHEPPARD
ORVILLE L. MAGELI
Pennwalt Corporation

D. Swern, ed., *Organic Peroxides*, Vols. 1–3, John Wiley & Sons, Inc., New York, 1970–1972.

PERSULFATES.   See Peroxides and peroxy compounds, inorganic.

PESTICIDES.   See Insect control technology; Poisons, economic.

PETROLATUM WAXES.   See Waxes.

PET AND LIVESTOCK FEEDS

Most of the nutritional needs of the U.S. population are met by foods of animal origin. As the world's population increases from 4×10^9 in 1975 to ca 6.35×10^9 in the year 2000, world food production must rise 90% with an estimated increase of 4% in arable land.

Animal foods vary from simple diets of grass and water to purified diets in which the essential nutrients are supplied in chemical form and in specific ratios. Modern animal production requires uniform nutrient supplies throughout the year.

The primary limitation to animal production is the adequate supply of high quality feed. As the world population increases, the demand for animal products as human food escalates. Some people believe that food designated for animals should be given to people and the animal link thus eliminated. However, much of animal food is derived from herbage plus industrial and agricultural by-products not acceptable to humans.

Feed efficiency has improved steadily. Today ca 600 kg feed is required to produce a 100-kg gain in cattle, 325 kg for a 100-kg gain in swine, 200 kg feed for a 100-kg gain in broilers, and 125 kg for 100-kg gain in fish.

Animals are capable of utilizing plants and grains generally not suited for humans. Most animal diets are mixtures of ingredients supplying proteins, amino acids, fats, carbohydrates, minerals, vitamins, and additives (see below). A beef-fattening ration, for example, may contain 22% alfalfa (grass hay, ground), 79% barley (dry, rolled), and 1% of a mixture of vitamins and trace mineral salts. Commercial feed companies and laboratories are spending large sums of money each year on animal research, and develop information on the precise nutritional needs of various animals.

Feed manufacturers purchase ingredients after they are harvested, and after drying grain to a 13–15% moisture content, store huge quantities in elevators. By-products, such as those obtained from the manufacture of paper pulp, sugar, and meat and fish processing are also used. Animal fats, containing energy and fatty acids including arachidonic and linolenic acids essential for cats, are supplied by the fat-rendering industry. Supplies are stored under carefully controlled conditions to minimize spoilage, including insect and fungal infestation. Ingredients are weighed, ground, and blended into complete or supplemental diets. Animal foods are regulated by the Association of American Food Control Officials (AAFCO).

During pelleting, finely ground material is forced through dies with small holes, producing strands of dense feed, which are cut into pellets. Animals generally prefer pellets to ground food.

Cats, dogs, and fish, however, prefer dry extruded food. Heat and pressure of the extrusion process converts starch into a form more easily digested by dogs and cats. Following extrusion, shaping, and cutting to size, moisture is reduced to 10% in huge gas-fired or steam-heated ovens to obtain products suitable for dry storage and sale.

Canning, which is expensive, is used extensively for dog and cat food. Mixtures containing 75–78% water are highly palatable.

Silage, produced on the farm from high moisture roughage, provides excellent nutritive feedstuffs; it is consumed *in situ*.

Animal food sold in the United States must have a label, that usually provides more information than labels on human food. The label includes product name, weight, guaranteed analysis, and a list of ingredients in descending order of prevalence. Feeding directions are often included.

Feeds may constitute complete diets or supplements that offer all or part of the essential vitamins, minerals, proteins (amino acids), fats (fatty acids), and carbohydrates. The chemical composition of a diet that produces good growth in kittens is given in Table 1. Cattle, grazing on pastures high in carbohydrates and vitamins, may only need protein and mineral supplements (see Mineral nutrients).

Over 80% of the animals produced for human consumption in the U.S. may have been given drugs. These drugs may have been injected, placed under the skin in the form of pellets, or added to food or drinking water. They are added for prevention and treatment of diseases, and improvement in weight gain and feed efficiency. Antibiotics, for example, increase the rate of weight gains in poultry and swine (see Veterinary drugs).

The hormone-like promoter, diethylstilbesterol (DES), although originally approved by the FDA in 1954, was banned in 1979 when traces were found in beef (see Hormones, nonsteroid estrogens).

The FDA Bureau of Veterinary Medicine (BVM) deals with problems caused by residues or metabolites of drugs remaining in animal tissues after slaughter.

Additives are subject to FDA approval; they are either on the GRAS or regulated lists. Additives include antioxidants, emulsifiers-stabilizers, clay, flavors, colors, antibiotics, fungistats, bacteriostats, antihelmintics, cocciostats, hormones, humectants, and buffers. A strict rule based on the Delaney Amendment excludes additives suspected of carcinogenicity (see Food additives). This rule has led to controversy, eg, sodium nitrite and saccharin (see *N*-Nitrosamines; Sweeteners). Instrumentation capable of detecting extremely small quantities of chemicals has confounded this problem.

Table 1. Chemical Constituents of Cat Food

Constituent	%	Constituent	%
Amino acids	18.85	*Minerals*	5.37
L-arginine.HCl	1.00	CaCo	0.3
L-histidine.HCl.H$_2$O	0.40	Ca$_3$(PO$_4$)$_2$	2.8
L-isoleucine	0.60	K$_2$HPO$_4$	0.9
L-leucine	1.20	NaCl	0.88
L-lysine.HCl	1.00	MgSO$_4$.7H$_2$O	0.35
L-methionine	0.45	MnSO$_4$.H$_2$O	0.065
L-tryptophan	0.15	ferric citrate	0.05
L-cystine	0.45	ZnCO$_3$	0.01
L-phenylalanine	0.50	CuSO$_4$.5H$_2$O	0.002
L-threonine	0.80	H$_3$BO$_3$	0.0009
L-tyrosine	0.50	Na$_2$MoO$_4$.2H$_2$O	0.0009
L-valine	0.60	KI	0.004
L-asparagine.H$_2$O	1.60	CoSO$_4$.7H$_2$O	0.0001
L-serine	0.80	Na$_2$SeO$_3$	0.00002
L-proline	1.60		
L-glycine	1.60		
L-glutamate	4.80		
L-alanine	0.80	*Vitamins (per kg diet)*	0.3
turkey fat	25.0	thiamine.HCl, mg	150
corn starch	24.27	niacin, mg	150
sucrose	24.28	riboflavin, mg	24
choline chloride	0.33	calcium pantothenate, mg	30
sodium acid carbonate	1.50	vitamin B$_{12}$, mg	0.03
taurine	0.10	pyridoxine.HCl, mg	9
ethoxyquin, 125 mg/kg diet			
		biotin, mg	0.9
		folic acid, mg	6
		inositol, mg	150
		p-aminobenzoic acid, mg	3
		menadione, mg	7.5
		ascorbic acid, mg	375
		retinyl acetate, I.U.	15,000
		cholecalciferol, I.U.	900
		α-tocopherol acetate, mg	30

JIM CORBIN
University of Illinois

W.G. Pond, R.A. Merkel, L.D. McGilliard, and V.J. Rhodes, *Animal Agriculture, Research to Meet Human Needs in the 21st Century*, Westview Press, Boulder, Col., 1980.

R.J. Webb and G.F. Cmarik, *Comparison of Roughages Fed to Wintering Steer Calves as Baled Hay, Chopped Hay, Hay Pellets, or Silage, Dixon Springs Station D.S. 40-239*, University of Illinois, 1957.

Association of American Feed Control Officials, Official Publication, Department of Agriculture, Room E-111, State Capitol Building, Charleston, W. Va., 1980.

We Want You to Know About Drugs for Food-Producing Animals, Department of Health, Education and Welfare Publication No. (FDA) 73-6009, FDA, Washington, D.C., 1973.

PETROLEUM

NOMENCLATURE IN THE PETROLEUM INDUSTRY

Most petroleum types are complex mixtures that are difficult to characterize in detail, and therefore, many definitions used by the exploration, production, and refining sectors of the industry to describe petroleum and its products lack precision. Traditionally, the unit of crude-oil production has been the barrel (bbl), equal to 0.159 m^3, 42 U.S. gal, or 5.61 ft^3. Today, however, petroleum reserves are often given in metric tons. Density may be reported in metric units but is more often given in degrees API. The relationship between specific gravity and °API, also called API gravity, is defined by

$$\text{sp gr (at 16°C)} = \frac{141.5}{(\text{°API} + 131.5)}$$

or

$$\text{°API} = -\frac{141.5}{\text{sp gr (at 16°C)}} - 131.5$$

Other terms relating to physical properties are determined under defined conditions established by ASTM.

Gas production is generally given in cubic meters or cubic feet (28.3 m^3 = 1000 ft^3) and reserves are often quoted in trillions of cubic feet (10^{12} ft^3 = 28.3 × 10^9 m^3). Dry gas contains mostly methane; it is called wet gas if it contains more than 4 liters per 100 m^3 of natural gas (see Gas, natural).

Cycloalkanes (or cycloparaffins) are commonly called naphthenes. Naphtha, however, is a distillate boiling below 204–260°C. Gas oil boils slightly higher than kerosene (232–426°C). Cylinder oil is a viscous lubricating oil. The term distillate is sometimes used to denote distillate fuel oil as opposed to residual fuel oil. Organic compounds are named according to the IUPAC system (see Hydrocarbons, nomenclature; Nomenclature).

COLIN BARKER
University of Tulsa

ORIGIN OF PETROLEUM

Petroleum is a naturally occurring complex mixture consisting predominantly of hydrocarbons, frequently with other hydrogen and carbon compounds containing significant amounts of nitrogen, sulfur, and oxygen as well as small amounts of nickel, vanadium, and other elements. It can occur in solid, liquid, or gaseous form as asphalt, crude oil, or gas, respectively.

Organic matter is incorporated into sediments as they are deposited and some biogenic gas may be generated; with increasing depth of burial, temperature converts the organic matter to petroleum-like material; part of this material migrates from the source rock through carrier beds to the reservoir; here, compositional changes may be produced by increasing temperature, water washing, and bacterial degradation.

Organisms produce a wide range of organic compounds, and the types incorporated into source rocks influence the type of petroleum generated. Much of the algal-derived amorphous matter (kerogen) is rich in hydrogen and poor in oxygen and generates oil, whereas the woody type of kerogen is rich in oxygen and poor in hydrogen and generates mainly gas. The generation of petroleum is nonbiological, induced by temperature and influenced by time. It follows essentially first-order kinetics and an increase of 10°C approximately doubles the rate.

An important exception to thermal generation is the bacterial formation of methane. The bacteria are anerobic, effective in sulfate-free anoxic conditions, and give shallow fields with gas that is isotopically light.

Movement of a separate crude-oil phase occurs when source rocks develop high internal pressures which squeeze oil droplets out and into adjacent permeable carrier beds. Here, buoyancy is the main driving force and migration distances can be in excess of 100 km. Oil may be remobilized after its initial accumulation in the reservoir which can lead to considerable compositional changes if both gas and oil are involved.

Petroleum composition can be changed by rising temperature or by contact with flowing water which may remove the more soluble components. If water brings bacteria and oxygen into contact with oil below 75°C, substantial degradation can result giving a higher-density, more sulfur-rich crude oil.

COLIN BARKER
University of Tulsa

J.M. Hunt, *Petroleum Geochemistry and Geology*, W.H. Freeman, San Francisco, Calif., 1979.

B.P. Tissot and D.H. Welte, *Petroleum Formation and Occurrence*, Springer-Verlag, New York, 1978.

C. Barker, *American Association of Petroleum Geologists Continuing Education Course Note Series No. 10*, 1979.

COMPOSITION

The term petroleum, literally, rock oil, is applied to the deposits of oily material found in the upper strata of the earth's crust, where they were formed by a complex series of chemical reactions from organic materials laid down in previous geological eras.

The elemental compositions of petroleum vary greatly from one deposit to another. Carbon and hydrogen make up the bulk, but other elements present include sulfur (trace to 8%), nitrogen (trace to 6%), oxygen (trace to 1.8%), and nickel and vanadium.

Modern techniques, such as gas chromatography and mass spectrometry and their combination (gc-ms), permit complete molecular characterization of the lower boiling range (through C_7) without prior separation. However, isolation and identification of individual compounds is not practical for the higher distillate range. The crude is usually distilled, followed by liquid–liquid chromatography to separate saturated, aromatic, and polar compounds. The vacuum residuum includes the heaviest compound types that do not distill under commercial operating conditions. Special techniques are required for characterization.

Saturated Hydrocarbons

n-Alkanes are found in all crudes and throughout the entire boiling range. n-Alkanes are present through n-$C_{44}H_{90}$ in the distillate range and have been identified as high as n-$C_{78}H_{158}$. Proportions and concentrations vary greatly; carbon number and concentrations have a strong effect upon the pour point, ie, the temperature at which crude or its products turn from a liquid to a solid. This is an important physical property often defining the necessary degree of processing.

All possible isoalkanes through C_8 have been identified in petroleum, together with many of the C_9 and some of the C_{10} compounds and a number of isoprenoids.

Cycloalkanes isolated from petroleum contain five-, six-, or seven-membered rings. Alkyl derivatives of cyclopentane and cyclohexane are commonly found; dicycloalkanes have also been reported, eg, *cis*- and *trans*-decahydronaphthalene and hexahydroindan. Cycloalkanes containing up to seven rings have been detected by mass spectrometry.

Analysis. All possible isomers with carbon number 7 or less and most of the C_8s can be separated by capillary gas chromatography. Hydrocarbon mixtures are analyzed by mass spectrometry according to the number of rings present in the molecule. These methods have been standardized and published by ASTM. However, the saturated compounds need to be separated from the aromatic compounds before analysis, usually by liquid chromatography over silica gel. Alkenes interfere with this analysis.

Unsaturated Hydrocarbons

Both alkenes (olefins) and aromatic compounds occur in petroleum. Alkenes are rare, whereas aromatic compounds of almost every known type have been found, eg, benzene and alkylbenzene derivatives (4.6–12.6%) through the 22 isomeric C_{10}-alkyl isomers, indan and tetrahydronaphthalenes and their methyl isomers, naphthalene and its 1- and 2-alkyl derivatives, biphenyl and derivatives, and many multiring condensed aromatic compounds. Instrumental ASTM methods permit rapid compound-type analysis of aromatic compounds.

Sulfur, Nitrogen, and Oxygen Compounds

Sulfur is found in all crudes in concentrations ranging from traces to 8.0%, mostly in the 0.5–1.5% range. It is present in aliphatic and aromatic compounds and also as elemental sulfur, hydrogen sulfide, and carbonyl sulfide (COS). Sulfur-specific flame photometric detectors in gas chromatography permit the rapid identification and measurement of sulfur compounds in the lower-boiling fractions. It is important to know the distribution of sulfur compounds within a crude. As the reserves of high quality, low sulfur crude diminish, crudes of poorer quality with high sulfur contents must be converted catalytically into low sulfur products.

Nitrogen concentrations are usually below 0.2%, but may be as high as 1.6%. Nitrogen compounds are concentrated in the high boiling fractions and in the residuum; very little is present in the fraction boiling below

300°C. Compounds containing oxygen as well as nitrogen are also found, eg, hydroxypyridines and hydroxyquinolines. The nitrogen compounds affect processing, particularly of the higher-boiling fractions Aromatic nitrogen-containing molecules are refractory and some are basic and may interact and adsorb onto acid catalysts and promote coke formation.

The oxygen content of most crudes is low. Oxygen is present mainly in the form of carboxylic acids, eg, fatty acids through stearic, branched-chain acids, alicyclic acids, and dicarboxylic acids.

Residuum

Specifications of a residuum fraction must always include the boiling range taken at the end of the distillate. Compound types are separated by chromatographic techniques. Unique to the residuum is solvent precipitation or deasphalting. The portion of the residuum insoluble in a large excess of n-heptane is known as asphaltenes (up to 25%), and the heptane-soluble fraction as maltenes. Saturated compounds may constitute 40 wt% of the residuum, usually much less.

Metals

In trace quantities, all metals through atomic number 42 (molybdenum) have been found, except rubidium and niobium; a few heavier elements have also been detected. Present in all crudes, nickel and vanadium are the most important. Analytical methods can detect metal concentrations in crudes at 10 ppb (10 ng/g).

A number of compounds, identified in petroleum, are referred to as biological markers or geochemical fossils. These are compounds whose basic carbon skeleton has survived chemical changes that occurred during petroleum genesis and whose structures are frequently found in terrestrial and marine life.

J.J. ELLIOTT
M.T. MELCHIOR
Exxon Research and Engineering Company

J.E. Dooley, D.E. Hirsch, C.J. Thompson, and C.C. Ward, *Hydrocarbon Process.* **53**(11), 187 (1974).

J.M. Hunt, *Petroleum Geochemistry and Geology*, W.H. Freeman & Co., San Francisco, Calif., 1979.

B.P. Tissot and D.H. Welte, *Petroleum Formation and Occurrence*, Springer-Verlag, New York, 1978.

RESOURCES

Petroleum resources are distributed widely in the earth's crust as gases, liquids, and solids. The products derived from these naturally occurring resources are used mainly as energy sources, although substantial volumes serve as feedstocks in the plastics, chemical, and other industries (see Feedstocks; Gas, natural; Liquefied petroleum gas; Tar and pitch; Oil shale).

Petroleum resources are classified as follows: The *resource base* is the total amount present in a specified volume of the earth's crust (also called in-place petroleum); *resources* represent the total amount of petroleum, including reserves, that is expected to be produced in the future; *reserves* constitute the petroleum that has been discovered and can be produced at the prices and technology that existed when the estimate was made; *proved reserves* are estimates of reserves contained primarily in the drilled portion of the fields; *indicated reserves* constitute known petroleum that is currently producible but cannot be estimated accurately enough to qualify as proved; *inferred reserves* are producible but the assumption of their presence is based upon limited physical evidence and considerable geological extrapolation; *subeconomic resources* constitute the petroleum in the ground that cannot be produced at present prices and technology but may become producible; *undiscovered resources* are estimated totally by geological speculation with no evidence by drilling available.

The world's proved petroleum reserves have increased from 62.2×10^9 m³ $(391 \times 10^9$ bbl) in 1966 to 91.7×10^9 m³ $(577 \times 10^9$ bbl) in 1978. More than 40% of this increase has been due to discoveries in the Middle

East, principally in Saudi Arabia and the United Emirates, which have by far the largest known petroleum reserves in the world.

The U.S. proved reserves of crude petroleum and NGL (natural gas liquids) together are ca 5.4×10^9 m³ $(34 \times 10^9$ bbl) and constitute about 6% of the world's proved reserves. The U.S. position in proved reserves has fallen substantially since 1966; among the states, only Alaska, because of the discoveries in Prudhoe Bay, recorded a significant increase in reserves.

Mexico has recorded a dramatic increase in discoveries and production. In 1978, it reported crude reserves of more than 4.5×10^9 m³ $(28 \times 10^9$ bbl), nearly 5% of the world total.

In South America, Venezuela continues to dominate in both proved reserves and production.

The increase in the petroleum reserves of Western Europe from 0.32×10^9 m³ $(2 \times 10^9$ bbl) in 1966 to more than 2.5×10^9 m³ (16×10^9) in 1978 is due to discoveries made in the UK and Norwegian sectors of the North Sea.

Proved reserves have almost doubled in the African countries between 1966 and 1980; Algeria, Libya, and Nigeria provide 87% of the reserves of 0.35×10^9 m³ $(2.233 \times 10^9$ bbl) and 84% of the production. Nearly half of the proved reserves of 2.5×10^9 m³ $(16 \times 10^9$ bbl) in the Far East are in Indonesia.

The communist nations reported proved reserves of 12.8×10^9 m³ $(80.4 \times 10^9$ bbl) in 1978; ca 72.6% were in the USSR, which in 1978 was the world's leading producer of crude petroleum. The People's Republic of China reported 3.2×10^9 m³ $(20 \times 10^9$ bbl) in 1978, when for the first time petroleum production was reported at ca 0.32×10^6 m³ $(2 \times 10^6$ bbl) per day.

The rate of discovery of large fields has declined sharply since the early 1950s, and current search efforts are mainly directed to the discovery of fields in the Arctic and deep-water offshore regions where the cost of exploration is far higher than in previously explored regions.

Domestic U.S. supplies may be increased by new techniques for extracting at least a portion of the crude petroleum left behind in reservoirs after conventional primary and secondary extraction techniques have been applied. These techniques include thermal methods and carbon dioxide miscible or chemical flooding.

Petroleum displaced coal as the principal source of energy in the United States in 1948, and in the world by 1965. Since then, world consumption has doubled and today petroleum supplies about half of the world's energy requirements. This spectacular growth is owing to the relative ease with which petroleum is discovered, produced, transported, and processed. The world is not running out of energy resources but most of the energy stored in the unconventional and conventional nonpetroleum resources is more difficult to extract and convert into readily usable forms.

L.J. DREW
U.S. Geological Survey

J.J. Schanz, *Oil and Gas Resources—Welcome to Uncertainty*, Special Issue No. 58, Resources For the Future, Washington, D.C., 1978.

DeGolyer and MacNaughton, *Twentieth Century Petroleum Statistics—1979*, Dallas, Texas, Nov. 1979.

International Petroleum Encyclopedia, Vol. 10, The Petroleum Publishing Co., Tulsa, Okla., 1977.

Oil Gas J. (Worldwide Issue) **77**(53), 67 (1979).

DRILLING FLUIDS

Drilling fluids and muds perform a variety of functions that influence the drilling rate, cost, efficiency, and safety of the drilling operation. They are pumped down a hollow drill string through nozzles in the bit at the bottom, and up the annulus formed by the hole or casing and the drill string to the surface. The bit is turned by rotating the entire string from the surface or by using a down-hole motor. After reaching the surface, the drilling fluid is passed through a series of screens, settling tanks or pits, hydroclones, and centrifuges to remove formation material brought to the surface. It is then treated with additives to obtain the desired properties. Once treated, the fluid is pumped back into the well and the cycle repeated. Since 1926, over 1000 patents have been issued for drilling-fluid materials. A recent product list of 87 suppliers offers ca 1500 trade names.

Gas-based muds, ranging from compressed dry air or natural gas to mist to stable foams (qv), are used mostly for the drilling of hard rock. About 90% of all drilling fluids is water-based systems, ranging from clear water without additives to high density muds containing clays and various organic additives. Offshore wells are frequently drilled with a seawater system, which needs more additives to achieve the desired flow and filtration properties.

Oil-based drilling fluids have diesel, a low-aromatic-content oil, or occasionally crude oil as a continuous phase with both internal-water and solid phases. Fluids with more than 5–10 vol% water are called invert oil-based muds or inverts. Oil muds are employed for high temperature wells where water-based systems may be unstable and where problems may arise from water-sensitive shale formations.

Properties

The density is adjusted with dissolved salts or powdered high density solids to provide a hydrostatic pressure against exposed formations in excess of the pressure of the formation fluids. In addition, the hydrostatic pressure of the mud column prevents collapse of weak formations into the borehole. Fluid densities may range from that of air to > 2500 kg/m³ (20.8 lb/gal). Most drilling fluids have densities > 1000 kg/m³ (8.33 lb/gal), the density of water. The hydrostatic pressure imposed by a column of drilling fluid is expressed as follows:

$$P = 0.098 L_{\rho_m} \left(0.052 L_{\rho_m} \right)$$

where P = hydrostatic pressure in kPa (psi), ρ_m = drilling-fluid density in kg/m³ (lb/gal), and L = column length or well depth in m (ft). Wells with bottom-hole pressures of 100 MPa (14,500 psi) at 5000 m are not unusual.

Viscosity and velocity must be high enough to remove drill cuttings and other formation material that may fall into the well bore. Low viscosity fluids are circulated at high rates and vice versa.

Density control. Mud density is controlled accurately by suitable weighting materials that do not adversely affect the other properties. Most important is the specific gravity of the weighting agent as well as its chemical inertness. Barite (barium sulfate, $BaSO_4$) meets these requirements; it is virtually insoluble in water and does not react with other mud constituents. The API specifications are given in Table 1. The barite content in drilling mud can be as high as 2000 kg/m³ (700 lb/bbl). Other weighting materials include galena, hematite, magnetite, ilmenite, siderite, celesite, dolomite, and calcite.

The drilling rate is improved by solids-free fluids, which are designed for work-over and completion operations to avoid formation damage. Weighted fluids without solids are provided by solutions of various water-soluble salts (see Table 2 for maximum densities). The incorporation of a limited amount of drilled solids is an economical way to increase mud density, but it also reduces penetration rates.

Viscosity. The type of solids and the solids content of water-based muds controls the viscosity. The reactive solids are composed of commercials clays, of which bentonite (montmorillonite), attapulgite, and sepiolite are the most important; organophilic clays, prepared from bentonite or attapulgite and aliphatic amine salts, are used for oil-based muds.

Table 1. API Specifications for Barite

Assay	Value
specific gravity	4.20 min
wet-screen analysis	
residue on U.S. sieve (ASTM no. 200) (74 μm)	3.0% max
residue on U.S. sieve (ASTM no. 325) (44 μm)	5.0% min
soluble alkaline-earth metals as calcium	250 ppm max

Table 2. Soluble Salts that Increase the Density of Drilling Muds and Work-Over and Completion Fluids

Salt	Formula	Specific gravity	Saturated brine, 20–25°C Wt%	Saturated brine, 20–25°C Density, g/cm³
potassium chloride	KCl	1.984	26.5	1.18
sodium chloride	NaCl	2.165	28.5	1.20
sodium carbonate	Na_2CO_3	2.532	22.5	1.24
calcium chloride	$CaCl_2$	2.150	46.1	1.46
potassium carbonate	K_2CO_3	2.428	52.9	1.56
sodium bromide	NaBr	3.203	53.7	1.65
calcium bromide	$CaBr_2$	3.353	58.7	1.70
zinc chloride	$ZnCl_2$	2.910	81.2	2.14
zinc bromide	$ZnBr_2$	4.201	81.7	2.65

Organic polymers are used to increase viscosity and also to control filtration rates. Water-soluble polymers include starch, guar gum, sodium carboxymethyl cellulose, hydroxyethyl cellulose, xanthan gum, polyacrylates, and polyacrylamides.

Viscosity is decreased by a reduction of solids, or deflocculation of clay-water suspensions (see Flocculating agents). Thinning is obtained by reducing plastic viscosity, yield point, or gel strength, or a combination of these properties. Typical mud-thinning chemicals are polyanionic materials that are absorbed on the positive-edge sites of the clay particles, thereby reducing the attractive forces between the particles without affecting clay hydration.

Filtration control. Filtration control is particularly important in permeable formations where the mud hydrostatic pressure exceeds the formation pressure. It reduces drill-string sticking, torque, and drag due to filter-cake development; minimizes damage to protective formations; and, in some areas, improves borehole stability. Materials for this purpose include clays, organic polymers, and lignite derivatives. The bentonite present in freshwater systems acts as the primary filtration agent; concentrations vary widely, but may range up to 100 kg/m³ (35 lb/bbl).

Although a combination of bentonite clay and lignosulfonate provides filtration control in many water-based muds, organic polymers are generally needed as filtration aids, especially starches in concentrations of 6–29 kg/m³ (2–10 lb/bbl). Fermentation by microorganisms is prevented by saturating with salt or adding a preservative, eg, paraformaldehyde or another biocide (ca 0.6–1.4 kg/m³ or 0.2–0.5 bbl). Carboxymethyl cellulose is most commonly used in fresh- to seawater fluids (1.4–5.7 kg/m³ or 0.5–2.0 lb/bbl).

Alkalinity control and removal of contaminants. Water-based drilling fluids are generally maintained at an alkaline pH by the addition of sodium hydroxide (3–14 kg/m³ or 1–5 lb/bbl), lime (6–57 kg/m or 2–20 lb/bbl), or magnesium oxide.

Drilling-fluid contaminants are defined as any material or condition that adversely affects fluid performance. Drill solids are removed mechanically by various combinations of screens, hydroclones, and centrifuges. Electrolytes are removed by dilution and chemical treatment.

Shale stabilization. Shales can be stabilized with an oil mud that prevents direct contact between the shale and the emulsified water. The most recent shale-protective agents are the polymer/potassium muds. The reaction of potassium ions with clay results in potassium fixation and formation of a less water-sensitive clay. Potassium chloride is generally used in combination with a polymer. Ammonium chloride, ammonium sulfate, and diammonium phosphate have also been used, along with a number of anionic polymers.

Surfactants perform various functions in drilling fluids (see Surfactants and detersive systems). Lignites and lignosulfonates act as emulsifiers (see Emulsions; Lignin), whereas aluminum stearate is used as a defoamer (see Foams).

To function properly a drilling fluid must be circulated through the well and back to the surface. Loss of fluid, owing to openings in the formation, can result in a loss of hydrostatic pressure at the bottom of the hole, influx of formation fluids, and possibly loss of well control.

Lost-circulation zones are sealed off by the addition of flakes (shredded cellophane or paper, laminated plastic), fibrous materials (wood fibers, saw-dust, straw), or granules (ground walnut shells, rubber, plastics, limestone); concentrations of these materials range from 14 to 143 kg/m³ (5–50 lb/bbl).

Lubricants and spotting fluids are added to overcome frictional resistance of the drilling string or sticking of the drill pipe. Blends of anionic and nonionic surfactants and natural products are used as well as gilsonite and air-blown asphalts. String and casing corrosion is generally caused by oxygen dissolved or entrained in the mud, and can be prevented by excluding or scavenging corrosive gases and by maintaining a high pH.

R.K. Clark
J.J. Nahm
Shell Development Corporation

G.R. Gray and H.C.H. Darley, *Composition and Properties of Oil Well Drilling Fluids*, 4th ed., Gulf Publishing Company, Houston, Texas, 1980.

API Bulletin on Oil and Gas Well Drilling Fluid Chemicals, API Bull. 13F, 1st ed., American Petroleum Institute, Dallas, Texas, Aug. 1978.

Standard Procedure for Testing Drilling Fluids, API Recommended Practice, API RP 13B, 8th ed., American Petroleum Institute, Dallas, Texas, April, 1980.

ENHANCED OIL RECOVERY

Crude oil exists in contact with gases and water in subterranean rock formations called reservoirs. Wells serve as pressure sinks that allow the oil to be brought to the surface. Normal depletion, called primary recovery, results from the driving forces provided by dissolved gases or an associated aquifer. Following primary production, fluid may be injected into a portion of the wells to provide a secondary energy source. This is called secondary oil recovery, eg, waterflooding or immiscible-gas flooding. Continued fluid injection, in time, results in ever-increasing production of injected fluid and eventually only injected fluid (ie, water) is produced, leaving unrecovered oil behind as a residual oil saturation.

Enhanced oil recovery (EOR), often referred to as tertiary oil recovery, is recovery of oil over and above that which can be recovered by primary and secondary methods (see Fig. 1). Estimates of the potential of EOR are highly speculative, but can be 5–10% of the oil discovered to date. In the United States, 7.2×10^{10} m³ (4.5×10^{11} bbl) of oil have been discovered so far, and worldwide 6.4×10^{11} m³ (4×10^{12} bbl).

Residual oil remains in a reservoir because of the high interfacial tension (IFT) between oil and water. Of importance is the interrelationship between capillary and viscous forces given by the capillary number:

$$\text{capillary number} = (\text{velocity} \times \text{viscosity})/\text{IFT}$$

where the velocity refers to the injected fluid and the IFT is measured between the oil and the injected fluid. High capillary numbers are required to get low residual oil saturations. For miscible fluids, the capillary number is infinite.

Because of the density differences between gas, oil, water, and injected fluids, oil may be trapped high in the reservoir and not contacted by an injected water-base fluid. Gravitational forces are expressed in terms of densities.

The economics of EOR methods are governed by the amount of oil that can be recovered and the cost of the injected fluids. A high initial investment is needed and oil production is delayed after initial injection of the fluid.

Miscible-Gas Drives

A gas can either miscibly or immiscibly displace oil, depending on the physical properties of the gas and the oil and the reservoir temperature and pressure. In a miscible-gas process, the residual oil is mobilized by an injected solvent (miscible gas) and the mobilized oil is displaced to a production well. A thorough understanding of the physical/chemical phenomena (displacement mechanisms) occurring within the oil-moving and solvent-moving zones is needed for a project design.

Miscible displacement may be controlled by either a first-contact or multiple-contact mechanism. In the former, the reservoir oil and injected solvent are mixed in all proportions at reservoir conditions and no phase separation occurs. In the latter, oil and solvent cannot be mixed in all proportions without phase separation upon equilibration. Miscibility is developed by either a condensation or a vaporization (extraction) process.

First-contact miscible solvents include liquefied petroleum gas fractions (LPGs) or enriched gas (see Liquefied petroleum gas). Multiple-contact miscible agents include methane plus LPG components, carbon dioxide, methane, nitrogen, and flue gas. Enriched-gas flooding is an example of miscibility developed by the condensation process where the LPG components condense into the oil.

Application. Miscible-gas floods are applied in sandstone or carbonate reservoirs of sufficient depth containing oil of moderate viscosity; ca 900–5400 m³ of gas per cubic meter oil recovered is required.

Micellar-Polymer Flooding

In micellar-polymer flooding, a bank of micellar fluid is injected to displace residual oil from a reservoir. The micellar fluid is pushed by a polymer-thickened water bank. The mechanism is based on reducing the IFT between crude oil and water. The fluid consists of a surfactant in a solvent in which the surfactant ions are arranged in oriented aggregates or micelles. As the fluid moves through the porous medium, mass transfer (qv) occurs between the injected fluid and the in-place fluids, and an oil-moving zone develops. The oil-moving zone creates an ultra-low IFT condition which results in residual oil being displaced. The desired mobility of the various banks can be adjusted, and the proper mobility keeps the integrity of the bank and maximizes sweep efficiency. This technique is still in the developmental stage.

Thermal Processes

Thermal processes utilize heat generated above ground or *in situ* to enhance oil recovery; they are applied primarily to heavy oils. The simplest process is steam injection. Steam is used as a driving fluid to displace oil from an injector to a producer, or in a cyclic-injection production sequence using the same well as both injector and producer with a soak period to heat the well, the so-called steam–soak or steam huff-and-puff process. Underground or *in situ* combustion processes utilize injected air or oxygen and hydrocarbons or coke within the reservoir as fuel.

H.R. Froning
D.D. Fussell
E.W. Heffern
Amoco Production Company

Enhanced Oil Recovery—An Analysis of the Potential for Enhanced Oil Recovery from Known Fields in the United States—1976–2000, National Petroleum Council, Washington, D.C., Dec. 1976 (LC 76-62528).

F.I. Stalkup, *Miscible Displacement*, SPE Monograph, Vol. 8, 1983.

W.B. Gogarty, *J. Pet. Technol.* **35**, 1168 (1983).

M. Pratts, *Thermal Recovery*, SPE Monograph, Vol. 7, 1982.

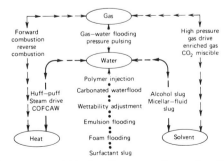

Figure 1. Improved methods for oil recovery (connecting lines indicate several combinations commonly in use). COFCAW is the combination of forward combustion and water flooding.

REFINERY PROCESSES, SURVEY

Petroleum products derived from crude oil are a convenient source of energy. Since petroleum liquids are easy to handle and store, they are well suited for transportation fuels, eg, for cars and airplanes. Other refinery products include lubricants, waxes, asphalt, solvents, and specialties such as liquefied petroleum gas (LPG), hydraulic fluids (qv), and others. Petroleum is the principal raw-material source for petrochemicals such as plastics, synthetic elastomers, certain alcohols, and other important products. The various fuel and chemical markets have their own product-quality requirements and it is the function of the refining operation to separate crude and other raw materials into fractions that are then processed to meet product specifications (see Petroleum products).

Crude petroleum contains a wide range of hydrocarbons from light gases to residuum that is too heavy to distill even under vacuum. Crude oil is primarily made up of paraffins, cycloparaffins (naphthenes), and aromatic compounds in varying proportions, some sulfur compounds, a small amount of nitrogen, but no appreciable amount of oxygen or olefins.

Refining processes can be grouped into three classes: separation, usually distillation to give the desired type of compounds; conversion, usually cracking, to change molecular weight and boiling point; and upgrading, eg, hydrotreating, to meet product-quality specifications.

In general, refineries are located near a large body of water, partly to supply cooling water but also for transportation. Large refineries process about 80,000 m³ (5×10^5 bbl) of crude per day, which corresponds to one supertanker load. Rail transportation would require 1600 tank cars per day to carry the same amount of crude.

Since the 1970s, the great increase in crude cost has been accompanied by greater emphasis on high value products at the expense of fuel products. For example, fuel oil previously used in large power plants is displaced by coal or nuclear fuel. Since 1975, no utility power plants have been built in the United States based on burning oil or gas as fuel. Increased attention to environmental aspects has led to demand for low sulfur products despite the trend toward higher sulfur crudes. Unleaded gasoline is another example of environmental concern.

Petroleum refining has shown a rapid growth and is now the largest manufacturing industry in the United States, where petroleum products amount to ca 10% of the GNP. Gasoline accounts for ca 43% of petroleum-product consumption, diesel and fuel oil for ca 20%. Imports amount to over 50×10^9 per year, causing a serious imbalance of foreign-trade payments as well as uncertainty of supply.

Efforts to decrease the gasoline consumption (km/L) of vehicles by reducing their size and weight have been successful. More efficient engines are in wide use, particularly diesel engines, in which the fuel is injected directly into the combustion chamber. Diesel engines have efficiencies of about 35% versus ca 25% efficiency for gasoline engines.

Refineries range in size from 1600 m³/d (10,000 bbl/d) to over 64,000 m³/d (4×10^5 bbl/d). Small refineries make only gasoline, diesel, and domestic heating oils. Large refineries include the manufacture of lubricating oils and greases. Refining is also the main source of raw materials for petrochemical manufacture. A large steam-cracking unit for 500,000 t/yr of ethylene may consume ca 2×10^6 t/yr of oil feed (40,000 bbl/d), which may be 10% of the crude used in very large refinery. Clearly, chemical and fuel-refining operations must be carefully coordinated.

Processing Steps

Desalting. Salt and clay or other suspended solids are removed by washing with water at 65–90°C to reduce viscosity. Typical salt content of crude may be 280 g/100 m³; desalting may remove over 90% without the loss of oil.

Distillation. The crude is separated in continuous-fractionation plate towers, as shown in Figure 1 (see Distillation). Primary distillation takes place at atmospheric pressure and the bottom temperature is limited to 370–400°C to prevent thermal cracking.

Naphtha, the fraction taken from the top, is mainly used for motor gasoline and processed further for octane improvement by catalytic reforming. The middle distillate includes diesel, and jet fuels, and heat-

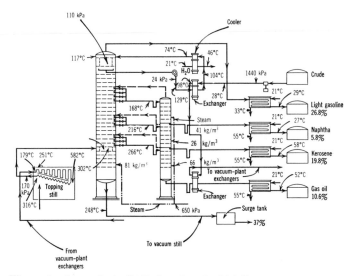

Figure 1. Atmospheric distillation of crude. Multidraw crude-oil topping plant. To convert kPa to psi, multiply by 0.145. Courtesy of McGraw-Hill Book Co., Inc.

ing oil. Kerosene and certain specialty solvents distill between these two fractions. The bottoms fraction can be used as fuel oil but is usually vacuum distilled in order to increase the yield of high value distillate oil for catalytic cracking.

Vacuum distillation provides low sulfur fuel oil by hydrosulfurizing vacuum gas oil, which is then blended back into untreated vacuum bottoms. In addition, various specialty materials are obtained, such as wax and lube fractions.

Hydroprocessing. Hydroprocessing improves the quality of various products or cracks heavy carbonaceous materials to lower-boiling, more valuable products. Mild hydrotreating removes sulfur, nitrogen, oxygen, and metals, and hydrogenates olefins. A fixed bed may be employed at 1.5–2.2 MPa (200–300 psig) and 350–400°C, without catalyst regeneration. Severe conditions are 7–21 MPa (1000–3000 psig) and 350–500°C with catalyst regeneration.

Hydrogen consumption increases with severity and depends on the amount of sulfur removed and the feed content of aromatic materials and olefins, which also consume hydrogen. Net consumption can range from 18 m³/m³ (100 ft³/bbl) feed for hydrofinishing to well over 180 m³/m³ (1000 ft³/bbl) feed in hydrocracking operations.

Hydrocracking. In hydrocracking, high molecular weight compounds are cracked to lower boiling materials. Severity is increased by operating at higher temperatures and longer contact time than in hydrotreating. Hydrocracking is used extensively on distillate stocks. It is of increasing importance in view of the trends to heavier crudes and the need for processing synthetic crudes.

Catalytic cracking. In catalytic cracking, heavy distillate oil is converted to lower molecular weight compounds in the boiling range of gasoline and middle distillate. Gasoline yield is high and so is the octane number. About half of the gasoline sold in the United States is obtained from petroleum by catalytic cracking, mostly by the fluidized-bed process where small particles of catalyst are suspended in upflowing gas to be handled like a liquid and circulated through pipes and valves between reaction and regeneration vessels (see Fluidization).

Catalyst circulation rates are over 50 t/min in a large plant. Temperatures range from 480–510°C in the reactor to ca 620°C in the regenerator using a synthetic silica-gel catalyst activated with 15–60% Al_2O_3. Temperatures throughout the fluidized bed vary by less than 5°C; pressures are 150–200 kPa (22–29 psi). The new zeolite catalysts can withstand higher temperatures and they are usually regenerated at 700°C. In addition, all CO is oxidized to CO_2; addition of a noble metal or other combustion catalysts in ppm concentrations assures complete combustion. With zeolite-type catalysts, 80–90% conversions are obtained. A feed rate of 16,000 m³/d (10^5 bbl/d) is not uncommon for large units having a 15-m-dia regeneration vessel and a 6.5-m-dia reactor; the structure may be 65 m high. Configurations of typical cracking units are shown in Figure 2, and typical yields in Table 1.

Coking. Both delayed and fluid coking are used on very heavy low value residuum feeds to obtain lower-boiling cracked products. Coking can be considered as high severity thermal cracking or destructive distillation. It is generally used on vacuum residuum (which is not a suitable feed for catalytic cracking) to generate lighter components. In delayed coking, a cyclic batch operation, the oil is held at 450°C and 35–70 kPa (5–10 psi) to deposit coke while cracked vapors are taken overhead. Fluid coking is a continuous process similar in design to fluid catalytic cracking.

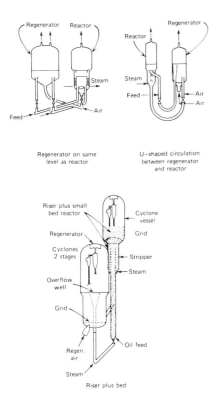

Figure 2. Typical fluid-catalytic-cracking-unit configurations. Courtesy of Exxon Research and Engineering.

Table 1. Catalytic-Cracking Operation, South Louisiana Gas Oil

	Operation	
	Maximum gasoline	Maximum middle distillate
Feed		
boiling range, °C	260–566	343–566
conversion, vol%	85	60
Yield[a]		
methane and C$_2$, wt%	2.8	2.4
propane, vol%	3.0	2.0
propylene, vol%	7.0	6.0
butane, vol%	8.8	5.2
butene, vol%	7.6	7.9
C$_5$ naphtha up to 221°C, vol%	69.5	47.1
heating oil vol%	10.0	35.0
slurry, vol%	5.0	5.0
coke, wt%	5.3	5.4
RON[b] clear	92	92
carbon on regenerated zeolite catalyst, wt%	< 0.20	0.35

[a] Because of expansion, vol% adds up to more than 100%.
[b] Research octane number.

The products include gas, naphtha, gas oils, and coke. Hydrotreated coker gas oil augments supplies of residential heating oil; heavier fractions are cracked catalytically.

The sulfur content of the coke may be as much as 50% higher than that of the feedstock, but the coke can be used as fuel if SO$_x$ control is provided. Needle coke, a premium product especially suited for graphite production, is made by delayed coking of low sulfur, highly aromatic heavy oil obtained from catalytic or thermal cracking. It is the world's main source of graphite for carbon electrodes (see Carbon).

In Flexicoking, high sulfur coke is gasified to clean fuel gas in a fluidized-solids circulating system. The coke is gasified with steam and either air or oxygen. The gas is scrubbed to remove sulfur compounds and impurities and thereby provides a cleaner fuel. Gasification is carried almost to completion while ash and metals build up to a high concentration in the purge solids. Yield pattern is given in Figure 3.

Visbreaking. Viscosity breaking is one of the few thermal-cracking processes still used in refining other than coking. It reduces the viscosity of heavy fuel oil and permits handling at lower temperature. Vacuum residuum can usually be burned directly but would require perhaps 200°C for proper atomization in burners. After visbreaking, a temperature of 100°C or less is adequate, which gives a fuel oil suitable for large commercial buildings, etc. In visbreaking, heavy oil is heated in a furnace to ca 480°C and held at this temperature in a soaking coil long enough to give the desired degree of cracking. The process yields 1–2% gas, 5–10% naphtha, and 20–30% distillate gas oil. In some cases, the bottoms are distilled under vacuum to give a pitch that can be used in asphalt (qv) or roofing tar.

Steamcracking. Steamcracking is used primarily to produce olefinic raw materials for petrochemicals manufacture. It plays such an important role in refining, however, that the planning and operation of a large steam-cracking plant is usually closely integrated with a large refinery. Steam cracking is carried out at ca 800–850°C and slightly above atmospheric pressure. Feedstocks range from ethane to vacuum gas oil; heavier feeds give higher yields of by-products such as naphtha, etc. Steam is added with the feed to reduce the hydrocarbon partial pressure, thereby giving more olefins. The amount of steam may be equal in weight to the oil feed.

Catalytic reforming. Naphtha recovered directly from crude by distillation is called straight-run or virgin naphtha. It is too low in octane (30–50 clear octane) to meet quality requirements for motor gasoline and is upgraded by catalytic reforming. The feed is a naphtha fraction boiling in the 80–230°C range. Lighter fractions are not amenable to reforming, whereas heavier fractions would give products that remain outside of the gasoline boiling range since reforming gives only a small reduction in boiling point. Typical operating conditions are 430–520°C and 1–6 MPa (145–870 psi). By-product hydrogen amounts typically to 36–45 m^3/m^3 of feed (202–253 ft^3/bbl) with a yield of C$_5$ and higher of 80–85 vol%.

For regeneration, the catalyst is burned in a low oxygen atmosphere; the on-stream run may last for one year or more before regeneration is needed. Currently, almost all catalysts are based on platinum or pal-

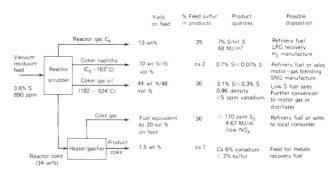

Figure 3. Flexicoker yield pattern, product qualities, and possible dispositions. To convert MJ/m^3 to Btu/SCF, multiply by 26.85. Courtesy of Exxon Research and Engineering.

ladium. They lose their activity in the presence of sulfur or nitrogen compounds and are poisoned by metals such as arsenic or lead.

The following reactions occur in reforming (see Fig. 4): dehydrogenation of cycloparaffins; isomerization of paraffins; dehydrocyclization of paraffins; hydrocracking of paraffins; and removal of sulfur, olefins, nitrogen, and oxygen.

Alkylation. In alkylation, olefins of 3, 4, and 5 carbon atoms react with isobutane to give higher molecular weight products of high octane. Thus, alkylation increases the high octane component in the gasoline pool. Having a 93–96 octane on either the ASTM scale (MON = motor octane number) or research scale (RON = research octane number), the product is especially valuable for unleaded gasoline. The octane number quoted in retail sales is an average of MON and RON.

Sulfuric or hydrofluoric acids are employed as catalysts. Sulfuric acid alkylation operates at 4–15°C and a pressure slightly higher than atmospheric. Feed consisting of olefins and isobutane is injected into the reaction emulsion through high velocity nozzles, and the mixture is agitated with mechanical stirrers. A large amount of sulfuric acid is consumed owing to the solubility of by-products and accumulation of water. With a hydrofluoric acid catalyst, the operating temperature is 25–45°C; the pressure of 0.7 MPa (101.5–145 psi) maintains the HF as a liquid phase.

Extraction. Extraction (qv) separates by chemical type rather than by boiling point and is much more expensive than distillation. It is used primarily for removal of aromatic material from naphtha fractions, upgrading middle distillates, and preparation of high quality lubricating oils.

Lubricants. Petroleum is the main source of lubricants for a wide range of uses (see Lubrication and lubricants). Lubricants are derived from distillates boiling above 370°C and they are separated by vacuum distillation and refined to meet specifications. Aromatic material is removed from the raw distillate by extraction with *N*-methylpyrrolidin-2-one, phenol, or furfural. Wax is crystallized by chilling and filtered. Various additives play an important role in meeting quality requirements.

Gasoline blending and treatment. Separate components are blended to give various grades. Elaborate instrumentation controls in-line blending and offers large savings compared to blending in tanks. Leaded gasoline of reduced lead content is still being made, but may be phased out by about 1990.

Naphtha-treatment processes, such as the Merox process, are used to extract sulfur compounds and to convert trace amounts of mercaptans to disulfides, which are much less objectionable.

Today's unleaded gasoline usually contains additives for octane improvement, eg, methyl *tert*-butyl ether and *tert*-butyl alcohol; gasohol contains ethyl alcohol (see Ethanol).

Hydrogen manufacture. The increase in hydrotreating has expanded the need for hydrogen beyond that available as a by-product from catalytic reforming. A common source is catalytic steam reforming of methane, refinery gas, or naphtha (see Hydrogen). Hydrocarbons treated with steam over a nickel-base catalyst at ca 2.8 MPa (400 psi) and 815°C give hydrogen and carbon oxides.

Sulfur recovery. Typical crude contains a large amount of sulfur. Thus, 10×10^6 t/yr of crude (200,000 bbl/d) with a sulfur content of 3% would yield 870 t/yr sulfur by-product. In refining, the sulfur compounds are converted to H_2S, which is removed by amine scrubbing. The H_2S is converted to pure sulfur in a Claus plant where one-third is allowed to react with the remaining H_2S in a series of catalyst beds. The reaction is exothermic and cooling is provided between beds together with separation of liquid sulfur. The final tail gas from a Claus plant is cleaned further before being released to the atmosphere.

Other refinery facilities. A complete refinery is a highly complex and integrated installation requiring many auxiliary facilities (so-called offsites), such as large storage capacities for feed and products, extensive port and dock facilities, and utilities including steam, electricity, cooling water and wastewater cleanup. Some utility requirements can be supplied from coal.

Conservation

Since 1975, conservation in refineries has produced energy savings of 15–25%. Efficiency of existing equipment has been increased by using less air in furnaces, and pumps with less excess capacity operate at more nearly optimum efficiency. Pressure energy in the flue gas from regeneration can be recovered, and heat pumps save distillation energy. Steam leaks, losses by evaporation, and leakage from valves have been greatly reduced. The overall efficiency of a refinery is ca 90%, individual processes ranging from 95% for distillation and catalytic cracking to about 80% for hydrocracking residuum (see Table 2).

Environmental Aspects

Environmental aspects of refining concern air, water, and solid-waste problems (EPA classification). Petroleum refining was one of the first industries to be covered by Federal regulations mandating ambient air-quality standards and allowable emissions (see Table 3). State standards are at least as strict; Japan has very strict standards.

Air-pollution control is directed toward the principal pollutants: SO_x, NO_x, particulates, CO, and hydrocarbons (see Air-pollution control methods). Control measures on furnaces include the use of low sulfur fuels. The EPA standards regulate emission of flue gases from furnaces and catalyst regenerators as well as hydrocarbon vapors generated by leaks, evaporation, and handling.

Water pollutants include oil, H_2S or other sulfur compounds, and ammonia, phenols, chromium compounds, and chlorine used to control corrosion and biological fouling. The main sources of wastewater are cooling systems that may contain additives to inhibit corrosion or fouling; low toxicity inhibitors are now available (see Corrosion and corrosion inhibitors). Process water containing H_2S, ammonia, and oil is termed sour water. It is stripped with steam, and treated in an API-type separator where the oil rises to the surface and is removed with the aid of air flotation.

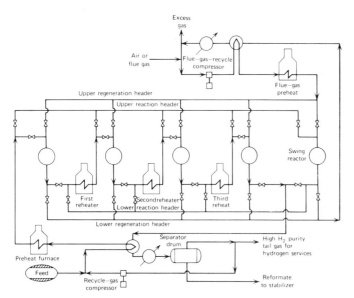

Figure 4. Fixed-bed cyclic catalytic reforming. Courtesy of Exxon Research and Engineering.

Table 2.   Thermal Efficiencies of Various Processes

Operation	Thermal efficiency, %
methanol from coal	50
hydrogen manufacture	
from coal	55
from methane	60
SNG from coal	65
coal liquefaction	70
oil from shale	80
fuel gas from coal	80–85

Table 3. National Primary and Secondary Ambient Air-Quality Standards

Pollutant	Type of standard	Averaging time	Concentration $\mu g/m^3$	Concentration ppm
carbon monoxide	primary and secondary	1 h	40,000	35
		8 h	10,000	9
hydrocarbons[a]	primary and secondary	3 h (6 to 9 a.m.)	160[b]	0.24[b]
nitrogen dioxide	primary and secondary	1 yr	100	0.05
photochemical oxidants[c]	primary and secondary	1 h	160	0.08
particulate matter	primary	24 h	260	
		24 h	75	
	secondary	24 h	150	
		24 h	60[d]	
sulfur dioxide	primary	24 h	365	0.14
		1 yr	80	0.03
	secondary	3 h	1,300	0.5
lead	primary	90 d	1.5	
ozone	primary and secondary	1 h	235	0.12

[a] Nonmethane.

[b] As a guide in devising implementation plans for achieving oxidant standards.

[c] Expressed as ozone by the Federal reference method.

[d] As a guide to be used in assessing implementation plans for achieving the annual maximum 24-h standard.

Solid wastes consist mainly of spent catalysts, and ash if coal was the refinery fuel. Federal regulations are officially promulgated in *The Federal Register*. The EPA updates on a regular basis the Standards of Performance for Stationary Sources of emissions.

Catalytic cracking units are subject to the following regulations: emission of particulates directly from regenerators should not exceed 1 kg/kg of coke burned; a stack opacity of 30% maximum; and a maximum of 0.05 vol% of CO in gases discharged to the atmosphere.

Plant Economics

Utilities and general off-site expenditures include utilities production (equipment to generate steam and electricity), tie-in lines for power and water; roads and motor-vehicle access; port and dock facilities; tanks for feed and products; product loading and shipping; pipelines and railroads; repair shops, spare parts, and maintenance; technical service and engineering support; analytical and testing laboratories, an instrument and controls shop; business and accounting offices; and housing and facilities for construction crews. Investment for these facilities may amount to 40% of the costs of a new plant in a large existing refinery and often doubles the costs for a new installation.

Operating costs are strongly related to investment and include maintenance and repair; utilities; chemicals, catalysts and additives; and labor. Costs have increased drastically since 1973, both for the increased investment and the need for a higher return in order to make additional investment attractive. Capital charges include taxes and insurance.

Natural petroleum sources are still the lowest in cost of liquid fuels, but liquid from oil shale, fuel gas by coal gasification, and methanol plus substitute natural gas (SNG) from coal are now nearly competitive in some situations (see Fuels, synthetic).

C.E. JAHNIG
Consultant

W.L. Nelson, *Petroleum Refinery Engineering*, 4th ed., McGraw-Hill Book Co., Inc., New York, 1958.

Hydrocarbon Processing 1982 Refining Handbook, Gulf Publishing Co., Houston, Tex., 1982.

Technical Data Book and *Basic Petroleum Data Book, 1980*, API Refining Department, Washington, D.C.

PETROLEUM PRODUCTS

The importance of petroleum products is reflected in the great consumption of crude oil, the starting material for these products. In 1979, world crude-oil consumption averaged 10.2×10^6 m^3/d (64×10^6 bbl/d), a 50% increase over 1969. United States consumption accounted for almost 2.9×10^6 m^3/d (18×10^6 bbl/d), a 30% increase since 1969.

Petroleum products include power fuels for passenger cars, trucks, aircraft engines, and gas turbines (gasoline, jet fuel, diesel fuel and LPG); heating oils; illuminating oils (kerosene); solvents (petroleum spirits, aromatic solvents); lubricants (engine oils, greases); building materials (asphalts); insulating and waterproofing materials; and other products, such as cutting oils, heat-treating oils, hydraulic oils, petroleum jelly, etc.

The amounts produced are influenced by several factors, especially feed costs, processing costs and market demands For example, a greater proportion of gasoline is produced in the United States as compared to western Europe, where distillate and residual fuels are more in demand. Distillate fuels usually denote burner fuels, diesel fuels, and similar intermediate-volatility products taken as condensed vapors from a distillation process. Gasoline is also a distillate product but is identified separately (see Gasoline and other motor fuels). Residual fuels are the unvaporized products from distillation and include asphalt (qv), petroleum coke, and heavy industrial and bunker fuels.

Because of U.S. Government regulations for the control of automobile emissions, new automobiles are equipped with catalytic mufflers for the reduction of unburned hydrocarbons emanating from engine exhausts (see Exhaust control, automotive). Since catalytic mufflers are rendered ineffective by leaded fuels, the production of unleaded gasoline has increased (see Table 1). High compression engines tend to develop knock, so that refiners developed processes to improve antiknock properties, even though they may reduce gasoline yields.

Table 1. Distribution of Gasoline Production, Vol%

Year	Unleaded	Leaded Regular	Leaded Premium
1978	33.3	54.4	12.3
1980[a]	49.4	43.2	7.5
1982[a]	63.6	32.2	4.2

[a] Estimated.

Table 2. Specifications for Automotive Gasoline (ASTM D 439-79)

Property	Volatility class A	B	C	D	E
octane number[a]	no limit specified				
distillation temperature, °C					
10% evaporated, max	70	65	60	55	50
50% evaporated, min	77	77	77	77	77
50% evaporated, max	121	118	116	113	110
90% evaporated, max	190	190	185	185	185
end point, max	225	225	225	225	225
temp for vapor–liq ratio of 20, min, °C	60	56	51	47	41
vapor pressure, max, kPa[b]	62	69	79	93	103
lead content, max, g/L					
unleaded grade	0.013	0.013	0.013	0.013	0.013
conventional grade	1.1	1.1	1.1	1.1	1.1
corrosion, copper strip, max, no.	1	1	1	1	1
gum, existent, max, mg/100 mL	5	5	5	5	5
sulfur, max, wt%					
unleaded grade	0.10	0.10	0.10	0.10	0.10
conventional grade	0.15	0.15	0.15	0.15	0.15
oxidation stability, min	240	240	240	240	240

[a] Measure of antiknock quality for spark ignition.

[b] To convert kPa to mm Hg, multiply by 7.5.

Table 3. Specifications for Diesel Fuel Oils (ASTM D 975-78)

Property	Grade		
	1-D	2-D	4-D
distillation (90%) point, °C	288 max	282–338	
flash point, min, °C	38	52	55
water and sediment, max, vol%	0.05	0.05	0.05
carbon residue on 10% bottom, max, %	0.15	0.35	
ash, max, wt%	0.01	0.01	0.01
viscosity at 40°C, kinematic, mm^2/s (= cSt)	1.3–2.4	1.9–4.1	5.5–24.0
sulfur, max, wt%	0.50	0.50	2.0
corrosion, copper strip, max, no.	3	3	
cetane number[a], min	40	40	30

[a] Measure of ignition quality for autoignition.

Aviation gasoline for commercial aircraft has been largely replaced by jet fuels, also called aviation turbine fuels (see Aviation and other turbine fuels).

Many petroleum products of commerce are blended, compounded, or otherwise modified from primary stock. Trade names are not necessarily those given by the refiner, and a product may have several trade names, depending on uses. Specifications for automotive gasoline are given in Table 2, and for diesel fuel in Table 3.

Distribution and Marketing

The marketing and distribution of petroleum products are large-scale operations involving delivery of 2.9×10^6 m^3/d (18×10^6 bbl/d) of a wide range of products from the refineries to storage and distribution points, and ultimately to the consumer.

Wholesalers are the intermediaries between the refinery and the customer. Home-heating oil is distributed by the wholesaler directly to the consumer, whereas gasoline is resold by the retail service station. Wholesalers use bulk terminals and bulk plants consisting of storage tanks plus handling and transportation equipment.

Gasoline retailers sell more than 80% of the gasoline purchased in the United States in about 170,000 primary service stations as well as a number of secondary outlets. About 85% of all service station are operated by independent, self-employed local businessmen. They buy the gasoline from a refiner–supplier or a jobber and sell it on their own account. Independent dealers may either lease their stations from an oil company or a distributor, or own their station outright.

HAROLD L. HOFFMAN
Gulf Publishing Company

Annual Book of ASTM Standards, American Society for Testing and Materials, Philadelphia, Pa., 1980, Pts. 23, 24 29.

PHARMACEUTICALS

The U.S. pharmaceutical industry has emerged since World War II as a significant component of the U.S. industrial complex, total global sales probably exceeding 20×10^9 annually by 1985. From about 1940 to 1965, the industry went through an impressive growth stage, wherein a large number of significant discoveries occurred.

In 1962, amendments to the Federal Food, Drug and Cosmetic Act promulgated regulations concerning the requirements for premarketing approval by the FDA. Requirements of proof of both safety and therapeutic efficacy and strict control of clinical testing have extended both the time and cost to introduce a new drug. It is estimated that the development of a new drug can take longer than seven years and cost more than 60×10^6. This has reduced the effectiveness of the patent period. Patent-coverage time available for recouping the research investment is shortened, and thus the generation of reasonable returns on investment is reduced. As trade-name drugs lose patent protection, they are produced as generic-name products by multi-sources.

The therapeutically active drug can be extracted from plant or animal tissue, or be a product of fermentation, as in the case of antibiotics (qv). Frequently, it is a synthetic compound, designed to correlate structure with therapeutic activity. Pharmacological activity is first tested on laboratory animals. Should the results be encouraging, physical and chemical properties are determined in the so-called preformation stage, and analytical procedures are developed for control. Biological characterization, drug product (ie, dosage form) development, patent application, clinical evaluation, packaging research, stability studies, and manufacturing procedures follow before filing the New Drug Application (NDA) in its entirety to the FDA.

The following are several organizations that represent the industry, pharmacists, and educational institutions: The Pharmaceutical Manufacturers Association, The National Pharmaceutical Council, The Proprietary Association, The American Association of Colleges of Pharmacy, The American Pharmaceutical Association, The National Association of Retail Druggists, and The National Pharmaceutical Association. Most national associations publish a journal as do most state pharmaceutical organizations. In addition, lists of recalled drugs are published by the APhA newsletter, *A Pharmacy Weekly*, the *FDA's Recall Report*, and the *Federal Register*.

Standards, sales, and distribution of prescription as well as over-the-counter (OTC) drugs are regulated by various Federal and state laws and regulations. The *United States Pharmacopeia* (USP) and the *National Formulary* (NF) are the recognized standards for potency and purity for most common drug products. They provide various chemical, physical, and biological tests and specifications. In the 1960s, a review of nonprescription drugs was begun for safety, efficacy and labeling, and some prescription drugs were changed to OTC status. As of 1984 this review had not been completed officially, and FDA regulations are yet to be formalized.

Manufacture

The common dosage forms of pharmaceuticals include liquid solutions, liquid dispersions, semi-solids (eg, creams, ointments), solids (powders, capsules, pills, troches, granules, compressed and coated tablets, and pellets), and suppositories (semi-rigid, plastic). Compressed tablets offer convenience, stability, accuracy, and in general, good bioavailability of active ingredients. Bioavailability is the amount of a drug that is absorbed from an administered dosage form at a certain rate by the body.

Compressed tablets are produced directly by compression of powder blends or granulations. In wet granulation, the powdered drug and diluent are blended with a dispersion of the binder excipient (eg, gelatin) to a consistency that can be screened to, eg, 840–1800 μm (10–20 mesh). The coarse granules are dried on trays in hot-air ovens or fluid-bed dryers. In slugging or dry granulation, the ingredients are blended and compressed on heavy-duty tablet presses. Direct compression is an excellent, relatively simple and time-saving method. The ingredients are blended and compressed directly into the tablet without a granulating procedure. However, not all substances can be compressed directly, and a granulation step is then required. Direct compression may well achieve great popularity in the future as manufacturers replace old equipment and/or build new units.

Tablet-compression machines (presses) are equipped with dies, which hold a measured volume of material to be compressed; upper punches, which exert pressure on the down stroke; and lower punches, which control the volume of the die fill and thus the tablet weight. The actual compression process is a cycle of die fill, compaction, and ejection. Compressed tablets that are composed of several layers require specially adapted presses with multiple feed hoppers.

Formulations for most types of dosage forms for drug products contain several types of inert adjuvant ingredients necessary for proper preparation and therapeutic performance. Effervescent tablets, for example, disintegrate by virtue of the chemical reaction occurring in water between several included ingredients (eg, sodium bicarbonate and citric or tartaric acid) to achieve release of carbon dioxide. Colors and flavors increase the elegance and acceptability of the product; colors also may be used for identification. Sugar or film coatings offer protection from moisture, oxygen, or light and mask unpleasant taste or appearance. Enteric coatings delay the release of the active ingredient from the dosage form so it does not occur in the stomach, when for various reasons (eg, gastric upset, acidic degradation) it is more appropriate for the drug to be released in the small intestine. Sugar coating is applied in rotating pear-shaped pans. Several layers are applied including shellac, sucrose, sucrose solutions (colored or noncolored) and a polish of waxy composition. Sugar coating increases the weight of the tablet, is time-consuming, and is a skilled operation. Film coating in pans with various polymers and plasticizers is a much quicker process, which can be automated or programmed more easily. In compression coating, an outer coating is compressed around a core tablet, producing a tablet within a tablet; this calls for a very specific and specialized pressure.

Capsules are made as hard gelatin capsules, where powders or granules are enclosed in rigid-gelatin shells, or as soft gelatin capsules that contain glycerol as well as gelatin to maintain plasticity of the outer shell. Powder, semi-solids, or certain liquids (ie, those that do not soften or dissolve the gelatin shell) can be enclosed; the former are made in two sections, cap and body, whereas the latter have their shell formed, filled, and heat-sealed, all within one operation, utilizing specially designed equipment.

The size system of the capsule is inversely related to the volume, ie, a No. 1 capsule is larger than a No. 2, and a No. 0 is larger than a No. 1. For human consumption, Nos. 0–2 are most common. Hard-shell gelatin capsules vary in size from those that contain 100 mg to those for veterinary use, which may contain several grams (see Veterinary drugs).

Prolonged or sustained-action dosage forms maintain safe and effective drug concentrations in the blood two to three times longer than can be obtained with regular tablets or capsules (see Pharmaceuticals, controlled release). Such dosage forms as capsules and tablets were first investigated and developed in the late 1950s and early 1960s. The earlier products depended, for example, on multicoatings which included tiny pellets, erosion from waxy cores, leaching from plastic matrices, and ion-exchange concepts for the slow, steady, and prolonged release of the active ingredient(s). Renewed interest in such prolonged (sustained) drug products has occurred in the last 5–6 yr period. Current developments include gradual release by transdermal transport diffusion to the blood system from externally applied "medicated patches" and osmotic-pressure-dependent release from ingested tablets with tiny, laser-cut openings, as examples. There will be much more research and development in the next 10–20 yr of such concepts for drug delivery, utilizing very advanced methodology and equipment.

Liquid-dosage forms are simple aqueous solutions, syrups, elixirs, and tinctures prepared by dissolution of solutes in the appropriate solvent system. Naturally occurring materials with therapeutic activity are prepared in such dosage forms by extractive processes, especially percolation and maceration. Solutions for external or oral use do not require sterilization, but generally contain antimicrobial preservatives. Ophthalmic and parenteral solutions require sterilization (see Sterilization techniques).

For the preparation of liquid dispersions, such as suspensions and emulsions, colloid mills and homogenizers, respectively, are used.

For semisolid forms, such as creams, ointments, and pastes, the base ingredients (eg, petrolatum, waxes) are melted together, the powdered drug components are added, and the mass stirred with cooling. Suppositories are semirigid plastic forms, designed to deliver a unit dose to a body cavity; they either melt at body temperature or dissolve in the fluids of the body cavity into which they are inserted.

Parenteral dosage forms are designed for injection into the body-fluid system subcutaneously (sc), intramuscularly (im), or intravenously (iv). Formulation requirements are sterility, freedom from pyrogens, isotonicity, and pH adjustment, as much as possible, to that of body fluids. Water for injection is prepared by reverse osmosis (qv) or distillation (qv). Fixed oils, eg, cottonseed or peanut oil or liquid esters, may also be used as solvent systems. Containers must be able to withstand the heat and pressure of sterilization and have traditionally been made of glass although plastics are being increasingly used (see Barrier polymers). Aseptic techniques must be followed scrupulously throughout production and packaging.

Pressurized containers to deliver drug products through appropriate systems of valves and actuators are used for the external application of lotions, creams, and some cosmetics and toiletries, and for the application of contraceptive foams (see Contraceptive drugs). Nasal applications of drugs from such products/devices is receiving some current research attention and may become a valid mode of drug administration.

PAUL ZANOWIAK
Temple University

A. Osol, ed., *Remington's Pharmaceutical Science*, 16th ed., Mack Publishing Co., Easton, Pa., 1980.

G.S. Banker and C. Rhodes, eds., *Modern Pharmaceutics*, Marcel Dekker, New York, 1979.

H.C. Ansel, *Introduction to Pharmaceutical Dosage Forms*, 3rd ed., Lea & Febiger, Philadelphia, Pa., 1981.

L.S. Goodman and A. Gilman, *The Pharmacological Basis of Therapeutics*, 4th ed., Macmillan, New York, 1970, pp. 29–35.

PHARMACEUTICALS, CONTROLLED RELEASE

Conventional dosage forms (tablets, capsules, drops, liquids, sprays, ointments, and injectables) usually produce high drug concentrations immediately following administration. The concentration declines gradually through the dosing interval until the next dose causes it to peak again. This peak-and-trough pattern is owing to a lack of control over the rate of drug release into the body. Such nondrug components that the product contains are simply to preserve the drug, prevent its contamination, facilitate its manufacture, and retard or facilitate dissolution after administration (see Pharmaceuticals).

In the 1970s, controlled-delivery pharmaceuticals were developed that incorporate technology to control the rate of drug release into the body. They also release the drug over much longer dosage intervals than ordinary dosage forms can provide. Control over drug release provides less fluctuation of drug levels in blood and tissues; as a result, toxic effects can be greatly reduced for some drugs. In addition, the prolonged drug delivery available from controlled-release pharmaceuticals can improve patient compliance by simplifying the regimen. Some of these products—which are also called rate-controlled or time-release pharmaceuticals—are applied and replaced by the user; others require administration by the physician. They differ from sustained release and similar formulations by the predictability of their drug release *in vivo* and *in vitro*.

The drug is chosen for its therapeutic effect, pharmacokinetic behavior, physiochemical characteristics, and half-life. A short biological half-life is preferable for some controlled-release pharmaceuticals because it permits quick reduction of drug concentration, if necessary. Controlled-release pharmaceuticals can be used for local or systemic therapy.

A drug- and biocompatible polymer module regulates the drug release. It consists of a drug reservoir, a rate controller, an energy source for drug transfer, and a delivery portal. A platform integrates the various elements and permits appropriate site placement. For example, the ocular therapeutic system delivering pilocarpine (Ocusert Pilo-20 system, ALZA) uses poly(ethylene-*co*-vinyl acetate) (EVA) as the rate-controlling membrane and platform. The unit is small enough to be worn in the *cul-de-sac* of the eye.

Several sites can be used for drug administration. The gastrointestinal tract is the usual site for systemic therapy because of the ease with which medication can be swallowed. Several oral commercial products release drugs more slowly than conventional tablets and capsules. For example, Searle Laboratories markets a theophylline capsule called Theo-24 containing hundreds of beads coated with theophylline and a chemical timing complex to control drug release.

Because of variations in epidermal thickness, sweating, and blood flow near the surface, the skin is limited as a portal for systemic therapy. However, a nitroglycerin ointment indicated for angina pectoris (Nitro-Bid, Marion Laboratories) provides relief despite variable skin permeability and the inherent lack of control over the drug quantity absorbed.

Another transdermal system is Transderm-Scop (CIBA), which supplies scopolamine for the prevention of motion sickness. It consists of a thin, small, adhesive-backed disk and is placed on the skin behind the ear. It provides a 75% reduction in the incidence of nausea and vomiting for three days (see Fig. 1) and reduced incidence of parasympatholytic side effects compared to the drug's oral or injectable forms. A similar transdermal system (Transderm-Nitro, CIBA), worn on the upper torso, has recently become available for nitroglycerin administration (see Cardiovascular agents); it is more convenient to apply than the previously mentioned ointment, and prevents dose-dumping by means of a rate-controlling membrane.

The uterus offers a route for localized administration of contraceptive agents. For example, the Progestasert intrauterine progesterone contraceptic system (ALZA) provides effective protection for at least one year without the side effects associated with systemic forms of hormonal contraception. It has a T configuration of solid cross arms and a hollow stem composed of EVA copolymer. The stem contains 38 mg progesterone dispersed in silicone oil, which acts as the thermodynamic diffusional energy source.

Several energy sources are being investigated for drug transfer. The newest developments for continuous iv or subcutaneous drug infusion are miniaturized electromechanical pumps attached to the patient, who might then become ambulatory. Examples are the Mill Hill Infuser and the Auto-Syringe, each weighing ca 400 g. Another wearable self-contained infuser (AR/MED Infuser, Travenol) provides rate-controlled systemic therapy by means of an elastomeric drug reservoir that, as it deflates, exerts steady pressure on a drug solution.

An electrical feedback system is also used in a cardiac pacemaker. It detects the atrial rate and utilizes this information to program electrical stimulation of the ventricles when needed.

For oral therapy, an osmotic pump has been developed that consists of a core containing the drug, surrounded by a semipermeable membrane with a delivery orifice (see Fig. 2). The system resembles a conventional tablet, but functions through the osmotic activity of the drug, which draws water from the gastrointestinal tract through the semipermeable membrane into the drug core. The drug solution leaves the interior of the tablet through the small orifice at the same controlled rate that the water enters by osmosis. The system maintains a constant delivery rate of drug until the latter is completely dissolved.

Delivery systems based on the same technology are available for research. The OSMET drug-delivery module for clinical studies is loaded with the drug of interest in solution or suspension. Drug delivery at a constant rate begins immediately after the patient swallows the module. For animal studies, ALZET mini-osmotic pumps are implantable units that provide continuous drug release for weeks at a time at rates chosen by the investigator.

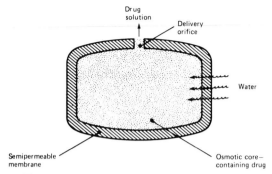

Figure 2. Cross section of an oral osmotic system.

In addition to inert polymers that control drug or water diffusion, polymers can be designed to dissolve, swell, or degrade in a controlled manner, thereby releasing the incorporated drug. It is, however, necessary that the polymer be transformed into a water-soluble product that evokes no limiting toxic response if the spent product is not to be reclaimed. The drug is locked into the polymer matrix (the drug reservoir) before its transformation. The surface area of the polymer–drug mass, the drug concentration and solubility characteristics, and the rate of polymer transformation affect the rate at which the drug is delivered. The polymer structure undergoes a phase change during which it or its by-products are removed or eliminated from the body, either during drug release or when most of the drug is deployed.

The polymers investigated for such systems include polyester, polyorthoesters, polyacids, hydrogels, celluloses, polypeptides, polyaminotriazoles, and albumin beads. Therapeutic agents investigated for delivery from polymeric matrices include narcotic antagonists (naloxone), steroids, antimalarials, insulin, enzymes, antibacterials, ophthalmic agents, vitamins, and anticarcinogens.

Encapsulation with liposomes promotes the passage of drugs across cell-membrane barriers, prolongs plasma lifetime of drugs with short biological half-lives, and directs drug disposition. These aqueous compartments bounded by bimolecular lipid layers carry the drug-containing platform closer to the target site, thus providing higher concentrations than the usual systemic therapy. The quantity of the drug or agent administered can, therefore, be reduced considerably.

HARRIET BENSON
BARBARA HARLEY
EDWARD E. SCHMITT
ALZA Corporation

J. Urquhart, ed., *Controlled-Release Pharmaceuticals*, American Pharmaceutical Association, Washington, D.C., 1981.

J.R. Robinson, ed., *Sustained and Controlled Release Drug Delivery Systems*, Marcel Dekker, Inc., New York, 1978.

K. Heilmann, *Therapeutic Systems, Pattern-Specific Drug Delivery: Concept and Development*, Georg Thieme Publishers, Stuttgart, FRG, 1978.

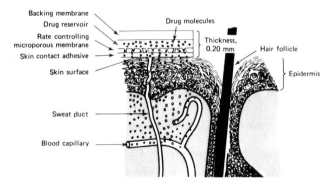

Figure 1. Diagram of the scopolamine transdermal therapeutic system (not to scale). The system is a thin (0.2 mm thick) unit, comprising a steady-state reservoir that contains the drug in a gel, sandwiched between an impermeable backing layer and a rate-controlling microporous membrane. On the epidermal side of that membrane is an adhesive gel layer that secures the system to the skin and contains the priming-dose scopolamine reservoir. After the patient places the system on the skin surface, drug diffuses through intact skin into capillaries within the dermis, where it is carried into the general circulation.

PHARMACEUTICALS, OPTICALLY ACTIVE

The biological activities of many compounds, eg, drugs, hormones, herbicides (qv), insecticides, and sweeteners (qv), are influenced by stereochemical factors (see Hormones; Insect control technology). For example, D-enantiomers of many amino acids (qv) are sweet, whereas the L isomers are bitter or tasteless. The fate of a drug is determined by the sum of the constants with which its chemical constitution has endowed it, but these constants can only be applied to the animal species in which they were determined. Optical activity, a function of the asymmetric or dissymmetric nature of a compound, represents one such parameter. However, optical antipodes, ie, enantiomers, are unique probes for the investigation of chemical–biological interactions or pharmacological mechanisms since, in the case of a single asymmetric center, physical and chemical characteristics for the two enantiomers are the same and enzyme systems are often highly enantioselective, ie, one enantiomer interacts to a greater extent with a receptor or active site on an enzyme. Generally, enantiomeric potency ratios increase with the activity of the more potent isomer.

Stereoisomers are substances with the same molecular formula but differing in their arrangement of atoms in space. Enantiomers are stereoisomers having a mirror-image relationship; they have the same physical and chemical properties except for their ability to rotate a plane of polarized light. The dextrorotary isomer (+ or d) rotates the plane of polarized light clockwise, whereas the levorotary isomer (− or l) rotates the plane of light counterclockwise. This property has no bearing on the interaction of drugs with the receptor surface. Rather, it is the absolute configuration of the enantiomers that is important in drug–host interaction. Although the letters D and L denote absolute configuration, there is no simple relationship between the sign of rotation (d or l) and the absolute configuration.

Generally, naturally occurring α-amino acids are designated as D (or R) or L (or S); however, for example, cysteine is D(S) or L(R) because of the sequence priority of CH_2SH over CO_2H.

The Role of Amino Acids

The precise role of D-amino acids in the metabolism of the producing organism is obscure. Biochemical studies have shown that ^{14}C-L-valine may be the precursor for both L- and D-labeled acids, that D-amino acid residues in microbial peptides may arise from the L isomer, and that the D isomers are not precursors for the L form.

Physical Methods for Investigating Drug-Receptor Interactions

These include the identification of stereospecific binding sites or receptors in animal brain and other organs by incubating radiolabeled drugs with organ homogenates; pharmacological receptors have been isolated and characterized. Computer graphics is a useful tool for modeling complex structural interactions of receptor sites and drugs. These interactions have also been studied with nmr spectroscopy. Thus, addition of chymotrypsin causes a wider signal broadening in the ^{1}H-nmr spectrum for D as opposed to L-tryptophan (see also Pharmacodynamics).

Preparation of Chiral Drugs from Synthons of Known Absolute Configuration

Chirality or other stereochemical activity relationships provide evidence concerning the specificity of chemical–biological interactions. The potency of structurally nonspecific compounds, eg, anesthetics (qv), hypnotics, and sedatives, is determined mainly by physiochemical rather than specific structural requirements (see Hypnotics, sedatives, anticonvulsants).

The pharmacology of several anticonvulsants prepared from amino acids of known absolute configuration has been investigated in .mice. (R)-Glutarimides (1) generally exhibit equal or more rapid onset of anticonvulsant action and possess greater neurotoxicity than their (S) isomers.

(1)

(R)-glutarimides

(S)-Viloxazine (2) is a novel antidepressant which exhibits stereoselective inhibition of norepinephrine uptake into mouse heart *in vivo*. It has the same geometric orientation of aryl, aminomethyl, and oxygen functions as in (R)-norepinephrine. The (S) isomers are at least ten times

more potent than the (R) isomers in reserpine-induced hypothermia and locomotor activity assays.

A number of neuroleptics exhibit stereoselective activity. Thus, (+)-(3S,4aS,13bS)-butaclamol is at least 100 times more potent than its (−) isomer in its ability to antagonize amphetamine-induced stereotyped behavior in rats and (S)-octoclothepin is 36 times more potent than its R-enantiomer in this test.

The dopaminergic tetralin derivative (3) exhibits enantioselectivity for binding, inhibition of prolactin secretion, and stimulation of adenylate cyclase and renal vasodilation; the most active isomer is (2R)-tetralin.

(2)

(S)-viloxazine

(3)

The cholinergic agonist (2S)-muscarine is 200–800 times more active than its enantiomorph; correct geometric configuration is important for this activity. The activity of the long-acting anticholinergic benzetimide, which has both central and peripheral effects, results exclusively from (R)-benzetimide. These enantiomeric potency differences may in part reflect differences in distribution, since the concentration of the (R) isomer, after subcutaneous administration of the racemate in rats, is greater in the brain, especially in the caudate nucleus; the (R) isomer is active for a longer time than its enantiomer.

Examples of stereostructure–activity relationships in drugs affecting the cardiovascular system include propanolol (4), whose (−) isomer is 100 times more potent as a β-adrenergic blocking agent and is utilized as an antihypertensive drug; α-methyl-DOPA (5), whose activity is in the (S) enantiomer; and captopril, whose antihypertensive activity is attributed to its binding to the active site of the angiotensin-converting enzyme.

(4)

propranolol

(5)

α-methyl-DOPA

Selective toxicity to various life forms may be observed using optical isomers. Although both isomers of tetramisole exhibit equal acute toxicity to the host, the (S) isomer, levamisole, exhibits activity toward nematodes and is useful as a broad-spectrum anthelmintic (see Chemotherapeutics, anthelmintic). For selective antistaphylococcal activity in the α-aminobenzylpenicillin (ampicillin) series, the (R) side chain affords analogues four times as potent as the (S) side chain. The adrenocorticoids and sex steroid hormones (qv) generally are highly stereoselective in terms of their activity.

Chirality and Drug Metabolism

Selective metabolism of one enantiomer may explain enhanced potency or toxicity of another enantiomer. Formation of diastereomeric complexes is to be expected from the interaction of a chiral macromolecular site with each enantiomer of a racemic drug. Substrate stereoselectivity describes the phenomenon whereby one enantiomer or diastereomer predominates during the metabolic process in which formation of one diastereomer predominates during the metabolic induction of a second asymmetric center. Product stereoselectivity defines the phenomenon whereby a prochiral center is preferentially and metabolically converted to one of two possible enantiomers. Sedative-hypnotic agents, eg, gluthimide and pentobarbital, also undergo stereoselective metabolism.

Methodology

Resolution involves formation of diastereomeric derivatives by means of an optically active resolving agent. Unlike enantiomers, diastereomers have different physical properties and may be separated by fractional crystallization, gas-liquid chromatography, thin-layer chromatography, and liquid chromatography. The separated diastereomer is the source of the pure enantiomer. There are rules for predicting which isomer of a diastereomeric salt will preferentially crystallize from a given medium.

Liquid chromatography has recently gained in popularity. Diastereomeric oxazolines have been separated by preparative, medium-pressure liquid chromatography from silica gel columns. The cost of the unit, which can analyze gram quantities, is moderate.

Optically active packing materials may be used in liquid chromatography. A novel approach uses silica gel coated with chiral polyethers (qv) to resolve racemic amino acid esters and primary amine salts.

Asymmetric synthesis. In asymmetric synthesis an achiral unit in an ensemble of substrate molecules is converted by a reactant into a chiral unit in such a manner that the stereoisomeric products are obtained in unequal amounts. A prochiral function serves as the precursor for a chiral product during the reaction. Optical purity is expressed in terms of enantiomeric excess, EE.

Reactions involving selected noncarbohydrate chiral reagents. Chiral amino acids and peptides are synthesized involving selected noncarbohydrate chiral reagents. For example, $(S)(+)$-alanine (**6**) is prepared in 63% EE by catalytic hydrogenation of the asymmetric imine (**7**) followed by hydrogenolysis.

A general synthesis of (S)-amino and (S)-methylamino acids in high optical yield utilizes (S)-proline methyl ester as a recoverable chiral reagent.

Reaction of (R,S)-oxazolone with (S)-amino methyl ester in the presence of triethylamine in THF at room temperature for 12 h yields (R,S)-peptides in ca 50% EE. Peptides of known absolute configuration can also be prepared in moderate EE using N-amino-(S)-proline as the starting material.

The synthesis of chiral alkanoic acids involves the use of synthetic chiral oxazolines. This method is highly stereoselective and the outcome of the asymmetric induction is predictable. When two different alkyl groups are introduced successively, the resulting acid is preferentially of the (S) configuration if the group of lower priority is introduced first; introduction in the reverse order gives the (R) configuration. Organolithium reagents or iodoethoxytrimethylsilane is used for the alkylation. Chiral (deuterated/tritiated) methyl groups, —CH^2H^3H, are of biochemical significance, eg, in chiral acetic acid.

In the synthesis of chiral ketones, asymmetric alkylation of cyclohexanone using chiral imines give low optical yields (26–37% EE) of 2-alkylcyclohexanones. However, 82–95% EE is achieved by using conformationally constrained metalloenamines.

The synthesis of chiral bicyclic ring systems is characterized by its potential application for the stereoselective synthesis of steroids and terpenes. Thus, cyclization of trione (**8**) in the presence of catalytic amounts of (S)-proline or another chiral amino acid in DMF or acetonitrile yields chiral aldol products (**9**) in high chemical and optical yields.

(8)   **(9)**

$n = 1$ or 2
$R = CH_3$ or C_2H_5
$R' = H, CH_3, (CH_2)_2CO_2CH_3, (CH_2)_2C_6H_4$-$m$-$OCH_3$

Reactions involving chiral carbohydrates. Carbohydrates serve as synthons and reagents for a multitude of asymmetric inductions. Protected carbohydrates with exposed carbonyl functions have been employed for the synthesis of chiral tertiary alcohols of known absolute configuration; the latter may serve as precursors for cholinergic agents.

Chiral metal – organic compounds and complexes. A versatile and commonly employed chiral hydroboration reagent is di-3-pinanylborane (**10**), available in both enantiomeric forms from naturally occurring $(+)$ α- or $(-)$ α-pinene (see Hydroboration). Relatively unhindered cis olefins yield trialkylboranes (**11**), which upon oxidation with alkaline hydrogen peroxide, yield (R)-alcohols at 76% EE from $(-)$-(**10**).

Chiral rhodium phosphine complexes are used as catalysts for asymmetric hydrogenation of substitutes enamides, α,β-unsaturated acids, and vinyl acetates. (S,S)-Chiraphos and (R)-prophos are two readily available phosphine ligands used for the synthesis of (R)- and (S)-amino acids, respectively.

(10)   **(11)**
$(-)$-di-3-pinanylborane

Donald T. Witiak
Muthiah N. Inbasekaran
Ohio State University

G.A. Crosby, G.E. Dubois, and R.E. Wingard, Jr. in E.J. Ariens, ed., *Drug Design*, Vol. VIII, Academic Press, Inc., New York, 1979, p. 261.

R. Bentley, *Molecular Asymmetry in Biology*, Vol. I, Academic Press, Inc., New York, 1969, p. 280.

A. Albert, *Selective Toxicity*, 6th ed., Methuen, London, UK, 1981.

PHARMACODYNAMICS

Pharmacology relates to all aspects of drug action, including synthesis, isolation, physiological and toxicological effects, therapeutic applications, and mechanisms of action. Drugs are chemical species that affect cellular functions (see Antibiotics; Chemotherapeutics; Hormones; Neuroregulators). The availability of a drug at its site of action is determined by absorption, metabolism, distribution, and excretion (see Fig. 1). These processes constitute the pharmokinetic aspects of drug action; the onset, intensity, and duration of action are determined by these factors as well as by the availability of the drug at its receptor site and events initiated by receptor activation. Pharmacodynamics is the study of drug action primarily in terms of drug structure, site of action, and the biochemical and physiological consequences of the action.

For a large number of drugs, the cell membrane is the locus of action. Some drugs, however, eg, the steroid and thyroid hormones, exert their effect intracellularly at the level of the genetic material.

The Receptor Concept

The structural requirements for drug interaction at the recognition site are elucidated by the study of structure–activity relationships

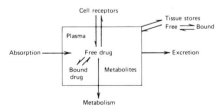

Figure 1. The interrelationship of the several processes that serve to regulate the concentration of a drug at its locus of action, ie, the cell receptors.

(SAR), in which the effects of systematic molecular modification of a parent structure are determined. It was found that two classes of agents act as receptors: those that bind to the receptor and initiate its specific response, ie, agonists, and those that bind but are unable to initiate response, ie, antagonists.

Structure–Activity Relationships

Many drug molecules are flexible structures and, although their conformations in the solution and solid states can be determined these bear no necessary relationship to that adopted at the receptor site. The SAR has been of great value in providing qualitative concepts of binding-site geometry, classifying receptors, providing evidence for the existence of new classes of receptor-specific drugs, and generating new therapeutically effective compounds.

Relatively unambiguous monotonic SARs occur where the activity depends upon the ionization of a particular functional group. A classic example is that of the antibacterial sulfonamides whose activity is exerted by competitive inhibition of the incorporation of p-aminobenzoic acid into folic acid (see Fig. 2).

Increasing attention has been paid to the generation of quantitative SARs, in which the effects of molecular substitution on pharmacological activity can be interpreted in terms of the physiochemical properties of the substituents. Biological activities may also correlate with electronic substituent factors alone, eg, the inhibition of acetylcholinesterase by diethyl phenyl phosphates. More commonly, however, multiparameter correlations can be made. Thus, for the relative sweetness of 4-nitro-2-aminobenzenes, the log relative sweetness = 1.03 σ^+ + 1.43 π + 1.58 ($n = 8$, $r = 0.97$), where σ^+ is a parameter describing the electronic effect, π is the hydrophobic constant, n is the number of compounds tested, and r is the correlation coefficient. Thus, linear free-energy correlations of biological activity with the physiochemical properties of molecules may be used to deduce the types of interaction involved in biological activity and to predict new compounds.

Quantum-mechanical approaches to the definition of SARs provide measurement of conformational energies, electron densities, and electrostatic potential maps of drug molecules.

Drug–Receptor Interactions

Pharmacological responses are generally graded and a relationship defined between the concentration (A) of a drug and the receptor response (R). The law of mass action can be applied to the basic equation of drug-receptor interaction:

$$A + R \rightleftharpoons AR \xrightarrow[\text{steps}]{\text{intermediate}} \text{response}$$

Assumptions can be made that response is proportional to the concentration of the drug-receptor complex and that maximum response occurs when all receptors are occupied. However, more complex interpretations are usually necessary.

A more direct receptor quantitation is achieved through measurement of the specific binding of drugs that are labeled to high specific activity with ^{3}H or ^{131}I to their receptors. Such binding has been demonstrated for many hormone systems.

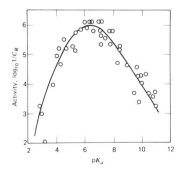

Figure 2. Relationship between antibacterial activity of sulfonamides (log 1/C_R) and pK_a of sulfonamide NH group. Courtesy of the *Journal of the American Chemical Society*.

$A + [\text{Receptor–channel}]_{\text{closed}}$

$\longrightarrow A - [\text{Receptor–channel}]_{\text{open}}$

(a)

$A + [\text{Receptor–cyclase}]_{\text{inactive}}$

$\longrightarrow A - [\text{Receptor–cyclase}]_{\text{active}}$

(b)

Figure 3. Two concepts of receptor activation. In (**a**), the receptor and ionophore are coupled directly and in (**b**), the response is generated through the intermediary step of cyclase activation.

Specific recognition and response initiation are the accepted attributes of drug–receptor interaction. In addition, target cells respond to changes in drug concentration by altering the number and affinity of their receptors. This regulatory process takes place with a wide variety of agents and is integral to drug–receptor interaction.

The function of the drug–receptor complex (D–R) are characterized by the action and the fate of the D–R complex. The sequence of action is: D–R is coupled to effector E, D–R–E initiates response, D–R–E desensitizes, and D–R–E dissociates to DR and E. The fate of the D–R complex may involve: D is released and R is reutilized; and D–R may be internalized resulting in the loss of receptors, degradation of the D–R complex, and mediation of intracellular effects.

Nonreceptor-Mediated Drug Action

It is assumed that general anesthetics (qv) do not owe their profound pharmacological effect to a specific receptor process but probably act by interference with neurotransmitter-release processes. Anesthetics may bind to hydrophobic sites on membrane proteins, thereby producing a direct perturbation of protein function.

Drug Action

The actions of neurotransmitters and many hormones are initiated at specific membrane receptors by two mechanisms: the drug-receptor interaction opens an ion channel resulting in alteration of ion flow across the membrane, or it leads to the production of a cyclic nucleotide, eg, cyclic 3',5'-adenosine monophosphate (c-AMP), which initiates the corresponding response (see Fig. 3). Receptor aggregation, driven by activator interaction with unoccupied mobile and homogeneously distributed receptors, may constitute a mechanism for initiating a membrane signal.

Ionic processes mediated by acetylcholine–receptor interaction occur at vertebrate skeletal muscle, vertebrate heart, cat cerebral cortex, and aplysia neurones. Acetylcholine receptor-mediated conductance increases in postsynaptic membranes of skeletal muscle involving an increase in permeability, primarily to Na$^+$ and K$^+$.

Epinephrine-mediated glycogenolysis is associated with the formation of c-AMP, and the actions of many agents are associated with the activation of adenylate cyclase and increases in terms of the intracellular concentrations of c-AMP, also called a second messenger. Most actions of c-AMP are mediated through activation of protein kinases and phosphorylation of substrate proteins. The discrete character of the recognition and catalytic components of hormone-sensitive adenylate cyclase has been demonstrated.

The action of steroids and thyroid hormones are exerted through interactions at specific intracellular receptors (see Thyroid and antithyroid preparations). Structure-activity studies indicate that the relationship of these receptors corresponds to biological response. Thyroid hormone may interact at a membrane receptor as a prerequisite for cellular entry.

D. J. TRIGGLE
State University of New York at Buffalo

A. Goldstein, L. Aronow, and S.M. Kalman, *Principles of Drug Action*, 2nd ed., John Wiley & Sons, Inc., New York, 1974.

A. Albert, *Selective Toxicity: The Physico-Chemical Basis of Therapy*, 6th ed., Chapman and Hall, London, UK, 1979.

J.W. Lamble, ed., *Toward Understanding Receptors*, Elsevier-North Holland, Amsterdam, Neth., 1981.

PHENAZINE, $C_{12}H_8N_2$. See Antioxidants and antiozonants; Azine dyes; Sulfur dyes.

PHENAZINE ANTIBIOTICS. See Antibiotics, phenazines.

PHENAZONE, $C_{11}H_{12}N_2O$. See Antipyrine in Analgesics, antipyretic and anti-inflammatory agents; Pyrazoles, pyrazolines, and pyrazolones.

β-PHENETHYL ALCOHOL, $C_6H_5CH_2CH_2OH$. See Benzyl alcohol and β-phenyl alcohols; Perfumes.

PHENETIDINES, $H_2NC_6H_4OC_2H_5$. See Analgesics; Antipyretics and anti-inflammatory agents.

PHENOBARBITAL. See Hypnotics, sedatives, and anticonvulsants.

PHENOL

Phenol is the name of monohydroxybenzene, C_6H_5OH, and the class of compounds containing one or more hydroxyl groups attached to an aromatic ring (see also Alkylphenols; Chlorophenols; Hydroquinone, resorcinol and catechol; (Polyhydroxy)benzenes).

Phenol is a white, crystalline compound with a characteristic odor. Some of its physical properties are given in Table 1.

The hydroxyl group imparts high reactivity, and the hydrogens ortho and para to it are highly reactive. Substitution leads first to 2- or 4-monoderivatives, then to the 2,4- or 2,6-derivative, and finally to the 2,4,6-derivative. Commercially most important is the condensation with formaldehyde; this reaction accounts for ca 40% of U.S. phenol consumption.

Table 1. Physical Properties of Phenol

Property	Value
congealing or freezing point, °C	40.91
boiling point at 101.3 kPa (atm)	181.84
d_4^{41} (liquid), g/cm³	1.0576
critical temperature, °C	419
critical pressure, MPa[a]	6.11
specific heat, J/(g·K)[b]	
at 22.7°C (solid)	1.41
viscosity at 60°C, mm²/s (= cSt)	2.47
heat of fusion, J/g[b]	122.2
heat of vaporization at bp, J/g[b]	487.9
heat of combustion, J/g[b]	−32,428
flash point, °C,	
closed cup	79

[a] To convert MPa to atm, divide by 0.101.
[b] To convert J to cal, divide by 4.184.

Manufacture

Phenol is mainly produced by the cumene hydroperoxide process. Benzene (qv) is alkylated to cumene (qv) which is oxidized by air oxidation to cumene hydroperoxide; cleavage gives phenol. Process temperature is 80–130°C; pressure and promoters, eg, sodium hydroxide or carbonate, are used. The cleavage reaction is conducted under acidic conditions with agitation at 60–100°C; nonoxidizing acids, eg, sulfur dioxide gas, are used. The mixture of phenol, acetone, and various by-products is neutralized with sodium phenoxide or another suitable base or with ion-exchange resins. Phenol yield is ca 93% based on cumene and 84% based on benzene. Approximately 0.40–0.45 kg acetone/kg cumene is produced as a coproduct; the ratio of phenol to acetone production is 0.6. Therefore, the economics of the process depend on the price obtained for the acetone (qv).

In the toluene–benzoic acid process, toluene is converted to benzoic acid (qv) by liquid-phase free-radical oxidation. The benzoic acid is oxydecarboxylated to phenol either as a melt or in a high boiling solvent at 220–250°C in the presence of steam and air and over a copper catalyst. The yield is 85% based on benzoic acid.

The economic advantage of the classic sulfonation process depends on the market of the main by-product, ie, Na_2SO_3. It is still used in Italy and the GDR, but no longer in the United States (see Sulfonation and sulfation). Other processes use monochlorobenzene or cyclohexanol/cyclohexanone.

Analysis and Storage

In the absence of other phenolic compounds, phenol is easily determined by the Kopeschaar method where phenol is brominated in aqueous solution to 2,4,6-tribromophenol; the consumed bromine is determined volumetrically. The colorimetric Gibbs method uses aminoantipyrine as reagent. Gas chromatography is widely used in industry for quality control.

In storage, phenol tends to become yellow, pink, or brown; the discoloration is promoted by water, light, air, and traces of iron or copper. When stored as a solid in nickel, glass-lined, or aluminum tanks, phenol remains colorless for a long time.

Health and Safety Factors

Phenol is toxic if absorbed through the skin and may be lethal; contact with the skin also causes burns. Phenol is a local anesthetic and no warning is given before it penetrates the skin. Spills must be washed off immediately with alcohol. Phenol fumes are irritating to eyes, nose, and skin. Phenol is a Class B poison, and containers must so be labeled.

Uses

The largest use is in phenolic resins (qv), which are produced by condensation with an aldehyde, usually formaldehyde. Bisphenol A is produced from phenol (2 mol) and acetone (1 mol). It is used for the production of epoxy and polycarbonate resins.

CARL THURMAN
Dow Chemical U.S.A.

Phenol: Properties, Uses, Storage, Handling, Dow Chemical Form 298-276-79, 1979.

PHENOLIC ALDEHYDES. See Aldehydes.

PHENOLIC ETHERS. See Ethers.

PHENOLIC FIBERS. See Novoloid fibers.

PHENOLIC RESINS

The largest-volume building blocks for phenolic resins are phenol and formaldehyde. Other important phenolic starting materials are the alkyl-substituted phenols; diphenols and bisphenol A are employed for applications requiring special properties (see Epoxy resins; Hydroquinone, resorcinol and catechol). Condensation products with acetaldehyde (qv) or furfuraldehyde are sometimes used.

Phenolic resins production has grown continuously for several decades and has about doubled since the 1960s in North America and western Europe.

Chemistry

The molecular structure and the physical properties are determined by the catalyst and the molar ration of the reactants. Because phenol is trifunctional and formaldehyde is difunctional, a formaldehyde-to-phenol ratio of 0.5–0.8 is used with an acid catalyst (sulfuric, p-toluenesulfonic, hydrochloric, phosphoric, or oxalic acids at pH 0.5–1.5) to produce a novolak. These products are thermoplastic, brittle solids, which react with various cross-linking substances that confer the desired properties on the resins. Higher formaldehyde-to-phenol ratios, ca 1.0–3.0, and basic catalysts give resoles, which are characterized by uncatalyzed advancement and cross-linking. Molecular weights are generally lower than those of novolaks. Molecular-weight advancement occurs until the glass temperature, T_g, of the product is sufficiently high that molecular mobility decreases and reactive end-group collisions subside. Resoles are most often used in liquid form, novolaks as solid products.

The use of certain metal salts, eg, zinc, magnesium, or calcium acetates, results in the formation of novolaks that have a disproportionally high concentration of o,o' repeat units. Consequently, more repeat units have open para positions than materials obtained with strong acid catalysts. The mechanism proposed for ortho direction involves chelation of the phenolic hydroxyl and methylene glycol molecules with metal ions.

Reactions conducted under basic conditions give predominantly methylol or methylene ether-containing resins, ie, resoles, the largest-volume phenolic resins. Resoles may be liquids, solids, or solutions; formaldehyde-to-phenol ratios vary, but are often in the 1.5 range. Catalysts include sodium, barium, or calcium hydroxides, sodium carbonate, and organic amines.

Low molecular weight phenolic resins must be advanced to a cross-linked network, usually by heat treatment. The limiting cured structure shown below results from a formaldehyde-to-phenol ratio of exactly 1.5 : 1.

100% cross-linked

Novolak resins are cured by the addition of a second component, such as a one-step resole resin, hexamethylenetetramine (8–15 wt%), paraformaldehyde, or trioxane.

Manufacture

A typical batch-production unit for novolak resins is shown in Figure 1. Resole resins can also be prepared in such equipment, if rapid discharge and proper cooling are provided. In a continuous novolak process, reactants and catalyst are prepolymerized in the first stage, and heated to 120–180°C under ca 700 kPa (101 psi) in the second stage. The product is recovered by flash devolatilization followed by separation of the residue into an aqueous and a resin phase. Phenol is recovered from the former, whereas the latter is dehydrated and recovered from the molten state using, eg, a belt flaker.

Carbon steel is the common material of construction of the equipment when light color is not necessary and/or low iron content is permissible. Otherwise, stainless steel is used. Strong acids promote an early and rapid exotherm and are added in increments for safety reasons. They are neutralized with lime or caustic before the dehydration step. Weaker acids give a milder reaction and a lighter-colored product.

High ortho novolak resins are manufactured in conventional equipment with a higher mole ratio of formaldehyde to phenol and different reaction conditions. Hydroxymethyl (methylol) or methylene ether groups are removed by heating to ca 110°C after removal of water. Yields of conventional novolaks of high molecular weight are 105–110% based on original phenol charge and a maximum of ca 5 wt% phenol in the final resin. Yields of high ortho resins are 90–100%. The latter have excellent cure speed and rheological properties.

Higher molecular weight resoles that are to be handled in solid form are made in relatively small reactors equipped with an agitator to prevent local gelation and circulate viscous fluids. Molten phenol is charged to the reactor, followed by a basic catalyst. Aqueous formaldehyde is added at a rate to maintain ca 60–70°C. Resole dehydration must be completed rapidly, preferably at < 100°C. Resins that are to be used in solid form generally display a 150°C hot-plate gel time of 20–110 s, as measured by the stroke test. When the desired end point is reached, the resin is discharged to a cooled surface and then refrigerated until used.

Most resoles are produced by a batch process of several steps. Initially, phenol and formaldehyde are heated at a controlled rate. Low molecular weight, water-soluble resoles are then cooled and neutralized. For advancement, partial condensation is accomplished in a second step, and water is removed by vacuum dehydration. The product is recovered as a syrup or in solution. Additives and modifiers are added at various stages. Liquid resoles are used mainly for plywood and insulation (qv) (see

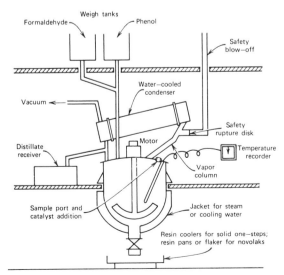

Figure 1. Schematic diagram of phenolic-resin production unit.

Laminated wood-based composites). They are generally made at a 1.25–1.40 formaldehyde-to-phenol ratio with a caustic catalyst. Reaction cycles are short, ca 45 min at 100°C or 3 h at 65°C; the products are recovered as 40–50% aqueous solutions.

Substituted phenols condensed with formaldehyde give specialty resins (ca 5% of U.S. phenolic resin production). Resorcinol increases cure speed and is predominantly used in water-soluble resole resins for tirecord adhesives (see Tire cord).

In the *in situ* manufacture of phenolic dispersions, the resins are prepared as stable, discrete, spherical particles in water as the continuous phase. A protective colloid allows *in situ* dispersion rather than postemulsification. The process involves a primary cycle under basic conditions at 90–105°C and a secondary cycle at 80–90°C under neutral conditions. These dispersions are used in pollution abatement. One-step solid resins and spray-dried resins are both redispersible in water.

Reaction at pH 4–7 under anhydrous conditions with a dual-catalyst system, ie, a divalent electropositive metal and an acidic material, gives high ortho novolaks with an o,o' content of 97 wt%.

Molding materials in pelleted form display great uniformity and low dust; they can be fed easily to various types of molding machines.

Analytical techniques include vapor-phase chromatography, ir spectroscopy for functional groups, thermal analysis, gel-permeation chromatography, nmr, ^{13}C nmr, and electron spectroscopy for chemical analysis (see Analytical methods).

Health and Safety Factors

No detrimental toxicological effects of cured phenolic resins have been reported. They are widely used in food applications, eg, can coatings, closures, and containers. In the uncured state, phenolic resins are skin sensitizers, and have to be handled with the usual precautions. Formaldehyde is a lacrimator and a suspected carcinogen and causes nasal irritation and coughing. Phenol itself is highly toxic, and production workers have to be protected. Air- and water-pollution control regulations must be observed.

Uses

Phenolic resins are used in wood composites where durability in high humidity is required, in fiber bonding, as laminates, in foundry resins, as abrasives, and in coatings and adhesives. Among the more recent developments are flame-resistant glass-reinforced phenolics (see Flame retardants), phenolic fibers, phenolic foams (see Foamed plastics), graphite-reinforced composites, and hydroxymethylfurfural/phenol resins. Macrocyclic compositions, ie, cyclic as opposed to linear condensation products (calixarenes), are also known.

GEORGE L. BRODE
Union Carbide Corporation

A. Knopf and W. Scheib, *Chemistry and Application of Phenolic Resins*, Springer-Verlag Publishing Co., New York, 1979.

G.L. Brode, J. Harding, T. Marrion, N.J. McCarthy, and S.W. Chow, *Synthesis of Phenolic Dispersions*, Proc. of 28th IUPAC Macromolecular Symposium, 1982.

N.J.L. Megson, *Phenolic Resins Chemistry*, Academic Press, Inc., New York, 1958.

PHENOLSULFONIC ACIDS.   See Sulfonic acids.

PHENOTHIAZINE.   See Haloalkylamines; Azine dyes; Psychopharmacological agents.

PHENYLACETIC ACID.   See Benzyl alcohol and β-phenethyl alcohol; Perfumes.

PHENYLENEDIAMINES AND TOLUENEDIAMINES.   See Amines, aromatic, phenylenediamines; Amines, aromatic diaminotoluenes.

PHORATE.   See Insect control technology.

PHOSGENE

Phosgene, Cl_2CO, is a highly toxic, colorless, low boiling liquid formed at elevated temperatures by oxidation of chlorinated solvents. It has been identified as the product of carbon tetrachloride metabolism in rat liver and aerobic incubation of hepatic microsomal fractions. Phosgene reacts with many classes of inorganic and organic reagents. It is soluble in hydrocarbons and organic acids and esters (see Table 1 for physical properties).

Manufacture

The manufacture of phosgene consists of the preparation and purification of carbon monoxide, preparation and purification of chlorine, metering and mixing of reactants, reaction of the mixed gases over activated charcoal, purification and condensation of phosgene, and recovery of traces of phosgene to assure worker and environmental safety. Continuous processing and a high degree of automation is required for condensation, purification, and storage.

Analytical and Test Methods

Trace quantities to a lower limit of 0.005 µg/L of air can be detected by uv spectroscopy. Both ir and gas chromatography have been used extensively to measure phosgene in air at 1 ppb–1 ppm. Special and multiple-column gas-chromatographic methods are used for more complex gas mixtures. Absolute determination of phosgene below 100 ppb has been reported using pulsed-flow coulometry. Quantities as small as 0.005 µg/L are measured with a tape (Universal Environmental Instruments). Phosgene dose-indicator badges for medical emergencies have been prepared.

Health and Safety Factors

Phosgene is highly toxic and may cause pulmonary edema; oxygen should be administered when breathing is restored. The permissible exposure TLV by volume in air is 0.1 ppm. Hazards can be minimized by the use of outdoor installations or extensive ventilation where phosgene must be employed indoors. Safety in handling depends mainly on the effectiveness of employee education, proper safety instrumentation, alert supervision, and safe equipment. Plant design should include facilities for neutralization, and water-fog equipment for emergencies.

Entrance into drains and sewers must be prevented. Phosgene waste is best removed by caustic scrubbing in packed columns. It should not be burned.

Uses

Phosgene is an important intermediate in the manufacture of plastics; practically all production is captive. The primary use is in the poly-

Table 1.   Some Physical Properties of Phosgene

Property	Value
melting point, °C	−127.84
boiling point (at 101.3 kPa = 1 atm), °C	7.48
density at 20°C, g/cm³	1.387
vapor pressure at 20°C, kPa[a]	161.68
latent heat of vaporization (at 7.5°C), J/g[b]	243
molar heat capacity of liquid (at 7.5°C), J/K[b]	100.8
molar heat of formation, kJ[b]	
from elements	218
from CO and Cl₂	108
molar entropy, J/K[b], at 25°C	284

[a] To convert kPa to psi, multiply by 0.145.
[b] To convert J to cal, divide by 4.184.

urethane (see Urethane polymers) industry, which consumes over 85% of the world's phosgene output. Polycarbonates (qv) consume ca 6%, and the remaining 9% is used for herbicides (qv).

EDGAR E. HARDY
San Diego State University

Chemical Safety, *Data Sheet SD-95*, Chemical Manufacturer's Association, Washington, D.C., Revised 1978.

H. Babad and A.G. Zeiler, *Chem. Rev.* **73**(1), 75 (1973).

W.F. Diller, *J. Occup. Health* **20**, 189 (Mar. 1978).

PHOSPHAMIDON.   See Insect control technology.

PHOSPHORIC ACIDS AND PHOSPHATES

Phosphates may be defined as compounds containing four phosphorus–oxygen (P–O) linkages. Compounds containing discrete, ie, monomeric PO_4^{3-} ions are known as orthophosphates or simply phosphates, linear condensed P–O–P chains as polyphosphates, cyclic rings as metaphosphates, branched polymeric materials and cage anions as ultraphosphates. Stoichiometrically, phosphates have been represented as combinations of oxides, eg, H_3PO_4 as $P_2O_5.3H_2O$ and Na_2HPO_4 as $P_2O_5.2Na_2O.H_2O$ (see Table 1).

Phosphoric Acid

Phosphoric acid, H_3PO_4, is the highest value inorganic acid marketed in the United States, and the second largest in terms of volume. Its main use is in the manufacture of phosphate salts, mainly for fertilizers (qv).

Phosphoric acid is a tribasic acid in which the first hydrogen is strongly ionizing ($K_1 = 7.1 \times 10^{-3}$), the second moderately weak ($K_2 = 6.3 \times 10^{-8}$), and the third very weak ($K_3 = 4.4 \times 10^{-13}$). Aside from its acidic behavior, phosphoric acid is relatively inert at room temperature. It is not reduced by strong reducing agents below 350–400°C; at elevated temperature it reacts with most metals and their oxides.

Pure phosphoric acid is a white crystalline solid (monoclinic), mp 42.35°C; the crystalline hemihydrate melts at 29.25°C. Properties of aqueous solutions are given in Table 2.

Phosphoric acid is manufactured either by the wet process or the furnace process. Over 90% of the U.S. production is made by the former, by digesting phosphate rock (apatite form) with sulfuric acid. The phosphoric acid is separated from the resulting calcium sulfate slurry by filtration. Chemical precipitation and solvent extraction are the main methods of purification; crystallization and ion exchange are also used.

Table 1.   Traditional Classification of Sodium Phosphates

Oxide ratio, R ($Na_2O + H_2O_{comp}$) : P_2O_5	Designation	General formula	Structure
> 3	phosphate + metal oxide (includes double salts and solid solutions)		mixtures
3	phosphate or orthophosphate, $n \le 3$	$Na_n H_{3-n} PO_4$	one phosphorus atom
3 > R < 2	mixture of ortho- and pyrophosphates		
2	pyrophosphate, $n \le 4$	$Na_n H_{4-n} P_2O_7$	two phosphorus atoms
2 > R > 1	polyphosphates, $n = 2, 3, 4, \ldots$	$Na_{n+2} P_n O_{3n+1}$	linear chains
1	metaphosphates, $n = 3, 4, 5 \ldots$	$Na_n(PO_3)_n$	cyclic or extremely long chains
1 > R > 0	ultraphosphates, $0 < x < 1$	$(x Na_2O)P_2O_5$	cross-linked chains and/or rings
0	phosphorus pentoxide	$(P_2O_5)_n$	P_4O_{10} or continuous structures

Table 2.   Physical Properties of Aqueous Solutions of Phosphoric Acid

Concentration, wt %		Density, 25°C, g/cm³	Bp, °C	Fp, °C	Viscosity, mPa·s (= cP) at		
H_3PO_4	P_2O_5				20°C	60°C	100°C
0	0	0.997	100.0	0	1.0	0.48	0.30
5	3.62	1.025	100.1	−0.8	1.1	0.54	0.33
10	7.24	1.053	100.2	−2.1	1.2	0.61	0.38
20	14.49	1.113	100.8	−6.0	1.6	0.78	0.48
30	21.73	1.182	101.8	−11.8	2.2	1.0	0.62
50	36.22	1.333	108	−44.0	4.3	1.8	1.1
75	54.32	1.573	135	−17.5	15	4.8	2.4
85	61.57	1.685	158	21.1	28	8.1	3.8
100	72.43	1.864	261	42.35	140	25	9.2
105	76.10	1.925	>300	16.0	600	70	19
115	83.29	2.044	>500			1500	250

The double-neutralization process is used to purify phosphoric acid during the production of large-volume detergent-builder phosphates.

In the furnace process, elemental white (yellow) phosphorus is burned in excess air; the resulting phosphorus pentoxide is hydrated, the heats of combustion and hydration are removed, and the phosphoric acid mist is collected. The processes are called wetted-wall, water-cooled, or air-cooled, depending on the protection of the combustion-chamber wall.

A potential environmental problem is created by the burning of phosphorus, which produces a persistent white cloud of phosphorus pentoxide and phosphoric acid droplets of such high obscuring power that it is used as a screening smoke by the military (see Chemicals in war). Neither precipitators nor Venturi scrubbers reduce the pentoxide content to a level acceptable in states with low plume-opacity regulations (see Air pollution); therefore, high efficiency mist eliminators are now standard practice.

Elemental phosphorus produced by the electrothermal process is a distilled product of high purity, which yields phosphoric acid pure enough for most industrial uses. The high heat of combustion of phosphorus [3.05 MJ/mol (730 kcal/mol)] can be recovered for process use such as evaporation of water from dilute phosphate solutions.

Furnace-grade phosphoric acid is used in metal treatment, refractories, catalysts, and food applications.

Phosphates

Both mono- and disodium phosphates are prepared commercially by neutralization of phosphoric acid with sodium carbonate or hydroxide. Trisodium phosphate (TSP) is crystallized from a wet-mix solution.

Monosodium phosphate is used as a pH buffer in acid-type cleaners, in boiler-water treatment, as a precipitant for polyvalent metal ions, and as an animal-feed supplement. Mixtures of mono- and disodium phosphates are used in textile and food processing (see Water, industrial water treatment), in the preparation of glazes and enamels, and in leather tanning, textile dyeing, and detergents. Trisodium phosphate is strongly alkaline and is a constituent of many heavy-duty cleaning compositions. The hypochlorite complex is used in disinfectants, scouring powders, and dishwashing formulations (see Bleaching agents).

In addition to the three simple potassium phosphates, the $K_2O–P_2O_5–H_2O$ system contains a number of crystalline hydrates and double salts.

The piezoelectric effect of monopotassium phosphate has led to its use in sonar systems and other electronic applications; it is also used in buffering systems and paper processing.

Tri- and dipotassium phosphates are marketed both as a solid and in a 50% solution; their main uses are as corrosion inhibitors in ethylene glycol antifreeze formulations and in coffee creamers.

Although thermally unstable, ammonium phosphates are present as solid phases in the $NH_3–P_2O_5–H_2O$ system from 0 to 75°C. Mono- and diammonium phosphates (MAP and DAP, respectively) are the world's leading phosphate fertilizers. Other applications are related to flame retardation and fire extinguishing (see Flame retardants).

The alkaline-earth phosphates are less soluble than those of the alkali metals. Apatite (calcium phosphate) ores supply the basic raw material for the production of phosphorus and its derivatives. Commercial calcium phosphates, produced from furnace-grade phosphoric acid, constitute the second largest volume phosphate salts. Most calcium phosphates are mixtures of several salts; their composition depends on the manufacturing conditions. Uses include fertilizers, animal feeds, food applications, and dentifrices.

The tertiary metal phosphates are of the formula MPO_4, where M = B, Al, Ga, Fe, and Mn. The boron and aluminum compounds are used as refractories.

Phosphate salts of heavy metals are insoluble in water. Zinc phosphate is used in dental cements (see Dental materials). Chromium and zinc phosphates provide corrosion protection and paint adhesion in some metal-treating applications. Mixed sodium–aluminum phosphates are utilized in some food applications.

Polyphosphoric Acids

The only clearly defined crystalline compositions of the H_2O–P_2O_5 system are three forms of phosphoric acid and their hemihydrates, pyrophosphoric acid, and crystalline P_4O_{10}. Amorphous condensed phosphoric acids are hygroscopic; they may be viscous, oily, gummy, and may consist of a mixture of glassy and crystalline materials.

Linear-polyphosphoric acids are strongly hygroscopic and undergo viscosity changes and hydrolysis to less complex forms when exposed to moist air (see Table 3). Upon dissolution in excess water, hydrolytic degradation to phosphoric acid occurs.

Pyrophosphoric (diphosphoric) acid, $H_4P_2O_7$, crystallizes in two forms (mp = 54.3 and 71.5°C); acid salts are known. Tripolyphosphoric (triphosphoric) acid, $H_5P_3O_{10}$, occurs in varying amounts as a component of condensed phosphoric acids containing more than ca 72% P_2O_5.

The largest use of polyphosphoric (superphosphoric) acids is as an intermediate in the production of liquid fertilizers. Condensed acids of 82–84% P_2O_5 are employed as catalysts in the petroleum industry. Polyphosphoric acid is also used as a dehydrating agent and in the production of phosphoric esters and agricultural chemicals.

Polyphosphates

The condensed phosphates are derived from phosphates by the loss of water. These materials range from simple diphosphates (pyrophosphates) to long-chain polymeric structures with molecular weights in the millions (10^6).

Polyphosphates are resistant to chemical attack, but are susceptible to hydrolysis. Short-chain polyphosphates hydrolyze without the concurrent formation of cyclic metaphosphates; in this respect they differ from the long-chain acids. Many polyphosphates form water-soluble complex ions, a phenomenon called sequestration (see Chelating agents). This property forms the basis for the water-treatment and detergent applications.

Pyrophosphates

The simplest linear condensed phosphates ($M_{n+2}P_nO_{3n+1}$, where n = 2, 3, 4 . . .) are pyrophosphates ($M_4P_2O_7$). A water molecule is eliminated from condensed orthophosphates:

$$2\,MH_2PO_4 \xrightarrow{\Delta} M_2H_2P_2O_7 + H_2O$$

$$2\,M_2HPO_4 \xrightarrow{\Delta} M_4P_2O_7 + H_2O$$

Insoluble pyrophosphates are obtained by treating a soluble salt of the desired cation with a sodium pyrophosphate solution.

Tetra- and disodium pyrophosphates (sodium acid phosphate) are prepared by thermal dehydration of di- and monosodium orthophosphate, respectively. Tetrasodium pyrophosphate is used as a builder in detergent and cleaning formulations (see Surfactants and detersive systems), in food applications (see Food additives), and as a deflocculant in drilling muds, dyes and inks. Sodium acid pyrophosphate is used as a leavening and chelating agent.

Calcium pyrophosphate exists in three polymorphic modifications; they form progressively upon dehydration of calcium hydrogen phosphate dihydrate. Calcium pyrophosphates are used primarily as abrasives in fluoridated toothpaste (see Dentifrices).

The tripolyphosphate anion, $P_3O_{10}^{5-}$, consists of triply condensed PO_4^{3-} tetrahedra:

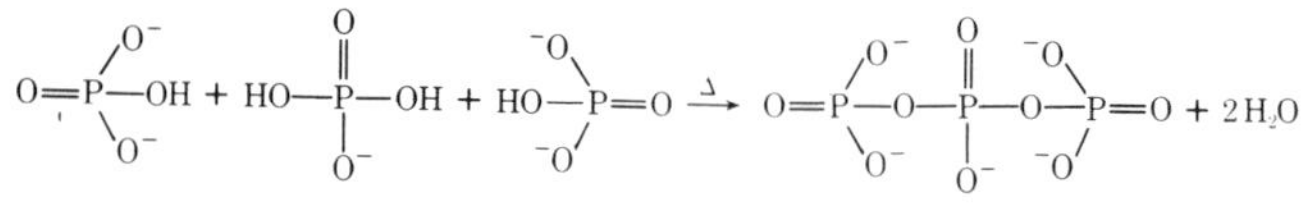

Sodium tripolyphosphate (STP, pentasodium triphosphate), $Na_5P_3O_{10}$, occurs as the anhydrous forms I (STP-I) (thermodynamically stable) and II (STP-II), the low temperature form; the transition temperature is 417 ± 8°C. They are differentiated by x-ray diffraction, ir, Raman spectra, and the temperature-rise test (ASTM D 501, 30). A hexahydrate forms by the addition of either anhydrous form to water or by the hydrolysis of sodium trimetaphosphate, $(NaPO_3)_3$. STP is produced commercially by calcination of a mixture of mono- and disodium phosphates.

The solubility and hydration behavior of STP are of particular importance in its industrial applications. As a builder for synthetic detergents, STP is the largest-volume phosphate salt for purposes other than fertilizers.

Thermal dehydration of monosodium phosphate gives rise to numerous condensed polyphosphates, eg, Graham's, Madrell's, and Kurrol's salts.

In the manufacture of phosphate salts, phosphoric acid is treated with a base, eg, carbonate, hydroxide, or ammonia, to form a solution or slurry. Most phosphates crystallize readily from solution and are separated by conventional techniques; the solutions or slurries are often evaporated to dryness.

The phosphate glasses are manufactured in refractory-type furnaces, where they are heated to 1000°C, then quenched rapidly to a solid glass; trade names are used for glasses of only slightly different composition.

Sodium tripolyphosphate (90–95% pure) is manufactured by drying and subsequent calcination of a solution or slurry with a Na_2O/P_2O_5 mole ratio of 1.67, corresponding to two mol disodium phosphate and one mole monosodium phosphate. Calcining conditions are less critical than drying conditions; combinations of dryer/calcining units are available.

Disodium phosphate, Na_2HPO_4, is marketed as the dihydrate and the anhydrous salt. Tetrasodium pyrophosphate, $Na_4P_2O_7$, is obtained by calcination of disodium phosphate or any of its hydrates. Commercial manufacture is similar to that of STP, except that the final Na_2O/P_2O_5 ratio is adjusted with NaOH; the same equipment can be used.

Chlorinated TSP, the second largest volume sodium phosphate salt, is a complex mixture approximating $(Na_3PO_4.11H_2O)_4.NaOCl$. It is made by the addition of sodium hypochlorite solution to a hot concentrated

Table 3. The Linear Polyphosphoric Acids of General Formula $H_{n+2}P_nO_{3n+1}$

Formula	Wt % P_2O_5	Structure	Prefix	Number of dissociated hydrogen atoms Strong	Weak
H_3PO_4	72.42		mono- (ortho)	1	2
$H_4P_2O_7$	79.76		di- (pyro)	2	2
$H_5P_3O_{10}$	82.54		tri- (tripoly)	3	2
$H_6P_4O_{13}$	84.01		tetra-	4	2
$H_{n+2}P_nO_{3n+1}$			poly	n	2

sodium phosphate solution, followed by cooling, crystallization, and granulation. Chlorinated TSP is unstable above 40°C and up to 20% of the available chlorine may be lost on heating to elevated temperatures during processing.

In the manufacture of ammonium phosphates, the high partial pressure of ammonia has to be considered, because it rises rapidly with the temperature or the $NH_3 : P_2O_5$ mole ratio. Phosphoric acid reacts quickly with ammonia vapor and is used as a scrubber fluid to prevent ammonia emissions and recover ammonia values. MAP and DAP fertilizers are made in granulation processes from ammonia and wet-process phosphoric acid. Fire retardancy is second to fertilizers in MAP consumption.

Most calcium phosphates are mixtures of several salts; their composition depends on the manufacturing conditions. For fertilizer and animal-feed uses, the primary concern is the CaO and P_2O_5 analysis. Calcium phosphates are used as food additives.

Environmental Considerations

Inorganic phosphates present little health hazard to humans and are essential to life processes. However, phosphates can create environmental problems, mainly because they increase the growth of algae in lakes and streams. Problems caused by sewage-borne phosphates are mostly localized to areas that employ lakes as receiving waters for sewage effluents. Average concentrations vary according to industrial input, storm waters, seasonal fluctuations, and population density.

MICHAEL J. DOLAN
ROBERT B. HUDSON
Monsanto Company

A.V. Slack, ed., *Phosphoric Acid*, Pts. 1–2, Marcel Dekker, Inc., New York, 1968.

J.R. Van Wazer, ed., *Phosphorus and Its Compounds*, Vols. 1–2, Interscience Publishers, Inc., New York, 1958.

E.J. Griffith and co-workers, eds., *Environmental Phosphorus Handbook*, John Wiley & Sons, Inc., New York, 1973.

A.D.F. Toy, *Phosphorus Chemistry in Everyday Living*, American Chemical Society, 1976.

PHOSPHORS. See Luminescent materials; Photocells and photodetectors.

PHOSPHORUS AND THE PHOSPHIDES

Phosphorus, the twelfth most abundant element, is fairly widely distributed in igneous and sedimentary rocks. The only important source is apatite, $Ca_5X(PO_4)_3$ (X = F, OH, or 0.5 carbonate). The coordination number of phosphorus (at. wt 30.98) is 3 or 4, but 1, 2, 5, and 6 are also known. There is one stable nuclide, ^{31}P, and four radioactive isotopes.

Properties

Phosphorus is a colorless or white waxy solid, d 1.28 g/cm³, which melts at 44.1°C to a clear colorless liquid (bp 280.5°C). Commercial white phosphorus (99.9%) is slightly yellowish. Red phosphorus forms from molten white phosphorus at ca 580°C. Amorphous or crystalline black phosphorus (d 2.70 g/cm³) is made from the white form under high pressure (> 980 MPa or 9670 atm) and at high temperature. The black modifications resemble graphite and are good conductors of electricity.

White phosphorus is more reactive than the red modification, which is more reactive than the black. White or liquid phosphorus ignites spontaneously in air and is usually protected from oxidation by a layer of water. Commercial red phosphorus is more stable and reacts only slowly with oxygen and water vapor. Spontaneous combustion, catalyzed by traces of iron, sometimes takes place in storage piles. Oxygen at sub-atmospheric pressures or moist air causes white phosphorus to emit a greenish glow.

Halogens, sulfur, and oxidizing acids give halides, sulfides, and oxoacids, respectively, of phosphorus. Phosphine, PH_3, is formed when phosphorus is heated with aqueous solutions of strong alkalies.

Manufacture

Elemental phosphorus is manufactured either in the electric or the blast furnace; both depend on silica as a flux for the calcium present in the phosphate rock.

$$4 \, Ca_5F(PO_4)_3 + 18 \, SiO_2 + 30 \, C \rightarrow 18 \, CaO{\cdot}SiO_2{\cdot}\tfrac{1}{9}CaF_2 + 30 \, CO\uparrow + 3 \, P_4\uparrow$$
fluoroapatite   silica   coke   slag

The present annual world production of ca 10^6 metric tons of elemental phosphorus is from a one-step process in an electric furnace; the furnace burden must be adequately porous to permit the gases to escape from the reaction zone near the bottom of the furnace.

Modern furnaces are three-phase units with potentials of 200–300 V. The outer shell is of welded steel sheets and is often cooled by a water spray. The floor is of monolithic carbon which extends a few meters up the side and above the surface of the molten slag. The upper walls are of firebrick or cast refractory cement and the roof is a cast monolithic structure. Operating data are given in Table 1; capacities of 50 MW are common.

Most furnaces have two tap holes, a lower one for ferrophosphorus, and an upper one for slag. From furnaces with only one tap hole, the molten stream runs into a catch basin from which the lighter slag overflows. The practice of selling granulated slag as a liming agent has been discontinued because of emission, albeit low, of radioactivity.

Hot gases are emitted at ca 100 m³/min; ca 93% is carbon monoxide and the remainder is primarily phosphorus with some silicon tetrafluoride and dust. The phosphorus is condensed in towers equipped with water sprays of 45–55°C temperature. The carbon monoxide is usually burned as fuel.

Red phosphorus is manufactured by a batch process, although continuous methods have been developed. White phosphorus is heated gradually in a steel or cast-iron vessel equipped with a condenser. The temperature is kept at 400°C for several hours. After cooling, the product is wet-ground and boiled with sodium carbonate solution to remove remnants of the white modification. The red phosphorus is sieved, washed, and vacuum dried. It can be stabilized by suspension in a 1 wt% solution of sodium aluminate, the slurry being aerated for several hours.

Phosphorus is poisonous and can cause bone necrosis and fatty degeneration of the viscera. It must be handled with special safety measures. The vapor is highly poisonous, and the established exposure limit in the United States is 0.1 mg/m³ air. Phosphine is much more lethal than elemental phosphorus, but it does not exhibit long-term effects.

Uses

Nearly all of the phosphorus production is converted to phosphoric acid or other phosphorus compounds (qv). White phosphorus is used in

Table 1. Operating Characteristics for a 15-MW Phosphorus Furnace

average potential between electrodes, V	300
power factor	0.97
raw materials consumed per kg of elemental phosphorus produced	
power, kW·h	14.3
phosphatic material, kg	10.0
silica material, kg	1.5
coke material, kg	1.5
products formed per kg elemental phosphorus produced	
slag, kg	4.0
ferrophosphorus, kg	0.30
carbon monoxide, kg	2.8
recovery, based on the element, of the charged phosphorus, %	87
temperature of off-gases, °C	370
temperature of slag at tapping, °C	1500

roach and rodent poisons. The military uses phosphorus to produce smoke clouds or to ignite gasoline bombs. Red phosphorus is used in safety matches and for fireworks (see Pyrotechnics).

Metallic Phosphides and Alloys

Phosphorus forms phosphides with most metals. Sodium phosphides (Na_2P, Na_3P, Na_3P_{11}) are reddish-brown to black materials thermally stable to 650°C; they react instantly with moisture to give phosphine. Magnesium phosphide, aluminum phosphide, or the mixed alloys are shiny gray-to-yellow crystalline materials which are stable in dry air but which decompose upon contact with water. They are used with an igniting agent in sea flares. Addition of ferrophosphorus to high strength, low alloy steel eliminates quenching or tempering. Phosphor copper is used for the deoxidation of copper and its alloys, and phosphor tin for deoxidation of bronzes and German silver. Phosphor bronzes are copper and tin alloys containing 1.25–11 wt% tin; 0.03–0.35 wt% phosphorus deoxidizes the alloy. Zinc phosphide, Zn_3P_2, is used as a rodenticide. Silver solder contains small amounts of silver phosphide.

JOHN R. VAN WAZER
Vanderbilt University

D.E.C. Corbridge, *Phosphorus, An Outline of Its Chemistry, Biochemistry, and Technology*, 2nd ed., Elsevier Scientific Publishing Co., Amsterdam, Neth., 1980, pp. 33–50.

Mellor's Comprehensive Treatise on Inorganic and Theoretical Chemistry, Vol. VIII, Suppl. III, Wiley-Interscience, New York, 1971, pp. 111–271; 289–363; 1129–1140; and 1230–1254.

J.R. Van Wazer, *Phosphorus and Its Compounds*, Vol. 1, Interscience Publishers, Inc., New York, 1958; Vol. 2, 1961.

PHOSPHORUS COMPOUNDS

Phosphorus exhibits oxidation states from -3 to $+5$. The largest group of phosphorus compounds contains oxygen, and the oxo acids form the basis for the most systematic nomenclature. However, the term phosphorous acid is often used to describe phosphonic acid, $H(H_2PO_3)$, which is commonly written as H_3PO_3 or $(HO)_3P$. The least ambiguous method for interpreting the nomenclature of phosphorus compounds is by referring to the structural formula. The phosphorus oxo acids are listed in Table 1.

Many phosphorus compounds are named as salts with phosphorus being the metallic or electropositive element, ie, phosphorus trichloride, PCl_3; phosphorus oxybromide (phosphoryl bromide), $POBr_3$; phosphorus triamide, $P(NH_2)_3$; and tetraphosphorus decasulfide (phosphorus(V) sulfide), P_4S_{10}. Some compounds are named as derivatives of the simple phosphorus hydrides, phosphine, PH_3, diphosphine, P_2H_4, and their oxides.

An unshared electron pair, characteristic of many trivalent phosphorus compounds, can be described in terms of the general formula

$$\begin{array}{c} R \\ R' - P: \\ R'' \end{array}$$

where R, R', and R'' can be hydrogen or alkyl, aryl, alkoxy, amino, halo, mercapto, or other groups.

The strength of a complex formed between a phosphorus donor and an electron acceptor depends on the nature of the acceptor. Strong acids provide stronger complexes than weak acids. Of phosphonium halides, only the iodide has a low dissociation pressure, which imparts stability to the compound under ambient conditions.

The unshared electron pair on phosphorus reacts with oxidizing agents, eg, hydrogen peroxide, sulfur, or halogens:

$$(RO)_3P: + H_2O_2 \rightarrow H_2O + (RO)_3P{=}O$$

$$R_3P: + S \rightarrow R_3P{=}S$$

$$:PCl_3 + Cl_2 \rightarrow PCl_5$$

Table 1. Phosphorus Oxo Acids

Formula	Structure	Name
H_3PO_4	$\begin{array}{c}HO \\ HO - P{=}O \\ HO\end{array}$	ortho-phosphoric acid
$H_2(HPO_3)$ H_3PO_3	$\begin{array}{c}HO \\ H - P{=}O \\ HO\end{array}$	phosphonic acid (frequently called phosphorous acid)
$H(H_2PO_2)$ H_3PO_2	$\begin{array}{c}H \\ H - P{=}O \\ HO\end{array}$	phosphinic acid (also called hypophosphorous acid)
H_3PO	$\begin{array}{c}H \\ H - P \\ HO\end{array}$	phosphinous acid
$H_4P_2O_7$	$\begin{array}{c}O \quad O \\ \parallel \quad \parallel \\ HOPOPOH \\ \mid \quad \mid \\ OH OH\end{array}$	pyrophosphoric acid
$H_3P_3O_{10}$	$\begin{array}{c}O \quad O \quad O \\ \parallel \quad \parallel \quad \parallel \\ HOPOPOPOH \\ \mid \quad \mid \quad \mid \\ OH OH OH\end{array}$	tripolyphosphoric acid
$H_6P_4O_{13}$	$\begin{array}{c}O \quad O \quad O \\ \parallel \quad \parallel \quad \parallel \\ HOPO(PO)_2POH \\ \mid \quad \mid \quad \mid \\ OH \ OH \ OH\end{array}$	tetrapolyphosphoric acid
$H_4P_2O_5$	$\begin{array}{c}O \quad O \\ \parallel \quad \parallel \\ HOPOPOH \\ \mid \quad \mid \\ H \ H\end{array}$	pyrophosphonic acid
HPO_3	$HO - P\begin{array}{c}{\nearrow}O \\ {\searrow}O\end{array}$	metaphosphoric acid
$H_3P_3O_9$	ring structure	trimetaphosphoric acid

Another reaction of the unshared electron pair is quaternization with alkyl halides:

$$R_3P: + R'X \rightarrow [R_3PR']^+ X^-$$

The tetrahedral structure is the most common coordination pattern for phosphorus. Frequently, there are distortions in a pure tetrahedral environment that result from steric and electronic effects. Pure tetrahedral coordination probably occurs only in species where there are four identical groups and no packing distortions.

Phosphorus Sulfides

Phosphorus and sulfur form the binary compounds tetraphosphorus trisulfide (phosphorus sesquisulfide), P_4S_3, mp 171–172.5°C, the most stable; tetraphosphorus pentasulfide, P_4S_5; tetraphosphorus heptasulfide, P_4S_7; and phosphorus(V) sulfide (tetraphosphorus decasulfide), P_4S_{10}. A stable oxysulfide, $P_4O_6S_4$, exists as a colorless, deliquescent crystalline solid, mp 102°C. Structures are shown in Figure 1, and physical properties in Table 2.

O,O'-Dialkyl or diaryl thiophosphoric acids, eg, $(C_2H_5)_2P(O)SH$, are obtained readily by alcoholysis of phosphorus(V) sulfide. They provide the basis for high pressure lubricants, oil additives, and flotation agents. The insecticides parathion and methylparathion are made from dialkyl thiophosphates.

The phosphorus sulfides are manufactured by direct union of the elements. The pentasulfide, P_4S_{10}, is purified by distillation. Exposed to

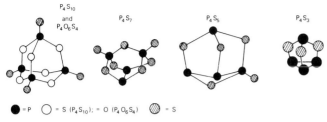

Figure 1. Structures of phosphorus sulfides.

Table 2. Physical Properties of the Phosphorus Sulfides

Property	P_4S_3	P_4S_5	P_4S_7	P_4S_{10}
mp, °C	171–172.5	170–220	305–310	286–290
bp, °C	407–408		523	513–515
sp gr at 17°C	2.03	2.17	2.19	2.09
color				
solid	yellow	sulfur yellow	almost white	yellow
liquid	brownish yellow		light yellow	reddish brown
solubility at 17°C, g/100 g CS_2	100	ca 10	0.029	0.222
action with cold water	scarcely attacked		fairly readily decomposed	slowly decomposed
stability of solid on standing in air	slowly oxidized		decomposes	slowly decomposed

moisture, it hydrolyzes, giving off H_2S; ventilation in storage avoids excessive H_2S build-up and fire and explosion hazards.

Phosphorus Halides

Phosphorus forms well-defined halogen compounds of the type PX_3, PX_5, POX_3, and PSX_3. All, except the pentaiodide and the oxy- and sulfoiodides are known. Physical properties of some halides are given in Table 3.

The trihalides, obtained by direct halogenation under controlled conditions, are Lewis bases. Examination by electron diffraction has confirmed pyramidal structures for the gaseous molecules.

Phosphorus trichloride is a clear volatile liquid with a pungent, irritating odor. Depending on the mole ratio of $H_2O:PCl_3$, hydrolysis produces phosphonic acid, a mixture of phosphonic and pyrophosphonic acids, or an unstable polymer of indefinite composition. Although it is nearly insoluble in water, PCl_3 hydrolyzes rapidly, and a polymer forms at low $H_2O:HCl$ ratios. It also forms upon storage in contact with moist air.

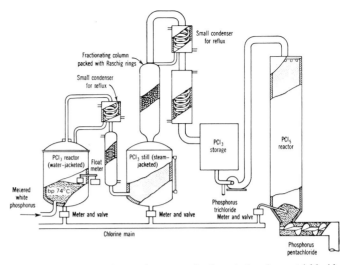

Figure 2. Plant for the continuous production of phosphorus trichloride and phosphorus pentachloride.

Phosphorus trichloride is made by combining the chlorine and phosphorus in the presence of a precharge of phosphorus trichloride which is refluxed continuously (see Figure 2).

Phosphorus oxychloride (phosphoryl chloride), $POCl_3$, is a colorless fuming liquid with a pungent disagreeable odor. It is made by heating PCl_5 with oxalic or boric acid. It is stable above 300°C, yields phosphoric acid upon hydrolysis, and is used extensively for the manufacture of alkyl and aryl orthophosphate esters.

Phosphorus pentachloride is a pale greenish solid with a pungent odor. It is made from PCl_3 and chlorine or by burning phosphorus in excess chlorine; reaction with water is violent. It serves as chlorinating agent and catalyst in organic synthesis.

Phosphorus halides are shipped in glass containers and must be handled with care. They are all irritating to skin, eyes, and mucous membranes; contaminated clothing must be removed immediately.

Phosphorus Oxides

Phosphorus forms five well-defined oxides: phosphorus(III) oxide, P_4O_6; phosphorus(V) oxide (phosphorus pentoxide), P_4O_{10}; phosphorus tetroxide, P_2O_4; phosphorus heptoxide, P_4O_7; and phosphorus nonaoxide, P_4O_9. They are obtained by direct oxidation of phosphorus.

The pentoxide is a very stable, white solid which exists in several crystalline, liquid and amorphous modifications. It is produced commercially by burning phosphorus in a stream of dry air. The usual plant

Table 3. Physical Properties of Some Phosphorus Halides

Compound	Mp, °C	Bp, °C	Specific gravity	Heat of vaporization, kJ/mol[a]	Critical temperature, °C	Color[b]	State at STP
PCl_3	−111.8 (−91)	75.5	1.57	30.5	285.5		liquid
PCl_5	167 (under pressure[c])	sublimes	1.6	64.9	372	white to pale yellow	solid
$POCl_3$	−1.2	106.5	1.68	34.7	331.8		liquid
PBr_3	−40	175.3	2.89	38.9			liquid
PBr_5	< 100 (dec)	dec				yellow	solid
PI_3	61	dec 120[d]				dark red	solid
PF_3	−151.5	−101.1	3.91	14.6	−2.05		gas
PF_5	−93.7	−84.5	5.84				gas
POF_3	−39.4	−39.8	4.65	21.1	73.3		gas

[a] To convert J to cal, divide by 4.184.

[b] Colorless unless otherwise noted.

[c] Sublimes at 101.3 kPa (1 atm) at 159°C; $\log P_{kPa} = 10.159 - 16{,}100/4.57$ K; ($\log P_{mmHg} = 11.034 - 16{,}100/4.57$ K).

[d] At 2.0 kPa (15 mm Hg).

consists of a phosphorus feed system, provisions for drying the air, a burning chamber, and the barn, which is a large room, cooled externally by air or water, where the gas produced is condensed.

Phosphorus pentoxide is treated as a flammable solid because of the explosive violence with which it reacts with water. It is sold in glass bottles which must be cushioned with incombustible packing material.

Phosphorus pentoxide is corrosive and irritates eyes, skin, and mucous membranes (1 mg/m^3 is suggested as TLV in air). It is an important drying and dehydrating agent and as such is used in the manufacture of methyl methacrylate resins (see Methacrylate polymers).

Phosphorous Acids

Phosphinic (hypophosphorous) acid, $H(H_2PO_2)$ or H_3PO_2, a deliquescent crystalline solid (mp 26.5°C, dec > 133°C), is manufactured by treating white phosphorus with a boiling slurry of lime. Calcium hypophosphite remains in solution, whereas insoluble calcium phosphite precipitates.

The reducing capacity of hypophosphorous acid and its salts is utilized in electroless-plating (qv) processes, in which nickel salts are chemically reduced to form a smooth adherent surface plate.

Phosphonic (phosphorous) acid, $H_2(HPO_3)$ or H_3PO_3, is a white deliquescent crystalline compound, mp 73.6°C, manufactured by the hydrolysis of phosphorus trichloride; the reaction can be violent.

Phosphonic acid and hydrogen phosphonates are used as strong but slow-acting reducing agents, eg, to precipitate heavy metals from solutions. Aminoalkylphosphonic acids are used in the production of herbicides (qv) and water-treatment systems (see Water, industrial water treatment).

Phosphazenes and other Phosphorus–Nitrogen Compounds

The phosphazenes $(NPX_2)_n$, are linear or cyclic compounds with alternating phosphorus and nitrogen atoms; they are toxic and irritating. The lower linear chlorophosphazenes are oils, the cyclic members are crystalline white solids. The latter can be polymerized to chains of highly controlled molecular weight. Chlorophosphazenes are prepared from phosphorus(V) chloride and ammonium chloride (see Inorganic high polymers).

Of the binary phosphorus–nitrogen compounds, only triphosphorus pentanitride has been obtained in a pure state by treating P_2S_5 with ammonia.

Phosphine and its Derivatives

Phosphine, PH_3, is produced in a number of ways; however, the inadvertant evolution of phosphine in an otherwise safe reaction is an element of hazard in many procedures involving phosphorus chemicals. Phosphine is conveniently produced by the hydrolysis of an active metal phosphide, eg, Ca_3P_2 or AlP, or from the acid- or base-catalyzed reaction of elemental phosphorus with water.

The addition of P—H bonds in phosphines across an olefinic double bond is an economical method to produce alkyl phosphines. Aryl groups are conveniently introduced by Friedel-Crafts reactions of phosphorus halides.

Aliphatic phosphines may be gases, volatile liquids, or oils. Aromatic phosphines are frequently crystalline, although many are oils.

Because of the bonding ability of phosphines to acceptors, eg, metals, they are widely used as ligands in catalysts (see Catalysis). A phosphine can block a specific site on a central metal under all conditions, or a site to some species but not to others, or simply make a metal more soluble and thus dispersible in a liquid medium (see Organometallics).

Phosphine, a central nervous system and liver toxin, is highly toxic and lethal to adults in a 0.5–1-h exposure of 0.05 mg/L. Transport on both passenger and cargo aircraft is forbidden.

I.A. BOENIG
M.M. CRUTCHFIELD
C.W. HEITSCH
Monsanto Company

J.R. Van Wazer, *Phosphorus and Its Compounds*, Interscience Publishers, Inc., New York, 1958.

M. Grayson and E. J. Griffith, *Topics in Phosphorus Chemistry*, Vols. 1–11, Wiley-Interscience, New York, 1964–1983.

G. Kosolapoff and L. Maier, *Organic Phosphorus Compounds*, Vols. 1–7, Wiley-Interscience, New York, 1972–1976.

J. Emsley and D. Hall, *The Chemistry of Phosphorus*, John Wiley & Sons, Inc., New York, 1976.

A. J. Kirby and S. G. Warren, *The Organic Chemistry of Phosphorus*, Elsevier, New York, 1967.

PHOTOCHEMICAL TECHNOLOGY

Most nonspontaneous chemical reactions are initiated, promoted, and controlled by the addition of energy. Photochemical energy most often causes reactions from the potential-energy surfaces of various electronically excited states.

Photochemical Activation

In order for a photochemical reaction to occur, a molecule must absorb a photon of light that has an energy greater than or equal to the energy difference between the ground state and an electronically excited state.

Laws

When a system is irradiated, the light may be transmitted, scattered, refracted, or absorbed. Only the light that is absorbed by a molecule can be effective in promoting photochemical change (first law of photochemistry). Consequently, not only should the absorption spectrum of the reaction and the spectral distribution of the lamp be known, but the absorption spectra of the solvent, product, and glassware between the lamp and the reactant should also be known.

According to the second law of photochemistry, the absorption of light by a molecule is a one-quantum process, and the sum of all primary-process quantum yields must be unity.

The quantum yield Φ of a reaction is defined as

$$\Phi = \frac{\text{moles of product obtained}}{\text{einsteins of light absorbed}}$$

and is measured in terms of monochromatic light.

Excitation Mechanism

The excited states have one electron in a π^*, ie, antibonding, molecular orbital and one in the π, ie, bonding, molecular orbital. If the angular-momentum or spin-quantum number s of the excited electron is unchanged, the total multiplicity S of the excited state $\Sigma s + 1$ is one and the state is designated a singlet state. If the spin of the excited electron is flipped, then $S = 3$, and a triplet state exists.

Electronic energy transfer from an electronically excited molecule to a substrate is shown schematically below.

$$A + h\nu \rightarrow A \text{ (excited)}$$

$$A \text{ (excited)} + S \text{ (ground state)} \rightarrow A \text{ (ground state)} + S \text{ (excited)}$$

where A = absorber and S = substrate. For such transfer to occur, the excited state of the substrate must be of lower energy than that of the absorber. The use of triplet-state photosensitizers in synthesis provides many examples of electronic energy-transfer reactions.

Quantum Yield

The quantum yield of a reaction Φ is measured at one wavelength by complicated actinometric systems. A synthetic quantum yield Φ' can be defined as

$$\Phi' = \frac{\text{moles product}/(\text{kW}\cdot\text{h})}{\text{einsteins}/(\text{kW}\cdot\text{h})}$$

Thus, reactions having very low synthetic quantum yields are not industrially feasible. A useful value of Φ' can be obtained over a spectral region rather than at one wavelength.

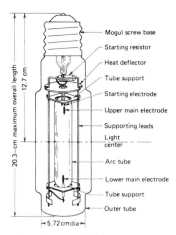

Figure 1. Construction of a typical high pressure mercury arc. Courtesy of General Electric Company.

Light Sources

A photochemical lamp must have the following characteristics: high intensity in the desired spectral region, long life, stability of output, ease of operation, proper physical dimensions for the process under consideration, and a minimum amount of necessary auxiliary equipment.

Mercury lamps best meet all these requirements. When an electric current passes between two electrodes that are separated by a gas or vapor, radiation of various wavelengths is generated. The intensity and wavelength of the light depends upon the gas, its pressure, the applied current, and the arc-tube diameter. The most useful vapor is mercury. The spectral distribution from a mercury arc is rich in uv light. Mercury-vapor lamps are designated as high, medium, or low pressure arcs. A very common high pressure lamp is a small-diameter, heavy-wall quartz tube enclosed in an outer glass jacket that serves as a filter, heat insulator, and shield in case of breakage (see Fig. 1). Operating pressures are 0.2–11 MPa (2–110 atm). At lower lamp pressures, air cooling is sufficient whereas at higher pressures, water cooling is necessary.

Reactors

In photochemical reactors, either the lamp, a long cylinder, is surrounded by a cooling jacket that is immersed in the reaction solution, or the reaction mixture is irradiated from outside the solution. Transmission limits range from 166 nm for a 10-mm quartz to 330 nm for 10-mm window glass.

Continuous reactors are used in high quantum-yield reactions in vapor-, liquid-, or mixed-phase reactions. Contact times are short and very high reaction rates are achieved.

Design and Scale-Up

The largest lamp with the greatest electrical efficiency in the region in which the starting materials absorb should be selected. Immersion reactors offer the greatest electrical efficiency because very little light is lost by reflection or scattering. The long arc lamp adds a great amount of heat to the solution in the immersion reactor and external cooling may be needed. Scale-up to full production is simply a matter of adding more lamps and reactors.

Economic aspects. Production costs for nonchain-reaction processes are more expensive than for thermal processes. If the products are of unusual structure and high economic value, photochemical synthesis may be economically feasible. Reactions of low product quantum yields are uneconomical. The relationship between the product quantum yield and the electrical cost of running a photochemical lamp is

$$\text{electrical cost per mole} = \frac{\text{cost}/(\text{kW}\cdot\text{h})}{\text{mol}/(\text{kW}\cdot\text{h})} = \frac{\text{cost}/(\text{kW}\cdot\text{h})}{\Phi\left(\text{einsteins}/(\text{kW}\cdot\text{h})_{\text{lamp}}\right)}$$

Uses

Photochlorinations utilizing free-radical reactions are frequently employed by industry because fewer side reactions occur at lower operating temperatures (see Chlorocarbons and chlorohydrocarbons). Sulfochlorination of n-paraffins produces almost random substitution. In direct nitrosation, cyclohexane is converted in one step to cyclohexanone oxime, an intermediate for caprolactam and nylon-6. Photochemical initiation of polymerizations is used in curing systems, where it eliminates air-pollution problems caused by solvent removal (see Initiators; Radiation curing). The synthesis of vitamin D_3 is a particularly good example of a low efficiency reaction requiring 80 (kW·h)/kg vitamin (see Vitamins (D)). Laser photochemistry is still very expensive but promises interesting applications (see Lasers).

JORDAN J. BLOOMFIELD
DENNIS C. OWSLEY
Monsanto Company

J.C. Calvert and J.N. Pitts, Jr., *Photochemistry*, John Wiley & Sons, Inc., New York, 1966, Chapts. 1, 4.

R.B. Woodward and R. Hoffmann, *Conservation of Orbital Symmetry*, Academic Press, New York, 1970.

U.S. Pat. 4,087,342 (May 2, 1978), J.J. Bloomfield (to Monsanto).

PHOTOCHROMISM.   See Chromogenic materials.

PHOTODETECTORS

Photodetectors are devices that convert electromagnetic radiation to electric signals which can be processed to obtain information inherent in the temporal and spatial variations of the radiation. For example, the image formed by a lens may be sampled by a small detector which scans the image plane. The electric signal, which is a measure of the photon density at each image point, is amplified and coupled to a television display which is synchronized with the image scan; in effect, the eye or photographic display is replaced by the detector. This simple technique offers advantages, eg, remote sensing, such as in satellite and planetary probes; sensitivity to uv, visible, and ir radiation, ie, heat sensing; and high speed detection or fast, automatic-response functions (see Infrared technology).

Detectors are classified as thermal or photon detectors, depending on the manner in which the incident radiation is utilized. Absorption of radiation in thermal detectors generates a rise in temperature which produces a measurable change in some properties. Photon detectors are greatly superior to thermal detectors in terms of sensitivity and speed of response. In thermal detectors, however, response to long-wavelength radiation can be obtained without cooling of the detector.

Photoabsorption is intrinsic if it is characteristic of the pure semiconductor and extrinsic if it results from an impurity or lattice defect. In the case of combined extrinsic and intrinsic photoconduction, the thermal-equilibrium conductivity of the detector is determined by shallow energy-donor or acceptor impurity levels which are fully ionized by lattice vibrations. Photons are detected by valence-band to conduction-band photon excitation of electrons and consequent lowering of conductivity.

Photovoltaic detectors utilizing $p-n$ junction diodes are often called photodiodes. Metal-insulator semiconductor (MIS) technology produces a photosensitive surface-potential.

Performance. Performance is described by noise, responsivity, detectivity, and response time. Related parameters include temperature, bias power, spectral response, background photon flux, noise spectra, impedance, and linearity. In addition, detector-element size, uniformity of response, availability of arrays, array density, reliability, cooling time, radiation tolerance, vibration and shock resistance, shelf life, and cost must be considered.

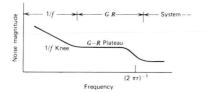

Figure 1. Noise power spectra for typical semiconductor photodetectors.

Semiconductor photodetectors exhibit sensitivity to variations in photon flux emitted from objects that are characterized by time and spatial variations in temperature and emissivity. A detector responds to variations in photon-flux density ϕ_s by producing a signal voltage V_s or current.

Systems parameters, eg, lens size, focal length, cold shielding of the detector element, detector size, bias power, etc, should be so arranged that detector-element noise dominates the system-input noise. The noise spectrum is characterized by $1/f$ noise at low frequency f, generation combination $g-r$ at midrange, and system noise at high frequency, as shown in Figure 1. At low frequency, noise voltage typically varies inversely as the square root of frequency and is a function of the surface-trap density and detector-bias current density.

A useful parameter in performance evaluation is the normalized responsivity-to-noise ratio D^*. The spectral detectivity D^*_λ in (cm· $Hz^{1/2}$)/W is a fundamental quantity. The noise-equivalent power (NEP) is the total radiant signal power absorbed by the detector to yield a signal-to-noise ratio of unity.

Cooling requirements. Free carriers can be excited by the thermal motion of the crystal lattice or phonons as well as by photon absorption. These thermally excited carriers determine the magnitude of the dark current and constitute a source of noise that defines the limit of the minimum radiation flux that can be detected. The dark-carrier concentration is temperature-dependent and decreases exponentially with reciprocal temperature at a rate that is determined by the magnitude of E_g or E_i for intrinsic or extrinsic material, respectively. Therefore, the smaller the value of E_g or E_i, the lower the temperature must be. The detector can be cooled by providing good thermal conductivity to a suitable cryogen, eg, solid CO_2 at 195 K, liquid nitrogen at 77 K, and liquid helium at 4.2 K; for temperatures above 200 K, a thermoelectric cooler may be used. The cooled detector is kept in a Dewar vessel.

Semiconductor Elements

Commercial photodetectors are made from one of nine basic semiconductor elements or compounds (see Table 1).

Compounds formed from Group-IIB and -IVA elements are grouped according to their characteristic band-gap energies. They are prepared by evaporation, spraying, settling, *in situ* reaction, or sublimation in the form of polycrystalline or amorphous thin films < 600 nm thick, as powders, as polycrystalline sintered layers, or as single crystals. The latter are grown from the melt by pulling, float-zone melting, zone melting, recrystallization, or liquid-phase or vapor-phase epitaxial growth. More recent methods include molecular-beam epitaxial growth, sublimation, reactive sputtering, and laser-beam annealing. The crystals are usually purified by zone refining (qv).

Photodetectors of high quality are made by junction formation, as well as by direct formation during bulk and epitaxial growth. The alloyed junction technique is difficult to control in small geometries and is used primarily for contacts. Performance may be improved by employing the epitaxial process wherein a semiconductor is grown on a substrate by chemical vapor deposition (see Film deposition techniques).

Germanium and silicon are doped uniformly with a selected impurity for operation as intrinsic detectors. Germanium single crystals can be grown with the Czochralski method. They must be of high purity before doping in order to avoid the presence of excess impurities of smaller ionization energies than those of the activator. In most cases, the activator impurity is incorporated during crystal growth. Mercury is added by modified zone leveling; concentration can be controlled by

Table 1. Typical Operating Characteristics of Some Photodetectors

Detector material	Operating mode[a]	Operating temperature, K	Response time, μs	Responsivity, V/W[b]	D^*_λ, (cm·$Hz^{1/2}$)/W[c]
CdS	PC	300	100,000	1×10^8	1×10^{14}
Si	PV	300	100	1×10^8	5×10^{13}
	CCD	300		1×10^5	1×10^{12}
Ge	PV	300	10	1×10^6	7×10^{11}
PbS	PC	300	300	4×10^3	1×10^{11}
	PC	200	1,000	5×10^4	5×10^{11}
PbSe	PC	200	50	5×10^4	2×10^{10}
	PC	77	150	1×10^6	2×10^{10}
InAs	PV	200	1	5×10^2	1×10^{11}
	PV	77	5	1×10^5	5×10^{11}
InSb	PV	77	1	2×10^4	1×10^{11}
HgCdTe					
$\lambda_c = 3.0\ \mu m$	PC	200	200	7×10^5	5×10^{11}
$\lambda_c = 4.5\ \mu m$	PC	200	10	2×10^5	1×10^{11}
	PV	200	10	1×10^5	1×10^{11}
$\lambda_c = 12\ \mu m$	PC	77	1	5×10^4	3×10^{10}
	PV	77	1	5×10^4	2×10^{10}
Ge–Hg	PC	30	0.1	5×10^5	3×10^{10}
Ge–Cu	PC	8	0.1	5×10^5	4×10^{10}
Si–Ga	PC	35	0.1	5×10^5	2×10^{10}
PbSnTe					
$\lambda_c = 12\ \mu m$	PV	77	1 or 0.1	5×10^4	3×10^{10} or 1×10^8

[a] PC, photoconductive; PV, photovoltaic; CCD, charge-coupled device.
[b] Peak spectral responsivity.
[c] Peak spectral detectivity for 300-K background and 180°C field of view (FOV).

adjusting the travel rate through the molten zone to ca 4 cm/h and the temperature to ca 300°C. All operations must be carried out in an inert or reducing atmosphere to prevent oxidation of the germanium.

Some impurities, eg, Cu, Ag, Au, Ni, Co, and Fe, have diffusion coefficients that are large enough to permit doping by solid-state diffusion well below the melting point of germanium. The impurity concentration should be as large as possible, within limits, in order to maximize the absorption coefficient.

Germanium and silicon form a continuous series of solid solutions in which the energy gap increases monotonically from Ge to Si; impurity-ionization energies increase as the Si content of the alloy is increased.

The only binary-compound semiconductors containing Group-III and Group-V elements useful as ir detectors are InAs, InSb, and mixtures of InAs with InSb and of GaSb with InSb.

The lead chalcogenides PbS, PbSe, and PbTe were among the first ir-detector materials to be investigated. They are prepared by sublimation or chemical deposition as polycrystalline films ca 1 μm thick, deposited upon glass or quartz substrates between gold or graphite electrodes. Large single PbSnTe crystals have been grown by the Bridgman technique with a free carrier concentration > 10^{19} cm^{-3}; they must be annealed before fabrication. Long-wavelength, intrinsic photoresponse has been observed in the PbTe–SnTe system.

Photoconductors

Photoconductor detectors have been fabricated from HgCdTe, doped germanium, doped silicon, CdS, PbS, and PbSe. The sensitivity of the latter three is due to a potential barrier-trapping mechanism. The photoconductive-detector element is prepared as a small slab of semiconductor; the electric contacts usually define the detector, as shown in Figure 2. Single-crystal detectors, eg, HgCeTe, Ge, and Si, must be prepared with great care. The Ge and Si detectors typically measure one or several millimeter, whereas HgCdTe may be as thin a 10 μm to increase sensitivity.

The photoconductive element, formerly called a photocell or cell, of resistance R is electrically biased, with current i being placed in series with a battery at voltage V_B and load resistor R_L. Because the voltage drop across the detector is

$$V = \frac{V_B R}{(R + R_L)} \qquad (1)$$

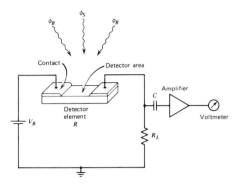

Figure 2. Schematic of photoconductive detector element which is exposed to background photon flux and signal flux and which is in a series circuit with battery and load resistor.

and given the usual condition, $R_L \gg R$, the constant current is

$$V = \frac{R}{R_L} V_B \tag{2}$$

The photoresponsivity is

$$R_V = \frac{\eta \tau V}{h\nu n\nu} \tag{3}$$

where ν is the detector volume, $h\nu$ is the photon energy which is absorbed with efficiency η, and τ is the lifetime.

For a well-designed photoconductor, the g–r noise is dominant and may be expressed in terms of a minority-carrier density p and a majority-carrier density n. Semiconductor noise analysis for the intrinsic photoconductor yields

$$\left(V_{g-r}\right)^2 = \frac{4p\tau V^2 \Delta f}{n(n+p)\nu} \tag{4}$$

when the ratio of electron-to-hole mobility is much larger than unity.

Saturation limits are imposed on the bias potential V in equations 3 and 4. The joule heating power P in a photoconducting element limits the bias voltage since $V = (PR)^{1/2}$ and the detector, upon excess bias, raises the static heat load, which warms the detector and results in lowered signal and detectivity.

The lead salt detectors, eg, PbS, PbSe, and PbTe, in the form of polycrystalline oxygen-activated thin films, are sensitive at wavelengths of 1–6 μm. Lead sulfide is the most sensitive material available for use at 2 μm; its response extends to ca 3 μm and it can be operated at room temperature.

The important, commercially available photoconductive cells for visible light are the CdS–CdSe devices, silicon field-effect devices, silicon n–p–n planar-junction phototransistors, germanium p–n–p alloy-junction phototransistors, and silicon p–n–p–n controlled rectifiers.

The CdS–CdSe devices are highly sensitive to visible light and can operate a relay directly without additional amplification. The photoconductive material may be the sulfide, the selenide, or a solid solution of the two. The maximum spectral responses for the sulfide and the selenide occur at 510 and 735 nm, respectively; the maximum of the spectral response for the solid solution is between these two values.

Both germanium and silicon can be manufactured as n–p–n or p–n–p photosensitive transistors which provide an inherently more sensitive cell, although the minimum detectable signal is not improved by the amplification. A silicon p–n–p–n device yields a light-activated silicon-controlled rectifier; light is used for the gating action. If no light falls on the device, it does not conduct and the load is not energized.

Extrinsic germanium and silicon detectors have been prepared from impurity-doped single crystals. Rectangular slabs of material, a few tenths to several millimeter thick, are sawed, lapped, and chemically polished. Resistivity is increased exponentially upon cooling, since thermal activation from the impurity state controls bulk conductivity. The rate of change upon cooling depends on the degree of compensation of the desired impurity level by residual impurities.

Photodiodes. Photovoltaic detectors generate a voltage when receiving radiation, and therefore, do not require an external voltage source for operation. A current flows when the terminals of the cell are connected to an electric circuit. For sufficiently low circuit resistance, the current is proportional to the intensity of radiation incident on the cell. The current generated by the photovoltaic cell is usually measured directly by a current meter and may be expressed in terms of the photon-generation rate ϕ as

$$I_\phi = \eta \phi q A \tag{5}$$

The quantum efficiency η can be determined with the aid of a calibrated photon source, eg, a blackbody, using the measured current. The detector area A should be measured by a spot-scanning apparatus, since the minority-carrier diffusion length effectively increases the diode collection area.

The most widely used photovoltaic devices are the selenium heterojunction cell, employed in photographic exposure meters, and the silicon homojunction cell used as a power source for space vehicles and for data processing by reading holes in punched cards and tape.

Silicon and germanium p–n junction detectors provide excellent coverage of the short-wave spectrum (ca 1.1 μm for Si and 1.8 μm for Ge); they can be operated at room temperature. Such detectors are prepared by diffusion at elevated temperature into wafers or single-crystal p-type Si or Ge with a p–n junction parallel to the surface at a depth of ca 1 μm. They can be operated in the photovoltaic mode, as reverse-biased photodiodes, or as avalanche photodiodes.

Silicon photovoltaic cells are available as single elements or in arrays measuring up to several centimeters on one side. The device can be used to read punched holes in computer cards; larger cells are used for solar-power conversion for space and terrestial applications (see Photovoltaic cells).

Gallium–arsenic and gallium–arsenic–phosphorus diodes are fabricated as photodiodes as well as light emitters (see Light-emitting diodes and semiconductor lasers). Spectral cutoff range extends from 500 to 900 nm, depending on the phosphorus content. These detectors are used for color discrimination without expensive interference filters.

Photodiodes are also available in PbSnTe and HgCdTe; their cutoff wavelengths are 5–12 and 2–12 μm, respectively. The HgCdTe photodiodes are fabricated by ion implantation (qv) into p-type bulk crystals or p-type epitaxial films.

Charge-transfer or charge-coupled devices (CCDs) offer exciting possibilities in focal-plane design since tens of thousands of detector elements can be incorporated in an area smaller than 1 cm^2. These devices utilize the semiconductor field-effect phenomenon, whereby an empty well is generated momentarily at the surface by a voltage pulse on a metal plate a small distance from the surface. Silicon CCDs have been made with as many as 640,000 picture elements for advanced applications in space telescopes.

The integrated-circuit industry has developed expanded-metal technology for the fabrication of CCDs. The semiconductor is prepared to ensure a nearly defect-free surface. In silicon CCDs, SiO$_2$ grows at elevated temperature; steam and oxygen are used for high quality insulator deposition and low surface-state density. Aluminum films are employed to protect the semiconductor outside the channels that are intended for detection and charge transport. The shift register is fabricated with two levels of metal to generate the overlapping four-phase structure and bus lines. The entire array is addressed with a single lead to each phase component. In this way, only five lead wires are required for a 640,000-element imaging array, one for each phase and a transfer gate at the output. A more recent version called virtual phase utilizes ion implant and only one phase electrode (see Electrical connectors).

An important CCD parameter is the charge-transfer efficiency (CTE), which is the fraction of the charge that is transferred compared to the charge that was in the well before. Typical values for silicon are greater than 0.9999 and have exceeded 0.999 for HgCdTe. Poor CTE results in image smearing since the image information is contained as the number of electrons in the charge packet.

Health and Safety Factors

Photodetectors are usually packaged hermetically in inert glass or plastics or in an evacuated container which affords environmental protection. Most detector materials are toxic. The usual safety rules have to be observed in fabrication; electrical hazards are minimal.

SEBASTIAN R. BORRELLO
Texas Instruments Incorporated

W.L. Wolfe and G.J. Zissis, eds., *The Infrared Handbook*, Environmental Research Institute of Michigan, Ann Arbor, Mich., 1978.

W.L. Eisenman, J.D. Merriam, and R.F. Potter, "Operational Characteristics of Infrared Photodetectors" in R.K. Willardson and A.C. Beer, eds., *Semiconductors and Semimetals*, Vol. 12, Academic Press, Inc., New York, 1977.

S.M. Sze, *Physics of Semiconductors*, Wiley-Interscience, New York, 1969.

PHOTOELECTROCHEMICAL CELLS.   See Solar energy.

PHOTOGRAPHY

White crystalline silver chloride turns violet when exposed to sunlight because of a photochemical reaction in which Ag^+ (silver ion in the ionic silver halide crystal) is reduced to Ag (elemental silver). As the photochemical reduction continues, elemental silver atoms aggregate and grow into clusters of a colloidal size sufficient to scatter light and produce hue shifts. Photography makes use of this photochemical property of silver halide to form images.

The AgBr, Ag(Br, I), and Ag(Br, Cl, I) microcrystals popularly used in modern photography are nominally sensitive only to electromagnetic radiation with wavelength shorter than 500 nm. Photographically, these crystals are said to be blue-sensitive but green- and red-insensitive. The blue sensitivity refers to the intrinsic sensitivity of the silver halide crystals. To reduce the number of blue photons required to produce a developable latent image, ie, a catalytic center, the silver halide crystals are treated with chemical compounds that adsorb to the crystal surfaces and may or may not react with them. Such compounds, called chemical sensitizers, usually contain sulfur or gold, and do not significantly alter the light-absorption properties of the silver halide crystals, but they do alter the efficiency with which the latent image is formed.

To achieve photographic sensitivity in the green (500–600 nm) and red (600–700 nm) regions, the silver halide crystals are sensitized with dyes, ie, dye molecules are adsorbed on the crystal surface (see Dyes, sensitizing).

Today, a broad range of photographic materials is available for amateur and professional uses including x-ray films, graphic-art films, microfilms, and complex multilayer coatings for color films (see Color photography).

The Photographic Crystal

The preparation of light-sensitive microcrystals is referred to as precipitation (see Fig. 1). These light-sensitive crystals are emulsified silver halide grains. Various techniques are applied during crystal growth to achieve the desired grain morphologies (shapes), size-frequency distributions, solid-state properties, light sensitivity, and catalytic activity. Chemicals are usually added to control crystal-growth rates, ripening characteristics, stability, and light sensitivity. During the crystal-growth process, reactant solutions containing a halide and a silver salt (usually silver nitrate) are mixed in the presence of a peptizing agent, preferably gelatin. This produces a suspended solid phase which separates in the form of microscopic crystals of silver halide. The peptizing agent adsorbs on the grain surface but does not inhibit continued growth; it prevents coagulation of the microcrystalline grains, maintaining a uniform dispersion.

The ultimate size-frequency distribution of these emulsion grains depends upon the rate of addition of reactant solutions, the temperature, and the presence of growth modifiers or ripeners. The latter are compounds that form water-soluble silver salts or complexes and preferentially dissolve the smallest grains from a given population of crystal sizes and enhance growth in the larger grains. Ripeners, eg, ammonia, sodium thiosulfate, or sodium thiocyanate, can be added before, during, or after the crystal-growth process.

The reaction to form a silver halide crystal is initiated by nucleation, for which the solution must be supersaturated in silver and halide ions to overcome adverse free-energy effects. As silver halide nuclei form, they provide a substrate or surface for continued growth. When there is a wide range of grain sizes within the reaction vessel, Ostwald ripening can occur. There are three stages in this ripening process: dissolution of small grains, ionic diffusion through the aqueous phase, and finally redeposition of ions on large grains. Ripening is enhanced by temperature increases.

For certain photographic systems, small amounts of inorganic impurities or dopants are added to the emulsion during precipitation to achieve desirable photographic responses. Dopants can have strong effects on solid-state properties of the grains as well as on light sensitivity, contrast, and developability.

When crystal growth is completed, the photographic emulsion is a dispersion of silver halide microcrystalline grains in an aqueous gelatin phase. Counter ions, ripeners, and other additives are also present. These by-products must be removed, usually by so-called noodle washing or by flocculation washing. In the former, more gelatin is added after the precipitation and the emulsion is solidified into a jelly after cooling. The more recent technique of flocculation washing requires less time and no additional gelatin and gives a more concentrated emulsion. The gelatin is coagulated by adjusting the pH or by adding salts. The gelatin floccules carry with them the silver halide grains and leave the water-soluble by-products in the supernatant.

Response Enhancement

Chemical sensitization. After the microcrystals are precipitated, but before they are coated on a support, they are chemically sensitized to enhance their native or intrinsic light sensitivity and spectrally sensitized to increase the range of the wavelength sensitivity. Chemical sensitization reduces the number of photons required to produce a developable latent-image center by as much as a factor of ten. Typical chemical sensitizers include thiourea and sodium thiosulfate, gold thiocyanate and potassium tetrachloroaurate, and reducing agents such as hydrogen, *tert*-butylamineborane, stannous ions, or hydrazine. Used in trace amounts (μmol/mol Ag) alone or in combination, chemical sensitizers optimize photographic properties and have beneficial effects on the solid-state phenomena during exposure without affecting amplification during development.

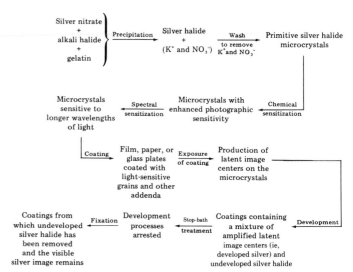

Figure 1. Flow chart of the photographic process.

Silver sulfide is generally considered to be the active chemical species resulting from sensitization with sulfur compounds. The reaction of thiosulfate with silver halide crystals to form adsorbed sulfide on the grain surfaces is activated thermally. If the reaction is allowed to continue too long before quenching or if too much thiosulfate is used, the grains become spontaneously developable (no exposure is required to induce catalytic activity), and image discrimination is lost. The sulfiding reaction is envisioned to be a two-step process involving adsorption followed by a thermally activated chemical reaction:

$$(AgBr)_n + S_2O_3^{2-} \rightleftharpoons (AgBr)_{n-1}[Ag(S_2O_3)]_{adsorbed}^- + (n-1)\,Br^-$$

$$(AgBr)_{n-1}[Ag(S_2O_3)]_{adsorbed}^- + Ag^+ + H_2O \rightleftharpoons (AgBr)_{n-1}Ag_2S + SO_4^{2-} + 2\,H^+$$

Sulfur sensitization is an effective and popular method for improving photographic sensitivity. Gold is often used in combination with sulfur, particularly to improve sensitivity to exposures of high light intensity. Gold enhances the catalytic activity of the latent-image center and reduces the number of photochemically formed silver atoms required for developability, thereby increasing the light sensitivity of the grain.

Spectral Sensitization

The intrinsic absorption, and therefore the intrinsic photographic sensitivity, of silver bromide and silver iodobromide microcrystals falls off rapidly for wavelengths greater than 500 nm. In fact, silver chloride crystals have almost no sensitivity in the visible region of the spectrum. The need to extend sensitivity into the green and red regions is obvious for color photography. Extending the wavelength sensitivity beyond the intrinsic region is called spectral sensitization. It is usually done after precipitation but before coating by adsorbing certain dyes to the crystal surfaces. Once the dye molecule is adsorbed, the effects of electromagnetic radiation absorbed by the dye are transferred to the crystal. The part of a dye molecule that enables the molecule to absorb visible or infrared light is called the chromophore. The resonance structures for three common chromophores are shown below.

Amidinium ion system

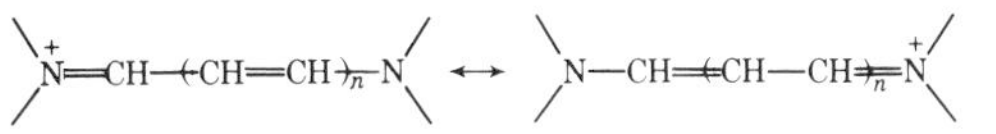

$$\overset{+}{N}=CH\text{---}(CH=CH)_{\overline{n}}\,N\ \longleftrightarrow\ N\text{---}CH=CH\text{---}CH)_{\overline{n}}\overset{+}{N}$$

Carboxyl ion system

$$O=CH\text{---}(CH=CH)_{\overline{n}}\,O^-\ \longleftrightarrow\ {}^-O\text{---}CH=CH\text{---}CH)_{\overline{n}}=O$$

Dipolar amidic system

$$\overset{+}{N}=CH\text{---}(CH=CH)_{\overline{n}}\,O^-\ \longleftrightarrow\ N\text{---}CH=CH\text{---}CH)_{\overline{n}}=O$$

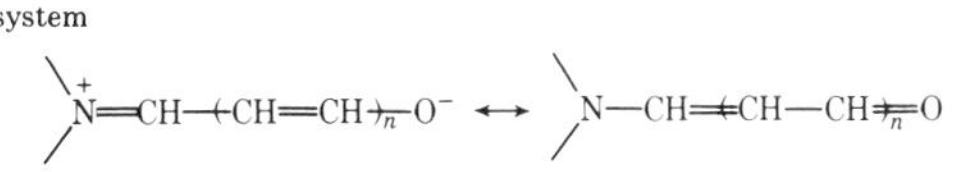

Coating Additives

Certain additives are used to facilitate coating operations (eg, surfactants), reduce spontaneous development in unexposed regions (eg, tetraazaindenes, mercaptotetrazoles), and reduce abrasion and permit high temperature processing (eg, aldehydes). Stabilizers include halide ions, benzimidazoles, benzotriazoles, and mercaptotetrazoles.

Gelatin cross-linking agents render the coated emulsion layers more resistant to abrasion during handling and improve the thermal stability and hardness of the gelatin.

Emulsion coating. For most applications, the sensitized emulsions must be coated on a base or support of glass, plastic, or paper for convenient handling. Supports are chosen on the basis of dimensional stability, low water permeability, lack of surface irregularities, compactness, cost, and safety. Today, clear plastic film supports made of cellulose esters or poly(ethylene terephthalate) (polyester) are most commonly used. These materials are safe and dimensionally and chemically stable. Paper supports are used for color or black-and-white print materials. Paper supports may be undercoated with barium sulfate in a gelatin matrix to improve smoothness and enhance whiteness, or waterproofed with impervious resin coatings of polyethylene whitened with titanium oxide.

Emulsion coatings must be uniform in thickness and composition and free of streaks. Multilayered coatings are often composed of more than

ten layers, each containing a variety of different chemicals. As light-sensitive materials, they must be coated in the dark. The individual coated, dried layer is generally 1–30 μm thick. The support is transported on rollers past a coating station where the liquid emulsion is delivered to the moving support by pumps or gravity flow from hoppers. In other systems, melted emulsion is held in troughs or trays and the moving support is brought into contact with the liquid emulsion.

Intermediate layers are coated between the emulsion and the base to facilitate spreading and improve the adhesion of the gelatin layer to the hydrophobic support. Such interlayers may also contain light-absorbing (antihalation) materials to prevent stray light from reflecting back into the emulsion layer during exposure. Antihalation material, eg, carbon particles, dyes, or colloidal silver, can degrade the appearance of the final photographic image and must be removed in processing. Overcoating the emulsion layers with a gelatin layer affords protection against image-degrading effects owing to pressure and abrasion.

Exposure

When the image is produced by uv, visible, or ir radiation, an optical-lens system is required which focuses the image on the emulsion layers of the sensitive coating. The degree of magnification is a function of the effective focal length of the optical system. In negative-working emulsion coatings, the photoelectrons react with silver ions to form clusters of silver metal on the grain surfaces. These clusters function as catalytic centers for amplification during subsequent development. Negative dye-scale images are produced when the developer molecules that have reacted with the catalytic center initiate process reactions. In positive working emulsion coatings, the density produced by developed silver or by dye decreases with increasing exposure.

Exposure is a measure of the total incident light energy and is therefore equal to the mathematical product of the light irradiance, I, and the exposure time, t. Positive photographic images can be produced by at least four different exposure-related phenomena: solarization, the photobleach effect, the Herschel effect, and the Clayden effect. Special development solutions or processing sequences are not required in these four cases.

Development. Developer solutions contain reducing agents, restrainers, and preservatives. The ability to discriminate between exposed and unexposed grains is a property of chemical reducing agents that possess the Kendall structure represented by $A\text{---}CH=CH$ and, where n may have zero or integral values and A and B may be hydroxyl, amine, or substituted amino groups. Most, but not all, chemical reducing agents are benzene derivatives with Kendall structures, eg, hydroquinones, catechols, aminophenols, p-phenylenediamines, and ascorbic acid. Phenidone developing agents and certain thiadiazoles also discriminately reduce silver ions. The activity of development solutions, such as hydroquinone (an important black-and-white developer), depends on hydrogen and halide ion concentrations.

Oxidation

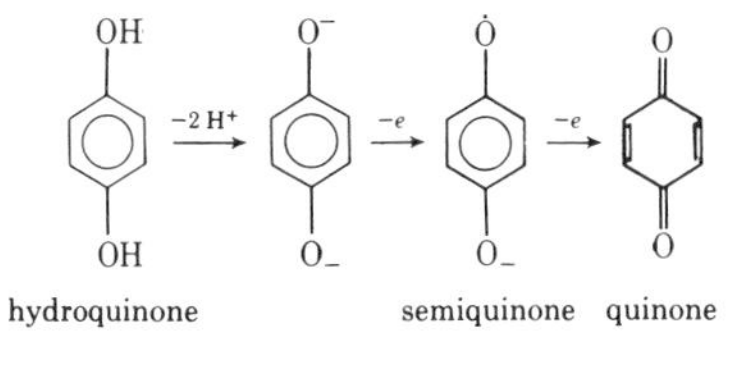

hydroquinone   semiquinone   quinone

Reduction

$$AgBr \xrightarrow{+e} Ag^0 + Br^-$$

The presence of bromide ions in the development solution restrains the conversion of Ag^+ to Ag^0 by the effect of its concentration on the electrochemical overpotential for the overall redox couple. Preservatives such as sodium sulfite are scavengers for oxidation products.

Development can be viewed as an electrochemical redox reaction

$$n\,AgBr + H_mD \rightleftharpoons D_{ox}^{(n-m)+} + n\,Ag^0 + n\,Br^- + m\,H^+$$

where D and $D^{(n-m)+}$, respectively, correspond to the reduced and oxidized forms of the reducing agent.

Development depends upon the diffusion rates of the developer components through the gelatin to the silver halide grain surfaces where the catalytic silver centers are located. Such diffusion processes are rate limiting, and once a reducing agent has diffused to a grain surface, nucleation and growth of an elemental silver phase can be initiated.

In general, the development rate increases with increasing temperature; many modern systems have been designed for high temperature processing which requires specially hardened coatings. Several photographic materials are based on special development techniques such as activator processing, thermal development, monobath processing, and silver-diffusion transfer.

Once satisfactory development has been achieved, the reaction is quenched by the rapid decrease in pH upon transferring the coatings from the alkaline developer solution to an acidic "stop" bath. For both the silver-diffusion-transfer films and the dye-transfer films used in instant photography, development is initiated by spreading a viscous alkaline reagent between an emulsion layer and an image-receiving layer. Recent integral films rely on the timed release of an acid to quench development.

Fixation

In conventional photography, undeveloped silver halide is removed with thiosulfate ions which convert the remaining silver to water-soluble complexes such as argentodithiosulfate and argentotrithiosulfate. Thiosulfate is stable, nontoxic, inexpensive and does not react with gelatin or with the developed silver.

In black-and-white photography, fixation is generally conducted under acidic conditions. Decomposition of thiosulfate is retarded by the addition of bisulfate. The rate of fixation is monitored in terms of clearing time, ie, the time required for the last visible opacity to disappear. At various post-development stages, the coatings are rinsed to reduce carry-over of chemicals; washing is essential for image permanence.

Environmental Aspects

Before being discharged, photographic effluents require special treatment, such as settling, biochemical degradation, aeration, or chlorination. Silver, usually bound in thiosulfate complexes, can be converted to silver sulfide and removed as a solid sludge (see Recycling).

The silver image. Humidity, temperature, chemistry of the environment, and particularly air oxidation, adversely affect image permanence. The subjective quality of a developed silver image depends upon the color tone, the brightness reproduction, and the perceived graininess and sharpness. The impression of crispness is achieved when the boundaries and edges of the objects are clear and well defined. The ability of photographic material to record fine details is a function of both development and optical effects.

DAVID J. LOCKER
Kodak Research Laboratories

C.B. Neblette in J.M. Sturge, ed., *Neblette's Handbook of Photography and Reprography*, 7th ed., Van Nostrand Reinhold Co., New York, 1977, p. 8.

M.A. Kriss in T.H. James, ed., *The Theory of the Photographic Process*, 4th ed., Macmillan Publishing Co., Inc., New York, 1977, pp. 592–635.

L. Wolfman, *1980–81 Wolfman Report on the Photographic Industry in the United States*, ABC Leisure Magazines, Inc., a subsidiary of American Broadcasting Co., Inc., New York, 1981.

PHOTOMULTIPLIER TUBES

The photomultiplier is a highly sensitive detector of radiant energy in the uv, visible, and near-ir regions of the electromagnetic spectrum. The basic sensor is a photocathode located inside a vacuum envelope. Photoelectrons are emitted and directed by an electric field to an electrode or dynode within the envelope. A number of secondary electrons are emitted at this dynode for each impinging primary photoelectron. These sec-

ondary electrons, in turn, are directed to a second dynode and so on until a final gain of perhaps 10^6 is achieved. The electrons from the last dynode are collected by an anode which provides the output-signal current.

Despite the solid-state revolution, the photomultiplier, because of its high gain and fast time response, continues to be manufactured on a large scale.

Design

Dynodes are shaped and positioned in such a manner that all the stages are properly utilized and no electrons are lost to support structures. The shape of the field should encourage the return of electrons to a center location on the next dynode. Magnetic fields may be combined with electrostat fields to provide the required electron optics, although most photomultipliers are electrostatically focused. For some applications, the time spread of the electron trajectories must be minimized and strong electric fields are provided at the surfaces of the dynodes to assure high initial acceleration of the electrodes. Regenerated effects must be avoided.

The original design was a circular array of photocathode and dynodes, as typified by the present 931A tube. In scintillation counting, the photocathode must be of the semitransparent type located on a relatively large flat glass surface through which the scintillating crystal is coupled. The electron-optical requirements for this case usually call for an increased photocathode-first-dynode spacing. Various dynode configurations are then utilized, including the circular cage, the box-and-grid and the venetian-blind for good electron-collection efficiency, the linear-array cage for short-time response, and the close-spaced mesh-dynode array for use in high magnetic field environments.

A recent innovation in front-end design is the so-called teacup photomultiplier, named after its large first dynode. Secondary electrons from the first dynode are directed to an opening in the side of the teacup and then to the second dynode. Fields between the photocathode region and the first dynode region are separated by a very fine grid structure (see Fig. 1). Sidewall photoemission results from the light that passes through the semitransparent photocathode. The increased photoemission and collection efficiency improve the pulse-height resolution in scintillating–counting applications.

Other recent designs include high speed tubes utilizing microchannel plates in proximity to the photocathode and anode. Time resolution or pulses initiated by single photoelectrons measured at full width at half maximum (FWHM) is less than 800 ps. Sensitivity to external magnetic fields is much reduced, a fact that is important in some nuclear-physics experiments. Even faster photomultipliers use crossed electrostatic and

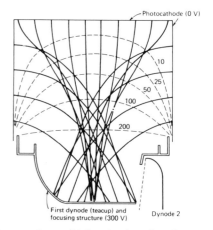

Figure 1. Teacup photomultiplier showing photoelectron paths directly from the photocathode and those initiated by light transmitted through the photocathode and striking the sidewall which also has an active photoemissive layer. Equipotential lines are indicated in the region between photocathode and first dynode.

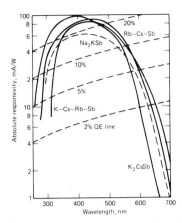

Figure 2. Typical spectral-response curves for various photocathodes used in scintillation–counting application. The variation in the cutoff at the low end is the result of different envelope materials.

magnetic fields to direct electrons to repeated stages of secondary emission.

The photoemitters used in these tubes are all semiconductors (qv). Light is absorbed by the valence-band electrons in the semiconductor only if the energy of the photon is at least equal to the band-gap energy E_G. If, as a result of light absorption, electrons are raised from the valence band into the conduction band, photoconductivity is achieved. For photoemission, an electron in the conduction band must have energy greater than the electron affinity E_A, and photoemission can occur only if the photon energy exceeds $E_G + E_A$.

Photoemission from semiconductors can be improved by modification of the energy-band structure. Reduction of the E_A permit the escape of electrons that have been excited into the conduction band at greater depths within the material. Indeed, if the electron affinity is reduced to less than zero, the escape depth may be as much as 100 times greater than for normal material. At less than zero, the vacuum level is lower than the bottom of the conduction bond and, a condition develops described as negative electron affinity (NEA). In secondary emission, the impact of primary electrons rather than of incident photons causes the emission of electrons.

Photocathodes. The most common photocathodes are cesium antimonide Cs_3Sb, multialkali or trialkali ($Na_2KSb:Cs$), and bialkali (K_2CsSb) derivatives. The bialkali cathodes have the greatest thermal stability. Recently, a rubidium–cesium–antimony (probably Rb_2CsSb) has been introduced that has excellent blue sensitivity. Typical spectral-response curves are shown in Figure 2.

Dynode materials are usually alloys, eg, copper–beryllium, silver–magnesium, or CsSb. Negative electron affinity materials, eg, GaP:Cs, have very high secondary-emission yields and are used to reduce statistical noise or reduce the number of stages.

Characteristics

When several secondary-emission stages are coupled, the total gain μ is given by $\mu = \delta^n$, where δ is the secondary emission per stage, and n is the number of stages. Photomultiplier gain is usually presented as a function of the applied voltage, and may vary from 10^4 at 500 V to 10^7 at 1200 V.

All photomultipliers are sensitive to external magnetic and electrostatic fields. The higher the voltage, the smaller the effect of these fields.

The anode current is proportional to the incident radiant flux. The limit of linearity occurs when space charge begins to form, usually between the last two dynodes.

Operating temperatures can be as high as 75°C although, with a Na_2KSb photocathode, 175°C can be tolerated for a short time, an important consideration in oil-well operations. Very low temperatures (below −40°C) may damage the tube because of stresses in the metal-to-glass seals. Operating stability depends on the magnitude of the average anode current; ≤ 1 μA is recommended if stability is of great

importance. Photomultipliers should be stored in the dark since light, especially blue or uv, increases dark emission.

At very low light levels, the limitation of detection and measurement is generally the signal-to-noise ratio determined by the fluctuation in the thermionic dark emission from the photocathode. Secondary emission contributes very little to the relative noise output of the tube. The limit to detection can be described by stating the equivalent-noise input or the noise-equivalent power.

Theoretical estimates of photoemission or secondary-emission times for metals or insulators are 0.01–0.1 ps. In NEA semiconductors, the lifetime of internal free electrons having quasi-thermal energy can be on the order of 0.1 ns.

Uses

Photomultipliers are recommended for the detection of very low light or radiation levels. They are also capable of operating over a wide range of light flux, and are used to detect or measure radiant fluxes (see Photodetectors). Direct applications include oil and gas exploration; blood-sample analysis; the detection and determination of air pollutants; radiation monitoring at nuclear plants; rapid sorting of food and manufactured goods; control of color-printing machines; gun-fire control; detection of pinhole flaws in steel and paper; laser ranging; photometry; spectrometry; radioimmunoassay; and thermoluminescent dosimetry. Other applications include scintillation counters and the gamma camera for locating tumors or other biological abnormalities. Photomultiplier tubes are used in CAT scanners and the positron camera (see X-ray technology); Recently, photomultiplier tubes having very large hemispherical photocathodes have been developed for use in proton-decay measurements.

RALPH W. ENGSTROM
Consultant, RCA

R.W. Engstrom, *Photomultiplier Handbook*, RCA Corporation, Lancaster, Pa., 1980.

A.H. Sommer, *Photoemissive Materials*, John Wiley & Sons, Inc., New York, 1968. (Reprinted by Robert E. Krieger Publishing Co., Huntington, N.Y., 1980).

Physics Today, **37**, 74 (April 1983).

PHOTOREACTIVE POLYMERS

The physical properties of photoresist coatings change when they are exposed to light. The change is usually one of solubility and results in solvent discrimination between exposed and unexposed areas. Photocross-linking and photoinitiated polymerization decrease solubility, whereas photomodification of functionality and photodegradation increase it. Exposure to light through a pattern results in solubility changes and, therefore, image boundaries that can be used to form resist images. These images can be produced by solvent development in either negative- or positive-working modes, as shown in Figures 1 and 2, respectively.

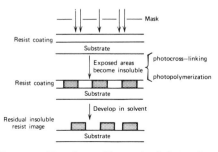

Figure 1. Negative-working mode of photoresists.

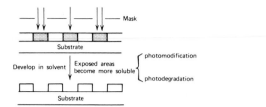

Figure 2. Positive-working mode of photoresists.

Photoresist polymer compositions must fulfill a number of physical-property-dependent requirements under practical working conditions; these depend on critical balancing of polymer and copolymer compositions. The most important and most difficult requirements to fulfill are image discrimination, thermal stability, and etch resistance. For image discrimination, the development solvent must remove the soluble portions of the exposed imagery without distorting or swelling the insoluble areas.

In the past, resist polymers were required only to withstand liquid aqueous etchants, eg, ammonium fluoride or hydrofluoric acid. However, present trends toward plasma etching require that the polymers resist the chemical attack of free radicals and ions. Stability to etching and sufficient instability to allow stripping are achieved through compromises in structure and fabrication conditions (see Plasma technology).

Chemistry

Photomodification of functionality depends on the modification of the solubility of polymer-bound chromophore units upon light absorption. A change in functionality modifies the solubility of appended groups and, thus, that of the polymer in selected solvents. For example, with sufficient exposure, a polymer-bearing diazoketone unit becomes hydrophilic.

The solubility and photochemical response of the copolymer can be modified by varying the ratios of the comonomers as well as the degree of substitution of the diazoketone or the other substituents, eg, benzoyl, on the amino group.

The photocross-linking of polymers for photoresist application is accomplished by means of bifunctional addenda or photoreactive chromophores. Bisazide addenda were believed to cross-link with polydiene chains by absorption of two photons per molecule of azide. More recently, however, investigation of the photoresist system of cyclized poly(cis-isoprene) and bis(4-azidobenzal)cyclohexanone using ir and uv spectroscopy, solvent extraction, and tlc showed that both azido groups of the bisazide are cleaved by absorption of a single photon.

The types of bisazides generally used for photocross-linking of polydienes are based on the condensation products of azidobenzaldehyde and 4-alkyl-substituted cyclohexanones:

2 N_3—⟨ ⟩—CHO + (cyclohexanone) $\xrightarrow{\text{base}}$ N_3—⟨ ⟩—CH= (ring) =CH—⟨ ⟩—N_3

azidobenzaldehyde

4-alkyl-2,5-bis(p-azidobenzal)cyclohexanones
λ_{max} 365 nm
spectral-sensitivity range: 340–420 nm

R = CH_3, C_2H_5, C_3H_9, etc

The spectral-response range of a bisazide-polydiene resist depends entirely on the absorption spectrum of the azide.

Attachment of photodimerizable groups to polymer chains readily provides photocross-linkable polymers, which, in conjunction with spectral-range-controlling dyelike sensitizers, offer versatility that has not been achieved by photomodification of bifunctional addenda. Photocross-linkable units can be made to function at 200–700 nm by judicious use of specific chromophores with specific triplet sensitization. Photocross-linkable units include the following groups, among others: cinnamates (absorption max 250 nm), chalcones (275 nm), p-azidophenyls (260 nm), azidophthalates (298 nm), and p-phenylenebis(acrylates) (320 nm).

Cinnamate moieties have been attached to various polymer backbones by almost every type of known organic structural link, eg, poly(vinyl alcohol) esters, urethanes, amides, vinyl ethers, and polystyryl groups. The polymer of choice for most commercial application is poly(vinyl cinnamate).

Some special uses require careful control of the physical properties, eg, adhesion, softening points, and solubilities. The properties of vinyl polymers are less easily modified than those of condensation polymers. However, the properties of polycarbonates (qv) and polyesters (qv) can be modified easily.

Compounds that act as sensitizers with poly(vinyl cinnamate) include aromatic nitro compounds, aromatic amines, aromatic ketones, aromatic hydrocarbons and various heterocyclic compounds.

Photopolymerization. Photocross-linking, photomodification of functionality, and photodegradation involve preformed, high molecular weight, photosensitive polymers. However, in photopolymerization, the result of the photoreaction is the polymerization of low molecular weight monomeric material to form a high molecular weight polymer. In general, the net effect of the light energy in photopolymerization is the production of initiating species. Thus, a photopolymerization system usually consists primarily of mono- or multifunctional polymerizable monomers, a photoinitiating system, and a preformed polymeric binder. The monomer can be of any type, although compounds bearing vinyl groups are most commonly used. The mode of polymerization is always addition.

Photoinitiators, free radical or ionic, are responsible for producing initiating species by the action of light energy. Free radicals are produced by intramolecular photocleavage or intermolecular proton abstraction. Compounds such as benzoin ethers, 4-$tert$-butyltrichloroacetophenone, 1-phenyl-1,2-propanedione-2-O-benzoyloxime, and α,α-diethoxyacetophenone undergo photocleavage.

Ionic photoinitiator systems are either anionic or cationic. Cationic initiators include aryldiazonium salts, diaryliodonium salts, triarylsulfonium salts, and triarylselenium salts.

A preformed polymeric binder is usually included in a photopolymerization system. In addition to binding the ingredients, sometimes in the form of a flexible film, it determines the ultimate physical properties of the system.

Uses

The most significant advantage of photopolymerization is the hardening speed, which permits rapid curing of coatings of millimeter thickness. In addition, it saves energy and reduces air pollution and occupational and health hazards.

In pigmented coatings, 2-chlorothioxanthone acts as a photoinitiator for titanium dioxide pigments. The polymerization of acrylate monomers is strongly inhibited by oxygen, and for thick coatings, excess initiator dilutes the effect of oxygen. Also, a combination of aromatic ketone and amine suppresses oxygen inhibition. The red-light photolysis of dissolved oxygen in the presence of a dye sensitizer and the scavenging of the resulting singlet oxygen are shown below.

$^1O_2^*$ + (1,3-diphenylisobenzofuran) → (1,2-dibenzoylbenzene) + $\frac{1}{2}$ $^3O_2$

1,3-diphenylisobenzofuran   1,2-dibenzoylbenzene

Another solution of the problem of oxygen inhibition is an oxygen-insensitive ionic-initiated process.

Photopolymerization can be panchromatically sensitized with photoreducible dyes and suitable reducing agents. Many activators have proved effective in the preparation of spectrally sensitized photopolymer imaging systems, eg, the trialkylbenzylstannates, which offer good dark stability.

Photoresists. The development of dry-film resists increased the efficiency and convenience of the production of large quantities of macrocircuits, where considerable labor savings are achieved in coating, drying, and inspection. However, dry-film resists are not used in the microimaging field, where liquid resists allow adjustment of thickness. Most coatings are made by spinning the substrate, usually a silicon wafer, after or while the resist solution flows onto the wafer. Recent developments are deep-uv, electron-beam (E-beam), x-ray, ion, and proton-beam resists. The annual market growth of photoresists for application in the semiconductor industry is expected to be 10–15%.

JACK L.R. WILLIAMS
MICHEL F. MOLAIRE
Eastman Kodak Company

W.S. Deforest, *Photoresists, Materials and Processes*, McGraw-Hill, New York, 1975.

C.G. Roffey, *Photopolymerization of Surface Coatings*, John Wiley & Sons, Inc., New York, 1982.

S.P. Pappas, ed., *UV Curing: Science and Technology*, Technology Marketing Corporations, Conn., 1978.

PHOTORESISTS. See Photoreactive polymers.

PHOTOVOLTAIC CELLS

Photovoltaic cells offer the following advantages: they capture sunlight, an inexhaustible and nonpolluting source, and convert it directly into electricity; the conversion requires no heat engines and produces no noise, waste, or pollution. Photovoltaic systems are modular and, therefore, can be adapted for a variety of applications. Solar cells function wherever sunlight is available and are therefore particularly useful in areas where power lines cannot be readily or cheaply installed (see Semiconductors; Solar energy).

Solar cells have been used extensively in space, where their high power-to-weight ratio and reliability are desirable. Terrestrial solar cells, however, are expensive; furthermore, sunlight has a comparatively low energy density, and the best solar cells convert sunlight to electricity with only limited efficiency.

The power density of sunlight is only ca 1350 W/m^2 (428 $Btu/(ft^2 \cdot h)$) at just above the earth's atmosphere, ie, where the air mass = 0 (AM0). Air mass defines the effect on the solar spectrum of passing through varying effective depths of atmosphere.

The DOE has developed a systems test-and-application plan emphasizing the establishment of the technical credibility of solar power systems; the identification and elimination of technical and institutional constraints on their widespread acceptance; the provision of data to permit the economic modeling of solar photovoltaic power systems; and the establishment of overall national energy goals.

Mechanism

When sunlight falls on a silicon solar cell, a voltage is induced and an electric current flows in an external circuit which is connected to the cell. Each atom in the crystal lattice is surrounded by and bound to four equidistant neighboring atoms. The outermost electron shell of each silicon atom contains four valence electrons, each of which is shared in a bonding orbital with an electron from one of its four neighbors. This electron-pair or covalent bond binds the crystal firmly. The energy required to break a covalent bond is the bond energy, also called the energy gap, E_g, which in silicon is ca 1.1 eV.

The absence of an electron from a covalent bond creates a hole which can be filled by a neighboring valence electron, thereby creating a hole in a new location. The new hole, in turn, can be filled by a valence electron from another covalent bond, and so on. Hence, a mechanism is established for electrical conduction that involves the motion of valence electrons, but not of free electrons. Because holes and electrons move in opposite directions under the influence of an electric field, a hole has the same magnitude of charge as an electron but is opposite in sign.

Junctions

If the hole and the electron are not kept apart, they recombine to produce a small amount of energy within the crystal and no net current flows. When the holes and electrons are kept apart, collected, and made to flow in a circuit outside the crystal, they produce a current in that circuit. Solar cells are equipped with a barrier or junction, which provides an internal field that segregates photogenerated electrons and holes.

A homojunction joins semiconductor materials of the same substance; a heterojunction is formed between two dissimilar semiconductor substances; a Schottky junction is formed when a metal and a semiconductor material are joined; and in a metal–insulator semiconductor junction (MIS), a thin oxide layer, generally less than 0.003 μm, is sandwiched between a metal and a semiconductor material.

Homojunctions are fabricated by diffusion, ion implantation (qv), chemical vapor deposition (CVD), vacuum deposition, and liquid-phase deposition; heterojunctions by CVD, vacuum deposition, and liquid-phase deposition; and Schottky and MIS junctions by vacuum deposition (see Film deposition techniques).

Efficiency

The most efficient silicon cells are based on $p–n$ homojunctions; they convert only ca 15% of the energy in incident sunlight. Such cells, made of expensive semiconductor materials, eg, InP, GaAs, and CdTe, with energy gaps of 1.2–1.4 eV, have maximum theoretical conversion efficiencies of ca 26%.

In commercial silicon cells of thicknesses of hundreds of micrometers, most of the solar radiation is absorbed in 20–30 μm. The homojunction is usually ca 0.5–1.0 μm from the surface. A comb or narrow metal bar is connected to a number of fingers and collects charge carriers from the side of the cell facing the sun (see Fig. 1). The total area of the fingers is so small that only a minimal cell area is in their shadow. Antireflection coatings are used over the silicon surface which, without coating, reflects ca 40% of incident sunlight.

The current method of producing wafers from ingots is an energy-consuming, costly, and wasteful process. The ingot is slowly pulled from a silicon melt, cut into thin wafers, and polished, which reduces more than 60% of the ingot to dust. A promising cost-reducing approach is to grow good quality crystalline sheets directly from molten silicon. Smoothly grown sheets, ca 100 μm thick, require little or no cutting and polishing and very little is wasted.

Optical systems, eg, lenses, concentrate the sunlight on the cells and reduce overall costs. Concentrator optics range from low ratio designs to much higher ratio systems based on parabolic mirrors or Fresnel lenses, which require precise, two-axis tracking.

Breaking up sunlight with tandem cells and filter reflectors or the use of thin films (20–30 μm) with good conversion efficiencies (> 10%) could also reduce costs.

Solar cells fabricated from plasma-deposited, amorphous, silicon-hydrogen films are reported to give a conversion efficiency of about 10% (see Plasma technology). The most efficient amorphous Si–H cells contain up to 15 wt% hydrogen.

Thin-film solar cells are also fabricated from Cu_xS/CdS materials. A conversion efficiency of ca 9.2% in sunlight has been reported.

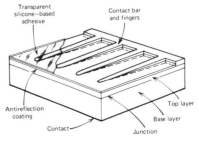

Figure 1. Basic design of a commercial silicon solar cell.

Gallium arsenide is a promising material for both high efficiency, single-crystal, and thin-film solar cells. Such cells could be much lighter than silicon cells of similar output because GaAs absorbs sunlight much more readily than silicon, and therefore, thinner layers of ca 1–2 μm could be used. Both GaAs homojunctions with a top window layer of $Ga_{1-x}Al_xAs$ and GaAs shallow homojunctions are being developed. The surface recombination effect is reduced by the formation of a thin layer of p-doped $Ga_{1-x}Al_xAs$ on the top surface of a p–n junction GaAs cell. The thin layer transmits almost all of the solar spectrum and forms an effective solar window. Single-crystal cells that are built in this manner produce conversion efficiencies as high as 21%. The structure is built around an n–p junction. The top layer, n, is less than 0.1 μm thick, thus avoiding the surface recombination effects.

Electrochemical Photovoltaic Cells

The main disadvantage of electrochemical photovoltaic cells is the instability of the semiconductor electrode, especially under sunlight, for extended operation. However, cells using GaAs, CdTe, and CdS have been made with 12, 7, and 16% efficiencies, respectively.

Although photovoltaic conversion is generally nonpolluting, environmental, health, and safety aspects must be considered, especially with regard to harmful emission and waste products. Proper encapsulation should minimize contamination.

JOHN C.C. FAN

Massachusetts Institute of Technology

J.C.C. Fan, *Technol. Rev.* **80**, 14 (1978).

E.A. Perez-Albuerne and Y.-S. Tyan, *Science* **208**, 903 (1980).

J.L. Smith, *Science* **212**, 1472 (June 26, 1981).

J.C.C. Fan, "Solar Photovoltaic Cells," in S.W. Yuan, ed., *Proceedings of the First U.S.-China Conference on Energy, Resources and Environment, Beijing, China, Nov. 1982*, Pergamon Press, New York, 1983, p. 480.

PHTHALIC ACIDS AND OTHER BENZENEPOLYCARBOXYLIC ACIDS

Physical and Chemical Properties

The physical properties of the benzenepolycarboxylic acids and of some anhydrides are given in Table 1.

The chemistry of these acids is generally the same as that of other carboxylic acids. They form esters, salts, acid chlorides, and anhydrides. They also undergo reactions characteristic of the benzene nucleus, such as halogenation or sulfonation.

Mixed diesters or triesters can be prepared by sequential reactions with different alcohols, acid chlorides by reaction with thionyl chloride, and amides by reaction with ammonia or amines.

Condensation polymerization is also possible:

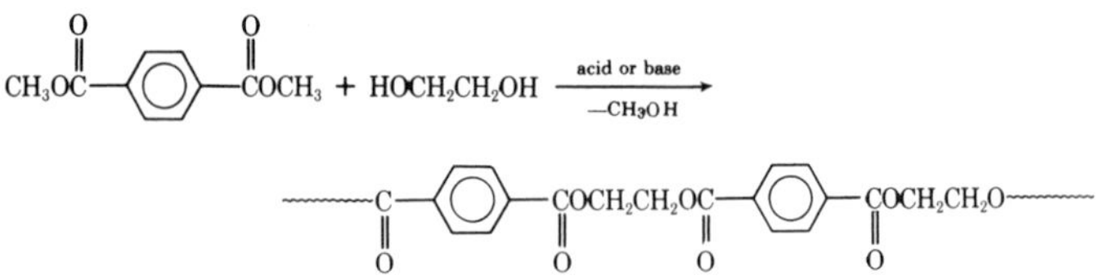

The dipotassium salts of phthalic and isophthalic acids can be rearranged thermally or catalytically to the dipotassium salt of terephthalic acid (Henkel reaction).

Condensation of phthalic anhydride with benzene gives anthraquinone (qv) and with phenol gives phenolphthalein. Phthalic anhydride reacts with urea and metal diacetates to form phthalocyanines (qv).

Phthalic Acid and Phthalic Anhydride

Manufacture. Until World War II, phthalic acid and, later, phthalic anhydride, were manufactured by liquid-phase oxidation of suitable feedstocks. Today, most phthalic anhydride is made by catalytic air oxidation of o-xylene using a fixed-bed, vapor-phase process. In this process, o-xylene is mixed with air and passed over a catalyst, typically vanadium oxide plus titanium oxide, in a multitubular reactor at about 400°C. Crude phthalic anhydride is recovered by condensation in switch condensers, heat treated, and then distilled to give the pure product.

Both fixed-bed and fluidized-bed reactors are used for vapor-phase oxidation of naphthalene with vanadium oxide-based catalysts. Yields are usually lower than those obtained from fixed-bed processes based on o-xylene.

In general, in vapor-phase oxidation processes, phthalic anhydride is recovered from the cooled reactor effluent by passing through automatically controlled switch condensers. Hot oil is charged through these condensers in order to melt the anhydride, which is drained into a tank. The vent gases, containing small amounts of by-products and traces of the anhydride, are scrubbed before being vented to the atmosphere. Scrubbing recovers maleic acid and small amounts of phthalic, benzoic, and citraconic acids.

The crude anhydride is purified by continuous distillation under vacuum in two columns which are connected in series. In the first column, maleic anhydride and benzoic and toluic acids are removed overhead; and in the second column, pure phthalic anhydride. High boiling residues are removed from the bottom of the second column.

Phthalic anhydride is also produced by liquid-phase air oxidation of o-xylene. Acetic acid is the solvent and a combination of cobalt,

Table 1. Physical Properties of Benzenepolycarboxylic Acids and Some Anhydrides

| Common name | Substitution | Melting point, °C | Dissociation constants in aqueous solution, 25°C | | | ΔH_f° at 25°C, kJ/mol[a] | Solubility, g/100 g water | | Anhydride, melting point, °C |
			pK_1	pK_2	pK_3		at 25°C	at 100°C	
phthalic	1,2-di	211 dec	2.95	5.41		−782	0.7	19.0	131
isophthalic	1,3-di	348 sub	3.62	4.60		−803	0.013	0.24	
terephthalic	1,4-di	> 300 sub	3.54	4.46		−816	0.0019	< 0.04	
hemimellitic[b]	1,2,3-tri	197 dec	2.80	4.20	5.87	−1160	v sol	v sol	196
trimellitic	1,2,4-tri	238 dec	2.52	3.84	5.20	−1179	2.1	60	168
trimesic	1,3,5-tri	350	2.12	3.89	4.70	−1190	0.24	6.4	
mellophanic	1,2,3,4-tetra	241 dec	2.06	3.25	4.73	−1562	sol	v sol	198 (di)
prehnitic	1,2,3,5-tetra	238	2.38	3.51	4.44	−1549			
pyromellitic	1,2,4,5-tetra	281	1.92	2.87	4.49	−1571	1.5	> 30	285 (di)
benzenepentacarboxylic		238	1.80	2.73	3.97	−1930	sol	v sol	
mellitic		288 dec	1.40	2.19	3.31	−2299	sol	v sol	320 dec (tri)

[a]To convert J to cal, divide by 4.184.

[b]Hemimellitic acid usually is handled as the dihydrate; formula wt 246.18, mp 191°C, decomposes.

manganese, and bromine salts is used as catalyst. After reaction, phthalic acid is filtered from the solvent and catalyst, dehydrated, and distilled. Overall yields are significantly higher than those from the vapor-phase process, but capital costs are higher.

The solidification point (131°C) is a good indication of purity, as is the molten color. The latter is determined initially at 250°C (APHA 10) and after being held for 2 h at this temperature (APHA 30).

Phthalic anhydride is shipped molten or as solid flakes.

Health and safety factors. Phthalic anhydride is a severe irritant to the eyes, respiratory tract, and skin, especially when moist, where it may cause burns. Flammability data must be consulted when working with phthalic anhydride. Phthalic acid dust forms explosive mixtures with air.

Uses. Phthalic anhydride is used mainly in plasticizers (qv) (diesters of a monohydric alcohol, eg, dibutyl phthalate, and mixed esters of two monohydric alcohols), unsaturated polyester resins, and alkyl resins. The largest-volume plasticizer is dioctyl phthalate (DOP) [di(2-ethylhexyl) phthalate]. Unsaturated polyester resins are usually condensation polymers of phthalic anhydride, an unsaturated dibasic acid or anhydride, and a glycol. A fully cross-linked product is made by polymerizing the unsaturated portion of the polycondensate with a vinyl monomer, usually styrene. Approximately 25% of the U.S. phthalic anhydride production is used for unsaturated polyesters. Alkyd resins, which are obtained from polybasic acids or anhydrides, polyhydric alcohols, and fatty oils and acids, consume ca 20%. These resins are used as binders for surface coatings. Phthalic anhydride is also used in the manufacture of dyes eg, 2-chloro-anthraquinone, phthalocyanine blues, quinone yellow, and anthracene brown (see Alkyd resins; Polyesters, unsaturated).

Terephthalic Acid and Dimethyl Terephthalate

Dimethyl terephthalate (DMT) and polymer-grade terephthalic acid are used as raw materials for the production of poly(ethylene terephthalate) (PET) and other polyesters (qv). Terephthalic acid is also produced in technical grades that are not suitable for the preparation of PET. The large commercial production of terephthalic acid/dimethyl terephthalate makes terephthalic acid the most important benzenepolycarboxylic acid.

In 1983, dimethyl terephthalate was the leading intermediate for the worldwide production of PET. However, polymer-grade terephthalic acid offers costs advantages to the PET producer, and the acid is expected to become the main intermediate by 1990.

Manufacture. Technical-grade terephthalic acid is produced by liquid-phase air oxidation of p-xylene. Although several processes have been commercialized, the technology developed by Amoco Chemicals Corporation is the most popular. In the Amoco process, acetic acid is used as a solvent, bromine is used as a renewable source of free radicals, and multivalent metals such as cobalt and manganese are used as catalyst. The product from the oxidation is recovered by centrifugation, and the crystals are subsequently washed and dried. Overall process yield is at least 90 mol%, and product purity is above 99 wt%.

Almost all polymer-grade terephthalic acid is produced from the technical-grade acid by using the purification technology developed by Amoco Chemicals. However, polymer-grade terephthalic acid can also be produced by hydrolyzing DMT and by modifying the processes that produce the technical-grade acid.

In the Amoco purification process, the technical-grade acid is dissolved in water at 250°C, and the solution is exposed to a hydrogenation catalyst, eg, a noble metal on a carbon support. The hydrogenation step converts various color bodies to colorless products, and it converts the main feedstock impurity, 4-formylbenzoic acid, to p-toluic acid. The product is then recovered by crystallization, centrifugation, and drying. The resulting polymer-grade terephthalic acid has a 4-formylbenzoic acid concentration below 25 ppm. The overall process yield is above 97%.

Several Japanese companies have modified their p-xylene oxidation processes to produce polymer-grade terephthalic acid, which is used captively by the producer. These modifications include changes in the p-xylene concentrations, the reaction times, the reaction temperatures, and the catalyst concentrations. Also, additional processing steps are used between the reactor and the product recovery section.

Most dimethyl terephthalate is produced by the Hercules/Dynamit-Nobel process which involves four steps that alternate between liquid-phase oxidation and liquid-phase esterification. First, p-xylene is air-oxidized with a cobalt catalyst to p-toluic acid, and esterification with methanol yields methyl p-toluate. A second oxidation gives monomethyl terephthalate which is esterified to the diester. Two reactors and two recycle streams are used and the product is purified by distillation. The overall process yield based on p-xylene feed is 87 mol%.

Dimethyl terephthalate is also produced by the liquid-phase esterification of technical-grade terephthalic acid. Metal catalysts such as zinc, molybdenum, antimony, and tin can be used with a large excess of methanol. Conversion to dimethyl terephthalate is limited by equilibrium, and overall process yields approach 96 mol%.

The combined domestic production of terephthalic acid and dimethyl terepthalate reached a peak of 3.3×10^6 t (DMT basis) in 1979. Both chemicals are sold under long-term contracts, and prices are greatly influenced by the price for p-xylene.

Terephthalic acid is shipped in hopper rail cars, hopper trucks, and 1000-kg polyethylene bags. Dimethyl terephthalate is shipped in molten form by road or rail and in pellet form using bags or bins.

Health and safety factors. Terephthalic acid has a low order of toxicity, but the normal precautions in handling industrial chemicals should be observed. Dimethyl terephthalate irritates the eyes and the respiratory system. If ventilation is inadequate, a toxic-dust respirator should be worn when handling either chemical. Molten dimethyl terephthalate burns when ignited, and vapors form explosive mixtures with air. The dust from either chemical can also form explosive air mixtures.

Uses. Almost all polymer-grade terephthalic acid/dimethyl terephthalate is used to make PET, which is the main polymer used to produce polyester fibers, polyester films, and bottle resins. Polyester fibers are used in textiles and in industrial applications such as tire cord (qv). Polyester films are coated with adhesives and emulsions to produce wrapping tapes, photographic films, and recording tapes. FDA approval makes PET the polymer of choice for food packaging, including beverage bottles (see Barrier polymers). Terephthalic acid and dimethyl terephthalate are also used to produce thermoplastic engineering resins and specialty fibers (see Engineering plastics).

Isophthalic Acid

Manufacture. Isophthalic acid is manufactured by liquid-phase air oxidation of m-xylene, similar to the Amoco process for oxidation of p-xylene to terephthalic acid; production facilities may be interchangeable. However, because isophthalic acid is more soluble than terephthalic acid, the crystallization step is more important.

Isophthalic acid is made containing different amounts of isomeric terephthalic acid. Grades range from 85 to 99% isophthalic acid. Isomeric purity depends upon the xylene feed. Product impurities also include 3-formylbenzoic acid and m-toluic acid. Some grades are purified in a similar way to that described for polymer-grade terephthalic acid.

The specifications include triethylene glycol color evaluated by heating isophthalic acid and triethylene glycol at 260°C for 60 min; the color of the resulting ester is measured on the APHA scale. Polyester color is evaluated by heating isophthalic acid, fumaric acid, ethylene glycol, and propylene glycol and measuring the resulting color on the Gardner varnish scale.

Health and safety factors. Isophthalic acid is a mild eye irritant and eye protection is recommended for handling. The autoignition temperature is > 650°C, and the dust forms explosive mixtures with air.

Uses. More than 50% of the U.S. isophthalic acid production is used for unsaturated polyester resins (isopolyesters) which are condensation polymers of isophthalic acid, an unsaturated dibasic acid, and a glycol. This polymer condensate is cross-linked with a vinyl monomer (eg, styrene). Special resin properties can be developed by the proper selection of reactants. Isopolyesters are more expensive than polyesters based on phthalic anhydride. However, they have greater flexural strength, higher heat distortion temperature, and better resistance to water, chemicals, and impact. They are used with glass-fiber reinforcement in bulk-molding compounds and filament windings.

Isophthalic alkyds are reaction products of isophthalic acid and other polybasic acids with polyhydric alcohols and fatty oils. These resins are formulated in solvent-based, water-based, or solventless coatings which are more expensive than those based on phthalic alkyds, but are faster drying, harder, and resist higher temperatures.

Trimellitic Acid and Trimellitic Anhydride

Manufacture. Trimellitic anhydride is the largest-volume benzenetricarboxylic acid. It is made by liquid-phase air oxidation of pseudocumene, followed by crystallization and filtration of the product trimellitic acid, which is dehydrated to the anhydride and then distilled. The oxidation step is similar to that used in the Amoco process for terephthalic acid.

Health and safety factors. Trimellitic anhydride causes severe respiratory irritation and sensitization. It should be handled as an extremely toxic substance, because exposure may result in a toxic reaction. Precautions should include effective ventilation and protective clothing and goggles. Flaked or molten trimellitic anhydride burns if ignited and the dust or vapor can explode in air.

Uses. The largest single use for trimellitic anhydride is in the manufacture of esters. The ester of largest volume is 2-ethylhexyl trimellitate, but other trimellitate esters of aliphatic alcohols of seven to twelve carbon atoms are common. All of the esters are used as vinyl plasticizers. Trimellitic anhydride is also used in producing wire enamels and coatings (see Plasticizers; Insulation, electric).

Other Benzenepolycarboxylic Acids

Trimesic acid. Commercial trimesic acid is a small-volume, synthetic chemical produced by the oxidation of mesitylene (1,3,5-trimethylbenzene) with potassium permanganate. It is used as a cross-linking agent for solid rocket-propellant fuel (see explosives and propellants).

Hemimellitic acid. Hemimellitic acid is available as the dihydrate for laboratory purposes. It is made by oxidizing hemimellitine (1,2,3-trimethylbenzene) or hemimellitol. Hemimellitic anhydride is obtained from the acid by thermal dehydration.

Pyromellitic acid and pyromellitic dianhydride. Pyromellitic acid is made by liquid-phase oxidation of durene (1,2,4,5-tetramethylbenzene) with nitric acid, chromic acid, or potassium permanganate; dehydration gives the dianhydride. Vapor-phase catalytic air oxidation of durene gives pyromellitic dianhydride directly. Using air as the oxidizing agent facilitates continuous operation, minimizes corrosion, and simplifies product isolation.

Both pyromellitic acid and its dianhydride irritate skin, eyes, and mucous membranes. Direct contact should be avoided; protective clothing is recommended. The principal commercial use for pyromellitic dianhydride is as raw material for polyimide resins (see Polyimides). These polypyromellitimides are condensation polymers of the dianhydride and aromatic diamines (eg, 4,4′-oxydianiline). They combine excellent thermal, electrical and physical properties and are processed as films, filled and unfilled molding compounds, precision machine parts, and coatings.

Mellophanic and prehnitic acids. Neither of these two acids is a commercial product. They are synthesized by oxidizing the corresponding 1,2,3,5- and 1,2,3,4-tetramethylbenzenes, respectively. Because mellophanic dianhydride has a relatively low melting point and good solubility in epoxy resins, it has been considered as a substitute for pyromellitic dianhydride as a cross-linking agent.

Benzenepentacarboxylic and mellitic acids. These two acids are synthesized by oxidation of the corresponding penta- and hexamethylbenzenes. Mellitic acid is available as a laboratory chemical. In the early 1970s, benzenepentacarboxylic and mellitic acids, their sodium salts, and other benzenetricarboxylic and benzenetetracarboxylic acids were considered as possible substitutes for phosphorus- and nitrogen-containing detergent builders. Although satisfactorily biodegradable, these chemicals are not as efficient as other builders and, consequently, have not been commercialized.

ALAN G. BEMIS
JOHN A. DINDORF
CECELIA SAMANS
Amoco Chemicals Corporation

O. Widemann and W. Gierer, *Chem. Eng.*, 62 (Jan. 1979).

P.H. Towle and R.H. Baldwin, *Hydrocarbon Process.* 43(11), 149 (1964).

PHTHALOCYANINE COMPOUNDS

The discovery of iron phthalocyanine and the elucidation of its structure led to the commercial application of copper phthalocyanine (**1**) as a pigment (see Pigments, organic). Phthalocyanine complexes of 63 metals have been prepared.

(1)

The phthalocyanine pigments account for ca 30% of U.S. organic pigment consumption. They dominate the blue- and green-pigment market because they are moderately priced, high performance products with good tinting strength.

Properties

Most phthalocyanine compounds do not melt; they sublime and vaporize only under greatly reduced pressure and above 500°C. Solubility in water and organic solvents is very low. They absorb in the visible region at 600–700 nm. Their color varies from dark blue to metallic bronze to green.

Although stable to atmospheric oxygen up to > 100°C, in the presence of moisture phthalocyanines are oxidized to phthalimides. The carbon atoms of the ring system and the central metal atom can be reduced. A metal phthalocyanine can be reduced reversibly during dye vatting.

Preparation and Manufacture

Metal phthalocyanines are prepared by the reaction of o-phthalonitrile with a bivalent metal, metal halide, or metal alcoholate, or by heating a mixture of a phthalic acid derivative and a metal salt in a high boiling solvent to 200–300°C.

Copper phthalocyanine is manufactured by heating o-phthalonitrile with cuprous chloride or by the phthalic anhydride–copper salt–urea–catalyst–solvent process. The crude product from either process is slurried, washed and dried, and stabilized with chlorine or a sulfonic acid. Pigment properties are improved with various additives.

The pigmentary forms of phthalocyanine, both blues and greens, are widely used as colorants (qv) for paints, printing inks, and plastics. They provide excellent resistance to heat and light, and acids and alkalies. The blues are broadly classified as alpha (reddish), beta (greenish), and epsilon crystals, and the greens as chlorinated (bluish) and brominated (yellowish) derivatives. They may be diluted with inert materials or incorporated with resins or aluminum benzoic acid salts to provide lakes. Water-soluble dyes include the sodium salts of the sulfonate derivatives, which are used for natural and synthetic textiles and paper.

Uses

In addition to their use as pigments, phthalocyanines are employed in many types of dyestuffs and as catalysts, eg, cobalt or vanadium phthalocyaninesulfonate in the air oxidation of mercaptans and other sulfur compounds in petroleum. Metal-free, chloroaluminum, vanadyl, or magnesium phthalocyanines are sufficiently soluble in organic solvents and show enough bleachable absorption at 694.3 nm to serve as repeated Q-switching elements for ruby lasers (qv). Iron phthalocyanine-sodium sulfonate solution has been used as a hemoglobin substitute, other phthalocyanines for staining biological specimens.

FRANK H. MOSER
Consultant

WILLIAM H. RHODES
BASF Wyandotte Corporation

F.H. Moser and A.L. Thomas, *The Phthalocyanines*, CRC Press Inc., Boca Raton, Fla., 1983, 2 vols.

F.H. Moser in T.C. Patton, ed., *The Pigment Handbook*, Vol. 1, John Wiley & Sons, Inc., New York, 1973, pp. 679–695, 699.

PICKLING OF STEEL.   See Metal surface treatments.

PIGMENTS

INORGANIC

Inorganic pigments are an integral part of decorative, protective, and functional coatings. They provide mass coloration for fibers, plastics, paper, rubber, elastomers, glass, cement, glazes, and porcelain enamels, and they are colorants in inks, cosmetics, and markers (see Coatings; Colorants; Paint).

Pigments can retard corrosion (see Corrosion and corrosion inhibitors) and act as fungistats and antistatic agents (qv) (see Fungicides).

In the late 1970s, pigment consumption was distributed as follows: coatings and paints, > 50%; paper, 20%; plastics, 10%; rubber, 3%; and ceramics, 2%.

Pigments can be colored, colorless, black, white, or metallic. Most pigments are insoluble solids. When dispersed in a binder or medium, most pigments remain insoluble.

Properties

Color is described quantitatively by the Munsell and CIE (Commission Internationale de l'Eclairage) systems (see Color). Color results from the pigment's selective absorption of visible light. Large pigment particles also may scatter light and thereby influence the opacity of the binder. Because inorganic pigments involve both transition and nontransition elements, several different theories explain how color is produced by these compounds.

The intense yellow color of lead chromate, $PbCrO_4$, is attributed to a charge-transfer process between oxygen and chromium. The presence of two different oxidation states for iron in iron blue, $Fe(III)NH_4[Fe(II)(CN_6)]$, results in charge-transfer and the associated intense blue color. The color of cadmium sulfide is based on its semiconductor properties and the charge transfer transitions between the valance band and the conductance band (see Semiconductors). The color of ultramarine blue results from electronic transitions within S_3^-.

Most inorganic pigments are crystalline. Polymorphism is displayed by pigments such as titanium dioxide (rutile, anatase, and brookite), lead chromate (light and primrose type) and calcium carbonate (calcite and aragonite).

The average ultimate particle size of most commercial pigments, excluding extenders, is 0.01–1.0 μm dia. (see Table 1). The hiding power of a pigment depends primarily upon the ability of the dispersed particles to scatter light. Factors that influence hiding power are refractive index and particle size. The effective particle size for most pigments should be ca one-half the wavelength of visible light.

Microscopy, gas absorption, and sedimentation techniques are used to determine particle size. Particle shape can be noted by transmission electron microscopy.

Another property of pigments is surface area. Carbon blacks have surface areas of 6–1100 m^2/g, whereas most other pigments have areas of 1–100 m^2/g. To achieve complete dispersion, the surfaces of the pigment particles must be properly wetted. Surface-area determination

can be made by the BET method, which is based on the absorption of nitrogen at extremely low temperatures and low pressures.

Manufacture

Many pigments are manufactured by precipitation, ie, lead chromate; or by precipitation and calcination, ie, cadium sulfide; or by calcination, ie, chromium oxide. Titanium dioxide is prepared by a vapor-phase reaction.

Care is taken in selection of raw materials and control of manufacturing parameters, ie, temperature, concentration, and pH excesses, and agitation. Other process steps in manufacture include chemical end treatment, filtration, washing, drying, grinding, and packaging.

Health and Safety Factors

Pigment usage in contact with food is regulated by the FDA, other usage by the CPSC. Pigments listed as nuisance dusts are calcium carbonate, kaolin, titanium dioxide, and zinc dust; recommended TLVs are 10 mg/m^3 total dust or 5 mg/m^3 respirable dust. Some pigments are listed as possible carcinogens, eg, lead and zinc chromates.

White Pigments

The principal white hiding pigment is titanium dioxide (see Titanium compounds); others are zinc oxide, zinc sulfide, lithopone, lead pigments, and antimony oxide (see Table 2).

Titanium dioxide. The stable crystal form of TiO_2 is rutile; anatase can be converted to rutile at ca 700–950°C (see Table 3). Titanium dioxide, mp > 1800°C, is insoluble in water, organic solvents, alkalies, and most acids. It is attacked by sulfuric and hydrofluoric acids after long contact and at high temperature. Titanium dioxides are not ideal whites because they do not reflect all wavelengths of visible light equally. They absorb below 430 nm in the far blue or violet end of the spectrum.

Titanium oxide provides whiteness, brightness, and opacity in paints; as a paper-filler pigment, it improves printability. Because of its chemical inertness and resistance to degradation by uv, it is used in plastic materials.

Table 1. Particle Size of Some Inorganic Pigments, μm

Pigment	Particle size, μm
iron blue	0.01–0.2
titanium dioxide	0.2–0.3
red iron oxide	0.3–4.0
natural crystalline silica	1.5–9.0
strontium chromate	0.3–20.0
hydrated aluminum oxide	0.4–60.0
micaceous iron oxide	5.0–100.0

Table 2. Characteristics of Some White Hiding Pigments

Pigment	Chemical composition	Refractive index[a]	Average particle size, μm
titanium dioxide, anatase	TiO_2	2.55	0.2
titanium dioxide, rutile	TiO_2	2.70	0.2–0.3
zinc oxide	ZnO	2.01	0.2–0.35
zinc sulfide	ZnS	2.37	0.2–0.3
lithopone	28% ZnS, 72% $BaSO_4$	1.84	0.2–0.3
lead carbonate, basic			
white lead	$2PbCO_3 \cdot Pb(OH)_2$	2.0	1.0
lead sulfate, basic	$PbSO_4 \cdot PbO$	1.93–2.02	0.8
antimony oxide	Sb_2O_3	2.1	1.0

[a] A common refractive index for a paint vehicle is 1.6.

Table 3. Typical Pigment Properties of Anatase and Rutile TiO_2

	Anatase	Rutile
density, g/cm^3	3.8–4.1	3.9–4.2
refractive index	2.55	2.76
oil absorption, g oil/100 g pigment	18–30	16–48
tinting strength, Reynolds method	1200–1300	1650–1900
particle size (av), μm	0.3	0.2–0.3

Zinc oxide. Zinc oxide is manufactured from sphalerite. In the indirect process, zinc metal is vaporized, followed by oxidation and collection of the zinc oxide powder. In the direct process, zinc sinter and coal are burned and the resulting zinc vapor is oxidized. Zinc oxide, together with sulfur and organic accelerators, is used in the vulcanization of elastomers. In addition, zinc oxide absorbs uv radiation and thereby protects organic binders from photodegradation and reduces chalking. Leaded zinc oxide is prepared by cofuming lead and zinc ores or by blending lead-free zinc oxide and basic lead sulfate or lead silicate. It commonly contains 12–55 wt% basic lead sulfate.

Zinc sulfide. Zinc sulfide is no longer manufactured in the United States but is imported from the FRG. It is softer than TiO_2 and does not have the yellowish undertone associated with TiO_2. It is produced from $ZnSO_4$ and Na_2S.

Lithopone. Lithopone is a mixed pigment which originally consisted of 28–30 wt% ZnS and 70–72% $BaSO_4$; grades with higher zinc sulfide contents (ca 60 wt%) have been developed. It is prepared by coprecipitation of ZnS and $BaSO_4$. Lithopones are used in water-based paints and in rubber and plastics.

Lead pigments. White lead is the oldest white hiding pigment. Today's usage of lead pigments, eg, basic lead sulfate, basic lead silicate, basic lead silicosulfate, and dibasic lead phosphite is limited because of restrictions on lead compounds.

Antimony oxide. Sb_2O_3 is an effective fire retardant controls chalking, and improves tint retention in exterior enamels. It is prepared by roasting stibnite, Sb_2S_3; its use has declined (see Flame retardants, inorganic).

Zirconium oxide and zircon. These are used in porcelain or vitreous enamels. Vanadium-doped zircon is a blue pigment used in glazes and enameling.

Extender Pigments

Extender pigments are inexpensive colorless or white pigments with refractive indexes < 1.7. Kaolin is a hydrous aluminosilicate mineral found in Georgia, South Carolina, and Texas. It is used as a filler in the paper and paperboard industry and in paper coatings. Calcined kaolins are white and hard; they are used in water-based and traffic paints.

Clays. Clays are graded according to particle size, ie, fine, intermediate, and coarse. Mined clay is slurried at the open-pit mine site and the slurry is pumped to a degritting station and then stored in tanks. Slurries are blended, and particle-size distribution is controlled through processing. Attapulgus clay is a crystalline, hydrated magnesium aluminum silicate (Fuller's earth) used in paints as a flattening agent and in sealants (see Clay).

Calcium carbonates. Natural calcium carbonate, a widely used pigment, is quarried or mined. Limestone is the raw material for precipitated grades. Synthetic calcium carbonates are available in several grades of calcite and aragonite. High calcium types are used in paints, plastics, rubber, paper, adhesives, and joint fillers (see Lime and limestone).

Barium sulfate. $BaSO_4$ can be prepared synthetically or from the mineral barite. It is used mainly as a weighting material in drilling muds. This application accounts for ca 90% of U.S. consumption (see Petroleum, drilling fluids). It is an excellent extender pigment because of its inertness and high refractive index. Blanc fixe is the precipitated form.

Silicates. Magnesium silicate (talc) is derived from natural sources such as soapstone, steatite, and asbestine. Talcs reduce chalking, blistering, and cracking of paint film (see Silicon compounds).

Calcium silicate is derived from wollastonite. The synthetic material is prepared hydrothermally from diatomaceous silicas and lime. It is used in coatings for water-thinned emulsion paints where hiding is achieved by air voids. Natural silicas are obtained from tripolitic ores. They are used as abrasives, reinforcing fillers (qv) for silicone rubber, and in caulking compounds and porcelain insulators.

Diatomaceous silica pigments are hydrous silicas obtained by open or strip mining. Diatomaceous silica controls gloss and sheen of flattening paints and is used in water-thinned emulsion flat wall paints, flat varnishes, primers, concrete, and stucco finishes (see Silca; Diatomite).

Precipitated synthetic silicas have large surface areas (100–300 m^2/g) and high oil absorption. Pyrogenic silicas are extremely pure SiO_2 prepared by high temperature reaction. They are used in rubber and plastics. Aerogel and hydrogel silicas have surface areas of 100–400 m^2/g and high porosity; they are used as flattening agents in coatings.

Micas. Micas (see Mica) are complex aluminum potassium silicates; muscovite is the most important. They are used in wallpaper and coated paper, primer and corrosion coatings, and in joint cements and lubricants.

Other extender pigments. Hydrated aluminum oxides, manufactured by seed nucleation of a saturated alumina solution, are fine white pigments. Pumice is a glossy volcanic lava used to impart a rough nonslip surface. Asbestos (qv), a fibrous magnesium silicate, functions as a viscosity-control agent. Calcium sulfate is used in composite pigments and as a filler in plastics and primer paints. Sodium silicoaluminate is an ultrafine synthetic pigment used as a filler in paper and as an extender of TiO_2. Sodium potassium aluminum silicate (nepheline syenite) is used in the glass and ceramic industries (see Aluminum compounds; Fillers).

Colored Pigments

Iron oxides are characterized by low chroma and excellent lightfastness; they are nontoxic and inexpensive. They are processed from hematite, limonite, siderite, and magnetite, and provide a range of reds, yellows, purples, browns, and blacks. Raw sienna is a dark yellow that is converted to burnt sienna, a deep brown, by calcination. Ochers are yellow iron oxides, lighter in color than siennas. Cyprus is the source of fine raw umber, commonly called Turkey umber. Burnt umber is produced by calcining raw umber. The flakelike crystalline structure of gray micaceous iron oxide resembles that of mica. It is used in metal-protective coatings.

Synthetic iron oxides are red, yellow, brown, and black. Advantages over the natural oxides include chemical purity, uniform particle size and size distribution, and the possibility of use in predispersed vehicle systems by flushing techniques.

Ferrous sulfate heptahydrate, $FeSO_4 \cdot 7H_2O$ (copperas) is the main source of iron for these pigments. For example, calcination gives copperas red.

$$FeSO_4 . 7H_2O \xrightarrow{\Delta} FeSO_4 . H_2O + 6\,H_2O$$

$$6\,FeSO_4 . H_2O \xrightarrow{\Delta} 2\,Fe_2O_3 + Fe_2(SO_4)_3 + 6\,H_2O + 3\,SO_2$$

$$Fe_2(SO_4)_3 \xrightarrow{\Delta} Fe_2O_3 + 3\,SO_3$$

Direct precipitation of red iron oxides involves growth of iron oxide particles on specially prepared nucleating particles or seeds of Fe_2O_3.

Synthetic red iron oxides are prepared in various grades from light to dark reds and are sold as Indian red, Turkey red, and Venetian red. The Penniman-Zoph process for the manufacture of yellow iron oxides involves the preparation of a seed or nucleating particle by alkali precipitation of ferrous sulfate. Magnetic gamma iron oxide, used in recording tape, is prepared from an acicular yellow iron oxide precursor (see Magnetic tape). Zinc and magnesium ferrites are tan pigments formed by the interaction of iron oxides with metallic oxides. The principal use of transparent iron oxides is in metallic automotive finishes. Chemically they are the same as their opaque counterparts but are much smaller in particle size.

Lead chromate and molybdate oranges. These pigments provide a gamut of hues from greenish yellow through orange to light red. Medium chrome yellows are essentially pure $PbCrO_4$ and provide the reddest yellow hue. Primrose chrome yellow and light chrome yellows are solid solutions of lead chromate and lead sulfate. Chrome oranges are $PbCrO_4 \cdot PbO$ and the variation from light to dark orange is a consequence of change in particle size. Molybdate oranges are solid solutions of lead chromate, lead molybdate, and lead sulfate. They display hues from orange to red. They are used in industrial finishes, where the risk of health hazards is minimal. Medium-yellow shades are used in traffic-paint formulations. Blends with organic reds and violets produce inexpensive,

Table 4. Typical Chemical Compositions and Related Hues for Cadmium Pigments[a]

	Compositions, wt %			
	CdS	ZnS	CdSe	HgS
cadmium yellow, concentrated				
primrose	79.5	20.5		
lemon	90.9	9.1		
golden	93.4	6.6		
deep golden	98.1	1.9		
cadmium sulfoselenide orange-reds, concentrated				
light orange	85.0		15.0	
light red	67.5		32.5	
medium light red	58.8		41.2	
medium red	51.5		48.5	
dark red	44.8		55.2	
maroon	35.0		65.0	
Mercadium orange-reds				
deep orange	89.0			10.9
light red	83.4			16.6
medium light red	81.0			19.0
medium red	78.5			21.5
dark red	71.6			23.9
maroon	73.5			26.5

[a] Cadmium pigments manufactured by CIBA-GEIGY.

durable automotive finishes. Heat-resistant types, eg, Kroler (DuPont) and Rampart HR (CIBA-GEIGY), are suitable as colorants in plastic materials.

Lead chromates are prepared by precipitation from soluble salts in aqueous media. Primrose and light chrome yellows require coprecipitation of lead chromate and lead sulfate. The light chrome yellows are precipitated hot and with excess lead. Chrome oranges are precipitated from alkaline solutions. Chrome yellows and molybdate oranges are soft textured. Their dispersibility results in proper pigment-grind development and high gloss suitable for topcoat finishes. Normal lead silicochromate is manufactured by coating a core of silica with a medium-yellow lead chromate. The pigment is used primarily for traffic paint.

Cadmiums. Cadmium pigments provide brilliant colors ranging from the greenish primrose yellow through orange, red, and maroon. They display clean, bright hues; heat stability up to and in some cases above 400°C; excellent bleed and alkali resistance; fair acid resistance; and good tint lightfastness when protected from moisture. They are used in artists' paints, heat-resistant coatings, printing inks, and latex paints.

Mercadmium pigments are based on cadmium sulfide–mercury sulfide solid solutions; variation in chemical composition yields a hue similar to those of the selenide oranges, reds, and maroons (see Table 4).

Synthetic mixed metal oxides. These pigments are classified by crystal structure, eg, mixed-phase rutile pigments or mixed-phase spinel pigments. They are produced by solid-state chemical reactions at high temperature. The dry or wet raw materials are mixed and blended and then calcined. These pigments show excellent heat stability as well as weather resistance and lightfastness. They are used in the glass and ceramics industries. Nickel titanates are colorants for PVC siding materials (see Vinyl polymers, poly(vinyl chloride)).

Iron blue. Iron blue, as first developed, was hard-textured and difficult to grind. The modern iron blues are ferric ammonium cyanides used for coloring paper, for bluing, and in printing inks. Their poor alkali resistance can be improved by treatment with nickel compounds. Preparation is based on the oxidation of Berlin white, $Fe(NH_4)_2[Fe(CN)_6]$, which is produced from sodium ferrocyanide, ferrous sulfate, and ammonium sulfate.

Ultramarine (Lapis lazuli blue) provides good brilliance, heat resistance, lightfastness, and alkaline stability. It is prepared from intimate mixtures of china clay, sodium carbonate, sulfur, silica, sodium sulfate, and charcoal or pitch. The intensity of the color depends on reactant concentrations, residence time, and temperature. Ultramarine is no longer manufactured in the United States.

Manganese violet and cobalt violet are prepared by precipitation. The former is used in cosmetics and is also used to tone white pigments.

Chrome oxide green is a calcined pigment prepared by reduction of sodium bichromate with sulfur or carbonaceous materials. Because it reflects irradiation and is similar in reflective properties to chlorophyll, it

is used extensively in camouflage coatings. Transparency permits formulations of polychromatic finishes. These pigments are manufactured by hydrolyzing a complex chromium borate obtained by heating sodium bichromate with boric acid.

Chrome greens are blends of variable composition of primrose chrome yellow and iron blue. They are used in paints, printing inks, flooring materials, plastics, and paper goods because of their high chroma and excellent hiding power.

Natural mercuric sulfide, HgS, can be derived from the mineral cinnabar. As the pigment vermilion, it occurs in a red crystalline or black amorphous form. Antimony vermilion, Sb_2S_3, varies in hues from orange to red. Van Dyke brown is a natural pigment, consisting of as much as 92 wt% organic matter, water, and traces of iron oxide, alumina, and alkali oxides. It is found in peat beds in the FRG.

Black Pigments

Carbon black, one of the oldest pigments, is used in plastics, paints, and printing inks (see Carbon).

Specialty Pigments

Specialty pigments include zinc yellow (corrosion-inhibitive primers), basic zinc chromate (a wash primer), strontium chromate (metal protection and corrosion resistance), red lead (protection of iron and steel surfaces), cuprous oxide (an antifoulant agent for marine paints and colorant for ceramic glazes and glasses), calcium plumbate (anticorrosion pigment for steel), basic lead silicochromate (electrodeposition of water-based automotive coatings), white molybdates (corrosion inhibitors), modified barium metaborate (mold and enzyme inhibition, corrosion and chalk resistance), zinc phosphate (corrosion-resistant steel undercoatings), and nacreous pigments (pearl essence for decorative effects).

Metallic pigments. Metallic pigments are prepared from metallic elements and their alloys, eg, aluminum, copper, bronze, and zinc. Aluminum flake pigment is sold in a paste form, which typically contains 65 wt% flake metal and 35 wt% volatile hydrocarbon. Bronze pigments or powders are manufactured from granular alloys of copper and zinc; they are used for coating paper and cardboard. Gray metallic zinc powder or dust is widely used on iron and steel because of its cathodic effect. Luminescent pigments are based on zinc sulfide and zinc–cadmium sulfide that have been doped with an activating material; they are used for indoor decoration, safety signs, television and electronics, and military coatings.

ROBERT C. SCHIEK
CIBA-GEIGY Corporation

T.C. Patton, ed., *Pigment Handbook*, Wiley-Interscience, New York, 1973.

H.P. Preuss, *Pigments in Paint*, Noyes Data Corporation, Park Ridge, N.J., 1974.

ORGANIC

Pigments are colored, colorless, or fluorescent particulate solids which are usually insoluble in and essentially physically and chemically unaffected by the vehicle or medium in which they are incorporated. They alter appearance either by selective absorption or by scattering of light. Pigments may be organic or inorganic chemicals. For application, they are usually dispersed in vehicles, eg, inks (qv) or paints (see Paint). In some cases, the substrate may serve as a vehicle, eg, in the mass coloration of polymeric materials (see Colorants).

In 1980, the principal U.S. producers of organic pigments included American Hoechst, BASF Wyandotte, CIBA-GEIGY, DuPont, and Sun Chemical Corp.

Color and structure. In organic compounds, color is associated with double bonds. The unsaturated, conjugated double bonds in chromophores contribute to the selective absorption of light. Chromophores include nitroso, nitro, carbonyl, thiocarbonyl, azo, azoxy, azomethine, and ethenyl groups, whereas auxochrome groups include amino, alkylamino, dialkylamino, methoxy, and hydroxy groups.

Properties

The characteristics that control the performance of a pigment include its chemical composition; chemical and physical stability; solubility; particle size and shape; degree of dispersion or aggregation; crystal geometry; refractive index; specific gravity; absorption; extinction coefficients; surface area and character; and the presence of impurities, extenders, or modifying agents.

The inherent strength of a pigment is controlled by its light-absorbing properties, which are related to molecular and crystalline structure. The intensity is a measure of brightness or cleanness, as opposed to dullness, or shade. Generally, a pigment having a molecular structure containing two or more chromophores is less intense than one containing a single chromophore.

Fastness defines the inherent ability of a pigment to withstand the chemical and physical factors to which it is exposed in its application. Durability defines the ability to withstand weather, light, water, gases, and industrial effluents.

Dispersibility is measured by the effort required to develop the full tinctorial potential of a pigment in a vehicle; it affects the maximum surface area and homogeneity of the pigment system. Texture describes the ease of dispersion.

The working properties include compatibility, oil absorption, contribution to rheology, ease of grinding, wettability, gloss, bronzing, hiding power, flocculation, etc.

Classification

Organic pigments are classified as azo and nonazo pigments. The former contain chromophore (—N=N—) groups and are subdivided into pigment dyes and precipitated azos. Pigment dyes are insoluble on formation, whereas precipitated azos include products containing salt-forming groups, principally sulfonic or carboxylic acids.

Azo pigments.   The Hansa Yellows are semiopaque, intense, monoazo pigments which are used in emulsion paints and paper coatings. The intense color of diarylide yellows provides greater tinctorial strength and resistance to bleeding and heat than is provided by the Hansa Yellows. Nickel Azo Yellow is a greenish yellow, chelated nickel azo pigment with high transparency and excellent durability.

Nickel Azo Yellow

Hansa Orange, Dinitraniline Orange, and Pyrazolone Orange all exhibit intense, high strength color with varying degrees of lightfastness and chemical resistance.

Azo reds and maroons.   Toluidine Red is one of the most popular red pigments for industrial enamels. It exhibits good bake and chemical resistance but poor lightfastness. Para Red is similar but darker and bluer. Chlorinated Para Reds are intense, very light yellowish red. Parachlor Red is one of the more lightfast azo-pigment dyes. Lithol Reds are intense reds of dark shades. They are formed by coupling diazotized Tobias acid to 2-naphthol and precipitated as the Ba, Ca, and Sr salts. Lithol Reds are inexpensive and are used widely where extreme durability is not required.

The BON Reds and Maroons derive their name from β-oxynaphthoic acid as the second component in the coupling of various diazotized amines containing salt-forming groups. Lithol Rubine exhibits a wide range of intense shades suitable for industrial enamels. Permanent Red 2B defines the calcium, barium, and manganese precipitations of the coupling from diazotized 2-chloro-4-aminotoluene-5-sulfonic acid with 3-hydroxy-2-naphthoic acid. They are used in printing inks, plastics, and

automotive finishes. Yellow BON Maroon is a lightfast, yellow maroon with superior solvent bleed resistance suitable for outdoor finishes. The manganese toner of Lithol Red 2G is intense in color and is used mainly in blends with inorganic molybdate orange for durable reds; the calcium toner is used in printing inks.

The pyrazolone reds are disazo pigments which provide high color intensity, excellent masstone lightfastness, high transparency and good bleed, bake, and chemical resistance (see Pyrazoles, pyrazolines, and pyrazolinones). The search for pigments with improved properties has resulted in a group of higher molecular weight disazo products.

The monoazo Naphthol Reds and Maroon pigments provide a wide range of colors from light reds to dark maroons.

Lakes are either dry toner pigments extended or reduced with a solid diluent or an organic pigment prepared by precipitation of a water-soluble dye on an adsorptive surface. Alizarine Red B or madder lake is a coordination complex of alizarin with alumina and calcium plus sulfated castor oil and a phosphate (see Dyes, anthraquinone). It has an intense, deep bluish-red of fair lightfastness. Helio Fast Rubine 4BL is a precipitated quinizarinsulfonic acid derivative used to shade pigments for organic coatings.

A basic-dye pigment is the precipitated product of the reaction between a basic dye and either complex inorganic heteropoly acids or precipitants, eg, tannic acid, tartar emetic, clay, or rosin soaps. Despite their high costs, these pigments are widely used for their high strength and intensity.

Phthalocyanines.   Phthalocyanines are characterized by excellent lightfastness, intensity, bleed and chemical resistance, and heat stability (see Phthalocyanine compounds). Copper Phthalocyanine Blue exists in a red-shade blue alpha form and the more stable green-shade beta form. Chlorination gives Copper Polychlorophthalocyanine Green which has excellent pigmentary properties. Copper polybromochlorophthalocyanines provide yellower shades of green than the polychloro compound.

Copper Phthalocyanine Blue

Quinacridone pigments offer outstanding fastness properties in the orange, maroon, scarlet, red, magenta, and violet ranges. An example of a violet with excellent fastness properties is Pigment Violet 23, a dioxazine pigment derived from 3-amino-N-ethylcarbazole.

Quinacridone

Vat pigments.   A number of vat dyes based on anthraquinone and substituted and/or condensed anthraquinones have found use as pigments because of fastness and durability characteristics. In general, vat dyes must be modified chemically and physically to develop pigmentary strength and color intensity.

The thioindigos provide a broad range of hues from a yellow shade of red to violet. The perinone pigments are diimides of naphthalene-1,4,5,8-tetracarboxylic acid formed by condensation of o-diamines with the acid or acid anhydride. The perylene pigments are diimides of perylene-3,4,9,10-tetracarboxylic acid and generally are stronger and more resistant to chemicals, heat, and solvent bleeding than the thioindigos.

Perylene Scarlet

Among the anthraquinone vat dyes, indanthrone is redder but less intense than the phthalocyanine blue pigments, but as durable as Copper Phthalocyanine Blue. It is used in automotive and other high quality finishes. Partially chlorinated indanthrone blues are greener than the unchlorinated but not as bake resistant. Other anthraquinone pigments include Isodibenzanthrone Violet, Dibromoanthranthrone Orange, Flavanthrone Yellow, and Anthrapyrimidine Yellow.

Indanthrone Blue

Pigments can be standardized only in terms of performance, color, durability, and working properties, and often only for the specific application or vehicle system for which they are intended. In general, organic pigments have no toxic effects.

Organic pigments are used for decorative and functional effects; they provide color, hiding power, and high visibility and contribute to durability. They contribute to countless applications involving coloration.

MILTON FYTELSON
Sandoz Colors & Chemicals

Colour Index, 3rd ed., The Society of Dyers and Colourists, Bradford, Yorkshire, UK, and American Association of Textile Chemists and Colorists, Lowell, Mass., 1982.

German Dyestuffs and Dyestuff Intermediates, Vol. 3 of Dyestuff Research, FIAT Final Report No. 1313, U.S. Dept. Comm. Office Tech. Serv. PB Report 85172, Feb. 1948.

T.C. Patton, ed., Pigment Handbook, Vols. 1–3, John Wiley & Sons, Inc., New York, 1973.

DISPERSED

A dispersed pigment concentrate is an extremely fine distribution of color pigment in any medium; it is suitable for supplying color to surface printing and coating or a complete mass coloring (see Coatings; Dispersions; Paint).

Dispersion

Reduction of the pigment to primary pigment particle size is necessary in order to develop the optimum visual performance and economic properties. The extent of the reduction is determined by the nature of the pigment, the dispersion system and processing equipment, product-use requirements, and economics.

The maximum aggregate size that adversely affects physical properties is a small fraction of the thickness of the film, coating, or thin mass-colored material. The optimum size to which a pigment should be reduced is the primary pigment particle. In dispersed pigments, the term primary pigment particle refers to individual crystals and tightly held aggregates. Agglomerates are larger associations of the primary pigment particles, ie, crystals and aggregates. Ideally, a dispersion consists mainly of primary

pigment particles and a minimum of loosely held aggregates. The system of wetted primary particles must be stabilized in order to avoid reversal of the dispersion process.

Flushing is the direct transfer of pigments in an aqueous phase to an oil or nonaqueous phase without intermediate drying. Because of the difficulties involved in flushing inorganic pigments, it is often more practical to disperse them in dry form. Flushing is therefore confined mostly to organic pigments, eg, Diarylide Yellow. It is usually carried out in a heavy-duty dough mixer and produces dispersions of better gloss, transparency, and strength than methods using dry pigments.

Diarylide Yellow

Equipment. Kneader dispersers process systems with extremely high plastic viscosity, ie, up to 10^7 Pa·s (10^8 P). The Banbury mixer is the most efficient mixer for preparing large amounts of concentrated pigment dispersions in rubber and thermoplastic carriers. The flusher or heavy-duty mixer is used primarily in the ink and paint industries. The two-roll mill can be considered a kneader or internal mixer; clearance between the rolls is large (up to 15 mm) and the mass to be processed is subjected to kneading just before it enters the nip itself. These mills are used to prepare dispersions in elastomers and thermoplastic resins.

Close-tolerance mills can be classified as cylindrical roll mills, rotating-disk mills, and cone mills. Cylindrical roll mills of several types are used for processing pigment dispersions in paste form for inks, paints, and other uses. Rotating-disk mills, cone mills, stone mills, and colloid mills essentially are rotors in the shape of disks or truncated cones which rotate at high speed (3000–5000 rpm) and at a small distance (10–100 μm) from a stationary surface.

Impeller and impingement high speed, fluid-energy mills are used for the preparation of fairly low viscosity mill bases for inks and paints.

Dispersion-process equipment employing a large number of dispersing surfaces, whose movement is not precisely controlled, are categorized into ball or pebble mills (media size > 3.0 mm) and bead, shot, and sand mills (media size 0.2–3.0 mm). The most common design of a ball or pebble mill is a horizontal cylinder, partially filled with balls or pebbles, which rotates on the horizontal axis at a rate that cascades the balls or pebbles rather than centrifuges them or allows them to slide. These mills operate more efficiently with small particle size media. They have been largely replaced by bead, shot, and sand mills, which are vertical or horizontal cylindrical containers filled with grinding medium under intense agitation.

Uses

A dispersed pigment concentrate must be suitably formulated for manufacturing the concentrate and converting it into an application product. High pigment concentrations provide maximum flexibility and economy. For heavy-duty mixers and impingement mills, the formulation used for the dispersion process must be modified. Ink films 0.002–0.01 mm thick require relatively high pigment concentration of ca 10–20 wt%. Manufacturers offer a wide variety of pigment concentrates in liquid, paste, and solid form.

In the United States, most letterpress and lithographic oil inks are manufactured from dispersed pigment concentrates, mainly in flushed form. Most gravure and flexographic liquid inks are made from dry pigment or presscake (see Table 1). Gravure inks are characterized by very fluid viscosities (ca 0.1 Pa·s 1 P) when press ready; these inks dry by solvent evaporation, leaving a film ca 10 μm thick. Flexographic inks, like gravure inks, are pigment-resin-solvent systems of low viscosity (ca 0.3 Pa·s = 3 P) and film thickness of ca 8 μm; they dry by solvent evaporation.

Environmental and safety considerations will affect future development in ink manufacture. Concern for organic-solvent vapor emission favors aqueous or water-based inks. Today, flexographic ink is the only one based on water as a solvent.

Table 1. Distribution of Usage of Paste Ink Colorants by Pigment Form, %

Type of Ink	Flush	Dry	Other
letterpress news	95	5	
heat-set letterpress	95	5	
glycol letterpress	0	70	30
other letterpress	90	10	
lithographic news	95	5	
heat-set web offset	95	5	
quick-set sheet offset	90	5	ca 5
uv offset	0	100	
conventional metal decorating	0	100	
dry offset metal and plastic	50	50	

Table 1. Typical Classes of Pilot Plants by Size

bench-top and microunits	< 1.5 m^2 area, typically located in a hood or on a bench top small tubing throughout (3–6 mm dia) limited automation	10–50
integrated pilot plant	10–25 m^2 area placed in open bay or over-pressure cell tubing and small (13–25 mm dia) pipe fairly heavily automated feed and product systems generally included	70–250
demonstration or prototype unit	> 250 m^2 area, usually in a dedicated building or area pipe used extensively heavily automated feed and product systems included	> 500

Coatings are generally heavier films, a fraction of to a few millimeters thick; they require a lower pigment content than inks. Inorganic pigments are hydrophilic and are easily dispersed in water-based systems. Organic pigments, on the other hand, are hydrophobic and require wetting agents for dispersion.

Dispersed-pigment or color concentrates for use in plastics vary widely in pigment concentration and physical form (see Colorants for plastics). Some are manufactured as liquids and high viscosity gels which are usually metered directly into the processing equipment. Pellet concentrates are the most popular form. The effect on the plastic is a consideration in pigment selection. Acetals, ABS, fluoropolymers, and nylon present special problems.

Mass coloration, ie, spin-coloration or dope-dyeing of synthetic fibers, particularly polyolefins, can be accomplished with pigment dispersions (see Olefin fibers). In this process the colorant is incorporated before the fiber is formed.

Pulp colorants are water pastes derived from pigments, extended pigments, or lakes. The principal applications are in paper and textile printing, leather and paint.

JOHN J. SINGER
BASF Wyandotte Corporation

T.C. Patton, *Paint Flow and Pigment Dispersion*, 2nd ed., John Wiley & Sons, Inc., New York, 1979.

G.D. Parfitt, *Dispersion of Powders in Liquids*, 2nd ed., John Wiley & Sons, Inc., New York, 1973.

PILOT PLANTS AND MICROPLANTS

The design of a plant and its construction and start-up represent a large investment in time and money. The rewards are great if a significant improvement is realized; the risks are also great if a multimillion (> 10^6) dollar project fails. To reduce risk, lengthy and expensive research programs are often undertaken involving extensive pilot-plant work.

A pilot plant is a collection of equipment constructed to investigate some critical aspects of a process operation or to perform basic research. It can range in size from a laboratory bench unit to almost a commercial-size unit, but it is usually intermediate. Its purpose is to provide solutions to engineering problems or to prove the economic feasibility of a process. The basic scientific discoveries underlying the process are generally investigated before construction.

Specific aims of pilot-plant work may include: confirming the feasibility of the proposed process; providing design data for the commercial plant; determination of optimum materials of construction; testing the operability of a control scheme; determination of plant-maintenance requirements; producing sufficient quantities of a product for market evaluation; or assessing process hazards or disposal problems. Objectives may include: obtaining kinetic data; screening catalysts; determination of long-term effects; testing areas of advanced technology; or providing data for scale-up.

Pilot plants are classified by size (Table 1) or automation, ie manual, limited automation (local control only), and heavy automation (local and remote control).

Since the goal of a pilot-plant operation is the commercialization of a new process, scale-up is an important aspect, and the commercial situation should be duplicated as closely as possible. Scale-up is considered successful if the plant produces the product at planned rates and cost, and of desired quality. Some processes are suitable for the pilot plant, but not for the commercial plant. In addition, larger equipment may perform differently than expected.

Statistical designs for experiments maximize information and reduce research time and costs, minimize the element of human bias, reduce the number of pilot-plant runs needed to define the effect of variables, and increase confidence in the experimental results for both the research engineer and management (see Design of experiments).

The costs of designing, building, and running a pilot plant range from $10^4 to $10^7, and are typically between 5×10^4 and 2×10^5, assuming a facility is available. Estimates may be developed from the cost of a similar unit, ratioed from a detailed estimate of the cost of the purchased materials or the estimated labor to construct, or both detailed material and labor estimates. This last method is fairly accurate but is time-consuming and requires a significant amount of information. Operating costs for pilot plants include raw feedstock, product removal, utilities, spare parts, maintenance, salaries, and support services (analytical, clerical, plant operation, etc). Most of these costs are usually fixed, but the number of operators, and their salaries, can be reduced by increased automation.

The space required for a pilot plant varies greatly according to size and type. A small unit may require only a section of a laboratory (5–10 m^2), whereas a large unit may require one or more buildings (500–2000 m^2). The average pilot plant needs 20–80 m^2 excluding extended feed or product storage. Special structures that can withstand high pressures in the event of a component failure (over-pressure cells) may also be needed.

The specific purpose for which the pilot plant is to be designed must be identified, and a decision reached as to whether to model a process or investigate a specific problem. In process modeling, the specific unit operations are reproduced on a small scale. When investigating a specific problem, on the other hand, the pilot plant may not even resemble the commercial operation, but may provide the means to solve the specific problem.

The time between project inception and data generation is usually 6–18 mo. However, construction may be started before the final design is completed.

Process-control instrumentation ranges from simple variable-voltage transformers to dedicated on-line computer systems and usually constitutes a significant part of the costs. Analytical instrumentation, off-line or on-line, provides information on the basis of which the success or failure of the process is evaluated. Data-gathering instrumentation includes dedicated recording instruments, data loggers, and computers (see Instrumentation and control).

A pilot-plant start-up is more difficult than the start-up of a process unit, and some companies maintain a specialized start-up group if start-up and shut-down occur frequently.

All safety rules issued for process plants must also be observed in pilot plants.

RAYMOND VAN SWERINGEN
DENNIS J. CECCHINI
RICHARD P. PALLUZI
Exxon Research and Engineering Company

W.G. Cochran and D.R. Cox, *Experimental Designs*, John Wiley & Sons, Inc., New York, 1965.

W.E. Biles and J.J. Swain, *Optimization and Industrial Experimentation*, John Wiley & Sons, Inc., New York, 1980.

C.R. Hicks, *Fundamental Concepts in the Design of Experiments*, Holt, Rinehart and Winston, New York, 1973.

PIPELINE HEATING.   See Pipelines.

PIPELINES

The general design of a pipeline system is similar to that of a rail transportation system including feeders, terminals, dispatchers, etc (see Transportation). A dispatcher can calculate how much crude oil or refinery product is delivered to any point by ascertaining how fast the fluid is being pumped through the pipeline. Communication is maintained along the entire system by telephone, teletype, and radio (see also Piping systems).

Pipeline systems for both crude oil and natural gas use smaller-diameter pipe to convey material from producing sites to a central point and then by larger-diameter transmission pipeline for long-distance transport.

Product pipelines carry the refined products from an oil refinery or natural-gas treatment plant to markets, storage, and shipping terminals. The products include gasoline, jet fuels, distillate fuel oil, and natural-gas liquids. Monthly shipments are reported regularly by the Bureau of the Census in its *Statistical Abstracts*.

Design

A pipeline is designed by mapping a preliminary route, starting at the source of the fluid to be carried and proceeding to its final destination via any additional sources and markets. Simultaneously, the mechanical design of the pipe can be created. Sizing, ie, the choice of a diameter, is based upon the volume and nature of the fluid, length of pipeline, design pressures, and economical spacing of compressor or pumping stations. Wall thickness governs the weight of the steel, which in turn, governs pipe costs, the main item of installed costs. Wall thickness is selected by its relation to desired operating pressure and yield strength of the pipe steel selected. That relationship is expressed in a code design formula based on the primary stress created in the pipe wall by the internal fluid pressure. By the addition of safety factors, the code design formula also recognizes the presence of secondary stresses caused by forces such as those produced by weight of trench backfill, bending, and thermal expansion or contraction against earth constraint.

The maximum allowable operating pressure of the pipeline can also be limited by the pipe-mill test-pressure level or the hydrostatic-strength test performed when construction is complete; both must be adequate.

Another task is the specification of fittings such as valves, branch tees, forged weld fittings, and others. Other specifications include external pipe coating, cathodic protection, and internal coating. The latter are usually epoxy paints applied to the inside surface at the pipe mill or coating yard prior to joining the lengths during construction.

Compressor or pumping stations are spaced at 80–160 km intervals along the pipeline in order to boost pressure lost to internal friction by the transported fluid. Compressors may be either engine-driven, reciprocating, or turbine-driven centrifugal designs.

Construction

Before construction, the necessary permits must be secured. In the United States, numerous government agencies are involved. From approval for the pipeline by the Federal Energy Regulatory Commission, through the Environmental Impact Statement, and innumerable river-, highway-, and railway-crossing permits, these preliminary steps take years to complete. In addition, easements to cross hundreds of privately owned parcels of land must be obtained, listed, and surveyed.

The *ASME Guide for Gas Transmission and Distribution Piping Systems* contains the Code of Federal Regulations plus guide material and appendixes that offer design recommendations, material references, and recommended practices.

The contractor is selected by competitive bidding based on specifications and drawings. In the case of a truly cross-country pipeline, the total project can be divided into 160-km segments that are separately bid, awarded, and constructed somewhat simultaneously.

Construction generally proceeds via the following steps, each with a specialized crew: right-of-way clearing and grading; ditching, also called trenching; pipe stringing along the right-of-way; pipe bending to fit the trench bottom; line-up of pipe lengths for welding; manual or automatic electric welding; cleaning and priming for exterior coating; coating and wrapping for cathodic corrosion protection; lowering the pipe into the trench; and backfill of earth over the pipeline. Finally, the pipeline is tested before being put into operation. Then it is put in service by tie-in to existing supply and distribution systems. Sign posts are installed at road, railroad, and river crossings as required by Federal standard.

Operation

Centralized control from a computerized communications office is common practice. Communications are transmitted by microwave or leased telephone lines to the central office. Some systems feature remote control of main-line valves to isolate known or threatening damage. Crews are assigned to one or several stations for operation, maintenance, and emergencies. Between stations, the pipeline is patrolled by aircraft about once every two weeks. Any visible leaks, damage or dangers are promptly reported and personnel dispatched to the site. In the case of natural gas, odorization of the usually colorless, odorless, and tasteless gas is mandatory in populated areas; sulfur compounds are used.

Economic aspects.   Transportation costs can amount to as much as 50–80% of the combined production–delivery costs for fuels. For the transport of natural gas, no other mode can compete with pipelines. Pipeline transport of crude oil and refinery products is generally the cheapest mode of moving liquids over land, although water transport on barges competes effectively. Pipeline costs are only 17% of rail costs and 3.5% of truck costs. For transportation of solids, however, the railroad unit train competes very effectively with slurry pipelines.

Table 1 gives a typical breakdown of construction costs. They are, of course, much higher for offshore than for onshore pipelines.

Offshore and Submarine Pipelines

Crude oil and natural gas are brought to shore from offshore locations in marine producing areas such as the Arabian Gulf, the Gulf of Mexico, and the North Sea in pipelines (submarine pipelines) positioned on the ocean floor. These are usually welded on a specially constructed "lay"

Table 1.   Distribution of Pipeline Construction Costs

Category	%
construction	40.8
line pipe and fittings	32.2
pump stations and equipment	17.1
land and right-of-way aquisition	3.2
miscellaneous	6.7

barge and lowered to the ocean floor. Closer to shore or on inland waters, they may be welded onshore, floated on pontoons, and towed into position where they are sunk or dragged into position by powerful ship- or barge-mounted winches. Underwater pipelines in shallower water are laid in trenches.

Slurry Pipelines

The transport of solids such as coal (qv) or minerals by slurry pipelines offers a low cost method of moving solids over both short and long distances. It is, however, vigorously opposed by the railroads, who fear the competition. Furthermore, many coal-slurry operations would require large quantities of water from the western United States where the proposed lines would originate and water supply is limited.

Chemical Pipelines

Wherever a large chemical or petrochemical complex exists, gases, liquids, or solid slurries are transferred by pipeline within or between plants. Caustic–chlorine plants supply nearby customers with either chlorine or caustic soda solutions via pipelines (see Alkali and chlorine products), and oil-refinery pipelines supply light hydrocarbons to nearby customers for conversion to plastics and other materials.

Ammonia pipelines have been constructed between plants in Texas and farming districts in Kansas, Nebraska, Iowa, and Minnesota. Storage costs are reduced and high volume, peak delivery is guaranteed. Carbon dioxide pipelines provide supplies for enhanced-oil-recovery flooding, and about 20% of all chlorine produced is shipped by pipeline either in liquid or gaseous form. Experience with long-distance piping of hydrogen is limited but interest has increased recently because of the potential of hydrogen as a renewable, nonpolluting gaseous fuel (see Hydrogen energy). Chemical pipelines may also be used for sulfur, oxygen, nitrogen, and helium.

The outlook for chemical-energy pipelines promises the transport of heat by reversible conversion of thermal to chemical energy either in a closed- or open-loop system.

Agricultural products, industrial raw materials, wastes, and other solids are transported over short distances by pneumatic pipelines or pneumo-capsule pipelines.

The encapsulation of material in plastic or aluminum allows transportation through a pipeline in a fluid whose density is similar to that of the material in the capsule.

District heating may utilize waste heat from a power plant or steel mill via twin insulated pipes which may be incorporated into the municipal services. This technique is used in Sweden for the distribution of hot water.

Environmental and Safety Factors

Once the pipelines are buried, their environmental effects are negligible, and pipelines provide the safest material transport known. The Natural Gas Pipeline Safety Act of 1968 provides for Federal safety regulation of facilities for gas storage and pipeline transmission.

THOMAS P. WHALEY
GEORGE M. LONG
Institute of Gas Technology

C.G. Segeler, ed., *Gas Engineers Handbook*, Industrial Press, Inc., New York, 1969.

Federal Rules and Regulations, *Title 49—Transportation; Chapt. 1—Hazardous Materials Regulations Board, Department of Transportation; Part 192*; "Transportation of Natural and Other Gas by Pipeline," Minimum Federal Safety Standard, Nov. 12, 1970, rev. Oct. 1, 1979.

A Primer of Pipeline Construction, 2nd ed., Petroleum Extension Service, University of Texas-Austin and Pipeline Contractors Association, Dallas, Tex., 1966.

PIPERAZINE, $C_4H_{10}N_2$.   See Amines, cyclic.

PIPERIDINE, $C_5H_{11}N$.   See Pyridine and pyridine and derivatives.

PIPERONAL, $C_8H_6O_3$.   See Aldehydes.

PIPERONYL BUTOXIDES.   See Insect control technology.

PIPING SYSTEMS

A piping system provides safe and economical transfer of fluid materials, often with provisions for controlling the rate of flow (see Hydraulic fluids; Pipelines).

Piping must be suitable for the temperature, pressure, and corrosivity of the fluid, the size must be adequate for the flow rate desired, and valving must be provided for flow control.

Materials

Piping materials are metallic, nonmetallic, or lined metallic. The most common and generally the least expensive material is carbon steel (see Steel).

The corrosivity and temperature of the fluid are of prime importance in material selection (see Corrosion and corrosion inhibitors). The corrosion rate is frequently temperature-dependent; perceptive process design and plant layout can give significant cost savings.

For moderate temperatures and pressures, and if corrosivity eliminates carbon steel, nonmetallic piping is an alternative to expensive alloys. Plastic and reinforced plastic piping is available in many different forms. In general, nonmetallic piping is limited to situations where fire risk is low. Glass piping can be used to ca 230°C. When product purity is of prime importance or when the fluids are highly corrosive, glass piping has no equal (see Glass; Laminated and reinforced plastics; Plastic building products).

With lined pipe, safety problems could arise owing to liner failure. Water jacketing provides warning of insulation failure and protection in high temperature systems. Piping must be protected against overpressure by safety relief valves or rupture disks.

Pipe Size

Piping systems may be divided into three classes: connecting equipment whose pressure is set by process requirements; piping systems whose pressure loss determines all or part of the pressure rise developed by pumps or compressors; and piping systems whose pressure loss contributes to setting the elevations of towers and drums.

The driving force causing circulation in thermosiphon reboilers is derived from the difference between the liquid in the piping from the tower to the reboiler and that of the liquid-vapor or vapor in the piping from the reboiler to the tower. A sufficient pressure difference must be developed to overcome the pressure loss in the reboiler and in the piping.

In order to select the pipe size, the pressure loss is calculated and velocity limits are established. For laminar flow (Reynolds number, $Re < 2000$), generally found only in circuits handling heavy oils or other viscous fluids, the Fanning friction factor $f = 166/Re$. For turbulent flow, f is dependent on the relative roughness of the pipe and on the Reynolds number. Occasionally, piping systems are designed to carry two-phase fluids (gas plus liquid), slurries (liquid plus solid), or non-Newtonian fluids.

Valves

Valves in piping systems are employed for on-off service (gate, plug, and ball valves), for controlling fluid-flow rate (globe, needle, angle, butterfly, and diaphragm valves), and for ensuring unidirectional flow (check valves). For any given application of any type of valve, temperature, pressure, and corrosivity must be considered. Temperature, pressure, and general-service limitations for valves are usually indicated in manufacturers' catalogues.

The operation of system valves and the starting and shutdown of pumps has significant effects on the transient fluid pressures in the piping system because of the acceleration and deceleration of the fluid as it changes its velocity. The basic wave velocity a is a function of the bulk modulus and density of the fluid:

$$a = \left(K/\rho \right)^{1/2}$$

where K = bulk modulus (kPa or psi $\times$ 6.895) and ρ = fluid density.

Pipe-Wall Thickness

The design code to which the piping system must conform is the ANSI B31 Code for Pressure Piping. The section of this code that must be followed is determined by the general service for which piping is intended. These sections specify allowable stress, joint efficiencies, and minimum allowance for threading, mechanical strength, corrosion, and thickness.

Supports and Restraints

In addition to wide ranges of movements, piping systems may be subject to temperature variations, unusual start-up conditions, or emergency shutdowns. Therefore, the support design must be adequate for all circumstances.

The spacing of supports is governed by the allowable stress of the piping materials; stability, in the case of large-diameter thin-wall pipe; deflection to avoid sagging or pocketing; and the natural frequency of the unsupported length to avoid susceptibility to undesirable vibration.

A restraint may be defined as a device preventing, resisting, or limiting the free thermal movement of a piping system; its effect on system movement and stress is established mathematically. Anchors provide full restraint against the three deflections and three rotations relative to the principle axis. They are usually subject to large loadings and must, therefore, be rigid.

Connections or devices attached directly to the surface of the pipe transmit thermal as well as dynamic or weight loads from the pipe shell to the restraining, bracing, or supporting fixtures. Pipe-attachment devices are either integral or nonintegral with the pipe shell. Nonintegral attachments are widely used for support. Integral attachments are frequently of special design (see Fig. 1) and are generally used as anchors.

Piping supports, guides, and anchors increase stresses at the point of attachment. Code-allowable stresses are conservative with respect to structural failure that occurs when the limit load is reached.

System Flexibility

The basic problem of the theoretical analysis of the three-dimensional rigid piping system is the analysis of a statistically indeterminate structure. Design charts and tables are used for noncomputerized hand calculations.

The systematic analysis of a piping system using elbows and straight pipe sections as building blocks was developed long before the advent of the digital computer. In the flexibility method, a basic anchor point is assigned and the entire system is worked in reference to this basic anchor. Most of the computer programs developed in the 1960s used this method.

In contrast, the stiffness method considers the displacement as unknown quantities in constructing the overall stiffness matrix. This method has a simple straightforward logic even for complex systems. The most recent developments in computational structural analysis are almost entirely based on the direct stiffness-matrix method, eg, piping-stress computer programs such as SIMFLEX, ADLPIPE, NUPIPE, and PIPESD.

Flexibility and stress-intensification factors. The flexibility factor k is defined as the ratio between the rotation per unit length of the part in question produced by a given moment to the rotation of a straight pipe (of the same size and schedule) produced by the same moment.

The solution of the problem of the rigid system is based upon the linear relationship between stress and strain. If this relationship is nonlinear, an elementary problem, such as a single-plane two-member system, can be solved only with difficulty. Using linear analysis in an apparently nonlinear problem is justified by the stress-range concept which is applied to piping systems.

Stress levels should be based upon fatigue considerations because of the cyclic nature of the piping-system stresses. The ANSI B31 Code for Pressure Piping, Chemical Plant, and Petroleum Refinery Piping assigns stress levels for the allowable stress range.

Semirigid and nonrigid systems. The calculated results of flexibility analysis with which the designer is concerned are the stress levels and movements at significant points with the piping system, and the magnitude and direction of forces and moments at the terminal equipment. When stresses or reactions exceed allowable limits, modifications are necessary. The simplest method of reducing stresses and reaction is to provide additional pipe in the form of loops or offset bends.

If a rigid system is impractical, the configuration may always be made more flexible with hinge joints, rotation joints, or translatory joints. Systems using such devices are commonly classified as semirigid or nonrigid. In nonrigid piping, all members are free of thermal forces and moments.

The movement-absorbing devices used in semirigid and nonrigid piping systems are called expansion joints. They are either of the packed type or of the packless or bellows type. The selection and application of an expansion joint is not as simple as a pipe or valve fitting and requires a sound understanding of the joint's capabilities and limitations. Bellows expansion joints require special attention to design and installation. They should always be inspected to ensure that temporary shipping bars are removed and flow liners are installed in the right direction.

S.E. HANDMAN
The M.W. Kellogg Company

R.P. Benedict, *Fundamentals of Pipe Flow*, John Wiley & Sons, Inc., New York, 1980.

R.C. King, *Sabin Crocker Piping Handbook*, McGraw-Hill, New York, 1973.

Standards of the Expansion Joint Manufacturers Association, Inc., 5th ed., EJMA, White Plains, N.Y., 1980.

PITCH. See Tar and pitch.

PIVALIC ACID, (CH$_3$)$_3$CCOOH. See Carboxylic acids.

PLANETARY CHEMISTRY. See Space chemistry.

PLANT-GROWTH SUBSTANCES

The use of plant-growth regulators may be the cause of the most important quantitative yields yet achieved in agriculture. The principal

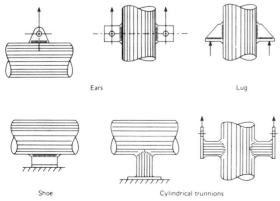

Ears   Lug

Shoe   Cylindrical trunnions

Figure 1. Integral pipe attachments.

Table 1. Plant-Growth Substances

Chemical name	Common name(s)	Structure	Trade name(s)	Oral LD$_{50}$, mg/kg[a]	Uses[b]
6-benzylaminopurine	benzyladenine			1,690 (mice)	dormancy breaker
N,N-bis(phosphonomethyl)glycine	glyphosine	$HOOCCH_2N(CH_2PO_3H_2)_2$	Polaris	3,925	sugarcane ripener
5-chloro-3-methyl-4-nitro-1H-pyrazole			Release		abscission stimulant for citrus
2-chloroethylphosphonic acid	ethephon		Ethrel, Cepha, Florel	4,000	stimulator of latex flow in rubber, sugarcane ripener, fruit ripener, flowering agent for pineapple, color enhancer
2-chloroethyltrimethyl-ammonium chloride	chlormequat chloride	$[(CH_3)_3NC_2H_4Cl]^+Cl^-$	Cyocel	670–1,020	height shortener in wheat and poinsettias, sugarcane ripener, lodging reducer in cereal grains
2-(3'-chlorophenoxy)propionic acid	3-CPA		Fruitone, CPA		stone-fruit thinner, size increaser for pineapple
2,4-dichlorophenoxyacetic acid	2,4-D		Weed-B-Gon, Plantgard, Agrotect	375	herbicide, fruit-drop controller
2,3-dihydro-5,6-dimethyl-1,4-dithiin 1,1,4,4-tetraoxide			Harvade	1,150	defoliant
6,7-dihydrodipyrido[1,2-a:2',1'-c]pyrazinedium dibromide	diquat		Reglone, Reglox, Pathchar	231	herbicide, sugarcane-flowering suppressant, desiccant
1,2-dihydro-3,6-pyridazinedione	maleic hydrazide		MH-30, Sprout Stop, Sucker Stuff	6,950	growth retardant, sprout-growth inhibitor, turf-growth inhibitor
1,1'-dimethyl-4,4'-bipyridinium dichloride	paraquat		Gramoxone, Dextrone X, Esgram	150	herbicide, sugarcane-flowering suppressant, desiccant pine-oleoresin stimulant
3-(2'-{3'',5''-dimethyl-2''-oxocyclohexyl}-2'-hydroxyethyl)glutarimide	cycloheximide		Acti-Aid	2	abscission agent for citrus
N,N-dimethylpiperidinium chloride	mepiquat chloride		Pix	1,420	cotton-growth regulant

Table 1. Continued

Chemical name	Common name(s)	Structure	Trade name(s)	Oral LD$_{50}$, mg/kg[a]	Uses[b]
6-furfurylaminopurine	kinetin	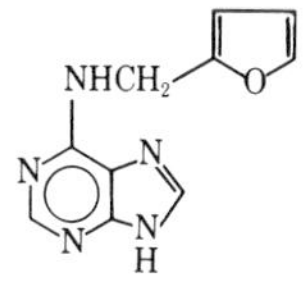			dormancy breaker
1-hydroxytriacontane	triacontanol	$C_{30}H_{61}OH$			general growth stimulant
3-indolylacetic acid	IAA	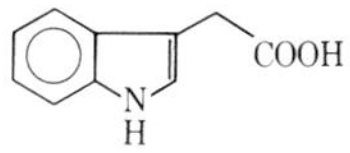		150 (ip) (mice)	plant-cell enlarger
2-methyl-4,6-dinitrophe-nol(4,6-dinitro-o-cresol)	DNOC	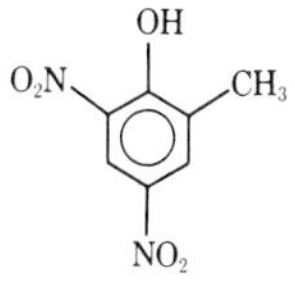	Chemset, Elgetol, Sinox, Trifocide	20–50	apple thinner, herbicide, fungicide, insecticide
3-methyl-5-(1′-hydroxy-4′-oxo-2′,6′,6′-trimethyl-2′-cyclohexen-1′-yl)-cis,trans-2,4-penta-dienoic acid	abscisic acid	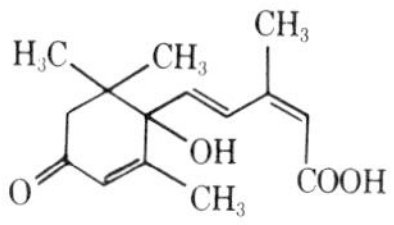	ABA, Abscisin		defoliant, growth inhibitor, antitranspirant
1-naphthaleneacetic acid	NAA	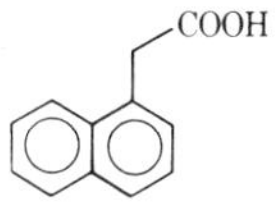	Tree-Hold, Phyomone, Rootone, Fruitone-N, Transplantone, Niagara Silk	1,000–5,900	fruit thinner, flowering agent for pineapple, tree-fruit thinner, preventer of preharvest fruit drop, root inducer
2-naphthoxyacetic acid	BNOA	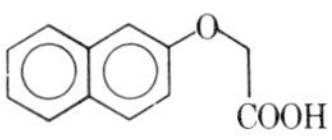	Betapal	600	enhancer of fruit set and fruit growth
N-1-naphthylphthalamic acid, sodium salt	naptalam (sodium salt)	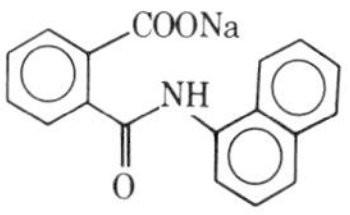	Alanap-3, Peach-Thin-322		fruit thinner
N-phenyl-N′-(1,2,3-thia-diazol-5-yl)urea	thidiazuron	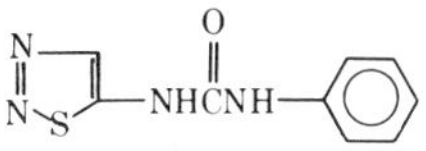	Dropp	> 4,000	cotton defoliant
sodium 2,3:4,6-di-O-isopropylidene-α-L-xyl-o-2-hexulofuranosonate	dikegulac (sodium salt)	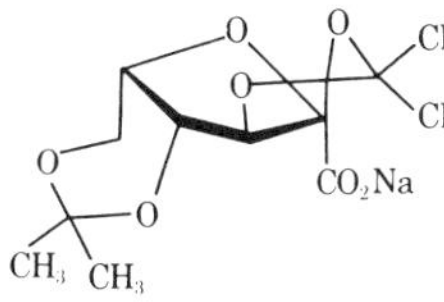	Atrinal	18,000–31,000	apical-dominance reducer, side-branch increaser, flower-bud increaser (chemical pincher)
succinic acid, 2,2-dimethylhydrazide	daminozide, SADH	$HOOCCH_2CH_2CONHN(CH_3)_2$	Alar, Kylar, B-Nine	8,400	growth retardant, multiple-flower stimulator
2α,3α,-22(R),23(R)-tetra-hydroxy-24(S)-methyl-B-homo-7-oxa-5α-chol-estan-6-one	brassinolide	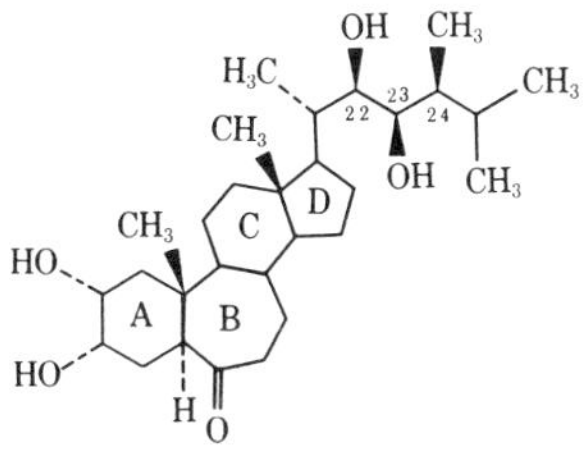			growth promoter

897

Table 1. Continued

Chemical name	Common name(s)	Structure	Trade name(s)	Oral LD$_{50}$, mg/kg[a]	Uses[b]
tributyl phosphorotrithio-ite	merphos	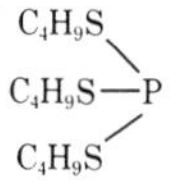	Folex	1,272	cotton defoliant
2,4 a,7-trihydroxy-1-methyl-8-methylene gibb-3-ene-1,10-dicarboxylic acid-1,4 a-lactone	gibberellic acid	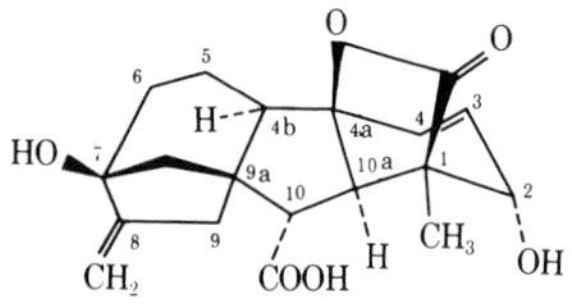	Pro-Gibb, Gib-Sol, Gib-Tabs, Gibrel, Activol, Barelex	> 15,000	seedless-grape enlarger, stimulator of amylase content of malting barley, flowering initiator, shoot-growth stimulator
2,3,5-triiodobenzoic acid	TIBA	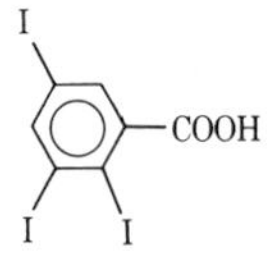	Regim-8	813	growth retardant, increaser of pod set in soybean

[a] For rats unless otherwise noted.
[b] In most cases, the uses are commercial; in some, they are experimental.

aim of the agrochemical industry has been to provide chemicals that control the competition to crops, ie, the weeds, insects, fungi, and nematodes that reduce yield or quality or that interfere with harvesting. Product performance has been judged in simple terms, eg, in the case of herbicides it is the death of the weed and adequate margins of safety for the crop and the farmer. Plant-growth substances or regulators are used to modify the crop by changing the rate or pattern, or both, of its response to the internal and external factors that govern all stages of crop development from germination through vegetative growth, reproductive development, maturity, and senescence or aging, as well as postharvest preservation (see Fungicides; Herbicides; Insect control technology). A list of plant-growth substances, their structures, trade names, related toxicity information, and uses is given in Table 1.

Louis G. Nickell
Velsicol Chemical Corporation

F.B. Abeles, *Ethylene in Plant Biology*, Academic Press, Inc., New York, 1973.

Agricultural Production Efficiency, National Academy of Sciences, Washington, D.C., 1975.

L.G. Nickell, *Plant Growth Regulators—Agricultural Uses*, Springer-Verlag, Berlin, FRG, New York, 1982.

L.G. Nickell, *Plant Growth Regulating Chemicals*, Vols. 1–2, CRC Press, Boca Raton, Fla., 1983.

PLANT LAYOUT

Plant layout involves developing a physical arrangement of equipment for a processing facility. The development must effect a balance of equipment spacing and integration of specific systems related to the facility as a whole.

Principles

Process. The sequence relationship of equipment is defined by the process-flow requirements. Process data specify temperatures, pressures, quantities, and flow media, all of which influence the layout of critical alloy-piping and plant-flexibility requirements.

Economics. The most economical plot arrangement in petrochemical plants is the in-line unit. The in-line plot arrangement consists of a central overhead pipe rack and pumps located in a row below or adjacent to the pipeway. An access road is provided through the entire unit. All towers, drums, and exchangers are in a single row on each side of the central pipe rack. A service road is provided on the opposite side of the equipment beyond the central pipe rack.

Client requirements. Many manufacturers and refiners prefer standardized features in operation and maintenance for convenience at the sacrifice of some economy.

Operation. The control house should be as conveniently close to the main operating equipment as safety considerations allow. It should not be located near vibrating equipment, noise, or continuous maintenance areas.

Sufficient access ways, platforms, stairways, ladders, and possible elevators must be provided for the convenience of the plant operator.

Erection and maintenance. The arrangement of all equipment for direct access from roads serves a two-fold purpose: it enables each piece of equipment to be erected as soon as it arrives on the job site rather than according to an erection sequence, which might be required with hidden equipment, eg, condensate pots, filters, and strainers; and it enables important maintenance work to proceed simultaneously on all pieces of equipment during a scheduled maintenance or turnaround period.

Safety. Plant layout must be developed with consideration of personnel, equipment, and public safety (see Plant safety).

Environment. Environmental factors are a principal consideration in plant-layout design; requirements of local, state, and national water-, noise-, and air-quality criteria must be considered because of the increasing concern for conservation and effects on nature.

Appearance. The finished plant serves as an advertisement for the plant-layout engineering group. Unless absolutely unavoidable, lines should drop straight down without any offsets from vertical vessels. The lines then can be readily supported and guided from the vessel to accommodate dead loads and wind loads. This also is good practice in terms of processing. Hose-connection risers on towers should be close to the vessel. Groups of equipment, eg, exchangers, always should be in line. Where possible, structures and equipment should be arranged for consistent appearance in heights. Excessive offsets and turns in piping should be avoided by, eg, making flat turns at the edges of a pipe alley. Lines should be self-draining to avoid pocketing of liquids during shut-downs. Tower manholes should be in the same verticle line, preferably facing toward the road or other access. Relief valve and tray drop-out areas should be maintained for accessibility and operation requirements.

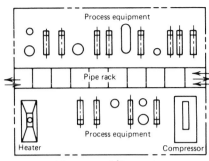

Figure 1. I-Shape plot.

Access ways must never be blocked by field-run items, eg, steam-tracing manifolds and instrument air manifolds.

Expansion.   Provisions for future expansion should be considered when developing the plant layout.

Application

In-line plant layouts are made in various arrangements, which often are referred to by letter designation, as in Figure 1. Generally, an I-shaped plot is used for small process units and an H-shape plot for larger units.

Prerequisites and objectives.   Design-feature characteristics of the specific units must be reviewed before the plant layout is developed. Certain information involving the plant, eg, existing site location, selection of plant site, surveys required, topography, etc, must be obtained from the client. A rough sequence of intended equipment arrangement is prepared, and the overall size of equipment should be established for towers, drums, shell-and-tube exchangers, air coolers, compressors, pumps, and tanks.

Site development. Prevailing wind.   If a prevailing wind can be established for the site, the office buildings, furnaces, and maintenance buildings should be positioned upwind from or to the side of process areas to reduce potential damage from fire or volatile vapors.

Consideration also should be given to the location of cooling towers in relationship to prevailing winds. Other development considerations include buildings, road systems, rail systems, cooling towers, pipe racks and sleepers, and product storage and tank farms.

Energy efficiency.   Energy efficiency for new or existing chemical plants can provide utility savings and increased production. Ideally, in a new plant design, the energy-efficiency equipment is incorporated in the layout of the plot to provide the space necessary to accommodate the heating and cooling equipment in a proper relationship. Upgrading existing units may be problematic because of insufficient space.

Ethanol and cogeneration projects.   Conservation of oil resources at plants is achieved by implementation of ethanol and cogeneration projects. The units consist of corn-based fermentation-derived ethanol plants in conjunction with coal-fired, cogeneration power plants (see Ethanol; Power generation).

F.V. ANDERSON
Stone & Webster Engineering Corporation

A.F. Waterland and W.L. Viar, *Power* (Energy Guidebook), 77 (1978).

J.M. Apple, *Plant Layout and Materials Handling*, 3rd ed., John Wiley & Sons, Inc., New York, 1977.

PLANT SAFETY

Safety assessments of entire processes began with quantification of overpressure potential and of flammability hazards by measurements of vapor pressure and of flash points and flammability limits, respectively. Process designers make use of data pertaining to reaction rates and energies for exothermic reactions and unstable chemicals; temperature limits beyond which explosive decompositions or other undesirable behavior can occur; rates of gas or vapor generation for proper design of emergency pressure-relief devices; recommended limits for exposure to toxic materials; radiation, noise, and heat; and strengths and corrosion rates of materials of construction (see Noise pollution; Corrosion and corrosion inhibitors). The application of fault-free analysis to chemical processes provides a means for quantitatively combining characteristics of process hazards with component and human failure rates to obtain a safety assessment of a process. It is likely that further quantification of process hazards, design standards, and risks and benefits to employees, consumers, and the public will be undertaken by industry, insurance companies, governmental authorities, and other groups and associations concerned with loss prevention and personnel safety (see also Product liability; Regulatory agencies).

In the United States, the Occupational Safety and Health Administration (OSHA) generally has limited its activity to enforcement of codes or consensus standards concerning toxicity and flammability; however, there has been a recent trend towards applying the "general duty" clause more broadly to management of hazardous chemical processes.

Occupational Safety and Health Act (OSHA)

The Occupational Safety and Health Act establishes standards for several types of occupational hazards, including toxicity, noise, equipment guarding, and protection against falling and electrical shock. It also promulgates other consensus standards for exit facilities and fire and explosion control. A recent development has been the availability of OSHA and NIOSH personnel for consultation, to identify, evaluate, and correct workplace hazards.

Threshold Value

The ANSI has published standards regarding the maximum acceptable concentration for certain gases and vapors in the air at work locations. A list of threshold limit values (TLVs), which is published annually by the ACGIH, provides the concentrations of dust, mist, or vapor believed to be harmless to most humans when exposed for five 8-h days per week.

Hazard identification of the contents of in-plant bulk storage tanks, warehouses, etc, may be achieved by a system developed by the NFPA. The system makes use of diamond-shaped symbols, which are marked with numbers, ie, 0, 1, 2, 3, or 4, indicating respective hazards of toxicity, fire, and reactivity.

Plant Site and Layout

The choice of a location for a chemical plant depends upon a number of factors and their effects on personnel in the plants and on the surrounding community. These factors include the assessment of hazards in the operation based upon the flammability of materials, reaction energy, and presence of toxic materials. Consideration then is given to the possible effect upon the community from the worst possible accident. An overcrowded plant can lead to damage or shutdown of adjacent units and may impede the movement of vehicles and materials in case of emergency. Reliable water supplies for process cooling, fire fighting, and safety showers are a vital necessity. The direction of the prevailing wind should also be considered. Open areas around the process units of the plant provide buffers to the surrounding community. Process units with significant hazard potentials should be segregated from operations that are nonhazardous, such as offices, laboratories, and warehouses.

Start-up

Often, key personnel from the design and construction organizations remain at the plant during start-up. Depending on the hazards of the process and materials involved, it may be advisable to use less hazardous materials, such as water or solvents, under working conditions before going ahead with the actual process materials.

Shutdown

Written procedures for normal shutdowns, as well as for emergency shutdowns, should be prepared, rehearsed, and kept up-to-date. Operating supervisors must be responsible for preparing plant equipment for maintenance work (see Maintenance).

Locking and Tagging

Safe maintenance requires that no one work on or be exposed to power-driven equipment without positively disconnecting the source of power beforehand. Written procedures should be prepared and thorough training given to carry them out.

Materials Handling

Piping. Liquids are usually moved through pipelines by a variety of pumps. Plastic pipe and liners, glass, and ceramics are widely employed in the chemical industry. Drainage valves should be provided at the low points of the plant system, and all piping should slope downward toward them. Valves and piping should be identified, and the operating instructions should specify the proper valving sequences.

Solids. Solid materials may be handled by belt or screw conveyers, air lifts, and a variety of other mechanisms. Common hazards associated with solids handling are dust explosions and the escape and dissemination of noxious or combustible dust. Two items must be considered in designing pneumatic conveying systems. First, if the material handled is combustible, inert gas should be used as the conveying fluid or blowout panels or vents should be provided to avoid explosion damage. Second, where noxious materials are being transported, special attention must be given to recovery of fines from the exit air.

Forklift trucks and other such means of transporting solid materials should be purchased only when full consideration has been given to any hazardous atmospheres in which they may be used, and they must be properly maintained to preserve the integrity of built-in safety devices.

Bulk Transport of Chemicals

Design and construction of tank cars have been controlled by regulations drafted by the Association of American Railroads, which is a private organization, and compliance with its regulations is voluntary. The DOT is responsible for regulation of rail transportation. The DOT regulations require placarding of over-the-road tank cars and trucks to alert the public and emergency personnel to the nature of their contents.

When flammable materials are loaded or unloaded, static bonding lines must be attached between the truck or car and the fixed piping system.

In the United States, regulation of barge and ship transport is by the U.S. Coast Guard. A special committee of the NAS is working with the Coast Guard to improve these regulations and to develop and disseminate information that will help increase the safety of shipping chemicals along waterways.

Labeling

The Federal Hazardous Substances Labeling Act requires that all containers sold to consumers be labeled with appropriate precautionary wording to protect the user and employees from injury resulting from contact with the chemical.

Sampling

The first consideration in sampling (qv) is protection of the sample collector with, eg, gloves, face and eye protection, and respiratory protection. Line sampling usually is carried out at a suitable valve tap. The collector should be aware of the possibility of high pressure or temperature when the valve is opened and operators should always be told that a sample is being taken.

Storage

A source of danger in bulk storage occurs when workers enter tanks, silos, or other storage spaces for maintenance or other duties. Fumes or atmospheres above the stored materials can cause intoxication, suffocation, and explosions.

In any warehousing operation, it is essential that incompatible substances are isolated to avoid a reaction in case of a spill or fire.

In design of warehouses in which flammable or combustible materials are to be stored, consideration should be given to the installation of fire walls, fire doors, and duct dampers. Automatic sprinklers are standard equipment in such locations. Fire-detection devices, such as flame-sensing or ionization types, operate much more rapidly than sprinkler heads and are used extensively as alarms and to activate fixed fire-extinguishing systems.

Disposal. Disposal of hazardous waste must be done according to precautions against fire and explosion hazards, severe corrosion or reactivity with water, and toxic effects.

Personnel Selection and Training

The quality of operating personnel is of paramount importance to the safe operation of a chemical plant. Operators must be well-trained and emotionally stable.

First aid and rescue. Thorough knowledge of first aid, as taught in courses by the American Red Cross or the U.S. Bureau of Mines, should be a major part of chemical-plant training programs. Rescue techniques also should be taught and practiced.

Medical Programs

Large chemical plants have at least one full-time physician who is at the plant five days a week and on call at all other times. Smaller plants either have part-time physicians or take injured employees to a nearby hospital or clinic by arrangement with the company compensation-insurance carrier.

Clinical tests can and should be made prior to employment or work assignment and at frequent intervals thereafter when employees could be exposed to toxic materials.

Fire Protection

Prevention of fire and extinguishment or control of fire are the two components of fire protection. Prevention involves study of material characteristics to determine appropriate hazard-avoidance methods.

Fire fighting. Automatic sprinkler or deluge systems are widely used for control of fire. Foam is a preferred extinguishing agent for large-scale chemical or petroleum fires. A principal foam is a protein-based material that is mixed with water and aerated at the nozzle.

Disaster Planning

Plant managers should recognize the possibility of natural and industrial emergencies and should oversee formulation of a plan of action in case of disaster. The plan should be well documented and made known to all personnel. Practice drills should be carried out to make sure that all personnel are accounted for, and that the participants know what to do in a major emergency.

Accident Investigation

A study of all accidents and injuries with the objective of determining the causes can lead to correction of the situation and prevention of recurrence.

RICHARD W. PRUGH
E.I. du Pont de Nemours & Co., Inc.

American Institute of Chemical Engineers, *Loss Prevention*, Vols. 1–14, New York, 1967–1981.

Institution of Chemical Engineers, *Chemical Process Hazards*, I—No. 7, 1960; II—No. 15, 1963; III—No. 25, 1967; IV—No. 33, 1972; V—No. 39a, 1974; VI—No. 49, 1977; VII—No. 58, 1980;

Process Industry Hazards, No. 47, 1976; *Major Loss Prevention in the Process Industries*, No. 34, New York, 1971.

PLASMA TECHNOLOGY

Plasmas consist of mobile, positively and negatively charged particles that interact because of Coulomb forces. They are either in the gaseous or condensed state. Synthetic gaseous plasmas are produced in diverse ways, eg, in fluorescent tubes and nuclear explosions. Some liquids and solids exhibit the collective behavior of charged particles that is characteristic of the plasma state.

Natural and synthetic plasmas are classified according to plasma temperature and density.

The temperatures and densities of synthetic plasmas, plasmas composed of carriers in metallic and semimetallic, and semiconducting liquids and solids, vary widely.

Gaseous plasmas. The distribution of equilibrated practice velocities for atoms and molecules in gases and for ions and electrons in plasmas is Maxwellian:

$$n(v) = 4\pi n \left(\frac{m}{2\pi kT}\right)^{3/2} v^2 \exp\left(-\frac{mv^2}{2kT}\right) \qquad (1)$$

where particles of mass m and density n (particles/cm^3) have temperature T (K). The most probable velocity is $v_0 = (2kT/m)^{1/2}$, and the mean velocity is $\bar{v} = (8kT/\pi m)^{1/2}$.

The important length parameter for plasmas is the Debye length λ_D, which is given by $\lambda_D = (kT/4\pi ne^2)^{1/2}$, where e is the charge on the electron.

A wide variety of special frequencies exists in plasmas, most notably the plasma frequency $\omega_P = (4\pi ne^2/m)^{1/2}$, which is a measure of the vibration rate of the ions and electrons relative to each other. For true plasma behavior, ω_P must exceed the particle-collision rate f.

Magnetic-field effects in plasmas are so important that, at one time, plasma physics and magnetohydrodynamics (MHD) were practically synonymous (see Coal conversion processes, MHD). Transverse electromagnetic waves propagate in plasmas if their frequency is greater than the plasma frequency. The presence of a static magnetic field within a plasma affects microscopic particle motions and microscopic wave motions. However, it does not alter the propagation of longitudinal electrostatic electron or ion waves if the propagation direction is parallel to the field. Propagation that is orthogonal to the field involves new frequencies that depend on the field strength. Magnetic fields introduce hydromagnetic waves, which are transverse modes of ion motion and wave propagation that do not exist in the absence of an applied field.

Gaseous plasmas are far from equilibrium and, therefore, exhibit microscopic or particle instabilities, and macroscopic or hydromagnetic instabilities. Microscopic instabilities are caused by departures from equilibrium Maxwellian distributions for the electrons or ions. Macroscopic instabilities produce the motion of the plasma as a whole and are caused by pressure or density gradients or magnetic-field curvature.

Matter and energy are necessary for the production of gaseous plasmas which serve as sources of matter and energy in their application. The molecules that are disassociated and the atoms that are ionized during plasma production can be in any state at the start.

The energy source may be internal, eg, the release of chemical energy in flames, or involve electric discharges through a plasma. Electromagnetic fields can be used to form plasmas that interact with or feed back into the source. Externally produced beams of photons, eg, laser beams or energetic particles, create a plasma by their impact and absorption independent of the source. Plasmas may also be produced by strong shock waves. Changes in the composition or energy of a plasma after it is formed often are desired.

Gaseous plasma can be restrained from its tendency to expand by compression or magnetic fields. Low density plasmas are confined magnetically by a variety of field configurations that are designed to prevent particle losses and overall fluid instabilities. Low temperature plasmas used for sputtering and high temperature, fusion-research plasmas are subjected to applied external magnetic fields. High density plasmas can be confined and compressed magnetically by fields produced by strong electric currents flowing in and heating the plasmas, as well as by externally applied fields. Inertial confinement depends on plasma heating outpacing plasma expansion (see Fusion energy).

Plasma diagnostics is the determination of conditions within gaseous plasmas, such as the identities and distributions of the various particle species (neutrals, electrons, and ions), and other characteristics (plasma flow velocities, turbulence, instabilities, and flow of energy). Diagnostic methods involve external probes of plasmas or rely only upon plasma self-emission. Diagnostic techniques that involve natural emissions are applicable to plasmas of all sizes and temperatures and, clearly, do not perturb the plasma conditions. Diagnostic methods based on x-ray emission are especially useful, since the x rays are preferentially emitted from the hottest and densest parts of plasmas.

Plasmas in condensed matter. In perfect semiconductors, there are no mobile charges at low temperatures. Temperatures or photon energies high enough to excite electrons across the band gap produce plasmas in semiconductors.

Impurity-produced plasmas in semiconductors do not have to be compensated by charges of the opposite sign. Plasmas in semiconductors generally are dilute. They are subject to many plasma effects including waves and instabilities.

Plasmas in gaseous and condensed states are related by more than the principles that govern them. For gaseous matter, there is a continuum of behavior from a low density, Maxwellian plasma to a high density, Fermion plasma. Densities less than the Fermi density, n_F, indicate classical nondegenerate behavior, whereas those above n_F imply quantum degenerate conditions, such as can be found in condensed-state plasmas.

Fundamental differences between gaseous and condensed plasmas include their states of excitation and characteristic lengths. Gaseous plasmas are produced by classical, ie, collisional, effects, whereas solid-state metallic plasmas are produced by quantum effects.

Uses of gaseous plasmas. Plasmas serve as electron and ion sources. Pulsed plasmas containing hydrogen isotopes can produce bursts of alpha particles and neutrons as a consequence of nuclear reactions. Intense neutron fluxes also are expected from thermonuclear fusion-research devices employing either magnetic or inertial confinement.

Plasmas are a source of incoherent and coherent, ie, laser, electromagnetic radiation. Plasma uv sources are used commercially for the production of microelectronics circuits by lithography.

The material and energy available in plasmas is used to excite materials and drive chemical reactions. The unique characteristics of plasmas, especially their abundance of energetic quanta, are exploited in many plasma chemical applications, eg, analysis (atomic emission or absorption spectroscopy).

Plasmas can accelerate reactions that are otherwise slow to the point of impracticality. Substances not producible by conventional means can be made with plasmas. Complex molecules that are exposed to nonequilibrium plasmas can be affected in terms of polymerization, rearrangements (isomerizations), elimination of constituent parts, and total destruction of the original molecules and the generation of atoms and ions. A cylindrical radial-flow reactor for deposition of silicon nitride is illustrated schematically in Figure 1.

Plasma materials production and modification embraces processes such as production of thin coatings, heat treatments, and the joining of materials. Plasmas are used extensively to melt materials for a variety of purposes.

The rapid burning of powder in a gun barrel produces relatively cold plasmas which eject the projectile on a ballistic trajectory. The chemical

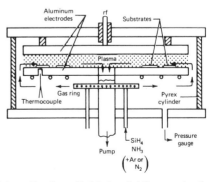

Figure 1. Schematic of a cylindrical radial-flow reactor for deposition of silicon nitride. The reactive gases SiH_4 and NH_3, mixed with Ar or N_2 carrier gases, flow from the gas ring, are excited by radio-frequency (r-f) energy, and exhaust to a pump. The temperature of the plate holding the substrates onto which the silicon nitride is deposited is monitored with a thermocouple.

Table 1. Physical Properties of Plastics[a]

Property	Plastics									
	HDPE	LDPE	PP	PVC	PS	ABS	Polyacrylic, glazing	Polycarbonate, glazing	Epoxy, mineral filled	Polyacetal
glass-transition temp, T_g, °C				75–105	85–105	90–120	90–105	105		
melting point, °C	135	115	168							
injection-molding temp, °C	150–260	150–230	200–290	150–215	225–250	200–275	165–260	290		195–250
injection-molding press, MPa[b]	70–140	55–200	70–140	70–270	70–200	55–170		70–140		70–140
tensile strength, MPa[b]	20–38	4–16	30–38	40–50	34–83	33–43	55–76	55–66	34–69	69
elongation, %	20–1300	90–800	200–700	40–80	1–2.5	5–70	2–7	100–130	16–21	25–75
Izod impact strength, J/m[c]	27–1070	no break	27–117	214–1070	13–21	27–267	16–21	640–690	16–21	75
heat-deflection temp, °C	60–90	45–50	50–60	75–80	80–110	100	70–100	130–135	125–200	125

	Foams			
	PU	Polyisocyanurate	PS board	PS expandable beads
density, kg/m³	14–42	24–56	24–80	13–15
maximum service temp, °C	80–170	150	75–80	75–80
thermal conductivity, W/(m·K)	0.0125–0.034	0.012–0.02	0.023–0.034	0.03
fire resistance	HF-1[d]		HF-1	HF-1

[a] HDPE = high density polyethylene; LDPE = low density polyethylene; PP = polypropylene; PVC = poly(vinyl chloride); PS = polystyrene; ABS = poly(acrylonitrile–butadiene–styrene); PU = polyurethane.

[b] To convert MPa to psi, multiply by 145.

[c] To convert J/m to ft·lbf/in., divide by 53.38 (see ASTM D 256).

[d] HF-1 = UL Standard 94 for Foamed Plastics.

reactions that occur during burning produce plasmas in rockets. Nuclear weapons and plasmas are intimately related.

The fusion or joining of two light nuclei, generally isotopes of hydrogen, leads to the formation of a heavier nucleus, eg, helium, with conversion of mass to energy. Net energy production by nuclear fission may occur in a plasma (see Fusion energy). High temperature is an important requirement for the attainment of fusion reactions in a plasma. Plasmas are an inevitable result of the detonation of fission and fusion weapons.

Most schemes that have been proposed to propel outer-space vehicles involve plasmas. They differ in the selection of matter for propulsion and the way it is energized for ejection.

D.J. NAGEL
Naval Research Laboratory

F.F. Chen, *Introduction to Plasma Physics*, Plenum Publishing Corp., New York, 1974.

G. Schmidt, *Physics of High Temperature Plasmas*, Academic Press, Inc., New York, 1979.

K. Miyamoto, *Plasmas Physics for Nuclear Fusion*, the MIT Press, Cambridge, Mass., 1979.

PLASTIC BUILDING PRODUCTS

The general advantages of plastics over other materials are that they provide a combination of properties and permit effects not possible with other materials; they provide lower costs, either of materials or in fabrication, or both; they give better performance in some critical respect; they are lighter in weight and have a greater strength-to-weight ratio; they provide a better appearance for a longer time and require less maintenance; and they may replace a scarce, often more expensive, material.

Preparation and Properties

The physical properties of the plastics that are of particular importance in building are the glass-transition or melt temperatures, ease of

processing, as shown by the temperatures and pressures needed for molding, heat-deflection temperature, tensile and impact strength, and elongation. For foams, the density, thermal conductivity, and fire resistance are important. The physical properties of the most important plastics are summarized in Table 1.

Exterior Uses

Plastic building products are used in roofing, siding, rainware, shutters and exterior trim, fascias, soffits, skirting, panels and curtain walls, door and window sashes, weather-stripping and sealants, thermal insulation, and solar heating.

Interior Uses

In interiors, plastics are used in wall and ceiling panels, decorative laminates, moldings and trim, plastic flooring, interior hardware, pipe, plumbing fittings and sanitary ware, and electrical applications.

Toxicity

The relative toxicity to laboratory mice of the fumes and smoke from burning plastics is generally less than that from wood.

LAWRENCE H. DUNLAP
Consultant

ROBERT DESCH
Armstrong World Industry

F.W. Billmeyer, *Textbook of Polymer Science*, 3rd ed., John Wiley & Sons, Inc., New York, 1984.

Modern Plastics Encyclopedia, Vol. 59, No. 10A, McGraw-Hill, Inc., New York, 1982–1983, pp. 6–423.

C.J. Hilado, H.J. Cumming, and C.J. Casey, *Mod. Plast.* **55**(4), 92 (1978).

PLASTICIZERS

A plasticizer is incorporated in a material to increase its workability, flexibility, or distensibility. Addition of plasticizer may lower the melt viscosity, the second-order transition temperature, or the elastic modulus of the plastic.

Organic plasticizers usually are moderately high molecular weight liquids or, occasionally, low melting solids. Most commonly, they are esters of carboxylic acids or phosphoric acid.

Mechanism of Plasticizer Action

Four general theories have been proposed to account for the effects that plasticizers have on certain resins. According to the lubricity theory, the rigidity or the resistance of a resin to deformation results from intermolecular friction. The plasticizer acts as a lubricant to facilitate movement of the resin macromolecules over each other and provides internal lubricity.

The gel theory, which was devised for amorphous polymers, contends that their resistance to deformation results from an internal, three-dimensional honeycomb structure or gel. This gel is formed by loose attachments, which occur at intervals along the polymer chains. The dimensional of the cells in a stiff or brittle resin are small because the points of attachment are close. Any attempt to deform the specimen cannot be readily accommodated by movement within the mass. Its limit of elasticity is low. A thermoplastic or even a thermosetting resin with widely separate points of attachment between macromolecules is flexible without plasticization, eg, unvulcanized and lightly vulcanized natural rubber.

The free-volume theory is related to the molecular theory, which assumes that there is nothing but free space between molecules, and this increases with the concentration of end groups and with increase in temperature.

Enough free volume permits freedom of movement, as if there were a hole for a nearby atom, molecule, or chain segment to move into, and polymer chains can move more readily at a given temperature. Thus, the normal result of adding more free volume to a polymer is that it is plasticized, T_g is lower, the modulus and tensile strength decrease, and elongation and impact strength increase. Yet, the freedom of movement afforded by the plasticizer also permits the polymer molecules, if it is their nature to do so, to associate tightly with each other. Sometimes, therefore, the effect observed after addition of small amounts of plasticizer is a reduction in freedom of chain motion and increased rigidity as the gel structure becomes tight, even though T_g is lower and the free volume is present in higher, local concentrations where it has not been excluded by resin–resin association; this is antiplasticization.

Processing aids versus end-use plasticizers. With the increasing use of increasingly higher temperature thermoplastic materials or engineering plastics (qv), more effort is being expended on the use of plasticizers as aids in processing. Easier processing can be achieved by better internal lubricity, reduced sticking of resin to metal equipment, lower hot-melt viscosity, or improved resin stability. At times, the proper choice of a plasticizer can aid processing in all of these ways simultaneously.

Volatile plasticizers also may be used as processing aids. A large portion is flashed during processing, but traces remain and provide some plasticization. In general, when a plasticizer is used only to aid in processing, best results are obtained when the plasticizer is removed from the point of action by volatilization, crystallization, localized incompatibility, or chemical change such as polymerization.

Plasticization. Correlation of the different steps that occur during plasticization have been described.

A hypothesis involving four steps has been proposed. Initially there is a rapid, irreversible plasticizer uptake as plasticizer molecules enter porous areas of the resin and adhere there. Then absorption occurs and the total volume of resin–plasticizer may decrease, although the resin particles swell slowly on the outside, causing strong internal strains at an energy of activation of absorption 21–207 kJ (5–50 kcal). The next step is characterized by higher activation energies [284–464 kJ (68–111 kcal)], during which more severe changes occur inside the particle, but with little or no overall volume change.

At this point, most or all of the plasticizer is incorporated in the resin and slight volume changes have taken place, but plasticization is not complete. Plasticizer is present as clusters of molecules between clusters or bundles of polymer segments or molecules. A mixture of PVC and plasticizer may be formed into a finished article which is hard and has high tensile strength and poor elongation compared with a properly fused piece of identical composition. As energy is applied to the system, a marked change occurs: the volume remains the same but the dielectric constant rises significantly, indicating that the polymer molecules are no longer held together rigidly but are free to move as if the polymer were rubbery rather than glassy. Thus, the plasticizer molecules have penetrated the clusters or bundles of polymer molecules and plasticization essentially is complete.

Antiplasticization. Sometimes, when plasticizers are added to polymers, they increase modulus and tensile strength and decrease elongation. Such antiplasticization results most frequently when small amounts of plasticizer are used. Although conventional plasticizers at low concentrations are effective processing aids for rigid PVC, the product is unacceptably brittle.

Compatibility

To those concerned with development and production of plastics materials, eg, resins, solvents, and plasticizers, compatibility is the ability of two or more substances to mix with each other to form a homogeneous composition of useful plastic properties. One solvent may dissolve a given resin, another only swell it, and a third leave the resin unchanged. Thermodynamic properties of polymer solutions are very dependent on the molecular weight of the polymer, and deviations from ideality in polymer solutions are much greater than those encountered in low molecular weight systems.

Solubility parameter. The cohesive energy density (CED) is one measure of the intensity of intermolecular interactions in a pure liquid or solid, and the strengths of solvent–solvent bonds and polymer–polymer bonds are related to it. Quantitatively, CED is the amount of energy at a given temperature that prevents 1 cm³ of molecules from separating to infinite distance. The solubility parameter, δ, equals $(CED)^{1/2}$.

If a number of assumptions, which are detailed in the literature and which are seldom or never fulfilled in real polymer solutions, are taken into account, the theory requires that a polymer is soluble in a solvent when the solubility parameters of each are the same or do not differ by more than ca ± 6.3 $(J/cm^3)^{1/2}$ $(\pm 1.5$ $(cal/cm^3)^{1/2})$.

None of the parameters so far mentioned is truly a measure of compatibility, but compatibility of a plasticizer with any resin is a function of its chemical makeup. The number, kind, and arrangement of atoms govern the forces involved. All influence the energy of vaporization, δ, dipole moment, hydrogen bonding, viscosity, etc, but not in the same proportions. Measurement of any one of these properties gives one viewpoint of the plasticizer molecule which may then be related to its compatibility with any given resin.

Interaction parameter. The theory of Flory and Huggins is based on a statistical-mechanical treatment of a lattice model of polymer solutions. According to theory, the free energy of mixing a polymer solution is given by

$$\Delta G = RT(n_1 \ln v_1 + n_2 \ln v_2 + \chi n_1 v_2) \tag{1}$$

where ΔG is the free energy of mixing, R is the gas constant, T is the absolute temperature, n_1 and n_2 are the number of moles of solvent and solute, respectively, v_1 and v_2 are the volume fractions of solvent and solute, respectively, and χ is the interaction parameter. If ΔG is negative, the polymer and solvent form a solution or, presumably, the polymer and plasticizer are compatible.

Primary and secondary plasticizers. It is not at all uncommon for plasticizers to have considerable but incomplete compatibility with a given resin. Plasticizers that are highly compatible with a given resin do not exude to form droplets or a liquid surface film nor bloom as a crystalline surface crust; these are primary plasticizers. Those that, on standing, do exude or bloom are secondary plasticizers and are used commonly in combination with primary plasticizers. The apparent compatibility of a plasticizer with a given resin may be influenced by factors such as pressure, temperature, humidity, and sunlight.

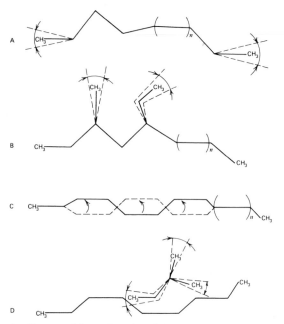

Figure 1. Sources of free volume in polymer systems. Types A, B, and C represent internal plasticization. Type D illustrates external plasticization.

Temperature dependence. The effect of temperature on the mechanism of plasticization has been explained according to the mechanistic theory. With polymer–polymer forces tending toward aggregation of polymer macromolecules and resultant syneresis or exudation of plasticizer, and with polymer–plasticizer forces tending toward disgregation and solvation, the system is in dynamic equilibrium at a given temperature. As the temperature changes, the magnitudes of the effects of these forces change but not to the same degree. With more energy, ie, higher temperature, it is easier to separate polymer–polymer bonds, but it also is easier to separate polymer–plasticizer bonds, which favor aggregation. At lower temperatures, both separations are more difficult.

External versus internal plasticization. The solvation–desolvation equilibrium is characteristic of all externally plasticized resins. Since plasticizer molecules are not attached to the resin by primary bonds, the plasticizer may be lost by evaporation, migration, or extraction. When side chains of about the same size as the plasticizer molecules attach to the resin, an internally plasticized resin results which does not suffer from the deficiency of plasticizer loss, but which does not have the efficiency of external plasticizers (Fig. 1).

Many such products have been developed, usually to be modified by external plasticizers; the copolymers of vinyl chloride and vinyl stearate are of this kind. The nitration or acetylation of cellulose is internal plasticization.

General-Purpose Plasticizers

Phthalate esters, particularly dialkyl phthalates, have dominated plasticizer technology since the 1930s. For many years, bis(2-ethylhexyl) phthalate (DOP) was the accepted industry standard for a general-purpose plasticizer for PVC and is a benchmark for comparison of other plasticizers.

bis(2-ethylhexyl) phthalate (DOP)

Specialty Plasticizers

Specialty plasticizers include benzyl phthalates, glycol benzoates, miscellaneous phthalates and terephthalates, trimellitates, hydrocarbons and petroleum products, flame-retardant plasticizers (chlorinated paraffins and phosphate esters), low temperature plasticizers (adipates, azelates, sebacates, and tris(2-ethylhexyl) phosphate), polymeric plasticizers, and epoxy plasticizers.

Miscellaneous Resins and Materials

The many amorphous polymers, eg, polystyrene, poly(vinyl butyral) and acrylics with no toughening supermolecular structure, do not respond to plasticization as PVC does. However, if they are to be used as surface coatings, adhesives, caulks, etc, they usually are improved by at least small amounts of plasticizers.

Poly(Vinyl Chloride)

Modification of PVC by plasticizers may be considered the standard for plasticization. PVC is unique in its acceptance of large amounts of plasticizers with gradual change in physical properties from a rigid solid to a soft gel or viscous liquid.

In general, plasticizers for PVC and poly(vinyl chloride-co-vinyl acetate) resins are esters of aliphatic and aromatic di- and tricarboxylic acids and organic phosphates. Secondary plasticizers include high molecular weight alkylaromatic hydrocarbons, chlorinated aliphatic hydrocarbons, and occasionally other types.

Methods of plasticization. There are four principal ways in which plasticizers are incorporated into resins such as PVC. Hot compounding involves mixing the plasticizer and resin powder with stabilizers, fillers, pigments, and lubricants as desired and fluxing on a two-roll mill at about 150–175°C, or in an internal batch-mixer with two sigma-shaped blades rotating in a heated chamber, or in a continuous mixer.

Plasticizers in moderate amounts may be totally absorbed into certain grades of resins after a few minutes' stirring at 50–80°C to yield a dry blend, which is barely distinguishable in outward appearance from the original resin.

Certain PVC resins, ie, paste resins, are formulated so that the plasticizer wets the resin particle at room temperature but only very slowly penetrates and solvates the resin. These stable, pourable dispersions of resin powders in plasticizers are called plastisols.

Poly(vinyl chloride) also dissolves in suitable solvents such as cyclohexanone, tetrahydrofuran, or mixtures of ketones and hydrocarbons. The solution can be mixed with a plasticizer and used to make a solvent-cast film, or it can be used as a surface-coating material.

Fusion. The effect of plasticizer on ease of fusion of resin–plasticizer blend is economically as well as technically important in determining the best processing conditions. The fusion effect is additive in that a plasticizer with rapid fusion characteristics can be added to a slower-fusing product to yield fusion results intermediate between the two. Through proper selection of a plasticizer, optimum fusion times and power requirements can be developed.

Fusion points of plastisols prepared from various plasticizers can be used to rank the plasticizers according to ease of fusion and can aid in choice of plasticizer for each industrial need. A list of plasticizers in order of decreasing ease of fusion, with exceptions, is phosphates, phthalates, isophthalates, terephthalates, adipates, azelates, and sebacates.

Vinyl foam. Vinyl foam can be produced by mechanical frothing, ie, incorporation of air or other gas in the system prior to fusion, or by chemical blowing, ie, decomposition of an additive by heat during processing to form gas bubbles. Both techniques require a balance of resins, plasticizers, and additives to yield a product with the desired density and open/closed cell ratio (see Foamed plastics).

Modifications of physical properties. The main variables in plasticizer properties and performance are the kind of plasticizer used and its concentration. In the usual range of concentration, ie, greater than 20 phr, addition of plasticizer softens the resin and makes it more flexible, reduces modulus and tensile strength, and provides better elongation. The reverse effects are observed at low concentrations (see section on Antiplasticization). For most resin–plasticizer combinations at room temperature, a plasticizer threshold concentration must be passed before the normal plasticizer effects on physical properties are observed.

Other physical properties vary with plasticizer type and concentration. Processing conditions of time, temperature, and method are additional variables. Thermal properties vary with increasing concentration of plasticizer: at ca room temperature, specific heat and thermal conductivity increase and thermal diffusivity decreases. The dissipation factor and the dielectric constant are sensitive to choice and concentration of plasticizer; the latter varies with plasticization from ca 3.0 to 8.0. Dielectric strength and volume resistivity decrease. Conductivity depends in part on the dielectric constant and efficiency of the plasticizer, and it increases with plasticizer concentration. Actual conductance values depend to a large extent on the purity of the plasticizer.

According to Leilich's rule, more viscous plasticizers are less efficient than less viscous ones.

Mixtures of plasticizers. In the modification of physical properties, a blend of two or three, or occasionally more, plasticizers provides effects that are essentially the weighted average of the effects produced by the individual plasticizers.

Effect on resin chemical properties. Heat stability. Incorporation of plasticizers into PVC tends to inhibit thermal degradation during processing (see Heat stabilizers).

Plasticizers containing the epoxy or oxirane group were among the first stabilizers to be used in PVC and related resins. They were added to react with the liberated HCl and retard its catalytic effect. Other plasticizers can aggravate thermal degradation of plasticized PVC, eg, chlorinated paraffins, phosphates, sulfonamides, and tartrates.

Oxidative stability. During processing, conditions are favorable for initiating autoxidation. Plasticizers with structures particularly susceptible to autoxidation, such as tertiary hydrogens, may contribute to early color development and other undesired effects. However, an antioxidant, eg, bisphenol A (4,4'-isopropylidenediphenol), is effective in controlling oxidation.

Light stability. In general, plasticizers improve light stability of PVC when they are used in plasticizing amounts.

Flammability. Because of its high chlorine content, PVC burns with difficulty. Although plasticizers add fuel value, flammability of plasticized PVC is heavily dependent upon type and concentration of plasticizer present. Plasticizers, eg, phthalates, adipates, and polyesters, contribute to burning. Both triaryl and alkylaryl phosphates inhibit burning of plasticized PVC and can be blended with other types of plasticizers to achieve an acceptable balance of flame resistance and physical properties.

Corrosion resistance. Low water permeability of a protective coating should accompany good corrosion resistance. With increase in plasticizer content, water permeability increases.

Corrosion of PVC is governed largely by the plasticizer. When films of plasticized PVC are exposed to twelve environments ranging in corrosiveness from distilled water through solutions of various acids, bases, and salts to ferric chloride, phthalate-plasticized resins are least attacked, the phosphates, adipates, and polymerics being less corrosion-resistant. Secondaries are sometimes beneficial.

Fungal and bacterial attack. Plasticizers may impart to an otherwise inert resin system their own biological properties. The extent of their influence is governed by their compatibility with the resin, ease of extraction, migration, and volatility in addition to their innate biological properties.

It is highly probable that there are no plasticizers that are completely free from fungal or bacterial attack. Some are utilized readily by microorganisms as sources of carbon. Others are sufficiently resistant to be useful in warm, moist, or wet environments.

Insects, termites, or rodents. Plasticization increases the probability of attack, particularly by termites and rodents (see also Poisons, economic). The susceptibility increases as the plasticizer amount increases.

Environmental effects. Exudation under pressure. A plasticizer that is compatible with PVC under ordinary conditions may exude under slight pressure. Proper compounding is required to avoid this and other compatibility problems.

Effect of humidity on compatibility. Many plasticizers that are compatible under normal conditions exude in humid and, especially hot and humid environments.

Light-promoted exudation. Plasticizer molecules can migrate to and from the surface. If the plasticizer contains carbon–carbon unsaturation, it may photooxidize at the surface and polymerize to compounds that are too polar and of too high a molecular weight to return into the plastic. Photooxidation is aggravated by the presence of epoxy plasticizers.

Volatility. Retention of plasticizer is necessary for good service life of a fabricated article. Plasticizer loss can occur through evaporation and is influenced by the structure and the molecular weight of the plasticizer. Antioxidants are frequently used to prevent oxidation and subsequent loss of the volatile by-products. In commercial practice, dialkyl phthalates of an alcohol chain length of greater than seven have been used as the sole plasticizer.

Extraction. Because the plasticizer is not held permanently to the resin, it may be extracted partially by oils, waxes, water, soapy water, and other agents. Resistance to extraction by any environmental agent can be controlled by a suitable plasticizer.

Migration to other materials. If plasticized PVC is in contact with other polymeric materials and if the resistance at the interfaces is not too high and the plasticizer is compatible with the second polymer, the plasticizer may migrate from the PVC to the other polymer. Such migration may soften or, in severe cases, destroy the surface coating.

Migration from other materials. Migration of trace amounts of other plasticizers from surface coatings or adhesives usually does not noticeably change plasticized PVC. Problems arise when the material migrating into the plasticizer of the PVC composition has an obviously undesirable property, eg, staining.

Other Thermoplastic Resins

Many resins other than PVC are modified for specific uses either with internal or external plasticizers, eg, cellulosics (cellulose nitrate, cellulose acetate, cellulose acetate–propionate and cellulose acetate–butyrate, ethyl cellulose, and 2-hydroxypropyl cellulose), polyacrylonitrile, polyamides (nylons), polycarbonate, linear polyesters, acrylics, polyolefins, polystyrene, and fluoroplastics.

Specialized Resin Systems

Adhesives. The main types of plasticized resin adhesives are emulsion, hot-melt, delayed tack, solution, pressure-sensitive, and adhesive primers or coatings. Plasticizers help control viscosity, wet tack, green strength, adhesion, hardness, and creep.

Rubber and Related Polymers

Natural rubber and synthetic elastomers consist of long-chain structures that are randomly zigzagged in the relaxed state but more or less straight in the stretched state (see Rubber; Elastomers, synthetic). An elastomer is cured subsequent to the introduction of a few cross-links between polymer chains; the cured product is tougher, and of increased resilience but lower elongation.

Natural rubber, because of its high initial molecular weight, is not easily fabricated into commercially useful articles. It is common practice to mill the raw rubber on a roll mill or in a Banbury mixer to lower the molecular weight by breaking the polymer chains. The product is more plastic and less elastic, or has less nerve, than raw rubber. The low molecular weight fragments, which are produced from milling, plasticize the remaining high polymer so that it processes easily and may be readily fabricated. Usually it is more economical to use a petroleum oil for at least part of the plasticizer rather than degrade more expensive rubber to produce the plasticizer *in situ*.

Peptizers or chemical plasticizers. Traces of free-radical acceptors in the absence of oxygen and oxidation promoters in the presence of oxygen increase the rate of plasticization during milling. Among commercial materials used for this purpose are zinc thiobenzoate, pentachlorothiophenol and its zinc salt, phenylhydrazine, and thio-β-naphthol. Such chemical plasticizers induce plasticization by chemical changes.

External plasticizers. Addition of external plasticizers during milling can reduce hot-melt viscosity and retard chain cleavage. Because most elastomers are comprised of long hydrocarbon segments and are largely amorphous, they accept heavy petroleum oils, coal tars, and other predominantly hydrocarbon products or extender oils, which are less expensive than the esters that commonly are used with PVC.

Effects of plasticization. The first noticeable effect of plasticizer addition is the reduction of viscosity.

Physical properties of the formulated, vulcanized rubber vary with the type and concentration of extender oil used. Correlations of the effect on physical properties with the type of oil are based on viscosity and specific gravity.

Thermosetting resins. As there is no sharp distinction between thermoplastic and thermosetting resins, there is no discontinuity in plasticizer technology between them. A resin that is highly cross-linked by many primary valence bonds is rigid, insoluble, and infusible; as such, it cannot be plasticized, but plasticizer can be added prior to extensive cross-linking. Among the more important effects of plasticizers on thermosets are increased potlife and improved setting and penetration of fillers, laminating papers, and fabrics, improved temperature release, postforming, and hot and cold punchability.

Polymerizable Plasticizers

Emphasis has been placed on the manufacture of a polymerizable plasticizer that would permit use of plastisol techniques to make rigid products. Polyglycol dimethacrylates, eg, MG-1, were among the first to appear promising, followed by 1,3-butylene glycol dimethacrylate and trimethylolpropane bis(methacrylate) [2-ethyl-2-hydroxymethyl-1,3-propanediyl dimethacrylate] and other cross-linkable esters. All of these can be used with paste PVC resins to yield satisfactory plastisols.

Plasticization by Fillers, Pigments, Salts, or Air

In a resin, nonreinforcing fillers, eg, precipitated calcium carbonate chalk, kaolin, silica, etc, usually are wet by the resin molecules, and secondary bonds form between them as do resin–resin and resin–plasticizer bonds (see Fillers). The normal effect of the filler is to stiffen the resin systems, ie, increase modulus and hardness. With increasing filler concentration, the tensile strength and ultimate elongation drops as the binder, ie, resin and plasticizer, becomes severely diluted with solid filler particles. However, small amounts of filler or pigments may behave abnormally, yielding a product that has been softened as if by a liquid plasticizer. Hydrophilic polymers may be plasticized by certain inorganic salts. Air incorporated into sheet-flooring resins has a noticeable plasticizing effect.

Miscellaneous Materials

Many materials not commonly considered to be plastics are frequently plasticized by suitable additives. These range from recognized natural and synthetic polymeric materials to high melting inorganic substances.

Economic Aspects

Approximately 450 plasticizers are available commercially. The rate of introduction into the market of new plasticizers has decreased significantly in the last two decades and is characteristic of maturing plasticizer developments. Of the 450 available plasticizers, perhaps 100 are of significant commercial value.

Health and Safety Factors, Toxicity

In plasticized PVC, the resin, its stabilizer, and the plasticizer must be essentially nontoxic when used, for example, in film or bottles for food packaging or in tubing for maple syrup, milk, or carbonated water. It also must be nonsensitizing to skin, eg, when used for disposable hospital aprons and sheets.

J.K. SEARS
N.W. TOUCHETTE
Monsanto Company

J.K. Sears and J.R. Darby, *The Technology of Plasticizers*, Wiley-Interscience, New York, 1981.

P.F. Bruins, ed., *Plasticizer Technology*, Vol. 1, Reinhold Publishing Corp., New York, 1965.

Modern Plastics Encyclopedia 1983–1984, Vol. 60, No. 10-A, McGraw-Hill, New York, 1983, pp. 156, 158, 161, 162, 636–645 (issued yearly).

PLASTICS, ENVIRONMENTALLY DEGRADABLE

Environmental degradation results from exposure to sunlight, heat, water, oxygen, pollutants, microorganisms, insects and animals, and from mechanical forces such a wind, sand, rain, wave action, traffic, etc.

Photodegradation refers to the degradation of polymeric substances and other organic compounds when exposed to sunlight and other uv-containing sources of light. Oxidative degradation is degradation through attack by oxygen and ozone. Biodegradation is degradation and assimilation by the action of living organisms.

With a few notable exceptions, synthetic plastics are resistant to microbiological degradation, but naturally occurring polymers are susceptible to fungi and bacteria. Biological degradation of polymers is facilitated by linearity, by the presence of unhindered aliphatic ester and peptide linkages, and by low molecular weights (below 1000 M_n).

Biodegradation of Synthetic Polymers

Several general methods have been developed to measure biodegradation. In one procedure, designated ASTM G 21-70 (before 1980: ASTM D 1924–63) the specimens are placed in or on a solid agar growth medium that is deficient in available carbon. After inoculation with the test microorganisms, the medium and samples are incubated for three weeks. Any growth of the colony is dependent on utilization of the polymer as a source of carbon.

In an elaborated variation, a polymer sample is deposited on a Petri dish and the dish is dried and weighed. Nutrient agar is poured over the polymer, which is then inoculated and incubated. After the test period (3–4 wk), the agar and culture are washed off and the plates dried and weighed again to determine the weight loss.

Alternatively, the polymer in finely powdered form is suspended in a nutrient-agar medium in Petri dishes. After inoculation and incubation of the hazy suspension, colonies of cells are observed growing on the gel if the polymer is being assimilated.

Samples retrieved from soil burial may be tested for weight loss or deterioration of mechanical properties, or they may be examined by scanning electron microscopy.

In the plate-count method, used only with bacteria, the polymer sample is finely ground, suspended in a shaker flask in nutrient broth which has been inoculated with the bacteria of choice, and incubated in the presence of air. Plate counts are performed on aliquots removed at intervals.

The first report of ^{14}C in polymer biodegradation was a study of ^{14}C-labeled polyethylene. The plastic was ground, mixed with soil, and incubated. Water-saturated air was passed through the pot containing the sample and then through a solution of 2 M KOH to absorb both $^{12}CO_2$ and $^{14}CO_2$ generated in the pot by microbial action. After 30 d, the KOH solution was titrated with 1 M HCl to pH 8.35, and the total amount of CO_2 formed was calculated.

Biodegradation depends on molecular size and structure, the microbial population, and environmental factors such as temperature, pH, humidity, and availability of nutrients.

Microbiological degradation is caused by enzymes produced by the organisms. The reaction rate of enzyme-catalyzed reactions increases with temperature until activity begins to decrease because of heat-activated denaturation.

Biodegradability of additives. Organic compounds are added to synthetic polymers as oxidation inhibitors, plasticizers, lubricants, colorants, slip agents, uv light stabilizers, and antistatic agents, and for other purposes. Some of these compounds are biodegradable, some are resistant, and a few are biocides.

Burial studies have demonstrated that blends of biologically inert polymers, such as polyethylene, with up to 50 wt% of biodegradable fillers such as starch or sugar, undergo removal of the biodegradable filler by microbial action.

Biodegradability of polymers. Biological oxidation of hydrocarbons to fatty acids has been established by several investigators. Cetyl palmitate is a product of the growth of a gram-negative coccus grown in hexadecane as the sole source of carbon. The formation of methyl ketones from the oxidation of hydrocarbons by *Pseudomonas methanica* and *Mycobacterium* species has also been reported. Among low molecular weight paraffins, the linear molecules are more easily biodegraded than the branched molecules. The ease of biodegradability of polyethylene decreases with increasing molecular weight.

Studies of high and low density polyethylene also show greater activity of lower molecular weight samples (see Table 1).

In a study of the rate of CO_2 evolution from a commercial polyethylene sample and a ^{14}C-labeled sample buried in soil enriched by composted garbage, it was found that when the ^{14}C-labeled polymer was protected from light, only traces of ^{14}C-labeled CO_2 were evolved, probably from low molecular weight oligomers originally present in the polymer.

Increasing polarity and hydrophilicity of polyethylene by adding short, polar side branches does not increase the ease of biodegradation. Copolymers of ethylene with unsaturated vegetable oils also give negative results. None of these copolymers is biodegradable.

Efforts to make polystyrene biodegradable by attaching biologically active end groups, by copolymerizing with other monomers, and by pyrolysis have been unsuccessful.

Poly(vinyl chloride) is resistant to fungi and bacteria and is widely used for outdoor sheathing, gutters, and window frames. It must be protected against sunlight with stabilizers and coatings. A number of commercial vinyl chloride copolymers and blends have been tested for biodegradability and found to be inactive.

Poly(*cis*-isoprene) is readily biodegraded, whereas the following synthetic rubbers are not susceptible to microbial attack: ABS rubber, styrene–butadiene block copolymer (thermoplastic elastomer), butadiene–acrylonitrile rubber, polyisobutylene, and chlorosulfonated polyethylene (Hypalon rubber).

Aqueous latex emulsions are subject to spoilage by microbial attack on additives in the formulation. The use of broad-spectrum biocides is recommended.

The following polymers were reported not to be susceptible to microbial attack: poly(methyl methacrylate) (Lucite), rubber modified poly(methyl methacrylate), poly(4-methyl-1-pentene) (TPX), poly(vinyl butyral), poly(vinyl ethyl ether), poly(vinyl acetate), and partially hydrolyzed poly(vinyl acetate). Poly(vinyl alcohol) is degraded

Most commercially available high molecular weight condensation polymers, such as nylons and aromatic polyesters, are resistant to biological attack. Aliphatic polyesters and polyurethanes derived from ester diols have been found to be susceptible to microbial degradation. High molecular weight crystalline polymers such as nylon-6, nylon-6,6, and nylon-12 have been found to be resistant to microbial attack.

Linear polyureas are crystalline and resistant to degradation, but substituted polyureas prepared from the methyl and ethyl esters of L-lysine support the growth of *Aspergillus niger*.

Polyesters containing significant amounts of aromatic constituents are generally found to be resistant to biodegradation. Poly(ethylene

Table 1. Molecular Weight Effects on Polyethylene Degradability

Density, g/cm³	Mol wt viscosity average	ASTM (G 21-70) growth rating
0.96	10,970	2
0.96	13,800	2
0.96	31,600	0
0.96	52,500	0
0.96	97,300	0
0.88	1,350	1
0.95	2,600	3
0.92	12,000	2
0.92	21,000	1
0.92	28,000	0

Table 2. Effect of Soil Burial on High Molecular Weight Poly(ε-Caprolactone)-PCL-700

Burial time, mo	Tensile strength, MPa[a]	Elongation at break, %	Weight loss, %
0	18 ± 0.7	369 ± 59	0
1.25	13 ± 1.5	9 ± 1.4	
2.0	11 ± 1.2	7 ± 2	8
4.0	3.06 ± 1.5	2.6 ± 1.1	16
6.0	0.7	negligible	25
12.0	negligible	negligible	42

[a] To convert MPa to psi, multiply by 145.

terephthalate) fibers show good resistance to biological attack after seven years exposure to seawater.

Aliphatic esters and polyesters have been reported as being biodegradable. The biodegradability of aliphatic polyesters was investigated using ASTM G 21-70 as the test method. A good correlation exists between biodegradability and melting point for a series of aliphatic polyesters, ie, a lower melting point indicates better biodegradability. Attaching biodegradable polymer molecules to nondegradable polymer molecules does not enhance the degradability of the inert component.

Polyether-linked polyurethanes are, in general, much less susceptible to attack by fungi than polyester-linked polyurethanes. Other polymers reported to be resistant to microbial attack are phenol–formaldehyde and urea-formaldehyde resins (in cross-linked form); cellulose triacetate, butyrate, and propionate; chlorinated polyethers; and silicone resins.

Using the quantitative Petri dish method and the organism *Pullularia pullulans*, an inverse dependence of weight loss per cm² on initial molecular weight of the polymer has been observed for poly(ε-caprolactone).

The effect of soil burial on bars molded from poly(ε-caprolactone) (ca 40,000 $\overline{M}_n$) has been determined. The bars were buried in a mixture of equal parts of New Jersey garden soil, Michigan peat moss, and builder's sand for periods up to 12 months. Samples were removed periodically to determine weight loss and measure tensile properties (see Table 2).

The biodegradability of water-soluble polymers is dependent on their molecular structure, the presence of specific enzymes, and in some instances, on the molecular weight. Water-soluble synthetic polymers that are resistant to biodegradation by the BOD test are poly(ethylene glycol)s, carboxyvinyl polymer (Carbapol), poly(ethylene oxide) (Polyox), polyvinylpyrrolidinone, poly(vinyl methyl ether), poly(acrylic acid) salts, polyacrylamide, and poly(vinyl alcohol). Most water-soluble cellulose derivatives are biodegradable.

Applications of biodegradable polymers. The three application areas which have received the most attention are medical, agricultural, and packaging. The advantage of biodegradable or absorbable plastics for medical implantation is that they obviate the need for surgical removal.

In vivo degradation involves hydrolytic, oxidative, and enzymatic processes. In some instances, polymers which are stable when buried in the soil are found to undergo degradation and erosion inside a human or animal body. Absorbable surgical sutures have been made by the extrusion of polylactide polymers and copolymers (see Sutures). Biodegradable polyesters have been studied as possible drug-delivery systems for contraceptives and narcotic antagonists implanted under the skin (see Pharmaceuticols, sustained release).

Perhaps the largest commercial application of degradable polymers is the use of urea–formaldehyde polymer in slow-releasing fertilizers, which permits the application of up to 30% nitrogen fertilizer without damaging the lawn ("burning"). Containerized tree seedlings grown in biodegradable containers have been planted extensively in Washington, New Mexico, several southern states, and British Columbia, Canada.

The low cost, high volume packaging plastics, such as polyethylene, polystyrene, polypropylene, and poly(ethylene terephthalate), are not biodegradable; efforts to make them so by blending with biodegradable fillers or other additives have not been successful.

Photodegradability of Synthetic Polymers

Most synthetic polymers are subject to weathering degradation, which is caused by the actions of sunlight, heat, oxygen, and moisture. The most direct way to evaluate the photodegradability of a given polymer is to subject samples to outdoor weathering and measure some physical property as a function of exposure time.

Degradable polymers become brittle upon exposure to actinic light and then disintegrate from the action of natural forces such as wind and rain. Photomicrography, scanning electron microscopy, and conventional photography are useful for documenting changes in plastic samples with exposure time.

Mechanisms. When exposed to uv radiation, polymers containing carbonyl or hydroperoxide groups have been observed to undergo photochemical chain cleavage (see also Photochemical technology).

Polymers containing ketone groups undergo chain cleavage by Norrish I and Norrish II mechanisms, as illustrated below.

Norrish type I reatcion

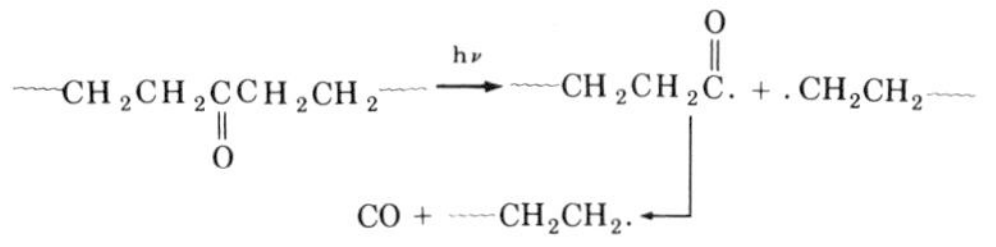

Norrish type II reaction

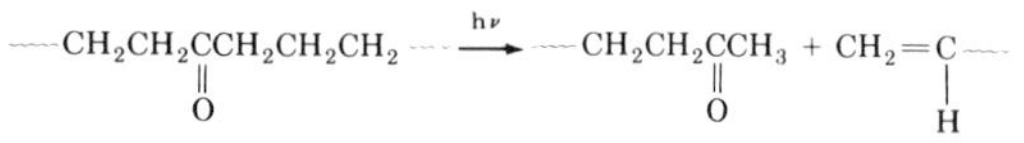

Quantum yield depends on chain length and on whether the carbonyl group is in a side chain or in the main chain.

Degradability can be enhanced by the addition of catalysts and prooxidants. Photooxidative degradation begins with the formation of a polymer-chain radical which forms polymer peroxides and hydroperoxides by reaction with oxygen. The hydroperoxides are highly susceptible to decomposition by heat and metal salts.

Antioxidants (qv) stabilize polymers by donating hydrogen atoms to polymer radicals. The resulting antioxidant radical reacts with polymer radicals by a chain-termination reaction.

Many organic compounds, such as aromatic ketones, diketones, quinones, nitroso compounds, and dyes, are known to sensitize the photoreactions of polymers that lead to degradation, cross-linking, and grafting. Whereas p-hydroxybenzophenones act as photodegradation sensitizers, o-hydroxy compounds act as photostabilizers.

The science and technology of polymer degradation and stabilization have developed to the point that almost any addition polymer could be formulated to degrade rapidly if necessary when exposed outdoors. Varying the concentration of ferric N,N-dibutyldithiocarbamate in low-density polyethylene affects the degradation rate of LDPE sheet. Soluble salts of cobalt, iron, manganese, cerium, zinc, lead, zirconium, and calcium catalyze the photoinduced decomposition of 0.922 density polyethylene containing 2% isotactic polypropylene.

Photooxidative degradation of polystyrene is accelerated by the presence of compounds that become activated by sunlight, such as benzophenone, acetophenone, and anthrone, and by metal salts that catalyze the decomposition of hydroperoxides.

Exposure of polypropylene to uv radiation in air can produce rapid photooxidation and loss of physical strength through the formation of hydroperoxide and carbonyl groups. Traces of transition-metal catalysts used in making the polymer contribute to its ease of photooxidation.

The alternative to enhance polymer degradability with additives is to incorporate photosensitive groups into the polymer molecule during polymerization. Copolymerization with carbon monoxide or a vinyl ketone is the usual method. Most patents pertaining to photodegradable polymers have been issued for addition polymers rather than condensation polymers.

Cellulose paper products, cellophane, viscose rayon, cellulose acetate, and cellulose nitrate are all subject to photodegradation and photooxidative degradation.

Table 3. Outdoor Weathering of Materials Used to Package Beverage Cans

Material	Thickness, mm[b]	Not exposed	1 mo	2 mo	5 mo
Plastic	*Ultimate elongation after exposure, %*				
low-density polyethylene	0.43	584	625	535	460
ethylene-CO copolymer[a]	0.43	550	10	0	0
polyethylene shrink wrap	0.05	563	380	40	350
polypropylene	0.25	527	380	40	0
Paper	*Ultimate tensile strength, MPa[c]*				
six-pack carton		43.2	18.9	24.5	8.07
corrugated box		5.36	5.49	4.07	1.16

[a] Used primarily for the six-pack loop carrier.
[b] To convert mm to mils, divide by 2.54×10^{-2}.
[c] To convert MPa to psi, multiply by 145.

Polymer films buried in the soil under aerobic conditions are subject to oxidative degradation of varying degrees depending on polymer structure and morphology, the prior history of thermal oxidative abuse, and the presence of additives.

Photodegradable plastics. Currently available photodegradable plastics can be divided into two categories: products, mostly packaging material, that are designed to degrade as rapidly as possible outdoors; and plastics that are designed to degrade only after having served a useful function outdoors for a definite time span. Only a few degradable packaging applications have been commercialized. The outdoor disintegration rates of the most commonly used packaging materials for beverage cans are shown in Table 3.

The consumption of mulch film in Europe and Japan is higher than in the United States because of more intensive farming methods practiced there. In order for photodegradable films to compete successfully with conventional agricultural mulch film, they must meet certain requirements: films must be stable in storage, films with different induction periods must be available for various vegetable crops, films must disintegrate into small pieces in order to reduce litter and avoid interference with soil preparation for a new crop, and disintegrated films must not release harmful substances. Poly(1-butene) plastics have received much experimental attention as photodegradable mulch films.

JAMES E. POTTS
Union Carbide Corporation

M. Alexander, *Microbial Ecology*, John Wiley & Sons, Inc., New York, 1971.
Proceedings of the Conference on Degradable Polymer Plastics, The Plastics Institute, London, UK, Nov. 1973.
J.E. Potts in H.H.G. Jellined, ed., *Aspects of Degradable and Stable Polymers*, Elsevier, New York, 1978, Chapt. 14.

PLASTISOLS. See Vinyl polymers, poly(vinyl chloride).

PLASTICS PROCESSING

Thermoplastic Resins

Almost 90% of the resins produced are thermoplastics, and over a dozen chemically different kinds of thermoplastics are used, of which there are two broad classes, ie, amorphous and crystalline. Crystalline resins are characterized by melting and freezing points, unlike the amorphous resins. Common crystalline plastics are high and low density polyethylenes (HDPE and LDPE), polypropylenes, acetal resins, nylons, and thermoplastic polyesters. Common amorphous thermoplastics are acrylonitrile–butadiene–styrene terpolymers (ABS), cellulose acetate, phenylene oxide-based resins, polycarbonates, poly(methyl methacrylate)

(PMMA), polystyrene, poly(vinyl chloride) (PVC), and styrene–acrylonitrile copolymers (SAN).

With few exceptions, thermoplastics are purchased in the form of cubic, cylindrical, or spherical pellets 3 mm in diameter.

Extrusion

Approximately 50% of all resins is converted into products by extrusion. An idealized extrusion line is shown in Figure 1.

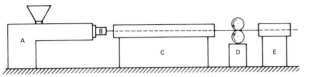

Figure 1. An extrusion line. A, extruder. B, die. C, sizer cooler (water trough). D, pull rolls. E, cutter (coiler). Courtesy of Society of Plastics Engineers.

The resin is melted in the extruder and then is pumped through a die. The die forms the melt into a shape closely related to that of the final product. The melt may be drawn to a thinner cross-section in an air gap. The melt then is cooled and its shape is maintained. The solid plastic enters a puller, which is characterized by a complementary shape to that of the plastic; it is the puller that draws melt from the die and through the cooling system. The final piece of equipment cuts the product to length or winds it into a roll or coil.

Blown film. Blown film is a plastic film that is extruded through a tube (see Fig. 2). Most blown film is made from polyethylene and is used for food and trash bags (see Film and sheeting materials).

Cast film. The cast-film process provides a film with gloss and sparkle and can be used with various resins.

Sheet. Sheeting thicknesses are 0.25–5 mm and widths are as great as 3 m. Usually sheets are cut to a specified length and are stacked. Cooling is controlled by a three-roll stack.

Profile extrusion. Profile extrusion or shape extrusion is the process of making siding, channels, gaskets, and decorative trim of irregularly shaped cross-sections. Shape extrusion is done most often with amorphous thermoplastics, eg, poly(vinyl chloride) and polystyrene.

Coating

Extrusion. Extrusion coating with polyethylene is applicable for paper, milk-carton stock, and aluminum foil (see Coating processes). Coated products are used primarily for packaging and the coating allows rapid assembly of packages by heat sealing. Coatings also may improve the barrier properties of the substrate. In contrast to most extrusion processes, extrusion coating involves a very hot melt, ca 340°C.

Wire. Usually an extruder with a special die is used to coat electric wire with insulation (qv). Wire is unwound from a reel, straightened, and sometimes heated before passing through the die. Coated wire is cooled in a long water trough and collected on a reel.

Molding

Injection molding. Injection molding is a process by which a molten thermoplastic is injected under high pressure into a steel mold (see Fig. 3). After the plastic solidifies, the mold is opened and a part the shape of the mold cavity is obtained.

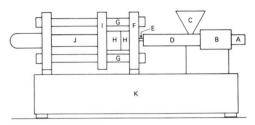

Figure 3. Injection-molding machine. A, hydraulic motor for turning the screw. B, hydraulic cylinder and piston which allows the screw to reciprocate about three diameters. C, hopper. D, injection cylinder (a single-screw extruder). E, nozzle. F, fixed platen. G, tie rods. H, mold. I, movable platen. J, hydraulic cylinder and piston. These are used to move the movable platen and to supply the force needed to keep the mold closed. K, machine base.

Structural foam. Structural-foam molding is a modified injection-molding process which is used to produce large parts with 20–40% reduction in density.

Blow molding. Blow molding is the most common process for making hollow parts, eg, bottles (see Fig. 2).

Thermoforming

Thermoforming is a process for converting a plastic sheet into parts, eg, a tray for packaging meat or a sign. If vacuum is used, the process is called vacuum-forming. Amorphous resins, eg, polystyrene, poly(vinyl chloride), and poly(methyl methacrylate), usually are used for thermoforming.

Cast Acrylic Sheeting

Cast acrylic sheeting has been made for over 40 yr by polymerization of methyl methacrylate in a cell assembled from two glass plates and a flexible gasket (see Acrylic ester polymers). The product is used for glazing and thermoforming.

Polystyrene

Expandable polystyrene molding. Molding of expandable polystyrene is used to make three kinds of foamed products: insulation board, shapes for packaging, and coffee cups.

Foamed polystyrene sheet extrusion. Foamed polystyrene sheet is used for thermoforming egg cartons, meat trays, and coffee cups.

Calendering

Calenders are used for making sheeting from rubber and for making similar products from semirigid and flexible PVC and ABS. Product thicknesses are 0.05–0.75 mm. A calender has four heavy, large steel rolls which usually are assembled in the inverted L configuration.

Thermosetting

Compression molding. Compression molding is the oldest process of the plastics industry. The equipment consists of a vertical hydraulic press with platens for mold attachment. A compressed cake of granular resin is placed in the mold, and the mold is closed. The resin melts and pressure forces the liquefied material to fill the cavity. Continued heating cures the resin within a minute or two and the part is removed from the mold.

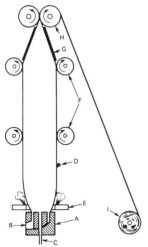

Figure 2. Extrusion of blown film. A, blown-film die. B, die inlet. C, air hole and valve. D, plastic tube (bubble). E, air ring for cooling. F, guide rolls. G, collapsing frame. H, pull rolls. I, windup roll. Courtesy of Society of Plastics Engineers.

Three important thermosetting resins are used in the compression-molding process: phenol–formaldehyde resins, urea–formaldehyde resins, and melamine–formaldehyde resins (see Amino resins; Phenolic resins).

Polyester. Most reinforced thermoset plastics are blends of polyester and glass fibers (see Laminated and reinforced plastics; Polyesters, unsaturated).

Polyurethanes. Polyurethane describes several kinds of products (see Urethane polymers) that are characterized by common chemistry: the reaction of an isocyanate with an alcohol to make a urethane. Products include flexible and rigid polyurethane foams.

Reaction-injection molding. Reaction-injection molding (RIM) is utilized for the production of partially foamed polyurethane moldings by rapid injection of metered liquid streams of polyol and an isocyanate into a mold.

PAUL N. RICHARDSON
E.I. du Pont de Nemours & Co., Inc.

P.N. Richardson, *Introduction to Extrusion*, Society of Plastics Engineers, Brookfield Center, Conn., 1974.

I.I. Rubin, *Injection Molding Theory and Practice*, John Wiley & Sons, Inc., New York, 1972.

Z. Tadmor and C.G. Gogos, *Principles of Polymer Processing*, John Wiley & Sons, Inc., New York, 1979.

J.L. Thorne, *Plastics Process Engineering*, Marcel Dekker, Inc., New York, 1979.

PLASTICS TESTING

Plastics standards and test methods define the properties of plastics and nonpolymeric additives and provide terminology, criteria, and procedures for accurate analysis and characterization. In the present context, a plastic is defined as a material that contains as an essential ingredient one or more organic polymeric substances of high molecular weight, is solid in its finished state, and at some stage in its manufacture or processing into finished articles can be shaped by flow (see Plastics processing). The properties of polymeric materials are determined by their chemical composition, structure and morphology, molecular weight parameters, and additives. Test methods that are capable of accurately measuring the chemical, physical, and mechanical properties of plastics serve a useful purpose in quality control, design, and research and development. Standard reference materials for the calibration of instruments and test methods are critical components of the measurement process.

Molecular Weight Measurements

Determination of molecular weight can be made by light-scattering, ultracentrifugation, chromatographic, osmetic, and other techniques.

Instrumental Analysis and Characterization

The development of instruments for the analysis and characterization of plastics has greatly facilitated the study of their composition, structure, molecular parameters, and performance (see Table 1) (see Analytical methods).

Additive Analysis

The large number of additives used in polymer technology complicates the problem of accurate identification and analysis. Separation techniques include column chromatography, thin-layer chromatography with Silica Gel G as the absorbent, and programmed-temperature gas–liquid chromatography.

Mechanical Tests

Mechanical properties that are tested include tensile strength and stress, elongation, impact strength, abrasion resistance, and tear strength.

Table 1. Instrumental Polymer Analysis and Characterization

Method	Determination
chromatography	
gel-permeation chromatography	molecular weight
	molecular weight distribution
gas chromatography	composition
	structure
light scattering	molecular weight
	molecular size and shape
	solvent parameters
ultracentrifugation	molecular weight
	molecular weight distribution
thermal analysis	
differential thermal analysis	glass-transition temperature
	melting temperature
	degree of crystallinity
	purity
thermogravimetric analysis	weight loss
	degradation mechanisms
	reaction kinetics
	activation energies
	thermal stability
nuclear magnetic resonance	structure
	configuration
	molecular dynamics
mass spectrometry	polymer composition
	polymer structure
	degradation mechanisms
	degradation products
infrared spectroscopy	structure
	configuration

Flammability and Fire Tests

Tests include ease-of-ignition test, flame spread, heat-release test, limiting-oxygen-index test, determination of the relative amounts of smoke produced by burning or thermal decomposition, analysis and evaluation of toxic gas emissions produced by pyrolysis or combustion of polymers, and fire-resistance tests (see Flame retardants).

Medical Applications

The performance of surgical-implant plastic materials in the body depends on the chemical, mechanical, and physical properties of the polymer and the conditions these materials are subjected to in service. There is a need to understand the relationship between performance and molecular weight, molecular weight distribution, purity, additives, and chemical reactivity. Surface charge, morphology, crystallinity, surface area, and pore size affect the suitability of a plastic material to serve as an implant as well as its interaction with blood and tissue. Polymers that have use in surgical-implant applications include polyolefins, polyurethanes, silicones, and epoxy resins and hydrogels (see Prosthetic and biomedical devices). Test methods are under development by the ASTM Committee F-4 on Medical and Surgical Materials and Devices and are published in the 1984 ASTM Book of Standards Section 13.

Miscellaneous Tests and Standards

The ISO Technical Committee 61 on Plastics has developed and promulgated more than 100 standards, many of which describe test methods for analysis and evaluation of plastic materials. Information may be obtained from the ANSI. Standards for plastics used in electrical applications are contained in the International Electrotechnical Commission catalogue of publications.

The World Index of Plastics Standards published by the Department of Commerce contains more than 9000 national and international standards on plastics and related materials. Department of Defense Index of Specifications and Standards (DODISS) lists unclassified Federal, military, and departmental specifications and standards (Group 9330) dealing with fabricated plastic materials, including resins.

EMANUEL HOROWITZ
The Johns Hopkins University

F.A. Bovey and F.H. Winslow, *Macromolecules, An Introduction to Polymer Science*, Academic Press, Inc., New York, 1979.

E.A. Collins, J. Bares, and F.W. Billmeyer, Jr., *Experiments in Polymer Science*, John Wiley & Sons, Inc., 1973.

C.D. Craver, ed., *Polymer Characterization, Advances in Chemistry Series No. 203, Proceedings 181st Meeting, American Chemical Society, Atlanta, 1981*, ACS, 1983.

1984 Annual Book of ASTM Standards, ASTM, Philadelphia, Pa., 1984, Section 8.

PLATINUM-GROUP METALS

The six members of the platinum group, ruthenium, rhodium, palladium, osmium, iridium, and platinum, exhibit distinctive properties, including resistance to chemical attack, excellent high temperature characteristics, the ability to catalyze chemical reactions, and stable electrical properties (see Table 1) (see Catalysis).

Sources and Production

The Republic of South Africa, the USSR, and Canada account for nearly all of the world's newly mined, or primary platinum-group metals.

Secondary platinum-group metals. Secondary metals are recovered by refining precious-metal scrap, used equipment, and spent catalysts. These metals can be of the same high purity as primary platinum-group metals.

Recovery Techniques

Concentration. Although alluvial deposits are the result of natural concentration processes involving weathering and gravity-separation phenomena, considerable additional treatment is required to yield a product suitable for marketing and refining. In South Africa, the dredges employing sluice boxes produce a rough concentrate which is later amalgamated to remove the gold. Finally, the residue is given a countercurrent wash to remove the black sand and leave a high grade platinum concentrate (see Gravity concentration).

In the Merensky horizon in South Africa, gravity and flotation (qv) methods produce a high grade concentrate and a sulfide fraction that also contains various base metals.

Refining. Most refining procedures take advantage of the ready solubility of gold, platinum, and palladium in aqua regia and the ease with which gold can be reduced to the metallic form from the chloride solution by the addition of ferrous salts or sulfur dioxide. However, in some refineries the gold is separated by solvent extraction methods (see also Gold).

Health and Safety Factors

The widespread use of platinum metals and their alloys in dental and medical devices and in jewelry would strongly indicate that the metals themselves are not toxic. When working with finely divided metals, inhalation should be avoided. Osmium tetroxide causes ocular disturbances, probably by reduction and subsequent deposition of lower-valent oxide on the cornea. It can cause dermatitis and ulceration of the skin on contact.

Uses

Chemical and laboratory applications. Platinum has many uses in chemical laboratories because of its resistance to chemical attack. It is also used in single-crystal growth, catalysts, for electrical and electronic applications, in the glass industry, in jewelry and decorations, and for dental and medical use. Currently, a large use is as automotive catalysts. A key use in the future is anticipated to be as a catalyst for fuel cells (see Exhaust control, automotive; Batteries).

Commercial forms. The commercial forms in which the platinum-group metals (PGM) are available depend on their ease of workability. Platinum and palladium can be worked with relative ease either hot or cold to a wide range of useful shapes and sizes. Rhodium and iridium are worked hot. Ruthenium has some limited hot workability, but it normally is fabricated by powder-metallurgy (qv) techniques. Osmium forms

Table 1. Physical Properties of the Platinum-Group Metals

Property	Platinum	Palladium	Iridium	Rhodium	Osmium	Ruthenium
crystal structure at (°C)	fcc (25)	fcc (20)	fcc (20)	fcc (25)	hexagonal (25)	hexagonal (20)
a, nm	0.39231	0.389	0.384	0.3803	0.27341	0.27056
specific gravity, g/cm³	21.45	12.02	22.65	12.41	22.61	12.45
linear coefficient of thermal expansion at 20°C per °C	9.1×10^{-6}	11.1×10^{-6}	6.8×10^{-6}	8.3×10^{-6}	6.1×10^{-6}	9.1×10^{-6}
specific heat at 0°C, J/g[a]	0.1314	0.244	0.13	0.247	0.13	0.2306
thermal conductivity, W/(m·K)	71.1	75.3	146.4	150.6		
vapor pressure at melting point, Pa[b]	0.0187	3.47	0.467	0.133	1.8	1.31
electrical resistivity, μΩ·cm						
at 0°C	9.85	9.93	4.71	4.1	8.12	6.80
at 20°C	10.6	9.96	5.11	4.5	9.66	7.4
temperature coefficient of electrical resistivity, 0–100°C, per °C	0.003927		0.00427	0.0046	0.0042	0.0042
mass susceptibility, cm³/g	9.0×10^{-6}	5.23×10^{-6}	6.6×10^{-6}	0.99×10^{-6}	0.052×10^{-6}	0.43×10^{-6}
work function, eV	5.32	4.99		4.80		
tensile strength, annealed, MPa[c]	137.9	165.5	1103	758.6		4.96[d]
Young's modulus of elasticity, GPa[e]	172.4	117.2	524	344.8	558.6	413.8
elongation, %	25–40	24–30		30		
Vickers hardness, annealed	38–40	37–39	240[f]	120	400	220
Poisson's ratio	0.39	0.39	0.26	0.26		
Emf vs Pt at 1000°C, mV		−11.491	12.750	14.12		9.760

[a] To convert J to cal, divide by 4.184.

[b] To convert Pa to μm Hg, multiply by 7.5.

[c] To convert MPa to psi, multiply by 145.

[d] Hot-swaged bar.

[e] To convert GPa to psi, multiply by 145,000.

[f] Annealed at 1000°C, 0.5-mm dia wire.

Table 2. Physical Properties of Commercial Platinum Alloys

Alloy	Sp gr	Electrical resistivity, $\mu\Omega$-cm		Temperature coefficient of electrical resistivity, 20–100°C	Brinell hardness[a]	Tensile strength,[b] MPa[c]	Elongation, %
		At 0°C	At 20°C				
crucible grade[d]	21.4	10.6	11.4	0.0038	50[e]	165	30
0.3% Ir–Pt	21.4		12.7		50	234	24
0.8% Ni–Pt	21.1	11.9	12.2	0.0034	75	207	23
3.5% Rh–Pt	20.8	15.7	16.6	0.0025	75	241	28
10% Rh–Pt	19.97	18.5	19.2	0.0018	90	310	35
20% Rh–Pt	18.74	20.2	20.8	0.0015	120	483	40
10% Ir–Pt	21.53	24.4	25	0.0013	130	379	25
25% Ir–Pt	21.66	32.5	33	0.00065	240	862	20
5% Ru–Pt	20.67	31	31.5	0.0009	130	414	34
10% Ru–Pt	19.94	42.4	43	0.0008	190	586	31
4% W–Pt	21.3	34.5	35.1	0.0008	150[e]	517	25

[a] Baby Brinell, 2-mm ball, 120-kg load.
[b] Annealed.
[c] To convert MPa to psi, multiply by 145.
[d] > 99.5% pure.
[e] Vickers hardness.

a toxic oxide at ambient temperatures and is not hot- or cold-worked. The PGM are coated on base metals.

Sheet, strip, ribbon, foil, and seamless tubing are produced in a broad range of materials, sizes, and thicknesses.

The PGM are available as blacks, powders, catalysts, chemical compounds, solutions, and colloids.

Platinum, Palladium, Iridium, Rhodium, Osmium, Ruthenium

Properties. The physical properties of the PGM and their alloys are shown in Tables 1 and 2, respectively.

RICHARD D. LANAM
EDWARD D. ZYSK
Engelhard Corporation

Supply and Use Patterns For the Platinum-Group Metals, Publication NMAB-359, National Academy of Sciences, Washington, D.C., 1980, p. 26–28.
E. Savitsky, V. Polyakova, N. Gorina, and N. Roshan, *Physical Metallurgy of Plantinum Metals*, trans. by I.V. Savin, MIR Publications, Moscow, USSR, 1978, pp. 30–45.

PLATINUM-GROUP METALS, COMPOUNDS

The compounds of the platinum-group metals may be divided into three classes: binary salts, oxides, etc; coordination compounds (qv); and organometallic compounds (see Organometallics). The importance of the latter two classes has risen dramatically in recent years with the application of complexes of the metals in homogeneous catalysis and cancer chemotherapy (see Catalysis; Chemotherapeutics, antimitotic). The compounds also are used in oxidation, staining, and electroplating. Physical properties of selected compounds are listed in Table 1.

Table 1. Physical Properties of Selected Platinum-Group Metal Compounds

Compound	Density, g/cm³	Mp, °C	Bp, °C	Color	Crystal form	Solubility
H_2PtCl_6	2.43	60		red to red brown		v sol H_2O, alcohol
$(NH_4)_2PtCl_6$	3.06	dec ≥ 380		yellow-orange	cubic	0.69 g/100 mL H_2O at 16°C; insol alcohol
$PtCl_2$	5.87	dec 581		olive green	hexagonal	insol H_2O, alcohol; sol HCl, NH_4OH
PtF_6	3.83	61.3	69.14	dark red-black	rhombic	
K_2PtCl_4	3.382	dec > 500		red	tetragonal	sol H_2O; insol alcohol
PtO	14.9	dec 560		black		insol H_2O, alcohol; sol aqua regia
cis-[$Pt(NH_3)_2Cl_2$]	3.738	dec 270		yellow		0.253 g/100 mL H_2O at 25°C
$PdCl_2$	4.0	680	dec > 680	red-brown	rhombic	insol H_2O[b]; sol HCl
$Pd(CH_3COO)_2$		205 (with dec)		orange-brown		insol H_2O, alcohol; sol $CHCl_3$, acetone, ether
PdO	8.3	dec > 700		black		insol H_2O, alcohol
$IrCl_3$	5.30	dec 763		red/brown[a]	rhombic/monoclinic[a]	insol H_2O[b], acids, alkalies
[$IrCl(CO)((C_6H_5)_3P)_2$]		215		yellow		insol H_2O, alcohol; sol benzene, toluene, $CHCl_3$
$RhCl_3$	5.38	dec 450		red	monoclinic	insol H_2O[b]; sol KOH, KCN
[$RhCl((C_6H_5)_3P)_3$]				maroon		sol $CHCl_3$, CH_2Cl_2, benzene, toluene
$Rh_2(SO_4)_3$		dec > 500		yellow/red[a]		
$OsCl_3$		dec > 450		dark gray	cubic	insol H_2O; sol HNO_3
OsO_2	11.4			dark blue-black	rutile	insol H_2O, acids
OsO_4	4.91	40.6	131.2	pale yellow	monoclinic	7.24 g/100 mL H_2O at 25°C, sol alcohol, CCl_4, ether
$K_2[OsO_2(OH)_4]$		$-H_2O$, 200		violet	rhombic	sol H_2O; insol alcohol, ether
$RuCl_3$	3.11	dec > 500		brownish-red	hexagonal	insol H_2O[b]
RuO_2	7.0	dec		dark gray-black	tetragonal	insol H_2O, acids; sol fused alkali
RuO_4	3.29	25.4	40	golden yellow	monoclinic	2.03 g/100 mL H_2O at 20°C, sol CCl_4
K_2RuO_4		$-H_2O$, 200		black with green luster	rhombic	v sol H_2O
$KRuO_4$		dec 400		black	tetragonal	sol H_2O

[a] Two forms.
[b] Hydrated form is water soluble.

Health and Safety Factors, Toxicology

Extremely small quantities of anionic salts and complexes of platinum may cause allergic reactions, ie, platinosis, in sensitive individuals.

The most prominent toxic effect of cis-$[Pt(NH_3)_2Cl_2]$ is damage to the renal tubules. Other important effects are bone-marrow depression, damage to the gastrointestinal epithelium, and damage to the spleen.

Compounds of the platinum metals generally are toxic to the kidneys. Other toxic manifestations that have been observed in animals are heart and liver damage, hemolysis, albuminurea, and diuresis (with $PdCl_2$), and peripheral vasoconstriction (with $PdSO_4$).

The tetroxides of ruthenium and osmium are extremely volatile; their vapors affect the eyes and lungs of experimental animals and humans. Ruthenium tetroxide has been less thoroughly studied but should be regarded as similarly hazardous.

The use of platinum metals in automobile exhaust catalysts appears to pose no demonstrable hazard to humans. OSHA exposure limits, which are based on 24 h exposure, are 2.0 $\mu g/m^3$ of airborne platinum salts and osmium tetroxide and 1.0 $\mu g/m^3$ of airborne soluble rhodium salts (see Exhaust control, automotive).

ALAN R. AMUNDSEN
ERIC W. STERN
Engelhard Corporation

F.R. Hartley, *The Chemistry of Platinum and Palladium*, John Wiley & Sons, Inc., New York, 1973.

W.P. Griffith, *The Chemistry of the Rarer Platinum Metals*, Wiley-Interscience, New York, 1967.

J.W. Hightower, H.E. Griffin, and co-workers, *Medical and Biological Effects of Environmental Pollutants—Platinum Group Metals*, National Research Council, Washington, D.C., 1977.

PLUTONIUM AND PLUTONIUM COMPOUNDS

Plutonium

Plutonium, Pu, is element number 94 in the periodic table. It is a member of the actinide series and is metallic. Isotopes of mass number 232 through 246 have been identified and all are radioactive. The most important isotope is plutonium-239; also of importance is plutonium-238 (see Actinides and transactinides).

The large energy release associated with the fission reaction is the most significant property of plutonium. The energy can be applied in electric-power generating reactors, industrial explosives, or military explosives. Large quantities of plutonium are produced in uranium-fueled power reactors (see Nuclear reactors). Approximately 200 kg Pu are produced per 1000 MW_e of electric power.

The isotope ^{238}Pu also is of technical importance because of the high heat of its constant radioactive decay. Such radiation is used as fuel in small terrestrial and space nuclear-power sources (see Thermoelectric energy conversion).

Plutonium is unique in that it is the first element to be synthesized.

Sources. From a practical viewpoint, plutonium does not occur in natural ores. All of the fifteen plutonium isotopes have been synthesized and all are radioactive (see Radioisotopes). Technologically, plutonium-239 is the most important isotope. It is characterized by a high fission reaction cross-section and is abundant in irradiated natural uranium.

Commercial electric-power generating reactors generally produce plutonium by irradiating uranium fields to a total neutron exposure of more than 5000 megawatt-days per ton ($MW \cdot d/t$). The recoverable plutonium contains a larger fraction of heavier isotopes. A large future source of plutonium will be from fast-neutron breeder reactors.

Physical Properties

Thermal. The expansion coefficient of α-plutonium is exceptionally high for a metal, whereas those of δ- and δ'-plutonium are negative. The net linear increase in heating a polycrystalline rod of plutonium from room temperature to just below the melting point is 5.5%.

Thermodynamic. The thermal conductivity of plutonium-242 is 0.084 and 0.155 $W/(m \cdot K)$ (0.020 and 0.037 $cal/(s \cdot cm \cdot °C)$) for the α- and β-phase, respectively.

Electrical and magnetic. The electrical resistivity of plutonium is high in all modifications as a result of the band structure in metallic plutonium.

Radioactive self-heating. Because of their radioactivity, all plutonium isotopes generate heat. The self-heating of a ^{239}Pu metal sphere has been determined: $(1.923 + 0.019) \times 10^{-3}$ W/g $(1.824 + 0.18$ Btu/$(s \cdot g))$.

Spectroscopic. Isotope shifts of the 238, 239, and 240 isotopes were obtained in the spark spectrum, reexamined with hollow-cathode spectral data, and supplemented by ^{241}Pu data. Based on the shifts of 20 lines of Pu I, and assuming the 238–240 interval as unity, the following shift values were obtained:

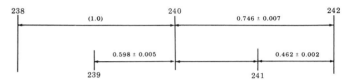

Chemical Properties

Plutonium is the fifth member of the actinide series. Its electronic structure is

$$1s^2 2s^2 2p^6 3s^2 3p^6 3d^{10} 4s^2 4p^6 4d^{10} 4f^{14} 5s^2 5p^6 5d^{10} 5f^{5-6} 6s^2 6p^6 6d^{0-1} 7s^2$$

Plutonium is an active metal and, as such, it can be made by the usual methods for active metals such as electrolysis in fused salts, reduction of its halides with an active metal such as Li, Ca, Ba, or reduction of the oxide with Ca in a $CaCl_2$ melt. Plutonium forms compounds with all the nonmetallic elements except the rare gases. The halogens and halogen acids form Pu halides, other chalcogens form chalcogenides, and CO forms a carbide. Nitrides are formed with NH_3 and N_2 and hydrides with H_2.

The metal dissolves readily in concentrated HCl, H_3PO_4, HI, $HClO_4$, or HSO_3NH_2.

The corrosion of plutonium has been summarized. Plutonium oxidizes very slowly in dry air, ie, $< 10^{-2}$ mm/yr; the rate is accelerated by water vapor.

Extractive metallurgy and conversion chemistry. The production methods used for the separation of Pu from U and fission products are liquid–liquid extraction and ion exchange (qv) (see Extraction).

Plutonium Compounds

Plutonium forms compounds with many of the metallic elements and all of the nonmetallic elements except the helium-group gases. The refractory compounds are of interest as potential fast-breeder reactor fuels. Commercially important compounds include the oxides, halides, and oxalates.

Health and Safety Factors

The principal hazards of plutonium are those posed by its radioactivity, its nuclear critical potential, and its chemical reactivity in the metallic state.

F. WEIGEL
University of Munich

Gmelin Handbuch der Anorganischen Chemie, System Nr. 71, Transurane, Springer-Verlag, Frankfurt, FRG, 1973–1979.

O.F. Wick, *Plutonium Handbook*, Gordon & Breach Science Publishers, New York, 1967.

W.T. Carnall and G.R. Choppin, *Plutonium Chemistry ACS Symposium Series No. 216*, ACS, Washington, D.C. 1983.

E.L. Christensen and W.J. Maraman, *Plutonium Processing at the Los Alamos Scientific Laboratory*, LA-3542 (April 1969).

F. Weigel in J.J. Katz, G.T. Seaborg, and L. Morss, eds., *The Chemistry of the Actinide Elements*, Chapman and Hall Ltd., London, UK, to be published in 1985, Chapt. 8.

PLYWOOD. See Laminated and reinforced wood.

POISONS, ECONOMIC

Economic posions are chemicals that are intended for the control, suppression, or destruction of plants or animals that are of economic significance as pests; pesticide has become a widely used synonym. The technology of pesticide development has resulted in the use of chemicals to control fungi (fungicides (qv)), chemicals to control insects (insecticides) (see Insect control technology), and chemicals to control weeds (weed killers or herbicides (qv)). In addition, there are other groups of specific pest-control agents, eg, molluscicides, nematocides, rodenticides, and chemicals for the control of certain obnoxious fish, birds, and predatory animals (see Repellents).

Invertebrate

Molluscicides. Terrestrial. Metaldehyde $(CH_3CHO)_3$, is used widely for the control of slugs and snails in crops. It is produced by the polymerization of acetaldehyde (qv) in ethanol in the presence of an acid. Its solubility at $17°C$ is 200 mg/L water, it is soluble in benzene and chloroform, and is moderately soluble in ethanol and diethyl ether. The most commonly used formulation is a bait containing 2.5–4.0 wt% metaldehyde in protein-rich bran or pellets.

The most commonly used carbamate insecticide is Methiocarb, 3,5-dimethyl-4-(methylthio)phenyl methylcarbamate, which is used in bait pellets at 200 g active ingredient per 10^4 m^2.

Aquatic. Copper sulfate, $CuSO_4$, Cuprobam, copper dimethylthiocarbamate, and the fungicide Ziram, zinc bis(dimethyldithiocarbamate), are effective against most aquatic snails at 3–5 ppm in water. The dinitrophenols also are effective at 3–5 ppm.

Other products include triphenyltin hydroxide, $(C_6H_5)_3SnOH$, triphenyltin acetate, $(C_6H_5)_3SnOC(O)CH_3$, Bayluscide, 5,2′-dichloro-4-nitrosalicylanilide, mp 199°C, and Frescon, N-tritylmorpholine, mp 174–176°C.

Nematocides. The first practical soil fumigant was chloropicrin, O_2NCCl_3, bp 112°C, d^{20} 1.651 g/cm³, which was evaluated as a nematocide in England in 1919 and was later shown to be a practical nematocide when injected 15–20-cm deep at 170–190 kg/ha. It is particularly useful and economically advantageous where both nematodes and fungi must be controlled.

Halogenated. Halogenated nematocides include D–D mixture, which consists of two parts of 1,3-dichloropropene, $ClCH=CHCH_2Cl$, bp 111°C, d^{20} 1.224 g/cm³, and one part of 1,2-dichloropropane, $ClCH_2CHClCH_3$, bp 95.4°C, d^{20} 1.159 g/cm³, 1.2-Dibromoethane, $BrCH_2CH_2Br$, ethylene dibromide (EDB), bp 131.6°C, d^{20} 2.172 g/cm³, and methyl bromide, CH_3Br, bp 4.5°C, d^0 1.732 g/cm³.

Others. Other nematocides include methyl isothiocyanate bp 88°C, d 1.520 g/cm³, compounds of the organophosphorus insecticides, and the carbamates Temik, 2-methyl-2-methylthiopropionaldehyde, (methylcarbamoyl)oxime, Furadan, 2,3-dihydro-2,2-dimethylbenzofuran-7-yl N-methylcarbamate, and Bydate, or N,N-dimethyl-2-methylcarbamoyloxymino-2-(methylthio)acetamide.

Vertebrate

Rodenticides. Acute. The classic acute rodenticides act within minutes or, at most, a few hours after a single ingestion. Acute rodenticides include zinc phosphide Zn_3P_2, LD^{50} in rats of 45.7 mg/kg, thallium sulfate, Tl_2SO_4, LD^{50} in rats of 22.5 mg/kg, Red Squill or scilliroside, LD^{50} in male and female rats of 0.7 mg/kg and 0.43 mg/kg, strychnine, and strychnidin-10-one.

strychnine

Antu,$α$-naphthylthiourea, LD^{50} in *Rattus norvegicus* is 6 mg/kg, fluoroacetamide, $FCH_2C(O)NH_2$, LD^{50} in *R. norvegicus* of 13 mg/kg, Crimidine or Castrix, 2-chloro-4-dimethylamino-6-methylpyrimidine, LD^{50} 1.25 mg/kg in rats, Norbormide, 3a,4,7,7a-tetrahydro-5-(hydroxyphenyl)-2-pyridylmethyl)-8-(phenyl-2-pyridinylmethylene)-4,7-methano-H-isoindole-1,3(2H)-dione, LD^{50} 52 mg/kg in rats, 11.5 mg/kg in *R. norvegicus*, and ca 10 mg/kg in *R. exulans*, and pyriminil, or Vacor, N-(3-pyridinylmethyl) N'-(4′-nitrophenyl)urea, LD^{50} in rats 4.75 mg/kg.

Chronic. Chronic poisons generally require several feedings to produce a lethal effect.

Anticoagulants. Difenacoum and brodifacoum are new hydroxycoumarin anticoagulants and are effective against rodents resistant to the earlier chronic toxicants. These compounds also are considerably more potent than earlier anticoagulants. Warfarin has a subacute oral LD^{50} of five daily dosages of 2 mg/kg in *R. norvegicus*. For difenacoum, the acute, single feed LD^{50} is 1.8 mg/kg and that for brodifacoum is only 0.4 mg/kg.

Calciferol, Vitamin D_2, is essential to normal development in all mammals but, in high doses, it is lethal. Death is caused primarily by hypercalcemia and generally occurs 3–10 days after the consumption of a toxic dose. This combination is semiacute; death follows feeding on 1–3 occasions; thus, this compound bridges the demarcation between acute and chronic rodenticides. Calciferol antidote procedure is symptomatic.

Miscellaneous

Sterilants. Diethylstilbestrol has been used successfully as a contraceptive for the coyote. As an anticholesterol agent, 22,25-diazacholestanol dihydrochloride prevents reproduction in pigeons for up to six months (see Contraceptive drugs; Hormones). Mestranol, which is an estrogen contraceptive, reduces rodent reproductive capacity when fed as a grain bait at 0.1–0.5 wt%.

Narcotics and immobilizing agents. In some instances, particularly where pest birds are concerned, it is necessary to narcotize a population so that pest species can be removed by hand and desirable species revived and released. $α$-Chloralose has been used for this purpose for many years in baits containing 1.5 wt% active ingredient.

Lampreycides. Lampreys are parasitic fish that cause severe losses in commercial fisheries. Two synergistic compounds, 3-trifluoromethyl-4-nitrophenol (TFM) and 5,2′-dichloro-4′-nitrosalicylanilide (DCN) or Bayluscide (see Molluscicides) provide effective control of sea lampreys without appreciable damage to other aquatic organisms.

Avicides. Avicides are poisons used to control bird pest species.

Narcotics. The use of narcotics such as $α$-chloralose and 2,2,2-bromoethanol has already been described.

Perch toxicants. Endrin and fenthion are both used to impregnate perches at 9.4 and 11%, respectively.

Repellents. Repellents (qv) typically are applied to a growing crop as sprays, or to seed as a dressing prior to planting. Repellents include 4-aminopyridine, methiocarb or Mesural, aldicarb or Temik, and anthraquinone or Morkit.

Predacides. Predatory carnivores can cause considerable losses of domesticated animals. The most common use of chemicals for predator control is in the form of poisoned baits. This technique requires skill and perseverance; poisoned meats represent a considerable hazard to all wildlife.

Sheep collars. In the United States, a selective predator-control technique has been developed whereby collars containing a toxicant are attached to target sheep. Coyotes typically grasp their prey at the neck and, in so doing, ingest the contents of the collar. Sodium cyanide, sodium fluoroacetate, and diphacinone have been evaluated as active ingredients for this technique.

Health and Safety Factors

All pesticides are, by definition, poisons and, to date, a totally safe pesticide has not been developed. All of the aforementioned compounds pose a degree of hazard to organisms other than the chosen pest target. All personal contact with the product should be avoided by the use of gloves, respirators, goggles, and overalls when handling the product. Strict personal hygiene such as washing after handling, and before eating or drinking, should be practiced in all cases. In the event that an accidental absorption of toxicant does occur, antidotes are available for many of the products listed but should only be administered under medical supervision.

MALCOLM R. HADLER
Sorex Limited

G. Thorne, *Principles of Nematology*, McGraw-Hill Book Co., Inc., New York, 1961.

H. Martin, *The Scientific Principles of Crop Protection*, Edward Arnold Ltd., London, UK, 1959.

H. Martin, "Guide to the Chemicals Used in Crop Protection," *Canadian Department of Ariculture Publ. 1093*, 4th ed., 1961.

A.C. Dubock and E.K. Kaukeinen, *Proc. 8th Vertebr. Pest Conf.*, Sacramento, *Calif.* **8**, 127 (1978); G.E. Connoly and co-workers, p. 197.

POLISHES

Polishes are used to produce or restore a glossy finish on various surfaces as well as prolong the useful lives of those surfaces. The appearance enhancement provided by polishes generally results from the presence in the polish of components that leave a glossy coating, and/or materials that smooth and clean surfaces. Floor, furniture, and shoe polishes rely on the deposition of a film. Applications of car-polish formulas leave glossy and protective films and contain abrasives to remove weathered paint and soils. Metal polishes are based on either abrasive smoothing and cleaning or tarnish-removing chemicals, and they sometimes deposit materials that retard future tarnishing (see Abrasives).

Furniture

There are five different types of marketed furniture polishes: liquid or paste solvent waxes, clear oil polishes, emulsion oil polishes, emulsion wax polishes, and aerosol or spray polishes (see Aerosols). Paste waxes contain ca 25 wt% wax, the remainder being solvent. Clear oil polishes contain 10–15 wt% oil, a small amount of wax, and the rest solvent. Aerosol or spray products may contain 2–5 wt% of a silicone polymer, 1–3 wt% wax, 0–30 wt% hydrocarbon solvent, and ca 1 wt% emulsifier; the remainder is water (see Waxes; Silicon compounds).

Floor

Aqueous, self-polishing, polymeric floor polishes contain two or three polymeric film formers, coalescents, leveling aids, plasticizers, zinc complexes, ammonia, and wetting and emulsifying agents. An aqueous formula may contain 0–12 wt% polymer, 0–12 wt% resin, 0–6 wt% wax, 0.3–1.5 wt% tris(butoxyethyl) phosphate, 1–6 wt% glycol ether, 0–1 wt% zinc, and the rest is water.

Automobile

Much of the automobile-polish market is represented by one-step products. One-step products generally contain four functional ingredients: abrasives, straight- and branched-chain aliphatic hydrocarbons, waxes, and silicones. A representative liquid emulsion product might contain 10–15 wt% abrasive, 10–30 wt% solvent, 2–12 wt% silicone, and 0–4 wt% wax; an emulsion paste product might contain 3–15 wt% wax and similar amounts of other ingredients.

Metal

Formulated metal polishes consist of fine abrasives similar to those involved in industrial buffing operations, ie, pumice and tripoli. Other ingredients include surfactants, chelating agents (qv), and solvents.

Many modern polishes contain inhibitors to prevent oxidation and tarnishing. Metal polishes may contain emulsifiers and thickeners for control of consistency and stabilization of abrasive suspension, and the product form can be solid, paste, or liquid. A representative liquid emulsion product may contain 8–25 wt% abrasive, 2–6 wt% surfactant,

0–5 wt% chelating agents, 0–25 wt% solvent, and the remainder water. The abrasive content in an emulsion paste product is greater than in a solvent product.

Shoe

Three general types of shoe polishes are produced: solvent pastes, self-polishing liquids, and emulsion creams. Solvent pastes represent ca 60% of the market.

Health and Safety Factors

Liquid polishes and waxes containing 1 wt% or more petroleum distillates must be contained in childproof packaging. General experience indicates that natural waxes and polyethylene waxes are nontoxic.

FRANCIS J. RANDALL
SEAN G. DWYER
S.C. Johnson & Son, Inc.

W.J. Hackett, *Maintenance Chemical Specialties*, Chemical Publishing Co., New York, 1972.

E.W. Flick, *Household and Automotive Chemical Specialties—Recent Formulations*, Noyes Data Corporation, Park Ridge, N.J., 1979.

POLLUTION. See Air, noise, water pollution.

POLYAMIDES

GENERAL

Polyamides are condensation products that contain recurring amide groups as integral parts of the main polymer chains. Linear polyamides are formed by condensation of bifunctional monomers. If the monomers are amino acids, eg, 6-aminohexanoic acid, the polymers are called AB types (A representing amine groups and B representing carboxyl groups). If the polymers are formed from condensation of diamines and dibasic acids, they are called AABB types. Polyamides are frequently referred to as nylons. Although they generally are considered as condensation polymers, polyamides also are formed by addition polymerization. This method of preparation is especially important for some AB polymers, in which the monomers are cyclic lactams, eg, ϵ-caprolactam, hexahydro-2H-azepin-2-one or 2-pyrrolidinone, γ-aminobutyrolactam.

Typical structural formulas of linear polyamides may be represented as

(1)

or, in the case of self-condensation of an amino acid

(2)

where R and R′ represent chains between functional groups in the reactants and n represents the degree of polymerization or number of recurring groups in the polymer chain.

Properties

Crystallinity. Symmetrical, hydrogen-bonded, linear polyamides invariably are highly crystalline and owe their excellent mechanical behavior to this property. Yield point, tensile strength, elastic and shear moduli, hardness, and abrasion resistance increase with increasing crystallinity, whereas moisture absorption and impact strength drop slightly.

Solubility. Aliphatic polyamides, eg, polyhexamethyleneadipamide, nylon-6,6, usually are soluble in phenols, formic acid, chloral hydrate, mineral acids, and similar substances at room temperature. At higher temperatures, alcohol–halogenated hydrocarbon mixtures, unsaturated alcohols, nitro alcohols, and calcium chloride–methanol mixtures often are solvents for aliphatic polyamides. Some copolymers, eg, nylon-6,6–6,10 (40:60), are soluble in methanol–chloroform; however, the respective homopolymers are not. Nylon-6,6 also is soluble in methanol under pressure. Aliphatic–aromatic polyamides are slightly less soluble than the aliphatic polyamides and are soluble in trifluoroacetic and sulfuric acids.

Hydrolysis. Most nylon fibers and plastics are unaffected by room temperature or boiling water; although at higher temperatures, and especially if the nylon is in the melt, hydrolysis and degradation occur.

Acidolysis and aminolysis. Polyamides undergo rapid acidolysis and degradation on heating with monobasic acids at elevated temperature, especially in the melt. Heating of nylon-6,6 with hexamethylenediamine results in rapid reduction of molecular weight, presumably by aminolysis.

Substitution of the amide hydrogen. Heating of polyamides in an autoclave with ethylene oxide, oxirane, results in the formation of hydroxyethylated polyamides, which are characterized by high elasticity and vapor permeability. The products are block copolymers. Ethylene carbonate, 1,3-dioxolan-2-one, also reacts with polyamides to form block copolymers by reaction with both amino and carboxyl end groups and with the amide NH groups. Nylons-6,6 and -6 yield water-soluble products.

Elastic *N*-methoxymethylated polyamide nylons have been prepared by treating nylon-6,6 with formaldehyde and methanol in the presence of an acidic catalyst.

Formaldehyde reacts with polyamides in the solid state or in the formic acid solution to yield *N*-methylol derivatives.

Grafting with unsaturated acids. Exposure of polyamides to high intensity radiation, eg, α rays, x rays, etc, leads to the formation of free radicals, which can be used as sites for polymerization with vinyl monomers, on the polymer chain:

$$-CH_2\overset{\overset{O}{\|}}{C}NCH_2- \longrightarrow -CH_2\overset{\overset{O}{\|}}{C}N\overset{\cdot}{C}H- \xrightarrow{CH_2=CHR} -CH_2\overset{\overset{O}{\|}}{C}NCH-\left[CHCH_2\right]_n$$
$$\qquad\ \underset{H}{}\qquad\qquad\quad \underset{H}{} \qquad\qquad\qquad\quad \underset{H}{}\ \ \underset{R}{}$$

(3)

Degradation. Thermal. The rate and type of thermal degradation depend on temperature, polymer structure, and whether the heating is carried out in the presence or absence of oxygen.

The order of thermal stability of aromatic polyamides is (4) < (5) < (6) < (7). Thermal degradation of aromatic polyamides in the absence of oxygen probably involves direct cleavage of the amine, ie, the C—N bond. Degradation in the presence of oxygen is more complex because the polymers cross-link rapidly, possibly by free-radical coupling of the aromatic rings.

(4)   (5)   (6)   (7)

Light. Exposure to sunlight causes a deterioration in the properties of many synthetic and natural textile materials. There is an accompanying reduction in the molecular weight and slight changes in chemical constitution for the polymer in individual fibers. Chain-breaking and cross-linking occur in the polymer, with the latter occurring chiefly at shorter wavelengths.

High intensity radiation. Although nylon is cross-linked primarily by ionizing radiation, scission must also play an important role since the degree of cross-linking saturates at a low radiation value. Differences in radiation effects have been attributed to the absence or presence of oxygen. Also, pressure affects the rate of diffusion of the oxygen and subsequent reaction with the radicals formed by irradiation.

Structure-property relationships. Attempts have been made to systematize the data associated with the physical properties of condensation polymers and to correlate polymer structure with properties. It is assumed that polymers of sufficiently high molecular weight are compared, so that differences in molecular weight cause negligible differences in the properties of the polymer types.

Linear polyethylene has been considered to be the parent polymer of all other classes of polymers. Insertion of various polar groups or ring structures in the hydrocarbon chain considerably changes the softening point from that of polyethylene, ie, ca 132°C. In general, introduction of urea, amide, and urethane groups yields polymers that melt at higher temperatures than polyethylene, whereas the aliphatic polyesters melt at considerably lower temperatures (see Olefin polymers).

Effects of rings in polyamide chains. When an aliphatic section of a polyamide chain is replaced by a ring segment, the flexibility of the chain decreases and the melt temperature increases because of the decrease in the entropy factor, ΔS. Introduction of aromatic rings generally causes decreased flexibility, solubility, and moisture absorption.

Copolymerization. The melt temperature of a polymer decreases when the regularity with which the monomer groups are spaced along the backbone chain decreases. This occurs in random copolymerization.

Aromatic polyamides. The aromatic polyamides of concern are aromatic in both the diamine and the dibasic acid. Many of the polymers decompose at ca 400°C prior to melting so that the effects of structure on melt temperature are difficult to determine. However, data show that para polymers are highest melting and that replacement of ortho by meta and meta by para raises the melt temperature by ca 50°C.

Manufacture

Processes include direct amidation, acid chloride reaction, and ring-opening polymerization.

Polyhexamethyleneadipamide, Nylon-6,6

Properties. Bulk nylon-6,6 normally is a tough, white, translucent crystalline material, mp 265°C. An amorphous, clearer form is obtained if the melt is quenched rapidly below room temperature, but this is unstable and crystallizes on standing at room temperature or if annealed. Physical properties of nylon-6,6 are listed in Table 1.

Poly(ε-caprolactam), Nylon-6

Properties. Typical properties for nylon-6 are shown in Table 1.

Other AB polyamides

In recent years AB polyamides other than polycaprolactam have resulted from improved lactam syntheses and new polymerization technology; none, however, has assumed the commercial importance of nylon-6. These polymers include polypropiolactams, nylon-3; poly(pyrrolidin-2-one), nylon-4; poly(ω-enanthamide), nylon-7; polycapryllactam, nylon-8; poly(ω-pelargonamide), nylon-9; poly(aminodecanoic acid), nylon-10; and poly(ω-undecaneamide), nylon-11.

Polyamides from Long-Chain Fatty Acids

Another class of polyamide resins is based on condensation of diamines or triamines, eg, ethylenediamine or diethylenetriamine, *N*-(2-aminoethyl-1,2-ethanediamine, with relatively high molecular weight dibasic acids or esters, eg, the dimer acid obtained from thermal polymerization of linoleic acid, 9,11-octadecadienoic acid.

The polyamides vary from tacky to high melting (175°C) resins and are used as heat-sealing adhesives and as moisture-barrier coatings for paper, wood, cellophane (see Barrier polymers), etc. The polymer coatings are applied as solutions, dispersions, or powders.

Table 1. Physical Constants of Nylon-6,6 and Nylon-6

Property	Nylon-6	Nylon-6,6
Melting point, °C		
at equilibrium	200–220	250–260
α-crystalline	231	
pressure dependence	260	< 270
Density, g/cm³		
crystalline		
α, monoclinic	1.24	
α, monoclinic	1.23	
α, monoclinic	1.21	
γ, hexagonal	1.13	
γ, monoclinic	1.17	
γ, pseudohexagonal	1.155	
α, triclinic		1.220
α, triclinic		1.24
β, triclinic		1.248
amorphous		
γ	1.09	
α	1.11	
Refractive index, n_D		
single crystals		
$α^a$		1.475
$β^a$		1.565
$γ^b$		1.58
moldings	1.53	1.53

[a] Calculated.
[b] Observed.

Partly Aromatic Polyamides

A typical example of an aliphatic–aromatic polyamide that is a more thermally stable polymer than an aliphatic–aliphatic polyamide and with better property retention at moderately high temperatures is nylon-6,T. However, none of the aliphatic–aromatic polyamides has yet assumed commercial importance, because the higher melting ones, eg, nylon-6,T, are difficult to prepare and fabricate and do not offer the long-term thermal stability of the all-aromatic polyamides.

Wholly Aromatic Polyamides

The need for fibers with a high degree of heat and flammability resistance was met in 1959 with the introduction of aramid fibers (qv). Aramid fibers are made from long-chain synthetic aromatic polyamides in which at least 85% of the amide linkages are attached directly to two aromatic rings.

Properties. In general, aromatic polyamides are colorless, high melting ($T_m > 300°C$), and crystalline. Many aromatic polyamides discolor on exposure to light; however, the discoloration apparently acts as an effective uv shield and tends to reduce the rate of strength loss.

Aromatic polyamide fibers, like their counterparts, are degraded by uv radiation. However, the resistance of wholly aromatic amides to ionizing radiation is greatly superior to that of nylon-6,6. Aramid fibers also are more resistant than nylon-6,6 fibers to degradation from gamma radiation and x-rays.

Extended-chain polyamides. The rigorous property requirements for space-age applications have been satisfied by newer polymers of the aramid family that can be spun into high strength, high modulus fibers. These fibers are spun from very highly crystalline rigid-chain polymers that form extended-chain crystals. Kevlar aramid has a higher modulus and a higher tensile strength on an equal weight basis than glass or steel. Extended-chain aramids also have higher moduli or stiffness and tenacity than glass or steel, and a lower density. Extended-chain aromatic polyamides are used in reinforcing composite structures, including tires, bulletproof vests, boat hulls, drilling-platform anchor cables, and where advantage can be gained by replacing 5 kg of wire or 2 kg glass, nylon, or polyester fibers with 1 kg of Kevlar aramid fiber (see Aramid fibers).

As intermediates. Aromatic polyamides with substituents that can react to form heterocyclic rings have been used as soluble precursors to intractable, thermally stable polymers.

Production and Economic Aspects

Manufacturers of nylon polymer or fiber include E.I. du Pont de Nemours & Co., Inc., Monsanto Co., Celanese Corp., Allied Chemical Corp., American Enka Corp., The Firestone Tire and Rubber Co., Badische Corp., Courtaulds North America, Inc. Nylons are used chiefly in the production of synthetic fibers.

Advantages of nylons over other natural and synthetic fibers are relatively low specific gravity, high strength, and good durability. The four largest uses of nylon fibers are in tires, carpets, stockings, and upholstery.

Although current commercial polyamides fulfill many use requirements, certain deficiencies in selected applications, eg, moisture effects in wash-wear applications, insufficient dimensional stability in industrial plastics, electrostatic-charge buildup in carpets, etc, have given reason for research efforts toward product improvement and discovery of even better polyamide types (see Polyamides, polyamide plastics).

RICHARD E. PUTSCHER
E.I. du Pont de Nemours & Co., Inc.

M.I. Kohan, *Nylon Plastics*, John Wiley & Sons, Inc., New York, 1973, p. 48.
H.K. Reimschuessel, *J. Polym. Sci. Macromol. Rev.* **12**, 65 (1977).
P.E. Cassidy, *Thermally Stable Polymers Synthesis and Properties*, Marcel Dekker, New York, 1980.

POLYAMIDE FIBERS

Polyamide fibers have monomer units joined by amide groups $[CONHRNHCOR']_n$ and are prepared from diamines and dicarboxylic acids, or in the case of $[RCONH]_n$, from lactams. If R and R' are aliphatic, alicyclic, or mixtures containing less than 85 wt% aromatic moieties, the polyamides usually are referred to as nylon. If more than 85 wt% of the repeating units are aromatic in structure, the fibers are called aramids (see Aramid fibers).

Nylon-6 and Nylon-6,6

Properties. Tensile. The main properties of representative nylon-6 and -6,6 yarns are listed in Table 1.

The stress–strain properties of nylons depend to a large extent on spinning speed and draw ratio (see Manufacture). The properties are controlled by the intimate morphology of the fibers, especially the crystalline orientation and the amorphous orientation.

Creep and recovery. Creep and recovery with respect to time for nylon yarns under different initial loads for different total elongations have been measured. Initially, recovery is rapid, and most of it occurs within a few minutes after release of the load. The outstanding elastic recovery of the nylons is responsible for the cling or fit in women's hosiery and for their extensive market acceptance.

Staple. The use of nylon staple rather than nylon continuous filament stems from a need to satisfy the demand for a soft and warm staple hand or sensation when fabric is touched with the hand, and for ease of

Table 1. Physical Properties of Nylon-6 and -6,6 Continuous-Filament Yarns[a]

Property	Normal tenacity	High tenacity
tenacity, N/tex[b]	0.4–0.6	0.75–0.84
wet	0.36–0.51	0.64–0.7
loop	0.40–0.49	0.57–0.69
knot	0.35–0.42	0.52–0.72
tensile strength, MPa[c]	580–635	855–952
elongation at break (conditioned), %	23–43	12–17
wet	28–41	14–21
tensile modulus (conditioned), N/tex[b]	2.2–3.1	2.9–4.1
average toughness, N/tex[b]	0.07–0.14	0.06–0.14
moisture regain at 21°C, %		
65% rh	4.5	3.3–4.5
95% rh	7.1–7.8	7.8

[a] Conditioned at 65% rh and 21°C.
[b] To convert N/tex to gf/den, multiply by 11.33.
[c] To convert MPa to psi, multiply by 145.

styling in carpets. Best performance in many types of fabrics usually requires blends of fibers. Since intimate blending only can be achieved in staple form, the combining of the blend components takes place early in the textile spinning process, ie, either before or immediately after carding. Therefore, the properties of nylon staple must be compatible with each of the other blend components, and the fiber must also be adaptable to processing on the respective spinning systems. The outstanding contributions of nylon staple to carpet, apparel, and upholstery fabrics are its abrasion resistance and its low moisture absorption. The former property increases the wear life of the textile structure, the latter contributes to faster drying.

Thermal and moisture. The thermal behavior of nylon fiber and the interrelated effects of moisture have a pronounced influence on the physical properties of the yarn and its products. They are of basic importance in fiber manufacture, in converting the yarn into fabric, and in ultimate use.

Thermal properties of fibers have not been extensively investigated, because much of the practical thermal behavior of textiles is strongly influenced by other factors.

Thermal and light-resistant properties are to a large extent related to the specific stabilizing substances used and to certain aspects of the processes by which the yarns are made. These vary from producer to producer and are changed from time to time by each producer.

Light and heat stabilization. Light. In the presence of light, titanium dioxide, which normally is used as a delusterant, is considered to react with oxygen to form peroxide autocatalytically, and the peroxide degenerates polyamides in the absence of stabilizers, eg, manganese salts (see Heat stabilizers; Uv stabilizers). Additives, eg, hypophosphorous acids, phosphites, phosphates, etc, also are used.

Heat and light. Water, carbon dioxide, and ammonia are the three principal gaseous products of nylon pyrolysis. The preferred stabilizers in nondelustered fibers from degradation by heat and light are copper salts.

Stabilization against high temperature strength loss at high loads. One of the significant factors that has made it possible to produce a nylon-6 tire yarn is the ability to increase the breaking temperature under high loads to essentially equal nylon-6,6-9,9. Dialkyldihydroacridine compounds are employed in polyamide stabilization at high temperature and high loads.

Stabilization against heat disorientation in liquids. The disorientation of polyamides as vulcanization temperatures of tires are raised becomes an increasingly significant problem. The problem can be alleviated partially by maintaining some degree of tension on the tire cord at the end of the pressure release after vulcanization. It can also be avoided partially by minimizing the moisture content of tire yarns prior to vulcanization.

Electrical. Electrical conductance of nylon is very low. Conductivity increases as moisture content rises.

Manufacture and processing. Essentially all polyamides other than the aromatic (aramid) types are melt-spun.

Texturing. In general, textured yarns are yarns that have greater apparent volume or are made more extensible by mechanical distortion of the filaments. The distortion may be produced by buckling the filaments under endload compression, by either bending them over an edge of small radius or twisting the strands as a whole. Bulked nylon yarns are either fine tex [1.7–22.2 tex (15–200 den)] for woven, knitted stretch, and textured fabrics in apparel applications or heavy tex [110–400 tex (1000–3600 den)] for carpet.

Staple. The production of staple is simpler than that of filament yarn, since the breaking of a filament in the spinning step usually is not a problem, and uniformity of filaments is not as critical because the staple is blended. Several hundred to several thousand filaments can be spun from one spinneret, and many of the yarns can be combined into a large tow for drawing, crimping, and cutting. Polymerization is usually coupled with spinning to lower processing cost. As for filament, the polymer is metered to spinnerets, the filaments are quenched, and finish is applied. Yarns then are combined to form a tow, which is drawn by being passed over hot rolls. More finish can be applied at this step. The tow may be crimped, then cut to staple.

Modified cross-sections. Conventional spinneret orifices are circular. Improved machining methods, advances in fiber-production technology, and the study of the effect of profiled fibers, ie, with noncircular-cross sections, on fabric characteristics, eg, luster, sparkle, opacity, air permeability, resistance to showing soil, and heat insulation, have given greater importance to the production of modified cross-section filaments.

Finishes. The main functions of a fiber finish are to provide surface lubricity and yarn cohesion.

Spin finishes are applied to the fiber during spinning. Yarn lubricant or coning oil defines all lubricants added subsequent to application of spin finish. Yarn lubricants are used in textile milling and processing.

The finish generally is an emulsion or a water-soluble mixture of one or more lubricants and an antistatic agent to avoid static charge, which would cause the filaments to repel each other, giving a yarn with poor cohesion and so flaring the filaments, which causes uneven processing of the yarn. Wetting agents usually are added to aid the spreading of the finish on the yarn. The concentration of finish on the yarn and after evaporation of water usually is 0.3–0.8 wt%.

Additives. A variety of substances usually are added to the reactive mixture before or during polymerization to change the properties and nature of the polymer.

Additives include delustering agents, spin-dyeing agents, antistatic agents (qv), and flame retardants (qv).

Dyeability. Fiber characteristics. Chemical structure, crystallinity, molecular orientation, and fabric preparation affect the rate and the extent of nylon dyeability.

Dyebath characteristics. In general, uniform dyeing is most easily achieved when dyeing conditions are selected so that dye, diffusing into and out of the fiber, establishes an equilibrium with an appreciable amount of dyestuff remaining in the bath. Under this condition, nonuniform fabrics eventually achieve a level shade. Disperse dyes are unexcelled for this purpose.

Modified Nylon Fibers

Modified polyamides include bicomponent and biconstituent fibers, block copolymers, and random copolymers.

Other Nylons

Other nylons include Qiana and dimensionally stable nylons and miscellaneous polyamides.

Uses

Principal commercial uses of nylon-6 and -6,6 are in carpets, apparel, tire reinforcement, and other industrial applications. Qiana is used primarily in apparel.

J.H. SAUNDERS
Monsanto Company

E.M. Hicks, Jr., and co-workers, *Text. Prog.* **3**(1), 1 (1971).

A.J. Hughes and J.E. McIntyre, *Text. Prog.* **8**, 18 (1976).

H.F. Mark, S.M. Atlas, and E. Cernia, eds., *Man-Made Fibers Science and Technology*, 3 vols., Wiley-Interscience, New York, 1967–1968.

POLYAMIDE PLASTICS

Polyamide plastics, under the generic name of nylon, were developed in 1938 as a result of a classic research effort by Wallace H. Carothers at DuPont. Nylon is a generic term for any long-chain, synthetic, polymeric amide in which recurring amide groups are integral to the main polymer chain. There is a wide choice of starting materials from which polyamides can be synthesized. The two primary mechanisms for polymer manufacture are condensation of a diamine and a dibasic acid or their equivalents or polymerization of monomeric substances.

The difference in numbers of carbon atoms between the amide groups results in a significant difference in mechanical and physical properties. Although the theoretical number of nylon types is very large, there are fewer commercially available nylons. The more common types of poly-

Table 1. Properties of Nylons—Dry, as Molded

Property	Nylon-6	Nylon-6,6	Nylon-11	Nylon-12	Nylon-6,9	Nylon-6,12
specific gravity	1.13	1.14	1.04	1.02	1.09	1.07
water absorption, wt%						
24 h	1.6	1.5	0.3	0.25	0.5	0.5
equilibrium at 50% rh	2.7	2.5	0.8	0.70	1.8	1.4
saturation	10.5	9.5	1.9	1.5	4.5	3.0
melting point, °C	215	265	194	179	205	217
tensile yield strength, kPa[a]	8.1×10^4	8.3×10^4	5.5×10^4	5.5×10^4	5.5×10^4	5.5×10^4
elongation at break, %	50–200	40–80	200	200	125	150
flexural modulus, kPa[a]	2.7×10^6	2.8×10^6	1.2×10^6	1.1×10^6	2.0×10^6	2.0×10^6
Izod impact strength, J/m[b] of notch	54	68	40	95	58	58
Rockwell hardness, R scale	119	120	108	107	111	114
deflection temp under load, °C						
at 455 kPa[a]	185	245	150	150	150	165
at 1850 kPa[a]	65	75	55	55	55	82
dielectric strength, kV/mm						
short-time	17	24	16.7	18	24	16
step-by-step	15	11		16	20	
dielectric constant						
at 60 Hz (= c/s)	3.8	4.0	3.7	4.2	3.7	4.0
at 10^3 Hz	3.7	3.9	3.7	3.8	3.6	4.0
at 10^6 Hz	3.4	3.5	3.1	3.1	3.3	3.5
starting material(s)	polycaprolactam	hexamethylene-diamine and adipic acid	11-aminoundecanoic acid	polylaurolactam	hexamethylene-diamine and azelaic acid	hexamethylene-diamine and dodecanedioic acid

[a] To convert kPa to psi, multiply by 0.145.
[b] To convert J/m to ft. lbf/in., divide by 53.38.

amide plastics, their properties, and starting models for manufacture are listed in Table 1. Nylon-6 and nylon-6,6 comprise ca 75–80% of the nylon molding-compound market.

Processing

Polyamides are most commonly processed in the molten state, either by injection molding or extrusion.

Additives and modifications. Plasticization increases the flexibility and impact strength of polyamide plastics. One of the most common plasticizers (qv) is caprolactam.

Modifiers. Nylon-6 and nylon-6,6 have undergone significant product development involving specially treated olefins and rubbers. These materials are not treated as plasticizers. As a result of the special treatment, a degree of grafting and copolymerization does take place and results in a product from which the additive is not leachable. The properties of nylon-6 and -6,6 can be altered dramatically by these techniques. These modifications result in significant improvements in cold-temperature impact strength.

Nucleation. The crystallinity of nylon can be controlled by seeding the molten polymer to produce a more uniform growth in the rate, number, type, and size of the spherulites. The result of seeding is formulation of smaller spherulites of uniform size.

Reinforcing fillers. The use of chopped glass fiber ca 3-mm long has become dominant in the area of fiber-glass reinforcement. The use of fiberglass in polyamide plastics increases tensile and flexural strengths and creep resistance and significantly reduces molding shrinkage and dimensional change resulting from moisture absorption.

Environmental Effects

Heat. Polyamides undergo molecular weight degradation and mechanical property loss when exposed to elevated temperatures.

Ultraviolet radiation. Exposure to ultraviolet light results in degradation of polyamide plastics. However, such degradation can be retarded significantly by the use of uv stabilizers (qv).

Moisture. Moisture has a plasticizing effect on nylons similar to that produced by temperature, ie, modulus decreases and impact strength increases.

Production

The principal United States producers of nylon resins, the common type(s) which they manufacture, and the trade names under which these resins are marketed are documented in Table 2.

Table 2. Manufacturers of Polyamide Plastics

Manufacturer	Nylon type	Trade name
United States		
Allied Corp.	6	Capron
American Hoechst	6	Fosta
Badische Corp.	6	Ultramid
Bemis Co.	6	CRI
Celanese Corp.	6,6	Celanese
E.I. du Pont de Nemours & Co., Inc.	6,6	Zytel
	6,12	
	6,6 (mineral-filled)	Minlon
	6,12 (mineral-filled)	
Firestone	6	Firestone
Monsanto	6,6	Vydyne
	6,9	
Rilsan Corp.	11	Rilsan
	12	

Uses

Injection-molded applications comprise 70% of the polyamide plastics consumed in the United States. A principal portion of this total is automotive and truck components and parts.

The consumer area is a rapidly growing one for polyamide plastics, especially because they can be colored to meet the aesthetic needs of this marketplace. Typical applications are power-tool housings, combs and brush backs, and bicycle wheels.

R.J. WELGOS
Allied Corporation

M.I. Kohan, *Nylon Plastics*, John Wiley & Sons, Inc., New York, 1973, pp. 14–29.

CAPROLACTAM

Caprolactam, 2-oxohexamethylenimine, hexahydro-2*H*-azepin-2-one, is one of the most widely used chemical intermediates. However, almost all of the annual production of 2.3×10^6 metric tons is consumed as the monomer for nylon-6 fibers and plastics (see Polyamides, polyamide plastics, polyamide fibers). Cyclohexanone, which is the most common organic precursor of caprolactam, is made from benzene by either phenol hydrogenation or cyclohexane oxidation (see Cyclohexanol and cyclohexanone). Reaction with ammonia-derived hydroxylamine forms cyclohexanone oxime which undergoes molecular rearrangement to the seven-membered ring ε-caprolactam.

Physical Properties

Caprolactam (mol wt 113.16) is a white, hygroscopic, crystalline solid at ambient temperature, with a characteristic odor. It is very soluble in water and in most common organic solvents and is sparingly soluble in high molecular weight aliphatic hydrocarbons. Molten caprolactam is a powerful solvent for polar and nonpolar organic chemicals. Selected physical properties and solubilities of caprolactam are listed in Table 1.

Table 1.   Physical Properties of Caprolactam $\overset{\frown}{CH_2(CH_2)_4CONH}$

Properties	Values
melting point, °C	69.3
density (at 77°C), g/cm³	1.02
bulk density, kg/m³[a]	600–700
boiling point at 101.3 kPa[b], °C	266.9
refractive index	
at 40°C	1.4935
at 31°C	1.4965
viscosity, mPa·s (= cP)	
at 70°C	12.3
at 80°C	8.5
specific heat, J/(kg·K)[c]	
solid	
at 25°C	1380
at 35°C	1420
liquid	
at 70°C	2117
at 110°C	2412
at 178°C	2608
vapor	
at 100°C	1640
thermal conductivity, W/(m·K)	0.169
heat of fusion, J/g[c]	108.86
heat of vaporization at 80°C, J/g[c]	511
heat of combustion (liquid at 25°C), J/g[c]	−31,900
heat of formation (liquid at 25°C), J/g[c]	−2,840
flash point (closed cup), °C	125
fire point, °C	140

[a] To convert kg/m³ to lb/ft³, divide by 16.02.
[b] To convert kPa to mm Hg, multiply by 7.5.
[c] To convert J to cal, divide by 4.184.

Reactions

Caprolactam is a cyclic amide and therefore undergoes the reactions of this class of compounds. It can be hydrolyzed, *N*-alkylated, *O*-alkylated, nitrosated, halogenated and subjected to many other reactions.

Health and Safety Factors

Caprolactam has a low order of toxicity and presents no appreciable health hazard if it is handled properly. Threshold limit values for caprolactam dust and vapor are 1 and 20 mg/m³, respectively (time-weighted averages).

WILLIAM B. FISHER
LAMBERTO CRESCENTINI
Allied Corporation

Technology Process Summary, *Eur. Chem. News* **28**(743), 24 (April 30, 1976).
C.S. Hughes, *Chemical Economics Handbook*, Marketing Research Report 625-2031A, SRI International, Menlo Par, Calif., Aug. 1983.

POLY(BICYCLOHEPTENE) AND RELATED POLYMERS

Bicyclo[2.2.1]hept-2-ene is commonly known as 2-norbornene. This highly strained molecule undergoes ring-opening polymerization in the presence of various catalysts. The main features of ring-opened poly-norbornene are the abilities to absorb oil, provide soft vulcanizates with good mechanical strength, vary the dynamic properties as desired, and be processed in fully compounded, dry-blend form.

The properties of 2-norbornene and poly(2-norbornene) are given in Table 1. 2-Norbornene is made by the Diels-Alder condensation of ethylene and cyclopentadiene.

Table 1.   Properties of 2-Nobornene and Poly(2-norbornene)

Property	Monomer	Polymer
CA name	bicyclo[2.2.1]hept-2-ene	poly(1,3-cyclopentylene-vinylene)
appearance	white crystalline solid	white amor-phous powder
structure		
	C_7H_{10}	$[C_7H_{10}]_n$ ($n => 20,000$)
molecular weight	94.15	> 2,000,000
melting point, °C	47	32 (T_g)
boiling point, °C	95–95	201–202 (dec)
density, g/cm³	0.950	0.977
refractive index	1.4475	1.534
solubility parameter, MPa$^{1/2\,a}$	18.9	19.8
heat capacity, J/(g·K)[b]	1.38	2.09
heat of vaporization, J/g[b]	400.4	
heat of formation, J/g[b]	968.7	218
free energy of formation, kJ/g[b]	2.165	
thermal conductivity, W/(m·K)		0.285

[a] To convert MPa$^{1/2}$ to (cal/cm³)$^{1/2}$, divide by 2.046.
[b] To convert J to cal, divide by 4.184.

Related Polymers

Related polymers include bicyclo[2.2.1]hepta-2,5-diene, polar derivatives of 2-norbornene such as 5-substituted esters or nitriles, and alkylated norbornene derivatives.

ROBERT OHM
CLAUDE STEIN
CdF Chimie S.A.

C. Stein and A. Marbach, *Rev. Gen. Caoutch. Plast.* **52**, 71 (1975).
R.F. Ohm, *Chemtech* **10**, 183 (1980).
S. Matsumoto, K. Komatsu, and K. Igaraski in T. Saegusa and E. Goethals, eds., *Ring Opening Polymerization*, ACS Symposium Seres 59, 1977, Chapt. 21, pp. 303–317.

POLYBLENDS

Polyblends are mixtures of structurally different homopolymers, copolymers, terpolymers, and the like (see also Copolymers). The copolymers, terpolymers, etc, may be random, alternating, graft, block, starlike, or comblike, as long as the constituent materials exist at the polymeric level. Figure 1 gives a classification of polyblends in terms of their method of preparation.

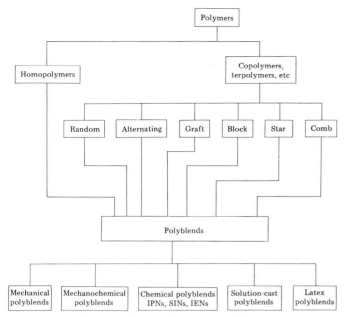

Figure 1. The classification of polyblends.

A mechanical polyblend is made by melt-blending the polymers either on an open roll or in an extruder or any other suitable intensive mixer. The processing temperature must be well above the glass-transition temperature (T_g) of each constituent for mixtures of amorphous polymers and above the melting temperature (T_m) of mixtures containing semicrystalline polymers, whichever is higher. Commercial polyblends are almost exclusively mechanical polyblends. Depending on the state of thermal stability of the poymers being mixed, the high processing shear could initiate degradation, resulting in free radicals. If the free radicals react with the other structurally different polymers present, resulting in true chemical graft or block copolymers, the mixture is called a mechanochemical polyblend. A chemical polyblend is made by *in situ* polymerization and cross-linking of the constituent polymers, giving an interpenetrating cross-linked polymer network of structurally different polymers. The three main categories are interpenetrating polymer networks (IPNs), simultaneous interpenetrating polymer networks (SINs), and interpenetrating elastomeric networks (IENs). In general, the IENs are made by mixing and coagulating two different polymer latexes; cross-linking the coagulum forms a three-dimensional mosaic structure. If the latex coagulum is not cross-linked, the resulting product is called a latex polyblend. This method is particularly useful in the preparation of rubber-toughened polyblends, which are mixtures of latex polymers (see Latex technology). In a latex, polymers are present as suspended microscopic particles. Interactions, ie, coalescing or coagulating of neighboring spheres, are prevented by the stabilizing surfactant medium. Even after extensive mixing of two different latexes containing different polymers, the mixture remains a random suspension of dissimilar particles, each unaffected by the other. During the coagulated state, the rate of flocculation generally depends entirely on the surfactant concentration, and not on the polymer characteristics; hence, the coagulated material contains a random intimate mixture of the polymers in the original latex. If, however, the polymers contain strong functional groups such as hydroxyl or carboxyl, the coagulated material may not be random. A melt-processing method is then employed for compounding and pelletizing the latex-blended material, and necessary precaution is taken to avoid degradation. Acrylonitrile–butadiene–styrene resins (ABS) are usually prepared in this manner. Solution-cast polyblends are prepared by dissolving the constituent polymers in a common solvent in such a way that the solutions have about the same viscosity; these solutions are mixed thoroughly. The resulting solution can be film-cast, coagulated, spray-dried, or freeze-dried to form the solution-cast polyblend. A melt-processing method can be used for compounding and pelletizing the solution-cast polyblends (see Plastics processing).

Most polymers are immiscible. Polymer mixtures that are thermodynamically miscible are characterized by a single thermal transition as well as a single amorphous phase. The physical properties of these polyblends are generally a compromise that, on balance, may be superior to the properties of the individual constituent polymers. Thus, a new set of products can be obtained without the usual significant capital investment.

Generally, blending technology rests on the premise of property additivity, although the additivity principle is not strictly valid for most polyblends.

The strongest reason for blending polymers is the cost : performance ratio. An expensive polymer whose property spectrum is much higher than is needed for a new application may be blended with an inexpensive polymer whose property spectrum is such that the resulting polyblend has a cost : performance ratio that makes it very attractive for the given application. Thus, the standard of performance demanded by the new application is satisfied by a mixture of commercially available polymers without the need to develop a new polymer or to invest in a new plant.

Miscible Versus Multiphase Polyblends

The state of immiscibility of a polyblend does not preclude its utility; multiphase polyblends have made respectable inroads into the polymer market. Because of their specific advantages, multiphase polyblends have utility in applications requiring improved physical and mechanical properties. Furthermore, antislip, antiblock, and coefficient of friction properties can be improved by the addition of a small-to-moderate amount of one polymer to another polymer, which is the continuous phase. The main advantages of miscible amorphous polyblends are that there is only one continuous phase and the principle of property additivity is obeyed, usually over the entire concentration range. Thus, the miscible polyblend is characterized by average or above average mechanical properties and absence of shear-induced phase separation and additive migration. Such polyblends have application in permanent plasticization and when it is necessary to increase or decrease the heat distortion of a polymeric system by addition of a suitable miscible polymer.

Compatibility Design

Multiphase polyblends have assumed far greater commercial importance than the miscible polyblends. Because of the success of these commercial polyblends, attempts are being made to modify the mechanical response of normally immiscible polyblends by the addition of agents that could guarantee good stress transfer between the components of the multicomponent multiphase polyblend.

Block or graft copolymers (A-B type) can be used to increase the compatibility of normally immiscible A and B homopolymers. The ternary blend of styrene–isoprene A-B block copolymer with the polystyrene and polyisoprene homopolymer is an excellent example of this technique. When block or graft copolymers are used in a ternary polyblend, the blend property is primarily dominated by the material that makes up the continuous phase; it also is affected by the degree of miscibility of the hard and soft blocks in the other constituent polymers. Pendant chemical groups have also been used to enhance compatibility.

During preparation, a mechanical, compatible multiphase polyblend forms a co-continuous phase if the viscosities and concentrations of the individual polymers are properly controlled during blending.

Probably the best method of enhancing the miscibility of polyblends is to introduce specific interactions. Miscibility almost always occurs when a strong specific interaction exists, and the concept of complementary dissimilarity appears to explain the miscibility of the better-known blends. The potentially useful specific interactions are random dipole–induced dipole, dipole–induced dipole, dipole–dipole, ion dipole, hydrogen bonding, acid–base, and charge-transfer interactions.

Characterization

Numerous techniques are available to determine the function of each polymer in the polyblend, achieve phase separation, identify the predominant phase, determine the character of the dispersed phase, and measure the interactions between the polymers. These techniques include optical methods, glass-transition-temperature methods, microscopic methods, scattering methods, and solution methods.

Miscible Polyblends

Experimental results in the past decade have cast serious doubt on the hitherto far-reaching conclusion that in polyblends miscibility is the exception and immiscibility is the rule. Numerous miscible polyblends have been discovered.

A large number of miscible polyblends are amorphous. Poly(vinyl chloride) (PVC), because of its specific interactions, hydrogen bonding, and charge-transfer capabilities, is the most miscible with structurally different polymers and copolymers. Many PVC polyblends have a much lower T_g than PVC and are used where flexibility and rubberlike properties are in demand. Several polymers miscible with PVC are used as permanent plasticizers for PVC. These are co- and terpolymers of ethylene as well as polyester oligomers in the 2000–4000 molecular weight range such as those made from butanediol, hexanediol, and ethylene glycol, and adipic, sebacic, azelaic, or succinic acids. Several polymers with higher T_g than PVC are used as T_g enhancers for PVC. The most important commercial miscible polyblend is the mixture of polystyrene, PS, with poly(2,6-dimethyl-1,4-phenylene oxide), PPO, sold under the trade name Noryl (see Polyethers, aromatic).

The semicrystalline polymer in a miscible polymer blend usually retains part or all of its crystallinity with its attendant solvent resistance, toughness, and weatherability. Consequently, only the amorphous phase contains the homogeneous mixture of the two polymers, and the polyblend in the solid state is essentially multiphase, consisting of amorphous, homogeneous, glassy polyblend interspersed with the crystals of the semicrystalline polymer.

A few polymers form cocrystalline polyblends. The phenomenon of cocrystallization or isomorphism is exhibited by different species that crystallize over a range of concentration with a common crystal lattice. The conditions (which need not be simultaneously satisfied) for isomorphism for polyblends are chemical similarity of the macromolecules, physical similarity (size) of the macromolecules, physical similarity of the backbone chains, practically equal cross-section dimensions, similarity of the forces of attraction, and similarity of conformation and configuration.

Polyblends that form complexes constitute a surprisingly large class. The best example is the mixture of aqueous solutions of poly(acrylic acid) and poly(ethylene oxide) (PEO). As soon as the solutions are mixed, a precipitate forms that is insoluble in water, the medium of its very formation. The main reason for complexation behavior has been ascribed to the presence of oppositely charged functional groups on the polymers being mixed; one usually slightly basic and the other slightly acidic.

Commercial Multicomponent Polymer Systems

Commercially available multicomponent polymer systems include copolymers, terpolymers, and physical mixtures of homopolymers, copolymers, and terpolymers. They are either rubber-toughened polymers including polymer blends, blocks, and grafts, or physical blends of any number of polymers including mixtures of copolymers with different comonomer unit ratios.

Uses

Polystyrene. The largest single market for impact polystyrene is packaging (see Styrene plastics).

ABS. ABS is used in applications where a high strength-to-weight ratio, high impact strength, and weatherability are required. The three largest markets are pipes and fittings (30%), appliances (20%), and automotive (13%) (see Acrylonitrile polymers).

Polypropylene. Toughened polypropylene is used in molded containers, caps and closures, film, covers, and sheets (see Olefin polymers).

Epoxy resins. The versatility and usefulness of epoxies are demonstrated by their varied applications, eg, as protective and decorative coatings, electrical and electronic components, adhesives, and structural and reinforced plastics (see Epoxy resins).

Poly(vinyl chloride). The PVC polyblends are used in electronic housings (smoke detectors, timers, calculators), power tool handles, and computer devices, among other products (see Vinyl polymers).

Polycarbonates. The ABS–PC polyblends are used in products that sustain high abuse and exposure to heat and electricity (see Polycarbonates).

OLAGOKE OLABISI
Union Carbide Corporation

O. Olabisi, L.M. Robeson, and M.T. Shaw, *Polymer-Polymer Miscibility*, Academic Press, Inc., New York, 1979.

D.R. Paul and S. Newman, eds., *Polymer Blends*, Vols. I and II, Academic Press, Inc., New York, 1978.

A. Noshay and J.E. McGraph, *Block Copolymers: Overview and Critical Survey*, Academic Press, Inc., New York, 1976.

POLYBUTYLENE TEREPHTHALATE.   See Polyesters, thermoplastic.

POLYCARBONATES

Polycarbonates (PC) are a special class of polyesters derived from the reaction of carbonic acid derivatives with aromatic, aliphatic, or mixed diols. They may be produced by the Schotten-Baumann reaction of phosgene with a diol in the presence of an appropriate hydrogen chloride acceptor (eq. 1) or by a melt transesterification reaction between the diol and a carbonate ester (eq. 2) (see Esters).

$$n \text{ HOROH} + n \text{ COCl}_2 + (2n + 1) \text{ NaOH} \longrightarrow$$
$$\text{Na} \underset{}{\overset{O}{\text{-(OCOR)}_n}} \text{OH} + 2n \text{ NaCl} + 2n \text{ H}_2\text{O} \quad (1)$$

$$n \text{ HOROH} + n(\text{R'O})_2\text{CO} \xrightarrow{\text{catalyst}} \text{R'} \underset{}{\overset{O}{\text{-(OCOR)}_n}} \text{OH} + (2n - 1) \text{ R'OH} \uparrow \quad (2)$$

Bisphenol A Polycarbonate

Properties and characterization. Solvents, solubility, and solvent resistance. Most of the commercial polymer is produced and characterized in solution. Methylene chloride is an excellent solvent (350 g polymer/L solvent at 25°C) having the advantages of low flammability and toxicity.

Bisphenol A polycarbonate exhibits high hydrolytic stability.

Molecular weight. Based on light scattering and osmometry data, a typical polymer with an intrinsic viscosity of 0.50–0.55 dL/g might have a weight-average molecular weight ($\overline{M}_w$) of 30,000 and a number-average of 11,000 to yield a dispersity ratio $\overline{M}_w/\overline{M}_n$ of 2.7.

Crystallinity and structure. The mechanical–optical properties of polycarbonates are those common to amorphous polymers.

Glass-transition temperature and dependent properties. Bisphenol A polycarbonate has a bulky structure, and packing at the pivotal linkages tends to restrict rotation. Furthermore, the repeat units along the chain are longer than those of most common polymers, which makes ordered arrangement difficult. These features play a key role in determining the physical properties.

The T_g at 159°C for bisphenol A polycarbonate is unusually high compared to that of polystyrene (100°C), poly(ethylene terephthalate) (69°C), nylon-6,6 (45°C), and polyethylene (−45°C). A high T_g indicates excellent dimensional stability and resistance to creep under load, both properties of bisphenol A polycarbonate.

Melt behavior. Bisphenol A polycarbonate becomes plastic at the reported melting range of 215–225°C and it may be shaped with sufficient pressure.

Thermal stability. Bisphenol A polycarbonate exhibits outstanding thermal stability. The dry polymer may be held for hours in the molten state up to 320°C and for short times up to 330–350°C with minimal degradation.

Light transmission. Polycarbonate resins exhibit water-white clarity with visible-light transmission of approximately 90%.

Mechanical properties. Whereas the room-temperature modulus and tensile strength of bisphenol A polycarbonate are in the same range as those of other amorphous thermoplastics at corresponding reduced temperatures, its impact strength and ductility are altogether exceptional.

Processing

The polycarbonates may be fabricated by all conventional thermoplastic processing operations, of which injection molding is the most common (see Plastics processing).

Health and Safety Factors

The toxicity of combustion products of plastic and wood construction materials has been rated. Unmodified polycarbonate was ranked 27th out of 30 (1 worst, 30 best) with respect to incapacitation and 30th with respect to lethality. No unusually toxic by-products were detected. The worst combustion component was carbon monoxide.

Uses

Extreme toughness, transparency, resistance to burning, and maintenance of useful engineering properties over a temperature range from -200 to $+140°C$ are the outstanding features of polycarbonates. This balance of properties qualifies polycarbonates for use in aeronautics, and impact-resistant and electrical products.

Other Polycarbonates and Related Technology

Copolycarbonates based on bisphenol A and tetrabromobisphenol A have been produced. The bromine content enhanced the inherent fire resistance of the bisphenol A polymer to yield products with excellent ratings. Furthermore, brominated bisphenol A did not detract significantly from the excellent balance of properties of the basic polymer. Polycarbonates from tetrachlorobisphenol A were explored but not commercialized, partly because the pure monomer was too expensive.

Polycarbonates and copolycarbonates derived from 2,2-bis(4-hydroxyphenyl)-1,1-dichloroethylene have impact strength equivalent to that of bisphenol A polycarbonate, high transparency, outstanding resistance to burning, very low smoke from forced combustion, and low toxicity of combustion products.

D.W. Fox
General Electric Company

H. Schnell, *Chemistry and Physics of Polycarbonates*, Wiley-Interscience, New York, 1964.

D. Fox and W. Christopher, *Polycarbonates*, Reinhold Publishing Corp., New York, 1962, pp. 161–177.

POLYELECTROLYTES

The term polyelectrolyte refers to a substance that contains polyions, which are macromolecules bearing a large number of ionizable groups (see Ionomers) (see Table 1). To preserve the electroneutrality of a polyelectrolye substance, the polyion charges must be compensated by counterions, typically ions of low molecular weight such as H^+ or Na^+. Unlike most uncharged polymers, polyelectrolytes usually are soluble in polar solvents, eg, water. With regard to their protonation equilibria in aqueous solution, they can be classified as polyacids, polybases, or if both acidic and basic groups are present, as polyampholytes. Like uncharged polymers, polyelectrolytes in solution can be classified according to the nature of their structure, ie, linear, branched, or cross-linked.

The presence of a large number of charges on the same macromolecule produces some striking effects that generally are not observed in solutions containing uncharged macromolecules of comparable chemical composition, or in solutions containing an equivalent concentration of unpolymerized charged monomers. As flexible chainlike polymers are ionized

Table 1. Typical Examples of Ionizable Chain Molecules

Acid	Formula
poly(acrylic acid)	$+CH_2CH+_n$ / COOH
poly(methacrylic acid)	CH_3 / $+CH_2C+_n$ / COOH
poly(vinylsulfonic acid)	$+CH_2CH+_n$ / SO_3H
poly(p-styrenesulfonic acid)	$+CH_2CH+_n$ / (ring) / SO_3H
poly(styrene-co-maleic acid)	$+CH_2CH-CH-CH+_n$ / C_6H_5 COOH COOH
poly(vinyl methyl ether-co-maleic acid)	$+CH_2CH-CH-CH+_n$ / OCH_3 COOH COOH
poly(acrylic acid-co-maleic acid)	$+CH_2CH-CH-CH+_n$ / COOH COOH COOH
poly(metaphosphoric acid)	O ‖ / $+O-P+_n$ / OH
polyvinylamine	$+CH_2CH+_n$ / NH_2
polyethyleneimine	$+CH_2CH_2N+_n$ / H
poly(4-vinylpyridine)	$+CH_2CH+_n$ / (pyridine ring) N
poly(4-vinyl-N-dodecylpyridiniumchloride)	$+CH_2CH+_n$ / (pyridinium ring) N^+ Cl^- / $C_{12}H_{25}$

by titration, their average configuration in solution becomes more extended as a result of the Coulombic repulsions among the polyion charges. The superposition of contributions from these charges produces a strong electric field, which can exert a long-range influence on other charged species in solution, especially the small mobile ions. The polyion field causes oppositely charged species to accumulate in the volume of solution immediately surrounding the polyion. This high local concentration of unbound ions affects any process in which the polyion charged groups are involved. Most of the distinctive features of polyelectrolyte solutions can be explained qualitatively in terms of the field arising from the polyion charges. In dilute solutions, the primary variables that determine the strength of this field are the density of structural charges on the polyion and the concentration(s) of added simple electrolyte(s). Measuring the physical properties of a polyelectrolyte solution as functions of these variables provides information for the development of quantitative descriptions.

The Shape of Flexible Polyions in Solution

The average shape of a flexible chainlike macromolecule in a sufficiently dilute liquid solution can be described in terms of the mean squared distance between the two ends of the molecule.

Models for polyion dimensions. If long-range interactions among segments of a flexible macromolecular chain are neglected, its unperturbed overall dimensions can be calculated theoretically from a model for the short-range interactions that determine the conformational distribution of adjacent segments. As ionizable groups attached to the backbone of a weak polyacid are titrated, the resulting Coulombic repulsions are assumed to affect the local conformational distribution and hence the unperturbed dimensions of the polyion.

All theoretical treatments of polyion expansion are based on idealized models that omit or oversimplify features of the real system. Models that consider the distribution of polyion chain conformations, each having a discrete charge distribution, are more realistic than models that assume that the polyion charge is smeared out over an equivalent volume.

Experimental determination of polyion dimensions. Polyion dimensions in solution usually are determined by measuring the concentration dependence of viscosity or the angular dependence of scattering intensities. The former method is more convenient and typically capable of higher precision, but its interpretation entails some uncertain assumptions. For either method, the analysis is simpler and theoretically more certain if the polyelectrolyte solution also contains some low molecular weight electrolyte.

Polyion Interactions With Small Ions in Solution

If the structural charge density on a polyion is high enough, interactions between the polyion field and the small ions present in solution cause their radial distributions to be highly nonuniform with respect to the polyion. A number of theoretical models describe the effects of long-range Coulombic interactions on the measurable properties of polyelectrolyte solutions. In many polyelectrolyte solutions, there is evidence for the close association of counterions with polyion charges. The molecular nature of this binding interaction is a topic of both theoretical and practical interest. The equilibrium properties of solutions of highly charged polyions frequently reflect both long-range electrostatic interactions and counterion association with the polyion; these two effects are often difficult to separate on the basis of experimental observations.

Theoretical models for long-range interactions in polyelectrolyte solutions. The distributions of small ions in polyelectrolyte solutions are determined primarily by their interaction with the electric field that arises from the charged groups on the polyion. For polyions whose structures are relatively rigid in solution, it is reasonable to assume a fixed charge configuration for the purpose of computing the surrounding field. For more flexible species, including most chainlike synthetic polyions, a wide distribution of chain configurations is expected. To arrive at an accurate calculation of the average polyion field resulting from the charge distributions associated with the different configurations, it would be necessary to consider how interactions of the small ions affect the configurational distribution in solution. No rigorous approach to this problem exists. Instead, in theoretical computations of properties that depend on small-ion distributions, it has generally been assumed that the average polyion field can be calculated from a single average configuration of structural charges. This average polyion charge distribution usually is a highly simplified representation of the true geometrical disposition of the structural charges. Rigid polyions with relatively compact structures, eg, globular proteins or polysoaps, have typically been modeled as impenetrable spheres having uniform, continuous surface-charge densities. Flexible chainlike polyions have been modeled either as penetrable spheres, consisting of a uniform continuum of charge density, or as impenetrable cylinders, bearing on their surface a charge distribution that is either uniform and continuous or discrete and regular. Further approximations employed in computing the polyion field are the neglect of interactions between polyion and the use of a spatially invariant dielectric constant, usually equated to that of the pure solvent.

Counterion binding in polyelectrolytes. The charge density of a polyion, and hence the strength of the surrounding electric field, can be decreased by the binding of counterions. The resulting effective charge density is an important parameter in models that have been proposed to account for the dimensions of polyions in solution and for their equilibrium and transport properties. Thus, information about the nature and extent of counterion binding to a polyion may contribute to the interpretation of measurements of its physical properties.

Protonation equilibria of weak polyelectrolytes. The binding of protons to weakly acidic or basic groups on a polyelectrolyte molecule is studied by potentiometric titration. This method has been applied extensively in investigations of relatively rigid polyions, eg, proteins, and of flexible-chain synthetic polyions.

CHARLES F. ANDERSON
University of Wisconsin

HERBERT MORAWETZ
Polytechnic Institute of NY

G.S. Manning, *Acc. Chem. Res.* **12**, 443 (1979).
C.F. Anderson and M.T. Record, *Ann. Rev. Phys. Chem.* **33**, 191 (1982).

POLYENE ANTIBIOTICS.   See Antibiotics, polyenes.

POLYESTER FIBERS

A polyester fiber is a manufactured fiber in which the fiber-forming substance is any long-chain synthetic polymer composed of at least 85 wt% of an ester of a dihydric alcohol (HOROH) and terephthalic acid (p-HOOCC$_6$H$_4$COOH). The most widely used polyester fiber is made from linear poly(ethylene terephthalate) (PET).

Although the rate of increase has decreased somewhat, polyester fibers constitute the fastest-growing market in the United States and worldwide.

Properties

Structural. Drawn polyester fibers may be considered to be composed of crystalline and noncrystalline regions. The structural unit or unit cell for PET has been deduced by means of x-ray diffraction techniques and is considered triclinic, with one repeating unit.

Mechanical. Typical physical and mechanical properties of polyester fibers are summarized in Table 1.

Table 1.   Physical Properties of Poly(Ethylene Terephthalate)

	Filament yarn		Staple and tow	
Property	Regular tenacity[a]	High tenacity[b]	Regular tenacity[c]	High tenacity[d]
breaking tenacity[e], N/tex[f]	0.25–0.50	0.5–0.86	0.2–0.5	0.52–0.63
breaking elongation, %	19–40	10–34	25–65	18–40
elastic recovery, %	88–93 at 5%	90 at 5%	75–85 at 5%	75–85 at 5%
initial modulus, N/tex[f]	6.6–8.8	10.2–10.6	2.2–3.5	4.0–4.9
specific gravity	1.38	1.39	1.38	1.38
moisture regain[g], %	0.4	0.4	0.4	0.4
melting temperature, °C	258–263	258–263	258–263	258–263

[a] Textile filament yarns for woven and knit fabrics.
[b] Tire cord and high strength, high modulus industrial yarns.
[c] Regular staple for 100 wt % polyester fabrics, carpet yarn, fiberfill, and blending with cellulosics or wool.
[d] High strength, high modulus staple for industrial sewing thread and blending with cellulosics.
[e] Standard measurements are conducted in air at 65% rh and 22°C.
[f] To convert N/tex to g/den, multiply by 11.33.
[g] The equilibrium moisture content of the fibers at 21°C and 65% rh.

Chemical. Polyesters show good resistance to most mineral acids, but dissolve with partial decomposition in concentrated sulfuric acid. Hydrolysis is highly dependent on temperature. Basic substances attack the fiber in two ways. Strong alkalies, eg, caustic soda, etch the surface of the fiber and reduce its strength. Ammonia and other organic bases, eg, methylamine, penetrate the structure initially at the noncrystalline regions and cause degradation of the ester linkages and a general loss in physical properties. The polyesters display excellent resistance to conventional textile bleaching agents and are resistant to cleaning solvents and surfactants. Most of these effects are conditioned by the structural morphology of the specimens.

Other. Poly(ethylene terephthalate) fibers display good resistance toward sunlight, and addition of modern light stabilizers and antioxidants enhances the polymer's performance (see Uv stabilizers; Antioxidants and antiozonants). The resistance to mildew, aging, and abrasion generally is excellent.

Manufacture and Processing

Terephthalic acid or its dimethyl ester reacts with ethylene glycol to form a diester monomer, which polymerizes to the homopolymer PET. The molten polymer is extruded or spun through a spinneret, forming filaments which are solidified by cooling in a current of turbulence-free air. The spun fiber is drawn by heating and stretching the filaments to several times their original length to form a somewhat oriented crystalline structure with the desired physical properties.

Waste recovery. Glycol and antimony catalysts are recycled into the polymerization system. Polymer and fiber waste can be separated into two chemical components, usually by methanolysis; the DMT that forms is purified and repolymerized. The waste also can be melted directly and converted to a lower quality fiber.

Texturing. Continuous-filament, 5.6–33-tex (50–300-den) yarn can be textured easily by several techniques, eg, false-twist, stuffer-box, knit-deknit, and air-jet methods. In texturing, filaments are crimped or looped at random to give the yarn greater volume or bulk; then they are heat-set for dimensional stability. The most common textured yarns are the false-twist textured yarns made with aggregates of tiny, high speed spindles, stacked friction disks, or cross-belt nips, all of which impart a high degree of twist.

Dyeing. Polyester fibers are dyed almost exclusively with disperse dyes because of the lack of reactive dye sites in the fiber (see Dyes, application).

Finishing. Poly(ethylene terephthalate) is a thermoplastic and, therefore, heat is applied to set the structural memory desired for maximum performance of PET fiber either alone or in blends. This important heat-setting technique known as no-cure-permanent-press, or durable press, provides the wrinkleproof, ease-of-care, and low laundering shrinkage properties for which polyester fabrics are noted.

Economic Aspects

Polyester accounts for almost half of all synthetic fibers and one sixth of all textile fibers.

Staple fiber represents the largest share (ca 60%) of the polyester market in the United States and worldwide.

Health and Safety Factors

Poly(ethylene terephthalate) is chemically stable and virtually nonbiodegradable.

Uses

Polyester fibers are used in staple, filament, and spun-bonded textile fabrics (see Nonwoven textile fabrics).

GERALD W. DAVIS
ERIC S. HILL
Fiber Industries, Inc.

H. Ludewig, *Polyester Fibers, Chemistry and Technology*, Wiley-Interscience, New York, 1971, p. 45.

G. Clayton and co-workers, *Text. Prog.* **8**(1), 31 (1976).

E.M. Hicks and co-workers, *Text. Prog.* **3**(1), 38 (1971).

R.W. Moncrieff, ed., *Man-Made Fibers*, 6th ed., Butterworth & Co., Ltd., London, UK, 1975, p. 434.

POLYESTERS, THERMOPLASTIC

Thermoplastic polyesters are condensation products that are characterized by many ester linkages distributed along the polymer backbone.

Properties

The properties of crystalline thermoplastic resins are listed in Table 1.

Unfilled and glass-filled poly(alkylene terephthalate) formulations. The widespread acceptance of semicrystalline thermoplastic polyesters generally is attributable to their physical, mechanical, and electrical properties. Both poly(butylene terephthalate) (PBT) and modified, glass-reinforced poly(ethylene terephthalate) (PET) exhibit fast mold-cycle times to give parts that maintain mechanical integrity almost to the crystalline melting points (see Processing, Additives). Also, they show excellent chemical resistance and good hydrolytic stability. Enhancement of mechanical properties may be achieved with a variety of additives. The additive most widely used, fiber glass, produces a two-to-six-fold increase in flexural modulus and a concommitant increase in heat-deflection temperature. The properties of virgin materials are compared with those of several fiber-glass-reinforced compositions in Table 1. Factors that favor polyesters over other plastics include surface and electrical properties, toughness, low water absorption, and thermal stability for continuous use at high temperatures.

Amorphous polyesters. The amorphous poly(1,4-cyclohexylenedimethylene terephthalate-*co*-isophthalate) is reported superior to other amorphous thermoplastics in solvent resistance but is lacking in this respect compared to the semicrystalline polyesters. Although such amorphous materials are used in injection molding, they have broader application in bottles and films because of their clarity, good barrier characteristics, and toughness.

Manufacture

Poly(ethylene terephthalate) for many years has been prepared commercially from ethylene glycol and dimethyl terephthalate. Since terephthalic acid of high purity is available through air oxidation of pure *p*-xylene, its direct esterification with ethylene glycol at high temperature and elevated pressure is the preferred synthesis route because of higher reaction rates, the elimination of methanol as a by-product, and reduced catalyst requirements (see Phthalic acids). Poly(butylene terephthalate) (PBT) is made in a similar manner from 1,4-butanediol and dimethyl terephthalate. Terephthalic acid is not used in a commercial PBT process.

Processing

Thermoplastic polyesters can be processed by injection molding and blow molding, and may be extruded as filament, film, or sheet.

Recycling. Economic mandates require that recycled or scrap polymer resin and parts be reclaimed. Scrap parts, eg, sprues and runners, are reground and blended with virgin resin at up to 25 wt% for injection-molding-grade polyester (see Recycling, plastics).

Additives. Additives include glass, minerals, flame retardants (qv), impact modifiers, colorants (qv), and stabilizers.

Health and Safety Factors (Toxicology)

Poly(ethylene terephthalate). In 1954, poly(ethylene terephthalate) film was sanctioned by the FDA for use in contact with food. In 1958, it was regulated as a food additive but only in the form of sheets, ie, film, or coatings. In 1973, the regulation was amended to permit poly(ethylene terephthalate) use in articles and containers that are intended for use in contact with food.

Poly(butylene terephthalate). Poly(butylene terephthalate) resin can be used for food-contact applications consisting of nonalcoholic foods when thickness, time, and temperature do not exceed 0.25 mm, 24 h, and 82°C, respectively.

Kodar PETG. Kodar PETG 6763, which is a modified poly(ethylene terephthalate) resin, can be used in contact with food and nonalcoholic beverages at a maximum temperature not to exceed 82°C and for storage temperatures not exceeding 49°C. Alcohol content may not exceed 8

Table 1. Properties of Crystalline Thermoplastic Polyesters

Properties	Glass-fiber concentration, wt%						
	Poly(butylene terephthalate)				Poly(ethylene terephthalate)		
	0	15	30	40	0	30	45
General							
specific gravity	1.31	1.41	1.53	1.60	1.37	1.56	1.69
water absorption at 23°C, %							
after 24 h	0.08	0.07	0.06	0.06	0.08	0.05	0.04
at equilibrium	0.34	0.30	0.26	0.36	0.60	0.45	0.45
mold shrinkage, %	2.0	0.8	0.6	0.5	2.0	0.2	0.2
Rockwell hardness	R117/M68	R118	R118/M90	R117	M106	M100	M100
Taber abrasion ((CS-17) 1kg), mg/1000 cycles	9	12	19	26	3	6	7
coefficient of friction							
against self	0.17	0.22	0.15			0.28	0.17
against metals	0.13	0.24	0.19			0.17	0.20
Strength and stiffness at 23°C							
tensile strength, MPa[a]	52	90	117	131	53	158	193
tensile elongation, %	300	5	4	3	300	3	2
flexural strength, MPa[a]	83	138	193	207	114	234	283
compressive strength, MPa[a]	90	103	124	124	128	172	179
shear strength, MPa[a]	53	55	61	59	59	79	86
flexural modulus, MPa[a]	2,340	4,620	7,580	8,970	2,830	8,960	13,790
Toughness							
Izod impact at 23°C, J/m[b]	53	59	96	117	43	101	128
unnotched	no break	530	800	960	no break	370	370
Thermal/flammability/electrical							
heat-deflection temperature, °C							
at 0.46 MPa[a]	155	210	216	216	115	250	250
coefficient of linear thermal expansion, $\times 10^{-5}$/°C	12.8	4.5	2.5	2.5	7.2	2.9	2.3
thermal conductivity, W/(m·K)[c]	0.16		0.21			0.29	0.31
oxygen index, %	20.6	18.5	18.2	18.5	21	20	20
chemical stability							
chemical resistance	excellent	excellent	excellent	excellent	excellent	excellent	excellent
stress-crack resistance	excellent	excellent	excellent	excellent	excellent	excellent	excellent
dielectric strength (short-term, 1.6 mm), kV/mm	23.2	23.2	24.8	26.4	23.6	29.6	28.8

[a] To convert MPa to psi, multiply by 145.

[b] To convert J/m to ft·lbf/in., divide by 53.38.

[c] To convert W/(m·K) to (Btu·in.)/(h·ft^2·°F), divide by 0.1441.

vol%. It is not lawful for use with carbonated beverages, beer, or packaged foods requiring thermal treatment in the container.

Uses

The thermoplastic polyesters can compete effectively with the thermoset resins in certain applications requiring good electrical properties, better impact strength, and superior processing with regard to cycle time and scrap generation. Thermoplastic polyesters also will be used more often in die-cast zinc and aluminum applications because of their ease of processing and the absence of a secondary finishing requirement.

Blow-molded poly(ethylene terephthalate). One of the fastest-growing markets for poly(ethylene terephthalate) is the soft-drink bottle (see Barrier polymers; Carbonated beverages).

DONALD B.G. JAQUISS
W.F.H. BORMAN
R.W. CAMPBELL
General Electric Company

I. Goodman in N. Bikales, ed., *Encyclopedia of Polymer Science and Technology*, Vol. 11, Interscience Publishers, a division of John Wiley & Sons, Inc., New York, 1969, pp. 62–128.

J.R. Caldwell, W.J. Jackson, Jr., and T.F. Gray, Jr. in N.M. Bikales, ed., *Encyclopedia of Polymer Science and Technology*, Supplement Vol. 1, Interscience Publishers, a division of John Wiley & Sons, Inc., New York, 1976, pp. 444–467.

Modern Plastics Encyclopedia, Vol. 60, McGraw-Hill Book Co., New York, 1983–1984, pp. 42–53.

POLYESTERS, UNSATURATED

Unsaturated polyesters are macromolecules with polyester backbones derived from the interaction of unsaturated acids or anhydrides and polyhydric alcohols (see Alcohols, polyhydric). The reaction normally proceeds at 190–220°C until a predetermined acid value–viscosity relationship has been achieved. Solutions of these polymers in vinyl monomers, eg, styrene, often are called polyester resins. They are compounded with fillers (qv) or fibers, or both, in the liquid stage and then are cured with the aid of free-radical initiators (qv) to yield thermoset articles. Market penetration, especially in the area of fiber-glass reinforcement, is greatly enhanced as a result of greater latitudes in compounding and processing than is possible with other polymeric systems. Unsaturated polyesters can be mass-cast, laminated, molded, pultruded, and made into gel coats in a variety of colors. Depending on the application, the physical and chemical properties of the product often can be met by judicious choice of polyester backbone ingredients and the type and amount of the diluent vinyl monomer.

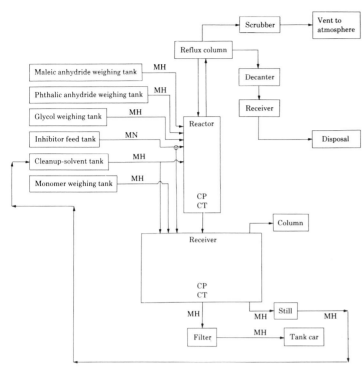

Figure 1. Typical flow diagram of unsaturated polyester by fusion method. MH = mechanical transfer, MN = manual transfer, CP = closed process, CT = cleaning operation.

Polyester resins are considered one of the three principal thermoset systems of economic importance. The other two are phenolics and epoxy resins (qv) (see Phenolic resins).

Properties

As the molecular weight of unsaturated polyesters increases, the cured products exhibit improvements in certain properties, eg, heat-deflection temperature (HDT), hardness, tensile strength, and flexural strength, until a constant value is reached. With an increase in molecular weight, the effect of the end groups becomes less pronounced. However polyesters that terminate with carboxyl groups generally are characterized by higher viscosities and improved physical properties compared to those with hydroxyl terminations.

The structure of the backbone of the polyester characterizes the physical properties of the cured resin. For most cured polyester resins, optimum physical properties are obtained when the degree of unsaturation, as represented by mol% fumarate, is 50–70%. Retention of physical properties at higher temperatures is not necessarily aided by higher cross-link density, although an increase in HDT may be observed.

Unsaturated polyester resins rarely are used alone without some type of filler or fibrous reinforcement. Fillers tend to increase the flexural modulus, HDT, and penetrating hardness and to degrade other physical properties. Fiber glass, however, transfers its strength to the matrix resin and improves most physical properties, depending on the length of the fiber and the amount of glass used (see Composite materials; Laminated and reinforced plastics).

Raw Materials

Raw materials include ethylenically unsaturated acids, aromatic and saturated aliphatic acids, glycols, monomers, inhibitors, accelerators, and initiators.

Manufacture

The manufacture of polyester is carried out in large stainless-steel lined kettles, which are provided with overhead fractionating columns to reduce the loss of components that are less volatile than water. The columns normally are packed and steam-jacketed. The most widely used manufacturing method is the direct fusion process (Fig. 1). The oxide method also is used.

Health and Safety Factors

Most of the concerns for safety and health associated with polyester resins pertain to worker exposure to styrene and its emission into the environment.

JOSEPH MAKHLOUF
PPG Industries

H.V. Boenig, *Unsaturated Polyesters: Structure and Properties*, Elsevier Publishing Company, Amsterdam, Neth., 1964.

E.E. Parker, *Ind. Eng. Chem.* **58**, 53 (1966).

POLYETHER ANTIBIOTICS.   See Antibiotics, polyethers.

POLYETHERS

AROMATIC POLYETHERS

The aromatic polyethers are characterized by their thermal stabilities and mechanical properties. The aromatic portion of the polyether contributes to the thermal stability and mechanical properties and the ether functionality facilitates processing but still possesses both oxidative and thermal stability (see also Ethers). With these properties and the ability to be processed as molding materials, many of the aromatic polyethers can be classified as engineering thermoplastics (see Engineering plastics).

Poly(Phenylene Oxide)s

The poly(phenylene oxide)s also are referred to as polyoxyphenylenes and poly(phenylene ether)s. Variations in the configuration of the ether group, ie, ortho, meta, or para, and in the extent and type of substitution, eg, alkyl, aryl, halo, etc, on the aromatic backbone give rise to the large number of possible homo- and copolymers. The polymers with para-oriented ethers have been studied most extensively and several of them have significant utility. In particular, poly(2,6-dimethyl-1,4-phenylene oxide) (DMPPO), made by the oxidative-coupling polymerization of 2,6-dimethylphenol, is marketed as a blend with polystyrene and impact modifiers under the trade name of Noryl thermoplastic resin.

Chemical properties. Phenolic end groups in poly(phenylene oxide)s react with oxidizing agents in a variety of ways; quinone formation and backbone redistribution can occur.

Poly(phenylene oxide)s undergo many substitution reaction. Reactions involving the aromatic rings and the methyl groups of DMPPO include bromination, displacement of the resultant bromine with phosphorus compounds and amines, and lithiation. Reactions at the open position of the aromatic ring include bromination, nitration, lithiation, and alkylation.

Syntheses. Synthetic processes include copper-catalyzed oxidative coupling of 2,6-disubstituted phenols and halogen-displacement polymerization of halophenols.

Copolymers. Copolymers of poly(phenylene oxide)s can be prepared in several ways. Oxidative coupling of mixtures of phenols usually provides random copolymers although, with a phenol that co-redistributes slowly, block copolymers can form. Another route is the oxidation of mixed dimers, which forms random copolymers. Chemical reactions involving the homopolymer, such as partial halogenation, can produce a copolymer.

Blends. Poly(2,6-dimethyl-1,4-phenylene oxide) and polystyrene make up one of the rare examples of compatible polymer systems. Blends of chemically distinct high polymers, however well-mixed, normally exist as separate phases, each composed almost exclusively of a single species; however, DMPPO and polystyrene are miscible in all proportions.

Noryl. The Noryl engineering thermoplastics are compatible polyblends formed by melt-blending DMPPO and high impact polystyrene

(HIPS) with proprietary stabilizers, flame retardants, impact modifiers, and other additives. Noryl is produced as sheet and for vacuum forming, but by far the greatest-volume use is in pellets for injection molding. Principal application areas are in water distribution, electrical-electronic appliances, business machines, and automobiles. General Electric is the only U.S. producer of Noryl resin.

Health and Safety Factors

The Noryl blends of DMPPO and HIPS and the polysulfones have FDA approval for re-use food appliations. Animal feeding studies of DMPPO itself have shown it to be nontoxic on ingestion. The solvents, catalysts and monomers that are used to prepared the polymers, however, should be handled with caution.

Polyethersulfones

The aromatic sulfone polymers are a group of high performance plastics of similar structure and properties (see Polymers containing sulfur, polysulfone resins). All are chemically polyethersulfones, ie, they have both aryl ether (ArOAr) and aryl sulfone (ArSO$_2$Ar) linkages in the polymer backbone. The simplest polyethersulfone consists of aromatic rings linked alternately by ether and sulfone groups:

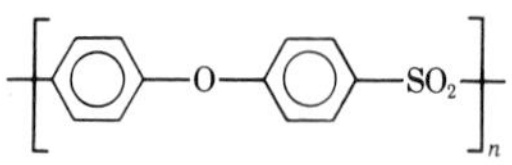

Syntheses. Aromatic polyethersulfones can be produced by two different routes: polyetherification and polysulfonylation.

Commercial products. The commercially available polyethersulfones, ie, Udel polysulfone, Victrex polyethersulfone, and Radel polyphenylsulfone, are rigid amorphous plastics, transparent in natural grades, and light-yellow to amber in color. All are hydrolytically stable and have good electrical properties which do not change over a wide range of temperature and frequency.

Polysulfone was the first of the sulfone polymers and its markets are developed more than those of the other polyethersulfones. Its dimensional stability, high use temperature, and good electrical properties suit it for a variety of electrical and electronic applications. The combination of hydrolytic stability, heat resistance, and FDA sanction for use in contact with foods and beverages is responsible for important applications in food processing and cookware, especially professional cookware. Applications of polyethersulfone are similar to those of polysulfone.

Union Carbide is the only producer of polysulfone and polyphenylsulfone.

Polyetheretherketones (Polyetheresters)

The polyetherification route to polyethersulfones can be adapted to the synthesis of polyethers containing strongly electron-withdrawing groups other than sulfone groups. Poly(1,4-oxyphenylenecarbonyl-1,4-phenylene) (PEEK) is produced by condensation of 4,4′-dichlorobenzophenone with 4,4′-dihydroxybenzophenone, or by the self-condensation of 4-chloro-4′-hydroxybenzophenone. It has a melting point of 367°C and a glass-transition temperature of 154°C. Although PEEK can be molded, its principal application areas are in coatings and electrical insulation for high temperature service.

Polyetherimides

Commercial aromatic polyetherimides are relatively small-volume commodities. Manufacture of one polyetherimide is based on the reaction of pyromellitic dianhydride and 4,4′-oxydianiline and the product is sold by DuPont as Kapton. A thermoplastic polyetherimide is sold by the General Electric Company as ULTEM resin. These materials possess excellent thermal stability, mechanical characteristics, and electrical properties.

Syntheses. The presence of the ether and imide functionalities in polyetherimides provides two general approaches for their syntheses. They can be prepared by a nucleophilic displacement polymerization similar to the halide displacement in polysulfone synthesis, or by a condensation of dianhydrides and diamines that is similar to normal polyimide syntheses (see Polyimides).

DWAIN M. WHITE
GLENN D. COOPER[‡]
General Electric Company

H.L. Finkbeiner, A.S. Hay, and D.M. White in C.E. Schildknecht and I.S. Skeist, eds., *Polymerization Processes, High Polymers*, Vol. 29, John Wiley & Sons, Inc., New York, 1977, p. 537.

W.J. MacKnight, F.E. Karasz, and J.R. Fried in *Polymer Blends*, Vol. 1, Academic Press., Inc., New York, 1978, p. 185.

D.M. White and co-workers, *J. Polym Sci. Poly. Chem. Ed.* **19**, 1635 (1981).

ETHYLENE OXIDE POLYMERS

Poly(ethylene oxide) resins are high molecular weight homopolymers produced by the heterogeneous catalytic polymerization of ethylene oxide (qv). The white, free-flowing resins are characterized by the structural formula:

$$-(CH_2CH_2O)_n-$$

The resins are available in a broad range of molecular weight grades, ie, as low as 100,000 to $> 5 \times 10^6$.

Physical Properties

Crystallinity. At molecular weights of 10^5–10^7, poly(ethylene oxide) forms a highly ordered structure. This has been confirmed by nmr and x-ray diffraction patterns and by the sharpness of the crystalline melting point (62–67°C).

Density. The density is 1.15–1.26 g/cm³.

Glass-transition temperature. The glass-transition temperature (T_g) of poly(ethylene oxide) has been measured over the molecular weight range of 10^2–10^7. These data indicate a rapid rise in the transition temperature to a maximum of -17°C for a molecular weight of 6000.

Solutions. Solubility. Poly(ethylene oxide) is soluble in water and several organic solvents, particularly chlorinated hydrocarbons (see also Resins, water soluble).

Concentration and molecular weight effects. The viscosity of aqueous solutions of poly(ethylene oxide) depends upon the concentration of the polymer solute, the molecular weight, the solution temperature, concentration of dissolved inorganic salts, and the shear rate. Viscosity increases with concentration and this dependence becomes more pronounced with increasing molecular weight.

Temperature effect. Near the boiling point of water, the solubility-temperature relationship undergoes an abrupt inversion. Over a very narrow temperature range, solutions become cloudy and the polymer precipitates; the polymer cannot dissolve in water above this precipitation temperature (98°C).

The viscosity of the aqueous solution is also significantly affected by temperature. In polymers of molecular weights $(1–50) \times 10^5$, the solution viscosity may decrease by one order of magnitude as the temperature is increased from 10°C to 90°C.

Effects of salts. The presence of inorganic salts in aqueous solutions of poly(ethylene oxide) reduces the upper temperature limit of solubility and viscosity.

Effect of shear. Concentrated aqueous solutions of poly(ethylene oxide) are pseudoplastic. The degree of pseudoplasticity increases as the molecular weight increases. Therefore, the viscosity of a given aqueous solution is a function of the shear rate used for the measurement.

Thermoplastic. Above the crystalline melting point of 63–67°C, high molecular weight poly(ethylene oxide) becomes thermoplastic. It can be molded, extruded, or calendered by means of conventional thermoplastic processing equipment (see Plastics processing).

[‡] Deceased.

Chemical Properties

Association complexes. The unshared electron pairs of the ether oxygens, which give the polymer strong hydrogen-bonding affinity, can also take part in association reactions with a variety of monomeric and polymeric electron acceptors. These include poly(acrylic acid), poly-(methacrylic acid), copolymers of maleic and acrylic acid, tannic acid, naphtholic and phenolic compounds, as well as urea and thiourea.

Oxidation. High molecular weight polymers of ethylene oxide are susceptible to oxidative degradation either in bulk, during thermoplastic processing, or in solution.

Manufacturing and Processing

Processes include heterogeneous catalytic polymerization, polymer suspension, thermoplastic processing, and cross-linked polymer processing.

Health and Safety Factors, Toxicology

Poly(ethylene oxide) resins are safely used in numerous personal-care applications and possess a very low degree of oral toxicity. Because of their high molecular weight, they are poorly absorbed from the gastrointestinal tract. They are relatively nonirritating to the skin and have a low sensitizing potential. The potential for eye injury also is slight. Flooding rabbits' eyes with an aqueous 5 wt% poly(ethylene oxide) solution causes only a trace of inflammation of the eye. Hence, both the dry resin as well as dissolved aqueous solutions can be handled quite safely.

Considerable interest has been shown in poly(ethylene oxide) for diverse applications in food, drug, and cosmetic products. Such uses fall within the scope of the Federal Food, Drug, and Cosmetic Act. The FDA has recognized and approved the use of poly(ethylene oxide) for specific food and packaging uses.

The WSR N and the WSR Polyox series are nontoxic, since the polymer chain common to both species is not broken down by metabolic process. Six separate toxicological studies were carried out on Polyox resins of molecular weights of ca $(1-40) \times 10^5$. The studies consisted of 2-yr and 90-d feeding studies and of metabolic studies. No toxic effects were noted.

Uses

Significant use properties of poly(ethylene oxide) are complete water solubility, low toxicity, unique solution rheology, complexation with organic acids, low ash content, and thermoplasticity.

Solutions are used in adhesives, acid cleaners, contact-lens fluids, detergents and lotions, drift control, foam stabilization, friction reduction, ink rheology, jet cutting, lubricants, oil-well flooding, and paint and coating thickener.

Solid mineral–polymer complexes are used as mineral-process handling aids, in groundwork fines retention, and for flocculation.

Solid polymer is used in retarding additives, in packaging film, hydrogels, monofilaments, coatings, and compound articles.

DAVID B. BRAUN
D.J. DELONG
Union Carbide Corporation

F.E. Bailey, Jr., and F.V. Koleske, *Poly(ethylene oxide)*, Academic Press, Inc., New York, 1976, p. 105.

POLYOX Water-Soluble Resins are Unique, Technical Bulletin, F44029C, Union Carbide Corporation, New York, 1981.

PROPYLENE OXIDE POLYMERS AND HIGHER 1,2-EPOXIDE POLYMERS

In this article, polymers are discussed that are made principally from propylene oxide or from propylene oxide with the addition of varied amounts of ethylene oxide, either as a random mixture or in blocks, as well as polymers of 1,2-butylene oxide. The polymers produced from these epoxides are applied in a wide range of industries. On the basis of chemistry and volume, the largest group is the polyisocyanates that are used for polyurethane articles such as rigid or flexible foam, coatings, elastomers, and adhesives (see Isocyanates; Urethane polymers). These polyethers range in molecular weight from about 250 to 7000. They are obtained from propylene oxide and up to 20% ethylene oxide which are added to a variety of initiator molecules.

Polyether diols are usually initiated with propylene glycol or dipropylene glycol. Polyether triols may be initiated with 1,2,6-hexanetriol, trimethylolpropane, or glycerol. Poly(oxyalkylene) derivatives of many other polyfunctional active hydrogen compounds also are used as initiators (see under Manufacture). These products are frequently called polyurethane chemicals or polyurethane polyols. Another large group of polyether products may be grouped by the terms industrial polyglycols or functional fluids. Included are surfactants, lubricants, hydraulic fluids, cosmetics, and pharmaceuticals, as well as many smaller-volume products, some for highly specialized uses.

Nonionic surfactants, another large family of polyether products, are usually molecules with a hydrophobic and hydrophilic portion. The latter is nearly always a block of oxyethylene units.

Physical Properties

The physical properties of these polyethers provide the basis for their utility. Physical properties are influenced by using mono-, di-, or polyfunctional initiator molecules and by the amount and type of alkylene oxide groups present in the polymer molecule. The properties thus affected are viscosity and viscosity index, water and oil solubility, and surface activity. Polyethers are liquids, except for the block polymers containing relatively long blocks of oxyethylene units. These are of interest mainly as surfactants.

Solubility. Polymers prepared from propylene oxide are soluble in almost all organic solvents.

Viscosity. The viscosity of the polypropylene glycols (diols) varies in the expected fashion, increasing with molecular weight.

Chemical Properties

The most important chemical property of these polymers in polyurethane applications is the hydroxyl end group. Its concentration, of course, varies inversely with the molecular weight of the polyether chain of which it is a terminal group. It may be either primary or secondary, that is, attached to either a primary carbon (methylene) group or a secondary carbon (methine) group. The secondary hydroxyl is considered to be less reactive than the primary in the reaction with isocyanate groups.

Antioxidants and other additives. As polyethers, these products have about the same susceptibility to oxidation as all ethers. To stabilize the product against normal handling conditions, only a small amount of antioxidant may be necessary (see Antioxidants and antiozonants).

Health and Safety Factors

High molecular weight polyethers derived from propylene oxide have a low acute oral toxicity and are not irritating either to eyes or skin. They can be stabilized by antioxidants, are noncorrosive, and have high flash points. Lower molecular weight products, especially the low hydroxyl equivalent-weight adducts used principally in rigid foams, may vary considerably in toxic hazard, depending upon the base compound and the amount of propylene oxide. Because of the wide variety of products and the range of possible additives, the manufacturer must be consulted for safety recommendations.

ROBERT A. NEWTON
Dow Chemical U.S.A.

N.G. Gaylord, ed., *Polyethers*, Vol. 1, Interscience Publishers, a division of John Wiley & Sons, Inc., New York, 1963.

Foams, 2nd ed., Desk Top Data Bank, The International Plastics Selector, Inc., San Diego, Calif., 1980.

J.H. Saunders and K.C. Frisch, *Polyurethanes Chemistry and Technology*, Vol. 1, Interscience Publishers, a division of John Wiley & Sons, Inc., New York, 1963.

TETRAHYDROFURAN AND OXETANE POLYMERS

The polymerization of THF and the oxetanes are classic examples of cationic ring-opening addition polymerization. Polytetrahydrofuran (PTHF) is a linear polymer consisting of a chain of methylene groups with an oxygen atom inserted after every fourth methylene group.

$$-(CH_2CH_2CH_2CH_2O)_n-$$
polytetrahydrofuran

The polymer is also referred to as poly(tetramethylene oxide) or polyoxytetramethylene. Substituted THFs generally do not polymerize and are not important. The structure of polyoxetane is closely related to that of PTHF:

$$-(CH_2CH_2CH_2O)_n-$$
polyoxetane

Polyoxetane differs only in that three methylene groups are between the oxygen atoms. In contrast to THF, however, substituted oxetanes polymerize readily. The oxetane of commercial significance is 3,3-bis(chloromethyl)oxetane (BCMO).

Polytetrahydrofuran

Physical Properties. Typical properties of PTHF are summarized in Table 1.

Chemical properties. End-group reactions. Polytetrahydrofuran can be prepared so that it has reactive end groups, eg, hydroxyl or amino, or relatively inert end groups, eg, alkoxy. Additional chemistry carried out on reactive end groups leads to polymers with transformed end groups.

The most commercially important PTHFs are probably the hydroxy-terminated polymers, ie, the poly(tetramethylene ether glycol)s, which are used to prepare polyurethanes and polyesters (see Urethane polymers; Polyesters, thermoplastic).

Main-chain reactions. Stability. The synthesis of PTHF involves the use of strong acid initiators. Thus, the polymer is quite stable to attack by bases but is somewhat acid sensitive. As a polyether, it is subject to oxidative attack and subsequent breakdown, much as is a monomeric aliphatic ether. The addition of common antioxidants, eg, amines or pyrocatechol, serves to inhibit these reactions and thereby imparts adequate stability to the polymer for commercial applications (see Antioxidants and antiozonants).

During polymerization. The PTHF chain is also subject to attack and reaction during the polymerization process. These reactions lead to randomization of the molecular-weight distribution or to formation of small amounts of macrocyclic oligomers.

Chemistry. The THF molecule is a planar five-membered ring with some internal strain resulting from repulsion of eclipsed hydrogens. Since the ring contains an oxygen atom with two unshared pairs of electrons, THF is a nucleophilic monomer in which steric interference to potential electron acceptors is low. A slightly negative free energy of polymerization of ca -3.35 kJ/mol (-800 cal/mol) at 25°C has been estimated from the experimentally derived heat of polymerization of -18.8 kJ/mol (-4.5 kcal/mol) and an entropy of polymerization of ca -75.3 J/(mol·K) (-18 cal/(mol·K)). Purity of reagents, dryness of apparatus, choice of initiator, atmosphere, monomer concentration, and solvent determine the amount and many properties of the polymer.

Copolymerization. Copolymerization of THF with other cyclic ethers is possible; even cyclic ethers that have little or no homopolymerizability copolymerize with THF.

Manufacture and processing of poly(tetramethylene ether glycol). Synthesis. The oldest and most readily visualized method of preparing PTHF glycol is initiation of the polymerization of THF with a protonic acid to form an OH head group and termination of the polymerization with water to produce an OH end group:

$$HX + n \, O\langle\rangle \rightarrow HO(CH_2)_4O)_{n-2}(CH_2)_4\overset{+}{O}\langle\rangle X^-$$
$$\downarrow$$
$$HO(CH_2)_4O)_nH + HX$$

Table 1. Typical Properties of Polytetrahydrofuran (PTHF) and Poly[3,3-Bis(Chloromethyl) Oxetane] (PBCMO)

Property	PTHF	PBCMO
melting temperature (T_m), °C	43[a], 58–60	186[a]
		180[b]
glass-transition temperature (T_g), °C	−86	7–32[c]
density		
amorphous (at 25°C), g/cm³	0.975	1.386
crystalline (at 25°C), g/cm³	1.07–1.08	1.47
compression modulus, MPa[d]		900
tensile strength, MPa[d]		41.4
high mol wt	29.0	
cured	16.8–38.3[e]	
elongation, %		35
high mol wt	820	
cured	400–740[e]	
modulus of elasticity, MPa[d]	97.0	
heat-distortion temperature, °C		
at 1.82 MPa[d]		85–93
at 0.48 MPa[e]		149
Izod impact (notched, at 22.8°C), J/m[f]		27
Shore A hardness	95	
Rockwell hardness		R100
thermal expansion coefficient (α)	4–7 × 10⁻⁴	< 7.5°C, 2 × 10⁻⁴
$= (1/V)(\partial V/\partial T)_p$, K⁻¹		7.5–40°C, 4 × 10⁻⁴
		> 180°C, 5 × 10⁻⁴
compressibility (β) $= (1/V)(\partial V/\partial T)_T$, kPa⁻¹[g]	4–10 × 10⁻⁷	
internal pressure (P_i), MPa[d]	281	
coefficient of expansion (dV_s/dT), cm³/(g·K)	7.3 × 10⁻⁴	
refractive index at (20°C)	1.48	
dielectric constant (k_c) (at 25°C)	5.0	3.0
solubility parameter (δ_p) (J/cm³)^{1/2}[h]	17.3–17.6	

[a] α-form.
[b] β-form.
[c] Depends on method of determination.
[d] To convert MPa to psi, multiply by 145.
[e] Depends on curing system.
[f] To convert J/m to ft·lbf/in., divide by 53.38 (see ASTM D 256).
[g] To convert kPa to psi, multiply by 0.145.
[h] To convert (J/cm³)^{1/2} to (cal/cm³)^{1/2}, divide by 2.045.

However, this method is full of severe problems which have led to the development of modified processes. One is a process that leads first to a diacetate intermediate from which the diprimary glycol is obtained by alcoholysis or hydrolysis.

Health factor. PTHF is a nohazardous material.

Oxetane Polymers

Physical properties. Extensive physical-property measurements have been reported for only the 3,3-bis(chloromethyl)oxetane polymer (PBCMO) (Table 1).

Chemical properties. The chemical considerations, as outlined for PTHF, also apply to most of the polyoxetanes. Linear low molecular weight oligomers of oxetanes have not received much attention until very recently. Functionally terminated copolyoxetanes are currently being explored for use as the soft segment of energetic polyurethane binders in solid-fuel rocket propellants.

Manufacture and processing. The chemistry of the polymerization of the oxetanes is much the same for THF.

P. DREYFUSS
The University of Akron

M.P. DREYFUSS
BF Goodrich Co.

P. Dreyfus, *Poly(tetrahydrofuran)*, Gordon & Breach, New York, 1982.

S. Penczek, *Makromol. Chem. Suppl.* **3**, 17 (1979).

T. Saegusa and S. Kobayaski, *Prog. Polym. Sci. Jpn.* **6**, 107 (1973).

P. Dreyfuss and M.P. Dreyfuss in C.H. Bamford and C.F.H. Tipper, eds., *Comprehensive Chemical Kinetics*, Vol. 15, Elsevier Scientific Publishing Co., Amsterdam, Neth., 1976, Chapt. 4.

POLYETHYLENE.   See Olefin polymers.

(POLYHYDROXY)BENZENES

Polyhydric phenols with more than two hydroxy groups, ie, the three positional isomers of benzenetriol, the three isomeric benzenetetrols, benzenepentol, and benzenehexol, are discussed below (see also Hydroquinone, resorcinol, and catechol).

Derivatives of these compounds or their corresponding quinones are of widespread occurrence in nature. They are abundant in plants and fruits as glucosides, chromones, coumarin derivatives, flavonoids, essential oils, lignins, tannins, and alkaloids (see Coumarin; Oils, essential; Lignin; Alkaloids). They also occur in microorganisms and animals. Many of these compounds have distinct properties and uses, eg, antibiotics, plant-growth factors, insecticides, astringents, antioxidants, toxins, sweeteners (qv), pigments and dyes, drugs, and many others (see Antibiotics; Plant-growth substances; Insect control technology; Antioxidants and antiozonants; Pharmaceuticals; Pigments; Dyes). The new and developing uses for the benzenepolyols and derivatives appear particularly valuable in the pharmaceutical and agricultural chemical areas.

The biochemical activity of the benzenepolyols is at least in part based on their oxidation–reduction potential.

Pyrogallol

Pyrogallol (1) which is of widespread occurrence in nature, is incorporated in tannins, anthocyanins, flavones, and alkaloids.

(1)

Properties. Pyrogallol forms colorless needles or leaflets, which gray on contact with air or light. Some of its other properties are mp 133–134°C and bp at atmospheric pressure with partial decomposition, 309°C. Pyrogallol is the strongest reducing agent among the benzenepolyols. Therefore, it is oxidized rapidly in air; its aqueous alkaline solution absorbs oxygen from the air and darkens rapidly. Sodium sulfite retards such oxidation.

Manufacture and synthesis. The commercial manufacturing process is based on Scheele's original procedure starting with crude gallic acid, which is extracted from nutgalls or tara powder. It proceeds according to the following equation:

$$C_6H_2(OH)_3COOH \rightarrow C_6H_3(OH)_3 + CO_2$$

The only U.S. manufacturers of pyrogallol are Mallinckrodt and BFC Chemicals.

Health and safety factors, toxicology. Pyrogallol is extremely poisonous. The principal symptom of poisoning attributable to pyrogallol is its effect on the red blood corpuscles, which break down and lose their hemoglobin.

Uses. The main commercial applications of pyrogallol are in pharmaceuticals and pesticides.

Derivatives. Derivatives of pyrogallol include gallic acid, propyl gallate, bismuth subgallate, alkyl gallates, gallic acid amides, gallein (pyrogallolphthalein; 4,5-dihydroxyfluorescein; tetrahydrofluoran; CI 45445), gallacetophenone (4-acetylpyrogallol, 2,3,4-trihydroxyaceto-phenone; Alizarin Yellow C), mescaline (2-(3,4,5-trimethoxyphenyl)ethylamine), and colchicine.

Hydroxyhydroquinone

Hydroxyhydroquinone (2) forms colorless plates from diethyl ether when freshly prepared. It occurs in many plants and trees in the form of ethers, quinoid pigments, coumarin derivatives, and complex compounds. It has strong reducing properties. Applications have been suggested in the synthesis of agricultural and photographic chemicals, drugs, and stabilizers.

(2)

Properties. Hydroxyhydroquinone forms plateles or prisms (mp 140.5°C). The compound is easily soluble in water, ethanol, diethyl ether, and ethyl acetate and is very sparingly soluble in chloroform, carbon disulfide, benzene, and ligroin.

Synthesis. The most convenient preparation of hydroxyhydroquinone is the reaction of *p*-benzoquinone with acetic anhydride in the presence of sulfuric acid or phosphoric acid. The resultant triacetate can be hydrolyzed to hydroxyhydroquinone.

Health and safety factors, toxicology. The LD_{50} of 1,2,4-trihydroxybenzene in mice after intracutaneous injection is 371 μg/g. Contact with hydroxyhydroquinone may blacken skin and fingernails.

Uses. Hydroxyhydroquinone has been used in hair and mordant dyes, healing plant wounds, and in corrosion inhibitors and adhesives.

Derivatives. Derivatives include scopoletin (6-methoxyumbelliferone), veriscolin (1,2,4-trihydroxy-3-methylbenzene), and precocene-2 (2,2-dimethyl-6,7-dimethoxy-2*H*-chromene).

Phloroglucinol

Phloroglucinol (3) is a colorless and odorless solid which is only sparingly soluble in cold water. Phloroglucinol occurs in many other natural products in the form of derivatives such as flavones, catechins, coumarin derivatives, anthocyanidins, xanthins, and glucosides.

(3)

There has been much interest in improved synthetic processes for phloroglucinol and in natural-product-derived food sweeteners, each of which are now characterized by a phloroglucinol nucleus in the structure.

Properties. Phloroglucinol forms odorless, colorless, sweet-tasting, rhombic crystals which tend to discolor on exposure to air or light. The dihydrate loses its water of crystallization at about 110°C (mp 113–116°C on quick heating); the anhydrous material melts at 217–219°C when heated rapidly.

Although most of the physical and chemical properties of phloroglucinol characterize it as a polyhydric phenol, in many cases it reacts in a tautomeric keto form or as the triketone, 1,3,5-cyclohexanetrione.

Manufacture and synthesis. The only commercial process in use in the United States through the 1970s involved the oxidation of 2,4,6-trinitrotoluene (TNT) with dichromate in sulfuric acid to 2,4,6-trinitrobenzoic acid. This was followed by the reduction of the nitro groups to amino groups with iron and hydrochloric acid and simultaneous decarboxylation to give 1,3,5-triaminobenzene. Acid hydrolysis at ca 108°C gave phloroglucinol in ca 75% yield.

Health and safety factors, toxicology. Phloroglucinol has low toxicity by ingestion. Prolonged, severe overexposure may disrupt the thyroid function. High dust concentration may cause respiratory irritation; the product is irritating to eyes and skin. Toxicity data include LD_{50} oral (rat) 5800 mg/kg.

Uses. Two of the principal commercial applications of phloroglucinol, ie, in the diazotype copying process and in textile-dyeing processes, are based on the ability of each mole of phloroglucinol to couple rapidly with 3 mol of diazo compound. Phloroglucinol also is used in resins and adhesives, as a plastics component or additive, as an intermediate for hydraulic fluids, as a rubber additive, as a photographic chemical, and as a starting material for priming compositions.

Derivatives. Derivatives include cotoin, conglomerone, griseofulvin, uvaretin, aflatoxin B, bioflavanoids (vitamin P complex), and hesperidin (hesperetin (7-rhamnoglucoside or 7-rutinoside)).

Benzenetetrols

1,2,3,4-Benzenetetrol. 1,2,3,4-Benzenetetrol (1,2,3,4-tetrahydroxybenzene, apionol) forms needles from benzene (mp 161°C). It is easily soluble in water, diethyl ether, ethanol, and glacial acetic acid and is sparingly soluble in benzene.

1,2,3,4-Benzenetetrol is best prepared by the hydrolysis of 4-aminopyrogallol hydrochloride.

Derivatives. The most important derivatives of 1,2,3,4-benzenetetrol are the ubiquinones.

1,2,3,5-Benzenetetrol. 1,2,3,5-Benzenetetrol (1,2,3,5-tetrahydroxybenzene) forms needles from water (mp 165°C). The compound is easily soluble in water, alcohol, and ethyl acetate and is insoluble in chloroform and benzene.

1,2,3,5-Benzenetetrol has been prepared by hydrolysis of 2,4,6-triaminophenol with dilute hydrochloric acid and by heating aqueous solutions of < 0.2 M 2,4,6-triaminophenol at > 130°C.

Derivatives. Derivatives include 3,6-dihydroxy-2,4-dimethoxyacetophenone and 3,4,5-trimethoxyphenol (antiarol). Many 1,2,3,5-benzenetetrol derivatives are used medicinally including khellin, eupatin and eupatoretin, and baicalein.

1,2,4,5-Benzenetetrol. 1,2,4,5-Benzenetetrol (1,2,4,5-tetrahydroxybenzene) forms leaflets from glacial acetic acid (mp 215–220°C). It is easily soluble in water, ethanol, and diethyl ether but is not quite as soluble in concentrated hydrochloric acid and glacial acetic acid.

1,2,4,5-Benzenetetrol is obtained by the reduction of 2,5-dihydroxy-1,4-benzoquinone.

Benzenepentol

Benzenepentol (pentahydroxybenzene) has been prepared by boiling 2,4,6-triaminoresorcinol diethyl ether in water, followed by ether cleavage with HI. The product is very soluble in water but very sparingly soluble in organic solvents.

Benzenehexol

Properties. Benzenehexol (hexahydroxybenzene), forms snow-white crystals when freshly prepared and collected in an inert atmosphere. Benzenehexol of good purity does not melt up to at least 310°C. It is sparingly soluble in water, ethanol, diethyl ether, and benzene.

Synthesis. Benzenehexol is made by the aeration of the glyoxal bisulfite addition product in sodium carbonate solution, isolation/acidification of the sodium salt of tetrahydroxybenzoquinone, and reduction of the quinone.

Derivatives. A considerable number of compounds that contain the benzenehexol structure possess therapeutic activity. Derivatives include tetroquinone (tetrahydroxy-p-benzoquinone), dipotassium salt of rhodizonic acid, inositols, myoinositol esters, and tetraaryloxybenzoquinones.

HANS DRESSLER
SAMUEL N. HOLTER
Koppers Company, Inc.

G.D. Clayton and F.E. Clayton, eds., *Patty's Industrial Hygiene and Toxicology*, 3rd ed., Wiley-Interscience, New York, 1982.

U.S. Pat. 4,182,912 (Jan. 8, 1980), J.W. Foley (to Polaroid Corp.).

U.S. Pat. 4,139,693 (Feb. 13, 1979), J.E. Schoenberg (to National Starch and Chemical Corp.).

POLY(HYDROXYBENZOIC ACID).   See Heat resistant polymers.

POLYIMIDES

Properties

Polyimides are characterized by the presence of the phthalimide structure in the polymer backbone.

The thermal stabilities of polyimides depend on their backbone compositions. The wholly aromatic systems, which in processed form are nearly all films, have similar properties; these are summarized in Table 1.

Synthesis

The synthesis that made polyimides usable is a two-step process. Dianhydrides react with diamines to form poly(amic acids)s, which are thermally or chemically cyclized to the polyimide. The advantage of this method is that the intermediate is soluble and can be processed. The processed form can then be cyclized, whereupon it becomes insoluble. This is the most common method for preparation of polyimides.

Table 1.   General Properties of Aromatic Polyimides

solubility	insoluble or soluble in conc H_2SO_4 or fuming HNO_3 with degradation; phenylated versions sol in $CHCl_3$
crystallinity	usually crystallizable
zero-strength temperature, °C	800–900
thermal stability in air, °C	> 500 by tga; 275–300 by isothermal methods
tensile strengths (film), MPa[a]	
at 25°C	172
at 500°C	28
elongation (film), %	
at 25°C	70
at 500°C	60
tensile modulus (film), MPa[a]	
at 25°C	2800
at 500°C	280
glass-transition temperature, °C	280–385[b]
dielectric constant at 1000 Hz	
at 23°C	3.1–3.7
dissipation factor at 100 Hz	
at 23°C	0.0013–0.002
volume resistivity (at 23°C and 50% rh), $\Omega \cdot cm$	
at 23°C	10^{17}–5×10^{18}
dielectric strength, V/μm	
at 23°C	180–270
color	colorless to deep red[c]
density, g/cm³	1.41–1.43
chemical resistance	varies widely with structure
radiation resistance	fair to ultraviolet, excellent to high energy (Van de Graaff 2 MeV and thermal neutrons)
flammability	self-extinguishing

[a] To convert MPa to psi, multiply by 145.
[b] Depending on method.
[c] Depending on diamine.

Uses

Uses have expanded from the early molded products to varnishes, laminating resins, molding powders, and adhesives.

PATRICK E. CASSIDY
Southwest Texas State Univ.

NEWTON C. FAWCETT
University of So. Mississippi

S.R. Sandler and W. Karo, *Polymer Synthesis*, Academic Press, Inc., New York, 1974, pp. 216–224; A.M. Wilson, D. Laks, and S.M. Davis, *Org. Coat. Plast. Prepr.* **42**, 470 (1980); G.M. Bower and L.W. Frost, *J. Polym. Sci.* **1**, 3135 (1963).

C. Arnold, *J. Polym. Sci. Macromol. Rev.* **14**, 265 (1979).

P.E. Cassidy, *Thermally Stable Polymers—Syntheses and Properties*, Marcel Dekker, Inc., New York, 1980, pp. 94–129.

POLYMERIZATION MECHANISMS AND PROCESSES

A polymerization reaction is the conversion of a particular compound to a large chemical multiple of itself (see Polymers). Classically, polymerization reactions that yield linear, high molecular weight products are classified into two main groups on the basis of a comparison of the structure of the repeating unit of the polymer with the structure of the monomer from which the polymer is derived. The two general divisions are addition polymerization and condensation polymerization.

An addition polymerization is a polymerization reaction yielding a polymeric product in which the molecular formula of the repeating unit is identical with that of the monomer, and the molecular weight of the polymer so formed is a simple summation of the molecular weights of all combined monomer units in the chain. For example, polystyrene (2) is obtained by the addition polymerization of styrene (1):

(1)   (2)

A condensation polymerization is a polymerization reaction yielding a polymeric product in which the repeating unit contains fewer atoms than the monomer or monomers and, necessarily, the molecular weight of the polymer so formed is less than the sum of the molecular weights of all the original monomer units that were combined in the reaction to form the polymer chain. For example, nylon-6,6 or poly(hexamethylene adipamide) (5) is obtained by the condensation polymerization of hexamethylenediamine (3) and adipic acid (4); the secondary product of the condensation is water (see Polyamides).

(3)   (4)   (5)

Reaction Types

Step-growth vs chain-growth polymerization. The general character of step-growth and chain-growth polymerization reactions is that each polymer chain grows at a relatively slow rate over a much longer period of time for the former than for the latter. In step-growth polymerization, there is generally only one type of reaction and the same basic mechanism is involved in the reactions of either two monomer units with each other or a monomer with the end group on a polymer chain, and/or two polymer-chain end groups with each other. Most condensation polymerizations occur by a step-growth mechanism. In contrast, a chain-growth polymerization generally consists of three different types of reactions: an

initiation reaction, which creates a highly active species; a propagation reaction, in which the only reaction possible is the addition of monomer to active polymer-chain end groups; and termination, in which the activity of the end groups is destroyed and the polymer chain can no longer add new monomer units. Most addition polymerizations occur by a chain-growth mechanism.

Copolymerization. More than one type of monomer may be involved in the polymerization reaction and more than one type of repeating unit may be generated in the polymer. The simultaneous polymerization of two or more monomers, ie, copolymerization, is a common occurrence and is very important to step-growth and chain-growth polymerizations.

In step-growth polymerization reactions, different monomers of the same general class with a given type of functional group generally show only minor differences in reactivity. In a broad series of diester monomers to be used for preparing polyesters, the difference in reactivities between ester functions is negligible compared to the differences in reactivities generally encountered for double bonds in a series of olefins. As a result, most copolymers prepared by step-growth copolymerization reactions contain essentially random placements of repeating units, the proportion of each monomer incorporated into the copolymer essentially being the same as the proportions in the original monomer mixture.

In contrast, strong selective effects often occur in chain-growth copolymerization reactions, and the compositions of the copolymers that are formed may be greatly different from the composition of the monomer mixture.

Stereochemistry

The chain-growth polymerization reaction of monomers containing carbon–carbon double bonds involves the formation of two new bonds at tetravalent carbon atoms as each new monomer adds onto the end of the growing polymer chain. The spatial arrangement of covalent bonds, or the configuration of the carbon–carbon bonds formed at the active terminal carbon atom of the growing polymer chain, is permanently established during the reaction, and two distinguishable diastereomeric configurations are possible. During the growth of the polymer, the regularity with which the two different configurations are established in the addition of each successive repeating unit has a marked effect on the long-range order in the macromolecule.

Most crystalline polyolefins have ordered chain structures consisting of either the triad (6) or the triad (7). A polymer containing principally repeating units with identical configurations, eg, triad (6), is an isotactic polymer, whereas a polymer containing principally units of exactly alternating configurations, eg, triad (7), is a syndiotactic polymer. A chain showing no regular order of repeating unit configurations is an atactic polymer (see also Olefin polymers, propylene).

(6)

(7)

Not all polymerization reactions involve the type of bond formation in which specific configurations are unequivocally established. Reactions in which only divalent or trivalent atoms are involved in bond formation generally do not necessarily lead to this stereochemical situation, and polymers formed by such reactions are free of the structural complexity that arises from the formation of dissymmetric centers occurring in the polymerization reaction.

Polymerization Processes

Most step-growth polymerization reactions are carried out in homogeneous systems by simple combination of two or more monomers in the melt in the absence of solvent. In contrast, a wide variety of methods is used experimentally and industrially for the preparation of chain-growth polymers and, in many cases, the reaction requires the presence of a solid catalyst or is complicated by the formation of a two-phase reaction system. In almost all cases of step-growth and chain-growth polymerizations, the reactions are conducted in an inert atmosphere.

Radical chain-growth polymerization reactions are considerably more important industrially than either cationic or anionic chain-growth reactions, but the latter two are used to some extent and their importance continues to grow. Similarly, chain-growth polymerization reactions involving the presence of solid catalysts, ie, heterogeneous polymerization reactions, have recently become very important industrially, particularly for the polymerization of olefin monomers to linear, high molecular weight polyolefins.

Radical chain-growth polymerization reactions of olefin monomers can be carried out in a number of different types of reactions systems, including bulk polymerization, solution polymerization, suspension polymerization, emulsion polymerization, and precipitation polymerization. Each of these methods is used industrially. Two other general methods of experimental interest only are gas-phase polymerization and solid-state polymerization.

Robert W. Lenz
University of Massachusetts

J. Brandrup and E.H. Immergut, eds., *Polymer Handbook*, 2nd ed., John Wiley & Sons, Inc., New York, 1975, Sect. II-5.

R.W. Lenz and F. Ciardelli, eds., *Preparation and Properties of Stereoregular Polymers*, D. Reidel Publishers, Dordrecht, Neth., 1980.

C.E. Schildknecht, ed., *Polymer Processes*, 2nd ed., Wiley-Interscience, New York, 1978.

POLYMERS

Polymers, or macromolecules, consist of both biological and nonbiological organic and inorganic materials. Biological polymers form the very foundation of life and intelligence, and provide much of the world's food supply (see Biopolymers; Carbohydrates; Proteins). Nonbiological organic polymers are primarily the synthetic materials used for plastics, fibers, and elastomers, but include a few naturally occurring polymers, eg, rubber (qv), wool (qv), and cellulose (qv). Today, these substances are truly indispensable and are essential for clothing, shelter, transportation, communication, and the conveniences of modern living.

A polymer (from the Greek *polys*, many; *meros*, part or unit) is a large molecule built up by the repetition of small, simple chemical units. In some cases the repetition is linear, much as a chain is built up from its links. In other cases the chains are branched or interconnected to form three-dimensional networks. The repeat unit of the polymer is usually equivalent or nearly equivalent to the monomer or starting material from which the polymer is formed.

The length of the polymer chain is specified by the number of repeat units in the chain. This is called the degree of polymerization (DP). The molecular weight of the polymer is the product of the molecular weight of the repeat unit and the degree of polymerization.

In contrast to the homopolymers, which are derived from a single repeat unit, macromolecules can exist as copolymers, which consist of chains made up of two or more chemically different repeat units joined in a somewhat regular sequence. The formation of copolymers is often accompanied by differences in chemical composition from molecule to molecule.

In contrast to the linear-chain polymers first mentioned, graft copolymers are an example of branched-chain polymers. Branching can also occur in homopolymers.

In commercial practice, cross-linking reactions may take place during the fabrication of articles made with thermosetting resins. The cross-linked network extending throughout the final article is stable to heat and cannot be made to flow or melt. In contrast, most linear polymers can be made to soften and assume new shapes by the application of heat and pressure. They are said to be thermoplastic.

Of particular interest are polymers with stereoregular configurations in which monomeric units follow one another along the chain in geometric arrangements according to some rule or tacticity.

Polymerization Processes

Polymerization processes can be divided into condensation and addition polymerization or, in more precise terminology, step-reaction and chain-reaction polymerization (see Polymerization mechanisms and processes).

Structure and Properties

In dilute solution, where the polymer chain is surrounded by small molecules, or in the melt, where it is in an environment of similar chains, the polymer molecule is in continual motion because of its thermal energy, and it assumes many different conformations in rapid succession. As a polymer melt is cooled, or as this molecular motion characteristic of polymers is restrained through the introduction of strong interchain forces, the nature of the polymer sample changes systematically in ways that are important in determining its physical properties and uses.

In addition to undergoing transition to a glassy state as the temperature is lowered, some polymers crystallize at temperatures below a crystalline melting point. The properties of crystalline polymers are highly desirable. Crystalline polymers are strong, tough, stiff, and generally more resistant to solvents and chemicals than their noncrystalline counterparts. These properties can be further improved by increasing intermolecular forces through the selection of highly polar polymers. By using inherently stiff polymer chains, crystalline melting points can be raised so that the desirable mechanical properties associated with crystallinity are retained at high temperatures. This has led to the development of engineering plastics (qv) capable of competing with metals and ceramics. In addition, the properties of crystalline polymers can be improved for materials in fiber form by the process of orientation or drawing. The result is the increased strength, stiffness, and dimensional stability associated with synthetic fibers (see Fibers, chemical).

Elastomers. All substances exhibiting a high degree of rubberlike elasticity contain long-chain structures. The restoring force that leads to elastic behavior results directly from the decrease in entropy associated with the distortion of a chain macromolecule from its most probable conformation. In order to maintain sufficient freedom of molecular motion to allow the distortions to take place rapidly, the polymer must be used at temperatures above its glass transition, ie, $> T_g$. In addition, it must be amorphous, at least in the undistorted state, and lightly cross-linked to inhibit flow (see Elastomers, synthetic).

Fibers. In contrast to elastomers, the requirements of high tensile strength and modulus-characteristic fibers are almost always obtained by utilizing the combination of molecular symmetry and high cohesive energy associated with a high degree of crystallinity. Usually, the fiber is oriented to provide optimum properties in the direction of the fiber axis.

General-purpose and specialty plastics. The wide range of plastics uses requires a variety of property combinations. In general, the properties of plastics are intermediate between those of fibers and elastomers, with much overlapping on either end.

Engineering plastics. In contrast to general-purpose and specialty plastics, the term engineering plastics (qv) or polymers is applied to materials that command a premium price, usually associated with relatively low production volume, because their outstanding balance of properties allows them to compete successfully with other materials (metals, ceramics) in engineering applications. They are strong, stiff, tough, abrasion-resistant materials capable of withstanding wide ranges of temperatures, and they are resistant to attack by weather, chemicals, and other hostile conditions. The value they contribute to the product justifies their higher price per kilogram.

Fred W. Billmeyer, Jr.
Rensselaer Polytechnic Institute

F.W. Billmeyer, Jr., *Textbook of Polymer Science*, 3rd ed., John Wiley & Sons, Inc., New York, 1984.

F.A. Bovey and F.H. Winslow, ed., *Macromolecules: An Introduction to Polymer Science*, Academic Press, Inc., New York, 1979.

H.-G. Elias, *Macromolecules*, Plenum Publishing Corp., New York, 1977.

POLYMERS, CONDUCTIVE

Three developments during the 1970s have contributed a resurgence of interest in polymers with conducting as opposed to insulating electrical-transport properties. First, polymeric materials are used as active elements in electronic devices, especially electrophotographic copying machines (see Electrophotography). Second, carbon- and metal-filled polymers are increasingly utilized as moldable semiconductors in the electronics industry. Third, modern studies of charge transport in polymers and the discovery that strong oxidizing, eg, AsF_5, and reducing, eg, Na, agents can render the electrical conductivity of certain unsaturated-backbone polymers such as polyacetylene initially semiconducting and ultimately metallic in character have motivated extensive current research into semiconducting polymers and synthetic metals (see Semiconductors, organic).

Two themes characterize modern research, development, and applications of conducting polymers. First, materials-development activities are guided by the goal of combining the desirable properties of polymers, eg, low cost, light weight, moldability, and mechanical flexibility, with acceptable electrical behavior characteristic of semiconductors or metals. Achievement of this goal is elusive because most polymers exhibit negligible densities of intrinsic bulk carriers and tend to trap injected charges. The second theme is the construction of microscopic models of the charge densities, nq, and mobilities, μ. Electronic conduction occurs via electron [$q_i = -e$, $e = 1.6 \times 10^{-19}$ C (5.8×10^{-16} A·h)], or holes ($q_i = e$) or both. An important goal of this activity is the establishment of sufficiently reliable relationships between molecular architecture and the electrical properties of polymers that the preparation of polymeric materials of specified conductivity and mobility can be systematized.

The focus of most current research is on the molecular doping of polymers. This technique for producing acceptable electrical properties has been utilized commercially for nearly a decade in photoconductors employed in electrophotographic copying machines.

Nature of Charges in Polymers

The polymeric solid state. The two central structural distinctions between polymers and familiar covalent crystalline semiconductors are the molecular character and the lack of periodic long-range order of polymeric materials.

The combined influence of molecularity and disorder can lead to profound differences between the fundamental physical phenomena that occur in traditional covalent network semiconductors and those that are characteristic of semiconducting organic materials. The molecular character of polymers makes electron motion along the individual macromolecules one-dimensional. This reduced dimensionality implies that, even if polymeric materials were perfectly periodic crystalline solids, their electronic properties could be governed by the occurrence of certain types of collective ground states, which characteristically occur in one-dimensional systems. The occurrence of disorder in polymers leads to the localization of injected charges and hence to the insulating behavior of most polymeric materials.

Relaxation. When a charge is injected into a polymer, it induces changes in the electronic charge density and atomic positions both on the molecular site that it occupies (intramolecular relaxation) and on neighboring molecular sites (intermolecular relaxation). This phenomenon is called relaxation and leads to a lowering of the energy of the composite system of added charge plus polymer by an amount called the relaxation energy.

Localization. The local or extended nature of charges in polymers is determined by a competition between fluctuations in the local site energies, which tend to localize the changes, and the hopping integrals

for intersite charge transfer, which tend to delocalize the charges. In practical materials, the fluctuations almost always win yielding localized charges and the electrical properties characteristic of insulators.

Insulator-metal transitions. *Disordered Semiconductor Model.* The molecular-doping approach to conducting polymers involves inserting molecular dopants into nominally semiconducting or insulating materials to form charge-transfer complexes with the macromolecular chains. Common dopants are *p*-tetracyanoquinodimethane (TCNQ), AsF_5, and I_2, all of which are acceptors, ie, they extract electrons from the host polymer to form a molecular anion, eg, TCNQ or I_3^-. The central issues in describing electrical transport in the doped polymers are the nature, density, and mobility of the charges produced by the charge-transfer process. It must be determined whether the mobile carriers are the excess electrons on the molecular anions, positive holes induced by the charge transfer on the otherwise undistorted macromolecular chains, or some type of structural defect associated with deformations of the polymer macromolecules induced by the charge-transfer process.

Collective ground states. Under certain circumstances, macromolecules exhibit collective ground states induced by the interaction of their valence electrons with each other or with the molecular motion of the macromolecular chains. These collective states may be either superconducting or semiconducting. Superconducting behavior has been observed only for polythiazyl, whereas many unsaturated backbone materials, eg, polyacetylene, $(CH)_x$, and the polydiacetylenes, are thought to exhibit collective semiconducting states.

The possibility of collective behavior in macromolecules is significant for two reasons. First, it greatly enriches the list of candidates for mobile carriers in phenomenological models. Also, the occurrence of such behavior has implications for the molecular design of polymeric metals. In particular, to achieve the goal of preparing synthetic metals it is necessary to suppress the natural tendency of polymers to become semiconductors by virtue of the formation of a collective state.

Preparation of Conducting Polymers

Pyrolysis. This approach to the preparation of conductive polymers consists of eliminating heteroatoms, eg, halogens, oxygen, and nitrogen, from the polymer by heating and forming extended aromatic structures approaching that of graphite (see Carbon, graphite).

The products of polymer pyrolysis can take the form of powders, films, or fibers, depending on the form and nature of the starting polymer and the pyrolysis conditions. Electrical characterization of the products includes measurement of the conductivity and its activation energy, the mobility, and the Seebeck coefficient to determine the sign of the majority carriers.

Molecular doping. Polymeric photoconductors. Polymeric photoconductors are polymers that have low dark conductivity but good mobilities [$\mu > 10^{-7}$ cm²/(V·s)] of carriers generated by the absorption of light. Since the carriers are generated by light rather than thermally, such materials are referred to as photoconductors rather than semiconductors.

The problem of constructing a polymer that combines high carrier mobilities, low intrinsic carrier densities, and high photogeneration efficiencies is a formidable one. Therefore, many polymeric photoconductors are multilayer devices in which the carriers are created from absorbed light in one layer, ie, the photogenerator layer, and drift across the sample in one or more transport layers. In both layers, the trapping of the photogenerated charge must be small. The photogenerator layer is designed to yield high photogeneration efficiencies, and the transport layers provide high carrier mobilities. Both layers may consist of molecularly doped polymeric materials.

Synthetic metals. The primary strategy in the design of molecularly doped polymers for synthetic metals is to increase the conductivity of conjugated polymer systems, which are characterized by conductivities of ca 10^{-12}–10^{-5} S/cm and, therefore, possess some charge carriers. The number of charge carriers increased by doping, and extension of the conjugation length, increases the mobility of the carriers. However, doping also increases the degree of disorder in the composite doped polymer. This disorder tends to localize the carriers, thereby reducing

Table 1.   Data for Typical Commercially Available Conductive Polymer Composites

Composite	Tensile strength, MPa[a]	Tensile elongation, %	Flexural strength, MPa[a]	Flexural modulus × 10^4, MPa[a]	Izod impact strength, J/m[b]	$\sigma_{25°C}$, S/cm
polycarbonate (PC)	65	94	82.1	2.5		10^{-16}
PC + 20 wt% Al flake	47	5	77.9	3.99	84.4	10^{-15}
PC + 30 wt% Al flake	41	3	74.5	4.54	107	1
PC + 25 wt% metallized glass	82.1	4	113	6.61	59.8	10^{-7}
PC + 40 wt% metallized glass	86.9	5	123	9.79	72.6	10^{-1}
PC + 10 wt% PAN carbon fiber	78.6	6	123	5.59	58.2	10^{-8}
PC + 40 wt% PAN carbon fiber	181	6	232	21.5	101	10^{-2}
PC + 25 wt% pitch carbon fiber	69.7	5	101	5.97	59.8	10^{-6}
PC + 40 wt% pitch carbon fiber	83.4	5	120	10.1	57.1	10^{-2}
nylon-6,6 (N-6,6)	52	78	68	2.56	52	10^{-14}
N-6,6 + 40 wt% metallized glass	87.6	4	132	9.79	55	10^{-7}
N-6,6 + 25 wt% metallized glass	71.0	4	97.2	6.22	33	10^{-13}
N-6,6 + 5 wt% carbon black	41	4	55	1.85	21	10^{-12}
N-6,6 + 15 wt% carbon black	26	8	27	1.19	23	10^{-3}
N-6,6 + 20 wt% pitch carbon fiber	61	3	170	0.7	69	10^{-4}
N-6,6 + 40 wt% pitch carbon fiber	100	5	167	13.7	32	10^{-2}
N-6,6 + 10 wt% PAN carbon fiber	86	8	135	5.46	28	10^{-7}
N-6,6 + 40 wt% PAN carbon fiber	190	4	280	23.9	72	5

[a] To convert MPa to psi, multiply by 145.

[b] To convert J/m to ft·lbf/in., divide by 53.38 (see ASTM D 256).

their mobility. Therefore, it is possible that the conductivities exhibit a maximum value of $\sigma_{25°C}$ a ca 100 S/cm at moderate doping levels. At lower doping levels, the carrier densities decrease; at higher levels, the mobilities can decrease faster than the carrier densities increase. Consequently, the structure of the doped polymeric solid state exerts a decisive influence on the achievable values of the conductivity.

Filled polymers.   The aim of obtaining modest conductivity values (σ = ca 0.001 S/cm) with good mechanical properties can be achieved by filling inert polymers with conductors, eg, metals or carbon black. The conductivity depends upon charge transfer between the particles of the conductor. If properly chosen, the filler can also improve the mechanical properties of the composite relative to those of the polymer.

Fillers.   The filler material must be characterized by high conductivity (see Fillers). Other criteria determine the cost of the material. The following materials are used in order of increasing current costs on a weight basis: carbon black, metals, metallized glass fibers, pitch-derived carbon fibers, and polyacrylonitrile-derived carbon fibers.

Polymer matrix.   The polymer matrix often determines the mechanical properties of the composite system. Therefore, the polymer is chosen to yield the mechanical properties required by the particular application.

Uses.   A number of companies produce and market conductive composites in the United States. Representatives systems are indicated in Table 1.

Conductive polymers are used for static-charge elimination, electromagnetic shielding, and video disks.

CHARLES B. DUKE
HARRY W. GIBSON
Xerox Corporation

R.B. Seymour, ed., *Conductive Polymers*, Plenum Press, New York, 1981.

A.R. Blythe, *Electrical Properties of Polymers*, Cambridge University Press, Cambridge, UK, 1979, pp. 123–132.

J. Mort and G. Pfister, eds., *Electronic Properties of Polymers*, Wiley-Interscience, New York, 1982.

POLYMERS CONTAINING SULFUR

POLY(PHENYLENE SULFIDE)

Although poly(*p*-phenylene sulfide), which frequently is referred to as poly(phenylene sulfide) or PPS, is classified as one member of a family of new engineering plastics (qv), it is by no means new to chemistry. It was prepared at the same time as the phenolic resins (qv), which generally are considered to be the precursors of all synthetic resins. Structurally, the polymer backbone of PPS is composed of a series of alternating aromatic rings and sulfur atoms.

poly(*p*-phenylene sulfide)

Properties

Virgin poly(phenylene sulfide) is isolated as an off-white powder and is a linear material of modest molecular weight. Although exact molecular weight determinations by conventional polymer-analysis techniques are precluded by insolubility characteristics, recent dilute-solution light-scattering studies, which were conducted in chloronaphthalene at 235°C, indicate that the molecular weight of virgin resin is ca 18,000, ie, n = ca 170. The polymer has a very low melt viscosity (3–5 Pa·s (30–50 P)) when measured at low shear rates. According to differential thermal analysis results, amorphous PPS has a broad and high melting point (T_m 285°C) and a moderate glass-transition temperature (T_g 85°C).

Curing effects.   Possibly the most important characteristic of the modest molecular weight version of PPS is its ability to undergo change upon heating in the presence of oxygen. This change generally represents alterations in properties and characteristics as a result of increasing molecular weight. Investigations based on model systems indicate that the chemistry involves varying degrees of oxidation, cross-linking, and chain extension. This change is referred to as curing.

Miscellaneous.   In addition to its thermosetting–thermoplastic attributes as an organic polymeric material, PPS possesses a unique set of characteristics including good thermal stability, unusual insolubility, resistance to chemical environments, and inherent flame resistance.

Crystallinity.   PPS, as prepared by the Phillips process, is a highly crystalline polymer (see Manufacture and Processing).

Manufacture and Processing

Polymer production.   PPS has been produced solely by Phillips Petroleum Company since 1973 in a commercial facility in Borger, Texas.

The steps involved in PPS production include the following: preparation of sodium sulfide from aqueous caustic and aqueous sodium hydrosulfide in a polar solvent, removal of water from this feedstock by distillation, production of polymer from the sodium sulfide stream and

Table 1. *m*- and *p*-Poly(Phenylene Sulfide) Copolymers

Wt% para	T_g, °C	T_m, °C	Solubility in tetrahydrofuran, wt%
100	83	284	< 0.1
75	68	205	< 0.1
50	49	a	> 20
25	27	a	42
0	15	a	0.5

a These polymers are amorphous and soften at ca 90–100°C.

p-dichlorobenzene at elevated temperature in the polar solvent, polymer recovery, washing to remove the sodium chloride produced as a by-product, drying, and packaging. The polymer produced in the above process is linear and of modest molecular weight and mechanical strength. It can be used directly in the production of coatings by slurry-coating procedures. However, the principal use of the virgin polymer is as a feedstock for the production of various molding-grade resins. PPS is used in injection molding, compression molding, free sintering, and coatings.

Health and Safety Factors (Toxicology)

Poly(phenylene sulfide) compounds have been approved for a number of applications where safety is of the utmost importance, and certain grades also comply with government regulations for contact with food.

Special precautions should be observed when molding or curing Ryton PPS compounds at or above 371°C because of the off-gases from decomposition. The gases, ie, sulfur dioxide and carbonyl sulfide, and products of decomposition are considered irritants to the mucous membranes. As with other plastic materials, adequate ventilation of the molding-shop area is recommended when injection-molding PPS compounds.

The acute oral toxicity test in albino rats shows that the resin is essentially nontoxic by ingestion and, when in contact with the skin and eyes, it produces a minimum of irritation.

Under the terms of the Federal health and safety regulations, exposure to inert dusts, eg, PPS, should not exceed time-weighted average concentrations of 5 mg/m³ of air, including a ceiling concentration of 10 mg/m³.

Uses

Because of its unique combination of properties and because it can be made to provide a wide variety of melt viscosities, poly(phenylene sulfide) is used in large, thick-walled mechanical parts as well as in tiny and delicate electrical components. By far the largest fabrication technique used in production is injection molding. Electrical, electronic, and mechanical applications comprise the largest markets for PPS compounds.

Copolymers and New Polymers

An entire family of arylene sulfide polymers can be prepared by substitution of other polyhalogenated aromatics for all or part of the *p*-chlorobenzene in the PPS polymerization process.

An illustration of the variations in polymer properties that can be achieved through copolymerization is the preparation of a series of copolymers based upon mixtures of *m*-dichlorobenzene and *p*-dichlorobenzene and sodium sulfide (see Table 1). As the meta content increases, the mp and T_g of the polymers decrease and, at or above 50 wt% meta content, the polymers are no longer crystallizable. Many other types of copolymers can also be prepared.

H. WAYNE HILL, JR.
D.G. BRADY
Phillips Petroleum Company

H.W. Hill, Jr., and J.N. Short, *Chemtech* **2**, 481 (1972).

H.W. Hill, Jr. and D.G. Brady, *Polym. Eng. Sci.* **16**, 831 (1976).

H.W. Hill, Jr., *Ind. Eng. Chem. Prod. Res. Dev.* **18**, 252 (1979).

POLYSULFIDES

The polysulfide polymers occur as solids and as thiol-terminated liquid polymers, which are convertible to the solid form. The original polysulfide polymers were solid elastomers and were first marketed in 1929; later versions of these are currently being marketed. The bulk of the polysulfide polymers are manufactured and sold as liquid polymers with thiol terminals. The general procedure for their synthesis and manufacture calls for the reaction of dichloro aliphatic compounds with an excess of aqueous sodium polysulfide to ensure complete polymerization (see Thiols).

The polysulfide polymers derive their utility from their unusually good resistance to solvents and to the environment, and to their good low temperature properties. The solid elastomers in the vulcanized cured state are used in printing rolls, paint-spray hose, solvent hose, gaskets, and gas-meter diaphragms. The liquid polysulfide polymers are used mainly in sealants (qv); the largest application is as sealants for double-pane insulating glass windows (see Insulation, thermal). Other applications are general sealants and high quality sealants for building construction, boat hulls and decks, printing rolls, aircraft integral fuel tanks, and aircraft bodies.

Properties

The polysulfide liquid polymers have a rather typical viscosity–molecular weight relationship; the viscosity of the polymers varies approximately as the cube of the molecular weight. This relationship seems to be true over a broad range of molecular weights and degrees of branching.

A number of liquid polysulfide polymers have been commercialized under the trade name LP (Morton Thiokol). The physical properties of these polymers in the uncured liquid form are listed in Table 1.

The most common way to convert the liquid polysulfide polymers to solid elastomers is to oxidize the terminal thiol groups to disulfides. This is usually carried out in a formulation along with fillers (qv), plasticizers (qv), and curing-rate modifiers, eg, stearic acid as in typical sealant compounds. The mole percent of cross-linking agent used in making the polymer affects the modulus, elongation, and hardness of the cured polymer.

Solvent resistance. A very important property of the polysulfide polymers is their outstanding resistance to solvents. The solvent resistance depends quite closely on the sulfur content of the polymer.

Low temperature properties. The use of bis(2-chloroethyl) formal as the predominant monomer in the manufacture of polysulfide liquid polymers results in the formation of polymers with very good low temperature properties. The glass-transition temperature T_g of these disulfide polymers is ca −59°C.

Stress relaxation. Depending on the degree of cure, the polysulfide polymers can undergo varying degrees of stress relaxation. The rate of stress relaxation depends on temperature, uv light, or catalysts, especially basic inorganic salts such as sodium monosulfide, sodium hydroxide, and amines.

High temperature properties. The bulk of the commercially available polysulfide polymers containing the repeating ethylene formal disulfide structure are generally useful up to ca 100°C. At substantially above 100°C in the cured and uncured state and depending on the type of cure, thermal degradation of the polymer may be observed as, for example, continued weight loss and a hardening of the polymer to a plasticlike material.

Manufacturing and Processing

Polymerization. Polysulfide polymerization belongs to the class of condensation polymerizations.

Copolymers. Random polysulfide copolymers can be formed by the reaction of a mixture of two monomers, eg, bis(2-chloroethyl) formal and ethylene dichloride.

Processing. The polysulfide elastomers Thiokol FA and Thiokol ST are compounded in standard rubber-processing equipment, eg, rubber mills or sigmoid-bladed mixers (see Rubber compounding).

Table 1.   Physical Properties of Commercial Polysulfide Liquid Polymers[a]

	LP-31	LP-2	LP-32	LP-3	LP-33	LP-8	LP-5	LP-12
viscosity (at 25°C), dPa·s (= P)	800–1400	350–450	350–450	7–12	14–16.5	2.5–3.5	100	400
molecular weight	7500	4000	4000	1000	1000	600	2500	4000
specific gravity	1.31	1.27	1.27	1.27	1.27	1.27	1.27	1.27
refractive index	1.57		1.5689		1.5689	1.557		
pour point, °C	45–50	45–50	45	−15	5–10	−25		
flash point (open cup), °C	235	232	235	216	204	182		
fire point (open cup), °C	246	246	252	241	241	204		
cross-linking agent, wt%	0.5	2	0.5	2	0.5	2	2	0.1

[a] LPs are reduced polymers of 1,2,3-trichloropropane, 1,1′-[methylenebis(oxy)]bis[2-chloroethane], and sodium sulfide (Na_2S_x).

Health and Safety Factors (Toxology)

In general, the polysulfide polymeric products are regarded as safe. Under the conditions of intensive exposure, the polymers may cause nasal irritation and, therefore, they should be used in an adequately ventilated environment. Acute oral toxicity tests of polysulfide liquid polymers have given results more than or equal to 3.5 g/kg, which is similar to table salt. As a precautionary measure, it is advised that protective gloves be used and that repeated contact with the skin should be avoided whenever possible.

S.M. ELLERSTEIN
E.R. BERTOZZI
Morton Thiokol Corporation

E.R. Bertozzi, *Rubber Chem. Technol.* **41**, 114 (Feb. 1968).

M.B. Berenbaum in N.G. Gaylord, ed., *Polyethers*, Pt. 3, Interscience Publishers, a division of John Wiley & Sons, Inc., New York, 1966, Chapt. XIII.

M.B. Berenbaum in N.M. Bikales, ed., *Encyclopedia of Polymer Science and Technology*, Vol. 11, Interscience Publishers, a division of John Wiley & Sons, Inc., New York, 1969, pp. 425–447.

POLYSULFONE RESINS

Polymers that have repeating units containing structures of the general formula

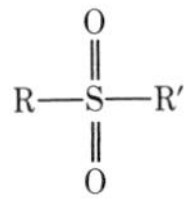

$$R-\overset{\displaystyle O}{\underset{\displaystyle O}{\overset{\displaystyle \|}{\underset{\displaystyle \|}{S}}}}-R'$$

are referred to as polysulfones. The aromatic thermoplastic variety has been made in several high molecular weight forms, which are being used increasingly in widely diversified applications. Emphasis in the following discussion is on the most commercially significant aromatic resins of the sulfone family.

The first of the sulfone polymers to be introduced commercially was Union Carbide Corporation's Bakelite polysulfone in 1965. Known as Udel polysulfone, it is by far the most prevalent of the four polysulfone resins. In 1967, the 3M Company introduced Astrel 360 polyarylsulfone, for which the manufacturing and marketing rights now belong to the Carborundum Company. Victrex polyethersulfone was first marketed by ICI Ltd. in 1972, and the most recent sulfone polymer, Radel polyphenylsulfone, was added to this resin family by Union Carbide Corporation in 1976. The characteristic feature of each is the highly resonant diaryl sulfone grouping. As a result of the sulfur atom being in its highest state of oxidation and the enhanced resonance of the sulfone group in the para position, these resins exhibit excellent oxidation resistance and thermal stability.

Udel Polysulfone

Properties. The physical properties of Udel polysulfone are summarized in Table 1. The resin's ability to retain a large proportion of its

Table 1.   Physical Properties of Udel Polysulfone[a]

Property	Value (average)
color	amber
clarity	transparent; 5% haze
refractive index	1.633
density, g/cm³ (lb/in.³)	1.24 (0.0447)
glass transition temp (T_g), °C	190
Rockwell hardness	M69, R120
water absorption, wt%	
after 24 h	0.3

[a] At 22°C unless otherwise noted.

physical, mechanical, and electrical properties over a wide range of temperatures (−101°C to >149°C) and in the presence of hot water or steam has led to most of its current applications.

Uses

The primary markets for polysulfone include appliances and cookware; chemical, food, and beverage processing; electrical and electronic components; medical products; and some aerospace and automotive components.

Astrel 360 Polyarylsulfone

A Friedel-Crafts reaction is used in the manufacture of Astrel 360 resin. An overview of the physical properties of Astrel 360 polyarylsulfone is provided in Table 2.

Table 2.   Properties of Astrel 360 Polyarylsulfone

Property	Value
specific gravity	1.36
T_g, °C	290
color	clear
water absorption (in 24 h), wt%	1.4
tensile strength, MPa[a]	90
tensile modulus, MPa[a]	2550
Izod notched impact strength, J/m[b]	106–212
Rockwell hardness	M110
thermal conductivity, W/(m·K)[c]	0.19
coefficient of linear expansion, mm/(mm·°C) × 10⁻⁵	4.7

[a] To convert MPa to psi, multiply by 145.
[b] To convert J/m to (ft·lbf)/in., divide by 53.38.
[c] To convert W/(m·K) to (Btu·in.)/(ft²·h·°F), multiply by 6.94.

The two most significant markets for polyarylsulfone are in the electric and electronics industry.

Victrex Polyethersulfone

As in the production of polysulfone, the polymerization of polyethersulfone is achieved by means of high temperature, nucleophilic substitution. A listing of many polyethersulfone properties is given in Table 3.

Table 3. Properties of Victrex Polyethersulfone

Property	Value
specific gravity	1.37
T_g, °C	ca 223
water absorption (in 24 h), wt%	0.43
tensile strength at 20°C, MPa[a]	83
tensile modulus, MPa[a]	2441
tensile elongation (at break), %	40–80
flexural strength, MPa[a]	129
Izod notched impact strength, J/m[b]	90
Rockwell hardness	M88
thermal conductivity, W/(m·K)[c]	0.18
coefficient of linear expansion, mm/(mm·°C) × 10⁻⁵	5.5

[a] To convert MPa to psi, multiply by 145.
[b] To convert J/m to (ft·lbf)/in., divide by 53.38.
[c] To convert W/(m·K) to (Btu·in.)/(ft²·h·°F), multiply by 6.94.

Table 4. Properties of Polyphenylsulfone

Property	Value
melt flow (at 400°C), 0.30 MPa[a], g/10 min	10
density, g/cm³	1.29
tensile strength (at yield and 22°C), MPa[a]	71.7
tensile elongation (at yield and 22°C), %	7.0
tensile modulus (at 22°C), MPa[a]	2140
tensile elongation (at break and 22°C), %	60
notched Izod impact (3.2 mm), J/m[b]	
at 22°C	640
dart impact (3.2 mm) J[c]	
at 22°C	> 136
coefficient of linear thermal expansion, mm/(mm·°C) × 10⁻⁵	5.5
specific heat, J/(g·°C)[d]	
20–200°C	1.17

[a] To convert MPa to psi, multiply by 145.
[b] To convert J/m to (ft·lbf)/in., divide by 53.38.
[c] To convert J to ft·lbf, divide by 1.36.
[d] To convert J/(g·°C) to Btu/(lb·°F), divide by 4.184.

The markets in which most PES resins are used are electric/electronics and aircraft. Victrex PES is used in the production of medical devices and has been cleared by the FDA for use in food-contact articles under 21 CFR 177.2440.

Radel Polyphenylsulfone

The specifics concerning raw materials and polymerization techniques for Radel polyphenylsulfone are proprietary. In terms of material properties and performance, polyphenylsulfone is in many ways an improvement over Udel polysulfone. A list of property data is presented in Table 4.

Outstanding among the characteristics of polyphenylsulfone is its resistance to degradation by heat and hydrolysis, its toughness, and its desirable combustion properties.

Radel polyphenylsulfone is used in place of metal in many applications. Its combustion characteristics allow it to be considered for a number of aerospace and safety-equipment applications.

NICOLAAS J. BALLINTYN
Union Carbide Corporation

E.J. Goethals in N.M. Bikales, ed., *Encyclopedia of Polymer Science and Technology*, Vol. 13, Interscience Publishers, a division of John Wiley & Sons, Inc., New York, 1969, pp. 448–477.

R.N. Johnson in N.M. Bikales, ed., *Encyclopedia of Polymer Science and Technology*, Vol. 11, Interscience Publishers, a division of John Wiley & Sons, Inc., New York, 1969, pp. 447–463.

R.N. Johnson, A.G. Farnham, R.A. Clendinning, W.F. Hale, and C.N. Merriam, *J. Polym. Sci. Part A-1* 5, 2375 (1967).

POLYMERS OF HIGHER OLEFINS. See Olefin polymers.

POLYMETHINE DYES

Polymethine dyes are colored substances in which a series of —CH= (methine) groups connect two terminal groups of a chromophore. In typical polymethine dyes and related polymethine radicals, the color can be varied throughout the visible region by changing the number of methines between X and Y (see also Color).

polymethine dyes $\left[X(CH=CH)_n CH=Y \right]$

polymethine radicals $\left[X(CH=CH)_n Y \right]$

Polymethine dyes are divided into simple polymethine dyes having nonheterocyclic terminal groups like aminopolymethines or 3H-indolopolymethines, cyanine dyes (qv) having conjugated heterocyclic groups linked by the polymethine chain, and polymethine radicals, which encompass generally less stable polymethines with either heterocyclic or nonheterocyclic groups. The simple polymethines are described in this article; the chromophore and terminal-group structures in Figure 1 are typical of dyes in this class. Polymethine and cyanine are often used interchangeably as generic terms for all polymethine dyes, and *Chemical Abstracts* uses cyanine as the main polymethine descriptor (see also Dyes and dye intermediates).

Synthetic polymethine dyes have been known since 1856 and are widely used as fabric dyes, filter dyes, biological stains, and to some extent as spectral sensitizers for photographic, xerographic, and photopolymer systems (see Dyes, sensitizing; Electrophotography; Photoreactive polymers). The symmetrical dye containing the 3.3-dimethyl-3H-indole nucleus was first prepared in 1924 (Astraphloxin FF (**1**) with R = methyl and *n* = 1). It was too unstable to be an important fabric dye, but related dyes from the 3H-indole nucleus were marketed as Astrazon dyes in the late 1930s. Substantial quantities of Astrazons are now made, and several thousand metric tons of Basic Yellow 11 and Basic Orange 21 is produced yearly. Their main uses are dyeing acetate rayon (brilliant hue, moderate stability) as well as polyacrylonitrile and polyacrylamide (improved dye stabilities). Other Astrazons and substituted Astraphloxins are used as biological stains. Indocyanine Green (**2**), patented in 1959 as an infrared tracer in blood and related heptamethine dyes are used as infrared laser dyes.

Properties

Absorption, x-ray crystal structure, and bonding. The polymethine dyes with structures as in Figure 1 have high extinction, electronic absorption bands in the ultraviolet, visible, or infrared regions of the spectrum. Extending the total length of the chromophore by successive —CH=CH— (vinyl) units shifts absorption maxima to longer wavelengths by about 100 nm per vinyl unit. Symmetrical polymethine dyes with terminal dialkylamino groups, known as streptocyanines, exhibit this shift regularly up to chromophore lengths of nine amethine units. However, the dimethylaminomerocyanines (structure (**b**), R = methyl) converge to a limiting wavelength as more methines are added, since the unsymmetrical nature of the chromophore leads to more polyenelike spectral characteristics.

The simple structural features of the streptocyanines permit more direct observation of the properties of the cyanine chromophore without interference from conjugated heterocycles. Fluorescence polarization indicates that the long-wavelength transition in the visible region is oriented strongly parallel to the methine chain. Proton magnetic resonance coupling constants for the —CH=CH— vinyl units indicate a trans orientation. X-ray crystal analyses confirm this trans geometry and show almost equivalent bond lengths for each —C—C— distance in symmetrical polymethines.

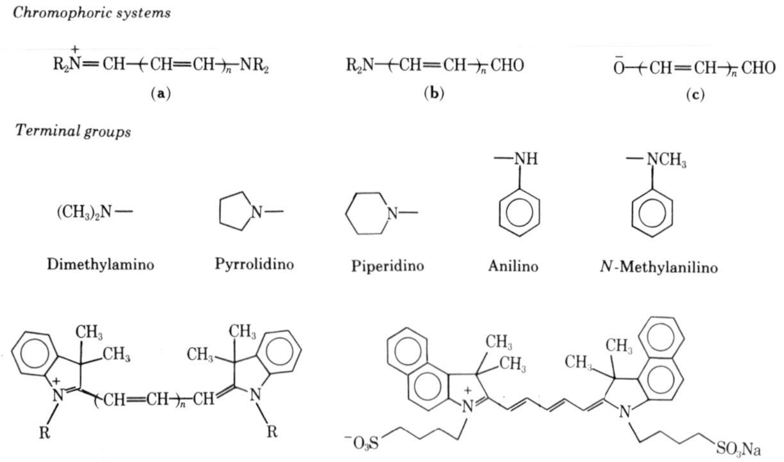

Chromophoric systems

(a)   (b)   (c)

Terminal groups

Dimethylamino   Pyrrolidino   Piperidino   Anilino   *N*-Methylanilino

(1)

(2) Indocyanine Green

Figure 1. Chromophoric systems and terminal groups for typical amino-polymethine and 3*H*-indolopolymethine dyes.

Synthesis and Reactivity

The aminopolymethine dyes have the general structure shown in Figure 1 and are most conveniently discussed in terms of the number of carbon atoms in the polymethine chain, ie, trimethine, pentamethine, etc. A typical dimethylaminotrimethine dye is prepared directly from propynal, dimethylamine, and dimethylammonium salts.

Anilinoacrolein anil hydrochloride, a trimethine dye that is also an important intermediate in cyanine-dye formation, is conveniently prepared from tetramethoxypropane and aniline.

A plethora of pentamethine dyes have been prepared readily by the reaction of substituted or unsubstituted *N*-(2,4-dinitrophenyl)pyridinium salts or *N*-cyanopyridinium salts with primary or secondary amines. These reactions are used to prepare pentamethine dyes with alkyl or aryl substitution at all five chain carbons. Heptamethine and nonamethine dyes with unsubstituted chains are synthesized from 2,4-hexadienal and 2,4,6-octatrienal, respectively, via the Vilsmeier formylation reaction.

The 3*H*-indole dyes are routinely prepared by use of the important intermediates Fischer's base (**3**) and Fischer's aldehyde (**4**) in combination with other reagents to give Astraphloxin FF, Astrazon Yellow 3G, and Astrazon Orange R.

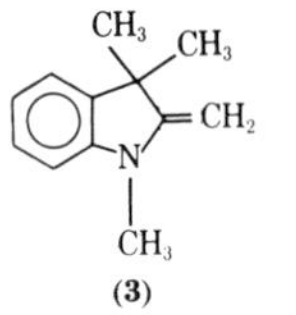

(3)
Fischer's base

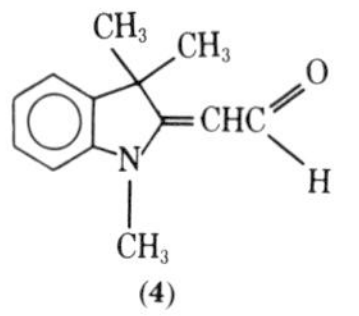

(4)
Fischer's aldehyde

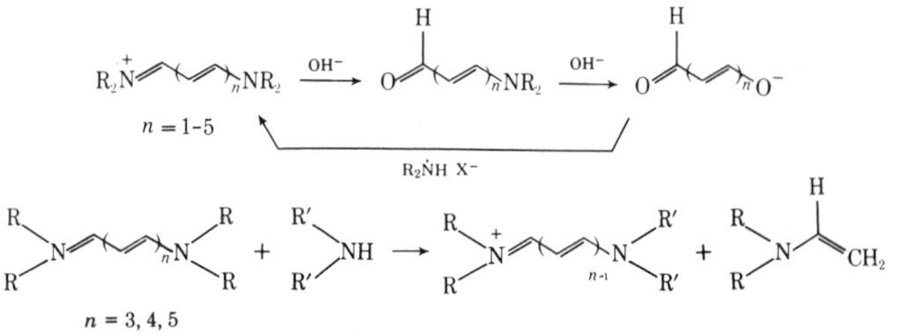

$n = 1-5$

$n = 3, 4, 5$

The chemical reactivity of the streptocyanines is well known. They are easily hydrolyzed in aqueous alkaline solution to the corresponding merocyanines and then to the oxonols. This reaction is reversible, ie, treatment of the oxonol salts with dialkylammonium salts regenerates the dialkylaminostreptocyanine. Merocyanines and oxonols resulting from longer-chain streptocyanines, eg, $n = 3-5$, are difficult to isolate

Table 1. Suppliers of Polymethine Dyes

Company	Location
Polymethine sensitizers, laser dyes	
Aldrich Chemical Company	Milwaukee, Wisc.
Fluka Chemicals	Basel, Switzerland
Japanese Institute for Photosensitizing Dyes	Okayama, Jpn.
Kodak Laboratory Chemicals	Rochester, N.Y.
Pfaltz and Bauer	Stamford, Conn.
Polymethine fabric dyes	
Allied Chemical Corporation	Morristown, N.J.
American Cyanamid Company	Bound Brook, N.J.
Aziende Colori Nazionali Affini A.C.N.A.	Milan, Italy
Bayer AG	Leverkusen, FRG
CIBA-GEIGY Corporation	Ardsley, N.Y.
Crompton and Knowles Corporation	Fair Lawn, N.J.
E.I. du Pont de Nemours & Co. Inc.	Wilmington, Del.
Fabricolor Chemical Corporation	Paterson, N.J.
GAF Corporation	New York
Hodogaya Chemical Co. Ltd.	Tokyo, Jpn.
L.B. Holliday and Co. Ltd.	Huddersfield, UK
Mitsubishi Chemical Industries Ltd.	Tokyo, Jpn.
Sumitomo Chemical Co. Ltd.	Osaka, Jpn.
Yorkshire Chemicals Ltd.	Leeds, UK

and purify. The longer-chain streptocyanines themselves are subject to aminolysis of the carbon–carbon linkages in the chain. For example, the nonamethine- and undecamethinestreptocyanines ($n = 4, 5$) are cleaved by strongly basic secondary amines even at room temperature; the heptamethine analogue requires high temperatures, whereas the pentamethine and trimethine dyes are stable.

Heterocyclic nucleophiles like Fischer's base (**3**) react readily with the more simple polymethine dyes, making these simpler dyes very useful synthetic intermediates for dyes. Reaction of anilinoacrolein anil hydrochloride with any of a variety of heterocyclic quaternary salts in the presence of acetic anhydride, suitable base, and suitable solvent is a routine preparation of dicarbocyanine dyes having an unsubstituted chain. Similarly, the dianilinopentamethine dye, glutacondialdehyde dianil hydrochloride, reacts with a variety of heterocyclic quaternary salts to form tricarbocyanine dyes. This reaction is severely limited for the corresponding chain-substituted glutacondialdehyde dianil hydrochloride salts, since the cis form of this chromophore is subject to an electrocyclic ring closure to form *N*-phenylpyridinium salts.

The chemical reactivity of polymethine dyes with organometallic reagents has also been investigated. Treatment of dialkylaminostreptocyanines with cyclopentadienyl salts results in the clean formation of a novel anionic chromophore. The reaction of the phenylmagnesium bromide Grignard reagent and a substituted tetramethinemerocyanine yields the enammonium salt of a substituted cinnamylideneacetaldehyde.

Suppliers of polymethine dyes are listed in Table 1.

DAVID M. STURMER
DONALD R. DIEHL
Eastman Kodak Company

S. Hünig and H. Quast in W. Jung, ed., *Optische Anregung Organischer Systeme* (Internationales Farbensymposium; Elmau, Switz.,), Verlag Chemie, GmbH, Weinheim, FRG, 1966, pp. 184–262.

S. Dähne, *Science* **199**, 1163 (1978).

F.M. Hamer, *The Cyanine Dyes and Related Compounds*, Interscience Publishers, a division of John Wiley & Sons, Inc., New York, 1964, pp. 200–213.

(POLYMETHYL)BENZENES

(Polymethyl)benzenes, PMBs, are aromatic compounds that contain a benzene ring and three to six substituted methyl groups. Included are the trimethylbenzenes (mesitylene, pseudocumene, hemimellitene), the tetramethylbenzenes (durene, isodurene, prehnitene), pentamethylbenzene, and hexamethylbenzene.

Properties

Structures and physical and thermodynamic properties of the PMBs are shown in Table 1.

The Koch Company is the only U.S. supplier of all PMBs (except hexamethylbenzene). The Koch process is based on isomerization, alkylation, and disproportionation conducted in the presence of a Friedel-Crafts catalyst.

Health and Safety Factors

The PMBs, as higher homologues of toluene and xylenes, are handled in a similar manner, even though their flash points are higher. Containers are tightly closed and use areas should be ventilated. Breathing vapors and contact with the skin should be avoided.

Table 2. Commercial Uses for PMBs

| PMB | Commercial derivative | |
	Via oxidation	Via other methods
pesudocumene	trimellitic anhydride	a few small-volume specialty products
mesitylene	trimesic acid	Ethanox
hemimellitene	none	musk intermediate
durene	pyromellitic dianhydride	none

Uses

The principal uses for the four commercial PMBs are shown in Table 2 (see Phthalic acid and other benzenepolycarboxylic acids).

H.W. EARHART
Koch Chemical Company

H.W. Earhart and H.E. Cier, *Advances in Petroleum Chemistry and Refining*, Wiley-Interscience, New York, 1964, Chapt. 6.

H.W. Earhart, *The Polymethylbenzenes*, Noyes Development Corporation, Park Ridge, N.J., 1969.

POLYMYXIN. See Antibiotics, peptides.

POLYPEPTIDE ANTIBIOTICS. See Antibiotics, peptides.

POLYPEPTIDES

Polypeptides are attractive model compounds for both fibrous and globular proteins. These macromolecules can be synthesized to include residues that are of interest for studies of protein structure. Such synthetic analogues of proteins allow one to investigate secondary, and even tertiary, structures, using techniques such as circular dichroism, ir spectroscopy, and nmr. With these procedures, transformations from ordered to disordered structures can be examined, and it becomes possible to elucidate the compositional factors that determine protein structure.

Table 1. Physical Properties of Polymethylbenzene

	Mesitylene	Pseudocumene	Hemimellitene	Durene	Isodurene	Prehnitene	Pentamethyl-benzene	Hexamethyl-benzene
systematic (benzene) name	1,3,5-trimethyl-	1,2,4-trimethyl-	1,2,3-trimethyl-	1,2,4,5-tera-methyl-	1,2,3,5-tetra-methyl-	1,2,3,4-tetra-methyl-	1,2,3,4,5-penta-methyl-	1,2,3,4,5,6-hexa-methyl-
molecular weight	120.186	120.186	120.186	134.212	134.212	134.212	148.238	162.264
boiling point, °C	164.72	169.35	176.08	196.80	198.00	205.04	231.8	263.8
flash point, °C	43.0	46.0	51.0	67.0	68.0	73.0		
specific gravity, 15.6/15.6°C	0.8696	0.8802	0.8987	0.8918	0.8946	0.9094	0.921	
freezing point, °C	−44.720	−43.80	−25.375	+79.240	−23.685	−6.25	+54.3	+165.5
refractive index, n_D at 25°C	1.49684	1.50237	1.51150	1.5093	1.5107	1.5181	1.525	
heat of vaporization at bp, kJ/mol[a]	39.0	39.2	40.0	45.5	43.8	45.0	45.1	48.2
heat of formation at 25°C, liquid, kJ/mol[a]	−63.52	−61.86	−58.63	−98.66	−98.49	−96.40	−135.1	−171.5
specific heat, C_p, liq., at 25°C, J/(mol·K)[a]	200.5	214.9	216.4		240.7	238.3		
surface tension, mN/m (= dyn/cm), at 20°C	28.83	29.71	31.27		33.51	35.81		

[a] To convert J to cal, divide by 4.184.

Synthesis

Polypeptides are usually synthesized by the polymerization of amino acid N-carboxyanhydrides (NCAs or 4-substituted-2,5-oxazoldiones) first discovered by Leuchs in 1906.

$$n \begin{array}{c} R \\ HN \end{array} \longrightarrow \left[NHCHC \atop R \right]_n + n\ CO_2$$

NCA

Copolymerization. An impressive number of copolymers of natural and synthetic amino acids have been prepared via NCA copolymerizations. In most cases, a random sequence distribution has been assumed in copolypeptides prepared via NCA copolymerization. However, at present the copolymerization process is incompletely understood in terms of reactivity ratios of the various monomers, and a complete reactivity scale of NCA monomers has not been determined.

Sequential polypeptides. During the past three decades, biologically active peptides have become the subject of intensive research activity. Large numbers of linear peptides containing 3 to > 30 residues play important metabolic and physiological roles.

Usually, preparation of these compounds is initiated by blocking or protecting the amino or carboxyl terminus of the amino acid in question. Dipeptide formation is accomplished by activation of the carboxyl terminus of an N-protected amino acid and addition of an amino acid ester as the nucleophile. The N-terminal blocking groups most widely used in peptide synthesis are the benzyloxycarbonyl (Z) and the t-butoxycarbonyl (Boc) groups. Chain elongation is continued by selective removal of protecting groups and additional coupling or polymerization reactions.

Conformational Analysis of Peptides

Polypeptide chains can fold into different ordered conformations. By far the most famous structure is the right-handed α-helix. This particular helical conformation, initially described by Pauling and Corey, is characterized by having 3.6 amino acid residues per turn, and by a translation of 15 nm along the axis of the helix (see Biopolymers).

Conformational Analysis of Polypeptides

Analytical methods used for the determination of conformation include optical rotation, ir spectroscopy, circular dichroism, and nmr spectroscopy.

Specific Structures

Under well-defined experimental conditions, synthetic homo- and copolypeptides fold into ordered structures typical of those found in proteins. Investigations of the conformational properties of high molecular weight polypeptides have determined the role of different amino acid residues in stabilizing given ordered structures. From these studies it has been possible to determine precisely the spectroscopic properties of various kinds of ordered conformations, including helices of different symmetry, collagenlike triple helices, parallel and antiparallel β-structures, β-turns, etc. By using the spectroscopic tools described above and minimum potential-energy calculations, it is possible to determine the conformational preference of amino acid residues and amino acid sequences in polypeptide chains. The α-helical conformation common to a

Figure 1. The α-helix.

number of synthetic polypeptides (Fig. 1) and other interesting native structures have been observed in sequential polypeptides.

It is likely that synthetic copolypeptides with predetermined amino acid sequences will continue to play an important role in elucidating the structure of natural biopolymers such as fibrous proteins and linear glycoproteins (see also Amino acids; Proteins).

MURRAY GOODMAN
University of California, San Diego

EVARISTO PEGGION
University of Padua

FRED NAIDER
The College of Staten Island

M. Bodanszky, Y.S. Klausner, and M.A. Ondetti, *Peptide Synthesis*, John Wiley & Sons, Inc., New York, 1976.

E. Blout, F. Bovey, M. Goodman, and N. Lotan, eds., *Peptides, Polypeptides and Proteins*, John Wiley & Sons, Inc., New York, 1974, p. 240.

M. Goodman, A.S. Verdini, N.S. Choi, and Y. Masuda in N.L. Allinger and E.L. Eliel, eds., *Topics in Stereochemistry*, Vol. 5, John Wiley & Sons, Inc., New York, 1970, pp. 69–166.

POLYPROPYLENE. See Olefin polymers.

POLYPROPYLENE FIBERS. See Olefin fibers.

POLY(PROPYLENE OXIDE). See Polyethers.

POLYSTYRENE. See Styrene plastics; Elastomers, synthetic.

POLYVINYL COMPOUNDS. See Vinyl polymers.

PORCELAIN. See Ceramics; Dental materials; Enamels, porcelain and vitreous.

POROMERIC MATERIALS. See Leatherlike materials.

POTASSIUM

Potassium and sodium compounds are present in nearly equal abundance throughout the earth's crust. They share the position of the seventh most abundant element. Commercial sources of potassium are usually limited to the natural deposits of sylvite, arcanite, or carnalite.

Physical Properties

Potassium is a soft, silver-colored metal at ambient conditions having physical and chemical properties similar to sodium (qv). Differences in the properties of the alkali metals reflect the variations in atomic size and weights.

The physical properties of potassium are, in general, those commonly associated with metals (see Table 1).

Chemical Properties

The chemical properties of potassium are quite similar to those of sodium. At higher temperatures, silicates, sulfates, and nitrates of heavy metals are reduced, often liberating the metal. Reaction with halogens is explosive. Carbonyls (qv) are formed with carbon monoxide and inter-

Table 1. Physical Properties of Potassium

Property	Value
atomic radius, nm	0.235
ionic radius, nm	0.133
Pauling electronegativity	0.8
crystal lattice	body-centered cubic
analytical spectral line, nm	766.4
viscosity, at 25°C, mPa·s (= cP)	0.258
melting point, °C	97.8
boiling point, °C	760
density, at 20°C, g/cm^3	0.86
specific heat, J/(g·K)^a	0.741
heat of fusion, J/g^a	0.598
heat of vaporization, kJ/g^a	2.075
electrical conductance, at 20°C, μs	0.23
surface tension, at 100°C, mN/m (= dyn/cm)	86
thermal conductivity, at 200°C, W/(m·K)	44.77

^{a} To convert J to cal, divide by 4.184.

laminar compounds are formed with graphite. In air, KO_2 forms, which is unstable in contact with molten potassium with which it reacts to give K_2O. Potassium is unique among the more readily available alkali and alkaline-earth metals in forming a superoxide in air. Rubidium and cesium also form superoxides in air. Organic alkoxides are produced easily; eg, potassium *t*-butoxide is used in organic syntheses (see Alkoxides, metal). Potassium is soluble in liquid ammonia, ethylenediamine, aniline, and mercury.

Preparation and Manufacture

Laboratory methods. Small quantities of potassium are prepared by chemical, thermal, or electrolytic processes.

The Greisheim process

$$2 KF + CaC_2 \rightarrow CaF_2 + 2 C + 2 K$$

takes place at 1000–1100°C at ambient pressure. Modifications utilize potassium carbonate or potassium silicate with silicone polymers to improve efficiency and operating conditions.

The reduction of potassium chloride with sodium, and the fractionation of the free metals, has been described for producing potassium or sodium–potassium alloys.

A continuous process currently in use depends upon the equilibrium that is rapidly established at high temperatures between sodium and potassium metals and sodium and potassium chlorides. The potassium is removed by distillation and, with a constant feed of molten sodium and molten potassium chloride, a continuous process results.

Health and Safety Factors

Potassium reacts with both oxygen and water. The explosive reactivity of potassium with water is similar to that of sodium. Inhalation and skin contact must be avoided. Safety goggles (with sideguards), gloves, and fire-resistant clothing are considered minimum safety equipment. Full-body-coverage suits of fire- and chemical-resistant materials are used when removing spills and fighting fires.

Potassium fires are best extinguished by powders, such as coated sodium chloride (Ansul's Metal-X), dry sand, or amorphous carbons.

Uses

Today, potassium is used mostly in the manufacture of potassium superoxide (KO_2) for life-support systems (see Oxygen-generation systems).

JOHN S. GREER
JOHN H. MADAUS
J.W. MAUSTELLER
MSA Research Corporation

J.W. Mellor, *Supplement III to Mellor's Comprehensive Treatise on Inorganic and Theoretical Chemistry*, Vol. II, John Wiley & Sons, Inc., New York, 1963.

O.J. Foust, *Sodium and Sodium–Potassium Engineering Handbook*, Vol. I, Gordon and Breach Science Publishers, Inc., New York, 1972, Chapt. 2.

J.W. Mausteller, F. Tepper, and S.J. Rodgers, *Alkali Metal Handling and Systems Operating Techniques*, Gordon and Breach Scientific Publishers, Inc., New York, 1967.

NaK and Potassium Technical Bulletin, Callery Chemical Company, Callery, Pa., 1980.

POTASSIUM COMPOUNDS

Occurrence

Potassium is the seventh most abundant element in the earth's crust and is about equal in abundance to sodium. Potassium minerals occur naturally in all parts of the world. Commercial production of potassium compounds is generally limited to the extraction of ores from underground deposits containing significant concentrations of soluble potassium salts. Exceptions include commercial potassium chemical operations on the Dead Sea and the Great Salt Lake (see Chemicals from brine). Canadian plants that came on stream in 1983 shifted the production leadership from the USSR to Canada.

Approximately 98% of the potassium recovered in primary ore and natural-brine refining operations is recovered as potassium chloride. The remaining 2% consists of potassium that is recovered from a variety of sources.

Physical properties of selected potassium compounds are listed in Table 1.

Refining. Process selection for the separation of potassium chloride as a relatively pure product from other constituents is based on the physical and chemical characteristics of a given ore. Ores amenable to treatment by the physical-separation methods that are commonly used in other nonmetallic minerals processing industries generally are chosen to recover the potassium chloride as a salable product. These methods include heavy-media and froth-flotation separations. Physical-separation processes are much less energy-intensive than fractional crystallization, which is the traditional method of producing potassium chloride.

Table 1. Physical Constants of Selected Potassium Compounds

Potassium compound	Formula	Mol wt	Form	Specific gravity	Melting point, °C
acetate	$KC_2H_3O_2$	98.14	white powder	1.8	292
bromide	KBr	119.01	cubic	2.75	730
carbonate	K_2CO_3	138.20	monoclinic	2.43	891
bicarbonate	$KHCO_3$	100.11	monoclinic	2.17	dec
chlorate	$KClO_3$	122.55	monoclinic	2.32	368
chloride	KCl	74.55	cubic	1.984	776
formate	$KHCO_2$	84.11	rhombic	1.91	168
hydroxide	KOH	56.10	rhombic	2.044	360.4 ± 0.7
iodide	KI	166.02	cubic	3.13	723
nitrate	KNO_3	101.10	rhombic, trig.	2.109	334
nitrite	KNO_2	85.10	colorless, prism	1.915	387
normal phosphate	K_3PO_4	212.27	rhombic	2.56	1340
monohydrogen phosphate	K_2HPO_4	174.18	amorphous		dec
dihydrogen phosphate	KH_2PO_4	136.09	tetragonal	2.338	252.6
sulfate	K_2SO_4	174.26	rhombic or hexagon	2.662	1076
bisulfate	$KHSO_4$	136.17	monoclinic, rhombic	2.24–2.61	210
sulfite dihydrate	$K_2SO_3 \cdot 2H_2O$	194.29	monoclinic		dec

Potassium Acetate

Potassium acetate, $KC_2H_3O_2$, is usually made from the carbonate and acetic acid. It is very soluble and is used in the manufacture of glass, as a buffer, a dehydrating agent, and in medicine as a diuretic (see Diuretics). It is deliquescent and is used as a softening agent for papers and textiles (see Papermaking additives).

Potassium Bromide

Potassium bromide, KBr, can be prepared by a variation of the process by which bromine is absorbed from ocean water; potassium carbonate is used instead of sodium carbonate:

$$3 K_2CO_3 + 3 Br_2 \rightarrow KBrO_3 + 3 CO_2 + 5 KBr$$

Potassium bromate is much less soluble than the bromide and can largely be removed by filtration.

Potassium bromide is extensively used in photography (qv) and engraving. It is the usual source of bromine in organic synthesis. In medicine, it is a classic sedative (see Hypnotics, sedatives, and anticonvulsants).

Potassium Carbonate

Except for small amounts produced by obsolete processes, potassium carbonate is produced by the carbonation of potassium hydroxide. It is available commercially as a concentrated solution containing ca 47 wt% K_2CO_3 or in granular crystalline form containing 99.5 wt% potassium carbonate. Impurities are small amounts of sodium and chloride plus trace amounts (< 2 ppm) of heavy metals as lead. Heavy metals are a concern, because potassium carbonate is used in the production of chocolate intended for human consumption.

In many heavy-chemical manufacturing operations requiring an intermediate alkaline-metal carbonate reactant, potassium carbonate and sodium carbonate can be used with equal effectiveness. Potassium carbonate possesses properties for some applications that preclude the substitution of sodium carbonate, eg, in television glass. Uses of potassium carbonate other than glass include applications in ceramics, chemicals, dyes and pigments, foods, cleansers and gas purification, among others.

Potassium Bicarbonate

Potassium bicarbonate, $KHCO_3$, is made by absorption of CO_2 in a carbonate solution. The bicarbonate is used in foods and medicine.

Potassium Formate

Potassium formate, HCO_2K, is made by the reaction:

$$CO + KOH \rightarrow HCO_2K$$

Potassium formate melts at 167°C and decomposes almost entirely to the oxalate at ca 360°C. Most of the formate produced is converted to oxalate.

Potassium Hydroxide

Potassium hydroxide is produced industrially by electrolysis of potassium chloride. Principal uses of KOH include chemicals, fertilizers, and other agricultural products; soaps and detergents; scrubbing and cleaning operations, eg, industrial gases; dyes and colorants; rubber chemicals; and others.

Potassium Iodide

Some potassium iodide, KI, is made by the iron and carbonate process described for the bromide. However, most U.S. production is absorption of iodine in KOH.

$$3 I_2 + 6 KOH \rightarrow 5 KI + KIO_3 + 3 H_2O$$

Approximately half the iodine consumed is used to make potassium iodide; production of the latter is almost 1000 t/yr. Its main uses are in animal and human food, pharmaceuticals, and photography.

Potassium Nitrate

Most of the potassium nitrate, KNO_3, produced commercially in the United States is based on the reaction of potassium chloride and nitric acid.

Potassium Phosphates

Phosphoric acid is the source of phosphate for the production of potassium phosphates (see Phosphoric acids and phosphates).

Condensed potassium phosphates are used as builders in liquid detergents. The compound commonly used is tetrapotassium pyrophosphate, $K_4P_2O_7$.

Potassium Sulfate

Compared with potassium chloride, potassium sulfate, K_2SO_4, and its complexes with magnesium sulfate play a minor role as sources of potassium in agriculture.

Potassium sulfate, however, is a well-established source of soluble sulfur, an essential element for plant growth. Complexes with magnesium sulfate supply water-soluble magnesium, an agronomically essential element. Although much less significant than potassium in terms of tonnage consumption, magnesium and sulfur are becoming increasingly important as essential fertilizer elements.

Health and Safety Factors

Potash mining and refining operations in the United States are strictly regulated by appropriate Federal and state agencies. Field studies conducted by NIOSH failed to disclose any evidence of predisposition of underground miners to any of the diseases, including lung cancer, evaluated in investigations.

W.B. DANCY
International Minerals & Chemical Corporation

World Survey of Potash Resources, 4th ed., The British Sulfur Corporation, Ltd., London, UK, 1983.

R.M. McKercher, ed., *Potash Technology: Mining, Processing, Maintenance, Transportation, Occupational Health and Safety, Environment*, Pergamon Press, New York, 1983.

POWDER COATINGS

Powder coating processes are considered fusion-coating processes, that is, at some time in the coating process the powder particles must be fused or melted. This is usually carried out in an oven.

In the fluidized-bed coating process, the coating powder is placed in a container with a porous plate as its base. Air is passed through the porous plate which causes the powder to expand and fluidize. In this state, the powder possesses some of the characteristics of a fluid. The part to be coated, usually metallic, is heated in an oven to a temperature above the melting point of the powder and is dipped into the fluidized bed where the particles melt on the surface of the hot metal to form a coating. Using this process, it is possible to apply coatings ranging in thickness from ca 250 to 2500 μm. It is difficult to obtain coatings thinner than 250 μm, and, therefore, fluidized bed-applied coatings are generally referred to as thick-film coatings, differentiating them from most conventional thin-film coatings applied from solution at thicknesses of 25–75 μm (see also Film deposition techniques).

In the electrostatic-spray process, the coating powder is dispersed in an air stream and passed through a high voltage field where the particles pick up an electrostatic charge. The charged particles are attracted to and deposited on the object to be coated which is usually at room temperature. The article is then placed in an oven where the powder melts and forms a coating. With this process, it is possible to apply coatings comparable in thickness to conventional solution coatings. The

Table 1. Physical and Coating Properties of Thermoplastic Powders[a]

Property	Vinyls	Polyamides	Polyester	Polyethylene	Polypropylene	Cellulosics
primer required	yes	yes	no	yes	yes	yes
melting point, °C	130–150	186	160–170	120–130	165–170	160–170
typical preheat/post-heat, °C	290–230	310–250	300–250	230–200	250–220	280–230
specific gravity, g/cm³	1.20–1.35	1.01–1.15	1.30–1.40	0.91–1.00	0.90–1.02	1.15–1.35
adhesion[b]	G–E	E	E	G	G–E	G–E
surface appearance	smooth	smooth	sl OP	OP	smooth	sl OP
gloss, Gardner 60° meter	40–90	20–95	60–95	60–80	60–80	80–90
hardness, Shore D	30–55	70–80	75–85	30–50	40–60	65–75
flexibility[c]	pass	pass	pass	pass	pass	pass
Resistance						
impact	E	E	G–E	G–E	G	G
salt spray	G	E	G	F–G	G	G
weathering	G	G	E	P	P	G
humidity	E	E	G	G	E	E
acid[d]	E	F	G	E	E	F
alkali[d]	E	E	G	E	E	P
solvent[d]	F	E	F	G	E	F

[a] E = Excellent, G = good, F = fair, P = poor, OP = orange-peel effect, sl OP = slight orange-peel effect.
[b] With primer where indicated.
[c] No cracking, 3-mm-dia mandrel bend.

Table 2. Physical and Coating Properties of Thermosetting Powders[a]

Property	Epoxy	Polyurethane[b]	Polyester[c]	Hybrid	Acrylic[b]
fusion range, °C	120–200	160–220	160–220	140–210	120–200
cure time at °C, min	1–30 at 240–135	10 at 200	10 at 200	8 at 190	10 at 200
storage temp, °C max	30	30	30	30	30
adhesion	E	G–E	G–E	G–E	G
gloss, Gardner 60° meter	5–95	20–95	40–95	20–95	80–95
hardness	H–4H	H–2H	H–2H	H–2H	H–2H
flexibility	E	E	E	E	F
Resistance					
impact	E	G–E	G–E	G–E	F
overbake	F–P	G–E	E	G–E	G
weathering	P	G–E	E	P–F	G–E
acid[d]	G	F	G	G	F
alkali[d]	G	P	F	G	P
solvent	G	F	F–G	F	F

[a] E = Excellent, G = good, F = fair, P = poor.
[b] Hydroxy functional-blocked isocyanate cure.
[c] TGIC (triglycidyl isocyanurate) cure.
[d] Inorganic; dilute.

electrostatic-spray process is the dominant process (ca 80%) of powder coating.

Compared with other coating methods, coating powders and powder-coating processes offer a number of significant advantages: they are essentially 100% nonvolatile; no solvents or other pollutants are given off during application or curing. They are ready to use; no thinning or dilution is required. They are easily applied by unskilled operators and automatic systems. Overspray powder is reused, ie, nothing is exhausted to the atmosphere.

Thermoplastic Coating Powders

The physical properties of polymeric materials improve with increasing molecular weight. However, as molecular weight increases, melt viscosity also increases. As a coating powder, a thermoplastic resin must melt and flow at the application temperature without any significant degradation. The principal polymer types in use today are based on plasticized PVC, polyamides, polyesters, and other specialty thermoplastics. Typical properties of coating powders based on these resins are given in Table 1.

Thermosetting Coating Powders

Thermosetting coating powders, with minor exceptions, are based on resins that are cured by addition reactions rather than condensation reactions. The resins are more versatile than thermoplastic resins in the following respects: many types are available in varying molecular weights ranges and with different functional groups; a variety of cross-linking agents is available, and physical, electrical, and chemical properties of the applied film can be modified; they possess a low melt viscosity during application and thin films can be applied; because of the lower melt viscosity, the quantities of pigments and fillers required to achieve opacity in thin films can be incorporated without adversely affecting flow; gloss, textures, and special effects can be produced by modifying the curing mechanism or through the use of additives; manufacturing costs are lower since compounding is carried out at lower temperatures; and the resins are friable and can be ground to a fine powder without using cryogenic techniques.

The properties of thermosetting coating powders are given in Table 2.

Manufacture

Coating powders are either melt-mixed or dry-blended (see Fig. 1).

Health and Safety

Increases in environmental regulations and legislative deadlines for compliance have contributed to the growth of powder-coating processes. They have been favorably evaluated by the EPA and recognized as one of several acceptable technologies to achieve the required level of volatile organic compound emissions.

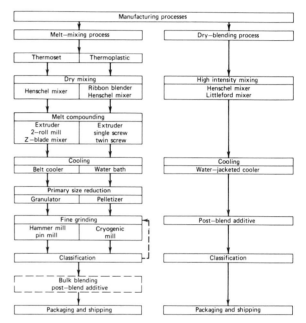

Figure 1. Flow diagram for powder-coating manufacture.

The most significant hazard in the manufacture and application of coating powders is the potential of a dust explosion. However, the explosive hazards of powders are considerably less than for solvent-based coatings.

Polyester resins and epoxy and acrylic coating powders have negligible toxicity, according to animal studies.

DOUGLAS S. RICHART
The Polymer Corporation

S.T. Harris, *The Technology of Powder Coatings*, Portcullis Press, London, UK, 1976.

M.W. Ranney, *Powder Coatings Technology*, Noyes Data Corporation, Park Ridge, N.J., 1975.

E.P. Miller and D.D. Taft, *Powder Coating*, Society of Manufacturing Engineers, Dearborn, Michigan, 1976.

POWDER METALLURGY

Powder metallurgy, or P/M, is formally defined as the production of metal powders and their utilization for the manufacture of massive

Table 1. Approximate Surface Area of Powder Particles Fabricated by Various Methods

Process	Particle shape	Approximate surface area[a]
carbonyl process	uniform spherical	πD^2
atomization	round irregular spheroids	$1.5-2\pi D^2$
reduction of oxides	irregular spongy	$7-12\pi D^2$
electrolytic process	dendritic	
mechanical comminution		
crushing	angular	$3-4\pi D^2$
ball milling	flakes or leaves	varies over wide range

[a]D = diameter.

materials and shaped objects. It encompasses both the total technology and the application of metals in powder form. It is an industry based on the metallurgical phenomenon called sintering, ie, the bonding of particles in a mass of metal powder by molecular (or atomic) attraction in the solid state through the application of heat below the melting point of the metal. Sintering causes the strengthening of the powder mass and normally results in densification and often recrystallization because of material transport (see Metal treatments).

Characteristics

Characteristics of individual particles. Characteristics include: size, particles smaller than 44 μm (-325 mesh) are called fines; shape, which depends to a large extent on the method of fabrication and is expressed as a deviation from a sphere of identical volume; density, which is not necessarily identical to the density of the material from which it is produced because of the particles' internal porosity; surface area (see Table 1), which shows that the particle surface area depends strongly on the method of production (which determines the particle shape); microstructure, ie, the crystal grain size, shape and orientation which depend also on the method of fabrication; surface oxide layer, the thickness of the oxide layer on an individual particle depending on the conditions under which oxidation occurs; and particle activity, which refers to the reaction of a powder particle with its environment and which determines the type and rate of reaction.

Characteristics of powder mass. The most important properties of a good molding-grade powder are flow rate, usually expressed by the time necessary for a specific amount of powder (usually 50 g) to flow through a specific orifice; particle size, the average particle size referring to a statistical diameter; size distribution (the distribution of the particles of various sizes in the powder mass); apparent density, the weight of a unit volume of loose powder (usually expressed in g/cm^3) which depends on the friction conditions between the powder particles which are a function of the relative surface area of the particles, the surface conditions, and the packing arrangement of the particles; green strength, the yield strength of a compacted but unsintered powder mass; compressibility, the density to which a powder may be pressed at any given pressure; and dimensional stability during sintering. There is a close relationship between particle size distribution and such factors as powder flow, apparent density and compressibility.

Manufacture

The primary methods for the manufacture of metal powders are atomization, in which a stream of molten metal is struck with air or water jets and the formed particles are collected, sieved, and annealed (this is the most common commercial method in use for all powders); the reduction of metal oxides to give so-called sponge iron or hydrogen-reduced mill scale; others include electrolytic deposition, mechanical comminution of metal masses, carbonyl decomposition, etc.

Processing

In general practice, metal powders or mixtures of various powders are fashioned into some form, a P/M gear for example, by first flowing them

at room temperature into a die cavity shaped as the finished gear, and then applying pressure from above and below to form a compact. After ejection from the die, the compact is heated (sintered) to form a coherent mass with the configuration of the original die and properties related to the metals used. A low density, highly porous compact such as a metal filter may be formed by pouring the powder into a die and then heating it without first applying pressure. The shape and size of the die cavity may be exactly as required by the finished part or it may be some intermediate configuration.

The pressing operation is usually carried out at room temperature, although warm or even hot pressing is sometimes used. The pressing operation consolidates the powder into a mass with sufficient strength to be handled without breaking after ejection from the die. Consolidation produces a coherent mass of definitive size and shape for further working, heat treating, or use as is. During pressing, a coherent mass is formed through the process of interparticle binding and interlocking.

The most frequently used technique for consolidation of powders is uniaxial pressing of metal powders in a die of specific dimensions and configurations. Other consolidation techniques include: cold isostatic pressing; hot isostatic pressing (HIP); vibratory consolidation; hot pressing; extrusion, swaging, or rolling; and injection molding.

The application of heat during pressing, or as a separate step, does not cause melting of the powder except in special instances involving liquid-phase sintering. The temperature to which the pressed compact is then heated is usually less than the melting point of the elemental metal or the solidus temperature of any alloy that may form. During heating, the bonding that was initiated in pressing is carried further by the solid-state movement of atoms across particle boundaries.

Liquid-phase sintering combines powder metallurgy and fusion metallurgy. In this procedure, part of the structure is liquid for part or all of the sintering time. The sintering temperature may be high enough to melt one constituent during the heating time, or the metal matrix may be infiltrated with another metal that melts at a lower temperature, such as the infiltration of copper in a sintered iron matrix.

After sintering, supplementary operations may be required. A sintered component may be worked by any number of metal-working procedures, such as swaging, rolling, or forging. Shaped items, such as machine parts and structural components, may require some additional machining, plating, or other finishing operations. The result is a P/M product, which is a metal shape equivalent in function, although usually of a lower density and often equivalent in physical and mechanical properties, to a wrought metal product but produced faster, automatically, and normally at a lower cost in terms of labor, materials, and energy.

Health and Safety

Metal powders possess a high ratio of specific surface area to volume. This can contribute to several potentially hazardous properties, such as pyrophoricity and explosiveness for metals such as aluminum or magnesium. The toxicity of metals such as lead and beryllium is exacerbated by the increased surface area.

Uses

Powder metallurgy is used to make structural parts from iron- and copper-base powders, and more recently, from aluminum for use in automobile engines, transmissions, business machines, power tools, tractors, etc; porous materials such as filters for separating combinations of solids, liquids and gases, surge dampeners and flame arrestors, metering devices and distribution manifolds; small, complex tool-steel parts; sintered friction materials classified in metal–nonmetal combinations (eg, clutch plates, brake blocks); electrical contact materials; permanent magnets (eg, Alnico, rare earths); iron powder cores; batteries; incandescent lamp filaments; molybdenum heating elements; roll-compacted glass sealing strips (Invar); cemented carbides; and in cermets.

P/M forgings produced by forging a heated P/M preform in a precision die are becoming commonly used because of the economies of production and superior mechanical and physical properties. Transmission components, automobile connecting rods, and ball-bearing cups are examples.

KEMPTOM H. ROLL
Metal Powder Industries Federation
American Powder Metallurgy Institute

W.D. Jones, *Fundamental Principles of Powder Metallurgy*, Edward Arnold, London, UK, 1960.

F.V. Lenel, *Powder Metallurgy—Principles and Applications*, Metal Powder Industries Federation, Princeton, N.J., 1980.

J.S. Hirschhorn, *Introduction to Powder Metallurgy*, American Powder Metallurgy Institute, Princeton, N.J., 1969.

Journal of Powder Metallurgy and Powder Technology, American Powder Metallurgy Institute, Princeton, N.J.

POWER GENERATION

The term power generation in the engineering sense implies the production of mechanical or electrical power from some other source of energy, eg, thermal, hydroelectric, or electrochemical.

Geothermal Power

Geothermal energy (qv), a vast potential source of power, is thought to result from natural decay of radioactive materials.

Solar Power

The supply of solar energy is so great that it can be considered virtually inexhaustible (see Solar energy). Solar energy is not expected to adversely affect the environment, but it is a diffuse source and only available part of the year.

Photovoltaic cells (qv) directly convert solar-radiation energy into electricity without the intervention of thermal cycles. Several materials have been utilized, including $p-n$ junctions of various semiconductors. The efficiency of the cells is wave-length dependent. Although employed in unusual small-scale applications, large-scale development of photovoltaic systems has not taken place because of extremely high cost and relatively low efficiency.

Wind Power

Interest in wind as an energy source has been revived because of the recent oil and gas shortages. The design of wind-energy conversion systems (WECS) is based on the fact that wind power is proportional to the cube of the wind velocity. The annual average wind speed at any site depends upon its geographical position, altitude, distance from trees, its exposure to prevailing wind, and the shape of the land in the immediate vicinity. The frequency of wind speed or its variability is important to determine the peak generation capacity of the installation and gives an assessment of the variation in power generation.

Wind turbines are categorized in terms of orientation of the axis of rotation relative to the wind speed, eg, the horizontal-axis wind turbine generator (HA-WTG) has its axis of rotation parallel to the direction of the wind and the earth's surface; the vertical-axis-wind-turbine generator (VA-WTG) has its axis of rotation perpendicular to both the horizontal earth's surface and the wind stream.

The main problems in wind-energy utilization are the intermittent nature and relatively low energy level of wind compared to conventional fuels.

Nuclear Power

A nuclear power plant uses heat obtained from the release of energy associated with fission of splitting of the ^{235}U into lighter elements (see Nuclear reactors).

Hydroelectric Power

Hydroelectric power plants may be classified depending upon the height of the water level in the reservoir above the plant into roughly high head (ca 160 m or more), medium head, and low head plants (< 16 m). High head plants require much less flow than low-head plants of the same capacity. Therefore, turbines and generators may be designed for higher speeds which permits smaller units and housings. This advantage is offset by the fact that long conduits or penstocks are necessary between dams and power plant.

Total U.S. hydroelectric capacity in 1979 was about 64 GW. Because of a lack of new hydroelectric sites as well as environmental and esthetic considerations, hydroelectric capacity is not expected to increase greatly.

Fuel Cells

A fuel cell converts the chemical energy of a fuel directly to d-c power without going through a thermal or combustion cycle in the process. It is composed of two electrodes and an electrolyte where the principal reactions create an electric current at the electrodes. Unlike a conventional lead-acid battery cell, the reactants in a fuel cell can be replenished externally, thereby enabling the fuel cell to operate continuously at a constant power output (see Batteries and electric cells, primary-fuel cells).

Internal-Combustion Engines

The power in an internal-combustion engine is produced by the expansion of the working substance against a piston that reciprocates inside a cylinder. The reciprocating motion of the piston is converted into rotary motion by means of a connecting rod, crankshaft, and flywheel; the latter serves as a device for smoothing out variable forces exerted against a crankshaft.

Internal-combustion engines can be built with one cylinder (as in smaller machines), or several cylinders (as in large diesel power plants, automobiles or aircraft engines). The cylinders may be arranged in the in-line and V-arrangements (the most popular), or the opposed-piston (used in diesel engines) and radial type (used in aircraft engines). Engine sizes range from small model-airplane engines to large units capable of delivering several megawatts. Rotating speeds are 50–2000 rpm, depending on the size and design.

Thermal efficiencies are usually high in well-designed units, ranging from ca 20% for automotive and aircraft engines to as high as 40% for diesel-locomotive and large stationary diesel power-plant engines.

The two most important cycles employed are the Otto-cycle engine, and the diesel cycle. They are called four-stroke cycles because four strokes of the piston (intake, compression, expansion, and exhaust) are required to complete the cycle.

Uses. Internal combustion engines operating on the Otto cycle are well adapted to applications where speed and power requirements vary over a wide range (eg, automobile and aircraft engines). Diesel engines are used widely for stationary and marine power plants and for railroad locomotives, as well as for trucks and heavy-duty excavating and road-building equipment where engines must be designed to maintain reasonably high efficiencies throughout wide ranges of speed and load.

Because of heightened concern over environmental problems, a great deal of effort has recently gone into designing automobile engines that are more fuel efficient and free from harmful exhaust gases (see Exhaust control, automotive).

Gas Turbines

The simplest gas-turbine cycle is essentially the Brayton cycle, shown in Figure 1. Its efficiency is computed easily from a temperature–entropy diagram assuming that air follows the ideal gas laws, that the specific heat of air at constant pressure is a constant, and that the weight of fuel supplied contributes a negligible additional weight to the heated air flowing through the turbine.

$$\eta_T = 1 - \frac{1}{r_p^{(k-1)/k}} \tag{1}$$

where η_T = ideal cycle efficiency and r_p = pressure ratio $P_2 : P_1$.

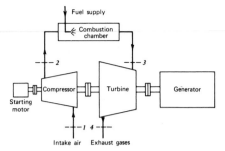

Figure 1. Simple open-cycle gas turbine.

In an actual gas turbine, both compressor and turbine have internal efficiencies of less than 100%, and the net energy delivered to the generator shaft of the turbine is substantially less than values computed by equation 1.

The efficiency of the simple open-cycle gas turbine can be improved considerably by employing a regenerator in which the exhaust gases are passed countercurrent to the air leaving the compressor and, in this manner, the air is heated by the exhaust gases before entering the combustion chamber, thereby reducing the amount of heat that must be supplied by the fuel.

Uses. The most important gas-turbine application is for jet-propulsion aircraft (see also Aviation and other gas-turbine fuels). It is also used for locomotive and marine applications, but only to a limited extent, partly because of the low efficiency of the single open-cycle gas turbine compared to the diesel engine. Stationary gas turbines are employed in the oil and gas industries to drive axial or centrifugal compressors. Gas turbines have innumerable applications in the petrochemical industry and other processing industries, and in the electric-utility industry where they are employed for peak load generation or for standby service.

Steam Engines

The reciprocating steam engine operates on the principle of the expansion of steam against a piston that moves back and forth inside a cylinder (see Steam).

Central-station power plants. Most central-station steam power-plant turbines are of capacities exceeding 100 MW. Turbines for fossil-fuel power plants use highly superheated steam with reheating. Typical steam conditions at the inlet for the newer units of 500 MW and above are pressures of 17–24 MPa (2400–3500 psi) and temperatures of ca 540°C. Most units are reheated once to 540–565°C, but some are reheated twice, each to ca 540–565°C.

Because of economic and technical limitations imposed by water-cooled nuclear reactors, turbines designed for pressurized-water and boiling-water nuclear power plants are limited to a maximum inlet pressure of ca 6.5 MPa (950 psi). Since a successful nuclear superheater has not yet been developed, steam is supplied to these turbines in the dry saturated state at ca 280°C. Because of these disadvantages, turbines employed in nuclear power plants are considerably larger and less efficient than those employed in fossil-fuel plants and require com-

plicated devices for extracting the moisture that is released from the steam as it expands through the turbine.

Steam Power Plants

Rankine cycle. The Rankine cycle (Fig. 2) is the basis from which all modern steam power plants have evolved. The work of the cycle is the difference between the work of the turbine and the work of the feed pump. The higher the initial pressure and temperature, the higher the cycle efficiency. This fact has led to a gradual trend toward higher pressures and temperatures in power-plant design.

Regenerative cycle. The regenerative cycle is similar to the Rankine cycle except that certain modifications increase the overall thermal efficiency. Steam is bled from the turbine at several points for the purpose of heating the condensate feedwater before returning it to the boiler. In the reheat–regenerative cycle, steam that has already expanded partially through the turbine is withdrawn and reheated at constant pressure to a higher temperature before it is returned to the turbine for completion of expansion. Reheating increases thermal efficiency.

Cogeneration. Because of the necessity and economic advantage of conserving fuel, governmental authorities have encouraged the generation of electric power as a by-product in plants built primarily for supplying process and other heat requirements. This type of operation is known as cogeneration. The real advantage of cogeneration is its more efficient use of fuel. Since ca 44% of the heat supplied to a conventional central power plant is lost to the cooling water in the condenser, this loss should be prevented by any means available. However, to be economically attractive, cogeneration requires a reasonably satisfactory balance of energy requirements for process or space heating in comparison to electric needs. Because of the wide variety of needs, there is no one preferred industrial cogeneration cycle. Any cycle that is adopted is essentially Rankine or the regenerative cycle having modifications for extracting steam at various pressures and in various amounts for process or other heating requirements.

Other Methods

Other methods include the following. Magnetohydrodynamics (MHD), in which electric power is produced by the movement of an electrically conducting gas through a magnetic field, requires extremely high temperatures to produce dc which must be converted to ac for suitable transmission and distribution (see Coal conversion processes, MHD). Electrogasdynamics (EGD) is also a method that directly converts the enthalpy of a working fluid to electric energy but does not require the extremely high temperatures of MHD. It can be incorporated into a power-generation scheme without the need for combining it with a steam plant of overall efficiencies comparable to that achievable with MHD; however, only low current, high voltage electric power is produced. Fusion power is a process that has received a great deal of interest recently (see Fusion energy; Plasma technology). Thermoelectric generators utilize the Seebeck effect, which operates on the same principle as a thermocouple, and have low efficiency and power output (see Temperature measurement). Thermionic generators operate on the principle that if a metal is heated to a sufficiently high temperature, electrons are emitted from its surface. If two parallel metal plates are separated by an ionized gas, and if one plate is heated and the other is kept cool, electrons flow through the ionized gas from the hot plate to the cold plate (see Thermoelectric devices).

Energy Storage

Principal energy-storage media are water, ie, pumped hydroelectric energy (pumped-hydro), compressed air, heat, and chemical batteries. Until recently, the only economic storage option available to electric utilities was the conventional pumped-hydro in which, during low demand period, water is pumped to large storage areas at higher elevation and released during peak demand periods to fall to the original elevation and, in the process, operates a hydraulic turbine that drives an electric generator. Another potentially attractive energy storage for peak use is thermal-energy storage in which thermal energy is extracted as steam or water and stored in hot rock, oil, or hot water and later is converted to

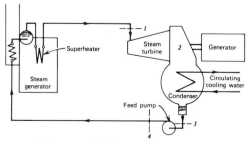

Figure 2. Rankine-cycle power plant.

steam by heat exchangers or flash evaporators for conversion to electricity.

SUSHIL K. BATRA
GIMRET International, Inc.

R.A. BUDENHOLZER
Illinois Institute of Technology

Evaluation of Phase 2 Conceptional Designs and Implementation Assessment Resulting from the Energy Conversion Alternatives Study (ECAS), NASA TM X-73515, April 1977.

J.A. Booth and D.H. Hall, *Turbomachinery International* 22(4), 13 (Apr. 1981).

A.W. Culp, Jr., *Principles of Energy Conversion*, McGraw-Hill, Inc., New York, 1979.

PRASEODYMIUM.   See Rare-earth elements.

PRESERVATIVES.   See Antioxidants and antiozonants; Coatings; Food additives; Paint; Wood.

PRESSURE MEASUREMENT

Measurement of pressure in the chemical industries and laboratories is of interest for a variety of reasons: differential pressure is the driving force in fluid dynamics; pressure is one of the fundamental terms in the ideal gas law and its corollaries and is one of the determining factors in vapor–liquid equilibria; and pressure is a safety consideration in the operation of process equipment (see High pressure technology; Vacuum technology).

Units of Measurement

Pressure is defined as force per unit of area. It can be defined in a wide variety of units, eg, the SI pressure unit, the pascal (Pa), defined as 1 N/m^2 (see also Units and conversion factors; Front matter to this volume).

Terminology

Atmospheric or barometric pressure is the pressure exerted by the column of air on the earth's surface. It varies with place, time, elevation, and weather conditions. These variables may be eliminated by establishing a normal atmosphere at 101,325 Pa. This is equal to the pressure exerted by a column of mercury 760 mm high at a temperature of 0°C, or 29.921 in. Hg, or 14.696 psi.

Absolute pressure is pressure measured from zero pressure. However, pressure gauges frequently measure from atmospheric pressure. Positive pressure measured in this manner is called gauge pressure. Gauge pressure is equal to the absolute pressure minus the atmospheric pressure and has often been reported specifically, ie, lbf/in^2 gauge, or psig. The expression psia has often been used to emphasize that the measurement is absolute pressure. In SI units, no official provision has been made to differentiate between gauge or absolute pressure. If the context does not clarify which pressure is meant, then a statement should be included to identify the pressure. A gauge that measures vacuum pressure reads the amount by which the pressure is less than atmospheric pressure. A compound gauge measures a pressure range that is above and below atmospheric pressure, ie, both gauge pressure and vacuum on the same scale.

Pressure-Measuring Devices

Pressure and vacuum are generally measured by directly actuated mechanical elements, particularly where they are monitored as opposed to being controlled. However, various electronic measuring devices continue to be introduced and are useful in certain areas. The mechanical elements are reliable and inexpensive. Liquid-filled indicators are used where vibration, pulsation, or atmospheric corrosion are a problem. Designs with blow-out backs are utilized for safety. Plastic mechanical movements on indicators exhibit good wear and are lightweight and corrosion resistant. The electronic devices are available in a large variety of forms. A broad choice of options, costs, and operating characteristics is offered. In plants where economies can be effected by extensive automatic instrumentation, and where precision of control, automatic data processing, and quick analysis of operations are important, electronic instrumentation meets these needs (see Instrumentation and control).

Diaphragm.   A diaphragm element is a thin flexible disk upon which pressure acts to create a force which causes a deflection of the diaphragm which, in turn, serves to move an indicator pointer, recorder pen, or other mechanism. Diaphragm elements are sensitive to small pressure changes and, therefore, are particularly useful in the measurement of low pressures.

Inverted bell.   An inverted-bell pressure element consists of two inverted bells that are partly immersed in oil, which provides a liquid seal. The bells are suspended from the opposite ends of a balance beam. A pressure can be introduced under each bell. This arrangement weighs the most minute difference between the two pressures. One of these pressures is atmospheric pressure and the other is the pressure to be measured. The bell with the higher pressure rises in the oil, and the beam tilts and moves a pointer on a scale and linkage to a controller. This element is sensitive to within 0.1 Pa (0.0005 in. H_2O).

Diaphragm-capsule.   A diaphragm-capsule pressure element is made up of two or more circular metal diaphragms that are welded at the inner and outer edges.

Bourdon tube.   A Bourdon tube is made from a flattened or elliptical tube, with one end sealed and the other open to the process pressure through connecting tubing. The final shape of the tube and the amount of flatness determine the trade name of the element and identify the overall shape and form. There are spiral-, helix-, and C-type Bourdon tubes. When the process pressure is applied through the connecting tube, the resulting force tends to uncoil or straighten the tubing. The rotating motion of the spiral or helix through a suitable linkage arrangement can actuate a pointer or pen arm.

Piezoelectric pressure-element designs.   Designs of piezoelectric pressure elements are based on the principle that a piece of quartz, when properly cut and oriented with respect to its crystallographic axes, generates a small electric charge on certain surfaces when stressed. In practice, a stack of properly cut quartz plates is mounted in a housing, which has a thin diaphragm at one end. The housing is usually designed to be mounted in the wall of a pressure vessel. The diaphragm is exposed to the pressure and deflects, thereby applying a compressive force to the quartz stack which, in turn, generates a charge directly proportional to the force. These are available in ranges as low as 0–69 Pa (0–0.01 psi) and as high as 0–830 MPa (0–120,000 psi). Response time is generally very fast. Frequency responses are as high as 500 kHz.

Piezoresistive sensors (integrated-circuit sensors).   Piezoresistive transducers convert a change in pressure into a change in resistance caused by strain. Pressure applied to one side of a silicon wafer strains resistors diffused into the wafer. The change in applied pressure causes a linear change in the resistance value, which can be converted and amplified to a usable output signal.

Vacuum measurement.   Weak vacuums or negative pressures can be measured by most of the pressure sensors described above. For measurement of high [0.13–1.3×10^{-4} Pa (10^{-3}–10^{-6} mm Hg)], very high [1.3×10^{-4}–1.3×10^{-7} Pa (10^{-6}–10^{-9} mm Hg)], and ultra high [$< 1.3 \times 10^{-7}$ Pa (10^{-9} mm Hg)] vacuums, thermoelectric and ionization sensors can be used. The thermoelectric sensor operates on the principal that the heat loss from a hot wire varies as the pressure of the gas or vapor surrounding the hot wire varies. This variation in heat loss with pressure is relatively large in high vacuum ranges for which it is used. The ionization sensor is based on the ability of electrons emitted from a hot filament to bombard the molecules of the residual gas in an evacuated system, thereby generating a current from the resulting ions. The magnitude of the current flow is directly proportional to the number of ions formed.

C. RAYMOND BRANDT
Honeywell, Inc.

B.G. Liptak in B.G. Liptak, ed., *Instrument Engineers Handbook*, Vol. 1, Chilton Book Company, Philadelphia, Pa., 1969, p. 249.

H.E. Soisson, *Instrumentation in Industry*, John Wiley & Sons, Inc., New York, 1975, p. 72.

R.P. Benedict, *Fundamentals of Temperature, Pressure and Flow Measurements*, John Wiley & Sons, Inc., New York, 1969.

PRESSURE VESSELS.   See High pressure technology.

PRIMING COMPOSITIONS.   See Explosives and propellants.

PRINTING INK.   See Inks.

PRINTING PROCESSES

The invention of modern printing using movable type is credited to Johann Gutenberg in ca 1440 A.D. Through the years, printing has developed into four main printing processes, as illustrated in Figure 1: relief or letterpress, intaglio or gravure, planographic or lithography, and stencil or porous printing. Letterpress printing traditionally excelled in the reproduction of text and pictorial matter, but improvements in the other processes have made it possible to produce reproductions equal to it in quality. In the production of all printing there are two physical areas: the printing (image) areas, and the nonprinting (nonimage) areas. In relief printing, the method by which typewriters, rubber stamps and letterpress printing operate, the image area is raised above the contact with the paper or other surface. Letterpress printing is used to print magazines, newspapers, advertising brochures, business forms, labels, packages, invitations, etc. Handset or machine-set cast-metal type can be used for direct printing but, for long printing runs, it is common to prepare printing plates or engravings from the type. Plates made on zinc, magnesium, or copper plated with nickel, are made of all illustration materials. When rubber or other elastomeric plates and water-base or solvent inks are used in printing, the process is called flexography. Letterset describes the use of relatively thin relief plates for printing by the offset principles.

In the intaglio process, the nonprinting area is at a common surface level and the printing area is recessed and consists of wells etched or engraved, usually to different depths. The most typical method of intaglio printing is the gravure process. Solvent inks with the consistency of light cream are transferred to the whole surface and a metal doctor blade is used to remove the excess ink from the nonprinting surface. Gravure printing is used for long-run magazines, mail-order catalogues, newspaper supplements, preprints for newspapers, plastic laminates, floor coverings, etc.

In the planographic or lithographic process, the image and nonimage areas are on the same plane and the difference between image and

Figure 1. The four chief printing processes: relief (letterpress), intaglio (gravure), planography (lithography), and porous (screen process). Courtesy of The Printing Industries of America.

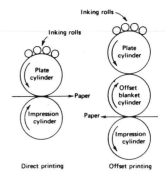

Figure 2. Printing cycle for direct and offset printing.

nonimage areas is maintained by the physiochemical principle that grease and water do not mix. The image area is grease-receptive and water-repellent; the nonimage area is water-receptive and grease-repellent. Ink adheres only to the image areas, from which it is transferred to the surface to be printed, usually by the offset method. This process is used for printing general commercial literature, books, catalogues, greeting cards, letterheads, business forms, checks, maps, art reproductions, labels, packages, etc.

In the stencil or screen printing process, a stencil representing the nonprinting areas is applied to a silk, nylon or stainless-steel fine-mesh screen to which ink with the consistency of paint is applied and transferred to the surface to be printed by scraping with a rubber squeegee. This process is used for printing displays, posters, signs, instrument dials, wallpaper, textiles, etc.

Direct printing is the transfer of the image directly from the image carrier to the paper. Most letterpress and gravure and all screen printing is done by this method. In indirect or offset printing, the image is transferred from the image carrier to an intermediate rubber-covered blanket cylinder, from which it is transferred to the paper (see Fig. 2). Most lithography is printed in this way and lithography is usually called offset printing. Letterpress and gravure can also be printed by the offset method (see Lithography).

Comparison of the Printing Processes

Lithography is an ideal process for text and pictorial reproduction for short and medium runs that are sheet-fed or web-fed up to ca 10^6 impressions; letterpress printing is good for text matter, flexibility, broad lettering, solid backgrounds and large expanses of color from short to long runs; gravure printing is best for long-run pictorial reproduction. Research in these and other processes, eg, flexography and waterless lithography, could result in improvements that could change these categories radically. Other printing processes, eg, ink-jet and electronic printing, could become competitive with lithography, letterpress, flexography, or gravure printing, when their reliability and quality are improved.

Prepress: The Preparatory Stage

Prepress consists of the operations involved in converting the original to be reproduced into a printing plate or image carrier. The starting point in all printing processes is the original or copy. This can be in many and varied forms, such as typed matter or text, and pictorial matter consisting of line drawings, art sketches, black-and-white and color photographs, paintings, etc. In some cases, the original can be produced directly on the plate surface, as in stone lithography, linoleum blocks, steel-die and copperplate engraving, and screen printing. In all other cases, the original is converted to the form for printing by photomechanical means.

Typesetting. A vast amount of printing begins with typesetting, which can be done by hand, machine, photography, or electronics. The second landmark in the history of printing was reached with the invention of the linotype machine (also called a line-casting machine because it casts a slug or line of type at a time). It is known as a hot-metal process and results in type metal, a raised cast-metal type consisting of an alloy of lead, tin, and antimony.

As distinguished from hot-metal type, there are cold-type forms of compositions, eg, strike-on composition, including typesetting for reproduction on special typewriters with carbon-paper ribbons and word-processing machines; and electronic typesetting, consisting of phototypesetting and computer typesetting. The third landmark in printing was reached in 1952 with the invention of the Photon typesetter. Photo- and computer typesetting have displaced more than 90% of cast-metal typesetting and the degree of complexity extends to the electronic setting of complete pages without any photographic film or handwork in the assembly of the images on the pages.

Copy assembly. All originals, consisting of text, pictures, and illustrations, must be photographed to convert them into the proper positive or negative films for the photomechanical plate processes by which they will be reproduced. On complicated jobs, eg, color advertising, magazines, books, etc, in which the copy can come from a number of sources, separate pieces of copy are photographed at the correct size and assembled in film form. This completed form is called a flat and the operation of film assembly is image assembly or stripping.

Process Photography

Process photography refers to the photographic techniques employed in the graphic arts processes. Many of the materials and techniques used in process photography are similar to those in photography (qv). Graphics arts involves highly specialized photographic equipment, methods, and materials. The photographic requirements of the various printing processes are diverse and exacting. The successful execution of any printing process involving photomechanical methods for making the image carriers depends to a great extent on the quality of its photographic components.

Letterpress, lithography, screen printing, and lateral-dot gravure are binary processes, ie, at one time or in one impression on the press, they can only print a solid color in the image area on the press and no color in the nonimage area. They cannot print intermediate tones or gradations of tone, ie, continuous tone. Most pictures or scenes to be reproduced have many intermediate tones between the shadows and the highlights. Such a picture is reproduced by a process involving a simulation of continuous tone called halftone. This is an optical illusion in which the tones are represented by solid dots, which are spaced equally but which vary in area (see Figs. 3 and 4).

Equipment. The most commonly used piece of equipment in process photography is the process camera, similar to an oversized enlarger.

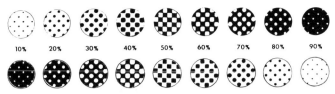

Figure 4. Diagram of halftone dots. The dots in the upper row are positive, those in the lower row are negative. Courtesy of Graphic Arts Technical Foundation.

Most are of the darkroom type for which the film back is mounted in a darkroom wall.

Process cameras use special lenses which are color-corrected, eg, apochromatic, and are relatively free of spherical and chromatic aberrations and distortion.

Materials. Three types of photographic emulsions are used in photomechanical processes. For line and halftone reproductions, high contrast orthochromatic or panchromatic emulsions of the lith type are needed. These are slow, thin, have a high silver content and fine grain, and the characteristic curves show little or no toe, very high gamma (ca 6–10), and maximum density of over 4.0 (see Photography).

Color Reproduction

Color theory. Like color photography (qv), process color printing is based on the Young-Helmholz theory of three-color vision. According to this theory, white light, which is a combination of all the wavelengths of light, consists of three primary colors: blue, green, and red. These are broad bands of color as distinguished from the physical concept of color in which each wavelength of light varies in color from every other wavelength (see Color). Any other color can be produced by appropriate mixtures of the three color bands. Covering the red area with a black sector and spinning produces a blue-green, which is called cyan. Cyan reflects blue and green light and absorbs red light. Covering the green area with the black sector produces a blue-red, which is called magenta and reflects blue and red light while absorbing green light. Covering the blue area produces a yellow, which reflects red and green light and absorbs blue. These are recognized as the colors of the printing inks used in four-color process reproductions. They are sometimes called complementary colors but are usually referred to as subtractive primaries. Each is a combination of two colors that are left when one primary color is subtracted from white light. Blue, green, and red light are additive primaries, as these colors of light add to form white light. When two subtractive primaries are printed over each other, they produce an additive primary. Overprinting yellow and cyan, yellow and magenta, and magenta and cyan forms green, red, and blue, respectively.

A typical good set of process color inks consists of dispersions of benzidine yellow, rhodamine Y (magenta), and phthalocyanine blue (cyan) pigments in the proper vehicles. For greater permanence or resistance to fading, rubine is often substituted for or mixed with the rhodamine. Rubine is a poorer pigment in blue reflectance than rhodamine so its use further affects color balance.

Color separation. Color reproduction is based on the three-color theory of vision in terms of duplicating the operations that the eye and brain perform when a color scene is viewed. The scene is usually photographed on color film and the transparency or a color print made from it is used as the original for reproduction. A negative is produced in the camera or enlarger by contact in a vacuum frame.

A negative made with the blue filter is a recording of all the blue light reflected from or transmitted, in the case of transparencies, through the copy. When a positive is made from this negative, it becomes a recording of the red and green colors in the original. In effect, the negative serves to subtract the blue from the original. The color that reflects red and green light is yellow. In color reproduction, the positive made from the blue filter or blue separation negative is printed with yellow ink. The negative made with the green filter is a recording of the green light, which is reflected or transmitted, and the positive made from this negative is printed in magenta ink. Likewise, the negative made with the

Figure 3. Halftone illustration and enlarged portion showing halftone dots. Courtesy of *Graphic Arts Manual*.

red filter records the red light from the original and the positive made from this is printed in cyan ink.

Because the printing-ink colors are not ideal, a reproduction made according to these simple principles lacks crispness, cleanness, and color purity and saturation. They must be color-corrected using photographic or electronic masks in the cases where color separation is done on electronic scanners, which are replacing many cameras for color reproduction. Electronic pre-press systems are in use that eliminate all photographic and image-assembly operations. Originals are scanned electronically, corrected, and composed into pages by computers, which are exposed directly onto plates using lasers.

Photochemical Methods of Platemaking

The photochemical method of platemaking is by far the most widely used and most important, and is a vital part of all other methods of plate production. Photomechanical methods involve exposure of photographic images to light-sensitive coatings that are either directly on the printing member or can be transferred to it. The distinguishing feature of the light-sensitive coatings is that, on exposure to light, they undergo changes in physical characteristics, usually solubility in water or in other chemicals, so that they can be developed to produce images that serve either as the printing images or as resists for producing the printing images. Where the light-exposed or hardened coatings are used as resists, eg, in deep-etch and bimetal plates, they must remain soluble in other chemicals so that they can be removed after the images have been produced (see Photoreactive polymers).

Light-sensitive coatings can be applied to the printing member by hand or machine in the processes where the printing plates are coated directly. Most plates are precoated or presensitized. A vacuum frame, such as is used for making contact exposures in photography, is usually used to expose the coated plate to the negative or positive. Until recently, the majority of the coatings used in the photomechanical processes were bichromated colloids; now diazo compounds, photopolymers (eg, cinnamic ester resins, polymethacrylate, polyamide, vinyl urethane systems, photosensitive synthetic rubber), silver halide photographic emulsions, and electrostatics have replaced them (see Photoreactive polymers; Epoxy resins; Dyes, sensitizing; Methacrylic polymers).

Lithographic Platemaking

Lithography was invented by Alois Senefelder, a Bavarian, in ca 1798. He discovered that if he drew characters on smooth Bavarian Solnhofen limestone with a greasy crayon and then dampened the surface with gum water, he could repeatedly ink the greasy design and pull impressions on paper. In 1906, the offset principle was introduced. Until then, all lithographic printing on paper involved a direct transfer of ink from stone or plate to paper. The rotary offset press embodied an additional cylinder covered with a rubber blanket between the plate and impression cylinders; thus, the ink was transferred first from the plate to the rubber blanket and the from the blanket to the paper. Offset press is characterized by the following advantages: the rubber printing surface conforms to irregularities in the paper surface, improving print quality; paper does not come into contact with the metal printing plate and thus the plate is less subject to abrasive wear; the speed of printing is increased; the image on an offset printing plate reads right instead of reverse or wrong-reading, facilitating both hand and photomechanical preparation; and less ink is required for total coverage.

Lithographic plates are of many types and can be made by many processes, including surface plates (diazo-presensitized, wipe-on, and photopolymer plates); deep-etched plates; bimetal plates; waterless plates, in which dampening in the lithographic process is eliminated; electrostatic plates, using xerography and electrofax (see Electrophotography) to produce the images; and laser plates, using lasers (qv) to expose images from paste-ups or computer memories onto film or plates. Other processes include collotype printing, screenless printing, and direct-image plates.

Health and Safety

New government regulations may severely restrict all solvents that are commonly used in printing inks from being emitted to the atmosphere without some control method, eg, incineration, solvent recovery, or electrostatic precipitation. Enforcement of such regulations may stimulate the development and use of radiation-cured, chemically reactive, or water-base inks (see Radiation curing).

MICHAEL H. BRUNO
Consultant

V. Strauss, *The Printing Industry*, Printing Industries of America, Washington, D.C., 1967.

R.F. Reed, *Offset Platemaking*, GATF, Pittsburgh, Pa., 1967, pp. 23–25.

M.H. Bruno, *Status of Printing in the U.S.A.—1983*, published biennially by New England Printer and Publisher, Salem, N.H.

PROCESS ENERGY CONSERVATION

Between 1973 and 1983, the ratio of energy price to capital price increased by a factor of 5–10. As a result, the old rules for optimum reflux, pressure drops, and temperature differentials must be adjusted. The increase in energy price has also increased the value of good engineering.

Energy Balance

Today, with increased energy cost, energy balance has become almost as important as material balance. On some studies, this is extended to an analysis of lost work. The key concept for this is exergy E, the potential to do work, ie, the maximum theoretical work a stream can deliver by coming into equilibrium with its surroundings.

$$E = (H - H_0) - T_0(S - S_0)$$

where H and S are the enthalpy and entropy, respectively, of the stream at its original conditions; H_0 and S_0 are the enthalpy and entropy, respectively, of the same stream at equilibrium with the surroundings; and T_0 = temperature of the surroundings (sink). Energy is sometimes called availability or work potential.

Lost work LW is irreversible loss in exergy. It occurs because of process driving forces or mixing material at different temperatures or compositions.

$$LW = E_{in} - E_{out}$$

Reactor Design

Often the greatest single contribution to reduced energy cost is increased yield. High yield reduces the amount of material to be pumped, heated, and cooled and simplifies downstream separation. (It also saves raw materials and the chemical industry uses almost as much energy in its raw materials as it does in purchases of electricity and fuel.)

Reaction temperature is usually dictated by yield considerations, but where possible in an endothermic reaction, temperature should be as low as practical, and in an exothermic reaction, as high as possible.

Continuous reactors use much less energy because of increased opportunities for heat interchange.

Separation

About one-third of the energy consumed by the chemical industry is used for separation.

Distillation. Distillation (qv) is the most common separation technique. Phase separation is clean; it is relatively easy to build a multistage, countercurrent device, and equilibrium is closely approached in each stage. Distillation is generally preferred for feed concentrations of 10–90%. It is probably a poor choice for feed concentrations less than 1%. Adsorption, absorption (qv), extraction (qv), chemical reaction, and ion exchange (qv) are chiefly used to remove impurity concentration < 1% (see Adsorptive separation).

The chief opportunity for improved practice in distillation is a reduction in the temperature differential used in reboilers and condensers. The economic optimum ΔT is typically under 15°C. Multieffect distillation is one way to reduce these ΔT's. Use of intermediate reboilers and condensers is a second way. (An intermediate reboiler moves the heat input location up the column to a slightly colder point and permits the use of waste heat when the bottoms temperature would be too hot.)

At today's energy prices, the optimum reflux ratio is generally below 1.15 minimum and often below 1.05 minimum. Operation at this low reflux often will require microprocessor or computer control with feed-forward capability. Experience has shown that this type of control can save 5–20% of a unit's utilities.

The penalty for column-pressure drop ΔP is an increase in temperature differential across the column

$$\Delta T = \left(\frac{dT}{dP} \right) \Delta P \qquad \frac{dT}{dP} = \frac{R}{\Delta H} \frac{T^2}{P}$$

The work penalty associated with this ΔT is approximately defined by the ratio:

$$\frac{\Delta T \text{ for pressure drop}}{T_{\text{reboiler}} - T_{\text{condenser}}} = \text{fraction of work potential for } \Delta P$$

This penalty is severest for close-boiling mixtures. Conventional packing can cut this penalty by a factor of four.

Relative volatility increases as pressure drops. For some systems, a 1% drop in absolute pressure can cut the required reflux by 0.5%.

Steam (or other stripping gas) and vacuum are largely interchangeable. Steam stripping allows more tolerance for pressure drop, but at the penalty of much higher energy use.

Other separation techniques. Absorption (extractive distillation) shares most of the advantages of distillation. In addition it separates by molecular type and hence can be tailored to obtain a high relative volatility. Extraction is conceptually similar, but because it purifies a liquid rather than a vapor, it is difficult to obtain high efficiency countercurrent processing (see Azeotropic and extractive distillation).

Adsorbents achieve still better selectivity than extraction. A fixed bed with thermal regeneration attains essentially 100% removal and carries little penalty for low feed concentration. The simulated moving-bed system has large-volume applications in paraffin and xylene separations.

Purification by reaction is used for low concentrations; for example, the hydrogenation of acetylene:

$$C_2H_2 + H_2 \rightarrow C_2H_4$$

Crystallization from a melt offers the advantage of a heat of fusion much lower than that of evaporation, but is rarely used because of the much greater difficulty of solids/liquids processing.

Liquid separation via membranes (reverse osmosis (qv)) is used in the production of pure water from seawater. Gas separation via membranes is used in hydrogen separation (see Membrane technology).

Heat Exchange

Heat exchangers use energy as frictional pressure drop and as the loss in ability to do work when heat flows from a hot temperature to a colder one.

$$\text{lost work} = QT_{\text{sink}} \left(\frac{1}{T_{\text{cold}}} - \frac{1}{T_{\text{hot}}} \right)$$

Heat-exchange network analysis is applied to crude still preheat trains, multistage exothermic reactors, and furnace-convection sections (see Heat-exchange technology, network synthesis).

ΔT is optimized by finding the point where savings in utility costs balances incremental surface costs. There are several cases of optimum ΔT. The simplest is the waste-heat boiler. In a waste-heat boiler, the optimum ΔT occurs when

$$\Delta T_{\text{approach}} = \frac{K_1}{K_v} \frac{1.33}{U}$$

where K_1 = annual cost per unit of surface, $/(m^2 \cdot yr)$; K_v = annual cost per unit of utility saved, $/(W \cdot yr)$; and U = heat transfer coefficient, $W/(m^2 \cdot K)$.

For most heat exchangers, there is also an optimum pressure drop. The total cost curve is fairly flat within $\pm 50\%$ of the optimum. The most important factor in setting optimum pressure drop is density, hence, the much lower optimum P for gases than liquids.

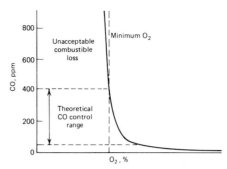

Figure 1. Relationship of CO concentration to O_2 concentration, in a fired heater.

Fired heaters. Unpreheated air in the combustion step is probably the biggest waste of thermodynamic potential in industry. The most common type of air preheater on new units is the rotating wheel. On retrofits, heat pipes or hot-water loops are often more cost-effective.

Limitations in the material of construction make it difficult to use the high temperature potential of fuel fully. This has led to the insertion of gas turbines into power-generation steam cycles, and even to their use to preheat air for ethylene-cracking furnaces.

Improved efficiency in fired heaters has tended to focus on heat lost with the stack gases, yet the losses for ΔT in the convection section are often twice as great.

Excess air can be controlled by measuring stack carbon monoxide (see Fig. 1). This is one of the most significant new energy technologies.

The much lower cost of coal has caused a rapid resurgence of coal firing for steam generation. Its direct use in process heaters has been negligible (see Power generation; steam).

Dryers. Drying (qv) can be viewed as both a separation and a heat-exchange step. When seen as a separation, the aim is to reduce the required work via reducing the water in the feed. When seen as a heat exchange, the aim is to reduce the enthalpy of the air purged with the evaporated water. This is achieved by using less hot inlet air as well as by heat recovery (see also Evaporation).

Pumping, Compression, and Vacuum Systems

In an optimized system, the annual cost for pumping power should be roughly one-seventh the cost of piping. Similarly, for an optimized heat exchanger, the annual cost for pumping should be one-third of the annual cost of the surface for the thermal resistance connected with that stream. These simple relations can be very useful in quick design checks.

The work of compression is typically compared with the isentropic minimum

$$\eta_{\text{comp}} = \frac{W_{\text{min}}}{E_{\text{out}} - E_{\text{in}}}$$

where η is the efficiency. Efficiencies for mechanical compressors should always exceed 0.6, and 1.00 is approachable in reciprocating devices.

A thermocompressor is a single-stage jet using a high pressure gas stream to supply the work of compression. The commonest application is in boosting waste-heat-generated steam to a useful level. Efficiency is generally below 30%.

Vacuum systems use vacuum jets and pumps. Because of the low jet efficiency, there is a range of vacuum above 13 kPa (100 mm Hg) where pumps will usually be more cost effective. As pressure falls, the capital cost of the vacuum pump rises more swiftly than the energy cost of the steam ejector, hence, there is also a range of vacuum starting at around 10 mm Hg below which the jet is usually more cost effective.

Refrigeration

The value of refrigeration (qv) is the work required to pump heat it to the sink temperature. Refrigeration is generally much more valuable than heat above ambient and justifies thicker insulation and lower ΔT in heat exchange.

Steam and Condensate Systems

Many process plants employ accounting systems where all steam is carried at the same price regardless of temperature or pressure. This may be appropriate in a polymer or textile unit where high temperatures are not required, but in a petrochemical plant it will generally lead to excess low pressure steam. A pricing system that depends on work content (energy) will function better. There are many areas for optimizing a steam system, among them: use of gas turbines upstream of the boiler, feedwater heating, condensate flash-steam recovery, and metering of all users.

Cooling-Water Systems

Cooling water can cost one-fifth as much as the primary fuel. Heat exchangers should be designed to use the available pressure drop and flow should be limited to that needed. If temperature requirements permit, the system will cost less to operate with exchangers in series.

Special Techniques

Heat pumps. There are refrigeration systems that raise heat to a useful level. Application depends primarily on low cost power relative to the alternative heating media.

Energy-management systems. Reduction in computing costs permits a wide range of routine monitoring and controlling, ranging from steam systems to building heating and cooling (see Energy management).

Existing Plants

Process needs are measurable and an energy balance is a mandatory first step for a retrofit program. In general, the existing plant will yield many attractive energy projects, as well as a large reduction in energy use achieved by tightening operations.

D. STEINMEYER
Monsanto Co.

D.E. Steinmeyer, *Chemtech*, 188 (Mar. 1982).

E.P. Gyftopoulos, *Industrial Energy Conservation*, 17 manuals, MIT Press, Cambridge, Mass., 1982.

PROCESS RESEARCH AND DEVELOPMENT

The aim of process research and development is to adapt a laboratory-scale procedure to a commercial process (see also Pilot plants and microplants; Operations planning). Further input may be needed to resolve problems that arise on start-up and for optimization of performance. During process development, a stage may be reached when the process is operational and the prime responsibility becomes a concern to the production management. Further research and development may then be transferred to a plant technical staff; the latter phase is called process improvement.

Process research and development combines experimental work with technical and economic calculations, which are guided largely by chemical and chemical engineering principles. Because the ultimate aim is an operating plant, the technical aspects are inseparable from the economic and legal (patent) ones. If these activities are not properly integrated, the process design will not be satisfactory (see Research management).

Early Stage

Research frequently provides the concept for the process, ie, the new chemistry, and always generates the necessary data. Research begins with an exploratory phase during which chemistry is at the center of the activities, followed by the process-research stage, which involves the experimental development of a processing scheme based on chemical engineering principles, and leads to process design.

Economic Evaluation

The economic evaluation (qv) is started long before the process research and development phase. Economic evaluation may be the impetus for technical personnel to investigate a certain reaction system. An important aspect is the preparation of conceptual process flow sheets based on reaction data, by-products and separation schemes, recycle streams, product purification, and materials of construction. Such flow sheets are the basis for cursory plant design, followed by estimates of capital investment, utilities, plot plan, and environmental safeguards (see Plant layout; Plant location).

The economic-evaluation group must be able to draw up conceptual flow sheets of alternative proposals and guide the experimental and computational development work to the best economics.

Project Management

A general outline of a research and development (R & D) organization is given in Figure 1. Some companies have centralized R & D organizations; in others, responsibilities are decentralized and distributed among various operating groups. Some companies emphasize a strict line organization, but others use the matrix system, where line functions and project functions interact, as shown in Figure 1.

Process Research

Reaction engineering. The reactors are the central concern in most chemical processes. Their performance determines the development of the process. Determination of the reaction regime is vital for a meaningful interpretation of reaction studies. In the case of reversible reactions, the equilibrium constant permits the investigator to outline the region of desirable reaction conditions.

In industry, mass-transfer or chemical-rate-limited reactions are more common than equilibrium-limited reactions. The energy of activation is an indication of the rate-limiting regime, since mass-transfer operations such as diffusion through catalyst pores or interfacial films are not highly temperature dependent, ie, they have low energy of activation (see Mass transfer).

The process-research engineers explore the catalyst system to the point where a model with a sufficiently wide range of variables has been developed and proved in a steady-state mode, including recycle streams (see Catalysis). This work includes the modeling of the rate equation.

The process control and safety system is based on process dynamics. The following data are needed for its design: a reaction-rate model; physical properties of all components; a residence time–distribution model; a heat-transfer model; response models for measuring elements such as analytical instruments, temperature, flow and pressure-controlled elements; and response models for process control, eg, control valves.

By-product identification and separation. After investigation of the reactions, the biggest problem in process research is the separation of by-products. By-products must be purged from the system. Economics usually determines the composition of the purge streams.

Materials of construction. Materials of construction must be evaluated throughout the process research and development. Like catalyst life-testing, corrosion tests, preferably under conditions of continu-

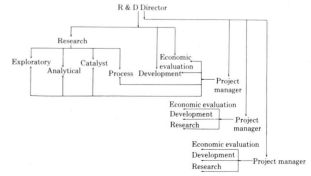

Figure 1. Typical matrix R & D organization.

ous operation, should be started as soon as stream compositions and temperatures are reasonably well known (see Corrosion and corrosion inhibitors).

Health and Safety

The process-research phase offers the opportunity to investigate any potentially dangerous features of the process, eg, potentially explosive gas mixtures (see Plant safety). Determination of explosive regions of the gas mixtures encountered in the process is necessary for safe equipment, process and control-system design.

Process Patents

Regardless of whether the chemistry of a new process is known and patented, there is generally room for further inventive improvement. Criticalities of certain variables or specific limits of the ranges of such variables may be discovered once a thorough study of the kinetics and thermodynamics of the reaction takes place. There is an endless variety of aspects in a process where inventions can be made and patented. Such patents may not establish a fundamental position in the particular technology; nevertheless, they can prevent other practitioners of similar technology from applying certain advantages.

ERNEST I. KORCHAK
Scientific Design Company
a division of The Halcon SD Group, Inc.

J.R. Fair, *AIChE Mongr. Ser.* **76**, (1980).

G. Jordan, *Chemical Process Development*, Interscience Publishers, a division of John Wiley & Sons, Inc., New York, 1968.

J.H. Perry, *Chemical Engineers' Handbook*, 5th ed., McGraw-Hill, Inc., New York, 1973.

PRO DRUGS. See Pharmaceuticals, controlled releases; Pharmacodynamics.

PRODUCER GAS. See Fuels, synthetic.

PRODUCT LIABILITY

Product liability is the legal term used to describe an action in which an injured party (plaintiff) seeks to recover damages for personal injury or loss of property from a seller (defendant) when it is alleged that the injuries resulted from a defective product.

Theories of Liability

The following three principles can be used in most states as the framework within which the plaintiff can bring an action in product liability: negligence, which tests the conduct of the defendant; strict liability and implied warranty, which test the quality of the product; and express warranty and misrepresentation, which test the performance of products against the explicit representations made by the manufacturer and sellers.

Negligence. To make out a cause of action in negligence, it is not necessary to establish that the defendant intended harm or acted recklessly. The defendant's conduct is deemed negligent when it falls below what a "reasonable person" would have done under similar circumstances. In every negligence action, the court must consider the probability of harm occurring; the gravity of harm, if it should occur, and the burden of precaution against the harm.

The negligence test focuses on the technology and information available at the time of manufacture. Since the determination of negligence arises on a case-by-case basis, it is possible that the same conduct may be judged both negligent and nonnegligent in different legal proceedings.

Strict liability and implied warranty. In 1964, the concept of strict liability emerged as an alternative avenue for establishing an action in product liability. The clearest expression is part of the Restatement (Second) of Torts, Section 402A (1965):

(*1*) One who sells any product in a defective condition unreasonably dangerous to user or his property is subject to liability for harm caused to user or his property if (a) the seller is engaged in the business of selling such a product, and (b) it is expected to and does reach the user or consumer without substantial change in the condition in which it is sold.

(*2*) The rule under (*1*) applies although (a) the seller has exercised all possible care and (b) the user has not bought the product from or entered into any contractual relation with the seller.

In deciding whether a product is unreasonably dangerous, the focus in strict liability is on the product and not on the conduct of the manufacturer. It is essential, as in the theory of negligence, to weigh the burden of precaution to protect against the harm together with the probability and gravity of the harm.

A guarantee of product quality, similar to that of strict liability, is given by the Uniform Commercial Code, Section 2-314. The code has been adopted in every state except Louisiana. Section 2-314 provides the same consumer protection as the Restatement requirement that a product be reasonably safe.

Express warranty and misrepresentation—implied warranty of fitness. An alternative theory for recovery does not require the establishment of a defect; ie, it is not necessary to establish that the product is unreasonably dangerous. Where a product fails to meet the seller's own warranty or representations, if the buyer can prove that injury resulted from the failure of the product to meet the warranty, liability is established. If a manufacturer markets beyond the capability of the product to perform, liability may result. An express warranty is a form of absolute liability.

Defect types. A production-defect case is one in which the product does not meet the manufacturer's internal design specifications. In almost all jurisdictions, courts impose liability if the product was defective without proof of negligence.

In contrast to the production-defect case, the design-defect case does not provide a built-in internal standard. The product meets the manufacturer's own standard. A balancing process must help determine the standard, which must be one of reasonableness.

Strict liability theory, unlike negligence, is not concerned with the conduct of the defendant. Whether a manufacturer should be liable for failing to design a product against risks, which were not discoverable at the time of manufacture, has been a source of conflict among courts and scholars.

Torts Concerning Toxic Substances

Toxic substances. Product liability law has dealt with products that have toxic qualities. Asbestos litigation spawned claims in such numbers that they threatened the financial solvency of the main producer, Johns-Manville Company. In a class action brought by veterans who served in Vietnam, the contention is that the companies that sold Agent Orange to the U.S. Government permitted contamination with dioxin.

Application of the Balancing Process

The case of Wilson v. Piper Aircraft Corp. is an illustration of how the "standard of reasonableness" is determined. The plaintiffs alleged that the airplane crash was caused by engine failure resulting from carburetor icing and that susceptibility to icing was inherent in the design. The court based its decision primarily upon the impracticality of requiring all aircraft to be manufactured with fuel-injection engines and that the plaintiffs had provided no evidence as to the effect of the proposed design alternative on such matters as "cost, economy of operation, maintenance requirements, over-all performance, and safety" of such planes. This case demonstrates that courts can take their policy-making role very seriously.

Causation

Even if the plaintiff establishes that the product was defective, a causal relationship between the defect and the malfunction must be established.

In some cases of defective products that cause injuries, the manufacturers do not bear the onus of liability. Products can be altered and misused and thereby cause injury. In some cases the injury is attributed to the misuse or alteration rather than the original defect.

In most personal-injury actions, if the plaintiff failed to act reasonably, the recovery will be reduced under the doctrine of comparative fault. In a minority of jurisdictions, the contributory fault of the plaintiff will prevent the recovery of any damages at all.

Developments in the field of product liability have fostered an atmosphere of crisis among manufacturers and many states have passed legislation seeking to protect manufacturers.

Aaron Twerski
Hofstra Law School

A.W. Weinstein, A.T. Twerski, W.D. Donaher, and H.P. Piehler, *Products Liability and the Reasonable Safe Product*, John Wiley & Sons, Inc., New York, 1978.

Beshada v. Johns–Manville Products Co., 90 N.J. 191, 447 Atlantic 2d 539 (1932).

282 Oregon 61, 577 Pacific 2d 1322 (1978).

PROGRAMMABLE POCKET COMPUTERS

There are two firms that market programmable computers: the Hewlett-Packard Corp. (HP) and the Texas Instrument Co. (TI), each of which has several models which vary in cost and capability. Several new, slightly longer, computers that are programmable in a higher level language, such as BASIC, rather than being step-programmable, have recently been introduced. There are other differences too numerous to be mentioned here. The HP and TI computers are all pocket size, weigh less than 0.5 kg, and are battery-powered. Alternating-current adapter cords are available and most models have rechargeable battery packs. The face of the computer has a single line-display area which consists of light-emitting diodes (LEDs) (qv) or liquid crystals (LCs) (qv) capable of displaying up to 8 or 10 decimal digits plus a two-digit exponent (one model can display up to 12 alpha characters), and a keyboard for data entry and to invoke the many function capabilities designed into the computer. Most keys have two or more functions associated with each of them, thus making it possible for a relatively small keyboard to access the computer's more than one hundred functions.

These computers contain a program memory, data-storage area, and some form of operational stack. The program memory can vary from a few hundred to several thousand program steps, depending on the model and manufacturer. The data-storage registers also vary in number and, in some models, the data-storage area can be increased or decreased as required by interchanging the data-storage and program-storage area, usually on the basis that eight program steps equals one data-storage register. Some models have a continuous-memory feature, which permits the computer to retain data and program information even when the computer is turned off. Also contained within some computers is a small electric motor, which drives a magnetic-card reader, thus permitting program and/or data to be both read from or written to small magnetic cards. It is this ability store programs and data on these tiny cards that contributes so much to the power, versatility, and general usefulness of these small machines. Some of these pocket computers are capable of being connected to a small thermal printer, which permits hard copy output to be obtained at the expense, however, of no longer being pocket size. Another feature available on some of the newer models is plug-in modules of read-only memory. These are preprogrammed to solve specific problems and permit very substantial increases in total memory capacity.

Computer Languages

The HP and TI pocket computers, although basically competitive with each other, are based on fundamentally different concepts insofar as programming is concerned. The latter is based on algebraic notation, and the former is based on Polish notation. When one writes a formula, for example, that includes the addition of two variables x and y, one generally writes $x + y$; that is to say, one places the addition operator between the two operands. This is generally called infix notation and is the essence of algebraic programming language. If instead of writing about the addition of two values, one performed the addition with pencil and paper, one would first write down the value of the first operand, then write down the value of the second operand, and then perform the addition. Thus, the operands precede the operator. This is postfix notation and in computer parlance is called Polish notation.

Algebraic notation involves infix notation for diadic operators, ie, those requiring two operands, and postfix notation for monadic operators, ie, those requiring one operand, and involves a hierarchy of operator precedences and the use of parentheses to alter the hierarchical precedence. The usual hierarchical precedence places monadic operators at the highest level, followed by exponentiation, followed by multiplication and division, which occupy the same hierarchical level, and finally by addition and subtraction, which also occupy the same level. Inner parentheses take precedence over outer parentheses and, at the same hierarchical level, operators are executed from left to right in the expression. Thus, to evaluate the expression:

$$A + B \div C^*(D - E)$$

one keys in the value of A, then presses the addition key, then keys in the value of B, then presses the divide key, then keys in the value of C, etc. Actually, no computation takes place until the multiply key is pressed, at which time B, which was pending, is divided by C, which was also pending. After continuing to key in the above expression, no further arithmetic is performed until the closing parenthesis is pressed, at which point the sum of D and E would be formed and multiplied by the quotient of B/C and the product is added to A. Thus, to implement algebraic notation, the computer must be able to store or stack the pending operators as well as all open parentheses. The size of these operand and operator stacks determines the degree of complexity of the algebraic statement that can be handled as well as the depth of nesting of parentheses that is permitted. Evaluation of a fifth-order polynomial expressed in the nested parenthetical form:

$$a_0 + x\big(a_1 + x\big(a_2 + x\big(a_3 + x(a_4 + x^*a_5)\big)\big)\big)$$

requires the stack as follows:

$$\text{Operand stack } a_0, x, a_1, x, a_2, x, a_3, x, a_4, x, a_5$$
$$\text{Operator stack } +, *, (, +, *, (, +, *, (, +, *, (, +, *, =$$

Thus, before the first arithmetic operation takes place, a stack of eleven pending operands and a stack of fifteen pending operators and parentheses have been constructed. With pressing of the equal key, all open parentheses close and statement evaluation proceeds from the innermost parentheses to the outermost ones.

Polish notation involves no parentheses nor rules of hierarchical precedence but involves postfix placement of the operator for both diadic and monadic operators. It does require a stack for pending operands and the number of pending operands is limited by the size of that stack. Using the same example:

$$A + B \div C * (D + E)$$

and assuming a four-position stack, one would key as follows:

$$A, \text{ENTER}, B, \text{ENTER}, C, \div, D, \text{ENTER}, E, +, *, +$$

Each arithmetic operation occurs when its operator key is pressed, ie, there is never any pending operator. At one point, the operand stack holds four pending operands. To evaluate the fifth-order polynomial

$$a_0 + x\big(a_1 + x\big(a_2 + x\big(a_3 + x(a_4 + x * a_5)\big)\big)\big)$$

one starts at the inner parenthesis and keys as follows:

$$a_5, \text{ENTER}, x, *, a_4, +, x, *, a_3, +, *, a_2, +, x, *, a_1, +, x, *, a_0, +$$

The stack never holds more than two operands and no operators are ever pending.

Programming

Programming the pocket computer simply means recording each keystroke in solving the problem. For example, to calculate the vapor–liquid equilibrium curve for a binary mixture with constant relative volatility using the Smoker equation:

$$y = \frac{\alpha x}{1 + (\alpha - 1)x} \qquad (1)$$

Suppose that the value of the relative volatility α was already stored in register 0 and that the value of the liquid-phase composition x was stored in register 1. The key sequence to solve for y using Polish notation:

$$RCL0, RCL1, *, RCL0, 1, -, RCL1, *, 1, +, \div$$

where RCL = recall. The sequence of key strokes is used either to solve the problem or to program its solution. Thus, to solve the problem is to program the problem.

Both the HP and TI operate clearinghouses of user-developed programs for their respective machines. These clearinghouses and their related newsletters provide a vast store of programs covering a wide variety of interests from computer games to programs that solve practical problems in diverse areas of human activity. Technical publications are also a fruitful source of program listings, eg, the *Oil and Gas Journal* publishes program listings related to petroleum technology. There also is a Professional Program Exchange (PPX) for the TI-59 with its own newsletter.

Data correlation. Data correlation, which encompasses all aspects of fitting a suitable equation to a set of data points, is of almost universal interest (see Engineering and chemical data correlation). The two programs presented in this category show the wide range of possibilities that exist.

Linear forms. The program in Listing 1 below is based on an expression of the form:

$$y = a + b * f(x) \qquad (2)$$

As coded, the program determines the coefficients a and b of equation 1 for seven different functions to other functions of x of the user's choice. If y increases with x and the data follow a trend that is concave upward, appropriate forms of $f(x)$ might be $f(x) = 10^x$, $f(x) = e^x$, $f(x) = x^2$, or $f(x) = x^n$ where $n > 1.0$. If, however, the trend of the data is concave downward, appropriate forms of $f(x)$ might be $f(x) = -1/x$, $f(x) = \ln(x)$, $f(x) = \log(x)$, $f(x) = \sqrt{x}$, or $f(x) = x^n$ where $0 < n < 1$. If y decreases when x increases, the above functions result in concavity in the direction opposite to that mentioned above.

Listing 1. Data correlation—linear forms
LBLA, CF1, STOB, Σ+, R↓, STOA, GSB1, LBLØ, DSPØ, P⇄S, ST+3, X², ST+2, RCLA, RCLC, ×, ST+1, RCLA, RCLD, Σ+, R↓, RCLE, P⇄S, ST+3, ×, ST+1, RCLE, X², ST+2, RCL9, R/S, F3?, GTO5, GTOØ, LBL5, F1?, GTOC, GTOA, LBLB, CLRG, P⇄S, CLRG, CLX, R/S, LBLC, SF1, STOB, Σ+, R↓, STOA, GSB2, GTOØ, LBLØ, DSP4, P⇄S, RCL7, RCL6, RCL9, ÷, STOI, RCL6, ×, −, STOC, GSB3, P⇄S, GSB3, R/S, LBLD, CF1, CF3, STOB, Σ−, R↓, STOA, GSB1, LBL4, P⇄S, ST−3, X², ST−2, RCLA, RCLC, ×, ST−1, RCLA, RCLD, Σ−, R↓, RCLE, P⇄S, ST−3, ×, ST−1, RCLE, X², ST−2, R/S, F3?, GTO6, GTOØ, LBL6, F1?, GTOE, GTOD, R/S, LBLE, SF1, F3?, STOB, Σ−, R↓, STOA, GSB2, GTO4, LBL1, RCLB, 1/X, STOC, RCLB, LOG, STOD, RCLB, √X, STOE, RCLC, RTN, LBL2, RCLB, e^x, STOC, RCLB, 10^x, STOD, RCLB, X², STOE, RCLC, RTN, LBL3, RCL8, RCLI, RCL4, ×, −, STOØ, RCL5, RCL4, X², RCL9, ÷, −, STOE, ÷, PRTX, RCL4, ×, RCL9, ÷, CHS, RCLI, +, PRTX, RCLØ, X², RCLE, ÷, RCLC, ÷, √X, R/S, RCL1, RCLI, RCL3, ×, −, STOØ, RCL2, RCL3, X², RCL9, ÷, −, STOE, ÷, PRTX, RCL3, ×, RCL9, ÷, CHS, RCLI, +, PRTX, RCLØ, X², RCLE, ÷, RCLC, ÷, √X, R/S, RTN, R/S

To use program, start by pressing key B for BEGIN, which clears all the data registers to zero. Next, key in the value of y_1, press ENTER, key in the value of x_1, and press key A if the functions $1/x$, $\log x$, or $\sqrt{x}$ are wanted, or press key C if the functions e^x, 10^x, or x^2 are wanted. Shortly, the computer stops, showing 1, in the display. For each additional data pair, key in Y_i, ENTER, x_i, R/S. Each time the computer stops, the display shows the total number of pairs already entered and is ready to receive the next data pair. When there are no more data pairs to be entered, press R/S and the computer proceeds with the least-squares fitting of the selected functions. The program displays the slope b for the equation $y = a = bx$ for 5 s, after which it continues execution, then displays the intercept a for 5 s, and finally displays the correlation coefficient and stops. Pressing R/S causes this display sequence to be repeated for the second function chosen, etc.

Suppose one or two of the data pairs appear to be erratic and the user wants to determine what the correlation would look like if those data pairs had not been included. The user can simply key in the y and x values for the mavericks and press key D. (If key C had been used instead of key A at the start of the data entry, key E now should be used instead of key D.) When the computer stops, other erratic data pairs can be deleted by repeating this procedure or by reevaluating the new slopes, intercepts, and correlation coefficients for all four functions by pressing R/S. Keys D or E can be used during data entry to delete a set of y and x values that may have been entered in error.

Example. Suppose one collected the data shown in Figure 1 and wanted to fit an equation to the data. Noting that the curvature appears to be concave downward, one would start to enter the data with key A, as described above. A few seconds after the last data point is entered, the computer displays the slope, then the intercept, and then the correlation coefficient for each of the four equation forms represented by key A. The results of the example are as follows and are shown in Figure 2.

Form	Equation	Correlation coefficient
linear	$y = 0.329 + 0.496x$	0.0856
reciprocal	$y = 2.317 - 0.6781$	0.756
logarithmic	$y = 1.005 - 12.586 \log(x)$	0.931
square root	$y = -0.907 + 1.715\sqrt{x}$	0.920

The correlation coefficient indicates that a better fit is obtained with logarithmic and square-root forms in this instance. Figure 2, which is a plot of these four equations and the original data points, shows that the square-root function fits the data better at low values of x but that the logarithmic function fits the data better at high values of x. It also appears from Figure 2 that the fifth data point ($y = 2.54$, $x = 1.91$) might be a maverick. To determine what the correlations are with this single point omitted, one keys in the y and x values for this point, then presses key D and, when the computer stops, presses R/S, after which the following results are displayed:

Form	Equation	Correlation coefficient
linear	$y = 0.073 + 0.523x$	0.938
reciprocal	$y = 2.209 - 0.654/x$	0.754
logarithmic	$y = 0.896 + 2.559 \log(x)$	0.961
square root	$y = -1.108 + 1.747\sqrt{x}$	0.977

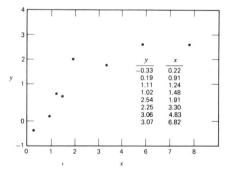

y	x
−0.33	0.22
0.19	0.91
1.11	1.24
1.02	1.48
2.54	1.91
2.25	3.30
3.06	4.83
3.07	6.82

Figure 1. Raw data, y vs x.

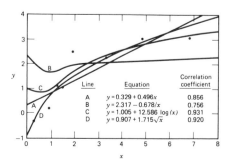

Line	Equation	Correlation coefficient
A	$y = 0.329 + 0.496x$	0.856
B	$y = 2.317 - 0.678/x$	0.756
C	$y = 1.005 + 12.586 \log(x)$	0.931
D	$y = 0.907 + 1.715\sqrt{x}$	0.920

Figure 2. Correlations developed by program 1 for data of Figure 1.

There has been an obvious improvement in the correlation coefficient for all but the reciprocal form, and the logarithmic and square-root forms have switched their relative positions, although both still show very high correlation.

Graphical solutions.   Another area, upon which programmable pocket computers can have a significant impact, involves problems that used to be solved graphically, such as the McCabe-Thiele solution of a binary-distillation problem. The process involved in a binary distillation is shown in Figure 3. In a typical design problem, the following question is asked: for a given feed composition and for given product compositions, how many separation stages are required? Usual input specifications include other parameters describing the feed conditions, reflux condition, stage efficiency, etc. The McCabe-Thiele solution technique results from material-balance equations, which are developed for both the top section and the bottom section of the column:

$$y_{n+1} = \frac{L}{V}x_n + \frac{Dx_D}{V} \tag{3}$$

$$y_{m+1} = \frac{\overline{L}}{\overline{V}}x_m + \frac{Bx_b}{\overline{V}} \tag{4}$$

where n and m refer to the stage indexes in the top and bottom sections of the column, respectively. Based on certain simplifying assumptions, which lead to L/V being constant in the upper or rectifying section of the column and $\overline{V}/\overline{L}$ being constant in the lower or stripping section of the column, these equations plot as straight lines on the equilibrium diagram and the number of ideal stages can be stepped off, ie, determined by drawing a series of horizontal and vertical lines. This graphical procedure is a little more complicated when a Murphree tray efficiency has been specified and when the count of real rather than ideal stages is wanted. When very pure products are required, ie, $x_B \rightarrow 0.0$, $x_D \rightarrow 1.0$, the graphical stepping-off process becomes very imprecise unless very large graph paper and very sharp pencils are used (see Distillation).

The program presented in Listing 2 below solves this problem in terms of the various ways in which the problem might be specified. One starts by keying in the values of q, α, the reflux multiplier, and the stage efficiency, followed by key E. The reflux multiplier is to be used with the minimum reflux ratio to obtain the actual reflux ratio L/D. If the actual reflux ratio is to be specified instead of the reflux multiplier, the former can be keyed in instead of the reflux multiplier; however, it should be entered as a negative number. The programmed response is to ignore the minimum reflux ratio and to take the absolute value of the number that was entered as the actual reflux ratio. Stage efficiencies are sometimes specified as overall efficiencies and sometimes as Murphree efficiencies. Here too, entering a positive number signals the former interpretation and a negative number signals the latter interpretation.

Listing 2. Graphical solutions: McCabe-Thiele distillation
LBLE, CF0, X<0?, SF0, ABS, STO3, R↓, X<0?, SF2, ABS, STOC, R↓, STO1, 1, −, ..., X⇌Y, X≠0?, GSBd, EEX, 4, CHS, +, LBLd, STO0, ×, STOA, R/S, LBLC, STO4, R↓, STO5, X⇌Y, STO6, −, RCL4, RCL6, −, ÷, STO8, GTOa, LBLB, STO8, 1, R↑, STC5, R↑, STO4, R↑, ×, −, X⇌Y, RCL8, −, ÷, STO6, GTOa, LBLD, 1, X⇌Y, −, STO8, R↓, STO5, X⇌Y, STO6, LSTX, ×, −, RCL8, ÷, STO4, LBLa, RCLC, F2?, GTO0, 1, RCL1, −, RCL0, RCL5, +, ×, RCL1, +, STOB, X², 4, RCLA, ×, RCL5, ×, +, √X, STOD, RCLB, −, 2, ÷, RCLA, ÷, STO9, RCL1, ×, RCL9, RCL1, 1, −, ×, 1, +, ÷, RCL6, −, RCL9, LSTX, −, ÷, STO9, 1, RCL9, −, ÷, RCLC, ×, LBL0, STO2, STOE, 1, +, STOB, RCL5, RCL2, ×, RCL0, RCL6, ×, +, RCL0, RCL2, +, ÷, STO9, CLX, STOI, RCL6, STO7, LBL0, ISZI, RCL7, RCL1, RCL1, 1, −, RCL7, ×, −, +, RCL2, ×, RCL6, +, RCLB, +, F0?, GTOA, GTOb, LBLA, RCL7, −, RCL3, ×, RCL7, +, LBLb, PRTX, STO7, RCL9, X≤Y?, GTO0, F2?, GTOe, SF2, RCL6, STOA, RCL2, 1, RCL8, −, STOD, ×, RCL0, +, STO2, RCL4, STO9, RCL8, ×, CHS, STO6, RCLB, 1, RCL8, −, ×, RCL0, 1, −, +, STOB, GTOc, LBLe, RCLI, F0?, GTOd, RCL3, ÷, INT, LBLd, RCLE, RCLD, RCLA, R↑, PRST, R/S

Computer-aided design.   Computer-aided design is an area generally not thought of as being within the realm of programmable pocket computers because of their limited memory capacity and because of the transient nature of the display. This example is presented to show that these disadvantages can be overcome and that the programmable pocket computer can make a significant contribution. The example chosen involves distillation. The design of a distillation column for separating a multicomponent mixture is so involved that various short-cut procedures

have been developed over the years to speed up preliminary sizing calculations. Three such developments are the Fenske equation for total reflux, the Underwood equation for minimum reflux ratio, and the Gilliland correlation for number of stages versus reflux.

The Fenske equation:

$$\left(\frac{X_i}{X_r}\right)_D = \alpha^N \left(\frac{X_i}{X_r}\right)_B \tag{5}$$

is used with an average value of the relative volatility α to determine the number of stages at total reflux. Once the number of stages N_m is determined, this same equation can be used to determine the distribution of all components between the overhead and bottoms products.

The Underwood equation:

$$\sum_{i=1}^{n} \frac{\alpha_i D_i}{\alpha_1 - \theta} = 1 - q \tag{6}$$

is first used to determine θ, after which another form of the Underwood equation:

$$R_m + 1 = \sum_{i=1}^{n} \frac{\alpha_i D_i}{\alpha_i - \theta} \tag{7}$$

can be solved for the minimum reflux ratio R_m.

The Gilliland correlation relates N vs R based on the knowns N_m and R_m. This correlation has been expressed as

$$\frac{N - N_m}{N + 1} = 0.75 - 0.75\left(\frac{R - R_m}{R + 1}\right)^{0.5668} \tag{8}$$

The above equations have been programmed using the algebraic notation of the TI-59 computer for a feed mixture containing up to eight components. This program is reproduced as Listing 3 below. To use this program, one must first store the input-feed composition in registers 11–18, the relative volatilities in registers 21–28, the bottoms-product compositions in registers 41–48, the overhead compositions in registers 51–58, and the feed condition q in register 50. The value of R_m can then be determined by pressing 2nd B' and the value of N_m can be determined by pressing 2nd C'. At this point, the desired R, which is greater than R_m, is entered, and 2nd D' is pressed to determine the actual number of plates N.

Listing 3. Fenske — Underwood — Gilliland distillation (Courtesy of *Chemical Engineering*, McGraw-Hill, Inc.).
Find heavy key
LBL, B, RCL, 40, STO, 01, STO, 19, STO, 20, LBL, EE, 2, 0, +, RCL, 01, =, STO, 35, 1, X⇌T, RCL, 01, −, 1, =, STO, 01, RC*, 35, INV, GE, EE, RCL, 01, +, 2, 0, =, STO, 35, RC*, 35, STO, 33, 0, STO, 01, GTO, 01, 26,

Calculate θ
LBL, A↑, STO, 10, RCL, 19, STO, 40, 0, STO, 32, CP, LBL, LNX, (, 2, 0, +, RCL, 40,), STO, 35, (, 1, 0, +, RCL, 40,), STO, 31, (, ST*, 35, RC*, 31, +, (, RC*, 35, −, RCL, 10,),), SUM, 32, (, RCL, 40, −, 1,), STO, 40, INV, EQ, LNX, (, RCL, 32, +, RCL, 50, −, 1,), STO, 34, RCL, 34, RTN, 1, ., 0, 0, 0, 5, PGM, 08, A, RCL, 33, PGM, 08, B, ., ., 1, PGM, 08, 0, ., ., 0, 0, 1, PGM, 08, D, PGM, 08, E, 0,

Calculate R_{min}
STO, 37, LBL, X², CP, 5, 0, +, RCL, 19, =, STO, 35, 2, 0, +, RCL, 19, =, STO, 36, RC*, 35, ×, RC*, 36, +, (, RC*, 36, −, RCL, 10,), =, SUM, 37, RCL, 19, −, 1, =, STO, 19, INV, EQ, X², RCL, 37, −, 1, =, STO, 29, RTN

Calculate N_{min}
LBL, C↑, 2, 0, +, RCL, 20, =, STO, 35, 1, X⇌T, RCL, 20, −, 1, =, STO, 20, RCL*, 35, INV, GE, C↑, RCL, 20, +, 5, 1, =, STO, 35, −, 1, =, STO, 31, RCL, 20, +, 4, 0, =, STO, 38, +, 1, =, STO, 39, RC*, 31, ×, RC*, 39, ÷, RC*, 38, =, LOG, ÷, RCL, 33, LOG, =, STO, 49, R/S

Calculate one of N, R, N_{min}, R_{min}
LBL, D', CP, RCL, 29, STO, 01, RCL, 49, STO, 03, LBL, SUM, RCL, 01, INV, EQ, √x, SBR, LOG, GTO, COS, LBL, LOG, RCL, 04, −, RCL, 03, =, +, (, RCL, 04, +, 1,), =, STO, 05, (, 1, −, 1, ., 3, 3, 3, ×, RCL, 05,), Y^X, 1, ., 7, 6, 4, 3, =, STO, 06, RTN, LBL, COS, RCL, 06, ×, (, RCL, 02, +, 1,), =, +/−, +, RCL, 02, =, R/S, LBL, √X, RCL, 02, INV, EQ, 1/X, SBR, LOG, (, RCL, 01, +, RCL, 06,), ÷, (, 1, −, RCL, 06,), =, R/S, LBL, 1/X, RCL, 03, INV, EQ, CE, SBR, CLR, GTO, SIN, LBL, CLR, RCL, 02, =, STO, 07, Y^X, −, 5, 6, 6, 8, =, x, ., 7, 5, +/−, +, ., 7, 5, =, STO, 08, RTN, LBL, SIN, RCL, 08, X, (, RCL, 04, +, 1,), =, +/−, +, RCL, 04, =, R/S, LBL, CE, SBR, CLR, RCL, 03, +, RCL, 08, =, +/−, ÷, (, RCL, 08, −, 1,), =, R/S

R.F. BENENATI
Polytechnic Institute of New York

Nomenclature

B = bottoms product flow rate
D = distillate product flow rate
F = feed flow rate
K = equilibrium constant = y/x
L = liquid flow rate
N = total number of stages in distillation
q = function defining thermal condition of distillation-column feed
R = reflux ratio = L/D
V = vapor flow rate
x = liquid composition
y = vapor composition
y^* = equilibrium vapor composition
α = relative volatility = $y/(1-y)/x/(1-x)$

Subscripts

0 = designates initial condition
b = bottoms product
D = distillate product
f = designates feed stream
i = designates any component
m = designates minimum reflux
m = plate number in stripping column
n = plate number in rectifying section of column

Texas Instruments Learning Center and Rensselaer Polytechnic Institute, *Sourcebook for Programmable Calculators*, McGraw-Hill, New York, 1979.

T.E. Croley, *Hydrologic and Hydraulic Computations on Small Programmable Computers*, Iowa Institute of Hydraulic Research, 1979.

J.E. Barnes and A.J. Waring, *Pocket Programmable Calculators in Biochemistry*, John Wiley & Sons, Inc., New York, 1980.

PROPYL ALCOHOLS

ISOPROPYL ALCOHOL

Isopropyl alcohol (2-propanol, dimethylcarbinol, *sec*-propyl alcohol) is the lowest member of the class of secondary alcohols. As one of the lower (C_1–C_5) alcohols, isopropyl alcohol is second in commercial production to methanol (qv) (see also Alcohols, higher).

Physical Properties

Isopropyl alcohol is a colorless, volatile, flammable liquid. Its odor is slight, resembling a mixture of ethyl alcohol and acetone. Unlike ethyl alcohol, it has a bitter, unpotable taste. The physical and chemical properties of isopropyl alcohol reflect its secondary hydroxyl functionality. For example, its boiling and flash points are lower than *n*-propyl alcohol, whereas its vapor pressure and freezing point are significantly higher. Thus, isopropyl alcohol boils only 4°C higher than ethyl alcohol and possesses similar solubility properties, which accounts for the competition between these two alcohols in many solvent applications. Table 1 contains some important physical constants of anhydrous and CBM (constant-boiling-mixture with water) isopropyl alcohol.

Chemical Properties

Most of isopropyl alcohol chemistry involves the introduction of the isopropyl or isopropoxy group into other organic molecules. The use of isopropyl alcohol for this purpose accounts for ca 60% of its production.

Isopropyl alcohol undergoes reactions typical of an active secondary alcohol. It can be dehydrogenated, oxidized, esterified, etherified, aminated, halogenated, or otherwise modified at this site more readily than primary alcohols, eg, *n*-propyl alcohol or ethyl alcohol (see Propyl alcohols, *n*-propyl alcohol; Ethanol). Manufacture of commercially important aluminum isopropoxide and isopropyl halides illustrates this reactivity. The former reaction involves replacement of the hydrogen atom of the hydroxyl group by aluminum metal with concomitant

Table 1. Physical Properties of Isopropyl Alcohol

Property	Grade	
	Anhydrous	91 vol %
boiling point (at 101.3 kPa), °C[a]	82.3	80.4
freezing point, °C	−88.5	−50
sp gr, 20/20°C	0.7861	0.8179
surface tension (at 20°C), mN/m (= dyn/cm)	0.0213	0.0214[b]
specific heat (liquid at 20°C), J/(kg·K)[c]	2510.4	
refractive index, n_D^{20}	1.3772	1.3769
heat of combustion (at 25°C), kJ/mol[c]	2005.8	
latent heat of vaporization (at 101.3 kPa[a,b]), kJ/mol[c]	39.8	
vapor pressure (at 20°C), kPa[a]	4.4	4.5
critical temperature, °C	235.2	
critical pressure (at 20°C), kPa[a]	4760	
viscosity, mPa·s (= cP)		
at 0°C	4.6	
at 20°C	2.4	2.1[b]
at 40°C	1.4	
solubility (at 20°C)		
in water	complete	complete
water in	complete	complete
flammability limit in air, vol %		
lower	2.02	
upper	7.99	
flash point, °C		
Tag open cup	17.2	21.7
closed cup	11.7	18.3

[a] To convert kPa to mm Hg, multiply by 7.50.
[b] At 25°C.
[c] To convert J to cal, divide by 4.184.

hydrogen evolution and, in the latter, the hydroxyl group is displaced, eg, with a halogen acid (see Alkoxides, metal).

Glycol ethers are also commercially important derivatives of isopropyl alcohol. These are readily prepared from isopropyl alcohol by reaction with olefin oxide eg, propylene or propylene oxide (see Glycols). Reaction is generally catalyzed by an alkali hydroxide.

$$(CH_3)_2CHOH + \overset{O}{\overset{\triangle}{CH_2CH_2}} \xrightarrow{KOH} (CH_3)_2CHOCH_2CH_2OH$$

2-isoproxyethanol
(isopropyl Cellosolve)

Higher alkoxylated products (oligomers) are formed by secondary reaction of oxide with the hydroxy group of the product.

$$(CH_3)_2CHOCH_2CH_2OH + \overset{O}{\overset{\triangle}{CH_2CH_2}} \xrightarrow{KOH} (CH_3)_2CHO(CH_2CH_2O)_xH$$

Glycol ethers have commercial utility.

Isopropyl alcohol also has significant industrial use in dehydrogenation, amination, and esterification reactions. Isopropyl alcohol is catalytically dehydrogenated by a wide variety of catalysts in high conversions (75–95 mol%) in an endothermic vapor-phase process to acetone. This was once the main commercial route to acetone, but it is now used to supplement the cumene–phenol process for acetone production (see Cumene). Amination by either ammonolysis in the presence of dehydration catalysts or reductive ammonolysis with hydrogenation catalysts is employed to produce two amines, isopropylamine and diisopropylamine.

$$(CH_3)_2CHOH + NH_3 \xrightarrow[\Delta,\,pressure]{catalyst} (CH_3)_2CHNH_2 + H_2O$$

$$(CH_3)_2CHOH + (CH_3)_2CHNH_2 \xrightarrow[\Delta,\,pressure]{catalyst} [(CH_3)_2CH]_2NH + H_2O$$

Isopropyl carboxylates are readily prepared by treatment of isopropyl alcohol with carboxylic acids in the presence of an acidic catalyst, eg, *p*-toluenesulfonic acid. The equilibrium reaction is typically carried out at 100–160°, and 101.3 kPa (1 atm) (depending on the carboxylic acid employed) and with an excess of alcohol.

Manufacture

There are two basic processes for the commercial manufacture of isopropyl alcohol, and each involves synthesis from propylene. The standard method since ca 1920 is the so-called indirect-hydration process (esterification–hydrolysis or the sulfuric acid method) which involves reaction of propylene with sulfuric acid. This is the only process used in the U.S. The other method, which is rapidly replacing indirect-hydration in Europe and Japan, involves catalytic hydration of propylene, ie, direct hydration, using superheated steam and high pressures. Availability of low purity propylene feedstock makes the indirect process more economical in the U.S.

Because low temperature and high pressure favor product formation in propylene hydration, efforts in developing this technology focused on a search for a catalyst that maximizes alcohol productivity at low temperatures within a reasonable time. Three basic processes evolved which are in commercial operation: vapor-phase hydration over a fixed-bed catalyst of supported phosphoric acid (Veba-Chemie), or silica-supported tungsten oxide with zinc oxide promoter (ICI); mixed vapor-liquid-phase hydration at low temperature (150°C) and high pressure [10.13 MPa (100 atm)] with a strongly acidic cation-exchange resin catalyst (Deutsche Texaco AG); and liquid-phase hydration at high temperature and high pressure [270°C, 20.3 MPa (200 atm)] in the presence of a soluble tungsten catalyst (Tokuyama Soda).

Health and Safety

Although alcohols as a class are not considered very toxic, isopropyl alcohol is about twice as toxic as ethyl alcohol. There is no systematic investigation of the effects of inhalation, eg, from aerosols, of isopropyl alcohol in humans. The known human toxicity is based on numerous cases of accidental ingestion or topical application. Toxic doses of ingested isopropyl alcohol, used as rubbing alcohol, may produce narcosis, anesthesia, coma, and death. The fatal dose for humans is ca 166 mL; death occurs from paralysis of the central nervous system.

Uses

Much of the production is for the manufacture of acetone, coatings, and other solvents, as well as agricultural chemicals, pharmaceuticals, and process catalysts.

ANTHONY J. PAPA
Union Carbide Corporation

L.F. Hatch and W.R. Fenwick, *Isopropyl Alcohol*, Enjay Chemical Company, New York, 1966.

F.C. Fielding in E.G. Hancock, ed., *Propylene and its Industrial Derivatives*, John Wiley & Sons, Inc., New York, 1973.

n-PROPYL ALCOHOL

n-Propyl alcohol, 1-propanol, $CH_3CH_2CH_2OH$, mol wt, 60.09, is a clear, colorless liquid having a typical alcohol odor and miscible in water, ethyl ether and alcohols. 1-Propanol occurs in nature in fusel oils and forms from fermentation and spoilage of vegetable matter.

Properties

Fp − 126.2°C; bp 97.20°C; vapor pressure at 20°C, 1.987 kPa (15.0 mm Hg); density at 20°C, 0.80375 g/cm³; refractive index (n_D^{20}, 1.38556; heat capacity, 141 J/(mol·K) (33.7 cal/(mol.°C); liquid at 25°C; electrical conductivity at 25°C, 2×10^{-8} S/cm.

Manufacture

1-Propanol has been manufactured by hydroformylation of ethylene followed by hydrogenation of propionaldehyde (see Oxo process) and as a by-product of vapor-phase oxidation of propane (see Hydrocarbon oxidation). Production by hydroformylation or oxo technology is a two-step process in which ethylene is first hydroformylated to produce propanal. The resulting propanal is hydrogenated to 1-propanol.

$$CH_2{=}CH_2 + CO + H_2 \xrightarrow[\Delta,\ pressure]{catalyst} CH_3CH_2CHO$$

$$CH_3CH_2CHO + H_2 \xrightarrow[\Delta,\ pressure]{catalyst} CH_3CH_2CH_2OH$$

Propane, 1-propanol, and heavy ends (the latter are made by aldol condensation) are minor by-products of the hydroformylation step. A number of transition-metal carbonyls, eg, Co, Fe, Ni, Rh, and Ir, have been used to catalyze the oxo reaction, but cobalt and rhodium are the only economically practical choices.

Health and Safety

Eye contact can cause irritation or burns. Repeated skin contact can result in dermatitis. Exposure to excessive vapor concentrations irritates the eyes and respiratory tract. Very high concentration has a narcotic effect. 1-Propanol gives negative results in the Ames test and in the Mouse Lymphoma Forward Mutation assay; TLV is 200 ppm.

Uses

1-Propanol is used mainly as a solvent and chemical intermediate. Historically, its chief use has been as a speciality solvent in flexographic printing inks (qv) (see Printing processes). n-Propyl acetate, a principal derivative of 1-propanol, is a powerful solvent and is used in nitrocellulose lacquers, cellulose esters and ethers, waxes (qv) and insecticide formulations (see Cellulose derivatives; Insect control technology).

J.D. UNRUH
L. SPINICELLI
Celanese Chemical Company, Inc.

J.A. Monick, *Alcohols, Their Chemistry, Properties and Manufacture*, Reinhold, New York, 1968, pp. 117–119.

R.C. Wilhoit and B.J. Zwolinski, *J. Phys. Chem. Ref. Data* **2**, Suppl. 1, 1–66 (1973).

R.L. Pruett in F.G.A. Stone and R. West, eds., *Advances in Organometallic Chemistry*, Vol. 17, Academic Press, Inc., New York, 1979, pp. 1–60.

PROPYLAMINES.   See Amines, lower aliphatic.

PROPYLENE

Propylene (propene, $CH_3CH{=}CH_2$) is perhaps the oldest petrochemical feedstock and is one of the main light olefins (see Feedstocks). It is used widely as an alkylation or polymer-gasoline feedstock for octane improvement (see Alkylation; Gasoline). In addition, large quantities of propylene are used in plastics as polypropylene, and in chemicals, eg, acrylonitrile (qv), propylene oxide (qv), 2-propanol, and cumene (qv) (see Olefin polymers, polypropylene; Propyl alcohols).

Physical Properties

Mol wt 42.081; fp 87.9 K; bp 225.4 K; critical temperature 365.0 K; critical pressure 4.6 MPa 45.5 atm; liquid density 0.612 g/cm³ at 223 K; refractive index n_D 1.3567; soluble to the extent of 44.6 mL gas/100 mL water (at 20°C and 101.3 kPa or 1 atm).

Chemical Properties

The chemistry of propylene is characterized both by its double bond and by its allylic hydrogen atoms. Carbon atoms 1 and 2 have a trigonal planar geometry identical to that of ethylene. Carbon atom 3 is tetrahedral, like methane. Hydrogen atoms attached to this carbon atom are allylic hydrogens.

The double bond is responsible for many of the reactions that are characteristic of alkenes. It serves as a source of electrons for electrophilic reactions; addition reactions typically are in this category. Simple examples are the addition of hydrogen or a halogen, eg, chlorine:

$$CH_3CH{=}CH_2 + H_2 \xrightarrow{\text{catalyst}} CH_3CH_2CH_3$$

$$CH_3CH{=}CH_2 + Cl_2 \xrightarrow{\text{catalyst}} CH_3CHClCH_2Cl$$

The allylic hydrogens in propylene often distinguish its chemistry from that of ethylene. For example, their presence causes cross-linked, gummy materials to form when propylene polymerizes with peroxide initiators. The effect of the allyl hydrogens on propylene reactions can be explained by the stability of allyl radicals and allyl carbocations. When an allylic hydrogen is abstracted from propylene, the sp^3 hybridized carbon of the methyl group changes to sp^2. The p orbital of this carbon can then overlap with the p orbitals that formed the π bond. This forms a new π bond, which overlaps all three carbon atoms. The electrons from the alkene π bond and the free-radical electron are delocalized over the entire molecule.

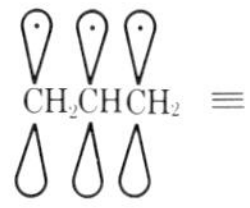

$$\overset{}{CH_2CHCH_2} \equiv CH_2{=}CHCH_2\cdot \leftrightarrow \cdot CH_2CH{=}CH_2$$

Manufacture

Steam cracking. In steam cracking, a mixture of hydrocarbon and steam is preheated to ca 870 K in the convective section of a pyrolysis furnace. Then it is further heated in the radiant section to as much as 1170 K where thermal pyrolysis reactions occur to form olefin products (see Ethylene; Petroleum, refinery processes). Steam reduces the hydrocarbon partial pressure in the reactor. The steam-to-hydrocarbon weight ratio is generally a function of the feedstock and ranges from ca 0.2 for ethane to ≥ 2.0 for gas oils. The amount of steam used is probably a compromise between yield structure (olefin selectivity), energy consumption, and furnace run length, which is limited by coking. The residence time in the radiant section varies from ca 1 s in older plants to as low as 0.1 s in some newer furnaces.

In the radiant section, the hydrocarbon mixture undergoes reactions involving free radicals. These mechanisms have been generalized to include the molecular reactions shown below:

Chain-initiation reactions: $R{-}R' \to R\cdot + R'\cdot$
Hydrogen-abstraction reactions: $R\cdot + R'H \to RH + R'\cdot$
Radical-decomposition reactions: $R\cdot \to RH + R'\cdot$
Radical-addition reactions to unsaturated molecules: $RH + R'\cdot \to R''\cdot$
Chain-termination reactions: $R\cdot + R'\cdot \to R{-}R'$
Molecular reactions: $RH + R'H \to R''H + R'''H$
Radical-isomerization reactions: $R'\cdot \to R''\cdot$.

The total number of reactions depends on the number of constituents present in the hydrocarbon feedstock. As many as 2000 reactions can occur simultaneously.

The yield of propylene produced in a pyrolysis furnace is a function of the feedstock and the operating severity of the furnace. Under practical operating conditions, ethylene yield increases with increasing severity of feedstock conversion. Propylene yield passes through a maximum. The economic optimum effluent composition for a furnace usually is beyond the propylene maximum.

In an olefins-plant separation train, propylene is obtained by distillation of a mixed C_3 stream, ie, propane, propylene, and minor components, in a C_3-splitter tower. Propylene is produced as the overhead distillation product, and the bottoms are a propane-rich stream. The size of the C_3-splitter depends on the purity of the propylene product.

Refinery production. Refinery propylene is formed as a by-product of fluid catalytic cracking of gas oils and, to a far lesser extent, of thermal processes, eg, coking. The total amount of propylene that is produced depends on the mix of these processes and the refinery product slate.

In fluid catalytic cracking, a partially vaporized gas oil is contacted with zeolite catalyst. Contact time is 5 s–2 min and pressure is 250–400 kPa (2.5–4 atm) depending on the design of the unit. Reaction temperatures are 720–850 K (see also Butylenes). Converted feedstock forms gasoline-boiling-range hydrocarbons, C_4 and lighter gas, and coke. Propylene yield varies, depending on reaction conditions, but yields of 2–5% based on feedstock are common.

Health and Safety

Propylene is a colorless gas under normal conditions, has anesthetic properties at high concentrations, and can cause asphyxiation. It does not irritate the eyes, and its odor is characteristic of olefins. Propylene is a flammable gas under normal atmospheric conditions. Vapor-cloud formation from liquid or vapor leaks is the main hazard that can lead to explosion. The autoignition temperature is 731 K in air and 696 K in oxygen.

Uses

Propylene is consumed in both oil refineries and chemical plants. In the refinery, propylene occurs in varying concentrations in fuel-gas streams. As a refinery feedstock, propylene is alkylated by isobutane or dimerized to produce polymer gasoline for gasoline blending. Commercial chemical derivatives include polypropylene, acrylonitrile, propylene oxide, isopropyl alcohol, and others. Polypropylene has been the largest consumer of propylene since the early 1970s and is likely to dominate propylene utilization for some time to come.

MORRIS R. SCHOENBERG
JOHN W. BLIESZNER
CHRISTOS G. PAPADOPOULOS
Amoco Chemicals Corporation

S.B. Zdonik, E.J. Bassler, and L.P. Hallee, *Manufacturing Ethylene*, The Petroleum Publishing Co., Tulsa, Okla., 1970.

D.J. Hadley, R.E. Saunders, and P.T. Mapp in E.G. Hancock, ed., *Propylene and Its Industrial Derivatives*, Earnest Benn Ltd., London, UK, 1973, pp. 416 ff.

PROPYLENE OXIDE

Physical Properties

Propylene oxide is a colorless, low boiling liquid. It is miscible with most organic solvents but forms a two-layer system with water. Propylene oxide can exist as two optical isomers; the racemic mixture is the commercial product. Except as noted, properties are of the racemic mixture. Physical properties are: bp at 101.3 kPa 34.2°C; $\Delta bp/\text{pressure}$ from 98.66 to 101.3 kPa (= 1 atm) 0.28 K/kPa; fp -112°C; dipole moment 6.61×10^{-30} C·m (1.98 debye); flash point (calculated) < -20°C; n_D^{25} 1.36322. Propylene oxide forms azeotropes with methylene chloride, ether, and several hydrocarbons.

The specific rotation $[\alpha]_D^{18}$ of the two optical isomers has been reported as $+12.72$° and -8.26° for (R) and (S)-propylene oxide, respectively.

Chemical Properties

Propylene oxide exhibits a high degree of reactivity, a result of the presence of the strained, three-membered oxirane ring. Reactions such as those with hydrogen halides and ammonia proceed at a satisfactory rate without a catalyst. Most reactions, however, require an acidic or basic catalyst; in both cases, the reaction proceeds through a nucleophilic substitution (S_N2) mechanism.

In base-catalyzed reactions, the nucleophilic reagent attacks the least-substituted oxirane carbon, and the primary product is the secondary alcohol.

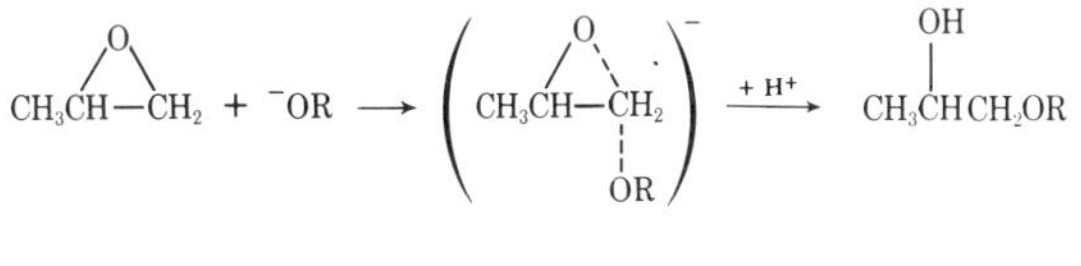

In the transition state of the acid-catalyzed reaction, the oxygen is protonated, and bond cleavage is more complete than bond formation. Both partial bonds are longer than usual, and a partial positive charge develops on the central carbon. The methyl group stabilizes the partial positive charge by electron release; however, this mechanism is also subject to steric hindrance from the methyl group. Thus, a mixture of products is obtained:

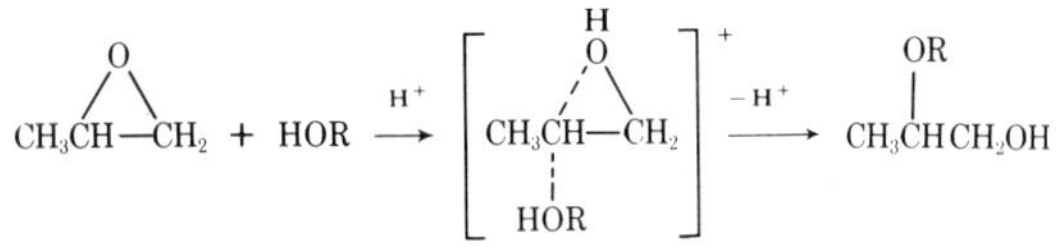

$$CH_3CH-CH_2 + HOR \xrightarrow{H^+} \left[CH_3CH \cdots CH_2 \atop HOR \right]^+ \xrightarrow{-H^+} CH_3CHCH_2OH \atop OR$$

Polymerization. The formation of polyether polyols is commercially the most important reaction of propylene oxide. A polyol is the product of reaction of an epoxide and compounds or initiators (eg, glycols, amines, acids, or water) that contain active hydrogens. Poly(propylene glycol), the simplest propylene oxide-based polyol, is prepared by the base-catalyzed polymerization of propylene oxide with propylene glycol as the initiator:

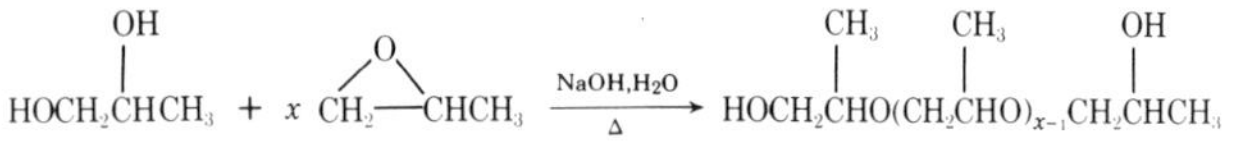

$$HOCH_2CHCH_3 + x\ CH_2-CHCH_3 \xrightarrow[\Delta]{NaOH, H_2O} HOCH_2CHO(CH_2CHO)_{x-1}CH_2CHCH_3$$

Such a polyol is commonly known as a polyol diol; a polyol triol results from the polymerization of propylene oxide initiated with glycerol.

Poly(propylene oxide) polymers with molecular weights of 100,000 or more can be prepared with a catalyst that consists of $FeCl_3$ and approximately five equivalents of propylene oxide. The addition of small amounts of toluene 2,4- and 2,6-diisocyanates greatly increases the molecular weights of the polymers obtained. Propylene oxide homopolymers can also be prepared with catalysts such as diethylzinc and trialkylaluminum compounds (see Polyethers).

Reactions. The reaction, or hydration, of propylene oxide with water to produce propylene glycol is utilized commercially.

Propylene oxide and ammonia with a small amount of water give isopropanolamine. Further reaction yields the di- and triisopropanolanimes. The ratio of primary, secondary, and tertiary amines obtained depends upon the molar ratios (see Alkanolamines).

Propylene oxide reacts with carbon dioxide to yield propylene carbonate which, in turn, can be hydrolyzed to propylene glycol. The reaction is catalyzed by potassium iodide, tetraalkylammonium bromides, calcium bromide, or magnesium bromide.

Carboxylic acids and propylene oxide give a mixture that contains the monoesters of the primary and the secondary alcohol groups of propylene glycol. The monoesters may then react with additional acid to form the glycol diester.

Alcohols or phenols and propylene oxide give monoethers of glycol. These glycol ethers may then react further to produce di-, tri-, and poly(propylene glycol) ethers. As the ratio of alcohol to epoxide in the reaction mixture is increased, the molecular weight of the products tends to decrease, ie, the yield of propylene glycol ether increases relative to that of the di-, tri-, and poly(propylene glycol) ethers.

Propylene oxide and hydrogen sulfide form 1-mercapto-2-propanol which may then react further to form the thiodiglycol, bis(2-hydroxypropyl) sulfide.

Starches, sugars, cellulose, and glycerol react with propylene oxide in the presence of alkaline catalysts to produce propylene glycol monethers and poly(propylene glycol) ethers.

In the presence of catalytic amounts of aluminum chloride, propylene oxide and aromatic compounds such as benzene or toluene give 2-arylpropanols (see Friedel-Crafts reactions).

Reaction with a Grignard reagent, RMgX, yields a secondary alcohol, $RCH_2CHOHCH_3$, as the principal product; a common by-product is the halohydrin $CH_3CHOHCH_2X$.

Propylene oxide isomerizes to allyl alcohol, propionaldehyde, and acetone. A 95% yield of allyl alcohol from epoxide may be obtained with a supported Li_3PO_4 catalyst.

Ketones and aldehydes give cyclic ketals and acetals (dioxolanes), respectively.

Hydrogenation of propylene oxide over nickel affords 1-propanol. With sodium amalgam, sodium in ammonia, lithium aluminum hydride, or lithium dimethylcuprate, the product is isopropyl alcohol.

Manufacture

Propylene oxide is produced either by the chlorohydrin or the hydroperoxide process.

Chlorohydrin process. The older of the two processes, chlorohydrin process has been used for both ethylene and propylene oxide. It involves the reaction of propylene, chlorine, and water to produce propylene chlorohydrin, followed by dehydrochlorination with lime or caustic to give propylene oxide and a salt. The main steps are chlorohydrination, epoxidation, wastewater-effluent treatment, and propylene oxide purification. The most frequently proposed mechanism for chlorohydration is as follows:

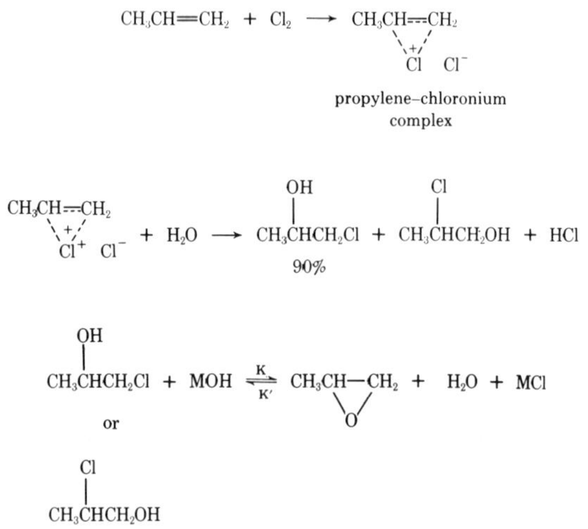

$$CH_3CH=CH_2 + Cl_2 \longrightarrow CH_3CH=CH_2 \atop Cl\ \ Cl^-$$
$$\text{propylene-chloronium complex}$$

$$CH_3CH=CH_2 \atop Cl^+\ Cl^- + H_2O \longrightarrow CH_3CHCH_2Cl \atop OH + CH_3CHCH_2OH \atop Cl + HCl$$
$$90\%$$

$$CH_3CHCH_2Cl \atop OH + MOH \underset{K'}{\overset{K}{\rightleftharpoons}} CH_3CH-CH_2 \atop O + H_2O + MCl$$
or
$$CH_3CHCH_2OH \atop Cl$$

Hydroperoxide process. In the hydroperoxide process, an organic hydroperoxide is used to epoxidize propylene; an organic alcohol is a coproduct. Currently, tert-butyl alcohol and α-methylbenzyl alcohol are produced this way. The former can be sold as a gasoline-octane booster or be dehydrated to isobutylene, whereas α-methylbenzyl alcohol is converted to styrene (qv). This process is based on the use of an organic hydroperoxide as the oxygen carrier to epoxidize propylene, and is based on the following reactions:

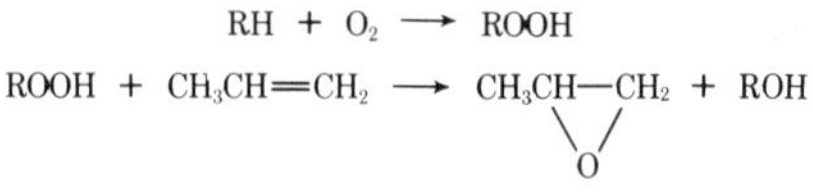

$$RH + O_2 \longrightarrow ROOH$$
$$ROOH + CH_3CH=CH_2 \longrightarrow CH_3CH-CH_2 \atop O + ROH$$

At present, ethylbenzene and isobutane are being used industrially as the starting materials; thus, isobutane is oxidized to tert-butyl hydroperoxide. The next step is the epoxidation of propylene in the presence of a metal catalyst. tert-Butyl alcohol can be used as is or dehydrated to isobutylene.

Health and Safety

The lower flammable (explosive) limit of propylene oxide is 2.3 vol% in air; the upper limit is 37 vol%. Dry chemical, alcohol foam, or carbon dioxide may be used to extinguish propylene oxide fires, whereas water is ineffective. Propylene oxide should not be stored in the presence of acids, bases, chlorides or iron, aluminum, and tin or peroxides of iron and aluminum, as any of these may cause violent polymerization.

Propylene oxide is a toxic substance with a permission exposure limit of 100 ppmv or 240 mg/m^3 at 25°C and 101.3 kPa (760 mm Hg). It is mutagenic in the Ames test, and a suspected animal carcinogen.

Uses

Propylene oxide is used as a package fumigant, to treat cotton fibers to improve moisture sorption (see Textiles), and also in admixture with other compounds as a stabilizer for methylene chloride.

RICHARD O. KIRK
T. JOHN DEMPSEY
The Dow Chemical Company

Alkylene Oxides, Product Bulletin, Form 110-551-77R, The Dow Chemical Co., Midland, Michigan, 1977.

Propylene Oxide, Product Bulletin F-41180, Union Carbide Corp., New York, Jan. 1965.

A.J. Gait, in E.G. Hanock, ed., *Propylene and Its Industrial Derivatives*, Halsted Press, a division of John Wiley & Sons, Inc., New York, 1973, p. 285.

PROSTAGLANDINS

Prostaglandins are a class of organic compounds that are biosynthetically derived from polyunsaturated fatty acids and that possess as a common feature the prostan-1-oic acid (1) skeleton. Arachidonic acid (2) is the predominant fatty acid precursor of the prostaglandins.

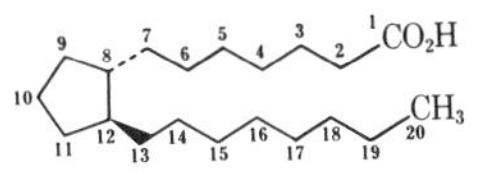

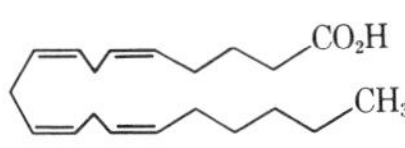

Prostaglandins exhibit a wide range of biological activities and are biosynthesized by most mammalian tissues. The investigation of the prostaglandins, thromboxanes, and leukotrienes (collectively referred to as eicosanoids) has led to important advances in synthetic organic chemistry, biochemistry, and physiology (see also Contraceptive drugs).

In 1960, the isolation from sheep prostate glands of the crystalline prostaglandins PGE$_1$ (3) and PGF$_{1\alpha}$ (4) was reported. Mass spectrometric analysis and chemical manipulation led to the structural elucidation of PGE$_1$, PGF$_{1\alpha}$, and PGF$_2$. By 1964, the structures of the three remaining primary prostaglandins, PGE$_2$, PGE$_3$ (5), and PGF$_{3\alpha}$ (6), were known.

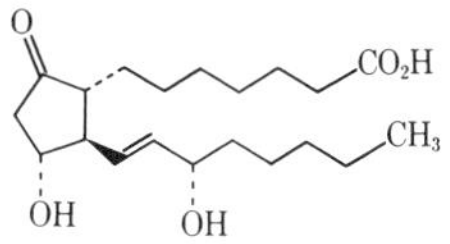

Arachidonic Acid Cascade

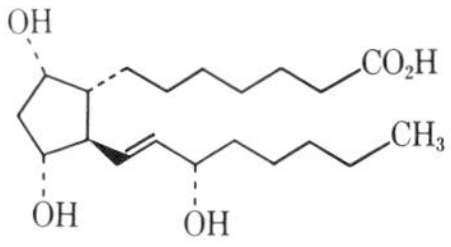

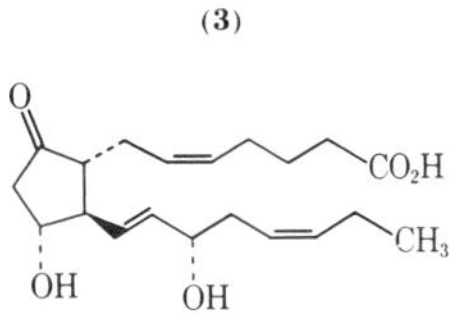

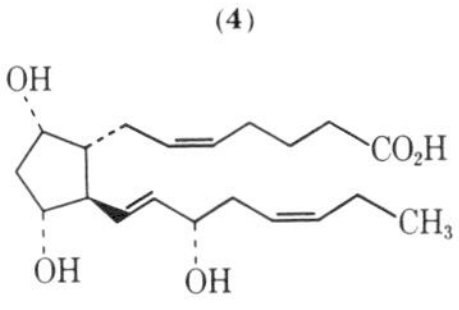

Arachidonic acid is the most abundant 20-carbon precursor fatty acid, which accounts for the predominance of the two-series prostaglandins, eg, PGE$_2$. The avenues of arachidonic acid metabolism are known as the arachidonic acid cascade. Various eicosanoids are the products.

The naturally occurring two-series prostaglandins, thromboxanes, and leukotrienes are not stored as such in cells but are biosynthesized on demand from arachidonic acid released from cell-membrane phospholipids. The precursor arachidonic acid is generally esterified at the 2 position of the phospholipids, eg, phosphatidylcholine (lecithin (qv)) (7), phosphatidylinositol, phosphatidylethanolamine, and phosphatidylserine. Arachidonic acid is released by the action of a phospholipase A$_2$ (PLA$_2$), which is specific for the 2 position of the phospholipid molecule.

In the case of (7), lysophosphatidylcholine (lysolecithin) (8) is the by-product of this enzymatic hydrolysis. Triglycerides have also been reported to be a source of precursor arachidonic acid.

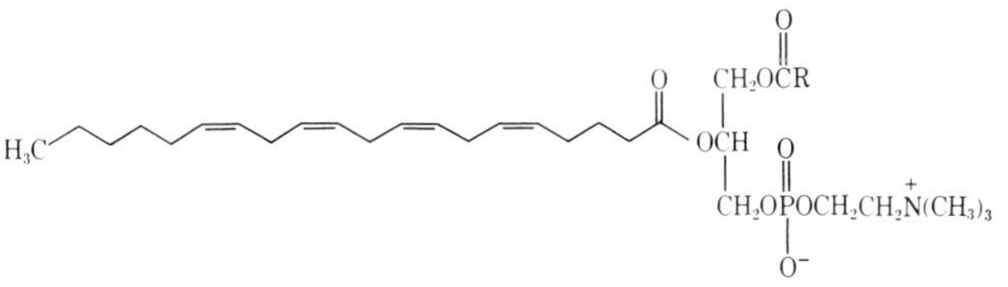

Arachidonic acid is converted into endoperoxides PGG$_2$ and PGH$_2$ by the enzyme complex prostaglandin cyclooxygenase. These endoperoxides are extremely important intermediates in the metabolism of arachidonic acid. In addition to having pronounced biological activity, eg, aggregation of blood platelets and constriction of vascular tissue, they serve as substrates for several other enzymes in the cascade.

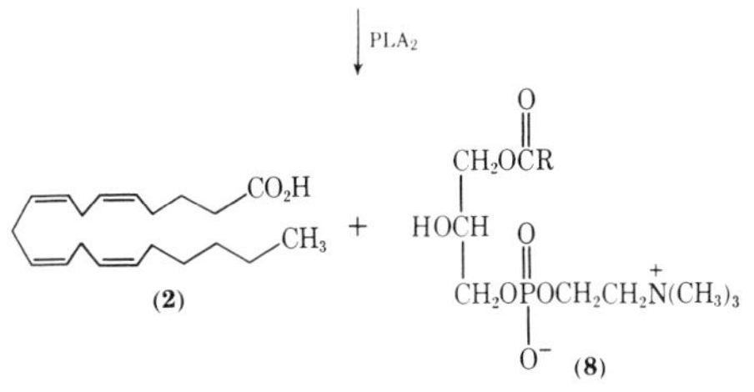

Arachidonic acid is also a subtrate for a number of lipoxygenases. The 5-lipoxygenase pathway of the arachidonic acid cascade has been the focus of intensive research.

There is increasing emphasis on the modulation of the arachidonic acid cascade. Nonsteroidal anti-inflammatory drugs, eg, aspirin and indomethacin, inhibit the cyclooxygenase which transforms arachidonic acid into PGG$_2$ or PGH$_2$, or both. Anti-inflammatory steroids (qv) are effective inhibitors of arachidonic acid release from phospholipids and thereby inhibit prostraglandin biosynthesis.

The naturally occurring prostaglandins and thromboxanes have a short biological half-life. These substances are subject to four major metabolic transformations as illustrated in Figure 1 for PGE$_2$.

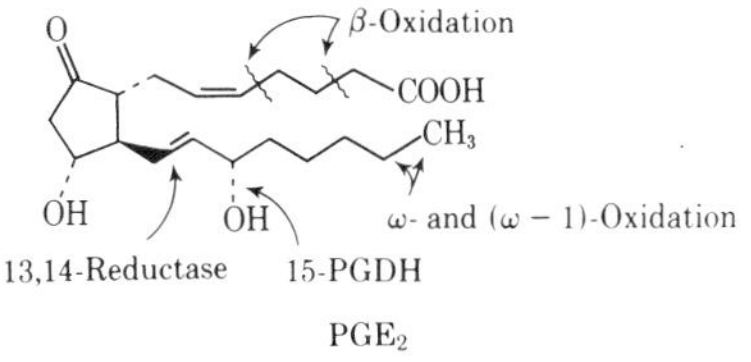

Figure 1. Principal metabolic transformations of PGE$_2$.

Table 1. Biological Activities Generally Associated with Naturally Occurring Eicosanoids

Eicosanoid	Associated biological activity
PGD$_2$	inhibitor of blood-platelet aggregation
PGE$_1$	inhibitor of blood-platelet aggregation; stimulates release of erythropoietin from renal cortex; relaxes bronchial and tracheal muscle (bronchodilator); inhibits basal rate of lipolysis from adipose tissue; inhibitor of allergic responses; and is a vasodilator
PGE$_2$	stimulates release of erythropoietin from renal cortex; relaxes bronchial and tracheal muscle (bronchodilator); contracts uterine muscle; inhibits gastric acid secretion; and protects gastric mucosal lining
PGF$_{2\alpha}$	contracts bronchial and tracheal muscle (bronchoconstrictor); contracts uterine muscle; luteolytic in some mammalian species; and is a vasoconstrictor
PGH$_2$	vasoconstrictor; inducer of blood-platelet aggregation; and is a bronchoconstrictor
PGI$_2$	inhibitor of blood-platelet aggregation; slight bronchodilator; vasodilator; and protects gastric mucosal lining
TXA$_2$	vasoconstrictor; bronchoconstrictor; and is an inducer of blood-platelet aggregation
LTB$_4$	stimulator of leukocyte behavior and function, eg, aggregation, chemokinesis, chemotaxis, and release of lysosomal enzymes
LTC$_4$/LTD$_4$	bronchoconstrictor; and stimulates airway mucus secretion and plasma leakage from venules

The molecular conformations of PGE$_2$ and PGA$_1$ have been determined in the solid state by x-ray diffraction, and special ^{1}H- and ^{13}C-nmr spectral studies of several prostaglandins have been reported.

Eicosanoids

The biological activities associated with naturally occurring eicosanoids are given in Table 1 (p. 963).

Total syntheses.   The synthesis of classical prostaglandins PGE$_2$ and PGF$_{2\alpha}$, developed by E.J. Corey and co-workers at Harvard University, is one of the most widely utilized procedures. It starts from thallous cyclopentadienide and includes ca 20 steps.

The Corey process is also useful for the synthesis of the one-series prostaglandins. Catalytic reduction of the bis-protected PGF$_{2\alpha}$ (9) results in selective saturation of the 5,6-double bond to afford (10). Subsequent transformations lead to PGE$_1$ (3) in 64% yield and PGF$_{1\alpha}$ (4) in 80% yield.

Intermediates in the Corey synthesis are useful for the synthesis of three-series prostaglandins. The key step is the Wittig reaction of aldehyde (11) with a β-oxidoylid to yield stereospecifically the lactone alcohol (12) with a trans-13,14-double bond. This lactone is subsequently converted to PGF$_3$ and PGE$_3$.

Another commercially important total synthesis starts with norbornadiene. In a third synthetic procedure for PGE$_2$ and PGF$_{2\alpha}$, racemic bicyclo[3.2.0]hept-2-en-6-one (13) is the starting material. An enantioconvergent approach is used, ie, both enantiomers (14) and (15) are utilized, which obviates the need for optical resolution and its associated loss of at least one half of the resolved product.

Another total synthesis includes the 1,4-addition of an organometallic reagent to an α,β-unsaturated ketone. The conjugate addition has been modified with the use of a cis-homocuprate, which gives higher yields of adduct than the corresponding trans-homocuprates. Furthermore, the 1,4-addition of cis-homocuprates occurs with a higher degree of stereoselectivity than with trans-homocuprates.

Prostaglandin D was first synthesized in 1973. The key step is the selective functionalization of the C-15 hydroxyl in the Corey intermediate (16).

In general, the classical prostaglandins of the A, B, and C series are prepared by partial syntheses from prostaglandins of the E series.

The endoperoxides PGH$_1$ and PGH$_2$ have been prepared by biosyntheses. The latter, its methyl ester, and PGG$_2$ have been prepared by total syntheses.

The first total synthesis of the PGI$_2$ methyl ester and the PGI$_2$ sodium salt (also known as epoprostenol sodium) was completed in 1976.

As of March 1984, the successful synthesis of thromboxane A$_2$ (TXA$_2$) had not been reported; however, several synthetic procedures for its stable hydrolysis product thromboxane B$_2$ (TXB$_2$) have been described. One of these starts from the 9,15-diacetate of the PGF$_{2\alpha}$ methyl ester (17).

The key intermediate in the first total stereospecific synthesis of leukotriene C$_4$ (LTC$_4$) is product (18), which was formed stereoselectively with a cis-11,12 double bond.

Partial syntheses.   The discovery that (15R)-PGA$_2$ and its 15-acetate, methyl ester, were present in relatively large quantities in the gorgonian *Plexaura homomalla* stimulated research aimed at the utilization of these products in the syntheses of PGE$_2$ and PGF$_2$. Synthetic procedures exist for the interconversion of prostaglandin families.

Prostaglandin E$_2$ can be selectively reduced to PGF$_{2\alpha}$ which can, in turn, be selectively reoxidized to PGE$_2$, whereas PGF$_{2\alpha}$ can be converted into thromboxane B$_2$ (TXB$_2$), PGI$_2$, or PGD$_2$. Prostaglandin E$_2$ can be dehydrated to PGA$_2$, which can be selectively isomerized to PGC$_2$. Alternatively, PGE$_2$, PGA$_2$, and PGC$_2$ can be directly isomerized to PGB$_2$ with aqueous base. Finally, PGA$_2$ can be converted to PGE$_2$.

Modified Eicosanoids

Naturally occurring eicosanoids derived from arachidonic acid, as well as from related unsaturated fatty acids, display a wide variety of biological activities. Since the metabolic oxidation effected by C-15 prostaglandin dehydrogenase (15-PGDH) is very rapid, numerous analogues that contain alkyl substituents at or near C-15 have been synthesized. These are less susceptible to the dehydrogenase. Susceptibility to

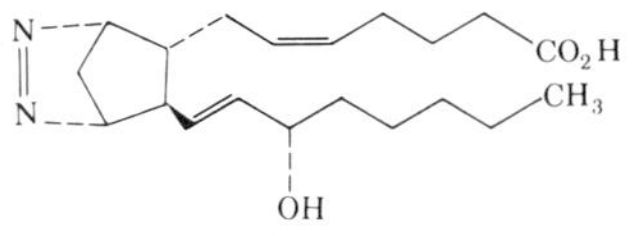

Figure 2. Chemically stable endoperoxide analogues.

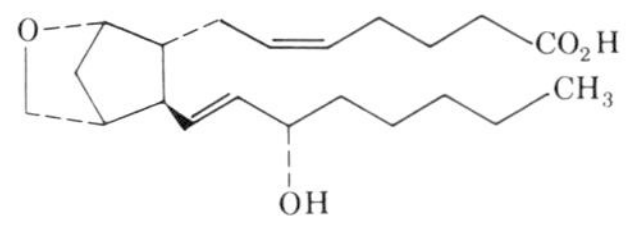

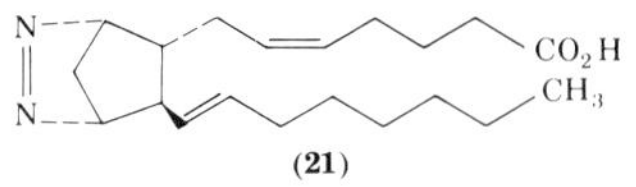

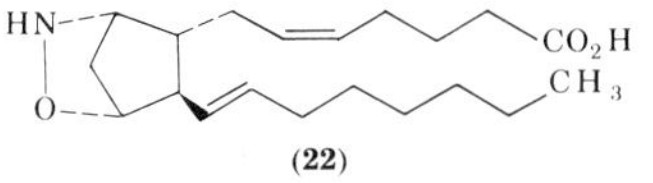

other metabolic enzymes is reduced also by other molecular modifications.

Endoperoxide analogues were designed to be more chemically stable than the naturally occurring PGG_2 and PGH_2. Azo analogue (**19**) and the epoxymethano analogue (**20**) appear to mimic the biological actions of the naturally occurring endoperoxides, whereas the 15-deoxy analogues (**21**) and (**22**) inhibit PGH_2-induced human platelet aggregation (see Fig. 2).

Chemical stability has also been one of the principal goals in the synthesis of thromboxane A_2 analogues. Compound (**23**) inhibits PGH_2-induced blood-platelet aggregation; however, it mimics TXA_2 in its ability to constrict vascular tissue. Totally carbon-substituted analogues, eg, (**24a** and **b**), have been prepared and their biological properties have been investigated.

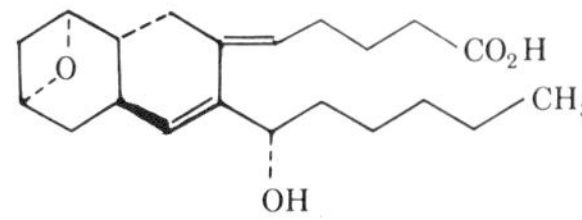

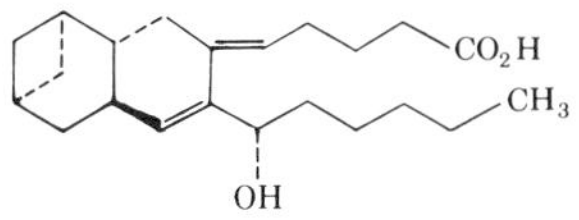

As with endoperoxides, PGG_2 and PGH_2, and TXA_2, the syntheses of stable PGI_2 (prostacyclin) analogues has been pursued. Notable among the chemically stable ones initially synthesized are the nitrilo analogue (**25**), the thia analogue (**26**), and the carbon analogue (**27**) shown in

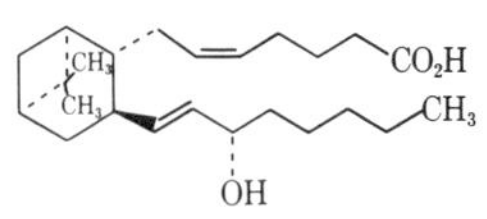

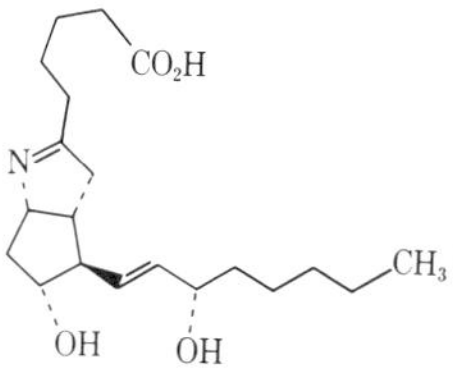

Figure 3. Chemically stable analogues of prostaglandin I_2 (prostacyclin).

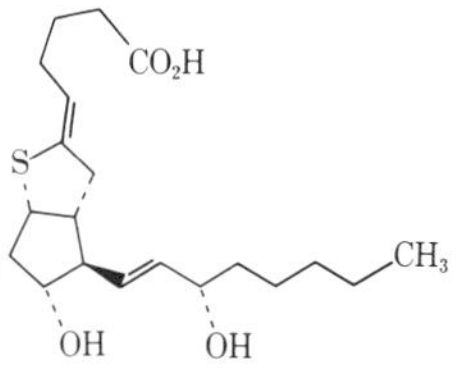

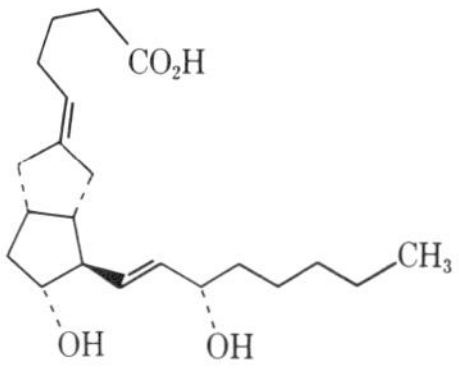

Figure 3. All of these inhibit adenosine-diphosphate-induced aggregation of human blood platelets *in vitro*.

DOUGLAS R. MORTON, JR.
The Upjohn Company

S.M. Roberts and F. Scheinmann, eds., *New Synthetic Routes to Prostaglandins and Thromboxanes*, Academic Press, Inc., New York, 1982.

Advances in Prostaglandin, Thromboxane and Leukotriene Research, Vols. 1–11, Raven Press, New York.

S. Moncada, R.J. Flower, and J.R. Vane in A.G. Gilman, L.S. Goodman, and A. Gilman, eds., *Goodman and Gilman's The Pharmacological Basis of Therapeutics*, 6th ed., MacMilland, New York, 1980, pp. 668–681.

PROSTHETIC AND BIOMEDICAL DEVICES

A prosthetic or biomedical device is an artificial part or device that replaces or augments a part of the human body. The area of prosthetic and biomedical devices is extremely broad and encompasses hundreds of specific devices.

The materials of construction are metals, ceramics (including glass), and polymers, including natural polymers such as collagen (see Biopolymers). Synthetic polymers are the principal materials. The glass-transition temperature T_g corresponds to the temperature above which a polymer is flexible. Thus, a polymer with a T_g below the normal body temperature of 37°C is flexible, whereas a polymer with a T_g above this value is rigid. For example, polydimethylsiloxane with a T_g of -123°C can be used as a soft-tissue replacement, whereas poly(methylmethacrylate) with a T_g of 105°C is used in dentures. Obviously no single material is suitable for all biomedical devices, because the specific requirements vary greatly. The range of requirements is so broad that it is necessary to delineate the specific requirements for each type of biomedical device and then find or develop a suitable material or combination for materials. This aim is difficult to achieve and many materials are not completely satisfactory.

The main requirements for biomedical materials are biocompatibility with the body tissues and fluids, blood compatibility, external environment compatibility (wear, discoloration, etc), and physical properties such as strength, flexibility, permeability, and degradation.

Biocompatibility

Biocompatibility refers to the manner in which a prosthetic or biomedical device interacts with body fluids and tissues. It depends to a great extent on where and how the device contacts the body. A device could be completely external, completely implanted under the skin (subcutaneous), protrude through the skin (percutaneous), or be deeply implanted in the body. Essentially, any foreign body contact with body tissues or fluids elicits some type of response. These reactions could be owing to the shape, design or movement of the device, its chemical nature, extractable materials (usually plasticizers, stabilizers, catalysts, or degradation products), and infections caused by improper sterilization. The reactions range from outright rejection to relatively benign tolerance. Biological rejection means that the body does not accept the implant and is manifested in several ways, including extrusion of the implant from the body, destruction of the implant by enzymatic, phagocytic, or other action, or encapsulation of the implant by fibrous tissue. The term biocompatibility can be defined in a variety of ways. If the definition is accepted that a biocompatible material has no effect on the surrounding tissues and has no adverse effect on the normal healing process in the body, then no material currently in use is completely biocompatible. Usually, a fibrous capsule grows around implanted materials, which is a definite response by the body. Even a completely external device may produce a rash or callous formation on the skin. The materials used today might better be termed biotolerable, but this term is not often used. In practice, materials have various levels or degrees of biocompatibility and an attempt is made to employ only those materials that most nearly approach the ideal.

Although suture materials, such as catgut, silk, and multifilament nylon, produce a considerable amount of adverse tissue, they are considered acceptable. In this article, the term biocompatibility describes various degrees of interaction of materials with the body. Thus, a completely biocompatible material shows no effect or interaction with the body, a satisfactory biocompatible material shows only slight interaction, and an incompatible material elicits a severe reaction. A number of screening tests have been devised, including: infrared spectroscopy on the pseudo-extracellular fluid (PECF) extracts of the material; tissue-culture methods such as agar overlay, direct contact, and suspension culture; animal implant studies; and utilization tests. In the case of blood contact, further tests involving blood are essential.

Blood Compatibility

Blood compatibility is much more complex than the compatibility of a biomaterial with other body fluids or tissues. The extent of the compati-

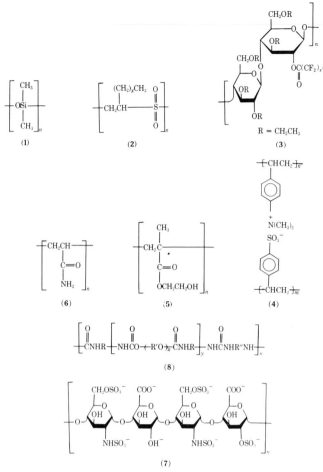

Figure 1. Structures of some thromboresistant polymers. (1) Polydimethyl-siloxane, (2) polyalkylsulfone, (3) perfluoroacyl ethyl cellulose, (4) Ioplex 101, (5) poly(2-hydroxyethyl methacrylate), (6) polyacrylamide, (7) heparin, (8) polyether polyurethane urea (PEUU).

bility of blood with a specific biomaterial depends on whether the blood is moving (as in a heart device or blood vessel) or static (as in a storage bag or bottle); whether the blood is arterial or venous; flow patterns and especially changes in flow patterns; and interactions with red cells, white cells, platelets, plasma proteins, and other blood components. Blood is a heterogeneous, non-Newtonian fluid consisting of ca 45% solids (red cells, white cells, platelets) and 55% plasma. The plasma contains a variety of inorganic ions and a series of soluble proteins which can be classified as albumins, fibrinogens, and globulins (see Blood, coagulants and anti-coagulants). Blood forms a clot or thrombus when injury occurs or when it is contacted by a foreign substance. This basic defense mechanism prevents fatal bleeding. Almost all biomaterials set off this clot-formation process and soon become coated with an irreversible clot of varying size that could have an adverse effect on the utility of the biomedical device and even be fatal to the patient. Biomedical and prosthetic devices that contact blood are either extracorporeal devices (tubing and membranes used in heart-lung machines, dialysis units, etc) or internal devices (replacement blood vessels, heart-assist devices, total artificial hearts, etc). Their main difference with regard to blood compatibility is the length of time of contact with the blood. In addition, anticoagulants and clot filters can be used with external devices, but not with internal devices.

A number of biomaterials have limited utility in various extracorporeal devices. Polydimethylsiloxane and other types of tubing can be used with heart-lung machines, dialyzers, and related devices if a suitable anticoagulant, such as heparin, is added to the blood. However, administration of heparin reduces or prevents the natural clotting of the

blood. Materials that have shown promise for use as membranes in blood oxygenators, usually with an anticoagulant, include polydimethylsiloxane, perfluoroacyl ethyl cellulose, polyalkylsulfones, and various heparinized polymers (see Fig. 1).

Several experimental polymer systems have shown promise for extracorporeal (or internal) use. These include the Ioplex materials and other hydrogels such as those based on 2-hydroxyethyl methacrylate or acrylamide.

Prostheses coated with low temperature isotropic pyrolytic carbon (LTI carbons) show excellent thromboresistance and good overall biocompatibility. These LTI carbon prostheses have excellent mechanical strength and wear resistance and have been used in heart valves since 1969 without failure of the biomaterial. Whether this material will prove as useful in other blood-contact applications, such as artificial hearts or blood vessels, remains to be seen (see Carbon and artificial graphite).

Tubular Prostheses, Plastic Surgery, and Related Devices

Tubular prostheses and devices. A wide variety of tubular prostheses and devices is used in medicine, including drains, catheters, cannula, shunts, and reconstructions or replacements of natural tubular-type organs (eg, windpipe or intestines). Some of these devices are lifetime implants; others are used only briefly. Some implants are internal; others protrude through the skin. Some are exposed to the blood and must be nonthrombogenic (see also Sutures). The materials used most frequently for tubular applications are polyethylene, polypropylene, polytetrafluoroethylene, poly(vinyl chloride), polydimethylsiloxane, and natural rubber.

A drain prosthesis removes liquid from one region of the body to the outside or to some other region. Drains are normally inserted following surgery to alleviate fluid buildup at the operation site; they are generally for short-term use. Other types are used for longer periods; these include drains for mucus secretions in sinus conditions, eustachian-tube drains, and the hydrocephalus shunt. A typical hydrocephalus shunt, made from silicone rubber, is shown in Figure 2a. In hydrocephalus treatment, the device is implanted for long-term use and must not clog at any point. Several one-way valves prevent fluid from rising in the tube and increasing pressure on the brain. Most of these devices are implanted in infancy and must be modified with growth. For this reason, the tubing is fitted with male and female joints, and a section can be replaced with a longer one.

A catheter is a tubular device for introduction into canals, passages, or tubes (including blood vessels). Such devices can be used for drainage, fluid sampling, or the introduction of fluids, medication, or devices such as an intraaortic balloon. A typical catheter, made of silicone rubber, is shown in Figure 2b. Catheters are usually temporary devices, but are sometimes used for many months.

A cannula is a tube inserted into a cavity or another tube in the body for various purposes, usually by percutaneous insertion into a blood

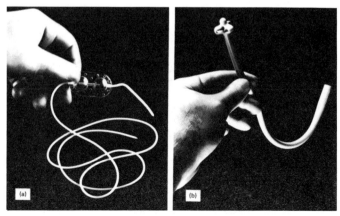

Figure 2. Some tubular devices. (a) Silastic hydrocephalus shunt. (b) Silastic malecot catheter. Courtesy of Dow Corning Corp.

vessel, and for long periods of time. Apart from blood compatibility, the main problem associated with such devices is to prevent motion, irritation, infection, and the formation of a sinus tract with weeping of fluid. These objectives can be achieved by anchoring the devices in various ways, eg, the attachment of material containing holes or fenestrations for tissue ingrowth or a textured surface on the implanted tube made of foamed material, or nylon or dacron velour. The texturized surfaces interact with the surrounding tissues and a bond is formed.

In a tracheotomy, a circular opening is cut through the skin into the trachea or windpipe, and a tube, usually made of silicone rubber or polyethylene, inserted to permit ready passage of air. Replacement or reconstruction of the trachea, when necessary, is effected with polytetrafluoroethylene tubes or polypropylene mesh.

Ear prostheses. The ear consists of an external part (auricula), a middle portion containing the tympanic membrane (eardrum) and the ossicular bones, and the inner ear, which contains the nerve connections. The semicircular canals, or labryinth, which is involved in balance control, is also situated in the ear. The auricula directs sound into the middle ear. At times it is necessary to replace this external ear because of congenital defects, disease, or accidents. The replacement may be a device that the patient can remove, or it may be a permanent implant. The replacement may be modeled from pictures of the ear before trauma or from the other ear. In the case of an externally attached ear, the pigmentation is closely matched to the patient. The materials used most commonly include natural rubber, plasticized poly(vinyl chloride), and polysilicone rubber.

Eye prostheses. The eyeball is generally replaced with a plastic globe, made from acrylic or silicone plastics, although glass had been used in the past. The patient normally does not have control of eyeball movement, although some experimental work has been done to connect the eye muscles to the prosthesis for better cosmetic effect.

Craniofacial reconstruction prostheses. Reconstruction of the head is necessary to correct deformations caused by heredity, disease, or accident. This reconstruction is complex because the head contains skin, soft tissue, cartilage and bone; each has very different physical characteristics that must be matched by the reconstruction biomaterials. When skin must be replaced, as in burn damage, it is usually grafted from another part of the body, if possible. Facial regions can be augmented or structurally modified by implantation or by affixing an external prosthesis. In the latter case, almost any stable material can be used that does not cause an adverse reaction in the surrounding or underlying surfaces and that is well matched to the surrounding facial material in texture, feel, and color. The materials most commonly used include silicone rubber, polyurethane, poly(vinyl chloride) and natural rubber.

Soft-tissue prostheses. A soft-tissue replacement or augmentation prosthesis takes the place of relatively soft muscle, fatty tissue, or connective tissue within the body.

Wall defect and hernia repairs are internal prostheses and must exhibit good biocompatibility and remain soft indefinitely. Certain soft tissues, such as breast, may be replaced by an external plastic device made from almost any material, eg, poly(vinyl chloride), polyurethanes, silicone rubbers, and natural rubber. These external devices are attached by friction, adhesives, or straps. Until fairly recently, the usable soft-tissue materials were transplanted from another part in the patient's body; cadaver transplants cause rejection problems.

Bone, Joint, and Limb Prostheses

Casts, braces, and splints. Most bone fractures can be repaired by resetting the break externally and restraining the limb, etc, in a cast, brace, or splint. Casts are generally made from plaster but plastics are much lighter and work at least as well. These consist of a polypropylene sheet molded thermally to match the patient's contours and normally lined with a polyester or polyurethane foam for comfort.

Internal bone fixation. In cases of severe bone fracture or destruction by disease, the bone must be repaired internally. Usually a metal plate is placed on the bone to bridge the fracture and give rigidity and strength during healing. Such plates are normally made from cobalt or titanium alloys or stainless steel. Unfortunately, the desired purpose may be

Figure 3. Hip-joint prostheses. Left and back: metal femoral ball-and-shaft portion; front and center: plastic acetabular socket.

defeated by resorption of the bone when insufficient stress is placed on it. In addition, most metals readily undergo fatigue fracture in physiological environments if the bone does not heal. Studies have shown that a more flexible bone plate promotes better bone regrowth.

Joint prostheses. Although the human body has many different types of joints, prostheses are mostly applied to the hinge-type (finger, elbow, or knee) or the ball-and-socket type (hip or shoulder). Freely movable joints (diarthoses) have several features in common. In these joints, also called synovial joints, the contacting surfaces are covered with cartilage and the entire joint is enclosed in a fibrous capsule. The viscous fluid, called synovial fluid, that lubricates the joint is secreted from a synovial membrane within the fibrous capsule. The motion of the joint is normally restricted by ligaments or tendons as well as the design. A joint prosthesis must meet the following requirements in order to function properly: maintenance of normal joint space; good, steady, natural joint motion; durable fixation; self-lubrication or the possibility of being readily lubricated; stress resistance; resistance to deterioration or erosion; simple, efficient design; ease of fabrication, implantation, and sterilization; biocompatibility with surrounding tissues; lack of leachable materials, and lack of carcinogenicity.

Finger-joint prostheses. The most successful device is the Swanson design prosthesis and related designs made from silicon rubber. This finger (metacarpal) prosthesis consists of two triangular rods of silicone rubber joined in the center by a concave hinge; these are available in several sizes. Recently, experimental finger-joint prostheses have been developed where the polymeric portion is poly(1,4-hexadiene) which has exceptional flexural strength and durability.

Hip prostheses. The natural hip joint consists of a ball on the femoral leg bone inserted into a socket in the acetabulum. Hip prostheses attempt to duplicate this basic design (see Fig. 3) using different acetabular–femoral combinations including metal–metal, plastic–plastic, plastic–metal, and ceramic–metal. In the United States, approximately 250,000 full or partial hip prostheses are implanted each year. The most successful designs are the plastic–metal and ceramic–metal systems; the former are more widely used.

Organ Replacement or Augmentation Prostheses

Many different artificial organs are in use at the present time, but all fall short of the ideal. Some are implanted (eg, heart valves), whereas others are solely extracorporeal devices (eg, the artificial kidney). Much research has been expended to develop a total artificial heart, whereas little attention is given to the development of an implantable artificial kidney, since a patient can be maintained on an extracorporeal dialysis unit. However, maintenance of life with a heart-lung machine does not permit the patient to approach a normal life style and would be a highly unsatisfactory approach. Ideally, any artificial organ is more satisfactory as an implanted device but some extracorporeal systems, such as the present artificial kidney, can be tolerated whereas others cannot.

Heart pacemakers. Cardiac pacemakers regulate the heartbeat and are employed to treat Stokes-Adams disease (heart block) in which the heart slows down to 20–30 beats per minute resulting in weakness, fainting spells, and sometimes sudden death. Pacemakers operate by electrical stimulation of the heart to beat faster. The plastic-coated electrodes are usually sewn into the myocardium of the heart and are powered by mercury or lithium cells.

Heart valves. The heart has four valves, ie, aortic, mitral, tricuspid, and pulmonary. Most replacements are made on the aortic (58.6%) or mitral (41.2%) valves. Many different replacement heart valves have been developed. Valve replacements are made either from polymers and metals or natural materials. A typical mitral valve prosthesis consists of a metal ring and cage (struts) made from Stellite (cobalt–chromium–molybdenum alloy) with a polymeric ball (silicone rubber) in the cage. The ring is covered with Teflon or Dacron mesh to anchor the prosthesis by endothelial ingrowth.

Blood-vessel replacement. Many materials have been investigated as blood vessel replacements. Currently, the knitted or woven Dacron prosthesis and the expanded polytetrafluoroethylene graft (Gore-Tex) are widely used.

CHARLES G. GEBELEIN
Department of Chemistry
Youngstown State University
Department of Pharmacology
Northeast Ohio Universities
College of Medicine

C.G. Gebelein and F.F. Koblitz, eds., *Biomedical and Dental Applications of Polymers*, Plenum Publishing Corp., New York, 1981.

W.J. Kolff, *Artificial Organs* **1**(1), 8 (1977).

L.L. Hench, *J. Biomed. Mater. Res.* **14**, 803 (1980).

PROTACTINIUM.   See Actinides and transactinides.

PROTEINS

Proteins are complex macromolecules that are fundamental to life. Much of the cellular content of plants and animals is protein, and metabolism is dependent on protein enzymes. Metabolic processes regulated by protein hormones (qv) and bioelectric processes at the cellular level are dependent on protein-ion interactions.

Plants can synthesize all their needs, including amino acids, but animals require certain (essential) amino acids. Human diets in developed countries contain much protein from animal sources (meat, milk, eggs, and fish), but poorer countries are heavily dependent on cereals and legumes.

Much of traditional technology is dependent on the properties of proteins. Baking of bread requires the properties of wheat gluten in forming an elastic network and thus holding moisture and gases. Brewing requires yeast protein enzymes to transform sugars to alcohol (see Beer), and the enzymes from lactobacilli are the basis of cheesemaking (see Milk and milk products). Flavor formation often depends on protein behavior during cooking.

Proteins are widely used in clothing (wool (qv), silk (qv), and leather (qv)). Adhesives are obtained from casein, blood albumin, collagen, and its breakdown product, gelatin (qv) (see Adhesives). Casein, soybean, and egg albumin have also been used for fiber production. The production of the protein fractions from blood for use in medicine is a further example of protein technology (see Blood). Food, however, constitutes the largest use of protein by far, and accordingly, is emphasized in this article.

Structure

Although several hundred amino acids are known, only about 20 occur in proteins (see Amino acids). The possible number of different proteins that could be produced from only 20 amino acids, even with a moderate chain length of 100 residues (20%), is so enormous that all protein molecules could be unique. However, the actual number of different proteins in an individual, although large (10^4–10^5), is only an infinitesimal fraction of the theoretical possibilities.

The molecular weights of the residues range from 57 to 186, the weighted average in a typical protein being about 110. Thus, a protein with a molecular weight of 33,000 contains ca 300 residues.

The amino acid composition of proteins may be expressed in many different ways, depending on the context, eg, nutritional considerations or composition and sequence data. Although composition can vary greatly, certain amino acids, eg, tryptophan, are generally present in much smaller amounts than others, eg, alanine.

Table 1.   Properties of Side Chains in Relation to Protein Structure

Amino acid	Abbreviation	Observation
glycine	Gly	Increases main-chain flexibility. No asymmetry at C-α.
alanine	Ala	Small nonpolar residue. Abundant, little preference for inside or surface of protein.
leucine	Leu	Branching nonpolar side chain allows for large stiffened side chains with limited flexibility which facilitate chain folding. Mainly inside molecules.
isoleucine	Ileu	
valine	Val	
phenylalanine	Phe	The presence of C-β methylene between C-α and the aromatic ring allows moderate side-chain flexibility. Without the methylene group, there would be severe steric hindrance at C-α. Mainly inside molecules.
tyrosine	Tyr	
tryptophan	Trp	
proline	Pro	Completely nonpolar, side chain curls back to main chain, produces rigid side chain. Fixes dihedral angle between C-α and the peptide nitrogen to a small range of $\pm 20°C$.
methionine	Met	Flexible side chain with one sulfur atom in a thioether bond, sulfur atom introduces an electric dipole.
cysteine	Cys	All polar and form hydrogen bonds. Cysteine has special role because of formation of cystine cross bridges. Hydroxyl groups (Ser, Thr) and amido groups (Gln, Asn) facilitate hydrogen-bond formation.
serine	Ser	
threonine	Thr	
asparagine	Asn	
glutamine	Gln	
tyrosine	Tyr	
histidine	His	Imidazole ring can be charged or uncharged in the physiological pH range. When uncharged, accepts H^+, thus often found in active center of enzymes.
aspartic acid	Asp	Most charged side chains are at the molecular surface. Long and flexible Lys and Arg residues increase solubility. Lys residues often react at the exposed ϵ-NH_2.
glutamic acid	Glu	
lysine	Lys	
arginine	Arg	
hydrophobic amino acids		Most amino acids except the hydrophilic charged Asp, Glu, Lys, and Arg are hydrophobic. The formation of a hydrophobic core strongly affects folding. Hydrophobicity can be evaluated by measuring free energy when amino acids are transferred from water to an organic solvent.
size of side chain		Amino acids can be arranged to reveal exchange groups wherein amino acids exchange preferentially with each other. In general, large and small residues find themselves in different groups; thus, it appears that size of side chain may be almost as important as the chemical nature.

Proteins, like nucleic acids, are derived from straight-chain molecules with one standard linkage. In peptides, all amino acid residues are of the α-type with L-configuration at the α-carbon atom. Differences and information are therefore restricted to the rather short side chains of the amino acids. The relation of amino acid properties to protein structure is shown in Table 1.

Separation and Classification

Proteins are amphoteric and function as acids or bases, depending on the conditions. The isoelectric point (pI) is the pH at which a protein does not migrate in an electric field. The minimum solubility of many proteins is at this point. Electrophoretic mobility is zero at the isoelectric point, and thus, electrophoresis at different pH can be used for protein separation and identification. These techniques are used especially for the separation of the blood-serum protein fractions into various components.

Solubility varies greatly, and the earliest systems of classification were based on solubility. Some proteins are water-soluble, and others, such as keratin, are completely insoluble.

Protein Sequence and Biosynthesis

As indicated above, the nature and function of a protein are determined by its primary structure (also called sequence), ie, the amino acids present and the order in which they are linked. The genetic code for translation of nucleotide triplets to amino acids in protein synthesis entails five principal steps.

(1) The amino acids are activated.

(2) The peptide chain is initiated, starting at the amino end.

(3) Elongation of the chain occurs because of codon-directed binding. (amino acid–tRNA + polysome polypeptide–polysome complex.)

(4) Termination requires guanosine triphosphate (GTP) and protein release factors that recognize the termination codons UAA, UAG, and UGA (U = uracil, A = adenine, and G = guanine). (Peptide–polysome complex → polypeptide + polysome.)

(5) Postribosomal modification (Fig. 1) occurs after synthesis on the mRNA–ribosome complex and involves many types of modification.

Nutrition

According to recent estimates, 70% of world protein production is derived from plants, and the remaining 30% from animal sources. In the developed countries, animal sources provide most of the dietary protein. The less-developed countries have argued that scarce grain reserves should be diverted from livestock to human consumption, but it is not clear how such a change would produce any long-term benefit.

Protein takes many forms and is fundamental to life; its biological relationships encompass all of biochemistry, and the elucidation of the biosynthesis of proteins has been a significant accomplishment. Mammalian protein metabolism overlaps with carbohydrate and lipid metabolism and is an important aspect of nutritional studies. The principal metabolic pathways are shown in Figure 2.

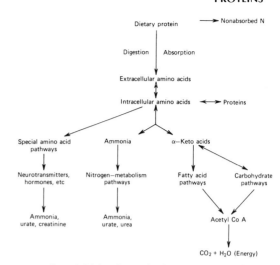

Figure 2.   Outline of chief pathways in the metabolism of dietary protein.

Quality of dietary proteins.   The capacity of a protein to meet the amino acid and nitrogen requirements of an organism depends not only on the amino acid composition and digestibility of the protein itself, but also upon the composition and adequacy of the diet and the health status of the consumer. These factors interact in a complex manner to modify the utilization of dietary protein. Furthermore, all animals, including humans, have some ability to reuse amino acids for the synthesis of proteins.

Protein requirements.   In general, recommended dietary allowances (RDA) protein are established as follows. The minimum requirement of high quality protein for maintenance of nitrogen equilibrium is estimated. It is adjusted to allow for the poorer utilization of proteins from normal mixed diets, as compared to diets containing only high quality proteins. The protein allowance is adjusted to allow for the extra needs of growth, pregnancy, and lactation. In Table 2, the safe levels of protein intake recommended by FAO/WHO are shown. These recommended levels are increased when the protein consumed is assumed to have a quality lower than that of milk or eggs.

Malnutrition.   There are two forms of protein-energy malnutrition (PEM). In the kwashiorkor syndrome, first described in the 1930s, edema, skin changes, and many biochemical abnormalities predominant features. In nutritional marasmus, the main features are those of starvation.

In general, there is cause for optimism in the decreasing prevalence of many nutritional-deficiency diseases. Curative and preventative programs based on purely nutritional measures such as increasing the supply of the missing nutrient by fortification or direct administration are continuing to produce results. Unfortunately, in spite of enormous research involvement, protein-energy malnutrition, especially of the nutritional marasmus type, remains a worldwide problem.

Food Technology

Extraction and processing.   Since cereals often contain more than 12% protein and are grown worldwide, they are a vast potential protein resource. Grain processing (wet and dry milling), with or without fractionation to improve protein quality, has been widely advocated as a new protein source. The protein content of legumes, especially soybeans, is much higher than that of cereals and the preparation of soy protein is an important industry in the United States. Soybeans are processed for both oil and protein. The residue from oil production is used as animal feed, but for human protein food the residue is extracted as hexane. Nutritionally, soy-protein products, unless supplemented with methionine or other proteins, are of poorer quality than animal products. Soy proteins are unlikely to displace animal proteins in consumer preference but will continue to supplement animal proteins (see Soybeans and other seed proteins).

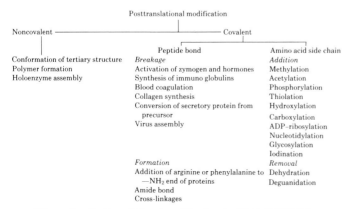

Figure 1.   Posttranslational modification reactions of proteins.

Table 2.   Levels of Protein Intake Recommended by FAO / WHO

Age group	Body weight, kg	Per kg per day		Per day		Adjusted level of protein of different quality, g per person per day		
		Protein, g	Nitrogen[a] millimoles	Protein, g	Nitrogen,[a] millimoles	Score 80	Score 70	Score 60
infants,[b] months								
6–11	9.0	1.53	17.5	14	160	17	20	23
children, yr								
1–3	13.4	1.19	13.6	16	183	20	23	27
4–6	20.2	1.01	11.5	20	229	26	29	34
7–9	28.1	0.88	10.1	25	286	31	35	41
adolescents, yr								
male								
10–13	36.9	0.81	9.3	30	343	37	43	50
13–15	51.3	0.72	8.2	37	423	46	53	62
16–19	62.9	0.60	6.9	38	434	47	54	63
female								
10–12	38.0	0.76	8.7	29	331	36	41	48
13–15	49.9	0.63	7.2	31	354	39	45	52
16–19	54.4	0.55	6.3	30	343	37	43	50
adults								
male	65.0	0.57	6.5	37	423	46	53	62
female	55.0	0.52	5.9	29	331	36	41	48
pregnancy, latter half				add 9	add 103	add 11	add 13	add 15
lactation, first 6 months				add 17	add 194	add 21	add 24	add 28

[a]Assumes protein = N × 6.25.

[b]Infants < 3 months 2.40 and 3–6 months 1.85 g protein/(kg/d).

Modification. Although chemical modification of food proteins is currently applied only to a limited extent, it offers opportunities for improving food proteins and extending their availability from nonconventional sources. Chemical modification of protein side chains most frequently involves the ε amino group of lysine and the sulfhydryl or disulfide group of cystine. Side-chain modification can improve physical state or nutritional quality by blocking deterioration. However, safety and consumer acceptability have to be considered. Chemical modification of soy proteins includes treatment with acids and alkalies, acylation, alkylation and esterification, and oxidation and reduction. In most instances, these reactions have been applied to heterogeneous protein mixtures that contain nonprotein impurities. Nevertheless, it is evident that protein functional properties can be altered significantly by such reactions (see Table 3).

Modification also includes enzymatic action, by which the peptide bonds are hydrolyzed (see Enzymes).

When followed by resynthesis, enzymatic degradation can improve both the nutritive value of food proteins and their functional quality. The mechanism of resynthesis (plastein reaction) is complicated but both condensation and transpeptidation reactions are known to be involved. Since it is possible to add amino acids, plasteins of improved essential amino acid composition can be formed.

PETER L. PELLETT
University of Massachusetts

Table 3.   Functional Properties of Proteins in Food Systems

Property	Function	Remarks
adhesion, cohesion	binding in meats and meat analogues	soy proteins enhance ability
coagulation and curding	gel structure in milk products: matrix in holding other components	rennin coagulates milk proteins in presence of Ca^{2+}
emulsification	stability and capacity of emulsions in sausages, mayonnaise, coffee whiteners	affected by proteases on meat and protein concentrates
elasticity and dough formation	bread doughs: proteins affect dough strength	soybean lipoxygenase oxidizes —SH groups and increases cross-linkage; ascorbic acid oxidase also affects dough quality
fat binding	sausages, meat products, doughnuts	soy proteins enhance ability
gel formation	gel structures act as a matrix for holding moisture, lipids and polysaccharides; cheese and yogurt	rennin and pepsin function in cheesemaking in association with coagulation and curding; proteases can reduce
solubility	allows selective precipitation to form structured products; meat analogues, cheese, yogurt	proteases and alkaline treatment increase solubility
viscosity	proteins increase viscosity in protein containing foodstuffs, soups, gravies, etc	proteases reduce viscosity by increasing solubility of proteins
water binding and absorption	prevents moisture loss in breads, cakes, etc	protein polar groups bind water by hydrogen bonding
whippability	foaming and air-holding capacity enhanced by soluble proteins (albumins); reduction of surface tension	proteases can increase foam volume but also decrease stability

V.R. Young and N.S. Scrimshaw in M. Milner, N.S. Scrimshaw, and D.I.C. Wang, eds., *Protein Resources and Technology: Status and Research Needs*, Avi Publishing Co., Westport, Conn., 1978, p. 136.

G.E. Schulz and R.H. Schirmer, *Principles of Protein Structure*, Springer-Verlag, New York, 1978, p. 314.

P.L. Pellett and V.R. Young, eds., *Nutritional Evaluation of Protein Foods*, United Nations University, Tokyo, Japan, WHTR-3/UNUP-129, 1980.

FAO/WHO, *Energy and Protein Requirements*, Report of a FAO/WHO Ad Hoc Expert Committee, Food and Agricultural Organization of the United Nations, Rome, Italy, 1973.

Committee on Dietary Allowances, Food and Nutrition Board, National Research Council, *Recommended Dietary Allowances*, 9th rev. ed., National Academy of Sciences, Washington, D.C., 1980.

R.E. Feeney and J.R. Whitaker, eds., *Food Proteins: Improvement Through Chemical and Enzymic Modification*, Advances in Chemistry Series No. 160, American Chemical Society, Washington, D.C., 1977.

PROTEINS FROM PETROLEUM.   See Foods, nonconventional.

PROTOZOAL INFECTIONS, CHEMOTHERAPY. See Chemotherapeutics, protozoal.

PSYCHOPHARMACOLOGICAL AGENTS

The need for psychotherapeutic agents is clear. It is estimated that in the United States, ca 1% of the population is schizophrenic, 1% is deeply depressed, and another 25% suffer from mild to moderate depression and anxiety. Mental disorders still account for more hospital admissions than any other single illness.

The discussion in this article is limited to the most used groups of psychotherapeutic agents, ie, the anxiolytics, the antipsychotics, and the antidepressants, and concerns itself mainly with the drugs marketed in the United States.

Anxiolytics

Anxiety is an emotion experienced to some degree by nearly everyone. Generally, anxiety is an unpleasant sensation that may play an important and necessary role in motivation and drive. It is well known, however, that excessive anxiety can interfere with normal function. Clinical anxiety consists of a pervasive, subjective feeling of apprehension and foreboding not related to a specific external threat. Clinical anxiety may also manifest itself through somatic signs, eg, palpitations, hyperventilation, and tics or twitching. Apprehension and somatic symptoms may appear together.

Anxiolytic drugs, such as meprobamate and various benzodiazepines, reduce both the feelings of apprehension and somatic signs not only in cases of clinical anxiety, where no definable external threat can be identified, but also in those cases where anxiety accompanies some threatening event, such as a physical illness. The majority of prescriptions is written for patients with a primary diagnosis of a physical disorder as opposed to emotional disorders. Benzodiazepines, particularly chlordiazepoxide and diazepam, are the most widely used anxiolytics.

Mechanisms of action. The benzodiazepine anxiolytics are now known to exert their therapeutic effects by interacting with specific receptor sites located on neuronal membranes in brain. These receptors are intimately associated and interact with GABA (gamma aminobutyric acid) receptors. Compounds have been identified that antagonize the pharmacological actions of the benzodiazepines by blocking the benzodiazepine receptors.

Chemistry. The most important anxiolytics in current use belong to two classes of compounds: propanediols and benzodiazepines.

Propanediols. Propanediols were studied in connection with the search for compounds that would be superior to mephenisin, a short-acting muscle relaxant and anticonvulsant. These studies culminated in the synthesis and introduction of the first tranquilizer, meprobamate, which became widely used under the trade names Miltown (1955, Wallace Laboratories) and Equanil (1957, Wyeth).

Its sales in the United States were highest in 1963. However, its use fell gradually with the appearance of the 1,4-benzodiazepines, which are the most widely used anxiolytic agents.

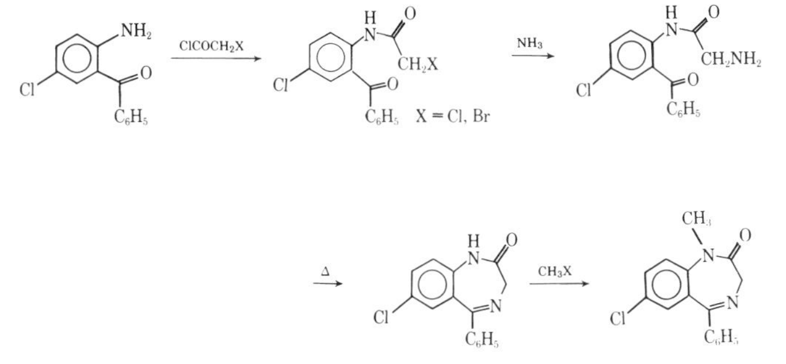

1,4-Benzodiazepines and related compounds. 1,4-Benzodiazepines have pronounced anxiolytic properties, combined with very low toxicity and a minimal effect on the autonomic nervous system. Of these, Librium (by Roche Laboratories in the United States and worldwide by F. Hoffmann-La Roche & Co., Ltd.) achieved broadest acceptance and was the most frequently used prescription drug in the United States until 1969, when sales of Valium (1963, Roche Laboratories) surpassed its sales.

The synthesis of Valium is depicted in Figure 1.

Side effects and toxicity. Anxiolytic drugs are relatively free of toxicity, addiction potential, and unpleasant side effects. The most common side effects experienced are ataxia, drowsiness, and dizziness and, at higher doses, some impairment of cognitive function may occur.

Antipsychotics (Neuroleptics)

Antipsychotic drugs reduce or reverse the fundamental symptoms of schizophrenia, such as thought disorder, blunted effect, and autistic withdrawal. Secondary symptoms, including hallucinations, paranoia, and belligerence, are also attenuated by antipsychotic therapy. Although the antipsychotics are often referred to as tranquilizers, their effects are not caused by simple sedation or tranquilization, but they appear specifically to inhibit or reverse psychotic symptoms.

Preclinical Pharmacology

The clinically active antipsychotics are well known for their ability to block postsynaptic dopaminergic receptors in the brain; thus, the various *in vivo* and *in vitro* tests employed for identifying new antipsychotic drugs are based on this property.

Mechanism of action. Drugs useful in the treatment of psychoses have many pharmacological properties, but the one property that they share is their ability to block dopaminergic transmission in brain. Reserpine accomplishes this by depleting stores of dopamine in presynaptic nerve endings, whereas the phenothiazines and butyrophenones inhibit postsynaptic dopamine receptors. It is widely believed that these antidopaminergic properties may account for both the antipsychotic effects and extrapyramidal influences of these drugs.

The ratios of antipsychotic to extrapyramidal activities vary among the antipsychotic agents. For example, thioridazine produces very few Parkinsonian effects at therapeutic doses. Thioridazine's unusual profile may be explained by its apparent selectivity for a specific dopaminergic pathway. There are two main dopaminergic pathways, which project forward from the midbrain. One originates in the substantia nigra and terminates in the striatum. The second originates in the ventral tegmental area and projects forward to the limbic forebrain and cortex. Blockade of the nigro-striatal pathway mediates extrapyramidal symptoms and inhibition of the tegmental-limbic pathway is believed to mediate antipsychotic actions.

Chemistry. Reserpine. Reserpine is an alkaloid obtained from the roots of the plant *Rauwolfia serpentina* (see Alkaloids). Its use as an antipsychotic drug decreased considerably after the development of the phenothiazine-type antipsychotics. Reserpine preparations are mostly used in the treatment of high blood pressure (see Cardiovascular agents).

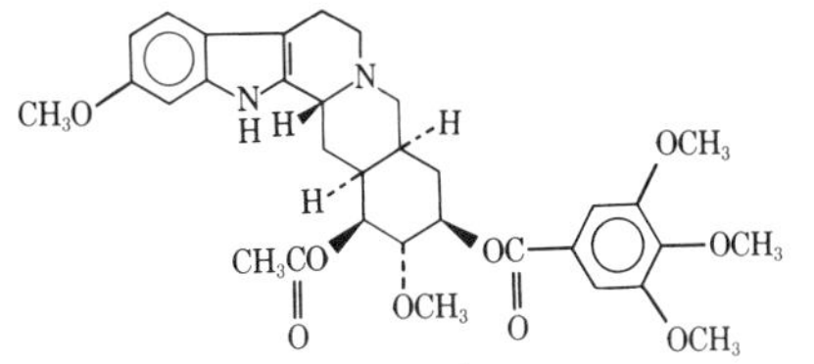

resperine

Phenothiazines. The starting materials for phenothiazine neuroleptics are the appropriately 2-substituted phenothiazines. Phenothiazines with Cl, CF_3, or SCH_3 substituents are prepared by heating the corresponding diphenylamines with sulfur. In most cases, mixtures of the 2- and 3-isomers are obtained. These can be separated quite readily by crystallization.

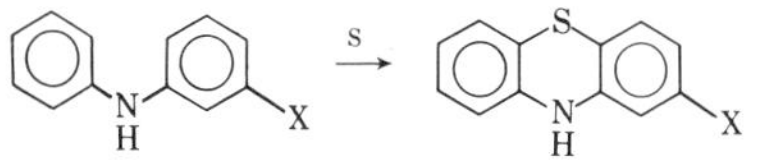

diphenylamines   phenothiazines

desmethyldiazepam   diazepam (Valium)

Figure 1. Synthesis of diazepam.

Those phenothiazines bearing an alkanoyl group in the 2-position are prepared from the unsubstituted phenothiazines by Friedel-Crafts acylation.

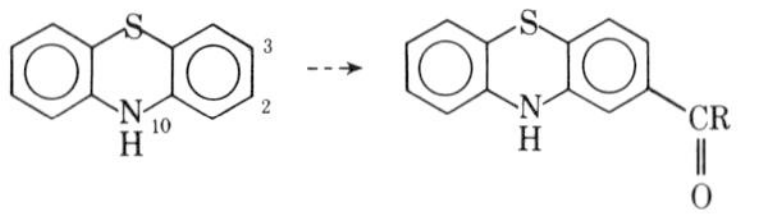

phenothiazine   R = CH$_3$, 2-acetylphenothiazine
                 C$_2$H$_5$, 2-propionylphenothiazine
                 C$_3$H$_7$, 2-butyrl phenothiazine

The substituent in the 10-position is introduced by treatment of the phenothiazines in the presence of a base, eg, NaNH$_2$ or LiH, with an appropriately substituted haloalkylamine:
Compounds bearing a hydroxyl group on the basic substituent can be esterified with fatty acid chlorides to yield esters, eg, fluphenazine enanthate and fluphenazine decanoate. These esters provide prolonged action on parenteral administration.

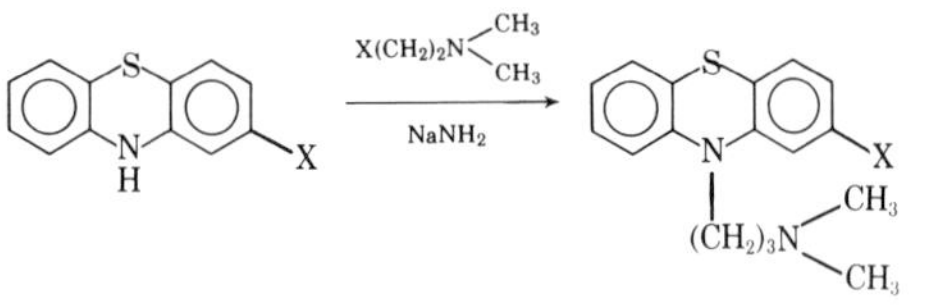

X = Cl: chlorpromazine
X = CF$_3$; salt = HCl:
trifluopromazine hydrochloride

Thioxanthenes. The thioxanthenes' antipsychotic properties are similar to those of the phenothiazines, but they never achieved the widespread popularity of the latter compounds.

In the United States, only two thioxanthene derivatives, chlorprothixene and thiothixene are marketed. The starting material is the appropriately substituted thioxanthenone, and the basic side chain is introduced via a Grignard reaction. The intermediate carbinol is dehydrated and the cis and trans isomers are separated by fractional crystallization of the bases or of their oxalate salts. In the case of chlorprothixene, the α or trans isomer is the biologically more active compound, whereas in the case of thiothixene, the cis compound shows higher activity.

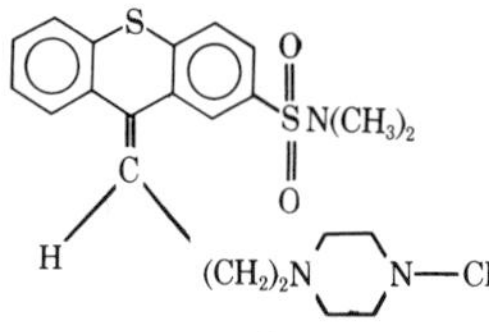

(cis)

Thiothixene (Navane, Roerig)

Butyrophenones. The butyrophenones have pronounced antipsychotic properties. The first compound of this series to be clinically tested was haloperidol, marketed in the United States as Haldol by McNeil Laboratories.

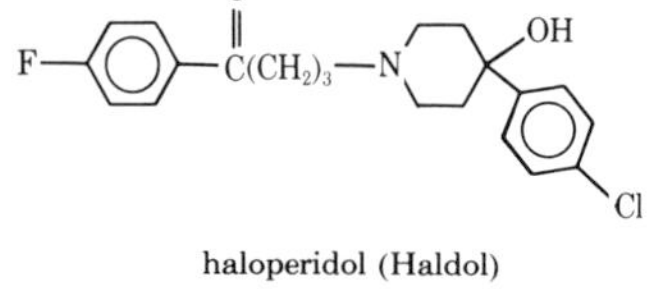

haloperidol (Haldol)

Side effects and toxicity. The phenothiazine antipsychotics influence a number of important neurotransmitters and, consequently, produce several undesirable side effects (see Neuroregulators). These include anticholinergic and antiadrenergic effects, eg, dry mouth, flushing, blurred vision, nasal congestion, constipation, and postural hypotension.

The development of tardive dyskinesia after extended exposure to antipsychotics is a severe and disturbing phenomenon for which there is no known treatment.

Antidepressants

Drugs employed in the treatment of depression belong to either of two classes: the tricyclic antidepressants or the monoamine oxidase inhibitors (MAOIs). The tricyclic antidepressants are by far the most frequently used agents in the treatment of depression and are effective in ca 70% of treated patients. The MAOIs are usually reserved for those patients who do not respond to treatment with tricyclics. Although the overall efficacy rate of the MAOIs is considerably less than for the tricyclics, there appears to be a subpopulation of depressed patients who respond particularly well to MAOI treatment.

Preclinical pharmacology. Tricyclic antidepressants influence as many as 20 different biochemical events in the brain and alter the activity of a number of important neurotransmitters, including acetylcholine, norepinephrine, dopamine, serotonin, and histamine. Which of these actions, if any, is responsible for their antidepressant properties is not known.

Prominent among the tricyclic actions is their ability to potentiate central adrenergic and serotonergic functions by blocking the re-uptake of these amines into presynaptic neurons. The re-uptake mechanism is believed to be the principal means of terminating the synaptic action of these neurotransmitters.

The limited use of MAOIs in antidepressant therapy has discouraged developmental work with this type of drug; however, two types of monoamine oxidases have been described. One type appears to be localized predominantly in neuronal tissues, and the other exists in nonneuronal tissues.

Chemistry. Tricyclics. Tricyclic antidepressants have a structural similarity to the tricyclic antipsychotics of the phenothiazine and thioxanthene group. The sulfur atom joining the two benzenoid rings in the antipsychotics is replaced by a two-carbon chain. Thus, instead of the (six-membered) thiazine or thiopyran ring, there is a (seven-membered) azepine or cycloheptadiene ring system. The basic substituent attached in the 5-position is usually a mono- or dimethylaminopropyl group.

The first tricyclic antidepressant to be used clinically was imipramine, which is synthesized by alkylation of a dibenzazepine. Imipramine hydrochloride was introduced in 1959 by Geigy under the trade name Tofranil and is widely used.

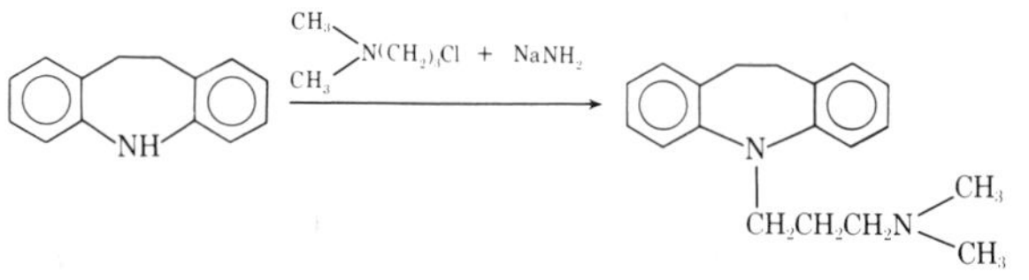

Dibenzazepine   Imipramine

The next tricyclic antidepressant to be marketed was amitriptyline, which is a dibenzocycloheptane derivative. Its valuable antidepressant properties led to the introduction in 1958 of amitriptyline hydrochloride under the trade name Elavil, by Merck in the United States.

Five other tricyclic antidepressants are marketed in the United States; two are analogues of amitriptyline, two are related to imipramine, and the fifth is a dibenzoxepine. These compounds differ in duration of action; side effects, eg, sedation, anticholinergic properties, etc; and potencies.

Monoamine oxidase inhibitors (MAOIs). Three MAOIs are marketed in the United States and a few others are marketed abroad. The simplest MAOI to be used as an antidepressant is phenethylzine, which is prepared by treatment of phenethyl bromide with hydrazine.

phenethyl bromide   phenethylzine

Another hydrazine–MAOI is isocarbazid.

A third MAO inhibitor, tranylcypromine, is not a hydrazine derivative.

isocarbazid

tranylcypromine

Side effects and toxicity. Antidepressant compounds exhibit a number of undesirable side effects and can be very toxic at high doses. The tricyclic compounds are well known for their potent inhibition of cholinergic, adrenergic (alpha), serotonergic, and histaminic receptors. In therapy, anticholinergic effects are particularly troublesome, causing dizziness, postural hypotension, tachycardia, dry mouth, and constipation.

The MAOIs produce many autonomic side effects like those of the tricyclics, ie, dizziness, dry mouth, postural hypotension, and constipation. In addition, the MAOIs are more prone than are the tricyclics to induce CNS (central nervous system) side effects, such as mania, psychosis, and disorientation.

LEO STERNBACH
DALE HORST
Hoffmann-La Roche

L.H. Sternbach, in R. Jucker, ed., *Progress in Drug Research*, Vol. 22, Birkhaüser Verlag, Stuttgart, FRG, 1978, p. 229.

L.H. Sternbach, *The Benzodiazepine Story*, 2nd rev. ed., F. Hoffmann-La Roche & Co., Ltd., Basel, Switz., 1983.

F. Hoffmeister and G. Stille, eds., *Psychotropic Agents*, Springer-Verlag, New York, 1981–1982.

E. Usdin, P. Skolnick, J.F. Tallman, D. Greenblatt, and S.M. Paul, eds., *Pharmacology of Benzodiazepines*, Macmillan Press, Ltd., London, UK, 1982.

R.S. Feldman and L.F. Quenzer, *Fundamentals of Neuropsychopharmacology*, Sinauer Assoc., Inc., Sunderland, Mass., 1984.

PULP

Pulp is the raw material for the production of paper (qv), paperboard, fiberboard, and similar manufactured products. In purified form, it is a source of cellulose (qv) for rayon (qv), cellulose esters (qv), and other cellulose-derived products. Pulp is obtained from plant fiber and is, therefore, a renewable source (see also Chemurgy).

As with most industries, the environmental and energy concerns of the 1970s effected large changes in the operation of pulp and paper mills as well as initiating much research effort to develop the most energy-efficient and cleanest methods of production. Recent trends have been the increasing use of high yield pulps by modifying the groundwood process to improve pulp quality, the use of more of the tree in harvesting and chipping, and the elimination or minimization of malodorous sulfur compounds in pulping and of the toxic and corrosive chlorine compounds from bleaching.

Wood is the original source of 99% of the pulp fiber produced in the United States. The common pulpwoods in the United States are listed in Table 1.

In terms of abundance and suitability for pulping, there are two chief botanical classifications of trees: the softwoods or evergreens, which are gymnosperms; and the hardwoods or broad-leaved deciduous trees, which are dicotyledon angiosperms. The chemistry and anatomy of wood varies somewhat with the species of tree, but there are gross similarities within the two classifications. The softwoods, which are preferred for most pulp products because of their longer fibers, generally contain a higher percentage of lignin (26–32% on an extractives-free basis) and a lower

Table 1. Pulpwood Species by Main U.S. Pulp-Producing Regions

Region	Softwoods		Hardwoods	
	Dominant	Secondary	Dominant	Secondary
Northeast	spruce	hemlock	oak	aspen
	fir	tamarack	hickory	poplar
		white pine	maple	
South	yellow pines	cypress	oaks	
			gums	
Northwest	douglas fir	true firs	red alder	
	hemlock	spruce		
Lake States	jack pine	white pine	red oak	birch
	red pine	tamarack	aspen	
			maple	

percentage of hemicellulose (14–17%) than the hardwoods, which contain 17–26% lignin and 18–27% hemicellulose (see Lignin).

Other fiber sources. A wide variety of plants can be used as a source of paper-making fibers. The only requirement is the ability of the fibers to bond to one another with sufficient strength so that a cohesive sheet is formed. However, there are several considerations that determine whether pulp from a particular plant source is suitable for the commercial production of paper or other fiber products. These include the characteristics of the fiber, supply, ease of storage, yield of desirable fibers, and wastes generated.

In countries where the wood supply is scarce, plants such as bamboo, rice, esparto, and sugarcane residues or bagasse (qv) are used to produce pulp (see Fibers, vegetable). Because of the increasing demand for paper and other wood products, alternative fiber sources, including less desirable wood species, annual plants specifically for fiber use, tropical woods, and agricultural residues, are being sought. Approximately 22% of the pulp produced in the United States in 1978 was secondary, eg, recycled fibers from newspapers, used corrugated boxes, computer printouts, etc (see Recycling).

Wood Preparation

Harvesting. The wood-processing operations of harvesting, topping and delimbing, barking, and chipping are combined into as few individual steps as possible, and all of them can take place on site in the woods. There are several factors that influence how much mechanization or which combination of procedures is used in a particular operation. These include land ownership, tree size, terrain, climate, required chip quality, and other uses for the wood.

Where the woods are located in small individually owned farms or lots, or where the terrain is mountainous or otherwise difficult to reach, or where thinning as opposed to clear-cutting is the preferred silvicultural practice, advanced mechanization is not as suitable.

Barking. Although the bark of some tree species, eg, mulberry, contains bast fibers which may be used in papermaking, the outer bark of trees usually does not contain fibrous material and, therefore, is a contaminant in the pulping and papermaking process. Under some circumstances, varying amounts of bark can be tolerated, as with thin-barked species and where the product, eg, fiberboard or corrugated medium, can accept a dirtier, lower quality pulp than can be used for fine papers.

Logs are barked in the woods where humus and nutrients are returned to the soil, or they are barked at the mill where the waste bark is used as fuel. Several types of barkers are in use in pulp mills. Drum barkers are large open-ended cylinders that are rotated on a sloped axis; this enables logs entering at one end to move towards and out the other. The bark, which is broken off by rubbing and pummelling between the logs themselves and between the logs and the cylinders, drops out through slots in the cylinder. Inline barking units, which remove the bark from one log at a time, are either of the hydraulic or mechanical-friction type. With the former, the bark is removed by jets of water at pressure in excess of 6900 kPa (1000 psi). With friction barkers, eg, the Cambio barker, special tools, which are pressed against the log, are rotated around it as it is passed through.

Chipping. The purpose of chipping for pulping is to reduce the wood to a size that allows penetration and diffusion of the processing chemicals without excessive cutting or damage to the fibers. The chips, which are ca 20 mm long, are fairly free-flowing and can be transported pneumatically or on belts and then stored in piles or bins.

Screening. The chipped wood is screened to remove large knots and oversized chips, which are separately reduced in size by mechanical means, and to remove fines, which are burned or pulped separately. Accepted chips are usually screened by length and width, but disk screens have been developed to give chips uniform thickness.

Mechanical Pulps

Groundwood. In the hydraulic magazine grinder, the logs are fed from above and are pressed against a grinding stone by a hydraulically operated pressure foot. The logs caught between the foot and the stone wear away and, when the foot reaches the end of its travel, it rapidly moves back, allowing more logs to drop. These logs are then pressed against the stone, and the cycle is repeated. Two magazines are provided on either side of the stone, and pulp from the first magazine is removed by showers ahead of the second magazine.

Groundwood pulp contains a considerable proportion (70–80 wt%) of fiber bundles, broken fibers, and fines in addition to the individual fibers. The fibers are essentially wood with the original cell-wall lignin intact. They are, therefore, very stiff and bulky and do not collapse like the chemical-pulp fibers.

The principal uses of paper-grade groundwood pulps are in newsprint, magazine papers, including coated publication grades, board for folding and molded cartons, wallpapers, tissue, and similar products. The paper has high bulk, excellent opacity, but relatively low mechanical strength.

Refiner mechanical pulping. The ground wood process requires bolts of roundwood as raw material. In the 1950s, the refiner mechanical pulping (RMP) process was developed, which produced a stronger pulp and utilized various supplies of wood chips, sawmill residuals, and sawdust. However, the energy requirement of RMP is higher, and the pulp does not have the opacity of stone-ground wood fibers.

The refiners are rotating-disk attrition mills. The disk plates have a construction of the type shown in Figure 1. The plates are paired face-to-face with a small interval between them. One disk rotates against a stationary disk or they both move in a counterrotating manner. The chips are fed through channels near the shaft in one of the disks and they move toward the periphery while undergoing attrition. The chips are first broken down into matchsticklike fragments by the action of the breaker bars, then into progressively smaller bundles as they move through the intermediate and fine bar sections.

Thermomechanical pulping. If the chips are presteamed to 110–150°C, they become malleable and do not fracture readily under the impact of the refiner bars. This modification is called thermomechanical pulping (TMP). A thermoplasticization of the wood occurs when it is heated above the glass-transition point of wet lignin. When these chips are fiberized in a refiner at high consistency, whole individual fibers are released; separation occurs at the middle lamella, and the same ribbon-like material is produced as with RMP.

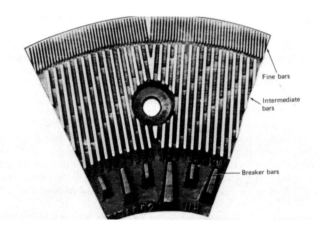

Figure 1. Refiner plate designs. Courtesy of Sprout-Waldron Division, Koppers Company, Inc.

Chemical Pulps

In chemical pulps, sufficient lignin is dissolved from the middle lamella to allow the fibers to separate with little, if any, mechanical action.

The concentration of the cooking liquor in contact with the wood influences the rate of delignification. Because of the time required for diffusion of the chemical through the wood structure and the depletion of the reagent concentration as it penetrates the chip, delignification proceeds more slowly at the center of the chip. This is particularly apparent in the case of oversize chips. In order to prevent overcooking of the principal portion of the pulp, digestion is normally halted before the centers of these larger chips are adequately delignified. The resultant pulp thus contains a portion of nondefibered wood fragments, which are separated by screening and returned to the digester or are fiberized mechanically.

Kraft. The dominant chemical wood-pulping process is the kraft or sulfate process. The alkaline pulping liquor or digesting solution contains about a 3:1 ratio of sodium hydroxide to sodium sulfide. The name kraft, which means strength in German, characterizes the stronger pulp produced when sodium sulfide is included in the pulping liquor, compared with the pulp obtained if sodium hydroxide alone is used, as in the original process.

Chemistry of delignification. In the presence of alkali, the acidic phenolic units in lignin are ionized. Above 120°C, quinonemethides form from the phenolic units, as shown below.

In the presence of sulfide or sulfhydryl anions, the quinonemethide is attacked with the formation of a benzyl thiol. The β-aryl ether linkage to the next phenylpropane unit is broken by neighboring-group attack by the sulfur with elimination of the aryloxy group as a newly reactive phenolate ion. If sulfide is not present, a principal reaction is formation of the very stable aryl enol ether, $ArCH=CHOAr$. A smaller amount of this product also forms in the presence of sulhydryl anion.

Under acid or alkaline catalysis, condensation reactions take place homolytically within the lignin polymer and very likely between lignin and carbohydrates or extraneous components. These reactions are undesirable in delignification, and prevention of such condensation accelerates pulping. Formaldehyde is generated in the formation of the arylenol ether, and it also causes condensation and cross-linking of phenylpropane units.

The very reactive quinonemethide intermediates are susceptible to attack by any carbanion. In kraft pulping, this type of condensation is minimized by the effective competition for the quinonemethide by the sulfhydryl anion.

Pulping. The chemical charge, liquor composition, time of heat-up, and time and temperature of reaction are functions of the wood species or species mix being digested and the intended use of the pulp. A typical set of conditions for southern pine chips in the production of bleachable-grade pulp for fine papers is active alkali, 18%; sulfidity, 25%; liquor-to-wood ratio, 4:1; 90 min to 170°C; and 90 min at 170°C. Hardwoods require less vigorous conditions primarily because of the lower initial lignin content.

Both batch and continuous digesting systems are in operation. The advantages of continuous digesters in addition to uninterrupted process flow include higher pulp yields and heat recovery, relative ease of automation, and in-line processing, eg, partial washing and blowline refining.

Modified soda pulping. Wood pulping with caustic soda solutions was the first chemical pulping process, but the beneficial effect of including sodium sulfide in the liquors was soon discovered. The soda process is used to advantage in pulping some hardwood species and nonwood plants, but wood pulping with sodium hydroxide was never widely used. The volatile malodorous sulfur compounds hydrogen sulfide, methyl mercaptan, dimethyl sulfide, and dimethyl disulfide are produced as undesirable by-products of kraft pulping. These compounds are not easily contained in the large-volume pulping process, and research efforts have been aimed at eliminating the use of sulfur. Two possibilities have been tested on a pilot-plant scale: use of soda–oxygen or soda–anthraquinone.

The first uses oxygen as an effective, readily available, and innocuous delignifying agent in aqueous alkaline solutions. The latter is an effective pulping accelerator in very small quantities and functions as a catalyst in the process.

Sulfite pulping. In the original sulfite pulping process, wood was pulped with an aqueous solution of SO_2 and lime. Calcium sulfite has very limited solubility above pH 2, and an excess of SO_2 gas was maintained in the digester to keep the pH below this level. Thus, the process was contrasted with the kraft or soda processes as being an acid process. Currently, bases other than calcium are used with SO_2 solutions, and sulfite pulping refers to a variety of processes in which the full pH range is utilized for all or part of the pulping. Magnesium, sodium, and ammonia are used as alternatives to calcium. Magnesium sulfite has decreasing solubility above pH 5, but sodium and ammonium sulfites are soluble at pH 1–14.

Semichemical Pulping

The distinctions between semichemical and high yield chemical processes are very small and are more a matter of gradation between the mechanical and full chemical procedures. A semichemical process is essentially a chemical delignification process in which the chemical reactions are stopped at a point where mechanical treatment is necessary to separate fibers from the partially cooked chips. Any chemical pulping process can be used to produce semichemical pulp. The pulps, although less flexible, resemble chemical pulps more than mechanical pulps because they are not as dependent on rupture of the fiber wall for bonding (see Figure 2). The yield is 60–85% with a lignin content of 15–20%. The lignin is concentrated on the fiber surface.

Bleaching

There are basically two types of bleaching operations: those that chemically modify the chromophoric groups by oxidation or reduction but remove very little lignin or other substances from the fibers, and those that complete the delignification and remove pitch and some carbohydrate material.

Mechanical pulps. The lignin-retaining type of bleaching is used with high yield mechanical and chemimechanical pulps in paper grades, eg, newsprint, where brightness stability is not critical. The initial brightness values of these pulps usually are 50–65% GE (General Electric

standard). If sodium bisulfite is added in a chemimechanical process, the pulps are a few points brighter.

The most effective bleaching agent for most groundwoods is hydrogen peroxide (qv).

Chemical pulps. If all of the lignin, pitch, carbohydrate degradation products, and other chromophores and uv-absorbing materials are removed, a very white (over 90% GE), highly color-stable pulp can be obtained. This condition is limited to full chemical pulps, because it is much less expensive and more efficient to remove most of the lignin with pulping chemicals. The reagents for full bleaching are mostly oxidative. Since the carbohydrates are also susceptible to oxidation, bleaching is accomplished under the mildest conditions possible.

One bleaching method is the five-stage sequence CEDED, an initial chlorination of the lignin under acidic conditions followed by alkaline hydrolysis and extraction of the chlorinated lignin, mild oxidation with chlorine dioxide followed by another alkaline extraction, and final brightening with chlorine dioxide. Other common sequences are C/DEDED, CEHDED, and OCEDED where H and O refer to hypochlorite and oxygen, respectively. Washing is performed between stages when necessary.

JAMES MINOR
U.S. Forest Products Laboratory

R.G. MacDonald and J.N. Franklin, eds., *Pulp and Paper Manufacture*, Vol. 1 of *The Pulping of Wood*, 2nd ed., McGraw-Hill Book Co., New York, 1969.

J.P. Casey, ed., *Pulp and Paper Chemistry and Chemical Technology*, Vol. 1, 3rd ed., John Wiley & Sons, Inc., New York, 1980, p. 301.

R.P. Singh, ed., *The Bleaching of Pulp*, 3rd ed., TAPPI Press, Atlanta, Ga., 1979, p. 6.

J.E. Huber, ed., *Kline Guide to the Paper Industry*, 4th ed., C.H. Kline & Co., Inc., Fairfield, N.J., 1980.

PULP, SYNTHETIC

Synthetic pulp generally defines very fine, highly branched, discontinuous, water-dispersible fibers made from plastics. The visual appearance and dimensions of synthetic pulps closely resemble those of cellulose pulps and asbestos (qv) (see Pulp). Synthetic pulps are not to be confused with extruded stable fibers, which are smooth rods of solid polymer. Nearly all current pilot, semicommercial, and commercial production units produce pulps based on either high density polyethylene or polypropylene, with or without inorganic fillers (see Olefin polymers). The exceptions to this are pulps based on aromatic polyamides, ie, aramids (see Aramid fibers).

Polyolefin synthetic pulps are designed to be blended in all proportions with wood pulp and glass fibers and made into papers and boards using conventional papermaking equipment (see Paper). The sheets so produced may or may not be heated to melt the polyolefin, depending on the intended application. In either event, the resulting materials differ significantly from the 100% plastic sheets or mats of continuous fibers made by spunbounding, extrusion, melt-blowing, and other processes (see Plastics processing; Nonwoven textile fabrics, spunbonded).

Typical characteristics of standard commercials synthetic pulps are listed in Table 1.

Manufacture

The main processes used to prepare synthetic polyolefin pulps are solution flash spinning, emulsion flash spinning, melt extrusion/fibrillation, and shear precipitation. Solution flash spinning is outlined in Figure 1. It consists of forming a true solution of polyolefin in a low boiling organic solvent at high temperature and pressure. The solution is passed into a specially designed spurting nozzle or spinneret, in which a small, controlled pressure drop is effected. Two liquid phases, one of which is polymer-rich and the other polymer-lean, result. This two-phase mixture exits through a small orifice at high shear into a chamber of low

Figure 2. Electron micrograph of neutral sulfite semichemical (NSSC) pulp handsheet.

Table 1. Typical Characteristics of Pulpex^a Polyolefin Pulps

Property	Pulpex E-A (polyethylene)	Pulpex P-AD (polypropylene)
melting range, °C	130–135	160–165
specific area, m^2/g^b	10–15	5–10
poly(vinyl alcohol) content, wt %	1–1.5	0.6–0.8
drainage value, %^{c}	ca 120	ca 80
average length, mm	0.8–1.2	0.8–1.2
maximum length, mm	2.5	2.5
fines content^d, %	<3	<10
brightness, %^{e}	>94	>94
scattering coefficient	>0.15	>0.15
diameter, μm	10–20	20–40
polymer density, g/cm^3	0.965	0.91
pulp density, g/cm^3	ca 0.4	ca 0.2

^{a} Trademark of Lextar (Hercules, Inc. company)
^{b} To convert m^2/g to yd^2/oz, multiply by 29.5.
^{c} Drainage of a 70:30 blend of synthetic pulp and bleached kraft at 500 CSF; expressed as a percentage of the drainage of cellulose alone.
^{d} Percent of pulp passing through a 74-μm (200-mesh) screen.
^{e} MgO reference.

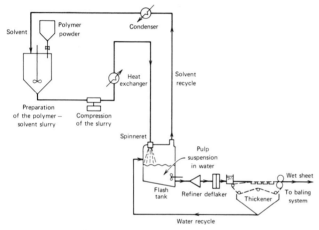

Figure 1. Flow sheet of flash spinning.

temperature and approximately atmospheric pressure to quickly and completely evaporate or flash the solvent in an almost explosive manner. Vaporization of solvent provides the energy to form the fibrous product, and the resulting cooling causes rapid crystallization of the polyolefin. The solvent is condensed and reused. With proper nozzle design and control of process variables, fine, discrete pulp fibers form at the last nozzle orifice. The pulp fibers are conveyed into water, passed through refiners or deflakers, or both, to control fiber length, if necessary, and then dewatered with conventional equipment.

The emulsion flash-spinning method involves initial formation of an emulsion at high temperature and pressure from a polymer solution and water. The emulsion is passed through a small orifice at high shear into a chamber at low temperature and approximately atmospheric pressure to partially evaporate the solvent. Solvent evaporation is completed by refining the resulting fibrous gel in hot water at atmospheric pressure or under slight vacuum. The refining step, which typically involves multiple passes through a disk refiner, is an essential step in the process because it provides fibrillation and adjustment of fiber length. Dewatering involves conventional technology.

Health and Safety Factors (Toxicology)

Both polyethylene and polypropylene in nonpulp form have been demonstrated to be practically nontoxic in animals and humans. The acute oral lethal dose for rats is greater than 500 mg/kg, which is the maximum feasible dose. Neither polymer is absorbed through the skin, they do not irritate the skin nor are they skin sensitizers.

Uses

Paper. Commercially established paper-related applications for synthetic pulps include wallpaper, teabags, wet-laid wipes, flooring felts (qv), battery separators, filters, and fibrous cement.

Nonpaper. Rheology control. Use of synthetic pulps in rheology-control applications (ie, for control of flow properties) generally, although not always, is to replace asbestos. However, polyolefin synthetic pulps can be used only below their melting points. Typical uses are solvent-cutback and emulsion-asphalt coatings and paints, caulks, mastics, and textured paints.

TERENCE RAVE
Hercules, Incorporated

V.M. Wolpert, *Synthetic Polymers and the Ppaer Industry*, Miller-Freeman Publications, San Francisco, Calif., 1977.

V.M. Wolpert, *Review of Synthetic Pulps and Papers*, 1976 International Synthetic Papers and Pulps Symposium, Mar. 31–Apr. 2, 1976.

PUMPS

Pumps are required in nearly every process plant, from the largest petroleum-refining unit or chemical complex to the smallest laboratory prototype operation.

Pumps are divided into dynamic and positive-displacement types. The dynamic type operates by increasing the velocity of the fluid as it passes through a rotating impeller. Positive-displacement types include pumps in which a chamber is filled with liquid to which pressure is imparted.

Operating conditions. Before a pump is selected, the duty conditions must be known, including type of liquid, density, temperature, viscosity, flow, inlet and outlet pressures, and presence of solids and corrosive or erosive material. Piping sizes must be determined. Pressure drops must be estimated. Most designers utilize some form of preprinted worksheet. This helps assure that, for a given system, all possible situations are considered. A helpful parameter for pump selection is the specific speed (N_s), a dimensionless term defined as:

$$N_s = \frac{N\sqrt{Q}}{H^{3/4}}$$

where N = rotating speed, Q = flow, and H = head.

Net positive suction head (NPSH) must be available in any pumping system. This is the net amount of head or pressure available at the inlet of the pump to overcome inlet losses and to allow liquid to flow into the pump.

Available NPSH (NPSH$_A$) is calculated as follows:

$$\text{NPSH}_A(\text{m}) = \left(P_s - P_{vp} - P_{frict} + P_{static} \right) \times \frac{102}{\rho}$$

where pressures are in kPa (101.3 kPa = 1 atm).

The pump, in turn, requires a certain NPSH (NPSH$_R$), which is a function of the pump design, the operating speed, and, for dynamic pumps, the flow rate. The NPSH$_A$ must exceed NPSH$_R$ to avoid cavitation. When the two values are equal, the liquid vaporizes as it flows into the pump because of entrance losses in the impeller eye. This causes formation of vapor bubbles, which soon collapse with such force as to damage most pumps. Most installations are not designed to tolerate this condition; the customary rule on simple systems is to provide at least 0.5 m greater NPSH$_A$ than the pump requires at the rated condition. On large, high energy pumps, or other specialized services where conditions may change, this margin should be increased.

System head exists for every pumping system. It is often desirable on complex systems to calculate the system-head curve, which can be overlaid on the proposed pump curve. From this comparison, the rated condition and operating conditions can be adequately estimated.

Applications and Selection

Centrifugal pumps. Efficiency and power are calculated for the rated point and all other operating points along the pump curve. Each centrifugal pump impeller and casing can be optimized for only one maximum efficiency. At the best operating point, efficiencies are ca 30–40% on very small, low specific speed pumps, and as much as > 90% on very large pumps. Impeller diameter must be individually determined for each application. For each impeller, there is a minimum diameter which is normally shown on the manufacturer's curve.

The shape of performance curves varies with the blade angle, impeller geometry, number of blades, and specific speed.

Parallel pumping occurs when two or more pumps operate in parallel in the same system. If the pumps are identical, they should share the duty equally. However, two supposedly identical pumps may not prove to be the same. In such a case, one pump may assume more of the load than the other.

Pumps handling viscous liquids and slurries perform differently than they would on water. The Hydraulic Institute Standards provide a method of derating a pump for a viscous liquid, based on experience. For slurries, the water performance of a pump must likewise be derated.

Dynamic pumps. Most dynamic pumps are centrifugal, and have either horizontal or vertical shafts. As sizes (specific speeds) increase, pumps are classified as mixed-flow and finally as axial-flow or propeller types. In very small sizes, there are speciality types.

The chemical process pump, or so-called ANSI type, is a single-stage, overhung bearing bracket design. External mounting dimensions, suction and discharge flange sizes, and orientations are standardized.

The refinery process pump, sometimes called the API-610 type, was developed for higher temperature liquids as in oil-refinery applications.

Other types include the horizontal-split-casing type, using a flat, unconfined casing gasket, a single-stage double-suction-type impeller, and between-bearing design. It is the standard for use on cooling water, circulating water, solution circulating, fire fighting, and in certain modified higher pressure versions.

Close-coupled, single-stage pumps are widely used for general service application. This type is mounted directly on an electric-motor frame and shaft, using the bearings of the motor for support. Alignment problems are minimized. Close-coupled pumps are not generally desired for hazardous or toxic fluids.

Self-priming pumps are made in a limited range of sizes and have applications where a horizontal pump to pull a suction lift is required.

Canned-rotor pumps are used where no leakage can be tolerated. Slurry pumps are single-stage, overhung bearing designs. Slurry services are generally severe. Horizontal multistage pumps are used when a single impeller cannot produce the required head. As many as 14 or more impellers may be mounted on a single shaft to form the rotor. Casings are split on the horizontal or vertical axis.

Bearings and seals are always to be considered in preparing the specification. Mechanical shaft seals are used almost exclusively in modern process pumps.

Cooling water is usually supplied for bearing brackets and stuffing-box jackets when pumping temperature is expected to be 150°C or higher. Above 300°C, the pump-support pedestals may also be water-cooled.

Vertical in-line pumps are mounted directly in the piping system with no additional elbows or changes in direction of flow. High speed versions of the vertical in-line type also exist, operating at speeds up to 20,000 rpm, and producing very high heads in a single stage.

The vertical-can pump is a vertical-shift, multistage pump which can be arranged to produce pumping heads exceeding 1200 m by operating at conventional 60 Hz speeds and utilizing as many stages as necessary.

Other vertical pumps include the single-stage vertical volute type, used from the smallest wet-pit sump-pump applications up to the very large wet- or dry-pit circulating pumps.

Vertical-turbine pumps are used in cooling-tower basins, river-and-lake water pumping, vertical firewater service, vertical process services, and for pumping water from a well. For deep-well pumps, the pumping element is either submerged in the water-well and connected to the driver on the surface by means of long vertical lineshafts, or it has a submersible motor.

The inertia pump is an oscillating type of dynamic pump. Other dynamic pumps include the regenerative turbine, pitot-tube, disk pump, etc.

Positive-displacement pumps. This class is divided into reciprocating, controlled-volume, and rotary pumps. Their formulas are like those for dynamic pumps, with the important distinction that positive-displacement pumps produce pressure, not head, and are essentially fixed-volume machines if operated at constant speed and stroke. The capacity can be changed by changing the speed.

Pulsating flow is inherent in reciprocating pumps as each piston or plunger alternates from zero flow to maximum. Most reciprocating pumps have two or more cylinders, and pistons or plungers are arranged about the crankshaft in such a way that the pulsating flow is cancelled to some extent by the alternate strokes of the multiple plungers or pistons. The $NPSH_A$ for reciprocating pumps is calculated much as it is for dynamic pumps.

Reciprocating pumps are either direct-acting or crank-driven. Direct-acting pumps have air, steam, or gas cylinders at one end of the pump, connected by piston rods and valve linkages to liquid cylinders at the other end. The crank-driven type has a crankshaft operating at low-to-medium speed (60–500 rpm) with connecting rods to pistons or plungers arranged either horizontally or vertically.

Controlled-volume pumps, also known as metering or proportioning pumps, are applied when a precise amount of fluid must be pumped. These are a variation of the reciprocating type, having the added feature of adjustable stroke length.

A rotary pump is defined in Hydraulic Institute Standards as "a positive-displacement pump, consisting of a chamber containing gears, cams, screws, vanes, plungers, or similar elements actuated by relative rotation of the drive shaft and casing, and which has no separate inlet and outlet valves."

The rotary pump is particularly useful in pumping viscous liquids, non-Newtonian fluids, liquid with entrained air or gas, and pumping in systems requiring suction lift. The rotary is often a better choice than a centrifugal pump when the liquid viscosity is higher than about 100 mPa·s (= cP) or when it is variable.

The engineer can make a choice by using the Hydraulic Institute method for rating a centrifugal pump on viscous liquids and comparing the results with the manufacturer's data on rotary-pump performance (see Table 1).

Table 1.   Pump Evaluation for Viscous Liquid

Service and selection	Charge solution	Centrifugal	Rotary
liquid			
pumping temperature, °C	30		
viscosity at pumping temperature, mPa·s (= cP)	110		
density at pumping temperature kg/m³	950		
capacity, dm³/s[a]	35		
differential pressure, kPa[b]	600		
pump differential head, m	64.4		
$NPSH_A$, m	5		
pump size			
API-610 type, cm		7.6 × 10.2 × 21.6	
gear type, mm			150
pump speed, rpm		3550	350
efficiency, %			
water		76	
viscous		63	80
efficiency correction		0.83	
power required, kW		33.3	26
end-of-curve power, kW		40	
power at relief valve setting, kW			30
probable motor size, kW		40	30
$NPSH_R$, m		4.3	2.5

[a]To convert dm³/s to GPM, multiply by 15.85.
[b]To convert kPa to psi, multiply by 0.145.

The rotary pump has become standard for lube-oil and seal-oil systems associated with large turbomachinery.

Drivers for pumps. Drivers used for pumps include all varieties of electric motors, as well as diesel and gas engines and steam or gas turbines. A steam-turbine-driven or engine-driven spare pump is often installed to take over during electric-power failure.

The optimum arrangement is to match the driver speed to the pump speed. When this is not possible, the use of speed-reducer or speed-increaser gears is common. In small sizes, V-belt or chain drives may be desirable.

Estimation of Pump Costs

Costs of pumps can be estimated in several ways, and at varying accuracy. A rough-order-of-magnitude (ROM) method has been offered in trade publications. Estimates do not include freight shipment to the site or installation.

Evaluation of Pump Performance

All pumps undergo certain tests before shipment. Hydrostatic tests of pressure-containing parts such as casings or cylinders should always be performed, at pressures of 150–200% of the pump design maximum.

Mechanical running tests are performed on even the smallest pumps to assure adequate assembly. Certain small pumps are tested in groups. Normally, for pumps in process service or those used as auxiliary equipment, the mechanical running test is not sufficient.

A performance test is therefore specified for most process-type pumps. Five or more performance points are usually required in a centrifugal-pump test. The test liquid is usually water. An NPSH test may be needed in certain applications to prove the $NPSH_R$ for the particular pump.

Field evaluation must often be made and when pumps have been in service for some time, it may be necessary to evaluate performance on actual duty conditions to determine whether maintenance is required. Among the revisions readily made in the field maintenance shop are replacement of pump impeller or wearing rings to restore original performance or to meet new requirements. When a larger impeller is being installed, the power requirements and thrust capabilities of the pump should be rechecked. In more complex revisions, it may be desirable to return the pump to the manufacturer for rebuilding and possibly for retesting. Replacement of seals, bearings, sleeves, shafts, etc, should be considered routine maintenance.

Nomenclature

$NPSH_A$ = net positive suction head available, m
$NPSH_R$ = net positive suction head required, m
P_s = pressure, suction, kPa (psi)
P_{vp} = pressure, vapor, kPa (psi)
P_{frict} = pressure loss due to friction, kPa (psi)
P_{static} = pressure, static, kPa (psi)
ρ = density of fluid, kg/m^3

RICHARD NEERKEN
The Ralph M. Parsons Company

I.J. Karassik, W.C. Krutzch, W.H. Fraser, and J.P. Messina, *Pump Handbook*, McGraw-Hill Book Co., New York, 1976.

Hydraulic Institute Standards, 14th ed., Hydraulic Institute, Cleveland, Ohio, 1983.

PTC 8.2 Power Test Code for Centrifugal Pumps and *PTC 7.1 Power Test Code for Displacement Pumps*, American Society of Mechanical Engineers.

PURGATIVES. See Gastrointestinal agents.

PURINES. See Alkaloids; Chemotherapeutics, antimitotic.

PYRAZOLES, PYRAZOLINES, AND PYRAZOLINONES

Pyrazoles

Pyrazoles (**1**) are stable, five-membered, heterocyclic compounds having two nitrogen atoms in a 1,2 relationship. They can be represented by several tautomeric forms, of which two are the most common. Pyrazoles are very scarce in nature, whereas the isomeric imidazoles, which are characterized by two nitrogens in a 1,3 relationship, occur in most living systems.

(1)

Physical properties. Pyrazoles are very stable compounds of high boiling points. Increasing substitution on the carbon atoms causes higher boiling points, but substitution on nitrogen lowers both melting and boiling points. For example, 3-methylpyrazole boils at 205°C, but 1-methylpyrazole boils at 127°C. These properties reflect an association of pyrazoles that have hydrogen on nitrogen.

Pyrazole and its lower homologues are soluble in water and most organic solvents. The specific gravities are 0.89–1.02 and the refractive indexes are 1.46–1.48. Pyrazoles that have hydrogen on nitrogen are weakly acidic and weakly basic. They form salts with strong acids, and with metals by replacement of the hydrogen on the nitrogen atom. The ^{1}H nmr spectra of pyrazoles have been extensively reviewed.

Chemical properties. Pyrazole is an aromatic molecule of six π electrons. In N-substituted pyrazoles, the 1-N has some cationic character and the 2-N retains two electrons and, therefore, has basic properties. The greater nuclear charges of the two nitrogen atoms reduce the charge density at positions 3 and 5, and this leads to a greater charge density at position 4, which leaves that position more prone to electrophilic attack. Conversely, nucleophilic attack is more likely to occur at the 5-position.

Chlorination with many reagents leads to the formation of 4-chloropyrazole.

Bromination of pyrazole introduces one to three bromine atoms. Iodination of pyrazoles is quite similar to bromination, substitution occurring at all positions. Sulfonation of pyrazole occurs only under drastic conditions and yields amphoteric compounds. Nitropyrazoles can be obtained by direct synthesis or by nitration of pyrazole derivatives.

The pyrazole ring is as resistant to oxidation as it is to reduction. Alkyl side chains can be oxidized by permanganate or chromic acid to carboxylic acid, without destruction of the pyrazole ring. Only ozonolysis or electrolytic oxidation destroys the ring.

Synthesis. There are over twenty methods for synthesizing pyrazole derivatives. The most common procedure is the reaction of β-dicarbonyl compounds with hydrazine. Unsymmetrical β-diketones with substituted hydrazines give a mixture of products. The reaction of α-acetylenic carbonyl compounds with substituted hydrazines, also leads to a mixture of pyrazoles (eq. 1).

$$\underset{O\quad\;\; O}{RCCH_2CR'} + R''NHNH_2 \rightarrow \text{(mixture of pyrazole products)} \quad (1)$$

$$\leftarrow \underset{O}{RCC{\equiv}CR'} + R''NHNH_2$$

A third procedure consists in the reaction of diazo derivatives with acetylenes; the reaction proceeds readily when the triple bond is activated. Finally, pyrazoles can be obtained from pyrazolines by dehydration, oxidation, or elimination.

Health and Safety

Pyrazoles that are substituted in the 4-position with alky or cycloalkyl groups strongly inhibit the activity of alcohol dehydrogenase in the human liver and may be used to combat the ill effects of alcohol abuse.

Uses. There are few registered products on the market. Among these are fungicides, herbicides, antibacterials, and hypnotics as well as dyes in the textile industry (see Antibacterial agents; Hypnoticas sedatives, and anticonvulsants).

Pyrazolines

Pyrazolines (2) are dihydropyrazoles and are less stable than pyrazoles. Three tautomeric forms are known; 2-pyrazolines are the most common.

(2)

Physical properties. Pyrazolines are colorless, high-boiling or low-melting solids. Substitution on nitrogen lowers the melting point. Low molecular weight pyrazolines are water soluble, but on an increase in molecular weight they are more soluble in organic solvents. Pyrazolines are weakly basic and can be protonated depending on the position of the double bond.

Chemical properties. A very facile reaction of pyrazolines is loss of nitrogen. Thus, 1-pyrazolines extrude nitrogen to yield mainly cyclopropanes; however, olefins may also be produced. Generally, *trans*-1-pyrazolines yield a mixture of *cis*- and *trans*-cyclopropanes.

The ratio depends on the nature of substituents and on the use of photochemical or thermal decomposition.

Pyrazolines can be oxidized to pyrazoles by means of a variety of reagents, eg, bromine or potassium permanganate; however, upon air oxidation, 1-pyrazoline undergoes ring cleavage to yield propylene and nitrogen. 2-Pyrazolines that are unsubstituted at the 1-position can be easily acylated, carbamoylated, alkylated, etc; rearrangement takes place on occasion.

Synthesis

The most common procedure for the synthesis of pyrazolines is the reaction of aliphatic or aromatic hydrazines with α,β-unsaturated carbonyl compounds. The reaction proceeds in an acidic or basic medium, with or without solvent, and at low or high temperatures. It proceeds through a hydrazone, although with aliphatic hydrazines its isolation before cyclization may be difficult. The second most important method for the synthesis of pyrazolines is the cycloaddition of diazoalkanes to carbon–carbon double bonds.

Uses. Pyrazolines are used as fluorescent optical whitening agents (see Brighteners, fluorescent).

Pyrazolinones

Pyrazolinones are oxygenated pyrazolines. They exist in three isomeric forms: 2-pyrazolin-5-one (3); 3-pyrazolin-5-one; and 2-pyrazoline-4-one. They display keto-enol tautomerism.

Physical properties. The pyrazolinones are usually crystalline and have poorly defined melting points. Generally, they are more soluble in polar solvents, including water. 5-Pyrazolones are both acidic and basic having pKa values of 6.2–11.0. The 3-pyrazolinones are more basic than the 5-pyrazolinones, and the former can form salts, eg, hydrochlorides.

Chemical properties. The 4-position in 5-pyrazolinones is highly reactive and undergoes all of the typical reactions of an active methylene group. With ketones and aldehydes, a condensation product and dimers are formed; but with formaldehyde, a hydroxymethyl moiety is introduced into the 4-position.

Pyrazolinones undergo C-alkylation at the 4-position; O- and N-alkylation can also occur and lead to a mixture of products. Introduction

of a halogen atom into the 4-position is also a facile reaction. Acylation, bromination, and carbamoylation at the 4-position are accomplished by the usual reagents. The Mannich reaction occurs in the 4-position, as does the introduction of the arylazo group. The latter is used in the preparation of azo dyes (qv). Strong oxidizing agents rupture the pyrazolinone ring, whereas catalytic reduction leaves the ring intact.

Synthesis. The most common procedure for the synthesis of 5-pyrazolones is the reaction of almost any nonsubstituted or monosubstituted β-ketoester with almost any monosubstituted hydrazine.

Health and safety. Pyrazolinones are suspected of being carcinogenic because their ease of nitrosation can lead to the formation of dimethylnitrosamine, a potent carcinogen.

Uses. Of the three classes of compounds, pyrazolinones have had the greatest commercial application, mainly as pharmaceuticals (qv) and dyes (see Dyes, application and evaluation).

GABRIEL KORNIS
The Upjohn Company

A.N. Kost and I.I. Grandberg in A.R. Katritzky and A.J. Boulton, eds., *Advances in Heterocyclic Chemistry*, Vol. 6, Academic Press, Inc., New York, 1966, pp. 347–429.

L.C. Behr, R. Fusco, and C.H. Jarboe in R.H. Wiley, ed., "Pyrazoles, Pyrazolines, Pyrazolidines, Indazoles and Condensed Rings," Vol. 22 of A. Weissberger, ed., *The Chemistry of Heterocyclic Compounds*, Wiley-Interscience, 1967.

J. Elguero in A.R. Katritzky and C.W. Rees, eds., *Comprehensive Heterocyclic Chemistry*, Vol. 5, Pergamon Press, Oxford, UK, 1984.

PYRIDINE AND PYRIDINE DERIVATIVES

Pyridine is the parent of a series of compounds that is important in medicinal, agricultural, and industrial chemistry. Although many polysubstituted pyridine compounds, like other heterocyclic compounds, are synthesized with their functional groups present from acyclic compounds, most derivatives are prepared by manipulation of pyridine and its simple homologues in a manner similar to the chemistry of the benzenoid aromatics. However, the simple pyridine compounds are prepared by cyclization of aliphatic raw materials.

Since pyridine has the symmetry of a monosubstituted benzene, there are three possible monosubstituted pyridine isomers, six compounds with two like substituents, etc (see Benzene). The three monomethylpyridines or picolines are 2- or α-picoline, 3- or β-picoline, and 4- or γ-picoline. Although pyridine and picolines dominate the commercially important chemistry of pyridine derivatives, 2-methyl-5-ethylpyridine (MEP or aldehyde collidine) also is important. Dimethylpyridines are called lutidines, and the 2,6- and 3,5-lutidines are readily available. The trivial name of trimethylpyridine is collidine, and the symmetrical 2,4,6-collidine is the most common. Pyridine chemistry has been comprehensively reviewed.

Properties

The physical properties of simple pyridine compounds (see Table 1) are largely controlled by the presence of the basic electronegative nitrogen atom in the ring.

Pyridine is a tertiary amine and all simple pyridine compounds are basic, but they are less basic in solution than typical aliphatic amines. The basicity of pyridine derivatives is increased by electron-donating substituents and decreased by electron-withdrawing substituents. Most alkyl pyridines form azeotropes with water.

Manufacture

Syntheses. The most important synthetic reactions for the manufacture of pyridine bases have been the reactions of aldehydes and ketones with ammonia.

Reaction of acetaldehyde (qv), preferably in the form of its trimer paraldehyde (1), with aqueous ammonia in the liquid phase, takes place

Table 1. Properties of Pyridine Derivatives

Compound	Freezing point, °C	Boiling point, °C	Density at 20°C, g/cm³	pKa thermodynamic in H₂O at 25°C	Solubility in H₂O at 20°C, g/100 g	Water azeotrope Bp, °C	Water azeotrope Wt% H₂O
pyridine	−41.6	115.3	0.9830	5.22	miscible	93.6	41.3
2-methylpyridine	−64	129.5	0.9462	5.96	miscible	93.5	48
3-methylpyridine	−18.3	143.9	0.957	5.63	miscible	96.7	63
4-methylpyridine	3.7	144.9	0.9558	5.98	miscible	97.4	63.5
2,3-dimethylpyridine	−15.5	161.5	0.9491	6.57	13.3^a		
2,4-dimethylpyridine	−64	158.7	0.9325	6.63	miscible^b		
2,5-dimethylpyridine	−15.7	157	0.9331	6.40	10.0^c		
2,6-dimethylpyridine	−6.1	143.7	0.923	6.72	miscible	93.3	51.5
3,4-dimethylpyridine	−10.6	179.1	0.9534	6.46	5.2		
3,5-dimethylpyridine	−6.6	172.7	0.944	6.15	3.3		
2,4,6-trimethylpyridine	−44.5	170.4	0.913		3.6		
5-ethyl-2-methylpyridine (MEP)	−70.9	178.3	0.9208		1.2	98.4	72
2-vinylpyridine		110 (20 kPa^d)	0.9746	4.98	2.75	97	62.0
4-vinylpyridine		121 (20 kPa^d)	0.988		2.91	98	76.6
piperidine	−11.0	106.3	0.8659 (15°C)	11.12	miscible	92.8	35

^{a} Miscible below 16°C.
^{b} Miscible below 23°C.
^{c} Miscible below 13°C.
^{d} To convert kPa to mm Hg, multiply by 7.5.

at 230°C and 5.6–20.8 MPa (800–3000 psig) with catalysis by ammonium salts, eg, ammonium acetate, to give MEP (**2**).

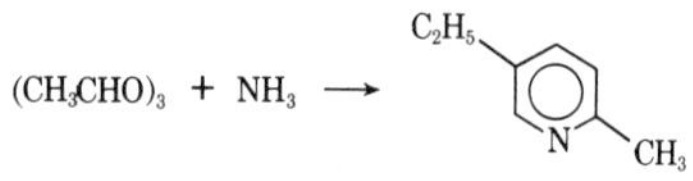

(CH₃CHO)₃ + NH₃ → (**2**)

(**1**)   (**2**)

In the vapor phase, acetaldehyde reacts with ammonia in the presence of heterogeneous catalysts to give 2- and 4-methylpyridine in about equal proportions; acetylene reacts similarly. The reaction is carried out at 350–550°C.

Under essentially the same vapor-phase reaction conditions, acetaldehyde, formaldehyde, and ammonia react to yield a mixture of pyridine (**3**) and 3-methylpyridine (**4**); methanol is included in the feed for increased yield.

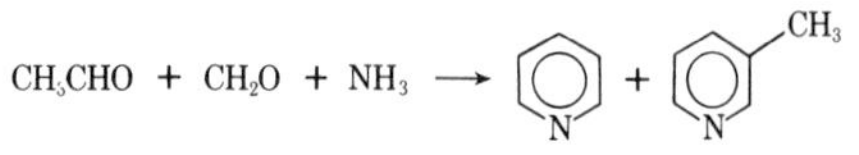

CH₃CHO + CH₂O + NH₃ → (**3**) + (**4**)

(**3**)   (**4**)

Toxicology

All pyridine chemicals should be handled according to the best standards of laboratory and manufacturing safe practice. Pyridine and the alkyl pyridines demand additional handling precautions to prevent the escape of vapors, because the compounds typically have intense odors that are detectable and disagreeable at very low concentrations. Contact of 2- or 4-vinylpyridine with the skin should be avoided, as temporary, painful skin burns frequently result.

Reactions and Uses of Derivatives

Reactions at nitrogen. *Quaternization.* Pyridine and most of its derivatives react with alkylating agents, typically alkyl halides or sulfates, to produce quaternary (quat) salts.

Quaternary salts of 2,2'- and 4,4'-bipyridyls are useful herbicides (qv).

***N*-Oxidation.** Most pyridine derivatives react with a peracid, usually peracetic acid, or hydrogen peroxide (qv) in the presence of a catalyst to give the corresponding *N*-oxides (see Amine oxides).

Reactions on the nucleus. *Amination.* In contrast to their general lack of useful reactivity in electrophilic substitution, some nucleophilic substitutions occur on pyridines in a synthetically useful fashion. Sodamide reacts with pyridine, typically in refluxing toluene or xylene

(the Chichibabin amination) to give, after hydrolysis of the intermediate sodium salt (**5**), 2-aminopyridine (**6**) in high yield.

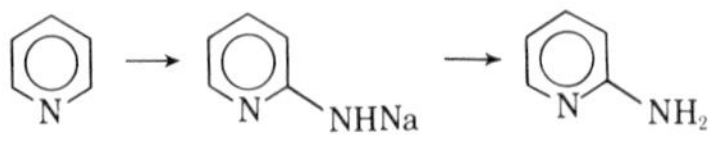

(**3**) → (**5**) → (**6**)

Chlorination. Direct chlorination of pyridine occurs in the vapor phase at over 300°C in the presence of a diluent to give 2-chloropyridine; 2,6-dichloropyridine is a by-product. Vigorous chlorination of pyridine or most of its alkyl derivatives eventually gives pentachloropyridine.

Reactions of side chains. Alkyl groups attached to the pyridine ring undergo many of the typical oxidation and substitution reactions of alkyl aromatic compounds. In addition, hydrogen on carbon atoms adjacent to the ring are particularly acidic, leading to additional kinds of reactions.

Ammoxidation of 3-methylpyridine followed by partial hydrolysis gives nicotinamide [niacinamide, (**7**)]. Oxidation of MEP (**2**) yields nicotinic acid [niacin, (**8**)]. Niacin or niacinamide are therapeutically equivalent forms of a vitamin required by all living cells.

Condensation of 2-methylpyridine with formaldehyde yields 2-vinylpyridine (**9**). A latex copolymer of styrene, butadiene, and 2-vinylpyridine is used as a tire-cord (qv) binder.

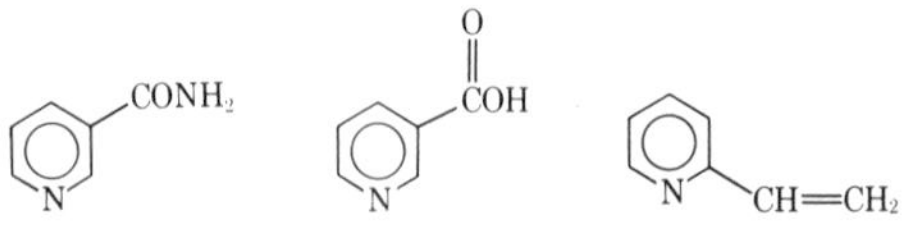

(**7**)   (**8**)   (**9**)

Reduction and reduced derivatives. Pyridine and most of its derivatives are easily hydrogenated at elevated temperatures and pressures with nickel, palladium, or ruthenium catalysts. Platinum is a very effective catalyst in acetic acid solution.

GERALD L. GOE
Reilly Tar & Chemical Corporation

E. Klingsberg, ed., *Pyridine and Its Derivatives*, Interscience Publishers, Inc., New York, 1960.

R.A. Abramovitch, ed., *Pyridine and Its Derivatives, Supplement*, John Wiley & Sons, Inc., New York, 1974.

M.H. Palmer in S. Coffey, ed., *Rodd's Chemistry of Carbon Compounds*, 2nd ed., Vol. IV, Pt. F, Elsevier Scientific Publishing Co., Amsterdam, Neth., 1976, pp. 1–26.

A.R. Katritzky and J.M. Lagowski, *Chemistry of the Heterocyclic N-Oxides*, Academic Press, Inc., New York, 1971.

PYRIODOXINE, PYRIDOXAL, AND PYRIDOXAMINE. See Vitamins.

PYRITE, FeS₂. See Iron; Pigments, inorganic; Sulfur; Sulfuric acid.

PYROCATECHOL (1,2-BENZENEDIOL), C₆H₄(OH)₂. See Hydroquinone, pyrocatechol, and resorcinol.

PYROGALLOL (1,2,3-BENZENETRIOL), C₆H₃(OH)₃. See Polyhydroxybenzenes.

PYROMETALLURGY. See Extractive metallurgy.

PYROMETRIC CONES. See Ceramics.

PYROMETRY. See Temperature measurement.

PYROTECHNICS

Pyrotechnics is the art of using chemically generated light, heat, or sound for entertainment, convenience, or war. Fireworks refers to the general discipline of civilian pyrotechnics which encompasses fireworks, matches (qv), and such devices as highway flares, gopher bombs, flashbulbs, automotive airbag inflators, thermitic welding kits, and items for theatrical effects. Military pyrotechnics includes a wide range of devices for illumination, signaling, incineration, and gas generation. Military devices are characterized by more rugged construction and greater resistance to adverse environmental conditions at concomitant higher cost, reliability, and safety than are civilian pyrotechnics (see Chemicals in war).

Civilian Pyrotechnics

Fireworks. Fireworks is historically the art of using black powder. Modern displays are based on mortar-fired aerial shells whose effects derive from a combination of devices containing colored starts. Many important facets of manufacture are trade secrets. Fireworks are divided into dangerous fireworks whose manufacture, transport, and display is specially regulated, "exempt fireworks" which are of utilitarian nature, and safe-andsane fireworks which are sold to the general public.

Fireworks fabrication is mainly a matter of hand labor and a great amount of skill. Both China and Japan have been in recent years the greatest sources of imported devices. Most complex are the spectacular bursting charges from which are propelled cascading stars and streamers. The bursting charge consists of potassium chlorate and charcoal (no sulfur) or of potassium perchlorate, charcoal and sulfur, or black powder. Around the central range are packed in close array up to three layers of stars which when burning form the "double petalled chrysanthemum". Stars are approximately 4.1 cm³ (0.25 in.³) any may be cut from a wet cake (American method) or pasted by gluing powder on core grains (Japanese method). Pasted stars can change color in flight. Stars are made from mixtures of accroides resin or shellac, potassium perchlorate and strontium carbonate (red), or sodium nitrate (yellow), or barium nitrate with chlorinated isoprene (green), or copper oxide and copper

carbonate with chlorinated isoprene (blue). Magnesium powder enhances the brilliance of the combustion. Addition of coarse aluminum flakes to stars results in luminous tails. Flash thunder is produced from mixtures of potassium perchlorate and aluminum flake, with or without sulfur. The quality of the effects is crucially dependent on the correct selection of ingredients for particle size, purity, sequence of compounding, and composition.

Roman candles are stars that are propelled from cardboard tubes, one at a time. Pin wheels are propelled by tubes filled with a black power composition. Waterfalls have black power mixed with coarse aluminum flakes. Toy pistol caps, "torpedoes," and "pull caps" contain minute quantities of potassium chlorate–arsenic monosulfide mixtures. Whistles are cardboard tubes containing mixtures of chlorates, perchlorates or nitrates with certain aromatic compounds such as potassium benzoate or gallic acid. Firecrackers contain mixtures of potassium perchlorate, sulfur, and flaked aluminum, loosely packed in cardboard tubes.

Model rockets and missiles are toys that are propelled by black powder that is pressed into cardboard tubes which are closed with clay nozzles.

Theatrical effects. Most theatrical effects for motion pictures or television, such as smoke, fog, lightning, and flame, are not pyrotechnic in origin but are produced from oil dispersions, dry ice, and arc lamps. Cannon and artillery fire is simulated by the use of special blanks. Blanks in muzzle-loading weapons contain small quantities of black powder which produce a better visual effect than do commercially produced blanks. Bullet effects or bullet hits are small, concealed, electrically initiated charges that have been built up to various degrees of intensity from various amounts of primary explosive.

Utilitarian devices. Utilitarian devices, eg, highway flares, are made chiefly of strontium nitrate mixed with sawdust, wax, sulfur, and potassium perchlorate and contained in a waterproof cardboard tube. A small quantity of a safety-match composition also is incorporated and, upon ignition, the device burns up to 30 min with a distinctive red flame.

Military Pyrotechnics

Military pyrotechnics generate light, eg, in flares, flash charges, and tracers; conceal or signal by means of smoke; generate heat, eg, in incendiaries; ignite and propagate pyrotechnic reactions, eg, with delays; and are sources of prime ignition, eg, in percussion primers and matches. Many aspects of military pyrotechnics are amenable to quantitative analysis through application of physical and chemical principles.

Infrared flares depend on the high degree of reflectance of natural terrain at long wavelengths. Recent developments in electro-optical devices, such as image intensifiers, for the purpose of detecting such images have enhanced performance in the red or near infrared spectral region. These pyrotechnic mixtures contain alkali nitrates other than sodium nitrate.

Smoke generators. Smoke generators are pyrotechnic devices for daytime obscuration and signaling. Concealment of troop movements and structures is not generally done pyrotechnically but through atomization of fog oil or release of titanium tetrachloride (FM smoke) or of sulfur trioxide. An exception is HC smoke, which is formed by the combustion of aluminum, hexachloroethane and zinc oxide. Signal smokes may be white or colored but not black, because black smoke is not sufficiently opaque to be distinguished against the background. Colored smokes derived their color from organic dyes which, because of their low combustion temperatures, evaporate and recondense.

Tracer munitions. Tracer bullets guide the direction of the fire, aid in range estimation, mark target impact, and act as incendiaries. Tracers serve by delay-train action to self-destruct munitions. Tracer munitions contain mixtures of magnesium, strontium nitrate, and potassium perchlorate. Daylight smoke tracers have been developed that produce colored trails by dissemination of a dry powder, by the sublimation of organic dyes, or by the combustion of phosphorus (white smoke) or of cadmium with sulfur (yellow smoke).

Ignition. Many methods for the initiation of pyrochemical reactions are not pyrotechnic but rely on primary explosives (qv). Others consist of materials and mixtures that ignite by the action of friction, shock, heat, electrical impulse, or by environmental action as from air, water, or chemical reagents. The selection of the best pyrotechnical technique for prime ignition depends on the desired functioning of time and the availability of suitable external stimuli. Typical ignition mixtures contain boron or magnesium and barium nitrate. Percussion primers are initiated by impact and these contain lead styphnate and barium nitrate, or potassium chlorate and antimony sulfide. Delay elements are fuses containing (typically) mixtures of manganese and lead and barium chromates. These burn without generating gas pressure and permit precise control of ignition sequences.

Health and Safety Factors

Safety concerns permeate all aspects of pyrotechnics. The number of fatalities and serious injuries experienced annually in the manufacture and handling of pyrotechnic devices far exceeds those resulting from any other pyrochemical activity. Death and injury are caused usually by burns and lung damage and, to a lesser extent, by shock and by impact of flying fragments. Mixtures of potassium chlorate with sulfur or red phosphorus are so dangerous that they must never be made.

ALEXANDER P. HARDT
Lockheed Missiles and Space Co., Inc.

R. Lancaster, *Fireworks Principles and Practices*, The Chemical Publishing Company, New York, 1972.

Pyrotechnics: Occasional Papers in Pyrotechnics, Vols. I–IX, 2302 Tower Drive, Austin, Tex., 1977–1984.

S.M. Kaye, ed., *Encyclopedia of Explosives and Related Items*, Vol. 7, ADA 019502, Vol. 8, ADA 057762, Vol. 9, ADA 097595, and Vol. 10, U.S. Army Research and Development Command, Dover, N.J., 1982.

PYRROLE AND PYRROLE DERIVATIVES

Physical Properties of Pyrrole

Pyrrole (**1**) is a colorless, slightly hygroscopic liquid and, if fresh, it emits an odor like that of chloroform. However, it darkens on exposure to air and eventually produces a dark-brown resin. It can be preserved by excluding air from the storage container, preferably by displacement with ammonia to prevent acid-catalyzed polymerization. Some physical properties of pyrrole are listed in Table 1.

Pyrrole has a planar, pentagonal (C_{2v}) structure and is aromatic in its reactions since it has an aromatic sextet of electrons. It is isoelectronic with the cyclopentadienyl anion. The π-electrons are delocalized throughout the ring system; thus, pyrrole is best characterized as a resonance hybrid with contributing structures (**1**)–(**5**). These structures explain its lack of basicity, which is less than pyridine, its unexpectedly high acidity, and its pronounced aromatic character.

The resonance energy is ca 100 kJ/mol (24 kcal/mol), or about two-thirds that of benzene. Its resonance energy is intermediate between those of furan and thiophene; thiophene has the higher value.

Table 1. Physical Properties of Pyrrole

Property	Value
melting point, °C	−18.5
boiling point, °C	130
critical temperature, °C	366
density, d_4^{20}, g/cm³	0.9698
refractive index, n_D^{20}	1.5085
dielectric constant (at 20°C), ϵ	8.00
flash point (closed-cup), °C	39

Pyrrole is freely soluble in alcohol, benzene, and diethyl ether, but is only sparingly soluble in water and in aqueous alkalies. It dissolves with decomposition in dilute acids. Pyrroles with substituents in the β-position are usually less soluble in polar solvents than the corresponding α-substituted pyrroles.

Syntheses of Pyrroles

Knorr. The Knorr reaction and its modifications are the most important and widely used methods for the synthesis of pyrroles. Since the α-aminoketone is subject to self-condensation, the condensation with a β-dicarbonyl derivative is usually carried out by generating the α-aminoketone *in situ* through reduction of an oximino derivative: zinc in glacial acetic acid is used as the reductant.

Knorr synthesis

The Knorr synthesis is not particularly sensitive to the nature of R and R‴, ie, they may be alkyl, acyl, aryl, or carbalkoxy without significantly affecting the yield. Similarly, good yields are obtained if R′ and R″ are acyl or carbalkoxy, but poor yields are obtained if they are alkyl or aryl.

Hantzsch and Feist. The Hantzsch synthesis of pyrroles involves condensation of an α-haloketone with a β-keto ester in the presence of ammonia or an amine. The Feist synthesis is similar to the Hantzsch method and involves condensation of acyloins, eg, with aminocrotonic esters in the presence of zinc chloride.

Pyrrolines and Pyrrolidines

The pyrrolines or dihydropyrroles can exist in three isomeric forms:

(**2**) 1-pyrroline   (**3**) 2-pyrroline   (**4**) 3-pyrroline

1-Pyrroline (**2**) (3,4-dihydro-2H-pyrrole) is an unstable material that resinifies upon exposure to air. 2-Pyrroline (**3**) (2,3-dihydro-1H-pyrrole) is even more unstable. Only 3-pyrroline (**4**) (2,5-dihydro-1H-pyrrole) is reasonably stable. It boils at 91°C and has a density d_4^{28} of 0.9097 g/cm³ and a refractive index n_D^{20} of 1.4664. Pyrrolidine (**5**), or tetrahydropyrrole, is a water-soluble, strong base having the usual properties of a secondary amine.

(**5**)

An important synthesis of pyrrolidines is the reaction of reduced furans with excess amine or ammonia over an alumina catalyst in the vapor phase at 400°C. However, if labile substituents are present in the tetrahydrofuran, pyrroles may form.

Reactions of Pyrroles

In keeping with its decidedly aromatic character, pyrrole is relatively difficult to hydrogenate, it does not ordinarily serve as a diene in Diels-Alder reactions, and it does not undergo typical olefin reactions. Electrophilic substitutions are the most characteristic reactions, and pyrrole has often been compared with phenol or aniline in its reactivity. Acids strong enough to form salts with pyrrole destroy the aromaticity and cause polymerization.

N-Acylation is readily carried out by reaction of the alkali-metal salts with the appropriate acid chloride. *C*-Acylation of pyrroles carrying negative substituents occurs in the presence of Friedel-Crafts catalysts. Pyrrole and alkyl pyrroles can be acylated noncatalytically with an acid chloride or an acid anhydride. The formation of trichloromethyl 2-pyrryl ketone is a particularly useful procedure, since the ketonic product can be readily converted to the corresponding pyrrolecarboxylic acid or ester by treatment with aqueous base or alcoholic base, respectively.

Halogenation reactions usually involve pyrroles with electronegative substituents. Mixtures are usually obtained and polysubstitution products, ie, tetrahalopyrroles, predominate.

Pyrrole can be reduced catalytically to pyrrolidine over a variety of metal catalysts, ie, Pt, Pd, Rh, and Ni. Of these, rhodium on alumina is one of the most active.

Pyrrole oxidizes in air to red or black pigments of uncertain composition.

Functional Derivatives

Hydroxypyrroles. Pyrroles with nitrogen-substituted side chains containing hydroxyl groups are best prepared by the Paal-Knorr cyclization. Pyrroles with hydroxyl groups on carbon side chains can be made by reduction of the appropriate carbonyl compound with hydrides, by Grignard synthesis, or by insertion of ethylene oxide or formaldehyde.

Pyrrolidinones. Because of the labile hydrogen on the nitrogen, 2-pyrrolidinone (6) is not as good a solvent as 1-methyl-2-pyrrolidinone (7). Nevertheless, moderate amounts are sold as a solvent and as a plasticizer and coalescing agent for polymer emulsion coatings. There is also continuing interest in (6) as a monomer for polypyrrolidinone and as a source of 4-aminobutanoic acid. The main use of (6) is as an intermediate for the manufacture of 1-vinyl-2-pyrrolidinone (8) (see Vinyl polymers, N-vinyl).

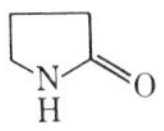

(6) 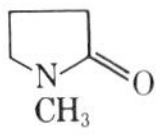   (7) 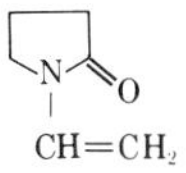   (8)

N-Methyl-2-pyrrolidinone (NMP) (7) is a dipolar aprotic solvent. It has a high dielectric constant and cannot donate protons for hydrogen bonding. All of its commercial uses involve its strong and frequently selective solvency. In recent years it has replaced other solvents of poorer stability, higher vapor pressures, greater toxicities, or more facile skin penetration.

The largest use of NMP is in extraction of aromatics from lube oils. In this application, it has been replacing phenol and, to some extent, furfural. Other petrochemical uses involve separation and recovery of aromatics from mixed feedstocks; recovery and purification of acetylenes, olefins and diolefins; removal of sulfur compounds from natural and refinery gases; and dehydration of natural gas.

Pyrrole esters. The pyrrole esters are important synthetically, since they stabilize the ring and may also act as protecting groups.

Condensed pyrroles. Pyrroles can be condensed to compounds containing two, three, or four pyrrole nuclei. These are important in synthetic routes to the tetrapyrrolic porphyrins, corroles, and bile pigments, and to the tripyrrolic prodigiosins. The pyrrole nuclei are joined by either a one-carbon fragment or a direct pyrrole–pyrrole bond.

Eugene V. Hort
L.R. Anderson
GAF Corporation

A.H. Jackson, *Compr. Org. Chem.* **4**, 275 (1979).

E. Baltazzi and L.I. Krinen, *Chem. Rev.* **63**, 511 (1963).

Q

QUATERNARY AMMONIUM COMPOUNDS

Quaternary ammonium compounds are usually tetrasubstituted ammonium salts. Originally, it was considered that the R groups were only hydrocarbon radicals attached to the nitrogen by a C–N bond, but now a large variety of substituents can be used, eg, nitrogen or oxygen atoms. The alkyl radicals may be substituted or unsubstituted, saturated or unsaturated, aliphatic or aromatic, or branched or normal chains. Also, there may be a great variety of substituents on the R group. In all cases, the nitrogen atom is pentavalent and is in the positively charged portion of the molecule. Thus, quaternary ammonium salts or hydroxides are cationic electrolytes.

Nomenclature

The quaternary ammonium salts are usually named as substituted nitrogen compounds. The expressions pyridinium and quinolinium are used for the corresponding quaternaries, for example,

is hexadecylpyridinium chloride (1).

Properties

Physical. The structure of the quaternary ammonium compound (quaternary) determines the physical properties of the material. The lowest molecular weight quaternary, ie, methylenedimethylammonium chloride, is very soluble in water and insoluble in nonpolar solvents such as ether, benzene, and aliphatic compounds. As the molecular weight of the quaternary increases, its solubility in polar solvents, including water, decreases and its solubility in nonpolar solvents increases.

Quaternary ammonium compounds have indefinite melting points.

Quaternary ammonium salts in which one of the alkyl groups contains twelve or more carbon atoms are often referred to as invert soaps, because the lipophilic portion of the molecule is cationic instead of anionic, as in sodium stearate.

Biological. One of the most important uses for quaternary ammonium compounds depends on their biological activity. Generally, optimum activity of completely aliphatic compounds is achieved if the higher aliphatic group contains a normal chain of 16–18 carbon atoms. The bactericidal activity of benzyl quaternary compounds is optimum if the higher aliphatic chain contains 14 carbon atoms. The anion has little influence except on solubility (see also Disinfectants and antiseptics).

The mechanism of the bactericidal action is closely related to the surface activity of the quaternary ammonium compound. Undoubtedly, interaction of the bactericidal agent with the cell wall interferes with the metabolic process of the organism, and this causes the inhibiting or killing action.

Quaternary ammonium compounds play an important part in biological functions. The vitamin B complex contains two components that have the quaternary nitrogen atom (see Vitamins, vitamin B_1, and vitamin B_{12}).

Preparation

The methods of preparation of quaternary ammonium compounds (2) are many and varied, depending on the structure of the final compound. The most convenient reaction is one in which the suitable tertiary amine reacts with an alkylating agent, which can be an alkyl ester.

$$RNR'' + R'''X \longrightarrow \underset{\underset{R'''}{|}}{\overset{\overset{R'}{|}}{RNR''}} + X^-$$

There are many variations in the product because of the large number of diverse starting amines and alkylating agents.

Toxicity

Because of their biocidal, algicidal, and fungicidal properties, quaternary ammonium compounds are toxic to some sewage systems.

Uses

Quaternary ammonium compounds have scores of uses because of their affinity for negatively charged surfaces. Their single largest market is as fabric softeners, involving three types of commercial product.

The second largest market for quaternary compounds is in the manufacture of organomodifed clays. The main use for compounds of this type is in the addition of organomodified clay to drilling mud to improve the lubricity and rheology of the systems (see Petroleum, drilling fluids).

RICHARD A. RECK
Armak Company

D.N. Eggenberger, F.K. Broome, R.A. Reck, and H.J. Harwood, *J. Am. Chem. Soc.* 72, 4135 (1950).

H.J. Hueck, D.M.M. Adema, and J.R. Wiegmann, *Appl. Microbiol.* 14 (3), 308 (May 1966).

Armak Quaternary Ammonium Salts, Armak Co., Chicago, Ill., 1980.

QUARTZ, SiO_2. See Silica.

QUENCHING OILS. See Petroleum products.

QUINHYDRONE. See Quinones.

QUININE. See Alkaloids.

QUINOLINE DYES. See Quinolines and isoquinolines.

QUINOLINES AND ISOQUINOLINES

Quinoline (1) and isoquinoline (2) are the two isomeric benzopyridines. They have the same relationship with pyridine that naphthalene has to benzene (see Pyridine and pyridine derivatives).

Quinoline

Physical properties. Quinoline is a colorless, highly refractive liquid with a pungent odor. It is very hygroscopic, is more soluble in hot than in cold water, and distills in steam. As a weak tertiary base (basic ionization constant, 8.9×10^{-10}, quinoline dissolves in acids and forms characteristic salts, eg, the sparingly soluble dichromate, $2C_9H_7N \cdot H_2Cr_2O_7$. Quinoline is soluble in ethanol, ethyl ether, acetone, carbon disulfide, and other common organic solvents. Some physical properties of quinoline and isoquinoline are listed in Table 1.

Reactions. Quinoline and quinoline derivatives exhibit the reactions common to benzene and pyridine. Electrophilic substitution occurs almost exclusively in the benzene ring. Nucleophilic substitution occurs in the pyridine ring. These substitutions are influenced by the ring nitrogen and the polarizabilty of different positions of the molecule by attacking reagents.

Table 1. Physical Properties of Quinoline and Isoquinolines

Table 1. Physical Properties of Quinoline and Isoquinolines

Property	Value	
	Quinoline	Isoquinoline
mp, °C	ca − 15	6.4
bp, °C	237.63	243.25
n_D	1.62928^a	1.62078^b
d^{30}, g/cm^3	1.08579	1.09101
K_a	$8.9 \times 10^{-10\,c}$	$2.5 \times 10^{-9\,d}$
viscosity (at 30°C), mPa·s (= cP)	2.997	3.2528
T_c	509	530

^{a}At 15°C.　　　　　　^{c}At 25°C.
^{b}At 30°C.　　　　　　^{d}At 20°C.

Nitration. In nitration of quinoline, the position takes the nitro group is determined by the conditions of the nitration. It has been shown that, in mixed acid, the nitronium ion attacks the protonated quinoline molecule. Under less acidic conditions of nitration, as when acetic anhydride is used with nitric acid or with dinitrogen tetroxide, the main product is 3-nitroquinoline.

Sulfonation. Sulfonation of quinoline at 220°C gives mainly 8-quinolinesulfonic acid which, when heated to 300°C, rearranges to 6-quinolinesulfonic acid.

1,2-Addition. Quinoline reacts with allylmagnesium chloride in THF in the absence of air to form 2-allyl-1,2-dihydroquinoline in 80% yield. This is a labile compound and isomerizes to 2-n-propylquinoline when heated to 170°C in an inert atmosphere.

Amination. Treatment of quinoline with barium amide in liquid ammonia affords an 80% yield of 2-aminoquinoline, which can be obtained together with 4-aminoquinoline by amination of quinoline with sodium or potassium amide in solvents, eg, xylene, toluene, or dimethylaniline.

Halogenation. Halogenation is analogous to nitration of quinoline in that 3-substitution involves electrophilic attack on the neutral quinolne molecule and 5- and 8-substitutions involve attacks on the protonated molecule.

Oxidation. Oxidation of quinoline electrolytically or by alkaline permanganate, or boiling concentrated sulfuric acid in the presence of selenium dioxide or by ozone produces quinolinic acid, which can be thermally decarboxylated to nicotinic acid (3-pyridinecarboxylic acid).

Quaternary salts. Quinoline forms quaternary salts with alkyl halides, dimethyl sulfate, and a wide variety of aliphatic and aromatic acid halides, anhydrides, or sulfonyl chlorides. The reaction with acid chlorides, eg, benzoyl chloride and potassium cyanide, gives quinoline intermediates that are 1,2-adducts of quinoline known as Reissert compounds.

Alkylation and arylation. Photochemical alkylation of quinoline by irradiation of a solution in benzene containing acetic acid gives 2-methylquinoline (20% yield), 4-methylquinoline (10% yield), and 2,4-dimethylquinoline (5% yield). A similar reductive alkylation occurs when benzyl radicals that are generated by oxidation of phenylacetic acid with $Na_2S_2O_8$, $Ti^{3+}-Na_2S_2O_8$, or $Ag^+-Na_2S_2O_8$ react reversibly with quinoline to give 2,4-dibenzyltetrahydroquinoline and other products.

Reduction. Catalytic hydrogenation of quinoline yields products in which the pyridine ring, the benzenoid ring, or both, are reduced, depending upon the absence or presence and nature of the substituents as well as upon reaction conditions. Reduction of quinoline over Raney Ni at 70–100°C and 6.1–7.1 MPa (60–70 atm) affords 70% yields of 1,2,3,4-tetrahydroquinoline, whereas an increase in temperature to 210–270°C affords 52–62% yields of decahydroquinoline.

Manufacture and synthesis. Quinoline is isolated from the chemical oil fraction of coal-tar distillates. After removal of tar acids by a caustic extraction, the acid-free oil is distilled to afford the methylnapthalene fraction (bp, 230–280°C). This fraction is washed with dilute sulfuric acid to remove the higher boiling tar bases. The tar bases are liberated from the sulfate salts with caustic and then are distilled. Properties of commercial refined quinoline are 90% min purity, distillation range of

2°C from 235 to 238°C, and sp gr at 15.5°C, 1.095. Typical composition (glc) of such a product is 92 wt% quinoline, 5 wt% isoquinoline, and 3 wt% others.

Isoquinoline, 2-methylquinoline, and 4-methylquinoline occur in the cruder fractions of tar bases that distill at 230–251°C.

Separation of quinoline and isoquinoline mixtures can be effected by chromatography or complex formation.

Quinoline and quinoline derivatives. *Skraup Synthesis.* The Skraup synthesis is a general reaction that can be used for the synthesis of many quinolines. It consists of heating primary aromatic amines with glycerol, concentrated sulfuric acid, and an oxidizing agent, eg, the nitro compound corresponding to the aromatic amine that is used, arsenic acid, or ferric chloride. The reaction proceeds through dehydration of glycerol to acrolein, addition of acrolein to the amine to form β-anilinopropionaldehyde (3), cyclization to 1,2-dihydroquinoline (4), and oxidation to quinoline (1) in 84–91% yields. The preparation of quinoline in 75% yield by the Skraup method and the use of methanesufonic acid as the dehydrating agent has been reported.

The Skraup method, which was originally designed for the synthesis of quinolines substituted in the benzene ring, has been extended to the synthesis of quinolines substituted in the heterocyclic ring. This is done by means of substituted acroleins instead of glycerol. Other syntheses follow.

Döbner-von Miller synthesis. The Döbner-von Miller synthesis is very closely related to the Skraup synthesis but is experimentally simpler and not nearly as violent. It consists of the reaction of one mole of an aromatic amine with two moles of acetaldehyde in the presence of hydrochloric acid or zinc chloride.

Conrad-Limpach-Knorr synthesis. The Conrad-Limpach-Knorr reaction involves the condensation of β-keto esters with aromatic amines. It has been applied to the synthesis of many quinoline derivatives, eg, 2- and 4-hydroxyquinolines, that are intermediates for chemotherapeutic agents (see Chermotheraputics).

Toxicology. Quinoline vapors are irritating to the eyes, nose, and throat and may cause headaches, dizziness, nausea, etc. The liquid is rapidly absorbed through the skin: it causes skin burns and eye irritation. Toxicity values are: oral LD_{50} (rat), 330–460 mg/kg; dermal LD_{50} (rat), 540 mg/kg.

Uses. *Antioxidants.* Most quinoline antioxidants are 1,2-dihydroquinoline derivatives. For example, 1,2-dihydro-6-decyl-2,2,4-trimethylquinoline, 1,2-dihydro-6-ethoxy-2,2,4-trimethylquinoline, and 1,2-dihydro-2,2,4-trimethylquinoline have been produced commercially in the United States as antioxidants, antiozonants, and stabilizers in rubber processing (see Antioxidants and antiozonants).

Polymers. Quinoline and quinoline derivatives are either added to, or incorporated in, polymers to impart ion-exchange capability (see Ion exchange).

Metallurgy. Quinoline and some of its derivatives are used in plating baths, extraction of metals from aqueous solutions, and separation of metals (see Electroplating; Extractive metallurgy).

Catalysts. Rigid foams (qv) have been prepared by treatment of unsaturated dicarboxylic acids with diols and decarboxylating the adduct at 170°C in the presence of 2 wt% quinoline.

Analytical reagents. As with the metallurgical applications, analytical applications of quinoline derivatives rely on their chelating properties. The determination of numerous metals can be performed by gravimetric methods with 8-hydroxyquinoline (see Analytical methods; chelating agents).

Isoquinoline

Isoquinoline (2) (2-benzazine, leucoline) is a heterocyclic compound formed by the fusion of a benzene and a pyridine ring with nitrogen in the 2-position. The structure is assigned on the basis of synthesis and degradation studies. Isoquinoline is oxidized to phthalic acid (5) and cinchomeronic (3,4-pyridinedicarboxylic) acid (6) on treatment with alkaline permanganate.

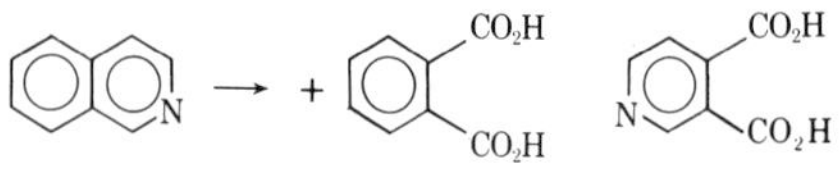

(2)   (5)   (6)

Physical properties. Isoquinoline has an odor resembling that of benzaldehyde, is a stronger base than quinoline, and reacts vigorously with alkyl halides to form quaternary salts. Selected physical constants are given in Table 1. Isoquinoline is sparingly soluble in water and volatile in steam. It dissolves readily in ethanol, ethyl ether, and the common organic solvents.

Reactions. In general, isoquinoline undergoes electrophilic substitution reactions in the 5-position and nucleophilic reactions in the 1-position. Nitration of isoquinoline with mixed acid at 0°C affords a 90/10 mixture of 5-nitroisoquinoline and 8-nitroisoquinoline and, at 100°C, the proportions are 85/15. Sulfonation of isoquinoline results in a mixture of products with isoquinoline-5-sulfonic acid as the principal product.

Direct bromination of isoquinoline hydrochloride in an inert solvent, preferably nitrobenzene, with excess bromine affords an 81% yield of 4-bromoisoquinoline.

Synthesis of isoquinoline and isoquinoline derivatives. Bischler-Napieralski reaction. The Bischler-Napieralski reaction consists of the cyclodehydration of N-acyl derivatives of β-phenethylamines to 3,4-dihydroisoquinolines with Lewis acids, eg, phosphorus pentoxide, phosphoryl chloride, polyphosphoric acid, or zinc chloride in a dry inert solvent.

Pictet-Spengler synthesis. The Pictet-Spengler synthesis involves the condensation of β-phenethylamines with carbonyl compounds in the presence of an acidic catalyst to give 1,2,3,4-tetrahydroisoquinolines.

Toxicology. Isoquinoline vapors are irritating to the eyes, nose, and throat and may cause headaches, dizziness, nausea, etc. The liquid is rapidly absorbed through the skin and it causes skin burns and eye irritation. Toxicity data include oral LD_{50}, (rat), 350 mg/kg; dermal LD_{50} (rabbit), 590 mg/kg.

Uses. Isoquinoline and isoquinoline derivatives are useful as corrosion inhibitors, antioxidants, pesticides, and catalysts, and they are used in plating baths and miscellaneous applications, such as photography, polymers, and azo dyes (qv). Numerous derivatves have been prepared and evaluated as pharmaceuticals. Isoquinoline is a main component in quinoline still-residue bases, which are sold as corrosion inhibitors and acid inhibitors for pickling of iron and steel.

Samuel N. Holter
Koppers Company, Inc.

G. Jones, "Quinolines," in A. Weissberger and E.C. Taylor, eds., *The Chemistry of Heterocyclic Compounds*, Vol. 32, Wiley-Interscience, New York, Pt. I, 1977, Pt. II, 1982.
G. Grenthe, "Isoquinolines, Pt. I," in A. Weissberger and E.C. Taylor, eds., *The Chemistry of Heterocyclic Compounds*, Vol. 38, Wiley-Interscience, New York, 1981.

QUINONES

Quinones are a class of cyclic enones whose best known example is 1,4-benzoquinone (1)

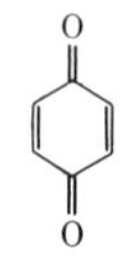

(1)

Physical Properties

Selected physical constants of various quinones are gven in Table 1.

Table 1. Physical Properties of Selected Quinones

Name	Melting point °C	Solubility Sol	Solubility Insol	Uv spectra Wavelength at max absorption (ϵ); solvent[a]	$-E_{1/2}$ 25°C,SCE[b], CH_3CN	$-E_{1/2}$ 25°C,SCE[b], $(C_2H_5)_4NClO_4$
1,2-benzoquinone	60-70 (dec)	ether, benzene	pentane	580 (30), 385 (1585); ether	0.31	0.90
3,4,5,6-tetrachloro-1,2-benzoquinone	133, 122–127				0.1	−0.71
1,4-benzoquinone	113, 116	alcohol, ether	water, pentane	243; CH_3OH	0.51	1.14
2-chloro-1,4-benzoquinone	57	water, alcohol		251 (7740); CH_3OH	0.34	0.97
2,5-dichloro-1,4-benzoquinone	161–162	ether, chloroform	water, alcohol	271 (5710); CH_3OH	0.18	0.81
2,3-dichloro-5,6-dicyano-1,4-benzoquinone	201–203 (dec)				−0.51	0.30
2,5-dimethyl-1,4-benzoquinone	125	ether, alcohol	water, alcohol	293 (1770), 251 (11650), 220 (4080); CH_3OH	0.67	1.27
2-methyl-1,4-benzoquinone	69	ether, alcohol	water	429 (19), 314 (589), 246 (13804), C_2H_5OH	0.58	1.12
2,3,5,6-tetrachloro-1,4-benzoquinone	290, 294	ether	water, ligroin	364 (248), 286 (12600); CH_3OH	−0.01	0.71
1,2-naphthoquinone	145–147	water, alcohol	ligroin	398 (1800), 336 (2280), 248 (20400); CH_3OH	0.58	1.18
3-chloro-1,2-naphthoquinone	172 (dec)	alcohol, benzene	water			
1,4-naphthoquinone	125, 128.5	alcohol, benzene	water, ligroin	330 (3020), 250 (19953), 246 (20417); CH_3CN	0.71	1.25
2,3-dichloro-1,4-naphthoquinone	193, 195	benzene, chloroform	water, alcohol	337,279, 252, 246; CH_3OH		
2-methyl-1,4-naphthoquinone	105–107	ether, benzene	water, alcohol	328, 264, 253, 249, 244; C_6H_{12}	0.77	1.28

[a] In many cases, more than one set of data is given.
[b] Saturated calomel electrode.

Chemical Properties

The quinones in biological systems play a variety of important roles. In addition to defense purposes eg, in insects, the vitamin K family members, which are based on 2-methyl-1,4-naphthoquinone, are blood-clotting agents (see Vitamins). Quinones of various degrees of complexity have antibiotic, antimicrobial, and anti-cancer activity, eg, (2) and (3) (see Antibiotics; Chemotherapeutics, antimitotic).

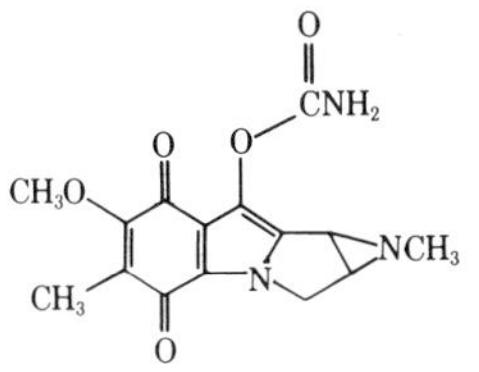

aziridinomitosene

(2)

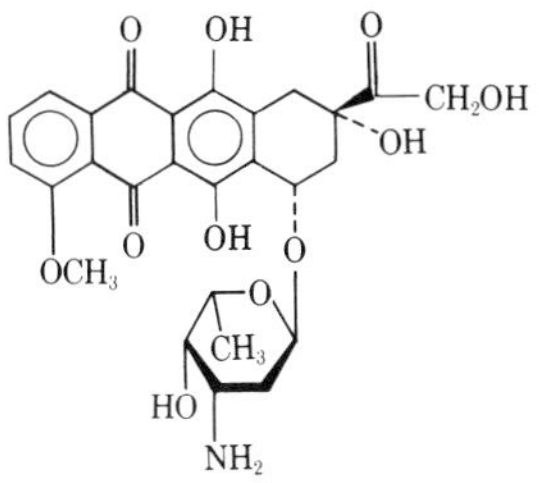

doxorubicin
(adriamycin)

(3)

The oldest and still important synthetic use of quinones is as dehydrogenation agents, especially for aromatization.

The use of 1,4-benzoquinone in combination with palladium (II) chloride converts terminal alkenes to alkyl methyl ketones in very high yield.

Quinones are extensively used in the dehydrogenation of steroidal ketones. Such reactions are marked by high yields and high selectivity.

Unlike most simple carbonyl compounds, the quinones do not yield bisulfite addition products but undergo ring addition.

A significant carbonyl reaction is the addition of tertiary phosphites under anhydrous conditions.

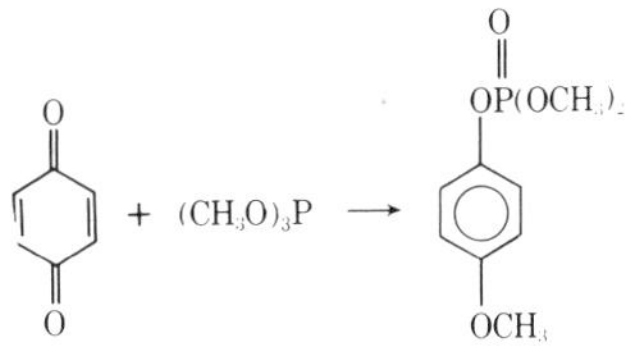

(1)

The ester product is easily hydrolyzed, and the reaction sequence provides an excellent synthesis of hydroquinone monoethers. The exact path of the reaction is uncertain.

A main area of quinone synthesis chemistry involves the nucleophilic substitution of labile groups. Most of this chemistry has involved amines. A smaller number of studies have been done with sulfur nucleophiles.

Syntheses

Syntheses of quinones often involve oxidation, since it is the only completely general method. Thus, in several instances, quinones are the reagents of choice for the preparation of other quinones. Oxidation has been especially useful with catechols and hydroquinones as starting materials. The preparative utility of these reactions depends largely on the relative oxidation potentials of the quinones.

For the preparation of ≤ 10 g of a quinone, the oxidation of a phenol with Fremy's salt (Teuber reaction) is perhaps the method of choice. A wide range of phenols has been used, including some with 4-substituents. The yield for simple phenols is frequently in excess of 70%, and some complex phenols show highly selective oxidation. With an occasional exception, substituents and side chains are not attacked by Fremy's salt. Thallium trinitrate oxidizes naphthols and hydroquinone monoethers to quinones or 4,4-dialkoxycyclohexa-2,5-dienones, respectively.

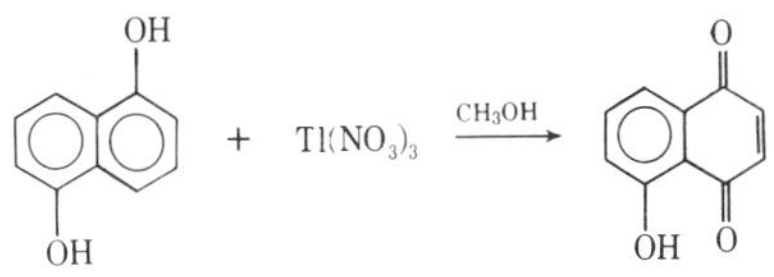

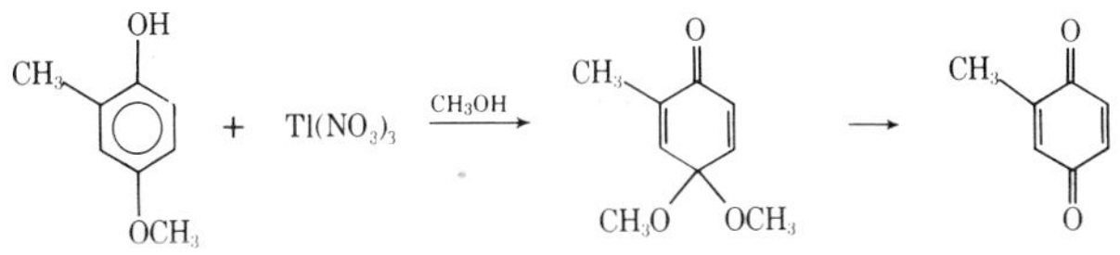

The oxidation of 4-bromophenols to quinones can also be accomplished with periodic acid.

The dimethyl ethers of hydroquinones and 1,4-naphthalenediols can be oxidized with silver(II) oxide or ceric ammonium nitrate.

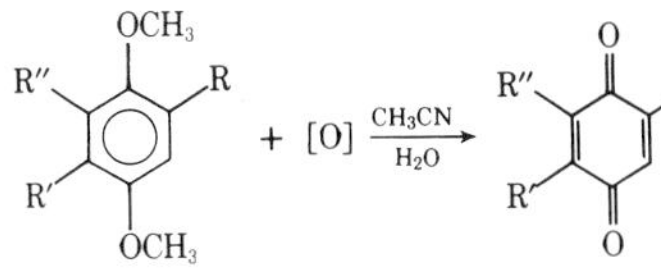

R = CH$_3$ R′,R″ = CH$_3$, H

[O] = AgO   yield = 91%
[O] = Ce(NH$_4$)$_2$(NO$_3$)$_6$ yield = 99%

The yields of quinones are excellent when pyridine- or pyrazinecarboxylic acids are used as catalysts.

Manufacture

With the exceptions of 1,4-benzoquinone and 9,10-anthraquinone, quinones do not have a substantial market, but a few of them are commercially available (see Anthraquinone). The few large-scale preparations involve oxidation of aniline, phenol, or aminonaphthols. In the case of 1,4-benzoquinone, the product is steam-distilled, chilled, and obtained in high yield and purity. The direct oxidation of the appropriate unoxygenated hydrocarbon has been described for a large number of ring systems, but it is generally utilized only for the polynuclear quinones.

Uses

Uses are listed in Table 2.

Table 2. Uses of Some Quinones

Quinone	Use
1,4-benzoquinone	oxidant, amino acid determination
2-chloro-, 2,5-dichloro-, and 2,6-dichloro-1,4-benzoquinones	bactericides
2,3-dichloro-5,6-dicyano-1,4-benzoquinone chloranils	oxidation and dehydration agent intermediate, oxidant
2-methyl- and 2,3-dimethyl-1,4-naphthoquinones	vitamin K substitutes, antihemorrhagic agents
2,3-dichloro-1,4-naphthoquinone	intermediate, fungicide

Health and Safety Factors

Because of the high vapor pressures of the simple quinones and their penetrating odor, adequate ventilation should be provided in areas where they are handled or stored. Quinone vapor can harm the eyes, and a limit of 0.1 ppm of 1,4-benzoquinone in air has been recommended. The solid or solutions can cause severe local damage to the skin and mucous membranes.

K. THOMAS FINLEY
State University College at Brockport

S. Patai, ed., *The Chemistry of the Quinoid Compounds*, Wiley-Interscience, New York, 1974.

R.H. Thomson, *Naturally Occurring Quinones*, 2nd ed., Academic Press, Inc., New York, 1971.

R.A. Morton, ed., *Biochemistry of Quinones*, Academic Press, Inc., New York, 1965.

R

RADIATION CURING

The use of electromagnetic radiation to alter the physical and chemical nature of a material is sometimes termed radiation-curing technology. The following discussion of radiation curing concerns processes that involve interaction of electromagnetic radiation with organic substrates to develop cross-linked or solvent-insoluble network structures. In general, radiation-curing technology involves consideration of at least four main variables: type of radiation source; type of organic substrate to be irradiated; kinetics and mechanisms of radiation energy; organic substrate interactions; and final chemical, physical, and mechanical properties of network formation.

Radiation and Electromagnetic Radiation Sources

Radiation curing, as applied to cross-linking of polymers or coating materials, involves the full spectrum of electromagnetic radiation energies to effect chemical reaction. These forms of radiation energy include ionizing radiation, ie, α, β, and γ rays from radioactive nuclei; x rays; high energy electrons; and nonionizing radiation such as are associated with uv, visible, ir, microwave, and radio-frequency wavelengths of energy (see Table 1).

Curing of Polymers with γ-Ray, X-Ray, and High Energy Electron Sources

Radiation curing of preformed polymers with ionizing-radiation processing equipment can result in two types of chemical change that are associated with cross-linking and degradation reaction mechanisms. Cross-linking reaction mechanisms on preformed polymer substrates usually involve removal of hydrogen atoms to form a macroradical intermediate. These macroradical intermediates can then couple to form a single molecule. This coupling results in an increase in the original average molecular weight of the starting polymer. If irradiation continues, the original polymer substrate is transformed into one gigantic molecule of infinite molecular weight with lower solvent solubility, higher melting points, and improved physical properties over the original material. Enhancement of cross-linking can be facilitated through the use of multifunctional vinyl monomers or oligomers which copolymerize and propagate much more rapidly than in a direct recoupling reaction to form greater amounts of gel or cross-linked material at lower dose rates and shorter reaction times.

High energy electron- and light-energy radiation-curable coatings generally are of multifunctional acrylic or methacrylic unsaturated polymers. They differ from conventional coatings in that the solvents for the polymers are high boiling, nonvolatile, and 100% corrective with themselves and with all of the other organic components in the film. The curing process for these coatings is a free-radical chain reaction. Ionizing radiation from the processing equipment is absorbed directly in the coating where the free radicals form uniformly in depth. Since electron energies of only 100 eV or less are required to break chemical bonds and to ionize or to excite components of the coating system, the shower of scattered electrons produced in the liquid coating leads to a uniform population of free radicals throughout the coating. These initiate the polymerization and the polymerization process results in a dry, three-dimensional cross-linked coating.

Electron beam. An electron-beam processing unit consists mainly of a power supply and an electron-beam acceleration tube. The power supply increases and rectifies line current and the accelerator tube generates and focuses the beam and controls the electron scanning.

Multiple planar-cathode processors. The design criterion for this electron-accelerator system is a planar array of concentrated cathode-control grid elements. The modular cathode construction allows for broad-beam (250 cm wide) processing materials with powers of 30 kGy (3 Mrad) at 300 m/min.

Curing with Ultraviolet, Visible, and Infrared Processing Equipment

Polymers. Upon direct absorption of uv or visible wavelengths of light, polymer substrates undergo chain scission and cross-linking. Cross-linking or curing of preformed polymeric materials, ie, of thermoplastics, or can be markedly enhanced through used of special photosensitive molecules that are mixed into the polymer matrix or that chemically attach to the backbone of the polymer chains. These special photosensitive molecules, when compounded into the preformed polymer matrix, can undergo light-induced radical abstraction or insertion reactions which result in coupling of the polymer chains and in network formation (see also Photoreactive polymers; Uv stabilizers).

Radiation curing of polymers with uv and visible-light energies is used widely in photoimaging and photoresist technology. Infrared processing is involved with thermoforming or heat-bonding of thermoplastic polymeric materials. These polymer heat-forming or melting processes do not usually cure the polymer but only cause physical changes and maintain original polymer thermoplastic characteristics.

Light source. The light source normally used in commercial photocuring reactions is the medium pressure mercury-arc lamp enclosed in a quartz or Vicor envelope. These lamps may contain electrodes for electrical to light-energy conversion or may be electrodeless, in which case a radio-frequency wave causes mercury-atom excitation and subsequent light emission (see Photochemical technology).

VINCENT D. MCGINNISS
Battelle Columbus Laboratory

V.D. McGinniss, L.J. Nowacki, and S.V. Nablo, *ACS Symposium No. 107*, 1979, pp. 51–70.

J.W.T. Spinks and R.J. Woods, *An Introduction to Radiation Chemistry*, John Wiley & Sons, Inc., New York, 1964.

V.D. McGinniss, "Light Sources," in S.P. Pappas, ed., *UV Curing Science and Technology*, Technology Marketing Corporation, Stamford, Conn., 1978, pp. 96–132.

A.F. Readdy, Jr., *Plastics Fabrication by Ultraviolet, Infrared, Induction, Dielectric and Microwave Radiation Methods*, Plastic Report R43, Plastics Technical Evaluation Center, Picatinny Arsenal, Dover, N.J., 1972.

RADIOACTIVE DRUGS

Radioactive drugs are useful as diagnostic or therapeutic agents by virtue of the physical properties of their constituent radionuclides. Thus, their utility is not based on any pharmacologic action. Today, most clinically used drugs of this class are diagnostic agents incorporating a gamma-emitting nuclide which, because of their physical or metabolic properties, localize in a specific organ after intravenous injection. Images reflecting organ structure or function are then obtained by means of a scintillation camera which detects the distribution of ionizing radiation emitted by the radioactive drug. The principal isotope used in clinical diagnostic nuclear medicine is reactor-produced metastable technetium-99m (^{99m}Tc). It either is injected directly as sodium pertechnetate ($NaTcO_4$) or is added to instant kits consisting of nonradioactive carrier molecules, to which it spontaneously binds to form the radiopharmaceutical product (see section on Kits).

Table 1. Electromagnetic Spectrum

Types of radiation	Wavelengths, nm	Frequency, Hz	Energy, eV
gamma ray	10^{-4}–10^{-2}	10^{19}–10^{22}	10^5–10^8
electron beam	10^{-3}–10^{-1}	10^{18}–10^{21}	10^4–10^7
x ray	10^{-2}–10	10^{16}–10^{19}	10^2–10^5
ultraviolet	10–400	10^{15}–10^{16}	5–10^2
visible	400–750	10^{15}	1–5
infrared	750–10^5	10^{12}–10^{14}	10^{-2}–1
microwave	$> 10^6$	10^{11}–10^{12}	$< 10^{-2}$
radio frequency	$> 10^6$	$< 10^{11}$	$< 10^{-2}$

Diagnostic Radioactive Drugs for Imaging

Design of diagnostic radioactive drugs requires the combination of low toxicity, specific biodistribution, low radiation dose, and radionuclidic emissions compatible with currently available instrumentation. Since the sensitivity of scintillation cameras, also called gamma cameras, is great, only trace amounts of the radioactive drug need be administered, and the potential for toxicity is thus reduced. Rapid excretion also minimizes radiation dose to the patient.

Reactor produced. Nuclear-reactor production of radioisotopes is limited to neutron reactions (see Nuclear reactors; Radioisotopes). Reactor-production processes of radiopharmaceutical interest are basically of two types: fission of a heavy nucleus upon interaction with a neutron to form two or more fission fragments of greater stability that are neutron-rich and that decay by beta emission, and reactions of the type (n, γ) which, for a monoisotopic or highly enriched target, yield mainly a single product.

Cyclotron produced. By accelerating charged particles of specific energy to collide with targets, cyclotrons produce nuclides on the proton-excess side of the region of stability in a proton–neutron configuration plot of the nuclides. In most commercial cyclotrons, protons bombard the targets, although deuterons, alpha and ^{3}He particles are also used in research machines.

Thallium (^{201}Tl) has become the most widely used cyclotron-produced clinical radioisotope by virtue of its tendency to localize in the heart and thereby reflect regional blood flow. Thallium-201 decays by electron capture with a half-life of 73 h and emits low abundance 135- and 167-keV gamma rays and high abundance (94.5%) imageable mercury K-x-rays of 69–83 keV.

Kits. Radiopharmaceutical kits are composed of sterile, nonpyrogenic, nonradioactive carrier materials that are configured so that they can be labeled by the aseptic addition of a radioisotope. All kits in use in North America are designed primarily for use in conjunction with the molybdenum–technetium generator and for labeling with its product, ie, sodium pertechnetate (^{99m}TcO$_4^-$) in sterile saline. In general, tin is the catalyst in the technetium labeling process. Although the exact chemistry of ^{99m}Tc-labeling is unknown in all cases, it is clear that Sn(II) acts to reduce TcO$_4$ to the Tc(IV) oxidation state, where labeling occurs. The carrier ligand competes for binding of Tc(IV), which tends to form spontaneously insoluble and unreactive TcO$_2$.

A typical kit is methylenediphosponate for bone imaging. It is available commercially as a sterile, nonpyrogenic lyophilized powder in vials suitable for reconstitution with Na^{99m}TcO$_4$ to form ^{99m}Tc–MDP which localizes the skeleton after injection. Kits contain 10 mg ^{99m}Tc–MDP and 0.85 mg stannous chloride dihydrate whose pH has been adjusted to pH 7.0–7.5. Vial contents are stored under nitrogen to inhibit stannous oxidation (see Medical diagnostic reagents).

Radioimmunoassay

Radioimmunoassay (RIA) is the generic term for systems of quantitative *in vitro* measurement based on the principle of saturation analysis, displacement analysis, or competitive protein binding. Since they are extremely sensitive and specific, RIA techniques are used widely for the determination of drug and hormone levels in biological fluids. The development of this field stems from the observation that unlabeled insulin displaces ^{131}I-labeled insulin from insulin antibody *in vitro*. With the antibody concentration and radioiodinated antigen held constant, the binding of the label is quantitatively related to the amount of unlabeled antigen that is added. Thus, in the insulin RIA, known insulin standards are used initially to prepare a plot of the fraction of bound ^{131}I-insulin against the concentration of insulin added.

In general, RIAs involve the separation of a labeled antigen of interest into bound and unbound fractions after its interaction with an antibody in the presence of an unknown quantity of unlabeled antigen. Homogeneous assays that do not require a separation step are becoming increasingly popular.

Allan M. Green
Irwin Gruverman
New England Nuclear Corporation

F.P. Castronovo in D. Rollo, ed., *Nuclear Medicine Physics, Instrumentation and Agents,* C.V. Mosby Company, St. Louis, Mo., 1977, pp. 560–636.

J.H. Howanitz and P.M. Howanitz in J.B. Henry, ed., *Clinical Diagnosis and Management,* 16th ed., W.B. Saunders Company, Philadelphia, Pa., 1979, pp. 385–401.

L.M. Freeman and P.J. Hohnson, *Clinical Scintillation Imaging,* 2nd ed., Grune and Stratton, New York, 1975.

RADIOACTIVE TRACERS

Properties

Any radioactive element can be used as a radioactive tracer, eg, chromium-51, cobalt-60, tin-113, and mercury-203, but the preponderance of use is with carbon-14, hydrogen-3, sulfur-35, phosphorus-32, and iodine-125. By far the greater number of radioactive tracers produced are based on carbon-14 and hydrogen-3 since these atoms exist in almost all known natural and synthetic chemical compounds.

Syntheses

Syntheses of radioactive tracers involve all of the classical biochemical and synthetic chemical reactions used in the synthesis of nonradioactive chemicals. There are, however, specialized techniques and considerations required for the safe handling of radioactive chemicals, strategic synthetic considerations in terms of their relatively high cost, and synthesis scale constraints governed by specific activity requirements.

Basic precursor materials, eg, carbon dioxide (^{14}C), benzene (^{14}C), methyl iodide (^{14}C), sodium acetate (^{14}C), sodium cyanide (^{14}C), etc, require vacuum-line handling in well-ventilated fume hoods. Tritium gas and methyl iodide (^{3}H), iodine, and tritium, which are the most difficult of the isotopes to contain, must be handled in specialized closed systems. Sodium sulfate (^{35}S) and sodium iodide (^{125}I) must be handled similarly in closed systems to avoid the liberation of volatile sulfur oxides (^{35}S) and iodine (^{125}I).

A multistep synthesis is strategically designed such that the labeled species is introduced as close to the last synthetic step as possible in order to minimize yield losses and cost. Use of indirect reaction sequences frequently maximizes the yield of the radioactive species at the expense of time and labor.

Biosynthetic techniques are ideally suited for the synthesis of many radiolabeled compounds. Plants, eg, potato and tobacco, when grown in an exclusive atmosphere of radioactive carbon dioxide, utilize the labeled chemical as their sole source of carbon. After a suitable period of growth, almost every carbon atom in the plant is radioactive. Thus, plants can serve as an available source of labeled carbohydrates.

The introduction of tritium into molecules is most commonly achieved by reductive methods, including catalytic reduction by tritium gas (^{3}H$_2$) of olefins, catalytic reductive replacement of halogen (Cl, Br, or I) by ^{3}H$_2$, and metal hydride (^{3}H) reduction of carbonyl compounds (see Deuterium and tritium).

Detection and Quantification

The methods for detection and quantification of radiolabeled tracers are determined by the type of emission (β or γ) the tracer affords, the energy of the emission, and the efficiency of the system by which it is measured. Detection of radioactivity can be achieved in all cases with the Geiger counter. However, in the case of the weaker-emitting isotopes, ie, ^{3}H, ^{14}C, and ^{35}S, large amounts of isotopes are required for detection of a signal. This is in most cases undesirable and impractical. Thus, more sensitive methods of detection and quantitation have been developed.

Liquid-scintillation counting is by far the most common method of detection and quantitation of β-emission. This technique involves the conversion of the emitted β-radiation into light by a solution of a mixture of fluors (the liquid scintillation cocktail), and the sensitive detection of this light by a pair of matched photomultiplier tubes in a dark chamber which is amplified, measured, and recorded with the liquid-scintillation counter.

Health and Safety Factors

Depending on the quantities used and type of operation, the more energetic emissions of ^{32}P (β-ray) and ^{125}I (γ- and x rays) may require appropriate shielding to minimize personnel exposure. The energy of the β-rays of ^{35}S, ^{14}C, and ^3H are weak enough to require no shielding. The use of closed systems, well-ventilated work areas (ie, with hoods and glove boxes), disposable gloves, disposable lab coats, etc, and neat work habits provide a safe working environment.

Uses

The detectability of minute quantities of a radiolabeled tracer makes possible the determination of micro quantities of substances. The most effective use of the radiotracer has been in biomedical research. A radiolabeled, nonmetabolized tracer for glucose, ie, 2-deoxyglucose (1-^3H), is administered to a test animal to identify areas of brain activity, ie, of glucose metabolism, corresponding to particular external stimuli. An external stimulus is given and the animal is sacrificed. The brain is frozen, sectioned, and exposed to x-ray film (autoradiography), and the location of the radioactivity is noted. In this way it is possible to relate the areas and to produce a brain map.

ROBERT E. O'BRIEN
New England Nuclear Corporation

Y. Wang, *Handbook of Radioactive Nuclide*, The Chemical Rubber Company, Cleveland, Ohio, 1969, pp. 16–63.

A. Murray and D. Williams, *Organic Synthesis with Isotopes*, Interscience Publishers, New York, 1958.

J. Shapiro, *Radiation Protection*, Harvard University Press, Cambridge, Mass., 1972.

RADIOACTIVITY, NATURAL

The spontaneous emission of penetrating energetic radiations by matter, now called radioactivity, was discovered accidentally in 1896 by Becquerel, who was seeking a connection between x rays, discovered the preceeding year, and phosphorescence. The photographic effect of a uranium salt caused by penetrating radiation was found to be independent of prior illumination and to be associated with the uranium, regardless of its chemical or physical state. In 1898 the Curies discovered polonium and radium. Marie Curie and Schmidt discovered independently in the same year that thorium is also naturally radioactive. Other primary radionuclides are potassium, rubidium, platinum, and samarium.

The four categories of natural radionuclides are primary, secondary, induced, and extinct. The primaries are those whose survival through geologic history is owing to their slow decay and long lifetimes (see Table 1). The secondaries owe their existence to continual generation by primaries. The induced natural radionuclides, eg, ^{14}C and ^3H, stem from cosmic-ray-induced nuclear reactions in the earth's atmosphere (see Table 2). The extinct nuclides are those whose lifetimes are too short for survival from their pre-solar-system production, yet long enough for them to have left observable effects in nature, especially in meteorites.

Others soon discovered a host of radioactive substances associated with uranium and thorium, including some that behaved chemically as inert gases, called emanations (now collectively called radon).

Following a prediction by Libby that cosmic-ray-induced nuclear reactions in the earth's atmosphere should produce ^{14}C and ^3H, he and his collaborators succeeded in detecting their natural occurrence. Carbon-14 and hydrogen-3 were the first of many induced natural radionuclides now known and provided the basis for the ^{14}C- and tritium-dating methods (see below).

Genetic relationships. The three natural radioactive series are those of ^{238}U, ^{235}U, and ^{232}Th. By 1903, Rutherford and Soddy had deduced that the radioactive emission of energetic radiations was an accompaniment of spontaneous transformation of an atom from one kind to another, generally of a different element. The α particles were deduced and promptly demonstrated to be helium ions, which explained the high abundance of helium in uranium and thorium minerals. The atomic weight of the transforming atom must then be reduced by about 4 units, the atomic weight of helium. The particles were found to be high speed electrons, whereas the radiation was found to be electromagnetic in nature; in neither case was the atomic weight appreciably affected. Soddy recognized in 1910 that, in several cases, two or more substances of different radioactive properties (radiations, half-lives) had identical chemical properties. In 1913, he introduced the concept of isotopes, ie, atomic species of the same element that differ in atomic weight. Concurrently, he formulated the displacement laws: α transformation results in a displacement two places downward in the periodic table, whereas β transformation results in a displacement upward by one place.

Table 1. Known Primary Natural Radionuclides

Nuclide	Disintegration modes (%)	Half-life, yr	Abundance, %	Element specific activity, Bq/g^a	Isotopic decay products (isotopic abundance, %)
^{235}U	$7\alpha + 4\beta^-$	$7.038 \times 10^{8\,b}$	0.720	5.685×10^2	$^{207}Pb^c$ (22.1)
	SF^d	3.5×10^{17}		1.14×10^{-6}	$FP^{e,c}$
^{40}K	β^- (89.5)^b	$1.250 \times 10^{9\,b}$	0.01167^b	28.3	$^{40}Ca^c$ (96.94)
	EC (10.5)^b			3.32	$^{40}Ar^c$ (99.60)
	β^+ (0.0010)			3.2×10^{-4}	$^{40}Ar^c$ (99.60)
^{238}U	$8\alpha + 6\beta^-$	$4.468 \times 10^{9\,b}$	99.275	1.2346×10^4	$^{206}Pb^c$ (24.1)
	SF^d	8.2×10^{15}		6.7×10^{-3}	$FP^{e,c}$
^{232}Th	$6\alpha + 4\beta^-$	$1.401 \times 10^{10\,b}$	100^f	4.07×10^3	$^{208}Pb^c$ (52.4)
	$SF^{d,g}$	$> 1 \times 10^{21}$		$< 6 \times 10^{-8}$	FP^e
^{176}Lu	β^-	3.6×10^{10}	2.61	5.5	$^{176}Hf^c$ (5.2)
^{187}Re	β^-	4×10^{10}	62.60	1.1×10^3	$^{187}Os^c$ (1.6)
^{87}Rb	β^-	$4.88 \times 10^{10\,b}$	27.83	8.8×10^2	$^{87}Sr^c$ (7.0)
^{147}Sm	α	1.06×10^{11}	15.1	1.25×10^2	$^{143}Nd^c$ (12.2)
^{138}La	EC (68)	1.1×10^{11}	0.089	0.52	^{138}Ba (71.7)
	β^- (32)			0.25	^{138}Ce (0.25)
^{190}Pt	α	6×10^{11}	0.013	1.5×10^{-2}	^{186}Os (1.58)
^{152}Gd	α	1.1×10^{14}	0.21	1.6×10^{-3}	^{148}Sm (11.3)
^{115}In	β^-	5.1×10^{14}	95.7	0.22	^{115}Sn (0.38)
^{174}Hf	α	2.0×10^{15}	0.16	6×10^{-5}	^{170}Yb (3.2)
^{144}Nd	α	2.1×10^{15}	23.8	1.0×10^{-2}	^{140}Ce (88.5)
^{148}Sm	α	8×10^{15}	11.3	1.2×10^{-3}	^{144}Nd (23.8)
^{113}Cd	β^-	9×10^{15}	12.2	1.6×10^{-3}	^{113}In (4.3)
^{82}Se	double-β^-	1.4×10^{20}	9.2	1.1×10^{-7}	$^{82}Kr^c$ (11.6)
^{130}Te	double-β^-	2×10^{21}	34.5	2×10^{-8}	$^{128}Xe^c$ (4.1)
^{128}Te	double-β^-	1.5×10^{24}	31.7	2.2×10^{-11}	$^{130}Xe^c$ (1.91)
all α emitters					$^4He^c$ (99.99986)

^{a} Bq = disintegration per second.

^{b} Value recommended for geo- and cosmochemistry by International Union of Geological Sciences, Subcommittee on Geochronology, 1977.

^{c} Decay product has been observed with enhanced isotopic abundance.

^{d} SF = spontaneous fission. Half-life is that which would result if SF were only disintegration mode.

^{e} FP = fission products, radioactive and stable.

^{f} In Th from natural sources containing U, some ^{230}U is also present.

^{g} Not observed.

Table 2. Observed Induced Natural Radionuclides

Nuclide	Half-life, d^a	Observation^b	Nuclide	Half-life, yr	Observation^b
$^{34m}Cl^c$	32.0 min	T	^{22}Na	2.60	T, M, L
^{38}Cl	37.3 min	T	^{55}Fe	2.7	M, L
^{39}Cl	56 min	T	^{60}Co	5.27	M, L
^{31}Si	2.62 h	T	^3H	12.33	T, M, L
^{38}S	2.8 h	T	^{44}Ti	47	M, L
^{24}Na	15.0 h	T	^{32}Si	108	T, M
^{28}Mg	21.0 h	T	^{39}Ar	269	T, M, L
^{198}Au	2.70	T	^{14}C	5370	T, M
^{199}Au	3.14	T	^{94}Nb	2.0×10^4	T
^{52}Mn	5.59	M	$^{239}Pu^d$	2.41×10^4	M
^{32}P	14.3	T, M	^{59}Ni	7.5×10^4	T, M, L
^{48}V	16.0	M, L	$^{233}U^d$	1.59×10^5	T
^{33}P	25.3	T, M	^{60}Fe	$\sim 2 \times 10^5$	M
^{51}Cr	27.7	M, L	^{81}Kr	2.1×10^5	M
^{37}Ar	35.0	T, M, L	^{36}Cl	3.0×10^5	T, M, L
^{59}Fe	44.6	M	^{26}Al	7.2×10^5	T, M, L
^7Be	53.3	T, M, L	^{10}Be	1.6×10^6	T, M, L
^{58}Co	70.8	M	$^{237}Np^c$	2.14×10^6	T
^{56}Co	78.8	M, L	^{53}Mn	3.7×10^6	T, M, L
^{46}Sc	83.8	M, L	^{40}K	1.25×10^9	M
^{35}S	87.4	T			
^{45}Ca	165	M			
^{57}Co	271	M, L			
^{54}Mn	312	M, L			
^{49}V	330	M, L			

^{a} Unless otherwise noted.

^{b} T = terrestrial materials; M = meteorites; L = lunar materials.

^{c} m = metastable.

^{d} Also radioactive precursors and descendants.

The concept of nuclear charge number and its identification with elementary atomic number Z was the result of Moseley's x-ray studies in 1913–1914. Mass-spectrographic studies in the 1920s permitted a distinction between mass number A and atomic weight. The discovery of the neutron in 1932 led to the concept of neutron number N, which is simply A minus Z. In terms of these quantities, the displacement laws assume their modern formulation:

α decay: $\Delta Z = -2$ $\Delta A = -4$ $\Delta N = -2$
β^- decay: $\Delta Z = +1$ $\Delta A = 0$ $\Delta N = -1$

The emission of γ radiation was recognized as an adjustment of energy content subsequent to an α or β transformation (an isomeric transition, IT).

γ emission: $\Delta Z = 0$ $\Delta A = 0$ $\Delta N = 0$
IT: $\Delta Z = 0$ $\Delta A = 0$ $\Delta N = 0$

Radioactive disintegration by positron (β^+) emission was discovered along with the phenomenon of artificial radioactivity by Frederic Joliot and Irene Curie in 1934. Four years later, the occurrence of electron capture, EC, was established. These processes follow identical displacement laws.

β^+ decay: $\Delta Z = -1$ $\Delta A = 0$ $\Delta N = +1$
EC: $\Delta Z = -1$ $\Delta A = 0$ $\Delta N = +1$

Temporal variations in activity. In 1902, Rutherford and Soddy formulated the law of radioactive decay, an exponential decay.

Radioactive disintegrations occur randomly, and the disintegration constant is actually a statistical probability that an atom disintegrates per unit time; this theoretically justifies the Rutherford-Soddy decay law. It follows that, when the number of atoms is essentially constant, the standard deviation of the number of disintegrations in a given time is the square root of that number.

Radiations. For details of the disintegration schemes and radiations of the heavy natural radionuclides, the 1978 *Table of Isotopes* should be consulted (see Isotopes).

The α-particle energies for ground-state-to ground-state transitions encountered among the heavy natural radionuclides range from 4.01 MeV (^{232}Th) to 8.78 MeV (^{212}Po). In general, they increase down the disintegration chains because of the special nuclear stability of $Z = 82$ (Pb) and $N = 126$ configurations. Because of the generally close spacing and large number of energy levels in heavy nuclei, many α emitters have complex spectra consisting of from two to many groups of particles of discrete energies. In addition to the overlapping, continuous components of the β^- particles of nuclear origin, the negative-electron spectra include conversion electrons of discreet energies from lull energy γ transitions and from high energy γ transitions. In addition to true γ rays of nuclear origin, there are x rays of atomic origin and weak continuous inner-bremsstrahlung components.

Primary Natural Radionuclides

All primary natural radionuclides whose radioactivity or transformation has been detected are listed in Table 1, along with decay-product particulars that are relevant to geochronology (see Radiometric Dating).

Induced Natural Radioactivity

There are two main sources of nuclear reactions that occur in nature in materials accessible for observation: α particles and cosmic rays. The particles emitted by heavy natural radionuclides, particularly by the short-lived end members, are sufficiently energetic to induce nuclear reactions in light elements; oxygen is the most abundant target. In addition, the spontaneous fission of ^{238}U yields neutrons directly. Uranium and thorium are the most abundant natural targets in n-rich regions, and the induced fission of ^{235}U enhances the neutron flux. Cosmic rays are protons and nuclei of helium and heavier elements accelerated in space or remote astrophysical sites to very high energies.

Production of Natural Radioactive Substances

Radium. After the discovery of radioactivity, the chief motivation for locating, mining, and processing uranium ores was their radium (^{226}Ra) content. Production was greatly accelerated during World War II. After the development of nuclear reactors and the widespread production of artificial radionuclides having a variety of properties and low costs, commercial radium production virtually ceased. Today, most commercial radium is recycled from hospitals and industrial users who have changed to other radiation sources.

The commercial production of radium followed the procedure developed by Marie Curie. Typically, ground ore is dissolved in a mixture of nitric and sulfuric acids with some barium. The insoluble sulfates of lead, barium, and radium and the siliceous gangue are removed by filtration, and $PbSO_4$ and some silica are leached out with boiling NaOH or NaCl. The residue is converted to barium and radium carbonates in an autoclave with Na_2CO_3, filtered, dissolved in HCl, and filtered from the remaining silica. The radium is separated from barium and purified by extensive fractional crystallization.

Radon. Sealed radon (^{222}Rn) sources are preferred to radium itself for many uses for reasons of economy, logistics, and small physical size. Such preparations are made at central facilities that serve both medical and industrial users.

Typically, ca 1 g radium is kept in solution, or preferably as a highly emanating solid. Semiautomatic devices have been developed for extracting and purifying the radon and compressing it into long capillary tubes. Glass capillaries can be sectioned by a fine torch and metal ones by a crimping cutter to yield sources of desired strength.

Other nuclides. Protactinium (^{231}Pa) and actinium (^{227}Ac) can be extracted from radium-process residues in which they are concentrated.

Highly concentrated ^{210}Pb is obtained by crushing or dissolving spent radon needles or radiographic sources and purifying the lead by radiochemical techniques. Either can be used for the extraction of polonium (^{210}Po).

The thorium extracted from high grade uranium minerals is predominantly ^{232}Th by mass but ionium (^{230}Th) by activity.

Radiometric Dating

Radiometric dating is based on the functional relationship between the rate of nuclear transformation, the time during which the transformation has occurred, and the extent of the transformation. If the rate and the extent can be determined, the time can be calculated.

Radioactive transformations occur at fixed rates expressed as disintegration constants (or half-lives). The extent is measured by mass spectral measurement of the amount of radioactive decay produced.

In all these methods, the transforming radionuclide is usually available for measurement. Geochronometric methods include the rubidium–strontium method, the uranium and thorium–lead methods, the helium method, the potassium–argon method, the samarium–neodymium method, the fission-track method, the potassium–calcium method, the rhenium–osmium method, and the lutetium–hafnium method.

Uses

Medical radiology. Bacquerel found that γ rays, like x rays, could produce absorption-contrast radiographs, but they never became competitive with the latter for medical-diagnostic radiography (see X-ray technology). Early workers established that radioactive radiations were damaging to living tissue and could be lethal to micro- and macro-organisms. Radioactive substances are very useful for medical therapy, particularly cancer treatment; however, many early experimenters, radiologists, and patients died of radiation-induced cancer.

Radium and radon are used mainly because of their radiations. Sealed radium sources develop equilibrium levels of radiation in about a month. Radon sources develop their maximum intensity within hours, then decay at the 3.823-d $t_{1/2}$. Implanted sources are often prepared to deliver the desired doses over the radon lifetime and left in place indefinitely.

Industrial radiography. Radiography for technological and industrial purposes was dominated by x rays until it was demonstrated that high energy radium-series rays could radiograph metallic objects of great thickness (see Nondestructive testing).

As with medical sources, radon offers practical advantages over radium, but both have been largely displaced by synthetic radionuclides, especially ^{60}Co, ^{137}Cs, ^{192}Ir, and by high energy electron accelerators.

Luminescent materials. An important use of concentrated α-emitting substances has been as radiation sources for phosphors—usually ZnS—for illuminating clock, watch, and instrument dials, for signs visible in the dark, and similar applications. This usage increased during World War II, but subsequently α emitters have been almost completely replaced by artificial β emitters (see Luminescent materials, phosphors). Radioactive tracers (qv) are especially useful when the radioactive emissions are great relative to the mass of the compound.

Ionization devices. Static electricity is a serious problem in many industrial operations. Ionization of the ambient air provides a conduction path for discharging static and preventing its buildup. Air can be ionized in a separate chamber or tube and blown over the devices or workspace. The preferred radiation source is ^{210}Po; ^{3}H is also used. The polonium is incorporated in ceramic microspheres bound to a tape with epoxy resin.

TRUMAN P. KOHMAN
Carnegie-Mellon University

C.M. Lederer and V.S. Shirley, eds., *Table of Isotopes*, 7th ed., John Wiley & Sons, Inc., New York, 1978.

G. Friedlander, J.W. Kennedy, E.S. Macias, and J.M. Miller, *Nuclear and Radiochemistry*, 3rd ed., John Wiley & Sons, Inc., New York, 1981.

E.K. Hyde, I. Perlman, and G.T. Seaborg in *The Nuclear Properties of the Heavy Elements*, Vol. II, Prentice-Hall, Inc., Englewood Cliffs, N.J., 1964, Chapt. 6, pp. 409–576.

RADIOCHEMICAL ANALYSIS AND TRACER APPLICATIONS. See Radioactive tracers.

RADIOCHEMICAL TECHNOLOGY. See Radiation curing.

RADIOGRAPHY. See X-ray technology.

RADIOIMMUNOASSAY. See Radioactive drugs.

RADIOISOTOPES

Radioactivity was discovered in 1896 by Becquerel following the discovery of x rays by Roentgen in the previous year. The penetrating radiation emitted was later shown to consist of the particulate α and β rays and the electromagnetic γ rays. By 1898 β rays were shown to be identical to electrons, and by 1909 α rays were shown to be identical to nuclei of the element helium.

In 1911 Soddy established the existence of chemically identical atoms with different atomic weights, for which he proposed the name isotopes (qv).

Decay Constants, Half-Lives

Radioisotopes decay spontaneously by the emission of particles or electromagnetic radiation. In the context of this article, the term radioisotope, whether a ground state or an isomeric state, is characterized by a probability of decay per unit time, λ, also called the total radioactive decay constant. For radioisotopes that decay by more than one mode, a partial decay constant, λ_i, is associated with each mode; thus $\lambda = \Sigma_i \lambda_i$. An example of such a nuclear species is ^{40}K, which decays by β-emission to ^{40}Ca ($\lambda_\beta = 4.86 \times 10^{-10}$ yr^{-1}), and by electron-capture decay to ^{40}A ($\lambda_\epsilon = 5.70 \times 10^{-11}$ yr^{-1}). All other partial-decay constants are zero. The total decay constant for ^{40}K is thus $\lambda = \lambda_\beta + \lambda_\epsilon = 5.43 \times 10^{-10}$ yr^{-1}.

The decay constant λ determines the rate of decay of a nuclear species. Given an initially large number of atoms, N(0) at a time $t(0) = 0$, of a radioisotope with decay constant λ, the number of atoms N, present at a later time t, is given by the well-known exponential relation

$$N = N(0)_e - \lambda t \qquad (1)$$

The number of disintegrations per unit time at time t, called the activity, is given by the product λN. From equation 1, the activity at time t is related to the activity at time $t(0) = 0$ by

$$\frac{dN}{dt} = -\lambda N = -N(0)\lambda e^{-\lambda t} \qquad (2)$$

Thus, both the number of radioactive atoms and the activity of these atoms decrease exponentially with time.

An assumption usually made concerning λ is that it is unaffected by external conditions such as temperature, pressure, or chemical state. This assumption is valid to a high degree; however, recent experiments have shown some variation in λ for a few radioisotopes under certain conditions. The measured effects are slight and unimportant for most chemical or biological uses.

It should be emphasized that the preceding discussion refers to neutral atoms. Dramatic changes in λ can occur for atoms that are stripped, even partially, of their orbital electrons. For example, a radioisotope that decays by electron capture only would, if fully stripped, become stable. Similarly, for an isomeric transition, λ would become zero for that fraction of the decays that would normally proceed by internal conversion. A special case of β^- decay, where orbital stripping dramatically affects λ, is ^{187}Re with a decay energy to ^{187}Os of only 2.6 keV. The atomic binding energies of ^{187}Re and ^{187}Os are such that a completely stripped ^{187}Re nucleus would have a mass less than that of ^{187}Os and would thus be stable against β^- decay. Although these effects can in principle be significant, in practice they are unimportant.

Rather than the decay constant λ, a more commonly used characteristic of a radioisotope is the half-life $t_{1/2}$, defined as the time interval during which the number of atoms of that radioisotope decreases by one-half. From equation 1, with $N = N(0)/2$:

$$t_{1/2} = \ln 2/\lambda = 0.693/\lambda \qquad (3)$$

Whenever a decay mode is energetically allowed, then $\lambda \neq 0$ for that decay mode and the nuclide in question is, by definition, radioactive. However, the association of a unique half-life with each radioisotope leads to an empirical definition of nuclear stability based on the upper limit of measurability of the quantity $t_{1/2}$. The present limit is about 10^{14} to 10^{19} yr for single-process decays, depending upon the type of emission. Longer $t_{1/2}$ values have been measured for a few nuclides that can decay by double β^- or double ϵ emission. Each element from hydrogen to bismuth, with the exception of technetium ($Z = 43$) and promethium ($Z = 61$), has at least one stable isotope. All isotopes above bismuth are unstable.

Naturally Occurring Radioisotopes

Some radioisotopes occur in nature. They are classified according to their origin (see Radioactivity, natural).

Primary radioisotopes. Primary radioisotopes have half-lives greater than about 10^9 yr, a value comparable to the age of the earth. They are assumed to have been present since the formation of the earth's crust (see Table 1).

Secondary radioisotopes. Secondary radioisotopes are decay products of the three chains beginning with the primary isotopes ^{235}U, ^{238}U, and ^{232}Th. Each chain proceeds by a series of α and β decays and ends in a stable isotope of lead.

Induced radioisotopes. Induced radioisotopes have half-lives much shorter than the age of the earth, but their loss owing to decay is compensated for by their continual production by nuclear reactions occurring in nature. High energy cosmic rays interacting with nuclei in the atmosphere can break up these nuclei into neutrons and residual fragments. Although some fragments may themselves be radioactive, the neutrons, n, give rise to most of the naturally occurring short-lived

Table 1. Naturally Occurring Radioisotopes

Radioisotope	Isotopic abundance, %	Half-life, yr[a]	Decay mode	Q-value, keV[a]	Daughter radioisotope
^{40}K	0.012	1.277×10^9 8	β^-	1311.6 5	^{40}Ca
			ϵ	1505.0 6	^{40}Ar
^{87}Rb	27.83	4.80×10^{10} 13	β^-	273.3 19	^{87}Sr
^{113}Cd	12.2	9.3×10^{15} 19	β^-	322 5	^{113}In
^{115}In	95.7	4.41×10^{14} 25	β^-	495 8	^{115}Sn
^{123}Te	0.89	$>10^{13}$	ϵ	52 23	^{123}Sb
^{138}La	0.089	1.35×10^{11} 16	β^-	1041 12	^{138}Ce
			ϵ	1749 5	^{138}Ba
^{144}Nd	23.8	2.4×10^{15} 3	α	1910.3 31	^{140}Ce
^{147}Sm	15.1	1.06×10^{11} 2	α	2310.5 15	^{140}Nd
^{148}Sm	11.3	8×10^{15} 2	α	1986.2 13	^{144}Nd
^{152}Gd	0.20	1.08×10^{14} 8	α	2206.2 34	^{148}Sm
^{174}Hf	0.16	2.0×10^{15} 4	α	2504 6	^{170}Yb
^{176}Lu	2.6	3.60×10^{10} 16	β^-	1186.5 22	^{176}Hf
^{180}Ta	0.012	$>10^{13}$	β^-	710 11	^{180}W
			ϵ	865 13	^{180}Hf
^{186}Os	1.58	2.0×10^{15} 11	α	2816.5 28	^{182}W
^{187}Re	62.60	5×10^{10}	β^-	2.64 4	^{187}Os
^{190}Pt	0.013	6×10^{11}	α	3243 20	^{186}Os

[a] Each one or two of the uncertainty digits (the numbers after the spaces) refers to the last one or two digits of the numerical value.

radioisotopes. The interaction of neutrons with nitrogen produce ^{3}H (tritium) and ^{14}C:

$$^{14}\text{N} + {'n} \begin{cases} ^{12}\text{C} + ^3\text{H} \\ ^{14}\text{C} + ^1\text{H} \end{cases}$$

whereas ^{239}Pu is formed by

$$^{238}\text{U} + n \rightarrow ^{239}\text{U} \xrightarrow{\beta^-} ^{239}\text{Np} \xrightarrow{\beta^-} ^{239}\text{Pu} \qquad (4)$$

and ^{242}Pu and ^{244}Pu are formed by multiple neutron capture on, for example, ^{238}U and ^{239}Pu.

Artificially Produced Radioisotopes

By far the largest group of radioisotopes is produced in the laboratory, either by charged-particle or neutron-induced reactions on targets of stable isotopes. Cobalt-60, for example, is commonly produced by the (n, γ) reaction on ^{59}Co (relative isotopic abundance 100%). Cobalt-58 is produced by the (p, n) reaction on ^{56}Fe (relative isotopic abundance 91.7%). Iodine-131 is a product of the neutron-induced fission of ^{235}U. Although nuclear reactions have also been carried out on a few radioactive targets, such as ^{210}Bi (isomeric state, $t_{1/2}$ 3.0×10^6 yr), these are specialized cases not suitable for commercial production. So far, more than 2100 radioisotopes have been artificially produced (see Actinides and transactinides).

The above methods are suitable for the production of radioisotopes that have half-lives long enough for production and use without significant loss of activity. Some short-lived radioisotopes can be obtained in radioisotope generators and separated, usually in an ion-exchange column, from the relatively long-lived parent.

Radiation Processes

The decay of a nuclear species to a different species can take place by the emission of β particles (positively or negatively charged electrons, β^+, β^-), α particles, neutrinos, and, less commonly, neutrons and protons. Accompanying these emissions, depending upon the type of decay, can be γ rays (photons), x rays, Auger electrons, and bremsstrahlung. For the isomeric branch of an isomeric state decay, the parent and daughter states are in the same nucleus and therefore, of the decays mentioned, only γ rays, internal-conversion electrons, x rays, and Auger electrons would be observed.

α-Particle transitions. For α decay between ground states of parent and daughter atoms, the maximum energy available for the particle is

$$E = Q_\alpha - E_R \qquad (5)$$

where Q_α is the difference, in energy units, between the mass of the parent atom and the sum of the masses of the daughter atom and a helium atom. The recoil energy E_R of the daughter atom, if M and M_α

are the masses of the daughter atom and the particle, respectively, is given by

$$E_R = Q_\alpha \frac{M_\alpha}{M_\alpha + M} \qquad (6)$$

For decay to a particular energy level E_L in the daughter nucleus from an excited level E^* in the parent nucleus, the maximum energy available for the alpha particle is $Q_\alpha + E^* - E_L - E_R$.

β-Particle transitions. β^- Decay. In β^- decay, an antineutrino $(\bar{\nu})$ and a negative electron (β^-) are emitted from the nucleus as a result of the process $n \rightarrow p + \beta^- + \bar{\nu}$. The decay increases the nuclear charge by one unit.

The energy released in a single β transition is divided between the β particle and the antineutrino in a statistical manner in such a way that, when a large number of transitions is considered, both the antineutrinos and β particles have energy distributions extending from zero up to some maximum value. For decay to a particular energy level, E_L in the daughter nucleus, the maximum energy available is $E_{max} = Q^- + E^* - E_L$, where Q^- is the atomic mass difference, expressed in energy units, between ground states of parent and daughter nuclides, and E^* is the excitation energy of the decaying level in the parent nucleus.

β^+ Decay. In β^+ decay, a neutrino (ν) and a positive electron (β^+) (positron) are emitted from the nucleus as a result of the process $p \rightarrow n + \beta^+ + \nu$. This decay decreases the nuclear charge by one unit.

As in β^- decay, the β^+ particles emitted in a transition to a particular level in the daughter nucleus have a continuous distribution of energies with a definite E_{max} and E_{av}. For positron decay,

$$E_{max} = Q^+ + E^* - E_L - 2mc^2 \qquad (7)$$

β^+ decay to a particular energy level, E_L, cannot occur unless

$$Q^+ + E^* - E_L > 2mc^2 \qquad (2mc^2 = 1022 \text{ keV}). \qquad (8)$$

Electron-capture decay. In electron-capture decay ϵ, an atomic electron is captured by a nucleus and a neutrino is emitted as a result of the process $p + e \rightarrow n + \nu$. This decay decreases the nuclear charge by one unit and leaves the daughter nucleus with a vacancy in one of its atomic shells. K-shell electron capture, for example, refers to a capture process where the final-state vacancy is in the K shell.

γ-Ray transitions. Electromagnetic radiation is emitted by a nucleus in a transition from a higher to a lower energy state. The energy of this γ ray, $E(\gamma)$, is equal to the energy difference between the two levels (except for a usually negligible amount of nuclear recoil energy).

$$E_r \text{ (keV)} \approx 5.4 \times 10^{-7} E(\gamma)^2 / A \qquad (9)$$

where A is the mas number and $E(\gamma)$ is in keV.

Internal-conversion-electron transitions (ce). An atomic electron from the cloud of electrons orbiting the nucleus can be emitted as an alternative to γ-ray emission in the transition of a nucleus from a higher to a lower energy state. In the internal-conversion process, the energy difference between the states is transferred directly to a bound atomic electron that is ejected from the atom.

X-ray and Auger-electron transitions. Whenever a vacancy is produced in an inner electron shell of an atom, the filling of this vacancy is accompanied by the emission of either an x ray (X) or Auger electron (e_A). Vacancies created by the filling of the initial vacancy produce, in turn, further x rays or Auger electrons. This cascade of radiations continues until all vacancies have been transferred to the outermost electron shell. Inner-shell vacancies are always produced in two types of nuclear decay, ie, electronic capture and internal conversion. Other processes that lead to electron vacancies following a nuclear decay occur rarely.

Bremsstrahlung. In addition to any monoenergetic x rays or γ rays that may be present in radioactive decay, every β or electron-capture decay produces continuous electromagnetic radiation called bremsstrahlung; two processes contribute to this continuous spectrum.

External bremsstrahlung is produced by collisions between β particles or conversion electrons and the atoms of the material surrounding the radiating atoms. The intensity of the external bremsstrahlung associated with a β group of maximum energy E_β (in keV) is given for the thick-target approximation as

$$E_{av} \approx 1.4 \times 10^{-7} Z E_\beta^2 \text{ keV per } \beta \tag{10}$$

Internal bremsstrahlung, originating within a decaying atom, is produced by the sudden change of nuclear charge that occurs in β^+, β^-, or ϵ decay. The average energy of the internal bremsstrahlung associated with a β group of maximum energy E_β (in keV) is given for $E_\beta \gg mc^2$ (511 keV) as

$$E_{av} \approx 1.5 \times 10^{-3} E_\beta \log \left(0.004 E_\beta - 2.2\right) \text{ keV per } \beta \tag{11}$$

In electron capture, the corresponding expression for a transition with energy E_ϵ (in keV) is

$$E_{av} \approx 1.5 \times 10^{-7} E_\epsilon^2 \text{ keV per capture} \tag{12}$$

The low average energy in both external and internal bremsstrahlung is a result of the low probability of the process. The actual spectrum of this electromagnetic radiation extends in energy up to the E_{max} for the β transition giving rise to it or, in the case of electron capture, up to E_ϵ.

Spontaneous fission (SF). This mode of decay has been established for several transuranic nuclides. In this process, a heavy nucleus decays into two lighter fragments with emission of several prompt neutrons. Each of the resulting fission fragments is neutron rich and undergoes several successive β^- decays before reaching a stable nucleus. In the case of ^{252}Cf, the two fragments are centered around A = 106 and A = 142, and the width at one-tenth maximum of each peak is approximately 27 mass units. The fragment with A = 106 decays as follows:

$$^{106}\text{Nb}\beta^- \rightarrow {}^{106}\text{Mo}\beta^- \rightarrow {}^{106}\text{Tc}\beta^- \rightarrow {}^{106}\text{Ru}\beta^- \rightarrow {}^{106}\text{Rh}\beta^- \rightarrow {}^{106}\text{Pd} \text{ (stable)} \tag{13}$$

whereas the fragment with A = 142 follows the chain:

$$^{142}\text{I}\beta^- \rightarrow {}^{142}\text{Xe}\beta^- \rightarrow {}^{142}\text{Cs}\beta^- \rightarrow {}^{142}\text{Ba}\beta^- \rightarrow {}^{142}\text{La}\beta^- \rightarrow {}^{142}\text{Ce} \text{ (stable)} \tag{14}$$

Hence, the radiations from spontaneous fission include the fission fragments themselves, neutrons, β particles, and prompt and delayed gammas. The energy released in spontaneous-fission decay from these radiations is about 200 MeV. Thus, when compared to the typical α energies of about 5 MeV, spontaneous fission can be the most important mode of decay in terms of total energy released, even though the SF branch may be small.

IT decay. The decay of an isomeric state can proceed by any of the modes discussed. The term IT decay, ie, isomeric transition decay, refers to those decay modes that can involve deexcitation of the isomeric state to states with lower energy in the same nucleus. Thus, there is no change in Z or A.

MURRAY J. MARTIN
Oak Ridge National Laboratory

L.H. Meyer, ed., *Applications of Radioisotopes*, Pt. 22 of *Nuclear Engineering*, American Institute of Chemical Engineers, New York, 1970.

G.F. Knoll, *Radiation Detectors and Measurements*, John Wiley & Sons, Inc., New York, 1979.

W.B. Mann, R.L. Ayres, and S.B. Garfinkle, *Radioactivity and Its Measurement*, Pergamon Press, New York, 1980.

RADIOPAQUES

Radiopaques are diagnostic agents that permit physicians to examine patients and diagnose abnormalities and impairment of organ functions (see also Medical diagnostic agents). By definition, radiopaques cause soft-tissue structures, such as the stomach, heart, and gallbladder, to become visible during x-ray examination by inhibiting the passage of x rays and producing a shadow of positive contrast. The term contrast media refers to radiopaques and also to materials that produce a shadow of negative contrast, that is, the part to be visualized is less dense than the surrounding tissue structures. The ideal contrast medium should permit rapid and adequate visualization of the organ under investigation, have no pharmacodynamic effect in the body, and be rapidly eliminated from the body. Almost all contrast media currently in use are radiopaques (see also Radioactive drugs).

Iodine is the most useful element for producing radiopaques with satisfactory properties (see Iodine and iodine compounds). It can be incorporated into a large number of organic compounds in which it forms a strong covalent bond. Barium sulfate is currently the only widely used radiopaque that does not contain iodine.

Urography and Angiography

Urography or pyelography is the visualization of the urinary tract by means of radiopaques. In excretion urography, the solution is given intravenously and the kidneys concentrate and excrete the radiopaque, whereas in retrograde urography, the solution is injected by catheter up the ureter into the pelvis of the kidney. Excretion urography is used more frequently and the drug may be administered by injection at a moderate rate or by drip infusion. The compounds developed for excretion urography are also employed for retrograde urography and angiography, which is the visualization of the blood vessels by means of radiopaques. Most of the urographic–angiographic agents are ionic, but recently nonionic agents were introduced for myelography and angiography. These nonionic media have lower osmolality and toxicity than the ionic media. Some urographic–angiographic agents and their iodine content, osmolality, and toxicity are given in Table 1.

Table 1. Some Urographic–Angiographic Agents

Name	Structure	Iodine, %	LD$_{50}$ in mice, g I/kg	Osmolality, osmol/kg at 280 mg I/mL
sodium acetrizoate	(1) Na salt	65.8	5.5	
sodium diatrizoate	(2) Na salt	59.9	8.4	1.51
sodium iothalamate	Na salt	59.9	8.0	1.69
sodium metrizoate	Na salt	58.5	9.1	1.66
sodium iodamide	Na salt	58.5	9.0	1.79
metrizamide	(3)	48.2	17.5	0.43
iopamidol		49.0	21.8	0.47
iohexol	(4)	46.4	23.4	0.62
sodium ioxaglate	Na salt	59.0	11.5	0.49

(1) R = H
(2) R = NHCOCH₃

(3)

(4)

Cholecystography

Visualization of the gallbladder with radiopaques is called cholecystography. In this procedure the administered substance or its metabolite is excreted by the liver into the bile and collected in the gallbladder. Iopanoic acid is a widely used oral cholecystographic agent. A number of other aliphatic acids or salts that contain the 2,4,6-triiodophenyl group are also employed as oral agents.

Myelography

Myelography is the visualization of the subarachnoid space of the spinal canal by means of a radiopaque. Both water-soluble agents and oils not miscible with water are used for myelography. Iophendylate became the preferred oil-type medium for this purpose.

Water-soluble media offer the advantage that the agent does not need to be removed from the patient after the procedure. These agents are excreted primarily in the urine after being absorbed from the cerebrospinal fluid.

The nonionic agent metrizamide (3) improved the safety of myelography. In spite of its high cost, this agent has almost completely replaced the other water-soluble media for this purpose. The nonionic agents iopamidol and iohexol also show promise in this field.

Bronchography, Lymphangiography, and Other Radiopaque Procedures

Bronchography is the x-ray visualization of the bronchial tree with the help of a radiopaque. The first bronchographic agent was iodinated poppyseed oil. Suspension-type agents and tantalum have also been used. The difficulties with the procedure and the introduction of computerized tomography (CT) have greatly diminished the application of bronchography.

The visualization of the lymphatic ducts by radiopaques is called lymphangiography. Water-soluble agents and oils usually are used for this procedure. Iodetryl, ethiodized oil, is the most commonly employed oil-type medium.

JAMES ACKERMAN
Sterling-Winthrop Research Institute

J.O. Hoppe in E.E. Campaigne and W.H. Hartung, eds., *Medicinal Chemistry*, Vol. 6, John Wiley & Sons, Inc., New York, 1963, pp. 290–349.

J.F. Sackett and C.M. Strothers, eds., *New Techniques in Myelography*, Harper & Row Publishers, Inc., Hagerstown, Md., 1979.

C.T. Peng in M.E. Wolff, ed., *Burger's Medicinal Chemistry*, Pt. III, 4th ed., John Wiley & Sons, Inc., New York, 1979, pp. 1139–1203.

RADIOPROTECTIVE AGENTS

The death, suffering, and permanent damage inflicted on many people by high levels of radiation incurred by the dropping of atomic bombs on Hiroshima and Nagasaki in 1945 stimulated research in several countries, notably the United States, the USSR, France, the United Kingdom, Canada, the Federal Republic of Germany, and Japan, to find chemical agents that would minimize the effects of radiation.

In an early experiment performed under the Manhattan Project, it was discovered that an irradiated sulfur-containing enzyme could be reactivated by the addition of cysteine; this suggested that radioprotection of biological systems was possible. Several thousand compounds, mainly sulfur-containing, have since been designed, synthesized, and tested in animals, mostly rodents. When administered in advance of irradiation, many of them show excellent ability to prolong the life of animals subjected to lethal radiation. In humans, some of the more promising radioprotective agents are being considered as adjuncts in cancer radiotherapy and chemotherapy.

Theories of protection. On the molecular level, reduced radiation damage in biological systems is theoretically approached through the following methods.

Radical scavenging. To the extent that radiolysis products of water play a role in the cause of cell injury in mammalian systems, the ability of radioprotective compounds to scavenge these mediators of the indirect effect must be considered a relevant mechanism in radioprotection. To account for the observed protection, some phenomenon must be responsible for concentrating the protector at the site of critical target macromolecules by approximately a factor of 100 above that predicted from a uniform distribution of the protector in cell water. Model studies involving DNA as the target molecule indicate that the most effective radical-scavenging radioprotectors form some type of complex with DNA and are then able to scavenge radicals at this presumably critical site. However, DNA binding is not a universal requirement for radioprotection and does not correlate with radioprotective efficacy in the N-heterocyclic aminoethyl disulfides. Sulfur-containing radioprotective compounds are also excellent scavengers of hydrogen atoms and hydrated electrons, which are two other significant water radicals.

Hydrogen transfer. If the initial damage to the target consists effectively of loss of a hydrogen atom, then its restoration constitutes instantaneous repair and, thus, protection. Although H-atom transfer has only been observed in model systems, studies of mammalian cells in tissue culture have, by analysis of the shape of survival curves, resulted in the identification of two types of protection: competitive and restitutive. As in the case of radical scavenging, some mechanism must be invoked to account for the localization of H-atom donors at critical sites.

Mixed disulfide hypothesis. In one explanation of radioprotectant localization, it is proposed that the aminothiol protectors form temporary mixed disulfides with —SH and —SS— groups within the cells. If the mixed disulfides were attacked by either direct- or indirect-radiation action, in at least half of such encounters radiolytic scission of the disulfide bond would restore the originally covered sulfhydryl moiety or allow restoration of a disulfide bond. Times for formation of mixed disulfides *in vivo* correspond well with times of observed optimal radiation protection, and effective protectors exhibit greater propensity for mixed disulfide formation. The most protective sulfur compounds are probably able to induce polarization of the temporary mixed disulfide bond and, thereby, increase the probability that the S atom receiving the radiation insult would be that contributed by the protector to the mixed disulfide bond, thus accounting for greater than 50% protection.

Endogenous nonprotein sulfhydryl compounds. A second hypothetical mechanism for localization of sulfur-containing radioprotective compounds at critical sites also involves the formation of temporary mixed disulfides between exogenous sulfhydryl radioprotective compounds and cellular disulfides. However, this hypothesis suggests that it is the released endogenous sulfhydryl that scavenges water radicals.

At the physiological-biochemical level, theories of protection include: a suggestion that the mechanism for protection by hypothermia is that during the period of lowered temperature, reduced metabolic rate permits repair of crucial radiation damage before the demand of normal metabolism returns.

Compounds related to histamine are thought to be capable of protection by preventing the reactions involved in the oxygen effect.

Administration of compounds that result in the release of interferon endogenously or of interferon itself has been reported to increase radioresistance of animals.

Chemical Radioprotective Agents

Thiols. Cysteine. The vast majority of antiradiation agents is aminoalkylthiols or derivatives thereof, the prototype of which is the sulfur-containing amino acid, cysteine (1).

$$HSCH_2CHCOOH$$
$$|$$
$$NH_2$$
(1)

This compound protects 75–89% of rats subjected to 8 Gy (800 rad) if it is administered 5 min prior to x irradiation at 175–575 mg/kg. In this study, 19% of the irradiated control rats survived. Cysteine is equally effective if given up to one hour before irradiation. Mice given 1000 mg/kg of cysteine iv (intravenously) are protected to the extent of 50%

from the effects of lethal radiation. Chromosome damage in irradiated human bone-marrow cells has been reduced 58% by cysteine.

A number of carboxylic esters of cysteine have been reported to give good protection, in terms of percent survival, to rats: cysteine methyl ester hydrochloride, 70%; cysteine ethyl ester hydrochloride, 55%; cysteine propyl ester hydrochloride, 100%; cysteine isopropyl ester hydrochloride, 40%; cysteine butyl ester hydrochloride, 60%; cysteine isobutyl ester hydrochloride, 100%; and cysteine isoamyl ester hydrochloride, 70%. The oxidized form of cysteine, namely, cystine, and its diethyl ester impart no protection.

2-Mercaptoethylamine. The decarboxylated form of cysteine, namely 2-mercaptoethylamine (2) (MEA, cysteamine, 2-aminoethanethiol, mercamine, Becaptan) is an even more promising antiradiation agent than cysteine.

$$H_2NCH_2CH_2SH$$
(2)

Because of its structural simplicity, MEA hydrochloride is one of the most studied antiradiation agents. It is the compound that not only serves as a model for the design of other agents, but generally is also the standard by which the activity of other agents is judged. The compound confers greater protection to mice irradiated with a single 8-Gy (800-rad) dose than to the mice given four 2-Gy (200-rad) doses at intervals of 7 d. It offers protection against at least 3 repeated lethal exposures, provided they are at 30-d intervals. When administered in the drinking water of mice, MEA did not protect against chronic radiation. The compound protects the gastrointestinal tract and bone marrow of mice.

The antiradiation properties of MEA are optimized in mice if it is given 10 min prior to radiation, whereas in rats, best results are obtained 45 min before radiation. 2-Mercaptoethylamine protects mouse and rat spermatozoa.

The radiation-prophylactic action of MEA has been ascribed to its ability to scavenge free radicals, to form mixed disulfides, to induce hypoxia, and to prevent cross-linking and DNA breakdown induced by radiation.

Other mercaptoalkylamines. The placement of amino and thiol functions at adjacent positions in an alicyclic system, eg, DL-*trans*-2-aminocyclohexanethiol (3) and *cis*- and *trans*-cyclobutyl derivatives (4)–(6), yields compounds with considerable antiradiation activity.

The *N*-phenethyl- and carboxymethyl-[cysteamine-*N*-acetic acid], derivatives of MEA have good antiradiation properties. The latter compound and its esters and salts are better tolerated and more effective than MEA.

Alkylthiols lacking an amino group. One of the most interesting compounds developed as an antiradiation agent is sodium 2,3-dimercaptopropanesulfonate (7) (Unithiol), which was first synthesized in the USSR.

$$CH_2CHCH_2SO_3Na$$
$$|\quad\ |$$
$$SH\ \ SH$$
(7)

It is more protective and less toxic (LD$_{50}$, 1400 mg/kg) in mice than MEA, and is protective in rats and dogs. Unithiol is structurally related to the heavy-metal antidote, 2,3-dimercaptopropanol (BAL). In rodents, it is an efficient chelating agent which, when complexed, is eliminated from mammalian systems in water-soluble form. Unithiol has been studied in the treatment of poisoning by mercury, arsenic, antimony, gold and cadmium, and mixtures of metals, eg, mercury, nickel, copper, and cadmium.

Disulfides and trisulfides. 2-Mercaptoethylamine can be oxidized to its disulfide, ie, bis(2-aminoethyl) disulfide (cystamine). The free base is a water-soluble liquid; however, it is usually administered as a solution of its crystalline dihydrochloride salt (8). The compound has lower acute toxicity, is as about as effective as MEA, and exhibits activity when administered orally to mice, rats, and guinea pigs.

$$(HCl.H_2NCH_2CH_2S)_2$$
(8)

The compound affords 60% survival at a dose of 146 mg/kg to rats subjected to lethal radiation and also protects antibody production in rats. A clue to its mechanism of action may be in the reduction of cystamine to MEA *in vivo* during irradiation. Other ideas are that cystamine protects DNA by complexing with it, thereby stabilizing the DNA helix, and that mobilization of endogenous catecholamines may be involved.

A limited number of aromatic mixed (unsymmetrical) disulfides, eg, o-(2-aminoethyldithio)benzoic acid hydrochloride, show possible activity.

The mixed disulfides of type (9) give good protection to mice at moderately low doses.

$$SCH_2CH_2NHAr$$
$$|$$
$$CH_3COS$$
(9)

Since unsymmetrical disulfides tend to disproportionate, especially under alkaline conditions, so as to give a mixture of symmetrical disulfides, it may be that under physiological conditions the combined effects of the symmetrical disulfides are observed.

An interesting class of antiradiation compounds bearing both disulfide and butanesulfinate groups but lacking a basic amino moiety has been developed. The most active of the class are (10) and its disproportionation product (11).

These compounds protect 93% and 73% of lethally irradiated mice at doses of 172 mg/kg and 200 mg/kg, respectively. The former also protects 100% of the mice if it is administered at a dose of 278 mg/kg po. The trisulfide corresponding to (11) protects 100% of irradiated mice when 300 mg/kg is given ip.

Organic thiosulfates (Bunte salts). In contrast to thiols, which are susceptible to air oxidation, organic thiosulfates (Bunte salts) are essentially unaffected by air. In addition, they can be solubilized by formation of their alkali salts. Furthermore, the latter react *in vitro* and *in vivo* with sulfhydryls to form mixed disulfides. Bunte salts usually have lower acute toxicity than the corresponding thiols; however, their antiradiation properties tend to be inferior to the latter, especially when they are administered orally.

2-Aminoethanethiosulfuric acid (12) increases the survival of irradiated mice.

$$H_2NCH_2CH_2SSO_3H$$
(12)

When administered at a dose of 150 mg/kg ip to lethally irradiated mice, (12) protects 73% of them.

Phosphorothioates. The most promising of the modified thiol groups to be incorporated into potential radioprotective agents is the phosphorothioate functionality. Compounds bearing this group do not undergo typical thiophilic displacements, to which disulfides or Bunte salts are subject, to give mixed disulfides. Phosphorothioates, however, are hydrolyzed very rapidly in the presence of acid to give the corresponding thiols and are enzymatically converted to thiol and orthophosphate by human erythrocytes, bovine brain, rat liver homogenates, and isolated acid or alkaline phosphatases.

Sodium 2-aminoethanephosphorothioate (13) (sodium 2-aminoethanethiol dihydrogen phosphate, WR 638, cystaphos) has excellent radioprotective action (> 95% survival in lethally irradiated mice) and is superior to MEA when given orally. In rats, the compound exerted its maximum radioprotective action when given 60–90 min prior to irradiation.

$$H_2NCH_2CH_2SPO_3HNa$$
(13)

A particularly valuable series of antiradiation agents consists of 2-(ω-aminoalkylamino)ethyl- **(14)** and -propyl- **(15)** dihydrogen phosphorothioates. High survivals and low toxicities characterize the former series when $n = 2$–6 and in the latter, when $n = 2, 3$.

$$H_2N(CH_2)_n NHCH_2CH_2SPO_3H_2 \qquad H_2N(CH_2)_n NHCH_2CH_2CH_2SPO_3H_2$$
(14)   **(15)**

Probably the most effective of all antiradiation agents is 2-(3-aminopropylamino)ethanephosphorothioic acid [WR 2721, amifostine (World Health Organization), gammaphos (USSR), YM-08310 (Japan)]. This compound protects mice, dogs, and rhesus monkeys against the effects of γ and x radiation.

$$H_2NCH_2CH_2CH_2NHCH_2CH_2SPO_3H_2$$
(16)

The compound protects 86% of irradiated mice at a dose of 300 mg/kg. WR 2721 promotes wound healing in irradiated rats and increases the resistance of the immune response to radiation injury.

Thioureas. There has been intense interest in aminoalkylthiopseudoureas and, particularly, 2-aminoethylisothiuronium bromide hydrobromide (AET) **(17)**. In aqueous solution, especially near neutrality, AET undergoes a rearrangement through an intermediate diaminothiazolidine to give 2-mercaptoethylguanidine **(18)** (MEG, 2-guanidinoethanethiol).

HBr·H₂NCH₂CH₂SC̈NH₂·HBr
(17)

(18)

Thiazolines

The condensation of MEA or an *N*-substituted MEA with an aldehyde or ketone yields a thiazolidine. Numerous compounds of this type possess antiradiation activity, probably because of their ability to hydrolyze slowly *in vivo* to form their constituent aminoalkylthiols.

Other radioprotective agents. Potassium iodide or iodine is used to prevent thyroid damage in humans exposed to high levels of radioiodine (^{131}I), which is attached to macromolecules, eg, an antibody or fibrinogen, during cancer therapy or diagnosis. Here, the body, particularly the thyroid gland, is saturated with nonradioactive iodine given as Lugol's solution at a dose of about 250 mg iodide per day before, during, and after administration of the radioiodinated macromolecule. Any radioiodine released from the macromolecule during metabolism or by autoradiolysis is then excreted with the excess iodide, rather than sequestered in an iodine-requiring organ, which would result in radiation damage.

Additional Uses

Cancer treatment. In clinical radiotherapy of malignant tumors, adjacent tissue is unavoidably damaged to some extent. It is, therefore, desirable to chemically protect normal tissue from radiation injury without affecting the radiosensitivity of the tumor. Sensitizers, eg, misonidazole, aid in achieving some selectivity. Modest success in the protection of normal tissue has been achieved with the use of several phosphorothioate antiradiation agents. The compounds WR 638 **(13)** and WR 2721 **(16)** are apparently less absorbed by solid tumors than by the surrounding tissue.

DANIEL L. KLAYMAN
Walter Reed Army Institute of Research

EDMUND S. COPELAND
National Institutes of Health

Z.M. Bacq, *Chemical Protection Against Ionizing Radiation*, Charles C. Thomas Publisher, Springfield, Ill., 1964.

V.S. Balabukha, ed., *Chemical Protection of the Body Against Ionizing Radiation*, Macmillan, Inc., New York, 1964.

W.O. Foye in W.E. Wolff, ed., *Burger's Medicinal Chemistry*, 4th ed., Part III, John Wiley & Sons, Inc., New York, 1980, p. 11.

RADIUM. See Radioactivity, natural.

RADON. See Helium-group gases.

RARE-EARTH ELEMENTS

The rare earths comprise a group of 17 elements in the periodic table and have similar properties in aqueous solutions. All are metals in the elemental state and all form salts that are strong electrolytes when dissolved in water. They ionize in this medium to give triply charged ions and, because of the high charge on these ions, react strongly with water dipoles to form a tight sheath of water molecules about them. Other ions in aqueous solutions only contact this sheath, giving rise to the similar properties of rare-earth cations in water.

The group consists of the following elements: scandium (^{21}Sc), yttrium (^{39}Y), lanthanum (^{57}La), all of which appear in the Group IIIB of the periodic table, and cerium (^{58}Ce), praseodymium (^{59}Pr), neodymium (^{60}Nd), promethium (^{61}Pm), samarium (^{62}Sm), europium (^{63}Eu), gadolinium (^{64}Gd), terbium (^{65}Tb), dysprosium (^{66}Dy), holmium (^6Ho), erbium (^{68}Er), thulium (^{69}Tm), ytterbium (^{70}Yb), and lutetium (^{71}Lu).

Occurrence

The rare earths are widely distributed in low concentrations throughout the earth's crust. They occur as mixtures in many massive rock formations, eg, basalts, granites, gneisses, shales, and silicate rocks, in which they are present in amounts of 10–300 ppm. They also occur in ca 160 discrete minerals, most of which are rare, but in which the rare-earth content, expressed as R_2O_3 or Ln_2O_3 (REO), can be as high as 60% (REO is an abbreviation used in industry for rare-earth oxide content). Ln_2O_3 is used when Y_2O_3 is not present in the ore or mineral. Approximately ten of these occur in sufficient quantities that they may furnish some REO to commerce, but more than 95% of the REO occurs in three minerals: monazite and bastnasite for the light rare earths, and xenotime for yttrium and the heavy rare earths. Xenotime occurs mixed with monazite in alluvial deposits.

Properties

Physical. The rare-earth metals alloy with most metals to form intermetallic compounds and occasionally solid solution. In rare-earth-rich alloys, other elements can change the properties of the pure metal by drastically lowering (or, more rarely, raising) the melting point by 200–300°C in some cases. Alloying with other elements can make the rare earth either pyrophoric or corrosion-resistant. Some properties of the rare earths are listed in Table 1.

Chemical. The chlorides, bromides, nitrates, bromates, and perchlorate salts are soluble in water and, when their aqueous solutions evaporate, they precipitate as hydrated crystalline salts. The acetates, iodates, and iodides are somewhat less soluble. The sulfates are sparingly soluble and are unique in that they become less soluble with increasing temperature. The oxides, sulfides, fluorides, carbonates, oxalates, and phosphates are insoluble in water. The oxalate, which is important in the recovery of highly pure Ln, can be calcined directly to the oxide.

Anhydrous rare-earth salts usually cannot be prepared by evaporating the water of crystallization.

The lanthanides can form hydrides of any composition up to LnH_3. Small amounts of hydrogen dissolve interstitially but, with increasing amounts of hydrogen, a second phase (LnH_2) appears. These alloys are metallic in their properties and lose hydrogen at relatively high temperatures.

Table 1. Some Properties of the Rare Earths

	Scandium	Yttrium	Lantha-num	Cerium	Praseo-dymium	Neodym-ium
atomic number	21	39	57	58	59	60
atomic weight	44.9559	88.9059	138.9055	140.12	14.9077	144.24
color	silvery	silvery	silvery	silvery	silvery	silvery
melting point, °C	1541	1522	918	798	931	1021
boiling point, °C	2836	3338	3464	3433	3520	3074
density, g/cm³	2.989	4.469	6.146	6.770	6.773	7.008
heat of fusion, kJ/mol^a	14.10	11.40	6.20	5.46	6.89	7.14

	Prome-thium	Samarium	Europium	Gadolin-ium	Terbium
atomic number	61	62	63	64	65
atomic weight	145	150.36	151.96	157.25	158.9254
color	silvery	silvery	silvery	silvery	silvery
melting point, °C	1042	1074	822	1313	1356
boiling point, °C	3000 (estd)	1794	1527	3273	3230
density, g/cm³	7.264	7.520	5.244	7.901	8.230
heat of fusion, kJ/mol^a	7.6 (estd)	8.62	9.21	10.05	10.79

	Dyspro-sium	Holmium	Erbium	Thulium	Ytterbium	Lutetium
atomic number	66	67	68	69	70	71
atomic weight	162.50	164.9304	167.26	168.9342	173.04	174.967
color	silvery	silvery	silvery	silvery	silvery	silvery
melting point, °C	1412	1474	1529	1545	819	1663
boiling point, °C	2567	2700	2868	1950	1196	3402
density, g/cm³	8.551	8.795	9.066	9.321	6.966	9.841
heat of fusion, kJ/mol^a	11.06	16.87	19.90	16.84	7.66	18.65

^{a}To convert J to cal, divide by 4.184.

The rare earths form many compounds with organic ligands; some of which are water-soluble. Chelating agents (qv), which are organic molecules that engulf rare-earth ions and displace some of the water of hydration, form complexes with rare-earth ions that have a much wider range in their formation constants than do those of ordinary mineral-acid salts.

Production and Processing

Monazite and bastnasite are the minerals used commercially to supply most of the rare-earth chemicals. Monazite is a brown, dense phosphate mineral. There are extensive deposits of enriched monazite sands on beaches in many parts of the world, eg, along the southwest coast of India and the east coast of Brazil; in the uplands of Australia, South Africa, the USSR; and in the United States in Idaho, South Carolina, and Florida. Such deposits are dredged, pulverized if necessary, and further enriched by flotation (qv) methods. Sometimes they are also subjected to cross-belt magnetic separation (qv) since they are weakly magnetic.

Liquid–liquid extraction. As the industrial demand for individual rare earths increased, some of the rare-earth companies developed liquid–liquid extraction systems. An organic stream immiscible with water is flowed countercurrent to the aqueous stream containing the rare-earth mixture. The organic phase may absorb neutral rare-earth molecules by complexing with them, eg, as does tributyl phosphate, or it may contain a complexing molecule or anion added to it for the same purpose. Also, complexing ligands may be added to the aqueous phase so as to effect a synergistic equilibrium between the rare-earth ions and the complexant. If the various equilibria between the organic and aqueous phases have overall equilibrium constants sufficiently different for various rare earths, separation occurs.

Health and Safety

The rare earths are considered only slightly toxic according to the Hodge-Sterner classification system and can be handled safely with ordinary care.

Uses

The rare earths are used in gasoline-cracking catalysts; in carbon arcs; as additives to steel and cast irons; as polishing compounds; to made optical glass; as both colorants and decolorants for glass, depending upon the rare earth element used; and in magnets. In the electronic area, the most important industrial rare-earth compounds are garnet-based materials, eg, yttrium iron garnet and gadolium gallium garnet, used in microwave devices (see Microwave technology), memory storage (see Magnetic materials), and oxygen sensors (Y_2O_3-stabilized ZrO_2). In nuclear reactors (qv), the rare earths are used in the form of oxides as absorbers of neutrons. They are also used as additions to superalloys, as hydrogen-storing materials (see Hydrogen; Hydrogen energy), and as synthetic gems (see Gems, synthetic).

F.H. SPEDDING
Ames Laboratory
Iowa State University

T. Moeller, *Inorganic Chemistry, A Modern Introduction*, John Wiley & Sons, Inc., New York, 1982.

E.J. McCarthy, J.J. Rhyne, and H. Silber, eds., *The Rare Earths in Modern Science and Technology: Papers Presented at the Meeting*, Plenum Publishing Company, New York, 1980.

K.A. Gscheidner, Director, The Rare Earth Information Center, Ames Laboratory of the USDOE, Ames, Iowa (a partial list of books and review articles on rare earths published in the past 10 years is available).

RASCHIG RINGS.   See Absorption; Distillation.

RAYON

Viscose Rayon

Most of the world's rayon is made by the viscose process. The term rayon designates a range of products that have widely varying properties, for the most part owing to the versatility of the viscose process. This versatility, which is the result of the many process steps and spinning changes that can be made, is a mixed blessing, since each stage of processing and spinning requires close attention to guarantee the desired product properties. The viscose process is a demanding process that requires continuous, year-long operation to prevent gelling of the system and to yield high quality products.

Process. The stages in the preparation of a satisfactory cellulose xanthate solution are given in Figure 1.

In order to have cellulose react with carbon disulfide to form the corresponding cellulose xanthate, cellulose must be converted to alkali

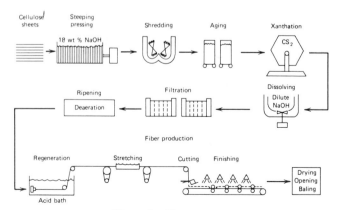

Figure 1. Viscose process.

cellulose (see Cellulose). Normally, this is done by placing cellulose-pulp sheets in a steeping press and filling the press with a closely controlled concentration of sodium hydroxide at a desired level of 18–20 wt%, depending on the type of cellulose used.

In addition to converting the alkali cellulose to a crumblike material, shredding serves additional functions, eg, the squeezing action of the shredder blades distributes the caustic more uniformly in the cellulose. Most alkali cellulose sheets are wetter at the edges than in the centers and, when the caustic solution drains, the bottoms of the sheets have more alkali. Thus, shredding gives a more uniform distribution of caustic in the alkali-cellulose crumb.

Aging is used to decrease and control the cellulose DP, ie, the alkali cellulose stands in covered containers in a temperature-controlled room. The alkali crumb should never become dry. In the continuous process, the alkali cellulose may be dried while conveyed in an air stream; therefore, humidified air must be recirculated to the conveying section. In the batch process, drying can occur during aging if the cans are not covered properly.

Xanthation is a critical part of the viscose process; many reactions proceed simultaneously during this step. Pure cellulose xanthate is white; any yellow color is owing to by-product formation, usually in the form of trithiocarbonate. The lower the xanthation temperature and the better the alkali-cellulose preparation, the lighter the color of the xanthate crumb and the less the hydrogen sulfide (H_2S) evolution during spinning.

Xanthate crumb is dumped into large, stirred tanks containing dilute caustic solution to dissolve the cellulose xanthate into a clear, honeylike viscous dope known as viscose. The caustic and water in the dissolver are measured to give the desired cellulose : alkali ratios. Many dissolvers have standpipes that force the mixtures upwards in a cascading fashion while the moving blades propel the mixtures in a circular manner.

When the cellulose xanthate is first dissolved, it is not ready for spinning because it will not coagulate readily. The xanthate groups attached to the cellulose molecule are distributed over the three hydroxyl positions of the anhydroglucose monomer units; more xanthate groups are on the kinetically favored 2 or 3 positions ($\gamma_{2,3}$) than on the thermodynamically favored 6 position (γ_6). In essence, the xanthate groups in the 2 and 3 positions act as wedges to keep the cellulose chains from approaching one another; these must be removed or relocated to the 6 positions before closer chain packing can be achieved. This redistribution of xanthate groups is the main function of the ripening stage.

Spinning. Once the viscose is ripened to the proper xanthate level, it is ready for spinning. Chief goals in spinning are control of coagulation vs regeneration rates and the maximal use of the differences in these rates to obtain maximum responses to stretching.

Stretching. An advantage of retarding regeneration is to make use of the densification of the extruded dope so as to obtain better molecular alignment on stretching. A fixed amount of stretching can occur; this is balanced between what is called jet stretch and godet stretch. Jet stretch is the ratio of the linear speed at the first godet roll to the average linear flow rate in the spinneret. Godet stretch is the stretching between the first and the last godet rolls. The higher the godet stretch, the less residual elongation is left in the final fiber. Therefore, the manufacturer balances fiber strength and residual elongation.

Crimped fibers. In the early 1970s, Rayonier patented a chemically crimped rayon trademarked Prima. The crimp is produced by selective control of the conditions in the primary and secondary regeneration baths wherein a gel-core fiber is allowed to leave the primary bath and is exposed to hot-acid conditions in a secondary bath. Under these conditions, significant tensions are established throughout the fiber, which causes permanent chemical crimping. These fibers also have high wet modulus and give cover, hand, bulk, and working performance in finished fabrics superior to that of ordinary rayon. When produced in higher deniers, Prima blends with polyester closely resemble wool fabrics.

Hollow fibers. Courtaulds has developed a hollow rayon fiber trademarked Viloft. This fiber is much lighter than regular rayon and therefore gives significantly more coverage per unit weight. Moisture-absorbing capacity is about 50% greater than ordinary rayons; this makes Viloft a candidate for use in towels, sheets, pillow cases, and garments requiring improved absorbency. It is produced by adding sodium cabonate to the viscose and then controlling the regeneration conditions to capture the released CO_2 gases to give the hollow fiber (see Hollow-fiber membranes).

Solvent-Spun Rayon

Several companies are exploring the possibility of using organic solvents with closed recycle-recovery loops to produce rayon-type fibers through solvent-spun processes because of the continued projected need for a fiber such as rayon. The American Enka Company has issued reports about the Newcell process, which appears to be based on spinning solutions of up to 20 wt% cellulose dissolved in hot aqueous N-methyl-morpholine N-oxide. This system appears to be amenable to closed-loop recovery-recycle and relatively rapid spinning speeds.

A new solvent system for cellulose has been developed that can give up to 16% solutions of 500 DP cellulose. This system, which does not cause cellulose degradation even on extended heating or air exposure, utilizes lithium chloride in dimethylacetamide or in N-methyl-2-pyrrolidinone to form solutions of cellulose that are stable in extended storage. This system permits total recycle and recovery of all components to give a nonpolluting process. The economics of the process appear to be competitive with a new viscose plant installation.

Cuprammonium Rayon

Essentially all of the rayon made today is produced by the viscose process; nevertheless, there are some uses where the cuprammonium process has retained an advantage because of specific performance factors. Fibers from cuprammonium rayon are significantly more supple than viscose fibers and are used where a very soft hand is desired. The use of films and hollow fibers from the cuprammonium process for making artificial kidneys is critically important in this branch of medicine (see Dialysis; Prosthetic and biomedical devices). At present, almost all artificial-kidney units use membranes prepared from such films and fibers. These films and fibers exhibit superior clearance performance for urea, creatinine, and metabolites, possess better dewatering characteristics, and cause less blood clotting as compared to any synthetics or corresponding products from the viscose process (see Membrane technology).

Dissolving. The ablity of selected cupric ion, ammonia and alkali mixtures to dissolve cellulose was originally reported in 1847. Such solutions, known as Schweitzer's reagent, are used not only for making cuprammonium rayon, but also for measuring dilute-solution viscosities that can be related to degrees of polymerization of cellulose.

Spinning. The spinning of cuprammonium yarns depends on the use of a funnel into which the coagulating-regenerating fluid flows and into which the dope is extruded.

As the following liquid travels down the funnel, its velocity increases and the entrained gelatinous cellulose fibers are stretched up to 400%. Subsequently, the fibers are washed as free as possible of occluded material, which is recycled, and the fibers enter a 5% H_2SO_4 bath where final removal of copper is achieved and where any remaining alkali or ammonia is converted to the corresponding sulfate.

Health and Safety

Ecological and pollution considerations. The cuprammonium and the viscose processes consume huge amounts of water for each kilogram of product; ca 420–750 L (110–200 gal) of water per kilogram of rayon is needed directly for processing and 8–10 times that amount of water must be handled to provide the plant supplementary-service facilities.

In the cases of the solvent spinning process, all recovery and recycle stages are considered to be closed loops with essentially none of the solvent being lost. This must be achieved because the solvent systems are too expensive to permit anything less than almost complete recycling.

JOHN LUNDBERG
ALBIN TURBAK
Georgia Institute of Technology

R.L. Mitchell and G.C. Daul in *Encyclopedia of Polymer Science and Technology*, Vol. 11, Interscience Publishers, a division of John Wiley & Sons, Inc., New York, 1969, pp. 810–847.

A.F. Turbak, *Amer. Chem. Soc. Symp. Ser.* **58**, (1977).

A.F. Turbak, R.B. Hammer, R.E. Davies, and H.L. Hergert, *Chemtech* **10**, 51 (Jan. 1980).

REACTOR TECHNOLOGY

A reactor consists of the vessels used to produce desired products by chemical means and is the heart of a commercial processing plant. Its configurations, operating characteristics, and underlying engineering principles constitute reactor technology. Besides stoichiometry and kinetics, reactor technology includes requirements for introducing and removing reactants and products, supplying and withdrawing heat, accommodating phase changes and material transfers, assuring efficient contacting among reactants, and providing for catalyst replenishment or regeneration.

Reactor Types and Characteristics

All reactors have in common selected characteristics of three basic reactor types: the well-stirred batch reactor, the continuous-flow stirred-tank reactor, and the tubular reactor (see Fig. 1). A reactor often may be represented by or modeled after one or a combination of the three types.

Batch reactor. A batch reactor is one in which a feed material is treated as a whole for a fixed period of time. The semibatch reactor is a modification characterized by the continuous addition of some reactant in addition to those initially placed in the reactor or by the continuous removal of one or more products.

Commercial reactors generally process continuously rather than in single batches, because overall investment and operating costs of continuous processes usually are less. Batch reactors often are used to develop continuous processes because of their suitability and convenient use in laboratory experimentation. The data, except for very rapid reactions, can be well-defined and used to predict performance of larger-scale, continuous-flow reactors.

Because uniform residence times can be achieved with both reactors, better yields and higher selectivities can be obtained than with continuous reactors.

Continuous-flow stirred-tank reactor. If reactants and products are continuously added and withdrawn from a well-stirred vessel, the resultant reactor is the continuous-flow stirred-tank reactor (CSTR). In practice, mechanical or hydraulic agitation is required to achieve uniformity of composition and temperature. The CSTR is the idealized opposite of the well-stirred batch and tubular plug-flow reactors. Analysis of judicious combinations of these reactor types can be useful in quantitatively evaluating more complex gas-, liquid-, and solid-flow behaviors.

Tubular reactor. The tubular reactor is a vessel through which flow is continuous, usually at a steady state, and configured so that process variables are functions of position within the reactor rather than of time. In the ideal tubular reactor, the fluids flow as if they were solid plugs or pistons, and reaction time is the same for all flowing material at any given tube cross section; hence, position is analogous to time in the well-stirred batch reactor. Tubular reactors also resemble batch reactors in providing initially high driving forces, which diminish as the reactions progress down the tubes.

Multiphase Reactors

The presence of more than one phase, whether it is fixed or flowing, further compounds analyses of reactor performance and increases the multiplicity of reactor configurations. Gases, liquids, and solids each flow in characteristic fashions, either dispersed in other phases or separately. Flow patterns in these reactors are complex and the phases rarely exhibit plug-flow behavior.

A fixed-bed reactor is one that is packed with catalyst. If a single phase is flowing, the reactor can be analyzed as a tubular plug-flow reactor or, essentially, as plug flow modified for uniform axial diffusion. If both liquid and gas or vapor are injected downward through the catalyst bed, or if substantial amounts of vapor are generated internally, the reactors are mixed-phase, downflow, fixed-bed reactors. If the liquid and gas rates are so low that the liquid flows as a continuous film over the catalyst, the reactors are called trickle beds. At higher total flow rates, particularly when the liquid is prone to foaming, the reactor is a pulsed column; this designation arises from the observation that the pressure drop within the catalyst bed cycles at a constant frequency. At high liquid flow rates, gas becomes the dispersed phase and bubble flow develops, with flow characteristics being similar to those in countercurrent packed-column absorbers. At high gas rates, spray and slug flows can develop. Downflow is preferred because the reactors are more readily designed mechanically to hold a catalyst in place and are not prone to inadvertent excessive velocities, which upset the beds. Upflow is used less often but has the advantage of optimum contacting between gas, liquid, and catalyst over a wider range of conditions.

Moving beds are fixed-bed reactors in which spent catalyst or reactive solids are slowly removed from the bottom and fresh material is added at the top. A fixed bed that collects solid impurities present in the feed or produced in the early reaction stages is a guard bed. If catalyst deposits are periodically burned or otherwise removed, the operation is cyclic, and the catalyst remaining behind the combustion front is regenerated.

In bubble columns, gas bubbles upward through a slowly moving liquid. The bubbles, which rise in essentially plug flow, draw liquid in their wakes, thereby back-mixing the liquid with which they have come in contact. These reactors are used for slow reactions where contact is not critical. In spray columns, liquid as droplets descends through a fluid, which is usually a gas. These are used for reactions where high interfacial areas between phases are desirable. If beds of solids are lifted by either gas or liquid flows, the reactors are termed fluid beds because the suspended solids appear to behave as liquids. The process is usually referred to as fluidization (qv). The most common fluid bed is the gas-fluidized bed.

Reactor Selection

Selection is often determined by economics, reliability, or availability of a proven system that is amenable to extension in a new service. For example, fixed beds and slurry reactors are favored at high pressures

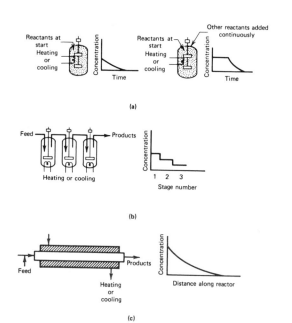

Figure 1. Three basic reactor types. (**a**) Batch. (**b**) Continuous-flow stirred-tank. (**c**) Tubular.

over fluid beds; fluidized systems are less likely to develop hot spots or be subject to temperature runaways; and downflow vapor-phase fixed-bed technology for desulfurization of naphthas and light gas oils has been extended to mixed-phased operations with higher-boiling gas oils and, subsequently, to residua.

Scale-Up

In the design and scale-up of reactors, key factors are the nature of the reaction sites, the specific reaction rates, and the mass- and heat-transport rates to and from the reaction sites. When reaction and transport rates are comparable, they affect each other and are said to be coupled. In these situations, increasing the reactor size alters mass- and heat-transport rates and, therefore, changes the apparent reaction rate. Effects on selectivity are also possible. In such situations, conversions are underestimated in the small reactor, but there is no general trend in selectivity. The effects of scale on selectivity must be determined by experimentation. If transport rates to and from the reaction sites are more than ca ten times greater than the specific reaction rate, the overall reaction rate will be uncoupled from the transport rates. Increasing reactor size, then, does not affect the apparent reaction rate.

Complex Flow Behavior

The concepts of well-mixed and plug flow become inadequate when flow patterns deviate significantly from ideal behavior. Departures from ideal behavior result in a range of residence times, as measured by the residence-time distribution (RTD). The RTD in a flow reactor can be determined by analyzing the effluent pattern resulting from a change in an inlet flow property, eg, the concentration of one component or of an added tracer.

The concept is applicable to more than fluids. Varying catalyst deactivations in a reactor as results from a loss of catalyst surface area or other aging mechanisms can be evaluated with equivalent particle-population-balance models.

Cold-Flow Models

Use of tracers to obtain RTD characteristics followed by inferential analysis of flow patterns establishes conceptual but not actual detailed flow characteristics of a system. This limitation often can be removed with the aid of cold-flow models, replicas of portions of reactors that operate at ambient conditions and designed for detailed observation of flow patterns.

Numerical Experimentation

Numerical experimentation offers prospects for reducing dependencies on idealized models. The reduced cost of computation and the increased cost of experimentation makes numerical experimentation increasingly attractive. Numerical experimentation involves combining modern quantitative representations of laminar and turbulent flows, new finite difference algorithms and other mathematics and advanced computer graphics. Associated with these efforts has been the development of computer programs for carrying out such calculations (see Simulation and process design).

Commercial Reactor Design and Use

Most reactors have evolved from concentrated efforts focused on one type of reactor. A few processes have emerged from parallel developments using markedly different reactor types. In most cases, the reactor selected for laboratory study has become the reactor type used commercially. Following are illustrative examples of reactor usage.

Batch reactors. The batch reactor is the preferred configuration for manufacturing plastic resins. Such reactors generally are 6–40-m^3 (ca 200–1400-ft^3), baffled tanks, in which there are blades or impellers that are connected from above by long shafts, and heat is transferred either through jacketed walls or by internal coils. Finger-shaped baffles near the top are used instead of full-length baffles. All resins, including polyesters, phenolics, alkyds, urea–formaldehydes, acrylics, and furans, can be produced in essentially the same way. Raw materials are held at temperatures of up to 275°C for ca 12 h until the polymerized liquid becomes sufficiently viscous. One plant has been designed to produce 200 different types and grades of synthetic resins in three 18-m^3 (ca 640-ft^3) reactors.

The cost savings of large-scale operations are possible for batch polymerizations Worldwide production capacity of poly(vinyl chloride) (PVC) is greater than 15×10^6 t/yr, but no continuous process reactor has been developed for PVC production.

Continuous-flow stirred-tank reactors. The synthesis of p-tolualdehyde (PTAL) from toluene and CO involves the use of CSTR equipment. p-Tolualdehyde is an intermediate in the manufacture of terephthalic acid. Hydrogen fluoride–boron trifluoride catalyzes the carbonylation of toluene to PTAL. In the commercial process, separate stirred tanks are used for each process step.

Thermal tubular reactors. Tubular reactors are used for the pyrolysis or thermal cracking of petroleum feeds to olefins, particularly ethylene. Given the current uncertainty in feed supply, the trend is towards processing a wide range of feeds, including high molecular weight gas oils in single large units, ca 250 t/h. Reactor configurations strongly depend on feed composition and the degree of flexibility must be weighed against the cost. There are limits to the range of feeds that can be processed in tubular reactors. The residua are difficult to crack and thus must be processed in other reactor types, eg, fluid beds.

Bubble and spray columns. Bubble columns in series have been used to establish the same effective mix of plug-flow and back-mixing behavior required for liquid-phase oxidation of cyclohexane, as obtained by staged reactors in series.

Tubular fixed-bed reactors. Bundles of downflow reactor tubes that are filled with catalyst and surrounded by heat-transfer media are tubular fixed-bed reactors. Such reactors are used most notably in steam reforming and phthalic anhydride manufacture. Steam reforming is the reaction of light hydrocarbons, preferably natural gas or naphthas, with steam over a nickel-supported catalyst to form synthesis gas, which is primarily H_2 and CO and some CO_2 and CH_4.

Fixed-bed reactors. Single-phase flow. Fixed-bed reactors supplied with single-phase reactants are used extensively in the petrochemical industry for catalytic reforming, ammonia synthesis, various hydroprocesses, eg, hydrocracking and hydrodesulfurization, and oxidative dehydrogenation. The feeds in these processes are gases or vapors. The reactors generally are of large diameter, operate adiabatically, and often house multiple beds in individual pressure vessels.

Multiphase flow. The most generally used mixed-phase reactor is the trickle bed. Special distributors are used to uniformly feed the two phase mixtures. Hydrodesulfurization, hydrocracking, hydrogenation, and oxidative dehydrogenation are carried out with high boiling feeds in such reactors. Pressures generally are higher than in their single-phase flow counterparts. Though some of these reactors may operate in the pulsed-flow regime, these reactors are called trickle beds because the same configurations are used.

Fluid beds and other fluidization reactors. The most well-known and one of the early applications of fluidization technology is catalytic cracking. In this process, gas oils are cracked at ca 500–525°C to produce gasoline and other light hydrocarbons, and fine silica–alumina-based catalysts averaging ca 50–60-μm dia are used. The reactors are either dense-phase fluid beds or dilute-phase transfer lines or combinations of the two fluidization regimes.

BARRY L. TARMY
Exxon Research and Engineering Company

G.F. Froment and K.B. Bischoff, *Chemical Reactor Analysis and Design*, John Wiley & Sons, Inc., New York, 1979.

O. Levenspiel, *Chemical Reaction Engineering*, 2nd ed., John Wiley & Sons, Inc., New York, 1972.

C.N. Satterfield, *Heterogeneous Catalysis in Practice*, McGraw-Hill Book Company, New York, 1980.

RECORDING DISKS

Magnetic Disks

Magnetic-disk systems record data by writing patterns of magnetization on layers of ferromagnetic material which is coated on disks. This information can be read, erased, or overwritten; thus, the data storage is stable but alterable. In some magnetic-disk systems, the disk is removable, and thereby provides a means of long-term or off-line storage or information interchange among systems. In the most recent high performance magnetic-disk systems, the disk or disks, ie, the pack, is/are permanently mounted within the machines or drives that move them.

Video and PCM (Pulse Code Modulation) Audio Disks

In the first of the stylus systems (the TED (Telefunken-Decca) system from Teldec), a stylus and pressure pickup were designed to scan a groove. This system, which was basically a miniaturization of a conventional audio playback system, suffered from poor performance and short stylus lifetime. A flat record is used in the more recent VHD (video high density) system for JVC (Japan Victor Company) and an electromechanical servo system guides the stylus. The laser optical systems read a flat record by means of an optical stylus in the form of a light spot produced by focusing the beam from a low powered laser. The spot is held in focus and on track by optoelectromechanical servos.

There has been much interest recently in the application of videodisk technology to audio recording by a PCM or digital encoding of the audio signal.

Optical Data Disks

Optical data disk systems have been developed to combine the rapid random access, local writing, and comprehensive error management and control characteristic of magnetic-disk systems with the extremely high capacities characteristics of optical video disk systems. Optical data disk systems are expected to play an important role in the development and proliferation of computer and communication systems that are designed to handle text as well as numeric information.

Properties

Audio master records are made by stylus-cutting the surface of appropriate master blanks. These blanks are made by spin-coating an acetate lacquer on both surfaces of a machined aluminum disk. The disks are available in standard diameters of 356, 301.6, and 254 mm, are 0.91 mm thick at an overall thickness with coatings of 1.3 mm, and are finished with a center hole 7.26 mm dia to fit the standard phonograph spindle.

The lacquer coating, typically 0.18 mm thick, must be very homogenous in order that the walls of the groove cut by the stylus be smooth.

Magnetic disks. Rigid magnetic disk platters look much like audio master blanks. They are aluminum disks that are spin-coated on both sides with layers of plastic. The plastic is a binder supporting fine ferromagnetic particles, called oxide, which are magnetized during the writing process. On some magnetic-disk platters, continuous-film (plated) media are applied to the aluminum disk.

Platters are used singly or stacked in packs of two to twelve platters. They are secured to hubs, which usually are cast aluminum, by mechanical clamping. The hubs in turn may be removable or may be permanently mounted to the spindles of the drives.

In most current rigid-disk drives, a servo keeps the head or heads on track despite disk eccentricity and vibration of the drive. The servo typically derives its guidance information from a pattern of servo tracks that exactly describe the desired data-track format and which are written on one surface of one platter of the pack.

Manufacture

Lacquer recipes for audio master blanks are proprietary; however, most current lacquers are based on a nitrocellulose resin mixed with plasticizers, fillers, lubricants, and dye. This mixture, carried in a volatile solvent, is directed onto a spinning aluminum disk to produce a smooth, thin coating. The oxide media on most magnetic disks are layers of acicular gamma ferric oxide particles suspended in a polymer binder. The oxide particles are typically 100 mm long and 5–10 mm across; the binder is typically an epoxy resin. The layer is coated from solution, usually by spinning or sometimes by spraying.

LEONARD LAUB
Vision Three, Inc.

J. Audio Eng. Soc. **25**(10/11), (1977).

L. Laub, *SMPTE J.* **85**(11), 881 (1976).

A. Bell and co-workers, *IEEE Spectrum*, **27** (Aug. 1978).

RECREATIONAL SURFACES

For the purposes of this article, the term recreational surfaces means man-made surfaces that provide a durable area of consistent properties for recreational activities. The characteristics of the playing surface may be selected to match natural surfaces under ideal conditions, or may provide special characteristics not otherwise available. In all cases, the intent is to provide desirable and durable playing characteristics. Recreational surfaces are used for football, soccer, field hockey, cricket, baseball, tennis, track, jumping, golf, wrestling, and general purposes. Included in the latter category are indoor–outdoor carpets, patio surfaces, and similar materials designed for low maintenance in light recreational service.

Types

Light-duty recreational surfaces. These are artificial surfaces intended for incidental recreational use and are designed primarily to provide a practical, durable, and attractive surface, eg, for swimming-pool decks, patios, and landscaping.

Specific athletic surfaces. These include running tracks, tennis courts, golf tees and putting greens, and other applications designed for a particular sport or recreational use. Specific performance criteria are important.

Multi-purpose recreational surfaces. The performance demands for artificial surfaces in this category control the design. A good example is the playing surface for professional football in the United States. The shock absorbency of the system affects player safety and long-term performance under very heavy use. The grasslike fabrics used for these applications are made from various pile materials, including polypropylene, nylon-6,6, nylon-6, and polyester (see Polyamides; Polyesters). The fabric may be woven, knitted, or tufted. The shock-absorbing underpad is derived from various materials, representing a compromise of performance properties.

Materials and Components

The principal parts of a recreational-surface system include the top surface material directly available for use and observation, backing materials that serve to hold together or reinforce the system, the fabric backing finish, the shock-absorbing underpad system, if any, and adhesives or other materials.

Surface materials. Pile materials used in grasslike surfaces are selected from fiber-forming synthetic polymers, such as polyolefins, polyamides, polyesters, polyacrylates, vinyl polymers, and many others (see Fibers, elastomeric). These polymers exhibit good mechanical strength in the necessary direction.

Backing materials. Any fiber-forming polymer of reasonable strength may be used in backing materials, including polyamides, polyesters, and polypropylenes. The backing provides strength and offers a medium to which the pile fibers may be attached. The backing usually is not visible in the finished product, nor does its presence contribute much to the characteristics of the playing surface. However, it provides dimensional stability and prolongs service life.

Backing finish. Usually, the backing material is consolidated with a pile ribbon (see Textiles). In tufting, for example, the tufts are locked to the backing medium. In weaving and knitting, the finish seals and stabilizes the product. Backing materials are usually applied as a coating that is subsequently heat-cured. For tufting, preferred choices are poly(vinyl acetate), poly(vinyl chloride), polyurethane resins, and various latex formulations. For knitted fabrics, poly(vinylidene chloride) and acrylics, or polystyrene–rubber latexes are used.

Underpads. Shock-absorbing underpad material is usually made of foamed elastomer, which provides good energy absorption at reasonable cost (see Foamed plastics). The foamed materials may be poly(vinyl chloride), polyethylene, polyurethanes, or combinations of these and other materials. Typical foam densities may range from 32–320 kg/m^3. Important criteria include a resistance to absorbing water, tensile strength and elongation, open-cell vs closed-cell construction, resistance to chemical attack, low cost, availability in continuous lengths, softness in energy-absorbing properties, and compression-set resistance.

Adhesives and joining materials. Grasslike surfaces are employed over substantial areas, and lengths of rolls must be joined, glued, or sewn together. Sewing is an excellent technique, but a variety of adhesives ranging from low cost poly(vinyl acetate) materials to cross-linked epoxy cements also are available (see Adhesives).

Fabrication and Installation

Tufting. The tufting process is frequently employed in construction of grasslike surfaces. The techniques are essentially those developed for the carpet industry and have economical high speed characteristics. In the tufting operation, pile yarn is inserted in one side of a woven or nonwoven fabric constituting the primary backing. Yarn is inserted by a series of needles, each creating a loop or tuft as the yarn penetrates the backing and forms the desired pattern on the other side. For artificial surfaces, the looped tufts that form in this process are cut to provide the desired individual blades in the playing surface. Cutting elements are incorporated in the tufting machine which severs the loops automatically in the process of forming the pile.

Knitting. The knitting process as applied to manufacture of artificial turf and related products provides a high strength, interlocked assembly of pile fibers and backing yarns. Pile yarn, stitch yarn, and stuffer yarn are assembled in the operation. The pile and stitch yarns run in the machine or warp direction, whereas the stuffer yarns interlock the wales, ie, rows, formed by the pile and stitch yarns, knotting the whole system together. Knitted fabrics typically possess high strength and high tuft bind.

Weaving. As a general rule, weaving is slower than tufting or knitting. The process consists of a two- or three-dimensional intermeshing of warp, pile, and fill yarns that may be of different types. In contrast to a knitted fabric, the yarns are not knotted together, but interwoven at right angles. The pile yarns are cut by a series of tires that are continuously assembled into and withdrawn through the fabric loops. A suitable finish further stabilizes the fabric.

Finishing. The artificial turf fabric is subjected to a finishing operation in which a suitable adhesive is applied to the back side, thus bonding the components and stabilizing the material. The finish may be applied with a knife or brush, in foam or paste form, followed by a heating and drying stage. The temperatures also affect the pile-ribbon properties.

Underlay. An installed artificial turf system may or may not include components between the fabric and the subbase. As mentioned earlier, such components are not a requirement for light-duty use, but are essential in attaining the shock-absorbing properties required by heavy-duty surfaces. The foam underpads used in shock-absorbing systems are made by incorporating a chemical blowing agent into the foam latex or plastisol.

Installation. In general, grass-like surfaces are glued down or otherwise affixed to a subbase. For light-duty purposes, it may suffice to tack the edges to the perimeter of the area to be covered. For heavy-duty systems, a solid subbase of asphalt or other materials is first installed. A recent variation is the use of permeable asphalt to drain rainfall. The shock-absorbing underpad is glued to this surface, followed by glueing the turf on top of the pad.

For artificial surfaces in the athletic category, eg, running tracks, the installation techniques are different. A poured-in-place or interlocking-tile technique may be employed; the latter is used for tennis courts. Adequate provision for weathering and water drainage is essential. In general, the resilient surfaces are installed over a hard base that contains the necessary curbs to provide the proper finished level. Out-of-doors, asphalt is the most common base, and indoors, concrete. A poured-in-place polyurethane surface is mixed on-site and cast from at least two components, an isocyanate and a filled polyol, of the polyether or polyester type (see Urethane polymers).

W.F. HAMNER
T.A. OROFINO
Monsanto Company

J.E. Nordale in N.M. Bikales, ed., *Encyclopedia of Polymer Science and Technology*, Vol. 15, Interscience Publishers, a division of John Wiley & Sons, Inc., New York, 1971, p. 490.

ASTM D 3574, 1977; *ASTM D 1667*, 1976; *ASTM D 624*, 1973; *ASTM D 2856*, 1976; *ASTM F 355*, 1978; *ASTM D 1175*, 1980; *ASTM D 1335*, 1972; *ASTM D 1682*, 1975.

RECYCLING

INTRODUCTION

Recycling is the recovery for reuse of the economic values of materials and energy from wastes that are usually destined for disposal. Although the term is applied most often to municipal wastes, it applies also to industrial or any other generated wastes or arisings. Many industrial waste streams, including most sources of industrial metal scraps, are never destined for disposal but for recycling (see Wastes, industrial).

Municipal solid waste comes from three main sources: construction and demolition; commercial and light industry; and households, ie, domestic waste. In addition, there are special categories, eg, pathogenic, hazardous, and radioactive wastes. Special wastes can sometimes be recycled, eg, in nuclear-waste processing (see Nuclear reactors, waste management).

Recycling is part of the total system of solid-waste management, which is shown diagrammatically in Figure 1. As a general rule, but subject to wide variations, the three main sources of municipal waste shown in Figure 1 each make up about one third of the total.

Methods

Scrap processing. The scrap industry generally processes obsolete capital equipment, eg, railroad cars, automobiles, ships, etc, to produce metals or other valuable materials to users' specifications. In addition, the scrap industry processes metal trim and like wastes from manufacturing processes, surplus odd lots of plastics, or various sources of waste papers, etc, to forms that meet users' specifications.

Mechanical and chemical processing. Mechanical processing includes unit operations, eg, size reduction (qv), screening, air classification, and magnetic separation (qv), to reduce the waste to a homogeneous mixture suitable for materials handling and separation. After these unit operations, several others, eg, froth flotation (qv) or optical sorting for glass recovery, or eddy-current or heavy-media separation for aluminum recovery, may be used (see Gravity separation). Such unit operations also prepare homogeneous refuse-derived fuels (RDF) for burning or feedstocks for pyrolysis or biological fermentation (see Fuels from waste).

Production and Economic Aspects

The daily capacity of municipal resource-recovery plants, but not of modular incinerators, constructed in the United States by 1981 was 27,110 t/d. The capacity of modular incinerators was 1416 t/d. The capital value of the plants was $1.5 × 10^9. The total worth of the industry is more because it includes industrial waste processing, the scrap industry, and other activities that cannot be estimated.

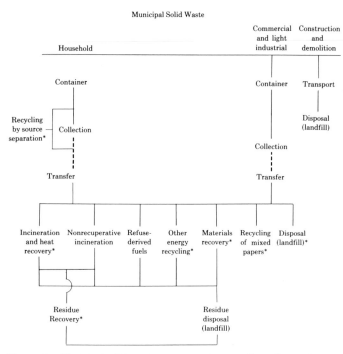

Figure 1. Municipal solid-waste management system. Recycling options are indicated by an asterisk (*).

Table 1. Total Ferrous-Scrap Consumption by Industry

Industry	Consumption, 10^6 metric tons	Year
iron and steel production	66	1978
iron foundries	13.6	1979
ferroalloy	0.45	1973
detinning	0.64^a	1974
copper precipitation	0.45	1974

^{a} Based on 4.5 kg of tin per metric ton of tinplate scrap.

The costs and economic risk of constructing and operating resource-recovery plants of different designs have been analyzed by several authors. The methods of analysis are not the same for all types of plants so that the reported costs must be used cautiously. One method has been structured to compute the tipping fee necessary for a given return. Another approach is based on the indifference value (at which an indifferent operator of resource recovery decides whether to use resource recovery or competitive disposal) of the fuel fraction that must be received in order for resource recovery to be competitive with landfill in a community.

Health and Environmental Factors

Two factors of principal concern are air emissions from burning solid waste or refuse-derived fuels (RDF) and the environmental health of workers in resource-recovery plants. The first factor has been addressed for both the United States and the UK. The overall conclusions, including those from several unpublished experimental trials using d-RDF–coal mixtures, are that the particulate and other priority-pollutant emissions do not increase, either beyond legal limits or those from coal alone, until an RDF-coal weight ratio of 60:40 is used.

HARVEY ALTER
Chamber of Commerce of the United States

J.L. Jones and S.B. Radding, eds., *Thermal Conversion of Solid Wastes and Biomass*, American Chemical Society, Washington, D.C., 1980.

P.A. Vesilind and A.E. Rimer, *Unit Operations in Resource Recovery Engineering*, Prentice-Hall, Inc., Englewood Cliffs, N.J., 1981.

H. Alter, *Materials Recovery from Municipal Waste, Unit Operations in Practice*, Marcel Dekker, Inc., New York, 1983.

FERROUS METALS

Ferrous scrap is consumed by industries other than iron-and-steel producers, iron foundries, and ferroalloy producers. The detinning and copper-precipitation industries use ferrous scrap. In terms of annual scrap consumption, however, iron and steel producers are the dominant force in the ferrous-scrap market, as suggested in Table 1.

Sources of Ferrous Scrap

Traditionally, ferrous scrap has been called either home scrap, prompt industrial scrap, or old scrap. Home scrap or revert scrap is generated during the production of steel or cast iron and is always recycled. Manufacturing industries produce prompt industrial scrap during the fabrication of various industrial, commercial, or consumer steel products. Prompt industrial scrap or new scrap is also widely recycled because its chemical and physical characteristics can be well documented. However, in some cases, additional processing, eg, detinning or compacting, must be carried out to process the scrap in a form suitable for recycling. The availability of prompt industrial scrap is directly related to the level of industrial economic activity. Prompt industrial-scrap producers usually cannot allow it to accumulate because of storage requirements and costs of inventory control. Thus, it is rapidly available at current prices to the steelmaker or ferrous-scrap industry.

All other ferrous scrap is included in the third category, called old scrap, obsolete scrap, or postconsumer scrap. This category includes all goods or products in which the iron content can at least in theory be recovered and recycled. The main types of old scrap recycled in the United States are railroad, machinery, and automotive, whereas in foreign countries with low labor costs, shipbreaking is a principal source of obsolete scrap. Ferrous scrap recovered from municipal solid waste (MSW), sometimes called MSW magnetics or municipal ferrous scrap (MFS), is a new nontraditional source of obsolete scrap.

Role of Ferrous Scrap in Steelmaking

The open-hearth furnace has the greatest flexibility in its consumption of scrap because a portion of the heat needed to melt the scrap is supplied by an external fuel. The furnace can process 100% hot metal, 100% scrap, or solid pig iron, or any combination thereof. Usually 40–60% scrap is charged with hot metal. However, the importance of the open-hearth process is rapidly declining and so flexibility in scrap usage is diminishing.

As a result of low capital costs and the availability of low cost oxygen after World War II, the basic-oxygen furnace (BOF) steelmaking process has surpassed the open-hearth process to become the dominant steelmaking process in the United States. However, scrap usage in the BOF is limited to about 33% (without preheating) because there is no external source of fuel.

The electric-arc furnace, like the open-hearth furnace, refines a material charge with any ratio of hot metal to solid metal, but typically operates with a solid charge of almost all scrap. The rapid increase in electric-furnace capacity in the last two decades has been paced by the growth of minimills, small capacity steel mills with only electric-arc furnaces based on 100% scrap, and expansion of electric-arc-furnace capacity by integrated steel mills that produce carbon steels. Growth in electric-furnace capacity has led to increased demand for ferrous scrap.

Recycling

Automobiles. The recycling of ferrous scrap from automobiles has two basic steps: dismantling the car to recover used parts of value, and processing the stripped car hulk to recover ferrous scrap.

Municipal solid waste (MSW). Solid waste can be separated into glass, ferrous, aluminum, and the organic or combustible fractions by a process called front-end separation. The sale of the materials would partially offset separation costs and disposal of the unsalable residue. Recycling of ferrous scrap from municipal solid waste, however, cannot be a main supplier of iron because the quantities of ferrous scrap cannot provide enough material to satisfy annual consumption. However, municipal ferrous scrap could supply up to 10% of annual scrap consumption in the United States.

JAMES EARLY
National Bureau of Standards

ASTM STP 592, Philadelphia, Pa., 1975.

Mineral Resources and the Environment, Supplementary Report: Resource Recovery from Municipal Solid Wastes, NAS, Washington, D.C., 1975.

M.B. Bever, *paper presented at the 7th Mineral Waste Utilization Symposium*, Chicago, Ill., 1980, pp. 174–183.

GLASS

Glass (qv) recycling involves the recovery of used glass (postconsumer glass) and reusing it as a raw material in useful products. The main repository for recovered glass is glass containers. Other applications, eg, fiber glass and aggregate for road building and construction products, are not significant because only very small quantities are used on an experimental basis. Recycled glass that is in bulk form and is suitable for melting is called cullet. Waste glass, ie, off-quality material and scrap from the manufacture of glass products, may also be called cullet but is not considered recycled because it has not been used by the consumer.

Approximately 10% of municipal refuse is glass, most of which is in the form of discarded containers.

Properties of Recovered Glass

Source-segregated glass is normally in the form of unbroken containers, which have a bulk density of 350–435 kg/m^3 (22–27 lb/ft^3). Glass recovered at a central resource-recovery plant is generally sized at less than 6 mm and has a density of 1300–1500 kg/m^3 (81–94 lb/ft^3). The color distribution of the glass in postconsumer municipal solid waste is ca 65% flint (colorless), 20% amber, and 15% green. A predominate proportion of the glass is soda-lime bottle glass with a composition of 66–75 wt% SiO$_2$, 1–7 wt% Al$_2$O$_3$, 9–13 wt% CaO and MgO, and 12–16 wt% Na$_2$O.

Processing

Source separation. Processing is usually necessary at some central point before the cullet can be used in the manufacture of glass containers. This intermediate processing may in it simplest form consist of hand sorting to segregate bottles by color and to remove nonglass containers. An intermediate processing operation designed to process glass with high contaminant levels might consist of manual sorting, magnetic separation, crushing, and screening.

Resource recovery. Several large solid-waste resource-recovery plants are equipped with glass-recovery modules. The glass-recovery system is generally combined with and inseparable from nonferrous-metal recovery equipment. Two basic final purification methods are employed by resource-recovery plants: the froth-flotation process and the optical-sorting process. Both approaches are, as yet, unproven.

The froth-flotation process is generally applied to a glass-rich fraction separated from dry shredded solid waste and produces a mixed colored product. The important preparation steps are magnetic separation of ferrous metal, separation of a heavy glass-rich product by screening and air classification, removal of aluminum with so-called aluminum magnets, and separation of residual organics and heavy metals by jigging and grinding (see Magnetic separation; Flotation).

The aluminum magnet operates on the principle that when metals pass through an electromagnetic field, eddy currents are generated in each piece. These currents have a magnetic moment, which is phased to repel the moment of the applied field. This force is sufficient to cause the metal to be thrown away from the nonmetallic particles.

The jig acts with a pulsing motion to separate solids in a liquid medium. The solids separate into layers of different apparent specific gravities. Organics move to the top and are skimmed. Heavy metals, eg, lead, zinc, and copper, form the bottom layer and are drawn off. A glass-rich middle layer is removed separately.

Froth flotation is a minerals-processing technique that has been adapted for use in waste-glass recovery. Treating an aqueous mixture of finely ground (≤ 850 µm) glass and mineral particles with a cationic fatty amine, which is selectively absorbed by the glass, causes the glass particles to float in a froth produced by aeration.

Optical sorters scan each particle of the waste feedstock. Opaque particles are removed in transparency sorters. When a light beam is broken, an air jet knocks the opaque particle out of the mixed colored feed stream. Color sorting is achieved by comparing each particle to a standard and rejecting off-color material.

Economic Aspects

Neither source separation nor the centralized resource-recovery processing techniques separate all of the glass present in solid waste. A recovery rate of ca 50% for source-separated refuse is very high. An average operation yields less than 25% of the glass present. Low recovery results from low public participation caused by indifference to, or ignorance of, the program. Recovery from central resource facilities is generally 40–50%. Glass is lost with the organic fraction and losses also are inherent in the recovery process.

PAUL MARSH
Marsh Eco-Service Co., Inc.

J.H. Heginbotham, *Proceedings of the Sixth Mineral Waste Utilization Symposium*, Illinois Institute of Technology, Chicago, Ill., 1978.

Multimaterial Source Separation in Marblehead and Somerville, Massachusetts, Composition of Source-Separation Materials and Refuse, EPA Report SW 823, EPA, Washington, D.C., 1979.

NONFERROUS METALS

The most widely used nonferrous metals are aluminum, copper-based alloys, lead, and zinc. Because of their high cost and current rate of recycling or limited use in products that would be discarded, or both, other nonferrous metals rarely occur in solid-waste streams. Prompt, old, or obsolescent scrap is recycled at a high rate from these other nonferrous metals, since they are much more valuable than the aforementioned metals.

Aluminum amounts to ≥ 2/3 of the typical 1% nonferrous metal content of municipal refuse. The remaining one third consists of varying quantities of heavy nonferrous metals, including copper-based alloys, zinc, small quantities of stainless steel, tin, lead in solders, and only traces of other metals.

Recovery of nonmagnetic metals from automotive-shredder residues is based on dense-media separation of aluminum and the subsequent recovery of heavy nonferrous metals as a residue (see Gravity separation). Heavy nonferrous metals are separated after aluminum separation by handsorting red and yellow metals as well as metals that are identifiable on the basis of shape, eg, thin strips of stainless-steel moldings. Residues from sorting can be processed in a sweat furnace to recover zinc, which can be purified by vacuum distillation. Copper-based metals are generally shipped to copper refiners, where they are processed in blast furnaces or open-hearth furnaces and then are electrolytically refined.

Sources of Nonferrous Metal Scrap

In 1982, scrap amounted to ca 31% of the supply to the aluminum industry (vs 23% in 1979). Of this, "old" scrap increased from ca 25% of scrap sources in 1950s to 47% in 1982 (up from 35% in 1979). "New" scrap made up the balance of the scrap supply. In 1982, old-scrap use was equivalent to ca 15% of industry shipment (nearly double the rate of 1979).

Projections of the container and packaging market through 1986 show that nearly 500,000 t of additional aluminum scrap will be available than in 1979 from these short-life-cycle products. It also is projected that for 1986 aluminum beverage cans will represent ca 75% of the aluminum tonnage in this market.

Another principal aluminum market, ie, transportation, also projects significant increases in aluminum use by 1990. Use of aluminum in automobiles is increasing, ca 61 kg being used in the average 1983 U.S. automobile; this is almost a 70% increase from the 1967 value.

The lead-acid battery is generally removed from every dismantled automobile for reasons of safety as well as to recover readily marketable lead. Recycled lead amounted to ca 57% of lead-industry consumption in 1982, most of which came from automobile batteries.

In 1982, scrap copper provided 56% of total domestic consumption. A large source of scrap is batteries. Copper radiators, which are stripped from automobiles, represent high quality material in that typically 8 kg Cu is recycled from one radiator.

Since many of zinc's markets are dissipated, used zinc scrap provided only 16% of domestic consumption in 1982. Old scrap in 1981 amounted to 31% of the scrap supply. Half of this old scrap was die castings from various sources.

Recovery Systems

Municipal refuse. Since there is no currently available nonferrous-metal recovery equipment analogous to the magnet used for ferrous metals, nonferrous metals are often by-products or residues from other unit operations. In municipal-refuse processing facilities, as in automotive-shredder-residue processing, heavy nonferrous metals are generally recoverable as by-products of aluminum-separation processes. They can be further separated by hand picking and/or selectively melting zinc and any residual aluminum in a sweat furnace. Sweat furnace residues can be processed by copper refineries.

Unit operations. There are several facilities in operation that use a rotary screen, or trommel, for processing raw refuse. Depending on the screen sizes used, a trommel can remove fine dirt and grit, glass, and wet putrescible material from the refuse and can recover a concentrate of metal containers, eg, beverage cans. The trommel oversize product, consisting mostly of paper and plastic, contains only a relatively small amount of the metallic and glass contaminants that are problematic in combustion systems. Further screening can separate most of the broken glass, rock, and fine putrescible material from the metal cans. The successive concentrating steps facilitate eddy-current separation, hand-picking, or the shredding of a can-rich fraction prior to shipment to a dense-media system. Most of the heavy nonferrous metals are present in the fine-screen fraction, mixed with glass and putrescibles.

For scrap copper, lead, and zinc, energy use for recycling varies widely, depending on the grade of scrap used. Energy use ranges from 4.4 to 49 MJ/kg (ca 1900 to 21000 Btu/lb) for copper-scrap recycling. Energy use is approximately 11 MJ/kg (4800 Btu/lb) for lead. It ranges from 2.3 to 23 MJ/kg (1000 to 10000 Btu/lb) for zinc depending on the type of scrap processed. Energy to produce virgin copper, lead, and zinc varies owing to ores and processes used.

Energy Savings

Energy saving is a main incentive for the recycling of aluminum and, to a lesser but significant extent, other nonferrous metals. Approximately 220 MJ/kg (ca 95,000 Btu/lb) is necessary to produce primary ingot. Approximately 10 MJ/kg (ca 4300 Btu/lb) is needed to convert aluminum scrap into molten metal comparable to that recovered in primary production facilities.

GILBERT BOURCIER
Reynolds Metals Co.

A Study to Identify Opportunities for Increased Solid Waste Utilization, Vols. II–III, and V, Batelle Columbus Laboratories, Columbus, Ohio, June 1972.

C.L. Kusik and C.B. Kenahan, *U.S. Bureau of Mines Information Circular IC 8781*, Washington, D.C., 1978.

First through *Seventh Mineral Waste Utilization Symposia*, U.S. Bureau of Mines and the Illinois Institute of Technology Research Institute, Chicago, Ill., 1968–1980.

OIL

The term oil includes a variety of liquid or easily liquefiable, unctuous, combustible substances that are soluble in ether but not in water and that leave a greasy stain on paper or cloth. These substances can include animal, vegetable, and synthetic oils, but usually the word oil refers to a mineral oil produced from petroleum (qv). An oil that has been used or contaminated, or both, but not consumed, can often be recycled to regain a useful material regardless of its origin.

Characteristics of Used Oils

Used petroleum oils to be recycled can be obtained from a variety of sources. These sources include automotive garages and service stations, truck and taxi fleets, military installations, individuals, industrial plants and manufacturing facilities of all types, and wastewater treatment plants. The main types of used petroleum oils that are being recycled are motor oils and hydraulic and industrial oils. The additives and contaminants that are typically present in these oils can cause both performance-related and environment-related problems.

Technology

Used oil recycled as burner fuel. Combustion of used oil as burner fuel has often been condemned because it destroys a valuable resource and it can cause substantial environmental pollution through widely dispersed distribution of metal oxides and stable organic contaminants.

Processing techniques for the recycling of used oil into fuel include pretreatment of the used oil to remove all or most of the contaminants which cause the environmental or operational concerns, or use of specialized facilities with acceptable environmental controls, ie, electrostatic precipitators, Venturi scrubbers, or a fabric-filter baghouse.

Reclaiming. Used oils can be reclaimed within the user facility or can be sent outside the facility to a commercial reclaimer. The primary difference between these two options is usually the level of treatment available. Often the in-plant reclaiming facility is limited to gravity purification or settling, filtration, centrifuging, and heating to remove volatiles and water. A commercial reclaimer can usually perform all of the above plus provide clay treatment, chemical treatment (acid–caustic), demulsification, distillation, and reformulation with additives if necessary. A general description of such processes may include any or all of the following: removal of solid particles by settling, centrifuging, or filtering; neutralization of acidic components with clay or alkalies, and removal of resulting soaps by washing; heating–distillation to remove volatile solvents, gasoline, and water; clay contacting to remove oxygenated components and spent additives or for decolorization; aeration and use of biocides to reduce bacterial levels; and replenishment of additives.

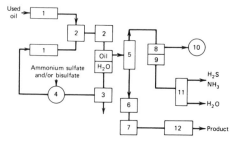

Figure 1. Flow chart of the Phillips PROP process (Phillips Petroleum, Bartlesville, Okla.) for refining used oil. 1, preheater; 2, reactor; 3, filter; 4, mixing; 5, flash; 5, adsorbent (clay); 7, filter; 8, gasoline; 9, water; 10, storage; 11, stripper; 12, hydrotreater.

Re-refining. Petroleum refining processes are employed for used lubricating oil to produce clean, high quality, lubricating base oil. These processes often include a pretreatment to reduce the impurity content, described in the previous section. This pretreatment is particularly important to protect catalyst systems if hydrotreating is used. Many modern re-refining processes use vacuum-distillation procedures in their systems.

One modern re-refining process is shown schematically in Figure 1.

DONALD A. BECKER
National Bureau of Standards

Proceedings of the Fifth International Conference on Used Oil Recycling, Association of Petroleum Re-refiners, Washington, D.C., 1983.

D.A. Becker, ed., *Proceedings, Measurements and Standards for Recycled Oil —IV*, NBS SP 674, National Bureau of Standards, Washington, D.C., 1982.

PAPER

Recycling of paper (qv) is a commercial practice involving the collection and processing of wastepaper for secondary fiber supply to the paper industry. In 1979, the world production of paper and paperboard was 171×10^6 metric tons, and recycled paper furnished 25% of this fiber supply. The use of recycled fiber will increase to help meet world paper demand, which is forecast to grow 3.5%/yr through 1990.

Recycled paper is a generic term for wastepaper that is first generated as scrap material during the original manufacture and conversion of paper and board products. No wastepaper grade is consciously produced as a raw material; the generation of wastepaper is a consequence of use. After consumer use, these products are discarded as part of the solid-waste stream. These sources of wastepaper are mostly available for recovery in concentrated population centers of developed countries.

Recycled fibers differ from virgin pulp and paper fibers in physical and mechanical properties because of the presence of contaminants and the changes within the fiber that result from drying, aging, and the recycling process (see Pulp). A recycled fiber has less strength value, lower specific volume, and often a greater mean specific surface area than its original virgin-fiber counterpart. The fiber lumen is dewatered and collapsed, which produces a flattened structure. The S-1 and S-2 layers are unraveled and separated at the fiber ends and often the fiber is split, broken, and shortened in average length. Its formation and interfiber bonding characteristics equal or exceed those of virgin fibers.

The recycled-fiber industry has developed distinct grades of wastepaper called paper stocks, which are defined by fiber type and degree of contamination. Amounts of pulp, brightness, and strength describe its general recycle use. The value of wastepaper is directly proportional to its economic substitution for virgin fiber. Paper-stock grades, eg, office waste, mixed paper, and old corrugated containers, have the lowest values because they are highly contaminated, are low in brightness, and require substantial processing. Other grades that have not been extensively converted receive a higher price. Converting adds contaminants in the form of coatings, films, inks, and adhesives. Uncirculated newspapers, ledgers, new corrugated cuttings, and publication trimmings are examples of unconverted, less contaminated paper stocks. These grades are suitable for deinking and production of newsprint, tissue, toweling, and printing papers. Direct pulp substitutes are the highest-valued paper-stock grades, are uncoated, clean, and high in brightness, and are made of strong chemical-pulp fibers. White envelope cuttings, unprinted ledgers, and blank newsprint are examples of direct pulp substitutes.

Paper and board mills that make use of recycled fiber require special equipment and facilities, as illustrated in Figure 1. The grade being produced determines the basic wastepaper furnish and the degree of required stock cleaning. Recent developments in pulping, cleaning, screening, and refining have permitted the increased use of wastepaper in higher-quality paper grades. Nonfibrous contaminants, eg, dirt, glass, sand, and paper clips, are removed by selective screening or centrifugal cleaning at various stock consistencies. Fine debris is removed by pressure screens with small perforated openings or fine flots. Low density

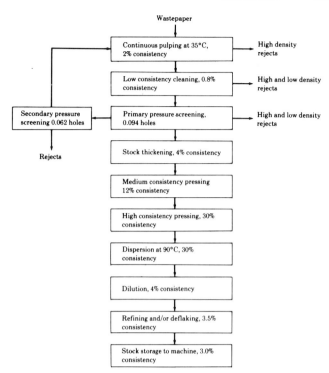

Figure 1. Typical process flow sheet of recycled fiber-stock preparation.

contaminants, eg, plastic, styrofoam, and coating films, are concentrated in stock-flow vortexes, which are generated in special screens and reverse-flow cleaners. Chemical contaminants are the most difficult to remove. These include pressure-sensitive adhesives, inks, colors, special coatings, and internal additives.

Recycled fiber-based grades of linerboard, newsprint, boxboard, printing papers, and tissue compete effectively in the market at competitive prices with acceptable product function and quality. Deinked, bleached, recycled fiber from colored ledger and printed bleached kraft has strength properties similar to those of bleached, softwood sulfite pulp.

The United States is the world's largest consumer and supplier or recycled paper.

JAMES H. ROBINS
JAMES R. GRANT
Mytec, Inc.

M. Bayliss and L. Haas, *Pulp Pap.* **54**(8), 64 (1980).

W.E. Crossland, *Recycle Fiber—Proceedings of the Technical Association of the Pulp and Paper Industry, 1980, Annual Meeting, Tappi Press*, Atlanta, Ga., 1980.

Official Board Markets, January–December 1981, 20 N. Wacker Dr., Chicago, Ill., 60606.

PLASTICS

Plastics are made from hydrocarbon feedstocks for the most part; therefore, they should be recycled as a means of energy conservation (see Plastics processing).

Many industrial waste-plastic streams are recycled, although some are difficult or impossible to recycle because of inherent limitations in the nature of plastics and because the plastics are often intimately combined with other nonplastic materials. Because of fabrication and service requirements, there are formidable technological difficulties in separating the plastics in forms and to specifications that are suitable for reuse.

Origin of Plastic Wastes

Home scrap is the waste and off-specification materials produced as by-products during polymer manufacture and compounding. Much of this is mixed, reformulated, or downgraded in specification and product application as a form of recycling.

Prompt scrap comes from fabrication and conversion operations as off-specification products and trim, eg, cutting sprues and gates. It often is recycled by the fabricator as regrind by reducing it in size, sometimes by pelletizing and further compounding, eg, by adding more heat stabilizer, and by mixing it with virgin material.

Obsolete scrap, ie, postconsumer wastes, is usually destined for disposal and, because these are often in the form of products or composites, they present the greatest challenge for separation and preparation for recycling. Typical postconsumer plastic wastes occur in mixed municipal waste, source-separated items such as plastic beverage containers, used agricultural mulch, or scrapping of large items such as obsolete automobiles.

Properties

Broadly, plastics are of two types: thermoplastic and thermosetting. Thermoplastic materials can be reheated and reformed, often several times. Thermosetting materials cannot; the initial heating and fabrication cause permanent chemical changes and subsequent reheating can cause degradation. Most plastics used either in durable goods or single-service items are thermoplastic. Thermosetting plastics are used mostly in durable goods and in much lower quantity than thermoplastics. Thermosetting plastic waste arises during manufacture and when obsolete items or capital goods, eg, automobiles or buildings, are scrapped. Thermoplastic waste is predominantly from packaging, and thermoplastic manufacturing waste is often recycled.

Processing

Froth-flotation separation. A method similar to froth flotation of minerals has been described for the separation of plastic mixtures from various industrial streams, eg, the plastic residues from chopping scrap wire and cable for recovery of the metal conductor or the residue base films after recovery of silver from obsolete x-ray plates (see Flotation). Such mixtures of waste plastics are chopped to pieces ca 0.5 cm dia and then passed through a series of froth-flotation cells containing proprietary aqueous mixtures of surfactants. At the same time, air is bubbled through the solutions to form a froth. Separation is believed to occur because of differences in densities of the plastic pieces and their wettability in the different baths. The degree of separation is high and the properties of some of the separated plastics indicate that they could be used on a commercial scale.

Mechanical methods. Several special extruders and similar devices have been developed to process mixtures of waste plastics for recycling. One such machine is the Mitsubishi Reverzer, which is intended to mix intimately a molten mixture of plastics so as to achieve acceptable physical properties. Several criteria for using this machine have been defined. Principally, the feed should not be widely dispersed mixtures of plastics. The mixture must be subjected for a short time to a high rate of shear at high temperature to achieve intensive mixing at a viscosity much lower than that of the normal plastic melt. After homogenization, the melt should be transformed directly into the finished product to avoid an additional processing step. For efficient production, the raw material for recycling must have a roughly constant composition.

Agricultural mulch. Several methods of recycling waste agricultural film, used as mulch, have been developed. For example, the film is shredded and then passed through a water wash to remove sand and other foreign matter; then it is shredded again and another water-wash tank of different design is used to separate polyethylene from poly(vinyl chloride). The plastics are mechanically dehydrated and dried.

Plastic beverage bottles. Plastic beverage bottles consist of a body made from poly(ethylene terephthalate) (PET), often with medium density polyethylene. The general method of separation is to chop the bottle into small pieces and to use float-sink techniques, air separation, or both (see Barrier polymers).

Several other plastics-recycling processes have been reported, such as for obsolete telephone equipment, scrap automobile upholstery, scrap nylon, and polyacrylonitrile fiber, and in a few cases the chemical conversion of waste plastics to other products. The latter may be accomplished by pyrolysis, acetylation, ozonation, or other reactions.

Uses

Once separated from a waste stream, the thermoplastic material must be heated to be reformed. Thermoplastics contain heat stabilizers (qv) so that the initial processing and subsequent forming does not cause excessive degradation or decomposition. Some plastics are more thermally labile than others and so require addition of more heat stabilizer prior to recycling.

HARVEY ALTER
Chamber of Commerce of the United States

A.J. Warner, C.H. Parker, and B. Baum, *Solid Waste Management of Plastics*, Manufacturing Chemists Association, Washington, D.C., 1970; B. Baum and C.H. Parker, *Plastics Waste Management*, Manufacturing Chemists Association, Washington, D.C., 1974.

International Research & Technology, Recycling Plastics, A Survey and Assessment of Research and Technology, Society of the Plastics Industry, New York, 1973; J.L. Holman, J.B. Stephenson, and M.J. Adam, "Recycling of Plastics from Urban and Industrial Refuse" in *Bureau of Mines Report of Investigations 7955*, U.S. Department of Interior, 1972.

RUBBER

Fuel Source

The use of scrap rubber for fuel is one of the best alternatives for reusing rubber if natural-gas and fuel-oil costs continue to increase. Whole tires and 2.5-cm tire chips are the most economical fuels because of shredding-cost savings. Tires contain more than 90% organic materials and have a heat value of ca 32.6 MJ/kg (ca 14,000 Btu/lb). Coal varies from 18.6 to 27.9 MJ/kg (ca 8000–12,000 Btu/lb). Shredded tire chips have been burned successfully in stoker-fired boilers. Uniroyal fired a 15% mixture of tire chips with coal and both General Motors and B.F. Goodrich burned a 10% tire-chip mixture with coal. The Japanese are using whole scrap tires to fuel portland-cement kilns. In a recent development, a cement company in California is planning to use scrap tire chips for coal fuel supplement after successfully completing preliminary tests.

Pyrolysis

Scrap-tire pyrolysis has been the subject of several research studies by rubber, oil, and carbon-black interests throughout the world (see Rubber compounding; Petroleum; Carbon, carbon black). The Tosco II process pyrolysis-research study was conceived to develop process equipment and to maximize quality carbon-black production. The Tosco II process is shown schematically in Figure 1.

The pilot-plant process was designed to handle 15 t of tires per day, and generally one metric ton of tires produces 0.5–0.6 m^3 (3–4 bbl) of oil, 1270–1540 kg of carbon black, 190–220 kg of steel, and 154–176 kg of fiber glass.

In Asphalt

The United States generates ca 2.2×10^8 scrap tires per year. Approximately 58 t of reclaimed rubber and 16,000 t of crumb rubber were produced in 1977. Approximately 4500 t of reclaimed and crumb rubber was used in asphalt–rubber compounds in 1980, which is less than 5% of the recycled rubber produced.

There are several methods for mixing and applying asphalt rubber to roadways. One conventional method is to mix the rubber and asphalt cement at ca 175–220°C for 1–2 h. The hot asphalt rubber is applied to the roadway and is covered with a layer of stone chips to form a chip seal. In addition to chip seals, rubber asphalt is used for waterproofing membranes, crack-and-joint sealers, hot-mix binders, and roofing materials (qv) (see Waterproofing and water repellency; Sealants). Rubber is also mixed with asphalt and sand to make running-track and tennis-court surfaces.

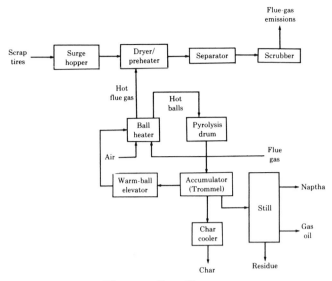

Figure 1. Tosco II process.

Reclaiming

Three basic processes are used in rubber reclaiming: digester, heater or pan, and the Reclaimator processes. Goodyear has devulcanized and reclaimed scrap rubber with microwaves. Tires are most commonly reclaimed by digesting. Two-roll corrugated mills or other grinding devices reduce whole tires to relatively uniform particle size. Fiber is mechanically separated from the rubber with hammer mills, blown into collectors, and baled. Instead of separating, metal chlorides are sometimes used to reduce tire fiber chemically during digesting. Reclaiming oils and processing aids are blended with the crumb rubber in ribbon blenders or similar mixers and are transferred to a digester, which is a steam-pressurized tank equipped with horizontal mixing paddles. The blend is mixed continuously at steam pressures of 1.03–1.7 MPa (10–17 atm) for 4–6 h. The pressurized digester batch is forced into a blowdown tank and is washed and dried. Compounding ingredients, ie, carbon black, clay, and other materials, are added to modify and maintain certain physical and chemical properties of the rubber for specific applications. Metal and other contaminants are strained from the digested rubber by screening through extruders. High friction refining mills smooth the digested rubber into sheets. The number of mill passes varies, depending on the desired smoothness and physical properties needed in the final product. The reclaim is then baled, extruded into pellets, or made into slabs for shipment.

Butyl and natural-rubber tubes and other fiber-free scrap rubbers are reclaimed by means of the heater or pan process. The scrap is mechanically ground, mixed with reclaiming agents, loaded into pans or devulcanizing boats, and autoclaved at steam pressures of 1.03–1.4 MPa (10–14 atm) for 3–8 h. The reclaim is refiner-milled, extruded, and finish-milled much the same as in the digester process.

The Reclaimator, a high pressure extruder, devulcanizes fiber-free rubber continuously with reclaiming oils and other materials. High pressure and shear forces between the rubber mixture and the extruder-barrel walls effectively reclaim the rubber mixture. Devulcanizing occurs at 175–205°C in 1–3 min.

Roughly 75% of the discarded tires is disposed of in landfills; 20% is retreaded; and 5% is reclaimed, burned for fuel, split, etc.

JOHN PAUL
Pedco Environmental, Inc.

M. Weintraub, A.A. Orning, and C.H. Schwartz, *Experimental Studies of Incineration in a Cylindrical Combustion Chamber*, U.S. Bureau of Mines RI 6908, U.S. Department of the Interior, Washington, D.C., 1967.

B.L. Schulman and P.A. White, *Pyrolysis of Scrap Tires Using the Tosco II Process—Progress Report*, rept. ACS Symp. Ser. 76 (1978).

REFRACTORIES

Refractories are materials that resist the action of hot environments by containing heat energy and hot or molten materials. The type of refractories that are used in any particular application depends upon the critical requirements of the process. There is no well-established line of demarcation between those materials that are and those that are not refractory, although the ability to withstand temperatures above 1100°C without softening has been cited as a practical requirement of industrial refractory materials (see also Ceramics).

Physical Forms

Refractories may be preformed (shaped) or formed and installed on site. Common forms include brick, setter tile and kiln furniture, fusion-cast shapes, cast and hand-molded refractories, insulating refractories, castables and gunning mixes, plastic refractories and ramming mixes, mortars, composite refractories, and refractory coatings.

Raw Materials

Today, high purity natural raw materials are increasingly in demand as well as synthetically prepared refractory grain made from combinations of high purity and beneficiated raw materials. The material produced upon firing raw as-mined minerals or synthetic blends is called grain, clinker, co-clinker, or grog. Raw materials include silica, fireclay, high alumina diaspore clays, calcium aluminate cements, zirconia, and other basic raw materials, eg, magnesite, dolomite, forsterite, chrome ore, silicon carbide, beryllia and thoria, and carbon and graphite.

General Properties

General properties of pure refractory materials are listed in Table 1.

Phase diagrams can be used to predict the reactions between refractories and various solid, liquid, and gaseous reactants. However, phase diagrams are derived from phase equilibria of relatively simple pure compounds and real systems are highly complex and may contain a large number of minor impurities that significantly affect equilibria. Furthermore, equilibrium between the reacting phases may not be reached under actual service conditions, and the physical environment of a product may be more influential to its life than the chemical environment.

Physical properties. The important physical properties of some refractories are listed in Table 2.

Mechanical properties. The physical properties of a particular refractory product depend upon its constituents and the manner in which they were assembled. The physical properties may be varied to suit specific applications; for example, for thermal insulation, highly porous products are employed, whereas dense products are used for slagging or abrasive conditions.

Thermal properties. Refractories, like most other solids, expand upon heating, but much less than most metals. The degree of expansion depends upon the chemical composition.

Manufacture

Processing. Initial processing may include an extensive survey of the raw material deposit, selective mining, stockpiling by grade, and beneficiation techniques such as weathering, grinding, washing, heavy-media separation, froth flotation, etc.

Further steps include crushing and grinding, screening, mixing, forming, drying, and burning.

Specialty refractories. Bulk refractory products include gunning, ramming, or plastic mixes, granular materials, hydraulic-setting castables, and mortars. These products are generally made from the same raw materials as their brick counterparts.

Health and Safety Factors

Because industrial refractories are by their very nature stable materials, they usually do not constitute a physiological hazard. This statement does not apply, however, to unusual refractories that might contain heavy metals or radioactive oxides such as thoria, urania, and plutonia, or binders or additives that may be toxic.

Table 1. Properties of Pure Refractory Materials

Material	Formula	Mp, °C	True specific gravity, g/cm^3	Mean specific heat J/(kg·K)^a	Mean specific heat Temp range, °C	Thermal conductivity, W/(m·K) at 500°C	Thermal conductivity, W/(m·K) at 1000°C	Linear thermal expansion coefficient per °C × 10^6, from 20–1000°C
aluminum oxide	Al$_2$O$_3$	2015	3.97	795.5	25–1800	10.9	6.2	8.6
beryllium oxide	BeO	2550	3.01	1004.8	25–1200	65.4	20.3	9.1
calcium oxide	CaO	2600	3.32	753.6	25–1800	8.0	7.8	13.0
magnesium oxide	MgO	2800	3.58	921.1	25–2100	13.9	7.0	14.2
silicon dioxide^b	SiO$_2$		2.20	753.6	25–2000	1.6	2.1	0.5
thorium oxide	ThO$_2$	3300	10.01	251.2	25–1800	5.1	3.0	9.4
titanium oxide	TiO$_2$	1840	4.24	711.8	25–1800	3.8	3.3	8.0
uranium oxide	UO$_2$	2878	10.90	251.2	25–1500	5.1	3.4	
zirconium oxide^c	ZrO$_2$	2677	5.90	460.6	25–1100	2.1	2.3	
mullite	3Al$_2$O$_3$.2SiO$_2$	1850^d	3.16	628.0	25–1500	4.4	4.0	4.5
spinel	MgO.Al$_2$O$_3$	2135	3.58	795.5		9.1	5.8	6.7
forsterite	2MgO.SiO$_2$	1885	3.22	837.4		3.1	2.4	9.5
zircon	ZrO$_2$.SiO$_2$	2340–2550^e	4.60	544.3		4.3	4.1	4.0
carbon	C		2.10	1046.7	25–1300	13.4	9.9	4.0
silicon carbide	SiC	3990^f	3.21	795.5	25–1300	22.5	23.7	5.2

^{a}To convert J to cal, divide by 4.184.
^{b}Silica glass.
^{c}Cubic, stabilized with CaO.
^{d}Congruent.
^{e}Incongruent.
^{f}Dissociates above 2450°C in reducing atmosphere and is readily oxidized above 1650°C.

Table 2. Typical Ranges of Physical Properties of Alumina, Silica, and Zirconia Refractory Brick

Type	PCE^a	Modulus of rupture MPa^b	Deformation under 172 kPa^b load, % linear change after 1½ hours Temperature, °C	Deformation under 172 kPa^b load, % linear change after 1½ hours Change, %	Linear reheat change^c at °C	Linear reheat change^c %	Bulk density, g/cm^3
silica		2.8–11.2	1650	0			1.60–1.80
fireclay							
semisilica	27–31	2.1–4.2	1450	0.1–2			1.80–2.10
medium duty	29–31	7.0–11.2	1450	1–6			2.11–2.20
high duty	31–33	2.8–21.0	1450	0.5–15	1600	0–1 S	2.13–2.30
super duty	33–34	2.8–24.0	1450	0–9	1600	0–1 S	2.28–2.48
high alumina							
50% Al$_2$O$_3$	34–35	7.0–11.2	1450	2–6	1600	0–1 E	2.27–2.43
60%	36–37	7.0–11.2	1450	1–4	1600	0–5 E	2.10–2.49
70%	37–38	7.0–11.2	1450	1–3	1600	1–7 E	2.20–2.66
80%	38–39	7.0–12.6					2.50–2.90
85%^{d}	38–39	21.0–35.0	1450	0.3–3	1600	0–2 E	2.70–2.92
90%	40–41	14.0–35.0	1700	0–1	1700	0–1 E	2.67–3.10
100%	41–42	12.6–21.0	1650	1–2	1700	0–0.5 S	2.84–3.10
zircon			1600	2–5	1540	0	3.77
molten cast							
Al$_2$O$_3$–ZrO$_2$–SiO$_2$							3.70–3.74
Al$_2$O$_3$–SiO$_2$							3.00–3.20
Al$_2$O$_3$ high soda							2.89
Al$_2$O$_3$ low soda							3.50

^{a}Pyrometer cone equivalent, as determined by ASTM C 24.
^{b}To convert MPa to psi, multiply by 145; for kPa, multiply by 0.145.
^{c}S = Shrinkage; E = expansion.
^{d}Phosphate bonded.

Inhalation of certain fine dusts may constitute a health hazard. OSHA regulations specify the allowable levels of exposure to ingestible and airborne particulate matter.

Selection and Uses

Any manufacturing process requiring refractories depends upon proper selection and installation. When selecting refractories, the environmental conditions are evaluated first, then the functions to be served, and finally the expected length of service. All factors pertaining to the operation, service, design, and construction of equipment must be related to the physical and chemical properties of the various classes of refractories.

By far the most common industrial refractories are those composed of single or mixed oxides of Al, Ca, Cr, Mg, Si, and Zr. These oxides exhibit relatively high degrees of stability under both reducing and oxidizing conditions. Carbon, graphite, and silicon carbide have been used both alone and in combination with these oxides. Refractories made from the above materials are used in ton-lot quantities in industrial applications. Other refractory oxides, nitrides, borides, and silicides are used in relatively small quantities for specialty applications in the nuclear, electronic, and aerospace industries.

Silica refractories. Uses include open-hearth roof linings, refractories for coke ovens, coreless-induction foundry furnaces, and fused-silica technical ceramic products.

Semisilica refractories. Uses include shapes for open-hearth stoves and checkers.

Fireclay refractories. Fireclay refractories are used in kilns, ladles, and heat regenerators, acid-slag-resistant applications, boilers, blast furnaces, and rotary kilns.

High alumina refractories. These refractories are used in kilns, ladles, and furnaces that operate at temperatures or under conditions for which fireclay refractories are not suited.

Chrome refractories. These refractories are used in nonferrous metallurgical furnaces, rotary-kiln linings, secondary refining vessels, such as argon–oxygen decarborizers (AODs) and glass-tank regenerators.

Magnesite refractories. These refractories are used in lining and maintenance of steelmaking and refining vessels and checkers.

Dolomite refractories. Dolomite refractories are primarily used in linings of BOF vessels and refining vessels, and in ladles and cement kilns.

Spinel refractories. Spinel refractories are used in cement kilns and steel-ladle linings.

Forsterite refractories. Uses include nonferrous-metal-furnace roofs and glass-tank refractories not in contact with the melt, ie, checkers, ports, and uptakes.

Silicon carbide refractories. Silicon carbide has a wide range of refractory uses including chemical tanks and drains, kiln furniture, abrasion-resistance linings, blast-furnace linings, and nonferrous metallurgical crucibles and furnace linings.

Zirconia refractories. The most common zirconia-containing refractories are made from zircon sand and are used mostly for glass-tank paver brick.

H.D. LEIGH
C-E Basic, Inc.

F. Singer and S.S. Singer, *Industrial Ceramics*, Chemical Publishing Company, Inc., New York, 1964.

F.H. Norton, *Refractories*, 3rd ed., McGraw-Hill Company, New York, 1949.

W.D. Kingery, *Introduction to Ceramics*, John Wiley & Sons, Inc., New York, 1960.

E. Ryshkewitch, *Oxide Ceramics*, Academic Press, Inc., New York, 1960.

REFRACTORY COATINGS

Refractory coatings denote those metallic, refractory-compound (ie, oxides, carbides, nitrides) and metal-ceramic coatings associated with high temperature service as contrasted to coatings used for decorative or corrosion-resistant applications. They also denote coatings of high melting materials that are used in other than high temperature applications. A coating may be defined as a near-surface region with properties that differ significantly from the bulk of the substrate (see Ceramics; Metallic coatings; Metal surface treatments).

The highest melting refractory metals are tungsten, tantalum, molybdenum, and niobium, although titanium, hafnium, zirconium, chromium, vanadium, platinum, rhodium, ruthenium, iridium, osmium, and rhenium may be included. Many of these metals do not resist air oxidation; hence, very few if any, are used in their elemental form for high temperature protection. However, bulk alloys based on nickel, iron, and cobalt with alloying elements such as chromium, titanium, aluminum, vanadium, tantalum, molybdenum, silicon, and tungsten are used extensively in high temperature service. Some modern high temperature oxidation- and corrosion-resistant coatings have compositions similar to the high temperature bulk alloys and are applied by thermal spraying,

evaporation, or sputtering. The protection mechanism for these high temperature alloy coatings is based on adherent impervious surface films of the Al_2O_3, SiO_2, CrO_2, or a spinel-type that grows on high temperature exposure to air.

Refractory coatings also include materials with high melting points, eg, silicides, borides, carbides, nitrides, or oxides, and combinations such as oxy-carbides, etc. In addition, mixtures of metals and refractory compounds (sometimes called metallides) of various microstructural configurations (ie, laminates, dispersed phases, etc) can also be classified as refractory coatings.

In some cases, a coating is a new material that is deposited onto the substrate by a variety of methods. It is then called a deposited or overlay coating. In other cases, the coating may be produced by altering the surface material to produce a surface layer composed of both the added and substrate materials. This is called a conversion coating, cementation coating, diffusion coating, or chemical-conversion coating when chemical changes in the surface are involved. Coatings may also be formed by altering the properties of the surface by melting and quenching, mechanical deformation, or other processes that change the properties without changing the composition.

All coating methods consist of three basic steps: synthesis or generation of the coating species or precursor at the source; transport from the source to the substrate; and nucleation and growth of the coating on the substrate. These steps can be completely independent of each other or may be superimposed on each other, depending on the coating process. A process in which the steps can be varied independently and controlled offers great flexibility, and a larger variety of materials can be deposited.

Numerous schemes can be devised to classify deposition processes. The scheme used here is based on the dimensions of the depositing species, ie, atoms and molecules, liquid droplets, bulk quantities, or the use of a surface-modification process (see Table 1).

The coating has to adhere to the substrate. The bonding may be mechanical as a result of the interlocking between the asperities on the surface and the coating. In diffusion or chemical bonding, the substrate and the coating material interdiffuse at the interface.

Residual stresses, which are always present in coatings, arise from thermal-expansion mismatch between coating and substrate, or growth stresses caused by imperfections in the coating that are built in during the process of film growth.

Coatings may be permeable either to the atmosphere or the substrate material. Diffusion of oxygen through a coating can result in gaseous products that may rupture the coating. The design of a coating that might equilibrate with its substrate during use is based on the phase diagram of the system.

Microstructure of Coatings

The microstructure of bulk coatings resembles the normal microstructure of metals and alloys produced by melt solidification. The micro-

Table 1. Coating Methods

Atomistic deposition	Particulate deposition	Bulk coatings	Surface modification
electrolytic environment	thermal spraying	wetting processes	chemical conversion
electroplating	plasma spraying	painting	electrolytic
electroless plating	D-gun	dip coating	anodization (oxides)
fused-salt electrolysis	flame spraying	electrostatic spraying	fused salts
chemical displacement	fusion coatings	printing	chemical–liquid
vacuum environment	thick-film ink	spin coating	chemical–vapor
vacuum evaporation	enameling	cladding	thermal
ion-beam deposition	electrophoretic	explosive	plasma
molecular-beam epitaxy	impact plating	roll bonding	leaching
plasma environment		overlaying	mechanical
sputter deposition		weld-coating	shot peening
activated reactive			
evaporation		liquid-phase epitaxy	thermal
plasma polymerization			surface enrichment
ion plating			diffusion from bulk
chemical-vapor environment			sputtering
chemical-vapor deposition			ion implantation
reduction			
decomposition			
plasma enhanced			
spray pyrolysis			

structure of particulate-deposited materials resembles a cross between rapidly solidified bulk materials with severe deformation and powder compacts produced by pressing and sintering. A special feature of particulate coatings is a significant degree of porosity (ca 2–20 vol%) which strongly affects the properties of the deposit.

The microstructure and imperfection content of coatings produced by atomistic deposition processes can be varied over a very wide range to produce structures and properties similar to or totally different from bulk-processed materials.

Applications

Coatings can be classified into six categories: chemically functional, mechanically functional, optically functional, electrically functional, biomedical, and decorative. In addition, there are some unique applications in the aerospace program, such as the ablative coatings of pyrolytic carbon and graphite- and silica-based materials for protection of nose cones and the space shuttle during reentry (see Ablative materials). another unique energy-related application is the coating of low-Z elements such as TiC for the first wall of thermonuclear reactors to minimize contamination of the plasma.

ROINTAN F. BUNSHAH
University of California, Los Angeles

H.H. Hausner, ed., *Coatings of High Temperature Materials*, Plenum Publishing Corp., New York, 1966.

R.F. Bunshah in *New Trends in Materials Processing*, American Society of Metals, Metals Park, Ohio, 1974, pp. 200–269.

D.H. Leeds in J.E. Hove and W.C. Riley, eds., *Ceramics for Advanced Technologies*, John Wiley & Sons, Inc., New York, 1965, Chapt. 7.

R.F. Bunshah, ed., *Techniques of Metals Research*, Vol. I, parts 1–3; Vol. VII, Pt. 1, John Wiley & Sons, Inc., New York, 1968.

REFRACTORY FIBERS

The refractory fiber defines a wide range of amorphous and polycrystalline synthetic fibers used at temperatures generally above 1093°C (see also Fibers, chemical; Refractories). Chemically, these fibers can be separated nito oxide and nonoxide fibers. The former include alumina–silica fibers (Fig. 1) and chemical modifications of the alumina–silica system, high silica fibers (> 99% SiO_2), polycrystalline zirconia, and alumina fibers. The diameters of these fibers are 0.5–10 μm (av ca 3 μm). Their length, as manufactured, ranges from 1 cm to continuous filaments, depending upon the chemical composition and manufacturing technique. Such fibers may contain up to ca 50 wt% unfiberized particles. Commonly referred to as shot, these particles are

the result of melt fiberization usually associated with the manufacture of alumina–silica fibers. The presence of shot reduces the thermal efficiency of fibrous systems. Shot particles are not generated by the manufacturing techniques used for high silica and polycrystalline fibers, and consequently, these fibers usually contain < 5 wt% unfiberized material. Refractory fibers are manufactured in the form of loose wool. From this state, they can be needled into flexible blanket form, combined with organic binders and pressed into flexible or rigid felts, fabricated into rope, textile, and paper forms, and vacuum-formed into a variety of intricate, rigid shapes.

The nonoxide forms, silicon carbide, silicon nitride, boron nitride, carbon, or graphite have diameters of ca 0.5–50 μm. Generally, nonoxide fibers are much shorter than oxide fibers except for carbon, graphite, and boron fibers, which are manufactured as continuous filaments. Carbon, graphite, and boron fibers are used for reinforcement in plastics in discontinuous form and for filament winding in continuous form. In addition to moderate temperature resistance, these fibers have extremely high elastic modulus and tensile strength. Carbon and graphite fibers cannot be accurately classified as refractory fibers because they are oxidized above ca 400°C. This is also true of boron fibers which form liquid boron oxide at approximately 560°C. Most silicon carbide, silicon nitride, and boron nitride fibers are relatively short, ranging in length from single-crystal whiskers of less than 1 mm to fibers as long as 5 cm. Diameters are ca 0.1–10 μm. These fibers are used mostly to reinforce composites of plastics, glass, metals, and ceramics. They also have limited applications as insulation in and around rocket nozzles and in nuclear-fusion technology requiring resistance to short-term temperatures above 2000°C. For inert atmosphere or vacuum applications, carbon-bonded carbon-fiber composites provide effective thermal insulation up to 2500°C.

Health and Safety Factors

Synthetic or man-made refractory fibers do not appear to pose the health hazards of naturally occurring mineral fibers like asbestos. Nonetheless, proper respirators should be worn in environments of excessive exposure to refractory fibers.

W.C. MILLER
Manville Corporation

H.W. Rauch, Sr., W.H. Sutton, and L.R. McCreight, *Ceramic Fibers and Fibrous Composite Materials*, Academic Press, New York, 1968, p. 24.

L. Olds, W. Miller, and J. Pallo, *Am. Ceram. Soc. Bull.* **59**(7), 739 (1980).

J. Leineweber, *ASHRAE J.* **3**, 51 (1980).

REFRIGERATION

The chemical industry uses refrigeration in installations covering a broad range of cooling capacities and temperature. The variety of applications results in a diversity of mechanical specifications and equipment requirements. Nevertheless, the methods for producing refrigeration are well-standardized.

Basic Principles

Thermodynamic principles govern all refrigeration processes (see Thermodynamics). Refrigeration is accomplished in a Rankine cycle by using a fluid that evaporates and condenses at suitable pressures for practical equipment designs. The vapor-compression cycle is illustrated by a pressure-enthalpy diagram, Figure 1.

Many of today's chemical plants use pumped recirculation of refrigerant rather than direct evaporation of refrigerant to service remotely located or specially designed heat exchangers. This technique provides the users with wide flexibility in applying refrigeration to complex processes and greatly simplifies operation. Secondary refrigerants or brines also are commonly used today for simple control and operation. Ice and brine storage tanks may be used to level batch cooling

Figure 1. Alumina–silica–chromia fiber after 120 h at 1426°C showing crystallization and sintering at contact points. ×5000.

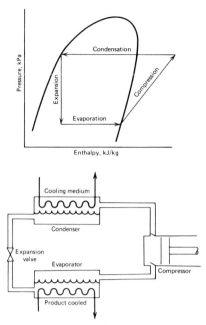

Figure 1. Basic compression cycle. To convert kPa to atm, divide by 101.3 To convert kJ/kg to Btu/lb, multiply by 0.4302.

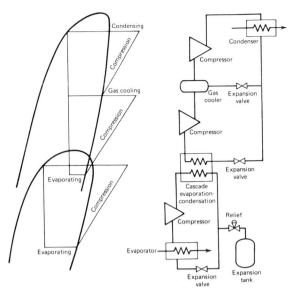

Figure 2. Cascade cycle.

loads and reduce equipment size. This approach provides stored refrigeration where temperature control is vital as a safety consideration to prevent runaway reactions or pressure buildup.

All mechanical cooling results in the simultaneous production, somewhere else, of a greater amount of heat. It is not always realized that this heat can be put to good use by applying the heat-pump principle. This requires provision to recover the heat normally rejected to cooling water or air in the refrigeration condenser. Recovery of this waste heat at temperatures up to 65°C is frequently employed in modern plants to achieve improved heat balance and operating economy (see Energy management).

If a steady supply of waste heat is available, absorption (qv) systems with water as the refrigerant and lithium bromide as the absorbent solution may be used for water-chilling service. For process applications that require chilled fluid < 7°C, the ammonia–water pair is used, ammonia serving as the refrigerant. Single-stage systems are most common at a generator heat input < 95°C. The coefficient of performance (COP) of a system is the cooling achieved in the evaporator divided by the heat input to the generator. The COP of a lithium bromide machine is generally 0.65–0.70 for water-chilling duty.

Historically, capacities of mechanical refrigeration systems have been stated in tons of refrigeration, a unit of measure related to the ability of an ice plant to freeze one short ton (907 kg) of ice in 24 h. Its value is 3.51 kW (12,000 Btu/h). Often a kilowatt of refrigeration capacity is identified as kW_{th} to distinguish it from the amount of electricity (kW_e) required to produce the refrigeration.

Refrigeration Cycles and System Overview

Refrigeration can be accomplished in either closed-cycle or open-cycle systems. In a closed cycle, the refrigerant fluid is confined within the system and recirculates through the processes in the cycle. The system shown at the bottom of Figure 1 is a closed cycle. In an open cycle, the fluid used as the refrigerant passes through the system once on its way to be used as a product or feedstock outside the refrigeration process. An example is the cooling of natural gas to separate and condense heavier components (see Gas, natural).

In addition to the distinction between open- and closed-cycle systems, refrigeration processes are also described as simple cycles, compound cycles (upper portion of Fig. 2) or cascade cycles (lower portion of Fig. 2). Simple cycles employ one set of components and a single refrigeration

cycle as in Figure 1. Compound or cascade cycles may be thought of as simple cycles that interact to accomplish cooling at several temperatures or to allow a greater span between the lowest and highest temperatures in the system than can be achieved with the simple cycle.

Refrigerant selection for the closed cycle. The operating fluid of choice is the one whose properties are best suited to the operating conditions. The factors below should be taken into account when selecting a refrigerant: discharge (condensing) pressure; suction (evaporating) pressure; standby pressure (at ambient temperature); critical temperature and pressure; suction volume; freezing point; theoretical power required for adiabatic compression of the gas; vapor density; liquid density; latent heat; and refrigerant cost.

Refrigerants. No one refrigerant meets all the ideal requirements for the wide range of temperatures and the multitude of applications required by modern chemical processing. However, the chemical industry normally uses low cost fluids such as propane and butane whenever they are available in the process These rediscovered hydrocarbon refrigerants, once thought of as too hazardous because of flammability, are entirely suitable for use in modern compressors and frequently add no more hazard than already exists in the oil refinery or petrochemical process. These low cost refrigerants are used in simple, compound, and cascade systems, depending on operating temperatures.

Most of the nontoxic, nonflammable refrigerants are halogenated hydrocarbons containing one or more of the halogens: fluorine, chlorine, and occasionally bromine (see Chlorocarbons and chlorohydrocarbons; Fluorine compounds, organic).

Refrigerant mixtures. Because the bubble-point and dew-point temperatures are not the same for a given pressure, nonazeotropic mixtures may be used to help control the temperature differences in low temperature evaporators.

Open-cycle refrigerant selection. Process gases used in the open cycle include chlorine, ammonia, and mixed hydrocarbons. Specifications should consider the following: composition, corrosion, dirt and liquid carryover, and polymerization.

Indirect refrigeration (brine). The process fluid is cooled by an intermediate liquid, water or brine, that is itself cooled by evaporating the refrigerant as shown in Figure 3. In the chemical industry, process heat exchangers frequently must be designed for corrosive products, high pressures, or high viscosities, and are not well-suited for refrigerant evaporators. Other problems preventing direct use of refrigerant are remote location, lack of sufficient pressure for the refrigerant-liquid feed, difficulties with oil return, or inability to provide traps in the suction line to hold liquid refrigerant. Use of indirect refrigeration simplifies the piping system; it becomes a conventional hydraulic system.

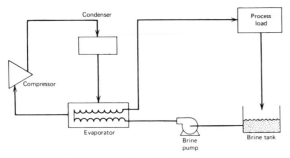

Figure 3. Secondary brine system.

The brines commonly used are brines having a salt base, brines with a glycol base, brines derived from an alcohol base, and brines for low temperature heat transfer.

Use of ice. Where water is not harmful to a product or process, ice may be used to provide refrigeration. Direct application of ice or of ice and water is a rapid way to control a chemical reaction or remove heat from a process.

System Components, Construction, and Operation

Compressors. Reciprocating or centrifugal compressors can be used singly or in parallel and series combinations for process compression and refrigeration applications (see High pressure technology).

Condensers. The refrigerant condenser is used to reject the heat of compression and the process heat load picked up in the evaporator. This heat can be rejected to cooling water or air, both of which are commonly used.

Evaporators. The broad range of chemical-plant cooling requirements has led to an extremely wide variety of refrigerant evaporator designs. There are special requirements for evaporators in refrigeration service that are not always present in other types of heat-exchanger design. These include problems of oil return, flash-gas distribution, gas-liquid separation, and submergence effects (see Evaporation).

System-Design Considerations

Chemical processes may be continuous or batch. Continuous processing is characterized by temperatures, pressures, flow levels, compositions, and other parameters that do not change with time. Associated with continuous operation are refrigeration start-up and shutdown conditions that invariably differ, sometimes widely, from those of the process itself. These conditions, although they occupy very little time in the life of the installation, must be properly accommodated in the design of the refrigeration system. Consideration must be given to the amount of time required to achieve design operating conditions, the need for standby equipment, etc.

In batch processing, operating conditions are expected to change with time, usually in a repetitive pattern. The refrigeration system must be designed for all extremes. Use of brine storage or ice banks can reduce equipment sizes for batch processes.

Refrigeration System Specifications

To minimize costly and time-consuming alterations owing to unexpected requirements, the refrigeration specialist who is to do the final design must have as much information as possible before the design is started. Usually, it is best to provide more information than thought necessary, and it is always wise to note where information may be sketchy, missing, or uncertain. Carefully spelling out the allowable margins in the most critical process variables and pointing out portions of the refrigeration cycle that are of least concern is always helpful to the designer.

A checklist of minimum information needed by a refrigeration specialist to design a cooling system for a particular application may be helpful.

It will include process flow sheets, basic specifications, instrumentation and control requirements, and off-design operation.

K.W. Cooper
K.E. Hickman
York Division
Borg-Warner Corporation

F.C. McQuiston and J.D. Parker, *Heating, Ventilating and Air Conditioning, Analysis and Design*, John Wiley & Sons, Inc., New York, 1977.

W.F. Stoecker, *Design of Thermal Systems*, Industrial Press, Inc., New York, 1968.

ASHRAE Handbooks, American Society of Heating, Refrigerating and Ventilating Engineers, Inc., Publications Department, Atlanta, Ga., 4 vols.: Fundamentals, Equipment, Systems, Applications.

REGULATORY AGENCIES

The U.S. government must provide for the welfare of its citizens; thus, the maintenance of a clean and healthy environment, especially as related to chemical engineering, is regulated by government agencies.

The two main Federal agencies involved in the protection of human health and environment are the Environmental Protection Agency (EPA) and the Occupational Safety and Health Administration (OSHA). The EPA's principal concern is the protection of the environment, especially as it affects the quality of life. The principal function of OSHA is the protection of people in the workplace.

Figure 1 shows the main Federal and state environmental permit and regulatory approvals.

Water

Water pollution is regulated in two ways. Water quality standards are set by the states based on stream-use classification and water quality criteria. The EPA issues Effluent Guidelines and Standards for 42 industrial categories limiting the discharge of pollutants. In order to ensure compliance, EPA developed the National Pollutant Discharge Elimination System (NPDES) permit program. Any source discharging to U.S. waters must obtain an NPDES permit.

Other EPA regulations protect wetlands, coastal areas and the ocean. The Army Corps of Engineers regulates water pollution as it affects navigation (see Water, water pollution).

Air

EPA regulates ambient air quality as well as air-pollutant emissions. The National Ambient Air Quality Standards (NAAQS) set primary and

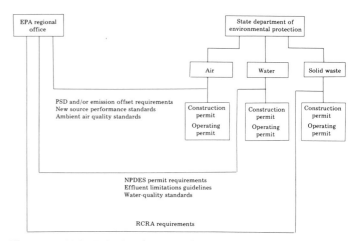

Figure 1. Main Federal and state environmental permits and regulatory approvals.

secondary standards for seven criterion pollutants; sulfur oxides, particulates, carbon monoxide, hydrocarbons, nitrogen oxides, ozone, and lead. The New Source Performance Standards (NSPS) limit emissions of specific pollutants from over 30 different industrial sources.

The EPA has also issued regulations for certain hazardous air pollutants. These include asbestos, benzene, beryllium, mercury, and vinyl chloride (see Air pollution).

Solid Waste

Control of pollution from solid-waste disposal is covered by the Resource Conservation and Recovery Act (RCRA). EPA regulates hazardous waste from generation to disposal under RCRA. The regulations include definitions of hazardous wastes, testing procedures, transportation requirements, storage methods, and disposal methods (see Wastes, industrial).

Chemicals

The control of the manufacture, use, and exposures to hazardous or toxic chemicals is mainly divided between the EPA and OSHA. In addition, the Food and Drug Administration (FDA) has control over chemicals in food, drugs, and cosmetics (qv). The Consumer Product Safety Commission (CPSC) is concerned with the safety of all consumer products, including child-resistant packaging regulations. The U.S. Department of Agriculture (USDA) maintains strict controls over chemicals in food.

Toxic Substance Control Act. In 1976, Congress passed the Toxic Substance Control Act (TSCA), which is administered by the EPA. The two main goals of TSCA are acquisition of sufficient information to identify and evaluate potential hazards from chemical substances and regulation of the production, use distribution, and disposal of these substances.

Occupational Safety and Health Act. OSHA has very broad responsibilities for protecting the workplace. The Occupational Safety and Health Act is administered by the Occupational Safety and Health Administration under the Department of Labor. The act covers all health and safety aspects of a worker's environment. Subpart Z of the Act, Toxic and Hazardous Substances, lists allowable employee exposure to many different chemical substances (29 CFR 1910.1000). These are given as ambient-air concentrations over a certain period, which usually is an 8-h TWA. Sometimes a ceiling concentration is given as well. Certain substances, eg, vinyl chloride, benzene, etc, are discussed in terms of necessary controls and limits.

Noise

Noise has long been considered a nuisance if not a definite health hazard (see Noise pollution). Nonetheless, until recently, its regulation has not been as vigorous as for other environmental concerns. The 1969 Amendment to the Walsh-Healy Public Contracts Act established a maximum sound-pressure level of 90 dBA (decibels measured on the A scale) for noise exposure in the workplace as an eight-hour TWA. This standard was incorporated under OSHA in 1970 (29 CFR 1910.95).

The jurisdictions of OSHA and EPA differs in that OSHA protects workers' hearing in the plant, whereas EPA is concerned with the surrounding community. EPA is developing standards controlling noise from specific equipment.

Conservation, Land Use, and Other Areas Subject to Regulation

A new facility should be located in an area that is economically suitable. At the same time, the social, environmental, and aesthetic effects must be considered. These decisions are part of land-use policies. At this time, there is no national land-use program. However, there are some laws that cover portions of this decision-making process and some states have such programs. The most important Federal law is the National Environmental Policy Act.

NANCY R. PASSOW
Lummus Crest Inc.

Chem. Eng., 81 (June 30, 1980); 55 (July 28, 1980).

J. Quarles, "Federal Regulation of New Industrial Plants: A Survey of Environmental Regulations Affecting the Siting and Construction of New Industrial Plants and Plant Expansion," *Environment Reporter*, Monograph No. 28, Bureau of National Affairs, Inc., Washington, D.C., 1979.

N.R. Passow, *Chem. Eng.*, 69 (Dec. 15, 1980).

RENEWABLE RESOURCES. See Chemurgy.

REPELLENTS

Repellents affect insects and other organisms and disrupt their behavior. Hostseeking and biting of humans and animals is interrupted to prevent the spread of disease.

Evaluation

Cloth test. Repellents are applied to a cloth at 33 g/m^2, usually as a 1% solution in acetone. Two hours later, the cloth is placed over an untreated nylon stocking on the arm of a subject, and the arm is exposed to 1500 female mosquitoes for one minute. If fewer than five bites are counted, the test is repeated until failure which is, by definition, five bites per minute.

Skin test. The chemical is dissolved in ethanol and spread over one forearm; N,N-diethyl-m-toluamide (DEET) (1) is applied to the other forearm. Each arm is then exposed and the results are compared.

DEET
(1)

Space-borne repellents. Air is drawn over a human arm through a 9.5-cm disk of cotton netting treated with repellent; the air is then drawn into an olfactometer cage containing 125 female *Aedes aegypti*. The number of days the repellent prevented > 10% of the mosquitoes from passing through the netting constitutes "effectiveness."

Skin-patch test. Small areas of forearm are treated with repellent. The patches are tested against small numbers of mosquitoes.

Repellents not using human bait. A treated strip of fabric and a control strip are lowered into a container of crawling arthropods. After a predetermined time, the strips are lifted and the animals remaining attached are counted.

Inanimate and animal attractants. Machines have been constructed to remove the factor of human odor in attempts to simplify the measurement of repellency. There is no standardized procedure. Numerous methods involve animals as attractants, followed by evaluation of the repellents as skin or cloth treatments.

Arthropod Repellents

Of 435 arboviruses (arthropod-borne viruses) found in insects, 100 cause diseases in humans, including yellow fever, dengue haemorrhagic fever, and many forms of encephalitis.

Mosquito repellents. Clothing impregnates may be applied by dipping into emulsions or manual application, or by spraying the surface with an aerosol. The standard clothing impregnate adopted by the U.S. Armed Services, M-1960, is intended for utilization in military laundries for protection from mosquitoes, fleas, ticks, and chigger mites.

Dimethyl phthalate (DMP) is used as a standard against *Ae. aegypti*, and is effective for 11–22 d on cloth. Rutgers 6-12, 2-ethyl-1,3-hexanediol, is exceptionally effective against *Ae. aegypti*, lasting 196 days on cloth. DEET has proved to be the best all-purpose repellent yet developed.

In a comparison of nine commercial repellents against *Ae. aegypti*, Stabilene and MGK Repellent 326 were inferior to DEET, dibutyl phthalate, Indalone (**2**), dimethyl phthalate, MGK Repellent 11 (**3**), 2-ethyl-1,3-hexanediol, and Citronyl (**4**).

Indalone
(**2**)

MGK Repellent II
(**3**)

Citronyl
(**4**)

In Florida, DEET averaged 28 times greater protection than an untreated control, whereas other materials averaged 190 times greater protection. DEET was considered an effective repellent when applied to exposed skin as a 25 wt% formulation, but four novel alicyclic piperidines (carboxamides) were more effective.

Mineral-oil preparations were shown to have a repellent effect when applied thickly to skin.

Dimethyl phthalate and pyrethrum cream were partially protective on skin against *Phlebotomus papatasi* for 6 h. DEET-treated bed nets gave complete protection. DEET-treated net jackets also provided good protection, but application of repellent to the face was necessary for maximum protection.

Tick and chigger repellents. Repellents are best impregnated into clothing. A 5% DEET solution gives 80–98% protection, depending on tick density.

Permethrin, available in an aerosol formulation called Permanone Tick Repellent, gives extremely effective protection against ticks. It acts more as toxicant than as repellent. The lethal barrier has been shown to give 100% protection (see Poisons, economic).

Cockroach Repellents

Although repellents may never completely stop the development of infestations, they may be helpful in preventing transport of cockroaches. Some uses of repellents are on cardboard cartons for food and soft drinks, on beer crates, and in vending machines.

Dibutyl succinate (Tabutrex) is formulated as an emulsion concentrate and an oil spray. Cockroaches were repelled from wooden beverage crates for 15 wk. Hexahydrodibenzofurancarboxaldehyde–butadienefurfural copolymer (MGK R-11) gave > 80% repellancy for 2 mo.

tert-Butyl *N,N*-dimethyldithiocarbamate (MGK R-55) is a rodent and insect repellent. It repels *Blattela germanica* from cartons for 90 d (at 2%) and for 63 d (at 1%). 2-Hydroxyethyl *n*-octyl sulfide (MGK R-874) tested against German cockroaches is marginally more effective than R-55 and lasts twice as long as R-11. Toxicity is low. The use of this repellent near food should not create a health risk.

Many toxicants are known to have repellent effects. Pyrethrins often are used on ships to flush cockroaches from harborages during a treatment with another less activating toxicant.

Diazinon (**5**) is the pesticide of choice for cockroach control.

Diazinon
(**5**)

Bird Repellents

Blackbirds, starlings, and sparrows cause serious damage to crops. Repellent devices such as propane cannons, scarecrows, metallic pinwheels, and recorded distress calls give temporary results at best. Chemical repellents are expensive. It seems practical to breed the ability to resist bird depredation into the genetic composition of the plants, and much effort has been directed to this end.

Some bird repellents are viscous, sticky materials that birds dislike. These are often based on incompletely polymerized isobutylene.

Intoxicating chemicals are not necessarily toxic, but operate as repellents whose mode of action is owing to a bad taste. Rejection of food is the desired result. These chemicals include condensed tannins obtained from sorghum and other cereal grains that have astringent properties. However, they are effective only if other foods are not available

Aldicarb (**6**) is classified as an insecticide, acaricide, and nematicide, and has been used as a bird repellent. Avitrol, 4-aminopyridine, has repellent–toxicant properties for birds and is classed as a severe poison and irritant. Methiocarb (**7**) (Mesurol) is classed as an insecticide and acaricide and is used as a slug and snail bait.

(**6**)

(**7**)

Anthrahydroquinones were patented as bird repellents, and anthraquinone is used widely in Europe as a spray to protect growing crops and as a wood dressing.

Deer, Rabbit, and Rodent Repellents

The feeding activity of deer has become a problem in the Pacific Northwest of the United States, especially where black-tailed deer and Roosevelt elk cause damage to Douglas fir seedlings. A pellet containing selenium embedded in the soil near a seedling causes the sapling to have "bad breath." Wild deer are repelled by the odor of dimethyl selenide. A fermented-egg product has been shown to be attractive to coyotes and repellent to deer.

Hinder or Repel is registered under the Federal Insecticide, Fungicide, and Rodenticide Act (FIFTA) to repel deer and rabbits. It is best applied as an aqueous spray before damage occurs.

Area repellents having a bad odor are intended to keep animals away from a broad area. They include lion scent (lion or tiger manure), blood meal, tankage, bone tar oil, rags soaked in kerosene or creosote, and human hair.

Dog repellents have been generally unsuccessful in laboratory tests, as have ultrasonic repellents used against mosquitoes and cockroaches.

Health and Safety Factors

Toxicological testing has been carried out on many of the older, widely used materials. Few of the newer compounds have been submitted for extensive testing.

D.A. CARLSON
University of Florida

C.E. Schreck, *Ann. Rev. Entomol.* **22**, 101 (1977).

M.D. Buescher and co-workers, *Mosq. News* **42**, 428 (1982).

P.B. Cornwell, *The Cockroach*, Vol. II, Associated Business Programmes, Ltd., London, UK, 1976, pp. 157–190.

REPROGRAPHY

Reprography is the art and science of reproducing documents. It includes copying, which is the reproduction of one or more copies by processes that involve the complete formation of an image for every copy, and duplicating, which is the multiple reproduction from an intermediate-imaged master. Copying by some methods is usually restricted to one or a few copies but, with the advent of high speed copiers, hundreds of copies are routinely produced on copier-duplicators (CDs). These machines are either high speed electrophotographic copiers or high speed integrated master makers and duplicators that produce copies with the same automatic operation as copiers. Duplicating overlaps the CD range: some processes are used for as few as a dozen copies and, in others, masters are used to print 10,000 copies or more (see Electrophotography printing processes).

Copying, duplicating, and printing have been profoundly affected by electronically generated, stored, and transmitted information. The production of documents through word and data processing has not substituted display for copies, but rather has expanded the use of most of the reprographic methods used to reproduce copies plus many new techniques to produce or compose copies and masters where no nonelectronic original exists.

Reprography and printing make up the graphic arts industry. Reprography is growing faster than printing; copying (including copying/duplicating) is growing faster than conventional duplicating. The fastest-growing segment is the reprographic production of copy on paper of digital information (ie, represented by a code or a sequence of discrete elements) from computer-base files, facsimile transmission and word-processing sources, and to a lesser extent, copies from microfilm produced directly from documents or from computers, ie, computer-output microfilm (COM).

Copying

Copying primarily provides copies of documents for office use. It produces legible copy, ie, hard copy on paper and other substrates, from documents and sources such as electronic output from computers, word processors, facsimiles, and microfilms. It can produce many copies conveniently and cheaply, or it can be used to produce image carriers or masters for duplicators and printing presses.

Silver halide photocopying. Until the 1960s, photocopying by methods based on silver halide photography was economically significant. Despite the development of methods that increased the convenience of using basically wet chemistry, first dry thermography and then the dry or at least nonaqueous electrophotography that yielded stable copies in increasingly convenient equipment replaced silver halide photocopying for office work. Before the rapid escalation in the cost of silver-based materials, silver halide was used primarily for some offset applications and micrographics.

The disadvantage of ordinary silver halide processing, which requires exposure, development, short stop, rinsing, fixing, washing, drying, exposure, development, short stop, rinsing, fixing, washout, and drying, arises from the necessity of using a negative and essentially repeating the steps to obtain a positive copy.

Silver halide photocopying processes include gelatin-dye-transfer process and diffusion-transfer reversal (DTR) (see Photography).

Thermal silver copying. Two thermographic processes involving silver and heat development were marketed by 3M in the 1960s. The direct dual-spectrum process was used in low volume photocopying but lost its market share to electrophotography. The dry-silver, negative-working process is used in computer-output microfilm and for phototypesetting.

Thermocopying. In thermocopying, an image is produced by heat. Radiation thermocopying with infrared-sensitive photomaterials, mainly the 3M Thermofax, was very popular in the 1950s and 1960s. It offered a dry, less expensive, and quick alternative to DTR copying. Other than producing inexpensive disposable copies, the thermal units are used to image infrared transparency films for overhead projection. Direct-working films are available in black, red, yellow, blue, and green and in negative or reverse films which give images with a black, red, or blue background. The same thermal-imaging equipment is used for rapid preparation of thermal stencils and thermal spirit masters (see Thermal Stencils; Facsimile Spirit Masters).

Thermorecording by heat conduction preceded the radiation process by several decades.

Plan-copy processes. The photochemical methods of blueprinting and brownprinting, which date from the first half of the 19th century, are obsolete and are only of interest historically to show the rapid product utilization of chemical discoveries. Diazotype expansion into the office market in the 1950s was tempered by the need for translucent originals and the inconvenience of aqueous solutions, ammonia, or almost-stable thermal systems, and was severely inhibited by the silver systems, Thermofax, and the emergence of electrophotography.

Blueprinting. The photosensitivity of iron salts and the blueprinting process were discovered in 1842. The mechanisms of blueprinting reac-

tions are incompletely understood in terms of modern photochemical principles. The basic photoreaction of blueprinting is a photoreduction requiring the presence of an electron donor:

$$Fe^{3+} + e + h\nu \rightarrow Fe^{2+} \tag{1}$$

Reaction 1 is the basic photoreaction of many iron processes which include blueprint processes and brownprint processes.

Diazo processes. The photosensitivity of diazo compounds $RN{=}NX$, where R is a hydrocarbon radical and X is any electronegative substituent, has given rise to many processes used in photography, printing, and reprography. In lightstruck areas of diazo coatings, the potential for coupling to form diazo dyes is lost. Immersing the exposed coating in a coupler solution develops in color the nonilluminated or image area giving a positive print. In 1920, a German monk put both diazo and the coupling component on the same paper, thereby conceiving the dry process. Diazotypy, diazocopy, and diazoreprography are often classified as autopositive photocopying processes.

The anhydrous photoreaction is ideally formulated as a photolysis of the diazo group.

$$ArN_2X + h\nu \rightarrow ArX + N_2 \tag{2}$$

Vesicular process. The submicroscopic bubbles of nitrogen gas given off by diazonium compounds during photo exposure and photolysis are expanded by heat to give light-scattering images by vesicular trapping in a suitable plastic matrix. The diazo compound, eg, p-(dialkylamino)benzenediazonium chlorozincate, p-(diphenylamino)benzenediazonium sulfate, sodium 4-diazo-3,4-dihydro-3-oxo-1-naphthalenesulfonate, or 4-(cyclohexylamino)-3-methoxybenzenediazonium p-chlorobenzenesulfonate, is coated on film base in a hydrophobic thermosetting or thermoplastic resin matrix, which protects the diazo compound from ambient moisture and extends shelf-life. After cold exposure with near-uv light, a reversal image is developed in a thermal process that expands the latent-image vesicles. Subsequent overall exposure stabilizes the image by decomposing the diazo in the background. The nitrogen produced by blanket exposure is lost from the matrix if heat is not applied quickly. This sequence yields a very stable, light-scattering bubble image with a clear background and is widely used to produce positive duplicates of the primary negative-silver microforms. A direct duplicate can be obtained by cold exposure with time allowed for the nitrogen to diffuse out of the film, and blanket exposure followed by heat development.

Washoff process. The washoff process is used for reproducing engineering drawings. Polyester film and waterproofed tracing cloth or paper may be coated with a layer of hydrophilic gelatin to improve adhesion of the photosensitive coating that is to be applied. This intermediate coating is called subbing. The photosensitive coating consists of a colloid of glue, gelatin, or poly(vinyl alcohol); a dichromate, ie, the photosensitizer and photohardener; and silver halide, which provides nonphotoacting image material, in this case in water. After the coating is dry, it is exposed under a photographic negative, which may be conventional or a brownprint. After exposure, the coating is washed with warm water, leaving a positive, photohardened image. The image is visible because some of the silver prints out. The printout image is developed to yield a black image on a clear background.

Electrophotography

The sharp increase of the service and white-collar segment of the work force meant an even greater need for and growth of copying and duplicating other than in the commercial print shop, which led to the popularity of electrophotography. By the 1980s, reprographic prints were produced at over 10^{12} copies, and well over 50% of the copies were being made directly on electrophotographic copiers or copier–duplicators or indirectly on duplicators with masters prepared by electrophotographic means.

Chemical technology contributed significantly to the development of electrophotographic materials, mainly photoreceptors and developers, but the overwhelming success of electrophotography came from the combination of equipment and materials that virtually eliminate any chemistry or any technical input by the user. The convenience of masterless duplicating blurred the traditional distinction between copying and duplicating, as the cost of making copies in an edition directly from an original became competitive with the cost of short-run and long-run duplicating.

Office copy machines. In the 1960s, the first automated xerographic machine, the Xerox 914, was the forerunner of a long line of increasingly automated, faster, and more versatile photocopying machines. The first machines were marketed by Xerox but, after a Federal Trade Commission order in July 1975, making the thousands of Xerox patents available for license, numerous manufacturers in the United States, Europe, and particularly Japan, introduced machines that were called PPCs (plain-paper copiers). Plain-paper copier better describes this type of equipment, for although they are often called xerographic because of the development and transfer steps, liquid development belies dry writing.

Photoreceptors. The first commercial photoreceptor used by Xerox for indirect or transfer electrophotography was selenium. Selenium that was modified with tellurium to enhance red response, as well as an arsenic alloy (ie, $As_2Se_{2.7}Te_{0.3}$) and arsenic triselenide, succeeded vitreous selenium. In 1980, more PPC models used selenium photoreceptors than any other type including those manufactured in Japan, only a third of which in 1976 had selenium-based photoreceptors.

Organic photoconductors (OPC) were introduced in higher-speed copier–duplicators first by IBM and later by Kodak. The IBM copiers had a roll with a coating of a 1:1 molar complex of polyvinylcarbazole and trinitrofluorenone.

Dye-sensitized ZnO coatings on paper were introduced commercially in the 1961 Apeco Electrostat, the first direct electrophotographic office copier. The blending of the dyes to get good color sensitivity with a fresh white appearance was the prime requisite for this coated photocopy paper.

In the 1970s, a ZnO-coated paper master for use in the photoreceptors in PPCs was developed by Japanese manufacturers. In 1976, almost a third of the Japanese PPCs had ZnO photoconductors, although by 1980 none of the new models used this approach. Changing from the single use of ZnO coating in the direct process to the multiple use in the PPCs changed the characteristic for both coating and the base and was analogous to the changes necessary for a good ZnO offset master.

Cadmium sulfide dispersed in an organic polymeric binder with a polyester overlayer having the extended spectral response as well as the surface, to make it suitable for both dry and liquid development, made it a more desirable substitute for selenium than the ZnO system.

Cadmium sulfide in lead-sealing glass binder has been developed to provide a smooth glass-enamel surface that would be very durable and resist damage from development, cleaning, and handling.

A sputtered, amorphous doped silicon (Si:H) photoconductive layer that is superior to ZnO binder in having low fatigue and high speed has excellent photosensitivity over the spectral range and is deposited by glow discharge of SiH_4 in multiplayers and monolayers.

Development. The first methods of development in both indirect and direct electrophotography automated equipment were in dry two-component toning systems: cascade in Xerox 914 and magnetic brush in the Apeco Electrostat. By the late 1960s, CPCs principally involved liquid-immersion development (LID) as it offered much simpler and smaller development and fixing equipment more suitable for low cost desktop machines than the magnetic brush with a radiant fuser. Coated-paper copiers in the 1970s were designed for single-component, magnetic-toner development with either cold-pressure roll or heat-roll fusing.

Fixing. The single-component toners in use are formulated for both cold-pressure and heated-roll fusing, permitting an instant-on feature, or a very short (less than one minute) warm-up time, which is important in convenience copiers.

Sublimation transfer. Based on the sublimation transfer of dyes to textiles, a copy process utilizing direct electrophotography with single-component toners containing one or more sublimable dyestuffs was developed. The sublimable dyes are monoazo, anthraquinone, quinophthalone, or styryl dyestuffs (see Azo dyes; Dyes, anthraquinone).

Plastic-Polymer Imaging

Plastic-Deformation Imaging. Thermal deformation of coatings for thermoplastic and photoplastic recording as well as the frost or thermo-plastic process is a significant reprographic technology, since the deformation images are really not legible.

Photopolymerization. The main use of photopolymerization in imaging is the production of printing plates for letterpress and lithography beyond the needs of duplication runs. Cross-linking printing-ink polymers make up another special case of uv-curing a surface coating.

Micrographic recording. Equipment for projection viewing and retrieval as aperture cards, microcards and microfiche was developed for storage and distribution. Silver halide, both conventional and dry-thermal, is used to record computer output on microfilm and microfiche by cathode ray tube (CRT) and laser beam. Kodak, 3M, Datagraphics, and Bell & Howell supply equipment and materials for computer output microfilm.

Ultramicrofiches with images reduced 75–200 times have been used for microform catalogues for mail-order houses and parts suppliers.

Microform duplication. Diazo and vesicular microforms on a polyester base are used extensively for duplicating silver-based microimages. Vesicular films have two advantages over diazo: greater image permanence than diazo type or silver print, and the ability to produce either direct print or reversal prints.

Updatable micrographics. The usefulness of microfiche is extended by technology that allows entry-posting for an active or open file. Micrographic recording makes it possible to maintain multiple access, record integrity, efficient retrieval, savings in space, and updated records. Technologies are available to add documents conveniently to a file on microfiche through special step-and-repeat camera processors.

Micrographic retrieval. The need for enlarged hard copy from microforms has led to the use of most of the reprographic technologies available for one-to-one copying. Kodak has transparent electrophotographic films that can be used for both microfilm and full-size copies. Plain-paper enlarger-printers have been developed by Xerox and others.

Electronic Printout

The printout of information generated and transmitted electronically with digital signals increased as computers expanded in the post-World War II period. The digital information can be recorded by impact printing with full type characters or by the formation of characters or graphics with a matrix of closely packed dots. The use of nonimpact methods to produce hard copy involves dot-matrix systems. The print quality and resolution depend on the matrix and the number of picture elements, or pels, describing the character.

Impact printing. Impact printers at first were an extension of the typewriter and became more advanced in step with computer input sources. The imaging of whole characters serially by a hammering member striking paper through inked ribbon involved typical typewriter print elements, ie, keys and ball impacters. The high speed, full-character daisy wheel has a circular series of flexible spokes at the ends of which are raised characters.

Much faster at 60–90 (as opposed to 30–55) characters per second are the dot-matrix printers. The printhead contains banks of wires arranged in matrix formats. A dot grouping or matrix in the desired character shape is formed by firing the wires at high speeds against an inked ribbon and paper. The matrix wires can be fired to form characters serially or to print a whole line on a line printer.

The disadvantage common to all impact printers is noise. A typical daisy-wheel printer produces over 80 dB, and the operating noise of the line printer often is well above the OSHA target of 55–60 dBs for offices (see Noise pollution).

Nonimpact printing. Other than the use of digital electronic information to form dot-matrix light images with lasers, light-emitting diodes (LEDs) (see Light-emitting diodes and semiconductor lasers), and CRTs and to print these images by electrophotographic techniques, electronic, magnetic, and thermal signals are used to form images directly. Compared to impact printers, all of these methods are virtually noiseless.

Ink-jet printing. In the 1970s, commercial ink-jet devices that form images by controlled projection of ink provided a nonimpact and noncontact means to produce high quality images at low to very high speeds. The use of ink-jet printing has grown rapidly as it is a reliable printing method for business forms, product coding, computer hard-copy printout, and chart recorders.

There are three basic systems: a high frequency, pressurized, continuous synchronous jet that is electrostatically deflected (Fig. 1); a low frequency, pressurized, electrostatically generated–gated, hence a synchronous or intermittent, jet; and a low-to-medium frequency, nonpressurized impulse jet, which is also known as ink-on-demand or drop-on-demand (Fig. 2).

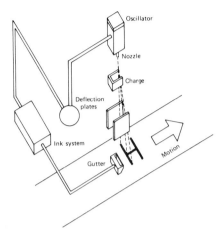

Figure 1. A continuous deflected-drop ink-jet printer. Courtesy of the Society of Photographic Scientists and Engineers.

Electrophotographic printing. The adaptation of systems of electrophotographic copiers to computer printout involves a variety of digitally controlled light inputs. Although a direct electrophotographic process could be used, the transfer plain-paper techniques are extensively adopted for computerized printing equipment. Three types of electrophotographic printing use laser-beam recording, cathode-ray-tube recording, and light-emitting-diode recording.

Intelligent copier. The intelligent copier is a concept that arose in the 1970s as the electronic impact printers roughly provided the quality and speed of electrographic and electrophotographic copiers.

Color Prints

Electrophotographic color. The Xerox 6500 was marketed in 1973 and this machine with both document and 35-mm-color-slide input has been principally marketed as a color-copy service through Xerox Reproduction Centers. In 1980, Canon demonstrated a full-color large-format [28 × 43 cm (11 × 17 in.)], indirect electrophotographic machine, ie, the Canon NP color, which produces 15 copies per minute. Konishiroku also announced the development of a full-color PPC.

Silver halide color copy. Using silver halide color technology, Ilford introduced the Ilford Cibachrome color copying system in 1981. This combined camera process can produce, per hour, 50 of either color prints on coated paper or color transparencies and does not require a darkroom or darkroom operator.

Nonphoto color prints. The use of computer intelligence rather than an optical imaging technique has been used to form color prints with color ink jets in the Applicon color plotter. Impulse ink-jet technology in the Print A Color Corporation IS 8001 and GP 1024 color ink-jet printer terminals with intelligence supplied by a computer gives the hard-copy

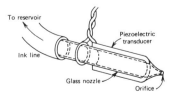

Figure 2. The Gould impulse ink jet. Courtesy of the Society of Photographic Scientists and Engineers.

form to the display on a color CRT terminal. The continuous ink jet with computer control of color shading and intensity reportedly produces very high quality color prints.

Duplicating

Duplication has grown rapidly in large part because of the use of high speed copiers as xerographic copier-duplicators producing hundreds and thousands of copies in a run.

Offset duplication. Offset lithography is by far the fastest growing sector of duplicating. It represents an advance in plate or master preparation and, with improved presses, it is the most prolific and least expensive high volume reprographic method. Stencil printing and spirit duplicating remain significant reprographic methods with continued modification maintaining their niche in low volume, low cost installation and in special applications.

Direct-image offset plates. In direct-image plates, a hydrophilic surface is imaged with oleophilic material, eg, the oily ink of a typewriter ribbon. On a lithographic press, the oleophilic image accepts the oily printing ink, and the unimaged hydrophilic surface, when wet with aqueous fountain solution, repels the hydrophobic ink. A direct-image plate is right-reading and hence is printed by offset. The direct plate has been supplanted by higher quality prints from copiers either by simply copying or making a master on a copier from typescript or another original.

Photolithographic plates include dichromate plates, presensitized photoplates (PSPs), and wipe-on plates.

Offset duplicators. Offset duplicators are manufactured by A.B. Dick Company, AM Multigraphics, ATF Davidson, Didde Graphics Systems Corporation, Gestetner Corporation, Hamada of America, Inc., Heidelberg USA, Itek Graphic Products, Rotoprint, Inc., Royal Zenith Corporation, Solna Corporation, and Ryobi. Web-fed offset machines can produce 3000 sheets per hour to 50,000 web copies per hour. The weight of the stock can be 7.5–208 g/m² (1.5–43 lb/ft²), and card stocks and envelopes can be printed on duplicators.

Unusual offset processes. Several offset processes print quite differently from those based on direct-image or common photoplates. These processes include collotype and continuous-tone lithography.

Camera-direct masters. Although the contact plates have the highest quality and longest run length on aluminum, camera-direct masters can be prepared very rapidly, do not require a darkroom, yield high quality line reproduction, and can be enlarged or reduced. They include silver halide camera masters, electrostatic camera masters, and copier-made masters.

Stencil printing. Stencil printing is an ancient form of duplication in which ink flows through openings on a porous sheet to form an image on a receiving surface; no ink can pass through unopened blocked areas. Screen-process printing, or silk screen, rotary screen, or seriography, is based on woven screens. In the other type of stencil printing, ie, mimeography, the stencil is coated on a special nonwoven medium.

Screen-process printing. Screen-process printing is for specialty printing on almost any surface and for fine-art printing. The screens are mainly of monofilament nylon, mono- and multifilament polyester, metalized polyester, and stainless-steel mesh fabric. Screen printing is commonly used for art prints, decals, nameplates, greeting cards, and program covers, and is particularly well adapted for printing on fabrics, felt, leather, glass, ceramic, and plastic materials.

Mimeograph duplicating. Mimeography first became popular in ca 1885 when the office typewriter provided a convenient means of punching through a wax covering on a fibrous-baselike Yoshino paper. The mimeograph was an A.B. Dick Company machine; the term has since become popular and generic. Electronic facsimile stencil cutters and thermal stencils have extended the range beyond the alphanumerics afforded by typing. Imaged stencils consist essentially of ink-impervious coatings, which are cut to expose a permeable sheet through which ink can pass. The stencils are wrapped around the rotating drum of a duplicator and the drum is perforated to permit brushed-on ink to pass through it to the stencil and through the images on the stencil to the image paper feeding against the stencil on the drum.

Spirit duplicating. In spirit duplicating, a dye image on a master is contacted to a surface that absorbs the dye. The surface gives up the dye where contacted with a moist copy sheet.

Hectography. In its earliest and simplest form, spirit duplicating was called hectography and was practiced using flat trays into which hot gelatin solution had been poured and allowed to set. Paper bearing methyl violet writing or typing ink was then contacted to the gelatin, flattened thereon, and allowed to rest until the gelatin had a substantial image pickup. Then successive moist copy sheets were pressed flat against the tray image to secure 10–20 copies. Tray hectography became popular with the invention of hectograph ribbon for typewriters, and when flatbed and rotary-type duplicators were introduced.

Spirit duplication. Spirit duplicators are rotary machines and are hand-operated or electrically driven and use a wraparound master. The master, prepared by typing or thermal means, consists of a manifold composed of a glazed paper sheet and a backing sheet coated with pigment on the side adjacent to the glazed sheet. Thus, typing on the glazed sheet produces a right-reading typed proof image on the front of the sheet and a wrong-reading transfer image on the back side of the glazed sheet. The take-off image, which is thus generated in the manner of a direct-image lithoplate image, is called a carbon image, and the backing sheet is called a carbon sheet or transfer sheet. As is usual for carbon paper, the dyed and/or pigmented wax or polymer breakaway coating is called carbon ink.

Facsimile spirit masters. Spirit masters are conveniently made from originals in an infrared copier. The original can be typed, printed, drawn, or handwritten. The original and a thermal spirit master construction of a master sheet and an ink-coated, 0.013-mm polyester film with a backup carrier sheet are passed through a Thermofax machine in a matter of seconds. The spirit ink transfers thermally in the image areas to the master sheet to form the wrong-reading master with a resolution of 20 lines per centimeter

Color and specialty spirit duplicating. Spirit duplicating has a unique capability for multicolor replication. The master can be part-imaged using one color of carbon, and can be imaged further with a second color-transfer sheet. As many different carbons as desired may be used to effect desired artwork. Although a multicolor original can be reproduced by a number of other processes, only spirit duplicating can do so with a single master on a single impression.

THOMAS J. KUCERA
Consultant, Evanston, Ill.

M.P. Doss, ed., *Information Processing Equipment*, Reinhold Publishing Corporation, New York, 1955.

Graphic Arts Manual, Arno Press, New York, 1981.

F. Eichenberg, *Lithography and Silkscreen*, Harry N. Abrams, New York, 1978.

RESEARCH MANAGEMENT

The term research and development (R & D) has largely replaced the more limited label of research in industrial usage. The term is being broadened to include the full range of activities required to introduce new technology into commercial application.

Changing Concerns

During the post-World War II period and extending up until the mid-1960s, the focus of R & D management attention was on the creation of the R & D function itself.

In about the middle of the 1960s, the attention of R & D management shifted to provide greatly increased effort on solving the problems of achieving commercial application of the results of R & D. More recently this effort has been broadened to include increased attention to the process of defining the goals, strategies, and management characteristics of the enterprise.

The attention of R & D management is focused today on the interface between research and development and the rest of society and between the research and development function and the rest of the corporation. Higher priority is being given to, ensuring that the mission and charter of the R & D organization are appropriate, that the objectives and strategies of the corporation make allowance for the contributions that can be obtained from new technology, and that R & D managers as well as the programs of the laboratory contribute to the formulation and realization of corporate objectives.

Innovation

Innovation is the introduction of change. Successful R & D management requires understanding and fostering the circumstances that lead to change, both on the part of the R & D managers and the managers of the enterprise or organization supporting the R & D effort.

The manager seeking to encourage an environment for successful research and development must recognize the inevitability of a considerable degree of conflict which cannot be managed away by careful planning and organizing. It is important to try to ensure that this kind of conflict does not become so disruptive that it unduly impedes effective progress. However, the manager should be equally and possibly more concerned with ensuring that those who are striving to introduce innovation are at least strong enough to be worthy opponents for those defending the *status quo* of conventional technology.

Technology

It is important to recognize the various ways in which technology can contribute to technological innovation, because the various dimensions of technology should be taken into account in defining the direction of the R & D organization. Technology can contribute to technological innovation by providing the basis for new and improved products that can lead to increased sales or improved profit margins. In addition to creating new products, there is a dimension that is becoming increasingly important: contributing to the internal efficiency and effectiveness of the enterprise. This can lead to improved productivity, greater flexibility in operations, and shorter response time in reacting to a competitive threat or in capitalizing on a new market opportunity.

Technology can be elevated to ensure the strength and growth of present businesses or to provide opportunities for growth. The fundamental strategic decision facing R & D management is to determine the balance between these two objectives.

Centralized vs Decentralized Organization

Another basic decision that affects the management of research and development is the determination of an appropriate organization. An excellent case can be made for the theory that all advanced technical work should be done as close to the user, ie, the production, engineering, and manufacturing operations, as possible. This approach encourages close coordination in planning, reduces communication barriers inherent in any technology-transfer situation, and fosters a team spirit in those involved.

On the other hand, organizational and geographic decentralization sacrifices the advantages of interdisciplinary synergism and the ability to assemble a critical mass of technical expertise at a single location. Centralized laboratories are less committed to the applications of conventional technology and thus are better adapted to creating and introducing the revolutionary new technology that can replace the technology presently being used. The choice of organization depends on the particular company, its size, the diversity of its business, the extent of its geographic dispersal, and the mode of its overall organization, ranging from strictly functional components to multiproduct divisions.

Program Planning and Evaluation

Irrespective of the mode of organization chosen, one of the critical tests of R & D management is in strategic program planning, which begins with the identification of the main areas of technology that are important to achieving the objectives of the enterprise. This work requires analysis of the various businesses of the company for their technical requirements and consideration of the external trends in science and technology that may provide opportunities for the company.

Strategic program evaluation requires addressing three questions: First, if the program is successful, what impact will it have on the company? Given a choice, one would prefer to work on those programs that have the largest potential impact. Second, what is the likelihood of achieving the desired and needed technical success on this program? Third, what level of effort is required to provide a rate of progress that has a reasonable chance of competitive R & D success?

The program evaluation procedure may be performed in two different ways. One is to establish a specialized program-planning-and-evaluation group to carry out all such analyses. The other is to insist that the analyses be prepared by the R & D managers; the latter is preferable.

Level of Effort

The third critical step in evaluation is the determination of the minimum size of effort necessary to provide a reasonable likelihood of competitive success. This evaluation is judgmental, but it is frequently made unrealistically. R & D managers confront many competing demands for their resources. They often try to take on too many programs in an attempt to satisfy many clients. In this situation, an independent appraisal from the planning component can be helpful in calling attention to the range of effort being applied in other organizations and in questioning the effectiveness of the proposed effort. Strenuous efforts are required by higher levels of R & D management to focus adequate resources on key programs and to stop programs of more limited potential.

Communication between Operations and Corporate Management

Effective communication in research and development almost always requires a two-way interchange of information. Research and development generates new information and creates opportunities for change. Consequently, the level of uncertainty in language and substance inherent in the communication is large compared with most management communication. The possibility of error or misinterpretation is large; therefore, communication must occupy a significant fraction of the time of the R & D manager if he or she is to ensure understanding of the program and support for the objectives.

Communication with operations involves three elements. The first is a flow, primarily from operations to the laboratory, of inputs regarding problems and opportunities as perceived by operations management. The second involves the collaborative planning of joint programs. And the third element of communication with operations involves selling the entire R & D program and function.

Transition to Commercial Application

Participation by operations in evaluating commercial potential is one way of effecting early participation of operations in research management. The transition process cannot be considered to have begun, however, until operations management commits its own resources. Obtaining this commitment of resources is the most important single step to commercial application. It requires creativity and persistence on the part of R & D management as well as flexibility in reshaping program objectives to reflect the wishes and perceptions of operations management. Eventual success, however, requires the creation of a truly joint program.

A.M. BEUCHE[‡]
L.W. STEELE
General Electric Co.

J.A. Morton, *Organizing for Innovation: A Systems Approach to Technical Management*, McGraw-Hill, New York, 1971.

L.W. Steele, *Innovation in Big Business*, Elsevier, New York, 1975.

[‡]Deceased.

RESINS, NATURAL

Natural resins, being organic glasses above their T_g, have historic and current uses in varnishes and diverse adhesives, sealants, and coatings. They are mainly residues from evaporation of most of the volatiles, such as turpentine, from the saps of trees and other plants, but also include lac exudation by insects and flakes in unique coal deposits in Utah.

Their diverse uses depend on degrees of hardness, chemical properties (see Table 1) and durability in the wide range from hard but scarce fossil amber used as gems to soft but abundant rosin from conifers in the United States, Portugal, and China. Annual volume of hand-gathered resins dwindles with time, although nearly one million metric tons of rosin is used in paints, paper size, etc.

Table 1.   Sources and Properties of Natural Resins

Resin	Class	Country of origin	Mp, °C	Acid no.	Saponi-fication no.	Iodine no.
accroides (yacca)		Australia	130		65	200
Congo	amber dark	Zaire	170	100	125	125
Congo	white pale	Zaire	150	110	125	120
damar	Batavia	Indonesia	105	28	34	104
damar	Singapore	Indonsia	115	30	38	113
East India	batu	Indonesia	174	19	33	81
	black	Indonesia	162	20	32	83
	pale, Macassar	Indonesia	140	18	40	103
	pale, Singapore	Indonesia	152	24	35	84
elemi, gum		Philippines	plastic	30	30	118
gilsonite		Utah, U.S.	170			
kauri	brown	New Zealand	160	70	90	120
	pale	New Zealand	130	70	90	140
Manila	Boea, Loba	Indonesia	130	123	160	130
	Macassar	Indonesia	121	136	180	121
	Philippine	Philippines	120	115	145	120
mastic, gum		Greece	76	65	75	100
pontianak		Indonesia	135	118	150	130
rosin	gum	U.S	70	165	174	220
	tall oil	U.S.	80	170	175	
	wood	U.S.	60	156	166	215
sandarac		Morocco	140	135	150	130
shellac		India				
Utah coal resin		Utah, U.S.	170	7		145

Traditional perfumes and medicines depended on the aromatic, volatile components of plant-exuded resins such as Balm of Gilead and Frankincense (see also Oils, essential).

JOHN C. WEAVER
Case Western Reserve University

C.L. Mantell and co-workers, *The Technology of Natural Resins*, John Wiley & Sons, Inc., New York, 1942.

C.L. Mantell, ed., *Natural Resins Handbook*, American Gum Importers Association, New York, 1939. (Now inquire of the O.G. Innes Corp., New York.)

F.J. Martinek in C.R. Martens, eds., *Technology of Paints, Varnishes and Lacquers*, Reinhold Book Corp., New York, 1968, Chapt. 15, pp. 258–272.

RESINS, WATER-SOLUBLE

Water-soluble resins are polymeric materials that are completely soluble or swell substantially in water. They include natural, modified-natural, and completely synthetic materials. Water-soluble resins are of commercial interest because they alter the properties of aqueous solutions or form films. Functions include dispersion, rheology control, binding, coating, flocculation, emulsification, foam stabilization, and protective-colloid action (see Flocculating agents; Emulsions).

Natural

Natural gums (qv) are produced principally in Africa or Asia and are obtained from seeds, seaweed, or exudates from trees (see also Resins, natural). All hydrocolloids from plants or microbes are polysaccharides. The hydrophilic groups can be either nonionic, anionic, or cationic. Chemically, the polysaccharides are composed of pentoses or, more commonly, hexoses. The stereochemistry of the hydroxyl groups on C-1 and C-2 of the glycoside ring is either cis (α) or trans (β), as shown in the following structures:

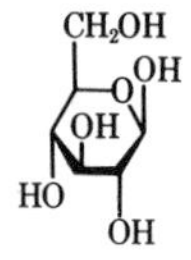

β-D-glucose 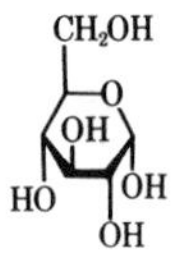 α-D-glucose

Table 1 lists the source, chemical composition, and principal uses of the important natural water-soluble resins. Collection, isolation, and purification methods vary with the material. In the United States, corn starch is by far the largest-volume, natural, water-soluble polymer produced (see Starch). Some chemically modified starches and celluloses are of commercial importance (see Table 2) (see Cellulose; Cellulose derivatives; Carbohydrates).

Water-soluble polymers from animal sources are proteins (qv), ie, amphoteric polyelectrolytes, which have the following general polypeptide structure:

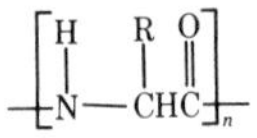

Industrially, the most important products are casein, gelatin (qv), and animal glue (see Glue).

Synthetic

Completely synthetic, water-soluble polymers are prepared by polymerizations in which chain growth results from breaking of ring structures, opening of double bonds, or condensation reactions involving elimination of a small molecule. Since many monomeric materials are available, a broad range of compositions may be synthesized, including homopolymers, copolymers, and copolymers with monomers whose homopolymers are not water-soluble. Synthetic water-soluble polymers include poly(vinyl alcohol), polyacrylamides, cationic resins, poly(acrylic acid) and derivatives, poly(ethylene oxide), poly(N-vinyl-2-pyrrolidinone), vinyl ether polymers, styrene–maleic anhydride copolymers, and ethylene–maleic anhydride copolymers.

Hydrophilic Gels

Hydrophilic gels are lightly cross-linked hydrophilic polymers that undergo substantial volume increase upon the absorption of water. These do not include soft contact lenses, which are heavily cross-linked to provide dimensional stability (see Contact lenses).

Various cross-linked products have been proposed for this use, eg, anionic cellulosics and starch polyacrylamides, polyvinylpyrrolidinone, and maleic or acrylic acid polymers. Commercially the most important materials are the acrylic polymers.

Polymeric Surface-Active Agents

Polymeric surface-active agents make up a special class of water-soluble polymers that is beginning to achieve commercial status in advanced emulsion polymerization and emulsification technology (see Emulsions).

Polymeric surfactants are surface-active, water-soluble polyelectrolytes (qv) in which the monomer units are not surface-active. The polymeric surfactants are random copolymers of water-insoluble and water-soluble ethylenically unsaturated monomers. The water-soluble

Table 1. Principal Natural Water-Soluble Resins

Resin	Chemical composition	Source	Uses
agar	polygalactose sulfate ester	seaweed extract	food, dentistry, medicine, microbiology
carrageenan	anhydrogalactose–galactose polymer sulfate ester	seaweed extract	food, pharmaceuticals, cosmetics
corn starch	poly(1,4-D-glucopyranose) (amylose) + amylopectin (branched)	grain extract	food, sizing
guar gum	mannose polymer with galactose branches on every 2nd unit	seed extract	food, papermaking, mining, petroleum production
gum arabic	highly branched polymer of galactose, rhamnose, arabinose, and glucuronic acid	plant exudate	food, pharmaceuticals, printing inks
gum karaya	polymer of galactose, rhamnose, and partially acetylated glucuronic acid	plant exudate	food-pharmaceuticals, paper binder, printing dyes
gum tragacanth	polymer of fucose, xylose, arabinose, and glucuronic acid	plant exudate	food, cosmetics printing pastes, pharmaceuticals
locust bean gum	mannose polymer with galactose branches on every 4th unit	seed extract	food, papermaking, textile sizing, cosmetics
potato, wheat, and rice starches	poly(1,4-D-glucopyranose) (amylose) + amylopectin (branched)	grain and root extract	food
tapioca	poly(1,4-D-glucopyranose) (amylose) + amylopectin	root extract	food

Table 2. Some Modified Natural Water-Soluble Polymers

Resin	Chemical composition, if not given as resin name	Preparation	Uses
cationic starch	aminoalkyl starch	starch + chloroethylamine or epoxy alkylamine	textiles, paper retention and dry strength, flocculant
dextran	poly(1,5-α-D-glucopyranose)	enzyme fermentation of sucrose	blood-plasma extender
hydroxyalkyl starches		starch + ethylene oxide or propylene oxide	paper coating, textile size adhesives, laundry starch
hydroxyethyl and hydroxypropyl cellulose		sodium cellulose + ethylene oxide or propylene oxide	coatings, petroleum production, cement, paper, textiles, adhesives, asphalt emulsions
methyl cellulose		sodium cellulose + methyl chloride	pharmaceuticals, cosmetics, adhesives, concrete, gypsum, polymerizations
sodium carboxymethyl cellulose		sodium cellulose + sodium chloroacetate	detergents, textiles, foods, drilling muds
xanthan gum	glucose, mannose, glucuronic acid polymer with acetyl pyruvic acid side chains	microbial fermentation	petroleum production, food, pharmaceuticals

monomer can be anionic or cationic. The molecular weight must be low to achieve surface activity. In some cases, the materials are used in-plant and in others the monomers instead of the polymeric surfactant are sold to the user (see Surfactants and detersive systems).

A.S. TEOT
Dow Chemical, U.S.A.

R.L. Davidson, ed., *Handbook of Water-Soluble Gums and Resins*, McGraw-Hill Book Co., New York, 1980.

W. Banks and C.T. Greenwood, *Starch and Its Components*, Edinburgh University Press, Edinburgh, UK, 1975.

Philip Moyneux, *Water-Soluble Synthetic Polymers: Properties and Behavior*, Vols. I–II, CRC Press, Inc., Boca Raton, Fla., 1983.

RESORCINOL.   See Hydroquinone, resorcinol, and catechol.

REVERSE OSMOSIS

In reverse osmosis, solutions of salt or other low molecular weight solutes are contacted with a membrane and subjected to pressure. A solution lower in solute concentration emerges from the other side of the membrane. The name reverse osmosis is derived from one aspect of the process. To reverse the normal osmotic flow from the low to the high concentration side, a pressure difference greater than the difference in osmotic pressures of the solutions adjacent to the surfaces of the membrane (or of the salt-rejecting layer of it) is needed.

Basic aspects are shown schematically in Figure 1. The membrane is in contact with a solution of solute concentration c_α circulated under pressure. The solution emerging from the membrane is of a lower concentration c_ω, and passes unchanged through the porous layer or layers underneath. The concentration at the brine–membrane interface c_α is higher than the average concentration on the high pressure side c_t because of build-up of rejected salt at the inteface. This concentration polarization penalizes performance by decreasing product quality because only a fraction of salt in contact with the membrane at the interface is rejected, and by lowering the concentration of slightly soluble components that can be tolerated without fouling by precipitation on the membrane. A velocity profile in a boundary layer frequently assumed in analysis of systems with turbulent flow is indicated.

For the permeabilities of present conventional membranes in systems of simple flow geometry, theory describes salt concentration polarization adequately. Fouling, especially by high-molecular-weight species, is more complex and unpredictable and frequently involves slow or irreversible reactions and perhaps specific interactions with the membrane surface.

Membranes

A thin homogeneous layer gives better production rates than thick membranes without sacrifice in rejection. Whatever mechanism of salt removal is assumed, the distances over which forces between molecular species are appreciable is so small that, if a membrane material is to discriminate between solution components, the passages through which fluid is forced must be small.

Thin membrane films can be attained by several approaches, eg, by casting procedures giving a tight layer on a porous mass (asymmetric), by casting or interfacial polymerization of thin homogeneous layers on porous supports (composite), or by deposition on a porous support of constituents in a solution (dynamically formed). Most present commercial membranes are asymmetric or composite and are composed of nonionic organic polymers such as cellulose acetate or polyamides. They can remove over 99% of the solute from seawater, but lower rejections in the interest of higher fluxes can be attained and are frequently advantageous. Dynamic membranes may be formed of organic or inorganic ion exchangers, as well as of uncharged materials (see Membrane technology).

Modules

Equipment for applications must include several features. The thin fragile layers have to be supported against process pressures, which range up to ca 10 MPa (100 atm). The ratios of pressurized volumes to membrane surface area should be kept low in order to minimize the cost of heavy-walled equipment. There should be a continuous flow path for brine past the membrane, open enough to allow flows sufficient to keep concentration polarization and fouling to an acceptable level. The last two features are contradictory to an extent and trade-offs are necessary for specific applications and different membranes.

At present, four configurations are offered commercially: plate-and-frame, tubular, spiral-wound, and hollow-fine-fiber types; the last two are dominant. A membrane equipment market of over 100×10^6 (see Hollow-fiber membranes) yr has grown up in the twenty-five years since the first research papers on reverse osmosis appeared. Capacities of over 1.4×10^6 t/d of product water were on line or under construction by 1980, most in the United States or Middle East. The largest markets are municipal supply and pure water for industry or for power production. Applications in pollution control and in concentration of industrial streams have so far developed only to a relatively small fraction of their apparent potential.

JAMES S. JOHNSON, JR.
Oak Ridge National Laboratory

U. Merten, ed., *Desalination by Reverse Osmosis*, MIT Press, Cambridge, Mass., 1966.

L. Dresner and J.S. Johnson, "Hyperfiltration (Reverse Osmosis)," in K.S. Spiegler and A.D.K. Laird, eds., *Principles of Desalination*, 2nd ed., Academic Press, Inc., New York, 1980, Chapt. 8.

Working Papers, First World Congress on Desalination and Water Reuse, Vols. 3 and 4, Florence, Italy, May 1983.

REYNOLDS' NUMBER.   See Fluidization; Fluid mechanics; Rheological measurements; Sedimentation.

RHENIUM AND RHENIUM COMPOUNDS

Rhenium

Rhenium, the seventy-fifth element in the periodic table, is the heaviest element in Group VIIB. Its congeners in this periodic group are manganese and technetium. Rhenium has an atomic weight of 186.2. The toxicities of most rhenium compounds have not been established, but users are advised to handle them with care.

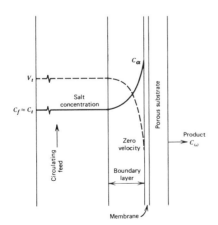

Figure 1.   Schematic salt and velocity distribution near membrane in reverse osmosis–turbulent flow. Key: C = conc at high pressure soln (membrane interface); C_f = feed conc; C_t = turbulent core conc; c_ω = conc at membrane–effluent interface; and v_t = velocity high enough to produce turbulence.

Occurrence. Rhenium is one of the least abundant of the naturally occurring elements. Various estimates of its abundance in the earth's crust have been made; the most widely quoted figure is 0.027 atoms per 10^6 atoms of silicon (0.05 ppm wt). However, this number has a high uncertainty; it is based on analyses for the most common rocks, ie, granites and basalts.

Isolation. Rhenium is obtained from molybdenite concentrates from porphyry copper ores. Rhenium, although present in small amounts in the original ore, is concentrated in the molybdenite by-products to as much as 2000 ppm.

Physical properties. Selected properties of rhenium are summarized in Table 1.

Chemical properties. Rhenium does not react with atmospheric oxygen at ambient temperatures, but at higher temperatures it burns, giving Re_2O_7. This volatile yellow crystalline compound dissolves in water to give perrhenic acid, $HReO_4$. Reactions of the metal occur with chlorine or bromine to produce Re_2Cl_{10} or Re_2Br_{10} and with fluorine to yield ReF_6 and ReF_7. Rhenium metal is not affected by water and hydrohalic acids, but it reacts quickly with HNO_3 to produce $HReO_4$. Fusing the metal with NaOH and doxidizing agents, eg, KNO_3 or Na_2O_2, produces perrhenate salts. The metal also oxidizes to $HReO_4$ in a reaction with 30 wt% H_2O_2; the rate of this reaction depends on the nature of the sample.

Uses. Certain properties of rhenium permit its use at high temperatures as filaments in electron tubes, light bulbs, and photoflash bulbs.

By far the largest use of rhenium is in bimetallic petroleum-reforming catalysts in the production of unleaded and low-lead gasoline.

Rhenium Compounds

A survey of known types of rhenium compounds is presented in Table 2.

PAUL M. TREICHEL
University of Wisconsin

F.A. Cotton and G. Wilkinson, *Comprehensive Inorganic Chemistry*, 4th ed., John Wiley & Sons, Inc., New York, 1980, p. 883.

R.D. Peacock, *The Chemistry of Technetium and Rhenium*, Elsevier Publishing Company, Amsterdam, Neth., 1966.

R. Colton, *The Chemistry of Rhenium and Technetium*, Interscience Publishers, a division of John Wiley & Sons, Inc., New York, 1965.

Table 1. Selected Physical Properties of Rhenium

Property	Value
mp, °C	3180
bp, °C	5926 est
density, g/cm^3	21.02
ΔH°_{subl}, kJ/mol^a	791
specific heat (at 20°C), J/(mol·K)^a	25.1
ionization potentials, kJ^a	757, 1597, 2502
electrical resistivity (at 20°C), $\mu\Omega\cdot$cm	19.3
thermal coefficient of electrical resistivity, °C^{-1}	3.95×10^3
modules of elasticity, Pa^b	0.46

^{a} To convert J to cal, divide by 4.184.
^{b} To convert Pa to psi, divide by 6895.

Table 2. Examples of Rhenium Compounds

Oxidation state	Electronic configuration	Coordination number	Examples^a
Re(−I)	d^8	5	$Na[Re(CO)_5]$
Re(0)	d^7	6	$Re_2(CO)_{10}$
Re(I)	d^6	6	$ReCl(CO)_5$
			$K_5Re(CN)_6$
			$Re(CO)_3(\eta\text{-}C_5H_5)$
Re(II)	d^5	6	$ReCl_2(diars)_2$
Re(III)	d^4	6	$ReCl_3(PR_3)_3$
			$[Re_3Cl_9]_x$
			$Cs_3[Re_3Cl_{12}]$
		5	$K_2Re_2Cl_8$
		7	$K_4[Re(CN)_7]\cdot2H_2O$
			$ReH_3(dppe)_2$
Re(IV)	d^3	6	K_2ReCl_6
			$ReCl_4$
			$K_4Re_2OCl_{10}$
			$ReCl_4(PR_3)_2$
Re(V)	d^2	5	Re_2Cl_{10}
			ReF_5
		6	$ReOCl_3[P(C_6H_5)_3]_2$
		8	$ReH_5[P(C_6H_5)_2C_2H_5]_3$
Re(VI)	d^1	5	$ReOCl_4$
			$ReO(CH_3)_4$
		6	ReF_6
			$Re(CH_3)_6$
		8	K_2ReF_8
Re(VII)	d^0	4	$KReO_4$
		4,6	$(Re_2O_7)_x$
		7	ReF_7
		9	$[(C_2H_5)_4N]_2(ReH_9)$

^{a} dppe = 1,2-bis(diphenylphosphino)ethane; diars = o-bis(dimethylarsino)-benzene; PR_3 = tertiary phosphine.

RHEOLOGICAL MEASUREMENTS

Rheology is the science of the deformation and flow of matter. It is concerned with the response of materials to mechanical force. That response may be irreversible flow, reversible elastic deformation, or a combination of the two (see Fluid mechanics). An understanding of rheology and the ability to measure rheological properties is necessary before rheology can be controlled, and control is essential for the manufacture and handling of a great many materials, eg, foods, cosmetics (qv), plastics, paints, drilling muds, etc (see Paint; Petroleum, drilling fluids).

The usual way of defining the rheological properties of a material is to use an index that provides a measure of the resistance to deformation, preferably in terms of the stress required to produce a unit deformation or rate of deformation. The index for flow is viscosity, which is the resistance to flow under mechanical stress. The index for elastic deformation is elastic modulus.

Viscosity

A liquid is a material that continues to deform as long as it is subjected to a tensile or shear stress. For a liquid under shear, the rate of deformation or shear rate is proportional to the shearing stress. The original exposition of this relationship is Newton's law, which essentially states that the ratio of the stress to the shear rate is constant. That constant is viscosity. Under Newton's law, viscosity is independent of shear rate. This is true for ideal or Newtonian liquids, but the viscosities of many liquids, particularly a number of those of interest to industry, are not independent of shear rate. These non-Newtonian liquids may be classified according to their viscosity behavior as a function of shear rate. Some exhibit shear thinning, whereas others give shear thickening. Some liquids at rest behave like solids until the shear stress exceeds a certain value, called the yield stress, after which point they flow.

The viscosity of a fluid is equal to the slope of the shear stress–shear rate curve ($\eta = d\gamma/d\dot\gamma$). The quantity $\tau/\dot\gamma$ is the absolute viscosity η for a Newtonian liquid and the apparent viscosity η_a for a non-Newtonian liquid. The kinematic viscosity is the viscosity coefficient divided by the density, $\nu = \eta/\rho$. The fluidity is the reciprocal of the viscosity, $\phi = 1/\eta$. The most common units for viscosity η are ((dyn·s)/cm^2 or g/(cm·s)), which are called poise and often are expressed as centipose cP. These units are being superseded by the SI units of pascal seconds (Pa·s) and mPa·s (1 mPa·s = 1 cP). In the same manner, the shear-stress units of dyn/cm^2 are being replaced by pascals (10 dyn/cm^2 = 1 Pa) and newton per square meter (N/m^2 = Pa). Units of shear rate are s^{-1} in both systems. The common units for kinematic viscosity ν are stokes (st) and centistokes (cSt), and the exactly equivalent SI units are cm^2/s and mm^2/s, respectively.

Many flow models have been proposed and are useful for treatment of experimental data or for describing flow behavior (see Table 1). It is probable that no given model fits the rheological behavior of a material over an extended shear-rate range. Nevertheless, these models are useful for summarizing rheological data and are frequently encountered in articles relating to the rheology of liquids.

Thixotropy and other time effects. In addition to the nonideal behavior described, many fluids exhibit time-dependent effects. Some fluids increase in viscosity (rheopexy) or decrease in viscosity (thixotropy) with time when sheared at a constant shear rate. These effects can occur in fluids with or without yield values. Rheopexy is a comparatively rare phenomenon, but thixotropic fluids are fairly common. Because of the decrease in viscosity with time as well as with shear rate, the up-and-down flow curves do not superimpose. Instead, they form a hysteresis loop, the so-called thixotropic loop. Since flow curves for thixotropic or rheopectic liquids depend on the shear history of the sample, different curves for the same material can be obtained depending on the experimental procedure.

One method for measuring time-dependent effects is to determine the decay of shear stress as a function of time at one or more constant shear rates. Another method for estimating thixotropy involves the hysteresis of the thixotropic loop. There are two different techniques. One method involves calculating or measuring the area of the thixotropic loop and has been shown to work well with printing inks. A variation is to determine the up curve on an undisturbed sample, shear the sample at high shear ($> 2000 \ s^{-1}$) for 30–60 s, then determine the down curve. The data are then plotted as Casson-Asbeck plots, $\eta^{1/2}$ vs $\gamma^{-1/2}$.

Dilute polymer solutions. The measurement of dilute-solution viscosities of polymers is widely used for polymer characterization. Very low concentrations reduce intermolecular interactions and allow measurement of polymer–solvent interactions. These measurements usually are made in capillary viscometers, some of which have provisions for direct dilution of the polymer solution in the viscometer. The key viscosity parameter for polymer characterization is the limiting viscosity number or intrinsic viscosity, η. It is calculated by extrapolation of the viscosity number, ie, reduced viscosity, or the logarithmic viscosity number, ie, inherent viscosity, to zero concentration.

Concentrated polymer solutions. Knowledge of the viscosity behavior of concentrated solutions is important to the manufacture and application of a number of commercial materials, eg, caulks, adhesives, inks, paints, and varnishes. This knowledge may be gained by a variety of methods including the use of simple capillary viscometers, extrusion rheometers, and rotational viscometers (see Flow measurement).

Melt viscosity. The study of the viscosity of polymer melts is important for the manufacturer who must supply suitable materials and for the fabrication engineer who must select polymers and fabrication methods. Thus, melt viscosity as a function of temperature, pressure, rate of flow, and polymer molecular weight and structure is of consider-

able practical importance. A number of experimental methods have been applied to measure the melt viscosity of polymers, but capillary extrusion techniques are very popular. Rotational methods are also used, and some permit the measurement of normal stress effects resulting from elasticity as well as of viscosity. Oscillatory shear measurements also are useful for measuring elasticity. A special rotational rheometer has been designed to measure both viscosity and elasticity of polymer melts.

Dispersed systems. Many fluids of commercial and biological importance are dispersed systems, eg, solids suspended in liquids, ie, dispersions, and liquid–liquid suspensions, ie, emulsions (qv). Most of them are non-Newtonian and they are best characterized through the use of rotational viscometers.

Extensional viscosity. In addition to the shear viscosity η, two other rheological constants can be defined for fluids. They are the bulk viscosity κ and the extensional or elongational viscosity η_e. The bulk viscosity relates the hydrostatic pressure to the rate of deformation of volume, whereas the extensional viscosity relates the tensile stress to the rate of extensional deformation of the fluid. Because of the difficulty in measuring these viscosities, relatively little has appeared in the literature concerning them. However, interest in extensional viscosity has been growing recently as a result of the discovery that it is important to a number of industrial processes and problems and the realization that shear properties alone are not sufficient for characterization of many fluids, particularly polymer melts.

Elasticity and Viscoelasticity

Elastic deformation is a function of stress and is expressed in terms of relative displacement or strain. Strain may be in terms of relative change in volume, length, or measurement depending on the nature of the stress. An ideal elastic body is a material that deforms reversibly and for which the strain is proportional to the stress (Hooke's law) with immediate recovery to the original volume and shape when the stress is released.

Mechanical models. Because the complex rheological behavior of viscoelastic bodies is difficult to visualize, mechanical models often are used to represent it. In these models the viscous response to applied stress is assumed to be that of a Newtonian fluid and is represented by a dashpot, ie, a piston operating in a cylinder of Newtonian fluid. The elastic response is idealized as a Hookean solid and is represented by a spring. The dashpot represents the dissipation of energy in the form of heat, whereas the spring represents a system storing energy. Stress–strain diagrams for the two elements are given in Figure 1. With the dashpot, the stress is relieved by viscous flow and is independent of strain. The result is a plot with a constant value for stress. With the spring there is a direct dependence of stress on strain and the ratio of the two is the modulus E (or G).

Dynamic behavior. Knowledge of how the mechanical models would behave if they underwent stress–strain and time-dependent measurements is important for the characterization of real materials, particularly to determine their response under conditions of processing or use. For this reason, stress–strain, creep, and stress–relaxation measurements are commonly made to define material properties.

Dynamic methods depend on measuring the response of a viscoelastic material to periodic stresses or strains. By subjecting a specimen to a sinusoidal stress at a given frequency and determining the response, both the elastic and viscous or damping characteristics can be obtained.

Table 1. Flow Equations for Various Flow Models[a]

Model	Flow equation
Newtonian	$\tau = \eta\dot{\gamma}$
plastic body or Bingham body	$\tau - \tau_0 = \eta\dot{\gamma}$
power law	$\tau = k\lvert\dot{\gamma}\rvert^n$
power law with yield value	$\tau - \tau_0 = k\lvert\dot{\gamma}\rvert^n$
Casson fluid	$\tau^{1/2} - \tau_0^{1/2} = \eta_\infty^{1/2}\dot{\gamma}^{1/2}$
Williamson	$\eta = \eta_\infty + \dfrac{(\eta_0 - \eta_\infty)}{1 + \dfrac{\lvert\tau\rvert}{\tau_{\text{rel}}}}$
Cross and extended Williamson	$\eta = \eta_\infty + \dfrac{(\eta_0 - \eta_\infty)}{1 + \alpha\dot{\gamma}^n}$

[a] Key: τ = shear stress; η = viscosity; $\dot{\gamma}$ = shear rate; T_0 = yield stress.

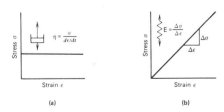

Figure 1. Stress–strain diagrams. (**a**), Dashpot of viscosity η; the intercept indicates the force resisting the motion and is proportional to the speed of testing. (**b**), Spring of modulus E; the slope is the modulus, which is independent of the speed of testing.

Normal stress (Weissenberg effect). Many viscoelastic fluids flow in a direction normal to the direction of shear stress in steady-state shear. This normal stress effect was first analyzed and linked to elasticity by Weissenberg. Manifestations of the effect include flour dough climbing up a beater, polymer solutions climbing up the inner cylinder in a concentric-cylinder viscometer, and paints forcing apart the cone and plate of a cone–plate viscometer. The normal stress effect has been put to practical use in certain screwless extruders designed in a cone–plate or plate–plate configuration, such that the polymer enters at the periphery and exits at the axis.

Viscosity Measurement

Whether a rheologist encounters a flow problem or just wishes to characterize a given fluid better, the decision of which instrument to use must be faced. There are so many commercial viscometers to choose from with such a variety of geometries and such wide ranges of viscosity and shear rates that it rarely is necessary to construct an instrument. However, in choosing a commercial viscometer, a number of criteria must be considered. One of the most important is the nature of the material to be tested: whether it is of high or low viscosity, whether it is elastic or not, the temperature dependancy of its viscosity, etc. Other important considerations are the accuracy and precision required and whether the measurements are for quality control or research.

There are three basic viscometer types: capillary (Fig. 2), rotational (Fig. 3), and moving body (Fig. 4). The choice depends on the particular requirements of the investigator and what can be afforded.

Capillary viscometers. There are three main classifications of commercially available capillary viscometers: glass (Fig. 2), orifice, and

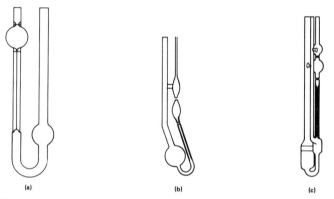

Figure 2. Various types of viscometers: (**a**), Ostwald glass capillary viscometer; (**b**) Cannon-Fenske viscometer; (**c**), Ubbelohde viscometer.

Figure 3. Brookfield Synchro-Lectric viscometer with Small Sample Adapter and water circulator. Courtesy of Brookfield Engineering Laboratories, Inc.

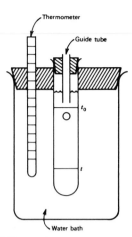

Figure 4. A simple falling-ball viscometer. Lines t_o and t are the timing lines for velocity determinations.

cylinder–piston (extrusion rheometers) types. They are used mainly for Newtonian or near-Newtonian fluids.

Rotational viscometers. Rotational viscometers consist of two basic parts separated by the fluid being tested. The parts may be concentric cylinders (cup-and-bob), plates, a low angle cone and a plate, or a disk, paddle, or rotor in a cylinder. One of the parts rotates relative to the other and produces a shearing action on the fluid. The torque required to produce a given angular velocity or the angular momentum force resulting from a given torque is a measure of the viscosity.

Moving-body viscometers. Moving-body viscometers are instruments or techniques by which the motion of a ball, bubble, plate, or rod through a material is monitored to determine viscosity.

Other viscometers. A number of other viscometers exist, many of which are custom-built for specific research or product applications. One design involves the use of ultrasonic vibrational techniques to measure viscosity. Since the rate of shear is not easily determined or changed, these instruments are best used for controlling or studying processes in which viscosity changes with time or temperature.

Extensional viscosity. Methods exist for measuring all three types of extensional viscosity: uniaxial, biaxial, and pure shear. There are very few commercial instruments available, however, and most measurements are made with improvised equipment.

Viscoelasticity Measurement

There are a number of methods for measuring the various qualities that describe viscoelastic behavior. Some of them require expensive commercial instruments, others depend on customized research instruments, and still others require only very simple devices. Even qualitative observations can be quite useful in the case of liquids, eg, polymer melts, paints, and resins, where elasticity may mean a bad batch or unusable formulation. Examples include observing die-swell of the material from a syringe with a microscope and noting the Weissenberg effect as seen in the forcing of a cone and plate apart during viscosity measurements or the climbing of a resin up the stirrer shaft during polymerization or mixing.

Creep. One of the simplest experiments to describe viscoelastic behavior is creep. Creep experiments involve measurement of deformation as a function of time after a given load has been applied. Such measurements may be made on samples in tension, compression, or shear.

Stress relaxation. In a stress-relaxation experiment, deformation is held constant and the resulting stress in the sample is measured as a function of time. The deformation of the sample produces an initial stress, which decays with time in the case of viscoelastic materials.

Penetration–indentation. Various penetration and indentation tests have long been used to characterize viscoelastic materials, eg, asphalt, rubber, plastics, and coatings. There are many variations but the basic test consists of pressing an indentor of prescribed geometry against the test surface. Most instruments have an indenting tip, ie, a cone, needle, hemisphere, etc, attached to a short rod that is held vertically. The load is controlled at some constant value and the time of indentation is specified. The size or depth of the indentation is then measured.

Tensile testing. One of the best and most common methods for measuring the viscoelastic properties of solids is the tensile tester or stress–strain instrument which extends a sample at constant rate and records stress. Creep and stress-strain measurements can also be made. A large number of commercial instruments of various sizes and capacities are available. They vary a great deal in terms of complication and degree of automation, ie, from hand operated to completely microprocessor controlled. Some have temperature chambers so that measurements can be made over a range of temperatures.

Dynamic measurements. Dynamic methods are required for investigating the response of a material to rapid processes, for studying fluids, or for examining a solid as it passes through a transition region (T_g, melting, etc.). The techniques impart cyclic motion to a specimen and measure the resultant response. Dynamic techniques are used to determine storage and loss moduli G' and G'' and the loss tangent $\tan \delta$. Some instruments are sensitive enough for the study of liquids and can be used to measure the dynamic viscosity η'. Measurements are made as a function of temperature, time, or frequency and results can be used to determine transitions and chemical reactions as well as the properties noted above. Dynamic-mechanical techniques for solids can be grouped into three main areas: free vibration (torsional pendulum and torsional-braid analyzer (TBA)), resonance-forced vibrations (the vibrating reed), and nonresonance-forced vibrations (Rheovibron (Imass, Inc.)).

Fluids. Rheometrics, Inc. supplies several devices designed especially for characterizing viscoelastic fluids (see also Viscosity Measurement). All of the instruments measure the response of a liquid to sinusoidal oscillatory motion to determine dynamic viscosity and storage and loss moduli. Some commercial rotational viscometers can be used to make such measurements or include a means for determining normal stresses.

Time–temperature superposition, master curves. Since the modulus of a viscoelastic material varies with time and temperature, it is necessary to make measurements over wide ranges of these variables for full characterization. A series of stress–relaxation curves measured at different temperatures can be shifted on the log time axis to give a single modulus–time master curve that covers a very wide time range. This shifting is possible because time and temperature have equivalent effects on modulus.

The rubber, plastics, inks, coatings, food, and oil industries use a wide variety of rheological measurements for research, quality control, and testing. The correlation of rheology with product and process performance requires much study and knowledge as well as careful measurement and appropriate measurement conditions.

Percy E. Pierce
Clifford K. Schoff
PPG Industries, Inc.

J.R. Van Wazer, J.W. Lyons, K.Y. Kim, and R.E. Colwell, *Viscosity and Flow Measurement, A Laboratory Handbook of Rheology*, Interscience Publishers, a division of John Wiley & Sons, Inc., New York, 1963.

R.W. Whorlow, *Rheological Techniques*, John Wiley & Sons, Inc., New York, 1980.

L.E. Nielsen, *Mechanical Properties of Polymers and Composites*, Vols. 1–2, Marcel Dekker, New York, 1974.

J.D. Ferry, *Viscoelastic Properties of Polymers*, 2nd ed., John Wiley & Sons, Inc., New York, 1970.

RHEOLOGY. See Rheological measurements.

RIBOFLAVIN. See Vitamins.

ROOFING MATERIALS

The bituminous types are by far the most common roof coverings in the United States. The built-up roofing membrane of bituminous materi-
als is estimated to cover over 90% of the commercial and industrial roof surfaces in North America. Asphalt-based materials predominate over coal-tar materials. Various types of roofing materials are employed for residential structures, both for new construction and reroofing, especially in the western area of the United States. Wood and miscellaneous materials account for 20% of residential roofing in the western region of the United States, but only for 5% in the northeast and central areas (see Wood); asphalt roofing accounts for the rest (see Asphalt).

Built-up roofings. Built-up roofing (BUR) is a continuous-membrane covering manufactured on site from alternate layers of bitumen, bitumen-saturated felts or asphalt-impregnated glass mats, saturated and coated felts, and surfacings. These membranes are usually applied with hot bitumens or by a cold process utilizing bituminous solvent-solution or water-emulsion cements (see Adhesives; Felts). Insulation is usually placed between the deck and the roof membrane. The use of fiber glass is increasing.

The deck may be nailable, eg, wood or light concrete, or not, eg, steel or dense concrete. The felts or mats may be organic (cellulose), asbestos, or fiber glass (see Cellulose; Asbestos; Glass). The ply adhesives may be hot-melt or cold-process bitumens (emulsion or cutback).

Protected-membrane roofs. In modern construction, insulation is needed that is unaffected by water or that can be kept dry in some manner and that stays in place over the roof membrane. In the United States, extruded polystyrene-foam insulation boards are commonly employed (see Insulation, thermal). They are placed over the roof membrane which is, usually the single-ply plastic type, and is attached to the decks. The insulation may or may not be bonded to the membrane. Gravel or slag at the rate of 48.8 kg/m² (1000 lb/100 ft²) holds the insulation in place and offers protection from the sun. The insulation joints are open and drainage must be provided. Various other materials, eg, patio blocks, mortar, and concrete slabs, are also used as surfacings and ballast. The extra weight imposes more exacting requirements on construction. Although of relatively recent design, the protected-membrane roof is finding a place in roof systems.

Single-ply plastics. Elastomeric and thermoplastic products are applied as single- or limited-ply roofing membranes. New materials have proliferated during the last decade; their claimed advantages include light weight, high chemical and weather resistance, high elasticity, and ease of application and repair. These new roofing systems depend primarily on the application of single-ply-sheet membranes giving continuous coverage (see Film and sheeting materials). Materials applied as liquids prepared from a variety of base materials are often used in combination with glass-mat or plastic reinforcements (see Plastic building products).

Other types. Asbestos-cement corrugated sheets (AC sheets) show excellent strength and durability properties in industrial and commercial roofing applications. The sheets are lapped and usually applied over steel purlins. The usual span is 1.37 m; special products are suitable for spans up to ca 3 m.

Metal roofs are used on industrial and farm buildings and are usually made of corrugated aluminum or galvanized steel.

Concrete-slab roofs that do not require a membrane save both on cost and maintenance. Cracks are prevented by post-tensioning the roof slabs during installation.

Fabric roofs are either air-supported or tensioned.

Plastic sheets have been installed in roof-support systems. Because of their excellent strength properties, polycarbonate sheets have been reported to be highly successful in severe environments.

Asphalt Shingles

For many years, organic-felt-base asphalt shingles have been the standard for residential roofing. In general, the individual three-tab strip shingle is 91.4 by 30.5-cm wide. Maximum exposure of 12.7 cm to the weather in application provides a head lap of not less than 18 cm to give double coverage. The typical organic-felt shingle has a felt base saturated with an asphalt of a 60.0–68.3°C softening point and a penetration of 3–5 mm. The use of fiber-glass shingles has increased. Organic-felt-base asphalt shingles are no longer predominant.

Other shingles and units. The preference for wood, slate, asbestos-cement, or tile roofing shingles and units varies in different parts of the

United States. Nationwide, the average is ca 87% asphalt, 7% wood, and 6% miscellaneous (tile, concrete, and metal).

Roofing Performance

Weathering. Bitumen films harden progressively owing to oxidation reactions and, to some extent, loss of plasticizing oils, especially with coal tars. Temperature changes may result in expansion and contraction of the roof covering. Other weather factors include moisture, ice, and hail. Wind affects shingles as well as BUR membrane performance.

Thermal effects. Temperature influences the oxidation rate, whereas temperature changes impart a mechanical stress to roof coverings.

Water effects. Water in its different forms (liquid, vapor, hail, and ice) profoundly influences the performance of roof coverings. Moisture migrations in roof insulation, eg, vapor that can accumulate and later liquefy, or water leakage through the roof covering, reduces insulating efficiency and leads to physical deterioration of the roofing material.

Fire and wind hazards. Weather resistance of roof coverings is not necessarily correlated to fire and wind resistance. Underwriter's Laboratories (UL) and the Factory Mutual Systems (FM) are nonprofit organizations that test and rate fire- and wind-hazard resistance.

Health and Safety Factors

Asphalt derived from petroleum differs from coal tar, which is a condensation by-product obtained from the carbonization of coal. The predominant emissions from asphalts are paraffinic or cycloparaffinic in character, whereas coal-tar emissions are predominantly aromatic in character. Low temperatures for handling are suggested as a precaution. In order to reduce worker exposure to emissions from roofing kettles, it was suggested that the kettles operate at the lowest possible temperature and that the lid remain closed.

ARNOLD J. HOIBERG[‡]
ERNEST G. LONG
Manville Corporation

M.C. Baker, *Roofs*, Multiscience Publications Ltd., Montreal, Can., 1980, pp. 1–6.

ROSANILINE.   See Triphenylmethane and related dyes.

ROSIN AND ROSIN DERIVATIVES.   See Resins, natural; Terpenoids.

RUBBER CHEMICALS

Accelerators of Vulcanization

The main applications of elastomers require that the polymer chains be cross-linked after being formed into a desired shape (see Elastomers, synthetic). After cross-linking of the polymer chains, which is called curing or vulcanization, the article is elastic. It deforms under stress but returns to the shape it had when vulcanization occurred if the stress is removed. The most common method of cross-linking elastomeric polymers is through the use of sulfur. Vulcanization was first discovered in 1839 by the observation of physical change in a mixture of natural rubber (NR), sulfur, and basic lead carbonate after heating (see Rubber, natural).

The principal elastomer applications require a variety of operations that produce and require heat, eg, mixing, extruding, calendering, molding, etc, during which cross-linking of the polymer to an appreciable extent cannot be tolerated. The need to prevent premature cross-linking has been met by the development of delayed-action accelerators. These materials are not accelerators initially but undergo chemical reactions during processing to produce the active accelerator species in a delayed manner. The main accelerators in use exhibit some degree of delayed action. When more delay than the accelerator furnishes is required, a vulcanization retarder can be incorporated.

A large portion of the vulcanization accelerators produced in the United States is comprised of derivatives of 2-mercaptobenzothiazole (MBT). There is a large variety of cross-linked polymers. They range from polymers having a high degree of unsaturation to those with very low levels or no unsaturation. Other polymers are cured through other functional groups, eg, halogens. This variety in polymers and the wide variety of vulcanizate physical-property and processing requirements has resulted in the industrial use of a large number of different accelerators and curing agents. A list of some commercially available materials is given in Table 1.

Mechanism of accelerated sulfur vulcanization. Although a good deal of effort has been directed at understanding the chemical mechanism of sulfur cross-linking, the exact steps are in contention. The main uncertainty is whether the steps are ionic or free radical in nature. The evidence indicates that the following general steps occur: the accelerator, which generally is an organic sulfur anion or radical, attacks S_8 to form a polysulfide adduct; this adduct becomes attached to the polymer chain through sulfur as a pendant polysulfide and is capped by the accelerator moiety; the cross-link forms through these pendant polysulfide groups by one of several possible mechanisms; and the average number of sulfur atoms in the cross-link decreases as the vulcanization process continues (this decline is probably a continuation of attack on the polysulfide groups by the accelerator).

Peroxide cures. In 1914, it was discovered that dibenzoyl peroxide cross-links rubber. The use of the more effective dialkyl peroxides to cross-link polymers started shortly after 1950. The cross-link is attained through free radicals formed by homolytic decomposition of the peroxide. This produces alkoxy radicals, which can abstract hydrogen from the polymer chain, producing alkyl radicals on the polymer chain. Coupling of these radicals leads to cross-links, as shown in the following simplified scheme, where R represents an alkyl group and P represents a polymer chain (see Initiation; Peroxides).

$$\text{ROOR} \xrightarrow{\text{heat or } h\nu} 2\,\text{RO·}$$
$$\text{RO·} + P\text{—H} \rightarrow \text{ROH} + P\text{·}$$
$$2\,P\text{·} \rightarrow P\text{—}P \text{ (cross-link)}$$

The carbon–carbon cross-links produced by this process have appreciably higher bond energies (343 kJ/mol or 82 kcal/mol) than sulfur–sulfur or carbon–sulfur bonds. Consequently, the cross-links are thermally more stable than typical sulfur vulcanizates and even more stable than an EV (efficient vulcanization) sulfur system. The physical properties of the resultant vulcanizates have unique features that are not attained in the EV sulfur system.

Other cure systems. A large variety of other types of cure systems has been described; of these, the most important are resin cures. Phenolic-resin curing systems are used extensively to cure latex tire-cord dips and butyl-curing bladders.

Toxicity of vulcanization systems. The long-term biological effects of most of the compounds listed in Table 1 are not known. Two areas of concern are the possible carcinogenicity of several of the thioureas and the presence of secondary alkyl nitrosamines in the air of certain rubber-processing plants. Many nitrosamines produce cancer in animals (see *N*-Nitrosamines).

Retarders

Retarders are chemicals that prevent the premature vulcanization of rubber compounds during mixing, calendering, and other processing steps. In the absence of the processing safety provided by retarders, scorched stocks, and consequently, waste results either during the processing steps or during the storage of the fully compounded green stocks. Retarders are often called antiscorching agents, scorch inhibitors, cure retarders, or prevulcanization inhibitors. Retarders having a sulfonamide group are called prevulcanization inhibitors, whereas such conventional retarders as salicylic acid, phthalic anhydride, and *N*-nitrosodiphenylamine (NDPA) are simply called retarders.

Conventional retarders include benzoic acid, phthalic anhydride, and NDPA. More recent ones include a sulfonamide derivative Vulkalent E (Mobay) and *N*-(cyclohexylthio)phthalimide (CTP), Santogard PVI and

[‡]Deceased.

Table 1. Selected Commercial Accelerators, Curing Agents, and Retarders

Compound	Structure	Acute oral LD$_{50}$, mg/kg	Uses	Trade name
Benzothiazoles 2-mercaptobenzo-thiazole (MBT)		3,000	primary accelerator for NR and synthetic rubbers	Akrochem MBTA[a] Nocceler M[b] Pennac MBT[g] Captax[l] Sanceler M[i] Soxinol M-G[j] Thiotax[d] Vulcafor MBT[m] Vulkacit Mercapto[e] Naugex MBT[k]
bis(2,2-benzothiazolyl) disulfide (MBTS)		> 5,000	primary and scorch-modifying secondary accelerator for NR and SBR	Akrochem MBTS[a] Altax[l] Nocceler DM[f] Pennac MBTS[g] Sanceler DM[i] Soxinol DM-G[j] Thiofide[d] Vulcafor MBTS[m] Vulkacit DM[e] Naugex MBT[k]
Benzothiazolesulfenamides N-*tert*-butyl-2-benzo-thiazolesulfenamide			delayed-action accelerator for NR and synthetic rubbers	Akrochem BBTS[a] Pennac TBBS[g] Santocure NS[d] Vanax NS[l] Vulkacit NZ[e] Delac NS[k]
N-cyclohexyl-2-benzo-thiazolesulfenamide		7,000	delayed-action accelerator for NR and synthetic rubbers	Akrochem CBTS[a] Durax[l] Nocceler CZ[f] Pennac CBS[g] Sanceller CM[i] Santocure[d] Soxinol CZ-G[j] Vulcafor CBS[m] Vulkacit CZ[e] Delac S[h]
2-(4-morpholinylthio) benzothiazole (MTB)		> 10,000	delayed-action accelerator for NR, SBR, IR, and SBR	Akrochem OBTS[a] Amax[l] Nocceler MSA[b] Sanceller NOB[i] Santocure MOR[d] Soxinol NBS-G[j] Vulcafor MBS[m] Vulkacit MOZ[e] Delac MOR[k]
2-(4-morpholinyl-dithio)benzo-thiazole		> 16,000	accelerator and sulfur donor from NR and synthetic rubbers	Morfax[l] Nocceler MDB[f]
Dithiophosphate zinc O,O-di-n-butyl phosphorodithioate			nonblooming, nonstaining accelerator for EPDM; used in combination with other accelerators	Vocol[d] Rhenocure TP/S[e] Royalac 136[k]
Guanidine 1,3-diphenylguanidine		375–850	secondary accelerator for thiazoles, sulfenamides, and thiurams	DPG[a,b,d] Nocceler D[f] Sanceler D[i] Soxinol D[j] Vanax DPG[l] Vulcafor DPG[m] Vulkacit D/C[e]

Table 1.   Continued

Compound	Structure	Acute oral LD$_{50}$, mg/kg	Uses	Trade name
Thiourea ethylenethiourea (2-imidazolidinethione)	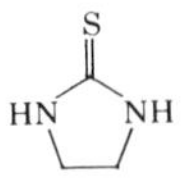	1,832 (rodent), carcinogenic in rats and mice	nonstaining accelerator for CR and epichlorohydrin	Akrochem ETU-22[a] Robac 22[h] Sanceller 22[i] Vulkacit NPV/C[e]
Thiurams tetramethylthiuram disulfide (TMTD)	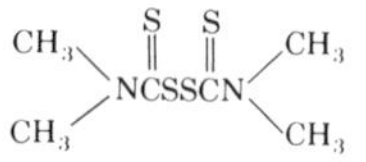	780–1,300; LD$_{Lo}$ (humans), 50	excellent for fast press cures; especially good for IIR and CR	Akrochem TMTD[a] Methyl Tuads[l] Metiurac 0[g] Nocceler TT[f] Robac TMT[h] Sanceler TT[i] Soxinol TTG[j] Thiurad[d] Vulcafor TMTD[m] Vulkacit Thiuram/C[e] Tuex[k]
tetramethylthiuram monosulfide		1,250–1,390	booster for thiazoles, especially in nitrile rubbers	Akrochem TMTM[a] Mono Thiurad[d] Nocceler TS[f] Pennac MS[g] Robac TMS[b] Sanceller TS[i] Soxinol TS-G[j] Unads[l] Vulcafor TMTM[m] Vulkacit Thiuram MS[e] Monex[k]
Thiocarbamyl sulfenamide N-oxydiethylenethio-carbamyl-N′-oxy-diethylenesulfen amide	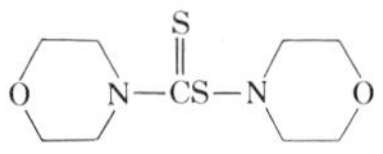		primary accelerator for NR and synthetic rubbers provides fast cure and scorch safety	Cure-rite[a, c]
Curing(vulcanizing) Agent 4,4′-dithiobismorpho-line	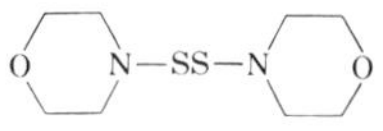	3,690	vulcanizing agent for NR and synthetic rubbers; provides excellent heat aging	Naugex SD-1[k] Sulfasan R[d] Vanax A[l] Vulnoc R[f]
Retarder N-(cyclohexylthio)phthal-imide (CTP)	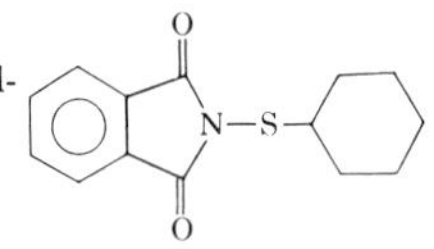		retarder for NR and synthetic rubbers; it works best with sulfenamide type accelerators	Santogard PVI[d]

[a]Akron Chemical.
[b]American Cyanamid.
[c]BF Goodrich.
[d]Monsanto.
[e]Molbay.
[f]Ouchi Shinko.
[g]Pennwalt.

[h]Uniroyal.
[i]Sanshin.
[j]Sumitomo.
[k]Uniroyal.
[l]Vanderbilt.
[m]Vulnax.

AK-8169 (Monsanto). Ca 80% of the annual consumption of these retarders is used in the manufacture of tires; the rest is used in miscellaneous mechanical goods, footwear, and sheet and calendered goods.

RAY TAYLOR
P.N. SON
BF Goodrich Company

J.F. Krymowski and R.D. Taylor, *Rubber Chem. Technol.* **50**, 671 (1977).

A.Y. Coran, "Vulcanization" in F.R. Eirich, ed., *Science and Technology of Rubber*, Academic Press, Inc., New York, 1978, Chapt. 7.

P.N. Son, *Rubber Chem. Technol.* **46**, 999 (1973).

C.D. Trivette, Jr., E. Morita, and O.W. Maender, *Rubber Chem. Technol.* **50**, 570 (1977).

RUBBER COMPOUNDING

Rubber is defined in ASTM D 1566 as a material that is capable of recovering from large deformations quickly and forcibly, and that can be or already is modified to a state in which it is essentially insoluble but can swell in boiling solvent, eg, benzene, methyl ethyl ketone, and ethanol–toluene azeotrope.

With a few exceptions, raw rubber in the dry state has few commercial applications. For the great majority of uses, the rubber must modified, usually by the addition of vulcanizing agents and other materials followed by vulcanization (see Rubber chemicals). The exceptions include such uses as crepe-rubber shoe soles; cements, eg, in rubber adhesives; and adhesives and masking tape (see Elastomers, synthetic; Rubber, natural).

Synthetic Rubbers

Synthetic rubbers include polybutadiene rubbers (BRs); polyisoprene rubbers (IRs); the 1000 series of styrene–butadiene rubber (SBR); nitrile rubbers (NBRs); butyl rubber (IIR); ethylene–propylene terpolymer (EPDM); silicone rubbers; types GN, GNA, GW, FB, and GRT neoprenes (CRs); polysulfides; polyacrylate rubbers; epichlorohydrin rubbers; fluoroelastomers (FDM); Hypalon, a chlorosulfonated polyethylene (CSM); halogenated butyl, eg, chlorobutyl (CIIR) and bromobutyl (BIIR); chlorinated polyethylene rubbers (CPEs); polyurethanes; and thermoplastic rubbers.

Compounding

The object of compounding is to select the most suitable combination of materials in their correct proportion and to determine the treatment that the chosen combination shall undergo in the processes of mixing, forming, and vulcanization so that the finished rubber product is of the required quality and is produced at the lowest possible cost. The operations of mixing, forming, and vulcanizing are the essential fabrication steps.

Basic recipes. The terms pure gum or gum compound are used for a rubber admixture containing only those ingredients necessary for vulcanization, processing, coloring, and resistance to aging. The simplest of these is natural rubber, 100 parts by wt and; sulfur, 8–10 parts by wt.

When it is desired to produce mixtures with lower sulfur ratios and shorter curing times, organic accelerators combined with metal oxides are used.

Because the proportion of naturally occurring fatty acids varies from one grade of rubber to another, it is common to add additional fatty acid to the composition when thiazole accelerators are used.

The other essential ingredient in a base recipe is the metallic oxide. Although the oxides of calcium, magnesium, and lead can be and occasionally are used, zinc oxide is by far the most widely used material.

Natural-rubber vulcanizates deteriorate during aging. In addition to the advantages obtained by the use of low sulfur ratios and good aging accelerators, antioxidants (qv) further improve resistance to deterioration during aging.

The basic compound can be the source of innumerable products simply by modifying the amount and kind of the ingredients or by adding reinforcing agents, fillers, colors, softeners, and extenders or other materials.

Reinforcing Agents and Fillers

Dry pigments other than those added to rubber as vulcanizing agents or vulcanizing aids are loosely classified either as reinforcing agents or fillers (qv). The former improve the properties of the vulcanizates and the latter serve primarily as diluents. There is no general agreement among rubber technologists as to what reinforcement is, how it can be measured, or precisely where the dividing line is that distinguishes the reinforcing materials from the fillers. There is general agreement that all grades of carbon black with the exception of N 880 (FT, fine thermal) and N 990 (MT , medium thermal) are reinforcing agents. The degree of

reinforcement increases with a decrease in particle size. There is more disagreement with respect to the nonblack fillers, but it is generally conceded that some reinforcement results from the use of fine-particle zinc oxides; fine, precipitated calcium carbonates; calcium silicates; amorphous, hydrated silicon dioxide; pure silicon dioxides; fine clays; and fine-particle magnesium carbonate.

Reinforcement has been defined as the incorporation into rubber of small-particle substances which give to the vulcanizate high abrasion resistance, high tear and tensile strength, and some increase in stiffness.

Softeners, Extenders, and Plasticizers

Softeners include a wide variety of oils, tars, resins, pitches, and synthetic organic materials and are used for a number of reasons, some of them having little or no relation to softness of either the raw, uncured stock or the vulcanizate. Some of these reasons are to decrease the viscosity and thereby improve the workability of the compound, to reduce mixing temperatures and power consumption, to increase tack and stickiness, to aid in the dispersion of fillers, to reduce mill and calendar shrinkage, to provide lubrication so as to aid extrusion and molding, and to modify the physical properties of the vulcanized compound. A few of these reasons are exact opposites, so it is important to be able to pick the type of softener that gives the desired effect. Extenders are materials possessing plastic or rubberlike properties and can be used to replace a portion of the rubber, usually with a processing advantage. Both softeners and extenders can be used simply as diluents. Thus the distinction between a softener or extender and a diluent is somewhat arbitrary. Certain materials may be used as both softeners and extenders on one hand and as extenders and diluents on the other.

Chemical plasticizers (qv), eg, pentachlorothiophenol (Renacit 7, Bayer), zinc salt of pentachlorothiophenol (Renacit 4, Bayer), 2,2'-dibenzamidodiphenyl disulfide (Pepton 22, American Cyanamid), and activated dithiobisbenzanilide (Pepton 44, American Cyanamid), are used in very low concentrations and their effect is chemical rather than physical. The chemical plasticizers or peptizers are generally used to lower the viscosity of the uncured compound. They function in the thermomechanical and thermooxidative breakdown of rubbers as oxidation catalysts at high temperatures but as radical acceptors at low temperatures. They have little effect on vulcanizate properties. The addition of certain compounding ingredients to the batch, eg, clay, carbon black, and sulfur, stops their plasticizing action.

Vulcanization Agents and Auxiliary Materials

When elemental sulfur is used as the vulcanizing agent, certain auxiliary materials must also be added to obtain desirable properties. The most important of these materials is the organic accelerator. The accelerator has a profound influence on the nature of the sulfur cross-link, which largely determines the thermal stability, flex-cracking, and aging resistance of the vulcanizate. Because of the use of organic accelerators, scorch of rubber compounds plays an important part in the preparation, storage, and further processing of the compounds. How various types of accelerators affect scorch properties as measured by the Mooney viscometer in both natural rubber and N 330 (HAF) black compounds is shown in Table 1.

Accelerators are frequently used in combination in order to produce a faster cure than can be obtained by either material separately. The most common combinations consist of an acidic accelerator, eg, mercaptobenzothiazole (MBT), or one that becomes acidic during vulcanization, eg, benzothiazyl disulfide (MBTS), combined with a basic type, eg, diphenylguanidine (DPG) or thiuram disulfide.

Occasionally retarders are used to lengthen the scorch time and to slow the cure rate of excessively fast accelerator combinations.

Styrene–butadiene polymers. Styrene–butadiene rubber is vulcanized by sulfur and can be vulcanized by dicumyl peroxide, *p*-quinone dioxime, trinitrobenzene, chloranil, and others.

Polyisoprene. The one manufacturer of polyisoprene rubbers in the United States uses an aluminum trialkyl–titanium tetrachloride catalyst, which gives a 96% cis-content polymer. Polyisoprene, on account of

Table 1.   Comparison of Mooney Scorch Times of Various Accelerators in Gum-and Carbon-Black-Loaded Compounds, min

| | Natural rubber (smoked sheets)[a] | | |
| | Gum | | Black |
Accelerator	121°C	138°C	121°C
N, N-diisopropyl-2-benzothiazolesulfenamide (DIBS)	62.7	19.0	21.3
N-tert-butyl-benzothiazolesulfenamide (TBBS)	> 60	34.6	22.8
2-benzothiazyl-N,N-diethylthiocarbamyl sulfide (EA)	26.1	10.6	14.8
N-cyclohexyl-2-benzothiazolesulfenamide (CBS)	> 60	28.4	19.5
benzothiazyl disulfide (MBTS)	72.8	23.8	11.8
mercaptobenzothiazole (MBT)	11.8		8.8
tetramethylthiuram monosulfide (TMTM)	25.2		11.5
tetraethylthiuram disulfide (TETD)	18.0		8.9
zinc dibutyldithiocarbamate (ZBDC)	4.9		4.0
zinc dimethyldithiocarbamate (ZMDC)	6.2		3.7
condensation product of butyraldehyde and aniline (BA)	19.3		5.1
diphenylguanidine (DPG)	16.0		9.6
tetramethylthiuram disulfide (TMTD)	13.2		6.4

[a] Natural-rubber recipe, in parts by wt: rubber, 100; N 330 (HAF) black, 0 (gum) or 50 (black); zinc oxide, 5.00; stearic acid, 3.00; sulfur, 3.00; accelerator, 1.00.

its chemical similarity to natural rubber (98% cis content), generally responds the same way to vulcanizing agents and accelerators.

Butyl rubber.   Butyl rubber can be vulcanized by sulfur and by certain oxidizing agents that vulcanize the other polymers.

Halogenated butyl rubbers.   Chlorobutyl and bromobutyl rubbers, which include either allylic chlorine or bromine and double bonds, can be cured by a number of systems. These include the normal systems for butyl rubbers, and by some systems related to neoprene.

Chlorosulfonated polyethylene.   Chlorosulfonated polyethylene (Hypalon) can be cross-linked by a variety of curing systems. Systems based on zinc oxide–magnesia as acid acceptors are most useful.

Polybutadiene.   The preferred vulcanizing agent for BR is sulfur, but the well-known nonsulfur curing agents can also be used.

Nitrile rubber.   Nitrile rubber is vulcanized by sulfur at 1.0–1.75 parts in carbon black compounds and 2.0–3.0 parts sulfur for nonblack compounds.

Chloroprene rubbers.   The general-purpose neoprenes are classified in three groups, the G, W, and T types. The oxides of zinc and magnesium are vulcanizing agents but also serve other functions.

Ethylene–propylene terpolymers.   Ethylene–propylene rubber can be vulcanized successfully by a variety of systems. The various third monomers that provide the unsaturation in the molecule allow EPDM to be vulcanized by the broad range of accelerator sulfur systems, p-quinone dioxime–dibenzoyl-p-quinone dioxime combinations, and resin systems. Vulcanization by peroxides and radiation is effective, though unrelated to molecular unsaturation.

Bonding Agents

In certain rubber articles designed to withstand considerable stresses in use, the rubber is reinforced with plies of comparatively inextensible textile, glass, or steel materials. Thus, rubber tires, hoses, and belts, etc, are more commonly reinforced with filamentary textiles, glass, or steel in the form of yarns, cords, or fabric. In such articles, it is important that the plies of reinforcing material be firmly adhered to the rubber and remain effective after the article has been subjected to repeated and varying strains in use. Thus, the article's durability and its ability to perform under increasingly severe operating conditions is directly linked to the adhesion of the ply-reinforcing material to its adjacent rubber surface.

A technique for obtaining adhesion is the dry adhesive system in which the bonding agents are mixed directly into the rubber compound. Bonding agents used in the dry adhesive system are as follows: 2.5 parts resorcinol (dihydric phenol) and 1.6 parts HEXA; 2.0 parts Bonding Agent R-6 (Uniroyal) (a resorcinol donor) and 1.0 part Bonding Agent M-3 (Uniroyal) (a methylene donor); 2.5 parts hexamethoxymethyl-melamine (Cohedur A, Mobay; Cyrez 963 resin, American Cyanamid) and 3.0 parts of the condensation product of resorcinol and formaldehyde which is deficient in formaldehyde (Penacolite Resin B-18, Koppers; SRF-1501, Schenectady Chemicals).

Antioxidants and Stabilizers

Natural rubber is different from SBR and other synthetic rubbers of the diene types in its need for protection. Latex as taken from the rubber tree contains natural antioxidants, proteins, and complex phenols, which protect it from deterioration during the coagulation and drying steps. These natural protectants are destroyed during vulcanization and thus synthetic protectants must be added to ensure an adequate service life of cured natural-rubber products. On the other hand, synthetic rubber of the diene types is quite unstable when freshly prepared and require synthetic protectants both during the flocculation and drying of the polymers and sometimes after vulcanization. The former protectants generally are referred to as stabilizers and the latter are called antioxidants or antiozonants (see Antioxidants and antiozonants).

Reclaimed Rubber and Ground Scrap

There are two basic factors that determine the type of reclaim. The first and most important is the type of scrap from which the reclaim is made. The second is the process by which the scrap is reclaimed (see Recycling, rubber). The most important properties of reclaimed rubber for compounding purposes are its viscosity and elasticity, its rubber hydrocarbon content, its freedom from lumps, and several less tangible properties, eg, tack, dryness, or stickiness and mushiness.

The choice of the relative proportions of new and reclaimed rubber to be used in a compound depends completely upon its intended use. The use of reclaim in tire treads always results in a sacrifice of quality, eg, abrasion loss, whereas in carcass and bead insulations, reclaim produces no sacrifice in quality.

The reinforcing effect of fillers in reclaimed rubber follows the same order of effectiveness as new rubber. The same amount and same type of softeners are used in reclaim as in new rubber. Reclaimed rubber needs less zinc oxide than new rubber. The same accelerators and vulcanizing agents are used with reclaimed rubber as with new rubber.

The use of grinding to reclaim rubber scrap has been increasing. The conventional grinding methods produce course, ie, 600-μm, powders. The coarse powders generally reduce the physical properties below satisfactory limits. The use of cryogenic grinding of cured rubber scrap for reuse in molded parts has been advocated. Use of liquid nitrogen produces fine particles because the rubbers are pulverized at temperatures below the rubber compound's glass-transition or embrittlement temperature.

Conditions of Cure

The combination of vulcanizing agents, accelerator, activator, type of polymer, kind and amount of filler loading, and other accessory materials chosen for a particular compound determines its curing rate. The curing rate is measured by the rapidity with which the physical properties of the rubber compound develops with the time of heating. The method of determining the curing rate is to use a cure meter to obtain a torque-vs-curing time curve. In Figure 1 three cure-meter curves are shown, illustrating the varying types of behavior encountered.

The cure to be used on a particular product depends on the size and shape of the product and the temperature of cure.

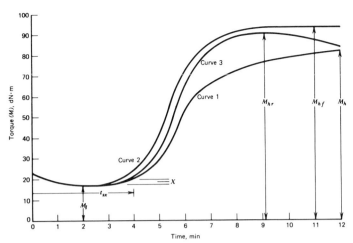

Figure 1. Curing curves. M = torque in dN·m (to convert dN·m to lbf·in., divide by 1.130); M_l = minimum torque, M_h = highest torque attained during specified period of time when no plateau or maximum torque is obtained (Curve 1), M_{hf} = maximum torque where curve plateaus (Curve 2), M_{hr} = maximum torque of reverting curve (Curve 3). T = temperature. t = time; t_{sx} = scorch time in minutes to increase x units above M_1, tX = cure time in minutes to $X\%$ of maximum torque, $t'X$ = cure time in minutes to $X\%$ of torque increase, ie, minutes to $M_1 + X(M_h - M_l)/100$ torque. The most commonly used values of X are 50 and 90.

Hard Rubber

Hard rubber, also called ebonite or vulcanite, is the hard, tough composition produced by prolonged vulcanization of rubber with a large proportion of sulfur. The ASTM classifies hard rubber as compositions having a combined sulfur-to-rubber hydrocarbon ratio in excess of 15%. The polymers used in compounding hard rubber are new or reclaimed natural rubber, polyisoprene, SBR, and nitrile rubber. Butyl, neoprene, EPDM, and Thiokol cannot be vulcanized to hard rubbers. Sulfur is the only vulcanizing agent for hard rubber.

Hard rubber has a unique combination of properties, which has caused it to be adopted for many varied uses. However, it is being challenged by many of the synthetic plastic resins. The properties of hard rubber are desirable physical properties, eg, toughness combined with an adequate degree of hardness; resistance to many chemicals; excellent electrical insulating properties; ease of machining; and attractive physical appearance, especially its high polish. There are some disadvantages associated with hard rubber. The material requires very long cures which are troublesome and costly. This is in contrast to the very short curing or molding cycles used in the plastic industry. The product is also subject to discoloration and loss of electrical properties after light and air exposure and during service at high temperatures, ie, > 70°C.

The physical properties of greatest interest in the development of hard-rubber material are the following: tensile strength and elongation, Shore D hardness, impact strength, distortion temperature, flexural strength, electrical properties, and stability on aging in air at elevated temperatures and on aging in light. Some typical applications for hard rubber include: withstanding corrosive chemicals in pipes and fittings, pumps, and tank linings; as a hard-rubber base for rolls and industrial solid tires; providing a means of adhering soft rubber to a metal core; forming sheet, rod, and tube for electrical insulating purposes, as storage-battery cases, dentures, and grinding wheels in which hard rubber is used to bond the abrasive particles; and in bowling balls, combs, fittings for drug sundries, and cellular gasoline floats.

Cellular Rubber

A cellular product is a two-phase system consisting of gas as the dispersed phase in a continuous phase of solid material, eg, rubber. The properties of such a cellular product are generally determined by: the physical properties of the rubber phase in which the gas is dispersed, the density of the material, the type of cell structure, whether open or closed cells, and the nature of the dispersed gas.

Chemical blowing agents are widely used in the manufacture of cellular rubber parts for consumer, automotive, and industrial applications, as well as in cellular plastics. There are two distinct classes of blowing agents: the inorganic and the organic types.

Cellular-rubber products prepared from the chemical blowing agents are becoming increasingly popular for technical and economic reasons. Their special features include heat- and sound-insulation properties, lightness and comfort in apparel, shock absorbency, improved textures and consumer appeal, and low weight-per-volume costs.

The type of cell structure in cellular materials is a chief factor in determining their properties and uses. Open-cell cellular material made from a dry polymer or polymers is called a sponge, but when it is made from a latex or latices it is called a foam. Closed-cell, expanded cellular materials have excellent thermal-insulation and sound-absorption properties. The water-vapor transmission and water-absorption properties of the closed-cell materials are very low.

Blowing. Blowing can be compared to baking a cake with yeast; the blowing agent decomposes and causes the compound to expand, and is followed by curing. An increase in volume by several hundred percent can be obtained, depending upon the amount and type of blowing agent used and several other components in the base compound. Cellular products prepared by the free-expansion method include sponge carpet underlay, direct-molded cellular soling, extruded and molded automobile sealing strips, pipe insulation, and sheet materials for flotation and insulation.

Expansion. In the expansion process, the uncured compound containing the blowing agent is placed in a mold. This method differs from the blowing process; instead of the uncured compound filling the mold only partially, enough compound is placed in the mold to fill it completely leaving 5% excess to seal the mold. The blowing agent decomposes during the cure, that takes place under considerable pressure; under high pressure, the nitrogen which is formed partly dissolves in the rubber. The press is opened before the cure of the rubber is complete. Owing to the high internal gas pressure, the precured compound expands and a material with very small closed cells is obtained.

A second curing operation, known as a postcure, is usually used to complete the cure and stabilize dimensions. Organic blowing agents are used almost exclusively for cellular rubbers prepared by this process, and consequently, a product with a predominantly closed-cell structure results. Molded closed-cell sheet material prepared by the expansion process is usually split into sheets of convenient thickness from which finished articles, eg, gaskets, shoe soles, etc, can be dried out.

Latex

The compounding technique for latex differs from that of dry rubber and is fundamentally much simpler. Latex compounding is done in the liquid state, and the compounded liquid latex is used directly to produce commercial articles without being first converted to a solid. For such use, the latex is usually concentrated to ca 60 wt% solids by one of several methods, the most common of which are centrifuging and creaming. As with dry rubber, most uses of latex require modification of the latex by addition of vulcanizing agents and other ingredients. The only exceptions are a few adhesives in which latex is used without modification by compounding.

The general principles of latex compounding are similar to those of dry-rubber compounding. The absence of mastication in the processing of latices saves time and power, and the resultant vulcanizate is inherently resistant to aging. A disadvantage of latex is the difficulty of obtaining any measure of reinforcement through the use of small-particle-size inorganic fillers. This failure of reinforcement is related to the lack of mastication. Another disadvantage is the difficult drying of latex deposits. This difficulty and the usual accompanying shrinkage rules out the use of latices in the production of solid articles of thick sections.

In order to obtain a homogeneous and stable latex compound, it is necessary that nonwater-soluble additives be dispersed or emulsified in water to about the particle size range of the latex particles. Thus the compounding ingredients must be capable of being readily dispersed.

Therefore, water-soluble ingredients are advantageous, since they can be either added directly to the latex or used in the form of an easily prepared water solution. Dispersions to be added to latex must have good storage stability, ie, the dispersions must not settle or change in pH.

Offsetting the limitations or difficulties in latex compounding is possible. It is possible to process latex compounds designed for fast cures at low temperatures by use of ultra-accelerators and very fast curing combinations. Some latex compounds can be cured simply by allowing the dried film or article to remain at room temperature for a sufficient length of time. It is possible to further increase cures by preparation of prevulcanized compounds in which, by heating the latex with the curing ingredients, the latex particles in the liquid latex compound are vulcanized so that, after formation of a deposit or article, it is necessary only to dry at low oven or room temperatures to obtain products of properties similar to those of a conventionally postcured article.

As in the case for many dry-rubber compounds, the addition of zinc oxide increases the rate of vulcanization and generally improves the physical properties and aging properties.

Fillers, eg, clays and whiting, are used to reduce cost or provide special properties. Fillers do not reinforce rubber deposited from latex. They are used to increase viscosity to give the latex compound suitable working properties for spreading.

The more active accelerators, particularly the dithiocarbamates, are used more widely in latex compounding than in dry-rubber compounding because scorching during processing is not a problem. Many articles from latex compounds are light in color and it is desirable to keep vulcanizing temperatures at or below 104°C to prevent darkening, and in many processes, it is very advantageous economically to reduce the time and temperature of vulcanization as much as possible.

The use of fatty acids in latex compounds follows the same principles as in dry rubber and, although natural-rubber latex contains some fatty acids or their soaps, it is sometimes necessary to add more, usually in the form of a soap. Antioxidants are used in latex compounding in a manner similar to dry-rubber compounding. Waxes, which form a protective coating by rising to the surface, are used in latex compounds as with dry-rubber compounds. Antiozonants can be used but they are very discoloring and staining.

Synthetic. The main types of elastomeric polymers commercially available in latex form from emulsion-polymerization systems are butadiene–styrene, butadiene–acrylonitrile, butadiene–styrene–acrylonitrile, and chloroprene (neoprene). There are also a number of specialty latices that contain polymers that are basically variations of the above polymers, eg, those in which a third monomer has been added to provide a polymer that performs some specific function. The most important of these are products containing either a basic, eg, vinylpyridine, or an acidic monomer, eg, methacrylic acid. These latices are specifically designed for tire-cord (qv) solutioning, paper coating, and carpet backsizing.

The basic constituents of all commercial emulsion-polymerization recipes are monomers, emulsifiers, and polymerization initiators. Other common components are modifiers, inorganic salts and free alkali, and shortstops.

Vinylpyridines. The vinylpyridine latices were developed specifically for use in adhering rubber stocks to fibers, particularly nylon. In general, the polymers are high diene types containing 10–15 wt% copolymerized 2-vinylpyridine and an approximately equal amount of styrene. There are three commercially available vinylpyridine latices, namely, Gen-Tac (General Tire and Rubber Co.), Hycar 2518 (BF Goodrich), and Polysar 781 (Polymer Corp.).

Carboxylated. One advance in latex technology has been the introduction of latices in which the polymer phases contain functional groups. The functional groups are derived from the use of unsaturated monomers containing carboxy groups in the polymerization system. Carboxylated styrene–butadiene latices have been used increasingly in carpet-backing applications because of their self-curing feature. Presumably, carboxylated latices obtain their strength from the polar carboxyl groups and the fairly high styrene content. Metal oxides and melamine–formaldehyde resins can be used as curing systems. In carpet-backing

applications, zinc oxide should be omitted from formulations since it can reduce tuft-retention values.

Types of latex compounds. For comparison with dry-rubber compounds, some examples of various latex compounds and the physical properties of their vulcanizates are given in Table 2. Table 3 includes similar data for an SBR latex compound.

Selection of Rubbers for Specific Applications

The selection of the type of rubber to be used in a particular product depends on the technical requirements of the product, the properties attainable by compounding, and economic factors. By far the largest volume of rubber is used in the manufacture of pneumatic tires for passenger cars, trucks, airplanes, and farm machinery. The polymers used for these applications are natural rubber, SBR, polybutadiene, polyisoprene, halogenated butyl, and some EPDM rubber. Butyl, EPDM, and natural rubber are also used for inner tubes. The essential requirements for the tread of a typical passenger tire include: resistance to abrasion and to the closely related phenomena of cutting and chipping; reduced rolling resistance; resistance to cracking and to crack growth; adequate flexibility at the lowest temperature encountered in service; a sufficiently high coefficient of friction between the tread and the road to minimize slipping and skidding; adequate adhesion of the tread to the carcass; sufficient stability of the material with time so that excessive deterioration does not occur in the normal life of the tire; and a moderate hysteresis so that excessive temperatures do not develop in service.

Table 2. Properties of a Natural-Rubber Compound[a]

Property		No antioxidant	Antioxidant 1.0 part (dry)
Unaged air-cured at 121°C			
3 min cure	700% modulus, MPa[b]	8.2	9.7
	tensile strength, MPa[b]	33.3	32.4
	elongation, %	> 970	> 910
15 min cure	700% modulus, MPa[b]	5.5	9.1
	tensile strength, MPa[b]	34.3	33.5
	elongation, %	> 990	> 920
After air-oven aging for 7 d at 70°C			
3 min cure	700% modulus, MPa[b]	1.9	2.1
	tensile strength, MPa[b]	35.6	37.1
	elongation, %	920	880
15 min cure	700% modulus, MPa[b]	1.6	1.3
	tensile strength, MPa[b]	32.8	34.2
	elongation, %	980	920
After oxygen-bomb aging for 96 h at 70°C			
3 min cure	700% modulus, MPa[b]	1.9	3.3
	tensile strength, MPa[b]	27.3	33.0
	elongation, %	900	810
15 min cure	700% modulus, MPa[b]	1.5	2.8
	tensile strength, MPa[b]	20.5	30.5
	elongation, %	940	880
After oxygen-bomb aging for 7 d at 70°C			
3 min cure	700% modulus, MPa[b]	2.9	4.4
	tensile strength, MPa[b]	20.4	30.1
	elongation, %	870	820
15 min cure	700% modulus, MPa[b]	1.5	2.5
	tensile strength, MPa[b]	11.0	16.8
	elongation, %	920	840

[a] Dry basis natural-rubber compound recipe, in parts by wt: natural latex (NC 356), 100.0; potassium hydroxide, 0.5; Nacconal 90F (an alkylarenesulfonate produced by Allied Chemical Company), 1.0; zinc oxide, 3.0; sulfur, 1.0; ZMBT, 1.0; zinc diethyldithiocarbamate (ZEDC) (trade names: Ethazate, Uniroyal, Inc.; Ethyl Zimate, R.T. Vanderbilt), 0.3; antioxidant, as indicated. Wet-basis natural-rubber compound recipe, in parts by wt: natural latex (NC 356), 167.9; potassium hydroxide, 2.5; Nacconal 90F, 5.0; zinc oxide, 5.45; sulfur, 1.65; ZMBT, 2.0; ZEDC, 2.0; antioxidant, as indicated. All films were poured from freshly mixed compounds, dried overnight in place, then lifted and dried one hour in air at 50°C before curing.
[b] To convert MPa to psi, multiply by 145.

Table 3. Properties of an SBR Compound[a]

Property	No antioxidant	Antioxidant, 1.0 part (dry)
Unaged, air-cured at 121°C		
10 min cure 500% modulus, MPa[a]	2.8	2.6
tensile strength, MPa[a]	18.4	15.2
elongation, %	790	790
30 min cure 500% modulus, MPa[b]	3.3	2.9
tensile strength, MPa[b]	17.3	13.3
elongation, %	750	710
After air-oven aging for 16 h at 130°C		
10 min cure 500% modulus, MPa[b]	4.0	3.7
tensile strength, % retained	27.4	37.1
elongation, % retained	58.5	76.0
30 min cure 500% modulus, MPa[b]	4.6	3.7
tensile strength, % retained	33.6	41.5
elongation, % retained	74.8	81.8
After oxygen-bomb aging for 96 h at 70°C		
10 min cure 500% modulus, MPa[b]	4.3	4.8
tensile strength, % retained	63.4	106.0
elongation, % retained	82.3	82.5
30 min cure 500% modulus, MPa[b]	3.7	3.5
tensile strength, % retained	67.0	89.2
elongation, % retained	89.5	97.2

[a] Dry basis SBR compound recipe, in parts by wt: SBR latex (type 2000), 100.0; Triton X-200 (sodium salt of alkylaryl polyether sulfonate, Rohm & Haas Company), 1.0; zinc oxide, 3.0; sulfur, 1.0; ZMBT, 1.0; ZEDC 0.3; antioxidant, as indicated. Wet-basis SBR compound recipe, in parts by wt: SBR latex (type 2000), 265.5; Triton X-200, 5.0; zinc oxide, 5.45; sulfur, 1.65; ZMBT, 2.0; ZEDC, 0.6; antioxidant, as indicated.
[b] To convert MPa to psi, multiply by 145.

Table 4. Qualitative Comparisons of Rubber[a]

	NR	SBR	IR	BR	IIR	EPDM	CR	NBR
Physical properties								
electrical resistance	B	B	B	B	B	B	C	X
flame resistance	X	X	X	X	X	X	B	X
gas permeation	C	C	C	C	A	C	B	B
heat resistance	C	C	C	C	B	B	B	B
cold resistance	B	B	B	B	C	B	C	C
Mechanical properties								
tensile strength (at 7 MPa[b])								
pure gum	> 21	< 7	> 21	< 7	> 10	< 7	> 21	< 7
reinforced	> 21	> 14	> 21	> 14	> 14	> 21	> 21	> 21
hardness, Shore A	30	40	30	40	40	30	40	40
	90	90	90	90	80	85	95	95
compression set	A	A	A	A	B	B	A	A
rebound, cold	A	B	A	A	X	B	B	B
rebound, hot	A	B	A	A	A	B	B	B
tear resistance	A	C	A	C	A	B	C	C
abrasion resistance	A	A	A	A	B	A	B	A
Chemical stability								
sunlight aging	C	C	C	C	A	A	A	C
oxidation resistance	B	B	B	B	A	A	A	A
ozone resistance	X	X	X	X	B	A	B	X
aliphatic hydrocarbon	X	X	X	X	X	X	B	A
aromatic hydrocarbon	X	X	X	X	X	X	C	B
chlorinated solvents	X	X	X	X	X	X	X	C
oxygenated solvents	B	B	B	B	B	B	C	C
petroleum, crude	X	X	X	X	X	X	B	A
natural gas	X	X	X	X	X	X	B	A
gasoline, fuel oil	X	X	X	X	X	X	B	A
lubricating oil	C	C	C	C	X	X	B	A
animal, vegetable oils	C	C	C	C	A	A	B	A
acids, dilute	B	B	B	B	A	B	A	B
acids, concentrated	C	C	C	C	B	B	B	C
sodium hydroxides	B	C	B	C	A	B	B	B
water swell resistance	B	A	B	A	A	B	B	B
specific gravity	0.92	0.93	0.92	0.94	0.91	0.86	1.23	1.00

[a] Ratings: A = excellent, B = good, C = fair, X = poor.
[b] To convert MPa to psi, multiply by 145.

In the thousands of other products made from rubber, the choice of polymer depends, as in the case of tires, on the properties that can be developed by compounding, on the requirements of the service, and on the cost of the polymer. In products that require a soft, flexible, resilient, highly extensible rubber, eg, for stationers' bands or golf-ball thread, natural rubber and polyisoprene are preferred. Another choice for such applications is neoprene. For solid golf balls, the excellent rebound characteristics of polybutadiene make it the choice of compounders. For products of low-to-moderate quality, eg, soles, heels, garden hose, and mats, SBR and natural rubber are competitive and can be used with high proportions of reclaimed rubber to reduce cost. For applications requiring excellent ozone resistance, EPDM, butyl, neoprene, chlorosulfonated polyethylene, polysulfides, nitrile–vinyl blends (Paracril Ozo), chlorobutyl rubber (CIIR), acrylics, polyurethanes, silicones, fluoroelastomers, and chlorohydrin rubbers can be used, the choice depending on what other properties are required by the service. Even the nonozone-resistant polymers, eg, SBR, BR, nitrile, natural rubber, etc, can be made resistant to ordinary ozone attack by the use of the proper blend of antiozonant and wax.

In Table 4, a general rating of the various polymers is shown with respect to certain important properties of their vulcanizates.

Fabrication

The recipe that prescribes definite proportions of rubber, fillers, softeners, vulcanizing agents, and certain accessory materials must be compounded or weighed, mixed, formed, and vulcanized to make a product. In some instances, certain materials require pretreatment before compounding. For example, it may be necessary to dry fillers to remove excess moisture or to pass them through screens to ensure adequate fineness and the removal of lumps. Also, the polymer in the case of natural rubber may require some processing, eg, washing or straining to remove foreign material, or it may require mastication to make the rubber more adaptable to succeeding processing steps.

Mastication. Probably the most important of these preliminary steps is the mastication of the rubber. Crude natural rubber as received from the plantation is high in viscosity, and for most uses it is advantageous to lower its viscosity prior to mixing.

The combination of heat and work on crude natural rubber produces a physical and chemical change that is measured in relative softness or plasticity. The amount of softness is prescribed by both the finished article use and the operation through which the compound will pass in its subsequent operations. Very high masticated or soft rubber is used in friction compounds, sponge stocks, and rubber solutions or cements. Medium soft rubber is used in calendering compounds, and lightly masticated rubber is used for stiff compounds.

Studies of mastication indicate that maximum softening effect on natural rubber is obtained by mastication either at < 54°C or > 132°C. Very little breakdown is obtained at 88–104°C. The breakdown of rubber is accomplished either on a roll mill, in an internal mixer, or in a screw plasticator, eg, the Gordon Plasticator. The roll mill functions in the low temperature region, whereas the plasticator and the Banbury operate at 149–177°C.

Mixing. The mixing operation is another very important stage through which the composition must pass. The processing steps subsequent to mixing depend on an adequate and uniform mix, and the quality of the final product is directly affected by the kind of mixing. The primary objectives in mixing include: attaining a uniform blend of all the constituents of the mix, ensuring that each portion of the resulting batch is of uniform composition; attaining an adequate dispersion of the fillers, ie, avoiding lumps or agglomerates of the fillers; and producing consecutive batches that are uniform both in degree of dispersion and viscosity (see Mixing and blending).

Roll mills and internal mixers, eg, the Banbury mixer, Intermix, and Bolling mixer, are used for batch operations.

The continuous mixers offer the following advantages: time savings since there are no separate loading and discharging operations; labor savings; lower capital investment because production rates are generally higher than with batch mixers; and probably, a more consistent mix overall than can be obtained in batch operations. A disadvantage of the continuous mixers is the relatively long mixing of a given compound. One such type of continuous mixer is the Shearmix extruder now known as the Transfermix. Other continuous mixers are the Farrel continuous mixer (FCM), Ko-Kneader, Plastificator, and Kneadermaster.

Forming. For the fabrication of most rubber products, the mixed compound must be formed in some way to prepare it for vulcanization. In some instances, the batch sheets as obtained from the mixing mills may not require processing other than to cut disks or rectangular pieces from the sheets, which are then suitable for charging into a mold. In most cases, however, the mixed compound must be processed into a form suitable for further fabrication. The most important of these processes are calendering and extrusion. Of less importance is the making of cements.

Assembly. Following the forming of the composition by calendering, extrusion, or deposition from a cement, some articles can be vulcanized directly without further treatment. For example, sheeted material suitable for gasketing, sheet rubber from which thread may be cut after cure, extruded shapes such as tubing, and irregularly shaped window strips need no further treatment before vulcanization. For other products, assembly steps, which vary from simple to complex, may be required.

Vulcanization. Vulcanization is an irreversible process during which a rubber compound, through a change in its chemical structure, eg, cross-linking, becomes less plastic and more resistant to swelling by organic liquids. The result is that elastic properties are conferred, improved, or extended over a wide range of temperature. The term vulcanization was originally employed to denote the process of heating rubber with sulfur, but has been extended to include any process with any combination of materials that produces this effect. Vulcanization can be carried out under numerous conditions. Methods include mold curing, injection cures, steam cures, hydraulic cures, air cures, and special and combination cures.

Latex mixes. The first step in latex fabrication is to bring the compounding ingredients into solution or dispersion form. In the case of water-soluble materials, it is necessary merely to dissolve in water, but most ingredients to be used are not water soluble, and it is necessary to emulsify the liquid ingredients and disperse the solid materials in water (see Latex technology).

Preparation of dispersions. The general procedure for preparing dispersions is to make a coarse slurry of the powder with water that contains small amounts of dispersing agent and stabilizer, and then grind the slurry in a suitable mill to give the desired particle size.

Preparation of emulsions. The general method for preparing an oil-in-water emulsion is first to make a coarse suspension of oil droplets in water and then subject this suspension to a refining process which may involve intense shearing and impact such as is delivered in colloid or ultrasonic mills or in a homogenizer. A homogenizer is a machine that forces the rough emulsion through a fine orifice under high pressure.

Mixing. The mixing of latex compounds is a simple operation and involves the following: weighing the proper amounts of the various solutions, emulsions, or dispersions, based on the recipe and the concentration of the various dispersions; then stirring these materials into the latex, usually in a large tank equipped with a mechanical agitator.

Forming. For the production of useful articles from latex compounds, it is necessary to convert them into solids of the desired form. A variety of methods are used to accomplish this. The straight-dip method is the simplest of any method used in making articles from latex. It is used for the manufacturing of very thin-walled dipped goods from which water can readily and quickly be removed by evaporation. Other processes include electrodeposition, coagulant deposition, the Kaysam process, and the heat-sensitizing process.

Thread. Although manufacturing of latex thread makes use of the general vulcanization principles and methods, it is unique with respect to the details of their application. The most widely used method is extrusion of latex compound through fine orifices into a coagulant bath which gels the thread, followed by mechanical handling of the thread through toughening, washing, drying, and curing operations. The coagulant bath is usually dilute acetic acid.

Latex foam. The flexible-foam market, especially in the furniture and transportation industries, has been taken over by the flexible urethane foams because of their wide range of physical properties. The high resilience (hr) type especially shows superior properties and lower energy demand over the conventional molded hot-cure foam. Latex foam, however, still constitutes an important application for synthetic and natural latexes.

Many different processes are patented for preparing this product but there are only two of commercial interest for preparing molded cushioning stock: the Dunlop (which is the most widely used) and the Talalay processes.

Back sizes. Backsizing defines the process of applying a film-forming adhesive material to the underside of carpets, rugs, or upholstery fabrics to achieve the following: anchor or lock the pile yarns to the basic weave or backing fabric, produce a firmer hand, increase weight, overcome raveling of cut edges, impart anti-skid properties, improve laying characteristics, which are especially important to wall-to-wall carpeting, and improve dimensional stability. In the cases of tufted carpet and pile fabric, the pile anchoring or tuft-locking function of the back size is critically important. A latex that accepts a high loading of filler for low costs and that at the same time develops the desired anchorage is required. The medium styrene–butadiene latices, and particularly, the so-called self-curing carboxy-modified SBR latices are most widely used.

Paper applications. In beater addition, the latex is mixed with the beaten paper pulp either by addition at the beater or to the stock chest at the wet end of the paper machine. A latex for this purpose must possess the proper balance between mechanical and chemical stability.

Adhesives. Both natural and synthetic latices are used in the preparation of combining adhesives for various applications. Latex adhesives are used for bonding sheets of fabric, paper, or leather, either to themselves or other materials. Essentially, there are two types of latex adhesives: dry combining and wet combining.

Tire-cord solutions. The latex solutioning of tire cord prior to building it into a carcass is a fabric adhesive application (see Tire cord).

Vulcanization. Vulcanization of articles made from latex compounds generally follows the same principles and methods as those used in vulcanizing articles made from dry-rubber compounds.

ROBERT R. BARNHART
Uniroyal Chemical

F.R. Eirich, ed., *Science and Technology of Rubber*, Academic Press, Inc., New York, 1978.

M. Morton, ed., *Rubber Technology*, 2nd ed., Van Nostrand Reinhold Publishing Corporation, New York, 1973.

W.M. Saltman, ed., *The Stereo Rubbers*, John Wiley & Sons, Inc., New York, 1977.

RUBBER, NATURAL

Natural rubber (NR) (*cis*-1,4-polyisoprene) occurs in over 200 species of plants, including dandelions and goldenrod. Only two species have had commercial significance: the *Hevea brasiliensis* tree and the guayule bush *Parthenium argentatum*. Guayule once was an important source of rubber and provided 10% of the world's supply in 1910. Today, the *Hevea* tree accounts for over 99% of the world's natural-rubber supply, but researchers are again studying guayule as a potential source of rubber.

Agriculture

Producing rubber trees grow to heights of 15–20 m and require 200–250 cm/yr of rainfall. They thrive at altitudes up to 300 m. Modern practice calls for the planting of trees at the rate of 400–500/ha (ca 160–200 per acre). Trees are commonly planted 3–4 m apart in parallel rows 6 m apart. Experiments are being conducted to determine the feasibility of higher planting densities. It takes up to 7 yr for a rubber tree to grow to maturity. The time of immaturity in the field can be reduced to 5 yr by growing the trees in plastic polybags in nurseries for up to 1 yr before transplanting to the field.

Rubber trees are subject to a variety of diseases, but the most devastating is the South American leaf blight caused by the fungus *Microcyclus ulei*.

Tapping

Latex is harvested from the tree by a process called tapping, which has been described as a controlled wounding of the tree. A specially designed tapping knife is used to remove shavings of bark from the surface of a groove made into the tree to a depth about 1 mm from the cambium. The groove is made from left to right at an angle of 30° to the horizontal across half the tree. This type of cut is the most common and is called a half spiral. Typical tapping intensities call for tapping the half spiral every second or third day.

Two types of raw material are brought to a processing factory: field latex and raw coagulum. Raw coagulum can be classified as either field or small-holder. Field coagulum is rubber that coagulates naturally in the field, ie, cup lump and tree lace. Small-holder coagulum is rubber coagulated either naturally or chemically by the small farmer who is unable to bring the rubber to the processing plant as latex. These raw materials are the basis for all types and grades of natural rubber.

Grades of Natural Rubber

Visually graded (conventional) rubber. Visually graded natural rubber is often referred to as conventional rubber, since visual inspection is the oldest method of grading rubber. The eight types of visually graded rubber, as classified by the source of raw material used in their manufacture, are latex grades: ribbed smoked sheet (RSS), white and pale crepes, and pure blanket crepes; and remilled grades: estate brown crepes, compo crepes, thin brown crepes or remills, thick brown crepes or ambers, and flat bark crepes. Latex grades are manufactured directly from field latex and remilled grades are produced from estate or small-holder coagulum. Compo crepes, thick blanket crepes (ambers), and pure smoked blanket crepes are not being produced in significant quantities.

Technically specified rubber. Technically specified natural-rubber (TSR) was introduced by the Malaysians in 1965 under the Standard Malaysian Rubber (SMR) scheme. Other rubber-producing countries soon followed with their own versions of TSR: Indonesia (SIR), Singapore (SSR), Thailand (TTR), and Sri Lanka (SLR). The introduction of TSR brought innovations in processing, packaging, and quality control to the natural rubber industry. In 1982 TSR accounted for ca 45% of worldwide dry natural rubber production.

Physical Properties

The physical properties of natural rubber vary slightly depending on the non-rubber content, degree of crystallinity, amount of storage hardening, and particular clone or blend of clones used as the raw material. Some average physical properties are listed in Table 1.

Chemical Properties

Commercial grades of *Hevea* natural rubber contain 93–95 wt% *cis*-1,4-polyisoprene. The remaining nonrubber portion is made up of moisture (0.30–1.0 wt%), acetone extract (1.5–4.5 wt %), protein (2.0–3.0 wt%), and ash (0.2–0.5 wt%). Natural rubber is soluble in most aliphatic, aromatic, and chlorinated solvents, but its high molecular weight makes it difficult to dissolve. The molecular weight is generally reduced by mastication on a mill or in an internal mixer, such as a Banbury, prior to dissolving the rubber. A chemical peptizing agent can be added during mastication to reduce the molecular weight further and to enhance the solubility of the rubber. The most important chemical reaction that natural rubber undergoes is vulcanization.

Other types of *Hevea* rubber include superior-processing (SP), technically classified (TC), air-dried sheet (ADS), skim, deproteinized natural rubber (DPNR), oil-extended natural rubber (OENR), heveaplus MG, and epoxidized natural rubber.

Table 1. Physical Properties of Natural Rubber

Property	Value
specific gravity	
at 0°C	0.950
at 20°C	0.934
refractive index	
at 20°C, RSS	1.5195
at 20°C, pale crepe	1.5218
coefficient of cubical expansion, °C^{-1}	0.00062
heat of combustion, J/g[a]	44,129
specific heat	0.502
thermal conductivity, W/(m·K)[b]	0.13
dielectric constant	2.37
power factor (at 1000 cycles)	0.15–0.20
volume resistivity, Ω·cm	10^{15}
dielectric strength, V/mm	3,937
cohesive energy density, J/cm^{3a}	266.5
glass-transition temperature, °C	−72

[a] To convert J to cal, divide by 4.184.
[b] To convert W/(m·K) to (Btu·in.)/(h·ft^2·°F), divide by 0.1441.

Natural rubber vs synthetic. Natural rubber's share of the world elastomer market has declined from well over 90% in the early 1940s to ca 32% in 1982. Synthetic-rubber usage increased during this period because of relatively inexpensive raw materials and the slow expansion of natural-rubber production.

Latex Concentrate

Natural-rubber latex concentrate is made by concentrating field latex with a dry-rubber content of 30–40 wt% to a minimum of 60 wt% dry rubber. Natural latex is used in a variety of applications: dipped goods, adhesives, latex thread, carpet backing, foam, and various miscellaneous uses, eg, reconstituted leather board and rubberized hair for upholstery.

Centrifuging is the most common method for concentrating field latex. Creaming is another method for concentrating *Hevea* latex. Cream latex has a dry rubber content of 64%.

High and low ammonia latices. Both centrifuged and creamed natural latices are produced in high and low ammonia versions. The high ammonia latices are preserved with about 0.75% ammonia and low ammonia latices are preserved with about 0.20% ammonia plus sodium pentachlorophenate (SPP), tetramethylthiuram disulfide (TMTD) and zinc oxide, or sodium dimethyldithiocarbamate (SDC) and zinc oxide.

The low ammonia latices are preferred by some consumers because of their lower odor and the elimination of a costly and time-consuming deammoniation step.

Balata and gutta-percha. Natural rubber (polyisoprene) exists in nature as both the cis and trans isomer. The rubber from the *Hevea* tree occurs only in the cis form, whereas both balata and gutta-percha occur as the trans isomer. Gutta-percha is obtained from trees of the family *Sapotaceae*, which are native to Malaysia, Borneo, and Sumatra. Balata is harvested from wild bushes and trees which grow primarily in the Surinana and Guiana regions on the northeastern coast of South America.

Guayule. Guayule (*Parthenium argentatum*) is a member of the sunflower family. The shrub grows to ca 0.5–1 m tall in its natural habitat of northern Mexico and the Big Bend area of Texas. In 1976, the Mexican Government built a pilot plant to extract rubber from wild guayule. The objective was to reduce Mexico's dependence on imported natural rubber and to create jobs for northern Mexico. However, at this time there is no commercial production of guayule rubber.

There are no significant structural differences between guayule and *Hevea* rubber. Guayule is more linear and is practically gel-free, which should result in better processing.

DAVID R. ST. CYR
The Goodyear Tire & Rubber Company

S.T. Semegen and Cheong Sai Fah, *The Vanderbilt Rubber Handbook*, R.T. Vanderbilt & Co., Inc., Norwalk, Conn., 1978, pp. 18–41.

A.T. Edgar, *Manual of Rubber Planting*, The Incorporated Society of Planters, Kuala Lumpur, Malaysia, 1958.

S.T. Semegen in M. Morton, ed., *Rubber Technology*, 2nd ed., Van Nostrand Reinhold Company, New York, 1973, p. 162.

RUBBER, SYNTHETIC.   See Elastomers, synthetic.

RUBBERY ACRYLONITRILE POLYMERS.   See Acrylonitrile polymers; Elastomers, synthetic, nitrile rubber.

RUBIDIUM AND RUBIDIUM COMPOUNDS

Rubidium (Rb) is an alkali metal in Group IA of the periodic table. Its chemical and physical properties generally lie between those of potassium and cesium (see Cesium and cesium compounds; Potassium; Potassium compounds). Rubidium is the sixteenth most prevalent element in the earth's crust. Despite its abundance, it usually is widely dispersed and is not found as a principal constituent in any mineral. At present, most of the rubidium produced is obtained from lepidolite containing 2–4% rubidium oxide. Lepidolite is found in Zimbabwe and at Bernic Lake, Canada.

Physical Properties

Rubidium is a soft, ductile, silvery-white metal and is the fourth lightest metallic element. Table 1 lists certain physical properties.

Chemical Properties

The reactions of rubidium are very similar to those of cesium and potassium, but rubidium is slightly more reactive than potassium.

Table 1.   Properties of Rubidium

Property	Value
atomic weight	85.47
melting point, °C	39.0
boiling point, °C	689
density, g/cm^3	
solid, 18°C	1.522
viscosity at 39°C, mPa·s(= cP)	0.6713
heat of fusion, J/g[a]	25.69
heat of vaporization, J/g[a]	887
specific heat J/(kg·K)[a]	
solid	331.37
thermal conductivity, W/(m·K), liquid	29.3

[a] To convert J to cal, divide by 4.184.

Rubidium burns with a violet flame in the presence of air and reacts violently with water, liberating hydrogen which spontaneously explodes if oxygen or air are present. Rubidium forms a mixture of four oxides: the yellow monoxide, Rb_2O; the dark brown peroxide, Rb_2O_2; the black trioxide, Rb_2O_3; and the dark orange superoxide, RbO_2. Rubidium is the second strongest Lewis base.

Health and Safety Factors

Few toxicity data are available regarding the response of humans to exposure to rubidium or its compounds. Localized ventilation of equipment and the use of approved dust respirators are recommended when handling dry rubidium salts.

Rubidium Compounds

Table 2 lists some properties of certain rubidium compounds.

Table 2.   Properties of Rubidium Compounds

| Compound | Formula | Solubility, g/100 cm^3 °C | | Mp, °C | Bp, °C |
		Hot water	Cold water		
acetate	$RbC_2H_3O_2$	86[44.7]		246	
aluminum sulfate dodecahydrate	$RbAl(SO_4)_2 \cdot 12H_2O$	43[80]	1.3[0]	99	
bromide	RbBr	205[113.5]	98[5]	682	
chloride	RbCl	139[100]	77[0]	715	1390
fluoride	RbF		130.6[18]	760	1410
iodide	RbI	163[25]	152[17]	642	1300
sulfate	Rb_2SO_4	82[100]	36[0]	1073	
chromate	$RbCrO_4$	95.7[60]	62[0]		
nitrate	$RbNO_3$	452[100]	19.5[0]		
carbonate	Rb_2CO_3	450[20]		837	740 (dec)
hydroxide	RbOH		180[15]	301	
perchlorate	$RbClO_4$	100[18]	0.5[0]		

FREDERICK B. WHITE, JR.
W.G. LIDMAN
Kawecki Berylco Industries, Inc.

C.A. Hampel in *Rare Metals Handbook*, 2nd ed., Reinhold Publishing Corp., New York, 1961, pp. 434–440.

J.W. Mellor, *Comprehensive Treatise on Inorganic and Theoretical Chemistry*, Vol. 2, Suppl. 3, John Wiley & Sons, Inc., New York, 1963, pp. 2136–2293, 2488–2505.

KBI Division of Cabot Corporation Product Data, *Metals and Alloys—Rubidium*, File No. 320-PD-1, Reading, Pa.

RUTHENIUM.   See Platinum-group metals.

RUTHERFORDIUM.   See Actinides and transactinides.

S

SABADILLA, SABADINE, SABALINE.   See Insect control technology.

SACCHARIN.   See Sweeteners.

SAFETY.   See Plant safety; Materials reliability.

SALICYLIC ACID AND RELATED COMPOUNDS

Compounds of the general structure

are commonly known as the monohydroxybenzoic acids mol wt 138.12. Of the three acids, the ortho isomer, salicylic acid, is by far the most important. The main importance of salicyclic acid and its derivatives lies in their antipyretic and analgesic actions (see Analgesics, antipyretics, and anti-inflammatory agents). Natural salicylic acid, which exists mainly as the glucosides of methyl salicylate and salicyl alcohol, is widely distributed in the roots, bark, leaves, and fruits of various plants and trees. As such, their use as preparations for ancient remedies is probably as old as herbal therapy.

Physical properties.   Salicylic acid is obtained as white crystals, fine needles, or fluffy white crystalline powder. It is stable in air and may discolor gradually in sunlight. The synthetic form is white and odorless. When prepared from natural methyl salicylate, it may have a lightly yellow or pink tint and a faint, wintergreenlike odor. m-Hydroxybenzoic acid crystallizes from water in the form of white needles and from alcohol as platelets or rhombic prisms. p-Hydroxybenzoic acid crystallizes in the form of monoclinic prisms. Various physical properties of hydroxybenzoic acids are listed in Table 1.

Reactions.   The hydroxybenzoic acids have both the hydroxyl and the carboxyl moieties and, as such, participate in chemical reactions characteristic of each. In addition, they can undergo electrophilic ring substitution. Reactions characteristic of the carboxyl group include decarboxylation, reduction to alcohols, and the formation of salts, acyl halides, amides, and esters. Reactions characteristic of the phenolic hydroxyl group include the formation of salts, esters, and ethers. Reactions involving ring substitution include nitration, sulfonation, halogenation, alkylation, and acylation.

Salicylic Acid

Manufacture.   Modern methods of commercial manufacture employ the Kolbe-Schmitt reaction in which alkali phenol and CO_2 are the reactants.

Health and safety factors (toxicology).   In the laboratory, salicylic acid should be handled so that its irritant properties do not become a problem. Avoidance of skin or eye contact, general good hygiene, and good housekeeping should be sufficient. If large quantities are to be handled, protective clothing, ie, long sleeves and gauntlets, and NIOSH-approved dust respirators should be used. Dust concentrations as low as 9 g/m^3 can ignite; therefore, "No Smoking" signs should be posted in or near the work area. Eye fountains and safety showers should also be in the vicinity of the work area, especially if large quantities are being handled.

The single-dose oral toxicity of salicylic acid is moderate. The LD_{50} in rats is 400–800 mg/kg.

Uses.   Approximately 60% of the salicylic acid produced in the United States is consumed in the manufacture of aspirin: this statistic has remained relatively constant for at least the last ten years. Approximately 10% of the salicylic acid produced is consumed in various applications, eg, foundry and phenolic resins, rubber retarders, dyestuffs, and other miscellaneous uses. The remaining 30% is used in the manufacture of its salts and esters for a variety of applications.

Salts.   Sodium salicylate is by far the most important salt of salicylic acid. Others include magnesium salicylate, basic bismuth salicylate (oxysalicylate, subsalicylate), and aluminum, ammonium, calcium, lead, lithium, mercury, potassium, and strontium salts.

Esters.   The esters of salicylic acid account for ca 25% of the salicylic acid produced. They include methyl salicylate, phenyl salicylate, benzyl salicylate, menthyl salicylate, isoamyl salicylate, and salicyl salicylate.

Other derivatives.   The derivatives of salicylic acid have been used in a wide variety of applications; however, the primary emphasis has been in the development of medicinal agents. Derivatives include p-aminosalicylic acid, methylene-5,5-disalicylic acid, salicylamide, salicylanilide, and 5-sulfosalicylic acid.

m-Hydroxybenzoic Acid

Of the three hydroxybenzoic acids, the meta isomer is of little commercial importance. It offers no outstanding points of chemical interest and is used industrially in small quantities in a limited number of applications.

Manufacture.   m-Hydroxybenzoic acid was first obtained by the action of nitrous acid on m-aminobenzoic acid. It is more conveniently prepared by the sulfonation of benzoic acid with fuming sulfuric acid.

Uses.   m-Hydroxybenzoic acid is reported as an intermediate in the manufacture of germicides, preservatives, pharmaceuticals, and plasticizers (qv).

p-Hydroxybenzoic Acid

p-Hydroxybenzoic acid is of significant commercial importance. The most familiar application is in the preparation of several of its esters, which are used as preservatives.

Manufacture.   Several methods have been described for the preparation of p-hydroxybenzoic acid. The commercial technique is similar to that for salicylic acid, ie, Kolbe-Schmitt carboxylation of phenol.

Uses.   There are many polymer and plastic applications for p-hydroxybenzoic acid. It has been converted to epoxy resins by the reaction with epichlorohydrin, used as a modifier for ethylene–propadiene copolymers, and employed in copolyether esters as fibers for use in radial tire cord (qv). One of the most recent applications is the development of a linear p-hydroxybenzoic acid polymer.

Acetylsalicylic Acid (Aspirin)

Acetylsalicylic acid was first synthesized in 1853 from the reaction of acetyl chloride and sodium salicylate.

Aspirin normally occurs in the form of white, tabular or needlelike crystals, or as crystalline powder. It melts at 135–137°C and decomposes at 140°C. The solubility of aspirin is ca 1 g/300 mL of water at 25°C, ca 1 g/5 mL of ethanol at 25°C, 29 g/100 g of acetone at 20°C, and 1 g/17 mL of chloroform at 25°C.

Table 1.   Physical Properties of Hydroxybenzoic Acids

Property	Value (isomer)		
	Ortho	Meta	Para
melting point, °C	159	201.5–203	214.5–215.5
boiling point, °C	211 sub		
density, g/cm³	1.443^{20}_4	1.473^{25}_{25}	1.497^{20}_{20}
refractive index	1.565		
flash point (Tag closed-cup), °C	157		
K_a (acid dissociation) at 25°C	1.05×10^{-3}	8.3×10^{-5}	2.6×10^{-5}
heat of combustion, mJ/mol[a]	3.026	3.038	3.035
heat of sublimation, kJ/mol[a]	95.14		116.1

[a] To convert J to cal, divide by 4.184.

Manufacture. Aspirin is manufactured by the acetylation of salicylic acid by acetic anhydride.

Production. Aspirin is produced in the United States by the Dow Chemical Company, Monsanto Company, Morton-Norwich Products, Inc., and Sterling Drug, Inc.

Uses. Aspirin has analgesic, anti-inflammatory, and antipyretic actions. It is used for relief of less severe types of pain, eg, headache, neuritis, acute and chronic rheumatoid arthritis, myalgias, and toothache.

Salicyl Alcohol

Salicyl alcohol (saligenin, o-hydroxybenzyl alcohol), $C_6H_4(OH)$-CH_2OH, crystallizes from water in the form of needles or white rhombic crystals. It occurs in nature as the bitter glycoside salicin, which is isolated from the bark of *Salix helix, S. pentandra, S. praecos*, some other species of willow trees, and the bark of a number of species of poplar trees, eg, *Populus balsamifera, P. candicans, and P. nigra.*

The alcohol has the following properties: mp, 86°C; d_{25}^{13}, 1.61 g/cm³; heat of combustion 3.542 MJ/mol (846.6 kcal/mol); solubility in 100 mL water at 22°C 6.7 g; very soluble in alcohol and ether; sublimes readily.

Manufacture. Numerous methods for the synthesis of salicyl alcohol have appeared in the literature. These involve the reduction of salicylaldehyde or of salicylic acid and its derivatives.

Uses. Saligenin has been used medicinally as an antipyretic and tonic.

Thiosalicylic Acid

Thiosalicylic acid (o-mercaptobenzoic acid) is a sulfur-yellow solid that softens at 158°C; has a mp of 164°C; sublimes; is slightly soluble in hot water and freely soluble in glacial acetic acid and alcohol; and yields dithiosalicylic acid upon exposure to air.

Manufacture. Thiosalicylic acid is prepared by heating o-chlorobenzoic acid with an alkaline hydrosulfide in the presence of copper sulfate and sodium hydroxide.

Uses. Thiosalicylic acid has been used as an anthelmintic (see Chemotherapeutics, anthelmintic), as a bactericide, and as a fungicide. It also has been used as a rust remover, a corrosion inhibitor for steel, and a polymerization inhibitor (see Corrosion and corrosion inhibitors). In photography (qv), it has application in print-out emulsions and as an activator for photographic emulsions.

STEPHEN H. ERICKSON
The Dow Chemical Company

M. Gross and L.A. Greenberg, *The Salicylates—A Critical Bibliographic Review*, Hillhouse Press, New Haven, Conn., 1948.

M.J.H. Smith and P. Smith, *The Salicylates—A Critical Bibliographic Review*, Interscience Publishers, a division of John Wiley & Sons, Inc., New York, 1966.

SAMPLING

The chemical industry produces many gaseous, liquid, and solid materials, ranging from basic chemicals to functional specialties. In addition, many processes require the use of intermediates in the form of gases, liquids, and solids. These various materials are sampled for the purposes of process control, product quality control, environmental control, and occupational health control. Sampling of some materials can be hazardous, particularly those involving toxic, unstable, and pressurized substances, which require special safety precautions. However, with the exception of these special circumstances, the problems encountered in sampling materials in the chemical industry are principally those of proper choice and handling of a sampling device.

Definitions and Problems

Sampling is the operation of removing a portion from a bulk material for analysis in such a way that the portion is representative of the physical and chemical properties of that bulk material. From a statistical point of view, sampling is expected to provide analytical data from which some property of the chemical may be determined with known and controlled errors and at the lowest cost.

For pure liquids and gases, sampling is relatively easy, but it becomes difficult when particulates are involved, and almost all samples taken in the chemical industry contain solids in a subdivided form. Some raw materials, intermediates, or products are particulates; others contain them as contaminants. Some exist in natural deposits, eg, strata, in a heap on the ground; in storage bins, tanks, pipes, or ducts; in railcars, drums, bottles, or bales; or in other containers that may or may not be subdivided easily into representative units. Furthermore, many systems containing particulates tend to segregate during handling and storage, and this introduces a sampling error in the form of bias. Although these systems introduce sampling difficulties, they do obey one rule of good sampling, which is that all samples should be taken when the system is in motion.

The design engineer must be extremely familiar with the potential problems that any sampling operation can impose. Rarely are chemical plants well-designed in terms of sampling capabilities, and rarely do they permit all rules of sampling to be practiced. Therefore, most sampling involves some compromise. Decisions must be made regarding several factors, including the use to which the subsequent analysis is to be put; the test procedure to be employed and the quantity of material to be taken; how representative is the quantity and quality of the sample; minimization of chemical and physical changes during sampling; analytical facilities available, whether manual or automatic; skill and experience of the sampling personnel; sample and analysis use correlation; and cost-benefit relationship.

Gases

By far the largest proportion of gas-sampling operations in industry is conducted in the environmental field, and the sampling methods have been well researched and well-documented. The preparation, precautions, and equipment requirements involved in the sampling of air-pollution sources are applicable to most other gaseous environments (see Air-pollution control methods).

Before a source-analysis program is undertaken, it is important to decide which information is really required. In the chemical industry, most operations include both simple and complex emission problems. Incorrect sampling can cause the greatest single error in analysis and can often incur the largest fractional cost. Sampling sites must be selected with care, as the choice of the sampling point can significantly influence accuracy and cost. For most purposes, measurement requires the determination of temperature, concentration, and characteristics of the gas contaminants. It requires the mass rates of emission of each contaminant and, therefore, concentration and volumetric-flow data need to be taken.

For acid mists, the Brink impactor is often used. The mist is first drawn through a cyclone to remove all particles greater then 3 μm. A five-stage impactor is used to classify and collect all mist particles of 0.3–3.0 μm dia.

Liquids

Unlike gases, liquids are contained and stored in a wider variety of containers, which require specific designs of samplers. In the chemical industry, liquids are sampled from process vessels, tanks, tank trucks, tank cars, ships, barges, pipelines, transfer lines, drums, carboys, cans, bottles, open lagoons, settling ponds, sewers, and open, flowing streams and rivers. Liquid and gaseous systems are difficult to sample representatively when particles are present. Conditions of isokinetic sampling, probe alignment, and sampling location are important to the design of the sampling procedure; however, with liquid systems, they are not as critical as with gases. Viscosity tends to dampen the effects of sudden changes in flow direction, and particles do not separate readily from streamlines unless the particle masses and velocities are large.

Solids

Solids occur in several forms in the chemical industry. Raw materials from natural deposits are compacted in the ground, and sampling is performed during the exploration stages. This type of material is typified by minerals and fossil fuels. Before use, such materials must be mixed, crushed, ground into particulate form, cleaned, and stored in a heap on the ground or in a silo, bin, or hopper. Products in particulate form are usually stored in drums, railcars, ships, barges, cans, bags, boxes, etc. During manufacture, they are transported by conveyors, pipes, and chutes and are packaged by the use of free-flowing streams, pneumatic conveyors, spouts, and gravity chutes. Sampling may be required before, during, or immediately after any one of these operations, and different methods are used for most of them.

Sampling Efficiency

Sampling of bulk solids from grinding circuits or chemical plants represents tons of material per day. A primary sampler generally removes 10–100 kg as a gross sample, which is further subdivided into 1–10-kg quantities by a secondary device. This secondary or laboratory sample is submitted for analysis. The sampling might be analyzed in its present form or it might be crushed prior to examination, depending on the type of analysis needed. Samples are required from this analytical sample in gram and milligram quantities. Because of the convenient size and reasonable quantities of test material, more has been reported on the efficiency of these devices than on commercial units. However, the findings are important to sampling design on a small and large scale.

REG DAVIES
E.I. du Pont de Nemours & Co., Inc.

T. Allen, *Particle Size Measurement*, 2nd ed., Chapman & Hall, Ltd., London, UK, 1975.

U.S. EPA Regulations on Standards of Performance for New Stationary Sources, 40 CFR 60, Appendix A, Reference Methods, Washington, D.C., 1980.

Handbook for Monitoring Industrial Wastewater, U.S. EPA, Washington, D.C., Aug. 1973.

SCANDIUM.   See Rare-earth elements.

SCREEN PRINTING.   See Dyes, Application and evaluation; Printing processes; Reprography.

SCREENS, SCREENING.   See Size classification; Size measurement of particles.

SCRUBBERS.   See Absorption; Gas cleaning.

SEALANTS

Although there is some overlap in application. Adhesives and sealants usually have different functions (see Adhesives). The former are selected for their ability to bind two materials, whereas the latter are selected as load-bearing elastic jointing materials that exclude dust, dirt, moisture, and chemicals and contain a liquid or gas. Sealants are also used to reduce noise and vibrations and to insulate or serve as space fillers (see Insulation, acoustic; Chemical grouts).

The sealant is a more modern term describing compositions based on synthetic resins as well as asphaltic and oil-based caulking compounds. These composites require little if any energy for conversion to solids and usually conserve energy when applied. The principal resins used as sealants are polysulfides, silicones, polyurethanes, acrylics, neoprene, butyl rubber, and sulfochlorinated polyethylene.

Bitumens

Bitumens or asphaltic materials have been used as sealants for many centuries (see Asphalt). Bituminous hot-melt compositions usually contain scrap rubber or neoprene, which acts as a flexibilizing agent. These dark-colored sealants are used for joints in highways and buildings. Sealants produced from asphalt emulsions have been used to prevent the escape of radon gas from storage bins of uranium-mill tailings.

Latex Caulking Compositions

Filled polymeric emulsions, which are related to latex paints, have been used as tub caulks and spackling compounds. Since these systems set by evaporation, their shrinkage is greater than hot-melt or prepolymer sealants (see Paint; Latex technology).

Oil-Based Caulks and Sealants

Oil-based caulks or putty have been used for centuries as glazing sealants. These 100% solids compositions are much more rigid than other sealants, but the rigidity is often overcome by the addition of elastomers, eg, neoprene.

Hot-Melt Sealants

In addition to butyl rubber and bitumens, several other polymers are used as hot-melt sealants. Among these are the copolymers of ethylene and vinyl acetate (EVA), atactic polypropylene, and mixtures of paraffin wax and polyolefins (see Olefin polymers).

Poly (vinly Chloride) Plastisols

Poly (vinyl chloride) (PVC) plastisols have been used widely as gap fillers and sealants in automobiles. These plastisols are dispersions of finely divided PVC in liquid plasticizers, eg, dioctyl phthalate (DOP) (see Vinyl polymers).

Polyesters

Fibrous-glass-reinforced unsaturated polyesters and silica- and calcium carbonate-filled polyester grouting compositions are strong, lightweight plastic cements that are used for making sewer pipe. Plastic concrete, trade name Polysil, is used for making utility poles, bathroom fixtures, and manholes (see Polyesters, unsaturated).

Epoxy Resins

Epoxy resins (qv) were first synthesized by the reactions of bisphenol A and epichlorohydrin. Most epoxy resin is used as an *in situ* polymerized plastic, in adhesives, flooring, protective coatings, and laminates (see Embedding).

Phenolic Resins

The two-package system consists of a silica or carbon filler containing an acid setting agent, eg, *p*-toluenesulfonic acid (PTA), and an A-stage resole phenolic resin. Since *p*-toluenesulfonic acid may cause dermatitis and formalydehyde may be carcinogenic, extreme care must be taken when applying these materials (see Phenolic resins).

Urea Resins

The disadvantage of the dark color associated with the phenolic resin cement was overcome by the use of a urea–formaldehyde prepolymer. Kaolin-filled, *in situ* cast-urea resinous composites had been used for sealing cracks in underground rocks and for the fabrication of irrigation pipes (see Amino resins).

Furan Resin

The lack of resistance of phenolic-resin cements to alkali was overcome by the development of comparable cements based on furan resins. These two-package resinous cements have been used extensively as mortars, grouts, and setting beds for brick and tile (see Furan compounds).

Health and Safety Factors

Care should be taken to prevent exposure to solvents, plasticizers, and fillers used with polysulfide sealants.

RAYMOND B. SEYMOUR
The University of Southen Mississippi

A. Damusis, *Sealants*, Reinhold Publishing Co., New York, 1967.

R.B. Seymour, ed., *Plastic Mortars, Sealants, and Caulking Compounds*, ACS Symposium Series 113, Washington, D.C., 1979.

H.H. Buchter, *Industrial Sealing Technology*, John Wiley & Sons, Inc., New York, 1979.

SEASONINGS. See Flavors and spices.

SEAWEED COLLOIDS. See Gums.

SEDIMENTATION

Sedimentation is defined as "the separation of a suspension into a supernatant clear fluid and a rather dense slurry containing a higher concentration of solid". The uses of sedimentation in industry fall into the following categories: solid–liquid separation; solid–solid separation; particle-size measurement by sedimentation; and other operations such as mass transfer, washing, etc.

In solid–liquid separations, the solids are removed from the liquid either because the solids or the liquid are valuable or because they have to be separated before disposal. If the primary purpose is to produce the solids in a highly concentrated slurry, the process is called thickening. If the purpose is to clarify the liquid, the process is called clarification (see Dewatering). Usually, the feed concentration to a thickener is higher than that to a clarifier. Some types of equipment, if correctly designed and operated, can accomplish both clarification and thickening in one stage (see also Extraction, liquid–solid).

In solid–solid separation, the solids are separated into fractions according to size, density, shape, or other particle property.

Sedimentation is also used for size separation, ie, classification of solids. It is one of the simplest ways to remove the coarse or dense solids from a feed suspension. Successive decantation in a batch system produces closely controlled size fractions of the product. Generally, however, particle classification by sedimentation does not give sharp separations (see Size measurement).

In particle-size measurement, gravity sedimentation at low solids concentrations (< 0.5 vol%) is used to determine particle-size distributions of equivalent Stokes diameters in the range from 2 to 80 μm. Particle size is deduced from the height and time of fall using Stokes' law, whereas the corresponding fractions are measured gravimetrically, by light, or by x rays. Some commercial instruments measure particles coarser than 80 μm by sedimentation when Stokes' law cannot be applied.

In addition to the above-mentioned applications, sedimentation is also used for other purposes. Relative motion of particles and liquid increases the mass-transfer coefficient. This motion is particularly used in solvent extraction in immiscible liquid–liquid systems (see Extraction, liquid–liquid). An important commercial use of sedimentation is in continuous countercurrent washing, where a series of continuous thickeners is used in a countercurrent mode in conjunction with reslurrying to remove mother liquor or to wash soluble substances from the solids.

Most applications of sedimentation are, however, in straight solid–liquid separation.

Equipment

Sedimentation equipment can be divided into batch-operated settling tanks and continuously operated thickeners or clarifiers (Figs. 1 and 2).

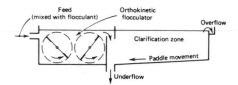

Figure 1. Schematic diagram of a rectangular basin clarifier.

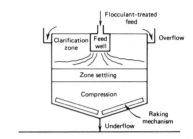

Figure 2. The circular-basin continuous thickener.

The operation of the former is very simple and their use has recently diminished. They are still used, however, when small quantities of liquids are to be treated, eg, in the cleaning and reclamation of lubricating oil (see Recycling, oil). Most sedimentation processes are operated in continuous units.

LADISLAV SVAROVSKY
University of Bradford

L. Svarovsky, ed., *Solid–Liquid Separation*, 2nd ed., Butterworths, London, UK, 1981.

J.M. Keane, *World Min.*, 44 (Nov. 1979).

SELENIUM AND SELENIUM COMPOUNDS

Selenium

Selenium (Se, at no. 34, at wt 78.96) is between sulfur and tellurium in Group VIA and between arsenic and bromine in period 4 of the periodic table. Its outer electronic configuration is $3d^{10}4s^24p^4$, and its three inner shells are completely filled. Strikingly similar to sulfur in most of its chemistry, its important oxidation states are -2, 0, $+2$, $+4$, and $+6$. As far as is known, the $+2$ state does not occur in nature. Selenium exists in various allotropic modifications and forms many inorganic and organic compounds, which are like those of sulfur and some of which are isomorphous. Selenium was discovered in 1817 and its name was derived from *selene* (moon, in Greek).

Physical properties. Some physical constants for selenium are given in Table 1.

Chemical properties. The chemical properties of selenium are intermediate between those of sulfur and tellurium. Selenium reacts with active metals and gains electrons to form ionic compounds containing the selenide ion Se^{2-}. Selenium forms covalent compounds with most other substances. The oxidation states in elemental and combined forms are as follows: Na_2Se, -2; Na_2Se_2, -1; Se_8, 0; Se_2Cl_2, $+1$; $SeCl_2$, $+2$; Na_2SeO_3, $+4$; and Na_2SeO_4, $+6$.

Selenium combines with metals and many nonmetals directly or hydrochemically. The selenides resemble sulfides in appearance, composition, and properties. Selenium forms halides by reacting vigorously with fluorine and chlorine, and less so with interhalogen compounds and bromine. Selenium combines with oxygen, yielding a number of oxides, the most stable being selenium dioxide. Under proper conditions, selenium forms selenides with hydrogen, carbon, nitrogen, phosphorus, and sulfur (see Inorganic Compounds).

Table 1. Physical Constants of Selenium

Property	Value
melting point, °C	217
boiling point, °C	ca 685
heat of fusion (trigonal liquid), J/mol[a]	6.224
heat of evaporation, J/mol[a]	95.90
heat of combustion (at 298 K), J/mol[a]	−236.8
heat capacity, J/(g·K)[a]	
trigonal	24.52
liquid	29.288
thermal conductivity, W/(m·K)	248.1
thermal expansion coefficient	3.24×10^{-5}–7.5×10^{-5}, depending on the form
viscosity, mPa·s (= cP)	
at 220°C	221
at 360°C	70
density, g/cm³	
trigonal at 298 K	4.819
liquid at 490 K	3.975
standard reduction potential, V	
$Se + 2 e \rightarrow Se^{2-}$	−0.78
$Se + 2 H + 2 e \rightarrow H_2Se$ (aq)	−0.36
surface tension (liquid), mN/m (= dyn/cm)	
at 220°C	105.5
at 310°C	95.2
electronegativity (Pauling scale)	2.4

[a] To convert J to cal, divide by 4.184.

Table 2. Physical Properties of Some Inorganic Selenium Compounds

Compound	Formula	Mp, °C	Bp, °C	Heat of formation (at 25°C), kJ/mol[a]
hydrogen selenide	H_2Se	−65.7	−41.3	85.7
carbon selenide	CSe_2	−40–45	125	155.2
selenium chloride	Se_2Cl_2	−85	127 (dec)	−43.7
selenium bromide	Se_2Br_2		227 (dec)	
selenium dichloride	$SeCl_2$			−40.6
selenium tetrafluoride	SeF_4	−13.2	101	
selenium tetrachloride	$SeCl_4$	305	196 (sub)	−188.3
selenium tetrabromide	$SeBr_4$	75 (dec)		
selenium hexafluoride	SeF_6	−34.6	−46.6 (sub)	−1029.3
selenium dioxide	SeO_2	340	315 (sub)	−238.5
selenium trioxide	SeO_3	118		−184
selenious acid	H_2SeO_3			−531.3
selenic acid	H_2SeO_4	60		−538.0
selenium oxyfluorides	$SeOF_2$	15	125–126	
	SeO_2F_2	−99.5	−8.4	
	$SeOF_6$	−54	−29	
selenium oxychloride	$SeOCl_2$	10.8	177.6	
selenium oxybromide	$SeOBr_2$	41.7	217	

[a] To convert J to cal, divide by 4.184.

Selenium remains unaffected by dilute sulfuric acid or hydrochloric acid, but it dissolves in a nitric–hydrochloric acid mixture, strong nitric acid, and concentrated sulfuric acid. It is oxidized by ozone and solutions of alkali-metal dichromates, permanganates, and chlorates and calcium hypochlorite.

Selenium also forms a large number of organic compounds. Of special interest is the oxidizing and reducing action of selenium and its compounds on many organic compounds.

Occurrence. At 0.68 atom per 10,000 atoms Si, selenium is the thirtieth element of cosmic abundance. Selenium is widely dispersed in igneous rocks, probably as selenide materials, in volcanic sulfur deposits with sulfides, in hydrothermal deposits, and in massive sulfide and porphyry copper deposits. In such sedimentary rocks as sandstones, carbonaceous siltstones, phosphorite rocks, and limestones, it is syngenetic.

Manufacture. Recovery. Practically all selenium is obtained as a by-product of precious-metal recovery from electrolytic copper refinery slimes.

Purification. Single-stage distillation of selenium at atomospheric pressure and at 680°C from cast iron or iron alloy retorts expels most of such occluded impurities as sulfur dioxide, water, organic substances, halogens, sulfuric acid, and mercury, and leaves the bulk of the nonvolatile impurities, eg, metals and tellurium, in the residue.

Health and safety factors. Commercial elemental selenium is relatively inert and can be handled without special precautions.

Industrial precautions include the common-sense measures of good housekeeping, proper ventilation, personal cleanliness, frequent change of clothing, provision of dust masks where needed, gloves, and either safety glasses or chemical goggles.

Inorganic Compounds

Selenium forms inorganic and organic compounds similar to those of sulfur and tellurium. The oxidation states are −2, 0, +4, and +6. The most important inorganic compounds are the selenides, halides, oxides, and oxyacids. Some important physical properties of selenium compounds are listed in Table 2.

Organic Compounds

The chemical properties of organosulfur, organoselenium, and organotellurium compounds are markedly similar. Because bond stability decreases with increasing atomic number of the element, thermal stability and stability on exposure to light of all wavelengths decrease and oxidation susceptibility increases. The compounds range from the simple COSe, CSSe, and CSe_2 to complex heterocyclic compounds, selenium-containing coordination compounds, and selenium-containing polymers.

Examples of the more simple ring compounds include selenophene, selenolo[2,3-b]selenophene, seleno[3,2-b]selenophene, seleno[3,4-b]selenophene, benzo[b]selenophene, selenathrene, selenazole, selenetane, and cyclic ethers 1,4-diselenane and 1,4-selenoxane.

Uses

Selenium and its compounds are used in photoelectric cells, photovoltaic cells, rectifiers, and xerography.

Metallurgical applications include the production of ferrous metals, copper and copper alloys, nickel–iron and cobalt–iron alloys, lead and lead alloys, and chromium plating.

Selenium and its compounds are also used in glass and ceramics, pigments, rubber, lubricants, organic synthesis and pharmaceuticals, and medicine and nutrition.

E.M. Elkin
Noranda Mines Limited

M. Schmidt, W. Siebert, and K.W. Bagnall, *The Chemistry of Sulphur, Selenium, Tellurium, and Polonium*, Pergamon Press Ltd., Oxford, UK, 1973.

R.A. Zingaro and W.C. Cooper, *Selenium*, Van Nostrand Reinhold Company, New York, 1974.

Proceedings of the Symposium on Industrial Uses of Selenium and Tellurium, Toronto, Canada, Oct. 1980, Selenium–Tellurium Development Association, Darien, Conn.

SEMICONDUCTORS

THEORY AND APPLICATION

Semiconductors are a class of materials exhibiting electrical conductivities intermediate between metals and insulators. The strong influence of impurities on the semiconductor conductivity facilitates the development of electron devices. One such device, the transistor, ushered in an era of explosive growth in the information and telecommunication technologies with its invention in 1947.

Solid State Theory

The band theory of solids predicts the existence of an energy gap which electrons must surmount in order to break free of their host atom and roam freely through the crystalline lattice. The escaping conduction electron leaves behind a vacancy or "hole" which may propagate, much like a bubble, through the lattice when neighboring bound electrons drop into the vacant site. The conductivity is governed by the density of conduction electrons and holes (which varies inversely with the magnitude of the energy gap) and their mobility μ (defined as the ratio of the carrier velocity to the inducing electric field). Energy gaps and mobilities are summarized in Table 1 for the important electronic semiconductors.

Impurities and crystalline defects may drastically alter the conductivity, either by releasing weakly-bound electrons and holes or by tapping available charge carriers. A semiconductor is n-type or p-type depending upon whether electrons or holes dominate the conduction process. In silicon, phosphorus or arsenic doping is typically used to produce n-type material whereas boron is used for p-type material. The product of electron and hole concentrations remains constant in equilibrium. Equilibrium is maintained by a balance between thermal generation of electron-hole pairs and electron-hole recombinations.

Charge carriers may move through a semiconductor device by random thermal motion (diffusion) or with the assistance of an electric field (drift). At low fields, the velocity is proportional to the field with proportionality constant μ. At high fields, the velocity saturates owing to interactions of the carriers with the lattice (or more specifically, with optical phonons). The saturation velocity poses fundamental limits on device switching speeds. Velocity-field behavior is illustrated in Fig. 1.

Device Physics

The rectifying diode is a two-terminal device which permits current flow in one direction only. It is formed at the junction of p-type and n-type material. Charge carriers diffuse across the junction creating a charge imbalance and hence an electric field. The field produces a drift

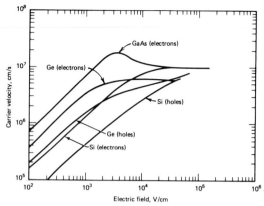

Figure 1. Electron and hole velocities vs. electric field for germanium, silicon, and gallium arsenide at 300 K.

current which balances the diffusion current. A voltage applied to the junction will either increase the field, thereby inhibiting current flow, or decrease the field, encouraging current flow. The current is limited by the rate at which charge carriers can recombine in the bulk of the semiconductor device, thereby preserving charge neutrality. The recombination rate also governs the switching speed of the device.

The transistor is a three-terminal device which may act as an amplifier or a switch. The bipolar transistor is formed from two back-to-back p–n junctions located in close proximity. Current flow between the outer regions (termed collector and emitter) is controlled by a small current in the central or base region which maintains charge neutrality by recombining with a small fraction of the charge in transit across the base region. The metal-oxide-semiconductor (MOS) transistor utilizes a thin oxide which separates a charging metal plate (gate) from the semiconductor material. A charge on the plate can attract and support a stable population of charge in an inversion layer at the oxide semiconductor interface. This charge is provided by an overlapping p–n junction, termed the source, and is collected by a second p–n junction, the drain. Current flow is controlled by minute changes in the amount of charge on the gate.

Device contacts are formed at metal–semiconductor junctions. Lightly doped junctions behave much like p–n junctions and are termed Schottky barrier diodes. Heavy doping results in a very high junction electric field, allowing electrons to tunnel through the energy gap into the conduction band and resulting in little resistance to current flow.

Applications

Several devices such as the solar cell and the photodiode rely on the production of charge carriers by light absorption. Semiconductors formed from columns III and V in the periodic table, such as GaAs and InP, display some remarkable and useful features not exhibited by silicon. In these materials, electron–hole recombination results most frequently in photon emission, thereby permitting the development of such optoelectronic devices as the light-emitting diode and the semiconductor laser (see Light-emitting diodes and semiconducting lasers). Negative resistance in these materials, ie, the reduction of velocity with field at high fields illustrated in Fig. 1, leads to useful instabilities, exploited in Gunn oscillators and IMPATT diodes.

Photographic and chemical processes are used to fabricate thousands of carefully arranged p–n junctions, metal contacts, and MOS devices on a small semiconductor slab to form an integrated circuit (IC). Several IC families have evolved that exploit the various characteristics of the bipolar and MOS technologies. The bipolar families (TTL, ECL) exhibit fast switching speed but consume much power. The MOS families (NMOS, CMOS) consume less power and allow greater circuit densities. IC development is slowly evolving in GaAs with much potential for high speed applications. The GaAs metal semiconductor field-effect transistor (MESFET) emulates the MOS transistor as it employs a Schottky gate. A comparison of semiconductor IC families is shown in Fig. 2 where the

Table 1. Properties of Important Semiconductors

Semiconductors		Band gap, eV		Mobility (at 300 K), $cm^2/(V \cdot s)$[a]			Effective mass[b], m^*/m		Relative dielectric, ϵ_s/ϵ_0
		300 K	0 K	Electrons	Holes	Band[c]	Electrons	Holes	
Element	C	5.47	5.48	1800	1200	I	0.2	0.25	5.7
	Ge	0.66	0.74	3900	1900	I	1.64^d	0.04^e	16.0
							0.082^f	0.28^g	
	Si	1.12	1.17	1500	450	1	0.98^d	0.16^d	11.9
							0.19^f	0.49^g	
	Sn		0.082	1400	1200	D			
IV–IV	α-SiC	2.996	3.03	400	50	I	0.60	1.00	10.0
III–V	AlSb	1.58	1.68	200	420	I	0.12	0.98	14.4
	BN	ca 7.5				I			7.1
	BP	2.0							
	GaN	3.36	3.50	380			0.19	0.60	12.2
	GaSb	0.72	0.81	5000	8500	D	0.042	0.40	15.7
	GaAs	1.42	1.52	8500	400	D	0.067	0.082	13.1
	GaP	2.26	2.34	110	75	I	0.82	0.60	11.1
	InSb	0.17	0.23	80,000	1250	D	0.0145	0.40	17.7
	InAs	0.36	0.42	33,000	460	D	0.023	0.40	14.6
	InP	1.35	1.42	4600	150	D	0.077	0.64	12.4
II–VI	CdS	2.42	2.56	340	50	D	0.21	0.80	5.4
	CdSe	1.70	1.85	800		D	0.13	0.45	10.0
	CdTe	1.56		1050	100	D			10.2
	ZnO	3.35	3.42	200	180	D	0.27		9.0
	ZnS	3.68	3.84	165	5	D	0.40		5.2
IV–VI	PbS	0.41	0.286	600	700	I	0.25	0.25	17.0
	PbTe	0.31	0.19	6000	4000	I	0.17	0.20	30.0

[a] The values for drift mobilities obtained in the purest and most perfect materials available to date.

[b] m^* = effective mass, m = rest mass.

[c] I = indirect, D = direct.

[d] Longitudinal effective mass.

[e] Light-hole effective mass.

[f] Transverse effective mass.

[g] Heavy-hole effective mass.

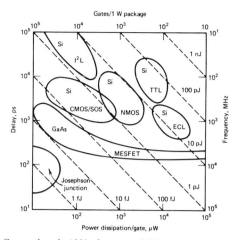

Figure 2. Comparison in 1980 of commercial Si technologies and laboratory GaAs and Josephson devices. The delay is the average switching time of a transistor or gate in a moderately complex circuit and is plotted against the average power dissipation per transistor in such a circuit. Opposing axes show the maximum switching frequency and the number of gates that can be accomodated on a chip with a 1 W power-dissipation limit, which is typical for Si. Dashed lines indicate energy dissipation.

switching or delay time of a single transistor (gate) is contrasted with its power dissipation (see Integrated circuits).

Outlook

Superconducting Josephson technology has excellent high speed potential but currently receives little research support owing to the promise of evolving Si and III–V technologies. New growth technologies such as molecular-beam epitaxy permit the development of structured band-gap materials in III–Vs with tremendous consequences for the emerging IC and lightwave technologies. Integrated optic circuits will combine optoelectronic, optical and IC components on a single semiconductor slab. Semiconductor memories based on large arrays of transistors or gates such as the random-access memory (RAM) and the charge-coupled device (CCD) are already approaching one million (10^6) gates/IC. Magnetic bubbles, amorphous semiconductor devices, and solid-state displays are in promising stages of development (see Amorphous magnetic alloys).

STEVEN A. SCHWARZ
Bell Communications Research, Inc.

S.M. Sze, *Physics of Semiconductor Devices*, 2nd ed., John Wiley & Sons, Inc., New York, 1981.

B.G. Streetman, *Solid-State Electronic Devices*, Prentice Hall, Englewood Cliffs, N.J., 1972.

A.S. Grove, *Physics and Technology of Semiconductor Devices*, John Wiley & Sons, Inc., New York, 1967.

FABRICATION AND CHARACTERIZATION

Silicon Technology

Discrete, monolithic, hybrid, integrated, power, and high voltage devices. Semiconductor devices made in the 1960s generally were of three types: discrete devices, monolithic circuits, and hybrid circuits. A discrete device contained just one element, eg, a diode, transistor, light-emitting diode (LED), etc (see Light-emitting diodes and semiconductor lasers). Monolithic circuits were made on a single chip of silicon by planar technology and contained many elements. Hybrid circuits were made by attaching monolithic circuits to substrates, which were usually ceramic and contained interconnects, resistors, and capacitors. Integrated circuits (qv) (ICs) were made by building up the circuit elements, eg, transistors, resistors, and interconnects simultaneously by use

of planar, thin film, or thick-film technologies. Monolithic and hybrid ICs were the most important types. Improvements in fabrication technology led from the first integrated circuits, ie, small-scale integration (SSI) in 1960 with tens of transistors, to medium-scale integration (MSI) in 1965 with ca 100 transistors, to the large-scale integration (LSI) in the 1970s with ca 10^4 transistors, to the present very large-scale integration (VLSI) in the 1980s for incorporating more than 10^5 transistors on a chip.

In parallel with these advances for low power devices were the developments for high power applications; these were the power transistor, ie, the *npn* transistor and its derivatives, and the thyristor, ie, the *pnpn* transistor and its derivatives.

Commercial manufacture of high voltage ICs began in 1980. These ICs can replace the costly and clumsy mechanical relays used in communication switching networks, because they have sufficient bandwidth, power, and voltage-blocking capability.

Integrated circuits: unipolar and bipolar. Bipolar transistors are the direct descendants of junction transistors; unipolars are field-effect devices, eg, field-effect transistors (FETs) and MOS transistors. Most bipolars are current-driven devices and have low input impedance. Most unipolars are voltage-driven devices; junction field-effect transistors (JFETs) generally have higher input impedance than bipolars, and MOS devices have the highest impedance. High input impedance is advantageous for large fanouts (number of devices that can be driven by a given transistor) and low power dissipations. Metal oxide semiconductor ICs have advantages over bipolars because of the former's lower power consumption and higher circuit density.

Organization of IC fabrication plants. The IC facility is composed of several interacting groups centered around the device-processing line. These are the design, the mask, the development, and the electrical testing groups, and representatives of product users.

Basic processing operations. The basic processing steps in IC production are cleaning; crystal growth, eg, of Si and compound semiconductors; oxidation (atmospheric and high pressure); lithography (photolithography, electron-beam, and x-ray); doping by diffusion, implantation, and molecular-beam expitaxy (MBE); deposition (electron-gun, sputter, chemical vapor deposition (CVD), low pressure chemical vapor deposition (LPCVD), and plasma); etching (wet chemical and dry or plasma); and electrical testing (see Photoreactive polymers; Ion implantation; Film-deposition techniques).

Compound-Semiconductor Devices

Compound semiconductors provide a variety of materials properties that permit the fabrication of devices that can not be made using silicon. The main difficulties with compound-semiconductor processing arise from the complexity of processing materials containing at least two elements and from the lack of a nearly ideal dielectric–semiconductor system, eg, SiO_2–Si. However, most of these difficulties originate from manufacturers' inexperience rather than any fundamental limitations. Because compound-semiconductor technology is less advanced than silicon technology, applications have been developed mostly in areas where silicon cannot compete, eg, microwave technology (qv) and photonics. The greatest difference between silicon technology and compound-semiconductor technology is the dominance of MOS and bipolar devices for silicon and the almost complete absence of these devices for compound semiconductors.

Many compound semiconductors are being studied, but only a few have attained commercial importance. These are GaP, GaAsP. and AlGaAs for light-emitting diode (LED) applications, GaAs for field-effect transistors (FETs), and GaAs and AlGaAs for lasers. Indium phosphide and indium gallium arsenic phosphide are being actively researched and are expected to become commercially important before 1985.

The most important field in compound-semiconductor technology is crystal growth. The methods used are Czochralski, Bridgeman, CVD, vapor-phase epitaxy (VPE), liquid-phase epitaxy (LPE), and molecular-beam epitaxy (MBE).

CHUAN C. CHANG
Bell Laboratories

J.V. DiLorenzo, ed., *GaAs FET, Principles and Technology*, Artech House Books, Dedham, Mass., 1982.

S.K. Ghandhi, *Semiconductor Power Devices*, John Wiley & Sons, Inc., New York, 1977.

H. Kressel and J.K. Butler, *Semiconductor Lasers and Heterojunction LED's*, Academic Press, Inc., New York, 1977.

AMORPHOUS

Although the modern theory of solids is intimately connected with crystalline materials, the vast majority of solids that is encountered in everyday experience is amorphous. Technically, an amorphous material exhibits no correlation between two atoms located more than ca 5 nm apart, so that the long-range periodicity of crystals is entirely absent. Nevertheless, amorphous solids ordinarily possess a great deal of short-range order, particularly correlations between nearest-neighbor atoms, which arises from their chemical interactions.

Since, historically, solid-state theory was restricted to periodic crystals, it was believed for many years that amorphous semiconductors did not exist. Consequently, it was a surprise when it was reported in the 1950s that many chalcogenide glasses, ie, amorphous alloys containing significant concentrations of one or more of the chalcogen elements and prepared by rapid cooling of the liquid, were semiconducting. The field of amorphous semiconductors attracted little scientific interest before 1968, when the existence of reversible switching effects in thin films of chalcogenide glasses was reported and a wide array of potential applications was proposed. This work spurred intensive investigations into the electronic properties of chalcogenides and amorphous analogues of the conventional, tetrahedrally coordinated, crystalline semiconductors, eg, silicon and germanium. Many amorphous-based semiconductor devices, from computer memories and television pickup tubes to solar cells, are commercially available.

Structure

Amorphous solids do not exhibit the long-range order of crystals, and this results in a freedom from periodic constraints that allows a wide range of potential compositions and structures. In this sense, an amorphous solid can be considered as a giant chemical molecule. Since small molecules can have many different structures with very different properties, it is clear that amorphous solids with the same exact composition are far from unique. Under ordinary conditions, a great deal of short-range order is expected, ie, each atom is most likely to have a local coordination consistent with its electron structure. Thus, neutral atoms in Groups I–IV should have 1–4 nearest neighbors, respectively, whereas neutral atoms in Groups V, VI, and VII should have coordination numbers of 3, 2, and 1, respectively, in accordance with the 8-N rule of chemical bonding. This arises because any deviation from the optimal coordination would ordinarily cause a large increase in total energy and thus would be suppressed if at all possible.

It is important to consider the ideal structure of an amorphous semiconductor. In amorphous alloys, different chemical bonds are possible, and the strongest bonds are favored. Each type of bond has an optimal length, and the large values of the bond-stretching frequencies indicate the presence of strong forces which tend to suppress significant bond-length deviations. Although considerably weaker, the bond-bending frequencies are usually sufficiently large that there are at most small bond-angle distortions under ordinary circumstances. The forces tending to impose third-neighbor, ie, dihedral-angle, constraints are generally very much weaker. It has been recently suggested that the optimal average coordination for an amorphous alloy is the one in which the number of constraints introduced by fixed bond lengths and bond angles is equal to the number of degrees of freedom, namely three. This calculation leads to the conclusion that the average coordination of an atom in an ideal glass is about 2.4.

Electronic. As indicated, the quantum theory of solids as originally formulated was entirely based on crystalline periodicity. In the absence of such periodicity, the problem of electronic structure becomes mathematically much more complex, and additional approximations are essential. However, several techniques have been developed to analyze this problem, and the electronic structure of amorphous semiconductors is generally regarded as at least qualitatively understood.

Doping

The fabrication of conventional semiconductor devices requires the ability to decouple the electrical activation energy from the optical gap (see Semiconductors, fabrication and characterization). In crystalline semiconductors, this is accomplished by substitutional doping, in which small concentrations of an impurity atom with a different valence from the host are introduced into the crystal. The periodic constraints then force a defect configuration, which leads to a large change in the Fermi energy, E_f. This would appear to be impossible in an amorphous semiconductor, in which no crystalline constraints exist. However, a(amorphous)-Si—H and a-Si—F—H alloys can be routinely doped with either phosphorus or boron, although the exact mechanism for this is still a matter of controversy.

Chemical Modification

Amorphous chalcogenide alloys possess a particularly low energy defect, called a valance-alternation pair (VAP). These consist of a positively charged overcoordinated center and a negatively charged dangling bond. When VAPs are present, the Fermi energy is effectively pinned and ordinary doping is impossible. Even if an impurity atom could be positioned in a nonoptimal chemical environment, the resulting release of excess electrons, for example, would only convert some of the positive centers to negative without moving E_f. To overcome this problem, one researcher introduced an ingenious technique for decoupling E_f, even in the presence of large VAP densities, thereby making possible the use of chalcogenides in conventional semiconductor devices. The technique of chemical modification requires both the use of large concentrations of one or more modifying elements to overcome the large VAP density and the introduction of the modifier in a nonequilibrium manner to preclude its incorporation in the network without any significant effect on electrical conductivity.

Transient Effects

Virtually all of the present electronic uses of amorphous semiconductors involve nonequilibrium processes introduced by the application of light or applied electric fields. Under these conditions, the nature and concentration of the defects are of the utmost importance, as they ordinarily completely control the kinetics of the return to equilibrium.

Despite the sharp contrasts in steady-state behavior between amorphous silicon alloys and chalcogenides, many of the unusual transient phenomena observed in chalcogenide glasses have also been observed in a-Si—H. These include dispersive transport, photoluminescence fatigue, light-induced creation of unpaired spins, and apparent photostructural changes.

Health and Safety Factors

The following compounds, which are involved in amorphous-semiconductor technology, should be handled with great care: silane, silicon tetrafluoride, phosphine, diborane, arsine, tellurium, and selenium.

Uses

The applications of amorphous semiconductors arise primarily from their low preparation costs and their resistance to deterioration under severe conditions rather than from their unique physical properties. The main application has been as the photoreceptor in the electrophotographic process, ie, xerography (see Electrophotography). Other important devices based upon the photoconductive properties of amorphous semiconductors include the use of chalcogenide alloys in television pickup tubes, ie, vidicons.

A second area of application results from the reversible structual changes induced by either an external electric field or light. Electrically alterable and programmable read-only-memories and an array of imaging films have been commercially available for several years, and a film for updatable microfiche applications has been developed.

Future applications are particularly promising in many areas. The reversible switching phenomena exhibited by chalcogenides in high electric fields should result in novel control functions not possible with conventional materials. A wide array of related computer applications is also being developed (see Computers).

Perhaps the most important application in the near future will be as solar cells (see Photovoltaic cells; Solar energy).

DAVID ADLER
Massachusetts Institute of Technology

D. Adler, *Scientific American* **236**(5), 36 (May 1977).

D. Adler, *J. Solid State Chem.* **45**, 40 (1982).

R.K. Willarson and A.C. Beer, eds., *Semiconductors and Semimetals*, Vol. 21, Academic Press, Inc., New York, 1984.

ORGANIC

Many families of organic compounds have thermally activated conductivity; however, actual working organic-semiconductor junction devices, eg, diodes, crude transistors, photovoltaics, etc, were not reported until 1978 (see Photovoltaic cells; Polymers, conductive). Other typical, low conductivity, high dielectric-constant semiconductor components, eg, electrolytic capacitators, were prepared in the mid-1960s but only have limited application. On the other hand, poly(vinylcarbazole–co-trinitrofluorenone) (PVK–TNF) has wide application in electrophotographic copiers (see Electrophotography). The commercial uses of organic semiconductors in terms of their electrical properties are extremely rare, and this article is concerned largely with their theoretical and experimental aspects.

Fundamental research on organic conductors has led from anthracene (resistivity of ca 10^{22} $\Omega\cdot$cm) in the early part of this century to the tetracyanoquinodimethane (TCNQ) salts (semiconductors, resistivities of ca 10^{-1} $\Omega\cdot$cm) in the 1960s to tetrathiafulvalenium (TTF) salts (organic metals, resistivities of 10^{-2}–10^{-3} $\Omega\cdot$cm) in the 1970s to tetramethyltetraselenafulvalenium (TMTSF) salts (metals, resistivities of 10^{-5} $\Omega\cdot$cm and superconductivity at 1.3 K) in the 1980s. Thus, in only two decades, there has been a revolution in terms of electrical conductivity, with a progression from ca 10 S/cm ($= 1/\Omega\cdot$cm) to infinite conductivity. Fundamental research on organic conductors has also led to profound contributions to theoretical solid-state physics, particularly with respect to low dimensional solid-state transitions. Theories of hypothetical one-dimensional metals were corroborated by the pseudo-one-dimensional organic solids, eg, TTF–TCNQ. The intimate relationship between Fermi-surface instabilities, eg, charge-density waves (CDWs) and metal-insulator phase transitions in low dimensional solids, were elucidated by studies of organic metals. Other collective modes, eg, spin-density waves (SDWs), occur (see Superconducting materials).

Theory

The theory of the conductivity of these materials is a fast-growing field marked by many points of contention. The more exotic transport mechanisms that have been suggested have played an important role in stimulating chemists to become involved in the synthesis of these materials. The excitonic superconductivity model provides the best example of this generation of interest. It has been suggested that the dipole fields set up by electronic transitions could replace the phonon field in supplying the necessary electron-pair coupling mechanism for the transition to a superconducting state, presumably at higher-than-usual temperatures. The verity of this mechanism has yet to be demonstrated.

Charge-transfer salts. Charge-transfer salts are composed of regular arrays, ie, stacks or columns, of individual molecules. They are conductors because one or both of the components possesses unpaired electrons that can participate in the formation of a partially filled energy band. The unpaired electrons are generated by electron transfer between the constituents, so that one component is termed an electron donor and the other an electron acceptor, and the whole is usually referred to as a charge-transfer salt. Many of the one-dimensional charge-transfer salts undergo Peierls transitions to semiconducting ground states.

Polymers. The principal difference between the charge-transfer salts and the polymers is the magnitude of the bandwidth. Whereas the former are narrow-band materials (≤ 1eV), the bandwidths of the polymers are of the order of those exhibited by inorganic metals and semiconductors (ca 10 eV). As a result, these materials are more susceptible to Peierls CDW transitions than the charge-transfer salts but are less prone to Coulomb localization.

Physical Properties

The physical properties of the organic conductors, which set them apart from normal solids, arise from their anisotropic electronic structure. The electrical properties are the focus of interest and provide the most useful materials classification.

Electrical. Charge-transfer salts. Class 1 (semi-) conductors. Class 1 conductors are characterized by room-temperature resistivities of 1–10^{10} $\Omega\cdot$cm and a monotonic decrease in conductivity as the temperature decreases.

Class 2 (intermediate) conductors. Class 2 conductors show room-temperature resistivities of 1–10^{-2} $\Omega\cdot$cm, and their conductivities gradually increase to a broad maximum as the temperature decreases, after which they decline rapidly.

Class 3 (metallic) conductors. The crystals of class 3 conductors are typified by room-temperature resistivities of less than 10^{-2} $\Omega\cdot$cm and a steady increase in conductivity as the temperature decreases.

Polymers. The polymers differ from the charge-transfer salts in that their bandwidths are about an order of magnitude broader, and as a result, the insulating transitions are of the CDW type. This accounts for the poor conductivity of the amorphous polymer $(CH)_x$ which, in the pure state, has a room temperature resistivity of 10^9 $\Omega\cdot$cm.

Magnetic. Charge-transfer salts. Class 1 (semi-) conductors. The Class 1 compounds are semiconductors by virtue of the band gap that is opened by coulomb localization, since the charge-transport process involves the intramolecular repulsion energy of two electrons or holes, ie, the half-filled-band case. The localized spins on adjacent molecules couple antiferromagnetically, and the paramagnetic susceptibility decreases strongly as the temperature decreases (see Magnetic materials).

Class 2 (intermediate) conductors. The intermediate conductors are characterized by incomplete charge transfer and there is a large antiferromagnetic contribution to the susceptibility at low temperatures.

Class 3 (metallic) conductors. The magnetic susceptibilities of the highly conducting charge-transfer salts show complex behavior, and no single model accounts for the temperature dependence of the magnetism at all temperatures. In TTF–TCNQ, three distinct regimes may be distinguished. Above 300 K, the magnetism is essentially temperature-independent and may be viewed as arising from Curie behavior or from an enhanced Pauli susceptibility. Below 300 K, the susceptibility decreases and, at temperatures above 60 K, this has been attributed to the formation of a pseudogap either as a result of CDW fluctuations or rehybridization effects, ie, interchain coupling. At 60 K, there is a break in the magnetic-susceptibility data, and at temperatures below this point, the magnetism rapidly approaches the core diamagnetism of the constituent molecules.

Polymers. In polythiazyl, the magnetic properties are characteristic of a normal semimetal with low carrier density.

Optical. Charge-transfer salts. Class 1 (semi-) conductors. The class 1 salts only exhibit one important electronic transition, apart from intramolecular excitonic features. In K TCNQ, this is a charge-transfer transition (energy, ca 1 eV) between adjacent pairs of $TCNQ^-$ molecules in the stack, ie, $2\,TCNQ^- \rightarrow TCNQ + TCNQ^{2-}$, and is a reflection of the electronic structure of these compounds: a half-filled band with a single unpaired electron per TCNQ molecule.

Class 2 (intermediate) conductors. Class 2 conductors are characterized by incomplete charge transfer and, in addition to the absorptions that occur in class 1 salts, there is a low energy mixed valence transition.

Class 3 (metallic) conductors. The most significant feature of the optical properties of the highly conducting charge-transfer salts is the metallic plasma edge in the optical reflectivity when light is polarized parallel to the conducting axis and the absence of this structure when light is polarized perpendicular to the same direction.

Polymers. With light polarized parallel to the highly conducting axis, the optical reflectivity of polythiazyl exhibits a well-defined, metallic plasma edge.

Chemical Properties

Electrochemistry of donors and acceptors. Charge transfer is an essential feature for conductivity in donor-acceptor salts and, thus, electrochemical oxidation and reduction potentials provide a useful index for evaluating likely combinations.

Donors. Empirically, it has been shown that sulfur- and selenium-based donors produce organic metals if their first oxidation potential ($E_{ox}^{1/2}$) in solution

$$D \rightarrow D^+ + e, \qquad E_{ox}^{1/2} \text{ (SCE)}$$

does not exceed 0.6 V, where $E_{ox}^{1/2}$ (SCE) is the half-cell oxidation potential measured against the standard calomel electrode (SCE).

Acceptors. Teracyanoquinodimethanide is the predominant progenitor of organic semiconductors and metals. The first reduction potential of an acceptor is given by

$$A + e \rightarrow A - E_{red}^{1/2} \text{ (SCE)}$$

Quinones are the oldest family of organic acceptors, but they have little use in the chemistry of organic semiconductors and metals, even though they exhibit a wide range of reduction potentials.

Stability. Thermal stability. The highest temperature to which an organic compound can be exposed in the absence of air is ca 500°C because of the nature of carbon–carbon, carbon–hydrogen, and carbon–heteroatom bonds. This temperature is reduced by ca 200°C when the materials are exposed to prolonged heating in the atmosphere.

Photostability. No in-depth studies on the photostability of TCNQ in the solid state have been reported. In solution, TCNQ rapidly decomposes upon uv irradiation.

Solubility. Most nitrogen- and sulfur-based donors are soluble in commonly employed solvents, with the exception of TTN and TTT. Of the selenium-based donors, HMTSF and TSeT are the least soluble. In general, the larger and more symmetrical molecules are the least soluble.

Synthesis and Manufacture

The compounds TTF, TCNQ, PVK, TNF, polyvinylpyridine, poly-(phenylene sulfide), phthalocyanines, $(CH)_x$, and many dyes are commercially available. The preparative literature of TTF derivatives has been reviewed. None of the selenafulvalenes are commercially available.

Organic metals. Organic metals are prepared by combining donors and acceptors or by electrolysis (see Crystal Growth).

Organic semiconductors. Charge-transfer organic semiconductors are prepared by the combination of a donor and acceptor in solution.

Crystal growth. With a few rare exceptions, all organic semiconductor and metal crystal growths are carried out in solution at ca RT. There are three main methods: diffusion, metathesis, and electrolysis.

Polymers. The usual polymer-fabrication techniques apply to PVK, polyvinylpyridine, poly(phenylene sulfide), etc, but not to $(CH)_x$. The latter is prepared from acetylene and a large excess of a Ziegler-Natta catalyst in toluene at low temperature.

Health and Safety Factors

Since most of the molecules employed in the preparation of organic semiconductors and metals exhibit facile oxidation and reduction processes, they should probably be used with care, because they could have profound effects on the biochemical electron-transport processes. Tetrathiafulvalene is toxic when administered intraperitoneally to mice of 200–400 mg/kg body weight, and it is not active against L-1210 lympoid leukemia and P-388 lymphocytic leukemia.

Uses

The practical applications of organic semiconductors are still largely experimental, and their inroads into commercial uses are extremely limited. They are being investigated for use in photovoltaic devices, thermometers, switches and diodes, reproduction and resist materials,

batteries, electrolytic capacitors and rectifiers, and electrochromic devices (see Chromogenic materials).

Robert C. Haddon
Martin L. Kaplan
Fred Wudl
Bell Laboratories

F. Gutmann and L.E. Lyons, *Organic Semiconductors*, John Wiley & Sons, Inc., New York, 1967.

J.S. Miller in A.J. Epstein, ed., *Synthesis and Properties of Low-Dimensional Materials*, Annals, Vol. 313, New York Academy of Science, New York, 1978.

M. Pope and C.E. Swenberg, *Electronic Processes in Organic Crystals*, Clarendon Press, Oxford, UK, 1982.

SEPARATIONS, LOW ENERGY

For industrial separation of a liquid mixture into its components, distillation is generally the preferred, lowest-cost technique. For complete separation of a liquid mixture into its components, the minimum work (W_{min}) of separation is given by:

$$W_{min, T} = -RT \sum_j x_{jF} \ln(\gamma_j x_{jF})$$

where γ_j is the activity coefficient for component j in the feed and x_{jF} the mole fraction of component j.

In both the chemical and petroleum-refining industries, steam is generated for a plant in a central boiler-plant facility at elevated pressures. This steam is piped around the plant at two or three pressure. Steam may also be generated in waste-heat boilers within the plant where excess process heat is available. The energy can be extracted from the steam before it is used in reboilers by decreasing its pressure from the maximum at which it is generated in the steam-boiler plant to the pressure at which it is utilized in the reboiler in the steam turbine.

Economic Considerations

The only certain approach to determining whether fuel is conserved by a particular option is to evaluate its impact on plant steam and fuel balances (see also Energy management; Economic evaluation).

Energy evaluation of plants applying cogeneration. The value of energy is assumed to be proportional to its availability, ie, to the amount of work that can be extracted from it so as to reduce it to dead-state conditions:

$$V = (\Delta B / \Delta H) \times C_p \times K$$

where ΔB is the availability of the stream or energy source relative to dead-state or ambient conditions, ΔH is the change in enthalpy in going from process conditions to the dead state, C_p is the cost of power, and K is a factor depending upon the type of energy source being considered.

For compression energy, K may be considered equal to 1.47. For thermal energy higher in temperature than dead-state conditions, K is equal to 1.0. However, a factor of less than 1 may be appropriate, since a mechanical device must be used extract work from the fluid or source.

For typical industrial payback and depreciation criteria, the following rule has been developed: if more than three dollars must be invested to save one dollar per year in energy costs, no net economic benefit accrues.

Distillation

Distillation (qv) has many advantages: the phases are fluids with relatively large density differences and large interfacial tensions; the low liquid- and vapor-phase viscosities lead to high diffusivities and efficient mass transfer; small increments in boiling temperature can produce vapor-pressure differences sufficient to cause vapor to flow through the equipment without need for blowers; throughputs are high, stage heights are low, and the equipment is reliable; and no mass-separation agents are used.

Distillation may not be suitable if relative volatilities are low; classes of compounds with a broad overlapping range of boiling points are to be separated; an impurity is to be separated from a more volatile liquid; a less-volatile organic compound is to be recovered from a water solution; extreme temperatures or pressures are required; heat-sensitive materials are to be recovered; and an azeotrope is present in the feed components.

More efficient separation. A basic source of energy waste in distillation processes is inadequate control over product purity. Improved column control reduces the extent of the overdesign. Factors that favor improved controls include high annual production volume, low relative volatility, and high distillate-to-product ratio. In material-balance control, product purity is controlled by direct manipulation of the product flow rate.

Feed-forward controls are normally used to account for flow variations by, for example, establishing a ratio of reboiler steam to feed flow rate.

Combined reflux and distillate controls have material-balance and reflux-boilup control systems for composition control. However, trained operators and maintenance personnel are required.

A retrofit with additional or more efficient trays conserves energy by virtue of the decreased column-reflux requirements, whereas in retrofits with lower pressure drop internals, energy is conserved because product degradation is reduced or vacuum-system energy requirements are lower.

Multicomponent separations generally require a sequence or train of distillation columns. A composition-node design method results in small but acceptable overdesign. Design heuristics for an ideal, three-component mixture ABC have been developed.

In many multicomponent separations, the withdrawal of side streams reduces the number of columns and the energy required. Side-stream removal can be advantageous where a mixture is to be separated into three or more products that need not be very pure or recovered in high yield, a stream contains both heavy impurities and a small quantity of relatively volatile light impurities, and the mixture to be separated contains an impurity which tends to concentrate at an intermediate location in a fractionating column.

If the feed contains a small concentration of impurities that are more volatile than the desired product, the distillate product is taken as a side cut and the impurity as the top product. The impurity concentration in the product decreases with increasing reflux.

Pasteurization is used industrially in the purification of ethylene, ethanol, methanol, propylene oxide, propylene glycol, acrylonitrile, and phenol, among others.

More efficient heat utilization. Efficient heat-transfer surfaces minimize condenser and reboiler temperature-difference driving forces and area requirements. Heat-pump, vapor-compression, reboiler-flashing, and heat-cascading techniques require increased energy and (increased heat-exchanger) temperature differences.

Enhanced heat-transfer surfaces include Linde's porous boiling surfaces (high flux tubes); a thin, metallic, microporous coating is applied to the inside or outside of heat-transfer tubing. The coating enhances boiling and condensing heat-transfer coefficients by a factor of 10–30. Enhanced heat-transfer surfaces can also be extended by fins or flutes.

Plate-and-frame and spiral-plate exchangers are examples of compact heat exchangers, which are being used increasingly between two process streams where close-approach temperature differences are important.

Air-cooled and evaporative condensers are more expensive and bulkier than water-cooled exchangers but discharge directly to the atmosphere, whereas cooling water involves intermediate exchange. Air-cooled condensers are less subject to fouling, require no water, and are less apt to generate environmental problems; utility costs are lower.

Single-column heat integration implies the possibility of exchanging heat between the feed and product streams of a column to reduce energy requirements. This technique can be applied in chemical and light-hydrocarbon plants because the bottoms product from one column is the feed to the next.

Heat-pumping reduces energy requirements as heat is pumped from the condenser to the reboiler. In order for heat to transfer from a low temperature to a high temperature reservoir, there must be an input of work. In a heat-pumping scheme, eg, external refrigeration, vapor recompression, and reboiler flashing, a compressor supplies the work. Only a change in the conventional column design is required. However, the form of energy used, ie, compressor work, is more expensive than the form it replaces, ie, reboiler or condenser heat.

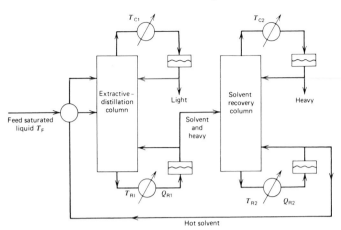

Figure 1. Typical extractive-distillation process scheme. T_{C1}, Temperature condenser 1; T_{C2}, temperature condenser 2; T_{R1}, temperature reboiler 1; T_{R2}, temperature reboiler 2; Q_{R1}, heat flux, reboiler 1; Q_{R2}, heat flux, reboiler 2; T_F, temperature feed.

Intermediate heat exchange can be based on a less-expensive energy source than that required by the condenser or reboiler, eg, hot water instead of steam or cooling water instead of refrigeration. Intermediate condensing of vapor is widely used in petroleum refining by means of pump-around loops. Crude distillation and catalytic cracking-fractionator towers incorporate intermediate condensing loops.

Mass-separating agents. In extractive distillation, a solvent is introduced near the top of a column to increase the relative volatility of the materials being separated (see Fig. 1). The solvent is less volatile than the materials being separated and flows down to the bottom of the column, where it is removed with the extracted or less volatile material. The solvent and extract mixture in the bottoms product is separated in another column by distillation or other means. A solvent flow from one to ten times the molar rate of feed to the column is required. The materials to be separated must not form an ideal solution and must differ sufficiently in chemical type (see Azeotropic and extractive distillation).

Extractive distillation is well-suited for separating materials that differ in polarity, eg, aromatics from paraffins. Desired solvent properties include high selectivity. The energy required for heating and cooling the recirculation solvent can be a main factor determining the energy required by the process.

In azeotropic distillation, a solvent is added, generally at the feed, to a distillation column to increase the relative volatility of the materials being separated. The solvent volatility is similar to that of the materials being separated. The solvent passes overhead with the distillate product, normally forming a minimum-boiling, frequently homogenous azeotrope. Azeotropic distillation is used principally in overhead distillation of a material that is present in the feed in low concentration and that is difficult to separate by straight distillation. The azeotropic agent should be easily recoverable from the overhead product, and should have a low heat of vaporization.

Nondistillation Techniques

Gas absorption. In gas absorption, a gas containing a mixture of components contacts a solvent which extracts one or more of the compounds to be separated. The adsorbed components are desorbed by pressure reduction or temperature increase, or both, and the solvent is recycled. Absorption (qv) and stripping are normally carried out in a countercurrent manner in tray or packed powers.

Absorption is widely used for the removal of acid gas from natural gas, synthesis gas, and refinery gas, and for the scrubbing of stack gases and acid gases.

Solvents may be either physical or chemical absorbents or a combination of the two. Desirable solvent properties include high selectivity, high capacity, low volatility or high water solubility, low viscosity, and low cost, toxicity, and corrosivity.

Energy is required to pump the solvent from the low pressure stripper to the high pressure absorber, to compress the solute to the product pressure, to compress the feed where necessary to the absorber pressure, and to compress the solute to the absorber pressure.

Liquid–liquid extraction. In liquid–liquid extraction, a liquid containing a mixture of components contacts a partially miscible solvent, which extracts one or more of the components to be separated. The process potentially is more selective than distillation. Despite the potential for energy conservation, it is not widely used because of the additional complexities and costs (see Extraction).

Supercritical-fluid extraction. In supercritical-fluid extraction (SFE), a liquid or comminuted solid contacts a supercritical fluid at temperatures and pressures near the critical point and thereby permits the extraction of one or more components to be separated. Pressures and diffusivities are high and the fluid's density and viscosity are low, enabling rapid extraction and phase separation. The extracted material is recovered by lowering the pressure or raising the temperature. The process is particularly well-suited for the separation of high boiling or heat-sensitive materials. For maximum solvent power, the fluid should have a critical temperature close to the extraction temperature. A compression ratio of three is roughly equivalent in energy requirements to a typical separation by distillation (see Supercritical fluids).

Adsorption. In adsorption, a gas or liquid mixture of components contacts a microporous solid, which selectively adsorbs certain components. Desorption can be effected by raising the temperature, decreasing the pressure, or displacing the adsorbed material with another material. The process is normally carried out in fixed beds.

Regenerative processes include thermal-swing adsorption, heatless adsorption, displacement regeneration, and parametric pumping.

Adsorption is an alternative to distillation where highly polar or high molecular weight trace impurities must be removed, molecular isomers must be separated, or molecules of different polarities need separation.

In purge-gas stripping, the gas may be a slipstream of the nonadsorbed product leaving an adsorber bed, or it may be a nonadsorbed gas separated from the process fluid.

For temperature-swing adsorption, (TSA), the energy requirements per kilogram of adsorbate are determined by the requirements to heat the bed:

$$Q_{TSA} = \left[(f/g)C_p\Delta T + \Delta H \right]/\eta$$

where f = the ratio of the heat capacity of the adsorbent bed plus vessel to that of the bed alone and g = the ratio of the weight or the adsorbent bed to the weight or adsorbate per cycle (see Adsorptive separation).

For fixed-bed adsorption of organic vapors from air, energy requirements are frequently ca 12,100 kJ/kg (2900 kcal/kg) of organic vapors.

Membrane gas separations. In gas separation by membrane processing, pressurized gas is introduced into a membrane module. Components with higher permeabilities are preferentially transported across the membrane. Separation is effected and the permeate is collected in a port connected to the lower pressure, downstream side of the membrane and the higher pressure residue is collected in a port connected to the upstream side. A continuous membrane column effects a clean separation by circulating the feed gas continuously and bleeding only a small fraction of the most permeable and least permeable gases at the top and bottom of the column, respectively. The energy requirements of this process are those involved in the recompression of the permeate.

Membrane processes are used in hydrogen purification and recovery in the petrochemical industry, in ammonia production, and to separate CO_2 and H_2S from sour natural-gas streams (see Membrane technology).

Crystallization. In fractional crystallization, a component is separated from a liquid mixture by freeze crystallization involving indirect heat exchange, direct-contact heat exchange, or vacuum to remove heat by vaporization. The crystals are separated and washed.

Crystallization (qv) is particularly applicable to a solid–liquid phase diagram of the eutectic type rather than the solid–solution type. Fractional crystallization is feasible when the melting points of the two products differ markedly. Since low temperatures are required, energy costs may be high.

Reverse osmosis. In reverse osmosis (qv), an applied pressure is used to reverse the normal osmotic flow of water across a semipermeable membrane. It has almost exclusively been applied to water desalination. High pressures are required to overcome osmotic pressure for even modest solute concentrations, and pumping-power requirements and membrane capital costs dictate single-pass operations. Operating pressures are typically 2.8–4.1 MPa (400–600 psi).

Selection of Energy-Conserving Options

Heuristic separation-sequence selection factors include the following guidelines: the most difficult separations should be left until last in a sequence; products should be recovered one at a time in order of decreasing volatility in the column overheads; preferred sequences are those that divide the feed to each column equally between the distillate and bottoms flows on a molal basis; and very pure products should be separated last in a sequence.

Tom Mix
Merix Corporation

G.E. Keller II in E.P. Gyftopoulos, ed., *Industrial Energy-Conservation*, Manual 9, the MIT Press, Cambridge, Mass., 1982, p. 9.

T.W. Mix and J.S. Dweck in E.P. Gyftopoulos, ed., *Industrial Energy-Conservation*, Manual 13, The MIT Press, Cambridge, Mass., 1982.

C.J. King, *Separation Processes*, 2nd ed., McGraw-Hill Book Co., Inc., New York, 1980.

SEPARATION SYSTEMS SYNTHESIS

The separation of chemical species (components) is required in most chemical processes. The selection, arrangement, interconnection, and energy integration of the separation operations is a complex task.

An example of the synthesis of a separation process is the recovery of butenes in a butadiene process (see Butadiene; Butylenes). The feed is a C_4 concentrate from the catalytic dehydrogenation of *n*-butane, which is to be separated into four fractions: a propane-rich stream containing 99% of the propane entering the separation system; an *n*-butane-rich stream that contains 96% of the entering *n*-butane, which is recycled; a stream containing a mixture of three butenes, at a 95% recovery, which is sent to a butenes-dehydrogenation reactor to produce butadienes; and an *n*-pentane-rich stream containing 98% of the entering *n*-pentane (see also Hydrocarbons, C_1–C_6).

An economical separation process for this problem, based on the availability of relatively inexpensive energy, is given in Figure 1. In the first separation, 1-butene and propane are removed from *n*-butane and the heavier components in a 100-tray distillation (qv) column (shown as two 50-tray columns), C-1. This separation is difficult because the relative volatility between 1-butene and *n*-butane is only about 1.2 and, therefore, a reflux ratio of 25:1 is needed in conjunction with 100 trays. In the second step, propane and 1-butene, which have a relative volatility of about 2.2, are easily separated by distillation in C-2 with 25 trays and a reflux ratio of about 12:1 on a small quantity of distillate. The pentane in the bottoms from C-1 is easily removed by distillation in the deoiler, C-3, with 20 trays and a reflux ratio of about 1:1. The fourth step is essentially impossible by distillation, because the relative volatility between *n*-butane and the lower-boiling 2-butene isomer is only about 1.03. However, in the presence of appreciable quantities of 96% furfural in water, this relative volatility is increased to about 1.17. Thus, extractive distillation in C-4 removes the *n*-butane (see Azeotropic and extractive distillation). Furfural solvent is recovered for recycle in the stripper, C-5 (see Furan compounds).

In a low cost adiabatic technique for superpurification of volatile liquids, which is said to be particularly useful for closely boiling liquids, a thin layer of liquid feed enters a horizontal vacuum chamber having a pressure slightly lower than the vapor pressure of the mixture. The impurity begins to vaporize, which causes the desired product to crystal-

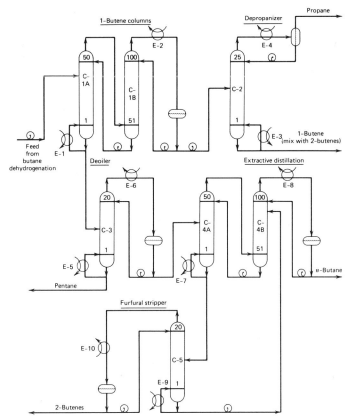

Figure 1. Process for butenes recovery. C = distillation column; E = heat exchanger.

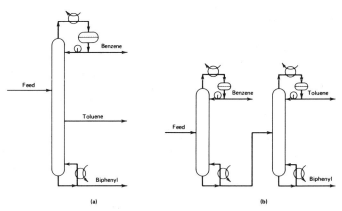

Figure 2. Distillation schemes for separation of a benzene–toluen–biphenyl mixture. (**a**) Single complex operation. (**b**) Sequence of two simple operations.

lize; it exits to a melt chamber for removal from the system. The vapor, which contains some product and essentially all of the impurities, is condensed to provide the vacuum for the system. Final purities over 99.99% are reported to be obtainable from 90% pure feed.

Types and Characteristics of Separation Operations

The feed to a separation process generally consists of a single phase, ie, vapor, liquid, or solid. However, in some cases a feed may be comprised of two or more coexisting phases. If so, phase separation by mechanical means, to the extent practical, should precede the species-separation process.

The large number of available species-separation techniques may be classified into three categories: interphase mass-transfer separation, intraphase mass-transfer separation, and chemical-reaction separation operations (see Mass transfer).

Selection of Criteria

The development of a separation process involves the selection of separation operations, including any necessary energy-separating agents (ESAs) or mass-separating agents (MSAs), types of equipment, and the sequencing of equipment for multicomponent separations.

If the feed has only two components, a single operation may suffice, provided that an ESA is used with either an interphase or intraphase mass-transfer separation operation. If an MSA is needed, at least one recovery operation is required, unless the MSA and associated material are expendable.

The selection of a particular separation operation is based on the nature of the feed components and any MSA used, as well as temperature, pressure, and phases. In general, interphase mass-transfer-separation operations are readily converted into efficient countercurrent cascades that produce very sharp separations. Thus, unless only a moderate degree of separation is required, certain intraphase mass-transfer-separation operations may not be feasible.

The reasons for a separation usually include purification of a species or group of species, removal of undesirable constituents, and recovery of constituents for subsequent processing or removal. In the case of purification, an MSA may prevent exposure to high temperatures that may cause decomposition. Removal of undesirable species containing a modest amount of desirable species may be economically acceptable. Likewise, in the recovery of constituents for recycle, a high degree of separation may not be necessary.

Equipment Selection

In general, equipment selection is based on stage- or mass-transfer efficiency, pilot-plant tests, scale-up feasibility, investment and operating cost, and ease of maintenance (qv) (see Pilot plants; Economic evaluation; Process development; Reactor technology).

Synthesis of Separation Sequences

In most industrial separations, multicomponent mixtures are separated into more than two products. Although one separator of complex design can often be devised to produce all the desired products, a sequence of two-product separators is more commonly used. For example, a mixture of benzene, toluene, and biphenyl can be conveniently separated by distillation. A single complex distillation operation, shown in Fig. 2**a**, is compared with the sequence of two simple distillation operations in Figure 2**b** giving the same products.

The number of possible sequences can often be greatly reduced by excluding certain separations. To synthesize the optimal or near-optimal sequences, all feasible sequences may be examined or synthesis techniques may be applied to find the best sequences with the least effort. These techniques include heuristic, evolutionary, and algorithmic methods.

Energy Integration

Before the energy crisis in the mid-1970s, energy (heat and work) integration in separation sequences was not common except at cryogenic or high temperature conditions. Rarely were condensers and reboilers coupled because of control problems. Today, consideration of energy integration is common at all levels, with and without phase change, and control problems are solved by digital process control.

J.D. SEADER
University of Utah

E.J. Henley and J.D. Seader, *Equilibrium-Stage Separation Operations in Chemical Engineering*, John Wiley & Sons, Inc., New York, 1981.

J.E. Hendry and R.R. Hughes, *Chem. Eng. Prog.*, **68**(6), 71 (1972).

D.W. Tedder and D.F. Rudd, *AIChE J.* **23**, 951 (1977).

SHALE OIL. See Oil shale.

SHAPE-MEMORY ALLOYS

The shape-memory effect is based on the continuous appearance and disappearance of martensite with falling and rising temperatures. This thermoelastic behavior is the result of transformation from a phase stable at elevated temperature to the martensite phase. A specimen in the martensite condition may be deformed in what appears to be a plastic manner but is actually deforming as a result of the growth and shrinkage of self-accommodating martensite plates. When the specimen is heated to the temperature of the parent phase, a complete recovery of the deformation takes place. Complete recovery in this process is limited by the fact that strain must not exceed a critical value which ranges from 2 to 4% for copper memory-effect alloys to 6–8% for the Ni–Ti system. A number of other characteristics associated with shape memory are referred to as pseudoelasticity or superelasticity, two-way shape-memory effect, martensite-to-martensite transformations, and rubberlike behavior.

Table 1 lists the martensite alloy systems that have been investigated with respect to thermoelasticity, pseudoelasticity, shape-memory effect, and two-way shape-memory effect. The bcc (β) phase is the dominant parent because of the comparatively large thermodynamic difference between the martensite transformation from a bcc parent to 3R, 2H, 18R, or 9R, and from an fcc parent to bcc or bct. The former is typical for nonferrous alloys, whereas the latter accounts for the hardening of steel.

Table 1.   Alloys Exhibiting Martensitic Effects

Thermoelastic	Pseudoelastic	Shape memory	Two-way shape memory
Ag–Cd	Ag–Cd	Ag–Cd	Cu–Al
Au–Cu–Zn	Au–Cd	Au–Cd	Cu–Zn–Al
Cu–Al–Ni	Au–Cu–Zn	Au–Cu–Zn	In–Tl
Cu–Zn	Cu–Al–Mn	Cu–Al	Ti–Ni
Cu–Zn[a]	Cu–Al–Ni	Cu–Al–Ni	
Fe–Pt	Cu–Au–Zn	Cu–Zn	
Ti–Ni	Cu–Zn	Cu–Zn–Al	
	Cu–Zn–Al	Cu–Zn–Ga	
	Cu–Zn–Sn	Cu–Zn–Si	
	Cu–Zn–X	Cu–Zn–Sn	
	Fe₃Be	Fe–Pt	
	Fe₃Pt	Fe–Ni	
	In–Tl	In–Cd	
	Ni–Ti	In–Tl	
	Ti–Ni	Ni–Al	
		Ni–Ti	
		304 stainless steel	
		Ti–Nb	
		Ti–Ni	

[a] With ternary additions of Ni, Ag, Au, Cd, In, Ga, Si, Ge, Sn, and Sb.

The behavior of shape-memory-effect (SME) alloys is exactly the opposite to that of normal metals in the following essential feature: as the temperature rises above A_s and martensite is increasingly converted to the β phase, the modulus of elasticity increases.

The Crystallographic Nature of Shape Memory

The martensitic memory or marmem effect occurs in alloys where both the parent and the martensite are ordered and exhibit crystallographically reversible, thermoelastic martensite transformations.

Applications

The first shape-memory alloy to be placed in commercial use was Nitinol employed as a high reliability coupling. Another use for the Ni–Ti memory alloys is in high reliability strip-chart pen-recorder drives.

Medical applications. The alloy Nitinol resulted from a successful research program intended to create a high strength metal resistant to corrosion by seawater. With these properties, it is not surprising that it has interesting applications in memory devices for medical implants requiring passive chemical behavior (see Prosthetic and biomedical devices). These alloys are being used in orthodontic devices, and are being studied for their application to the treatment of cardiovascular problems, orthopedic therapy, and prosthetic devices, among others.

Mechanical devices. As the Ni–Ti alloys have dominated the biomedical field, so have the Cu–Zn–Al family of alloys taken a com-

manding lead in the field of thermally actuated mechanical devices. These include circuit breakers and electric relays, thermostatic valves, window openers, climate-control devices for buildings and cars, safety devices, aerospace-antenna systems, and various fastener and connector devices.

L. McDonald Schetky
International Copper Research Assoc.

L. Delaey, R.V. Krishman, H. Tas, and H. Warlimont, *J. Mater.* 9, 1521 (1974).

J. Perkins, ed., *Proceedings of the First International Conference on Shape Memory*, Plenum Press, Toronto, Canada, 1975.

L.M. Schetky, *Scientific American* 241, 74 (1979).

SHELLAC

Shellac is the purified product of the hardened resinous secretion (lac) of an insect that is parasitic on selected trees and bushes of India, Burma, and Thailand. This tiny insect, *Kerria lacca* (formerly *Laccifer lacca*) of the family Coccoidea secretes the lac as a protective covering for its larva. Lac is the only known commercial resin of animal origin. Its outstanding properties and versatility as a resin are attested to by its continued widespread use in industry (see also Resins, natural).

Shellac is a hard, tough, nontoxic, amorphous resin that produces films of good water resistance and exceptional gloss. Its chemical formula has eluded chemists even after many years of investigation, but it is generally believed to be a mixture of two resins secreted simultaneously by the lac insect. These resins are composed of aliphatic polyhydroxy acids present in the form of lactones, lactides, and intermolecular esters. Associated with the secreted resin are the water-soluble dye, laccaic acid, a water-insoluble dye, erythrolaccin, and a wax, also produced by the insect.

Physical Properties

Some properties are listed in Table 1.

Table 1.   Physical Properties of Shellac

Property	Value
melting point °C	77–90
softening range, t_f–t_q °C	30.5–56.5
specific gravity at 15.5 °C	1.11–1.22
molecular weight	964–1,100
refractive index, n_D^{20}	1.514–1.524
abrasion resistance, sand[a], mm⁻¹	2.3–2.8
ultimate tensile strength at 20°C, MPa[b]	13
modulus of elasticity, MPa[b]	
by sound transmission at 20°C	1094
adhesion, MPa[c]	
to glass	7.6
copper	22.8
an optically plane surface	44.1
heat of fusion, J/g[a]	52.7
specific heat at 10–40°C, J/(g·°C)[c]	1.51–1.59
thermal conductivity, mW/(cm·K)	
at 35°C	2.42

[a] ASTM D 968-76.
[b] To convert MPa to psi, multiply by 145.
[c] To convert J to cal, divide by 4.184.

Solubility.   The best solvents for shellac are the lower alcohols, methyl and ethyl, followed by amyl alcohols, glycols, and glycol ethers.

Chemistry

Chemical properties of orange and bleached shellacs are given in Table 2 as well as the properties of hard- and soft-resin components separated by solvent extraction.

Table 2. Chemical Properties of Shellac

Property	Orange, mol wt 1006	Bleached, mol wt 949	Hard resin, mol wt 1900–2000	Soft resin, mol wt 513–556
acid value, mg KOH/g	68–79	73–91	55–60	103–110
saponification value, mg KOH/g	220–232	185–260	218–225	207–229
ester value, mg KOH/g	155–167	103–155	163–165	104–119
hydroxyl value, mg KOH/g	250–280	230–260	116–117	235–240
iodine value, mg I/g				
Wijs (1 h)	1.3–1.6	0.7–1.0	1.1–1.3	5.0–5.5
carbonyl value				
sodium sulfite method	7.8–27.5		17.6	17.3

Laccaic acid and erythrolaccin. Laccaic acid, the water-soluble dye associated with the crude lac, is a mixture of several components, one of which contains nitrogen. The components are anthraquinone derivatives. Other investigators, however, believe that the dye remains associated with a protein and belongs to the monochrome group of pigments.

Erythrolaccin, the water-soluble dye, $C_{15}H_{10}O_6$, melts with decomposition at 314°C.

Lac wax. The lac insect secretes the wax in the form of thin white filaments along with the lac resin. These filaments are embedded in the resin and thus form an essential, although minor, constituent of between 3.5 and 5.5% associated with the resin. A melting point of 72–82°C, acid value of 12.0–24.3, and saponification value of 79–126 have been reported.

Polymerization. When heated above their melting points, lacs behave as thermoplastic materials for short periods of time. After a given interval, which varies with the temperature, polymerization begins.

Esterification. When dissolved in low molecular weight alcohols such as methanol or ethanol, shellacs tend to esterify slowly, even in the absence of a catalyst.

Processing

After the life cycle of the insect has been completed, the lac is ready for harvest. The cultivator (ryotat) cuts the coated twigs and scrapes off the encrustation or chops up the twigs into small pieces, called stick lac, which he takes to the market. There it is bought and transferred to refining centers.

The stick lac is scraped to remove the lac or crushed to separate the lac resin from the sticks. At this stage the crushed lac contains a mixture of resin, insect remains, dye, twigs, and other impurities. The large sticks are removed by screening and the material, known as bueli, is washed with water to remove the water-soluble dye and small sticks and insect bodies. After a few washings, the crushed lac is dried; this product is known as seed lac.

The shellacs of commerce are made from seed lac and are grouped under three processes: handmade, machine-made, and bleached.

Shellac is used in the electrical-insulating industry, varnishes, pharmaceutical tablets and pellets, printing inks, coatings of candies, shellac-bonding grinding wheels, and hat stiffening.

JAMES MARTIN
William Zinsser & Co.

J.W. Martin in *Treatise on Coatings, Film-Forming Compositions*, Vol. 1, Pt. II, Marcel Dekker, Inc., New York, 1972, p. 442.

G.S. Misra and S.C. Sengupta in N.M. Bikales, ed., *Encyclopedia of Polymer Science and Technology*, Vol. 12, Interscience Publishers, a division of John Wiley & Sons, Inc., New York, 1970, p. 419.

SHORTENINGS AND OTHER FOOD FATS. See Fats and fatty oils; Vegetable oils.

SHRIMP MEAL. See Pet and livestock feeds; Aquaculture.

SIDERITE. See Pigments.

SIENNA, BURNT. See Pigments.

SIEVES. See Size measurement of particles.

SIGNALING SMOKES. See Chemicals in war.

SILANES; SILANOLS. See Silicon compounds.

SILICA

INTRODUCTION

The term silica denotes the compound silicon dioxide, SiO_2. In technological usage, this designation includes several crystalline forms of SiO_2, as well as various amorphous forms of the parent compound which are hydrated or hydroxylated to a greater or lesser degree, eg, types of colloidal silica and silica gel.

Silicon dioxide is the most common binary compound of silicon and oxygen, the two elements of greatest terrestrial abundance. It constitutes ca 60 wt% of the earth's crust, occurring either alone or combined with other oxides in the silicates. It is thus a ubiquitous chemical substance and, owing to its rich chemistry, is of great importance geologically and to ceramic science. Commercially it is the source of elemental silicon and is used in large quantities as a constituent of building materials. In its various amorphous forms it is used as a desiccant, adsorbent, reinforcing agent, filler, and catalyst component (see Drying agents; Fillers). It has numerous specialized applications, eg, piezoelectric crystals, in vitreous-silica optical elements and glassware. Silica is a basic material of the glass, ceramic, and refractory industries and an important raw material for the production of soluble silicates, silicon and its alloys, silicon carbide, silicon-based chemicals, and the silicones (see Carbides; Ceramics; Glass; Refractories).

Structure and Bonding

Silicon shares with the other elements of Group IVA of the periodic system the property of forming an oxide of formula MO_2. All these oxides show acidic properties, most distinctly in the cases of CO_2 and SiO_2. Like the dioxides of germanium, tin, and lead, silicon is a solid of high melting point, although it differs from the former in having its common form a three-dimensional lattice based on four-coordinate silicon, whereas the heavier analogues possess the more ionic structures of the rutile type (except for a high temperature form of GeO_2 which has the four-coordinate structure).

The basic structural unit of most of the forms of silica and of the silicate minerals is a tetrahedral arrangement of four oxygen atoms surrounding a central silicon atom. The structures in which SiO_4 tetrahedra share all four oxygen atoms lead to the principal forms of silica.

At ordinary temperatures, silica is chemically resistant to many common reagents. However, it undergoes a wide variety of chemical transformations under appropriate conditions, particularly at high temperatures or when volatile products escape from the reaction. Reactivity is strongly dependent upon the form, pretreatment, and state of subdivision of the particular sample investigated.

Common aqueous acids do not attack silica, except for hydrofluoric acid which forms fluorosilicate anions, eg, SiF_6^{2-}. Silica is reduced to silicon at 1300–1400°C by hydrogen, carbon, and a variety of metallic elements. Gaseous silicon monoxide is also formed. Of the halogens, only fluorine attacks silica readily, forming SiF_4 and O_2.

The acidic character of silica is shown by its reaction with a large number of basic oxides to form silicates. The reactions of silica with organic and organometallic compounds result in compounds containing Si—C and Si—O—C bonds.

An important aspect of silica chemistry concerns the silica–water system. The interaction of the various forms of silica with water has geological significance and is applied in steam-power engineering where the volatilization of silica and its deposition on turbine blades may occur, in the production of synthetic quartz crystals by hydrothermal processes, and in the preparation of commercially important soluble silicates, colloidal silica, and silica gel.

When a solution of $Si(OH)_4$ is formed (as by acidification of a solution of a soluble silicate) at a concentration greater than the solubility of amorphous silica (100–200 ppm), the monomer polymerizes to form dimers and higher molecular weight species.

Forms of Silica

Forms of silica include crystalline silica (quartz, tridymite, cristobalite, keatite, coesite, stishovite, silica W, microcrystalline silicas), and the noncrystalline forms of silica (bulk vitreous silica and a variety of other amorphous types).

Health and Safety Factors

The principal hazard has been associated with inhalation of dust over long periods, particularly of crystalline silica, which is the main cause of a disabling pulmonary disease known as silicosis. Federal standards are in effect for work-place exposures.

T.D. COYLE
National Bureau of Standards

R.K. Iler, *The Chemistry of Silica*, John Wiley & Sons, Inc., New York, 1979.

E.M. Levin, C.R. Robbins, and H.R. McMurdie, *Phase Diagrams for Ceramists*, American Ceramic Society, Columbus, Ohio, 1964, and supplements 1969, 1975, 1981.

W.D. Kingery, H.K. Bowen, and D.R. Uhlmann, *Introduction to Ceramics*, 2nd ed., John Wiley & Sons, Inc., New York, 1976.

AMORPHOUS SILICA

The word amorphous, when used to describe silica, denotes a lack of crystal structure, as defined by x-ray diffraction. Some short-range organization may be present and is indicated by electron-diffraction studies but this ordering gives no sharp x-ray diffraction pattern. Silica, SiO_2, can be either hydrated (up to ca 14%) or anhydrous. The chemical bonding in amorphous silica is of several types, including siloxane (—Si—O—Si), silanol (—Si—O—H), and at the surface, silane (—Si—H) or organic silicon (—Si—O—R or —Si—C—R).

Amorphous silica can be broadly divided into three categories: vitreous silica or glass made by fusing quartz; silica M made by irradiating either amorphous or crystalline silica with high speed neutrons; and microamorphous silica. Silica M is a dense form of amorphous silica; it is thermally unstable and converts to quartz at 930°C after 16 h. Microamorphous silica includes sols, gels, powders, and porous glasses, all of which are composed of ultimate particles or structural units < 1 μ. These silicas have high surface areas, generally > 3 m^2/g. Microamorphous silica can be further divided into microparticulate silica, microscopic sheets and fibers, and hydrated amorphous silica. The microparticulate silicas are the most important group commercially and include pyrogenic silicas and silicas precipitated from aqueous solution. Pyrogenic silicas are formed at high temperature by condensation of SiO_2 from the vapor phase, or at lower temperature by chemical reaction in the vapor phase followed by condensation. Properties of silica sols, silica gels, precipitated silica, and pyrogenic silica are given in Table 1.

Heterogeneous Reactions

Dissolution. Amorphous silica dissolves (or depolymerizes) in water to a limited extent (1.4–2.2 mm/kg at 25°C) in water, according to the following equation:

$$SiO_2(s) + 2 H_2O(l) \rightarrow H_4SiO_4(aq)$$
$$\text{silicic acid}$$

Below pH 9 the solubility of amorphous silica is independent of pH. Above pH 9 the solubility increases because of greater ionization of the weak acid (silicic acid).

Molecular Precipitation

Amorphous silica is precipitated from a supersaturated solution. Supersaturation is obtained by concentrating an undersaturated solution, cooling a hot saturated solution, or generating $Si(OH)_4$ by hydrolysis of a silica ester, SiH_4, SiS_2, $SiCl_4$, or Si.

Organic reactions. Silicon forms chelate-type bonds with some oxygen- and nitrogen-containing organic compounds; silicon is then hexacoordinated. Silicon may also occur as a chelate in humic compounds. Polysilicic acid can undergo esterification with alcohols to form products with the Si—O—C—R structure. Organosilicon compounds have the form $R_y SiX_z$ in which X = halogen or hydroxyl ions and $y + z = 4$.

Silica Sols and Colloidal Silica

Properties. A silica sol is a stable dispersion of discrete, colloid-size particles of amorphous silica in aqueous solutions. Silica sols do not gel or settle even after several years of storage. Sols may contain up to 50% silica, and particle sizes of up to 300 nm are possible, although particles larger than 70 nm slowly settle.

Preparation. To produce sols that are stable at relatively high concentration, particles must be grown to a certain size in weakly alkaline aqueous dispersion.

Silica sols are purified by ion-exchange, dialysis, electrodialysis, or washing.

Modifications and alterations. Coagulation, flocculation, or gelling and drying of silica sols produces amorphous silica powders. The sol particles are extremely small, so particulate silica obtained from sols is composed of aggregates or porous particles that have a much higher specific surface area than estimated from apparent size.

Silica Gel

Properties. Silica gel is a coherent, rigid, continuous three-dimensional network of spherical particles of colloidal silica. Silica gels are classified into three types: regular-density gel, intermediate-density gel, and low-density gel. Silica powder can be made by grinding or micronizing dried gels, which decreases the size of the gel fragments but leaves the ultimate gel structure unchanged. Gels and powders are characterized by the density, size, and shape of the particles, particle distribution, and by aggregate strength or coalescence.

When silica is used as an absorbent, the pore structure determines the gel-adsorption capacity. Pores are characterized by specific surface area, specific pore volume (total volume of pores per gram of solid), average pore diameter, pore-size distribution, and the degree to which entrance to larger pores is restricted by small pores.

Surfaces can be categorized as fully hydroxylated in which the surface consists solely of silanol (Si—O—H) groups, as in a siloxane (Si—O—Si), or as an organic surface.

Preparation. Silica gels can be prepared by several methods. Most commonly, a sodium silicate solution is acidified to a pH less than 10 or 11.

Table 1. Properties of Different Forms of Amorphous Silica

Property	Silica sols	Dry silica gels	Silica precipitated from solution	Pyrogenic silica
SiO_2, %	10–50	96.5–99.6	80–90	99.7–99.9
CaO, %	na	na	0.1–4	na
Na_2O, %	0.1–0.8	0–1	0–1.5	na
wt loss, %				
at 105°C	50–80	na	5–7	0.5–2.5
at 1200°C	50–90	2–17.5	10–14	0.5–2.5
ultimate particle size, nm	5–100	1–100	10–25	1–100
aggregate particle size, μm		3–25	1–10	2–3
surface area, m^2/g	50–700	200–700	45–700	15–400
pH, aqueous suspension	3–5, 8–11	2.3–7.4	4–9	3.5–8
apparent or bulk density, g/cm^3	1.2–1.4	0.1–0.8	0.03–0.3	0.03–0.12
true density, g/cm^3	2.2–2.3	2.22	2.0–2.1	2.16
refractive index, n_D	1.35–1.45	1.35–1.45	1.45	1.45
oil absorption, g/g		0.9–3.15	1–3	0.5–2.8

Modification. Once a gel structure is formed, it can be modified in the wet state to strengthen the structure or enlarge the pore size and reduce surface area. Dried gels can be sintered to modify surface characteristics.

Precipitated Silica

Properties. Precipitated silica (also called particulate silica) is composed of aggregates of ultimate particles of colloidal size that have not become linked in a massive gel network during the preparation process.

Preparation. Precipitated silicas are either formed from the vapor phase (fumed or pyrogenic silicas) or by precipitation from solution.

Silica can be precipitated from a sodium silicate solution using a lower concentration than in gel preparation. Precipitated silica powders have a more open structure with higher pore volume than dried pulverized gels.

Naturally Occurring Amorphous Silica

Naturally occurring forms of amorphous silica include biogenic silica, opal, diatomaceous earth, chert, amorphous silica of volcanic origin, geothermally deposited silica, and silicified biogenic materials.

Health and Safety Factors

The only health hazard reported is so-called transient dermatitis, which causes skin dehydration and loss of skin oils.

Uses

Amorphous silica, depending on its form and purity, is used mainly as a filler and reinforcing material in rubber and plastics; to improve ink retention on paper; as a pigment and filler in paints and coatings; as an abrasive, adsorbent, desiccant, and catalyst base; and in electrical insulation (see Fillers; Abrasives; Drying agents).

JOAN D. WILLEY
University of North Carolina at Wilmington

R.K. Iler, *The Chemistry of Silica*, John Wiley & Sons, Inc., New York, 1979, pp. 21–28.

G. Bendz and I. Lindquist, eds., *Biochemistry of Silicon and Related Problems*, Plenum Press, New York, 1977.

S. Aston, ed., *Silica Geochemistry and Biogeochemistry*, Academic Press, London, UK, 1983.

J.G. Falcone, ed., *Soluble Silicates*, American Chemical Society Symposium Series 194, Washington, D.C., 1982.

VITREOUS SILICA

Vitreous silica is a glass composed essentially of SiO_2. It has been the subject of considerable study for two reasons. First, it is a material with many unique and useful properties, eg, low thermal expansion, high thermal shock resistance, high ultraviolet transparency, good refractory qualities, dielectric properties, and chemical inertness. A second reason is the simplicity of its chemical constitution. It is one of the relatively few binary oxide glasses and consequently has been investigated by countless chemists, physicists, spectroscopists, and materials scientists. However, vitreous silica is actually a very complicated material whose properties vary with, among other things, raw material, method of manufacture, and thermal history (see also Glass).

The question arises as to why, if vitreous silica has such outstanding properties, it is not used even more extensively. The answer is the high cost of manufacture as compared to most glasses, caused by very high viscosity, small temperature coefficient of viscosity, and volatility at forming temperatures. This means that even at 2000°C the melt is very stiff and difficult to shape, particularly by mass production methods, although there are significant research and development efforts aimed at circumventing some of these obstacles.

Vitreous silica can be transparent or nontransparent. The nontransparent fused material contains a large number of microscopic bubbles that create a milky appearance caused by the scattering of light. This material, sometimes called translucent fused silica, is more economical to produce than the transparent type and is often used where optical properties are not important. Another nontransparent type is opaque and is formed by sintering powdered vitreous silica.

Structure

X-ray, neutron, and electron-diffraction studies, carried out by numerous investigators, have created several areas of agreement: The x-ray diffraction pattern is typical of amorphous material, having broad diffuse rings with no indication of crystallites of significant size. The radial distribution function indicates that the separation between bonded silicon and oxygen atoms is in the 0.159–0.162 nm range. The oxygen-to-oxygen distances similarly calculated are 0.260–0.265 nm, whereas Si–Si distances are 0.305–0.322 nm. The —SiOSi— angles range from 120–180°, with an apparent max ca 150°. The —OSiO— (tetrahedral) angle appears normal (109.5°).

The infrared and Raman spectra and the x-ray and neutron diffraction patterns of vitreous silica were compared with the patterns observed for the various crystalline polymorphs of silica. Thus, in addition to cristobalite, tridymite and β-quartz have been offered as structural models. Vibrational frequencies observed for vitreous silica are in agreement with the frequencies calculated for the β-quartz model. As with many other properties of vitreous silica, the thermal history of the sample should be taken into account. The Raman spectra of Type III-vitreous silica change when the samples are stabilized at 1000°C instead of 1300°C.

Devitrification

Devitrification of vitreous silica at atmospheric pressure occurs as cristobalite formation from 1000°C to the cristobalite liquidus at 1723°C with a maximum growth rate at ca 1600°C. Trace amounts of water and metallic impurities can significantly increase devitrification rates.

Physical Properties

Vitreous silica has many exceptional properties, a number of which are abnormal when compared to other glasses and even other solids. The following anomalous properties are more or less interdependent: the expansion coefficient is negative below ca −80 to −100°C and is positive and very small above these temperatures. The elastic moduli increase with increasing temperatures above ca −190°C. Young's modulus increases linearly with applied longitudinal stress at −196°C, whereas that of soda glass decreases. The compressibility increases as the pressure increases in the low to moderate range, ie, to ca 3 GPa (< 30,000 atm). The equilibrium density decreases with heat treatment in the transformation range, in contrast to that of most glasses. The bulk modulus shows a negative pressure dependence. There is a divergence from diffusion-controlled permeation of hydrogen at elevated temperatures. The temperature coefficient of sound velocity is positive over the range of 0–800°C.

Radiation Effects

Depending upon the type of radiation, significant structural changes are generally due to high energy particle radiation, eg, a neutron stream, whereas electronic changes are usually the result of ionizing radiation, such as electron beams, x rays, α rays, or protons. Damage by ionizing radiation manifests itself in the formation of optical absorption centers (also called color centers or defect centers). Though the predominant effect of neutron radiation is structural, electronic changes can also occur and, if the ionizing radiation is of very high intensity, structural defects may be produced.

Chemical Properties

Stoichiometric vitreous silica contains two atoms of oxygen for every one of silicon, but it is extremely doubtful if such a material really exists. In general, small amounts of impurities derived from the starting materials are present; water is incorporated in the structure as —OH.

In the presence of water vapor at high temperature, the following reaction can take place:

$$\text{—SiOSi— + H}_2\text{O} \rightarrow 2\text{ —SiOH}$$

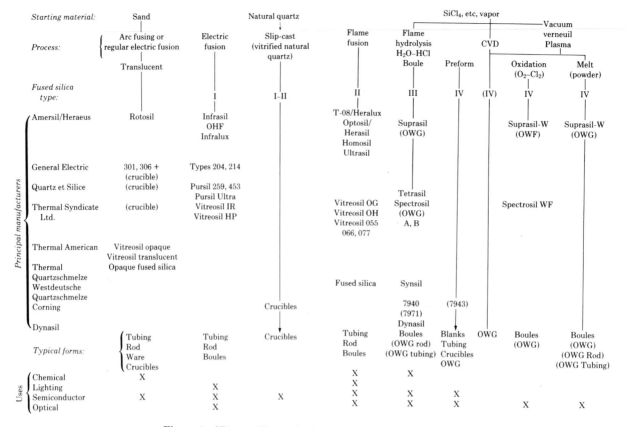

Figure 1. Vitreous silica production. OWG = optical waveguides.

Reduction to a certain degree also takes place, depending on the method of manufacture. In the presence of carbon, the following reaction may occur:

$$SiO_2 + C \rightarrow SiO + CO$$

The resistance of nontransparent vitreous silica to chemical attack is slightly less than the resistance of transparent vitreous silica. This difference is primarily due to the higher surface area of the former caused by the presence of a large number of bubbles. Most data in the literature are on the transparent material.

Metals do not generally react below 1000°C or their melting point, whichever is lower. Exceptions are magnesium, alkali metals, and aluminum; the latter readily reduces silica at 700–800°C.

Fused basic salts and basic oxides react at elevated temperatures. Reaction with alkaline-earth oxides takes place at ca 900°C. Halides may tend to dissolve vitreous silica at high temperatures; fluorides are the most reactive. Fused borates react appreciably.

Dry halogen gases do not react below 300°C. Reaction with hydrogen is very slight at 1000°C, but at much higher temperatures reduction takes place.

Attack of dilute basic solutions is very slight at room temperature. In 5% NaOH at 95°C, surface erosion is ca 10 μm in 24 h; however, crazing may occur. At higher caustic concentrations and temperatures, the reaction rate increases significantly.

Vitreous silica does not react with water or steam at moderate temperatures and pressures.

Manufacture

A summary of the important manufacturing methods, manufacturers, and product names and forms is given in Figure 1.

PAUL DANIELSON
Corning Glass Works

J. Wong and C.A. Angell, *Glass-Structure by Spectroscopy*, Marcel Dekker, Inc., New York, 1976, pp. 436–442.

R. Bruckner, *J. Non-Cryst. Solids* **5** 123 (1970); 177 (1971).

SYNTHETIC QUARTZ CRYSTALS

Silicon dioxide, SiO_2, exists in both crystalline and glassy forms. In the former, the most common polymorph is α-quartz (low quartz). All

Table 1. Properties of α-Quartz[a]

Property	Value
structural	
crystal class, space group	32[b]
optical	
indexes of refraction, Na D line	
n_o	1.5442
n_E	1.5533[c]
electrical	
resistivity, Ω-cm	10^{15}
dielectric const	
ϵ_1^T	4.58
piezoelectric coupling coefficient, %	10
piezoelectric const, FC/N[d]	
d_{11}	-23.12
mechanical	
hardness, Mohs	7
thermal conductivity, W/(m·K)	6.69–12.13
acoustic Q	0.1×10^6–3×10^6

[a] Many properties are directionally dependent; therefore, the values listed are indicative only.

[b] Trigonal trapezohedral class of the rhombohedral subsystem.

[c] Birefringent; $n_E - n_o = 0.0091$.

[d] To convert FC/N to stat C/dyn, divide by 333×10^8.

commercial applications use α-quartz which is stable only below ca 573°C at atmospheric pressure. Some of the properties of α-quartz are listed in Table 1.

Quartz is mainly used in electronic applications for which it must be free of electrical and optical twinning, voids, inclusions of foreign minerals and liquids, and must be large enough for convenient processing. The principal source of electronic-grade natural quartz is Brazil, but today manufacturers use synthetic quartz for electronic devices.

Synthesis

The first known successful attempt to grow quartz crystals hydrothermally was reported in 1905. All successful quartz-growth processes depend upon the supersaturation produced by dissolving small particles of quartz nutrient in a hot region of the high pressure system and crystallizing it onto α-quartz seeds in a cooler part of the system. Thus, it is necessary to employ a solvent in which quartz is the stable solid phase with reasonable solubility and in which the dependence of solubility upon temperature produces an appropriate supersaturation (ΔS) with an appropriate temperature differential (ΔT) between the dissolving and the growth zones. All commercial processes use either NaOH or Na_2CO_3.

Health and Safety Factors

The principal consideration in quartz synthesis is the safe management of the high pressures required for growth.

R.A. LAUDISE
E.D. KOLB
Bell Laboratories

R.A. Laudise in F.A. Cotton, ed., *Progress in Inorganic Chemistry*, Vol. III, John Wiley & Sons, Inc., New York, 1962, pp. 1–47.

A.A. Ballman and R.A. Laudise in J.J. Gilman, ed., *The Art and Science of Growing Crystals*, John Wiley & Sons, Inc., New York, 1963, pp. 231–251.

R.A. Laudise and J.W. Nielsen in F. Seitz and D. Turnbull, eds., *Solid State Physics*, Academic Press, Inc., New York, 1961, pp. 149–222.

SILICA, BRICK.   See Refractories.

SILICATES.   See Silicon compounds.

SILICIDES.   See Silicon and silicon alloys.

SILICON AND SILICON ALLOYS

PURE SILICON

Silicon (from the Latin *silex, silicis* for flint) is the 14th element of the periodic series (at wt 28.083). Elemental silicon does not occur in nature; however, as a constituent of various minerals, eg, silica and the silicates, it accounts for ca 25% of the earth's crust. There are three stable isotopes that occur in nature as well as several that are artificially prepared and radioactive. Silicon has a gray, metallic luster and may appear irridescent if it has a thin oxide covering. It is a brittle material with a hardness slightly less than that of quartz. Small single-crystal filaments are very strong and exhibit breaking strengths of up to ca 1.4 GPa (200,000 psi). In more massive pieces, substantially smaller values are observed because of difficulties in removing the effect of stress enhancement at small surface cracks. Below 800°C there is very little plastic flow. In thin sections, single crystals cleave primarily along (111) planes, but in larger pieces, fracture is conchoidal.

Crystal Structure

At atmospheric pressure, silicon has a diamond cubic structure, ie, two interpenetrating face-centered cubes displaced 1/4, 1/4, 1/4 from each other.

Physical Properties

Values for various thermal and mechanical properties are given in Table 1.

Table 1.   Thermal and Mechanical Properties

Property	Value
at wt	28.085
atomic density (atoms/cm³)[a]	5.0×10^{22}
melting point,°C	1410
boiling point,°C	2355
vapor pressure, Pa[b]	
1000°C	1.33×10^{-5}
density, g/cm³ at 25°C	2.329
critical temp, °C	4886
critical pressure, MPa[c]	53.6
hardness, Mohs/Knoop	6.5/950
heat of fusion, kJ/g[d]	1.8
heat of vaporization at mp, kJ/g[d]	16
volume contraction on melting, %	9.5

[a] Calculated.
[b] To convert Pa to mm Hg, multiply by 0.0075.
[c] To convert MPa to atm, divide by 0.101.
[d] To convert J to cal, divide by 4.184.

Electrical properties.   Silicon is a semiconductor with a band gap E_g of 1.12 eV at ca 25°C; E_g is the amount of energy required to raise an electron from the valence band to the conduction band.

Radiation Effects

Gamma radiation produces free carriers much as does visible light. High energy electrons and protons produce deep-level defects that reduce minority-carrier lifetime according to the equation

$$\frac{1}{\tau_f} = \frac{1}{\tau_o} + k\phi$$

where ρ_f is the lifetime after irradiation with a fluence ϕ, τ_o is the original lifetime, and k is a radiation-damage constant. Neutrons produce deep-level defects that not only degrade the lifetime but also increase resistivity through the removal of carriers.

Chemical Properties

Silicon, carbon, germanium, tin, and lead comprise the Group IVA elements of the periodic system. Silicon and carbon form silicon carbide which, although most widely known as an abrasive and heating element, is also a semiconductor. Germanium and silicon are isomorphous and thus are mutually soluble in all proportions. Molten tin and lead are immiscible with molten silicon. Molten silicon dissolves most materials and no container material has been found that is not noticeably dissolved.

The elements in the adjacent columns (IIIA and VA) form compounds with silicon and also enter substitutionally in small amounts into the lattice of a silicon crystal.

Oxygen forms strong bonds with silicon. There are two oxides, numerous silicates, and almost endless variations of silicones.

At elevated temperatures, silicon reacts with the halogens as well as with their anhydrous acids to form $SiCl_4$, $SiBr_4$, SiI_4, and SiF_4. In addition, hydrogen bonding gives rise to a series of hydrides (SiH_4, Si_2H_6, etc) and halosilanes, eg, $SiHCl_3$, SiH_2Cl_2, SiH_3Cl, and numerous organosilanes.

At ca 25°C, silicon appears relatively inert, partly because of the rapid formation in air of the protective oxide coating. Silicon is insoluble in acids but can be dissolved in a two-stage operation in which the surface is first oxidized and the oxide is then removed. For this operation, an aqueous solution of HNO_3 and HF is generally used.

Manufacture

Reduction. Semiconductor-grade silicon is prepared by careful purification of some easily reduced silicon compounds such as $SiCl_4$ followed by reduction with an equally pure reducing agent (usually hydrogen).

Electrolytic deposition and metallurgical methods. Silicon is deposited electrolytically from molten mixtures such as K_2SiF_6—LiF—KF or SiO_2—Na_3AlF_6 and, with less success, from $SiCl_4$ or $SiHCl_3$ in an organic solvent. However, the purity obtained to date has not been sufficient to serve the semiconductor market.

Silicon can also be purified by various metallurgical methods, eg, directional freezing or zone refining (qv).

Crystal growth. Because most devices require single-crystal semiconductors for best performance, the growth of single crystals of silicon has been studied extensively since the 1950s. A number of processes are available, depending on the desired properties. The most common and least expensive is the Czochralski process, first described in 1918, also called the Teal-Little method in honor of G.K. Teal and J.B. Little who first applied the process to semiconductor materials. The principal disadvantage of this method is the difficulty of maintaining the reactive molten silicon free of contaminants for hours at a time. Thus, in order to provide very high resistivity material, alternative methods such as float-rezoning are sometimes used. Other growth processes include edge-defined growth and vapor-phase growth.

Health and Safety Factors

Elemental silicon is inert; in air, it is only classified as a nuisance particulate with a TLV of 10 mg/m^3. However, crystalline-quartz dust is considered hazardous.

Uses

Silicon is used widely in metallurgical applications, both as constituent of various alloys and as an oxidizer in steelmaking (see Steel). It is also basic to the silicone industry, which is built around compounds with long oxygen-silicon chains (see Silicon compounds, silicones). However, silicon is best known as the material from which transistors and integrated circuits (qv) are made, although the quantities consumed are small compared to metallurgical uses.

W.R. RUNYAN
Texas Instruments, Inc.

S.M. Sze, *Physics of Semiconductor Devices*, 2nd ed., John Wiley & Sons, Inc., New York, 1981.

J.W. Mellor, *A Comprehensive Treatise on Inorganic and Theoretical Chemistry*, Vol. 4, Longmans, Green & Co., Inc., New York, 1957.

A.S. Berezhnoi, *Silicon and Its Binary Systems* (trans. from Russian), Consultants Bureau, New York, 1960.

Semiconductor Silicon, The Electrochemical Society, Princeton, N.J., 1969.

METALLURGICAL

The most important and most widely used method for making metallurgical silicon and silicon alloys is the reduction of oxides or silicates with carbon in an electric furnace. In the reduction process, metal carbides usually form first because of lower temperature requirements. As silicon is formed, it displaces the carbon, because silicon alloys and silicides have a higher heat of formation than metal carbides.

Silicon

Silicon is soluble in aluminum in the solid state to a maximum of 1.65 wt% at 577°C. It is soluble in silver, gold, and zinc at temperatures above their melting points.

Silicon metal is used extensively by the nonferrous metal industry, mainly in the production of aluminum and copper alloys. In aluminum, silicon improves castability, reduces shrinkage and hot-cracking tendencies, and increases corrosion resistance, hardness, tensile strength, and wear resistance. Silicon metal added to copper produces silicon bronzes.

The silicon improves fluidity, minimizes dross formation, and enhances corrosion resistance and strength.

Silicon Alloys and Silicides

Metal silicides form well-defined crystals with a bright metallic luster; they are usually hard and high melting.

Alloys and silicides include ferrosilicons, boron-bearing ferrosilicons, calcium silicons, ferrochrome–silicon alloys, magnesium ferrosilicons, rare-earth silicides, aluminum–silicon alloys, silicon–aluminum alloys, manganese–silicon alloys, strontium–silicon, titanium–silicon, vanadium–silicon, and zirconium–silicon.

Health and Safety Factors

The principal health hazard that may be associated with silicon is caused by the oxide. Silica, the natural oxide of silicon, is both the raw material and main polluting effluent in the electric-furnace production of metallurgical silicon and silicon alloys. Silica in its crystalline form is the chief cause of a disabling pulmonary fibrosis, ie, as silicosis.

Uses

In the iron and steel industry, silicon alloys (silicides) are used for alloying, deoxidizing, and reducing other alloying elements such as manganese, chromium, tungsten, and molybdenum. Metallurgical-grade silicon is used for the production of high purity silicon crystals required by the electronics industry and of monocrystalline and polycrystalline material for solar cells. Substantial quantities of silicon metal are used in the production of silicones, which was greatly expanded in the 1970s.

RONALD F. SILVER
Elkem Metals Company

H. Moissan, *The Electric Furnace*, 2nd ed., trans. by V. Lenker, Chemical Publishing Co., Easton, Pa., 1920.

M. Hansen, *Constitution of Binary Alloys*, 2nd ed., McGraw-Hill Book Co., New York, 1958.

D.N. Matter, *Silicon: Its Alloys and Their Use*, Ohio Ferro-Alloys Corp., Canton, Ohio, 1961.

SILICON COMPOUNDS

SYNTHETIC INORGANIC SILICATES

Naturally occurring silicate minerals make up roughly 90% of the earth's crust. These minerals are slightly soluble and are generally in dynamic chemical equilibrium with the mineral components of the aquasphere in the timeless process of mineral breakdown and reformation. Because of this slight solubility, dissolved silica is usually found at concentrations of 10–100 ppm. These minerals may be viewed as the natural analogs of the synthetic silicates, both soluble and insoluble.

The soluble silicates of commerce have the general formula:

$$M_2O.mSiO_2.nH_2O$$

where M is an alkali metal; m has been called either the ratio or modulus of the silicate. The most common soluble silicates are sodium silicates. Potassium silicate and lithium silicate are manufactured to a limited extent for use in special applications. Commercial forms of these materials are generally manufactured as a glass that dissolves in water to form viscous alkaline solutions. The values of m for commercial materials generally are 0.5–4.0. The most common form of soluble silicate, sometimes called waterglass, has an m value of 3.3.

Silicate glasses. Synthetic silicates and silica are made up of oligomers of the basic silicate building block, SiO_4^{4-}. In the glass state, a complex distribution of silicate anions is envisioned as well as an equally complex distribution of alkali cations. The physical and chemical properties of these glasses are quite sensitive to the modulus of the glass and the ion size or coordination number of the modifying cation.

Silicates in solution. The distribution of silicate species in solution has long been of interest because of the wide variations that are found for the typical properties of these solutions when the ratio is varied. The silica in these solutions exists as a complex mixture of silicate anions, of varying degrees of polymerization in a dynamic equilibrium. Species equilibrium attainment is rapid in alkaline solution above a pH value of ca 10, unless the solutions contain particles of colloidal dimensions. Colloidal particles begin to take on greater importance at ratio values greater than 2.0. Because of their polymeric nature, compositionally equivalent solutions of soluble silicates may possess different physical properties and chemical reactivity.

Crystalline soluble silicates. The most common crystalline soluble silicates are of the metasilicate family $Na_2O.SiO_2.nH_2O$. The anhydrous sodium metasilicate, Na_2SiO_3, contains SiO_2 chains, and the hydrates, written as $Na_2H_2SiO_4.xH_2O$ (where $x = n - 1$), contain silicate monomer. Only the anhydrous and so-called pentahydrate ($x = 4$) forms are of general commercial importance. There are five known mineral sodium silicates, and all appear to have layer structures and exhibit some degree of inner crystalline reactivity.

Dissolution of Soluble Silicates

The dissolution of soluble silicates is of great commercial importance. The rate of solution is generally thought to be dependent on the glass ratio, concentration, temperature, pressure and glass particle size.

The dissolution of silicate glass involves a two-step mechanism:

Ion exchange

$$\equiv\!\!Si\text{–}ONa + H_2O \rightleftharpoons \equiv\!\!Si\text{–}OH + Na^+OH^-$$

Network breakdown

$$\equiv\!\!Si\text{—}O\text{—}Si\!\!\equiv + OH^- \rightleftharpoons \equiv\!\!Si\text{–}O^- + HO\text{–}Si$$

Thus the removal of silica from the glass trails the alkali.

Polymerization of Silicate Species in Solution

The complex silanol condensation process may be represented empirically as:

$$\equiv\!\!Si\text{–}OH + \equiv\!\!Si\text{–}O^- \underset{k_2}{\overset{k_1}{\rightleftharpoons}} \equiv\!\!Si\text{—}O\text{—}Si\!\!\equiv + OH^-$$

It appears that this condensation occurs most readily at a pH value equal to the pK_a of the participating silanol group. This empirical representation becomes less valid at pH values greater than ca 10, where the rate of the depolymerization reaction (k_2) becomes significant and at very low pH where H^+ exerts a catalytic influence. Silicate polymerization in dilute solutions at pH values up to 10 is sensitive to pH and other factors that generally influence colloidal systems, eg, ionic strength, dielectric constant and temperature. If SiO_2 concentration is sufficiently high (approx 1%), interparticle aggregation and ultimately network formation (gelation) occur, yielding a continuous structure throughout the medium. This structure initially encompasses the whole system and appears quite uniform, but in time further condensation occurs with gel shrinkage and water release (syneresis).

Chemical Activity

Silicate polymer/metal ion interactions in solution. The reaction of metal ions in solution with polymeric silicate species may be thought of as an ion-exchange process, and it is expected that the silicate species as a ligand would exhibit a range of reactivity toward cations in solution because of variations of silanol acidity with degree of polymerization. Soluble silicates, at very high degrees of association, appear to interact with metal ions in solution in a manner analogous to silica gel and it also appears that, as the degree of polymerization decreases, these silicates species exhibit decreased interaction with cations.

Effect of soluble silicates on oxide/water interfaces. The adsorption of ions at clay mineral and rock surfaces is an important factor in many natural and industrial processes. Silicates are adsorbed on oxides to a far greater extent than would be predicted from their concentration. Soluble silicates not only specifically adsorb onto oxide surfaces, but also play a significant role in maintaining a negative surface charge on oxide surfaces in the presence of cations that could reverse the surface charge.

General Characteristics

General characteristics that are relevant to various uses are the pH behavior of solutions, rate of water loss from silicate films, and the dried film strength.

Table 1. Typical Commercial Sodium and Potassium Silicates

Commercial silicates	Wt ratio[a], $SiO_2 : M_2O$	Modulus[a], $SiO_2 : M_2O$	Flow point[b], °C	H_2O, wt %	d_{20}^{20}, g/cm³	Viscosity (at 20%C), Pa·s[c]	pH
anhydrous glasses							
sodium silicates	3.22	3.33	840				
	2.00	2.06	760				
potassium silicates	2.50	3.92	905				
hydrated amorphous powders							
sodium silicates	3.22	3.33		18.5			
	2.00	2.06		18.5			
solutions							
sodium silicates	1.60R	1.65			1.68	7.00	12.8
	2.00	2.06			1.69	70.00	12.2
	2.50	2.58			1.41	0.06	11.7
	2.88	2.97			1.49	0.96	11.5
	3.22	3.32			1.39	0.18	11.3
	3.75	3.86			1.32	0.22	10.8
potassium silicates	2.50	3.93			1.259	0.04	11.30
	2.20	3.45			1.261	0.01	11.55
	2.10	3.30			1.381	1.05	11.70
	1.80	2.83			1.490	1.30	12.15
crystalline solides							
sodium orthosilicate		0.50		9.5			
anhydrous sodium metasilicate		1.00		2.0			
sodium metasilicate pentahydrate		1.00		42.0			
sodium sesquisilicate		0.67		38.1			

[a] M represents Na or K.
[b] Viscosity reaches 10 kPa·s (10^5P).
[c] To convert Pa·s to P, multiply by 10.

Manufacturing and Processing

The soluble silicate glasses are manufactured, for the most part, in oil- or gas-fired open-hearth regenerative furnaces. The glass is obtained by reaction of quartz sand and sodium carbonate (soda ash) at a temperature sufficient to provide a reasonable quartz dissolution rate in the molten batch and manageable melt viscosity. The glass can be dissolved in pressure dissolvers or ground for sale as fine powders. Solutions can be spray dried to produce "hydrous silicates" containing roughly 18% water, which are stable enough to be handled commercially, yet they dissolve significantly faster than their ground-glass counterparts. Crystalline metasilicates are manufactured by processing high solids solutions of 1.0-ratio sodium silicate. Potassium silicates are manufactured in a similar manner by the reaction of K_2CO_3 and sand.

Commercial Products

The average composition and relevant properties or characteristics of soluble silicates generally available commercially are shown in Table 1.

Health, Safety, and Environmental Aspects

The moderate to strong alkalinity of commercial soluble silicates is their primary hazard. Contact-exposure effects can range from irritation to corrosion, depending on the concentration of the soluble silicate, its silica to alkali ratio, the sensitivity of the tissue exposed, and the duration of exposure.

Uses

The largest single use for soluble silicates as a functional additive is in soaps and detergents followed closely by uses in the manufacture of derivative materials. Other applications include water treatment, mineral beneficiation, adhesives and binders, enhanced oil recovery, bleach stabilization, deflocculation, and slurry thinning.

Derivatives

Derivatives include precipitated silicas, silica sols, silica gels, and synthetic insoluble silicates. Examples of the latter materials are hectorite, zeolite NaA, and various amorphous metal-ion silicates.

JAMES S. FALCONE, JR.
The PQ Corporation

R.K. Iler, *The Chemistry of Silica*, John Wiley & Sons, Inc., New York, 1979, p. 172.

J.G. Vail, assisted by J.H. Wills, *Soluble Silicates*, 2 vols., ACS Monograph No. 116, Reinhold Publishing Corp., New York, 1952.

J.S. Falcone, Jr., ed., *Soluble Silicates*, ACS Symposium Seres 194, ACS Book, Washington, D.C., 1982.

SILICON HALIDES

Despite the extensive research in silicon halides, only two of these chemicals are produced on a large industrial scale (excluding organohalosilanes). These are tetrachlorosilane, $SiCl_4$, and trichlorosilane ($HiCl_3$).

Physical Properties

The physical properties of silicon tetrahalides are listed in Table 1; those of the halohydrides are listed in Table 2.

Chemical Properties

Silicon halides are typically tetravalent compounds. The silicon–halogen bond is very polar; thus, the silicon is susceptible to nucleophilic attack. This in part accounts for their broad range of reactivity with various chemicals. Furthermore, their reactivity generally increases with the atomic weight of the halogen atom.

Manufacturing Processes

The silicon halides can be easily prepared by the reaction of silicon or silicon alloys with the respective halogens.

Table 1. Properties of Silicon Tetrahalides

Compound	Mp,°C	Bp,°C	Density, g/cm³ (°C)	Bond energy, kJ/mol[a]
SiF₄		−90.3	1.66 (−95)	146
SiCl₄	−68.8	56.8	1.48 (20)	381
SiBr₄	5	155.0	2.81 (29)	310
SiI₄	124	290.0		234

[a] To convert J to cal, divide by 4.184.

Table 2. Properties of Silicon Halohydrides

Compound	Mp,°C	Bp,°C	Density, g/cm³ (°C)
H₃SiF		−99.0	
H₂SiF₂	−122.0	−77.8	
HSiF₃	−131.2	97.5	
H₃SiCl	−118.0	−30.4	1.145 (−113)
H₂SiCl₂	−122.0	8.3	1.42 (−122)
HSiCl₃	−128.2	31.8	1.3313 (25)
H₃SiBr	−94.0	1.9	1.531 (20)
H₂SiBr₂	−70.1	66.0	2.17 (0)
HSiBr₃	−73.0	111.8	2.7 (17)
H₃SiI	−57.0	45.4	2.035 (14.8)
H₂SiI₂	−1.0	149.5	2.724 (20.5)
HSiI₃	8.0	111.0[a]	3.314 (20)

[a] At 2.9 kPa (21.8 mm Hg).

Silicon tetrachloride. A substantial percentage of commercially available silicon tetrachloride is made as a by-product from industrial processes. The two primary sources are the production of metal halides, particularly $ZrCl_4$ and $TiCl_4$, and of semiconductor-grade silicon by thermal reduction of trichlorosilane.

Trichlorosilane. Trichlorosilane, like other organochlorosilanes, is produced exclusively by the direct reaction of hydrogen chloride gas with silicon metal in a fluid-bed reactor.

Health and Safety Factors

Halosilanes should only be handled in areas that are equipped with adequate ventilation, eye-wash facilities, and safety showers. It is recommended that personnel handling halosilanes wear rubber aprons and gloves and chemical safety goggles. Furthermore, all personnel handling halosilanes should be thoroughly trained in safe handling procedures, hazardous characteristics of halosilanes, and procedures for all foreseeable emergencies.

Uses

Silicon tetrachloride. Although there is a broad range of industrial applications for $SiCl_4$, the vast majority of $SiCl_4$ is used in the manufacture of fumed silica.

Trichlorosilane. There are essentially only two large industrial applications for trichlorosilane. These are the synthesis of organotrichlorosilanes and the production of semiconductor-grade silicon metal.

WARD COLLINS
Dow Corning Corp.

A.G. MacDiarmid, *Organometallic Compounds of the Group IV Elements; Volume 2, The Bond to Halogens and Halogenoids*, Marcel Dekker Inc., New York, 1972.

W. Noll, *Chemistry and Technology of Silicone*, Academic Press, Inc., New York, 1968.

E. Hengge in V. Gutman, ed., *Inorganic Silicon Halides in Halogen Chemistry*, Vol. 2, Academic Press, Inc., New York 1967.

SILANES

Silanes are compounds containing a hydrogen–silicon bond. They also are referred to as silicon hydrides. Silane, SiH_4, is the simplest hydride and provides the basis of nomenclature for all silicon chemistry. Compounds are named as derivatives of silane with the substituents prefixed, eg, trichlorosilane, $HSiCl_3$; disilane, H_3SiSiH_3; methyldichlorosilane, $CH_3SiH(Cl_2)$; methylsilane, CH_3SiH_3; diethylsilane, $(C_2H_5)_2SiH_2$; and triethylsilane, $(C_2H_5)_3SiH$. Two or more substituents are listed alphabetically substituted organic moieties being named first, followed by simple organic fragments. Alkoxy substituents are named next, followed by acyloxy, halogen, and pseudohalogen groups; for example, ethylmethylethoxysilane, $C_2H_5(CH_3)SiH(OC_2H_5)$, and (3-chloropropyl) methylchlorosilane, $ClCH_2CH_2CH_2SiH(CH_3)Cl$ (see also Hydrides).

Only a few of the thousands of silane compounds reported have any commercial significance. These include inorganic silanes, organic silanes, and polymeric siloxanes. Despite the small number of compounds, a wide range of applications has developed, including high purity and electronic-grade silicon metal, epitaxial silicon deposition, selective reducing agents, monomers, and elastomer intermediates. Not least is the use of these materials as intermediates for production of other silanes and silicones.

Inorganic Silanes

The inorganic silanes of commercial importance include silane, dichlorosilane, and trichlorosilane. The last, trichlorosilane, is preponderant. It is not only the preferred intermediate for the first two, but it is also used in the production of high purity silicon metal and as an intermediate for silane adhesion promoters, coupling agents, silicone resin intermediates, and surface treatments. Other silanes that appear to have potential in solar electronics are monochlorosilane, disilane, and some silylmetal hydrides. Additionally, siloxene, $(H_6Si_6O_3)_x$, an inorganic polymer containing silicon hydride bonds, is of interest as a catalyst.

Table 1 contains selected physical properties of inorganic silanes.

Reactions. Oxidation. All inorganic silicon hydrides are readily oxidized.

Water and alcohols. Silanes do not react with pure water or slightly acidified water under normal conditions.

Silane reacts with methanol at room temperature to produce methoxymonosilanes of types $Si(OCH_3)_4$, $HSi(OCH_3)_3$, and $H_2Si(OCH_3)_2$, but not H_3SiOCH_3.

Halogens, hydrogen halides, and other covalent halides. Most compounds containing Si—H bonds react very rapidly with the free halogens.

Metals and metal derivatives. Silane dissolved in various solvents reacts with alkali metals (potassium has been the most commonly studied), forming as the chief product the silyl derivative of the metal, eg, $KSiH_3$.

Silanes react with alkyllithium compounds, forming various alkylsilanes.

Manufacturing and processing. Four fundamental methods of production of compounds containing a Si—H bond are noteworthy. Silicides of magnesium, aluminum, lithium, iron, and other metals react with acids or their ammonium salts to produce silane and higher binary silanes. The method was generally abandoned in favor of methods involving reduction of silicon halides.

Treatment of calcium silicide with HCl–ethanol or glacial acetic acid yields the complex polymer called siloxene.

The reduction of chlorosilanes by lithium aluminum hydride, lithium hydride, and other metal hydrides offers the advantages of higher yield and purity and the flexibility in producing a range of silicon hydrides comparable to the range of silicon halides.

Direct synthesis is the preparative method that accounts for most of the commercial silicon hydride production. Trichlorosilane is produced by the reaction of hydrogen chloride with silicon, ferrosilicon, or calcium silicide in the presence of a copper catalyst.

Health and safety factors, toxicology. At low concentrations, chlorosilanes affect nasal and pulmonary membranes. Only a minimal amount of toxicity information is available on these materials. The LD_{50} for trichlorosilane is 1050 mg/kg.

Organic Silanes

The organosilane of greatest commercial importance is methyldichlorosilane. Careful hydrolysis of this material with water affords polymethylhydrosiloxanes, which are used in the textile industry to waterproof and improve the wear resistance of fabrics. It is also used as a waterproofing agent in the leather industry, in the paper industry for sizing, in electronic applications, and in construction. Methyldichlorosilane is also used captively by silane and silicone producers in thermal condensation and addition reactions to produce vinyl, phenyl, alkyl, and cyanoalkyl precursors to silicone fluids. The addition reaction of methyldichlorosilane to fluorocarbon alkenes has enabled production of methyltrifluoropropyl silicone fluids, gums, and rubbers. Trialkoxysilanes, eg, triethoxysilane and trimethoxysilane, are being used to prepare a number of organic coupling agents utilized by the plastics industry as adhesion promoters (see Adhesives). Organosilanes containing one or more Si—H bonds have excellent reducing capabilities.

Physical properties. The physical properties of organosilanes, as with inorganic silicon hydrides, are determined largely by the properties of the silicon atom (see Table 2).

Chemical properties. Organohydrosilanes undergo a wide variety of chemical conversions. The Si—H bond of organohydrosilanes reacts with elements of most groups of the periodic system, especially Groups VIA and VIIA. There are no known reactions where the Si—H bond is replaced by stable bonds of silicon with elements of Groups IIA, IIIA, and VIII.

Reactions include oxidation and hydrolysis.

Alcohols, phenols, silanols, and carboxylic acids. The catalyzed reaction of organosilanes with hydroxyl-containing organic compounds affords organoalkoxy- and organoaryloxysilanes, usually in high yields.

Amines and phosphines. As in reactions of alcohols and acids with organosilanes, reaction of the Si—H bond with amines and phosphines proceeds only under catalysis. Alkali-metal amides or phosphines are the catalysts of choice and effect replacement of the Si—H bond with Si—N or Si—P bonds, respectively. Catalytic activity of the alkali metals for these reactions is K > Na > Li.

Halogens and halogen compounds. The reaction of organosilanes with halogens and halogen compounds usually proceeds in good yield through cleavage of the Si—H bond and formation of the silicon–halogen bond.

Metals and organometallic compounds. There are no reports of the direct reaction of the Si—H bond in organosilanes with magnesium, zinc, mercury, aluminum, and other elements of the Groups IIA, IIB, and IIIA metals. The Group IA alkali metals, ie, sodium, potassium, and their alloys, react with arylsilanes in amines or ammonia to produce the arylsilyl derivatives of these metals.

Addition of organosilanes to olefins. The addition of organosilanes to olefins is commonly termed hydrosilylation (see Silicon compounds,

Table 1. Properties of Inorganic Silanes

Property	SiH_4	H_2SiCl_2	$HSiCl_3$	H_3SiSiH_3
mp,°C	−185	−122	−126.5	−132.5
bp,°C	−111.9	−8.2	−31.9	−14.5
ΔH vaporization, kJ/mol[a]	12.5	25.2	26.6	21.2
ΔH fusion, kJ/mol[a]	0.67			
critical temperature, °C	−3.5	176		109
critical pressure, MPa[b]	472	455		
ΔH formation, kJ/mol[a]	32.6			
dipole moment, $C \cdot m \times 10^{-30}$[c]			3.913	3.24
density, g/cm³	0.68 at −185°C	1.22	1.34	0.69 at −15°C

[a] To convert J/mol to cal/mol, divide by 4.184.

[b] To convert MPa to psi, multiply by 145.

[c] To convert C·m to debye, divide by 3.336×10^{-30}.

Table 2. Properties of Commercial Organosilanes, Siloxanes, and Silazanes

Compound	Mol wt	Bp, °C$_{kPa}$[a]	Mp, °C	d$_4^{20}$	n$_D^{20}$
CH_3SiH_3	46.1	−57	−157	0.6277[b]	
$C_2H_5SiH_3$	60.2	−14	−180	0.6396[c]	
$C_6H_5SiH_3$	108.2	120		0.8681	1.5125
$C_{18}H_{37}SiH_3$	284.6	195–196$_2$			
$CH_3(C_6H_5)SiH_2$	122.2	139		0.889	1.506
$(C_6H_5)_2SiH_2$	184.3	100–101$_4$		0.9964	1.5756
$CH_3SiH(Cl)_2$	115.0	41–42	−93	1.105	1.422
$CH_3SiH(OC_2H_5)_2$	134.3	94.5		0.829[e]	1.372[d]
$CH_3SiH[N(CH_3)_2]_3$	176.4	112–113			
$CH_3SiH(C_2H_5)Cl$	108.6	67–68		0.8816	1.4020
$(CH_3)_2SiH(Cl)$	94.6	36	−111	0.851	111
$(CH_3)_2SiH(C_2H_5)$	88.2	46		0.6681	1.3783
$(CH_3O)_3SiH$	122.2	86–87			1.3687
$(C_2H_5)_2SiH(Cl)$	122.7	100		0.889	1.4152
$(C_2H_5)_3SiH$	116.3	107–108	−157	0.7318	1.4119
$(C_2H_5O)_3SiH$	164.3	131.5	−170	0.875	1.337
$(C_6H_5)_2SiH(Cl)$	218.8	143$_{1.3}$		1.118	1.581
$C_6H_5SiH(CH_3)Cl$	156.7	113$_{13}$	1.054	1.571	
$(C_6H_5)_3SiH$	260.4	160–165$_{0.4}$	42.44		
$(C_8H_{17})_3SiH$	368.8	163–165$_{0.02}$		0.821	1.454
$[(CH_3)_2SiH]_2O$	134.3	70–71		0.757	1.370
$[(CH_3)_2SiH]_2NH$	133.3	99–100			
$[(CH_3)_3SiO]_2SiH(CH_3)$	222.5	141–142			1.3815
$[CH_3SiHO]_4$	240.5	134–135	−69	0.9912	1.3870

[a] To convert kPa to mm Hg, multiply by 7.5.
[b] At −58°C.
[c] At −14°C.
[d] At 25°C.

silylating agents). It is of commercial importance in the silane and silicone industry for the production of organofunctional coupling agents, low-temperature-vulcanizing (LTV) silicone rubbers and elastomers, and specialty monomers.

Organosilanes as reducing agents in organic synthesis. These reactions are based on the hydridic character of the Si—H bond to prepare C—H bonds. A catalyst is usually required to promote reduction. Selective reduction, ie, chemoselective and regioselective, can sometimes be obtained by careful selection of catalyst.

Photolysis. Irradiation of 2,2-bis(2,4,6-trimethylphenyl)-1,1,1,3,3,3-hexamethyltrisilane in hydrocarbon solution yields tetramesityldisilene, which can be isolated as a yellow-orange solid that is stable to 20°C and above in the absence of air.

Manufacture and processing. Processes include the direct process, reduction with metal hydrides, disproportionation, reaction with Grignard reagents, and addition to olefins.

BARRY ARKLES
WILLIAM R. PETERSON, JR.
Petrarch Systems, Inc.

V. Bazant, V. Chvalovsky, and J. Rathovsky, *Organosilicon Compounds*, Vol. 1, Academic Press, Inc., New York, 1965.

Proceedings of the XV Silicon Symposium, Duke University, Durham, N.C., March, 1981.

SILICON ETHERS AND ESTERS

Silicon esters are silicon compounds that contain an oxygen bridge from silicon to an organic group, ie, SiOR. The oldest reported silicon compounds contain four oxygen bridges and are often named as derivatives of orthosilicic acid, $Si(OH)_4$. With the advent of organosilanes that contain silicon–carbon bonds (Si—C), an organic nomenclature was developed by which compounds are named as alkoxy derivatives.

The applications for alkoxysilanes range broadly. They are classified roughly by whether the Si—OR bond is expected to remain intact or be

hydrolyzed in the final application. The susceptibility to hydrolysis, volatility, and other properties of alkoxysilanes predicate their particular applications. Applications in which the Si—OR bond is hydrolyzed include binders for foundry-mold sands used in investment and thin-shell castings, binders for refractories, resins, coatings, low heat glasses, cross-linking agents, and adhesion promoters. Applications in which the Si—OR bond remains intact include lubricant, heat-transfer, hydraulic, dielectric, and diffusion-pump fluids. In general, lower molecular weight compounds, eg, tetraethoxysilane and tetramethoxysilane, are used in reactive applications whereas such compounds as tetrabutoxysilane and hexakis(2-ethylbutoxy)disiloxane are associated with mechanical applications. Tetraethoxysilane and its polymeric derivatives account for > 90% of all production.

Properties

The alkoxysilanes possess excellent thermal stability and a broad temperature range of liquid behavior which widens with length and branching in the substituents. The physical properties of the silane esters, particularly the polymeric esters containing siloxane bonds, ie, Si—O—Si, are often compared to the silicone oils. They have low pour points and similar temperature–viscosity relationships. The alkoxysilanes generally have sweet, fruity odors that become less apparent as molecular weight increases. With the exception of tetramethoxysilane, which can be absorbed into corneal tissue, causing eye damage, the alkoxysilanes generally exhibit low levels of toxicity.

Preparation

The preferred method of production is described by Von Ebelman's 1846 synthesis:

$$SiCl_4 + 4\,C_2H_5OH \rightarrow Si(OC_2H_5)_4 + 4\,HCl$$

The reaction is generalized to:

$$R'_{4-n}SiCl_n + n\,R'OH \rightarrow R'_{4-n}Si(OR')_n + n\,HCl$$

Process considerations must not only take into account characteristics of the particular alcohol or phenol to be esterified, but also the self-propagating by-product reaction that results in polymer formation.

BARRY ARKLES
Petrarch Systems, Inc.

R.C. Mehrotra, V.D. Gupta, and G. Srivastava, *Rev. Silicon Germanium Tin Lead Compd.* **1**, 299 (1975).

V. Bazant, V. Chvalovsky, and J. Rathousky, *Organosilicon Compounds*, Academic Press, Inc., New York, 1965 (1st serial); Marcel Dekker, Inc., New York, 1973 (2nd serial); Institute of Chemical Process Fundamentals, Prague, Czechoslovakia, 1977 (3rd serial).

W. Noll, *Chemistry and Technology of Silicones*, Academic Press, Inc., New York, 1968, pp. 639–662.

SILICONES

The name silicone denotes a synthetic polymer

$$\left(R_n SiO_{(4-n)/2} \right)_m$$

where $n = 1$–3 and $m \geq 2$. A silicone contains a repeating silicon–oxygen backbone and has organic groups R attached to a significant proportion of the silicon atoms by silicon–carbon bonds. Silicone has no place in scientific nomenclature, although it was originally introduced under the supposition that compounds of the empirical formula RR′SiO were analogous to ketones. Several decades later it was used to describe related polymers. In commercial silicones, most of the R groups are methyl; longer alkyl, fluoroalkyl, phenyl, vinyl, and a few other groups are substituted for specific purposes. Some of the R groups in the polymer can also be hydrogen, chlorine, alkoxy, acyloxy, or alkylamino, etc. These polymers can be combined with fillers, additives, and solvents to make products that are loosely classed as silicones.

Silicones have an unusual array of properties. Chief among these are thermal and oxidative stability and a relatively mild dependence of

Table 1. Formulas and Symbols for Silicones

Formula	Functionality	Symbol
$(CH_3)_3SiO_{0.5}$	mono	M
$(CH_3)_2SiO$	di	D
$(CH_3)SiO_{1.5}$	tri	T
$(CH_3)(C_6H_5)SiO$	di	D'
$(C_6H_5)_2SiO$	di	D'
$(CH_3)(H)SiO$	di	D'
SiO_2	quadri	Q

Table 2. Properties of Silane Monomers

Compound	Boiling point, °C	Density, g/cm^3	Refractive index, n_D
$HSiCl_3$	32	1.3298[25]	1.3983[25]
$(C_2H_5)_2SiCl_2$	129	1.0472[25]	1.4291[25]
$(C_2H_5)SiH(Cl)_2$	74.5	1.0926[20]	1.4148[20]
$(C_6H_5)_2SiCl_2$	305	1.218[25]	1.5765[26]
$(CH_3)(C_6H_5)SiCl_2$	205	1.174[25]	1.5180[20]
$(CH_3)(C_6H_5)_2SiCl$	93	1.085[20]	1.5742[20]
$(CH_3)(C_2H_3)SiCl_2$	93	1.085[25]	1.4200[25]
$(CH_3)(CF_3CH_2CH_2)SiCl_2$	122	1.211[20]	1.3817[25]
$(CH_3)(CNCH_2CH_2)SiCl_2$	215	1.187[25]	1.4564[20]
$(CH_3)Si(OCH_3)_3$	103.5	0.955[25]	1.3687[25]
$(CH_3)_3Si(OCH_3)$	56.5	0.7537[25]	1.3678[20]
$(CH_3CO_2)_3SiCH_3$	95	1.1677[25]	1.407
$(NH_2CH_2CH_2CH_2)Si(OC_2H_5)_3$	217	0.943[25]	1.4190[25]

physical properties on temperature. Other important characteristics of these materials include a high degree of chemical inertness, resistance to weathering, good dielectric strength, and low surface tension. As the general formula implies, the molecular structure can vary considerably to include linear, branched, and cross-linked structures. These structural forms and R groups can provide many combinations of useful properties that lead to a wide range of commercially important applications. Silicones include fluids, resins, and elastomers. Many derived products, eg, emulsions, greases, adhesives, sealants, coatings, and chemical specialties, have been developed for a large variety of uses.

There are four basic manufacturers of silicones in the United States: Dow Corning, General Electric, Union Carbide, and Stauffer-Wacker Silicones. There are also large producers in the UK, France, the FRG, the GDR, Japan, and the USSR and small manufacturers in Belgium, Italy, Czechoslovakia, and elsewhere.

Nomenclature

Polymer nomenclature is inherently complex and difficult to use and, as a result, that of silicones is simplified by the use of the letters M, D, T, and Q to represent monofunctional, difunctional, trifunctional, and quadrifunctional monomer units, respectively. Primes, eg, D', are used to indicate substituents other than methyl. Equivalent symbols are shown in Table 1.

Silane Monomers

Silane monomers are the precursors of silicone polymers. Siloxane compositions are obtained from these silane monomers by hydrolysis. Organic radicals, attached to silicon by hydrolytically stable linkages, survive these hydrolyses and thereby become a part of the siloxane product. The widespread commercial usefulness of silicone products is the direct result of the discovery of economic routes for the manufacture of silane monomers.

Methylchlorosilanes. Methylchlorosilanes are the starting materials for methyl silicones. They are made industrially by the copper-catalyzed exothermic reaction of methyl chloride with silicon at ca 300°C.

Other chlorosilanes and derivatives. The same kind of direct reaction involving an organic chloride and silicon can also be used to make ethyl and phenylchlorosilanes.

Properties of several industrially important silane monomers are listed in Table 2.

Uses. As silylation agents, silane monomers are being used increasingly in synthetic and analytical chemistry.

Silicone Polymers

Chlorosilane hydrolysis. Conversion of chlorosilane monomers to useful polymer products generally involves 2–4 processes: hydrolysis plus cleanup and acid reduction of the hydrolysate; in some cases, conversion of the hydrolysate to cyclic oligomers; polymerization, equilibration, or bodying of the hydrolysate or cyclic oligomers; and stripping, devolatilization, or solvent removal. The hydrolysis can be carried out in a continuous system, by batch processes, or especially in the case of silicone resins, by hydrolysis of a solvent solution of the chlorosilanes.

The production of silicone fluids and elastomers is largely based on the hydrolysis of dimethyldichlorosilane. Batch or continuous processes are used, but the continuous process is preferred.

Equilibration and polymerization. Very important to the manufacture of silicone fluids and elastomers are the siloxane rearrangement reactions, which occur in the presence of acids or bases. An example is the reaction of hexamethyldisiloxane with octamethylcyclotetrasiloxane in the presence of sulfuric acid, which gives a mixture of linear and cyclic polymers.

$$MM + D_4 \overset{H_2SO_4}{\rightleftharpoons} MD_nM + D_m$$

This reaction involves numerous equilibria.

The rate of alkali-catalyzed polymerization is proportional to the concentration of the cyclic siloxane and to the square root of the concentration of alkali.

Branched polymers can be made by the introduction of T or Q units (see Table 1), or by irradiation of linear polymers.

Copolymerization involving different siloxane units is important, because many useful fluids and gums are copolymers containing groups, eg, methylvinyl-, methylphenyl-, or diphenylsiloxane as well as dimethylsiloxane units.

In addition to these reactions, several other reactions are important in the manufacture and use of commercial silicones. The Si—H bond is easily solvolyzed by water or alcohols, easily oxidized, and can be added across carbon–carbon multiple bonds to form Si—C bonds. Siloxanes containing SiH can be cross-linked or modified by such reactions.

Silicone–organic copolymers. Copolymers of silicones and organic polymers have the advantages of low cost and physical strength and they retain the durability and surface properties of silicones.

Copolymers in which the silicon is present to modify properties or to provide room-temperature cure for organic polymer systems can in principle be made with almost any organic polymer. Several such compounds have been reported, for example, polyolefins, polyurethanes, and polyethers.

Methyl silicone polymers. Linear polydimethylsiloxanes have been extensively studied. Because intermolecular forces are weak, the polymers have low melting points and second-order transition temperatures.

Effects of substitution on properties. Physical properties of several oligomers with various substituents are listed in Table 3. A great many such oligomeric compounds have been prepared and described, and several are used industrially as intermediates in the preparation of polymers.

Silicone Fluids

Dimethyl silicone fluids are made by catalyzed equilibration of dimethyl silicone stock, ie, the crude fluid or distilled cyclic polymers, with a source of the chain terminator, $(CH_3)_3SiO_{0.5}$.

Some properties of silicone fluids are listed in Table 4.

Derived products. Several types of greases are made from mixtures of silicone fluids and fillers. For insulating and water-repellent greases, silica filler is used, and the fluid may be dimethyl silicone or dimethyl silicone copolymerized with methylphenyl, methyltrifluoropropyl, or

Table 3. Properties of Siloxane Oligomers

Compound	Boiling point, °C$_{kPa}$[a]	Density d^{20}, g/cm^3	Refractive index, n_D^{20}	Melting point, °C
$[(C_6H_5)_2SiO]_4$	335$_{0.13}$			200
$[(CH_3)(C_6H_5)SiO]_4$	237$_{0.13-0.67}$	1.183	1.5461	99
$(CH_3)_3SiOSi(C_6H_5)_2OSi(CH_3)_3$	172$_{2.4}$	0.984	1.4927	
$[(CF_3CH_2CH_2)(CH_3)SiO]_4$	134$_{0.4}$	1.255	1.3724	
$[(CH_2{=}CH)(CH_3)SiO]_4$	111$_{1.3}$	0.9875	1.4342	−43.5
$(CH_3)_3Si[OSiH(CH_3)]_2OSi(CH_3)_3$	177	0.8559	1.3854	
$[(CH_3)(H)SiO]_4$	134	0.9912	1.3870	−69

[a] To convert kPa to mm Hg, multiply by 7.5.

methylalkyl silicone. For lubricating grease, the fillers are generally lithium soaps and the preferred fluids are methylphenyl, chlorinated phenylmethyl, or methyltrifluoropropyl silicones. The electrical properties of silica-filled greases are good. Such dielectric compounds are workable at low temperatures and do not have a dropping point, as soap-filled greases do. Lubricating greases vary in properties, depending on the type of fluid and thickener used. They have wide service-temperature ranges, ie, from −70 to 230°C; dropping points of 200–260°C, depending on the soap used; low bleed; and low evaporation rate and weight loss in vacuum. Those based on fluids with good lubricating properties are themselves good lubricants, capable of prolonged performance up to 200°C.

Emulsions of silicone fluids in water are made for convenience in applying small amounts of silicone to textiles, paper, or other surfaces.

Silicone Resins

Silicone resins are highly cross-linked siloxane systems, and the cross-linking components are usually introduced as trifunctional or tetrafunctional silanes in the first stage of manufacture. The general effects of the most commonly used monomers on the properties of a film are listed in Table 5.

Dipping or impregnating varnishes based on silicone resins are used to bond and insulate electrical coils and glass cloth and are usually supplied as 50 wt% or 60 wt% silicone resins in an organic solvent.

Silicone laminating resins are used first to coat glass cloth, and this coating is partially cured to a nontacky stage. Stacks of cloth so treated are pressed at ca 7 MPa (1000 psi) and for high pressure laminates are heated to cure or are bag-molded or vacuum-formed at ca 0.7 MPa (100 psi) or less for low pressure laminates before cure.

Table 5. Effect of Monomers on the Properties of Silicone Resin Films

Property	CH_3SiCl_3	$C_6H_5SiCl_3$	$(CH_3)_2SiCl_2$	$(C_6H_5)_2SiCl_2$	$CH_3(C_6H_5)SiCl_2$
hardness	increase	increase	decrease	decrease	decrease
brittleness	increase	great increase	decrease	decrease	decrease
stiffness	increase	increase	decrease	decrease	decrease
toughness	increase	increase	decrease	decrease	decrease
cure speed	much faster	some increase	slower	much slower	slower
tack	decrease	some decrease	increase	increase	increase

Pressure-sensitive adhesives (qv) are made by compounding silicone elastomer gums with silicone resins which are not completely compatible with each other.

The cure of silicone resins usually occurs through the formation of siloxane linkages by condensation of silanols.

Silicone resins change little on exposure to weather.

Silicone Elastomers

Elastomers are an extremely important genus of silicone products. Silicone polymers of appropriate molecular weight must be cross-linked to provide elastomeric properties. Fillers are used in these formulations to increase strength through reinforcement. Extending fillers and various additives, eg, antioxidants, adhesion promoters, and pigments, can also be used to obtain specific properties (see also Elastomers, synthetic). The usual reinforcing fillers (qv) for silicone elastomers are finely divided silicas made by either the fume process or the wet process. Where physical strength is not required of the finished product, nonreinforcing fillers are used.

Different silicone elastomers are conveniently distinguished by their cure-system chemistries and can be categorized by the temperature conditions needed to cure them properly, eg, room-temperature-vulcanizing rubbers and heat-cured rubbers.

Properties. The properties of silicone rubber change with temperature; for example, Young's modulus decreases from ca 10,000 to 200 MPa (14.5×10^4 to 2.9×10^4 psi) from −50°C to RT and then remains fairly constant to 260°C. Resistivity decreases; electric strength does not change greatly; dielectric constant increases at 60-Hz current and decreases at 10^4-Hz current and above; and the power factor increases considerably. Tensile strength decreases from ca 6.9 MPa (1000 psi) at 0°C to 2.1 MPa (300 psi) at 300°C. Thermal conductivity of silicone rubber usually is ca 1.5–4 W/(m·K) and increases with increased filler loading.

Table 4. Properties of Silicone Fluids

Type of fluid		Viscosity at 25°C, mm^2/s (= cSt)	d^{25}, g/cm^3	n_D^{25}	Pour point, °C	Flash point, °C	Surface tension at 25°C, mN/m (= dyn/cm)	Electric strength, kV/μm
$(CH_3)_2SiO$, mol %	Copolymer silicone							
100	none	10	0.940	1.399	−73	210	20.0	1.4
100	none	100	0.968	1.4030	−55	302	20.9	1.4
100	none	1,000	0.974	1.4035	−50	315	21.1	1.4
100	none	10,000	0.975	1.4035	−47	315	21.3	1.4
100	none	100,000	0.978	1.4035	−40	315	21.3	1.4
50	$CH_3(C_6H_5)SiO$	125	1.07	1.495	−45	302	24.7	1.3
91.2	$CH_3(C_6H_5)SiO$	50	0.99	1.425	−73	282	25.0	1.3
91.2	$CH_3(C_6H_5)SiO$	100	0.99	1.425	−73	293	24.1	1.3
95.6	$(C_6H_5)_2SiO$	100	1.00	1.421	−73	302	24.0	1.4
>90	tetrachlorophenyl siloxane	70	1.045	1.428	−73	288	21	
0	$CH_3(H)SiO$	25	0.98	1.397			20	
0–10	$CF_3CH_2CH_2(CH_3)SiO$	300	1.25		−48		26	
92	$(CH_3)SiO_{1.5}$	50	0.972	1.403	−84	315	21.0	1.4

Silicone rubber (gum) films are permeable to gases and hydrocarbons and in general are ca 10–20 times as permeable as organic polymers. Solvents diffuse into silicone rubber and swell, soften, and weaken it.

Heat alone can cause depolymerization and volatilization. Permanent deformation occurs when rubber is compressed or stretched at high temperature.

Health and Safety Factors

Methyl silicones are biologically inert. They do not react with body fluids, cause coagulation of blood, adhere to body tissues, and do not show irritating or toxic effects. These properties no doubt result from the molecular weight, incompatability, and stability of the polymers. These qualities of the methyl polymers are usually also true of phenyl-, alkyl-, and fluoroalkyl-substituted polymers. Trifluoropropyl silicones, however, form toxic materials when heated above 280°C (see Biomedical and prosthetic devices).

Formulated silicones contain ingredients that may be irritating or toxic. For example, some of the metallic or organometallic catalysts in resins or RT-vulcanizing (RTV) rubbers are irritating or toxic, but they are used at very low levels in these products. By-products of elastomer cure may be irritating. Emulsifiers in silicone emulsions or bactericides used to keep these or silicone elastomers from deterioration are potential hazards. Many such products, however, are formulated to meet specific safety standards. Silicone resins are sold in solvents, and these may be hazardous with regard to health or fire.

Methylchlorosilanes are flammable and corrosive because HCl is liberated on hydrolysis; other chlorosilanes are less flammable, but all are hazardous chemicals. Volatile oligomeric siloxanes are somewhat flammable and can be generated by decomposition of siloxane polymers. Compounds containing SiH liberate hydrogen on hydrolysis or alcoholysis, and this is a hazard both in terms of flammability and generation of pressure.

A few other silicon compounds are toxic. For example, methyl and ethyl orthosilicates are somewhat toxic. Materials of this type may be used in formulating silicone products. It should not be assumed that because silicone polymers are physiologically inert, all silicones and silanes are similarly benign, although this is usually the case.

Silicone rubber is generally resistant to bacterial or fungal growth, but bacterial growth has been noted in a few cases. This is probably caused by nonsilicone components in the composition, eg, fatty acids, but the net result is some ability to support growth.

A few low molecular weight silicone oligomers show varying biological activity. Hexamethyldisiloxane has toxic properties similar to that of many solvents and the methyltrifluoropropyl and methylphenyl trimers and tetramers are biologically active.

Uses

The range of silicone applications is extremely large and a compilation of general product types and example application areas is presented in Table 6.

BRUCE B. HARDMAN
ARNOLD TORKELSON
General Electric Company

R.R. McGregor, *Silicones*, McGraw-Hill Book Company, New York, 1954.

E.G. Rochow, *An Introduction to the Chemistry of the Silicones*, John Wiley & Sons, Inc., New York, 1951.

W. Noll, *Chemistry and Technology of Silicones*, Academic Press, Inc., New York, 1968.

M.D. Beers, in I. Skeist, ed., *Handbook of Adhesives*, D. Van Nostrand Company, New York, 1977.

W.J. Bobear in M. Morton, ed., *Rubber Technology*, 2nd ed., D. Van Nostrand Company, New York, 1973.

SILYATING AGENTS

Silylation of Organic Compounds

Silylation is the displacement of active hydrogen from an organic molecule by a silyl group. The active hydrogen is usually OH, NH, or SH, and the silylating agent is usually a trimethylsilyl halide or a nitrogen-functional compound. A mixture of silylating agents may be used.

Derivatizing an organic compound for analysis may require only a few drops of reagent selected from silylating kits supplied by laboratory supply houses. Commercial synthesis of penicillins requires silylating agents purchased in tank cars from the manufacturer.

Typical commercial silylating agents are listed in Table 1.

All of the silylating agents are classified by the Department of Trans-

Table 6. General Product Types and Example Application Areas

Fluid applications

plastic additives	greases
hydraulic fluids	coagulants
vibration damping	particle and fiber treatments
release agents	cosmetic and health-product additives
antifoamers	heat-transfer media
dielectric media	polishes
water repellency	lubricants
surfactants	

Resin applications

varnishes	electrical insulation
paints	pressure-sensitive adhesives
molding compounds	laminates
protective coatings	release coatings
encapsulants	adhesives
junction coatings	

RTV rubber applications

sealants	encapsulants
adhesives	electrical insulation
conformal coatings	glazing
gaskets	medical implants
foams	surgical aids
molding parts	mold making

Heat-cured rubber applications

tubing and hoses	auto-ignition cable and spark-plug boots
belting	extruding
wire–cable insulation	medical implants
surgical aids	laminates
fuel-resistant rubber parts	electrically conducting rubber
penetration seals	fabric coating
molded parts	foams
embossing–calendering rollers	

Table 1. Methyl Silylating Agents

Chemical name	Formula
trimethylchlorosilane (TMCS)	$(CH_3)_3SiCl$
dimethyldichlorosilane (DMCS)	$(CH_3)_2SiCl_2$
hexamethyldisilazane (HMDZ)	$(CH_3)_3SiNHSi(CH_3)_3$
chloromethyldimethylchlorosilane (CMDMS)	$ClCH_2(CH_3)_2SiCl$
N,N'-bis(trimethylsilyl)urea (BSU)	$[(CH_3)_3SiN]_2CO$
N-trimethylsilyldiethylamine (TMSDEA)	$(CH_3)_3SiN(C_2H_5)_2$
N-trimethylsilylimidazole (TSIM)	
N,O-bis(trimethylsilyl)acetimide (BSA)	$(CH_3)_3SiN=C(CH_3)OSi(CH_3)_3$
N,O-bis(trimethylsilyl)trifluoroacetimide (BSTFA)	$(CH_3)_3SiN=C(CF_3)OSi(CH_3)_3$
N-methyl-N-trimethylsilyltrifluoroacetamide (MSTFA)	$(CH_3)_3SiN(CH_3)COCF_3$
t-butyldimethylsilylimidazole (TBDMIM)	
N-trimethylsilylacetamide (MTSA)	$(CH_3)_3SiNHCOCH_3$

portation as flammable liquids. The chlorosilanes are clear liquids that react readily with water to form corrosive HCl gas and liquid. Liquid chlorosilanes and their vapors are corrosive to the skin and extremely irritating to the mucous membranes of the eyes, nose, and throat. They should be treated as strong acids. The nitrogen-functional silanes react with water to form ammonia, amines, or amides. Since ammonia and amines are moderately corrosive to the skin and very irritating to the eyes, nose, and throat, silylamines should be handled like organic amines.

The techniques of silylation and their application in analysis have been reviewed as have the intermediate steps in organic synthesis.

Silylation of Inorganic Compounds

Silicate modifications. A method has been described in which silicate minerals are simultaneously acid-leached and trimethylsilyl end-blocked to yield specific trimethylsilyl silicates with the same silicate structure as the mineral from which they were derived.

Silylation of Inorganic Surfaces

Alkyl silylating agents. Alkyl silylating agents convert mineral surfaces to water-repellant, low energy surfaces useful in water-resistant treatments for masonry, electrical insulators, packings for chromatography, and in noncaking fire extinguishers. Methylchlorosilanes react with surface water or hydroxyl groups of the surface to liberate HCl and deposit a very thin film of methylpolysiloxane, which has a very low critical surface tension and is therefore not wetted by water (see Waterproofing).

Organofunctional Silylating Agents

Whereas alkylsilylating agents provide low energy surfaces designed to prevent adhesion, a series of organofunctional silylating agents is offered commercially as adhesion promoters. Their prime application has been as coupling agents in mineral-filled organic resin composites.

Organofunctional silylating agents are used in liquid-crystal alignment, ion removal, metal oxide electrodes, antimicrobials, polypeptide synthesis and analysis, immobilized enzymes, immobilized metal-complex catalysts, and reinforced composites.

EDWIN P. PLUEDDEMANN
Dow Corning Corporation

K. Balu and G. Kind, eds., *Handbook of Derivatives for Chromatography*, Heyden, London, UK, 1977.

E.P. Plueddemann, *Silane Coupling Agents*, Plenum Press, New York, 1982.

D.E. Leyden and W. Collins, eds., *Silylated Surfaces*, Gordon and Breach Science Publishers, London, UK, 1980.

SILK

Silk is the solidified viscous fluid excreted from special glands (or orifices) by a number of insects and spiders. It is a polymer consisting of amino acids; glycine and alanine are the main components. The only significant source for textile usage is the silk-moth caterpillar, commonly referred to as the silkworm. Several varieties are known; the most valuable is the caterpillar of the moth *Bombyx mori*, which was domesticated centuries ago. A somewhat different product known as wild silk, sometimes called tussah, is produced by species of moths that are not domesticated.

Silk is available for textiles (qv) as a continuous filament and as staple yarn. Although not truly continuous, a filament may be 300–1200 m long. The strand produced by the silkworm consists of two filaments encased in a protein gum. The silkworm uses it to form the cocoon in which it encases itself for metamorphosis. If the moth is allowed to emerge alive, the filaments are ruptured. These and other damaged filaments, called silk waste, are used in producing staple yarns, which are formed by twisting short lengths of filament together.

Because of the enormous amount of manual labor required in the production of silk, it has always been expensive. Its properties have

made it highly sought after and revered for apparel and furnishing fabrics for centuries. In recent years, man-made fibers have been produced that rival silk's desirable properties of luster, hand, and drape at a far lower cost (see Fibers, chemical). Consequently, the demand for silk has declined, although it is still used for specialty and high quality luxury items.

Physical Properties

Throughout history, silk has been sought after because of its unique fiber characteristics. It is both strong and elastic and has a tensile strength of 0.34–0.39 N/tex (3.9–4.5 gf/den) and an elongation at break of 20–30%. Its natural crease resistance is due to good resilience and it recovers readily from deformation. A highly hygroscopic fiber (moisture regain of 10–15%), it is a poor conductor of heat and electricity. It has the desirable tactile properties associated with light weight, warmth, and good drapability. It is a smooth, translucent fiber with a triangular cross-section. Its cross-sectional dimensions correspond to ca 0.17 tex (1.5 den) per filament of fibroin. The continuous filament has a high luster or sheen that contributes to its aura of luxury.

Chemical Properties

Raw silk consists primarily of the proteins sericin and fibroin. The bulk is fibroin (fiber), which is coated with 15–25% sericin (silk gum). The principal amino acid constituents are glycine, alanine, serine, and tyrosine.

Processing

Steps include reeling, spinning, scouring and bleaching, weighting, dyeing, and finishing.

CHARLES D. LIVENGOOD
North Carolina State University

W.F. Leggett, *The Story of Silk*, Lifetime Editions, New York, 1949, pp. 70–73.

M.S. Otterburn in R.S. Asquith, ed., *Chemistry of Natural Protein Fibers*, Plenum Press, Inc., New York, 1977, p. 56.

E.R. Trotman, *Dyeing and Chemical Technology of Textile Fibers*, 5th ed., Charles Griffin and Co., Ltd., London, UK, 1975, p. 261.

SILLEMANITE, Al₂O₃SiO₂. See Refractories.

SILLEMANITE, $Al_2O_3SiO_2$. See Refractories.

SILVER AND SILVER ALLOYS

Silver, along with copper and gold, its colorful neighbors in the periodic table, has been used throughout recorded history.

There are 55 silver minerals. Ores in which silver is the main component are associated with igneous rocks of intermediate felsic composition. The Coeur d'Alene mining district of Idaho is the principal silver-producing area in the United States. The formerly rich silver ores in Mexico, Peru, and Bolivia are now largely exhausted, and the ores from which silver is produced today in Peru, Mexico, and Canada are mostly by-product ores.

Properties

The primary valence of silver is 1, although divalent silver oxide and other higher-valent compounds exist; two stable isotopes, ie, ^{107}Ag and ^{109}Ag, and 25 radioactive isotopes exist. Tensile-strength and hardness values vary widely. Electrodeposited silver (HV = 100) has a higher electrical resistivity than wrought silver. Properties are given in Table 1.

Molten silver dissolves in ca 10 times its own volume of oxygen (ca 0.32 wt%) just above its melting point and ejects it violently before solidification; at ca 810°C, silver dissolves hydrogen.

Table 1. Properties of Silver

Property	Value
melting point, °C	961.9
boiling point, °C	2163
density, g/cm^3	
at 20°C, hard drawn,	10.43
annealed	10.49
at 960.5–1300°C (liq)	9.30–9.00
thermal conductivity, W/(m·k)^a	
at 20°C	428
at 450°C	356
mean specific heat, J/(kg·K)^b	
at 25°C	235
at 961°C	297
at 961–2227°C (liq)	310
electrical resistivity, R	
at 0°C, $\mu\Omega\cdot$cm	1.59
elastic modulus, GPa^c	71.0
Poisson's ratio	0.39 (hard drawn)
0.37 (annealed)	
latent heat of fusion J/g^b	104.2
latent heat of vaporization, kJ/g^b	2.636

^{a} To convert W/(m·k) to (Btu·in.)/(h·ft^2·°F), divide by 0.1441.

^{b} To convert J to cal, divide by 4.184.

^{c} To convert Pa to psi, divide by 6895; to convert MPa to psi, multiply by 145

Substances that corrode silver are chlorine, mercury, and sulfur; chromic, nitric, and sulfuric acids; alkali-metal cyanides; hydrogen peroxide; hydrogen sulfide; ferric sulfate; and permanganates. Sulfur tarnishes silver, coating it with a light-colored-to-black film. The reaction with nitric acid is used in the production of chemical compounds.

$$4\ Ag + 6\ HNO_3 \rightarrow 4\ AgNO_3 + NO + NO_2 + 3\ H_2O$$

Mining and Processing

Silver is mined by open-pit methods and subsurface shafts and drifts. In the 1970s, improvements in the cyanidation of low grade gold–silver ores, etc, and the development of the activated-carbon stripping of the pregnant solutions led to the use of this process.

In the United States, gravitation was replaced by flotation (qv). The latter as well as crushing, grinding, classification, thickening, and filtering, increases the concentration.

Silver is recovered electrolytically from the sludges obtained from copper, lead, and zinc concentrates.

Secondary silver recovery. Since 1951, the recovery of silver from scrap has been necessary to meet demand. Photographic film contains 5–15 g Ag/m^2, almost all of which is salvaged, as well as other photographic materials (see Recycling).

Standards and Specifications

Commercial fine silver contains a minimum of 999.0 ppt (parts per thousand) silver as measured by the Gay-Lussac-Volhard method. This was the standard prior to 1964, when ASTM issued a tentative specification for 99.90 wt% Ag, which was modified in 1966 to include a higher grade (see Table 2).

Analysis

Fire assaying is used on low grade, voluminous materials, and the silver is determined spectrographically or by atomic absorption. The gravimetric method is used for silver alloys containing no interfering base metals and ≤ 90 wt% Ag. Gay-Lussac-Volhard titration is used for silver alloys containing ≥ 90 wt% Ag.

If a sample is volatilized in an electric arc, the elements in the sample emit characteristic spectra by which they can be identified. Most precious metals, however, cannot be identified below 28 g/t in ores and concentrates. X-ray fluorescence analysis is simple, rapid, reliable, and

Table 2. Specifications for Silver, wt%

Method	Ag, min	Cu, max	Pb, max	Fe, max	Bi, max
Gay-Lussac-Volhard	99.90	0.08	0.025	0.002	0.001
ASTM	99.95	0.04	0.015	0.001	0.001

accurate for most noble metals. However, each type of sample, ie, ore, concentrate, alloy, etc, must be individually calibrated. Atomic absorption is fast, accurate, and economical.

Health and Safety Factors

A problem that may be encountered in the handling of silver is argyria, an irreversible, gray, or purple pigmentation of the skin, which is apparently harmless, although disfiguring.

Uses

Conventional photographic emulsions contain fcc AgBr and AgCl and may contain up to 10 wt% hexagonal AgI.

Silver is used in electrical contacts because of its high electrical and thermal conductivity and oxidation resistance.

Silver paints are used for producing electrically conducting surfaces on nonconductors, solder bonding and for decorating porcelain or glass.

Silver batteries include silver oxide–potassium hydroxide–zinc and silver chloride–seawater–magnesium primary batteries (qv).

With respect to modern brazing alloys, silver is present in five of the seven copper–phosphorus alloys, all of the seventeen silver–copper alloys, and four of the six silver–copper vacuum-tube alloys.

Silver catalysts are used most frequently in oxidation reactions. Mirrors are silver-coated by reducing AgNO$_3$ solutions.

Dental amalgam is the intermetallic silver–tin phase, that contains 67–70 wt% Ag, 25.3–27.7 wt% Sn, 0–5.2 wt% Cu, and 0–1.2 wt% Zn, plus mercury (see Dental materials). Silver has a bactericidal effect.

Lead-coated, silver-plated steel bearings are used in aircraft engine bearings (see Bearing materials).

GEORGE H. SISTARE
Consultant

S.C. Carapella, Jr. and D.A. Corrigan in D. Benjamin, ed., *Metals Handbook: Pure Metals*, 9th ed., Vol. 2, American Society for Metals, Metals Park, Ohio, 1979, pp. 794–796.

A. Butts and C.D. Coxe, *Silver—Economics, Metallurgy and Use*, D. Van Nostrand Co., Inc., Princeton, N.J., 1967.

F.A. Shunk, *Constitution of Binary Alloys*, 2nd Suppl. (Hansen), McGraw-Hill Book Co., Inc., New York, 1969.

SILVER COMPOUNDS

Silver is a white, lustrous metal, slightly less malleable and ductile than gold. It has high thermal and electrical conductivities. Most silver compounds are made from silver nitrate, which is prepared from silver metal.

Silver(I) Compounds

Silver acetate, CH$_3$CO$_2$Ag, is prepared from silver nitrate and acetate ion. Silver azide, AgN$_3$, is prepared by treating a solution of silver nitrate with hydrazine or hydrazoic acid. It is shock-sensitive. Silver acetylide or silver carbide, Ag$_2$C$_2$, is prepared by bubbling acetylene through an ammoniacal solution of silver nitrate. Silver carbonate, Ag$_2$CO$_3$, is produced by the addition of an alkaline carbonate solution to a concentrated solution of silver nitrate. Silver chromate, Ag$_2$CrO$_4$, is prepared by treating silver nitrate with a chromate salt. Silver cyanide, AgCN, forms when silver nitrate and a soluble cyanide are mixed.

Silver halides and other halogen salts. Silver chloride, AgCl, precipitates when chloride ion is added to a silver nitrate solution. Silver bromide, AgBr, is formed by the addition of bromide ions to a solution of silver nitrate. Silver iodide, AgI, precipitates as a yellow solid when iodide ion is added to a solution of silver nitrate. Silver fluoride, AgF, is prepared by treating a basic silver salt with hydrogen fluoride. All silver halides are reduced to silver by treating an aqueous suspension with more active metals.

Silver chlorate, $AgClO_3$, bromate, $AgBrO_3$, and iodate, $AgIO_3$, have been prepared. Silver perchlorate, $AgClO_4$, and periodate, $AgIO_4$, are well known, but the perbromate, $AgBrO_4$, has only recently been described. Silver tetrafluoroborate, $AgBF_4$, is formed from silver borate and sodium borofluoride or bromine trifluoride.

Other silver(I) salts. Silver nitrate, $AgNO_3$, is prepared by the oxidation of silver metal with hot nitric acid. The nitrite, $AgNO_2$, is prepared from silver nitrate and a soluble nitrite.

Slightly soluble or insoluble silver salts are precipitated when carboxylic aliphatic acids or their anions are treated with silver nitrate. Silver oxide, Ag_2O, a dark brown-to-black material, is formed when an excess of hydroxide ion is added to a silver nitrate solution.

Silver permanganate, $AgMnO_4$, is a violet solid formed when a potassium permanganate solution is added to a silver nitrate solution. Silver phosphate, Ag_3PO_4, is a bright yellow material formed by treating silver nitrate with a soluble phosphate or phosphoric acid. Silver selenate, Ag_2SeO_4, is prepared from silver carbonate and sodium selenate.

Silver sulfate, Ag_2SO_4, is prepared by treating metallic silver with hot sulfuric acid. Silver sulfide, Ag_2S, forms as a finely divided black precipitate when solutions or suspensions of most silver salts are treated with an alkaline sulfide solution. Silver sulfite, Ag_2SO_3, is obtained as a white precipitate when sulfur dioxide is bubbled through a solution of silver nitrate. Silver thiocyanate, AgSCN, is formed by the reaction of silver ion and a soluble thiocyanate. Silver thiosulfate, $Ag_2S_2O_3$, forms as an insoluble precipitate when a soluble thiosulfate reacts with an excess of silver nitrate.

Silver(I) complexes. In the presence of excess ammonia, silver-ion forms the complex ammine ions $Ag(NH_3)_2^+$ and $Ag(NH_3)_3^+$. Insoluble silver cyanide, AgCN, is readily dissolved in an excess of alkali cyanide. The predominant species in such solutions is $Ag(CN)_2^-$ with some $Ag(CN)_3^{2-}$ and $Ag(CN)_4^{3-}$. Silver halides form soluble complex ions, AgX_2^- and AgX_3^{2-}, with chloride, bromide, and iodide. Silver ion forms complexes with olefins and many aromatic compounds. Silver compounds other than the sulfide dissolve in excess thiosulfate.

Other Oxidation States

Silver(II) is stabilized by coordination with nitrogen heterocyclic bases. Silver(II) fluoride, AgF_2, is a brown-to-black, hygroscopic material obtained by the treatment of silver chloride with fluorine gas. Silver(II) oxide, AgO, is prepared by persulfate oxidation of Ag_2O in basic medium at 90°C or by the anodic oxidation of solutions of silver(I) salts.

No simple silver(III) compounds exist. When mixtures of potassium or cesium halides are heated with silver halides in a stream of fluorine gas, yellow $KAgF_4$ or $CsAgF_4$, respectively, is obtained.

Analytical Methods

The classic method for the identification of silver is precipitation as silver chloride. Silver in solution is determined by titration with thiocyanate. Gravimetrically, silver is determined by precipitation with chloride, sulfide, or 1,2,3-benzotriazole. Silver can be precipitated as the metal by electrodeposition or chemical reducing agents. A colored silver diethyldithiocarbamate complex is used for the spectrophotometric determination of silver complexes.

Health and Safety Factors

Silver compounds may cause adverse health effects. Chronic exposure to silver and silver compounds seems to affect only the skin. Argyria and argyrosis have resulted from therapeutic and occupational exposures to silver and its compounds. These disorders are characterized by deposition of a silver–protein complex in parts of the body. The affected areas show a blue-gray discoloration. The gradual accumulation of 1–5 g may lead to generalized argyria.

In 1980, the ACGIH adopted a TLV for airborne silver particles of 100 $\mu g/m^3$, and proposed a TLV of 10 $\mu g/m^3$, as silver, for airborne soluble silver compounds. The U.S. Public Health Service adopted an interim primary drinking standard of 50 $\mu g/L$.

Environmental Impact

In a standardized, acute aquatic bioassay, fathead minnows were exposed to various concentrations of silver compounds for a 96-h period, and the concentration of total silver lethal to half of the exposed population was determined. For silver nitrate, the value obtained was 16 $\mu g/L$.

Free ionic silver forms soluble complexes with substances present in natural waters. Because of the direct relationship between the availability of free silver ions and adverse environmental effects, the ambient freshwater criterion for the protection of aquatic life is expressed as a function of the water hardness. Silver discharged to secondary waste-treatment plants or into natural waters is present as a complexed or insoluble species.

Uses

The ability of silver ion to form sparingly soluble precipitates with many anions has been applied in quantitative determination (see Analytical methods).

Primary (nonrechargable) batteries containing silver compounds have gained in popularity in miniaturized electronic devices (see Batteries and electric cells, primary). Silver and silver compounds are widely used as catalysts for oxidation, reduction, and polymerization reactions (see Catalysis).

Silver iodide can initiate ice-crystal formation because, in the β-crystalline form, it is isomorphic with ice crystals. Cloud seeding with AgI is used for weather modification.

Most silver-plating baths employ alkaline solutions of silver cyanide. The use of silver mirrors results in a highly reflective coating.

Silver nitrate is used in medicine in the form of a stick containing 1–3% silver chloride, or in solutions of varying concentrations. A procedure still in use is the drop of 1% silver nitrate required in many states to be placed in the eyes of newborn infants as a prophylactic against ophthalmia neonatorum.

The largest single user of silver compounds is the photographic industry, where silver nitrate and a halide salt of an alkali metal or an ammonium halide give a light-sensitive silver halide, which can account for up to 30–40% of the total emulsion (see Photography). Photochromic glass contains silver chloride and silver molybdate (see Chromogenic materials).

HAINES B. LOCKHART, JR.
Eastman Kodak Company

I.C. Smith and B.L. Carson, *Silver*, Vol. 2 of *Trace Metals in the Environment*, Ann Arbor Science, Ann Arbor, Michigan, 1977, p. 12.

F.A. Cotton and G. Wilkinson, *Advanced Inorganic Chemistry*, Interscience Publishers, a division of John Wiley & Sons, Inc., New York, 1962, p. 642.

SIMULATION AND PROCESS DESIGN

Simulation is the use of one system to imitate another, ie, to simulate it. Simulation may be important because the second or subject system does not yet exist (the design problem), because the system is too time-consuming, dangerous or costly to operate itself to develop the necessary information (the cost problem), or because not enough is yet known about it (the information problem). Although simulation in theory embraces the use of any kind of physical system to imitate the one in question, the term has come almost universally to mean the use of a digital computer in exercising a mathematical model of the system under study.

A Description of Simulation

The principal steps in any simulation study are as follows: (*1*) A precise formulation of the mathematical model of the system to be studied; (*2*) conversion of the mathematical model into a computer program; (*3*) validation of the model and its representation in the computer program; (*4*) exercise of the model on the computer including the design of the experimental program; (*5*) use of the results to achieve the purpose of the study; and (*6*) recording of results.

Simulation makes it possible to investigate and experiment with the complex internal interactions of the system being studied. It allows investigation of the sensitivity of a system to small changes in parameters. Simulation affords complete control over time since a phenomenon may be speeded up or slowed down at will. In addition, the process of modeling the desired system for a simulation may often be more valuable than the simulation itself.

For certain types of stochastic or random-variable problems, the sequence of events may be of particular importance. Statistical information about expected values or moments obtained from plant experimental data alone may not be sufficient to completely describe the process. In these cases, computer simulations with known statistical inputs may be the only satisfactory way of providing the necessary information. At the same time, drawbacks or disadvantages of a simulation-based study should be kept in mind. For example, creation and exercise of a simulation model can be very expensive in terms of manpower and computer time.

Many systems of interest cannot be studied or evaluated adequately by standard mathematical or operations-research techniques (see Operations planning). These problems may arise for any number of reasons. Of course, direct experimentation on the physical system itself could perhaps supply the information needed, but only with the appearance of other difficulties.

The Development of Computer Simulation

Simulation was originally carried out with physical models. With the development of the analogue computer in the late 1940s and early 1950s, these computers became the primary medium for such simulation studies, particularly for aircraft and missile systems. However, the difficulty of programming the analogue computer resulted in their almost total replacement by digital computers by the 1970s.

The heart of every simulation study is the mathematical model. Since simulation studies must often be carried out on dynamic systems, such simulations require the solutions of sets of differential or difference equations. These form the mathematical model of the system being studied. Static or steady-state systems, on the other hand, are generally modeled and simulated as sets of algebraic equations.

Simulation models are in a sense input-output models which are run rather than solved to obtain the desired results. Thus, simulation is not a theory but a methodology of problem solving.

Dynamic Simulation Techniques

The models developed for simulation studies, particularly those for dynamic system studies, may be divided in two different ways, according to whether the output is completely determined by the input and initial values of the system or whether there is some inherent randomness in the system.

Continuous systems are those whose dependent variables can assume any value over a range or interval. Discrete systems are those whose variables can assume only integer values.

Discrete systems are idealized as network flow systems and are characterized by the following three factors: the system contains components, each of which performs prescribed functions; items flow through the system, from one component to another, requiring the performance of a function at a component before the item can move on to the next component; and components have a finite capacity to process the items.

The main objective in studying discrete systems is to examine their behavior and determine the capacity of the system. The analytical techniques used to solve such problems are queueing theory and stochas-

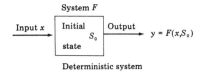

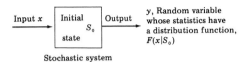

Figure 1. Comparison of deterministic and stochastic systems.

tic processes. Figure 1 illustrates the difference between a deterministic and a stochastic system. Any particular system may contain both deterministic and stochastic elements.

The single most vital step of the simulation study is model validation in order to assure credibility of the results. Another problem is that of determining when the simulated or the real system is in a true equilibrium or steady-state regime in its behavior.

The importance of dynamic simulation is reflected in the fact that over 20 special simulation languages for dynamic systems are listed in a recent survey. In addition the two classes of computer programming languages, continuous and discrete change, others are themselves self-contained methods of analysis and modeling. These languages are composed of groups of subroutines programmed in a common base language such as FORTRAN. With such wide possibilities, the choice of the best language for a particular simulation task becomes difficult indeed.

Dynamic simulations are used in the chemical processing industry for several types of problems. The kinetic parameters for mathematical models of complex chemical reaction systems have been determined through application of parameter-identification techniques against the composition-change history of experimental runs of real reactions.

Steady-State Simulations

A principal use of simulation in chemical engineering has been for process design and operation studies. Although these have involved the use of both steady-state and dynamic simulations, the former has predominated. The great importance of these types of computations to chemical engineering has resulted in the development of special languages or computer programs for such use.

The most widely used of the steady-state simulation programs is probably FLOWTRAN, made widely available to the educational community through the CACHE (Computer Aids for Chemical Engineering Education) Committee. FLOWTRAN has been greatly expanded in the ASPEN program.

Problem Solution — The Experimental Phase

Once the model and computer program have been validated, the investigator's work has often only just begun. The design of the necessary simulation experimental program is vital. The design of a computer simulation experiment is essentially a plan for covering the range of parameter values of interest in its most expeditious or inexpensive manner while assuring statistical validity of the results. Such designs are also valuable for process-optimization studies since they allow closure of the optimization in minimum time (see Design of experiments).

Differences between experimentation and simulation. Although the underlying objectives of computer simulation experiments are essentially the same as for physical experiments, some differences must be considered. Among the more important of these are the difficulties in defining a single datum point or sample; the ease with which experimental conditions can be repeated or reproduced; the ease of stopping and resuming experimentation; the presence or absence of correlation between subsequent data points; and the ability to control stochastic variability which in physical experiments is beyond the control of the experimenter.

The ease with which experimental conditions can be repeated or reproduced in a computer model is often a distinct advantage of computer simulation over physical experimentation.

In most complex simulation studies, the number of possible combinations of factors and factor levels of interest is almost infinite; hence, a large number of design trade-offs must be made to stay within the resource constraints.

THEODORE J. WILLIAMS
Purdue University

C.L. Johnson, *Analog Computer Techniques*, McGraw-Hill Book Company, Inc., New York, 1956.

W.L. Luyben, *Process Modeling*, McGraw-Hill Book Company, Inc., New York, 1973.

D.D. Pegden and A.A.B. Pritsker, *Introduction to Simulation and SLAM*, John Wiley & Sons, Inc., New York, 1979.

J.D. Seader, W.D. Seider, and A.C. Pauls, *FLOWTRAN Simulation, An Introduction*, 2nd ed., CACHE Committee, Ulrich's Bookstore, Ann Arbor, Mich., 1977.

SIMULTANEOUS HEAT AND MASS TRANSFER

Heat transfer and mass transfer occur simultaneously whenever a transfer operation involves a change in phase or a chemical reaction. Of these two situations, only the first is considered here.

Interphase transfer occurs when one or more components change phase in the presence of inert or less active components. For the condensation of a single component from a binary gas mixture, the gas-stream sensible heat and mass-transfer equations for a differential condenser section take the following forms:

$$G \cdot C_p \frac{dT_G}{dA} = - h_g \cdot \left(T_g - T_s \right) \frac{\epsilon}{e^\epsilon - 1}$$

$$\frac{dV}{dA} = - k_g \cdot \left(p_g - p_s \right)$$

Latent and sensible heat transfer to the surface from the condensing vapor affects the coolant temperature:

$$L \cdot M_L \cdot C_w \frac{dT_w}{dA} = \pm h_o \cdot \left(T_s - T_w \right)$$

These basic relations have been solved for a wide range of cooler-condenser conditions and for different complexities of systems to permit condenser design. A procedure based on the assumption that the mixture is saturated throughout the condensation process has been developed. The same approach extended to superheated mixtures has been used to develop the following equation for calculating T and partial pressures during condensation:

$$\frac{dp_g}{dT_g} = \frac{p - p_g}{(Le)^{2/3} \cdot p_{BM}} \cdot \frac{p_g - p_s}{T_g - T_s} \cdot \frac{e^\epsilon - 1}{\epsilon}$$

This relation was tested experimentally for water condensation in various gases and was found to be acceptable.

In considering the effect of mass transfer on the boiling of a multicomponent mixture, both the boiling mechanism and the driving force for transport must be examined. The boiling mechanism can conveniently be divided into macroscopic and microscopic mechanisms. The macroscopic mechanism is associated with the heat transfer affected by the bulk movement of the vapor and liquid. The microscopic mechanism is that involved in the nucleation, growth, and departure of gas bubbles from the vaporization site.

Gas–Vapor Systems

In engineering applications, the transport processes involving heat and mass transfer usually occur in process equipment involving vapor–gas mixtures where the vapor undergoes a phase transformation such as condensation to or evaporation from a liquid phase. The system of

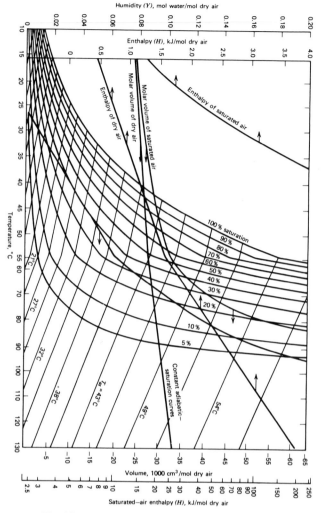

Figure 1. Humidity chart for the air–water system, molal quantities. To convert kJ to Btu, divide by 1.054; to convert cm³ to ft³, multiply by 35.31×10^6.

primary interest is that of a condensable vapor moving between a liquid phase, mostly pure, and a vapor phase in which other components are present.

For the air–water system, the humidity is easily measured with a wet-bulb thermometer. The balance between heat transfer to the wick covering the thermometer bulb and the energy required by the latent heat of the mass transfer from the wick at steady state gives:

$$- k_Y \cdot A \cdot \left(Y_l - Y_w \right) \cdot \lambda_w = \left(h_c + h_r \right) \cdot A \cdot \left(T_l - T_w \right)$$

$$T_l - T_w = \frac{kY\lambda_w}{\left(h_c + h_r \right)} \left(Y_w - Y_l \right)$$

From this relationship, a humidity diagram (Fig. 1) for the air–water system can be constructed and used. Often, such a diagram is extended to include such properties as specific volume and enthalpy.

For systems other than air–water vapor or for total system pressures different from 101.3 kPa (1 atm), humidity diagrams can be constructed if basic phase-equilibria data are available. The simplest of the phase equilibrium relations is Raoult's law:

$$P \cdot y_s = p_i^\circ \cdot x_i$$

which for a two-component system, where one component exists only in the vapor phase, reduces to

$$y_s = \frac{H_s}{1 + H_s} = \frac{p_i^\circ}{P}$$

Calculations for Humidification and Dehumidification Processes

The temperature of the vapor phase is important to the designer of humidifiers, either as one of the variables specified or as an important indicator of fogging conditions in the column. Such a condition would occur if the gas temperature equaled the saturation temperature, ie, the interface temperature.

If a steady state is obtained, measurement of air- and water-flow rates, water-inlet and -outlet temperatures, and air-inlet and -outlet wet- and dry-bulb temperatures comprises all the information needed to evaluate the unit performance and to check for fogging or other malfunctions. Overall coefficients of heat and mass transfer are frequently available (see also Air conditioning).

Humidification and Dehumidification Equipment

The addition of a condensable component to, or its removal from, a noncondensable gas can be accomplished by direct contact between the vapor and the gas in a countercurrent or cross-flow tower. The direction of transfer depends upon the relative temperatures of the two streams. If the air is relatively dry, liquid evaporates into it, both humidifying the air and cooling the liquid. Spray ponds can also be used. In these, a grid of nozzles sprays liquid, usually water, into the gas phase, usually air. Humidification processes also occur in spray contactors used to scrub minor components from a gas stream. Here the gas passes through successive sprays of liquid.

Water-cooling towers. By far the most common large-scale mode of humidification processing is in water-cooling towers. As supplies of cooling water become more strained, and as discharge-water temperatures are more closely controlled, water cooling and recirculation become a necessity more often. Typically, the tower consists of a set of louvres and baffles over which the water falls, breaking into films and droplets. Air flow may be across this cascading liquid or countercurrent to it. The thermal design of cooling towers follows the same general procedures presented above (see also Water, industrial water treatment).

Cooling-tower design is usually done by finite-difference solutions to the mass- and heat-transfer equations presented above, plus fundamental equations of fluid mechanics and transfer rates.

Though approximate methods are no longer needed for design work, they are still useful for rapid estimates of the effects of changing conditions on performance. Using Merkel's approximation and knowing the desired thermal performance, the flow rates and transfer coefficient can be quickly calculated.

In a natural-draft cooling tower, the driving force for the air is provided by the buoyancy of the air column in a very tall stack. Such movement results from fans and mechanical-draft towers.

Natural-draft cooling towers are extremely sensitive to air-inlet conditions, owing to the effects on draft. As the air-inlet temperature approaches the water-inlet temperature, the allowable heat load decreases rapidly. For this reason, natural-draft towers are unsuitable in many regions of the United States.

Until recent years, natural-draft cooling towers were rare in the United States, since ample water supplies were available for power-plant cooling, and natural-draft towers are best suited for large heat loads. However, there has been a dramatic change, and almost all new, large power plants require cooling towers. In the southern United States, mechanical-draft towers dominate the field.

When all the expenses involved in using wet cooling towers in a power plant are considered, it appears that a mechanical-draft cooling tower system may raise the cost of generating electricity ca 3%, and a natural-draft tower ca 6% over the generating cost of a direct river-cooled power plant.

An important consideration in the acceptability of either a mechanical-draft or a natural-draft tower cooling system is the effect on the environment.

Not only may the cooling-tower plume be a source of fog, which in some weather conditions can ice roadways, but it carries salts from the cooling water itself. These salts may come from salinity in the water, or they may be added by the cooling-tower operator to prevent corrosion and biological attack in the column.

The natural-draft tower is much less likely to produce fogging than the mechanical-draft tower. Nonetheless, it is desirable to devise techniques for predicting plume trajectory and attenuation.

Much work has been done on the modeling of wet cooling-tower plumes with the aim of determining their effect on the environment. Plume modeling owes much to meteorology and especially to the theory of cumulus clouds.

The recirculation of cooling water via a cooling tower ultimately removes process heat by evaporating water rather than by warming it, as could be the case with once-through systems. When water is especially scarce, it may be necessary to cool process water by transferring the heat to air through indirect heat transfer. This requires dry cooling towers, which, though they are relatively expensive, have been built in a few dry regions of the United States.

Nomenclature

A = interfacial area
C_p = heat capacity, at constant pressure
C_w = heat capacity of water
G = mass flow rate of gas phase, $kg/(h \cdot m^2)$
H = enthalpy
h = heat transfer coefficient
h_c = convective heat transfer coefficient
h_r = coefficient for heat transfer by radiative mechanism
k_g = gas phase mass transfer coefficient in partial pressure driving force units
k_Y = mass transfer coefficient in gas phase mole ratio units
L = liquid stream molar flow rate
Le = Lewis number, Pr/sc
M = molecular weight
P = total pressure
p = partial pressure of condensable components
$p°$ = vapor pressure
T = temperature
V = gas phase molar flow rate
x = mole fraction in liquid phase
Y = mole ratio
y = mole fraction in gas phase
ϵ = Ackerman correction term, $\epsilon = m_i C_{p_i}/h_g$
λ = latent heat of vaporization

Subscripts
BM = mean value for noncondensing component
g = gas phase
i = interface condition
L = liquid phase
o = at reference condition
s = at saturation
w = for water, or at wet-bulb temperature

LEONARD A. WENZEL
Lehigh University

R.E. Treybal, *Mass Transfer Operations*, 2nd ed., McGraw-Hill, Inc., New York, 1968, pp. 176–220.

D.W. Green, ed., *Perry's Chemical Engineers Handbook*, 5th ed., McGraw-Hill, Inc., New York, 1984.

A.S. Foust and co-workers, *Principles of Unit Operations*, 2nd ed., John Wiley & Sons, Inc., New York, 1980, Chapt. 17, pp. 420–453.

SIZE ENLARGEMENT

Size-enlargement processes bring together fine powders into larger masses in order to improve the powder properties. Modern high volume processing requires consistent feeds with good flow properties, requirements that for powders can often only be met through some form of agglomeration.

Particle-Bonding Mechanisms

Based on statistical-geometrical considerations, Rumpf developed the following equation for the mean tensile strength of an agglomerate in which bonds are localized at the points of particle contact:

$$\sigma_T \approx \frac{9}{8}\left(\frac{1-\epsilon}{\epsilon}\right)\frac{H}{d^2}$$

where σ_T is the mean tensile strength per unit section area, Pa; ϵ is the void fraction of the agglomerate; d is the diameter of the (assumed) monosized spherical particles, m; and H is the tensile strength, N, of a single particle–particle bond. (To convert Pa to psi, multiply by 0.145×10^{-3}).

In practice, simple and quick test methods are used to assess the quality of bonding and other desirable properties of product agglomerates. Compression tests, in which agglomerates are crushed between parallel platens, are probably most universal. For approximately spherical agglomerates, compression strength is calculated as follows:

$$\sigma_C = L/(\pi D^2/4)$$

where L is the compression force at failure in N, and D is the agglomerate diameter in m. Several other tests of agglomerate quality are done routinely, the details depending on the practice accepted in a specific industry.

Classification of size-enlargement methods reveals two distinct categories. The first comprises forming-type processes in which the shape, dimensions, composition, and density of the individual larger pieces formed from finely divided materials are of importance. The second comprises processes in which creation of a coarse granular material from fines is the objective and the characteristics of the individual agglomerates are important only in their effect on the properties of the bulk granular product. Only the latter methods are considered in detail here.

Tumbling and Other Agitation Methods

When fine particles, usually in a moist state, are brought into intimate contact through agitation, binding forces come into action to hold the particles together as an agglomerate. Agglomerate growth can occur through coalescence, crushing and layering, layering of fines, and abrasion transfer. More than one mechanism may operate at the same time. To survive and grow in an agitated system, agglomerates must be able to withstand the destructive forces generated by the moving charge of powder.

Although a wide variety of agitation equipment is used industrially to produce agglomerates, rotary drums or cylinders and inclined disks or pans are the most important.

Drum agglomerators consist of an inclined rotary cylinder powered by a fixed- or variable-speed drive (see Fig. 1). The drums used for fertilizer granulation range from 1.5 m in dia by 3-m long up to 3-m dia by 6-m long. A typical drum used to ball iron ore is 3 m in dia by 9.5 m long. An inclined-disk agglomerator consists of a tilted rotating plate equipped with a rim to contain the charge.

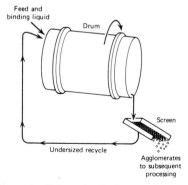

Figure 1. Schematic of a drum agglomerator.

Operation and control of tumbling agglomerators is affected primarily by the character of the feed powder, its optimum liquid content for agglomeration, and the retention time.

A most important feature of the inclined-disk agglomerator is its size-separating ability. The inherent classifying action of inclined disks offers an advantage in applications that require accurate agglomerate sizing.

Advantages claimed for the drum compared with the disk agglomerator are greater capacity, longer residence time for difficult materials, and less sensitivity to upsets due to the damping effect of a larger recirculating load.

Mixer agglomerators. Various industries use mixers in which, in contrast to drum or disk equipment, internal agitators of several designs provide a positive rubbing and shearing action to accomplish both mixing and size enlargement. Pug mixers (blungers, pugmills, paddle mixers) have been widely used in the granulation of fertilizer materials. These mixers consist of a horizontal trough with a rotating shaft to which mixing blades or paddles of various designs may be attached. An intensive countercurrent pan mixer can be used to homogenize feed powders while adding binding liquid to help pregranulate extreme fines before pelletizing. Shaft mixers operating at very high rotational speeds are also used to granulate fines, such as clays and carbon black, which may be highly aerated when dry, and plastic or sticky when wet.

Powder clustering. Many applications of size enlargement require only relatively weak, small, cluster-type agglomerates to improve behavior of the original powder in flow, wetting, dispersion, or dissolution. Tableting feeds in pharmaceutical manufacture, detergent powders, and "instant" food products are examples. Continuous-flow mixing systems are commonly used in the agglomeration of powdered food products.

Pressure Compaction and Extrusion Methods

Compression techniques produce agglomeration by applying suitable forces to particulates held in a confined space. In extrusion systems, the particulates undergo shearing and mixing as they are consolidated while being forced through a die or orifice under the action of a screw or roller. Tableting, pressing, molding, and extrusion operations are commonly used to produce agglomerates of well-defined shape, dimensions, and uniformity.

The compaction process of void reduction may be considered to occur by two mechanisms. The first is the filling of the holes of the same order of size as the particles. The second process consists of the filling, by plastic flow or fragmentation, of voids that are substantially smaller than the original particles. The success of the compaction operation depends partly on the effective utilization and transmission of applied forces and partly on the physical properties and condition of the mixture being compressed.

Roll briquetting and compacting machines. In roll presses, particulate material is compacted by squeezing as it is carried into the gap between two rolls rotating at equal speed. This is probably the most versatile method of size enlargement. Roll presses consist of a frame, two rolls, and the associate bearing, reduction gear, and fixed or variable-speed drive. For fine powders that tend to bridge or stick and are of low bulk density, some form of forced feed must be used.

Pellet mills. Unlike roll briquetting and compacting machines, pellet mills compress and form particulates into agglomerates by extrusion through a die rather than by squeezing. The roller-and-die assembly produces a shearing and mixing action which yields a plastic mix that is pushed through the die.

Heat Reaction, Fusion and Drying Methods

These methods of size enlargement depend on heat transfer to accomplish particle bonding. Agglomeration may occur by supplying heat, as in the drying of a concentrated slurry or paste, the fusion of a mass of fines, or chemical reaction among particles at elevated temperatures. Alternatively, heat may be removed from the material to cause agglomeration by chilling. A wide variety of preagglomeration equipment is used to preform powders and pastes into agglomerates suitable for drying, firing, or chilling.

Table 1. Product Characteristics and Capacity Data for Some Materials Treated in Belt Cooling Systems[a]

Product	Thickness, mm	Feed temp, °C	Discharge temp, °C	Capacity, kg/(h·m²)
phenolic resins	1.2–1.3	138	33	277
sulfur	6.4	143	66	269
asphalt	3.2	218	52	90
urea	2.4	191	60	190
ammonium nitrate	1.6	204	71	439
chlorinated wax	1.6	149	38	303
sodium acetate	3.2	82	38	183
hot-melt adhesive	11.1	166	39	70
wax blend	0.6	132	29	129
epoxy resin	1.0	177	38	195

[a] Courtesy of Sandvik Conveyor, Inc.

In extractive metallurgy, sintering and pelletizing processes have been developed to allow processing of fine ores, concentrates, and recyclable dusts. Four separate sequential processes take place during these high temperature operations: drying, preheating, firing or high temperature reaction, and cooling.

Drying and solidification on surfaces. In this type of equipment, granular products are formed directly from fluid pastes and melts, without intermediate preforms, by drying or solidification on solid surfaces. Surfaces formed by single or double drums are common. Drum dryers consist of one or more heated metal rolls on which solutions, slurries, or pastes are dried in a thin film. Molten materials can also be cooled to solid products on endless-belt systems (see Table 1), (see Drying).

Suspended-particle techniques. In these methods, granular solids are produced directly from a liquid or a semiliquid phase by dispersion in a gas to allow solidification through heat and mass transfer. In spray drying (see Sprays), the largest particles produced are normally ca 1 mm in dia. The pharmaceutical and ceramics industries employ this technique to produce granular dried peanuts.

Spray cooling or solidification, more commonly known as prilling, is similar to spray drying; the liquid feed is dispersed into droplets at the top of a chamber. However, the liquid droplets are produced from a melt which solidifies primarily by cooling in the chamber with little, if any, drying.

Agglomerations from Liquid Suspensions

Size enlargement of particles contained in liquids is a frequent aid to other operations such as filtration, dewatering, settling, etc. Flocculation procedures are the traditional means used to promote such size enlargement (see Flocculating agents).

Fine particles in liquid suspension can readily be formed into large dense agglomerates of considerable integrity by adding a second or bridging liquid under agitation.

The sol-gel process is a related technique for the preparation of spherical oxide fuel particles, up to ca 1 mm in dia, for nuclear reactors.

Flocculation improves settling, drainage, and filtration characteristics of the aggregates. On the other hand, flocculated particles form a loose, bulky layer with relatively large pores retaining a larger proportion of suspending medium than untreated particles. Pellet flocculation eliminates this problem.

C.E. CAPES
National Research Council of Canada

D.F. Ball, J. Dartnell, J. Davison, A. Grieve, and R. Wild, *Agglomeration of Iron Ores*, Heinemann, London, UK, 1973.

H. Rumpf in W.A. Knepper, ed., *Agglomeration*, Interscience Publishers, a division of John Wiley & Sons, Inc., New York, 1962, pp. 379–414.

C.E. Capes, *Particle Size Enlargement*, Elsevier, Amsterdam, Neth., 1980.

T.P. Hignett in V. Sauchelli, ed., *Chemistry and Technology of Fertilizers*, Reinhold, New York, 1960, Chapt. 11, pp. 269–298.

SIZE MEASUREMENT OF PARTICLES

Particle size influences the combustion efficiency of powdered coal and sprayed liquid fuels, the setting time of cements, the flow characteristics of granular materials, the compacting and sintering behavior of metallurgical powders, and the hiding power of pigments. The size of a particle is best expressed in terms of a single, linear dimension, eg, a $50-\mu m$ particle, but a problem arises in selecting the appropriate dimension. The common solution is to characterize particle action in terms of an equivalent simple shape, generally that of a sphere.

Data Representation

Systems composed of identical particle sizes are extremely rare. Normally, a distribution of individual sizes is encountered. Mean diameters are commonly employed in terms of the number of particles having a particular diameter. Descriptive designations involve two and only two parameters of a particle system (see Table 1).

Presentation of particle-size information in useful tabular form requires data classification. It may be sufficient to group the measurements in linear intervals, and then list the intervals as a function of the corresponding percent, or fraction, of the whole that each interval represents.

Grouping into linear size intervals results in a serious loss of information and is unsatisfactory when the range of sizes is large because the difference between 1 and 2 μm, for example, is much more significant than the difference between 100 and 101 μm. In this case, data are best classified on a geometric scale. Although a tabular presentation offers the ultimate in precision since all data are included, it is inconvenient to compare tables; a graph usually offers advantages.

Mathematical. The maximum of useful information is revealed when particle-size data are summarized by a mathematical expression. Its application is simplified if it can be linearized.

The Gaudin-Schuhmann distribution often gives a remarkably good fit to data generated for crushed minerals by sieving. If $Y_m(x)$ is the weight of particles finer than size x, this distribution function is expressed by:

$$Y_m(x) = \frac{x^b}{k}$$

where k is a size modulus and b a distribution modulus.

The normal or Gaussian distribution equation is very commonly utilized in statistical applications. It is appropriate for particle-size data only in the case of very narrow distributions or where a sharp cut has been extracted from a wider range of sizes. Computer programs have been written to aid the fitting of data to both normal and log-normal distribution functions.

The logarithmic distribution function is expressed by:

$$y(x) = \frac{2.303}{x\sqrt{2\pi}\log\sigma_g} \exp\left[-\frac{(\log x - \log\bar{x})^2}{2\log\sigma_g}\right]$$

where σ_g is the geometric standard deviation and $\bar{x}$ is the geometric mean size. All other diameters are readily calculated once any geometric

Table 1. Mean Diameters

Descriptive designation	Common designation	Mathematical expression[b]
number–length	arithmetic, number	$\Sigma nd/\Sigma n$
number–surface	surface diameter	$(\Sigma nd^2/\Sigma n)^{1/2}$
number–volume	volume diameter	$(\Sigma nd^3/\Sigma n)^{1/3}$
length–surface	length	$\Sigma nd^2/\Sigma nd$
length–volume	volume diameter	$(\Sigma nd^3/\Sigma nd)^{1/2}$
surface–volume	surface	$\Sigma nd^3/\Sigma nd^2$
volume (or weight)–moment mean[a]	volume	$\Sigma nd^4/\Sigma nd^3$

[a] Value is identical for volume and weight if all particles in system are of identical density.
[b] n = number; d = diameter.

mean diameter and the geometric standard deviation are established for a particle system.

The log-probability function frequently fails to describe adequately a particle-size distribution because of deviations near the upper extreme of the size range.

Methods

Different methods have been developed for particle-size measurement. *In situ* analysis is sometimes required, whereas in other situations measurement a distance away from the point of origin is acceptable.

Sieving. the most widely employed sizing method determines particle size by the degree to which a powder is retained on a series of sieves of different dimensions. A typical sieve is a shallow pan with a wire mesh bottom or an electroformed grid. Opening dimensions in any mesh or grid are generally uniform within a few percent. Dry-sieving is typically performed on a stack of sieves having openings diminishing in size from the top downward. The lowest pan has a solid bottom to retain the final undersize. Wet-sieving is performed on a stack of sieves in a similar manner except that water or another liquid that does not dissolve the material is applied to facilitate particle passage. A detergent is frequently added.

Microscopic examination. Examining particles one at a time by the aid of magnification is an obvious means for assessing size. Several magnifications are usually required.

Optical microscopy is used for particles larger than ca 1.0 μm in diameter. The particles need to be spread evenly on a microscope slide with considerable free space between. The field of view should contain a maximum of 30 particles. Electron microscopy is applicable to particles of diameters from ca 0.002–15 μm; the upper limit is set by the viewing field. Microscopy is applied when particle identification and, perhaps, shape evaluation are important in addition to size. Such uses include pollution or contamination assessment and forensic studies (see Forensic chemistry).

Sedimentation. Measurement of the rate at which particles move under gravitational or centrifugal acceleration in a quiescent liquid provides a most important means for determining particle sizes. In a simple sedimentation method, a pan is suspended from a sensitive balance. The particles accumulate in this pan as a function of time as they settle under gravity from a well-dispersed suspension. The concentration of particles remaining in suspension is readily calculated from density measurements made with a hydrometer.

Using a collimated beam of x rays permits particle-concentration detection as a function of mass without disturbing the suspension. Another version employs light instead of x-ray transmission to determine concentration. The technique is generally referred to as turbidity or photoextinction. X-ray sedimentation analysis gives quick well-defined results.

Centrifugal force in sedimentation analysis permits evaluation of smaller diameters but adds mechanical complexity. Disk centrifuges operate with both a vertical and horizontal axis of of rotation. The disturbances arising from acceleration have largely been eliminated in a centrifuge that employs a single cylindrical chamber rotated with its axis at right angle to and about three chamber heights away from the axis of rotation.

Sensing-zone methods. Sensing-zone methods produce size information as a result of particles being directed, one at a time, into a region where their presence affects a measurable condition, such as electrical conductivity; magnetic flux; light extinction, diffraction, and scattering; acoustic response; and thermal stability. Such systems require calibration.

The number and size of particles in an electrically conducting liquid may be determined by passing the particles through a small, short, nonconducting aperture, on either side of which is an immersed electrode.

The sensing-zone method is employed most widely in medical applications involving blood cells, bacteria, bacteriophages, etc; it is also extensively used by the food, beverage, and pharmaceutical industries.

Analysis of the light absorbed, diffracted, refracted, and scattered from a restricted beam of white light, or from a laser beam as either

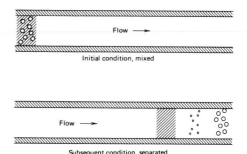

Figure 1. Hydrodynamic chromatography effect in a capillary tube.

liquid- or gas-borne particles pass through, provides the basis for another type of particle-sizing instrumentation.

Instruments designed to distinguish size on the basis of the light extinguished by particles passing one at a time through a small lighted orifice are particularly useful for the determination of particle contamination in lubricating oils and hydraulic fluids. Expressions for light scattering from particles that are not spherical have been worked out only for special cylinders, disks, and ellipsoids. True size data are only obtained when the complete scattering pattern of a particle is analyzed in detail.

Other sensing-zone techniques include the momentary emission of visible light when an airborne particle encounters an intense laser beam.

Fluid dynamic. A noteworthy hydrodynamic unit incorporates a helical path in which radial stratification of particles occurs. Other devices utilize a series of small hydroclones. Devices employing these techniques are found primarily in mineral dressing and deep-well operations. Discrete particle-size separation results from a transport phenomenon generally described as hydrodynamic chromatography (see Fig. 1). Analysis of particles of 1–5 μm dia appears feasible with capillaries, whereas packed columns extend the range from ca 1 μm into colloidal dimensions.

Another broad class of size-separation techniques, termed field-flow fractionation, may also result in commercial instrumentation. Submicrometer particle separation is achieved by applying a field of force at right angles to a liquid migration carrying the particles.

In other devices, the average diameter of submicrometer particles is deduced from their Brownian motion. The frequency of this motion is inversely proportional to particle size.

Ultrasonic-energy attenuation provides particle-size information in a device that utilizes two pairs of receiving and detecting transducers; one pair operates at a frequency to emphasize scattering loss (see Ultrasonics).

CLYDE ORR
Micrometrics Instrument Corp.

J.K. Beddow, *Particulate Science and Technology*, Chemical Publishing Co., Inc., New York, 1980.

T. Allen, *Particle Size Measurement*, 3rd ed., Chapman and Hall, London, UK, 1981.

SIZE REDUCTION

Size reduction, or comminution, is the process whereby particulate materials are mechanically reduced in size. Size reduction to produce a material suitable for use is an important operation in many process industries, eg, cement, grain, stone, fertilizer, and coal. In others, eg, mineral ores, size reduction is an intermediate step in the separation of valuable constituents from waste materials (see Extractive metallurgy; Flotation; Gravity concentration).

Comminution processes are energy-intensive and size-reduction devices are inefficient and costly to operate (frequently $\geq$ 25% of total raw-material processing costs). It has been estimated that < 1% of the energy input to tumbling mills, for example, manifests itself as new surface. Equipment for a particular application must be chosen with regard to energy efficiency. The choice is dictated largely by the nature of the material to be comminuted and by the size desired. In the mineral industries, coarse size reduction (from large sizes to 1 cm) is normally achieved by explosive shattering followed by crushing, where individual pieces are broken by compression or impact. Fine size reduction (from 1 cm to 10 μm) is typically achieved by grinding in tumbling mills, where breakage occurs by probabilistic impact in a loosely tumbling mass of grinding media. Ultrafine size reduction ($< 10 \mu$m) is achieved mainly by pulverizing, where size reduction results from abrasion and attrition.

Theory

Solids may be homogeneous or heterogeneous, crystalline or amorphous, hard or soft, brittle or plastic. However, most of the commercially important materials are heterogeneous, flawed but essentially brittle solids that fail catastrophically when stressed to a sufficient degree.

Most comminution machines work through compression or impact. However, when bodies are subjected to compression, tensile forces are set up at right angles to the imposed compressive stress.

Orders of precedence for the type of fracture likely to occur in a homogeneous material can be developed on the relative magnitudes of its ultimate tensile, compressive, and shear strengths. Although very few of the materials subjected to comminution are homogeneous, most naturally shaped particles break under tension or shear. Ultimate shear and tensile strengths are usually very similar, and it is not particularly important which predominates from an energy viewpoint.

The dissipation of energy as the material fractures can take the form of kinetic energy, sound, potential strain, heat, light, and the energy of the new surfaces generated by fracture. Of these various forms, only the increase in surface energy bears a direct relationship to the ultimate objective of comminution, namely, the reduction of particle size.

The only way to determine exactly the degree of size reduction that can be achieved in a particular comminution device is to carry out full-scale tests on a representative sample of the material. Nevertheless, considerable success has been achieved in empirically predicting the performance of full-scale equipment from laboratory and pilot-scale tests. More recent methods of analysis show promise of developing into true representations of comminution processes. However, these have not as yet supplanted the empirical laws, principally because of an insufficient data base.

Of the various comminution theories proposed since the mid-1800s only that of Bond is widely used today. Bond's "third theory of comminution," based on a detailed analysis of a large number of operating and laboratory data, states that:

$$W = W_i \left(\frac{10}{\sqrt{P}} - \frac{10}{\sqrt{F}} \right)$$

where W is the required work input, P and F are, respectively, the 80% passing sizes of the product and feed particles, and W_i is the "work index," representing the power required to comminute an infinitely large particle to 80% passing 100 μm (in kW$\cdot$h/t). This relationship is widely used for the design of grinding circuits because a very large body of information on W_i values exists. In addition, experimental laboratory methods have been developed to determine work indexes for a particular material and comminution device. This parameter can then be used, along with equation 1, to size ball and rod mills. It can also be used for crusher specifications, although these machines are more typically selected from manufacturers' catalogues.

Comminution Mechanics

The principal limitation of the energy relationships discussed above is that they fail to account for the mechanics of the comminution device. Any such device must provide a means for moving material into a stress-application zone, applying stress to the material and then permitting the material to exit. Thus, a detailed study of a given comminution device must include the breakage process and machine dynamics.

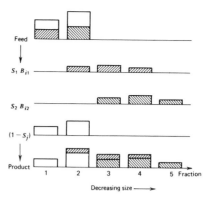

Figure 1. Schematic representation of the comminution process.

The breakage process. Comminution can be treated as a process of transforming particles from a given size interval into a mass distribution spread throughout a set of finer size intervals (see Fig. 1). The parameters of the process can then be defined as S_j, the specific rate, with respect to time or energy input, at which particles in size class j are broken; and b_{ij}, the mass fraction of the fragments produced by breaking size j, which fall into size class i.

Population-balance models developed to describe the different machines generally utilize a first-order hypothesis which states that the rate of breakage of any size is proportional to the mass of that size present in the comminution zone of the particular device. There is no self-evident reason why this should be so, but the first-order assumption is valid to a first approximation for a wide variety of practical systems.

So far these models have been of limited use in the design and operation of industrial comminution systems because of an inadequate data base and a lack of good scale-up laws. However, because of significant progress, it seems likely that this approach will supplant traditional empirical approaches within the decade, especially for devices other than ball mills where the empirical approach has not been very successful.

Machine dynamics. A commercial comminution device must perform both material breakage and material transport and the capacity of a given machine may be controlled by either. Generally, the relative energy consumption for material transport increases as the complexity of this function increases.

The relative efficiency of comminution equipment tends to decrease as the size of material decreases because of the difficulties of effectively applying stress to fine materials. Operating ranges and specific energy requirements for a variety of common comminution devices are given in Figure 2.

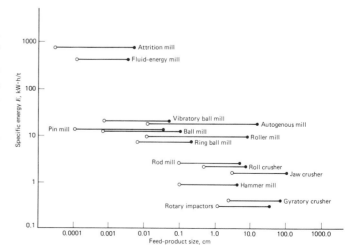

Figure 2. Average energy requirements for several size-reduction devices. ●, typical feed size; ○, typical product size.

Equipment

Equipment can be categorized on the basis of primary comminuting action, namely, compression or nipping, blow or impact, tumbling or projection, cutting or shredding, and attrition. Compression or nipping devices include disk pans, roller mills, and jaw, gyratory, and roll crushers. These devices are subdivided into those in which the gap between the crushing surfaces is controlled (eg, the jaw crusher) and the force is allowed to vary, and those in which a fixed force is used (eg, the roller mill) and the gap between the crushing surfaces is permitted to vary.

Impact mills operate by subjecting particles to sudden high stress through impact. The kinetic energy generated by each impact may also give rise to breakage of particles against the walls of the mill. Particles are broken in vibratory mills by high speed compression between two surfaces (balls).

Tumbling mills, the most important comminution devices, are a special subclass of impact devices where particles are broken by the action of a tumbling mass of loose grinding media. The mechanics of breakage in these devices is still not fully understood. Cutting and shredding devices operate through the action of a knife blade that can be moving or stationary. Attrition mills are generally used for ultrafine grinding.

Several other specialized methods of comminution are used or have been proposed but cannot be classified conveniently in the above categories. Some of these methods are used on a large scale, whereas others are not fully developed. These techniques include weathering, thermal shock, ultrasonics, electrical methods, pressure alteration, chemical methods, and explosion.

Selection criteria. The selection of comminuting equipment relies on many elements. The size of the feed material often determines the choice of the primary comminuting device. The top size determines the feed opening required, whereas the size distribution determines the flow through the comminuting machine and affects the reduction ratio. The latter relates feed size to product size and is usually specified as $R_R = F_{80}/P_{80}$. All comminuting equipment operates within a range of possible reduction ratios. The rate at which a material is to be comminuted obviously influences equipment selection.

Compression devices. Jaw crushers, used for hard, abrasive materials, crush virtually any mineral. Crushing is performed by compression without rotary motion. Jaw crushers are rated according to their receiving area, which is the width of the plates times the gape (the distance between the plates).

Gyratory crushers consist of a solid cone set on a revolving shaft located within a hollow shell. The cone revolves eccentrically. Gyratory crushers are high capacity devices and can be used as primary, secondary, or tertiary crushers.

Roll crushers can consist of a single or six or more rolls. The rolls can be smooth, corrugated, grooved, or beaded; they can have intermeshing teeth, fingers, or lugs; and they may be fluted, waffled, or serrated. Roll crushers produce a very narrow size distribution of material; they also exhibit small reduction ratios, typically around four. They can handle wet and sticky material.

Ring-roll and ring-ball mills consist of a number of rolls or balls pressed by mechanical or gravitational means onto a ring which is driven. An air sweep which entrains the finer particles is always incorporated, thereby removing them from the comminuting area. Closed-circuit grinding is often employed, ie, the mill discharge is classified and the coarse material is returned to the mill. Roller mills are usually restricted to materials with a Mohs hardness of ≤ 5.

Impact mills. Impact mills feature a fast-moving part that transfers a portion of its kinetic energy to the material by contact. This transfer of energy sets up stresses that cause the material to shatter. Secondary breakage occurs when the material strikes a breaker plate or stationary surface. Numerous types of mills can be classified as impact devices, eg, hammer, sledge, vibration, pin, centrifugal, cage, swing, and stamp mills. Impact mills are generally divided into those that rotate and those where the energy is transferred by other means, such as vibration and stamping.

Some hammer mills have adjustable cages, a feature that permits changing the product size. The tighter the clearance between the screen bar and hammers, the smaller the particle size of the product.

Vibration mills are also classified as impact devices. They have either a torus-shaped or cylindrical-shaped shell. Loose bodies are contained within the shell and caused to vibrate; the medium and the material impact against one another and the shell, causing breakage.

Tumbling mills. Tumbling mills are characterized by a cylinder that contains grinding medium, rotating about a horizontal, or nearly horizontal, axis. The medium can be rods of one dimension several times the other, balls, or the material to be ground itself. Tumbling mills are commonly classified according to the medium used to effect breakage, eg, rod mills, ball mills, or autogenous mills. Compeb mills have two or more compartments containing balls, pebbles, or rods and combine two or more grinding stages in one unit. These are popular in the cement industry.

Attrition mills. In attrition machines, size reduction is effected by particles breaking each other after having acquired energy from a solid or fluid impeller. The principal task of attrition mills is to produce very fine material ($< 10 \mu m$). They permit control of the product within a very narrow size range.

Other methods. Thermal size reduction is achieved by raising or lowering the temperature of a material causing differential thermal expansion and fractures. In vacuum comminuting, particles are subject to sudden pressure changes. Resonant vibration causes particles containing mixed compositions to be excited or to vibrate at different frequencies.

Comminution Circuit Design

A sizing-classification device separates an input particle stream into fine and coarse particles. Closed-circuit comminution systems can save energy. Comminution-classification circuits are often equipped with environmental-control systems.

Applications in the Chemical Process Industries

Size reduction in the chemical process industries is used mainly in the preparation of mineral feed stocks, eg, limestone plus lime, phosphate rock, coal (as a chemical feedstock), silica plus feldspars, etc. With impure minerals, the size-reduction step may be incorporated into a mineral-processing plant to upgrade the ore to meet user specifications. Lesser but still important applications exist in coal pulverization of fuel and in the preparation of suitably sized products for sale (see Coal). Size reduction is rarely an integral part of a chemical process. More typically, it provides an appropriately sized intermediate, whether for increased reaction rates, for more rapid heat transfer, or to improve ease of handling, eg, the cutting of nylon polymer into small chips for blending before spinning.

Cement. The materials required for portland-cement manufacture are lime, alumina, and silica obtained from limestone, shale, clay, or cement rock. Run-of-mine stone is first crushed, then blended in appropriate proportions and ground to 75–90% passing 75 μm in wet or dry circuits. Modern plants prefer dry grinding in order to minimize fuel requirements for drying before calcination. The material is dried in a separate dryer or in a mill-classifier closed circuit, employing waste heat from the cement kiln (see Cement). Cement clinker from the kiln must be dry-ground to fine sizes (10–50 μm) to produce a product having surface areas, as determined by the Blaine air-permeability analysis, of 3200–6000 cm^2/g.

Pulverized coal is used increasingly in suspension firing in power-plant installation because of its high thermal efficiency. Small, self-contained equipment, eg, ball mills and ring-roller and bowl mills are used.

Dispersion of powders in liquids. In the paint industry, size-reduction processes are used in the preparation of finely divided pigments and in the dispersion process itself (see Dispersion of powders in liquids). A good example is titanium dioxide, the most important white pigment. It is manufactured from ground ilmenite or rutile via sulfate digestion or chlorination, then reground in pebble or roller mills with air classification. In the paint process itself, the pigment is mixed with various tints, thinners, resins, and oils and then dispersed in tumbling, vibratory, or planetary ball mills, attrition mills, sand mills, or specialized roll mills. The action of all these mills is one of breakdown of agglomerates rather than particle size reduction, though this occurs also (see Paint).

B.P. FAULKNER
H.W. RIMMER
Allis Chalmers Corporation

D.W. Green, ed., *Perry's Chemical Engineers' Handbook*, 6th ed, McGraw-Hill, Inc., New York, 1984.

A.F. Taggart, ed., *Handbook of Mineral Dressing*: *Ores and Industrial Minerals*, John Wiley & Sons, Inc., New York, 1945.

G.C. Lowrison, *Crushing and Grinding*, CRC Press, Cleveland, Ohio, 1974.

A.L. Mular and G.V. Jergensen, eds., *Design and Installation of Comminution Circuits*, SME/AIME, New York, 1982.

SIZE SEPARATION

Size separation devices fall into two general categories: those that separate by forces of fluid dynamics, and those involving passage through an aperture.

Screening devices are used to make coarse separations with fine products having 95% passing ca 100-mm to 50-μm size. Dry-screening devices have a lower recommended size of ca 500 μm.. Wet-screening devices that produce 95% passing ca 500–50 μm size are continually being improved.

Fluid-dynamic devices are used to make fine separations; fine products having 95% passing ca 1000- to 1-μm size. However, different devices have different separating ranges. For example, hydraulic-settling classifiers produce fine streams in the range of 95% passing ca 1000- to 100-μm size. Hydraulic-cyclone classifiers, or hydrocyclones, produce fine streams in the range of 95% passing ca 500- to 5-μm size. Generally, as the size becomes finer, the capacity of the separating device decreases.

Evaluation of Separating Devices

The mass-flow relationships of a size separation are usually expressed on a relative rather than an absolute basis. The ratio of the product stream to the feed stream is known as the yield. Recovery is used as a measure of the quality of the separation.

Classification efficiency is defined as a corrected recovery efficiency. Classification efficiency is the same as separation efficiency.

Quantitative efficiency is the ratio of the sum of the amount of material, less than the cut size in the product stream, plus the amount of material greater than the cut size in the other stream, to the feed rate.

The separation efficiency depends upon the feed properties and hence cannot be used to compare different operating situations unless the same feed is used. Therefore, the AIChE has selected an independent criterion, based upon the fractional recovery of material to the coarse stream, to evaluate size separations. Three parameters, the apparent bypass, the cut size, and the sharpness index, are used for evaluation.

The apparent-bypass value does not affect the size makeup of the fine stream. It does, however, affect the size makeup of the coarse stream.

Screening

Screening is a process whereby particles are presented to a series of apertures that allow undersize particles to pass through and oversize particles to be retained.

By deriving a simple theory that views the screening process as a rate process, observations can be made concerning the rate at which particles pass through the apertures. The rate depends upon the open area of the screen, the total area of the screen, the quantity of material on the screen, the presentation per aperture·second, the density and shape of the material, and the fraction of oversize and near-size particles on the screen.

Screening devices can have stationary or moving-screen decks. Moving-screen decks can be rotating cylinders (trommels) or vibrating surfaces. The most popular screen-deck design is the inclined vibrating screen. The formula for selecting screen-deck size determines the area of screening surface needed to remove the undersize material from the feed before it is discharged from the screen surface.

Undersize (percent of material less than one-half of the screen opening in size) in the feed is expected to be 40%.

Oversize (percent of material greater than the screen opening in size) in the feed is expected to be 25%.

Efficiency (the ability to remove undersize material from a given feed) is expected to be "commercially perfect" or 95%.

Wet screening (spraying the material on the screen deck with water in order to remove and screen the finer-size particles) requires less screening surface for screen sizes < 50 mm.

Percent open area of the screen is a function of the screen opening.

Aperture geometry (the shape of the screen openings) decreases the required screening surface for rectangular openings and increases it for circular openings.

Mixed results have been reported in screening slurries with small-aperture (100-μm) vibrating screens.

Stationary screens, known as cross-flow screens, used mainly for dewatering or desliming, have been applied to slurry sizing.

A recent screening design developed to handle feed materials with extensive damp, clayey fines content, is the rotating probability screen (Ropro) (see Fig. 1).

Hydraulic Classification

The settling-pool group of classifiers consists of a rectangular tank with an inclined floor [25 (min)–35 (max) cm/m, ie, 14–19°] that creates a pool. Feed slurry is introduced at the side of the tank and the overflow of fine particles and water exits through an overflow weir-and-box arrangement. The length of time the particles stay in the pool determines the distance they settle in the pool. When the settling type of classifiers were used to produce finer size separations, the pool volume was increased (see Fig. 2).

Increasing the gravitational force by developing centrifugal force decreases the settling time of smaller particles. A centrifugal force can be imposed by rotating the slurry of particles. The classifying hydrocyclone is designed to rotate the slurry of particles by introducing it tangentially into a cylinder. As the fluid rotates in the hydrocyclone, forming an air core in the center, three velocities are of interest: the tangential, the radial, and the axial velocity. The diameter of the air core varies with the feed volumetric flow rate. A particle entering the cyclone finds a point where its velocity, with respect to the fluid, is equal to the radial velocity and hence is at rest with respect to radial movement.

Pneumatic Classification

Pneumatic classification can conveniently be partitioned into coarse, intermediate, and fine. Pneumatic classification, like hydraulic classification, balances the force of gravity with drag forces (counterflow) in order

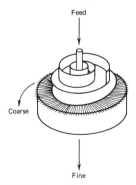

Figure 1. Rotating probability screen (Ropro).

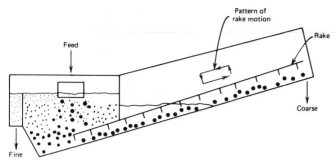

Figure 2. Settling-pool classifier.

to bring about a separation. The simplest example of counterflow classification would be an expansion classifier.

The counterflow principle can be extended to movement in a centrifugal field, improving the selectivity of the classification process and the fineness of the product. The fixed-vane centrifugal counterflow principle is termed a free vortex and has been extended to fine classification. An alternative is the forced-vortex design in which the vanes are rotated.

Another pneumatic classification design is the balance of inertial forces with drag forces (transverse flow) in order to bring about a separation.

Another classifier design incorporates both the gravitational and centrifugal counterflow principle.

The concept of two-stage classification has been very popular in both hydraulic and pneumatic-classifier installations.

Most solids, when finely divided, suspended in air in the proper concentration, and ignited, can produce violent explosions. The only safe materials are those that are fully oxidized or hydrated. Before dusty materials are handled, the flammability characteristics of the dust must be determined.

Prediction of Size Separation

Separation can be predicted if the appropriate fractional recovery values can be estimated. The fractional recovery values, $r(x_i)$, are given by:

$$r(x_i) = (1 - a - b) c(x_i; \kappa, d_{50}) + a$$

where a is the apparent bypass fraction of the feed to the coarse stream; b is the apparent bypass fraction of the feed to the fine stream; $c(x_i; \kappa, d_{50})$ is the corrected curve function with characteristic parameter d_{50}, the equiprobable size, κ, the sharpness index; and x_i is the upper size of ith size interval.

PETER LUCKIE
The Pennsylvania State University

T. Allen, *Particle Size Measurement*, 3rd ed., Chapman & Hall, New York, 1981.

Particle Size Classifiers, A Guide to Performance Evaluation, AIChE Equipment Testing Procedure, Equipment Testing Procedure Committee, AIChE, 1980.

D. Bradley, *The Hydrocyclone*, Pergamon Press, New York, 1965.

SLAGCERAM. See Glass ceramics.

SLIMICIDES. See Industrial antimicrobial agents.

SMOKES, FUMES, AND SMOG. See Air pollution.

SOAP

Soaps are alkali-metal salts of long-chain monocarboxylic acids.

Raw Materials

The fats and oils used in soapmaking are triglycerides. A triglyceride is converted to soap by saponification.

$$\begin{array}{l} RCO_2CH_2 \\ | \\ RCO_2CH \\ | \\ RCO_2CH_2 \end{array} + 3\ NaOH \rightarrow 3\ RCO_2Na + \begin{array}{l} CH_2OH \\ | \\ CHOH \\ | \\ CH_2OH \end{array}$$

Tallow is the principal animal fat in soapmaking. It is obtained by rendering the body fat from cattle and sheep. Lard (rendered hog fat) can be used as a replacement for tallow after partial hydrogenation. Coconut oil is obtained by crushing and extracting the dried fruit (copra) of the coconut palm. Palm oil often serves as a partial substitute for tallow. Palm kernel oil is extracted from the center nuts of the fruit cluster that yields palm oil, which can be used as a partial substitute for coconut oil. Relatively small quantities of olive, castor, and babassu oils may be used.

The crude soap or "foots" obtained from refining edible oils (cottonseed, soybean, etc), as well as the foots obtained by the soapmaker in refining fats and oils, may be used in limited amounts. A modern trend is to hydrolyze oils to fatty acids and glycerol (qv) prior to saponification. The fatty acids may then be vacuum-distilled to improve the resulting soap. Rosin, tall oil (qv), and cycloparaffinic (naphthenic) acids are used in the preparation of laundry or textile-scouring soaps.

Aqueous sodium hydroxide solution (50 wt%) is commonly used for the preparation of hard soaps, flakes, or beads. Potassium hydroxide is used for the preparation of soft soaps.

Processing of Fats and Oils

Crude fats or oils may be refined and bleached before saponification. Coloring matter can be removed by agitating with bleaching clays or activated carbon. The soapmaking qualities of highly unsaturated triglycerides can be improved by hydrogenation.

Soap Phases

Different phases can coexist in the three-component system of a soap kettle and equilibria between the phases are governed by the phase rule. The most important two-phase separations in kettle boiling are the kettle wax–lye and neat–nigre separations.

The product of the boiling and continuous procedures, starting with either fat or fatty acid, is neat soap of ca 69% soap and 30% water. Hence, the compositions that are finished bar, flake, or bead products are essentially two-component systems of soap and water.

In solid hydrous soaps, four different phases have been characterized by different x-ray-diffraction patterns. Three of these phases are encountered in samples at room temperature of commercially important soap–water compositions. Phase transformations are induced by temperature or composition changes. The temperature of ready solubility (T_s) is defined as the temperature at which soap shows a sudden sharp increase in water solubility.

Batch Processes

Full-boiled kettle method. Much of the world's soap production still begins in open steel kettles or pans capable of processing batches of 1–1000 kg. Saponification comprises a slow incubation period, a rapid exothermic stage, and gradual completion. During the slow incubation period, the caustic must be added slowly since a large excess of unreacted alkali renders the oil and water phases even less miscible. When enough soap has formed to start emulsification, the exothermic stage begins and caustic is consumed rapidly. The last, small amounts of caustic soda are added slowly with testing for alkali consumption. The product is washed several times. Improvement in color can be obtained by adding sodium hydrosulfite or sodium hypochlorite. If rosin is to be incorporated, it is added to the pan just prior to the fit.

Cold-process saponification. The fat charge is melted in a vessel equipped with a mechanical stirrer, and caustic soda is added with vigorous stirring. After emulsification and thickening, the mass is poured into frames where saponification is completed during cooling and solidification. In semiboiled saponification, a higher temperature is used to speed the process.

Fatty acids obtained from natural oils by hydrolysis or by acidulation of foots are readily neutralized with caustic.

Continuous Processes

The Procter & Gamble process is a continuous process that converts raw fats to finished soaps in a matter of hours.

The Sharples system converts fats directly to soaps and uses centrifuges to separate lyes for washing and glycerol recovery.

The Mon Savon process continuously converts fats and oils directly into soap. Washings and lye are removed by countercurrent flows of crude soap and hot brine solutions through a multistage washing tower.

In the De Laval Centripure process, the fats and caustic soda solution are proportioned countercurrently into a vertical reactor.

Automated continuous Mazzoni systems are available for the saponification of fats. J. Crosfield and Sons, a Unilever subsidiary in the UK, uses continuous production of soap for spray-dried products.

The Armour Company produces soap by an automated continuous process combining units of the Mazzoni and the De Laval systems.

The Unilever process for the jet saponification of fats conserves steam by withholding it during the exothermic stage of the reaction.

The high caustic–high solids saponification process can result in production of 86% soap, thereby eliminating drying.

Finishing Operations

Incorporation of various additives, eg, builders, into the soap may be accomplished using vessels called crutchers. Bar soaps can be produced by allowing neat soap to cool in rectangular frames. The solidified soap is then cut into cakes. The 30% water content of neat soap must be reduced to 10–15% before the soap can be shaped into bars and to 5–10% before it is cut into flakes. The drying is generally accomplished in cabinet, flash, or vacuum-drying units, or by vacuum spray drying (see Drying).

Pigments, dyes, perfumes, germicides, antioxidants, etc, may be added to bar and flake soap products. These are generally worked into pellets or flakes in a batch mixer equipped with a helical agitator.

Mill ribbons or refiner pellets are fed through a hopper into a plodder where an internal screw forces the soap into a compression area. The log is then cut, cooled, stamped, wrapped, and cartoned. All of these operations are conducted continuously by machines capable of producing 300 bars/min.

In the manufacture of flakes, soap from the mixer is passed through finishing rolls adjusted to a close tolerance. The flake shapes are marked by rotating cutters and stripped from the final roll by a knife.

In the manufacture of spray-dried powders, additives are suspended or dissolved in hot neat soap in a crutcher before drying.

Heavy-duty soap powders containing inorganic builders have been almost completely replaced in the United States by heavy-duty synthetic detergents.

Specialty Soaps

Modern antibacterial agents have supplanted the phenol and cresylic acid bar-soap formulations and are in great demand as deodorant soaps. Many toilet soaps are superfatted with 2–10% excess of unsaponified oil, fatty acid, or lanolin. Liquid soaps are formulated using more soluble potassium, ammonium, or triethanolamine soaps. The original floating soaps were made by beating air into molten neat soap. These have been improved by newer methods of incorporating air under pressure. Marbleized soap bars are produced either by mixing different-colored soap pellets in a plodder or by injecting a liquid into the soap. Additions of alcohol, sugar solution, and glycerol to hot soap promote a glossy, transparent condition. Slightly abrasive soaps contain a finely powdered insoluble material such as pumice. Soaps for shaving are generally formulated by saponifying a mixture of coconut oil and stearic acid with caustic potash and soda.

Analysis

The American Oil Chemists' Society publishes testing methods. For total fatty acids, the sample is hydrolyzed with acid and the fatty acids are extracted with ether, which is evaporated, and the residue is weighed. Color is compared with standard colors. For free alkalinity, a sample is dissolved in alcohol and titrated. Salt is determined by titration with silver nitrate. The glycerol content is determined by oxidation with potassium dichromate or sodium periodate.

Health and Safety Factors

In the soap industry, the handling of caustic soda solutions is probably the greatest hazard. Goggles and protective clothing are worn. High dust concentrations may prove irritating to the mucous membranes.

Uses

The principal use of the soluble alkali soaps is in toilet bars. Soaps play a key role in emulsion–polymerization processes. Soaps are widely used in the cosmetic industry. Sodium and lithium soaps are extensively used to thicken mineral oil in the manufacture of lubricating greases. Soaps are used as wetting and spreading agents to improve the dispersal of insecticidal and fungicidal components of agricultural sprays.

FREDERICK S. OSMER
Lever Brothers Co., Inc.

W.H. Mattil, *Oil Soap* **21**, 198 (1944).

D. Swern, ed., *Bailey's Industrial Oil and Fat Products*, 4th ed., Vol. 1, John Wiley & Sons, Inc., New York, 1979.

SODA. See Alkali and chlorine products.

SODIUM AND SODIUM ALLOYS

Sodium, an alkali metal, is the second element of Group IA of the periodic table, at wt. 22.9898. Sodium occurs naturally as a component of many complex minerals and of such simple ones as sodium chloride, sodium carbonate, sodium sulfate, sodium borate, and sodium nitrate. Sodium-23 is the only naturally occurring isotope.

Properties

Sodium is a soft, malleable solid which is readily cut with a knife or extruded as wire. Physical properties are given in Table 1. Sodium is

Table 1. Physical Properties of Sodium

Property	Value
ionization potential, V	5.12
melting point, °C	97.82
heat of fusion, kJ/kg[a]	113
volume change on melting, %	+2.63
boiling point, °C	881.4
heat of vaporization at bp, MJ/kg[a]	3.874
density of the solid at 20°C, g/cm^3	0.968
density of the liquid at mP, g/cm^3	0.927
viscosity at 100°C, mPa·s (= cP)	0.680
surface tension at mP, mN/m (= dyn/cm)	192
specific heat, kJ/(kg·K)[a]	
solid at 20°C	2.01
liquid at mp	1.38
electrical resistivity, μΩ·cm	
solid at 20°C	4.69
liquid at mp	9.64
thermal conductivity, W/(m·k)	
solid at 20°C	1323

[a] To convert J to cal, divide by 4.184.

paramagnetic. The vapor is chiefly monatomic. Sodium is easily dispersed in inert liquid hydrocarbons. Dispersions may contain as much as 50 wt% sodium. In this form, it is easily handled and reacts rapidly.

Liquid sodium readily wets many high surface solids. This property may be used to provide a highly reactive form of sodium without contamination by hydrocarbons.

Sodium forms unstable solutions in liquid ammonia where a slow reaction takes place to form sodamide and hydrogen. Iron, cobalt, and nickel catalyze this reaction. At high temperature, sodium and its fused halides are mutually soluble.

Sodium is soluble in ethylenediamine, but solubility in other amines such as methyl- or ethylamine may require the presence of ammonia. Sodium reacts with naphthalene in dimethyl ether to form a soluble, dark green, reactive comlex. The solution is electrically conductive.

Sodium forms well-defined compounds with a number of metals; some of these alloys are liquid below 300°C.

The reaction of sodium and water according to the equation below is very rapid:

$$Na + H_2O \rightarrow NaOH + \tfrac{1}{2} H_2 + 141 \text{ kJ/mol}(98.95 \text{ kcal/mol})$$

The liberated heat melts the sodium and frequently ignites the evolved hydrogen if air is present.

Hydrogen and sodium do not react at room temperature, but at 200–350°C sodium hydride is formed. There is little evidence for the formation of sodium carbide from the elements, but sodium and graphite form lamellar intercalation compounds. Nitrogen and sodium are reported to form the nitride or azide under the influence of an electric discharge.

Sodium reduces the oxides of Group IA elements except lithium, but not oxides of Group IIA elements except mercury, cadmium, and zinc. Many other oxides are reduced by metallic sodium.

Sodium reacts with many organic compounds, particularly those containing oxygen, nitrogen, sulfur, halogens, carboxyl, or hydroxyl groups. Organosodium compounds are prepared from sodium and other organometallic compounds or active methylene compounds by reaction with organic halides, cleavage of ethers, or addition to unsaturated compounds. Sodium amalgam or sodium and alcohol are employed for organic reduction.

Manufacture

The earliest commercial processes were based on the carbon reduction of sodium carbonate or sodium hydroxide.

Electrolysis. The first successful electrolytic production of sodium was achieved with the Castner cell:

Cathode	$4 Na^+ + 4 e \rightarrow 4 Na$	
Anode	$4 OH^- - 4 e \rightarrow 2 H_2O + O_2$	

The Castner cell was so simple that over the years only minor changes were made.

For the more efficient electrolysis of fused sodium chloride, the widely used Downs cell has been most successful. It consists of three chambers. The upper chamber is outside the chlorine dome and above the sodium-collecting ring. The other two chambers are the chlorine-collecting zone inside the dome and diaphragm, and the sodium-collecting zone outside the diaphragm and under the sodium-collecting ring. This arrangement prevents recombination of the sodium and chlorine. The cell bath in early Downs cells consists of about 58 wt% calcium chloride and 42 wt% sodium chloride. Salt substantially free of sulfate and other impurities is the cell feed. This grade may be purchased or made on site by purification of rock salt. Cell life is determined by the loss of graphite from the anodes. A dimensionally stable anode consisting of an electrically conducting ceramic substrate coated with a noble metal oxide has been developed (see Metal anodes).

Sodium in the form of amalgam as made by the electrolysis of brine in mercury cathode cells is much cheaper than any other form of the metal but commercial use is restricted largely to production of caustic soda (see Alkali and chlorine products).

Searching for a method for using sodium and sulfur as reactants in a secondary battery, the Ford Motor Company developed a polycrystalline β-alumina ceramic material that selectively transports sodium cations when subjected to an electric field. This ceramic is useful as a diaphragm in a two-compartment cell. The sodium is in contact with the ceramic, which connects it electrochemically with the liquid electrolyte. Sodium of exceptional purity is produced at satisfactory operating conditions. However, because a ceramic of predictable properties and long service life has not yet been developed, these cells have not been commercialized.

Specifications, Shipping

Sodium, generally about 99.95% Na, is available in two grades:

	Calcium, wt%	Chlorides, wt%
regular	0.040	0.005
reactor	0.001	0.005

Sodium is usually shipped in 36- or 54-t tank cars. Smaller amounts are shipped in tank trucks.

Analytical Methods

Sodium is identified by the intense yellow color that sodium compounds impart to a flame, or spectroscopically, by the characteristic sodium lines. Gravimetric procedures employ precipitation as the uranyl acetate of sodium–zinc or sodium–magnesium. Quantitative determination without separation is frequently possible by emission or atomic-absorption spectrometric techniques. Metallic sodium is determined with fair accuracy by measuring the hydrogen liberated on the addition of ethyl alcohol. Calcium in commercial sodium is usually determined by permanganate titration of its oxalate.

Health and Safety Factors

With properly designed equipment and procedures, sodium is used in large and small applications without incident. Contact with the skin can cause deep, serious burns. Goggles, face shield, gloves, and flame-retardant protective clothing are recommended when working with sodium.

Perhaps the greatest hazard presented by metallic sodium stems from its extremely vigorous reaction with water. Another hazard arises from the reaction of air and sodium. Fires that cannot be extinguished by excluding air may be quenched by large quantities of dry salt or other dry, cold, inert powder. Dry soda ash is excellent.

In the laboratory, sodium is best handled in a glove box filled with nitrogen or another inert gas. Contact with air should be kept to a minimum, since moisture in the air reacts rapidly with sodium. Most reactions of sodium are heterogeneous, occurring on the surface of solid or liquid sodium. Dispersion of the sodium in a suitable medium accelerates the reactions.

Uses

The production of tetraethyllead and tetramethyllead antiknock agents for gasoline is the largest outlet for sodium in the United States (see Lead compounds; Organometallics).

The manufacture of refractory metals such as titanium, zirconium, and hafnium by sodium reduction of their halides is growing. A typical overall reaction is the following:

$$TiCl_4 + 4 Na \rightarrow Ti + 4 NaCl$$

Potassium–sodium alloys are made by the reaction of sodium with fused KCl or KOH.

Sodium is a catalyst for many polymerizations. Sodium hydride, made from sodium and hydrogen, is employed as a catalyst or reactant. Sodium is employed as a reducing agent in the manufacture of dyes, herbicides, pharmaceuticals, high molecular weight alcohols, perfume materials, and isosebacic acid.

Sodium as an active electrode component of primary and secondary batteries offers the advantages of low atomic weight and high potential. Electrical cables are made of sodium sheathed in polyethylene. Sodium is used as a heat-transfer medium in liquid–metal fast-breeder power reactors (see Heat-transfer technology; Nuclear reactors). Sodium vapor lamps contain only a few milligrams of sodium.

Sodium Alloys

Sodium is miscible with many metals in the liquid phase and forms alloys or compounds. The brittleness of metals is frequently increased by the addition of sodium.

Alloys of lead and sodium containing up to 30 wt% sodium are obtained by heating the metals together in the desired ratio, allowing a slight excess of sodium to compensate for loss by oxidation. Sodium–lead alloys that contain other metals, eg, the alkaline-earth metals, are hard even at high temperatures, and are thus suitable as bearing metals (see Bearing materials).

Up to ca 0.6 wt% sodium dissolves in mercury to form amalgams which are liquid at room temperature. Sodium does not form alloys with aluminum but is used to modify the grain structure of aluminum–silicon alloys and aluminum–copper alloys.

CHARLES H. LEMKE
E.I. du Pont de Nemours & Co., Inc.
University of Delaware

O.J. Foust, ed., *Sodium—NaK Engineering Handbook*, Gordon and Breach, Science Publishers, Inc., New York, 1972.

R.E. Robinson and I.L. Mador in N.M. Bikales, ed., *Encyclopedia of Polymer Science and Technology*, Vol. 1, Interscience Publishers, a division of John Wiley & Sons, Inc., New York, 1970, pp. 639–658.

T.P. Whaley, *Sodium, Potassium, Rubidium, Cesium and Francium*, Chapt. 8 in A.F. Trotman-Dickenson, ed., *Comprehensive Inorganic Chemistry*, Pergamon Press, Oxford, UK, 1973.

SODIUM CARBONATE.   See Alkali and chlorine products.

SODIUM COMPOUNDS

SODIUM HALIDES, SODIUM CHLORIDE

Sodium chloride, NaCl, here referred to as salt, supplies sodium and chloride ions, both essential to animal life. It is one of the principal raw materials used in the chemical industry and the source of almost all industrial compounds containing sodium or chlorine (see Alkali and chlorine products; Chemicals from brine).

Most of the world's salt is contained in the oceans. In North America, huge quantities of rock salt are found in the Silurian basin, covering parts of Michigan, Ontario, Ohio, Pennsylvania, and New York.

The salt dome, a large vertical column of salt, is one of the most unusual types of rock-salt deposits. The top of a dome used for salt production is usually within 100 m of the earth's surface. The principal impurity in naturally occurring rock salt is calcium sulfate, generally 1–4%, with small amounts of calcium chloride and magnesium chloride. Salt also occurs as sylvinite, which consists of mixed crystals of potassium and sodium chloride.

Pure sodium chloride is colorless. Physical properties are given in Table 1.

Processing.   Salt is mined similarly to coal. A shaft is sunk into the rock-salt deposit, and with well-established techniques of undercutting, side-shearing, drilling, blasting, loading, and transporting, the salt is removed for further processing.

Rock salt is mined by the room-and-pillar method. As the salt is removed, large empty spaces (rooms) are formed; pillars are left for support.

Solution mining.   Salt brine is obtained by pumping water into a rock-salt deposit. The salt is dissolved and the brine brought to the surface. Solution mining produces essentially saturated brine. Its prin-

Table 1.   Properties of Sodium Chloride

Property	Value
mp, °C	800.8
bp, °C	1465
density, g/cm^3	2.165
hardness	2.5
refractive index, n_D^{20}	1.544
specific heat, $J/(g \cdot °C)$[a]	0.853
heat of fusion, J/g[a]	517.1
critical humidity at 20°C, %	75.3
heat of solution in 1 kg H_2O at 25°C, kJ/mol[a]	3.757

[a] To convert J to cal, divide by 4.184.

cipal impurity is calcium sulfate. For ordinary grades of salt, no brine treatment is required other than hydrogen sulfide removal, if necessary, and settling to remove solids.

The multiple-effect crystallizing evaporator system, or vacuum-pan system, is the most common technique for the evaporation of brine to produce salt (see Evaporation). Salt produced in this manner is called evaporated, or granulated, salt.

Evaporators are either the calandria (internal heating-surface) type, or the forced-circulation (external heat-exchanger) unit. Untreated brine may cause scaling of heating surfaces with calcium sulfate. This problem is controlled by maintaining a suspension of very fine calcium sulfate seed crystals in the evaporators.

With the grainer process, a type of salt called flake salt is obtained. The Alberger process produces a salt that is a mixture of flake salt and fine cubic crystals. The recrystallizer process was developed primarily to convert rock salt to evaporated salt.

Solar evaporation.   Solar evaporation is the oldest method of salt recovery. Solar-salt production is limited to areas that combine favorable meteorological conditions, availability of land, and accessibility to markets. Solar evaporation is basically a fractional crystallization process that uses the sun as the energy source.

Solar-salt operations, with their large open ponds, depend upon the weather; a particularly rainy year, or even one hurricane, may wipe out most of an entire season's production. As an alternative to solar evaporation, an ion-exchange-membrane process produces a salt-rich brine from seawater by means of electrodialysis (see Membrane technology); the salt is then crystallized in evaporators.

Prevention of caking.   A water-insoluble free-flow agent is generally added to table salt and some industrial grades. The most common agents are finely divided adsorbents that are approved food additives, eg, sodium silicoaluminate, tricalcium phosphate, magnesium carbonate, calcium polysilicate, and silicon dioxide.

Analytical Methods

The most common impurities are calcium sulfate, calcium chloride, magnesium chloride or magnesium sulfate, sodium sulfate, and water-insoluble material. Surface moisture is determined by drying, material insoluble in water by weighing, calcium and magnesium by ethylenediaminetetracetic acid (EDTA) titration, and sulfate gravimetrically.

Health and Safety Factors

In every mammal, sodium regulates the volume of blood and maintains the balance of fluids and pressure inside and outside the cells. The daily human requirement for sodium is difficult to establish because the need fluctuates, depending upon sweating and waste elimination. There has been recent concern about excessive sodium in the U.S. diet with regard to hypertension. High blood pressure can be reduced by a diet severely restricted in sodium. However, the value of low sodium diets has been questioned.

Except in the case of infants, acute oral toxicity of salt is hardly meaningful. The oral LD_{50} is 3,000 mg/kg in rats and 4,000 mg/kg in mice, indicating a toxic dosage of 200–280 g for a 70–kg adult.

Environmental Concerns

Mainly because of its large-scale usage, salt creates significant environmental problems. In the production of granulated salt, impurities concentrate in the brine. The waste brine must eventually be disposed of, usually by injection into deep wells.

The general public is mainly concerned with deicing salt for roads. Shade trees that are more salt resistant are now planted along roadways. Coverings over deicing salt piles and improved salt-spreading equipment have minimized wastage. Many substitutes for deicing salt have been suggested, but most are too expensive.

Uses

Historically, salt has been used to flavor and preserve meat and fish. In ham, bacon, hot dogs, sauerkraut, and sausage, salt promotes the natural development of color. In sausage and other processed meat, salt promotes formation of a binding gel. It improves the tenderness of cured meats, and imparts a smooth, firm texture. Salt develops an even consistency in bread. In production of pickles, sauerkraut, summer sausage, cheese, and bread, salt controls the rate of fermentation.

More than 50% of the table salt sold in the United States is iodized. Iodine is used in the body for formation of thyroxine, an essential hormone. Iodine deficiency occurs in areas where excessive run-off has leached naturally occurring iodine compounds from the soil. The only iodizing agent approved for table salt in the United States is potassium iodide (see Hormones).

In some countries, table salt is used as a carrier of fluoride ion for prevention of dental caries.

Many important industrial chemical processes are based on sodium chloride. A growing use is for regeneration of cation-exchange resins in water softeners (see Ion exchange).

Although most commercial salt substitutes are based on potassium chloride, no other compounds seem to impart the salty flavor characteristic of sodium chloride. A wide variety of spices, herbs, and seasonings is used in low sodium diets to counteract blandness.

JOHN F. HEISS
Diamond Crystal Salt Co.

EUGENE J. KUHAJEK
Morton Salt

D.W. Kaufmann, *Sodium Chloride*, Reinhold Publishing Corporation, New York, 1960.

D.S. Kostick in *Bureau of Mines Minerals Yearbook* 1980–1981, U.S. Department of the Interior, Washington, D.C.

SODIUM HALIDES, SODIUM BROMIDE

Sodium bromide, NaBr, is a colorless compound with a somewhat bitter salty taste, which crystallizes in the cubic system; properties are given in Table 1.

Preparation

Sodium bromide is obtained by neutralizing sodium carbonate or hydroxide with hydrobromic acid, followed by evaporation and crystallization, or by formation from bromine, sodium hydroxide or carbonate, and a reducing agent in water.

Health and Safety Factors

Sodium bromide dust and solutions are moderately irritating to eyes and skin; prolonged exposure should be avoided. Internally in large doses, it causes depression of the central nervous system. Continued intake may lead to acne and mental deterioration.

Uses

Sodium bromide serves generally as a source of bromide ion. It is used in photographic processes for the preparation of light-sensitive silver

Table 1. Properties of Sodium Bromide

Property	Value
mp, °C	747
refractive index, n_D^{25}	1.6141
sp gr$_4^{25}$	3.200
heat of fusion, J/g^a	253.6
heat capacity at 25°C, J/(kg·K)^a	498
liquid at mp.	870

^{a}To convert J to cal, divide by 4.184.

bromide emulsions and as a restrainer in developers. It is also used as a catalyst for the partial oxidation of hydrocarbons and other organic compounds. Sodium bromide is widely employed as a sedative, hypnotic, and anticonvulsant (see Hypnotics, sedatives, and anticonvulsants).

V.A. STENGER
Dow Chemical, U.S.A.

"Bromine" in *Minerals Yearbook*, Bureau of Mines, U.S. Dept. of the Interior, Washington, D.C., latest edition.

SODIUM HALIDES, SODIUM IODIDE

Sodium iodide crystallizes in the cubic system; physical properties are given in Table 1. Sodium iodide is soluble in methanol, ethanol, acetone, glycerol, and other organic solvents.

Table 1. Physical Properties of Sodium Iodide

Property	Value
mp, °C	651
bp, °C	1304
d_4^{25}, g/cm^3	3.667
specific heat, J/(kg·K)^a	
at 0°C	350
at 50°C	360

^{a}To convert J to cal, divide by 4.184.

Production is based on the reaction of sodium carbonate or hydroxide with an acidic iodide solution. After removal of impurities, the solution is filtered and concentrated.

The principal use of sodium iodide is as an expectorant in cough medicines. It has been used in cloud seeding.

P.H. MERRELL
E.M. PETERS
Mallinckrodt, Inc.

J.C. Bailar, Jr., H.J. Emelius, R. Nyholm, and A.F. Trotman-Dickenson, eds., *Comprehensive Inorganic Chemistry*, Pergamon Press, Inc., Elmsford, N.Y., 1973, Vol. 1, p. 402; Vol. 2, p. 1107.

The United States Pharmacopeia XX (USP XX–NF XV), The United States Pharmacopeial Convention, Inc., Rockville, Md., 1980, p. 732.

SODIUM NITRATE

Sodium nitrate, $NaNO_3$, occurs in nature associated with sodium and potassium chlorides, potassium nitrate, sodium sulfate, magnesium chloride, and other salts. Chilean nitrate is the only inorganic nitrogen fertilizer produced today. Commercially valuable Chilean deposits of

Table 1. Selected Properties of Sodium Nitrate

Property	Value
crystal system	trigonal, rhombohedral
mp, °C	308
refractive index, n_{D}^{20}	
trigonal	1.587
rhombohedral	1.336
density (solid), g/cm^3	2.257
solubility in H$_2$O, molality ($\pm 2\%$)	
at 0°C	8.62
at 80°C	17.42
specific conductivity (at 300°C), S/cm	0.95
viscosity^a, mPa·s (= cP) at 590 K	2.85
heat of fusion, J/g^b	189.5
heat capacity, J/g^b	
solid at 0°C	1.035 ± 0.005
liquid at 350°C	1.80 ± 0.02

^{a} Measurement method: capillary.
^{b} To convert J to cal, divide by 4.184.

sodium nitrate are found in the northern part of the country. Properties are given in Table 1.

Manufacture and Processing

In the Shanks process, the ore was crushed and loaded into vats. The leaching solution consisted of water and a mother-liquor brine with ca 450 g/L sodium nitrate. The Shanks process was discontinued in the early 1960s.

Guggenheim process. The Guggenheim nitrate process demonstrated that caliche ores as low as 7 wt% in nitrate content could be economically exploited. The mined ore is transferred to crushing units where it is crushed in three stages. The crushed ore is leached in 8–10 vats, of which four are used at any given time at an average temperature of 40°C. The leaching cycle lasts 40 h, the total vat cycle ca 168 h.

The fines produced during crushing are mixed in the filter plant with liquor containing 150–200 g/L of nitrate, at a mixing ratio of two tons of solids per cubic meter of liquor; the leaching is carried out at 40–50°C. The rich liquor from the leaching vats containing 450 g/L of sodium nitrate is sent to crystallizers where it is chilled to 0–5°C. At the exit of the last crystallizers, the mother liquor and the crystals are pumped out as a slurry to continuous centrifuges.

The sodium nitrate crystals are transferred to the graining plant, which consists of large, oil-fired, reverberatory-type furnaces, where the crystals are heated to 315–325°C. The molten sodium nitrate is pumped to large spray chambers where the spray solidifies as prills. The Guggenheim process recovers 80–85 wt% of the NaNO$_3$ contained in coarse fractions at a total fuel consumption of 0.1 metric ton per metric ton of sodium nitrate produced.

In the modified Guggenheim process, the ore is leached countercurrently with water in vats. From the weak sodium nitrate brines, sodium sulfate is extracted by a chilling process that yields Glauber's salt. Rich sodium nitrate solution is passed to the crystallizers and mother liquor to the solar-pond system. Continuous evaporation in the ponds results in concentration increases of potassium nitrate and iodine. Potassium nitrate is removed through crystallization.

Synthetic sodium nitrate is produced by neutralization of nitric acid with soda ash or caustic soda. The sodium nitrate is melted and prilled.

Health and Safety Factors

The acceptable daily intake by adults for nitrates (suggested by WHO) is 5 mg/kg in addition to naturally occurring nitrates; large doses are lethal. Accidental ingestion of ca 8–15 g or more causes severe abdominal pain. Poisonings from the ingestion of meat containing sodium nitrate and nitrite have occurred (see Meat products; *N*-Nitrosamines).

Uses

Sodium nitrate is used primarily in agriculture as a fertilizer. The sodium in sodium nitrate corrects soil acidity and liberates phosphates,

and is a partial substitute for potassium. The nitrogen from sodium nitrate is available faster than nitrogen from ammonia.

The main industrial use is in the manufacture of explosives, eg, dynamite. Sodium nitrate is also used in the manufacture of glass, fiber glass, enamels (qv), and porcelain as an oxidizing and fluxing agent. It is an ingredient in the production of charcoal briquettes. Sodium nitrate has application in the production of certain antibiotics and pharmaceuticals. In curing beef, bacon, and other meats, it is used as a preservative.

SAMUEL MAYA
Beecham Products

MARTIN LABORDE
CODELCO

V. Sauchelli, ed., *Fertilizer Nitrogen, Its Chemistry and Technology*, Reinhold Publishing Corporation, New York, 1964.

International Critical Tables, McGraw-Hill Book Co., Inc., New York, 1928.

SODIUM NITRITE

Sodium nitrite, NaNO$_2$, is supplied commercially in two forms, dry and liquid, and three grades, technical, USP, and food. It is hygroscopic and very soluble in water, but relatively insoluble in most organic solvents. Aqueous solutions are mildly basic.

Manufacture

Methods for the preparation of sodium nitrite can be divided into two groups: reduction of sodium nitrate and absorption of oxides of nitrogen in alkali. Industrial sodium nitrite is prepared by contacting oxides of nitrogen with aqueous alkaline solutions. Nitrous gases can be produced by the arc process of nitrogen fixation, the air oxidation of ammonia, or as a waste gas in the production of nitric acid and oxalic acid.

Storage and Handling

Dry sodium nitrite is usually shipped in multiwall bags or fiber drums. It should be stored in a tightly closed container in a cool, dry place away from combustible materials. Spills should be kept away from acids.

Health and Safety Factors

Although small quantities are used in food processing, sodium nitrite is moderately toxic. Inhalation of large volumes of dust or direct ingestion can result in death. Irritation of the mucous membranes and the respiratory tract may be caused by ingestion or inhalation. Gloves, safety glasses, and protective clothing are recommended when handling sodium nitrite.

Uses

Sodium nitrite is a convenient source of nitrous acid in the nitrosation and diazotization of aromatic amines.

Sodium nitrite is important in the manufacture of rubber-processing chemicals (see Rubber chemicals), eg, accelerators, retarders, and antioxidants (qv).

Combinations of sodium nitrite, sodium nitrate, and potassium nitrate are used in the preparation of molten salt baths and heat-transfer media. The most frequently used heat-transfer mixture is the triple eutectic 40 wt% NaNO$_2$, 7 wt% NaNO$_3$, and 53 wt% KNO$_3$, which has a melting point of 143°C.

Sodium nitrite is used in metal treatment and finishing operations. It is an anodic inhibitor that forms a tightly adhering film over steel and prevents the dissolution of metal at anodic areas. It is used to reduce corrosion of steel reinforcing bars in concrete.

Sodium nitrite is a curing and preserving agent in bacon, ham, frankfurters, etc. The maximum allowable quantity of sodium nitrite in the brines and the residual remaining in the meat is under the jurisdiction of FDA and USDA (see Food additives). Other applications of sodium nitrite include synthesis of insecticides, herbicides, synthetic caffeine, drugs, flavors, and aromatic fluorine compounds.

JOHN KRALJIC
Allied Corporation

"Natrium" in *Gmelins Handbuch der Anorganischen Chemie*, System 21, Vol. 3, Verlag Chemie, Weinheim, FRG, 1966.

J.W. Mellor, *Supplement to Mellor's Treatise on Inorganic and Theoretical Chemistry*, Vol. VIII, Supp. II, Pt. II, John Wiley & Sons, Inc., New York, 1967.

Sodium Nitrite—1981, Allied Corporation, Morristown, N.J.

SODIUM SULFATES

The sulfates of sodium are commonly sold in four forms: anhydrous sodium sulfate, Na_2SO_4; technical-grade sodium sulfate, known as salt cake; sodium sulfate decahydrate, $Na_2SO_4 \cdot 10H_2O$, known as Glauber's salt; and sodium hydrogen sulfate, $NaHSO_4$, known as niter cake.

Sodium sulfate deposits are found throughout the world as a result of the evaporation of inland seas and lakes. In cold climates, eg, Canada and the USSR, deposits of mirabilite, the decahydrate, occur, whereas in warmer areas, eg, South America, India, Mexico, and the western United States, anhydrous thenardite is found.

Properties

Properties are given in Table 1. Reactivity varies widely with temperature. In the solid state and at ambient temperatures, Na_2SO_4 is relatively inert but forms acid compounds in the presence of sulfuric acid. Only $NaHSO_4$ is of industrial importance. At higher temperatures and particularly in the molten state, Na_2SO_4 is very active.

Manufacture and Processing

Nearly all producers purify the sulfate by crystallizing it as Glauber's salt. The latter is obtained by cooling a natural brine, a solution obtained by solution-mining a sulfate deposit, or cooling a process stream. Glauber's salt is crystallized by cooling in ponds, surface cooling in crystallizers using polished tubes, and vacuum cooling. Dewatering is accomplished by draining or filtration.

Glauber's salt is processed in a variety of ways. The anhydrous salt is obtained by evaporation in a double-effect evaporator, by dehydration in a rotary dryer, or by melting followed by salting out. Energy requirements for the different processes vary widely depending on local conditions.

The principal U.S. process for $NaHSO_4$ involves the reaction of sulfuric acid and salt cake under anhydrous conditions.

Health and Safety Factors

Most by-product sulfate producers have been forced to recover Na_2SO_4 because of regulations concerning discharge into waterways. Paper producers have been similarly restricted. Acid rain containing sulfuric acid can react with a sodium-containing base to produce Na_2SO_4. In general, however, Na_2SO_4 does not appear to be environmentally dangerous.

Uses

In the United States, 67% of the total sodium sulfate production is used in paper-pulp manufacture; 26% for detergents; and 7% for glass and all others. Consumption in the paper industry is declining, and uses for detergents are increasing. In some kraft or sulfate paper processes, a mixture of sodium sulfide and sodium hyroxide is replacing salt cake to digest wood chips. This process can be used for a wider range of wood types than the competing pulping processes.

The high temperature properties of Na_2SO_4 are advantageous in glassmaking. Sodium sulfate improves the boiling and working properties of high silica glasses. Household laundry detergents contain as much as 75 wt% Na_2SO_4 as a diluent and builder. Anhydrous Na_2SO_4 is preferred because of its purity and whiteness.

Interest in solar energy has created renewed interest in the high heat of crystallization of Glauber's salt as a means of storing energy.

As a solid, $NaHSO_4$ provides a convenient source of acid for such household uses as toilet-bowl cleaning, automobile-radiator cleaning, and the adjustment of pH in swimming pools. Industrially, it is used for metal pickling, as a dye-reducing agent, for soil disinfecting, and as a promoter in hardening certain types of cement.

THOMAS F. CANNING
Kerr-McGee Chemical Corporation

"Sodium Sulfate," *Chem. Mark. Rep.* **213**(5), 9 (Jan. 30, 1978); **207**(8), 9 (Feb. 24, 1975).

J.A. Kent, ed., *Riegel's Handbook of Industrial Chemistry*, 7th ed., Van Nostrand Reinhold, New York, 1974, pp. 132–134.

F.A. Lowenheim and M.K. Moran, eds., *Faith, Keyes and Clark's Industrial Chemicals*, 4th ed., John Wiley & Sons, Inc., New York, 1975.

SODIUM SULFIDES

Sodium sulfide (Na_2S, mol wt 78.05), sodium hydrosulfide (NaHS, mol wt 56.06), and sodium tetrasulfide (Na_2S_4, mol wt 174.24) have valuable industrial uses.

Sodium Sulfide

Pure sodium sulfide is a white, crystalline solid (mp 1180°C, sp gr 1.856). The oldest manufacturing process for Na_2S is not used in the United States but was part of the LeBlanc soda ash process. Sodium sulfate was reduced by powdered coal in a furnace at 900–1000°C.

Currently, two processes are practiced. The reduction of baryte ore ($BaSO_4$) with coal in a rotary kiln at ca 800°C gives a crude black ash. Leaching dissolves barium sulfide, which then is converted to barium carbonate and sodium sulfide by addition of sodium carbonate.

In the other process, sodium hydrosulfide, obtained from hydrogen sulfide and caustic soda, reacts with caustic soda to yield Na_2S.

$$H_2S + NaOH \rightarrow NaHS + H_2O$$

$$NaHS + NaOH \rightarrow Na_2S + H_2O$$

Hydrogen sulfide can be obtained by the reaction of hydrogen and sulfur, as a by-product of the carbon disulfide-from-methane process, or from desulfurizing petroleum.

Sodium sulfide is marketed as 30–34 wt% fused crystals and 60–62 wt% flakes.

Sodium Hydrosulfide

Pure sodium hydrosulfide (sodium sulfhydrate, sodium hydrogen sulfide, sodium bisulfide) is a white, crystalline solid (mp 350°C, sp gr 1.79).

Processes to make NaHS are closely related to the source of the hyrogen sulfide which reacts with caustic soda as shown above. Sodium hydrosulfide is marketed as 70–72 wt% flakes and 44–60 wt% liquor in the high purity grades and as 10–40 wt% liquor. Use of sodium hydrosulfide are listed in Table 1.

Table 1. Properties of Sodium Sulfates

| Properties | Sodium sulfate | | Sodium hydrogen sulfate |
	Anhydrous	Decahydrate	
melting point, °C	882	32.4	315
specific gravity	2.664	1.464	2.435
refractive index	1.464, 1.474, 1.485	1.394, 1.396, 1.398	1.459, 1.479
crystalline form	rhombic, monoclinic, or hexagonal	monoclinic	triclinic
heat of formation[c], MJ/mol[a]	−1.384	−4.324	−1.126
heat of solution (at ∞ dilution), MJ evolved/mol[a]	1.17	−78.41	7.28
heat of crystallization from saturated soln, J/mol[a]	−8.8$_{50°C}$	74.98$_{25°C}$, 75.52$_{11°C}$	

[a]To convert J to cal, divide by 4.184.

Table 1. Estimate of Sodium Hydrosulfide Uses in the United States in 1980

Use	%
mining (flotation)	33
leather depilatory	20
chemicals and dyestuffs	19
export	18
other	10

Polysulfides

Sodium tetrasulfide, Na_2S_4, is prepared by the reaction of sodium sulfide with sulfur.

$$Na_2S + 3\,S \rightarrow Na_2S_4$$

Other polysulfides have been studied from di- through pentasulfide, but there are questions as to their actual structures. No commercial polysulfide of significance is produced other than the tetrasulfide.

Analysis

A double-end-point, acid-base titration can be used to determine Na_2S and either NaHS or NaOH. Analysis of sodium tetrasulfide requires sulfide determination, liberation of hydrogen sulfide, and determination of free sulfur by a gravimetric procedure.

Health and Safety Factors

Sodium sulfides are corrosive on animal tissues. Contact irritates and burns the eyes, skin, and respiratory tract.

CHARLES DRUM
PPG Industries, Inc.

Sulfur Chemicals, PPG Industries, Inc., Pittsburgh, Pa., 1979.

Sulfur Products, *Chemical Economics Handbook*, 780.4000A-H, Stanford Research Institute International, Menlo Park, California, Dec., 1983.

SOIL CHEMISTRY OF PESTICIDES

In the 1970s in the United States, pesticides came under scrutiny. Soils were of special interest since they represent the ultimate reservoir for most pesticides (see Insect control technology; Poisons, economic).

Insecticides

The principal classes of insecticides are the chlorinated hydrocarbons, organophosphates, carbamates, and synthetic pyrethroids.

In 1971, registration of DDT, dichlorodiphenyltrichloroethane, for most uses in the United States was canceled, and environmental residues have been declining steadily since then. The two processes that structurally alter DDT are microbial metabolism and photodecomposition. The course of degradation in soils is not fully established.

Aldrin (1) is oxidized to dieldrin (2) and heptachlor (3) to heptachlor epoxide (4) in soils. Both epoxides are insecticidal and prolong biological activity. Dieldrin is degraded by soil microorganisms.

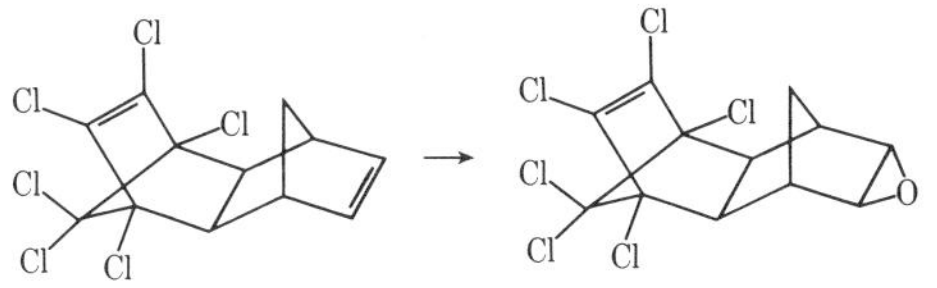

(1)   (2)

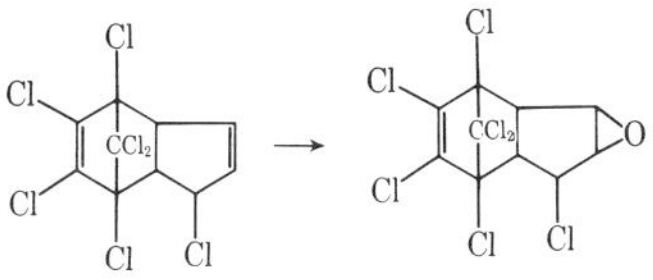

(3)   (4)

Lindane $1\alpha,2\alpha,3\beta,4\alpha,5\alpha,6\beta$-hexachlorocyclohexane, is less persistent in soil than DDT. It degrades to γ-pentachlorocyclohexene.

Several *methylcarbamates* are sold as broad-spectrum insecticides. Of these, carbaryl, 1-naphthyl methylcarbamate, is the most extensively used. Little is published on the fate of these insecticides in soils; however, this may relate to their brief persistence.

Like the carbamates, the *organic phosphates* are cholinesterase inhibitors. Parathion, diazinon, and phorate are used to control soil-borne insects. Organophosphorus compounds degrade fairly rapidly. Malathion is rapidly metabolized by a soil fungus and a bacterium; the latter is isolated from soils receiving heavy applications of the insecticide.

The insecticidal activity of the extracts of pyrethrum plants has been exploited commercially for over a century. The active principles are the *pyrethrins* and related compounds. Modifications of the structure led to more effective compounds that are economical and convenient to synthesize and more stable toward light. The degradation of pyrethroids in soil is rapid.

Herbicides

More than 180 herbicides (qv) are manufactured in the United States and more than 6000 formulations are registered. Principal chemical classes include the phenoxyalkanoic acids, s-triazines, phenylcarbamates, phenylureas, dinitrobenzenes, benzoic acids, thiocarbamates, anilides, dipyridyls, chlorinated aliphatic acids, and uracils.

The phenoxy herbicides are related by the following structure:

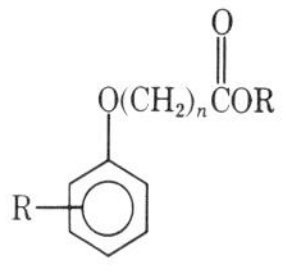

The phenoxyalkanoic acids decompose in aqueous solution in sunlight. The main reactions occurring are fission of the ether bond and replacement of halogen by hydroxyl. Microbial metabolism is the principal mechanism of breakdown in soils.

Herbicidal *s*-triazines are 2-substituted derivatives of the 4,6-bis-(alkylamino)-*s*-triazines, where X may be Cl, SCH_3, or OCH_3 and R and R' are alkyl groups.

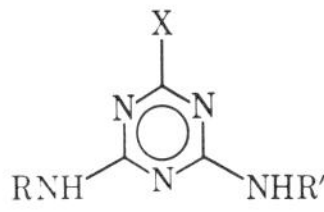

Chemical and biological reactions play an important role in altering these herbicides in soils. The principal degradation product is the corresponding 2-hydroxy-4,6-bis(alkylamino)-*s*-triazine.

Most phenylureas contain the 1,1-dimethyl-3-phenylurea moiety, where X and Y may be H, Cl, CF_3, or OC_6H_5.

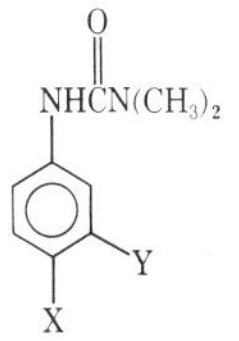

Degradation occurs under conditions conducive to microbial activity, ie, warm, moist, soils with high organic content. Urea herbicides readily undergo photolysis.

Phenylcarbamates are metabolized by microorganisms, as shown below for chlorpropham.

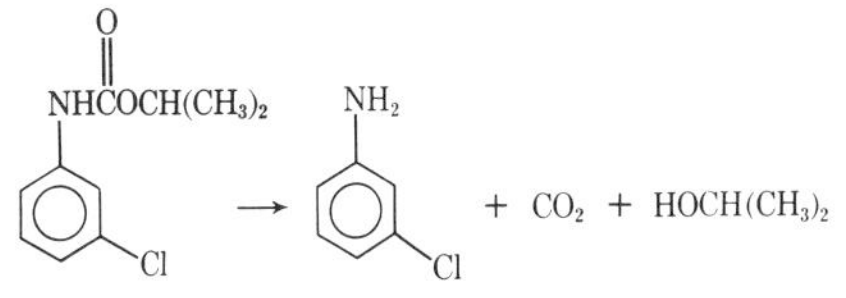

The *acylanilide herbicides* closely resemble the phenylcarbamates. Propanil, *N*-(3,4-dichlorophenyl)propanamide, is the principal herbicide used for weed control in rice. Soil microorganisms cleave propanil to 3,4-dichloroaniline and propionic acid.

The *chlorinated aliphatic acids*, TCA (**5**) and dalapon (**6**), are used extensively as herbicides. Microorganisms rapidly metabolize most members of this family.

$$CCl_3COOH \qquad\qquad CH_3CCl_2COOH$$
$$(5) \qquad\qquad\qquad (6)$$

Amitrole, 3-amino-1*H*-1,2,4-triazole, has been used for a variety of industrial and agricultural weed-control applications, but in 1971, uses on food crops were canceled. Several adducts have been identified as plant metabolites.

The *dinitroanilines* trifluralin, nitralin, and bensulide, are used in cotton and soybean cultivation. Photodecomposition, volatilization, and soil microorganisms contribute to their disappearance. Degradation proceeds by removal of the *N*-alkyl groups.

The *dipyridyls* paraquat and diquat are desiccants and contact herbicides. In soils they are inactivated almost immediately.

Arsenical Pesticides

Arsenicals have been used in agriculture for many years as insecticides, herbicides, soil sterilants, and silvicides. The inorganic arsenicals have largely been replaced by the less toxic organic arsenicals, eg, DSMA (disodium monomethanearsonate), $CH_3AsO(ONa)_2$, and cacodylic acid (dimethylarsinic acid), $(CH_3)_2AsO(OH)$. Rice is particularly sensitive to arsenic residues in soil where they exist in water-soluble form, and iron, aluminum, calcium, and nonextractable arsenate salts.

Fungicides

Fungicides are used at low concentrations as seed protectants and disinfectants on most crop areas. On small areas, however, fairly high concentrations are used to control diseases mainly on citrus and deciduous fruits.

The inorganic fungicides, including sulfur and Bordeaux mixture, $CuSO_4 + Ca(OH)_2$, are among the oldest known pesticides. In soils, sulfur oxidation is rapid when initiated by microbes.

Copper is a natural component in soils, but continuous use of copper sulfate as a fungicide has caused phytotoxicity to citrus trees.

Mercurous chloride and the organic mercurials are used as seed disinfectants. After 30–50 d, a large portion of soil-applied organic mercurial is unaltered.

Thiocarbamate fungicides may be classified into thiuram disulfides, $(R_2NCS_2)_2$; metallic dithiocarbamates, R_2NCS_2M; and ethylenebis (dithiocarbamates), EBDC, $(CH_2NHCS_2)_2M$. Nonvolatile fungicides are inactivated nonbiologically in soils, whereas others are metabolized by soil microorganisms.

Processes Affecting Pesticides

Adsorption. Adsorption influences pesticide movement, photodecomposition, microbial or chemical decomposition, and plant uptake. Pesticide adsorption is usually presented in the form of adsorption isotherms, ie, graphs relating equilibrium solution concentration to the amount of solute sorbed per unit mass of adsorbent. Among mechanisms identified or postulated to account for pesticide-soil adsorption are weak van der Waals forces, hydrogen bonding, ligand exchange, charge transfer, entropy generation or hydrophobic bonding, strong ion exchange (qv), and strong chemisorption. The amount of organic matter and the amount and type of clay are usually well correlated with the amount of pesticide sorption to soil; secondary properties, eg, surface area and cation-exchange capacity, are also related to sorption.

Characteristic functional groups of pesticides that seem to be associated with enhanced adsorption include: R_3N^+, $CONH_2$, OH, NHCOR, NH_2, OCOR, and NHR. Protonation of any functional group increases adsorption, as does increasing molecular size.

Leaching. Leaching refers to the transport of solutes within soil. Although movement of most pesticides is probably restricted to less than 1 m within soil, residues of at least twelve pesticides have appeared in groundwater.

Adsorption of pesticides has been correlated with restricted mobility. Organic matter is the most significant soil component with respect to pesticide retardation, but clay is more important for organic cations and perhaps for chemicals that may coordinate with mineral cations.

The chemical characteristics that affect adsorptive behavior are related to mobility. Organic cations are strongly sorbed despite very high water solubility; at the other extreme, very poorly soluble pesticides are also immobile. Among the more mobile chemicals are organic anions. Degradation is an important attenuating factor. Thus, dalapon is highly mobile in laboratory leaching tests but is not a problem in use because it degrades rapidly. Leaching is usually evaluated in the laboratory by soil column or soil thin-layer chromatographic methods.

Photodecomposition. Decomposition of pesticides by light occurs not only through direct absorption of light, but also by reaction with products of a photochemical process. Photosensitized reactions may accelerate decomposition of pesticides. Studies of the photochemistry of pesticides on silica reveal that the acidic nature of silica changes the adsorption spectrum and it is therefore difficult to extrapolate results of experiments that are not conducted in the soil environment. The effect of soil and silica environments on photolysis has been compared, and the latter may favor photodecomposition of adsorbed materials because it allows passage of uv energy to a much greater extent than soil. The complexity of the natural environment has restricted photochemical studies primarily to solutions or solid films.

Photooxidation usually transforms amines and phenols into dark materials of high molecular weight. In aqueous solution, halogenated phenoxyaliphatic acids generate polyhydroxyphenols. Predictions based on extrapolation from the photochemical behavior of a model system are of limited value until the effects of complicating factors can be studied in depth.

Metabolism and chemical degradation. Microbiological and chemical reactions affect pesticide persistence in soils. The principal reactions associated with pesticide decomposition by soil microorganisms include dehalogenation, dealkylation, amide or ester hydrolysis, oxidation, reduction, ether fission, aromatic ring hydroxylation, and ring cleavage.

Oxidation, an important reaction of chlorinated insecticides containing an isolated double bond, leads to formation of epoxides. Purely chemical reactions are probably less well understood than the biochemical reactions mediated by soil microorganisms.

Plant absorption and translocation of residual pesticides used in a current or previous cropping system may represent one possible avenue of human exposure to pesticides in the diet. The most important property of the pesticide influencing plant contamination is water solubility. Under actual field conditions, it is impossible to separate contamination by root uptake and translocation from contamination by dust, drift, splashing, or volatilization.

Problems

A number of unanticipated problems arose in the 1970s. The main object of concern is dioxin, 2,3,7,8-tetrachlorodibenzo-*p*-dioxin (TCDD), which is extremely toxic (LD_{50} 6 µg/kg in male guinea pigs); it is fetotoxic and teratogenic to laboratory animals. It is uncertain whether toxic effects in laboratory animals can be translated to man. Chloracne (a severe form of acne) is the most common sympton observed in industrial workers exposed to TCDD.

Interest continues in the exposure of military personnel to TCDD in Agent Orange, a defoliant used in Vietnam. TCDD is very insoluble in water (0.2 ppb), which governs many of its properties in soils. In recent studies, TCDD was detected in fish and bird samples collected in the Great Lakes region. The source of these residues is the subject of much debate.

Ethylenethiourea (ETU) is a manufacturing, processing, and metabolic product of EBDC fungicides. It became an environmental issue when toxicological studies indicated that it may be goitrogenic, tumorogenic, and teratogenic to laboratory animals. In soils ETU is rapidly oxidized.

The highly effective nematicide DBCP, 1,2-dibromo-3-chloropropane, was synthesized following the discovery that a halogenated hydrocarbon mixture improved pineapple growth. However, it causes temporary sterility among male production-plant workers and was identified as a potential carcinogen. Use of DBCP in California was suspended in 1977 and in all states by 1984, partly as a result of detection of DBCP in some well water.

Consumption and Regulatory Actions

Although the regulatory actions of EPA changed the pattern of compounds used for pest control, the industry grew substantially during 1970–1980. U.S. consumption peaked in 1975, whereas exports nearly doubled and imports increased almost eightfold. The single most important event affecting the use of pesticides in the United States was the establishment of the EPA in 1970 (see Regulatory agencies). Most of the important chlorinated hydrocarbon insecticides have been discontinued, suspended, or otherwise restricted, as have been many metal-containing pesticides.

P.C. KEARNEY
C.S. HELLING
J.R. PLIMMER
U.S. Department of Agriculture

R.J. Hance, ed., *Interactions between Herbicides and Soil*, Academic Press, Inc., New York, 1980.

C.S. Helling, P.C. Kearney, and M. Alexander, *Adv. Agron.* **23**, 147 (1970).

P.C. Kearney and D.D. Kaufman, *Herbicides: Chemistry, Degradation and Mode of Action*, 2d ed., Vols. 1 and 2, Marcel Dekker, Inc., New York, 1976.

SOLAR ENERGY

Solar-energy technology comprises two distinct categories: thermal conversion and photoconversion. Thermal conversion takes place through direct heating, ocean waves and currents, and wind. Photoconversion includes photosynthesis, photochemistry, photoelectrochemistry, photogalvanism, and photovoltaics.

The collection of solar energy is illustrated in Figure 1. The United States is making appreciable use of solar energy in the form of hydroelectric power, annually approximately equal to the contribution of nuclear power plants. Additional energy is derived from burning biomass.

Thermal Energy

The sun radiates energy at an effective surface temperature of ca 6000 K and terrestrial solar concentrators capture significant amounts of this thermal energy.

The energy in solar radiation is not a function of ambient air temperature but depends on the clarity of the sky and the angle at which the sun's rays are received. Even on an overcast day, 50% of maximum solar radiation may reach the surface. A well-designed passively heated home may receive all the heat it needs from the sun. Surplus heat is stored in floors, walls, water tanks, or rock bins. Trombe walls add effectiveness to passive designs. Passive cooling techniques include cross-ventilation, insulation, shading, regenerative rock bed cooling, evaporation, and radiation to the night sky.

In active solar heating, the heat-exchange medium is moved from a solar collector to the space that is to be heated. The most common solar collectors are used for domestic water heating. Water systems are much more common than air systems. They offer better heat-exchange performance but have the disadvantage of leaks and susceptibility to freezing. The antifreeze loop seems to offer the best protection. Much effort has been spent on developing selective surfaces by coating or other treatments to obtain high absorptance but low emittance.

Current solar air-conditioning projects include solar-powered refrigeration systems of both Rankine-cycle vapor-compression types and absorption systems. Other cooling systems include solar-regenerated desiccant cooling. Open-cycle adsorption systems seem to offer the best prospects. Desiccant cooling is best suited for regions with about equal heating and cooling loads and high humidity.

Industrial process heat. Industrial and agricultural applications are often amenable to active solar heating. Food dryers and car washes represent the low temperature end of this spectrum, which extends upward to very hot water and other liquids in the food, textile, brewing, and other industries, and even to steam. The best flat-plate collectors produce temperatures up to ca 100°C, but at considerable loss of efficiency at the high end. Thus, concentrating collectors are preferred at medium temperatures. Temperatures produced by various solar collector systems are given in Table 1.

Power generation. It is still widely believed that solar energy is too diffuse for use as a power source; however, an area of one square meter intercepts ca 1 kW of solar radiation in clear weather. The south-facing portion of a roof of ca 80 m² (860 ft²) receives far more energy than the house consumes for heating, cooling, light, and appliances. A case can also be made for central solar power, and a DOE 10-MW pilot plant is now being operated in southern California. Costs on the order of $2.5/W are projected for commercial solar power plants, competitive with most fossil fuels. In the U.S., solar power plants function best in the Southwest, where solar radiation is abundant. Typical average direct normal radiation is ca 7 (kW·h)/(m²·d).

Other proposed projects include the orbiting solar-power satellite (SPS), comprising arrays of photovoltaic cells for converting solar radiation to electricity. The electric-power output would be converted continuously to microwave frequencies and beamed to rectifying antenna arrays on earth. Here, the microwave radiation would be converted to alternating current at frequencies compatible with utility grids.

Solar ponds. In Israel, the first man-made, reverse-gradient solar pond was completed and a specially designed turbine for producing electric power from water at 90°C was installed. In 1979, a 150 kW generator began operation at a 7500 m² (ca 81,000 ft²) solar pond. The Israeli Ormat team is building a 5-MW solar pond at the Dead Sea and a 48-MW solar pond for Southern California Edison in California. Solar ponds operate in a horizontal plane, thus suffering some loss of intercepted radiation as compared to a tilted or tracking solar collector. However, construction costs are relatively low.

A nonconvecting solar pond typically comprises a thin, convecting top layer that serves as a buffer for wind; a middle, salt-gradient, nonconvecting layer; and a convecting bottom layer that stores most of the heat.

Shallow convecting ponds have black, heat-absorbing bottoms, insulated sides, and plastic-glazed tops (see Fig. 2). Water is heated during the day in the ponds and stored at night in underground tanks. Such installations are nearly competitive with petroleum.

Materials and equipment. Materials may be categorized as glazing, heat absorbers, plumbing, insulation, supporting structures, reflective materials, heat-transfer fluids, and sealants.

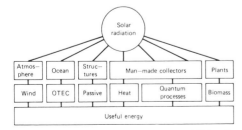

Figure 1. Natural and man-made solar collectors. OTEC = Ocean Thermal Energy Conversion.

Table 1. Temperatures Produced by Various Collectors

Collector type	Temperature, °C
solar ponds/flat plate	100
evacuated glass-tube collectors	50–175
line-focus concentrators	65–300
point-focus concentrators	200–800
central receiver	800–1400

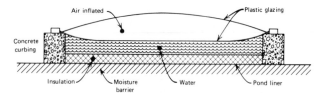

Figure 2. Shallow solar pond. Courtesy of Lawrence Livermore Laboratories.

Glazing is the first component of a solar collector to receive radiation. Glass is the traditional material, and plastic substitutes have yet to match its stability, weatherability, and resistance to uv radiation. Reflective materials for heliostats or concentrators include silvered glass, polished metals, and reflective foils and films. Absorber plates in solar collectors vary from blackened plastics for low temperature use, such as pool heating, to metals with selective surface coatings. Heat-transfer media include liquids and gases. Some solar collectors utilize air; some use chlorofluorocarbons, including Freons. Collector boxes are made from wood, plastic, and metal.

Insulation prevents loss of collected heat energy. Polystyrene foam may be used in a solar swimming-pool heater, which has an upper temperature limit of ca 75°C. However, a glazed, medium temperature solar collector can attain sufficiently high temperatures that fiber-glass insulation or an equivalent is necessary. Seal and sealant materials must withstand thermal cycling and undergo minimum outgassing. Elastomers are choice materials.

Heat Storage

Insolation varies diurnally as well as seasonally, and it is generally necessary to provide some means of storing heat for nights and cloudy periods.

Heat can be stored in water at ca 4 kJ/(kg·K) [1 Btu/(lb·°F)] increase in temperature. Rock stores heat, too, although not as much for a given volume. Both these media store sensible heat:

$$Q = MC_pT$$

where Q = heat stored, M = mass of storage material, C_p = specific heat of storage material, and T = temperature change of storage material. Sensible heat storage in residences is common in Europe. Typical systems use refractory brick.

A heat-storage system, called Annual Cycle Energy Storage (ACES), was developed at Oak Ridge National Laboratories; water is stored in the basement of a house. In winter, a heat-pump system heats the house, drawing on the stored water. By spring, the water is frozen and then used for cooling the house in summer.

Latent heat storage requires a phase change. Water is a common example. Melting ice absorbs great quantities, but the phase-change temperature is not suitable for storing heat. Material is needed that changes phases at a higher temperature. Sodium sulfate decahydrate (Glauber's salt) was one of the first materials used. The problems encountered in phase-change heat storage stem from crystallization and gradual loss of ability to change from liquid to solid after a number of phase-change cycles.

Solar power and industrial process-heat systems call for higher temperature heat storage, ie, > 250°C, than residential or commercial applications. The insulating material must not only insulate the heat-transfer medium but also survive the elevated temperature.

Heat-storage materials used for reversible thermochemical reactions include hydrated, ammoniated, or methanolated salts, sulfuric acid, hydrogenated metals, and hydrated zeolites. Thermochemical systems are extremely complex and thermochemical storage is a long way from commercial use.

Ocean Energy

More than 70% of the solar radiation reaching the earth falls on the ocean. The flux in the upper layer of ocean water influenced by winds is estimated at ca 100 EJ/yr (95 × 10^15 Btu/yr). However, the prospect for converting more than a very small fraction of the potential is slim. Japan, the United Kingdom, the United States, Canada, and Norway have for some time sought methods to produce large amounts of power from waves.

Ocean gyres could be tapped for power. Driven by prevailing global wind systems and modified by the Coriolis effect, the currents rotate clockwise in the northern hemisphere and counterclockwise south of the equator. Of the most interest to the United States is the Gulf Stream. Many schemes have been proposed for tapping ocean currents. These include huge undersea turbines, as well as large parachutes attached to long cables on a system of pulleys.

Ocean thermal energy conversion (OTEC). The idea of exploiting thermal gradients was expanded to a heat engine. The OTEC plant as a heat engine is limited in efficiency by the Carnot cycle:

$$\text{Eff} = \frac{T_{in} - T_{out}}{T_{in}}$$

where T = temperature in K. The OTEC plant operates on a temperature difference of ca 30°C at best. Assuming surface water at 27°C (300 K) and at 1000 m depth at 4°C (277 K), the maximum theoretical efficiency is

$$\frac{300 - 277}{300} \times 100\% = 7.67\%$$

This is a theoretical value which is reduced to ca 2–3% in practice. Nevertheless, the OTEC scheme is attractive because the ocean heat source is virtually limitless and offers baseload power.

A commercial OTEC plant would be ca 100 m across and would dangle a cold-water pipe (20–60-m dia) hundreds of meters below it. In addition to metals and plastics, concrete could be used for the plant and cold-water pipe. It is felt that the state of the art is capable of fulfilling OTEC needs.

An OTEC system raises nutrient-laden cold water which increases fish populations in the vicinity of the plant. It has been suggested that minerals and chemicals might be recovered profitably as an adjunct to the production of power.

The hybrid-cycle design combines features from both closed and open cycles. Warm seawater is flash-vaporized in partially evacuated evaporators. The vapor produced is then used to evaporate a second working fluid as in a closed-cycle system. Variations use hydraulic effects rather than expansion to produce mechanical power.

Marine biomass. The ocean is the largest but least cultivated pasture on earth, with ca 10^10 metric tons of carbon per year fixed by photosynthesis. In the mid-1970s, the Ocean Food and Energy Farm (OFEF) project was funded by several public and private organizations. A huge, open-ocean kelp farm was planned. A test module was installed off Laguna Beach, California. Although many technical and economical problems remain to be solved, research on a full-scale project continues.

Wind Energy

Wind is not generally understood as the solar-thermal process it is; however, a small portion of solar energy in the atmosphere is transformed into wind. Factors involved in this transfer of heat energy to the atmospheric cycle include unequal heating of land and air masses, resulting atmospheric pressure changes, the rotation of the earth, and the diurnal and seasonal changes in solar-energy input. The atmosphere is indeed a huge heat engine fueled by solar radiation. A study at Princeton University suggests that a wind resource of 8% of the area of the contiguous American states would provide 220 GW of electric power.

Wind velocities of 5.4–17.9 m/s are attractive for existing wind machines. Wind power varies with the cube of velocity (see Fig. 3); thus, higher speeds are desirable. The rated output of a given machine should be at ca 1.5 to 2 times the measured average wind.

Wind machines, rotating about a vertical or a horizontal axis, extract power from wind by converting its kinetic energy to rotary or other mechanical motion. Large wind machines are designed and built by large industrial firms, eg, General Electric, Bendix, Hamilton Standard, and

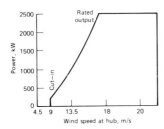

Figure 3. Typical cut-in/cut-out speed regime for wind machine. Courtesy of *Solar Age*.

Boeing. In addition to wood, steel, and aluminum, blade materials include cement, plastics, fiber glass, and even carbon and boron fibers. Cost projections show large wind machines competitive with conventional electric power.

Because of the Public Utilities Regulatory Policies Act, utilities are required to purchase electric power from small power producers such as operators of wind machines. For this reason, the operator may not need to provide storage. Instead, surplus electricity can be sold to the utility and bought during periods of little or no wind. Large wind machines are obviously of greater interest to utilities geared to power-plant operation.

Regions of high wind-power density include the Midwest, Pacific Northwest, Texas coast, Northeast, and Northeast seacoast. Hill crests and narrow canyons usually create stronger-than-normal winds. Careful monitoring is desirable for a specific site.

Wind power presents a number of problems, including icing of blades, blade or tower failure, and alteration of surface wind regime. Noise and interference with TV reception are particularly objectionable in urban areas and have created legal obstacles to the installation of windmills. High winds also pose problems. Sizable wind machines have been damaged or destroyed by severe gusts.

Development of wind technology is taking place on two fronts. New machines provide power of up to 100 kW in rural areas and in small communities; in addition, machines of much higher capacity are producing power for electric utility grids. About 300 MW of wind electric power was fed into utility grids in early 1983.

Photoconversion

In the context of solar-energy conversion, photoconversion refers to the direct production of fuels and chemicals from sunlight and water, carbon dioxide, nitrogen, and simple organic compounds. Typical products are hydrogen, methane, alcohols, ammonia, and organic nitrogen compounds.

The natural photosynthetic process produces large amounts of biomass useful for fuel, but only at modest conversion efficiencies.

Photobiological systems require the physical separation of dark reactions from light-driven reactions, as has been demonstrated for both bacteria and green plants, eg, by immobilization of the photosynthetic mechanism which produces charge separation. The immobilized system can be coupled to a photoelectrochemical cell.

The production of hydrogen by various phototropic organisms occurs in the presence of light with photosynthetic bacteria, cyanobacteria, or algae. *In vivo* cultures of photosynthetic bacteria offer the best prospect for efficient photobiological hydrogen production. *In vitro* hydrogen generation, employing isolated spinach chloroplasts, ferredoxin, *Chromatium* hydrogenase, and cysteine, was first reported in 1961. Since then, considerable improvements have been made.

Photochemical reactions are carried out in either homogeneous or heterogeneous media. Liquid-phase processes are preferred. Synthetic chloroplasts have been produced analogous to the natural chloroplasts in green plant cells. These structures split hydrogen from water and form carbohydrates by combining hydrogen with carbon from carbon dioxide in the atmosphere.

Photoelectrochemistry is based on the properties of photoactive semiconductor electrodes in contact with liquid electrolytes. Cells can be designed to produce either chemicals and fuels or electricity. In the former case, the different oxidation and reduction reactions result in a net change in the electrolyte that creates fuels or chemicals.

Storage of electricity is possible in an electrochemical photovoltaic cell with the help of a third storage electrode or the use of redox electrolytes in a redox battery. Experimental solar-powered, electrochemical, photovoltaic storage cells have been operated for months, providing an electric current day and night.

Liquid-junction solar cells offer several advantages over conventional solid-state photovoltaic cells (qv). A redox electrolyte has a Fermi level or redox potential and thus can take the place of a *p*-type semiconductor or a metal to form the necessary junction. Liquid-junction cells are much larger and heavier and may require sealing if the electrolyte solution is affected by air.

Photoelectrosynthetic cells that drive reactions thermodynamically uphill are of particular interest since solar energy is thus stored as chemical energy. The photolytic splitting of water, called photoelectrolysis, is of exceptional interest and research is in progress worldwide.

Photogalvanic processes involve absorption of incident light by dye molecules in the cell electrolyte. The photoexcited dye molecules drive redox reactions in the electrolyte, and these redox products then transfer charge to the cell electrodes to produce an electric current.

DAN HALACY
Solar Energy Research Institute

A.R. Hoffman, *Domestic Policy Review of Solar Energy*, Solar Energy Policy Committee, U.S. Department of Energy, Washington, D.C., Aug. 25, 1980.

A.F. Clark and W.C. Dickinson, *Solar Technology Handbook*, Part A, Marcel Dekker, Inc., New York, 1980, p. 377–402.

SOLDERS AND BRAZING ALLOYS

Solders

Soldering is used for making a mechanical or electrical connection. Common solder alloys are given in Table 1.

Joints. A good solder joint should provide visual inspectability, electrical conductivity, mechanical strength, ease of manufacture, and simplicity of repair. The metals to be joined must be compatible with the solder, and the surfaces must be solderable. Parts being joined must be clean.

Fluxes. A flux is essential. It removes surface compounds, reduces surface tension of the molten solder alloy, and prevents oxidation. Most inorganic fluxes are a combination of salts and acids dissolved in water with a wetting agent. They should not be used in electrical or electronic soldering, because of corrosive residues.

Most fluxes are organic water-insoluble rosins or water-soluble organic acids. Rosin flux is inert, noncorrosive, and nonconductive in the cold solid state, but active in removing tarnish when hot. Water-soluble organic fluxes are used increasingly in electromechanical and electronic soldering.

Most solder alloys are combinations of tin and lead made in many shapes and forms, eg, wire, bar, foil, spheres, ring, or paste.

Small runs are generally heated manually. All parts to be soldered are heated to ca 50°C above the melting point of the solder. Most popular is the electric soldering iron.

Table 1. Common Solder Alloys

Sn	Pb	Cd	Bi	Ag	Sb	Melting range, °C	Use
63	37					183	eutectic solder for electronic application
60	40					183–190	high quality solder
50	50					183–216	general-purpose solder, plumbing
40	60					183–238	wiping solder, radiator solder
30	70					183–255	machine and torch soldering
20	80					183–277	automotive-body solder
95					5	235–240	refrigeration soldering
62	36			2		179	soldering silver surfaces
1	97.5			1.5		309	high temperature soldering
15.5	32		52.5			90	fusible links
13	27	10	50			70	low melting solder

In wave soldering, the heat is applied to the mass soldering of printed wiring assemblies on a plastic-laminated board. Hundreds of connections can be soldered in a few seconds. In oven soldering, the solder is placed on the assembly and the units are placed in an oven for soldering. Vapor-phase reflow soldering is a special technique for special mass soldering applications.

After soldering, flux residues and oxides have to be removed. Flux residues may be corrosive and cause malfunction. Flammable solvents, although excellent solvents for rosins and activators, are not used for safety reasons. Chlorinated solvents remove rosin but not polar activators. Combinations of alcohols and chlorinated solvents are preferred. Water-soluble organic flux residues are removed with water.

Solders present no safety or health hazard under proper working conditions.

Brazing Alloys

In welding, similar components are fusion-bonded at or just below their melting points. In most brazing and almost all soldering, the components are bonded well below their melting points. Alloying may or may not take place, but extensive alloying should be avoided.

Brazing is simple. To produce good joints, the surfaces must be prepared, and fluxing, assembling, jigging, heating, and cleaning are required. Butt joints are used where lap joints cannot be accommodated by the shape of the parts. Oil, grease, dirt, oxides, and scale must be removed. Fluxes prevent oxidation. Protective atmospheres reduce the requirement for flux. Fluxes used with Al–Si and Mg–Al alloys function at 370–675°C and 480–650°C, respectively; they contain chlorides and fluorides.

In assembling the parts to be joined, the filler metal should be preplaced. Thin sheets or washers may be used between flat surfaces and rings made of rectangular strip or round wire on tubular members. Joints should be designed to be as self-supporting as possible. The jig should be simple and light with a minimum of contact between it and the assembly, which is heated broadly and uniformly just above the flow point of the brazing alloy. Chloride and borate fluxes must be removed immediately after brazing.

Health and safety factors. Zinc and cadmium volatilize during brazing and inhalation of their vapors must be avoided. Cadmium presents more hazards than zinc and cadmium brazing alloys should not be used in food applications.

Alloys. Copper is a furnance brazing alloy that is extremely fluid on iron and iron alloys. Its brazing temperature (1093–1194°C) is fairly high and its service temperature is low (ca 200°C). It is used in alloy combinations with zinc, silver, and tin. Some copper–phosphorus alloys are very fluid on stainless steel but erode it far less than silicon–boron alloys. Silver–copper alloys include pure metals and eutectics. They are used in the step brazing of high grade electron tubes and electronic devices.

Gold alloys are vacuum-tube step-brazing alloys. Nickel alloys are brittle and are only available as powder, powder paste, and transfer tapes. Cobalt brazing alloy is compatible with cobalt alloys and is used in diffusion brazing of jet engine parts. Aluminum–silicon alloys are the conventional aluminum brazing alloys. The furnace, dip, or torch brazing of magnesium–aluminum alloys must be precisely controlled.

GEORGE SISTARE
Consultant

FREDERICK DISQUE
Alpha Metals, Inc.

H.H. Manko, *Solders and Soldering*, 2nd ed., McGraw-Hill Book Company, New York, 1979.

Welding and Brazing, Vol. 6 of *Metals Handbook* 8th ed., American Society for Metals, Metals Park, Ohio, 1971, pp. 593–702.

Materials and Processing Databook "81," *Metals Progress*, American Society for Metals, Metals Park, Ohio, June 1981, pp. 150–151.

SOLVENT RECOVERY

Processes employing valuable solvents generally include solvent-recovery systems as part of the initial installation. Furthermore, solvents that have little value are also commonly recovered in order to conform with governmental or industry standards.

This article deals with the recovery of nonaqueous solvents employed in the formation, deposition, and drying of polymers; solvent extraction of oils and fats, of fossil fuels, and of metallic compounds from treated ores; solvent refining of oils; extractive and azeotropic distillations; absorption; degreasing; dry cleaning; solute chemical reactions and precipitations.

The solvent flow in a typical recovery system is given in Figure 1. Fresh and recovered solvents join together and enter an industrial process. The products are separated from recycle streams and a stream containing the bulk of the solvent for recovery. The last stream is recycled after purification. Solvent that is not recycled leaves the system as a component of the product, by-product, or waste streams.

Solvent-Recovery Systems

Solvent-recovery units may be classified according to the method used to make the initial separation between solvent and product streams, ie, mechanical separation, fractional distillation, or drying in the presence of air or inert gas followed by condensation, absorption, or adsorption.

Techniques

Draining of liquids from solids is a common operation. The solids are retained by stationary or moving screens, perforated plates, baskets, belts, or chains. Filtration is indicated where solids are present in small particle sizes. For flammable or noxious solvents, filter closures should be pressure tight or at least fumeproof. Continuous pressure-tight filters are sometimes used in the solvent refining of lubricating oils.

Settling followed by decantation is an obvious way to separate immiscible liquids, solids, and gases. Solutions of materials occurring in nature or of synthetic polymers may produce relatively stable foams, which must be counteracted. Heat, energy, and chemicals may be required for the phase separation.

Centrifugal filters, perforated basket centrifuges, and solid-bowl centrifuges use centrifugal force to increase the efficiency of filtering, draining, and settling operations, respectively (see Centrifugal separation).

Extraction. Solvent and product are sometimes separated by washing with water or another solvent. Extraction is sometimes accompanied by a chemical reaction between a component of the washing solution and the impurity (or product) being extracted. The extraction step may be preceded by a chemical reaction in which the material to be extracted is converted to a more easily extractable compound.

Chemical reaction techniques are particularly used in solvent-recovery systems associated with the extraction of transition-metal compounds from ores (see also Extractive metallurgy). Extraction with chemical reaction is performed in the same type of equipment as extraction without chemical reaction, but the design calculations are more complex.

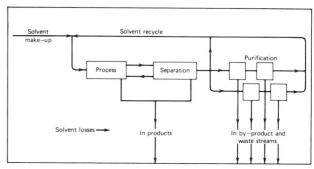

Figure 1. Solvent flow in generalized solvent-recovery system.

Evaporation. Solvents are often recovered by evaporation and condensation. In dry cleaning, for example, grease and dirt must be removed from the solvent. Package stills, complete with condenser, feed preheater, and semiautomatic controls are available for this purpose. For high boiling Stoddard solvent the operation may be under vacuum, whereas for lower-boiling chlorinated solvents it may be at atmospheric pressure. Multiple-effect evaporation and vapor-reuse distillation conserve energy.

The solvent power of high pressure intermediate-density supercritical fluids can be changed by varying the density. Less energy is required to separate solute from solvent than with conventional vaporization. Possible uses include secondary and tertiary petroleum recovery.

Direct steam is used to steam distill solvent from the surfaces or pores of solids.

Fractional distillation. Solvents are separated from products, water, or other solvents by fractional distillation. A series of distillation operations may be required. Sieve-plate or bubble-cap columns are generally specified, but packed columns and other designs are also used. Operation is normally continuous, except for very small quantities (see Distillation).

Drying. Drying solids by vaporizing the solvent in the absence of air or inert gas reduces explosion hazards, eliminates mechanical air circulation, and simplifies solvent recovery. Heat is usually supplied by surfaces heated with steam or other media or occasionally radiant lamps.

Drying in the presence of air or a mixture of nitrogen and CO_2 (deoxygenated air) permits vaporization of solvent at lower temperatures than in the absence of air. In addition, hot circulated air provides good heat transfer and does not require contact of product with hot surfaces. Recovery of solvent from the air may be by adsorption, absorption, or condensation (see Drying).

Condensation. Vaporized solvents may be liquefied either in shell-and-tube surface condensers or direct-contact condensers. Direct contact with water is largely restricted to solvents with very low water solubilities, but direct contact with cold solvent may be used in any case.

Solvents may be recovered from air or gas streams by scrubbing with a suitable liquid, frequently water, in sieve-plate or bubble-cap columns with moderate pressure-drop designs. The temperature of absorption determines the amount of scrubbing liquid required.

Adsorption. The predominant method for removing solvent vapors from air streams is adsorption on activated carbon. Steaming causes the solvent to be vaporized, and the latent heat is supplied by a portion of the steam that condenses into the bed. Unlike water scrubbing, adsorption can be applied to water-insoluble solvents. Unlike condensation, it can reduce the solvent content of the air as much as desired.

Some volatile compounds react chemically on the surface of activated carbon. The success of carbon adsorption depends to a large extent upon proper control of the operation cycle. At constant rate of solvent recovery, automatic time-cycle controllers are often used. Continuous contacting is feasible on a large scale. Today completely automatic control systems are available and a single piece of equipment may be used for a whole series of operations (see Adsorptive separation).

Solvent-recovery decisions are based mainly on economic analysis. The recovery system should be designed early in the development of a new process. Even with efficient recovery, make-up solvent is always required. Investment and operating costs increase as the quantity of solvent increases, but not proportionally. In small-volume solvent applications, for which individual recovery and disposal systems would not be economical, solvent effluents can be collected, stored, and periodically recovered.

Health and Safety Factors; Government Regulations

Accepted practice with regard to the handling of flammable solvents is contained in UL codes and those of the National Fire Protection Association. They form the basis for municipal safety laws and insurance inspection requirements. Mixtures of air and solvent must be kept below the explosive range. Permissible vapor concentrations in work places are specified by OSHA standards.

Because of new legislation and regulations, innocuous solvents and systems giving essentially complete recovery are increasingly preferred.

Application

Processes for the manufacture of synthetic fibers and sheets, impregnated articles, and related products use large amounts of solvents as volatile vehicles for the deposition of polymeric materials. Industry is expected to substantially reduce emissions of volatile organic solvents.

In solvent extraction of vegetable oils and other solid-liquid extraction processes in the United States, more than 2×10^6 t/yr of solvent is recovered. Solvent refining of lubricating oils and other solvent operations in the petroleum industry entail the recovery of solvents in quantities far exceeding all other applications.

C.M. COOPER
Michigan State University

R.E. Treybal in R.H. Perry and C.H. Chilton, eds., *Chemical Engineers' Handbook*, 5th ed., McGraw-Hill Book Co., New York, 1973, Sec. 15.

L.A. Robbins in P.A. Schweitzer, ed., *Handbook of Separation Techniques for Chemical Engineers*, McGraw-Hill Book Co., New York, 1979.

J.W. Drew in P.A. Schweitzer, eds., *Handbook of Separation Techniques for Chemical Engineers*, McGraw-Hill Book Co., New York, 1979.

SOLVENTS, INDUSTRIAL

The term industrial solvents is generally applied to organic compounds used on an industrial scale to dissolve, suspend, or change the physical properties of materials. These applications include production of dissolved or suspended materials for the ink or coatings industries, reaction solvents for the process industry, and cleaning agents for the metal cleaning or dry-cleaning industries. Most organic liquids can act as solvent on an industrial scale. Generally, these compounds are aromatic or aliphatic hydrocarbons, alcohols, aldehydes, ketones, amines, esters, ethers, glycols, glycol ethers, or alkyl or aromatic halides; boiling points range from 75 to 220°C.

Selection as well as the development of new solvents is aided by knowledge of the solvent–solute interaction. Solubility parameters are useful, particularly in the coatings industry, because they characterize systems for experimental evaluation. Solubility parameters must be verified experimentally.

Solubility

Solution of one compound by another is based on the concept that the process must be spontaneous. This ideal is expressed by the equation for the free energy, ΔG, of mixing:

$$\Delta G = \Delta H - T\Delta S$$

where ΔH is the enthalpy and ΔS is the entropy. The controlling term for the spontaneous process is the enthalpy. If it is negative or a small positive number, the process could be considered spontaneous.

Single-component solubility parameter. An expression was developed for the heat of mixing:

$$\Delta H_m = \frac{x_1 x_2 V_1 V_2}{x_1 V_1 + x_2 V_2} \left[\frac{a_1^{1/2}}{V_1} - \frac{a_2^{1/2}}{V_2} \right]^2$$

where x_1 and x_2 are mole fractions, α_1 and α_2 are interaction constants, and V_1 and V_2 are volumes. Cohesive energy of a mole of liquid mixture could be expressed as:

$$\Delta E_m = (x_1 V_1 + x_2 V_2) \left[\left(\frac{\Delta E_1^v}{V_1} \right)^{1/2} - \left(\frac{\Delta E_2^v}{V_2} \right)^{1/2} \right]^2 \phi_1 \phi_2$$

where ΔE^v is the energy of vaporization and ϕ_1 and ϕ_2 are volume fractions. The term $\Delta E^v/V$, the energy of vaporization per unit volume, can be taken as a measure of the internal pressure. It is often called the solubility parameter δ. In other words,

$$\left(\frac{\Delta E^v}{V} \right)^{1/2} = \delta = \frac{a^{1/2}}{V}$$

is the expression most commonly used for the solubility parameter, and δ has the units of $(J/m^3)^{1/2}$ [$= 4.89 \times 10^{-4}$ $(cal/cm^3)^{1/2}$]. Therefore, the free energy of mixing is given by the expression

$$\Delta G = V_t [\delta_1 - \delta_2] \phi_1 \phi_2 + RT(x_1 \ln x_1 + x_2 \ln x_2)$$

and solution should be assured as δ_1 approaches δ_2. According to this theory, two substances should mix when the solubility parameters are equal. The solubility parameter for a blend is given by

$$\delta_{blend} = \delta_1 \phi_1 + \delta_2 \phi_2 + \delta_3 \phi_3 \ldots \delta_n \phi_n$$

where ϕ represents the volume fractions.

Two-component solubility parameter. Polar molecules, or molecules that hydrogen-bond, have interactions that influence both the enthalpy and entropy terms. The correction for the heats of mixing caused by polar effects has been calculated and an additional term included:

$$\frac{\Delta E^v}{V} = \delta^2 + \omega^2$$

where δ signifies the dispersion forces and ω the dipole forces. The g factor is introduced to account for the effect of the dipole moment on the nearest neighbors:

$$g = 1 + z \int \cos \gamma \cdot \exp(-W/kt) d\omega$$

where γ is the angle between the dipole moments and W is the potential of the average torque hindering their rotation. Values of g are usually near one for dilute solutions, but at high concentrations they approach the value for the pure polar solute.

Three-component solubility parameters. Three-component parameters have been developed in order to extend the solubility parameter concept to polar and hydrogen-bonding systems. It is assumed that the cohesive energy, $-E$, per unit volume is given as follows:

$$-\frac{E}{V} = -\frac{E_d}{V} - \frac{E_p}{V} - \frac{E_h}{V}$$

where $-E_d$ is the dispersion interaction, $-E_p$ the polar interaction, and $-E_h$ the hydrogen-bonding interaction. This gives the following solubility parameter:

$$\delta_o^2 = \delta_d^2 + \delta_p^2 + \delta_h^2$$

where δ_o^2 is the total solubility parameter for the three-component system.

Regulations

At least five Federal agencies and many more state and local agencies regulate the transportation, use, and disposal of chemicals, including industrial solvents (see Regulatory agencies). In addition, state and local fire marshals regulate the storage and handling of flammable liquids. Current listings for the specific compounds should be consulted for an accurate assessment of the potential hazard of any given solvent (see Industrial hygiene and toxicology).

Cleaning. Dry-cleaning solvents are generally perchloroethylene, mineral spirits, or 140 flash naphtha. The impact of environmental regulations on chlorinated and chlorofluorinated hydrocarbons is not yet clear, but regulations designed to help attain the ozone standard may further restrict the use of petroleum solvents. Owing to a potential fire and explosion hazard, and a decrease in solvency, petroleum solvents are not likely to again become the primary solvents for the dry-cleaning industry.

Cold cleaning applications include degreasing of metals, cleaning of parts, maintenance, stripping of paints and varnishes, and printing-ink solvents. The solvency requirements can be different from those for coatings. In some cases it is not desirable to dissolve the polymeric materials but only to cause them to soften and swell so they can be removed easily. Vapor degreasing is a very efficient method of cleaning metal parts, and generally eliminates residual solvent-soluble soil that could cause production problems. The part to be cleaned is suspended in the vapors of the solvent. The refluxing solvent bathes the part in fresh distilled solvent.

Coatings. Solvents in paints and varnishes dissolve resins that provide protective coatings and support pigment and resin on the surface. Formulation of the solvent system controls ease and method of application, drying time, and nature of the resin film. New regulations from EPA and OSHA have caused many reformulations. This has resulted in a shift toward water-based rather than solvent-based systems (see Coatings, industrial; Paint).

Staining and wood treatment.. Treatment with solutions of insecticides and fungicides is helpful when wood must be protected from weathering and pest damage. Pigmentation and resin can be added to the solvent to produce wood stains as well as protective coatings (see Stains, industrial).

Other applications. Solvents used in printing inks must evaporate very fast. They are usually petroleum products. Insecticides or herbicides are dissolved in solvent systems that contain emulsifiers. Many chemical systems use solvents as inert reaction media. Solvents are also used in purification steps for extraction or recrystallization.

CLYDE F. PARRISH
3M Company

Reviewed by
HUGH FARBER
Dow Chemical Company

J.H. Hildebrand and R.L. Scott, *The Solubility of Nonelectrolytes*, 3rd ed., Dover Publications, Inc., New York, 1964.

J.G. Kirkwood, *J. Chem. Phys*, **7**, 911 (1939).

H. Burrell, *Polymer Handbook*, 2nd ed., Wiley-Interscience, New York, 1975.

SORBIC ACID

Sorbic acid (*trans,trans*-2,4-hexadienoic acid) is a white crystalline solid. Its antimicrobial effect was discovered in 1940, and its applicability as a food preservative was approved in 1953. It is widely used in moist foods below pH 6.5 where control of bacteria, molds, and yeasts is essential to obtain safe and economic storage life.

sorbic acid

Properties

Physical properties are given in Table 1. The chemical reactivity of sorbic acid is determined by the carboxyl group and the conjugated double bonds. Sorbic acid is brominated faster than other olefinic acids.

Sorbic acid is oxidized rapidly in the presence of molecular oxygen or peroxide compounds. The double bond farthest from the carboxyl group

Table 1. Physical Properties of Sorbic Acid and Potassium Sorbate

Properties	Values	
	Sorbic acid	Potassium sorbate
mol wt	112.13	150.22
melting point, °C	134.5	270 dec
boiling point, at 101.3 kPa[a], °C	228	
density, g/cm³	1.204$_{19}$	1.363$_{20}$
flash point, °C	126–130	
specific heat, J/(g·K)[b]	1.84	
latent heat of fusion, kJ/mol[b]	13.6	
heat of neutralization, kJ/mol[b]	6.07	
vapor pressure at 130°C, kPa[b]	1.3	

[a] To convert kPa to mm Hg, multiply by 7.5.
[b] To convert J to cal, divide by 4.184.

is oxidized. More complete oxidation leads to acetaldehyde, acetic acid, fumaraldehyde, fumaric acid, and polymeric products. Sorbic acid dimerizes and undergoes Diels-Alder reactions with many dienophiles. Polymerization catalyzed by free radicals occurs.

Sorbic acid forms salts, esters, amides, and acid chlorides. The most important compound is the potassium salt because of its stability and high solubility.

Synthesis and Manufacture

Sorbic acid was first synthesized from crotonaldehyde and malonic acid. Currently, commercial methods involve the condensation of ketene and crotonaldehyde in the presence of metal catalysts. The condensation adduct has been identified as:

$$\left[\begin{array}{c} \overset{O}{\overset{\|}{-CHCH_2CO-}} \\ | \\ CH=CHCH_3 \end{array}\right]_n$$

Acidic decomposition gives high yields of pure trans,trans-2,4-hexadienoic acid, whereas the pyrolysis in the presence of alkali or amine catalysts gives a mixture of isomers which must be converted to the pure trans,trans form. Food-grade specifications are met by further purification in the form of carbon treatments and recrystallization. A USSR method is based on the reaction of crotonaldehyde and acetone followed by oxidation.

Quality requirements are defined by the U.S. Food Chemicals Codex. The presence of isomers other than the trans,trans form causes instability and affects the melting point.

Sorbic acid is marketed in the United States in dust-free crystalline form for food use or as powder for feed. Potassium sorbate is marketed as powder or granules.

Analysis

Sorbic acid is normally assayed spectrometrically by absorbance at ca 260 nm. In the presence of interfering compounds, the sample should be extracted with suitable solvents.

Health and Safety Factors

The extremely low toxicity of sorbic acid enhances its desirability as a food preservative. The WHO has allowed for it the highest daily intake of all food preservatives, ie, 25 mg/kg body weight.

Uses

Sorbic acid and its potassium salt, collectively called sorbates, are used primarily in a wide range of food and feed products and to a lesser extent in certain cosmetics, pharmaceutical, and tobacco products.

Sorbates have inhibitory activity against a wide spectrum of yeasts molds, and bacteria, including most foodborne pathogens. As bacterial inhibitors, sorbates are least effective against lactic acid bacteria. The activity of the sorbates at a higher pH is one distinct advantage over benzoic and propionic acids. A number of enzyme systems in fungi and bacteria have been designated as sites of sorbate inhibition. Some molds can metabolize sorbates.

Sorbates are classified as GRAS in the United States and have no upper limit set for foods that are not covered by Standards of Identity. Compared with other antimicrobial preservatives, sorbates can be used in higher concentrations without affecting the flavor of foods.

Sorbates are applied by direct addition, dipping in or spraying with an aqueous solution, dusting with sorbate powder, or impregnating food-packaging material.

Sorbates and benzoates are used in margarine, sorbates being the more effective. Sorbic acid is used in table wines to prevent secondary fermentation of residual sugar. The largest commercial consumer is the dairy industry where it is used mostly in processed cheeses. Sorbates extend the shelf life of fresh and processed seafood products.

Sorbates are used in yeast-raised and chemically leavened bakery products. In cooked cured sausages, ie, beef, pork, and chicken-frankfurter emulsions, sorbic acid delays germination and growth of *Clostridium botulinum*.

For fresh poultry, a potassium sorbate dip reduces the total number of viable bacteria and doubles the refrigerated shelf life. Sorbates prolong the shelf life of pet foods.

Sorbic acid also improves the feed utilization and weight gain of chickens. Similar effects have been observed in swine feeds.

C.L. KELLER
S.M. BALABAN
C.S. HICKEY
V.G. DiFATE
Monsanto Co.

E. Lück, *Sorbinsaure*, Band 1, 1969, and *Sorbinsaure Chemie–Biochemie–Mikrobiologie–Technologie*, Band 2, B. Behr's Verlag, Hamburg, 1972, p. 21.

M.A. Ikrima and V.D. Simonov, *Sorbic Acid and Its Derivatives*, Khimiya, Moscow, 1977.

SORBITE. See Steel.

SORBITOL, CH_2OH $(CHOH)_4CH_2OH$. See Alcohols, polyhydric.

SORGHUMS. See Wheat and other cereal grains.

SOYBEANS AND OTHER OILSEEDS

The four principal oilseed crops in the United States are soybeans, cottonseed, peanuts, and sunflowers. Some are consumed directly as foods, and all serve as sources of edible oils. After removal of these oils, the resulting meals are rich in proteins and are used mainly for animal feeds (see also Vegetable oils).

Physical Characteristics

Plants and seeds of the four oilseeds vary in growth habit, size, shape, and other features. Soybeans grow on erect, bushy annual plants, 0.3–2.0 m high with hairy stems and trifoliolate leaves. Seeds are produced in pods. Cotyledons contain protein and lipid bodies.

Cotton grows as an annual or perennial herb or shrub. The seeds are produced in leathery capsules (boll), covered with fibers.

Peanuts grow on annual herbaceous plants. The seeds are produced in pods containing two or three seeds.

Sunflowers are grown in two types. The varieties grown for oilseed production are generally black-seeded. Nonoilseed varieties have striped seed coats and thick hulls. The sunflower is an erect annual; the seed is four-sided and flattened.

Chemical Composition

The four oilseeds all have a high seedcoat or hull content except soybeans (see Table 1). The proteins found in the four oilseeds are complex mixtures. Some fractions are considered to be storage proteins. In addition, a variety of minor proteins including trypsin inhibitors, hemagglutinins, and enzymes are present.

Cottonseed, peanut, and sunflower oils are classified as oleic–linoleic acid oils because of their high content (> 50%) of these fatty acids. Soybean oil is called a linolenic acid oil. In addition to the triglycerides, the four oilseeds contain phosphatides.

Sterols are present in four forms: free, esterified, nonacylated glucosides, and acylated glucosides. Soybeans contain a total of 0.16% of these sterol forms in the ratio of ca 3:1:2:2 (see Steroids). Oilseeds contain soluble mono- and oligosaccharides and insoluble polysaccharides. Minor components, eg, phytic acid, affect feed and food applications.

Table 1. Compositions[a] of Oilseeds, Percent

Oilseed	Hulls	Oil	Protein[b]	Ash	Protein in dehulled, defatted meal[c]
soybean	8	20	43	5.0	52
cottonseed					
acid delinted	36	21.6	21.5	4.2	
kernels		36.4	32.5	4.7	63
peanut	20–30				
kernels	2–3.5[d]	50.0	30.3	3.0	57
sunflower					
arrowhead variety, low oil type	47	29.8	18.1		67
armavirec variety, high oil type	31	48.0	16.9		60
kernels		64.7	21.2		

[a] Approximate; moisture-free basis.
[b] N × 6.25.
[c] Data vary with efficiency of dehulling and oil extraction, variety of seed, and climatic conditions during growth.
[d] Red skins or testa.

Harvesting and Storage

The U.S. soybean crop normally is harvested in September or October. Soybeans at ≤ 12% moisture can be stored for 2 yr or more in concrete silos. Aflatoxin contamination is not as much a problem as it is with cottonseed and peanuts (see Food toxicants).

In the United States, cotton harvesting begins in late July and is usually completed in December. After picking, the cotton is processed in gins. Cottonseed is stored in Muskogee-type warehouses equipped with aeration and temperature-monitoring systems. High temperatures cause rapid deterioration.

When the peanut kernels are fully developed, the plants are dug mechanically, shaken to remove the soil, and inverted into windrows to dry (cure) and mature completely. Ideally, the peanuts are left for several days until the moisture content drops to ca 10%; then they are harvested mechanically. The cured peanuts are stored or shelled. Shelled peanuts should have a moisture content of ca 7%, which is critical to maintain quality.

In the northern United States, sunflower seeds are harvested in late September and October. Harvesting is frequently delayed until after a killing frost to speed drying. A moisture content of 9.5% is safe for short-term storage, but ≤ 7% is recommended for long-term storage without aeration.

Processing

Cottonseed is converted into oil and meal by hydraulic pressing, screw-pressing, prepress-solvent extraction, or direct solvent extraction. For processing, the seed is cleaned to remove sticks, stones, leaves, and other foreign materials. The cotton fibers remaining after ginning are removed mechanically (delinting). The hulls are split and removed, and the separated kernels are passed between smooth rolls to form flakes, which are extracted with hexane to remove the oil. A metric ton of cottonseed yields ca 91 kg linters, 247 kg hulls, 162 kg oil, and 455 kg meal.

Only 10–15% of the U.S. peanut crop is converted into oil and meal. Processing is carried out by screw-pressing or prepressing followed by solvent extraction. A metric ton of peanuts yields ca 317 kg oil and 418 kg meal; the remainder is shells and foreign matter.

Virtually all soybeans processed in the United States are solvent-extracted. The beans are cleaned, dried, and cracked to loosen the seed coat or hulls. Then they are dehulled, flaked, and extracted with hexane. A metric ton of soybeans yields ca 180 kg oil and 790 kg meal.

Like cottonseed and peanuts, sunflower seeds are processed by screw-pressing, solvent extraction, or both. For prepress solvent extraction, expeller meal is conditioned, flaked, and extracted with hexane. When processed without dehulling, a metric ton of sunflower seed yields ca 400 kg oil and 550 kg high fiber meal.

Nutritional Properties and Antinutritional Factors

Oil. Because of their high linoleic acid contents, unhydrogenated and partially hydrogenated cottonseed, peanut, soybean, and sunflower oils are good sources of this essential fatty acid. Hydrogenation imparts high temperature stability to cooking oil, extends shelf life, and improves flavor stability and physical and plastic properties. Although studies have failed to reveal toxic effects on ingestion or partially hydrogenated soybean oil, this complex problem is under active investigation.

Heating and oxidation of fats, especially under severe conditions, result in the formation of a variety of compounds including hydrocarbons, cyclic hydrocarbons, alcohols, cyclic dimeric acids, and polymeric fatty acids. Some of these compounds are toxic, but the present consensus is that an oil such as soybean oil is safe and nontoxic when used under normal cooking conditions.

Proteins and meals. Nutritional properties of the oilseed protein meals and their derived products are determined by the amino acid compositions and the biologically active proteins and nonprotein constituents found in the defatted meals. Cottonseed proteins are low in lysine, threonine, isoleucine, and leucine. These deficiencies can be minimized by blending cottonseed flour with cereal flours (Incaparina). The FDA limits the content of free gossypol in edible cottonseed flour to 450 ppm.

Like cottonseed proteins, peanut proteins tend to be low in lysine, threonine, isoleucine, and leucine. The nutritive value of cornmeal is improved by blending with peanut flour and still more by soy flour. Raw peanuts contain ca one-fifth the trypsin inhibitor found in raw soybeans, but this concentration is high enough to cause hypertrophy of the pancreas in rats. Wet-heating inactivates the inhibitor but has little effect on the nutritional value of peanut flour.

In soybean proteins, methionine is in greatest deficit for meeting the nutritional requirements of a given species. It is common practice to add synthetic methionine to broiler feed to compensate for this deficiency. Sunflower proteins are low in lysine and leucine and are borderline in threonine and isoleucine.

Oilseed Products and Uses

Soybeans supply ca 50% of the total world oilseed production, followed by cottonseed, peanuts, and sunflower seed. Soybeans are the most important oilseeds in international trade; the United States, Brazil, and Argentina are the main suppliers. Although soybeans contribute about one-half the world's production of oilseeds, they supply only one-third the total edible vegetable fats and oils because of their relatively low oil content.

Oil. Most oil obtained from oilseeds is processed and converted into edible products such as salad and cooking oils, shortenings, and margarines. Partial hydrogenation increases stability to oxidation and reduces flavor deterioration. Soybean oil can be hydrogenated under selective or nonselective conditions to increase its melting point and produce hardened fats.

Vegetable oils are utilized in a variety of nonedible applications, but only ca 6% of U.S. soybean oil production is used for such products.

Protein products. Most of the meal obtained in processing of oilseeds is used as protein supplements in animal feeds. Because of its gossypol content, high fiber, and low lysine content, cottonseed meal is used primarily for beef and dairy feeds.

Only defatted peanut and soybean flakes are converted into edible-grade products. Defatted soybean flakes gives flours and grits (50% protein); protein concentrates (70% protein); and protein isolates (90% protein).

Oilseed proteins are used in foods at concentrations of 1–2 to nearly 100%. Textured soy flours and concentrates serve as meat substitutes. The use of some oilseed proteins in foods is limited by flavor, color, and flatus effects. Raw soybeans, for example, taste grassy, beany, and bitter.

In the United States, only soybean protein isolates are used for industrial applications, eg, as adhesives for clays used in coating of paper and paperboard.

Food products. Peanuts are processed into peanut butter, candy, salted nuts, and roasted-in-the-shell peanuts. Peanut butter is made and consumed primarily in the United States.

Small amounts of soybeans are roasted and salted for snacks. Nut substitutes for baked products and confections are also manufactured from soybeans. Soybeans are used for oriental food products, such as soy milk, tofu, miso, tempeh, and soy sauce. The latter is made by fermentation or acid hydrolysis.

Sunflower seeds are roasted in the shell and in the dehulled form. Small amounts are sold through health-food stores.

WALTER J. WOLF
U.S. Department of Agriculture

A.E. Bailey, *Cottonseed and Cottonseed Products*, Interscience Publishers, Inc., New York, 1948.

A.K. Smith and S.J. Circle, *Proteins*, Vol. 1 of *Soybeans: Chemistry and Technology*, Avi Publishing Co., Inc., Westport, Conn., 1972.

J.G. Woodruff, ed., *Peanuts: Production, Processing, Products*, 3rd ed., Avi Publishing Co., Inc., Westport, Conn., 1983.

J.F. Carter, ed., *Sunflower Science and Technology*, American Society of Agronomy, Inc., Madison, Wisconsin, 1978.

SPACE CHEMISTRY

The abundance of the elements, their present chemical forms, and their distribution within the universe are the result of lengthy and complex evolutionary processes. Ninety-nine percent of the atoms in existence were created in a single event, ie, the origin of the universe. This synthesis produced most of the atoms, but essentially none of the elements were heavier than lithium. The heavier elements were created later within stars from the earlier-produced hydrogen and helium. The origin of the universe is referred to as the big bang, and the reality of this event is widely accepted in the astrophysics community.

In the initial stages of the big bang, the temperature of the universe was $> 10^{13}$ K and the density was $> 10^{15}$ g/cm³. After a few tens of minutes, element production associated with the origin of the universe was finished. The primary reactions that occurred during this period and produced most of the atoms in the universe are as follows:

$$^1H + {}^1H \rightarrow {}^2H + \beta^+ + \nu$$

$$^1H + {}^2H \rightarrow {}^3He + \gamma$$

$$^3He + {}^3He \rightarrow {}^4He + 2\,{}^1H + \gamma$$

$$^3He + {}^4He \rightarrow {}^7Be + \gamma$$

$$^7Be + \varepsilon \rightarrow {}^7Li + \nu$$

The elements up to iron, to a certain extent, are produced by fusion reactions in stars.

Beyond the iron peak elements, elements form primarily by neutron-capture reactions. When the available neutron flux is low, as is the case in evolved but stable stars such as red giants, the dominant process is the s (slow) process. In the s process, sequential neutron captures build successively more neutron-rich isotopes of a given element until a radioactive isotope is reached and it decays creating the next element in the periodic table (see Radioactivity, natural).

In violent stellar events, such as supernova explosions, intense neutron fluxes are produced and the r (rapid) process neutron capture occurs. Successive neutron captures on a given element produce highly unstable nuclei which decay to produce neutron-rich isotopes.

Cosmic Abundances of the Elements

Mass-loss processes and catastrophic explosions of stars recycle newly created elements from stars back into interstellar gas and dust where they are eventually incorporated into new stars. This mixing and reworking of material has not only increased the number of heavy elements in the universe over time, but it has also led to a degree of homogenization. The elemental composition of most stars in our galaxy is very close to that of the Sun, and in general, most galaxies appear to have compositions similar to our own.

The solar abundances of H, C, N, and O are obtained by spectroscopy, whereas those of the noble gases are derived from solar wind studies and a variety of nucleosynthetic criteria. Abundances are also determined from meteorites. On the average, an element with an even atomic number is approximately ten times as abundant as its odd-numbered neighbors in the periodic table. Besides the odd–even effect there are other peaks in the abundance curve associated with high binding energy.

Interstellar Gas and Dust

Interstellar gas and dust are both the raw materials from which stars form and the materials returned to the interstellar medium when stars lose mass gradually or explode violently. Interstellar material typically is very tenuous; the gas density in the Galaxy averages less than one atom per cubic centimeter. The total abundances in the interstellar medium are thought to be close to solar.

Complex gas molecules usually have lifetimes < 100 yr before they are destroyed by photodissociation. A very important exception, however, is the case of molecular clouds where the high dust density acts as a shield blocking external radiation.

The Solar System

The solar system consists of the Sun, nine planets, the planetary satellites, and an enormous number of asteroids and comets. The planets include the inner or terrestrial planets and outer or giant planets.

Terrestrial planets. Mercury, Venus, Earth, and Mars are Earth-like objects that differ greatly from the outer planets. Roughly 90% of the mass of each terrestrial planet is composed of magnesium, silicon, iron, and oxygen. Oxygen is the most abundant element. The variation of Fe^{2+}–Fe^0 among the terrestrial planets is probably owing to the oxidation state of iron in the dust grains from which the planets originally accreted.

All of the terrestrial planets are differentiated bodies; their compositions vary radially in their interiors. The separation of immiscible fluids and the separation of materials of different densities produced planetary regions that differ significantly in composition from their original cosmic abundances. The principal effect was the gravitational separation of an immiscible metal phase to form iron cores. Core formation concentrates a planet's metallic iron at its center and depletes the rest of the planet in siderophile (iron-loving) elements. In contrast to the siderophiles that concentrate in planetary cores, planetary processing enriches lithophile elements in the crusts. Lithophile elements include calcium, aluminum, lithium, rubidium, strontium, potassium, scandium, titanium, barium, and vanadium.

Venus is essentially a twin of the Earth in that its mass, density, and distance from the Sun are very similar. The surface of Venus, perpetually covered by clouds, was mapped to a spatial resolution of 30 km by radar altimetry and radar imaging on the Pioneer Venus mission. Recent imaging by Soviet Venera spacecraft have produced large-scale images with a resolution of 1 km. The data show that although the total range in elevations on Venus is similar to that on the Earth, the distribution is quite different.

Mars is a small planet with only 10% the mass of the Earth. Unlike Mercury and the Moon, Mars has been active geologically through much of the planet's lifetime. Although water may exist on Mars, any organic matter would likely be destroyed by photocatalytic reactions, leaving little chance of biological activity there.

The Earth's atmosphere is anomalous because its composition is determined by biology and not by chemistry alone. With the exception of argon, the gases in the atmosphere are controlled and rapidly recycled by biological processes. The most remarkable aspect of this control is the maintenance of a large oxygen abundance above a planetary surface that would rapidly react with practically all atmospheric oxygen in the absence of life. The "normal" composition of the atmosphere of a terrestrial planet is nearly pure CO_2 (see Table 1).

Outer planets. The outer planets include Pluto and the giant planets Jupiter, Saturn, Uranus, and Neptune. Very little is known about Pluto, but it appears to be small, even in comparison with the inner planets, and it clearly is quite different from the other outer planets. Jupiter and Saturn have interiors composed primarily of hydrogen. The composition

Table 1. Atmospheric Compositions of the Terrestrial Planets

Planet	Relative pressure	Principal gases, %	Other gases, ppm
Mercury	10^{-15}	He, ca 98 H, ca 2	
Venus	90	CO_2, 96 N_2, 3.5	H_2O, ca 100; SO_2, 150; Ar, 70 CO, 40; Ne, 5; HCl, 0.4; HF, 0.01
Earth	1	N_2, 77 O_2, 21 H_2O, 1 Ar, 0.93	CO_2, 330; Ne, 18; He, 70; Kr, 1.1 Xe, 0.087; CH_4, 1.5; H_2, 5; N_2O, 0.3 CO, 0.12; NH_3, 0.01; NO_2, 0.001 SO_2, 0.002; H_2S, 0.0002; O_3, 0.4
Mars	0.007	CO_2, 95 N_2, 2.7 Ar, 1.6	O_2, 1300; CO, 700; H_2O, 300 Ne, 2.5; Kr, 0.3; Xe, 0.08; O_3, 0.1

of Uranus and Neptune is probably similar to the bulk composition of comets.

Asteroids and comets. The asteroids, which apparently formed between Mars and Jupiter, seem to be small planetesimals that did not accrete to form a planet. They are widely believed to be the main, and possibly the sole, source of meteorites.

Comets resemble asteroids in that they too are small early solar-system objects that escaped incorporation into planets. The comets differ, however, in that they must have accumulated much farther out in the solar system in the regions where conditions were cooler and volatile ices could survive as grains.

Satellites. Satellites are primarily a phenomenon of the outer planets. The only satellites around the inner planets are Phobos and Deimos, the two tiny satellites of Mars, and the Earth's moon. The Moon, anomalous in many ways, is the only satellite about which a great deal is known. Its average density, 3.3 g/cm^3, implies that, unlike the Earth and Venus, it cannot have a significant iron core. Carbon and water are essentially nonexistent on the Moon.

The Voyager revealed an incredible range in the physical character of the satellites. Some are stony and could be thought of as terrestrial. Most satellites, however, contain varying amounts of icy materials and are compositionally different from the terrestrial planets.

Near-Earth Extraterrestrial Resources

If large-scale manned operations are ever conducted in space, the utilization of extraterrestrial materials may become a necessity. The energy required to remove material from selected near-Earth bodies is, of course, less than that required to lift materials from the Earth into orbit. Various forms of extraterrestrial resources could be used. The potential suppliers of near-Earth space resources are the Moon and the Apollo (Earth-crossing) asteroids.

The future of asteroid mining is a question of economics, not technology. It seems possible that, with sufficient funding, near-Earth asteroids in the 100-m to several-kilometer size could be moved from solar orbit to Earth orbit. Asteroids might also be brought through the atmosphere to the Earth's surface, but their greatest value is as a resource of raw materials in orbit.

D.E. BROWNLEE
University of Washington

Special Voyager issues, *Science* **204**, 945 (1979); **206**, 925 (1979); **212**, 159 (1981).

B. Mason, *Handbook of Elemental Abundances in Meteorites*, Gordon and Breach, New York, 1971.

W.H. Arnold, S. Bowen, K. Fine, D. Kaplan, M. Kolm, H. Kolm, J. Newman, G. O'Neill, and W.R. Snow, in *Space Resources and Space Settlements*, *NASA SP-428*, 1979, p. 87.

SPANDEX AND OTHER ELASTOMERIC FIBERS. See Fibers, elastomeric.

SPRAYS

A spray is a liquid-in-gas dispersion in the form of a multitude of drops.

Characteristics

Sprays are produced by the breakup of liquid filaments or sheets. In an early analysis, it was predicted that a filament would break up into essentially spherical drops with a uniform diameter equal to 1.89 times the orifice diameter. Studies have shown this prediction to be valid for low viscosity liquids. For more viscous liquids, a modified diameter ratio appears appropriate:

$$\frac{D}{d_o} = 1.89 \left[1 + \frac{3We_L^{1/2}}{Re_L} \right]^{1/6}$$

where the two dimensionless groups, the liquid Weber number and the liquid Reynolds number, are defined as follows:

$$We_L = \frac{u_o^2 \rho_L d_o}{\sigma}$$

$$Re_L = \frac{d_o u_o \rho_L}{\mu_L}$$

The typical spray pattern consists of a range of droplet sizes. A useful function for evaluating spray distribution has the log-normal form:

$$\frac{dn}{dD} = f(D) = \frac{1}{Ds_g\sqrt{2\pi}} e^{-(\ln D - \ln D_g)^2 / 2s_g^2}$$

Diameters. Median diameters divide the spray into two equal portions by number, surface area, or volume. Mean diameters are defined as follows: arithmetic mean (simple weighted average based on all the individual droplets in the spray); geometric mean (simple average based on the logarithm of the diameters of all droplets); surface mean (diameter of a droplet whose surface area, if multiplied by the total number of droplets, equals the surface of all particles in the spray); volume mean (diameter of a droplet whose volume, if multiplied by the number of droplets, equals the total volume of the sample); and sauter mean (diameter of a droplet whose ratio of volume to surface area is equal to that of the entire spray).

Techniques for the measurement of drop-size distribution vary greatly in accuracy, cost, time for data reduction, and compatibility with the system geometry.

Atomizers

The geometry of the atomizing device is based on the characteristics of the spray desired.

Centrifugal pressure nozzles. In the swirl chamber (hollow cone), a circular orifice outlet is preceded by a chamber in which the liquid is given a swirl. As pressure drop across the nozzle increases, liquid capacity increases and the dispersion becomes finer.

Droplet size data supplied by nozzle manufacturers are usually based on water dispersed in air. For conversion to liquids other than water, the following relationship is suggested:

$$\frac{(D_{vs})_1}{(D_{vs})_2} = \left(\frac{\rho_{L,2}}{\rho_{L,1}} \right)^{0.35} \left(\frac{\mu_{L,1}}{\mu_{L,2}} \right)^{0.15} \left(\frac{\sigma_1}{\sigma_2} \right)^{0.20} \approx \frac{D_{v,1}}{D_{v,2}}$$

In a similar relationship, the liquid density, viscosity, and surface tension ratios have exponents of 0.3, 0.2, and 0.5, respectively; a viscosity ratio exponent of 0.32 has been proposed.

In the solid-cone nozzle, a special core or axial jet fills the center of the conical pattern. The resulting full-volumetric coverage enhances rates of mass and heat transfer between the spray liquid and gas passing through the cone.

Fan-spray nozzles. These are used in the coating industry and where a narrow elliptical pattern is more appropriate than a circular pattern. The spray is formed by discharge from an elliptical or circular orifice impinging on a curved surface.

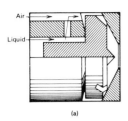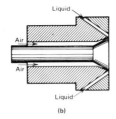

Figure 1. Two-fluid atomizers. (**a**) Internal-mixing type. (**b**) External-mixing type.

Two-fluid atomizers. In these devices, a separate gas flow promotes atomization of the liquid, and by adjustment of gas-to-liquid ratios of flow and pressure, extremely fine dispersions can be produced. The gas and liquid may be mixed within the nozzle before discharge through a single-outlet orifice. Alternatively, gas and liquid may discharge through separate orifices arranged in such a manner that the gas impinges on the liquid at or just outside the orifice. The two-fluid atomizer produces a finer spray than pressure nozzles, but consumes more energy (see Figure 1).

The sonic atomizer is a special type of two-fluid device. Gas (normally air) is accelerated within the device to sonic velocity. The sound waves produced impinge on a plate or annular cavity (resonation chamber) and are reflected to the path of the emerging liquid sheet.

Rotary atomizers. These are disk- or cup-shaped devices that are rotated by a separate power source and are capable of handling slurries and other materials which might clog the narrow passages of pressure and two-fluid nozzles. The devices may be smooth or vaned, flat or bowl shaped, and may have multiple tiers. The mechanism of droplet formation is similar to that for pressure nozzles.

Sprays can form by flashing liquid jets, and an orifice through which a high pressure liquid expands with partial vaporization can produce a spray.

Chemical Applications

A number of processes require sprays. Though a small component in most systems, the atomizer must be chosen carefully according to the nature of the process, and the characteristics of the sprayed products.

Nomenclature

d_O = orifice diameter
D = droplet diameter
D_g = number of geometric mean diameter
D_v = volume (or mass) mean diameter
D_{vs} = volume/surface, or Sauter, mean diameter
n = number of droplets
Re = Reynolds number (dimensionless)
s_g = geometric standard deviation
u_O = superficial liquid velocity
We = Weber number (dimensionless)
μ = viscosity
ρ = density
σ = surface tension
L = liquid phase

JAMES FAIR
University of Texas

D. Steinmeyer in D.W. Green, ed., *Perry's Chemical Engineers' Handbook*, 6th ed., McGraw-Hill Book Co., New York, 1984, pp. 48–57.

D.A. Lundgren and co-eds., *Aerosol Measurement*, University of Florida Press, Gainesville, Fla., 1979.

W.R. Marshall, *Atomization and Spray Drying*, Chemical Engineering Progress Monograph Series No. 2, American Institute of Chemical Engineers, New York, 1954.

M.J. Ashley, *Chem. Eng.* (*London*), 368 (June 1974).

SPUTTERING. See Film deposition techniques; Metallic coatings.

STAINS, INDUSTRIAL

A stain is a solution or dispersion of colorants designed primarily to impart color rather than to form a protective coating. Stains are used as architectural finishes and for concrete, paper products (paper plates), plastics, textiles, and numerous wood products.

Types

Water stains are 1–3% solutions of water-soluble dyes. They raise the grain of wood to such an extent that sanding is required.

Non-grain-raising stains (NGR) are solutions of acid or spirit-soluble dyes in a nonaqueous vehicle, eg, methanol and toluene. These stains give bright, transparent, light-resistant effects. They are generally used as the base stain to attain a specified undertone. Spirit stains are generally 0.5–12% solutions of dyes in alcohols. Oil stains are 1–3% solutions of oil-soluble dyes in hydrocarbons.

Pigmented, solvent-borne stains are the most widely used stains in the furniture industry. Occasionally, waterborne resins and colorants are used in stains and glazes. Wiping stains are composed of pigments usually ground in an alkyd resin or drying oil. Nonwipe spray stains are low solids wiping stains. They do not impart the depth of color or enhance the wood grain like a wiping stain. Solvent-borne dip stains differ from nonwipe-spray stains only in solvent balance. Glazes are slow-drying wiping stains formulated for brushing over a seal coat. Wood fillers are used for filling open-pore woods (80–90 wt% solids). Blond or simulated bleached finishes are achieved with toners, which offer white as a coloring aid. The toner offers the means of subtracting color from the wood without bleaching.

Ingredients

Dyes and pigments enhance or obscure the surface to which they are applied. Alkyd resins are polyesters generally modified with oils. The oils used in wood stains are linseed, soybean, tung, and oiticica. Rosin is used mainly as a resin modifier.

Hydrocarbons used as solvents include petroleum naphthas, toluene, xylene, aromatic petroleum, solvents, turpentine, and dipentene. Esters, ketones, alcohols, and glycol ethers are also used as solvents.

Application Techniques

Stains are applied by brush, spray, dip, flow coat, padding, and tumbling. Formamide mixed with an equal volume of water is the most effective penetration promoter when applied to veneered panels. Frequently, articles to be stained are first exposed to a vacuum to remove air and moisture.

Spattering is a method of obtaining spotted effects. Differential or contrast staining is used for wood that includes hard and soft tissues. Surfaces are often colored by simply pouring on a liquid stain and letting the excess drain off.

Graining requires great skill. The varnish stain is applied with special tools in transparent films of uneven thickness over a light, opaque background to simulate the grain effect of wood. Transfers and decalcomanias impart stainlike effects, even though they are not strictly classifiable as stains.

Stains are most effective on lumber or plywood, but are also satisfactory on smooth surfaces. Solid-color stains obscure the natural wood color and grain. Semitransparent penetrating stains are moderately pigmented and do not totally hide the grain.

Health and Safety Factors

Safe handling of industrial stains requires adequate ventilation. Solvent vapor concentrations are expressed as TLV and vary between 5 and 1000 ppm.

ROBERT S. BAILEY
Lilly Industrial Coatings, Inc.

Colour Index, 3rd ed., 1975, and *Supplement* to Vols. 1–4, and Vol. 6, 1975, Society of Dyers and Colourists, Bradford, UK, and AATCC, U.S., rev. 1981.

STARCH

Starch, $(C_6H_{10}O_5)_n$, is a mixture of linear (amylose) and branched (amylopectin) polymers of α-D-glucopyranosyl units. It is the principal reserve polysaccharide in plants, and constitutes a substantial portion of the human diet.

Properties

Starch occurs in plants in the form of granules. Rice starch has the smallest granules and potato starch the largest. Undamaged starch granules are insoluble in cold water but imbibe water reversibly and swell slightly. In hot water, irreversible swelling occurs, producing gelatinization. The property of forming thick pastes or gels is the basis of many starch uses.

Most common starches contain two types of D-glucopyranose polymers. Amylose is essentially a linear polymer of α-D-glucopyranosyl units linked $(1 \rightarrow 4)$. Amylopectin is a highly branched polymer of α-D-glucopyranosyl units containing $1 \rightarrow 4$ links with $1 \rightarrow 6$ links at branch points (see Fig. 1).

Starch is heterogeneous not only in relation to polymer structure but also in relation to molecular weight. For amylose, osmotic pressure measurements give a range of 10,000–60,000. With an anerobic isolation technique to prevent oxidative degradation, a range of $(1.6–7.0) \times 10^5$ was obtained. Amylopectin is a much larger molecule, ranging in mol wt between 50,000 and 10^6.

Hydrolysis of starch is an important industrial reaction which is accomplished by acid, enzymes, or both. The action of acids produces glucose; other products may also be formed.

α-Amylase hydrolyzes starch to a mixture of D-glucose, maltose, and a dextrin obtained from amylopectin. β-Amylase removes maltose units successively until, in the case of amylose, the reducing end of the molecule is encountered or, in the case of amylopectin, an α-$(1 \rightarrow 6)$ branch point is met. Glucoamylase hydrolyzes both amylose and amylopectin, yielding D-glucose.

Oxidation of the hydroxyl groups of starch, eg, with hypochlorite, gives aldehydes, ketones, or carboxylic acids.

Starch gives a characteristic blue complex with iodine which is used as a test for starch in various systems.

Manufacture

Milling of corn, *Zea mays*, provides corn starch, which is extensively used in food and nonfood applications. Steeped corn is coarsely ground in an attrition mill to break loose the germ. The mill gap during this step must be adjusted to maximize the amount of germ freed, but minimize rupture of the germ. Germ is removed from the aqueous slurry in a cyclone separator. The germ fraction is processed for corn oil. The cyclone underflow is milled a second time for complete release of the starch granules. Following the second milling, the kernel suspension contains starch, gluten, and fiber. The collected fiber is combined with gluten for feed use.

The starch–gluten suspension, commonly known as mill starch, is concentrated by centrifugation. The concentrated starch is passed through hydroclones to remove the last traces of protein. Starch suspension may be processed dry and marketed as unmodified corn starch, modified by chemical or physical means, gelatinized and dried, or hydrolyzed to corn syrup.

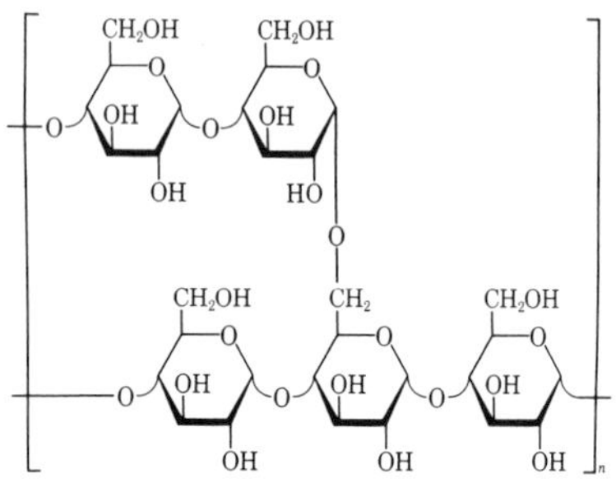

Figure 1. Branch point structure in amylopectin.

Chemical modification. Acid-treated starches exhibit decreased intrinsic viscosity, decreased gel strength, and increased gelatinization temperature. In industrial production, a 40% slurry of corn or waxy maize starch is acidified with hydrochloric or sulfuric acid at 25–55°C. When the desired degree of thinning is attained, the mixture is neutralized.

In the manufacture of oxidized starch, a slurry of starch granules is treated with alkaline hypochlorite, neutralized, washed, and dried. Oxidation results in lower pasting temperatures, decreased thickening power, and lower paste setback.

In the manufacture of pyrodextrins, dried starch is sprayed with an acid and dried to 1–5% water content. The acidified starch is hydrolyzed and reverted by heating. Chemical changes produced by acid-heat treatment lead to starches with lower water content, decreased solution viscosity, and greater solubility in hot water.

Uses

Unmodified corn starch is used in nonfood applications in the mining, adhesives, and paper industries.

Acid-modified or thin-boiling starches are employed mainly in the textile industry as warp sizes and fabric finishes. Oxidized starches are used in paper coatings and adhesives. Starch pyrodextrins and British gums have extensive use as adhesives for envelopes, postage stamps, and other products.

Ethanol, isopropyl alcohol, *n*-butanol, acetone, 2,3-butylene glycol, glycerol, and fumaric acid are produced from starch by fermentation.

Food uses. Unmodified starch is used in food preparations requiring thickening, gelling, or similar properties. Pregelatinized starch is used where thickening is required but cooking is to be avoided. Acid-modified starches are used for gum candies.

Derivatives

Starches, as polyhydroxy compounds, undergo reactions of alcohols, including esterification and etherification. Such modifications may produce useful polymers. Hydroxyethyl starches are used as paper coatings and sizes.

The commercial cationic starches are the tertiary and quaternary aminoalkyl ethers. Cationic starches are used on paper for fiber and pigment retention. They have been employed as emulsifiers for water-repellent paper sizes.

Starch monophosphates are useful in food applications because of their freeze-thaw stability. Degree of substitution is generally low (DS < 0.15). Starch phosphate diesters contain ester cross-links and are used as thickeners and stabilizers in baby foods, salad dressings, fruit pie filling, and cream-style corn.

Starch acetates are used in food because of their clarity and stability. Applications include frozen fruit pies and gravies, baked goods, instant puddings, and pie fillings.

ROY L. WHISTLER
JAMES R. DANIEL
Purdue University

R.L. Whistler, E.F. Paschall, and J.N. BeMiller, eds., *Starch: Chemistry and Technology*, Academic Press, Inc., New York. 1984.

W. Banks and C.T. Greenwood, *Starch and Its Components*, Edinburgh University Press, Edinburgh, UK, 1975.

J.A. Radley, ed., *Examination and Analysis of Starch and Starch Products*, Applied Science Publishers, Ltd., London, UK, 1976.

STEAM

Steam can be generated by evaporation of water at subcritical pressures, by heating water above the critical pressure, and by sublimation of ice. It is generated and used saturated (or dry saturated), wet, or superheated. Saturated steam has no moisture or superheat, wet steam contains moisture, and superheated steam has no moisture and its temperature is above the saturation temperature. Thermodynamic and physical properties of steam are well established.

Properties

The properties of steam can be divided into thermodynamic, transport, physical, and chemical properties. In addition, molecular structure and chemical composition are of interest. Because of the universal use of water and steam, the need for international research cooperation and for property formulations was recognized as early as 1929, when the First International Steam-Table Conference was held in London. In 1972, the International Association for the Properties of Steam (IAPS) was formed. Today, four IAPS Working Groups are studying equilibrium, transport, and other properties, and the chemical thermodynamics of power cycles. Steam properties in the form of graphs, tables, and theoretical and empirical equations are widely used in design and service analysis of steam engines, power systems, and heat-transfer and -process equipment.

The molecular structure of steam is not as well understood as that of ice and water. There are indications that even in the steam phase some H_2O molecules are associated into small clusters of two and more molecules.

If temperature T and pressure p are independent variables, the dependent basic properties are volume v, internal energy u, entropy s, and enthalpy (formerly called heat content) $h = u + pv$. The thermodynamic potentials are the Helmholtz function, $f = u - Ts$, and the Gibbs function, $g = h - Ts = f + pv$. With given values of these properties and their derivatives, a thermodynamic surface of a substance can be constructed. If a surface is defined, all other properties follow.

Dynamic and kinematic viscosities and thermal conductivity are the transport properties. The derived parameters, thermal diffusivity $(\lambda/\rho C_p)$ and the Prandtl number $(\mu C_p/\lambda)$, are two transport parameters used in engineering calculations. In the above formulas, λ is the thermal conductivity, ρ the density, C_p the specific heat at constant pressure, and μ the dynamic viscosity.

Surface-tension data are important in evaluations of condensation, droplet formation and transport, surface wetting, and emulsions. The static dielectric constant of saturated and superheated subcritical steam is between 1 and 3.

Chemical properties. The ionization constant has a fundamental influence on equilibria and is therefore the most important characteristic of an electrolyte.

Solubility of chemicals in steam reflects the capacity to transport them. When concentration exceeds solubility, chemicals precipitate. Dynamic solubility in rapidly expanding steam may be much higher than the equilibrium solubility.

Investigations of solubility in steam and of vaporous carryover (the distribution of substances between water and vapor) have shown that solubility in superheated steam rapidly decreases as the pressure decreases or the specific volume increases. The lowest solubility, and therefore lowest tolerance to impurities, is in the lowest pressure areas, just before the saturation line.

Distribution of substances between water and vapor (also called vaporous carryover, distribution ratio constant or coefficient, and partition coefficient) is one factor governing the concentration of substances with low vapor pressure transferred into steam. For constant concentrations, compositions, and pH, and in the absence of chemical reaction, the distribution ratios depend only on the ratio of water and steam densities:

$$K = \left(\frac{\rho_w}{\rho_s}\right)^n$$

where n is a constant for each compound and concentration.

Compounds ionized in water (strong electolytes) are hydrated and not as easily transferred into steam as the weakly hydrated molecules of weak electrolytes, such as Al_2O_3, B_2O_3, and SiO_2. Superheated steam can hydrolyze salts. Hydrolysis of $NaCl$ has been observed at temperatures as low as 250°C.

Steam Generation

Steam below the critical pressure is normally generated in boilers.

Fuel-fired boilers. Both process-steam generating boilers and central heating boilers normally operate at 0.2–10.4 MPa (15–1500 psig). Most fire-tube boilers currently in service are of the small-package type, completely instrumented. Most modern boilers are of the water-tube type, in which the water is circulated through tubes that are exposed to radiant or convection gas heating. Steam generators used for power generation in central power stations can be designed for either subcritical or supercritical pressures (see Furnaces, fuel-fired).

In natural or controlled-circulation drum-type boilers, means must be provided within the drum to separate the steam formed in the generating tubes from the remaining boiler water. The continuing effort to increase the thermal efficiency of fossil-fuel steam power generating plants has led to the development of steam at pressures above the critical point.

Steam generators known as package or shop-assembled units are used in units under 50 t/h in single boilers.

Fuel efficiency can be increased by a combination of a steam turbine and one or more large gas turbines with a water-tube heat-recovery steam generator utilizing heat from the exhaust gases (combined cycle). Large industrial users of steam are combining generation of electric power by high pressure steam and combined cycles with production of low pressure process steam (cogeneration).

In nuclear-fission systems, steam is generated from water heated either directly in the reactor core or in steam generators by heat-transfer medium (see Nuclear reactors).

Many concepts for generation of steam using solar energy (qv) have been proposed, but no commercial steam-generating plants are in operation today.

Process-energy demand can be significantly reduced by a recovery of residual heat and heat generated in the chemical processes. The recovered heat is used in the process itself, for heating feedwater in steam generation, or for generation of steam in recovery or waste-heat boilers (see Fig. 1).

Steam generation in heat-recovery systems can be classified into four categories: removal of heat from vessels in which exothermic chemical reactions proceed at isothermal temperature conditions; cooling of process stream and/or removal of latent heat of condensation at temperatures capable of generating steam; combustion of burnable materials in which the gases produced are capable of generating steam; and combustion of fuel in which by-product flue gases are capable of generating steam.

The generation of steam as a means of controlling the temperature at which an exothermic reaction takes place is of considerable interest in the chemical industry. The reaction may require careful temperature control in order to avoid production of undesired by-product materials or, perhaps, in order to avoid short life or physical breakdown of the catalyst.

A good example of high pressure steam generation as a by-product of the heat of reaction of a chemical process is a large synthetic-ammonia plant where steam is generated in tube bundles inserted in the reaction vessel.

Geothermal steam. Geothermal steam comes from wells as steam or is produced by flash evaporation of hot water or brine. The feasibility of producing steam by pumping water into molten rock or inserting a conventional closed heat exchanger was demonstrated in April 1981, at

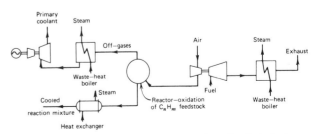

Figure 1. Waste-heat recovery system in a chemical plant.

the Kilauea Iki Lava Lake in Hawaii. Magma temperature is 850–1100°C, and the magma bodies considered for steam production are within 10 km of the earth's surface (see Geothermal energy).

Steam accumulators. Some process plants require steam at a very high rate for a short period of time and then at a low rate for a considerable length of time. The problem can be solved by the use of accumulators. These are insulated high pressure reservoirs containing water into which steam is charged at high pressure during times of low steam demand. At the time of peak withdrawal from the system, the rated output of the boiler, augmented by the flow of flash steam from the accumulator, is released to the low pressure flow line.

Steam and Water Chemistry

High purity water is generally used for steam generation. Combinations of purification processes reduce impurity content to low ppb levels.

High system pressure and heat fluxes require highly purified water. Wet-steam systems can tolerate higher concentration of impurities than superheated-steam systems. Although harmful impurities are removed, other chemicals are added to protect the system against corrosion and scale formation.

Precipitation and deposition occur when the concentration of an impurity exceeds its solubility in superheated steam. When the specific volume of steam is rapidly increasing, any component is prone to deposit. Impurities also deposit from boiling-water films. A sampling of steam is needed because its composition cannot be accurately predicted from the composition of feedwater and boiler water. After condensation, impurities are analyzed by continuous instruments or chemical methods. For most constituents, a detection limit of 1 ppb is desirable.

Solids precipitation results from steam impurity; more than 120 compounds and elements have been identified in turbine, boiler, and piping deposits.

Uses

Steam is used by all principal manufacturing industries in the United States. The approximate breakdown is as follows:

Distribution	Percent
process steam	40.6
electric drive	19.2
electrolytic process	2.8
direct process heat	27.8
feedstock for chemicals	8.8
other	0.8

Carnot efficiency η_c is a theoretical maximum efficiency of a perfect cycle (without losses):

$$\eta_c = \frac{T_1 - T_2}{T_1} \times 100\%$$

where T_1 is the absolute temperature at the inlet and T_2 is the absolute temperature at the exhaust. Cogeneration and combined cycle systems increase T_1; using condensing cycles decreases T_2. The energy balance of a real cycle is illustrated in Figure 2.

The maximum thermal efficiency of a steam cycle is the Carnot efficiency, which is always less than 100%. The development of boiler-tube material more resistant to creep and oxidation permitted raising pressures and temperatures and increased efficiencies.

Steam turbines. The steam turbine is the simplest and most efficient engine for converting large amounts of heat energy into mechanical work. Steam turbines employ three principles of action: impulse, reaction, and shear torque. The impulse principle involves a stationary nozzle and moving blades, which absorb the mechanical energy from the steam as it flows over the blades. In the reaction turbine, the nozzles themselves are attached to the shaft. In a shear-torque turbine, drag (friction force) is exerted by high velocity steam (usually wet) on rotating disks; there are no blades. A special type of turbine is the turboexpander (expansion turbine) used to recover power from steam and other hot gases and to reduce steam pressure.

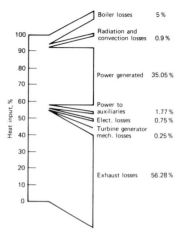

Figure 2. Energy distribution in a simple cycle of a steam power plant using a boiler, turbine, and condenser.

Turbines have few moving parts. The rotor with disks and blades is usually supported by two radial bearings; axial forces are supported by an axial-thrust bearing.

For the synchronous operation of the electric-power generator, utility turbines are operated at constant rpm. Turbines for mechanical drives, such as marine turbines and compressor and blower drives, operate at variable speeds.

Steam turbines are very efficient at high load ratings, and are normally considered as drive units if more than 37 kW (50 hp) is required. Steam turbines operate very effectively at high speeds (3,000–10,000 rpm) and thus lend themselves to large-power output, high speed drives.

Generation of electric power. Steam cycles for generation of electric power are designed for maximum efficiency and reliability. Because of energy conservation and other factors, the U.S. growth rate of electric energy has been decreasing (see Power generation).

Wet and saturated steam has a definite pressure for each fixed boiling or condensing temperature. Control of the desired temperature for any process-heating requirement may be fixed by choosing the steam pressure. Installation of steam-heating systems in a cascade system, eg, multiple-effect evaporators, permits the recovery of heat at successive levels by reducing pressure at each stage. When close temperature control is required to prevent overheating or to ensure a high heating density, steam is generally used.

Steam is used to supply heat to most evaporation and distillation processes, such as sugar-juice processing and alcohol distillation.

In the steam-reforming process, light hydrocarbon feedstock reacts with steam over a nickel-containing catalyst to produce hydrogen (above 700°C), methane (below 550°C), and carbon oxides. Steam-methane reforming is used in the production of ammonia.

In the coal gasification process, coal reacts at high temperature with steam and air or oxygen to produce a mixture of gases, typically carbon monoxide, carbon dioxide, hydrogen, and methane. In coal liquefaction, steam is used to produce the hydrogen required for the dissolution of the coal.

Steam is being used more and more in tertiary oil recovery by flooding oil wells (see Petroleum, enhanced oil recovery).

In the production of synthetic rubber by solution polymerization, steam is used for dewatering and the removal of solvents. Steam distillation separates water-immiscible liquid mixtures of components of varying volatility (see Fig. 3). High pressure steam is effective in removing dirt and scale from solid surfaces. Low pressure turbine steam is often extracted for low pressure distillation of salt water to produce drinking water.

Corrosion

The use of metals in hot steam is limited by their oxidation rate, mechanical strength, and creep resistance. Corrosion rates in pure steam are about the same as in high purity deoxygenated water, except for gray

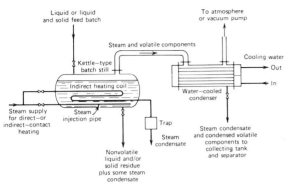

Figure 3. Single-stage batch-type steam distillation or stripping system.

iron, nickel, lead and zirconium, which corrode faster in steam. More than 160 compounds have been found in steam and steam-borne deposits. To minimize corrosion, impurities must be controlled, acid impurities neutralized, oxygen minimized, and metal surfaces kept clean and covered with protective oxide (see also Corrosion and corrosion inhibitors).

OTAKAR JONAS
Westinghouse Electric Corp.

Reviewed by
KEES A. BUETERS
C-E Power Systems

F. Franks, ed., *Water, A Comprehensive Treatise*, Plenum Press, New York, 1972.

Steam—Its Generation and Use, 39th ed., Babcox and Wilcox Company, New York, 1978.

P.M. Goodall, ed., *The Efficient Use of Steam*, IPC Science and Technology Press, Guilford, Surrey, UK, 1980.

J.G. Singer, ed., *Combustion: Fossil Power Systems*, 3rd ed., Combustion Engineering, Inc., Windsor, Conn., 1981.

STEARIC ACID, $CH_3(CH_2)_{16}COOH$. See Carboxylic acids.

STEATITE. See Talc.

STEEL

Steel is the generic name for a group of metals composed principally of iron which, because of their abundance, durability, versatility, and low cost, are the most useful metallic materials known. In a typical year, over 7×10^8 metric tons of raw steel are produced throughout the world. The most useful classification is by chemical composition into carbon, alloy, and stainless steels.

Modern steelmaking processes are either acidic or basic processes. Carbon, manganese, and silicon are removed with relative ease by both processes. Oxidation is employed to convert a molten bath of pig iron and scrap, or scrap alone, into steel. The oxygen combines with the unwanted elements to form oxides which either leave as gases or enter the slag.

In general, steels with similar chemical compositions have similar mechanical and physical properties, no matter by which process they are made. The basic oxygen process has become the leading steelmaking method in the United States.

Open-Hearth Process

The open-hearth furnace is a large structure comprised of several parts, all constructed primarily of refractory bricks. The charge of steel scrap and liquid blast-furnace iron or hot metal is placed on the bottom of an elongated tunnel-like furnace, and fuel is injected. Preheated air burns the fuel, thereby heating the charge until it melts. Excess carbon and silicon are removed by oxidation.

Electric-Furnace Processes

In the electric-arc furnace and the induction furnace, steel is made that is different in composition and shape from the starting material which is usually steel scrap or direct-reduced iron (DRI). The steel used for consumable-electrode melting and electroslag remelting resembles the desired steel ingot which is subsequently rolled or forged. The last two processes are employed for very high quality steel for applications with extremely strict requirements.

Electric-arc furnace. The chemical reactions taking place in the electric-arc furnace are similar to those in other steelmaking processes. When the carbon, phosphorus, and sulfur concentrations are decreased and the temperature has been raised to the desired level, the steel is tapped into a ladle at the same time that ferromanganese, ferrosilicon, and other agents are added to deoxidize the steel and obtain the desired composition. These furnaces are lined with acidic or basic refractories.

In the basic process, a slag containing ca 50% CaO is used for steels of the highest quality with low sulfur and phosphorus contents. With the power turned off, solid scrap and other components of the charge are placed in the furnace. Alloying materials that are not easily oxidized are charged in the furnace prior to melting. Excess carbon is used in the melt bath, permitting some carbon to be removed by ore additions or oxygen injection. This excess carbon is removed as carbon monoxide gas, which bubbles out of the liquid steel bath, thereby stirring the bath. This action equalizes composition and temperature. After charging has been completed, the arcs are struck and the charge melted as quickly as possible. As oxidation progresses, the temperature is raised to remove carbon and increase fluidity.

Acidic electric furnaces operate with partial oxidation, complete oxidation with a single slag, or with silicon reduction and double-slag technique. Charging and melting are similar to the basic process, except for the necessity of using scrap with a low content of phosphorus and sulfur which cannot be removed in the acidic process. After the bath is covered with the proper oxidizing slag and the carbon content of the metal is high enough, the temperature is raised to bring the bath to a boil. The amounts of carbon and oxygen in the bath should be sufficient to maintain boiling for at least 10 min. When the carbon has reached the desired concentration, silicon and manganese in the form of ferroalloys are added to the bath. The furnace should be tapped soon as these have melted and diffused.

The high frequency, coreless induction furnace is used in the production of complex, high quality alloys such as tool steels. It consists of a refractory crucible surrounded by a water-cooled copper coil through which an alternating current flows. Most commonly, the constituents of the charge melt smoothly and mix. The charge is selected to produce the composition desired in the finished steel with a minimum of further additions.

In vacuum and atmosphere melting, a high frequency, coreless induction furnace of the type described above is enclosed in a container or tank which can be either evacuated or filled with a gaseous atmosphere of any desired composition or pressure.

Consumable-electrode melting produces special quality alloy and stainless steels by casting or forging the steel into an electrode that is remelted and cast into an ingot in a vacuum. The furnace consists of a tank above ground level that encloses the electrode and a water-cooled copper mold below ground level (see Fig. 1).

Electroslag remelting has the same general purpose as consumable-electrode melting; a conventional air-melted ingot serves as a consumable electrode and no vacuum is employed.

Oxygen Steelmaking Processes

In the basic oxygen process (BOP), pure oxygen is mixed with hot metal, causing oxidation of the excess carbon and silicon and producing steel.

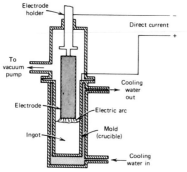

Figure 1. Schematic representation of a consumable-electrode furnace for melting steels in a vacuum.

The top-blown basic oxygen process is conducted in a cylindrical furnace similar to a Bessemer converter. Gaseous oxygen is blown onto the surface of a bath of molten pig iron at the bottom of the furnace by a vertical water-cooled retractable pipe or lance inserted through the mouth of the vessel. With the furnace tilted toward the charging floor, scrap is dumped by an overhead crane into the mouth of the furnace. The crane then moves away from the furnace and another crane carries a ladle of molten pig iron to the furnace and pours the molten pig iron on top of the scrap.

The bottom-blown basic oxygen process is conducted in a furnace similar to a Bessemer converter consisting of a bottom and a barrel. In Europe, this process is called the OBM process. It is generally used with high phosphorus (1.5–2.7%) hot metal which does not permit the use of the catch-carbon technique.

Another bottom-blown basic oxygen process is the Q-BOP process, which is used with low phosphorus (usually < 0.2%) hot metal permitting application of the catch-carbon practice described below.

Catch-carbon technique. In many countries, including the United States, the iron ore used in blast furnaces to make molten pig iron or hot metal contains less than ca 0.2% phosphorus. With low phosphorus hot metal, medium carbon (0.20–0.40%) and high carbon (0.45–1.00%) steels are made by the catch-carbon technique in which oxygenation is stopped when the desired carbon content is reached. Because it is advantageous to oxygenate a heat of steel only once, it is desirable to charge the correct amounts of hot metal, scrap, and oxygen, giving the steel the desired temperature and carbon content at the end of the oxygenation. This aim is accomplished in more than 50% of the heats made by the catch-carbon technique.

Determination of Chemical Composition

The chemical composition of a steel is generally specified by the customer within narrow limits. It is followed closely at various stages since the carbon content dictates the course of refining. Instruments are stationed near the furnaces for rapid analysis.

Scrap as Raw Material

Scrap consists of the by-products of steel fabrication and discarded articles containing iron or steel. Every ton of scrap consumed in steelmaking is estimated to conserve 3.5–4 t of natural resources. The BOP uses 15–30% scrap (up to 45% with preheating), open-hearth processes use 35–60%. Electric-arc furnaces are charged almost entirely with scrap.

Unsalable products are termed home or revert scrap. These include pit scrap; ingots too short to roll; rejected ingots; crop ends from slabs, blooms, and billets; shear cuttings from trimming flat-rolled products; pieces damaged in handling and finishing; ends cut from bars, pipe, and tubing, and so on. In general, ca 3×10^7 t of home scrap would result from the manufacture of 10^8 t of raw steel and the processing of this steel into finished products. Since 5.5×10^7 t of scrap is normally used to produce 10^8 t of raw steel, 2.5×10^7 t of purchased scrap is used to supplement 3×10^7 t of home scrap.

Purchased (dormant) scrap comprises obsolete, worn-out, or broken consumer products. It requires careful sorting to prevent contamination

of steel in the furnace. Junked automobiles represent enormous resources of steel scrap.

Prompt industrial scrap may consist of unwanted portions of plate or sheet, trimmings from stamping and pressing, machine turnings, rejected products scrapped during manufacture, short ends, flashes from forgings, and other types. Scrap should be free of unknown and unwanted elements referred to as tramp alloys. Analysis of samples from individual lots is sometimes employed in scrap classification.

Scrap can be a valuable source of alloying elements, and full advantage is taken of this source in the production of alloy steels in electric furnaces. However, since most open-hearth production consists of carbon and low alloy steels, alloying elements in scrap are generally a source of trouble.

Addition Agents

In steelmaking, various elements are added to the molten metal to effect deoxidation, control of grain size, improvement of the mechanical, physical, thermal and corrosion properties, and other specific results. Ferroalloys are a special class of addition agents.

Addition agents may be added to the charge in the furnace, the molten bath near the end of the finishing period, the ladle, or the molds. Admixture at ambient temperature to liquid steel lowers the temperature (chill effects) (see Table 1).

Ladle Metallurgy

Most steels are made without ladle metallurgy treatment. After the liquid steel has been tapped from the basic oxygen or other steelmaking furnace into the ladle and the desired addition agents added, the steel is usually continuously cast or cast into ingots. Most ladles hold all the steel produced in one furnace heat.

To meet higher quality specifications, liquid steel is subjected to one of a variety of ladle metallurgy treatments. These treatments accomplish one or more of the following objectives: removal of sulfur, oxygen, hydrogen, or carbon; addition of ferroalloys with very high recoveries; and decrease or increase of temperature to meet specifications for continuous casting.

To facilitate consideration of the many ladle metallurgy treatments, they have been here classified into the following groups: argon treatments, vacuum processes, and electric-arc processes.

Argon treatment. In argon stirring (CAS, SAB, and AOD processes) argon gas is passed through liquid steel in order to mix ferroalloys with the steel, homogenize the steel with respect to chemical composition and temperature, accelerate cooling, and remove oxide and sulfide inclusions.

Vacuum processes. In stream degassing, a bottom-pour ladle containing the molten steel to be degassed is set upon the evacuated tank; the bottom of the ladle and the top of the tank are equipped with mating seals to exclude air. When the stopper rod of the tapping ladle is raised, molten metal flows through the nozzle, melts a metal diaphragm that seals the opening to the tank, and passes into the ladle (or mold) in the vacuum tank. As the molten metal enters the evacuated space, it breaks up into tiny droplets, exposing an enormous surface to vacuum degassing.

In the DH (Dortmund-Hoerder) process, a refractory-lined chamber with one hollow leg or pipe extending from the bottom is inserted into

Table 1. Chill Values of Various Addition Agents

Addition agent	Chemical composition, wt %	Chill value, %C for 1 kg per metric ton steel
high carbon ferrochrome	68 Cr–25 Fe–5 C–1.5 Si–0.5 Mn	+2.50
cobalt	100 Co	+1.40
high carbon ferromanganese	76 Mn–16 Fe–7 C–1 Si	+2.28
nickel	100 Ni	+1.30
silicomanganese	61 Mn–17 Si–14 Fe–2C	+1.78
ferrophosophorus	71.5 Fe–24 P–3 Mn–1.5 Si	+2.60
ferrosilicon, 75%	75 Si–23 Fe–2 Al	−0.52
steel scrap	100 Fe	+1.64
sulfur	100 S	−1.46

the liquid steel in a ladle. After the chamber is evacuated, liquid steel is moved back and forth between the ladle and the chamber by moving them together and apart.

In the RH (Ruhrstahl-Heraeus) process, a refractory-lined chamber similar to the DH degasser is used, but with two hollow legs instead of one. Liquid steel moves continuously from the ladle, up through one leg into which argon gas is injected, into the vacuum chamber, down the other leg, and back into the ladle.

The RH-OB (Ruhrstahl-Heraeus oxygen-blowing) process combines the RH vacuum degassing process with oxygen blowing. Oxygen is injected into the liquid steel in the vacuum chamber through double tuyeres.

The VOD (vacuum-oxygen-decarburization) process was originally developed for producing stainless steel but has also been used to refine carbon steels. The liquid steel is tapped into a ladle into which argon is injected through a porous plug in the bottom.

Electric-arc processes. In the LF (ladle-furnace) process, the liquid steel is refined by argon stirring with a synthetic slag and heated by an electric arc; vacuum is not used.

In the VAD (vacuum-arc degassing) process, the liquid steel is refined by heating under vacuum while argon is injected through a porous plug in the ladle bottom.

In the ASEA-SKF process, named for two Swedish companies, the liquid steel is tapped into a ladle and exposed to vacuum while argon is injected through a porous plug.

Ingots

The ladle is carried by an overhead crane to a pouring platform where the steel is teemed into a series of molds, where it solidifies to form large castings called ingots. Most steel produced today in the United States is cast into ingots. The molds are made of cast iron with square, rectangular, or round cross sections. The mold cavity is tapered to facilitate removal of the ingot.

In all except killed steels, the evolution of gas produces cavities of roughly cylindrical shape (skin or honeycomb blowholes) or spherical shape (located deeper in the ingot). Blowholes serve a useful purpose in diminishing or preventing the formation of pipe and increasing the amount of usable steel. In properly made ingots, the gas evolution is so controlled that a skin of adequate thickness covers the blowholes closest to the surface.

When a liquid consisting of more than one component freezes, the solid is seldom uniform in composition; this is termed segregation. When liquid steel freezes, it generally exhibits positive segregation in which the part that freezes first is highest in iron content and the part that freezes last is lowest. Some elements tend to segregate more than others; sulfur segregates most.

Applications. The thick skin of relatively clean metal or rimmed-steel ingots makes them desirable for rolling products where the surface of the finished products is most important. Capped steel has a thin-rimmed zone that is relatively free from blowholes, and a core zone that is less segregated than that of a rimmed ingot of the same volume. It is used when the carbon content is above 0.15%. Semikilled steel has wide application in structural shapes, plates, and bars. Killed steel is generally used when a homogeneous structure with strength and toughness is required.

Casting

Although continuous casting appears deceptively simple in principle, many difficulties are inherent in the process. For example, when molten steel comes into contact with a water-cooled mold, a thin solid skin forms on the wall. However, because of the physical characteristics of steel, and because thermal contraction causes the skin to separate from the mold wall shortly after solidification, the rate of heat abstraction from the casting is so low that molten steel persists within the interior of the section some distance below the bottom of the mold; eventually, the whole section solidifies. Continuous casting gives a higher yield than ingot casting and avoids the cost of rolling ingots into slabs. Continuous-cast slabs are only ca 20–25 cm thick and exhibit much less segregation than slabs rolled from ingots. Consequently, there is usually less variation in steel composition from one heat to the next than there is from the top to the bottom in a single ingot.

In bottom-pressure casting, a ladle filled with molten steel is placed in a pressure vessel covered with a lid in which a pouring tube has been inserted that dips into the molten steel almost to the ladle bottom. This method gives better yields than ingot casting and produces slabs and billets with surfaces that require little conditioning.

Plastic Working of Steel; Heat Treatment

Plastic working is the permanent deformation accomplished by applying mechanical forces to a surface. The primary objective is usually the production of a specific shape or size. Plastic deformation of steel can be accomplished by hot working or cold working.

The great advantage of steel as an engineering material arises from the fact that its properties can be controlled by heat treatment. Thus, if steel is to be formed into some intricate shape, it can be made very soft and ductile by one heat treatment. After forming, the steel can then be given high strength by subjecting it to a different heat treatment.

Iron – Iron Carbide Phase Diagram

The iron–iron carbide phase diagram (Fig. 2) shows the ranges of compositions and temperatures in which the various phases are present in slowly cooled steels. The only changes occurring on heating or cooling pure iron are the reversible changes at ca 910°C from bcc α iron to fcc γ iron and from fcc δ iron to bcc γ iron at ca 1390°C. Eutectoid steels contain 0.8% carbon. At and below 727°C the constituents are α ferrite and cementite.

Hypoeutectoid steels contain less carbon than eutectoid steels. If the steel contains more than 0.02% carbon, the constituents present at and below 727°C are usually ferrite and pearlite. The behavior on heating and cooling hypereutectoid steels (containing > 0.80% carbon) is similar to that of hypoeutectoid steels, except that the excess constituent is cementite rather than ferrite.

Grain Size

A significant aspect of the behavior of steels upon heating is the grain growth that occurs when the austenite, formed upon heating above A$_3$ or A$_{cm}$, is heated even higher; A$_3$ is the upper critical temperature and A$_{cm}$ the temperature at which cementite begins to form (see Fig. 2). The microscopic grain size of steel is customarily determined from a polished plane section. Austenitic-grain growth may be inhibited by undissolved carbides or nonmetallic inclusions. Steels of this type are commonly referred to as fine-grained steels, whereas steels that are free from grain-growth inhibitors are known as coarse-grained steels.

Phase transformations. At equilibrium, that is, with very slow cooling, austenite transforms to pearlite when cooled below the A$_1$ temperature. The faster the cooling, the lower the temperature at which transformation occurs.

The main factors affecting transformation rates of austenite are composition, grain size, and homogeneity. An isothermal transformation diagram illustrates the structure formed if the cooling is interrupted and the reaction completed at a given temperature.

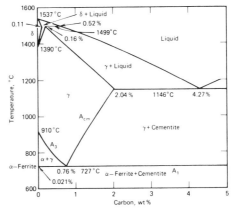

Figure 2. Iron–iron carbide phase diagram.

The transformation behavior of austenite is best studied by observing the isothermal transformation at a series of temperatures below A_1. In carbon and low alloy steels, transformation over the temperature range of ca 700–540°C gives pearlite microstructures of the lamellar type. Transformation to bainite occurs over the temperature range of ca 540–230°C. Transformation to martensite occurs almost instantly during cooling. The temperature at which transformation to martensite starts on cooling is designated as the M_s temperature.

Phase properties. Pearlites are softer than bainites or martensites. However, they are less ductile than the lower temperature bainites and, for a given hardness, far less ductile than tempered martensite. In a given steel, bainite microstructures are generally found to be both harder and tougher than pearlite, although less hard than martensite, the hardest and most brittle microstructure obtainable in a given steel. The hardness of martensite increases with increasing carbon content up to the eutectoid composition. Martensite is tempered by heating to a temperature range of 170–700°C for 30 min to several hours.

Hardenability

Hardenability refers to the depth of hardening or to the size of a piece that can be hardened under given cooling conditions, and not to the maximum hardness that can be obtained in a given steel. The maximum hardness depends almost entirely upon the carbon content, whereas the hardenability (depth of hardening) is far more dependent upon the alloy content and grain size of the austenite. Although the critical cooling rate can be used to express hardenability, cooling rates ordinarily are not constant. Hardenability is measured by a test in which a steel sample is subjected to a continuous range of cooling rates.

Heat-Treating Processes

In heat-treating processes, steel is usually heated above the A_3 point and then cooled at a rate that gives the desired properties. In austenitization, the temperature is usually high enough to dissolve the carbides completely and to take advantage of the hardening effects of the alloying elements. The temperature should not be so high as to produce pronounced grain growth.

The purpose of quenching is to cool rapidly enough to suppress all transformation above the M_s temperature.

Tempering, ie, heating below the lower critical temperature A_1, relieves stresses and improves ductility, although at some expense of strength and hardness. In order to minimize cracking, tempering should follow quenching immediately.

Martempering minimizes the high stresses created by the transformation to martensite during the rapid cooling characteristic of ordinary quenching. It is ordinarily carried out by quenching in a molten salt bath just above the M_s temperature.

Lower bainite is generally as strong as and somewhat more ductile than tempered martensite. Austempering, an isothermal heat treatment that results in lower bainite, offers an alternative heat treatment for obtaining optimum strength and ductility.

In normalizing, steel is heated above its upper critical temperature (A_3) and cooled in air in order to refine the grain and obtain a carbide size and distribution favorable for carbide solution.

Annealing relieves cooling stresses induced by hot or cold working and softens the steel to improve its machinability or formability. Annealing to coarse pearlite can be carried out isothermally by cooling to the proper temperature for transformation to coarse pearlite and holding until transformation is complete. Coarse pearlite microstructures are too hard for optimum machinability in the higher carbon steels. Such steels are customarily annealed to develop spheroidized microstructures. Process annealing is subcritical annealing of cold-worked materials. It involves heating at a temperature high enough to cause recrystallization of the cold-worked material and soften the steel.

In carburizing, low carbon steel acquires a high carbon surface layer by heating in contact with carbonaceous materials.

In the nitrogen case-hardening process, termed nitriding, machined and heat-treated parts are subjected to the action of a nitrogenous medium, commonly ammonia.

Carbon Steels

Plain carbon steels represent by far the largest volume steel produced. Properties are governed principally by carbon content and microstructure and can be controlled by heat treatment. Most plain carbon steels, however, are used without heat treatment.

Carbon steels with relatively low hardenability are predominantly pearlitic in the cast, rolled, or forged state. Cast steel is generally coarse-grained, since austenite forms at high temperature and the pearlite is usually coarse.

Many carbon steels are used in the form of as-rolled finished sections. The microstructure and properties of these sections are determined largely by composition, rolling procedures, and cooling conditions.

The manufacture of wire, sheet, strip, and tubular products often includes cold working. The most pronounced effects are increased strength and hardness and decreased ductility.

Residual elements. In addition to the carbon, manganese, phosphorus, sulfur, and silicon which are always present, carbon steels may contain small amounts of gases, such as hydrogen, oxygen, or nitrogen, introduced during the steelmaking process; nickel, copper, molybdenum, chromium, and tin, which may be present in the scrap; and aluminum, titanium, vanadium, or zirconium, which may be introduced during deoxidation.

Dual-Phase Sheet Steels

Dual-phase steels derive their name from their unique microstructure of a mixture of ferrite and martensite phases. Normally, high strength hot-rolled sheets are manufactured by hot rolling and cooling on a hot-strip mill, which produces a microstructure of ferrite and pearlite. On heating to ca 750–850°C, a microstructure of ferrite and austenite is produced, and by cooling at an appropriate rate, the austenite is transformed to a very hard martensite phase contained within the soft, ductile ferrite matrix.

Alloy Steels

Alloy steels may be defined as having enhanced properties owing to the presence of one or more special elements or larger proportions of elements (such as silicon and manganese) than are ordinarily present in carbon steel. They are classified into low alloy, intermediate alloy, and high speed tool steels, and may contain up to ca 50% of alloying elements, which enhance the properties.

High strength low alloy steels (HSLA) acquire superior mechanical properties by the addition of alloying elements (other than carbon), singly and in combination. HSLA steels may be substituted for structural carbon steel with a weight decrease.

Alloy steels are classified into low alloy, intermediate alloy, and high speed tool steels.

Steels that contain $\geq 4.00\%$ chromium are included by convention among the special types of alloy steels known as stainless steels, which are resistant to rusting and staining. The effect of chromium may be enhanced by additions of molybdenum, nickel, and other elements. Martensitic stainless steels are iron–chromium alloys that are hardenable by heat treatment. Ferritic stainless steels are iron–chromium alloys that are largely ferritic and not hardenable by heat treatment. Austenitic stainless steels are iron–chromium–nickel alloys not hardenable by heat treatment and predominantly austenitic.

High-temperature service and heat-resisting steels are used for high temperature equipment, eg, steam boilers and turbines, gas turbines, cracking stills, tar stills, hydrogenation vessels, heat-treating furnaces and fittings for internal-combustion engines. A class of quenched and tempered low carbon constructional alloy steels, referred to as low carbon martensites, has been extensively used in pressure vessels, mining and earth-moving equipment, and large steel structures (see High temperature alloys).

A group of high nickel martensitic steels, called maraging steels, contain so little carbon that they are referred to as carbon-free iron–nickel martensites. These are soft and ductile and become hard and strong upon aging at 480°C. Iron–carbon martensite is hard and brittle in the as-quenched condition and becomes softer and more ductile when tempered.

Silicon steels are characterized by high permeability, high electrical resistance, and low hysteresis loss when used in magnetic circuits.

Health and Safety Factors

The hazards associated with steelmaking have been sharply reduced by industry-wide efforts. Over 60% of serious accidents in steel mills involved machinery, cranes, railroad equipment, and motor vehicles. The industry has recognized potential hazards and is resolving problems by good work practices and engineering controls.

ROBERT J. KING
United States Steel Corporation

H.E. McGannon, ed., *The Making, Shaping, and Treating of Steel*, 9th ed., U.S. Steel Corporation, Pittsburgh, Pa., 1971.

W.C. Leslie, *The Physical Metallurgy of Steel*, Hemisphere Publishing, McGraw-Hill, New York, 1981.

J. Walker, *Scientific American*, **250**, 148 (May 1984).

STERILIZATION TECHNIQUES

Sterilization technology is important in industries as diverse as food processing (qv) and space exploration. Generally, however, it is associated with health care.

Kinetics and Thermodynamics

The rate of destruction of microorganisms is logarithmic, ie, first order with respect to the concentration of microorganisms. The process is described by the expression

$$\frac{N_0}{N_t} = e^{-kt}$$

in which N_t = the number of organisms alive at time t, N_0 = the initial number of organisms, and k = the rate constant. Since N_t approaches zero as t approaches infinity, absolute sterility is impossible to attain. As a practical matter, total kill is assumed to have taken place when culturing results in no growth. The rate of kill is expressed in terms of a decimal reduction rate or D value. The D value represents the time of exposure required for a tenfold decrease in the viable population. The D value concept simplifies the design of sterilization cycles. A 6-D reduction from the last survivor or a 10^{-6} concentration of microorganisms is generally regarded as an acceptable criterion for sterility.

The Eyring equation can be applied to the sterilization process:

$$k = \frac{k_B T}{H} e^{(\Delta S/R)(-\Delta H/RT)}$$

where k_B = Boltzmann's constant, h = Planck's constant, and ΔS and ΔH are the standard entropy and enthalpy changes, respectively; ΔH and ΔS are usually between 167 and 335 J/mol (40–80 cal/mol).

The relationship between the D value and k can be derived by considering the meaning of D:

$$D = \frac{t_2 - t_1}{\log_{10} N_1 - \log_{10} N_2}$$

where $N_1 = 10 N_2$. Substituting into the first expression gives:

$$D = \frac{2.3}{k}$$

Testing and Monitoring

Testing for sterility is destructive, and the product is rendered useless for food and medical purposes. Indirect methods rely on sampling or process monitoring.

Biological indicators are preparations of microorganisms resistant to the sterilization process they are intended to monitor (see Table 1). The bioload, ie, the number of organisms present on or in an entire chamber load, is determined first. The carriers containing the biological indicators

Table 1. Typical Performance Characteristics of Biological Indicators

Culture	Sterilization process	Approximate D-value
Bacillus subtilis spores	ethylene oxide at 50% rh and 54°C	
	600 mg/L	3 min
	1200 mg/L	1.7 min
Bacillus stearothermophilus spores	saturated steam at 121°C	1.5 min
Bacillus pumilus spores	gamma radiation	
	wet preparations	2×10^{-6} Gy[a]
	dry preparations	1.5×10^{-6} Gy[a]
Bacillus subtilis spores	dry-heat at 121–170°C	1–60 min
Clostridium sporogenes spores	saturated steam at 112°C	0.7–3.5 min

[a] To convert Gy to rad, multiply by 100.

are retrieved following exposure, transferred to sterile culture media, and incubated. If no growth is observed while viability control displays growth, sterilization was successful.

Electromechanical instrumentation. All sterilizers are equipped with gauges, sensors, and timers; some are computerized. It is not possible to locate sensors inside packages to be sterilized. Electromechanical instrumentation therefore cannot detect inadequate sterilization conditions inside packages.

Chemical monitoring. Chemical indicators combine the functions of biological indicators and electromechanical instrumentation. Sensors can be located inside packages, and the results are observable immediately without the need for incubation.

Dry-heat sterilization. Dry-heat sterilization is generally conducted at 160–170°C for > 2 h. At higher temperatures, exposure times are shorter. Chemical indicators are available in the form of pellets enclosed in glass ampules or paper strips containing a heat-sensitive ink.

Steam sterilization. This process is usually carried out with saturated steam in an autoclave at ≥ 121°C. Constant presence of moisture is essential.

The gravity-displacement type autoclave relies on the nonmiscibility of steam and air to allow the steam to rise to the top of the chamber, and fill it by pushing the air out through the drain at the bottom. In the prevacuum autoclave, air is eliminated by creating a vacuum.

The critical parameters of steam sterilization are temperature, time, air elimination, steam quality, and the absence of superheating. Temperature and time are interrelated. Success depends on direct contact with steam; air interferes. The term steam quality refers to the amount of dry steam present relative to liquid water. Superheated steam results when steam is heated to a temperature higher than that which would produce saturated steam.

Cycle-development studies may be conducted with full loads that produce fractional kill. The results are plotted as D values of the load vs time of exposure, and the logarithmic plot is extrapolated to the desired degree of safety.

Chemical indicators can be classified into those that integrate the time and temperature of exposure, those that determine whether a specific temperature has been achieved, those that determine if uniform steam penetration has been achieved, and those that distinguish packages that have been processed from those that have not.

Gas Sterilization

Certain articles, particularly those used for health care and space exploration, cannot withstand the temperatures and moisture of steam sterilization or exposure to radiation. Gaseous sterilants that function at low temperatures offer an attractive alternative. Ethylene oxide (qv) is the most frequent choice. The critical parameters of ethylene oxide sterilization are time, temperature, gas concentration, and humidity. Cycle development is similar to that for steam.

Ionizing Radiation

Radiation sterilization employs electron accelerators or radioisotopes (qv). Gamma-radiation sterilization permits deeper penetration and usually employs ^{60}Co and occasionally ^{137}Cs as the radiation source.

Filtration and Liquid Sterilization

Filtration depends on a filter membrane or similar medium (see Ultrafiltration).

Formalin has sterilizing properties, as does glutaraldehyde. Careful precleaning is required and the exposure time is critical. Liquid sterilants corrode metal. However, glass or certain corrosion-resistant alloys can safely be processed. In the United States, the manufacture and sale of chemical sterilants and disinfectants are regulated by the EPA.

Ultraviolet radiation has sterilizing properties but cannot penetrate opaque materials. It is used usually for fluids, especially air.

Packaging

Packaging protects sterility. When an article is placed in its protective container and subsequently sterilized, the process is called terminal sterilization. When the article is first sterilized and then placed in a sterilized container, the process is called sterile filling.

Related Techniques

Procedures less thorough than sterilization may be used for food and medical supplies, if judged acceptable for the purpose by a qualified professional.

Pasteurization destroys potentially harmful organisms, such as *Microbacteria tuberculosis*, *M. bovis*, or *M. avium*. Pasteurization is carried out at 62°C for 30 min or at 72°C for 15 s; it is not equivalent to sterilization (see Milk and milk products).

Sanitization is a cleaning procedure that reduces microbial contaminants to safe levels, usually with hot water and detergents. Sanitization is not safe for articles inserted into the body.

Decontamination and sterilization are similar, except that in the former the bioburden is higher.

Sterilization in the Food Industry

Foods may act as nutrients for and carriers of pathogenic organisms and may be spoiled by certain organisms or enzymatic action. The most widely used sterilization method in the food industry is moist heat. Cooking and sterilization can frequently be combined. Acidic foods, such as fruits, tend to retard microbial growth and resist certain types of contamination. In the United States, the FDA has regulatory responsibility over the preparation, sterilization, and distribution of foods.

Food is sterilized by one of the following procedures: batch or continuous terminal sterilization in filled containers; aseptic filling following batchwise cooking in a retort; and aseptic filling in a continuous system.

THOMAS A. AUGURT
Propper Manufacturing Co., Inc.

S.S. Block, *Disinfection, Sterilization, and Preservation*, 2nd ed, Lea and Febinger, Philadelphia, Pa., 1977.

G. Sykes, *Disinfection and Sterilization*, 2d ed., Spon, London, 1965.

S. Turco and R. E. King, *Sterile Dosage Forms, Their Preparation and Clinical Application*, Lea and Febinger, Philadelphia, Pa., 1974.

STEROIDS

Steroids are a large class of marine and terrestrial organic compounds that have the perhydro-1,2-cyclopentenophenanthrene ring system (**1**) as a common structural feature.

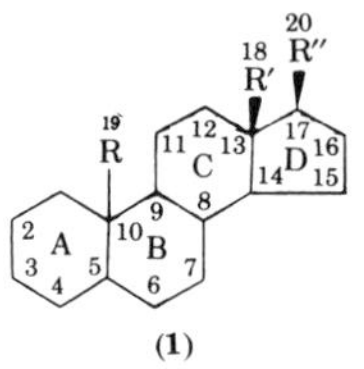

(1)

The vast diversity of the natural and synthetic members of this class depends primarily upon variations in the side chains, R, R′, and R″, as well as on nuclear substitution, degree of unsaturation and stereochemical relationships at the ring junctions. Included under the designation steroid are the naturally occurring and synthetic substances, eg, marine, myco, phyto-, and zoosterols; adrenocortical hormones; steroidal plant and animal alkaloids; antibiotics; antihypertensive drugs; bile acids; cardiac aglycones; contraceptive drugs (qv); insect hormones; insect moulting hormones; sapogenins; male and female sex hormones; toad poisons; and vitamin D.

The term steroid is derived from the isolation of the initial solid crystalline secondary alcohols from the nonsaponifiable lipid extracts of plants and animals called sterols. Study of steroids has led to significant advances in synthetic and mechanistic chemistry, as well as in biochemistry, biology, and the medical sciences. Pharmaceutical research has developed a wide variety of novel steroid-based medicines with sales of ca 10^9.

The nomenclature rules approved by the International Union of Pure and Applied Chemistry in 1972 are based on IUPAC 1957 rules as revised. Compounds are systematically named as derivatives of the parent hydrocarbons. Substituents attached to the plane of the steroid ring system from above are called β and designated by a solid line; those attached below the plane are called α and are shown by a dotted line; substituents of unknown configuration are indicated with a wavy line and designated ξ. Many, if not most, of the more important steroids are designated by trivial names.

Biosynthesis

The formulation of cholesterol from acetyl coenzyme A involves ten enzymes from acetyl–CoA to squalene and at least 19 enzymic steps from squalene to cholesterol. The enzymic conversions from acetate through the formation of farnesyl pyrophosphate, with the exception of HMG (hydroxymethylglutaryl)-CoA reductase, occur in the cytoplasm of the cell. In subsequent steps leading to cholesterol, both enzymes and substrates are microsomally bonded. In mammalian systems, cholesterol is the prime progenitor of steroids. A simplified relationship showing the main biosynthetic pathways is shown below.

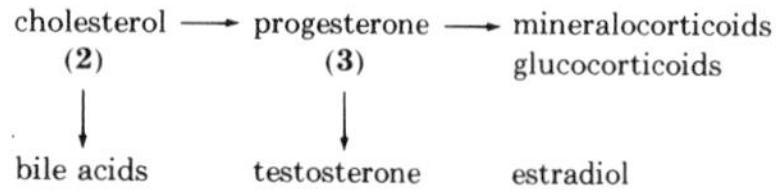

Structural Types

Sterols. The natural sterols are crystalline C_{26}–C_{30} alcohols containing an aliphatic side chain at C-17; they are widely distributed and occur both free and as esters and glycosides.

Cholesterol (**2**) is present in all mammalian tissue either free or esterified, and serves an important biochemical function in membranes. It functions as a metabolic precursor for the bile acids and mammalian steroid hormones as well as for the cardenolides, ecdysones, and many other steroid types.

The D vitamins are steroids which are part of the hormonal system that regulates calcium homeostasis. Rickets is a disease caused by faulty calcium deposition in bones. It was observed that patients responded favorably to sunlight or cod liver oil. Provitamin D_2 was identified as ergosterol (**3**), which is a characteristic sterol of yeast and other fungi, and provitamin D_3 was identified as 7-dehydrocholesterol. Irradiation yields vitamin D_2 and D_3, respectively.

The plant sterols are exemplified by stigmasterol (**4**) and sitosterol. Stigmasterol occurs chiefly in soybean oil as an approximately 1:4 mixture with sitosterol. It serves as starting material for the preparation of the sex hormone progesterone (**5**) and corticoids through oxidation, ozonolysis, and oxidation of the subsequently formed enamine with air.

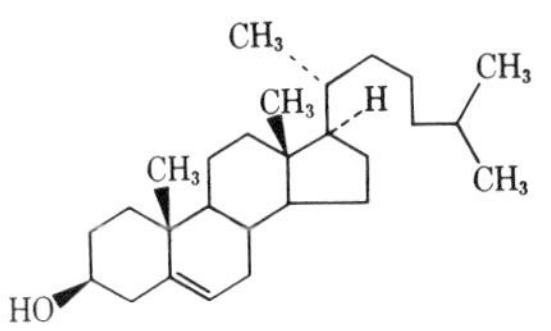

(2)

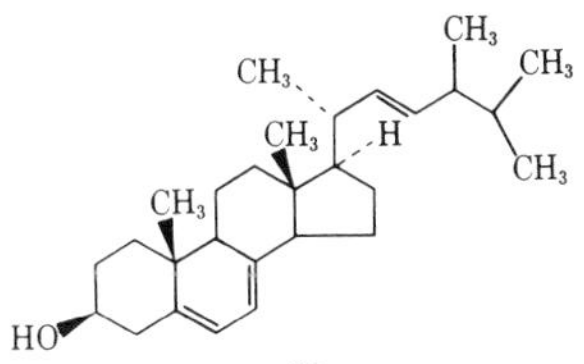

(3)

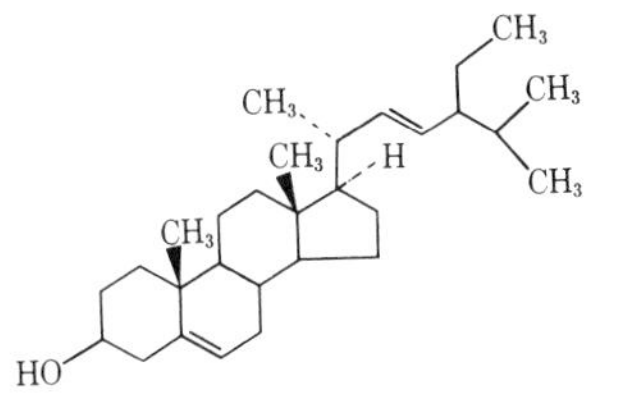

(4)

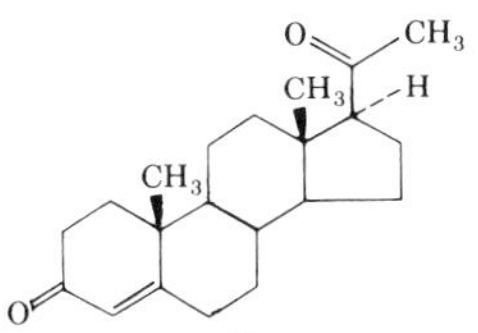

(5)

Marine sterols generally have a carbon range of C_{26}–C_{30}. The carbon variation occurs almost exclusively in the side chain. The steroid gorgosterol has a cyclopropane in the side chain. This unusual structural feature is characteristic of a class of marine steroids.

The steroidal antibiotics of the fusidane series are fungal metabolites with antibiotic properties; they are used against staphylococcus infections. They exhibit an unusual trans-syn-trans-anti-trans stereochemistry which forces ring B into the boat configuration.

Bile acids. The active constituents of bile are the bile salts, which are either glycine or taurine conjugates of a polyhydroxylic steroidal acid. The principal bile acids of mammals are the hydroxy derivatives of 5 β-cholan-24-oic acid. Ox bile is the most important commercial source and contains primarily cholic acid with lesser amounts of deoxycholic acid. The bile acids are synthesized in the liver from cholesterol.

Saponins and sapogenins. Steroidal saponins are plant glycosides that give soapy solutions; the crude plant extracts have been used as detergents. The saponins cause hemolysis and have been used as fish poisons but are not toxic to mammals, probably because they are not absorbed from the gut. Saponins form molecular complexes with several classes of compounds.

The saponins are not easily isolated in a pure state. The complete structures of several have been elucidated and in general the sugar is linked to the steroid through its 3-hydroxyl group. Thus digitonin consists of the aglycone digitogenin and two glucose, one xylose, and two galactose, moieties. Diosgenin (6) and its 5,6-dihydro-12-oxo derivative hecogenin are important raw materials in the steroid industry.

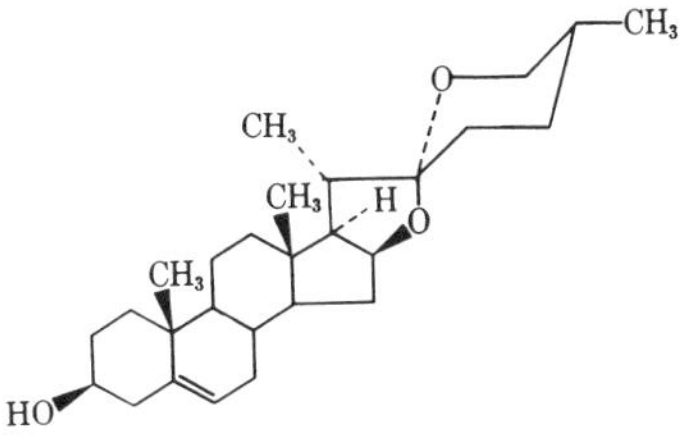

(6)

Steroidal alkaloids. The steroidal plant alkaloids are widespread. Normal steroidal alkaloids are divided into five groups, all of which are derived from 5 α-pregnane and differ in the number and position of the amino groups.

The modified plant steroidal alkaloids are of two types: the C-nor-D-homo steroids, which have a modified ring system where ring C has been contracted to a five-membered ring and ring D has been expanded to a six-membered ring; and the buxus alkaloids, which usually have a cyclopropane ring fused into a normal steroid nucleus at positions 9 and 10.

Salamander venom is secreted in skin glands. Its main constituent is the animal steroidal alkaloid samandrine which causes convulsions followed by respiratory paralysis. This steroid is toxic to all animals.

For centuries, the skin secretions of the small, vividly colored frog *Phyllobates amotaenia* have been used by the Indians in the Colombian rain forest to prepare poisoned darts. The active constituents are 20α esters of the steroidal alkaloid batrachotoxin A with various pyrrole carboxylic acids. Batrachotoxin, the 20α ester of batrachotoxin A with 2,4-dimethylpyrrole-3-carboxylic acid, is among the most lethal substances known, with an LD_{50} of 2 μg/kg subcutaneously in mice as compared to an LD_{50} of 500 μg/kg for strychnine.

Steroidal lactones. Digitalis and other glycosides have a powerful action on the myocardium and are invaluable in the treatment of congestive heart failure. Many plant extracts containing cardiac glycosides have been used as arrow and ordeal poisons. The aglycone with the simplest chemical structure is digitoxigenin (7), although more complicated structures, eg, ouabagenin (8), are more common. Synthetic approaches to the construction of cardenolides have centered on the construction of the lactone ring and introduction of the 14 β-hydroxyl group.

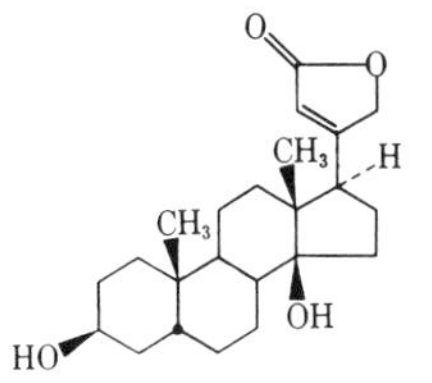

(7)

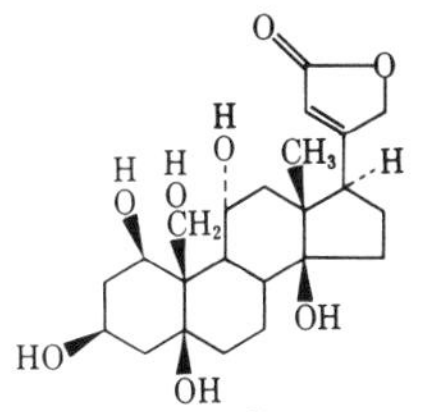

(8)

Except for a 5-substituted 2-pyrone at the 17β-position, the bufadienolides resemble the cardenolides in respect to functional groups, stereochemistry, and biological activity. They occur in several plant families and in the venom of certain Asian toads.

Sea cucumbers secrete saponins which are toxic to fish and have hemolytic and oncolytic properties. Acidic hydrolysis gives holothurigenins, eg, seychellogenin.

The withanolides occur in plants of the family *Solanaceae*. They constitute a series of C_{28} lactones of the ergostane type with a pyran ring in the side chain. Interest in these compounds is due to their reported anticancer activity.

Insect moulting hormones. Moulting processes, essential to the development of insects, are under hormonal control. The moulting hormone was long considered to be secreted from the prothoracic glands; however, experiments demonstrate hormone biosynthesis outside this gland. In 1954, 25 mg of the crystalline moulting hormone of the silkworm, *Bombyx mori*, was isolated from 500 kg of silkworm pupae. Chemical studies and an x-ray analysis of the hormone α-ecdysone (9) established its steroidal nature. The closely-related β-ecdysone containing a 20R-hydroxyl group also occurs in the silkworm.

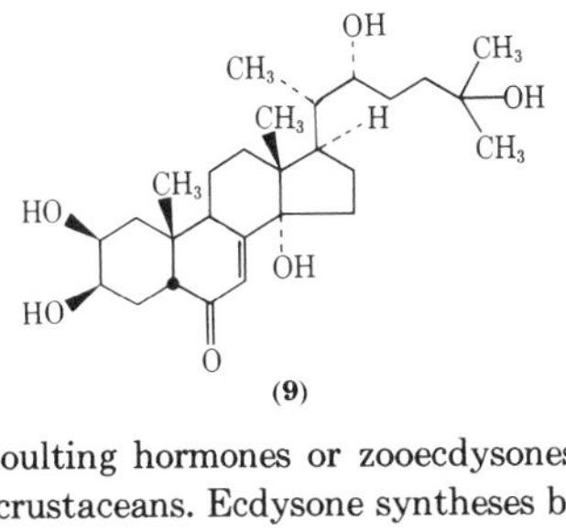

(9)

A variety of moulting hormones or zooecdysones have been isolated from insects and crustaceans. Ecdysone syntheses based on stigmasterol and diosgenin have been reported.

Manufacture and Synthesis

The starting materials for steroids may either be of petrochemical (total synthesis) or animal or vegetable (partial synthesis) origin. Approximately two-thirds of the raw material for chemical synthesis of the steroid hormones produced has depended on diosgenin (6) obtained from the plant *Dioscorea compositae*, which grows in hot and humid mountainous areas. In the United States, stigmasterol (4) serves as raw material for the synthesis of corticoids, whereas in Europe, the abundant

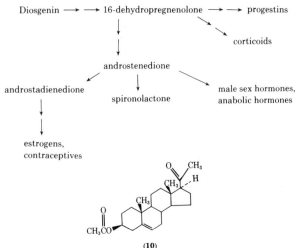

Figure 1. The central role of 16-dehydropregnenolone in the manufacture of steroid products. This critical intermediate was prepared commercially from diosgenin (**6**) obtained from Mexican yams.

sapogenin hecogenin and the bile acids are also used. The two most commonly available sterols are cholesterol (**2**) and β-sitosterol. Their nonfunctionalized side chains are cleaved by microbiological oxidation.

Partial synthesis. The main method for steroid synthesis for over two decades was the degradation of diosgenin (**6**) to 16-dehydropregnenolone acetate (**10**) (see Fig. 1).

The increasing use of 17-keto steroids as starting materials is due to a breakthrough in the microbial degradation of the two commonest steroids, cholesterol and β-sitosterol. Because of the shift to 17-keto steroids as starting materials, new methods have been developed for introducing the corticoid dihydroxyacetone side chain.

Intramolecular functionalization of the nonactivated *C*-10 methyl groups has been used to prepare 19-norsteroids. With the methodology to prepare 6β,19-ethers and 19-hydroxy steroids, fermentation routes have been developed to estrone (**11**) and the enone ether (**12**).

An 18-methyl group in progestins strongly enhances biological activity and, although these compounds are made commercially by total synthesis, several partial syntheses have been described.

Aldosterone (**13**), obtained from adrenal extracts, is necessary for survival by maintaining fluid homeostasis in the human body. Commercially available corticosterone acetate is converted into its nitrite ester, irradiation of which generates an oxime, which is oxidatively deoximated to yield aldosterone. An alternative synthesis starts from 11β-hydroxy progesterone.

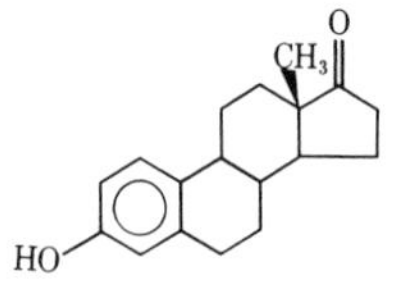

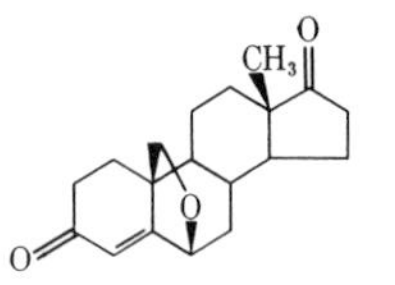

Because of the interest in vitamin D and its metabolites, the switch in sterioidal starting materials from diosgenin to androstenedione, and the need for synthetic comparison samples for sterols obtained from natural sources, much work has been done on the construction of *C*-17 side chains (see Vitamins, Vitamin D).

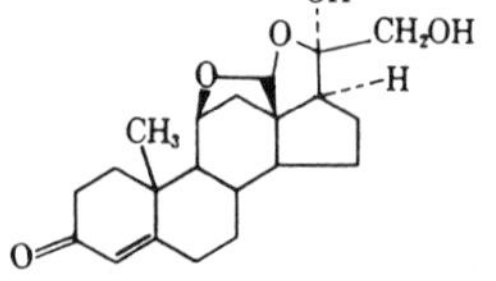

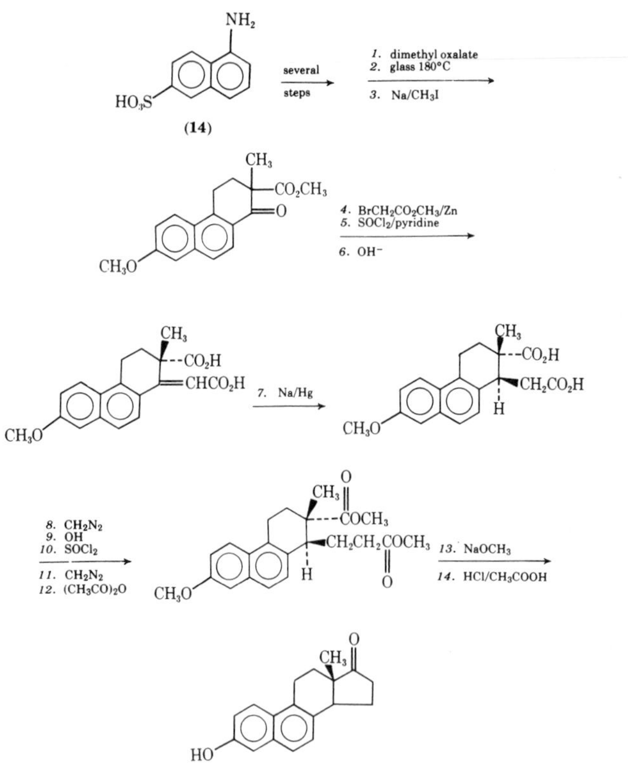

Figure 2. The classical approach to the total synthesis of the estrane steroid, equilenin (**15**).

Total synthesis. The earliest attempts at total synthesis focused on the A-ring aromatic steroids because of their lower stereochemical complexity. The AB-aromatic steroid equilenin (**15**) has only two asymmetric centers, estrone (**11**) has four, and cholesterol (**2**) has eight. Equilenin is the simplest representative of the natural steroids and was synthesized from 5-aminonaphthalene-2-sulfonic acid (**14**) in 20 steps with an overall yield of 2.7% (see Fig. 2).

Estrone (**11**) was the second natural steroid to be synthesized and was prepared from 1-(2-bromoethyl)-3-methoxybenzene via (**16**) in 0.1% yield in 18 steps.

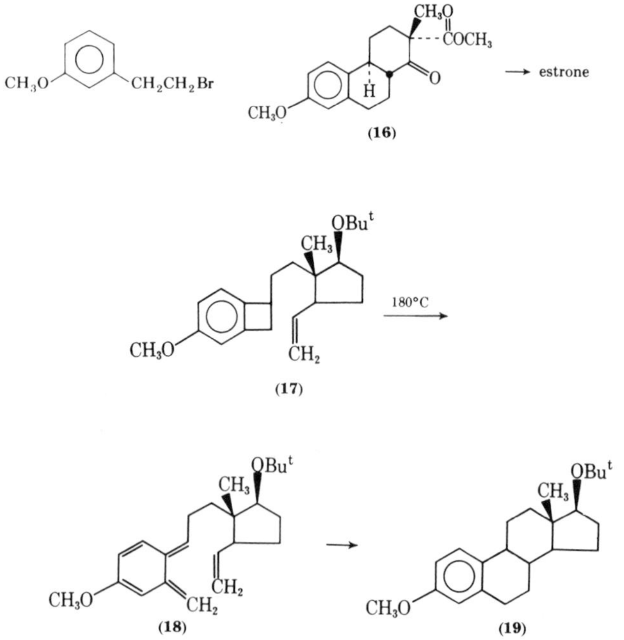

Figure 3. Final steps in the total synthesis of aromatic steroids.

An interesting new approach to A-ring aromatic steroids involves ring opening of a benzocyclobutene to an *o*-quinodimethane and a subsequent intramolecular Diels-Alder reaction. The optically active benzocyclobutene (17) is synthesized and then undergoes ring opening at 180°C, forming the quinodimethane (18), which immediately cyclizes to the estradiol derivative (19) (see Fig. 3).

The main economic interest has been in the preparation of saturated 19-norsteroids for use in oral contraceptives. Most 19-norandrostane oral contraceptives are produced by total synthesis. In a synthesis of 19-nortestosterone (22), 6-methoxytetralone is converted by isoxazole formation, methylation, and Stobbe condensation to a tricyclic compound. A series of steps gives the secosteroid (20). Hydrolysis of the chlorine in (20) gives the ketone (21) which, after an intramolecular aldol condensation and dehydration, is hydrolyzed to 19-nortestosterone (22).

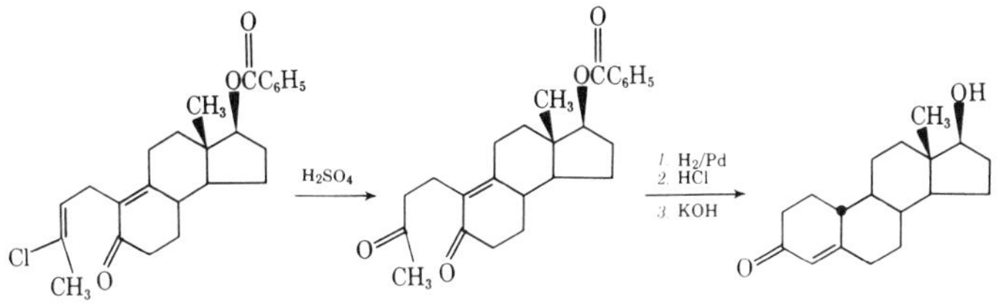

(20) → H₂SO₄ → (21) → 1. H₂/Pd 2. HCl 3. KOH → (22)

(20)   (21)   (22)

The strategy for the total synthesis of steroidal natural products previously discussed has involved, for the most part, step-by-step annelations, ie, the rings are formed one at a time. In the biomimetic approach, a number of rings are synthesized in a single step by ring closure of an acyclic chain having 1,5-placed trans-olefinic bonds. This process is analogous to the biogenetic conversion of squalene to lanosterol.

Modified steroids. Steroids containing a heterocyclic ring attached to the tetracyclic nucleus have been investigated following the discovery that fusion of a pyrazole ring to the 2,3-positions of the steroid nucleus enhances biological activity in androstane derivatives. For example, stanazolol (23) is 10 times as potent an anabolic agent as 17α-testosterone but has only twice the androgenic activity with no lasting adverse side effects. In the corticoid series, a 2'-phenylpyrazole ring substantially enhances the antiinflammatory activity of the parent steroid. However, moving the phenyl ring to the 1'-position results in a sharp drop in activity.

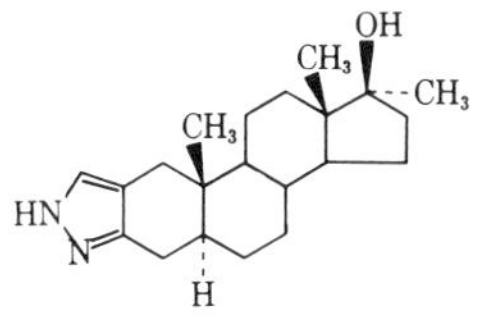

(23)

The highly potent activity of the pyrazoles instigated the syntheses of other heterocyclic fused steroids including thiazoles, pyridines, pyrimidines, pteridines, oxadiazoles, pyrroles, indoles, triazoles, and isoxazoles. One of the most potent of these heterocycles is androisoxazole (24), which exhibits an oral anabolic: androgenic ratio of 40. The 17α-ethynyl derivative danazol is devoid of estrogenic side effects and has been evaluated as an oral contraceptive in women. In men, danazol functions as an antifertility agent by reducing sperm concentration.

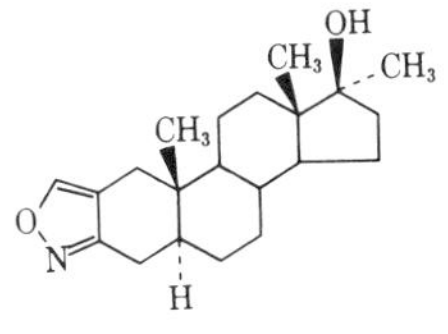

(24)

Pancuronium (25) is a member of a series of bis-quaternary ammonium steroids which cause neuromuscular block. Studies have shown that it is five times as potent as *d*-tubocurarine as a competitive neuromuscular

blocking agent, with minimal cardiovascular and little or no histamine-releasing or hormonal actions.

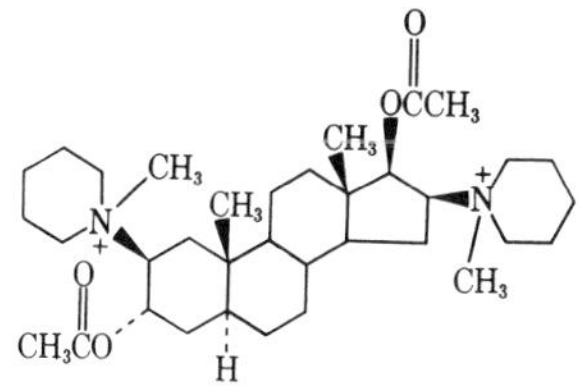

(25)

Research on steroids containing a heteroatom has led only to modest medicinal chemical success. In a synthesis of a series of azacholesterols, one or two of the carbon atoms in the side chain were replaced by nitrogen; these compounds displayed interesting hypocholesterolemic activity. The compound 22,25-diazacholesterol (26) acts by blocking the biosynthetic conversion of demosterol into cholesterol. When tested in humans, it lowered cholesterol and serum triglyceride levels but accumulated desmosterol and caused reversible myotonia. However, it subsequently found use as an oral contraceptive for pigeons.

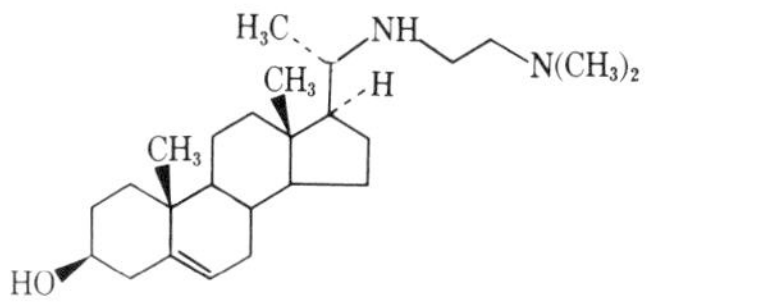

(26)

One of the advantages of introducing a heteroatom into the A-ring of a steroid is the blocking of the metabolic aromatization of ring A to estrogens. It also permits the separation of desired biological activities from undesired ones. Although the 4-oxadihydrotestosterone derivatives have no demonstrable biological activity, the isomeric 2-oxa steroids are potent anabolic agents with reduced androgenic activity. The same type of synthetic approaches used for the construction of the carbocyclic skeleton are for the most part amenable to the formation of heterocyclic steroids.

Cyclopropyl steroids were for many years a laboratory curiosity and were chiefly used for protecting the 3β-hydroxy-5-ene groupings during reactions involving other portions of the molecule. Solvolysis of the 3β-tosylate in methanol yields the methoxycyclopropane, termed an isosteroid.

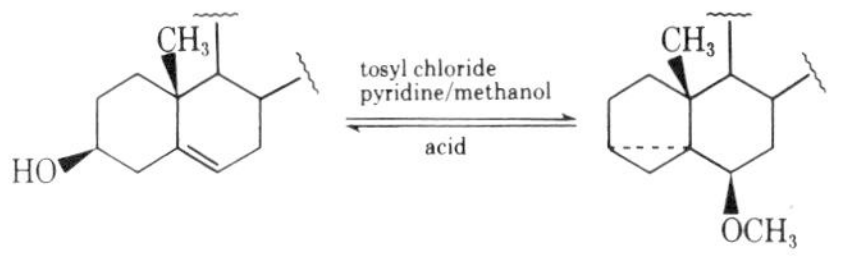

tosyl chloride / pyridine/methanol ⇌ acid

Subsequently, a novel synthesis of a cyclopropane ring system involves the photochemical cyclization of the conjugated diene system in compound (27) into a bicyclo[1.1.0] bridged ring system, which contains two cyclopropanes and solvolysis of one of the cyclopropane rings to give the ether (28).

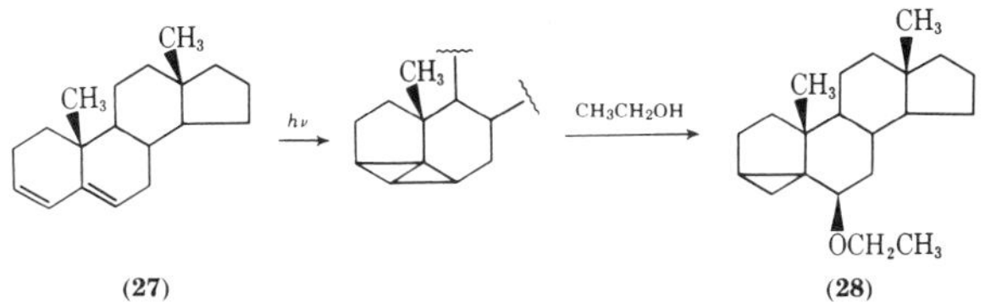

(27) → hν → → CH₃CH₂OH → (28)

(27)   (28)

This reaction is restricted to the unsubstituted diene. A general way of introducing the 1α,2α-cyclopropane is the conjugate addition of dimethyloxosulfoniummethylide to a 3-oxo-1-ene steroid. To introduce a β-oriented cyclopropane in the 1,2-position, the Simmons-Smith reaction, which involves an electrophilic organozinc reagent, is used. The introduction of 6,7-cyclopropane into a 3-oxo-4,6-diene is accomplished via the methylide reagent. A 6β,7β-cyclopropylprogesterone derivative and 6β,7β-cyclopropylspironolactone have been prepared in this way. Both are potent antimineralocorticoids.

Steroids having an inverted configuration at C-9 and C-10 are termed retrosteroids. These compounds were originally prepared through a series of reactions based on studies of vitamin D. Ergosterol (3) was converted photochemically into its 10α-epimer. Oxidation and selective reduction then yielded the enone (29), which was converted by standard methods into retroprogesterone (30). Conversion to the conjugated dienone (31) affords the potent, orally active progestin dydrogesterone. In contrast to progesterone, compound (31) is nonthermogenic. A wide variety of retrosteroids in both the sex hormone and corticoid series have been synthesized and evaluated biologically.

(29)   (30)

Δ^6

(31)

Homo-, nor-, and secosteroids are steroids that have been modified by expansion or contraction or simultaneous expansion and contraction of one or more of the carbocyclic rings in the tetracyclic nucleus. The driving force for the preparation of these modified steroids has been the enhancement or separation of biological activities. Ring expansion is accomplished by the Tiffeneau-Demjanov rearrangement, whereby a ketone is converted to a cyanohydrin, reduced to the amino alcohol, diazotized, and solvolyzed to the homoketone.

Ring expansion can also be accomplished with diazomethane on saturated ketones or the homologation of α,β-unsaturated ketones with diazomethane in the presence of boron trifluoride. The main product is formed by the insertion of a methylene group between the carbonyl group and the unsaturated α-carbon to give a $\beta\gamma$-unsaturated ketone.

Formation of A-norsteroids can be effected by a retropinacol reaction of triterpenoid 4,4-dimethyl-3β-ols. 3β-Hydroxy-4,4-dimethyl steroids are usually rearranged with phosphorus pentachloride in inert solvents. B-Norsteroids are generally prepared by chromium trioxide oxidation of 3β-acetoxy-5-ene steroids. Although the B-nor derivatives of most biologically interesting steroids have been synthesized, their biological activity has been disappointing. Similarly, D-norsteroids are not of biological significance; they are prepared by photochemical ring contraction of a diazoketone.

Simultaneous ring expansion and contraction reactions have been extensively used in the preparation of modified steroids. A general method for A-nor-B-homosteroids is the photolysis of α,β-epoxy ketones. Fission of the carbon–oxygen bond of the epoxide is followed by a 1,2-alkyl or hydrogen shift. Most C-ring contractions are accompanied by simultaneous D-ring expansion.

Analytical Methods

Ultraviolet and visible spectroscopy allows the detection and identification of chromophores within a molecule. Optical rotatory dispersion and the complementary circular dichroism have been extremely useful in solving stereochemical and configurational problems with steroids containing a chromophore or in those steroids where a chromophore can be attached to a nonchromophoric functional group. An empirical analysis of the circular dichroism of ketones has allowed the estimation of the numerical contributions of the rings and substituents to the intensity of the absorption.

Proton nuclear magnetic resonance has allowed the direct observation of hydrogen atoms in their steroidal molecular environment. It is possible to obtain routine nmr spectra on trace amounts of material by means of pulse and Fourier transform nmr. A more fundamental advance has resulted because of the commercial availability of ^{13}C nmr spectrometers to observe individual carbon atoms in their molecular environment.

Extensive x-ray crystal-structure determinations have been done on a wide variety of steroids.

GEORGE R. LENZ
Searle Laboratories

R.F. Witzmann, *Steroids: Keys to Life*, Van Nostrand Reinhold, New York, 1981.

L.F. Fieser and M. Fieser, *Steroids*, Reinhold Publishing Corporation, New York, 1959.

Terpenoids and Steroids, The Chemical Society, London, UK, annual.

A.S. Mackenzie, S.C. Russell, G. Eglington, and J.R. Maxwell, *Science* 217, 491, (1982).

STILBENE DERIVATIVES.   See Brighteners, fluorescent; Stilbene dyes.

STILBENE DYES

Stilbene dyes are mixtures of indeterminate structure resulting from the condensation of 5-nitro-2-toluenesulfonic acid (1) with itself or with other aromatic compounds. The chemical moiety 2,2'-stilbenedisulfonic acid (2), either as the free acid or in salt form, is common to all stilbene dyes.

(1)   (2)

According to the complexity of these condensation reactions, stilbene dyes are classified as follows: self-condensation products of (1) or its derivatives and further products after oxidation; condensation products of (1) and its derivatives with phenols, naphthols, or aminophenols; condensation products of (1) and its derivatives with aromatic amines; condensation products of (1) and its derivatives with aminoazo compounds; and azostilbene dyes formed by diazotization of a condensation product containing primary amino groups and coupling.

Manufacture

When 5-nitro-2-toluenesulfonic acid (1) is heated with aqueous sodium hydroxide, a variety of yellow dyes is obtained depending on the concentration of the caustic, temperature, and reaction time. These are collectively grouped as Direct Yellow 11. The mixture is generally heated at 70–80°C in aqueous sodium hydroxide (initially 4.5 wt%); the concentration is increased to 17.5 wt% over 5 h. Continued heating gives (3).

(3)

If a reducing agent is added in the initial condensation or as aftertreatment, additional yellow and orange dyes are produced, eg, Direct Orange 15. Its reduction product, called Diamine Orange D, is used in the dyeing of paper. If formaldehyde is the reducing agent, Direct Yellow 6, also a paper dye, is the product.

Oxidation of (3), Direct Yellow 6, with nitric acid, sodium hypochlorite, or chromic acid, gives Mikado Yellow G. The sodium salt of (1) can be used for self-condensation.

Condensation of (1) or 4,4'-dinitrostilbene-2,2'-disulfonic acid (4) with primary amines leads to a wide range of orange-to-brown dyes. Such condensations yield azo dyes (eg, (5)) without the diazotization/coupling reaction sequence.

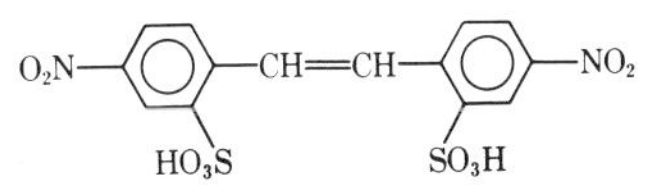

(4)

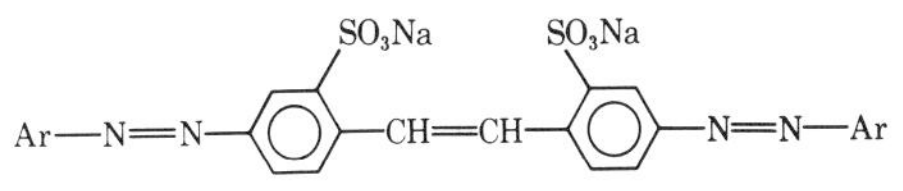

(5)

Condensation of *o*-oxy-*o*'-methoxyaminoazo compounds with (1) under caustic alkaline conditions with subsequent metallization gives olive, brown, and gray dyes with excellent lightfastness.

The dyes described thus far are obtained primarily from caustic alkaline condensation reactions giving mixtures and products of indeterminate structure. However, if 4,4'-dinitro-2,2'-stilbenedisulfonic acid (4) is reduced, the resulting 4,4'-diamino-2,2'-stilbenedisulfonic acid (6) may be tetrazotized and coupled to a variety of aromatic systems to yield azo dyes (eg, (7)), which can be characterized with more exact structural certainty (see Fig. 1). The important dyes Brilliant Yellow and Chrysophenine are made this way.

When (6) is tetrazotized and coupled to two moles of *N,N*-dialkyl-substituted aniline-type couplers, disazo dyes ranging from yellows to violets with excellent water solubility and good fastness properties on polyamide fibers are obtained.

Dyes that yield reddish-black to black shades on paper have been reported by coupling (6) to 2-amino-8-hydroxynaphthalene-6-sulfonic acid, which is then coupled to resorcinol. A series of dyes that produces dark blue shades on paper has been made by incorporating the 2-substituted-4-phenyl-3,1-thiazole moiety. Unsymmetrical molecules can be obtained from the reaction of two different couplers to tetrazotized (6).

When 4,4'-dinitro-2,2'-stilbenedisulfonic acid is reduced with sodium sulfhydrate under carefully controlled conditions, 4-nitro-4'-aminostilbene-2,2'-disulfonic acid is obtained. Unsymmetrical products, eg, Direct Green 36, can be prepared by diazotizing the 4'-amino group, coupling to one mole of a coupling component, reducing the 4-nitro group, diazotizing the newly formed 4-amino group, and coupling to one mole of a different coupling component.

A similar series of reactions is involved in the production of Direct Green 23; *N*-acetyl-*S*-acid is the final coupling component. A variation on this reaction involves the formation of aminostilbenenaphthotriazines, such as Direct Green 34. Recent patents claim such stilbenenaphthiotriazines as excellent paper dyes with bright shades, especially for dyeing alum or rosin-sized paper.

Unreduced 4-nitro-4'-amino-substituted stilbene-2,2'-disulfonic acids are excellent dyes, particularly because of their superior water solubility. Diazotized 4-amino-4'-nitrostilbene-2,2'-disulfonic acid coupled to H-acid and *p*-nitroaniline yields a dark green dye on cotton.

Fiber-reactive dyes incorporate a cyanuric halide moiety into a stilbene system. Such dyes are used for leather, paper, and cotton (qv) where excellent fastness properties and high tinctorial strength are required.

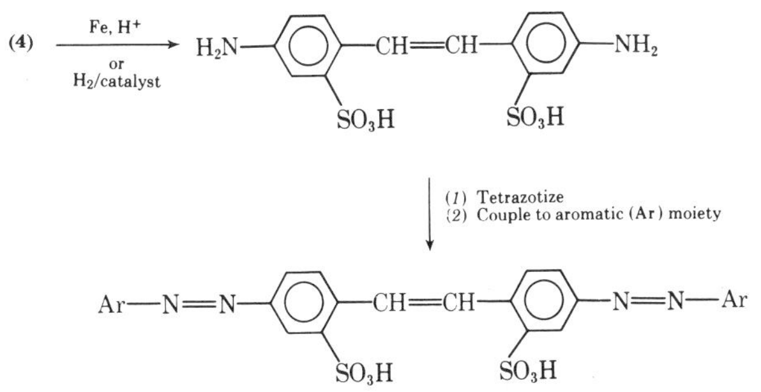

Water solubility. Stilbene dyes are used in the dyeing of paper and paper pulp where an aqueous medium is required. Powder form improves cold-water solubility. Lithium or lithium–sodium salts increase solubility, as does the addition of urea.

Health and Safety Factors

As a class, stilbene dyes have not exhibited health or safety properties warranting special precautions. Some nitro intermediates may cause skin or eye irritation.

RUSSELL E. FARRIS
Sandoz Colors & Chemicals, Inc.

Colour Index, 3rd ed., Vol. 4, The Society of Dyers and Colourists, Bradford, Yorkshire, UK, 1971, pp. 4212–4214, 4365–4373; Vol. 6, 1975, p. 6400.

K. Venkataraman, *The Chemistry of Synthetic Dyes*, Vol. 1, Academic Press, Inc., New York, 1952, pp. 628–635.

STIMULANTS

Therapeutic agents that stimulate the central nervous system (CNS) include natural and synthetic compounds and are characterized according to structure, locus of action, and pharmacologic or clinical effects. Therapeutically, they are used to elevate low levels of physiologic activity to normal. For the purposes of this discussion, the term includes analeptics, psychostimulants, sympathomimetics, monoamine oxidase inhibitors, and antidepressants (see Fig. 1).

Analeptics

Analeptics or restoratives are compounds capable of restoring depressed medullary and cerebral functions. The compounds are used for the

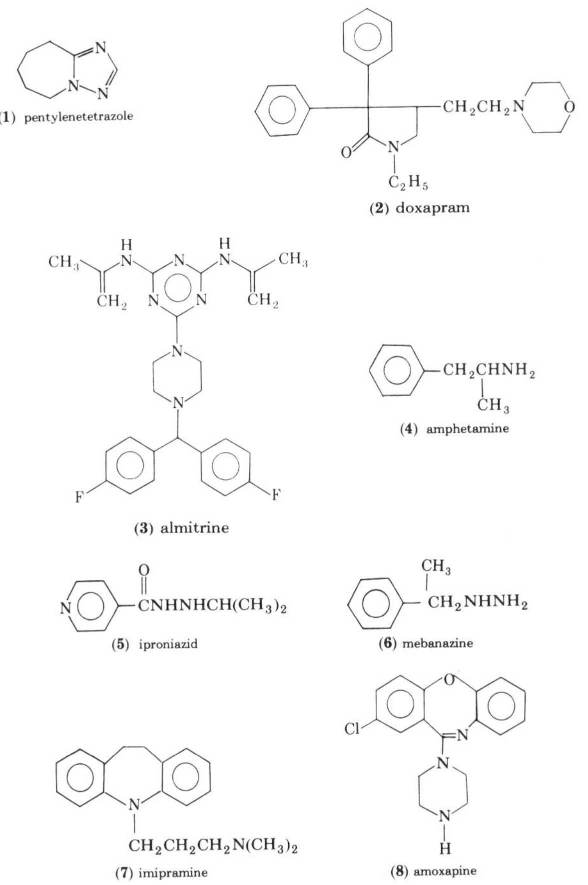

Figure 1. Representative stimulants of different classes.

Figure 1. Preparation of stilbene azo dyes.

treatment of barbiturate intoxication, arousal after general anesthesia, and treatment of pulmonary dysfunction. Overdoses may produce extreme CNS excitation, including restlessness, hyperexcitability, skeletal muscle hyperactivity, and convulsions.

Pentylenetetrazole (1) was one of the first totally synthetic analeptics, prepared by the reaction of cyclohexanone with hydrazoic acid. It is used to enhance mental and physical activity in elderly patients and as an EEG activator in epilepsy.

Flurothyl, $CF_3CH_2OCH_2CF_3$, is a volatile analeptic with strong convulsant properties that has been used for chemical shock therapy.

Ideally, analeptics used to enhance depressed respiration should selectively stimulate respiratory centers without affecting other motor centers in the CNS. Several synthetic analeptics are sufficiently selective to be clinically useful. Doxapram (2) is by far the leading respiratory stimulant marketed in the United States. Almitrine (3) is a more recent analeptic marketed in Europe.

Psychostimulants

Compounds with specific cerebral-stimulant properties are classified as psychostimulants or psychoanaleptics, eg, deanol, marketed in the United States as the acetamidobenzoate, and pemoline. Both are used primarily in the treatment of children with attention-deficit disorder and hyperkinetic syndrome (see also Psychopharmacological agents).

Caffeine is a mild psychostimulant, which is by far the most widely used psychoactive substance on earth (see Alkaloids).

Sympathomimetics resemble the neurotransmitters epinephrine and norepinephrine (qv) in their actions and, to some extent, chemically. They have wide-ranging effects including, in some cases, profound CNS excitatory action, and are used primarily as anorexigenic agents. Sympathomimetics, eg, amphetamine (4), belong to the most abused classes of drugs marketed in the United States. Continued use often leads to habituation and tolerance. The manufacture, distribution, and use of sympathomimetics are controlled by the Drug Enforcement Agency.

Antidepressants

Depressive disorders, often called affective disorders, are divided into several categories. The most common is reactive depression (also called secondary or neurotic depression) and is usually precipitated by a serious adverse experience. A more serious type is endogenous or primary depression, probably of genetic origin. Evidence indicates that endogenous depression is associated with abnormally low levels of the neurotransmitter norepinephrine, and possibly serotonin at central nerve synapses.

A spectacular breakthrough in the treatment of depression was achieved in the 1950s with the development of the monoamine oxidase (MAO) inhibitors and the tricyclic antidepressants. Since the introduction of these compounds, there has been a significant decline in the number of patients hospitalized for depression. Three MAO inhibitors are marketed in the United States for the treatment of depression: isocarboxazid (5), phenelzine, and tranylcypromine. Nialamide and mebanazine (6) are marketed in Europe. The use of MAO inhibitors for the treatment of depression is severely restricted because of potential side effects, the most serious of which is hypertensive crisis.

Tricyclic antidepressants. Imipramine (7) is one of many useful psychoactive compounds derived from systematic molecular modifications of the antihistamine promethazine. Related compounds were prepared and evaluated, eg, amitriptyline and doxepin. The latter has an oxygen atom in the bridge between the two aromatic rings.

A recent addition to the U.S. antidepressant market is amoxapine (8), the desmethyl derivative of the antipsychotic drug loxapine. Other tricyclic antidepressants marketed outside the United States are noxiptilin, butriptyline, clomipramine, and dibenzepin.

Although tricyclic antidepressants have proved much safer than MAO inhibitors, side effects can occur, such as blurred vision, dry mouth, constipation, and urinary retention, which are associated with anticholinergic properties. More serious, however, are effects on the cardiovascular system.

In the correlation of structural features and physical properties of antidepressant drugs with pharmacologic effects, two of the most im-

portant factors are the relative orientation of the aromatic rings and the distance between the basic nitrogen and one of the aromatic rings. Maximum activity is usually achieved when the angle between the two aromatic rings is large.

At present, more than thirty antidepressants are undergoing clinical trials worldwide. Many have unique, nontricyclic structures.

Maprotiline and nisoxetine are effective antidepressants that are claimed to be specific inhibitors of norepinephrine reuptake. Other compounds, eg, trazodone, fluoxetine, and fluvoxamine, preferentially inhibit serotonin reuptake. Mianserin and iprindole substantially reduce the responsiveness of post-synaptic beta-receptors. It has been postulated that beta-adrenergic receptor desensitization may be an effect common to all antidepressants though brought about by several different mechanisms.

WILLIAM J. WELSTEAD, JR.
A.H. Robins Company

S.C. Wang and J.W. Ward in J.G. Widdecombe, ed., *International Encyclopedia of Pharmacology and Therapeutics*, Vol. 104, Pergamon Press, Oxford, UK, 1981, p. 85.

B.K. Koe in S. Fielding and H. Lal, eds., *Antidepressants*, Futura Publishing Co., Mt. Kisco, N.Y., 1975, p. 143.

S.J. Enna, J.B. Malick, and E. Richelson, eds., *Antidepressants: Neurochemical, Behavioral, and Clinical Perspectives*, Raven Press, New York, 1981.

K.A. Nieforth and M.L. Cohen in W.O. Foye, ed., *Principles of Medicinal Chemistry*, Lea and Febiger, Philadelphia, Pa., 1974, p. 275.

STOICHIOMETRY, INDUSTRIAL. See Simulation and process design.

STONEWARE. See Ceramics.

STORAX. See Resins, natural.

STOUT. See Beer.

STREPTOMYCIN AND RELATED ANTIBIOTICS. See Antibiotics.

STRONTIUM AND STRONTIUM COMPOUNDS

Strontium is in Group IIA of the periodic table between calcium and barium. Strontium is present at 0.02–0.03% concentration in the earth's crust and is the fifth most abundant metallic ion in seawater. Strontium rarely forms independent minerals in igneous rocks but usually occurs as a minor constituent of rock-forming minerals.

Strontium is a hard, white metal with a melting range of 768–791°C, a boiling range of 1350–1387°C, and a density of 2.6 g/cm³, crystallizing in the fcc system. Strontium-90 is a radioisotope produced by nuclear fission; four stable isotopes of strontium are known.

Strontium is produced by reduction of the oxide with aluminum *in vacuo*. There are no commercial uses of strontium metal. It has been used to remove traces of gas from vacuum tubes.

Strontium Compounds

Although many strontium compounds are known, only a few have commercial importance; the carbonate and nitrate are made in large quantities.

The principal strontium mineral is celestite, $SrSO_4$. The carbonate is found as the mineral strontianite; no economically workable deposits are known. Almost all celestite is used to make strontium carbonate, which is sold in granular and powder grades. Celestite ore is imported from Mexico; quantities and values are reported annually by the U.S. Bureau of Mines.

Strontium compounds are remarkably free of toxic hazards. Strontium nitrate should not be stored in areas of potential fire hazard.

Strontium acetate, $Sr(CH_3CO_2)_2$, is a white crystalline salt soluble to the extent of 36.4 g in 100 mL of water at 97°C. When heated, it decomposes.

Strontium carbonate, $SrCO_3$, is a colorless or white crystalline solid. Ground celestite ore is reduced in a kiln to strontium sulfide, known as black ash. Treatment of a saturated solution of black ash with soda ash or CO_2 precipitates strontium carbonate crystals. The largest use of $SrCO_3$ is in the manufacture of glass faceplates for color-television tubes.

Strontium chromate, $SrCrO_4$, is used as a yellow pigment and as an anticorrosive primer for zinc.

Strontium hexaferrite, $SrO.6Fe_2O_3$, is used for magnets in small electric motors.

Strontium halides are used in medicine as replacements for other bromides and iodides. The bromide forms white, needlelike crystals, which are very soluble in water; the chloride is similar to calcium chloride but less soluble in water; the fluoride is insoluble in water but soluble in hot HCl; and the iodide forms colorless crystals which decompose in moist air.

Strontium nitrate, $Sr(NO_3)_2$, is a colorless crystalline powder classified as an oxidizer by the DOT. The principal use of strontium nitrate is in pyrotechnics (qv), where it imparts a characteristic, brilliant crimson color to a flame.

Strontium oxide, SrO, is a white powder which with water gives the hydroxide, $Sr(OH)_2$. The latter is a white deliquescent solid used to make strontium greases. Strontium peroxide, SrO_2, is used in pyrotechnics and medicines.

Strontium sulfate forms colorless or white rhombic crystals.

Strontium titanate, $SrTiO_3$, is used in the form of disks as electrical capacitors in television sets, radios, and computers.

ANDREW F. ZELLER
FMC Corporation

Metals Handbook, 9th ed., Vol. 2, American Society for Metals, Metals Park, Ohio, 1979.

Strontium, A Chapter from Mineral Facts and Problems, 1980 Edition, Preprint from Bulletin 671, U.S. Bureau of Mines, 1981.

Mineral Commodity Summaries, 1981, U.S. Bureau of Mines, U.S. Department of the Interior, Washington, D.C., 1981.

STRUCTURE – ACTIVITY RELATIONSHIPS. See Pharmacodynamics.

STYRENE

Styrene, $C_6H_5CH{=}CH_2$, is the most important aromatic monomer. It is used extensively for the manufacture of plastics, including crystalline polystyrene, rubber-modified impact polystyrene, acrylonitrile–butadiene–styrene terpolymer (ABS), styrene–acrylonitrile copolymer (SAN), and styrene–butadiene rubber (SBR) (see Styrene plastics; Acrylonitrile polymers).

Commercial manufacture began on a small scale shortly before World War II. Since that time, annual production in the United States has grown rapidly.

Ethylbenzene dehydrogenation remains the dominant manufacturing technique, although coproduction of styrene with propylene oxide accounts for a part of world production (see Xylenes and ethylbenzene).

Properties

The physical properties of styrene are given in Table 1. Polymerization is the only reaction of commercial significance, although styrene also undergoes other reactions of a typical unsaturated compound.

Table 1. Physical Properties of Styrene Monomer

Property	Value			
boiling point (at 101.3 kPa = 1 atm), °C	145.0			
freezing point, °C	−30.6			
flash point (fire point), °C				
Tag open-cup	34.4 (34.4)			
Cleveland open-cup	31.1 (34.4)			
autoignition temperature, °C	490.0			
explosive limits in air, %	1.1–6.1			
refractive index, n_D^{20}	1.5467			
	at 0°C	60°C	100°C	140°C
viscosity, mPa·s (= cP)	1.040	0.470	0.326	0.243
surface tension, mN/m (= dyn/cm)	31.80	29.01	27.15	25.30
density, g/cm^3	0.9237	0.8702	0.8346	
heat of formation (liquid) at 25°C, ΔH_f, kJ/mol[a]	147.36			
heat of polymerization, kJ/mol[a]	74.48			

[a] To convert J to cal, divide by 4.184.

Manufacture

The two commercial routes are based on ethylbenzene produced by alkylation of benzene with ethylene by the Friedel-Crafts reaction.

In the Union Carbide/Badger process, catalyst complex, benzene, and recycled higher alkylated benzenes are agitated in a brick- or glass-lined reactor at low pressure and temperature. After the reaction is completed, the organic phase is separated and the ethylbenzene distilled. The system is sensitive to catalyst poisoning. However, fresh catalyst is fed continuously.

Vapor-phase alkylation at moderate to high pressure has been practiced commercially with solid acid catalysts as early as 1942. During World War II, the U.S. government financed the construction of a plant with a fixed-bed catalyst of alumina deposited on silica gel. Temperatures over 300°C and pressures over 6000 kPa (870 psi) were required.

Solid phosphoric acid catalysts developed by Universal Oil Products were used by the El Paso Natural Gas Company. The reaction was carried out in the liquid phase with dilute ethylene feedstock; conversions of 90–98% were obtained.

The Mobil/Badger ethylbenzene process developed during the 1970s is based on a synthetic zeolite catalyst, which combines high activity with good resistance to coke formation. In the reaction section, fresh and recycled benzene are heated, vaporized, combined with an alkyl–aromatics recycle stream and fresh ethylene, and fed to a reactor. Two reactors are included in the system; one is onstream while the other is being regenerated. Reactor effluent vapor is passed to a prefractionator for recovery of unreacted benzene. Condensed prefractionator distillate is recycled to the reactor, and uncondensed light components, which are vented from the system, can be used as fuel gas. A small regeneration heater is provided for the catalyst-decoking cycle. The distillation section consists of the benzene-, ethylbenzene-, and polyethylbenzene-recovery columns.

The Alkar process, commercialized in 1960, is based on alumina activated with boron trifluoride.

Recovery of ethylbenzene from mixed C_8 aromatics by superfractionation is a process with high capital costs. Several commercial superfractionation units were built in the United States, Europe, and Japan during the 1950s and 1960s. However, the supply of mixed xylenes has been growing at a much slower pace than the demand for styrene, and there is little need for new superfractionation capacity.

Dehydrogenation of ethylbenzene accounts for ca 90% of the world styrene capacity. Peroxidation and reaction with propylene followed by dehydrogenation of α-phenylethanol to styrene accounts for the remaining 10%.

In Halcon International's process, styrene and propylene oxide are produced. First, ethylbenzene is oxidized to ethylbenzene α-hydroperoxide:

$$C_6H_5CH_2CH_3 + O_2 \rightarrow C_6H_5CH(OOH)CH_3$$

The reaction takes place in the liquid phase, and no catalyst is required. At constant temperature, this process is more selective toward the production of by-product acids, acetophenone, and α-phenylethanol; pressure is not critical. The hydroperoxide product reacts with propylene to form propylene oxide and α-phenylethanol.

$$C_6H_5CH(OOH)CH_3 + CH_3CH{=}CH_2 \rightarrow C_6H_5CH(OH)CH_3 + CH_3\overset{O}{\overset{\diagup\diagdown}{CHCH_2}}$$

The catalysts are molybdenum, tungsten, or vanadium compounds. The alcohol can be dehydrated to styrene or reduced to ethylbenzene.

Although all processes share the catalytic dehydrogenation of ethylbenzene to styrene in the presence of steam, the reaction can proceed either by an adiabatic or isothermal process. As the dehydrogenation reaction is endothermic, heat must be supplied. In the adiabatic process, steam superheated to 800–950°C is mixed with heated ethylbenzene feed prior to exposure to the catalyst. The isothermal process takes place in a tubular reactor and reaction heat is provided by indirect heat exchange between the process fluid and a heat-transfer medium, eg, flue gas (BASF technology).

Conversion of ethylbenzene is a function of temperature. An increase in conversion can be obtained by increasing the reactor temperature. The main difference between the isothermal and adiabatic processes is in the way the endothermic reaction heat is supplied. Use of the isothermal process on a commercial scale has been limited.

The most important factor in establishing the economics of the process is the catalyst. All commercial catalysts are formulated around an iron oxide. The most widely used additives are chromia (Cr_2O_3) and potassium oxides (as K_2CO_3).

Crude styrene effluent must be fractionated and undesirable by-products must be removed. Polymerization is minimized by low-temperature vacuum distillation and inhibitors. In the Cosden/Badger purification, the styrene recovery section consists of three vacuum distillation columns. The liquid hydrocarbon phase from the settler is fractionated in the benzene–toluene column. The benzene–toluene overhead cut is sent to a splitter to recover benzene. The separation of ethylbenzene and styrene is difficult because of their close boiling points; high efficiency, low pressure drop distillation trays are used.

Specifications for a typical polymerization-grade styrene are shown in Table 2.

Analysis

For high purity products, gas chromatography has superseded the more conventional wet methods. Color is determined with a spectrophotometer. The polymer content is determined by measurement of turbidity produced by the addition of methanol.

Health and Safety Factors

Styrene is mildly toxic and flammable. It forms explosive mixtures with air and can polymerize violently. However, these hazards are not severe, and it is considered a safe chemical. Oral toxicity is low.

Uses

Styrene consumption and distribution is governed by polystyrene demand.

Table 2. Specifications for Typical Polymerization-Grade Styrene Monomer Product

Assay	ASTM test method
purity, wt % 99.6	D 3962
color, APHA <10; 10 max (Pt–Co scale)	D 1209
polymer, ppm by wt <10	D 2121
C_8, ppm by wt 400–800	D 3962
C_9, ppm by wt 500–1000	D 3962
aldehydes (benzaldehyde), ppm by wt <50	D 2119
peroxides (hydrogen peroxide), ppm by wt <30	D 2340
inhibitor (TBC), ppm by wt 10–50	D 2120
chlorides (chlorine), ppm by wt <10	

Table 3. Typical Physical Properties of Styrene Derivatives

Property	Vinyltoluene	α-Methyl-styrene	Divinyl-benzene[a]
molecular weight	118.17	118.17	130.08
refractive index, n_D^{20}	1.5422	1.53864	1.5326
viscosity, mPa·s (= cP), 20°C	0.837	0.837	0.833
surface tension, mN/m (= dyn/cm), 20°C	31.66	32.40	30.55
density, g/cm³, 20°C	0.8973	0.9106	0.8979
boiling point (at 101.3 kPa (= 1 atm)), °C	172	165	180[b]
freezing point, °C	−77	−23.2	
flash point (Cleveland open-cup), °C	60	57.8	57
explosive limits, % in air	1.9–6.1	0.7–3.4	1.1–6.2
heat of polymerization, kJ/mol[c]	+66.9 ± 0.2	39.75	
solubility, wt% in H_2O at 25°C	0.0089	0.056	
H_2O in monomer at 25°C	0.47	0.010	

[a] DVB-22.
[b] Calculated.
[c] To convert J to cal, divide by 4.184.

Polystyrene. Uses include: packaging and disposable service ware associated with food packaging; housewares, furniture, appliances, television cabinets, and related consumer products; electrical and electronic equipment; and industrial molding products.

ABS. ABS is used in drain, waste, and vent pipes, automobile parts, appliance parts, casings for business machines, recreational goods, luggage, toys, and hobby materials.

SB latexes. Uses include paper coatings, carpet-back coatings, adhesives, latex paint, and similar uses.

SBR. SBR is used in tires and related products.

Derivatives

Vinyltoluene, methylstyrene, and divinylbenzene have been restricted to specialty applications because of their high cost. They are all infinitely soluble in acetone, carbon tetrachloride, benzene, and ethanol and other organic solvents. Physical properties are given in Table 3.

Vinyltoluene. The toxicological properties of vinyltoluene are similar to those of styrene. As a copolymer with styrene, vinyltoluene increases the temperature range of paints, coatings, and varnishes. The higher reactivity reduces drying times.

p-Methylstyrene. p-Methylstyrene (PMS) is ca 97 wt% p-vinyltoluene and 3 wt% m-vinyltoluene. PMS analogues of crystalline and high impact polystyrene offer improved high temperature properties. The use of PMS in reinforced polyesters offers potential for high temperature applications and reduced volatility of the monomer in curing operations. The methyl group causes reduction of polymer densities and an increase in the glass-transition temperature.

α-Methylstyrene. α-Methylstyrene (AMS) is a specialty monomer produced as a by-product of cumene–phenol operations. It is much less reactive than styrene. As a copolymer in ABS and polystyrene, it increases the heat-distortion temperature of the product.

Divinylbenzene. Divinylbenzene (DVB) is an unusual specialty monomer because of its bifunctionality. Its commercial value is in a styrene copolymer. Cross-linking results in reduced solubility, increased heat-distortion temperatures, increased surface hardness, and improved strength. It is made by dehydrogenation of diethylbenzene, similar to the dehydrogenation of ethylbenzene. Toxicology is similar to that of styrene.

P. J. LEWIS
C. HAGOPIAN
P. KOCH
The Badger Company

Styrene-Type Monomers, Technical Bulletin No. 170-151B-3M-366, The Dow Chemical Company, Midland, Michigan.

R.H. Boundy and R.F. Boyer, eds., *Styrene, Its Polymers, Copolymers, and Derivatives*, Reinhold Publishing Corp., New York, 1952.

K.R. Coulter, H. Kehde, and B.F. Hiscock, in E.C. Leonard, ed., *Vinyl Monomers*, Interscience Publishers, a division of John Wiley & Sons, Inc., New York, 1969.

STYRENE–BUTADIENE SOLUTION COPOLYMERS. See Elastomers, synthetic.

STYRENE PLASTICS

Polystyrene (PS), the parent of the styrene plastics, is a high molecular weight linear polymer, $[CH(C_6H_5)CH_2]_n$. The commercially useful form is amorphous and highly transparent. Under ambient conditions, the polymer is stiff and clear, whereas above the glass-transition temperature it becomes a viscous liquid which can be easily fabricated. Monomeric styrene is a worldwide commodity chemical. Polystyrene, a product family with as many as 30 members, although not a commodity in the same sense, is produced on a very large scale. Butadiene-based rubbers increase impact resistance, and copolymerization of styrene with acrylonitrile produces heat-resistant and solvent-resistant plastics. Packaging applications are the largest use for styrene plastics (see Acrylonitrile polymers).

Properties

Considerable differences in performance can be achieved by using the various styrene plastics. For molecular weights above 5000, many properties are independent of chain length. The densities of amorphous and crystalline PS are 1.04–1.065 and 1.12 g/cm³, respectively; T_m = 240–250; refractive index, n_D = 1.60; and the compressive modulus = 3000 MPa (435,000 psi).

The strain energy, derived from the area under the stress-strain curve shown in Figure 1, indicates the toughness of a polymer. High impact polystyrene (HIPS) has a higher strain energy than ABS plastics, although the latter are usually tougher. Tensile strengths of styrene polymers vary with temperature. Increased temperature lowers the strength.

The molecular orientation of the polymer in a fabricated specimen can significantly alter the stress-strain data, as compared with an isotropic specimen. For example, tensile strengths as high as 120 MPa (18,000 psi) have been reported for PS films and fibers. Polystyrene tensile strengths below 14 MPa (2000 psi) have been obtained in the direction perpendicular to the flow.

Creep tests involve the measurement of deformation as a function of time at constant stress. At room temperature, styrene and its copolymers show low elongation with only small variation with stress.

Stress-relaxation measurements, where stress decay is measured as a function of time at constant strain, have also been used to predict the long-term behavior of styrene-based plastics.

Fatigue is of considerable interest, and the inherent fatigue resistance of the material is tested and the relationship between specimen design and fatigue failure is determined.

Polystyrene and styrene copolymers are brittle. Rubber-modified styrene polymers are more impact resistant. A brittle fracture of a styrene polymer can be brought about by producing uniaxially oriented moldings. Tough moldings can be obtained through the introduction of balanced, multiaxial orientation. Embrittlement of tough rubber-modified styrene polymers occurs through aging.

Polystyrene and copolymers. Polystyrene, $\overline{M}_w$ = 2–3 × 10⁵, is a crystal-clear, hard, rigid thermoplastic free of odor and taste. In addition, PS materials have excellent thermal and electrical properties. When lubricants are added, easy-flow materials are produced.

Standard polystyrenes are carefully prepared and characterized materials available from the National Bureau of Standards and the Pressure Chemical Company. Monodisperse polystyrene latices are produced by Dow Chemical as calibration for scientific measurements.

Isotactic polystyrene can be obtained by polymerization with stereospecific catalysts of the Ziegler-Natta type. It can be crystallized and has a threefold helix-chain conformation.

Crystalline polystyrene has a high melting temperature indicating a first-order transition temperature of ca 240°C. It is insoluble in most polystyrene solvents. The density of the 100% crystalline polymer is calculated to be 1.12 g/cm³ from x-ray data. Although highly isotactic polystyrene has been prepared, only partially crystalline polymers have been obtained and generally with less than 50% relative crystallinity. The lack of commercial interest in isotactic polystyrene may result in part from its low degree of crystallinity and its slow crystallization rate. Syndiotactic polystyrene is unknown.

Polymers containing flame retardants (qv) have been developed. The addition of flame retardants does not make a polymer noncombustible, but increases its resistance to ignition and reduces the rate of burning with minor fire sources.

Antistatic polystyrenes have been developed. For styrene-based polymers, alkyl and/or aryl amines, amides, quaternary ammonium compounds (qv), anionics, etc, are used.

Acrylonitrile, butadiene, α-methylstyrene, methyl methacrylate, and maleic anhydride have been copolymerized with styrene to yield commercially significant copolymers.

Butadiene copolymers are rubbers. Many latex paints are based on styrene–butadiene. Most block copolymers prepared in the presence of anionic catalysts, eg, butyllithium, are elastomers; some are thermoplastic rubbers.

Methyl methacrylate copolymers with styrene are clear materials which, when properly stabilized, are similar in their light stability to poly(methyl methacrylate). Maleic anhydride copolymers with styrene tend to have alternating structures. Accordingly, equimolar copolymers are normally produced, corresponding to 48 wt% maleic anhydride.

Some polymers from styrene derivatives meet specific market demands. For example, monomeric chlorostyrene is useful in glass-reinforced polyester recipes.

Rubber is incorporated into polystyrene to impart toughness. The resulting materials are called high impact polystyrenes (HIPS). The rubber is dispersed in the polystyrene matrix in the form of discrete particles.

Acrylonitrile–butadiene–styrene (ABS) polymers are also two-phase systems in which the elastomer component is dispersed in the rigid styrene–acrylonitrile (SAN) copolymer matrix. They are rigid at room temperature, and have excellent notched impact strength. This combination makes ABS polymers suitable for demanding applications. Several ABS polymers exhibit a minimum tendency to orient or develop mechanical anisotropy during molding. Accordingly, uniform tough moldings are obtained. In addition, ABS polymers exhibit good ease of fabrication and produce moldings and extrusions with excellent gloss, which can be decorated by many techniques. Some of the high rubber compositions are excellent impact modifiers for poly(vinyl chloride) (PVC).

Glass-reinforced styrene polymers. Glass reinforcement of PS and SAN markedly improves their mechanical properties; strength, stiffness, and fracture toughness are generally at least doubled. Creep and relaxation rates are significantly reduced and creep rupture times are increased. The coefficient of thermal expansion is reduced by more than one half, and response to temperature changes is minimized. Glass-rein-

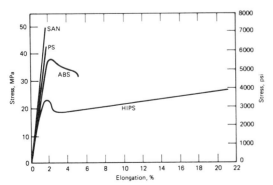

Figure 1.   Stress-strain curves for styrene-based plastics.

forced parts are produced with a preblended, reinforced molding compound; blending of reinforced concentrates with virgin resin; a direct process, in which the glass is cut and weighed automatically and blended with the polymer at the molding machine; and general implant compounding.

Degradation. Styrene plastics are susceptible to degradation by heat, oxidation, uv radiation, high energy radiation, and shear. Degradation by heat *in vacuo* is detectable only above 250°C. Even though polystyrene does not absorb radiation above 300 nm, terrestrial sunlight causes rapid yellowing and embrittlement. The chromophore appears to be a π-complex of oxygen (perhaps in the singlet state) with the aromatic ring. Photolysis at 254 nm results in hydrogen generation, cross-linking, embrittlement, and discoloration. Polybutadiene-modified polymers degrade more rapidly. Ionizing radiation has little or no effect on polystyrene until very large doses are given.

Polymerization

Styrene and most of its derivatives are among the few monomers that can be polymerized by free-radical, anionic, cationic, and Ziegler mechanisms. Styrene–butadiene block copolymers are made with anionic chain carriers, and low molecular weight polystyrene by a cationic mechanism.

The styrene family of monomers are almost unique in their ability to undergo spontaneous or thermal polymerization merely by heating to ca 100°C. Styrene and most styrene derivatives act as their own initiators. Styrene undergoes slow Diels-Alder dimerization.

(1)

Of the two stereoisomers of (1), only one reacts further.

(1a)

These two radicals then initiate chain growth. An important consequence of thermal initiation is the formation of styrene dimers and trimers. It is often advantageous to add free-radical initiators.

Chain-transfer agents are occasionally added to reduce the molecular weight of the polymer. The agents of commercial significance include α-methylstyrene dimer, terpinolene, dodecane-1-thiol, and 1,1-dimethyldecane-1-thiol. Chain transfer to styrene monomer has been reported, but recent work strongly suggests that this reaction is negligible and transfer with the Diels-Alder dimer is the actual transfer reaction.

In some cases, inhibition of polymerization can be regarded as a special type of chain transfer. This is of importance in commercial operations involving styrene storage for extended periods. Most inhibitors are of the phenol/quinone family. These function as inhibitors only in the presence of oxygen. Other inhibitors include sulfur and sulfur compounds, picryl hydrazyl derivatives, carbon black, and some soluble transition-metal salts. Both inhibition and acceleration have been reported for styrene polymerized in the presence of oxygen.

Styrene polymerization can proceed through a positively or negatively charged species. The reaction is more sensitive to impurities than the free-radical system. Two mechanisms are known for anionic initiation: (a) direct attack of a base B, eg, butyllithium, on the monomer, and (b) electron transfer from an active donor molecule:

(a) $B\colon + CH_2{=}CH \longrightarrow BCH_2{-}\bar{C}H$
       $\qquad\qquad\quad | \qquad\qquad\qquad |$
       $\qquad\qquad\quad C_6H_5 \qquad\qquad\quad C_6H_5$

(b) $e + CH_2{=}CH \longrightarrow \cdot CH_2{-}\bar{C}H$
       $\qquad\qquad\quad | \qquad\qquad\qquad |$
       $\qquad\qquad\quad C_6H_5 \qquad\qquad\quad C_6H_5$

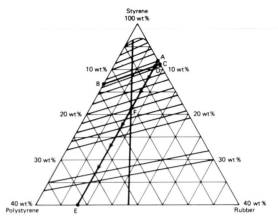

Figure 2. Ternary phase diagram for the system styrene–polystyrene–polybutadiene rubber.

Dispersions of alkali metals initiate anionic polymerization by an electron-transfer process.

Cationic polymerization of styrene can be initiated either by strong acids or by Friedel-Crafts reagents with a proton-donating activator. The solvent plays an important role, and chain-transfer reactions are very common where the reactants are polymer, monomer, solvent, and counterion. High molecular weights are more difficult to achieve and molecular weight distributions are often comparable to those obtained from free-radical polymerizations. Cationic styrene polymerization is used commercially only where low molecular weight polymers are desired.

Copolymerization

Copolymers with butadiene are an important family of rubbers. Methyl methacrylate–styrene copolymers are used where improved resistance to weathering and light are required. Divinylbenzene copolymers with styrene are supports for the active sites of ion-exchange resins and in biochemical synthesis. Polymerization of styrene in the presence of polybutadiene rubber yields a much tougher material than polystyrene. Although the material is usually opaque and has a somewhat lower modulus, it is used widely because of its increased impact resistance. The graft copolymer of polystyrene on polybutadiene acts as an emulsifier, stabilizing the dispersion against coalescence. The polymerization can be followed with the ternary phase diagram shown in Figure 2.

Commercial Processes

Problems encountered in the manufacture of polystyrene include the removal of the heat of polymerization and handling a partially converted polymer syrup with a viscosity of ca 10^8 mPa·s (= cP). Two types of reactors are used for continuous solution polymerization: the linear-flow reactor (LFR), approximating in the ideal case a plug-flow reactor, and the continuous-stirred tank reactor (CSTR), which ideally is isotropic in composition and temperature. The linear-flow reactors usually involve conductive heat transfer to many tubes through which a heat-transfer fluid flows. Multiple temperature zones can easily be achieved in a single reactor; agitation is provided by a rotating shaft with arms down the central axis of the tubular vessel. Reactors of this type operate for long periods and can handle partial polymers of very high viscosity.

A few manufacturers of rubber-modified polystyrene use a mass-suspension method. Acrylonitrile–butadiene–styrene plastics are sometimes manufactured by a similar process, although the favored method involves preforming submicrometer, cross-linked polybutadiene particles in an emulsion process, followed by grafting with styrene–acrylonitrile to the desired rubber content or blending in preformed SAN copolymer.

Fabrication. Injection-molding machines are equipped with a reciprocating screw or a screw preplasticator. Their simple design, uniform melt temperature, and excellent mixing characteristics make them preferable for injection molding. Large solid moldings include automotive dash panels, television cabinets, and furniture components.

Injection molding is the forcing of melted polymer into a cool mold where it freezes and is removed. Injection molding of styrene-based plastics is usually carried out at 200–300°C. For ABS polymers, the upper limit may be lower. To obtain satisfactory moldings with good appearance, contamination must be avoided.

Extrusion of styrene polymers is one of the most convenient and least expensive fabrication methods, particularly for obtaining sheet, pipe, irregular profiles, and films. Many rubber-modified styrene plastics are fabricated into sheet by extrusion primarily for subsequent thermoforming operations.

Thermoforming of HIPS and ABS extruded sheet is of importance in several industries. In the refrigeration industry, for example, large parts are made by vacuum-forming extruded sheet. A plastic sheet is heated above its softening point and forced against a mold by applying vacuum, air, or mechanical pressure. Several modifications have been developed, such as vacuum snapback forming, drape forming, and plug-assist pressure-and-vacuum forming. Thermoforming is perhaps the process with the lowest unit cost.

Blow molding is a multistep process for manufacturing hollow symmetrical objects. Styrene-based plastics are used in blow molding but not as much as linear polyethylene and PVC. Polystyrene or copolymers are used extensively in injection blow molding. Even though polystyrene is stable, compounds are sometimes added to give extra protection for a particular application. For food-contact applications, the additives must be approved by the FDA.

Characterization

Chemical analyses, mechanical characterization, morphology, and rheology are employed for characterization. Chemical analysis has three components: determination of the molecular weight; the additives, ie, plasticizer and mold-release agent; and the residuals remaining from the polymerization process. Particle-size distribution is usually measured with a Coulter counter or directly from electron photomicrographs. Rheological studies are made with a modified tensile tester with capillary rheometer (ASTM D 1238–79). For product specifications, a simple melt-flow test is used (ASTM D 1238 condition G).

Health and Safety Factors

Styrene-based plastics have very low toxicity. Under normal conditions they should pose no problems. Heating usually results in the release of vapors. Adequate ventilation should be provided. Styrene polymers burn under the right conditions, even when modified by addition of ignition-suppression chemicals. Combustion products should be considered toxic.

Uses

Polystyrene foams are used for extruded planks and boards, for low-temperature thermal insulation, buoyancy, floral display, novelty packaging, and construction purposes. Batch and automated continuous molding are employed. Polystyrene is the dominant material in rigid thermoplastic foams.

Extruded rigid foam is used for insulations. In addition to such uses in buildings, there is a new area in the form of highway underlayment to prevent frost damage.

Expandable polystyrene (EPS) has been in wide use in packaging since its introduction. The basic resin for EPS is in the form of foaming-in-place (FIP) spherical beads that are expanded to a desired density before molding. The Dow Company pioneered an elongated shape, called Pelaspan-Pac, which expands into wormlike shapes ideally suited as loose fill for packaging.

Lightweight concrete is made from prefoamed EPS beads, portland cement, and organic binders. Precast shapes provide structural strength, thermal insulation, and soundproofing.

Polystyrene foamed sheet is used for foamed trays, egg cartons, disposable dinnerware, and packaging.

Although polystyrene is normally considered brittle, biaxial orientation imparts extremely desirable properties, particularly in regard to an increase in elongation.

Polystyrene film contains no plasticizers, absorbs negligible moisture, and exhibits exceptional dimensional stability.

A.E. PLATT
T.C. WALLACE
The Dow Chemical Company

R.H. Boundy, R.F. Boyer, and S. Stoesser, *Styrene, Its Polymers, Copolymers and Derivatives*, American Chemical Society, Monograph No. 115, Reinhold Publishing Corporation, New York, 1952.

R.F. Boyer, H. Keskkula, and A.E. Platt in N.M. Bikles, ed., *Encyclopedia of Polymer Science and Technology*, Vol. 13, John Wiley & Sons, Inc., New York, 1970, pp. 128–447.

C.A. Brighton, G. Pritchard, and G.A. Skinner, *Styrene Polymers: Technology and Environment Aspects*, Applied Science Publishers Ltd., London, 1979.

SUBERIC ACID, $HOOC(CH_2)_6COOH$.   See Dicarboxylic acids.

SUCCINIC ACID AND SUCCINIC ANHYDRIDE

Succinic acid, $C_4H_6O_4$, is a constituent of almost all plant and animal tissues. Succinic anhydride (1) (3,4-dihydro-2,5-furandione) was first prepared by dehydration of the acid; hydrogenation of maleic anhydride gives nearly quantitative yields.

Properties

The acid occurs both as colorless triclinic prisms (α form) and as monoclinic prisms (β form). Succinic anhydride forms rhombic pyramidal or bipyramidal crystals. Physical properties are given in Table 1.

Succinic acid undergoes reactions characteristic of dicarboxylic acids. In addition, the methylene groups are responsible for many reactions.

On heating, succinic acid forms an internal anhydride with a stable ring structure. Further heating gives the dilactone of γ-ketopimelic acid (2). This reaction can occur with explosive violence when molten succinic anhydride is distilled in the presence of alkali ions.

When heated with bromine in a sealed vessel at 100°C, succinic acid yields meso-2,3-dibromosuccinic acid almost quantitatively. Succinic anhydride, when heated with an equivalent of bromine, gives mainly the monobromo derivative (bp, 130–133°C); two moles of bromine give dl-2,3-dibromosuccinic anhydride (mp, 118–119°C).

Succinic esters condense with aldehydes and ketones in the presence of bases to form half esters of alkylidenesuccinic acids. This reaction,

Table 1. Physical Properties of Succinic Acid and Succinic Anhydride

Property	Succinic acid	Succinic anhydride
mp, °C	188.1	119.6
bp, °C	dehydrates at mp	261
sublimation pt at 267 Pa[a], °C	156–157	90
specific gravity	1.552–1.577	1.572
solubility, g/100 g soln		
in water at 100°C	121	
heat of formation at 298.15 K, J/mol[b]	940.35 ± 0.54	
heat capacity at 298.15 K, J/(mol·K)[b]	152.9	
dielectric constant at 3–97°C, 5 kHz	2.29–2.90	

[a] To convert Pa to μm Hg, multiply by 7.5.
[b] To convert J to cal, divide by 4.184.

known as the Stobbe condensation, is specific for succinic esters. Dialkylidenesuccinic acids and anhydrides are formed in a manner similar to aromatic aldehydes. Succinic acid and its anhydride are readily esterified by the usual methods.

Ketones containing reactive methyl or methylene groups form diketones with succinates in the presence of sodium hydride.

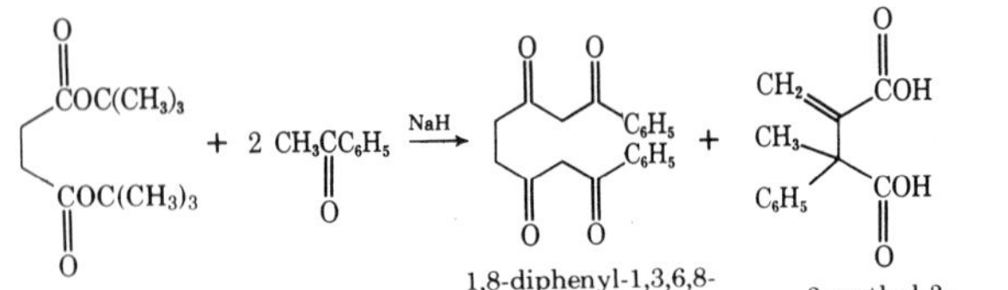

1,8-diphenyl-1,3,6,8-octantetraone

3-methyl-3-phenylitaconic acid

Under Friedel-Crafts conditions, succinic anhydride forms alkylbenzoylpropionic acids with alkylbenzenes.

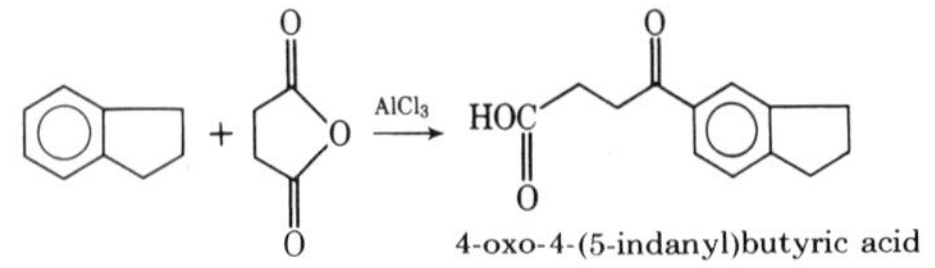

4-oxo-4-(5-indanyl)butyric acid

(1)

Succinimide (mp, 126°C) can be prepared from succinic acid or its anhydride with ammonia, urea, adipamide, or an isocyanate. Succinimide derivatives are used as anticonvulsants.

N-Bromosuccinimide (mp, 176–177°C), formed by the addition of bromine to a cold, aqueous solution of succinimide in sodium hydroxide, is a bromination and oxidation agent.

Ethylenediamine heated with two moles succinic anhydride gives N,N'-ethylenedisuccinimide (3).

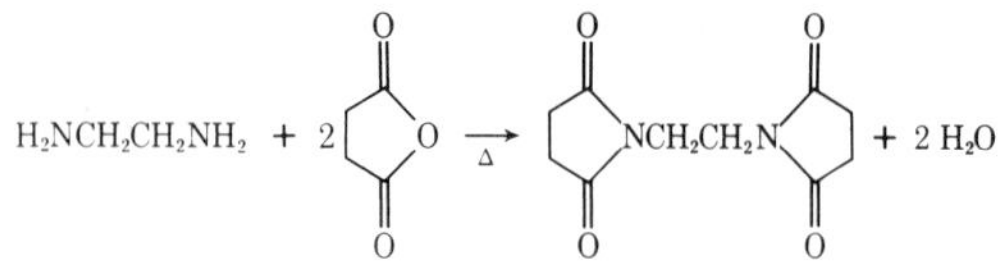

(2)   (3)

Succinamide, $(-CH_2CONH_2)_2$, (mp, 268–270°C) is obtained from succinyl chloride with ammonia or by the partial hydrolysis of succinonitrile.

Diethyl or diphenyl succinate react with potassium hydrogen sulfide to form dipotassium dithiosuccinate. Acidification gives thiosuccinic anhydride with the evolution of hydrogen sulfide.

Succinic anhydride is hydrated practically instantaneously in boiling water. Potassium permanganate oxidizes succinic acid to oxalic acid or a mixture of malic and tartaric acids. Sodium perchlorate yields 3-hydroxypropionic acid.

Catalytic hydrogenation of succinic acid or anhydride yields 1,4-butanediol, γ-butyrolactone, tetrahydrofuran, or mixtures.

Manufacture

Numerous processes are available for the production of succinic acid and its anhydride. Oxidation of paraffins C_{12}–C_{32} gives mixtures of acids from which succinic acid can be separated. Oxidation of naphthenic acid with nitric acid in the presence of a vanadium catalyst gives a mixture containing mostly succinic acid. Cycloolefin oxidation with ozone or ozone followed by hydrogen peroxide gives better selectivities for succinic acid.

Other routes starting with carbon monoxide include the reaction of acetylene with carbon monoxide and the oxidative carbonylation of ethylene in the presence of methanol and a catalyst to give methyl esters of succinic acid.

A U.S. patent describes a vapor-phase, catalytic process in which 2-butene is oxidized to give a mixture of 45% fumaric acid and 55% succinic acid.

The preparation of succinic acid by fermentation has been studied extensively in Japan. Yields of succinic acid from glucose as high as 51% based on glucose have been reported.

Hydrogenation of maleic acid or anhydride continues to be a preferred method. Palladium on carbon, palladium on $CaCO_3$, and carbon-carrier catalysts containing rhodium and ruthenium have been cited as catalysts.

Manufacture of the acid from the anhydride is much simpler. Sales-grade anhydride is dissolved in boiling water. The solution is cooled, and the crystals formed are separated and dried.

Analytical Methods

Titration in conjunction with mixed melting points is often sufficient. Gas chromatography has been recommended for plant tissues and fruit juices. Mass spectrometry and a combination of gas chromatography with mass spectrometry are used for the analysis of succinic acid in urine.

Health and Safety Factors

The GRAS status of succinic acid has been affirmed by the FDA as a flavor enhancer, as a pH control agent in condiments and relishes, and for meat products. A carbon respirator is recommended when handling hot succinic acid or anhydride (see Food additives).

Uses

Succinic acid is used in medicinals; in agricultural applications; as preservative and flavoring agent; in plating baths; as corrosion inhibitor; in gas scrubbing; in textile finishing; in photography; in cosmetics; as plastic and resin additive; in detergents and emulsifiers; and in catalysts.

LEON O. WINSTROM
Consultant

Technical Data Bulletin D55, Allied Chemical Corp., New York, 1966.

Technical Data Bulletin AA5, Allied Chemical Corp., Plastics Division, Morristown, N.J., Oct. 1965.

SUGAR

PROPERTIES OF SUCROSE

Sucrose (α-D-glucopyranosyl-β-D-fructofuranoside), $C_{12}H_{22}O_{11}$, is a disaccharide composed of D-glucosyl and D-fructosyl moieties.

Sucrose is extracted commercially from sugarcane, sorghum, and sugar maple. The cultivation of sugarcane is restricted to tropical and semitropical regions; the sugar beet is more suited to temperate zones as are sorghum and maple. Beets provide sugar for Europe, central Canada, and the central and western United States, whereas cane supplies the rest of the world.

Properties of Sucrose

Sucrose crystallizes as monoclinic, hemimorphic (sphenoidal) crystals. Its melting point has been reported between 160 and 200°C, with 188°C generally accepted. The structure is shown in Figure 1.

Sucrose is soluble in water to the extent of 2.07 g/g H_2O at 25°C, and is readily soluble in aqueous protic solvents such as methanol and ethanol. It is insoluble in ether and chloroform and anhydrous ethanol and glycerol. It is moderately soluble in DMF, pyridine, and dimethyl sulfoxide. Sweetness is affected by total acidity, pH, viscosity, temperature, and other constituents present.

Derivatives

A variety of sucrose esters has been prepared. They are used in baking, as emulsifiers and viscosity modifiers, and in drug and cosmetic formulations. Ethers have similar properties and uses. Detergents formulated with sucrose are biodegradable.

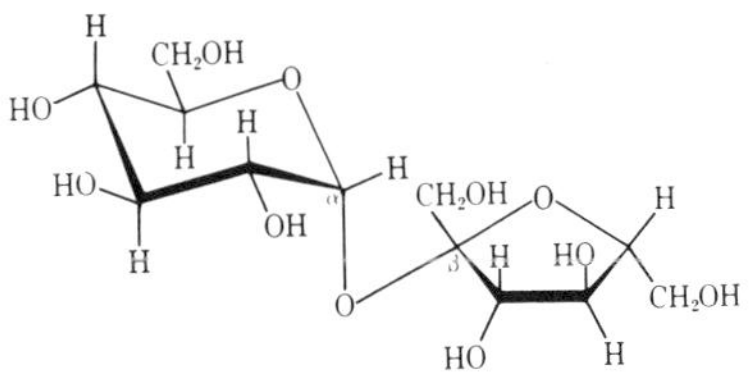

Figure 1. Structure of sucrose.

Reactions of Sucrose

Sucrose reacts with alkali- and alkaline-earth metal salts to form complexes. Sucrose complexes with calcium form the basis for separating sucrose from beet molasses (Steffen process). Similar processes based on strontium and barium have been employed.

Under mild acid catalysis, usually dilute HCl, sucrose is easily hydrolyzed and inversion occurs (change in the sign of polarization from + 66.5 to a negative value, owing to the large negative rotation of fructose). A small amount of the fructose formed is converted to D-fructose dianhydrides. A small degree of inversion is detectable under alkaline conditions.

Sucrose is hydrogenated with Raney nickel to a mixture of sorbitol and mannitol; under more drastic conditions, glycerol and propylene glycol are produced. Mild oxidation gives oxalic and tartaric acids, as well as acidic materials designated as saccharic acids.

Degradation of dry sucrose at temperatures of 90–200°C begins with cleavage of the glycosidic bond followed by condensation and formation of water. At 170–210°C, this process is referred to as caramelization; the mixture of products formed is commercially called caramel. At high temperatures carbon–carbon cleavage occurs. Degradation reactions are sensitive to impurities. Thermal degradation in solution produces 5-hydroxymethylfurfural and colored condensation products.

R.M. Sequeira
Amstar Corporation

R.A. McGinnis, ed., *Beet Sugar Technology*, 3rd ed., Beet Sugar Development Foundation, Fort Collins, Colo., 1982.

G.P. Meade and J.C.P. Chen, *Cane Sugar Handbook*, 10th ed., John Wiley & Sons, Inc., New York, 1977.

W. Mauch and E. Farhoudi, *Sugar Technol. Rev.* **7**, 87 (1979–1980).

SUGAR ANALYSIS

The authority for sugar analyses is the International Commission for Uniform Methods of Sugar Analysis (ICUMSA), which convenes every four years.

Physical Methods

The concentration of a pure sugar solution is determined by measurement of polarization (optical rotation), refractive index, and density. The saccharimeter is a polarimeter reading directly in percent sucrose. However, polarimeter scales are not inherently stable and must be calibrated frequently using quartz plates. Direct polarization can give the value for sucrose only when no other optically active substances are present. When interfering substances are present, the correct sucrose value can be obtained by the Clerget method, which is a double polarization, before and after inversion of the sucrose.

Density, measured by the standard hydrometer or spindle, is widely used to determine the sugar concentration of syrups, liquors, and juices. Hydrometers graduated in sucrose concentration are called Brix hydrometers or Brix spindles; the readings are called spindle Brix. The degrees Brix equals the wt% of sucrose in the solution. Hydrometers graduated in °Baumé are used for molasses and corn sugar. The relation between °Baumé and density d, in g/cm³, is

$$°\text{Baumé} = 145(1 - 1/d)$$

Refractive index is another measurement of sugar concentration in solution. Many refractometers have Brix scales. The readings are called refractometer Brix or refractometer dry substances (RDS).

Chemical Methods

Reducing action is the basis for methods for the determination of aldoses and ketoses. The alkaline copper tartrate reagent, Fehling's solution, is used in a number of methods. All have the problem that the reaction is not stoichiometric, and the procedure must be followed exactly.

In the Lane and Eynon method, the boiling reagent is titrated with sugar solution and the point must be reached in exactly 3 min.

The Munson and Walker method is convenient for an occasional analysis. The reduced cuprous oxide is determined gravimetrically.

The Ofner method is used for up to 10% invert sugar in sucrose. The reduced cuprous oxide is treated with excess standardized iodine, which is then back-titrated. Small amounts of reducing sugars in sucrose are determined by the Knight and Allen method. After the sugar reacts with the copper reagent, the excess unreduced copper is determined with EDTA. Very small amounts of reducing sugars are determined by the Emmerich method. It is based on the formation of a colored complex of 3,6-dinitrophthalic acid and invert sugar.

Enzymatic methods can also be used. Glucose oxidase, for example, is specific for the oxidation of glucose.

Sucrose content is also obtained by measuring reducing sugars before and after inversion.

Micromethods

These methods include reduction of ferricyanide to ferrocyanide; reduction of cupric sulfate to cuprous oxide; and the development of color with phenol in strong sulfuric acid.

Chromatographic Methods

Paper chromatography gives good separation and, with highly specific sprays, the individual spots can be identified.

Electrophoresis uses a high voltage gradient. The hydroxyl groups on sugars readily form complexes with borate and other ions. A charge is thus imparted and electrophoresis can be applied.

For gas chromatography, the sample is vaporized. Although sugars are not very volatile, their methyl esters and silyl derivatives are. Liquid chromatography offers the advantage that no derivatives are needed and the solvent can be water.

Purity. Purity denotes sucrose content as a percentage of total solids. It is calculated as pol/Brix.

Ash. Ash is determined by incineration. The residue is called carbonate ash. Bubbling and swelling of the incinerating mass can be avoided by adding concentrated sulfuric acid. The ash components are converted to sulfates and the ash is called sulfate ash. Electrical conductivity gives a conductivity ash that is simple and quick.

Color. The standard conditions for colorimetry are 50 Brix concentration and pH 7.0. The light transmission T is measured at 420 nm. Sugar color is expressed as:

$$a^* = -\log T/bc$$

where b is the cell depth in cm; c is the sugar concentration in g/mL; and a^* is the attenuation index. The number given by the above equation is multiplied by 1000 to give sugar-color units. Raw sugars have a color of 1000 or more, refined sugar, 35.

Frank G. Carpenter
United States Department of Agriculture

F. Schneider, *Sugar Analysis ICUMSA Methods*, International Commission for Uniform Methods of Sugar Analysis, 1979.

Official Methods of Analysis of the AOAC, 13th ed., Association of Official Analytical Chemists, Washington, D.C., 1980.

Standard Analytical Methods, 6th ed., Corn Refiners Association, Washington, D.C., 1980.

CANE SUGAR

Cane sugar is the name given to sucrose, a disaccharide produced from the sugarcane plant. The other principal source of sucrose is the sugar beet. Refined sugars from the two sources are practically indistinguishable. However, the trace constituents are different, notably raffinose and odor constituents. In addition, since the two plants have different photosynthetic pathways, the carbon isotope ratio in the sugars can be used to identify the source. The other important sweetener is corn sugar (glucose), derived from corn starch (see Sweeteners).

Sugar is used almost entirely for food. In the United States, only ca 1% is used for nonfood purposes. Sugar is one of the purest substances produced in large volume.

The principal by-product of cane sugar production is molasses, which is the residual plant juice concentrate from which no more sugar can be economically recovered.

Cultivation

Sugar cane, a tropical grass, grows very tall, 3–5 or even 7 m, and has a thick stalk (2–3 cm) in which the sugar is stored in solution (in the juice). There are many varieties and all are hybrids. A continuing breeding program is essential because all varieties suffer a gradual decline.

The experimental station at Canal Point, Florida, has a collection of 30,000 varieties for cross breeding, from which ca one million (10^6) cross breeds are made each year.

Cane is grown by planting stalks called seed cane or setts. The growing time is about one year, although in some areas it grows two or more years. Upon harvesting, the stalks are cut off even with the ground and the next day more sprouts start from the buds just below ground level. Good agricultural practice requires close attention to soils, fertilizers, weeds, pests, disease, and water supply. The sugarcane plant stores sugar at all times, and the harvesting season is therefore long. Sugarcane produces the heaviest yield of all crops in weight of both biomass and useful product per unit area of land. Table 1 gives sugarcane yields from various regions.

Raw Sugar Manufacture

Sugar cane cannot be stored more than a few hours after it is cut, so the raw sugar mills are located in the cane fields. The cane is washed and broken into pieces. The juice is extracted either by milling, in which the cane is pressed between heavy rolls, or by diffusion, in which the cane is leached with water. The sugar recovered is nominally 10 wt% of the cane.

Milling achieves ca 95% extraction of the sucrose in the cane, diffusion achieves ca 97% extraction. Diffusion juice contains fewer suspended solids. The plant costs are lower and less energy is required to run it. The fiber remaining is bagasse (qv) and is burned in the boilers that power the mill.

Clarification. The first step in the process is to add lime, which raises the pH, stops inversion, and helps to settle the suspended matter. Clarification is by means of a Dorr-type clarifier. Although clarification removes most of the mud, the resulting juice is not necessarily clear. Supplemental schemes complete the clarification.

Evaporation. The clarified cane juice is only 15° Brix and a large amount of water must be removed. Multiple effect evaporators are used; about three or four are used in series in which the first operates at atmospheric pressure and subsequent effects are at higher vacuum.

Table 1. Yield of Sugarcane

Region	t/(ha·yr)
Louisiana	60
Florida	90
Hawaii	100
Philippines	65
Puerto Rico	75
Australia	100
Iran	110

Evaporation is carried to a final Brix of ca 65–68. The juice, after evaporation, is called syrup and is almost black and a little turbid.

Crystallization. Crystallization from the concentrated syrup is traditionally a batch process. After seeding, the evaporation and feeding of the syrup are balanced for the fastest possible rate of crystal growth.

Vacuum pans. Crystallization is done under a vacuum such that the water boils at about 65°C in order to minimize thermal destruction of sucrose. The vacuum pan has a very large discharge opening, typically 1 m dia. At the end of a strike, the massecuite contains more crystals than syrup and is very viscous. The large opening is required in order to empty the pan in a reasonable time.

Centrifuging. The massecuites from the vacuum pans enter a holding tank called a mixer that has a slowly turning paddle to prevent the crystals from settling. The centrifuge is fed from the mixer. In batch-type centrifuges, the mother liquor is separated from the crystals in batches of ca 1 t. A centrifugal force of about one thousand G is applied. The starting and stopping of the centrifugal batch consumes a great amount of energy, much of which can be saved by the use of a continuous centrifuge consisting of a conical basket with a screen.

Boiling systems. In raw-sugar manufacture, the first strike is called the A strike, and the mother liquor obtained from the centrifuges is called A molasses. The so-called B sugar from the second strike is only half as pure as that obtained from the first strike. Usually, only three strikes are obtained from cane juice.

Final strikes are not sent to the centrifuges directly but to crystallizers, which are holding tanks where the crystals grow, over a period of four d, and as much sucrose as possible is extracted from the final molasses.

Raw sugar. The A and B sugar from the centrifuges comprise the raw sugar of world commerce. Raw sugar can be stored for several years. If raw sugar is washed slightly in the centrifuge, and then dried and cooled, the resulting sugar has the very high pol (% sucrose) of 99.4%. This material is free flowing and keeps well.

Direct-consumption sugar. Many raw-sugar mills produce sugar for direct consumption (off-white) by performing a supplemental clarification on the cane juice, or by filtering it, followed by decolorization. From these improved syrups one strike of sugar for direct consumption can be obtained; later strikes are treated as raw sugar.

Refining

Sugar refineries are located in large cities, near harbors. The first step in refining is to remove the molasses film from the raw sugar crystals by a washing process known as affination. The resulting syrup is called affination syrup. It is estimated that 90% of refining takes place in this first step.

In the recovery house, sugar is recovered from the affination syrup. The recovered sugar is called remelt and is sent back to processing. The residual syrup is refiners' molasses.

The washed sugar is melted in hot water and the pH is adjusted with lime. The melted liquor is strained through a screen and then clarified.

The object of clarification is the complete removal of all particular matter by filtration, carbonatation, or phosphatation. The most straightforward process is filtration, which requires a filter aid, such as diatomaceous earth or kieselguhr. With a tight, efficient filter aid, flow is very slow. Filtration, as a sole means of clarification, is being replaced by other methods.

To make liquor more suitable for clarification, substances that form precipitates and coagulate the impurities are added. Both processes used today, carbonation and phosphatation, employ lime. In the course of the precipitation of the calcium carbonate, insoluble lime salts are coprecipitated. The final step is filtration. In phosphate clarification, lime and phosphoric acid are added simultaneously. The calcium phosphate precipitate forms a floc. At the same time, air is injected. The precipitate floats to the surface as a scum and is scraped off without filtration. The clarified liquor from either process is brilliantly clear but dark.

The key process in sugar refining is decolorization. Color is the property that distinguishes refined sugar from raw sugar. The color is measured by light transmission at 420 nm. Traditionally, carbon adsorbents have been used for decolorization. However, they adsorb everything, including sugar, with little selectivity. Bone char and granular carbon behave similarly. Bone char is made by charring cattle bones. Granular carbon is made from selected mixtures of coal steam-activated by heating to ca 1000°C (see Carbon). Process variations include sending liquors of increasing color to the adsorbent.

Ion exchange (qv) is also used for decolorization. Its advantages are *in situ* generation without heat, short contact time, and small equipment.

The Tafloc process takes advantage of the anionic properties of the sugar colorant as in ion exchange. The colorants are precipitated with a quaternary ammonium compound, and the precipitate is removed with a phosphate clarifier.

Crystallization. The color of the washed, clarified, and decolorized liquor proceeding to crystallization ranges from water white to slightly yellow. The same vacuum pans are used as described above. They must be operated carefully to produce crystals of the desired size.

Packaging, storing, and shipping. The present trend is away from consumer-sized (< 50 kg) packages and toward bulk shipment.

Products

Sugar marketing is primarily local. Each refinery serves its own area. Refined granulated sugar is the usual output of a refinery. By far, the largest production is in this class. Large-grain specialty sugars are used for candy and cookies. Fine-grain sugar consists of small crystals obtained by screening; it dissolves quickly. Powdered sugar is made by grinding granulated sugar and adding 3% cornstarch to prevent caking. Cubes are made by mixing a syrup with granulated sugar to form cubes. Reagent-grade sugar is sold by chemical supply houses but its quality is not higher than that of regular sugar. Brown sugars, known in the trade as soft sugars, can be made either by coating white sugar with syrup or by a special boiling technique that produces conglomerates. Liquid sugar is a solution of sucrose and invert sugar. Syrups have a high percentage of invert. High test molasses is partially inverted cane syrup. Molasses is the final syrup left after all the sugar that can be economically removed has been crystallized.

Frank G. Carpenter
United States Department of Agriculture

M. Patureau, *By-Products of the Cane Sugar Industry*, Elsevier Publishing Co., Amsterdam, Neth., 1969.

A.C. Barnes, *The Sugar Cane*, 2nd ed., John Wiley & Sons, Inc., New York, 1974.

G.P. Meade and J.C.P. Chen, *Cane Sugar Handbook*, 10th ed., John Wiley & Sons, Inc., New York, 1977.

BEET SUGAR

Sugarbeets are often profitable in areas where few other crops, except potatoes, can be grown successfully. Sugar beets are an excellent crop to fit into a crop-rotation scheme, as their culture tends to improve the soil. Beets are deep-rooted and bring up plant food from a considerable depth, which is then made available to shallow-rooted crops. It is possible to plant them in populated areas.

Sugar Beet Composition

The composition of sugar beets is given in Table 1. Of the nitrogenous compounds, ca 58% is protein. Of the other nitrogen compounds, betaine constitutes 40%, which is carried through to the molasses. The remainder consists of amino acids, purines, amides, and ammonium salts. The principal acid is citric acid which, together with oxalates, is eliminated as insoluble calcium salts in the purification stage.

Manufacture

The manufacture of beet sugar is mainly confined to the time during which the beets are harvested. In most areas of the United States and Canada, the harvest lasts from the beginning of October until the end of November. It is customary to build storage piles of beets and to continue factory operations with these beets for about two more months.

Table 1. Composition of Sugar Beets

Component	Percent
juice	ca 92
insoluble matter[a]	ca 5
water, chemically bound	ca 3
soluble solids	11–25
sucrose in solids[b]	87.5
sucrose in beet	10–22
nonsucrose substances soluble in juice	
organic nitrogen compounds	ca 44
nitrogen-free organic compounds	36
inorganic compounds	20

[a]So-called marc.

[b]Expressed as purity of the juice.

Diffusion process. The sugar in the sugarbeet is contained within the parenchyma cells. On heating to 70–80°C, diffusion of dissolved molecules takes place. The slow extraction of the small sodium and potassium ions is due to the electrostatic retarding effect of the high molecular weight polymeric anions.

The sugar beets are moved to the processing area in water flumes equipped with rock and stone removers, vegetation and trash catchers, and beet washers. At the factory, the beets are sliced into cossettes (long thin strips of V-shaped or square cross section). Typical cossettes may be 2–3 mm thick and up to 15 cm long. The sucrose is extracted from the cossettes by passing them through a diffuser, countercurrent to hot leaching water. The juice leaving the diffuser is screened and may contain ca 14.5 wt% total solids, which, with a sucrose content of 12 wt%, has a purity of ca 83 wt%.

The slope diffuser is a covered, sloping trough, ca 4–7.3 m wide and 15.8–19 m long, depending on the capacity. Today, two other types of diffusers are preferred. The RT-4 consists of a horizontal rotating cylinder built on the screw principle. A tower diffuser consists of a vertical tank with a rotating vertical central shaft equipped with conveying arms that move the cossettes upward against a downward flow of juice.

The dried pulp leaving the diffuser, having a water content of ca 95%, is pressed in screw-type presses to 74–80% moisture. The pressed pulp is enriched by the addition of molasses or concentrated Steffen filtrate, dried, cooled, and stored. Molasses-dried beet pulp is excellent cattle feed.

Purification with lime and carbon dioxide. Treatment with lime gives precipitates and soluble products; 60–70% of the nonsucrose products removed by precipitation are acids that form insoluble lime salts, eg, oxalates, hydroxycitrates, citrates, tartrates, phosphates, and some sulfates. The other nonsucrose products are removed through coagulation.

Passing carbon dioxide through the limed mixture precipitates calcium carbonate on which nonsucrose materials are adsorbed. The calcium carbonate crystals and the adsorbed impurities are removed by sedimentation and filtration.

Many modifications of liming and carbonation procedures have been devised, such as predefecation or preliming and main liming in advance of carbonation. In preliming, about one tenth of the total lime to be used is gradually added to the diffusion juice; many nitrogenous substances are precipitated at their isoelectric points and are not redispersed during main liming. In the Dorr-system purification, the hot juice is simultaneously limed and carbonated with the aid of recirculation, and high alkalinities are avoided.

Alternatives or supplements are treatments with activated carbon or ion-exchange resins. Deliming resins reduce the lime salts content of the thin juice.

A small quantity of sulfur dioxide is added to the thin juice to inhibit the Maillard or browning reaction between reducing sugars and amino acids.

Evaporation and fuel economy. For fuel economy, all high pressure steam from the boilers is expanded to exhaust-steam pressure through turbines; all exhaust steam is used in the first effect of the evaporator; all possible evaporation takes place in the evaporator to reduce the load on the vacuum pans; and for heating of vacuum pans, heaters, or melters, vapors are employed from one of the evaporator effects at the lowest possible pressure. The beet-end juice is heated gradually, using portions of the later-effect vapors to start, and finishing with first-effect vapors (see Evaporation).

Crystallization. Sufficient feed liquor in the vacuum pan permits its concentration to supersaturation. Crystallization (qv) may be induced by adding a small amount of fine crystals of sugar. The fully grown sugar crystals are spun free of syrup in centrifugal separators and washed briefly with hot water. The spun-off syrup is known as green syrup (although it is brown). It provides the feed syrup for the vacuum pan for the second or intermediate boiling. The spun syrup from the second boiling serves as feed syrup for the third (raw) boiling. The syrup spun from the third boiling is termed molasses. It may be used as cattle food, or for production of chemicals by fermentation, or may be treated for removal of more sucrose, as in the Steffen process. The sugar crystals produced in the second and third boilings are normally dissolved in the feed syrup for the first boiling.

Steffen process for desugaring molasses. In North America, ca 25% of the factories are using this process in which about 95% of the sucrose is recovered.

Molasses, diluted with water to ca 6% sucrose, is treated with lime and the calcium saccharate is filtered off. The filter cake is slurried in thin juice, and the so-called saccharate milk is carbonated, yielding free sucrose and calcium hydroxide. Galactosidase may be used to hydrolyze the raffinose.

Granulated sugar is dried, screened, cooled, and bagged or stored in bulk. It is also used for the manufacture of liquid and powdered sugars. Beet-sugar products as manufactured in the United States are essentially identical with refined cane sugar products. Partial or complete computer control is anticipated by the industry.

Material Balance and Quality Control

Sucrose is relatively fragile in solution. Sucrose in sliced beets is determined first by weighing the beet cossettes, and then by analyzing samples of cossettes for sucrose. The grower is paid on this basis.

Sucrose in molasses is determined by analyzing a sample.

Yields and losses statements are made for intervals of one day to one month. Losses of sucrose in the beets between the time of purchase and the time of slicing, so-called outside-factory losses, are partly due to metabolic changes. Total inside-factory losses are found by subtracting the weight of granulated sugar produced from the amount of sucrose in the beets sliced. Extraction denotes the percentage of granulated sucrose recovered of the total sucrose in the beets.

Analyses are based on the use of the refractometer or Brix hydrometer for measuring the total solids in solution; on the polariscope for the determination of sucrose; and on copper-reduction methods for invert sugar; pH is determined throughout.

R.A. McGINNIS
Consultant

R.A. McGinnis, ed., *Beet-Sugar Technology*, 3rd ed., The Beet-Sugar Development Foundation, Fort Collins, Colo., 1982.

F. Schneider, ed., *Technologie des Zuckers*, 2nd ed., Verein der Zuckerindustrie, M. & H. Schaoer, Hannover, FRG, 1968.

P.M. Silin, *Technology of Beet-Sugar Production and Refining*, Pishchepromizdat, Moscow, USSR, 1958; Israel Program for Scientific Translation, Jerusalem, Isr., 1964.

SUGAR DERIVATIVES

Sucrose

The focus of interest on sucrose (**1**) as a chemical raw material has given rise to the name sucrochemistry for the field of carbohydrate chemistry concerned specifically with sucrose and its derivatives. Sucrose can be used as a starting material in degradative reactions, or in syntheses, in which case the carbon framework is retained and the chemistry of sucrose is that of a polyhydric alcohol.

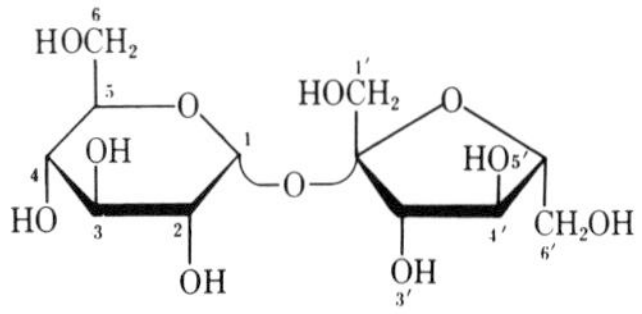

(1) Sucrose

Degradative reactions. Above its melting point (180°C), sucrose decomposes rapidly with the formation of a complex mixture of volatile compounds and an involatile residue, known as caramel, widely used as a coloring agent in foods and drinks (see Colorants for foods, drugs, and cosmetics).

In acidic aqueous solution, sucrose is hydrolyzed to an equimolar mixture of glucose (**2**) and fructose (**4**) which, in alkaline solution, are interconvertible through their common intermediate enediol form (**3**) (Fig. 1). Under acidic conditions at moderate temperatures, hexoses undergo progressive loss of three molecules of water with the formation of 5-hydroxymethylfurfural. At high temperatures, in the presence of strong acids, sucrose gives a mixture of levulinic and formic acids. Under alkaline conditions, sucrose is relatively stable, but glucose and fructose decompose rapidly, forming highly colored polymers and a complex mixture of decomposition products in which saccharinic acids or lactic acid predominate according to the conditions.

Sucrose heated in solution with an excess of lime gives predominantly lactic acid, which can be isolated in high yield. However, this is not an economically favorable route to lactic acid compared with fermentation or synthesis from acrylonitrile.

Sucrose is not readily oxidized by gaseous oxygen, unless it is first hydrolyzed, as in strongly alkaline solution. Glucose is oxidized to arabonic acid when the alkaline solution is shaken with oxygen under pressure at ambient temperature, although in the presence of a palladium catalyst gluconic acid is obtained.

Sucrose is not reducible, but gives a mixture of sorbitol and mannitol upon hydrolysis and electrolytic reduction or hydrogenation in the presence of a nickel or palladium catalyst.

Sugars, when heated in solution with aqueous ammonia, undergo a sequence of reactions in which a wide range of nitrogen heterocyclic compounds form, including derivatives of imidazole, pyrazine, piperazine, and pyridine. This reaction is rarely of synthetic value, although 4-hydroxymethyl-1H-imidazole and 2-methylpiperazine can be made by this route.

The reaction of reducing sugars with amino acids (qv), known as the Maillard reaction, is involved in the nonenzymic browning of foods during cooking, processing, and storage.

Synthetic derivatives. Sucrose undergoes reactions characteristic of alcohols, giving, for example, esters, ethers, acetals, and urethanes.

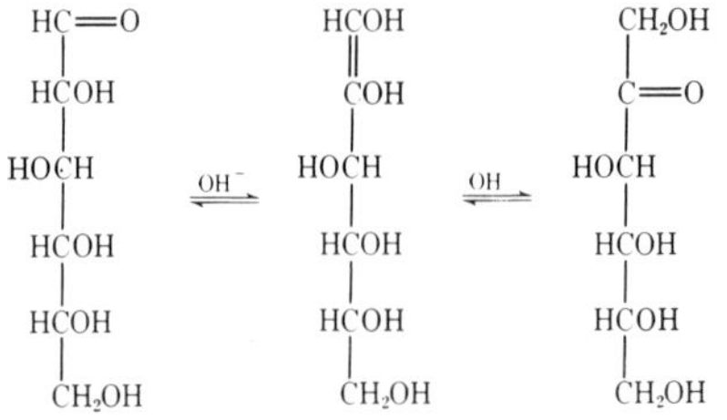

Figure 1. The interconversion of glucose and fructose.

Sucrose is readily esterified to the octaester in high yield by reaction with excess of the anhydride or chloride of a sterically unhindered organic acid in the presence of pyridine or other suitable base.

An important alternative route to sucrose esters is by transesterification between sucrose and an alkyl ester or triglyceride, whereby an acyl group is transferred to sucrose in the presence of a basic catalyst, eg, potassium carbonate, at temperatures > 100°C.

Sucrose reacts with a reactive alkyl halide in the presence of a base to form an ether, eg, the octabenzyl, trimethylsilyl and methyl ethers. Ethers of sucrose are also formed by addition to a reactive ethylenic bond, such as that of acrylonitrile or dihydropyran.

Reaction of sucrose with ethylene oxide gives the 1-hydroxyethyl ethers. Sucrose under appropriate conditions will form cyclic acetals despite unfavorable steric factors. For example, benzylidene bromide in pyridine reacts with sucrose to give 4,6-O-benzylidenesucrose, which is isolated as the hexaacetate in 35% yield.

Sucrose reacts readily with an isocyanate to give the corresponding urethane; a diisocyanate, eg, toluene diisocyanate, giving a polyurethane. Such products are brittle, and ethoxylated or propoxylated sucrose are the preferred cross-linking polyols in the manufacture of polyurethane forms (see Urethane polymers; Isocyanates).

Sucrose can form internal ether bridges, creating multiple ring systems. The epoxide (oxirane), unlike the stable ether rings of sucrose anhydrides, is readily opened by nucleophilic displacement. Sucrose anhydrides are formed on treatment of a primary alkyl- or aryl-sulfonyl ester of sucrose with sodium methoxide in methanol, ring formation taking place with the appropriate secondary hydroxyl group at the 3,3' or 4' position.

The alkyl- and arylsulfonyl esters of sucrose undergo nucleophilic displacement of the sulfonyl groups with the formation of the corresponding substituted deoxysucrose derivative, on treatment with the appropriate anion in a suitable solvent. This reaction provides a valuable synthetic route to deoxysucrose derivatives, eg, azidosucrose. Sucrose can be chlorinated directly with an excess of sulfuryl chloride in pyridine, or preferably with carbon tetrachloride–triphenylphosphine reagent in pyridine. Deoxysucroses are usually obtained by the reductive dehalogenation of the corresponding halogen derivatives.

Enzymic conversion. Under the action of certain microbial enzymes, sucrose undergoes transglycosylation, a process by which a tri- or higher saccharide is formed by the transfer of a sugar residue. Simple enzymic rearrangement of the hexose linkages of sucrose may also occur as, for example, in the formation of the isomeric reducing disaccharide palatinose (6'-O-(α-D-glucopyranosyl)-D-fructose) by *Erwinia rhaptonici*.

Palatinose has only one third the sweetness of sucrose, which it closely resembles in physical properties, and would be suitable for use in the manufacture of food products for human or animal consumption where the sweetness of sucrose is undesirable.

Sucrose monoesters of long-chain fatty acids have considerable potential as nonionic surfactants in detergent formulations. They are particularly suited for use in cosmetics (qv) and shampoos, and also in food applications as emulsifying and dispersing agents.

Trichlorogalactosucrose, which is 650 times sweeter than sucrose itself, is being developed as a noncaloric, high intensity sweetener.

A potentially large market for sucrose and its derivatives is in synthetic resins and plastics, eg, polyurethane foams. Flame-retardant properties can be introduced into the resin by copolymerizing with halogen or phosphorus derivatives of sucrose.

K.J. PARKER
Tate & Lyle, Ltd.

V. Kollonitsch, *Sucrose Chemicals*, The International Sugar Research Foundation, Inc., Bethesda, Md., 1970.

J.L. Hickson, ed., *Sucrochemistry*, ACS Symposium Series 41, American Chemical Society, Washington, D.C., 1977.

M.R. Jenner in C.K. Lee, ed., *Developments in Food Carbohydrate-2*, Applied Science Publishers, Ltd., London, UK, 1980, pp. 91–143.

SPECIAL

Fructose

D-Fructose (fruit sugar) is a monosaccharide constituting one half of the sucrose molecule. The sweetness of fructose is 1.3–1.8 times that of sucrose, which makes it an attractive alternative for sucrose. When stored in solution at high temperatures, it browns rapidly and polymerizes to dianhydrides.

Because of its sweetness, fructose has been the object of commercial production for decades. Recent technologies involve ion-exchange separation from glucose in a mixture obtained by the isomerization of glucose. In food formulations, fructose is used for special dietary purposes, eg, to lower the calorie (kcal or 4184 J) intake. Fructose-based foods, then, may be helpful in the diets of diabetics and to control blood-sugar levels.

Maltose

Maltose (malt sugar) occurs occasionally in plants and fruits. It is a structural component of starch.

4-O-α-D-glucopyranosyl-D-glucose
maltose

Purification can be achieved by way of the β-maltose octaacetate. Commercial maltose typically contains 5–6 wt% of the trisaccharide maltotriose with traces of glucose. Maltose is about half as sweet as sucrose and has been used in intravenous feeding.

The maltose in malt syrups is important in brewing. High maltose syrups from starch typically contain ca 8–9 wt% glucose and 40–52 wt% maltose; the remainder is higher saccharides. Such syrups are used in the preparation of confections, preserves, and other foodstuffs. Hydrogenation gives maltitol.

Lactose

Lactose (milk sugar) is the only sugar available commercially that is derived from animal rather than plant sources. It is prepared commercially from whey and is available in a fermentation grade (98 wt% pure); lactose, a reducing sugar, reacts with amines and undergoes browning.

Lactose has been used as a nutritional sugar, flavor enhancer, texture controller and color retainer, and as a carrier for synthetic sweeteners (qv).

G.N. BOLLENBACK
The Sugar Association, Inc.

W. Pigman and D. Horton, eds., *The Carbohydrates*, Vol. IA, 2nd ed., Academic Press, New York, 1970.

C.A.M. Hough, K.J. Parker, and A.J. Vlitos, eds., *Developments in Sweeteners —1*, Applied Science Publications, Ltd., London, 1979.

SULFAMIC ACID AND SULFAMATES

Sulfamic acid, HSO_3NH_2, is produced and sold in the form of water-soluble crystals and granules.

Properties. The acid is highly stable and may be kept for years without change in properties. Physical properties are listed in Table 1. Corrosion rates are low in comparison to other acids. Concentrated solutions that are heated in closed containers can generate sufficient internal pressure to cause container rupture. An ammonium sulfamate, 60 wt% aqueous solution, exhibits runaway hydrolysis when heated to

Table 1. Selected Physical Properties of Sulfamic Acid

Property	Value
mol wt	97.09
mp, °C	205
decomposition temperature, °C	209
density (at 25°C), g/cm^3	2.126
refractive indexes, 25 ± 3°C	
α	1.553
β	1.563
γ	1.568
solubility, wt %	
in water, °C	
0	12.80
20	17.57
40	22.77
60	27.06
80	32.01
in formamide, °C	
25	0.1667
in methanol, °C	
25	0.0412
in ethanol (2% benzene), °C	
25	0.0167
in acetone, °C	
25	0.0040
in ether, °C	
25	0.0001

200°C at pH 5 and to 130°C at pH 2. Alkali metal sulfamates are stable in neutral or alkaline solutions even at boiling temperatures.

Nitrous acid reacts rapidly and quantitatively with sulfamic acid:

$$HSO_3NH_2 + HNO_2 \rightarrow H_2SO_4 + H_2O + N_2$$

This reaction can be used for the quantitative analysis of nitrites.

Chlorine, bromine, and chlorates oxidize sulfamic acid to sulfuric acid and to nitrogen, whereas chromic acid, permanganic acid, and ferric chloride do not attack sulfamic acid.

Sodium and potassium add to both the amido and sulfonic portions to give salts, eg, $NaSO_3NHNa$. Sodium sulfate and sulfamic acid form the complex $6HSO_3NH_2 \cdot 5Na_2SO_4 \cdot 15H_2O$.

Primary alcohols react with sulfamic acid to form alkyl ammonium sulfate salts. Sulfation by sulfamic acid has been used in the preparation of detergents from dodecyl, oleyl, and other higher alcohols and in sulfating phenols and phenol–ethylene oxide condensation products.

Amides react in certain cases to form ammonium salts of sulfonated amides. Ammonium sulfamate or sulfamic acid and ammonium carbonate dehydrate liquid or solid amides to nitriles.

Aldehydes form addition products with sulfamic acid salts. These are stable in neutral or slightly alkaline solutions but are hydrolyzed in acid and strongly alkaline solutions. The N-alkyl and N-cyclohexyl derivatives of sulfamic acid are comparatively stable; the N-aryl derivatives are very unstable and can only be isolated in salt form; thiazolylsulfamic acids have been prepared.

Sulfamates. Sulfamates form readily by the reaction with the metal or its oxide, hydroxide, or carbonate. Sulfamates prepared from weak bases form acidic solutions, whereas those prepared from strong bases produce neutral solutions. Both the ammonium and potassium sulfamates liberate ammonia at elevated temperatures and form imidodisulfonates. Inorganic sulfamates are water-soluble except for the basic mercury salt.

Manufacture

Sulfamic acid is manufactured by the exothermic reaction of urea, sulfur trioxide, and sulfuric acid.

The most important commercial salt, ammonium sulfamate, is made by adding anhydrous ammonia and the acid to ammonium sulfamate mother liquor from a preceding crystallization operation to form a hot concentrated solution, and repeating the crystallization step.

Analytical Methods

Sulfamic acid is determined by titration with sodium nitrite solution. An automatic titrator is used with a silver-to-platinum polarized electrode system using potassium bromide as the electrometric indicator.

Health and Safety Factors

Sulfamic acid and its solutions cause eye burns and irritate nose, throat, and skin. Goggles are recommended. A number of regulations govern the use of sulfamic acid in the cleaning of food-processing equipment and in the manufacture of food-packaging materials. Sulfamic acid is approved for use in all departments under the Meat, Poultry, Rabbit, and Egg Products Inspection Program of the USDA.

Uses

Sulfamic acid is particularly well-suited for scale-removal operations and chemical cleaning in a large variety of applications.

In the manufacture of dyes and pigments, sulfamic acid removes excess nitrite from diazotization reactions and is also used for pH adjustment. In paper-pulp chlorination and hyprochlorite bleaching stages, sulfamic acid reduces pulp degradation (see Pulp).

Sulfamic acid has been recommended as a reference standard in acidimetry. Sulfamic acid can be regarded as sulfation and sulfamation agent, eg, for the preparation of sodium cyclohexylsulfamate, a synthetic sweetener.

Sulfamates. A number of flame retardants (qv) used for cellulosic materials are based on ammonium sulfamate. Ammonium sulfamate is highly effective in nonselective products to control weeds, brush, stumps, and trees (see Herbicides).

Nickel and other sulfamates are used in the plating industry. Ferrous sulfamate is used in nuclear-fuel processing solutions.

ELMER B. BELL
E.I. du Pont de Nemours & Co., Inc.

Reviewed by
J.H. BOSTWICK
E.I. du Pont de Nemours & Co., Inc.

K.K. Anderson, "Sulfamic Acid and Their Derivatives," *Compr. Org. Chem.* **3**, 363 (1979).

L.P. Bicelli, "Structure and Properties of Sulfamic Acid, Its Compounds, and Aqueous Solutions," *Symp. Sulfamic Acid, Its Electrometall. Appl.*, Milan, Italy, Discussion 28, 19 (1966).

SULFANILAMIDE. See Antibacterial agents, synthetic, sulfonamides.

SULFANILIC ACID, p-$H_2NC_6H_4SO_3H$. See Sulfonation and sulfation; Amines, aromatic—aniline and its derivatives.

SULFATED ACIDS, ALCOHOLS, OILS, ETC. See Sulfonation and sulfation; Surfactants and detersive systems.

SULFATION. See Sulfonation and sulfation.

SULFIDES. See Sulfur compounds.

SULFITE PROCESS. See Pulp.

SULFITES. See Barium compounds; Sulfur compounds; Sulfuric and sulfurous esters.

SULFOALKYLATION. See Sulfonation and sulfation.

SULFOCHLORINATION. See Sulfonation and sulfation; Sulfurization and sulfur chlorination.

SULFOLANES AND SULFONES

Sulfolane is a colorless, highly polar, water-soluble compound; for properties, see Table 1. Chlorine can be added to sulfolane by a uv-initiated process to give 3-chloro-, 3,4-dichloro-, and 3,3,4-trichlorosulfolane. Bromination gives 2-bromosulfolane, which is further brominated to dibromosulfolanes. The sulfolane ring can be cleaved by sodium or potassium. The addition of tributyltin chloride and olefins is catalyzed by alkali metals.

Table 1. Physical Properties of Sulfolane

Property	Value
molecular weight	120.17
boiling point, °C	287.3
melting point, °C	28.5
specific gravity, 30/30°C	1.266
100/4°C	1.201
density (at 15°C), g/cm³	1.276
flash point, °C	165–178
viscosity, mPa·s at 30°C (= cP)	10.3
refractive index, n_D (at 30°C)	1.48
heat of fusion, kJ/kg[a]	11.44
dielectric constant	43.3

[a] To convert J to cal, divide by 4.184.

Sulfolane and its alkyl homologues react with ethylmagnesium bromide to give sulfolanyl 2-mono- and 2,5-dimagnesium bromides. Sulfolane complexes with Lewis acids, eg, boron trifluoride and phosphorus pentafluoride.

Sulfolane is made by hydrogenating 3-sulfolene

When administered intraperitoneally, sulfolane is excreted both unchanged and as 3-hydroxysulfolane.

Sulfolane causes minimal and transient eye and skin irritation. Inhalation of vapors is not considered biologically significant.

Uses

Sulfolane is used principally as a solvent for extraction of benzene, toluene, and xylene from mixtures with aliphatic hydrocarbons (see BTX processing). The sulfolane extraction unit consists of extractor, extractive stripper, extract recovery column, and water-wash tower.

The urea-adduction method for separating normal and branched aliphatic hydrocarbons can be carried out in sulfolane. Overall recovery by this process is 85% with an n-paraffin purity of 98%.

Sulfolane exhibits selective solvency for fatty acids and fatty acid esters. It is a suitable extractive-distillation solvent for separating close-boiling alcohols.

Sulfolane removes acidic components and mercaptans from sour-gas streams (Sulfinol process).

Sulfolane can be used alone or with a cosolvent as a polymerization solvent. Sulfolane polymer solutions have been patented for fiber-spinning processes. Sulfolane is a polymer plasticizer.

Sulfones

3-Sulfolene is an intermediate in the synthesis of sulfolanyl ethers, which are used as hydraulic-fluid additives. 3-Sulfolene or its derivatives are used in cosmetics (qv) and slimicides.

Other sulfones of commercial potential include dimethyl sulfone, diiodomethyl p-tolyl sulfone, and 4,4′-dihydroxydiphenyl sulfone.

MERLIN LINDSTROM
RALPH WILLIAMS
Phillips Research Center

J.A. Reddick and W.E. Bunger, *Organic Solvents*, Vol. 2, 3rd ed., Wiley Interscience, a division of John Wiley & Sons, Inc., New York, 1970, pp. 467–468.

Technical Information on Sulfolane, Bulletin 524, Phillips Chemical Company.

SULFONATION AND SULFATION

Sulfonation and sulfation are chemical methods for introducing the SO_3 group into organic entities; they are closely related and usually treated jointly. Although the terms sulfonation and sulfation are often used interchangeably, the two methods differ chemically.

In sulfonation, an SO_3 group is introduced into an organic molecule, to give a product with $—SO_3—$ linkages (sulfonates) in the form of a sulfonic acid ($—SO_3H$), salt ($—SO_3Na$), or sulfonyl halide ($—SO_3X$). Sulfonic acids are generally produced directly by reaction of an aromatic hydrocarbon with sulfuric acid, sulfur trioxide, or chlorosulfuric acid. Sulfonation of unsaturated hydrocarbons with metal sulfites or bisulfites produces the metal sulfonate salts. These reactions are referred to as sulfitation or bisulfitation. In instances where the sulfur atom, at a lower valence, is attached to a carbon atom, the sulfonation process entails oxidation. Thus, the reaction of a paraffinic hydrocarbon with sulfur dioxide and oxygen is referred to as sulfoxidation, and the reaction of sulfur dioxide and chlorine as chlorosulfonation. Sulfoalkylation and sulfoarylation reactions involve the addition of a sulfoalkyl or sulfoaryl group to an organic molecule.

Sulfation is defined as any process of introducing an SO_3 group into an organic compound in which the reaction product (sulfate) exhibits the characteristic $—OSO_3—$ molecular configuration. Unlike the sulfonates, which show remarkable stability even after prolonged heating, sulfated products are unstable toward acid hydrolysis. This difference forms the basis of an analytical method to identify mixtures of these two species.

In sulfamation, also termed N-sulfonation, compounds of the general structure R_2NSO_3H are formed as well as the corresponding salts, acid halides, and esters. The reagents are sulfamic acid (amidosulfuric acid), SO_3–pyridine complex, SO_3–tertiary amine complexes, SO_3–aliphatic amine adducts, and chlorine isocyanate–SO_3 complex.

Sulfonation and sulfation processes are important tools for the organic chemist in the design of specific molecules. Sulfonation, sulfation, and sulfamation processes have led to the commercial development of new sulfonated and sulfated products.

Detergent sulfonates or sulfates for use in household detergent and personal care products must have light color, little unreacted material or free oil, low inorganic salt content, and negligible odor. Sulfonation, sulfation, and sulfamation products include surfactants, dyes, pigments, medicinals, pesticides, and cyclic intermediates.

Sulfonation and sulfation reagents are listed in Table 1.

Sulfonation

The sulfonation of benzene and short-chain alkylbenzenes, such as toluene, the xylenes, cumene, and ethylbenzene, is an important commercial process carried out in a number of ways using sulfuric acid, oleum, or sulfur trioxide. The sulfur trioxide route presents potential fire and explosion hazards and produces by-product sulfones that impair product

Table 1. Reagents for Sulfonation and Sulfation

Reagent[a]	Chemical formula	Physical form	Principal uses	Relative usage	Reactivity
sulfur trioxide	SO_3	liquid	with solvents to modify reactivity	very limited	extremely reactive
		gas	for organic compounds	significantly	highly reactive; generally mol per mol; instantaneous
from sulfur + air		gas	same as above; optional source of SO_3	moderate	same as above
oleum, 20%, 30%, and 65%	$H_2SO_4 \cdot SO_3$	liquid	for alkylaryls for detergents, dyes	most widely used sulfonating, sulfating agent	high reactivity
chlorosulfuric acid	$ClSO_3H$	liquid	for alcohols, dyes	moderate	high reactivity; mol per mol reaction
sulfuryl chloride	SO_2Cl_2	liquid	sulfonation of acetylene groups, lab mainly	low; mainly lab research tool	moderate
sulfamic acid	NH_2SO_3H	solid	sulfation of alkyl phenols, some specialties	limited	relatively low
sulfuric acid (96–100%)	H_2SO_4	liquid	to sulfonate aromatics	significant	low
sulfur dioxide plus chlorine gases	$SO_2 + Cl_2$	gas mixture	chlorosulfonation of paraffinic hydrocarbons	very limited	low
sulfur dioxide and oxygen gases	$SO_2 + O_2$	gas mixture	sulfoxidation of paraffinic hydrocarbons	very limited	low
sodium sulfite	Na_2SO_3	solid	sulfonation of alkyl chlorides	substantial	low
sodium bisulfite	$NaHSO_3$	solid	sulfation of conjugated olefin species (succinates, etc), sulfonation of lignins	substantial	low
hydroperoxide-bisulfite (O_2 + $NaHSO_3$)	$NaHSO_3$ + O_2	gas and solid	sulfonation/sulfation of paraffinic petroleum fractions	very limited	low

[a] Reagents listed in the order of generally descending reactivity.

solubility. Precise process control is essential, and the sulfones are removed by extraction.

Sulfonated toluene, xylene, and cumene, primarily in the form of ammonium and sodium salts, are important hydrotropes or coupling agents in the manufacture of liquid cleaners and surfactant compositions, and serve as crisping agents in drum and spray-drying operations.

Sulfation

Sulfation involves the reaction wherein a —COS— linkage is formed by the action of a sulfating agent on an alkene, alcohol, or phenol. Concentrated sulfuric acid forms a $-COSO_3H$ moiety with an alkene, whereas, SO_3 yields a $-COSO_3H$ group with an alcohol or a phenol. Unlike the sulfonates, which exhibit excellent hydrolytic stability, the alcohol sulfates are readily susceptible to hydrolysis in acidic media. Sulfation of fatty alcohols and polyalkoxylates has produced a substantial body of commercial detergents and emulsifiers.

Linear ethoxylates are the preferred raw materials for the production of ether sulfates used in detergent formulation because of their uniform-

ity, high purity, and biogradability. The alkyl chain is usually in the C_{12}–C_{15} range at a molar ethylene oxide: alcohol ratio of 3:1. In 1977, ca 18,200 metric tons of ether sulfates were used in emulsion polymerization and textile processing. Propoxylates, ethoxylates, and mixed alkoxylates of aliphatic alcohols or alkylphenols are sulfated for use in specialty applications.

Sulfamation

Sulfamation is the reaction of an amine in chloroform with SO_3, an adduct, chlorosulfuric acid, or sulfamic acid

Sodium cyclohexylsulfamate

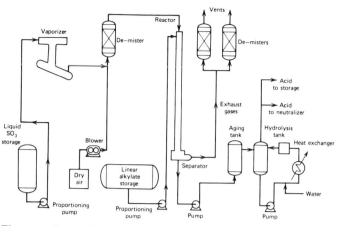

Figure 1. Process flow scheme for continuous SO₃ film sulfonation process.

Sodium cyclohexylsulfamate (sodium cyclamate) is a sweetener. Because biological tests indicated a potential health hazard, the product has been removed from the U.S. market except under prescription (see Sweeteners).

Heating an amine with sulfamic acid is an alternative method of sulfamation.

Other processes make use of SO₃ or its adducts. Thus, refluxing cyclohexylamine with SO₃–triethylamine in methylene chloride, followed by treatment with aqueous NaOH, produced a 90% yield of sodium cyclohexylsulfamate. Direct addition of SO₃ to an amine is a highly exothermic reaction that can be used to prepare a sulfamate.

Sulfamic acid is prepared from urea and oleum. Sulfamation is little used commercially but sulfamic acid has various applications, eg, as dyeing aid, for bleaching wood pulp, and in metal cleaning; cigarette paper is treated with ammonium sulfamate to reduce the hazard of tumor formation from tobacco smoke.

Industrial Processes

Industrial processes include oleum sulfonation and sulfur trioxide sulfonation, eg, liquid SO₂ as solvent, gaseous SO₃, and continuous SO₃ processes (see Fig. 1).

E.A. KNAGGS
M.L. NUSSBAUM
A. SHULTZ
Stepan Chemical Company

E.E. Gilbert, *Sulfonation and Related Reactions*, Interscience Publishers, a division of John Wiley & Sons, Inc., New York, 1965.

A. Lanteri and G. Mazzoni, *Sulfonation and Sulfation Technology*, World Conference on Soaps and Detergents, Montreux, Switzerland, Oct. 1977.

Sulfur Trioxide Detergent Process Equipment, Technical Bulletin, Chemithon Corp., Seattle, Washington, 1980.

SULFONIC ACIDS

The sulfonic acids are any of numerous acids that are characterized by the sulfonic group —SO₃H (or —SO₂OH) and may be regarded as derived from sulfuric acid by replacement of a hydroxyl group by either an inorganic anion or univalent organic radical (see Sulfuric acid and sulfur trioxide).

Physical Properties

The nature of the R group has a significant influence upon the physical and chemical properties of the specific acids. In general, they are strong acids comparable to sulfuric acid and are, for the most part, hygroscopic, nonvolatile, soluble in water and polar solvents, and chemically stable. Only in select instances can the sulfonic acids be distilled either at atmospheric pressure or even *in vacuo*, because they undergo thermal decomposition at elevated temperatures.

Chemical Properties

The sulfonic acids are strong acids like sulfuric acid. They ionize in water and in many polar organic solvents. Unlike the organic sulfates, they are moderately stable to hydrolysis and not prone to decomposition on heating. Classical chemical reactions of sulfonic acids include transformation of the sulfonic acid group and cleavage of the carbon–sulfur bond.

Arene- and Alkylarenesulfonic Acids

One of the simplest products, *p*-toluenesulfonic acid, made by sulfonation of toluene with 100% sulfuric acid, is used commercially to effect aqueous solubility of disperse dyes (see Dye carriers). Azosulfonic acids are potentially suitable as dyes for lipstick. Both *p*-toluenesulfonic acid and 2-naphthalenesulfonic acid are used as initiators (qv) for the catalytic polymerization of caprolactam.

Alkylarenesulfonic acids are used for the preparation of agricultural and industrial fungicides (see Fungicides).

Arenesulfonic acids are suitable for stabilization of maleic anhydride–olefin copolymers. Another interesting application of alkylarenesulfonic acids is in the compounding of wax-flux compositions for low temperature soldering processes.

Organic Sulfonic Acid Oligomers

Organic disulfonic acids are produced by heating a sulfonate monomer at above 110°C in the absence of water.

The salts of oligomers of C₁₄–C₂₀ sultones, after undergoing saponification processing, are useful surfactants in heavy-duty-detergent compositions. The disulfonic acids, as unneutralized oligomers, are commercially important as the foaming agents for the production of preformed acidic circulation fluids for oil wells (see Petroleum, drilling fluids).

Fluorinated and Chlorofluorinated Sulfonic Acids

Trifluoromethanesulfonic acid is useful in a number of commercial applications. It is used as a catalyst with sulfuric acid for the liquid-phase alkylation of low molecular weight isoparaffin hydrocarbons with olefin hydrocarbons. In another petroleum feedstock application, trifluoromethanesulfonic acid is used as a superacid catalyst (Magic acid) in combination with antimony pentafluoride (SbF₅) for liquid- or vapor-phase hydroisomerization of C₄–C₁₂ paraffin hydrocarbon to give products useful in motor fuels (see Friedel-Crafts reactions).

Of the alkenylamidoalkanesulfonic acids, 2-acrylamido-2-methylpropanesulfonic acid has considerable commercial utility and is marketed by Lubrizol Corporation under the trademark of AMPS monomer. Some of the applications of the AMPS monomer are as surfactants for aqueous solutions, brass electroplating-bath additives, for dye receptivity in fibers and films, as textile sizes and permanent press and soil-release agents, in printing inks, and as clear antifog coatings.

Acrylimidoaminoethanesulfonic Acids

Acrylimidoaminoethanesulfonic acids and their derivatives are prepared by the reaction of acrylonitrile sequentially with SO₃, 2,2-dimethylpropane, and NH₃ at low temperatures (−30 to −75°C). The products are useful as flocculating agents (qv) and in the preparation of polymers.

Amidoalkanesulfonic Acids

Amidoalkanesulfonic acids are prepared by the reaction of nitriles with olefins and concentrated sulfuric acid or oleum. They are useful as dispersing agents for calcium soaps.

Perfluoroisoalkoxyalkanesulfonic Acids

Perfluoroisoalkoxyalkanesulfonic acids have a terminal polyfluorinated branched-chain- or cyclic-perhaloisoalkoxy group linked through an ether or oxygen to a perfluoromethylene substituent.

The perfluoroisoalkoxyalkanesulfonic acids are used as strong acid catalysts for organic reactions, as ingredients of dry-powder fire-extinguishing compositions (see Fire-extinguishing agents; Plant safety), and as intermediates in the manufacture of other polyfluoroisoalkoxyal-kanesulfonyl-substituted compounds (see also Fluorine compounds, organic).

3,3'-Disubstituted Benzidine Derivatives

The 3,3'-disubstituted benzidine derivatives have in each of the 3,3'-positions the substituent

$$-O_n-AS-OW$$

where W is hydrogen or an alkali metal, and n is zero when A is a divalent branched-chain alkylene of 2–7 carbon atoms or one when A is a straight-chain divalent alkylene with three or four carbons.

Alkanesulfonic Acids

Methanedisulfonic acid and methanetrisulfonic acid are used as catalysts for the alkylation of phenols.

Health and Safety Factors, Toxicology

See Table 1.

Hydrolysis of Oils and Fats

An important commercial application for sulfonic acids involves hydrolysis of oils and fats, which are glycerol esters of saturated and unsaturated fatty acids.

ANDREW SHULTZ
Stepan Chemical Company

C.J.M. Stirling, *The Chemistry of the Sulphonium Group*, Parts 1 and 2, John Wiley & Sons, Inc. New York, 1981.

SULFOXIDES

Sulfoxides are compounds that contain a sulfinyl group covalently bonded at the sulfur atom to two carbon atoms. They have the general formula RS(O)R′, ArS(O)Ar′, and ArS(O)R. Sulfoxides represent an intermediate oxidation level between sulfides and sulfones. The naturally occurring sulfoxides often are accompanied by the corresponding sulfides or sulfones. The only commercially important sulfoxide is the simplest

Table 1. Properties, Toxicity, and Uses of Sulfonic Acids

Compound	Bp, °C	Mp, °C	Toxicity	Uses
benzenesulfonic acid	110.6	50–51	highly irritating to skin, eyes, and mucous membranes; oral LD_{50} (rat) 890 mg/kg versus 3,800 mg/kg for benzene	manufacture of phenol by fusion with NaOH
m-nitrobenzenesulfonic acid		70	oral LD_{50} (rat) of sodium salt 11,000 mg/kg versus 640 mg/kg for nitrobenzene	organic syntheses; sodium salt is a protective antireduction agent
m-aminobenzenesulfonic acid			oral LD_{50} (rat) 12,000 mg/kg versus 440 mg/kg for aniline	manufacture of azo dyes; synthesis of certain sulfa drugs
p-toluenesulfonic acid	140 at 2.7 MPa[a]	106–107	highly irritating to skin and mucous membranes; oral LD_{50} (rat) 2,480 mg/kg versus 5,000 mg/kg for toluene	manufacture of dyes and oral anti-diabetic drugs; conversion to sodium and ammonium salts for manufacture of hydrotropes
3,5-dimethylbenzenesulfonic acid (m-xylene-m-sulfonic acid)			intraperitoneal LD_{50} (mouse) 500 mg/kg	
dodecylbenzenesulfonic acid			oral LD_{50} (rat) 890 mg/kg versus 34,000 mg/kg for dodecylbenzene	electronic cleaning chemicals, pickling baths, and detergent manufacture
1-naphthalenesulfonic acid		90[b]	oral LD_{50} (rat) 420 mg/kg versus 1,780 mg/kg for naphthalene	manufacture of α-naphtholsulfonic acid and α-naphthylaminesulfonic acid; solvent for phenol in the manufacture of disinfectant soap
2-naphthalenesulfonic acid		91[c]	oral LD_{50} (rat) 400 mg/kg versus 1,780 mg/kg for naphthalene	manufacture of beta-naphthol, beta naphtholsulfonic acid, beta-naphthylaminesulfonic acid, and other intermediates
2-naphthol-6-sulfonic acid		170 dec	oral LDL_0 (mouse) 250 mg/kg	preparation of azo dyes
1-naphthylamine-4-sulfonic acid		170 dec	intraperitoneal LD_{50} (mouse) 300 mg/kg	therapeutic category; Na salt as hemostatic
fluorosulfuric acid	165	−87	highly irritating to skin and mucous membranes; aquatic toxic rating, TL 100–110 ppm after 96 h	fluorinating agent, catalyst in alkylation, acylation, polymerization, and condensation reactions; hydrofluorination of olefins; production of substituted pyridines
fluoromethanesulfonic acid	162		oral LD_{50} (mouse) 112 mg/kg	alkylating catalyst; methylation of carbamates, octaethylporphyrin, and octaethylchlorine

[a] To convert MPa to psi, multiply by 145.
[b] Dihydrate crystals.
[c] Monohydrate, mp = 124–125°C.

member, dimethyl sulfoxide (DMSO, sulfinylbismethane). Sulfoxides occur widely in small concentrations in plant and animal tissues.

Properties

For the most part, sulfoxides are crystalline, colorless substances, although the lower aliphatic sulfoxides melt at relatively low temperatures (see Table 1). The lower aliphatic sulfoxides are water-soluble, but as a class the sulfoxides are not soluble in water. They are soluble in dilute acids and a few are soluble in alkaline solution. Dimethyl sulfoxide is a colorless liquid and its properties are listed in Table 2. Dimethyl sulfoxide generally undergoes typical sulfoxide reactions and is used as an illustrative example in the following sections.

Thermal stability. Dimethyl sulfoxide decomposes slowly at 189°C to a mixture of products that includes methanethiol, formaldehyde, water, bis(methylthio)methane, dimethyl disulfide, dimethyl sulfone, and dimethyl sulfide. The decomposition is accelerated by acids, glycols, or amides.

Sulfoxides undergo oxidation, reduction, carbon–sulfur cleavage and Pummerer reactions.

Methylsulfinyl carbanion (dimsyl ion). Strong bases, eg, sodium hydride or sodium amide, react with DMSO producing solutions of methylsulfinyl carbanion, which are synthetically useful.

Methoxydimethylsulfonium and trimethylsulfoxonium salts. Alkylating agents react with DMSO at the oxygen. For example, methyl iodide gives methoxydimethylsulfonium iodide as the initial product. The alkoxysulfonium salts are quite reactive and, upon continued heating, decompose to carbonyl compounds or rearrange to the more stable trimethylsulfoxonium salts.

Complexes. The sulfoxides have a high dipole moment (ca 4), which is characteristic of the sulfinyl group and a basicity about the same as

that of alcohols, and they are strong hydrogen-bond acceptors. They would be expected, therefore, to solvate ions with electrophilic character, and a large number of DMSO complexes of metal ions have been reported.

Synthesis and Manufacture

The sulfoxides are most frequently synthesized by oxidation of the sulfides.

Dimethyl sulfoxide. Dimethyl sulfoxide is manufactured from dimethyl sulfide (DMS), which is obtained by processing spent liquors from the Kraft pulping process or by the reaction of methanol or dimethyl ether with hydrogen sulfide.

Health and Safety Factors

Dimethyl sulfoxide is a relatively stable solvent of low toxicity. The LD_{50} for single-dose oral administration to rats is ca 17,400–28,300 mg/kg. Dimethyl sulfoxide has received considerable attention as a useful agent in medicine.

Uses of Dimethyl Sulfoxide

DMSO is used as a polymerization and spinning solvent, solvent for displacement reactions, solvent for base-catalyzed reactions, extraction solvent, solvent for electrolytic reactions, cellulose solvent, pesticide solvent, and clean-up solvent.

W.S. MacGregor
J.V. Orle
Crown Zellerbach Corp.

Dimethyl Sulfoxide, Technical Bulletin, Crown Zellerbach Corp., Chemical Products Division, Vancouver, Washington, 1982.

T. Durst in E.C. Taylor, H. Wynberg, eds., *Advances in Organic Chemistry: Methods and Results*, Vol. 6, Interscience Publishers, a division of John Wiley & Sons, Inc., New York, 1969, pp. 285–388.

D. Martin and H.G. Hauthal, *Dimethyl Sulfoxide*, Halsted Press, a division of John Wiley & Sons, Inc., New York, 1975.

SULFUR

Sulfur, S, a nonmetallic element, is the second element of Group VIA below oxygen and above selenium. In massive elemental form, sulfur is often referred to as brimstone. Sulfur is one of the most important raw materials, especially in the fertilizer industry (see Fertilizers). Its consumption is generally regarded as a measure of a nation's industrial development and economic activity (see also Sulfur compounds; Sulfur recovery; Sulfuric acid).

Sulfur recovered as a by-product, accounts for a larger portion of world supply than the mined material. By-product sulfide is obtained from hydrogen sulfide, which evolves when natural gas, crude petroleum, tar sands, oil shales, and coal are desulfurized. Other sources include metal sulfides, eg, pyrites; sulfate materials, eg, gypsum; and elemental sulfur in native and volcanic deposits mined in the traditional manner.

Sulfur constitutes ca 0.052 wt% of the earth's crust. The forms in which it is ordinarily found include elemental or native sulfur in unconsolidated volcanic rocks, in anhydrite over salt-dome structures, and in bedded anhydrite or gypsum evaporate basin formations; combined sulfur in metal sulfide ores and mineral sulfates; hydrogen sulfide in natural gas; organic compounds in petroleum and tar sands; and a combination of pyritic and organic compounds in coal.

Properties

Sulfur occurs in a number of different allotropic modifications, that is, in various molecular aggregations which differ in solubility, specific gravity, crystalline form, etc. Like many other substances, sulfur exhibits dynamic allotropy, ie, the various allotropes exist together in equilibrium

Table 1. Melting and Boiling Points of Sulfoxides

Name	Formula	Melting point, °C	Boiling point, °C
sulfinylbismethane	$(CH_3)_2SO$	18.55	189.0
1,1'-sulfinylbisethane	$(C_2H_5)_2SO$	15	88–90 (at 2.0 kPa[a])
1,1'-sulfinylbispropane	$(n\text{-}C_3H_7)_2SO$	18	
1,1'-sulfinylbisbutane	$(n\text{-}C_4H_9)_2SO$	32	
1,1'-sulfinylbis(2-chloroethane)	$(ClCH_2CH_2)_2SO$	110.2	
1,1'-sulfinylbisbenzene	$(C_6H_5)_2SO$	70.5	340 (slowly decomposes)
methylsulfinylbenzene	$C_6H_5S(O)CH_3$	30–30.5	139–140
phenylmethylsulfinylbenzene	$C_6H_5S(O)CH_2C_6H_5$	125.5	
1,1'-sulfinylbis(methylenebenzene)	$(C_6H_5CH_2)_2SO$	135	

[a] To convert kPa to mm Hg, multiply by 7.5.

Table 2. Properties of Dimethyl Sulfoxide

Property	Value
boiling point, °C	189.0
conductivity (at 20°C), S/cm	3×10^{-8}
dielectric constant (at 25°C), 10 MHz	46.7
dipole moment, C·m[a]	1.4×10^{-29}
entropy of fusion, J/(mol·K)[b]	45.12
free energy of formation (g) (at 25°C), kJ/mol[b] C_{graph}, $S_{2(g)}$	115.7
freezing point, °C	18.55
refractive index, n_D^{25}	1.4768
flash point (open cup), °C	95
viscosity, mPa·s (= cP)	1.996^{25} 1.396^{45} 0.68^{100}

[a] To convert C·m to D, divide by 3.336×10^{-30}.

[b] To convert J to cal, divide by 4.184.

Table 1. Physical Constants of Sulfur

Property	Value Ideal	Value Natural
freezing point of solid phase, °C		
rhombic, S	112.8	110.2
monoclinic	119.3	114.5
boiling point, °C		444.6
density of solid phase (at 20°C), g/cm³		2.07
rhombic		1.96
monoclinic		1.92
amorphous		
density of liquid, g/cm³		
125°C		1.7988
140°C		1.7865
density of vapor at 444.6°C and 101.3 kPa (= 1 atm), g/L		3.64
refractive index, n_D^{110}	1.929	
vapor pressure, p = Pa, T = K^a		
rhombic (20–80°C)	$\log p = 16.557 - 5166/T$	
monoclinic (96–116°C)	$\log p = 16.257 - 5082/T$	
liquid (120–325°C)	$\log p = 19.6 - 0.0062238T - 5405.1T$	
(325–550°C)	$\log p = 12.3256 - 3268.2/T$	
surface tension, mN/m (= dyn/cm)		
120°C		60.83
critical temperature, °C		1040
specific heat, J/(kg·K)^b		
rhombic (24.9 to 95.5°C)	$Cp = 468 + 0.814T$	
monoclinic (−4.5 to 118.9°C)	$Cp = 465 + 0.908T$	
liquid (Sλ) (118.9 to 444.6°C)	$Cp = 706 + 0.65T$	
gas (S) (25 to 1727°C)	$Cp = 709 - 0.034T - 3.5 \times 10^6 T^{-2}$	
gas (S₂) (25 to 1727°C)	$Cp = 558 + 0.018T - 5.2 \times 10^6 T^{-2}$	
thermal expansion of rhombic sulfur (linear)		
0–13°C	4.567×10^{-5}	
50–78°C	8.633×10^{-5}	
98–110°C	103.2×10^{-5}	
latent heat of vaporization, J/g^b	L^c	L^d
200°C	308.6	
400°C	286.4	278.0
440°C	290.1	274.6
electrical resistivity, ohm·cm		
20°C	1.9×10^{17}	

^{a} To convert $\log P_{Pa}$ to $\log P_{psi}$, subtract 3.8384 from the constant.
^{b} To convert J to cal, divide by 4.184.
^{c} Including heat of dissociation to S₂ present in vapor.
^{d} Minus heat of dissociation to S₂ in vapor.

in definite proportions, depending on the temperature and pressure. The formula for the molecules of various allotropes is S—S$_n$, where n is a very large but unidentified number ($n > 10$)^6.

Sulfur crystallizes in the rhombic and monoclinic forms. The molecular constitution of liquid sulfur undergoes significant and reversible changes with temperature variations. These changes are evidenced by the characteristic temperature dependence of the physical properties of sulfur. In most studies of liquid sulfur, some striking changes in its physical properties are observed at ca 160°C.

The physical constants of sulfur are presented in Table 1.

Elemental Sulfur

Extraction processes include the Frasch process, the hydrodynamic process, distillation, flotation, autoclaving, filtration, solvent extraction, or a combination of several of these processes.

Sulfide Ores

The metal sulfides are the most important source of elemental sulfur. The potential for their recovery exists, although at present they are less attractive economically and technologically than other sources of brimstone.

Pyrometallurgical processes. Pyrometallurgical recovery processes include the Orkla, Noranda and Outokumpu processes.

Hydrometallurgical processes. Hydrometallurgical processes include the CLEAR process and the Sherritt-Cominco (S.C.) copper process.

By-Product Sulfur

When sulfur is the principal and often the only product, it is called voluntary sulfur. When recovered as a by-product, it is termed involuntary sulfur (see Sulfur recovery).

Sulfates

The largest untapped source of sulfur occurs in the ocean as dissolved sulfates of calcium, magnesium, and potassium. The average sulfur concentration in seawater is 880 ppm; thus, 1 km³ of seawater contains ca 0.86×10^6 t of elemental sulfur in the form of sulfates. Natural and by-product gypsum, $CaSO_4 \cdot 2H_2O$, and anhydrite, $CaSO_4$, rank second only to the oceans as potential sources of sulfur. Mineral deposits of gypsum and anhydrite are widely distributed in large quantities. Gypsum is a by-product waste material from several manufacturing processes, most notable in the manufacture of phosphoric acid from phosphate rock and sulfuric acid.

Extraction. One extraction process being investigated is thermal reduction of gypsum.

Phosphogypsum. A significant technical achievement would be the economic recovery of sulfur from phosphogypsum, a by-product produced in tremendous quantities in the manufacture of phosphatic fertilizers.

Bacteriological sulfur. Anaerobic, sulfate-reducing bacteria burn hydrocarbons as a source of energy. Several experimenters have tried to develop this area into a controlled process for producing sulfur from gypsum or anhydrite. This process requires a strain of sulfate-reducing bacteria, an organic substrate whose hydrocarbons provide food for the bacteria, and close control of environmental conditions in order to obtain maximum sulfur yields.

Health and Safety Factors

Solid elemental sulfur is nontoxic and can be taken internally without injury.

Sulfur becomes a liquid at ca 100°C. The primary hazards when handling liquid sulfur include its relatively low ignition point, the possible presence of hydrogen sulfide, and the fact that it is a hot liquid capable of producing severe burns. Hydrogen sulfide is not only toxic, but it also has a low explosive limit. The maximum acceptable concentration of hydrogen sulfide averaged over a normal working day and recommended by the ACGIH is 10 ppmv. The OSHA standard is a ceiling concentration of 20 ppm and a peak concentration of 50 ppm for 10 min.

Uses

Sulfur is used as a chemical reagent rather that as a component of a finished product. Its predominant use as a process chemical generally requires that it first be converted to an intermediate chemical product prior to its initial use in industry. In most of the ensuing chemical reactions between these intermediate products and other minerals and chemicals, the sulfur values are not retained in the final product. Rather, they are discarded as the component of a waste product.

Sulfuric acid is the most important of these intermediate products. More than 85% of sulfur consumed in the world is either converted to sulfuric acid or produced directly as such. Worldwide, over half of the sulfuric acid production is used in the manufacture of phosphate and ammonium sulfate fertilizers. The sulfur source may be voluntary elemental, ie, from the Frasch process; recovered elemental from the natural gas or petroleum; or sulfur dioxide from smelter operations.

In recent years, the largest demand for sulfur has been for agricultural purposes, the principal use being for processing of phosphate fertilizers. Other uses have been in petroleum refining; leaching of copper and uranium ores; production of organic and inorganic chemicals, paints and pigments, pulp and paper, and synthetic materials; and in many industrial uses.

D.W. Bixby
H.L. Fike
The Sulphur Institute

J.E. Shelton
U.S. Bureau of Mines

T.K. Wiewiorowski
Freeport Minerals Co.

B. Meyer, *Sulfur, Energy, and Environment*, Elsevier Publishing Co., New York, 1977.

W.N. Tuller, ed., *The Sulphur Data Book*, McGraw-Hill Book Co., New York, 1954.

U.S. Bureau of Mines Yearbook, Washington, D.C., 1982.

SULFUR COMPOUNDS

Carbon–Sulfur Compounds

The only commercial carbon sulfide is carbon disulfide (qv), CS_2. There are several unstable carbon sulfides.

Carbon subsulfide, C_3S_2, is a red liquid (mp $-0.5°C$, bp 60–70°C at 1.6 kPa (12 mm Hg)) produced by the action of an electric arc on carbon disulfide.

Carbon monosulfide, CS, is an unstable gas produced by the decomposition of carbon disulfide at low pressure in a silent electrical discharge or photolytically.

Incompletely characterized carbon sulfides include a black solid known as carsul. It occurs as a residue in sulfur distillation or as a precipitate in molten Frasch sulfur.

Carbonyl sulfide. *Physical properties.* Carbonyl sulfide (carbon oxysulfide), COS, is a colorless gas that is odorless when pure; however, it has been described as having a foul odor. Physical constants and thermodynamic properties are listed in Table 1.

Chemical properties. Carbonyl sulfide is a stable compound which can be stored under pressure in steel cylinders as compressed gas in equilibrium with liquid. At ca 600°C, carbonyl sulfide disproportionates to carbon dioxide and carbon disulfide; at ca 900°C, it dissociates to carbon monoxide and sulfur. It turns with a blue flame to carbon dioxide and sulfur dioxide. Carbonyl sulfide reacts only slowly with water to form carbon dioxide and hydrogen sulfide. Much technology has been developed for hydrolysis of carbonyl sulfide in gas streams to improve the removal of the sulfur content as hydrogen sulfide.

Occurrence and preparation. Carbonyl sulfide is formed by high temperature reactions of carbon compounds with donors of oxygen and sulfur. A principal route is the following reaction:

$$CO + S \rightarrow COS$$

Carbonyl sulfide occurs as a by-product in the manufacture of carbon disulfide. It is an impurity in some natural gases, in manufactured fuel and refinery gases, and in combustion products of sulfur-containing fuels.

Carbonyl sulfide is the most abundant sulfur-containing compound in the earth's atmosphere: 430–570 parts per trillion (10^{12}).

Health and safety factors. Carbonyl sulfide is dangerously poisonous, more so because it is practically odorless when pure. It is lethal to rats at 2900 ppm. Recent study shows an LD_{50} (rat, intraperitoneal) of 22.5 mg/kg.

Uses. There may be some captive use of carbonyl sulfide for production of certain thiocarbamate herbicides (qv).

Thiophosgene. *Physical properties.* Thiophosgene (thiocarbonyl chloride), $CSCl_2$, is a malodorous, red-yellow liquid (bp 73.5°C, d_{20}^{15} 1.509 g/cm³, n_D^{20} 1.5442). It is only slightly soluble in water (with decomposition), but it is soluble in ether and various organic solvents.

Preparation. Thiophosgene forms from the reaction of carbon tetrachloride with hydrogen sulfide, sulfur, or various sulfur-containing reducing agents at elevated temperatures.

Health and safety factors. Thiophosgene has an LD_{50} (rat, oral) of 929 mg/kg and an LC_{50} (inhalation, rat) of 370 mg/m³. It has both irritant and systemic toxic properties.

Trichloromethanesulfenyl chloride. *Physical properties.* Trichloromethanesulfenyl chloride (perchloromethyl mercaptan), a misnomer but used as the common commercial name for CCl_3SCl, is a strongly acrid, pale-yellow liquid; boiling at 149°C with some decomposition at atmospheric pressure, 68°C at 6.93 kPa (52 mm Hg), and 25°C at 0.8 kPa (6 mm Hg); sp gr$_4^{20}$ 1.6996; n_D^{23} 1.541. It slowly hydrolyzes and is soluble in most organic solvents.

Manufacture. Trichloromethanesulfenyl chloride is made commercially by chlorination of carbon disulfide with the careful exclusion of iron or other metals, which catalyze the chlorinolysis of the C—S bond to produce carbon tetrachloride.

Health and safety factors. Trichloromethanesulfenyl chloride is extremely toxic, with an LC_{50} (mouse) of 9 ppm and a TLV of 0.1 ppm (provisional value). The primary hazard is from inhalation of vapor.

Uses. The principal commercial application for trichloromethanesulfenyl chloride is as an intermediate for the manufacture of fungicides, the most important being captan, N-(trichloromethylthio)-4-cyclohexene-1,2-dicarboximide), and folpet, N-(trichloromethylthio)phthalimide (see Fungicides).

Hydrogen Sulfide

Hydrogen sulfide is present in the gases emanating from volcanoes, sulfur springs, undersea vents, swamps, and stagnant bodies of water. Bacterial reduction of sulfates and bacterial decomposition of proteins give hydrogen sulfide. Of greater importance as a source of sulfur are sour gases, which occur in large amounts in several locations.

Hydrogen sulfide is a by-product of many industrial operations, eg, coking and the hydrodesulfurization of crude oil and of coal. Hydrodesulfurization is increasing in importance as the use of high sulfur crude oil becomes more and more necessary (see Petroleum, refinery processes). A large source of hydrogen sulfide may result if coal liquefaction attains commercial importance.

Physical properties. Hydrogen sulfide, H_2S, is a colorless gas with a characteristic rotten-egg odor. The physical properties of hydrogen sulfide are given in Table 2.

Chemical properties. Although hydrogen sulfide is thermodynamically stable, it dissociates at very high temperatures. Hydrogen sulfide is oxidized by a number of oxidizing agents.

Certain of the above reactions are of practical importance. The oxidation of hydrogen sulfide in a flame is one means for producing the sulfur

Table 1. Physical and Thermodynamic Properties of Carbonyl Sulfide

Property	Value
mol wt	60.074
mp, °C	−138.8
bp, °C	−50.2
ΔH, fusion at 134.3 K, kJ/mol[a]	4.727
ΔH, vaporization at 222.87 K, kJ/mol[a]	18.57
density at 220 K, 101.3 kPa (= 1 atm), g/cm³	1.19
sp gr gas at 298 K (air = 1)	2.10
critical temperature, °C	105
critical pressure, kPa[b]	5946
critical volume, cm³/mol	138
autoignition temperature in air, °C	ca 250
solubility in water at 101.3 kPa (= 1 atm), vol%	
0°C	0.356
20°C	0.149

[a] To convert J to cal, divide by 4.184.
[b] To convert kPa to mm Hg, multiply by 7.5.

Table 2. Physical and Thermodynamic Properties of Hydrogen Sulfide

Property	Value
mol wt	34.08
mp, °C	−85.60
bp, °C	−60.75
ΔH fusion, kJ/mol[a]	2.375
ΔH vaporization, kJ/mol[a]	18.67
density at −60°C, g/cm³	0.993
sp gr, gas (air = 1)	1.19
critical temperature, °C	100.4
critical pressure, kPa[b]	9020
critical density, g/cm³	0.3681
$S°$ formation at 25°C, J/(mol·K)[a]	205.7
$C_p°$, J/(mol·K)[a]	34.2
autoignition temperature in air, °C	ca 260
vapor pressure, kPa[b]	
0°C	1033
40°C	2859
solubility in water at 101.3 kPa (= 1 atm)	
total pressure, g/100 g soln	
0°C	0.710
20°C	0.398

[a] To convert J to cal, divide by 4.184.
[b] To convert kPa to mm Hg, multiply by 7.5.

dioxide required for a sulfuric acid plant. Oxidation of hydrogen sulfide by sulfur dioxide is the basis of the Claus process for sulfur recovery (qv).

Anhydrous gaseous or liquid hydrogen sulfide is practically nonacidic, but aqueous solutions, ie, hydrosulfurous acid, are weakly acid.

Anhydrous hydrogen sulfide does not react at ordinary temperatures with metals, eg, mercury, silver, or copper.

Hydrogen sulfide causes the precipitation of sulfides from many heavy metal salts. It reacts with molten sulfur and depresses the viscosity of the latter, particularly at 130–180°C.

Hydrogen sulfide reacts with olefins under various conditions forming mercaptans and sulfides.

Manufacture. Small cylinders of hydrogen sulfide are readily available for laboratory purposes, but the gas can also be easily synthesized by action of dilute sulfuric or hydrochloric acid on iron sulfide, calcium sulfide, zinc sulfide, or sodium hydrosulfide.

Recovery from gas streams. The crude oil refined in the United States contains varying amounts of sulfur, eg, 0.04 wt% in Pennsylvania crude to ca 5 wt% in heavy Mississippi crude. Hydrodesulfurization is becoming increasingly important as a refinery operation. Hydrogen sulfide is recovered extensively from sour gas streams by scrubbing with various means and is usually fed to a Claus unit for conversion to sulfur.

Corrosivity. Anhydrous hydrogen sulfide has low corrosivity towards carbon steel, aluminum, Inconel, Stellite, and 304 and 316 stainless steels. However, hard steels, if highly stressed, are susceptible to hydrogen embrittlement by hydrogen sulfide. Wet hydrogen sulfide is quite corrosive to carbon steel; corrosion rates are as high as 2.5 mm/yr. High strength steels may suffer stress-corrosion cracking.

Health and safety factors. Hydrogen sulfide has an extremely high acute toxicity and has caused deaths both in the workplace and in areas of natural accumulation, eg, cisterns and sewers. Brief exposure to hydrogen sulfide at a concentration of 140 mg/m³ causes conjunctivitis and keratitis (eye damage), and exposures at above ca 280 mg/m³ cause unconsciousness, respiratory paralysis, and death.

OSHA's acceptable ceiling concentration is 20 ppm. Protective measures involve prompt detection and adequate ventilation.

Uses. Most of the hydrogen sulfide recovered as a by-product or scrubbed out of sour-gas streams is converted to elemental sulfur by the Claus process or to sulfuric acid (qv) where a market for the acid is near the source of the hydrogen sulfide. Hydrogen sulfide is also used to prepare various inorganic sulfides, notably sodium sulfide and sodium hydrosulfide, which are used in the manufacture of dyes, rubber chemicals (qv), pesticides, polymers, plastics additives, leather (qv), and drugs.

A large amount of sodium hydrosulfide or sodium sulfide is used and recycled in kraft pulping; hydrogen sulfide can be used for replenishing the sulfide content (see Pulp). An important application in organic synthesis is the reaction of hydrogen sulfide to produce thiols (qv) (mercaptans).

Hydrogen polysulfides (sulfanes). Individual hydrogen polysulfides have been characterized from H_2S_2 up to at least H_2S_8. They are of no commercial utility, although sodium and calcium polysulfides, which are made by addition of sulfur to the corresponding monosulfides, are used commercially. The atmospheric boiling point of H_2S_2 is 70.7°C and the boiling point of H_2S_3 is 69°C at 0.3 kPa (2 mm Hg).

Sulfur Halides and Oxyhalides

Sulfur forms several series of halides with all halogens except iodine. The fluorides, including the commercially important sulfur hexafluorine, are discussed elsewhere (see Fluorine compounds, inorganic).

Sulfur monochloride. *Physical properties.* Sulfur monochloride, S_2Cl_2, is a yellow-orange liquid with a characteristic pungent odor. The typical commercial sulfur monochloride has a melting point of −76°C and a boiling point of 137.8°C.

Manufacture. Sulfur monochloride is made commercially by direct chlorination of sulfur and usually in a heel of sulfur chloride from a previous batch.

Health and safety factors. Sulfur monochloride is highly toxic and irritating by inhalation, and corrosive to skin and eyes. The OSHA-permissible exposure limit is 1 ppm (6 mg/m³).

Uses. The principal commercial uses of sulfur monochloride are in the manufacture of lubricant additives and vulcanizing agents for rubber (see Lubrication and lubricants; Rubber chemicals).

Sulfur dichloride. *Physical properties.* Sulfur dichloride, SCl_2, is a deep red, fuming liquid, that decomposes in moist air with the evolution of hydrogen chloride. Pure sulfur dichloride is unstable and is supplied commercially as a 73.1–79.4 wt% SCl_2 mixture, with sulfur monochloride comprising the remaining percentage. The melting point is reported in the range of −121.5 to −61°C, and the boiling point (with decomposition) is 59°C.

Manufacture. The manufacture of sulfur dichloride is similar to that of sulfur monochloride, except that the last stage of chlorination proceeds slowly and must be conducted at temperatures below 40°C.

Uses. Sulfur dichloride is used as a chlorinating agent in the manufacture of parathion insecticide intermediates (see Insect control technology). It is also useful in the rapid vulcanization of rubber, eg, in the preparation of thin rubber goods by coating molds or fabrics with rubber latex.

Thionyl chloride. *Physical properties.* Thionyl chloride, $SOCl_2$, is a colorless fuming liquid with a choking odor. Selected physical and thermodynamic properties are listed in Table 3.

Manufacture. Thionyl chloride may be made by any of the following reactions:

$$SCl_2 + SO_3 \rightarrow SOCl_2 + SO_2$$
$$SCl_2 + SO_2 + Cl_2 \rightarrow 2\,SOCl_2$$
$$SCl_2 + SO_2Cl_2 \rightarrow 2\,SOCl_2$$

The sulfur dichloride can be fed as such or produced directly in the reactor from chlorine and sulfur monochloride.

Health and safety factors. Thionyl chloride as a reactive acid chloride can cause severe burns to the skin and eyes and acute respiratory-tract injury upon vapor inhalation.

Uses. A principal use of thionyl chloride is in the conversion of acids to acid chlorides, which are employed in the syntheses of herbicides, surfactants, drugs, vitamins, and dyestuffs (see Surfactants and detersive systems; Vitamins). Possible applications are also in the preparation of engineering thermoplastics of the polyarylate type made from iso- and terephthaloyl chlorides, which can be made from the corresponding acids plus thionyl chloride (see Engineering plastics).

Sulfuryl chloride. *Physical properties.* Sulfuryl chloride, SO_2Cl_2, is a colorless liquid with an extremely pungent odor. Physical and thermodynamic properties are listed in Table 4.

Table 3. Physical and Thermodynamic Properties of Thionyl Chloride

Property	Value
mol wt	118.97
mp, °C	−104.5
bp, °C	76
density at 25°C, g/cm^3	1.629
latent heat of vaporization, kJ/mol^a	31.3
viscosity at 0°C, mPa·s (= cP)	0.801
refractive index, n_D^{20}	1.517
dielectric constant at 20°C	9.25
vapor pressure, kPa^b	
0°C	4.5
20°C	11.6

^{a}To convert J to cal, divide by 4.184.
^{b}To convert kPa to mm Hg, multiply by 7.5.

Table 4. Physical and Thermodynamic Properties of Sulfuryl Chloride

Property	Value
mol wt	134.968
mp of last crystal point, °C	−54
bp, °C	69.5
density at 25°C, g/cm^3	1.6570
latent heat of vaporization, kJ/mol^a	27.97
surface tension at 23.5°C, mN/m (= dyn/cm)	35.26
viscosity at 0°C, mPa·s (= cP)	0.918
refractive index, n_D^{20}	1.443
vapor pressure at 0°C, kPa^b	5.45
coefficient of expansion, 0–38°C, °C^{-1}	0.0012
electrical conductivity, S/cm	3×10^{-8}

^{a}To convert J to cal, divide by 4.184.
^{b}To convert kPa to mm Hg, multiply by 7.5.

Manufacture. The preparation of sulfuryl chloride is carried out by feeding dry sulfur dioxide and chlorine into a water-cooled steel vessel containing a catalyst, eg, activated charcoal.

Health and safety factors. Sulfuryl chloride is corrosive to the skin and toxic upon inhalation. The TLV suggested by the manufacturer is 1 ppm.

Uses. Uses of sulfuryl chloride include the manufacture of chlorophenols, eg, chlorothymol for use as disinfectants. It is also used in the manufacture of alpha-chlorinated acetoacetic derivatives, eg, CH$_3$CO-CHClCOOC$_2$H$_5$, which are precursors for important substituted-imidazole drugs, phosphate insecticides, and heterocyclic fungicides. A large captive use may be in the chlorination of thermoplastics and elastomers.

Sulfur Nitrides

Although no commercial applications have as yet been developed for these compounds, interest has been stimulated by the discovery that polymeric sulfur nitride is electroconductive (see Polymers, conducting; Inorganic high polymers).

Tetrasulfur tetranitride. Tetrasulfur tetranitride, N$_4$S$_4$, is a bright orange crystalline solid (mp 178° with subl). It is explosive and detonates when struck. It decomposes slowly in water.

The usual satisfactory synthesis of tetrasulfur tetranitride involves saturating a solution of sulfur monochloride in carbon tetrachloride with chlorine and treating it with dry ammonia. However, this method is dangerous because of the risk of explosion.

Disulfur dinitride. Disulfur dinitride, N$_2$S$_2$, is a colorless liquid which, when kept at 20°C, gradually polymerizes to polythiazyl (NS)$_x$.

Polymeric sulfur nitride. Polymeric sulfur nitride (polythiazyl), (NS)$_x$, is a bronze, crystalline substance with metallic luster. It exhibits anisotropic electrical conductivity, which is highest along one crystal axis. At temperatures approaching absolute zero (< 0.26 K), superconductive properties are observed (see Superconducting materials).

Sulfur Oxides

Numerous oxides of sulfur have been reported and those that have been characterized are SO, S$_2$O, SO$_3$, and SO$_4$, Among them, SO$_2$ and SO$_3$ are of principal importance.

Sulfur dioxide. *Physical properties.* Sulfur dioxide, SO$_2$, is a colorless gas with a characteristic pungent, choking odor. Its physical and thermodynamic properties are listed in Table 5.

Manufacture. For most chemical process applications requiring sulfur dioxide gas or sulfurous acid, sulfur dioxide is prepared by burning sulfur or pyrite, FeS$_2$. Large amounts of sulfur dioxide are produced and discharged to the atmosphere from the combustion of coal from smelting operations.

Corrosivity. Almost all common materials of construction are resistant to commercial dry liquid sulfur dioxide, dry sulfur dioxide gas, and hot sulfur dioxide gas containing water at above the dew point.

Health, safety, and environmental factors. Sulfur dioxide is an irritant; concentrations above 20 ppm have a marked irritant, choking, and sneeze-inducing effect. Chronic exposure to sulfur dioxide has been widespread in certain industries, notably smelting and paper manufacture. Sulfur dioxide occurs in industrial and urban atmospheres at 1 ppb–1 ppm and in remote areas of the earth at 50–120 parts per trillion (10^{12}).

Green plants are far more sensitive to sulfur dioxide than people and animals. Injury to vegetation may occur at ground level concentrations below 1 ppm. Sulfur dioxide and nitrogen oxides are converted to sulfuric and nitric acids, respectively, by oxidation processes in the atmosphere. Acidity of rain contributed to by sulfuric and nitric acid may have caused environmental damage, particularly to aquatic life in freshwater lakes; however, the magnitude of the problem, the sources of the pollutants, and their quantitative trends are much disputed as are the preventative measures to be taken.

Uses. The dominant use of sulfur dioxide is captively in the production of sulfuric acid, and there is also substantial captive production in the pulp and paper industry for sulfite pulping. Sulfur dioxide is used as an intermediate for on-site production of bleaches, eg, chlorine dioxide or sodium hydrosulfite (see Bleaching agents). In food processing, sulfur dioxide has a wide range of applications as a fumigant, preservative, bleach, and steeping agent for grain.

Table 5. Physical and Thermodynamic Properties of Sulfur Dioxide

Property	Value
mol wt	64.06
mp, °C	−72.7
bp, °C	−10.02
ΔH fusion, kJ/mol^a	7.40
ΔH vaporization at −10.0°C, kJ/mol^a	24.92
vapor density at 0°C, 101.3 kPa (= 1 atm), air = 1	2.263
liquid density at −20°C, g/cm^3	1.50
critical temperature, °C	157.6
critical pressure, kPa^b	7911
dielectric constant at −16.5°C	17.27
dipole moment at 25°C, C·m^c	3.87×10^{-30}
vapor pressure, kPa^b	
10°C	230
30°C	462
solubility in water at 101.3 kPa^b, g/100 g H$_2$O	
0°C	22.971
20°C	11.577
40°C	5.881

^{a}To convert J to cal, divide by 4.184.
^{b}To convert kPa to mm Hg, multiply by 7.5.
^{c}To convert C·m to D, divide by 3.336×10^{-30}.

In water treatment, sulfur dioxide is often used to reduce residual chlorine from disinfection and oxidation. This technology is applied in potable-water treatment, sewage treatment, and especially industrial waste-water treatment (see Water, industrial water treatment; Water, municipal water treatment; Water, sewage).

In petroleum technology, sulfur dioxide, usually as sodium sulfite, is used as an oxygen scavenger. Another application in oil refining is as a selective extraction solvent in the Edeleanu process.

In mineral technology, sulfur dioxide and sulfites are used as flotation depressants for sulfide ores.

Sulfur Oxygen Acids and Their Salts

Sulfuric acid, H_2SO_4, is the most important commercial sulfur compound (see Sulfuric acid and sulfur trioxide). Peroxymonosulfuric acid (Caro's acid), H_2SO_5, is discussed elsewhere (see Peroxides and peroxy compounds, inorganic). The lower-valent sulfur acids are not stable species at ordinary temperatures. Dithionous acid, $H_2S_2O_6$, sulfoxylic acid, H_2SO_2, and thiosulfuric acid, $H_2S_2O_3$ are unstable species.

Sodium sulfite. *Physical properties.* Anhydrous sodium sulfite, Na_2SO_3, is an odorless, crystalline solid; the commercial grades are colorless or off-white. It melts with decomposition. The specific gravity of the pure solid is 2.633 (15.4°C). Sodium sulfite is soluble in water.

Manufacture. In a typical process, a solution of sodium carbonate is allowed to percolate downward through a series of absorption towers, through which sulfur dioxide is passed countercurrently. The solution leaving the towers is chiefly sodium bisulfite of, typically, 27 wt% combined sulfur dioxide content. The solution is run into a stirred vessel where aqueous sodium carbonate or sodium hydroxide is added to the point where the bisulfite is fully converted to sulfite. The solution may be filtered if necessary to attain the required product grade. A pure grade of anhydrous sodium sulfite can be crystallized above 40°C, since the solubility decreases with increasing temperature.

Health and safety factors. Although sodium sulfite has no detectable odor, its dust and solutions are irritating to the skin, eyes, and mucous membranes.

Uses. Sodium sulfite is used in pulp manufacture, water treatment, and photography.

Sodium bisulfite. Sodium bisulfite, $NaHSO_3$, is not stable in the solid state. The sodium bisulfite of commerce consists chiefly of sodium metabisulfite, $Na_2S_2O_5$. Aqueous sodium bisulfite solution with 26–77 wt% SO_2 (sp gr 1.36), is a commercial product.

Sodium metabisulfite. *Physical properties.* Sodium metabisulfite (sodium pyrosulfite, sodium bisulfite (a misnomer)), $Na_2S_2O_5$, is a white granular or powdered salt (sp gr 1.48), storable when kept dry and protected from air.

Manufacture. Aqueous sodium hydroxide, sodium bicarbonate, or sodium sulfite solution are treated with sulfur dioxide to produce sodium metabisulfite solution.

Health and safety factors. Sodium metabisulfite is nonflammable but, when strongly heated, releases sulfur dioxide. It has a low degree of acute toxicity; the LD_{50} (rat, oral) is 0.5–5 g/kg. Large doses can cause colic, diarrhea, and death.

Uses. Sodium metabisulfite is extensively used as a food preservative and bleach in the same applications as sulfur dioxide.

In tanneries, sodium bisulfite is used to accelerate the unhairing action of lime. It is also used as a chemical reagent in the synthesis of surfactants.

The reversible addition of sodium bisulfite to carbonyl groups is used in the purification of aldehydes. Sodium bisulfite also is employed in polymer and synthetic fiber manufacture.

Sodium dithionite. *Physical properties.* Sodium dithionite (sodium hydrosulfite, sodium sulfoxylate), $Na_2S_2O_4$, is a colorless solid, soluble in water to the extent of 22 g/100 g of water at 20°C. The commercial product is an anhydrous white powder, assaying 88–90 wt% $Na_2S_2O_4$, with the remainder comprising sodium metabisulfite, sodium thiosulfate, and in some cases sodium formate and sodium carbonate.

Manufacture. Dithionites can be manufactured by the reduction of sulfites, bisulfites, or sulfur dioxide with sodium or zinc.

Health and safety factors. Dry sodium dithionite, exposed to moist air, can ignite spontaneously.

Sodium dithionite is considered only moderately toxic. As a food additive, sodium dithionite is generally recognized as safe (GRAS).

Uses. Dyeing consumes ca 50% of the U.S. production of sodium dithionite. Bleaching of mechanical pulp and thermomechanical pulp comprises ca 20% of U.S. production. Miscellaneous uses include reductive bleaching of clay, glue, gelatin, soap, oils, and food products.

Zinc dithionite. Zinc dithionite, ZnS_2O_4, is a white, water-soluble powder. Although it exhibits greater stability in aqueous solution than sodium dithionite at a given temperature and pH, its use has declined sharply because of regulatory constraints on pollution of water by zinc.

Sodium and zinc formaldehydesulfoxylates. Although free sulfoxylic acid, H_2SO_2, has not been isolated and its salts are in doubt, organic derivatives, which may be viewed as adducts of sulfoxylic acid, are commonly made. The latter are mainly sodium formaldehydesulfoxylate, $HOCH_2SO_2Na$ (commercially sold as the dihydrate), and zinc formaldehydesulfoxylate. These compounds are water-soluble reducing agents with uses similar to the dithionites but are more stable. They can be used in reducing and bleaching applications at lower pH values and at somewhat higher temperatures than the dithionites.

Thiocyanic Acid and Its Salts

Free thiocyanic acid, HSCN, can be isolated from its salts but is not an article of commerce because of its instability; dilute solutions can be stored briefly. Commercial derivatives are principally ammonium and sodium thiocyanates as well as several organic thiocyanates.

Ammonium thiocyanate. *Physical properties.* Ammonium thiocyanate, NH_4SCN, is a hygroscopic crystalline solid which deliquesces at high humidities. It melts at 149°C with partial isomerization to thiourea. It is soluble in water to the extent of 65 wt% at 25°C and 77 wt% at 60°C.

Manufacture. The principal route used in the United States is the reaction of carbon disulfide with aqueous ammonia, which proceeds by way of ammonium dithiocarbamate.

Health and safety factors. Ammonium thiocyanate has a low acute toxicity; the LD_{50} in mammals is 500–1000 mg/kg. Exposure of skin, eyes, and mucous membranes should be avoided.

Uses. Ammonium thiocyanate is a chemical intermediate for the synthesis of several proprietary agricultural chemicals, mainly herbicides.

Sodium and potassium thiocyanates. *Physical and chemical properties.* Sodium thiocyanate, NaSCN, is a colorless deliquescent crystalline solid (mp 323°C). It is soluble in water to the extent of 58 wt% NaSCN at 25°C and 69 wt% at 100°C. It is highly soluble in methanol and ethanol, and moderately soluble in acetone.

Potassium thiocyanate, KSCN, is a colorless crystalline solid (mp 172°C) soluble in water to the extent of 217 g/100 g of water at 20°C, and in acetone and alcohols. Much of the chemistry of sodium and potassium thiocyanates is that of the thiocyanate anion.

Manufacture. In the United States, sodium and potassium thiocyanates are made by adding caustic soda or potash to ammonium thiocyanate, followed by evaporation of the ammonia and water.

Uses. The largest use for sodium thiocyanate is as 50–60 wt% aqueous solution as a component of the spinning solvent for acrylic fibers (see Acrylic and modacrylic fibers; Acrylonitrile polymers). Other textile applications are as a fiber swelling agent and a dyeing and printing assist. Sodium thiocyanate and other thiocyanate salts are used to prepare organic thiocyanates.

EDWARD D. WEIL
Stauffer Chemical Co.

Gmelins Handbuch der Anorganischen Chemie, 8th ed., System No. 9, 14 (D4–D6), Springer-Verlag, Berlin, Heidelberg, 1953–1977.

A. Senning, ed., *Sulfur in Organic and Inorganic Chemistry*, Vols. 1–4, Marcel Dekker Inc., New York, 1971–1982.

SULFUR DIOXIDE, SO$_2$.   See Sulfur compounds.

SULFUR DYES

Sulfur dyes are used mainly for dyeing textile cellulosic materials or blends of cellulosic fibers with synthetic fibers such as polyacrylates, polyamides, and polyesters. They are also used for silk and paper in limited quantities for specific applications. The use of solubilized sulfur dyes on certain types of leather is gaining favor.

From an application point of view, the sulfur dyes are between vat, direct, and fiber-reactive dyes in importance. They give good to moderate lightfastness and good wetfastness at low cost and rapid processing.

Traditionally, these dyes are applied from a dyebath containing sodium sulfide. However, over the years, development in dyeing techniques and manufacture has led to the use of sodium sulfhydrate, sodium polysulfide, sodium dithionite, thiourea dioxide, and glucose as reducing agents. In the reduced state, the dyes have affinity for cellulose and are subsequently exhausted on the substrate with common salt or sodium sulfate and fixed by oxidation.

Little is known about the structure of sulfur dyes, and therefore, they are classified according to the chemical structure of the starting materials.

Sulfurization

The process of sulfurization is usually carried out by a sulfur bake, in which the dry organic starting material is heated with sulfur between 160 and 320°C (Table 1); a polysulfide bake, which includes sodium sulfide (Table 2); a polysulfide melt, in which aqueous sodium polysulfide and the organic starting material are heated under reflux or under pressure in a closed vessel (Table 3); or a solvent melt, in which butanol, Cellosolve, or dioxitol are used alone or together with water. In the last two methods, hydrotropes may be added to enhance the solubility of the starting material. The hydrotropes improve yield and quality of the final dyestuff.

Application

Sulfur dyes are applied in leuco form. In this form, the dye has affinity for the fiber. After the dye is completely absorbed by the fiber, it is reoxidized *in situ*. In dyes, such as the bright blues, which contain quinonimine groups, further reduction takes place in a manner similar to the reduction of the keto group in vat dyes. Sodium sulfide or sodium sulfhydrate are the most widely used reducing agents and the dyeings are normally oxidized chemically. Sodium or potassium bichromate mixed with acetic acid is the traditional oxidation method, but because of ecological restrictions on bichromates, other oxidizing agents are being

Table 1.   Sulfur-Bake Dyes

Intermediates	Shade	CI designation Name	Number
(structure)	orange	Sulfur Orange 1	53050
(structure)	yellowish brown	Sulfur Brown 12	53065
(structure)	reddish yellow	Sulfur Yellow 1	53040

Table 2.   Polysulfide-Bake Dyes

Intermediate(s)	Shade	CI designation Name	Number
(structure)	yellow	Sulfur Yellow 9	53010
(structure)	dull green	Sulfur Green 9	53005
(structure) (+ CuSO$_4$)	olive / dull green	Sulfur Green 11 / Sulfur Green 1	53165 / 53166
(structure)	dull reddish brown	Sulfur Brown 7	53275

Table 3.   Polysulfide-Melt Dyes

Intermediates	Shade	CI designation Name	Number
(structure)	greenish black	Sulfur Black 1	53185
(structure)	reddish blue to bluish violet	Sulfur Blue 7	53440
(structure)	green	Sulfur Green 3	53570
(structure)	Bordeaux	Sulfur Red 6	53720
(structure)	blue to reddish navy	Vat Blue 42	53640
(structure)	bluish black	Sulfur Black 11	53290
(structure)	Bordeaux brown	Sulfur Red 10 Leuco Sulfur Brown 96	53228
(structure)	dull bluish red	Sulfur Red 7	53810

used such as hydrogen peroxide or sodium perborate and potassium iodate or sodium bromate mixed with acetic acid.

Additional aftertreatments include resin finishes, which improve fastness properties, and dye-fixing agents of the epichlorhydrin-organic amine type. These agents react with the dye to give products that are not water soluble and hence more difficult to remove.

Commercial forms of sulfur dyes include powders, prereduced powders, grains, dispersed powders, dispersed pastes, liquids, and water-soluble brands.

Health and Safety Factors

During the last 20 yr, much more emphasis has been placed on health and safety aspects within the chemical and dyestuff industries. As a consequence, benzidine and β-naphthylamine, which are known carcinogens, have been banned in many countries. The handling of hazardous chemicals, such as nitroanilines, dinitro- and diaminotoluenes, nitro- and dinitrophenols, and chlorodinitrobenzenes is kept to a minimum. In all cases, suitable protective gear is recommended.

The effluent from the dye manufacture and textile-dyeing industry is usually treated by aeration in large tank farms, although small dyehouses may utilize spent flue gases or bleach liquors to oxidize residual sulfides.

Uses

Sulfur dyes are widely used in piece dyeing of traditionally woven cotton goods, such as drill and corduroy fabrics (see Textiles). The cellulosic portion of polyester/cotton and polyester/vicose blends is dyed with sulfur dyes. The fastness matches that of the disperse dyes on the polyester portion, especially when it is taken into account that these fabrics are generally given a resin finish. Yarn is dyed with sulfur dyes, although raw-stock dyeing has declined in recent years. The dyeing of knitted fabrics, both 100% cotton or blends of cotton with synthetic fibers, is increasing, but continuous dyeing of piece goods by pad-steam methods is still one of the principal outlets for sulfur dyes today.

R.A. GUEST
W.E. WOOD
James Robinson & Company, Ltd.

O. Lange, *Die Schwefelfarbstoffe, Ihre Herstellung und Verwendung*, 2nd ed., Spamer, Leipzig, 1925.

J.F. Thorpe and M.A. Whiteley, eds., *Thorpe's Dictionary of Applied Chemistry*, 4th ed., Vol. 11, Longmans, Green & Co., London, 1954.

K. Venkataraman, *The Chemistry of Synthetic Dyes and Pigments*, Vol. 2, 1952, Chapt. 35; Vol. 3, 1970, Chapt. 1; Vol. 7, 1974, Chapt. 1, Academic Press, London and New York.

SULFURIC ACID AND SULFUR TRIOXIDE

Sulfuric acid, H_2SO_4, is a colorless, viscous liquid with a specific gravity of 1.8357 and a boiling point of 270°C. Its anhydride, SO_3, is also a liquid with a specific gravity of 1.857 and a boiling point of 44.8°C. Sulfuric acid is by far the largest volume chemical commodity produced and is sold or used commercially in a number of different concentrations including 78 wt% (60°Bé), 93 wt% (65.7°Bé), 96 wt% (66°Bé), 98–99 wt%, 100 wt%, and as various oleums (fuming sulfuric acid, $H_2SO_4 + SO_3$). Stabilized liquid SO_3 is also an item of commerce.

Sulfuric acid has several desirable properties that lead to its use in a wide variety of applications. It typically is less expensive than any other acid; it can be readily handled in steel or common alloys at normal commercial concentrations; and it is available and readily handled at concentrations > 100 wt% (oleum). Sulfuric acid is a strong acid; it reacts readily with many organic compounds to produce useful products, and forms a slightly soluble salt or precipitate with calcium oxide or hydroxide, the least expensive and most readily available base. Concentrated sulfuric acid is also a good dehydrating agent and under some circumstances functions as an oxidizing agent. The physical properties of sulfur trioxide are given in Table 1.

Table 1. Properties of Sulfur Trioxide

Property	Value
critical temperature, °C	217.8
critical pressure, kPa[a]	8,208
critical density, g/cm³	0.630
normal boiling point temperature, °C	44.8
melting point (γ phase), °C	16.8
liquid density (γ phase at 20°C), g/cm³	1.9224
solid density (γ phase at −10°C), g/cm³	2.29
liquid coefficient of thermal expansion (at 18°C), per °C	0.002005
liquid heat capacity (at 30°C), kJ/(kg·°C)[b]	3.222
heat of formation of gas (at 25°C), (MJ·kg)/mol[b]	−395.76
free energy of formation of gas (at 25°C), (MJ·kg)/mol[b]	−371.07
heat of dilution, MJ/kg[b]	2.109
heat of vaporization (γ liquid), MJ/kg[b]	0.5843
diffusion in air (at 80°C), m/s	0.000013
liquid dielectric constant (at 18°C)	3.11
electric conductivity	negligible

[a] To convert kPa to psi, multiply by 0.145.
[b] To convert J to cal, divide by 4.184.

Manufacture

The contact process produces sulfuric acid from a wide range of sulfur-bearing raw materials by several different process variants, depending largely on the raw material used. In some cases sulfuric acid is made as a by-product of other operations, primarily as an economical or convenient means of minimizing air pollution or disposing of unwanted by-products. Economics favor production of sulfuric acid close to the point of use, since truck- or rail-freight costs are substantial compared to production costs.

Generation of sulfur dioxide gas. *Sulfur burning.* With the trend to very large single-train plants, current practice is to use horizontal, brick-lined combustion chambers with dried air and atomized molten sulfur introduced at one end. Atomization typically is accomplished by pressure-spray nozzles or by mechanically driven spinning cups. Because the degree of atomization is a key factor in producing efficient combustion, sulfur nozzle pressures have reached 1.03 MPa (150 psi) in recent years. Sulfur burners are typically designed as proprietary items by companies specializing in acid-plant design and construction. Some designs contain baffles or secondary air inlets to promote mixing and effective combustion.

Spent-acid or H_2S burning. Burners for spent acid or hydrogen sulfide are generally similar to those used for sulfur with a few critical differences. Special types of nozzles are required both for H_2S, a gaseous fuel, and for corrosive and viscous spent acids. In a few cases, spent acids may be so viscous that only a spinning cup can satisfactorily atomize them.

Ore roasting, sintering, or smelting. Generation of SO_2 at nonferrous metal smelters is determined primarily by the needs of various metallurgical processes and only incidentally by the requirements of the sulfuric acid process. In general, three main types of equipment are used to generate SO_2 gas from metallic sulfides, ie, roasters, sintering machines, and converters. Reverberatory furnaces are also widely used in the copper industry, but the SO_2 gas they generate is too weak and low in oxygen to be handled economically in sulfuric acid plants (see Extractive metallurgy).

Process Details

The following equation presents the stoichiometric relation between reactants and products for the contact process.

$$SO_2 + 1/2\,O_2 \rightleftharpoons SO_3$$

There are three important characteristics of this reaction: it is exothermic and reversible, and shows a decrease in molar volume in the direction of the desired product. To improve the equilibrium or driving force of the reaction, the sulfuric acid industry has attempted one or a combination of the following process-design modifications: increasing concentration of SO_2 in the process-gas stream; increasing concentration of O_2 in the process-gas stream by air dilution or oxygen enrichment; increasing the number of catalyst beds; removal of the SO_3 product by interpass absorption (double-absorption process); lowering catalytic-converter inlet-operating temperatures (better catalysts); and increasing the catalytic-converter operating pressure (pressure plants).

The double-absorption process at relatively low pressure is widely used today to minimize emissions of free SO_2. Many older plants or plants located where SO_2 emissions are not regulated still utilize the single-absorption process. Figure 1 illustrates the single-absorption sulfur-burning process. The double-absorption process is similar, except that gas heat exchangers and another absorptive tower and acid circulation system are provided to remove SO_3 at an intermediate stage during catalytic oxidation of SO_2 to SO_3.

Oleum manufacture. Production of fuming sulfuric acid (oleum) is accomplished by absorbing sulfur trioxide in one or more special absorption towers irrigated by recirculated oleum.

Sulfur trioxide. The anhydride of sulfuric acid, SO_3, is a strong organic sulfonating and dehydrating agent which has some specialized uses (see Sulfonation and sulfation). Its principal applications are in production of detergents and as a raw material for chlorosulfuric acid (qv) and 65% oleum. In recent years, SO_3 gas has been added to cooled combustion gases at many coal-burning power plants to improve dust removal in electrostatic precipitators (see Air pollution control methods). Liquid SO_3 is usually produced by distilling SO_3 vapor from oleum and condensing it.

Energy-efficient plants. The dramatic increases in energy costs in the pass decade have led to significant plant-design changes in the sulfuric acid industry. The current trend is to reduce power consumption of the process, increase heat recovery in the steam system, and use steam-driven turbines to generate electric power. To reduce the power consumption of a sulfuric acid plant, three main options are available: increased efficiency of rotating equipment; reduced plant pressure drop on the gas side; and reduced gas volume processed per ton of acid produced.

Health and Safety Factors

Sulfuric acid is injurious to the skin, mucosa, and eyes. Dangerous amounts of hydrogen may develop in reactions between dilute acid and metals. Sulfuric acid at high concentrations reacts vigorously with water, organic compounds, and reducing agents. Oleums and liquid SO_3 frequently react with explosive violence, particularly with water.

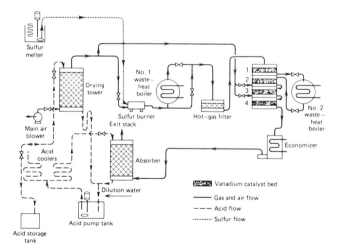

Figure 1. Typical flow sheet for air-dilution sulfur-burning acid plant. The following table of optimum catalyst temperatures at the inlet and exit of each pass is considered normal for ca 9% SO_2 gas when the plant and catalyst are operating efficiently at rated load. If greater or smaller temperature differentials occur, it indicates that the plant or catalyst may not be functioning properly.

Converter pass	1	2	3	4
inlet, °C	415–425	440–450	435–440	425–430
outlet, °C	600	500	450	430–435
ΔT, °C	175–185	50–60	10–15	5

Sulfuric acid, oleum, and sulfur trioxide are classified as corrosives for labeling and freight purposes and must carry a DOT corrosive label as well as a product label. Small bottles, pails, or drums must be placed inside outer boxes or containers as specified by government regulations.

Sulfuric acid toxicity. Studies of prolonged exposure to sulfuric acid fumes have been performed on workers in plants manufacturing lead–acid batteries. From these studies it is well documented that prolonged exposure to mineral acid fumes causes the teeth of the exposed subject to deteriorate. Overexposure to sulfuric acid aerosols results in pulmonary edema, chronic pulmonary fibrosis, residual bronchiectasis, and pulmonary emphysema.

The threshold limit value (TLV) of sulfuric acid mist for humans agreed upon by the ACGIH, OSHA, and NIOSH is 1 mg/m³ air. Sulfuric acid aerosols below the TLV are commonly not detected by odor, taste, or irritation. A TLV of 1 mg/m³ is recommended by the ACGIH to prevent pulmonary irritation and injury to the teeth at particle sizes likely to occur in industrial situations.

Uses

It has been said that consumption of sulfuric acid is a good measure of a country's degree of industrialization and that it also serves as an index of general business conditions. Until recently, this statement was valid, but current heavy use of sulfuric acid by the phosphate fertilizer industry has changed matters. Of total U.S. sulfuric acid production, over two-thirds now goes into phosphate fertilizers, whereas other uses have either grown slowly or declined. This trend will undoubtedly continue.

Other uses of sulfuric acid are in the production of inorganic pigments, textile fibers, explosives, petroleum products, alcohols, pulp and paper, detergents, other chemicals, and as a leaching agent for ores, a pickling agent for iron and steel, and a component of lead storage batteries. Use in the production of pigments has been relatively static or declining because of a trend to chloride processes for titanium oxide pigments. Iron and steel-pickling usage is also declining because of the increased use of hydrochloric acid, which can be more easily regenerated on site at large pickling installations (see Metal surface treatments).

J.R. DONOVAN
J.M. SALAMONE
Monsanto Enviro-Chem Systems, Inc.

J.R. Donovan, R.D. Stolk, and M.L. Unland in B.E. Leach, ed., *Applied Industrial Catalysis*, Vol. 2, Academic Press, New York, 1983, pp. 245–286.

W.W. Duecker and J.R. West, eds., *The Manufacture of Sulfuric Acid*, Reinhold Publishing Co., New York, 1959; reprinted by Robert E. Krieger Publishing Co., Huntington, N.Y., 1971.

A.I. More, ed., *Making the Most of Sulfuric Acid, Proceedings of the British Sulphur Corporation's Fifth International Conference*, London, Nov. 1981.

Sulfuric/Phosphoric Acid Plant Operations, CEP Technical Manual, AIChE, New York, 1982.

SULFURIC AND SULFUROUS ESTERS

Sulfuric and sulfurous acids form a series of esters analogous to those from other acidic materials. The hydrogen of the acid is replaced by a carbon-containing group. Since two hydrogens are present in the sulfur-based acids, there are two series in esters. Replacement of one hydrogen results in an acid ester. If both hydrogens are replaced, symmetrical and unsymmetrical diesters form. The two series are represented by the following general formulas:

$$ROS(O)_2OR' \quad ROS(O)OR'$$

where R is a carbon group and R' is a carbon group, hydrogen, or metal cation.

In the acid ester series, there are compounds in which another group, eg, chlorine or amine, is present in place of the hydroxy group. These compounds can be used as intermediates in making diesters. The carbon groups most commonly present are those from short- and long-chain alcohols and hydroxyaromatic and heterocyclic compounds. In this review, the nomenclature is primarily that used in commercial practice, rather than the more rigorous *Chemical Abstracts* nomenclature by which these esters are called sulfuric and sulfurous acid organic ester. Table 1 illustrates the variety of known compounds and some of their properties. Of these, dimethyl sulfate, diethyl sulfate, and long-chain monoalkyl alkali metal sulfates are the compounds of practical interest.

Chemical Properties

Sulfates. The chemistry of alkyl sulfates is dominated by reactions with nucleophiles and reactions as acids. Reaction with nucleophiles results in alkylation.

Sulfites. Reactions involving alkylation are similar to those of sulfates. Sulfites also undergo elimination, transesterification, and isomerization. The last two reactions parallel those of phosphites.

Manufacture

Monoester hydrogen sulfates. Hydrogen sulfates are prepared by the action of a sulfating agent on the corresponding alcohol or phenol. The following reagents are used for sulfation: sulfur trioxide, sulfuric acid, chlorosulfuric acid, sulfur trioxide–amine complexes, and sulfamic acid (see Sulfonation and sulfation).

Diorgano sulfates. Dialkyl sulfates up to octadecyl can be made from the alcohols via the following reactions:

$$ROH + SO_2Cl_2 \rightarrow ROSO_2Cl + HCl$$

$$2\ ROH + SOCl_2 \rightarrow (RO)_2SO + 2\ HCl$$

$$ROSO_2Cl + (RO)_2SO \rightarrow (RO)_2SO_2 + RCl + SO_2$$

Table 1. Some Sulfuric and Sulfurous Acid Esters

Ester	Bp$_{kPa}{}^a$, °C	Mp, °C
Sulfates, (RO)$_2$SO$_2$		
R		
methyl	$188.8_{101.3}, 69.70_{1.33}$	
ethyl	$208, 89_{1.20}$	
n-propyl	$95_{0.670}$	
n-butyl	$103_{0.200}$	
chloromethyl	$97_{1.87}$	
phenyl	$194.6_{2.66}$	
ethylene^b		99
1,3-propylene^b		63
methylene (dimer)^b		155
Sulfites, (RO)$_2$SO		
R		
methyl	$126–127_{101.3}$	
ethyl	$159–160_{101.3}$	
n-propyl	$82_{2.00}$	
n-butyl	$110_{1.73}$	
ethylene^b	$80_{3.72}$	
1,2-propylene^b	$84_{3.92}$	
pentaerythritol, di^b		154
Organo hydrogen sulfates, (RO)SO$_3$H		
R		
methyl	$130–140_{101.3}$ (dec)	
n-decyl	(liquid)	
n-dodecyl		25–27
phenyl		
Organo halosulfates, (RO)SO$_2$X		
R, X		
methyl, chloro	$134–135_{101.3}, 42_{2.13}$	
methyl, fluoro	$92_{101.3}, 45_{21.4}$	
Organo halosulfites, (RO)SOX		
R, X		
methyl, chloro	$35_{8.00}$	
ethyl, chloro	$550–53_{8.00}$	

^{a}To convert kPa to mm Hg, multiply by 7.5; 101.3 kPa = 1 atm.
^{b}Cyclic.

Mixed esters are synthesized by the reaction of one alkyl chlorosulfate with a different sodium alkoxide or a different dialkyl sulfite as follows:

$$ROSO_2Cl + NaOR' \rightarrow ROSO_2OR' + NaCl$$

$$ROSO_2Cl + (R'O)_2SO \rightarrow ROSO_2OR' + R'Cl + SO_2$$

Diorgano Sulfites

Symmetrical or mixed dialkyl sulfites are prepared by the stepwise reaction of thionyl chloride with two molecules of an alcohol or the stoichiometric quantities of two alcohols in pyridine.

Halosulfates and Halosulfites

A general method for the preparation of alkyl halosulfates and halosulfites is the treatment of the alcohol with sulfuryl or thionyl chloride at low temperatures while passing in inert gas through the mixture to remove hydrogen chloride.

Health and Safety Factors

The most commonly used dialkyl sulfate is dimethyl sulfate. This is most hazardous in liquid and vapor forms. The hazard arises from its toxicity, high reactivity, and to some extent, combustibility. Dimethyl sulfate is corrosive and poisonous, and its effects may be acute or chronic. Because it has an analgesic effect on many body tissues, even severe exposures may not be immediately painful. Dimethyl sulfate is particularly dangerous to the eyes and respiratory system. It causes severe burns, but symptoms may be delayed. Exposed workers must be immediately and properly treated to avoid permanent injury to eyes and lungs, or even death. Skin burns can also be severe. Ingestion causes convulsions and paralysis, with later damage to the kidneys, liver, and heart. Genetic effects have been extensively reviewed, and its mutagenicity is correlated with carcinogenicity.

According to OSHA, exposure to dimethyl sulfate shall not exceed an 8-h time-weighted average of 1 ppm in air. Since both liquid and vapor can penetrate the skin and mucous membranes, control of vapor inhalation alone may not be sufficient to prevent absorption of an excessive dose. The ACGIH recommends a time-weighted average threshold limit value of 0.1 ppm.

Uses

The sulfuric acid esters as compared to the sulfurous esters are the most widely used. In nature they appear as solubilizing groups in detoxification-excretory mechanisms and as sulfated carbohydrate groups in modified proteins. The significant commercial uses are alkylation, formation of long-chain alcohol monosulfates as surfactants, and formation of intermediates in preparation of some lower alcohols. Alkylation involves primarily dimethyl and diethyl sulfates in the preparation of a wide variety of intermediates and products, especially dyes, agricultural chemicals, drugs, and other specialties.

Ethylenesulfonate salt formation and ethyleneimine (aziridine) formation are the most significant uses of the esters as intermediates.

W.B. McCormack
B.C. Lawes
E.I. du Pont de Nemours & Co., Inc.

Dimethyl Sulfate, Properties, Uses, Storage and Handling, bulletin E.I. du Pont de Nemours & Co., Inc., Jan. 1981.

E.E. Gilbert, *Sulfonation and Related Reaction*, Wiley-Interscience, New York, 1965, Chapt. 6.

C.M. Suter, *Organic Chemistry of Sulfur*, John Wiley & Sons, Inc., New York, 1944, Chapt. 1.

SULFURIZATION AND SULFURCHLORINATION

Sulfur reacts with unsaturated compounds forming products that are usually dark and often viscous cross-linked mixtures of mono-, di-, and polysulfides. Sulfur monochloride, S_2Cl_2, also reacts with unsaturated

Table 1. Chemical and Physical Properties of Sulfurized and Sulfurchlorinated Fats and Oils[a]

Product designation	Sulfur, wt %	Active sulfur, wt %	Copper strip tarnish test, 10 wt % in oil	Viscosity mm²/s (= cSt) 40°C	Viscosity mm²/s (= cSt) 100°C	Density (at 25°C), g/cm³	Total acid number, mg KOH/g	Pour point, °C
Base 10-L[b]	10.0	0	1b	1100	80	0.98	25	18
Base 14-L[b]	13.5	3	3b	1700	100	0.99	26	27
Base A-92[b]	16.5	6	4b	625	50	1.00	18	21
Base L-66[b,c]	6.0	0	1b	1000	100	0.99	5	10
Base 10-SE[d]	8.5	0	1b	20	4	0.94	5	10
Base 12-SE[d]	13.0	2	4a	30	6	0.96	4	10
Base 18-SE[d]	16.5	6	4a	40	9	0.98	7	10
Base 99[c,d]	8.0	0	2b	75		1.02	8	10
Sul-perm 10[e]	10.0	0	1b	400	45	0.98	8	16
Sul-perm 12[e]	12.0	1	1b	500	50	0.98	12	18
Sul-perm 18[e]	16.5	6	4a	600	55	1.02	12	16
Sul-perm 88[c,e]	6.5	0	2b	4000	330	0.99	7	18
recommended ASTM procedure	D 129	D 1662	D 130	D 445	D 445	D 1298	D 974	D 97

[a] The products listed in the table and their properties have been submitted by Keil Division of Ferro Corp.
[b] Based on lard.
[c] Sulfurchlorinated fats and oils have chlorine %s similar to their sulfur %s.
[d] Based on methyl ester of lard.
[e] Based on mixture of animal and vegetable oils plus synthetic fatty esters.

materials forming products that are cross-linked by sulfur but which also contain chlorine. For many years, sulfurized (sulfurated) and sulfurchlorinated, unsaturated fatty oils have been added to mineral oils to make lubricants that provide antiwear, low friction, increased load-carrying ability, and improved oxidation resistance (see Lubrication and lubricants).

Properties

Properties of typical sulfurized fatty oils are listed in Table 1.

Manufacture

Sulfurization. Manufacture of a sulfurized product is normally carried out in a mild steel, batch-reaction vessel equipped with an agitator.

Sulfurchlorination. Manufacture of sulfurchlorinated products is carried out preferably in a glass-lined reaction vessel equipped with a sparging system and a cooling jacket or coils. The process is schematically similar to that of sulfurization except that a separate vessel for air blowing is normally not used.

Health and Safety Factors

Sulfurized and sulfurchlorinated fatty oils and esters have a long history of manufacture and usage without incident of hazard to workers.

KARL KAMMANN
GUY E. VERDINO
Keil Chemical Division
Ferro Corporation

J.P. Friedrich in E.H. Pryde, ed., *Fatty Acids*, American Oil Chemists Society, Champaign, Ill., 1979, Chapt. 30, pp. 591–607.

C.L. Hermann and J.J. McGlade, *J. Am. Oil Chem. Soc.* **51**, 88 (1974).

J. Wisniak and H. Benajahu, *Ind. Eng. Chem., Prod. Res. Dev.* **14**(4), 247 (1975).

SULFUR RECOVERY

Sulfur is generally categorized as elemental sulfur (brimstone) or sulfur in other forms. Elemental sulfur is further classified by its method of production. Frasch sulfur is produced from salt domes or sedimentary deposits by injecting superheated water into the formation, melting the sulfur in place, and pumping out the molten sulfur. Native sulfur is produced by conventional mining methods, and recovered sulfur is generally produced as a by-product of sour natural gas and petroleum by which hydrogen sulfide is stripped from the hydrocarbon and converted to elemental sulfur by the Claus process (see Gas, natural; Petroleum, refinery processes).

Production of recovered sulfur has increased rapidly in the last 25 yr and accounts for about one-half of the world production of sulfur in the elemental form. Discovery and development of large sour natural-gas fields in the United States, Canada, France, the FRG, the Middle East, and the USSR have been main factors in this rapid growth. Increased processing of sour crude oil and more strict pollution control has forced most petroleum refineries to recover the sulfur content of the crude in their refineries. Based on the nature of the process or the economics of a particular situation, sulfur is recovered also as a salable sulfur compound.

Processing of Sour Natural Gases and Sour Crude Petroleums

Recovery of acid gases. In the processes in which the reagent is regenerated and the hydrogen sulfide is recovered, amine reagents in aqueous solutions are the most widely used. Three amines are available industrially: monoethanolamine (MEA), diethanolamine (DEA), and triethanolamine (TEA) (see Alkanolamines). The first two are used widely in gas treating. Other amines, as well as nonamine compounds, are also used (see also Carbon dioxide). A simplified flow diagram of the conventional MEA cycle is illustrated in Figure 1.

Other recovery processes include the hot carbonate process, the Sulfinol process, the Selexol process, the methyldiethanolamine process, the tripotassium phosphate process, the sodium phenolate process, the Alkazid process, the vacuum carbonate process, and water washing.

Processes for direct recovery include the dry-box process, the Thylox process the Giammarco-Vetrocoke process, the Stretford process, the Takahax process, the molecular-sieve process, processes involving iron cyanide complexes, and the Perox process.

Claus process. The basic Claus process was invented in 1883. Its large-scale use began in the 1950s in the United States, and there are two principal process variations. The first is the straight-through process, in which all of the air and the acid gas pass through the combustion zone. The second is the split-flow process, in which all of the air and at least one-third of the acid gas pass through the combustion zone with the remaining acid gas being sent to the first catalytic reactor.

A simplified flow diagram of a typical Claus sulfur-recovery plant is shown in Figure 2.

Claus plant tail-gas treatment. Most Claus sulfur plants cannot meet existing or proposed air-pollution regulations without addition to the original design or inclusion of some method of reducing or eliminating the sulfur content of the gas leaving the plant and entering the atmosphere. Tail-gas clean-up processes have been proposed and some are in use.

Recovery from Metallurgical Operations

In early metallurgical operations, sulfur recovery was not practiced. Sulfur dioxide in the air may be helpful at times to farmers but can also cause ecological damage and political controversy. In locations where the economics are favorable, sulfur is recovered as sulfuric acid in metallurgical sulfuric acid plants (see Sulfuric acid and sulfur trioxide). Newer metallurgical processes have been and are being developed. One aim of these processes is to recover sulfur as the element or as a useful

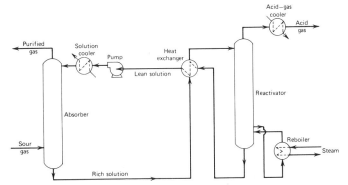

Figure 1. Flow diagram of standard amine process.

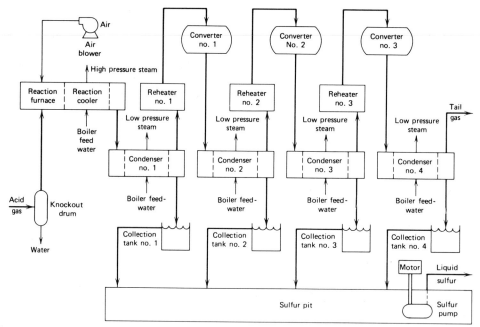

Figure 2. Flow diagram of typical Parsons Claus sulfur-recovery plant. Courtesy of The Ralph M. Parsons Co.

compound, eg, ammonium sulfate. To avoid the formation of SO_2 in winning, ie, recovering, metals from sulfide ore concentrates, some investigators turned to hydrometallurgy.

Recovery from Stack Gases

Scrubbing of flue gases with lime or limestone, which produces a calcium sulfite–sulfate waste for dumping, has dominated the flue-gas desulfurization (FGD) systems market. Other ways to meet air-pollution requirements are by selection of low sulfur fuels, conversion of high sulfur coal to gas or oil with removal of the sulfur during the processing, removal of the SO_2 in the boiler, and use of tall stacks.

Recovery from Coal

Coals high in sulfur content can be burned if the sulfur is removed or recovered from the flue gases or during the combustion process, the coal is desulfurized before combustion, or the coal is converted to gas or oil (see Coal conversion processes, desulfurization).

Recovery from Oil Shale

Recovery of oil from shale containing organic matter has been investigated and, in some countries, carried out (see Oil shale). Generally, oil from shale can be recovered by mining and processing the shale on the surface or by *in situ* processing. The sulfur in the shale passes into the oil vapors and into the gas.

JAMES R. WEST
Texasgulf Inc.

P.A. Ferguson, *Hydrogen Sulfide Removal from Gases, Air, and Liquids*, Noyes Data Corporation, Park Ridge, N.J., 1975.

The Ralph M. Parsons Co., *Sulphur Recovery Plants and Tail Gas Purification Units*, Bulletin No. 210, New York, Dec. 1976.

J.B. Pfeiffer, ed., *Sulfur Removal and Recovery from Industrial Processes*, American Chemical Society, Washington, D.C., 1975.

A.V. Slack and G.A. Hollinden, *Sulfur Dioxide Removal from Waste Gases*, 2nd ed., Noyes Data Corporation, Park Ridge, N.J., 1975.

SUPERCONDUCTING MATERIALS

Superconductivity, discovered by Onnes in 1911, is a phenomenon occurring in many metals and alloys at very low temperatures, near the boiling points of liquid helium and hydrogen. The term superconductivity was derived from its primary characteristics, namely, the complete absence of electrical resistance below the superconducting transition temperature T_c.

Superconductivity was first applied on a small scale in 1964 when quantum interference effects were observed between two Josephson junctions located in a superconducting ring. The device potential of this phenomenon was quickly realized and soon a very sensitive magnetometer known as a superconducting quantum interference device (SQUID) was fabricated. Josephson junctions have since been used in mixers, detectors, and as circuit elements for digital computers. The material problems are somewhat different in this case. For example, fabrication procedures for computer-circuit elements must be compatible with integrated-circuit technology. Common to both large- and small-scale applications is the necessity of finding materials with higher superconducting transition temperatures and good mechanical properties (see Integrated circuits).

Materials

At present, the largest commercial applications of superconductors rely upon the property of zero electrical resistance in the superconducting state. Practical materials for large-scale applications must also possess good mechanical properties. The commercially available superconducting materials can be roughly divided into ductile solid-solution alloys such as those of niobium and titanium, and intermetallic compounds, such as Nb_3Sn or V_3Ga, which are relatively brittle. At this time, NbTi, containing 45–50 wt% niobium, is the most widely used ductile material for superconducting magnet windings.

Consideration of the configuration in which these materials are used clarified the processing procedure. Between the upper and lower critical fields, type-II superconductors contain flux-line vortices. The passage of a current through the superconductor results in a Lorentz force on the flux lines and may lead to their motion, thereby making the conductor resistive. To prevent this development and improve the current-carrying capability of the wire, the flux lines are pinned by microstructural defects, eg, dislocations and grain boundaries. In spite of the pinning,

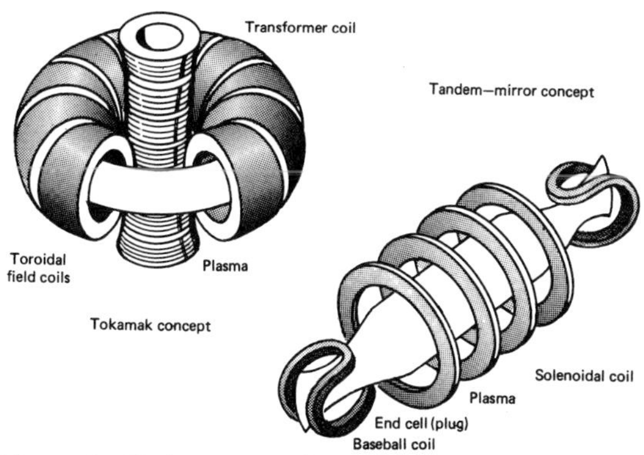

Figure 1. Basic concepts of magnetic confinement systems.

disturbances such as a small heat pulse caused by friction or a sudden change in the field can result in the motion of the vortices known as flux jumps, which leads to resistance. In the earliest superconducting wires, these flux jumps would cause the magnet to revert to normal, often in a catastrophic fashion. To stabilize the magnet against damage caused by flux jumps, the conductor must be modified. The methods of stabilization depend upon the close proximity of the superconductor to some other material of high conductivity, eg, copper or aluminum. Methods include cryostatic stabilization, adiabatic stabilization, and dynamic stabilization.

Materials for tunnel junctions. Since the discovery of superconducting tunneling and the Josephson effect, various electronic applications of superconductivity have been developed. Measuring devices employing the SQUID have been available for some time, and the tunnel junctions as detectors of infrared radiation have been successfully employed in radioastronomy. However, the potentially largest electronics application of Josephson junctions depends upon their ultrafast, low power-switching characteristics in digital circuits. An extensive effort, at IBM and other laboratories, has been made to develop materials for Josephson devices compatible with the thin-film integrated-circuit technology of modern computers (qv). Until recently, IBM emphasized the development of Pb alloys for thin-film devices. Persistent problems with mechanical stability of the Pb-base materials, however, have led to renewed efforts to prepare junctions with refractory Nb alloys (see Magnetic materials, thin film).

New materials. New materials include Chevrel-phase superconductors and amorphous superconductors.

Applications. Applications include magnets for high energy physics, rotating machinery (synchronous generators, homopolar d-c machines), fusion magnets (Fig. 1) magnetohydrodynamic generators, and magnets for nmr imaging.

FRANK P. MISSELL
Universidade de São Paulo

BRIAN B. SCHWARTZ
Brooklyn College, CUNY

S. Foner and B.B. Schwartz, eds., *Superconducting Machines and Devices: Large Systems Applications*, Plenum Press, New York, 1974.

S. Foner and B.B. Schwartz, eds., *Superconductor Materials Science: Metallurgy, Fabrication, and Applications*, Plenum Press, New York, 1981.

B.B. Schwartz and S. Foner, eds., *Superconductor Applications: SQUIDS and Machines*, Plenum Press, New York, 1977.

SUNSCREEN AGENTS. See Cosmetics.

SUPERACIDS. See Antimony compounds.

SUPERALLOYS. See High-temperature alloys.

SUPERCRITICAL FLUIDS

Supercritical-fluid extraction (*SFE*) utilizes the advantages of both distillation and liquid extraction. Slight changes in the temperature and pressure in the critical region cause extremely large changes in the solvent density, and thus its dissolving power. Heavy nonvolatile substances dissolve in supercritical fluids typically 2–7 orders of magnitude in excess of the amount based on the ideal gas law.

The two types of SFE consist of an extraction stage and a separation stage. In the pressure-controlled type, the compressed solvent dissolves the solute in an extraction vessel. The solution is expanded in the separation stage to precipitate the extract, and finally, the solvent is recompressed for recycle. The temperature-controlled type involves the same kind of extraction stage, except that the extract is precipitated by heating the solution to lower the solvent density. The density is increased for recycle by isobaric cooling. This type is highly energy efficient because the heat is transferred directly between the heating and cooling stages, and the nearly isobaric conditions minimize compression energy.

The density of a fluid is extremely sensitive to pressure and temperature near the critical point ($P_r = 1$, $T_r = 1$). In addition to the density, the viscosity and diffusivity for typical supercritical fluids are intermediate between those of a liquid and a gas as shown in Table 1. Supercritical solvents are superior to liquids for penetrating the micropores of a solid structure such as coal.

The advantages of SFE compared to distillation (qv) and liquid extraction are as follows: SFE offers many options for achieving and controlling the desired selectivity, which is extremely sensitive to variations in pressure, temperature, and choice of solvent. The extract is virtually free of residual solvent. Supercritical fluids can be used to vaporize thermally labile nonvolatile substances at moderate temperatures. Nontoxic, nonhazardous supercritical carbon dioxide can be used in the food and pharmaceutical industries without contaminating the product. The lower viscosity and higher diffusivity provide advantages in transport rates. The extract may be fractionated into numerous components, even if they have similar volatilities. Additional components can be used for further manipulation of the phase behavior. The ability of the supercritical fluid to vaporize nonvolatile compounds at moderate temperatures reduces the energy requirements compared to distillation (see Extraction).

Supercritical-Fluid Phase Behavior

The classifications of phase boundaries for binary systems are usually based on pressure–temperature (P–T) projections of mixture critical curves and three-phase equilibria lines. The experimental data are obtained by the simple synthetic method in which the pressure and temperature of a homogeneous solution of known concentration are manipulated to precipitate a visually observed phase. Unfortunately, this method does not usually give the composition of this phase, and the complementary pressure–composition (P–x) diagrams are not obtained. Six classes of binary P–T diagrams are identified, each one located along a continuum, which traverses the degree of dissimilarity in the intermolecular forces.

Binary systems. Miscibility behavior is influenced primarily by the volatility, size, and polarity rather than the structure, aromaticity, or degree of saturation. The solubility of solid hydrocarbons in solvents

Table 1. Physical Properties of a Typical Gas, Supercritical Fluid, and Liquid

Property	Gas	Supercritical fluid	Liquid
density, g/cm³	10^{-3}	0.3	1
viscosity, mPa·s (= cP)	10^{-2}	0.1	1
diffusion coefficient, cm²/s	0.1	10^{-3}	5×10^{-6}

such as ethylene, ethane, and carbon dioxide has been measured primarily using dynamic experimental techniques to develop a pressure–temperature–composition (P–T–x) data base for modeling SFE. Compared to conventional gas chromatography, SFC offers lower separation temperatures for thermally labile substances and the alternative of pressure programming to achieve fractionation instead of temperature programming.

Ternary systems. A near-supercritical solvent can be used to modify the mutual miscibility of a binary mixture.

Ternary systems near the critical point of the supercritical solvent may display regions of three-phase behavior, which are analogous to the binary case. For the system ethylene–methyl ethyl ketone–water, small loadings of ethylene below its critical pressure greatly reduces the mutual miscibility, yielding a selectivity of 705.

The combination of a supercritical as well as a liquid solvent provides flexibility in utilizing the advantages of both types of solvents. In a system of two solids in equilibrium with a supercritical fluid, the solubility of a given solid may be significantly enhanced when the second solid also dissolves in the fluid. These benefits resulting from supercritical solvents with or without entrainers should lead to a wide variety of new separation processes.

Theory of Supercritical-Fluid Phase Behavior

Quantitative correlation and, to a lesser degree, prediction of supercritical-solution phase behavior are beginning to emerge, based on recent advancements made in high pressure solution thermodynamics, using computers. The development of thermodynamic models presents two problems: these mixtures are often highly asymmetric, in terms of the size and energy differences between the components reflected, eg, in the critical properties; and the isothermal compressibility of the fluid is ca 1000 times that of a triple-point liquid. Dissolution of a solid in the highly compressible fluid leads to strong attraction that causes the mixture to contract with very large negative values of the partial molar volume of the solute.

Thermodynamic models for SFE are based on the principle that the fugacities, f_i of component i are equal for all phases at equilibrium at constant temperature and pressure. In the supercritical-fluid phase, the fugacity can be represented by an expression for an expanded liquid or a dense gas.

Most recent investigators have chosen the dense-gas approach utilizing two or three parameter cubic equations of state that are suitable for compressible mixtures. These models are used to calculate the fugacity coefficient of the solute, which is typically on the order of 10^{-4}, and is the primary contributor to the enhancement factor.

Environmental Considerations

The safety features of carbon dioxide make it an attractive solvent for the food and pharmaceutical industries. Generally, SFE does not leave residues in the product.

As the ASME Code is limited to 20.7 MPa (3000 psi), most of the designs for SFE are restricted below this pressure, if possible, to avoid further regulations. At these pressures, the stored energy is much larger for highly compressible supercritical fluids compared with incompressible liquids.

Applications

Carbon dioxide can be used to extract thermally labile food components at near-ambient temperatures. A process for the removal of caffeine from coffee beans using supercritical carbon dioxide was patented in the United States in 1974. One process uses water to wash the caffeine from this mixture, and another uses activated carbon to adsorb the caffeine. The third and most simple method involves charging a vessel containing the beans and activated carbon to supercritical conditions to cause the caffeine to migrate to the fluid phase where it is subsequently adsorbed by the carbon. Carbon dioxide is used to extract α-acids from hops. The USDA has been testing supercritical carbon dioxide for extraction of oils, particularly triglycerides, from soybean flake and corn germ. Supercritical fluids are used for the extraction of flavors and fragrances in natural products (see Coffee; Fats and fatty oils; Flavors and spices; Perfumes).

The energy costs for distillation or liquid solvent extraction have increased tenfold during the last decade; therefore, the less-energy-intensive SFE can be attractive if the capital costs for high pressure equipment are not prohibitive. Several feedstocks, including atmospheric and vacuum-distillation residues, were converted to cat-cracker feed and lubricating-oil stocks using supercritical pentane. Supercritical-fluid extraction can be used to fractionate low-vapor-pressure oils.

Supercritical fluids are utilized in coal processing in the critical solvent deashing process. In the coal-SFE process, a mixture of bicyclic aromatic/naphthenic hydrocarbons is used to dissolve coal and to stabilize the intermediate products from the thermal decomposition.

Supercritical-fluid extraction has been proposed to derive alternative sources of fuel from tar sands (qv), lignite (qv), wood (qv), and oil shale (qv).

Other applications involve the separation of chemicals from aqueous solutions, separations of specialty chemicals, environmental control, crystal growth, solids settling, and polymer processing.

Although a number of companies are known to be developing supercritical fractionation and processing methods for polymers, the specific applications are proprietary.

KEITH JOHNSTON
University of Texas, Austin

L.G. Randall, *Sep. Sci. Technol.* **17**, 1 (1982).

M.E. Paulaitis, V.J. Krukonis, R.T. Kurnik, and R.C. Reid, *Rev. Chem. Eng.* **1**, 179 (Jan.–Apr. 1983).

M.E. Paulaitis, J.M.L. Penninger, P. Davidson, and R.D. Gray, eds., *Chemical Engineering at Supercritical-Fluid Conditions*, Ann Arbor Science Publishers, Ann Arbor, Michigan, 1983.

SUPEROXIDES. See Peroxides and peroxy compounds.

SUPERPHOSPHATE. See Fertilizers.

SURFACTANTS AND DETERSIVE SYSTEMS

The term surfactant is a contraction of the longer, more awkward term surface-active agent. Coined in 1950, surfactant has become universally accepted to describe organic substances with certain characteristic features in structure and properties. The term detergent is often used interchangeably with surfactant. As a designation for a substance capable of cleaning, detergent can also encompass inorganic substances when these do in fact perform a cleaning chore. More often, detergent refers to a combination of surfactants with other substances, organic or inorganic, formulated to enhance functional performance, specifically cleaning, over that of the surfactant alone. It is so used in this article.

Surfactants are characterized by the following features:

Amphipathic structure. Surfactant molecules are composed of groups of opposing solubility tendencies, typically an oil-soluble hydrocarbon chain and a water-soluble ionic or polar group.

Solubility. A surfactant is soluble in at least one phase of a liquid system.

Adsorption at interfaces. At equilibrium, the concentration of a surfactant solute at a phase interface is greater than its concentration in the bulk of the solution.

Orientation at interfaces. Surfactant molecules and ions form oriented monolayers at phase interfaces.

Micelle formation. Surfactants form aggregates of molecules or ions called micelles when the concentration of the surfactant solute in the bulk of the solution exceeds a limiting value, the so-called critical micelle concentration (CMC), which is a fundamental characteristic of each solute–solvent system.

Functional properties. Surfactant solutions exhibit combinations of cleaning (detergency), foaming, wetting, emulsifying, solubilizing, and dispersing properties.

Surfactants classification depends on the charge of the surface-active moiety, usually the larger part of the molecule. In anionic surfactants, this moiety carries a negative charge. In cationic surfactants, the charge is positive. In nonionic surfactants there is no charge on the molecule. Finally, in amphoteric surfactants, solubilization is provided by the presence of positive and negative charges in the molecule.

In general, the hydrophobic group consists of a hydrocarbon chain containing ca 10–20 carbon atoms. The chain may be interrupted by oxygen atoms, a benzene ring, double bonds, and amide, ester, and other functional groups. A propylene oxide hydrophobe can be considered as a hydrocarbon chain in which every third methylene group is replaced by an oxygen atom. In some cases, the chain may carry substituents, most often halogens. Siloxane chains have also served as the hydrophobe in some surfactants developed in recent years.

Hydrophilic, solubilizing groups for anionic surfactants include carboxylates, sulfonates, sulfates, and phosphates. Cationics are solubilized by amine and ammonium groups. Ethylene oxide chains and hydroxyl groups are the solubilizing groups in nonionic surfactants. Amphoteric surfactants are solubilized by combinations of anionic and cationic solubilizing groups.

The molecular weight of surfactants ranges from a low of ca 200 to a high in the thousands for polymeric structures.

In the application of surfactants, physical and use properties, precisely specified rather than chemical homogeneity are of primary concern.

Anionic Surfactants

Carboxylate, sulfonate, sulfate, and phosphate are the polar, solubilizing groups found in anionic surfactants. In dilute solutions of soft water, these groups are combined with a C_{12}-chain hydrophobe for best surfactant properties. In neutral or acidic media, or in the presence of heavy metal salts, the carboxylate group loses most of its solubilizing power.

Of the cations (counterions) associated with polar groups, sodium and potassium impart water solubility, whereas calcium, barium, and magnesium promote oil solubility. Ammonium and substituted ammonium ions provide both water and oil solubility. Triethanolammonium is a commercially important example. Salts of these ions are often used in emulsification.

Higher ionic strength of the medium depressed surfactant solubility. To compensate for the loss of solubility, shorter hydrophobes are used for application in high ionic-strength media.

Carboxylates. Soaps represent most of the commercial carboxylates. The general structure of soap is $RCOO^-M^+$, where R is a straight hydrocarbon chain in the C_9–C_{21} range and M^+ is a metal or ammonium ion. Interruption of the chain by amino or amido linkages leads to other structures, which account for the small volumes of the remaining commercial carboxylates.

Large volumes of soap are used in industrial applications as gelling agents for kerosene and paint driers and as surfactants in emulsion polymerization (see Soap; Driers and metallic soaps; Emulsions). Soap usage in consumer products is principally in person washing products, ie, soap bars.

Carboxylates include polyalkoxycarboxylates and *N*-acylsarcosinates.

Acylated protein hydrolysates. These surfactants are prepared by acylation of protein hydrolysates with fatty acids or acid chlorides. The hydrolysates are variable in composition, depending on the degree of hydrolysis. Collagen from leather processing is a common protein source. Acylated protein hydrolysates are mild surfactants recommended for personal-care products (see also Cosmetics).

Sulfonates. The sulfonate group, $-SO_3M$, attached to an alkyl, aryl, or alkylaryl hydrophobe, is a highly effective solubilizing group. Sulfonic acids are strong acids and their salts are relatively unaffected by pH. They are stable to oxidation and, because of the strength of the C—S bond, also to hydrolysis. They interact moderately with the hardness ions Ca^{2+} and Mg^{2+}, significantly less so than carboxylates. Modification of the hydrophobe in sulfonate surfactants, by introduction of double bonds or ester or amide groups into the hydrocarbon chain or as

substituents, gives surfactants that offer specific performance advantages.

Because the introduction of the SO_3H function is inherently inexpensive, eg, by oleum, SO_3, SO_2, Cl_2, or $NaHSO_3$, sulfonates are heavily represented among the high-volume surfactants (see also Sulfonation and sulfation).

Sulfonates include alkylbenzenesulfonates; alkylarenesulfonates, short-chain lignosulfates; naphthalenesulfonates; α-olefinsulfonates; petroleum sulfonates; and sulfonates with ester, amide, or ether linkages.

Sulfates and sulfated products. The sulfate group, $-OSO_3M$, which can be viewed as the sulfuric acid half-ester of an alcohol, is more hydrophilic than the sulfonate group because of the presence of an additional oxygen atom. Attachment of the sulfate group to a carbon atom of the hydrophobe through the C—O—S linkage limits hydrolytic stability, particularly under acidic conditions.

Usage of sulfated alcohols and sulfated alcohol ethoxylates has expanded dramatically during the last decade as the detergent industry reformulated consumer products to improve biodegradability and lower phosphate content.

Other sulfates include alkylphenols, (ethoxylated and sulfated); sulfated acid, amides and esters and sulfated natural oils and fats.

Phosphate esters. The mono- and diesters of orthophosphoric acid

$$\begin{array}{ccc} HO & & RO \\ RO-P{=}O & \text{and} & HO-P{=}O \\ HO & & RO \end{array}$$

and their salts are useful surfactants.

In contrast to sulfonates and sulfates, the resistance of alkyl phosphate esters to acids and hard water is poor. Calcium and magnesium salts are insoluble. In the acidic form, the esters show limited water solubility; the alkali metal salts are more soluble. The surface activity of phosphate esters is good, although in general it is somewhat lower than that of the corresponding phosphorus-free precursors. Thus, a phosphated nonylphenol ethoxylated with 9 mol of ethylene oxide is less effective as a detergent in hard water than its nonionic precursor. At higher temperatures, however, the phosphate surfactant is significantly more effective.

Because of their high cost and the limitations noted above, phosphate surfactants find application as specialty surfactants in situations where such limitations are of no concern. As specialty surfactants, phosphate esters and their salts are remarkably versatile. Applications include emulsion polymerization of vinyl acetate and acrylates; drycleaning compositions where solubility in hydrocarbon solvents is a particular advantage; textile mill processing where stability and emulsifying power for oil and wax under highly alkaline conditions is called for; and industrial cleaning compositions where tolerance for high concentrations of electrolyte and alkalinity is required. In addition, phosphate surfactants are used as corrosion inhibitors, in pesticide formulations, in papermaking, and as wetting and dispersing agents in drilling-mud fluids.

Nonionic Surfactants

Unlike anionic or cationic surfactants, nonionic surfactants carry no discrete charge when dissolved in aqueous media. Hydrophilicity in nonionic surfactants is provided by hydrogen bonding with water molecules. Oxygen atoms and hydroxyl groups readily form strong hydrogen bonds, whereas ester and amide groups form hydrogen bonds less readily. Hydrogen bonding provides solubilization in neutral and alkaline media. In a strongly acidic environment, oxygen atoms are protonated, providing a quasi-cationic character. Each oxygen atom makes a small contribution to water solubility. More than a single oxygen atom is therefore needed to solubilize a nonionic surfactant in water. Nonionic surfactants are compatible with ionic and amphoteric surfactants. Since a polyoxyethylene group can easily be introduced by the reaction of ethylene oxide with any organic molecule containing an active hydrogen atom, a wide variety of structures can be solubilized by ethoxylation.

Polyoxyethylene surfactants (ethoxylates). Ethoxylates are moderate foamers and do not respond to conventional foam boosters. Foaming shows a maximum as a function of ethylene oxide content. Low foaming nonionic surfactants are prepared by terminating the polyoxyethylene chain with less soluble groups such as polyoxypropylene groups and methyl groups. Ethoxylates can be prepared to attain almost any hydrophilic–hydrophobic balance. For incorporation into powdered products, they suffer from the disadvantage of being liquids or low melting waxes, which complicates the manufacture of free-flowing, crisp powders. Solid products are manufactured with ethoxylates of high ethylene oxide content. The latter, however, are too water-soluble to provide optimum surface activity.

Ethoxylation. Base-catalyzed ethoxylation of aliphatic alcohols, alkylphenols, and fatty acids proceeds in two stages: formation of a monoethoxy adduct and addition of ethylene oxide to the monoadduct to form the polyoxyethylene chain. Polyoxyethylene surfactants include alcohol ethoxylates and alkylphenol ethoxylates.

Carboxylic acid esters. In the carboxylic acid ester series of surfactants, the hydrophobe, a naturally occurring fatty acid, is solubilized with the hydroxyl groups of polyols or the ether (and terminal hydroxyl) groups of ethylene oxide chains. Carboxylic acid esters include glycerol esters; polyoxyethylene esters; anhydrosorbitol esters; ethoxylated anhydrosorbitol esters; natural fats, oils and waxes; and ethoxylated and glycol esters of fatty acids.

Carboxylic amides. Carboxylic amide nonionic surfactants are condensation products of fatty acids and hydroxyalkyl amines. They include diethanolamine condensates, monoalkanolamine condensates, and polyoxyethylene fatty acid amides.

Polyalkylene oxide block copolymers. The higher alkylene oxides derived from propylene, butylene, styrene, and cyclohexene react with active hydrogens in a manner analogous to the reaction of ethylene oxide. Because the hydrophilic oxygen constitutes a smaller proportion of these molecules, the net effect is that the oxides, unlike ethylene oxide, are hydrophobic. The higher oxides are not used commercially as surfactant raw materials except for minor quantities that are employed as chain terminators in polyoxyethylene surfactants to lower their foaming tendency. The hydrophobic nature of propylene oxide units, $-CH(CH_3)CH_2O-$, has been utilized in several ways in the manufacture of surfactants.

Block copolymer nonionic surfactants are not strongly surface active but exhibit commercially useful surfactant properties. Aqueous solutions characteristically foam less than those of other surfactant types. They act as detergents, wetting and rinsing agents, demulsifiers and emulsifiers, dispersants, and solubilizers. They are used in automatic dishwashing-detergent compositions, cosmetic preparations, spin-finishing compositions for textile processing, metal-cleaning formulations, papermaking, and other technologies.

Cationic Surfactants

As the name implies, the solubilizing group of a cationic surfactant carries a positive charge when dissolved in an aqueous medium. The positive charge resides on an amino or quaternary nitrogen. A single amino nitrogen is sufficiently hydrophilic to solubilize a detergent-range hydrophobe in dilute acidic solution; eg, laurylamine is soluble in dilute hydrochloric acid. For increased water solubility, additional primary, secondary, or tertiary amino groups can be introduced or the amino nitrogen can be quaternized with low molecular weight alkyl groups such as methyl or hydroxyethyl. Quaternary nitrogen compounds are strong bases that form essentially neutral salts with hydrochloric and sulfuric acids. Most quaternary nitrogen surfactants are soluble even in alkaline aqueous solutions. Polyoxyethylated cationic surfactants behave like nonionic surfactants in alkaline solutions and like cationic surfactants in acid solutions.

Cationic surfactants are widely used in acidic aqueous and nonaqueous systems as textile softeners, dispersants, emulsifiers, wetting agents, sanitizers, dye-fixing agents, foam stabilizers, and corrosion inhibitors. To some extent, the usage pattern mirrors that of the anionic surfactants in neutral and alkaline solutions. Positively charged cationic surfactants are adsorbed more strongly than nonionic surfactants on a variety of substrates including textiles, metal, glass, plastics, minerals, and animal and human tissue, all of which often carry a negative surface

charge. Substantivity of cationic surfactants is the key property in many applications in which they are uniquely effective. In general, they are incompatible with anionic surfactants. Reaction of the two large, oppositely charged ions gives a salt that is insoluble in water. Ethoxylation moderates the tendency to form insoluble products with anionic surfactants.

Many benzenoid quaternary cationic surfactants possess germicidal, fungicidal, or algicidal activity. Solutions of such compounds, alone or in combination with nonionic surfactants, are used as detergent-sanitizers in hospital maintenance. Since biocidal products are classified as economic poisons, their labeling is regulated by the EPA.

Amines. *Oxygen-free.* Aliphatic mono-, di-, and polyamines derived from fatty and rosin acids make up this class of surfactants. Primary, secondary, and tertiary monoamines with C_{18} alkyl or alkenyl chains constitute the bulk of this class. The products are sold as acetates, naphthenates, or oleates. Principal uses are as ore-flotation agents, corrosion inhibitors, dispersing agents, wetting agents for asphalt, and as intermediates for the production of more highly substituted derivatives.

Oxygen-containing amines, except amides. This group includes amine oxides, ethoxylated alkylamines, 1-(2-hydroxyethyl)-2-imidazolines, and alkoxylates of ethylenediamine. It is steadily increasing in economic importance.

2-Alkyl-1-(2-hydroxyethyl)-2-imidazolines. 2-Alkyl-1-(2-hydroxyethyl)-2-imidazolines are used in hydrocarbon and aqueous systems as antistatic agents, corrosion inhibitors, detergents, emulsifiers, softeners, and viscosity builders. They are prepared by heating the salt of a carboxylic acid with (2-hydroxyethyl)ethylenediamine at 150–160°C to form a substituted amide; 1 mol water is eliminated to form the substituted imidazoline with further heating at 180–200°C.

Substituted imidazolines yield three series of cationic surfactants: by ethoxylation to form more hydrophilic products; by quaternization with benzyl chloride, dimethyl sulfate, and other alkyl halides; and by oxidation with hydrogen peroxide to amine oxides. Amide amines are dark-colored liquids that find application in corrosion inhibitors, petroleum demulsifiers, and metal-processing formulations.

Quaternary ammonium salts. The quaternary ammonium ion is a much stronger hydrophile than primary, secondary, or tertiary amino groups, strong enough to carry into solution a hydrophobe in the surfactant molecular weight range, even in alkaline media. The discrete positive charge on the quaternary ammonium ion promotes strong adsorption on negatively charged substrates, such as fabrics, and is the basis for the wide-spread use of these surfactants in domestic fabric-softening compositions (see Quaternary ammonium compounds).

Amphoteric Surfactants

Amphoteric surfactants contain both an acidic and a basic hydrophilic group. These ionic functions may be any of the anionic or cationic groups described in the preceding sections. In addition, ether or hydroxyl groups may also be present to enhance the hydrophilicity of the surfactant molecule.

Amphoteric surfactants are generally considered specialty surfactants. They do not irritate skin or eyes; they exhibit good-surfactants properties over a wide pH range; and they are compatible with anionic and cationic surfactants.

Imidazolinium derivatives. Amphoteric imidazolinium derivatives are prepared from 2-alkyl-1-(2-hydroxyethyl)-2-imidazolines and sodium chloracetate. Imidazolinium derivatives are recommended as detergents, emulsifiers, wetting and hair-conditioning agents, foaming agents, fabric softeners, and antistatic agents. There is some evidence that in cosmetic formulations, certain imidazolinium derivatives reduce eye irritation caused by sulfate and sulfonate surfactants present in these products.

Uses

Household and personal products. Detergency is the primary function of household and personal products. In recent years, a secondary function, such as softening in combination with detergency in laundry detergents or conditioning in combination with detergency in shampoos, has been offered as an additional product benefit. In general, products have tended toward functional specialization.

Industrial uses. Surfactants are widely used outside the household for a variety of cleaning and other purposes. Often the volume or cost of the surfactant consumed in industrial processes is small compared to the benefit. Although usage of surfactants in industrial applications is smaller than in the large household products market, expansion of enhanced-oil-recovery activities could consume sufficient surfactant to tip the balance in favor of industrial uses.

Detergency and Detersive Systems

The cleaning of a solid object, ie, the removal of unwanted foreign matter from its surface, is accomplished by a variety of methods ranging from simple mechanical separation such as blotting or abrasion to removal by solution or selective chemical action. The term detergency is restricted to cleaning systems in which the following conditions prevail: A liquid bath is present and is the main cleansing constituent of the system. The action of the bath involves more than simple solution or hydraulic dislodging of soil although both of these processes inevitably occur and make some contribution to cleaning. The enhanced cleaning effect is caused primarily by the presence of a special solute, the surfactant, which acts by altering interfacial effects at the various phase boundaries within the system (see also Drycleaning and laundering).

In the cleaning or washing process in a typical detersive system the soiled substrate is immersed in or brought into contact with a large excess of the bath liquor. Sufficient bath is used to provide a thick layer over the whole surface of the substrate. During this stage, air is displaced from soil and substrate surfaces, ie, they are wetted by the bath. The system is subjected to mechanical agitation, either rubbing or shaking, which provides the necessary shearing action to separate the soil from substrate and disperse it in the bath. Agitation also aids the mass-transfer processes in the system, just as in any other heterogeneous chemical reaction. The fouled bath carrying the removed soil is drained, wiped, squeezed, or otherwise removed from the substrate. The substrate is rinsed free of the remaining fouled bath. This rinsing step determines the final cleanliness of the substrate. The cleaned substrate is dried or otherwise brought to the desired finished state.

A discussion of detergency requires a definition of clean. In a practical sense, a surface is clean if it has been brought to desired prespecified state with regard to foreign matter present, as judged by some set of agreed-upon tests or standards.

Components of detersive systems. *Substrates.* Solid objects to be cleaned by detersive processes, vary widely in chemical composition and surface configuration.

Soils. Soils vary greatly in composition. They may consist of a single solid or liquid phase but usually two or more phases are present, intimately and randomly mixed and irregularly dispersed over the substrate surface.

Baths. The baths discussed here are aqueous solutions.

Formulation. Detergents are formulated to satisfy the requirements of the soiled substrate to be cleaned and the expected range of washing conditions.

Surfactants. The most important components of detersive systems are, of course, the surfactants described earlier.

Builders. Builders are substances that augment the detersive effects of surfactants. Most important is their ability to remove hardness ions from the wash liquor and thus to prevent them from interacting with the surfactant.

In general, builders supply alkalinity to the wash liquor and thus function also as alkalies. In addition, they can exert a suspending effect and are therefore effective in keeping detached soil suspended in the wash liquor, particularly builder ions with multiple charges. Builders include phosphates, sodium carbonate, silicates, zeolites, clays, nitrilotracetic acid, alkalies, and neutral soluble salts.

Organic additives. Certain nonsurfactant organic additives improve cleaning performance and exhibit other desirable properties. Such additives are usually present in low concentrations and serve one or more of the following specific functions: reduced redeposition of soil from the

detergent bath onto the substrate; increased whiteness or appearance of cleanliness; enhanced cleaning effect on specific types of soils and stains; promotion or inhibition of foaming power and foam stability; increased solubility or other modification of the physical form of the detergent composition; sequestering of heavy-metal ions, both in the concentrated detergent and in the diluted cleansing bath; reduced injurious effects which the detergent may have on the substrate or the washing machine, such as tarnishing of silverware or etching of glassware, corrosion of metals, or irritation of skin in toilet and cosmetic applications.

Factors influencing detergency. Detergency is mainly affected by the concentration and structure of surfactant, hardness and builders present, and the nature of the soil and substrate. Other important factors include wash temperature; length of time of the washing process; mechanical action; relative amounts of soil, substrate, and bath, generally expressed as the bath ratio, ie, the ratio of the bath weight to substrate weight; and rinse conditions.

Mechanisms. It is evident from the preceding discussion that even the simplest detersive system is surprisingly complex and heterogeneous. It can nevertheless be conceptually resolved into simpler systems that are amenable to theoretical treatment and understanding. These simpler systems are represented by models for substrate–solid soil and substrate–liquid soil. In practice, many soil systems include solid–liquid mixtures. However, removal of these systems can generally be analyzed in terms of the two simpler model systems. Although these two systems differ markedly in behavior and structure and require separate treatment, there are certain underlying principles that apply to both.

The first principle is that soil systems can be regarded and treated as classical systems of colloid and surface chemistry. A second principle applying to these model systems is derived from their colloidal nature. With the usual thermodynamic parameters fixed, the systems come to a steady state in which they are either agglomerated or dispersed. No dynamic equilibrium exists between dispersed and agglomerated states. In the solid-soil systems, the particles (provided they are monodisperse, ie, all of the same size and shape) either adhere to the substrate or separate from it. In the liquid-soil systems, the soil assumes a definite contact angle with the substrate, which may be anywhere from 0° (complete coverage of the substrate) to 180° (complete detachment). The governing thermodynamic parameters include pressure, temperature, concentration of dissolved components, and electrical conditions.

A final consideration in resolving practical detersive systems into their simpler components relates to soil removal versus redeposition. Superficially, it would appear that the redeposition phenomenon contradicts the all-or-nothing concept that the system must exist in either the agglomerated or the dispersed state. Keeping in mind both the composite nature and the kinetics of a practical system, it is readily shown that no such contradiction exists.

Solid-soil detergency. *Adsorption.* Adsorption of bath components is a necessary and possibly the most important and fundamental detergency effect. Adsorption is the mechanism whereby the interfacial free energy values between the bath and the solid components (solid soil and substrate) of the system are lowered, thereby increasing the tendency of the bath to separate the solid components from one another. Furthermore, the solid components acquire electrical charges that tend to keep them separated or acquire a layer of strongly solvated radicals that have the same effect.

Mass transfer near the substrate surface. Mechanical action has a great effect on soil removal probably by influencing mass transfer, ie, the diffusion of soluble material away from the immersed fibers. In a Terg-O-Tometer investigation of mass transfer, using only water-soluble substances, the transfer coefficient was found to be directly proportional to agitator speed and stroke angle, inversely proportional to the water-holding capacity of the cloth load, and independent of bath volume.

Colloidal stabilization. Surfactant adsorption reduces soil–substrate interaction and facilitates soil removal. For a better understanding of these interactions, a consideration of collidal forces is required.

There are two general theories of the stability of lyophobic colloids, or, more precisely, two general mechanisms controlling the dispersion and flocculation of these colloids. Both theories regard adsorption of dissolved species as a key process in stabilization. However, one theory is based on a consideration of ionic forces near the interface, whereas the other is based on steric forces. The two theories complement each other and are in no sense contradictory. In some systems, one mechanism may be predominant and in others both mechanisms may operate simultaneously. The fundamental kinetic considerations common to both theories are based on Smoluchowski's classical theory of the coagulation of colloids (qv).

Oily-soil detergency; roll-up. The principal means by which oily soil is removed is probably by roll-up. The applicable theory is simply the theory of wetting.

Solubilization. The role of micellar solubilization (as the term is used in the physical chemistry of surfactants) in oily-soil removal has been debated for many years. The amount of oily soil that could be present in a normal wash load could not all be removed and held in micellar solution by anionic surfactants. On the other hand, nonionic surfactants could do so, because of their greater solubilizing ability. High solubilizing power is definitely linked with good detergency.

Phase changes at the soil–bath interface. Closely related to solubilization is a phenomenon that involves polar organic soils and surfactant solutions. If a complete phase diagram is plotted for a ternary system containing sodium dodecyl sulfate (or another representative water-soluble surfactant), a polar organic compound (eg, glyceryl oleate), and water, several important and unusual features are noted: (*1*) A large area represents a liquid phase consisting of a microemulsion, where the dispersed particles are so small that the system is isotropic, like the familiar soluble oils. (*2*) Over another large area, a liquid-crystalline phase is formed, containing all three components. This liquid-crystalline phase flows like a liquid, at least in one direction. Flow perpendicular to the oriented planes is accomplished by folding the planes cylindrically, but the physical flow is still of the purely viscous type, with no yield point evident. These two phases, particularly the liquid-crystal phase, play an important part in detergency. Furthermore, liquid-crystal formation lowers interfacial tension. Although this phenomenon was demonstrated in tertiary oil recovery, the principles could also apply to oily-soil detergency (see Petroleum, enhanced oil recovery).

Measurement of detergency. The measurement of detergency in the laboratory requires the following components: a means for measuring or estimating the amount of soil of the substrate or the degree of cleanness before and after washing; satisfactory substrates and soiling compositions; a means for applying soil to substrate in a realistic manner; and a realistic and reproducible cleaning device. These fundamental requirements apply regardless of the particular type of substrate that is being cleaned.

Fabric detergency. Reflectance is commonly used to measure the whiteness (cleanness) of fabrics, most often on a Gardner colorimeter or a Zeiss Elrepho reflectometer. Special filters are available to prevent masking of results by the effects of fluorescent whitening agents. Color effects, such as fabric yellowing, can also be measured.

No single artificial soil can fully represent the variety of natural soils encountered in fabric washing. Nonetheless, useful though limited correlations with field testing can be obtained with artificial soil cloths, particularly those soiled with a mixture of particulate and oily (sebum) soils.

For practical laboratory detergency measurements, the Terg-O-Tometer (U.S. Testing Co.), a small-scale washing machine, is most commonly employed.

Performance of various detergent formulations can be compared on the basis of reflectance R_f of washed soil cloths. However, it is generally more useful to express cleaning as percentage detergency, $\% D$:

$$\% D = \frac{R_f - R_i}{R_o - R_i} \times 100$$

where R_i is the initial reflectance of soil cloth prior to washing and R_o is the reflectance of unsoiled cloth.

Redeposition can be assessed simultaneously with detergency by measuring the loss of reflectance of a clean cloth added to the wash load.

Table 1. Rat Oral LD$_{50}$ Values of Surfactant

Type of compound	Oral LD$_{50}$ (rat)
alkylbenzenesulfonates	700–2,480 mg/kg
alcohol ethoxylates	1,600 to greater than 25,000 mg/kg
sulfated alcohol ethoxylates	7,000 to greater than 50,000 mg/kg
alcohol sulfates	5,000–15,000 mg/kg

Manufacture. *Liquid products.* The manufacture of liquid detergent products is a straightforward process requiring simple equipment with provisions for metered addition of individual ingredients, agitation, and if needed, heating.

Spray-dried powders. The manufacture of powdered products is more complicated. In the last three decades, high-pressure spray-drying of an aqueous slurry has replaced the earlier process in which a solidified cake of the product was broken up mechanically.

Dry-blended powders. In addition to lower capital outlay, dry blending requires considerably less processing energy. Dry blending of powders provides little control over product density. Moreover, particle size and product densities are less constant than in spray-dried powders (see Mixing and blending).

Agglomerated powders. The process of agglomeration is intermediate between spray-drying and dry-blending. Process-water concentrations are between 35–45% in a crutcher slurry and essentially zero in dry blending. In agglomeration, a spray of water or other liquid is aimed at dry powder under agitation.

Health and safety factors; environmental considerations. Under conditions of normal use, detergent products are not hazardous to users. Nonetheless, surfactants possess some toxicity and are mild irritants. Particularly under conditions of misuse, such as accidental ingestion or spillage, they can produce irritation and discomfort in the form of nausea and vomiting as well as irritation to skin and eyes. The long-term effects, however, are minimal. Ranges of surfactant LD$_{50}$ values are shown in Table 1.

Environmental considerations. The introduction of surfactant products into the environment, after use by consumers or as part of waste disposed during manufacture, is regulated by the Clean Water Act, the Clean Air Act, and the Resource Conservation and Recovery Act. In this respect, surfactants are subject to the same regulations as chemicals in general. There are, however, two areas of specific relevance to surfactants and detergent products, ie, biodegradability and eutrophication.

ARNO CAHN
Consultant

JESSE L. LYNN, JR.
Lever Brothers Company

A.W. Adamson, *Physical Chemistry of Surfaces*, 4th ed., John Wiley & Sons, Inc., New York, 1982.

McCutcheon's Detergents & Emulsifiers, 1981 North American Edition, MC Publishing Company, Glen Rock, N.J., 1981.

M.J. Rosen, *Surfactants and Interfacial Phenomena*, Wiley-Interscience, New York, 1978.

SUTURES

Surgical sutures are sterile filaments used to hold tissues together until they heal adequately for self-support or to join tissues with implanted prosthetic devices. They are normally attached to needles for stitching the edges of wounds or surgical incisions. As ligatures they are used, generally without a needle, to tie off ends of severed tubular structures such as blood vessels and ducts to prevent bleeding or fluid leakage.

Sutures are characterized according to type of material, physical form, biodegradability, size, surgical use, etc. Among the materials utilized for

Figure 1. Scanning electron photomicrograph of a braided, a monofilament, and a catgut suture (from top to bottom).

sutures are natural products, such as surgical gut or silk (qv), many commonly known synthetic fibers, and some fibers made from polymers that have been synthesized specifically for suture applications (see Fibers, chemical). Sutures may be fabricated as monofilaments or multifilaments; the latter are generally braided, but sometimes twisted or spun, and may be coated with waxes, fluorocarbons, silicones, and other polymers to decrease capillarity and improve handling or functional properties.

A typical braided suture, a monofilament, and catgut are shown in Figure 1.

An absorbable suture is one that degrades in body tissues and disappears from the implant site, usually within two to six months. A nonabsorbable suture is resistant to biodegradation and becomes encapsulated in a fibrous sheath or is otherwise incorporated by the host, and remains in the tissue as a foreign body unless it is surgically removed (eg, skin sutures) or extruded from the site by the host.

Suture-Tissue Interactions

As with all surgical implants, some tissue response to sutures is inevitable. The initial tissue reaction of ca 5-d duration is essentially the same for all sutures. Beyond this point, the tissues respond to the foreign body by absorbing or encapsulating it in an envelope of fibrous connective tissue; the difference in the degree, intensity, and type of cellular reaction depends on the nature of the material. Catgut and its related collagens evoke the strongest reaction, followed by silk, cotton, and other natural fibers. The synthetic fibers currently used result in relatively bland reactions as do metallic wires.

Mechanical Properties

Tensile strength and knot security are perhaps the two most important mechanical properties of sutures. Since sutures are usually knotted in use, surgeons are concerned with knot-pull rather than straight-pull tensile strength. This property is generally measured with a tensiometer by pulling on the free ends of a surgeon's knot or by pulling apart a loop tied with sufficient squared throws to prevent knot slippage. As might be expected, metallic sutures are the strongest; natural fibers including linen, silk, catgut, collagen, and cotton are the weakest; and synthetic materials are of intermediate strength.

Sterilization

Suture materials must be sterile when used, and it has become standard practice for manufacturers to sterilize their products before marketing, essentially eliminating a procedure that had mostly been done in the hospital (see also Sterilization techniques).

Needles

Surgical needles used for stitching may be considered as integral parts of sutures. They are generally made from heat-treatable (400 series) stainless steel or from carbon steel, which may be nickel-plated to retard corrosion.

Colorants

Although some surgical procedures require sutures that are not colored, dyed or pigmented sutures are used more frequently to enhance their visibility while surgery is in progress. Only dyes or pigments are utilized for which Color Additive Petitions have been approved by the Division of Color Technology, Bureau of Foods of the FDA (see Colorants for foods, drugs, and cosmetics).

JAMES B. MCPHERSON
American Cyanamid Company

D.E. Clarke, *Contemporary Surgery*, **17**, 33 (1980).

N.A. Swanson and T.A. Tromovitch, *Intern. J. Dermatology*, **21**, 373 (1982).

G.V. Yu and R. Cavaliere, *J. Am. Podiatry Assoc.*, **73**, 57 (1983).

SWEETENERS

Saccharin

The taste of saccharin (**1**), 3-oxo-2,3-dihydro-1,2-benzisothazole-1,1-dioxide (*o*-sulfobenzimide or *o*-benzosulfimide), was discovered in 1879 during an investigation of the oxidation of *o*-toluenesulfonamide. The compound is acidic and not very soluble in water. Therefore, it is normally employed as the sodium or calcium salt.

saccharin
(**1**)

The potency of saccharin relative to sucrose, as with other synthetic sweeteners, is inversely related to sucrose concentration, and ranges anywhere from 200 to 700 times that of sucrose. As normally used, saccharin has a potency of ca 300. Although sweet, saccharin has a bitter, metallic aftertaste which is very difficult to mask.

Saccharin is manufactured by one of two processes. The first is based on the original discovery, whereas the second is designed without *o*-toluenesulfonamide, a suspected carcinogen.

In recent years, saccharin has come increasingly under fire as a result of a number of toxicological studies.

Cyclamic Acid

The taste of cyclamate was discovered in 1937 at the University of Illinois. Cyclamic acid (**2**), cyclohexanesulfamic acid, has a sweet-sour flavor and a very acidic pH of ca 1. It is readily soluble in water at ca 10% and is normally used in the form of its neutral sodium salt, called sodium cyclamate or simply cyclamate. The calcium salt is used in low sodium diets. Cyclamic acid and its sodium salt were patented by DuPont in 1940. Abbott Laboratories introduced cyclamate in 1950 and the product enjoyed a steady increase in popularity and sales until it was banned in the United States in 1970.

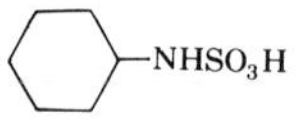

cyclamic acid
(2)

Cyclamate is about 30 times as sweet as sucrose, but has an off-flavor or unpleasant aftertaste which can be detected if the concentration is high enough.

Aspartame

In 1965, the taste of aspartame (3), APM, L-aspartyl-L-phenylalanine methyl ester, was discovered accidentally. The compound was an intermediate in the synthesis of the C-terminal tetrapeptide of gastrin; the latter was used as biological standard in connection with an antiulcer project. Aspartame is a white crystalline compound which may be obtained anhydrous or, more usually, as a half hydrate. When heated, it cyclizes to the corresponding diketopiperazine (4).

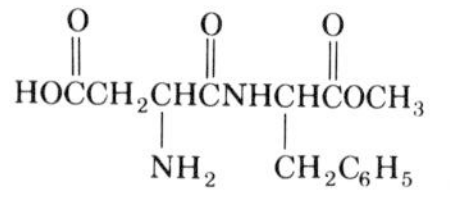

aspartame (APM)
(3)

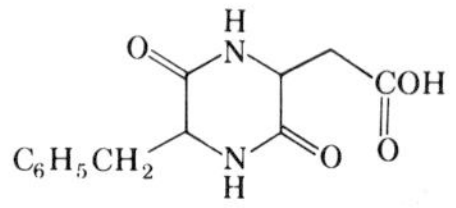

3,6-dioxo-5-benzyl-2-piperazineacetic acid
(4)

Formation of the diketopiperazine (4) also takes place when APM is dissolved in aqueous solution; the rate is a function of temperature and pH.

The taste of APM could not have been predicted from its constituent amino acids (qv). L-Aspartic acid has a flat taste, whereas L-phenylalanine is bitter. When the two are linked in the proper way and the phenylalanine carboxyl is converted to a methyl ester, the resulting product has a totally unexpected taste. Aspartame is very much like sucrose, and some tasters in paired comparison studies have preferred it to sucrose.

Aspartame is a suitable sweeteer for a wide variety of foods because of its sucrose-like taste and its ability to blend well with other food flavors. Because of interactions with other flavors, the replacement of sucrose by aspartame requires reformulation and not simple substitution of ingredients.

Aspartame is metabolized into its constituent parts, which are metabolized by the usual body pathways.

Other Sweeteners

β-Neohesperidine dihydrochalcone. Conversion of the bitter flavanone, naringin, to the dihydrochalcone, β-neohesperidine dihydrochalcone, at the Western Regional Research Laboratories of the USDA resulted in a sweet-tasting product.

The sweet taste of β-neohesperidine dihydrochalcone is lingering, with a cooling or menthol-like aftereffect. These properties would sharply restrict any possible commercial application.

Acesulfam-K. A new class of sweeteners was reported in 1973 from the laboratories of Farbwerke Hoechst A.-G. The structure is reminiscent of saccharin and, in fact, the saccharin homologue containing the oxathiazinone dioxide ring is sweet but also bitter. Acesulfam, 3,4-dihydro-6-methyl-1,2,3-oxathiazine-4-one-2,2-dioxide, is slightly soluble in water but, as a strong acid, has a simultaneous sweet and sour taste which impedes its usefulness. The neutral salts, like Acesulfam-K (5), are stable and allow suitable taste comparisons.

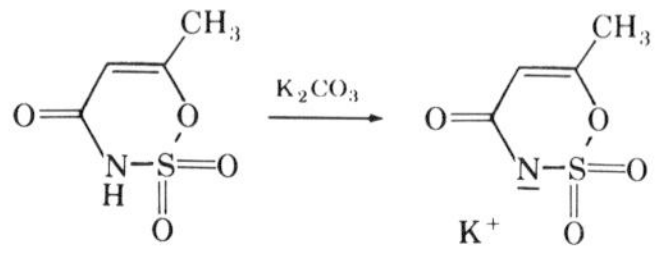

acesulfam   acesulfam-K
(5)

ROBERT MAZUR
GD Searle and Co.

I. Remsen and C. Fahlberg, *Am. Chem. J.* **1**, 426 (1879); C. Fahlberg and I. Remsen, *Ber.* **12**, 469 (1879).

U.S. Pat. 319,082 (June 2, 1885), C. Fahlberg (one-half to A. List, Leipzig, Germany), *Chem. Eng. News* **41**, 76 (1963).

A.I. Bakal, *Chem. Ind.*, 700 (1983).

SYNTHETIC AND IMITATION DAIRY PRODUCTS

Definitions

Imitation dairy products are referred to as imitations, simulates, substitutes, analogues, and mimics and are associated with terms such as filled, nondairy, vegetable nondairy, and artificial milk, cheese, etc. There are no universally accepted definitions of imitation dairy foods. The products can be divided into three types: those in which an animal or vegetable fat has been substituted for milk fat; those that contain a milk component, eg, casein; and those that contain no milk components (see Milk and milk products). The first two types make up most of the substitute dairy products.

Ingredients

The characteristics and stability of imitation dairy products depend largely on the characteristics of the main ingredients, ie, fat, protein, and carbohydrates, and of the minor functional ingredients that stabilize the fat and protein systems. The ingredients generally include fats or oils and the stabilizing emulsifiers, proteins, and their stabilizing gums, salts, and carbohydrates (see Fats and fatty oils; Vegetable oils; Gums).

Fats and oils. A comparison of the characteristics of milk fat and three different fats that are used in acceptable filled and imitation milks are listed in Table 1 and those that comprise whipped topping, in Table 2.

Emulsifiers. The physical form of the food relates to the selection of the emulsifier. For example, coffee whiteners are marketed as liquids, frozen products, or dry powders. The optimum emulsifier differs for each system. For liquid products, the emulsifier must impart good stability to the emulsion so that it remains in a uniform state on standing after preparation and prior to sale. It must also prevent oiling off (emulsion breaking) and feathering (formation of unstable white flocs on top that resemble feathers) when used in coffee (qv). For the frozen product, the emulsifier must also contribute to the freeze–thaw stability of the emulsion. In dry products, the emulsifier must contribute to the free-flowing properties and to the dispersal of the product in coffee.

Proteins. Proteins contribute to a number of functions in an imitation food, which include emulsification, gelation, melting, and whipping. The desired functionality of proteins in different types of imitation dairy food are shown below:

Imitation	Desired functionality of protein
cheese	emulsification, melting
coffee cream	emulsification, whitening, buffering
cultured product	emulsification, gelation
ice cream	emulsification, whipping
milk-based pudding	emulsification, gelation
whipped topping	emulsification, whipping

Table 1. Typical Fat Characteristics for Imitation and Filled Milks

Characteristics[a]	Milk fat	Fat products A	B	C
SFI, %				
at 10°C	34	25	15	15
at 21°C	13	13	11	9
at 27°C	8	9	10	8
at 33°C	0.5	4	6	2.5
iodine value	33	80	95	105
polyunsaturated fatty acid, wt% (cis–cis)	1.6	21	24	36
saturated fatty acids, wt%	92.1	21	20	18

[a]SFI = solid-fat index.

Table 2.   Typical Fat Characteristics for Imitation Dairy Products

| Characteristic | Ice cream and whipped topping | | | | Coffee whiteners | |
	A	B	C	D	Fluid	Dry
SFI, %						
at 0°C	38	30	24	25	25	61
at 21°C	23	20	13	18	14	50
at 27°C	20	19	10		9	46
at 33°C	12	14	6	4	4	28
at 41°C	4	8	2			
iodine value	68	82	83	85	80	67

a Ref. 2.

Table 3.   Gums and Associated Functions in Imitation Dairy Products

Gum	Imitation product use	Function
alginates	ice cream	aeration, reduce whip time
	cheese	texture modifier, prevent oil separation
carrageenan	milk puddings	gelation
	ice cream	prevent wheying off
	infant formula	stabilize proteins to heat
	evaporated milk	stabilize proteins to heat
locust bean gum	ice cream	smooth melt down, freeze–thaw resistance
	cheese spread	texture modifier
guar gum	ice cream	water binding, body control
carboxymethyl cellulose	ice cream	prevent ice crystal growth
	whipped topping	aeration, protective colloid
xanthan gum	sour cream	prevent wheying off

Table 4.   Recommended Chemical Composition for Milk Substitute

Property	Fluid, sterilized	Dry
moisture, wt% max	89.0	3.5
fat, wt% min	2.0	18.0
nonfat solids, wt% min	9.0	
protein (N × 6.25), wt% min	3.5	28.0
essential fatty-acid content, expressed as linoleic acid as wt% of fat, min	2.0	2.0
carbohydrates (sucrose or malt sugar–dextrin), wt% min	4.0	51.5
calcium, mg/100 mL or /100 g, min	100	800
vitamin A, µg/100 mL or /100 g, min	50	450
vitamin D, IU/100 mL or /100 g, min	40	360
iron, mg/100 mL or /100 g, min	0.4	4.0
thiamine, mg/100 mL or /100 g, min	0.15	1.0
riboflavin, mg/100 mL or /100 g, min	0.15	1.3
niacin, mg/100 mL or /100 g, min	1.0	8.0
folic acid, µg/100 mL or /100 g, min	5	50.0
pantothenic acid, µg/100 mL or /100 g, min	200	1600
vitamin B_{12}, µg/100 mL or /100 g, min	0.5	5.0
pyridoxine, µg/100 mL or /100 g, min	200	1600
ascorbic acid, mg/100 mL or /100 g, min	3.0	25.0
total ash, wt% max		7.0
acid-insoluble ash, wt% max		0.05

Casein and caseinates are the most widely used proteins in imitation dairy products, primarily because of their superior functionality and flavor as compared to most vegetable proteins.

Soybean protein isolates are the only vegetable proteins that are widely used in imitation dairy products.

Gum stabilizers. The principal gums and their functions in imitation dairy products are listed in Table 3.

Stabilizing salts. Citrates and phosphates are used in imitation dairy products for one or more of the following purposes: alter buffering capacity of the system, improve the stability of the protein to calcium ions, improve the heat stability of the protein, minimize the age gelatin of ultrahigh temperature (UHT)-processed imitation products, serve as emulsifying salts in imitation-cheese manufacture, modify the water-binding capacity of proteins, and improve solubility.

Composition

The composition of imitation dairy products is highly variable and generally represents the least-cost formulation consistent with consumer acceptance of the product. In the absence of Standards of Identity the imitation products invariably have lower fat and protein levels than the dairy products that they are made to resemble.

Processing

The processes used for imitation dairy products are essentially the same as for their dairy product counterparts. The processes common to all fluid products include: ingredient blending, pumping, pasteurization or sterilization, homogenization, cooling, and packaging.

Margarine. Margarine or oleomargarine was originally marketed as an imitation butter; however, it now has a recognized identity of its own. The proportion of the fat blend and other ingredients varies with the type of margarine and with the country of manufacture.

Cheese. Imitation cheese began with the substitution of milk components in the development of filled and nondairy cheese and the development of a synthetic cheese based on the Chinese food (tofu), which is based on soybean curd.

Filled and caseinate-based cheese. Filled cheese was made traditionally by adding vegetable fat to skim milk, homogenizing at low pressure to make a uniform emulsion, followed by traditional cheesemaking procedures. At present most imitation cheese is based on calcium caseinate

Coffee creams/whiteners. Coffee-cream substitutes have been given the generic name of coffee whiteners and are available in liquid, frozen,

and dry forms. The emulsifiers and proteins are key stabilizing factors with respect to the stability of the product in coffee.

Ice cream and frozen desserts. Imitation frozen desserts include ice cream, sherbets, and specialty products, eg, milkshake bases. Filled ice cream, called mellorine in the United States, is the dominant product in the world market and is sold both as a soft-serve and as a hardened product. The composition and processing is as for ice cream, except for the substitution of vegetable fat for milkfat.

Milk. Imitation milks fall into three broad categories: filled products based on skim milk, buttermilk, whey, or combinations of these; synthetic milks based on soybean products; and toned milk based on the combination of soy or groundnut (peanut) protein with animal milk (see Table 4).

Soy-based beverages. Soybean-based beverages may be classified as follows: (1) traditional soy milks, unfermented, starting with uncomminuted full-fat beans, full-fat or defatted soy flours, or soy protein concentrates; (2) traditional fermented yogurtlike products; or (3) simulated milks based on soy-protein concentrates and isolates, ie, of the dairy type; fluid single-strength, concentrated, or sterile nonfat milk replacers as dry blends of soy protein isolates containing dry whey; infant-feeding beverages simulating human milk; and fermented yogurt-like products.

Infant formulas. Infant formulas are either milk-based or vegetable-protein based.

Whipping creams and whipped toppings. Whipped topping is the generic term used for imitation whipping cream. Products are available as liquids to be whipped, frozen prewhipped products, liquids in aerosol cans that form whips upon release, and dry powdered products that are reconstituted and whipped. Toppings were first made from nonfat dry milk or sodium caseinate as the protein source and vegetable fat as the source of fat.

Miscellaneous. Miscellaneous products with current or potential market significance include sour cream, chip dips, milkshake bases, puddings, and yogurt. These products have been formulated from caseinates and soybean proteins.

Regulatory Aspects

The legal status of imitation dairy products varies widely among countries. The FDA has held that filled products should be nutritionally equivalent to the products they resemble. A section of the Food, Drug, and Cosmetics Act defines misbranded foods and the FDA has set up standards of identity for food under this part of the law.

W. James Harper
Ohio State University

A.H.M. van Gennip, *International Dairy Federation Document 107*, International Dairy Federation, Brussels, Belgium, 1978.

F. Winklemann, *Imitation Milk and Milk Products*, Food and Agricultural Organization of the United Nations, 1974.

M.T. Gillies, *Shortenings, Margarines, and Food Oils*, Noyes Data Corp., Park Ridge, N.J., 1974.

SYNTHETIC LUBRICANTS.   See Lubricants and lubrication.

SYRUPS

Dextrose and corn syrups are hydrolysis products of starch (qv) and are generally referred to as corn sweeteners. Although distinct commercial products, they have in common the raw material source, general methods of preparation, and many properties and applications. Dextrose is the common or commercial name for D-glucose, the pure crystalline solid recovered from almost completely hydrolyzed starch. Corn syrups are clear, colorless, viscous liquids prepared by hydrolysis of starch to solutions of dextrose, maltose, and higher molecular weight saccharides. Alternatively, the enzymatic isomerization of dextrose hydrolysate yields high fructose corn syrup (HFCS). Maple syrup, like corn syrup, is a nutritive sweetener produced as a concentrated solution of carbohydrate (sucrose) and is used in food applications. Molasses is a syrup produced as a by-product of sugar manufacture (see Sweeteners; Sugar, special sugars).

Dextrose

Dextrose (D-glucose, corn sugar, starch sugar, blood sugar, grape sugar) is by far the most abundant sugar in nature and occurs either in the free state (monosaccharide form) or chemically linked with other sugar moieties.

Properties.   Physical properties of the three crystalline forms of dextrose are listed in Table 1.

Dextrose shows the reactions of an aldehyde, a primary alcohol, a secondary alcohol, and a polyhydric alcohol (see Carbohydrates).

Metabolism.   Dextrose is the common intermediary metabolite in carbohydrate metabolism, since other utilizable monosaccharides are converted to dextrose before they are further metabolized. Starch, glycogen, and the common monosaccharides are hydrolyzed enzymatically in the alimentary canal. Dextrose is normally absorbed into the portal-vein blood, by which it is transported first to the liver and then circulated to all parts of the body.

Manufacture.   Dextrose is manufactured from a variety of starches. Corn starch is used almost exclusively in the United States, whereas starches from wheat, potato, sweet potato, and tapioca, as well as corn are used in other countries (see also Wheat and other cereal grains).

Table 2.   U.S. Distribution by Industry of Dextrose, HFCS, and Corn Syrup in 1980, 1000 metric tons

	Baking	Beverage	Canning	Confectionery	Dairy	Total[a]
Dextrose[b]	51	66	4	55	2	513
HFCS[c]	365	1039	235	15	140	2659
Corn syrup[d]	196	384	126	446	241	2201

[a] Includes applications not listed.
[b] Monohydrate basis.
[c] 71 wt% solids basis.
[d] Sp gr 1.42 (43°Bé, 80.3 wt% solids).

Health factors.   Dextrose as well as corn syrups are substances that are presumed to be GRAS by the FDA.

Uses.   The main use of dextrose is in the food-processing industry, where it is of value for its physical, chemical, and nutritive properties. Distribution of dextrose to various industries in the United States is shown in Table 2.

Dextrose is also used in the pharmaceutical industry for intravenous feeding as well as for tableting and other formulations. In fermentation, dextrose is a raw material for biochemical synthesis and, chemically, it is a raw material for sorbitol, mannitol, and methyl glucoside production. Yeast fermentation of dextrose to ethanol is of current importance in the production of gasohol. The production of other bulk chemicals from dextrose by chemical or biochemical means appears to have promise in future applications. Reaction of dextrose with sorbitol and citric acid produces a low calorie dextrose polymer, which is used as a bulking agent in reduced-calorie foods.

High Fructose Corn Syrups

High fructose corn syrups (HFCS, isosyrup, isoglucose) are concentrated solutions containing primarily fructose and dextrose with lesser quantities of higher molecular weight saccharides. HFCS is produced by partial enzymatic isomerization of dextrose hydrolysates followed by refining and concentration. Commercial products contain 42, 55, or 90% fructose on a dry weight basis.

In nature, fructose (levulose, fruit sugar) is the main sugar in many fruits and vegetables.

Properties.   Fructose, a ketohexose monosaccharide, crystallizes as β-D-fructopyranose; it has a molecular weight of 180 and a melting point of 102–104°C.

Because of the presence of fructose, HFCS is sweeter than conventional corn syrups, although intensity of sweetness results from many factors, eg, temperature, pH, and concentration.

Manufacture.   HFCS containing 42 wt% fructose is produced commercially by column isomerization of clarified and refined dextrose hydrolysate with immobilized glucose isomerase. Enriched syrup, containing 90 wt% fructose, is prepared by chromatographic separation and blended with 42 wt% HFCS to obtain 55% HFCS.

Uses.   High fructose corn syrup is used as a partial or complete replacement for sucrose or invert sugar in food applications to provide sweetness, flavor enhancement, fermentables, or humectant properties.

Corn Syrups

Corn syrups are defined on the basis of reducing sugar content as one of four types having DE (dextrose equivalent) of 20–99.4. Lower DE products are classified as maltodextrins and higher DE products as dextrose. Principal uses of corn syrups are shown in Table 2. Health and safety aspects are the same as for dextrose. Corn syrup is produced by continuous or batch hydrolysis of starch.

Maple Syrup

Maple syrup is prepared by evaporating sap from the maple tree to a concentrated solution containing predominantly sucrose. Its characteristic flavor and color are formed during evaporation. Syrup is clarified, graded as to color, flavor, and density, and finally packaged in small containers for retail sale as table syrup. Typically the product contains

Table 1.   Physical Properties of D-Glucose

Property	α-D-Glucose	α-D-Glucose hydrate	β-D-Glucose
molecular formula	$C_6H_{12}O_6$	$C_6H_{12}O_6 \cdot H_2O$	$C_6H_{12}O_6$
mp, °C	146	83	150
solubility (at 25°C), g/100 g soln	62 → 30.2 → 51.2[a]	30.2 → 51.2[a,b]	72 → 51.2[a]
$[\alpha]_D^{20}$	112.2 → 52.7[a]	112.2 → 52.7[a,b]	18.7 → 52.7[a]
heat of soln (at 25°C), J/g[c]	−59.4	−105.4	−25.9

[a] Equilibrium value.
[b] Anhydrous basis.
[c] To convert J to cal, divide by 4.184.

88–99 wt% sucrose and 0–12 wt% invert sugar. Maple sugar is also used in candy manufacture by blending with sucrose. Other applications include addition to cookies, cake, ice cream, baked beans, baked ham, and baked apples.

Molasses

Molasses, another type of syrup, is a by-product of the sugar industry; it is the mother liquor remaining after crystallization and removal of sucrose from the juices of sugar cane or sugar beet and is used in a variety of food and nonfood applications (see Sugar).

Composition. Molasses composition depends on several factors, eg, locality, variety, soil, climate, and processing. Cane molasses is generally at pH 5.5–6.5 and contains 30–40 wt% sucrose and 15–20 wt% reducing sugars. Beet molasses is ca 7.5–8.6 pH, and contains ca 50–60 wt% sucrose, a trace of reducing sugars, and 0.5–2.0 wt% raffinose.

Uses. The primary use of molasses is in animal feed. Molasses provides a carbohydrate source, as well as salts, protein, vitamins, and palatability, and may be used directly or mixed with other feeds (see Pet and livestock feeds).

The second main use for molasses is in various fermentation processes as an inexpensive source of carbohydrate. Molasses is the basic raw material for rum production and is also used for production of yeast (qv) and citric acid (qv) (see Fermentation).

Food applications utilize first and second molasses in baking (bread, cakes, cookies) for the molasses flavor. Molasses is also used in curing of tobacco and meats in confections such as toffees and caramels, and in baked beans and glazes.

R.E. HEBEDA
CPC International

R.A. McGinnis, ed., *Beet Sugar Technology*, 3rd ed., Beet Sugar Development Foundation, Fort Collins, Colo., 1982.

Nutritional Sweeteners from Corn, 2nd ed., Corn Refiners Association, Inc., Washington, D.C., 1979.

C.O. Willits and C.H. Hills, eds., *Maple Syrup Producers Manual, Agricultural Handbook No. 134*, Agricultural Research Service, USDA, Washington, D.C., 1976.

T

TACK. See Rubber compounding.

TACONITE. See Iron.

TALC

The term talc covers a wide range of natural minerals, most of which are high magnesium silicates. It is a hydrous magnesium silicate, $Mg_3SiO_{10}(OH)_2$, theoretically 31.7% MgO, 63.5% SiO_2, and 4.8% H_2O. The mineral talc is usually, although not always, a main constituent of mineral mixtures offered commercially as talc.

Talc deposits were probably formed by hydrothermal alteration or contact metamorphism of preexisting rocks. The whiter, purer talcs derive chiefly from dolomite, dolomitic limestone, and magnesite. Those from ultrabasic igneous rocks usually contain intermediate minerals, eg, serpentine.

Generally speaking, the purest talcs have the greatest commercial value. Those used for cosmetics, pharmaceuticals, selective adsorption, and electrical ceramics approach theoretical purity and are also characterized by a white color.

Talc, although not as common as clay, is found in many parts of the world (see Clays). The United States is the world's largest producer.

Properties

Pure talc mineral is characterized by softness (Mohs scale 1), hydrophobic surface properties, and slippery feel. The crystal form may be foliated, lamellar, fibrous, or massive.

Crude talcs range in color from white to green and brown. The refractive index of talc is 1.54–1.59, and the specific gravity is 2.7–2.8. Pure talc is heat-stable up to 900°C. Talc is inert in most chemical reagents, although it exhibits a marked alkalinity (typically pH 9.0–9.5). It is, however, soluble in hot concentrated phosphoric acid.

Mining and Processing

Mining methods vary from straightforward open-pit operations to complicated, carefully timbered underground operations. Where color and chemical purity are important, screening, washing, and selection by hand or electron beam may be used.

Health and Safety Factors, Environmental Considerations

Talc is subject to regulation for respirable dust concentration in work places by OSHA and the Mine Safety and Health Administration. The threshold limit value (TLV) for pure talc is 2 mg talc per m^3 air.

The TLVs for commercial talc containing deleterious impurities such as free silica are lower. Those containing fibrous asbestos impurities, eg, chrysotile, anthophyllite, and tremolite, are regulated as asbestos with a TLV of two fibers of $> 5 \mu m$ in length per cm^3 air (see also Asbestos).

All open-pit mines require dust-suppression measures such as wet drilling and haul-road watering. Mine-equipment operators are customarily protected from dust by enclosed air-conditioned cabs.

Uses

The paper industry is the world's largest consumer of talc. The ceramics industry has been the largest consumer of talc-group minerals in the United States for more than three decades. The next most important application is in protective coatings. Other uses of talc are as insecticide carriers, rubber-dusting and textile-filling materials, and as an additive in asphalt roofing compounds. Polypropylene replacements for metal in domestic applications represents another important new market for talc (see Engineering plastics; Fillers).

H.T. MULRYAN
Cyprus Industrial Minerals

R.A. Clifton, *Mineral Facts and Problems*, Bulletin 671, U.S. Bureau of Mines, Washington, D.C., 1980, pp. 1–12.

The Economics of Talc and Pyrophyllite, 3rd ed., Roskill Information Services Limited, London, 1981.

J.A. Pask and M.F. Warner, *J. Am. Cer. Soc.* **37**(3), 118 (1954).

TALL OIL

Tall oil, a natural product of pine trees, is isolated by means of the Kraft pulping process. Current applications are increasingly based on specific chemical functionality of refined rosin and fatty-acid cuts produced by fractional distillation of crude tall oil.

Tall oil is composed of the ether-extractable nonlignin, noncellulosic portion of the pine tree. The two main chemical structures represented are rosin acids and fatty acids. The rosin acids are diterpene carboxylic acids based on an alkyl-substituted perhydrophenanthrene ring structure (see Terpenoids). These acids are nonfood-chain acids and are involved with wound-recovery mechanisms of the tree. The fatty acids (see Carboxylic acids, fatty acids from tall oil) are predominately C_{18} straight-chain mono- or diunsaturated fatty acids; oleic and linoleic acids predominate.

The tall-oil fatty acids exhibit the chemistry expected of fatty acids derived from any vegetable-oil source. These linear, long-chain, unsaturated acids are pale yellow liquids at ambient temperature. They are insoluble in water and soluble in most organic solvents. Their chemistry is centered in the aliphatic carboxyl group and their ethylenic unsaturation.

Government Regulations

Environmental requirements defined by the Air Quality Act, Toxic Substances Control Act, and the Resources Conservation Act have been met by the tall oil industry. Although a cost factor, these regulations do not create technological problems (see Air pollution control methods).

Uses of Fatty Acids, Rosin, and Refinery By-Products

A principal use of tall-fatty acids is for the production of dimer acids; the ratio of unsaturated acids and the low saturated-acid content makes it an optimum raw material. Other uses of these fatty acids are varied with substantial amounts being sold to the mining, rubber, surface-coating, plasticizer, and oil-production industries.

The by-products produced during tall-oil refining are tall-oil pitch, tall-oil heads, and distilled tall oil. Tall-oil pitch is a dark thermoplastic material used in asphalt emulsions, low-performance mastics, and as a fuel (41.8 MJ/kJ or 18,000 Btu/lb). Tall-oil heads contain lower boiling fatty acids generally with ca 16 carbons. Heads can be used as inexpensive fatty acids where carboxyl functionality is required or as a raw-material source for the production of palmitic acid. Distilled tall oil has many uses where its liquid nature and its mixed rosin and fatty acid content result in favorable product characteristics.

HERBERT G. ARLT, JR.
Arizona Chemical Company

Tall Oil Fatty Acids, Composition and Characteristics, Arizona Chemical Co., Wayne, N.J., 1980.

L.G. Zachery, H.W. Bajak and F.J. Eveline, *Tall Oil and Its Uses*, F.W. Dodge Co., a division of McGraw-Hill, Inc., New York, 1965.

TANTALUM AND TANTALUM COMPOUNDS

Tantalum, Ta (at no. 73, at wt 180.948, sp gr 16.6) appears in Group VA of the periodic table directly below niobium, Nb, with which it is closely associated in nature and to which it is very similar in properties. Tantalum has only one natural isotope, ^{181}Ta, but several artificial radioactive isotopes have been made. The most common valence of the element is 5. Tantalum is best known as a refractory metal with a combination of unique properties that make it useful in a number of unrelated commercial applications. It is very ductile and can be worked cold into fine wire or thin foil. It is completely inert to strong acids below their boiling points and yet is very reactive to almost all substances at high temperatures.

Tantalum does not occur naturally in the free state. The most important tantalum-bearing minerals are tantalite and columbite, which are modifications of $(Fe, Mn)(Ta, Nb)_2O_6$. Other sources include slags from tin smelting and recycled scrap.

Physical and mechanical properties of tantalum are listed in Table 1.

Alloys

Tantalum has been alloyed with tungsten to improve its strength with little change in corrosion properties (see Table 2). Two high-tantalum alloys with exceptional creep strength have had limited use in the handling of hot liquid metals in nuclear power systems:

	Alloy	
Composition	T-111	T-222
tungsten, wt%	7–9	9.6–11.2
hafnium, wt%	1.8–2.4	2.2–2.8
carbon, ppm	< 40	80–170
tantalum	balance	balance

Table 1. Typical Physical and Mechanical Properties of Tantalum

Property	Value
lattice type	body-centered cubic
lattice constant at 20°C, nm	0.33026
density (at 20°C), g/cm^3	16.6
melting point, °C	2996
boiling point, °C	5425 ± 100
specific heat at 20°C, J/(kg·K)^a	150.7
linear coefficient of expansion, °C^{-1}	6.5×10^{-6}
thermal conductivity (at 20°C), W/(m·K)	54.4
electrical resistivity, μΩ·cm	
at 20°C	13.5
at 1500°C	71
temperature coefficient of electrical resistivity (at 0 – 100°C), °C^{-1}	0.00382
thermionic work function, eV	4.12
magnetic susceptibility (cgs)	0.93×10^{-6}
tensile strength, MPa^b	
at room temp	241–483
at 1000°C	90–117
Young's modulus, GPa^b	
at room temp	186
at 1000°C	152
Poisson's ratio	0.35

^{a} To convert J to cal, divide by 4.184.
^{b} To convert MPa to psi, multiply by 145; for GPa, multiply by 145,000.

Table 2. Properties of Tantalum-Tungsten Alloys

Alloy	Elongation, %	Tensile strength, MPa^a	Corrosion resistance in sulfuric acid, μm/yr	
			175°C	200°C
tantalum	42	262	5.08	48.3
tantalum–2.5 wt % tungsten	36	375	5.84	25.4
tantalum–5 wt % tungsten	32	455	5.33	27.9
tantalum–10 wt % tungsten	28	676	5.59	43.2

^{a} To convert MPa to psi, multiply by 145.

Tantalum is also used to increase the high temperature strength of nickel- and cobalt-based superalloys, which are primarily used in aircraft gas turbines.

Manufacture and Processing

The manufacture of tantalum metal is accomplished by extraction of tantalum from the ore or tin slag, separation of the extract from other metals present, formation of a pure tantalum compound (usually potassium fluorotantalate, K_2TaF_7, and reduction of the compound to metal powder, usually by means of sodium metal added to the molbeu salt. The metal powder can be used for electrolytic capacitor applications or it can be consolidated to solid tantalum metal by powder metallurgy (qv) or melting methods. These consolidated metal forms are generally further fabricated to mill products.

Toxicology

Tantalum is inert and does not appear to have detrimental effects on the human body, in which it is used in surgical implants (see Prosthetic and biomedical devices). Like other reactive metals, tantalum burns, particularly if in a finely divided state, and when exposed to heat, flame, or a chemical oxidizer. The recommended threshold limit value (TLV) for exposure in workroom air is 5 mg/m^3 of air. Tantalum powder and some tantalum compounds have been suspected of causing skin irritation and mild fibrosis of the lungs.

Uses

About 55% of the world's annual production of tantalum is used in capacitors, and 27% is used as the carbide, TaC, in cemented carbide cutting tools for steels and special alloys where it increases hot hardness and resistance to deformation (see Carbides; Refractory coatings). Its use as a mill product, primarily in corrosion applications, is ca 10%, and ca 8% is used as an alloying addition to high temperature superalloys. A smaller amount is used in other special applications.

Tantalum Compounds

The refractory nature of tantalum prevents the formation of many ordinary simple compounds of this element. Of those that have been prepared, only a very few are considered important, eg, aluminum tantalide, Al_3Ta (sp gr 7.02, mp ca 1400°C); tantalum boride, TaB_2 (sp gr 11.15, mp ca 3000°C); tantalum carbide, TaC (sp gr 13.9, mp 3880°C). Tantalum halides include tantalum pentachloride, $TaCl_5$ (sp gr 3.68, mp 216.0°C, bp 242°C); tantalum oxytrichloride, $TaOCl_3$; tantalum tribromide, $TaBr_3$; tantalum pentabromide, $TaBr_5$ (sp gr 4.67, mp 265°C, bp 348.8°C); tantalum pentaiodide, TaI_5 (mp 496°C, bp 543°C); tantalum pentafluoride, TaF_5 (sp gr 4.74, mp 96.8°C, bp 229.5°C); potassium fluorotantalate (potassium tantalum fluoride), K_2TaF_7 (sp gr 5.24, mp 740 ± 10°C). Tantalum hydride, TaH, has a specific gravity of 15.1, and tantalum nitride, TaN, has a specific gravity of 13.80 and a melting point of 2800 ± 50°C). Tantalum oxides and tantalic acid include tantalum (II) oxide, Ta_2O_4; tantalum pentoxide Ta_2O_5 (sp gr 8.2, mp 1800°C); tantalic acid $HTaO_3$ or $Ta_2O_5 \cdot nH_2O$. Another compound is tantalum sulfide, Ta_2S_4.

ROBERT E. DROEGKAMP
MORTIMER SCHUSSLER
JOHN B. LAMBERT
DONALD F. TAYLOR
Fansteel, Inc.

G.L. Miller, *Tantalum and Niobium*, Academic Press, Inc., New York, 1959.

F.T. Sisco and E. Epremian, eds., *Columbian and Tantalum*, John Wiley & Sons, Inc., New York, 1963.

L.D. Cunningham, "Cb and Ta in 1983," U.S. Bureau of Mines, Washington D.C.

TAR AND PITCH

Most organic substances, other than those of simple structure and low boiling point, when pyrolyzed, eg, heated in the absence of air, yield dark-colored, generally viscous liquids termed tar and pitch. The differentiation between these terms is not precise. When the by-product is a liquid of fairly low viscosity at ordinary temperature, it is regarded as a tar; if of very viscous, semisolid, or solid consistency, it is designated as a pitch.

Large amounts of tar or pitch by-products are produced by industrial processes. The distillation of crude petroleum yields a pitch-like residue termed bitumen or asphalt.

Wood Tar

Production. The pyrolysis or carbonization of hardwoods, eg, beech, birch, or ash, in the manufacture of charcoal yields, in addition to gaseous and lighter liquid products, a by-product tar in ca 10 wt% yield. Dry distillation of softwoods, eg, pine species, for the production of the so-called DD turpentine yields pine tar as a by-product in about the same amount.

Composition, processing and uses. There are no statistics available for the amount of wood tar processed, but almost all of it is burned. The commercial by-products from wood carbonization are limited to methanol, denatured methanol, methyl acetate, and acetic acid.

Small amounts of the sedimentation tar, ie, the separated organic layer from the condensed wood-carbonization vapors, are distilled, first at atmospheric pressure to give wood spirit, crude acetic acid, and light-wood oils.

Chemically, wood tar is a complex mixture that contains at least 200 individual compounds, among which the following have been isolated: guaiacol (2-methoxyphenol), 4-ethylguaiacol (2-methoxy-4-ethylphenol), creosol (4-methylguaiacol), 2,6-xylenol, butyric acid, crotonic acid, acetol (1-hydroxy-2-propanone), butyrolactone, maltol (3-hydroxy-2-methyl-$4H$-pyran-4-one), tiglaldehyde (2-methyl-2-propenal), methyl ethyl ketone, methyl isopropyl ketone, methyl furyl ketone, and 2-hydroxy-3-methyl-2-cyclopenten-1-one.

Coal Tar

By far the largest source of tar and pitch is the pyrolysis or carbonization of coal and, generally, the terms tar and pitch are synonymous with coal tar and the residue obtained by its distillation (see Coal, coal-conversion processes, carbonization). Today, coal tar is chiefly a source of anticorrosion coatings, wood preservatives, feedstocks for carbon-black manufacture, and binders for road surfacings and electrodes.

Physical properties. The physical properties of crude tars vary over a wide range. Investigation has been mainly concerned with establishing correlations between the more readily determined chemical and physical properties of the distillate oils and residual pitch, and other properties. Based on the correlations, other properties can be predicted with an accuracy sufficient for such purposes as plant design (see Table 1).

Viscosity of coal-tar pitch; change with temperature. Since pitch is mainly used as a hot-applied binder or adhesive, the viscosity and its change with temperature are important in industrial practice.

Among a number of equations for absolute viscosity (η) or kinematic viscosity, the absolute viscosity of a straight-run or fluxed-back pitch can be calculated from the Rt_s (ring-and-ball) softening point:

$$\log \eta_t = -4.175 + \frac{711.8}{86.1 - t_s + t} \tag{1}$$

Chemical composition. The tars recovered from commercial carbonization plants are not the primary products of the thermal decomposition of coal. The initial products undergo a complex series of secondary reactions.

Table 2 gives the average properties of various types of tar and Table 3 lists the amounts of components which have current or potential industrial possibilities.

Manufacture and processing. In the future, crude low temperature tar may be supplied as by-product of synthetic natural gas (SNG) and syncrude from coal substitute. A number of the more advanced SNG

Table 1. Correlations for Predicting the Physical Properties of Tars and Tar Products

Property	Applicable to	Correlation expression
density, d, at 20°C, g/cm^3	coke-oven dry tars and tar oils	$d_{20} = 1.877 \times 10^{-3}M^a + 0.808$ $d_{20} = 7.337 \times 10^{-4}t_b{}^b + 0.890$
viscosity, η, mPa·s ($= cP$)	tar oils	$\log \eta_{20} = 0.0078t_b - 1.123^c$
specific heat capacity, C, $kJ/(kg·K)^d$	tar oils	$C_t\{(0.7360 + 0.8951 \times d_{20} + 0.00360\,t)/d_{20}\} + (0.00904 - 0.0000221 \times t_b) \times$ T.A.^{e}
	pitches	$C_t = \dfrac{3.665}{d_{20}} - 1.729 + 0.00389 \times t$
thermal conductivity, K, $W/(m·K)$	tar oils	$K = (1.34 - 0.00084t \pm 0.084) \times 10^{-5}$
	pitches	$K = (1.423 \pm 0.084) \times 10^{-5\,f}$
surface tension, S, mN/m ($= dyn/cm$)	dry tars, tar oils, pitches	$S = 93.8 \times d_{20} - 0.0496 \times t_b - 47.5$ $S_t = 18.4 \times d_t^4$
		$S_t = \dfrac{S_{20}d_t}{d_{20}^4}$
latent heat of vaporization, L, kJ/kg^b	tar oils	at the average boiling point, $L = d_{20}$ $(486.1 - 0.599t_b)$

^{a}Av molecular weight.
^{b}Av boiling point defined as the mean of the temperatures in °C at which 10%, 20% ...80%, 90% by volume distills in a standard flask distillation.
^{c}At other temperatures, $\log \eta$ varies linearly with the absolute temperature; at t_b the viscosity of any tar oil is approximately 0.25 mPa·s ($= cP$).
^{d}To convert J to cal, divide by 4.184.
^{e}%Tar acids defined as the percentage by volume extracted by 10% aqueous caustic soda.
^{f}No significant variation over range 25–105°C.

Table 2. Properties of Coal Tars

	Coke-oven tars, av			CVR^b tars, UK av	Low temperature tars, UK av	Lurgi tars, UK av
	UK	FRG	U.S.			
yield, L/t	33.6	26.8		70.9	95.5	12.7
density at 20°C, g/cm^3	1.169	1.175	1.180	1.074	1.029	1.070
water, wt%	4.9	2.5	2.2	4.0	2.2	2.8
carbon, wt%	90.3	91.4	91.3	86.0	84.0	84.2
hydrogen, wt%	5.5	5.25	5.1	7.5	8.3	7.7
nitrogen, wt%	0.95	0.86	0.67	1.21	1.08	1.09
sulfur, wt%	0.84	0.75	1.2	0.90	0.74	1.39
ash, wt%	0.24	0.15	0.03	0.09	0.10	0.02
TI^a, wt%	6.7	5.5	9.1	3.1	1.2	0.7

^{a}Toluene-insoluble components.
^{b}Continuous vertical retort.

processes such as Lurgi, Bi-Gas, and Cogas employ low temperature pyrolysis of coal and yield a by-product tar.

Primary distillation. Today, 99% of the tar produced in the UK and FRG and 75% of U.S. production is distilled. Most of the crude tar regarded as being burned in the United States is first topped in simple continuous stills to recover a chemical oil, ie, a fraction distilling to 235°C that contains the bulk of naphthalene and phenols. Primary distillation of crude tars is carried out in continuous pipe stills of 100–700 t/d capacity.

Secondary processing of tar distillate oils. Light oils, refining of benzene and naphtha. The only processing that light oils might receive at the refinery would be a fractional distillation into crude benzene (formerly called benzol or benzole, now obsolete) distilling up to 150°C, a naphtha fraction distilling from 150 to 190°C, and a creosote residue (see BTX processing; Benzene).

Crude benzene from coke-oven gas. The crude benzene, together with that extracted from the carbonization gases, is refined to pure

Table 3. Constituents of Coal Tar

Component, wt% of dry tar	Coke-oven tars, av		
	UK	FRG	US
benzene	0.25	0.4	0.12
toluene	0.22	0.3	0.25
o-xylene	0.04		0.04
m-xylene	0.11	0.2	0.07
p-xylene	0.04		0.03
ethylbenzene	0.02		0.02
styrene	0.04		0.02
phenol	0.57	0.5	0.61
o-cresol	0.32	0.2	0.25
m-cresol	0.45	0.4	0.45
p-cresol	0.27	0.2	0.27
xylenols	0.48		0.36
high boiling tar acids	0.91		0.83
naphtha	1.18		0.97
naphthalene	8.94	10.0	8.80
α-methylnaphthalene	0.72	0.5	0.65
β-methylnaphthalene	1.32	1.5	1.23
acenaphthene	0.96	0.3	1.05
fluorene	0.88	2.0	0.64
diphenylene oxide	1.50	1.4	
anthracene	1.00	1.8	0.75
phenanthrene	6.30	5.7	2.66
carbazole	1.33	1.5	0.60
tar bases	1.77	0.73	2.08
medium-soft pitch	59.8	54.4	63.5

benzene, nitration-grade toluene, and mixed xylenes at the coke-oven benzole refinery, or, in the case of continuous vertical retort and low temperature light oils, to a product suitable as a solvent or gasoline additive.

Coumarone-indene resins. These should be called polyindene resins (see Hydrocarbon resins). They are derived from a close-cut fraction of a coke-oven naphtha free of tar acids and bases.

Pyridine bases. Formerly, pyridine bases were recovered from coal-tar light oils but in recent years synthetic pyridine and picolines have replaced the coal-tar products.

Carbolic oils and low temperature tar middle oil, tar acids. The fractions of CVR tars and, in some cases, coke-oven tars, distilling in the range of 180–240°C, and the middle oil fraction (180–310°C) from low temperature tars are treated for the recovery of tar acids, which may be further refined to give phenol, o-cresol, m/p-cresol mixtures, and higher boiling cresylic acids.

Naphthalene oils. Naphthalene is the principal component of coke-oven tars and the only component that can be concentrated to a reasonably high content on primary distillation. Naphthalene oils from coke-oven tars distilled in a modern pipe still generally contain 60–65% naphthalene. They are further upgraded by a number of methods.

The modern process adopted in the UK and some European plants is based on crystallization of the primary naphthalene oil, which is diluted with lower-crystallizing material to give a feedstock crystallizing point at 55°C. This material is cooled in closed, stirred tanks to 30–35°C and the resultant slurry of naphthalene crystals and mother liquor is centrifuged, washed, and spun dried. These operations are automatically timed and controlled.

Wash oils. No tar chemicals can be extracted commercially from tar oils distilling in the range of 250 to 300°C. Although the wash-oil fraction of coke-oven tars, distilling mainly in the 250–280°C range, is employed at coking installations to scrub benzene from coal gas, most oils in this boiling point range are used in creosote blends.

Anthracene oils. In the UK and Europe, but not in the United States, crude anthracene (40–45%) is isolated from coke-oven anthracene oils.

Health and safety factors. The volatile components of coal tar, ie, mononuclear aromatic hydrocarbons, phenols, and pyridine bases, are toxic when ingested, inhaled, or absorbed through the skin and the usual precautions must be taken when crude benzene or tar light oils are

handled. Most polynuclear aromatic compounds are primarily skin and eye irritants but are tolerated internally.

The carcinogenic hazard of tar and tar products. The main health hazard associated with coal tar products is cancer of the skin (usually the hands, face or scrotum) or of the oesophagus due to frequent long term contact with, or inhalation of, finely divided pitch. Inhalation of the fumes from heated tar or pitch does not involve any risk of lung cancer and the use of tar products in road surfacing, preservation of telegraph and transmission poles with creosote or the use of metal pipes internally coated with coal tar enamel for portable water supply constitute no danger whatever to the general public.

Uses. Coumarone-indene resins have outlets in paints, as tackifiers in rubber compounding, and as adhesives in the manufacturing of flooring tiles (see Hydrocarbon resins).

Cresylic acids. The higher boiling cresylic acids are mixtures of cresols or xylenols with higher boiling phenols (see Phenol: Alkylphenols). Their main uses are in phenol-formaldehyde resins, solvents for wire-coating enamels, as metal-degreasing agents, froth-flotation agents, and synthetic tanning agents (see Flotation; Leather).

Naphthalene. Today, only ca 21.3% of the phthalic anhydride made in the United States is derived from naphthalene: in Western Europe, ca 10%; and in Japan, ca 30%.

The main outlet for naphthalene is now the production of β-naphthol and dyestuff intermediates such as H-acid and J-acid.

Creosote. This is the term given to blends of tar oils to meet certain specifications. The current most important outlet is as a feedstock for carbon-black manufacture. Next in importance is the use of creosote oils for the preservation of timber (see Wood) and minor outlets are for horticultural winter wash oil and for the production of disinfectant emulsions.

Pitch. The principal current outlet for coal-tar pitch is as a binder for the electrodes used in aluminum smelting.

Other important uses are as a binder in the manufacture of coal briquettes and formcoke, for the manufacture of pitch-fiber pipes and as both a binder and impregnant in the production of refractories.

Fluxed pitches and refined tars; Road tars. In the United States, which has a large supply of bitumen, tar is little used in road construction or maintenance, but in Europe road binders still constitute an important, though declining, market for tar bulk products.

Surface coatings. Tar-based surface coatings range from the so-called black varnishes, which consist of a soft pitch fluxed back to brushing or spraying consistency with coal-tar naphtha, to pipe-coating enamels and pitch-polymer coatings.

DONALD McNEIL
Technical Consultant
Bradford, UK

D. McNeil, *Coal Carbonization Products*, Permagon Press, Oxford, 1966.

D. McNeil in M.A. Elliot, ed., *High Temperature Coal Tar*, John Wiley & Sons, Inc., New York, 1981.

A.W. Goos and A.A. Reiter, *Ind. Eng. Chem.* **38**, 132 (1946).

H-G. Franck and C. Collin, *Steinkohlenteer: Chemie Technologie and Verwandung*, Springer-Verlag, Berlin, 1968.

TAR SANDS

Tar sands (also known as oil sands or bituminous sands) are sand deposits impregnated with dense, viscous petroleum. Tar sands are found throughout the world, often in the same geographical areas as conventional petroleum. The largest deposits are in the Athabasca area in the northeastern part of the province of Alberta, Can., and in the Orinoco region of east central Venezuela.

Recoverable Reserves

The two conditions of vital concern for the economic development of tar-sand deposits are the concentration of the resource, or the percent

Table 1. Bulk Properties of Tar Sands

| Property | Canada, Lower Cretaceous | | | | Venezuela, Jobo | United States, various deposits |
	Athabasca, Wabiskaw/ McMurray	Cold Lake A, Grand Rapids	Cold Lake B, Clearwater	Wabasca B, Wabiskaw		
proved in-place reserve, m³	132 × 109	26,000	6,400	8,000	1,400	
depth, m	0–610	300	410	200–670	1,030–1,170	0–975
av pay thickness[a], km	27	11	11	7	14–52	4–150
reservoir temperature, °C	10		13		54–64	
reservoir pressure, kPa[b]			3,030			
initial gas–oil ratio, m³/m³						
initial oil saturation, %	44–98	76	73	64	76–88	35–82
porosity, %	39	34–40	38–46	19–32	30–31	17.5–35
permeability, $\mu m^{2\,c}$	≤ 500		≤ 300		150	1.2–600

[a] 3% bitumen cutoff for Canadian fields.
[b] To convert kPa to psi, multiply by 0.145.
[c] To convert μm^2 to darcy, multiply by 1.013.

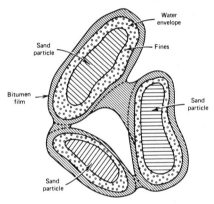

Figure 1. Typical arrangement of tar-sand particles. Courtesy of Research Council of Alberta.

bitumen saturation, and its accessibility, usually measured by the overburden thickness. Recovery methods are based either on mining combined with some further processing or operation on the oil sands *in situ*. The mining methods are applicable to shallow deposits, characterized by an overburden ratio (ie, overburden depth to thickness of tar-sand deposit) of ca 1.0.

Properties

Bulk. Tar sand has been defined as sand saturated with a "highly viscous crude hydrocarbon material not recoverable in its natural state through a well by ordinary production methods." Technically, the material should perhaps be called bituminous sand rather than tar sand since the hydrocarbon is bitumen (ie, a carbon disulfide-soluble oil).

Tar sand is a mixture of sand, water, and bitumen (see Fig. 1).

Average bulk properties of samples from Canadian locations, Venezuelan locations, and U.S. deposits are given in Table 1. The large number and wide distribution of these deposits result in a very wide range of property values.

Tar-sand minerals. Usually, > 99% of the tar-sand mineral is composed of quartz sand and clays. In the remaining 1%, more than 25 minerals have been identified, mostly calciferous or iron based.

Bitumen. Although there are wide variations both in the bitumen saturation of tar sand (0–18 wt% bitumen) and in the particle-size distribution of the tar-sand mineral, the chemical composition of Athabasca bitumen is relatively constant over a wide geographical area. Density at 15°C is slightly greater than that of water. About 50% of the bitumen is distillable without cracking. Elemental analyses are relatively constant; sulfur and nitrogen averages are 4.5–5.0 and 0.4–0.5 wt %, respectively.

Table 2. Properties of Synthetic Crude Oil Refined from Athabasca Bitumen

| Property | Source of samples | | | |
	GCOS[a] Commercial	Shell Canada, Ltd.	Direct fluid coking, Bitumount	Feed for hydro-treating tests
sp gr	0.833	0.861	0.961	0.965
(°API)	(38.3)	(32.8)	(15.8)	(15.5)
Distillation temperature, °C				
initial bp	72		83	
5%	105		211	
10%	123		250	
30%	209	204	310	
50%	264	282	343	
90%	324	474		
end pt	379			
recovery, %				
Elemental analysis, wt%				
carbon				84.4
hydrogen				11.2
nitrogen[b]		0.09		0.24
oxygen				0.10
sulfur[b]	0.022	0.29	3.9	4.1
heavy metals[c], ppm				
Hydrocarbon type, wt%				
asphaltenes				
resins				
aromatic oil				
saturated oil	79			

[a] Mildred-Ruth Lakes area.
[b] Can be removed to ppm concentrations.
[c] Vanadium, nickel, and copper.

Refined products. Properties of synthetic crude oils made from Athabasca bitumen by various processes are given in Table 2.

Recovery Technology

The tar sand may be mined and transported to a processing plant where the bitumen is extracted and the sand is discharged. Alternatively, the bitumen may be separated *in situ*. The latter process is similar to tertiary recovery of crude oil.

In situ processes. *In situ* processes include combustion, steam stimulation, steam drive, and the emulsion–steam-drive process.

Mining. The alternative to *in situ* processing is to mine the tar sands, transport them to a processing plant, extract the bitumen value, and dispose of the waste sand.

Bitumen recovery. Recovery processes include direct coking, anhydrous solvent extraction, cold-water separation process, hot-water process, and froth cleanup.

Bitumen Upgrading

Bitumen is a hydrogen-deficient oil that is upgraded by carbon removal (coking) or hydrogen addition (hydrocracking) (see also Petroleum, refinery processes).

Health and Safety Factors

Health and safety factors in *in situ* operations are associated with high temperature, high pressure steam, or high pressure air. The environmental considerations relate to air and water quality and surface reclamation. In some environmentally sensitive areas such as the oil-sands deposits in Utah, environmental considerations may make development unfeasible.

DONALD TOWSON
Petro-Canada

D.A. Redford and A.G. Winestock, eds., *The Oil Sands of Canada—Venezuela*, The Canadian Institute of Mining and Metallurgy, Montreal, Can., 1978.

Oil Sands Fuel of The Future, Canadian Society of Petroleum Geologists, Calgary, Alb., Can., Sept. 1974.

TARTARIC ACID. See Hydroxy dicarboxylic acids.

TEA

The term refers to the plant *Camellia sinensis*; the dried, processed leaf manufactured from it; extracts derived from the leaf; and beverages prepared from leaf or extract of this species. Tea is the most widely consumed of all beverages aside from water. Apart from its extensive usage, it is of interest to the food technologist as well as to the biochemist and organic chemist, because of the dependence of the product characteristics on the unusual chemical composition of the starting material, and because of the complex series of biochemical and organochemical reactions that occur during processing (see Food processing).

During the last few decades, tea has been primarily propogated from leaf cuttings. Planting material for the establishment of new clones is selected on the basis of beverage quality, yield, ease of establishment, pest resistance, frost resistance, and whatever other criteria local considerations might dictate. Selection for yield has greatly improved productivity.

Composition of the Fresh Leaf

Like all plant-leaf matter, the tea leaf contains all the enzymes, biochemical intermediates, and structural elements normally associated with plant growth and photosynthesis. In addition, it contains substances that are known to be responsible for the unique properties of tea.

Flavanols. The outstanding characteristic of tea leaf is its very high concentration of polyphenolic material. Although amounts vary depending on the factors influencing plant metabolism, such as light, rainfall, temperature, nutrient availability, leaf age and genetic make up, a small group of flavanols usually constitutes 20–30% of the dry matter. The flavanols of tea belong to the group known as catechins. (see Fig. 1).

The tea flavanols are water-soluble, colorless substances with an astringent taste. They oxidize readily in the pure state and form complexes with many other substances. Tea beverage quality is positively correlated with fresh-leaf flavanol concentration, especially gallated flavanols which decrease as the leaf ages. As discussed later, this group of compounds plays the most significant role of all substances in tea manufacturing. Tea flavanols occur in the cytoplasmic vacuoles of leaf cells.

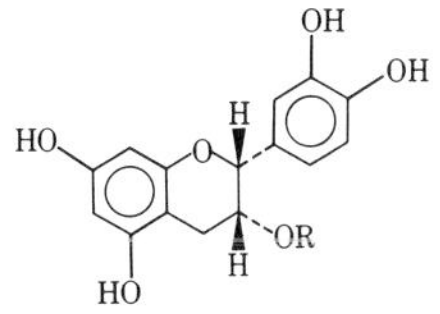

Figure 1. The flavanols occurring in tea.

(1) (−)-epicatechin; R = H; 1–3% of dry wt
(2) (−)-epicatechin gallate; R = 3,4,5-trihydroxybenzoyl; 3–6% of dry wt
(3) (−)-epigallocatechin; R = H; 3–6% of dry wt 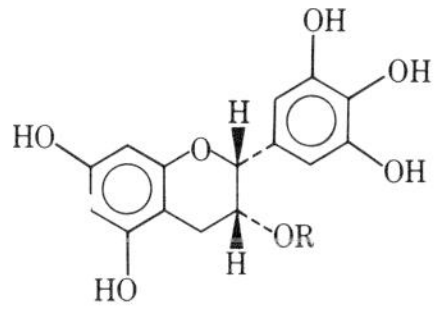
(4) (−)-epigallocatechin gallate; R = 3,4,5-trihydroxybenzoyl; 9–13% of dry wt
(5) (+)-Catechin; 1–2% of dry wt 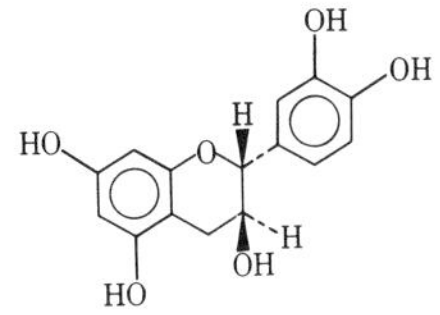
(6) (+)-Gallocatechin; 3–4% of dry wt 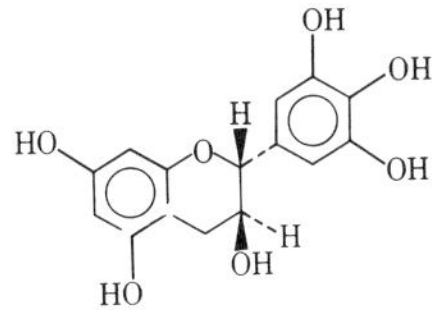

Other polyphenols. These include gallic acid; flavanols, such as quercetin, kaemferol, myricetin, and their glycosides; and depsides such as chlorogenic acid, *p*-coumarylquinic acid, and theogallin (7) (unique to tea) (see Fig. 2).

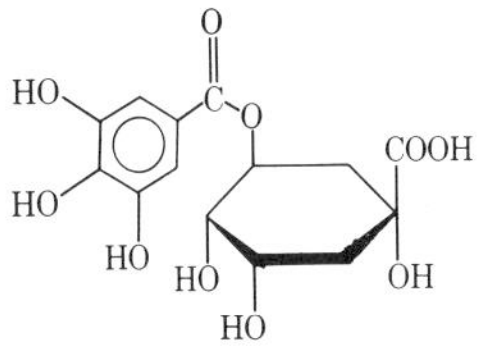

(7) theogallin

$$HO_2CCH(NH_2)CH_2CH_2CONHC_2H_5$$
(8) theanine

Figure 2. Other tea components.

Other important components are caffeine and other xanthines, theanine (8) (unique to tea) (see Fig. 2) and other amino acids (qv), carotenoids, chlorophyll, volatiles, lipids, and enzymes (qv) (see Alkaloids).

Manufacture

The manufacturing process converts freshly harvested leaf to products of commerce. Black tea is the most widely consumed form of tea. It is the result of promoting the oxidation of fresh leaf catechins by atmospheric oxygen through catalysis by tea polyphenolase. This is accomplished by macerating withered leaf to breakdown cell structure and allowing for intimate contact between enzyme and substrate. Over a period of 90–180 min catechins are oxidized to a series of quinones that react further with other tea leaf components or condense with each other to produce an extremely complex mixture of highly colored substances known as theaflavins, theaflavic acids, and thearubigens. The latter are not well characterized. These substances in combination with caffeine provide the basic black tea taste. In addition, new volatile compounds are produced during the oxidation step and the final firing (drying) process. This mixture of several hundred volatile compounds constitutes tea aroma. The oxidized product is dried.

Green tea, consumed mostly in Japan, the People's Republic of China, North Africa, and the Middle East, is processed in a manner designed to prevent the enzymic oxidation of catechins before drying. Oolong tea is partially oxidized tea manufactured primarily in the People's Republic of China and the Republic of China. Instant tea, usually a powder, is prepared by the aqueous extraction of black tea followed by concentration and drying. It is also possible to make instant tea from green tea or oxidized leaf before the drying step.

Black tea used in the United States has an average caffeine content of about 3%. The average cup of tea contains 30–35 mg caffeine.

Decaffeination

Decaffeinated tea has begun to appear on the market and is of growing significance. Extraction of caffeine with supercritical CO_2 is the preferred process.

HAROLD GRAHAM
Thomas J. Lipton, Inc.

T. Eden, *Tea*, 3rd ed., Longman Group, Ltd., London, 1976.

R.L. Wickremasinghe in C.O. Chichester, ed., *Advances in Food Research*, Vol. 24, Academic Press, New York, 1978, pp. 229–286.

G.W. Sanderson in V.C. Runeckles, ed., *Structural and Functional Aspects of Phytochemistry*, Academic Press, New York, 1972.

TECHNETIUM. See Radioactive tracers.

TECHNICAL SERVICE

Definitions

The meaning of the term technical service can be interpreted in several ways. A study of the impact of communication factors of R & D performance divides R & D functions into basic and applied research, development, and technical service. The study defines the latter as tasks meeting the specific local service needs of an organization, such as improving performance or reducing cost in existing products or operations. Opening new markets for existing products is included, and emphasis is placed on applying current knowledge.

Functions

A number of functions are basic to technical service; these are provided by one person, a group, or a department and include: development of specific technical recommendations for customers; information on products (including use, properties, application, etc); trial or start-up of products at a customer's location; assistance in training customer's or the companys' own sales personnel in product application and monitering; solving application problems; laboratory and testing service; investigation of customer complaints; the gathering and preparing of technical data and information regarding the product line; and information concerning government regulations regarding the sale and use of chemicals.

Technical Service and the Chemical Business

From the viewpoint of technical service in the chemical industry, the level of service provides the distinction between commodity chemical and chemical specialities. Inherently, commodities like acetone or sodium phosphate require minimal technical-service support; generally, only standard information is required by a customer. Technical-service effort in these areas usually focuses on cost reduction or process improvement. Differentiated products, on the other hand, require varying degrees of technical-service support. These materials are generally made by different suppliers to performance specifications rather than to content specifications. Technical service demonstrates the superior capability of one supplier versus another.

Organization

The first approach to a formal technical-service function, whatever its designation, might be within the sales or marketing organization of which customer contact is an integral part. This arrangement offers the best opportunity for intimate knowledge of the customer. Although under any structure, contact through the sales force is primary, a sales orientation to the technical-service group ensures rapid response; up-to-date knowledge of all matters that could affect the customer's reception of the technical-service effort; awareness of the sales situation; and priority setting within the department itself. However, laboratory and other interdepartmental contacts and activities to support a technical-service effort have decreased; there is pressure on the type, amount, and method of data to be presented; and sales details and activities are attended to rather than technical functions.

Technical-service activities within a laboratory organization are more prevalent in the chemical industry. In the past, emphasis was on autonomous laboratories, working both with sales and marketing personnel and with R & D staff. Today, the technical-service responsibility appears to reside with development or applications group personnel within R & D. These groups are usually organized around a product or product line and are occasionally subdivided by industry classifications.

Computer Application

Computer usage can replace experienced personnel who normally would travel extensively to customer locations. Today, data and information are obtained, transmitted to a central location, and analyzed. Results or recommendations are returned almost instantly.

ROBERT J. ZIEGLER
HILLEL LIEBERMAN
Betz Laboratories

T.J. Allen, D.M.S. Lee, and M.L. Tushman, *IEEE Trans. on Eng. Mgmt.* **EM-27** (1), 2 (Feb. 1980).

Betz Handbook of Industrial Water Conditioning, 8th ed., Betz Laboratories, Inc., Trevose, Pa., 1980.

The Versatile Ingredient VEEGUM, Bulletin No. 15, R.T. Vanderbilt, Norwalk, Conn., undated booklet.

TELLURIUM AND TELLURIUM COMPOUNDS

Tellurium, Te, at no. 52, at wt 127.61, is a member of the sixth main group (VIA) of the periodic chart located between selenium and polonium and between antimony and iodine. The configuration of its six outer electrons is $5s^2 5p^4$, and its four inner principal shells are completely filled. Tellurium is more metallic than oxygen, sulfur, and selenium, yet it resembles them closely in most of its chemical properties. Whereas oxygen and sulfur are nonmetals and electrical insulators, selenium and tellurium are semiconductors, and polonium is a metal. Tellurium forms inorganic and organic compounds superficially similar to the corresponding sulfur and selenium compounds, and yet dissimilar in properties and behavior. The valance states assigned to the central atom in tellurium compounds are $-2, 0, +2, +4, +6$.

Physical properties. At least 21 tellurium isotopes are known, with mass numbers from 114 to 134. Of these, eight are stable (120, 122–126, 128, 130). The others are radioactive with lifetimes from 2 min to 154 d; the heaviest six (131m, 131, 123, 133m, 133, 134) are fission products (see Radioisotopes). Physical properties are given in Table 1.

Chemical properties. Although chemically tellurium resembles sulfur and selenium, it is more basic, more metallic, and more strongly amphoteric. It behaves as an anion or a cation, depending on the medium.

Tellurium forms ionic tellurides with active metals, and covalent compounds with other elements. Tellurium reacts with concentrated, but not with dilute, sulfuric acid to form tellurium sulfite.

Elemental tellurium reduces chlorides such as $AsCl_3$, $AuCl_3$, and $PbCl_4$ to the element; it reduces $FeCl_3$ to $FeCl_2$ and SO_2Cl_2 to SO_2. Oxidation of metals by tellurium gives metallic tellurides.

The stability of organic chalcogen compounds decreases mostly in the order sulfur > selenium > tellurium.

Recovery. The yield of tellurium recovered from copper ore is indeed small. About 90% is lost in flotation concentration and from 20 to 60% in each metallurgical operation, eg, roasting, smelting, converting, fire-refining, and slimes treatment (see also Copper). Losses are also high in the metallurgical treatment of lead and zinc ores, by-products of which are sometime sent to a copper smelter.

Most tellurium of commerce is recovered from electrolytic copper refinery slimes, in which it is present from a trace to 8% (see also Selenium).

Table 1. Physical Properties of Tellurium

Property	Value
specific gravity[a] at 18°C	
crystalline	6.24
precipitated	6.0–6.2
hardness, Mohs[b]	2.0–2.5
modulus of elasticity, MPa[c]	4140
Poisson ratio at 30°C	0.33
heat capacity at 25°C, kJ/mol[d]	25.70
entropy at 25°C, J/K[d]	49.70
heat of fusion, kJ/mol[d]	17.87
mp, °C	450
viscosity at mp, mPa·s (= cP)	1.8–1.95
bp, °C[e]	990
heat of formation[f], kJ/mol[d]	171.5
heat of vaporization, kJ/g[d]	46.0
thermal conductivity at 20°C[g], W/(m·K)	0.060

[a] Increases under pressure.
[b] Anisotropic.
[c] To convert MPa to psi, multiply by 145.
[d] To convert J to cal, divide by 4.184.
[e] Extrapolated.
[f] Te(g) atom to Te₂(g) molecule.
[g] Polycrystalline material; in single crystals, it is anisotropic and affected by impurities and lattice imperfections.

Commercial products include tellurium dioxide, sodium tellurate, ferrotellurium, and tellurium diethyldithiocarbamate.

Health and safety factors. Elementary tellurium and the stable tellurides of heavy nonferrous metals are relatively inert and are not thought to be a health hazard. All other tellurium compounds, however, should be handled with caution, including reactive tellurides and volatile and soluble compounds, such as hydrogen telluride, tellurium hexafluoride, and the organic compounds.

As with selenium, industrial precautions include the common sense measures of good housekeeping, adequate ventilation, personal cleanliness, and frequent changes of clothing. No ambient air standard has

Table 2. Isolated Organic Tellurium Compounds

Compounds	Formula	Mp, °C	Bp, °C
tellurols or tellanes	RTeH		> 90
ethanetellurol	C_2H_5TeH		
benzenetellurol	C_6H_5TeH		
alkyl, aryl, and cyclic tellurides			
dimethyl telluride (methyl telluride)	$(CH_3)_2Te$		82
diphenyl telluride	$(C_6H_5)_2Te$	53.4	182
tetraphenyl telluride	$(C_6H_5)_4Te$	104–106	
tetrahydrotellurophene	$TeCH_2(CH_2)_2CH_2$		
ditellurides	R_2Te		
diphenyl ditelluride	$C_6H_5Te–TeC_6H_5$	53–54	
alkyltellurium trihalides	$RTeX_3$		
methyltellurium tribromide	CH_3TeBr_3	dec 140	
dialkyltellurium dihalides	R_2TeX_2		
dimethyltellurium dichloride	$(CH_3)_2TeCl_2$	92(α), 134(β)	
tellurium salts	R_3TeX		
trimethyltellurium iodide	$(CH_3)_3TeI$		
telluroxides	RTeO		
diphenyl telluroxide	$(C_6H_5)_2TeO$	185	
tellurones	R_2TeO_2		
dimethyl tellurone	$(CH_3)_2TeO_2$		
telluroketones	R_2CTe		
dimethyl telluroketone	CH_3CTeCH_3	63–66	
tellurinic acid	RTeO.OH		
phenyltellurinic acid	$C_6H_5Te.OH$	211	
heterocyclic compounds			
1,4-oxatellurane	$OCH_2CH_2TeCH_2CH_2$		
3,5-telluranedione	$CH_2COCH_2TeCH_2CO$		

been established in the United States; the accepted TLV is 0.01–0.1 mg/m³.

Inorganic Compounds

Tellurium forms inorganic and organic compounds very similar to those of sulfur and selenium. The most important tellurium compounds are the tellurides, halides, oxides, and oxyacids.

Tellurides include hydrogen telluride, H_2Te; tellurium sulfide, TeS; carbon sulfotelluride, CSTe; carbonyl telluride, COTe; and tellurium nitride, Te_3N_4. Tellurium halides include tellurium tetrafluoride, TeF_4; tellurium decafluoride, Te_2F_{10}; tellurium hexafluoride, TeF_6; tellurium dichloride, $TeCl_2$; tellurium tetrachloride, $TeCl_4$; tellurium dibromide, $TeBr_2$; tellurium tetrabromide, $TeBr_4$; tellurium tetraiodide TeI_4; tellurium oxychlorides, $Te_6O_{11}Cl_2$ and tellurium oxydibromide. Tellurium oxides, oxyacids, and salts include tellurium trioxide, TeO_3; tellurous acid, H_2TeO_3; orthotelluric acid H_6TeO_6; polymetatelluric acid, H_2TeO_4, and tellurates.

Organic Compounds

Organotellurium compounds range from the simple carbon sulfotelluride to complex heterocyclic compounds and organotellurium ligands. Various types of tellurium compounds and specific examples are listed in Table 2.

Uses

Tellurium and tellurium compounds are used in free-machining steels, chilled castings, copper alloys, lead alloys, metal coatings, pigments and glass, catalysts, lubricants, rubber, explosives, thermoelectric materials, semiconductors, and photography and thermography.

E.M. ELKIN
Noranda Mines Ltd.

K.W. Bagnall, *The Chemistry of Selenium, Tellurium, and Polonium*, Elsevier Publishing Co., Amsterdam, 1966; *Comprehensive Inorganic Chemistry*, Vol. 2, Pergamon Press, Oxford, UK, 1973, Chapts. 23–24.

W.C. Cooper, ed., *Tellurium*, Van Nostrand Reinhold Co., New York, 1971.

B. Mason, *Principles of Geochemistry*, 3rd ed., John Wiley & Sons, Inc., New York, 1971.

TEMPERATURE MEASUREMENT

Thermodynamic Kelvin Temperature Scale

Temperature is a measure of hotness. For a measure to be rational, there must be agreement on a scale of numerical values for hotness and upon devices for realizing and displaying these values. The only temperature scale with an absolute basis in nature is the thermodynamic Kelvin temperature scale (TKTS), which is based on functions that can be deduced directly from the first and second laws of thermodynamics (qv).

Accurate measurements of the TKTS present substantial experimental difficulties. Most easily used thermometers are not based on functions of the first and second laws of thermodynamics. Thus, most thermometers depend upon some function that is a repeatable and single-valued analogue of temperature, and they are used as interpolation devices of practical and utilitarian temperature scales which are themselves artifacts. The main purpose of the study of the TKTS is to establish relationships between the thermodynamic scale of nature and the practical temperature scales of the laboratory, so that measurements made on nonthermodynamic scales can be translated into TKTS terms.

Practical Temperature Scales

Many properties of materials change with temperature, eg, the expansion and contraction of solids and liquids, the electrical properties of conductors and semiconductors, and the color and brilliance of light emitted from a very hot source. Any of these properties can be used to make a thermometer; all nonthermodynamic properties require the con-

Table 1. Defining Fixed Points of the IPTS, 1968

Defining fixed point[a]	°C	K
tp e H$_2$[b]	−259.34	13.81
bp e H$_2$, 2.5/7.7 MPa[b,c]	−256.108	17.042
nbp e H$_2$[b]	−252.87	20.28
nbp Ne	−246.048	27.102
tp O$_2$	−218.789	54.361
nbp O$_2$	−182.962	90.188
ice point		
tp H$_2$O	0.01	273.16
nbp H$_2$O	100	373.15
fp Sn	231.9681	505.1181
fp Zn	419.58	692.73
nbp S		
fp Ag	961.93	1235.08
fp Au	1064.43	1337.58

[a]tp = triple point, bp = boiling point, nbp = normal
  boiling point, and fp = freezing point.
[b]eH$_2$ = eka H$_2$.
[c]To convert MPa to atm, divide by 0.101.

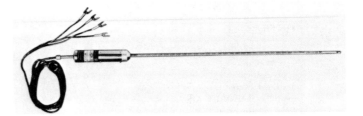

Figure 1. Laboratory standard platinum resistance thermometer (SPRT).

struction of a concensus scale to relate the properties of a prescribed interpolation device to temperature.

Although only one fixed point is necessary to define the TKTS, practical temperature scales require the specification of a sufficient number of fixed points to allow the coefficients of the characteristic equations of the stipulated interpolation devices to be determined.

In 1968, an extensive revision of the International Practical Temperature Scale (IPTS) was made. The motivation for this revision was to extend the scale downward from the normal boiling point of oxygen to 13.81 K (259.34°C) and to bring it into closer conformance with the then best-known values of fixed points on the TKTS. The interpolation instruments of the scale did not change, except for the requirement that the platinum-resistance-thermometer ratio be not less than 1.39250. The platinum-resistance-thermometer interpolation equations were changed to accommodate changes in the values of all fixed points except those of the triple point and boiling point of water. Table 1 shows the values assigned to the defining fixed points of the IPTS of 1968.

Another revision of the IPTS is considered for 1987. One change which will almost certainly be made is an increase in the range of the platinum-resistance thermometer.

Fixed Points of the Scale

The fixed points are the basis of the scale and they are of two kinds: defining and secondary or reference fixed points. Defining fixed points are two and three-phase equilibria of very pure materials to which temperature values are assigned. Secondary or reference fixed points are two and three-phase equilibria of very pure materials, whose values of equilibrium temperature are not assigned but are established by measurements made with interpolation instruments that have been calibrated at the defining fixed points.

The triple point of water is the most important fixed point because it is the one point that the TKTS and the IPTS have in common and because it can be determined with great accuracy.

Metal freezing points are an important class of fixed points. Their pressure dependence is of small order (for tin, for example, it is 3.0×10^{-8} K/Pa (0.003 K/atm)) and unlikely to be a problem in all but the most critical measurements. The metals used in fixed-point cells are usually better than 99.9999% pure.

Liquid-vapor fixed points or boiling points are the least satisfactory fixed points because of their pressure dependence.

Resistance Thermometers

Resistance thermometers are of three types: high precision platinum, industrial-grade metallic, and nonmetallic. High precision platinum resistance thermometers are suitable for use as interpolation instruments of the IPTS and are frequently referred to as SPRTs (standard platinum

resistance thermometers). These thermometers are generally useful over the entire platinum thermometer range of the IPTS, ie, from −182.962 to 650°C. A special capsule form of this thermometer can be used from −259.34 to 232°C. A typical SPRT is shown in Figure 1.

Industrial-grade metallic resistance thermometers are wound of metal wire, usually high purity platinum, but copper, nickel, and alloys are also used. The temperature-sensitive element is typically protected by a covering of ceramic tubing or ceramic cement.

Nonmetallic resistance thermometers or thermistors are generally temperature-sensitive semiconducting ceramics (see Ceramics as electrical materials). The variety of sizes, shapes, and performance characteristics is large. The thermistor material is usually a metal oxide, eg, manganese oxide. Dopants, eg, nickel oxide or copper oxide, may be added to obtain a variety of resistance and slope characteristics. The material can be sintered into a disk or bead with integral or attached connecting wires.

Instrumentation. The principal considerations for instrumentation to measure the resistance at a certain temperature are adequate sensitivity, elimination of the effects of lead-wire resistance, and avoidance of the effects of self-heating.

Seebeck-Effect Thermometers

Seebeck-effect thermometers or thermocouples are junctions of two wires of dissimilar materials that accomplish a net conversion of thermal energy into electrical energy with the appearance of an electric current (see Thermoelectric energy conversion). Unlike resistance thermometers, whose output is proportional to temperature, the output of thermocouples is proportional to temperature difference; that is, the magnitude of the current is proportional to the difference of the temperatures of the two junctions; Figure 2 illustrates this principle.

Standard (IPTS) Thermocouple

At 630–1064.43°C (the freezing temperatures of antimony and gold, respectively), the specified interpolation standard of the IPTS (1968) is a thermocouple, one conductor of which is pure platinum and the other an alloy of platinum with 10% rhodium. The thermal electromotive force (emf) required for such a thermocouple is stipulated in the scale. The coefficients of the characteristic quadratic of this thermocouple are determined by measurements at the two fixed points, which are the high and low ends of its range, and at one intermediate fixed point, the freezing point of silver, 961.93°C. In a system for designating thermocouple pairs by letter, the Pt-90%/Pt-10% rh thermocouple is called type S.

Type-S thermocouples intended for use as IPTS interpolation standards are usually protected by an enclosure that minimizes contamination and provides an oxidizing atmosphere.

Working Thermocouples

Properties of various types of working thermocouples are shown in Table 2.

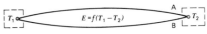

Figure 2. Basic thermocouple circuit. A and B are wires of different materials.

Table 2. Properties of Various Thermocouple Pairs

ASTM type	Materials	Range[a], °C	Seebeck[b] coefficient, μV/C°
J	iron–constantan	0–760	64.3
K	chromel–alumel	0–1260	36.5
R	Pt–87 Pt–13 Rh	0–1450	13.8
S	Pt–90 Pt–10 Rh	0–1450	11.8
T	copper–constantan	−183–375	53.0
E	chromel–constantan	0–875	78.5
B	70 Pt–30 Rh–94 Pt–6 Rh	870–1700	11.6

[a] Wire of small diameter may require closer range.
[b] Emf is given for example at the high temperature end of the range; many thermocouples are non-linear.

Measurement of Emitted Radiation

Temperatures of materials that are too hot to be placed in physical contact with a thermometer, either because the temperature to be measured would be affected by such contact or because the thermometer's characteristics would be changed, may be determined by measuring the radiation emitted by the materials. Instruments include the optical pyrometer and the strip lamp.

Other Electrical Thermometers

Many special-purpose electrical thermometers have been developed, either as practical thermometers or as research devices for study of temperature and the temperature scale. They are research thermometers based on thermal noise and thermometers based on the change in the speed of sound as a function of absolute temperature.

A novel and useful thermometer based on the change in the resonant frequency of a quartz crystal is produced by Hewlett-Packard.

Yokogawa Electric Works has developed a thermometer based on the nuclear quadrupole resonance of potassium chlorate and usable over the range from −183 to 125°C.

Nonelectrical Thermometers

Liquid-in-glass. The thermal expansions of liquids are reliable indicators of temperature. Liquids are suitable for this use between their freezing and boiling points; mercury and colored alcohol are the most common fluids. The typical liquid-in-glass thermometer includes a thin-walled glass bulb, which is attached to a capillary stem partially filled with a visible liquid and sealed against the environment.

Miscellaneous. Other nonelectrical thermometers are bimetal, filled-system, and pyrometric-cone thermometers.

HENRY E. SOSTMAN
Yellow Springs Instrument Co., Inc.

R.P. Benedict, *Fundamentals of Temperature, Pressure and Flow Measurement*, John Wiley & Sons, Inc., New York, 1969.

H.B. Sachse, *Semiconducting Temperature Sensors and Their Applications*, John Wiley & Sons, Inc., New York, 1975.

J.F. Schooley, ed., *Temperature, Its Measurement and Control in Science and Industry*, Vol. 5, Instrument Society of America, 1982.

T.J. Quinn, *Temperature*, Academic Press, New York, 1983.

TERBIUM. See Rare-earth elements.

TEREPHTHALIC ACID. See Phthalic acid and other benzenepolycarboxylic acids.

TERGITOL. See Surfactants and detersive systems.

TERNE PLATE. See Metallic coatings.

TERPENES AND TERPENOIDS. See Terpenoids.

TERPENOIDS

Terpenes are widely distributed in nature, and occur in nearly all living plants. They are generally regarded as derivatives of isoprene, wherein the isoprene units are arranged in a head-to-tail fashion, although there are some exceptions to this arrangement. The terpenes are therefore classified according to the number of isoprene units in their carbon skeletons, with a single terpene unit being regarded as two isoprene units.

The terpenes are further categorized as acyclic (open chain), monocyclic (one ring), bicyclic (two rings), tricyclic (three rings), etc. Although the term terpenes might seem to be appropriate only for unsaturated hydrocarbons, it is generally recognized to apply not only to isoprene oligomers, but also to their saturated or partially saturated isomers, as well as to their derivatives, eg, alcohols, aldehydes, esters, etc. The term terpenoids properly designates such derivatives. Terpenoids, especially the oxygenated derivatives, are important flavor and perfume materials. In general, their mammalian toxicity is relatively low. Many are listed as GRAS as food additives and flavorings. Properties of selected terpenes and terpenoids are listed in Tables 1 and 2.

In nature, the two most abundant natural sources of terpenes are turpentine and other essential oils. Nearly all commercially important terpenes were obtained from these sources until relatively recently (see Oils, essential). Since about the late 1950s, synthetic methods have been developed for manufacturing most industrially important monoterpenes, and these synthetics have taken over a large share of the market. Essential oils are still used in substantial quantities for their flavor and aroma characteristics, which largely depend on their terpene content.

Monoterpenes

From turpentine. The southeastern United States is the largest turpentine-producing region in the world. Nearly all industrial syntheses

Table 1. Properties of Selected Monoterpene Hydrocarbons and Cineoles

Name	Bp, °C at 101.3 kPa[a]	Bp, °C at 13.33 kPa[a]	Mp, °C	Density, d^{20}, g/cm³	n_D^{20}[b]	[α]$_D$	Gas-chromatography retention time relative to α-pinene on Carbowax 4000
tricyclene	152	85	+65			0	0.96
α-pinene (1)	156	89	−50; −75, (±)	0.8595	1.4658	±51	1.00
α-fenchene	157	91.5		0.8697	1.4740	±44	1.22
camphene (2)	158	91	+49			±108	1.28
β-pinene	165	98	−50; −61, (±)	0.8722	1.4790	±22	1.58
myrcene (3)	167			0.7880[b]	1.4680^{25}	0	1.70
cis-pinane	168	101	−53	0.8575	1.4629	±23	1.25
cis-p-menthane	172	105	−90	0.8002	1.4431	0	1.10
1,4-cineole	172	105.5	−46	0.8986	1.4446	0	2.09
1,8-cineole	174	108	+1	0.9245	1.4574	0	2.57
limonene	176.5	110	−74; −89, (±)	0.8411	1.4730	±124	2.29
p-cymene	77	110	−73	0.857	1.4905	0	3.15

[a] 101.3 kPa = 1 atm.
[b] Unless otherwise indicated.

α-pinene (1)   camphene (2)   myrcene (3)

Table 2. Properties of Selected Oxygenated Terpenes

| Name | Bp, °C$_{kPa}$^{a} | | Mp, °C | Density, d^{20},^{b} g/cm^3 | n_D^{20}^{b} | $[\alpha]_D$ |
	at 101.3 kPa^a	at 13.33 kPa^a				
fenchone	193	122	+ 5	0.9452	1.4628	± 70
linalool (4)	199			0.8607	1.4616	± 22
α-fenchol	200	133	48; 39, (±)	0.935^{40}	1.4734	± 12.5
citronellal	203			0.851	1.4467	± 12
camphor	209		179			± 45
trans-β-terpineol	209	139	33	0.919	1.4712^{30}	0
trans-menthone	210	138	− 6	0.8903^{25}	1.4500	± 29
terpinen-4-ol	211	123$_{6.66}$		0.9259	1.4762	± 29
neomenthol	212		− 22; 53, (±)	0.8917^{30}	1.4604	± 20
borneol	212		209; 210, (±)			± 38
isoborneol	214		212 dl			± 34
menthol	216		43; 38, (±)	0.8911^{30}	1.4615	± 50
citronellol (5)	224			0.8550	1.4559	± 7
nerol (6)	225	128$_{3.33}$		0.8735^{25}	1.4736^{25}	0
geraniol (6)	230	131$_{3.33}$		0.8770^{25}	1.4756^{25}	0
carvone (7)	231	157		0.9550^{25}	1.4990	± 62
hydroxycitronellal		116$_{0.67}$		0.9220	1.4494	± 10.5
terpin (ordinary)	258		105; 117, hydrate			0
α-ionone	258			0.9309	1.4971	± 400
β-ionone	271		− 35	0.9461	0.5202	0

^{a}101.3 kPa = 1 atm.
^{b}Unless otherwise indicated.

linalool (4)   citronellol (5)   geraniol /nerol (6)   (−)-carvone (7)

of monoterpenes from turpentine involve sulfate turpentine rather than gum or wood turpentine because of its availability and low price, although some wood turpentine is used in the production of dipentene, camphene, and terpineol (pine oil). Industrially, turpentine is separated into its components by high efficiency vacuum fractionation.

From acetylene and acetone. Whereas the synthesis of oxygenated monoterpenes from turpentine involves transformations of molecules already containing 10 carbon atoms, the so-called acetylene–acetone route represents a total synthesis of monoterpenes. Industrially, acetone reacts with sodium acetylide in liquid ammonia, producing 2-methyl-3-butyne-2-ol (methylbutynol). Methylbutynol is partially hydrogenated over a Lindlar catalyst to obtain methylbutenol (2-methyl-3-buten-2-ol). The acid-catalyzed reaction of methyl isopropenyl ether with methylbutenol gives β-methylheptenone (6-methyl-5-hepten-2-one), which on subsequent reaction wth sodium acetylide produces dehydrolinalool (3,7-dimethyl-6-octen-1-yn-3-ol). Dehydrolinalool is readily dehydrogenated to linalool.

From isobutylene. Another route to the key intermediate methylheptenone is a direct one-step process. Isobutylene, formaldehyde, and acetone under high pressure and relatively high temperature form methylheptenone.

From isoprene. The first step is the production of prenyl chloride (1-chloro-3-methyl-2-butene) by continuous hydrochlorination of isoprene by either gaseous or aqueous HCl. Prenyl chloride next reacts with acetone in the presence of a quaternary ammonium salt catalyst and solid sodium hydroxide at modest temperatures to produce β-methylheptenone. Dehydrolinalool is obtained by a reaction of the ketone with acetylene.

α-Pinene. *Reactions.* Historically, the chief industrial reactions of α-pinene (1) have been acid-catalyzed reactions to produce pine oil and camphene. Recently, however, two additional large-scale uses have been developed commercially. These are the isomerization of α-pinene to β-pinene and the conversion of α-pinene to linalool, geraniol, and nerol.

Manufacture. α-Pinene can be obtained readily by isomerization of β-pinene over a variety of catalysts with retention of optical activity.

Uses. α-Pinene is an important starting material for large-scale manufacture of a variety of other terpenic hydrocarbons, alcohols, aldehydes, and ketones. It also is used in the manufacture of terpene resins. High purity α-pinene is used in flavor and perfume applications.

β-Pinene. *Reactions.* β-Pinene undergoes many of the same reactions as α-pinene but is seldom used for these reactions when the less costly and more plentiful α-pinene is available.

Manufacture. β-Pinene is obtained mainly by fractionation of turpentine and also by the isomerization of α-pinene.

Uses. The two most important uses of β-pinene are as a starting material for manufacturing aroma and flavor compounds and as monomer for terpene resins (see Terpene resins).

3-Carene. (+)-3-Carene, like the pinenes, is obtained by fractional distillation of turpentine. 3-Carene has little industrial use other than as a solvent.

Camphene. *Reactions.* One of the most important commercial reactions of camphene (2) is chlorination. The second most important commercial reaction depends on the readiness with which camphene rearranges to the isobornyl structure.

Manufacture. Camphene is produced commercially by treatment of α-pinene with acidic catalysts in the absence of water.

Uses. The most important uses of camphene are in the production of toxaphene, isobornyl acetate, isoborneol, and camphor.

p-Menthadienes and p-cymene. *Manufacture.* The p-menthadienes are obtained as by-products from the manufacture of synthetic pine oil and camphene and from extraction of Southeastern pine stumps. Most p-menthadienes are sold as mixtures called dipentene.

Uses. In addition to being an excellent solvent for paints, varnishes, and enamels that contain synthetic resins, particularly phenolic resins, dipentene serves as antiskinning agent and as wetting agent in the dispersion of pigments (see Pigments). The solvency of dipentene for rubber and its swelling and softening properties make it useful in rubber reclaiming and in the processing of natural and synthetic rubbers. Dipentene is also formulated into a variety of cleaners similar to pine-oil cleaners. It is sometimes used as a diluent for pine oil in flotation (qv) applications.

Myrcene. *Reactions.* Hydrohalogenation is the single most important reaction of myrcene (3), since it is the first step in the industrial synthesis of the large-volume terpene alcohols geraniol, nerol, and linalool.

Manufacture. Myrcene occurs in small amounts in many essential oils, but the only important commercial source is its manufacture by pyrolysis of β-pinene at 550–600°C.

Uses. By far the biggest use of myrcene is as an intermediate in the commercial production of the terpene alcohols, geraniol, nerol, and linalool.

3,7-Dimethyl-1,6-octadiene. *Manufacture.* 3,7-Dimethyl-1,6-octadiene is produced by the pyrolysis of pinane, preponderantly in the cis isomer, which can be made by hydrogenation of either α- or β-pinene or mixtures thereof.

Uses. The chief use for 3,7-dimethyl-1,6-octadiene is as a starting material for the synthesis of acyclic perfumery aldehydes and alcohols.

Alloocimene. *Manufacture.* Alloocimene is produced by the thermal isomerization of α-pinene with an approximately equal quantity of dipentene and smaller quantities of α- and β-pyronenes.

Alloocimene is useful in resins and epoxy modifiers, as a scavenger for decomposition products of halogenated solvents, and as a chemical intermediate for perfumery alcohols.

Oxygenated Monoterpenes

Geraniol and nerol. In general, geraniol/nerol (6) undergo the same reactions, although sometimes at different rates and under slightly different conditions. The most important industrial reactions are hydrogenation, dehydrogenation (oxidation), and rearrangement.

Manufacture. The chief natural source for geraniol is citronella oil, wherein it occurs with citronellal and citronellol and from which it is separated by fractional distillation. Citronella oil contains only small amounts of nerol. Much larger amounts of geraniol and nerol are produced synthetically than are isolated from natural sources.

Uses. Geraniol and nerol are used in perfumery for their rosy scents. By far their largest use is as starting materials for the synthesis of citral, citronellol, cirtonellal, hydroxycitronellal, the ionones, vitamins A and E, and carotenoids.

Linalool. *Reactions.* Linalool (4) can be esterified to produce linalyl acetate by reaction with acetic anhydride.

Manufacture. The most important natural source of linalool is *bois de rose* oil from which it is separated by fractionation, but much larger quantities are produced synthetically.

Uses. Linalool is used in multitonnage quantities for the production of geraniol/nerol and as an intermediate in the synthesis of vitamin E and vitamin A. Substantial quantities of linalool and its acetate are used in perfumery.

Dehydrolinalool. Dehydrolinalool is not obtainable from natural source but is manufactured commercially.

Uses. The chief uses of dehydrolinalool are as a precursor for the manufacture of linalool, geraniol/nerol, pseudoionone, and citral, all of which are important intermediates in the production of vitamins A and E, the ionones, methylionones, and other perfumery compounds.

Citral. *Reactions.* The most important commercial reactions of citral depend on base-catalyzed aldol condensation with ketones.

Manufacture. Lemongrass oil was the chief source of citral with *Litsea cubeba* oil more recently becoming commercially significant.

Uses. Multitonnage quantities of pseudoionone are produced each year from citral for use in the manufacture of vitamins A and E. Appreciable quantities of pseudoionone and the pseudomethylionones are used for producing perfumery ionones and methylionones.

Ionones and methylionones. *Reactions.* The only important large-scale commercial reactions of β-ionone are those associated with the manufacture of vitamin A and carotenoids. α-Ionone can be isomerized to β-ionone under acid conditions.

Manufacture. Pseudoionone (ionone precursor) is generally made industrially from citral. A commercially significant synthesis from mesityl oxide and prenyl chloride has been developed.

Uses. All ionones are used extensively in perfumery with α-ionone and α-isomethylionone being the most highly valued. β-Ionone is manufactured in the largest volume of any of the ionones because of its use in vitamin-A manufacture.

Citronellol. *Manufacture.* Citronellol (5) is manufactured on a commercial scale by the hydrogenation of geraniol/nerol or citronellal.

Uses. The chief use of citronellol is in fragrances for the soap and detergent industries for its natural rosy scent. It also serves as an intermediate in the synthesis of certain fragrance and flavor compounds.

Citronellal. *Manufacture.* The most common natural sources of citronellal are citronella oil and *Eucalyptus citriodora*. It is produced commercially by rearrangement of geraniol/nerol or by dehydrogenation of citronellol.

Uses. Principal uses of citronellal are in the production of hydroxycitronellal, methoxycitronellal, and menthol.

Hydroxy- and methoxycitronellal. *Reactions.* An important reaction of methoxycitronellal is its aldol condensation with dialkyl-3-methylglutaconates to produce an intermediate for the synthesis of certain juvenile hormone mimics, eg, methoprene.

Manufacture. Hydroxycitronellal is produced by hydration of citronellal under strong acid conditions. Methoxycitronellal is produced by the same reactions, except that in the hydrations step, methanol is used instead of water.

Hydroxycitronellal is valued for its delicate lily-of-the-valley fragrance by the perfumery industry, which uses large volumes of this compound. Methoxycitronellal is used as a starting material for the synthesis of juvenile hormone mimics (see Insect-control technology).

Myrcenol and dihydromyrcenol. *Reactions.* The most important chemical reaction of myrcenol has been the Diels-Alder reaction with acrolein to produce a mixture of the two isomeric aldehydes, 3- and 4-(4-hydroxy-4-methyl-pentyl)-3-cyclohexene-1-carboxaldehyde, which are used in perfumery.

Manufacture. One of the oldest methods for making myrcenol involves initial formation of an adduct of myrcene by its reaction with sulfur dioxide. The adduct is hydrated with aqueous sulfuric acid, and the hydrated product is thermally decomposed to produce myrcenol.

Dihydromyrcenol can be synthesized by hydrochlorination of 3,7-dimethyl-1,6-octadiene, followed by treatment with dilute caustic.

Uses. The primary use of myrcenol is in the production of the aldehyde mixture discussed above.

Menthol. *Reactions.* Menthol undergoes all of the typical chemical reactions of a cyclic secondary alcohol.

Manufacture. Of the menthol isomers, only (−)-menthol and (±)-menthol are of large-scale commercial importance. The (−)-product is becoming increasingly popular. All (±)-menthol is obtained by synthetic methods.

Uses. (−)-Menthol is used primarily for its physiological cooling effect and minty characteristics.

Carvone. *Manufacture.* (−)-Carvone (7) is the most abundant constituent of spearmint oil from which it can be separated by efficient fractionation. (+)-Carvone is the most abundant component of caraway- and dill-seeds oils from which it can be isolated by similar procedures.

A substantial amount of (−)-carvone is produced synthetically from (+)-limonene.

Uses. Most (−)-carvone is used in compounding or enhancing mint flavors and aromas for use in dentifrices, confections, pharmaceuticals, and odorants.

Camphor. *Manufacture.* Camphor is obtained both naturally and synthetically. Natural camphor is obtained from the wood of the camphor tree (*Cinnamomum camphora*), which grows in the People's Republic of China, the Republic of China, and Japan. The camphor is isolated by steam distillation, filtration, distillation, and sublimation. Most synthetic camphor is produced from camphene made from α-pinene.

Uses. Camphor is used as an ingredient in pharmaceutical preparations, as a moth repellent (see Repellents), a religious incense, and an odorant-flavorant in a variety of consumer products.

Pine Oil

There are three types of pine oil: steam-distilled, sulfate, and synthetic. By far the largest use of pine oil is in the manufacture of cleaners and disinfectants.

Terpene Resins

Large quantities of terpene hydrocarbons are used in the manufacture of terpene resins. The cationic polymerization is typically accomplished by reaction in a solvent, eg, toluene, and catalyzed by a Lewis acid, most commonly $AlCl_3$. The largest use of terpene resins is in hot-melt and pressure-sensitive adhesives.

Sesquiterpenes

Sesquiterpene hydrocarbons contain 15 carbon atoms usually comprised of three isoprene units, although these are frequently rearranged. There are acyclic, monocyclic, bicyclic, tricyclic, and tetracyclic sesquiterpenes. Their structures can be simple or complex.

Reactions. The two most important commercial reactions of sesquiterpenes are esterification and oxidation to obtain aroma products; sesquiterpene acetates are especially valued for this application. Acetylation of longifolene and hydroxylation of caryophyllene yield products useful in perfumery.

Manufacture. With a few exceptions, the limited number of commercially available sesquiterpenes have been isolated from natural sources, primarily essential oils to which they impart characteristic organoleptic properties. Isolation is usually accomplished by extraction, fractionation, crystallization, or a combination thereof.

Sesquiterpenes are used almost exclusively in perfumery.

Higher Terpenes and Terpenoids

Diterpenes. Diterpenes contain 20 carbon atoms. Commercially, the resin acids and vitamin A are the most important diterpenic substances. Gibberellic acid is produced by fermentation processes and is used as a growth promoter for plants, especially seedlings (see Plant growth substances). It is approved by the FDA for increasing the enzymatic activity of malt in brewing applications.

Sesterterpenes. Sesterterpenes contain 25 carbon atoms. They are a rare group of natural products with complex structures and are not commercially significant.

Triterpenes. The triterpenes contain 30 carbon atoms and are widely distributed in nature, especially in plants, both in the free state and as esters or glycosides. A smaller but important group, including lanosterol, occurs in animals. Squalane has a rather wide usage as a skin lotion and fixative. Squalane is produced synthetically by the coupling of two molecules of geranyl acetone with diacetylene, followed by dehydration and complete hydrogenation.

Tertraterpenes. Carotenoids comprise the most important group of C_{40} terpenes and terpenoids, although not all carotenoids contain 40 carbon atoms. They are widely distributed in plant, marine, and animal life.

Carotenoids were usually isolated from natural sources by solvent extraction. The extracts were used primarily for coloring foods. An important function of certain carotenoids is their provitamin A activity.

Carotenoids are used as pigments and dietary supplements in animals and poultry feedstuffs (see Pet and livestock feeds). They are added to pharmaceutical products to provide a control during manufacturing and to distinguish one product from another; they also enhance the appearance of the products.

β-Carotene is prescribed in the treatment of the inherited skin disorder, erythropoietic protoporphyria (EPP), to reduce the severity of photosensitivity reactions in such patients.

JOHN M. DERFER
MARIAN M. DERFER
SCM Corporation

S. Arctander, *Perfume and Flavor Chemicals* (*Aroma Chemicals*), Vols. I–II, published by the author, Montclair, N.J., 1969.

P.Z. Bedoukian, *Perfumery and Flavoring Synthetics*, 2nd rev. ed., Elsevier Publishing Co., New York, 1967.

J.L. Simonsen and co-workers, *The Terpenes*, Vols. 1–3, 2nd ed. rev., Cambridge University Press, Cambridge, New York, 1953 and 1957.

A.F. Thomas and Y. Bessière in J. ApSimon, ed., *The Total Synthesis of Natural Products*, Vol. 4, John Wiley & Sons, Inc., New York, 1981, pp. 451–591.

TETRACYCLINES. See Antibiotics, tetracyclines.

TEXTILES

SURVEY

Textile materials are among the most ubiquitous in society. They provide shelter and protection from the environment in the form of apparel, and comfort and decoration in the form of household textiles, such as sheets, upholstery, carpeting, drapery, and wall covering, and

they have a variety of industrial functions, eg, tire reinforcement, tenting, filter media, conveyor belts, insulation, etc. Textile materials are produced from fibers (finite lengths) and filaments (continuous lengths) by a variety of processes to form woven, knitted, and nonwoven (felt-like) fabrics. In the case of woven and knitted fabrics, the fibers and filaments are formed into intermediate continuous-length structures known as yarns, which are interlaced by weaving or interlooped by knitting into planar flexible sheetlike structures known as fabrics. Nonwoven fabrics are formed directly from fibers and filaments by chemically or physically bonding or interlocking fibers that have been arranged in a planar configuration (see Nonwoven textile fabrics; Tire cords).

Textile fibers may be classified into two main categories and into a number of subcategories, as indicated in Table 1.

Man-made fibers are formed by extrusion processes known as melt, dry, or wet spinning. The spinning or extrusion of filaments is normally followed by the operation known as drawing. In this step, the newly formed filaments are irreversibly extended and stabilized by setting or crystallization processes.

With the exception of silk (qv), naturally occurring fibers have finite lengths and generally require several cleaning and purification steps prior to processing into yarns and fabrics.

The manufacture of textile fabrics from fibers is based on processes that were developed to accommodate the special geometric and physical properties of naturally occurring fibers. Thus, there is a cotton system of yarn manufacture, a wool system, and a worsted system.

LUDWIG REBENFELD
Textile Research Institute

W. von Bergen, ed., *Wool Handbook*, Vols. I–II, John Wiley & Sons, Inc., New York, 1963, 1968.

H.F. Mark, S.M. Atlas, and E. Cernia, eds., *Man-Made Fibers Science and Technology*, Vols. 1–3, Interscience Publishers, a division of John Wiley & Sons, Inc., New York, 1967, 1968, and 1969.

M.L. Rollins and D.S. Hamby, eds., *The American Cotton Handbook*, John Wiley & Sons Inc., New York, 1965, 1966.

FINISHING

Finishing of textiles includes, in its broadest sense, any process used to improve a knitted, woven, or bonded textile fabric for apparel or other home or industrial use.

Some of the more important mechanical finishing processes are compressive shrinkage (Sanforized process) and calender finishes, which include schreinering (to produce a silklike appearance), chintz finishing (usually a glazed fabric with bright prints), and embossing. These finishes are semidurable but are made more durable, especially if the base fabric is cotton or rayon, by including a treatment with a conventional N-methylol agent (see Amino resins).

Chemical Treatments

Generally, chemical treatments are divided into processes involving topical treatments and those involving formation of a chemical bond with the substrate. Most chemical treatments of synthetic fibers are topical because of the general inertness of the substrate. Because the cellulosics and wool are reactive, these fibers may readily form chemical bonds with the finish, as well as act as substrates for nonreactive topical treatment.

Conventional chemical finishing processes involve three steps commonly referred to as pad-dry-cure (Fig. 1).

By the mid or late 1950s, most textile chemists had decided that cross-linking of cellulose and not polymer formation within or about the cellulose fiber was primarily responsible for the desired smooth-drying properties. However, even today the role of polymer formation with aminoplast resins in promoting smooth-drying properties is not completely discarded.

The appreciable loss of strength and abrasion resistance when cellulosics are cross-linked can be minimized, but this is usually accompa-

Table 1.   Classification of Textile Fibers

Naturally occurring fibers
    vegetable (based on cellulose), cotton, linen, hemp, jute, ramie
    animal (based on proteins), wool, mohair, vicuna, other animal hairs, silk
    mineral, asbestos
Man-made fibers
    based on natural organic polymers
      rayon, regenerated cellulose
      acetate, partially acetylated cellulose derivative
      triacetate, fully acetylated cellulose derivative
      azlon, regenerated protein
    based on synthetic organic polymers
      acrylic, based on polyacrylonitrile (also modacrylic)
      aramid, based on aromatic polyamides
      nylon, based on aliphatic polyamides
      olefin, based on polyolefins (polypropylene)
      polyester, based on polyester of an aromatic dicarboxylic acid and a dihydric alcohol
      spandex, based on segmented polyurethane
      vinyon, based on polyvinyl chloride
    based on inorganic substances
      glass
      metallic
      ceramic

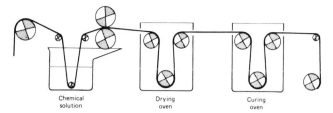

Figure 1. Pad–dry–cure textile finishing.

nied by a loss in smooth-drying performance. One solution to this problem of loss of strength and abrasion resistance is to blend polyester fibers into the yarns that make up the fabric. The cellulosic portion of the fabric still loses strength because of cross-linking, but the effect is masked by the polyester.

Chemistry of N-methylol cross-linking systems. The urea-formaldehyde (UF) reactant system is not dimethylolurea, as depicted in eq 1, but generally a methylolated urea or a partially methylated methylolurea in which the ratio of formaldehyde to urea is 1.3–1.8 : 1. This ratio provides a reasonably stable pad bath, and the reactant, in the presence of a catalyst, is capable of cross-linking cellulose, polymerization, or both. Improved treatments have been developed because finishes based on UF suffer from hypochloride damage, lack of durability, and excessive release of formaldehyde.

$$HOCH_2NHCONHCH_2OH + 2\ CellOH$$
$$\rightleftharpoons CellOCH_2NCONHCH_2OCell + 2\ H_2O \tag{1}$$

Reaction mechanisms for formation and hydrolysis of finishes. In the early 1960s research on these reaction mechanisms, as applied to chemical finishing of cellulosic fabrics, commenced. The formation and hydrolysis (acid) reactions were shown to pass through the same intermediate immonium–carbonium ion.

$$-CONHCH_2OH + H^+ \rightleftharpoons -CONHCH_2\overset{H}{\underset{+}{O}}H \tag{2}$$

$$-CONHCH_2\overset{H}{\underset{+}{O}}H \rightleftharpoons -CONHCH_2^+ + H_2O \tag{3}$$

$$-CONHCH_2^+ \leftrightarrow -CONH\overset{+}{=}CH_2 \tag{4}$$

$$-CONHCH_2^+ + CellOH \rightleftharpoons -CONHCH_2OCell + H^+ \tag{5}$$

Curing catalysts. Curing catalysts are mineral or organic acids and latent acids, such as ammonium salts, amine salts, and metal salts. Metal salts, such as zinc nitrate and magnesium chloride, have been the most widely used.

Melamines and other amino-s-triazines. These finishes are unlike UF finishes in that hypochlorite bleaching during laundering does not cause fabric degradation, but yellows the fabric. The N-chloromelamines are thermally stable, but a dilute solution of sodium bisulfite removes color from the fabric. Conventional melamine agents are poor from the standpoint of loss of methylol groups, and even a fully methoxymethylated melamine or amino-s-triazine resins develops sites for color formation after processing and laundering.

Alkyleneureas, triazones, and urons. These cyclic ureas were the first commercial N-methylol products offered as cross-linking agents for reducing or eliminating the hypochlorite bleaching problem. They are still available commercially but have only limited use.

Delayed cure and permanent press. With emphasis on high performance or permanent press, the curing step may be delayed. Fabric is padded as before with a cross-linking formulation, dried, and fabricated into the garment, which is then cured. In a popular variation of this process, fabrication of the garment follows curing, and creases are formed by high pressure, high temperature presses. The presses are curing the cross-linked fabric a second time by breaking cross-links formed in the initial cure of the flat fabric and allowing them to reform in the fabric in the press. If the fabric in the press is creased, the reformed cross-links keep the fabric in this state.

Sources of formaldehyde in finished fabrics. The release of formaldehyde from a chemically treated fabric has been reduced significantly. A urea–glyoxal–formaldehyde adduct, 1,3-dimethylol-4,5-dihydroxyethyleneurea (DMDHEU), has been the primary cross-linking agent used in the United States since the combination of decreased formaldehyde release, as measured by the AATCC Test Method 112, and an emphasis on durable properties, ie, hydrolysis resistance, essentially eliminated the other standard agents from general use. Another finish, based on N,N-dimethylolcarbamates, is the only other durable low-formaldehyde treatment used industrially, and its use is only about one-fifth that of DMDHEU. 1,3-Dimethyl-4,5-dihydroxyethyleneurea has been used almost exclusively since 1974 for nonformaldehyde finishing.

Miscellaneous Finishing Processes

The finishing of synthetic fibers is limited primarily to heat setting, soil release, hand or comfort effects, and antistatic agents (qv).

Obtaining properties of synthetic fibers in cotton. Cotton and cotton–polyester blends have been chemically modified to accept disperse dyestuffs and thereby rival polyester and other synthetics as a substrate in transfer printing. Successful treatments include the use of glycols with cross-linking agents, of melamine–formaldehyde resins, and acetylation. For achieving durable-press performance, no elastomer equals the conventional N-methylol cross-linking reactants.

Elastic properties needed for stretch textile fabrics are inherent or readily developed in most synthetic, thermoplastic fibers.

Outdoor fabrics. Outdoor fabrics must withstand the combined effect of mildew and rot, sunlight, airborne acids, alternate wetting and drying, and abrasion. Treated cotton has been widely used. The lighter, stronger, microbial-resistant synthetic fabrics have made large inroads into markets formerly held by cotton, even though the service life of outdoor cottons has been significantly increased since 1960 (see also Fungicides; Industrial antimicrobial agents).

The microorganisms that grow on cotton textiles cause several undesirable effects. Loss of strength is the most serious effect, but they may also stain the fabric.

Three different types of agents or processes are used to protect cotton against microorganisms. Certain phenols, organometallic compounds, and inorganic salts are toxic to microorganisms. Many of the highly effective organometallic compounds (particularly those of mercury) are highly toxic to humans, and other agents have been developed. Copper salts have been widely used, particularly copper 8-quinolinolate, but because of its color, its use is limited (see also Industrial antimicrobial agents).

Other processes used to protect cotton against microorganisms include a blocking mechanism and chemical modification of cellulose.

Flame retardants. Flame-retardant finishes provide protection to cotton during weathering, and there has been a trend to develop flame-retardant standards for outdoor fabrics (see Flame retardants for textiles).

Tris is an effective flame retardant on some synthetic fibers but not in finishes for cellulosic fabrics. A number of other phosphorus compounds have been used effectively to flameproof cellulosics for use in garments. Currently, the finishes based on tetrakis(hydroxymethyl)phosphonium chloride (THPC) or the sulfate (THPS) appear to be the most widely used.

Tentage fabrics are generally treated with an antimony oxide-halocarbon topical finish. A number of states have enacted flame-retardant standards for tents, and the Canvas Products Association International has been successful in its efforts to obtain uniformity.

Finishing of wool. Wool (qv) competes for markets primarily with the noncellulosic, synthetic fibers where warmth and wrinkle recovery are desired. Its disadvantages are fuzzing, shrinkage, and difficulty in creasing. Chlorination, modification of the disulfide cross-links, and polymer treatments appear to be chemical reactions of choice.

Hand, comfort, and bioactivity. Factors involving the thermal and moisture-transparent aspects of comfort are possibly the simplest to evaluate and are important in the finishing of textiles, especially of synthetics. Stiffness of a fabric is also of primary concern.

A variety of chemical products and finished fabrics said to have antibacterial finishes are on the market. Two promising new finishes of this type are based on an organosilicon quaternary ammonium chloride compound or on peroxide-containing complexes of zinc acetate.

Health and Safety Factors

The principles of toxicology for textile dyeing and finishing have been reviewed. Hundreds of substances have been examined; the LD_{50}s for common finishing chemicals range from 800 for formaldehyde to 7070 for glyoxal. Many polymeric materials used to coat textiles are durably bonded to the fiber and are essentially inert even if removed.

The Consumer Product Safety Commission (CPSC) has been active in a number of textile-related matters involving potential carcinogenicity. The most notable case was the ban on Tris and the loss of formaldehyde from durable-press and other fabrics.

Contact textile dermatitis has been traced to dyes, formaldehyde, unreacted monomer from synthetic fibers, and an early flameproofed cotton fabric. Skin irritations have also been related to the harshness of the fiber.

SIDNEY L. VAIL
Southern Regional Research Center, USDA

H.F. Mark, S.M. Atlas, and E. Cernia, eds., *Man-Made Fibers*, Vol. 3, Interscience Publishers, a division of John Wiley & Sons, Inc., New York, 1968, p. 499.

H.F. Mark, N. Woodling, and S.M. Atlas, eds., *Chemical Aftertreatment of Textiles*, Interscience Publishers, a division of John Wiley & Sons, Inc., New York, 1971, p. 135.

J.T. Marsh, *An Introduction to Textile Finishing*, 2nd ed., Chapman and Hall Ltd., London, 1966; early and current methods of chemical and mechanical finishing.

TESTING

The properties of any textile structure, whether made by weaving, knitting, or nonwoven technology, are determined by variables related to the selected fiber, yarn structure, fabric structure, and chemical, thermal, or mechanical finish (see Fibers, chemical; Nonwoven textile fabrics). The tests performed on textile structures relate to a measure of use performance and can be categorized as objective or subjective tests. The first type relates to physical performance and includes strength, elastic behavior, shrinkage in care, ironability, colorfastness, and resistance to tear, abrasion, pilling, and degrading media, eg, sunlight, heat, and various chemicals. Subjective test measurements relate more to textile esthetics than to engineering performance and include fabric hand, appearance following laundering or dry cleaning, luster, and comfort.

The properties of the finished textile structure or fabric are an accumulation of the properties of the fiber (or fibers if it is a blend), yarn configuration, construction, and the selected finish. Specific properties are, for fibers: diameter (tex), length, cross-sectional shape, crimp (staple plus textured filament), chemical composition, and fiber structure (molecular orientation and crystallinity); for yarn: fiber properties, fineness or count (number of fibers per cross-sectional area), twist and migration of fibers throughout the bundle cross section, compactness, surface character (hairy or smooth), texture for textured filament yarns, and blend ratio and uniformity of mix (intimate blended yarns); and for fabrics: fiber properties, yarn properties, design (threads per unit length in warp and weft of woven fabric or courses and wales of knitted structure), tension during fabrication, compactness of weave or knit structure, and mechanical or chemical finish. The properties of the textile structure are dictated by interactions of all of the parameters.

The textile industry and ASTM have developed test standards which, through much experience of relating test properties to actual performance, can be used as guidelines for predicting satisfactory performance for many consumer textile items. Textile test data are also useful to the buyers and sellers of such products.

ROBERT W. SINGLETON
University of Connecticut

Book of ASTM Standards, Parts 32 and 33, ASTM, Philadelphia, Pa., 1980.

AATCC Technical Manual, American Association of Textile Chemists and Colorists, Durham, N.C.

R.W. Singleton in H.F. Mark, S.M. Atlas, and E. Cernia, eds., *Man-made Fibers—Science and Technology*, Vol. 3, Interscience Publishers, a division of John Wiley & Sons, Inc., New York, 1968, pp. 571–634.

THALLIUM AND THALLIUM COMPOUNDS

Thallium Metal

Thallium belongs to Group IIIA of the periodic table with boron, aluminum, gallium, and indium. ^{203}Tl (29.5%) and ^{205}Tl (70.5%) are the two stable naturally occurring isotopes.

Thallium is not a particularly rare metal and its abundance in the earth's crust is ca 0.3 ppm. It occurs not only in oxide minerals but also as a chalcophilic element. The metal commonly occurs in potash minerals and in a number of other thallium-containing minerals; crookesite and lorandite are the most important ones.

Properties. Thallium is a grayish-white, heavy, soft metal. The physical properties are given in Table 1.

Production. Thallium is obtained as a by-product in the roasting of zinc blende and lead sulfide ores for the production of sulfuric acid (see Sulfuric acid and sulfur trioxide).

Uses. Thallium has limited commercial applications because of its toxic nature. A number of binary, ternary, and quaternary eutectic alloys are known with very low coefficients of friction and good resistance to acids. These alloys can be used in bearings. The most important alloy is the mercury–thallium alloy which forms a eutectic at 8.7 wt% Tl with a melting point of −60°C. It can be used as a substitute for mercury in switches and in seals for equipment used in polar regions or the stratosphere.

Thallium Compounds

The properties of the more important thallium compounds are listed in Table 2.

Organometallics. Organothallium compounds have attracted a great deal of interest in recent years, mostly because of their applications in organic synthesis. Stable compounds occur in both Tl(I) and Tl(III) oxidation states.

Organothallium(III) derivatives can be classified into three types: R_3Tl, R_2TlX, and $RTlX_2$. The trialkyl derivatives are reactive, unstable compounds, whereas the dialkyl derivatives are among the most stable and least reactive organometallic compounds. The monoalkyl compounds are unstable and often cannot be isolated. They are important intermediates in some Tl(III)-promoted organic reactions (see Organometallics).

Table 1. Physical Properties of Thallium

Property	Value
atomic weight	204.37
melting point, °C	303
boiling point, °C	1457
density, g/cm^3	11.85
thermal conductance, W/(cm·K)^a	0.39
specific heat (at 20°C), J/g^b	0.13
heat of fusion, J/g^b	21.1
heat of vaporization, J/g^b	795
Brinell hardness	2
linear coefficient of expansion	28×10^{-6}
electrical resistivity, μΩ·cm	18
tensile strength, MPa^d	9.0

^{a} To convert W/(cm·K) to (cal·cm)/(s·cm^2·°C), divide by 4.184.

^{b} To convert J to cal, divide by 4.184.

^{c} To convert MPa to psi, multiply by 145.

Table 2. Properties of Thallium Compounds

Compound	Formula	Mp, °C	Density, g/cm³	Solubility, g/100g, water (°C)
thallous carbonate	Tl₂CO₃	272	7.16	5.2 (18), 22.4 (100)
thallous formate	TlO₂CH	101	4.967	500 (10)
thallous acetate	TlO₂CCH₃	131	3.765	very soluble
thallous fluoride	TlF	327	8.23	78.6 (15)
thallic fluoride	TlF₃	550 dec	8.65	insol
thallous chloride	TlCl	430	7.0	0.32 (20), 2.38 (100)
thallic chloride	TlCl₃	155		very soluble
thallous bromide	TlBr	456	7.5	0.05 (20), 0.25 (60)
thallic bromide	TlBr₃			sol
thallous iodide	TlI	440	7.29	0.0006 (20)
thallous oxide	Tl₂O	300	9.52	dec
thallic oxide	Tl₂O₃	7.7	10.11	insol
thallous hydroxide	TlOH	139 dec	7.44	25.9 (0), 52 (40)
thallous nitrate	TlNO₃	206	5.55	8 (15), 594 (104.5)
thallic nitrate	Tl(NO₃)₃			dec
thallous sulfate	Tl₂SO₄	632	6.77	4.87 (20), 18.45 (100)

Toxicology

Based on industrial experience, 0.10 mg/m³ of thallium in air is considered safe for a 40-h work week. The lethal dose for man is not definitely known, but 1 g of absorbed thallium is considered a lethal dose for an adult; however, 10 mg/kg body weight has been fatal to children.

BENJAMIN C. HUI
Alfa Products

A.G. Lee, *The Chemistry of Thallium*, Elsevier Publishing Co., Amsterdam, 1971.

A. McKillop and E.C. Taylor in F.G.A. Stone and R. West, eds., *Advances in Organometallic Chemistry*, Vol. 2, Academic Press, New York, 1973, p. 147.

Charles Pfizer and Co., Inc., *Thallium Poisoning*, *Spectrum* (*N.Y.*) **6**, 558 (1958).

THERMAL POLLUTION BY POWER PLANTS

An important by-product of most chemical technologies, in the broad sense, is heat. Concern over heat rejection arose when the quantities at localized sites rose dramatically as the electric utility industry shifted to water-cooled, thermal-electric generating stations of high unit capacity in the 1950s, particularly in the UK and the United States. Concern was further heightened by a planned shift to nuclear-fission reactors as the energy source, which entailed both a further increase in localized generating capacity because of economics of scale and a higher percentage of rejected heat compared to useable electrical energy. The term thermal pollution took on fearsome portents among aquatic scientists, fishery managers, and eventually water-pollution-control agencies (see Water, water pollution). Direct lethal effects of high temperatures on aquatic life were predicted and, where sublethal temperatures were maintained, effects on reproductive cycles, growth rates, migration patterns, and interspecies competition were hypothesized based on the pervasive nature of temperature as a factor which controls ecological processes.

Initially, the source of environmental risk from cooling water was assumed to be the pollutant discharged, ie, heat, in the form of the elevated temperature of the water released from the condensers. Heat is now recognized as only one of several potential risks of power-station cooling. Generally unheralded until the early 1970s was the physical entrapment and impingement of fish on cooling-water intake screens.

Biocides, principally chlorine used periodically (0.5 h/d per condenser) for condenser cleaning, were identified as toxic risks for organisms in the cooling circuit at the time and for those in the vicinity of the discharge where the biocide dissipates (see Industrial antimecrobial agents; Water, industrial, water treatment).

Another risk derives from combined damages, ie, thermal, physical, and chemical, sustained by small organisms, especially young fish, which are pumped through the cooling system, ie, entrainment.

Questions also arose over the risks to aquatic life from changes in gas content of the water as a by-product of temperature change; warmer water holds less gas in solution.

Physical changes in habitats near the cooling water intake and discharge structures of power stations were also identified as posing some risk or at least potential for change to segments of aquatic communities.

Risks from cooling towers were identified as comparative ecological analyses of cooling-system alternatives became part of the assessment process.

Minimizing the Risks

Selection of the best system involves matching engineering options to the local aquatic system potentially at risk. General principles of aquatic ecology and of the life histories and environmental requirements of species represented locally can be adapted to local water-resource goals and with detailed understanding of the local aquatic setting to achieve site-specific risk prevention.

Thermal effects. Despite the immensity and complexity of known and suspected roles of temperature in aquatic ecosystems, certain thermal criteria have been especially useful in minimizing risks from thermal discharges, at least those short-term ones which have been most thoroughly investigated. More data have been organized at the physiological level than at higher levels of organization, so that such data have factored prominently in current approaches.

Preventing mortality. Upper and lower temperature tolerances of aquatic organisms have been well conceptualized over the past 30 yr, and standardized methods are available for determining a species' tolerance ranges under different conditions of thermal history so that brief exposures to potentially lethal temperatures are not actually lethal. Graphical and mathematical representations of these data for important species at a site allow a design engineer to tailor temperature elevations and duration of exposure in a plant's piping or in the effluent mixing zone to maximize organism survival. This is usually accomplished with detailed mathematical models of cooling-water-effluent dispersion and heat dissipation in the near field where temperatures are highest.

Preventing stressful high temperatures over long periods. Standards for upper limits on water temperatures for particular water bodies over periods of about one week or more can be based on species-specific growth rates. With an inventory of important species and life stages in the area during the warm season, the analyst can ascertain the upper temperature limit which does not stress those in the desired aquatic assemblage. Hydrothermal models of heat dissipation in the far field beyond the zone of effluent mixing are important for estimating the zones that may present a long-term risk from elevated temperatures.

Recently, the preferred and avoided temperatures in a gradient have been used as surrogates for optimum growth and upper danger levels for fish.

Preserving reproduction cycles. Reproduction success depends in part on the preservation of an annual temperature pattern, although the precise timing is usually not critical. An analyst can obtain dates and temperatures for the spawning of many important aquatic species, and thermal discharges can be designed, usually with the help of mathematical models, to assure the necessary thermal periodicity.

Maintaining ecosystem structure and function. Thermal heterogeneity of water bodies is an important structural feature of the environment and plays a large role in determining the composition and functioning of most aquatic systems.

Careful planning of thermal additions, including creation of new cooling reservoirs, can yield thermal structures which enhance rather than damage desirable aquatic species. Knowledge of the thermal niches of these species and their potential competitors or predators permits special provisions for thermal refuges, eg, cool summer zones in a heavily heated cooling pond. From a different perspective, aquatic species introduced to waters used for power-station cooling can be selected so that their thermal niche matches the thermal structure that the facility creates.

Impingement.   Current perspective on minimizing impingement risks focuses on site-specific analyses of potentially vulnerable species and selection of engineering designs which, within acceptable cost limits, keep impingement deaths few.

Biocides.   Chlorine and other biocides are used occasionally in cooling water to kill and dispose of organic growths on heat-exchange surfaces and on piping where water flow could be hampered by such growth. Of necessity, organisms passing through the cooling circuit or residing in the effluent area during periodic chlorine injections experience the potentially lethal exposures. The objective of conscientious plant operators is, however, to maximize the intended kill and minimize extraneous damages, particularly in the receiving water. Methods for accomplishing this goal have been developed progressively over the past several years.

Entrainment.   Stresses to small nonscreenable organisms, eg, fish larvae, during passage through the cooling circuit come from a combination of thermal shock, physical abuses, and periodic injections of biocide.

A principal frustration in attempts to minimize entrainment damages has been the contradictory demands of thermal and physical stresses. Thermal stresses can be quantitatively predicted based on dose-response data and minimized by increasing water-flow volumes, which dilute the fairly constant supply of rejected heat. The added volume of cooling water, however, includes proportionately more planktonic organisms, which are subjected to physical stresses.

The assumption of high percentage mortality resulting from physical stresses has recently been criticized. It appears that ameliorations of thermal stress with flow increases generally are not cancelled by additional physical mortalities except in cases of exceedingly high flows. An optimization procedure, such as that suggested by the Committee on Entrainment, appears fruitful for identifying on a site-specific and seasonally varying basis the most appropriate cooling-water flow regime.

Gas balance.   Damaging supersaturation of dissolved gases has occurred in some cooling-water discharges. The practice of winter increase in temperature rise across condensers by cutting back on pumping capacity has either ceased in general or the immediate discharge areas have been engineered to prevent long-term residence by susceptible biota. These remedial measures can be completely effective.

Cooling-tower chemicals.   The new risk posed by changing power-station systems from traditional once-through or open-cycle cooling to cooling towers is from chemicals added to the recirculating water. Blowdown can be treated for removal of chemical constituents, eg, chromates, with the additional benefit of chemical recycling and, thus, cost recovery. Chemical-laden sludges become a more long-term disposal problem; current practices include ponding and landfills. Chemical-recovery processes are also available for sludge treatment.

Human pathogens.   Most studies of the risks to humans from pathogens stimulated by environmental conditions in cooling systems, eg, amoebae, *Legionella*, are preliminary. There is, however, considerable indication that proper designs and good maintenance practices can reduce the potential risks to very low levels.

Maximizing the Benefits

Increasing attention is being given to finding productive uses for power-plant waste heat. The waste of approximately two thirds of the fuel energy during conversion of fossil or nuclear fuels to electricity is increasingly viewed as extravagant. Methods are thus being developed for converting the thermal pollution from power stations into useful heat resources. There are potential physical applications of power-plant rejected heat, eg, industrial heating, and biological applications, such as fish culture, soil warming, and heating greenhouses and livestock shelters.

Aquaculture.   Culturing of some aquatic species in essentially unmodified thermal effluents of power stations has been attempted both experimentally and commercially (see Aquaculture). Use of rejected heat to prolong optimal growth temperatures in cool months can significantly increase the sizes attained in a year.

Open-field agriculture.   Use of warmed water for field crops and orchards has been tested in several studies. Buried pipes can convey thermal-discharge heat to soils where warming aids plant growth and extends the growing season.

Greenhouse agriculture.   Greenhouse agriculture is well known for its many advantages over open-field agriculture for certain crops. One drawback is the high expense for heating in winter, which can be the most costly part of greenhouse operation. The use of waste heat from steam–electric power plants therefore appears promising as a source of low cost heat for greenhouses.

Animal shelters.   The advantages of temperature control for maximizing weight gain and avoiding animal losses in livestock and poultry are well known. Low grade heat from power-station cooling offers the possibility of low cost heating of animal shelters.

Space heating.   A large percentage of the energy requirements of most countries in temperate zones is for heating and cooling of living and working spaces and for hot water. The historical use of dual-purpose power plants for electricity generation and central district heating in the United States and their current extensive use in such countries as the USSR, Sweden, and the FRG suggests that expansion of this form of waste-heat utilization can contribute significantly to energy conservation and control of concentrated thermal discharges worldwide.

Industrial process heat.   Many industries use process heat at $77-110°C$. Much of this heat is supplied by combustion of oil and natural gas. Equipment manufacturers are developing industrial heat pumps to capture free industrial-plant waste heat and regenerate it to the desired process-heat temperature, thereby greatly reducing energy costs associated with direct heating.

Cooling reservoirs.   The most extensively developed productive use for power-plant cooling is in multiple-purpose cooling reservoirs. Small impoundments built especially for heat dissipation have been managed for recreational uses as well.

CHARLES C. COUTANT
Oak Ridge National Laboratory

W. Majewski and D.C. Miller, eds., *Predicting Effects of Power Plant Once-Through Cooling on Aquatic Systems*, Technical Papers in Hydrology 20, United Nations Educational, Scientific and Cultural Organization (Unesco), Paris, 1979.

J.R. Schubel and B.C. Marcy, Jr., eds., *Power Plant Entrainment—A Biological Assessment*, Academic Press, New York, 1978.

Water Quality Criteria 1972, Report No. R-73-033, National Academy of Sciences—National Academy of Engineering, U.S. Environmental Protection Agency, Washington, D.C., 1973, pp. 151–171 and appendix.

THERMODYNAMICS

The science of thermodynamics deals with energy and its transformations in macroscopic systems, especially in systems for which temperature is a characteristic coordinate. Every physical and chemical process can be examined thermodynamically.

Thermodynamics is built around two fundamental laws that restrict the behavior of material systems. The first law of thermodynamics affirms the principle of energy conservation; the second law states the principle of entropy increase. In the formal application of these laws, attention is focused on a particular object, quantity of matter, or region of space as the system. All else then constitutes the surroundings. A system may be open or closed to the exchange of matter with its surroundings, or it may be isolated from its surroundings, in which case it can exchange neither matter nor energy. Once isolated, a system is independent of its surroundings and can only change toward a final equilibrium state characterized by properties that have no further tendency to change. Systems not in equilibrium states undergo processes during which the properties of the system change; moreover, the system and surroundings may exchange energy in the forms of heat Q and W. A process that proceeds so that the system is never displaced more than differentially from an equilibrium state is said to be reversible, because such a process can be reversed in direction at any time by an infinitesimal change in external conditions.

First and Second Laws of Thermodynamics

The four basic principles of thermodynamics are as follows: (*1*) Associated with systems in equilibrium is an intrinsic property called internal energy U. For closed systems at rest, changes in this property are given by

$$dU = dQ - dW \tag{1}$$

(*2*) First law of thermodynamics: The total energy change of any system together with its surroundings is zero. (*3*) Associated with systems in equilibrium is an intrinsic property called entropy S. For reversible processes, changes in this property are given by

$$dS = \frac{dQ}{T} \tag{2}$$

(*4*) Second law of thermodynamics: The total entropy change of any system with its surroundings is positive and approaches zero for reversible processes. Thus:

$$\Delta S_{\text{total}} \geq 0 \tag{3}$$

The first and third axioms are prerequisite to the other two, which state the laws of thermodynamics. Each asserts the existence of a thermodynamic property, and each provides an equation connecting the property with measurable quantities. These are not defining equations; they merely provide a means to calculate changes in each property.

For a closed system at rest, the first law for a finite process is expressed as

$$\Delta U = Q - W \tag{4}$$

where ΔU is the energy change of the system and Q and W are heat and work, respectively, passing the boundary between system and surroundings. Equation 1 is the differential form of this equation. These equations require the numerical value of Q to be positive for heat transfer to the system and negative for heat transfer from the system. The opposite convention applies to W: numerical values are positive for work done by the system and negative for work done on the system.

Heat Engines and Heat Pumps

The following are Carnot's equations, and they apply to all reversible heat engines (Carnot engines) operating between a heat source at temperature T_H and a heat sink at temperature T_C

$$\frac{|Q_C|}{|Q_H|} = \frac{T_C}{T_H} \tag{5}$$

and

$$\frac{|W|}{|Q_H|} = 1 - \frac{T_C}{T_H} \tag{6}$$

Here, Q_H is the heat applied to the engine at T_H, and Q_C is the heat discarded from the engine at T_C. Equation 6 yields the thermal efficiency of a Carnot heat engine, ie, the fraction of the heat taken in that is converted into work. A Carnot engine is reversible and, thus, can be run backward as a heat pump or refrigerator, as indicated in Figure 1.

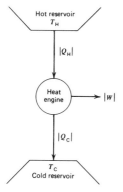

Figure 1. Schematic representation of a Carnot heat engine.

Fundamental Property Relations for Constant-Composition Fluids

The systems of interest in chemical technology are usually comprised of fluids not appreciably influenced by surface, gravitational, electrical, or magnetic effects.

For one mole of a homogeneous fluid of constant composition, thermodynamics provides a fundamental relationship giving the internal energy U as a function of entropy S and volume V:

$$dU - TdS - pdV \tag{7}$$

Alternative forms of this fundamental property relation follow from the definitions:

enthalpy $H = U + PV$

Helmholtz function $A + U - TS$

Gibbs function $G = U + PV - TS$

Taking the differentials of these defining equations and eliminating dU by equation 7 gives

$$dH = TdS + VdP \tag{8}$$

$$dA = -SdT - PdV \tag{9}$$

$$dG = -SdT + VdP \tag{10}$$

Since these are exact differential expressions, the Maxwell equations can be written by inspection; the two most useful ones are derived from equations 9–10:

$$\left(\frac{\partial S}{\partial V} \right)_T = \left(\frac{\partial P}{\partial T} \right)_V \tag{11}$$

$$\left(\frac{\partial S}{\partial P} \right)_T = -\left(\frac{\partial V}{\partial T} \right)_P \tag{12}$$

Enthalpy and Entropy as Functions of *T* and *P*

For a homogeneous fluid of constant composition:

$$H = H(T, P) \tag{13}$$

and

$$S = S(T, P) \tag{14}$$

Therefore:

$$dH = \left(\frac{\partial H}{\partial T} \right)_P dT + \left(\frac{\partial H}{\partial P} \right)_T dP \tag{15}$$

and

$$dS = \left(\frac{\partial S}{\partial T} \right)_P dT + \left(\frac{\partial S}{\partial P} \right)_T dP \tag{16}$$

By definition

$$C_P = \left(\frac{\partial H}{\partial T} \right)_P \tag{17}$$

is the heat capacity at constant pressure. If equation 44 is divided by dT and restricted to constant P:

$$\left(\frac{\partial H}{\partial T} \right)_P = T \left(\frac{\partial S}{\partial T} \right)_P \tag{18}$$

With equation 17, this becomes

$$\left(\frac{\partial S}{\partial T} \right)_P = \frac{C_P}{T} \tag{19}$$

When equation 8 is divided by dP and restricted to constant T:

$$\left(\frac{\partial H}{\partial P} \right)_T = T \left(\frac{\partial S}{\partial P} \right)_T + V \tag{20}$$

In view of equation 12, this becomes

$$\left(\frac{\partial H}{\partial P} \right)_T = V - T \left(\frac{\partial V}{\partial T} \right)_P \tag{21}$$

Combination of equations 15, 17, and 21 gives

$$dH = C_P dT + \left[V - T \left(\frac{\partial V}{\partial T} \right)_P \right] dP \tag{22}$$

Similarly, combination of equations 12, 16, and 19 gives

$$dS = \frac{C_P}{T} dT - \left(\frac{\partial V}{\partial T} \right)_P dP \tag{23}$$

For an ideal gas:

$$PV = RT \tag{24}$$

and

$$\left(\frac{\partial V}{\partial T}\right)_P = \frac{R}{P} = \frac{V}{T} \tag{25}$$

Thus, for an ideal gas, equations 22–23 reduce to

$$dH = C_P \, dT \tag{26}$$

and

$$dS = \frac{C_P}{T} \, dT - \frac{R}{P} \, dP \tag{27}$$

Equations 22 and 23 provide basic relationships that allow calculation of numerical values for the thermodynamic properties H and S from experimental heat capacity and PVT data.

Energy Equations for Steady-Flow Processes

Industrial chemicals are largely produced by continuous processes in which the rates of inflow and outflow of mass are constant and equal. Moreover, conditions at all points in the process are maintained constant with time. A simple steady-flow process is represented in Figure 2. A process occurs within the fixed control volume between points 1 and 2 that changes the properties of a fluid element as it flows from point 1 to point 2. The system is the fluid in the control volume plus the fluid element of mass δm, (Fig. 2a) that enters the control volume at point 1 during time δT. After time interval $\delta\tau$, the system appears as in Figure 2b and consists of fluid in the control volume plus the fluid element of mass δ_{m2} that has left the control volume. For a steady-flow process, $\delta m_2 = \delta m_1$ and the properties of the fluid in the control volume are unchanged. The fluid elements δm_1 and δm_2 have properties as measured at points 1 and 2, and these include a velocity, u and an elevation above a datum level z. Thus, these masses have kinetic and potential energy as well as internal energy, and the general energy balance for a closed system, as given by equation 4, must be changed to

$$\Delta U + \Delta E_K + \Delta E_P = Q - W \tag{28}$$

where Δ indicates the change from the initial to the final state of the system.

Partial Molar Properties

Since the macroscopic intensive properties of homogeneous fluids in equilibrium states are functions of T, P, and composition, it follows that the total property of a phase, say nM, can be expressed functionally as

$$nM = m(T, P, \, n_1, n_2, n_3, \dots) \tag{29}$$

The total differential of nM is, therefore,

$$d(nM) = \left[\frac{\partial(nM)}{\partial T}\right]_{P,n} dT + \left[\frac{\partial(nM)}{\partial P}\right]_{T,n} dP$$
$$+ \sum \left[\frac{\partial(nM)}{\partial n_i}\right]_{T,P,nj} dN_i \tag{30}$$

The derivatives in the summation are called partial molar properties and are denoted by $\overline{M}_i$; thus, by definition:

$$\overline{M}_i \equiv \left[\frac{\partial(nM)}{\partial n_i}\right]_{T,P,nj} \tag{31}$$

Equation 30 can now be written

$$d(nM) = n\left(\frac{\partial M}{\partial T}\right)_{P,x} dT + n\left(\frac{\partial M}{\partial P}\right)_{T,x} dP + \sum \overline{M}_i \, dn_i \tag{32}$$

Important equations follow from this result:

$$dM = \left(\frac{\partial M}{\partial T}\right)_{P,x} dT + \left(\frac{\partial M}{\partial P}\right)_{T,x} dP + \sum \overline{M}_i \, x_i \tag{33}$$

and

$$M = \sum x_i \overline{M}_i \tag{34}$$

Differentiation of equation 34 yields

$$dM = \sum x_i \, d\overline{M}_i + \sum \overline{M}_i \, dx_i \tag{35}$$

If equation 33 and equation 35 are both valid:

$$\left(\frac{\partial M}{\partial T}\right)_{P,x} dT + \left(\frac{\partial M}{\partial P}\right)_{T,x} dP - \sum x_i \, d\overline{M}_i = 0 \tag{36}$$

This result is known as the Gibbs-Duhem equation. It imposes a constraint on how the partial molar properties of any phase may vary with temperature, pressure, and composition.

Fugacity, Fugacity Coefficient, and Activity Coefficient

The fugacity and fugacity coefficient are auxiliary functions useful in phase- and chemical-equilibrium calculations; they are defined in relation to the Gibbs function. For a constant-composition mixture, the fugacity f and fugacity coefficient ϕ are defined by the equations:

$$dG = RT d \ln f \qquad (\text{constant } T, x) \tag{37}$$
$$\lim_{P \to 0} (f/P) = 1 \tag{38}$$

and

$$\phi \equiv f/P \tag{39}$$

These equations apply as well to pure species i, a special case of a constant-composition mixture, as indicated by subscripts: G_i, f_i, and ϕ_i. For species i in solution, $\hat{f}_i$ and $\hat{\phi}_i$ are defined by a similar set of equations:

$$d\overline{G}_i = RT d \ln \hat{f}_i \qquad (\text{constant } T) \tag{40}$$
$$\lim_{P \to 0} (\hat{f}_i / x_i P) = 1 \tag{41}$$
$$\hat{\phi}_i \equiv \hat{f}_i / x_i P \tag{42}$$

The calculation of values for f and ϕ for a constant-composition mixture or a pure species depends on equation 10:

$$dG = V dP \qquad (\text{constant } T, x) \tag{43}$$

Eliminating dG in equation 37:

$$d \ln f = \frac{V}{RT} dP \qquad (\text{constant } T, x) \tag{44}$$

By equation 39:

$$d \ln \phi = d \ln f - \frac{dP}{P} \tag{45}$$

Combining the last two equations and rearranging results in

$$d \ln \phi = \left(\frac{PV}{RT} - 1\right)\frac{dP}{P} = (Z - 1)\frac{dP}{P} \tag{46}$$

where the compressibility factor Z is defined as $Z \equiv PV/RT$. Integration from $P = 0$, where $\phi = 1$, to $P = P$ gives

$$\ln \phi = \int_0^P (Z - 1)\frac{dP}{P} \qquad (\text{constant } T, x) \tag{47}$$

An analogous equation allows calculation of $\ln \hat{\phi}_i$:

$$\ln \hat{\phi}_i = \int_0^P (\overline{Z}_i - 1)\frac{dP}{P} \qquad (\text{constant } T, x) \tag{48}$$

By definition, the activity coefficient is

$$\gamma_i = \frac{\hat{f}_i}{RT} \tag{49}$$

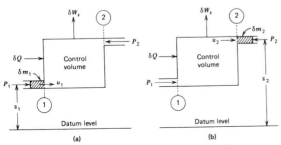

Figure 2. Schematic diagram of a simple steady-flow process.

It is related to the Gibbs function by the equation

$$\ln \gamma_i = \frac{\overline{G}_i^E}{RT} \qquad (50)$$

where

$$\overline{G}_i^E = \overline{G}_i - G_i - RT \ln Y_i \qquad (51)$$

Therefore, $\ln \gamma_i$ is a partial molar property with respect to G^E/RT. Because of the primary role played in γ_i in vapor–liquid equilibrium calculations, this relationship is important. However, for it to be useful, an expression G^E/RT as a function of composition is needed. Of the various empirical expressions that have been proposed to represent this dependence, the modern local-composition equations are of greatest practical use. Notable among these are the Wilson equation, the NRTL equation, and the UNIQUAC equation. All are inherently applicable to multicomponent systems. Two group-contribution methods for the prediction of activity coefficients have developed from local-composition equations and are commonly used. They are the *UNIFAC* method and the *ASOG* method.

Generalized Correlations and *PVT* Equations of State

The volumetric properties of fluids are conveniently represented by generalized correlations or by analytical equations of state. Modern generalized correlations are based on the extended theorem of corresponding states, which in its simplest, three-parameter form asserts that

$$Z = Z(T^r, P^r, \pi) \qquad \text{(all fluids)} \qquad (52)$$

Here, Z is a universal function of the reduced temperature $T^r (\equiv T/T_C)$, the reduced pressure $P^r(\equiv P/P_C)$, and a third corresponding-states parameter π.

The PVT equations of state may be variously classified; the most popular are virial equations and cubic equations.

The virial equations are series representations of Z, with either molar density $\rho(\equiv V^l)$ or pressure P taken as the independent variable of expansion. They are unsuitable for the calculation of properties of liquids or dense gases. The simplest expressions appropriate (in principle) for such applications are equations that are cubic in molar volume. These equations, inspired by the van der Waals equation of state, may be represented by the general formula:

$$P = \frac{RT}{V - b} - \frac{\theta(V - \eta)}{(V - b)(V^2 + \delta V + \epsilon)} \qquad (53)$$

where parameters, b, θ, δ, ϵ, and η can each depend on temperature and composition.

Vapor-Liquid Equilibria (VLE)

A criterion for vapor-liquid equilibrium is

$$\hat{f}_i^l = \hat{f}_i^v \qquad \text{(all } i) \qquad (54)$$

Effective use of this general equation requires explicit introduction of the compositions of the phases. This may be done through the activity coefficient $\hat{\phi}_i$ or the fugacity coefficient $\hat{\phi}_i$. When activity coefficients are used for the liquid phase and fugacity coefficients for the vapor phase, equation (54) becomes

$$x_i \gamma_i f_i = y_i \hat{\phi}_i P \qquad \text{(all } i) \qquad (55)$$

Chemical-Reaction Stoichiometry

Consider a system containing N chemical species, any or all of which can participate in r chemical reactions. The r reactions can be represented schematically by the algebraic equations:

$$0 = \sum_i^N \nu_{i,j} A_i \qquad (j = \text{I}, \text{II}, \dots, r) \qquad (56)$$

where the A_i represent chemical species and $\nu_{i,\text{j}} j$ is the stoichiometric number for species i in reaction j. Each ν_{ij} has a magnitude and a sign:

$$\text{sign} (\nu_{i,j}) = \begin{cases} - \text{ for a reactant species} \\ + \text{ for a product species} \end{cases}$$

If species i does not participate in reaction j, then $\nu_{ij} = 0$.

The stoichiometric numbers provide relationships among the changes in mole numbers of chemical species which occur as the result of chemical reaction. Thus, for reaction j:

$$\frac{\Delta n_{1,j}}{\nu_{1,j}} = \frac{\Delta n_{2,j}}{\nu_{2,j}} = \cdots = \frac{\Delta n_{N,j}}{\nu_{N,j}} \qquad (57)$$

Since all of these terms are equal, they can be equated to the change of a single quantity ϵ_j, called the reaction coordinate for reaction j, thereby giving

$$\Delta n_{i,j} = \upsilon_{i,j} \Delta \epsilon_j \begin{Bmatrix} i = 1, 2, \dots, N \\ j = \text{I}, \text{II}, \dots, r \end{Bmatrix} \qquad (58)$$

Now the total change in mole number n_i is just the sum of the changes $\Delta n_{i,j}$ resulting from the various reactions. Thus, by the last equation:

$$\Delta n_i = \sum_j^r \Delta n_{i,j} = \sum_j^r \nu_{i,j} \Delta \epsilon_j \qquad (i = 1, 2, \dots, N) \qquad (59)$$

If the initial number of moles of species i is n_{i_0} and if the convention is adopted that $\epsilon_j = 0$ for each reaction in this initial state, then

$$n_i = n_{i_0} + \sum_j^r \nu_{i,j} \epsilon_j \qquad (i = 1, 2, \dots, N) \qquad (60)$$

Equation 60 is the basic equation of material balance for a closed system in which r chemical reactions occur.

Criterion for Chemical-Reaction Equilibrium

The equilibrium state of a chemically reactive system is characterized by the r derivative conditions:

$$\left(\frac{\partial G^t}{\partial \epsilon_j} \right) T, P, \epsilon_k = 0 \qquad (j = \text{I}, \text{II}, \dots, r) \qquad (61)$$

subject to the material-balance constraints of equation 60.

Consider the case of a single-phase, multicomponent system undergoing just a single chemical reaction. The total Gibbs function is

$$G^t = n^t G = \sum_i n_i \mu_i \qquad (62)$$

and is minimized subject to the constraints

$$n_i = n_{i_0} + \nu_i \epsilon \qquad \text{(all } i) \qquad (63)$$

Equation 61 requires that

$$\left(\frac{\partial G^t}{\partial \epsilon} \right)_{T, P} = 0 \qquad (64)$$

If equation 64 is applied to equation 62, with $\partial n_i / \partial \epsilon = \nu_i$, then:

$$\sum_i \nu_i \mu_i = 0 \qquad (65)$$

which is the familiar algebraic criterion for single-reaction equilibria.

The approach just illustrated is easily extended to an arbitrary number of independent reactions r and produces the expected generalization of equation 65:

$$\sum_i \nu_{ij} \mu_i = 0 \qquad (j = \text{I}, \text{II}, \dots, r) \qquad (66)$$

Nomenclature

A = Helmholtz function
C_P = heat capacity at constant pressure
C_V = heat capacity at constant volume
E_K = kinetic energy
E_P = potential energy
f = fugacity of a mixture
$\bar{f}_i$ = fugacity of species i in solution
G = Gibbs function
H = enthalpy
M = intensive thermodynamic property of a solution (M can represent a variety of thermodynamic functions, including U, C_V, H, C_P, S, A, G, or Z)

M_i = intensive thermodynamic property of pure species i
$\overline{M}_i$ = partial molar property of species i in solution
m = mass
n = number of moles
n_i = number of moles of species i
P = absolute pressure
Q = heat
R = universal gas constant
S = entropy
T = absolute temperature
U = internal energy
V = volume
W = work
W_s = shaft work for a flow process
x_i = liquid-phase mole fraction of species i
y_i = vapor-phase mole fraction of species i
Z = compressibility factor
z = elevation above a datum level

Superscripts

E = excess
l = liquid phase
r = reduced
v = vapor phase

Subscripts

C = cold
H = hot

Greek letters

γ_i = activity coefficient of species i in solution
μ_i = chemical potential of species i
ν_{ij} = stoichiometric number of species in reaction j
π = corresponding-states parameter (eq. 52)
ϕ = fugacity coefficient of a mixture
ϕ_i = fugacity coefficient of pure species i
$\hat{\phi}_i$ = fugacity coefficient of species i in solution
τ = time

HENDRICK C. VAN NESS
MICHAEL M. ABBOTT
Rensselaer Polytechnic Institute

M.M. Abbott and H.C. Van Ness, *Schaum's Outline of Theory and Problems of Thermodynamics*, McGraw-Hill, New York, 1972.

J.M. Smith and H.C. Van Ness, *Introduction to Chemical Engineering Thermodynamics*, 3rd ed., McGraw-Hill, New York, 1975.

H.C. Van Ness and M.M. Abbott, *Classical Thermodynamics of Nonelectrolyte Solutions: With Applications to Phase Equilibria*, McGraw-Hill, New York, 1982.

THERMOELECTRIC ENERGY CONVERSION

Thermoelectric energy conversion is the interconversion of heat and electrical energy for power generation or heat pumping and is based on the Seebeck, Peltier, and Thomson effects.

The technique of direct energy conversion is characterized by the independence of size vs efficiency, the absence of moving parts, high reliability, quietness, lack of vibration, low maintenance, simple startup, and absence of pollution problems. Thermoelectric generators have been used increasingly in specialized applications in which combinations of their desirable features outweigh their high cost and low generating efficiencies, which are typically ca 3–7%. The most advanced thermoelectric systems are the radioisotope thermoelectric generators (RTSs), which have been developed for military and space systems under the aegis of DOE.

Thermoelectric heat pumping, like thermoelectric power generation, has had increased application in those areas where the advantage of the thermoelectric conversion process, ie, small space, light weight, high reliability, no noise or pollution, and simple temperature control, can be utilized. Thermoelectric cooling devices have been developed for a variety of military and commercial applications (see Refrigeration). These include submarine air-conditioning systems, small refrigerators and recreational cooler chests, cooling for electronic components, temperature-control units, laboratory instruments, and cooling for electrooptical systems. The state of the art is characterized by individual couples have pumping capacities of 1–4 W. Since the early 1970s, the trend in thermoelectric cooling has been increasingly for commercial applications.

The conversion efficiency of a thermoelectric generator and the coefficient of performance of a thermoelectric refrigerator depend upon the properties of the thermoelectric materials as expressed by their figure of merit. To date, three material technologies are established; these are based on bismuth telluride, lead telluride, and the Si–Ge alloys. The development of solid-state materials with enhanced figures of merit have not been realized. Therefore, thermoelectric energy conversion provides a unique solid-state technology that complements rather than replaces existing technologies.

Properties

The Seebeck coefficient, electrical conductivity, and thermal conductivity are properties of materials that can theoretically be related to the atomic structure of the materials. However, any real material is much too complicated to analyze completely, although various simplified models of the electronic and thermal processes in metals and semiconductors have been made, which are useful for describing the various material parameters. The thermoelectric properties of some metals and semiconductors at room temperature are given in Table 1. It may be observed that for a metallic system the highest figure of merit is $6 \times 10^{-4}\,\mathrm{K}^{-1}$. Metallic systems are not good sources of thermoelectric power where high power and efficiency are important, whereas semiconductors, since they have significantly higher efficiencies, do offer the potential for greater device utility.

Semiconductors. The optimum thermoelectric material should be a semiconductor doped to 10^{19}–10^{20} carriers/cm^3 and have a Seebeck coefficient of ca 200 V/°C. For power generation, it should have a large bandgap (ca 1 eV for operation at 800–1000°C), a high melting point, a low vapor pressure, and good mechanical strength, and should be chemically stable (see Semiconductors).

Thermoelectric materials. The chemical, physical, and thermoelectric properties of many materials have been examined for possible utility in thermoelectric devices. These materials include elements, alloys, compounds, and mixtures of compounds. The properties of the best thermoelectric materials are given in Table 2.

Fabrication of Thermoelectric Devices

Thermoelectric materials are prepared by hot pressing, cold pressing, and sintering, polycrystalline-casting, or single-crystal techniques.

Power generation. A thermoelectric generator converts heat directly from a heat source into electrical energy and is basically a power-limited, low voltage, high current d-c device. Most devices require multivolt inputs, and d-c-to-d-c converters are used in these instances to raise the voltage to an acceptable level. Therefore, the essential components of a thermoelectric generator are a heat source, a thermoelectric module, and

Table 1. Thermoelectric Properties of Metals and Semiconductors at Room Temperature

Substance	Seebeck coefficient, μV/°C	Electrical conductivity, S/cm	Thermal conductivity, W/(cm·K)	Figure of merit, K^{-1}
Cu	2.5	5.9×10^5	3.96	9.3×10^{-7}
Ni	18	1.5×10^5	0.87	5.6×10^{-5}
Bi	75	8.6×10^3	0.08	6.0×10^{-4}
Ge	200	1000	0.636	6.3×10^{-7}
Si	200	500	1.133	1.8×10^{-5}
InSb	200	2000	0.17	4.7×10^{-4}
InAs	200	3000	0.315	3.8×10^{-4}
Bi$_2$Te$_3$	220	1000	0.02	2.3×10^{-3}
ZnSb	170	556	0.03	5.8×10^{-4}

Table 2. Properties of Thermoelectric Materials

Material	Mp, °C	Type	Z^a max, K^{-1}	Z^a max, T^b	T max^c, °C
Bi_2Te_3	585	n or p	2.0×10^3	0.6	250
$BiSb_4Te_{7.5}$		p	3.3×10^{-3}	0.99	250
Bi_2Te_2Se		n	2.3×10^{-3}	0.69	250
PbSnTe + MnTe	904	n or p	1.5×10^{-3}	1.00	400
$Si_{78}Ge_{22}$	1215	n	0.63×10^{-3}	0.70	1000
$Si_{78}Ge_{22}$	1215	p	0.80×10^{-3}	0.91	1000
TeSbGeAg	576	p	1.40×10^{-3}	0.56	400
$Cu_{1.97}Ag_{0.03}Se$		p	1.23×10^{-3}	0.95	900
$GdSe_x$		n	1.25×10^{-3}	1.59	1000

[a] Z = figure of merit.
[b] Product of maximum figure of merit and temperature K.
[c] Maximum operating temperature.

a d-c-to-d-c converter. Since ca 1960, thermoelectric generators have been built using oceanic temperature gradients, and solar, fossil-fuel and nuclear-energy heat sources.

ALFRED C. GLATZ
General Foods Corporation

D. Benson and T. Jayadev, *Third Int. Conf. Thermoelectr. Energy Convers.* University of Texas at Arlington, Dallas, Texas, Mar. 12-14, 1980.

G. Bennett, J. Lombardo, and B. Rock, *Proc. Intersoc. Energy Convers. Eng. Conf.* 16, 362(1981).

G. Guazzoni, A. Herchakowski, and J. Angello, *Proc. Intersoc. Energy Convers. Eng. Conf.* 13, 1978 (1978).

THIAZOLE DYES

DIRECT DYES

Dyes are classified according to their mode of application, eg, acid dyes, direct dyes, disperse dyes, etc, or by their chemical structure, eg, azo dyes (qv), anthraquinone dyes, xanthene dyes (qv). The thiazole nucleus is a moiety common to basic, direct, disperse, cationic, and vat dyes.

(1)

Although these dyes once enjoyed wide usage, they have declined commercially in recent years; only four products were reported by the U.S. International Trade Commission in 1980 as being manufactured in or imported into the United States (see also Dyes and dye intermediates):

Basic Yellow 1	(CI 49005)
Direct Yellow 27	(CI 13950)
Direct Yellow 28	(CI 19555)
Vat Yellow 2	(CI 67300)

Direct dyes are commercially important because of their excellent substantivity to cellulose attributed to the ability of the dye molecule to form hydrogen bonds with the cellulosic substrate. These bonds form most readily when the dye molecule assumes a linear and coplanar conformation. The basic building block for this type of dye is dehydrothio-*p*-toluidine [2-(*p*-aminophenyl)-6-methylbenzothiazole], prepared by the fusion of *p*-toluidine with sulfur.

Basic Dyes

Basic dyes are synthesized from dehydrothio-*p*-toluidine by methylation in hydrochloric or sulfuric acid at 160–170°C under pressure.

Among the vat anthraquinone dyes, only Vat Yellow 2 is of commercial importance.

Health and Safety Factors

As a class, dyes derived from the thiazole nucleus have not exhibited properties warranting special health and safety precautions; however, standard chemical labeling instructions are required.

RUSSELL E. FARRIS
Sandoz, Inc.

K. Venkataraman, *The Chemistry of Synthetic Dyes*, Vol. 1, Academic Press, Inc., New York, 1952, pp. 622–627.

(BIOS) (*British Intelligence Objectives Subcommittee*) *Final Report 1153*, 1946, pp. 349–352 and *Final Report 987*, 1946, p. 59.

H.E. Fierz-David and L. Blangey, *Fundamental Processes of Dye Chemistry*, Interscience Publishers, Inc., New York, 1949, pp. 333–335.

DISPERSE DYES

Many disperse dyes contain a thiazole nucleus. The term disperse describes the manner in which this class of dyes is applied to textile fibers, ie, from a dispersion in water with the aid of a dispersing agent, eg, sodium lignosulfonate (see Dispersants; Dye carriers). The thiazole dyes in this class are azo compounds derived from 2-aminothiazole and 2-aminobenzothiazole derivative, insoluble or sparingly soluble in water (see Azo dyes). Since 1950, their main use has been for polyester fibers.

Azo dyes with improved lightfastness on acetate, triacetate, and polyester fibers can be obtained from 2-aminobenzothiazole with electron-withdrawing substituents instead of electron-donating substituents.

Cationic Dyes

The cationic thiazole dyes are characterized by a positive charge on the dye molecule. Acrylic and modacrylic fibers and the more recent basic dyeable polyester and polyamide fibers are produced with acid sites. During the dyeing process, the colored cation of the dye molecules associates with the fiber's acidic sites, the anion.

Health and Safety Aspects

The treatment of effluent from dye and dye intermediate factories and the toxicity of disperse dyes has been described (see Dyes and dye intermediates).

J.G. FISHER
G.T. CLARK
Eastman Kodak Company

D.R. Baer in K. Venkataraman, ed., *The Chemistry of Synthetic Dyes*, Vol. 4, Academic Press, Inc., New York, 1971, pp. 188–200.

J.M. Straley in K. Venkataraman, ed., *The Chemistry of Synthetic Dyes*, Vol. 3, Academic Press, Inc., New York, 1970, p. 425.

THIOCYANATES. See Sulfur compounds.

THIOGLYCOLIC ACID

Thioglycolic acid (2-mercaptoacetic acid), $HSCH_2COOH$, is the sulfur analogue of glycolic acid. It is a colorless liquid with a distinct sulfurous odor. Its primary use in the form of its salts and esters as a component of hair-waving and depilatory compositions. The metal mercaptide derivatives of its esters are used as stabilizing additives to prevent deterioration of plastics during processing and use (see Hair preparations; Heat stabilizers).

Properties

Pure thioglycolic acid freezes at −16.5°C and distills unchanged under reduced pressure. Reported constants include: bp 123°C at 3.9 kPa (29 mm Hg) and 90°C at 0.8 kPa (6 mm Hg); d_4^{20} 1.325 g/cm^3; and heat of combustion 1446 kJ/mol (345.6 kcal/mol). The acid is miscible with water, methanol, ethanol, acetone, ethyl ether, chloroform, and benzene, but insoluble in aliphatic hydrocarbons.

Reactions.   Thioglycolic acid is stable at room temperature in $\leq 70\%$ aqueous solutions. At higher concentrations, self-esterification takes place to some extent.

One of the most widely used reactions of thioglycolic acid is that with disulfides, particularly cystine, and in proteins, eg, wool (qv) and hair.

Manufacture

Thioglycolic acid is manufactured by the reaction of sodium or potassium chloracetate with alkali metal hydrosulfide in an aqueous medium under controlled conditions of pressure, temperature, and pH to give a high yield of thioglycolate salt in solution and to minimize the formation of such by-products as the salts of thiodiglycolic acid, dithiodiglycolic acid, and glycolic acid. The reaction mixture is acidified to liberate thioglycolic acid, which is extracted from the solution into an organic solvent and then purified by vacuum distillation or converted directly into derivatives.

Health and Safety Factors, Toxicology

Thioglycolic acid is a stronger acid than acetic acid and must be handled with precautions appropriate to corrosive liquids. Neutral or slightly alkaline solutions of thioglycolates are less irritating than solutions of the free acid.

Derivatives

Derivatives include thioethers; esters; amides; mercaptoacetylated amino acids, peptides, and proteins; and metal complexes.

OTTO S. KAUDER
Argus Chemical Corporation

W.W. Edman and co-workers, *Cosmet. Toiletries* **94**(4), 35 (1979); B. Bach, *Cosmet. Toiletries* **94**(4), 43 (1979).

W.H. Starnes, Jr., in D.L. Allara and W.L. Hawkins, eds., *Stabilization and Degradation of Polymers*, American Chemical Society, Washington, D.C., 1978, pp. 309–323; L.R. Brecker, *Pure Appl. Chem.* **53**, 577 (1981).

THIOLS

Thiols are the sulfur analogues of alcohols, ie, the oxygen in the alcohol is replaced with sulfur. Until 1930, these compounds were named as mercaptans and, particularly in commerce, the name mercaptan is still widely used.

Occurrence

Naturally occurring compounds containing thiol groups and related disulfide (RSSR) linkages are important in biological processes. Cysteine and cystine are two common amino acids which can be isolated from protein hydrolysates. They are generally considered to be interconvertible *in vivo*:

$$\underset{\substack{| \\ NH_2 \\ \text{cysteine}}}{HSCH_2CHCOOHO_2} \xrightarrow[\text{H}_2]{\text{O}_2} \underset{\substack{| \\ NH_2 \\ \text{cystine}}}{[SCH_2CHCOOH]_2}$$

Although there are only small amounts of cysteine and cystine in living tissue, most of the thiol and disulfide groups in nature come from combined forms of these two acids.

Properties

Thiols are notorious for their obnoxious odors; this is particularly true for the lower alkanethiols. The odors can be detected at very low concentrations, eg, ethanethiol can be detected by humans in air at less than 0.5 ppb (0.5×10^{-9}). The odor thresholds of various thiols have

Table 1.   Physical Constants of Some Alkanethiols

Compounds	Boiling point,°C$_{kPa}{}^a$	Freezing point,°C	Refractive index, n_D^{20}	d_4^{20}, g/cm³
methanethiol	5.9	−122.97		0.8665
ethanethiol	35.0	−147.89	1.43105	0.8391
2-propanethiol	52.6	−130.54	1.43554	0.8143
1-propanethiol	67.8	−113.13	1.43832	0.8415
2-methyl-2-propanethiol	64.2	1.11	1.42320	0.8002
2-butanethiol	85.0	−140.14	1.43673	0.8299
2-methyl-1-propanethiol	88.5	−144.86	1.43877	0.8343
1-butanethiol	98.4	−115.67	1.44298	0.8416
1-pentanethiol	126.5	−75.7	1.44692	0.8421
1-hexanethiol	152.6	−80.49	1.44968	0.8424
1-heptanethiol	176.9	−43.23	1.45215	0.9431
1-octanethiol	199.1	−49.2	1.4540	0.8433
1-decanethiol	240.6	−26	1.4569	0.8443
1-dodecanethiol	142.5$_2$	−9.2	1.4589	0.8450
1-hexadecanethiol	123–8$_{0.07}$	18–20		
1-octadecanethiol	188$_{0.1-0.2}$	28	14645	0.8475
cyclohexanethiol	158		1.4921	0.9782

^{a}To convert kPa to mm Hg, multiply by 7.5.

Table 2.   Boiling and Melting Points of Various Thiol Types

Compound	Boiling point, °C$_{kPa}{}^a$	Melting point, °C
1,2-ethanedithiol	146	−41.2–49.1
2,2-propanedithiol	57$_{13}$	4–6
benzenethiol	169.1	−14.9
p-benzenedithiol		97.5–98.5
pyridine-2-thiol		130–132
thiophene-2-thiol	54$_{0.7}$	

^{a}To convert kPa to mm Hg, multiply by 7.5.

been measured. The physical constants of various alkanethiols are shown in Table 1, and of a few other thiols and dithiols in Table 2.

Thiols undergo oxidation, addition to carbon-carbon multiple bonds, addition to carbonyls, and formation of mercaptides.

Preparation

The bulk of the alkanethiols manufactured on a commercial scale are prepared by the reaction of hydrogen sulfide with olefins or with alcohols.

Laboratory methods.   There are a number of other synthetic routes used primarily for small-scale preparation. The first syntheses of thiols were achieved by the reaction of an alkyl sulfate with a metal hydrosulfide. The similar reaction of an alkyl halide or *p*-toluenesulfonate with sodium or potassium hydrosulfide has been used extensively:

$$RX + NaSH \rightarrow RSH + NaX$$

Hydrolysis of thiol esters is also a method of general utility, particularly in making olefin-derived thiols from thioacetates.

Dithiols of the type $HS(CH_2)_nSH$ have been made by the procedures used for monothiols

Health and Safety Aspects, Toxicology

In general, most thiols are considered moderately toxic compounds.

Uses

One of the main uses for aliphatic thiols is as polymerization modifiers in emulsion-polymerization systems (see Emulsions; Initiators Polymerization mechanisms and processes).

Lower molecular weight alkanethiols, eg, methanethiol, ethanethiol, propanethiol, and 1-butanethiol, are used as intermediates in the manufacture of various agricultural chemicals, including insecticides, acaricides, herbicides (qv), and defoliants (see Insect control technology).

Lower molecular weight thiols are used as odorants in natural and liquefied petroleum gas (LPG).

2-Mercaptobenzothiazole and derivatives are an important class of rubber vulcanization accelerators.

Polysulfide polymers with terminal thiol groups, prepared from dihalo compounds and sodium polysulfides, are used in sealant and adhesive applications (see Sealants; Polymers containing sulfur).

Medicinal applications include the use of o-mercaptobenzoic acid for the preparation of ethylmercurithiosalicylate (sodium salt), a sterilizing agent known as Merthilate, and ethanethiol for the preparation of 2, 2-bis(ethylsulfonyl)propane (sulfonal) related hypnotics (see Hypnotics, sedatives and, anticonvulsants). 2-Mercaptoethylamine imparts some protection to animals against the effects of ionizing radiation, and has been the model structure for government-sponsored research into more effective and less toxic antiradiation drugs (see Radioprotective agents).

JOHN NORELL
R.P. LOUTHAN
Phillips Petroleum Co.

E.E. Reid, *Organic Chemistry of Bivalent Sulfur*, Chemical Publishing Co., Inc., New York, 1958–1962.

E. Block, *Reactions of Organic Sulfur Compounds*, Academic Press, New York, 1978.

S. Oae, ed., *Organic Chemistry of Sulfur*, Plenum Press, New York, 1977.

C.J.M. Stirling, ed., *Organic Sulphur Chemistry*, Butterworths, Woburn, Mass., 1975.

THIOPHENE AND THIOPHENE DERIVATIVES

Thiophene, C_4H_4S, a 5-membered heterocyclic compound (1), is a colorless liquid, bp 84°C, with an odor similar to benzene, immiscible with water, and soluble in most organic solvents.

(1)

It is highly flammable and moderately toxic and should be handled with proper safety precautions in a well-ventilated area free from sources of sparks and flames. It is available in the United States from Pennwalt Corp. and in Europe from Croda Synthetic Chemical, Ltd. and Pennwalt-Holland B.V. It is used primarily as an intermediate to pharmaceuticals, dyestuffs, agricultural chemicals, and polythiophene semiconductor films. Physical properties are given in Table 1.

Reactions

Reactions include oxidation, reduction, and hydrodesulfurization; halogenation, alkylation, and acylation; sulfonation, nitration, metallization, and polymerization.

Manufacture

In the preferred laboratory synthesis of thiophene, powdered, anhydrous sodium succinate is heated with phosphorus trisulfide to high (gas flame) temperatures under a stream of carbon dioxide. The crude thiophene from the effluent gas stream is condensed, steam distilled, and dried; the yield is 25–30%. Commercially it is produced from four-carbon hydrocarbons and furan and a sulfur source.

Health and Safety Factors

Based on the toxicological data available, thiophene is considered to be moderately toxic. An IC_{50} inhalation value of 9.5 mg/L for mice exposed for a period of two hours and an oral LD_{50} value of 420 mg/kg for mice and 1400 mg/kg for rats have been reported.

Table 1. Physical Properties of Thiophene

Property	Value
freezing point, °C	−38.3
boiling point at 101.3 kPa[a], °C	84.16
flash point, °C	−1.11
d_4^{25}, g/cm³	1.057
refractive index, n_D^{25}	1.52572
viscosity, abs at 25°C, mPa·s (= cP)	0.621
surface tension at 20°C, mN/m (= dyn/cm)	31.34
vapor pressure, kPa[a]	
temp, °C	
0	2.86
50.1	31.16
84.16	101.3
95.9	143.3
Critical constants	
temp., °C	306.2
volume, mL/mol	219
density, g/cm³	0.385
heat of formation at 298.16 K, kJ/mol[b]	
liquid	81.67
gaseous	116.4
heat of combustion[c] at 101.3 kPa[a], 25°C, kJ/mol[b]	−2791.5
dielectric constant at 20°C	2.7
dipole moment at 25°C in benzene, C·m[d]	1.73×10^{-30}

[a] To convert kPa to mm Hg, multiply by 7.5.
[b] To convert J to cal, divide by 4.184.
[c] $C_4H_4S(l) + \frac{13}{2} O_2(g) + 0.7 H_2O(l) \rightarrow 4 CO_2(g) + [H_2SO_4 \cdot 1.7H_2O](l)$.
[d] To convert C·m to D, multiply by 3×10^{29}.

Uses

The principal application for thiophene today is as an intermediate for the manufacture of pharmaceuticals.

BERNARD BUCHHOLZ
Pennwalt Corporation

H.D. Hartough, *Thiophene and Its Derivatives*, Interscience Publishers, Inc., New York, 1952.

C.D. Hurd, *Q. Rep. Sulfur Chem.* 4(2–3), 75 (1969).

THIOSULFATES

The thiosulfate ion, $S_2O_3^{2-}$, is a structural analogue of the sulfate ion where one oxygen atom is replaced by one sulfur atom. The two sulfur atoms are not equivalent and the unique chemistry of the thiosulfate ion is dominated by the sulfide-like sulfur atom which is responsible for the reducing properties and complexing abilities of thiosulfates. Their ability to dissolve silver halides through complex formation is the basis for their commercial application in photography (qv).

Physical Properties

Thermodynamic properties. The heat of formation of the thiosulfate ion is −5.75 kJ/g (−1.37 kcal/g). The standard free energy of formation is −4.58 kJ/g (−1.09 kcal/g). The partial molal entropy is 62.8 ± 25.1 J/K (15 ± 6 cal/K).

Electrochemical properties. The oxidation potential for the reaction

$$2 S_2O_3^{2-} = S_4O_6^{2-} + 2 e$$

ranges from 0.2 to 0.4 V in neutral solution, depending on the method of measurement.

Electrolytic reduction with a mercury or platinum electrode produces equimolar amounts of sulfide and sulfite:

$$S_2O_3^{2-} + 2\,e \rightarrow S^{2-} + SO_3^{2-}$$

Chemical Properties

Thiosulfuric acid is relatively unstable and thus cannot be recovered from aqueous solutions.

Pure thiosulfuric acid has been prepared in liquid CO_2 at $-50°C$ or in diethyl ether at $-78°C$. It decomposes at $-30°C$ to $H_2S_3O_6$ and H_2S, and rapidly at higher temperatures to H_2O, SO_2, and sulfur.

The ammonium, alkali metal, and alkaline earth thiosulfates are soluble in water. Neutral or slightly alkaline solutions containing excess base or the corresponding sulfite are more stable than acid solutions.

Acidification of thiosulfate with strong acid invariably leads to decomposition with the formation of colloidal sulfur and sulfur dioxide.

In dilute aqueous solution, the following equilibrium is established:

$$S_2O_3^{2-} + H^+ \leftrightharpoons HSO_3^- + S \qquad K = 0.013 \text{ at } 11°C$$

Reactions

Catalytic amounts of arsenic, antimony, or tin salts promote the formation of pentathionate. Mild oxidizing agents such as hydrogen peroxide in acid solutions produce tetrathionates and trithionates. The presence of Fe^{2+} promotes oxidation to the sulfate. The reaction with iodine in neutral or slightly acid solution is the basis for volumetric analysis.

Stronger oxidizing agents such as chlorine, bromine, permanganate, chromate, or alkaline hydrogen peroxide oxidize thiosulfate quantitatively to sulfate. Thiosulfates are reduced to sulfides by metallic copper, zinc, or aluminum.

Thiosulfate reaction with cyanide to give thiocyanate is the basis for the use of thiosulfate as an antidote in cyanide poisoning.

Corrosion. The preferred material of construction for pumps, piping, reactors, and storage tanks is austenitic stainless steels such as 304, 316, or Alloy 20. The corrosion rate for stainless steels is $< 440 \text{ g}/(\text{m}^2 \cdot \text{yr})$ at $100°C$ (see also Corrosion and corrosion inhibitors).

Preparation

Thiosulfates are normally prepared by the reaction of sulfur and sulfite in neutral or alkaline solution:

$$S + SO_3^{2-} \rightarrow S_2O_3^{2-}$$

Polysulfides react similarly. Sulfides react with sulfur dioxide, sulfite, or bisulfite. These three methods are employed commercially.

Sodium Thiosulfate

Sodium thiosulfate, as the anhydrous salt, $Na_2S_2O_3$, or the crystalline pentahydrate, is commonly referred to as hypo or crystal hypo.

Selected physical properties of sodium thiosulfate pentahydrate are shown in Table 1.

Health and safety factors. The LD_{50} of anhydrous sodium thiosulfate for mice is $7.5 \pm 0.752 \text{ g/kg}$. Sodium thiosulfate pentahydrate is affirmed as a GRAS indirect and direct human food ingredient under the Federal Food, Drug and Cosmetic Act (see Food additives).

Uses. The principal use for sodium thiosulfate continues to be as a fixative in photography to dissolve undeveloped silver halide from negatives or prints (see Photography). It also is used in leather tanning, paper, and textiles, and flue-gas desulfurization.

Ammonium Thiosulfate

Ammonium thiosulfate, $(NH_4)_2S_2O_3$, commonly referred to as ammo hypo, has displaced sodium thiosulfate in photography (qv). It is normally sold in the United States as the aqueous solution.

Health and safety factors (toxicology). The toxicological properties of ammonium thiosulfate are generally considered to be the same as those of sodium thiosulfate and thiosulfates in general.

Table 1. Physical Properties of Sodium Thiosulfate Pentahydrate

Property	Value
refractive index, n_D^{20}	1.4886
density, d_4^{25}, g/cm^3	1.750
heat of solution in water at $25°C$, J/g^a	-187
heat of formation, kJ/g^a	-10.48
heat of fusion, J/g^a	200
specific heat solid, $\text{J/(g} \cdot \text{K)}^a$	1.84
dissociation pressure at $20°C$, kPa^b	0.796
vapor pressure of saturated solutions, at $33°C$, kPa^b	1.33

^{a} To convert J to cal, divide by 4.184.
^{b} To convert kPa to mm Hg, multiply by 7.5

Uses. The use distribution of ammonium thiosulfate in 1981 was estimated to be for photography, 48%; agricultural applications, 50%; and others, including dechlorination, 2%.

Other Thiosulfates

Many other metal thiosulfates, eg, magnesium thiosulfate and its hexahydrate, have been prepared on a laboratory scale, but with the exception of the calcium, barium, and lead compounds, they are of little commercial or technical interest.

Complexes and Organic Thiosulfates

Gold thiosulfate complexes of the $MNa_3[Au(S_2O_3)]_2.H_2O$ are prepared by the addition of gold trichloride to concentrated sodium thiosulfate solution.

Organic thiosulfate are usually prepared by the reaction alkyl chlorides with sodium thiosulfate. Sodium ethyl thiosulfate is also known as Bunte's salt after its discoverer. Bunte salts have bacterial, insecticidal, and fungicidal properties, and are also used as chelating agents or surfactants. Bunte salts have been tested for preirradiation protection for mammals exposed to lethal radiation doses (see Radioprotective agents).

JAMES W. SWAINE, JR.
Allied Corporation

T. Moeller, *Inorganic Chemistry, A Modern Introduction*, John Wiley & Sons, Inc., New York, 1982.

Gmelins Handbuch der Anorganischen Chemie, 8th ed., Schwefel, Part B, No. 2, Verlag Chemie, G.m.b.H., Weinheim/Bergstrasse, 1960, p. 868.

THORIUM AND THORIUM COMPOUNDS

Thorium, element number 90, has an atomic weight of 232.038. Chemically, thorium is less basic than the rare earths, but there are insoluble thorium fluorides, carbonates, hydroxides, oxalates, and phosphates.

Refractory thorium dioxide, when heated to incandescence by a weak yellow gas flame, emits a brilliant white light. The best effect is achieved with material containing ca 1 wt% cerium oxide. Commercial mantles are made up of a mesh hood of oxide surrounding a flame and are used in decorative exterior lighting, gasoline lanterns for camping, etc. In the 1940s, it was discovered that the product of neutron absorption by thorium was ^{233}U, which could undergo fission by slow neutrons (see Radioactivity, natural). The work of the Manhattan Project during World War II on the fission chain reaction created tremendous commercial interest in thorium (see Nuclear reactors, fast-breeder reactors). This is still the main interest in the element, although small-scale applications have been developed in electronics and catalytic processes. A larger use of thorium that has growth potential is that of a strengthening alloying element in metallic magnesium (TD Mg); this application is being extended to alloys such as TD Ni, TD NiCr, and TD NiCrAl.

Table 1. Physical Properties of Thorium

Property	Value
atomic diameter, nm	0.3596
melting point, °C[a]	1750
boiling point, °C	ca 3800
density, x ray, g/cm^3	11.72
heat of fusion, kJ/mol[b]	< 19.2
heat of vaporization, kJ/mol[b]	ca 586
electrical resistivity, $\mu\Omega\cdot$cm	14
elastic constants	
Young's modulus, MPa[c]	7.24×10^4
shear modulus, MPa[c]	2.76×10^4
Poisson's ratio	0.27

[a] Higher values have been reported.
[b] To convert J to cal, divide by 4.184.
[c] To convert MPa to psi, multiply by 145.

Thorium is mildly radioactive, decaying by emission of alpha particles at a half-life of 1.41×10^{10} yr.

Occurrence

Thorium is not a very rare element; it comprises 0.001–0.002 wt % of the earth's crust and is distributed widely at up to 15 ppm in soils. The significant mineral source for thorium is the anhydrous rare-earth phosphate, monazite, from which thorium is currently obtained as a by-product. Thorium contents of monazite are commonly 5–30 wt % (expressed as oxide).

Recovery from Ores

Chemically stable monazite can be disintegrated successfully by strong acid or strong alkali. Strong acid dissolves monazite by converting phosphate ion to $H_2PO_4^-$ and H_3PO_4. The thorium in the solution can then be separated from rare earths or lanthanons by differential precipitation.

When the monazite is treated with strong alkali, both thorium and lanthanon hydroxides remain in the insoluble solids, and the phosphate is removed in the supernate. The final acid solution of the thorium is purified by extraction with an immiscible organic solvent.

Thorium Metal

Properties. Pure thorium is a bright, silvery metal and is dense, very high melting, weaker than the structural metals, softer than steel, and quite reactive chemically. See Table 1 for properties.

Compounds

Nonmetallic thorium is in the +4 oxidation state exclusively; Th^{4+} is one of the least hydrolyzed tetrapositive ions having a first acid dissociation constant of ca $10^{-3.6}$. With a small ionic radius (0.095 nm), it has a strong tendency to hydrate and to coordinate electron donors. Its coordination numbers are 11, 10, 9, 8, and occasionally 6. In general, the halides and some strongly binding oxygen compounds favor the lower coordination numbers, whereas weakly basic hydrated compounds favor the higher ones.

Complexes are formed with fluoride, iodate, bromate, nitrate, chlorate, chloride, sulfate, sulfite, carbonate, phosphate, pyrophosphate, and molybdate ions. Complexes form also with most anions of organic acids, eg, formate, acetate, chloracetates, oxalate, tartrate, malate, citrate, salicylate, sulfosalicylate, etc. Chelates form very readily, some of which lend themselves especially well to the solvent extraction process.

Health and Safety Factors

There are a few potential hazards relating to thorium, although no significant incidents have been reported. One potential hazard is the reactivity of finely divided metal and of the hydrides, with respect to oxygen and halogens. Such a reaction can have explosive or inflamma-

tory results. Th^{4+} is a heavy element capable of being hazardous when introduced into the body.

LEONARD I. KATZIN
Consultant

R.J. Callow, *The Industrial Chemistry of the Lanthanons, Yttrium, Thorium, and Uranium*, Pergamon Press, Inc., New York, 1967.

L.I. Katzin in J.J. Katz, G.T. Seaborg, and L. Morss, eds., *The Chemistry of the Actinide Elements*, Chapman-Hall, London, in press.

G.T. Seaborg and L.I. Katzin, *Production and Separation of U^{233}: Survey*, National Nuclear Energy Series, Div. IV, Vol. 17A, 1951, TID-5222, AEC.

J.F. Smith, O.N. Carlson, D.T. Peterson, and T.E. Scott, *Thorium: Preparation and Properties*, Iowa State University Press, Ames, Iowa, 1975.

THULIUM. See Rare-earth elements.

THYROID AND ANTITHYROID PREPARATIONS

The main role of the thyroid gland is the production of the thyroid hormones (iodinated amino acids) which are essential for growth, development, and energy metabolism. Thyroid underfunction is a frequent occurrence that can be treated with thyroid preparations. In addition, the thyroid secretes calcitonin (thyrocalcitonin), a polypeptide that lowers calcium blood levels. Thyroid hyperfunction can be treated with antithyroid drugs.

Thyroid Function and Malfunction

Human life without thyroid hormones is possible but of minimal quality. The two principal thyroid hormones, L-thyroxine (T_4) and L-triiodothyronine (T_3), are produced by the thyroid gland and secreted into the blood stream (see Fig. 1). The minute amounts secreted are regulated by a complex system that originates in the CNS (Fig. 2).

Thyroid hormones affect growth and development by stimulating protein synthesis. In mature animals, the main action is their calorigenic effect, which is caused by an increase in basal metabolic rate. Thyroid underfunction results in cretinism, if present in a fetus or an infant, and in myxedema in an adult. If the hypothyroidism is due to insufficient iodine intake, it is known as a simple goiter. Thyroid hyperfunction occurs as diffuse toxic goiter, also known as Graves' disease. Other forms of hyperfunction are thyrotoxicosis and toxic nodular goiter (Plummer's disease).

Thyromimetic Compounds

The main thyroid hormone is the 5'-desiodo analogue of thyroxine, triiodothyronine or T_3. The structural requirements for thyromimetic activity are: two aromatic rings isolated electronically from each other, which constitute a central lipophilic core; substitution at the 3 and 5 positions with alkyl or halogen; an acidic side chain at position 1; a small substituent capable of forming hydrogen bonds in the 4' position; and one lipophilic substituent ortho to the 4'-OH group. Iodine is not indispensable for thyromimetic activity.

Biosynthesis, metabolism, and synthesis. Iodine, a trace element in the environment and the diet, is extracted from the blood by the thyroid

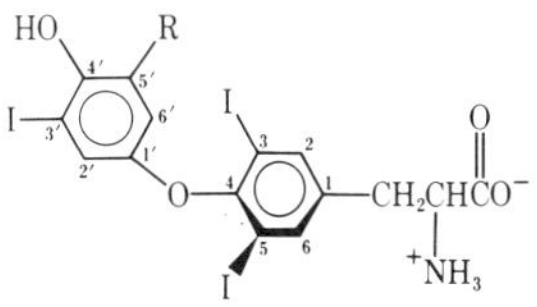

(1) R = I: L-thyroxine, 3,5,3',5'-tetraiodo-L-thyronine, T_4
(2) R = H: 3,5,3'-triiodo-L-thyronine, T_3

Figure 1. The principal thyroid hormones.

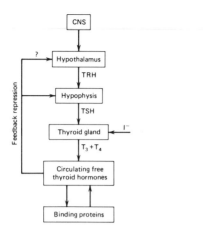

Figure 2. Mechanisms controlling free thyroid-hormone levels.

gland via an active transport system and converted to di-, tri-, and tetraiodothyronines.

The most widely employed synthetic route for T_3 and T_4 is the so-called Glaxo method. It has also been used with a variety of phenols and thiophenols to establish structure-activity relationships; L-tyrosine is the starting material.

Assays. The main problem in the chemical analysis of thyroid hormones is their separation. Paper or high performance liquid chromatography, as well as radioimmunoassay, methods are employed. *In vivo* and *in vitro* bioassays determine the potency of thyroid preparations and new compounds. The rat antigoiter assay is the most common test for thyromimetic activity; *in vitro* tests use binding to various macromolecules.

Antithyroid Substances

Administration of large doses of iodide ion results in a transient inhibition of synthesis and release of thyroid hormones by the so-called Wolff-Chaikoff effect. The selective uptake of iodide ion by the thyroid gland is the basis of radioiodine treatment of hyperthyroidism. Thiocyanate ion inhibits formation of thyroid hormones by blocking the iodination of tyrosine residues in thyroglobulin. A large group of compounds incorporating thionamide or thiourea moieties are potent antithyroid agents, but only four are used clinically and only two are accepted by the USP XX. Lithium salts are not used in the treatment of hyperthyroidism.

The relatively long duration of action of the thyroid hormones makes compounds desirable which are capable of blocking them competitively at their site of action. A large number of thyroid hormones analogues have been tested for this effect, but no clinically useful peripheral antagonists have been found.

The thyroid gland is the source of the hypocalcemic hormone calcitonin, whose effects oppose those of the parathyroid hormone. It is used clinically in various diseases in which hypercalcemia is present, eg, Paget's disease.

Commercial Preparations

The four basic preparations for hypothyroidism are thyroxine (sodium levothyroxine), triiodothyronine (sodium liothyronine), thyroglobulin, and desiccated thyroid extracts (*Glandulae thyroideae siccatae*). They are supplied under various trade names by Armour, Warner-Chilcott, Merck, Henning, and others.

Antithyroid drugs include propylthiouracil and methimazole.

PEDRO A. LEHMANN F.
Instituto Politecnico Nacional, Mexico

E.C. Jorgensen in M.E. Wolff, ed., *Burger's Medicinal Chemistry*, Pt. III, John Wiley & Sons, Inc., New York, 1981, pp. 10–3–145.

E.C. Jorgensen, *Thyroid Hormones and Analogs*, Vol. VI of C.H. Li, ed., *Hormonal Proteins and Peptides*, Academic Press, Inc., New York, 1978, pp. 57–105, 107–204.

TIN AND TIN ALLOYS

TIN AND TIN ALLOYS

Tin is one of the world's most ancient metals and has been associated with the growth of civilization. Of the nine tin-bearing metals found in the earth's crust, only cassiterite, SnO_2, is of importance. Low grade alluvial or eluvial placer deposits are scattered throughout Southeast Asia and complex tin sulfide minerals are found in Bolivia and Cornwall; no workable tin deposits have been found in the United States.

Mining methods depend on the character of the deposit. Gravel pumping is widely used in Southeast Asia and probably accounts for 40% of the world's tin production. Hydraulic and open-pit methods employ gravity separation with water in palongs; dredging is mining with a floating dredge on an artificial pond in a placer. Access to lode deposits in Bolivia and Cornwall is by shaft sinking or adits (passages driven into the side of the mountain). Tin concentrates obtained from lode deposits contain 40–60% tin and must be further upgraded.

Properties

Some physical properties are given in Table 1. Although the pure metal has a silvery-white color, it may have a yellowish tinge when cast; when polished it is highly reflective. Tin exists in β (white, body-centered tetragonal) or α (gray, diamond cubic) allotropic form.

Tin occurs in the form of ten isotopes; it is amphoteric and reacts with both acids and bases. A reversal of potential of the tin–iron couple occurs when tin-coated steel (tin plate) is in contact with acid solutions in the absence of air. The tin coating, acting as anode, is slowly attacked, but the steel is not. Thus, tin protects the appearance and flavor of canned goods.

Processing

The crude tin obtained from slags and smelting ore concentrates is refined by further heat treatment, eg, liquation or sweating and boiling. Iron, copper, arsenic, and antimony are readily removed by such pyrometallurgical processes. Large quantities of lead or bismuth are removed by electrolysis or vacuum refining.

Economic Aspects

Tin, a strategic material, is stockpiled by the General Services Administration. Since the mid-1950s, its price has been regulated by an agreement between producing and consuming nations. Under the International Tin Agreement, the International Tin Council seeks to deal with shortages or surpluses as they arise, and prevents excessive price fluctuations. World trading occurs mostly at Penang, London, and New York; most of the world's tin is produced in southeast Asia.

Table 1.   Physical Properties of Tin

Property	Value
mp, °C	231.9
bp, °C	2625
sp gr	
α-form (gray tin)	5.77
β-form (white tin)	7.29
liquid at mp	6.97
thermal conductivity at 20°C, W/(m·K)	65
coefficient of linear expansion, $\times 10^{-6}$	
at 0°C	19.9
at 100°C	23.8
Brinell hardness, 10 kg, 5 mm, 180 s	
at 20°C	3.9

Table 2. ASTM B 339-72 Classification of Pig Tin

Grade designation		Description[a]		
ASTM	Commercial	Class	Tin, % min	General applications
AAA	electrolytic	extra high purity	99.98	analytical standards and research
AA	electrolytic	high purity	99.95	research, pharmaceuticals, fine chemicals
A	A	standard	99.80	food containers, foil, collapsible tubes, unalloyed (block) tin products, electrotinning, tin-alloyed cast iron, high grade solders
B	B	general purpose	99.80	less exacting than A; general purpose
C	C	intermediate grade	99.65	general purpose; alloys
D	D	lower intermediate grade	99.50	general purpose; alloys
E	E	common	99.00	cast bronze, bearing metal, general-purpose solders, lead-base alloys

[a]A more complete description of these grades is given in the full ASTM Standard B 339.

The United States is by far the largest consumer of tin, followed by Japan and the FRG. Tinplate provides an outlet for over one-third of the primary tin used in the United States where large quantities of tin are recovered from scrap.

Specifications and Analysis

ASTM classifications are available (see Table 2) as well as ASTM methods for determination of tin in alloys. The tin content of ores, concentrates, ingot metal, and other products is determined by fire assay, fusion, and volumetric wet analysis.

Considerable quantities can be consumed without ill effects. Small amounts are present in most liquid canned products.

Uses

Tin ingots are cast into anodes for plating. Sheet tin is used to line distilled-water tanks. Mixtures of tin and copper powder are used for bronzing. The development of tinplate was associated with the need for a safe and reliable packaging material for foods. The tin coating (0.0025–0.05-mm thick) is applied by electroplating or from a bath of molten tin. It may also be applied by hot-dipping the article or, when a very thin film is required, by immersion. The latter technique is based on chemical displacement from a solution of tin salts.

The manufacture of tin requires less energy than the manufacture of other container materials, eg, aluminum or glass.

Tin alloys. Tin–copper electrodeposited coatings have the appearance of 24-carat gold and provide a bronze finish. Tin–lead coatings, such as terne plate, are used for roofing, automobile fittings, and radio and TV equipment. Terne plate is low carbon steel, coated by a hot-dip process with an alloy of tin and lead (7–25%). Tin–nickel electrodeposited coatings (65% Sn) provide a bright, corrosion-resistant, decorative finish. A new use for tin–nickel is for printed circuit boards. Solderable tin–zinc coatings are used for electronic components. A tin–copper–lead coating is a standard overplate for automotive bearings. Tin–cadmium coatings are used in aviation.

Solder. Tin and lead combine to form soft solders employed for low temperature joining. The important constituent is tin because it wets the base metal by alloying with it. Lead-free solders are available for special applications. Solder is the second largest use of tin after tinplating. Fusible alloys (mp 20–176°C), mostly eutectic, are two-, three-, four-, or even five-component mixtures of tin, lead, bismuth, cadmium, indium, and gallium (see Solders).

Copper–tin alloys, called bronzes, can be wrought, sand cast, or continuously cast. Bell-metal bronze is known for its tonal quality.

Bearing metals. Metals used for casting or lining bearing shells are classed as white bearing alloys, but are known commercially as babbitt (see Bearing materials). Included are high tin alloys (> 80% Sn) and high lead alloys (> 70% Pb, 12% Sn). Both exhibit the characteristic structure of hard compounds in a soft matrix. Bearing metals also include bronzes and aluminum–tin (6.5% Sn) alloys.

Modern pewter may have a composition of 90–95% Sn, 1–8% Sb, and 0.5–3% Cu. Lead should be avoided because it causes the surface to blacken with age. The annual U.S. production exceeds 1100 t. The

printing trade formerly required large amounts of lead-based alloys containing 10–25% Sb and 3–13% Sn.

Tin-alloyed flake and nodular cast irons are widely used. Tin-inoculated iron has uniform hardness and improved machinability and wear resistance; on heating, shapes are retained. Alloys of tin with niobium, titanium, and zirconium have been developed. The single-phase alloy Nb_3Sn has the highest transition temperature of any known superconductor (see Superconducting materials).

D.J. MAYKUTH
Tin Research Institute, Inc.

C.L. Mantell, *Tin: Its Mining, Production, Technology, and Applications*, 2nd ed., Reinhold Publishing Corp., New York, 1949.

C.J. Faulkner, *The Properties of Tin*, Publication 218, International Tin Research Institute, London, UK, 1965.

S.C. Pearce in J. Cigan, T.S. Mackey, and T. O'Keefe, eds., *Proceedings of a World Symposium on Metallurgy and Environmental Control at 109th AIME Annual Meeting*, The Metallurgical Society of AIME, Warrendale, Pa., 1980, pp. 754–770.

DETINNING

Detinning refers to the mechanical and chemical processing of tinplate scrap for the recovery of tin and the base metal, which is normally steel. The scrap is generated by fabricators of tin cans and other articles. Used tin cans are not utilized because of the unfavorable economics of collection and processing.

The alkaline electrolytic process permitted the direct detinning of small batches of tinplate. However, control was difficult because of changing current conditions. In the chlorination process, dry chlorine reacts with tin below 50°C to produce anhydrous stannic chloride without attack on the steel substrate. In the alkaline chemical process, which is preferred today, the scrap is treated with hot caustic soda containing an oxidizing agent to dissolve the tin as sodium stannate. The tin may be recovered by electrolysis or after neutralization as tin oxide.

There are no safety measures that are exclusive to the detinning industry. The usual precautions must be observed when handling heavy equipment. Spent electrolyte is the only liquid waste.

WILLIAM GERMAIN
H.P. WILSON
V.E. ARCHER
Vulcan Materials Co.

C.L. Mantel, *Tin*, 2nd ed., Reprint Hafner Publishing, New York, 1970, pp. 519–529.

Solid Waste Processing Facilities, American Iron and Steel Institute, Washington, D.C., Sept. 1981.

Minerals Yearbook, U.S. Bureau of Mines, U.S. Government Printing Office, Washington, D.C., yearly editions.

TIN COMPOUNDS

Tin has valences of +2 and +4 and forms stannous, ie, tin(II) compounds and stannic, ie, tin(IV) compounds. The important deposits of cassiterite, SnO_2, are in southeast Asia, whereas the complex sulfidic ores are found in Bolivia. Tin compounds are also present in natural waters, in soil, in marine organisms, and in meteorites.

Inorganic Tin Compounds

Tin reacts with strong acids and bases. A thin oxide film forms on tin exposed to oxygen. Sulfides are formed by a vigorous reaction when tin and sulfur are heated.

Halides. Properties of tin chlorides are given in Table 1. They are prepared by the reaction of chlorine with tin metal. Anhydrous stannous

Table 1. Physical Properties of Tin Chlorides

Property	$SnCl_2$	$SnCl_2 \cdot 2H_2O$	$SnCl_4$	$SnCl_4 \cdot 5H_2O$
mol wt	189.60	225.63	260.50	350.58
mp, °C	246.8	37.7	−33	ca 56 (dec)
bp, °C	623		114	
density (at 25°C), g/cm³	3.95	2.63	2.23[a]	2.04

[a] At 20°C.

chloride, a water-soluble white solid, is used in redox and plating reactions. Solutions are widely employed as reducing agents. Stannous chloride is also used as a food additive, for which it has FDA GRAS approval. Stannous chloride dihydrate is prepared by treatment of granulated tin with hydrochloric acid, followed by evaporation and crystallization or by reduction of a stannic chloride solution.

Stannic chloride is available as pentahydrate or as anhydrous colorless fuming liquid. It is soluble in water and organic solvents. Its main uses are as a raw material for other tin compounds and in the surface treatment of glass (qv) and other nonconductive materials. It is also used as a catalyst in Friedel-Crafts reactions (qv). The pentahydrate, a white, crystalline, deliquescent, water-soluble solid, is used in place of the anhydrous chloride where anhydrous conditions are not required.

Stannous fluoride (opaque white water-soluble crystals) is used in toothpaste and other dental preparations (see Dentifrices).

Oxides. Stannous oxide (stable, blue-black crystalline product (dec > 385°C)) reacts readily with acids, which accounts for its primary use in the manufacture of other tin compounds. Stannic oxide is prepared by blowing hot air over molten tin. It is used in binary catalyst systems and in the ceramics and glass industries as an opacifier. Hydrated stannic oxide is obtained by the hydrolysis of stannates or stannic chloride.

Metal stannates. Many stannates of the formula $M_nSn(OH)_6$ are known. Both potassium and sodium stannates are colorless, water-soluble crystals; they are used for alkaline tin electroplating. Insoluble metal stannates are prepared by metathetic reactions. They are used as additives for ceramic dielectrics.

Salts. Stannous sulfate is a white crystalline powder prepared from sulfuric acid and granulated tin at 100°C. It is used in tin plating, as is stannous fluoroborate. Solutions of the latter have good throwing and covering power. Stannous pyrophosphate is used in toothpaste and in x-ray technology.

Health and safety factors. Tin compounds are generally low in toxicity because of poor absorption and rapid excretion. Tin is present in all animals and in the human body in small amounts in all organs.

Organotin Compounds

Mono-, di-, tri-, tetra-, and hexaorganotin compounds are known. Most important are those where the organic radical is methyl, butyl, octyl, chclohexyl, phenyl, or β,β-dimethylphenethyl (neophyl).

Tetraorganotin compounds. These compounds are insoluble in water but soluble in organic solvents. Their most important reaction is the Kocheshkov redistribution reaction, by which organotin halides are prepared.

$$R_4Sn + SnCl_4 \rightarrow 2\ R_2SnCl_2$$
$$R_2SnCl_2 + R_4Sn \rightarrow 2\ R_3SnCl$$
$$3\ R_4Sn + SnCl_4 \rightarrow 4\ R_3SnCl$$
$$R_4Sn + 3\ SnCl_4 \rightarrow 4\ RSnCl_3$$

Tetraalkyl- and tetraaryltin compounds are prepared from stannic chloride and Grignard reagents or organoaluminum compounds. Organolithium or organosodium reagents can also be used. Functional groups must be blocked. Functional tetraorganotin compounds are prepared by tin hydride addition (hydrostannation) to functional unsaturated organic compounds.

The main use for tetraorganotin compounds is as intermediates for the tri, di, and monocompounds. Application as Ziegler-Natta-type catalysts has been reported (see Olefin polymers).

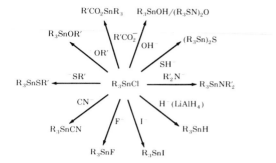

Figure 1. Reactions of triorganotin chlorides.

Triorganotin compounds. Triorganotin halides are soluble in organic solvents; only $(CH_3)_3SnCl$ is soluble in water. The utility of triorganotin chlorides results from the ease of nucleophilic displacement, as shown in Figure 1. They are generally prepared by Kocheshkov redistribution. Tribenzyltin chloride, however, is prepared directly from the halide and tin:

$$3\ C_6H_5CH_2Cl + 2\ Sn \xrightarrow[\text{reflux}]{H_2O} (C_6H_5CH_2)_3SnCl + SnCl_2$$

Physical properties of triorganotin compounds are given in Table 2.

Uses. Triorganotin compounds are widely used as biocides, agricultural chemicals, wood preservatives, and marine antifoulants. The most useful biological control agents are the tributyl-, triphenyl-, and tricyclohexyltin compounds. The fungicide triphenyltin acetate (Brestan) controls phytophthora (late blight) on potatoes and cercospora on sugar beets in applications of very low concentrations. Triphenyltin hydroxide (Du-Ter) has similar activity. Tricyclohexyltin hydroxide (Plictran) and hexaneophyldistannoxane (Vendex) are other important industrial chemicals. Bis(tributyltin) oxide is widely used in Europe for wood preservation. Triorganotin compounds are used as marine antifouling agents, alone and in combination with cuprous oxide. Preferred for this application are tributyltin fluoride, triphenyltin hydroxide, and triphenyltin fluoride. Eroding antifouling paints are based on tributyltin acrylate or methacrylate copolymers (see Coatings, marine). Triorganotin compounds have been used experimentally to control the snail vector in schistosomiasis and to control mosquitos.

Diorganotin compounds. Diorganotin dichlorides are the intermediates for the preparation of all commercial diorganotin compounds (for properties, see Table 3). Dibutyltin dichloride is manufactured by Kocheshkov redistribution from tetrabutyltin and stannic chloride.

$$(C_4H_9)_4Sn + SnCl_4 \rightarrow 2\ (C_4H_9)_2SnCl_2$$

Many organic halides, especially alkyl bromides and iodides, react with tin metal at elevated temperatures. Treatment of molten tin with methyl chloride gives good yields of dimethyl tin dichloride, an important intermediate in the manufacture of dimethyltin-based PVC stabilizers.

Table 2. Physical Properties of Commercially Important Tiorganotin Compounds

Compound	Mp, °C	Bp, °C	n_D^{20}	d^{20}, g/cm³
$[(C_4H_9)_3Sn]_2O$	< −45	$210–214_{1.3\ kPa}$[a]	1.488	1.17
$(C_4H_9)_3SnF$	218–219 (dec)			1.27[b]
$(C_4H_9)_3SnOCOC_6H_5$		$166–168_{0.13\ kPa}$[a]	1.5157	1.1926
$(C_4H_9)_3SnOCOCH_3$	80–85			1.27
$(C_6H_5)_3SnOH$	118–120 (dec)			1.552[b]
$(C_6H_5)_3SnF$	357 (dec)			1.53

[a] To convert kPa to mm Hg, multiply by 7.5.
[b] At 25°C.

Table 3. Physical Properties of Diorganotin Compounds

Compound	Mp, °C	Bp, °C	n_D^{20}	d^{20}, g/cm³
$(CH_3)_2SnCl_2$	107–108	185–190		
$(C_4H_9)_2SnCl_2$	41–42	$140–143_{1.3 kPa}{}^a$		
$(C_4H_9)_2SnBr_2$	21–22	$90–92_{0.04 kPa}{}^a$	1.5400	1.3913^b
$(C_4H_9)_2SnI_2$		$145_{0.8 kPa}{}^b$	1.6042	1.996^b
$(C_6H_5)_2SnCl_2$	42–44	$180–185_{0.7 kPa}{}^a$		
$(CH_3OC(O)CH_2CH_2)_2SnCl_2$	132			
$(CH_3)_2Sn(SC_4H_9)_2$		$81_{0.013 kPa}{}^a$	1.5400	1.280
$(C_4H_9)_2Sn(OCH_3)_2$		$126–128_{7 Pa}{}^a$	1.4880	

^{a} To convert kPa to mm Hg, multiply by 7.5.
^{b} At 25°C.

The direct reaction of metallic tin with higher haloalkanes is less satisfactory, even with catalysts, except for alkyl iodides. However, tin reacts directly with activated organic halides, eg, allyl bromide and benzyl chloride.

Uses. The largest industrial application for organotin compounds is in the stabilization of PVC; dialkyltin compounds are the best general-purpose PVC stabilizers. In the building industry, rigid PVC is stabilized with diorganotin carboxylates. In addition, dibutyltin compounds are used as catalysts in the preparation of rigid foams (see Urethane polymers). Dibutyltin as well as monobutyltin compounds are used as catalysts in the manufacture of organic esters for plasticizers, lubricants, and heat-transfer fluids. Other uses include the application of dibutyltin dilaurate as a coccidiostat in fowl. Dimethyltin dichloride forms a thin coating of stannic oxide on glass to improve abrasion resistance and bursting strength of bottles.

Monoorganotins. Monoorganotin trihalides are strong Lewis acids and resemble acid chlorides. The halides are easily replaced. Monoorganotin halides are raw materials for triorganotin compounds and are generally prepared by Kocheshkov redistribution. Although less effective as PVC stabilizers than the dialkyl derivatives, monoalkyltin compounds added to the dialkyltin compounds (5–20%) have a synergistic effect. Butylthiostannoic acid anhydride is used in the FRG as sole stabilizer for certain grades of PVC.

Compounds with tin–tin bonds. Ditin compounds are prepared by reductive coupling of a triorganotin halide with sodium in liquid ammonia:

$$R_3SnCl + 2 Na \rightarrow R_3SnNa + NaCl$$

$$R_3SnNa + R_3SnCl \rightarrow R_3SnSnR_3 + NaCl$$

Hexaorganoditin compounds with short-chain aliphatic groups are colorless liquids of little commercial importance.

Health and safety factors. Most toxic are the lower trialkyltin compounds. The toxicity is strongly dependent on the organic groups. Most triorganotin compounds are eye and skin irritants. The diorganotin compounds are less toxic, and the monoorganotin compounds present no special problems. They show the familiar trend of decreasing toxicity with increasing alkyl length but lower than the diorganotin compounds.

The current OSHA TLV for exposure to organotin compounds is 0.1 mg (as tin)/m³ air over an 8-h work shift.

MELVIN H. GITLITZ
MARGUERITE K. MORAN
M & T Chemicals, Inc.

W. Neumann, *The Organic Chemistry of Tin*, John Wiley & Sons, Inc., New York, 1970.

A.G. Davies and P.J. Smith, *Adv. Inorg. Chem. Radiochem.* **23**, 1 (1980).

A.F. Trotman-Dickenson, ed., *Comprehensive Inorganic Chemistry*, Vol. 2, Pergamon Press, Oxford, 1973, pp. 43–104.

TIRE CORDS

A pneumatic tire is a remarkable engineering achievement; many of its properties are derived from the high strength cords it contains. Tire cords give a tire its size and shape, load-carrying capacity, and bruise and fatigue resistance.

Cotton, the first tire-cord material, was replaced by rayon which, in turn, was partially replaced by nylon. Actually, polyester replaced rayon, whereas the use of glass and steel increased as the materials for belted-bias and radial tires. Rayon still retains a good share of the European tire market whereas in the U.S. its market has become much smaller. Steel consumption continues to grow in both the carcass and belts of truck and heavy-duty radial tires. The auspicious entry of aramid into the tire cord market has so far been cut short, but prospects for further growth are good as its price becomes more competitive and more efficient use is made of its high tenacity in tires.

Processing

A simplified processing train is shown in Figure 1 which indicates that textile yarns are twisted into tire cords, woven and treated with an adhesive prior to shipment to the tire plants. Proper storage and packaging are essential to preserve the quality of the fabric. After calendering, the rubberized fabric is cut to the required angle and width. Then the pieces are butt-sliced together automatically into a continuous sheet from which the "green" tire is built. These tires are cured in automatic press molds at 165–180°C and under 1.48–2.86 MPa (200–400 psig).

Steel tire cords are made from finely drawn high carbon (0.67 wt%) steel and brass plated just before the last draw. Then these wires are twisted together in a closing operation by either cabling or bunching, thus forming the appropriate lay length (length/turn). The brass coating helps lubricate the drawing operation and promotes adhesion to most rubber compounds. Moisture and heat aging promote adhesion degradation, which is resisted better by thin brass coatings than by thick coatings. Adhesion degradation, especially in the presence of moisture, is also reduced by improvements in rubber compounds and wire design.

Tire Performance Related to Tire Cords

A tire must safely support a specified load under dynamic conditions with a minimum of power loss, overcome minor obstacles, and provide a reasonable endurance life. In addition, a tire should provide a long wear life, a smooth and quiet ride with good cornering, adequate skid resistance and traction under various road conditions. A tire is also expected to have a pleasing appearance to complement the vehicle.

Endurance life is difficult to determine. It is measured by running test tires to failure in as short a time as possible. The accelerated endurance life of a tire is usually measured in combinations of overloads up to 150% of the recommended loads, inflation pressures as low as 75% of the recommended pressures, and speeds > 161 km/h.

The burst strength of a tire is related to the tensile strengths of its tire cords. The burst strength is lowered by weak or insufficient cords, inefficient cord placement, elevated tire temperatures and permanent

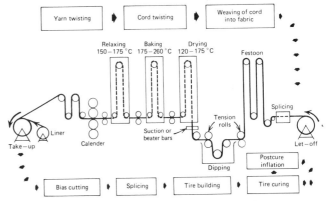

Figure 1. A schematic flow diagram for processing tire yarn into tires.

degradation from tire usage. Therefore the usual safety factor of a new tire is > 10, eg, a tire normally inflated cold to 274 kPa (25 psig) resists more than 1.82 MPa (250 psig) before burst.

Bruise resistance is tested by measuring the energy required to penetrate a 19-mm cylindrical plunger at a speed of 51 mm/min into the crown of an inflated tire at room temperature. Although high speed tests on warm tires are more realistic, low speed tests are simpler and faster and therefore used for the quality control of tires.

Separations are usually owing to poor adhesion between the cord and the carcass rubber caused by elevated temperature, moisture, oxygen, strain-stress cycling, and fatigue failure. At speeds > 145 km/h or under loads 20% above those recommended, or under inflation pressures 34.5 kPa (5 psi) below those recommended, fatigue failure can occur in the sidewall near the tire-tread shoulder or near the bead where the cord plies end. Separations are also caused by excessive stress concentrations in these regions, especially at high speeds.

In a bias tire, 30–40% of the total power loss, a significant portion of the heat generated, is caused by the carcass cords. Since power loss is synonymous with heat-generation rate, similar tires made from different cords should show equilibrium-operating temperatures corresponding to the heat-generation rates of these cords. Although it is almost impossible to make truly equivalent tires from different cords, a good experiment performed by Allied Chemical compared nylon-6 and nylon-6,6 in bias truck tires. They showed that under slight overload conditions the tires made with nylon-6 operate 10–25°C cooler than those made of nylon-6,6.

Radial tires tend to generate less heat and therefore less power loss than bias tires at normal speeds of 113 km/h or less. However, at higher speeds the normal radial passenger tire begins to go into resonance much sooner than bias passenger tires and its relative power loss advantage is soon lost.

Tread wear partially depends upon load distribution and movement (scuffing) of the tread elements in the tire foot print. Belted radial tires generally are more wear resistant than bias tires.

When tires are subject to excessive growth, the tread grooves are under strain and small cracks begin which eventually expose the rubber cords and allow deterioration of these tire cords.

Tire flatspotting can occur when thermoelastic tire cords, such as nylon and polyester, are heated above their T_g (glass temperature) and shrink. Upon cooling below T_g, the cord length freezes and a flatspot or out-of-roundness remains where the cords shrank. Reheating can reverse this process.

Test Methods

Yarn-to-cord conversion efficiency is usually measured by relating the ultimate tensile strength of the untwisted yarn to the ultimate tensile strength of the cord. The length of stability of a cord in the presence of moisture is measured by the degree of shrinkage. Length stability to heat is measured by heating the cord under minimal load and noting the length change. Simple creep tests can be related to expansion, groove cracking, and nonuniformity. Fatigue is measured by bare-cord and in-rubber tests. Adhesion tests are either peel or pull-out tests. The flatspot and "uniformity" tests are based on simulation procedures. Stiffness is measured by the angle of recovery after folding under a standard load.

Tire-Cord Status

Nylon and polyester cords are very strong, but rayon becomes comparable to nylon at elevated temperatures and high speed impact. Polyester resists high speed impact but loses modulus and strength faster than rayon at 150°C. Nylon and polyester retain their tensile strengths better than rayon when highly twisted. Very high modulus aramid, glass, and steel lose even more tensile strength when highly twisted. Under constant strain, nylon generates the least heat and is therefore used to reinforce most high load and high speed tires. Although polyester is dimensionally stable to moisture, it is subject to both adhesion loss and/or tensile loss by hydrolysis and aminolysis when at elevated temperatures. Costly aramid fibers (qv) have a very high tensile strength but their fatigue life remains inadequate for bias-type tires

unless they are highly twisted—which drastically reduces tensile strength. Each tire cord represents a different combination of properties which have to be optimized to fit the tire's performance requirements, including costs.

LEONARD SKOLNIK
BFGoodrich

T. Takeyama, J. Matsui, and M. Hijiri in S.K. Clark, ed., *Mechanics of Pneumatic Tires*, 2nd ed., Chapt. 2, U.S. Government Printing Office, Washington, D.C.; J.D. Walter and S.K. Clark, Chapt. 3.

TITANIUM AND TITANIUM ALLOYS

Titanium is a metallic element of Group IVB, that is known as a space-age metal because of the commercial development of metallic titanium and its use in aerospace applications. The properties of this element that account for its aerospace usage are its inertness and high strength:weight ratio. Its principal use, however, is as TiO_2 as a paint filler (see Paint; Pigments). The whiteness and high refractive index of TiO_2 are unequaled for whitening paints, paper, rubber, plastics, and other materials.

Titanium occurs in nature as ilmenite ($FeTiO_2$), rutile (tetragonal TiO_2), anatase (tetragonal TiO_2), brookite (rhombic TiO_2), perovskite ($CaTiO_3$), sphene ($CaTiSiO_5$), and geikielite ($MgTiO_3$). Ilmenite is by far the most common, although rutile has been the most important raw material source. Titanium ore bodies occur either as hard-rock deposits, magnetic in origin, or as secondary placer deposits. The latter contain ca 10^8 metric tons titanium as ilmenite and 15×10^6 t as rutile. Hard-rock deposits contain $> 2 \times 10^8$ t, excluding the oil sands of Alberta, which contain $> 2 \times 10^9$ t.

Alloys

Alloy development has been aimed at elevated temperature aerospace applications, strength for structural applications, and aqueous corrosion resistance. The principal effort has been in aerospace applications to replace nickel–cobalt-based alloys in the 500–900°C ranges. The useful strength and corrosion resistance temperature limit is ca 550°C.

Alloying elements alter the α–β transformation temperature. Elements that raise the transformation temperature are called α stabilizers, whereas β stabilizers depress it. The β stabilizers are divided into β-isomorphous and β-eutectic types. The α-stabilizing alloying elements include aluminum, tin, zirconium, and the interstitial alloying elements, ie, elements that do not occupy lattice positions, such as oxygen, nitrogen, and carbon. Small quantities of interstitial alloying elements, generally considered to be impurities, greatly affect strength and ultimately result in embrittlement at room temperature (see Table 1). These elements are always present and are difficult to control. Nitrogen, which has the greatest effect, is specified in commercial titanium alloys to be < 0.05%, whereas carbon is specified at 0.08% max. For cryogenic service, oxygen content is specified to be < 1300 ppm. Alloys with low interstitial content are identified as ELI (extra-low interstitials).

The most important alloying element is aluminum, an α stabilizer, which increases the mechanical strength of titanium. However, above 7.5% Al, the alloy becomes difficult to fabricate and embrittles. The important β-stabilizing alloying elements are the bcc elements vanadium, molybdenum, tantalum, and niobium of the β-isomorphous type, and manganese, iron, chromium, nickel, copper, and silicon of the β-eutectic type. Alloys of the β type respond to heat treatment, are denser than pure titanium, and are easily fabricated. The most important commercial β-alloying element is vanadium.

The aerospace alloys can be divided into three categories: all-α structure, a mixed α-β structure, and all-β structure. Most of the ca 100 commercial alloys are of the α-β structure type. The most important one is Ti-6 Al-4 V. It can be age hardened and is moderately ductile, and has an excellent record of successful applications. The only α alloy of

Table 1. Effects of O, N, and C on the Ultimate Tensile Strength[a]

Concentration of impurity, wt%	Oxygen[b,c]		Nitrogen[b,c]		Carbon[b,c]	
	UT, MPa[d]	Elong., %	UT, MPa[d]	Elong., %	UT, MPa[d]	Elong., %
0.025	330	37	380	35	310	40
0.05	365	35	460	28	330	39
0.1	440	30	550	20	370	36
0.15	490	27	630	15	415	32
0.2	545	25	700	13	450	26
0.3	640	23	embrittles		500	21
0.5	790	18			520	18
0.7	930	8			525	17

[a] Tests were conducted using titanium produced by the iodide process.
[b] UT = ultimate tensile stress.
[c] Elongation on 2.54 cm.
[d] To convert MPa to psi, multiply by 145.

commercial importance is Ti–5 Al–2.5 Sn. The commercial near-α alloys are Ti–8 Al–1 Mo–1 V and Ti–6 Al–2 Sn–4 Zr–2 Mo. Commercial development of β alloys has not been successful. The nonaerospace applications of titanium alloys are mainly as vessels, tanks, and heat exchangers used to contain industrial fluids. There is one grade (ASTM Grade 2) which accounts for most of this usage, and it is an unalloyed grade titanium.

Other alloys include the aluminides (TiAl and Ti_3Al), the superconducting alloys (Ti–Nb type), the shape-memory alloys (Ni–Ti type), and the hydrogen-storage alloys (Fe–Ti) (see Superconducting materials; Shape-memory alloys). Titanium alloyed with niobium exhibits superconductivity and a lack of electrical resistance below 10 K. Nickel alloys exhibit a memory effect in response to temperature changes. Titanium alloyed with iron is a leading candidate for solid-hydride energy storage material for automotive fuels.

Properties

The physical properties of titanium are given in Table 2; most important is the ratio of its strength (ultimate strength > 690 MPa or 100,000 psi) to its density 4.507 g/cm^3. Because of its high melting point, titanium can be alloyed to maintain strength well above the useful limits of magnesium and aluminum alloys as well as other light elements. This property gives it a unique position in applications between 150–550°C where the strength:weight ratio is the sole criterion.

Corrosion resistance. Titanium is immune to corrosion in all natural environments. It resists decomposition because of a tenacious protective oxide film, which is insoluble, repairable, and nonporous. When this film is broken, corrosion is very rapid. The presence of a small amount of water is sufficient to repair the damaged oxide film. Titanium is resistant to corrosion in oxidizing, neutral, and inhibited reducing conditions, but

Table 2. Physical Properties of Titanium

Property	Value
melting point, °C	1668 ± 5
boiling point, °C	3260
density, g/cm^3	
α phase at 20°C	4.507
β phase at 885°C	4.35
thermal conductivity at 25°C, W/(m·K)	21.9
electrical resistivity at 20°C, nΩ·m	420
magnetic susceptibility, mks[a]	180 × 10^{-6}
modulus of elasticity, GPa[b]	
tension	ca 101
compression	103
shear	44

[a] mks = meter-kilogram-second.
[b] To convert GPa to psi, multiply by 145,000.

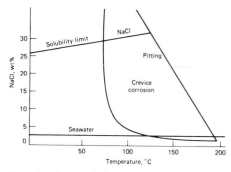

Figure 1. Corrosion characteristics of titanium in aqueous NaCl solution.

it corrodes rapidly in HCl, H_2SO_4, or acid fluoride environments, as well as in hot caustic solutions. Titanium is susceptible to pitting and crevice corrosion in aqueous chloride environments (see Fig. 1), and to failure by hydrogen embrittlement. In galvanic coupling, titanium is usually the cathode metal and consequently not attacked.

Manufacture

High grade ore (> 85% TiO_2) is refined to pigment-grade TiO_2 by chlorination. Lower grade ore is processed via the sulfate route. The chlorination process produces a better quality pigment, requires less energy, and has less waste.

In the sulfate process, ilmenite ore is treated with sulfuric acid at 150–180°C:

$$5\,H_2O + FeTiO_3 + 2\,H_2SO_4 \rightarrow FeSO_4 \cdot 7H_2O + TiOSO_4$$

Heating to 90°C hydrolyzes titanyl sulfate to insoluble titanyl hydroxide,

$$TiOSO_4 + 2\,H_2O \rightarrow TiO(OH)_2 \downarrow + H_2SO_4$$

which is calcined at 1000°C to TiO_2. Environmental problems have forced the industry to either shut down sulfate plants or install expensive pollution-control equipment.

In the chloride process, a high grade titanium oxide ore is chlorinated in a fluidized-bed reactor in the presence of coke at 925–1010°C.

$$TiO_2 + 2\,C + 2\,Cl_2 \rightarrow 2\,CO + TiCl_4 \qquad \Delta G_{1300°C} = -125\ kJ\ (30\ kcal)$$

The $TiCl_4$ is oxidized at 985°C to TiO_2. For reduction to the metal, the $TiCl_4$ is further purified.

Tetrachloride reduction. In the sodium reduction process, $TiCl_4$ is first reduced to the dichloride and the trichloride at ca 230°C. The lower chlorides are then reduced to the metal by adding more sodium to the salt bath at ca 1000°C. Titanium sponge forms in the center of the reaction vessel encased by sodium chloride from which it is separated.

In the magnesium reduction (Kroll) process, titanium tetrachloride gas is metered into a reaction vessel containing liquid magnesium in a helium or argon atmosphere. A 25% excess ensures that the lower chlorides ($TiCl_3$ and $TiCl_2$) are reduced to metal. The product, so-called titanium sponge, is further processed to remove unreacted titanium chlorides, magnesium, and magnesium chlorides. The recovered magnesium chloride is recycled.

Other methods of reducing $TiCl_4$ include electrolytic reduction, hydrogen reduction, plasma reduction, and aluminum reduction. Titanium dioxide can be reduced to metal with aluminum, calcium, or carbon, but acceptable purity of the metal product is impossible to attain.

To consolidate the sponge into ingot, it is blended with alloying elements or other sponge. Consumable electrodes are produced by welding 90-kg blended sponge compactions together in an inert atmosphere. These electrodes are then double-vacuum arc remelted (VAR). The double melt eliminates porosity, yields a smooth surface ingot, and removes residual volatiles (Mg, $MgCl_2$, Cl_2, and H_2).

Titanium casts are produced by precision or investment casting methods. Powder-making processes include the hydride–dehydride process, the electron-beam rotating-disk method, the Crucible Research Center Colt Titanium process, the pendant-drop process, and the rotating-electrode process.

The ingots are further processed by forging, hot- and cold-rolling, extrusion, etc. Minimum heating time at the lowest practical temperature under an inert gas attains the best mechanical properties, minimizes contamination and oxidation, and avoids excess grain growth. The surface is conditioned by lathe turning, grid blasting, belt grinding, centerless grinding, and caustic or acid pickling. Sheet, thin plate, welded tubing, and small bars are manufactured by conventional cold-working techniques. Fabrication techniques are similar to those used for austenitic steel (see Steel). Welding requires protection by an inert gas. Titanium cannot be fusion welded to other metals because of the formation of brittle intermetallic phases in the weld zone. For some applications, titanium can be clad to steel (Detaclad) to reduce cost.

About 5% of titanium mined is used a metal, 93% as pigment-grade TiO_2, and 2% as ore-grade rutile for fluxes and ceramics.

Specifications and Standards

The compositional specifications of titanium alloys are tabulated by ASTM; other material and quality standards have been set by ASME and AMS. Military specifications are found under MIL-T-9046 and MIL-T-9047, and aerospace material specifications for bar, sheet, tubing, and wire. Elements controlled by specifications include carbon, iron, hydrogen, nitrogen, and oxygen, and for more stringent specifications, yttrium.

Titanium and its corrosion products are nontoxic. A safety problem exists because of pyrophoric grindings, turnings, and some corrosion products.

Uses

Titanium is primarily used in the form of high purity titanium oxide in paint pigments. Other uses are in plastics, paper, and rubber. The use in aircraft is about equally divided between engines and frames. Corrosion-resistant applications include heat-exchange pipes and tubing for the power industry and marine and desalination applications. Titanium metal is utilized in environments of wet chlorine gas and bleaching solutions in the chlor-alkali and pulp and paper industries. In oil and gas refinery applications, titanium is used in environments of H_2S, SO_2, CO_2, NH_3, caustic solutions, steam, and cooling water.

DONALD KNITTEL
Cabot Corporation

R.A. Wood, *The Titanium Industry in the Mid-1970's*, Batelle Report MCIC-75-26, Battelle Memorial Institute, Columbus, Ohio, June 1975.

S.G. Glazunov in N.P. Sazhin and co-workers, ed., *Titanium Alloys for Modern Technology*, NASA TT F-596, National Aeronautics and Space Administration, Washington, D.C., March 1970, p. 11.

Titanium for Industrial Brine and Sea Water Service, Titanium Metal Corporation of America, Pittsburgh, Pa., 1968.

TITANIUM COMPOUNDS

INORGANIC

Titanium has four valence electrons; Ti(IV) is the most stable valence state. The most important commercial forms are titanium dioxide and titanium metal. Titanium dioxide is amphoteric, giving rise to a series of titanates as well as salts of Ti(IV), which are readily hydrolyzed in aqueous solution. This is the basis of the commercial process for the manufacture of titanium dioxide pigments (see Paint; Pigments, inorganic).

Titanium–Hydrogen System

The absorption of hydrogen by titanium is reversible above 400°C to a maximum composition of $TiH_{1.7}$. A compound with the stoichiometric formula TiH_2 has been claimed to have been prepared by the reduction of pure titanium dioxide with calcium hydride; its existence has not been proved. Titanium hydride is a light gray powder of metallic appearance.

It decomposes at ca 600°C; the hydrogen thus liberated is a convenient source of pure hydrogen. The titanium–hydrogen system offers the possibility of thermal storage (see Hydrogen energy).

Titanium hydride is the starting material for the preparation of titanium borides, nitrides, and silicides; the evolving hydrogen provides a convenient reducing atmosphere. Titanium hydride is also used for glass- or ceramic-to-metal seals and for titanium coatings on copper or copper-plated metals.

Titanium–Boron System

The equilibrium diagram of the titanium–boron system shows the presence of five phases: TiB_2, Ti_2B, TiB, Ti_2B_5, and TiB_{12}. Only TiB_2, a gray crystalline solid, is important; owing to its hardness and resistance to oxidation it is employed in the aircraft industry and in cutting and grinding. It may be made by direct combination of titanium or titanium hydride and boron at 2000°C.

Titanium–Carbon System

Pure titanium carbide, TiC, is difficult to prepare; some impurities, eg, oxygen and nitrogen, are usually present. It is manufactured by reduction of titanium dioxide with carbon. An intimate mixture of the two substances is heated to 1900–2100°C in an electric arc furnace or a graphite tube or in a tungsten tube under a hydrogen atmosphere. Titanium carbide is light gray when fractured but can be polished to a silver gray. The maximum melting point in the Ti–C system is 3067°C for $TiC_{0.8}$; the boiling point of TiC is ca 4800°C. It is one of the hardest pure carbides known (9–10 Mohs). It is a good conductor of electricity and is not attacked by acids or aqueous alkali. It is used extensively for cutting tools.

Titanium–Nitrogen System

Titanium nitride. Titanium nitride, TiN, melts at 2950°C. It is not attacked by acids except boiling aqua regia, and it is decomposed by boiling alkalies with the evolution of ammonia. It is a better conductor of electricity than the metal. It is prepared by direct synthesis at 1000–1400°C or from the dioxide and nitrogen in the presence of carbon at 1600°C. Titanium nitride is a brown-yellow powder which, after pressing and sintering, can be polished to a golden-yellow mirror.

Titanium nitrate. Prolonged reaction of liquid nitrogen dioxide and titanium tetrachloride at −60 to −20°C produces $2TiCl_4.3N_2O_4$, which decomposes to nitrosyl chloride and titanium nitrate, $Ti(NO_3)_4$. Aqueous solutions are readily prepared by dissolving hydrated titanium oxide in nitric acid.

Titanium-Oxygen System

A wide range of compounds is possible in this system. Irregularities in the crystal lattice are of great importance to the behavior of the dioxide as a pigment. For example, TiO_2 is white, whereas $TiO_{1.9995}$ is blue. The monoxide, sesquioxide, and pentoxide are made by heating a compressed stoichiometric mixture of titanium powder and the dioxide at 1600°C under vacuum. The pentoxide is also made by hydrogen reduction of the dioxide or monoxide at 700–1100°C. The properties of the monoxide and sequioxide are given in Table 1.

Hydrated titanium oxides. Addition of an alkali-metal hydroxide to a solution of a Ti(II) or Ti(III) salt precipitates the hydroxides of Ti(II) (black) or Ti(III), respectively. They readily oxidize in air to the hydrated titanium dioxide. The precipitate obtained from an alkali-metal hydroxide and a solution of a Ti(IV) salt is essentially hydrated titanium

Table 1. Properties of Titanium Monoxide and Titanium Sesquioxide

Property	TiO	Ti_2O_3
color	golden yellow	violet
density, g/cm³	4.888	4.486
melting point, °C	1750	1900
structure	fcc	hexagonal

Table 2. Crystallographic Properties of Anatase, Brookite, and Rutile

Property	Anatase	Brookite	Rutile
crystal structure	tetragonal	orthorhombic	tetragonal
optical	uniaxial, negative	biaxial, positive	uniaxial, positive
density, g/cm^3	3.9	4.0	4.23
hardness, Mohs scale	$5\frac{1}{2}$–6	$5\frac{1}{2}$–6	7–$7\frac{1}{2}$
unit cell	$D_4a^{19} \cdot 4TiO_2$	$D_2h^{15} \cdot 8TiO_2$	$D_4h^{12} \cdot 2TiO_2$

oxide; the exact composition depends on conditions. In view of its ready interchange with other compounds, hydrated titanium dioxide is used as an ion-exchange medium (see Ion exchange). It is also the starting material for other titanium compounds.

Titanium dioxide. Titanium dioxide occurs in nature in three crystalline forms: anatase, brookite, and rutile (see Table 2). They have been prepared synthetically, but only rutile, the thermally stable form, has been obtained in the form of large transparent single crystals. Titanium dioxide is used extensively in the electronics, plastics, and ceramics industries because of its electrical properties; rutile is a semiconductor (qv). Titanium dioxide is thermally stable (mp 1855°C) and highly resistant to chemical attack. It is reduced partially by hydrogen and carbon monoxide; at ca 2000°C under vacuum, it is reduced by carbon to titanium carbide. Chlorination is only possible in the presence of a reducing agent.

Chemically pure titanium dioxide is prepared from titanium tetrachloride that has been purified by repeated distillations. The tetrachloride is hydrolyzed in aqueous solution giving a precipitate of hydrated titanium oxide, which is calcined at 800°C. Titanium dioxide can also be precipitated from titanium tetrachloride as hydrated titanium dioxide, which is converted to the double oxalate, $(NH_4)_2[TiO(C_2O_4)_2]$; recrystallization from methanol and subsequent calcination gives highly pure TiO_2.

Nonpigment uses include applications in the vitreous enamel and electronics industries. Synthetic gems have been made from rutile and strontium titanate. Titanium dioxide is also used a catalyst, sometimes impregnated with precious metals.

Titanium dioxide pigments. The high refractive index, lack of absorption of visible light, ability to be produced in the correct size range, and the stability and nontoxicity of TiO_2 made it the predominant white pigment in the world.

Titanium dioxide pigments are produced as anatase or rutile either by the sulfate or the chloride process. In the former, an acid solution of titanyl sulfate is hydrolyzed, followed by calcination of the hydrous precipitate. In the latter, titanium tetrachloride is burned in oxygen to yield the dioxide and chlorine (see Pigments, inorganic). The pigment particles may be coated with silica, alumina, titania, or zirconia to improve durability and reduce yellowing, which occurs in certain types of paint. The coating process involves dispersion of the pigment from the calciner, or from the oxidation stage in the chloride process, in water. This may be done in a ballmill or sand mill; sodium silicate is frequently used as the dispersing agent. The dispersion is allowed to settle to remove oversized particles. Coatings of two or more hydrous oxides may be applied to a pigment, either separately by successive operations or simultaneously in one operation. Once the coating is formed, the pigment is washed and dried before being ground in a micronizer or a fluid-energy mill.

Pigments designed for plastics may be given an organic surface treatment in addition to the inorganic coating. Vesiculated bead systems, eg, water vesicles in rigid polymer beads or polymer beads containing air and TiO_2, promote opacity. The pigment color can be modified by the incorporation of guest ions into the rutile host lattice; thus, Titanium Nickel Yellow, also called Titanate Yellow or Sun Yellow, contains nickel and antimony. This pigment is very stable to acids, alkalies, and oxidizing and reducing agents, and insoluble in all solvents and has excellent heat stability; it is, however, tinctorially weak compared with organic yellows (see Pigments, organic).

Titanium dioxide pigments are frequently handled in the form of slurries, which amount to ca 20% of TiO_2 sales in the U.S. Pigment manufacturers have their own specifications; national and international specifications are also available.

Peroxidic compounds. The action of freshly precipitated hydrogen peroxide or hydrated titanium(IV) oxide or the hydrolysis of a peroxide, such as $K_2[TiO_2(SO_4)_2]$, gives a solid which, on dehydration with acetone, gives a yellow solid that behaves as a true peroxide.

Inorganic titanates. Titanium forms a series of compounds with other metals that must be considered as multiple oxides, such as meta $(M(II)TiO_3)$, ortho $(M(II)_2TiO_4)$, and poly$(M(I)_2Ti_2O_5)$. These compounds are prepared by heating stoichiometric amounts of titanium dioxide or the hydrated dioxide with the metal oxide, hydroxide, or carbonate at the correct temperature. Lithium titanate is prepared by heating lithium carbonate and titanium dioxide to ca 950°C, whereas potassium titanate is made by heating hydrated titanium dioxide and potassium hydroxide to 160–170°C. Titanates are insoluble in water, and are generally stable but are decomposed by acid.

Most alkali titanates are white. The polytitanates of potassium can be manufactured in fibrous form. Commercial applications include use as a reinforcement for plastics, filtration medium, friction-brake component, and insulating and reflective material.

Alkaline-earth titanates exhibit electrical characteristics that make them well suited for electrical applications. Polarized barium titanate, $BaTiO_3$, is ferroelectric and piezoelectric. It is used in phonograph pickups, underwater detection equipment, and for the generation of ultrasonic waves. Both strontium and calcium titanates are used as additives to barium titanate in electronic components. Magnesium titanate, Mg_2TiO_4, is used as an additive to ceramic dielectric components (see Ferroelectrics).

Ferrous metatitanate, $FeTiO_3$, occurs in nature as the mineral ilmenite, which is used extensively in the manufacture of pigments. Ferrous orthotitanate, Fe_2TiO_4, is prepared by fusing ferrous oxide and titanium dioxide. Ferrous dititanate, $Fe_2Ti_2O_5$, forms by heating ilmenite with carbon at 1000°C. Ferric titanate, Fe_2TiO_5 (pseudobrookite), is found in nature. It is prepared from Fe_2O_3 and TiO_2 in a sealed tube at > 1000°C.

Lead titanate, $PbTiO_3$, a yellow solid that had been used as a pigment, has been superseded by more durable rutile pigments. Zinc orthotitanate, Zn_2TiO_4, is obtained by heating the two oxides at 1000°C. Nickel antimony titanate, is a yellow pigment (Pigment Yellow 53).

Titanium–Halogen Compounds

The halides and oxyhalides are given below.

Oxidation state	Fluoride	Chloride	Bromide	Iodide
II	TiF_2	$TiCl_2$	$TiBr_2$	TiI_2
III	$TiF_3, TiOF$	$TiCl_3, TiOCl$	$TiBr_3$	TiI_3
IV	TiF_4	$TiCl_4, TiOCl_2$	$TiBr_4, TiOBr_2$	TiI_4

Fluorides. The trifluoride is a blue crystalline solid made by the action of gaseous HF on titanium or titanium hydride or by an exchange reaction between HF and $TiCl_3$. The tetrafluoride is made from HF and $TiCl_4$. Fluorotitanic acid is obtained by dissolving titanium dioxide in HF.

$$TiO_2 + 6\ HF \rightarrow H_2TiF_6 + 2\ H_2O$$

Although the acid is stable in solution, it has not been isolated. A well-defined series of salts includes the important potassium fluorotitanate, K_2TiF_6.

Chlorides. Titanium dichloride, a black solid, is prepared by the thermal disproportionation of $TiCl_3$ or the reduction of $TiCl_4$. The trichloride, a dull purple, crystalline solid, is obtained by the hydrogen reduction of $TiCl_4$ at 700°C. It is used as a polymerization catalyst. The

Table 3. Physical Properties of Titanium Tetrachloride

color	colorless
density (at 20°C), g/cm^3	1.70
freezing point, °C	−24
heat of fusion, kJ/mol^a	9.37
boiling point, °C	135.8
vapor pressure, kPa^b	
at 20°C	1.33
at 50°C	5.52
at 100°C	35.47
heat of vaporization, kJ/mol^a	
at 25°C	38.1
specific heat (at 20°C), J/g^a (J/mol^a)	0.81 (153.1)
refractive index, n_D^{20}	1.6985
magnetic susceptibility	$−0.287 \times 10^{-6}$
dielectric constant (at 20°C)	2.79

[a] To convert J to cal, divide by 4.184.

[b] To convert kPa to mm Hg, multiply by 7.5.

Table 4. Structure and Physical Properties of Titanium Silicides

Property	TiSi	TiSi$_2$	Ti$_5$Si$_3$
structure	rhombic	orthorhombic	hexagonal
sp gr	4.34	4.39	4.31
mp, °C	1760 (incongruent)	ca 1540	2120
hardness, 100-g load	1039	870	986
hardness, Mohs		4–5	
resistivity, $\mu\Omega \cdot cm$		123	

trichloride hexahydrate, owing to its reducing properties, is used as a stripping or bleaching agent in the dyeing industry, particularly where chlorine must be avoided.

Titanium tetrachloride is the starting material for titanium dioxide pigments, titanium metal, and many titanium compounds. It is also used as catalyst. Physical properties are given in Table 3. Titanium tetrachloride is very susceptible to hydrolysis and the liquid fumes strongly when in contact with moist air. Owing to its strong affinity for water, $TiCl_4$ is an excellent desiccating agent. It is soluble in organic solvents, but forms addition compounds with a number of inorganic substances and with solvents that contain a donor atom, eg, ketonic oxygen, nitrogen, and sulfur. The reaction with anhydrous ammonia yields amidochlorides, $Ti(NH_2)_xCl_{4-x}$. Reduction of $TiCl_4$ gives the metal. Manufacture is by chlorination of rutile with carbon as the reducing agent. However, the increasing cost of rutile has led to efforts to upgrade ilmenite. Fluidized-bed methods include the chlorination of titanium slags.

The acid H_2TiCl_6 forms when dry hydrogen chloride dissolves in $TiCl_4$. The yellow ammonium, potassium, rubidium, and cesium salts are known. An oxychloride, $TiOCl$, is prepared by the action of TiO_2, Fe_2O_3, or oxygen on $TiCl_3$. The compound $TiOCl_2$ is prepared from chlorine monoxide or ozone and $TiCl_4$.

Bromides. Titanium dibromide, a black powder, is made from the elements or by disproportionation of titanium tribromide at 400°C. The tribromide, a blue-black crystalline powder, is prepared by reduction of the tetrabromide with hydrogen or titanium.

$$2\ TiBr_4 + H_2 \rightarrow 2\ TiBr_3 + 2\ HBr$$

$$3\ TiBr_4 + Ti \rightarrow 4\ TiBr_3$$

The tetrabromide, a lemon-yellow solid forms addition compounds with NH_3, PH_3, SO_2, H_2S, and some organic compounds. It is prepared by double decomposition of titanium tetrachloride and HBr, by direct combination of the elements, or by bromination of carbon and titanium dioxide at 650–700°C.

Iodides. Titanium diiodide, triiodide, and tetraiodide are made by direct combination of the elements. The tetraiodide is used as a catalyst in organic syntheses and as an intermediate in the production of pure titanium.

Titanium–Silicon System

The three silicides $TiSi_2$, $TiSi$, and Ti_5Si_3 are made by direct synthesis. Their properties are given in Table 4. They are used in the preparation of abrasion and heat-resistant refractories. A mixture of Ti_5Si_3 with diamond and titanium carbide, hot pressed at 1450°C and 6 GPa (9×10^5 psi), forms a cutting tip with a much longer life than conventional tools. Titanium silicides are also used in electrical resistance

materials and electrically conducting ceramics (see Ceramics as electrical materials).

Titanium–Phosphorus Compounds

Titanium monophosphate is prepared by heating phosphine with $TiCl_4$ or titanium sponge, or titanium powder with phosphorus in a sealed tube. Titanium(III) phosphate, $Ti(PO_4)_3$, is made by adding a soluble phosphate to a solution of titanous chloride or sulfate. Titanium(IV) phosphate, a white compound, made by adding a soluble phosphate to titanium(IV) sulfate solution, is used in the dyeing and tanning industries. Titanium(IV) bis(hydrogen phosphate) dihydrate, $Ti(HPO_4)_2 \cdot 2H_2O$, is an ion-exchange medium. The pyrophosphate, $TiH_2P_2O_7$, a white powder, is an uv-reflecting pigment and is made by heating stoichiometric amounts of TIO_2 and phosphoric acid to 900°C.

Titanium–Sulfur Compounds

Titanium subsulfide, Ti_2S, forms as a gray solid when titanium monosulfide is heated with titanium at 1000°C in a sealed tube. Both the monosulfide, TiS, and the sesquisulfide, Ti_2S_3, are prepared by direct combination or by hydrogen reduction of the disulfide at high temperature.

Titanium(IV) sulfate, $Ti(SO_4)_2$, is a white hygroscopic solid. Upon heating it gives titanyl sulfate, $TiOSO_4$, and then TiO_2. Titanium sulfate is made by the reaction of $TiCl_4$ with SO_3 dissolved in sulfuryl chloride. Titanyl sulfate is an intermediate for titanium pigments.

Analytical Methods

Titanium is determined in solution by reduction methods, eg, with Jones or Nakazono reductors or by heating with the reducing agent, and titrating with ferric iron solution.

Titanium dioxide pigment samples can be dissolved in a boiling mixture of concentrated sulfuric acid and ammonium sulfate or hydrofluoric acid. The solution is cooled and diluted, and the titanium is determined titrimetrically as described above.

Small quantities of titanium are determined by adding hydrogen peroxide solution to the acidic sample solution. A yellow solution containing the peroxide–titanium complex is obtained; its intensity is measured spectrophotometrically.

Health and Safety Factors

There is no evidence that titanium is essential for life, nor is there any evidence that it is toxic. Titanium compounds are used in medical and food products without ill effects. Titanium dioxide pigment is classified as a nuisance dust because of its small particle size.

J. WHITEHEAD
Tioxide Group PLC

J. Barksdale, *Titanium, Its Occurrence, Chemistry and Technology*, 2nd ed., The Ronald Press, Co., New York, 1966.

P. Pascal, *Nouveau Traitè de Chimie Minerals*, Tome IX, Masson et Cie, Paris, France, 1963.

R.J.H. Clark, *The Chemistry of Titanium and Vanadium*, Elsevier Publishing Co., Amsterdam, Neth., 1968.

ORGANIC

Organic titanium compounds contain a covalent bond between titanium and another atom that is bonded to a carbon-containing group. Titanium tetrachloride, manufactured from ilmenite or rutile, is the basic raw material. It is readily converted to titanate esters, in particular tetraisopropyl titanate (TPT), by the Nelles process. TPT is converted by alkoxy exchange to a wide variety of commercial tetraalkoxytitanium compounds. True organometallic compounds with a titanium–carbon bond are prepared from titanium tetrachloride by reaction with organometallics such as organomagnesium, sodium, or lithium reagents. Most simple organometallic derivatives are unstable except bis(cyclopentadienyl)titanium dichloride (titanocene dichloride), Cp_2TiCl_2, and its analogues (see Organometallics).

Organometallics and complexes with trivalent titanium are stable at room temperature but are attacked by oxygen and moisture. Organic derivatives of divalent titanium are less common. They, as well as titanium(0) compounds, are potent reducing agents. Hydrotitanium derivatives are involved in olefin and acetylene isomerizations and polymerizations (see Olefin polymers; Polymerization mechanisms and processes).

Alkoxides (Titanate Esters)

The most important titanium alkoxide is TPT made from $TiCl_4$ and isopropyl alcohol. Alkylene oxides react stepwise with $TiCl_4$ forming 2-chloroalkoxides, eg, $Ti(OCH_2CH_2Cl)_4$ (see Alkoxides, metal). Higher alkoxides are prepared by alcohol interchange in benzene or cyclohexane. The affinity of an alcohol for titanium decreases in the order: primary > secondary > tertiary, and unbranched > branched.

Properties and Reactions

Physical properties of a few Ti(IV) compounds are given in Table 1. Organic titanates have a strong tendency to associate. Titanium(IV) achieves coordination number 6 by sharing electron pairs from oxygen atoms, a tendency that is opposed by steric crowding. Association may involve single or double bridges. In crystalline tetramers, each titanium atom is octahedrally coordinated to six oxygens; edge sharing of four tetrahedra gives the tetramer:

The lower titanium alkoxides, except the methoxide, are rapidly hydrolyzed by moist air or water giving a series of condensed titanoxanes (—TiOTiO—). Methoxides, aryloxides, and higher alkoxides with long chains hydrolyze slowly. Complete hydrolysis to TiO_2 is difficult to achieve. Titanoxane formation takes the following path:

Titanoxanes are also prepared from alkoxides and anhydrides. Dimers are made by oxidation of Ti(III) alkoxides. Oligomers are prepared from $TiCl_4$ and alcohols containing enough water to yield the desired degree of polymerization; the reaction is driven to completion by the addition of ammonia.

The tendency of titanium(IV) to reach coordination number 6 accounts for the rapid exchange of alkoxy groups with alcohols.

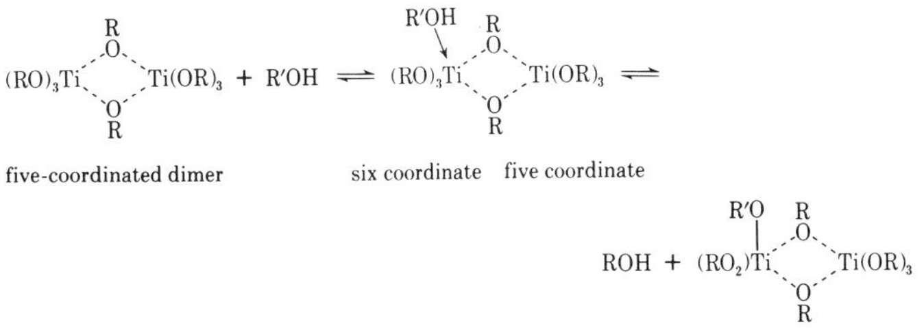

For preparative purposes, the equilibrium must be shifted with excess alcohol or by distilling the more volatile lower alcohol.

Ester interchange catalyzed by titanates is an important industrial reaction employed in the manufacture of methacrylates from dialkylaminoethanols and methyl methacrylate.

Organic acids form acylates when heated with titanium alkoxides. Best results are obtained with one or two moles of acid; a higher acid concentration gives polymers.

Lower alkoxides are stable and can be distilled quickly at atmospheric pressure. Prolonged heating promotes polymerization. Thermolysis is utilized in the coating of glass and other surfaces with a film of titanium dioxide. When a lower alkoxide, eg, TPT, vaporizes in a stream of dry air and is blown onto hot glass above 500°C, a thin transparent protective coating of TiO_2 is deposited.

Alkoxides are manufactured by a modification of the Nelles process in which $TiCl_4$ in an inert solvent is treated with a monohydric lower alcohol. If the hydrogen chloride formed is expelled by heating or sweeping with dry nitrogen, a dialkoxydichlorotitanate is obtained. Addition of an acid acceptor, eg, ammonia, is required to attain the tetraalkoxide stage; the insoluble NH_4Cl is filtered or centrifuged. Higher alkoxides are readily prepared from TPT or tetrabutyl titanate (TBT) by alcohol interchange with the removal of 2-propanol or n-butanol by distillation. Chelates are made by mixing the chelating agent with TPT or another tetraalkoxide. Since most titanates are hygroscopic, they should be handled under dry nitrogen.

Alkoxy Halides

Titanium alkoxy halides have the formula $Ti(OR)_n X_{4-n}$, where R = alkyl, aryl, or acyl and X = F, Cl, or Br but not I. The fluorides and chlorides are hygroscopic colorless or pale yellow solids or viscous liquids that darken on standing, especially when exposed to light. Bromides are yellow crystalline, hygroscopic solids. Aryloxytitanium halides are orange-to-red solids. Physical properties of some halides are given in Table 2. The alkoxy halides are intermediates in the preparation of alkoxides from a titanium tetrahalide (except the fluoride) and an

Table 1. Properties of Titanium(IV) Tetraalkoxides and Tetraaryloxides

Compound	Formula	Mp, °C	Bp, °C$_{Pa}$[a]	Other properties
titanium methoxide	$Ti(OCH_3)_4$	210	$170_{1.3}$ sublimes	dipole moment, 5.37×10^{-30} C·m[b]
titanium ethoxide	$Ti(OC_2H_5)_4$	< −40	103_{13}	n_D^{35}, 1.5051;
titanium 2-ethylhexyloxide	$TiOCH_2(C_2H_5)CHC_4H_{94}$	< −25	$248–249_{1467}$	d_4^{35}, 1.107 g/cm^3
titanium nonyloxide	$Ti(OC_9H_{19})_4$		$264–265_{200}$	n_D^{35}, 1.4750
titanium isopropoxide	$Ti OCH(CH_3)_{24}$	18.5	$49_{0.1}$	n_D^{20}, 1.4785
titanium m-methylphenoxide	$Ti(OC_6H_4CH_3)_4$		$323–325_{40}$	red crystalline solid
titanium 1-naphthyloxide	$Ti(OC_{10}H_7)_4$		does not distill	

[a] To convert Pa to mm Hg, divide by 133.

[b] To convert C·m to D, divide by 3.336×10^{-30}.

Table 2. Properties of Alkoxyhalides and Aryloxyhalides

Titanium Compound	Formula	Mp, °C	Bp, °C$_{Pa}$^{a}	Other properties
ethoxytrifluoride,	$Ti(OC_2H_5)F_3$	220		
ethoxytrichloride,	$Ti(OC_2H_5)Cl_3$	decomposes 80–81	185–186	
ethoxytribromide,	$Ti(OC_2H_5)Br_3$	indefinite		
diethoxydibromide,	$Ti(OC_2H_5)Br_2$		95–105$_{67}$	red solid
triethoxychloride,	$Ti(OC_2H_5)_3Cl$		176$_{240}$	dipole moment, 9.57×10^{-30} C·m^b
triisopropoxyfluoride,	$Ti(OC_3H_7\text{-}i)_3F$	3–85	140–150$_{80}$	
diphenoxydichloride,	$Ti(OC_6H_5)_2Cl_2$	116		reddish-brown solid

^{a}To convert Pa to mm Hg, divide by 133.
^{b}To convert C·m to D, divide by 3.336×10^{-30}

alcohol or phenol. Heating with excess primary alcohol replaces only two chlorine atoms and the yields are poor. Tetraalkoxides are cleaved by hydrogen chloride in an inert solvent. The dialkoxytitanium dichloride is obtained as an alcoholate.

$$Ti(OR)_4 + 2 HCl \rightarrow Ti(OR)_2Cl_2 \cdot ROH + ROH$$

Cleavage by acetyl chloride or bromide replaces one, two, three, or four alkoxy groups. The products, such as $Cl_2Ti(OR)_2$ and $ClTi(OR)_3$, disproportionate on vacuum distillation.

A principal use for alkoxytitanium halides is the reaction with organolithium or organomagnesium compounds to give compounds with carbon–titanium bonds.

Chelates

Titanium chelates are formed from tetravalent titanium alkoxides or halides with bi- or polydentate ligands. Primary diols react by alkoxide interchange yielding cross-linked polymers.

$$(RO)_4Ti + HOGOH \longrightarrow (RO)_3TiOGOTi(OR)_3 \longrightarrow$$

where G = H$_3$C—CH(CH$_3$)—CH$_2$—CH$_3$ ... CH$_3$

Where the glycol contains one or two secondary or tertiary hydroxyls, the products are more soluble and some are monomeric cyclic chelates.

Silanediols, eg, $(C_6H_5)_2Si(OH)_2$, yield four- and six-membered rings with titanium alkoxides. Pinacols and cyclic 1,2-diols form chelates rather than polymers.

Titanate α-hydroxy acid complexes are prepared in aqueous solutions and are stable in water over a wide pH range. Their structures are uncertain and probably depend on pH and concentration. The alkoxytitanate solution in acetone or THF is added to a solution of the α-hydroxy acid in the same solvent, or an aqueous titanium(IV) chloride, sulfate, or nitrate is treated with an α-hydroxy acid; chelate formation is detected spectroscopically. Oxalic acid behaves as an α-hydroxy acid and yields crystalline ammonium or potassium salts from aqueous titanium(IV) solution or tetraalkoxytitanium compounds. Succinic and adipic acids, however, do not dissolve titanic acid.

β-Diketones, reacting as enols, readily form orange-red chelates with titanium alkoxides which are soluble in common solvents. Since they are coordinately saturated (coordination number 6), they are more resistant to hydrolysis than the parent alkoxides (coordination number 4). Similar chelates are prepared from TiCl$_4$ and β-diketones or β-ketoesters.

$$TiCl_4 + 2\ Hacac \longrightarrow 2\ HCl + Cl_2Ti(acac)_2 \equiv$$

Alkanolamine chelates are used primarily in cross-linking water-soluble polymers in thixotropic paints, in hydraulic fracturing, and in the drilling of oil and gas wells. Their preparation is given below.

$$[(CH_3)_2CHO]_4Ti + 2\ N(CH_2CH_2OH)_3 \rightarrow [(CH_3)_2CHO]_2Ti\left[\begin{array}{c}O\\N\\(CH_2CH_2OH)_2\end{array}\right]_2 + 2\ (CH_2)_2CHOH$$

Alkanolamine titanates react rapidly with aqueous solutions of polyhydric polymers to form gels. Soluble cellulose derivatives, quar gum, and poly(vinyl alcohol) are thus cross-linked. The reactions of simple alkanolamines with titanium alkoxides are not completely understood. Ethanolamine reacts with the lower titanium alkoxides to give insoluble white solids.

Acylates. Titanium acylates are prepared from TiCl$_4$ or a tetraalkoxide; eg TiCl$_4$ and acetic acid give dichlorotitanium diacetate as an HCl-free white powder when passed preheated (136–170°C) into a heated chamber. Trichlorotitanium monoacylates form by the thermal decomposition of TiCl$_4$-ester complexes. Tetraacylates are prepared from TiBr$_4$ and excess carboxylic acid in an inert solvent. The usual products from reactions of alkoxides and acids are dialkoxytitanium diacylates.

Polymeric acyltitanoxanes have been prepared from (RO)$_4$Ti and R′COOH in various mol ratios; equimolar amounts of TiCl$_4$ and RCOOH react in the presence of an amine, followed by the addition of 1–2 mol H$_2$O to give

Alkoxytitanium acylates react with alcohol resulting in the exchange of alkoxy groups. The acyl group is unaffected.

Titanium(IV) Complexes with Other Ligands

The d^0-titanium(IV) atom is hard, ie, not very polarizable, and would be expected to form its most stable complexes with hard ligands, eg, fluoride, chloride, oxygen, and nitrogen. Soft or relatively polarizable ligands containing second- and third-row elements or multiple bonds should give less stable complexes. The stability depends on the coordination number of the titanium.

Titanates may influence reactions of organic peroxides; thus, t-butyl hydroperoxide plus TPT epoxidize olefins. Titanates trigger peroxide-initiated curing of unsaturated polyester to give noncolored products. Hydrogen peroxide produces an intense yellow with Ti(IV) in aqueous solution; this reaction provides a qualitative test for Ti.

Titanium amides, Ti(NR$_2$)$_4$, are typically prepared from a titanium chloride and LiNR$_2$. Amides have been prepared from both primary and secondary amines by these procedures and by their reaction with TiS$_2$. The properties of the tetrakisdialkylamides resemble those of the alkoxides. They are hydrolyzed rapidly by water, are sensitive to oxygen, and undergo reversible amine exchange with other secondary amines. The unstable compounds formed with β-diketones decompose to enamines. Amides of the formula TiN(R)COX are prepared by the insertion of a titanium alkoxide into an isocyanate:

This reaction occurs also with cyclopentadienyl alkoxides $CpTi(OR')_3$. No reaction occurs with the heterocumulenes CS_2 and $ArNCS$. With carboniimides, a 2 : 1 adduct forms regardless of the molar proportions of the reagents.

$$Ti[OCH(CH_3)_2]_4 + 2\,ArN{=}C{=}NAr \longrightarrow [(CH_3)_2CHO]_2Ti\overset{\overset{\displaystyle Ar}{\displaystyle |}}{[NC[OCH(CH_3)_2]{=}NAr]_2}$$

The nitrogen of a Schiff base unit ($>C{=}N{-}$) in a polydentate ligand coordinates readily. Azo ligands are encountered in metallized azo dyes (qv) when other groups, eg, phenol, thiophenol, or amine functions, provide the primary bonds. They are more common with lower-valent titanium.

Simple sulfur ligands bond to titanium. Mercaptoacetic esters displace the alcohol from TPT. Bidentate ligands with two coordinating sulfurs are represented by xanthates and dithiocarbamates.

$$\underset{NaSCNR_2}{\overset{\overset{\displaystyle S}{\displaystyle \|}}{}} + TiCl_4 \xrightarrow{CH_2Cl_2} Cl_{4-n}Ti{\left(\overset{S}{\underset{S}{\diagup\diagdown}}CNR_2\right)}_n$$

Dithiocatechol reacts with TPT in the presence of tertiary amines yielding the 3 : 1 adduct as a dianion:

$$3\,R{-}\overset{SH}{\underset{SH}{\diagup\diagdown}} + [(CH_3)_2CHO]_4Ti + 2\,R'N(C_2H_5)_2 \longrightarrow$$

$$\left[Ti{\left(\overset{S}{\underset{S}{\diagup\diagdown}}\,R\right)}_3\right]^{2-}\;2\,R'\overset{+}{\underset{H}{N}}(C_2H_5)_2$$

Seleninate bonds to titanium through both oxygens:

$$TiCl_4 + 4\,RSeO_2Na \xrightarrow{CH_3CN} Ti(OSeR)_4\overset{\displaystyle O}{}$$

Titanium borohydrides are unknown because BH_4^- reduces Ti(IV); however, $Ti(BH_4)_3$ has been reported.

Lower Valent Titanates

Titanium tetralkoxides are reduced to isolable $Ti(OR)_3[R = C_2H_5,$ $i\text{-}C_3H_7, n\text{-}C_4H_4]$ by potassium. A family of Ti(III) derivatives roughly parallels those of Ti(IV). Titanium(III) chelates are known, eg, trisacetylacetonate prepared in benzene from titanium trichloride, acetylace-

Table 3. Organotitanium Compounds of Valence < 4

Formula	Type	Appearance	Mp, °C	Other properties
$Ti(CH_3)_3$	Ti(III) trialkyl	not isolated; green in THF solution		solutions give positive Gilman test; decompose above $-20°C$
$(C_5H_5)_2TiCl$	Ti(III)Cp$_2$ halide	green crystals	279–281	
$C_5H_5TiCl_2$	Ti(III) Cp halide	violet	sublimes in vacuo, 150	insoluble in hydrocarbons; very sensitive to oxygen; blue solution in acetonitrile
$Ti(C_6H_5)_2$	Ti(II) diphenyl	black solid		pyrophoric; gives phenylmercury chloride with $HgCl_2$
$C_5H_5TiC_6H_5$	Ti(II) cyclopentadienyl phenyl	black solid		sensitive to air and moisture; thermally stable to 170°C
$(C_5H_5)_2Ti$	Ti(II) dicyclopentadienyl	dark green	200	pyrophoric; catalyst for polymerization of olefins and acetylenes

tone, and ammonia. This deep-blue compound is soluble in benzene but insoluble in water. Titanium(III) β-diketonates are prepared by reduction of the Ti(IV) chelate. Physical properties of some low-valent Ti compounds are given in Table 3.

A titanous oxalate, prepared in water from $TiCl_3$ and oxalic acid, precipitates upon addition of ethanol as a yellow solid. It forms double salts with metal oxalates, $MTi(C_2O_4)_2{\cdot}H_2O$. A group of violet titanium(III) acylates has been prepared from $TiCl_3$ and alkali carboxylates; they are strong reducing agents.

Photoreduction of aqueous Ti(IV)-containing alcohols or glycols, but not ethylene glycol, yields Ti(III) and the aldehyde or ketone corresponding to the alcohol. Titanium(IV) citrate can be reduced polarographically to Ti(III) citrate; this procedure has been advocated as a method for determining citric acid.

Organometallics

In classical organometallic chemistry, Grignard reagents or organolithium compounds react with halides of less active metals forming new C—M bonds. These reactions fail uniformly with titanium halides because many titanium alkyls are unstable thermally and to moisture and air. Thermal stability is enhanced in chelates; thus, the compound is much more stable than $(CH_3)_3Ti[OCH(CH_3)_2]_2$.

$$(CH_3)_2Ti\overset{O}{\underset{O}{\diagup\diagdown}}\overset{CH_3}{\underset{CH_3}{\diagdown\diagup}}CH_3$$

The general synthesis of covalent non-Cp compounds, R_nTiX_{4-n} (where R = alkyl or aryl and X = halogen, alkoxy, or amido) involves the reaction of a lithium, sodium, or magnesium organometallic compound with a titanium–halogen compound in an inert atmosphere in ether or hydrocarbon solvents.

The reactions of numerous alkyltitanium compounds resemble those of alkyllithium compounds and alkylmagnesium halides. They are protolyzed by water and alcohols, they insert oxygen, and they add to carbonyl groups:

$$R{-}Ti{\diagup}\;+\;O{=}C{\underset{CH_3}{\overset{CH_3}{\diagup\diagdown}}}\;\longrightarrow\;{\diagup}TiOC{\underset{\underset{CH_3}{\displaystyle CH_3}}{\overset{\displaystyle R}{\diagup\diagdown}}}$$

Titanium alkyls do not add to esters, nitriles, epoxides, or nitroalkanes at low temperature, but do add exclusively 1,2 to unsaturated aldehydes.

Olefin isomerization is catalyzed by titanium, eg, in the conversion of vinylnorbornene to ethylidenenorbornene.

Cyclopentadienyltitanium compounds. The structure of Cp_2TiCl_2 has been shown by x-ray diffraction to be a distorted tetrahedron. Changes in the structure are imposed by bridging the Cp rings with —$(CH_2)_n$— or other groups. Although $CpTiCl_3$ forms stable sublimable complexes with ditertiary amines or arsines, the less basic $O, S,$ and P analogues do not form stable complexes. Synthesis involves a salt of cyclopentadiene and a titanium halide, usually in an ether solvent. Different Cp groups can be introduced in two steps.

Cyclopentadienyltitanium trichloride and, particularly, Cp_2TiCl_2 react with RLi or $RA{<}$compounds to form one or more R—Ti bonds. Methyl and aryl groups are most commonly used; higher alkyltitanium compounds tend to decompose by β-elimination. Both vinyl and ethynyl groups can be attached to the Cp_2Ti frame, giving compounds that are more stable thermally and to air and moisture. Alkyl and aryl groups are cleaved by iodine, but Cp groups are not affected. The following reaction has been proposed for the quantitative determination of such compounds.

$$Cp_2TiR_2 + 2\,I_2 \rightarrow Cp_2TiI_2 + 2\,RI$$

Photolysis of Cp_2TiAr_2 in benzene yields titanocene and a variety of aryl products derived both intra- and intermolecularly. Dimethyltitanocene, $Cp_2Ti(CH_3)_2$, photolyzed in hydrocarbons yields methane.

Pyrolysis of solid $Cp_2Ti(CD_3)_2$ yields CD_3H but not CD_4. Pyrolysis of $(C_5D_5)_2Ti(CH_3)_2$ yields CH_3D. These results show that the radical attacks the Cp rings.

Cyclopentadienyltitanium halides undergo displacement with nucleophiles. Amides are formed with amines, whereas hydroxylic reagents cleave Ti—R bonds:

$$2\ Cp_2TiCH_3Cl + H_2O \rightarrow 2\ CH_4 + (Cp_2TiCl)_2O$$

In Cp–titanium acylates, the carboxylate ligands are unidentate, not bidentate; they are generally prepared from the halide and a silver acylate. Organometallic ligands can be attached to titanium by displacement reactions; both chlorines are also displaceable:

$$Cp_2TiCl_2 + LiOC[Co_3(CO)_9] \rightarrow Cp_2ClTi\text{-}OCCo_3(CO)_9$$

$$Cp_2TiCl_2 + LiOC(R)Cr(CO)_5 \rightarrow Cp_2ClTi\text{-}OC(R)Cr(CO)_5$$

Insertion into the CpTi–R bond with sulfur dioxide yields solfones and ultimately sulfinates. Isocyanides insert to yield imines:

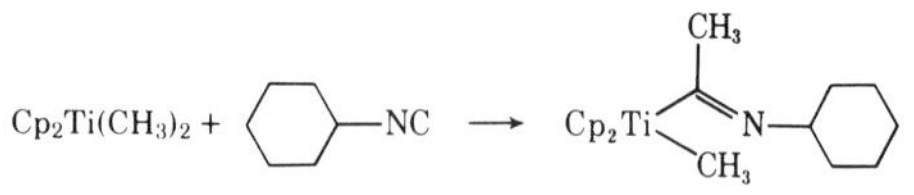

Organic isocyanates and isothiocyanates and nitric oxide insert similarly. Carbon monoxide inserts to yield highly stable acyltitanium compounds:

Transition metal-catalyzed polymerizations of β-propiolactone and $CH_2\!=\!CHOCH_2CH_2Cl$ are markedly accelerated by Cp_2TiCl_2.

Health and Safety Factors

The tetraalkoxides have a low acute oral toxicity (LD_{50} 7,500–11,000 mg/kg in rats). Because of their rapid hydrolysis, they can cause severe eye damage. Chelates possess the added toxicity of the chelating agent.

Uses

Titanates with a carbon–titanium bond are extensively used in Ziegler-Natta polymerization of olefins and other catalytic applications (see Catalysis).

Glass coating. A thin, transparent film of TiO_2 imparts scratch resistance to glass and reduces fragility. The lower alkoxides, particularly TPT, are preferred for this application. They are applied undiluted or in a solvent. The films are also applied for decorative purposes. A film ca 150 nm thick (one-quarter wavelength of visible light) is reflective toward infrared radiation. Window glass with such coatings is used in hot climates to reduce the effect of the sun (see also Ceramics; Glass). The bonding properties of $(TiO_2)_x$ have been used for size reinforcing of glass fibers.

Paints. Pyrolysis of organic titanates, obtained by controlled hydrolysis of $(C_4H_9O)_4Ti$ furnishes adherent films of nearly inorganic $(TiO_2)_x$. This application may be used in heat-resistant paints. Corrosion-resistant titanate coatings are used for tinplate, steel, and aluminum. Since titanates promote hardening of epoxy resins, they are used in epoxy-based paints. Water-based emulsion paints contain water-soluble, hydrolysis-resistant titanates. Titanium chelates in thixotropic emulsion paints improve rheology and give a water-repellent durable paint film.

Adhesives. Titanates bond on metals, nonmetal oxides, and polyethylene and fluorinated polymers. Packaging film, such as Mylar polyester or aluminized films, are coated with a titanate to give a film that is not tacky and can be rolled and stored.

Ester interchange. Ester interchange is strongly promoted by titanates. Transesterification is the classical preparative method for polyesters (qv). Titanate catalysts are much more active than traditional catalysts such as Pb_3O_4; $(TiO_2)_x$ is also active, especially at higher

temperatures. Thus, condensation of dimethyl terephthalate with p-$C_6H_4(COOCH_2CH_2OH)_2$ begins with titanium lactate and is completed with TiO_2. Oligomers from titanium lactate with ethylene glycol or $[(CH_3)_2CHO]_2Ti(acac)_2$ with polyester prepolymers yield high molecular weight polyesters. Polyurethanes react with titanium acylates to give artificial leather (see Leatherlike materials). An unsaturated polyester filled with crushed marble is cross-linked with triethanolamine titanate and cumene hydroperoxide; the product is a moldable artificial marble.

Epoxy cross-linking is catalyzed by TPT and TBT, alone or with piperidine, and by triethanolamine. Organic peroxides coordinate with titanates, then dissociate to radicals capable of initiating vinyl polymerization. Polyols, such as a natural polysaccharides and poly(vinyl alcohol), are cross-linked by titanates.

Cellulose (qv) or starch xanthate cross-linked by titanates adsorbs uranium from seawater. Carboxymethylcellulose cross-linked with isopropyl tristearoyl titanate is a bonding agent for clay, talc, wax, and pigments to make colored pencil leads.

Guar gels. Water under a pressure of 69–172 MPa (10,000–25,000 psi) is forced down well bores to fracture the rock and to open channels through which the oil can flow more rapidly. The water is thickened with a soluble polymer which is cross-linked with titanates to form a gel which will suspend sand. Titanium and zirconium are effective and environmentally innocuous cross-linking agents.

Addition of titanates or titanic acid improves the wet strength and ink acceptance of paper. Poly(vinyl alcohol), used as a paper size alone or with dyes and pigments, is rendered insoluble by cross-linking with titanates.

Complex titanates suspend particles of metals, metal and nonmetal oxides, clays, talc, and carbon black. When titanated particles are blended with polymers, the mixtures are less viscous and better adapted to extrusion, molding, or coating on substrates.

The numerous uses of titanium dioxide are discussed under Titanium compounds, inorganic.

CHRISTIAN S. RONDESTVEDT, JR.
E.I. du Pont de Nemours & Co., Inc.

D.C. Bradley, R.C. Mehrotra, and D.P. Gaur, *Metal Alkoxides*, Academic Press, Inc., New York, 1978.

P.C. Wailes, R.S.P. Coutts, and H. Weigold, *Organometallic Chemistry of Titanium, Zirconium, and Hafnium*, Academic Press, Inc., New York, 1974.

R.J.H. Clark in J.C. Bailar, H.J. Emeleus, R.S. Nyholm, and A.F. Trotman-Dickenson, eds., *Comprehensive Inorganic Chemistry*, Vol. 3, Pergamon Press, London, UK, 1973, Chapt. 32, pp. 355–417.

Gmelin, *Handbuch der Anorganischen Chemie*, 8th ed., Syst. No. 41, Springer Verlag, Berlin, FRG, 1977.

TOBIAS ACID. See Naphthalene derivatives.

TOCOPHEROLS. See Vitamins, Vitamin E.

TOILET PREPARATIONS. See Cosmetics.

TOLIDINES. See Benzidine and related biphenyldiamines.

TOLU BALSAM. See Perfumes.

TOLUENE

Toluene, methylbenzene, C_7H_8, is a colorless, mobile liquid with a distinctive aromatic odor somewhat milder than that of benzene. Its main source before World War I was coke ovens (see Coal conversion processes). Petroleum became the source of toluene with the advent of catalytic reforming and the need for large quantities used in aviation fuel during World War II. Toluene is generally produced along with benzene (qv), xylenes (qv), and C_9 aromatics by the catalytic reforming of C_6–C_9 naphthas. About 90–95% of the ca 31×10^6 metric tons (9.4×10^9 gal) produced annually in the United States is not isolated but blended directly into the gasoline pool as a component of reformate and pyrolysis gasoline (see Gasoline).

Properties

Physical and thermodynamic properties are given in Table 1. Toluene forms azeotropes with many hydrocarbons and most alcohols that boil in a similar range; all are minimum-boiling azeotropes. Toluene, water, and alcohols frequently form ternary azeotropes.

Because of the high electron density in the aromatic ring, toluene behaves as a base both in the formation of charge-transfer π complexes and in the formation of sigma complexes. When only π electrons are involved, toluene behaves much like benzene and xylene. When σ bonds and complexes are involved, toluene reacts much faster than benzene and much slower than xylenes.

Derivatives are formed by substitution of the hydrogen atoms of the methyl group, by substitution of the hydrogen atoms of the ring, and by addition to the double bonds. Substitutions on the methyl group are generally high-temperature, free-radical reactions. Thus, chlorination at ca 100°C, or in the presence of uv or other free-radical initiators, successively gives benzyl chloride, benzal chloride, and benzotrichloride (see Benzaldehyde; Benzoic acid). With oxygen in the liquid phase, particularly in the presence of a catalyst, good yields of benzoic acid are obtained. In the presence of alkali metals, toluene is alkylated. With a lithium catalyst and a chelating compound, telomers are obtained with ethylene.

$$\underset{\text{CH}_3}{\bigcirc} \xrightarrow[\text{110 °C}]{\text{C}_2\text{H}_4,\ \text{Li},\ \text{TMEDA}} \underset{\text{CH}_2 \text{--(C}_2\text{H}_4\text{)}_n\text{--CH}_2\text{CH}_3}{\bigcirc}$$

Additions to the double bonds results from both free-radical and catalytic reactions, eg, chlorination at 0°C and hydrogenation. Usually, all three double bonds react. Substitution of the ring hydrogen atoms by electrophilic attack takes place with the same reagents that react with benzene. Some of the common groups with which toluene can be substituted directly include

$$\text{—Cl, —Br, —CCH}_3, \text{ —SO}_3\text{H, —NO}_2, \text{—(C}_n\text{H}_{2n+1}\text{), and —CH}_2\text{Cl.}$$

Under the same conditions, toluene reacts more rapidly than benzene.

Table 1. Physical and Thermodynamic Properties of Toluene[a]

Property	Value
freezing pt, °C	−94.965
boiling pt, °C	110.629
density at 25°C, g/cm³	0.8623
heat of combustion at 25°C constant pressure, kJ/mol[a]	3910.3
heat of vaporization, kJ/mol[a]	
at 25°C	37.99
at bp	33.18
heat capacity, J/(g·K)[a]	
ideal gas	1.125
liquid at 101.3 kPa[b]	1.970
free energy of formation, ΔF_f°, kJ/K[a]	
gas	93.00
liquid	114.1

[a] To convert J to cal, divide by 4.184.
[b] 101.3 kPa = 1 atm.

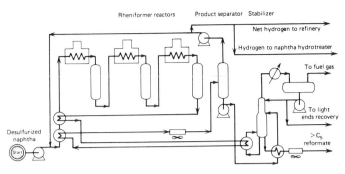

Figure 1. Chevron Research Co. Rheniforming process. Courtesy of Guld Publishing Co.

These reactivities and the related selectivity to the ortho and para positions can be explained in terms of the inductive effect of the methyl group. Toluene requires substitution by strongly negative groups, such as NO_2, to react with anions.

Manufacture

The principal source of toluene is catalytic reforming of refinery streams. This accounts for ca 87% of the total production. An additional 9% is separated from pyrolysis gasoline produced in steam crackers during the manufacture of ethylene and propylene. The reactions taking place in catalytic reforming, in order of decreasing rate, are dehydrogenation or aromatization of cyclohexanes, dehydroisomerization of substituted cyclopentanes, and the cyclodehydrogenation of paraffins. Hence, feeds rich in cycloparaffins are most easily reformed.

Because catalytic reforming is endothermic, most units comprise three reactors with furnaces in between to minimize kinetic and thermodynamic limitations caused by decreasing temperature. Three types of operation are used: semiregenerative, operating at low severity and requiring catalyst regeneration only every six to 24 mo; cyclic, operating at high severity and using a swing reactor regenerated as frequently as every 24 h; and continuous, in which catalyst is removed, regenerated externally, and returned to the reactor. Semiregenerative operation (see Fig. 1) is most common.

The predominant feeds are straight-run naphthas from crude stills. Proper choice of feedstocks and severe operating conditions produce streams high enough in toluene to be directly usable for hydrodemethylation to benzene.

Specifications and Analysis

Toluene is marketed mostly as nitration and industrial grades. The standards are given by ASTM D 841 and D 362. Purity of toluene as well as the impurities and by-products are determined by standard gas-chromatographic and spectrographic methods.

Health and Safety Factors

Toluene is classified as a flammable liquid by the ICC. Properties related to safety are: explosive limits in air 1.27–7.0 vol% and flash point (closed cup) 4.4°C. Exposure limits as set by OSHA are 752 mg/m³ (average for 8 h, TWA) and 376 mg/m³ as set by NIOSH.

Uses

Uses are summarized in Table 2. About 90% of crude toluene generated by catalytic reforming is blended directly into gasoline. Toluene itself has a blending octane number of 103–106, which is exceeded only by oxygenated compounds (see Table 2). About 17% of the total toluene production is isolated in extraction units; about half of the isolated material is used in chemicals production, and half is returned to the gasoline pool for blending to improve the octane number of premium fuels. During recent years, the use in fuels has increased, whereas the use for chemicals has slightly declined. The largest use of toluene for chemicals is the demethylation to benzene, which supplies 25–39% of the total benzene demand.

Table 2. Blending Octane Number, (R + M)/2[a] for Selected Components in Unleaded Gasoline

Component	(R + M)/2
methanol	120, 117
ethanol	119, 113, 117
methyl *tert*-butyl ether	108, 106, 111
tert-butyl alcohol	97.5, 94.5, 96
toluene	106, 103.5, 102.9
C_8 aromatics	105.5
unleaded regular	88.0
unleaded premium	93.0

[a] R = Research-method octane rating, ASTM D 2699; M = Motor-method octane rating, ASTM D 2700.

The second largest use of toluene is as solvent in surface coatings and other formulations. This use accounts for ca 25% of the total U.S. toluene demand for chemicals.

Toluene diisocyanate, the basic raw material for the production of flexible polyurethane foams, is made from toluene by nitration, followed by reduction and treatment with phosgene (see Foamed plastics; Urethane polymers).

Benzoic acid (qv) is manufactured from toluene by air oxidation in the liquid phase over a cobalt catalyst. Yields are generally > 90% and product purity is > 99%. Benzaldehyde is a by-product. Benzyl chloride is manufactured by free-radical chlorination of toluene at high temperatures.

Other uses of toluene are for disproportionation to benzene and xylene, used when excess toluene is available at a plant; manufacture of vinyl toluene for use as a resin modifier; and sulfonation to *p*-toluenesulfonic acid used mainly for conversion to *p*-cresol but also used as a condensation and esterification catalyst.

Processes to form styrene directly from toluene, for example by reaction with methanol, are reported in the literature. Although these processes avoid the intermediate conversion to benzene, none are competitive with production from benzene and ethylene. A more promising approach being commercialized by Mobil Chemical is to manufacture *p*-methylstyrene by selective alkylation of toluene with ethylene followed by dehydrogenation.

M.C. Hoff
Amoco Chemicals Corporation

"Toluene" in *Chemical Economics Handbook*, SRI International, Menlo Park, Calif., July 1979; update, Jan. 1982.

H.A. Wittcoff and B.G. Reuben, *Industrial Organic Chemicals in Perspective*, John Wiley & Sons, Inc., New York, 1980.

L.F. Hatch and S. Mater, *Hydrocarbon Process*, 189 (Jan. 1979).

TOLUENEDIAMINES. See Amines, aromatic, diaminotoluenes.

TOOL MATERIALS

Machining of materials with a cutting tool is a common operation in which the unwanted material is removed in the form of chips. A successful cutting tool must resist severe conditions of high temperature, high pressure, and chemically reactive surfaces, and provide a sufficiently long tool life (see also Ceramics; Carbon, diamond; Boron compounds, refractory boron compounds; High temperature alloys).

A tool material must meet stringent requirements. Both deformation energy and frictional energy are converted into heat, and consequently, tool temperatures are very high (ca 1000°C).

A wide range of materials is available with a variety of properties, performance capabilities, and costs. High speed steels (HSS) and ce-

mented carbides (coated and uncoated) are currently the most extensively used materials. Ceramics, diamond, and cubic boron nitride (CBN) are used for special applications. Guidelines for selection for different cutting operations are given in Table 1.

Carbon Steels and Low–Medium Alloy Steels

Low–medium alloy tool steels have been largely superseded by other materials, except for some low speed applications. Low–medium steels contain molybdenum (ca 0.25%) and chromium (0.5–0.75%) for hardenability, and tungsten (0.5–1.75%) and molybdenum for wear resistance. These materials, however, lose their hardness rapidly with temperature. Hence, low-medium alloy steels are used in relatively inexpensive tools for low speed cutting applications.

High Speed Steels

High speed steels (HSS) contain significant amounts of tungsten, molybdenum, cobalt, vanadium, and chromium, in addition to iron and carbon. These elements strengthen the matrix beyond the tempering temperature and increase hot hardness and wear resistance.

Tool steels are broadly classified as T-type or M-type, depending whether tungsten or molybdenum is the principal alloying element. Cutting performance of these types is similar. In general, M-type steels are more popular (ca 80% of all tool steels) and are ca 30% cheaper. High speed tools are available in cast, wrought, and sintered forms.

A new processing technique involves atomization of the prealloyed molten alloy into fine powder, followed by consolidation under hot isostatic pressure (HIP). Tool steels made in this manner grind more easily, exhibit more uniform properties, and perform more consistently (see also Powder-metallurgy). By an appropriate increase of Mo–W or V content, hot hardness can be maintained without cobalt, which has been short in supply recently and increasing in cost (see Table 2).

Cast-Cobalt Alloys

Cast-cobalt alloys, known as Stellite tools, were introduced for cutting applications about the same time as HSS. Though comparable in room-temperature hardness, cast-cobalt alloy tools retain their hardness to a much higher temperature and can be used at higher (25%) cutting speeds than HSS tools. They are generally cast to shape and finished to size by grinding. They are available only in simple shapes.

Cemented Carbides

Cemented carbides contain a large volume fraction (> 90%) of fine-grain, refractory carbides in a metal binder produced by cold pressing, followed by liquid-phase sintering. Carbides are 2–3 times stiffer than HSS.

There are at least four different classifications of cemented carbides. The U.S. system is based on relative performance, the U.K. system on properties (wear resistance, toughness or shock resistance, and crater resistance), and the USSR system on composition; the fourth system, used widely in Europe and supported by the ISO, is based on application and chip form.

The C-classification (C-1 to C-8) for cemented carbides, used unofficially in the United States for machining applications, was originally developed by the automobile industry to obtain a relative performance index of tools made by different manufacturers. It is by far the simplest system. Grades C-1 to C-4 are recommended for machining cast iron, nonferrous alloys, and nonmetallic material, whereas C-5 to C-8 are recommended for carbon steels and alloy steels. An attempt was made in 1953 by the ISO to arrive at a universally accepted classification confined to applications.

In selecting a carbide grade for a given application, the following guidelines should be followed: the grade with the lowest cobalt content and the finest grain size consistent with adequate strength to eliminate chipping should be chosen; straight WC grades can be employed if cratering, seizure, or galling is not experienced and for work materials other than steels; to reduce cratering and abrasive wear when machining steels, TiC is preferred; and for heavy cuts in steel where high temperature and high pressure deform the cutting edge plastically, a multi-

Table 1. Guidelines for Tool Materials

Tool materials	Work materials	Machining operation and cutting-speed range	Limitations
carbon steels	low strength, softer materials, nonferrous alloys, plastics	tapping, drilling, reaming; low speed	low hot hardness, limited hardenability and wear resistance, low cutting speed, low-strength materials
low–medium alloy steels	low strength–soft materials, nonferrous alloys, plastics	tapping, drilling, reaming; low speed	low hot hardness, limited hardenability and wear resistance, low cutting speed, low-strength materials
HSS	all materials of low–medium strength and hardness	turning, drilling, milling, broaching; medium speed	low hot hardness, limited hardenability and wear resistance, low to medium cutting speed, low- to medium-strength materials
cemented carbide	all materials up to medium strength and hardness	turning, drilling, milling, broaching; medium speed	not for low speed because of cold welding of chips and microchipping, not suitable for low speed application
coated carbides	cast iron, alloy steels, stainless steels, superalloys	turning; medium to high speed	not for low speed because of cold welding of chips and microchipping, not for titanium alloys, not for nonferrous alloys since the coated grades do not offer additional benefits over uncoated
ceramics	cast iron, Ni-base superalloys, nonferrous alloys, plastics	turning; high speed to very high speed	low strength and thermomechanical fatigue strength, not for low speed operations or interrupted cutting, not for machining Al, Ti alloys
CBN	hardened alloy steels, HSS, Ni-base superalloys, hardened chill-cast iron, commercially pure nickel	turning, milling; medium to high speed	low strength and chemical stability at higher temperature, but high strength, hard materials otherwise
diamond	pure copper, pure aluminum, aluminum–Si alloys, cold-pressed cemented carbides, rock, cement, plastics, glass–epoxy composites, nonferrous alloys, hardened high carbon alloy steels (for burnishing only), fibrous composites	turning, milling; high to very high speed	low strength and chemical stability at higher temperature, not for machining low carbon steels, Co, Ni, Ti, Zr

[a] Depth of cut line.

Table 2. Nominal Wt% Compositions of Representative HSS Tool Grades

HSS Grade	C	W	Mo	Cr	V	Co	W_{eq}[a]
General Purpose Grades (No Cobalt)							
T1	0.7	1.8		4	1		18
M1	0.8	1.5	8	4	1		17.5
M2	0.85	6.0	5	4	2		16
M10	0.9		8	4	2		16
Higher V Content Grades (No Cobalt)							
M3	1.2	6	5	4	3		16
M4	1.4	5.5	4.5	4	4		14.5
M7	1.0	1.75	8.75	4	2		19.25
Co HSS Grades							
TS	0.8	18		4	2	8	18
M34	0.9	2	8	4	2	8	18
M35	0.8	6	5	4	2	5	16
M36	0.8	6	5	4	2	8	16
Higher V and Higher Co Grades							
T15	1.5	12		4	5	5	12
M43	1.25	1.75	8.75	3.75	2	8.25	19.25

[a] $W_{eq} = 2(wt\%\ Mo) + wt\%\ W$.

carbide grade with low binder content and containing W–Ti–Ta should be used.

Composition, microstructure, and performance depend on cobalt content, grain size, and type of carbides. A high speed carbide material, based on TiC cemented with a nickel–molybdenum binder, has been developed for machining steels (see Table 3).

Cemented carbide tools are available in insert form in squares, triangles, diamonds, and rounds. They can be brazed or clamped to the tool shank, and are recommended for higher speed cutting operations (45–180 m/min). To conserve strategic materials and reduce costs, the W, Co, and Ta in so-called throwaway inserts are separated and recycled.

Coated tools. Although rapid advances in coated cemented carbide technology took place during the last decade, coating technology for HSS is lagging. Only a coating that requires heating below the HSS transformation temperature can be applied to HSS.

An analysis of the cutting process indicates that the material requirements at or near the surface of the tool are different from those of the body. A thin, chemically stable, hard, refractory binderless coating often satisfies these requirements. Refractory coatings include TiC, Al_2O_3, HfN, or HfC, and multiple coatings of Al_2O_3 or TiN over TiC. These are generally deposited by chemical vapor deposition (CVD). Multiple coatings prolong tool life and provide protection. Coated tools give a significantly better performance than uncoated tools, and 30–50% of the carbide tools in use are coated.

Table 3.   Composition and Properties of Some Representative Grades of Cemented-Carbide Tools

Grade	Composition, wt% WC	TiC	TaC	Co	Grain size	Density, g/cm³	HRA[a]	TRS[b] MPa[c]	Elastic modulus E, GPa[d]	Compressive strength, MPa[c]	Tensile strength, MPa[c]	Relative abrasion resistance, vol loss/cm³
Nonsteel grades[e]												
roughing	94			6	coarse	15.0	91	2210	640	5170	1520	15
general purpose	94			6	medium	15.0	92	2000	650	5450	1950	35
finishing	97			3	fine		92.8	1790	610	5930	1790	60
Steel grades[f]												
roughing	72	8	11.5	8.5	coarse	12.6	91.1	1720	560	5170		8
general purpose	71	12.5	12	4.5	medium	12.0	92.4	690	570	5790		7
finishing	64	25.5	4.5	6	medium	9.9	93.0	130	460	4900	480	5

[a] Rockwell hardness A scale.
[b] Transverse rupture strength.
[c] To convert MPa to psi, multiply by 145.
[d] To convert GPa to psi, multiply by 145,000.
[e] C-1 to C-4.
[f] C-5 to C-8.

Table 4.   Composition and Properties of Steel Grades of Cemented Titanium Carbide

Grade[a]	Composition, wt% TiC	Ni	Mo	HRA[b]	Transverse rupture strength, MPa[c]	Young's modulus, GPa[d]	Density g/cm³
roughing	67–69	22	9–11	91	1900	413	5.8
general purpose	72–74	17	9–11	92	1620	431	5.6
finishing	77–79	12	9–11	92.8	1380	440	5.5

[a] C-5 to C-8.
[b] Rockwell hardness A scale.
[c] To convert MPa to psi, multiply by 145.
[d] To convert GPa to psi, multiply by 145,000.

Ceramics

The newest class of tool materials are ceramics, which are used for a wide range of high speed, high removal-rate finishing operations. Ceramics are predominantly alumina based, although silicon nitride-based materials have attractive features for certain applications. A comparison of the physical properties of ceramic tools and carbide tools is given in Table 4. Toughness of ceramics with smaller grain size can be improved by the introduction of a more ductile second phase.

Recently, pure alumina and alumina–TiC dispersion-strengthened ceramics have attracted attention. The latter contain ca 30% TiC and small amounts of yttria as a sintering agent, resulting in a density 99.5% of theoretical. An Al_2O_3–ZrO_2–W alloy (Cer Max 460) ceramic has been introduced by General Electric. A similar material performs exceptionally well as an abrasive in heavy-stock grinding operations, such as cutoff and snagging. The three popular compositions contain 10, 25, and 40% ZrO_2; the remainder is alumina.

Another interesting material is based on silicon nitride with various additions of aluminum oxide, yttrium oxide, and titanium carbide. These ceramics were originally developed for automotive gas turbines and other high temperature applications. They are produced by sintering a mixture of Si_3N_4, AlN, Al_2O_3, and Y_2O_3. Because of its high toughness and excellent thermal shock resistance, square-, triangular-, and diamond-shaped tools of this material may be used for machining superalloys in the intermediate-speed range (150 m/min) where only round tools are used currently.

Certain ceramic tools, especially those based on alumina, are not suitable for machining aluminum, titanium, and similar materials because of a strong tendency to react chemically.

Diamond

Diamond is the hardest of all known materials (Knoop hardness ca 78.5 GPa or 8000 kgf/mm²). Both the natural and synthetic forms are used in cutting-tool applications. Natural diamond is used for special applications where it outperforms all other materials and where a long life justifies the cost. Examples include tools for finishing of copper-front surface mirrors and microtome knives.

A less expensive alternative is synthetic diamond produced under a pressure of ca 5 GPa (50,000 atm) at ca 1500°C in the presence of a suitable catalyst-solvent, which led to the development of polycrystalline sintered diamond tools (see Carbon, diamond synthetic). The latter are metallurgically bonded to a cemented-carbide base, which provides an elastic support. Sintered diamond tools are fabricated in an assortment of shapes and sizes from round blanks. They are much more expensive than conventional cemented carbide tools because of the high cost of processing. They are, however, economical for certain applications because of increased productivity. Sintered diamond tools are used for applications similar to those of the lower quality industrial diamonds.

Cubic Boron Nitride (CBN)

Next only to diamond in hardness, CBN is a remarkable material that does not exist in nature (Knoop hardness 46.1 GPa or 4700 kgf/mm²). Similar to diamond, it is produced by a high temperature, high pressure process. Sintered CBN tools are fabricated in the same manner as sintered diamond tools and are available in the same sizes and shapes. The two predominant wear modes are DCL (depth of cut line); notching and microchipping. These tools are used for heavy interrupted cutting and for milling white cast iron and hardened steels. They are not, however, recommended for very low or very high speed cutting applications.

Health and Safety Factors

The TLVs for the components of cemented carbides and tool steels range from 0.1 mg/m³ for cobalt to 5 mg/m³ for tungsten; no fire or explosion hazard is generally involved. Eye protection is recommended during machining, and respirators when metal fumes or dust are produced. Precautions must be taken against flying fragments, sharp edges, and rotating parts.

R. KOMANDURI
J.D. DESAI
General Electric Company

E.M. Trent, *Metal Cutting*, Butterworth and Co., Ltd., London, UK, 1977.

G.A. Roberts, J.C. Hamaker, Jr., and A.R. Johnson, *Tool Steels*, 3rd ed., American Society for Metals, Metals Park, Ohio, 1962.

P. Schwarzkopf and R. Keiffer, *Cemented Carbides*, The MacMillan Company, New York, 1960.

A.G. King and W.M. Wheildon, *Ceramics in Machining Processes*, Academic Press, Inc., New York, 1966.

TOOTHPASTE. See Dentifrices.

TORPEX. See Explosives and propellants.

TOXAPHENE. See Insect control technology.

TRACE AND RESIDUE ANALYSIS

Trace analysis, the determination of substances comprising $< 0.01\%$ of a sample, has increased in importance in recent years, partly as a consequence of the development of highly sensitive methods, and partly as the result of discoveries concerning the importance of trace substances in living organisms and materials of commerce. Environmental toxicology, in particular, is an area where the determination of residues and traces of pesticides and other toxic substances is frequently employed.

Instrumental methods much used in trace analysis include atomic and molecular spectrometry, microscopy, x-ray examination, activation analysis, radioimmunoassay, potentiometry, polarography, voltammetry, and gas and liquid chromatography (often coupled with mass spectometry) (see also Analytical methods).

Sampling and Sample Preservation

The method of collecting a sample should introduce a minimum of contamination and should not cause the loss of any substance of interest. A random sample may be sufficient for a particular trace analysis, but often a more systematic sampling protocol is necessary. To characterize a bulk quantity of material (eg, a field, a lake, a car load, etc), a number of samples must be taken with a specified spatial distribution, depending on available information and assumptions about the homogeneity of the material. If the material is a flowing stream of air, liquid, or even solid particles, automatic samplers can be used to collect composite or sequential samples. Common operations include separation of solids, dissolved gases, and volatiles from solution and separation of solids into size fractions.

Analytical samples must be representative. Solids may be ground, liquids and slurries stirred or shaken to provide a uniform distribution of particulate matter. Relatively large portions are required, ie, 0.1 g to 1 kg of solids, 1 mL to 2 L of liquids, and 1 cm^3 to 2000 m^3 of gases.

Pretreatment and Preconcentration

Contamination from the atmosphere and the equipment must be kept to a minimum and blanks carried through the entire procedure to correct for unavoidable impurities. The water used must be highly purified and the purity of the reagents must be known. Loss of sample constituents or products should be minimized. Quality assurance involves the use of reference samples or standards, control charts, and round-robin analyses.

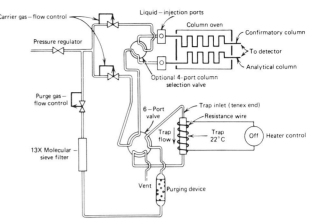

Figure 1. Purge and trap system (purge–sorb mode) for trapping volatile organics prior to desorption and gc detection. Note: all lines between trap and gc should be heated to 80°C.

Unless a direct method of determination is available, the procedure starts with the preparation of a sample solution. This may involve extraction with water or a nonaqueous solvent or chemical attack. Sorbed or dissolved substances are selectively removed by thermal volatilization and inorganic compounds by selective leaching. Reagents for selective leaching include water, electrolyte solutions, complexing agents, weak acids, strong mineral acids, and oxidizing or reducing agents. Complete dissolution may require attack by concentrated acids, fusion, and a combustion of organic matter.

Separation and preconcentration methods (batch or continuous), based on the phase changes involved, include volatilization (see Fig. 1), extraction (see Table 1), and sorption-precipitation. Volatilization efficiency depends on temperature control. Extraction of metals is often accomplished by the addition of complexing agents. Emulsions are broken by adding salt, using mixed solvents or wetting agents, or filtering through glass wool. Sorption efficiency increases with decreasing size of solid sorbent particles.

Table 1. Extraction Solvents and Regents

Compound[a]	Dielectric constant	Liquid range mp to bp, °C	Density, g/mL (20°C)	P'[b]
water	78.0	0 − 100	0.998	10.2
acetonitrille[c]	36.0	− 45–82	0.777	5.8
methanol[c]	32.6	− 98–65	0.792	5.1
acetone[c]	20.7	− 95–56	0.791	5.1
methyl isobutyl ketone (MIBK)	13.1	− 85–118	0.801	ca 4.5
ethanol[c]	24.3	− 117–78	0.789	4.3
chloroform[d]	4.8	− 64–61	1.489	4.1
methylene chloride	8.9	− 97–40	1.336	3.1
diethyl ether	4.3	− 116–35	0.708	2.8
benzene[d]	2.3	5.5–80	0.879	2.7
toluene	2.38	− 95–111	0.868	2.4
carbon tetrachloride[d]	2.24	− 23–77	1.594	1.6
n-hexane	1.91	− 94–69	0.659	0.1
isooctane	1.94	− 107–99	0.692	0.1

[a] Liquid CO_2 (triple pt − 56°C, critical pt 31°C at 7.4 MPa (73 atm), density 0.914 g/cm^3 at 0°C) is an inexpensive, nontoxic, nonflammable solvent used under moderate pressure as an extractant for organic compounds; also, if used as a mobile phase in lc, fractions are self-freezing as they are collected, then the CO_2 sublimes leaving pure solutes (perhaps useful for lc–ms interface).

[b] P' = solvent polarity, from liquid chromatography theory; decreasing polarity represents increasing power to extract hydrophobic organics.

[c] Miscible with water; used to extract solids or modify aqueous phase.

[d] Carcinogen; use with caution.

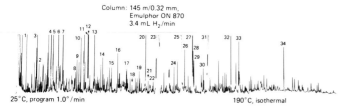

Column: 145 m/0.32 mm,
Emulphor ON 870
3.4 mL H₂/min

25°C, program 1.0°/min   190°C, isothermal

Figure 2. Capillary gc of middle fraction of cigarette smoke. Courtesy of K. Grob and *Chromatographia*, Pergamon Press, Elmsford, N.Y.

Methods of Determination

Recent developments that have expanded the scope of trace analysis, and in many cases lowered limits to the subnanogram range, include automated analyzers; semiconductor detectors for gamma- and x-ray spectrometry; improved sample-excitation methods for atomic spectometry; high intensity, stable excitation sources for spectrofluorimetry; new ion-selective electrodes (qv); pulse polarographic circuitry high performance liquid chromatography (hplc, normal and reverse phase); temperature programming and open tubular capillary columns for gas chromatography (gc) (see Fig. 2); and sensitive detectors for both gc and hplc. Computerized gc–ms methods have revolutionized the identification of trace substances.

Standard Reference Materials

About 450 standard reference materials (SRMs), certified for the concentrations of trace substances are available from the NBS, such as metals (ferrous, nonferrous, and high purity), solid inorganic materials, environmental and biological standards, river sediments, gases (in metals, nitrogen, and air), clinical laboratory standards, hydrocarbon blends, nuclear materials, isotopic reference standards, and spectrophotometer calibration standards. In addition, the EPA offers small samples of standard pesticides and organic pollutants. Other governmental agencies, such as the FDA, and many professional associations prepare and distribute standard materials.

Applications

Methods dealing with the composition of agricultural materials, foods, drugs, and household products are published by the AOAC. The *Official Methods* are used by the FDA to monitor compliance with government regulations. Tolerances from zero to 110 ppm have been established by the EPA for over 300 pesticides. The CPSC issues regulations concerning trace hazardous substances in consumer products.

Other countries have agencies comparable to the FDA, and the UN agencies, FAO and WHO, issue standards and methods for foods and drugs as well as pollutants.

Research in soil science, agronomy, and animal husbandry often requires analyses for trace chemicals and biochemicals in soils, nutrient media, feeds, tissues, seeds, and crops.

Trace analysis is a routine procedure in hospital laboratories and clinics, eg, radioimmunoassay and automated calorimetric determinations on blood serum. The American Association of Clinical Chemistry publishes standard methods.

Environmental studies. In the United States, the EPA has broad responsibilities for controlling and monitoring contaminants in the air, water, and soil. Methods of ambient air sampling and analysis have been developed by an intersociety committee representing thirteen organizations. Some of these methods are applicable to workplace and pollution-source monitoring.

The determination of trace substances in air generally starts with an air-sampling or preconcentration step. About half of the standard methods (as of 1982) were colorimetric and 25% chromatographic. For the determination of solid particulates, the so-called hi-vol sampler is often used. A cascade-impactor sampler determines particle-size distribution. Dichotomous samplers were used to characterize particulates in atmospheric haze. It is important to know the composition of particles below 2-μm dia because they penetrate deeply into the lungs.

Direct methods are used for determining some air pollutants, especially when continuous monitoring is required. Examples of such methods include particle concentration by light scattering in an integrating nephelometer; gc determination of C_1–C_5 hydrocarbons; flame-ionization detection of total hydrocarbons; long-pathlength in determination of CO; and flame-photometric detection of SO_2.

Standard methods for analysis of water and wastewater have been adopted by the EPA. These methods apply to drinking water as well as to surface waters; industrial wastewaters may require special procedures. Most of these methods are trace methods to determine a few ppm or less. Preconcentration is usually necessary. Biological as well as chemical determinations are generally required, especially in the reuse of treated wastewater (see also Water, water pollution).

Trace analysis is frequently vital in assessing the environmental impact of disposal methods, as in mapping the plume of groundwater contamination from a hazardous-waste dumpsite. Sewage sludge has to meet certain EPA and state guidelines to be approved for use on farmland. Trace analysis can be crucial in toxicity testing, eg, in classifying chemicals under TSCA. A number of bioassay tests have been devised to screen compounds for mutagenic and teratogenic potential.

Trace analysis is used in the metal and alloy industries and many standard reference samples are available. Advances in high purity electronic materials have been made with the benefit of ultratrace analysis, which is used, for example, to monitor impurity concentrations in doped semiconductor materials. In the field of organic chemicals and solvents, trace analysis after preconcentration is common, usually by gc–ms determination.

In the United States, OSHA sets the permissible levels of toxic and hazardous substances encountered in the workplace. Recommendations are made by NIOSH. Increasingly, battery-powered personnel monitors are being used in setting these standards.

BRUCE McDUFFIE
State University of New York at Binghamton

G.H. Morrison, ed., *Trace Analysis: Physical Methods*, Interscience Publishers, a division of John Wiley & Sons, Inc., New York, 1965.

J.F. Thompson, ed., *Manual of Analytical Methods for the Analysis of Pesticide Residues in Human and Environmental Samples*, U.S. Environmental Toxicology Division, Research Triangle Park, N.C., June 1977 rev.

H.H. Willard, L.L. Meritt, Jr., J.A. Dean, and F.A. Settle, Jr., *Instrumental Methods of Analysis*, 6th ed., Van Nostrand, New York, 1981.

TRACERS. See Radioactive tracers.

TRADEMARKS AND COPYRIGHTS

TRADEMARKS

A trademark, as defined in the Federal statutes, includes any word, symbol, or device, or any combination thereof, adopted and used by a manufacturer or merchant to identify goods and distinguish them from those manufactured or sold by others. Related to trademarks are service marks, certification marks, and collective marks. Ownership of a trademark confers the exclusive right to use it or to authorize its use in connection with the goods of its owner, or goods made by others to the owners' standards.

Patents, which afford an exclusive right to use an invention, are acquired by disclosing an invention in an application duly filed and prosecuted in accordance with the patent laws. Copyrights, which in most cases afford the right to control reproduction, adaptation, public performance, or public display of literary or artistic works, arise automatically upon creation of a work, but are dependent on Federal statute.

Selection

In selecting a trademark, it is essential that a mark be distinct from other trademarks for similar merchandise. A mark for which exclusive protection is sought cannot be purely descriptive, but should be fanciful or arbitrary. Before adoption of a mark, a search should be made in the field of commerce in which its use is contemplated.

The standards used to determine whether a prospective mark conflicts with an existing mark are twofold: similarity of the marks themselves; and the relationship among the kinds of merchandise for which they are intended. It is also necessary to consider whether a proposed new mark has such a descriptive character as to render the mark incapable of distinguishing the goods from such goods in general. A generic term cannot be preempted against the right of the public to use it. However, if a term, though descriptive, is used long enough that the public understands it as a trademark, it may function and be protected as such. Names of persons, especially surnames, are in the same class as descriptive terms that can nevertheless function as trademarks, eg, Ford or DuPont. Marks that are not descriptive but suggest the character, quality or function of the goods are much favored, eg, Talon for slide fasteners and Ivory for soap. Marks that are wholly arbitrary may be words having no relationship to the goods or they may be synthetic terms having no significance. These marks have the advantage of being least likely to lose their status by becoming the generic name of the product.

The functions of trademarks may be characterized as serving as an identification of origin and a guarantee of quality, and as an advertising device.

Registration

A right to trademark protection arises through exclusive use of the mark in connection with the goods of the owner. It is protected by common law independent of registration. Registration is a statutory creation, that affords a means of publicizing a trademark right and facilitating its protection. Registration procedure involves filing an application stating the kind of commerce in which the mark is in use, the date of first use and of first use in commerce, the register (principal or supplemental) on which registration is sought, the goods on which the mark is used, and the manner in which it is applied to the goods. The application is submitted to the Patent and Trademark Office accompanied by the filing fee. After examination, any objections are communicated in an office action requiring response within a limited period. The response may involve an amendment, an argument, or a demonstration of further facts. Final rejection may be appealed to the Trademark Trial and Appeal Board. When the application is found allowable, it is published in the *Official Gazette* of the Patent Office, and within 30 days any person may file an opposition.

Marks registered under the Trademark Act of 1946 are subject to cancellation unless an affidavit or declaration under penalty of perjury is filed within the sixth year of registration showing that the mark is still in use in commerce.

The registrant of a mark under the Federal statutes is afforded a remedy by civil action in the courts against unauthorized use of a reproduction, counterfeit, copy, or colorable imitation of the mark in connection with any goods or services for which such use is likely to cause confusion or mistake, or to deceive. In addition, importation of merchandise that copies the registered mark is prohibited. Unlike trademarks, trade names are not registerable.

Transfer

A trademark and its registration may be assigned but only together with the good will of the business in which the mark is used. A trademark may be licensed, but special rules apply to such licensing by reason of the nature and function of the trademark right.

Proper Use

Proper use of a trademark by the owner or the licensee under a valid license is essential to preservation of the exclusive trademark right. The 1946 statute provides, for example, that nonuse of two years constitutes *prima facie* abandonment of the trademark. Various rules have been devised to reduce the danger of a mark degenerating into a generic term. These rules are particularly significant with respect to chemical products.

State and Foreign Registrations

Registration of a trademark under state law may be appropriate when the mark is used primarily in intrastate commerce. The trademark laws of foreign countries differ in many fundamental respects from U.S. laws. Under the International Convention for Protection of Industrial Property, a trademark for which application for registration has been made in one member country can be registered in another member country with priority dating from the application in the first country if the later application is filed within six months and convention priority is claimed.

WALTER G. HENSEL
Patent Attorney

Reviewed by
JON A. BAUMGARTEN
CHARLES H. LIEB
Paskus, Gordon & Hyman

United States Trademark Association, *Trademark Management*, Clark Boardman, Ltd., New York, 1981, pp. 1–54.

Trademarks, The Who What Where and How of Black & Decker Trademarks, The Black & Decker Manufacturing Company, Towson, Md.; D. Fey, *The Practical Lawyer* **24**(8), 75 (1968), rev. and reprinted by U.S. Trademark Association, New York, 1978.

COPYRIGHTS

Copyright is the right of the creator of a work to control certain uses of that work by others. Works that are the subject of copyright encompass a wide variety of products of intellectual endeavor. In the United States, two agencies exercise limited administrative authority with respect to copyright: the U.S. Copyright Office, a department of the Library of Congress, and the Copyright Royalty Tribunal, an independent entity located in the legislative branch of the government.

Congress is empowered to enact copyright legislation by the Constitution in order to "...promote the Progress of Science and useful Arts..." Although both copyright and patent law derive from the same constitutional provision, they are quite different systems. Patents are granted by the Government; copyrights are not.

Subject Matter

The categories of subject matter protected by copyright law in the United States include literary works (including compilations of data, computer programs, and data bases); dramatic and musical works; pantomimes and choreographic works; pictorial, graphic, and sculptural works; motion pictures and other audiovisual works; and sound recordings. In order to be protected by copyright, the work must be fixed in a concrete mode of expression and it must be original. Copyright protection pertains only to the expression of a work and not to its underlying idea.

Acquisition

When the new copyright law became effective on January 1, 1978, the United States adopted a principle of automatic copyright, which means that works created on and after that date are copyrighted under Federal statute automatically upon their creation. However, certain statutory conditions (so-called formalities) pertaining to registration and copyright notice remain important.

Registration and Notice

Although registration with the U.S. Copyright Office is not a condition of copyright protection, the statute provides a number of advantages to prompt registration, eg, the ability to proceed rapidly against infringement to obtain certain remedies, and the evidentiary effects of a certificate of copyright registration.

Registration is accomplished by submitting an application for registration to the U.S. Copyright Office, together with an appropriate fee (in 1984 generally $10 per work) and a deposit of the work. Whenever a work is published, ie, made available to the public, a specific marking in conformity with the statute (copyright notice) should be placed on copies.

Duration. Copyright in works created on or after January 1, 1978 generally endures for the life of the author and 50 years thereafter. Different rules apply to many works created before January 1, 1978. A renewal for works copyrighted before January 1, 1978 is required in order to secure a second, or renewal, term.

Ownership

Copyright belongs in the first instance and upon creation of the work to its author. In the case of a work made for hire, however, copyright belongs to the employer for whom the work was written or created. The author has no rights in such works except as may be agreed upon by the employer.

Rights

The principal rights embraced under copyright are the right to reproduce; the right to prepare derivative works; the right to distribute copies; the right to perform; and the right to display the work publicly. The statute specifies a number of instances in which the use of a work is totally excluded from the copyright owner's rights. For example, the doctrine known as fair use permits limited, reasonable, and customary uses of prior works in connection with such activities as research, scholarship, criticism, comment, teaching, and news reporting. A compulsory license is a privilege given by the Copyright Act that permits certain other uses of a work without the consent of the copyright owner, but only upon compliance with certain conditions, including payment of royalties.

Effect of Technology

Tensions have developed between copyright and new technologies, which have affected copyright in a variety of ways, principally by eroding owners' rights to control or secure compensation for the use of their works. An example is the off-air video recording in private homes. With the aid of audio- or video-recording devices or photocopying machines, the consumer is now capable of serving as the publisher or creator of copies. Although Congress has made several attempts to accomodate copyright and new technology in the new copyright statute, the task is far from completed.

International Copyright

Works originating in the United States are also subject to copyright protection in many foreign countries. The phrase international copyright is, however, a misnomer, for there is no single law or code of international copyright. Two principal multilateral copyright treaties are the Universal Copyright Convention, to which the United States is a party, and Berne Convention, to which the United States does not belong.

Although treaties and conventions do impose certain minimal requirements on member states with respect to the type of protection to be accorded to foreign works, in each case the protection accorded by a foreign country is substantially determined by its own law.

JON A. BAUMGARTEN
CHARLES H. LIEB
Paskus, Gordon & Hyman

A. Latman, *The Copyright Law*, Bureau of International Affairs, Inc., New York, 1979.

M.B. Nimmer, *Nimmer on Copyright*, Matthew Bender, New York, 1981 (4-vol. treatise).

J.A. Baumgarten, *U.S.–U.S.S.R. Copyright Relations Under the Universal Copyright Convention*, Practising Law Institute, New York, 1973.

TRAGACANTH. See Gums.

TRANQUILIZERS. See Psychopharmacological agents.

TRANSISTORS. See Semiconductors.

TRANSPORTATION

The transportation of chemicals and related products is unusual in that substantial quantities are moved in packages as well as in bulk. Other materials, such as coal, grain, and ore, are transported in bulk but seldom in packaged form. Both the supply and pricing of transportation services are governed by Federal and state regulations.

The chemical industry is one of the largest users of commercial transportation. Its cost has an important effect on marketability, and its accessibility is often a significant consideration in the choice of locating production facilities. In 1980, railroads carried 28.9% of all intercity tonnage, trucks 36.4%, oil pipelines 16.7%, water 17.9%, and air transport 0.1%.

Railroads

Tank cars and other special-purpose rail cars, such as covered hopper cars, must generally be furnished by the shippers or receivers, who usually purchase or lease such equipment.

The so-called trip lease of private (shipper-furnished) tank cars, used primarily to avoid the assessment of railroad demurrage, is a unique feature of rail transport. Demurrage is a charge made by carriers for the detention of transport conveyance beyond a period considered sufficient for loading or unloading.

At many plants and manufacturing facilities, railroad tracks are constructed wthin the plant. Such tracks, called industry or private tracks or sidetracks, connect directly with the tracks of the railroad serving the plant. The storage of private cars on private tracks of like ownership is generally not subject to demurrage.

Motor carriage. Motor carriage is an essential part of the U.S. transportation system. Improved highways, including the interstate system, have led to the development of a transportation network competitive with railroads in both rates and service. Individual carriers tend to specialize in the type of service they offer. Transport of general commodities is frequently limited territorially. Thus, the services of common carriers are limited, in the case of trucking companies, to those they are willing or permitted by their governmental franchise to perform. Motor carriers who confine their services to particular shippers are known as contract carriers or house carriers.

Proprietary or private carriage is conducted in furtherance of a primary business other than transportation. Recent legislation has reversed a historic rule, that held that a parent corporation that performed transportation for compensation on behalf of a subsidiary, or vice versa, was engaged in for-hire transportation and, therefore, was subject to regulation.

A wide variety of highway vehicles, including trucks, tractors, trailers, tank vehicles, hopper vehicles, low-boys, vans, and others, are used by common, contract, and private motor carriers. Highway tractors used for long, continuous journeys are usually equipped with sleeper cabs.

Waterborne transport. Water carriers transport large quantities of bulk chemicals in barges between inland or coastal ports. Water carriage of bulk chemicals in the United States is essentially unregulated. Barges are available in a wide variety of types, sizes, and capacities. On inland waterways, they are usually unmanned and without power but pushed by towboats, which can move several barges in a group or tow. Deepwater or ocean-going barges sometimes carry a crew and are self-propelled. In recent years, ocean transportation has been dominated by container ships designed to load and carry large, trailer-sized containers. The amount of time spent in port is greatly reduced, thereby increasing the number of voyages possible in a given period and reducing operating costs. Other types of ocean vessels include tankers and dry-bulk ships.

Other services. Pipeline transportation is predominantly limited to hydrocarbons. Commercial petroleum lines are considered to be common carriers available to all customers.

Air transport. Small quantities of chemicals are transported by air; economic and safety considerations impede the development of this form of transportation of chemicals.

Domestic freight forwarders consolidate multiple small shipments which they forward in carload or truckload lots to a central location for subsequent distribution.

Warehousing is an integral part of the distribution system for goods. Although employed primarily for storage, warehousing may be used to assure timely deliveries. Warehouses may be owned and operated by private companies for their own purposes or they may be available to the public.

Shipping

Terms. Although frequently referred to as shipping terms, terms such as fob (free on board), fas (free alongside ship), and cif (cost, insurance, freight) are actually terms of sale. The selection of shipping terms has a material effect on the sales contract. The party with the risk of loss must decide whether to insure against such risk and must prepare and file a claim against the carrier when goods are lost or damaged in transit.

Domestics. The bill of lading serves as a receipt for goods delivered to a carrier as well as a contract of carriage. Bills of lading may be negotiable and, as such, constitute evidence of title or the right to possession of the goods. In international trade, the ocean bill of lading serves the same purposes. Where a shipper's freight occupies the whole or a substantial portion of a particular vessel, the document used may be a voyager charter, which provides for use of the vessel for a single voyage. The dock receipt is evidence that the goods have been delivered to a dock, pending transfer to an ocean-going vessel. A freight bill is an invoice issued by a carrier requesting payment for transportation. A delivery receipt is frequently a copy of the freight bill that has been signed by the consignee as evidence of delivery.

Interstate and Intrastate Commerce

The applicability of Federal and state transportation laws and regulations depends on whether the transportation constitutes interstate or intrastate commerce. Except in rare instances, transportation that requires physical movement across state boundaries is interstate commerce. However, transportation within a single state is not necessarily intrastate commerce, since such transportation may be a portion of a continuous movement in interstate commerce.

Economic Regulation

In the United States, Federal and state governmental regulations are directed at safety or at economic concerns such as discrimination in rates and services or excessive competition. The pertinent agencies at the Federal level are the Interstate Commerce Commission (ICC), the Federal Maritime Commission (FMC), the Civil Aeronautics Board (CAB), and the Federal Energy Regulatory Commission (FERC). Recent relaxation of controls has resulted in substantial changes in the transportation industry. Entry into the motor-carrier business has been liberalized. Railroads are now permitted to enter into contracts with individual shippers, and the ICC can grant exemptions from railroad regulation. Antitrust immunity has been removed to a large extent and more reliance has been placed on market competition.

Regulated carriers are required to publish and file with the ICC tariffs or schedules of their rates and charges, although many motor contract carriers have recently been exempted from such requirements. Contracts between shippers and railroads must be filed with the ICC but are treated as confidential.

Freigh rates and allowances. The establishment of a transportation price structure embraces virtually all articles of commerce, in a multitude of packages and quantities, via numerous routes, and between numerous locations and includes services such as storage or reconsignment. A somewhat standard nomenclature identifies products shipped in commerce. Chemical and freight nomenclatures, however, are frequently different. Many thousands of tariffs are filed with the regulatory agencies and numerous changes and revisions become effective daily.

Carload or truckload rates are applicable to minimum carload or truckload volumes. Multiple-car rates are applicable when a specified number of carloads is tendered for a single shipment. Annual (or periodic) volume rates are applicable to individual shipments that are part of an aggregate tonnage that a shipper has agreed to ship in a specified period. Accessorial charges are for services that are ancillary to line-haul transportation.

Freight loss and damage. The Interstate Commerce Act codifies liability for railroads, motor common carriers, and domestic freight forwarders, and provides that such carriers are liable for the full, actual loss or damage to the goods. Contract carriers are not held to the same standard of liability as common carriers since they are considered ordinary for-hire bailees. The liability of water carriers is established under the principles of traditional admiralty law, which generally reflects the fundamental concept of liability for negligence.

Safety Regulations

Upon the establishment of DOT in 1967, the Federal Aviation Administration (FAA) and Coast Guard were transferred to that department and the safety functions formerly administered by ICC were assumed by DOT.

DOT safety regulations fall into two categories: The first pertains to the qualifications and hours of service of carrier employees and the safety of transport operations and equipment. The second, of special concern to the chemical industry, pertains to the transportation of hazardous materials and related commodities. The National Transportation Safety Board is responsible for investigating serious accidents by all modes of transportation. The Hazardous Materials Transportation Act of 1974 consolidated the authority of DOT with respect to safety.

Many industrial or professional organizations, eg, ASME and the Bureau of Explosives, publish standards for containers, materials of construction, tests, and similar matters that are frequently incorporated in the regulations. Application of the hazardous materials regulations is greatly facilitated through the use of the Hazardous Materials Tables at 49 CFR, 172.101. The proper selection of chemical shipping descriptions and the determination of the hazard class require chemical expertise and familiarity with DOT definitions. The packaging requirements of the hazardous materials regulations (49 CFR part 173) are extremely detailed.

The recent adoption by the Materials Transportation Bureau (MTB) of a numbering system for more specific identification of hazardous materials reflects a worldwide effort to improve response to transportation emergencies. Although accidental release of hazardous materials in transit is relatively rare, the potential for significant harm is of constant concern to the public and industry and is magnified by the fact that many public emergency-response agencies have had little, if any, training or experience in dealing with chemical emergencies. In an effort to provide immediate and reliable information to carriers and public officials at the scene of an emergency, the Chemical Manufacturers' Association established the Chemical Transportation Emergency Center (CHEMTREC) in Washington, D.C. Since its formation in 1971, CHEMTREC has responded to thousands of emergency calls, providing information from its files containing data on more than 16,000 chemical products. Similarly, the Chlorine Institute has organized a mutual aid program, called CHLOREP, which offers assistance at the scene of emergencies involving chlorine. Industrial response teams are usually available for assistance in connection with clean-up of spills which may be hazardous to the public or the environment (see also Plant safety).

Enforcement of the hazardous materials rules may result in compliance orders, injunctions, or the assessment of civil or criminal penalties as high as $25,000 and imprisonment for five years for each violation.

The proliferation of local regulations has tended to burden the free flow of hazardous commodities in commerce with diverse requirements which are often inconsistent with Federal rules. An increasing number of lawsuits challenge the validity of individual state or local enactments.

STANLEY HOFFMAN
Union Carbide Corporation

J. Guandolo, *Transportation Law*, 3rd ed., Wm. C. Brown Co., Dubuque, Iowa, 1979.

W.J. Augello, *Freight Claims in Plain English*, Shippers National Freight Claim Council, Inc., Huntington, N.Y., 1979.

G. Gilmore and C.L. Black, Jr., *The Law of Admiralty*, 2nd ed., The Foundation Press, Inc., Mineola, N.Y., 1975.

L.W. Bierlein, *Red Book on Transportation of Hazardous Materials*, Cahners Books International, Inc., Boston, Mass., 1977.

TRIAZINETRIOL.   See Cyanuric acid and isocyanuric acids.

TRICRESYL PHOSPHATE.   See Plasticizers; Phosphorus compounds.

TRIETHANOLAMINE.   See Alkanolamines.

TRIETHYLENE GLYCOL DINITRATE.   See Explosives and propellants.

TRIMENE BASE.   See Rubber chemicals.

TRIMETHYLOLETHANE.   See Alcohols, polyhydric.

TRIMETHYLOLETHANE TRINITRATE.   See Explosives and propellants.

TRIMETHYLOLPROPANE.   See Alcohols, polyhydric.

TRIOXANE.   See Formaldehyde.

TRIPENTAERYTHRITOL.   See Alcohols, polyhydric.

TRIPHENYLMETHANE AND RELATED DYES

The triarylmethane dyes are of brilliant hue, exhibit high tinctorial strength, are relatively inexpensive, and may be applied to a wide range of substrates. However, they are deficient in fastness to light and washing and their use on textiles has decreased. On polyacrylonitrile fibers, they are readily absorbed and show surprisingly good light- and washfastness. The most important black dye for acid-modified fibers is a mixture of malachite green and fuchsine because of high tinctorial strength and low cost. This is, however, the only remaining major textile use for triarylmethane dyes.

Structures

Chemically, the triarylmethane dyes are derivatives of the colorless compounds triphenylmethane and diphenylnaphthylmethane. One or more primary, secondary, or tertiary amino groups or hydroxyl groups in the para positions to the methane carbon atom are required to give the spectral absorption characteristics of the dyes. Additional substituents such as carboxyl, sulfonic acid, or halogen may be present on the aromatic rings. The number, nature, and position of these substituents determine both the hue or color of the dye and the application class to which the dye belongs. When only one amino group is present, the shade is a weak orange-yellow. Additional amino groups increase the intensity of absorption and result in a strong bathochromic shift. Derivatives of commercial value contain two or three amino groups, eg, Doebner's violet.

Doebner's violet

A further bathochromic shift is observed as the basicity of the primary amines is increased by *N*-alkylation, eg, in malachite green, or *N*-phenylation.

Chemical Properties

Triarylmethane dyes are usually destroyed by strong oxidizing agents, eg, sodium hypochlorite, which further limits their use as textile dyes. Triarylmethane dyes are extremely sensitive to photochemical oxidation, which accounts for their poor lightfastness on natural fibers. The mechanism of degradation is believed to involve *N*-dealkylation, a general phenomenon in dye photochemistry, which is also observed with thiazine and rhodamine dyes and *N*-methylaminoanthraquinones.

Triarylmethane dyes are readily reduced to leuco bases. Reduction with titanium trichloride (Knecht method) is used for rapid assay:

$$Ar_3COH + 2\ TiCl_3 + 2\ HCl \rightarrow Ar_3CH + 2\ TiCl_4 + H_2O$$

The $TiCl_3$ titration is carried out to a colorless endpoint, which is usually very sharp.

Sulfonation of alkylaminotriphenylmethane dyes gives mixtures. Sulfonic acid groups are preferably introduced by sulfonation of the intermediates from which the dyes are manufactured.

Dyes containing alkylated amino groups are prepared from alkylated intermediates. However, 4,4′,4″-triaminotriphenylmethane (pararosaniline) may be *N*-phenylated with excess aniline and benzoic acid.

Manufacture

The methods of manufacture of triarylmethane dyes are grouped according to the source of the central methane carbon atom. In the aldehyde method the central carbon atom is derived from an aromatic aldehyde or a substance capable of generating an aldehyde during the course of the condensation.

For example, Malachite Green, CI Basic Green 4, is prepared by heating benzaldehyde under reflux with a slight excess of dimethylaniline in aqueous acid.

In the ketone method, a diarylketone is prepared from phosgene and a tertiary arylamine. This is then condensed with another mole of a tertiary arylamine (the same or different) in the presence of phosphorus oxychloride or zinc chloride. The dye is produced directly without an oxidation step. This method is useful for the preparation of unsymmetrical dyes.

In the diphenylmethane base method, the central carbon atom is derived from formaldehyde which is condensed with two moles of an arylamine to give a substituted diphenylmethane derivative. This is then oxidized to a benzhydrol derivative. Condensation with arylamines gives a leuco dye which is oxidized in the presence of acid. For example, Crystal Violet Lactone, used in carbonless copying paper, is prepared from Michler's ketone and *m*-dimethylaminobenzoic acid.

In the benzotrichloride method, the central carbon atom of the dye is supplied by the trichlormethyl group from *p*-chlorobenzotrichloride. Both symmetrical and unsymmetrical triarylmethane dyes suitable for acrylic fiber are prepared by this method.

Uses

Application of triarylmethane dyes is confined mainly to nontextile purposes. Substantial quantities are used for organic pigments for printing inks and paper printing where low cost and brilliance are more important than lightfastness. Triarylmethane dyes and their colorless precursors, eg, carbinols and lactones, are used extensively in high speed photoduplicating and photoimaging systems.

Related Dyes

The diphenylmethane dyes are usually classed with the triarylmethane dyes, although they have little in common. Only Auramine O and CI Basic Yellow 37 of the series are of commercial significance. The indicators phenolphthalein and phenol red are usually considered to be triphenylmethane derivatives.

DEREK BANNISTER
JOHN ELLIOTT
CIBA-GEIGY Corporation

K. Venkataraman, ed., *The Chemistry of Synthetic Dyes*, Vol. 2, Academic Press, Inc., New York, 1952; N.R. Ayyanger, D.B. Tilak, and D.R. Baer in Vol. 4, 1971; J. Lenoir in Vol. 5, 1971; E. Gurr and co-workers in Vol. 7, 1974; N.A. Evans and I.W. Stapleton in Vol. 8, 1978.

H.A. Lubs, ed., *The Chemistry of Synthetic Dyes and Pigments*, American Chemical Society Monograph Series, Reinhold Publishing Corp., New York, 1955.

H.A. Lubs, ed., *The Chemistry of Synthetic Dyes and Pigments*, American Chemical Society Monograph Series, Reinhold Publishing Corp., New York, 1955.

TRYPSIN. See Enzymes.

TRYPTOPHAN. See Amino acids.

TUADS. See Rubber chemicals.

TUNG OIL. See Fats and fatty oils; Drying oils.

TUNGSTEN AND TUNGSTEN ALLOYS

Tungsten (wolfram), at. wt 183.85 (five stable isotopes), is a silver-gray metallic element appearing in Group VIB below chromium and molybdenum. It has the highest melting point of any metal, 3695 K, a very low vapor pressure, and the highest tensile strength above 1650°C of any metal. The pure metal was first used as a filament for electric lamps at the beginning of this century.

Of the more than 20 tungsten-bearing minerals only four, ie, ferberite (iron tungstate), huebnerite (manganese tungstate), wolframite (iron-manganese tungstate), and scheelite (calcium tungstate) are of commercial importance. Deposits occur in association with metamorphic rocks and granitic igneous rocks throughout the world. Reserves are estimated at 6.8×10^6 metric tons.

Properties

Physical properties are given in Table 1. The oxidation states of tungsten range from +2 to +6; compounds with zero oxidation state also exist. Above 400°C, tungsten is highly susceptible to oxidation. At 800°C, the oxide sublimes with decomposition. Very fine powders are pyrophoric. Above 600°C, the metal reacts vigorously with water to form oxides. Tungsten is stable in nitrogen to > 2300°C. Fluorine attacks the metal at room temperature.

Manufacture

Tungsten mines are generally small, producing less than 200 t of raw ore per day. Worldwide, there are only 20 mines producing over 300 t/d. Owing to their high specific gravity, tungsten minerals can be beneficiated by gravity separation, usually tabling. Flotation is used for scheelite but not for wolframite ores.

Table 1. Physical Properties of Tungsten

Property	Value
density at 298 K, g/cm^3	19.25
melting point, K	3695 ± 15
boiling point, K	5936
entropy at 298 K, J/mol[a]	32.66
heat of fusion, kJ/mol[a]	46.0
heat of sublimation, 298.1 K, kJ/mol[b]	859.8
vapor pressure, 2600–3100 K, Pa[b]	$\log P_{Pa} = \dfrac{-45395}{T} + 12.8767$[c]
thermal conductivity at K, W/(cm·K)	
10	97.1
500	1.46
3400	0.90

[a] To convert J to cal, divide by 4.184.
[b] To convert Pa to mm Hg, multiply by 0.0075.
[c] To convert $\log P_{Pa}$ to $\log P_{mm\ Hg}$, subtract 2.1225.

In extractive metallurgy, an impure ore concentrate is converted into a high purity compound, usually ammonium paratungstate (APT). The metal powder is obtained from APT by stepwise reduction with carbon or, preferably, hydrogen.

Tungsten is usually processed by powder metallurgy (qv). Small quantities of rod are produced by arc or electron-beam melting. Ductility increases with working. Tungsten, sintered or after full recrystallization annealing, is as brittle as glass at room temperature, and is initially worked at very high temperatures. Swaging at 1500–1600°C is employed for the manufacture of lamp wire (Coolidge process). Rolling at 1600°C is employed for large-diameter rod or plate. Forging is also used; hammer forging is preferred over press forging.

Analytical Methods

Tungsten is usually identified by atomic spectroscopy. In a wet method, the ore is fired with sodium carbonate and then treated with hydrochloric acid; addition of zinc, aluminum, or tin produces a blue color. For quantitative determination, the ore is digested with acid and the tungsten is complexed with cinchonine, purified, ignited, and weighed. More commonly, x-ray spectrometry is employed and its accuracy enhanced by using tantalum as internal standard. Plasma spectroscopy determines concentrations of 0.1 ppm in solutions and 10 ppm in solids.

Health and Safety Factors

Although there are no documented cases of tungsten poisoning, numerous cases of pneumoconiosis have been reported in the cemented carbide industry but whether the cause is WC or cobalt has not been established. The exposure limits for insoluble tungsten compounds are 5 mg/m^3 and for soluble compounds 1 mg/m^3.

Uses

Tungsten is used in four forms: tungsten carbide, as an alloy additive, essentially pure tungsten, and tungsten chemicals. The carbide is used for cutting tools, abrasion-resistant surfaces, and forming tools. About 16% of tungsten usage is as an alloy additive. In steels, tungsten forms a dispersed WC phase which imparts a finer grain structure and increases high temperature hardness. For hot-work tool steels, up to 18% tungsten is added. In nickel- and cobalt-base superalloys, tungsten imparts high temperature strength and wear resistance (see High temperature alloys).

Metallic tungsten offering the advantages of a high melting point and low vapor pressure, accounts for 15% of usage. In lamp filaments, potassium, silicon, and aluminum dopants are added to the tungsten oxide. As an electron emitter, tungsten can be used at very high temperatures. Addition of thoria reduces the work function and improves emission.

Alloys of tungsten with various combinations of iron, nickel, and copper are called heavy alloys. They have the high density of tungsten but in a more machinable form. Alloys made by infiltrating porous tungsten with copper or silver are used as electrical contact material and rocket nozzles. Composite materials with barium and strontium compounds are used in electron-emitting devices.

Nonmetallurgical uses include brilliant organic dyes and pigments. Tungstates are used as phosphors in fluorescent light, cathode-ray tubes, and x-ray screens. Tungsten compounds are used as catalysts in petroleum refining.

JAMES A. MULLENDORE
GTE Product Corporation

G.D. Rieck, *Tungsten and Its Compounds*, Pergamon Press, London, UK, 1967.

C.J. Smithells, *Tungsten*, Chemical Publishing Co., New York, 1953.

S.W.H. Yih and C.T. Wang, *Tungsten*, Plenum Press, New York, 1979.

TUNGSTEN COMPOUNDS

Tungsten, a Group VIB element, has the valence states of 0, +2, +3, +4, +5, or +6. Its most stable valence state is +6. Tungsten complexes vary widely in stereochemistry and oxidation states.

Tungsten hexacarbonyl, $W(CO)_6$, may be prepared in yields > 90% by the aluminum reduction of tungsten hexachloride. A colorless to white solid, it decomposes without melting at ca 150°C.

Halides and oxyhalides. Tungsten forms binary halides for all oxidation states between +2 and +6; oxyhalides are only known for oxidation states +5 and +6. The hexachloride and hexafluoride are the starting materials for the chemical vapor deposition of tungsten, which is an important process technique for coatings and free-standing parts (see Film deposition techniques).

Oxides, acids, and salts. Tungsten oxides form a series of well-defined ordered phases to which precise stoichiometric formulas can be assigned (see Table 1). Their composition may vary without change in crystalline structure. The homogeneity ranges are represented by $WO_{2.95-3.0}$, $WO_{2.88-2.92}$, $WO_{2.664-2.766}$, and $WO_{1.99-2.02}$. Each tungsten atom is octahedrally surrounded by six oxygen atoms.

Tungsten bronzes are intensely colored, ranging from golden-yellow to bluish-black; in crystalline form, they exhibit a metallic sheen. They are inert to most acids but dissolve in basic reagents. Sodium tungsten bronzes serve as promoters for the catalytic oxidation of carbon monoxide and reformer gas in fuel cells. They are prepared by electrolytic reduction, vapor phase deposition, or solid-state reaction.

Tungstic acid, H_2WO_4 or $WO_3 \cdot H_2O$, is precipitated from hot tungstate solutions with strong acids. Ammonium tungstate, $(NH_4)_2WO_4$, is prepared by the addition of hydrated tungstic acid to liquid ammonia. Tungstates are of particular interest in electronic and optical applications.

Polytungstates. In acid solution, tungstate ions form condensed complex ions of isopolytungstates. Metatungstates of the alkali, alkaline-earth, rare-earth, and transition metals have been reported. The most important is ammonium metatungstate, $(NH_4)_6H_2W_{12}O_{40}$. The paratungstates are crystallized from slightly basic solutions; the most important is ammonium paratungstate, $(NH_4)_{10}H_{10}W_{12}O_{46}$. Heteropoly tungsten compounds are classified according to the ratio of the number of central atoms to tungsten.

Sulfides. Tungsten disulfide, WS_2, is prepared by heating tungsten powder with sulfur at 900°C. It forms an adherent soft film on a variety of surfaces and is a good lubricant. The trisulfide, a chocolate-brown powder, is prepared by treating an alkali-metal thiotungstate with HCl. Tungsten forms thiotungstates corresponding to tungstates, where one, two, three, or all oxygen atoms are replaced by sulfur.

Interstitial compounds. Tungsten forms hard, refractory, chemically stable interstitial compounds with nonmetals, particularly C, N, B, and Si. These compounds are used in cutting tools and structural elements in kiln, gas turbines, and other refractory applications. The two carbides, WC and W_2C, both melt at ca 2800°C and have a hardness approaching that of diamond. Their most important application is in hard metals. Approximately 67% of tungsten production is used for WC.

The nitrides resemble the carbides; they are prepared by heating tungsten in ammonia. The borides, W_2B, WB, and W_2B_5, are extremely hard and exhibit almost metallic electrical conductivity. Tungsten silicides, W_2Si_3 and WSi_2, form a protective oxide layer over tungsten to prevent oxidation at elevated temperatures.

Compounds of tungsten with acid anions, other than halides and oxyhalides, are few in number and are known only in the form of complex salts, eg, $M(I)_2(W_2F_4)$, $M(I_4)(W(CN)_8)$, and $M(I_3)(W_2Cl_9)$.

Toxicity

Considerable differences in the toxicity of soluble and insoluble tungsten compounds have been reported. In view of the degree of systemic toxicity of soluble tungsten compounds, a threshold limit of 1 mg tungsten per m^3 of air is recommended. A threshold limit of 5 mg/m^3 is recommended for insoluble compounds.

Uses

Tungsten compounds, especially the oxides, sulfides, and heteropoly complexes, are used as catalysts for a variety of chemical processes, eg, for petroleum. The hexachloride is the starting material for metathesis catalysts, which form double and triple bonds with carbon. Sodium tungstate is used in the manufacture of color lakes. Calcium tungstate is used in lasers, fluorescent lamps, high voltage sign tubes, and oscilloscopes for high speed photographic processes.

The trioxide is the principal source of tungsten metal and carbide powders. Because of its bright yellow color, it is also used as a pigment in oil and water colors (see Pigments). The tungstates are good corrosion inhibitors and are used in antifreeze solutions.

M.B. MACINNIS
T.K. KIM
GTE Products Corporation

S.W.H. Yih and C.T. Wang, *Tungsten*, Plenum Press, New York, 1979.

G.D. Rieck, *Tungsten and Its Compounds*, Pergamon Press, London, UK, 1967.

C.J. Smithells, *Tungsten*, Chapman and Hall Ltd., London, UK, 1952.

TURBIDITY AND NEPHELOMETRY.   See Analytical methods.

TURKEY RED OIL.   See Castor oil.

TURPENTINE.   See Terpenoids.

TYPE METAL.   See Lead alloys.

TYROCIDINE.   See Antibiotics, peptides.

TYROTHRICIN.   See Antibiotics, peptides.

Table 1.   Tungsten Oxides

Oxide	Phase	O:W, av	Theoretical density, g/cm³	Color
WO_3	α	3.00	7.29	yellow
$W_{20}O_{58}$	β	2.90	7.16	blue-violet
$W_{18}O_{49}$	γ	2.72	7.78	reddish-violet
WO_2	δ	2.00	10.82	brown
W_3O	(β-W)	0.33	14.4	gray

U

ULTRAFILTRATION

Ultrafiltration is pressure-driven filtration on a molecular scale (see also Dialysis; Filtration; Hollow-fiber membranes; Membrane technology; Reverse osmosis). Typically, a liquid is separated from small dissolved molecules, colloids, and suspended solids.

Ultrafiltration separations range from ca 2 to 20 nm. Above ca 20 nm, the process is known as microfiltration. Below ca 2 nm, interactions between the membrane material and the solute and solvent become significant. That process, called reverse osmosis or hyperfiltration, is best described by a solution-diffusion mechanism.

Membrane-retained components are collectively called concentrate or retentate. Materials permeating the membrane are called filtrate, ultrafiltrate, or permeate.

Media

Most ultrafiltration membranes are porous, asymmetric, polymeric structures produced by phase inversion, ie, the gelation or precipitation of a species from a soluble phase. Membrane structure is a function of the materials and the mode of preparation. Commonly used polymers include cellulose acetates, polyamides, polysulfones, poly(vinyl chloride-co-acrylonitrile)s, and poly(vinylidene fluoride).

Process

Pore-flow models most accurately describe ultrafiltration processes. Other membrane-transport mechanisms include dialysis, osmosis (anomalous and reverse), electrodialysis, piezodialysis, electroosmosis, Donnan effects, Knudsen flow, thermal effects, chemical reactions, and active transport.

When pure water is forced through a porous ultrafiltration membrane, Darcy's law states that the flow rate is directly proportional to the pressure gradient:

$$J = \frac{V}{A \cdot t} = \frac{K_m \Delta P}{\mu} \tag{1}$$

where J is the permeate flux of volume V per membrane area A, at time t; K_m is the membrane hydraulic permeability coefficient; ΔP is the membrane pressure drop between retentate and permeate; and μ is the fluid viscosity; K_m is a function of the pore size, tortuosity, and length, and any resistance in the substructure. These parameters change with the membrane pressure history.

The addition of small membrane-permeable solutes to the water affects permeate transport as follows: solvent–solute interactions change the viscosity of the permeating fluid; solute adsorption reduces the apparent membrane-pore diameter; the interfacial charge between the membrane-pore wall and the liquid affects permeate transport when the Debye screening length approaches (ca 10%) the membrane pore size; high surface tension on hydrophobic membranes forces water molecules to form large clusters in the pores; and solvents, swelling agents, and plasticizers that diffuse into the polymer structure can change the apparent pore size or increase the rate of long-term compaction.

If the solute size is greater than the pore dimensions, the solute is retained by mechanical sieving, forming a gel-polarization layer. This gel layer has a hydraulic permeability coefficient of K_g which controls the permeation rate and reduces the effective membrane pore size. Equation 1 becomes the following:

$$J = \frac{\Delta P}{\mu \left(\dfrac{1}{K_m} + \dfrac{1}{K_g} \right)} = \frac{\Delta P}{\mu \left(R_m + R_g \right)} \tag{2}$$

where R_m and R_g are the hydraulic resistances of the membrane gel.

At steady rate, flux varies with the concentration of retained species according to:

$$J = K \ln \frac{c_g}{c} \tag{3}$$

where K is the mass-transfer coefficient, c_g is the gel concentration, and c is the bulk concentration of all retained species.

Fouling. If the gel-polarization layer is not in hydrodynamic equilibrium with the fluid bulk (ie, K_g is not constant), the membrane is fouled. Fouling is caused either by adsorption of species on the membrane surface or consolidation of an existing gel layer. It is controlled by selection of proper membrane materials, pretreatment of feed and membrane, and operating conditions. When fouling is present, the ultrafiltration process is usually operated at high liquid shear rates and low pressure.

Fouling films are removed by chemical and mechanical methods. Dissolved material may pass into the membrane pores. Reprecipitation upon rinsing must be avoided. Certain applications require that the equipment meet FDA and USDA sanitary standards to ensure that the products are not contaminated by extractables or microorganisms (see Sterilization techniques).

Designs

Since the theoretical model cannot predict flux rates from physical data, plant-design parameters are obtained from laboratory tests, pilot-plant data, or operating performance. Flux is maximized when the upstream concentration is minimized. For any specific task, therefore, the most efficient (minimum membrane area) configuration is an open-loop system where the retentate is returned to the feed tank. These systems have long residence times which may be detrimental if the retentate is susceptible to degradation by shear or microbiological contamination. The feed-bleed or closed-loop configuration is a one-stage continuous-membrane system (see Fig. 1).

Electroultrafiltration combines forced-flow electrophoresis with ultrafiltration to control or eliminate the gel-polarization layer. Placing an electric field across an ultrafiltration membrane facilitates the transport of retained species away from the membrane surface and the retention of partially rejected solutes can be dramatically improved (see also Electrodialysis). Electroultrafiltration has been demonstrated on clay suspensions, electrophoretic paints, protein solutions, oil-water emulsions, and other materials.

In diafiltration, water is added to the concentrate and permeate is removed. The two steps may be simultaneous or sequential. Diafiltration improves the degree of separation between retained and permeable species. Constant-volume batch diafiltration is the most efficient process mode. Sequential-batch diafiltration is a series of dilution-concentration steps. Continuous diafiltration practiced in one or more stages of a cascade system has the same volume-turnover relationship for overall recoveries as sequential-batch diafiltration.

Membrane Equipment

Cartridges are inserted in series into square pipes. Retentate is pumped between the plates while permeate passes into the plates and out through a manifold.

Plate-and-frame systems are also comprised of plates with membranes on both sides. A rigid frame on the plate perimeter functions as a pressure vessel when the plates are stacked. At least one hole near the perimeter of each plate connects the flow channels from one side of the plate to the other.

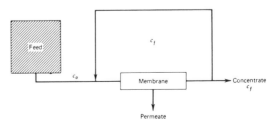

Figure 1. Closed-loop system (feed bleed); c_f = final concentration.

A spiral-wound cartridge has two flat membrane sheets (skin side out) separated by a flexible, porous permeate drainage material. Spiral-wound cartridges are inserted in series into cylindrical pressure vessels.

Supported tubular membranes are cast in place (integral with the support tube), cast externally and inserted into the tube (disposable linings), or are dynamically formed. Depending on the membrane material and operating pressure, self-supporting tubes have inside diameters less than 2 mm; tubes with inside diameters as small as 0.4 mm are commercially available.

The devices described here are assembled by connecting the modules into combinations of series or parallel-flow paths, or both. These are connected to pumps, valves, tanks, heat exchangers, instrumentation, and controls to provide complete systems.

Applications of ultrafiltration include electrophoretic paint, dairy wheys (protein concentration, purification, diafiltration), milk (cheese and yogurt manufacture, yield improvement), oil-water emulsions, effluents of wool and yarn scouring (lanolin recovery, pollution abatement), concentration and purification of enzymes, biological reactors (manufacture of antibiotics, alcohol fermentation, sewage treatment), extraction of vegetable proteins, latex concentration, production of pure water, pulp and paper manufacture (separation of lignosulfonates from spent liquor), and fractionation and purification of blood.

P.R. KLINKOWSKI
Dorr-Oliver, Inc.

S. Hwang and K. Kammermeyer, *Membranes in Separations*, John Wiley & Sons, Inc., New York, 1975.

M.C. Porter in P.A. Schweitzer, ed., *Handbook of Separation Techniques for Chemical Engineers*, McGraw-Hill, New York, 1979.

ULTRAMARINE.   See Pigments, inorganic.

ULTRASONICS

HIGH POWER

The object of high power ultrasonic (macrosonic) application is to bring about permanent physical change. The power density may range from less than one watt to thousands of watts per square centimeter. The piezoelectric sandwich-type transducer driven by an electronic power supply is the most common source of ultrasonic power. Ultrasonic-power transducers produce sinusoidal motion and can be coupled to the load directly or through intermediate resonant members. In most power applications, ultrasonic motion is perpendicular to the plane of the transducer-load interface, and compressional (longitudinal) waves are imparted to the load.

Applications in Liquids

In a strong ultrasonic field, most of the energy supplied by the source to the liquid is dissipated in cavitation. Practically all high power uses of ultrasound in liquids depend on cavitation and its secondary effects. The work performed depends on the energy of imploding bubbles as well as on the number of bubbles per unit volume. Cavitation is more intense at lower frequencies and ceases altogether in the high MHz range.

Industrial ultrasonic cleaning is utilized in a large variety of products and industries. Ultrasonic cleaning works best on hard materials, such as metal, glass, and plastic, which reflect rather than absorb vibration. It is particularly useful when applied to parts of complex geometry with inaccessible areas.

Ultrasonic cleaning agents should combine chemical action with the ability to cavitate. Aqueous solutions are the most common and include both alkali and acid. Ultrasonic cleaning is often combined with other cleaning operations such as degreasing and vapor rinsing.

An ultrasonic cleaning tank (5–150 L) is energized by transducers attached to its bottom; a resonant pattern forms in the liquid. Parts can

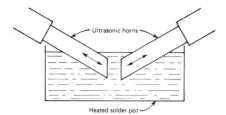

Figure 1. High intensity ultrasonic solder pot for soldering and tinning small parts.

be dipped in the tank directly or in perforated baskets. Small, inexpensive cleaners incorporate the tank and the electronic power supply in one unit for use in laboratories and small shops.

Cavitational implosions at or near the surface of a material can cause erosion of the hardest substance. Resistance to erosion is related to fatigue strength and hardness and can be tested with ultrasonics. Ultrasonic soldering allows tinning without flux and has been of particular interest for aluminum (see Fig. 1). Ultrasonic deburring is applied to delicate precision parts.

Ultrasonic disintegration of biological cells is used to extract active antigens, as well as alkaloids, glucosides, and other components from plant tissues. It is a safe procedure, with extraction times ranging from a few seconds to tens of minutes.

Ultrasonic emulsification produces emulsions without surfactants or with a reduced surfactant content. Coal powder mixed with oil in concentrations up to 40% can be burned as a substitute for oil, but continuous agitation is needed to prevent settling of coal particles. Addition of a small amount of water gives a stable coal-oil suspension suitable for storage and transportation.

Clusters of solid particles can be dispersed in liquids with ultrasound. Fine uniform dispersions that are free of flocculation result. Ultrasonic dispersions of magnetic oxides used in the manufacture of audio, video, and computer tapes and disks improve the quality of the coatings. Ultrasonic treatment of slurries reduces soaking time and improves yields in flotation and separation processes. Sterilization rates are improved by ultrasonics as is the treatment of sewage by ozone.

Diffusion of liquids through porous media can be enhanced by ultrasound. Ultrasonic agitation increases filtration rates, reduces the need for cleaning the filter, and produces drier deposits.

Ultrasonic atomization produces droplets of predictable, uniform size. It was first used for nebulizers in inhalation therapy. Ultrasonic atomization may be of value in the drying of fabric, where it works best on thin, nonabsorbing materials. Fine crystals of uniform size can be produced by atomizing a supersaturated solution and freezing the droplets.

In a strong ultrasonic field, nucleation in supersaturated solutions can be accelerated and growth of large crystals suppressed. In grain refinement of metals, exposure to ultrasound during solidification results in a finer grain and improved physical characteristics.

Macromolecules, such as high molecular weight polymers, can be degraded by high intensity ultrasound. In an ultrasonic field, gases dissolved in a liquid form cavitational bubbles which grow and coalesce into larger bubbles and rise to the surface. Industrial applications include treatment of beer and carbonated beverages, and degassing of photographic solutions. In molten glass or metals, ultrasonic degassing reduces porosity.

Applications in Solids

Ultrasonic welding of thermoplastics causes minimal distortion and material degradation and is used extensively. Power density is high, ca 500 W/cm^2 of weld area (see Fig. 2). Plastic films and synthetic fabrics can be welded continuously. Many applications involving ultrasonic plastic welding are actually forming or reforming operations where metals and thermosets are mechanically interlocked with or enclosed in thermoplastics. For metals, ultrasonic welding is limited to lap welds and is used when other processes present problems.

Brittle materials, difficult to machine by conventional methods, can be machined with ultrasound, eg, ceramics, glass, ferrites, and gem stones.

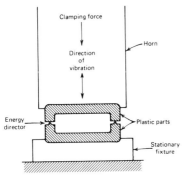

Figure 2. Cross section of an ultrasonic joint.

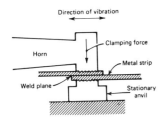

Figure 3. Ultrasonic shear metal welding.

In rotative abrasive machining, axial ultrasonic vibration is superimposed on the conventional rotary motion of the tool. Diamond-impregnated or diamond-plated core drills are used, cooled by a liquid through the center. Most applications involve drilling, but other operations are possible. Impact grinding is performed by ultrasonic vibration alone. An abrasive slurry is fed between the tool and the workpiece and the negative of the horn face is imparted to the work. The principle of ultrasonic metal welding is illustrated in Figure 3. The horn vibrates parallel to the plane of a stationary anvil and causes the upper part to slip with respect to the lower part, while it applies clamping pressure to keep the parts together. Ultrasonic shear motion breaks up and disperses the oxides and other contaminants at the interface, and the exposed plasticized-metal surfaces form a bond under pressure. The process is relatively cold, and parts usually can be welded below their melting temperatures, which reduces embrittlement due to recrystallization.

Ultrasonic fatigue testing allows a drastic reduction in testing time. Endothermic reactions can be accelerated by the heat induced in the volume of a solid under ultrasonic stress. High acceleration present at ultrasonic frequencies reduces apparent friction and produces a slippery effect as if caused by lubrication. An ultrasonically vibrated knife cuts with less effort through a spongy, fluffy, or sticky material. Ultrasonic chutes and sieves promote the flow of sticky powders.

Airborne Applications

Foams can be broken with high intensity ultrasound, but the effect is fairly local. Ultrasonic defoaming does not require any contact with the product, which is an advantage for sanitary reasons. Drying (qv) is accomplished at lower temperatures in the presence of high intensity airborne sound.

In medical applications, power ultrasound is used for dental-descaling, cataract removal, and some surgical operations. Other applications are under investigation.

Health and Safety Factors

Absorption of ultrasonic power at levels normally encountered in industry is negligible and the only area warranting consideration is hearing. Soundproof enclosures and ear-protective devices are available (see Insulation, acoustic).

ANDREW SHOH
Branson Sonic Power Company

L.D. Rosenberg, ed., *Physical Principles of Ultrasonic Technology*, Vols. 1–2, Plenum Press, New York, 1973.

B. Brown and J.E. Goodman, *High Intensity Ultrasonics*, Iliffe Books Ltd., London, UK, 1965.

Ultrasonic Plastics Assembly, Branson Sonic Power Company, Danbury, Conn., 1979.

LOW POWER

Low power ultrasound is used in many fields of science, engineering, and medicine as a nondestructive, noninvasive testing and diagnostic technique. Ultrasonic waves can be generated inside a test material with an ultrasonic transducer. Typical waveforms are shown in Figure 1. The sound-wave speed of a material is proportional to the square root of the ratio of the material's stiffness to density. In many materials, wave speeds correlate with physical properties.

As a sound wave propagates through a material, its amplitude is reduced by absorption, scattering, and beam spreading. Beam spread is caused by interference between portions of the sound wave, scattering occurs at inhomogeneities in the material, and absorption is due to viscoelastic effects.

Display Techniques

The A-mode, or amplitude mode, provides a display of echo amplitude versus time. A B-mode display provides a two-dimensional cross section of the material perpendicular to the inspected surface. A C-mode display gives a two-dimensional section of the part parallel to the inspected surface. In M-mode presentations, echo positions as a function of time are recorded.

Nondestructive Testing

Ultrasound is used to detect flaws in materials and to measure thickness (see Nondestructive testing). Flaws can be detected by reflected signals (pulse-echo) or by a decrease in the transmitted signal (through-transmission). Ultrasonic thickness measurement can be used when only one side of the part is accessible.

In contact scanning, the transducer is placed directly on the part with the sound coupled through a thin film of liquid between the two. In immersion testing, the part to be tested and the transducer are immersed in a tank of water and the transducer is scanned over the part. The angle at which the ultrasound waves enter the part can be varied and the path of the sound beam in the part can be determined using Snell's law to find the location of the echo reflectors.

Materials can be characterized with ultrasound. Changes in sound propagation parameters can be used to measure changes in other properties. Acoustic emission is another method of nondestructive testing. Propagating cracks emit noise which can be detected by ultrasonic transducers. Triangulation can be used to locate the source of the noise. This technique is particularly useful in proof-testing structures.

Process monitoring. Ultrasonic signals can be used as the sensing mechanism in a number of process-control and monitoring techniques. Pulse-echo equipment senses liquid levels in tanks either by measuring the length of the liquid column or the distance from the tank top to the top of the surface of the liquid. Doppler-shift measurement or differential time-of-flight can be used to measure flow rate in liquids being pumped through pipes. In electronic technology, low power ultrasound is

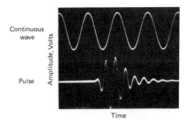

Figure 1. Ultrasonic waveforms.

used for signal processing either in resonators or in traveling-wave devices.

Medical Applications

Low power pulse-echo ultrasound has a role in imaging in diagnostic medicine. Two-dimensional section imaging applying B-scan techniques is used to image anatomy, and has also become a valuable aid in cardiology. The nonionizing nature of the energy and the lack of biohazards offer great advantages in obstetrics and gynecology and pregnancy monitoring. B-scanners, with their mechanically articulated arm, give a static or snapshot view of the anatomy. M-mode instrumentation, less expensive than two-dimensional imaging equipment, has become a ubiquitous tool for the cardiologist. Doppler measurements may be used to gather information about blood flow and areas of vascular stenosis or restriction. Doppler techniques are applied to the screening of stenotic carotid arteries and fetal monitoring. Current medical ultrasonic imaging makes primary use of the amplitude of the returning echo.

Health and Safety Factors

Sound intensities used in diagnostic and in industrial ultrasonic testing are typically in the range of 20–100 mW/cm^2 and are well below the levels needed to cause cavitation. Adverse effects have not been demonstrated.

CRAIG S. MILLER
RON E. MCKEIGHEN
Krautkramer-Branson, Inc.

JOSEPH ROSE
Drexel University

J. Krautkramer and H. Krautkramer, *Ultrasonic Testing of Materials*, 2nd English ed., Springer-Verlag, New York, 1977.

J.L. Rose and B.B. Goldberg, *Basic Physics in Diagnostic Ultrasound*, John Wiley & Sons, Inc., New York, 1979.

W.P. Mason and R.N. Thurston, eds., *Physical Acoustics: Principles and Methods*, Academic Press, New York, 1964 to present.

ULTRAVIOLET ABSORBERS.   See Uv stabilizers.

UMBER.   See Pigments, inorganic.

UNADS.   See Rubber chemicals.

UNITS AND CONVERSION FACTORS

In 1790, the French National Assembly requested the French Academy of Sciences to work out a system of units suitable for adoption by the whole world. This system was based on the meter as a unit of length and the gram as a unit of mass. Industry, commerce, and especially the scientific community benefited greatly. In 1893, the United States adopted the meter and the kilogram as the fundamental standards of length and mass, respectively. In 1954, the 10th General Conference on Weights and Measures (CGPM) added the degree Kelvin as the unit of temperature and the candela as the unit of luminous intensity. In 1960, this new system with six base units was formalized with the title International System of Units (SI). Each quantity is expressed in one and only one unit. Derived units are defined by simple equations relating two or more base units.

The International System of Units

SI rests on seven base units, two supplementary units, and a number of derived units. A list of these units is given in the front matter.

Table 1.   Units in Use with SI

Unit	Symbol	Value in SI unit
minute	min	1 min = 60 s
hour	h	1 h = 60 min = 3600 s
day	d	1 d = 24 h = 86 400 s
degree	°	1° = (π/180) rad
minute	′	1′ = (1/60)°
		= (π/10 800) rad
second	″	1″ = (1/60)′
		= (π/648 000) rad
hectare	ha	1 ha = 1 hm^2 = 10^4 m^3
liter	L	1 L = 1 dm^3 = 10^{-3} m^3
metric ton	t	1 t = 10^3 kg

The base units are the meter, kilogram, second, ampere, kelvin, mole, and candela; the supplementary units are the radian and steradian. The largest class, the derived units, consists of a combination of base, supplementary, and other derived units according to the algebraic relations linking the corresponding quantities. In addition, 16 prefixes are directly attached to form decimal multiples and submultiples of the units. The prefixes indicate the order of magnitude, thus eliminating nonsignificant digits and providing an alternative to powers of ten; eg, 45,300 kPa is 45.3 MPa, and 0.0043 m is 4.3 mm.

A number of non-SI units also used in SI are given in Table 1. Other non-SI units include the kilowatt-hour (eventually to be replaced by the megajoule), barn, bar (although the pascal and its multiples are preferred), curie, roentgen, and rad. These will be used until the International Committee on Weights and Measures (CIPM) considers their use no longer necessary. Except for the non-SI units referred to here, other metric units should be avoided.

The basic unit for force is the newton. It is best to avoid the term weight. It should be determined whether mass (kg) or force (N) is intended, and then the correct unit used. The degree centigrade was dropped in 1948 in favor of the degree Celsius (°C), which is used for both temperature and temperature interval in addition to the kelvin. The derived unit for energy (or work) is the joule (1 J = 1 N·m).

Certain quantities, eg, refractive index and relative density (formerly specific gravity) are expressed by numbers only.

Style and usage. If the advantages of SI are to be realized, the following rules must be observed: unit symbols are always in roman type; a space is required between number and unit; a period is not used after a symbol; the plural form of a symbol is the same as the singular; the prefixes for 10^6 and above are capitalized, as are symbols derived from proper names; the product of two or more symbols is indicated by a center dot; a solidus indicates the quotient of two unit symbols and "per" the division of two unit names; a prefix is not used in the denominator of a compound unit (except for the base unit kg); an exponent attached to a symbol containing a prefix indicates that the multiple of the unit is raised to the power expressed by the exponent; compound prefixes are not used; a comma should not be used to separate groups of digits; the prefix "giga" is preferred over the term "billion"; and when using powers with a unit name, the modifier "squared" or "cubed" is used after the unit name except for areas and volumes.

Conversion of quantities should be handled with careful regard to the implied correspondence between accuracy and the number of digits. Excellent tables of conversion factors are given by ASTM Standard E 380-82 for Metric Practice.

ROBERT P. LUKENS
American Society for Testing and Materials

Standard for Metric Practice E 380-82, ASTM, Philadelphia, Pa., 1982.

The International System of Units (SI), NBS Special Publication 330, Superintendent of Documents, U.S. Government Printing Office, Washington, D.C., 1981.

Metric Editorial Guide, 3rd ed., American National Metric Council, Bethesda, Md., 1981.

UNSATURATED POLYESTERS. See Polyesters, unsaturated.

URANIUM AND URANIUM COMPOUNDS

Uranium, at no. 92, at wt 238.03, is a member of the actinide series of transition elements (see Actinides and transactinides). The importance of uranium as feed material for the generation of nuclear energy caused a search for new supplies during the late 1940s and 1950s that was unparalleled in mining and metallurgical history (see also Nuclear reactors). The demand for uranium as a primary source of power will doubtless continue to increase (see Table 1).

Table 1. Annual Natural-Uranium Requirements for Fuel-Cycle Strategies for Noncommunist World, Thousand Metric Tons

Reactor strategy[a]	Nuclear power growth	Year		
		1980	2000	2025
once-through LWR[b]	high	32	175–200	430–590
	low	28	120–135	190–260
once-through HWR[c]	high	32	170	360–480
	low	28	115	160–220
large-scale FBR[d]	high	32	145	50–240
	low	28	100	50–70
LWR with recycle	high	32	125–140	320–420
	low	28	85–95	140–180
HWR with recycle	high	32	160–175	290
	low	28	110–115	130

[a] Tails assay: 0.2% ^{235}U.
[b] Light-water reactor.
[c] Heavy-water reactor.
[d] Fast-breeder reactor.

Occurrence

Uranium is present in the earth's crust at ca 2 ppm. Acidic rocks with a high silicate content, such as granite, have a uranium content above average, whereas the contents of sedimentary and basic rocks, such as basalts, are below average. However, 90% of the world's known uranium sources are contained in conglomerates and sandstone. Primary uranium minerals, crystallized from low-melting rocks, are generally black and contain uranium in a valence lower than six. Secondary minerals are yellow, greenish-yellow, bright green, or orange, and contain uranium in the valence +6.

Uraninite and pitchblende differ in physical form; their composition ranges from $UO_{2.0}$ to $UO_{2.67}$. Uraninite is well crystallized; pitchblende is either amorphous or consists of very fine crystals. Gummite, $UO_3.nH_2O$, and becquerelite, $2UO_3.3H_2O$, are typical hydrated oxides. They are fairly common, but commercially unimportant. Carnotite, $K_2O.2UO_3.V_2O_5.3H_2O$, and tyuyamunite, $CaO.2UO_3.V_2O_5.8H_2O$, are the commercially important minerals.

Phosphates are represented by autunite, $Ca(UO_2PO_4)_2.8H_2O$, and torbernite, $Cu(UO_2PO_4)_2.12H_2O$. Uranium arsenate minerals are found in European deposits. Numerous organic complexes are found in sedimentary deposits, eg, thucholite and carburan.

Resources. Reasonable assumed resources (RAR) contain deposits of such size, grade, and configuration that recovery is within the given production cost ranges with the current mining and processing technology. This type of resource is estimated at 1.8×10^6 metric tons. Estimated additional resources (EAR) occur as extensions of well-known deposits, little-explored deposits, or undiscovered deposits; they are estimated at 2.6×10^6 t. High grade resources include sandstone-type deposits found in the United States, Gabon, and Niger. Precambrium quartz-pebble conglomerates host the large uranium deposits of the Elliott Lake area in Canada and the gold uranium deposits in South Africa. In the Lake Athabasca region of Canada and the Alligator River region in Australia, high grade deposits of up to 1.5×10^5 t of uranium have been found (see also Tar sands).

Table 2. Thermodynamic Properties of Uranium Metal

Function or parameter	Value or equation
entropy of α-U, J/K[a]	50.21 ± 0.12
entropy of U(g), J/K[a]	199.6
enthalpy of fusion, kJ/mol[a]	8.326 ± 0.54
enthalpy of sublimation, kJ/mol[a]	1062.73
free energy of vaporization, G, kJ/mol[a]	
solid to gas	525.3–0.137
liquid to gas	487.6 ± 0.11
normal (extrapolated) boiling point, K	3818
vapor pressure of liquid[b]	$\log p \text{ (kPa)} =$ $-\dfrac{(25{,}230 \pm 370)}{T}$ $+ (7.72 \pm 0.17)$

[a] To convert J to cal, divide by 4.184.
[b] To convert kPa to mm Hg, multiply by 7.5.

Properties

Uranium is a dense, lustrous metal that resembles iron; it is ductile, malleable, and weakly magnetic. In air, it tarnishes rapidly, and in a very short time, even a polished surface becomes coated with a dark-colored layer of oxide. In the solid state, uranium exists in three allotropic modifications. The thermodynamic properties have been determined with great accuracy (see Table 2). Thermal conductivity ranges from 0.251 W/(cm·K) at 309 K to 0.326 W/(cm·K) at 673 K. Spectroscopic properties have been studied in great detail and more than 30,000 lines of the arc-and-spark-emission spectra have been catalogued.

Of the four oxidation states, +3, +4, and +6 are stable enough to be of practical importance; +5 is of minor importance. The alternation between +4 and +6 states is utilized in the extraction of uranium from ores and in purification.

Isotopes

There are fifteen known isotopes of uranium, not counting the isomeric states; ^{234}U, ^{235}U, and ^{238}U exist in nature (see Radioisotopes). The natural isotope composition usually varies. In addition to ^{235}U, an artificial isotope, ^{233}U, was found to be fissionable. Today, it is an industrial product produced in kilogram quantities by the thermal breeding reaction.

$$^{232}\text{Th} + n \xrightarrow{\gamma} {}^{233}\text{Th} \xrightarrow{\beta^-} {}^{233}\text{Pa} \xrightarrow{\beta^-} {}^{233}\text{U}$$

The product is isolated by a radiochemical process.

Extraction from Ore

Conventional ore-dressing techniques have not been successful in the preconcentration of uranium minerals. Gravity separation is sometimes possible (see Gravity concentration). Electrostatic methods generally give low recoveries in low concentrations. Flotation (qv) usually gives satisfactory concentrations.

A high temperature roasting or calcining operation before leaching is frequently desirable, and the characteristics of many ores are thus improved.

Treatment with acids or alkalies converts the uranium contained in the ore to water-soluble species. Most mills use acid leaching. Alkaline or carbonate leaching is used for ores with high lime content, which would require excessive consumption of acid.

The crude uranium isolated in the leach liquors requires additional purification. Direct or selective precipitation has not been commercially successful in acid leach systems. However, precipitation may be used to enrich the uranium in side streams. Anionic sulfato or carbonato complexes may be absorbed from leach liquors on anion-exchange resins. The uranium is eluted with a saline or acid solution.

Some uranium ores exhibit extremely poor filtering and settling characteristics after leaching. To avoid large liquid–solid separation equipment, the ion-exchange process has been modified to extract uranium directly from the leach pulp (resin-in-pulp process).

Solvent extraction is widely used in the recovery of uranium from ores. Contrary to ion exchange, solvent extraction can be operated in a continuous countercurrent flow. The preferred extractants are di(2-ethylhexyl) phosphate (D2EHPA) and dodecyl phosphate (DDPA). The product of the extraction process is a purified uranium solution. Uranium is usually precipitated directly from solutions resulting from carbonate leaching. From acid solutions, uranium is precipitated by neutralization with ammonia or magnesia. With the former, a precipitate of composition $(NH_4)_2(UO_2)_2SO_4(OH)_4 \cdot nH_2O$ is obtained. So-called yellow cake with a higher uranium content is obtained with magnesia.

Conversion and Purification

The crude product from the refinery is purified for nuclear applications. Yellow cake is dissolved in nitric acid and extracted with tributyl phosphate. The uranium is back-extracted with recycled concentrated acid containing $< 1\%$ HNO_3 or with deionized water. Evaporation gives uranyl nitrate hexahydrate, which is pyrolyzed to UO_3. Reduction with hydrogen gives the dioxide, which is converted to UF_4 by treatment with HF. Direct fluorination of the tetrafluoride, also called green salt, gives the hexafluoride.

For reactor applications, enriched UF_6 from an isotope-separation plant is converted to other compounds or the metal.

Manufacture of the Metal

Uranium metal is produced by reduction of the tetrafluoride in a bomb at ca 700°C (Ames process).

$$UF_4(s) + \begin{cases} 2\ Ca \\ 2\ Mg \end{cases}$$
$$\rightarrow U + \begin{cases} 2\ CaF_2 & \Delta H° = -560.5\ kJ/mol\ (-134\ kcal/mol) \\ 2\ MgF_2 & \Delta H° = -349.4\ kJ/mol\ (-83.51\ kcal/mol) \end{cases}$$

Most nuclear reactors built for the generation of electric power are based on uranium fuel enriched in ^{235}U. Reactors using natural uranium, such as the Hanford production reactors or some power reactors in the U.K., do not require enriched uranium. Purified natural uranium fabricated in rods (slugs) may be used.

Only the gaseous-diffusion process is used for the separation of ^{235}U and ^{238}U on an industrial scale. Highly purified gaseous uranium hexafluoride is pumped through a series of diffusion aggregates arrayed in cells in a cascade pattern. Enormous amounts of electric power are required.

The high capital cost and power consumption of gaseous-diffusion plants have led to the investigation of centrifugal separation of ^{235}U and ^{238}U. A gas-centrifuge enrichment plant (GCEP) is under construction at the site of the Portsmouth, Ohio, gas-diffusion plant.

Electromagnetic separation, one of the most powerful methods, was developed at the University of California Radiation Laboratory and employed on an industrial scale at Oak Ridge using so-called calutron (California University Cyclotron) separators. The calutron technique has been used to separate pure samples of ^{234}U, ^{236}U, and stable isotopes of many other elements. The Oak Ridge calutron plant was shut down in 1980.

Other methods for isotope separation were developed, but none advanced beyond the pilot-plant stage. Laser excitation appears very promising. Uranium hexafluoride vapor is ionized by means of a tunable rhodamine-B laser in such a manner that ^{235}U atoms are selectively excited to form UF_5 molecules which are collected on a sonic impactor (MLIS-process).

Isolation of specific isotopes. The uranium isotopes ^{232}U, ^{233}U, and ^{234}U are the daughters of the actinide isotopes ^{232}Pa, ^{233}Pa, and ^{238}Pu, respectively, which may be obtained pure. For the production of ^{232}U, proactinium-231 is bombarded with neutrons. The ^{232}U can be separated by an ion-exchange process. Uranium-233 is obtained by bombardment of thorium-232 with slow neutrons. Uranium-234 is the daughter product of plutonium-238 which, in turn, is produced by neutron bombardment of neptunium-237.

Uranium Compounds

Uranium metal, heated to 150–200°C in a hydrogen atmosphere, gives uranium trihydride, UH_3, a black powder. Rare isotopes of hydrogen, ie, tritium, can be stored in the form of hydrides. Deuterium or tritium are absorbed on uranium turnings heated to 200°C and then may safely be stored as solids. Upon heating to 500°C in vacuum, the isotope is released as a highly pure gas.

Fluorides. Uranium forms seven binary fluorides: UF_3, UF_4, U_4F_{17}, U_2F_9, α-UF_5, β-UF_5, and UF_6. The trifluoride is prepared from stoichiometric amounts of UF_4 and uranium metal. The tetrafluoride is prepared by hydrofluorination of UO_2 with excess gaseous HF at ca 550°C. The uranium (IV, V) fluorides, U_4F_{17} and U_2F_9, are obtained from the tetrafluoride and the hexafluoride at 200°C and 2.36 kPa (17.7 mm Hg). These two compounds are observed in gas centrifuges as decomposition products. The grayish-white α-pentafluoride is obtained from the tetra- and hexafluorides by reduction with HBr at 80–100°C; at 150–200°C, the yellowish-white β-pentafluoride is obtained. The hexafluoride is prepared by direct fluorination of the tetrafluoride; its properties are given in Table 3. It is produced on a large scale as feed for the gaseous-diffusion, separation-nozzle, and gas-centrifuge processes.

A large number of ternary uranium fluorides are composed of a binary fluoride and an alkali or alkaline-earth fluoride (see Table 4).

Chlorides. The trichloride, an olive-green solid, is prepared from the trihydride and HCl at 250–300°C. The tetrachloride, a dark green solid, is best prepared from UO_3 and hexachloropropene.

$$UH_3 + 3\ HCl \rightarrow UCl_3 + 3\ H_2$$

$$UO_3 + 3\ CCl_3CCl{=}CCl_2 \rightarrow UCl_4 + 3\ CCl_2{=}CClCOCl + Cl_2$$

The pentachloride is prepared by treatment of the trioxide with carbon tetrachloride. The hexachloride, a dark green solid, is obtained by disproportionation of the pentachloride in vacuum.

Bromides and iodides. The tribromide, a reddish-brown crystalline material, is prepared from the trihydride and HBr or from the elements. The dark brown tetrabromide is prepared by heating uranium turnings in a stream of nitrogen with bromine vapor. The pentabromide is unstable. The tri- and tetraiodides are made from the elements; the product depends on the temperature and the pressure of the iodine.

Oxides. The uranium–oxygen system is extremely complicated. Alterations of the oxidation state are significant in refining ore concentrates. The dioxide, mp ca 2800°C, is very stable and occurs in nature as pitchblende. The tetrauranium enneaoxide, U_4O_9, forms black crystals and occurs in three modifications. The triuranium octoxide, U_3O_8, is a component of a complicated phase system.

The trioxide, UO_3, exists in six polymorphic modifications that differ in crystallographic properties and color. It is readily obtained from the thermal decomposition of various uranyl compounds. The monoxide, UO, forms very slowly above 2000°C in mixtures of UO_2 and uranium. The hydrated peroxide, $UO_4 \cdot xH_2O$, is utilized in purification technology. It may be precipitated from uranyl nitrate solution with H_2O_2. Several hydrated oxides have been investigated, eg, $UO_3 \cdot 2H_2O$, $UO_3 \cdot H_2O$, and $UO_3 \cdot 0.5H_2O$.

Table 3. Properties of Uranium Hexafluoride

Property	Value
triple point at 151 kPa[a], °C	64.052
sublimation point, °C	56.4
density	
solid, g/cm³	5.09
liquid, g/mL	6.63
heat of formation, solid at 25°C, kJ/mol[b]	−2158.9
heat of vaporization at 64.01°C, kJ/mol[b]	28.899
heat of fusion at 64.01°C, kJ/mol[b]	19.196
heat of sublimation at 64.01°C, kJ/mol[b]	48.095

[a] To convert kPa to mm Hg, multiply by 7.5.
[b] To convert J to cal, divide by 4.184.

Table 4. Uranium–Fluorine Complexes

Formula	Metal
M_4UF_8	NH_4
M_3UF_7	Li, Na, K, Rb, Cs
M_2UF_6	Na, K, Rb, Cs, NH_4
$M_7U_6F_{31}$	Na, K, Rb, NH_4
MUF_5	Li, Rb, Cs
MU_2F_9	Na, K
MU_3F_{13}	K, Rb
MU_4F_{17}	Li
MU_6F_{25}	K, Rb, Cs
$M_2U_3F_{14}$	Cs
MUF_6	Ca, Sr, Ba, Pb

Other compounds. Uranyl nitrate hexahydrate, $UO_2(NO_3)_2.6H_2O$, may be prepared by evaporating a neutral solution. It is soluble in organic solvents, a property that is utilized in most uranium refining processes.

When phosphoric acid or arsenic acid are added to uranyl salt solutions, $HUO_2PO_4.4H_2O$ or $HUO_2AsO_4.4H_2O$, respectively, precipitate. The hydrogen can be exchanged for alkaline or alkaline-earth bivalent metal ions. The resulting salts are identical with a number of natural minerals.

Uranyl sulfate monohydrate and trihydrate are stable. This property is utilized in the high temperature sulfatization of uranium ores.

Uranium forms three nitrides, UN, $UN_{1.5}$, and $UN_{1.75}$. Only the mononitride is stable $> 1300°C$.

The compounds UC, U_2C_3, and UC_2 have been identified. Because they melt $> 2350°C$, they are considered as ceramic reactor fuel materials. They are obtained from the elements by arc melting, sintering, and other techniques.

Uranyl carbonates occur in high-lime ores, which may be leached by the carbonate process. Tetrasodium uranyl tricarbonate, $Na_4UO_2(CO_3)_3$, is obtained when ores are pressure leached at elevated temperature with soda-ash solutions. Tetraammonium uranyl tricarbonate, $(NH_4)_4$-$UO_2(CO_3)_3$, forms a series of solid solutions with the hexavalent plutonium carbonate, $(NH_4)_4PuO_2(CO_3)_3$, which may be decomposed in hydrogen-donor mixtures to form $(UPu)O_2$. These solutions are, therefore, of great importance in the production of fuel element.

The alloys of uranium are utilized in reactor technology, and numerous phase diagrams have been established.

Health and Safety Factors

Uranium is not only toxic because of its radiation, but also chemically toxic to the same degree as, for example, arsenic. However, the toxicity of uranium compounds varies:

Highly toxic: $UO_2(NO_3)_2.6H_2O$, UO_2F_2, UCl_4, UCl_5
Moderately toxic: UO_3, $Na_2U_2O_7$, $(NH_4)_2UO_2O_7$
Nontoxic: UF_4, UO_2, UO_4, U_3O_8

Radiation toxicity. The toxicity of uranium caused by radiation depends upon the isotopes present. Isotopes emitting fairly strong γ-radiation should be handled in a hot cell; for others a glove box is sufficient. ^{235}U and ^{238}U can be handled on an open laboratory bench. The laboratory should be equipped as an α-laboratory.

Finely divided uranium metal, some alloys, and uranium hydride are pyrophoric.

Large quantities of ^{233}U or ^{235}U may be the source of an unexpected critical excursion, an extremely dangerous phenomenon, and all possible precautions should be taken to prevent it from occurring.

FRITZ WEIGEL
University of Munich

O. Hahn and F. Strassmann, *Naturwissenschaften*, **27**, 11 (1939).

Gmelin Handbuch der Anorganischen Chemie, Erganzungswerk zur 8 Auflage, System-Nr. 55, Verlag-Chemie, Weinheim, FRG, 1982, Vol. A5.

E. Cordfunke, *The Chemistry of Uranium*, National Nuclear Energy Series Div. VIII, Vol. 5, McGraw-Hill Book Co., New York, 1952.

J.J. Katz and E. Rabinowitch, *The Chemistry of Uranium*, National Nuclear Energy Series Div. VIII, Vol. 5, McGraw-Hill Book Co., New York, 1952.

UREA

Urea can be considered the amide of carbamic acid, NH_2COOH, or the diamide of carbonic acid, $CO(OH)_2$. Properties are shown in Table 1. At atmospheric pressure and at its melting point, urea decomposes to ammonia, biuret (**1**), cyanuric acid (qv) (**2**), ammelide (**3**), and triuret (**4**). Biuret is the principal and least desirable by-product present in commercial urea.

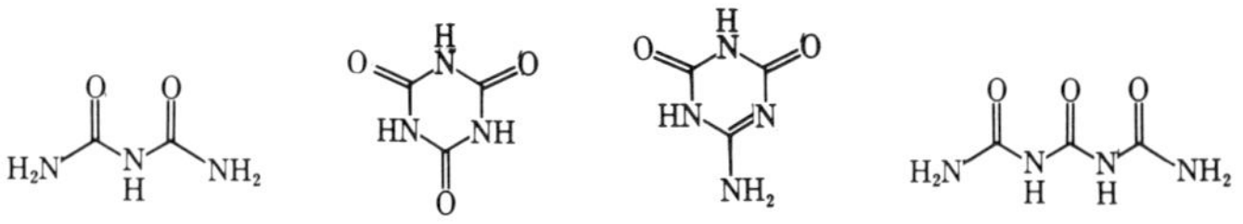

(1)	(2)	(3)	(4)
biuret	cyanuric acid	ammelide	triuret

Urea acts as a monobasic substance and form salts with acids. With alcohol urethanes (see Urethane polymers). Urea and malonic acid give barbituric acid, an important compound in medicinal chemistry (see also Hypnotics, sedatives, and anticonvulsants).

Manufacture

Urea is produced from liquid ammonia and gaseous CO_2 at high pressure and temperature. The formation of ammonium carbamate and the dehydration to urea take place simultaneously, for all practical purposes.

Ammonium carbamate is a white crystalline solid that forms by passing ammonia gas over dry ice. The conversion to urea begins below $100°C$. The maximum conversion is attainable at $185°C$, ca 53%. The ammonia and CO_2 gases recovered from the effluent mixture in several

Table 1. Properties of Urea

Property	Value
melting point, °C	135
index of refraction, n_D^{20}	1.484, 1.602
density, d_4^{20}, g/cm^3	1.3230
crystalline form and habit	tetragonal, needles or prisms
free energy of formation, at 25°C, J/mol^a	−197.150
heat of fusion, J/g^a	251^b
heat of solution in water, J/g^a	243^b
heat of crystallization, 70% aqueous urea solution, J/g^a	460^c
bulk density, g/cm^3	0.74
specific heat, $J/(kg \cdot K)^a$	
at 0°C	1.439
100	1.887

^{a} To convert J to cal, divide by 4.184.
^{b} Endothermic.
^{c} Exothermic.

pressure-staged decomposition sections are absorbed in water and recycled.

In the Mitsui-Toatsu total recycle process, the reactor is operated at ca 25 MPA (246 atm) and 195°C at an NH_3-to-CO_2 overall mol ratio (fresh feed plus recycle) of ca 4:1. A relatively high conversion of carbamate to urea per pass is reported (67–70%). Many urea plants with capacities of up to 1800 metric tons per day are using this process.

In the Montedison urea process, the reactor is operated at ca 20–22 MPa (197–217 atm) and an overall NH_3-to-CO_2 mol ratio of ca 3.5:1 (fresh feed plus recycle). A 62–63% conversion of carbamate is reported. In an improvement, based on the new isobaric double-recycle (IDR) technology, the reactor effluent is first stripped with NH_3 gas and then with CO_2 gas. Considerable reduction in steam consumption is reported.

The UTI heat-recycle process offers the following new concepts: An isothermal reactor is provided with an internal open-ended coil for countercurrent heat transfer from the strongly exothermic process of carbamate formation to the endothermic formation of urea; ca 40–50% of the makeup CO_2 feed is injected into the medium-pressure carbamate decomposition section, and more than 70% of the exothermic heat of carbamate formation in the medium-pressure absorption system is exchanged with relatively colder streams within the process.

High pressure gas stripping, developed in the mid-1960s, is based on high pressure CO_2 gas stripping at reactor pressure and relatively high temperature. The unconverted carbamate is decomposed to NH_3 and CO_2 by the stream of gaseous CO_2 passed through the reactor effluent solution at reactor pressure and condensed and recycled. Because of its energy efficiency, the stripping process accounts for almost half of the world's urea production.

In the Stamicarbon CO_2 stripping process, reactor, carbamate decomposer (stripper), and carbamate condenser each operate at ca 14 MPa (140 atm) at an $NH_3:CO_2$ mol ratio of ca 2.8:1.

In the Snamprogetti NH_3 stripping process, a synthesis loop is operated at ca 15 MPa (150 atm) and an $NH_3:CO_2$ overall mol ratio of 3.8:1. Carbamate conversion to urea per pass is reported to be ca 65–67%.

Wastewater Treatment

Under the pressure of government regulations with regard to residual NH_3 and urea in wastewaters, the fertilizer industry made an effort to improve wastewater treatment. It is, however, difficult to reduce the NH_3 and urea content to below 100 ppm. In a fairly efficient method, the urea is hydrolyzed to ammonium carbamate, and CO_2 and NH_3 are stripped off. The Stamicarbon system uses a hydrolyzer for urea and a separate dual-desorption system to strip the ammonia. The residual content of NH_3 and urea is reported to be reduced to about 70 and 80 ppm, respectively.

The UTI hydrolyzer stripper (see Fig. 1) consists of a single stainless-steel tower in which urea is hydrolyzed and NH_3 stripped simultaneously by means of steam and CO_2 stripping.

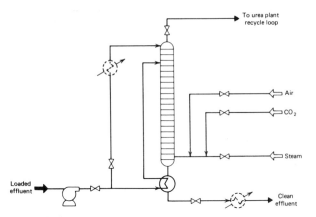

Figure 1. Hydrolyzer-stripper cleanup process. Courtesy of Urea Technologies, Inc.

Finishing Processes

Urea processes provide an aqueous solution containing 70–80% urea. This solution can be used directly for nitrogen-fertilizer suspensions or solutions. The water is evaporated from the steam-heated solutions under reduced pressure with or without the addition of hot air as drying agent.

A combination of crystallization and remelting removes biuret, which is detrimental in the technical-grade urea used for the manufacture of plastics. The solid urea thus obtained is relatively pure and well-suited for such applications.

Until recently, most solid urea was produced by prilling. Molten urea obtained either by evaporation or by crystal melting is sprayed as droplets from the top of a tall cylindrical tower (50–60 m high) and allowed to fall countercurrently through a stream of cold air emanating from a fluid-bed cooler at the base of the tower (see Fluidization). Prill towers require large quantities of air, and pollution abatement is very expensive. In addition, granulation offers far more flexibility, and therefore the U.S. industry is shifting to this technique.

In granulation, urea is concentrated to ca 99.5% and sprayed through nozzles onto a combination of falling granules and a cascading bed of recycled fines in a rotating cylindrical granulation drum. The physical characteristics of the granules are good, with usually < 0.15% H_2O; sizes range between 6.7–4 mm or 3.4–1.7 mm, and the granules are hard and spherical. Biuret content is 1–2%. In pan granulation, a concentrated urea solution is sprayed onto a cascading bed of recycled fines in an inclined rotating pan. The falling-curtain evaporative cooling process is based on TVA sulfur-coated urea technology in combination with some novel approaches to granulation. The process is extremely energy efficient. A full-scale plant of ca 608 t/d is expected to use ca 22.1 kW·h/t of product. Steam consumption, including concentration of scrubbing solution, is expected to be about 75 kg/t.

In the Nederlandse Stikstof Maatschappij N.V process, granulation is accomplished in a fluid bed divided into several chambers. Pneumatic atomizing nozzles are mounted just above the air-distribution plate and oriented to spray a 95–96% urea solution at 130–135°C upward into the active bed. Seed particles are introduced into the first chamber. The particles grow by accretion when the atomized droplets of urea solution strike them and the water evaporates as the concentrated urea crystallizes; biuret content increases only by 0.03%.

In the Mitsui-Toatsu and Toyo engineering process, concentrated urea solution is sprayed onto the surface of particles in a fluidizing granulation process; the heat released by solidification of the urea feed is removed by a combination of cooled recycle, fluidizing air, and some water evaporating. Hardness may be increased with additives. Granules with a 3-mm diameter are reported to have a crushing hardness of 20 N (2 kgf). A production capacity of 100 t/d per granulator is estimated to be the optimum size.

Uses

Solid urea containing 0.8–2.0% biuret is primarily used for direct application to the soil as nitrogen-release fertilizer. Urea is also used as feed supplement for ruminants, where it assists in the utilization of protein. Urea is a raw material for urea–formaldehyde resins (see Amino resins). Reagent-grade urea is needed in some pharmaceutical preparations.

Clathrates

Urea has the remarkable property of forming crystalline complexes or adducts with straight-chain organic compounds (see Clathration). These complexes consist of a hollow channel, formed by the crystallized urea molecules, in which the hydrocarbon is completely occluded. Urea clathrates are used in petroleum refining for the production of jet fuels and for the dewaxing of lubricating oils.

IVO MAVROVIC
Consultant

A. RAY SHIRLEY, JR.
Applied Chemical Technology

Fertilizer Manual, International Fertilizer Development, Muscle Shoals, Ala., Dec. 1979.

UREA – FORMALDEHYDE RESINS.   See Amino resins and plastics.

URETHANE POLYMERS

Polyurethanes contain carbamate groups, —NHCOO—, also referred to as urethane groups, in their backbone. They are obtained by the reaction of a diisocyanate with a macroglycol, a so-called polyol, or with a combination of a macroglycol and a short-chain glycol extender. The macroglycols are based on polyethers (qv), polyesters, or a combination of both. A linear polyurethane polymer has the structure (1), whereas a linear segmented copolymer obtained from a diisocyanate, a macroglycol, and ethylene glycol has structure (2).

$$\left[\text{ROCNHR'NHCO}\right]_n \qquad \left[\text{ROCNHR'NHCOCH}_2\text{CH}_2\text{OCNHR'NHCO}\right]_n$$

(1)   (2)

In addition to the linear thermoplastic polyurethanes obtained from difunctional monomers, branched or cross-linked thermoset polymers are made with higher functional monomers. Linear polymers have good impact strength, good physical properties, and excellent processability, but owing to their thermoplasticity, thermal stability is limited. Thermoset polymers, on the other hand, have higher thermal stability but lower impact strength. The higher functionality is obtained with higher functional isocyanates, so-called polymeric isocyanates, or with higher functional polyols (see Isocyanates; Glycols; Polyesters). Urethane network polymers are also formed by trimerization of part of the isocyanate group. This is used in the formation of rigid urethane-modified isocyanurate foams with structure (3).

(3)

Properties

The physical properties of polyurethanes are derived from their molecular structure and are determined by the choice of building blocks as well as the atomic interaction between the chains. Melt viscosity depends on the weight average molecular weight M_w and is influenced by chain length and branching. Thermoplastic polyurethanes are viscoelastic, ie, they behave like a glassy, brittle solid, an elastic rubber, or a viscous liquid, depending on the temperature and time scale of measurement. The melt temperature, T_m, is important for processability. For linear amorphous polyurethane elastomers, the T_g is ca -50 to $-60°C$. The pseudo-links generated by the interaction of the hard segments are reversed by heating or dissolution.

The physical properties of rigid urethane foams are usually a function of foam density. Strength is influenced by the catalyst, surfactant, polyol, isocyanate, and mixing. The thermal conductivity (K-factor) is greatly influenced by the blowing agent, cell size, cell content, and foam density. The density of insulation foams, usually blown with Fluorocarbon-11 (FC-11, CCl_3F), is 0.032–0.048 g/cm^3. Temperature variations affect the dimensional stability, whereas pressure causes distortion in closed-cell foams. Low density foams (d < 0.0032 g/cm^3) shrink as the polymer structure collapses, since internal pressure no longer maintains the cell structure (see Foamed plastics).

Manufacture

Isocyanates. The commodity toluene diisocyanates (TDI, 80 : 20 of 2,4 and 2,6 isomer, respectively) and PMDI (a polymer isocyanate obtained by the phosgenation of aniline–formaldehyde-derived polyamines) are the most widely used isocyanates in the manufacture of urethane polymers; a coproduct with the latter is 4,4′-methylenebis-(phenyl isocyanate), MDI.

The manufacture of TDI involves the nitration of toluene, catalytic hydrogenation to the diamines, and phosgenation. The basic raw material for PMDI and MDI is benzene. Nitration and reduction give aniline; reaction with formaldehyde in the presence of HCl gives a mixture of oligomeric amines, which are phosgenated to yield PMDI. The coproduct MDI is obtained by vacuum distillation.

More expensive specialty products include 1,5-naphthalene diisocyanate (NDI) and bitolylene diisocyanate (TODI); they are used in high quality cast elastomers.

Urethane polymers derived from aromatic diisocyanates are slowly oxidized by air and light; this causes a discoloration which is unacceptable in some applications. Polyurethane products derived from aliphatic isocyanates are color stable.

Difficulties encountered in handling and shipping volatile aliphatic diisocyanates have prompted the development of so-called blocked isocyanates, in which all isocyanate groups are blocked with a suitable blocking agent containing an active hydrogen, eg, caprolactam, phenol, and acetone oxime.

Polyether polyols. Polyether polyols are derived from cyclic ethers. Polymerization is usually initiated by KOH. In polyether polyols used for flexible foam, the viscosity increases with chain length of the polyether branches. In polyether polyols used in rigid foams, the type and functionality of the initiator determine the viscosity. For highly resilient flexible-foam and thermoset RIM (reaction-injection molding) elastomers, so-called graft or polymer polyols are used.

Polyester polyols. Initially, polyester polyols were the preferred raw material for polyurethanes. Today, the less expensive polyether polyols dominate the market. Recently, inexpensive aromatic polyester polyols have been introduced in rigid-foam applications. They are obtained from residues of terephthalic acid production or by transesterification of dimethyl terephthalate (DMT) or poly(ethylene terephthalate) (PET) scrap with glycols.

Polyurethane formation. The key to the manufacture of polyurethanes is the reactivity of diisocyanates toward nucleophilic additions. The polyaddition reaction is influenced by the structure and functionality of the monomers. Processing characteristics and properties are affected by the catalyst employed. Generally, an increase in base strength in tertiary amines increases the catalytic strength. Metal compounds catalyze the reaction of isocyanates with macroglycols; for example, di-n-butyltin diacetate is 2400 times more reactive than triethylamine. The catalysis of the reaction of isocyanates with water is especially important in the formation of water-blown cellular products. Highly resilient or cold-cure flexible foams are usually amine catalyzed. Most traditional catalysts promote the release of carbon dioxide at the same time as added fluorocarbon blowing agents vaporize. Efficient trimerization catalysts have been developed for urethane-modified isocyanurate foams.

Flexible foam. Flexible slab or bun foam is typically poured by multicomponent machines on continuous bun lines at rates > 45 kg/min (see Fig. 1). A typical formulation for furniture-grade foam with a density of 0.024 g/cm^3 includes a polyether triol (100 parts), TDI (50 parts), water (4 parts), catalyst (1 part), surfactant (1.5 parts), and FC-11 (3 parts). A high rate of block-foam production (150–220 kg/min) is required in order to obtain large slabs to minimize cutting waste.

Most flexible foams are based on polyether. These foams have excellent cushioning properties, are flexible over a wide range of temperatures, and resist fatigue, aging, chemicals, and mold growth. Polyester-based foams are superior in resistance to dry cleaning and can be flame bonded to textiles.

Rigid foams. Rigid polyurethane foam is mainly used for insulation. It is produced in slab or bun form on continuous lines similar to flexible

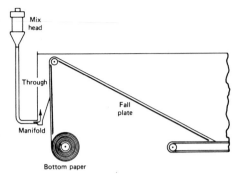

Figure 1. Flat-top bun line for flexible-foam production.

foam. The raw material is usually PMDI; flame retardants are sometimes added. The blowing agent is FC-11, (bp 23.8°C). Tertiary amines and certain organotin compounds, eg, dibutyltin dilaurate, are employed as catalysts.

Premixed machine-ready components for spray-foam applications for roof and tank insulation are sold by suppliers of polyurethane-foam systems. The polyols are preblended with surfactant and catalyst and sometimes fire retardants. The other component is the isocyanate. The catalyst can be added to either component. Recently developed urethane-modified rigid polyisocyanurate foams exhibit superior thermal stability and combustibility characteristics.

Polyurethane elastomers. Millable polyurethane elastomers are produced by chain extension of linear or slightly branched polyester or polyether polyols with aromatic diisocyanates, eg, TDI and MDI, ca 50,000 mol wt. Cast elastomers are usually slightly cross-linked thermoset polymers based on TDI and MDI. They are prepared by the one-shot method from MDI prepolymers or partial prepolymers. Recent developments include graft polymers and the use of aromatic amine glycols and amine extenders to give higher melting hard segments. Thermoplastic polyurethane elastomers are segmented linear polymers based on MDI, polyester, or polyether polyols, and glycol extenders. The desired properties and processing characteristics of the polymers are obtained by adjusting the amounts of the components. In the one-shot method, isocyanates and polyols are mixed before reacting. This method is used in the high pressure RIM (reaction-injection molding) casting of automotive polyurethane isomer parts.

Segmented elastomeric polyurethane fibers (spandex fibers) based on MDI were developed in the United States in the 1960s. Their elastic behavior is the result of the alternate arrangements of soft segments consisting of a macroglycol, such as a polyester polyol (polyadipate), a polyether polyol (polytetramethylene glycol), or a polycaprolactone, and hard segments or blocks containing an aromatic semicarbazide, urea, or urethane groups.

Recycling. Flexible polyurethane foam generated in production or recovered after use as packaging material can be recycled by shredding and mixing with a polyurethane binder to produce carpet underlay. Isocyanate-derived foams can be recycled after hydrolysis with steam, pyrolysis, or glycolysis, ie, transesterification of the carbamate groups with the glycol solvent (see Recycling).

Health and Safety Factors; Environmental Aspects

Fully cured polyurethanes present no health hazard; they are chemically inert and insoluble in water and most organic solvents. However, dust inhalation should be avoided. Because of their inertness, polyurethanes are the polymers of choice in biomedical applications. Some of the chemicals used in their production must be handled with caution. Isocyanates, for example, which are relatively nontoxic, must not be exposed to atmospheric moisture; formulation of insoluble ureas is the sign of improper handling. Thermal degradation of isocyanates occurs above 100–120°C.

The OSHA exposure limits for TDI and MDI have been set at 20-ppb (8-h TWA).

Polyurethane or polyisocyanurate dust may present an explosion risk. Polyurethanes are combustible and at no time should exposed foam be used in building construction. An approved fire-resistant thermal barrier must be applied over foam insulation on interior walls and ceilings.

Uses

The largest market for flexible foam is in the furniture and bedding industry. Most cushioning is made of polyurethane foam. Flexible and semi-rigid polyurethane foam products are also used in engineering packaging, gaskets, and linings, eg, weather stripping.

The bulk of the rigid polyurethane and polyisocyanurate foam production is used for insulation (qv) in the form of board or laminate or as liquid systems for *in situ* applications (pour-in-place or spray foam). Ships transporting LNG are usually insulated with rigid polyisocyanurate foam laminates providing temperature stability from −180 to 150°C.

Polyurethane elastomers are used where toughness, flexibility, strength, abrasion resistance, and shock-absorbing qualities are required. RIM elastomers are used in auto fascia, bumper, and fender extension as well as industrial and agricultural parts, such as oil-well plugs and grain buckets. Millable gums are used in gaskets, seals, and conveyor belts, elastomeric spandex fibers in hosiery, girdles, swim wear, and stretch pants (see Fibers, elastomeric).

Polyurethane surface coatings impart abrasion resistance, skin flexibility, fast curing, good adhesion, and chemical resistance. Polyurethane-modified alkyds are made from isocyanates and partially solvolyzed oils. They are used as coatings and foundry-core binders. Moisture-cured polyurethane coatings are applied as architectural finishes. Polyurethanes are also used in the formulation of water-based coatings. Linear polyurethane elastomers dissolved in organic solvents are used in surface coatings on fabrics and leather. Blocked isocyanates are used for baking enamels and wire and powder coatings.

A special application of polyurethane coating is in synthetic leather products. These materials are produced from textile length fiber mats impregnated and finished with polymeric compositions. Permeability to moisture vapor is the key property needed in synthetic leather, which is produced in Japan and Europe (see Leatherlike materials).

HENRI ULRICH
The Upjohn Company

H. Ulrich, *Modern Plastics Encyclopedia*, McGraw-Hill Publications Company, New York, 1983–1984, p. 76.

H. Saunders and K.C. Frisch in *High Polymers*, Vol. XVI, Pts. I and II, Wiley-Interscience, New York, 1962 and 1964.

URIC ACID

Uric acid, 7,9-dihydro-1H-purine-2,6,8(3H)-trione; 8-hydroxyxanthine was the first purine isolated from natural sources. It is widely distributed in nature and found in seeds and other plant parts. It forms the principal product of nitrogenous metabolism in birds and reptiles, and occurs in the blood and urine of all carnivores. Uric acid may be considered a minor pollutant in domestic sewage (0.2–1.0 mg/L); it is biodegradable by oxidation and hydrolysis.

Most commercial uric acid is obtained from natural sources by extraction from reptilian or bird excreta (guano) with sodium hydroxide, precipitation with ammonium chloride, and treatment of the precipitate with mineral acid.

Properties

Uric acid exists as white, colorless, tasteless crystals in the tautomeric forms (**1a**) and (**1b**).

(1a)   (1b)

It forms primary or monobasic salts and secondary or dibasic salts. It is stable in both acid and base under moderate conditions but decomposes to ammonia, carbon dioxide, and glycine by vigorous treatment with hydrochloric acid. Uric acid is easily oxidized with potassium permanganate to yield allantoin (2), which upon treatment with HI gives hydantoin (3).

(1b) uric acid   (2) allantoin   (3) hydantoin

Synthesis

Reaction of urea with acetoacetic ester gives 6-methyluracil, which upon treatment with nitric acid and more urea gives uric acid. Urea and malonic acid can also be used as starting materials. The condensation of urea and ethyl 2-chloro-3-diethoxyacrylate gives an imidazolone which undergoes further ring closure by treatment with base.

Analytical Methods

In the murexide test, the sample is treated with concentrated nitric acid and evaporated to dryness. Development of a red color upon the addition of ammonium hydroxide indicates the presence of uric acid or any other purine. For detection in biological fluid, the fluid is filtered through a cellulose membrane, treated with tungstic and mineral acids to remove proteins, and chromatographed. The fraction containing uric acid is detected at 220–315 nm. Reduction of ferric o-phenanthroline by uric acid may be automated and does not require protein removal.

Uses

Uric acid is used in cosmetic preparations because of its stability, inertness in solution, and ability to absorb uv radiation. It is a corrosion inhibitor and has been used as a leveling agent in textile dying.

Derivatives

Allantoin (2) 5-ureidohydantoin is a product of purine metabolism. Alloxan, 2,4,5,6-(1H,3H)-pyrimidinetetrone, is prepared by oxidation of uric acid. Alloxantin, 5,5′-dihydroxy-5,5′-bibarbituric acid is prepared by oxidation of alloxan monohydrate. Parabanic acid, imidazolidinetrione, is obtained by treatment of uric acid with 30% hydrogen peroxide.

L.G. SYLVESTER
CIBA-GEIGY Corporation

Thorpes Dictionary of Applied Chemistry, 4th ed., Vol. 9, Longman, Green & Co., London, UK, 1954, p. 802.
Rodd's Chemistry of Carbon Compounds, Vol. IV, Pt. 1, Elsevier Scientific Publishing Co., Amsterdam, Neth., 1980, p. 36.

UV STABILIZERS

Ultraviolet stabilizers are substances that protect light-sensitive materials from degradation by uv (see also Vitamins, vitamin E). Light-initiated oxidation is frequently responsible for most of the light-induced damage of polymeric materials. Photooxidation can be represented by a sequence of reactions comprising initiation, propagation, and chain branching. Initiation can be retarded by additives that function by photophysical mechanisms such as uv absorption and quenching of photoexcited chromophores. Oxidation can be retarded by antioxidants that scavenge free radicals or destroy hydroperoxides to yield nonradical species.

Many commercial stabilizers function by more than one mechanism. Thus 2-hydroxybenzophenones and 2-(2′-hydroxyphenyl)benzotriazoles contribute to stability by absorbing uv, trapping free radicals, and quenching photoexcited chromophores.

UV Absorbers

An ideal uv absorber should absorb the radiation between 290 and 400 nm while transmitting all visible light. Substituted 2-(2′-hydroxyphenyl)benzotriazoles approach this ideal closely. Compounds with little or no absorbance beyond 400 nm are preferred for most polymer applications. The absorber itself must be light stable. The mechanism by which 2-(2′-hydroxyphenyl)benzotriazoles dissipate the energy of absorbed radiant energy involves the tautomeric structures (1) and (2).

(1)   (2)

In the ground state, the phenolic structure (1) is preferred since the electron density on the oxygen atom is much greater than on the triazole nitrogen. In the photoexcited state, the electron density on the oxygen atom is greatly decreased, favoring structure (2).

For optimum activity, the absorber should be molecularly dispersed. Incomplete solubility results in a lower absorbance than the theoretical value calculated from the Beer-Lambert law.

Ultraviolet absorbers with extremely low volatility are required for polymers that are extruded at high temperatures, such as polycarbonate and poly(ethylene terephthalate). Low volatility is also required for applications such as automotive paints where the stabilizer must suffer only minimal losses during oven drying and outdoor exposure (see Paint). Ultraviolet absorbers should be chemically inert to other additives.

Ultraviolet absorbers provide a higher degree of stabilization that can be anticipated from light absorption alone. This unexpected activity can be explained either by the quenching of photoexcited chromophores or by inhibiting oxidative processes initiated by radiation.

Antioxidants. The development of hindered-amine light stabilizers (HALS) represents the most important advance in light-stabilizer technology since the introduction of uv absorbers. These compounds function primarily as light-stable antioxidants (qv).

Nickel chelates contribute to stabilization of polymeric substances by decomposing hydroperoxides, scavenging free radicals, absorbing uv radiation, and quenching photoexcited chromophores. However, the color of nickel chelates limits their utility for white or colorless polymers. Esters of 3,5-di-t-butyl-4-hydroxybenzoates function by scavenging free radicals.

Although thermal antioxidants based on hindered phenols and esters of trivalent phosphorus are not classified as light stabilizers, some markedly improve the light stability of photosensitive polymers.

Testing for Light Stability

Outdoor weathering is the most reliable method for determining light stability. Exposure sites are usually at southern locations. For meaningful comparisons, specimens are exposed at the same site at the same time. Years of exposure may be required before significant degradation takes place. Artificial exposure devices decrease testing time and permit stabilization of exposure conditions. Accelerated-weathering devices are used primarily as screening tools.

Health and Safety Factors

Safety is assessed by animal tests. A number of light stabilizers have been cleared by the FDA for use in polymers that come in contact with food (see Food additives).

Uses

Selection of uv stabilizers depends upon the composition of the substrate, thickness of sample, processing conditions, and expected service life. Ultraviolet absorbers, HALS, and phosphites are generally used at concentrations of ca 0.25–1.0%. In some applications, where clear film is desired to protect a light-sensitive substrate, concentration may be as high as 3%. The concentration of thermal antioxidants is generally 0.1–0.5%.

MARTIN DEXTER
CIBA-GEIGY Corporation

B. Ranby and J.F. Rabek, *Photodegradation, Photooxidation and Photostabilization of Polymers*, John Wiley & Sons, Inc., New York, 1975.

J.F. McKellar and N.S. Allen, *Photochemistry of Man Made Polymers*, Applied Science Publishers, Ltd., London, UK, 1979.

F. Gugumus in R. Gächter and H. Müller, eds., *Taschenbuch der Kunstuff-Additive*, 2nd ed., Carl Hanser Verlag, Munchen and Wien, 1983, pp. 101–192.

V

VACCINE TECHNOLOGY

A vaccine, live or killed, is a preparation that is used to prevent a specific disease by inducing immunity. Vaccines licensed in the United States are used either in the general population or in special populations. They are regulated by the Office of Biologics Research and Review of the FDA (see also Immunotheropeutic agents).

Vaccines for the General Population

In this category are vaccines recommended for the routine immunization of children or adults. These vaccines provide protection against poliomyelitis, diphtheria, tetanus, pertussis (whooping cough), measles (rubeola), mumps, and rubella (German measles).

Poliomyelitis. Currently, two vaccines are licensed for the control of poliomyelitis in the United States. The live, attenuated oral poliovirus vaccine (OPV) is recommended by the Advisory Committee on Immunization Practices (ACIP) for primary immunization of children (see Table 1). The killed or inactivated vaccine (IPV) is recommended for immunization of adults at increased risk of exposure.

Diphtheria, tetanus, and pertussis (DTP). Diphtheria and tetanus toxoids combined with pertussis vaccine are routinely used for active immunization of infants and young children (see Table 1). Diphtheria and tetanus vaccines contain purified antigens or toxoids, which are prepared by inactivation with formaldehyde. The pertussis component of DTP is killed whole cells of *Bordella pertussis* bacteria. Potency standardization of diphtheria and tetanus toxoids relies on *in vitro* antigenic tests. The U.S. Standard Pertussis Vaccine is used to determine the potency of pertussis vaccine.

Measles, mumps, and rubella. Live, attenuated vaccines are used for simultaneous or separate immunization. The combined vaccine is a mixture of the three live, attenuated viruses. The concentration of the virus constitutes the measure of potency. All titrations are run with a U.S. Reference Virus as a control.

Vaccines for Special Populations

Vaccines for special populations include influenza, pneumococcal polysacchlarides, hepatitis, rabies, yellow fever, adenovirus, smallpox, meningococcal, cholera, typhoid, and plague.

Table 1. Schedule for Immunization of Normal Infants and Children

Age	Immunization
2 mo	DTP[a], TOPV[b]
4 mo	DTP, TOPV
6 mo	DTP[c]
15 mo	measles, mumps, rubella[d]
18 mo	DTP, TOPV
4–6 yr	DTP, TOPV
14–16 yr	Td[e], repeat every 10 yr

[a] DTP, diphtheria and tetanus toxoids combined with pertussis vaccine.
[b] TOPV, trivalent oral poliovirus vaccine.
[c] TOPV is optional but may be given in areas where poliomyelitis is endemic.
[d] May be given as measles–rubella or measles–mumps–rubella combined vaccines.
[e] Td, combined tetanus and diphtheria toxoids (adult type), which contains a smaller amount of diphtheria antigen.

Influenza. The ACIP recommends annual influenza vaccination for all persons who are at risk from infections of the lower respiratory tract and for all older persons. Influenza viruses A and B are responsible for periodic outbreaks of febrile respiratory disease. Hemagglutinin and neuraminidase are the main immunogenic components of influenza virus. Standardization of potency relies on the quantitation of hemagglutinin using single radial immunodiffusion (SRID).

Pneumococcal polysaccharide. This vaccine consists of a mixture of purified capsular polysaccharides from 23 pneumococcal types that are responsible for at least 90% of serious pneumococcal disease in the world. Standards for these polysaccharides have been established by the Office of Biologics Research and Review.

Hepatitis. The vaccine is prepared by isolation of the viral antigen from infected patients, and standardized to obtain a specific amount of hepatitis B surface antigen. Administration of three doses of vaccine is recommended.

Vaccines in Development

Despite spectacular advances since the late 1960s, many diseases are endemic in many parts of the world.

Gonorrhea is the most commonly reported communicable disease in the United States. An increasing number of strains are becoming resistant to penicillin. Studies are being conducted on various structural components of the gonococcal bacterium as vaccine candidates.

The *Haemophilus influenza* serotype b is the most common cause of meningitis in infants in the United States. Because of the importance of the capsule polysaccharide of this organism in pathogenesis, vaccine studies have been mainly oriented toward raising an antibody to this bacterial component.

Hepatitis is caused by three distinct types of virus: type A (infectious hepatitis), type B (serum hepatitis), and type C (non-A, non-B hepatitis). Type B is of special interest since chronic carriage is possible. Generic recombinant vaccines are being pursued as second generation vaccines to replace the current product.

Immunological methods in relation to pregnancy have been examined. A vaccine has been developed that contains the *beta* subunit of human gonadatropic hormone (HCG) linked to tetanus toxoid. This preparation is immunogenic in humans and produces antibodies to HCG six to eight weeks after vaccination. These antibodies seem to disappear after six months.

Malaria infection occurs in over 30% of the world's population and almost exclusively in developing countries. Vaccine development has been pursued for many years. Vaccines used in monkeys and mice have been shown to induce immunity, but only to the sporozoite stage.

The development of a vaccine against Herpes simplex virus is hampered by the fact that recurrence of the disease is common.

Future Technology

Genetic engineering (recombinant DNA technology) involves preparation of DNA fragments coded for the substance of interest, inserting the DNA fragments into vectors, and introducing the recombinant vectors into living host cells. Genetic engineering (qv) offers new and, in some instances, safer and more effective methods for the development and production of vaccines against viral, bacterial, mycotic, and parasitic infection.

Monoclonal antibodies are produced by a culture derived from a single cell. These antibodies recognize a single determinant or structure of a given antigen. This technology provides a practically unlimited supply of uniform, highly specific antibodies. Immunoaffinity chromatography and recombinant-DNA technology will greatly enhance vaccine development.

Economic Aspects

Costs of vaccine manufacture vary. Live vaccines are generally less expensive. The number of strains of organisms or antigens also affect the price. Packaging is very expensive; the smaller the number of doses supplied in each container, the higher the cost. The cost-benefit relationship between preventive vaccination and disease treatment shows generally high cost savings.

In any immunization program, there are a small number of recipients that develop associated injury. The manufacturer can be held liable for failure to warn of possible adverse reactions. This liability has to be included in the economic aspects of vaccine technology.

V.A. JEGEDE
K.J. KOWAL
W. LIN
M.B. RITCHEY
Lederle Laboratories
American Cyanamid Company

W.K. Joklik, H.P. Willett, and D.B. Amos, *Zinsser Microbiology*, 17th ed., Appleton-Century-Crofts, New York, 1980.

B. Davis, R. Dulbecco, H. Eisen, and H. Ginsberg, *Microbiology*, 3rd ed., Harper and Row, New York, 1980.

V.A. Fulginiti, ed., *Immunization in Clinical Practice*, J.B. Lippincott Co., Philadelphia, Pa., 1982.

VACUUM TECHNOLOGY

Vacuum technology concerns the means to predict, effect, and control subatmospheric environments (vacuum). Vacuum environments must be safe and cost, energy, and materials effective. Vacuum systems can be grouped into crude, rough, controlled, highly controlled, and ultracontrolled categories. These categories, in some instances, correspond to the traditional categories of low, medium, high, ultrahigh, and beyond ultrahigh vacuum, ie, no single parameter is adequate to characterize a vacuum environment.

Within vessels, vacuum environments comprise gaseous molecular phases in contact but not necessarily in equilibrium with condensed molecular phases. Nonmolecular species, including radiant quanta, electrons, holes, and phonons may interact with the molecular environment. By species and concentration, molecules that are needed in one application may be anathema in another.

Vacuum Dynamics

In the gaseous as well as the condensed phase in each application, molecular concentration weighted by molecular species is of prime importance; thus, good and bad results may depend upon species at small concentration. By convention however, total base pressure in a Maxwellian gas is used as though it indicates the quality of the vacuum and as though Maxwellian gases were the rule rather than the exception.

Interaction between gaseous and condensed phases. In a closed vessel of volume V containing a nonionized, unexcited molecular gas with total number of molecules N, the change in equilibrium pressure P in the gas may not be predicted if the steady-state absolute temperature T of the walls and the gas is changed to another steady, constant level

$$PV = NkT$$

where k = the Boltzman constant.

In other words, the kinetic theory of gases is a valuable tool for vacuum technology, provided that the unmodified kinetic theory not be applied when the gas interacts significantly with itself or the molecular phases that bound it.

By convention, all vacuum environments are characterized in terms of one parameter, ie, pressure in the gaseous phase. However, when costs, energy, safety, hazardous waste, and other requirements are taken into account, each system must be characterized by a host of considerations that may include the molecular species under dynamic conditions, electrical breakdown, film contamination from the bulk phase, contamination drawbacks or benefits, and cold emission of electrons.

Action of vacuum on spacecraft materials. For service beyond the atmosphere, the vacuum environment allows materials to evaporate or decompose under the action of various forces encountered. The action of the space environment on materials and spacecraft can be predicted if simulation and measurement of a source–sink relationship in a vacuum environment is established under free molecular conditions. (It is not necessary to achieve the gas concentration of space.)

Electronic vacuum tube. In special electronic vacuum-diode tubes, high gas concentrations of some types of molecules are beneficial to the operation of the tube under proper control.

Pump-Down

Initially, the vessel is filled with ambient air. Any given macrosample of air may contain at least 1600 substances. The most important species is usually water. Gross cracks and voids are generally lined with microstructures that hold water. Such structures can be removed effectively by chemical and other methods.

Rough pumping. An oil-sealed mechanical pump in good condition, with vented or trapped exhaust, is gas purged of gas by running for several hours. A liquid-nitrogen (LN) trap that can be refilled automatically is positioned between pump and vessel. As the pressure falls, the composition of the gas in the vessel begins to change. At a pressure of ca 13 Pa (0.097 mm Hg), water is the predominant species in the gas phase.

Diffusion-pump (DP) system. After the roughing pump line has been shut off, a suitably large valve is opened slowly enough that the mass flow of gas from the chamber through the valve into the oil-diffusion pump system does not disrupt the action of the top jet of the diffusion pump.

Pumping-speed efficiency depends on trap, valve, and system design. The maximum contamination rate from the pump set for routine service can be $< 10^9$ molecules/$(m^2 \cdot s)$ for molecular weights > 44.

Leaks

A vacuum system can be stalled by gas leaks which may be real or virtual. A real leak refers to permeation processes or cracks or holes that allow external gas (air) to seep into the vacuum environment. Virtual leaks refer to gases that originate from within, eg, from trapped volumes, gauges, pumps, and bulk and surface species. Proper instruments readily distinguish real from virtual leaks.

Molecular Transport

Molecular transport concerns the mass motion of molecules in condensed and gaseous phases. The mass motions are driven primarily by temperature. As time progresses, the initial mass motion results in concentration gradients. In the condensed phase, flow along concentration gradients is described by Fick's law. No noble gas permeates a metal but hydrogen does. The least permeable material for hydrogen is carbon.

Gas transport. Initially, in a vessel containing air at atmospheric pressure, mass motion takes place when temperature differences exist and especially when a valve is opened to a gas pump.

The free molecular gas regime is illustrated in Figure 1. A duct of maximum transverse dimension D and length L connects two chambers, each of minimum interior dimension $\gg D$. Free molecular transport (Knudsen flow) is often sufficiently approximated when $\lambda > D$. In free molecular flow at steady state, the temperature of the gas entering a duct determines the rate of passage through the duct, not the temperature of the duct itself.

Work is performed by a pump but not by a conductant. The average velocity, v, of a free molecular gas, the conductants, C, probabilities of directional passage, $W_{1 \to 2}$ and $W_{2 \to 1}$, and areas, A_1 and A_2, are related by $C = \frac{1}{4} v A_1 W_{1 \to 2}$ and $A_1 W_{1 \to 2} = A_2 W_{2 \to 1}$. Under free molecular flow, the volumetric rates of transport in the gas phase are independent of the pump being on or off; bends in a duct hardly alter the probability of passage over a straight duct with the same axial length.

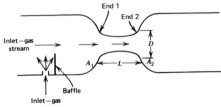

Figure 1. Duct connecting two volumes of dimension $\gg D$.

Wall geometries. Rougher-than-rough wall geometries can reduce transmission probabilities in Knudsen flow by as much as 25% compared to so-called rough walls. Therefore, conductance calculations that claim accuracy beyond a few percent may not be realistic.

The probability of a random gas entering a duct is not a random function but is proportional to the cosine of the angle between the molecular trajectory and the normal to the entrance plane of the duct. The latter assumption is consistent with the second law of thermodynamics, whereas assuming a random distribution entry is not.

The probability of passage is independent of the entrance velocity of free molecules and the subsequent velocity ($V = 0$) of these molecules within the tube. It depends upon the entering angular and wall-refraction distributions of the molecules. For engineering surfaces and gases at room temperature, reasonable results within $\pm 10\%$ are obtained by assuming that a statistical number of molecules impinging on a surface exhibits a cosine distribution upon reflection from the surface. In free molecular flow, all tubes of similar shape have the same probability of free molecular passage for the same entering gas distribution.

Combining conductances. Combining short conductances may be difficult because, if a free molecular gas that is Maxwellian in steady state enters conductance 1 (length = 0), the gaseous distribution is no longer Maxwellian at exit 1. This corresponds to the so-called beaming effect. The overall conductance can be estimated if the probabilities of passage of the individual components are known and if the juxtaposed components do not vary more than about a factor of two in cross-sectional areas.

Pumping speed. If the standard formulas for gas flow in vacuum are applied, it is assumed that a Maxwellian free molecular gas is entering the pump.

The operational speed of the pump is a systems effect. All current pumps perform more than one function. Pump are sinks and at the same time sources for molecules.

Instrumental Measurements

Gauges. At present, there is no way to evaluate a molecular vacuum environment except in terms of its use. Readings related to gas-phase concentration are provided by diaphragm, McCleod, thermocouple, Pirani gauges, and hot and cold cathode-ionization gauges (manometers).

Residual-gas analyzers. A gaseous molecular phase is analyzed with a mass spectrometer (see Analytical methods). If heat is delivered to the condensed phase or if electrons are caused to strike surfaces, molecular description alters the mass-spectrometer analysis.

Ultrasound frequencies can be introduced into the walls of the vacuum system. If a source of ultrasound is placed on the wall of an ultrahigh vacuum system, a large hydrogen peak is observed (see Ultrasonics).

Vacuum Systems and Equipment

Glass is often the material of choice for small laboratory systems and sealed systems in commercial practice. Glass exhibits high compressive strength but relatively low tensile strength, requiring careful selection of glass type and design. Evacuated glass tubes, such as photomultiplier tubes, are exposed to temperatures as high as 720 K. In some applications, high alumina ceramic is used (see Ceramics). Demountable joints using stainless steel and rubber gaskets are widely used.

In a properly designed and used vacuum brazing furnace, contamination can be negligible even with rapid cycling from one work load to the next.

Liquid-Nitrogen Traps

Cold traps are frequently inefficient in preventing the passage of oil or mercury because of warming of the trap and its internal filling lines when LN is added or as LN depletes. In some designs, however, this problem is eliminated. A well-designed LN trap provides a pumping speed of at least 100 m³/s per m² of system entrance area for water vapor, and oil contamination is reduced to negligible rates.

In general, uncracked oil from a DP is completely inhibited from creeping by keeping the surface temperature below 223 K. The effectiveness of an LN trap is confirmed by the absence of pressure pips on an ionization gauge when the LN is replenished in the reservoir.

Process Equipment

Crude vacuum ranges from 101 kPa to ca 100 Pa (760–0.75 mm Hg), and rough vacuum from 100 to ca 0.1 Pa (0.75–0.00075 mm Hg). The chemical engineer is likely to work in the crude-vacuum region in which distillation, evaporation, drying, and filtration are normally conducted. The rough-vacuum range is employed in molten-metal degassing, molecular distillation, and freeze-drying.

Steam ejectors. Ejectors are simple vacuum pumps. They have no moving parts; compression is accomplished through fluid-momentum transfer.

Liquid-ring pumps. In these pumps, the rotor is the only moving part. The liquid ring performs all the functions normally performed by mechanical pistons or vanes.

Rotary-piston pumps. These are positive-displacement, oil-sealed machines that isolate a specific volume of gas with each revolution. The piston revolves and traps the aspirated gas ahead of it by closing the inlet port. The gas is compressed, the discharge valve opens, and the gas is exhausted to the atmosphere. Compression ratios can be of higher than $10^6:1$ for a single-stage pump.

Rotary-vane pumps. These are positive-displacement machines with spring-loaded vanes that contact the inside of the pump casing. Gas entering the pump is trapped between adjacent blades and forced out through the discharge port.

Rotary-blower pumps. These pumps employ two interlocking rotors to trap and compress gases. The rotors are prevented from touching one another, and there is no sealing liquid in the pump. This type of pump is limited to small compression ratios, but can be designed for higher throughput than any other mechanical pump.

NORMAN MILLERON
Neven Corporation

N. Milleron in J.A. Dillon, Jr., and V.J. Harwood, eds., *Experimental Vacuum Science and Technology*, Marcel Dekker, Inc., New York, 1973.

J.F. O'Hanlon, *A User's Guide to Vacuum Technology*, Wiley-Interscience, New York, 1980.

G.L. Weissler and R.W. Carlson, eds., *Vacuum Physics and Technology*, Vol. 14 of Marton, ed., *Methods of Experimental Physics*, Academic Press Inc., New York, 1979.

J.L. Ryans and S. Croll, *Chem. Eng.*, 73 (Dec. 14, 1981).

VANADIUM AND VANADIUM ALLOYS

Vanadium, a member of Group VB of the periodic system and of the first transition series, is a gray bcc metal. It occurs in the uranium-bearing minerals of Colorado, in copper, lead and zinc vanadates of Africa, and with certain phosphatic shales and phosphate rocks in the western United States. It is a constituent of titaniferous magnetites with large deposits in the USSR, South Africa, Finland, the People's Republic of China, and Australia. There are more than 65 known vanadium-bearing minerals.

The United States dominated world production of vanadium until the late 1960s when several other countries, notably the USSR, expanded production significantly. At about the same time, the United States shifted from being a net exporter to a net importer.

Properties

In pure form, vanadium is a soft, ductile metal, which is hardened and embrittled by oxygen, nitrogen, carbon, and hydrogen. Addition of selected metals results in alloys of higher strength. Physical properties are given in Table 1.

Vanadium exists in the oxidation states $+2, +3, +4,$ and $+5$. When heated in air, it is oxidized to a brownish black trioxide, a blue-black tetroxide, or a reddish orange pentoxide, depending on the temperature. With chlorine at 180°C, it readily forms VCl_4, and with carbon and nitrogen at high temperatures, it forms VC and VN, respectively.

Table 1. Physical Properties of Vanadium Metal

Property	Value
melting point, °C	1890 ± 10
boiling point, °C	3380
density, g/cm^3	6.11
specific heat (at 20–100°C), J/g^a	0.50
latent heat of fusion, kJ/mol^a	16.02
latent heat of vaporization, kJ/mol^a	458.6
thermal conductivity (at 100°C), W/(cm·K)	0.31
electrical resistance (at 20°C), $\mu\Omega\cdot$cm	24.8–26.0

^{a}To convert J to cal, divide by 4.184.

Vanadium exhibits corrosion resistance in alkali solutions and in acids and liquid metals.

Manufacture

The vanadium-bearing ores are generally crushed, ground, screened, and mixed with sodium chloride or carbonate. The mixture is roasted at ca 850°C and the oxides are converted to the water-soluble sodium metavanadate, $NaVO_3$. The vanadium is extracted by leaching with water and precipitates at pH 2–3 as sodium hexavanadate, $Na_4V_6O_{17}$, a red cake, by addition of sulfuric acid. The hexavanadate is fused at 700°C to yield a dense, black product, which is sold as technical-grade vanadium pentoxide.

Ferrovanadium. Ferrovanadium, an important additive to steel, is produced by the reduction of vanadium ore, slag, or technical-grade oxide with carbon, ferrosilicon, or aluminum. Another steel additive is vanadium carbide, which is produced by the solid-state reduction of vanadium oxide with carbon in a vacuum furnace. A silicon reduction process has been developed by the Foote Mineral Company.

In the aluminothermic process, technical-grade vanadium oxide, aluminum, iron scrap, and a flux are charged into an electric furnace. The temperature of the highly exothermic reaction is controlled by the particle size and the feed rate.

In the thermite reaction, vanadium and iron oxides are reduced together by aluminum granules in a magnesite-lined steel vessel or in a water-cooled copper crucible.

Pure Vanadium

Vanadium metal is prepared either by the reduction of vanadium chloride with hydrogen or magnesium, or by the reduction of vanadium oxide with calcium, aluminum, or carbon. In the calcium reduction, the exothermic reaction is carried out adiabatically in a sealed vessel or bomb using $CaCl_2$ as a flux for the CaO slag. The vanadium metal is recovered in the form of droplets or beads. A massive ingot or regulus is obtained by using iodine as both a flux and a thermal booster.

In the aluminothermic process, vanadium pentoxide is heated with high purity aluminum in a bomb; a massive regulus of vanadium–aluminum alloy forms. Proprietary additives are used to increase the temperature, decrease the melting point of the slag or metal, or increase the fluidity of the two phases. A metallic regulus forms, which can be used as fuel-element cladding material following purification to 99% purity using electron-beam melting (see Nuclear reactors). In addition to electron-beam purification, vanadium can be refined by iodine refining (van Arkel-deBoer process), electrolytically in a fused salt, or by electrotransport.

Alloys

Consolidation by the consumable-electrode electric-arc melting technique is used extensively for the preparation of ingots. An electrode consisting of carefully weighed portions of each constituent is prepared, which is welded in vacuum or under an inert gas. Multiple-arc melting for a minimum of two melts ensures a homogenous ingot.

Forging or extrusion is used for primary or initial fabrication. To avoid oxidation, the machined ingot is clad and sealed in a steel container. Intermediate and final recrystallization is performed at 650–1000°C in vacuum or under an inert gas. Fabrication of most vanadium alloys is more difficult than the pure metal because of increased strength and decreased ductility, especially at low temperatures.

Vanadium alloys are used for fuel cladding and other structural components in liquid metal-cooled fast-breeder reactors. Selection is based on neutron considerations, corrosion resistance, ductility, and strength. Some alloys with room-temperature strength of 1.2 GPa (175,000 psi) exhibit strengths of up to 1 GPa (145,000 psi) at 600°C. Beyond this temperature, most alloys lose tensile strength rapidly. Weld ductility of the alloys is usually not as good as that of the pure metal.

Health and Safety Factors

Vanadium metal and its alloys pose no particular health or safety hazard. Dust or fine powder present a moderate fire hazard. Vanadium compounds may irritate the conjunctivae and respiratory tract.

Uses

Vanadium as an alloying element in steel increases grain refinement and hardenability. Vanadium alloys are used in dies or taps and for cutting tools. The titanium 6–4 alloy (6% Al, 4% V) is used in aircraft where strength-to-weight ratio is important. Because of its low capture cross section for fast neutrons as well as its resistance to corrosion by liquid sodium and its high temperature creep strength, vanadium is being developed as a fuel-cladding element for fast-breeder reactors.

A potential use is in the field of superconductivity. The compound V_3Ga exhibits a critical current of 20 T (20×10^4 G) at 20°C, which is one of the highest of any known material (see Superconducting materials).

EDMUND F. BAROCH
Consultant

Metals Handbook, 9th ed., Vol. 2, American Society for Metals, Metals Park, Ohio, 1979, p. 822.

R.W. Buchman, Jr., *International Metals Reviews*, 158 (1980).

G.A. Morgan, "Vanadium" in *Mineral Facts and Problems*, Bureau of Mines Bulletin 671, U.S. Bureau of Mines, Washington, D.C., 1980.

R. Rostoker, *The Metallurgy of Vanadium*, John Wiley & Sons, Inc., New York, 1965.

VANADIUM COMPOUNDS

Vanadium is widely dispersed in the earth's crust at an average concentration of ca 150 ppm. Deposits of ore-grade minable vanadium are rare. Vanadium is ordinarily recovered in the form of the pentoxide, but sometimes as sodium and ammonium vanadates. For metallurgical uses (ca 90% of consumption), the oxides are prepared for conversion to master alloys by fusion and flaking to form glassy chips (see Vanadium and vanadium alloys).

Properties

Properties of vanadium compounds are given in Table 1. Vanadium, a typical transition element, displays well-characterized valence states of 2–5 in solid compounds and solutions. All compounds of vanadium having unpaired electrons are colored, but a specific color does not necessarily correspond to a particular oxidation state.

The chemistry of vanadium compounds is related to the oxidation state. Thus, V_2O_5 is acidic and weakly basic, VO_2 is basic and weakly acidic, and V_2O_3 and VO are basic. Coordination compounds are mainly based on six coordination, in which vanadium has a pseudooctahedral structure.

Interstitial compounds. Vanadium is capable of taking atoms of nonmetals into its lattice. Carbides, hydrides, and nitrides so formed are called interstitial compounds. Nonmetals having large atoms, eg, Si, Ge, P, As, Se, and Te, form compounds with vanadium that are intermediate between interstitial and intermetallic.

Oxides. Vanadium pentoxide, V_2O_5, is intermediate in behavior and stability between the highest oxides of titanium, TiO_2, and chromium,

Table 1. Physical Properties of Some Industrial and Other Selected Vanadium Compounds

Compound	Formula	Appearance	Mol wt	Density, g/cm^3	Mp, °C,	Bp, °C
vanadic acid, meta	HVO$_3$	yellow scales	99.95			
ammonium metavanadate	NH$_4$VO$_3$	white–yellowish or colorless crystals	116.98	2.326	200 dec	
potassium metavanadate	KVO$_3$	colorless crystals	134.04			
sodium metavanadate	NaVO$_3$	colorless, monoclinic prisms	121.93		630	
vanadium carbide	VC	black cubic	62.95	5.77	2810	3900
vanadium(IV) tetrachloride	VCl$_4$	red-brown liquid	192.75	1.816	28 ± 2	148.5
vanadium(V) oxytrichloride	VOCl$_3$	yellow liquid	173.30	1.829	−77 ± 2	126.7
vanadium(III) oxide	V$_2$O$_3$	blue crystals	149.88	4.87	1970	
vanadium(V) oxide	V$_2$O$_5$	yellow-red rhombohedra	181.88	3.357	690	1750 dec

CrO$_3$. Vanadium(IV) oxide, VO$_2$, is a blue-black solid with a distorted rutile structure. The sesquioxide, V$_2$O$_3$, is a black solid with corundum structure. Vanadium(II) oxide is a nonstoichiometric material with a black-gray metallic luster and metallic-type electrical conductivity.

Vanadates. Ammonium metavanadate, NH$_4$VO$_3$, and to a lesser extent, potassium and sodium vanadates, are of commercial interest. The pure compounds are colorless crystals.

Halides and Oxyhalides. Only the tetrachloride, VCl$_4$, and vanadium(V) oxychloride are of commercial importance. Most of the possible halides and oxyhalides are known; they exhibit a wide range of colors.

Vanadium(V) oxytrichloride is prepared by chlorination of the pentoxide with charcoal at red heat. The tetrachloride is prepared by chlorination of crude metal at 300°C. The trichloride is prepared by heating VCl$_4$ in a stream of CO$_2$ or by treating vanadium metal with HCl.

Sulfates. Sulfate solutions derived from leaching vanadium ores with sulfuric acid are important in the recovery of vanadium from its sources. Vanadium in quadrivalent form may be solvent extracted as the oxycation (VO)$^{2+}$. Vanadyl sulfate pentahydrate, VOSO$_4$.5H$_2$O, is an ethereal blue solid. Vanadium(III) sulfate, V$_2$(SO$_4$)$_3$, is a powerful reducing agent.

Manufacture

Primary industrial compounds produced directly from vanadium raw materials are mainly 98% fused pentoxide, air-dried (technical-grade) pentoxide, and technical-grade ammonium vanadate. The ore is ground to pass through a 1.2-mm (14-mesh) screen, mixed with 8–10% NaCl and roasted at 800–850°C for 1–2 h. The hot calcine is quenched with water or cooled in air before being lightly ground and leached with water. The solution has a pH of 7–8 and a vanadium content of ca 30–50 g V$_2$O$_5$/L. The preferred routes for recovering the vanadium from the leach solution are solvent extraction and precipitation of ammonium vanadate or vanadic acid.

Direct acid leaching is used mainly for vanadium–uranium ores and less extensively for processing spent catalyst, fly ash, and boiler residues. A calcareous carnotite ore in Australia is ill-suited for salt roasting or acid leaching. Dissolution of uranium and vanadium by leaching in sodium carbonate solution is being tested on a pilot-plant scale. Commercial production is expected for the late 1980s.

The bulk of vanadium production is derived as a by-product or coproduct from the processing of iron, titanium, and uranium ores, and to a lesser extent, from phosphate, bauxite, and chromium ores and the ash or coke from burning or refining petroleum. South Africa, the world's largest producer, has three operating firms.

The energy required to produce fused pentoxide from a vanadium–uranium ore containing 1.3% V$_2$O$_5$ and 0.20% U$_3$O$_8$ is ca 360 MJ/kg (155,000 Btu/lb).

Analytical and Test Methods

A sensitive qualitative test is the formation of brownish-red pervanadic acid upon addition of hydrogen peroxide to a vanadate solution. X-ray absorption spectroscopy is convenient for identifying traces of vanadium in coal.

Health, Safety, and Environmental Aspects

Toxic effects have been observed from exposure to airborne concentrations of vanadium compounds of several milligrams or more per cubic meter of air. OSHA threshold limits in the workplace are 0.5 mg/m^3 for dust and 0.05 mg/m^3 for fumes. Oral toxicity in humans is minimal.

Uses

Conversion of fused pentoxide to alloy additives is by far the largest use of vanadium compounds. The dominant single use of vanadium chemicals is in catalysts. Minor uses include oxides in refractories and as coloring agents.

JOE B. ROSENBAUM
Consultant

A.F. Trotman-Dickenson, ed., *Comprehensive Inorganic Chemistry*, Vol. 4, Pergamon Press, Ltd., Oxford, UK, 1973.

A.E. Martel, ed., *Coordination Chemistry*, Vol. 1 ACS Monograph 168, American Chemical Society, Washington, D.C., 1978.

J.B. Rosenbaum, *Vanadium Ore Processing, Meeting of High Temperature Metal Committee*, preprint A71-52, AIME, New York, 1971.

Medical and Biological Effects of Environmental Pollutants—Vanadium, National Academy of Sciences, Washington, D.C., 1974.

"Vanadium" in *Minerals Yearbook*, U.S. Bureau of Mines, Battelle Columbus Laboratory, Columbus, Ohio, 1982.

VANILLIN

Vanillin, 3-methoxy-4-hydroxybenzaldehyde, occurs in nature as a glucoside, which hydrolyzes to vanillin and sugar. The best-known natural source is the vanillin plant. Vanilla beans are grown in Mexico, Madagascar, Java, Reunion, and Tahiti.

Properties

Vanillin crystallizes from water and organic solvents in monoclinic prismatic needles. Physical properties are given in Table 1.

Vanillin undergoes reactions of the aldehyde group, the phenolic hydroxyl, and the aromatic nucleus. The aldehyde group undergoes typical aldehyde condensation reactions that allow various substitutions; it can be partially or completely reduced. However, as a *p*-hydroxybenzaldehyde, vanillin does not undergo some very common aromatic aldehyde reactions, such as the Cannizzaro reaction, the benzoin condensation, and oxidation to vanillic acid. If the hydroxyl group is protected, oxidation to vanillic acid occurs readily. As a phenol, vanillin forms esters and ethers, and the nucleus is easily substituted by halogen and nitro groups. Vanillin is noted for its stability. When exposed to light in an alcoholic solution, vanillin slowly dimerizes to dehydrovanillin.

Vanillin is not considered toxic. However, it can cause irritation on skin and mucous membranes.

Table 1. Physical Properties of Vanillin

Property	Value
melting point, °C	81–83
boiling point, °C	
at 101.3 kPa (1 atm)	284 (dec)
at 133 Pa (1 mm Hg)	127
density, d_4^{20}, g/cm³	1.056
heat of soln in H_2O at infinite dilution, J/mol[a]	21.76
heat of neutralization[b], J/mol[a]	38.74
heat of combustion[c], J/mol[a]	3.83
solubility, wt %	
water, 14°C	1
water, 75°C	5
glycerol, 25°C	4.5
propylene glycol, 25°C	23.0

[a] To convert J to cal, divide by 4.184.
[b] When 1 mol vanillin in 30 L H_2O is neutralized with 0.2 N NaOH.
[c] Constant pressure and volume.

Manufacture

Vanillin is mainly produced from natural sources, but 10–20% of production is made synthetically from guaiacol. The lignin processes are based on alkaline air oxidation at 160–175°C and 1.1–1.2 MPa (150–160 psig) of a fermented spent-waste liquor from a sulfite pulp mill (see also Lignin). The crude sodium vanillinate obtained is treated with SO_2, filtered to remove acetovanillone and other phenolic impurities, and acidified to precipitate the vanillin, which is filtered, vacuum distilled, and crystallized.

Ultrafiltration has been investigated to upgrade the quality of the spent sulfite liquor by removing the undesirable low molecular weight compounds that cannot be oxidized to vanillin. Ultrafiltration can increase plant capacity by as much as 25% through yield improvement and by reducing the consumption of caustic and other raw materials.

Synthetic Routes

Catechol is alkylated to guaiacol, which is converted to vanillin by a modification of the Riedel process. Glyoxalic acid is condensed with guaiacol, followed by air oxidation; acidification gives vanillin.

Guaiacol (o-methoxyphenol) is obtained principally by the alkylation of catechol and by the destructive distillation of wood and coal. As starting material, it offers numerous routes to vanillin. The aldehyde group can be introduced with a Reimer-Tieman reaction ($CHCl_3$ plus KOH), with HCN (Gatterman synthesis), by a Fries rearrangement of guaiacol acetate, or by treatment with formaldehyde (Sandmeyer reaction).

Oil of cloves contains 85–90% eugenol. Upon treatment with alkali, it rearranges to isoeugenol, which is oxidized to vanillin.

Grades and Specifications

Both USP and FCC grades must contain not less than 97.0% vanillin. The USP grade should not lose more than 1% and the FCC grade not more than 0.5% of its weight on drying for 4 h. Limits for arsenic and heavy metals are 3 and 10 ppm max, respectively. The technical grade contains foreign odors and color.

Analysis

Vanillin is identified by the white to slightly yellow precipitate formed upon addition of lead acetate to aqueous solutions; with ferric chloride solution, a blue color appears. Most quantitative determinations are made by gc or hplc.

Uses

The use of technical-grade vanillin as an intermediate in the production of pharmaceuticals has surpassed its application in perfumes and as a flavoring agent in foods (see Flavors and spices; Perfumes). The manufacture of drugs such as Aldomet, L-dopa, and Trimethaprim consumes ca 40% of vanillin production.

Derivatives

Ethylvanillin forms white crystalline needles. It is a synthetic flavoring and fragrance agent three to four times stronger than vanillin.

Vanillic acid forms colorless needles (mp 207°C) from water, which sublime on heating. It cannot be obtained from oxidized lignin liquors. Esters are easily prepared in high yields. Acetovanillone, 4-hydroxy-3-methoxyacetophenone, is present in commercial vanillin. It crystallizes from water as colorless prisms (mp 115°C, bp 295–300°C). Veratraldehyde, obtained by methylating vanillin, is used as a metal brightener in the plating industry. Isovanillin, 4-methoxy-3-hydroxybenzaldehyde, crystallizes from water (mp 115–117°C).

J.H. VAN NESS
Monsanto Company

U.S. Pat. 4,151,207 (Apr. 24, 1974), H. Evju (to Booregaard Co.).

P.Z. Bedoukian, *Perfumery and Flavoring Synthetics*, Elsevier Publishing Co., New York, 1967, pp. 344–363.

VAPOR-LIQUID EQUILIBRIA. See Absorption; Distillation.

VARNISH. See Insulation, electric (properties); Resins, natural.

VEGETABLE OILS

Vegetable oils are obtained as a tree crop or from the seeds of annually grown crops. They include most of the fatty acid esters of glycerol, commonly called triglycerides, which provide the world with its supplies of edible oils (liquids) and fats (solids) (see Fats and fatty oils).

Of the extensive range of plants known to have oil-rich fruit or seeds, fewer than twenty are exploited commercially on a significant scale, and > 90% of world vegetable-oil production is accounted for by nine oils (Table 1), of which soybean oil has for many years contributed the largest volume (see Soybeans).

Oleic (cis-9-octadecanoic) acid and stearic (octadecanoic) acid are the most common fatty acids in vegetable oils, but the range present in significant amounts extends to erucic (cis-13-docosenoic) acid. Unsaturation occurs primarily, but not exclusively, in acids with an 18-carbon chain. Oils are classified according to their principal fatty acids, eg, lauric (coconut, palm kernel oils), palmitic (palm oil), oleic (olive, groundnut, rapeseed, sesame oils), linoleic (medium, soybean, cottonseed, corn oils; high, sunflower, safflower oils), linolenic (linseed oil), and ricinoleic (castor oil) (see Carboxylic acids).

Free fatty acids must be removed in order to make the oil acceptable for edible purposes. Color is primarily due to the presence of carotenoids

Table 1. World Production and Exports of Vegetable Oils, 1000 Metric Tons

oil	1969–1970 Production	1969–1970 Export	1974–1975 Production	1974–1975 Export	1979–1980 Production	1979–1980 Export
Edible oils						
cottonseed	2,345	341	3,114	410	3,063	440
groundnut	2,933	862	3,037	699	14,054	6,854
soybean	6,990	2,998	8,208	3,417	5,705	1,567
sunflower	3,430	700	3,802	687	3,308	1,091
rape	1,559	415	2,391	686	3,308	1,091
sesame	632	103	673	89	676	114
olive	1,380	237	1,552	205	1,559	245
corn	342	30	426	50	632	131
safflower	171	47	228	38	332	36
palm	1,433	704	2,648	1,695	4,405	2,752
palm kernel	424	310	474	385	587	406
coconut	2,289	1,038	2,729	1,305	2,944	1,300
miscellaneous	653	138	782	122	932	400
Total	*24,581*	*7,923*	*30,064*	*9,788*	*41,239*	*16,060*
Nonedible oils						
linseed	1,020	433	691	252	868	419
castor	337	244	432	153	339	204

(< 50 ppm) and chorophyll and its derivatives (10 ppm max). Metals such as copper (< 1 ppm) and iron (< 10 ppm) are frequently present and have an adverse effect on quality. Sulfur interferes with processing and must also be removed. Pesticides are found in low concentration (see Trace and residue analysis).

When spoilage occurs as a result of hydrolysis or oxidation, the oils become unsuitable for human consumption. Steps to reduce the danger of spoilage include destruction or inactivation of microorganisms, preservation of natural antioxidants, removal of oxidation promoters, and the exclusion of oxygen during processing and of water during storage.

Properties

Density, viscosity, and melting behavior are most important to processers and users. Densities and viscosities of most oils lie within a narrow band with the notable exception of castor oil (qv), which is characterized by much higher viscosity.

Dilatometric and nmr determinations are used to measure the amount of crystallized material in a fat. Because the refractive index is a function of molecular weight and unsaturation, refractometry is used in quality and process control. A more direct measure of unsaturation is the Iodine Value, a long-established analytical method that is, however, slowly being replaced by chromatography.

Production

The harvested oil seeds must be stored under carefully controlled moisture and temperature conditions with good ventilation. Cottonseed must be delinted before dehulling, and sunflower seed is dehulled by most processers in order to reduce the wax content of the extracted oil. Dehulling of soybeans mainly serves to increase the protein content of the meal.

Mechanical pressing as a means of oil recovery is virtually obsolete except in combination with solvent extraction. Most extractors used are based on percolation rather than total immersion. A combined percolation and immersion process, using intermediate desolventizing and flaking, is sometimes employed for material containing > 40% oil. Industrial hexane is the preferred solvent. Olive oil is recovered from the fruit in hydraulic presses operated at up to ca 39 MPa (5700 psi) (see Extraction).

Processing

Refining removes free fatty acids, phosphatides, pigments, and volatile compounds as well as artifacts introduced by the process itself, eg, soap formed in alkaline neutralization of free fatty acids.

Degumming, ie, removal of phosphatides, is usually combined with alkali refining. The phosphatides not precipitated by water can be precipitated by reaction with concentrated phosphoric acid or citric acid before alkaline neutralization. Iron content of the oil is also reduced by this treatment. The need to remove waxes (winterization) applies only to a few oils, especially sunflower oil.

Batch or continuous deacidification with alkali (alkali refining) is widely practiced. Oils of poor quality suffer serious losses of neutral oil, and large quantities of effluent are generated. Physical refining or deacidification by semicontinuous or continuous steam stripping are the alternatives to alkaline refining. The plants are frequently the same as those used for deodorization, and up to 5% of stripping steam may be used. The stripped fatty acids are recovered by means of a partial condenser or in a scrubber. Carotenoids decompose rapidly at the stripping temperature, but chlorophyll must be removed by adsorption before stripping is carried out.

Adsorptive bleaching has been traditionally carried out with natural or activated bleaching clay (0.5–2.0%) at 90–110°C and preferably under vacuum. Automatic centrifugal-discharge filters are increasingly used for the removal of the clay and are available in sizes up to 100 m^2 filtration area.

The volatile components present in the deacidified, bleached oil are removed by deodorization at 0.2–1.2 kPa (1.5–9 mm Hg) and 175–270°C. Deodorization is carried out batchwise by sparging steam into the oil through a distributor in the lower part of a cylindrical vessel. A semicontinuous deodorizer consists of a series of stages similar in principle to the batch deodorizer, the oil being held for a fixed time in each. Deodorization is the most energy-intensive operation in the refining sequence, and reduction of the energy consumption has been effected, principally by heat exchange between incoming and processed oil. In semicontinuous deodorizers, heat recovery of up to 70% has been claimed, and ever higher recovery can be expected in continuous operations.

Modification

Hydrogenation, interesterification, and fractional crystallization provide means of modifying the relationship between solids content and the temperature of oils and their blends to give products of specific functional properties.

Hydrogenation. Hydrogenation reduces the degree of unsaturation but may cause isomerization. Reduction in unsaturation raises the melting range and enhances resistance to oxidation and flavor deterioration. The principal catalyst is nickel on kieselguhr. Copper (copper chromite) is more selective but requires more stringent refining.

Batch hydrogenation is more flexible than continuous operation and is believed to offer better control over quality. The high rate of reaction in the early stages requires adequate cooling to remove a heat of reaction of ca 3.6 MJ (ca 3400 Btu) per metric ton of oil per unit change of Iodine Value. In continuous hydrogenation, the heat can be used more effectively; steam is required only during startup.

The trans isomer of oleic acid, ie, elaidic acid, is the main isomer present in hydrogenated fat; lesser amounts of isomerized dienes also occur, eg, *trans,trans*-linoleic acid.

Interesterification. Interesterification rearranges the fatty acid groups in the triglyceride mixtures. Sodium alcoholates, sodium metal, and sodium–potassium alloys are the catalysts. Batchwise esterification is carried out at 90–110°C, whereas continuous esterification is performed in stirred tanks and tubular reactors at 130–160°C.

Fractional crystallization. This procedure modifies the melting properties without resorting to chemical change. Crystallization from the melt (dry fractionation) produces a coarse separation between saturated and unsaturated glycerides. Crystallization from a solvent gives sharper separations but is far more energy-intensive and entails higher capital expenditure. In dry fractionation the slurry produced may be separated by filtration or by transfer of the solids to an aqueous phase. In solvent crystallization a filter is used for the solid–liquid separation in most cases.

Margarine

In most countries, the manufacture of margarine is subject to strict control; total fat content is generally required to be at least 80%, and water content must not exceed 16%. The addition of additives, eg, emulsifiers, flavor components, vitamins, color, preservatives is also controlled. Emulsifiers (0.2–0.5%) facilitate aeration and, in margarine, create and stabilize the initial water/oil emulsion.

In processing, scraped-surface heat exchangers (Votator A-units) remove sensible and latent heat. Storage at a temperature above that at which the product is packed can lead to deterioration. Quality control is carried out by sensory assessment and microbiological control, oxidative deterioration being detected under the former.

Applications

The edible-oil market is dominated by vegetable oils that have been refined and modified. Olive oil is substantially used as a salad oil in the unrefined state. Fats or blends of fats and oils are used for margarine and related spreads, shortening, and specialties. Vegetable butters having sharp melting characteristics are used in confectionary formulations.

Nonedible oils, mainly linseed, castor, and tung oils, account for < 5% of total vegetable-oil production. They are used in the form of crude or refined triglycerides or as fatty acids or fatty acid derivatives, in surface coatings, paints and varnishes, and in detergent formulation.

W. HAMM
Unilever Research
Colworth Laboratory, UK

D. Swern, ed., *Bailey's Industrial Oil and Fat Products*, 4th ed., Wiley-Inter-science, New York; Vol. 1, 1979; Vol. 2, 1983.

J. Baltes in *Grundlagen und Fortschritte der Lebensmitteluntersuchung*, Vol. 17, Verlag Paul Parey, Berlin, FRG, 1975.

A.J. Haighton, *J. Am. Oil Chem. Soc.* **53**, 397 (1976).

VELVETEX.   See Surfactants and detersive systems.

VERMICULITE.   See Insulation, thermal.

VERMILION.   See Pigments, inorganic.

VETERINARY DRUGS

Compounds developed for use in veterinary medicine are subjected to evaluations concerning both animal and human toxicity, efficacy claims, stability, metabolism, manufacturing, possible environmental effects and labeling information.

Antimicrobial Agents

The selection of an antimicrobial agent depends on an accurate diagnosis and identification of the offending organism(s). The antibacterial spectrum of activity is further described in terms of activity against gram-positive and/or gram-negative organisms; ie, staining characteristics using a blue primary stain of crystal violet with iodine and a red counterstain, usually safranin. Consideration is also given to whether an agent kills the organism or inhibits its growth. Many antimicrobials have both therapeutic and growth-promoting properties (see Antibacterial agents; Antibiotics, chemotheropeutics).

Sulfonamides.   The sulfonamides are derivatives of *p*-aminobenzene-sulfonamide. They are active against a broad spectrum of gram-positive and gram-negative organisms. Their mode of action is by competitive antagonism of *p*-aminobenzoic acid, a folic acid precursor. Administration is usually oral or intravenous and if absorbed they are distributed widely in the body and are excreted in the urine. Mammalian toxicity is low.

Penicillins.   The penicillins are most effective against actively multiplying organisms. The spectrum of activity may be narrow or broad, depending on the analogue structures. Mammalian toxicity is low, but allergic phenomena following sensitization may occur.

Aminoglycosides.   The aminoglycosides (streptomycin, kanamycin, gentamycin) have a hexose nucleus joined to two or more amino sugars. They are rapidly bactericidal by inhibiting protein synthesis. Toxicity following exaggerated or prolonged dosage schedules is characterized by ototoxicity or renal failure.

Tetracyclines.   The tetracyclines are fermentation products with a broad antibacterial spectrum. The mechanism of action is by protein synthesis inhibition. Toxic reactions are rare.

Antifungal agents.   Superficial fungal infections may respond to sunlight, administration of vitamin A, topical application of antifungal agents, eg, nystatin, cuprimixin, or oral griseofulvin. Systemic infections frequently require protracted treatment and often are refractory to any therapy.

Parasiticides

Organophosphates and carbamates inhibit cholinesterase enzymes. They are rapidly absorbed following inhalation or oral, parenteral or topical administration. Metabolism is by hydrolysis or oxidation. Concurrent exposure to more than one agent results in cumulative effects, and the incidence of toxicity is relatively high. Atropine and/or pralidoxime chloride are antidotal.

The avermectins are fermentation products derived from *Streptomyces avermitilis*. They have a very broad spectrum of insecticidal and anthelmintic activity. Levamisole and the benzimidizoles also used as anthelmintics against a wide variety of enteric nematodes.

Coccidiosis is ubiquitous in the poultry industry, with serious consequences. Anticoccidial agents are fed routinely throughout the life of broiler chickens. Although resistance by the coccidia has often developed rapidly in the past, this has not yet been observed against ionophores, the most commonly used agents today.

Anti-Inflammatory Agents

Aspirin is well tolerated and widely used in dogs and horses, but is relatively toxic to cats. It can be used in prolonged therapy in chronic inflammatory diseases such as arthritis.

Pyrazolone derivatives, eg, phenylbutazone, are particularly useful in managing lameness in horses and controlling inflammation after trauma or surgery. Most widely used are the corticosteroids and their synthetic analogues, which, in addition to anti-inflammatory action, can adversely affect the rate of healing and cause imbalances in a variety of other physiological processes (see Steriods).

Hormones

Hormones (qv) and releasing or regulating agents are used to adjust the reproductive processes and as therapeutics for hormonal imbalances and physical or physiological abnormalities. Additionally, estrogens, testosterone, or compounds that mimic their effects have shown utility in accelerating weight gain and decreasing the amount of feed required to produce these gains in food-producing animals.

Tranquilizers and Anesthetics

Tranquilizers are used in the management of excitement in individual animals and allow the practitioner to examine the frightened or injured patient with less chance of further damage. They are also useful in the management of animals during shipping (see Psychopharmacological agents).

Anesthetics (qv) are generally either injectable agents or inhalants. Specific products are chosen dependent upon the length of the procedure to be carried out, the patient's health status, and the depth of anesthesia necessary.

Cancer Chemotherapy

In the veterinary patient, as in humans, metastatic neoplasms are often widely disseminated throughout the body. Surgery and irradiation have serious limitations, and chemotherapy is used with increasing frequency in cancer management. Because of the expense and time involved, such treatment must be restricted to animals for which the risk-benefit evaluation is favorable and treatment seems appropriate. In general, such treatment must be viewed as palliative, not curative.

RODERICK B. DOUGHERTY
Pfizer Inc.

DAVID M. PETRICK
American Cyanamid Company

D.C. Blood and J.A. Henderson, *Veterinary Medicine*, 5th ed., Lea and Febinger, Philadelphia, Pa., 1979.

Feed Additive Compendium, Miller Publishing Co., Minneapolis, Minn., 1982, published annually.

J.R. Georgi, *Parasitology for Veterinarians*, 3rd ed., W.B. Saunders Co., Philadelphia, Pa., 1980.

L.R. Soma, *Textbook of Veterinary Anesthesia*, The Williams and Wilkins Company, Baltimore, Md., 1971.

VETIVER.   See Oils, essential.

VINEGAR

Vinegar is used as a preservative in pickling and to give a sharp or sour taste to foods. Most table vinegars derive from the acetic acid-bacterial fermentation of wine or cider (see Acetic acid). These are produced by alcoholic fermentation (qv) of dilute sugar solutions, such as grape juice, cider, or malt (see Table 1).

A number of factors govern the composition of vinegar, including the nature of the raw material, the substances added to promote alcoholic fermentation and the growth and activity of *Acetobacter*, the procedure used for the acetification, and finally the aging, stabilization, and bottling operations.

Manufacture

Vinegar is produced by surface or submerged-culture oxidation. Production is accelerated by increasing the surface of the liquid and bacteria exposed to oxygen by passing the liquid through a column, tank, or vat packed with inert material on which the bacteria are absorbed.

Starch hydrolysis and alcoholic fermentation. Yeasts cannot utilize starch as carbon source for growth, and the starch must be hydrolyzed to sugar. Malt vinegars are made from malted barley which may be mixed with other starches or grains. The enzymes present hydrolyze the starch to glucose and maltose, which are readily fermented by *Saccharomyces* yeast. In Japan, vinegars are made from rice.

In alcoholic fermentation, glucose from hydrolyzed starch or the glucose and fructose from fruits and honey are converted by yeast enzymes to ethyl alcohol and carbon dioxide (see Ethanol; Sugar). Both the fermentation of the sugars and the oxidation of ethanol are exothermic processes. The excess heat must be removed during the oxidation. Practical yields of acetic acid are 77–85%.

In the Orleans process, the wine oxidizes slowly in a barrel where it is covered with a film of *Acetobacter*. Holes in each barrel head, covered with screens, permit access to atmospheric oxygen. Wines with 10–12% ethanol give vinegars of 8–10% acetic acid concentration. Orleans vinegars have a relatively high concentration of ethyl acetate, detected by its characteristic odor.

In generator processes, generators are filled with beech wood shavings which provide a loose packing that allows open spaces for the free flow of liquid and air; cooling coils are provided. In submerged-culture generators, a mechanical system keeps the bacteria in suspension in the liquid in the tank and in intimate contact with fine bubbles of air. The Frings acetator uses a bottom-driven hollow rotor turning in a field of stationary vanes; the air which is drawn in is intimately mixed with the liquid throughout the bottom area of the tank. Submerged-culture oxidizers are smaller and more flexible in operation. They are usually operated on a semicontinuous basis. Foam production impedes bacterial growth and is minimized by keeping nutrient, ethanol, and acetic acid concentrations in the optimum ranges (see Defoamers). Dead cells promote foaming, which is usually broken by centrifugation.

In the tower fermenter the liquid is held without packing on a porous plate by the pressure of air introduced below.

Concentrated vinegars. U.S. regulations require at least 4 g acetic acid per 100 cm³ vinegar. Submerged-culture oxidizers can give concentrations of 10–13 g/100 cm³. Continuous aeration and carefully stepwise addition of ethanol seem to be the keys to successful operation.

Although aesthetically undesirable, nematodes, known as vinegar eels, consume dead bacteria, prolong the operation, and make nutrients more readily available to *Acetobacter*. "Mother of vinegar" is the cellulosic slime coating the bacterial cells. It may block passageways in packed-tank generators.

Processing

Raw vinegars vary widely in stability and may contain materials that form cloudiness or deposits. Vinegars that carry a high and cloudy suspension of bacterial cells are clarified and stabilized with bentonite and similar agents or activated carbon. Membrane filtration can be combined with aseptic bottling to provide a product free of all microorganisms. Wine, cider, and malt vinegars benefit from aging, which improves the character of the product. Vinegars bottled for table use or pickling usually are pasteurized at 77–78°C.

A.D. WEBB
University of California, Davis

H.A. Connor and R.J. Allegier, *Adv. Appl. Microbiol.* **20**, 81 (1976).

G.B. Nickol in H.J. Peppler and D. Perlman, eds., *Microbial Technology*, 2nd ed., Vol. 2, Academic Press, New York, 1979, pp. 5–72.

VINYLBENZENE. See Styrene.

VINYL CHLORIDE. See Vinyl polymers.

VINYL ETHER. See Anesthetics; Vinyl polymers.

VINYLIDENE CHLORIDE AND POLY(VINYLIDENE CHLORIDE)

Poly(vinylidene chloride) and its copolymers exhibit low permeability to gases and vapors, but thermal instability at melt-processing temperatures. In the United States, Saran is a generic term for high vinylidene chloride polymers. A letter indicates the comonomer, eg, Saran A for the homopolymer, B for the copolymer with vinyl chloride (VC), and F for copolymers with acrylonitrile (AN). The homopolymer is not used extensively because of fabrication difficulties. Of commercial importance are vinylidene chloride–vinyl chloride copolymers, vinylidene chloride–alkyl acrylate and methacrylate copolymers, and vinylidene chloride-acrylonitrile copolymers.

Vinylidene Chloride

Vinylidene chloride (VDC, 1,1-dichloroethylene) is a colorless, mobile liquid with a characteristic sweet odor. Its properties are given in Table 1.

Manufacture. Vinylidene chloride is prepared commercially by the dehydrochlorination of 1,1,2-trichloroethane with a slight excess of lime or caustic.

$$2\ CH_2ClCHCl_2 + Ca(OH)_2 \xrightarrow{90°C} 2\ CH_2{=}CCl_2 + CaCl_2 + 2\ H_2O$$

Commercial grades contain 200 ppm of the monomethyl ether of hydroquinone (MEHQ) as inhibitor. It is removed by distillation or washing with 25% caustic. For many polymerizations, MEHQ need not be removed. Uninhibited vinylidene chloride should be kept at −10°C in the dark under a nitrogen atmosphere.

Vinylidene chloride polymerizes by ionic or free-radical reactions. The latter are carried out by solution, slurry, suspension, and emulsion methods.

Solution polymerization in a medium that dissolves both monomer and polymer has activation energies and frequency factors in the normal range for free-radical polymerizations of olefinic monomers. Spontaneous polymerization of VDC at room temperature is caused by peroxides. The heterogeneous nature of bulk polymerization is apparent from the rapid

Table 1. Raw Materials and Fermentation Products

Raw material	Products
tropical fruits	wines, vinegars
mango waste	syrups, wines, vinegars
whey	vinegar
rejected bananas	vinegar
palm sap	vinegar
dates	vinegar
white soy sauce	rice vinegar
enzyme preparation	rice vinegar

Table 1.   Properties of Vinylidene Chloride Monomer

Property	Value
color (APHA)	10–15
solubility of monomer in water, at 25°C, wt%	0.25
solubility of water in monomer, at 25°C, wt%	0.035
normal boiling point, °C	31.56
freezing point, °C	−122.56
flash point (tag closed-cup), °C	−28
flash point (tag open-cup), °C	−16
flammable limits in air (ambient conditions), vol%	5.6–16.0
autoignition temperature, °C	513[a]
latent heat of vaporization, ΔH_v°, kJ/mol[b] at 25°C	26.48 ± 0.08
latent heat of fusion (at freezing point), ΔH_m°, J/mol[b]	6514 ± 8
heat of polymerization (at 25°C), ΔH_p°, kJ/mol[b]	−75.3 ± 3.8
heat of formation, at 25°C, ΔH_f°, kJ/mol[b]	−25.1 ± 1.3
heat capacity, at 25°C, C_p°, J/(mol·K)[b]	111.27
critical temperature, T_c, °C	220.8
critical pressure, P_c, MPa[c]	5.21
critical volume, V_c, cm³/mol	218
liquid density, at 20°C, g/cm³	1.2137
index of refraction, at 20°C, n_D	1.42468
absolute viscosity, At 20°C, mPa·s (= cP)	0.3302

[a] Inhibited with MEHQ.
[b] To convert J to cal, divide by 4.184.
[c] To convert MPa to atm, divide by 0.101.

development of turbidity caused by the presence of minute PVDC crystals. Heterogeneous polymerization is characteristic of a number of monomers, including vinyl chloride and acrylonitrile. Emulsion and suspension reactions are doubly heterogeneous; the polymer is insoluble in the monomer and both are insoluble in water.

The instability of PVDC is one reason why ionic initiation of VDC has not been used extensively. Many catalysts either react with the polymer or catalyze its degradation.

The importance of VDC as a monomer is due to its ability to copolymerize with other vinyl monomers. Bulk copolymerizations yielding high VDC copolymers are usually heterogeneous. During copolymerization, one monomer may add to the copolymer more rapidly than the other. Batch reactions carried to completion yield polymers of broad composition distribution, which is usually not desirable.

Health and safety factors.   Vinylidene chloride is highly volatile and, when free of decomposition products, has a mild, sweet odor. Even short exposure to a high concentration of vapor causes intoxication. Vinylidene chloride is hepatoxic, but does not appear to be a carcinogen. Skin contact should be avoided.

In the presence of air or oxygen, uninhibited VDC forms a violently explosive peroxide at temperatures as low as −40°C. Peroxides may be removed by washing with sodium hydroxide or sodium bisulfite solution.

Poly(Vinylidene Chloride) (PVDC)

The polymer chain is made up of vinylidene chloride units added head-to-tail:

$$-CH_2CCl_2CH_2CCl_2CH_2CCl_2-$$

High crystallinity indicates that no significant head-to-head addition or branching are present. This is confirmed by ir and Raman spectra.

Molecular weights are determined directly by dilute solution measurements. Viscosity studies indicate degrees of polymerization (DP) ranging from 100 to > 10,000.

Crystal studies indicate a unit cell with four monomer units. The calculated density is higher than the experimental values of 1.80–1.94 g/cm³ at 25°C. The melting point is independent of molecular weight above DP = 100. The properties (see Table 2) are usually modified by copolymerization.

The glass-transition temperatures of Saran copolymers increase with the comonomer content at low comonomer concentration, even when the T_g of the other homopolymer is lower. A maximum T_g is observed at

Table 2.   Properties of Poly(Vinylidene Chloride)

Property	Best value	Reported values
mp, °C	202	198–205
T_g, °C	−17	−19 to −11
transition between T_m and T_g, °C	80	
density (at 25°C), g/cm³		
amorphous	1.775	1.67–1.775
unit cell	1.96	1.949–1.96
crystalline		1.80–1.97
refractive index (crystalline), n_D	1.63	
heat of fusion (ΔH_m), J/mol[a]	6275	4600–7950

[a] To convert J to cal, divide by 4.184.

intermediate compositions. Plasticization lowers T_g and decreases crystallization induction times. Copolymers with lower T_g tend to crystallize more rapidly. Maximum crystallization rates of the common crystalline copolymers occur at 80–120°C. The maximum rate for PVDC is probably at 140–150°C, but observation is difficult because of degradation. Orientation or mechanical working accelerates crystallization and has a pronounced effect on morphology.

The highly crystalline particles of PVDC precipitating during polymerization are aggregates of thin, lamellar, highly branched crystals.

Melting points of as-polymerized powders are high (198–205°C). Although the T_g is not well defined because of high crystallinity, a sample can be melted at 210°C and quenched rapidly to an amorphous state at < −20°C. The amorphous polymer has a T_g of −17°C. Once melted, PVDC does not regain its as-polymerized morphology when subsequently crystallized.

Solubility and solution properties.   Poly(vinylidene chloride) does not dissolve in most common solvents at ambient temperatures. Copolymers, particularly those of low crystallinity, are much more soluble. However, one of the outstanding characteristics of Saran polymers is resistance to solvents and chemical reagents. The insolubility of PVDC results less from its polarity than from its high melting point. Above 130°C, it dissolves readily in a wide variety of solvents. Poly(vinylidene chloride) dissolves readily in certain solvent mixtures containing a sulfoxide or N,N-dialkylamide. Effective cosolvents are less polar and have cyclic structures.

Mechanical properties.   Because of the difficulty of fabricating PVDC into suitable test specimens, few direct measurements of its mechanical properties have been made (see Table 3). The performance of a given specimen is very sensitive to morphology. Incorporation of VC units in the polymer structure results in a drop in dynamic modulus because of reduced crystallinity. However, the T_g is raised; the softening effect observed at room temperature is therefore accompanied by increased brittleness at lower temperatures.

In cases where the copolymer has a substantially lower T_g, the modulus decreases with increasing monomer content, which is accompanied by an improvement in low temperature performance.

Vinylidene chloride polymers are more impermeable to a wider variety of gases than other polymers. This is a consequence of high density and high crystallinity. An increase in either reduces permeability (see Barrier

Table 3.   Mechanical Properties of Poly(Vinylidene Chloride)

Property	Range
tensile strength, MPa[a]	
unoriented	34.5–69.0
oriented	207–414
elongation, %	
unoriented	10–20
oriented	15–40
softening range (heat distortion), °C	100–150
flow temperature, °C	>185
brittle temperature, °C	−10 to 10
impact strength, J/m[b]	26.7–53.4

[a] To convert MPa to psi, multiply by 145.
[b] To convert J/m to ft·lbf/in., divide by 53.38 (see ASTM D 256).

polymers). Permeability is affected by the type and amounts of comonomer as well as crystallinity. A more polar comonomer, eg, an AN comonomer, increases the water-vapor transmission more than VC. All VDC copolymers are impermeable to aliphatic hydrocarbons. Plasticizers increase permeability.

Degradation. Poly(vinyl chloride) is thermally unstable and when heated above 125°C, evolves HCl. At lower temperatures the polymer is degraded by radiation or treatment with alkalies, reactive metals, and Lewis acids. Thermal decomposition in the solid state (T < 200°C) is a two-step process:

formation of a conjugated polyene:

$$+CH_2CCl_2+_n \xrightarrow[\Delta]{fast} +CH=CCl+_n + n\ HCl$$

and carbonization:

$$+CH=CCl+_n \xrightarrow[\Delta]{slow} 2\ n\ C + n\ HCl$$

The polymer discolors gradually from yellow to brown and finally to black.

The stability of VDC copolymers depends on the comonomer. Copolymers with VC and the acrylates degrade slowly. Acrylonitrile copolymers degrade more rapidly and release HCN as well as HCl. Degradation in solution in nonpolar solvents is much slower than the solid-state reaction. In both cases, a free-radical mechanism is evident.

Stabilization. The ideal stabilizer system should adsorb or combine with evolved HCl irreversibly under conditions of use, but not strip HCl from the polymer chain; act as a selective uv absorber; contain reactive dienophilic molecules capable of preventing discoloration; possess antioxidant activity; and be able to chelate metals in order to prevent the formation of degradation catalysts.

Polymerization and processing. Emulsion and suspension polymerizations are the preferred processes (see Polymerization mechanisms and processes). Both processes are carried out in a closed, stirred reactor, glass-lined and jacketed for heating and cooling. The reactor must be purged of oxygen and the reactants must be free of metallic impurities.

Emulsion polymerization produces high molecular weight polymers within reasonable reaction times, especially vinyl chloride copolymers. Initiation and propagation can be controlled. Monomer can be added during the polymerization to control composition. The disadvantage is the high concentration of additives needed, which affect water sensitivity, electrical properties, and heat and light stability.

Suspension polymerization is used for molding and extrusion resins. Fewer additives are needed than for emulsions, and stability is increased and water sensitivity decreased. However, longer reaction times are required and high molecular weight polymer are difficult to prepare.

Uses. PVDC is used in molding and extrusion resins for filaments, films, rods, tubing, and pipe. Thermal degradation and streamlined flow must be considered and the materials of construction must be corrosion resistant nickel alloys.

Saran resins are used for multilayer film and sheet. Solid-phase forming is ideally suited for the fabrication of thermally unstable polymers, which are used as preformed sheets, briquettes, or compressed powder. Other applications of PVDC include lacquer resins, latex, foams, and flame retardants (qv).

DALE S. GIBBS
R.A. WESSLING
Dow Chemical U.S.A.

R.A. Wessling, *Polyvinylidene Chloride*, Gordon and Breach Science Publishers, New York, 1977.

R.A. Wessling and F.G. Edwards in N.M. Biklaes, ed., *Encyclopedia of Polymer Science and Technology*, Vol. 14, John Wiley & Sons, Inc., New York, 1971, p. 590.

L.G. Shelton, D.E. Hamilton, and R.H. Risackerly in E.C. Leonard, ed., *Vinyl and Diene Monomers, High Polymers*, Vol. 24, Interscience Publishers, a division of John Wiley & Sons, Inc., New York, 1971, pp. 1205–1282.

VINYLIDENE POLYMERS, POLY(VINYLIDENE FLUORIDE) ELASTOMERS.   See Fluorine compounds, organic.

VINYL POLYMERS

POLY(VINYL ACETAL)S

Poly(vinyl acetal)s are resins made from the reaction products of poly(vinyl alcohol) (qv) and aldehydes. Monomeric acetals are formed from one molecule of aldehyde and two molecules of alcohol in the presence of acid.

$$RCHO + 2\ R'OH \xrightarrow{H^+} RCH(OR')_2 + H_2O$$
aldehyde   alcohol   acetal

Similarly, poly(vinyl acetal)s are prepared by the reaction of an aldehyde and two hydroxyl groups on the poly(vinyl alcohol) chain. The poly(vinyl alcohol) is in turn prepared from poly(vinyl acetate) by hydrolysis or alcoholysis.

The conditions of the acetal formation are closely controlled to form a poly(vinyl acetal) containing a predetermined proportion of acetate groups, hydroxyl groups, and acetal groups. The poly(vinyl acetal) may be represented as shown in Figure 1 where the three basic units are randomly distributed along the molecule. The poly(vinyl acetal) unit is a cyclic acetal because the hydroxyl groups are attached to the same chain.

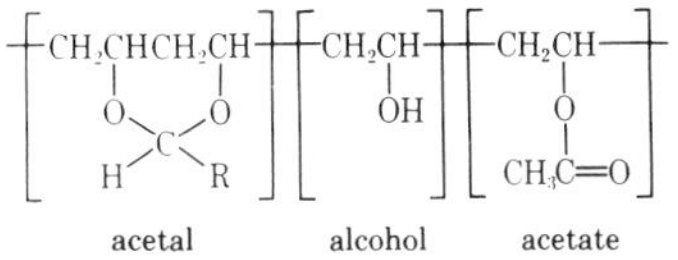

Figure 1.   A poly(vinyl acetal) structure.

Properties

Both poly(vinyl formal) and poly(vinyl butyral) are soluble in certain blends of polar and nonpolar solvents, eg, in 40:60 ethyl alcohol (95 wt%):toluene by weight. The low temperature flexibility of films, coatings, adhesives, etc, made from poly(vinyl acetal) resins can be improved by incorporation of plasticizers (qv).

Although the poly(vinyl acetal) resins are thermoplastic and soluble in various solvents, the secondary hydroxyl groups permit cross-linking with a variety of thermosetting resins. Incorporation of even small amounts of the poly(vinyl acetal) resin into thermosetting compositions markedly improves the toughness, flexibility, and adhesion of the cured composition. Alternatively, incorporation of smaller quantities of thermosets improves the balance of properties of the poly(vinyl acetal)s. Figure 2 illustrates the probable mechanism of cross-linking with epoxy groups.

Increasing the length of the acetal side chain increases the flexibility. Poly(vinyl formal) is very tough, whereas poly(vinyl butyral) is less tough and more flexible. Ketones can be used in place of aldehydes, but the resulting poly(vinyl ketal)s have been made only in the laboratory.

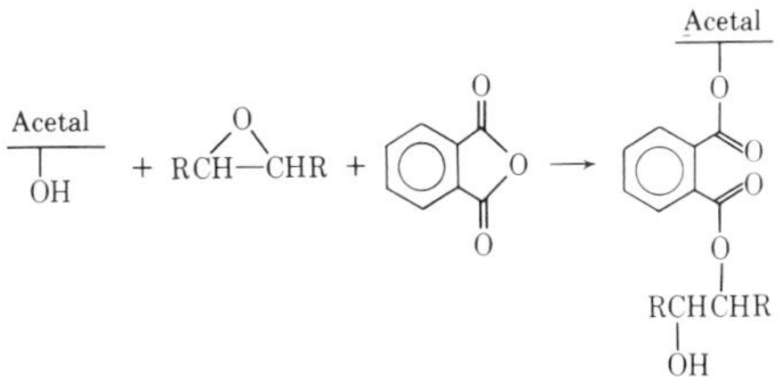

Figure 2.   Reaction with epoxy groups (anhydride cure).

Manufacturing and Processing

In the simultaneous or one-stage reaction, the hydrolysis of the poly(vinyl acetate) and acetalization of the resulting poly(vinyl alcohol) are carried out at the same time in the same kettle with an acid catalyst. In the sequential or two-step reaction, the alcoholysis and acetalization reactions occur separately. In the semisequential process, the aldehyde is added after the hydrolysis has progressed to a certain stage but without separation of the hydrolysis product.

Commercial poly(vinyl formal) (Formvar) is manufactured by a simultaneous process in acetic acid. Some properties are adjusted by varying the charge composition and the reaction conditions. However, molecular weight is controlled during the polymerization, not during the simultaneous process. These polymers contain the three basic units, ie, hydroxyl, acetate, and acetal.

Commercial poly(vinyl butyral)s are manufactured by sequential processes. The bulk density, grain structure, solvent and plasticizer absorption rates, and even transparency can be controlled by the conditions at precipitation. These resins contain essentially only two basic units, ie, hydroxyl and acetal. The acetate unit generally comprises < 2.5 wt%.

Aqueous poly(vinyl butyral) dispersions. In addition to poly(vinyl acetal)s in powder form, aqueous dispersions containing poly(vinyl butyral) have been developed commercially. They are usually produced by a mixer process similar to that used for making rubber latex from reclaimed stock. Films of poly(vinyl butyral) dispersion exhibit the toughness, flexibility, and transparency for which the resin is noted. The tensile strengths of unplasticized dispersions are 41–48 MPa (6000–7000 psi), whereas those containing 40–50 phr plasticizer break at ca 14 MPa (2000 psi). Elongations at break show a similar variation. Reactive thermosetting resins, eg, water-soluble or dispersible phenolics and melamines, can be incorporated to modify properties further. Poly(vinyl butyral) dispersions are widely used in the textile industry to impart abrasion resistance, durability, strength, and slippage control, and reduce color crocking, ie, color transfer when rubbed.

Health and Safety Factors

Butvar poly(vinyl butyral) resin is practically nontoxic. Butvar resin is slightly irritating to the eyes. Curable coatings containing Formvar or Butvar resin can be formulated to meet the extractibility requirement of the FDA. Butvar resins can be used in accordance with CFR regulations as ingredients of can enamels, adhesives, and components of paper and paperboard in contact with foods.

Uses

Poly(vinyl formal) and poly(vinyl butyral) are used in primers and surface-coating formulations for metal, wood, plastic, concrete, and leather substrates. The films, mostly in combination with other resins, can be air dried, baked, or cured at room temperature. As protective coatings in wash primers or metal conditioners, they prevent corrosion (see Corrosion and corrosion inhibitors).

Poly(vinyl acetal) resins can be compounded and formed into baked coatings, which offer chemical resistance and withstand postforming.

Plasticized poly(vinyl butyral) resin can be extruded into sheet, and used as the interlayer in safety-glass laminates which have excellent transparency, penetration resistance, and permanence (see Laminated materials, glass).

Poly(vinyl acetal)s are used in high performance thermosetting adhesives and in hot-melt formulations. Combinations of poly(vinyl formal) or poly(vinyl butyral) with phenolic resins are used for the structual bonding of metals.

Foams from the acetalization of poly(vinyl alcohol) yield tough, resilient, and soft synthetic sponges. Semipermeable membranes for reverse osmosis (qv) or ultrafiltration (qv) and ion-exchange membranes are another application.

Poly(vinyl formal) resins are used in electrical insulation for magnet wire and in wire enamels.

Poly(vinyl acetal)s are used in direct electrostatic recording, and in printed circuits.

Poly(vinyl butyral) is used in flexographic, letterpress, and gravure inks and in paper, textiles, ceramics, mold additives, cements, and refractories.

EDWARD LAVIN
JAMES A. SNELGROVE
Monsanto Company

Butvar, Poly(Vinyl Butyral) and Formar, Poly(Vinyl Formal), Technical Bull. No. 6070D, Monsanto Co., St. Louis, Mo., June 1977.

C.E. Schildknecht, *Vinyl and Related Polymers*, John Wiley & Sons, Inc., New York, 1952, pp. 323–385.

E. Lavin and J.A. Snelgrove in I. Skeist, ed., *Handbook of Adhesives*, 2nd ed., Van Nostrand Reinhold Co., New York, 1977, Chapt. 31.

POLY(VINYL ACETATE)

Vinyl Acetate Monomer

Vinyl acetate (VA) is a colorless, flammable liquid with an initially pleasant odor which quickly becomes sharp and irritating. Its only use is in polymerization.

Properties. Physical properties are listed in Table 1.

The most important chemical reaction of vinyl acetate is free-radical polymerization.

$$n \; CH_2=CHOCCH_3 \longrightarrow \begin{matrix} O \\ \| \\ CH_3CO \\ | \\ +CH_2CH+_n \end{matrix}$$

Halogens or hydrogen halides add readily giving the 1,2-dihaloethyl acetate or 1-haloethyl acetate, respectively.

Manufacture. Vinyl acetate is produced by the oxidative addition of acetic acid to ethylene in the presence of a palladium catalyst. Liquid-phase and vapor-phase processes are used commercially:

$$CH_3COH + H_2C=CH_2 + \tfrac{1}{2} O_2 \longrightarrow CH_3COCH=CH_2 + H_2O$$

In the liquid-phase process, a mixture of ethylene and oxygen at ca 3 MPa (30 atm) is fed into the single-stage reactor, which contains acetic acid, water, and catalyst at 100–130°C. The products, ie, vinyl acetate and acetaldehyde, are separated from the exiting gas stream in a series of distillation columns.

The catalyst solution contains palladium salts and copper salts; chloride ion is also necessary; overall yields are 90% based on ethylene and 95% based on acetic acid. In the catalytic vapor-phase process very little acetaldehyde is formed. Capital costs are ca 50% higher than required by the acetylene vapor-phase process and energy consumption is higher. However, acetylene is more expensive.

A catalytic vapor-phase process is also used for vinyl acetate production based on acetic acid addition to acetylene. It operates at 180–210°C with an acetylene-to-acetic-acid molar feed ratio of (4–5):1. The catalyst is zinc acetate.

Table 1.   Physical Properties of Vinyl Acetate

Property	Value
boiling point, °C	72.7
melting point, °C	−100, −93
specific gravity, 20/20	0.9338
refractive index, n_D^{20}	1.3952
viscosity 20°C, mPa·s (= cP)	0.42
flash point, tag open cup, °C	0.5–0.9
heat of vaporization at 72°C, J/g[a]	379.1
heat of combustion, kJ/g[a]	24.06
heat of polymerization, kJ/mol[a]	89.12

[a] To convert J to cal, divide by 4.184.

The liquid-phase process for the acetic acid–acetylene route to vinyl acetate employs a mercuric salt catalyst, which precipitates in the reaction medium upon addition of sulfuric acid, oleum, or phosphoric acid.

Acetaldehyde and acetic anhydride react in the presence of a catalyst at an elevated temperature, forming ethylidene diacetate. This product is passed to a cracking tower where vinyl acetate and acetic acid form. In addition, vinyl acetate can be made in excellent yields by the reaction of vinyl chloride and sodium acetate in solution at 50–75°C in the presence of palladium chloride.

Specifications, storage, and analysis. Vinyl acetate monomer is supplied in three grades which differ in inhibitor content. Vinyl acetate is commonly stored in carbon-steel tanks. Baked phenolic-coated steel, aluminum, glass-lined tanks, and stainless steel are also suitable. Inhibited vinyl acetate has good storage stability below 30°C.

Gas chromatography is an excellent method for determining vinyl acetate and its volatile impurities. Appropriate ASTM specifications and test procedures are D 2190, D 2191, D 2193, and D 2083.

Health and safety factors. Vinyl acetate is only moderately toxic if ingested or if absorbed through the skin. However, prolonged or repeated contact with the skin should be avoided.

Vinyl Acetate Polymers

With increasing molecular weight, poly(vinyl acetate)s (PVAcs) vary from viscous liquids and low melting solids to tough, horny materials. They are neutral, water-white to straw colored, tasteless, odorless, and nontoxic. Physical properties are listed in Table 2. Aging qualities of PVAc are excellent because of its resistance to oxidation and its inertness to uv-vis radiation.

The chemical properties of PVAc are those of an aliphatic ester. Hydrolysis produces PVA and acetic acid.

Poly(vinyl acetate) emulsion films adhere well to most surfaces and have good binding capacity for pigments and fillers. Plasticized films are strong and flexible.

Polymerization. Vinyl acetate has been polymerized industrially in bulk, solution, suspension, and emulsion. Perhaps 90% of the material identified as poly(vinyl acetate) or copolymers are made by emulsion techniques. An emulsion recipe, in general, contains monomer, water, protective colloid or surfactant, initiator, buffer and, perhaps, a molecular weight regulator. Poly(vinyl acetate) emulsions can be made with a surfactant alone or with a protective colloid alone, but a combination of the two is commonly used. The greater the quantity of emulsifiers in a recipe, the smaller the particle size of the emulsion.

The initiators or catalysts employed are the familiar free-radical types. Buffers are frequently added. The pH of commercial solutions is 4–6. Addition of a chain-transfer agent controls the molecular weight.

Industrial polymerizations are carried out to over 99% conversion. Most poly(vinyl acetate) emulsions contain a maximum of 0.5 wt% unreacted vinyl acetate. Emulsion processes are operated at atmospheric pressure in conventional glass-lined or stainless-steel kettles or reactors.

On the molecular scale, vinyl acetate polymerizations are generally understood as free-radical polymerizations characterized by extensive chain transfer. Vinyl acetate polymerizes chiefly head-to-tail but some of the monomers orient themselves head-to-head and tail-to-tail as the chain grows. The degree of grafting of PVAc or PVA during polymerization strongly affects latex properties, eg, viscosity, rheology, and polymer solubility.

The kinetics of vinyl acetate emulsion polymerization in the presence of nonylphenyl–polyethanol surfactants of various chain lengths indicate that part of the polymerization occurs in the aqueous phase and part in the particles.

Block copolymers of vinyl acetate with methyl methacrylate, acrylic acid, acrylonitrile, and vinylpyrrolidinone have been prepared by copolymerization in viscous, poor solvents for the vinyl acetate macroradical.

Specifications. Borax stability is an important property in adhesive, paper, and textile applications. Other emulsion properties include tolerance to specific solvents, surface tension, minimum filming temperature, dilution stability, freeze-thaw stability, and percent soluble polymer.

Table 2. Physical Constants of Poly(Vinyl Acetate)

Property	Value
absorption of water at 20°C for 24–144 h, %	3–6
coefficient of thermal expansion, K^{-1}	
cubic	6.7×10^{-4}
linear, below T_g	7×10^{-5}
above T_g	22×10^{-5}
compressibility, $cm^3/(g \cdot kPa)^a$	17.8×10^{-6}
decomposition temperature, °C	150
density at 20°C, g/cm^3	1.191
200°C	1.05
dielectric constant at 50°C (at 2 MHz)	3.5
dielectric dissipation factor at 50°C (at 2 MHz), tan δ	150
dielectric strength at 30°C, V/L	0.394
dipole moment at 20°C, $C \cdot m^b$ per monomer unit	2.30
elongation at break (at 20°C and 0% rh), %	10–20
glass-transition temperature, T_g, °C	28–31
hardness (at 20°C), Shore units	80–85
heat capacity (at 30°C), J/g^c	1.465
heat of polymerization, kJ/mol^c	87.5
refraction index, n_D	
at 20.7°C	1.4669
142°C	1.4317
interfacial tension, mN/m (= dyn/cm)	
at 20°C with polyethylene	14.5
at 20°C with polystyrene	4.2
internal pressure, MJ/m^{3b} (= MPa)	
at 0°C	255
60°C	418.7
modulus of elasticity, GPa^d	1.275–2.256
notched impact strength, J/m^e	102.4
surface resistance (Ω/cm)	5×10^{11}
surface tension, mN/m (= dyn/cm)	
at 20°C	36.5
180°C	25.9
tensile strength, MPa^d	29.4–49.0
thermal conductivity, $mW/(m \cdot K)$	159
Young's modulus, MPa^d	600

^{a}To convert kPa to atm, divide by 101.3.
^{b}To convert $C \cdot m$ to debye, divide by 3.336×10^{-30}.
^{c}To convert J to cal, divide by 4.184.
^{d}To convert MPa to psi, multiply by 145; GPa to psi, multiply by 145,000.
^{e}To convert J/m to lbf/in., divide by 53.38.

Film properties include clarity, gloss, light stability, water resistance, flexibility, heat-sealing temperature, specific gravity, and bond strength.

Homopolymer resin specifications usually include viscosity grade, volatiles, acidity as acetic acid, and softening point.

Poly(vinyl acetate) is nontoxic and is approved by the FDA for food-packaging.

Uses. The main areas of poly(vinyl acetate) adhesive use are packaging and wood gluing. Homopolymers adhere well to porous or cellulosic surfaces, and are suitable adhesives for high speed packaging.

Copolymers wet and adhere well to nonporous surfaces and form soft, flexible films, in contrast to the tough, horny films formed by homopolymers. Copolymer emulsions tend to wet slick surfaces better than homopolymer emulsions because of the extra mobility and softness given to the polymer particles by the plasticizing comonomer.

The setting speed of an adhesive is the time during which the bond becomes permanent. Before the setting of an emulsion adhesive occurs, inversion of the emulsion must take place; that is, it must change from a dispersion of discrete polymer particles in an aqueous, continuous phase to a continuous polymer film containing discrete particles of water. The rapid inversion possible with PVAc emulsions and their low viscosities allow them to be compounded into adhesives that not only set rapidly but also machine easily at high speeds. The viscosity of an adhesive influences its penetration into a substrate; as the viscosity increases, the penetrating power decreases.

Poly(vinyl acetate) emulsions are excellent for water-resistant paper adhesives for use in bags, tubes, and cartons.

Solvents are frequently used in emulsion adhesives where they promote adhesion to solvent-sensitive surfaces, increase the viscosity of the emulsion and intensify the tack of the wet adhesive, and improve the coalescing properties of the film. Tackifiers increase the tackiness and the setting speed of adhesives. They are usually rosin or its derivatives or phenolic resins.

Poly(vinyl acetate) dry resins and EVA copolymers are used in solvent adhesives which can be applied by typical industrial techniques, eg, brushing, knife-coating, roller-coating, spraying, or dipping.

Poly(vinyl acetate) latex paints are the first choice for interior use, and are also widely used as exterior paints. Their durability, particularly their resistance to chalking, far surpasses that of any conventional oleoresinous paints. The toughest paint films are formed from latex polymers having the highest molecular weights. Special vinyl acetate copolymer paints have been developed with greatly improved resistance to blistering or peeling in water.

Poly(vinyl acetate) emulsions and resins are used as binder in coatings for paper and paperboard. Emulsions used in paper coatings must meet special requirements: the particle size must be small (ca 1 μm) and its distribution rather narrow.

Poly(vinyl acetate) emulsions are widely used as textile finishes because of their low cost and good adhesion to natural and synthetic fibers. The use of vinyl acetate copolymers as binding agents for nonwoven fabrics has grown rapidly.

WILEY DANIELS
Air Products and Chemicals, Inc.

M.K. Lindemann in G.E. Ham, ed., *Vinyl Polymerization*, Vol. 1, Marcel Dekker, Inc., New York, 1967, Pt. 1, Chapt. 4.

M.K. Lindemann in N.M. Bikales, ed., *Encyclopedia of Polymer Science and Technology*, Vol. 15, John Wiley & Sons, Inc., New York, 1971, p. 636.

International Trade Commission Reports, Washington, D.C., 1960–1980.

POLY(VINYL ALCOHOL)

Poly(vinyl alcohol) (PVA) is a polyhydroxy polymer and, consequently, a water-soluble synthetic resin. It is produced by the hydrolysis of poly(vinyl acetate); the monomer, CH_2=CHOH, does not exist.

It is dry, white-to-cream colored solid, available in granular or powdered form. PVA forms tough, clear films that have high tensile strength and abrasion resistance. The main uses in the United States are in textile and paper sizing, in adhesives, and as an emulsion-polymerization aid.

Properties

The physical properties (see Table 1) of PVA are controlled by molecular weight and the degree of hydrolysis. The PVA product matrix has four molecular-weight ranges and three degrees of hydrolysis. Above 100°C, PVA gradually discolors; it darkens rapidly above 150°C and decomposes above 200°C.

All commercial grades are soluble in water, which is the only practical solvent for PVA. Solubility is influenced by surface area, molecular weight, and crystallinity.

The viscosity of a PVA solution is controlled by molecular weight, concentration, and temperature. Viscosity is proportional to degree of hydrolysis at constant molecular weight. Viscosity, rather than solubility, limits the concentration of PVA solutions.

The tensile strength of unplasticized, fully hydrolyzed PVA is 55–69 MPa (8,000–10,000 psi) at 50% rh. Tensile elongation is extremely sensitive to humidity and ranges from $\leq$ 10% when completely dry to 300–400% at 80% rh.

PVA is well known for its exceptional adhesion to cellulosic surfaces and its binding power in cement formulations. Gelation of PVA with boric acid reduces adhesive penetration into porous surfaces and produces aqueous solutions that exhibit excellent wet tack.

Table 1. Physical Properties of Poly(Vinyl Alcohol)

specific gravity	
of solid	1.27–1.31
of 10 wt% sol at 25°C	1.02
refractive index (film) at 20°C	1.55
thermal conductivity, W/(m·K)[a]	0.2
electrical resistivity, ohm·cm	$(3.1–3.8) \times 10^7$
specific heat, J/(g·K)[b]	1.5
melting point (unplasticized), °C	230 for fully hydrolyzed grades; 180–190 for partially hydrolyzed grades
T_g, °C	75–85
storage stability (solid)	indefinite when protected from moisture
flammability	burns similarly to paper
stability in sunlight	excellent

[a] To convert W/(m·K) to (Btu·in.)/(h·ft^2·°F), divide by 0.1441.
[b] To convert J to cal, divide by 4.184.

As little as 0.1% borax, based on solution weight, can cause thermally irreversible gelation. Congo red is an excellent gelling agent where colored films or coatings can be tolerated.

PVA is virtually unaffected by greases, petroleum hydrocarbons, and animal or vegetable oils.

The oxygen-barrier properties at low humidity are unsurpassed by any other synthetic resin. The gas barrier performance rapidly diminishes below the 98% hydrolysis level.

Unplasticized PVA is not considered a thermoplastic, because the degradation temperature is below the 230°C melting point for fully hydrolyzed grades. Glycerol is one of the most widely used plasticizers, as are poly(ethylene glycol)s.

The surface tension of PVA solutions varies linearly with degree of hydrolysis from 88 to 98%.

Poly(vinyl butyral) is formed by the reaction of PVA with butyraldehyde. Other reactions include the formation of esters, cyanoethylation with acrylonitrile, and reaction with ethylene oxide to form hydroxyethyl groups.

PVA can be grafted with other monomers. The reaction is usually carried out in aqueous solution with free-radical or ionic catalysts.

PVA can be readily cross-linked with many chemical additives, eg, glyoxal, urea–formaldehydes, and melamine–formaldehydes.

Manufacture

Conversion of poly(vinyl acetate) to PVA is generally accomplished by base-catalyzed methanolysis; sodium hydroxide is the usual base. The degree of hydrolysis during alcoholysis is controlled and is independent of molecular weight.

Waste disposal. Poly(vinyl alcohol) can be biodegraded in acclimatized, activated sludge wastewater systems. It does not appear to interfere with the treatment of other biodegradable materials.

Specifications and Handling

The PVA product matrix has four molecular-weight ranges (see Table 2) and three hydrolysis levels. The hydrolysis levels are described as partial (88%), intermediate (96%), and full (98%) according to the mole% of hydroxyl groups present.

Storage life in dry form is unlimited. Aqueous solutions are stable but must be protected from rust and bacterial growth. Granular PVA forms an explosive mixture with air with a low severity rating of 0.1 on a scale in which coal dust has a rating of 1.0. Handling in bags poses no significant explosion risk.

PVA solutions are prepared by dispersing the solids in water under good agitation prior to heating. Water should never be added to the dry

Table 2. Molecular Weight of Main Commercial Poly(Vinyl Alcohol) Grades[a]

Viscosity grade	Nominal M_n	4% solution viscosity, mPa·s (= cP)[b]
low	25,000	5–7
intermediate	40,000	13–16
medium	60,000	28–32
high	100,000	55–65

[a] Courtesy of Air Products and Chemicals, Inc.
[b] Measured at 20°C with Brookfield viscometer.

Table 1. Physical Properties of Vinyl Chloride

Property	Value
melting point, °C	−153.8
boiling point, °C	−13.4
specific heat, J/(kg·K)[a]	
vapor at 20°C	858
liquid at 20°C	1352
critical temperature, °C	156.6
critical pressure, MPa[b]	5.60
critical volume, cm³/mol	169
dipole moment, C·m[c]	5.0×10^{-30}
latent heat of fusion, J/g[a]	75.9
latent heat of evaporation, J/g[a]	330
standard enthalpy of formation, kJ/mol[a]	35.18
standard Gibbs energy of formation, kJ/mol[a]	51.5
vapor pressure, kPa[b]	
−30°C	50.7
0°C	164
viscosity, mPa·s (= cP)	
−40°C	0.3388
−10°C	0.2481
explosive limits in air, vol%	4–22
self-ignition temperature, °C	472
flash point (open-cup), °C	−77.75
liquid density (at −14.2°C), g/cm³	0.969

[a] To convert J to cal, divide by 4.184.
[b] To convert MPa to psi, multiply by 145; to convert kPa to mm Hg, multiply by 7.5.
[c] To convert C·m to D, divide by 3.336×10^{-30}.

solid. After dispersion, the temperature should be increased by direct steam injection to prevent PVA buildup on steam coils and heated surfaces.

PVA is not a hazardous material, according to the American Standard for Precautionary Labeling of Hazardous Industrial Chemicals (ANSI Z129.1-1976). Short-term inhalation of PVA dust has no known health significance but can cause discomfort and should be avoided.

Uses

Poly(vinyl alcohol) has been approved by the FDA for food-contact and cosmetic applications. The largest U.S. application is in textile sizing. Other uses include adhesives, paper coatings, joint cements, water-soluble films, nonwoven fabric binders, and binders for phosphorescent pigments in television picture tubes.

PVA is an excellent textile warp size because of its superior strength, adhesion, flexibility, and film-forming properties with a wide variety of fibers. Machine conditions and fabric construction determine the proper PVA grade required for good weaving performance.

Partially hydrolyzed PVA is widely used as an emulsifier and protective colloid for the polymerization of emulsion adhesives. PVA improves wet tack and can reduce adhesive penetration into porous surfaces.

As a paper size, PVA improves strength and solvent resistance, and reduces porosity. Paper coated with PVA is used in many printing applications, and for packaging foods and chemicals.

PVA is widely used in the formulation of cement coating and finishes, especially in dry-wall joint cements.

PVA fiber has good strength characteristics and an attractive feel in fabrics. Japan is the only noncommunist country that widely uses PVA fiber in textiles.

PVA film can be produced by solution casting or extrusion. PVA is useful as a temporary protective coating for metals, plastics, and ceramics. The emulsifying, thickening, and film-forming properties of PVA are utilized in many cosmetic applications.

DAVID CINCERA
Air Products and Chemicals, Inc.

C.A. Finch in C.A. Finch, ed., *Polyvinyl Alcohol*, John Wiley & Sons, Inc., New York, 1973, pp. 183–202.

J.G. Pritchard, *Poly(Vinyl Alcohol) Basic Properties and Uses*, Gordon and Breach, Science Publishers, Inc., New York, 1970.

R.L. Davidson, ed., *Handbook of Water-Soluble Gums and Resins*, McGraw-Hill Book Co., New York, 1980, Chapt. 20.

VINYL CHLORIDE

Vinyl chloride, $CH_2=CHCl$, by virtue of the wide range of applications for its polymers, is one of the largest commodity chemicals in the United States. On an energy-equivalent basis, rigid poly(vinyl chloride) (PVC) is one the most efficient construction materials available (see Engineering plastics). The physical properties of vinyl chloride are listed in Table 1.

Reactions

Vinyl chloride is generally considered to be inert to nucleophilic replacement compared to alkyl halides. However, chlorine substitution does occur in the presence of palladium and other transition metals.

Vinylmagnesium chloride (Grignard reagent) can be prepared directly from vinyl chloride. A useful vinyllithium compound can be formed from vinyl chloride by means of a lithium dispersion containing 2 wt% sodium at 0–10°C.

The oxidation of vinyl chloride by a chlorine-sensitized reaction yields 74 vol% ClCHO and 25 vol% CO in the gas phase with 30–32% conversion.

Chlorination proceeds by an ionic or a radical path. In the liquid phase and in the dark, 1,1,2-trichloroethane forms when a metal catalyst ($FeCl_3$) is used; above 250°C, unsaturated chloroethylenes are produced. Hydrogen halides give the 1,1-adduct. Various vinyl chloride adducts form under Friedel-Crafts conditions.

Vinyl chloride is more stable than chloroalkanes to pyrolysis. Heating to 450°C gives small amounts of acetylene.

Manufacture

Vinyl chloride was first produced commercially in the 1930s by the reaction of hydrogen chloride with acetylene. After ethylene became plentiful in the early 1950s, other commercial processes were developed, including direct chlorination of ethylene to produce 1,2-dichloroethane (ethylene dichloride, EDC) and pyrolysis of EDC to vinyl chloride.

The development of ethylene–oxychlorination technology in the late 1950s encouraged new growth in the vinyl chloride industry. In this process, ethylene reacts with HCl and oxygen to produce ethylene dichloride. Combining direct chlorination, EDC pyrolysis, and oxychlorination provided the so-called balanced process for the production of vinyl chloride with no net consumption or production of HCl.

Direct chlorination	$CH_2=CH_2 + Cl_2 \rightarrow ClCH_2CH_2Cl$
Oxychlorination	$CH_2=CH_2 + 2 HCl + \frac{1}{2} O_2 \rightarrow ClCH_2CH_2Cl + H_2O$
Ethylene dichloride pyrolysis	$2 ClCH_2CH_2Cl \rightarrow 2 CH_2=CHCl + 2 HCl$
Overall reaction	$2 CH_2=CH_2 + Cl_2 + \frac{1}{2} O_2 \rightarrow 2 CH_2=CHCl + H_2O$

Direct chlorination of ethylene to ethylene dichloride (EDC) is conducted by mixing ethylene and chlorine in liquid EDC. The reaction may be run with a slight excess of ethylene or chlorine. The heat of reaction is removed by cooling with water or by operating the reactor at the boiling point of ethylene dichloride and allowing the product to vaporize.

In the oxychlorination process, ethylene reacts with dry hydrogen chloride and air or oxygen to produce EDC and water. In general, the reaction is carried out in the vapor phase in a fixed- or fluid-bed reactor containing a modified Deacon catalyst. Fluidized-bed reactors typically are cylindrical vessels equipped with internal cooling coils for heat removal and external or internal cyclones to minimize catalyst carry-over. Fluidization of the catalyst assures intimate contact between feed and product vapors, catalyst, and heat-transfer surfaces and results in a uniform temperature within the reactor. Fixed-bed reactors resemble multitube heat exchangers with the catalyst packed in vertical tubes.

In the air-based oxychlorination process with fluid- or fixed-bed reactors, ethylene and air are fed in slight excess to ensure high conversion of HCl.

The use of oxygen instead of air in fixed- or fluid-bed reactors permits operation at lower temperatures and results in improved efficiency and yield.

Dichloroethane used for pyrolysis to vinyl chloride must be of purity greater than 99.5 wt% because the cracking process is exceedingly susceptible to inhibition. It must also be dry to prevent corrosion.

Direct chlorination generally produces EDC with a purity greater than 99.5 wt%; except for removal of $FeCl_3$, little purification is necessary.

Ethylene dichloride from the oxychlorination process is generally less pure than when obtained by direct chlorination; it is usually washed with water and caustic solution.

Thermal cracking of EDC to vinyl chloride and hydrogen chloride occurs as a homogeneous, first-order free-radical chain reaction. The cracking is relatively clean at atmospheric pressure and 425–550°C. Commercial operations, however, generally operate at gauge pressures up to 2.5–3.0 MPa (360–435 psig) and temperatures of 500–550°C. Rapid cooling is essential. The hot effluent gases are normally quenched and partially condensed by direct contact with cold EDC.

Disposal of by-products from vinyl chloride manufacturing processes involves a number of methods because a variety of gaseous, organic liquid, and aqueous streams must be handled. Chemical treatment, scrubbing, sorption, incineration, or catalytic combustion is employed.

Vinyl chloride emissions in the balanced process occur from a number of sources, but mainly from the vinyl chloride purification system vents and the product-loading facility vents. Because of the toxicity of vinyl chloride, the EPA promulgated emission standards in 1975.

Recent developments in commercial vinyl chloride processes based on the balanced ethylene-feedstock route, which is used worldwide, include boiling-liquid reactors for direct chlorination, a trend toward oxygen-based oxychlorination, and efforts to improve conversion and minimize by-product formation in ethylene dichloride (EPC) pyrolysis.

Vinyl chloride processes based on acetylene or mixed acetylene–ethylene streams are limited by the availability and relative costs of such feedstocks (see Chlorocarbons; Acetylene-derived chemicals).

Health and Safety Factors

Current OSHA regulations require that no employee be exposed to vinyl chloride concentration greater than 1.0 ppm over any 8-h period, or 5.0 ppm averaged over any period not exceeding 15 min. Wherever exposure is above the OSHA limit, respirators are required.

Vinyl chloride is flammable. Large fires are very difficult to extinguish. Vapors represent a severe explosion hazard. Because of possible peroxide formation, vinyl chloride should be transported or handled under an inert atmosphere.

Uses

Vinyl chloride has gained worldwide importance as the precursor to poly(vinyl chloride). It is also used in a wide variety of copolymers. The inherent flame-retardant properties, wide range of plasticized compounds, and the low cost of the polymers from vinyl chloride have made it an important industrial chemical (see Flame retardants, halogenated fire retardants).

J.A. COWFER
A.J. MAGISTRO
BF Goodrich Co.

S. Patai, *Chemistry of the Carbon–Halogen Bond*, Pts. 1–2, John Wiley & Sons, Inc., New York, 1973.

R.W. McPherson, C.M. Starks, and G.J. Fryar, *Hydrocarbon Process.*, 75 (March 1979).

M. Sittig, *Vinyl Chloride and PVC Manufacture, Process and Environmental Aspects*, Noyes Data Corp., Park Ridge, N.J., 1978, p. 75.

Vinyl Chloride Monomer—Handling and Properties, 3rd ed., PPG Industries Brochure, Pittsburgh, Pa., Sept. 1977.

POLY(VINYL CHLORIDE)

Poly(vinyl chloride) (PVC) has a sales volume between polyethylene and polystyrene. The Stanford Research Institute predicts that by the year 2000 in the United States, PVC will be the leader, with an annual volume of 17×10^9 metric tons. This widespread use arises from a high degree of chemical resistance and a truly unique ability to be mixed with additives to give a large number of reproducible PVC compounds with a wider range of physical, chemical, and biological properties than any other plastic material.

Poly(vinyl chloride) is produced by the free-radical polymerization of vinyl chloride and has the following basic structure.

$$\left[CH_2CH \atop \underset{Cl}{|} \right]_n$$

where n varies from 300 to 1500.

Structure

Although PVC is available in latex form, dispersion resins and general-purpose resins account for nearly all sales of PVC resins. Dispersion resins are characterized by a primary particle size ($< 1 \mu$m) and an agglomerate size (see Fig. 1). The primary particles form the agglomerates which may be well over 20 μm dia. General-purpose resins are made by either the mass or suspension process and have a particle size of 80–200 μm. The individual 80–200 μm particles are porous and commercial PVC particle porosities ranging from about 0.2 to 0.45 cm^3/g.

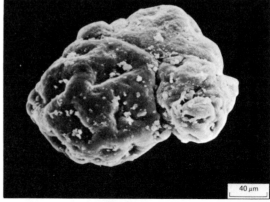

Figure 1. (Top) An electron micrograph of a typical PVC dispersion resin; (bottom) electron micrograph of a commercial PVC suspension-resin grain.

Addition of the vinyl chloride monomer units during polymerization can occur either in head-to-tail fashion, resulting in 1,3 positions for the chlorine atoms,

$$\text{+CH}_2\text{CHClCH}_2\text{CHCl+}_n$$

or head-to-head, tail-to-tail, placing the chlorine atoms in 1,2 positions.

$$\text{+CH}_2\text{CHClCHClCH}_2\text{+}_n$$

Saturated end groups are formed by chain transfer to monomer and polymer and by termination through disproportionations. Unsaturated chain ends result from termination by disproportionation and chain transfer to monomer. Because of the high transfer activity of the monomer, ca 60% of the polymer molecules are estimated to have unsaturated end groups. Long-chain branching can be caused by the incorporation of the terminal double bond of a polymer molecule into a growing chain:

$$\text{---CH}_2\dot{\text{C}}\text{HCl} + \text{---CH=CHCl} \rightarrow \text{CH}_2\text{CHClCH}\dot{\text{C}}\text{HCl}$$

or by intermolecular chain transfer to polymer:

$$\text{---CH}_2\dot{\text{C}}\text{HCl} + \text{---CH}_2\text{CHCl---} \rightarrow \text{---CH}_2\text{CH}_2\text{Cl} + \text{---CH}_2\dot{\text{C}}\text{Cl}$$

For the determination of total branching, the polymer is hydrogenated, and the chlorine removed with lithium aluminum hydride. The ratio of methyl to methylene groups is determined by ir spectroscopy. In conventional resins polymerized in bulk or suspension at 50–90°C, there are 0.2 to 2 branches per 100 carbon atoms; at lower temperature, the number of branches decreases. PVC is partially crystalline and the degree of crystallinity can be determined from x-ray patterns and correlates with the degree of syndiotacticity.

Properties

The molecular weight is commonly determined by measuring the intrinsic viscosity which is defined as follows:

$$[\eta] = \frac{\ln \eta_r}{c} \lim c \rightarrow 0$$

where η_r (relative viscosity) is the viscosity of the solution t divided by that of the pure solvent t_o. The quantity $\ln \eta_r/c$ is known as the inherent viscosity.

The glass-transition temperature T_g depends upon the polymerization temperature t_{pol} of the resin.

$$T_g = 93.4 - 0.20 \times t_{pol}$$

The glass-transition temperature of pure commercial PVC resins is ca 81°C. In the case of copolymers, T_g can be approximately determined by a weighted average of the transition temperatures, T_{g1} and T_{g2} of the homopolymer components

$$\frac{1}{T_g} = \frac{w_1}{T_{g1}} + \frac{w_2}{T_{g2}}$$

where w_1 and w_2 represent the weight fractions.

The melting point T_m of PVC cannot be measured directly because of the thermal instability of the resin. It is usually determined from the melting temperature T_m^* of solutions of the polymer in plasticizers. The melting-point temperatures increase linearly with decreasing polymerization temperature t_{pol}:

$$T_m = 257.5 - 0.81 \times t_{pol}$$

The linear coefficients of expansion of PVC are $(6-8) \times 10^{-5}$/°C above T_g. The specific heat values at constant pressure, c_p, in these two ranges are 1.046–1.255 and 1.757 J/(g.K) (0.25–0.30 and 0.42 cal/(g·K)), respectively. Thermal conductivity at 20°C is 0.1588 W/(m·K).

Above 100°C, PVC begins to decompose. Hydrogen chloride is evolved and the resin becomes discolored, brittle, and finally insoluble. The rate of HCl evolution at a given temperature increases with decreasing number-average molecular weight of the starting resin.

Networks are formed by elimination of HCl between chains. In the presence of oxygen, additional reactions take place, eg, formation of peroxides and β-chloroketones, as well as chain scission and network formation via radical mechanisms.

The frequency distribution of the polyene sequences can be determined by uv spectroscopy.

The primary objectives of chemical modification of PVC include increased solubility in inexpensive solvents, increased heat-distortion temperature, increased resistance to hot melt flow, introduction of ion-exchange capacity, preparation of polymer structures not otherwise available, improved stability to light and heat, and improved melt flow.

Chlorination can be carried out in the dry state or in an organic medium, eg, carbon tetrachloride, at moderate temperatures under the influence of uv irradiation. Chlorine content can be raised to ca 73 wt%, corresponding to the introduction of an additional chlorine atom per monomer unit. The T_g of a given resin increases with increasing degree of chlorination; it follows approximately the relation (in °C), as shown below:

$$T_g = (3.8 \times \text{wt\% Cl}) - 136$$

The structure of chlorinated PVC can be elucidated by chemical means, ir, nmr, and pyrolysis-gas chromatography. Chlorine can be incorporated into the polymer chains in the following ways:

$$\text{---CH}_2\text{CHCl---} + 1/2\,\text{Cl}_2 \begin{cases} \text{---CHClCHCl---} \\ \text{---CH}_2\text{CCl}_2\text{---} \end{cases}$$

Nucleophilic displacement of chlorine using thiol compounds gives products with enhanced physical properties.

Polymerization

The polymerization of vinyl chloride by a mass or bulk procedure, where a free-radical initiator is added to the liquid monomer, is normally difficult to control because of the heterogeneous reaction products (see also Polymerization mechanisms and processes).

In a two-stage process, the first stage is carried out in a prepolymerizer, a vertical reactor equipped with a flat-blade turbine stirrer and baffles, where the monomer is polymerized for ca 1–1.5 h with 7–10% conversion. In the second stage, the mixture from the prepolymerizer together with more monomer and initiator is transferred into a horizontal or vertical autoclave. The reaction proceeds through the liquid stage and, at ca 25% conversion, liquid monomer disappears and a powder is obtained; considerable heat is evolved.

The mass process has low utility costs with minimum water consumption, no drier operation, and few raw materials. However, suspension polymerization offers more flexibility.

In suspension polymerization, the monomer is first finely dispersed in water by vigorous agitation. Polymerization is started by means of monomer-soluble initiators. Suspension stabilizers and other suspending agents minimize coalescence of the growing grains by forming a protective coating.

The molecular weight of the resulting polymer is independent of the concentration of the initiator and exhibits only a slight increase with increasing conversion. Molecular weight decreases with increasing temperature, which is therefore controlled.

Suspension polymerization is used for an estimated 82% of U.S. production. Most current reactors are water-jacketed and lined with glass or stainless steel to minimize polymer buildup on the walls. Ratios of water to vinyl chloride are (1.2–4):1.

Suspending agents include maleic anhydride–vinyl acetate copolymers, partially hydrolyzed vinyl acetate polymers, precipitated inorganic carbonates, phosphates and silicates, water-soluble cellulose derivatives, and methacrylic acid copolymers.

Initiators, eg, peroxydicarbonates or peroxyesters, are employed in concentrations of 0.03–0.1% based on weight of monomer. The selection of initiators is important from the productivity point of view, where combinations of initiators may be employed to achieve more uniform and linear reaction rates. Typical polymerization times are 4–10 h, depending upon the mol wt of the resin being prepared, as well as the heat-removal capacity of the reactor system.

Mass and suspension kinetics and mechanisms. The suspension droplet can be simply considered to be a mass polymerization on a small scale. The polymer precipitates when the chain has reached 10–20 monomer units in length. The precipitated polymer is swollen by monomer, and the reduced termination rate in the swollen gel phase may be considered responsible for the observation that with increasing conversion, the polymerization rate increases. Polymerization is accompanied by ca 35% shrinkage.

The presence of two phases, a monomer-swollen polymer phase and the dilute-liquid monomer phase, forms the basis for all kinetic model descriptions. The polymerization rate is slower in the dilute phase because of lower mobility and decreased termination rate of the growing polymer chain.

Chain transfer to monomer is the predominant transfer reaction affecting the molecular-weight distribution. The chain-transfer constant C_m, defined as the ratio of the rate coefficient of transfer to monomer to that of chain propagation, is 6.25×10^{-4} at 30°C and 2.38×10^{-3} at 70°C.

In emulsion polymerization, vinyl chloride is emulsified in water by means of surface-active agents. Water-soluble initiators are added, and polymerization begins when a radical enters a monomer-swollen micelle. Chain termination takes place within each latex particle by the usual radical–radical interaction. This technique offers the possibility of obtaining high molecular weight polymers at very rapid rates. It is postulated that under certain assumptions, the overall reaction rate is proportional to the number N of latex particles per mL of emulsion.

Residual soaps affect the polymer clarity, viscosity, electrical resistivity, water absorption, and heat stability. Typical initiators are hydrogen peroxide, organic peroxides, peroxydisulfates, and redox systems. Oxygen is excluded, and the pH of the mixture is maintained at 6–8.

Vinyl chloride can be copolymerized with a large variety of monomers (M_2), such as other unsaturated halogenated hydrocarbons, styrene and its halogenated derivatives, vinyl esters and ethers, olefins, dienes, esters and various heterocompounds.

The most important product is the copolymer of vinyl chloride and vinyl acetate (3–20 wt%); 75% of U.S. production is manufactured by the suspension process, the remainder by emulsion. Molecular weight of the product may be controlled by the addition of chain-transfer agents, such as trichloroethylene.

A vinyl chloride–propylene copolymer was first introduced for a glass-clear plastic-bottle application (see Barrier polymers). Olefins act as chain-transfer agents by transfer of allylic hydrogens on the methyl groups. Propylene is reported to have a chain-transfer activity ten times greater than vinyl chloride.

Other commercial copolymers made by emulsion or suspension techniques are acrylonitrile, 2-ethylhexyl acrylate, vinylidene chloride, and vinyl isobutyl ether.

Compounding

Poly(vinyl chloride) is always mixed with other ingredients before being processed. A thermal stabilizer is nearly always required. Stabilizers are based on metallic salts of inorganic and organic acids and phenols, organometallics (eg, tin and antimony), epoxy compounds, and phosphates. Organotins are efficient heat stabilizers (qv) and provide good initial color, long-term stability, and excellent product clarity. Lead stabilization systems are used in wire and cable applications. Stabilizers approved by the FDA are required in PVC compounds intended for food- and beverage-packaging applications.

Lubricants are another essential part of PVC formulations that are extruded or injection molded. External lubricants reduce the tendency to stick to the hot metal surfaces of the processing machinery, whereas internal lubricants increase the flow of the individual PVC resin particles. Typically, lubricants are used at 0.1–4 phr.

Plasticization. Solutions of PVC prepared at elevated temperatures with high boiling solvents are flexible and elastic and exhibit a high degree of chemical inertness and solvent resistance. This behavior is due to unsolvated crystalline regions that act as cross-links and allow the PVC to accept large amounts of solvent. Thus, the rigid PVC is transformed to a rubberlike material with stable properties over a wide temperature range. The solvents are known as plasticizers. They are added at 15–20 phr for semirigid compounds and at > 100 phr for soft flexible compounds.

A plasticizer must be compatible with the resin in order to resist migration and extraction by liquids, such as water. It should be nonvolatile and boil above 400°C. Other desirable properties include nonflammability, good heat and light stability, lack of toxicity, and low cost. Primary plasticizers can be used alone and are highly compatible with PVC at concentrations as high as 150 phr. They are commonly esters of alcohols containing 8–10 carbons, eg, phthalate esters. Secondary plasticizers, because of limited compatibility, cannot be used alone. They are used to confer some special property on the compound or to reduce cost. Examples are chlorinated paraffins, esters of fatty acids, and alkylated benzene and petroleum fractions (see Plasticizers).

Both organic and inorganic pigments can be added to PVC in concentrations from 0.1 phr of a toner for clarity up to 20 phr of a pigment for weatherability. Dyes are used at low concentrations in transparent products.

Fillers are used both to reduce cost and gloss. The preferred filler is calcium carbonate.

Although PVC itself and most rigid PVC compounds are resistant to attack by microorganisms, flexible PVC products can require the addition of a biocide.

In rigid applications, toughness can be increased with impact modifiers, generally materials of low modulus with limited compatibility used at concentrations below 15 phr; they are, however, expensive.

Processing aids, eg, styrene–acrylonitrile (SAN) copolymers, increase the melt strength of rigid PVC compounds during processing.

Flexible PVC compounds that contain > 30 phr of plasticizer may require flame retardants (qv).

Cellular PVC in both open- and closed-cell forms may be made by a variety of techniques, such as whipping air into a plastisol or dispersion of fine-particle PVC in plasticizer; incorporation of a gas under pressure in a vinyl extrusion system near the die while the compound is in the melt state; or use of chemical blowing agents (see Foamed plastics).

Applications

Latex. PVC latices prepared by emulsion polymerization are true colloidal dispersions of submicrometer particles ($< 0.2 \ \mu m$) in water, stabilized by a surfactant system. Vinyl latices offer solvent-free methods of coating, impregnating, or saturating a variety of materials, eg, paper and woven and nonwoven fabrics. Treatment with vinyl latex improves abrasion resistance, toughness, and flame retardance and gives resistance to oils, water, and many chemicals. Vinyl latices are typically sold with 50–70% solids and a specific gravity of ca 1.09–1.18, surface tension of 33–45 mN/m (= dyn/cm), and viscosity < 50 mPa·s (= cP). Additives include thickeners, protective colloids, surfactants, plasticizers, antifoaming agents, pigments, and fillers.

Dispersion resins. Dispersion resins are mixed with plasticizer to form colloidal dispersions known as plastisols. In some applications, the plastisol is diluted with a volatile solvent that evaporates during the fusion process. Such a diluted plastisol is known as an organosol. Plastisols and organosols may be applied by spread coating or molding.

About 20–25% of U.S. dispersion-resin manufacture is used for resilient vinyl flooring. Dispersion-resin is also used for vinyl-coated fabrics. Metals are first coated with a solvent-based adhesive primer which must be flash-baked or partially cured before the application of the plastisol or organosol. Dip-coating and various molding techniques are also used. Both organosols and plastisols can be sprayed; coating thicknesses from 0.05 to > 1.5 mm can be obtained. In addition, plastisol can be extruded, a method that offers a convenient way for making extremely soft products.

General-purpose resins. General-purpose resins comprise ca 80% of all the PVC resin produced and are used chiefly to make so-called 100% vinyl products by a variety of molding and extrusion techniques. The

Table 1. Basic General-Purpose Flexible PVC Formulation

Component	Parts by weight
PVC resin, high mol wt[a]	100.0
plasticizer	
dioctyl phthalate	30.0–80.0
processing aid, epoxidized soybean oil	5.0
stabilizer, barium–cadmium	3.0
filler, calcium carbonate	≤30.0
lubricant, stearic acid	0.5
pigment	≤3.0

[a] Inherent viscosity, >0.95.

resin is mixed with the various compounding ingredients required (see Table 1). The mixture may be a dry free-flowing powder similar in consistency to the original resin, melted and fused cubes (0.256 cm³), or pellets of various dimensions. Powder mixing is a batch process with a cycle time of typically 2–4 min for high speed mixers that can handle up to 550 kg.

Melt mixers can be of either continuous or batch type. Typical batch-melt mixers or intensive internal mixers, as they are often called, are the Banbury and the Intermix.

In continuous mixers, similar to extruders, the powder is compressed, fused, and melted, and then transferred to a cooling line. Typical examples are the twin-rotor Farrel continuous mixer (FCM), the twin-screw Kombiplast, and the Ko-Kneader.

Most PVC products made from general-purpose resins are extruded for melt preparation by forcing the melt through a die. In sheet, profile, and pipe extrusion, the product is shaped by the die. In injection molding, the cooled mold is filled by a screw-type extruder.

Rigid pipe is extruded by forcing the PVC melt through an annular die with the hole shaped by an internal mandrel supported and centered by a device called a spider. For large diameter pipe, a floating plug seal is attached by means of a rope to the center of the spider. Rigid pipe is available in sizes of 0.95–45.72 cm, and is pressure-rated for ca 0.34–2.07 MPa (50–300 psi).

Rigid sheet is extruded through electrically heated dies and cooled on three stacked temperature-controlled rolls. It is generally 0.254–3.175 mm thick and has application in a variety of architectural uses. It is produced in widths up to 190.5 cm.

Other applications include rigid profiles, flexible wire and cable insulation, and flexible film. PVC is readily chlorinated, and PVCs with chlorine contents > 70% (CPVC) have been made.

Environmental Aspects

As PVC products enter the disposal stage, they are easily collected and can be disposed of by landfill, incineration, or recycling. In a sanitary landfill, PVC wastes do not contribute to instability and do not give off toxic decomposition products. Proper incineration operation and design can control HCl and other acidic emissions to eliminate these as air-pollution hazards. Solid waste can be reground and reprocessed.

JOHN A. DAVIDSON
KEITH L. GARDNER
B.F. Goodrich

R.N. Burgess, ed., *Manufacture and Processing of PVC*, Macmillan Publishing Co., Inc., New York, 1982.

L.I. Nass, ed., *Encyclopedia of PVC*, Marcel Dekker, Inc., New York, 1976.

VINYL ETHER MONOMERS AND POLYMERS

Monomers

Vinyl ethers have been used to convert hydroxyl groups to acetals, which are stable under neutral or alkaline conditions and are easily hydrolyzed with dilute acid. Several patents describe vinylation of alcohols in systems similar to those used in the Wacker oxidation of ethylene. Another method that gives very high yields is vinyl exchange with vinyl acetate.

$$\text{ROH} + \text{CH}_2\text{=CHOCCH}_3 \xrightarrow{\text{PdCl}_2, \text{Na}_2\text{WO}_4} \text{ROCH=CH}_2 + \text{CH}_3\text{COH}$$

Polymers

Poly(alkyl vinyl ether)s with broad molecular weight distributions are obtained from iodine-initiated polymerizations of alkyl vinyl ethers in polar solvents.

Physical characteristics of the homopolymers range from viscous liquid, through sticky liquids and rubbery solids, to brittle solids. Poly(methyl vinyl ether) (PVM) is soluble in water in all proportions at room temperature because of hydrogen bonding. The glass-transition temperatures of the amorphous straight-chain alkyl vinyl ethers decrease with increasing length of the side chain (see Table 1).

The only vinyl ether homopolymer manufactured in the United States (by GAF Corporation) is PVM. Dry PVM polymer or aqueous solution are not a primary irritant or sensitizer.

Homopolymerizations. Cationic initiators, eg, Lewis acids, are the preferred catalysts for preparing alkyl vinyl ether homopolymers. Atactic alkyl vinyl ether homopolymers are readily prepared by initiation wth Friedel-Crafts type catalysts, eg, boron fluoride, aluminum trichloride, stannic chloride, etc, either in bulk or in inert dry solvents.

Vinyl ether monomers with highly branched alkyl groups generally exhibit greater reactivity and greater tendency to form stereoregular polymers than straight-chain monomers.

Molecular sieves (qv) and vanadium pentoxide, V_2O_5, are heterogeneous initiators for the atactic homopolymerization of vinyl ether monomers at room temperature. Initiators that have been shown to give stereoregular vinyl ether homopolymers are listed in Table 2.

The significant element in steric control of alkyl vinyl ether cationic polymer chain growth is the degree of association of the growing cation and counterion. Polymerization of the α-methyl vinyl alkyl ethers tends to yield principally syndiotactic products even in nonpolar solvents.

Table 1. Glass-Transition Temperatures of Amorphous Poly(Vinyl Ether)s and Melting Points of Crystalline Poly(Vinyl Ether)s

Poly(vinyl ether)	T_g, °C	Mp, °C
methyl	− 34	144
ethyl	− 42	
isopropyl	− 3	191
n-butyl	− 55	
isobutyl	− 19	170
t-butyl		238
2-ethylhexyl	− 66	
n-hexyl	− 77	
n-octyl	− 80	

Table 2. Initiator Systems for the Stereoregular Polymerization of Alkyl Vinyl Ethers

Monomer[a]	Initiator	Polymerization solvent	Temperature, °C	Conversion, %	Tacticity
allyl VE	SbCl₅	C₆H₅CH₃	− 10	100	isotactic
IBVE	I₂	(C₂H₅)₂O	− 78	71	syndiotactic
IBVE	POCl₃ + Al(C₂H₅)₃	C₆H₆	30	28	isotactic
IBVE	SOCl₂ + Al(C₂H₅)₃	C₆H₆	0	58	isotactic
IBVE	V₂O₅ + Al(C₂H₅)₃	C₆H₆	30	20	isotactic
IBVE	C(channel black)	CCl₄	20	84	[b]
IBVE	Fe₂O₃	C₆H₅CH₃	25	43	isotactic
ethyl vinyl ether	Fe₂O₃	C₆H₅CH₃	− 20	97	[b]
MVE	Fe₂O₃	C₆H₅CH₃	− 10	60	[b]

[a] IBVE = isobutyl vinyl ether; MVE = methyl vinyl ether.
[b] Not specified.

Heterogeneous initiator systems frequently provide alkyl vinyl ether polymers having greater stereoregularity than homogeneous initiators, especially at ambient temperatures.

Divinyl ether and the divinyl ethers of ethylene glycol diethylene glycol (DEGDVE), and 1,4-butanediol (B1DDVE) can be polymerized in dilute solution with either free-radical or cationic initiators to give soluble polymers with highly cyclized, low molecular weight structures.

homopolymer of divinyl ether

Cationic-initiated cyclopolymerization of the divinyl ethers of aliphatic diols yields cyclic polymeric products, which are different structurally from those produced by free-radical-initiated cyclopolymerization.

Irradiation of vinyl ether monomers by ^{60}Co γ rays results in a small number of free ions that do not undergo geminate recombination. These can initiate the cationic polymerization of vinyl ether monomers by a free-ion mechanism. Vinyl ether polymerization propagation rates decrease dramatically as solvent polarity increases.

Uses

Applications for the poly(alkyl vinyl ether)s include adhesives (qv), surface coatings, lubricants, greases, elastomers, molding compounds, fibers, films, and chemical processing.

The principal commercial applications for PVM are for viscosity control and dry-film flexibility in uv photoresist coating solutions; as a tackifier and adhesion promoter for acrylic pressure-sensitive adhesives used in tapes, labels, and decals; and as a semipermeable membrane for reverse osmosis.

Interpolymers with maleic anhydride are the most significant commercial class of vinyl ether copolymers. Poly(methyl vinyl ether-co-maleic anhydride) (PVM–MA) is a white, fluffy powder soluble in ketones, ester, lactams, and aldehydes. It is used for latex paints and cosmetics (qv) and in hair sprays, textile sizes, thickeners, and adhesives; inclusion in food-packaging adhesives has been approved by the FDA. PVM–MA is not irritating or sensitizing.

Addition of potassium dichromate and the disodium salt of PVM–MA to oil-well drilling muds prevents loss of water from the mud to the surrounding formation and reduces swelling of clay cuttings from the drilling operation. Dilute (1 wt%) solutions of PVM–MA in dimethyl sulfoxide supply unusual photochromic compositions that are colored by heat or diffuse daylight and decolorized by uv or double-bond-cleaving reagents (see Chromogenic materials).

A copolymer of maleic anhydride with octadecyl vinyl ether is available as a toluene solution under the trade name Gantrez AN-8194. The waxy film coatings of this copolymer are effective as antiblocking agents and release coatings (see Abherents).

A lightly cross-linked version of PVM–MA copolymer is available for use in textile print-paste systems. This copolymer (Gaftex PT) belongs to the class of water-soluble polymers known as microgels. Impact-resistant plastics with high flexural strength, high softening points, and good gas and vapor resistance result from the copolymerization of butyl vinyl ether with acrylonitrile and indene in the presence of acrylonitrile–butadiene latex.

Copolymerizations. Cationic initiation of the polymerization of a mixture of different vinyl ether monomers or a mixture of a vinyl ether monomer with another electron-donating monomer does not always produce true copolymers. Frequently, a homopolymer forms first. The free-radical-initiated copolymerization of the vinyl ethers with monomers bearing strong electron-withdrawing substituents gives uniform, high molecular weight, high conversion copolymers, which are of great commercial interest.

Completely cyclized, alternating copolymers containing a divinyl ether–maleic anhydride ratio of 1:2 have been obtained. Radical copo-

lymerization of 1,4-butanediyl divinyl ether with maleic anhydride proceeds very rapidly and produces insoluble gels.

Eugene V. Hort
R.C. Gasman
GAF Corporation

R.W. Aben and H.W. Scheeren, *J. Chem. Soc. Perkin Trans.* **1**, 3132 (1979).

E.C. Leonard, ed., *Vinyl and Diene Monomers*, Pt. 1, Wiley-Interscience, New York, 1970.

N.M. Bikales in N.M. Bikales, ed., *Encyclopedia of Polymer Science and Technology*, Vol. 14, Wiley-Interscience, New York, 1971, pp. 511–521.

N-VINYL MONOMERS AND POLYMERS

Monomers

Vinylamines. Attempts to isolate vinylamine have been unsuccessful, although its presence as a transient intermediate has been detected by spectroscopic methods. Polyvinylamine is prepared by modifying another polymer.

Dialkylvinylamines are moderately stable at low temperatures in the absence of oxygen. They are prepared by treatment of a secondary amine with acetaldehyde or acetylene (see Acetylene-derived chemicals).

$$CH_3CHO + (C_2H_5)_2NH \rightarrow (C_2H_5)_2NCH=CH_2 + H_2O$$

Trialkylamines react with acetylene to give quaternary vinyltrialkylammonium compounds.

$$(CH_3)_3N + HC\equiv CH + H_2O \rightarrow (CH_3)_3\overset{+}{N}CH=CH_2\ OH^-$$

$$(CH_3)_3\overset{+}{N}H\ Cl^- + HC\equiv CH \rightarrow (CH_3)_3\overset{+}{N}CH=CH_2\ Cl^-$$

Weakly basic aromatic secondary amines, such as pyrrole, indole, and carbazole, give relatively stable vinylamines. Disubstituted N-vinylamines are more stable than the monosubstituted. The only commercially available vinylamines are N-vinylcarbazole and 1-vinylimidazole.

N-Vinyl isocyanate. Vinyl isocyanate is prepared by heating acryloyl chloride with sodium azide, by combining hydrocyanic acid and acetylene, or by cracking carbamoyl chlorides.

N-Vinylamides and N-vinylimides. Amides and imides are readily vinylated with acetylene by dehydration of hydroxyethyl substituents, by pyrolysis of ethylidenebisamides, or by vinyl exchange.

N-Vinyl-2-pyrrolidinone. N-Vinyl-2-pyrrolidinone, commonly called vinylpyrrolidone or VP, is mainly used to prepare the homopolymer, poly(N-vinyl-2-pyrrolidinone) (PVP), but is also used as a comonomer. It is readily polymerized with cationic (BF$_3$) or anionic (KNH$_2$) initiators, and alkali metals and their oxides. Physical properties are given in Table 1.

Catalytic hydrogenation gives N-ethyl-2-pyrrolidinone. Hydroxyl compounds add to the double bond. Alcohols require acid catalysis, but phenols react exothermically without a catalyst. Amides add across the vinyl double bond in the same fashion as hydroxyl groups. Fluoroacetic acid catalysts at 85–90°C gives a > 90% yield of the dimer, 1,1'-(3-methyl-1-propene-1,3-diyl)bis-2-pyrrolidinone.

The vinylation of 2-pyrrolidinone is carried out under alkaline catalysis analogous to the vinylation of alcohols. N-Vinyl-2-pyrrolidinone is

Table 1.   Properties of N-Vinyl-2-Pyrrolidinone

Property	Value
freezing point, °C	13.5
boiling point, °C$_{kPa}$[a]	$46_{0.1333}$
	$88_{1.33}$
	$147_{13.3}$
	$219_{101.3}$
density, g/cm^3 at 25°C	1.04
refractive index, n_D^{25}	1.511
viscosity at 25°C, mPa·s (= cP)	2.07
flash point, open cup, °C	98.4
fire point, °C	100.5

[a] To convert kPa to mm Hg, multiply by 7.5.

Table 2.   Specifications of Technical PVP Grades

Designation	Form	K range	Water, % max	Ash, % max	Residue monomers, % max
PVP K-15	powder	12–18	5	0.02	1.0
PVP K-30	powder	26–35	5	0.02	1.0
PVP K-60	aqueous solution	50–62	55	0.02	1.0
PVP K-90	aqueous solution	80–100	80	0.02	1.0
PVP K-90	powder	80–100	5	0.02	1.0
Polyclar AT	powder	cross-linked	5		

Table 3.   Reactivity Ratios (r) for Free-Radical Copolymerization of N-Vinyl-2-pyrrolidinone (M_1)

Comonomer (M_2)	r_1	r_2
acrylonitrile	0.06 ± 0.07	0.18 ± 0.07
allyl alcohol	1.0	0.0
allyl acetate	1.6	0.17
allylidene diacetate	0.92	0.94
crotonic acid	0.85	0.02
maleic anhydride	0.16 ± 0.03	0.08 ± 0.03
methyl methacrylate	0.005 ± 0.05	4.7 ± 0.05
trichloroethylene	0.54 ± 0.04	< 0.01
tris(trimethylsiloxy)vinylsilane	4.0	0.1
vinyl chloride	0.38	0.53
vinyl cyclohexyl ether	3.84	0.0
vinyl phenyl ether	4.43	0.22
vinylene carbonate	0.4	0.7

available in tank cars and trailers and in drums. Shipping containers are normally steel or stainless steel.

Contamination of N-vinyl-2-pyrrolidinone with strong acids must be avoided to prevent polymerization. N-vinyl-2-pyrrolidinone is neither a skin sensitizer nor a primary irritant.

The principal use of N-vinyl-2-pyrrolidinone is as a monomer for the preparation of poly(N-vinyl-2-pyrrolidinone) (PVP) homopolymer and various copolymers (see also Acetylene-derived chemicals).

Polymers

N-Vinyl monomers are enamines (or derivatives thereof) and possess high electron density at the β-carbon. The N-vinylimides and N-vinyl-amides, especially the cyclic analogues, where one or two electron-withdrawing moieties are attached to the nitrogen atom, are relatively stable and readily polymerized.

N-Vinylcaprolactam polymerizes readily. It has been copolymerized with vinyl formate, vinyl acetate, methyl vinyl ether, and isopropenyl methyl ether.

Poly(N-vinylcarbazole) can be cast into films from chloroform, toluene, methylene chloride, and ethylene chloride.

Poly(N-vinyl-2-pyrrolidinone). PVP derives its commercial success from its biological compatibility, low toxicity, film-forming and adhesive characteristics, unusual complexing ability, and inert behavior toward salts, acids, and thermal degradation. In the United States, PVP is sold in pharmaceutical and technical grades (see Table 2). It was widely used in Germany as a blood-plasma substitute during World War II.

Poly(N-vinyl-2-pyrrolidinone) consists of linear N-vinyl-2-pyrrolidinone groups of varying degrees of polymerization.

Molecular weight is determined by osmometry, ultracentrifugation, light-scattering photometry, and solution-viscosity techniques. Different molecular weights are distinguished by the K value, which is determined at 1% wt/vol of a given PVP sample in aqueous solution. Solubility in water is limited only by viscosity.

Films of PVP are clear, transparent, glossy, and hard. They can be cast from water, methyl alcohol, chloroform, or ethylene dichloride. A number of synthetic and natural resins can be combined with PVP to yield clear solutions and films.

The single most attractive property of PVP is its binding capability. This has permitted utilization in numerous commercial applications. PVP reduces the binding of albumin to salicylic acid.

Poly(N-vinyl-2-pyrrolidinone) is manufactured by bulk, solution, or suspension polymerization under free-radical catalysis. N-Vinyl-2-pyrrolidinone has been copolymerized with a variety of comonomers in both solution and emulsion systems (see Table 3). Poly(N-vinyl-2-pyrrolidinone) has been used to form graft copolymers.

Poly(N-vinyl-2-pyrrolidinone) is not a skin or eye irritant, or a skin sensitizer. Apparently, it is not absorbed from the gastrointestinal tract. The lower molecular weight material is readily excreted through the kidneys.

Poly(N-vinyl-2-pyrrolidinone) and its copolymers are widely used in hair and skin-care. It acts as a protective colloid and silver halide suspending agent and is useful in oil recovery. Incorporation of PVP into hydrophobic fibers such as polyacrylonitrile polyesters, nylon, and cellulosic material greatly increases their dyeability. Poly(N-vinyl-2-pyrrolidinone) has been formulated with borax in a washing formulation. The ability of PVP to complex with certain polyphenolic compounds (tannins and others) has led to its use in the clarification and chillproofing of fruit beverages. The lack of toxicity and high solubility of poly(N-2-vinyl-pyrrolidinone) have made it ideally suited for a number of pharmaceutical applications. As a blood extender, it is nonantigenic and requires no cross-matching; infections are avoided.

An important application is derived from the ability to complex and solubilize poorly soluble drugs and hence increase their bioavailability and efficiency. Highly cross-linked PVP, generically termed Crospovidone, NF, and sold in the United States as Polyplasdone XL (GAF), is a tablet excipient.

EUGENE V. HORT
B.H. WAXMAN
GAF Corporation

C.E. Schildknecht, *Vinyl and Related Polymers*, John Wiley & Sons, Inc., New York, 1952, pp. 658–659.

USP XX Second Supplement, The United States Pharmacopeial Convention, Inc., Rockville, Md., 1981, p. 122.

Poly Vinyl Pyrrolidone, Technical Bulletin 2303-112, GAF Corporation, Wayne, N.J., 1982.

VINYL POLYMERS, POLY(VINYL FLUORIDE). See Fluorine compounds, organic.

VINYLTOLUENE. See Styrene.

VIRAL INFECTIONS, CHEMOTHERAPY. See Chemotherapeutics, antiviral.

VISCOMETRY. See Flow measurement; Rheological measurements.

VISCOSE. See Rayon.

VISCOSITY. See Flow measurement; Rheological measurements.

VISCOSITY BREAKING, VISBREAKING. See Petroleum, refinery processes.

VITAMIN K. See Blood, coagulants, and anticoagulants; Prostaglandins.

VITAMINS

SURVEY

A list of the vitamins and their vitamers is shown on Table 1.

Solubilities

Vitamins are classified by solubility characteristics into the fat-soluble group and the water-soluble group containing the vitamins of the B series, including niacin, pantothenic acid, folic acid, biotin, and vitamin C. The fat-soluble vitamins are derivatives of partially cyclized isoprenoid polymers (A, E, K) or of sterol derivatives (D).

So-Called Vitamins

Several substances do not fit even the broad definition of the term vitamin and are not generally recognized as such, although they are called vitamins U, T, P, and L.

Production

The commercial production of vitamins may be by chemical synthesis, isolation from natural sources, or microbial fermentation.

PETER W. FRYTH
Hoffmann-LaRoche, Inc.

Table 1.   Vitamins, Vitamers, and Their Characteristics

| | | Vitamers | | |
Vitamin	Other names	Natural	Synthetic	Characteristics
A	vitamin A_1, retinol, axerophthol	α-carotene, β-carotene, γ-carotene vitamin A esters	vitamin A acid vitamin A esters	skin and mucosal integrity, maintenance of vision
C	ascorbic acid antiscorbutic vitamin	dehydroascorbic acid	6-desoxyascorbic acid isoascorbic acid	antioxidant, supports capillary integrity, collagen formation
D	antirachitic vitamin	vitamin D_2 (ergocalciferol), vitamin D_3 cholecalciferol	irradiated vitamin D_4	regulation of calcium and phosphorus metabolism. bone growth
E	α-tocopherol	β-tocopherol, γ-tocopherol,	racemic α-tocopherol, tocopherol esters	biological antioxidant, cell respiration
K	vitamin K_1, phylloquinone, coagulation vitamin	vitamin K_2	menadione, menadiol	blood clotting
B_1	thiamine, aneurin	thiamine pyrophosphate, cocarboxylase, thiamine orthophosphate	thiamine disulfide, acetylated thiamine	carbohydrate metabolism coenzyme in pyruvate metabolism, antineuritic
B_2	riboflavin, vitamin G, lactoflavin, hepatoflavin, ovoflavin, verdoflavin	riboflavin mononucleotide, FMN; riboflavin dinucleotide, FAD	7, 8, and/or 10-methyl or ethyl compounds	coenzyme in respiratory enzyme systems' cellular redox systems
B_6	pyridoxine, pyridoxol	pyridoxal, pyridoxamine, pyridoxal phosphate		coenzyme in some phases of amino acid metabolism
B_{12}	cobalamin, cyanocobalamin	hydroxycobalamin, aquocobalamin	nitrocobalamin	coenzyme in lipid, protein nucleic acid synthesis antipernicious extrinsic factor
biotin	vitamin H, anti-egg-white-injury factor	desthiobiotin	biotinsulfoxide	coenzyme for some carboxylations for pyruvic oxidase
folic acid	folacin, pteroylglutamic acid, antianemia factor	tetrahydrofolic acid pteroyltriglutamic acid, citrovorum factor, leucovorin	pteroic acid xanthopterin	synthesis of nucleic acid, coenzyme in purine, pyrimidine metabolism, antianemia requirement
niacin	nicotinic acid, antipellagra factor, vitamin B_3^a	niacinamide, NAD (CoI), NADP (CoII)	niacin esters, nikethamide, niacinamide	coenzyme in dehydrogenase and other systems, pellagra prevention, vasodilation
pantothenic acid	chick antidermatitis factor		pantothenic acid esters	component of coenzyme A in carbohydrate acetyl transferase system

[a] Vitamin B_3 has also been used to name pantothenic acid.

R.J. Kutsky, *Handbook of Vitamins and Hormones*, 2nd ed., Van Nostrand Reinhold Co., New York, 1981.

Vitamin–Mineral Safety, Toxicity, and Misuse, National Nutrition Consortium, Inc., The American Dietetic Association, Chicago, Ill., 1978, pp. 1-20.

C.S. Sodano,*Vitamins Synthesis, Production and Use*: *Advances Since 1970*, Noyes Data Corp., Park Ridge, N.J., 1978.

ASCORBIC ACID

Ascorbic acid (L-ascorbic acid, L-*xylo*-ascorbic acid, L-*threo*-hex-2-enonic acid γ-lactone) (1) is the name recognized by the IUPAC-IUB Commission on Biochemical Nomenclature for vitamin C. The name implies the vitamin's antiscorbutic properties, namely, the prevention and treatment of scurvy. L-Ascorbic acd is widely distributed in plants and animals. Humans, other primates, guinea pigs, and some bats, birds, and fish lack a liver enzyme, L-gulono-γ-lactone oxidase, cannot synthesize ascorbic acid and require the vitamin from exogenous sources to survive. The pure vitamin ($C_6H_8O_6$, mol wt 176.13) is a white crystalline substance derived from L-gulonic acid, a sugar acid, and synthesized both biologically and chemically from D-glucose. L-Ascorbic acid and its main derivatives are officially recognized by regulatory agencies and included in compendia such as the *United States Pharmacopeia* (USP) and the Food Chemicals Codex (FCC). The most significant characteristic of L-ascorbic acid (1) is its oxidation to dehydro-L-ascorbic acid (2) (L-*threo*-2,3-hexodiulosonic acid γ-lactone).

(1) L-ascorbic acid   (2) dehydro-L-ascorbic acid

Distribution

Ascorbic acid exists as a pool of ascorbate, which is distributed throughout the body with high concentrations in specific tissues. It is also present in vegetables and fruit.

Physical Properties

The physical properties of both synthetic L-ascorbic acid and natural vitamin C have been examined, and the data were comparable for both substances. Table 1 contains a summary of L-ascorbic acid's physical properties.

Table 1.   Physical Properties of L-Ascorbic Acid

Property	Characteristics
appearance	white, odorless, crystalline solid with a sharp acidic taste
formula; mol wt	$C_6H_8O_6$; 176.13
crystalline form	monoclinic; usually plates, sometimes needles
mp, °C	190–192 (dec)
density, g/cm³	1.65
optical rotation	$[\alpha]^{25}$ + 20.5° to +21.5° (c = 1 in water)
	$[\alpha]^{23}$ + 48° (c = 1 in methanol)
pH	
5 mg/mL	3
50 mg/mL	2
pK_1	4.17
pK_2	11.57
solubility in water, g/mL	0.33
Spectral properties	
uv	pH 2: $A_{1cm}^{1\%}$ 695 at 245 nm (nondissociated form)
	pH 6.4: $A_{1cm}^{1\%}$ 940 at 265 nm (monodissociated form)

Chemical Properties

The reversible oxidation–reduction with dehyro-L-ascorbic acid is L-ascorbic acid's most important chemical property and the basis for known physiological activities and stabilities. Dehydro-L-ascorbic acid has been prepared by uv irradiation and by oxidation with halogens, ferric chloride, hydrogen peroxide, 2,6,-dichlorophenol, indophenol, neutral potassium permanganate, selenium oxide, and many others. It has been reduced to L-ascorbic acid by hydrogen iodide, hydrogen sulfide, etc, without affecting the lactone ring. Solid L-ascorbic acid is stable when dry and gradually darkens on exposure to light.

Biochemistry. Ascorbic acid is an important reducing agent that functions primarily as an electron carrier. By giving up two electrons, it is converted to dehydro-L-ascorbic acid (2). Loss of one electron by interaction with oxygen, metals, or both, produces the reactive monodehydro-L-ascorbate free radical which is reduced to L-ascorbic acid by various enzymes in animals and plants. In some model systems, reactions occur because of its redox potential and not its coenzyme activity.

Biosynthesis. As in animals, L-ascorbic acid is also the product of hexose phosphate metabolism in plants.

Metabolism. Studies with labeled L-ascorbic acid in rats and guinea pigs disclosed that the vitamin is oxidized to respiratory CO_2.

Structure Determination and Synthesis

Chemical constitution. The elucidation of the sterochemistry of L-ascorbic acid played a large role in its structure determination and in the development of syntheses and manufacturing processes. During the early studies, the configurational assignment of ascorbic acid to the L-series was confirmed by synthesis from L-xylose. Subsequently, ascorbic acid was synthesized from D-glucose because the chiral centers at C-2 and C-3 were in the correct configuration to become C-5 and C-4, respectively, of L-ascorbic acid.

Manufacture

Although natural and synthetic vitamin C are chemically and biologically identical, in recent years, a limited amount of commercial isolation from vegetable sources, eg, rose hips, persimmon, citrus fruit, etc, has been carried out to meet the preference of some persons for vitamin C from natural sources. L-Ascorbic acid was the first vitamin to be produced in commercial quantities, and its principal manufacture is based on Reichstein's second synthesis from *D*-glucose. Many chemical and technical modifications have improved the efficiency of each step, enabling this multistep synthesis to remain the principal, most economic process. Other routes, eg, fermentative oxidation of L-sorbose and reduction of 2,5-dioxo-D-gluconic acid as intermediate steps, have been reported in the literature. L-Ascorbic acid is produced in large, integrated, and automated facilities, involving both continuous and batch operations.

Environmental issues. The environmental concerns of an ascorbic acid manufacturing facility are typical of a chemical processing plant. Its operating design must be patterned to conform to environmental protection regulations. Measures must be taken to contain solvents and to keep emissions within official guidelines. Special condensers, continuous instrumental monitoring, and emergency containment and cleanup systems are required. Wastewater-treatment facilities may have to be provided to remove by-product organics and inorganics from the effluent streams before disposal.

Biochemical Functions

L-Ascorbic acid deficiencies involve metabolic effects that can be related to subclinical stages of classical scurvy. Abnormal effects are observed in collagen formation, fatty-acid metabolism, brain function, drug detoxication, infection, and fatigue. In plants, L-ascorbic acid is involved in cellular respiration, growth, and maintaining the carbon balance. Definitive hypotheses on its activities have not yet been formulated. It plays important roles in mixed-function oxidation reactions involving molecular oxygen, whereby only one of the two oxygen atoms is incorporated into the substance. L-Ascorbic acid appears to play a classical cofactor-like role as part of the active site of hyroxylating enzymes, whereas other functions with hydroxylases in collagen bio-synthesis suggest a protective role. Its relation to transition-metal ions, its reducing properties, and its reaction with free-radical derivatives of oxygen may be central to its biological functions.

Uses

Vitamin C is an essential nutrient for health maintenance. The first uses of the vitamin were to prevent and treat scurvy. Subsequent uses developed from biochemical studies which delineated its extra-anti-corbutic activities related to maintenance of good general health. A daily intake of 60–100 mg of L-ascorbic acid is considered adequate, depending on body weight and rate of metabolism.

Important uses of vitamin C relate to the prevention of megalolastic anemia in formula-fed infants and other macrocytic anemias by impro-ving iron metabolism.

Vitamin C requirements for wound repair, normal healing processes, and trauma are based upon its role in collagen synthesis and fiber cross-linking.

Stimulation of the immune system by vitamin C accounts for its use in the prevention and treatment of infections.

The use of ascorbic acid in ameliorating and speeding recovery from the common cold is based on immunostimulation and is recommended by Pauling.

Vitamin C's possible role in cancer prevention, therapy, and manage-ment is being studied intensively. It has been related to immunostimula-tion, collagen encapsulation of tumors, deficiencies in tissue, activity as an antioxidant, and interaction with free radicals.

The role of vitamin C in lipid metabolism and its possible usefulness in cardiovascular disease were recently reviewed. Experimental, clinical and epidemiological evidence demonstrated L-ascorbic acid medition in lowering the incidences of hypercholesterolemia and atherosclerosis. Vitamin C deficiencies have been reported in people subject to various stresses, such as alcohol, smoking, emotional, environmental, etc; recom-mendations have been made that allowances be increased for these people.

Industrial uses of L-ascorbic acid relate to its antioxidant properties. L-Ascorbic acid is widely accepted as a preservative and is GRAS for this application.

Derivatives and Analogues

Various derivatives and analogues of ascorbic acid have been prepared in attempts to find substances with increased activity (see Table 2). Only salts and C-6 substituted esters have full activity; they readily form ascorbic acid in the body and are reversibly oxidized to dehydroascorbic acid (2).

Derivatives, eg, 6-deoxy-L-ascorbic acid, 3-O-methyl ether of L-ascorbic acid, 2-amino-2-deoxyl-L-ascorbic acid, etc, have been prepared and their respective activities compared to L-ascorbic acid are 1/3, 1/25–1/50, and 0.

It appears that the highest vitamin C activity correlates with the D-configuration for the C-4 hydroxyl group, ie, lactone ring formation, at least a two-carbon substituent on the C-4 carbon, two enolic groups, L-configuration for the C-5 hydroxyl group, and a carbon chain no longer than six carbons.

A bound form of ascorbic acid, ascorbinogen, occurs in cabbage, and it has been identified as a β-substituted indole derivative of L-ascorbic acid. Ascorbic acid polymer, which is a reaction product of polymerized ascorbic acid and polymerized formaldehyde and is known as P-2 Viru-

cide, has been used with some success as an antiviral agent for bovine leukemia.

GERALD M. JAFFE
Hoffmann-La Roche, Inc.

J.N. Counsel and D.H. Hornig, eds., *Vitamin C (Ascorbic Acid)*, Applied Science Publishers, London, UK, 1981.

S. Nobile and J.M. Woodhill, *Vitamin C. The Mysterious Redox-System—A Trigger of Life.* MTP Press, Boston, Mass., 1981.

W.H. Selrell and R.S. Harris, ed., *The Vitamins, Chemistry, Physiology, Pathology, Methods,* 2nd ed., Vol. 1, Academic Press, New York, 1967.

P.A. Seib and B.M. Tolbert, ed., *Ascorbic Acid: Chemistry, Metabolism and Uses, Adv. Chem. Ser.* 200 (1982).

L. Pauling, *Vitamin C and the Common Cold,* W.N. Freeman Co., San Francisco, Calif. 1970.

BIOTIN

Biotin is one of the water-soluble B vitamins. In bound form, it is distributed widely as a cell constituent of animal and human tissues. The main sources of biotin are liver, kidney, pancreas, egg yolk, yeast, and milk. A high content of biotin in cow's milk occurs early in lactation. It is in different plant materials, especially in seeds, pollen, molasses, rice, mushrooms, fresh vegetables, and some fruits. Most fish contain biotin in small amounts. Biochemically, biotin functions as a cofactor for enzymes principally to carboxylation reactions. These reactions are involved in important biochemical processes, eg, gluconeogenesis and fatty-acid synthesis.

(1) D-biotin

Any valuable synthesis of biotin requires an efficient and stereocon-trolled formation of three chiral centers in the all-cis configuration. This was first accomplished in the Sternbach synthesis of optically active biotin, which later was known as the Hoffmann-La Roche industrial synthesis of biotin.

In 1975, Sumitomo chemists replaced the optical resolution-reduction sequence of the Sternbach synthesis by an efficient asymmetric conver-sion of the prochiral cis-acid to the optically active lactone. The acid reacts with the optically active dihydroxy amine to give quantitatively the chiral imide. Sodium borohydride reduces stereoselectivity in the pro-R carbonyl group to give, after recrystallization, the optically pure hydroxy amide. Hydrolysis then yields the lactone.

Biosynthesis

A number of fungi and bacteria synthesize biotin from pimelic acid by a metabolic pathway, whose last step is the conversion of dethiobiotin to biotin.

Biotin Deficiency

Because of biosynthesis by intestinal flora, a deficiency of biotin seldom occurs in humans.

In poultry, biotin is an essential vitamin for normal growth, feed conversion, and reproduction as well as healthy skin, feathers, and bones. Biotin deficiency causes dramatic symptoms in swine, eg, reduced growth rate, impaired feed conversion, dermatitis, excessive hair loss, furry tongue, foot lesions, stiff-legged gait, squatness, and hind-leg spasms. Deficiencies are corrected by using biotin as a feed additive for poultry and swine (see Pet and livestock feeds).

MILAN R. USKOKOVIĆ
Hoffmann-La Roche, Inc.

Table 2. Isomeric Ascorbic Acids

Substance	Structure	Mp, °C	$[\alpha]_D$ (H$_2$O)	Activity
L-ascorbic acid	(1)	192	+24°	1
D-xyloascorbic acid (D-ascorbic acid, D-*threo*-hex-2-enonic acid γ-lactone)	(33)	192	−23°	0
D-araboascorbic acid (erythorbic acid, D-*erythro*-hex-2-enonic acid γ-lactone)	(22)	174	−18.5°	1/20–1/40
L-araboascorbic acid (L-*erythro*-hex-2-enonic acid γ-lactone)	(34)	170	+17°	0

P. Confalone, G. Pizzolato, D. Lollar, and M.R. Uskoković, *J. Am. Chem. Soc.* **102**, 1959 (1980).

F. Frappier, J. Jouany, A. Marquet, A. Olesker, and J.C. Tabet, *J. Org. Chem.* **47**, 2257 (1982).

R.B. Guchait, S.E. Polakis, D. Hollis, C. Fenselau, and M.D. Lane, *J. Biol. Chem.* **249**, 6646 (1974).

INOSITOL

The name inositol may be applied to the nine possible isomers of 1,2,3,4,5,6-hexahydroxycyclohexane, of which only one, *myo*-inositol, has been associated with vitaminlike properties.

1,2,3,5-*cis*-4,6-*trans*-hexahydroxycyclohexane
(*myo*-inositol)

Occurrence

Plants. Free *myo*-inositol is widely distributed in many types of plants. However, it is most commonly present in higher plants as the hexakis-*O*-phosphate, phytic acid (see Food toxicants, naturally occurring).

Animals. In its free state, *myo*-inositol occurs in significant amounts in brain, testes, semen, secretory tissues, and fetal blood, and the milk of rodents and humans contains appreciable levels of free *myo*-inositol.

Common foods. The greatest amounts of *myo*-inositol are in fruits, beans, grains, and nuts. Also, various domestic and foreign wines have relatively high levels of *myo*-inositol (see also Wine).

Properties

myo-Inositol is a highly stable white crystalline powder. Its anhydrous molecular weight is 180.16, but crystals are commonly obtained as the dihydrate. Other chemical and physical properties are listed in Table 1.

Manufacture

The main source of *myo*-inositol for commercial production is from plant phytin by degradative processes.

Advances in the understanding of the nutritionally related aspects of *myo*-inositol have chiefly involved its lipotropic role and its precursor relationship to the phosphoinositides. However, significant studies have been reported on the biological function of these phospholipids in various systems. The main effect of external stimuli on selected target tissues is rapid and enhanced phosphatidylinositol metabolism.

myo-Inositol Metabolism and Disease

In recent years, *myo*-inositol tissue levels have been associated with various disease states. A correlation has been observed between decreased peripheral motor and sensory nerve conduction velocities and decreased free *myo*-inositol in the nerves from experimentally diabetic rats. Dietary supplementation of 1 wt% *myo*-inositol increases plasma and nerve levels of free inositol and significantly improves motor nerve conduction velocities in these animals. However, when 3 wt% dietary *myo*-inositol is provided, a decreased motor conduction velocity occurs in both normal and diabetic rats.

WILLIAM W. WELLS
Michigan State University

Table 1. Physical Properties of *myo*-Inositol

Property	Value
melting range, °C	225–227
density (anhydrous crystals at 15°C), g/cm^3	1.752
optical rotation	inactive
solubility in water at 25°C, g/100 mL in water at 25°C	14

R.S. Clements, Jr., and B. Darnell, *Am. J. Clin. Nutr.* **33**, 1954 (1980).

T. Posternak in *The Cyclitols*, Holden-Day, Inc., San Francisco, Calif., 1965, pp. 127–143.

B.J. Holub in H.H. Draper, ed., *Advances in Nutritional Research*, Vol. 4, Plenum Publishing Corp., New York, 1982, pp. 107–141.

NICOTINAMIDE AND NICOTINIC ACID (B$_3$)

Nicotinamide (**1**) is a member of the B-vitamin complex. Other commonly used names include niacinamide, nicotinic acid amide, nicotinamidum, vitamin B$_3$, vitamin PP (pellagra preventing), and 3-pyridinecarboxamide.

(**1**) nicotinamide
3-pyridinecarboxamide

Both nicotinamide and nicotinic acid are building blocks for coenzyme I (Co I), nicotinamide–adenine dinucleotide (NAD), and coenzyme II (Co II), nicotinamide–adenine dinucleotide phosphate (NADP).

Today, acute avitaminoses are seen only rarely. The symptoms of vitamin B$_3$ deficiency are evident, above all, in the skin, digestive tract, and nervous system.

Occurrence

Nicotinamide and nicotinic acid occur in nature almost exclusively in bound form. Free nicotinic acid is more prevalent in plants, whereas in animals nicotinamide predominates. Almost all of the nicotinamide found in humans and animals is fixed as building blocks of NAD and NADP.

Properties

Nicotinamide is a colorless, fine crystalline solid with a bitter taste. At room temperature, 1 g is soluble in ca 1 mL water, 1.5 mL 95% ethanol, or 10 mL glycerol. The compound is soluble in butanol, amyl alcohol, ethylene glycol, acetone, and chloroform, but only slightly soluble in ether or benzene. The physical properties are given in Table 1.

Nicotinamide has been utilized in numerous reactions. Among the complex salts of nicotinamide, the bis(pyridine-3-carboxamide)–PdCl$_2$ complex can be used via a Mannich reaction for the synthesis of *N*-aminomethylated pyridine-3-carboxamides in good yields.

Important reactions of the ring nitrogen include quaternization (**2**) and *N*-oxide (**3**) formation.

(**2**) R = alkyl, aryl
X = halogen

(**3**) nicotinamide *N*-oxide

Preparation and Manufacture

The synthesis of nicotinamide using the nicotinonitrile route involves the following steps: synthesis, from aliphatic starting materials, of an alkylpyridine, primarily 3-methylpyridine or 2-methyl-5-ethylpyridine (MEP); ammoxidation or ammonodehydrogenation of the alkylpyridine to nicotinonitrile (in the case of 2-methyl-5-ethylpyridine, part of the carbon chain is oxidatively cleaved in this step); and hydrolysis of the nitrile to nicotinamide (**1**).

Ammoxidation of 3-methylpyridine is carried out on a commercial scale in the gas phase using heterogenous catalysts; oxides of antimony, vanadium, and titanium, or antimony, vanadium, and uranium are highly effective.

Table 1. Physical Properties of Nicotinamide

Property	Value
molecular weight	122.12
melting point, °C	
stable modification	129–132
boiling range, °C (0.067 Pa)[a]	150–160
sublimation range, °C	80–100
density of crystals, g/cm³	1.401
true dissociation constants in H₂O at 20°C	
K_{b1}	2.24×10^{-11}
K_{b2}	3.16×10^{-14}
specific heat, kJ/(kg·K)[b]	
solid, 55°C	1.30
75°	1.39
liquid, 135°C	2.18
heat of solution in H₂O, kJ/kg[b]	−148
density of melt at 150°C, g/cm³	1.19

[a] To convert Pa to mm Hg, multiply by 0.0075.

[b] To convert J to cal, divide by 4.184.

Biochemical Functions and Requirements

Nicotinamide is a building block of the hydrogen-transfer coenzymes NAD and NADP. These enzymes catalyze numerous hydrogenation and dehydrogenation reactions in intermediary metabolism. They are always bound to a substrate-specific apoenzyme in the form of freely dissociable complexes. These two coenzymes take part in most hydrogen-transfer reactions involved in metabolism.

A deficiency of vitamin B₃ gives rise to disturbances in the glycolysis function, the citric acid cycle, and the respiratory chain. Disturbances in synthetic processes, eg, fatty-acid synthesis, also occur. NAD can be endogenously synthesized from the amino acid L-tryptophan, provided the tryptophan requirement for protein synthesis is fulfilled.

Requirements. The best estimate of the average requirement for an adult is 15–20 mg per day. The recommended dietary allowance (RDA) for adults, expressed as vitamin B₃ equivalents, is 6.6 mg per 4184 kJ (1000 food calories) and not less than 13 mg for an intake of less than 8368 kJ (2000 food calories).

Toxicity

Toxicity after either single or repeated ingestion of nicotinamide (as well as nicotinic acid) is only observed at high concentrations or doses. The acute toxicity of nicotinamide and nicotinic acid is minimal.

In humans, a distinction should be made between single or repeated exposure, while handling nicotinamide at the workplace, and ingestion of nicotinamide, either daily in food (in which it may be either a natural component or added in an enrichment process) or therapeutically in medications. It has been reported that handling nicotinamide in the workplace does not cause the flush effect which is known to occur in handling nicotinic acid.

In the United States, nicotinamide is certified as a nutrient and dietary supplement. It has GRAS status in accordance with the CFR.

Uses

It is common practice in many parts of the world to enrich the human diet with nicotinamide or nicotinic acid (see also Food additives).

Nicotinamide and nicotinic acid have a stabilizing effect on the pigmentation of cured meats. Moreover, they are reported to contribute to the formation of light-stable, red pigments (hemochrome) during the curing of meats and sausages (see also Food processing).

Nicotinamide is administered to patients for prevention and treatment of hypovitaminosis. In animal nutrition, particularly in modern, intensive livestock raising, nicotinamide, nicotinic acid, and other vitamins are added to feed as concentrates and premixes. Nicotinamide also promotes the growth of plants.

Nicotinamide is used as a brightener in electroplating baths at concentrations of 1–10 g/L.

Nicotinic Acid

Nicotinic acid (**4**) (niacin, acidum nicotinum, vitamin B₃, vitamin PP, 3-pyridinecarboxylic acid) can be used instead of nicotinamide in some applications since it enters into the NAD biosynthesis quite effectively and is amidated (Preiss-Handler route).

(**4**) nicotinic acid 3-pyridine carboxylic acid

Nicotinic acid is found principally in plants. In cereals, such as corn and wheat, it is bound to polysaccharides and peptides and is inactive as a vitamin.

Properties

Nicotinic acid is a fine crystalline compound with needle-shaped crystals. It has a sour taste and a weak odor. Its physical properties are given in Table 2.

Compared to other pyridine carboxylic acids, nicotinic acid is stable. It is amphoteric and forms numerous salts with both acids and bases. Chemical reactions may involve the nitrogen atom, and carboxyl group, or the heterocyclic ring. The nitrogen atom is subject to quaternization, N-oxidation, and N-imine formation.

Preparation and Manufacture

Nicotinic acid is obtained by oxidizing 2-methyl-5-ethylpyridine (MEP) or 3-methylpyridine with the nitric acid or air. The reaction may be carried out in either the liquid or gas phase; catalysts are sometimes employed.

Most nicotinic acid is produced by the liquid-phase oxidation of MEP.

Biochemical Functions

Nicotinic acid is formed in animal metabolism by biosynthesis from tryptophan or by deamidation of nicotinamide.

Toxicity

The acute toxicity of nicotinic acid is low.

In the workplace, single or repeated contact with nicotinic acid can lead to a flush effect on the skin, including the face. Nicotinic acid is certified in the United States as a nutrient and dietary supplement with GRAS status according to the CFR.

Uses

Nicotinic acid, like nicotinamide, is used for vitamin enrichment of cereal products (eg, wheat flour, bread, macaroni, corn products). Nicotinic acid is utilized as a meat additive. It is often used instead of

Table 2. Physical Properties of Nicotinic Acid

Property	Value
molecular weight	123.11
melting point, °C	236–237
sublimation range, °C	≥ 150
density or crystals, g/cm³	1.473
true dissociation constants in H₂O at 25°C	
K_a	1.50×10^{-5}
K_b	1.04×10^{-12}
isoelectric point in H₂O at 25°C, pH	3.42
pH of saturated aqueous solution	2.7
solubility, g/L	
water, 0°C	8.6
ethanol, 96%, 0°C	5.7
methanol, 0°C	63.0
of sodium nicotinate hemihydrate	
water, 0°C	95.0

nicotinamide as a vitamin supplement for animal nutrition. In the area of industrial applications, nicotinic acid can replace nicotinamide as a brightener for zinc, cadmium, and palladium in electroplating baths.

Derivatives

Nicotinic acid and its ester and amide derivatives have widespread medical use as antihyperlipidemic agents (antilipemic agents) and peripheral vasodilators.

HERIBERT OFFERMANNS
AXEL KLEEMANN
HERBERT TANNER
HELMUT BESCHKE
HEINZ FRIEDRICH
Degussa Company

W.J. Darby, K.W. McNutt, and E.N. Todhunter, *Nutr. Rev.* **33**, 289 (1975).

H. Beschke and H. Friedrich, *Chemiker Zeitung*, **101**, 377 (1977).

Hearing Draft: Evaluation of the Health Aspects of Niacin and Niacinamide as Food Ingredients, Contract No. FDA 223-75-2004/1979, Life Sciences Research Office, Bethesda, Md., 1978.

PYRIDOXINE (B$_6$)

Pyridoxine refers specifically to 5-hydroxy-6-methyl-3,4-pyridinedimethanol (1). The older name for compound (1), pyridoxol, is no longer used.

The six pyridine derivatives that exhibit significant vitamin B$_6$ activity are pyridoxine (1), pyridoxal (2), pyridoxamine (3), and their respective 5'-phosphates (4), (5), and (6). The vitamin B$_6$ group also includes minor components, including various isomers and metabolites.

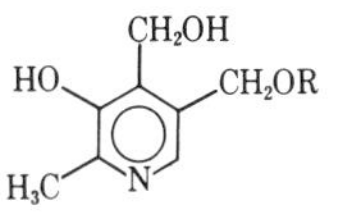

(1) R = H, pyridoxine
(4) R = OPO$_3$H$_2$, pyridoxine-5'-phosphate

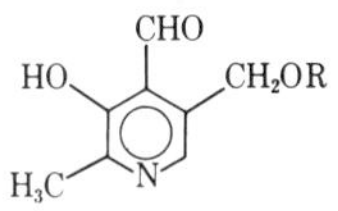

(2) R = H, pyridoxal
(5) R = OPO$_3$H$_2$, pyridoxal-5'-phosphate

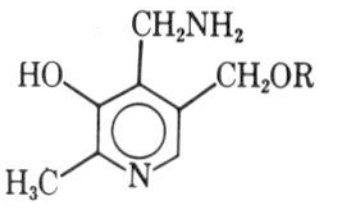

(3) R = H, pyridoxamine
(6) R = OPO$_3$H$_2$, pyridoxamine-5'-phosphate

Occurrence

Vitamin B$_6$ occurs in the tissues and body fluids of virtually all living organisms. This ubiquity of the vitamin reflects its essential role in amino acid biochemistry. Almost all foods in a typical human diet contain detectable amounts.

Function

As enzyme cofactors, pyridoxal-5'-phosphate (5) and pyridoxamine-5'-phosphate (6) are required for several important biotransformations of amino acids. These include decarboxylation, transamination, deamination, racemization, and the metabolism of sulfur-containing amino acids.

Properties

Physical properties of pyridoxine, pyridoxamine, pyridoxal, and their derivatives are listed in Table 1.

The chemical properties of pyridoxine that are related to its role as a vitamin are centered on the 4-hydroxymethyl group.

Synthesis

Since pyridoxine (1) is commercially the most important of the six forms of vitamin B$_6$, only the syntheses of pyridoxine are discussed. Moreover, pyridoxal (2) is best prepared by oxidation of pyridoxine and pyridoxamine (3), in turn, is obtained by catalytic reduction of pyridoxal oxime.

Condensation. Condensation reactions are the classical routes to pyridoxine. Typically, they involve formation of the pyridine ring by a Knoevenagel-type condensation of cyanoacetamide with a β-diketone, followed by a sequence of functional group modifications of substituents on the ring.

From 2-(α-aminoethyl)furans. 2-(α-Aminoalkyl)furans can be transformed into 3-hydroxypyridines. With appropriate substituents on the furan ring, the reaction sequence, which consists of oxidative addition of methanol to the ring followed by hydrolytic rearrangement, affords pyridoxine.

Diels-Alder reactions of oxazoles. Fundamental to vitamin B$_6$ production technology are processes based on the Diels-Alder reaction of oxazoles.

DAVID L. COFFEN
Hoffmann-La Roche, Inc.

E.E. Snell in J.E. Leklem and R.D. Reynolds, eds., *Methods in Vitamin B-6 Nutrition, Analysis and Status Assessment*, Plenum Press, New York, London, 1981.

M. Brin in *Human Vitamin B$_6$ Requirements*, National Academy of Sciences, Printing and Publishing Office, Washington, D.C., 1978.

H. König and W. Böll, *Chem. Ztg.* **100**, 105 (1976).

Table 1. Properties of Pyridoxine, Pyridoxamine, Pyridoxal, and Their Derivatives

Substance	Empirical formula	Mol wt	Form	MP, °C,	Solubility in water	pK values
pyridoxine (1)	C$_8$H$_{11}$NO$_3$	169.18	needles, acetone	160	sol	5.00, 8.96
pyridoxine hydrochloride	C$_8$H$_{12}$ClNO$_3$	205.64	colorless platelets or rods, ethanol/acetone	204–206 (dec)	22 g/100 mL	
pyridoxine-5'-phosphate (4)	C$_8$H$_{12}$NO$_6$P	249.16	colorless needles, water	212–213 (dec)	sol	
pyridoxamine (3)	C$_8$H$_{12}$N$_2$O$_2$	168.20	crystals	193	sol	3.31–3.54, 7.90–8.21, 10.4–10.63
pyridoxamine dihydrochloride	C$_8$H$_{14}$Cl$_2$N$_2$O$_2$	241.12	deliquescent white platelets	226–227 (dec)	50 g/100 mL	
pyridoxamine-5'-phosphate (6)	C$_8$H$_{13}$N$_2$O$_5$P	248.18			sol	< 2.5, 3.25–3.69, 5.76, 8.61
pyridoxamine-5'-phosphate dihydrate	C$_8$H$_{13}$N$_2$O$_5$P.2H$_2$O	284.21	rhombic plates			
pyridoxal (2)	C$_8$H$_9$NO$_3$	167.16		oxime: 225–226 (dec)	50 g/100 mL	4.2, 8.68, 13
pyridoxal hydrochloride	C$_8$H$_{10}$ClNO$_3$	203.63	rhombic crystals	165 (dec)	50 g/100 mL	
pyridoxal-5'-phosphate (5)	C$_8$H$_{10}$NO$_6$P	247.14		oxime:229–230 (dec)		
pyridoxal-5'-phosphate monohydrate	C$_8$H$_{10}$NO$_6$P.H$_2$O	265.16	needles		sol	< 2.5, 4.14, 6.20, 8.69

RIBOFLAVIN (B₂)

Riboflavin (1) (vitamin B_2, vitamin G, lactoflavin, ovoflavin, lyochrome, hepatoflavin, uroflavin) has the chemical name 7,8-dimethyl-10-D-ribitylisoalloxazine, $C_{17}H_{20}N_4O_6$, mol wt 376.37.

(1) riboflavin

In the free form, riboflavin occurs in the retina of the eye, and in whey and urine. Principally, however, riboflavin fulfills its metabolic function in a complex form. In general, riboflavin is converted into flavin mononucleotide (FMN, riboflavin-5'-phosphate) and flavin-adenine dinucleotide (FAD), which serve as the prosthetic groups (coenzymes), ie, they combine with specific proteins (apoenzymes) to form flavoenzymes, in a series of oxidation–reduction catalysts widely distributed in nature.

As a coenzyme component in tissue oxidation–reduction and respiration, riboflavin is distributed in some degree in virtually all naturally occurring foods. Liver, heart, kidney, milk, eggs, lean meats, and fresh leafy vegetables are particularly good sources of riboflavin. It does not seem to have long stability in food products.

Riboflavin is widely used in the pharmaceutical, food-enrichment, and feed-supplement industries. Riboflavin USP is administered orally in tablets or by injection as an aqueous solution, which may contain nicotinamide or other solubilizers.

Properties

Riboflavin forms fine yellow to orange-yellow needles with a bitter taste. It melts with decomposition at 278–279°C (darkens at ca 240°C). The solubility of riboflavin in water is 10–13 mg/100 mL at 25–27.5°C, and in absolute ethanol 4.5 mg/100 mL at 27.5°C.

Riboflavin is stable against acids, air, and common oxidizing agents such as bromine and nitrous acid (except chromic acid, $KMnO_4$, and potassium persulfate), zinc in acidic solution, or catalytically activated hydrogen, riboflavin readily takes up two hydrogen atoms to form the almost colorless 1,5-dihydroriboflavin, which is reoxidized by shaking with air.

Riboflavin forms a deep-red silver salt.

Chemical Synthesis

In 1935, Karrer and Kuhn independently proved that riboflavin was 7,8-dimethyl-10-D-ribitylisoalloxazine by total synthesis. These syntheses are essentially the same and involve a condensation of 6-D-ribitylamino-3,4-xylidine with alloxan in acid solution.

Microbial Synthesis

Biosynthesis. Riboflavin is produced by many microorganisms, including

Ashbya gossypii	*Hansenula* yeasts
Asperigillus sp	*Pichia* yeasts
Eremothecium ashbyii	*Azotobactor* sp
Candida yeasts	*Clostridium* sp
Debaryomyces yeasts	*Bacillus* sp

The mechanism of riboflavin biosynthesis has been deduced from data derived from several experiments involving a variety of organisms. Included are the conversion of a purine such as guanosine triphosphate (GTP) to 6,7-dimethyl-8-D-ribityllumazine, and the conversion of 6,7-dimethyl-8-D-ribityllumazine to riboflavin.

Fermentation. Throughout the years, riboflavin yields obtained by fermentation have been improved to the point of commercial feasibility. A suitable carbohydrate-containing mash is prepared and sterilized, and the pH adjusted to 6–7. The mash is buffered with calcium carbonate, inoculated with *Clostridium acetobutylicum*, and incubated at 37–40°C for 2–3 d. The yield is ca 70 mg riboflavin per liter (see Fermentation).

Deficiency, Requirements, and Toxicity

Riboflavin is essential for mammalian cells. A lack in the human diet causes well-defined syndromes, such as angular stomatitis; glossitis (magental tongue); reddened, shiny, and denuded lips; seborrhoeic follicular keratosis of the nasolabial folds, nose, and forehead; and dermatitis of the anogenital region (scrotum and vulva). An adult requires ca 1.5–3 mg riboflavin daily.

Derivatives

Derivatives include riboflavin-5'-phosphate, riboflavin-5'-adenosine diphosphate, covalently bound flavins, 6-hydroxyriboflavin, 8-nor-8-hydroxyriboflavin, roseoflavin, and 5-deazariboflavin.

FUMIO YONEDA
Kyoto University

R.S. Rivlin, *Riboflavin*, Plenum Press, New York and London, 1975.

T. Wagner-Jauregg in W.H. Sebrell, Jr., and R.E. Harris, eds., *The Vitamins*, Vol. V, Academic Press, Inc., New York, 1972, p. 19.

C. Walsh, *Enzymatic Reaction Mechanisms*, W.H. Freeman and Company, San Francisco, 1979, p. 362.

THIAMINE (B₁)

Thiamine (1), thiamin, vitamin B_1, aneurine, 3-[(4-amino-2-methyl-5-pyridmidinyl)methyl]-5-(2-hydroxyethyl)-4-methylthiazolium chloride, $C_{12}H_{17}$–N_4OSCl, is a member of the vitamin-B complex group. The terms thiamine, thiamin, and vitamin B_1, usually denoting the hydrochloride (2), are used interchangeably.

Thiamine is associated with carbohydrate metabolism. It is converted in the body to the pyrophosphate (3) (cocarboxylase) which acts as a coenzyme in the decarboxylations of α-keto acids and several other reactions. A deficiency of thiamine in the diet leads to the beriberi syndrome characterized by deterioration of the nervous system. Thiamine is used for prophylactic and therapeutic purposes and as an enrichment of foodstuffs. The commercially available forms of thiamine are the chloride hydrochloride (thiamine hydrochloride) (2) and the mononitrate (4).

(1) R = H, X = Cl
(2) R = H, X = Cl, Y = HCl
(3) R = —$P_2H_3O_6$ (ie, —PO(OH)OPO(OH)₂)
(4) R = H, X = NO_3

Thiamine is found in the tissues of animals, plants, and microorganisms. The predominant form in animal tissue is the pyrophosphate (3), which exists largely as a protein complex bound to an enzyme. The most abundant form in plant tissue is free thiamine (1).

In animals, including humans, the organs with high thiamine concentration (μg/g of moist tissue) are the heart (2.8–7.9), kidney (2.4–5.8), liver (2.0–7.6), and brain (1.4–4.4); lesser amounts are found in the spleen, lung, adrenals, and muscle. Normal human blood contains ca 90 ng/mL, although the reported values vary considerably. A value below 40 ng/mL may be indicative of thiamine deficiency.

The occurrence of thiamine in foodstuffs is of nutritional importance. Considerable amounts are likely to be destroyed in cooking processes owing to diverse factors such as heat, metals, chlorine in the water, and reactive organic compounds present in the food. Therefore, naturally occurring thiamine in cereals is often supplemented by synthetic thiamine or its derivatives (see also Food additives).

Properties

Thiamine consists of two heterocyclic moieties, pyrimidine and thiazole, connected by a methylene group. It forms mono- or diacid salts.

Degradation. Although an aqueous solution of thiamine hydrochloride (**2**) is stable under normal conditions, it decomposes to 4-amino-2-methyl-5-pyrimidinemethanol and 4-methyl-5-thiazole-ethanol when heated in a sealed tube at 140°C.

Oxidation. Thiamine is susceptible to oxidation under alkaline conditions. An oxidation product, thiochrome, is used for the qualitative and quantitative analyses of thiamine.

Reduction. Thiamine in the thiazolium form undergoes reduction of the thiazolium ring by lithium aluminum hydride, or sodium trimethoxyborohydride to afford tetrahydrothiamine via the intermediate dihydrothiamine.

Synthesis

The practical synthetic methods for thiamine are represented by several pathways. In the first, 4-amino-5-bromomethyl-2-methyl pyrimidine hydrobromide and 5-(2-hydroxyethyl)-4-methylthiazole are coupled to yield thiamine bromide hydrobromide, followed by the exchange of the bromide and chloride ions by treatment with silver chloride or ion-exchange resins.

The second method is characterized by the extension of the side chain at the 5-position of 4-amino-2-methyl-5-pyrimidinylmethylamine followed by cyclization into the thiazole ring. 5-Acetoxy-3-chloropentan-2-one is condensed with *N*-(4-amino-2-methyl-5-pyrimidinyl) methyl thioformamide, prepared from 4-amino-2-methyl-5-pyrimidinylmethylamine by formylation with formic acid and reaction with phosphorus pentasulfide. Hydrolysis of the resulting thiamine acetate with hydrochloric acid completes the synthesis. An alternative industrial method involves the condensation of 4-amino-2-methyl-5-pyrimidinylmethylamine, carbon disulfide, and 3-chloro-5-hydroxypentan-2-one to give a 4-thiazoline-2-thione derivative (thiothiamine), followed by oxidation with hydrogen peroxide.

Salts and Derivatives

Salts and derivatives include thiamine hydrochloride, thiamine mononitrate, thiamine pyrophosphoric acid ester, thiamine triphosphoric acid ester, and thiamine disulfide, as well as a number of derivatives of thiol-form thiamine which are used as thiamine prodrugs.

YOSHIKAZU OKA
Takeda Chemical Industries, Ltd.

W.H. Sebrell and R.S. Harris, *The Vitamins*, 2nd ed., Vol. V, Academic Press, Inc., New York, 1972, pp. 98–133.

F.A. Robinson, *The Vitamin Co-Factors of Enzyme Systems*, Pergamon Press, Oxford, UK, 1966, pp. 6–142.

A.F. Wagner and K. Folkers, *Vitamins and Coenzymes*, Interscience Publishers, a division of John Wiley & Sons, Inc., New York, 1964, pp. 17–45.

H.Z. Sable and C.J. Gubler, eds., *Ann. N.Y. Acad. Sci.* **378**, 1 (1982).

VITAMIN A

The substance first designated vitamin A, and subsequently vitamin A_1 (**1**), is the primary isoprenoid polyene alcohol (all-*E*)-3,7-dimethyl-9-(2,6,6-trimethyl-1-cyclohexen-1-yl)-2,4,6,8-nonatetraen-1-ol. *Chemical Abstracts* prefers the base name retinol for (**1**) and adheres to a numbering system for the carotenoid to show the relationship between the two. As a consequence, in *Chemical Abstracts* vitamin A aldehyde is retinal (**2**); vitamin A acid, retinoic acid (**3**); vitamin A_2, 3,4-didehydrovitamin A, 3,4,-didehydroretinol.

(**1**) R = CH₂OH
(**2**) R = CHO
(**3**) R = COOH

Vitamin A is indispensable for normal development and functioning of the ectoderm (ie, the cutaneous, mucosal, and epithelial tissues) and for normal growth, reproduction, and vision.

Biosynthesis

Practically all vitamin A found in nature is derived from carotenoids. The biosynthesis of carotenoids follows the usual pattern of terpenoid synthesis.

Stereochemical aspects. The four double bonds in the vitamin A side chain theoretically permit 16 possible geometrical isomers. The double bonds in the polyene chain or carotenoids and vitamin A are of three types: those for which the adoption of a cis configuration involves very little steric hindrance between two hydrogen atoms (type a), and those for which a cis configuration leads to a serious clash between a methyl group and either a hydrogen atom (type b) or a methyl group (type c).

Physical Properties

Vitamin A is a pale-yellow crystalline solid (mp 63–64°C) which can be recrystallized from ethyl formate or petroleum ether at low temperature. It is insoluble in water but dissolves readily in most organic solvents. Physical properties of various all-trans vitamin A compounds are listed in Table 1. Physical properties and vitamin A potencies of various stereoisomers of retinol and retinal are shown in Table 2.

Chemical Properties

Vitamin A is oxidized readily by atmospheric oxygen, especially in the presence of light and heat. It readily esterifies, reacts with maleic anhydride and gives an intense color in the Carr-Price test.

Synthesis

Although it is possible to isolate vitamin A from natural sources, practically all the vitamin A used today is obtained by total synthesis. The key intermediate in all industrial vitamin A synthesis is β-ionone.

Cis Isomers of Vitamin A

Properties. Some physical properties of the stable *cis*-retinols and *cis*-retinals are shown in Table 2. The biological potency is, in all cases

Table 1. Physical Properties of All-*Trans* Vitamin A Compounds

Compound	Molecular formula	Mp, °C	Absorption in ethanol	
			λ_{max}, nm	$A_{1\ cm}^{1\%}$
vitamin A (**1**)	$C_{20}H_{30}O$	62–64	325	1832
vitamin A acetate	$C_{22}H_{32}O_2$	57–58	328	1550
vitamin A palmitate	$C_{36}H_{60}O_2$	28–29	328	975
vitamin A aldehyde	$C_{20}H_{28}O$	61–62	381	1530
vitamin A acid	$C_{20}H_{28}O_2$	179–180	350	1510
vitamin A acid, methyl ester	$C_{21}H_{30}O_2$	56/72–73	354	1415
3,4-didehydroretinol	$C_{20}H_{28}O$	63–65	350	1455
3,4-didehydroretinal	$C_{20}H_{26}O$	78–79	401	1470
3,4-didehydroretinoic acid	$C_{20}H_{26}O_2$	183–184	370	1395

Table 2.   Physical Properties of Stereoisomers of Vitamin A

	mp, °C	λ_{max}	$A_{1\,cm}^{1\%}$	Vitamin A potency
all-*trans*-vitamin A				
alcohol (**1**)	62–64	325	1832	100%
aldehyde	57/65	381	1530	91%
13-*cis*-vitamin A				
alcohol	58–60	328	1686	75%
aldehyde	77	375	1250	93%
11-*cis*-vitamin A				
alcohol	oil	319	1220	23%
aldehyde	63.5–64.7	376.5	878	48%
11,13-di *cis*-vitamin A				
alcohol	86–88	311	1024	15%
aldehyde	oil	373	700	31%
9-*cis*-vitamin A				
alcohol	81.5–82.5	323	1477	22%
aldehyde	64	373	1270	19%
9,13-di *cis*-vitamin A				
alcohol	58–59	324	1379	24%
aldehyde	49/85	368	1140	17%

Table 3.   Properties of Provitamin A Compounds

Provitamin A	Molecular formula	Mp, °C	Absorption in petroleum ether		Biological activity
			λ_{max}, nm	$A_{1\,cm}^{1\%}$	
β-carotene	$C_{40}H_{56}$	180	273	383	100%
			453	2592	= standard
			481	2268	

The provitamins A are insoluble in water and glycerol, poorly soluble in ethanol, and fairly soluble in chloroform and benzene.

Biological Aspects

The normal daily requirement of vitamin A for adults is about 5000 IU (or 1500 RE, retinal equivalents). Although vitamin A is known to play a vital role in general metabolism, the only biological function in which its action is clearly understood is vision.

Application

The synthetic vitamin A acetate, propionate, and palmitate are the derivatives most generally used. They are offered in a number of application forms depending on the use in pharmaceuticals, food fortification, or animal feeds. Stabilizers (antioxidants) are commonly added but only those that are completely acceptable for food or pharmaceutical use.

The provitamin A, β-carotene, or apo-carotenoids are used when a yellow color is required or desirable, as in the case of margarine, butter, cheese, salad dressing, fruit juices, and carbonated beverages.

OTTO ISLER
FRANK KIENZLE
F. Hoffmann LaRoche & Co.

A.E. Asato, A. Kini, M. Denny, and R.S.H. Liu, *J. Am. Chem. Soc.* **105**, 2923 (1983).

O. Isler, ed., *Carotenoids*, Birkhäuser, Basel, Switzerland, 1971.

T.W. Goodwin, *The Biochemistry of the Carotenoids*, 2nd ed., Chapman & Hall, London, 1980.

J.C. Bauernfeind, ed., *Carotenoids as Colorants and Vitamin A Precursors*, Academic Press, New York, 1981.

F. Kienzle and O. Isler in K. Venkataraman, ed., *The Chemistry of Synthetic Dyes*, Vol. 8, Academic Press, New York, 1978, p. 389.

except the 13-cis compound, considerably less than that of (**1**). All 16 possible isomers of vitamin A are now known.

Synthesis.   There are two basically different ways of obtaining stereoisomers of vitamin A. One consists of isomerization of the all-trans compound through heating, sometimes in the presence of a catalyst, to an equilibrium mixture and then isolation of the various stereoforms. The other method uses syntheses specifically designed to give exclusively or predominantly one stereoisomer.

With just one exception, various cis isomers of vitamin A have no commercial significance. The exception is 13-*cis*-retinoic acid, which has been introduced for the treatment of cystic acne as isotretinoin, Roaccutane (Roche).

Vitamin A₂

Vitamin A₂, *all-trans*-3,4-didehydroretinol (Table 1), shows reactions similar to those of vitamin A.

Retinoids

All the compounds related to either the natural forms or synthetic analogues of vitamin A have been named retinoids. These molecules are characterized by a polyene chain connected at one end to a cyclic end group, and at the other to a polar group such as a carboxyl, ester, amide, or alcohol group. They play a role in controlling cell proliferation and differentiation and show inhibitory effects on the development of epithelial cancer.

all-trans-Retinoic acid (Tretinoin, Airol, Roche) is being used for the treatment of acne.

The Provitamins A

In a broad sense, any compound that can be converted in the animal body to vitamin A should be called a provitamin A. Usage, however, has confined the term to those naturally occurring carotenoids that are transformed *in vivo* to vitamin A such as α-carotene, β-carotene, γ-carotene, and cryptoxanthin. Certain synthetic analogues also show vitamin A activity. Approximately 40 carotenoids are currently regarded as having vitamin A activity.

Properties.   In Table 3, the properties of β-carotene are listed. β-Carotene is the most important provitamin A. It is widely distributed in nature and has been found in the leaves of practically all green plants.

The biological conversion of a carotenoid precursor to vitamin A can be accomplished by either symmetrical or asymmetrical fission or terminal oxidation. The metabolism of provitamins A takes place mainly during absorption in the intestinal mucosa.

Most carotenoids are unstable in the presence of oxygen and are converted to colorless products.

VITAMIN B₁₂

Vitamin B₁₂ is a red, crystalline cobalt complex synthesized by microorganisms. It belongs to a group of compounds named corrinoids. All corrinoids contain four reduced pyrrole rings joined into a macrocyclic ring by links between the α positions; three of these links are formed by methylidene =C— groups and the fourth by a C-C bond. The corrinoids act as catalysts in certain carbon-skeleton rearrangements. Vitamin B₁₂ is essential for normal blood formation, certain fundamental metabolic processes, neural functions, and human, animal, and microbial growth and maintenance. Human requirements are extremely small, ca 1 μg daily.

Vitamin B₁₂, like a number of other vitamins, occurs not singly but as a family of closely related compounds, most of which lack biological activity in animals.

Vitamin B₁₂ is present in food in very small amounts. The main dietary sources are of animal origin. Plant material, in general, contains little or no B₁₂.

Vitamin B₁₂ is sold under the trade names Bevatine-12, Berubigen, Betalin-12 crystalline, β-Twelv-Ora, Depinar, Dodecavite, Dodex, Endoglobin, Hepcovite, Normocytin, Poyamin, Rubramin PC, Sytobex, Vibalt, Vitron-C-Plus, Vi-Twel, and Tulag.

Structure

Vitamin B₁₂ is a large, octahedral cobalt complex [a cobalamin (Cbl)] consisting of a porphyrinlike nucleus with cobalt in the middle, a benzimidazole (or a purine) nucleotide, and a cyano or another group attached to cobalt.

Properties

Vitamin B_{12} crystallizes from water or water–acetone as red prisms. It darkens at 210–220°C, but does not melt below 300°C. The air-dried crystals contain 10–12% water of crystallization. Vitamin B_{12} dissolves only in highly polar solvents such as water (solubility ca 1.2%), methanol, lower aliphatic acids, phenols, dimethylformamide, dimethyl sulfoxide, and liquid ammonia. It is not soluble in acetone, ether, or benzene. The other cobalamins crystallize similarly and have similar thermal stability.

Cobalamin-Transport Proteins

Specific extracellular, membrane-bound, and intracellular proteins are needed for absorption, storage, transport, and metabolism of cobalamins in blood and tissues. The extracellular transport proteins are the intrinsic factor (IF), R protein (cobalophilin, non-IF), and transcobalamin (TC). The membrane-bound transport proteins are the receptors for both the IF–Cbl complex, located in the ileum, and the TC–Cbl complex, located in the tissues. The intracellular Cbl binders are probably identical with the enzymes methionine synthase and methylmalonyl–CoA mutase.

Biochemical Functions

There are three mammalian essential enzymatic reactions requiring Cbl. The methyl analogue is involved in transmethylation. Another mammalian reaction requiring Cbl is the isomerization of (R)-methylmalonyl-CoA (MM–CoA) to succinyl–CoA in the presence of methylmalonyl–CoA mutase. The third mammalian reaction requiring Cbl is the interconversion of α- and β-leucine.

Medical aspects. The amounts of Cbl stored in the body are relatively extensive, on the order of 2–5 mg, with ca 30–60% located in the liver and ca 30% in muscle. The plasma Cbl concentration is 200–900 pg/mL. It takes several years of total Cbl deprivation before the body stores are depleted and symptoms of Cbl deficiency become manifest. True nutritional deficiency in humans is rare but may occur in vegetarians. Deficiency in humans has two effects: an interference with the normal maturation of dividing cells such as those of the bone marrow (megaloblastic bone marrow) and all types of mucosa; and morphological and functional abnormalities of the nervous system.

WILHELM FRIEDRICH
Universität Hamburg

D. Dolphin, ed., *Vitamin B_{12}*. Vol. I, *Chemistry*; Vol. II, *Biochemistry and Medicine*, John Wiley & Sons, Inc., New York, 1982.

B. Zagalak and W. Friedrich, eds., *Vitamin B_{12}*, *Proceedings of the Third European Symposium on Vitamin B_{12}*, Zurich, 1979, Walter de Gruyter, Berlin, 1979.

J. Kirschbaum in K. Florey, ed., *Analytical Profiles of Drug Substances*, Vol. 10, Academic Press, New York, 1981, p. 183.

VITAMIN D

Vitamin D developed in the twentieth century as a dietary supplement to treat and prevent rickets, a disease in which the organic matrix of new bone is not mineralized. Natural vitamin D is made in the skin of animals during exposure to sunlight, and a lack of exposure causes a need for vitamin D as a dietary additive to supplement the natural vitamin D production in humans and animals (see Pet and livestock feeds).

Vitamin D referred to a substance that possessed antirachitic activity. Research during the 1970s revealed that vitamin D is better defined as those natural or synthetic substances that are converted by animals into metabolites that control calcium and phosphorus homeostasis in a hormonal-like manner. Vitamin D_2 and vitamin D_3 are the two economically important forms. The other D vitamins have relatively little biological activity and are only of historical interest. Vitamin D_2 (ergocalciferol) is active in humans, other mammals, such as cattle, swine, and dogs but is inactive in poultry. It is prepared (opening of ring B) by the uv irradiation of ergosterol (1), a plant sterol.

Vitamin D_3 (cholecalciferol) is active and occurs naturally in all animals. It is produced (opening of ring B) by the irradiation of 7-dehydrocholesterol (2). The D vitamins are fat-soluble.

(1) ergosterol
(24-methyl-cholesta-5,7,22-triene-3β-ol; provitamin D_2)

(2) 7-dehydrocholesterol
(cholesta-5,7-diene-3β-ol; provitamin D_3)

Physical Properties

The physical properties of the provitamins and vitamins D_2 and D_3 are listed in Table 1. The values are listed for the pure substances, although these are not usually isolated in normal production.

Chemical Properties

Provitamins. β-Hydroxy steroids, that contain the 5,7-diene system and can be activated with uv light to produce vitamin D compounds, are called provitamins. The two most important provitamins are ergosterol (1) and 7-dehydrocholesterol (2). These are produced in plants and animals, respectively, and 7-dehydrocholesterol is produced synthetically on a commercial scale.

Table 1. Physical Properties of Provitamins and Vitamins D_2 and D_3

Properties	7-Dehydro-cholesterol	Ergosterol	Vitamin D_2	Vitamin D_3
melting point, °C	150–151	165	115–118	84–85
color and form	solvated plates from ether-methanol	hydrated plates from alcohol; needles from acetone	colorless prisms from acetone	fine colorless needles from dilute acetone
optical rotation ($[\alpha]_D^{20}$), °				
acetone			82.6	83.3
ethanol			103	
chloroform	−113.6	−135	52	51.9
ether			91.2	
petroleum ether			33.3	
benzene	−127.1			
coefficient of rotation per °C in alcohol			0.515	
uv max, nm	282	281.5	264.5	264.5
specific absorption, E_{max} (at 1% conc)	308		458.9 ± 7.5	473.2 ± 7.8
potency[a], IU/g biological activity			40×10^6 in mammals	40×10^6 in mammals and birds
chicken efficacy, %			8–10[b]	100
solubility, g/100 mL				
acetone at 7°C			7	
acetone at 26°C			25	
absolute ethanol at 26°C	sl sol	0.15	28	
ethyl acetate at 26°C			31	
water	insol	insol	insol	insol

[a] The international standard for vitamin D is an oil solution of activated 7-dehydrocholesterol. The IU is the biological activity of 0.025 µg of pure cholecalciferol.

[b] Recent studies have claimed an efficacy as high as 10%.

Vitamin D. The irradiation of the provitamins to produce vitamin D as well as several isomeric substances was first studied with ergosterol (**1**).

The irradiation process, which converts 7-dehydrocholesterol (**2**) to vitamin D, occurs in the skin of animals if sufficient sunlight is available.

Biochemistry. Vitamin D is introduced into the bloodstream either from the skin after natural synthesis by the irradiation of 7-dehydrocholesterol (**2**) in the epidermis or by ingestion and absorption through the gut wall of vitamin D_2 or vitamin D_3.

The vitamin D is converted into polyhydroxylated forms which help to carry out the function of calcium and phosphorus homeostasis. These metabolites behave in a hormonal-like fashion and the several different forms are controlled in delicate balance within the endocrine system.

Synthesis

The total synthesis of vitamin D involves many steps and is not an economical route to the production of commercial material. The syntheses are useful for the preparation of molecules containing isotopes of various atoms for radioactive tracer work or for making derivatives of vitamin D.

Important synthetic developments have been made over the last several years in the synthesis of naturally occurring hydroxylated forms of vitamin D as well as derivatives that are not naturally occurring.

Manufacture

Most of the vitamin D produced in the world is made by photochemical conversion of 7-dehydrocholesterol (**2**).

Health and Safety Factors

Disease states. Although rickets is the most common disease associated with vitamin D deficiency, many other diseases can occur and are related to vitamin D. These can involve a lack of the vitamin, deficient synthesis of the metabolites from the vitamin, deficient control mechanisms, or defective organ receptors.

In the treatment of diseases where the metabolites are not being delivered to the system, synthetic metabolites or active analogues have been successfully administered.

Toxicology. The vitamins D are toxic when doses exceed 1000–3000 IU/kg body weight per day over extended periods of time.

Uses

Most of the vitamin D sold is synthetic. Vitamin D_2 as a concentrate or in microcrystalline forms is used in many pharmaceutical preparations, although vitamin D_3 is preferred by many. The swine and poultry industries are the largest markets for vitamin D_3. Swine account for 25.6% and poultry for 43.3% of vitamin D consumption in animal agriculture. The beef and dairy industries account for 30.3% of vitamin D consumption, of which 22% is by dairy calves. The remaining 0.7% of total animal consumption of vitamins is largely in prepared pet foods.

Crystalline vitamin D is used for medicinal preparations and formulations in the pharmaceutical industry as well as for the fortification of fresh and evaporated milk and nonfat dry milks.

Dietary requirements. Vitamin D is tremendously important for growth and maintenance of good health. According to the NRC, the vitamin D requirement for optimum health is ca 400 IU/d, regardless of age.

Arnold L. Hirsch
A.L. Laboratories, Inc.

A.W. Norman, *Vitamin D—The Calcium Homostatic Steroid Hormone*, Academic Press, Inc., New York, 1979, p. 19.

C.E. Bells in W.H. Sebrell and R.S. Harris, eds., *The Vitamins*, 1st ed., Vol. 2, Academic Press, New York, 1954, p. 132.

D.E.M. Lawson, *Vitamin D*, Academic Press, Inc., New York, 1978.

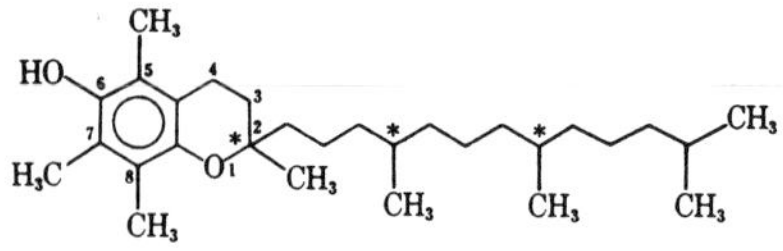

(**1**) α-Tocopherol (5,7,8-trimethyltocol) (*RRR*) [*59-02-9*], *all-rac* [*2074-53-5*]

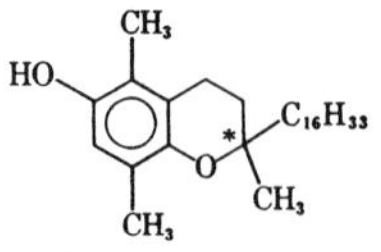

(**2**) β-Tocopherol [*148-03-8*] (5,8-dimethyltocol)

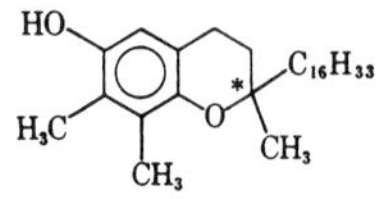

(**3**) γ-Tocopherol [*54-28-4*] (7,8-dimethyltocol)

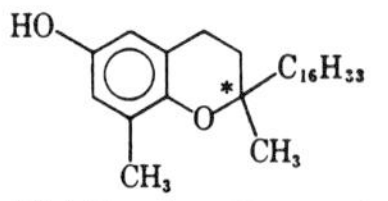

(**4**) δ-Tocopherol [*119-13-1*] (8-methyltocol)

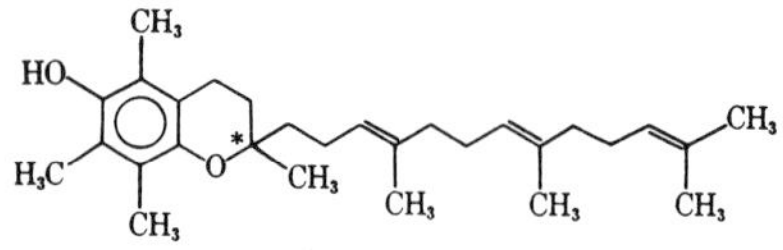

(**5**) α-Tocotrienol [*1721-51-3, 6829-55-6*] (5,7,8-trimethyltocotrienol)

Figure 1. The four naturally occurring tocopherols and α-tocotrienol. Asterisks denote asymmetric centers.

VITAMIN E

Eight naturally occurring compounds possessing vitamin E activity have been isolated and identified. Four have been designated tocopherols (**1**)–(**4**) (Fig. 1). One example of the tocotrienols (**5**) is also shown in Figure 1.

Although the tocopherols and tocotrienols are closely related chemically, they have widely varying degrees of biological effectiveness. α-Tocopherol (**1**) $C_{29}H_{50}O_2$, formula weight 430.72, having a completely methylated aromatic ring and a saturated side chain, has the highest activity. Because of the high potency of α-tocopherol and because it is the predominant form in animal tissues, the term α-tocopherol is now used widely instead of vitamin E. Because tocopherols can exist as several stereoisomers, the specific isomer form of α-tocopherol should always be indicated. The term vitamin E should be reserved as the generic descriptor for all tocol and tocotrienol derivatives having qualitatively the biological activity of α-tocopherol.

Both α-tocopherol and α-tocopheryl acetate are clear, odorless, viscous, slightly yellow oils. α-Tocopheryl acetate is the principal commercial form of vitamin E in food fortification, dietary supplements, and medicinals, and for domestic animals as a source of vitamin E activity. The tocopherols are used as dietary supplements and in food technology as antioxidants (qv) to retard the development of rancidity in fatty materials (see also Food additives).

Tocopherols are widely distributed in foods in an unesterified form and occur in the highest concentration in cereal-grain oils.

The per capita daily intake of α-tocopherol in the United States has been estimated from data on total purchases of food to be ca 15 mg.

Table 1. Some Physical Properties of Commercial Vitamin E Products

Product	Mp, °C	Sp gr$^{25°C}_{25°C}$	n^{20}_D	$A^{1\%}_{1cm}$ in ethanol
(*RRR*)-α-tocopheryl acetate	25	0.950–0.964	1.4940–1.4985	40–44 at 284 nm
(*RRR*)-α-tocopheryl acetate concentrate[a]	liq	0.920–0.950	[b]	[b]
(*RRR*)-α-tocopheryl hydrogen succinate; white, crystalline	73-78			35–40 at 284 nm
mixed tocopherols concentrate[c]	liq	0.920–0.950	[b]	[b]
all-rac-α-tocopherol	liq	0.947–0.958	1.5030–1.5070	71–76 at 292 nm
all-rac-α-tocopheryl acetate	liq	0.950–0.964	1.4940–1.4985	40–44 at 284 nm

[a]Contains ≥ 40% (*RRR*)-tocopheryl acetate.
[b]Varies with potency of concentrate.
[c]Contains ≥ 50% total tocopherol, at least half of which is (*RRR*)-α-tocopherol.

Most is derived from margarine, salad oils, and shortening. Appreciable losses of vitamin E may occur during the processing of foods, including cooking and baking.

Properties

Physical properties of commercial vitamin E products are given in Table 1.

Biochemical Function, Clinical Evaluation, and Toxicology

The exact mechanism whereby α-tocopherol functions in the body as vitamin E is not known. It probably acts as an antioxidant controlling redox reactions in a variety of tissues and organs, particularly in protecting the cellular membrane from free radicals generated during peroxidation of unsaturated lipids. That this is the exclusive role has been challenged, however, and the high isomeric specificity of the α-tocopherol structure in physiological reactions suggests other roles as an integral part of the biological machinery.

α-Tocopherol has been administered orally in large doses to a number of species and is generally well tolerated.

Uses

Both α-tocopherol and its esters are constituents of multivitamin and single-dose nutrient capsules or other liquid dietary supplements. Stable α-tocopherol acetate is used for human infant formulas and foods, and for animal feeds.

DAVID C. HERTING
Eastman Kodak Company

L.J. Machlin, ed., *Vitamin E: A Comprehensive Treatise*, Marcel Dekker, Inc., New York, 1980.

W.H. Sebrell, Jr., and R.S. Harris, eds., *The Vitamins*, Vol. 5, Academic Press, Inc., New York, 1972, Chapt. 16.

R. Porter and J. Whelan, eds., *Biology of Vitamin E*, The Pitman Press, London, UK, 1983.

W

WALLBOARD.   See Laminated and reinforced wood.

WAR GASES.   See Chemicals in war.

WARFARIN.   See Poisons, economic.

WASTES, INDUSTRIAL

Industrial wastes contribute at least two thirds of the organic matter entering the watercourses in the United States and, in order to improve this situation, a thorough investigation of the problems and production processes is needed. The EPA is currently making progress in this area. Each industry was studied by a separate contractor in order to determine treatments and acceptable levels of residual wasteloads at acceptable costs. The EPA established minimum wasteloads of key contaminants for each main wet industry. State and local governments may require more treatment in cases where local water quality is not maintained by the industrial treatment and effluent guidelines. Therefore, both effluent and stream standards are of importance in the planning of industrial waste treatment.

Some solid wastes produced by industry may be similar to municipal refuse, eg, from paper mills, plastic plants, and food-processing plants. Such industrial waste comprises ca 0.5 kg per capita per day of all municipal solid wastes collected (ca 25% of that generated). It is rapidly increasing in quantity as industrial production expands.

Airborne industrial effluents consist mostly of carbon monoxide, particulates (suspended visible matter), sulfur oxides (SO_x), hydrocarbons, and nitrogen oxides (NO_x). These contaminants originate from several sources including transportation, stationary sources of fuel combustion, industrial processes, solid-wastes disposal, and other miscellaneous sources of operation (see Air pollution).

In the United States, nine waste-treatment firms handle more than 50% of all commercial waste disposal and treatment: Browning-Ferris Industries, CECOS International, Chem-Clear, Chemical Waste Management, Conversion Systems, IT Corporation, Rollins Environmental Services, SCA Chemical Services, and US Ecology. Data on the various methods are given in Table 1.

The cost of waste treatment depends upon many factors such as plant location, environmental-control regulations, and type of wastes produced. It usually ranges from 1 to 10% of production costs. Overall costs can be minimized by utilizing the professional services of highly qualified and experienced environmental engineers. Information about such persons and establishments can be obtained from the EPA in Washington, D.C., the local agency in charge of environmental control, or the American Association of Air Pollution Engineers and the American Academy of Environmental Engineers.

Anaerobic wastewater treatment uses less energy and generates less sludge than aerobic treatment. Today, when energy and sludge disposal are so costly, anaerobic treatment is increasingly applied.

Air Contaminants

Treatment of waste gas is usually not complicated from an engineering standpoint, and a number of methods are available, eg, condensation of vapors, solvent extraction, odor control, incineration, and dispersion.

Liquid Wastes

Effects on streams.   Streams and rivers assimilate a certain quantity of waste contaminants before reaching a polluted state (see Water, water pollution). A polluted stream contains an excessive amount of specific contaminants relative to its specific use (see Table 2).

Effects on sewage plants.   Industry often presumes that its waste can be disposed of in the public sewer system, where it is accepted by the municipality. However, waste discharge should not be permitted into the sewer system without disclosure of its contents, the system's ability to handle them, and the effects upon the system. A sewer ordinance, restricting the types and concentrations of waste admitted, is a protection (see Water, sewage).

Characteristics of municipal and industrial sewage are compared in Table 3.

The BOD of industrial waste affects mainly the biological or secondary treatment units of a sewage-treatment plant.

Stream Protection

The quality of receiving waters can be preserved by stream standards or effluent standards. A polluter might be required to discharge only a certain quantity of contaminant or a certain concentration with a certain volume of wastewater. Many state agencies determine the best usage of a stream and assign certain quality standards to that water use. Any polluter violating these standards is cited and must abate the pollution. More recently, some agencies established the receiving-water quality desired and attempted to maintain the quality by controlling each waste discharge to the minimum contaminant units per unit of production or per capita. The latter is generally determined on an industry-wide basis from an analysis of effective treatment potential on an economically feasible basis.

Methods of stream protection include volume reduction, reduction of waste concentration, neutralization, equalization and proportioning, removal of suspended solids, removal of colloidal solids, removal of dissolved inorganic material, removal of dissolved organic material, disposal of sludge solids, joint treatment with municipal sewage, and reuse by other compatible industrial plants.

Table 2.   Industrial Contaminants of Water

Contaminant	Source	Effects
inorganic salts	oil refineries, desalination plants, munitions manufacture, pickle curing	interference with industrial usage, municipal water, and agriculture
acids, alkalies	chemical manufacturing	corrosion of pipelines and equipment; destruction of aquatic life
organic matter	tanneries, canneries, textile mills, etc	food for bacteria; oxygen depletion
suspended matter	paper mills, canneries, etc	suffocation of fish eggs; stream deterioration
floating solids and liquids	slaughterhouses, refinery oil	objectionable appearance, odor; interference of oxygen transfer
heated water	cooling waters from most industries and power plants	acceleration of bacterial action, reduction of oxygen saturation
color	textile, metal-finishing, and chemical plants	objectionable appearance
toxic chemicals	munitions manufacture, metal plating, steel mills, etc	alteration of stream biota and animal and human predators
microorganism	tanneries, municipal–industrial sewage plants	unsafe for drinking
radioactive particles	nuclear power plants, chemical laboratories	concentration in fish; unsafe for drinking
foam producers	glue manufacture, slaughterhouses, detergent manufacture	objectionable appearance

Table 3.   Municipal Sewage and Industrial Wastes

Waste	BOD, ppm	Suspended solids, ppm
municipal sewage	250	300
pulp and paper mills		
no chemical pulping	150	350
with chemical pulping	5000	2500
tannery	1200	2000
cannery	150–1000	250–1500

Table 1.   Waste-Disposal Methods

Waste management	Capacity, wet, 1000 t			Volume received, wet, 1000 t		
	1981	1980	Change, %	1981	1980	Change, %
land fill	37,372	25,672	46	1,965	2,182	−10
land treatment and solar evaporation	1,400	1,447	−3	282	300	−6
chemical treatment	1,305	1,105	18	734	544	35
deep-well injection	1,095	1,095	0	475	475	0
resource recovery	341	341	0	83	83	0
incineration	102	102	0	80	85	−6
Total	*41,615*	*29,762*	*40*	*3,610*	*3,669*	*−1*

Solid Wastes

The quantities involved are overwhelming, and the problem of solid-waste disposal is most serious. Only a few methods are available, eg, incineration, composting, sanitary landfill, and recycling of products and energy (see Incinerators; Recycling; Fuels from waste).

NELSON L. NEMEROW
Consulting Engineer

E.J. Middlebrooks, *Industrial Pollution Control. Vol. I. Agro-Industries*, John Wiley & Sons, Inc., New York, 1979.

R.A. Conway and R.D. Ross, *Handbook of Industrial Waste Disposal*, Van Nostrand Reinhold Co., New York, 1980.

N.L. Nemerow, *Streams, Lakes, Estuaries, and Ocean Pollution*, Van Nostrand Reinhold Co., New York, 1984.

WATER

SOURCES AND UTILIZATION

Sources

Sources include the hydrologic cycle (see Fig. 1) and world's water supply (see Table 1). Principal water environments of the earth are listed in Table 1.

Utilization

Withdrawal uses involve the diversion or withdrawal of water from its source, and in nonwithdrawal uses, the water is used within its natural setting. Diverted or withdrawn water used more than once by an industrial plant, farm, or other user is termed recycled. Withdrawn water that is not consumed in the course of its use may be returned to streams, lakes, groundwater reservoirs, or other natural sources, and thereby made available for subsequent repeated withdrawal.

Water Planning and Management

Most decisions relating to water apportionment and management are intricate, involving social, political, technical, and economic issues.

Table 1. Estimated Volumes of Water in Storage and Average Time of Residence in the Earth's Environment

Environment	Volume, km³[a]	Average residence time
atmospheric water	1.30×10^4	8–10 d
oceans and open seas	1.37×10^9	> 4000 yr
freshwater lakes and reservoirs	1.25×10^5	from days to years
saline lakes and inland seas	1.04×10^5	
river channels	1700	2 wk
swamps and marshes	3600	years
biological water	700	1 wk
moisture in soil and unsaturated zone (zone of aeration)	6.5×10^4	2 wk to 1 yr
groundwater	$(4-60) \times 10^6$	from days to tens of thousands of years
frozen water (glaciers and ice caps)	3.0×10^7	tens to thousands of years

[a] To convert km³ to mi³, divide by 4.16.

Computer-supported methods of systems analysis and simulation modeling have been devised to supplement traditional empirical and analytical methods for evaluation and resolution of developmental and management questions. Methods for coping with uncertainties inherent in the available information, in synthesis and projection of data, and in hydrologic and management techniques and methodologies, have been incorporated into the analytical process.

Extending water supplies. Methods of extension include interbasin transfer of water, desalination, recycling, weather modification, and artificial recharging of groundwater.

Community and rural water supplies. Trends in community and rural water-supply and sanitary-system development are indicators of cultural and economic progress. The magnitude of domestic water use reflects living standards since modern kitchen and sanitary conveniences and ornamental vegetation require water.

Industrial water supplies. Virtually all industrial processes are dependent on water for cooling and condensing, cleansing, conveyance of wastes (including heat), drinking and sanitary purposes, incorporation in products, and numerous other purposes.

Approximately one half to two thirds of the water used in most industrially developed nations is for power generation and cooling.

Agricultural water supply. Irrigation of commercial crops is the predominant use of water in agriculture. Additional agricultural applications include stock watering, gardening, cleaning, and other farm purposes.

Utilization in the United States. The gross water resources available within the United States greatly exceed demands for them. The average annual precipitation of ca 0.77 m, of which ca 4.5 km³/d (1.2×10^{12} gal/d) is shed as natural runoff, exceeds nationwide withdrawal usage ca threefold. However, precipitation is not evenly distributed over the nation, nor are the other natural conditions that determine water-supply availability. In addition, development of water geographically is not uniform, and levels of demand are uneven.

Water Quality

Water quality and quantity are interdependent, interacting elements of water systems. The interrelations of flow conditions and chemistry are relevant in the case of a stream receiving waste-liquid effluents: both the stream-flow characteristics and the chemical nature of the stream water govern the effectiveness of stream processes in the dilution and assimilation of the waste liquid.

Natural water. The exposure of water to earth materials and its participation in natural biological, hydrological, and geochemical processes alters its chemical and physical character. Solutes contained in water are derived from many sources, including products of biological activity, of the erosion and weathering of soil and rocks, and of the

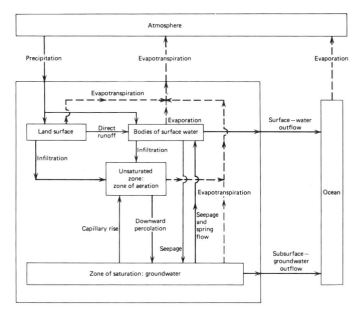

Figure 1. Components of the hydrologic cycle and directions of flow of water under natural conditions. Heavy lines represent principal flow paths; thin lines, minor flow paths; solid lines, flow of liquid water; and dashed lines, flow of gaseous water.

solution and precipitation reactions occurring during the flow of water on and through earth materials.

Practical management of water quality limits the term water pollution to chemical or physical changes of sufficient magnitude to hamper utility of the water for designated purposes and to changes caused only by human activities. For management and regulation of water resources, localized sources of pollutants or point sources must be distinguished from sources of more widespread occurrence or nonpoint sources. Polluting substances are chemical, biological, or physical (see Water, water pollution; Water, industrial water treatment; Water, municipal water treatment).

Criteria and regulatory standards. The term water quality refers to the level of suitability of water for specified purposes. Criteria and standards for water vary widely among nations. In 1978, the EPA published guideline criteria for 52 of the most common constituents and characteristics of water influencing its quality. Substances considered to be most significant to human health protection and to freshwater and marine aquatic environments, including natural and polluting chemicals, are discussed. The criteria are not intended for regulatory use, but rather for guidance in the development of water-quality management programs.

GERALD MEYER
U.S. Geological Survey

R.L. Nace, ed., *Scientific Framework of World Water Balance*, No. 7 of *Technical Papers in Hydrology*, United Nations Educational, Scientific, and Cultural Organization, New York, 1971, Table 2.

W.B. Solley, E.B. Chase, and W.B. Mann IV, *U.S. Geol. Surv. Circ.* **1001**, (1982).

U.S. Water Resources Council, *The Nation's Water Resources, 1975–2000, Second National Water Assessment*, Vol. 1, Washington, D.C., 1978, p. 22.

PROPERTIES

Natural-water systems contain numerous minerals and often a gas phase. They include a portion of the biosphere and organisms, and their abiotic environments are interrelated and interact with each other. The distribution of chemical species in water is strongly influenced by an interaction of mixing cycles and biological cycles.

Structure

The water molecule. Bond formation has little effect on the $1s^2$ electrons of the oxygen, but remaining eight electrons of the water molecule form four hybrid orbitals, two of which contain bonding-pair electrons and are directed along the O—H bond axes. The other two are nonbonding, ie, they contain lone-pair electrons. They are symmetrically located above and below the H—O—H molecular plane and form roughly tetrahedral angles with the bond hybrids (Fig. 1). These lone-pair electrons are responsible for the molecule's large induced dipole moment.

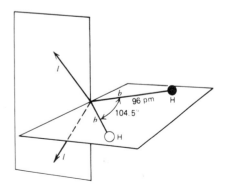

Figure 1. The structure of the isolated water molecule. *b*, Hybrid orbital with bonding-pair electrons; *l*, hybrid orbital with lone-pair electrons.

Thermodynamics and Physical Properties

The heat capacity of liquid water is ca twice that of ice at 0°C or steam at 100°c. This thermodynamic quantity of the liquid is nearly constant but exhibits a slight minimum near 35°C. Properties of water vapor, liquid water, and ice I are given in Tables 1–3.

Natural Waters

Hydrological cycle and water reservoirs. In hydrological studies, the transfer of water between reservoirs is of primary interest. The oceans hold ca 76% of all the earth's water. Most of the remainder (21%) is contained in pores of sediments and sedimentary rocks. A little more than 1% (or 73% of freshwater) is locked up in ice. The other freshwater reservoir of significant size is groundwater. Lakes, rivers, and the atmosphere hold a surprisingly small fraction of the earth's water.

The mean residence time for a water molecule in the atmosphere is ca 10 d.

Rocks and the course of weathering. The chemical characteristics of natural waters must be viewed in the light of the environmental history of the water and the chemical reactions of the rock-water-atmosphere systems.

The sedimentary rocks, oceans, and the atmosphere have probably been formed by the reaction of primary igneous rocks (bases) with excess volatiles (acids) originating from the earth's interior and concentrating at the surface by distillation. Seawater is one product of these reactions.

Biochemical cycle and oxidation-reduction processes. The maintenance of life in aquatic ecosystems results directly or indirectly from the steady impact of solar energy (qv). Photosynthesis is carried out mostly by algae and water plants; it may be conceived as a disproportionation of water into an oxygen reservoir and hydrogen, which forms high energy bonds with C, N, S, and P compounds that are incorporated as organic matter in the biomass. Various organisms catalytically decompose the

Table 1.   Selected Properties of Water Vapor at 101.3 kPa (1 atm)

Property	Value
molecular weight	18.015
heat of formation, kJ/mol at 100°C	242.49
viscosity, mPa·s (= cP) at 20°C	96×10^{-6}
velocity of sound, m/s at 100°C	405
diffusion coefficient, cm^2/s at 100°C in air	0.380
specific volume, cm^3/g at 100°C	1729.6
specific heat, J/(g·K)^a at 100°C	2.078
thermal conductivity, W/(cm·K) at 110°C	2.44×10^{-4}

^{a}To convert J to cal, divide by 4.184.

Table 2.   Selected Properties of Liquid Water at 101.3 kPa (1 atm)

Property	Value
heat of formation, kJ/mol^a at 25°C	285.890
ionic dissociation constant, M^{-1} at 25°C	10^{-14}
heat of ionization, kJ/mol^a at 25°C	55.71
apparent dipole moment, C·m^b	6.24×10^{-30}
viscosity, mPa·s (= cP), at 25°C	0.8949
velocity of sound, m/s at 25°C	1496.3
density, g/cm^3	
at 25°C	0.9979751
at 0°C	0.99987
freezing point, °C	0.0
boiling point, °C	100.0
specific heat at constant volume, J/(g·K)^a at 25°C	4.17856
thermal conductivity, W/(cm·K) at 20°C	0.00598
temperature of maximum density, °C	3.98
dielectric constant at 17°C and 60 MHz	81.0
electrical conductivity, S/cm at 25°C	$< 10^{-8}$

^{a}To convert J to cal, divide by 4.184.
^{b}To convert C·m to D, divide by 3.336×10^{-30}.

Table 3.   Selected Properties of Ice at 101.3 kPa (1 atm)

Table 3.   Selected Properties of Ice at 101.3 kPa (1 atm)

Property	Value
heat of formation, kg/mol at 0°C	292.72
Young's modulus of elasticity, MPa[a] at −10°C	967
density, g/cm^3 at 0°C	0.9168
coefficient of cubical thermal expansion, cm^3/(g·°C) at 0°C	120×10^{-6}
coefficient of linear thermal expansion, °C^{-1} at 0°C	52.7×10^{-6}
specific heat, J/(g·K)[b] at 0°C	2.06
thermal conductivity, W/(m·K)	210
dielectric constant at $^-$1°C and 3 kHz	79

[a] To convert MPa to psi, multiply by 145.
[b] To convert J to cal, divide by 4.184.

unstable products of photosynthesis through energy-yielding electron-transfer reactions (reduction–oxidation processes, respiration). These organisms use this source of energy for their metabolic needs.

Chemical composition.   The concentrations of biologically regulated components (ie, C, N, P, Si) vary with depth and are markedly influenced by the growth, distribution, and decay of phytoplankton and other organisms. The concentrations of other constituents, especially the salts (Cl$^-$, SO$_4^{2-}$, Mg^{2+}, and Na$^+$), are remarkably constant and are different from those in fresh waters. For most elements, a mass balance appears to exist between input into the sea (mostly by rivers) and removal (mostly by sedimentation). Thus, the oceans are often assumed to be at steady state.

Chemical variety.   The term species refers to the actual form in which a molecule or ion is present in solution. The inorganic fraction of suspended particulate matter consists mainly of minerals formed by weathering of terrestrial rocks. The organic particulate matter is primarily composed of living organisms, their decay, and metabolic products. Inorganic trace elements may be sorbed or bonded onto them.

Pollution

The products of human activities find their way into the environment and disturb ecosystems. Pollution has altered the surroundings to the detriment of humanity. In the last decades, the pollutional load increased, and its character has changed. Aquatic pollutants consist of salts (Cl$^-$, SO$_4^{2-}$, Na$^+$, Mg^{2+}, K$^+$, Ca^{2+}), nutrients (compounds intimately involved with the life cycle), and trace constituents (heavy metals, synthetic organic compounds, xenobiotic substances).

CHRISTINA MATTER-MULLER
WERNER STUMM
Swiss Federal Institute of Technology

D. Eisenberg and W. Kauzmann, *The Structure and Properties of Water*, Oxford University Press, London, 1969.

H.D. Holland, *The Chemistry of the Atmosphere and Oceans*, Wiley-Interscience, a division of John Wiley & Sons, Inc., New York, 1978.

Water-Related Environmetal Fate of 129 Priority Pollutants, Vols. I and II, EPA-440/4-79-029a, Environmental Protection Agency, Washington, D.C., 1979.

POLLUTION

Water is omnipresent on the earth. Constant circulation from the ocean to the atmosphere (evaporation) and from the atmosphere to land and the oceans (precipitation, runoff, etc) is generally known as the hydrologic cycle. Within the hydrologic cycle, there are several minor and local subcycles where water is used and returned to the environment.

Freshwater is withdrawn from various sources (rivers, lakes, groundwater, etc) and used many times before its discharge to the ocean. Water uses can generally be classified as follows: public water supply (domestic); industrial; commercial and institutional, eg, restaurants, schools; agricultural; and livestock.

The largest consumers of water in the United States are thermal power plants (eg, steam and nuclear power plants) and the iron and steel, pulp and paper, petroleum refining, and food-processing industries. They consume > 60% of the total industrial water requirements (see also Power generation; Food processing; Pulp; Petroleum, refinery processes).

Water-Quality Management

Quality criteria.   Whether the water of a given source is suitable for a specific use depends on the criteria or standards for that use.

The formation of the EPA in 1970 initiated specific water-pollution control programs for restoring national water quality. Amendments in 1973 (PL 92-500) established specific goals and targets to make the nation's waters suitable for fishing and swimming and achieve a zero discharge of pollutants by 1985. In addition, the National Pollutant Discharge Elimination System (NPDES) was established to provide the basic regulatory mechanism.

All waters must be free from substances attributable to wastewater or other discharges that settle to form objectionable deposits; float as debris, scum, oil, or other matters to form nuisances; produce objectionable, color, odor, taste, or turbidity; injure, are toxic, or produce adverse physiological responses in humans, animals, or plants; and produce undesirable or nuisance aquatic life.

Pollution Control

In past years, municipal wastewaters were treated to improve their appearance and bacteriological safety. Treatment included reduction in biochemical oxygen demand (BOD), suspended solids, pathogens, and inorganic dissolved solids.

The post-World War II growth in industrial activity has significantly altered the composition of wastewaters in urban treatment facilities. Pollutants that are resistant to biological oxidation have become predominant (eg, synthetic detergents, petrochemicals, synthetic rubber, etc), requiring the development of new nonbiological processes and approaches to water-pollution control. Today, the industrial-wastewater engineer must be familiar with the manufacturing process and the chemistry of the raw materials, products, and by-products.

Primary investigation.   The basic systems-engineering approach is the most suitable method for developing a solution to industrial wastewater management. The preliminary investigation includes a plant survey and the characterization of the wastewater source.

In-plant waste control.   Pollution can be reduced or eliminated by process modification, chemical and raw materials substitution, or recovery of by-products. In addition, process modification generally increases product yield by incorporating control devices.

Wastewater treatment.   The technologies developed for the treatment of municipal wastewaters generally apply to the removal of conventional pollutants, eg, dissolved organics, suspended and floating materials. A fixed or suspended-growth biological system generally provides the basis of the treatment process combined with physical-separation units (clarification-sedimentation) for the removal of grit, oil, grease, and biological solids. These process systems can remove > 90% of BOD$_5$ (5-d BOD) and suspended solids.

Toxic organic materials.   The term toxic organics includes synthetic organic compounds such as pesticides, herbicides, PCBs, and chlorinated hydrocarbons, usually produced by the manufacturers and formulators of these products. Because these compounds persist over a long period of time in a natural environment, the most effective treatment technology at present is incineration (see Incinerators).

Membrane separation.   Reverse osmosis (qv) or ultrafiltration (qv) can be used to concentrate toxic organic substances, depending on the type of compounds and the stability of the membrane against chemical attack (see also Membrane technology).

Chemical treatment.   Some organic compounds are attacked by chemical reagents such as potassium permanganate, sodium hydroxide, calcium hypochlorite, and ozone.

Adsorption. Organic compounds are adsorbed on activated carbon and synthetic resins (eg, XAD-2 and XAD-4, Rohm and Haas Co.). This technique depends on the properties of the compound being removed and the regenerative capability of the adsorbent.

Heavy metals. Heavy metals of particular concern in the treatment of wastewaters include copper, chromium, zinc, cadmium, mercury, lead, and nickel. They are usually present in the form of organic complexes, especially in wastewaters generated from textiles finishing and dye chemicals manufacture.

Inorganic heavy metals are usually removed from aqueous waste systems by chemical precipitation in various forms (carbonates, hydroxides, sulfide) at different pH values. Other methods, including activated carbon, ion exchange, and reverse osmosis, can be used to concentrate waste streams and remove the heavy metals.

MUTHIAH RAMANATHAN
Roy F. Weston, Inc.

D.K. Todd, *Groundwater Hydrology*, John Wiley & Sons, Inc., New York, 1980.

R.K. Linsley and J.B. Franzini, *Water Resources Engineering*, McGraw-Hill Book Company, New York, 1979.

M. Ramanathan and W.E. Verley, *Evaluation, Design and Startup of an Innovative and Cost-effective Wastewater Treatment Plant at Concord, New Hampshire*, paper presented at the 36th Annual Meeting, Virginia Water Pollution Control Association, Inc., Charlottesville, Va., 1982.

ANALYSIS

Since 1970, new analytical techniques, eg, ion chromatography, have been developed, and others, eg, atomic absorption and emission, have been improved. Detection limits for many chemicals have been dramatically lowered. Many wet chemical methods have been automated and are controlled by microprocessors which allow greater data output in a shorter time. Perhaps the best known continuous-flow analyzer for water analysis is the Autoanalyzer system manufactured by Technicon Instruments Corp. (Tarrytown, N.Y.).

Although simple analytical tests often provide the needed information regarding a water sample, recent discoveries, such as the formation and presence of chloroform and other organohalides in drinking water, require some very specialized methods of analysis. The separation of trace metals into total and uncomplexed species also requires special sample handling and analysis (see Trace and residue analysis).

A list of all water analyses would be extremely long since, under some conditions and with enough time, water can solubilize everything to some extent. Fortunately, a great deal can be learned about a water supply by carrying out a few physical and chemical tests. These simple tests might be all that are needed to characterize a water supply for many purposes, and it is usually the purpose for which the water is to be used that determines the type and extent of testing. The methods described in this review are intended primarily for freshwater analysis and may not be suitable for the analysis of saline water.

Physical Properties

An analysis of physical properties would include temperature, specific conductance, color, turbidity, taste and odor, dissolved solids, suspended solids, and pH.

Principal Mineral Constituents and Gases

An analysis of principal mineral constituents and gases would include measures of alkalinity, acidity, hardness (calcium and magnesium), sodium and potassium, chloride, sulfate, nitrate and nitrite, fluoride, phosphate, boron and borates, and oxygen.

Minor Mineral Constituents and Gases

An analysis of minor mineral constituents and gases would include measurements of metals, nonmetals (arsenic, selenium, cyanide, bromide and iodide), and gases (hydrogen sulfide and ammonia).

Organic Materials

The following organic factors and materials would be measured: biochemical oxygen demand, chemical oxygen demand, organic carbon, detergents, oil and grease, pesticides, and trihalomethanes.

Radioactive Materials

Radioactivity in environmental waters can originate from both natural and artificial sources. The natural or background radioactivity usually amounts to ≤ 100 mBq/L. The development of the nuclear power industry as well as other industrial and medical uses of radioisotopes (qv) necessitates the determination of gross alpha and beta activity of some water samples.

Bacteria

A bacteriological examination of water is primarily carried out to determine the possible presence of harmful microorganisms. Testing is actually done to detect relatively harmless bacteria, the *colon bacilli*, commonly called the coliform group, which are present in the intestinal tract of humans and animals. If these organisms are present in a water in sufficient number, then this is taken to be evidence that other harmful pathogenic bacteria may also be present.

R.B. SMART
West Virginia University

K.H. MANCY
University of Michigan

J.K. Kopp and G.D. McKee, eds., *Methods for Chemical Analysis of Water and Wastes*, 3rd ed., EPA/600/4-79-020, Washington, D.C., 1979.

Standard Methods for the Examination of Water and Wastewater, 15th ed., American Public Health Association, Washington, D.C., 1981.

Instrumentation for Environmental Monitoring, rev., Vol. II, Water LBL-1, Technical Information Department, Lawrence Berkeley Laboratory, University of California, Berkeley, Calif., 1980.

SUPPLY AND DESALINATION

Of the surface of the earth, 71% (3.60×10^8 km^2 or 1.39×10^8 mi^2) is covered by oceans; their average depth is 6 km and their volume is 8.34×10^8 km^3 (ca 2×10^8 mi^3). Unfortunately, much of this huge quantity of water is unsuitable for human use. Water with over 1000 ppm salt is usually considered unfit for human consumption, but in some parts of the world, people and land animals are forced to survive with much higher concentrations of salts. It has been recorded that when the native population in parts of Mexico was given desalinated potable water, the people refused at first to drink it as its taste was strange.

Freshwater is generally considered to be potable water at less than 500 ppm or 0.05% dissolved solids. Rain is the source of freshwater, and its precipitation of over 1.3×10^{15} L/d (3.4×10^{14} gal/d) over the earth's surface averages about 1.05 m/yr. Extremes range from almost zero in Northern Chile's desert bordering the Pacific Coast to > 25.4 m in some tropical forests and on some high slopes where the high, cold mountains condense floods from the clouds.

The oceans hold about 97% of the earth's water. More than 2% of the total water and over 75% of the freshwater of the world is locked up as ice in the polar caps. Of the remaining 1% of total water that is both liquid and fresh, some is groundwater at depths of > 300 m and therefore impractical to obtain, and only the very small difference, possibly 0.06% of the total water of this planet, is available for human use as it cycles from sea to atmosphere to land to sea. Only recently have humans been able to regulate that cycle to their advantage, and then only infinitesimally in a few isolated places.

Wells produce groundwater, stored from previous rains. However, the fact that in recent years wells have had to be made deeper to reach water shows that groundwater is being used faster than it is being replenished. Water lying in deep strata for millions (10^6) of years is being mined like other minerals, never to be replaced.

Transport of Freshwater

Initiated ca 120 yr ago, New York City's water system is truly an engineering marvel. Farsighted action in the late nineteenth century also gave the city extensive upstate watershed rights. This system, having a storage capacity of 2×10^9 m^3 (5.3×10^{11} gal), can safely furnish the city with an average of 5×10^6 m^3/d (1.3×10^9 gal/d). The system is not adequate today, but it has served well except in years of serious drought. Water enters the city via two tunnels, one built in 1917 and the other in 1937. The total cost of a third tunnel is now estimated at $\$3.5 \times 10^9$ in 1981 dollars, but could be as high as $\$11 \times 10^9$ by the year 2009, the date scheduled for its completion.

The 1974 Safe Drinking Water Act put all public water supplies in the United States under Federal supervision, and the 94th Congress authorized a six-state water study, the High Plains Study Council, to develop plans for increasing water supply in the area. If its recommendations are accepted, the government will move huge amounts of water from the Missouri and Arkansas Rivers to the high plains. It could take as long as 9 yr to design, 25 yr to build, and is estimated by the Army Corps of Engineers to cost $\$(6-25) \times 10^9$.

The problem of bringing water to southern California continues to be one of much controversy. The project to build the so-called Peripheral Canal, a 67-km long, 120-m wide channel to carry water from the Sacramento River delta to an existing aqueduct and then to southern California, was authorized by California's legislature in 1980.

In some places and under certain conditions, freshwater can be obtained by desalination of seawater more cheaply than by transporting water. This is true when all costs of the tremendous investments in dams, reservoirs, conduits, and pumps to move the water are considered.

The Water Problem

An analysis of water use in the United States and an estimate for the next 50 yr predicts withdrawals will increase by 400% and consumptive uses by > 100%. Most of this increase must be met by reuse; ie, in manufacturing, recycling must rise from 1.3 times reuse to 6.3 times reuse by the year 2020.

Water is far from evenly distributed in the United States, and principal shortages occur in some very populous areas. California and the Southwest have always been watershort, but in the first half of the 1960s and the late 1970s, the Northeast, too, experienced water shortages. For the United States as a whole, the demand for water will become equal to the total supply before 1985. Some areas will be desperately short much sooner. Water shortages are acute in years of low rainfall, as in 1957 when over 1000 communities in 47 of the 48 contiguous states restricted the use of water. Restrictions of water use in one or more states have become virtually annual occurrences.

Some of the most attractive areas of the world, particularly islands and beaches, are almost devoid of freshwater. This living space, and space for resort hotels, is lost. The biggest and one of the fastest growing of the industries of the world is tourism. It is a particularly attractive industry to developing countries, and in some of these it may be almost the only nonagricultural industry. Tourism accounts for a substantial use of the available freshwater, and may be stifled or entirely prevented in otherwise attractive places if there is insufficient freshwater. To conserve available water supplies, some hotels might use double water systems, ie, seawater for flushing, but they still must provide on the order of 400–600L/d (106–159 gal/d) per tourist to assure comfort to guests unaccustomed to water shortages. Production of these quantities of water by desalination techniques has become an important expense to the hotels. Today, the cost of production of desalinated water can be as high as $\$4.00-5.25$ per thousand liters ($\$15-20/1000$ gal), and in some locations even higher.

Saline water for municipal distribution. Only a very small amount of potable water is actually taken internally, and it is quite uneconomical to desalinate all municipally piped water, although all distributed water must be clear and free of harmful bacteria. Most of the water piped to cities and industry is used for little more than to carry off extremely small amounts of waste materials or waste heat. In many locations, seawater can be used for most of this service. If chlorination is required, it can be accomplished by direct electrolysis of the dissolved salt. Against the obvious advantage of economy, there are also several disadvantages. Use of seawater requires different detergents; sewage-treatment plants must be modified; the usual metal pipes, pumps, condensers, coolers, meters, and other equipment corrode more readily; chlorination could cause environmental pollution; and dual water systems must be built and maintained.

Water in industry. Freshwater for industry can often be replaced by saline or brackish water, usually after sedimentation, filtration, and chlorination (electrical or chemical) or other treatments.

Water for agriculture. Two liters (ca 0.5 gal) of water in some form is the daily requirement of the average human, depending on many personal and external conditions. However, at least several hundred liters per day is required for the growing of the vegetables, fruit, and grain that make up the absolute minimal daily food ration for a vegetarian.

The one seventh of the world's crop lands that are irrigated produce one quarter of the world's crops. Irrigations main losses result directly from seepage and evaporation from the open water-carrying channels and the soil. Only a small fraction of the water withdrawn from the irrigation ditch or pipe is absorbed by the plants. Plastic films, as ground covers through which the plants protrude, prevent some losses but at great expense for film and labor. Cheaper systems are necessary to assure better water utilization by plants. Other possible goals would be food plants with membranes capable of separating freshwater from brackish water to give a nonsalty crop. Progress has been made in both of these directions, and some plants have been developed that accumulate salt from the ground.

Desalination Development Programs and Associations

Office of Saline Water (OSW)-Office of Water Research and Technology (OWRT). The U.S. Department of the Interior, long active in many areas of water resources and management, was authorized by Congress in 1952 to organize the OSW with a $\$2 \times 10^6$ budget for 5 yr to develop economical processes for desalination. This budget was repeatedly expanded to a total of almost $\$318 \times 10^6$ authorized through fiscal 1980 for research and development efforts, including the construction of demonstration plants to prove the technical and economic feasibility of desalination processes.

Manufactured Freshwater

The possibility of producing freshwater from seawater or brackish water by separation of the salts opens a new dimension in the supply of freshwater areas bordering the sea would have an available raw material without limit or cost of transportation to the water facility. The successful realization of desalination by the combined effort of chemists, chemical engineers, mechanical engineers, and metallurgists, as opposed to the search for and transport of existing freshwater, gives new hope for adequate water in many, but not all, cases.

Freshwater may be obtained from saline solution by many processes, but all fall into one of two classes: freshwater is separated from seawater, which becomes more concentrated (water from saline solution); or salt or more concentrated brine is separated from seawater, which becomes less concentrated until an acceptable level is reached (salt from water). Usually the first type of process, water from saline solution, is the easier.

Materials for desalination plants. In addition to metals and alloys, a wide variety of plastic and polymeric materials and different concrete compositions have become an integral part of materials in construction of desalination plants. The plastics and polymers are used in pipes, sheets, membranes, ion-exchange resins, gaskets, coating and linings, etc. Other materials are also becoming part of the operations, among them various solvents, heat-transfer reagents, refrigerants, and similar thermodynamic reagents. Chemicals that are strictly expendable are those used for treating raw seawater and brackish water such as sulfuric acid, lime, algicides, antifoams, and antiscalants of the conventional type, eg, polyphosphates or the new high temperature polymeric variety such as Belgard (CIBA-GEIGY). Smaller amounts of other chemicals for treating product water are also used, such as chlorine, ozone, activated carbon or higher alcohols to stop evaporation in storage basins.

Evaporation Processes for Desalination

Evaporation with surface for boiling and for condensing. In such processes, heat is transferred from primary steam coming from a boiler in single-purpose plants, or more often, as is the case in dual-purpose plants, from the exhaust or some intermediate stage of a turbine. The steam is condensed on one side of the first or heated metal surface, the evaporating surface, to boil the seawater contained in a suitable vessel on the other side of the tubular surface. Obviously, the required area of heat-transfer surface (hence the amount of metal required for a given duty) is reduced by increasing the coefficient of heat transfer through the surface. For standard tubes, the heat-transfer coefficient, reported in W/(m·K) [0.578 Btu·ft/(ft^2·h·°F)], is usually considerably below 1.73 kW/(m·K) [1000 Btu·ft/(ft^2·h·°F)] but values up to 4.33 (2500) have been reached by Westinghouse, General Electric, and others. If these high transfer rates can be obtained, it will be possible to reduce the heat-transfer area, and hence the material cost, provided no other considerations intervene.

The vapors from the evaporation pass to a second metal surface where they are condensed to give freshwater on the tubular condensing surface. This condensing surface is cooled, for example, by circulating cooling water on the other side.

Today, at least four different vapor-compression systems are commercially utilized. Spray-film vapor compression is extensively used for hotels, small industrial plants, and power stations. Its capacity is usually 10,000–120,000 L/d [(2.6–32) × 1000 gpd]. The vertical tube flow vapor-compression unit has a similar range. The steam heat ejector vapor-compression unit operates in sizes up to 1.4×10^5 L/d (3.7×10^4 gpd) and the low temperature all-aluminum vapor-compressor unit operates at temperatures below 55°C.

Conventional multiflash evaporation. Flash evaporation (Fig. 1) is the vaporization, often violent, of a part of the water in a stream of hot seawater as it passes into a low pressure chamber. The vapors are formed adiabatically. The sensible heat given up by the cooling liquid as it comes to equilibrium with the lower vapor pressure is equated to the latent heat of the vapors formed. A high ratio of flash temperatures and pressures, as in the discharge of condensate from a stream trap to atmospheric pressure, causes an approach to equilibrium with almost explosive violence. The vapors formed are withdrawn and condense as they preheat the raw seawater passing the system of condensing tubes.

Freezing. The concept of the freezing desalination process is based upon thermodynamics and the phase diagram of an NaCl–H$_2$O solution. As the temperature of the solution is reduced below the freezing point of water, pure water solidifies, leaving a more concentrated, lower freezing brine. A freezing desalination process must have a freezer, a means of washing the ice crystals formed on their surface, and a melter to process the ice formed.

Reverse osmosis (RO). Reverse osmosis (qv) is a process that consists of allowing the solution flow under pressure through an appropriate porous membrane and withdrawing membrane-permeated product, generally at atmospheric pressure and ambient temperature. In this process, the permeate or product is enriched in one or more of the constituents of the mixture, and the solution on the high pressure side of the membrane is left at higher or lower concentration. The reverse-osmosis membrane need not be treated, and no phase change is involved in the process.

Reverse osmosis is a process of wide-ranging and continuously growing applications. Its application to the desalination of sea and brackish waters is the single largest use of this emerging technology.

Desalination by renewable energy sources. The sun, ocean, geothermal energy, waves, and direct labor are all renewable energy sources that have been tapped for desalination projects. The exploration of desalination using renewable energy sources has been prompted by the desire to conserve the standard fuels, the difficulties of obtaining standard fuels in remote regions, and especially, their high cost. Basic solar stills have been in use at least since the seventeenth century, and recently there have been serious studies of a diversity of concepts for advanced solar-powered desalination systems. The exploration of the other sources, although of notable potential, has seen less than either serious or consequential investment of effort to effect their practical realization, and even solar distillation remains almost unused (see Solar energy).

ROBERT BAKISH
Bakish Materials Corporation

O.K. Buros, *Desalination Manual*, U.S. Agency for International Development, International Desalination and Environmental Association, Teaneck, N.J., 1981.

K.S. Spiegler and A.D.K. Laird, *Principles of Desalination*, 2nd, ed., Academic Press, New York, 1980.

INDUSTRIAL WATER TREATMENT

Industrial water treatment involves the production from available water supplies of boiler feedwaters, cooling waters, or process waters that have the composition or properties required for each water-using system or process. A number of the processes discussed in this section are used for the treatment of waters discharged from industrial water systems or from municipal or industrial wastewater treatment plants in order to make these waters reusable for industrial applications (see Water, municipal water treatment; Water, reuse).

Water-Caused Problems

Water-caused problems in industry affect production costs and cause unscheduled shutdowns, which can result in extremely costly loss of production. Water-caused problems may reduce heat transfer, reduce water flow, cause premature equipment deterioration or failure, and reduce product quality or yield.

Treatments

Ideally, the water-quality and treatment requirements for each industrial plant and water use should be evaluated beginning with early design to anticipate possible water-caused operating problems and to provide the most cost-effective means for preventing or minimizing these problems during the life of the plant. The evaluation should take into consideration the quantity and quality of available water sources, including their ranges of variations; quantitative and qualitative requirements for each plant water use; possible cascading or sequential use of water for multiple-plant applications; in-plant recycling of water, alternative treatment processes capable of meeting the various water-quality requirements; space requirements for treatment processes; applicable effluent-discharge regulations; and quality and quantity of operating personnel.

External. External treatment processes are used for the reduction or removal of suspended or colloidal solids, dissolved solids, or dissolved gases.

Suspended or colloidal solids can be reduced or removed by rough screening, sedimentation (qv), centrifugal separation (qv), straining, filtration (qv), coagulation, flocculation, magnetic separation (qv), or combinations of these processes (see Flocculating agents).

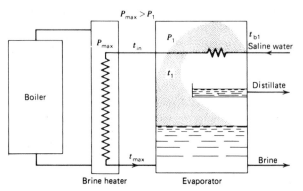

Figure 1. Flash distillation (brine heater and first stage). Courtesy of Longman Group Ltd., London.

Internal. In contrast to external treatment, most internal water-treatment processes are directed toward modification of the water quality within a system in order to minimize deposit formation, corrosion, and biological growths rather than toward the reduction or removal of constituents that cause these problems. One of the few exceptions is the treatment of boiler feedwater with sulfite or hydrazine for removal of residual dissolved oxygen by chemical reduction.

Internal treatment for deposit control is accomplished by means of a variety of chemical additives, which act as precipitants, chelating agents (qv), dispersants, or scale inhibitors (see Dispersants).

Corrosion control by internal treatment involves one or more of the following techniques: chemical reduction of dissolved oxygen, pH adjustment, protective film formation with corrosion inhibitors, and dispersion of suspended solids (see Corrosion and corrosion inhibitors).

Biological growths are usually controlled in industrial water systems by treatment of the water with oxidizing or nonoxidizing biocides, although several mechanical procedures are used, particularly for cooling-water systems. The latter procedures include automated, direct cleaning of condenser or heat-exchanger tubes for removal of microbiological slimes, installation of opaque covers on head basins of cooling towers to prevent algae growth, and periodic reversal or discontinuance of water flow to raise water temperatures high enough to kill organisms. Chlorine, an oxidizing agent, is by far the most widely used water-system biocide (see Industrial antimicrobial agents; Water, industrial water treatments).

Programs. Water treatment for an industrial plant frequently includes various external and internal treatment processes. The total plant water supply may be clarified and filtered. A part of the treated water may be used without further treatment for general plant uses. Another part may be used as make-up water to systems, eg, cooling systems, within which it is treated internally for the control of deposit formation, corrosion, and biological growths. Still other portions may be softened, deionized, or otherwise treated, externally or internally, to meet the quality or performance requirements for use as boiler feed or in specific processes.

Health and Safety Factors

Almost every chemical used for water treatment can be harmful under some circumstances. Therefore, operating personnel who handle the full-strength treatment chemicals should be fully informed of potential hazards, provided with necessary personal protective equipment, and trained in the safe handling of each chemical used.

Many chemicals added for internal treatment remain in the water when the latter is ultimately discharged from the system and from the industrial plant in which it is used. Therefore, consideration must be given to the effects of the chemicals of choice on the operation of the plant's wastewater-treatment system and their environmental acceptability in the plant's wastewater discharge.

S. Sussman[‡]
Olin Water Services

Cooling Tower Manual, Water Chemistry, Cooling Tower Institute, Houston, Texas, Jan. 1981.

Betz Handbook of Industrial Water Conditioning, 8th ed., Betz Laboratories, Inc., Trevose, Pa., 1980.

Drew Principles of Industrial Water Treatment, 1st ed., Drew Chemical Corp., Boonton, N.J., 1977.

MUNICIPAL WATER TREATMENT

Although the human physiological need for water is less than two liters per day (0.5 gal/d), the average production and distribution of potable water through public water supplies in the United States is ca 600 liters per capita per day (158 gal/d). This amount is used for household and sanitary needs as well as urban, industrial, and municipal requirements (see Water, industrial water treatment).

Water Quality

The principal sources of water are surface supplies, eg, streams, rivers, lakes, reservoirs, and groundwater supplies, ie, those obtained from deep wells. The general properties of the two sources are listed in Table 1. Selection of a supply source is based on the specific properties of the supplies in question and local availability.

Until the passage of the Federal Safe Drinking Water Act (SDWA) (Public Law 93-523), the states regulated the quality of the water furnished to water utility customers. The SDWA required that the EPA promulgate interim water-quality regulations, which would be subject to revision upon recommendation of the NAS. The Act requires the EPA to establish recommended maximum contaminant levels (MCLs) at which no known or anticipated adverse effects on the health of persons occur and which allow an adequate margin of safety. The interim regulations were to protect health by means of generally available technology and were based partly on cost and other factors (see also Regulatory Agencies).

Treatment Methods

The choice of which alternative sources of water to use for municipal consumption must be based on such factors as available quantity, continuity of availability, treatment and capital costs, and assurance that the chosen treatment process will provide water that is biologically safe, noncorrosive and nondamaging to plumbing fixtures and water heaters and containers, and aesthetically acceptable. The choice of treatment is a function of the raw-water quality. Most surface waters must be treated for the removal of suspended solids. These solids include clay, protozoan cysts, and organic debris resulting from partial breakdown of leaves and other plant materials. Some surface waters contain organic color, which is usually removed for aesthetic reasons. In addition, many utilities must either remove the natural organic compounds, which are precursors for trihalomethanes (THMs); use a disinfectant other than chlorine; remove the THMs formed; or modify their treatment scheme to reduce THM formation. Groundwater supplies also contain such precursors. Waters obtained from groundwaters or impounded supplies may contain reduced iron and manganese, which must be removed to prevent precipitation of the oxidized forms in the distribution system or at the point of use.

A complete sequence of treatment for a municipal plant includes application of copper sulfate to a reservoir or lake for algal control; addition of activated carbon and chlorine at the head of the plant for taste, odor, and bacterial control; addition of a coagulant in rapid-mix tanks for turbidity removal; passage through a flocculation tank to promote floc growth; passage through a horizontal or vertical sedimentation tank; pH adjustment; and gravity filtration (see Flocculating agents; Filtration). The water from the filters, after the addition of chlorine, is usually stored in a clearwell until it is pumped to a service or storage tank. Some or all of the preceding methods may be used in any given case for a turbid surface water. If the water is also hard or is hard rather than turbid, it can be softened by lime, lime-soda, or an ion-exchange process in addition to or instead of the coagulation–flocculation–sedi-

Table 1. General Properties of Natural-Water Supplies

Property	Surface	Groundwater
mineralization	low	high
dissolved oxygen	usually saturated	very low
H_2S	absent, unless polluted	may be present
pollution	common	common
color	common	uncommon
turbidity	common	uncommon
Fe and Mn	uncommon	common
quantity	variable unless impounded	constant
organics	variable	variable

[‡] Deceased.

mentation steps prior to filtration, chlorination, and distribution (see Ion exchange).

Since the SDWA was passed in 1974, emphasis in treatment has been on the control of trace organics in public water supplies. The discovery that chloroform and the other bromo and chloro haloforms form as a result of disinfection of water with chlorine led to consideration of treatment-process modification, alternative disinfectants, increased emphasis on the removal of natural organics, and removal of the haloforms after they form. Various of these alternatives have been included in treatment schemes (see also Disinfectants).

Recent attention has been focused upon the removal of industrial organic solvents from groundwaters. The effective treatment technologies have been air stripping and/or adsorption by granular activated carbon.

Fluoridation. The practice of adding fluoride ion to domestic water supplies has been increasing since the first experiments in the city of Grand Rapids, Michigan, in 1945. It has been shown that dental decay is reduced significantly by maintaining a fluoride residual of ca 1 mg/L (3.8 mg/gal) in the public water supplies. The concentration recommended by the U.S. Public Health Service is 0.7–1.2 mg/L depending upon the annual average maximum daily air temperature, since more water is ingested at higher temperatures.

J.E. SINGLEY
Environmental Science & Engineering, Inc.

Drinking Water and Health, National Academy of Health, Washington, D.C., 1977, Part I, p. I-3. *Fed. Regist.* 40, 59566 (Dec. 24, 1975).

M.N. Baker, *The Quest for Pure Water*, American Water Works Association, New York, 1949.

J. Arboleda-Valencia, *Teoria, Diseno y Control de los Procesos de Clarificacion del Aqua*, Pan American Center for Sanitary Engineering and Environmental Sciences, Pan American Health Organization, Lima, Peru, 1973.

SEWAGE

Sewage is the spent water supply of a community. Because of infiltration of groundwater into loose-jointed sewer pipes, the total amount of water treated may exceed the amount consumed. Sewage contains about 99.95% water and ca 0.05% waste material.

Strength of sewage is expressed in terms of BOD, total solids, suspended solids, fixed solids, volatile solids and filterable solids. The BOD is a measure of the load placed on the oxygen resources of the receiving waters. Treatment efficiency is usually evaluated on the basis of BOD removal by the plant. Unless otherwise stated, BOD signifies the biochemical oxygen demand for five days at 20°C (BOD$_5$). The significant parameters of wastewater are evaluated according to the analytical methods set forth by the U.S. Public Health Association.

Sewage Treatment and Disposal Systems

Sewage works were originally constructed for reasons based primarily on public-health concepts. However, prevention of disease is not the sole purpose of modern water-sanitation practice; it is also a means to protect the oxygen resources of the receiving water. An effluent that does not significantly reduce the oxygen concentration of the receiving water into which it is discharged can be expected to have a low concentration of food for microorganisms that deplete the dissolved oxygen. Thus, protection of the oxygen resources of the receiving water prevents the spreading of disease and satisfies aesthetic considerations.

Private and rural disposal systems. In areas not served by sewers, human and other water-carried wastes are disposed of in primitive privies, cesspools, or septic tanks.

Small communities. Small communities and recent subdivision additions to larger communities, which have not yet been connected to municipal collection systems, must have a means of waste disposal. Septic tanks are a possibility, but require periodic servicing and cleaning. Furthermore, the soil is not always suitable for accepting the effluent. An alternative is the package plant. These plants furnish primary treatment

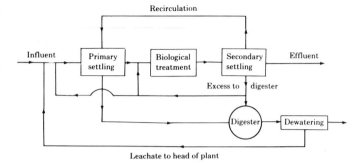

Figure 1. Sewage-treatment operations.

and some secondary treatment, and require only minimal operating supervision. Capacity can be varied as needs dictate. In general, public-health authorities prefer such installations instead of septic tanks.

City disposal systems. The operations necessary for treating sewage are outlined in Figure 1. In primary treatment large particles are removed by screening and sedimentation; 15 to 60% of the BOD can be removed in this manner. Collodial and dissolved substances are removed by secondary, or biological, treatment where the microorganisms of the process utilize the waste material for food. A commonly accepted definition of tertiary treatment is the use of any process or operation for further removals.

Sludge Digestion and Disposal

Digestion. Digestion reduces the volume and the pathogenic organisms. Digested sludge is black and granular and has a tarry color. Sludge is withdrawn periodically from the settling tanks and allowed to flow by gravity to a collection well. From there it is pumped to a digester where it is thoroughly mixed. Sludge digestion can be aerobic or anaerobic.

Disposal. Digested sludge is reasonably inert but has a high water content. It can be dewatered by filtration or by drying on open or covered sand beds. The dried sludge may be incinerated or used for landfill or fertilizer.

Ocean disposal. Disposal of raw or treated sludge by barging to sea was practiced for many years by some coastal cities, but today is highly controversial, and it appears that this method will be no longer economically feasible.

Chlorination of Effluent

It is common practice to chlorinate wastewater effluents for bacterial control. Regulations vary from state to state, but all require chlorination to specified residual concentrations.

Other Disposal Problems

Other disposal problems include watercraft-waste disposal, surfactants and phosphates.

Health and Safety Factors

Wastewater-treatment plants have numerous hazards to be expected in a chemical-process plant. Worker safety is covered by applicable OSHA, state and local standards, but two hazards require special notice: the potential for infection by pathogenic organisms is always present, and plant workers require innoculation against the common waterborne diseases. In addition, since wastewater-treatment plants utilize deep water-filled tanks, provision must be made against drowning.

JAMES R. PFAFFLIN
Consultant

G.M. Fair, J.C. Geyer, and D.A. Okun, *Water and Wastewater Engineering*, Vols. 1 and 2, John Wiley & Sons, Inc., New York, 1968.

C. MacInnis, "Municipal Wastewater," in *The Encyclopedia of Environmental Science and Engineering*, Vol. 1, Gordon and Breach Science Publishers, New York, 1976, pp. 587–600.

T.D. Reynolds, *Unit Operations and Processes in Environmental Engineering*, Wadsworth, Inc., Belmont, Calif., 1981.

REUSE

Formerly, water was accepted by a second user for reuse while it was still under control of the first user. Today, the used water is treated in such a manner that it can be used again before ultimate disposal. Furthermore, a distinction can be made between direct reuse, where the water is reclaimed without dilution or natural purification, and indirect use, where treated used water is returned to the environment for subsequent utilization as a raw water supply.

Water can seldom be reused directly. The treatment required depends on the intended second use. Disposal costs of the wastewater must be included in any economic analysis, and additional treatment for reuse may be justified when this expense is included. Costs of reclamation depend on the location, water scarcity, availability of public water supplies, and the intended reuse.

Conventional secondary wastewater treatment does not produce an effluent suitable for direct reuse, and a tertiary treatment step is required.

Tertiary Treatment

Tertiary treatments include chlorination, chemical precipitation, filtration, microstrainers, effluent polishing, foam separation, activated-carbon adsorption, ion exchange, oxidation ponds, reverse osmosis, electrodialysis, and vapor-compression evaporation and waste-heat evaporation (see Evaporation).

JAMES R. PFAFFLIN
Consultant

Guidelines for Water Reuse, EPA 600/8-80-036, Environmental Protection Agency, Washington, D.C., 1980.

T.D. Reynolds, *Unit Operations and Processes in Environmental Engineering*, Wadsworth, Inc., Belmont, Calif., 1982.

W.W. Eckenfelder, Jr., *Principles of Water Quality Management*, CBI Publishing, Boston, Mass., 1980.

TREATMENT OF SWIMMING POOLS, SPAS, AND HOT TUBS

Swimming Pools

Most swimming pools are of the recirculating type in which the water is changed only infrequently. Through continuous treatment by mechanical filtration and chemical additions, the same water can be recycled for many years before the pool need be drained and refilled. Sanitizing chemicals must be added regularly to kill and control disease-carrying bacteria introduced by swimmers and dirt entering the water. It is also necessary to destroy algae whose spores are carried into the water by wind and rain. Unchecked algal growth results in discolored water, unsightly growth on the walls and bottom of the pool, clogging of filters, and provides a breeding ground for bacteria. The pH of pool water must be maintained within the proper range for swimmer comfort and optimal effectiveness of chlorine sanitizers. In order to control the corrosive or scaling tendencies of pool water, it is also necessary to maintain a proper balance between pH, alkalinity, and hardness. Undesirable trace metals such as iron, manganese, or copper are sometimes found in feedwater or formed by corrosion of pool equipment. Unless removed, these metals discolor pool water and cause stains, especially damaging to plaster pool surfaces. Filtration of pool water is necessary for removal of suspended solids which otherwise cloud the water and interfere with the disinfection process. Good circulation of water is important for proper filtration and dispersal of sanitizers and other chemicals added to pool water.

Sanitizers. Chlorine and its compounds have been and continue to be the foremost chemicals for disinfecting swimming-pool water (see also Disinfectants). Chlorine gas, sodium hypochlorite solution, calcium hypochlorite, and chlorisocyanurates provide free available chlorine (FAC),

Table 1. Chlorine Sanitizers Used in Swimming Pools

Compound	Formula	Typical form	FAC, %
chlorine	Cl_2	liquefied gas	100
trichloroisocyanuric acid		tablets, sticks	89–91
calcium hypochlorite	$Ca(OCl)_2$	granules, tablets, briquettes	65 or 70
sodium dichloroisocyanurate		granules	62–63
sodium dichloroisocyanurate dihydrate	$2H_2O$	granules	55–56
lithium hypochlorite	$LiOCl$	granules	35
sodium hypochlorite	$NaOCl$	solution	10–15

ie $HOCl + ClO^-$, and are the most commonly used swimming-pool sanitizers (see Table 1) (see also Bleaching agents).

Stabilizers. Cyanuric acid has been successfully used to stabilize chlorine derived from chlorine gas, and hypochlorites or chloroisocyanurates against decomposition by sunlight.

Superchlorination. Superchlorination or shock treatment of pool water is necessary since accumulation of organic matter and nitrogen compounds consumes free available chlorine and impedes the process of disinfection.

Algicides. Algal growth in pools is unsightly, a safety hazard to swimmers, and usually a result of poor pool maintenance. It can cause slipperiness, development of odors, cloudy and discolored water, chloramine formation, increased chlorine demand, bacterial growth, and stubborn stains. Low FAC, high temperatures, sunlight, and certain mineral nutrients promote algal growth. Such growth can be prevented by continuous pumping to ensure efficient sanitizer dispersal and by maintaining the proper pH range and free available chlorine content, supplemented by periodic superchlorination or shock treatment.

pH control. The optimal range for bather comfort and efficiency of disinfection of chlorine (Cl^+) sanitizers is pH 7.2–7.6 where the biocidal agent $HOCl$ represents 47–69% of FAC. The pH of pool water is readily controlled with inexpensive chemicals. Hydrochloric acid solution or sodium bisulfate lowers it, whereas sodium carbonate raises it.

Alkalinity. In swimming-pool water at its normal pH range, the so-called carbonate alkalinity is due primarily to bicarbonate and a very small contribution from carbonate. Because of its buffering capacity, alkalinity resists changes in pH when sanitizing chemicals are added to pool water. Since cyanuric acid is 77% neutralized at pH 7.4, cyanurate ion also contributes to alkalinity. Therefore, in stabilized pools the alkalinity determination is corrected for 0.3 × ppm cyanuric acid in the normal pH range (7.2–7.6). Sodium bicarbonate is generally added to increase alkalinity, muriatic acid (HCl) or sodium bisulfate ($NaHSO_4$) to reduce it.

Hardness. Hardness (qv) is defined as the soap-consuming capacity of water. In swimming-pool water, the hardness is caused primarily by Ca^{2+}. The concentration of Ca^{2+} must be at or near the $CaCO_3$ saturation value in order to prevent deterioration of the pool and its equipment. Hardness is raised with $CaCl_2$ and is lowered by draining some of the pool water and adding water of lower hardness or by passing the pool water through a softener or demineralizer (see Ion exchange).

Balanced water. Materials of construction used in pools are subject to the corrosive effects of water, eg, iron and copper equipment can corrode owing to the presence of dissolved O_2 and anions such as Cl^- and SO_4^{2-}, whereas concrete and plaster can undergo simple dissolution. Maintaining proper water balance minimizes and may even prevent such corrosion by deposition of a protective layer of crystalline $CaCO_3$ (calcite) using the natural Ca^{2+} and alkalinity in pool water.

Other chemical treatments. Pool water may occasionally contain metallic impurities such as copper, iron, or manganese which enter the pool with the make-up water or by corrosion of metallic parts in the circulation system. Superchlorination oxidizes soluble Fe^{2+} and Mn^{2+} to the highly insoluble $Fe(OH)_3$ and MnO_2 which can be removed by filtration or allowed to settle to the bottom of the pool from where they are removed by vacuuming. For dispersed colloids, coagulation may be necessary.

Water-soluble, high molecular weight polymers can be used as flocculating agents (qv) to increase the settling rate or strength of a chemical floc, thereby acting as a filter aid to control the depth of floc penetration.

Test kits. Proper pool management requires routine analysis for free and combined chlorine and pH and, less frequently, alkalinity, hardness, and cyanuric acid. These analyses can conveniently be carried out at the poolside with simple test kits which are available at moderate cost.

Filtration. Efficient filtration for removal of suspended particles is essential for pool water. Filters may be of the fixed-bed, precoat, or cartridge type and operate under vacuum or pressure. The most common filter media are sand, diatomaceous earth (DE), anthracite, and paper or cloth cartridges.

Feeders (dispensers). Feeders dispense the chemicals in gaseous, liquid, and solid (both granular and compacted) forms.

Spas and Hot Tubs

The basic principles of treating swimming-pool water apply also to spas and hot tubs. However, spas and tubs are not miniature swimming pools but are unique in treatment requirements because of use patterns and a high ratio of bather to water. Currently, there are no national standards governing sanitation of spas and hot tubs. Standards proposed by the National Spa and Pool Institute (NSPI) are like those for swimming pools. Public spas and tubs are under the jurisdiction of local health departments.

Spa and hot-tub sanitation is dominated by chlorine- and bromine-based disinfectants.

In addition to replacing spa or tub water because of cyanuric acid build-up from chloroisocyanurate sanitizers, the NSPI recommends that the water be replaced at least monthly or more frequently when often used because bathers contribute microorganisms, perspiration, body oils, lotions, etc, which affect water chemistry, total dissolved-solids concentration, water surface tension (resulting in foaming), and sanitizer demand.

Chemicals such as defoamers, sequestrants, and flocculating agents are used in spas and in tubs. In addition, water tints and fragrances are added, further complicating the water chemistry. These chemicals are of concern in spa and tub sanitation since they might reduce the bacterial kill efficiency by lowering the FAC because of chlorine demand or by interfering in the disinfection process.

Higher calcium concentrations are required to maintain a pH-balanced water condition. As in swimming pools, the efficiency of the chlorine sanitizers in spas and tubs is dependent on pH, which should be maintained in the range 7.2–7.6.

Health and Safety Factors

In spas and tubs, diseases can be transmitted by contact with and ingestion of the water, inhalation of micoorganisms contained in aerosols produced by the aerated water, and infection through close personal contact.

Chlorine-based swimming-pool and spa and hot-tub sanitizers irritate eyes, skin, and mucous membranes and must be handled with extreme care. The toxicities are as follows: for chlorine gas, TLV = 1 ppm; acute inhalation LC_{50} = 137 ppm for 1 h (mouse).

Swimming-pool sanitizers should be kept in closed containers in a cool, dry area, segregated from other nonpool materials (such as paints, solvents, etc), and should never be mixed with each other or with other materials.

JOHN A. WOJTOWICZ
J. PHILIP FAUST
FRANK A. BRIGANO
Olin Corporation

G.C. White, *Handbook of Chlorination*, Van Nostrand Reinhold Co., New York, 1972, Chapt. 8.

D.G. Thomas, *Swimming Pool Operators Handbook*, National Swimming Pool Foundation, Washington, D.C., 1972.

WATER GLASS.   See Silicon compounds.

WATERPROOFING AND WATER / OIL REPELLENCY

Principles of Repellency

When a drop of liquid rests on a solid surface, the shape of the drop depends on the equilibrium among three forces, as shown in Figure 1. The contact angle θ depends on three surface tensions: the liquid–air interface γ_{LA}; the solid–air interface γ_{SA}; and the solid–liquid interface γ_{SL}. Wetting of the solid by the liquid results from reduction of the contact angle so that the liquid spreads easily. A surface is made repellent by raising the contact angle, so that spreading and migration into capillaries do not occur.

Textiles

The history of waterproofing textiles includes a long use of impermeable coatings followed by treatments that make the individual fibers of fabrics repellent without loss of permeability to air and water vapor.

According to the American Association for Textile Technology, Inc., the textile industry uses the following definitions for fabric treatments that repel water.

Water-repellent fabrics resist wetting or repel waterborne stains; they pass AATCC Test Method 22.

Shower-resistant fabrics protect against water penetration during a light or brief shower and pass AATCC Test Methods 22 and 42 (shower).

Rain-resistant fabrics protect against water penetration during a rain of moderate intensity and pass AATCC Test Methods 22 and 25 (rain).

Storm-resistant fabrics protect against water penetration during a heavy rain and pass AATCC Test Methods 22 and 35 (storm).

Waterproof is normally used to describe plastic, plastic-coated, or nonbreathable fabrics. However, the term has been used for some chemically coated fabrics that allow penetration of air.

Durable finishes maintain a high level of performance after laundering, dry cleaning, or both.

Oil-repellent fabrics resist wetting with oil and repel oilborne stains. The level of performance of such fabrics is judged by AATCC Test Method 118.

Waterproof finishes. Waterproofing results from coating a fabric and filling the pores with film-forming material, eg, varnishes, rubber, nitrocellulose, waxes, tar, and plastics.

Repellent finishes. The following are six classes of repellent textile finishes in order of increasing use by weight: pyridinium compounds, organometallic complexes, waxes and wax–metal emulsions, resin-based finishes, silicones, and fluorochemicals.

Fabric construction for water repellency. Fabric construction affects the performance of water repellents. The twist, ply, and coarseness of yarn affect performance. The tightness and the nature of the weave, eg, twill vs poplin, have some effect on repellency, tight weaves and poplins being more repellent. Some reports indicate that differences in fabric roughness affect repellency. Roughness reduces repellency, especially with low-to-moderate concentrations of repellent. Mechanical action on fabrics, even after treatment, can reduce repellency if the action increases fiber roughness or exposes fibers that have little repellency.

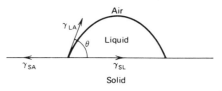

Figure 1.   Contact angle for a liquid on a surface, $\cos \theta = \dfrac{\gamma_{SA} - \gamma_{SL}}{\gamma_{LA}}$.

Modifications of textile fibers. The reaction of hydrophobic chemicals with textile fibers offers the possibility of permanent repellency without alteration of the other physical properties of fibers. The disadvantage is the difficulty of carrying out chemical reactions on fibers in commercial textile-mill operations. Most studies of fiber modification for repellency have involved attempts to achieve permanent water repellency of cotton (qv). The etherification and esterification of cellulose have been most effective in terms of achieving durable water repellency.

Commercial treatment of carpet fibers. The treatment of nylon fibers with fluorochemical finishes to provide durable soil, water, and oil repellency is an important part of carpet-fiber manufacture. The threadline application of fluorochemicals during fiber manufacture allows the carpet manufacturer to purchase treated fiber and makes topical treatments of carpets unnecessary. Fiber treatments are more durable to wear and cleaning than topical carpet treatments. Carpets containing such treated fibers benefit from a finish that covers the entire length of the fibers, in contrast with spray-applied fluorochemicals on carpets, which are deposited only on the upper portions of the fibers.

Health and safety factors. The toxicological properties of repellents vary widely. Only the suppliers of the products can provide more than general information. Most products have low levels of toxicity, but suppliers urge caution in use. Many suppliers assure the safety of their repellents, not only as shipped but also as used on textiles and perhaps next to human skin. The products as supplied should be assumed to be skin and eye irritants. Operators should use gloves, goggles or spectacles, and clothing sufficient to minimize physical contact.

Paper and Paperboard

Paper that repels water may be designated waterproof or water-resistant. Waterproof paper is paper or paperboard that is resistant to both water and water vapor. Chemical modifications of paper fibers include processes that develop water repellency. However, apparently none has been commercialized. Water resistant or water-repellent paper or paperboard is one of two kinds. One kind is analogous to water-repellent textiles, that is, it repels water because of a treatment of individual fibers without formation of a film impermeable to water and water vapor. The second kind is pigment-coated paper or paperboard (see Paper; Papermaking additives).

Waterproof. Waterproof paper or paperboard is coated or laminated with a molten thermoplastic material or coated with an aqueous latex or solvent solution of a polymer. Among the commercially important thermoplastic products are asphalt, wax, polyethylene, poly(ethylene-*co*-methacrylic acid), ethylene–vinyl acetate copolymers, poly(ethylene terephthalate), poly(vinylidene chloride), and combinations of thermoplastic synthetic resins and waxes.

Water resistant. Developing the resistance of paper or paperboard that is not pigment coated to the penetration of liquids, especially water and aqueous solutions, is called sizing. Repellents against water or aqueous solutions are hydrocarbon derivatives or siloxanes. However, repellents against oils, greases and alcohols are fluorochemicals or mixtures of fluorochemicals and hydrocarbon derivatives.

Pigment coating. Grades of pigment-coated paper and paperboard that may contact moisture during use must have water resistance or wet-rub resistance. The choice of adhesive used in a pigment coating determines the water resistance of a coated paper or paperboard. Casein, soybean protein, and synthetic resins produce the highest water resistance.

Leather

Water-resistance treatments are useful on leather garments, shoes, and boots to reduce the tendency of leather to becomes stiff and uncomfortable after wetting and drying.

Commonly used repellents for leather are silicones, chrome complexes of long-chain fatty acids, and fluorochemicals.

Concrete and Masonry

Waterproof. Waterproofing may be achieved by the application of membranes to the surface of a concrete structure. The sheet materials include such different products as roofing felt, poly(vinyl chloride), polyethylene, butyl rubber, neoprene, and sheet lead (see Film and sheeting materials).

Waterproofing may also be achieved by the application of sealants (qv), eg, tar, asphalt (qv), some solvent-based or latex paints, or mastics (qv) (see also Coatings, resistant).

Water repellant. Three techniques used for water repellency are modification of cement by the addition of waterproofers, use of repellent additives to the concrete mix, and surface treatment of concrete structures with repellents. The modification of portland cement by the addition of calcium stearate, aluminum stearate, or gypsum treated with tannic acid and oils yields proprietary cements (see Cement).

Admixtures are sometimes used to decrease capillary flow and the permeability of liquids in concrete. These additives may be pore-filling materials, eg, chalk, Fuller's earth, or talc. They may also be repellents, eg, mineral oil, asphalt emulsions, wax emulsions, or salts of fatty acids, especially stearates.

The third class of water repellents consists of materials applied to the surface of concrete, eg, oils, waxes, soaps, resins, and silicones. The most widely used repellents are silicones.

MASON HAYEK
E.I. du Pont de Nemours & Co., Inc.

J.J. Keavney and R.J. Kulick in J.P. Casey, ed., *Pulp and Paper Chemistry and Chemical Technology*, 3rd ed., Vol. III, John Wiley & Sons, Inc., New York, 1981, pp. 1547–1592.

J.L. Moillet, ed., *Waterproofing and Water-Repellency*, Elsevier Publishing Co., Amsterdam, The Netherlands, 1963.

W.H. Taylor, *Concrete Technology and Practice*, McGraw-Hill Book Co., New York, 1977.

WATTLE BARL. See Leather.

WAXES

Wax usually refers to a substance that is a plastic solid at ambient temperature and, on being subjected to moderately elevated temperatures, becomes a low viscosity liquid. Because it is plastic, wax usually deforms under pressure without the application of heat. The chemical composition of waxes is complex: they usually contain a broad variety of molecular weight species and reactive functional groups, although some classes of mineral and synthetic waxes are totally hydrocarbon compounds.

Insect and Animal Waxes

Beeswax. White and yellow beeswax have been known for over 2000 yr, especially as used in the fine arts. Beeswax is secreted by bees, eg, *Apis mellifera*, *A. dorsata*, *A. flores*, and *A. Indica*, and used to construct the combs in which bees store their honey. The wax is harvested by removing the honey and melting the comb in boiling water; the melted product is filtered and cast into cakes. The yellow beeswax cakes can be bleached with oxidizing agents, eg, peroxide or sunlight, to white beeswax, a product much favored in the cosmetic industry. Generally, beeswax produced in the United States is not suitable for bleaching.

Beeswax typically has a melting point of 64°C, a penetration (hardness) of 2.0 mm at 25°C and 7.6 mm at 43.3°C, a viscosity of 1470 mm²/s (= cSt = 67 SUs) at 98.9°C, an acid number of 230, and a saponification number of 84.

Beeswax is used widely in applications associated with the human body because it is safe to ingest in limited quantities and to apply to the skin.

Vegetable Waxes

The aerial surfaces of almost all multicellular plants are covered by a layer of wax. With the advent of more sensitive analytical tools, eg, gas–liquid chromatography and mass spectrometry, investigations of the character of epicuticular wax of many species has been undertaken. However, only a very few extant species, primarily those in semiarid climates, produce waxes in such quantities that commercial recovery is economically feasible.

Vegetable waxes include candelilla, carnauba, japan wax, ouricury wax, Douglas-fir bark wax, rice-bran wax, jojoba, castor wax, and bayberry wax.

Mineral Waxes

Mineral waxes include montan wax, peat waxes, ozokerite and ceresin waxes, and petroleum waxes.

A paraffin wax is a petroleum wax consisting principally of normal alkanes. Paraffin, microcrystalline, and semicrystalline waxes may be differentiated using the refractive index of the wax and its congealing point as determined by ASTM D 938. Typical physical properties of petroleum waxes are listed in Table 1.

Petroleum wax is outstanding as a cost-effective moisture and gas barrier; thus, food packaging is the largest market for petroleum waxes in the United states, where over 50% of the wax is used in this application. Much of the petroleum wax produced is food-grade quality, although such quality may well be used in non-food-grade applications to simplify inventorying.

Synthetic Waxes

Synthetic waxes include polyethylene waxes, Fischer-Tropsch waxes, chemically modified hydrocarbon waxes, and substituted amide waxes.

Polyethylene Waxes

Low molecular weight (less than ca 10,000) polyethylenes having waxlike properties are used in conjunction with petroleum waxes in

Table 1. Typical Physical Properties of Petroleum Waxes

Property	Paraffin	Microcrystalline
flash point, °C	204, min	260, min
viscosity at 98.9°C, mm²/s (SUs)	4.2–7.4 (40–50)	10.2–25 (60–120)
melting range, °C	46–68	60–93
refractive index at 98.9°C	1.430–1.433	1.435–1.445
average mol wt	350–420	600–800
carbon atoms per molecule	20–36	30–75
other physical aspects	friable to crystalline	ductile-plastic to tough-brittle

Table 2. Typical Properties of Selected Synthetic Hydrocarbon Waxes

Wax	Mp[a], °C	Penetration[b], 0.1 mm 25°C	Penetration[b], 0.1 mm 60°C	Viscosity at 149°C, mPa·s (= cP)	Density at 23°C, g/cm³
Allied A-C 6 polyethylene	106	4	20	220	0.92
BASF A polyethylene	108	3	15	450	0.92
Ciech WP-2 polyethylene	110	3	15	200	0.93
Epolene N-12 polyethylene	110	1	9	220	0.94
Hoechst PA 130 polyethylene	125	1	4	320	0.93
Leuna LE 114 polyethylene	115	2	7	260	0.93
Paraflint H-1 polymethylene	108	2	15	7	0.94
Polywax 500 polyethylene	86	7	61	3	0.94
Polywax 2000 polyethylene	125	1	2	50	0.96
Veba A227 polyethylene	108	2	10	260	0.93

[a]ASTM D 127.
[b]ASTM D 1321.

food-packaging applications. These polyethylenes are made either by high pressure polymerization, low pressure (Ziegler-type catalysts) polymerization, or controlled thermal degradation of high molecular weight polyethylene.

Low molecular weight polyethylenes are available with a variety of properties (see in Table 2). In addition to homopolymers of ethylene, copolymers of ethylene, propylene, butadiene, and acrylic acid with waxlike properties are also produced (see Olefin polymers).

C. SCOTT LETCHER
Petrolite Corporation

"Waxes" in *Chemical Economics Handbook*, SRI International, Menlo Park, Calif. March 1984.

H. Bennett, *Industrial Waxes*, Chemical Publishing Co., Inc., New York, 1975.

WEED KILLERS. See Herbicides.

WEIGHING AND PROPORTIONING

Weighing is the operation of determining the weight of material in one or more objects or in a definite quantity of bulk material.

Proportioning is the weighing and controlling of two or more materials to a specific formula to make a definite blend of the materials for a mixed product or for a chemical process.

The four most common applications of weighing in industry are the measuring of incoming material by weighing received shipments; controlling ingredients to the proper proportions; putting the product into packages of uniform weight (either by packaging directly on a scale or by using a scale to check the performance of other filling equipment); and measuring outgoing shipments for billing and transportation-charge purposes.

Types of Equipment

A scale is a device or machine used to perform a weighing operation. There are many types of scales and several principles of operation used. The type of scale to be used depends upon the conditions of the operation and the performance required.

In mechanical scales, the load is measured either by comparing it with a known weight or by directly measuring the distortion of a spring caused by the load.

In beam-type scales, the force due to the load, reduced by the multiplication of the lever system, is measured by the position of a known weight (the poise) on a graduated beam.

In pendulum-type dial or automatic-indicating scales, the known weight (the pendulum) is moved by rotation of a sector or cam until it counterbalances the force from the load, and the amount of movement is measured and indicated by an indicator which is driven by a rack and pinion.

In spring-type automatic-indicating scales, the deflection of a spring is measured and indicated by an indicator.

Weighing can also be done by electrical means (see Figs. 1 and 2). Electronic scales consist of a load-receiving element, an electrical transducer, ie, a load cell, and a digital indicator. Electronic scales have existed since 1952. Electronic scales have brought a new capability to check weighters, enabling them to keep track of individual and total weights as well as weight deviations with greater accuracy than previously possible. One method is to measure the change in length of a steel column or the bending of a steel beam, by measuring the change in resistance of grids of wire or foil bonded to the steel member.

Mass can be determined directly by measuring the absorption of γ or β rays owing to the presence of the unknown quantity of material (see Radioisotopes).

Counting scales have been developed from digital scales. A counting scale, usually a top dash-loading or platform-balance type, allows multi-

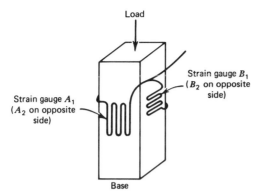

Figure 1. Strain-gauge load cell, column type.

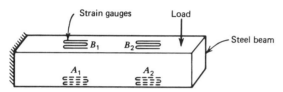

Figure 2. Strain-gauge load cell, beam type.

ples of the same item to be weighed. The result is placed in the scale memory and displayed digitally. The scale can then be used for counting unknown quantities.

Batch weighing is widely used in the chemical and food industries. The weighing system provides automatic weighing of a single ingredient at the start of the weighing process.

Factors Affecting Weighing

Factors include variations in the force of gravity, buoyant effect of the air, and moisture content.

Choice of Scale

Since the error in a scale, as in many other instruments, is approximately the same throughout its indicating capacity, for best accuracy a scale should be chosen so that most of the weighings to made on it are in the upper range of its indicating capacity. The capacity of a scale can be extended by the use of drop weights for a mechanical scale or range steps for a load-cell scale. This permits greater accuracy in weighing through a wide range of loads.

Forms of Scales

Scales of various types are available in different forms to facilitate the weighing operation. These include bench scales, portable scales, floor scales, motor-truck scales, railroad-track scales, and on-the-fly weighing systems.

Measuring Definite Quantities

For the measurement of bulk materials, which may be handled in anything from small packages to railroad cars, the weighing operation can consist of measuring the gross weight of material and container and subtracting the tare weight of the container. When the tare weight is known, its weight can be offset on the scale so that the net weight is indicated.

Controlling Quantity

The most common application for the controlling of quantities of materials is to fill containers to predetermined weights. One method is the net-weighing process, in which material is fed into a fixed container such as a hopper or tank permanently attached to the scale. The tare weight of the individual container is offset on the scale, and if this weight varies, the tare weight must be adjusted for each container to maintain a uniform net weight. In manual operation, the tare is adjusted

by the operator and the filling operation is controlled by the operator. In the gross-weighing method, the shipping container is placed directly on the scale and filled.

Continuous Weighing and Controlling

Sometimes it is required to determine the weight per centimeter or meter of material in sheet or strip form as it is being transported or in the process of manufacture. The material can pass over a roll which is weighed or over a power-driven belt conveyor mounted on a scale. Direct mass measurement with a nuclear scale can also be used for this application.

Weighing and controlling bulk material that is continuously conveyed is often necessary for optimizing the performance of such devices as grinders or pulverizers, for adding materials to a process at a continuous rate, or for controlling additives, such as to water supplies. A belt-conveyor scale can be installed in a belt conveyor, or a short belt feeder can be mounted on a scale (see also Conveying).

Proportioning

There are several methods of proportioning ingredients to obtain the desired formulation. Selection of the proper method depends upon the processing equipment and also upon the accuracy required. Where a definite quantity of a mixture is to be injected into a process, such as being placed in a batch mixer, or where the greatest accuracy is required, the batch system is generally used.

When the manufacturing process is continuous, consideration can be given to special forms of weighing equipment that deliver a continuous controlled stream of material. Equipment to perform this operation may be of the continuous-feeder type or of the loss-in-weight type.

Scales and Computers

Scales can be modified to perform some types of computing directly, eg, counting scales.

Feeding equipment suitable for weighing operations. For best performance, careful consideration must be given to the choice of proper equipment for the feeding of materials to scales for such operations as filling containers or for proportioning ingredients to a definite formula. The feeder should deliver material at as uniform a rate as possible and should respond quickly to control signals from the scale. When the feeder stops, there is a column of material in the air that falls into the weigh hopper. The cutoff point can be preset to allow for the average weight of this column, but deviations from the average weight will result in errors. For best accuracy, the weight in the column after cutoff should be small in proportion to the total weight. When the weighing operation is to be performed relatively fast, such as in filling the container in less than a minute, or when the rate of flow changes from such causes as a variation in head of material above the feeder, the feeder should be capable of being slowed down before the desired weight is in the scale.

Discharging the scale. When material is weighed into a hopper or tank on a scale, as in the net-weight method of filling containers or in a batch system, the hopper or tank can be emptied completely by means of a gate or valve.

Natl. Bur. Stand. Handb. **44** (1983).

Scale Men's Handbook of Metrology, National Scale Men's Association, Libertyville, Ill., 1980.

"Control System in a Small Box Has a Big Impact on the Operating Efficiency of Jumbo Air Freighters," *Mater. Handl. Eng.*, 64 (Feb. 1979).

WELDING

Welding comprises a group of processes whereby the localized coalescence of materials is achieved through application of heat and/or pressure. Today the American Welding Society recognizes over sixty methods of welding and more than twenty allied processes employing thermal cutting, thermal spraying, and adhesive bonding, some of which are identified in Table 1.

Table 1. Welding Processes

Process	Abbreviation	Process	Abbreviation
Arc welding		*Resistance welding*	
shielded-metal arc welding	SMAW	resistance spot welding	RSW
gas–tungsten arc welding	GTAW	projection welding	RPW
plasma arc welding	PAW	resistance seam welding	RSEW
gas–metal arc welding	GMAW	flash welding	FW
flux-cored arc welding	FCAW	upset welding	US
submerged-arc welding	SAW	*Solid-state welding*	
Oxyfuel-gas welding	OFW	diffusion welding	DFW
oxyacetylene welding	OAW	explosion welding	EXW
oxyhydrogen welding	OHW	friction welding	FRW
pressure–gas welding	PGW	ultrasonic welding	USW
Soldering and brazing		*Other processes*	
dip soldering	DS	electron-beam welding	EBW
wave soldering	WS	electroslag welding	ESW
resistance brazing	RB	laser-beam welding	LBW
furnace brazing	FB	thermit welding	TW

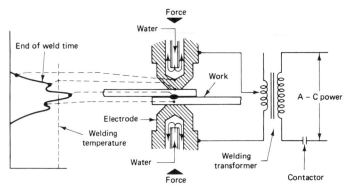

Figure 1. The resistance spot welding process.

Arc Welding Processes

In arc welding processes, the coalescence of metals is achieved through the intense heat of an electric arc which is established between the base metal and an electrode. The six arc processes listed in Table 1 are differentiated by various means of shielding the arc from the atmosphere.

Welding Systems

The various welding processes result in systems of varying complexity. They include at least the electrode and a device for holding or feeding it, the work piece, the power source, and heavy-duty cabling to provide a complete electrical circuit. Provisions for supply and control of gas and control of wire feed and movement of the electrode assembly are required, depending on process type and degree of automation.

Welding systems are generally classified as manual, semiautomatic, and automatic. In manual welding, the operator must maintain the arc, feed in filler metal, and provide travel and guidance along the joint. In semiautomatic welding, the welding machine maintains the arc and feeds filler metal, while the operator controls joint travel and guidance. In automatic welding, the machine assumes all of the preceding functions.

Other Important Processes

Oxyfuel-gas welding. This process, commonly called gas welding, uses the heat of combusting gases to melt and coalesce base metals. Although several different fuel gases, eg, propylene, hydrogen, or methane, can be added to the oxygen, the oxyacetylene flame is the most widely used because its high flame temperature (3100°C) is needed to weld steel.

Resistance welding. As noted in Table 1, resistance welding comprises several processes; the most widely used is resistance spot welding (RSW). The principles are quite different from the processes previously described (see Fig. 1).

The workpieces are firmly clamped between copper electrodes and an electric current is passed through the assembly. Heat is generated by the electrical resistance of the components, with maximum heat at the interface between the workpieces. A nugget of metal at the interface region is melted, at which moment the current is shut off and the clamping force on the electrodes released.

Electroslag welding. In this process, the heat of molten slag coalesces the base and filler metals. Electric current flows from a consumable electrode through a molten metallurgical slag into the molten weld metal. The electric resistance of the slag provides the heat to maintain the slag molten and to melt the weld and base metal.

Electron-beam welding. This welding process achieves the heat necessary for coalescence by bombarding the base metals with a concentrated stream of electrons (see Fig. 2).

Laser-beam welding. The heat of coalescence is produced by focusing the beam of light (photons) from a high power laser on the base metal (see Lasers).

Brazing and soldering. In brazing and soldering processes, a molten filler metal flows into the joint of the base metal at a temperature below the melting point of the base metal (see Solders and brazing alloys).

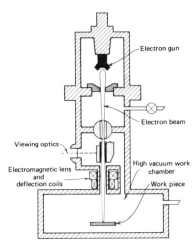

Figure 2. Electron-beam welding system.

Physics and Metallurgy

In most welding processes, the local regions of adjoining base metals and any added filler metal are melted and resolidified. Welding differs from conventional casting by the speed of the solidification process. Only a local region of material is melted in welding, and the surrounding, low temperature base metal acts as a large heat sink, producing rapid heat flow from, and solidification of, the weld zone. This produces a complex metallurgical microstructure and physical properties not typical of casting.

Metallurgy. Welding metallurgy deals with the interactions of base and filler metals and the interactions of these with various chemicals injected into the weld via gases, electrode coverings, fluxing and sagging agents, and surface contaminants. Oxygen, nitrogen, water vapor, and carbon dioxide are gases that react with ferrous metals to yield products harmful to the metallurgical properties of a weld. The nature of slag-metal reactions that occur in the molten state strongly depends on the composition of the flux or the electrode coating. Flux chemistry may be altered to control removal of specific weld-metal impurities, such as addition of manganese or silicon to provide strong deoxidizing action, or additions to enhance slag removal, with all of these influencing the final metallurgical characteristics of the weld.

Base and Filler Materials

Properties. The properties of materials are ultimately determined by the physics of their microstructure. For engineering applications, however, materials are characterized by various macroscopic physical and mechanical properties. Among the former, the thermal properties of materials are particularly important in welding, including melting temperature, thermal conductivity, specific heat, and coefficient of thermal expansion. The latter property greatly influences structual distortion that may occur in welding. The electrical conductivity of a material is important in any welding process where base or filler metal is a part of the welding electrical circuit. Material density is also of interest.

Base metals. Base metals include carbon steels, low alloy steels, stainless steels, aluminum and aluminum alloys, magnesium alloys, nickel and nickel alloys, copper and copper alloys, and reactive and refractory metals.

Filler metals. Filler metals are added to a weld by melting a consumable electrode or a separate wire fed into the weld pool. In the first category, the filler metal is part of the welding electrical circuit and may be in the form of short lengths of covered wire, as in shielded-metal arc welding, or in the form of continuous reels of wire used in semiautomatic and automatic welding processes. Solid wire is used in the gas-metal and submerged-arc welding processes, whereas a hollow, flux-filled wire is used in flux-cored arc welding. More filler metal in the form of iron powder is sometimes added to the electrode coating or the flux used in submerged arc welding. In the second category, the filler metal may be in the form of short lengths of bare, solid wire, as used in gas welding or manual gas-tungsten arc welding, or in continuous reel form, used in automatic gas-tungsten arc welding.

Design

Welded joints. The weld joint is the geometric arrangement between two pieces of base metal brought together for purposes of welding. There are only five recognized welding-joint configurations: butt, corner, tee, lap, and edge joints.

Weld types. The fillet weld joins corners and tees. In the plug weld, the two pieces are joined by weld metal deposited in a prepared hole in the overlying piece. For a spot weld, heat is appled to the overlying plate, creating fusion at the interface. Resistance welding is generally used for the spot weld and the seam weld. The groove weld joins base metals in a butt joint. For the backing weld, the root of the original weld is first removed by chipping or gouging. Surface welds are used to build up the surface of parts with special materials or to replace worn material. Flange welds are applied to edge joints.

Stress and distortion. The forces acting on a structure are transmitted through the welded joints; that is, the joint is subjected to simple tension (or compression), bending, shear, or torsional stresses, or to combinations of these stresses owing to combined loading situations. Weldments must be of proper size, length, and location to withstand the loads imposed during service.

Health and Safety Factors

Welders are subject to the same hazards as all other workers in the metalworking trades. Specific additional hazards of welding include electrical shock, arc radiation, fumes and gases, fires and explosions, compressed gases, cutting and chipping operations, and high noise levels.

KARL F. GRAFF
The Ohio State University

Welding Handbook, 7th ed., Vol. 1, American Welding Society, Miami, Fla., 1976; see also other volumes in this series.

H.B. Cary, *Modern Welding Technology*, Prentice-Hall, Inc., Englewood Cliffs, N.J., 1979.

Welding, Brazing and Soldering, Vol. 7 of *Metals Handbook*, 9th ed., American Society for Metals, Metals Park, Ohio, 1983.

WETTING AGENTS.   See Surfactants and detersive systems.

WHEAT AND OTHER CEREAL GRAINS

Origins and History

The word cereal derives from the name of the Roman goddess, Ceres, in whose honor a spring festival, The Cerealis, was celebrated. This indicates the antiquity of these foods and the reverence with which they

were esteemed. Cereals are still an important dietary ingredient. As the world population continues to grow, cereals will become an increasing fraction of the diets for more and more people. That is recognized by the use of an ear of wheat as the symbol of the FAO. Below the wheat is the Latin inscription, *Fiat panis* (Let there be bread).

Botanically, cereals are simple, one-seed fruits. They include wheat, rice, corn, (also called maize in some parts of the world), rye, barley, oats, sorghum, and millet. All cereals contain large amounts of starch and little fat; the fat is associated primarily with the germ and scutellum (single cotyledon or the first leaf). In most cases, the lipids in cereals contain a high concentration of unsaturated fatty acids, which are protected from oxidation by the presence of tocopherols (see Vitamins, vitamin E). One of the main purposes of milling, in addition to producing an acceptable flour, is to remove the wheat germ in as intact a state as possible. Thereby, the concentration of the lipids is reduced, the development of oxidative rancidity decreased, and the shelf life of the flour extended.

Finally, cereals grow in a wide variety of climatic and soil conditions, and they successfully compete with weeds for the limited amounts of nutrients and water where these plant factors are in short supply. This is an important reason why cereals have played, and continue to play, such an important role in the development of the human race.

Nutritional Value of Cereals

Deficiency diseases.   Not only did cereals make an important contribution to improving the general status of humankind, but they also were important dietary components of some groups of people who showed certain nutritional deficiencies. This observation led to the discovery of some of the vitamins and the role played by certain substances in human health. These conditions have been most prominently associated with the use of rice, corn, and wheat. They include beriberi (thiamine deficiency), pellagra (niacin deficiency), zinc deficiency, calcium adsorption, and lysine concentration.

Health Advantages of High Cereal Consumption

Health advantages include lowered blood urea levels and improved kidney function, prevention of osteoporosis, and prevention of dental caries. Cereals play an important role in prevention and treatment of colonic cancer, diabetes, and cardiovascular diseases.

Health Problems Associated with the Consumption of Cereals

Health problems include celiac disease, disturbances associated with flour-aging agents, and consumption of ergot.

Enrichment of Cereal Products

On the basis of a number of dietary surveys and the results of some biochemical tests, the Food and Nutrition Board of the NAS in 1974 suggested increasing both the levels and number of nutrients added to cereals. The rationale for this suggestion was that both the dietary and biochemical data indicated a significant risk of deficiency of vitamin A, thiamine, riboflavin, niacin, vitamin B_6, folacin, iron, calcium, magnesium, and zinc among significant segments of the population. On that basis, the amounts of nutrients in Table 1 were suggested for wheat, corn, and rice products where they were technically achievable.

Triticale

Until the development of triticale, no new cereal had been added to food crops for thousands of years. The ability of rye to grow under relatively adverse climatic and environmental conditions and the recognition that its genetic profile was not too different from that of wheat suggested the possibility of crossing these two cereals.

Although triticale has some advantages over wheat and rye, certain problems must be overcome before this plant begins to compete with either of its parents. For one, the seed is shriveled to such an extent that it can be milled with modern machinery only at the expense of losing large amounts of the endosperm with the bran.

There is still some question as to whether triticale flour, by itself, makes an acceptable loaf of bread. However, there are reports that triticale is satisfactory for use in such products as bread, pastas, tortillas, chapattis, and infant food. An increasing amount of farmland is being devoted to triticale, especially in Ethiopia and some South American countries.

Cereal Production

More than two thirds of the world's cultivated area is planted with cereal grains; in the less-developed countries, the fraction is even larger. Wheat, rice, and corn account for almost three fourths of the world's production. Over the past 50 yr, except for oats and barley, there has been a steady increase in world production of all cereal grains. Part of this increase reflects the improvements in yield brought about by the Green Revolution. Wheat varieties grown in the United States are listed in Table 2.

Storage of Cereals

One feature of cereal grains that makes them important as a year-round source of food for both people and animals is that they can be stored for long periods of time. The storage capability of the grain depends on low moisture content. It is reduced at harvest, almost always by solar energy. For most cereals, a maximum moisture content of ca 10% protects the grain during storage.

Milling of Cereals

Wheat. In 1979, the United States exported 62% of its wheat crop. Of the remainder, about two thirds was processed into human food of which

Table 1. Proposed Enrichment of Cereals[a]

Nutrient	mg/100 g
vitamin A (retinol equivalent[b])	0.48
thiamine	0.64
riboflavin	0.40
niacin	5.29
vitamin B_6	0.44
folic acid	0.07
iron	8.81
calcium	198.20
magnesium	44.10
zinc	2.20

[a]*Proposed Fortification Policy for Cereal Grain Products*, Food and Nutrition Board, NRC, NAS (Washington, D.C.), (1974).
[b]See Vitamins, vitamin A.

Table 2. Wheat Varieties in the United States

Variety	Where grown	Primary use
Triticum aestivum (*common wheat*)		
hard red spring	Minnesota North and South Dakota Montana	flour for yeast breads
hard red winter	Kansas Nebraska Oklahoma Texas	flour for yeast breads
soft red winter	Ohio Indiana	flour for chemically leavened baked goods; all-purpose flour
white wheat	Pacific Northwest	flour for pastries; all-purpose flour
Triticum durum		
red durum	North Dakota Minnesota	poultry and livestock feed
amber durum	North Dakota Minnesota	macaroni; semolina

flour was the principal component. Wheat that is unsuitable for high grade flour is often used for a number of other purposes, mainly animal feed. Some is malted and, as such, some is added to high grade flour to increase its diastasic activity; the rest is used primarily in making alcoholic beverages. Another product is starch (qv) used in making adhesives for, among other things, wallpaper and plywood. In the preparation of starch, the proteins of the endosperm are separated as gluten which has some use in the preparation of adhesives (qv), emulsifiers (see Emulsions), polishes (qv), and the meat-flavoring agent, monosodium glutamate (see Amino acids).

Wheat is difficult to mill because the large, deep crease that extends the length of the kernel precludes simply pearling the grain to obtain a white kernel for grinding. To achieve the separation of endosperm from bran, flour millers subject the wheat, after it has been cleaned, to a breaking process that opens the kernel, releasing its contents so the endosperm particles can be separated from the bran. The wheat kernels are passed between a pair of spirally fluted, chilled iron rolls. The two rolls are run at different speeds to produce a shearing action which separates the endosperm from the bran particles. The endosperm particles are separated by screening, usually through silk. At the first break, the largest particles consist primarily of bran and are sent to the second set of rolls or the second break. The next smaller particles form semolina which is a mixture of clean endosperm and pieces of bran, some with adhering endosperm. The semolina is transferred to a purifier which is a slightly inclined sieve through which air currents are drawn. The meshes of the sieve become coarser down the slope. By that means, the dense, uncontaminated endosperm particles are separated from the lighter particles of endosperm with adhering bran; the lightest particles are pure bran which move on with the air blast. From the purifier, the semolina that contains no bran goes to the reduction rolls, which reduce the size of the particles. The reduction rolls have smooth surfaces and the two rolls at each stage move at about the same speed, crushing and grinding the endosperm particles. The latter pass through the second set of reduction rolls, after which the screening process is repeated. There may be as many as five reduction steps. At each one, the amount of flour secured becomes progressively less. Furthermore, the flour from the fourth or fifth reduction stage contains an increased amount of fine bran particles, which color the flour. During the reduction process, the germ is flattened but not disintegrated; at the end of the reduction process, the germ is separated from the bran by screening.

Toward the end of the operation, the flour is treated with a bleaching and aging agent (chlorine dioxide in the United States and the UK) and then the vitamin–mineral enrichment mixture is metered into the flour. After that, the flour is sifted and then bagged or put into a tank truck for bulk distribution to large bakeries. The flour distributed in bulk is delivered through large flexible pipes (see Bakery products; Bleaching agents).

Corn. Of the corn produced in the United States in 1980, about 62% was not used on the farms where it was grown. Only about 4% of the harvested corn, but nevertheless, a very large amount on the basis of previous production figures ($> 6 \times 10^6$ t), is subjected to dry milling. This involves increasing the moisture content to 21–24%, degerminating the corn, and breaking the endosperm with rollers similar to those used in milling wheat. The endosperm particles are dried to a moisture content of 14–16% and then are passed through a hominy separator consisting of cylindrical sieves that separate particles according to size. These endosperm fractions are passed through series of reduction rolls. The material screened at the various reduction stages is used for different purposes. The coarsest particles are used in the preparation of corn flakes, the slightly smaller particles make up grits, and the next two groups constitute coarse and fine corn meal, respectively. Corn flour is composed of the smallest particles.

Corn starch is becoming an increasingly important product both as a food and as a starting material used in many industrial processes. It is prepared from corn kernels that have been steeped in dilute sulfurous acid for a few hours. The acid treatment softens the outer layers of the grain and dissolves certain soluble substances in the corn. The resulting solution, called corn steep liquor, is the primary nutrient for some of the molds and microorganisms involved in the production of antibiotics, primarily penicillin. The moist corn kernels are cracked in a special mill, which separates the intact germ from the rest of the kernel. Since the germ contains about 20% fat, it can be separated by water flotation; the rest of the corn kernel falls to the bottom. The kernels are removed from the bottom of the pan and powdered in a burr mill. The hulls are removed by a series of siftings. The remaining mixture of starch and protein is separated from a slurry which slowly run into long shallow channels. There, the starch settles while the protein is carried to the end of the channel where it is recovered and used with the hulls to form animal feed. The starch is recovered from the channels, washed, and dried.

Rice.   The rice kernel is covered with a tightly adhering hull which is relatively high in silicon, thus rendering it unsuitable for consumption. To remove the hulls, the cleaned rice is put into a shelling machine which consists of two horizontal circular steel plates or rubber rollers. The inner surface of the steel plates is coated with coarse carborundum embedded in cement. One plate remains stationary while the other rotates at such a height that the rice grains can assume a vertical position between the plates. The pressure thus exerted on the rice grain splits the hull and frees the endosperm with its adhering bran layers and germ; at that stage, it is known as brown rice. A blast of air is used to remove the light hulls from the brown rice and the remaining intact rice. The latter two are separated in a paddy machine where the lighter, unhulled rice rises to the top during shaking while the heavier endosperm moves downward. The unhulled rice emanating from the top of the paddy machine passes through a second shelling machine, whose plates are set a little closer together than in the first machine. Separation of hulls from the brown rice and intact grains is again accomplished by a blast of air, after which the rice is again put through the paddy machine. The brown rice is separated from the bran layers in a cylindrical, rotating machine where the rice kernels scour each other and thus remove the outer bran layers and the germ. Final polishing of the rice occurs in a brush machine which is a vertical cylinder containing soft leather strips on its outer surface. This cylinder revolves within another stationary cylinder made of wire screen. As the brown rice flows down between these two cylinders, it is polished by the moving leather strips and the wire screen. The rice that emerges from the brush machine is screened to separate the intact endosperm from the broken pieces. By this procedure, 66–70% of the initial rice is recovered as white rice.

In the United States and some other countries, there is an additional step before the rice can be packaged that involves the addition of vitamins and minerals since practically all rice sold in the United States is enriched. Techniques had to be devised for enriching the grains without changing their shape.

Oats.   Only 3–4% of the oat crop is destined for human consumption and, for that reason, oat milling is performed by only a few companies. The first step is the removal of foreign matter and the very lightest oat grains. To facilitate milling, the oats are dried to a moisture content of 6%; which increases the brittleness of the hulls and permits easier separation of the groats (the dehulled grain). The dried oats are separated into five or six groups on the basis of grain length. Each group of oats is conveyed to separate hullers which resemble those used in hulling rice. They consist of a lower, flat stationary "stone" and an upper one which is slightly conical and rotates rapidly. As in rice milling, the space between the two "stones" is adjusted to be slightly less than the length of the intact oat grain and slightly more than that of the groat. This exerts pressure on the oat grains, which assume a vertical position as they migrate from the center of the stones to the periphery. The pressure shatters the hull, thus releasing the groat. The mixture of unhulled oats, hulls, groats, and oat flour is sifted to remove the flour. The hulls are then removed by an air stream. Groats are separated from the unhulled grain in "apron" machines that have openings into which the smaller groats can pass while the unhulled oats pass through the machine for reprocessing. For human use, the groats are treated with steam which partially cooks then and increases their moisture content.

Rye.   Only about 20% of the rye crop in the United States is milled; about two thirds of the crop is used as stock feed. Rye milling is done in

somewhat the same way as that for wheat. The principal difference arises from the tenacity with which rye bran adheres to the endosperm. Partially for that reason, the whole flour is removed early in the milling process. It represents about 50–60% of the grain. As milling proceeds, the flour removed at successive stages becomes darker.

OLAF MICKELSON
Consultant

D.K. Mecham in Y. Pomeranz, ed., *Wheat, Chemistry and Technology*, American Association of Cereal Chemists, Inc., St. Paul, Minn., 1971.

W.H. Leonard and J.H. Marin, *Cereal Crops*, The Macmillan Co., London, 1963.

R.W. Schery, *Plants for Man*, Prentice-Hall, Inc., Englewood Cliffs, N.J., 1972.

WHEAT-GERM OIL.   See Fats and fatty oils.

WHEY.   See Pet and livestock feeds.

WHISKEY.   See Beverage spirits, distilled.

WHITE LEAD.   See Pigments, inorganic.

WHITENING AGENTS.   See Brighteners, fluorescent.

WHITING, $CaCO_3$.   See Pigments, inorganic.

WINE

The word wine was possibly first applied to the fermentation product of the sugars in the juice of grapes; this is its primary meaning. However, the fermented juices of many fruits are now called wine, and the term is also sometimes incorrectly applied to the alcoholic fermented juice of various plant materials containing sugars, eg, rice wine. It is this used in contrast to fermented liquids made from starch-containing materials, such as beer (qv) and related beverages (see Beverage spirits, distilled).

Standards

There are two types of standards for wines in the United States and most other countries. One is based on taxes. In the United States, the tax is levied according to the alcohol content:

alcohol, %	≤ 14	> 14– ≤ 20	> 21–24
tax, ¢/L	4.5	17.7	63.4

Classification

Wines may be classified in many ways: by alcohol (ethanol) content, place of production, color, method of production, or variety of grape or fruit. Wines with $\leq 14\%$ alcohol are defined as table wines since they are normally consumed with meals, especially if they do not taste sweet. Wines with $> 14\%$ are called dessert or aperitif wines since they are usually consumed after meals if sweet or before meals if not sweet.

Legal Restrictions

Not only is the composition of the wines offered for sale subject to legal restrictions, but every producer of U.S. wines, except those produced in limited quantities for home consumption, must secure a Federal permit and take out a bond before beginning operations. Since this basic permit may be canceled for willful violation of Federal laws or regulations, the government possesses a powerful tool to deter violations. In most countries, the sale of wines is subject to permits, regulations, and taxes.

Production

The special character of the fermentation of the various types of wines depends to a considerable extent on the composition of the fruit juice fermented (see also Fermentation).

Composition of grapes. Grapes contain ca 15–25% sugar; partially dried grapes contain 30–40%. The percentage of sugar in the grapes, the extent of the fermentation, and the losses or additions of alcohol during treatment and storage determine the percentage of alcohol in the finished product.

Only a small amount of nitrogenous material (0.3–1.0%) is found in grapes. However, this material is of considerable significance for yeast nutrition, bacterial stability, and flavor development, presumably because of the many amino acids present (see Amino acids).

The pigments of grapes and fruits are usually located in the epidermal cells. A few varieties have red juice as well. During alcoholic fermentation, the epidermal cells are killed and the colored pigments are released. By separating the skin of red grapes from the juice before fermentation, it is possible to produce a white or nearly white wine.

The pectins of some fruits and grapes are a source of difficulty in juice clarification. They are rather insoluble in alcohol and precipitate during alcoholic fermentation.

The inorganic constituents are not of critical importance as they are usually present in sufficient amounts to catalyze yeast or enzyme functioning; excess iron or copper may cause turbidity.

Microorganisms. Wines are normally produced by fermentation with the yeast *Saccharomyces cerevisiae*, sometimes with *S. bayanus* or *S. oviformis*. Taxonomists are still reclassifying the genus *Saccharomyces*. These and other yeasts, found in grapes or fruit, multiply rapidly in the sweet juice, eventually causing fermentation. To prevent growth and competition or undesirable organisms, 50–150 mg sulfur dioxide are usually added per liter ca 2 h before the pure yeast culture is added.

Equipment. Wine production can be very simple, but for large-scale operation, specialized equipment has been developed.

The grapes are crushed in combined stemmers and crushers.

Fermentation tanks may be of wood, concrete, stainless steel, or iron-lined with epoxy resins or a thin layer of stainless steel; they may be open or closed, and possibly, temperature-controlled.

After fermentation, the residue, ie, stems and skins, often called pomace or marc, must be transferred from the fermentor to the press. The oldest type of press still in use is the screw-type basket press, usually operated vertically, either from above or below. It has been largely replaced by the horizontally operated hydraulic press or other continuous-type presses. These presses are more expensive to operate but produce a relatively clear juice from both red pomaces or white musts.

White table wines. White wines of < 14% alcohol are designated as white table wines. They are usually made from white grapes. Occasionally, a white wine is made from red grapes (as in the Champagne district of France) by separating the skins from the juice immediately after crushing. They include dry and sweet wines.

Red table wines. Most of the world's wine is red table wine of < 14% alcohol. In France and Italy, particularly, they constitute an important part of the daily caloric intake. Red wines are relatively easy to produce compared to white table wines, and they are less subject to spoilage or clouding during aging.

The grapes should be harvested when they are sufficiently ripe to produce 11–13% alcohol. A sugar content of ca 21.5–23.5% is best. The grapes should be transported from the vineyard to the crusher without delay to prevent development of spoilage bacteria. Prompt crushing and destemming are essential.

Sparkling wines. Wines containing a permanent visible excess of carbon dioxide are called sparkling wines. The nomenclature of the sparkling wines is somewhat complicated. The most famous name is champagne, originally produced only in the region of that name in France where this appellation may be used only for wines produced and fermented in that region. However, the name is used in some countires as well, although local names are also used in some countries, eg, Sekt in Germany and spumante in Italy. In the United States, the term champagne is used for most sparkling wines produced by a secondary fermentation of sugar in closed containers. If fermented in tanks, it must be so stated on the label.

In Europe, many wines are bottled with a slight residual sugar content. During aging, this sugar may ferment, and the wines become slightly gassy; wines of Alsace, the Loire region, and Switzerland are frequently of this type. A certain amount of yeast growth is necessary to produce this gassiness, but the yeast deposit is often surprisingly small. At slightly higher temperature, this type of wine loses the gassiness which is one of its chief attractions; furthermore, all the residual sugar may ferment, thereby producing a large yeast deposit which affects the clarity and sometimes the odor.

Gassiness also results when the malic acid of the wine ferments, giving lactic acid and carbon dioxide. However, a high malic acid content is necessary for such fermentation. The vinho verde wines of Portugal are of this type, as are some gassy northern Italian wines.

Dessert wines. Less than 15% of all the Californian wines are fortified as dessert wines at an alcohol content of 17–21%. Both red and white dessert wines are produced; the white may be either dry or sweet. Worldwide, probably less than 10% of the wines are dessert types.

White dessert wines. Muscatel, Angelica, white port, and sweet sherry are the most important sweet dessert wines produced in the United States; dry sherry is the only dry type and it is often at least slightly sweet. Famous European dessert wines include Malaga (Spain), Marsala (Sicily), and Madeira (Madeira Islands).

Red dessert wines. Port and port-type wines are the main red U.S. dessert wines, although a small amount of red muscatel is also produced. The prototype port is produced in a delimited district along the Douro river east of Oporto in Portugal. A number of varieties are used, ie, the darker, stronger flavored wines are used for vintage or ruby port, whereas the lighter-colored wines are aged for tawny port.

Fruit and berry wines. The production of fruit wines is similar to that of grape wines, except that the sugar must always be added because of the low sugar content of most berries and fruit.

Naturally flavored wines. In the United States, these wines were legally recognized in 1958. Natural herbs, spices, fruit juices, aromatics, essences, and other natural flavorings may provide the base, and sugar and caramel may be added; apple wine may also be added. These wines are made both as table (< 14% alcohol) and dessert wines. All are sold with proprietary names, eg, Bali Hai, Silver Satin, and Thunderbird.

Vermouth. Vermouths are nearly dry or sweet fortified wines to which herbs or herb extracts are added. The nearly dry or French type is used straight or for martini cocktails. The sweet or Italian type is used for Manhattan cocktails, but in Europe is more often consumed as a dessert wine or aperitif. The herbs in vermouth should be easily detectable, but the odor of no single herb should be allowed to predominate (see also Flavors and spices).

Finishing. Finishing steps include filtration, fining, refrigeration, and, occasionally, pasteurization.

Packaging. Automatic bottling, corking or capping, labeling and casing lines are employed in large wineries. These are less labor-intensive and give better results than manual procedures. The bottles are packed in cardboard boxes for shipment.

Spoilage

Wines are relatively immune to spoilage because of their alcohol content and low pH. Contact with air may result in acetification, which can be avoided by keeping the containers full or by pasteurization. Low acid wines may be spoiled by lactic acid bacteria which can be prevented by low concentrations of sulfur dioxide. Yeast infection may be a problem. Iron and copper originating from production equipment occasionally cause cloudiness in white table wines.

M.S. AMERINE
University of California at Davis
Consultant, San Francisco Wine Institute

A. Lichine, *New Encyclopedia of Wines and Spirits*, 2nd ed., Alfred A. Knopf, Inc., New York, 1974.

M.A. Amerine and V.L. Singleton, *Wine: An Introduction for Americans*, 2nd ed., University of California Press, Berkeley and Los Angeles, Calif., 1977.

E. Peynaud, *Connaissance et Travail du Vin*, Dunod, Paris, France, 1981.

WINTERGREEN OIL.   See Oils, essential.

WITHERITE, BaCO₃.   See Barium compounds.

WOLFRAM AND WOLFRAM ALLOYS.   See Tungsten and tungsten alloys.

WOOD

Wood is one of the most important natural resources, and one of the few that is renewable. It supplies material for many objects necessary to everyday living, ranging from homes and furniture to bridges and railroad ties. Wood yields fiber for pulp, paper, and fiberboard products and material for plywood, particleboard, pallets, and rayon textile fibers. It is also a source of energy and industrially important chemicals (see also Laminated wood-based composites; Paper; Pulp).

Structure

The anatomical structure of wood affects strength properties, appearance, resistance to penetration by water and chemicals, resistance to decay, pulp quality, and the chemical reactivity of wood. To use wood most effectively requires not only a knowledge of the amounts of various substances that make up wood, but also how those substances are distributed in the cell walls.

Woods are either hardwoods or softwoods. Hardwood trees (angiosperms, ie, plants with covered seeds) generally have broadleaves, are deciduous in the temperate regions of the world, and are porous, that is, they contain a vessel element. Softwood trees (conifers or gymnosperms, ie, plants with naked seeds) are cone bearing, generally have scalelike or needlelike leaves, and are nonporous, that is, they do not contain vessel elements. The terms hardwood and softwood have no direct relation to the hardness or softness of the wood.

Many mechanical properties of wood, such as bending and crushing strength, and hardness, depend upon the density of wood; denser woods are generally stronger. Wood density is determined largely by the relative thickness of the cell wall and by the proportions of thick-walled and thin-walled cells present.

The cells that make up the structural elements of wood are of various sizes and shapes and are firmly bonded together.

Just under the bark of a tree is a thin layer of cells, not visible to the naked eye, called the cambium. Here, cells divide and eventually differentiate to form bark tissue outside of the cambium and wood or xylem tissue inside. This newly formed wood contains many living cells and conducts sap upward in the tree, and hence is called sapwood.

Composition

Wood is a complex polymeric structure consisting of lignin (qv) and carbohydrates (qv) (cellulose (qv) and hemicellulose), which form the visible lignocellulosic structure of wood. Also present, but not contributing to wood structure, are minor amounts of extraneous organic chemicals and minerals. The organic chemicals are extractable from the wood with neutral solvents, and are therefore called extractives. The minerals constitute the ash residue remaining after ignition at a high temperature.

Wood-Liquid Relationship

Adsorption.   Wood is highly hygroscopic. The amount of moisture adsorbed depends mainly on the relative humidity and temperature, as shown in Figure 1.

Shrinking and swelling.   The adsorption and desorption of water in wood is accompanied by external volume changes. At moisture contents below the fiber-saturation point, the relationship may be a simple one, merely because the adsorbed water adds its volume to that of the wood, or the desorbed water substracts its volume from the wood. The relationship may be complicated by the development of stresses causing changes in volume or shape.

Permeability.   Although wood is a porous material (60–70% void volume), its permeability (ie, flow of liquids under pressure) is extremely variable. This is owing to the highly anisotropic shape and arrangement of the component cells and to the variable conditions of the microscopic channels between cells.

Transport.   Wood is composed of a complex capillary network through which transport occurs by capillarity, pressure permeability, and diffusion.

Drying.   There are a number of important reasons for drying: it reduces the likelihood of stain, mildew, or decay developing in transit, storage, or use; the shrinkage that accompanies drying can take place before the wood is put to use; wood increases in most of its strength properties as it dries below the fiber-saturation point (30% moisture content); the strength of joints made with fasteners, such as nails and screws, is greater in dry wood than in wet wood dried after assembly; the electrical resistance of wood increases greatly as it dries; dry wood is a better thermal insulating material than wet wood; and the appreciable reduction in weight that accompanies drying reduces shipping costs (see Drying).

Ideally, the temperature and relative humidity during drying should be controlled; if wood dries too rapidly, it is likely to split, check, warp, or honeycomb because of stresses.

Air drying is a process of stacking sawmill products outdoors to dry. Kiln drying is a controlled heated drying process widely used for drying both hardwoods and softwoods.

Structural Material

Strength and related properties.   In the framing of a building or the construction of an industrial unit, where wood is used because of its unique physical properties, strength and stiffness are primary requirements. Different species of wood have different mechanical properties that relate to the amount of wood substance per unit volume, ie, its specific gravity. The strength of a piece of lumber depends also upon its grade or quality. The strength values of a lumber grade depend upon the size and number of such characteristics as knots, cross grain, shakes, splits, and wane. Wood free from these defects is known as clear wood.

The mechanical properties of wood tend to deteriorate immediately when it is heated and improve when it is cooled. In addition, high temperatures result in permanent degradation. This irreversible loss in

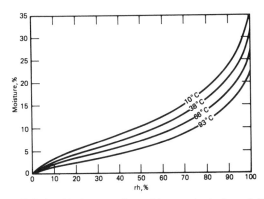

Figure 1.   Relationship between the moisture content of wood (% of dry wood) and relative humidity at different temperatures.

mechanical properties depends upon moisture content, heating medium, temperature, exposure period, and to some extent, species.

The effect of absorption of various liquids upon the strength properties of wood largely depends on the chemical nature and reactivity of the absorbed liquid. In general, neutral, nonswelling liquids have little if any effect upon the strength properties. Any liquid that causes wood to swell causes a reduction in strength.

Wood preservatives are applied either from an oil system, such as creosote and petroleum solutions of pentachlorophenol, or a water system. Preservatives applied in petroleum oils are practically inert to wood and have no influence on strength.

Inorganic salts used in fire-retardant treatments are reactive with wood and have a severe effect on mechanical properties of wood.

Heat and fire resistance. Wood in its untreated form has good resistance or endurance to fire penetration when used in thick sections for walls, doors, floors, ceilings, beams, and roofs. This endurance is owing to low thermal conductivity which reduces the rate at which heat is transmitted from the fire-exposed surface to the interior of the wood member.

To improve the fire performance of wood products, fire-retardant chemicals such as ammonium phosphate, ammonium sulfate, zinc chloride, dicyandiamide–phosphoric acid, borax, and boric acid, are often used alone or in combinations. Solutions of these fire-retardant formulations are impregnated into wood under a full cell-pressure treatment to obtain dry-chemical retentions of 65–95 kg/m^3; the treatment can greatly reduce flamespread and afterglow (see Flame retardants).

Resistance to chemicals. Different species of wood vary in their resistance to chemical attack. The significant properties are inherent to the wood structure, which governs the rate of ingress of the chemical, and the composition of the cell wall, which affects the rate of action at the point of contact.

Wood is widely used as a structural material in the chemical industry because it is resistant to a large variety of chemicals. Its resistance to mild acids is far superior to that of steel but not as good as some of the more expensive acid-resistant alloys.

Alkaline solutions attack wood more rapidly than acids of equivalent concentrations, whereas strong oxidizing chemicals are harmful. Wood is seldom used where resistance to chlorine and hypochlorite solutions is required.

Because traces of iron reduce the brilliance of many dyes, wood tanks have long been preferred to steel tanks in the manufacture of dyes.

Resistance to chemical attack is generally improved by resin impregnation, which protects the underlying wood and reduces movement of liquid into the wood. Resistance to acids can be obtained by impregnating with phenolic resin, and to alkalies by impregnating with furfural resin (see Furan compounds; Phenolic resins).

Biodeterioration. Wood may be attacked by fungi, bacteria, insects, or marine borers.

Decay fungi break down wood components enzymatically to forms they readily assimilate. Decay can be prevented by keeping wood either too dry (below 20% moisture content) or too wet (lumens filled with water) by using naturally decay-resistant species, or by heating with preservatives.

Termites are the most destructive insects attacking wood. Their attack can be prevented, or lessened, by using naturally resistant wood, or by treating wood with preservatives.

Modified Wood

In addition, to preservation or fire protection, wood is modified to reduce shrinking and swelling under conditions of fluctuating relative humidity. Resin impregnation is a successful method of adding bulking agents, provided the resin can permeate the lumens and penetrate the cell walls.

The mechanical properties of resin-impregnated wood are generally improved. Heat resistance can be improved by resin impregnation. The largest industrial application of resin-impregnated wood is in die models for automobile body parts and other model dies.

It is possible to add an organic moiety to the hydroxyl groups in wood cell-wall components. This type of treatment also bulks the cell wall with

a permanently bonded chemical. Many compounds modify wood chemically. Best results are obtained by the hydroxyl groups of wood reacting under neutral or mildly alkaline conditions below 120°C.

Steam or chemical bending is another treatment process. Above 80°C, or in the presence of certain chemicals wood becomes deformable. On cooling to room temperature and drying under restraint, the new shape persists. This is the basis for commercial bending of wood to various shapes.

Chemical Raw Material

The gasification of wood to produce methanol (qv) has received considerable recent attention (see Alcohol fuels). Because wood consists of two-thirds carbohydrates, considerable attention has also been given to the potential of wood residues as a raw material for conversion to ethanol (qv).

The principal chemical industry based on wood is, of course, the pulp and paper industry which produces over 65×10^6 t of fiber products ranging from newsprint to pure cellulose. Most of the over 43×10^6 t of organic chemicals removed from wood during pulping are burned for their energy content and to recover inorganic pulping chemicals. A few organic chemicals are recovered from this waste stream. The most valuable chemical by-products isolated at pulp mills are sulfate turpentine and tall oil (a mixture of fatty acids and rosin).

Wood is the raw material of the naval stores industry. Naval stores, so named because of their importance to the wooden ships of past centuries, consist of rosin (diterpene resin acids), turpentine (monoterpene hydrocarbons), and associated chemicals derived from pine (see Terpenoids).

Hydrolysis

In the acid hydrolysis process, wood is treated with acid to produce a lignin-rich residue and a liquor containing sugars, organic acids, furfural, and other chemicals. The process is adaptable to all species and all forms of wood waste. The liquor can be concentrated to a molasses for animal feed, used as a substrate for fermentation to ethanol or yeast, or dehydrated to furfural and levulinic acid (see Furan compounds).

Fuel Properties

The fuel properties of wood can be summarized by ultimate and proximate analyses and determination of heating value. The analytical procedures are the same as those for coal but with some modifications. The high heating value of wood does not vary between species except that values for softwoods are generally higher than hardwoods (see Fuels from waste).

Charcoal Production

Charcoal is produced by heating wood under limited access of oxygen. When wood is heated slowly to ca 280°C, an exothermic reaction occurs. In the usual carbonization procedure, heating is prolonged to 400-500°C in the absence of air. The term charcoal also includes charcoal made from bark. Charcoal is produced commercially from primary wood-processing residues and low quality roundwood in either kilns or continuous furnaces (see Carbon).

THEODORE H. WEGNER
A.J. BAKER
B.A. BENDTSEN
J.J. BRENDEN
W.E. ESLYN
J.F. HARRIS
J.L. HOWARD
R.B. MILLER
R.C. PETTERSEN
J.W. ROWE
R.M. ROWELL
W.T. SIMPSON
D.F. ZINKEL
U.S. Department of Agriculture

R.M. Rowell. ed., *The Chemistry of Solid Wood, Advances in Chemistry Series 207*, American Chemical Society. Washington, D.C., 1984.

D. Fengel and G. Wegener, *Wood Chemistry, Ultrastructure, Reactions*, Walter de Gruyter, New York, 1984.

A.J. Panshin and C. deZeeuw, *Textbook of Wood Technology: Structure, Identification, Uses, and Properties of the Commercial Woods of the United States*, 4th ed., McGraw-Hill Book Co., Inc., New York, 1980.

Wood Handbook, Agriculture Handbook 72, U.S. Forest Products Laboratory, U.S. Department of Agriculture, Madison, Wisc., revised 1974.

WOOD-PLASTIC COMPOSITES.   See Laminated wood-based composites.

WOOD PULP.   See Pulp.

WOOL

Wool, the fibrous covering from sheep, is by far the most important animal fiber used in textile manufacture. It commands a price premium over most other fibers because of its outstanding natural properties of soft hand (the feel of the fabric), good moisture absorption (and hence comfort), and good drape (the way the fabric hangs) (see Fibers, chemical; Textiles).

The three main distinctions of wool are fine, medium, and long (Table 1). Raw wool from sheep contains other constituents considered contaminants by wool processors. These can vary in content according to breed, nutrition, environment, and position of the wool on the sheep. The main contaminants are a solvent-soluble fraction called wool grease; protein material; a water-soluble fraction (largely perspiration salts collectively termed suint); dirt; and vegetable matter, eg, burrs and seeds from pastures.

Fiber Growth

The skin of the sheep has two layers, an inner dermis and an epidermis. The fiber grows out of a tubelike structure in the epidermis known as the follicle. The two types of wool follicles are the primary, which develop first in the outer skin of the unborn lamb, and the secondary, which develops later.

Fiber Morphology

The two types of cells in fine–medium wool fibers are cortical cells, which make up the center of the fiber, and cuticle cells which form the outer protective cover around the cortex. The cuticle cells overlap like the tiles on a roof, the protruding tips of the scales pointing toward the fiber tip. The orientation of the scales in this way leads to a directional friction effect. This effect is thought to be the main cause of wool felting (see Felts).

Between the cuticle and the cortical cells, and separating the cortical cells themselves, is a cell membrane complex. This complex, ca 25 nm thick, basically "cements" the cells, and has been called the intercellular cement.

Physical Properties

That wool is still used as a textile fiber is to a large extent the result of its unique physical properties. In particular, it absorbs large volumes of moisture and hence is comfortable to wear. The absorption of moisture yields heat. The bulk (volume occupied by fibers in a yarn) given to it by its bilateral structure (and the resulting fiber crimp) adds to warmth as a greater volume of air is trapped than in a yarn of the same weight but of other fibers. Wool's unusual elastic properties give it outstanding resistance to and recovery from wrinkles, and outstanding drape. The principal physical properties of the fiber are given in Table 2.

Chemical Structure

Wool belongs to a family of proteins (qv) called the keratins. However, as explained above, morphologically, the fiber is a composite, and each of the components differ in chemical composition. Principally, the compo-

Table 1. Main Types of Wool

Type	Breed	Average length, cm	Average diameter, μm^a	Grade or count[a]
fine	merino	3.7–10	10–30	90s–58s
medium	southdown hampshire dorset cheviot	5–10	20–40	60s–46s
crossbred	corriedale polwarth targhee columbia	7.5–15	20–40	60s–50s
long	lincoln	12.5–35	25–50	50s–36s

[a] Wool fineness is now largely described as the "micron," ie, micrometer. Traditionally, it has been expressed as grade or count which, in the worsted trade, represents the number of hanks, each 560 yd (1512 m) long, of the finest possible yarn that might be spun satisfactorily from 1 lb (0.454 kg) of the wool concerned. The higher the grade or count, the lower is the "micron."

Table 2. The Principal Physical Properties of Wool Fibers at 25°C

Property	Regain,[a] % 0	5	10	15	20	25	30	33
relative humidity, %								
absorption	0	14.5	42	68	85	94	98	100
desorption	0	8	31.5	57.5	79	91.5	98	100
specific gravity	1.304	1.3135	1.3150	1.325	1.304	1.2915	1.2765	1.268
volume swelling, %	0	4.24	9.07	14.25	20.0	26.2	32.8	36.8
length swelling, %	0	0.55	0.93	1.08	1.15	1.17	1.18	1.19
radial swelling, %	0	1.82	4.00	6.32	8.88	11.69	14.57	16.26
heat of complete wetting, kJ/kg wool[b]	100.9	64.5	38.1	20.5	8.8	4.1	1.13	0
heat of complete absorption of liquid water, kJ/kg water[b]	854	624	431	276	142	100	42	33
relative Young's modulus	1.00	0.96	0.87	0.76	0.66	0.56	0.44	0.38
rigidity modulus (torsion), GPa[c]	1.76	1.60	1.26	0.90	0.50	0.28	0.16	0.1
electrical resistivity, $10^6 \, \Omega \cdot cm$			4×10^4	800	40	6		
dielectric constant at 10^4 Hz	4.6	5.1	6.2	8.3	12.8			

[a] Moisture content.
[b] To convert J to cal, divide by 4.184.
[c] To convert GPa to psi, multiply by 145,000.

nents are proteinaceous, but cleaned wool (after successive extraction with light petroleum, ethanol, and water) contains small amounts of lipid material and inorganic ions equivalent to an ash content of ca 0.5–1% after combustion of the fiber.

Wool Processing

The conversion of raw wool into a textile fabric or garment involves a long series of separate processes. There are two main processing systems, ie, worsted and woolen, although an appreciable volume of wool is also processed on the short-staple (cotton) system or on the semiworsted system for carpet use. The main stages in the woolen and worsted systems are shown in Figure 1.

Wool Grease

In wool scouring, the contaminants on the wool, mainly grease, dirt, suint, and protein material, are washed off the fiber and remain in the wastewaters either in emulsion or suspension (grease, dirt, protein) or in solution (suint). Centrifugal extraction of the wastewaters produces a grease contaminated with detergent and suint. This product is called wool grease.

Lanolin is wool grease that has been refined to lighten its color and reduce its odor and free fatty acid content. Wool wax is the pure lipid material of the fleece, extractable with usual fat solvents such as diethyl ether and chloroform. Wool grease is a mixture of compounds that are classed as waxes (qv).

Chemical composition.   Wool wax is a complex mixture of esters of water-insoluble alcohols and higher fatty acids with a small proportion of hydrocarbons.

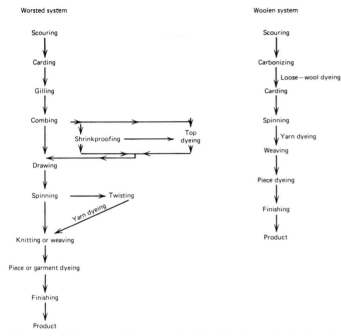

Figure 1. Main stages in the processing of wool by the worsted and woolen systems.

Wool-grease recovery. The principal recovery process in use involves centrifuging in a cream-separator type of centrifuge modified by the addition of peripheral jets or other mechanical devices to remove dirt.

Grease refining and fractionation. The refining process most commonly used involves treatment with hot aqueous alkali to convert free fatty acids to soaps followed by bleaching, usually with hydrogen peroxide, although sodium chlorite, sodium hypochlorite, and ozone have also been used. Other techniques include distillation, steam stripping, neutralization by alkali, liquid thermal diffusion, and the use of active adsorbents (eg, charcoal and bentonite) and solvent fractionation.

Uses of wool grease. The uses of wool grease, lanolin, and lanolin, derivatives are wide, ranging from pharmaceuticals and cosmetics (qv) to printing inks (qv), rust preventatives, and lubricants (see Lubrication and lubricants).

W.S. BOSTON
CSIRO

W. von Bergen, ed., *Wool Handbook*, 3rd ed., Interscience Publishers, a division of John Wiley & Sons, Inc., New York, Vol. 1, 1963.

J.A. Maclaren and B. Milligan, *Wool Science: The Chemical Reactivity of the Wool Fibre*, Science Press, Sydney, Australia, 1981.

J.D. Leeder, *Wool—Nature's Wonder Fibre*, Australasian Textiles Publishers. Geelong, Australia, 1984.

WORKING FLUIDS. See Heat-exchange technology.

XANTHATES

The salts of the *O*-esters of carbonodithioic acids and the corresponding *O, S*-diesters are xanthates. The free acids decompose on standing.

Properties

The free xanthic acids are unstable, colorless or yellow oils, and might decompose with explosive violence. They are soluble in the common organic solvents and are slightly soluble in water: methyl xanthic acid at 0°C, 0.05 mol/L; ethyl xanthic acid at 0°C, 0.02 mol/L; and *n*-butyl xanthic acid at 0°C, 0.0008 mol/L. Values for the dissociation constant for ethyl xanthic acid are $(2.0-3.0) \times 10^{-2}$. Potentiometric determinations for C_1-C_8 xanthic acids show a decreasing acid strength with increasing molecular weight.

The alkali-metal salts, in contrast to the free acids, are relatively stable solids, pale yellow when pure, and have a disagreeable odor.

When exposed to air, the sodium salts tend to take up moisture and form dihydrates. The alkali-metal xanthates are soluble in water, alcohols, the lower ketones, pyridine, and acetonitrile (see Table 1). They are not particularly soluble in the polar solvents, eg, ether or ligroin.

The heavy-metal salts, in contrast to the alkali metal salts, have lower melting points and are more soluble in organic solvents. They are slightly soluble in water, alcohol, aliphatic hydrocarbons, and ethyl ether.

Alkalies stabilize xanthate solutions somewhat and the solutions readily decompose at acidic pHs.

Reactions. The chemistry of the xanthates is essentially that of the dithio acids.

Preparation and Manufacture

The alkali-metal xanthates are generally prepared from the reactions of sodium or potassium hydroxide with an alcohol and carbon disulfide.

Many of the heavy-metal xanthates have been prepared from aqueous solutions of the alkali-metal xanthates and the water-soluble compound of the heavy metal desired.

Table 1. Solubilities of Some Alkali-Metal Xanthates

Xanthate	Solvent	Solubility, g/100 g soln	
		0°	35°C
potassium *n*-propyl	water	43.0	58.0
	n-propyl alcohol	1.9	8.9
sodium *n*-propyl	water	17.6	43.3
	n-propyl alcohol	10.2	22.5
potassium isopropyl	water	16.6	37.2
	isopropyl alcohol		2.0
sodium isopropyl	water	12.1	37.9
	isopropyl alcohol		19.0
potassium *n*-butyl	water	32.4	47.9
	n-butyl alcohol		36.5
sodium *n*-butyl	water	20.0	76.2
	n-butyl alcohol		39.2
potassium isobutyl	water	10.7	47.7
	isobutyl alcohol	1.6	6.2
sodium isobutyl	water	11.2	33.4
	isobutyl alcohol	1.2	20.5
potassium isoamyl	water	28.4	53.3
	isoamyl alcohol	2.0	6.5
sodium isoamyl	water	24.7	43.5
	isoamyl alcohol	10.9	15.5

Health and Safety Factors, Toxicology

The xanthates are low in acute and oral toxicity. Potassium amyl xanthate causes extensive pain and slight corneal injury to the eye and may burn the skin on prolonged contact.

The alkali-metal xanthates are fairly safe to handle. The standard precautions of rubber gloves, dust mask, and goggles are sufficient when handling the solid or the solution.

Under regulations for the enforcement of the Federal Insecticide, Fungicide, and Rodenticide Act, products containing over 50 wt% sodium isopropyl xanthate (**1**) must bear the label: "Caution. Irritating dust. Avoid breathing dust; avoid contact with skin and eyes." Rubber goods in repeated contact with food may contain diethyl xanthogen disulfide (**2**) not to exceed 5 wt% of the rubber products.

$$(CH_2)_3CHOC(S)SNa \qquad (C_2H_5OC(S)S)_2$$
$$(1) \qquad\qquad (2)$$

Uses

Outside of the importance of cellulose xanthates in the manufacture of rayon and cellophane, the primary use for the alkali metal xanthates is as collectors in the flotation (qv) of metallic sulfide ores.

Derivatives

The principal derivatives are of three types: the dixanthogens, the mixed anhydrosulfides, and dialkyl thionocarbamates.

GUY H. HARRIS
Consultant

Xanthate Handbook, American Cyanamid Company, Mining Chemicals Department, Wayne, N.J., 1972.

G. Gattow and W.R. Bahrendt, *Topics in Sulfur Chemistry, Vol. 2: Carbon Sulfides and Their Inorganic and Complex Chemistry*, Georg Thieme, Stuttgart, 1977.

J. Leja, *Surface Chemistry of Froth Flotation*, Plenum Press, New York, 1982.

XANTHAN GUM. See Gums; Microbial polysaccharides.

XANTHENE DYES

Xanthene dyes are those containing the xanthylium (**1**) or dibenzo-γ-pyran (**2**) nucleus as the chromophore with amino or hydroxy groups meta to the oxygen as the usual auxochromes.

(1) xanthylium   **(2) xanthene**

They have brilliant hues in the shade range of greenish-yellows to dark violets and blues, and they exhibit fluorescence but usually with inferior lightfastness compared with other chromophoric systems. They are used for the direct dyeing of wool and silk and mordant dyeing of cotton. Paper, leather, woods, food, drugs, and cosmetics are dyed with xanthenes (see Dyes, application and evaluation—application). Brilliant insoluble lakes are used in paints and varnishes.

Xanthene dyes are classified into three groups according to the nature of the aromatic substitution: amino derivatives, hydroxy derivatives, and aminohydroxy derivatives.

Amino Derivatives

Amino derivatives include pyronines, succineins, rhodamines, rosamines, and saccharein.

Hydroxyl derivatives

The building block of most hydroxyl-substituted xanthenes, or fluorones, is fluorescein (3).

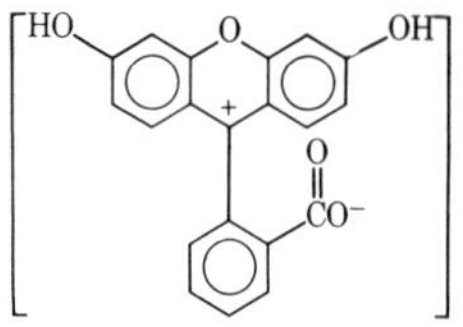

(3) fluorescein

Amino Derivatives

The rhodamines are economically the most important amino-substituted xanthene dyes. Phthalic anhydride is used in the condensation reaction with *m*-dialkylaminophenol to produce a triphenylmethane analogue, a 9-phenylxanthene. Rhodamine B (Basic Violet 10, CI 45170) (4) is usually manufactured by the condensation of two moles of *m*-diethylaminophenol with phthalic anhydride.

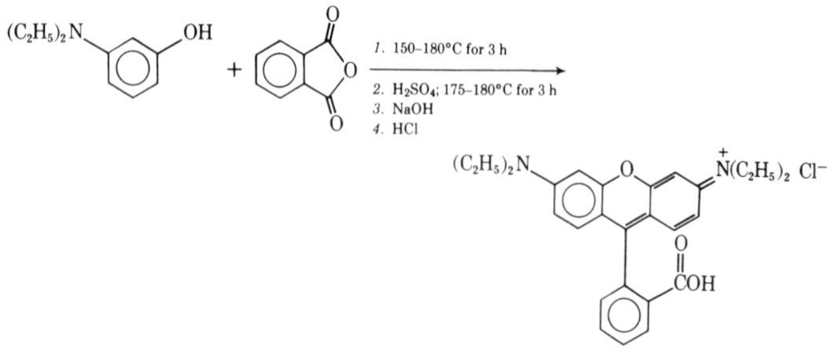

(4) Rhodamine B

Aminohydroxy Derivatives

Aminohydroxy-substituted xanthenes are of little commercial importance.

Table 1.　Toxicological Properties of Selected Xanthene Dyes

Compound	Property	Value, mg/kg
fluorescein, (3)	LD_{Lo} (rat), intraperitoneal	600
	LD_{Lo} (mouse)[b]	600
	LD_{Lo} (rabbit), oral	2,500
	LD_{Lo} (rabbit), intravenous	300
	LD_{Lo} (guinea pig)[b]	400
Rhodamine B, (4)	LD_{Lo} (rat), oral	500
	TD_{Lo} (rat), subcutaneous	3,825
	LD_{50} (rat), intravenous	89,500
	LD_{Lo} (mouse), intraperitoneal	128

Health and Safety Factors, Toxicology

Xanthene dyes have not exhibited health or safety properties warranting special precautions (see Table 1); however, standard chemical labeling instructions are required.

RUSSELL E. FARRIS
Sandoz Colors & Chemicals

K. Venkataraman, *The Chemistry of Synthetic Dyes*, Vol. 2, Academic Press, Inc., New York, 1955.

H.A. Lubs, *The Chemistry of Synthetic Dyes and Pigments*, R.E. Krieger Publishing Company, Huntington, N.Y., 1955.

Colour Index, 3rd ed., The Society of Dyers and Colorists, Bradford, Yorkshire, UK, Vol. 4, 1971, pp. 4417–4430; Vol. 6, 1975, p. 6401.

XENON.　See Helium-group gases.

XENON COMPOUNDS.　See Helium-group gases.

XEROGRAPHY.　See Electrophotography.

X-RAY ANALYSIS.　See X-ray technology.

X-RAY TECHNOLOGY

X-ray photons are electromagnetic radiations characterized by wavelength or energy, which are manifested in two forms, continuous radiation or characteristic radiation. Continuous radiation is produced when a high energy electron beam decelerates as it approaches the electron clouds that surround the atomic nucleus. Characteristic radiation is produced following the ejection of an inner-orbital electron by high energy particles and subsequent transition of atomic-orbital electrons from states of high to low energy. When a monochromatic beam of x-ray photons falls onto a given specimen, three basic phenomena may result, namely, absorption, scatter, or fluorescence. The coherently scattered photons may undergo subsequent interference leading in turn to the generation of diffraction maxima. These three phenomena form the bases of three important x-ray methods: the absorption technique is the basis of radiographic analysis, the scattering effect is the basis of x-ray diffraction, and the fluorescence effect the basis of x-ray fluorescence spectrometry (see Analytical methods).

In x-ray radiography, the sample being examined is placed between an x-ray source and a film, and an exposure will produce an absorption shadowgraph of the object being irradiated. In x-ray fluoroscopy, the film is replaced by some form of x-ray transducer which will convert the transmitted x-ray signal to a voltage. Both of these techniques are familiar since they are used in many medical applications. In normal radiography, one seeks to keep the image as sharp as possible during an exposure, and to this end, relative movement of source, object and detection medium, during exposure, should be kept at an absolute minimum. In x-ray tomography, suitable relative movement is introduced in all planes of the object except for the one of interest. This renders all nondesirable planes blurred except for the plane of interest which is kept sharp.

X-ray imaging tests are among the methods most widely used to examine interior regions of metal castings, fusion weldments, composite structures and brazed components. Radiographic tests are made on pipeline welds, pressure vessels, nuclear fuel rods and other critical materials and components that may contain three-dimensional voids, inclusions, gaps, or cracks aligned so that the critical areas are parallel to the x-ray beam (see Nondestructive testing). The thickness of a solid homogeneous material is also readily measurable by x-ray absorptiometry, and this method can be applied both for the estimation of bulk thickness as well as for the measurement of layer thickness. More recently, x-ray absorption methods have found a special application in security screening and today provides the means of high speed examination of luggage.

The basis of the x-ray fluorescence technique lies in the relationship between the wavelengths of the x-ray photons emitted by an irradiated sample element and the atomic number of that element. X-ray fluorescence spectrometers use either the diffracting power of a single crystal to isolate narrow wavelength bands, or a proportional detector to isolate narrow energy bands from the polychromatic beam-characteristic radiation excited in the sample. The first of these methods is called wavelength-dispersive spectrometry and the second, energy-dispersive spectrometry. Because of the known relationship between emission wavelength and atomic number, isolation of individual characteristic lines allows the unique identification of an element to be made and elemental concentrations can be estimated from characteristic line intensities.

Thus, this technique is a means of materials characterization in terms of chemical composition.

X-ray powder diffractometry makes use of the fact that a specimen in the form of a single-phase microcrystalline powder will give a characteristic diffraction pattern. A diffraction pattern is typically in the form of a graph of diffraction angle vs diffracted line intensity. A pattern of a mixture of phases is made up of a series of superimposed diffractograms, one for each unique phase in the specimen. The powder pattern can be used as a unique "fingerprint" for a phase, and a large file of about 40,000 standard single-phase patterns is now available. Analytical methods based on manual- or computer-search techniques are now available for unscrambling patterns of multi-phase materials, offering a unique method of qualitative and quantitative multi-phase identification. Special techniques are also available for the study of stress, texture, topography, particle size, high and low temperature phase transformations, and miscellaneous other research applications.

X-ray lithography is the application of x-ray microscopic techniques to the fabrication of electronic microcircuits. During the past several years, there has been a tremendous growth in the use of micro-electronic circuits. Contemporary designs have features that are about two orders of magnitude smaller than can be seen with the naked eye and the trend to even smaller devices seems likely to continue, giving structural dimensions down to the level of about 10 nm. At this size of component, optical magnifying devices have limited use and the use of x-ray and uv-based lithographic techniques have successfully provided the fine-line instrumentation required for the manufacture of these devices. X-ray lithography is used for the manufacture of the replicate pattern then chemical etching is used to reproduce this pattern on the required substrate (see also Photoreactive polymers).

RONALD JENKINS
Philips Electronic Instruments, Inc.

R. Jenkins, R.W. Gould, and D.A. Gedcke, *Quantitative X-Ray Spectrometry*, Marcel Dekker, New York, 1981.

H.P. Klug and L.E. Alexander, *X-Ray Diffraction Procedures*, 2nd ed., John Wiley & Sons, Inc., New York, 1974.

XYLENES AND ETHYLBENZENE

Xylenes and ethylbenzene are C_8 benzene homologues having the molecular formula C_8H_{10}. The three xylene isomers are o-xylene, m-xylene, and p-xylene, which differ in the positions of two methyl groups on the benzene ring. The term mixed xylenes describes a mixture of ethylbenzene and the three xylene isomers. Mixed xylenes are largely derived from petroleum.

p-xylene   m-xylene   o-xylene   ethylbenzene

Ethylbenzene is always present, except in the small amount of xylenes produced by toluene disproportionation. Ethylbenzene is a diluent which can accumulate in recycle processing schemes and, hence, has a strong impact on the separation of the individual xylene isomers. Ethylbenzene is commercially important as a precursor to styrene; it is usually made from benzene and ethylene (see Styrene).

Gasoline contains ca 24% aromatic material. Thus, the demand for mixed xylenes for petrochemicals use is strongly influenced by the demand for gasoline. By-product credits have a decisive impact on the production economics of the individual isomers.

U.S. mixed-xylenes production is distributed roughly as follows: p-xylene, 50–60%; to gasoline blending, 10–25%; o-xylene, 10–15%; solvents, 10%; ethylbenzene, 3%; and m-xylene, 1%.

Table 1. Physical Properties for C_8-Aromatic Compounds

Property	p-Xylene	m-Xylene	o-Xylene	Ethylbenzene
density at 25°C, g/cm³	0.8610	0.8642	0.8802	0.8671
boiling point, °C	138.37	139.12	144.41	136.19
freezing point, °C	13.26	−47.87	−25.18	−94.98
refractive index at 25°C	1.4958	1.4971	1.5054	1.4959
surface tension, mN/m (= dyn/cm)	28.27	31.23	32.5	31.50
dielectric constant at 25°C	2.27	2.367	2.568	2.412
dipole moment of liquid, C · m[a]	0	0.30	0.51	0.36
critical properties				
critical density, mmol/cm³	2.64	2.66	2.71	2.67
critical volume, cm³/mol	379.0	376.0	369.0	374.0
critical pressure, MPa[b]	3.511	3.535	3.730	3.701
critical temperature, °C	343.1	343.9	357.2	343.1
thermodynamic properties				
C_s at 25°C, J/(mol · K)[c]	181.66	183.44	188.07	185.96
S_s at 25°C, J/(mol · K)[c]	247.36	253.25	246.61	255.19
$H_o - H_o$ at 25°C, J/mol	44.641	40.616	42.382	40.219
$-(G_s - H_o/T)$ at 25°C, J/(mol · K)[c]	97.633	117.03	104.46	120.29
heats of transition, J/(mol · K)[c]				
vaporization at 25°C	42.04	42.04	43.413	42.23
fusion	17.11	11.57	13.6	9.17
vapor pressure, Antoine equation[d]				
A	6.1155	6.1349	6.1239	6.0821
B	1453.430	1462.266	1474.679	1424.255
C	215.307	215.105	213.686	213.206

[a] To convert C · m to D, divide by 3.336×10^{-30}.
[b] To convert MPa to psi, multiple by 145.
[c] To convert J to cal, divide by 4.184.
[d] $\log P_{kPA} = A - B/(C + t)(\log P_{mm\ Hg} = \log P_{kPa} + \log 7.50)$; to convert kPa to mm Hg, multiply by 7.51.

Physical Properties

Because of their similar structure, the three xylenes and the isomeric ethylbenzene exhibit similar properties (see Table 1).

Chemical Properties

Reactions involving the position of the alkyl substituents. These reactions include isomerization, disproportionation, and dealkylation.

Reactions of the alkyl groups. p-Xylene is oxidized to either terephthalic acid or dimethyl terephthalate which is then condensed with ethylene glycol to form polyesters (qv). Oxidation of o-xylene yields phthalic anhydride, another important commodity chemical. m-Xylene is oxidized to isophthalic acid, which is also converted to esters and eventually used in plasticizers and resins (see Phthalic acids and other benzenepolycarboxylic acids).

Reactions of the aromatic ring. The reactions of the aromatic ring of the C_8-aromatic isomers are generally electrophilic substitution reactions (see Friedel-Crafts reactions).

Complex formation. All four C_8-aromatic isomers have a strong tendency to form several different types of complexes. Complexes with electrophilic agents are utilized in xylene separation.

Manufacture

A schematic of a process for benzene, o-xylene, and p-xylene would be the following: The 65–175°C cut from a straight-run petroleum fraction or from an isocracker is used as feed to a reformer. This is followed by heart-cutting and extraction. If benzene is a required product, its concentration can be increased via toluene recovery and dealkylation, and recovery of benzene by distillation. o-Xylene and m-xylene are separated, and o-xylene is recovered by rerunning H (see BTX processing). p-Xylene is separated, and the mother liquor or raffinate is sent to the unit for reequilibration with the xylenes. Any toluene and lighter materials made in this process are rejected and sent to toluene dealkylation. Since p-xylene and o-xylene recovery is followed by reequilibration of xylenes, a recycle system or loop is established. Under a constant set of conditions, each stream eventually equilibrates both in volume and composition. However, impurities or diluents accumulate. In xylene processing, ethylbenzene accumulates unless a reaction with greater selectivity for ethylbenzene than xylenes occurred, or ethylbenzene was rejected by distillation.

Separation Processes for p-Xylene

Crystallization processes for p-xylene. Until the development of molecular-sieve adsorption, crystallization (qv) was the only practical method for p-xylene production.

Adsorption processes for p-xylene. In the late 1960s, UOP introduced the Parex process. By 1981, sixteen Parex units were on-stream, and production exceeded 10^6 t/yr. To date, fifteen plants are either being designed or are under construction. p-Xylene recovery per pass is 90–95% compared to 60–70% for crystallization. The Parex process is based on the principle of continuous selective adsorption in the liquid phase, employing a fixed bed of solid adsorbent.

Isomerization. An enriched xylenes stream receives considerable processing before entering the xylenes loop. Therefore, a once-through procedure for xylenes recovery is generally not economic. The introduction of the isomerization step to reequilibrate the xylenes enables efficient utilization of the enriched xylenes stream.

Dual-function processes. The Octafining process was developed by Engelhard Industries and Atlantic Richfield Company and has been employed successfully on a commercial scale for many years. The key to the Octafining process is its ability to convert ethylbenzene to xylenes and to interconvert the isomeric xylenes.

Amorphous silica-alumina catalysts. Amorphous silica-alumina catalysts have long been used for xylene isomerization, eg, in the Chevron, Maruzen, and ICI processes. The advantage of these processes is their simplicity. No hydrogen is required, and the only side reaction of any importance is disproportionation. However, in the absence of hydrogen, coke buildup on the catalyst is rapid, and regeneration is required every 3-30 d.

Zeolite-based xylene isomerization. In the 1970s, research performed at Mobil Oil Corporation on a versatile series of zeolite materials (see Molecular sieves), referred to as ZSM, initiated a new era in xylene isomerization. From this work emerged three xylene isomerization processes called LTI (low temperature isomerization), MVPI (Mobil's vapor phase isomerization), and MLPI (Mobil's low pressure isomerization).

Process choice. Currently, the MGCC process is preferred for m-xylene production. Crystallization processes are used for the production of p-xylene. The Parex process is well established and would appear to have advantages over crystallization because the higher p-xylene recovery results in less loop traffic.

A p-xylene separation process would certainly be accompanied by an isomerization process. If fresh feed to the loop is expensive or limited, and the need exists to convert it efficiently to p-xylene or p- and o-xylene, the choice might be Octafining II. In other cases, the choice might be MVPI or MLPI, the latter appearing to have some economic advantages.

Health and Safety Factors

The xylene isomers are designated as flammable liquids and as such should be stored in approved closed containers with a red label out of doors and away from heat or open flames. Limits for transportation by air are one liter (passenger planes) and 40 L (cargo planes).

The xylenes are not very toxic. They are mild skin irritants, and skin protection and cannister-type face masks are recommended. Prolonged (8 h) exposure by humans should be limited to 200 ppm.

Uses

The bulk of xylenes, which are mostly produced by reforming petroleum fractions, is used in gasoline (see Gasoline and other motor fuels). Mixed xylenes, which are obtained under severe reforming conditions or more likely, from reformate extracts, are used as solvents in the paint and coatings industry (see also Solvents, industrial). This use is likely to decline.

Ethylbenzene is mainly used as a precursor to styrene (qv).

o-Xylene is used almost entirely as feedstock for phthalic anhydride manufacture.

m-Xylene is used for the manufacture of isophthalic and to a lesser extent, isophthalonitrile.

Commercially, p-xylene is the most important isomer. Almost all is converted via terephthalic acid or dimethylterephthalate and reaction with ethylene glycol to poly(ethylene terephthalate) for use in fibers, films, or resins.

DEREK L. RANSLEY
Chevron Research Company

N.E. Ockerbloom. *Hydrocarbon Process*, 101 (Feb. 1972).
D.B. Broughton, *Chem. Eng. Prog.* **73**(10), 49 (1977).

XYLYLENE POLYMERS

The p-xylylene monomers are considered prototypes of the Chichibabin hydrocarbons and can be represented by a quinoid (**1**) or benzenoid (**2**) structure. If the structure exists in the quinoid form, it would have diamagnetic properties, whereas the benzenoid form would have a diradical structure with paramagnetic properties. It was established that the monomer is diamagnetic and, therefore, the p-xylylene probably exist in the quinoid form (**1**).

Because of their unique properties, the xylylene polymers are used as special coatings or films. These polymers are highly crystalline, straight-chain organic compounds having excellent dielectric characteristics and molecular weights of ca 500,000.

The poly(p-xylylene) polymers are usually prepared by the pyrolysis of di-p-xylylene. A patent has been issued that describes the electrolytic preparation of poly(p-xylylene).

Synthesis

The starting material for the synthesis of the commercial xylylene polymers is a dimer, di-p-xylylene or 2,5'-dichlorodi-p-xylylene.

Polymerization. Di-p-xylylene is vaporized at ca 133 Pa (1 mm Hg) and 200°C and the vapors are conducted through a pyrolysis or cleavage zone at 600°C to yield monomeric p-xylylene diradicals. At ambient temperature, these stable p-xylylene diradicals instantaneously polymerize quantitatively upon contact with any solid surface without the formation of any intermediate phase. This is often referred to as the Gorham process.

Parts to be coated are rotated in the vacuum chamber by a controlled variable-speed motor to allow even distribution of the polymeric coating.

Properties

Molecular weight. Measurement of the soluble polymer fraction with a high temperature viscometer gave a molecular weight of 20,000. This was confirmed by an osmotic-pressure measurement of the chloroform solution giving a value of ca 24,000.

The high molecular weight of poly(p-xylylene) polymers is demonstrated by the fact that the stretched polymer yields highly oriented fibers. Other evidence includes the swelling of sulfonated poly(p-xylylene) in water and the failure of the polymer solution to pass through a sintered-glass membrane which is designed to prevent diffusion of polymers of molecular weight above 20,000.

Contrary to early reports, poly(p-xylylene) is a linear, high molecular weight polymer that is completely soluble at high temperatures. Paramagnetic studies reveal a molecular weight above 2×10^5. Similar values were obtained when the polymer was prepared by the Wurtz reaction.

Table 1. Mechanical and Thermal Properties of Two *p*-Xylylene Polymers

Properties	Poly(*p*-xylylene)	Poly(chloro-*p*-xylylene)
tensile properties at room temperature[a]		
tensile strength, MPa[b]	46.9	69.0
tensile modulus, MPa[b]	2414	3172
elongation at break, %	10–15	220
thermal properties[c]		
crystalline melting point, °C	420	290
glass-transition temperature, °C	80	80
tensile modulus at 200°C, MPa[b]	172.4	172.4

[a] Measured on 0.025–0.050 mm films in Instron tensile tester at 10% strain/min.

[b] To convert MPa to psi, multiply by 145.

[c] From secant modulus–temperature curve. Melting points also obtained from x-ray data and from melting behavior in sealed melting-point tubes.

Table 2. Properties of Commercial Xylylene Polymers

Properties[a]	ASTM method or conditions	Poly(chloro-*p*-xylylene)	Poly(*p*-xylylene)[b]
secant modulus, MPa[c]	D 882-56T at 1% strain	2759	2414
tensile strength, MPa[c]	D 882-56T at 10% strain/min	69.0	41.4–75.9
yield strength, MPa[c]	D 882-56T at 10% strain/min	55.2	42.1
elongation to break, %	D 882-56T at 10% strain/min	200	20–250
yield elongation, %	D 882-56T at 10% strain/min	2.9	2.5
density, g/cm³	D 1505-57T	1.289	1.10–1.12
index of refraction, n_D^{23}	Abbe refractometer	1.639	1.661
water absorption, 24 h, %	D 570-57T	0.06	0.01
Rockwell hardness	D 785-65	RS0	R88
coefficient of friction	D 1894-63		
static		0.29	0.25
dynamic		0.29	0.25

[a] Measured on films 0.025–0.075 mm thick, except where specified.

[b] Properties depend on deposition conditions.

[c] To convert MPa to psi, multiply by 145.

Structure. The high birefringence of β-poly(*p*-xylylene) reveals that the molecules are all parallel to the *c* axis and are distributed nearly isotropically about it.

Molecular orientation. Early wide-angle x-ray diffraction patterns of films > 1 μm in thickness, and electron-diffraction patterns of thin films (0.01–0.02 μm), presented a typical randomly oriented powder pattern. This occurred when the beam of radiation was directed perpendicular to the plane of the film. However, when the beam was inclined at an angle other than 90° to the film, the continuous Debye rings broke up into arcs, establishing the presence of a preferred orientation.

Solubility. Poly(*p*-xylylene) polymers are resistant to solvents, acids, and bases. Poly(*p*-xylylene) and poly(chloro-*p*-xylylene) dissolve in high boiling liquids such as α-chloronaphthalene or benzyl benzoate above 200°C.

Reactivity. Chlorination with a pyridine solution of sulfuryl chloride under uv radiation results mainly in chlorinated ethylene groups.

Concentrated nitric acid at 50°C for a prolonged period gives poly(dinitro-*p*-xylylene), an explosive material with the sensitivity of pentaerythritol tetranitrate and the power of trinitrotoluene.

Boiling the polymer for 24 h in a chromic acid–acetic acid solution gives terephthalic acid as the main product.

Oxidation takes place slowly at high temperatures in the presence of atmospheric oxygen.

Both vacuum and oxidative pyrolysis of the vapor-deposited polymer gives low molecular weight products.

Physical and mechanical properties. The physical and mechanical properties of two xylylene polymers are given in Table 1 and Table 2.

Electrical properties. The poly(*p*-xylylenes) exhibit good electrical properties; dielectric strength and volume and surface resistivity are high. The dielectric constants and dissipation factors, however, are very low and better than most epoxies, silicones, or polyurethanes, but not as good as for Teflon and some other fluorocarbons.

High temperature and related properties. The high crystalline melting points of the poly(*p*-xylylenes) suggest the possibility of a useful service life at elevated temperatures. However, the thermal endurance in air is not exceptional.

In measurement of weathering properties by accelerated weather tests, 0.05 mm films of poly(*p*-xylylene) and poly(chloro-*p*-xylylene) become brittle in less than 100 h.

Optical. Both poly(*p*-xylylene) and poly(chloro-*p*-xylylene) absorb very little in the visible region and are, therefore, transparent and colorless. Films of poly(*p*-xylylene) develop a pronounced birefringence during stretching.

Health and Safety Factors

The hazardous products of thermal decomposition and burning are carbon dioxide, carbon monoxide and, for the chlorinated derivative, hydrogen chloride. Hazardous polymerization does not occur. Waste disposal by burial rather than incineration is recommended. The usual respirators may be employed as needed when working with these compounds.

Applications

Applications include surface preparation and adhesion enhancement; film stabilization; electrical and electronic applications (circuit boards, capacitors, thermistors, surface passivation, solid-state relays, hybrid circuits, discrete core memories, field-effect transitors, nichrome resistors, electrets, plastic-packaged microelectronics, transformers and coils, and transducers); biomedical applications (eg, long-term implantation devices); particle encapsulation; pellicles, thin membranes, and electro-optical applications; lubricants; and optical devices.

STUART M. LEE
Ford Aerospace and Communications Corp.

W.F. Gorham and W.D. Niegisch in N.M. Bikales, ed., *Encyclopedia of Polymer Science and Technology*, Vol. 15, Wiley-Interscience, New York, 1971, p. 99.

S.M. Lee, *Polymeric Films for Semiconductor Passivation*, NASA Contract NAS 12-2011, March 1969.

W.E. Loeb and C.E. White, *Proceedings of the Technical Program*, National Electronic Packaging Convention, NEP/CON '68 West, Los Angeles, Calif., Jan. 30–Feb. 1, 1968; NEP/CON '68 East, New York, June 4–6, 1968, pp. 228–240.

U.S. Pat. 3,399,124 (Oct. 24, 1967), H.G. Gilch (to Union Carbide Corp.)

Y

YEASTS

The genus *Saccharomyces* is of greatest practical and economic importance for the baking, beer (qv), and wine (qv) industries, as well as for the production of biomass (see Bakery processes and leavening agents, yeast-raised products). Other yeasts participate in alcoholic fermentations and occur as food-spoilage organisms or pathogens (see Beverage spirits, distilled; Fermentation).

Morphology, Reproduction, and Life Cycles

The shape of a single yeast cell is usually spherical to ellipsoidal but may be cylindrical, ogival, pyramidal, or apiculate. Often the shape results from specific sites of bud formation.

The yeast cell is surrounded by a strong and mechanically refractory cell wall which may account for 20–30 wt% of the cell solids. The cell wall consists of alkali-soluble β-glucan, alkali-insoluble glucan, and glycoproteins. The cell wall surrounds the plasma membrane, which regulates the transport of chemical compounds into and out of the cell either by simple diffusion or by active transport. The large nucleus of yeast cells is surrounded by a membrane, or tonoplast, which has many pores with an average diameter of ca 0.085 μm. The nucleus contains the genetic material of the cell.

Vegetable reproduction in yeasts occurs mostly by budding and, in the instance of *Schizosaccharomyces* and *Endomycopsis*, by fission. In *Ascomycetes*, sexual reproduction is through the formation of spores in a cell which serves as an ascus or spore sac.

Genetic Aspects

Yeasts have translation and transcription mechanisms similar to those of higher eukaryotic organisms. Accordingly, higher eukaryotic genes are expressed in yeasts when they are inserted by transformation. However, most of the work on genetic improvement of industrial yeast strains is performed by the more traditional methods of hybridization and mutation (see also Genetic engineering).

Fermentative and Respiratory Metabolism

Many yeasts are strict aerobes or ferment weakly. Strongly fermenting yeasts, eg, *Saccharomyces cerevisiae*, grow fairly well under anaerobic conditions but with a lower yield. The yield of yeast based on the weight of sugar consumed is greater in the presence of oxygen, and oxygen inhibits the rate of glucose utilization. However, under aerobic conditions, alcohol is produced if the concentration of glucose exceeds a threshold value of 0.1–0.2 wt%. This is the reverse Pasteur or Crabtree effect and is also called glucose inhibition or catabolite repression. In the presence of higher concentrations of glucose, synthesis of respiratory enzymes, eg, cytochromes, is inhibited.

Composition, Nutrients, and Growth Rate

The elemental and vitamin compositions of some representative yeasts are listed in Table 1.

The specific growth-rate constant for exponential growth μ is defined by the equation $\mu \times dt = dM/M$, where M is the mass of yeast and t is time.

Yeasts that have been grown in clear media, eg, clarified molasses or brewers' worts, can be recovered by centrifuging or filtration followed by pressing. Yeasts that have grown in distillers' mashes or grape musts, which contain insoluble particles, cannot be recovered economically.

Yeast-Fermented Foods and Beverages

Table 2 shows U.S. production of alcoholic beverages, baked goods, and yeast biomass and the production of industrial fuel alcohol by fermentation.

Table 1. Composition of Yeast (Dry Mass)

Component	Bakers' yeast	Brewers' yeast	*Candida* sp
C, wt %	47.0		45.94
H, wt %	6.0		6.72
N, wt %	8.5		7.31
O, wt %	32.5		32.08
ash, wt %	6.0	6.4	7.75
Ca, wt %		0.13	0.57
Fe, wt %		0.01	0.01
Mg, wt %		0.23	0.13
P, wt %		1.43	1.68
K, wt %		1.72	1.88
Na, wt %		0.07	0.01
Co, mg/kg		0.2	
Cu, mg/kg		33.0	13.4
Mn, mg/kg		5.7	38.7
Zn, mg/kg		38.7	99.2
dry matter, wt %		93.0	93.0
crude fiber, wt %		3.0	2.0
ether extract, wt %		1.1	2.5
protein (N × 6.25), estd wt %		44.6	48.3
protein, digestible, wt %		38.4	41.5
thiamine, mg/kg		91.7	6.2
riboflavin, mg/kg		35.0	44.4
nicotinic acid, mg/kg		447.5	560.3
vitamin B₆, mg/kg		43.3	29.5
biotin, mg/kg			1.1
pantothenic acid, mg/kg		109.8	82.9
folic acid, mg/kg		9.7	23.3
choline, mg/kg		3885.2	2910.6

Table 2. 1981 U.S. Production of Yeast-Fermented Foods, Beverages, and Fuel Alcohol and Production of Yeast

Product	Food, beverage, and fuel production, 10⁶ m³ (except where noted)	Production of yeast[a], 1000 metric tons
beer	18.52	40
wine	1.61	4
baked goods (yeast-raised)	7.5 × 10⁶ t	60
yeast biomass (inactive)		7–9
distilled alcohol beverages (50 vol % ethanol)	2.5	25
fuel and industrial alcohol (100 vol % ethanol)	1.5	30

[a] Estimates based on production of 2–2.5 kg of yeast solids per cubic meter of the alcohol beverage (at 8–12 vol % ethanol).

Bakers'-yeast production. Bakers' yeast is grown aerobically in fed-batch fermentations and under conditions of carbohydrate limitation. This assures the growth of yeast and prevents or minimizes the formation of ethanol. Such yeasts have excellent dough-leavening ability. All bakers' yeast strains are *S. cerevisiae*.

Bakers' compressed yeast. Compressed yeast is received by the baker in 0.45-kg cakes or as crumbled material in 22.7-kg bags and is refrigerated while stored.

Active dry yeast (ADY). In several applications, ADY has replaced compressed yeast, eg, where storage stability and convenience are key considerations.

Brewers' yeast. The basic raw materials for the production of beer are sweet worts formed by enzymatic hydrolysis of various cereal starches. The principal cereal is barley which, after malting, is also the source of enzymes that hydrolyze starches, glucans, and proteins.

Wine yeasts. Wine yeasts include several genera that occur naturally in grape musts and that participate in spontaneous fermentation in wine. The number of yeast cells on intact young grape berries is not large; however, it increases with increasing maturity of the grapes, and there are many yeasts on injured grapes where the grape juice is exposed. Nonetheless, most yeast cells that participate in spontaneous fermentations are probably derived from yeasts adhering to crushers, presses, and other cellar equipment. Wine yeasts include *Saccharomyces*, *Kloeckera*, *Metschnikowia*, *Torulopsis*, *Pichia*, and *Hansenula*. Active dry wine yeasts of the genus *Saccharomyces* are now used widely for direct inoculation of musts.

Distilled beverages. Distilled alcoholic beverages are made by fermentation of sugars from grains or fruits followed by distillation.

Distillers' yeast is made up of selected strains of *S. cerevisiae*. Many distillers use their own, proprietary strains. Commercial bakers' yeast may be used for the production of grain neutral spirits.

Microbial Biomass and Single-Cell Protein

In all fermented foods, microbes contribute to the food supply by their preservative action, eg, by lowering pH and producing ethanol, or by making such foods more palatable. However, the use of microbes as a direct source of food started with the drying of spent brewers' yeast in ca 1910 and with the production of *Candida utilis* yeast during World War II. There are, however, some notable exceptions: mushrooms (macrofungi) have always been used as food; algae have been used as part of the human diet by the Aztecs of Central America; and small beer, ie, the sediment of beer, has been used as a vitamin supplement for infants (see Foods, nonconventional).

During the 1960s and 1970s, the world protein shortage stimulated the production of microbial biomass from traditional substrates, eg, carbohydrates, and from alternative raw materials, eg, *n*-paraffins, ethanol, and methanol. Present industrial processes for the production of biomass use yeast cultures (see Fuels from biomass).

Substrates and nutrient requirements. Traditional carbon sources for the production of biomass are fermentable sugars. The preferred source is beet or cane molasses because of their low cost and because they contribute some nitrogenous nutrients, minerals, and vitamins. In addition, yeasts require large amounts of nitrogen, phosphorus, potassium, calcium, and magnesium as well as trace minerals.

Fermentation. Yeast biomass need not be produced under conditions of complete sterility. Most yeast fermentations are carried out at pH ≤ 4.5, which limits bacterial growth. The most common contaminants are lactic acid-producing bacteria and vinegar organisms. All of the processes designed specifically for yeast growth are highly aerobic and thus require costly heat removal. The supply of oxygen in the form of air is also costly, since oxygen-transfer rates from the gas to the liquid are low.

The simplest fermentor is a cylindrical vessel equipped with air-sparger tubes at the bottom, internal cooling coils, and perhaps an agitator.

Nutritional value. Microbial biomass is either recovered as a by-product of a food fermentation, eg, from beer, or grown specifically as a food or feed supplement. In either case, its value is mainly in its contribution to protein nutrition. There is an additional contribution by minerals and vitamins, but this is less important. Yeasts contain 7.5–9 wt% nitrogen, ie, N × 6.25 = 47–56% crude protein; 6–12 wt% nucleic acids; 5–9 wt% ash; and 2–6 wt% lipids.

Production. Microbial biomass is frequently referred to as single-cell protein (SCP). However, it is best to refer to the entire microbial cell mass as microbial biomass and to reserve the designation SPC for the proteins isolated from it. The most widely available biomass is a by-product of the brewing industry.

Bakers' inactive dried yeast is also used widely in the health-food industry.

Candida utilis yeast is grown on waste sulfite liquor in western countries, on molasses in Cuba and Republic of China, and on cellulose acid hydrolysates in the USSR and other Eastern European countries.

Cheese whey contains 70–75 wt% milk sugar, which can serve as the carbon source for the growth of lactose-fermenting yeasts; *Kluyveromyces fragilis* is generally used for this purpose.

Of the petrochemical substrates, both *n*-alkanes and gas oil can be used as carbon and energy sources. The yeasts that have been used commercially are *C. tropicalis* and *C. lipolytica*.

One of the most promising substrates for future production of microbial biomass and single-cell protein is cellulose (qv) from agricultural residues, eg, wood pulp, sawdust, feed-lot waste, corn stover, rice hulls, nut shells and bagasse. All of these materials contain cellulose as the principal carbon source.

Enzymatic hydrolysis with cellulase and cellobiase is the subject of intensive research. The most commonly used enzymes are cellulases and cellobiases derived from *Trichoderma reesei*.

Processing. Recovery of microbial biomass from the fermentor requires centrifugation, washing, concentration, and drying.

Health and safety factors. The presence of nucleic acids in yeast is one of the main obstacles to their use in human foods. The ingestion of purines by humans and some other primates increases plasma levels of uric acid, which may cause gout. No more that two grams of nucleic acids should be in a daily human diet. This limits the daily intake of inactive dry yeasts by humans to ca 20 g/d.

Uses. Inactive dried yeasts are used as ingredients in many formulated foods, eg, baby foods, soups, gravies, and meat extenders; as carriers of spices and smoke flavor; and in baked goods. Inclusion in dough formulas generally results in greater extensibility of the dough because of the reducing action of sulfhydryl groups leached from the dried yeasts. Usually, yeasts used in the health-food industry are fortified with higher concentrations of B vitamins, principally thiamine, riboflavin, and niacin.

Yeast-Derived Products

Yeast derived products include enzymes, yeast extracts, and nucleotides.

Food Preservation and Food Spoilage

Many foods are preserved by fermentation, including alcoholic beverages, pickles, cheese, and fish sauce. Usually, spontaneous fermentations by mixed populations of yeasts and bacteria are involved. The preservative effect results from a lowering of the pH or the formation of ethanol. Yeasts do not produce any potent antibiotics.

Yeasts also act as spoilage organisms. Jams, jellies, and honey can be fermented by osmophilic yeasts, eg, *S. mellis* or *S. rouxii*. Wild yeasts may also spoil wines or beers. Film-forming yeasts, eg, *Pichia membranefaciens*, may grow on the surface of fermented vegetables, such as sauerkraut or pickles. *Kloeckera fragilis* and other lactose-fermenting species frequently occur in milk products. Butter and oleomargarine are occasionally spoiled by *C. lipolytica*. Such spoilage always results in the development of undesirable flavors with serious economic losses.

Pathogenic Yeasts

Few yeasts are pathogenic. However, *C. albicans* is the best known of the potentially pathogenic species.

GERALD REED
Amber Laboratories, Inc.
Present affiliation
Universal Foods

H.J. Phaff, M.W. Miller, and E.M. Mrak, *The Life of Yeasts*, Harvard University Press, Cambridge, Mass., 1978.

G. Reed, ed., *Prescott and Dunn's Industrial Microbiology*, AVI Publishing Co., Westport, Conn., 1982.

H.J. Rehm and G. Reed, eds., *Food and Feed Fermentations*, in Vol. 5 of *Biotechnology*, Verlag Chemie, Deerfield Beach, Fla., 1983.

A.H. Rose, ed., *Microbial Biomass*, Vol. 4 of *Economic Microbiology*, Academic Press, London, 1979.

YOUNG-HELMOLTZ COLOR-VISION THEORY. See Color.

YTTERBIUM. See Rare-earth elements.

YTTRIUM. See Rare-earth elements.

Z

ZEOLITES. See Molecular sieves.

ZIEGLER-NATTA CATALYSTS. See Catalysis; Olefin polymers; Organometallics.

ZINC AND ZINC ALLOYS

Zinc is a relatively active metal and its compounds are stable. Since it is not found free in nature, it was discovered much later than less-reactive metals such as copper, gold, silver, iron, and lead.

The main application of zinc is to protect iron and other metals from corrosion. Another important use is in alloys for die casting. These alloys are used extensively because of their high quality and low cost. Brass and bronze products account for the third largest usage (see Copper alloy).

Occurrence

Zinc ores are widely distributed throughout the world; 55 zinc minerals are known. However, only those listed in Table 1 are of commercial importance. Of these, sphalerite provides ca 90% of the zinc produced today.

In the United States, the richest zinc district is the Mississippi Valley; however, Canada leads the world in estimated reserves and the United States is second.

Zinc minerals tend to be associated with those of other metals; the most common are zinc–lead or lead–zinc, and depending upon the dominant metal, zinc–copper or copper–zinc, and base metals such as silver.

Physical Properties

Zinc is a lustrous, blue-white metal, that can be formed into virtually any shape by the common metal-forming techniques such as rolling, drawing, extruding, etc. The hexagonal close-packed crystal structure governs the behavior of zinc during fabrication. Physical properties are given in Table 2.

Chemical Properties

The most significant chemical property of zinc is its high reduction potential. Zinc, which is above iron in the electromotive series, displaces iron ions from solution and prevents dissolution of the iron.

In batteries, a zinc anode undergoes the oxidation reaction,

$$Zn \rightarrow Zn^{2+} + 2\ e\ (+0.763\ V)$$

to provide a flow of electrons to the external circuit.

The capacity of zinc to reduce the ions of many metals to their metallic state is the basis of important applications.

Table 1. Common Zinc Minerals

Name	Composition	%Zn
sphalerite[a]	ZnS	67.0
hemimorphite[b]	$Zn_4Si_2O_7(OH)_2.H_2O$	54.2
smithsonite	$ZnCO_3$	52.0
hydrozincite	$Zn_5(OH)_6(CO_3)_2$	56.0
zincite	ZnO	80.3
willemite	Zn_2SiO_4	58.5
franklinite	$(Zn,Fe,Mn)(Fe,Mn)_2O_4$	15–20

[a] Zinc blend, wurtzite.
[b] Calamine.

Zinc hydrosulfite (zinc dithionite) is a powerful reducing agent used in bleaching paper and textiles. Another hydrosulfite reducing agent is zinc formaldehyde sulfoxylate, $Zn(HSO_2.CH_2O)_2$ (see Bleaching agents).

Processing

Processing steps include roasting (flash roasting, fluidization roasting, and sintering), and reduction (electrolytic and pyrometallurgical processes).

Secondary recovery. Zinc is recovered as metal, dust, and chemicals (including oxide) from secondary sources, mostly scrap (see Recycling). So-called old scrap originates from die castings and engraver's plates, whereas new scrap, such as drosses, skimmings, flue dust, clippings, and residues, originates in various processes.

Health and Safety Factors; Environmental Aspects

Zinc is not toxic to humans: ca one gram per day may be ingested without ill effects. Recommended dietary allowance is 15 mg/d for adults. Inhalation of fresh zinc oxide may cause a temporary illness called metal fume fever (see also Zinc compounds).

Zinc in the diet is necessary for growth and wound healing in animals and human beings. Zinc and cadmium intake are generally related, especially in contaminated areas, and the interaction of the two in the body is complex. The effects of zinc and cadmium vary greatly, but some of the adverse effects of the latter can be mitigated by sufficient zinc (see Mineral nutrients).

Hazards of production. In most zinc mines, zinc is present as the sulfide and coexists with other minerals, especially lead, copper, and cadmium. Therefore, the escape of zinc from mines and mills is accompanied by these other, often more toxic, materials. Mining and concentrating, usually by flotation (qv), does not present any unusual hazards to personnel. Atmospheric pollution is of little consequence at mine sites but considerable effort is required to flocculate and settle fine ore particles which could be deposited in receiving waters.

Table 2. Physical Properties of Zinc

Property	Value
ionic radius, Zn^{2+}, nm	0.074
covalent radius, nm	0.131
metallic radius, nm	0.138
ionization potential, eV	
first	9.39
second	17.87
third	40.0
density	
solid, g/cm^3	
at 25°C	7.133
at 419.5°C	6.830
liquid, g/mL	
at 419.5°C	6.620
at 800°C	6.250
melting point, °C	419.5
boiling point, °C	907
heat of fusion at 419.5°C, kJ/mol[a]	7.387
heat of vaporization at 907°C, kJ/mol[a]	114.8
coefficient of thermal expansion, mm/(m·K)	
volume	8.9
linear, polycrystalline	39.7
thermal conductivity, W/(m·K)	
solid at 18°C	113.0
liquid at 419.5°C	60.7
electrical resistivity, nΩ/m	
polycrystalline	$R = 54.6 (1 + 0.0042t)$[b]
liquid at 423°C	369.55
heat capacity, J/(mol·K)[a]	
solid	$22.39 + 10^{-2}T$[c]
liquid	31.39
gas	20.80

[a] To convert J to cal, divide by 4.184.
[b] $t = 0$–100°C.
[c] $T = 298$–692.7 K.

Table 3. Properties of High Strength Zinc Foundry Alloys

Property	Alloy No. 8, Permanent-Mold Cast	Alloy No. 12 Sand Cast	Alloy No. 12 Permanent Mold Cast	Alloy No. 27 Sand Cast[a]	Alloy No. 27 Sand Cast H.T.[b]
physical					
density, g/cm³	6.37	6.03	6.03	5.01	5.01
melting or solidification range, °C	375–404	377–432	377–432	375–487	375–487
mechanical					
tensile strength, MPa[c]	221–225	276–310	310–345	400–441	310–324
yield strength, 0.2% MPa[c]	207	207	214	365	255
elongation, %	1–2	1–3	4–7	3–6	8–11
Brinell hardness[d]	85–90	92–96	105–125	110–120	90–100

[a] Primary purpose.
[b] Heat treated.
[c] To convert MPa to psi, multiply by 145.
[d] 500 kg for 30 s.

In recent years, zinc plants have expended large sums to control in-plant dusts and environmental pollution. Liquid effluents are limed and settled to precipitate metals as hydroxides. Flocculants are used to reduce the total suspended solids and, in some instances, filtration of thickener overflow is practiced.

Uses

Zinc is used in metallic coatings, die-casting alloys, foundry alloys (eg, high strength alloys, see Table 3; slush alloys; and forming-die alloys), rolled zinc, and zinc dust and powder.

By-Product Metals

Ores exploited primarily for zinc invariably contain one or more other valuable metals. Cadmium and mercury are usually recovered in sep- arate processes at the zinc plant. The others are shipped as enriched residues to plants that specialize in their recovery.

Thomas B. Lloyd
Gulf and Western National Resources Group

Walter Shawak
The New Jersey Zinc Company, Inc.

Zinc, National Research Council, Subcommittee on Zinc, University Park Press, Baltimore, Md., 1979.

J.D. Nriagu, ed., *Zinc in the Environment*, John Wiley & Sons, Inc., New York, 1980, Parts 1 and 2.

E. Gervais, H. Levert, and M. Bess, *84th Casting Congress and Exposition of the American Foundrymen's Society*, St. Louis, Mo., April 21–25, 1980, American Foundrymen's Society, Des Plaines, Ill.

ZINC COMPOUNDS

Zinc usually occurs as the sulfide but significant quantities of the oxide, carbonate, silicate, and basic compounds of the latter two are also mined (see Zinc and zinc alloys). Table 1 lists properties and uses of zinc compounds.

Thomas B. Lloyd
Gulf and Western Natural Resources Group

Table 1. Properties, Prices, and Uses of Zinc Compounds

Zinc compound	Formula, synonym	Sp gr	Mp, °C	Solubility[a], g/100 g solvent Water	Solubility[a], g/100 g solvent Other	Uses
acetate	$Zn(C_2H_3O_2).2H_2O$	1.735	237	40[25°C] 67[100°C]	3[25°C] alcohol	wood preservative, mordant, antiseptics, catalyst, waterproofing
ammonium chloride	$ZnCl_2.2NH_4Cl$	1.88	150 dec	66[0°C] 69[30°C]		galvanizing, solder flux, adhesives
diborate	$ZnO.B_2O_3.2H_2O$	3.64		0.007[25°C]	sl sol HCl	fireproofing, ceramics, fungicide
dodecaborate	$2ZnO.3B_2O_3 3.5H_2O$	4.22	980	insol		fire retardant
bromide	$ZnBr_2$	4.21	394	471[25°C] 675[100°C]	sol alcohol, ether	photographic paper, catalyst, batteries
carbonate	$ZnCO_3$	4.40	−CO_2 at 300	0.001[15°C]	insol alcohol	ceramics, rubber, astringent (lotions)
chloride	$ZnCl_2$	2.91	275	432[25°C] 614[100°C]	sol ether	textiles, adhesives, flux, wood preservative, antiseptic, astringent
cyanide	$Zn(CN)_2$	1.85	800 dec	0.005[20°C]	sol alkali, CN^-	electroplating, gold extraction
dithiocarbamates dibutyl	$Zn[S-SCN(C_4H_9)_2]_2$	1.21	106	insol	sol C_6H_6, CS_2, $CHCl_3$	vulcanization accelerator, lube oil
diethyl	$Zn[S-SCN(C_2H_2)_2]_2$	1.48	176	insol	sol C_6H_6 CS_2,	vulcanization accelerator
ethylenebis				insol	sol C_6H_6, CS_2, $CHCl_3$	fungicide, insecticide
dimethyl	$Zn[S-SCNHCH_2]_2$, zineb	1.71	249	0.0065[25°C]	sol CS_2, acetone, alkali	vulcanization accelerator, fungicide
	$Zn[S-SCN(CH_3)_2]_2$, ziram					
2-ethylhexanoate	$Zn(C_8H_{16}O_2)_2$	0.90		insol	sol hydrocarbon	paint drier, silicone rubber cure
fluoroborate	$Zn(BF_4)_2.6H_2O$		−H_2O at 60°C	> 100[25°C]	sol alcohol	plating, bonderizing, textile resin cure
fluoride	ZnF_2	4.95	872	1.6[18°C]	sol hot acid, NH_4OH	ceramics, impregnating wood, galvanizing
formaldehyde sulfoxylate	$Zn(HSO_2.CH_2O)_2$		90 dec	60[25°C]	insol alcohol	reducing agent, drying, polymerization

Table 1.   Continued

Zinc compound	Formula, synonym	Sp gr	Mp, °C	Solubility[a], g/100 g solvent			Uses
				Water		Other	
hydrosulfite	$ZnS_2O_4.2H_2O$		200 dec	$40^{20°C}$			bleach, especially textile, paper, reducing agent
iodide	ZnI_2	4.70	446	$432^{18°C}$	$510^{100°C}$	sol alcohol	medicine, photography
2-mercaptobenzo-thiazolate	$Zn(SC_6H_4NCS)_2$	1.70	300 dec	insol	insol	sol $C_6H_6CS_2$, $CHCl_3$	vulcanization accelerator for latex
naphthenate	$Zn[(C_2H)_5CHCOO]_2$			insol	insol	sol hydrocarbon, acid	paint film improver, rot proofer
nitrate	$Zn(NO_3)_2.6H_2O$	2.065	−18	93	$900^{70°C}$	sol alcohol	textiles as resin catalyst, mordant latex coagulant
oxide	ZnO	5.47, 5.61	1800 sub-limes	$0.00042^{18°C}$		sol acid, alkali, NH_4OH	vulcanization accelerator, mildewstat, pigment, supplement in feed and fertilizer, catalyst ceramics, intermediate
peroxide	ZnO_2	1.57	212 ex-plodes	insol		sol acid	cosmetic powders as antiseptic
phosphate	$Zn_3(PO_4)_2$	4.00	900	$2.6^{25°C}$	insol	sol NH_4OH, insol alcohol	metal coatings, dental cement
potassium chromate	$4ZnO.K_2O.4CrO_3.3H_2O$, zinc yellow	3.36–3.46		$0.24^{25°C}$			rust-inhibiting pigment
resinate	$Zn(C_{20}H_{30}O_2)_2$	1.24	205	insol	insol	sol hydrocarbon	inks, paint drier
selenide	$ZnSe$	5.42	> 1100	insol		sol acid	phosphor
silicofluoride	$ZnSiF_6.6H_2O$	2.10	100 dec	$77^{10°C}$	$93^{60°C}$		laundry sour, wood preservative, plaster additive
stearate	$Zn(C_{17}H_{35}COO)_2$	1.09	120	insol	insol	sol hydrocarbon	lubricant, mold release, vinyl stabilizer, anticake, water repellant
sulfate	$ZnSO_4$	3.54	680 dec	$41.9^{0°C}$	$91^{70°C}$	sol glycerol	rayon bath, agriculture, zinc plating, intermediate floatation, mordant
sulfate	$ZnSO_4.H_2O$	3.28	238 dec	$101^{70°C}$	$87^{105°C}$		rayon bath, agriculture, zinc plating, intermediate floatation, mordant
sulfide	ZnS	3.98, 4.10	1185 sub-limes	$0.0007^{18°C}$	insol	sol acid	phosphor, white pigment dental materials
tetroxychromate	$4Zn(OH)_2.ZnCrO_4$	3.87–3.97		$0.01^{25°C}$			wash primer
undecylenate	$Zn[CH_2=CH(CH_2)_8COO]_2$		115	insol	insol	sol hydrocarbon	dermal fungicide

[a] Insol = insoluble, v sol = very soluble, dec = decompose.

M. Farnsworth and C.H. Kline, *Zinc Chemicals*, Zinc Institute, Inc., New York, 1973.

H.E. Brown, *Zinc Oxide, Properties and Applications*, International Lead Zinc Organization, Inc., New York, 1976.

Zinc-Medical and Biologic Effects of Environmental Pollutants, Subcommittee on Zinc, National Research Council, University Park Press, Baltimore, Md., 1979.

ZIRCONIUM AND ZIRCONIUM COMPOUNDS

Zirconium is classified in subgroup IVB of the periodic table with its sister metallic elements titanium and hafnium. Zirconium forms a very stable oxide. The principal valence state of zirconium is +4, its only stable valence in aqueous solutions. The naturally occurring isotopes are given in Table 1. Zirconium compounds commonly exhibit coordinations of 6, 7, and 8. The aqueous chemistry of zirconium is characterized by the high degree of hydrolysis, the formation of polymeric species, and the multitude of complex ions that can be formed.

Zirconium occurs naturally as a silicate in zircon, as an oxide in baddeleyite, and in other oxide compounds. Zircon is an almost ubiquitous mineral, occurring in granular limestone, gneiss, syenite, granite, sandstone, and many other minerals, albeit in small proportion, so that zircon is widely distributed in the earth's crust. The average concentration of zirconium in the earth's crust is estimated at 220 ppm, about the same abundance as barium (250 ppm) and chromium (200 ppm).

Zirconium is used as a containment material for the uranium oxide fuel pellets in nuclear power reactors (see Nuclear reactors). Zirconium is particularly useful for this application because of its ready availability, good ductility, resistance to radiation damage, low thermal-neutron absorption cross section of 18×10^{-30} m^2 (0.18 barns), and excellent corrosion resistance in pressurized hot water up to 350°C. Zirconium is used as an alloy strengthening agent in aluminum and magnesium, and as the burning component in flash bulbs. It is employed as a corrosion-resistant metal in the chemical process industry, and is accepted as a pressure-vessel material of construction in the ASME Boiler and Pressure Vessel Codes.

Table 1.   Naturally Occurring Zirconium Isotopes

Isotope	Occurrence, %	Thermal-neutron capture cross section, 10^{-28} m^{2a}
^{90}Zr	51.45	0.03
^{91}Zr	11.32	1.14
^{92}Zr	17.49	0.21
^{94}Zr	17.28	0.055
^{96}Zr	2.76	0.020

[a] To convert m^2 to barns, multiply by 10^{28}.

Occurrence and Mining

Zircon occurs worldwide as an accessory mineral in igneous, metamorphic, and sedimentary rocks. Weathering has resulted in segregation and concentration of the heavy mineral sands in layers or lenses of placer deposits in river beds and ocean beaches. All commercial sources of zircon are derived from the mining of these ancient, unconsolidated beach deposits, the largest of which are in Kerala State in India, Sri Lanka, the east and west Coasts of Australia, on the Trail Ridge in Florida, and at Richards Bay in the Republic of South Africa.

The deposits, usually ca 4% heavy minerals, are mined with front-end loaders or sand dredges. Typically, the overburden is bulldozed away, the excavation is flooded, and the raw sand is handled by a floating sand dredge capable of dredging to a depth of 18 m. Initial wet concentration using screens, Reichert cones, spirals, and cyclones removes the coarse sand, slimes, and light-density sands to produce a 40 wt% heavy-mineral concentrate. The tailings are returned to the back end of the excavation and used for rehabilitation of worked-out areas. The concentrate is dried and iron oxide and other surface coatings are removed; various combinations of gravity separation (qv), magnetic separation (qv), and electrostatic separation yield individual concentrates of rutile, ilmenite, leucoxene, zircon, monazite, and xenotime.

Physical Properties

Zirconium is a hard, shiny, ductile metal, similar to stainless steel in appearance. Physical properties are given in Table 2.

Table 2.   Physical Properties of Zirconium

Property	Value
atomic weight	91.22
density at 298.15 K, g/cm^3	6.5107
crystal structure	
αZr	
close-packed hexagonal space group	P6$_3$/mmc
a, nm	0.3231
c, nm	0.5146
c/a	1.5927
βZr	
body-centered cubic space group	Im3m
α–β transition temperature, K	1136 ± 5
melting temperature, K	2125 ± 10
boiling temperature, K	4577 ± 100
vapor pressure, $T < 2125$ K, $\log_{10} P_{kPa}{}^a$	
βZr	$8.956 \pm 0.080 - (30810 \pm 240)T^{-1}$
liquid Zr	$8.547 \pm 0.080 - (29940 \pm 240)T^{-1}$
heat of transition, kJ/mol^b	3.89 ± 0.08
heat of melting, kJ/mol^b	18.8 ± 2.1
heat of boiling, kJ/mol^b	573.2 ± 4.6
heat of sublimaton at 298 K, kJ/mol^b	600.8
heat capacity, $T = 298$–1136, J/(mol·K)^b	
αZr	$22.857 + 8.970 \times 10^{-3}T - 0.69 \times 10^5 T^{-2}$
βZr	$2.493 + 6.586 \times 10^{-3}T + 36.718 \times 10^5 T^{-2}$
entropy at 298.15 K, J/(mol·K)^b	38.99 ± 0.21
thermal expansion	
single crystal	
perpendicular to c-axis, $L_{TC} = L_{0°C}$	$1 + 5.145 \times 10^{-6}T$
parallel to c-axis, $L_{TC} = L_{0°C}$	$1 + 9.213 \times 10^{-6}T - 6.385 \times 10^{-9}T^2 + 18.491 \times 10^{-12}T^3 - 9.856 \times 10^{-15}T^4$
volumetric, $V_{TC} = V_{0°C}$	$1 + 19.756 \times 10^{-6}T - 7.023 \times 10^{-9}T^2 + 19.146 \times 10^{-12}T^3 - 9.890 \times 10^{-15}T^4$
polycrystalline	
linear, for random orientation, $L_{TC} = L_{0°C}$	$1 + 6.499 \times 10^{-6}T - 2.096 \times 10^{-9}T^2 + 6.108 \times 10^{-12}T^3 - 3.259 \times 10^{-15}T^4$
thermal conductivty, W/(cm·K)^c	
0–10 K	$1.08189 \times 10^{-1}T + 1.65524 \times 10^{-3}T^2 - 2.63239 \times 10^{-4}T^3$
10–25 K	$-9.48650 \times 10^{-1} + 3.44316 \times 10^{-1}T - 1.79663 \times 10^{-2}T^2 + 2.82821 \times 10^{-4}T^3$
25–80 K	$1.80754 - 5.50950 \times 10^{-2}T + 7.61839 \times 10^{-4}T^2 - 3.72006 \times 10^{-6}T^3$
80–500 K	$5.18009 \times 10^{-1} - 2.36738 \times 10^{-3}T + 6.28905 \times 10^{-6}T^2 - 5.58159 \times 10^{-9}T^3$
500–1900 K	$2.44486 \times 10^{-1} - 1.7982 \times 10^{-4}T + 2.27218 \times 10^{-7}T^2 - 6.24923 \times 10^{-11}T^3$
electrical resistivity, Zr, at 5°C, Ω·cm	$43.74 \pm 0.08 \times 10^{-6}$
temperature coefficient, 0–200°C, per °C	42.5×10^{-4}
Poisson's ratio	0.33
Brinell hardness number, HB^d	90–130

^{a}To convert kPa to mm Hg, multiply by 7.5.

^{b}To convert J to cal, divide by 4.184.

^{c}Tabulated selected experimental values were regressed to give the equation quoted. Estimated values tabulated were not used in the regression. All equations represent tabulated, nonestimated values within ± 2% or better.

^{d}At good Kroll-process purity, lower for iodide zirconium.

Chemical Properties

Zirconium forms anhydrous compounds in which its valence may be 1, 2, 3, or 4, but the chemistry of zirconium is characterized by the difficulty of reduction to oxidation states less than four. In aqueous systems, zirconium is always quadrivalent. It has high coordination numbers, and exhibits hydrolysis which is slow to come to equilibrium, and as a consequence, zirconium compounds in aqueous systems are polymerized.

Zirconium is a highly active metal that, like aluminum, seems quite passive because of its stable, cohesive, protective oxide film which is always present in air or water.

Corrosion resistance. Zirconium is resistant to corrosion by water and steam, mineral acids, strong alkalies, organic acids, salt solutions, and molten salts (see also Corrosion and corrosion inhibitors).

Processing

Decomposition of zircon. Zircon is a highly refractory mineral as shown by its geological stability; the ore is cracked only with strong reagents and high temperatures. Methods include the carbothermic reduction, caustic fusion, fluorosilicate fusion, chlorination, and thermal dissociation.

Separation of hafnium. Zirconium and hafnium always occur together in natural minerals and, therefore, all zirconium compounds contain hafnium, usually about 2 wt% Hf/(Hf + Zr). However, the only applications that require hafnium-free material are zirconium components of water-cooled nuclear reactors. Today, the separation of zirconium and hafnium by multistage counter-current liquid-liquid extraction is routine (see Extraction, liquid-liquid extraction).

Reduction. The Kroll Process is the most common method of reduction, in which zirconium tetrachloride vapor is reduced by molten magnesium in a protective argon atmosphere.

Refining. Zirconium sponge produced by the Kroll process has adequate purity and ductility for most uses. For applications requiring extremely soft metal and for research studies on the properties of the pure metal, it can be further purified by the van Arkel-de Boer (iodide-bar) process using a selective vapor transport.

Health and Safety Factors

Zirconium is generally nontoxic as an element or in compounds.

The oral toxicity is low; OSHA standards for pulmonary exposure specify a TLV of 5 mg zirconium per m^3.

Finely divided zirconium is classified as a flammable solid and shipping regulations are prescribed accordingly. Metal powder finer than $74 \mu m$ (270 mesh) is limited to 2.26 kg per individual container.

Uses

The largest use is foundry sand, where zircon is used as the basic mold material, as facing material on mold cores, and in ram mixes.

Zirconium oxide is used in the production of ceramic colors or stains for ceramic tile and sanitary wares (see Colorants for ceramics).

Zirconium oxide increases the refractive index of some optical glasses, and is used for dispersion hardening of platinum and ruthenium.

Yttria or calcia-stabilized cubic zirconias are used extensively as the solid electrolyte in oxygen sensors in automotive and boiler exhausts, and oxygen-content probes for molten copper or iron in smelters.

Zirconate compounds exhibit several interesting properties. Lead zirconate–titanate compositions display piezoelectric properties which are utilized in the production of FM-coupled mode filters, resonators in microprocessor clocks, photoflash actuators, phonograph cartridges, gas ignitors, audio tweeters and beepers, and ultrasonic transducers (see Ferrites). Lanthanum-modified lead zirconate-titanate ceramics have been studied for photoferroelectric image storage (see Ferroelectrics). Alkaline-earth zirconate dielectrics are used in ceramic capacitors.

The uses of other zirconium compounds result primarily from the ability of zirconium to complex with carboxyl groups to form an insoluble organic compound.

Zirconium metal is marketed in three forms: zirconium-containing silicon–manganese, iron, ferrosilicon, or magnesium master alloys; com-

mercially-pure zirconium metal; and hafnium-free pure zirconium metal.

Silicon–manganese–zirconium, ferrozirconium, and ferrosilicon-zirconium (and some pure zirconium) are used in the steel industry for deoxidizing. Magnesium–zirconium is added to magnesium and aluminum alloys for grain refining and strengthening. Most magnesium alloys used at elevated temperatures contain zirconium.

Pure zirconium is being increasingly used as a corrosion-resistant material of construction for chemical process industry equipment.

Hafnium-free zirconium alloys containing tin or niobium are used for tubing to hold uranium oxide fuel pellets inside water-cooled nuclear reactors.

Compounds include zirconium hydride, ZrH_2, zirconium carbide, ZrC, zirconium nitride, ZrN, zirconium diboride, ZrB_2, zirconium dodecaboride, ZrB_{12}, phosphides, ZrP_3, ZrP_2, and $ZrP_{0.6}$, chalcogenides, ZrS_3, $ZrSe_3$, and $ZrTe_3$, zirconium dioxide, zirconium silicate, $ZrSiO_4$, zirconium tetrafluoride, ZrF_4, potassium hexafluorozirconate, K_2ZrF_6, zirconium tetrachloride, $ZrCl_4$, zirconium tetrabromide, $ZrBr_4$, and zirconium tetraiodide, ZrI_4, zirconium trichloride, $ZrCl_3$, tribromide, $ZrBr_3$, triiodide, ZrI_3, zirconium monochloride, $ZrCl$, hydrous zirconium oxide hydrate, $ZrO_2 \cdot nH_2O$, zirconium oxide dichloride, $ZrOCl_2 \cdot 8H_2O$, anhydrous zirconium oxide chloride, $ZrOCl_2$, anhydrous zirconium tetranitrate, $Zr(NO_3)_4$, zirconium carbonate, $2ZrO_2 \cdot CO_2 \cdot xH_2O$, zirconium sulfate, $Zr(SO_4)_2 \cdot 4H_2O$, zirconium bis(monohydrogen phosphate), $Zr(HPO_4)_2 \cdot H_2O$, zirconium alkoxides, $ZrX_{4-n}(OR)_n$, zirconium hydroxy carboxylates, zirconium dichlorobis(dimethylamide), $ZrCl_2[N(CH_3)_2]_2$, and zirconium arylamines.

Organometallic compounds. Certain zirconium organometallic compounds are highly reactive toward low molecular weight unsaturated molecules. Some of these compounds are useful in various syntheses; others function effectively as catalysts for polymerization, hydrogenation, or isomerization. Compounds include hydrides, carbonyl complexes, dinitrogen complexes, alkyl and aryl complexes, and mixed-metal systems.

Catalysts. Several types of zirconium organometallic compounds are useful catalysts. In addition to the catalytic properties of the molecules, the fact that they can be bound to a relatively inert substrate increases their utility.

RALPH H. NIELSEN
JAMES H. SCHLEWITZ
HENRY NIELSEN
Teledyne Wah Chang Albany

O. Kubaschewski, ed., *Zirconium: Physico-Chemical Properties of its Compounds and Alloys*, International Atomic Energy Agency, Vienna, 1976.

A.M. Alper, ed., *High Temperature Oxides*, Academic Press, New York, 1970.

A. Clearfield, *Inorganic Ion Exchange Materials*, CRC Press, Boca Raton, Fla., 1982.

ZONE REFINING

Zone refining is one of a class of techniques known as fractional solidification, in which a separation is brought about by crystallization of a melt without solvent being added (see also Crystallization). A massive solid is formed slowly with a sizable temperature gradient imposed at the solid–liquid interface.

Zone refining can be applied to the purification of almost every type of substance that can be melted and solidified, eg, elements, organic compounds, and inorganic compounds. Because the solid–liquid phase equilibria are not favorable for all impurities, zone refining often is combined with other techniques to achieve ultrahigh purity.

The high cost of zone refining has thus far limited its application to laboratory reagents and valuable chemicals such as electronic materials. The cost arises primarily from the low processing rates, handling, and high energy consumption owing to the large temperature gradients needed (see Semiconductors).

Equipment and Techniques

Containers. The ideal container for zone melting should neither contaminate the melt nor be damaged by the melt or subsequent contraction of the solid. For organic materials, borosilicate glasses are especially suitable, although metals and fluorocarbons and other polymers have also been successfully employed. For many metals and semiconductors, fused silica (often erroneously called quartz) is ideal. High melting inorganic salts and oxides are conveniently, albeit expensively, treated in noble-metal containers: platinum, rhodium, and iridium. The container may either be in the form of a tube or a horizontal boat.

Drive mechanisms. Either the heaters or the sample may be moved. The optimal zone-travel rates are typically rather slow, ie, ca 1 cm/h. Thus, electric-motor drives require gearing systems, resulting in an undesirably jerky stick-slip movement. This is avoided by using double-rod supports with linear ball bearings and with low backlash gears, where tension on the moving piece push in the direction of motion.

Heating and cooling. Heat must be applied to form the molten zones, and this heat must be removed from the adjacent solid material. The most common method is to place electrical resistance heaters around the container.

Heat is often removed by simply allowing it to escape by convection, radiation, and conduction. However, such uncontrolled escape can lead to very large temperature fluctuations. It is better to surround the entire container, heaters and all, with a controlled-temperature cold chamber.

Floating-zone melting. No completely satisfactory container material exists for many high melting materials. In such cases, floating-zone melting may be employed. The primary application for floating-zone melting is crystal growth rather than purification.

Stirring. Stirring increases the optimal zone-travel rate by lowering the film thickness and aids significantly in removal of foreign particles.

WILLIAM R. WILCOX
Clarkson University

M. Zief and W.R. Wilcox, *Fractional Solidification*, Marcel Dekker, Inc., New York, 1967.

W. Keller and A. Mühlbauer, *Floating-Zone Silicon*, Marcel Dekker, Inc., New York, 1981.

ZYMURGY.   See Beer; Fermentation; Wine.

SUPPLEMENT

ANALYTICAL METHODS

Analytical chemistry is concerned with the determination of chemical composition (elemental, ionic, or molecular). Increasingly, analytical requirements are for the determination of trace constituents at the ppm, ppb, and lower concentration levels, species information in which determinations are made of trace organic molecular species or chemical form and valence states of elements, and overall chemical profile of the sample, and finally, local compositional information (compositional maps) for complete characterization.

Chemical analysis must be undertaken with a clear perspective of the information sought. In most cases, more than one method can provide the needed information. In general, determinations of concentrations are relative and not absolute.

Perhaps the most significant developments are combined techniques, such as gas chromatography–mass spectrometry and gas chromatography–Fourier transform infrared spectrometry. These hyphenated techniques afford a multidimensionality that is particularly well-adapted to the analysis of complex mixtures.

Advances in electronics and computer technologies have led to new generations of automated analytical instruments. These developments have in turn fostered a new area called chemometrics which is concerned with the use of mathematical and statistical methods to design or select optimal measurement procedures and experiments and to extract maximal information from the analysis of chemical data.

The use of lasers has become commonplace, and methods that have been improved through incorporation of lasers include atomic fluorescence, fluorometric analysis, polarimetry, and Raman spectroscopy. New methods include laser-enhanced ionization and resonance ionization mass spectrometry.

Flow-Injection Analyses

Flow-injection analysis (fia) utilizes converging streams of sample and reagent which mix and undergo chemical or physical change that is monitored by means of a flow-through detector. Flow-injection analysis is adaptable to a wide range of problems. For example, Mn(II) can be determined spectrophotometrically by the reaction with formaldoxime in alkaline medium. An important variant is the so-called stopped-flow technique where intermittent pumping lengthens the reaction time without significantly increasing the sample zone. An important application of the stopped flow technique is enzyme-based rate-reactions analysis. Advantages of fia are the reproducibility of experimental conditions and high sample throughput.

Inelastic-Electron-Tunneling Spectroscopy

Inelastic-electron-tunneling spectroscopy is a branch of vibrational spectroscopy based upon the measurement of the energy losses sustained by electrons tunneling from one metal to another through an insulation barrier containing the sample to be analyzed. Experimentally, current is monitored as a function of bias voltage. Inelastic-electron-tunneling spectroscopy is used to investigate metallic corrosion, surface orientation, radiation damage, sequencing of biomolecules, and mechanisms of catalysis.

Extended X-Ray Absorption Fine Structure

Extended x-ray absorption fine structure (exafs) is a technique in which the x-ray absorption spectrum of a material is measured over an energy range that extends above the absorption edge of a selected element. This technique is especially well-suited to the analysis of materials which have short-range order but lack long-range order, such as disordered solids and amorphous materials. It is unique in its ability to determine nearest-neighbor distances, coordination numbers, and mean-square relative displacements. Applications include investigation

of metal-on-support catalysts, ions in solution, impurities in solids and at surfaces, and structures of metalloenzymes.

Photoacoustic Spectroscopy

Photoacoustic spectroscopy (pas) is based on the fact that optically excited species can lose excess energy through nonradiative processes. This technique is highly sensitive for the analyses of species in gas, liquid, and solid phases. It has been used to study physical properties of materials, depth profiling, surface or subsurface microscopic imaging, low-probability transitions and overtone absorptions, energy-level splittings caused by the Zeeman effect, energy-transfer processes, and transitions from excited states.

Inductively Coupled and Direct-Current Plasma-Emission Spectrometry

Flame-atomic-emission spectrometry (faes) and flame-atomic-absorption spectrometry (faas) have been recognized as extremely useful techniques for elemental analysis in a wide variety of matrices. However, the relatively low temperatures of combustion flames (usually $< 3000°C$) limit the sensitivy of faes. The development of plasma sources with temperatures approaching $10,000°C$ has increased the potential for sensitive multielement analysis. Although many plasma sources have been developed, including inductively coupled plasma (ICP), direct-current plasma (DCP), capacitively coupled microwave plasma (CMP), microwave-induced plasma (MIP), and laser-induced plasma (LIP), only ICP and DCP are used extensively. Plasma techniques provide a wide linear dynamic range, increased sensitivity, and sequential/simultaneous multielement determinations.

Laser-Enhanced Ionization (lei)

A pulsed laser is used to pump a dye laser with the wavelength tuned to a resonance electronic transition of the element of interest. The analyte, aspirated as a solution into the flame, absorbs the energy provided by the laser and is raised to an excited state. Additional energy (thermal, kT) is supplied by the flame, and the sum of energies provided results in atom ionization producing an ion and an electron. This changes the current in an external circuit applying voltage across the flame and is proportional to the elemental concentration. The technique is very selective and should be applicable to all species currently determined by atomic spectrometry with greater sensitivity and selectivity in most cases.

Resonance-Ionization Mass Spectrometry (rims)

This combination technique is derived from resonance-ionization spectroscopy (ris) and mass spectrometry (ms). To date, its main use has been the determination of rare-earth and actinide elements. Excitation schemes have been developed for more than 50 other elements and it is anticipated that reports on quantification of these elements will follow soon.

Laser-Microprobe Mass Spectrometry

This technique, culminating in the production of the laser-microprobe mass analyzer (LAMMA) and laser-induced ion-mass analyzer (LIMA), has added a new dimension to analytical chemistry, providing unparalleled potential for yielding compositional data from H to U on samples with $1-2$ μm spatial resolutions at sensitivities from 10^{-18} to 10^{-20}g. Samples can range from thin films (< 1 μm thick) to macrosamples (> 20 μm). This technique has been used in the determination and identification of species in a broad variety of biomedical areas and in the analysis of particles, aerosols, polymers, soils, and wood.

Solid-State Nmr Spectroscopy

High power proton dipolar decoupling, magic-angle spinning (mas), and cross polarization (cp) increase the sensitivity and resolution of ^{13}C solid-state spectra. The barrier of low sensitivity for ^{13}C solid-state nmr spectra is removed by cp techniques, and it is now possible to obtain information on molecular and local nuclear motions and interactions leading to determinations of relaxation parameters, chemical-exchange processes, crystal structures, and the presence of reactive intermediates in the solid state.

Biomolecules as Analytical Reagents

Enzymatic and immunochemical analyses have brought analytical chemists out of the instrumental revolution back to sensitive and selective wet chemical methods. Enzymes are used to catalyze a specific reaction and, in general, the rate of the reaction is proportional to the concentration of the analyte. A problem encountered with enzymes is their instability when removed from a living organism. In some instances, this problem has been circumvented by attaching the enzyme to a polymeric support.

In immunossay procedures, antibody molecules have binding sites that react with molecular species called antigens having specific molecular structures. These techniques are selective and sensitive to as little as 10^{-16} g of a complex mixture. Biomolecules have been used in medicine, clinical and environmental chemistry, forensic science (see Forensic chemistry), pharmacology, and food and nutrition.

CURT W. REIMANN
RANCE A. VELAPOLDI
National Bureau of Standards

B.R. Kowalski, *Chemometrics: Theory and Applications*, ACS Symposium Series 52, Washington, D.C., 1977.

G.M. Hieftje, J.C. Travis, and F.E. Lytle, *Lasers in Chemical Analysis*, The Humana Press, Clifton, N.J., 1981.

J. Ruzicka and E.H. Hansen, *Flow Injection Analysis*, John Wiley & Sons, Inc., New York, 1981.

B.K. Teo and D.C. Joy, eds. *EXAFS Spectroscopy: Techniques and Applications*, Plenum Press, New York, 1981.

C.A. Fyfe, L. Bemi, R. Childs, H.C. Clar, D. Curtin, J. Davies, D. Drexler, R.L. Dudley, G.C. Gobbi, J.S. Hartman, P. Hayes, J. Klinowski, R.E. Lenkinski, C.J.L. Lock, I.C. Paul, A. Rudin, W. Tchir, J.M. Thomas, F.R.S., and R.E. Wasylishin, *Philos. Trans. R. Soc. London* **A305**, 591 (1982).

E.M. Chait and R.C. Ebersole, *Anal. Chem.* **53**(6), 682A (1981).

AZEOTROPIC AND EXTRACTIVE DISTILLATION

An azeotrope is a liquid mixture that exhibits a maximum or minimum boiling point relative to the boiling points of the components of the mixture and that distills without change in composition. The maximum or minimum boiling point is caused by negative or positive deviations, respectively, from Raoult's law. Simple distillation results in the removal of the azeotrope in the overhead or bottoms for a minimum or maximum boiling-point azeotrope, respectively. For the separation of an azeotropic mixture, the properties of the azeotrope can be altered by pressure adjustment, addition of a high boiling-point solvent (extractive distillation), or formation of a heterogeneous ternary azeotrope (conventional azeotropic distillation). Other processes have been considered, from the use of membranes to extractive distillation based on the salt effect.

For a given temperature and pressure, an azeotrope defines the condition in which the composition of the vapor phase is the same as that of the liquid phase in equilibrium. If only one liquid phase is present, the azeotrope is homogeneous. For example, ethanol and water form a homogeneous azeotrope at atmospheric pressure and 78.15°C and at a mole fraction of ethanol of 0.8943. If there are two liquid phases, the azeotrope is heterogeneous. *n*-Butanol and water form a heterogeneous azeotrope at atmospheric pressure and 92.25°C and an average mole fraction of *n*-butanol of 0.25.

Self-Entrained Systems

The separation of heterogeneous azeotropes is considerably easier than that of homogeneous azeotropes; the vapor is in equilibrium with each of the two liquid phases. Once the azeotrope has been split into its two liquid phases, the components may be obtained by simple distillation. Such systems are said to be self-entrained.

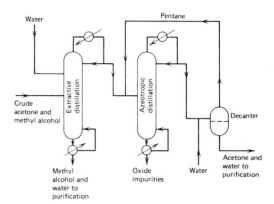

Figure 1. Separation of acetone and methanol by azeotropic and extractive distillation.

Industrial Processes

Unfortunately, very few industrially important systems form heterogeneous binary azeotropes, and generally, azeotropic or extractive distillation is used to separate close-boiling mixtures. Thus, the separation of ethanol and water by conventional distillation is not possible because of the presence of a homogeneous azeotrope. However, they can be separated by extractive and azeotropic distillation.

In both extractive and azeotropic distillation, a third component, called a solvent in the former or an entrainer in the latter, is added to increase the difference in volatility between the key components. The main difference between extractive and azeotropic distillation is in the different methods used to secure appreciable concentrations of the mass-separating agent. There is no general answer to the question which of the two processes is more efficient. When a suitable entrainer is available, azeotropic distillation usually is preferred.

Combinations of azeotropic and extractive distillation. In certain cases, separation cannot be accomplished by azeotropic or extractive distillation. In these cases, both processes are considered.

In the separation of acetone and methanol, the impurities consist primarily of tetramethylene oxide, which forms an azeotrope with acetone. Acetone and methanol also form an azeotrope. It is necessary, then, to separate two azeotropes. A polar solvent and a nonpolar entrainer are used to break multiple azeotropes. The process is shown in Figure 1.

A combination of azeotropic and extractive distillation has been suggested for the purification of methyl ethyl ketone. A nonpolar entrainer, hexane, removes the oxide impurities in an azeotropic tower. The entrainer is recovered in a water-extraction column; a second water-extraction column removes the acetal impurities. Methyl ethyl ketone and water are finally separated in a pentane-extraction column.

Design of Towers

The basic calculation procedures for extractive-distillation towers are the same as for conventional-distillation towers. Perhaps the single most important issue is the ability to predict the vapor–liquid and liquid–liquid equilibria. There are many sets of experimental data in the literature. The high degree of nonideality of azeotropic mixtures make simple models for the Gibbs free energy inadequate. For a successful design of an azeotropic-separation sequence, it is necessary to predict the vapor–liquid–liquid equilibrium of the system.

Most methods for the simulation of conventional-distillation towers are suitable, in principle, for azeotropic towers. However, steep composition profiles cause difficulties in many algorithms.

GEORGE PROKOPAKIS
Columbia University

C.D. Holland, S.E. Gallun, and M.J. Locket in J.J. McKetta, ed., *Encyclopedia of Chemical Processing and Design*, Vol. 16, Marcel Dekker, Inc., New York, 1982, pp. 96–132.

C.J. King, *Separation Processes*, McGraw-Hill Book Co., New York, 1971.
C. Black, *Chem. Eng. Prog.* **76**, 78 (1980).
L.H. Horsley, *Adv. Chem. Ser.* **116**, (1972).

CARDIOVASCULAR AGENTS

Cardiovascular agents are drugs that affect the circulatory system.

Calcium-Channel Blocking Agents

Calcium-channel blocking agents or calcium entry blockers alter the availability of calcium by affecting its entry into its receptor sites. The prototype drugs of this group of compounds, such as verapamil, nifedipine, and diltiazem, have been utilized widely for a long time for the treatment of angina pectoris and certain peripheral vascular diseases.

Calcium acts on sites associated with the control of the cytoplasmic-calcium concentration. Vascular smooth-muscle contraction can be initiated by the opening of the calcium-influx channel.

The calcium-channel blocking agents interfere with the entry of calcium through the membrane channel and therefore prevent the intracellular calcium from reaching the critical concentration necessary to initiate contraction. They have been used mainly for the treatment of angina and arrhythmias. Most of them are well absorbed after oral administration.

Verapamil exerts potent negative inotropic effects in isolated cardiac preparations. However, little change in contractility occurs in patients. Electrophysiological effects are selective for atrial conducting tissues. The onset of action is 30 min and peak effects occur 4–5 h later.

Nifedipine has more potent negative inotropic effects than verapamil, but little effect on cardiac contractility. Side effects include hypotension, facial flushing, headache, dizziness, nausea, vomiting, edema, and epigastric pressure.

Diltiazem has a slightly negative inotropic effect on isolated myocardial tissues.

Perhexiline exhibits a slightly negative inotropic effect in isolated cardiac preparation. Upon prolonged administration, it produces serious adverse effects including hepatitis, peripheral neuropathy, and weight loss.

Lidoflazine has a slightly negative inotropic effect in isolated cardiac preparations.

Nicardipine exhibits a moderate degree of negative inotropic effect in isolated cardiac preparations.

Nimodipine blocks the vasoconstrictor effect of a number of agonists in isolated vascular tissues. It dilates smaller arteries and increases cerebral and coronary blood flow and cardiac output.

Niludipine. In anesthetized dogs, the intravenous administration of this drug significantly decreased coronary artery resistance, heart rate, and systemic blood pressure. These changes were associated with an increase in coronary blood flow.

Nisoldipine. Experiments in dogs have shown that this compound effectively increases coronary blood flow and cardiac output and decreases total peripheral resistance and systemic blood pressure.

Other compounds described as having calcium-channel blocking activity are nitrendipine, flunarizine, cinnarizine, bepridil, and prenylamine.

The Renin-Angiotensin System and Blood-Pressure Regulation

Antihypertensive agents with different mechanisms of action are now available. The introduction of β-adrenergic receptor-blocking agents, such as propranolol, has provided highly effective control of hypertension.

Angiotensin-converting enzyme is identical to kininase II, which is responsible for the inactivation of bradykinin. In the anesthetized guinea pig, the angiotensin-converting enzyme inhibitors potentiate bradykinin-induced bronchoconstriction which can be blocked by indomethacin (a cyclooxygenase inhibitor).

Captopril is an efficacious antihypertensive agent in essential hypertensive patients. This effect is greatly increased by concomitant treatment with diuretics. Captopril may impair neurogenic vasoconstriction by interfering with the pre- and postjuctional actions of angiotensin on adrenergic neuroeffector systems or by a nonangiotensin-dependent mechanism. Captopril potentiates the release of a bradykinin-induced prostacyclin-like substance whose action is blocked by indomethacin. Captopril significantly increases plasma-renin activity and decreases aldosterone concentration. It inhibits angiotensin-converting enzyme *in vivo* and *in vitro*. Few side effects are observed.

Enalapril effectively lowers blood pressure in hypertensive patients with an associated increased plasma-renin activity. Absorption after oral administration is rapid. Adverse effects are rare.

CI 906. This compound inhibits plasma-angiotensin-converting enzyme in both rats and humans.

Pivlopril has a rapid onset but short duration of action and is less potent than captopril or enalapril.

SA 446. This compound reduces blood pressure in the renal and genetic spontaneously hypertensive rat.

Nitroglycerin

Glyceryl trinitrate, nitroglycerin, was introduced for the treatment of angina pectoris in 1879. Transdermal nitroglycerin-delivery systems address many problems of nitrate therapy, such as the ease of handling for control of efficacy and side effects, adequate absorption, onset and duration of action, and maintenance of potency. The mechanism of action is not clearly understood. The basic pharmacologic action is relaxation of smooth muscles. Nitroglycerin may release prostacyclin from the vascular endothelial lining and may be effective by redistributing coronary blood flow.

Sublingual preparations are rapidly absorbed and peak blood concentrations and pain relief occur within 1–4 min. Once nitroglycerin is absorbed, it is bound to plasma protein. The principal side effects are headache, nausea, flushing, dizziness, and weakness.

In addition to remaining the drug of choice for angina pectoris, nitroglycerin has been successful in the treatment of congestive heart failure, myocardial infarction, peripheral vascular disease, and mitral insufficiency.

Transdermal delivery systems. Marketing approval has been granted by the FDA for three transdermal nitroglycerin systems. These are patches that consist of several layers.

Transderm-Nitro consists of an impermeable, aluminized plastic outer layer that holds the drug reservoir, consisting of nitroglycerin on lactose, in a viscous silicone oil.

Nitro-Disc is a system in which a removable outer impermeable foil layer holds the drug reservoir, which is microsealed and contains nitroglycerin dispersed in a silicone rubber polymer matrix (see Fig. 1).

Nitro-Dur, available in four sizes or strengths, is slightly more complex than the other transdermal units. The outer layer is an impermeable polyethylene film that contains an absorbent bandage to trap moisture. The inner layer contains a diffusion matrix bonded to an occlusive aluminum foil base-plate support which is attached to the bottom of the adhesive tape of the outer layer. The diffusion matrix contains nitroglycerin on lactose in equilibrium with nitroglycerin in a glycerol–water phase.

Cardiotonic Agents

The most important use of drugs that produce positive inotropic effects is in acute or chronic congestive heart failure. At present, the digitalis glycosides and the β-adrenoceptor-agonist drugs are used as inotropic agents in heart failure.

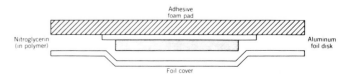

Figure 1. Schematic of the Nitro-Disc unit.

Amrinone is a novel nonglycosidic, non-β-adrenoceptor agonistic positive inotropic agent. Given intravenously or orally, it has been shown to increase cardiac output, stroke volume, and stroke work (indexes of inotrophy), and to increase exercise tolerance in patients with severe congestive heart failure, with or without ischemic heart disease. The positive-inotropic effects in animals and humans were not accompanied by arrhythmias. Although the margin of safety is better than for digitalis, the drug has been reported to produce thrombocytopenia and nephrogenic diabetes insipidus.

Peter Cervoni
Fong M. Lai
Lederle Laboratories
American Cyanamid Company

J.F. Dasta and D.R. Geraets, *Am. Pharm.* **22**, 29 (1982).

R.L. Talber and H.J. Bussey, *Clin. Pharm.* **2**, 403 (1983).

A. Zanchetti and R.C. Tarazzi, *Am. J. Cardiol.* **49**, 1381 (1982).

A.A. Alousi and H.P. Dobreuc, in A. Scriabine, ed., *New Drugs Annual*, Raven Press, New York, 1983, p. 259.

COAL CHEMICALS

INTRODUCTION

Development of the Coal Chemicals Industry

Initially, coal was the main raw-materials base of the chemical industry and remained so until well into the twentieth century. The synthesis of dyestuffs, for example, was based on aromatics, eg, benzene, naphthalene, and anthracene, obtained from bituminous coal tar.

Today, coal-derived products play only a secondary role. Petroleum and natural gas proved to be superior raw-materials sources. They are richer in hydrogen and easier to transport and process. Since the transition to oil and natural gas as the main chemical raw materials, worldwide production of chemicals has risen enormously, especially of plastics and fertilizers.

Phenol (qv), the starting material for many resins, occurs in coal tar only in small amounts, and synthetic processes for it were also developed. Large amounts of ethylene, as well as propylene and butadiene are easily obtained from petroleum.

By-products of coking, particularly tar and benzene, have maintained their commercial importance. Coal remains an important raw material in reduction processes for the recovery of phosphorus, sodium sulfide, silicon, and other elements and compounds.

Outlook

With decreasing supplies of oil, coal is expected to be used again more and more as a raw-material source. All basic chemicals can be manufactured from coal, though in most cases with greater difficulty than from petroleum. Coal hydrogenation gives mixtures of hydrocarbons containing large proportions of aromatic compounds.

Acetylene chemistry may become important, particularly if the costs of the carbide process decrease, or other techniques, eg, the thermo-oxygen or the plasma-based cracking process are developed (see Acetylene-derived chemicals).

George Kölling
Bergbau-Forschung GmbH

J. Falbe, ed., *Chemical Feedstocks from Coal*, John Wiley & Sons, Inc., New York, 1982.

H. Harnisch, R. Steiner, and K. Winnacker, eds., *Chemische Technologie*, 4th ed., Vol. 5, C. Hanser, Munich, FRG, and Vienna, Austria, 1981.

A. Stratton, *Energy and Feedstocks in the Chemical Industry*, Ellis Horwood Limited, Chichester, UK, 1983.

GASIFICATION

The Lurgi, Koppers-Totzek, and Winkler gasification processes are referred to as coal-gasification processes of the first generation. All three have been used in large-scale plants. The main product is chemical-feedstock synthesis gas, primarily a mixture of CO and H_2, which is converted to intermediates and products. The gasification temperature greatly influences the formation of chemical feedstocks. In general, higher reaction temperatures increase carbon conversion and thereby raise the yield of synthesis gas.

The Lurgi process is the main source of chemical feedstocks from the by-products of gasification, whereas the Koppers-Totzek and the Winkler gasification processes provide, in addition to carbon monoxide and hydrogen, only elemental sulfur as a final product. Coal dust and tars are removed from the raw gas which leaves the gasifier via a scrubbing cooler and is cooled further in a waste-heat boiler. Compared to the by-product potential of the Lurgi moving-fixed-bed gasification process, the two other primary-general coal-gasification processes are insignificant except for the recovery of sulfur. Carbon dioxide is obtained from all processes; it can be recovered from the scrubbing gases.

Syngas Chemistry

Second-generation coal-gasification processes, with the exception of the Lurgi process, generally yield synthesis gas rich in CO, especially when coal fines and high temperatures and pressures are used.

Synthesis gas from coal is used in many countries for the oxo synthesis, methanol production, hydrogen-based ammonia production, and the manufacture of motor fuels and chemical feedstocks.

If pure hydrogen is the desired product, the synthesis gas is first converted catalytically:

$$CO + H_2O \rightarrow CO_2 + H_2$$

In high temperature shift conversion, an iron–chromoxide catalyst is used at gas inflow temperatures of 280–350°C. For low temperature shift conversion, a copper catalyst, which requires sulfur-free feed gas, is used. Sulfur-resistant shift catalysts based on cobalt and molybdenum permit the conversion of untreated, sulfur-containing synthesis gas in one heating. For many applications, the COS must be converted to H_2S.

Carbon monoxide can be separated from hydrogen and methane in the liquid phase at low temperatures by condensation and distillation or by selective absorption. Carbon monoxide (97–99 wt%) that contains small amounts of hydrogen, methane, or nitrogen generally meets the quality demands of chemical processes. Low temperature separation, liquid methane washing, and absorptive methods can be used for separation. Low temperature processes for carbon monoxide recovery from synthesis gases involve partial condensation or liquid-methane scrubbing. In the latter, the gas mixture is scrubbed at a temperature just above the freezing point of methane (ca −180°C). The methane is recovered by distillation, and the carbon monoxide is flashed overhead.

In the first technical application of a second-generation coal-gasification process, in which the resulting synthesis gas is used as a chemical feedstock, acetic acid is produced from coal (see Fig. 1).

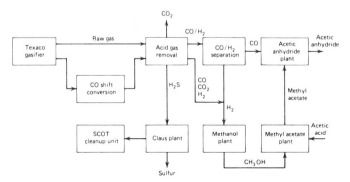

Figure 1. Texaco coal gasification process: gas processing and associated synthesis in the production of acetic anhydride from coal. SCOT = Shell Claus off-gas treating.

Table 1. Raw-Gas Composition of Second-Generation Coal-Gasification Processes

| | Fluidized-bed techniques | | | Entrained-bed techniques | | | | |
	U-gas[a]	Westinghouse	High-temperature Winkler	Texaco	Shell	Saarberg Otto	Combustion Engineering	Others KHD[b]
pressure, MPa[c]	0.4	1.6	1.0	8.0	2.0	2.5	0.1[d]	0.1
Typical gas composition, vol %								
CO	34	54	52	55	65	61	8–30	65–70
H_2	41	27	35	33	31	29	5–9.5	25–30
CO_2	19	10	9	11	3	7	3–12.5	0.3
CH_4	3.5	8	3	< 0.1	< 0.10.2	≤ 3	0.1	
N_2 + Ar	0.9	0.5	0.7	0.6	0.7	3	58–74	
sulfur compounds[e]	1.0	< 0.02	< 0.1	0.3	0.2	< 0.1	< 0.1	< 0.01
C_nH_{2m}				µg range		≤ 1		
NH_3, ppm				3	500			

[a] Utility Gas Process, developed by Institute of Gas Technology.
[b] Kloeckner-Humboldt-Deutz, FRG.
[c] To convert MPa to psi, multiply by 145.
[d] Airblown.
[e] H_2S + COS.

Recent Developments

Coal gasification. Second-generation coal-gasification processes require high gasification temperatures, which results in high carbon conversion, pure syngases (low environmental impact), possible slagging, and decomposition of water, which may serve as an oxygen source. Second-generation processes are aimed at maximum yields of CO and H_2 and minimum by-products by applying of the highest reaction temperatures possible. Compositions of raw gases obtained by various processes are given in Table 1.

The Texaco second-generation coal-gasification process (TCGP) provides synthesis gas that is practically free of by-products. Second-generation coal-gasification processes are also potential sources of CO_2, which can be further processed.

Syngas chemistry. Synthesis gas can be used as a raw material and energy source, for chemical feedstocks, and as a component in chemical syntheses.

In recent years, methane, produced by hydrogenation of carbon monoxide, has been used increasingly as a source of energy and as a raw material. The hydrogeneration is catalyzed by many metals. The heat of reaction must be removed. Cooled or adiabatic fixed-bed reactors mostly operate by a multiple-reactor system.

In the one-stage liquid-phase process, a light-hydrocarbon liquid stream is flushed through a granular catalyst. The pressures used depend on the composition of the synthesis gas.

Fischer-Tropsch synthesis. The catalytic hydrogenation of carbon monoxide with catalysts containing iron, cobalt, or ruthenium produces hydrocarbons and, therefore, provides an alternative to the direct hydrogenative liquefaction of coal. The Fischer-Tropsch synthesis provides a wide variety of hydrocarbons ranging from methane to waxes with an average molecular weight of ca 1000 (see Fuels, synthetic).

The conversion of synthesis gas on rhodium or cobalt catalysts gives mixtures of oxygen-containing C_2 compounds. The reaction takes place at 250–300°C at pressures of 2–10 MPa (290–1450 psi) in the gas phase.

On rhodium catalysts, the hydrogenation of carbon monoxide at increased pressure in the liquid phase gives a mixture of mono- and polyvalent alcohols, in which ethylene glycol and methanol are the main components (see Oxo reactions).

The reaction of alcohols or aliphatic carboxylic acids with synthesis gas leads to homologues:

$$RCH_2OH + CO + 2\,H_2 \rightarrow RCH_2CH_2OH + H_2O$$
$$RCOOH + CO + 2\,H_2 \rightarrow RCH_2COOH + H_2O$$

Homologation occurs at 25–50 MPa (3625–7250 psi) and 150–150°C on cobalt or ruthenium catalysts, which require a halogen cocatalyst.

Outlook

Future research will concentrate on processes that can use CO/H_2 without any other reactants as building blocks, those involving CO/H_2 and employed to make additional use of compounds derived from synthesis gas, those that use a derivative based on CO/H_2, eg, methanol, those that make syngas for use in combined-cycle plants, and those that use the carbon monoxide for subsequent syntheses.

J. FALBE
C.D. FROHNING
B. CORNILS
Ruhrchemie AG

H. Hiller and co-workers in *Ullmann's Encyklopädie der Technischen Chemie*, Vol. 14, Verlag Chemie, 1977, p. 357.

Proceedings of the Symposium "Ammonia from Coal," Tennessee Valley Authority, Muscle Shoals, Ala., May 1979.

J.A. Lacey and J.E. Scott, *paper presented to the 2nd Annual EPRI Contractors Conference on Coal Gasification*, Palo Alto, Calif., Oct. 1982.

HYDROGENATION

Processes

Hydrogenation. Direct coal hydrogenation (German technology) (see Fig. 1) is a development of the Bergius-Pier process (IG process). The reaction products are separated by distillation, and the oil that is used for the slurry preparation consists of asphaltene-free middle- and heavy-oil distillates in a 40:60 ratio; the residue is used for the production of hydrogen. Operating pressure is reduced from 70 to 30 MPa (10,200 to

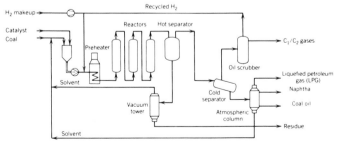

Figure 1. German technology flow diagram.

4,400 psi) using an inexpensive iron oxide catalyst. Specific coal throughput is increased by 50%, and heat recovery is improved.

The Exxon donor-solvent process (EDS), a development of the Pott-Broche process, uses a process-derived solvent and additional hydrogen; pressures are 10.5–17.5 MPa (1520–2540 psi) and the temperature is 450°C. The solvent is hydrogenated by means of a solid-bed catalyst, thus recovering its donor properties.

The H-coal process, developed by Hydrocarbon Research, Inc., is based upon the H-oil process for the desulfurization of mineral-oil distillation residues. Coal hydrogenation takes place in an ebullated-bed reactor with an extruded Co–Mo catalyst. The spent catalyst is replaced during the process, thus ensuring a continuous operation.

Gulf solvent-refined coal (SRC) process. The Gulf SRC II process was developed from the Gulf SRC I process, in which a solid and nearly ash- and sulfur-free extract is produced. In the SRC II process, part of the reacted slurry is recycled. This raises the catalytic mineral content in the slurry, which increases coal conversion during hydrogenation.

The LC fining process is a two-stage hydrogenation process. Short-contact-time hydrogenation is achieved by means of a catalytically hydrogenated solvent. The products are LPG, naphtha, and light, middle, and heavy distillates.

Japanese technology includes extractive coal liquefaction, direct hydroliquefaction, solvolysis coal liquefaction, and brown-coal liquefaction. The last technology was initiated in Australia, where as of early 1983, four coal liquefaction projects were under development. These projects are contracted to private companies, whereas the cost of research is underwritten by the government.

Large-scale plant projects. The construction of commercial plants for direct coal hydrogenation is possible, but a large capital is required and profits are low. Coal-hydrogenation technologies could be developed with a plant processing 10^6 t/yr.

Upgrading Oil from Coal

The oil produced by hydrogenation of coal is upgraded in a way similar to that for crude oil. Generally, the reaction conditions are more severe in order to remove sulfur and nitrogen (see Petroleum, refinery processes).

In the upgrading to gasoline, the light distillate, ie, naphtha (bp 185°C), is treated with the naphtha cut obtained from the middle distillate hydrotreatment. Heteroatoms are reduced to a concentration of < 1 ppm. Hydrotreating with a cobalt–molybdenum catalyst is sufficient.

For the production of heating oil,, the two alternatives differ in the degree of hydrogenation. Under relatively mild hydrotreating conditions and in a single-step process, middle distillates can be upgraded to a storable product. For the production of light heating oil most of the aromatic compounds must be hydrogenated by an additional process step.

Coal oil is an excellent feedstock for the production of some chemicals, eg, phenols, benzene, toluene, xylenes, ethylbenzene, and naphthalene-tetralin.

JOSEF LANGHOFF
ALFONS JANKOWSKI
Ruhrkohle Oel und Gas GmbH

J. Langhoff and K. Dohms, *paper presented at the 29th International Congress of Pure and Applied Chemistry*, Cologne, June 5–10, 1983.

W. Döbler, U. Graaser, J. Hallansleben, and A. Jankowski, *Erdoel Kohle* **36**, 370 (1983).

CARBONIZATION AND COKING

Industrialized countries with significant steel production have a combined output of metallurgical coke of ca 450×10^6 metric tons per year. It is used as the reducing agent in the manufacture of pig iron in blast furnaces. The present technology should provide a good economic basis in the future, since liquid by-products from coking and carbonization of coal are likely to expand in volume.

A typical coking coal produces an average of 80% coke, 12% coke-oven gas, 3% tar, and 1% crude benzene in high temperature oven-coking. Tar and crude benzene are separated by cooling the gaseous products, first to ca 100°C to precipitate 60–70% of the tar contained in the crude gas, followed by a secondary precipitation with indirect water condensers and electrofilters.

Refining

High temperature coke-oven coal tar. Refining operations for high temperature coke-oven coal tar include primary distillation and further separation of fractions, crystallization to obtain pure aromatic hydrocarbons, nonpolar heterocyclic compounds and filtered oils. Extraction with base and acid is applied to recover phenols and nitrogen-containing bases, catalytic polymerization to produce indene–coumarone resins, thermal polymerization to form electrode pitch and coking of pitch to produce low ash electrode coke.

Crude benzene. After separation of phenols, pyridine bases, and indene–coumarone resins, the remaining tar/light oil is refined together with the crude benzene obtained from the coke-oven gas. Sulfuric acid or oleum is used to sulfonate thiophene and polymerize unsaturated hydrocarbons. Sulfur compounds are converted to hydrogen sulfide and olefins are hydrogenated at 350°C and 2–6 MPa (20–60 atm) pressures over a molybdenum–cobalt–alumina catalyst.

The principal aromatic hydrocarbons produced by coke-oven benzene refining are benzene and toluene. In addition, ethylbenzene and xylene isomers are recovered.

Low temperature tars. Sources for low temperature tars include as by-products of the low temperature carbonization of selected bituminous coals as used to produce solid, smokeless fuels in the United Kingdom, and the liquid by-products of the Lurgi pressure carbonization–gasification process in South Africa. The main product of refining low temperature coal tars consist of phenolic compounds, as opposed to the aromatic hydrocarbons obtained from high temperature coal tars.

Lignite tars contain considerable amounts of paraffinic compounds. They are usually refined by continuous distillation, following pretreatments such as catalytic hydrogenation.

Chemicals and feedstocks made from carbonization and coking are important intermediates for many key chemicals.

Health and Safety Factors

Tars contain some compounds that have proven carcinogenic in animal testing. The acute toxicity of tar itself is low.

G. COLLIN
Rütgerswerke AG

G. LÖHNERT
Ruetgers-Nease Chemical Company

H.G. Franck and G. Collin, *Steinkohlenteer-Chemie, Technologie und Verwendung*, Springer Verlag, New York, 1968.

G. Collin in L.E. St. Pierre, ed., *Proceedings of World Conference, Future Sources Org. Raw Materials Chemrawn I*, Toronto, 10., 13.7.1978, Pergamon Press, New York, 1979, pp. 283–297.

D. McNeil, *Coal Carbonization Products*, Pergamon Press, London, 1966.

CARBIDE PRODUCTION

Calcium Carbide

For the manufacture of calcium carbide, lime is treated with a carbon-containing material:

$$CaO + 3 C \rightarrow CaC_2 + CO + 466 \text{ kJ/mol (441 Btu/mol)}$$

Industrial processes are carried out at 1800–2100°C.

Calcium carbide owes its importance to two reactions:

$$CaC_2 + N_2 \rightarrow CaCN_2 + C$$

$$CaC_2 + 2 H_2O \rightarrow C_2H_2 + Ca(OH)_2 + 129.7 \text{ kJ/mol (123 Btu/mol)}$$

The first reaction fixes the nitrogen contained in the air for use as fertilizer. The acetylene obtained according to the second reaction was used as an illuminant around the turn of the century. Later, it was used in welding, and more recently it became an important starting material for chemicals (see Acetylene-derived chemicals).

Calcium carbide has been used increasingly as a desulfurizing agent in the steel industry:

$$CaC_2 + [S] \rightarrow CaS + 2\,C$$

Manufacture. Calcium carbide is made in the electrothermal low-shaft furnace. The furnace vessel is made of welded or riveted iron and reinforced against distortions. The bottom is covered with carbon bricks or fireproof bricks. Most furnaces are operated by three-phase-current electric power supplied by three self-baking Soderberg electrodes. In open furnaces, the CO gas burns at the surface of the furnace contents and has to be purified. Modern furnaces having a power consumption of up to 60 MW are completely closed and for this reason can make use of the CO gas that forms. The carbide furnace has to be controlled in such a way that the production proceeds continuously and evenly. The open furnace is easiest to control.

The lime is usually quicklime with a grain size up to 50 mm. The carbon source is generally coke obtained from hard coal. The $Ca(OH)_2$ that forms when carbide is gasified is burned and recycled.

The liquid carbide from the furnace is collected in cast-iron crucibles. In some plants, the blocks are allowed to cook in the crucibles, in others they are removed after 2–4 h and then allowed to cool. Alternatively, the liquid carbide leaving the tap hole is directed onto a rotating drum. However, this procedure does not give the granulation required by many standards.

Calcium carbide is stable almost indefinitely in the absence of moisture. In contact with water, acetylene, which is flammable, forms. The explosive limit in air is 2.4%, and therefore, handling, storage, and transport are subject to special safety regulations (see Acetylene).

W. PORTZ
Hoechst AG

S.A. Miller, *Acetylene*, Vol. 1, Ernest Benn Ltd., London, 1965.

K. Hartmann, W. Schirmer, and M.G. Slinko, *Probleme der modernen chemischen Technologie*, Akademie-Verlag, Berlin, 1980.

GENETIC ENGINEERING

Recombinant DNA and Associated Technologies

Genetic engineering has generally been equated with recombinant DNA technology. It depends, however, on or, at least, exploits many other technologies. For the purpose of this article, genetic engineering is considered to be the direct operations performed on genes, ie, the isolation and amplification, the regulable expression, and the directed modification or mutagenesis of DNA segments encoding specific genes.

Gene cloning. Restriction enzymes recognize and cut sequence-specific sites in DNA. Many leave short single-stranded ends on the resulting fragments. If fragments taken from different sources but generated by cutting with the same purified restriction enzyme are mixed, they can reassociate by virtue of the so-called sticky ends. These hybrids can be covalently closed by treatment with DNA ligase. If one of the DNAs used in the process is, for example, an *Escherichia coli* phage or plasmid, the second DNA segment can be amplified by replicating the hybrid molecule in *E. coli*. Many variations have now been devised for generating and recombining DNA fragments *in vitro*.

The technical bottleneck tends to be identification and authentication of the desired hybrid. This is typically achieved by nucleic acid hybridization of the hybrid clones with a related and previously characterized DNA segment that has been radioactively labeled. Procedures depending on an antibody directed against the protein encoded in the DNA or on the activity of the protein have also been devised. Ultimately, it is

necessary to correlate the cloned DNA's sequence with the activity of the element whose DNA has been putatively cloned.

Expression of cloned genes. Besides incorporating the appropriate DNA fragment into the vector, it is necessary that the proper transcription and translation signals be juxtaposed to the gene. *E. coli* has been the most commonly used host–vector system for overexpression. The yeast *Saccharomyces* has been the second most commonly studied microbial system, especially in the fermentation industry. It secretes proteins, but is not likely to perform accurate glycosylations of mammalian proteins. It will undoubtedly be useful for some purposes. *Bacillus subtilis* is also used in fermentation and can secrete proteins. However, its phage and plasmids have not been easily incorporated into vector constructions. Mammalian cell systems are under extensive study.

Site-directed mutagenesis. Site-directed mutagenesis provides increased genetic diversity and a route to improved protein and cell traits. There are three general routes: a cloned fragment can be subjected to traditional methods of chemical or genetic mutagenesis, one of the strands of the fragment can be nicked at random or at specified sites and subjected to chemical or enzymatic mutagenesis at the site, or a synthetic oligonucleotide encoding the desired change can be used as a primer to direct the mutagenesis. When the desired change can be defined, the last route is the most direct way to synthesis.

Cell transformation. The development of shuttle vectors has been particularly important in permitting the directed and controlled transformation of higher eukaryotic cells. The shuttle vector carries regulatory elements and a selectable marker that permit its replication and selection in the control host, typically *E. coli* or *Saccharomyces*. The vector also carries a set of elements and selectable markers to permit its replication or integration and selection in the second host. Vectors that can be moved directly from *E. coli* to the higher eukaryotic cell have been devised and exploited in animal cells. There have been efforts to introduce genes into somatic and germ cells and to return them into the animal. A spectacular example was the demonstration of enhanced growth in mice that had developed from embryos into which a rat- or human-growth-homone gene had been introduced. Plant-cell and resulting-plant transformation has also been achieved with model vector systems.

Applied Biotechnology and Genetics

Chemical processes represent one of the largest opportunities for the future application of recombinant DNA technology. Evaluations have focused on new approaches to the manufacture of existing chemicals, reduced energy requirements, and waste-disposal problems.

The production of chemicals by biotechnological means and specifically by genetic engineering requires an analysis of the type of compounds that can be produced by such methods. Table 1 gives some of the commercial enzymes that are being or could be produced using the recombinant DNA technology (see Enzymes, industrial).

The prospects for the application of biological processes to the manufacture of high volume commodity chemicals appear poor with the exception of a few products, eg, ethanol and some short-chain organic acids. Prospects appear much better for the manufacture of higher

Table 1. Some Commercial Enzymes Whose Production Can Be Based on Recombinant DNA Technology

Enzyme	Applications
α-amylase	starch hydrolysis, manufacture of alcohol, textile desizing, and human digestive aid
ficin	wound debridement
glucose oxidase	diagnostic for diabetes
glucose isomerase	high fructose corn syrup (HFCS) manufacture
lipase	flavor production in cheese
papain	meat tenderizer
pectinase	clarification of beverages
penicillinase	treatment of penicillin allergic reaction
protease	leather batting
rennin	cheese manufacture

priced specialty chemicals or those with no practical alternative route, eg, alkaloids (qv) and other natural products.

Waste and industrial pollution. The most widely publicized applications of modern biotechnology concerned the "oil-eating" bacteria. This discovery led to the landmark Chakrabarty decision by the Supreme Court in 1980 permitting the patenting of new forms of microbial life developed through genetic manipulation. Although no significant commercial results have yet been achieved, the experiment demonstrated the feasibility of engineering microorganisms to degrade potentially toxic chemicals.

Pharmaceuticals and health care. The application of recombinant DNA technology and genetic engineering has had its greatest impact on the health care and pharmaceutical industry. Several genetically engineered proteins are being tested. Insulin was the first product to reach the market. Human growth hormone (somatotropin) is well along in its clinical trials. Massive efforts involving many companies are being directed to the clinical evolution of the interferons and immune system modulator interleukin 2. In many cases technology may prove more valuable in providing research clues to new small chemical products and in producing proteins which serve as products themselves.

Agriculture. Plant genetic engineering currently involves introduction of genes into plant cells cultured *in vitro* and the regeneration of whole plants from these cells. Research is also proceding on the genetic multiplication of microorganisms that inhibit and enhance the plant rhizosphere. Several animal-growth factors and health products including bovine-growth hormone have been produced and are now in testing.

Food processing. The food industry has begun to apply recombinant DNA techniques to the production of single-cell proteins, enzymes, amino acids, and synthetic sweeteners. The genetics of yeast are being studied in the baking, brewing, and wine industries. Some of the Federal constraints imposed on the food industry may possibly be circumvented by transferring desired genes into organisms that meet FDA standards.

Energy. Biomass fuel sources have advantages over fossil fuels. Genetic-engineering applications would involve the development of enzymes to break down lignocellulose to small molecules that can be metabolized to products such as ethanol.

Mining. Potential applications of genetic engineering to mining are in tertiary recovery of oil and recovery of minerals associated with rocks.

Regulations and Property Rights

The Asilomar Conference in 1975 and subsequent discussions resulted in the elaboration of a set of NIH *Guidelines for Research Involving Recombinant DNA Molecules.* Property rights of biological systems and their manipulation are protectable by patent.

ERNEST JAWORSKI
DAVID TIEMEIER
Monsanto Company

T. Maniatis, E.F. Fritsch, and J. Sambrook, *Molecular Cloning: A Laboratory Manual,* Cold Spring Harbor Laboratory, New York, 1982.

Impact of Applied Genetics, Office of Technology Assessment, U.S. Government Printing Office, Washington, D.C., 1981.

R.H. Zaugg and J.R. Swarz, *Assessment of Future Environmental Trends and Problems: Industrial Use of Applied Genetics and Biotechnologies,* Contract Grant No. 68-02-3192, EPA No. 600/8-81-020.

National Technical Information Service, U.S. Department of Commerce, Springfield, Va., 1981.

TOXICOLOGY

In the twentieth century, has developed significantly the discipline of toxicology and it has been establshed as a professsional activity. Toxicologists have made substantial advances in defining the nature and mecha-

nism of toxic injury, in determining factors that may influence the expression of a toxic effect, in developing methodologies, and in obtaining information on the toxicity of a multitude of substances.

Toxicology is a study of the interactions between chemicals and biological systems in order to quantitatively determine the potential for such chemicals to produce injury which results in adverse health effects in intact living organisms, and to investigate the nature, incidence, mechanism of production, and reversibility of such adverse effects.

Investigations are carried out using a variety of biological systems, including *in vivo* or *in vitro* studies. Dose–response relationship studies are of considerable value (see Pharmacodynamics).

A toxicological investigation should define the nature of the harmful effects, the incidence and severity of the effects as functions of the exposure dose, and the mechanisms by which the effects are produced. The study should include detection of the effects, ie, the development of methodologies for the specific recognition and quantitation of the toxic effects, and their reversibility.

Classification of Toxic Effects

In order to cause tissue injury, a substance must come into contact with an exposed body surface (local effects). However, material may be absorbed from a contaminated site and be disseminated by the circulatory system to various body organs and tissues (systemic effects). Many materials may cause both local and systemic toxicity.

The nature of a toxic effect, and the probability of its occurring, is often related to the number of exposures. Acute exposures involve a single exposure. Short-term repeated exposures involve consecutive daily exposures over a few days or a few weeks. Subchronic exposures involve consecutive daily exposures for a period amounting to 10–15% of the lifespan of the test species. Chronic exposures involve consecutive daily exposure over the lifespan of the test species. Some materials of low acute toxicity may have a significant potential for producing harmful effects by repeated exposure, and *vice versa.* Any given material may produce more than one type of toxic effect. Examples of toxic effects produced by different chemicals are shown in Table 1.

The Nature of Toxic Effects

Listed below are some of the more significant and frequently encountered types of injury or toxic response.

Inflammation describes the local and immediate response to injury characterized by increased blood flow, leaking of blood plasma into the tissues, and migration of particular blood cells to the affected area. Inflammation may be acute or chronic.

Degeneration is a description for structurally abnormal changes visible by microscopy. Necrosis describes the circumscribed death of tissue.

Table 1. Examples of Differing Types of Toxic Effect Classified According to Time Scale for Development and Site Affected

Time scale	Site	Effect	Chemical
acute	local	lung damage	hydrogen chloride
	systemic	hemolysis	arsine
	mixed	lung damage methemoglobinemia	oxides of nitrogen
short-term	local	sensitization	ethylenediamine
	systemic	peripheral neuropathy	methyl-*n*-butyl ketone
	mixed	respiratory irritation kidney injury	pyridine
chronic	local	bronchitis	sulfur dioxide
	systemic	liver angiosarcoma	vinyl chloride
	mixed	emphysema kidney damage	cadmium
latent	local	pulmonary edema	phosgene
	systemic	neuropathy lung fibrosis	organophosphates paraquat

The immune system protects against invasion by foreign materials. Such materials (antigens) stimulate immune mechanisms. In some instances, hypersensitivity develops.

Since a primary function of the immune system is protection against pathogenic materials, any substance capable of producing a suppression of immune function will have a deleterious effect on such protective mechanisms (immunosuppression) (see also Immunotherapeutic agents).

Neoplasms are abnormal masses of cells in which growth control and divisional mechanisms are impaired, resulting in aberrant proliferation. Neoplasms are classified as benign or malignant.

Chemically induced mutagenesis involves an interaction between the causative agent and cellular constituents, that produces deoxyribonucleic acid (DNA) damage that is heritable. Mutations can occur in somatic or germ cells and may be reflected in altered structure or function. It is generally conceded where genotoxic carcinogens are concerned that an early irreversible stage in the complex process of carcinogenesis is likely to involve a mutagenic event.

Some materials produce toxic effects by inhibition of biologically vital enzyme systems, leading to an impairment of normal biochemical pathways.

Uncoupling agents, such as dinitrophenol, interfere with the synthesis of high energy phosphate molecules, resulting in the liberation of energy as heat.

Lethal synthesis is a process in which the toxic substance has a close structural similarity to normal substrates. As a result, the material may be incorporated into the biochemical pathway and metabolized to an abnormal and toxic product.

Teratogenic effects result in the development of a structural or functional abnormality in the fetus or embryo.

Although not strictly a toxic effect, peripheral sensory irritation is important in many occupational health considerations since it produces discomfort and distraction.

Factors Influencing Toxicity

Some toxic effects are produced by a single exposure, whereas others require multiple exposures. The magnitude of the exposure influences the likelihood of an effect and its severity.

There may be marked differences between species with respect to the potency of a given material.

The time of day, or season of the year, may influence the toxic response, as may the formulation of a material. Impurities may modify the toxic response.

Routes of exposure. In order to induce a toxic effect, the causative material must first come into contact with an exposed body surface. The route of uptake may have a significant influence on the metabolism and distribution of a material.

A swallowed material may cause local effects, such as inflammation, and carcinogenic materials may induce tumor formation in the alimentary tract. The skin is a principal route of exposure in the industrial environment. Local effects include inflammation, allergic reactions, and neoplasia. The skin may also absorb systemically toxic materials.

The potential for adverse effects from materials dispersed in the atmosphere depends on physical state, concentration, and time and frequency of exposure.

Adverse effects on the eye may be produced by splashes of liquids or solids, and by materials dispersed in the atmosphere.

In most practical situations, simultaneous exposure to multiple chemicals can occur and thus a potential for complex biological interactions may result. Exposure to combinations of chemicals does not always produce clearly distinguishable interactions.

Fate of Absorbed Chemicals

The induction of systemic toxicity may involve complex interrelationships between the absorbed material and any conversion products, and their concentration and distribution in body tissues.

Materials may be absorbed by a variety of mechanisms, eg, passive diffusion, filtration processes, facilitated diffusion, active transport, and the formation of microvesicles for the cell membrane (pincocytosis). Metabolism may form a product of lower toxicity than the parent, ie, detoxification has occurred. In other cases, the result is a metabolite of greater toxicity than the parent, ie, metabolic activation has occurred. The kidney is an important organ for the excretion of toxic materials.

With respect to environmental exposure, the probability of adverse effects depends on many factors, such as magnitude, duration, and number of exposures.

Dose-Response Relationships

An observation that is fundamental to the interpretation of toxicological information is the variation in susceptibility to potentially harmful chemicals by individual members within a given population. Dose-response relationships are useful for many purposes, and by appropriate considerations of the dose–response data, it is possible to make quantitative comparisons and contrasts between materials or between species. Caution is necessary in making quantitative comparisons between different materials.

Since the largest proportion of deaths is distributed around the 50% mortality level, this forms a convenient reference point, and is referred to as the median lethal dose (LD_{50}). It is calculated from data obtained by using small groups of animals, and usually, for only a few dose levels. Therefore, an uncertainty factor is involved.

Testing Procedures

In general toxicology studies, animals are exposed to a test material and examined for all signs of toxicity. Specific studies are those in which exposed animals are monitored for a defined effect. Studies may be carried out by single or repeated exposure. It is usual to proceed in sequence from acute to multiple-exposure studies.

Animals are inspected to discover any departure from normal appearance and behavior. A decrease in the rate of gain in body weight may be one of the earliest indications of toxic effects. Changes in food and water consumption may indicate a toxic potential. The functional status of blood and blood-forming tissues can be assessed by red- and white-blood-cell counts, platelet counts, clotting time, coagulation tests, and examination of bone marrow.

Chemical pathology or clinical chemistry involves the measurement of the concentration of certain materials in the blood, or certain enzyme activities in serum or plasma. Urine is collected and examined.

Animals are examined macroscopically at autopsy. Measurement of the weight of organs removed at autopsy and microscopic examination of tissues is an integral part of most toxicology studies.

Acute toxicity studies are often dominated by considerations of lethality, including calculation of the LD_{50}, but it is important to note signs of toxicity. Short-term repeated studies should give information about cumulative toxicity. Short-term repeated-exposure studies may not be relevant for assessment of hazard over a longer period. Most types of repeated-exposure toxicity are detected by subchronic exposure conditions.

Many procedures are available to detect specific organ toxicity. Primary irritancy studies determine local inflammatory effects. Allergenic materials may produce hypersenitivity reactions by skin contact or inhalation.

When a material may produce structural or functional damage to the nervous system, special methods should be incorporated into general toxicology studies. Methods for assessing sensory irritation by inhalation and contamination of the eye are available.

Most studies conducted to determine the teratogenicity of materials are aimed primarily at assessing structural defects of development. Reproductive studies cover a wide spectrum of developmental biology.

Metabolism is concerned with the biotransformation of the parent material. Pharmacokinetic studies should allow an assessment of the relationship between the environmental-exposure conditions, the absorbed dose, and the target organ dose.

Studies to determine the potential for mutagenic events may be conducted *in vitro* and *in vivo*. The Ames procedure is based on the ability of mutagenic chemicals to cause certain bacteria to regain their ability to grow in media deficient in an essential amino acid.

The interpretation of toxicology studies is a matter requiring experience. The study should allow decisions on whether injury is a direct result of toxicity or secondary to other events.

Hazard is the likelihood that the known toxicity of a material will be exhibited under specific conditions. Toxicity is but one of many considerations to be taken into account in hazard-evaluation procedures.

In some instances, it may be economically impossible to conduct a complete spectrum of toxicology testing. The relevance and credibility of a toxicological study can be no better than its study design and conduct allow.

BRYAN BALLANTYNE
Union Carbide Corporation

D.T. Plummer in J.W. Gorrod, ed., *Testing for Toxicity*, Taylor and Francis, London, 1981.

T.A. Loomis, *Essentials of Toxicology*, Lea and Febiger, Philadelphia, Pa., 1974.

G.D. Clayton and F.E. Clayton, eds., *Patty's Industrial Hygiene and Toxicology*, Vols. 2A, 2B, and 2C, John Wiley & Sons, Inc., New York, 1981.

INDEX